NEWTON's TELECOM DICTIONARY

23rd Edition

Harry Newton

New York

NEWTON's TELECOM DICTIONARY

Published in the United States by
Flatiron Publishing
50 Central Park West
New York, NY 10023
email: Harry@HarryNewton.com
www.FlatironBooks.com

Distributed by
National Bank Network
4501 Forbes Boulevard, Suite 200
Lanham, MD 20706

Orders
Phone toll-free 1-800-462-6420
Direct 1-717-794-3800
Fax 1-800.338.4550
custserv@nbnbooks.com

ISBN Number 0-9793873-0-2
March 2007
Twenty Third Edition
Steven Schoen, Contributing Editor
Gail Saari, Layout and Production Artist
Saul Roldan and Damien Casteneda, Cover design

Stay In Touch

For suggestions, corrections, updates, special offers, please send an email to
Harry@HarryNewton.com.

I promise you I won't give your name to anybody. Nobody. Promise.

Harry Newton

Dedication to a Great Visionary

LeRoy T. Carlson

LeRoy T. Carlson founded, and is today Chairman Emeritus of Telephone and Data Systems, Inc. (TDS), one of the largest independent telephone companies in America still controlled by its founding family. TDS serves seven million happy customers in 36 states with 11,800 enthusiastic employees and enjoys over $4 billion in annual revenues.

Carlson, the son of Swedish immigrants, grew up on Chicago's South Side. He helped support his family during the Great Depression with a door-to-door produce business he created. He funded his college and graduate school studies by selling magazine subscriptions, used furniture, and laundry services to his classmates. Carlson served in the U.S. Navy, S.C., Lt. Jr. Gr. and the General Motors Overseas Corp. during World War II.

He and his wife, Margaret, have four children and ten grandchildren. Carlson, now 90, comes to work every day. He is among the oldest-living and hardest-working of the original telephone industry pioneers. I salute his great sense of humor and his zest for life. I am proud to dedicate this 23rd edition of my dictionary to LeRoy T. Carlson, a great American.

March, 2007

Contents

WHY I WRITE THIS DICTIONARY

How You Can Help

by Harry Newton

This 23rd edition is slightly smaller in pages than last year's. That's not because there are fewer definitions. In fact, there are many more. But a genius called Gail Saari, a great artist, figured some creative ways to squeeze more words on each page without affecting readability. One of her little tricks was to remove one blank line at the bottom of the page – between the text and the page number. Gail loves doing this dictionary: she finds it endlessly amusing and incredibly useful. Thank you Gail.

This dictionary is a labor of love. It began on a rainy weekend many eons ago, I said to my wife, "What this industry really needs is a dictionary. I'll write one this weekend." That was nearly 25 years ago. I'm still writing this dictionary. Every day I add another definition, or two, or three.

I wrote this dictionary to teach myself my twin loves – telecom and technology. I find that writing a definition gets it clear in my brain. Writing helps me understand the complexities.

Writing takes enormous work. Anyone can write in techno mumbo-jumbo. There's oodles of it out there. I write this dictionary in non-technical business language. That means distilling the geek-speak and the techno mumbo-jumbo into English normal people (like me) can understand. People use geek-speak to make themselves feel important. That annoys me. I believe (perhaps naively) that anyone in business – whether engineer or not – should be able to understand complex technical terms. I'm not an engineer. That's good and bad. It takes me longer, but I do get it eventually into English.

This dictionary is different from any other. Some of my definitions are short. Some are long – very long. They can be mini-essays. My objective is to explain what the technology means, what it does, how you use it, what its benefits are and what the pitfalls of using it are.

I also have this crazy section – The Best Money Saving Tips. I've collected the best ones over 25 years. I used to publish a telecom magazine called Teleconnect. The most popular section was Money Savings Tips. Here are the best. You'll easily pay off the cost of this dictionary by using only one of the tips.

Sadly, I'm not all knowing. I need your help. Sometimes my definitions are unclear, old, wrong, don't make sense or aren't up to date. Please send me an email. I'm good about responding. But, I'm totally obsessed about making this the best dictionary ever written.

Please help. Send me an email.

Harry Newton
New York, NY
Harry@HarryNewton.com

THE HOTTEST TELECOM OPPORTUNITIES

"I Love Telecom. Where Should I Work?"

By Harry Newton

Here are some booming telecommunications areas:

Broadband. Broadband. Broadband. The world is going broadband. Will it be cable TV, DSL, Wi-Fi, EvDo, FiOS or WiMax, satellite, or long-range wireless? Depends. In the U.S. cable TV seems to be winning, though I'm impressed with Verizon's portable Broadband Access service and its new fiber-based FiOS. Overseas, the phone company's DSL is working magic – largely by providing fast bandwidth (much faster than the U.S.) and low prices. Long-range wireless (such as WiMAX) could become a contender if prices drop and providers get better funded. Fact is, broadband Internet is our industry's fastest-growing, most popular new service. Broadband Internet is to economic growth today what the interstate highway system was in the 1950s. It's the engine of economic growth.

VoIP. Voice over the Internet. I have it. I use it everyday. Slowly, they're working the bugs out of it. It'll never work as well as normal circuit switched calling – unless you control all aspects of it (e.g. on a private network). This means you'll never get perfect quality on the Internet. But VoIP will get pretty close. And the price is right – basically free for each additional call. Today's VoIP quality is about as good as decent cell phone service, which means you get dropped and volume ebbs and flows. Adding features to VoIP is where the action is going to be – like voice mail you can read on your BlackBerry.

Adding value to phone calls, voice mail and faxes. There are serious computer telephony services out there that do creative things. I like the services that accept your incoming faxes, convert them to PDFs and them attach them to an email to you. That's super because it means you can receive faxes wherever you are – in New York or Sydney, Australia. I like the services that receive your incoming voice mails and attach them to emails to you. There's even one service called Simulscribe that transcribes your voice mails into text messages and attaches a wave file of the message – in case you want to check they got it correctly. What's nice about this is you don't have to write down all the numbers in the voice mail. The computer has already done it for you. Saves huge time.

Home entertainment. It used to be standalone. A separate TV. A separate radio, etc. Now everything is connected by Ethernet cables and Wi-Fi wireless to a home entertainment server and through that to the Internet for downloading movies, songs, photos and programs for your TiVo, known now as a PVR (Personal Video Recorder). This is a super vision that is selling. Right now, putting the whole thing together is a little "hard" for the average consumer. But it is getting easier. More and more of my friends are putting complex home entertainment networks into their home. I'm getting one also. I have a wired Ethernet and wireless network running around the house. I have TiVo recording TV shows, which I can play back on any of three TV sets. We have a sound system blasting into many places. Often I carry my central collection of songs on my iPod to the appropriate input connection. Today, my "switching" between devices is manual. But many of my friends have a whole low voltage network of switches. There's huge money in this business.

Better TV. DVDs are better TV. They look and sound better. High definition TV (HDTV) is even better. Once you've seen it, you won't want to go back to normal broadcast or analog cable TV. Sadly, it's hard to get HDTV in most parts of the world. But it's coming. The "playback" devices – plasma, LCD, DLP, projection TV – are getting by the day.

The Internet as a giant shopping mall. Many of us developed great skills in running telephone call centers. Now it's time to transfer those skills to running great web sites and providing great service to Internet customers. Too many web commerce designers, especially those coming out of IT (Information Technology), make their web sites according to the "service" they get out of computer vendors – like endlessly waiting for answers to questions, like little real information and like 48-hour answers to customer emails. In short, most web sites that peddle stuff could sell a lot more and be a lot more profitable if their web sites were designed to include customer service (and people) not replace them. I wrote a piece on this subject at the end of this introduction, called "Why is it so hard to buy?"

Remote healthcare. Millions of people with chronic health problems such as diabetes, heart, kidney or circulatory problems will be able to have their conditions automatically monitored, and their problems addressed, as they go about their daily lives. This will save lives, reduce the cost health care and create huge opportunities for telecom service, equipment and maintenance.

Location based services. GPS (Global Positioning Service) combined with telecom. Triangulation with cell phones. America has six million heavy trucks. Every wasted hour costs big bucks. $100 an hour is not uncommon. By combining location information with smart telecom networks and intelligent software, you can seriously improve the productivity, and thus the profitability, of these trucks. Imagine what location services could do for agriculture. Cell service and GPS could tell the bulldozer precisely where to go, to tell the harvester where and how to harvest most efficiently. Ditto for the construction industry.

Security telecom. Whenever there's a disaster, the nation's cell phones and landlines collapse. There's huge money being spent on emergency communications systems. The battlefield for the war on "terrorism" is focused heavily on the Internet. That's where our enemies do their recruiting, their planning and their call to arms. Corporations likewise are spending fortunes to secure their networks, their web sites, their wireless communications and their databases.

Web services. As broadband Internet access becomes more common, more people will find it convenient to draw their computing power and services from remote sites, especially as those sites draw their material from multiple sites and combine it in unique and creative ways. See my definitions for Web 2.0 and Web 3.0.

Brazil, Russia, India and China – also called BRIC in the investing world. Every major North American company now has operations in China and/or India. Most companies are selling into China and India. A 50% per year growth is common. China is the world's fastest-growing large economy. By 2008, in time for the Olympics, China will fiber optic up its entire country, spending at least $8 billion. The smallest village to the largest city will be served by the most advanced fiber optic telecommunications system in the world. And now India is exploding, especially with English-speaking call centers answering service phone calls from customers in the U.S. Neat useful business. If I were younger, I'd be working in China or India helping them with telecom.

Outsourcing. As services get outsourced to China, India and other places, the need for decent, reliable telecommunications and the management of it becomes paramount.

Web servers and web server farms are increasingly among the key assets of most major corporations. The care and protection of these servers and the telecom lines that feed them are hugely important jobs.

And finally the old standby, Refunds. Over 80% of the phone bills presented to corporations are wrong and usually wrong in the carriers' favor. This means that, if you know something about telecom bills, there's a handsome business in auditing corporate phone bills and securing hands refunds.

I wrote this book for those of us turned on to telecom, for those of us trying to keep up, for those of us new to this wonderful industry and for those of us who simply want a respite from life for a few moments.

Give my dictionary to your workmates, your employees, your users, your customers, your boss. Give it to your kids to let them understand what you do and why you're turned on. They'll understand why you, too, have no life.

Arthur Clarke once commented "Any sufficiently advanced technology is indistinguishable from magic." I hope the year 2007 brings lots of magic into your life. If you want to talk about Vision for our industry, send me an email or give me a call. Or ask me to give a talk at your user group.

Thank you.

Harry Newton

Harry@HarryNewton.com
March, 2007

THE BEST MONEY SAVING TIPS

How to Save on Telecom, PCs, Internet, Airline, and Investing

Expanded for 2007

by Harry Newton

The most popular section in any magazine I published was a section called "Dollar Saving Tips." Every month I'd write tips that saved money or time, or both. One day, we decided to put the best tips into this dictionary. The logic? Perhaps we'll sell more dictionaries? If readers can save many times the cost of this dictionary, they'll feel they got the definitions for free. Weird way to sell a dictionary. That's the entire rationale for including these Dollar Savings Tips in this edition. I believe you'll save the price of this book several hundred-fold with these tips. Please do not underestimate how much work has gone into compiling this collection. It's 20+ years of work. As you read them, you'll notice eight themes:

1. Check. Check. Check. You can always find it cheaper, faster, better elsewhere, especially now we have the Internet, Google, Froogle, Kayak and others.

2. Ask. Propose. Negotiate. Everything is negotiable. End of story.

3. Don't believe the "experts." You can become an instant expert on the Internet, at your local library and by calling people on the phone, or all three.

4. You should argue with "brain surgeons." Pick a professional. Bargain with him. He'll do a good job, even if he agrees to do it for slightly less. That's why he's a professional.

5. Things are not worth what you paid for it. Just because it costs more doesn't mean it's worth more, though my wife often believes it is.

6. When God closes a door, She opens a window. This means there's a always a better option or a better opportunity than the one you just lost. This is another way of saying, "always be positive." This applies to investments, houses, jobs, etc.

7. Never stop learning. The minute you do, the world will take you over. The world we live in is immensely complicated. The only weapon you have against it is to study.

8. Get everything you can in simple writing. Telecom salespersons often promise one thing. But their company's billing system can't bill it. You need simple paperwork to prove you're right.

Here are my best tips. They are in no particular order.

MOST PHONE BILLS ARE WRONG -- I

Over 80% of telecommunications bills contain errors, usually in the favor of the telecom service provider, local or long distance. Telecommunications rates and service contracts are complex and different deals are struck with different customers. What you have in writing is different to what I have in writing. The weakest part of any telecom carrier's operation is its billing system. Figure at least 20 years out of date. Different computer systems often exist within the same telecom company (think of all the mergers), and records of customer billing often conflict with the actual services. That's because most phone companies still have not integrated their billing and operations databases. In short, your corporate phone bills are wrong and definitely not in your favor.

MOST PHONE BILLS ARE WRONG -- II

In addition to incorrect rates on your telecom bills, you may also be billed equipment and for services such as outside lines that simply do not exist. You may have ordered a line that was never delivered, but billing was started and continues. Or you may continue to be billed for lines that you had in place, but then disconnected. If you can find these mistakes and document how far back the mistakes go, you can get a refund (plus interest) for all those years. Different rules apply to the refund process and may be dictated by State Public Utilities Commissions or the FCC. While you can probably get back some of the money yourself, professional telecommunications bill auditors are likely to get back more. Though you'll have to pay them a percentage of the refund, you are still likely to end up with more dollars in your pocket.

MOST PHONE BILLS ARE WRONG -- III

Audit all of your telecommunications at least once a year to be sure your billing is correct. You may consider using a professional telecommunications bill auditor. Auditors have the benefit of looking at many bills and contracts, so they know where to look for the errors. The best telecommunications bill auditor I know is DIgby 4 Group, Inc., 212-883-1191, www.digby4.com. Jane Laino, president, is a dear friend.

GET INTEREST ON YOUR REFUND

If you're successful getting a refund from your local telephone company, be sure you get all the interest you're entitled to. Interest rates vary from state to state and can run as high as 18% per year (in New York.) Not all states pay interest (such as New Jersey). State regulations also vary in terms of how far back in time the refund will cover.

SENDING A FAX FROM YOUR PC

How to send faxes from your email. Now you're on broadband and no longer on dial-up, how do you send faxes from your PC? There are a bazillion services. But all want to charge a monthly fee. I found one which charges by usage. It's Fax1.com. I tested it. You can send attachments with graphics, charts and spreadsheets. It works fine. The price is right – 12 cents a page through the U.S., higher to other countries.

NEED A SPARE CHARGER?

Need a spare charger for your cell phone, your BlackBerry or your iPod? Next time you check into a hotel, tell them you left you charger there on your last visit. They'll bring you a gigantic box of chargers and adapters their previous guests left. Take your pick. You'll be doing them a favor. Every day they accumulate more from forgetful guests.

HOW TO MAKE PDFs

PDF stands for portable document format. It's a universal computer file standard. This means if you send someone a PDF file, they should be able to view it on their computer – PC or Apple – and on some cell phones like BlackBerries and Treos. Most computers these days come with Adobe Reader, a simple free program to read PDF files. If you don't have one, go here – www.adobe.com/products/acrobat/readstep2.html. Some software allows you to save documents as PDFs. Microsoft's new Office 2007 has an available plug-in which lets you save documents as PDF. Look under www.microsoft/downloads. Search for "Save as PDF."

Making PDF documents is easy. There probably 500 programs. None do a perfect job. Which means they often mess up complex diagrams, shadowing, etc. The best way of making perfect PDF documents - i.e. ones that look exactly like the original – is to use an expensive program like QuarkXpress or Adobe InDesign and have it save your document in PDF format. Don't forget, there are many flavors of PDF. This dictionary goes to the printer in "high definition" PDF, which is very different to what you see every day.

Among the many PDF software makers I've tested, the best value for the money is a piece of software called PDF Printer Driver. You can download a trial version of the software, but it will put a small plug for itself on every page it converts to PDF. Or you can pay $9.95 (which I did) and buy the real thing. To use it, you simply "print" your material to a file, to a place on your hard drive, not to a piece of paper. It couldn't be easier.

If you rarely need to convert a document to a PDF, there's an another way. Email your document to PDFOnLine.com. They'll convert it for you and return your new PDF document as an attachment to an email. If you need to convert paper with a signature or drawings to PDF, the easiest way is to use www.CallWave.com. You fax them your document. They return it as a PDF attached to an email.

There is one other issue with PDFs. If someone sends me a PDF file – say for a prospectus – I like being able to excerpt words, mark up the document with yellow highlighting or add a little note with questions to myself. For that I've been using Adobe Acrobat. Its only problem is that it's expensive – $300. If you're curious you can get a 30-day free trial.

I've recently discovered something called Jaws PDF Editor, which seems to have everything Adobe Acrobat has and a little more, namely the ability to pull out, insert, delete and shuffle pages around. Best of all, it costs only $43. There are undoubtedly many others, equally as good.

RECEIVING FAXES ON YOUR PC

How to receive faxes on your PC. The best service is www.Callwave.com. They give you a phone number. People send faxes to that number. A minute or two later you receive an email containing your fax and cover sheet. They send your fax in PDF format. Callwave charges $3.95 a month to receive and send you an unlimited number of faxes each month. The faxes come in as a PDF attachment to an email.

BEWARE OF KEYLOGGING SOFTWARE

Keylogging software, also caller a keylogger, is a piece of software that sits on your PC and records every keystroke you type – including passwords and email messages. Then it secretly mails out copies to whoever planted the keylogger. Key logging software could be planted by your boss to see if you visit pornographic sites. It could also be installed when you surf a corrupt Web site and/or download applications such as file-sharing programs. Such keylogging software could easily be programmed to send its stolen information to remote databases offshore, where thieves sift through it for passwords, user names and account numbers. Essentially keylogging is automated identify theft via telecommunications. To protect yourself, don't download any stuff from web sites you're not familiar with. And run all the virus protection and adware removal software you can.

UPS WITH AUTOMATIC VOLTAGE REGULATION

When buying an Uninterruptible Power Supply, always buy one with Automatic Voltage Regulation (AVR). It can be the best friend your platform has. AVR eliminates the peaks and valleys of power surges and brown outs, providing you with clean, reliable power.

HOW TO BUY A FLAT SCREEN MONITOR

1. Only buy a digital monitor with DVI (digital video input).

2. 19" monitors are perfect. Especially several of them hooked to one computer.

3. Don't put yourself in a corner and circle screens around you. Better to have the screens in a straight line and angled slightly.

4. if you use a laptop, as I do, put your monitors on a 9 1/2 inch wooden stand. It should be 10 inches deep and at least 70 inches wide. That way you can put three 19" monitors up.

5. Look for monitors with adjustable stands.

6. I run three external 19" monitors with my laptop, including its own 14" screen. I do this with the aid of VillageTronics' PC card VTBook and a Toshiba docking station. The VTBook powers two of the monitors. The docking station powers the third and runs a DVI digital connection (as against a VGA analog connection – which is not as clear).

TIRED OF VOICE MAIL? TRY SIMULSCRIBE

Simulscribe converts your voicemail into emails and sends them directly to your inbox – which can be on your BlackBerry and/or your laptop. There are huge advantages to reading voice mail.

1. You don't have to keep checking your voice mail for the important one you're waiting for.

2. It's easier to read than listen. There are places you can't listen, but can read, like meetings.

3. They attach a wave recording of the voice mail. You can always listen to the message if you think the transcription screwed up.

4. It's great for traveling. Easier than calling the U.S. to find if you have a message, or not.

5. You can email a response to a voice mail.

6. You don't have to write down all the numbers and other information in your voice mail. As an email, it's there and indexable by your PC's search engine. You can find the information easily.

7. It's easier to scroll through emails than it is to scroll through voice mails. To get to an important voice mail, you have to listen to all your messages. Yuch.

My son, Michael, turned me onto Simulscribe. As a test, I called him this morning and left a voice mail with numbers. A minute later, my voice mail came in. The transcription from my weird accent to text was flawless. Simulscribe had even improved my miserable grammar. The message made more sense than the way I had spoken it. No kidding. Try Simulscribe for a week for free. Click here. www.simulscribe.com/products_index.php

TRAVEL WITH USB FLASH MEMORY STICK

I travel with a laptop. When I need something printed, I go to the local Staples or my hotel's business center. To give them my file, I save it to a small USB flash memory stick. I then hand the stick to the people at Staples who stick it into their computer and print my documents or photos. Saves huge time and is much more convenient that loading up printer drivers on my laptop and printing directly.

I'M GOING TO EUROPE. HOW DO I STAY IN TOUCH?

For voice in Europe:

Buy yourself a cheap unlocked GSM phone. Buy Nokia if you can. It's the biggest seller in Europe and thus easiest to get parts for. There are plenty on eBay. Buy cheap prepaid SIM cards in EACH country you visit and use them in only that country. When you visit another country, buy another prepaid wireless GSM card. Don't waste your money on global phones from the U.S. Their per minute charge will Put you into the poor house. Leave a message on your voice mail in the U.S. directing your callers to your European phone number.

When you need to call the U.S., do not use your cell phone. The cheapest way to call back to the U.S. is to use Skype and talk to a computer equipped with Skype in the U.S. That call will be free. You can also use Skype to call out of Skype to a normal phone number. That will cost you a couple of pennies a minute.

If you don't have a broadband connection or your laptop and hence you can't use Skype, go to one of the many web sites that sell prepaid calling cards to the U.S. – e.g. www.CallingCardsPlus.com. Before you use the card to dial from your hotel room, check with the hotel if they charge anything (or what they charge) for calling in essence a European toll-free 800 number.

For data and emails in Europe:

For quick messages, SMS works best. You can SMS (short message service) from a phone in the U.s. or from a PC. For your emails and surfing the Internet, the easiest is to find an Internet cafe – they're ubiquitous in Europe – and access your email via the Web. You can also take your laptop to Europe. I do. Most Internet cafes will hook you up for a few dollars. Some have an RJ-45 Ethernet cable available specifically for laptops. Others you'll need to unplug such cable from one of their networked PCs and plug it into your laptop. Assure them that you know what you're doing. Hotels in Europe are iffy. Sometimes they claim to have Wi-Fi "throughout" the hotel. But it rarely works. You might find it works in the lobby. The best solution remains an Internet cafe – or your friends. Often they have broadband access at their homes and will let you use it. Be careful, some European DSL providers charge you by megabytes transmitted. Some even charge by time.

If you use dial-up in Europe, never leave your computer plugged in and powered on. My friend did that once for three days and got hit with a $4,000 phone bill – many times the cost of his hotel bill.

"MY COMPUTER IS SLOWING"

As computer guru to all my friends, I hear "slowing computer" regularly. There are several solutions. These apply only if you're running Windows.

1. Add more memory. My laptops have three gigabytes of memory.
2. Back up all your data instantly. Best, back it up two places.
3. Run an up-to-date anti-virus program.
4. Run an anti-spyware program. Be careful not to erase needed cookies.
5. Delete all the files in the subfolders temp and prefetch.
6. Uninstall all software you're no longer using, especially any you recently installed. You do this by going to Control Panel/Add or Remove Software. Be ruthless. If you haven't use the software in the past three months, remove it.
7. Run MSCONFIG and stop programs starting which you don't need on startup. By unclicking the software in MSCONFIG, you are not removing it. You can still use the software any time. What you are doing is removing the little programs that run constantly and put major burden on your PC – e.g. software that checks for updates, software that loads printers you're not using, software that reminds you to register your software, software that loads wireless drivers you're not using, etc.
8. Run Disk Defragmenter. Right click on My Computer. Right click on Drive C: and go to Tools.
9. When all this fails to speed up your computer (as it may), it's time to start from scratch. Format your hard disk. Install Windows and your various software programs from scratch. When you do, look skeptically at all your software. Ask yourself, "Do I really need this program?" If not, don't install it. The more software you install, the more your computer will struggle and eventually slow.

One BIG trick: When you first buy your computer and get it set up with the minimum number of programs you need, make a clone of its hard drive. And put the clone away. That way when your working hard drive starts slowing down, you can move all your working data over to the clone. And then clone the slow disk with the fast one.

The BIG key is not to load needless software. What I do is to have a second "play" machine. Anytime I see some software I might like, I load it up on my play machine and play with it on that machine. If I really, really like it, I load it up on my good machine. But 90% of the stuff that looks good based on its marketing hype turns out not to be very useful – what I might call "clutterware."

If all this fails, get a Mac. More and more IT professionals are.

HIGH FEE FUNDS

Research shows that mutual funds with high fees do not outperform those with low fees. Personally, I prefer index funds from Vanguard for painless investing – especially if you have a day job.

TRAVELING WITH A LAPTOP

First, slow speed dial-up – perish the thought:

You'll need an ISP provider to allow you to dial into the Internet. But you'll need one who serves the places you're going to. You're looking to make a local call. If you're forced to make a long distance, life can become expensive – especially if you're calling from a hotel. If you sweet-talk them, the hotel will allow you to use the ISP it uses. This way you typically only make a local call. I've used this technique in hotels in the the Austrian and Swiss alps.

Don't leave your laptop plugged into a hotel phone line. A friend left his laptop plugged in for his entire four day visit to France. He came back to a $4,000 phone bill. All European hotels bill phone calls by time and distance. Some make more on phone calls than they do on renting the rooms. (At least that's my theory.)

Second, a broadband connection. There are two types: corded and wireless. For corded, you can plug directly into an Ethernet connection. You'll need to carry an RJ-45 male-to-male cord. For wireless (such as at Starbucks or at some airports), you'll need a Wi-Fi card. Most laptops these days come with one built-in. Typically you plug in, turn on and start your Internet Explorer. You'll then be taken to a page which asks for money or for your ID. Most places charge $10 a day, which seems to me reasonable if you're going to get some serious work done. But there are many places with "free" Wi-Fi – i.e. someone's unprotected network. Turn on. Search. You'll be surprised. There are even handheld gadgets out there whose only job is to ferret out available networks. Your laptop can do the same job, but a little less conveniently. Recently, I was in a midtown Manhattan lawyer conference room. I asked the boss lawyer did he have wireless? He said he didn't. When I turned on my laptop, it detected six available Wi-Fi networks. I latched onto one, picked up and sent my email. Some "public" Wi-Fi have proxy servers or other gadgetry that won't let you send SMTP email – in which case it's easier to use your own SMTP server. Key: In Outlook you need to check the box that says "My outgoing server (SMTP) requires authentication" and the box that says "Log on using Secure Password Authentication (SPA)."

Verizon runs a service calls Broadband Access. It's not cheap – $60 a month. And it's not as fast as DSL or a cable modem. But you can use it in your car, on trains, in offices that won't let you in on their LAN. And it works splendidly in most major cities around the U.S. and a heck of a lot of smaller cities. I'm very impressed with the service. It doesn't work in Europe. But I believe that Sprint or Cingular's portable broadband service might work in Europe.

WHAT IS RSS AND WHY IT'S USEFUL

RSS stands for Real Simple Syndication. You may also hear the term Web Feed, XML, RDF or ATOM. You use RSS to save you the time of revisiting web sites constantly. Basically you go to your favorite web sites, and subscribe to their RSS feed. From then on, you receive regular RSS "emails." Here's how:

1. Get a News Aggregator, also called a News Reader. Here are some.
http://reader.google.com/
http://www.sharpreader.net/
http://www.pluck.com/
http://www.feedreader.com/
2. Subscribe to Content
Go to your favorite web sites and look for an orange button that says RSS. Drag the orange RSS button into your News Reader. Drag the URL of the RSS feed into your News Reader.
3. That's it. Now you have personalized "E-Newspaper."

HOW TO GET BROADBAND ACCESS TO THE INTERNET

There are more ways than ever to get access to the Internet. In order of speed, the fastest is Verizon's FiOS fiber optic service. The second fastest is a cable modem service. The third fastest is a phone company's DSL service. The fourth fastest is the various cell phone providers 3G broadband access service. I prefer Verizon's service because of its broad geographic coverage. If you primarily use a laptop, and do some traveling, it makes sense to simply get one of the broadband access service and use it at home, at your office and traveling.

THE GEEK SQUAD IS A GOOD IDEA

Having trouble with your PC or laptop? Call The Geek Squad. They're expensive, but they usually do the job. They're part of Best Buy. Most Best Buy stores have members of its Geek Squad available for asking for questions. www.geeksquad.com/ or 1-800-433-5778. Apple has its "geniuses" who work in its stores. They will also answer questions and help you fix problems with your Apple equipment.

TIPS ON TRAVELING WITH A LAPTOP

Only stay at hotels with broadband Internet service – wired or wireless. Call in advance to make sure the service is working.

Carry one of the broadband access cards for your laptop. I use Verizon.

Check your POP3s and your SMTPs to make sure they work outside of your present broadband service. See the previous tip. Better yet, if you pay money to a hotel for broadband service, ask for its preferred SMTP server.

Take RJ45 and RJ11 extension cords.

Take a small triple adapter for AC power.

Take your cell phone charger and laptop power supplies.

Take spare batteries for your laptop, your cell phone and your digital camera.

Take a plane/car power adapter – the one that works in cigarette lighters. Make sure your adapter thing puts out sufficient power to drive your laptop. Do not plug your power adapter into the adapter on a private aircraft. Though the adapter may look the same as the 12 volt one on your car, you'll find the private aircraft one may be 28 volts. Plugging it in will destroy your adapter. Trust me. It's happened to me.

Take plug adapters for foreign countries if you're traveling overseas. Radio Shack has a selection for around $10.

Take a USB hub and a USB flash memory stick – for getting business centers, Staples, Kinkos and the like to print your documents.

DEALING WITH MICROSOFT ACTIVATION CODES

Each time you load Windows (now called) Vista and Office, Microsoft demands that it "activate" your installation – even though you have entered the correct product key. Activation means that Microsoft tells your computer it's OK to run the software. If you have installed your software (and thus entered the product key more times than Microsoft thinks is kosher, Microsoft will, over the Internet, deny your use of the software. There are two keys to solving this problem:

First, make sure you keep the product keys for Windows (Vista) and Office in a safe place. That often means copying them from some weird place – like the bottom of your laptop, or some scrap of paper that comes with all the other junk when you buy your PC. You can use your friends numbers – for example, if you've lost yours.

Second, try activation by the Internet first. If that doesn't work, call Microsoft on the phone and explain your plight, which is probably caused by your having to reformat your hard and reload all your software in order to clean your hard disk of sundry viruses and malware.

SHOULD WE UPGRADE OUR SOFTWARE?

There is an old IT expression: "If it works, don't mess with it." Software vendors constantly pressure you to upgrade your installation. They do this to earn fees. Most upgrades tend to be bloatware. They add more features than you need. The extra features clutter the software, slow it down and often make it more unreliable. I am a great believer in having a "test" machine and running new and upgraded software on it. This gives me the opportunity to see if I really want those new features and, more importantly, if the software is reliable. I certainly would never upgrade any software I like and am using without testing it. I would also not upgrade several systems simultaneously – like over a long weekend. That's only courting disaster.

THE CARE AND FEEDING OF OUR COMPUTERS

First, the computers. We expect our computers and our Internet to turn on and work as reliably as a light. Sadly, No. Most of us haven't a clue how this computer magic works. When it fails – as it will – we're distraught. Our files are gone. Our Outlook email lifeline is dead, etc.

I hate to suggest this. But you can't leave your computer to your 15-year old pimply computer whiz. You must understand enough to back up, troubleshoot and restore your computer. You must also understand how your network is connected, to where and how.

My basic strategy is:

Have all my working files in a folder called c:allharry. That often means telling software to save its files in a new place. You can do with Outlook by simply copying the file Outlook. pst to a new place, e.g. c:allharryemail. Once you've done that (with Outlook closed), next time you start Outlook, it will say it can't find your files. Show it where its new location is.

Daily backup of all my working files. I use FileSync to back my files onto external hard drives. These days external hard drives are cheap. For $150 you can buy an excellent Iomega or Fantom.

Clone my PC's hard disk before I update, change or add software. I always assume that new software I load will barf and screw everything up. I use my backup to revert to the old reliable disk. Windows XP has a "go back" system. I don't trust it.

I always have two connections to the Internet – a DSL or cable modem (depending on where I am) and Verizon's Broadband Access card.

I have two identical laptops. Should one crash, I can pop the original or the replacement hard disk in and keep working.

There are some things you can't backup or duplicate. That includes the idiot who puts up your web site and runs your email. (Hopefully, you're not stupid enough to do that yourself.) The best you can do is to find a reliable provider (if there is such a thing) and pray. I like ICDSoft.com.

HOW TO PROTECT YOURSELF AT WIRELESS HOT SPOTS

This is good advice, which Computerworld magazine published on January 5, 2007 and which I've excerpted (and fixed) here:

Wi-Fi hot spots in airports, restaurants, cafes and even downtown locations have turned Internet access into an always-on, ubiquitous experience. Unfortunately, that also means always-on, ubiquitous security risks. Connecting to a hot spot can be an open invitation to danger. Hot spots are public, open networks that practically invite hacking and snooping. They use unencrypted, insecure connections, but most people treat them as if they are secure private networks. This could allow anyone nearby to capture your packets and snoop on everything you do when online, including stealing passwords and private information. In addition, it could also allow an intruder to break into your PC without your knowledge.

Disable ad hoc mode. You don't need a hot spot or wireless router to create or connect to a wireless network. You can also create one using ad hoc mode, in which you directly connect wirelessly to another nearby PC. If your PC is set to run in ad hoc mode, someone nearby could establish an ad hoc connection to your PC without you knowing about it. They could wreak havoc on your PC and steal your files and personal information. The fix is simple: Turn off ad hoc mode. Normally it's not enabled, but it's possible that it's turned on without your knowledge. To turn it off in Windows XP. When you're connected to a wireless network do the following:

1. Right-click the wireless icon in the System Tray.
2. Choose Status.
3. Click Properties
4. Select the Wireless Networks tab.
5. Select your current network connection.
6. Click Properties, then click the Association tab.
7. Uncheck the box next to "This is a computer-to-computer (ad hoc) network."
8. Click OK, and keep clicking OK until the dialog boxes disappear.

In Windows Vista, there's no need to do this, because you have to take manual steps in order to connect to an ad hoc network; there's no setting to leave it turned on by default.

Turn off file sharing: Depending on the network you use at work or at home, you may use file sharing to make it easier to share files, folders and resources. That's great for when you're on a secure network. But when you're at a hot spot, it's like hanging out a sign saying, "Come on in; take whatever you want."

1. Right click on My Computer
2. Left click on Explore.
3. Right click your C: drive.
4. Click on the Sharing and Security.
4. Unclick the obvious boxes.

TEACH YOURSELF NEGOTIATION

Negotiation is one of the most valuable skills you'll ever learn. You can attend courses on negotiation. You can read books on it. But the best way to learn it is to practice it. The key element in negotiation is information. You must know your own strength. You got to know where the other guy is coming from. And you have to know what's out there in the real world – comparables, as they call them in the real estate business.

HOW TO BOOST YOUR CELL PHONE

Let's say your cell phone service is not good at your home. Here are some ways to solve the problem.

1. For incoming calls, use your cell phone's call forwarding to forward your cell calls to your land line. This assumes you have actually a landline.

2. For incoming calls, give people your VOIP phone number. You can have your VOIP provider ring your incoming calls to both your cell phone number and your land line number.

3. Use a passive antenna. There's one for $34.95 passive antenna called Freedom Antenna. You have to have an antenna port on your cell phone which you can plug into. Nokia and BlackBerry owners are out of luck.

4. Use a wired indoor antenna booster. This requires having a cable connect from the phone to an indoor antenna. For $299 there's the Spotwave Z1900 amplified antenna. It works only with 1900 MHz cell services, which include Sprint, T-Mobile and some Cingular areas. You can check if you're covered by punching your ZIP into Spotwave's Web site, www.spotwave.com.

There's also the $399 zBoost from Wi-Ex. This thing works at both 800 Mhz and

1900 MHz and thus works for most US cellphone companies, including Verizon, but not Nextel. www.Wi-Ex.com. Don't expect magic from all three antennas. You will get an improvement. If your house gets lousy reception, these gadgets are for you. For the amplified boosters to work, you should be speaking from a cell phone close to their indoor antennas.

CELLULAR COST SAVINGS TIPS

My friend John Perri runs a firm called ComView – the communications billing experts. He tells me of horrendous mistakes in cell phone billing. I asked him for his top tips on getting your (personal and corporate) cell phone bills under control:

Consumer + Monitor your minutes. Minutes should be at 70% or more of your plan minutes. If usage is consistently below 70%, change your plan. If you continually go over the purchased minutes, upgrade to a plan with more minutes.

Look for non-listed plans. When deciding on a new plan or carrier, ask about plans that are not listed. Also ask about new plans, equipment purchase discounts and promotions due to be released in the next 30 days.

Look at your average minute cost: Divide the total cost of the plan (excluding taxes) by the minutes used to calculate the cost per minute. If the cost per minute is above 11 cents, search for a new plan or service.

When purchasing data plans, consider the type of usage. If your usage is mostly e-mail, (low bandwidth) a measured plan may be better.

Don't believe the salesman. When buying new phones/plans, don't take the salesperson's word. They are usually on commission and may forget to mention the more cost effective plans.

Check organizations you belong to including your employer. Some have affinity cellular discounts.

If you regularly call other countries, ask your carrier for discounts to those countries. You can get lower rates by paying a few dollars for an International plan.

Buy phone cards to use when your minutes exceed your plan. If you have a basic plan, use a cellular calling card to place long distance calls. Cards can be used for US and international calls.

Do not buy insurance unless you're a total idiot – and have a tendency to break or lose your phone.

Get free options: Ask for as many free options as are available: mobile-to-mobile, nights and weekends, toll-free, caller ID, voicemail, etc.

Check you're not paying extra for detail billing when you could print the information you require via the Internet?

When buying new equipment, ask your carrier for a loyalty discount.

Carriers will negotiate termination fees.

Carriers will negotiate equipment prices.

Company: + Make sure all discounts are applied to all plans.

If pooling, make sure total usage is at 90% of total pooling minutes. Optimize plans for users with lower usage.

Compare old plans against new. You may be able to gain minutes by updating plans.

Make sure you are not charged for services that carriers now offer for free such as mobile-to-mobile.

Carriers will negotiate almost anything these days. Nothing is off the table even if the rep says different.

Make sure individual users' annual expirations are coterminous with your master agreement.

JUMP COMMERCIALS ON YOUR TIVO AND YOUR PVR

You can program a button on your TiVo to fast forward by 30 seconds. This is most useful in jumping over commercials. Go here: www.technologyinvestor.com/login/2004/Dec16-05.php

DEBIT PHONE CARDS ARE NICE

For cheap long distance and international calling, phone cards are often the best. Not only are they cheap on a per minute basis, but you get to skip all of the add-on taxes that show up on your regular monthly phone bill. The only tax you pay is the sales tax when you purchase the card. Some places like Wal-mart and Costco sell cheap cards. On the Internet, you'll find them at www.CallingCards.com and www.NobelCom.com. Because these cards use 800 toll-free numbers to begin their calling, you can use them with cell phones and make international calls much much cheaper than you can on your cell phone – if you're

even allowed to make calls on your cell phone.

DO YOUR PRICE HOMEWORK

Some telecommunications contracts provide specific rates for every type of telephone call (intra-state, interstate, etc.). Other contracts reference published tariffs or service guides and say that the rates will be discounted a certain percentage from the published rates (which are never provided as part of the contract.) Always get specific rates quoted for every type of type of call you will make written into your contract. These rates are always negotiable. Put together your requirements (number of minutes usage for each type of call) and get quotes from three separate service providers. You may get lower rates from a lesser known company, but check references to be sure the service is good.

VoIP CALLING ON THE INTERNET MAKES SOME SENSE

I pay $20 a month for unlimited local, long distance and much of my international calling. For the most part, quality is OK. But it can sound pretty awful on some international calls and for some reason it doesn't like certain cell phones. So the moral of this story is always keep a landline or cell phone handy – just in case. And make sure those landlines are on cheap long distance and international plans.

HOW TO STOP PROGRAMS FROM STARTING IN WINDOWS XP

Run msconfig. That program will show you which programs loaded at startup. You can selectively turn off programs. If you're running Windows 2000, you can find a program called msconfig.com on www.pcmag.com. The more programs that load when you start your machine, the slower your machine will be. You can turn off all the programs listed under the StartUp tab – except for ones you recognize and want.

GET RID OF LOAD COILS FOR BETTER DSL SERVICE

Load coils are also known as impedance matching transformers. Load coils are used by the telephone companies on long analog POTS (Plain Old Telephone Service) lines to filter out frequencies above 4 kHz, using the energy of the higher frequency elements of the signal to improve the quality of the lower frequencies in the 4 kHz voice range. Load coils are great for analog voice grade local loops, but must be removed for digital circuits to function. Load coils must be removed for DSL loops, as the frequencies required are well above 4 kHz. Today many phone companies offer broadband service, but often tell their customers that they can't get the service because "you live too far from the telephone company's office." Tell the company to remove the loading coils and any bridging taps on your local loop and it will work. If they balk, offer to pay for the removal of the loading coils. If that doesn't work, offer to buy commercial ADSL service.

THE INTERNET HAS THE BEST PRICES

Before you buy anything, check www.eBay.com, www.PriceGrabber.com, www.PriceWatch.com, www.google.com and, of course, http://Froogle.Google.com. There's always a cheaper price, somewhere.

DON'T UPGRADE TO A NEW OPERATING SYSTEM

Don't upgrade your existing computer to a new operating system. Start from scratch. Reformat. Install from scratch. Before you do any of this, make sure you have originals of all the software you'll need and all the drivers you'll need for all the hardware in your PC, and most important, copies of all your data – at least two copies.

NEVER GIVE YOUR SOCIAL SECURITY NUMBER OUT

Never give your social security number to anyone. Never. Ever. Your social security number is the key to your identity, which can then be stolen. My son's was. It was painful and time-consuming to fix the mess. They had him buying a house and taking a mortgage on that house. He was only 18 at the time. The bank must have lost all its marbles to give "him" the money.

BEFORE YOU BUY A HOUSE

Get a house inspector to inspect it thoroughly. Make sure he checks everything – from the dishwasher to the opening and closing of the windows. Then he should give you a report and you should insist that the seller correct all the faults before you move in. This will save you a lot of time, money and aggravation at finding things not working.

LOOK FOR YOUR TV's SCHEDULE

Try www.excite.com/tv/grid.jsp. Another way is to sign up for http://my.yahoo.com.

YOUR MOTHER IS PRECIOUS

Never post your recent family history on the web or in an unprotected PC system. Many systems use your mother's maiden name. Many people post their family history on geneal-ogy websites. Most of these are public data or open to anyone willing to join the service and steal your information. Identity theft is a huge and growing business.

SHREDDERS DO MAKE SENSE

Anything that has your name, address, or any kind of account information on it – be it a bill, or even a magazine subscription – shred it with a good heavy duty cross-cut shredder before disposing of it. Additionally – for those with fireplaces, this makes very good kindling to further completely dispose of your shreds. Also – every new kitchen trash bag you put into the kitchen trash can, toss into the bottom whatever shreddings you have and it will absorb any liquids you will throw away, reducing any leaks the trash bag develops during use. This has the added advantage of further obscuring any information on the clippings a dumpster diver might try to piece together on you or your family.

BLUE SCREEN OF DEATH

BSOD stands for blue screen of death, an affectionate name for the screen displayed when Microsoft Windows encounters an error so serious that the operating system cannot con-tinue to run. Windows may also display information about the failure and may perform a memory dump and an automatic system restart. The Macintosh equivalent is a blank screen with a text box containing a graphic of a bomb with a lit fuse.

There are two typical causes of a BSOD – a bad driver or bad memory. You get a new driver from the software company whose driver was bad. You get new memory from your memory company. The problem is figuring out which driver and which memory. Using a program call Alexander SPK from www.Alexander.com may help. Memory is a different problem. You may never figure out which memory is bad. You do it by trial and error. Intermittent BSODs without being traced to a specific program are likely memory problems. Don't bother trying to figure out which band of memory is bad. You never will. (I've been there, done that.) Get them all replaced, asap.

MUST-HAVE PLUGINS FOR FIREFOX

The Mozilla firefox is a far better browser than Microsoft's Internet Explorer. Trust me on this one. Pick yourself up a free copy, www.mozilla.com/en-US/firefox/. Firefox is open software, which means that users (like you and me) write add-ons to give firefox those few extra features we want and we make those add-ons free for the rest of us. Here are my favorite add-ons:

Firefox showcase. Hit F12 and you'll little pictures they're all the web sites you cur-rently have open. i typically open 14 web sites each morning – including the wall street journal, businessweek, forbes, the new york times business, the economist, yahoo! finance here and in australia, etc. click on one and you're right there. huge time saver. download here: https://addons.mozilla.org/firefox/1810/

session manager saves and restores the state of all windows and tabs - either when you want it automatically at startup or after crashes. it lets you reopen (accidentally) closed windows and tabs. if you're afraid of losing data while browsing – this extension allows you to relax…. download here: https://addons.mozilla.org/firefox/2324/

ie tab:some web sites insist on internet explorer. ie tab runs the internet explorer engine. ie tab is useful. you can easily check what your web site looks in firefox and internet explorer – often very different. download here. https://addons.mozilla.org/firefox/1419/

MeasureIt. This tool is useful if you're like me rushing to get out a web site. all it does is put a ruler around whatever you want and give you the measurements in pixels: https://addons.mozilla.org/firefox/539/

InFormEnter. The web is replete with forms to enter. firefox, internet explorer and cook-ies try to help you, but usually half fail. i've tried roboform and ewallet. too complicated. the easiest i've found is informenter. it doesn't fill forms out automatically. shucks. but you can choose an answer from a list. download here. https://addons.mozilla.org/firefox/673/

THE EASIEST PASSWORD

The worst thing you can do is to have the same userid and password for everything – your bank, your credit card, etc. but having the same one sure saves time. For your userid, I recommend a your first and last name – e.g. JohnDoe. And for your password, you need a combination of alpha and numbers – something like handsome321. The reason for this is that many sites won't accept something simple, like Handsome. They insist on one or two numbers thrown in. So, create one like Handsome321 that satisfies everyone. This way you can remember it easily.

HARD DRIVES DIE

There are only two things on PCs these days which move and therefore wear out – hard drives and keyboards. I replace each every six months. Both are cheap. Their cost is cheap insurance.

CHECK ADD/REMOVE PROGRAMS

Sometimes, even if you're running antivirus and firewall programs, malicious code can still find its way onto your machine. It's a good idea to check your Add/Remove Programs (in the Control Panel window) regularly, which shows most of your installed applications. If you see anything you don't use or need, simply click on Remove.

YOUR CURSOR IS DRIFTING

Some laptops have pointing sticks in the middle of their keyboards. I love these gadgets. Sometimes the cursor moves all by itself. This drifting is a function of adaptive software in the pointing stick which checks occasionally to make sure the strain gauges in the trackpoint are centered. If it happens to check while you are moving the cursor, it adjusts itself to take this as non-movement. When you stop moving the cursor, the pointing stick thinks the cur-sor is still when it is in fact moving the opposite direction. The solution is to relax and let it go until it checks centering again (a few seconds). If you catch it early and relax, it corrects very quickly. If you keep fighting it, it will just get worse. Think of it as zen training.

CALLING INTERNATIONAL

Never call international from a cell phone. Never call international from a phone that doesn't have an international "plan." Item: "Harry, last summer I made an 18-minute call to Hong Kong on my home landline phone via MCI. The bill? $85. MCI said that was the going rate since I did not have (or know about) their international calling plan. I don't usually make international calls. Their solution? Sign up for the plan, make the call(s), and then cancel." The cheapest way of calling international is with a a broadband VoIP line. Pre-paid calling cards are the second cheapest.

DON'T SPRAY LAPTOP SCREENS

Never spray anything directly onto a laptop screen. Never use Windex. Most manufacturers recommend water. Use a tinge of detergent if your screen is filthy. We use a spray called "Invisible Glass", sprayed onto a non-abrasive cloth. You can also use eyeglass cleaner. Never spray it onto the screen as it can drip down and short out the display. Displays are the most expensive part of any laptop.

WHEN MOVING, ALWAYS MOVE

When moving, move all of your subscriptions and catalogs with you and then cancel the ones you don't want. Old accounts and account cancellation notices often contain informa-tion someone could take illegal advantage of.

WHEN GETTING YOUR CAR SERVICED

When getting your car serviced, remove all remote control devices and telephone devices before releasing the car. The easiest way to penetrate someone's home security is to clone the security code of a garage door remote.

COVER YOUR CAR'S SERIAL NUMBER

There's talk that thieves can copy your car's VIN number which is located on top of the dashboard and take that number to a car dealer and have spare keys made to your car. I don't know if this is true. But it makes sense to be careful.

BE WARY OF SURFING WITH YOUR CELL PHONE

Cellular providers are encouraging us to surf the net using the data transmission abilities on their cell phones. Beware: such transmissions can be very very expensive. Some carriers charge by the thousand bytes. Others charge by time. Others charge hefty for providing the data transmission in areas or in countries they don't service – i.e. "roaming." By contrast, Wi-Fi is much more economical and a lot faster.

WANT TO SEND A MESSAGE TO A CELL PHONE?

Email your message to yourcellphonenumber@teleflip.com. If you want to send me a message, send it to 19172971818@teleflip.com.

HOW TO LENGTHEN YOUR LAPTOP'S BATTERY LIFE

1. Turn down the LCD brightness. The screen is your laptop's biggest consumer of power.
2. Buy a laptop with a Pentium M (Mobile) chip – sometime called Centrino (if it also has Intel Wi-Fi).
3. Use built-in power management.
4. Turn off wireless and wired Ethernet when not in use.
5. Set screen blanking to 1 or 2 minutes.
6. Add RAM memory to maximum capacity. This saves on disk access.
7. Close unused programs.
8. Fully drain the internal batteries at least once a month.
9. Remove unused PC Cards.
10. Remove second hard drives.
11. Removed unused notebook-powered USB devices.
12. Don't watch DVDs, and don't play graphics-heavy games on battery, without plugging into AC.

WHEN REMOVING SOFTWARE FROM YOUR PC

After you've removed software from your PC, make sure you re-start your computer before installing any new software.

WHEN INSTALLING NEW HARDWARE

When installing older hardware on your PC – e.g. a scanner, a printer – don't plug the device in. First, install the software. Then reboot. Then plug your equipment in. Then turn your PC on. If your computer doesn't immediately recognize your new hardware, go to Control Panel and add new hardware. If your equipment is new, you may try plugging it. Perhaps, with luck, your PC might recognize the new equipment and find the software you've already installed. Installing new hardware is often hit or miss. Sometimes you have to keep installing the software until it finally works.

CLOSE DOWN CREDIT CARDS

A friend has a habit of, every few years, closing down all his credit card accounts and starting over. This creates a security gap for you and causes a clean break that is easily tracked. There is no earthly way to stop fraud in its tracks more effectively than occasionally dumping all your cards and starting over clean. This is very much like "cold booting" your credit security. This is also a great way to weed out services and subscriptions you no longer use that might still have your credit card number on file.

SHAREWARE AND FREEWARE DO MAKE SENSE

There are hundreds of awesome and free software applications out there that save dollars. Try www.download.com, www.pcmag.com and www.nonags.com. Install free software on a spare PC before you put it on your working machine. Most free software is fine, but, unless you use it regularly, it's a waste of space and resources on your PC.

BUY IN THE QUARTER'S LAST WEEK

The cheapest time to buy anything is the last week of each quarter. Companies are desperate to close any deal they can to make that quarter look good to Wall Street.

PRICELINE IS GREAT FOR HOTELS

Priceline.com is cheapest for hotels, but you can't name the precise hotel you want to stay at. You choose a hotel by location and by stars (a dubious measure at best). And you have to take what they give you. But you usually get a hotel at a great price. Just make sure you're 100% certain you're going. Cancellation is a pain. And penalties are high. Hotels. com is often easier.

WHEN CONNECTING TO A NETWORK

When connecting your laptop to the Internet, you'll connect with DHCP (most common) OR with specific IP addresses, specific subnets (mostly 255.255.255.0) and a specific default gateway (mostly 10.1.1.1). If you're having trouble connecting, check that you need DHCP or you must have a static IP address.

OPTIONS FOR DIRECTORY ASSISTANCE CALLS

From Jon Giberson at Connections (www.CallAccounting.com):
How are you dealing with directory assistance (DA) 411 costs? When we ask people how they control this expensive service, they give us a variety of answers. Here are the most common:

1. Ignore it and just pay the phone bill. This is the most expensive solution. Local and long distance carriers charge up to $2.99 per call. On top of that, they offer direct connect service that can add a considerable amount to the cost of the call.

2. Use a "free" directory assistance service. These require that you listen to a commercial first - controversial because of unknown advertisers. It's time consuming, too.

3. Block 411 calls in your PBX and force employees to use the Internet. This can be a productivity thief. Internet databases typically come from compiled sources, and can be out of date. In addition, the data itself is not generally standardized and optimized for retrieval. Therefore, it is often likely a user will spend more time than necessary searching for a listing that does not exist. If they can't find the number, they may end up calling 411 by going through the company operator, further increasing unproductive time.

4. Use an alternative DA provider - This is the one we suggest. Third-party providers handle your calls at a savings of 50% or more. Cost and quality of this service varies, based on provider. Costs are determined by the volume and level of service you choose. Other variables that determine cost include: Live operator or automated answer, offshore operators, volume of calls, and whether you use a toll-free number or you pay for the call.

YOU CAN BARGAIN WITH BRAIN SURGEONS

I used to believe you didn't bargain with "brain surgeons." I didn't believe in bargaining with dentists, lawyers, doctors, accountants and consultants. My theory was they'd do a much worse job for me. I was wrong. "Brain surgeons" are professionals. They'll do the best job they can for you – so long as you organize their lower fees up front. These days, most professionals will cut their fees. Or better yet, negotiate a flat fee for your particular job. The idea is to get them off per-hour billing. With that sort of billing, they have no incentive to finish your job.

FIBER OPTIC AND COAXIAL CABLING PROBLEMS

Most fiber optic and coaxial cabling problems can be traced to faulty or contaminated connectors. The longer the connection without connectors, the better signal you'll get.
BAD DRIVERS My computer (a Compaq laptop with a D-Link wireless card) crashes with a blue-screen "serious error" message when I try to use a hot spot outside my house. I keep having to reboot.
That's the classic symptom of badly written driver software that, instead of failing gracefully when it encounters some unexpected situation, short-circuits the entire computer. The only cure for this situation is improved software from the original hardware vendor. Stop by D-Link's Web site and see if any bug fixes are available for you to download.

NATIONAL DO NOT CALL REGISTER:

Call 1-888-382-1222 from the number you wish to register. Or go to www.donotcall.gov. You can register landlines and cell phone numbers.

BIRTHDAY AND CHRISTMAS PRESENTS NOT TO BUY

Don't ever buy anyone technology – unless you feel 100% comfortable that they'll be able to use without your involvement. You are not your brother's computer guru/geek.

HOW TO SET UP A NEW LAPTOP

I'm now working on my 21th laptop. Each one has been faster and better than the previous one. But the length of time it has taken to switch to a new laptop has always stayed the same – about 11 hours. Here's my thinking on this whole process:

First, ask yourself, do you really need a new machine? For what? A teeny bit more speed? What about fixing up your present laptop? Put the most memory into it you can. Change out the word keyboard. Buy yourself a new battery. Replace the hard disk. All these changes will cost you $400 or so. But your old laptop will magically feel new. Then CHKDSK your old machine and defragment its hard disk. It will really feel new.

Check out the new laptop, is there something strange about it you really hate? My Toshiba M1 has a really beautiful keyboard. The new Toshiba M2 has a really horrible keyboard. Why do I need the extra speed if all I'm doing is typing? I'm thinking seriously going back to my old laptop – just because of the keyboard.

In short, do you really need a new laptop? if you decide you do, here are the keys to making the process of switching over less painful:

Start with your old machine:

1. Write down the name of all the software and hardware you want to install on your new machine. You'll find the names of most of it on your old machine. But look seriously at each. Do I really want to install it? Do I ever use it? Remember the less software and hardware you install, the more reliable your new machine will be.

2. Check that you have a pile of all the CDs of all the software you want to install.

3. Write down all the software you want to install and go to the web site of each manufacturer and download to a subdirectory called "downloads" all the drivers you will need.

Now move to the new machine.

1. The first step is to get Ethernet networking working on it. You want to be able to access the Internet. And you want to do be able to do it through a Ethernet hub, to which you've connected your old laptop. Try opening Internet Explorer, if it gets you to your favorite web site, great. If not, you may have to change your IP settings from DHCP to a static IP address. You do by rightclicking My Network and opening Internet Protocol adding your numbers in. You'll need to use a static IP address if you're using a proxy server or a router to get to the Internet.

2. The second step is to get Windows networking working. You want to be able to "see" and access all the files on all the other machines on your network. Using Windows Explorer (right click on Start), can you "see" your old laptop and open its hard drive? You probably won't be able to. That will make you made because it was so easy under previous versions of Windows – until XP. Here's the Harry Newton tip, you now must install Then you need to install NWLink NetBIOS and NWLink IPX/SPX/NetBIOS Compatible Transport Protocol. You do this by rightclicking on My Network, right clicking on the Local Area Connection (the wired one) and installing the NetBIOS protocol. It's all very easy to do. But you won't find any mention of the necessity of doing it in any Microsoft literature. Don't ask me why? I have no clue. Just I know my technique works. Theirs doesn't.

3. Now the next step is a discipline we need to establish. The folder "My Documents" will henceforth be the only folder into which dump our work, our data. Now we can set up as many subfolders under My Documents as we wish. But we will not scatter our work throughout the hard disk. The reason for this is simple: We want to be able to quickly and regularly back up our work to an external hard disk or another computer. There are many programs for doing this. The easiest backup I've found is called FileSync.

4. OK. Now to software. Remember we don't install software we don't need – like RealPlayer or Quicktime. And we'll try to avoid installing "free" software (e.g. Microsoft AntiSpyware) since it often comes with software that sneakily loads on startup and keeps insisting that we buy the upgraded version – which we don't want and which slows our machine down. OK. So now we take our installation disks and install the software we need. When installing the software we need, we'll need to bring over certain files. To wit:

Internet Explorer can export cookies and bookmarks from your old machine. And you can import them into the new IE on your new laptop.

Microsoft Office has a Save My Settings Wizard which will bring over in one file all your preferences and dictionaries.

Outlook has a file called Outlook.pst which contains all your contacts, your inbox and your calendar. You'll need that file. You can also tell Outlook to save it anywhere on your hard disk.

Most other software have settings or ini (i.e. initialization) files. When installing software, I always try to tell the program to put those files in a separate folder under My Documents. Sometimes it works (it works with Outlook). Sometimes it doesn't. If it doesn't, I have to remember to manually copy those files over.

5. You'll need to install your printers. With many simple and new printers, Windows XP will already have drivers. The moment you hook your printer to your PC, Windows XP will recognize the printer and install it. For older machines you'll need the original or updated software. Download it from the manufacturer. For networked printers, i.e. those connected to your Ethernet network, you'll need to install a "local" network port, usually an IP Port, or with some HP printers, you'll to install a jet port. I hate installing HP networked printers. I always have trouble.

6. Write down everything you do, step-by-boring step. That will make it faster next time. If there is a next time.

7. Use MSCONFIG and Windows XP Services (My computer/Manage/Services and Applications/Services) to close down virtually all the software on your machine that starts automatically. Much software that is set up by software makers to start automatically contributes absolutely nothing to your PC – like checking for updates, setting up "quick" starts,

loading stuff you don't want (like Windows Messenger). But it runs all the time, slows your machine down and, worse, makes it more buggy. Clean and simple are best. Take out most of the software that's under the Startup and Services tag by simply unchecking it. If you find you've stopped something you really want, you want always check it and reboot.

FOLDERS TO KILL

Make your Windows PC run faster, close down all programs, go to Start/Run and type %temp%. Delete all the temporary files. Also find the prefetch folder and kill everything in it.

KEEP YOUR ORIGINAL DISKS

You need your original PC computer disks when (1) you're reinstalling software that's broken down and (2) switching to a new, faster computer.

IMPACTED FOOD

You must dental floss after every meal. If you don't, food becomes impacted between your teeth, and then pushed further and further in each time you eat. It's not pleasant and can cause serious gum problems. I know. I go every three months to the dental hygienist for serious below-gum cleaning.

CRITICAL UPDATES FROM MICROSOFT ARE ALWAYS BOGUS

Emails from "Microsoft" and others advising that the accompanying software patch be installed are 100% bogus. The attached "patch" contains a nasty virus which will infect your computer. Avoid opening email attachments you don't expect or are not 100% comfortable with.

BEWARE THE NIGERIAN SCAM

"We have a temporary crisis in our family. Our money is tied up. Please send money to enable us to get it out. When we free the money up, we'll send you your money back plus 50%." Or words to that effect. Be careful. The whole plea is a gigantic scam. It usually comes with a Nigerian or other African address. It's called a 419 Fraud" (Four-One-Nine) after the relevant section of the Criminal Code of Nigeria. Typically the scam promises the recipient thousands of dollars. All you have to do is to send the Nigerians $500 for them to "normalize" your banking relationship with them. I don't know what normalize means.

CHAPSTICK IS GREAT FOR CUTS

ChapStick medicated is best for cuts. Also excellent: New Skin. The best cream for dry hands is No-Crack Super Hand Cream made by the Dumont Company, Lacrosse, Wisconsin. I cannot live without this cream in Winter.

CREDIT CARD SCAM

I received the following note from a friend. I checked with Visa. Visa Security tells it doesn't ask for numbers on stuff that should already be in its files. Here's how the scam works:

"My husband was called on Wednesday from "VISA" and I was called on Thursday from "MasterCard". It worked like this: Person calling says, this is <name> and I'm calling from the Security and Fraud department at VISA. My Badge number is 12460. Your card has been flagged for an unusual purchase pattern, and I'm calling to verify. This would be on your VISA card issued by <namebank>. Did you purchase an Anti-Telemarketing Device for $497.99 from a marketing company based in Arizona?

When you say "No". The caller continues with, "then we will be issuing a credit to your account. This is a company we have been watching and the charges range from $297 to $497, just under the $500 purchase pattern that flags most cards. Before your next statement, the credit will be sent to (gives you your address), is that correct?" You say "yes". Caller continues..."I will be starting a fraud investigation.

If you have any questions, you should call the 800 number listed on your card 1-800-VISA and ask for Security. You will need to refer to this Control "#". Then gives you a 6 digit number. "Do you need me to read it again?" Caller then "needs to verify you are in possession of your card. Turn card over. There are 7 numbers; first 4 are your card number, the next 3 are the security numbers that verify you are in possession of the card. These are the numbers you use to make Internet purchases to prove you have the card. Read me the three numbers".

"Then the person says "That is correct. I just needed to verify that the card has not been lost or stolen, and that you still have your card. Do you have any other questions? Don't hesitate to call back if you do.""

"You actually say very little, and they never ask for or tell you your credit card number. But after we were called on Wednesday, we called back within 20 minutes to ask a question. Are we glad we did! The REAL VISA security department told us it was a scam and in the last 15 minutes a new purchase of $497.99 WAS put on our card. Long story made short, We made a real fraud report and closed the VISA card and they are reissuing us a new number. What the scam wants is the 3-digit PIN number. By the time you get your statement, you think the credit is coming, and then it's harder to actually file a fraud report. The real VISA officials reinforced that they will never ask for anything on the card that they already know. What makes this more remarkable is that on Thursday, I got a call from "Jason Richardson of MasterCard" with a word-for-word repeat of the VISA scam. This time I didn't let him finish. I hung up. We filed a police report, as instructed by VISA. The police said they are taking several of these reports daily, and to tell friends, relatives and coworkers."

RUN MULTIPLE MONITORS ON YOUR LAPTOP OR PC

My laptop is now running four screens, three external and the internal one. I can drag windows from one screen to another. I can keep a screen to watch incoming emails. I have a screen to watch my stock portfolio. And I have two screens for doing my work, like researching and writing this dictionary. Windows XP on a laptop lets you run two screens – the internal monitor and an external monitor. A product called VTBook from VillageTronics allows you to run two more. For more information see www.VTBook.com and www.realtimesoft.com/multimon.

HOW TO MAKE CELL PHONE CALLS IN EUROPE

There are many ways of making and receiving cell phone calls in Europe or Asia. The cheapest is to rent a phone or a SIM card for a phone (more about that in a moment) in the country you'll be staying in. There are so many different charges for cell phone calls that it's impossible to give generic advice, except:

1. Always ask the price. You can easily see a five for one difference. The cheapest call can be one-fifth the most expensive. The BIG price difference is what it costs to call across borders, especially back to the U.S.

2. Pre-paid calling cards or pre-paid SIM cards tend to be much cheaper. In other words, paying for a bundle of minutes before you use them tends to be much cheaper than paying for the minutes AFTER you've used them.

3. Paying for them in the country you'll be using them tends to be cheaper than buying the minutes in the United States or in another country.

4. If you have a GSM phone that works on overseas frequencies (typically called a dual or triple mode phone), take it with you on trip. Sometimes, it's cheaper to simply buy or rent a prepaid SIM card overseas. Example: my friend wrote me "I use Sprint and they were going to charge me $6.00 a minute to call back to the U.S. Just by chance the phone did not work, so I went into a local cell phone store in Prague and they took out the Sprint "smart chip" (i.e. the SIM card) from the phone and inserted their own. And for about $40 prepaid I got to make the calls to the U.S. at 35 cents a minute (and no charge at all for incoming calls – Sprint was going to charge $1.50 a minute for incoming). They also gave me a telephone number to use while I was in Europe (actually gave me a choice of numbers!)"

Important to know: Cell phones in Europe and most of the world outside the U.S. typically work on GSM. Basic to all GSM phones is a SIM card – a small metal card that slides into a cell phone and tells it who you are and to whom the calls are to be billed. You can place your SIM card into any GSM phone and it will work as yours. You can also place another SIM into your phone and if it's working on the right frequencies, it will make and receive calls. Do not be confused, however. There are several carriers in the United States who bill themselves as "GSM." They are, but some work on frequencies different to overseas. So their "GSM" phones won't work overseas.

DON'T EVER DIAL AREA CODES 809, 284 AND 876

Don't respond to emails, phone calls, or web pages which tell you to call an "809" phone number. Below is a scam that can cost you $2400. It's difficult to avoid unless you are aware of it. Here's how it works:

You will receive a message on your answering machine or your pager, which asks you to call a number beginning with area code 809. The reason you're asked to call varies. It can be to receive information about a family member who has been ill, to tell you someone has been arrested, died, to let you know you have won a wonderful prize, etc. In each case, you are told to call the 809 number right away. Since there are so many new area codes these days, people unknowingly return these calls.

If you call from the US, you may be charged $2425 per-minute. You'll probably get a long recorded message. They will try to keep you on the phone as long as possible to increase the charges. Unfortunately, when you get your phone bill, you'll often be charged more than $24,100.00.

Here's why it works: The 809 area code is located in the British Virgin Islands (The Bahamas). The 809 area code can be used as a "pay-per-call" number, similar to 900 numbers in the US. Since 809 is not in the US, it is not covered by U.S. regulations of 900 numbers, which require that you be notified and warned of charges and rates involved when you call a pay-per-call" number. There is also no requirement that the company provide a time period during which you may terminate the call without being charged. Further, whereas many U.S. homes that have 900 number blocking to avoid these kinds of charges, do not work in preventing calls to the 809 area code.

We recommend that no matter how you get the message, if you are asked to call a number with an 809 area code that you don't recognize just disregard the message. Be wary of email or calls asking you to call an 809 area code number. It's important to prevent becoming a victim of this scam, since trying to fight the charges afterwards can become a real nightmare. That's because you did actually make the call. If you complain, both your local phone company and your long distance carrier will not want to get involved and will most likely tell you that they are simply providing the billing for the foreign company. You'll end up dealing with a foreign company that argues they have done nothing wrong.

PC MAINTENANCE TIPS

Courtesy PC World Magazine:
Every day:
 - Update your virus definitions.
 - Do an incremental backup of files that have changed since last time.
 - Reboot when programs crash.
Every week:
 - Perform a full virus scan
 - Do a complete backup.
 - Update your Windows operating system.
 - Run an spyware and adware removal program. (I use Lavasoft's Ad Aware and SpyBot's Search and Destroy.)
Every month:
 - Update your software and programs. Use www.VersionTracker.com
 - Check for new drivers.
 - Use a one-click utility checkup program.
Every year:
 - Vacuum and clean out your PC case.
 - Make a full backup. Then use your restore CD to return your system to its original state.
 - Do a full diagnostic check.

BEST CAR RENTAL TIPS.

1. Book early. There are typically no cancellation penalties.
2. Keep shopping around.
3. Five days may be far more expensive than weekly rates.
4. Reserve cheap and upgrade for free.
5. Renting on airport can add 12% or more to the cost of renting a car. But off-airport sites often close early.
6. Rent from smaller companies. Orbitz.com offers an extensive listing of rental companies.
7. Rent counter-intuitively. Big cars cost more during the week. Trucks and SUVs cost less on weekends.
8. Check your rental car for dents and dings. Make sure they're recorded before you leave the lot.
9. You don't need the insurance rental cars try to lumber you with. The credit card you use to rent the car covers most of your car rental insurance needs.

HOW TO AVOID IDENTITY THEFT

Don't give your social security number to anyone. The Federal Reserve has a publication detailing identity theft and how to avoid it. It is available online, at the Bank's public web site www.bos.frb.org/consumer/identity/index.htm.

BEST OUTLOOK TIP - I

Microsoft's popular Outlook software crams all your information – emails, contacts, calendar, tasks and journal – into one file, typically called outlook.pst. Find the latest *.pst file on your PC and make sure you regularly back it up to another machine or another drive. You cannot afford to lose your Outlook outlook.pst file. Believe me.

BEST OUTLOOK TIP - II

If your Outlook locks loading, or starts very slowly, find the file scanpst.exe on your hard disk. It's Microsoft's Inbox Repair Tool. Run it on your .pst file. This works with all Office variations – just takes a while. When it says it's found minor errors and asks do you want to fix them? say Yes. Don't bother asking it to make a backup. Any errors can mess up your Outlook.

BEST OUTLOOK TIP - III

You must regularly reduce the size of your Outlook .pst file. Delete or save all the attachments still in your inbox. Empty all your deleted files. Archive all your sent items. Then compact your pst file. Right click on Outlook Today. Go to Properties, Advanced, then Compact Now. That should substantially reduce the size of your .pst file.

BEST OUTLOOK TIP - IV

To automatically get rid of spam or porn emails, use Outlook 2003 and regularly update your spam definitions for free from the Microsoft.com web site. Ditto for the new Outlook 2007.

BEST OUTLOOK TIP - V

It makes sense to use Microsoft Word as your email editor. You get instant spellchecking and you can add a return address to each email you send out. To add your own personalized return address, close Outlook. Open Word. Tools/Options/E-mail options/E-mail signature.

Wi-Fi SECURITY

1. Install the latest encryption software.
2. Set up an "invite list" specifying which machines can connect to the network.
3. Build a virtual private network, protected by passwords.

HOW TO MAXIMIZE YOUR LAPTOP'S BATTERIES

At least once a month, disconnect the computer from a power source and operate it on battery power until the battery pack fully discharges. A main battery can be recharged many times. Gradually, over time, it loses its ability to hold a charge. If you will not be using the system for more than a month, remove the battery pack. Disconnect the AC adaptor when the battery is fully charged. Overcharging makes the battery hot and shortens life. Replace your laptop's main battery once a month.

DON'T OPT OUT

The "opt-out" option at the bottom of many spam messages often results in you getting more spam. It all depends on which spammer you're dealing with. Some honor the "unsubscribe" request, but man y don't. Tech glitches are also a factor. A 2002 study by the Federal Trade Commission found that two-thirds of all "unsubscribe" links fail.

GET RID OF SPYWARE

I recommend two programs that can clean out most spyware and adware from your PC. Two are free: Ad-Aware SE Personal, available at www.lavasoftusa.com1; and Spybot Search & Destroy, which you can get at www.download.com2, and typing "spybot" into the search box. Microsoft has a program called AntiSpyware, which may or may not work – but probably will eventually. I've tried it. It keeps running in the background. That annoys me. You have to turn that feature off.

THE UNIVERSAL FIX-IT -- REBOOTING.

The universal fix-it – rebooting, also called resetting. It works with anything with a micro-processor (i.e. small computer) in it. Which means practically every gadget you and I own. That includes computers, phone systems, cable modems, DSL boxes, Apple iPods, cable TV boxes, satellite TV boxes, your new car. Simply unplug the device. Count to 30. Pray a little. Plug it back in. Turn it on. Bingo, it now usually works. If you have several devices talking to each other – e.g. a cable modem, a broadband router, and a firewall – you may need to reboot the devices in some sort of order – from the back forward, or from the forward to the back.

HOW TO FIX YOUR CABLE MODEM

For homes and home offices, the fastest broadband service is typically a cable modem – unless you're in Europe or Asia, where DSL lines may be faster. Lessons from three days without broadband (cable modem) Internet:

Every time a coaxial cable – the type carrying cable modem and TV – splits, it loses signal strength. Hence, the fewer splits, the better.

Decent signal strength is absolutely critical to a cable modem, less so to a TV picture.

Eliminate as many splitters and as many joins as you can. Splitters can actually go bad, even if they're inside. If you have a problem – especially if it's intermittent cable modem service – remove and replace anything and everything. Tighten all joins. If you have open ends on your splitters, cap them or use a smaller splitter. If you have splitters, which are splitting your signal strength into more splits than you need, get a smaller splitter.

Some splitters send more signal strength out on one path than another. If you're losing signal, switch them. Send your cable modem a stronger signal. The cleaner and stronger the signal it receives is, the faster your Internet connection.

The big key to great cable modem service is to bring a clean, heavily-shielded (preferably quad-shielded) cable directly from the street into your house. Split it just before your modem. One side to the modem. One side to your TV sets.

Your cable modem service will fail from one day to the next. Water may seep into a splitter, especially if it's outside, as many are. Joins may become loose. Stuff happens.

When you're building and the walls of your house are still open, run at least three coaxial cables to all the important places – your home office, your mail living room, etc. That way, when one cable gives trouble, you can easily switch cables, without opening walls or running the cable on the outside of the wall. Also you may need a second cable in order to move TiVoed shows around between various TV sets.

Keep all joins tight, and sealed from water. This means you may have to add insulation and tape after your cable man has done his quick install.

A cable can fail from one day to the next. Keep a dial-up backup handy. Have your wireless connection working. In a pinch you can always go to Starbucks and work there. Find something to do as you wait for the cable technician.

HARRY'S TWOFOLD BACKUP STRATEGY

1. I keep all my working files in c:AllHarry. I back up all changed files in c:AllHarry to a second and third hard drive once a day. I carry a backup hard drive with me. I leave one backup drive at home, i.e. away from the office. I use software called filesync to copy only changed files. www.fileware.co.uk.

2. I make a clone of my main hard drive every week. I use software called EZ-Gig Transfer Utility and EZ-Gig II. www.apricorn.com.

CORRODED COAXIAL CABLE CONNECTORS

Does your cable TV picture sometimes freeze for seconds? If so, your coaxial connections have probably corroded and need replacing.

NEVER LEAN ON YOUR LAPTOP

The weakest point of any notebook computer is the top side of your machine, i.e. the screen. Don't lean on it. Don't put anything on top of it. Don't drop anything on it. Replacing your laptop's screen may cost more than your laptop originally cost you and usually a busted screen isn't covered by warranty. Or the warranty expired the day before you broke the screen.

SAVING MONEY ON FOOD AND OTHER SUPERMARKET PURCHASES

To save money on your food bill, look down. Less costly items are often on bottom shelves, whereas more expensive ones are placed at eye level.

Often bigger containers cost more than smaller containers on a per ounce basis.

BUYING ON-LINE? NEED COUPONS?

Buying on line? Need coupons, try www.naughtycodes.com, www.currentcodes.com and www.edealfinder.com.

HEARTBURN?

Suffering occasional heartburn while sleeping? Try sleeping on your other side. If it gets worse see your doctor.

INTERMITTENT PC PROBLEMS?

Intermittent PC problems? First, check what you did most recently. Did you just install some new memory? Is the memory seated correctly? Never do two things to your PC at once. If you do, you'll never figure out which action caused the mess.

INVIOLATE EMAIL RULE

Never ever write anyone a rude or angry email. Emails should be compliments. Emails should shower flattery. They should be funny and they should be tear-jerking. But they shouldn't be angry. If angry, calm down first, then call the person, or better, go visit them. Face to face always works better. Rude emails simply aggravate the problem. Moreover, emails can be subpoenaed in law suits. And employers do read their employees' emails.

WASH YOUR HANDS OFTEN

Cruise passengers have been advised that they can reduce the risk of infection by washing their hands frequently and keeping their hands out of their mouths to avoid ingesting viruses they may have picked up from touching doorknobs and railings.

BATTERIES

All rechargeable batteries need to be drained and recharged occasionally. Nickel cadmium (NiCad) need the most discharging. Lithium Ion ones less so.

THE BIGGEST DIFFERENCES BETWEEN LAPTOPS

The quality of the screen is the biggest difference between cheap and expensive laptops. The design, layout and feel of their keyboard is the second biggest difference between laptops. The third biggest difference between laptops is the quality of manufacture. Some are simply made better than others. Some models simply don't stand up to the rough and tumble of travel.

HOSPITALS ARE DANGEROUS PLACES

Hospitals are dangerous places. If you ever have to go to one, NEVER go without someone who will personally supervise everything about your presence there, the whole time you're there. A skeptical spouse works best.

BEST POWERPOINT TIP

When someone interrupts your PowerPoint presentation, press the letter B key (or the period key) to put up a blank black slide. Don't hold your finger on the key. Now your audience can now focus on you, instead of being distracted by your dumb slide.

BEST MICROSOFT WORD TIP

Microsoft Word is buggy. It locks frequently. Save often. If it crashes, close it. Open it again before rebooting. You may recover your document. Done for the day? Don't close every document you have open. Press Shift as you open the File menu. You'll discover that the Save command now says Save All. I do not use Word to write this dictionary. It's too unreliable. I use a simple program called the Semware Editor. It's available from www. semware.com.

GOING AWAY FOR A WEEK OR TWO?

Unplug your DVDs, VCRs, TVs, cordless phones, etc. Switching off these devices doesn't necessarily stop their power drain while they're plugged in. It just places them on standby, ready to spring to life when you touch the remote. The amount of electricity these devices devour while napping is staggering. Alan Meier of the Lawrence Berkeley National Laboratory has found that VCRs and DVD players draw 93% of their total power usage while inactive. Answering machines and cordless phones are worse – 98%. The Federal Government does not require manufacturers to label standby power usage on appliances, but the laboratory has created a web site listing devices that draw less than one watt (i.e. very little) of standby power – click here. In 2001, President Bush issued an executive order mandating that federal agencies buy only low-standby appliances. For more, http://standby.lbl.gov/index.html.

INTERNATIONAL CALLBACK WORKS

You can save 70% plus if you use international callback when calling the U.S. from overseas. One source: Premiere WorldLink 1-888-729-0107

BEST WEB TIP - 1

www.Google.com is the best search engine. Better yet, install the Google toolbar. Go to http://toolbar.google.com/install. Second best tip: http://maps.google.com. Best mapping site on the Internet.

BEST WEB TIP - 2

Always hit your browser's Refresh button when you alight on a site. A remarkable number of sites are cached by someone somewhere. Which means they often feed you yesterday's site. Get today's site by refreshing.

BUY YOURSELF A SHREDDER

Buy yourself a crosscut shredder and shred all your personal documents. A crosscut shredder makes shreds that are harder to tape together.

THREE OR FOUR DIGIT EXTENSIONS?

If you use 2-digit extension numbers in your company and are putting in an automated attendant, think how long before you outgrow it and need a three digit scheme. If you think you'll ever outgrow 3-digits (and you will), change now. Save the confusion later.

REPLACE HANDSET CORDS

Change console handset cords every three months – whether they need it or not. Replace all handset cords on all telephones once a year. If your phone crackles, change its handset cord.

10 WAYS TO KEEP YOUR CAR OUT OF THE SHOP

Stacey L. Bradford published these tips on SmartMoney.com on March 26, 2003. Here are excerpts of his article:

1. Start It Up Properly. "Most of the wear on an engine happens when you start the car... Make sure extra accessories, such as headlights or climate control, aren't on when you turn the ignition... An even more destructive habit: revving the engine. Believe it or not, this won't help warm up the car. In fact, revving can do serious damage to the engine and significantly shorten its life.

2. Come to a Complete Stop. Ever pull out of a parking spot and pop your car into drive while it's still coasting backward? If the answer is yes, you can kiss your transmission goodbye. By shifting into drive while the car is still in reverse, you're asking the transmission to do the work of the brakes – and that will wear down your gears.... A transmission driven properly should last more than 100,000 miles, one owned by an aggressive or impatient driver will give out long before that.

3. Don't Run the Needle Down to Empty. Sediment collects at the bottom of the fuel tank over time. You never want to do anything to unleash that dirt into your fuel system. You can stir up sediment by driving with less than a quarter of a tank of gas.

4. Turn Gently. Cutting the steering wheel too far to the right or left – something nearly everyone is guilty of while parallel parking – can also do damage. By turning the wheel to the point where it can move no further, you're putting 50 times more wear and tear on your steering pump than normal.

5. Check Your Tires... Too little or too much pressure can cause all sorts of problems. If you drive with too little air, your tires will get quite hot; excessive heat will wear out the tread much faster than normal driving. If you drive with too much air, less rubber will hit the road than necessary. Next thing you know, the outside portion of your tire will be doing all the work and wearing out faster than the inside.

6. Don't Ride the Clutch... Some of us use the clutch to avoid rolling backwards while on a hill... Better use the emergency brake. Another clutch-preserving tip: Pop the car into neutral and take your foot off the clutch while sitting in traffic.

7. Listen to Your Car. Every little sound your car makes is a cry for help. Ignore it, and a small problem could turn into a huge one. The most common sound people ignore is squeaky brakes. Steering systems need attention, too. Once you hear the whine of the steering wheel, you need to have it checked.

8. Heed Those Warning Lights. Don't ignore the warning lights on your dashboard... By waiting even a few days, you can turn a simple problem into a disaster. In some cases, you need to stop your car immediately and get it towed. Here are two such situations. The first is when you see your engine is overheated. "If the temperature gauge crosses into the red zone, it is as dangerous to your car as a heart attack or stroke is to the human body. The

second is you'll also ruin your engine if you drive after the loss of oil pressure. The engine oil light will pop up if this happens. Oil is the lifeblood of your engine. Drive without it, and you'll cause catastrophic damage.

9. Drive More. Many people don't realize it, but not driving your car can be as damaging as driving it too hard. The worst thing you can do with a machine is not to use it.

10. Neglect Will Cost You. The single most important thing you can do for your car is take it in for regular maintenance. Most important: the oil change. Regular oil changes every 3,000 or 4,000 miles can double the life of your car. But you also need to have filters changed, belts checked (and sometimes replaced) and all fluids inspected.

THIRTEEN TIPS FOR BETTER IVR SCRIPTS

1. Place strict limits on the number of levels, and the number of choices at each level. In most cases, four levels with four choices each is an appropriate limit.

2. At each level, tell callers how many choices they have, before telling them what the choices are. "You now have three choices. For sales, press ." That encourages callers to listen to the entire menu before making a selection, and reduces misdirected calls.

3. Place the most frequently selected choices at the beginning of each menu.

4. Instead of "Press one for customer service," say "For customer service, press one." With the first version, some callers will press one before the sentence is completed.

5. Have your menus recorded professionally. Use one voice throughout the system, for all menus and announcements.

6. Allow frequent callers to override spoken menus and jump directly to their frequent choices.

7. If the caller doesn't make a selection within a preset time (10 or 15 seconds) repeat the choices.

8. Provide an easy way to (a) reach a real person or (b) return to the main menu or (c) back up one level.

9. Use consistent commands. If one prompt requires the caller to "press the pound key" after entering data, then every data-entry prompt should do so. Similarly, always use the same keys for backing up or reaching a human.

10. Use consistent vocabulary. Always describe similar actions and options using the same words and phrases.

11. Keep promotional messages short, and don't inflict them on callers without warning. (It's one thing to hear an ad after you "press 1 for more information." It's quite another to hear one you didn't ask for and can't bypass.)

12. If a caller makes an invalid entry, say so politely and repeat the choices. If the same invalid entry is repeated, provide additional information and guidance.

13. Always let the caller dial zero and reach a real human being.

Source: Telemanagement, March 2003 www.Angustel.CA

DON'T STORE THINGS IN BLACK GARBAGE BAGS

Several of our young people stored their laundry in large black plastic garbage bags. The bags were thrown out by over-zealous cleaners. One bachelor, in need of extensive dry cleaning, brought his entire wardrobe of clothes into our office. It was thrown out that night. He now owned only the clothes he wore, and they were dirty.

KEEP GOOGLE NEWS GOING

Neat new service: http://news.google.com/

LEARN HOW TO SAY NO

NO is the hardest word in the English language. NO is also the kindest word in the English language. Saying NO often does the person you're saying it to a big favor. It may save them doing something foolish. It may even put them out of their misery. I'm learning to say "NO" more politely than I have in the past. The key to saying NO is not to explain why – unless you're really pushed. Most people don't want to hear the reason because it implies a criticism of them.

HTML OPTIMIZATION TIPS

Files that are too large move slowly across the Net and create data bottlenecks. Here are some ways to "optimize" your web content and make downloading your site much faster.

1. Fewer Colors in Images. If an image is simple and has few colors like a button or plain text, try to convert it to use less colors. Use 16 colors (4 bit) if you can, and 256 (8 bit) for everything else. Avoid 16 bit and higher because this will dramatically increase image size. Reduce the size of your images. Then convert them to GIF.

2. Use height and width Tags in IMG Tags. Always use height and width tags in all of the IMG tags. Web browsers try to lay out the entire webpage first and then insert the images whenever they get downloaded.

3. Code HTML Files by Hand. If at all possible, if you have the experience, code your HTML files by hand using a text editor. Don't rely entirely on a WYSIWYG editor, such as Macromedia's Dreamweaver. Coding by hand will give you greater control over your files and will keep them small and fast, just what visitors want. Some editors, e.g. Dreamweaver, put redundant duplicated new commands at the beginning of each paragraph, bloating your code for no reason.

PRICEGRABBER.COM IS GREAT FOR SHOPPING

www.PriceGrabber.com may be the best site for online shopping yet. It lets you search dozens of online stores for the items you want and pay the lowest price possible. It's brilliant.

SEATGURU.COM

Looking for the best seat on your next airline trip? Go to www.SeatGuru.com. Make sure you slide your mouse across the maps of the seats in the plane. The resulting "mouseover" reveals useful information.

BEWARE OF FAKE CASHIERS CHECKS

Scott Fripp, telecom dealer (sfripp@optiontelecom.com), writes: "To all telecom dealers: "Some of you may be aware that this September my company was robbed of over $40,000 worth of Norstar telephone equipment by a criminal posing as an employee of a company in the greater LA area. The company was fictitious, and the ship to address was to an apartment complex. We used FedEx as the shipping carrier and sent out the equipment via COD Secured Funds. The FedEx driver accepted a poorly drafted fraudulent "Cashier's Check" as the payment (incidentally, FedEx denies any responsibility for training their drivers to distinguish a real bank check from a fake one, and they claim they are not liable in any way for damages resulting from their services).

"If you go to Dunn & Bradstreet, you can look up a company by name, for free www.dnb.com.

"If you do not see this man's company there, or if the address does not match, then you have probably talked to a dangerous criminal who has already stolen more than $150,000 worth of merchandise from myself and others that we know of."

FAX MACHINES NEED PROGRAMMING

Check the programming on your fax machine for correct day, date time, and your fax number. Fax machines send back their fax number to the sending fax – but only the number programmed into their memory – not the actual phone number they're attached to. It can seriously confuse senders if they think they're sending to the wrong machine or the wrong phone number.

COOL YOUR PILLS

All drugs lose their potency with age. The loss of potency doubles its speed for every 10 degree increase in temperature, and halves its speed for every ten degree drop in temperature. Best advice: Put all your pharmaceuticals in the refrigerator. You'll extend their potency and their life.

HOW TO REMEMBER NAMES

A person's name is their favorite word. Repeat it many times when you're talking to them. "Catherine, it's nice to meet you. Is that Catherine with a 'C' or Katherine with a "K?" Try also "It's been nice meeting you, Catherine." Write something on her card that will bring back an image of her.

TROUBLESHOOTING MEMORY PROBLEMS

When you have a problem with computer memory, the cause is usually one of three things:

Improper Configuration: You have the wrong part for your computer or did not follow the configuration rules.

Improper Installation: The memory may not be seated correctly, a socket is bad, or the socket may need cleaning.

Defective Hardware: The memory module itself is defective.

Many intermittent computer problems are memory problems. But being intermittent makes troubleshooting difficult. For example, a problem with the motherboard or software may produce a memory error message.

The following basic steps apply to almost all situations:

Make sure you have the right memory part for your computer.

Confirm that you configured the memory correctly. Many computers require module installation in banks of equal-capacity modules. Some computers require the highest capacity module to be in the lowest labeled bank. Other computers require that all sockets be filled; still others require single-banked memory. These are only a few examples of special configuration requirements.

Re-install the module. Push the module firmly into the socket. In most cases you hear a click when the module is in position. To make sure you have a module all the way in the socket, compare the height of the module to the height of other modules in neighboring sockets.

Swap modules. Remove the new memory and see whether the problem disappears. Remove the old memory, reinstall the new, and see whether the problem persists. Try the memory in different sockets. Swapping reveals whether the problem is a particular memory module or socket, or whether two types of memory aren't compatible.

Clean the socket and pins on the memory module. Use a soft cloth to wipe the pins on the module. Use a PC vacuum or compressed air to blow dust off the socket. Do NOT use solvent, which may corrode the metal or prevent the leads from making full contact. Flux Off is a cleaner used specifically for contacts. You can purchase it at electronics or computer equipment stores.

Update the BIOS. Computer manufacturers update BIOS information frequently and post revisions on their Web sites. Make sure you have the most recent BIOS for your computer. This applies especially when you have recently installed new software or you are significantly upgrading memory.

When in doubt, send all your memory modules back to their manufacturer and ask for a complete replacement.

BUY A NEW KEYBOARD

New keyboards on laptops and desktops are the cheapest way of making your computer feel like new again. Keyboards are easy to install on laptops. Swap your manual mouse for an optical one. The optical ones work much better.

HOW TO MAKE YOUR PCs LAST LONGER

1. Buy quality voltage regulators/surge suppressors and connect them.
2. Turn them off at night.
3. Unplug your computers during thunderstorms.
4. Keep the computer case intact. When you install a new internal component, make sure to save the spacer you just removed so you can cover the slot again. When engineers design computer cases, they also design the airflow in the system. Removing a spacer will disrupt that carefully designed airflow and cause the heat inside the case to rise, damaging sensitive computer parts.
5. Check internal fans for dust and clean them. Vacuum cleaners work well.
6. Keep the power-supply fan clean and areas around it clean.
7. Keep the keyboard free of debris.
8. Clean the keys on the keyboard.
9. Clean your monitors. Clean the monitor screen with a nonabrasive glass cleaning solution. Be sure to spray the solution onto the rag and not directly onto the monitor to prevent any liquid from getting inside the high-voltage monitor.
10. Keep mice clean. To keep the mouse and mouse-ball clean, you need a can of compressed air and some alcohol wipes. Once a month, remove the mouse ball from the mouse and clean it with the alcohol wipes. Then use the compressed air to remove any dirt and debris that may have found its way into the mouse. Better yet, buy an optical mouse.
11. Clean the inside of computers.
12. Keep CDs clean to prevent the drive from getting dirty. Work from the inside of the disk to the outside using circular motions.
13. Change drives once a year whether they're still working or not. They're electromechanical. They wear out.

DRINK LOTS OF WATER ON PLANES

Don't drink alcohol on planes. Drink water. Lots of it.

HOW TO INSTALL A NETWORK PRINTER

The key to installing a network printer is to first install a LOCAL port – which seems stupid.

It means you install a "local" printer on a TCP/IP network port, then install the printer's software. Steps include:

1. Add a LOCAL printer.
2. Create a new port.
3. Standard TCP/IP port. Insert port ID, e.g. 10.1.1.XXX
4. Have disk. Point to CD.

Unless you're installing an Hewlett Packard printer, when you might need to install a JetDirect Port. You still need to start by adding a local printer, even though you're trying to install a network printer. HP's installation routines don't make sense. Their tech support is non-existent. Epson printers are easier to install.

COLLECTING MONEY

Phone calls are more effective than letters in collecting amounts due. During a collection call: Avoid small talk. Don't apologize for bothering the debtor or asking for your money. Don't ask for reasons why the bill hasn't been paid. Make a closing statement indicating what will happen if the bill isn't paid.

WEEKEND PREPARATIONS

Last thing Friday night, fill all your fax machines with paper. Be a shame to miss all the nice orders that might flow in over the weekend.

DON'T OPEN EMAIL ATTACHMENTS

Don't open email attachments. Call the person sending the attachment and ask if it's OK. Never open attachments with extensions .vbe and .exe and double extensions, such as .jpg.exe.

Don't ever open an email that says "Hi!. How are you? I send you this file in order to have your advice. Thanks." This email contains a virus.

Follow Newton's rule: The more friendly and intriguing the email is in its subject line or its first few lines, the most likely it is to be spam, porn, junk or contain a virus.

If you're really worried, install Microsoft's many security patches for Outlook and install Outlook 2003. It deletes all .exe, .vbe and .bat files before it delivers your email. The only way for friends to send these files to you now is to zip them or change their extension.

SAVE ON SHOULDER SURFING FRAUD

Shoulder surfing happens when the guy behind you in the payphone line copies your credit card number as you dial it and then steals it. One solution: Be wary. Another: Dial your credit number and then keep dialing extra digits. The system doesn't care how many extra numbers you dial. Just don't press the # key. The extra numbers may confuse the thieves.

FAX MACHINES ARE QUIRKY

When fax machines break (i.e. run out of supplies or memory) they will not busy your phone line. This means if the broken fax machine is the first in your hunt group, nothing else will get through. This is not a manufacturer design fault. The FCC will simply not allow fax machines to busy out phone lines. Thus, you must check fax supplies – especially paper – at least twice a day. You don't want your fax machines to be empty.

TIVO AND REPLAY

The two best gadgets for watching TV are digital video recorders made by TiVo and the general purpose digital video recorders supplied by some of the cable companies. They let you record your favorite programs whenever they come on and on whichever channel they appear. Among the multitude of advantages: zapping through commercials (automatic with Replay); ease of programming; knowledge of when your favorite programs are playing (they dial into TiVo and pick up the database of your programs). I couldn't live without my TiVo. www.TiVo.com.

CAR INSURANCE

There are a thousand ways of saving on your car insurance. To save time, find a broker you can trust. I trust Mardee Mieske at MetzWood Harder Insurance on 800-635-7170 and 518-392-5161 in Chatham, NY. Mardee is magnificent.

NEW SOFTWARE RELEASES

It doesn't make sense to be the first user on the block to use a new release of software. All software comes with bugs. Best to wait until the first service pack or service release comes out. New versions of software typically end in zero, e.g. 4.00. An upgrade (also called a fix) is 4.01.

AMERICAN EXPRESS PLATINUM

American Express platinum credit card will give you a two business class tickets for the price of one on many airlines. It's a great deal – if you have to travel business class and if you have to travel long distances. We used the credit card to go to South Africa. It was great.

HOLD BACK AT LEAST 10%

If you hire a contractor (any contractor), hold back at least 10% of the monies you owe him for at least three months to make sure you're completely happy and your Punch List has been completed satisfactorily.

FIRING SOMEONE

Firing someone? Cut their access off immediately to your company's computer network. Change all passwords immediately. Physically remove the person's PC. They may willfully harm their computer and the company's network if they still have access. It happened to us. Walk your ex-employee out of the building immediately on firing them.

THE BEST QUALITY COAXIAL CABLE

Wire your home with only the best quality quad-shielded coaxial cable you can get. Ask for RG6U or better. Install at least three sets of coaxial cable to every outlet in every room so that you can use one for cable TV, one for your cable modem, and one spare one to slot in when one of the other two degrades over time (as it will).

HOW TO AVOID INTERNAL THEFT

This tip from Chuck Whitlock, Scambusters. www.ChuckWhitlock.com

An ex-employee might get into your payroll files and arrange for checks to be cut twice a month to nonexistent employees. Or he might issue payments to a phantom supplier for hard-to-confirm services, such as consulting or cleaning, that were never provided. Your firm is less of a target if certain safeguards are in place:

When someone leaves, immediately discontinue computer passwords.

Institute cross-training procedures. Require each employee to make a procedure manual for his/her job. Everyone in your company should make at least one colleague familiar with his accounts and files. This makes it less likely that any employee will think he can get away with something.

Check the books carefully when someone in payroll or accounting declines to take a vacation. Embezzlers may be hesitant to have someone else look at their books.

Put firewalls around your computer system. Use codes with both numerals and letters to protect against cyber-criminals outside your firm. Do not let employees use names or birth dates as codes.

No security system is foolproof, but you can make it harder for high-tech criminals. The right computer security program depends on your system and the sensitivity of your data. I use a program from McAfee Software. www.mcafee.com.

Create written procedures for purchasing, expense reimbursement and payroll functions. These should include supervision by more than one employee.

WIRELESS LANS WORK (SORT OF)

Wireless LANs are not a panacea. A wired LAN will always work better. 802.11b wireless Wi-Fi LANs work well within a confined area – perhaps 100 feet in a place with few walls; perhaps 50 feet in a place with many walls and ceilings. There are two keys to getting your laptop on a wireless LAN. First you should set up another profile, in addition to the one you already have. Second, you should know the new LAN's name. You typically don't need an IP address. The wireless LAN should assign you one. The best part of wireless LANs: so many are around that you can now get easily onto the Internet in coffee shops, boardrooms, airline clubs, etc. Just pick a different profile. Be careful: few wireless LANs use encryption. So be careful what you send. See next tips.

10 TIPS FOR IMPROVING YOUR WIRELESS NETWORK by Tony Northrup and published on Microsoft's site on February 8, 2005:

If Microsoft Windows XP ever notifies you about a weak signal, it probably means your connection isn't as fast or as reliable as it could be. Worse, you might lose your connection entirely in some parts of your home. If you're looking to improve the signal for your wireless network, try some of these tips for extending your wireless range and improving your wireless network performance.

1. Position your wireless router (or wireless access point) in a central location.

When possible, place your wireless router in a central location in your home. If your wireless router is against an outside wall of your home, the signal will be weak on the other side (the further side) of your home.

2. Move the router off the floor and away from walls and metal objects (such as metal file cabinets).

Metal, walls, and floors will interfere with your router's wireless signals. The closer your router is to these obstructions, the more severe the interference, and the weaker your connection will be.

3. Replace your router's antenna.

The antennas supplied with your router are designed to be omni-directional, meaning they broadcast in all directions around the router. If your router is near an outside wall, half of the wireless signals will be sent outside your home, and much of your router's power will be wasted. Most routers don't allow you to increase the power output, but you can make better use of the power. Upgrade to a hi-gain antenna that focuses the wireless signals only one direction. You can aim the signal in the direction you need it most. Standard antenna and hi-gain antenna

4. Replace your computer's wireless network adapter.

Wireless network signals must be sent both to and from your computer. Sometimes, your router can broadcast strongly enough to reach your computer, but your computer can't send signals back to your router. To improve this, replace your laptop's PC card-based wireless network adapter with a USB network adapter that uses an external antenna. In particular, consider the Hawking Hi-Gain Wireless USB network adapter, which adds an external, hi-gain antenna to your computer and can significantly improve your range. Laptops with built-in wireless typically have excellent antennas and don't need to have their network adapters upgraded.

5. Add a wireless repeater.

Wireless repeaters extend your wireless network range without requiring you to add any wiring. Just place the wireless repeater halfway between your wireless access point and your computer, and you'll get an instant boost to your wireless signal strength. Check out the wireless repeaters from ViewSonic, D-Link, Linksys, and Buffalo Technology.

6. Change your wireless channel.

Wireless routers can broadcast on several different channels, similar to the way radio stations use different channels. In the United States and Canada, these channels are 1, 6, and 11. Just like you'll sometimes hear interference on one radio station while another is perfectly clear, sometimes one wireless channel is clearer than others. Try changing your wireless router's channel through your router's configuration page to see if your signal strength improves. You don't need to change your computer's configuration, because it'll automatically detect the new channel.

7. Reduce wireless interference.

If you have cordless phones or other wireless electronics in your home, your computer might not be able to "hear" your router over the noise from the other wireless devices. To quiet the noise, avoid wireless electronics that use the 2.8GHz frequency. Instead, look for cordless phones that use the 5.8GHz or 900MHz frequencies.

8. Update your firmware or your network adapter driver.

Router manufacturers regularly make free improvements to their routers. Sometimes, these improvements increase performance. To get the latest firmware updates for your router, visit your router manufacturer's Web site. Similarly, network adapter vendors occasionally update the software that Windows XP uses to communicate with your network adapter, known as the driver. These updates typically improve performance and reliability. To get the updates, visit Windows Update, and then click Select Optional Hardware. Install any updates relating to your wireless network adapter. It wouldn't hurt to install any other updates while you're visiting Windows Update, too.

9. Pick equipment from a single vendor.

While a Linksys router will work with a D-Link network adapter, you often get better performance if you pick a router and network adapter from the same vendor. Some vendors offer a performance boost of up to twice the performance when you choose their hardware: Linksys has the SpeedBooster technology, and D-Link has the 108G enhancement.

UNLEASH YOUR HOME WI-FI NET

Larry Armstrong gave me "a few tweaks" on how to boost Wi-Fi performance. "Does your wireless home network need more oomph? Maybe you hoped the signal would extend beyond your living room so you could check your e-mail or surf the Web from a chair on the patio or your office over the garage. The dirty little secret of wireless networks is that they work best when all of the antennas are within line of sight of the base station. But many homes have concrete block or plaster-and-metal-lath walls that can obstruct the signal. And because Wi-Fi shares the airwaves with other appliances in your home, such as cordless phones and microwave ovens, interference can cause the signal to fade. The result? You're probably going to have places where you get a slow or unreliable connection, or where you get no connection at all. Some tips to minimize those dead spots are to move your base station out into the open and 10 feet away from other wireless devices. If that doesn't work, change the channel, just like you would for your cordless phone. If you're still having problems, Linksys, the leading maker of networking gear, sells a wireless signal booster that attaches to the top of its most popular wireless models. The booster actually works, though it's not a universal panacea. It's a marginal improvement.

GUARDING YOUR WIRELESS NETWORK AT HOME

Q: How secure are home networks?
A: Most have such a minimal level of security they're practically Swiss cheese.
Q: What should you do when you get your new router home?
A: The very first thing is change the default password: Every hacker on the Web knows it. Also, turn off the remote administration feature so that changes can be made only from your computer. If it's a wireless router, turn on encryption. That way your neighbors won't get into your computer via your network.
Q: How serious is the threat?
A: Pretty darn serious. Your Quicken files and other intimate details are stored on your computer. When you shop on the Internet, your address and credit-card number are also kept in files. That information can be used to steal your identity. Even if no one is interested in stealing your stuff, your computer can be used to distribute illegal software, pornography, or attack other computers.
Q: How often does this happen to home users?
A: We don't have any hard data on individual attacks, but we have lots of data on attacks on businesses. In 1988, there were only six reported attacks. In 2003, there have been more than 82,000. Unfortunately, the consumer is often clueless, especially when it comes to computer viruses.
Q: How can I prevent an attack?
A: Install a firewall. Hardware firewalls are better than software firewalls. But software firewalls are better than no firewalls.

INSTALLING NEW HARDWARE AND NEW SOFTWARE - 1

You're installing a new scanner, a new digital camera, a new printer on your PC running Windows. Sometimes you hook up the device first and then install the software. Sometimes you install the software first, then hook up the device. Before you do anything, check the instructions. When in doubt, install the software first and then install the equipment.
INSTALLING NEW HARDWARE AND NEW SOFTWARE - 2
Never install the software that comes in the package. Go the maker's web site and download the latest drivers and latest software. The software that came in the package is months out of date.

HOW TO HOOK A DIGITAL CAMERA TO YOUR PC

Digital cameras come with cables to transfer their information to PCs. Those cables get lost. Much easier is to slide the memory card out of the camera and slide it into your laptop. All you need is a PC card adapter that slides into one of your laptop's PCMCIA (also called PC Card) slots. You can leave the empty card in your laptop. That way you won't lose the adapter.

WHEN IT RAINS, THE CABLE TV GOES OUT

The biggest problem with the coaxial cable that serves your home TVs and cable modem is that rain seeps into the cable and, over time, your picture starts to looks increasingly "snowy." Finally it craps out altogether. The only solution is to replace the cable and its connectors. Always remember to tighten the connectors using a spanner. The newer connectors contain a rubber o-ring which keeps the water out. Ask for it. Also the newest cable contains four metallic shields – very important to cut down on ghosting when you're laying cable to houses in clear line of sight to local TV transmitters. The easiest way to check your cable for water is to cut the connector off and tap the cable on a surface. If water appears, you have a problem. Wrap all your outside connections with oodles of insulation tape. Keep all connections out of the rain.

CHOOSING YOUR PASSWORD

Insist users of your network choose unique passwords – ones that they don't use for their personal email or their bank account, etc. Explain that the use of identical passwords makes them vulnerable.

UNSOLICITED CREDIT CARDS

Return all unsolicited credit cards. Don't just throw them out. Return them. We hear, from InfoWorld Magazine, if a Chase card, after six months, "is not activated but the customer doesn't complain about not getting the card, Chase activates it." This means that someone could now be using your credit card!

HOW TO GET THROUGH A CALL CENTER

Some call centers will answer callers faster who don't push the buttons, as in "dial 1 for English...". There's a "logic" to this. It's all about keeping waiting callers busy. If you keep call busy pushing buttons, they will not objecting to waiting. Better, don't push buttons. You'll often get faster service.

HOW NOT TO HAVE PROBLEMS WITH YOUR CD PLAYERS

To prevent accidental damage to your CD player, don't use any of the following CDs:
Discs with stick-on labels.
Excessively thick discs.
Small or irregularly shaped (triangle, rectangle, arrow, etc.) discs.
Any discs that are scratched or dirty.
Discs that are bent, chipped or warper.
Cleaning discs.
Discs fitted with adaptors or protectors.

DON'T MAKE INTERNATIONAL CALLS ON YOUR MOBILE

Don't use your cell phone to make international calls. They can be very, very expensive. Calling from your landline can easily be one-tenth the per minute price of calling from your cell phone.

POWER FAILURE TRANSFER IS CHEAP DISASTER RECOVERY

Many PBXs have power failure transfer. This means if your PBX crashes, you still have some single line telephones to make and receive calls. In most companies, no one knows if power transfer failure works or where the single line phones are. Be good to check now. Ditto for checking obvious things: Is there gasoline in your emergency generator? Is there water in your wet call emergency batteries, etc.?

DON'T LOCATE UNDER BATHROOMS

Never, ever put your phone system or computer room below a bathroom. A friend's PBX got rained on the other night. The toilet upstairs overflowed. It wasn't pleasant. No one wanted to do the cleanup.

CLOSE THE TOILET SEAT

Before you flush, please close the toilet seat. It's healthier for you not to breath the foul air.

BEST AIRLINE TIP - 1. CHECK THE METASEARCH ENGINES

Most people know the Internet is the place for their travel needs. Most people make many visits to many sites before they find what they want. The problem is you won't find the discount airlines such as JetBlue, Southwest and Independence on sites other than their own. The latest solution is something called a "metasearch" engine. These sites search all the sites on the web that you would check. They do your work for you. The sites you want are SideStep.com, Mobissimo.com, Kayak.com and Farechase.com. I can't vouch for any of them, since I haven't used any. But I did check that they actually do exist.

BEST AIRLINE TIP - 2. THEN CHECK YOUR CHOICES

Once you know your choices, you can bargain shop via Orbitz.com, Expedia.com, Cheap-Tickets.com, Travelocity.com, even Priceline.com, your local travel agent and the airline directly on the phone. Never discount the idea of calling the airline directly and asking for a "cheaper" fare. Don't use Priceline if you're traveling on business for a very short trip. They're useless for short trips.

BEST AIRLINE TIP - 3. CHECK DISCOUNT AIRLINES

Discount airlines have the best deals. They include Midwest Express, Sun Country, JetBlue,

ATA, Independence Air and Southwest. Be careful, their fares are often not listed on the major Internet travel sites. Fortunately, all have their own excellent ticket booking Internet sites. www.Ata.com, www.JetBlue.com, www.Southwest.com, www.MidwestExpress.com, www.flyi.com and www.SunCountry.com. And they all have "Internet specials," which means they discount their fares even more.

BEST AIRLINE TIP - 4. BACK-TO-BACK FARES ARE GREAT

You can get almost four back-to-back New York-Ottawa airfares for the price of one weekday roundtrip fare. In virtually all cases, two back-to-back tickets are far less than the price of one weekday roundtrip ticket. Some airline routes are "monopoly" routes and exceptionally expensive. New York to Ottawa, Canada is an example. $932 for up one night, back the next. But it drops to $162 if you buy in advance and stay over Saturday night. The obvious solution: Buy two roundtrip tickets and throw one half of each away. They're called "back-to-back" tickets. If you're worried about getting caught (you won't), buy each ticket on a different airline. Here's how it works. Buy one cheap roundtrip/over Saturday night ticket starting in New York. Buy one starting in Ottawa two days later. Use the first half of the ticket to go and the first half of the other ticket to return. Throw the second halves of both tickets away – or, even better, use them for a "free" trip a couple of weeks later. Sometimes the airlines will "catch" you. But, be aware, it's not illegal.

BEST AIRLINE TIP - 5. HIDDEN CITY TICKETS

Going from New York to Dallas? Buy a ticket to Austin through Dallas. Going to Chicago? Buy a ticket to Minneapolis through Chicago. Going from New York to Detroit? Buy a ticket to Akron, Ohio through Detroit. Going to Atlanta? Buy a ticket to Miami through Atlanta. Going to London? Buy a ticket to Paris through London. Hidden city tickets work best with promotions, e.g. Florida is on "sale" during the summer when no one wants to go to Florida. Hidden city tickets are "illegal" in economy, i.e. you could be denied boarding on the way back (but rarely will be). Hidden city tickets work 100% legally in first class and business class. That's assuming you want to travel first or business. You cannot be denied boarding in first or business class.

BEST AIRLINE TIP - 6. GETTING ON SOLD OUT FLIGHTS

Want to get on a sold-out flight? Call at the stroke of midnight. That's when most ticket reservations with a "hold" expire, potentially freeing up seats.

BEST AIRLINE TIP - 7. FRAGILE WORKS

Mark your baggage "fragile." Most airlines load fragile baggage last, potentially saving you the damage wrought by dumping heavy bags on top of yours.

TELEZAPPER WORKS

Walt Mossberg of the Wall Street Journal tested a device called TeleZapper which stops telemarketing calls dialed by computers and sends back a signal telling the dialing computer your phone is kaput (thus getting you off the list). He raved about the $50 device device, made by Privacy Technologies.

YOUR SOCIAL SECURITY NUMBER IS PRECIOUS

There are all sorts of sneaky ways via email and the phone that people use to get your social security number – saying they need "to check on your account," "make sure you're getting your refund," etc. Don't give anyone your social security number.

USED TELECOM EQUIPMENT MAKES SENSE

Because of slow development times, buying used corporate phone systems makes huge sense. Reasons: dollar savings, more responsive dealers (they're smaller), the inventory is in stock and sometimes you can wangle longer warranties than from the original maker. Tips on buying used:
1. Is it the latest model? If not, does it matter?
2. Am I buying from a reputable dealer? Check references.
3. Is it in stock? Can it be shipped today? Some secondary dealers advertise stuff they don't have, then go and try to buy it from someone else.
4. Ask for "advance replacement." If the thing (e.g. a card) you get sent doesn't work when you get it, your supplier should be prepared to overnight you another one before they get your broken one back. But you'd better demonstrate some intelligence about checking the card you got. Secondary dealers tell me that most of the "bad" cards they get back really aren't bad. They're just installed wrongly.

5. Does your supplier understand the technical aspects of your phone system so they can match appropriate revision and software levels.
6. What about software? Do you get it or do you have to buy it from the manufacturer? Manufacturers often use their software "upgrades" as a bargaining chip to keep their customers away from the secondary market.
7. Does your supplier offer technical assistance over the phone? When you get your stuff and it doesn't work, will they "talk you through" until you finally get it working?
8. Do they refurbish? To what level? Some secondary dealers clean their stuff and do nothing else. Some clean it, test it and add small but important things, like new designation strips, instruction manuals, etc.

COMPROMISED BY A VIRUS?

When a computer system is compromised by a virus or worm, the only way to truly clean it is to back up the data and reinstall the operating system, including any software patches issued since the computer was purchased.

DON'T USE FILE-SHARING NETWORKS

Files you download from file-sharing network (often called darknets) are often infected by viruses.

TAP YOUR PHONE'S HANDSET

If you hear static, tap your phone's handset. This won't work on an electronic handset. But it will work on a carbon mike. You'll recognize carbon ones because the handset is round. Carbon handsets are older but they often work better.

BEST PC TIP - 1. ALWAYS SAVE ALL THE TIME

No Windows or Mac software nor any computer operating system is 100% stable (i.e. won't lock up). Always save your work every few minutes. And ALWAYS save your work before you Alt-Tab to another application or start another application. Any version of Windows software can blow up any time, without logic.

BEST PC TIP - 2. HOW TO SAVE YOUR LOCKED-UP APPLICATIONS

If Windows (any version) locks up, hit Ctrl/Alt/Delete and close down open applications one by one. As you do, you may be able to save your work. Windows typically locks up because it runs out of memory. Watch for Microsoft Word. I find it buggy, though the newest version, Word 2003 is more stable. Outlook is now the buggiest Microsoft program, especially when its *.pst (the one in which all your data is stored) file gets too large. With Outlook, permanently delete all deleted and all junk email. Then compact the Outlook *.pst file. Then SCANPST the PST file for errors. There will be plenty.

BEST PC TIP - 3. KEEP AN OLD MACHINE

Before you install new software on your primary PC – the one you rely on every day – install the software on some other junk (i.e. old) PC. You may not like the new software or it won't work. Installing it on your good machine will only clutter your machine with junk. Most "uninstall" programs don't uninstall everything. They leave clutter. Sometimes they even swap good software with bad software and mess your machine up long-term. And sometimes you can't uninstall them at all.

BEST PC TIP - 4. UNINSTALL YOUR OLD SOFTWARE

Before you install a new version (not an upgrade) of software running under Windows, uninstall the old version completely. Use the uninstaller that comes with the software first. If there isn't one, use Windows' own uninstaller – Control Panel/Add or Remove Programs. Don't follow this rule on upgrades. Then you must leave the existing software. After uninstalling the software, reboot your machine.

BEST PC TIP -- 5. THE BEST SPEED ENHANCER

Extra memory is the best and cheapest way to speed up your PC. Put the most memory in you can. When I doubled the RAM on my laptop running Windows XP Professional, it felt like I'd bought a brand-new machine. The memory cost $100. A new machine would have cost $3,000.

BEST PC TIP -- 6. CHECK. CHECK. CHECK

I asked that my laptop be delivered with 512 megabytes of RAM. It was. Except that my laptop came with two 256 Meg memory slots. One wasn't working. The memory, however, was fine. I didn't check that the memory was actually working for several months. I was lucky I discovered the problem in time. To fix the second slot, and thus to move up from 256 Megs to 512 Megs required a new motherboard – an expensive fix. Except I caught the problem while the warranty was OK. Had I not checked I probably would have been $1,500 out of pocket for a new motherboard.

BEST LAPTOP CONFIGURATION TIPS

Things you need to consider when you're buying a new laptop:

1. Physical screen size and screen pixels. Screen size affects portability. The bigger the more cumbersome. Pixels affect viewability. Easiest to read is a 1024 x 768 screen, But laptop screens also come in 1400 x 1050 and in all sorts of wide screen shapes. Some screens are more readable than others. You must check beforehand, because you cannot change screen resolutions on LCD) monitor.

2. Weight. Light weight means flimsy and lack of bits and pieces. For example, some lightweight machines don't come with CD drives.

3. Erector set ability. Can you swap a CD drive out for a second battery or a second hard disk? I personally find that being able to put a second drive inside a laptop is great for backing my work up – especially when I'm on the road. Toshiba Tecras come with a "Select Bay" – you can slide in an extra battery or an extra hard disk, etc.

BEST LAPTOP TIP -- 1. GET THE LONGEST LIFE

To get the longest life out of your laptop's battery, turn your screen's brightness down to its lowest level. Your screen is the biggest consumer of power on your laptop.

BEST LAPTOP TIP -- 2. CLONE YOUR HARD DISK

Before you leave home for a trip, clone your laptop's entire hard disk on an identical hard disk, which you leave at home. If your laptop is stolen, you'll have a backup. Best software for cloning is called EZ-Gig Transfer Utility, from www.Apricorn.com.

BEST LAPTOP TIP -- 3. DON'T USE HIBERNATION

Never put your laptop into a power saving mode called "Hibernation." Such mode puts information directly onto your hard disk. When you next start, your laptop searches for that file. If it doesn't find it because the file is corrupted, it's all over. You cannot cold boot your machine. It will always look for that file. You may have to send your laptop's hard disk off to the hard disk recovery factory. On the other hand a mode called "standby" is OK. If you lose it, you can always reboot.

BEST INTERNET BROWSING TIP

Browsing the Internet is fastest when you have BOTH a high speed connection to the Internet (T-1, Cable Modem or DSL) and a fast PC. You need the line to get the information fast to you. And you need a fast PC to quickly decode the incoming bit stream into pretty pages you can read. The fast PC makes a huge difference. Do not underestimate its importance. Do not get a broadband connection without a fast PC.

CREDIT CARDS GO ONLINE

You can find how much you have owe on your credit cards at any time by checking them on the web. Register with your credit card issuers. For example, if you have a Visa from Citibank, go to www.citicards.com. Good news: Citibank downloads its information to Quicken – as do most large on-line stock trading systems.

BEST CELLULAR TIP - 1. BUY CELL PHONES CHEAP

Independent retail cell phone dealers will always give you a better deal than dealing directly with the service provider, e.g. AT&T or Sprint.

BEST CELLULAR TIP - 2. FINDING THE BEST COVERAGE

Verizon has "the best national coverage." But it may not have the best coverage in your area. AT&T Wireless has the second best coverage. Check.

BEST CELLULAR TIP - 3. DON'T CHANGE YOUR CELL PHONE SERVICE PLAN

Be wary of changing your cell phone service plan. Every time you do, you may have "signed" up for another year of service. They don't tell you this. Ask. If you do extend your service contract, ask for a new, better, smaller cell phone. They'll usually give you one for free.

HOW TO BUY LONG DISTANCE AND INTERNATIONAL SERVICES

1. You must be on a plan.

2. The small carriers often have better deals than the bigger carriers.

3. Most carriers will match another carrier's cheaper plan.

4. Most carriers will bargain – especially if you offer all your incoming or outgoing traffic, sufficient volume and agree to sign a one or two year contract.

5. All long distance carriers will charge much less if you can figure a way of delivering your traffic directly to them, bypassing the local monopoly phone company. That way is usually a leased T-1 or T-3 line.

6. You must check your bills every month. You'll probably be billed for calls you didn't make and lines you didn't have.

7. It's good idea to negotiate a service level agreement between you and your carrier.

8. Don't ever buy long distance from your local phone company. If you have a dispute with your local phone company about your long distance bill, they most likely won't solve your long distance problems, but they will threaten to cut off your local phone lines if you don't pay your long distance bill.

9. The cheapest long distance is usually with a VoIP phone, such as one from Vonage. Remember: Your ignorance is their victory. Your squeaky wheel is their nightmare.

MY FAVORITE FIREWALL

My favorite firewall is the Watchguard SOHO Firebox. It protects all the PCs at our house which sit behind our cable modem. This box is obsolete. You can often buy it on eBay. Watchguard makes newer firewall hardware that's excellent, but more pricey.

CHECK YOUR FIREWALL

Check out www.grc.com. Wait a minute for the page to load. Select "Shields Up" and "Probe My Ports." It's a great way to check the security of your firewall. I just tested my home network and it passed with flying colors. It is protected by a Watchguard SOHO Firebox.

BLOCK CALLS TO 900, 976, 970 etc.

Block calls to "Dial-it" services and area codes where you don't do business. Most businesses routinely block to the area code – 900. Be careful with the 212 area code which has expensive exchanges – 394, 540, 550, 970 and 976.

PHONE SYSTEM PROTECTION

You need to protect your phone lines from the possibility of AC coming up them and blowing out all or part of your phone system. It happened to us. Someone touched 240 volts to two of our incoming 800 lines. That blew one of our PBX's trunk cards. It cost us $500 to repair the damage. We could have protected ourselves with $40 worth of surge protection inserted into our 66 block.

Lightning strikes can blow most phone systems. But high voltage spikes don't crash phone systems; they just cause them to act funny. They may knock calls off the air. They may ring extensions when there's no one calling. They may send screwy error messages to the console. Make sure your system is backed by AC surge arrestors.

If in doubt, reboot your phone system. Rebooting for your phone system just as it works for your PC.

ASK WHAT THE FEES ARE

Never buy mutual fund from a stockbroker. Buy directly from the fund. Always negotiate the price. Mutual funds charge huge fees – some upfront, some ongoing and some at the end. Mutual funds kick back ongoing fees to stockbrokers. Basically, most mutual funds don't perform well. My advice: buy your own stocks.

GET MANY EMAIL ADDRESSES

Get many email addresses. Use one for signing to services you're dubious about. Use one to buy things online. Only use your best one for emailing your business acquaintances and family. Getting rid of spam and porn will thus be incredibly easy. Just progressively kill your email addresses. Good idea: Use a second, older PC to receive emails from your "junk" email addresses.

SECOND CREDIT CARDS MAKE HUGE SENSE

Get a second credit card. Use this second credit card for subscribing to services that are difficult to cancel. It's easier to pay off the card and close it down than closing down some services, e.g. AOL.

CHECK YOUR CONSULTANT's HONESTY

Telecom consultants can be great for checking bills and doing jobs you don't want to. But many are crooks. They may get paid under the table by the vendors whose equipment and services they recommend. There are some ways to check you won't get ripped off:

1. Ask your consultant which company won his last 10 recommendations. Not any ten, his last ten recommendations. If one vendor keeps popping up, be careful. You may have a "consultant" who's on the take from the vendor. Typical take is 10%.

2. Ask the three leading telephone equipment vendors in your community if they will bid on a proposal your proposed telephone consultant specifies. If any of the three says he won't, ask why and be very careful.

3. Don't let your consultant get all the proposals that come in for your phone system without you seeing them also.

4. Make sure the potential vendors know it's your company they are selling the equipment to. Not the consultant.

5. Before accepting your consultant's recommendation, check out the approximate cost of a telephone system of the size and complexity you're thinking about. You should know roughly what your system is going to cost you. That way you can see if your consultant's recommended bid is way out of line.

6. Check out your consultants' qualifications with his last five clients. Call the clients and ask them very specific questions, such as equipment recommended, cost of services, services contracted for, and services performed, etc. Don't ask them one dumb question, like "Were you happy with the consultant?"

7. Bid out your consulting firm's services as exhaustively as they will bid out a phone system for you.

KEEP YOUR PHONE SYSTEMS COOL

Phone systems, like computers, work better and longer when they're cool. Sadly, most companies hide them away in cramped closets and unventilated rooms.

CABLES ARE ALWAYS BAD

If in doubt, assume the cable is bad, the connections are corroded, you're connected to the wrong plug or you're using the wrong drivers. Electronics, chips and PC cards rarely are bad.

CHECK ALL YOUR LINES -- MORNING AND NIGHT

Check each of your phone lines first thing every day. Check them by dialing in on each line. Check that lines which are meant to hunt, do. Check that no lines in your hunt group are broken and are preventing calls from jumping to the next line. Can your callers get through? Check them again once you put your PBX on night answer. Can you get through with Night Answer?

PREDICTIVE DIALERS ARE RECOMMENDED

Predictive dialers are useful and easy to cost justify. A predictive dialer automatically dials phone numbers fed it by a computer. It paces calls according to agent workload (thus "predictive"). If the number dialed is not answered or gets a busy, the predictive dialer redials that number or another. The idea is to spare agents the agony of dialing and the frustrations of ring-no answer, busy tones, wrong numbers, answering machines, intercept messages, fax machines, etc. The idea is to get as many calls connected as possible and keep the agents busy talking to customers (or whatever) instead of dialing.

HOW TO HIRE A SALESPERSON

Advertise for the person. Ask your respondents to call into your voice mail and leave a two minute explanation of why they want the job. It's easy it is to ferret out suitable candidates.

DON'T LEAVE YOUR BATTERIES CHARGING

Don't leave your nicad batteries charging forever. You'll overcharge them and burn them out.

HOTEL PHONE TIP - 1. DON'T LET IT RING

If you must dial out of a hotel using "Dial 9," never let the phone ring and ring. Here's a

sign we found in one honest hotel, but it applies to many hotel phones: "Charges on all calls begin to accrue after six rings. If you're not connected within six rings, hang up and redial to avoid charges."

HOTEL PHONE TIP - 2. KEEP USING THE

If dialing through a hotel to AT&T, MCI or another carrier, use the # button to make multiple calls while you're on the same call. That way you'll save multiple dollar local calls.

EMAIL AS A MARKETING TOOL - 1. INCOMING EMAILS

Print your customer service email address on all your stationery, business cards, sales literature, etc. Make sure someone is designated to answer incoming emails within the hour, 24 hours a day seven days a week.

EMAIL AS A MARKETING TOOL - 2. OUTGOING EMAILS

Your customers want to receive emails from you IF they contain offers of interest to them. But only if they contain offers of interest to them.
EMAIL AS A MARKETING TOOL - 3. BLIND COPIES
Be careful when sending blind copies on your e-Mail. If the person who is blind-copied decides to click "Reply All" and responds, the other recipients of the e-mail will receive the response and may easily deduce that there was a "blind copy-ee" on the e-mail. This could be embarrassing!

BE CAREFUL WITH YOUR PORTABLE DEVICES

2,900 laptops, 1,300 PDAs and over 62,000 mobile phones were left in London's licensed taxi cabs in the six months to August, 2001. That's the only survey I have, but it's pretty indicative of the problem.

FAXING ON PAPER IS DUMB. CALLWAVE IS GREAT.

Receiving faxes on paper is dumb. More than 56% of faxes are junked. The paper and toner waste is huge. Moreover, you can't receive a fax if you're not near your fax machine. Get yourself a free fax number from www.Callwave.com or www.efax.com, which will encode your faxes and instantly send them as an email to wherever you are in the world. I prefer Callwave. Their faxes are easier to read.

PLUGGING INTO OTHERS' WIRED LANs

You want to connect your laptop to someone else's LAN. Right click Network Neighborhood. Scroll down to TC/IP settings. Click once on it. That highlights it. Then click Properties. The easiest way to connect is to highlight "Obtain an IP address automatically." If that doesn't work, you'll need five sets of numbers: an IP Address, a Subnet Mask, a Gateway address and two DNS addresses. You'll get these five numbers from the LAN boss. The Subnet Mask will look like this: 255.255.255.0. All the others will look like this 101.10.105.109. Don't settle for getting fewer than five numbers. You'll likely need them all.

PLUGGING THROUGH A PROXY SERVER

Some LANs connect their users to the Internet through a Proxy Server. One reason for doing this is to control (i.e. limit) access to their Internet. If your company connects through a Proxy Server, you'll need the proxy server's address – it will look like this 10.10.0.26 – and you'll need a port number, like 80. You add a proxy server through settings for Internet Explorer: Tools. Internet Options. Connections. LAN Settings. Click on Use a Proxy Server. Then add in the numbers you get from the local LAN boss.

INSTALL PLENTY OF OUTLETS

You can never install sufficient power, phone (RJ-11), data (RJ-45) and coaxial TV outlets in your offices and your home. It's a lot cheaper to fill the walls with cable when the walls are open than after you've closed them, spackled them, painted them and decorated them.

HOW TO WIRE YOUR HOUSE FOR TELECOM

You need a wiring closet – somewhere clean, dry, ventilated and accessible. It should be large enough to take a small phone system, a cable box or satellite box and a punchdown block (for incoming phone lines and for lines to be distributed around the house).

You need at least two wiring outlets in each room – they're the same size as the electrical box. You buy modules for data, voice and video. The Siemon company has good stuff. I recommend them. www.homecabling.com.

You need three sets of wiring to every room, including the kitchen:

a. Category three cabling for phone calls. At least three pairs. Two should be punched down. One is for spare. I recommend using four pairs wired T568A allowing cables to be split outside of the outlet if need be. Also allows up to four phone lines to be wired on the same jack.

b. Category six cabling for Ethernet data. Should be at least four pairs. Two pairs should be punched down. Two pairs should be kept for spares. (TIA/EIA 570A Standard recommends Category 5e cabling, and also that all 4 pairs be terminated allowing 10Base-T to 100Base-T to operate over cable)

c. Three wires of 75 OHM standard coaxial cable. One wire is for DirectTV, local CATV or distribution from a central video server. One wire is for a cable modem. One wire is spare.

All wiring should be home run from the wiring closet to the various rooms. We do NOT want loop wiring. This wiring does not need to be punched down entirely. It needs to be run into the walls - before they are closed. It is much too expensive to run cabling after walls are closed. When in doubt, run more and better cabling. When running cables that are not terminated make sure to leave enough slack for future re-location within the same room.

WATER ALERT

There are safety devices which sit on your phone or computer room floor. If they sense water, they will squeal for hours. Highly recommended. One maker, Dorlen Products. 414-282-4840. www.Dorlen.com.

YOUR EMAIL ON THE INTERNET

When buying anything on the Internet, make sure you uncheck the boxes that say "Please notify me of updates and specials." That way you keep off their spam list.

CHECK FOR BAD TRUNKS

Always run a trunk summary report from your call data recorder (CDR) at the end of each week. You will be able to spot any outgoing trunk malfunctions (no traffic). One bad trunk could mess up an entire hunt group and thoroughly degrade your incoming trunks, thus messing up your service to your poor customers.

LOCK YOUR SWITCHROOM

Elementary security will stop most casual thieves. Lock your switchroom. Put different locks on your PBX's cabinets. The original installers still have keys. The factory-installed keys are the same on all the cabinets they ship. Change your programming passwords from Admin.

NEGOTIATE YELLOW PAGE ADVERTISING

Don't hesitate to negotiate price on your yellow pages advertising. It's not tariffed. No one oversees the pricing. The price is flexible. When you sign up for a yellow pages ad, find out if you are paying a special promotional price which will jump in the second year. If you are, you may not want to make a big investment in artwork which may not be used if you decide to reduce the size of your ad next year. Remember, more and more people are now going on-line to search the yellow pages, so these people may never see your ad. On the other hand, there are some neat deals with the on-line yellow page providers. Sadly, there are many of them. Check out Google, which now provides a much better alternative to yellow page listing.;

WANT THE PHONE NUMBER YOU'RE CALLING FROM?

Call your cell phone. Or dial 958 in New York. Dial 114 in some parts of California. There's a similar number in most cities. Wait a few seconds for the switch to tell you your calling number. Call 700-555-4141 to find out the name of the carrier your phone is equal accessed to.

STOP BROKER CALLS

Yo get off junk phone call lists, write a letter "Requesting DMI and directory exclusion" as per general instruction 5G-12 to

Dun & Bradstreet Corporate Headquarters One Diamond Hill Road Murray Hill, NJ 07974-1218.

FAKE TELEX, FAX AND BUSINESS DIRECTORIES

One big scam in the phone business is sending companies an invoice for an appearance in a fax, telex or businesses directory that is never published. The "bill" often comes from Europe, often from Switzerland.

DIRECTORY ASSISTANCE HAS BECOME EXPENSIVE

Public Utilities Commissions now allow most local telephone companies to offer both local and long distance Directory Assistance by dialing 411. In New York when you dial 411 and ask for a local number you can easily be charged 80 cents. In California, when you dial 411 and ask for a local number, PacBell charges at least 460xF5. When you dial 411 and ask for a long distance number, PacBell charges at least $1.10. Charges vary by state, but can be up to $1.90 for local or long distance numbers.

Check your Long Distance Telephone Bills. When you dial XXX-555-1212 and ask for a long distance number, your long distance carrier charges you up to $1.99. MCI just announced a rate increase to $2.49!

All state rates are different and IXC rates vary widely, so the only way to know what you are spending is to look carefully at both your local phone bill and your long distance bill! Look carefully, the charges are not easy to find.

USE OLD PHONE WIRE?

Sometimes you move into a new space and find old phone wiring. "Don't worry," says the landlord, "you can save money by re-using all that wonderful wire that's already in the walls." Don't believe him. Go for new wiring.

PLAN FOR NETWORK FAILURE NOW

Make sure you have some plan for network failure. Each of the major long distance companies – AT&T, MCI and Sprint – will suffer major network failures. Never exclusively use one carrier. But be careful. Even though you're using multiple vendors, you may find they're all riding on the same cables they have leased from the same local vendor, namely your local phone company. Better: see if you can negotiate service from multiple central offices and have the wires come into your building from different directions.

HOW TO GET A HOTEL ROOM

When a convention tries to tie up all the hotels in the town and charge too much per room, call Ready Reservations – 1-800-999-9144. Hotels have more room rates than airlines have prices. These days talk to the reservation clerk at the hotel, ask for cheap prices. Ask for rates for AARP, AAA, airlines, AT&T, etc.

BELONG TO AAA

Belong to AAA (Triple A, the automobile people). You get 10% discounts on Amtrak and as much as 30% on hotels. One reader writes "AAA gets you all kinds of discounts you never think of. I now routinely ask anywhere I go traveling. I've got discounts at museums, cinemas, even Disney, although you now have to go to the AAA office in Kissimee to buy discounted tickets to Disney. My wife feels embarrassed when we travel because after I ask for the price, the next thing I ask for is what is the AAA discounted price?. You'll be surprised on the answer."

RADIO OVER THE INTERNET IS GREAT

Find it and dozens of Internet radio stations on http://windowsmedia.com/radiotuner/MyRadio.asp.

SEARCH BY AUTHOR. IT'S EASIEST

Looking for an article on the web site of the Wall Street Journal or New York Times, etc.? It's often more reliable to search by author. I always search by author first.

GET BETTER DEAL ON A CAR

The Internet is a great place to check out the best place to buy your next car. Start with www.AutoBuyTel.com. My son saved over 10% on his latest car.

INSTALL SCANMAIL ON YOUR CORPORATE EMAIL

Your corporate email needs virus protection. A great program is ScanMail for Microsoft Exchange. ScanMail checks mail for viruses and disposes of them before they can wreak havoc. www.AntiVirus.com.

A GREAT TELECOM EQUIPMENT CATALOG

Want an inline adapter that connects one handset cord to another? Want a secret butt-set? What about sharing phone lines between phones and modems? Mike Sandman runs a unique direct mail operation that specializes in hard to find tools, test equipment, telephone repair parts and the training needed to use these products. Highly recommended. Mike Sandman. www.Sandman.com. 630-980-7710.

STAY HEALTHY. HERE'S THE BEST INFO ON THE WEB

You are your own best doctor. Well, not really. But knowledge is comforting and a great supplement to visiting a doctor. For example, you get to ask the right questions. The best health sites on the web are www.HealthScout.com, www.InteliHealth.com; www.MedicineNet.com, www.MerckHomeEdition.com, www.WedMd.com and http://Consumer.pdr.net.

HOW TO TRUMP BACK PAIN

The principle to keep in mind is that movement trumps back pain. Stretching and exercising will do more to keep your back pain-free than all the painkillers in the pharmacy.

CHECK. CHECK. CHECK.

IBM had THINK as its corporate motto. I have CHECK. It means to check everything. It means to check that your pharmacist gave you the right pills. It means to check that all your phone lines work. It means checking that you're not being overcharged on your phone bills. It means checking the accuracy of each of my definitions. It means checking that I'm going to the right airport. It means checking everything in your daily life.

LOOKING FOR PHONE NUMBERS?

Looking for phone numbers of people or companies? The Web has many sites that will give numbers. Some will also let you do reverse lookups – you type the phone number; it tells you who owns it. The sites vary in quality. There is no "best one." Check several. Good web sites are www.AnyWho.com, www.TheUltimates.com, www.BigYellow.com, www.SuperPages.com, www.WhoWhere.com, www.SuperPages.com, www.WhitePages.com and www.freeality.com.

WHO OWNS A WEB SITE?

To find out who owns a web site, go to www.netsol.com/cgi-bin/whois/whois. It's also a fine way to find out contact information on a company which doesn't list anything on its web site. It will give you an address, a phone number and an email address.

HOW TO FIND DRIVERS FOR OLD HARDWARE

1. Go to the manufacturer's site.
2. Go to the site of the manufacturer which bought your manufacturer.
3. Go to www.Gnutella.com.

eBAY CAN SAVE MONEY

eBay now carries a remarkable number of telephone systems, networking and computing equipment. Best eBay bidding tip: Wait five minutes before the end of the auction to put in your bid. Watch for sellers who don't have a record, who are selling product at ridiculous prices and worse, who are based overseas. Best trick: Ask them to send the product by Fedex Collect. Have a bank check ready.

RECORD A PHONE CONVERSATION

Can you record a telephone conversation? Yes...but it is common courtesy to let the person on the other end of the line know you'll be recording. You may have heard you need to sound a beep-tone every 15 seconds or so. That is not necessary. Verbal notification at the beginning of the taped conversation should be adequate. Local regulations vary, so you should check with your local public service commission.

PBX DIGITAL EXTENSIONS CAN HARM

Digital PBX extensions work on much higher electrical voltage than normal analog lines. These high-voltage lines may damage your laptop's modem. Is the phone line is digital? Check underneath the phone. If it "complies with part 68, FCC Rules" and has a Ringer Equivalence Number (REN), the phone is analog.

YOU MUST INSTALL TELEMANAGEMENT SOFTWARE

Telemanagement software is a no-brainer for any corporation. The full suite of software includes call accounting, cost allocation, bill reconciliation, traffic analysis, network optimization, inventory management, cable and wire management, work order management, trouble ticket management and directory management.

Telemanagement software works with any PBX, or with multiple PBXs in a multi-site

enterprise. It can save a ton of money, since it helps you manage network assets that cost a ton of money. Call accounting and cost allocation are the applications on which most companies focus first because they yield immediate hard-dollar savings. Call accounting lets you run reports, either scheduled or on demand, that identify toll fraud and toll abuse...and stop it. Cost allocation lets you identify which individuals, work groups, divisions or other cost centers are costing you the most.

If those costs are out of line, you now have the information to bring them back under control. You can charge those costs against departmental budgets. You can bill clients for the costs you incur working on their project.

BEST STOCK MARKET ADVICE - 1

Never buy anything without placing a 15% stop loss order. When your stock rises (some do), lift your stop loss order. With stop loss orders, you lock in your gains and you limit your losses and bring some discipline to your investing.

BEST STOCK MARKET ADVICE - 2

Bad news always follows bad news. Or said another way, the first bad news is not the last bad news. I call this my Cockroach Theory. Once you find one cockroach, you know you'll find more – though it may take days.

BEST STOCK MARKET ADVICE - 3

Management, management, management should be the investors' main slogan, just as location, location and location is the real estate industry's slogan.

BEST STOCK MARKET ADVICE - 4

Wall Street's advice stinks. With rare exceptions, Wall Street's analysts, strategists and money managers (etc.) are salespeople, not analysts. Their job is to sell the stocks they recommend.

BEST STOCK MARKET ADVICE - 5

Todd Kingsley is my broker with Smith Barney in Washington, D.C. (202-857-5459) who always wanted to be a financial journalist. His best tips:

Every day your portfolio is cash. If you wouldn't buy a specific company at today's price (assuming tax issues) than you should not own it. Every day you own a stock it's as if you bought it that day! You must ask yourself "Would I buy this company today with fresh money?"

When the reason you bought a company (i.e. good earnings, takeover speculation, good momentum) ceases to exist, get out. Do not hang around on a prayer.

Better to be wrong and OUT than wrong and IN."

If you stick to 15% stop losses you will NEVER average down. Or conversely, never catch a falling knife. Don't keep buying companies till they become penny stocks.

Look at the charts. they don't lie!

When an investment sounds too good to be true, it is.

Diversify. Diversify. Diversify.

Don't ever buy on margin.

YOUR HARD DISK SHOULD BE THROWN OUT

Hard disks should be thrown out once a year. Hard disks are mechanical devices that wear out. Before yours wears out completely, clone it and then throw it out, or keep it as a remote backup.

SOME INVOICES ARE NOT INVOICES

Many organizations send "invoices" that aren't invoices. They're solicitations. But they look like invoices. They usually come from newsletter publishers or phone directory or telex publishers. Often they come from overseas. Be careful. Don't pay them.

PUT YOUR WALLET ON YOUR SCANNER OR COPIER

Place the contents of your wallet on a photocopy machine and copy both sides of each license, credit cards, etc. You will know what you had in your wallet and all of the account numbers and phone numbers to call and cancel, in case your wallet is lost or worse, stolen. If your wallet gets stolen, cancel your credit cards immediately. The key is having the toll free numbers and your card numbers handy so you know whom to call. File a police report immediately in the jurisdiction where it was stolen, This proves to credit providers you were diligent, and is a first step toward an investigation (if there ever is one).

But here's what is most important: Call the three national credit reporting organizations immediately to place a fraud alert on your name and social security number. The alert means any company that checks your credit knows your information was stolen and they have to contact you by phone to authorize new credit. There are records of all the credit checks initiated by the thieves' purchases.

The credit card database companies' numbers are:
Equifax: 1-800-525-6285
Experian (formerly TRW): 1-888-397-3742
Trans Union: 1-800-680-7289
Social Security Administration (fraud line): 1-800-269-0271

GET AN 800 FAX NUMBER FOR INCOMING ORDERS

Many companies take orders via fax. It's easy to get connect your fax machine to an 800 line or two. That way you'll give your customers a greater incentive to reach you.

BYPASSING PHONE TREES

The Wall Street Journal of November 22, 2005 carried this story:

The Problem: Getting a person on the line when calling customer support.

The Solution: Consumers looking for shortcuts to reach customer-service representatives can check the Interactive Voice Response cheat sheet (named for the technology in automated calling systems) at www.paulenglish.com/ivr. The site, started this year by Paul English, co-founder of travel search engine Kayak.com, has phone numbers for 110 organizations and tips on bypassing the touch-tone or voice-recognition process (such as saying "representative" when calling FedEx, or hitting "0" many times quickly for Ikea.) The site is constantly updated with numbers suggested by the site's supporters. A handful, including one for Best Buy, are unpublished internal numbers given anonymously by employees. For numbers not on the list, try a search engine. Entering "Amazon customer service," for example, brings up a Web page by ClicheIdeas.com, dedicated to contacting customer representatives at Amazon.com. Mr. English also suggests a common trick: Don't press any keys and wait until a person picks up the call.

THE SQUEAKY WHEEL WORKS BEST

When things go awry, the squeaky wheel gets the most attention.

DON'T EVER GIVE YOUR REAL EMAIL ADDRESS OUT

Get yourself at least two email addresses – your real one and others for giving out on the web and elsewhere. Answer your real email address on your real PC. Answer your other email addresses on another spare PC. This way you automatically prioritize your emails – just glancing at the unimportant messages.

TRAVELING FOR A TIME?

Spending a week or two in a nice resort? Call the local phone company up early and see if they have "guest dial-up Internet service." It won't be as fast as at your office. But it will probably work on a local number and consist of unlimited usage. When I wrote this I was in Hilton Head, North Carolina. The local phone company is providing me a guest Internet account for $15 a week – unlimited Internet access and accessible via a local phone line where all local calls are free. If you're staying a week or longer, it may even be cheaper to install your own phone line. I did this once in a hotel. It proved much cheaper and more convenient than using the hotel's system. I even installed my own portable answering machine.

CHECK CHECK CHECK

I ordered Internet service at the vacation resort. It didn't work when I got there. The password and userID were all in upper case. CHECK. I didn't ask if my money manager charged 1% for managing cash on which I was earning 1.75%. He did. CHECK. I bought 512 megs of RAM for new laptop. Half of it never worked. The motherboard was broken. CHECK.

THE UNIVERSAL CODE

Let's say you attach something on your LAN, like a router, and you need to get inside it. Open your browser and type 192.168.111.1, or 192.168.11.1, 192.168.1.1 or 10.1.1.1. You'll either get to the device's menu screen or you'll see a screen that says used ID and password. Type admin into both spaces. You should get into the device's menu.

THE RESET BUTTON SOLVES ALL

Let's say you've just messed up the installation of a wireless broadband router, as I did. Turn the thing around or flip it upside down and search carefully for the tiny button labelled "Reset." With the power on, hold it in for 20 seconds. That will typically restore all the factory's defaults – which means all your messups will be eliminated and exchanges for the factory's. They're usually better than yours.

WHAT'S THE TIME?

What's the time in your location? www.TheTimeNow.com. Cell phones and cable TV boxes also have accurate time.

UPDATE ON DOLLAR SAVINGS TIPS

For an update on and for more of these tips, please visit my investing web site, www.TechnologyInvestor.com, also called www.InSearchOfThePerfectInvestment.com.

Harry Newton

March 2007

WHY IS IT SO HARD TO BUY?

How You Should Orient Your Precious Telecom Budget in Today's Tight Times

I watched TV this morning. They advertised a product I wanted. I called the company. It was closed.

I sent an email to one of those sales@theircompany email addresses. I got back an automated reply. I'd hear from them within 72 hours. I never did.

I bought a scanner from the world's largest scanner company, Hewlett-Packard. My son lost the disk. The wouldn't let me download the installation software over the Internet. The ONLY way I could get it was to pay them $10 and wait three weeks for a CD. The software had no commercial value. Its only value was for running their scanner.

I call innumerable companies for information on their products only to get totally tangled up in automated attendant machines – "Dial 1 for this, Dial 2 for that..." Never did I hear an option that's for me. In desperation, I hit zero – only to be told that I've just made "an invalid selection."

Fred Goodwin, the CEO of the Royal Bank of Scotland, says, "Customers hate call centers with a passion because it verges on the miraculous if they are ever able to get back to the same operator" (and by definition get what they want done).

If the vast number of companies make life so difficult for their customers and potential customers, what are we investing so much money in all this fancy telecom technology?

Making the customer's life better is where I always start when I implement technology. How can what I do bring me more happier customers? How can what I do allow me to hold my customers tighter to my bosom? How can what I do allow me to sell my customers more of my products and services? To my tiny brain, the only reason we go to work in the morning is to please our customers.

Thinking this way puts telecom and computing technology into perspective. Let's try asking ourselves a few simple questions:

1. Can my customers reach my company 24 hours a day, seven days a week? By voice and by Internet?

2. When they reach a representative from my company, will they reach someone who wants to and can help? Or will they reach someone who is hog-tied by time-consuming, stupid rules – like having to transfer the call into another queue to reach the "engineer," forcing me to wait 35 minutes on eternity hold, or not being able to make a minor concession, like dropping the shipping and handling charge – a concession to the aggravation of waiting.

3. Does the web site have a phone number and an email address? Is someone around who can answer my email – without making me wait 72 hours? Is that person empowered to help me, like negotiating small items, like sending me a replacement disk.

4. What happens if the salesperson at your firm needs to bring someone else in? Can they quickly reach someone else? Does your phone system conference well? Do they understand the importance of getting back fast to customers?

5. Is my web site easy and logical to get around? Is there information on the web site that helps the customer understand how to use the products and how to buy new ones.

6. Can someone easily find the products our company makes? Are all our specifications and prices detailed? In contrast to a paper advertisement, the Web is of unlimited size. It costs virtually nothing to fill a web site with extensive specifications and great photographs on all the company's products. Better yet, the more information on the Web, the fewer dumb questions my expensive people have to answer. It's amazing how most companies ignore the simplest information on their products – like their full technical specifications, the cost, what colors they come in, what the availability is... How most web products pictures are too small.

7. Is my web site logical? For example, if you ask a leading airline for all its flights from New York to Los Angeles, it won't tell you about flights from Newark or those to John Wayne or Long Beach Airports. Why? Who knows? Too much fun technology. Not enough thinking about the customer. Fortunately, some airlines use the code NYC to mean "all New York area airports" – JFK, La Guardia and Newark.

8. Can someone buy something from me on my web site? Or do they have to go to another web site – for example a distributor, whose web site is no longer working? What about spare parts? Are they readily available? does my web site have exploded diagrams of the products I sell and thus the parts that may need fixing or replacing?

8. Do I provide something of value on my web site that my customers can use? For example, if I'm a credit card company, do I allow my customers to find out how much they owe me? Will I allow them to download their accounts into their own Quicken software? Ditto for brokerage companies, which are probably the worst.

9. Do I regularly check with my customers for their reactions to how well my computing and communications systems work? Microsoft tested one of its Office releases on 25,000 people. How often do you ask your customers, "How are we doing? What can we do better?"

10. How easy is it for my largest and most important customers to reach people at my company? Do my salespeople carry cell phones, BlackBerries, laptops, etc. Do my engineers hide behind voice mail or do they solve my customers' problems? Why do emails to salespeople take so long to be answered?

**11. How much can I shift the burden and expense of

my people selling you, the customer, to you buying yourself – self-service over the web. Airlines are giving discounts if I buy my ticket over the Internet and print my boarding pass on my home printer. They share their employee savings with their customer. Some companies take my credit card and download me my software. Saves me waiting three weeks for UPS. Saves me disposing of all that packaging. Gives me instant gratification.

How can I use computing, networking and communications to give instant gratification. Customers like buying directly on a web site if it's easy and fun to use. Check www.stables.com and www.llbean.com. They're two of the easiest sites to use. I also like dealing with Fromuth. I call up and say "send me a carton of tennis balls." Two days later I have my balls. Fromuth retains my address, phone number and my credit card. I don't have to repeat all my information. I prefer it that way.

There's a huge amount of really neat customer relationship management software, which helps agents solve customer problems. The Web is evolving some really neat customer-friendly tools. Here are some of my favorite "customer-pleasing" tools that some (very few) web sites have:

A "permanent" Shopping Cart, also called basket. My Amazon will always be there - even if I have to quickly close my laptop and go out to dinner.

I like stored profiles. I really like web sites that keep my name, address, phone number, credit card, expiration date, etc. Saves me enormous time entering the information the next time I want to buy something. Really and truly keeps me coming back to that web site.

A reminder list of what I bought last time. As I go to check out at Staples.com, it brings up a list asking me if I'd like to re-order some of the things I bought last time and the time before. It reminds me I need more toner for the copier and more paper for the fax machine.

A reminder emailing. When Zappos gets new Puma sneakers in, it tells me. It also includes a link so I can easily buy more Puma sneakers I don't need.

A "continue shopping" button that takes me back to the place I just left. Often if I want to buy two of something, perhaps in a different color, I have to spend huge time schlepping from my shopping cart all around the web site until I finally end up in the place I started.

I like "Wish Lists," also called "wishlists." When I get rich, I'll buy them. Meantime, it's great to have my list kept for me. Maybe my kids will buy them for me.

"It's out of stock, but when it comes in, we'll send you an email." This is extremely useful, for obvious reasons.

Customer reviews of products are often useful. Most web sites are afraid of what their customers might say about their products. But a great customer review will sell more than a great seller review.

I like to see the full specifications. That includes size, voltage needs, etc. I had to email PCConnection.com to find out the size of a printer they were selling. The printer fit my bill. But I didn't know if it fit my space. It took PCCConnection two days to respond to my email. Which brings me to:

Your web site should have an easy way customers can email you and ask a simple question. They should get an email response within five minutes - not 48 hours as many of today's web sites say.

The search engines on most web sites are pretty miserable. To find something, you often have to enter it in many "creative" ways before you actually find what you want. For example, I wanted to buy a Canon SD800 camera. But I didn't know if it was spelled SD800, SD-800 or SD 800. Searching all three ways on camera retailer sites turns up remarkably different results.

Some sites are awfully slow. I've complained, with no avail, to Vanguard.com about their awfully slow site.

Some sites are "all over the place." In some you'll find buttons on the top, on the left side, and on the bottom. My favorite is the site that puts in the logon button at the top and the logoff button at the bottom.

I like sites that include maps.

I like sites that have do-it-yourself things, like boarding passes, movie tickets, etc. I have a printer. If using it saves me standing in line, I'm all for it.

In short, lots of work. Lots of great opportunities.

Harry Newton

Harry@HarryNewton.com

DISASTER RECOVERY PLANNING

How To Protect Your Computing and Telecom Resources

By Ray Horak

Disasters are a fact of life. We know they will happen, but we hope they will not happen to us. They are completely unpredictable and totally predictable. There's nothing we can do about them and everything we can do about them. How well we minimize their impact on our businesses and our lives depends on logic, planning, effort and, ultimately, money.

Disaster planning stretches across a broad continuum. At one end is doing absolutely nothing. At the other end is complete real-time duplication of everything we do and every system we use, in two or three separate physical places.

We know something will happen to threaten our business at some time. It could be a natural disaster – earthquake, hurricane, fire or flood. It could be an unnatural force such as a power failure, a terrorist strike or a hacker attack.

Let's take a look at Disaster Recovery Planning, also known as Business Continuity Planning. Planning is the operative word, for it makes the difference between catastrophe and continuity. The planning process comprises risk assessment, criticality assessment, loss assessment and cost assessment.

- Risk Assessment is the process of assessing the risk of failure of a network and its subnetworks. You assess the risk down to the level of individual network elements that comprise them. It includes all forces that might cause such a failure, whether the result of forces of nature, man or machine. Risks can be categorized from low to high. The risk of a system failure often can be quantified in terms of MTBF (Mean Time Between Failures) and the duration of the failure in terms of MTTR (Mean Time To Repair). The failures themselves can be categorized as either hard or soft. A hard failure is a total, or catastrophic, system failure. A soft failure is a performance failure, or degradation in performance, that falls short of total failure. Risk assessment can be reduced to a set of mathematical probabilities of occurrence.
- Criticality Assessment is the assessment of the importance of individual computing and communications resources, including computer systems and their resident databases, telecommunications systems, and voice and data networks and subnetworks. This assessment also includes groups of users, the applications on which they rely and the business functions in which they are engaged. Levels of criticality might include critical or essential, important, and non-critical or non-essential. Or, the levels of criticality might be defined as very high, high, moderate, and low. Further assessment of criticality might establish an acceptable recovery window, i.e., the length of time that the business can survive or thrive in the event of the total failure of a specific resource or function. Utmost in this overall assessment must be the very nature of the core business. Call centers, which deal with customer sales, for example, cannot tolerate even short-lived failures of telecommunications systems or networks. Similarly, financial institutions cannot tolerate even short-lived failures of computer systems or networks. Airlines and courier services cannot tolerate either.

- Loss Assessment is the process of assessing the monetary cost to the business of the total failure of a specific resource or function. Losses clearly are sensitive to the criticality of the resource or function, and generally are sensitive to the length of the failure, or recovery window. For example, some critical functions might withstand a failure of 15 minutes, and some non-critical functions might withstand a failure of up to 30 days, which might be equivalent to a billing cycle. Losses can be external or internal. An external loss might take the form of the failure of an e-commerce web site can result in the immediate loss of revenues and profits. Such a failure also can result in the loss of customer confidence and market fidelity, which translates into the long term loss of revenues and profits. An internal loss associated with the same failure might take the form of the lost productivity and increased stress on employees, the loss of employee confidence in the systems and, ultimately, increased employee turnover.

Once the organization has assessed risk, criticality and loss, it's in a position to consider measures designed to prevent failures, and measures designed to recover from failures in the event that the preventative measures fail. All of this is in the context of cost, of course.

Barriers are designed to prevent disasters. Unfortunately, barriers are few in number, being limited to things like earthquake bracing, mechanical and electronic locks and dead-bolts, electrical surge protectors, sump pumps and software firewalls.

Backups are designed to assist in the recovery from failures. Redundancy translates into resiliency, which allows a business to snap back from either a hard or a soft failure. A hard failure forces a business to exercise a backup. A soft failure, on the other hand, affords the organization the choice of limping along for a period of time while the problem is diagnosed and corrected, or exercising the backup at any time.

Cost assessment, of course, is the final step in the development of a disaster recovery plan. The costs of implementing alternative business continuity plans must be considered in the context of the risks of failure, the criticality of various resources, and the potential losses arising from such failures over time. Striking this balance is the essence of optimization, and business continuity plans must be optimized. There are other constraints that also must be considered, such as the lack of trained personnel and the time available to put a solution in place, but cost is the one that usually gets our attention.

Math exercise

All of this can be expressed in a set of relatively simple mathematical formulas:

$$P \times STI = ESTE$$

P: Probability of occurrence (%)

STL: Short Term Loss ($)

ESTE: Expected Short Term Exposure ($)

ESTE - STC = NC
ESTE: Expected Short Term Exposure ($)
STC: Short Term Cost of Solution ($)
STNC: Short Term Net Cost ($)
P x LTI = ELTE
P: Probability of occurrence (%)
LTL: Long Term Loss ($)
ELTE: Expected Long Term Exposure ($)
ESTE - STC = NC
ELTE: Expected Long Term Exposure ($)
LTC: Long Term Cost of Solution ($)
LTNC: Long Term Net Cost ($)

If both the Short Term Net Cost and Long Term Net Cost are positive numbers (which sounds strange, because losses are never really positive), then the solution clearly should be implemented. The bigger the number, the greater the emphasis that should be placed on addressing that particular potential failure. When this process of evaluation is completed for all identified failures, it becomes a simple matter to rank them.

Continuity Checklist

Here are some of the things we can do to protect our businesses from disasters and to recover from them quickly. It's an abbreviated checklist. Entire books have been written on the subject.

Electrical continuity is critical, as most, if not all, network elements are electrically powered. The criticality of reliable electrical power became quite clear in the U.S. as a result of the power shortages during the summer of 2001. Surge protection is absolutely essential to protect network elements from voltage spikes caused by unclean power and lightening Grounding is very basic, but it is worth noting that a great many system failures, both hard and soft, are due to improper electrical grounding. Power supplies are redundant in fault tolerant computers and in carrier class switches and routers. UPS systems are always a good idea, and may comprise both battery backup and backup power generators. At the very least, a UPS system provides enough time to shut systems down gracefully. At the very most, backup generators may provide power indefinitely. Grid diversity, perhaps the ultimate in electrical continuity, involves access to multiple electrical grids, often through multiple utility companies.

System continuity clearly is critical. Fault tolerant computers and carrier class switches and routers routinely are highly redundant at the component level. Clearly, there are wide ranges of redundancy at the system level and wide ranges in associated costs.

Data continuity is extremely important. Application programs and files should be backed up routinely to ensure the continuity of the business, itself. Contemporary backup media options run the range from floppy disks to magnetic tapes and CD-ROMs. SANs (Storage Area Networks) routinely make use of highly redundant storage systems including RAID (Redundant Array of Inexpensive Disks). The ultimate in data backup involves storing backed up programs and files at a separate site.

Access continuity ensures that network access is available on a highly reliable basis. Access continuity includes loop diversity and media diversity. Loop diversity involves multiple levels. Entrance diversity involves multiple points of loop entrance to a building. Pair diversity involves access via diverse pairs in a multi-pair cable system. Cable diversity entails access via multiple cables, which may be UTP (Unshielded Twisted Pair) or fiber in nature. (Note: SONET standards for fiber optic systems specify as many as four fibers.) Path diversity requires that the local loops follow multiple, diverse physical paths between the network edge and the customer premises. Aside from the inherent redundancy of SONET systems, loop diversity is unusual in all but the most critical applications scenarios. Media diversity involves the use of several media. In the event that the primary access medium fails, the backup medium can be initialized, perhaps with little, if any, disruption in service. Wireless media (e.g., microwave and infrared) routinely are used as backups for wired media (e.g., UTP and optical fiber). Some service providers offer microwave systems as a backup to infrared (Free Space Optics, or FSO), and others offer do just the reverse.

Transport continuity ensures that continuity of connectivity is maintained within the core of the service provider's network. This level of continuity typically is supported through the use of optical fiber transport systems based on SONET standards. SONET, as I mentioned above, specifies redundant fibers. In the core of the carrier networks, this generally is in the form of a 4FBLSR (Four Fiber Bidirectional Line Switched Ring). Carriers generally provide connectivity assurances in the form of guaranteed service restoral windows, as stated in SLAs (Service Level Agreements). Frame Relay connectivity can be protected through the use of backup PVCs (Permanent Virtual Circuits), which generally are available at discounted cost. SVCs (Switched Virtual Circuits), while inherently redundant, generally are not available from service providers. The Internet is inherently highly redundant and IP's connectionless nature uses that redundancy to the fullest.

Carrier diversity involves the use of multiple carriers for a given service. Access to the circuit-switched PSTN (Public Switched Telephone Network), for example, might make use of a LEC (Local Exchange Carrier) for local calling purposes, and an IXC (InterExchange Carrier) for long distance. In the event of a failure in the LEC network, the IXC typically can be used for local calling, and vice versa. Carrier diversity is relatively easily accomplished in the contemporary competitive environment, but is all too rarely employed.

Network diversity involves the use of multiple network services, ideally through diverse carriers. For example, ISDN routinely is used as a backup for Frame Relay. Dial up modem access through the PSTN routinely is used as a backup for Frame Relay and other data network services. Cell phones back up landlines and vice versa. The options are so numerous and the costs so reasonable these days that it is irresponsible not to have some level of network diversity in place.

Site diversity takes several forms. Distributed vs. Centralized Operations: Distributed operations at multiple sites may involve additional costs, but certainly is less susceptible to catastrophic failure than is centralized operations at a single site. Mirrored Operations involve maintaining an exact copy of the center at a remote backup site for standby purposes. Critical data centers and call centers, for example, often are mirrored. Such an exact copy includes systems, applications, files, and networks. A hot standby is always powered up, and ready instantaneously, meaning that data is processed and files are maintained at the hot standby site concurrently (i.e., in parallel) with the operational sites. A warm standby is ready to fire up in a short period of time, perhaps once files are updated. A cold standby may require that some applications be updated, for example, which takes a bit longer. A number of companies are in the business of providing mirrored data centers, which sometimes are shared by multiple companies with the same general system and network configurations. Also, some companies provide backup call center capabilities on an outsourced basis.

Think of business continuity planning as a form of business insurance. Most businesses wouldn't even consider operating without medical insurance, liability insurance, fire and flood insurance, or vehicle insurance. Neither should they even consider doing without loss of business insurance, most of which policies require the existence of a well developed disaster recovery plan. It's not uncommon for a company to spend as much as 10% of its annual IT budget on the development and implementation of such a plan. It's just one more cost of doing business, and it is money well spent. In these times when it seems as though both Mother Nature and mankind have gone crazy, a little insurance can go a long way.

RULES FOLLOWED IN THIS DICTIONARY

How To Figure Our Ordering of Definitions and Our Spellings

by Harry Newton

For 20-plus years, I've started all my definitions with capital letters – e.g. Address Filtering, instead of address filtering. And for 20 years my readers have been emailing me. "You got it all wrong, idiot. You should be like other dictionaries. Go lower case. Not all words begin with capital letters." OK. OK. We've gone through the entire dictionary and changed the capitalization to the way it's meant to be. Or at least the way it seems to be meant to be.

The problem – and there's always a problem – is that it's not always clear which should be capitalized and which shouldn't be. Obviously proper names should stay capitalized. But what about words that spell an acronym – e.g. Multipurpose Internet Mail Extension which spells out MIME. When we first started de-capitalizing, we decided to leave in phrases that spelled out an acronym in order to make the acronym clear. But, after mulling more, we thought that didn't look right, so we dropped the capitalizations. So, now you'll find local area network as all lower case, even though it spells a common acronym, i.e. LAN (in capital letters). I suspect that we didn't catch all and there may be some small inconsistencies remaining. Suffice, this is a a work in progress. Let me know if you have thoughts (or a rule) on this.

ASCII Order.

Most dictionaries put their words in alphabetical order. Dah! I do, too. But, it's not that easy with a technical dictionary. Some "words" contain non-letters – like @, #, / and numbers like 1, 2, 3. So, I made a decision that we'd put my dictionary in ASCII order. What this means is that terms beginning with letters are in alphabetical order. Terms with non-letters are in ASCII. Here is the order you'll find in this dictionary:

Blank Space	= ASCII 32
!	= ASCII 33
#	= ASCII 35
& (Ampersand)	= ASCII 38
- (Hyphen or dash)	= ASCII 45
. (Period)	= ASCII 46
/ (Forward slash)	= ASCII 47
0 (zero)	= ASCII 48
1	= ASCII 49
2	= ASCII 50
3	= ASCII 51
4	= ASCII 52
5	= ASCII 53
6	= ASCII 54
7	= ASCII 55
8	= ASCII 56
9	= ASCII 57
: (colon)	= ASCII 58
; (semi colon)	= ASCII 59
A (capital A)	= ASCII 65
Capital letters to ASCII 90	
(back slash)	= ASCII 92
Lower case letters start with a	= ASCII 97

American Spelling

This dictionary conforms to American spelling. To convert American spelling to British and Canadian spelling typically requires adding a second "L" in words like signaling and dialing (they're American) and changing "Z" in words like analyze to analyse. Center in American is Center. In Britain, Europe, Australia and Canada, it's Centre. This dictionary contains more British, Australian and European words (and their correct spellings in those languages) than previous editions – a result of several overseas lecture tours. I was born in Australia. When I came to the U.S., I thought all the local spellings to be funny. Now I think the Australian spellings are funny. It all depends on what you're used to.

All high-tech industries make up new words by joining words together. They typically start by putting two words next to each other. Later, they join them with a hyphen. Then, with age and familiarity, the hyphen tends to disappear. An example: Kinder garten. Kindergarten. and now Kindergarten. A word that's evolving is electronic mail. At first, it clearly was electronic mail. Then it moved to e-mail and now it seems to be morphing to email, i.e. it's becoming a real word. The hyphen "rule" is not a rule. It's determined by time, and thus not easy to be "right." For example, all the literature refers to "C Band" in the radio world and C-Band in the optical world. It may very well be a construct to make the distinction, much like we use "frequency" in the electrical and radio domains, and "wavelength" in the optical domain. Some of this stuff doesn't have anything to do with rules of English. Rather, it comes down to the preferences and objectives of the standards bodies and others.

Sometimes it's a matter of personal choice. Some people spell database as one word. Some as two, i.e. data base. I prefer it as one, since it has acquired its own logic by now. Sometimes it's a matter of how it looks. I prefer T-1 (T-one), not T1, simply because T-1 is easier to recognize on paper. I define co-location as co-location. Websters spells it collocation, with two Ls, one more than mine. I think mine is more logical. And since Mr. Webster is dead, he can't argue with me. I like email. But readers have told me they prefer e-mail. You'll probably find it spelled both ways in this dictionary, since it's not easy to be thoroughly consistent in an industry (and dictionary) changing so fast.

Plurals

Plurals give trouble. The plural of PBX is PBXs, not PBX's. The plural of PC is PCs, not PC's, despite what the New York Times says. The Wall Street Journal and all the major computer magazines agree with me. The plural possessive is PBXs' and PCs', which looks a little strange, but is correct. In this dictionary, I spell the numbers one through nine. Above nine, I write the numbers as arabic numerals, i.e. 10, 11, 12, etc. That conforms to most magazines' style.

Sometimes the experts don't even get it right. Take something as common as 10Base-T. Or is it 10BaseT? 10Base-T is an IEEE standard. So you'd think the IEEE would know. Forget it. Go to their web site, www.ieee.org. You'll find as many hits for 10Base-T as for 10BaseT. I checked every known and unknown expert in the Western world. We now believe the correct spelling is 10Base-T.

Sometimes, I don't simply know. So I may list the definition twice – once as two separate words and once as one complete word. As words and terms evolve, I change them in each edition. I try to conform each new edition to "telecomese" and "Internetese" as it's spoken and written at that time.

Bits and Bytes per Second

Telecom transmission speed has confused many of my readers. I hope this will help:

The telecom and computer literature is loaded with references to Bps and bps. You'll see them as Kbps, kBps or KBps, or Kbits/sec. You'll see them as Mbps or MBps. You'll see them as Gbps or GBps. There is no consistency in the industry's "literature," i.e. brochures, articles, spec sheets, etc. Let me explain:

First, k means kilo or a thousand; m means mega or one million. And g means giga, which is a thousand million, or 1,000,000,000. The term kbps means a thousand bits per second. And that's a telecom transmission term meaning that you're transmitting (and/or receiving) one thousand bits in one second. The term mbps means one million bits per second. Note the k and the m are small, i.e. non-capital letters.

There is an exception to this neat rule. Fibre Channel and other transmission systems used in SANs (Storage Area Networks) measure transmission speeds in Bps (Bytes per second). Here, the terminology is driven by the application, which is the transfer of data between storage systems.

Now to computing speeds: The term KBps (with a big K and a big B) means one thousand bytes per second. MBps (with a big M and a big B) means one million bytes per second. They refer to speeds inside the computer, e.g. from your hard disk to your CPU (central processing unit – your main microprocessor). There's a big difference between a bit and a byte. A byte is typically (but not always) eight bits.

That's the way it's meant to be. But, there's a lot of sloppy writing out there. You'll see MBps or MB/s also meaning one million bits per second as a telecom transmission speed. You really have to figure out if the writer means telecom transmission – i.e. anything outside the computer – or whether the writer is referring to speed insider the the computer in which case it's bytes and a computer term. You can usually tell from the context.

Measuring the speed of a communications line is not easy. And tools to measure lines are still very primitive. The Internet added a whole new dimension to complexity. Since the Internet is a packet switched network, every transmission goes a different way. So, one transmission that might be one million bits per second might, a minute later, be 800,000 bits per second. There are sites on the Internet that measure your connection speed, by sending you a big file, and then waiting for you to send back some message that you've received the file. But they report numbers all over the space from one moment to another. About the only certain thing you know is that the speed of a circuit is always measured by the slowest part of the circuit. Look at the Internet. You might be getting horribly slow downloads, despite being on a T-1. That might be due to a horribly overloaded server at the other end or it might be due to the fact that your T-1 is overloaded with other users at the office, also downloading. These days with faster lines what's often a gating factor is the speed of your PC. It may be simply not be fast enough for your PC's browser to keep up with the speed of your incoming bits. In which case you need a faster PC. It happened at our home when we got fast new cable modem. We all had to upgrade to faster PCs.

Virtually all telecom transmission is full duplex and symmetrical. This means if you read that T-1 is 1,544,000 bits per second, it's full duplex (both ways simultaneously) and symmetrical (both directions the same speed). That means it's 1,544,000 bits per second in both directions simultaneously. If the circuit is not full duplex or not symmetrical, this dictionary points that out. For now, the major asymmetrical (but still full duplex) circuit is the xDSL

family, starting with ADSL, which stands for asymmetric, which means unbalanced. The DSL "family" no longer starts with "A," and most of it (but not all of it) is still asymmetrical. The one major exception, SDSL (Symmetrical Digital Subscriber Line) is clearly symmetrical. Cable modems and satellite Internet connections are typically asymmetrical, also.

There's one more complication. Inside computers, they measure storage in bytes. Your hard disk contains this many bytes, let's say sixty gigabytes (thousand million bytes). That's fine. But they're not bytes the way we think of them in internal computer transmission terms. They're different and they have to do with a way computer stores material – on hard disks or in RAM. They're what I call "storage bytes." When we talk about one KB of storage bytes, we really mean 1,024 bytes. This comes from the way storage is actually handled inside a computer, and calculated thus: two raised to the power of ten, thus 2 x 2 x 2 x 2 x 2 x 2 x 2 x 2 x 2 x 2 = 1,024. Ditto for one million, two raised to the power of twenty, thus 1,048,576 bytes.

Which Words Get Defined?

Which words get defined in this dictionary? These are my rules: All the important terms in the field are defined. No proprietary products, i.e. those made by only one firm, are defined. No proprietary terms are defined. I am the first to admit that my rules are not precise. Writing a dictionary is very personal. I read over 100 magazines a month. I study. I mull. I try to understand. Eventually, my wife calls, "Enough with the words, already. It's 2:00 AM. Time to sleep."

A or An? Here's The Logic

I admit my fallibility. This edition of this book is riddled with "a" when it should be "an" and "an" when it should be "a." I've never been confused. I always believe "an" is used before vowels, and "a" before consonants. Not so, says my friend, Jay Delmar, who edits technical documentation. Here's his explanation.

Concerning the problem of what article ("a" or "an") should be used with a word or an acronym, it all depends on how the acronym is pronounced, that is, whether it's pronounced as a string of letters or as a word. In some cases, the article would be the same. In others, the form would have to switch. Usually "an" is used before vowels, but some consonants require it as well, and some vowels require an "a." It all depends on the sound. Whether a letter is intrinsically a vowel or a consonant doesn't really matter; what matters is if it's pronounced as a vowel or a consonant in the particular context.

If an acronym is pronounced as a string of letters, the following shows the appropriate article to use with the first letter of the acronym:

An A	An H	An O	A V
A B	An I	A P	A W
A C	A J	A Q	An X
A D	A K	An R	A Y
An E	An L	An S	A Z
An F	An M	A T	
A G	An N	A U	

If an acronym is pronounced as a word, the article might need to change:
An RS-232, but a RAM (pronounced "ram")
An STP, but a SRDM (pronounced "sardem") and a SLC (pronounced "slick")
An FTP, but a FAIC (pronounced "fackey")
An HIC, but a HICUP (pronounced "hiccup")

According to The New York Public Library Writer's Guide to Style and Usage, "The article a is used before all consonant sounds, including a sounded h, a long u, and an o with sound of w (as in one). The article an is used before all vowel sounds except a long u and before words beginning with a silent h." This definition has never helped me because I've never really understood why in "an STP" the "s" sound is a vowel sound and in "a SRDM" the s sound is a consonant sound. Basically, I rely on my ear.

The real trouble, of course, is that unless one is really, really familiar with the acronym, one doesn't know how it's actually used: pronounced as a string of letters or as a word. I thought SIPL would be "an SIPL" (an ess-eye-pea-ell) until fairly recently. I didn't know it was pronounced "a sipple"-or "a sighpull" (I've heard it both ways). Jay makes sense. I'm going to try to be more in line with his concepts in upcoming editions. But this one may have a few inconsistencies.

Thank Yous for Help On This Dictionary

Among the manufacturers, special thanks to Anixter, Aspect Communications, AT&T, Intel/Dialogic, Ecos Electronics, General Cable, Intel, Lucent, Micom, Microsoft, Worldcom, NEC, March Networks, Mitel, New York Telephone (now Verizon), Northern Telecom (now Nortel), Racal Data, Ricoh, Sigma Designs, Sharp and Teknekron. Among the magazines I borrowed (or stole), the best were PC Magazine and Teleconnect, Call Center, Computer Telephony (now Communications Convergence) and Imaging Magazines. Special thank yous also to internetworking expert, Tad Witkowicz at CrossComm, Marlboro MA.; to Russ Gundrum, Network Engineering Manager - Transport at Southwestern Bell in Bellaire, TX; to Stephan Beckert of The Strategis Group in Washington, D.C.; to Ken Guy erstwhile of Micom, Simi Valley (near LA); to Robert M. Slade, who does a wonderful job reviewing books (including this one); Frank Derfler of of the US Air Force, then PC Magazine, then just by himself, was especially helpful; Chris Gahan of 3Com; to Bob Rich of Boeing's System Engineering Group, who's studying for his MCSE; to Glenda Drizos, Enhanced 911 Project Leader at Sprint PCS, Overland Park, Kansas. Special thanks to Jeff Deneen, erstwhile of the Norstar Division of Northern Telecom in Nashville; Stephen Doster erstwhile of Telco Research in Nashville; bugging expert Jim Ross of Ross Engineering, Adamstown, MD; wiring experts John and Carl Siemon of The Siemon Company, Watertown CT; to Jim Gordon and Parker Ladd at TCS Communications, Nashville, TN, the people who do workforce management software for automatic call distributors; to Jun Sun of Cisco, the people who make the Internet routers; to Judy Marterie and the electricity wiring, grounding and test experts at Ecos Electronics Corporation in Oak Park, Il; to Brian Newman of MCI, who understands wireless; to John Perri of SoftCom, NYC; to John Taylor of GammaLink, a Sunnyvale, CA company which produces beautiful fax products (but which is now owned by Dialogic); to Charles Fitzgerald at Microsoft and Herman D'Hooge at Intel who jointly helped created Windows Telephony; to everyone else at Microsoft (including Mark Lee, Toby Nixon, Bill Anderson, Lloyd Spencer and Mitch Goldberg) and Waggener Edstrom (Microsoft's PR agency) who produce such great White Papers and keep pushing the state of telecommunications standards further; to Bill Flanagan who's written fine books on T-1 and voice and data networking.

Special thanks also RFIDJournal.com for their help on RFID definitions, to Jane Laino of Digby 4 Group, NYC, the best telecom consultants in the world; to Jeffrey Welch, consultant of Fenton Michigan, Jon L. Forsyth, Manager at Cambridge Strategic Management Group, Cambridge, MA.; to Henry Baird of Seattle consultants Baird & Associates; to Sharon O'Brien formerly of Hayes Microcomputer Products in Norcross (Atlanta); to Howard Bubb, John Landau, Jim Shinn, Nick Zwick and Sam Liss at leading voice processing component manufacturer, Dialogic Corporation of Parsippany, NJ; to David Perez and Nick Nance of COM2001.com, San Diego, CA; to Al Wokas of MediaGate, San Jose, CA; to Nayel Shafei of Qwest Communications; to Alison Golan of networking company, Interphase Corporation in Dallas, which allowed me to steal some of the definitions from their excellent booklet, "A Hitchhiker's Guide to Internetworking Terms and Acronyms;" to Rusty Powell, a very talented Senior Editor at Alcatel USA in Plano, TX.

Special thanks also to Ian Angus at the Angus TeleManagement Group in Ajax, Ontario, who embarrassed me into expanding my Canadian coverage; to Bruce Watson of Sprint Canada, Toronto Ontario, to Andrew Reichman, who works in E911 data processing in the Pacific Northwest; to Glenn Estridge, one of the world's leading experts on dense wave division multiplexing and the whole wonderful world of fiber optics; to John Arias, a seriously good technician Bell Atlantic who came to fix a busted line and left educating me on the intricacies of cable naming at his company; to Ed Margulies at Miller Freeman, New York, NY who's written so many fine books on computer telephony; to Charlie Peresta, P.E., PMP, Telecommunications Manager, Intellisource who helped me with some of the telecom energy terms.

Extra special thanks to Lee Goeller, an electrical engineer who, after a lifetime in telecommunications at Bell Labs, RCA and his own consulting firm, Communication Resources, has edged into retirement. During the 1980s and early 1990s he wrote the BCR (Business Comunications Manual) of PBXs and taught BCR's PBX seminars. If you want to argue with him he can be reached at leegoeller@aol.com.

I'm very grateful to Dan Thomas, VP Marketing of Telemobile Inc., Torrance, CA who helped me a lot with wireless local loop definitions.

I'm very grateful to The ATM Forum of Mountain View, CA (www.atmforum.com) for allowing me to use many of their definitions from their really well-done ATM Forum Glossary. I'm also grateful to Donovan Bezer, who works as a law clerk at the NJ Ratepayer Advocate (consumer counsel for phone customers) and who helped me understand the government/judicial treatment of the term ISP. Thank you also to Kevin Allaway, who has 20+ Years of IT connectivity experience. He's a muliti-certified talented engineer.

I'm very gateful to the unbelievably talented people at UBS Warburg – Pip Coburn, Faye Hou, Qi Wang, David Bujnowski, Weiyee In, Boris Markovich, Sean Debow and Rafael Volet. These people put out the best research on telecom and technology on Wall Street. They also published a small, but great dictionary, called Telexicon, which I've consulted. As they say in my business, if you steal from one person, it's called plagiarism. If you steal from many, it's called research. I do research. And their Telexicon was most useful in my research. Thank you guys.

One informed person on satellite communications is Michael Brady, an American living in Norway. He went to Norway on a Fulbright Grant, fell in love with the place and now lives and consults out of there. He deals mostly in high-tech documentation and has completed assignments for NERA Satcom and for the Norwegian Space Center. Typically, he works with R&D engineers to compile final documentation as systems/hardware/software are being developed, which cuts the time spent on and the cost of user documentation. This affords a significant competitive advantage, because user documentation often is required along with FAT (factory acceptance testing). He holds three University degrees in electronics from George Washington University, MIT and Stanford. He is now principal of own company, M Brady Consultants A.S. of Oslo. He has written and translated 20 books. Michael contributed many of the satellite definitions in this dictionary. Thank you, Michael.

I'm very grateful to Muriel Fullam, my assistant, and Gail Saari, layout artist.

Huge thank you to Gavin Wedell, who did a major polishing job on the 19th edition. His resume says it all "Highly motivated, fast learning and creative individual with a thorough understanding of the business and technology paradigm. Strengths in writing, analysis and oral communication." Fortunately he also knew a lot about technology since he had recently graduated from the University of Technology in Sydney, Australia and was on a respite before plunging back into another degree, this time on psychology and philosophy. Big thanks also to Buck Sexton, who helped the capitalization problem and who plays great tennis.

A big "Thank You" to the dozens of people and dozens of companies who helped. If I left you out, I apologize.

I wrote this dictionary on a series of ever-newer, ever-faster Toshiba Tecra laptops (very reliable machines) using The Semware Editor, a very beautiful text editor, which Sammy Mitchell of Marietta, GA wrote, and which I wholeheartedly recommend – www.semware. com. The Toshiba laptop for this dictionary was the wonderful Tecra M5. It's a 2.33 GHz machine with three gigabytes of RAM and two hard drives – a main 60 gigabyte drive and a removable (i.e. backup) 120 gigabyte drive. Thanks to Howard Emerson and Craig Marking from Toshiba for all their help.

To design and lay out this book, Gail Saari used a Macintosh Pro with a 2.66 GHz Dual-Core Intel Xeon processor and Mac OS Version 10.4.8. Software was Adobe Creative Suite 2 (including InDesign, Photoshop, Illustrator, and Adobe Acrobat 8 Professional). For automatically producing the dictionary-style running heads with absolutely no effort, Script Programming for InDesign was provided by Gavriel Harbater. Many thanks!

HARRY NEWTON
The Author

Harry Newton
50 Central Park West
New York, NY 10023
Tel: 212-712-2833 Fax 777-254-3491
www.HarryNewton.com
Email: Harry@HarryNewton.com
Investing web site:
www.InSearchOfThePerfectInvestment.com

Harry Newton keeps busy writing this dictionary, writing a daily column on investing (www.InSearchOfThePerfectInvestment.com), writing a book on investing (called "In Search of the Perfect Investment") and being an angel investor in technology startups. He does some public speaking on technology, telecommunications and investing. He really likes speaking and he seems to be popular since he keeps being invited back. In an earlier life, Newton and his brillant partner, Gerry Friesen, co-founded six successful magazines – Call Center, Computer Telephony, Imaging, LAN (now Network Magazine), Teleconnect and Telecom Gear. They also co-founded the immensely successful shows Call Center Demo and Computer Telephony Conference and Exposition, which at its peak attracted 26,000 people to the Los Angles Convention Center. They also published over 40 books on networking, imaging, telecom and computer telephony. Friesen and Newton sold their publishing company to Miller Freeman (now part of CMP) in September 1997. Friesen retired to California. Newton tried to retire, but failed. Newton is always willing to listen to a new idea for a business, but says he does a mean due diligence, which means he says NO a lot. Here are his criteria: 1. It should be a "hard" technology. that means it should be hard and expensive for someone else to duplicate. Sound patents are key. 2. The management team had better be incredible. That means they should have broad skills, integrity and a serious desire to succeed and make money. 3. They had better be obsessed with sales and marketing. What the world doesn't need is another technology playpen – a place angels and venture capitalists can endlessly dump money so the entrepreneurs can invent more "cool" products. 4. The valuation ought to be reasonable, i.e. to allow for considerable upside. 5. The company ought to believe in regular and open reporting to its shareholders. No one expects instant success. There will be stumbling blocks along the way. If those stumbling blocks are hidden away they become insurmountable. Open, they become surmountable.

STEVE SCHOEN
Contributing Editor

Steve Schoen
Senior Manager Market Research
Hawaiian Telcom
PO Box 37427
Honolulu, HI 96837
Phone/fax: 808-735-6971
Email: steven.schoen@gmail.com

Steve Schoen began working in the telecommunications industry in 1984, the same year that the first edition of Newton's Telecom Dictionary was published. The dictionary has been his trusted guide to telecommunications terminology ever since. He keeps a copy of the latest edition in his office and another copy at home, for ready reference.

Steve has worked in a variety of telco positions over the years, first at GTE, then at Verizon Communications, and now at Hawaiian Telcom. He started out working for one of GTE's local phone companies in Hawaii, and in 1989 moved over to one of GTE's international business units, also located in Hawaii, which became part of Verizon Communications following GTE's merger with Bell Atlantic. He is now at Hawaiian Telcom, which was formed when Verizon sold off some of its Hawaii operations.

On Saturday mornings, since the late 1980s, Steve has also been teaching hands-on computer classes to working adults at Honolulu Community College. And since the 1990s he has been teaching C, SQL, database management, telecommunications and e-business classes, both on-ground and online, at colleges and universities in Hawaii and on the mainland, including, among others, UC Berkeley, University of Maryland University College, University of Phoenix, and Chaminade University. In prior existences, he taught math in England, worked for a federally funded project that developed educational materials in native languages of Micronesia, and spent four years in the Army.

Since 1988 Steve has also had a modestly successful software company, but these days he refers all new project opportunities to former students, to help them grow their own businesses.

For recreation and relaxation he likes coffee breaks and lunch with friends and colleagues, sports, spicy food, juggling, reading, and listening to audiobooks, podcasts, broadcast radio and public radio.

DATES

1024 China issues the first paper money.

1389 Serbs defeated by the Turks on the Field of Blackbirds. Prince Lazar, who led the Serbs, was reported to have said, "It is better to die in battle than to live in shame." The Serbian Church later made Lazar a saint.

1453 Johannes Gutenberg, a goldsmith from Mainz, Germany, prints his Mazarin Bible, which is believed to be the first book printed with metal movable type, i.e., his famous Gutenberg Press. Movable type is best defined as printing with individual letters than can be composed into texts, printed, then disassembled and reused. It took Gutenberg two years to compose the type for his first bible. But once he had done that he could print multiple copies. It took three years of constant printing to complete Johannes Gutenberg's famous Bible, which appeared in 1455 in two volumes, and had 1,284 pages. He reportedly printed 200 Bibles, of which 47 still exist. Before Gutenberg, all books were copied by hand. Monks residing in scriptoriums, usually did the copying. They seldom managed to make more than one book a year. The Gutenberg press was a major advance. Before Gutenberg, there were only 30,000 books on the continent of Europe. By the year 1500, there were nine million. They covered the areas of law, science, poetry, politics and religion. Some people (including me) have likened the invention of the Internet to the Gutenberg Press. Johannes Gutenberg lived from 1397 to 1468.

1517 Martin Luther nails his 95 theses which criticize papal "indulgences" to a church door. The Reformation, splitting western Christendom, is on its way. See Indulgence.

1553 The Muscovy Company of London issues the first equity shares.

1639 Japan cuts itself off from the outside world. It re-opened in 1853.

1659 It is illegal to celebrate Christmas in Massachusetts.

1660 Clocks made before 1660 had only one hand – an hour hand.

1666 The year 1666 was much feared throughout Western Europe because of its triple sixes, "666," which represent the "Number of the Beast." While the world did not end, London was nearly destroyed by the Great Fire.

1687 Isaac Newton publishes his "Principia" which sets out the laws of motion.

1753 Benjamin Franklin invents the lightning rod. It was the first practical victory of science over a natural phenomenon. Two years later, when Lisbon, Portugal, was destroyed by an earthquake and a tidal wave, some ministers in Boston proclaimed it was a punishment for the sacrilege of using lightning rods to avert the wrath of God.

1758 Kamehameha the Great is born on the Big Island (i.e., the island of Hawaii).

1776 James Watt improved steam engine first installed. See 1785. The steam engine becomes to the Industrial Revolution what the computer is to the Information Revolution – its trigger.

The United States declares itself independent of Britain and all men are created equal.

1778 April 2, 1778. The Banda Islands suffer an earthquake, a tidal wave, a volcanic eruption and a hurricane. This collection of disasters effectively ended the Dutch monopoly on nutmeg. See Nutmeg.

1785 James Watt's improved steam engine is first applied to an industrial operation – the spinning of cotton.

1787 According to a bill for a celebration party thrown September 15, 1787, the 55 framers of the U.S. Constitution drank 54 bottles of Madeira, 60 bottles of claret, 8 bottles of whiskey, 22 bottles of port, 8 bottles of cider, 12 bottles of beer, and 7 large bowls of spiked punch big enough "that ducks could swim in them." Sixteen players provided the background music for the bash. This would appear to explain why the Constitution was signed on the 17th, and not the 16th, of September.

Catherine the Great tours Crimea. See Potemkin Village.

1789 In Britain the law is changed to make hanging the method of execution. Before then, burning was the modus operandi. The last female to be executed by burning in England was Christian Bowman. Her crime was counterfeiting coins.

In 1789, the entire U.S. federal government debt was $190,000.

1790 King Kamehameha secures control of the island of Hawaii.

1791 April 27, Samuel Finley Breese Morse born.

1793 The Chappe brothers established the first commercial semaphore system between two locations near Paris. Napoleon thought this was a great idea. Soon there were semaphore signaling systems covering the main cities of France. Semaphore signaling spread to Italy, Germany and Russia. Thousands of men were employed manning the stations. Speed: about 15 characters per minute. Code books came into play so that whole sentences could be represented by a few characters. Semaphores weren't very successful in England because of the fog and smog caused by the Industrial Revolution. Claude Chappe headed France's system for 30 years and then was "retired" when a new administration came into power. There were semaphore systems in the U.S., especially from Martha's Vineyard (an island near Cape Cod) and Boston, reporting to Boston's Custom House on the movement of sailing ships. This was also true around New York City and San Francisco. Samuel F.B. Morse, the inventor of the electric telegraph, reportedly saw the semaphore system in operation in Europe. The last operational semaphore system went out of business in 1860. It was located in Algeria.

1795 Leading an armada of over 1,000 war canoes and 10,000 soldiers, King Kamehameha easily conquers Maui and Molokai. His armada then moves on to Oahu, landing at Waikiki and Waialae. Fierce fighting ensues, culminating in a decisive battle on the Pali. Kamehameha now controls Oahu, and all of the islands to its east and south. Prior to Kamehameha's unification of the islands, different parts of the island chain were ruled by different rulers.

1800 First battery invented by Alessandro Volta, an Italian physicist.

1801 Oil discovered in Red Fork, near Tulsa, Oklahoma.

1810 After unsuccessful attempts to conquer Kauai in 1796 and 1803, King Kamehameha assembles the largest armada Hawaii has ever seen and sets out for Kauai again. This time the ruling chief of Kauai knows that resistance is futile and negotiates terms of surrender instead. He becomes Kamehameha's vassal on Kauai. Kamehameha gains control of Kauai and Niihau, thereby becoming sole sovereign of a unified Hawaiian island chain.

1811 New York State passes the first limited liability law.

1832 The Scottish surgeon Neil Arnott devised the water bed as a way of improving patients' comfort.

1833 Analytical engine by Charles Babbage.

1835 Elisha Gray (born in Barnesville, Ohio, on Aug. 2, 1835, died Newtonville, Mass., on Jan. 21, 1901) would have been known to us as the inventor of the telephone if Alexander Graham Bell hadn't got to the patent office one hour before him. Instead, he goes down in history as the accidental creator of one of the first electronic musical instruments - a chance by-product of his telephone technology. Gray accidentally discovered that he could control sound from a self vibrating electromagnetic circuit and in doing so invented a basic single note oscillator. The 'Musical Telegraph' used steel reeds whose oscillations were created and transmitted, over a telephone line, by electromagnets. Gray also built a simple loudspeaker device in later models consisting of a vibrating diaphragm in a magnetic field to make the oscillator audible.

1837 Telegraphy by Samuel F. B. Morse. Morse invents American Morse Code.

1840 Samuel Morse patents the telegraph. Congress was asked to provide funding for a semaphore system running from NYC to New Orleans. Samuel Morse, it is said, advised against funding of this system because of his work on developing the electric telegraph.

1841 Punahou School is founded. Among its famous alumni are Chinese revolutionary Sun Yat-sen, Steve Case (founder of AOL), Ron Higgins (founder of Digital Island, which was later purchased by Cable & Wireless and is now part of SAVVIS), Dave Guard and Bob Shane (two members of the Kingston Trio), Barack Obama (U.S. senator from Illinois), Hollywood actors Buster Crabbe and Kelly Preston, over 20 Olympic athletes (including Buster Crabbe in 1928 and 1932), and golf phenom Michelle Wie.

1843 First commercial test of Morse's telegraph. The US Government paid for a telegraph line between Baltimore and Washington, D.C. It worked.

First successful fax machine patented by Scottish inventor, Alexander Bain. His "Recording Telegraph" worked over a telegraph line, using electromagnetically controlled pendulums for both a driving mechanism and timing. At the sending end, a style swept across a block of metal type, providing a voltage to be applied to a similar stylus at the receiving end, reproducing an arc of the image on a block holding a paper saturated with electrolytic solution which discolored when an electric current was applied through it. The blocks at both ends were lowered a fraction of an inch after each pendulum sweep until the image was completed. Bain's device transmitted strictly black and white images.

1844 Samuel Morse sends first long distance public telegraph message to Baltimore from the chambers of the Supreme Court in Washington, DC. The message, "What hath God wrought?" comes from Numbers 23:23 and marks the beginning of a new era in communication. Morse's first telegraph line between Washington and Baltimore opens in May. The telegraph enabled the first instantaneous transmission of information over vast distances. Its wide diffusion, by the 1870s and 1880s permitted more efficient operations of businesses and railways. With the transoceanic telegraph (1866 and afterwards) continents became linked and the telecommunication industry was born. Several telecommunication giants of today (such as American Telephone and Telegraph, AT&T), started out as telegraph service providers.

1845 First rotary printing press by Richard M. Hoe.

1847 First telegraph company offices opens. Boston and New York first joined by a telegraph line. Alexander Graham Bell is born in Edinburgh, Scotland.

1849 An optical telegraph signal station is built on a high hill in San Francisco. The hill, originally named Loma Alta, is later renamed Telegraph Hill.

1850 The first telegraph cable is laid on the floor of the English Channel, connecting Dover, England with Calais, France. It fails after one day, due to insufficient armoring to protect it from the elements.

1851 A second telegraph cable is laid on the floor of the English Channel, connecting England and France. Properly armored to protect it from the elements, the cable is a success. See 1850.

The Continental (more commonly called the International) Morse Code is adopted for European telegraphs, but American telegraphers reject it. See Morse Code.

There are 51 telegraph companies in operation.

1853 Commodore Matthew Perry drops anchor in Tokyo Bay and effectively forces Japan to open itself to the outside world after three centuries of running a closed society.

Electric telegraph service begins in San Francisco, connecting Point Lobos with Merchant's Exchange, in order to inform local merchants of ships arriving in San Francisco Bay.

Three telegraph cables are laid on the floor of the North Sea, between England and Holland.

1854 George Boole develops a system of mathematics called Boolean algebra, which uses binary operations. Today, programmers still think and work in binary.

1856 Western Union formed by six men from Rochester, N.Y. They start an acquisition spree.

California becomes a state. The State had no electricity. The State had no money. Almost everyone spoke Spanish. There were gun fights in the streets. Some wags claim not much has changed.

1857 Joseph C. Gayetty of New York City invents toilet paper in 1857.

1858 The first transatlantic telegraph cable is completed, and messages begin to flow between the shores of America and Europe. But the cable fails after 26 days because the voltage is too high.

Burglar Alarm - Edwin T. Holmes of Boston begins to sell electric burglar alarms. Later, his workshop will be used by Alexander Graham Bell as the young Bell pursues his invention of the telephone. Holmes will be the first person to have a home telephone.

1859 Darwin publishes his "Origin of Species."

1860 Telegraph service is established between San Francisco and Los Angeles.

Pony Express formed to carry mail to the Wild West. Pony Express lasted 18 months before the telegraph took over. See Pony Express for the full story.

On June 16 the US Congress passes, and President James Buchanan signs, the Pacific Telegraph Act of 1860, authorizing the Secretary of the Treasury to accept bids for the construction of a telegraph line from Missouri to San Francisco, thereby connecting the western United States with the eastern half of the country. Construction of the line begins later that year and is completed on October 24, 1861. See wigwag.

The US Army's Signal Corps is established.

1861 Construction of the telegraph line from Missouri to San Francisco is completed on October 24, 1861. This also signals the end of the Pony Express.

Pony Express disbanded. The telegraph took over.

1862 Telegraph service is established between New York and San Francisco.

1865 First commercial fax service started by Giovanni Casselli, using his "Pantelegraph" machine, with a circuit between Paris and Lyon, which was later extended to other cities.

Abraham Lincoln assassinated.

J.C. Maxwell mathematically predicts the propagation of electromagnetic waves through space.

The International Telegraph Union (ITU) is founded by 20 member European states in order to facilitate cross-border telegraphy. Up until that point, telegraph lines did not cross national borders and different countries used different systems. Even different regions within a country sometimes used different, incompatible systems. If a telegraph message needed to be sent from Paris, for example, to Geneva, it could only go as far as a French telegraph station on the border between France and Switzerland, where the message would then be transcribed and hand-carried over to a Swiss telegraph station on the other side of the border, where a Swiss telegraph station would then transmit it over the Swiss telegraph network to its final destination.

1866 First experimental wireless by Mahlon Loomis.

Two successful transatlantic submarine telegraph cables (one eastbound, one westbound) are laid by Cyrus Field between Valencia, Ireland and White Stand Bay, Newfoundland, Canada.

1867 Christopher Sholes, a Milwaukee newspaper editor, invents the typewriter.

First internal combustion engine built.

The first Atlantic cable, promoted by Cyrus Field, was layed on July 27th.

1868 The first telegraph cable between Florida and Cuba is completed.

1869 Elisha Gray and Enos Barton form small manufacturing firm in Cleveland, OH.

1870 1. Thomas Edison invents multiplex telegraphy.

2. The 11,000 kilometer Indo-European Telegraph Line, which connects London and Calcutta via Prussia, Russia, and Persia, is completed.

1871 First British submarine telegraph cable laid in Hong Kong.

Bell arrived in Boston to start his work in the teaching of the deaf.

1872 Western Union buys the telegraph equipment manufacturing firm, Gray & Barton, and renamed it Western Electric.

1874 April 25, 1874, Guglielmo Marconi born in Bologna, Italy.

King David Kalakaua ascends to the throne of the kingdom of Hawaii. He reigns until 1891.

1875 February, 1875. Alexander Graham Bell signs an agreement with two partners (one is his father-in-law) to start a company to oversee his patents. The deal covered the young man's telegraphic inventions, but also included "further improvements," one of them later turned out to the transmission of human voice.

June 2 - Bell's theory of the telephone confirmed by experiment. First words transmitted by telephone.

The director of the United States Patent Office sent in his resignation and advised that his department be closed. There was nothing left to invent, he claimed.

1876 Many inventors were working on transmitting speech over wire in the 1870s, but no one had ever produced a working model. Then, on February 14, 1876, Alexander Graham Bell's father-in-law submitted Bell's patent for "Improvements in Telegraphy" just hours before Elisha Gray applied for a patent caveat, outlining his own device. Months later, Gray was among the scientific dignitaries assembled to witness Bell's first public demonstration of the telephone, at the Centennial Exposition in Philadelphia. Gray's patent caveat (an announcement of an invention he expected soon to patent) described apparatus 'for transmitting vocal sounds telegraphically.' It was later discovered, however, that the apparatus described in Gray's caveat would have worked, while that in Bell's patent would not have.

March 7, Telephone patent issued to 29-year old Boston University professor, Alexander Graham Bell. The patent was number 174,465. Three days later he sent the landmark message, "Mr. Watson, come here. I want you." The telephone has become the most profitable invention in the history of mankind. Bell successfully defended himself against all 600 lawsuits claiming rights to his invention. See 1877.

March 10, First complete sentence of speech transmitted by telephone in Boston.

Western Union issues its famous internal memo which contains the incredibly wonderful words, "This 'telephone' has too many shortcomings to be seriously considered as a means of communications. The device is inherently of no value to us."

Braving a hostile ocean, the men of the Faraday, a steam-driven ship with three masts, laid the first transatlantic cable between Ireland and America. The cable was made by Siemens. It could carry 22 telegraph messages at one time. It carried the world into a new era of communications.

June 25, Bell exhibited the telephone to the judges at the Centennial Exposi-tion, Philadelphia.

October 9, Bell conducted the first successful experimental two-way talk over the telephone between Boston and Com-bridgeport, Mass., distance of 2 miles.

First complete sentence transmitted by telephone. First conversation by overhead line, 2 miles – Boston to Cambridgeport.

Edison invents the electric motor and the phonograph.

1877 First telephone in a private home. First telephone in New York City.

Phonograph invented by Thomas Edison. The phonograph is reputed to be Thomas Edison's most brilliant invention.

Western Union turns down a chance to buy the patent rights to the invention of the telephone for $100,000. Western Union believed the telegraph superior technology. They were flat out wrong. It was clearly one of the dumbest decisions made in American business history.

Bell Telephone Company formed, with Alexander Graham Bell as "electrician" and Thomas Watson as "superintendent."

1878 Theodore N. Vail begins his career with the Bell System as general manager of the Bell Telephone Company. In 1985, he became the first president of the American Telephone & Telegraph Company. He left AT&T two years later. After pursuing other interests for 20 years, he returned as president of AT&T in 1907, retiring in 1919 as chairman of the board. Vail believed in "One policy, one system, universal service." He regarded telephony as a natural monopoly. He saw the necessity for regulation and welcomed it.

President Rutherford Hayes has a telephone installed in the White House. His first call is to Alexander Graham Bell. It's long lost in history what President Hayes actually said. There are two theories. Hayes' first words were "Please speak more slowly" or "This is an amazing invention, but who would ever want to use one?"

The New Haven Telephone Company publishes the first telephone directory. It had one page of 50 listings. In 1996, some 6,200 telephone directories were published in the United States, generating about $10 billion in advertising revenues. I look up all my addresses and phone numbers now on the Internet. There are dozens of sites, including www.bigyellow.com.

The first female telephone operator was Emma M. Nutt, who started working for the Telephone Dispatch Company in Boston, on September 1, 1878. Prior to that, all operators were men.

The first two telephones in Hawaii go into service. They are at each end of a phone line that Maui shopkeeper C.H. Dickey installs between his store in Wailuku and his house 12 miles away in Haiku.

1879 Edison invents the electric light bulb – the first successful carbon-thread lamp. See Incandescent.

1880 Alexander Graham Bell develops the photophone which uses sunlight to carry messages. It was never commercially produced.

There are 30,872 Bell telephone stations in the United States.

On March 31, 1880, the good people of Wabash, Indiana (population 320), launched a technological revolution. Atop the town's courthouse dome, they mounted two crossarms with a 3,000-candlepower carbon-arc bulb at both ends of each. They then fired up a threshing-machine steam engine to generate electricity, and at 8 p.m. sharp, flipped a switch. Sparks showered, and Wabash became the first electrically lit city in the world. "The strange, weird light, exceeded in power only by the sun, rendered the square as light as midday," one witness reported. "Men fell on their knees, groans were uttered at the sight, and many were dumb with amazement. We contemplated the new wonder of science as lightning brought down from the heavens." Excerpted from July, 2003 Discover Magazine.

Hawaiian Bell Telephone Company is founded.

1881 First long distance line, Boston to Providence.

American Bell purchases controlling interest in Western Electric and makes it the manufacturer of equipment for the Bell Telephone companies.

Mr. Eckert who ran a telephone company in Cincinnati said he preferred the use of females to males as operators. "Their service is much superior to that of men or boys. They are much steadier, do not drink beer nor use profanity, and are always on hand."

Bell Telephone company purchases Western Electric Company.

1883 Brooklyn Bridge, New York, completed.

King Kalakaua grants a charter to Mutual Telephone Company to enter the telephone business in Hawaii.

1884 Paul Nipkow obtains a patent in Germany for TV, using a selenium cell and a mechanical scanning disk. First long distance call: Boston to NYC.

September 4, Opening of telephone service between Boston and New York, 235 miles.

Conversation by overhead line (hard-drawn copper), 235 miles - Boston to New York.

1885 Theodore N. Vail becomes the first president of the American Telephone & Telegraph Company. See 1878 for more.

The Bell Telephone Company formed a new subsidiary, American Telephone & Telegraph (AT&T).

Incorporation of American Telephone and Telegraph Company, New York City.

1886 Heinrich Rudolf Hertz proves that electricity is transmitted at the speed of light.

1887 AT&T (American Telephone & Telegraph Co.) starts business.

Heinrich Hertz shows that electromagnetic waves exist.

1888 Heinrich Hertz produces radio waves.

1889 A. B. Strowger invents the telephone switch, dial telephone.

Punch card tabulating machine invented by Herman Hollerith.

Wall Street Journal first published.

1890 Congress passes Sherman Act.

Herman Hollerith gets a contract for processing the 1900 census data using punched cards. His firm was eventually named IBM in 1924.

There are 211,503 Bell telephone stations.

1891 First underseas telephone cable, England to France.

Invention of 1,000 line switch with disc bank having ten concentric rows of line contacts. Not used commercially. Formation of Strowger Automatic Telephone Exchange.

Queen Liliuokalani ascends to the throne and reigns over Hawaii until 1893.

1892 Almon Strowger, the St. Louis undertaker, became upset on finding that the wife of a competitor was a telephone operator who made his line busy and transferred calls meant for him to her husband. "Necessity is the mother of invention" so Strowger developed the dial telephone system to get the operator out of the system. He forms a Chicago firm, Automatic Electric, to manufacture step-by-step central office equipment (which is now owned by GTE). The first automatic C.O. was installed in LaPorte, Indiana. I discovered in Ralph Meyer's book, Old Time Telephones, that actually, in 1879, Connelly, Connelly and McTighe patented an automatic dial system, although they did not commercialize it.

October 18, Opening of long distance telephone service, New York to Chicago, 950 miles.

Conversation by overhead line, 900 miles-New York to Chicago.

First commercial Strowger installation; LaPorte, Indiana, USA. Used switcher with 100 line disc-type banks.

The minimum age for marriage of Italian girls was raised by law to 12 years.

1893 An early form of broadcasting was started in Budapest over 220 miles of telephone wires serving 6000 subscribers who could listen at regular schedules to music, news, stock market prices, poetry readings and lectures.

The world's first Ferris Wheel was built for the 1983 Expo in Chicago.

Hawaii's monarchy is overthrown. Supporters of the coup d'etat form a provisional government which rules from 1893 to 1894.

1894 Basic telephone patents expire; period of intense competition begins.

Invention of gear-driven switch with "zither" (piano wire) line banks. Not used commercially. 200-line "zither" board with ratchet drive installed at LaPorte, Indiana, USA.

Mutual Telephone Company acquires Hawaiian Bell Telephone Company.

The Republic of Hawaii is established.

1895 Guglielmo Marconi of Italy invents wireless telegraph. See 1897.

When the X-ray was discovered by Wilhelm Roentgen in 1895, some journalists were convinced that the primary user of the revealing shortwave radiation would be the "peeping Tom." The titillating publicity led to a law introduced in New Jersey forbidding the use of 'X-ray opera glasses' and to merchants in London selling X-ray-proof underwear for modest ladies."

Third installation at LaPorte, Indiana. Earliest use of switch with semi-cylindrical bank and shaft with vertical and rotary motions. Invention of earliest dial-type calling device.

1896 The shortest war on record, between Britain and Zanzibar in 1896, lasted 38 minutes. The winner of the war was the manufacturers who made the guns, rifles, cannons, gunpowder and other implements of destruction.

Invention of selector trunking; first use of dial telephones in large exchange (Augusta, Georgia, USA).

Marconi patents wireless telegraph.

1897 German physicist Dr. Karl Ferdinand Braun created the world's first cathode ray tube (CRT) – the technology at the heart of every television set.

Guglielmo Marconi sends wireless signals across the Bristol Channel, achieving a new record distance of 8.7 miles (14km).

1898 Earliest use of relays for switch control instead of direct operation of magnets over line wires. First die cast switch frame.

Hawaii was formally annexed as a US territory. Self-government was granted by Congress. The territorial government ruled from 1900 to 1941, and again from 1944 to 1959.

1899 When using the first pay telephone, a caller did not deposit his coins in the machine. He gave them to an attendant who stood next to the telephone. Coin telephones did not appear until 1899.

Magnetic voice recorder by Vlademar Poulsen.

AT&T, created in 1885, takes over American Bell Telephone and becomes parent to Western Electric and the Bell System companies.

Strowger Automatic goes abroad (Berlin, Germany). Earliest use of automatic trunk selection with busy test.

1900 John J. Carty, Chief Engineer of NY Tel (and later AT&T), installs loading coils, invented by Michael Pupin, to extend range and utilizes open wire transposition to reduce crosstalk an inductive pickup from ac transmission lines. AT&T paid Pupin $255,000 for the use of his patent. There are now about 20,000 telcos in business. There are now 856,000 telephones in service.

676,733 Bell telephone stations owned and connected.

Basic trunking principles established for large exchanges. Bank terminals molded in plaster of Paris.

Wireless telegraph service begins in Hawaii, on the island of Oahu. The first transmission takes place between a wireless telegraph station in Kaimuki and one in downtown Honolulu. Ship-to-shore wireless telegraph service begins soon thereafter.

1901 Formation of Automatic Electric Company to take over Strowger Automatic Telephone Exchange. Installation at Fall River, Mass., used line banks with fiber insulators and aluminum fillers. First use of "slip multiple."

Guglielmo Marconi sends first transatlantic wireless signals from Cape Code, December 12.

Inter-island wireless telegraph service begins in Hawaii.

1902 First conversation by long distance underground cable, 10 miles - New York to Newark.

First installation in Chicago begun. Earliest use of measured service in automatic exchanges.

Poulsen-Arc Radio Transmitter invented.

1903 Large Strowger installations placed in service in Grand Rapids, Dayton, Akron, Columbus.

AIEE Committee on Telegraphy and Telephony formed.

Nikola Tesla, a Yugoslavian scientist/inventor, patents electrical logic circuits called "gates" or "switches".

Orville and Wilbur Wright take to the air.

New York Stock Exchange building opens in New York City. It contained over 500 telephones – a record for any one building.

An undersea telegraph cable connecting San Francisco, Hawaii, Guam and the Philipines is put into service.

1904 First use of multi-office trunking, and connections between automatic and manual offices (Los Angeles, Califonia).

John Ambrose Fleming invents the two-element "Fleming Valve".

1905 While working as an examiner in the Swiss Patent Office, Albert Einstein discovers the Theory of Relativity, which he publishes in his doctoral dissertation at the University of Zurich.

Earliest extended use of party lines and reverting calls. First system using common battery talking (South Bend, Indiana).

Marconi patents his directive horizontal antenna.

1906 Motion picture sound by Eugene Augustin Lauste.

Lee deForest invents the vacuum tube.

Conversation by underground cable, 90 miles-New York to Philadelphia.

Invention of Keith Line Switch, resulting in enormous reduction in cost of automatic boards. First used at Wilmington, Delaware.

Dr. Lee de Forest reads a paper before an AIEE meeting on the Audion, first of the vacuum tubes that would make long distance radiotelephony possible. Reginald Fessenden broadcasts Christmas Carols on Christmas Eve from Brant Rock, MA.

1907 States start to regulate telcos. Mississippi was among the first. (The idea of regulation goes back several centuries, when in England, innkeepers were required to post their charges to prevent gouging. (I wish it applied to plumbers.) "Common carrier" regulation refers to government approval of tariffs filed by railroads, truck lines, telcos, etc which provide the terms and conditions whereby the public can make use of their services.

Theodore Vail returns as President of AT&T (and Western Union). He is responsible for the concept of "end-to-end" service that guided AT&T and other telcos in providing the C.O., transmission systems, and CPE that lasted until the Carterphone and Specialized Common Carrier Decisions.

First installation in Canada (Edmonton, Alta.). Invention of small dial and two-wire system eliminating ground at subscriber's station.

The world's first transatlantic commercial wireless services is established by Marconi with stations at Clifden, Ireland and Glace Bay, Nova Scotia.

1908 Henry Ford introduces the Model T.

First two-wire system (large dial) installed at Pontiac, Illinois. Earliest use of automatic, intermittent ringing. Installation at Lansing, Michigan. Features use of small dial, secondary line switch, and 200-point selectors and connectors.

1909 Paris' best-known monument, the Eiffel Tower, was saved from demolition because there was an antenna, of great importance to French radio telegraphy, mounted at the top of the nearly 1000-foot-high structure.

The first airline, DELAG, was established on October 16, 1909, to carry passengers between German cities on Zeppelin airships. By November 1913, more than 34,000 people had used the service.

Western Union and AT&T are closely locked.

Invention of out-going secondary line switch, resulting in economy of inter-office trunks. First used at San Francisco.

Marconi shares the Nobel Prize in Physics, with Karl Ferdinand Braun for their work in the development of wireless telegraphy.

Geronimo, the Apache Indian chief, dies at nearly 90. As the leader of the warring Apaches of the Southwestern territories in pioneering days, Geronimo gained a reputation for cunning and cruelty never surpassed by that any other American Indian chief. For more than 20 years, he and his men were the terror of the country always leaving a trail of bloodshed and devastation.

1910 Peter DeBye in Holland, develops theory for optical waveguides. He was a few years ahead of his time. Interstate Commerce Commission starts to regulate telcos.

The Mann-Elkins Act enacted, putting interstate communications under the purview of the Interstate Commerce commission (ICC)

5,142,692 Bell telephone stations owned and connected.

Strowger system introduced in Hawaii and Cuba. Earliest use of dialing over toll lines. Introduction of revertive ringing tone.

The first commercial radios are sold by Lee de Forest's Radio Telephone Company.

1911 Multiplying and dividing calculating machine invented by Jay R. Monroe.

IBM was incorporated in the State of New York on June 15, 1911 as the Computing - Tabulating - Recording Co. (C-T-R), a consolidation of the Computing Scale Co. of America, The Tabulating Machine Co., and The International Time Recording Co. of New York. In 1924, C-T-R adopted the name International Business Machines.

Conversation by overhead line., 2,100 miles – New York to Denver.

Formation of Automatic Telephone Manufacturing Co., Ltd. For production of Strowger system in England.

Using loading coils properly spaced in the line, the transmission distance for telephone reaches from New York to Denver.

1912 First Strowger installation in England (Epsom "Official Switch").

1913 The Kingsbury Agreement. Mr. Kingsbury was an AT&T vice president. In his famous letter to the U.S. Government, AT&T agrees to divest its holdings of Western Union, stop acquisition of other telcos, and permit other telcos to interconnect.

The Kingsbury Commitment precludes un-approved expansion, and permits connections to network.

The U.S. Justice Department filed its first antitrust suit against Bell, charging an unlawful combination to monopolize transmission of telephone service in the Pacific Northwest.

Conversation by overhead line, 2,600 miles–New York to Salt Lake City. Conversation by underground cable, 455 miles-Boston to Washington.

Strowger system introduced in Australia and New Zealand. Development of key-type impulse sender, and Simplex dialing on toll lines.

1914 Congress passes Clayton Act.

Underground cables link Boston, NYC and Washington.

February 26, Boston-Washington under-ground telephone cable placed in commercial service.

Automatic Switches used as traffic distributors in manual exchanges (Indianapolis, Indiana and Defiance, Ohio).

The last pole of the transcontinental telephone line is placed in Wendover, Utah, on the Nevada-Utah state line.

1915 Vacuum tube amplifiers used the first time in coast-to-coast telco circuits. In opening the service, Bell, in New York, repeated his famous first telephone sentence to his assistant, Mr.Watson, who was in San Francisco, "Mr. Watson, come here, I want you." Watson replied, "If you want me, it will take me almost a week to get there." E.T. Whitaker develops the sampling theorem that forms the basis of today's PCM and TCM technologies.

January 25, Opening of First Trans-continental telephone line, New York to San Francisco, 3600 miles.

October 21, First transmission of speech across the Atlantic by radiotelephone, Arlington, Va., to Paris.

Development of modern covered switch with horizontal relays - used at St. Paul and Minneapolis. First use of cast iron switch frame at Hazelton, Pennsylvania.

Direct telephone communications opened for service at 4pm, EST. Alexander Graham Bell, in NY, greets his former assistant, Thomas Watson, in San Francisco, by repeating the first words ever spoken over a telephone, "Mr. Watson, come here I want you". Mr. Watson would reply that it would take him a week to get there.

Nora Bayes records "Hello Hawaii, How Are You?" a song inspired by the introduction of radiotelephone service in Hawaii that year.

1916 Earliest community automatic exchange network installed in Wisconsin.

1917 First transatlantic radio by Guglielmo Marconi.

Rapid expansion in the use of private automatic branch exchanges. Development of remote alarm equipment for unattended exchanges.

Nora Bayes records "Hello Hawaii, How Are You?" a song inspired by the introduction of radiotelephone service in Hawaii in 1915.

1918 First installation using rotary primary line switches (Elyria, Ohio).

Edwin Armstrong develops a receiving circuit - the superheterodyne.

1919 First Strowger board manufactured for Bell System (Norfolk, Virginia).

Radio Corporation of America (RCA) is formed.

1920 January 16. The 18th Amendment goes into effect at midnight. Alcohol is banned as an illegal substance. Organized crimes takes over distribution.

Bell introduces its own step-by-step offices that were previously acquired from Automatic Electric. G. Valensi develops the time domain multiplexing concept.

July 16, World's first radiotelephone service, between Los Angeles and Santa Catalina Island, opened to the public.

There are 11,795,747 Bell telephone stations owned and connected.

Beginning of wide-spread adoption of Strowger equipment for metropolitan areas both in the U.S. and abroad. First installation of call-indicator equipment for automatic-manual connections in multi-office areas.

The first regular commercial radio broadcasts begin when AM station KDKA of Pittsburgh delivers results of the Harding-Cox election to its listeners. Radio experiences immediate success; by the end of 1922, 563 other licensed stations will join KDKA.

Westinghouse Radio Station KDKA is established (2 November)

November 6, First commercial AM radio broadcast in the U.S. KDKA, Pittsburgh, sending the Harding-Cox election bulletins. The first words ever carried by a commercial radio station were, "We shall now broadcast the election returns."

1921 Facsimile technology (Wirephoto) from Western Union.

The Willis-Graham Act allows telcos to merge with permission of the States and the Interstate Commerce Commission.

April 11, Opening of deep sea cable, Key West to Havana, Cuba, 115 miles.

First conversation between Havana, Cuba, and Catalina Island by submarine cable, overhead and underground lines and radio telephone-distance 5,500 miles. Extension of Boston - Philadelphia cable to Pittsburgh - total distance 621 miles. President Harding's inaugural address delivered by loud speaker to more than 100,000 people. Armistice Bay exercises at burial of unknown soldier delivered by means of Bell loud speaker and long lines to more than 150,000 people in Arlington, Va., New York and San Francisco.

Wirephoto - The first electronically-transmitted photograph is sent by Western Union. The idea for a facsimile transmission was first proposed by Scottish clockmaker Alexander Bain in 1843.

First radio broadcast of a sporting event (Dempsey/Carpentier Heavyweight Championship Prize Fight, 2 July).

Radio Shack is founded in Boston, Massachusetts by two brothers, Theodore and Milton Deutschmann. The company starts out as a retail and mail-order operation, selling radio equipment to maritime radio officers and amateur radio operators. The name, Radio Shack, was picked because it was the nautical term at the time for a ship's radio room. Radio Shack is also the term that ham operators use for the room where their radio equipment is set up.

1922 First dial exchange in New York City – PE-6, derived from Pennsylvania 6.

Ship-to-shore conversation by wire and wireless between Bell telephones in homes and offices and the S. S. America 400 miles at sea in the Atlantic.

Introduction of improved steel wall telephones and improved desk stands (Type 21).

Alexander Graham Bell dies at his summer home in Beinn Breagh, near Baddeck, Cape Breton Island, Nova Scotia (August 2). Telephone service is suspended for one minute (6:25pm-6:26pm) on the entire telephone system in the United States and Canada during the funeral service (4 August). British Broadcasting Corporation (BBC) is formed. (It would receive its Royal Charter in 1927).

1923 June 7, Radio broadcasting networks had their beginning with a hook-up of four radio stations by long distance telephone lines.

December 22, Opening of Second Trans-continental telephone line, southern route.

There are 14,050,565 Bell telephone stations owned and connected. Successful demonstration of transoceanic radio telephony from a Bell telephone station in New York City to a group of scientists and journalists in New Southgate, England.

First British Post Office announces adoption of Strowger system (with Director) for London.

Meetings at New York and Chicago of the American Institute of Electrical Engineers (AIEE) are linked by long distance lines connected to loudspeakers so that both meetings could follow the same program (14 February).

Mutual Telephone Company (now known as Hawaiian Telcom) finishes replacing its fleet of 25 horse-drawn service wagons with motorized trucks. Horses and donkeys are still used, however, to pull telephone poles along trails in remote areas.

1924 The "A&P Radio Hour" is the first nationally broadcast radio program in the United States.

Thomas J. Watson renames Computing-Tabulating-Recording (CTR) the International Business Machines Corporation. Mr. Watson, president, is successor to the company's founder, Herman Hollerith, who invented a method of assembling databases and making computations using a system of paper cards with holes punched in them.

The work of Herbert Ives at Bell Labs on the photoelectric effect leads to the first demonstration of the transmission of pictures over telephone.

Strowger exchange installed throughout Canal Zone. First Strowger "Directors" installed in Havana.

Directive short wave antenna is developed by Professor Hidetsugu Yagi and his assistant, Shintaro Uda.

1925 IBM begins selling punch-card machinery in Japan.

Bell Laboratories is created from the AT&T and Western Electric engineering department, which had been combined in 1907. Frank B. Jewett becomes the first president of Bell Labs. 1.5 million dial telephones in service out of 12 million phones in service.

AT&T's Long Lines Department offers the press an early facsimile service between New York, Chicago, and San Francisco. The technology takes decades before it reaches a mass market.

October 1, Opening of long distance telephone cable, New York to Chicago.

Introduction of the Monophone - first hand set telephone of modern type.

The Combined Line and Recording (CLR) method of handling toll calls over long distances (100 miles or more) is introduced experimentally by Bell Systems. It reduces the handling of toll calls from 13 minutes (in 1920) to 7 minutes.

1926 The Knights of Columbus' Adult Education Committee recommends that the following questions be discussed in group meetings: "Does the telephone make men more active or more lazy? Does the telephone break up home life and the old practice of visiting friends?"

AT&T Bell Labs invents sound motion pictures.

First public test of transatlantic radiotelephone service – between New York and London.

Robert Goddard, the father of the space age, launches the first liquid-fuel rocket from an aunt's farm in Worcester, Mass. He developed a general theory of rocket action. When captured German rocket scientists were brought to the U.S. after World War II and were questioned about rocketry and its development, they asked with incredulity why the U.S. did not already have the answers from Goddard. The U.S. had to admit that it had neglected Goddard. He died in 1945 before the neglect could be corrected.

Warner Brothers announces a system developed by Bell Laboratories and Western Electric to allow synchronized voice and music in the movies. The next year, "The Jazz Singer," with Al Jolson, dazzled the United States.

Baird in Scotland and Jenkins in the U.S. demonstrate TV using neon bulbs and mechanical scanning disks. P.M. Rainey at Western Electric patents the PCM methodology.

Introduction of the Type 24 Dial - modern, quiet-running, long-life calling device. Strowger system adopted by Japan.

The first public test of radiotelephone service from New York to London.

1927 On January 17, 1927, transatlantic telephone service between London and New York opened, charging $25 a minute, or 15 English pounds for three minutes. When Time Magazine later reported the event, it said "Walter Sherman Gifford, president of the American Telephone and Telegraph Co. picked up a telephone receiver in Manhattan. Said Gifford into the transmitter, 'Good morning, Sir. This is Mr. Gifford in New York.' Sir George Evelyn Pemberton Murray, Secretary of the General Post Office of Great Britain in London replied, 'Good morning, Mr. Gifford. Yes. I can hear you perfectly. Can you hear me?' The distinction of talking to London on the first day of transatlantic service was also taken by Adolph S. Ochs, publisher of the New York Times, who let it be known that he was the first private speaker with editor Geoffrey Dawson of the London Times.

May 21, Charles A. Lindbergh successfully flies his monoplane "the Spirit of St Louis" from New York to Paris, ushering in a new transport technology and obsoleting the competitors – ships and blimps – and obsoleting the 200 tons of concrete and steel that was installed on the Empire State Building and which was meant to act as a downtown landing.

December. The first talking movie, "The Jazz Singer," starring Al Jolson is released.

Secretary of Commerce Herbert Hoover spoke over TV from Washington, D.C. to an audience in New York. Praising the invention, he said, "What its uses may finally be, no one can tell."

April 7, First public demonstration of television by Bell System engineers, by wire and radio.

First "Director" installation in London. Introduction of line switch with self-aligning plunger.

Television - Philo Farnsworth demonstrates the first television for potential investors by broadcasting the image of a dollar sign. Farnsworth receives backing and applies for a patent, but ongoing patent battles with RCA will prevent Farnsworth from earning his share of the million-dollar industry his invention will create.

First public demonstration of long distance transmission of television. Formal opening of telephone service between the US and Mexico, and also, Mexico- London, via New York.

1928 Mickey Mouse is born in Walt Disney's first cartoon, "Steamboat Willie." See 1937.

The Galvin Manufacturing Corporation is founded in Chicago by Paul Galvin and his brother, Joseph Galvin. The company's first product is a battery eliminator, a device that lets battery-powered radios run on standard household electric current.

Zworykin files patents on electronic scanning TV using the iconoscope.

First extended use of Strowger 200-point Line finder. Introduction of improved Monophone designs.

A joint meeting of the AIEE and the British IEE is held over radiotelephone channels, with the respective groups assembled in New York and London.

1929 February 14, The infamous "St. Valentine's Day Massacre" gangland hit is ordered by Al Capone against North Side boss Bugs Moran.

April 7, 1929, First public demonstration of long distance TV transmission. Moving black and white pictures were sent over telephone wires between Secretary of Commerce Herbert Hoover in Washington DC and AT&T executives in New York. They went at 18 frames per second. Further development of this technology led to the creation of TV.

Harold S. Black's negative feedback amplifier cuts distortion in long distance telephony. Black is at Bell Labs.

June 27, Bell Laboratories makes the first U.S. public demonstration of color television in New York. Images are roses and a U.S. flag.

Coaxial cable invented in Bell Telephone Laboratories; Herbert Hoover first president to have phone installed on his desk.

Bell Laboratories and Western Electric introduced the Sound Newsreel Camera. It used an AT&T "Light Valve" to record sound directly on the film as it passed through the camera. It was the first single system sound camera.

October 29, "Black Tuesday." The beginning of the Great Crash. The stock market crashes with 16 million shares sold. On November 13, the prices reach their lowest point for the year and $30 billion in stock values are wiped out. The crash, along with negative factors in the U.S. and world economics decisively brings an end to the Roaring Twenties and brings on the Great Depression, the rise of fascism in Europe and ultimately, the Second World War.

Commercial ship-to-shore telephone service opened.

U.S. Navy begins use of Strowger equipment. Monophones made available in color.

Not until Herbert Hoover was U.S. president, in 1929, did the U.S. chief executive have a private telephone in his office. (The telephone had been invented 53 years earlier.) The booth in a White House hallway had served as the president's private phone before one was installed in the Oval office.

1930 Galvin Manufacturing Corporation develops a car radio and names it the Motorola.

Opening of transoceanic telephone service to Argentina, Chile and Uruguay and subsequently to all other South American countries.

Development of new small switchboards of unit type. Networks of small Strowger exchanges installed in Italy.

1931 Nevada legalized gambling in 1931. At that time, the Hoover Dam was being built and the federal government did not want its workers (who earned the princely sum of 50 cents an hour) to be involved with such diversions, so they built Boulder City to house the workers, making gambling illegal in town. To this day, Boulder City is the only city in Nevada where gambling is illegal.

Empire State Building opened.

Development of Strowger Remote Toll Board. First installed in Elyria, Ohio.

Radio Astronomy - While trying to track down a source of electrical interference on telephone transmissions, Karl Guthe Jansky of Bell Telephone Laboratories discovers radio waves emanating from stars in outer space.

AT&T inaugurates the Teletypewriter Exchange Service (TWX) November 21.

1932 Development of unattended private automatic branch exchanges. Two-line Monophones introduced.

1933 Karl G. Jansky at Bell Labs discovers radio waves from the Milky Way. His discovery leads to the science of radio astronomy.

Bell Labs transmits first stereo sound, a symphony concert, over phone lines from Philadelphia to Washington.

FM radio invented by Edwin H. Armstrong.

1934 Federal legislation which established national telecommunications goals and created the Federal Communications Commission to regulate all interstate and international communications.

Congress passes Communications Act of 1934, with a goal of universal service at reasonable charges as its key tenet. The FCC was formed.

Introduction of new self-contained desk Monophone molded in bakelite (Type 34A3).

On Aug. 19, a plebiscite in Germany approved the vesting of sole executive power in Adolf Hitler as Fuhrer.

1935 First telephone call around the world. About 6700 telcos in operation.

April 25, First around-the-world telephone conversation by wire and radio.

New "all positions" transmitter. New bakelite wall Monophone (Type 35A5).

Western Union's "Telefax" begins operating. Telefax sent telegrams, manuscripts, line drawings, maps and page proofs for magazines.

1936 First TV broadcast by the BBC in Great Britain.

Invention of coaxial cable is announced at a joint meeting of the American Physical Society and the IRE (April 30).

1937 Galvin Manufacturing Corporation begins developing and selling radios and phonographs for the home. Galvin enhances its car radios by adding push-button tuning, fine-tuning and tone control.

PanAm makes first commercial flight across the Pacific. PanAm much later goes broke.

Walt Disney releases his first full-length animated film, "Snow White and the Seven Dwarfs."

German giant dirigible, the Hindenberg, explodes as it is being moored in Lakehurst, NJ on May 6, thus effectively ending the era of blimp travel across the Atlantic and sealing the fate of the Empire State Building as no longer an "airport" for downtown Manhattan – the purpose of the 200 tons of concrete and steel that so beautifully adorn the top of that building.

Bell introduces the Model 300 improved handset.

December 8, Opening of Fourth Transcontinental telephone line.

Seven-hour radio broadcast of the coronation of King George VI and Queen Elizabeth of England.

1938 Xerography invented by Chester Carlson.

Bell introduces crossbar central office switches.

The power of radio is demonstrated by Orsen Wells with the broadcast of "War of the Worlds". This causes telephone traffic to peak in nearly all cities and on long distance lines.

1939 Western Union introduces coast-to-coast fax service.

John Atanasoff and Clifford Berry invent the first electronic computer at the University of Iowa. In 1973 a judge ruled in a patent infringement suit that their research was the source of most of the ideas for the modern computer.

The Golden Gate Exposition (San Francisco) and New York Worlds Fair are opened. These exhibit the newest technologies, including the Voder (synthesized speech) and television. FM is used by Bell Laboratories in a radio altimeter that uses signal reflections from the surface of the earth.

For those Amerians wealthy enough to own a TV in 1939, Franklin Delano Roosevelt's speech at the opening session of the New York World's Fair of 1939 was beamed right into their homes – in glorious black and white.

1940 February 19, DuPont introduces "nylon," an artificial silk billed as a formidable rival to natural Japanese silk. Nylon is technically described as "synthetic fiber-forming polymeric amides." Its basic materials are coal, air and water. DuPont announces that nylon will be put to many other uses besides stockings: non-cracking patent leather, weatherproof clothing and flexible window panes.

June 24, Television transmitted over coaxial cable from Convention Hall in Philadelphia to television studio in Radio City, New York.

FM Police Radio Communications begin in Hartford, CT.

AT&T lays its first trans-continental coaxial cable link across the USA.

1941 Konrad Zuse in Germany develops the first programmable calculator using binary numbers and boolean logic.

The Japanese attack on Pearl Harbor affects the telephone system of the United States by causing tremendous traffic peaks in all cities, and an increase from 100 to 400 percent in long distance telephoning – which already is at a record high of 3 million messages. (The United States would again experience this phenomenon in 2001, during the After Pearl Harbor (1941) the US military ruled the islands until 1944. September 11, 2001 attacks.)

Radar successfully detects the attack on Pearl Harbor, but the warnings are ignored.

July 1. The first American television commercial was broadcast in New York during a baseball game at Ebbets Field between the Dodgers and the Phillies. The game was interrupted by an image of a Bulova watch face, superimposed on the screen and accompanied by a voice-over announcing, "America runs on Bulova time." Bulova paid $9 for the spot.

1942 Harry Newton born Sydney, Australia on June 10.

December 21, Opening of first all-cable transcontinental telephone line with completion of buried cable, connecting existing cable systems of East and West.

1943 The U.S. Army Ordnance Department commissions the ENIAC (Electronic Numerical Integrator Analyzer and Computer) to help produce missile trajectory tables for use in World War II. Even though it doesn't arrive until 1945 and misses the war, ENIAC can perform 5,000 additions per second and is later used in artillery calculations. ENIAC weighs 30 tons and is 100 feet long, 8 feet high, and contains 17,468 vacuum tubes. Programming ENIAC is anything but user friendly. It takes two days to set up problems that ENIAC solves in two seconds.

Philadelphia is the last city to have telephone service supplied by different local carriers (until the recent deregulatory moves by Congress and the FCC.) Western Union and Postal Telegraph permitted to merge.

August 22, First equipment for the dialing of called telephone numbers in distant cities directly by the operator placed in service in Philadelphia.

Construction of a telephone line from Calcutta, India to Kunming, China, along Stilwell Road, begins at Ledo, Assam.

1944 A telephone submarine cable is laid across the English Channel.

Howard Aiken's Mark I is the last great electromechanical calculator.

1945 AT&T lays 2000 miles of coax cable. Arthur C. Clarke proposes communications satellites.

Western Union installs the first commercial radio beam system.

1946 The first general purpose computer, Electronic Numerical Integrator and Computor (spelled with an O) is built at the University of Pennsylvania. It was called ENIAC. It was 30ft x 50ft. and weighed 30 tons. It included 18,000 vacuum tubes, 6,000 switches and 3,000 blinking lights. Although it did not store programs, it could multiply two 6-digit numbers in half a second and could hold an astounding 200 bytes of memory. The ENIAC was turned off for the last time on October 2, 1955.

February 12, New York-Washington co-axial cable circuits opened for television transmission on an experimental basis.

Dr. Robert N. Metcalfe, co-inventor of Ethernet, born Brooklyn, New York on April 7. Dr. Metcalfe now writes a prestigious column for IDG's InfoWorld weekly newspaper.

June 17, Opening of experimental mobile radiotelephone service in St. Louis.

Ray Horak born Niagara Falls, New York on November 10.

FCC's Recording Devices Docket required telcos to furnish connecting arrangements for conversation recorders. The use of "beep tones" required when conversations are recorded.

1947 Transistor invented at AT&T's Bell Labs in New Jersey. Reporting on the transistor's invention, the Wall Street Journal wrote, "In December 1947, William Shockley and his team of scientists came up with a solution to a problem vexing their employers at AT&T Corp. At the time, voice was amplified over long distances using vacuum tubes. The bulky tubes often overheated and broke down, making long-distance calling expensive. Working at AT&T's Bell Labs in Murray Hill, N.J., two members of the team, Walter Brattain and John Bardeen, came up with a device made from a paper clip, two slivers of gold foil, and a slab of germanium on a crystal plate. The contraption made it easier to transmit sounds with clarity over a long distance. It was later dubbed the transistor. More than any other technology, the invention unleashed the information revolution of the late 20th century. Over time, it also carried the seeds of AT&T's demise. The transistor, and its subsequent improvement and miniaturization on silicon chips, made it possible to store and distribute ever-greater amounts of information. That drove the development of computers, satellites, space exploration and much of modern communications and electronics. Dr. Shockley moved to California and founded a company that played a role in spawning Silicon Valley. The transistor invention won the Bell Labs team the Nobel Prize in 1956."

Galvin Manufacturing Corporation changes its name to Motorola.

Telcos install nationwide numbering plan.

Opening of commercial telephone service for passengers on certain trains running between New York and Washington, D.C.

Nov. 13, Opening of New York-Boston radio relay system for experimental service.

Invention of the point contact transistor by Brattain and Bardeen. (December 23). Demonstration of mobile telephone equipment from a United Airlines plane to ground stations.

August 15. India and Pakistan became independent after some 200 years of British rule. Larry Collins and Dominique Lapierre later write a brilliant book on the history of India's independence and partition. The book is called "Freedom at Midnight."

1948 May 11, Birth of the International Communications Association, among the larger groups of telecommunications users in North America.

Claude E. Shannon announces the discovery of information theory, the cornerstone of current understanding of the communication process. Shannon was a Bell Labs employee. See Shannon's Law.

Xerography introduced by Chester F. Carlson. Xerography copied documents with carbon paper and ink and totally revolutionized office work. It was introduced by the Haloid Company of Rochester, New York, later renamed the Xerox Corporation.

Invention of the junction transistor

The Hush-A-Phone case had its beginning. The Hush-A-Phone Corp. had developed and was marketing a cup-like device placed on a phone's mouthpiece to increase privacy of conversations. The Bell System complained to the FCC about this "foreign attachment."

1949 October 1, Mao Ze-dong or Mao Tse-tung (1893-1976) declares victory in China's civil war in Tiananmen Square, Beijing (Peking). Mao was the leader of of the Long March in the Chinese Civil War (1934), he became the first president of Communist China (1949-1967).

The volume of telephone calls reaches 180 million a day!

Justice Department files antitrust suit against AT&T. The Department wanted Bell to divest Western Electric, and to separate regulated monopoly services and unregulated equipment supply, among other actions.

An optical telegraph signal station is built on a high hill in San Francisco. The hill, originally named Loma Alta, is later renamed Telegraph Hill.

1950 Television network facilities extended to include 72 television stations in 42 cities, making television available to one half the population of the nation.

1951 First direct distance calling in North America. Phone users can dial long distance without an operator.

Sony unveils the first transistor radio.

The Eckert and Mauchley UNIVAC (Universal Automatic computer) was delivered to the U.S. Census bureau. The cost of constructing the first UNIVAC was close to one million dollars.

1952 IBM introduces the Defense Calculator, one of its first major entries in the computer business.

The first database was implemented on RCA's Bizmac computer. Reynold Johnson, an IBM engineer, developed a massive hard disk consisting of fifty platters, each two feet wide, that rotated on a spindle at 1200 rpm with read/write heads. These were called "jukeboxes".

Hawaii's first TV broadcast takes place, by KONA (now KHON) TV.   Like many TV stations, KONA started out in life as a radio station.

1953 May 9, Edmund Hillary and Tenzing Norgay reach the summit of Mt. Everest, the first climbers to do so.

Digital voltmeter (DVM) invented by Non-Linear Systems.

John Pierce proposes deep space communication.

1954 The solar cell invented by Gerald L. Pearson, Daryl M. Chapin and Calvin S. Fuller at Bell Labs.

William Shockley leaves Bell Labs to pursue the commercial opportunities offered by his invention of the transistor.

IBM brings out the 650, the first mass-produced computer. IBM management states that there will never be a need for more than six mainframes in the entire world. The 650 turns out to be a great success, with 120 installations in its first year. Gene Amdahl developed the first computer operating system for the IBM 704.

Sony introduces the first transistor radio that sold for $49.95.

Raytheon introduces the transistor for hearing aids replacing its line of subminiature tubes. Zenith's highly successful hearing aids using subminiature tubes, about the size of a pack of cigarettes with a separate battery pack sold for about $25.00. The new transistor hearing aids reduced the size of the electronic package to about the size of a box of matches with an internal battery and sold for about $100. The first in-the-ear hearing aids appeared about 1955-1956.

Mutual Telephone Company changes its name to Hawaiian Telephone Company.

1955 Bill Gates, founder and chairman of Microsoft, born October 28, 1955.

1956 The first transatlantic telephone cable was inaugurated on September 25. It was hailed as a major breakthrough in telecommunications.   It was designed to link both the US and Canada to the UK, with facilities for some circuits to be leased to other West European countries, giving them direct communication with US and Canada.   It provided 30 telephone circuits to the US and 6 to Canada, as well as a number of telegraph circuits to Canada.   Most were for communication with the UK, the remainder were permanently connected through London to give direct circuits to Germany, France, the Netherlands, Switzerland, and a circuit for Denmark which also carried American traffic with Norway and Sweden.   The whole project took 3 years to complete, at a cost of ú120 million, during which time the system had to be planned, manufactured and installed, requiring the development of new machinery and techniques for placing the cable in deep waters.

First modem was invented by AT&T Bell Laboratories, according to AT&T.

Videotape recorder invented by Ampex.

AT&T signs a consent decree with the federal government that allowed it to keep its structure under which it sold both phone service and telephones themselves. In exchange, AT&T promised to stay out of other businesses and license its patents freely. AT&T's equipment arm, Western Electric, had to withdraw from selling sound equipment for film producers and movie theaters – giving up experience in a competitive market that would have proved useful later, according to the Wall Street Journal. The decree also put the transistor patents in the public domain. As a result, while AT&T used transistors to improve the reliability and quality of the phone calls it relayed, it played little role in developing the integrated circuit. That fame and fortune went largely to Texas Instruments and Fairchild, which independently figured out a way in the late 1950s to fabricate and embed multiple, miniaturized transistors on a tiny silicon chip. Those advances led to the birth of the microprocessor, the engine of personal computers, and the birth of Intel, which invented the first microprocessor in 1971. See 1971.

AT&T signs consent decree limiting Western Electric to manufacturing equipment for the Bell system and the U.S. government.

John Bachus and his IBM team invent FORTRAN, the first high-level programming language.

First hard drive introduced by IBM. It was the size of a refrigerator and weighed a ton. It used 50 platters each measuring 24 inches in diameter. It stored five megabytes.

The 1956 Nobel Prize in Physics is awarded to the inventors of the transistor: Dr. Walter H. Brattain, Dr. John Bardeen and Dr. William Shockley.

1957 The U.S.S.R. launched Sputnik on October 4, 1957. It embarrassed the U.S. Government into a frenzy of space investments, culminating in the U.S. being the first country to have people walk on the moon.

President Eisenhower signs a bill to authorize the construction of interstate highways in the United States. See also Telecommunications Act of 1996.

1958 Integrated circuit invented by Jack S. Kilby at Texas Instruments.

United States forms the Advanced Research Projects Agency (ARPA).

AT&T introduces datasets (modems) for direct connection.

Jack Kilby, Texas Instruments, developed the first integrated circuit. TI introduces the silicon-based transistor which soon eclipsed germaninum devices in production volume.

Seymour Cray at Control Data Corporation develops the first transistorized computer, Model 1604. He later uses liquid nitrogen to enhance the speed of CDC's line of supercomputers.

1959 September 28, A desk-sized machine that reproduces documents on ordinary instead of specially treated paper was introduced by Haloid Xerox, Inc. Fixing dry ink permanently onto paper, the Xerox 914 turns out reproductions at the rate of six a minute.

Robert Noyce of Fairchild Semiconductors seeks a patent for a new invention: the integrated circuit.

Paul Baran begins to think about ways to make America's communications infrastructure resistant to a nuclear attack. He proposed using a system called "distributed adaptive message block switching", known today as packet switching. This involves breaking digital information into small chunks, or packets, and sending them separately, thus doing away with centralized switching centers and enabling the network to work even when partly destroyed. His idea was initially ignored and was only given its first proper test in 1969, when it was used as the basis for ARPANET, an experimental computer network run for the Department of Defense that later grew into the Internet. See Arpanet and Internet.

On March 12 the US Congress passes a bill to make Hawaii the 50th state of the Union. President Dwight Eisenhower signs the bill into law (Public Law 86-3) on March 18. A plebiscite in Hawaii on June 21 shows Hawaii residents in favor of statehood by a 17-1 margin. Hawaii officially is admitted into the Union on August 21.

1960 Western Union sends its last Morse Code encoded telegram.

First test of an electronic telephone switch.

MITI creates the Japan Electronic Computer Corporation to promote its domestic computer industry.

Laser invented by Theodore Maiman of the U.S. Laser stands for Light Amplification by the Stimulated Emission of Radiation.

2,000 computers are in use in the United States.

Digital Equipment Corporation introduces its first minicomputer, the PDP-1, priced at a relatively modest $120,000 (modest compared to the price of mainframe systems).

The first integrated circuits reach the market, costing $120. NASA selects Noyce's invention for the on-board computers of the Gemini spacecraft.

There are now 3,299 telephone companies in the United States.

ECHO I communications satellite is launched on 12 August. It provided the first satellite television broadcast in 1962.

1961 Leonard Kleinrock publishes the first details packet switching, the critical technology of the Internet.

T-1 transmission system created.

1962 LEDs – Light Emitting Diodes – invented.

AT&T Telstar I satellite was launched on July 10th, and later that same day transmitted the first live television images from the United States to France. The concept of a communications satellite was first proposed in 1945 by engineer-turned-science-fiction-writer Arthur C. Clarke., who now lives in Sri Lanka, once called Ceylon. See Clarke, Arthur C.

August 31, 1962 President Kennedy signs Communications Satellite Act.

Semiconductor laser invented.

Ross Perot forms EDS with a reputed $1,000. In 1984, he sold it to General Motors for $2.6 billion.

Wal-mart opens first store in Rogers, Arkansas. By 2000, it had grown to among America's top five companies, as measured by revenue.

EEST Electronic Switching Systems is introduced.

1963 President John F. Kennedy assassinated.

Touch Tone service introduced.

Audio cassette tape introduced by Philips.

C. Kumar N. Patel at Bell Labs develops the carbon dioxide laser now used around the world as a cutting tool in surgery and industry.

MCI started as a two-way radio company. Its original name is Microwave Communications, Inc. It was originally organized as a microwave carrier to allow truckers to speak to their home base. They were to communicate via antennas on top of microwave towers between Chicago and Springfield, Illinois.

Paul Baran of RAND publishes "On Distributed Communications Networks," outlining the operations of packet-switching networks capable of surviving node outages.

NASA announces that the new Syncom II communications satellite has been used successfully to transmit voices live between the U.S. and Africa. At the time of the conversations, Syncom II hovers 22,000 miles over Brazil. The satellite is the first successful synchronous satellite. This mean that the satellite's revolution matches the daily revolution of the earth about its axis, so that the satellite seems to remain "stationary" over the same earth location.

August 30, a telephone hotline connects Soviet and American leaders.

1964 Prototype of the first video phone made by the Bell System shown at The World's Fair in Queens, New York City. Pictures were black and white and the technology was very expensive. It was called the Picturephone.

IBM showed the first word processor.

IBM announces System 360, a family of computers that can be used for science and business, and share the same software printers, and tape drives.

The first Local Area Network (LAN) is developed at Lawrence Livermore Labs.

BASIC is developed at Dartmouth College by John Kemeny and Thomas Kurtz.

The first integrated circuit sold commercially is used in a Zenith hearing aid.

June 22, An improved stock ticker tape machine (designed, developed and manufactured by Teletype Corporation) is placed into service at the New York Stock Exchange. The ticker, which transmits stock prices to brokerage houses nearly twice as fast as the previous system, has a capacity of ten million shares a day without incurring delays.

The Beatles made their American debut on the Ed Sullivan Show.

1965 PDP-8 minicomputer introduced by Digital Equipment Corporation. Its speed, footprint (about the size of a small refrigerator), and reasonable cost (a mere $18,000) made the DEC PDP-8 the first successful minicomputer.

First trial offers for reversing telephone charges. Telephone bills start to go awry.

Moore's Law predicts that the number of components on an integrated circuit would double every 18 months and the cost of computers would be cut in half.

Audio tape cassette first introduced. Sony Walkman introduced in 1979.

K. C. Kao and G. A. Hackham publish influential paper on fiber optics.

April 6, The first commercial communications satellite, Early Bird, later named Intelsat 1, is launched into orbit from Cape Kennedy. The 85-pound satellite is a synchronous satellite, matching the earth's rotation to hover over the same spot all the time.

April 23, The Soviet Union launches its first communications satellite and carried out transmissions of television programs. The satellite is named "Molniya 1", which translates to "Lightning 1".

Digital rolls out the PDP-8, the first minicomputer.

1966 June 15, Lawrence G. Roberts of MIT publishes "Towards a Cooperative Network of Time-Shared Computers" which outlines the ARPANET plan. Worldwide direct telephone dialing has its first public demonstration, a call from Philadelphia to Geneva, Switzerland.

October, the Electronic Industries Association issues its first fax standard: the EIA Standard RS-328, Message Facsimile Equipment for Operation on Switched Voice Facilities Using Data Communications Equipment. The Group 1 standard, as it later became known, made possible the more generalized business use of fax. Transmission was analog and it took four to six minutes to send a page.

Suggestions made by Kao and Hockham that optical fiber could be used for long distance transmission.

In anticipation of direct distance dialing, Hawaiian Telephone Company begins converting Hawaii's 5-digit phone numbers to 7 digits.

1967 First 800 call made in the United States.

Electronic handheld calculator introduced by Texas Instruments.

Larry Roberts at the Advanced Research Projects Agency publishes a paper proposing ARPANET.

Bell Laboratories announces a new solid-state source of high frequency radio waves. The "LSA diodes" emitted millimeter waves, a part of the radio frequency range that could carry about nine times more telephone calls than all lower frequencies combined. An LSA diode and its power supply is about as large as a deck of cards. (February 15).

June 30, An experimental cordless extension telephone is introduced by Bell Laboratories.

GTE acquires Hawaiian Telephone Company.

1968 In a landmark decision, the FCC for the first time allows non-AT&T equipment to attach to the Bell System. The FCC rules that equipment which is privately beneficial, but not publicly harmful, is OK for connection. The Carterfone device connected two-way radios to the phone network – See Carterfone and Carterfone Decision.

Fiber optics for communications invented by Robert Maurer.

Intel Corporation is founded in Santa Clara, California, by Fairchild veterans Robert Noyce and Gordon Moore. Andy Grove is employee number four.

Packet switching network presented to ARPA.

AT&T starts development of the Integrated Digital Services Network (ISDN).

Gary Englehart at Stanford Research Institute demonstrates the first combination of a keyboard, keypad, mouse, windows and word processor.

Dan Noble, IBM, developed the 8-inch floppy disk. Its capacity increased from 33K in 1971 to 1200K in 1977. AT&T starts 56 Kbps service. Pieter Kramer (Philips) invents the compact disk.

FCC starts proceeding to set aside spectrum for land mobile communications.

Bell System adopts the use of "911" as a nationwide emergency telephone number. Huntington, Indiana became the first U.S. city served by the Bell System to receive the new universal emergency telephone number "911". (March 1).

1969 Ken Thompson and Dennis Ritchie, computer scientists at AT&T Bell Laboratories, create the Unix operating system.

September 2, Internet born with the installation of ARPAnet's first Interface Message Processor (IMP) in Professor Leonard Kleinrock's lab at the University of California at Los Angeles. IMPs were packet-switching minicomputers, pre-Cisco routers, developed at Bolt, Bernanek and Newman (BBN) in Cambridge, Mass.

ARPAnet introduced by the Advanced Research Projects Agency of the U.S. Defense

Department, comprising a 50 kilobit per second backbone and four computer hosts. UCLA, UCSB, Stanford Research Institute and University of Utah set up first four nodes on ARPAnet. See also 1959.

Traffic Service Position System replaces traditional cord switchboards. The system automates many operator functions for the first time.

In a landmark decision, Federal Communications Commission authorizes MCI Communications Corporation to be the first long distance company allowed to compete with AT&T in the U.S. long distance market. The route chosen for the competition is Chicago to St. Louis. The route was originally Chicago to Springfield, Illinois. But that meant it was an intrastate route and the Illinois Commerce Commission, which had jurisdiction, hinted it would not rule in MCI's favor, and suggested to MCI it would fare better if it moved the venue to Washington, D.C. by extending the route to St. Louis in Missouri.

July 20, an American spacecraft lands the first two men on the moon.

IBM develops GML for tagging content in documents for law offices.

1970 AT&T introduces customer dialing of international long distance calls, initially between Manhattan and London.

Optical fiber for long-distance communications developed.

Relational database invented by Dr. E. F. "Ted" Codd at IBM.

Gilbert Chin creates a new type of magnetic alloy now used in most telephone handset speakers.

Xerox creates the Palo Alto Research Center (PARC) to investigate what they called the "architecture of information" and make computers easy enough for anyone to use. PARC came up with black-on-white screens, a bitmapped display, icons, pointers, laser printers, word processors, and networks (notably Ethernet).The Xerox Star and the Alto were two PARC-created computers that embodied all of these groundbreaking ideas, but they were never successfully marketed. See PARC.

Hawaiian Telephone Company finishes converting Hawaii's 5-digit phone numbers to 7 digits.

1971 Ted Hoff at Intel invents the microprocessor – a single chip that contained most of the logic elements used to make a computer. Intel's twin innovations with the device were to put most of the transistors that make up a computer's logic circuits on to a single chip and to make that chip programmable. Here, for the first time, according to Robert X. Cringely's book "Accidental Empires," was a programmable device to which a clever engineer could add a few memory chips and a support chip or two and turn it into a real computer you could hold in your hands. Intel's first microprocessor, the 4004, a four bit machine, was released in November, 1971. See 4004. See also Microprocessor.

Floppy disk invented by IBM. It was designed originally to carry the latest IBM mainframe software to mainframe computers, each of which had a floppy drive for the sole purpose of uploading new software. Once the floppies arrived and their software uploaded, they were physically thrown away. It was only later that someone figured you could use floppy disks for permanent storage.

February 8, Trading begins on Nasdaq. Nasdaq stands for National Association of Securities Dealers Automated Quotation or "NASDAQ" System. It started with 2,500 over the counter stocks.

Ray Tomlinson develops a program for sending messages between computer systems. He designates the @ symbol to separate the user name from the computer name in the address. In short, email is invented.

The NAS Report recommended that an equipment certification program could be established to prevent harm to the network caused by hazardous voltages, excessive signal power, improper network control signaling and line imbalance. FCC establishes the PBX and Dialer and Answering Devices Committees to recommend certification standards based on the NAS Report.

Gary Starkweather, Xerox, patents first laser printer. A couple of years later HP and Canon jointly introduce the first commercial laser printers.

FCC establishes the PBX Advisory Committee and the Dialer and Answering Devices Committee and were terminated on the approval of Part 68. The PBX Committee's report was turned over to EIA where it eventually as a voluntary standard, 470. The Dialer and Answering Devices meetings were so contentious that no report was published.

The Intelsat IV communications satellite goes into commercial operation. Initially it has 830 circuits in service and linked ground stations in 15 countries. DUV (Data Under Voice) is introduced. It permits signals to "hitch-hike" on existing microwave radio systems by using the lower end of the frequency band not normally used for voice.

1972 IEEE Communications Society is established on 1 January.

Microwave Communications, Inc, later called MCI, wins an FCC license to transmit calls between Chicago and St Louis.

June 2, Blyth & Co. issues a red herring prospectus for MCI, offering three million shares at $10 a share, giving MCI a market capitalization of $30 million. The company began trading on Nasdaq on June 23, 1972. In 1998 WorldCom bought MCI for $40 billion.

First commercial video game (Pong) introduced by Nolan Bushnell at Atari.

Email introduced on Arpanet, precursor to the Internet.

IBM announces SNA.

Intel introduces the 8008 microprocessor.

Life, the magazine, unable to compete with TV dies. It followed the death of Look and The Saturday Evening Post.

First computer to computer chat takes place at UCLA.

Jon Postel writes the specifications for Telnet.

The landmark pop-porn film "Deep Throat" starring Linda Lovelace debuts. In December 2002, The New York Times wrote: You don't need to have seen a frame of the movie to be familiar with its ingenious, if ludicrous, premise: a woman is unable to achieve orgasm until a sympathetic doctor discovers that her clitoris resides in her throat. His "treatment" called upon Lovelace to perform hitherto unimaginable feats of fellatio – the more pleasure her character wants to feel, the more oral satisfaction she has to provide. Having been taught by her husband and manager, Chuck Traynor, to suppress her gag reflex, Lovelace, then in her early 20s, was shown in repeated close-ups following the doctor's advice to lengths that did not seem humanly possible. It was the stuff of which undomesticated male fantasy is made – the slavish gratification of a carnal desire that had previously been associated only with extramarital license. Small wonder that the 62-minute XXX-rated film, shot in six days by a Brooklyn hairdresser, went on to become astonishingly lucrative. Having been produced with what was, by porn standards, an astronomical budget of $25,000 (courtesy of Mob backers), the movie racked up at least $600 million in sales. Lovelace was paid $1,200, which she handed over to Traynor.

1973 The first ARPANET nodes outside of the United States are established, one at University College of London, the other at the Royal Radar Establishment in Norway.

Computerized Axial Tomography (CAT Scan) invented by Allan Cormack and Godfrey N. Hounsfield.

Ethernet invented by Dr. Robert N. Metcalfe on May 22, 1973 at the Xerox Palo Alto Research Center (PARC). Dave Boggs (Dr. David R. Boggs) was the co-inventor. Metcalfe and Boggs (in that order) were the authors of THE Ethernet paper, published July 76 in CACM. CACM is the Communications part of the ACM. ACM is the Association for Computing Machinery.

Vinton Cerf, computer scientist, invents the basic design of the Internet – the intermediate level gateways (now called routers), the global address space and the concept of end/end acknowledgement.

Gerhard Sessler and James E. West of Bell Labs receive a patent for their unidirectional microphone that improves hands-free telephone conversations.

ARPAnet, the predecessor to the Internet, has 2,000 users. Electronic mail represents three-quarters of its traffic.

Robert Metcalfe invents Ethernet at Xerox PARC. Ethernet uses a cable rather than a radio channel as the transmission medium.

The "Touch-a-matic" telephone is introduced. It can automatically dial a call anywhere in the U.S. at the touch of a single button. Its solid-state memory allows dialing up to 32 pre-coded telephone numbers.

The File Transfer Protocol (FTP) is introduced making it easier to transfer data information.

Reuters introduces the Reuters Monitor, a terminal that allowed banks to display newly deregulated foreign currency rates to traders all around the world. Banks paid to "contribute" their rates to the Reuters screens, then paid again to buy the screens themselves and rent an information feed. Reuters added market-moving news. The Reuters Monitor was phenomenally successful.

October 6, Egyptian and Syrian forces launch a surprise attack on Israeli positions in the Gaza Strip and the Golan Heights. Two weeks later King Faisal of Saudi Arabia cuts off his country's oil exports to the United States, triggering an oil crisis, long lines at the gas pumps and a recession.

1974 AT&T introduces Picturephone, a two-way color videoconferencing service at 12 locations around the country. Businesses rented meeting rooms equipped with the technology.

Hewlett-Packard introduces the first programmable pocket calculator, the HP-65.

Structured Query Language (SQL) invented by Don Chamberlain and colleagues at IBM Research.

Cerf and Kahn publish "A Protocol for Pcket Network Inconnection" detailing TCP.

SMCL invented by Charles F. Goldfarb.

First domestic satellites in operation.

AT&T introduces the digital subsriber loop.

The Department of Justice files its antitrust suit against AT&T. The Consent Decree, resulting therefrom, required AT&T to divest itself of the 24 Bell Operating Companies by 1984.

Value-added (packet-switched networks) come on the scene.

The term "Internet" is used for the first time.

1975 Bill Gates and Paul Allen co-found Micro-soft. Later the hyphen was quietly dropped.

First PCs on sale. MITS Altair 8800 personal computer kit from MITS. The Altair was the first commercially available personal computer kit. It was on the cover of the Popular Electronics January 1975 issue.

Live TV satellite feed. Ali-Frazier fight from HBO (Home Box Office).

Sony introduces Betamax, which doesn't do as well as the Video Home System (VHS) introduced later by Matsushita/JVC.

Telephones go mobile. The FCC reallocates a swath of the radio spectrum for mobile communications. Cellular is born.

April 30, Final day of Vietnam War. Saigon falls.

Bolt, Beranek and Newman (BBN) opens Telenet, the first public packet data service.

Viking is launched. Lands on Mars in 1976 and sends back data to Earth.

Transmission testing begins on the T4M, highest-capacity, short-haul digital transmission system in the U.S. The new system, linking Newark, NJ to New York City, transmits 274 million "bits" of information per second over a single coaxial tube.

1976 First digital electronic central office switch installed.

Apple Computer founded in a Cupertino, California garage by Steve Jobs and Steve Wozniak.

Unix-to-Unix Copy (UUCP) developed at AT&T Bell Labs.

Digital radio and time division switching introduced.

Alan Shugart, IBM, introduced the 5.25-in floppy. (Much later, in 1987, SONY introduced the 3.5" floppy). Floppies were first introduced with IBM's PCs when they first came on the market in 1981.

The telephone companies support "The Consumers Communications Reform Act of 1976" H.R. 12323, which was endorsed by more than 90 members of the House. This proposed legislation would have retained the telephone companies' monopoly. The FCC counters with its Docket 20003, Economic Implications and Interrelationships Arising from Policies and Practices Relating to Cusotmer Interconnection, Jurisdictional Separations and Rate Structures .Resale and sharing of carrier services permitted. Other Common Carriers (OCCs) now have access to telco Foreign Exchange (FX) and Common Control Switching Arrangement (CCSA) private network facilities.

Centennial of the Telephone. IEEE establishes the Alexander Graham Bell Medal to commemorate of the centennial of the telephone's invention and to provide recognition for outstanding contributions in telecommunications. Amos Joel, William Keister and Raymond Ketchledge are the first recipients.

COMSTAR is launched and begins commercial service. It is in permanent orbit over the Galapagos Islands.

The Cray-1 supercomputer looks like a C-shaped piece of furniture.

AT&T launches the Picturephone Meeting Service, an in-house network that allowed videoconferencing among the company's various offices. The idea was to use AT&T's service as a promotion. Other companies would see it and buy the service. The service never took off within AT&T, let alone other outside companies.

Computer engineers Vint Cerf and Robert Kahn develop IPv4. See 1983.

1977 First lightwave system installed and begins operation. It's under the streets of Chicago.

Interactive cable system (Qube) installed by Warner Cable.

Commodore PET was among the hot PCs of 1977.

Hayes introduces 300 bit per second modem for $280.

Datapoint introduces Arcnet, a 2.5 megabit per second local area network (LAN) that, at one stage, was the world's largest selling LAN.

Queen Elizabeth II becomes the first head of state to send email.

Voyager spacecraft is launched. Sends back signals from Jupiter (1979-1980), Saturn (1981), Uranus (1986) and Neptune (1989).

The Apple II, Commodore Pet and RadioShack TRS-80 debut. They are the first ready-out-of-the-box "microcomputers."

The US Supreme Court allows lawyers to advertise.

1978 Bell Labs invents cellular technology.

ITU comes out with Group 2 recommendation on fax.

Intel introduces the 8086 chip, with 29,000 transistors and processing 16 bits of data at one time. A variation of this chip, the 8088, introduced in 1980, caught IBM's eye and IBM used it in its first PC.

Before enactment of the 1978 law that made it mandatory for dog owners in New York City to clean up after their pets, about 40 million pounds of doggie souvenirs were deposited on the streets of New York City every year.

Airline industry in the United States is de-regulated. Before 1978 the Civil Aeronautics Board set fares, generally allowing the airlines to pass along their costs and to become lazy in the process.

TCP split into TCP and IP.

Usenet established between Duke and UNC by Tom Truscott, Jim Ellis and Steve Bellovin.

TAT-1, the world's first transoceanic telephone cable was retired.

Apple II introduced.

1979 Chapter 11 Federal bankruptcy provision introduced. Chapter 11 is court-supervised reorganization. It's put a moratorium on debts, allows the company work a deal with the debtors, try to get the company back on its feet. Chapter 7 is total liquidation. Close the company, sell the assets.

CompuServe Information Service starts and goes on-line.

Gordon Matthews invents corporate voice mail. See VMX.

A Federal Communications Commission inquiry restricts AT&T from selling enhanced services except through an AT&T subsidiary, American Bell, which begins operations in 1983 and closes down shortly thereafter.

July, Sony introduces its first Walkman, a name which it trademarks. The Walkman plays on cassette, but cannot record. The original Walkman came with two headsets.

Dan Bricklin and Bob Frankston introduce VisiCalc, a spreadsheet program that becomes the PC's first "killer app," or killer application. The spreadsheet, for the first time, showed a practical, business use for the desktop machines. Visicalc's VC.com file occupies slightly less than 28 kilobytes, far smaller than most gif images you see on the Internet.

Gavin Wedell born Sydney, Australia on April 22.

Oil prices rises and economy slides in a recession. See also 1974.

1980 ITU comes out with Group 3 recommendation on fax. Group 3 machines are much faster than Group 2 or 1. After an initial 15-second handshake that is not repeated, Group 3 machines can send an average page of text in 30 seconds or less.

Supreme Court of the United States rules that patents for software can be issued.

May 1980, CompuServe becomes a wholly-owned subsidiary of H&R Block, Inc.

The Osborne I makes portable computing practical - for those with strong backs.

Discount airline People Express starts.

1981 IBM introduces its first personal computer August 12 and soon has 75 percent of the market. Its PC uses a Microsoft disk-operating system called PC-DOS (for PC-Disk Operating System). Microsoft is intelligent and keeps the right to a virtually identical operating system called MS-DOS and competitors quickly develop lower-priced PC "clones" running on MS-DOS, which Microsoft sells much cheaper than PC-DOS. In 2004, IBM put its PC business up for sale.

First portable computer, by Osborne.

National electronic phone directory (minitel) starts in France.

3Com introduces the first 10 megabit per second Ethernet adapter. It cost $950. In 1999, you could buy 10 megabit Ethernet adapters for under $30.

France Telecom starts to deploy Minitel.

November. Gerry Friesen and Harry Newton launch Teleconnect Magazine. They sell it in September, 1997 to Miller Freeman, which later becomes CMP Media LLC, which closes the magazine in August 2001. Subscribers receive a magazine called Communications Convergence, the new name for Computer Telephony, a magazine also started by Friesen and Newton.

The Telecommunications Act of the U.K. is passed. It is the first step towards liberalizing the telecommunications market in the U.K. and has four main consequences:

- The General Post Office (the ernstwhile monopoly provider of telecommunications services in the U.K.) was divided into two separate entities: The Post Office and British Telecommunications (BT), which retained the monopoly over existing telecommunications networks.

- It determined that a duopoly would be created as a first step towards the introduction of competition in telecommunications.
- The Secretary of State for Trade and Industry was empowered to license other organizations to be known as Public Telecommunications Operators (PTOs), to operate public telecommunications networks (including cellular networks) in the U.K.
- It paved the way for the gradual deregulation of equipment supply, installation and maintenance which had previously been the monopoly of the GPO.

Following the Act, Mercury Communications, majority-owned by Cable & Wireless was created to compete with British Telecommunications. See also Telecommunications Act of 1981 and 1984.

Warner Amex and CompuSereve offer electronic mail to cable TV users in Columbus, Ohio.

Bell Telephone Labs design of a network-embedded database of Personal Identification Numbers (PINs) for calling card customers to be accessed by public telephones over Signaling System 7. (Today, improved architectures of this kind underlie all Intelligent Network services.)

First cellular mobile telephone service is offered, in Saudi Arabia and Scandinavia.

A new telephone service, DIAL-ITr allowed a caller to listen to the voice communications between the Space Shuttle Columbia and the ground command center.

1982 The US Postable Service begins an electronic mail service, E-Com, allowing messages to be sent by computer for postal delivery. It's abandoned in 1985.

January 8, the consent decree to break up AT&T into seven regional holding companies and what was left (long distance and manufacturing) is announced. The divestiture takes place two years later on January 1, 1984.

March 3, the FCC formally approves the startup of cellular phone services. The FCC indicated that it would accept applications for licenses in the top 30 markets 90 days after procedures were published and for smaller markets, 180 days after publication. The FCC subsequently gave one license in each market to the local phone company (the "wireline") and one for a competitor (the "non-wireline") carrier.

October 21, the FCC awards the first construction permit for a cellular radio license to AT&T's Ameritech subsidiary (this was prior to the AT&T breakup).

The Katharine Gibb Secretarial Schools teach electronic mail techniques in work processing classes.

The first full-color two-way video teleconferencing service is offered.

The development of TFM (Time Frequency Multiplexing).

The first compact disk players are available for sale in the U.S.

1983 The Federal Communications Commission set guidelines for the rollout of cellphone service. The local incumbent (i.e. monopoly) telephone company would get one license in each market while the second would be up for grabs. In early 2005, the Wall Street Journal wrote, "At the time, AT&T was about to be split up, and it could have demanded that the new AT&T long-distance company be given the incumbent licenses. But Charles Brown, then chief executive, decided that the cellphone was largely a local business. "He felt that it was logical that the cellular business should go to the Baby Bells," says Sheldon Hochheiser, AT&T's former historian, referring to the local phone companies spun off in the 1984 breakup. A study at the time by McKinsey & Co. predicted that by the year 2000 there would be 900,000 cellphone users in the U.S. The actual number was more than 100-fold greater than the prediction. The licenses that Mr. Brown had decided not to seek turned out to be worth many billions of dollars. Eventually AT&T realized it had missed the boat on cellphones. It bought a cellphone provider in 1993 but later spun it off as an independent company called AT&T Wireless. That company was acquired last year by Cingular Wireless, which is 60%-owned by SBC, and the AT&T Wireless name was retired.

Novell introduces its first local area network software called NetWare. It was originally introduced to allow a handful of personal computers to share a single hard disk, which at that stage was a costly and scarce resource. As hard disks became more available, the product evolved to allow the sharing of printers and file servers.

The Department of Defense project connecting a select group of academics and researchers, adopts an addressing system, IPv4, so that computers connected to the Internet could each have a unique identity for recognizing and communicating with each other. The addressing scheme, which uses a series of four decimal values, each of which can be a number from 0 to 255 (also known as 32-bit addressing), has a total of 4.3 billion possible addresses.

Nintendo introduces Famicom, a computer turned video game.

Cellular radio in the United States gets its first subscriber.

Sony introduces the Camcorder.

October 13, Ameritech turns on its new cellular radio system in Chicago, the first in the nation.

IBM introduces the PC XT, the first IBM PC to contain a hard disk.

Bill Gates of Microsoft announces Windows at November's Comdex.

IEEE approves 802.3 – Ethernet local area network.

DNS (Domain Name System) introduced on the Internet. DNS is a hierarchical database containing records that describe the name, IP address and other information about hosts. The database residents in DNS servers scattered throughout the Internet and private intranets.

TCP/IP protocol suite mandated for use in the Internet.

Name server developed at the University of Wisconsin.

April, the Cleaved Coupled-Cavity (C3) laser was introduced. The single frequency tunable laser emitted a light so pure that over a billion bits of information per second could be sent through a glass fiber.

Murray Waldron and William Rector sketch out a plan to create a discount long-distance provider called LDDS (Long-Distance Discount Service), later to become WorldCom.

1984 January 2, 1984. The breakup of the Bell System. AT&T gave up its local operating phone companies, which got formed by the Judge into seven, roughly equal holding companies. In turn, AT&T got the Justice Department off its back for an antitrust suit and got the right to get into industries other than telecommunications. Its chosen industry was the computer industry for which it felt it had unique skills and which it later lost around $8 billion, buying and selling NCR and fiddling around trying to build and sell its own Unix computers.

The Dot-Com is born. The Domain Name Systems is introduced, classifying network addresses by extensions, like .com.

January 24, Apple Computer Inc.'s Steve Jobs introduces the first Macintosh computer. It was the machine that changed the world of PC computing. Mr. Jobs often described the little machine as "insanely great."

March, Motorola introduces the DynaTAC 8000X, the first portable cellular phone. It listed for $3,995 and weighed two pounds.

Ken Oshman sells Rolm to IBM for $1.26 billion. It was not one of IBM's better investments. Rolm is now part of Siemens, which understands telecommunications.

Prodigy Information Service, a service of IBM and Sears, starts.

1984 Telecommunications Act passed in the U.K. See Telecommunications Act 1981 and 1984.

Local area signaling service is introduced. The service is used to trace nuisance calls, transfer calls, and provide other advanced calling services. (May 20).

AT&T and NASA space shuttle Discover launch its second Telstar 3 satellite.

1985 Early investor Bernard Ebbers becomes chief executive officer of LDDS, later to become WorldCom, later to go bust, later to end up in jail for perpetuating the largest fraud in American history.

CD-ROM introduced by Philips and Sony.

Steve Jobs driven from Apple Computer by John Sculley.

John Sculley, head of Apple Computer, licenses the copyrights protecting the "look and feel" of the Apple Macintosh operating system to Microsoft in order to get Microsoft to write more applications for the Mac. The license allowed Microsoft to launch its hugely successful Windows operating system in November of 1985 and to defend itself against a lawsuit brought by Apple alleging Windows was so similar to the Mac that it violated Apple's copyrights.

IBM introduces four megabit per second token ring local area network.

Symbolics.com is assigned on March 15 to become the first registered domain.

Symbolics registers the first commercial domain name, Symbolics.com. MITRE registers the first organization domain name, Mitre.org. Carnegie Mellon (cmu.edu), Purdue (purdue.edu), Rice (rice.edu), and UCLA (ucla.edu) register the first educational domain names. Domain names can be registered for free.

1986 Novell's SFT NetWare, first fault tolerant local area network operating system.

McDonalds becomes first commercial customer to trial ISDN. It's provided by Illinois Bell.

William G. McGowan, MCI's founding chairman, and a pioneer in the competitive United States telecom scene, has a massive heart attack and a heart transplant. He dies six years later.

MCI Mail, pioneered by Vinton G. Cerf, one of the founders of the Internet, links its mail service with CompuServe, creating a network with half a million subscribers. The services are linked to the Internet in 1989.

SGML becomes an official international standard. See SGML.

Network News Transfer Protocol (NNTP) proposed by Brian Kantor and PHil Lapsley.

Mail Exchange (MX) records developed by Craig Partridge.

The National Science Foundation introduces its 56kbps backbone network.

William Esrey elected CEO of of United Telecom, later to become Sprint. In 2003 Esrey was fired as a result of poor performance and amid a scandal of unpaid taxes on hefty options awarded by the board.

Microsoft IPO.

1987 The 290-square-mile Caribbean island of Dominica becomes the first country in the world to operate a fully digital national telephone system. To the north of Dominica is Guadeloupe and to the south Martinique and St Lucia.

The Year of the Fax Machine, according to the New York Times.

Prozac, the popular antidepressant, introduced.

October, 1987, the one-millionth cellular subscriber signs up for service in America.

Cable TV reaches the halfway mark, penetrating 50.5% of U.S. homes.

George Forrester Colony of Forrester Research is believed to have coined the term "client-server" computing. See Client-Server.

First InterOp trade show, Monterey, California.

Synoptics ships the first Ethernet hub. They progressively get cheaper and cheaper.

Archived email is retrieved from the National Security Council computer system in the investigation of the Iran-contra affair.

Perl released by Larry Wall.

Bellcore introduces the Asymmetric Digital Subscriber Line (ADSL) concept which has the potential of multimedia transmission over the nation's copper loops. SONY introduces the 3.5-in floppy. Philip Estridge, IBM, developed the first hard drive for PCs. It held 10MB. N.J. Bell is the first to implement Caller ID.

Superconductivity is discovered - the transmission of electricity without resistance through low temperature material.

TDD (telecommunications device for the deaf) is initiated.

1988 First transatlantic optical fiber cable.

Robert (rob) E. Allen takes over as CEO of AT&T.

The European Union chooses GSM as the general standard for mobile communications, ensuring that even though the 15-nation EU's electrical plugs and TV sets were largely incompatible, at least the cellular phone system across all those nations would be compatible.

IBM speeds up its token ring local area network to 16 megabits per seconds. Wags call token ring a "virtual engagement present."

Computer Emergency Response Team (CERT) formed by DARPA.

Internet Relay Chat (IRC) developed by Jarkko Oikarinen.

U.S. Congress passes the Telecom Trade Act of 1988 in response to alleged dumping of telecom systems in the U.S. by foreign manufacturers. One aspect was the requirement of all imported telecom equipment to comply with all applicable FCC requirements. Enforcement is by U.S. Customs.

1989 LDDS (later to become WorldCom) becomes public through the acquisition of Advantage Companies.

Fiber to the home field trial, Cerritos, CA.

Novell releases NetWare 3.0, the first 32-bit network operating system for Intel 80386/486-based servers.

Panasonic's household-size video phone with moving color images debuts in Tokyo.

IETF established. See IETF.

Berlin Wall falls, spelling the effective end of communism, except in North Korea. it came down on November 9, i.e. 11/9. See also 2001.

English computer scientist called Tim Berners-Lee invents the World Wide Web while working at a European nuclear research laboratory.

MP3 is patented.

Sprint begins transatlantic long-distance service.

Tim Berners-Lee proposes a scheme to enable electronic documents to link to other documents stored on other computers. This idea, which later grew into the world wide web, started out as a program called Enquire, which Mr Berners-Lee wrote for his own use while working at CERN, the European Particle Physics Laboratory in Geneva, Switzerland. He went on to write the first web browser and web server, both of which he gave away on the internet, along with details of the protocols to describe and transmit web pages. See also Berners-Lee.

1990 Demonstration of 2,000 kilometer fiber optic link using optical amplifiers without repeaters.

MVIP formed and first product shipped.

Arpanet officially called the Internet.

Sharp introduces the first LCD designed for a laptop and thus the laptop business is launched.

The world (world.std.com) becomes the first commercial provider of Internet dial-up access.

HTTP 1.0 specification published.

The first HTML document published by Tim Berners-Lee.

Archie file-indexing system developed.

AT&T filed a petition to strengthen DID rules for prevention of toll fraud.

EIA filed a petition to require digital security coding for cordless phones to prevent random dialing that interfered with 911 operations.

Docket 90-313 requiring hotels/motels and coin phones to provide equal access to competing long distance carriers was resolved in 1992.

Digital audiotape (DAT) makes its debut.

August 2. Iraq invades Kuwait.

The Berlin Wall comes down, reuniting East and West Germany.

1991 AT&T, under chairman Bob Allen, buys NCR for a gigantic $7.4 billion and soon renames it AT&T Global Information Solutions. Later AT&T took hundreds of millions of dollars in restructuring and other charges related to the fact that NCR lost pots of money after AT&T bought it. Part of the problem, according to analysts, was that AT&T bought NCR right at the time NCR was making the transition from traditional mainframe computers to so-called massively parallel computers powered by collections of small, cheaper processors run in tandem. NCR also got hit by a decline in its traditional cash register business as low-margin PCs came in. The skinny around the industry at the time AT&T bought NCR was that AT&T bought NCR to disguise the fact that its own computer operations at that time were losing so much money. And the senior management of AT&T at that time wanted to retire with the glories of booming long distance revenues and not lousy computer results.

Motorola introduces the lightest cellular phone yet, the MicroTAC Lite for about $1,000 retail.

The Electronic Industries Association approves and publishes on July 9, 1991, the Commercial Building Telecommunications Wiring Standard, the most important wiring standard ever published in the history of telecommunications.

Scott Hinton at Bell Labs heads a team that builds the first photonic switching fabric, bringing light-based switching technology in telecommunications networks closer to reality.

Wiltel introduces frame relay service.

Linus Torvalds, a student at the University of Helsinki, invents Linux, a computer operating system that would later go on to become one of the most successful computer operating systems ever created. Torvalds positioned Linux as a "free" operating system, with inputs from programmers all over the world – but with him as the ultimate decision maker as to which code finally made it to the latest release. Linux rhymes with cynics. See Linux for a much bigger explanation.

World Wide Web created by Tim Berners-Lee at CERN in Switzerland.

Gopher released by Paul Lindner and Mark P. McCahill from the University of Minnesota.

On August 29, 1991, the Supreme Soviet, the parliament of the U.S.S.R., suspended all activities of the Communist Party, bringing an end to the institution. The Soviet Union is basically dissolved.

1992 AT&T introduces VideoPhone 2500 marketed as the first home-model color video phone which works on normal dial-up analog phone lines. It meets cool reception because of poor image quality and its high price, namely $1,500. It is later withdrawn from the market.

LDDS (later to become WorldCom) merges in an all-stock deal with discount long-distance service provider Advanced Telecommunications.

Microsoft Windows 3.1 and IBM's OS/2 2.0 operating systems introduced. Windows NT (32-bit operating system) debuts in beta form.

Wang files for Chapter 11. Later it emerges from Chapter 11, but not the company it was before.

MCI introduces VideoPhone for normal dial-up analog phone lines. It retails for $750. It is not compatible with the AT&T video phone. It and the MCI phone promptly bomb and are withdrawn from the market.

The cellular industry signs its ten millionth subscriber on November 23, 1992. At least that's what the press releases said. Some carriers claimed that at year end, cellular subscribers in the United States had actually hit 11 million, way ahead of all predictions.

Apple, EO and others introduce the PDA, the Personal Digital Assistant. Later they called it the Newton.

RMON ratified by IETF.

William G. McGowan, MCI's founding chairman, dies. he suffered a serious heart attack in 1986 and received a heart transplant. He lived for six years with the heart transplant, an absolute miracle of modern medicine. For those of who loved the man, his death was particularly sad.

Congress allows commercial use on the Internet.

Veronica, a Gopherspace search tool, is released by the University of Nevada.

The World Wide Web is born - the brain child of CERN physicist Tim Berners-Lee. Congress required all agencies to metricize their rules. A major impact was on Part 68 plug and jack drawings.

Will G. McGowan, chairman and visionary of MCI Communications, dies.

United Telecommunicatons changes its name to Sprint.

1993 FCC announces its intention to auction off a chunk of spectrum larger than that used in 1993 for cellular radio. The new airwaves will be used for new types of wireless communications, including portable digital communications devices from phones to laptops, palmtops and PDAs equipped to receive and transmit data of all types, including faxes and video.

LDDS (later to become WorldCom) acquires long-distance providers Resurgens Communications Group and Metromedia Communications in a three-way stock and cash transaction that creates the fourth-largest long-distance network in the United States.

Sprint merges with Centel, a local and wireless phone company.

Microsoft releases Windows NT.

Registration of domain names begins to cost $35 per year.

January, Marc Andreessen and Eric Bina introduce first graphical Web browser called Mosaic. It's from the National Center for Supercomputing Applications.

February 25, McCaw Cellular announces North America's first all-digital cellular service, in Orlando, Florida.

February 26, a bomb explodes in the garage of New York's World Trade Center, killing six people and injuring more than 1,000 others.

March 17, The Clinton administration urges Congress to eschew comparative hearings and institute a lottery for awarding new radio spectrum. See PCS.

August, AT&T agrees to buy McCaw Cellular for $12.6 billion. The idea is to help get AT&T back into local phone service. The company is later spun off as AT&T Wireless.

October, Bell Atlantic sets $21.7 billion merger with TCI, cable TV giant. The assumption is that cable TV and telephone networks are "converging" into an information highway for transporting video, voice and data. The deal later fell apart after the FCC cut cable TV rates and TCI's profitability fell part.

December 31, Thomas J. Watson dies, age 79.

AT&T introduces the AT&T EO Personal Communicator 440, based on the Bell Labs-developed Hobbit microprocessor. This hand-held device combines the features of pen-based personal computers, telephones and fax machines. The device is later withdrawn from the market because of poor demand.

Louis V. Gerstner takes over a stumbling IBM and later turns it around.

InterNIC created by NSF.

Telecom Relay Service (TRS) available for the disabled.

The National Science Foundation (NSF) network backbone jumps from T-1 to T-3.

The world's first 64-bit home console video game system released. Made by IBM, the 'Jaguar' offered high-speed action, CD-quality sound, and polygon graphics processing beyond most other machines available at the time.

The first wireless headset portable CD player is marketed.

1994 The Year of the Internet, according to the New York Times.

NCR name changed to AT&T Global Information Services (GIS).

AT&T acquires McCaw Cellular, renames it AT&T Wireless.

LDDS (later to become WorldCom) acquires domestic and international communications network IDB Communications Group in an all-stock deal.

GO-MVIP, Inc. formed. Trade association for developers and manufacturers of MVIP computer telephony products.

AT&T pays $12.6 billion for McCaw Cellular Communications Inc. Robert E. Allen is CEO of AT&T. This is his second expensive purchase, the previous one being NCR.

Bill Gates marries Melinda. Later they have children and form the Bill and Melinda Gates Foundation, the largest in the world.

Hughes Satellite starts DirecTV, a direct-broadcast satellite service that beams 175 channels to a home satellite antenna dish 18 inches in diameter. It snags 1.3 million subscribers in less than a year.

April 4, Netscape Communications Corp which will go on to create the Navigator version of a browser, is founded. Many date the beginning of the World Wide Web as a serious tool of international commerce to this time.

July 17, Microsoft signs a consent decree with the Justice Department agreeing to give computer manufacturers more freedom to install programs from other companies. As a result, Microsoft slightly alters its licensing contracts.

September 12, Netscape ships its first Internet / Web browser.

October 8, A team of six programmers and a veteran Microsoft software developer begin writing the code that will become Internet Explorer version 1.0.

Microsoft licenses technology from Spyglass to help it quickly develop a Web browser.

North American Free-Trade Agreement signed by America with Mexico and Canada.

Internet is pretty much world-wide with the exception of most of the African interior, Pakistan, Mongolia, Cuba and some areas in South America and Southeast Asia.

Real Audio introduced to Internet which allows one to hear in near real time.

Radio HK, the first 24-hr Internet only radio station, starts broadcasting.

At the Winter Olympics in Lillehammer, Norway, fiber optics transmitted the first ever digital video signal.

1995 The World Wide Web gets major traction with the creation of Netscape, the first browser of the modern generation. Also Netscape does its IPO – Initial Public Offering. See also Mosaic.

The first digital video (DV) camcorders are sold. Digital cameras hit the market.

Internet traffic grows tenfold in 1995.

IBM buys Lotus for $3.5 billion, the main attraction being Lotus Notes. One of the key attractions of Lotus Notes is that it saves on phone bills by substituting electronic messaging for calling.

LDDS acquires voice and data transmission company Williams Telecommunications Group (WilTel) for $2.5 billion cash and changes its name to WorldCom.

May 26, William H. Gates sends his famous "The Internet Tidal Wave" memo to Microsoft's top executives, making the Internet the company's top priority.

Sun introduces Java.

August 24, Windows 95 finally ships. It contains computer telephony features, including TAPI, VoiceView and Fax-on-demand. See TAPI 3.0.

August, Disney agrees to buy Capital Cities/ABC for $19 billion. The idea of the merger was, according to the New York Times, that "biggest is best". The idea was to combine the most profitable TV network with a name-brand family entertainment empire.

August, CBS accepts Westinghouse's $5.4 billion takeover offer. According to the New York Times, the idea of the takeover is that "even an ailing Big 3 TV network is worth owning, if it comes with the collection of radio and TV stations reaching one-third of U.S. households."

September, Time Warner agrees to buy Turner Broadcasting for $7.5 billion.

September 20, AT&T announces it will split itself into three companies – long distance, equipment manufacturing, and computers. Wall Street applauds the decision and in one day lifts the price of AT&T's stock by 10%, or about $6 1/2 billion. Meantime, AT&T announces that it will substantially reduce the size of its failed computer activities, which were called AT&T Global Information Solutions. See 1991.

November, Microsoft ships Internet Explorer 2.0.

December 7, Microsoft publicly unveils its Internet strategy.

December 8, Digital Versatile Disk (DVD) is announced. DVD is a specification announced by nine companies for a new type of digital videodisk, similar to CD-ROMs but able to store far more music, video or data in a common format. DVDs will be 5 inches in diameter and will be able to store 4.7 gigabytes on each side, equivalent to 133 minutes of motion picture and sound, or enough to hold most feature-length movies. The companies announcing DVD were Philips, Toshiba, Matsushita Electric Industrial, Sony, Time Warner, Pioneer Electronic, the JVC unit of Matsushita, Hitachi and Mitsubishi Electric.

AT&T announces its plan for restructuring into three separate, publicly traded companies: a services company that will retain the name AT&T; a systems and technology company (Lucent Technologies) composed of Bell Labs, Network Systems, Business Communications Systems, Consumer Products and Microelectronics; and a computer company, which recently returned to the NCR name.

Netscape IPO.

1996 AT&T spins off its products and systems manufacturing arm, later named Lucent Technologies.

AT&T also spins off its computer company called NCR which it had bought in 1991. It does this by distributing all 101,437,174,698 common shares of NCR to AT&T shareowners, completing the spinoff of its computer arm and bringing back the legendary 113-year-old "cash register" company,along with its original name and stock market ticker symbol, to the New York Stock Exchange – a scenario not seen before on the Big Board.

WorldCom merges with MFS Communications Company (MFS), which owned local network access facilities via digital fiber optic cable networks in and around major U.S. and European cities, and UUNet Technologies, an Internet access provider for businesses.

September 10, 1996: Time Warner Cable launches Road Runner, its cable broadband service, in Akron, Ohio, following initial testing in Elmira, New York in 1995.

January, After 10 years of trying, Congress finally passes a bill deregulating most segments of the communications industry. Telephone companies, broadcasters and cable operators are all free to enter each other's markets. It's called The Telecommunications Act of 1996. It turns out later to be one of the worst written pieces of Federal legislation in the U.S. See Telecommunications Act of 1996.

The browser wars begin. On August 12, Microsoft launches Internet Explorer 3.0. One week later Netscape releases Navigator 3.0, an upgrade of its existing browser that previously held a 75% market share.

Internet traffic doubles again. It had doubled in 1995. One year it doubled in three months. Everybody starts to believe that it will continue doubling every three months forever. The fact that it didn't lead to the tech wreck of 2000, 2001 and 2002.

February, US West signs $5.3 billion deal for Continental Cablevision. The assumption, according to the New York Times, is that it will now have a wire into many homes, by dominating local phone service in 14 states and reaching 16.3 million cable subscribers.

April, SBC Communications (the name for the holding company owning Southwestern Bell Telephone) buys Pacific Telesis (the holding company for Pacific Bell) for $16.7 billion. The assumption of this merger, according to the New York Times, is that regional telephone companies can become national players by combining 30 million phone lines in five states with potential coast-to-coast cellular market of 80 million people.

April, Bell Atlantic buys NYNEX (the holding company for New York Telephone and New England Telephone) for $22.1 billion. The new company will be called Bell Atlantic.

IP Telephony introduced.

Late in the year, Microsoft releases Internet Explorer 3.0. Many reviewers praise its vastly improved nature.

September, Rockwell announced a 56 kbps modem chip set designed for Internet applications. 56K download (PCM); 33.6 upload (analog). Technical committees start development of standards for this new technology. Ccntroversy erupts over the fact the modulation technology limits the theoretical speed to about 53K because of Part 68's signal power limitation requirements to prevent crosstalk to third parties. Actually, because of line impairments the fastest practical speed is around 42 to 44K.

December 30. AT&T spins off NCR, the firm it had bought in 1991 for $7.4 billion.

The PalmPilot 1000 launches a handheld revolution.

Phone companies began to install 2.5 gigabit fiber-optic equipment.

1997 Amazon.com goes public.

IP Telephony starts to become a reality. Microsoft announces TAPI 3.0, whose cornerstone is IP telephony. See TAPI 3.0.

WorldCom Inc. buys MCI for $30 billion in its own stock for MCI. The offer for $41.50 an MCI share was at a 41% premium over MCI's stockmarket price before the bid was announced. WorldCom later reports accounting irregularities which change later into massive fraud. Along the way it goes bankrupt (Chapter 11) and changes its name – irony of irony –to MCI.

The Asian Financial Crisis happens. The Asian markets dropped and most of the Southeast Asian "miracle" economies suffered dramatic reversals. Those economies included Korea, Malaysia, Singapore, Hong Kong and Thailand.

June 12, The U.S and the E.U. reach agreement on mutual recognition of product testing and approval requirements covering everything from lawnmowers, pharmaceuticals, recreational craft to telecom equipment.

October 20, Justice Department files suit against Microsoft alleging that the company violated its 1995 consent decree – one section of which banned Microsoft from tying the licensing of one product to the acceptance of another.

December 11, Judge Thomas Penfield Jackson issues a temporary restraining order against Microsoft. Microsoft must at least temporarily halt its practice of requiring PC vendors to bundle Internet Explorer with Windows 95 while the case is being decided. Judge Thomas Penfield Jackson also appointed a special master to gather additional evidence.

However, Jackson declined to hold Microsoft in contempt – a move which could have cost Microsoft as much as $1 million a day. He also turned down a Justice Department request to strike down non-disclosure agreements between Microsoft and PC vendors that the department claimed had hampered its attempts to solicit testimony against the software vendor.

Congress mandates that most broadcasters convert to digital signals by 2006 and grants them an estimated $70 billion worth of new television spectrum to do so. But the promise of digital TV – sharper pictures, better sound, more channels and interactive capability – has been slow to materialize, with broadcasters and TV makers blaming each other for the sluggish pace of the changeover.

November 1, C. Michael Armstrong is elected chairman and CEO of AT&T, succeeding Robert E. Allen who has been chairman and CEO since 1988.

DVD comes to market.

On August 31, Diana, the Princess of Wales, is killed in an automobile accident in a tunnel by the Seine in Paris.

1998 January 1, 1998. The market for fixed telecommunications services in the EU (European Union) is opened to all competitors. European countries see a proliferation of long distance phone companies.

January 1, 1998. Local service competition becomes a reality in Canada. Local Number Portability (LNP) to be in Canada soon.

February, V.90 56K standard was approved ending months of difficult negotiations and modem wars.

WorldCom completes three mergers: with MCI Communications ($40 billion) - the largest in history at that time - Brooks Fiber Properties ($1.2 billion) and CompuServe ($1.3 billion).

May 18, The Justice Department and 20 states file antitrust suits against Microsoft.

May, Galaxy IV satellite messes up, causing 45 million pagers to be shut down and many credit card transactions to cease. The problem was fixed within a few days, but not after lots of publicity.

SBC Communications, Inc. the new name for the regional Bell holding company, Southwestern Bell, agrees to buy Southern New England Telephone Company – commonly called SNET. See Southern New England Telephone Corporation.

MCI gets bought by Worldcom for $40 billion. New company is called MCI Worldcom. Worldcom paid for the acquisition by issuing paper script it printed itself. And you thought the U.S. government had a monopoly on printing money.

May, Microsoft sued by the U.S. Justice Department for antitrust violations. The Justice Department is later joined by 20 states. A settlement is announced in the fall of 2001.

November 24, American Online agreed to buy Netscape for $4.2 billion. It also made a side agreement to jointly develop technology with Sun Microsystems and Netscape.

Digital Millennium Act Copyright Act passes, making it illegal to circumvent copyright publishing protection technologies.

China tried Lin Hai for plotting against the state by providing 30,000 email addresses to an American magazine. He gets a two-year jail term.

The first digital televisions are sold in the U.S. The DVD-Audio format is agreed upon. Recordable DVD formats emerge. The first portable DVD player is introduced.

1999 January 1, The new European currency called the euro is officially introduced. Eleven European countries have pegged their currency to it. Those countries are Austria, Belgium, Finland, France, Germany, Ireland, Italy, Luxembourg, Netherlands, Portugal and Spain. The euro started at around $US1.17.

WorldCom and Sprint agree to merge. WorldCom shares peak at more than $64.

Vice President Al Gore admits that he created the Internet.

October 4, In the largest corporate takeover deal ever, MCI Worldcom announces that it will acquire Sprint in a stock swap valued at $115 billion.

October 4, Rupert Murdoch sells TV Guide for $9.2 billion. He had bought it for $3 billion in 1987. That was an annual appreciation of about 9.5%.

November 5, Judge Thomas Penfield Jackson issues "findings of fact" showing he believes Microsoft to be a market-stiffling monopoly. No penalty is ordered. That comes later.

Portable MP3 players begin to sell. DVD Audio and SACD players are introduced. Personal video recorders (PVR) are first introduced.

Sprint and WorldCom announce plans to merge in a deal valued at $129 million. The merged was called off the following year in light of regulatory scrutiny.

GTE Wireless, Bell Atlantic Mobile and Vodafone AirTouch combine to form Verizon Wireless.

SBC and BellSouth merge wireless operations to form Cingular.

2000 A proposed merger between MCI WorldCom and Sprint is barred on antitrust grounds.

AT&T Wireless is spun off into separate company.

Nasdaq tops 5000.

Verizon Communications is formed by merger between Bell Atlantic and GTE.

U.S. and European regulators block the proposed merger with Sprint; WorldCom and Sprint terminate their merger agreement.

The year of potential computer apocalypse caused by the Millennium Bug, which was meant to bring down computers systems because they had encoded dates at two digits, i.e. 99, instead of four 1999. The crisis never happened. See Y2K for a full explanation.

March 16, 2000, when the Dow Jones industrial average rose nearly 500 points.

April 4. Judge Thomas Penfield Jackson finds that Microsoft violated antitrust law.

June 7. Judge Jackson orders that Microsoft be broken up.

October 25, AT&T announces plans to create four publicly held companies from its current business units: 1. AT&T Consumer is built around residential long-distance and WorldNet Internet access businesses. It is the U.S.'s largest consumer communications and marketing company. 2. AT&T Wireless is a cell phone company. 3. AT&T Broadband is the largest cable TV and broadband services company. It includes AT&T's investments in TCI, MediaOne, and Excite@Home. 4. AT&T Business combines AT&T's global investments in business communications and services.

DVD recorders begin to sell in the U.S.

November 16, 2000: ICANN, the authority that governs global Internet domain name system, has approved seven new domains extensions, in addition to .com, .net and .org. These new extensions are the first new global Internet domains approved by ICANN in over a decade.

1) .biz
2) .info
3) .name
4) .pro
5) .aero
6) .coop
7) .museum

2001 The Year of Wireless, according to the New York Times.

WorldCom merges with Intermedia Communications, a provider of data and Internet services to businesses.

September 11 (or 9/11), New York's World Trade Center is destroyed. The Berlin Wall came down on 11/9/1989. Note the similarity of

Digital satellite radio enters the market. XM Satellite Radio begins transmitting.

VoiceStream buys Deutsche Telekom; renamed T-Mobile in 2002.

2002 January 1, Euro bills and coins go into circulation. Twelve European nations – Austria, Belgium, Finland, France, Germany, Greece, Ireland, Italy, Luxembourg, the Netherlands, Portugal and Spain – begin using the new Euro. The currency is also accepted in places like Britain.

January 3, Cable Labs publishes OCAP 1.0 specification.

March 14, The FCC classifies cable modem service as an information service, eliminating "forced access" of ISPs on cable operators.

April 30, WorldCom's Bernard (Bernie) Ebbers is out as CEO as questions about $366 million in his personal loans from the company, a federal probe of its accounting practices and concerns about its finances mount. WorldCom would later turn out to have perpetuated the largest accounting fraud in American business history.

Microsoft begins work on Vista, called Longhorn at the time. It finally releases Vista in January, 2007 – five years later.

June 26, Adelphia Communications files for Chapter 11. It was the nation's sixth largest cable TV provider.

July 1, National bills and coins of those countries with the Euro as currency are no legal tender. Only the euro is now legal tender. See 1999.

FCC requires the installations of off-air DTV (digital TV) tuners in nearly all new U.S. television sets by 2007.

October 10, FCC declines to approve DirecTV/Echostar merger.

November 14, FCC approves $47.5 billion Comcast/AT&T Broadband merger.

December 19, Cable and consumer electronics industries reach consensus on how to build "unidirectional, plug and play" cable-ready digital TVs.

Digital satellite radio expands. Sirius Satellite Radio begins transmitting. The previous year XM Satellite Radio had started transmitting.

Dow Jones Industrials and Nasdaq fall for the third year running.

Since the telecom downturn started in the year 2000, 500,000 telecom workers in the United States have lost their jobs. The industry, at the end of 2002, had run up debts of $1 trillion globally and has a huge glut of excess capacity. By some estimates, no additional facilities for distributing telephone calls, data and video around the world (excluding local loops, i.e. the last mile) will be needed until at least 2007.

Here are the most significant events of 2002 in the history of WorldCom:

March 11 - WorldCom receives a request for information from the U.S. Securities and Exchange Commission relating to accounting procedures and loans to officers.

April 3 - WorldCom says it is cutting 3,700 jobs in the U.S. or 6% of WorldCom group's staff, 4% of WorldCom's overall work force.

April 22 - Standard & Poor's cuts WorldCom's long-term and short-term corporate credit ratings.

April 23 - Moody's Investors Service cuts WorldCom's long-term ratings. Fitch cuts the company's ratings, saying it expects WorldCom's revenue to deteriorate during 2002, with prospects for recovery in 2003 uncertain.

April 30 - WorldCom CEO Bernard Ebbers resigns amid slumping share prices and SEC probe of the company's support of his personal loans. Vice Chairman John Sidgmore takes over.

May 9 - Moody's cuts WorldCom's long-term debt ratings to junk status, citing the company's deteriorating operating performance, debt and expectations for further weakness.

May 10 - Standard & Poor's cuts WorldCom's credit rating to junk status.

May 13 - Standard & Poor's removes WorldCom from its S&P 500 Index.

May 15 - WorldCom says it would draw down a $2.65 billion bank credit line as it negotiates for a new $5 billion funding pact with its lenders.

May 21 - WorldCom says it will scrap dividend payments and eliminate its two tracking stocks, one that reflects its main Internet and data business and a second that reflects its residential long-distance telephone business.

May 23 - WorldCom secures $1.5 billion in new funding to replace a larger, $2 billion credit line.

June 5 - WorldCom says it will exit the wireless resale business and will cut jobs to reduce expenses and pare massive debts.

June 25 - WorldCom fires its chief financial officer after uncovering improper accounting of $3.8 billion in expenses over five quarters starting in 2001. The company also says it will cut 17,000 jobs, or 20% of its work force.

June 26 - SEC files civil fraud charges against WorldCom and seeks an order to prevent the company from disposing of assets, destroying documents and making extraordinary payments to senior officers. The U.S. Justice Department says it is probing the matter. Nasdaq market halts trading in its two tracking stocks, WorldCom Group and MCI Group. Shares of WorldCom fall as low as 9 cents before the halt.

June 27 - The U.S. House Financial Services Committee subpoenas top current and former WorldCom executives, Ebbers, Sidgmore and Sullivan, as well as Salomon Smith Barney analyst Jack Grubman to testify on July 8. The House Energy and Commerce Committee requests documents for its own probe.

July 1 - WorldCom says in a sworn statement to the SEC that its audit committee is reviewing its financial records for 1999 through 2001 regarding "certain material reversals of reserve accounts." The company receives notice from some of its lenders saying they could demand immediate repayment for defaulted loans. The company's shares are resumed on the Nasdaq, opening at about 8 cents. The Bush administration says it is reviewing existing government contracts with WorldCom and could deny the company new business.

July 2 - Sidgmore holds news conference, apologizing for the scandal and says WorldCom is working on funding proposals with its lenders to stave off bankruptcy. New York State comptroller files suit for losses in its pension funds.

July 3 - U.S. District Judge Jed Rakoff appoints former SEC Chairman Richard Breeden to prevent possible shredding of key documents and unwarranted payouts to top officers. Rakoff sets March 31 for WorldCom to go on trial for alleged fraud.

July 8 - Former WorldCom CEO Ebbers tells the U.S. House Financial Services Committee he did nothing wrong and refuses to answer questions. Ex-CFO Sullivan also refuses to testify. Salomon Smith Barney analyst Jack Grubman says he attended WorldCom board meetings but denied having inside information about the woes. Sidgmore blames Andersen and says turnaround plans are coming together, some of which include bankruptcy.

WorldCom says in a revised statement filed with SEC that Sullivan tried to delay an internal audit that discovered the transfers of expenses to capital spending accounts.

July 9 - Sidgmore says the company expects to decide within three weeks whether to pursue bankruptcy or some other financial reorganization and is seeking $3 billion in funding, less than the previously sought $5 billion.

July 10 - U.S. attorney in Mississippi is removed from investigation of WorldCom because of a conflict of interest and case is taken over by New York's U.S. attorney's office.

July 11 - WorldCom says it will not pay the $71 million second quarter dividend to shareholders of its MCI Group long-distance tracking stock.

July 12 - SEC wins a stay blocking WorldCom plans to convert MCI Group tracking stock into WorldCom stock for 10 business days while the agency reviews the conversion.

July 15 - Rep. Billy Tauzin says Congress interviews of witnesses indicate WorldCom's accounting errors may go back to 1999 and says company e-mails show efforts in March 2001 of top executives trying to manipulate earnings to meet Wall Street expectations. He also says company memos show executives had discussions in 2000 about accounting for expenses over a longer period of time than allowed.

July 16 - WorldCom lines up $2 billion in financing to keep operating if lenders force the company into bankruptcy protection, sources say. A lawsuit by 25 banks trying to limit WorldCom's use of $2.5 billion in loans was moved to federal court from state court. Three California pension funds sue WorldCom for allegedly misleading them about the company's financial health in a 2001 bond offering. WorldCom missed $79 million in interest payments, according to sources. FCC Chairman Michael Powell says he does not expect imminent service halts or disruptions by WorldCom.

July 17 - The company agrees to freeze some assets for 80 days in exchange for a temporary halt to legal efforts by a group of banks to recover $2.5 billion in loans.

July 18 - Sources say WorldCom plans to file for bankruptcy protection as early as July 21.

July 21 - WorldCom CEO Sidgmore says the company will file for Chapter 11 bankruptcy protection later in the day, but plans to emerge within 9 to 12 months. The company will have access to up to $2 billion in funding but does not plan to tap all of it. Additionally the company will hire a restructuring expert.

2003 The number of mobile phones in the world overtook the number of fixed line phones. There will be 1.47 billion mobile phones, but only 1.41 billion landlines.

Computer experts, according to the New York Times, called 2003 "The Year of the Worm." For 12 months, digital infections swarmed across the Internet with the intensity of a biblical plague. It began in January, when the Slammer worm infected nearly 75,000 servers in 10 minutes, clogging Bank of America's A.T.M. network and causing sporadic flight delays. In the summer, the Blaster worm struck, spreading by exploiting a flaw in Windows; it carried taunting messages directed at Bill Gates, infected hundreds of thousands of computers and tried to use them to bombard a Microsoft Web site with data. Then in August, a worm called Sobig.F exploded with even more force, spreading via e-mail that it generated by stealing addresses from victims' computers. It propagated so rapidly that at one point, one out of every 17 e-mail messages traveling through the Internet was a copy of Sobig.F. The computer-security firm mi2g estimated that the worldwide cost of these attacks in 2003, including clean-up and lost productivity, was at least $82 billion (though such estimates have been criticized for being inflated).

May. Time Warner Cable officially launches its VoIP service, Digital Phone, in Portland, Maine.

2004 AT&T announces plans to stop marketing consumer long-distance service.

Cingular acquires AT&T Wireless.

Sprint agrees to acquire Nextel.

MCI (formerly WorldCom) emerges from bankrupcy.

IBM puts its PC manufacturing business, which it started in 1981, up for sale, marking the end of an era. It eventually sells the money-losing company to a Chinese company called Lenova for around $1.5 billion. Several observers couldn't figure why anybody would pay that much money for a company that had been losing money for several years and had dubious prospects at best.

2005 SBC Communications announces it will buy AT&T for $16 billion, virtually all in SBC stock. Within days of the merger announcement, the companies predictably announced hujge layoffs of 13,000 to justify the deal. As is usual on Wall Street, most mergers benefit senior management of both companies, but not the employees or the shareholders.

Verizon announces it will buy MCI Corp. for $6.7 billion, virtually all of it in Verizon stock. MCI is the old WorldCom emerged out of bankruptcy.

Hewlett Packard's board fires Carly Fiorina, whose early training had been in AT&T and its spin-off, Lucent. She didn't do well there, either. During her 5 1/2 years running HP, HP's stock fell by roughly half, while she pocketed $188.6 million in salary, bonuses, stock options and severance pay.

IBM agrees to sell its money-losing PC making and selling company to China Lenovo Group for around $1.5 billion.

Alltel buys Western Wireless.

WorldCom's Bernie Ebbers is found guilty of securities fraud, conspiracy and filing false documents with regulators. He is sentenced to 25 years.

eBay buys Skype for $1.3 billion and 32.4 million shares of eBay stock in a transaction worth around $4.2 billion.

Verizon Communications sells its telephone operations and directories business in Hawaii to a group of investors led by the Carlyle Group. The Hawaii company's name is changed from Verizon Hawaii to Hawaiian Telcom.

2006 Early 2006: SBC (earlier called Southwestern Bell) bought AT&T. After the buyout, SBC changed its name to AT&T. Earlier, Southwestern Bell (SBC) bought two of the RBOCs, Pacific Telesis and Ameritech. It also bought Southern New England Telephone (SNET), which wasn't an RBOC (Regional Bell Operating Company). See December 29, 2006.

January 27. Western Union sends its last telegram. After more than 150 years, Western Union is out of the telegram business. Western Union will now dedicate itself to money transfers and other financial transactions. Parent company First Data recently said it planned to spin off Western Union. Founded in 1855, Western Union was the first company to stretch telegraph wire across the continent. It also originated the stock-ticker system. It made the famous decision not to buy the patent to the telephone for $100,000. See Divestiture.

May 18. Sprint Nextel spins off its local wireline operations. The new wireline company is named Embarq. Spinning off its local landline business enables Sprint Nextel to focus on its fast-growing wireless business.

July 17. Alltel spins off and merges its wireline business with Valor Communications. The new company is named Windstream. Alltel owns 85% of the new company, while Valor stockholders own 15%. Spinning off its landline business enables Alltel to focus on its fast-growing wireless business.

November 30. French telecommunications company Alcatel acquires Lucent, and with it the remaining part of Bell Labs. The new company is named Alcatel-Lucent, at least initially, and is headquartered in Paris. The stock is listed in Paris and New York. Its symbol is ALU.

December 29. The FCC approves AT&T's $86 billion buyout of BellSouth, the largest telecommunications takeover, so far, in U.S. history. The acquisition received unconditional approval in October from the Department of Justice's Antitrust Division, the same division that successfully fought for the breakup of AT&T in the early 1980s. AT&T now consists of the former AT&T Long Lines plus four of the seven regional Bell Operating Companies (Pacific Telesis, Southwestern Bell, Ameritech, and BellSouth) that were created by the breakup of AT&T on January 1, 1984.

December 15. The FCC eliminates the requirement that applicants for an amateur radio license pass a Morse code proficiency test.

2007 January 1, 2007 - The world's oldest newspaper, Sweden's Post-och Inrikes Tidningar, discontinues its paper edition and now exists only as an online publication. The newspaper was founded in 1645 by Sweden's Queen Kristina. You can read it online at https://poit.bolagsverket.se.

January 15. AT&T begins phasing out the Cingular brand and replacing it with the AT&T brand, the latter being a brand which, ironically, Cingular, itself, had once phased out. If this sounds confusing, here's a quick recap of events. Cingular Wireless, which was founded by SBC and BellSouth in 2000, purchased AT&T Wireless from AT&T Corp. in 2004 and replaced the AT&T Wireless name with Cingular Wireless. In 2006, SBC, one of the owners of Cingular, bought AT&T and adopted the AT&T brand for its entire portfolio of services, including its Cingular-branded wireless services. Then on the last day of 2006, AT&T bought BellSouth, the other owner of Cingular. This set the stage for replacing the Cingular brand in former BellSouth areas, i.e., the only areas where the Cingular brand had continued to exist.

2038 On January 20, 2038 the 32-bit signed time_t integer which clocks seconds in UNIX systems will expire. UNIX-based applications and embedded chips, and some other systems based on a 32-bit architecture, count seconds from midnight January 1, 1970, which is the "UNIX Epoch Start Date." On roughly January 20, 2038, the integer will roll over from a zero followed by 31 ones to a one followed by 31 zeros; the system will interpret the date as January 1, 1970. As most application software queries the OS for date information, rather than calculating it internally, the impact of this oversight could be very significant, indeed. Actually, the effect could be felt much earlier. For instance, 30-year mortgages may not be calculated correctly beginning in the year 2008. On the other hand, a UNIX-based system in a restaurant may calculate your check based on the cost of a meal in 1970 dollars – a pleasant thought, although an unlikely result.

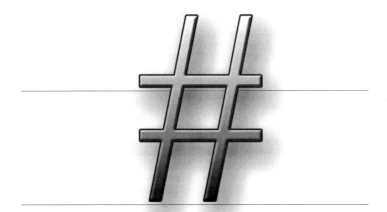

" 1. To computers, double and single quotes are often the same thing.

2. Put double quote marks around something and it typically means you want to search for it in its entirety. This has become less important with sophisticated search engines like Google.

3. Double quotation marks. Typically used to signify something your computer should print (to screen, disk or paper), as in PRINT "Thanks for being a good guy." Some programs allow you to use single quotation marks interchangeably with double ones. Some programs don't. Try one or the other if in doubt.

4. A " refers to inches. So 8" is the same as saying eight inches. See also U.

In the 1960s, Bell Labs were working on ways to get telephones to "talk" to computers and invented what is now called touch-tone dialing. This needed two additional special keys on telephones. One of these is the * symbol, usually known as the asterisk but which Bell Labs decided to call the star key. The other was the # symbol.

The character on the bottom right of your touchtone keypad, which is also typically above the 3 on your computer keyboard. The # sign is correctly called an "octothorpe," but sometimes it is also spelled without the "e" on the end. There is even an International Society of Octothorpians who maintains a web page at http://www.nynews.com/octothorpe/home.htm. The octothorpe is commonly called the pound sign, but it's also called the number sign, the crosshatch sign, the pound key, the tic-tac-toe sign, the enter key, the octothorpe (also spelled octathorp) and the hash. Musicians call the # sign a "sharp."

On some phones the # key represents an "Enter" key like the Enter key on a computer. On some phones it represents "NO." And on others it represents "YES." MCI, AT&T and some other long distance companies use it as the key for making another long distance credit card call without having to redial. Hold down the # key for at least two seconds before the person at the other end has hung up, you'll get a dial tone, punch in your phone number and you can make another long distance call – without having to punch in your credit card number again. (This service is often called Call Reorigination or just plain Next Call.)

The # key is used in the paging industry – national and local. When you dial a phone number which represents someone's beeper, you will typically hear a double beep. At that point you punch in your phone number, ending it with a #. At that point, the machine hangs up on you and sends out the numbers you punched in to the pager you just dialed. Many digital phone pagers allow people to send actual text messages to pagers. Many use the # sign to signal the use of certain digits (c, f, i, l, o, s, v, y) as well as to signal the end of the transmittal. The # character is also the comment character used in UNIX files (configuration, code, etc.). See also Octothorpe.

The third story has the merit of being documented, since ralph carlsen of bell Laboratories wrote a memorandum about it just before his retirement in 1995. He records that in the early 1960s a Bell Labs engineer, Don Macpherson, went to instruct their first client, the Mayo Clinic, in the use of a new telephone system. He felt the need for a fresh and unambiguous name for the # symbol. He was apparently at that time active in a group that was trying to get the Olympic medals of the athlete Jim Thorpe returned from Sweden, so he decided to add thorpe to the end. (Jim Thorpe, a native American who has been described as the greatest athlete of the twentieth century, had won medals in the decathlon and the pentathlon at the 1912 Olympics in Sweden, but had later been disqualified because he was found to have accepted money for playing baseball three years earlier, so making him a professional. His medals were finally returned in 1983.) See touchtone dialing.

$100 billion What cell phone operators worldwide spent in the late 1990s to buy third generation (3G) licenses from their various governments. By the time I wrote this entry in mid-2003, most of the operators had written off most of the money, since few could see any future for 3G cell phone service. See also 3G.

& The "and" sign. Its real name is an "ampersand."

* Asterisk. Also known as a "splat," "Nathan Hale," or the "star" sign. The term "splat" comes from the poor quality of some early printers, since "*" often looked like a splat of ink. "Nathan Hale" refers to his allegedly saying at his hanging: "I only regret that I have only one asterisk (i.e. one life) to give for my country." On IVR systems it typically means "No." In computer languages, it often means a multiplication sign. It's also used to represent a wild card or a joker. For example, the DOS command ERASE JOHN.* will erase all the files in that directory beginning with JOHN, e.g. JOHN.TXT, JOHN.NEW, JOHN.OLD, JOHN.BAK, etc. The asterisk was also called amplicon by the old Bell Telephone system, but I don't know why.

*57 Customer-Originated Trace. The North American universal dialing code which you touchtone in immediately after receiving a harassing, obscene or annoying phone call. By touchtoning that number in, you have alerted your central office to "tag" that phone call. Should a law enforcement agency get involved in investigating your annoying calls, they would be able to go into your records and find the phone number from which the annoying call was made. See also Trap and Trace, and Vertical Service Code.

*66 Usual code in North America to access Automatic Callback.

*67 Customer Number Delivery Blocking, or Call Block Code. Dial *67 before you make a call in North America and the person you call won't see your Caller ID, i.e. the number from which you're calling. See also Vertical Service Code.

*69 Automatic Recall. Dial *69 after you have received a call in North America which you didn't answer and it will dial that calling number for you. See also Vertical Service Code.

*70 The usual code in North America to disable call waiting for an individual call.

*82 The CNI (Calling Number Identification) Call Display Code. This code temporarily overrides the CNIR (Calling Number Identification Restriction) option. You dial it before you place an outgoing call when you want to override your profile restriction, and let the called party see your telephone number when you call them.

++ In C programming, the expression i++ means use the current variable i and add 1 to it. In more contemporary usage, ++ has come to mean an expansion, an improvement, an upgrading, etc. So, C++ is an improved version of the C programming language. MAE-East++ is an expansion of MAE-West located nearby. See MAE-East.

- 1. The hyphen. In typewriting, two of them together are called a dash. in Microsoft Word, two hyphens typed together are usually magically transformed by the program into a dash, which is called an m-dash, because it consumes the width of a lower-case m. Often we take two words and join them with a hyphen into a new word. As the word becomes more and more common, we remove the hyphen and the double word now becomes a single word.

2. The minus sign.

.NET Microsoft .NET is Microsoft's software platform for XML-based Web Services. It provides a distributed software environment where interactions between computer systems use Web Services as the communications mechanism. .NET spans client, server, service and development tools. Microsoft talks about .NET Services as service-based building blocks for developers, including .NET Passport, .NET Alerts, etc. See also SOAP, UDDI and Web Services.

/ The forward slash. Lotus made it famous. UNIX uses it as a directory separator. You see the use of forward slashes in Web addresses, e.g. www.ctexpo/index.html.

// The double forward slash. Filenames or other resource names that begin with the string // mean that they exist on a remote computer. The convention // is commonly seen on the Internet. See \.

**** The backslash. Used for designating directories on your MS-DOS / Windows PC. This dictionary is located in D:WorkDiction. That means it's in the "diction" subdirectory of the "work" directory, also called folder. If it were under Windows, the file name could be as long as 256 characters and could say something more descriptive, like d:workharry's great dictionary.

**** The double backslash. Filenames or other resource names that begin with the string \ mean that they exist on a remote computer. The double backslash is more common in the NT world, while the // (double forward slash) is more common in the Unix world, and thus on the Internet. See also //.

^ The character typically above the 6 on your keyboard. In some languages like French, it is a circumflex and it changes the way you pronounce the letter with the circumflex over it. In PC computer language, when you see it on paper, it means the Control (Ctrl) key. In typography, the ^ symbol is used to show where something is to be inserted. The ^ symbol also is used to indicate the power to which a number is raised when the text processing program does not support superscript character formatting. For example, 10^6 means 10 to the 6th power, which equals 10x10x10x10x10x10=1,000,000. As another example, 2^8 means 2 to the 8th power, which equals 256. The latter is particularly important in the digital world, as it is based on bits, each of which has one of two possible values – a one or a zero. Since most data fields are byte-length with most bytes comprising eight bits, the number of unique values that can be represented by such a field is 256. In most applications, the effective number of values is considered to be 255, as all zeroes value (i.e., 00000000) is considered "nothing."

~ This character is called a tilde. It is used as the UNIX shortcut for "home directory for this account. In Spanish, it tells you how to pronounce the n in senor. According to William Safire, it's a Spanish word from the Latin term for a tiny diacritical mark used to change the phonetic value of a letter.

0+ Calls Called "oh plus." 0+ calls are calls made by dialing zero plus the desired telephone number. Calls made this way may be interrupted by a live operator requesting billing information, or a recorded announcement requesting the caller to enter the billing information.

0- Calls Called "oh minus." 0- calls are operator-assisted calls. The caller dials zero and waits for the operator to pick up the line and talk to the caller.

0B+D An ISDN BRI circuit missing the voice B channels, and only provisioned with the signalling channel. This is common for automated teller machines, travel agency terminals and authorization services. This uses the D channel for a 9600 baud X.25 connection to a provider instead of a leased line. Monthly cost starts around $20. Some providers offer low cost LATA wide packet services for $1-$4 per month, with worldwide X.25 costs starting about $20+ per month usage fees. Anywhere a low speed data connection is needed, this is usually a low cost alternative. All it takes is an NT-1 that trips off the D channel and supports a serial port for your hardware.

00+ or 00- Dialing Double zero dialing. Allows a caller to get an IEC Assistance

Operator in areas where dialing only one zero would connect the caller with the local operator; occurs as a result of the division of services into Intra- and InterLATA.

010+ The access code for operator-assisted intercontinental calls in North America; after the user inputs the required code and number, an operator is signaled to come on the line, as in domestic "0+" dialing.

011 The prefix you use in the United States to dial a number to another country, except Canada and most countries in the Caribbean. Must be followed by a country code and the area code and the local phone number.

0345 Numbers A British Telecom LinkLine service in England where the caller is charged at the local rate irrespective of the distance of the call. The subscriber pays installation and rental charges in addition to a charge for each call. Additional numbers prefixes in England include 0645 and 0845 numbers.

0645 See 0345 Numbers.

0800 Numbers A British Telecom LinkLine service in England where the caller is not charged for the call. Similar to the North American 800 IN-WATS service.

Ericsson has 0800-type service on cellular systems it's put in. Ericsson describes it as a "network-oriented service – based on time of day, day of week, or special day – which allows calls to be redirected to other numbers."

0845 See 0345 Numbers.

0891 and 0898 Numbers A British Telecom Premium rate service in England where the caller is charged at a premium rate for the call. The calls are normally made to receive information or a service. The service provides revenue for the information provider who receives part of the call charge.

0839 and 0881 Numbers Mercury's premium rate service numbers in England.

1+ Pronounced "one plus." In North America, dialing 1 as the first digit has come to signal to your local phone company that the phone number you are dialing is long distance, i.e. is designed to reach a long distance number in the United States, Canada, or several of the Caribbean islands (including Bermuda, Puerto Rico, the Virgin Islands, Barbados and the Dominican Republic). The number 1 will typically be followed by an area code and then seven digits. For example, to dial me from outside New York City, you would dial 1-212-691-8215. To reach other international countries, from the United States, you dial the international access code "011." This "1+" will work with your local phone company, signaling it that you want to reach another local area code. (In New York City, calling from Manhattan to Brooklyn means dialing 1+718+the seven digit Brooklyn number.) It also will work with the long distance company you have presubscribed to, through the process known as equal access. To reach another long distance carrier and route calls over its network, you will need to dial 1-0XXX and then the area code and number.

1+1 A method of protecting traffic in which a protection channel exists for each working traffic channel. For optical systems, the protection channel fibers can be routed over a path separate from the working fibers. The traffic signal is bridged to both the working and protection transmitters so the protection signal can be selected quickly if the working channel fails.

1.544 Mbps The speed of a North American T-1 circuit. 1,544,000 bits per second. See T Carrier and T-1.

1.9 GHz The radio spectrum between 1850 MHz and 1990 MHz used in broadband Personal Communications Services (PCS).

1/4/80 The strange starting date embedded in the original IBM PC. April 1, 1980. Someone's idea of a joke.

1:1 redundancy A form of redundancy whereby for each piece of hardware there is a backup that can take over non-disruptively.

1:n redundancy A form of redundancy whereby for each n identical pieces of hardware or lines, there is a single backup that can take over non-disruptively in the case of a single one's failure. Only one traffic channel can be switched to the protection channel at any given time.

1U See U.

10 Baggers Venture capital jargon for companies returning 1000% on their investment.

10 Base X See 10Base X below, e.g. 10Base-2, 10Base-T.

10 Base T See 10Base-T. T stands for Twisted pair.

10 Digit Trigger A 10 digit trigger is used when porting a number from the ILEC to the CLEC.

10 Gea See 10 Gigabit Ethernet and 10 Gigabit Ethernet Alliance.

10 Gigabit Ethernet 10GbE.In many ways, 10 Gigabit Ethernet is the same as the original 10 million bit per second Ethernet that we have today in our offices and in our homes. The only difference is that it's faster – much faster. It has three more zeros of speeds. It's 10,000,000,000 bits per second versus only 10,000,000 bits per second for normal ten-meg Ethernet. 10 Gigabit Ethernet has the same header. It still has the same header format, the same 8-byte preamble, and the same minimum (64 byte) and maximum (1,518 byte) frame sizes. The biggest change is that CSMA/CD (Carrier Sense Multiple Access/Collision Detection) has been eliminated because 10 Gig will be implemented in full-duplex mode only, meaning that collision detection is turned off. This will make our life easier by eliminating the duplexity mismatches that have plagued some Fast Ethernet and Gigabit Ethernet installations. Another difference: It will only run on fiber. The seven types of physical interfaces, or PHYs, are all fiber. There's no IEEE working group focusing on a copper standard. If 10 Gig ever does run on twisted pair, distances would be very limited. Each PHY comprises a PCS (Physical Coding Sublayer), which is responsible for controlling the transmitted bit patterns, and a PMD (Physical Media Dependent) layer, which is responsible for converting bits into light signals. The PMD is sometimes referred to as the "optics." These layers were designed to be independent of one another. With Gigabit Ethernet, you have just two types of standardized fiber interfaces to keep straight: those that support multimode fiber and those that support single-mode fiber. The major difference between single mode and multimode is the light frequencies supported and the corresponding difference in range. Longer wavelengths running on single mode provide more distance. The other major change is that there are now LAN and WAN PHYs for each PMD. Multiplying three optics by two PHYs gives six unique interfaces. The seventh interface, sometimes referred to as the LX4 interface, is a LAN PHY and uses light frequencies in the 1,310-nm range. The main difference is that, though the other PMDs convert bits to light in a serial manner, the WWDM (Wide Wavelength Division Multiplexing) interface uses WDM technology to multiplex the bits across four light waves. This interface is the most versatile because it supports both 62.5 micron multimode fiber for short distances (300 meter) and single-mode 9-micron fiber for long-range (10 kilometers) connections. If you're wondering why there are so many different versions, you're not alone, especially when you consider that there is overlap between LX4 and some of the other standards. A lot of the variation is based on cost, range and the desire to take advantage of existing technologies and installed fiber. For example, the 850-nm optics that drive multimode fiber short distances are less expensive to build than the optics for single-mode fiber going longer distances. The thinking is, Why pay for what you don't need? This makes sense, so distance for the 850-nm PMD is limited to only 26 meters for existing (62.5 micron) fiber. Going 65 meters will require 50-micron fiber, which is much less common. If you're connecting switches and servers with patch cords in a data center, this isn't a big deal–aside from the burden of keeping track of different fiber types. But for cable installed in structured-wiring plans, it's a different ball of wax. One thing is clear: If you're running fiber today, you should pull some single mode (9 micron) fiber, especially if the runs are more than 300 meters. It will cost you a bit more initially, but the labor cost to add it down the road could be even higher. The newer, higher grades of fiber will increase distances even further. This definition owes much to a wonderful article by Peter Morrissey in the August 5, 2002 issue of Network Computing Magazine.

10-code A short numeric code beginning with the number 10, which is used in radio transmissions by the police, fire departments, truckers, and ham radio operators. The most well known of these ten-codes (also known as 10-codes) by the general public is 10-4, meaning "OK, message received."

10-Net An original local area network invented by a company called Fox Research, Dayton, OH. 10-Net is a baseband, Ethernet CSMA/CD peer-to-peer LAN running on one twisted pair at one megabit per second. It is easy to install and has many advantages. It's also slow. Very slow. And it's no longer being made. See Ethernet.

10/100 An Ethernet/Fast Ethernet designation, referring to having both 10Mbps and 100 Mbps on the same port.

100 Base T See 100Base-T. T stands for Twisted pair.

100 Test Line A Nortel Networks' switching term. The 100 test line, also known as a quiet or balanced termination, is used for noise and loss measurements. The S100 provides a quiet termination for noise measurements only. In this 100 family, there are the T100, S100 and N100 tests. The N100, a more recent version of the test, also includes a milliwatt test (i.e., a 102 test line) and therefore can be used for far-to-near loss measurements. The T100 is used when the equipment at the terminating office is unknown. When the T100 test line is performed, a two-second time-out is introduced to detect the presence or absence of a milliwatt tone. If the T100 test detects the milliwatt tone, it executes the N100 version of the test; otherwise, the S100 version is initiated. If the version of the distant office test line is known, then that version of the test line can be performed directly,

and thus the two-second delay per trunk of the T100 test line test is eliminated.

The 101 test line is used to establish two-way communications between the test position and any trunk incoming to the system. The connection to the 101 test line is established through the switching network.

The 102 test line, also known as the milliwatt line, applies a 1004 Hz test tone towards the originating office to facilitate simple one-way or automatic transmission loss measurements. The test tone is applied for a timed duration of nine seconds during which answer (off-hook) signal is provided. Then an on-hook signal followed by a quiet termination is transmitted to the originating end until the connection is released by the originating end.

The T103 is used for the overall testing of supervisory and signaling features on intertoll trunks. The test is performed to the far end to check overall supervisory and signaling features of the trunk. If the test fails or if a false tone signal is detected, the test is abandoned and the condition indicated.

The T104 test, used for two-way transmission loss measurements, far-to-near noise measurements and near-to-far noise checks is normally used in testing toll trunks.

When a 105 test line at a far-end office is called and seized, timing functions are initiated and an off-hook supervisory signal and test progress tone are returned to the originating office. If the responder is idle, the test line is connected to the responder and test progress tone is removed. Transmission tests are then initiated.

The T108 test provides far-end loop-around terminations to which a near-end echo suppression measuring set is connected for the purpose of testing echo suppressors.

100Base-FX IEEE standard (802.3u) for 100 Mbps Ethernet implementation over fiber, with the F denoting fiber. 100Base-FX runs on optical fiber cable, making the specification best suited for a backbone or long cable segment. 100Base-TX and 100Base-T4 as well as existing 10Base-T hubs can all be connected to a fiber backbone using appropriate hardware such as a bridge or router. 100Base-FX also supports full duplex operation. The bit encoding scheme used is 4B5B for purposes of Clock/Data Recovery (CDR). See 100Base-T for fuller explanation of 100 Mbps Ethernet LANs. See also 4B5B and CDR.

100Base-T 100 Mbps, Baseband, Twisted pair. In sort, 100Base-T is a 100 Mbps LAN (Local Area Network) standard known by the generic name of Fast Ethernet, and running over UTP (Unshielded Twisted Pair) copper cable. There are three basic implementations of Fast Ethernet – 100Base-FX, 100Base-T and 100Base-T4. Each specification is identical except for its interface circuitry, which determines on what type of cabled medium it runs. As a result, the technologies currently aren't interchangeable; each must be connected to its own type of hub. For example, a 100Base-T4 NIC must be connected to a 100Base-T4 interface; likewise, a TX NIC must be connected to a TX interface at the other end. 100Base-T is the most popular and cost-effective high-speed LAN technology because it is designed to integrate with existing Ethernets with minimal disruption, as all are based on the conventional Ethernet MAC (Media Access Control) standard. Essentially an extension of 10Base-T, 100Base-T achieves increased throughput by decreasing the latency period between bits, which effectively increases the packet speed by a factor of 10. The standard adheres to the 802.3u MAC specification, which builds on the 802.3 Ethernet standard to ensure compatibility with existing 10Base-T installations. As a result, an upgrade from 10Base-T to 100base-T is invisible to users, the NOS and current network management applications. 100Base-T uses the 4B5B Manchester encoding scheme for purposes of Clock/Data Recovery (CDR).

My own experience with upgrading to 100Base-T was not 100% overwhelmingly positive. First, we tried upgrading our Macintosh network to 100 Mbps (i.e., Fast Ethernet), but found zero improvement in the speed of transferring a file from the server to a Mac. Problem? Allegedly, the Macintosh network operating system simply won't work that fast. Then we started to upgrade some of our PC workstations to 100 Mbps, but found it made more sense to break our big PC LAN apart into smaller LANs, each run by an Ethernet switch and then have each Ethernet switch join to each other by 100 Mbps over fiber running 100Base-FX. That has worked wonderfully. The point of all this? Designing networks is not trivial. The speed of the individual links – 10 Mbps or 100 Mbps – may not be the gating factor. Check before you spend the money. See 4B5B, 100Base-FX, 100Base-T4, 100Base-FX, and CDR.

100Base-T2 A LAN that transmits data at 100 megabits per second over copper cabling. It is a half-duplex version of 100Base-T that uses two twisted pairs of category 3, 4 or 5 UTP cable. Officially called 802.3y. See 100Base-T.

100Base-T4 100 Mbps Ethernet implementation using four-pair Category 3,4, or 5 cabling. 100Base-T4 runs over four pairs, with three pairs used to transmit data at 33 MHz per pair in half-duplex mode. The fourth pair is used for CSMA/CD collision detection.

100Base-T4 uses the 8B6T encoding scheme for purposes of Clock/Data Recovery (CDR). See also 8B6T, 100Base-T, and CDR.

100Base-TX 100-Mbps Ethernet implementation over Category 5 twisted pair cabling. The MAC layer is compatible with the 802.3 MAC layer. 100Base-TX requires Category 5 UTP (Unshielded Twisted Pair) cabling, the type used in almost all new network installations. 100Base-TX uses two pairs of wire. One pair is used for transmission in half-duplex mode, and using the 4B5B Manchester encoding scheme used for data transmission. The other pair is used for receiving, and for signaling and control purposes. Upgrading a network from 10 megabits per second Ethernet to 100Base-TX requires 100Base-TX NICs and hubs as well as the Category 5 wiring to connect them. 100Base-TX is the best choice for interconnecting servers, hubs, switches and routers because it supports full duplex operation, meaning that it can simultaneously send and receive data. In addition, the cost and effort of upgrading these connections to Cat 5 is minimal since servers are often located very near these devices. See 100Base-T for a general discussion of this technology. See also 4B5B for discussion of the encoding technique.

100Base-X 100-Mbps baseband Fast Ethernet specification that refers to the 100BaseFX and 100BaseTX standards for Fast Ethernet over fiber-optic cabling. Based on the IEEE 802.3 standard. See also 100BaseFX, 100BaseTX, Fast Ethernet, and IEEE 802.3.

100VG-AnyLAN 100-Mbps Fast Ethernet and Token Ring media technology using four pairs of Categories 3, 4, or 5 UTP cabling. This high-speed transport technology, developed by Hewlett-Packard, can operate on existing 10BaseT Ethernet networks. Based on the IEEE 802.12 standard. See also IEEE 802.12.

1000Base-CX A standard for Gigabit Ethernet (GE) connectivity. 1000 means 1,000 megabits per second; Baseband means single-channel transmission; C means Copper; and X is the generic "whatever." The current specification calls for a very specialized copper cable in the form of an electrically balanced, shielded 150-ohm twinaxial cable which is limited to 25 meters in distance; the distance can be extended to 50 meters with a single repeater. Both conductors share a common ground in order to minimize concerns about safety and interference which could be caused by voltage differences. This cable is intended for use as a short jumper to interconnect clustered GE switches in wiring closets or computer rooms. Over time, the IEEE intends to develop a standard for 1000BaseCX connectivity over distances as long as 200 meters, but likely involving a much better form of copper; hence the X "whatever." See also 1000Base-LX, 1000Base-SX, Gigabit Ethernet, Twinax and UTP.

1000Base-F A 1-Gbps IEEE standard for Ethernet LANs.

1000Base-LX A standard for Gigabit Ethernet (GE) connectivity. 1000 means 1,000 Mbps (Megabits per second); Baseband means single-channel transmission; L means Long-wave laser; and X is the generic "whatever." Long-wave lasers are more expensive than short-wave lasers, but can transverse longer distances as they use low-energy lasers over single-mode fiber at a wavelength of approximately 1,300nm (nanometers). For example, a long-wave laser system can transmit reliably over distances of approximately 3 kilometers through fiber with an inner core diameter of 5 microns; long-wave laser transmission also is supported over multi-mode fiber, and generally over longer distances than short-wave laser. See also 1000Base-CX, 1000Base-SX, Gigabit Ethernet and Fiber Optics.

1000Base-SX A standard for Gigabit Ethernet (GE) connectivity. 1000 means 1,000 Mbps (Megabits per second); Baseband means single-channel transmission; S means Short-wave laser; and X is the generic "whatever." Short-wave lasers are less expensive than long-wave lasers, but cannot transverse the same long distances as they use high-energy lasers over multi-mode fiber at a wavelength of approximately 850nm (nanometers). For example, a short-wave laser system can transmit reliably over distances no farther than 550 meters through multi-mode fiber with an inner core diameter of 50 microns; short-wave laser transmission is not supported over single-mode fiber. See also 1000Base-CX, 1000Base-LX, Gigabit Ethernet and Fiber Optics.

1000Base-X A standard for Gigabit Ethernet (GE) connectivity. 1000 means 1,000 Mbps (Megabits per second); Baseband means single-channel transmission; and X is the generic "whatever," which refers to the range of current and future transmission media. The IEEE 802.3z task force developed 1000BASE-X, which defines MAC (Media Access Control) changes, a Gigabit Media Independent Interface (hence the "X"), management and general requirements for Ethernet operation at 1000 Mbps, and a set of physical layer interfaces based on the original Fibre Channel technology. See also 1000Base-CX, 1000Base-LX, 1000Base-SX, Fibre Channel, and Media Access Control.

100VG-AnyLAN A 100 megabit-per-second LAN (Local Area Network) standard established by the IEEE 802.12 committee in 1996, and originally known simply as AnyLAN. A joint development of AT&T Microelectronics (now Lucent), Hewlett-Packard and IBM,

VG=Voice Grade, meaning that voice-grade UTP (Unshielded Twisted Pair) generally is used as the transmission medium. AnyLAN means that LAN networking and internetworking can be accomplished, accommodating both Ethernet and Token Ring. 100VG-AnyLAN provides media flexibility, including Cat 3 (Category 3) UTP in a 4-pair configuration over distances up to 100 meters, Cat 4 UTP in a 4-pair configuration over distances up to 100 meters, and Cat 5 in a 4-pair configuration over distances up to 200 meters. Also accommodated are Level 1 STP (Shielded Twisted Pair) in a 2-pair configuration over distances up to 100 meters, and fiber optic cable (62.5 microns) over distances up to 2 kilometers. In a four-pair UTP configuration, all pairs are used for transmission in half-duplex mode, with the signal being split across the pairs at 25 MHz each. Data frames are encoded using the 5B6B encoding mechanism. 100VG-AnyLAN is deployed in a star topology, with up to five repeating hubs being tolerated. DPMA (Demand Priority Media Access) provides access priority and collisionless transmission, which is an improvement over Ethernet. 100VG-AnyLAN, however, requires equipment upgrade, which does not position it well relative to 100Base-T, also known as Fast Ethernet. See 100Base-T and FDDI for explanation of the competing 100-Mbps LAN standards. See also 5B6B.

1010XXX See 101XXX.

101B Closure Housing used to protect service wire splices.

101XXXX The Feature Group D Carrier Identification Code (CIC). To connect with a long distance carrier in the United States other than your preselected carrier, as of July, 1998, you must now dial 101XXXX, where X is any number between 0 and 9. Each long distance carrier in the United States now has a unique four digit code represented by XXXX – at least every carrier with Feature Group D access. That code is called a CIC code, which stands for Carrier Identification Code. AT&T's code is 0288. Thus, to reach AT&T you dial 101-0288, or as they advertise "Ten-Ten ATT." MCI's is 101-0222. There are two reasons for wanting to use 101XXXX to dial a different long distance to the one you're subscribed is simple – to save money, or to use an other carrier when your preselected carrier is experiencing an overload or a network failure. You can dial some carriers (e.g. 101-0457) and receive the bill for your phone calls on your monthly bill from your 101XXXX LEC (local exchange carrier). Others you have to contact in advance and set up an account. Dialing via 101XXXX is a little complicated. Let's say you want to call my sister, Barbara, in Sydney, Australia from the U.S. You would dial 101-0457-011-612-9-663-0411 (the first seven digits are the CIC). As of July 1, 1998, this 101XXXX code replaced the original 10XXX. The reason 101XXXX replaced 10XXX is clearly that 10-XXX lets you dial only 1000 long distance carriers, while 101XXXX allows you to dial 10,000 long distance carriers. Deregulation and competition simply resulted in more carriers than could be accommodated by the old dialing scheme. In a very forward thinking move, the FCC has already planned for the next expansion, whenever needed, to 10XXXXX. This allows for up to 100,000 long distance carriers, while preserving the current CICs exactly as they are now. For more – www.fcc.gov/Bureaus/Common_Carrier/FAQ/cic_faq.html. Kevin Ross, KevinR@seedberry.com helped on this definition. Thank you. See also 950-XXXX.

1080p If you buy an LCD or plasma or other TV with 1080p, you've bought the top of the class of high definition TVs available today (February 2007). The number 1080 stands for 1080 lines of vertical resolution, while the letter p stands for progressive scan or non-interlaced. The term 1080p usually assumes a widescreen aspect ratio of 16:9, versus the present NTSC TVs of 4:3. The 1080p standard implies a frame resolution of 1920 x 1080 or over two million pixels. High-definition is continually evolving and improving. 1080p is currently the digital standard for filming digital motion pictures. Directors such as George Lucas (in Star Wars Episode III: Revenge of the Sith) shoot their digital films in this high definition mode to be shown in theaters equipped with 1080p digital projectors.

10Base-2 10Base-2 is the implementation of the IEEE 802.3 Ethernet standard on thin coaxial cable. It's commonly called thin Ethernet or thinnet because the cable is half the diameter of 10Base-5 Ethernet cable. 10Base-2 LANs, which run at ten million bits per second, have their PCs daisy-chained along a terminated bus topology. The maximum segment length is 185 meters. Connectors are typically BNC. 10Base-2 uses RG58A/U 50-Ohm cable. Also called thinwire Ethernet.

10Base-5 A transmission medium specified by IEEE 802.3 that carries information at 10 Mbps in baseband form using bus topology, using 50-ohm coaxial cable and using AUI connectors. 10Base-5 was specified by the original Ethernet standards and is sometimes called ThickWire Ethernet. The maximum segment length (i.e. without a repeater) is 500 meters.

10Base-F Standard for Fiber optic Active and Passive Star based Ethernet segments. Described in IEEE 802.1j-1993 (not in an 802.3 supplement, as you might expect). 10Base-F includes the 10Base-FL standards.

10Base-FB Part of the new IEEE 802.3 10Base-F specification, "Synchronous Ethernet" which is a special-purpose link that links repeaters and allows the limit on segments and repeaters to be enlarged. It is not used to connect user stations. See 10Base-F.

10Base-FL 10 Mbps Baseband (single-channel transmission)-Fiber Link. A part of the IEEE Base-F specification that covers Ethernet over fiber over a distance of as long as two kilometers. 10Base-FL is interoperable with FOIRL. See 10Base-F.

10Base-T An Ethernet local area network which works on twisted pair wiring (T stands for twisted pair) that looks and feels remarkably like telephone cabling. In fact, 10Base-T was invented to run on telephone cable. 10Base-T Ethernet local area networks work on home runs in which the wire from each workstation goes directly to the 10Base-T hub (like the wiring of a phone system), just like a phone system. 10Base-T cards which fit inside PCs typically cost the same as those for Ethernet running on coaxial cable. The advantages of 10Base-T (which has become the most commonly installed local area network in the world) are twofold – namely if one machine crashes, it doesn't bring down the whole network (coax Ethernet LANs are typically in one long line, looping from one machine to another. One crash. Every machine goes down.); and secondly, a 10Base-T Ethernet network is easier to manage because the 10Base-T hubs often come with sophisticated management software. Though 10Base-T is designed to work on "normal" telephone lines, no one in their right mind would install "normal" phone wiring. The preferred method of installing 10Base-T networks is to use new Category 5 wiring. If you're forced (because you don't want to open up a pretty wall), then connect it to old phone cabling. It will probably work. Remember 10Base-T uses two pairs. Most phones need only one pair. 10Base-T's maximum segment length is 100 meters running on unshielded twisted pairs. See 802.3 10Base-T.

10Broad36 An IEEE 802.3 Ethernet LAN specification. 10Broad36 means ten million bits per second (10Mbps), Broadband, 3600 meter maximum segment length. In the LAN context, "broadband" means multichannel, which is accomplished over a coaxial cable system through FDM (Frequency Division Multiplexing). As FDM implies, the transmission method is analog, and the devices attach to the cable through modems, which almost defies the imagination for intensive LAN data transmission. However, analog transmission allows multiple communications channels (i.e., multiple transmissions) to be supported simultaneously. That is an advantage. The maximum segment length of 3600 meters is a real advantage, as well. The thick coax cable is a real disadvantage. 10Broad36 is seldom used. See 10Base-T, which is much more popular.

10GBase-ER See 802.3ae.

10GBase-EW See 802.3ae.

10GBase-LR See 802.3ae.

10GBase-SR See 802.3ae.

10GBase-SW See 802.3ae.

10GBase-LW See 802.3ae.

10GBase-LX4 See 802.3ae.

10GBase-T See 802.3an.

10GbE 10 Gigabit Ethernet. Positioned as a high-speed technology for MAN (Metropolitan Area Network) applications, 10 GbE is a developing IEEE 802.3ae standard that will enable networks to scale from the traditional 10 Mbps beyond the common 100 Mbps and increasingly common 1 Gbps, up to 10 Gbps. 10GbE will enable MSPs (MAN Service Providers) to create very high-speed links between collocated equipment (e.g., switches and routers) at very low cost. 10GbE retains the basic protocol structure of Ethernet, including frame format, minimum and maximum frame sizes, and MAC (Media Access Control) protocol. However, 10GbE operates only in full-duplex mode, thereby making the CSMA/CD (Carrier Sense Multiple Access with Collision Detection) MAC (Media Access Control) mechanism unnecessary, especially over longer distances.

10GbE employs an interface known as XAUI (pronounced "Zowie"), with "X" denoting the Roman numeral for 10, implying 10 Gbps, and "AUI" derived from Ethernet Attachment Unit Interface. The XAUI is an interface extender for the XGMII (10 gigabit Media Independent Interface), a 74-signal wide interface comprising one 32-bit wide data path for the transmit direction and one for receiving direction used to attach the Ethernet MAC (Media Access Control Layer) to the PHY (PHYsical Layer). The XAUI is a self-clocked serial bus evolved directly from the GbE 1000BASE-X PHY. The XAUI interface speed is 2.5 Gbps, and makes use of four serial lanes that operate in parallel over a WWDM optical fiber in order to achieve aggregate transmission speed of 10 Gbps. As is the case with 1000BASE-X, 10GbE employs the 8B/10B transmission code in order to ensure signal integrity through the copper circuitry of PCBs (Printed Circuit Boards).

10GbE operates only over optical fiber media, unlike GbE, which operates at 1 Gbps over either copper or fiber. 10GbE will operate over SONET/SDH fiber, at the OC-192 speed of 9.953 Gbps, which is close enough to 10 Gbps to be acceptable. In a SONET/SDH mode, GbE makes use of SONET framing conventions and some inherent overhead functions. Some of the more costly aspects of SONET are unnecessary and will be avoided, including TDM support, performance monitoring, and certain network management functions. This "thin SONET" approach has been dubbed PES (Packet over Ethernet over SONET). 10GbE also will operate directly over optical fiber transmission systems based on WWDM (Wide Wavelength Division Multiplexing), a intermediate variation on the theme of WDM and DWDM (Dense WDM). WWDM specifies four serial optical channels that run in parallel. This WDM-based approach is known as PEW (Packet over Ethernet over WDM). Another approach is that of 10GbE over POF (Passive Optical Fiber), which has been dubbed PEF (Packet over Ethernet over Fiber).

Here's some additional detail on the PMD (Physical Media Dependent) sublayer of the PHY. MMF is intended to support 10GbE in serial mode over target distances up to 65m running at a wavelength of 850nm, and using WWDM at up to 300m running at 1310nm. SMF is intended to support 10GbE using either serial mode or WWDM at up to 10km running at 1310nm, and up to 40km in serial mode running at 1550nm.

10GbE is intended for backbone applications in the MAN domain, although it is considered extendible to the WAN domain, where it will be used to interconnect 10GbE MANs. It will be considered for application in the LAN domain, perhaps even directly to end-user systems (e.g., servers) in applications such as NAS (Network-Attached Storage) and SANs (Storage Area Networks). Ultimately, 10GbE may extend even to the desktop, although the latter requirement seems remote at this time. See also 10GEA, 1000BASE-X, CSMA/CD, DWDM, Ethernet, GbE, GE, MAC, MMF, PEF, PEW, PHY, POF, SMF, SONET, WDM, WWDM.

10GEA 10 Gigabit Ethernet Alliance. Formed in January, 2000, the 10 Gigabit Ethernet Alliance was organized to facilitate and accelerate the introduction of 10 Gigabit Ethernet into the networking market. It was founded by networking industry leaders: 3Com, Cisco Systems, Extreme Networks, Intel, Nortel Networks, Sun Microsystems, and World Wide Packets. Additionally, the Alliance will support the activities of IEEE 802.3 Ethernet committee, foster the development of the 802.3ae (10 Gigabit Ethernet) standard, and promote interoperability among 10 Gigabit Ethernet products. See 10 Gigabit Ethernet. www.10gea.org.

10XXX Calling The original access code that you dialed in North America to reach carrier that you had not equal access to. On July 1, 1998, the 10XXX access code was changed to 101XXXX. You must now dial the full seven digits. See 101XXXX for a full explanation.

110-type Connecting Block The part of a 110-type cross connect, developed by AT&T (now Lucent), that terminates twisted-pair wiring and can be used with either jumper wires or patch cords to establish circuit connections.

110-type Cross Connect A compact cross connect, developed by AT&T (now Lucent), that can be arranged for use with either jumper wires or patch cords. Jumper wires, used for more permanent circuits, must be cut down to make circuit connections. Patch cords allow ease of circuit administration for frequently rearranged circuits. The 110-type cross connect also provides straightforward labeling methods to identify circuits.

119 Japan's equivalent of the United States' emergency 911 number.

1149.1 See JTAG.

12-Pack Coax Cable A bundle of 12 50-ohm coaxial cables that often run from a SONET carrier to a Digital Cross-Connect System (DCS), They carry a STS-1 (synchronous transport signal-1).

1284 IEEE standard for connecting a PC's parallel port to a printer. Buy one that's advertised as bi-directional.

13 The average married woman in 17th-century America gave birth to 13 children.

136 The TV channel which Manhattan Cable uses to deliver cable modem services.

1394 Also called FireWire, IEEE 1394, 1394 and i.Link. 1394 is an IEEE standard for an data transport bus between a host computer and its peripherals. 1394 runs at speeds of 100, 200 and 400 Mbps, with increases planned up to 2 Gbps. A single 1394 port can support up to 63 peripherals, and a single host computer can support up to 1023 buses. The cable length can be up to 4.5 meters, although as many as 16 cables can be daisy-chained to extend the length to as much as 72 meters. A tree configuration also is acceptable. The cables comprise six conductors, and can supply up to 60 watts of power, allowing low-power peripherals to operate on a line-powered basis (i.e., without separate power). Much like USB (Universal Serial Bus), although running at much higher speeds and costing much more, 1394 supports plug-and-play on a hot-swappable basis. In other words, peripherals can be plugged and unplugged while the computer is turned on, and the computer will automatically discover and configure the link between itself and the peripheral device. Example peripherals include high-density storage devices, and high-resolution still and video cameras. 1394 supports both asynchronous and isochronous data transfer. Since 1394 is not a particularly exciting name, manufacturers have come up with their own names. For example, Apple Computer calls it FireWire. See also USB.

Walt Mossberg of the Wall Street Journal, wrote:

FireWire, 1394, i.Link: These are three different, confusing terms for the same thing: a very fast connector on some PCs that can rapidly suck in and pump out large volumes of data from external devices, like camcorders, external hard disks and high-capacity portable music players. Apple, which helped invent the technology and has marketing in its bones, calls this connector "FireWire." The Windows PC makers use the geeky term "1394." To make matters a little more confusing, Sony calls the same connector "i.Link." This is worth buying only if you plan to import video or use a high-capacity music player like Apple's iPod.

The IEEE-1394 High Performance Serial Bus is a versatile, high-speed, and low-cost method of interconnecting a variety of personal computer peripherals and consumer electronics devices. The IEEE-1394 bus began life in 1986 as Apple Computer's alternative to the tangle of cables required to connect printers, modems, external fixed-disk drives, scanners, and other peripherals to PCs. The proposed standard (P1394) derived from Apple's original FireWire design, was accepted as an industry standard at the December 12, 1995 meeting of the Institute of Electrical and Electronics Engineers (IEEE) Standards Board. The official name is IEEE 1394-1995 Standard for a High Performance Serial Bus. The 1394 Trade Association was formed in 1994 to accelerate adoption of the Bus by personal computer and consumer electronic manufacturers. The 1394 Trade association has dubbed IEEE-1394 the MultiMedia Connection. Adaptec has licensed Apple's FireWire technology, trademark, and logo; FireWire is used interchangeably with IEEE-1394 in these pages.

The primary advantages of FireWire over other current and proposed serial buses are:

Versatility: FireWire provides a direct digital link between up to 63 devices without the need for additional hardware, such as hubs. Digital Video (DV) camcorders, scanners, printers, videoconferencing cameras, and fixed-disk drives all share a common bus connection not only to an optional PC, but to each other as well. FireWire is a candidate for the "Home Network" standard initiated by VESA (Video Electronic Standards Association) and other industry associations.

High speed: The present implementation of IEEE-1394 delivers 100 Mbps (Megabits per second) or 200 Mbps of data (payload) and control signals (overhead). Future versions that support 400 Mbps are in the development stage, and a 1.2 Gbps (Gigabits per second) version of IEEE-1394 has been proposed. Isochronous data transmission lets even the lowest-speed implementation support two simultaneous channels of full-motion (30-frame-per-second), "broadcast quality" video and CD-grade stereo audio.

Low cost: The cost of the integrated circuits and connectors to implement FireWire is often less than the cost of the connectors and circuitry it replaces. FireWire uses a flexible, six-conductor cable and connectors derived from Nintendo's Gameboy to interconnect devices. (A four-conductor version of the standard cable is used to interconnect consumer audio/video components.) Use of FireWire for consumer electronics gear, such as camcorders and VCRs, will provide the high-volume market needed to achieve low-cost implementation of FireWire on PCI adapter cards and PC motherboards. Ease of installation and use: FireWire extends Plug and Play features far beyond the confines of the personal computer. When you add a new device, FireWire automatically recognizes the device; similarly, on disconnect FireWire automatically reconfigures itself. The standard FireWire cable provides up to 1.5 amps of DC power to keep remote devices "alive" even when they're powered down. You don't need a computer to take advantage of FireWire; as an example, a VCR can act as a FireWire controller for camcorders, TV sets, receiver/amplifiers, and other home theater components.

An IEEE data transport bus that supports up to 63 nodes per bus, and up to 1023 buses. The bus can be tree, daisychained or any combination. It supports both asynchronous and isochronous data. 1394 is a complementary technology with higher bandwidth (and associated cost) than Universal Serial Bus. Intel told me in summer of 1996 that it was supporting USB for most devices that attach to PC up through audio and video conferencing. Intel told that they are "supporting IEEE 1394 as the preferred interface for higher bandwidth applications such as high quality digital video editing, and connection to new digital consumer electronics equipment." according to Zayante, a company setting itself up as a testing lab for 1394 products, "The IEEE 1394 multimedia bus standard is the "convergence bus" bringing together the worlds of the PC and digital consumer electronics. It is already the digital interface of choice for consumer digital audio/video applications,

providing a simple, low-cost and seamless plug-and-play interconnect for clusters of digital A/V devices, and it is being adopted for PCs and peripherals. The original specification for 1394, called IEEE 1394-1995, supported data transmission speeds of 100 to 400 Mbits/second. Most consumer electronic devices available on the market now support either 100 or 100/200 Mbits/second, meaning that plenty of headroom remained in the 1394 specification. But more devices are added to a system, and improvements in the quality of the A/V data (i.e., more pixels and more bits per pixel), lead to a need for greater bandwidth. The 1394a specification, currently in the final stages of approval (early 2000), offers efficiency improvements including support for very low power, arbitration acceleration, fast reset and suspend/resume features. The 1394b specification extends the 1394-1995 and 1394a efforts in three primary ways: It increases the speed to 800 Mbits/second and 1.6 Gbits/second, while adding the architectural infrastructure to support 3.2 Gbits/second and beyond; it specifies alternative media that allow 1394 products to be connected at distances of up to 100 meters (up from the 4.5 meters of the current specification); and it is overall more efficient, lower cost and easier to manage." See USB.

144-line Weighting In telephone systems, a noise weighting is used in a noise measuring set to measure noise on a line that would be terminated by an instrument with a No 144-receiver, or a similar instrument.

1588 IEEE 1588 is a precision clock synchronization protocol for networked measurement and control systems. The protocol enables precise synchronization of clocks in networked, local computing, and distributed-object environments. The protocol enables heterogeneous systems that have clocks of various degrees of precision, resolution and stability to synchronize, and to do so with sub-microsecond accuracy while consuming minimal network and local clock computing resources to achieve that accuracy.

16-bit An adjective that describes systems and software that handle information in words that are 2 bytes (16 bits) wide.

16-bit Computer A computer that uses a central processing unit (CPU) with a 16-bit data bus and processes two bytes (16 bits) of information at a time. The IBM Personal Computer AT, introduced in 1984, was the first true 16-bit PC.

16-CAP An ATM term. Carrierless Amplitude/Phase Modulation with 16 constellation points: The modulation technique used in the 51.84 Mb mid-range Physical Layer Specification for Category 3 Unshielded Twisted-Pair (UTP-3).

16:9 Four to three (4:3) is the ratio of width to height in a traditional TV set. High definition TV sets have a 16:9 ratio. That makes them much more rectangular. Sometimes expressed as 16x9 or 16 by 9 (known as 1.78:1 in the film world); the standard DTV (digital TV) wide-screen television screen size.

16450/8250A Found in most current PCs, these older UART chips use a 1-byte buffer that must be serviced immediately by the CPU. If not, interrupt overruns will result. See 16550 and UART.

16550 An enhanced version of the original National Semiconductor 16xxx series UART, which sits in and controls the flow of information into and out of virtually every PC serial port in the world. The older version contains only a one-byte buffer. This can slow down the transmission of high-speed data especially when you're using a multitasking program, like Windows. The "solution" is to get a serial card or port containing the 16550. This chip contains two 16-byte FIFO buffers, one each for incoming and outgoing data. Also new is the 16550's level-sensitive interrupt-triggering mechanism, which controls the amount of incoming data the buffer can store before generating on interrupt request. Together, these features help reduce your CPU's interrupt overhead and thus speed up your communications. See 16450/8250A and UART.

173 The number of words in the Ten Commandments.

179,584 Finland has the greatest number of islands of any other country in the world: 179,584.

1822 Historic term which refers to the original ARPANET host-to-IMP interface. The specifications for this are in BBN report 1822.

193rd Bit The frame bit for a T-1 frame. See Robbed Bit Signaling.

1A AT&T's first generation of standardized key telephony system equipment based on a variety of interconnected phone-line-powered relays. Prior to 1A1, key systems were often patched together from a variety of non-standard parts, with varying wiring schemes, making repairs and upgrades very difficult.

1A-ESS An analog central office, made by AT&T and widely deployed by the Bell Operating Companies prior to divestiture.

1A1 AT&T's second generation of standardized KEY TELEPHONE SYSTEM equipment. Unlike the phone-line powered 1A1, it used commercial AC power for added features such as illuminated buttons to indicate line status.

1A2 AT&T's third generation of standardized KEY TELEPHONE SYSTEMS. It was distinctive for its use of plug-in circuit cards, making it much easier to add features or diagnose and cure problems.

1A3 A cute term for an historic TIE electronic key system that provided advanced features, but was priced competitively with 1A2 electromechanical key systems.

1Base-5 Defined in IEEE 802.3, 1Base-5 was the first LAN standard to make use of UTP (Unshielded Twisted Pair). Running at one megabit per second with Manchester encoding, this Ethernet standard operates on a baseband basis, providing for a single transmission at a time. Connection to the centralized hub is accomplished over UTP of 22, 24 or 26 gauge at distances of up to 500 meters in a star topology. Two pairs are used, with one providing upstream connectivity to the hub and the other providing downstream connectivity. AT&T's StarLAN adhered to the 1Base-5 standard, which has long been eclipsed by 10/100Base-T. The maximum segment length for 1Base-5 is 1,640 ft. (500 m). See also 10Base-T and Manchester encoding.

1BL One Business Line, a term used by Bell Canada for a single business phone line.

1FB One Flat rate analog Business phone line. A phone line for which you pay a single monthly charge for and on which you may make as many local phone calls as you wish during that month. A 1FB is an increasing rarity in the United States. See also 1MB.

1FL One Family phone Line, used by Bell Canada to refer to a residential user's phone line.

1FR One Flat rate residential phone line. A phone line for which you pay a single monthly charge for and you on which you may make as many local phone calls as you wish during that month. See also 1MB.

1G Mobile Network First generation mobile network. Refers to the initial mobile wireless networks that use analog technology only and did not carry data. Advanced Mobile Phone Service (AMPS) is an example of a 1G mobile network standard. See 2G.

1K Pooling See Number Pooling.

1MB One Message rate Business phone line. A phone line for which you pay a single monthly charge. That charge typically allows you to make a small number of local calls for free. But each additional local call will cost you, either by the minute and/or by the distance, or just by the call. See 1FB.

Slang for a T-1 line. This derives from the fact that a T-1 data line will deliver 1.544 million bits per second.

1TR6 First National ISDN Signalling System used in Germany. This system is being phased out and will be completely replaced by Euro-ISDN.

1xEVDO 1xEVDO is a data-only wireless network that is separate from the traditional voice cellular carrier. Verizon runs one such network in the U.S. It is known as an asymmetrical network. Typically it produces download speeds of 400 Kbps to 500 Kbps and upload speeds of around 80 Kbps. Verizon calls it their Internet broadband acccess service. To use it, Verizon equips you with a PC card for your laptop – one not dissimilar to one you might use to access Wi-Fi in your home. Verizon's Internet broadband service is presently available (December, 2005) in the top 60 largest U.S. cities. See also 1xRTT.

1xRTT The first phase of CDMA2000 technology designed to double voice capacity and support data transmission speeds up to 144 Kbps, or 10 times the speed commonly available today. 1xRTT is compatible with today's IS-95A and IS-95B. IS-95 is Interim Standard 95. IS-95 is a TIA standard (1993) for North American cellular systems based on CDMA (Code Division Multiple Access), and is widely deployed in North America and Asia. IS-95a defines what generally is known as cdmaOne, which supports voice and 14.4 Kbps data rates. IS-95b supports data rates up to 115 Kbps. See also GPRS.

2-6 Code A 2-6 code is an alphanumeric designation that Verizon uses to identify a trunk group. It consists of two letters and six numbers. An example would be AB-123456.

2-line Network Interface. Old type interchangeable lightning protectors. The top is painted white to indicate gas type instead of carbon type.

2-way Trunk A trunk that can be seized at either end.

2-wire Facility A 2-wire facility is characterized by supporting transmission in two directions simultaneously, where the only method of separating the two signals is by the propagation directions. Impedance mismatches cause signal energy passing in each direction to mix with the signal passing in the opposite direction. See 4-WIRE FACILITY.

214 Licence Licence from the FCC (Federal Communications Commission) which lets you offer international communications services to customers in the United States.

218-219 MHz Also known as Interactive Video and Data Service (IVDS). A short-distance communication service designed for licensees to transmit information, product, and service offerings to subscribers and receive interactive responses within a specified service area. Mobile operation is permitted. Federal Communications Commission (FCC)

rules permit both common carrier and private operations, as well as one- and two-way communications. Potential applications include ordering goods or services offered by television services, viewer polling, remote meter reading, vending inventory control, and cable television theft deterrence. Until September 15, 1998, the 218-219 MHz Service was known as the Interactive Video and Data Service (IVDS). Many Commission documents still refer to the 218-219 MHz Service by its former name. The FCC has issued a warning that members of the public may receive solicitations to invest in enterprises that hold or plan to acquire licenses in the 218-219 MHz Service. These may still be represented as "IVDS licenses." Entrepreneurs may attempt to use the 218-219 MHz Service licensing process to deceive and defraud unsuspecting investors. Promotional material may promise unrealistic profits, and sales representatives may represent the investment as the "chance of a lifetime." The promotional material may include actual copies of FCC releases, or quotes from FCC personnel, giving the false appearance of FCC approval or knowledge of the solicitation. The FCC does not endorse any individual investment proposal, nor does it provide a warranty with respect to the license being allocated. Obtaining a license is not a guarantee of success in the marketplace. Between 1996 and 1999, many 218-219 MHz Service licenses cancelled automatically due to nonpayment under the terms of the FCC installment payment rules. Furthermore, under the terms of the Report and Order, certain existing licensees may elect to return their licenses to the FCC. You should be aware of these facts when evaluating investment and sales offers in the 218-219 MHz Service. Eighteen 218-219 MHz Service licenses were issued by lottery (random selection) in 1993 so the licenses for 9 of the top 10 MSAs were awarded. Two licenses per market were offered for auction at the same time, with the highest bidder given a choice between the two available licenses, and the second highest bidder winning the remaining license.   In 1994, the FCC conducted an auction for the remaining MSA licenses that weren't issued by lottery. No RSA licenses have been issued to date. The FCC has no set date for the auction of unallocated 218-219 MHz Service licenses. All licenses are issued for a ten-year term.

23B+D An easy way of saying the ISDN Primary Rate Interface circuit. 23B+D has 23 64 Kbps (kilobits per second) paths for carrying voice, data, video or other information and one 64 Kbps channel for carrying out-of-band signaling information. ISDN PRI can be derived from (i.e. channelized out of) a North American T-1 line. In ISDN 23B+D, the one D channel is out-of-band signaling. In T-1, signaling is handled in-band using robbed bit signaling. Increasingly, 23B+D is the preferred way of getting T-1 service since the out-of-band signaling is richer (delivers more information – like ANI and DNIS) and is more reliable than the in-band signaling on the older T-1. One good thing about PRI: You can now organize with your phone company to deliver the signaling for a bunch of ISDN PRI cards on one D channel. Thus your first line would have 23 voice channels. Your second would have 24 voice channels, etc. Several of the more modern voice cards will accept the signaling for up to eight ISDN PRI channels on the D channel of the first one. See ISDN PRI, Robbed Bit Signaling and T-1.

23 Skiddoo The famed New York expression, "Twenty-three skiddoo" came to be because the wind drafts created by the height of the skyscraper raised women petticoats, and constables had to "skiddoo" the men who came to peek. See Flatiron Building.

24-bit Mode The standard addressing mode of Apple Macintosh's System 6 operating system, where only 24 bits are used to designate addresses. Limits address space to 16MB (2 to the 24th power), of which only 8MB is normally available for application memory. This mode is also used under System 7 (the Mac's more modern operating system) if 32-bit addressing is turned off. See 32-BIT.

24-bit Video Adapter A color video adapter that can display more than 16 million colors simultaneously. With a 24-bit video card and monitor, a PC can display photographic-quality images.

24-Hour Format Sometimes known as military time. Using 24 hours to designate the time of day, rather than two, 12 hour segments.

24/365 Apple used 24/365 to show that an Apple Store is open 24 hours, 365 days of the year. I guess they close on February 29 during leap years.

24/7 Twenty four hours a day, seven days a week. When your company is open 24/7, it means it's open all the time.

24th Channel Signaling See 2G mobile network.

24x7 When you see 24x7, it means you're getting something (e.g. service) for 24 hours, seven days of the week. This means you're getting it all the time. 24x7 (pronounced "24 by 7") is a more sexy way of saying "all the time."

25 Cities Program A shorter name for the "25 High-Risk Metropolitan Areas Interoperability Project." It is an initiative by the Department of Justice, with support from the Department of Homeland Security and the Treasury Department, to provide a federal wireless network to support first-responder communications in 25 American cities that were identifed as having the highest risk for terrorist incidents after the 9/11 attacks. The program's objective is to provide federal law enforcement and homeland security agencies with intersystem communication in emergency situations and an ability to communicate with key local fire, police, and emergency medical services entities. The cities in the program are Atlanta, Baltimore, Boston, Charlotte, Chicago, Dallas, Denver, Detroit, District of Columbia, Hampton Roads/Norfolk, Honolulu, Houston, Jacksonville, Los Angeles, Miami, New Orleans, New York, Philadelphia, Phoenix, Portland, San Diego, San Francisco, Seattle, St. Louis, and Tampa. See also radio cache.

2500 The 2500 set is the "normal" single-line analog touchtone desk telephone. It has replaced the rotary dial 500 set in most – but not all – areas of the United States and Canada. No one seems to know why the addition of a "2" in front of a model number came to denote touchtone in the old Bell System. Colin Neal, a reader, suggests that the 2 could perhaps have come from 2 (dual)-tone as in DTMF. See DTMF.

255 See ^.

258A Adapter A device about 12 inches long and six inches wide and two inches deep that is used to connect a 25-pair Amphenol cable to RJ-45 patch cords.

2600 Tone Until the late 1960s, America's telephone network was run 100% by AT&T and used 100% in-band signaling, whereby the circuit you talked over was the circuit used for signalling. For in-band signaling to work there needs to be a way to figure when a channel is NOT being used. You can't have nothing on the line, because that "nothing" might be a pause in the conversation. So, in the old days, AT&T put a tone on its vacant long distance lines, those between its switching offices. That tone was 2600 Hertz. If its switching offices heard a 2600 Hz, it knew that that line was not being used. At one point in the 1960s, a breakfast cereal included a small promotion in its cereal boxes. It was a toy whistle. When you blew the whistle, it let out a precise 2600 Hz tone. If you blew that whistle into the mouthpiece of a telephone after dialing any long distance number, it terminated the call as far as the AT&T long distance phone system knew, while still allowing the connection to remain open. If you dialed an 800 number, blew the whistle and then touchtoned in a series of tones (called MF – multi-frequency – tones) you could make long distance and international calls for free. The man who discovered the whistle was called John Draper and he picked up the handle of Cap'n Crunch in the nether world of the late 1960s phone phreaks. Since then, in-band signaling has been replaced by out-of-band signaling, the newest incarnation being called Signaling System 7. See 2600, Captain Crunch, Multi-Frequency Signaling and Signaling System 7.

266 The number of words in the Gettysburg Address.

271 Section 271 of The Telecommunications Act of 1996 describes the conditions by which a Bell Operating Company (BOC) may enter the market to provide interLATA services, long distance in particular, within the region where they operate as the dominant local telephone service provider. The Act mandates that BOCs must open their local telephone markets to competition as a precondition to entry into the long distance market. The term 271 has come to be used as shorthand for referring to the strategic efforts of the BOCs to prove competition exists, and thereby gain FCC approval to provide interLATA long distance service. Although final authority to approve a BOC's entry into the long distance market is given to the FCC, Congress provided in Section 271 a checklist to guide the FCC's assessment of local market competition. The checklist points are (summarized): Interconnection for any requesting telecommunications carrier with the BOC's network that is at least equal in quality to that provided by the BOC to itself. Non-discriminatory access to network elements. Nondiscriminatory access to the poles, ducts, conduits, and rights-of-way owned or controlled by the BOC at just and reasonable rates. Local loop transmission from the central office to the customer's premises, unbundled from local switching or other services. Local transport from the trunk side of a wireline local exchange carrier switch unbundled from switching or other services. Local switching unbundled from transport, local loop transmission, or other services. Non-discriminatory access to 911, directory assistance and operator call completion services. White pages directory listings for customers of the other carrier's telephone exchange service. Nondiscriminatory access to telephone numbers for assignment to the other carrier's telephone exchange service customers. Nondiscriminatory access to databases and associated signaling necessary for call routing and completion. Telecommunications number portability. Nondiscriminatory access to services or information to allow the requesting carrier to implement local dialing parity (the ability to complete a connection without the use of additional access codes). Reciprocal compensation arrangements. Telecommunications services available for resale. See 271 Hearings.

271 Hearings The incumbent phone company goes before the local public service commission and begs to be allowed to sell long distance phone service. Such hearings are often used by CLECs, who agree to appear and say favorable things, in order to bargain better operating arrangements out of the local phone company. See 271 and CLEC.

2780 A batch standard used to communicate with IBM mainframes or compatible systems.

2B+D A shortened way of saying ISDN's Basic Rate Interface, namely two bearer channels and one data channel. A single ISDN circuit divided into two 64 Kbps digital channels for voice or data and one 16 Kbps channel for low speed data (up to 9,600 baud) and signaling. Either or both of the 64 Kbps channels may be used for voice or data. In ISDN 2B+D is known as the Basic Rate Interface. In ISDN, 2B+D is carried on one or two pairs of wires (depending on the interface) – the same wire pairs that today bring a single voice circuit into your home or office. See ISDN.

2B1Q Two Binary, One Quaternary. A line encoding technique used in ISDN BRI in the US, and used extensively in the U.S. in first-generation HDSL systems. 2B1Q is a four-level PAM (Pulse Amplitude Modulation) technique which maps two bits of data into one quaternary symbol, with each symbol comprising one of four variations in amplitude and polarity over a circuit. As the resulting signaling rate is half the bit rate, the efficiency of transmission is doubled. In other words, at a given frequency, this 4-level PAM approach allows two bits to be sent per baud (i.e., Hertz, or sine wave). As the rate of signal loss increases as the carrier frequency increases, higher frequency signals attenuate (lose power) more rapidly than do lower frequency signals. Therefore, a method such as 2B1Q allows you to send more data per second using a relatively low frequency. Through the use of 2B1Q, a good-quality twisted-pair local loop can support an ISDN BRI total bit rate of 144 Kbps over a distance of up to 18,000 feet, and without signal repeaters. ISDN BRI operates at a maximum rate of 40 KHz, therefore running at 80 KBaud, which supports a total signaling rate of 160 Kbps. A total of 128 Kbps is used for user payload (two B channels at 64 Kbps each), 16 Kbps for the D channel, and 16 Kbps for framing and synchronization. ISDN BRI uses echo canceling to support full-duplex operation. 2B1Q is described in ANSI T1.601 and ETR 080, Annex A. The line coding technique used in Europe is 4B3T. See also 4B3T, BRI, HDSL, Hertz, ISDN and PAM.

2FR A flat-rate party line with two subscribers.

2FSK Two-level Frequency Shift Keying. See FSK for an explanation.

2G 2G is the most common type of wireless telephone communication today. It allows slow data communication, but its primary focus is voice. See 2G mobile network.

2G mobile network Second generation mobile network. Refers to the second generation cellular phones that introduced digital technology and carried both voice and data conversations. CDMA, TDMA, and GSM are examples of 2G mobile networks. GSM is used throughout the world. CDMA and TDMA are used primarily in the Americas. See GSM and 2.5G.

2G+ mobile network Second generation plus mobile network. Refers generically to a category of mobile wireless networks that support higher data rates than 2G mobile networks. GPRS is an example of a 2G+ mobile network standard.

2W Two-Wire. See 2-Wire Facility.

2.5G Second-and-a-half generation wireless. Refers to the additional features and functionality added to digital cellular phones, such as Internet access and messaging. The main feature added to 2.5G is GPRS, a mobile data communications service running at the speed of a dial-up landline. See GPRS and 3G.

3 Computer hackers often use the number 3 as a substitute for e.

3.5G A faster version of 3G wireless networks. 3.5G uses a technology called HSDPA, which is a significant enhancement to W-CDMA and can achieve speeds of up to 14.4 Mbps. HSDPA stands for High Speed Downlink Packet Access. It is a packet based data service feature of the WCDMA standard which provides a downlink with data transmission up to 8-10 Mbps (and 20 Mbps for MIMO systems) over a 5MHz bandwidth in WCDMA downlink. The high speeds of HSDPA is achieved through techniques including; 16 Quadrature Amplitude Modulation, variable error coding, and incremental redundancy. HSDPA is a technology upgrade to current UMTS networks. See also UMTS.

3:2 Pull-down A method for overcoming the incompatibility of film and video frame rates when converting or transferring film (shot at 24 frames per second) to video (shot at 30 frames per second).

30B+2D In Europe, the equivalent of the North American T-1 line is called an E-1. And instead of carrying 1.544 million bits per second, it carries 2.048 million bits per second. This is the rate used by European CEPT carriers to transmit 30 64 Kbps digital channels for voice or data calls, plus a 64 kilobits per second (Kbps) channel for signaling, and a 64 Kbps channel for framing (synchronization) and maintenance. Thus the expression 30B+2D – 30 B channels for voice (i.e. user information) and 2 D channels for housekeeping

– signaling and maintenance. See E1, E2, E3, and T-1.

311 The Services Code now available for non-emergency access to police, fire and other governmental departments. The FCC instructed the North American Numbering Plan Administration to make 311 available in order to relieve the load on the 911 emergency number. See also 911 and 711.

3172 IBM's network controller. It connects to the mainframe channel on one end and the LAN media (Ethernet, Token Ring, FDDI) on the other.

3174 IBM's cluster controller. It connects to terminals and other I/O devices on one end, and a mainframe channel on the other.

32-bit An adjective that describes hardware or software that manages data, program code, and program address information in 32-bit-wide words. What is the significance of 32-bit? With 32-bit memory, each program can address up to 4 gigabytes (2 to the 32nd power) of memory, i.e. four billion bytes. This is in contrast to Windows 3.x where programs are limited to 16 MB of memory. Possibly more significant than the amount of memory that is available to a 32-bit application is how that memory is accessed. Under Windows 3.x, memory is accessed by using two 16-bit values that are combined to form a 24-bit memory address. (24-bits is the size of the memory addressing path of the Intel 80286. The 80286 is the architecture that Windows 3.x was designed for.) The first 16-bit value (selector) is used to determine a base address. The second 16-bit value (offset) indicates the offset from the base address. One of the side effects of this architecture is that the maximum size of a single chunk of memory is 64 KB. Windows 95 and Windows NT are 32-bit operating systems. Windows 95 and Windows NT developers can address memory with a single 32-bit value. Such an addressing scheme allows developers to view memory as one flat, linear space with no artificial limits on the size of a single segment. No longer are programmers concerned about selectors and offsets and the 64 KB segment limit. Also, Windows 95 and Windows NT take full advantage of the protection features of the Intel 80386 microprocessor. 32-bit applications are given their own protected address space which tends to prevent applications from inadvertently overwriting each other.

32-bit Addressing See 32-BIT.

32-bit Computer A computer that uses a central processing unit (CPU) with a 32-bit data bus and central processing unit (CPU) which processes four bytes (32 bits) of information at a time. Personal computers advertised as 32-bit machines – such as Macintosh SE, and PCs based on the 80386X microprocessor – aren't true 32-bit computers. These computers use microprocessors (such as the Motorola 68000 and Intel 80386SX) that can process four bytes at a time internally, but the external data bus is only 16 bits wide. 32-bit microprocessors, such as the Intel 80386DX, the Pentium and the Motorola 68030, use a true 32-bit external data bus and can use 32-bit peripherals. See also 64-bit processor.

3270 IBM class of terminals (or printers) used in SNA networks.

3270 data stream A data stream developed by IBM for communication between mainframe applications and 3270 terminals. The data stream includes control characters that tell the receiving terminal how to format or display information. The control characters allow the application to use parts of or the entire screen display for input and/or output, instead of just a single command line.

3270 Gateway An electronic link which uses 3270 terminals to handle data communications between PCs and IBM mainframes.

3270SNA A specific variation of IBM's System Network Architecture for controlling communications between a 3270 terminal connected to an IBM mainframe.

3274 IBM series of Control Units or Cluster Controllers provide a control interface between host computers and clusters of 3270 compatible terminals.

327X Belonging to IBM's 3270 collection of data communications terminals.

3299 A communications device for an IBM mainframe computer.

347x The type of fixed function computer terminals used with IBM mainframe computers.

370 Block Mux Channel See Block Multiplexer Channel.

3745 IBM's communications controllers, often called front-end processors. 3745 devices channel-attach to the mainframe and support connections to LANs and other FEPS.

3780 A batch protocol used to communicate with an IBM mainframe or compatible system.

3B2-400 A UNIX-based minicomputer, manufactured by AT&T and widely deployed by the Bell Operating Companies prior to divestiture.

3D API 3D Application Programming Interface. This generic term refers to any API that supports the creation of standard 3D objects, lights, cameras, perspectives, etc. APIs include Argonaut's BRender and Microsoft's Reality Lab.

3DES An encrypting algorithm that processes each data block three times, using a unique key each time. 3DES is much more difficult to break than straight DES. It is the most secure

of the DES combinations, and is therefore slower in performance. See also Data Encryption Standard (DES).

3DGF 3-D Geometry File. A platform independent format for exchanging 3-D geometry data among applications. Developed by Macromind.

3FR A flat-rate party line with three subscribers.

3G Third Generation Mobile System. An upgrade to 2G, 3G promises higher data speeds – up to 2 megabits per second, or roughly in line with cable modems and DSL lines. 3G is the generic term for the third generation of wireless mobile communications networks. Most commonly, 3G networks are discussed as graceful enhancements of the GSM cellular standards. Thereby, existing GSM networks can be upgraded on a non-disruptive basis. The enhancements include greater bandwidth, more sophisticated compression techniques, and the inclusion of in-building systems. 3G networks will transmit data at 144 kilobits per second, or up to 2 megabits per second from fixed locations. This planned evolution of GSM is an integral part of the ITU-T's vision of IMT-2000 (International Mobile Telecommunications for the year 2000), which clearly will miss the target date of 2000. 3G will, theoretically, standardize three mutually incompatible standards: FDD, TDD, and CDMA2000. FDD and TDD are extensions of GSM architecture using CDMA technology in the air interface. CDMA2000 is the extension of IS95 air interface for wideband data applications. The three standards will compete with each other in the marketplace. In addition to the technical differences between the standards, there is a strong political background in the competition. FDD and TDD were proposed by European firms, and will be promoted worldwide as the heirs to GSM systems. CDMA2000 was the standard championed by Qualcomm, the USA based company with many patents in CDMA technology. See 3GPP, 4G, CDMA2000, FDD and TDD.

3G data card Also spelled 3G datacard. A small card fitting into the PCMCIA slot of a laptop computer and giving the user access to a cell phone carrier's 3G wireless network for highspeed data communication – typically access to the Internet. The card performs the same function that Wi-Fi cards perform – provide access to a network, in this case, a Wi-Fi network.

3g2 A file with a .3g2 extension is most likely a video clip taken with a mobile phone or some sort of portable video device. You can play the file with most media players, including Apple's QuickTime and RealPlayer. The format belongs to the 3rd Generation Partnership Project (3GPP) and its Project 2 (3GPP2), standards developed for multimedia files used on the newer high-capacity wireless phone networks. The networks themselves are often referred to as third generation, or 3G. They are designed to carry data and voice communications easily.

3GPP Third Generation Partnership Project. The Organizational Partners comprise ARIB (Association of Radio Industries and Businesses) of Japan, CWSI (China Wireless Telecommunication Standard group), ETSI (European Telecommunications Standards Institute), T1 (Standards Committee T1 Telecommunications) sponsored by ATIS (Alliance for Telecommunications Industry Solutions) and ANSI (American National Standards Institute), TTA (Telecommunications Technology Association) of Korea, and TTC (Telecommunication Technology Committee) of Japan. The partners have agreed to co-operate for the production of Technical Specifications for a 3rd Generation Mobile System based on the evolved GSM core networks and the radio access technologies that they support. The TSG (Technical Specification Group) addressing the GSM evolution is known as GERAN (GSM/EDGE Radio Access Network). www.3gpp.org and www.etsi.org. In November 2000, 3GPP announced working relationships with the International Telecommunication Union (ITU) and the Internet Engineering Task Force (IETF). See also 3g2, EDGE, GPRS, and GSM.

3GSM World Congress Europe's primary mobile phone gathering forum.

3xRTT The second phase of cdma2000 technology expected to offer speeds up to 384 Kbps for mobile and 2 Mbps for stationary applications. 3xRTT offers greater capacity (i.e., three 1.25 MHz carriers) than current CDMA systems (one 1.25 MHz carrier).

4 Million English Pounds Four million English pounds was the amount King Fahd of Saudi Arabia is expected to spend per day while on holiday in the summer of 2002 at his palace on the Costa del Sol in Spain. according to the magazine, Money Week. The king brought with him three planes full of helpers and hospital equipment in addition to his customized 747. A local florist is to supply the palace with 1,000 English pounds of fresh flowers per day. 500 mobile phones have been ordered and a local department store has set up a direct line of credit and will stay open round the clock to satisfy any whims. Some of the money is being spent in London, where an escort agency has been contracted to supply blond companions to make the visitors feel at home.

4-wire Facility A 4-wire facility supports transmission in two directions, but isolates the signals by frequency division, time division, space division, or other techniques that enable reflections to occur without causing the signals to mix together. A facility is also called 4-wire if its interfaces to other equipment meet this 4-wire criteria (even if 2-wire facilities are used internally), as long as crosstalk between the two transmission directions, as measured at the interface, is negligible. See 2-WIRE FACILITY.

4.9% No regional Bell Operating Company is presently allowed to own more than 4.9% of the stock of a telecommunications manufacturing company. See Divestiture.

4:3 Four to three is the ratio of width to height in a traditional TV set. The newer high definition TV sets have a 16:9 ratio. That makes them much more rectangular.

4004 The world's first general-purpose microprocessor (computer on a chip). The 4004 was made by Intel, was 4-bit, was released on November 15, 1971 and contained 2,300 transistors. It executed 60,000 instructions per second. The tiny 4004 had as much computing power as the first electronic computer, ENIAC, which filled 3,000 cubic feet with 18,000 vacuum tubes when it was built in 1946. The 4004 found a home in desktop calculators, traffic lights and electronic scales. Despite its power, its 4-bit structure was too small to process all the bits of data at one time to handle all the letters of the alphabet. It was followed by the 8-bit 8008. See also 1971 and 1978.

40G A shortened way of saying 40 gigabit per second transmission technology. See SONET.

404 When a page on a web site is removed, the server will automatically generate a 404 error message when a visitor attempts to view that page.

411 The local number dialed for local directory assistance (we used to call it Information) in many, but not all, cities in North America. Sometimes you have to dial 555-1212. Sometimes you have to dial 1-555-1212. For long distance directory assistance, you would dial 1-213-555-1212 (for information in the 213 area code). See also N11.

41449 AT&T's specifications for its ISDN PRI (Primary Rate Interface). It is different from the ANSI standard T1.607.

419 Scam A fraud, particularly one originating in Nigeria, in which a person is asked for money to help secure the release of a much larger sum and is promised a piece of the larger sum. The classic 419 scam asks the victim to help some hapless relative of a deposed despot get millions of dollars out of the con artist's country. The con artist offers the victim a percentage of this illusive pot of gold, hoping to suck the victim into paying all sorts of fees to get trunks of money out of Nigeria, Sierra Leone, the Philippines or whatever exotic locale the con artist chooses. In the end, the money is never sent, but the victims are often out thousands of dollars. 419 refers to the relevant section of the Criminal Code of Nigeria.

419A A famous old Bell System tool which many installers found very convenient to hold a diminishing marijuana cigarette (called a 'roach') in the 1960s.

42A An early terminal block. The Model 42A is a plastic mounting base about two inches square with four screws and a cover. Before modular connections became widespread, the 42A was used to connect a phone's line cord to the wire inside a wall or running around the baseboard. Adapters, such as the No. 725A made by AT&T and Suttle Apparatus, can be used to convert a 42A into a 4-conductor modular jack. See also Terminal Block.

456 The area code, or Numbering Plan Area (NPA), used to identify certain carrier-specific services. The specific carrier is identified by the succeeding NXX, which is the next three digits. This number is used to ensure proper routing of inbound international calls destined for these services into and between North American Numbering Plan (NANP) countries. Current NXX assignments include 226 or Teleglobe Canada, 228 and 229 for AT&T, 624 for MCI, and 640 for Sprint.

46-49 In North America, most cordless phones operate within the band 46-49 MHz. That band contains only 10 channels and is horribly overcrowded. Recently, the FCC authorized a new frequency range – 905-928 MHz – for use by, amongst other things, cordless phones. The 900 Mhz contains 50 channels.

480p A better TV picture, the video image is "painted" on the screen without the flicker commonly associated with regular television. 480p is the native resolution of DVDs. P stands for progressive. You can also get 480i, where the i stands for interlaced, which is the common

488 IEEE 488 is the most widely-used international standard for computer-to-electronic instrument communication. It is also known as GPIB and HPIB.

4A The last generation of "telco-quality" add-on speakerphones, with separately-housed microphone and speaker; made by both Western Electric (AT&T) and Precision Components, Inc.

4B/5B Local Fiber 4-byte/5-byte local fiber. Fiber channel physical media used for FDDI and ATM. Supports speeds up to 100 Mbps over multimode fiber. See also TAXI 4B/5B.

4B3T 4 Binary 3 Ternary. A line coding technique used in Europe and elsewhere to support ISDN BRI (Basic Rate Interface). A "block code" that uses "Return-to-Zero" states on

the line, 4B3T combines 4 bits to represent one ternary (i.e., one of three) signal state on the line. Therefore, 4B3T supports a total signaling rate of 160 Kbps at a baud rate of 120 KBaud. The three signaling states presented to the ISDN BRI line are a positive pulse (+), a negative pulse (-), and a null pulse (zero-state, or 0). 4B3T is defined in ETR 080, Annex B, and various national standards. The corresponding line coding technique used in the US to support ISDN BRI is 2B1Q. See also 2B1Q and BRI.

4B5B 4 Bits 5 Bits. A Manchester data encoding/decoding scheme that encodes four data bits into a 5-bit transmission sequence. 4B5B is used in 25 Mbps ATM, as well as 10Base-T and certain 100Base-T (100Base-TX and 100Base-FX) implementations. With 4B5B, the data octets within the data frames to be transmitted over the serial (i.e., one bit at a time) link are divided into 4-bit nibbles (Note: A 4-bit value is known as a nibble, which is exactly half of an 8-bit byte. I kid you not.). Each nibble is then scrambled, using a standard scrambling algorithm, and is mapped into a 5-bit sequence prior to transmission. The 5-bit sequence includes bits for delineation of the data sets, and control indicators for clock recovery (i.e., synchronization at the receiving end of the transmission) and data recovery (i.e., error control). As 4B5B uses a NRZ (Non Return to Zero) unibit (i.e., one bit per baud, or one bit per Hertz) signaling scheme, a 100Base-T application requires aggregate bandwidth of 125 MHz to support transmission of 100 Mbps. See also 5B6B, 8B6T, 8B10B, 8B10T, 10Base-T, 100Base-T, ATM, and Manchester.

4ESS A digital central office switching system made by Lucent. It is typically used as an "tandem switching office, switching long distance phone calls. See 5ESS.

4FR A flat-rate party line with four subscribers.

4FSK Four-level frequency shift-keying. See FSK for an explanation.

4G 4G is what the next, next generation cellular might be. The idea is simple – universal high-speed Internet access. The thinking is Wi-Fi Internet access (at up to 10 megabits per second) with blanket coverage and fewer base stations than are needed in today's cell phone networks. Firms including IPWireless, Flarion, Navini, ArrayComm and Broadstorm offer just such a blend. But these are proprietary solutions, so far. Such proposed 4G wireless-broadband systems can be seen in two ways: as a rival to coffee shop Wi-Fi or as a wireless alternative to the cable modem and digital subscriber line (DSL) technologies that now provide broadband access to homes and offices. IPWireless sees their system as a fast internet connection that follows you around. Navini calls it "nomadic broadband"; ArrayComm's term is "personal broadband." Mike Gallagher of Flarion, a firm backed by Cisco, likens Wi-Fi to cordless phones that work within a limited range of a base-station, whereas 4G is akin to mobile phones that work anywhere. Advocates of 4G technology argue that, unlike with 3G and Wi-Fi, the business case for 4G is sound. Nobody is sure how commercial Wi-Fi hotspots will make money. The number of connections per day at most hotspots is still tiny. But 4G is being priced like fixed-line broadband, a service for which millions of users worldwide are already willing to pay about $50 a month. 4G networks may be built initially in regions where cable and DSL are unavailable, in order to capitalize on pent-up demand for broadband. Some cell phone companies are said to be considering skipping 3G altogether in favor of 4G.

4GL Fourth Generation Language.

4W Four wire.

4WL-WDM Four Wavelength Wave Division Multiplexing, also called Quad-WDM. MCI announced this technology in the Spring of 1996 as a method of allowing a single fiber to accommodate four light signals instead of one, by routing them at different wavelengths through the use of narrow-band wave division multiplexing equipment. The technology allowed MCI to transmit four times the amount of traffic along existing fiber. At that time MCI's backbone network operated at 2.5 gigabits per second (2.5 billion bits) over a single strand of fiber optic glass. Using Quad-WDM the same fiber's capacity will rise to 10 gigabits – enough capacity to carry 64,500 simultaneous transmissions over one single strand of fiber.

500 Service A non-geographic area code specifically assigned for Personal Communications Services (PCS) as originally defined – in other words, not necessarily the cellular-like PCS we hear so much about. 500 numbers provide for follow-me services, which allow the subscriber to define a priority sequence of telephone numbers which the network will use to search for him. For instance, the search might begin at your business phone, progressing to your cellular/PCS phone, then to your home phone, and then to your voice mailbox, assuming that you can't be found or don't want to be found. Options might include distinctive ringing for pre-defined callers of significance such as your spouse, significant other, or boss. Further options might include billing, such as caller pays any long distance charges (hopefully, with pre-notification), call blocking, and selective call blocking. 500 Service promises to offer a single telephone number which can find you anywhere, for life. All available 500 numbers were assigned in 1995; plans exist to expand 500 Service area codes, to include area codes such as 520 and 533. 500 Services, clearly, are network-based. CPE solutions recently have emerged, as well. See also Wildfire.

500 Set The old rotary dial telephone deskset. The touchtone version was called a 2500 set.

501(c)(3) 501(c)(3) refers to the specific section of the Internal Revenue Code that designates a tax-exempt organization. Such organizations are required to be nonprofit and must reinvest their revenues back into the organization. They are also, for the most part, public entities subject to oversight by the IRS (Internal Revenue Service) to ensure compliance with very stringent regulations. If you give money to an organization which has 501(c)(3) status, you can be assured that your donation will be allowed by the IRS as a deduction from your income. To be 100% sure, you need to make sure that the entity has the IRS approval. And the way to do that is to get a copy of the letter from the IRS authorizing grant of the 501(c)(3) exemption status to the charity you're giving money to.

511 FCC designated local phone number in the United States for traveler information.

5250 IBM class of terminals for midrange (System 3x and AS/400) environments.

5250 Gateway An electronic link which uses 5250 terminals to handle communications between PCs and IBM minicomputers.

555 Central office prefix numbers used to access a wide variety of information services. 555 numbers are in the format 555-XXXX. For example, you dial dial 411 for local Directory Assistance (we used to call it Information) in many, but not all, cities in North America. Sometimes you have to dial 555-1212. Sometimes you have to dial 1-555-1212. If you want to double-check to make absolutely sure of the identity of your IXC (Interexchange Carrier), to make sure that you haven't been "slammed," you can dial 1-700-555-4141. Many Hollywood movies use phone numbers like 213-555-5678, knowing that they won't be some real person's number. See also 700, and Slamming.

56 Kbps A 64,000 bit per second digital circuit with 8,000 bits per second used for signaling. Sometimes called Switched 56, DDS (Digital Data Service) or ADN (Advanced Digital Network). Each carrier has its own name for this service. The phone companies are obsoleting this service in favor of the more modern ISDN BRI, which has two 64 Kbps circuits (called Bearer circuits) and one 16 Kbps packet circuit. See 56 Kbps Modem, K56flex and ISDN.

56Flex See K56flex.

5B6B 5 Bits 6 Bits. A data encoding/decoding scheme that encodes five data bits into a 6-bit transmission sequence. 5B6B is used in 100VG-AnyLAN, which is standardized as IEEE 802.12, and which supports both Ethernet and Token Ring LANs. With 5B6B, the data frames to be transmitted over the serial (i.e., one bit at a time) link are divided into 5-bit data quintets. Each quintet is then scrambled, using a different scrambling mechanism for each of the four channels (i.e., wire pairs) in order to randomize the bit patterns on each channel and, thereby, to reduce radio frequency interference (RFI) and the resulting crosstalk between the pairs. At that point, the each quintet is encoded, or mapped, into a predetermined 6-bit "symbol," which process creates a balanced data pattern comprising equal numbers of 1's and 0's. The expanded symbol (five bits become six bits) provides both clock synchronization between transmitter and receiver, and error-detection capability. As there exist only 16 balanced 5-bit symbols (i.e., data patterns) available, and as there exist 32 unique five-bit data combinations (Note: 2 to the fifth power equals 32.), 16 of the data quintets require expression in the form of two 6-bit symbols. The pattern of one and two 6-bit symbols is used alternately to maintain DC balance. Then, each symbol is prepended with a preamble and starting delimiter, and is appended with an ending delimiter. Finally, and in a process known as "quartet channeling," the quintets are distributed sequentially over each of the four channels, with each channel being in the form of a wire pair in a four-pair configuration. This channel definition is based on the "lowest common denominator" assumption that Cat 3, 4, or 5 UTP (Unshielded Twisted Pair) is used for connectivity between the workstation and the hub. If either 2 pairs of STP (Shielded Twisted Pair) or two optical fibers (of 62.5 microns) are used, the 6-bit symbols are multiplexed before being transmitted. See also 5B6B, 8B6T, 8B10B and 100VG-AnyLAN.

5C Refers to the five founding companies of the Digital Transmission Content Protection (DTCP) technology. Sony, Matsushita, Intel, Toshiba and Hitachi. Also used to refer to 5C digital certificates.

5ESS A digital central office (also called a public exchange) switching system made by Lucent. It is typically used as an "end-office," serving local subscribers. But it is also used by some GSM cellular operators as transit switches connecting their MSCs (mobile switching centers).

5x5 See Five By Five.

5XB 5 X-Bar central office equipment.

606-A A loose standard that specifies a uniform administration approach and consistent cable labeling system for telecommunications cabling systems that supports multi-product and multi-vendors. The 606-A standard comes in four flavors:

Class 1: for systems with a single building with one Telecommunications Room (TR) that all workstation cables for that system run to.

Class 2: for systems within a single building that are served by multiple TRs.

Class 3: for a system that spans multiple buildings, called a campus environment.

Class 4: for systems that span multiple campuses. This is also called a multi-site system.

611 Phone number used by many carriers (including cell phone providers) in North America for telephone company repair service. You dial it to report problems or ask service-related questions. See also N11.

613 Mitzvah in Jewish can be literally translated as a commandment. Mitzvot is the plural form for Mitzvah and it means the 613 commandments, including the first ten, which are more well known as The Ten Commandments.

62 Sixty-two degrees Fahrenheit is the minimum temperature required for a grasshopper to be able to hop.

62 miles The beginning of space is 62 miles above the surface of the earth.

64 bit architecture A 32-bit processor can handle up to 4 billion bytes of memory, an amount rarely approached in ordinary desktop computing today. By comparison, a 64-bit chip can address 18 billion bytes.

66 Block The most common type of connecting block used to terminate and cross-connect twisted-pair cables. It was invented by Western Electric eons ago and has stood the test of time. It's still being installed. Its main claims to fame: Simplicity, speed, economy of space. You don't need to strip your cable of its plastic insulation covering. You simply lay each single conductor down inside the 66 block's two metal teeth and punch the conductor down with a special tool, called a punch-down tool. As you punch it down, the cable descends between the two metal teeth, which remove its plastic insulation (it's called insulation displacement) and the cable is cut. The installation is then neat and secure. 66 blocks are typically rated Category 3 and as such are used mostly for voice applications, although Category 5 66 blocks are available. 66 blocks are open plastic troughs with four pins across, and the conductors tend to be more susceptible to being snagged or pulled than the conductors terminated on 110, Krone or BIX. Why is it called 66 block? According to AT&T, all these things were developed and named by Bell Labs. They just started with "number 1" on whatever system they were working on. TD1 radio, TD2 radio, etc. Whenever there was a "hole" in the sequence, that meant that the labs had worked on something, but it didn't pan out for some reason. I guess 1 through 5 didn't pan out.

64 Kbps 64,000 bits per second. The standard speed for V.35 interface, DDS service, and also the effective top speed of a robbed-bit 64 Kbps channel. A 64 Kbps circuit (DS0). "Clear Channel" is 64 kbps where entire bandwidth is used. See also ISDN.

64-bit See 64 bit.

64-cap An ATM term. Carrierless Amplitude/Phase Modulation with 64 constellation points.

64b/66b 64 bits 66bits. A data encoding/decoding scheme that encodes 64 bits of data into a 66 bit physical layer signal. The eight bits of data to be transmitted over the serial (i.e., one bit at a time) line are converted into a 66-bit code group prior to transmission with the extra two bits of "special characters" serving such signaling and control functions as indicating the start of the data frame, the end of the data frame, and the link configuration. The primary purpose of the additional bits in the transmission is for the detection of bit errors in the data stream. 64b/66b offers a much more efficient coding mechanism than the 8B/10B encoding mechanism previously used in Gigabit Ethernet (IEEE 802.3z). With the 8B/10B coding mechanism, a Gigabit Ethernet switch generated a 1.25Ghz physical layer signal (1Gbps * 10 / 8) which forces a Metropolitan Ethernet service provider to provision an STS-24c SONET circuit to support wire-speed Gigabit Ethernet performance. The 64b/66b encoding mechanism transports a 10 Gigabit Ethernet signal at 10.3Ghz (10Gbps * 66 / 64). The same 10Gbps would have required 12.5Ghz with the 8B/10B encoding mechanism (10Gbps * 10 / 8). The IEEE 802.3ae 10 Gigabit Ethernet standard calls for the 64b/66b encoding mechanism. See also Gigabit Ethernet.

64QAM 64-state quadrature amplitude modulation. This digital frequency modulation technique is primarily used for sending data downstream over a coaxial cable network. 64QAM is very efficient, supporting up to 28-Mbps peak transfer rates over a single 6-MHz channel. But 64QAM's susceptibility to interfering signals makes it poorly-suited to noisy upstream transmissions (from the cable subscriber to the Internet). See also QPSK, DQPSK, CDMA, S-CDMA, BPSK and VSB.

66-type Connecting Block A type of connecting block used to terminate twisted-pair cables. All wires are manually cut down with a special tool to terminate or connect them. See 66 Block.

66-type Cross Connect A cross connect made up of the 66-type connecting blocks and jumper wires for administering circuits. All wires, including jumper wires, must be cut down (or punched down) and seated with a special tool. See 66 BLOCK.

6611 IBM's multi-protocol router, which supports APPN in addition to TCP/IP, DECnet, AppleTalk, IPX, NetBIOS, and other protocols.

6bone The Internet's experimental IPv6 network. 6bone is an informal collaborative project designed as a testbed backbone network for IPv6 (Internet Protocol version 6), commonly known as IPng (Internet Protocol next generation). The 6bone is a virtual network layered on top of portions of the IPv4-based Internet. In the core of the 6bone are production-level (versus beta test-level versions) routers running the IPv6 protocol suite, connected to workstation-class machines also running native IPv6. The islands of IPv6 are connected to the current IPv4-based Internet through edge routers running both the IPv4 and IPv6 protocol suites. The plan is that, over time, IPv6 will gradually work its way throughout the Internet, replacing IPv4 in the process. See also IPng, IPv4, and IPv6.

6to4 A tunneling mechanism in an IPv6 network that encapsulates IPv4 endpoint information inside IPv6 source and destination addresses. 6to4 is useful when two hosts need to exchange IPv6 traffic over network segments that support only IPv4.

7-bit ASCII The standard code for text in which a byte (eight bits) holds the seven ASCII digits that define the character plus one bit for parity.

700 Service A non-geographic area code reserved for the provisioning of special IXC services, 700 numbers were created in 1983, with the intent that interexchange carriers could use them to create and implement new services quickly. AT&T originally marketed 700 Service in the form of Easyreach, which allows your calls to follow you in the same fashion as would 500 Service. Some carriers (who will remain unnamed) once used 700 numbers for user access to the network for purposes of intraLATA long distance calling, such as those from Manhattan to Westchester County – the cost presumably was less than the cost of the same call through the serving LEC. This practice was illegal and no longer is necessary, as local competition now is in place in most states. 700 Service is still evolving, with each carrier having the right to create whatever services it wants with its 700 numbers. Currently, 700 Service commonly is used in both voice and data VPNs (Virtual Private Networks). See also 500 Service and VPN.

701 1. The 701 was one of the first manual switchboards, also called a cordboard. it was made by Western Electric. A manual switchboard required the operator to insert a cord into the slot on a board that represented the location of a department or person receiving a call.

2. IBM's first computer, introduced in 1953. It was also known as the Defense Calculator.

709.1 EIA/ANSI 709.1 is an open, device networking and control communications protocol. In most fields, it's known as the LonWorks Platform. It's designed to connect all manner of devices to the Internet – from electricity meters to subway doors. See also www.Echelon.com.

711 711 is the abbreviated dialing code for accessing Telecommunications Relay Services (TRS) from anywhere in the United States. Dialing 711 will connect you to the relay service in the state you are calling from. TRS allows a person who has a hearing or speech disability to communicate via the telephone with whomever they wish to call. See also 911 and 311.

73 One of Western Union's 92 codes, meaning "best regards." Its use continues today among amateur radio operators.

75 ohm cable 75 ohm cable is the most common cable used in the cable TV industry to transmit video to houses and businesses. It refers to coaxial cable. A 75ohm video cable typically consists of a solid inner copper and one or more braided wire shielding around the core, separated by a plastic sleeve. A 75 ohm cable will act as a conductor, picking up TV signals and causing ghosting on the TV screen (i.e. multiple images). The shielding/s are necessary to keep the ghosts out. The most shields, the more effective the cable at keeping out ghosts. In Manhattan, New York City, the local cable company typically uses 75 ohm cable with four braided wire shields. 75 ohm cable works well with several caveats: Corroded connections often cause your TV picture to freeze. The solution is to replace the connectors. Second, it's hard to replace 75 ohm connectors yourself. To do it properly you need heavy duty equipment – the sort of stuff your local cable TV guy carries, not the stuff sold at your local Radio Shack.

8 PSK Eight phase shift keying.

8.3 Under the MS-DOS naming structure, a file's name can be eight letters in front of the period and three after it, e.g. LAZARUS8.TXT.

8.3 minutes. The time it takes for light to travel from the Sun to the Earth.

8-bit Computer A computer that uses a central processing unit (CPU) with an 8-bit data bus and that processes one byte (8 bits) of information at a time. The first microprocessors used in personal computers, such as the MOS Technology 6502, Intel 8080, and Zilog Z-80, were installed in 8-bit computers such as the Apple II, the MSAI 8080, and the Commodore 64.

800 The first "area code" for what AT&T originally called In-WATS service. See 800 Service and 8NN.

800 MHz rebanding An FCC-ordered relocation of traditional public safety radio channels, currently located in the 800 MHz band, to a different part of the frequency spectrum. This will put more separation between public safety frequencies and those used by cell phones, thereby avoiding harmful interference.

800 Portability 800 Portability refers to the fact that you can take your 800 number to any long distance carrier. A case example, once I had 1-800-LIBRARY. For many years, that number was provided by AT&T. When portability came along, we were able to change it from AT&T to MCI and still keep 1-800-LIBRARY, which is 800-542-7279. 800 Portability is provided by a series of complex databases the local phone companies, under FCC mandate, have built. 800 Portability started on May 1, 1993. See 800 Service.

800 Service A toll free call paid for by the called party, rather than the calling party. A generic and common term for In-WATS (Wide Area Telecommunications Service) service provided by a phone company, whether a LEC (Local Exchange Carrier) or an IXC (Inter-eXchange Carrier). In North America and in order of their introduction, all these In-WATS services have 800 (1967), 888 (1996), 877 (1998), 866 (2000), or 855 (2001) as their "area code." (Note: Future 800 numbers will follow the convention 8NN, where NN are specific numbers which are identical. Such 800 service is typically used by merchants offering to sell something such as hotel reservations, clothes, or rental cars. The idea of the free service is to entice customers to call the number, with the theory being that if the call was a toll call and therefore cost the customer something, he or she might be less inclined to call. Suppliers of 800 services use various ways to configure and bill their 800 services.

800 Service works like this: You're somewhere in North America. You dial 1-800, 1-888, 1-877, 1-866 or 1-855 and seven digits. The LEC (Local Exchange Carrier, i.e., the local phone company) central office sees the "1" and recognizes the call as long distance. It also recognizes the 8NN area code and queries a centralized database before processing the call further, with the query generally taking place over a SS7 (Signaling System 7) link. The centralized database resides on a Service Management System (SMS), which is a centralized computing platform. The database identifies the LEC or IXC (InterExchange Carrier) providing the 8NN number. Based on that information, and assuming that the toll-free number is associated with an IXC, the LEC switch routes the call to the proper IXC. Once the IXC has been handed the call, it processes the 800 number, perhaps translating it into a "real" telephone number in order to route it correctly. Alternatively, the IXC translates the 800 number into an internal, nonstandard 10-digit number for further routing to the terminating Central Office (CO) and trunk or trunk group.

As a real-life example, the publisher of this book has an 800 number, 800-LIBRARY (or 800-542-7279). When you call that number, MCI routes that number to the first available channel on the dedicated T-1 circuit which leased from MCI's, and connecting the MCI New York City POP (Point Of Presence) to the CMP New York City office.

Because 800 long distance service is essentially a database lookup and translation service for incoming phone calls, there are endless "800 services" you can create. You can put permanent instructions into the company to change the routing patterns based on time of day, day of week, number called, number calling. Some long distance companies allow you to change your routing instructions from one minute to another. For example, you might have two call centers into which 800 phone calls are pouring. When one gets busy, you may tell your long distance company to route all the 800 inbound phone calls to the call center, which isn't busy. See Eight Hundred Service and One Number Calling for more, especially all the features you can now get on 800 service.

In May of 1993 the FCC mandated that all 800 (and by extension all 8NN) numbers became "portable." That means that customers can take their 800 telephone number from one long distance company to another, and still keep the same 800. See also 800 Portability.

800 Services are known internationally as "Freefone Services." In other countries the dialing scheme may vary, with examples being 0-800 and 0-500. Such services also go under the name "Greenfone." In June 1996, the ITU-T approved the E.169 standard Uni-

versal International Freefone Number (UIFN) numbers, also known as "Global 800." UIFN will work across national boundaries, based on a standard numbering scheme of 800, 888 or 877 plus an 8-digit telephone number. See also UIFN and Vanity Numbers.

802 See 802 Standards.

802 Standards The 802 Standards are a set of standards for LAN (Local Area Network) and MAN (Metropolitan Area Network) data communications developed through the IEEE's Project 802. IEEE stands for Institute of Electrical and Electronic Engineers. The two most important standards are 802.11b and 802.11a. The standards also include an overview of recommended networking architectures, approved in 1990. The 802 standards follow a unique numbering convention. A number followed by a capital letter denotes a standalone standard; a number followed by a lower case letter denotes either a supplement to a standard, or a part of a multiple-number standard (e.g., 802.1 & 802.3). The 802 standards segment the data link layer into two sublayers:

A Media Access Control (MAC) layer that includes specific methods for gaining access to the LAN. These methods – such as Ethernet's random access method and Token Ring's token passing procedure – are in the 802.3, 802.5 and 802.6 standards.

A Logical Link Control (LLC) Layer, described in the 802.2 standard, that provides for connection establishment, data transfer, and connection termination services. LLC specifies three types of communications links:

An Unacknowledged Connectionless Link, where the sending and receiving devices do not set up a connection before transmitting. Instead, messages are on a "best effort" basis, with no provision for error detection, error recovery, or message sequencing. This type of link is best suited for applications where the higher layer protocols can provide the error correction and functions, or where the loss of broadcast messages is not critical.

A Connection-Mode Link, where a connection between message source and destination is established prior to transmission. This type of link works best in applications, such as file transfer, where large amounts of data are being transmitted at one time.

An Acknowledged Connectionless Link that, as its name indicates, provides for acknowledgement of messages without burdening the receiving devices with maintaining a connection. For this reason, it is most often used for applications where a central processor communicates with a large number of devices with limited processing capabilities.

802.1 IEEE standard for overall architecture of LANs and internetworking. See all the following definitions.

802.11 Wireless specifications developed by the IEEE. It details a wireless interface between devices to manage packet traffic (to avoid collisions, etc.) Some common specifications include the following:

802.11a 802.11a is an updated, bigger, better, faster version of 802.11b (also called Wi-Fi), which is now commonly installed in offices, airports, coffee shops, etc. Many laptops now come with 802.11b built-in. The newer 802.11a, also an IEEE standard for wireless LANs, supports speeds up to 54 Mbps. 802.11a runs in a 300-MHz allocation in the 5 GHz range, which was allocated by the FCC in support of UNII (the Unlicensed National Information Infrastructure). Specifically, 200 MHz is allocated at 5.15-5.35 MHz for in-building applications, and 100 MHz at 5.725-5.825 MHz for outdoor use. This allocated spectrum is divided into three working domains. At 5.15-5.25 MHz, maximum power output is restricted to 50mW (milliWatts), 5.25-5.35 to 250mW, and 5.725-5.825 to 1 Watt. 802.11a has been dubbed Wi-Fi5 (Wireless Fidelity 5 MHz) by the Wireless Ethernet Compatibility Alliance (WECA).

802.11a uses Coded Orthogonal Frequency Division Multiplexing (COFDM) as the signal modulation technique. COFDM sends a stream of data symbols in a massively parallel fashion, with multiple subcarriers (i.e., small slices of RF, or Radio Spectrum, within the designated carrier frequency band. Each carrier channel is 20 MHz wide, and is subdivided into 52 subcarrier channels, each of which is approximately 300 KHz wide; 48 of the subcarrier channels are used for data transmission, and the remaining four for error control. Through the application of a coding technique, each symbol comprises multiple data bits. The specified coding techniques and data rates specified, all of which must be supported by 802.11-compliant products, include BPSK (Binary Phase Shift Keying) at 125 Kbps per channel for a total of 6 Mbps across all 48 data channels, QPSK (Quadrature Phase Shift Keying) at 250 Kbps per channel for a total of 12 Mbps, and 16QAM (16-level Quadrature Amplitude Modulation) at 500 Kbps per channel for a total of 24 Mbps. The standard also allows more complex modulation schemes, that offer increased data rates. Currently, the most complex and fastest is 64QAM (64-level QAM), at 1.125 Mbps per channel for a total of 54 Mbps.

The symbol rate is slowed down enough that each symbol transmission is longer than the delay spread. The delay spread is the variation in timing between receipt of the signals associated with a given symbol, with the delay spread caused by multipath fading. Multipath fading is the phenomenon whereby the RF signals carrying a given data symbol arrive at the receiver at slightly different times. This is because the signal spreads out from the transmitter, with certain portions of the signal reaching the receiver more or less directly, while other portions of the signal bounce around off of walls, furniture, your co-worker's pointy head, and such. Now, each of the symbols contains multiple bits, which are imposed on it through the coding processes identified above. As the multiple symbols reach the receiver, they are sorted out and decoded, with the decoding process providing some additional time for the receiver to adjust for the delay spread and to get ready to receive the next symbol. Both 802.11a and 802.11b are designed to be compatible with Ethernet LANs, using the MAC (Media Access Control) technique of CSMA/CA (Carrier Sense Multiple Access with Collision Avoidance).

If this sounds great, that's because it is great. If this sounds too good to be true, that's because it gets a little more complicated. While the 5 GHz spectrum is pretty clear in the US, it's not so readily available elsewhere. Military and government installations use portions of this band overseas. In Japan, only the 5.15-5.25 MHz spectrum is available. In Europe, the 5.725-5.825 MHz spectrum is already allocated for other uses. In Europe, ETSI (European Telecommunications Standards Institute) requires that two additional protocols be used in conjunction with 802.11a in order to protect incumbent applications and systems running over previously allocated shared spectrum. DFS (Dynamic Frequency Selection) allows the 802.11a system to dynamically shift frequency channels and TPC (Transmission Power Control) reduces the power level. In combination, these protocols serve to eliminate interference issues with incumbent signals. See also 802.11b, 802.11g, BPSK, CSMA/CA, MAC, OFDM, QAM, QPSK, WECA, Wi-Fi and (for a comparison of wireless standards) Wireless.

802.11b 802.11b is now the most common wireless local area network. 802.11b is now installed in offices, airports, coffee shops, hotels, boardrooms and homes. Many laptops now come with 802.11b wireless transmit and receive electronics built-in. 802.11b is also called Wi-Fi or WiFI (Wireless Fidelity). 802.11b is a low power wireless system so the closer you are to a transmitter, the faster it will be. This is roughly what you'll get: Wireless operating range (indoors): 100 feet at 11 Mbps, 165 feet at 5.5 Mbps, 230 feet at 2 Mbps, 300 feet at 1 Mbps. Wireless operating range (outdoors): 500 feet at 11 Mbps, 885 feet at 5.5 Mbps, 1,300 feet at 2 Mbps, 1,500 feet at 1 Mbps. 802.11b operates at the same frequency as some cordless phones, garage door openers, walkie-talkies, etc. So there's a real chance for interference in big cities, like New York. An 802.11b basestation is often attached to a local area network, which is then attached to the Internet and/or the corporate network. This means that you use 802.11b to surf the Internet or get to corporate databases, etc.

802.11b defines both the Physical (PHY) and Medium Access Control (MAC) protocols. Specifically, the PHY spec includes three transmission options – one Ir (Infrared), and two RF (Radio Frequency). 802.11b uses DSSS (Direct Sequence Spread Spectrum) modulation for digital communication. DSSS involves the transmission of a stream of one's and zero's, which is modulated with the Barker code chipping sequence. Barker code is an 11-bit sequence (e.g., 10110111000) that has advantages in wireless transmission. Each bit is encoded into an 11-bit Barker code, with each resulting data object forming a "chip." The chip is put on a carrier frequency in the 2.4 GHz range (2.4-2.483 GHz), and the waveform is modulated using one of several techniques. 802.11 systems running at 1 Mbps make use of BPSK (Binary Phase Shift Keying). Systems running at 2 Mbps make use of QPSK (Quaternary Phase Shift Keying). Systems running at 11 Mbps make use of CCK (Complementary Code Keying), which involves 64 unique code sequences, which technique supports six bits per code word. The CCK code word is then modulated onto the RF carrier using QPSK, which allows another two bits to be encoded for each 6-bit symbol. Therefore, each 6-bit symbol contains eight bits. Power output is limited by the FCC to 1 watt EIRP (Equivalent Isotropically Radiated Power). At this low power level, the physical distance between the transmitting devices becomes an issue, with error performance suffering as the distance increases. Therefore, the devices adapt to longer distances by using a less complex encoding technique, and a resulting lower signaling speed, which translates into a lower data rate. For example, a system running at 11 Mbps using CCK and QPSK, might throttle back to 5.5 Mbps by halving the signaling rate as the distances increase and error performance drops. As the situation gets worse, it might throttle back to 2 Mbps using only QPSK, and 1 Mbps using BPSK. Also to be considered in this equation is the fact that the 2.4 GHz range is in the unlicensed ISM (Industrial, Scientific and Medical) band, which is shared by garage door openers, microwave ovens, bar code scanners, cordless phones, Bluetooth LANs, and a wide variety of other devices. As a result, this slice of spectrum can be heavily congested at times, and performance can drop considerably. 802.11 divides the available spectrum into 14 channels. In the US, the FCC allows the use of 11 channels. Four channels are available in France, 13 in the rest of Europe, and only one in Japan. There also is overlap between adjacent channels (e.g., channels one and two), which fact further affects performance; therefore, any given system must maintain maximum channel separation from other systems in proximity.

Both 802.11a and 802.11b are designed to be compatible with Ethernet LANs. 802.11b uses a variation of the MAC (Media Access Control) technique of CSMA/CA (Carrier Sense Multiple Access with Collision Avoidance), which is used in some wired Ethernets, as well. A device seeking to transmit over the shared medium (in this case, a shared RF channel) listens to the network. If it senses no activity over the carrier frequency for a minimum period of time known as the DIFS (DCF (Distributed Coordinated Function) Inter-Frame Spacing), it requests access by first transmitting a RTS (Request To Send) packet. The RTS packet includes both the source (i.e., transmitter) and destination (i.e., intended receiver) addresses, the duration of the intended session (i.e., transmission), and the ACK (ACKnowledgement) associated with it. If the network is available, the destination device responds with CTS (Clear To Send), repeating both the duration and the ACK. All other devices back off the network until the session is concluded. If the network, on the other hand, is busy, the device waits a period of time equal to the DIFS, plus a random number of slot times, as calculated with several back-off timers. The "listening" process takes several forms. CAM (Constant Access Method), the default method, involves constant monitoring of the network. Since CAM creates a power consumption issue for battery-powered devices, PAM (Polled Access Mode) can be substituted. PAM calls for all client devices to go into sleep mode, all awaking at regular intervals, at the exact same time, to listen for network activity. On January 3, 2000 the 802.11 technology got another boost when Microsoft and Starbucks announced that they were to join forces to offer wireless access, using 802.11b among other standards, in most of Starbucks' coffee outlets over the next two years. The deal, some analysts say, is a further sign that 802.11b could become a serious competitor to better known wireless technologies such as Bluetooth, HomeRF, or even next-generation cellular networks. Apple was the first to launch an 802.11b product line (called AirPort). All Apple computers now include a built-in antenna which, in conjunction with a networking card, can exchange data with a small base station plugged into a broadband Internet connection up to 45 metros (150 feet) away. Although some PC laptops now come pre-equipped with wireless hardware, most users buy a PCMCIA card, or PC card, that serves as a wireless modem and antenna. See also 802.11a, 802.11g, Bluetooth, BPSK, Chip, CSMA/CA, DSSS, EIRP, Ethernet, HomeRF, MAC, QPSK, Spread Spectrum, WECA, Wi-Fi and (for a comparison of wireless standards) Wireless.

802.11e IEEE's Quality of Service (QoS) standard. It is designed to guarantee the quality of voice and video traffic. It will be particularly important for companies interested in using Wi-Fi phones.

802.11g Similar to 802.11b, but this standard supports signaling rates of up to 54 megabits per second. It also operates in the heavily used 2.4-GHz ISM band but uses a different radio technology to boost overall throughput. Compatible with older 802.11b. Two optional modulations are allowed if systems manufacturers chose to add them: CCK-OFDM and CCK-PBCC. The new standard 802.11g is important because 802.11a and 802.11b are incompatible, but 802.11g devices will work with both 802.11a and 802.11b devices.   For a comparison of wireless standards, see Wireless.

802.11i The IEEE standard for Robust Security Network for WLANs (Wireless Local Area Networks). 802.11i includes a strong message integrity check, allows for authentication using 802.1X and uses AES (Advanced Encryption Algorithm) for message encryption. WPA (Wi-Fi Protected Access) is an industry standard based on an early version of 802.11i. See also 802.1X, AES and WPA.

802.11j Standard for wireless LAN Medium Access Control (MAC) and Physical Layer (PHY) Its specifications are 4.9 GHz –5 GHz operation in Japan. In August 2002, the Japanese Government published new rules to use 4.9 and 5 GHz bands in hot spot (indoor), fixed (outdoor), and nomadic (mobile) modes using Wireless LAN technology. IEEE 802.11j amends IEEE 802.11 to deliver a standard method of supporting these capabilities with new technologies such as the ability to change channel widths and dynamically modify radio capabilities.

802.11k In December 2005, the IEEE 802.11 Working Group passed a major milestone in the development of IEEE 802.11k, "Wireless LAN Medium Access Control (MAC) and Physical Layer (PHY) Specifications: Radio Resource Management of Wireless LANs," by voting to accept a draft radio resource measurement document as a baseline for the final standard. Once completed, IEEE 802.11k will allow enhanced measurements and diagnostics for IEEE 802.11 wireless local area networks (WLANs) that operate in the

unlicensed 2.4GHz (ISM), 4.9GHz (Japan), and 5GHz (UNII) bands. This amendment to the IEEE 802.11 base standard will enable more accurate and efficient operation of WLANs in governmental, residential, enterprise and metropolitan settings. "Next generation video streaming, wireless VOIP and dense WLAN deployments present new challenges that call for more precise WLAN measurements," says Stuart Kerry, IEEE 802.11 Working Group Chair. "IEEE 802.11k will help optimize these radio environments so more devices can coexist even as it reduces wireless network traffic congestion. Final approval of this amendment is targeted for January 2007."

802.11n This will be the next, faster standard in wireless local area networks. Eventually it will supersede Wi-Fi. For now (February 2007), the standard is not finalized. This means that products bearing the "802.11n" word might be called Pre-N or draft. By introducing 802.11n products, the manufacturers are guessing what the final standard will be. Does this mean you should steer clear of 802.11n products? The answer depends on the product. If it conforms to earlier wireless standards including 802.11a, 802.11b and 802.11g, then you know the thing will work, albeit at speeds slower than the more than 500 million bits per second that 802.11n is scheduled to reach. If you stick with routers and cards from the same manufacturer, you can feel pretty comfortable that they will work together at the higher 802.11n speed – not at their maximum but higher. As to the standard, it is believed that 802.11n will run in both the 2.5 GHz and 5 GHz bands in order to be backward compatible with 802.11a, 802.11b and 802.11g. The 802.11n standard is based on MIMO (Multiple Input/Multiple Output), which will support theoretical transmission rates of up to 540 Mbps and will extend distances up to 200 meters – though you won't get 540 Mbps at 200 meters. 802.11n also adds Orthogonal Frequency Division Multiplexing (OFDM) for improved throughput. See also MIMO, Pre-N and OFDM.

802.11r Expected to be ratified in mid to late 2006, the 802.11r Fast Roaming standard will address maintaining connectivity as a user moves from one access point to another. This is especially important in applications that need low latency and high quality-of-service standards such as voice-over-WLAN (wireless LAN).

802.11s This standard will deal with mesh networking. It is predicted to be ratified in mid-2008.

802.12 Standard for 100VG-AnyLAN. Addresses 100 Mbps demand-priority access method physical-layer and repeater specifications. Approved in 1995.

802.15.1 The 802.15.1 is an IEEE standard for wireless personal area networks. Basically it is an enhanced Bluetooth v1.1 specification and it is fully compatible with the Bluetooth v1.1 specification. Bluetooth technology defines specifications for small-form-factor, low-cost wireless radio communications among notebook computers, personal digital assistants, cellular phones and other portable, handheld devices, and connectivity to the Internet. "The new standard gives the Bluetooth specification greater validity and support in the market and is an additional resource for those who implement Bluetooth devices," says Ian Gifford, IEEE 802.15 Working Group Vice Chair. "This collaboration is a good example of how a standards development organization and a special industry group (SIG) can work together to improve an industry specification and also create a standard. The IEEE standard also added a major clause on Service Access Points, which includes an LLC/MAC interface for the ISO/IEC 8802-2 LLC, a normative annex that provides a protocol implementation conformance statement (PICS) pro forma, and an informative, high-level behavioral ITU-T Z.100 specification and description language (SDL) model for an integrated Bluetooth MAC Sublayer. This SDL model offers an extensive overview (more than 500 pages long) of a significant portion of the Bluetooth protocols e.g., Baseband, LMP, L2CAP, and the Link Manager (using the host controller interface (HCI)). The IEEE-SA plans to further develop the 802.15.1 SDL model source to support the standard. The SDL code, which will be available on CD-ROM, will include a computer model for use with any SDL tool that supports the SDL-88, SDL-92 or SDL-2000 update of ITU-T Recommendation Z.100. The IEEE 802.15.1 Working Task Group used the SDL (Specification and Description Language) to translate the natural language of the Bluetooth Specification into a formal specification that defines how the Bluetooth protocols react to events in the environment that are communicated to a system by signals.

802.15.3 A standard developed to meet the requirements of portable consumer imaging and multimedia applications – what the IEEE refers to as the standard for High Rate Wireless Personal Area Networks (WPANs). (The IEEE defines High Rate as 20 Mbps or better.) It is based on a centralized and connection-oriented ad-hoc networking topology. The standard also supports peer-to-peer connectivity and isochronous and asynchronous data. 802.15.3 is optimized for low-cost, small-form factor, and low-power consumer devices, enabling multimedia applications that are not optimized by existing wireless standards. The current technology will operate in the unlicensed 2.4 GHz band and supports five selectable data rates; 11, 22, 33, 44, and 55 Mbps, as well as three to four non-overlapping channels. The standard is secure, implementing privacy, data integrity, mutual-entity authentication and data-origin authentication for consumer applications. Task Group 3a (TG3a) has been charged to define an enhancement amendment for even higher speeds.

802.15.4 The IEEE 802.15 TG4 (Task Group 4) is chartered to investigate a low data rate solution with multi-month to multi-year battery life and very low complexity. Intended to operate in an unlicensed, international frequency band, the standard will support sensors, interactive toys, smart badges, remote controls and home automation. This is an important standard which will extend wireless communications to low-power, low-cost, low-speed devices – like sensors and switches for industrial and residential use to smart tags and badges, interactive toys, inventory tracking and much more. 802.15.4 provides for low-data-rate connectivity among relatively simple devices that consume minimal power and typically connect at distances of 10 meters (30 feet) or less and operating at data rates of 10 to 250 kbps. It allows devices to form short-range ad hoc networks within which they can interact directly. "This is an enabling standard," says Pat Kinney, Chair of IEEE 802.15 Task Group 4. "It builds a framework so existing low-end wired devices can participate in wireless networks and also creates a path for many new applications. The potential uses have several things in common. They all involve relatively simple, low-speed wireless links that need so little power that a set of AA batteries might last three to five years or even longer. "We believe a host of new applications will be based on the standard. These might include motion sensors that control lights or alarms, wall switches that can be moved at will, meter reader devices that work from outside a house, game controllers for interactive toys, tire pressure monitors in cars, passive infrared sensors for building automation systems, and asset and inventory tracking devices for use in retail stock rooms and warehouses." See ZigBee.

802.16 Officially known as the WirelessMAN Air Interface for Broadband Wireless Access (BWA), 802.16 is an umbrella standard specification for what generically is known as a Wireless Local Loop (WLL). Specifically, 802.16 was released in 2001 as a means of standardizing Local Multipoint Distribution Services (LMDS). 802.16 focuses on frequencies in the 10-66 GHz range and requires Line of Sight (LOS). Since the initial release, 802.16 has evolved considerably through the 802.16a, 802.16d and 802.16e extensions. The first BWA standard to be released by an accredited standards body, 802.16 features a protocol-independent core, supports high-bandwidth on-demand environments and hundreds of users per channel, and can handle either continuous or bursty traffic. In combination, the IEEE 802.16 standards and related standards from the European Telecommunications Standards Institute (ETSI) have served as the basis for a set of solutions known as Worldwide Interoperability for Microwave Access (WiMAX), which is promoted by the WiMAX Forum. See WiMAX for a fuller explanation. See also 802.16a, 802.1d, 802.16e and LMDS.

802.16a An extension (2003) of the 802.16 standard, 802.16a is based on Multichannel Multipoint Distribution Services (MMDS) and the European HiperMAN system. 802.16a operates in the 2-11 GHz range, which includes both licensed and license-exempt bands. It is designed for both point-to-point and point-to-multipoint topologies, and usually requires Line of Sight (LOS). See also 802.16.

802.16d Also known as 802.16-2004, this is a compilation and modification of previous versions and amendments 802.16a, b and c. It operates in the 2-11 GHz range and was designed for point-to-point, point-to-multipoint and meshed topologies. 802.16d operates best with Line of Sight (LOS), but does not require it. Included is support for indoor Customer Premises Equipment (CPE). See also 802.16.

802.16e Finalized in December 2005, 802.16e adds hand-off capability to 802.16 and its extension, thereby extending portability and mobility. This extension operates in the 2-11 GHz range and does not require Line of Sight (LOS). IEEE calls this new standard the Mobile WirelessMAN. According to the IEE, the mobile WirelessMAN standard will facilitate the global development of mobile broadband wireless access (BWA) systems. The standard amends and extends the IEEE 802.16 WirelessMAN standard, which addressed Wireless Metropolitan Area Networks for broadband wireless access but previously supported only fixed (stationary) terminals. The amended standard specifies a system for combined fixed and mobile BWA supporting subscriber stations moving at vehicular speeds in licensed bands under 6 GHz. See also WiMAX.

802.17 IEEE 802.17 optimizes packet transmission at multi-gigabit rates for local, metropolitan and wide area networks. In issuing the standard in July 2004, the IEEE said, "The protocol offered in this standard will enable the fiber optic rings widely deployed in local, municipal and wide area networks to carry more data, voice, and video content with greater reliability, efficiency, and economy.   In addition to a new Media Access Control (MAC) method, the standard includes appropriate physical layer specifications and promotes multi-vendor interoperability. Its potential base of users includes telecommu-

nication carriers, multi-service cable operators, carrier-neutral service providers, data centers, metropolitan facilities-based service providers, municipal and utilities owned networks, and large enterprise networks." 802.17 is really a technological specification being developed by the IEEE for Resilient Packet Ring (RPR). 802.17 is intended to optimize Ethernet-based metropolitan ring networks for packet transport with resiliency matching or exceeding that of SONET rings. RPR will carry voice and other TDM (Time Division Multiplexed) traffic with the QoS (Quality of Service) and resiliency of SONET and ATM combined, while supporting LAN traffic with the efficiency of Ethernet.

802.1b Standard for LAN/WAN management, approved in 1992; along with 802.1k, became the basis of ISO/IEC 15802-2.

802.1d IEEE standard for interconnecting LANs through MAC bridges (specifically between 802.3, 802.4, and 802.5 networks). The standard was approved in 1990, and was incorporated into ISO/IEC 10038. Works at the MAC level. The original version of 802.1d was invented by Digital Equipment Corporation, that used it to prevent bridging loops by creating a spanning tree. The algorithm is now documented in the IEEE 802.1d specification, although the Digital algorithm and the IEEE standard 802.2 handles errors, framing, flow control, and the Layer 3 service interface. See spanning tree.

802.1e IEEE standard for LAN and MAN load protocols. Approved in 1990, formed the basis for ISO/IEC 15802-4.

802.1f Standard for defining network management information specified in 802 umbrella standards. Approved in 1993.

802.1g An IEEE standard for remote bridging at the MAC layer.

802.1h IEEE practices recommended for bridging Ethernet LANs at the MAC layer. Approved in 1995.

802.1i IEEE standard for using FDDI (Fiber Distributed Data Interface) as a MAC-layer bridge. Approved in 1992, the standard was incorporated into ISO/IEC 10038.

802.1j IEEE standard for LAN connectivity using MAC-layer bridges. A supplement to 802.1d, it was approved in 1996.

802.1k IEEE standard for the discovery and dynamic control of network management information. Approved in 1993. In conjunction with 802.1B, was the basis for ISO/IEC 15802-2.

802.1m A conformance statement for 802.1e, it addresses definitions and protocols for system load management. Approved in 1993, it was incorporated into ISO/IEC 15802-4.

802.1n See 802.11n.

802.1p 802.1p is an IEEE standard for providing quality of service (QoS) in 802-based networks. 802.1p uses three bits (defined in 802.1q) to allow switches to reorder packets based on priority level. It also defines the Generic Attributes Registration Protocol (GARP) and the GARP VLAN Registration Protocol (GVRP). GARP lets client stations request membership in a multicast domain, and GVRP lets them register into a VLAN. 802.1p is an IEEE extension of 802.1D. It is the specification for the use of MAC-layer bridges in filtering and expediting multicast traffic. Prioritization of traffic is accomplished through the addition of a 3-bit, priority value in the frame header. Eight topology-independent priority values (0-7) are specified, with all eight values mapping directly into 802.4 and 802.6. Switches that support 802.1P and 802.1Q provide a framework for bandwidth prioritization. Essentially what all these words mean is that you can assign a priority to the type of traffic with IEEE 802.1p class-of-service (CoS) values and these allow network devices along the way to recognize and deliver high-priority traffic in a predictable manner. When congestion occurs, QoS drops low-priority traffic to allow delivery of high-priority traffic. See also 802.1q.

802.1q An IEEE standard for providing VLAN identification and quality of service (QoS) levels. Four bytes are added to an Ethernet frame, increasing the maximum frame size from 1518 to 1522 bytes. Three bits are used to allow eight priority levels (QoS) and 12 bits are used to identify up to 4096 VLANs. 802.1q is the IEEE specification for implementation of VLANs (Virtual Local Area Networks) in Layer 2 LAN switches, with emphasis on Ethernet. Similar to 802.1P, prioritization of traffic is accomplished through an additional four bytes of data in the frame header. Most data fields in this addition to the header are specific to VLAN operation. Also included is a field which provides the same 3-bit priority flag specified in 802.1P's priority-mapping scheme. In addition to conventional data traffic, 802.1Q supports voice and video transmission through Ethernet switches. In short, the 802.1Q specification provides a 32-bit header for VLAN frame tagging. Each 802.1Q tag sits in an Ethernet frame between the source address field and the media access control (MAC) client type/length field. An important feature of 802.1Q is the ability to share multiple subnets across a high-speed link. This capability not only reduced the number of lower speed links needed for physical separation, but it also allowed for asymmetrical traffic management so that different speed links could be managed more easily. With IEEE 802.1p and 802.1Q, we saw the introduction of some important concepts that have been carried forward for further QoS (Quality of Service) development. These 802.1 features also can be mapped into higher layer protocols like IP and ATM.

802.1w The Task Group developing IEEE P802.11w is focused on improving the security of IEEE 802.11 management frames, including but not limited to action management frames and deauthentication and disassociation frames. The IEEE described their proposed 802.1w as "Amendment to Standard for Information Technology–Telecommunications and Information Exchange between systems–Local and Metropolitan Area networks–Specific requirements–Part 11: Wireless LAN Medium Access Control (MAC) and Physical Layer (PHY) specifications: Protected Management Frames," will provide enhancements to the IEEE 802.11 Medium Access Control layer to make available mechanisms that enable data integrity, data origin authenticity, replay protection, and data confidentiality for selected IEEE 802.11 management frames. See also 802.1x.

802.1x 802.1x is an authentication standard for local area networks (LANs), both wired and wireless. 802.1x attempts to introduce serious security checking by making sure that users of the LANs are clean and honest folks who are authorized to use the LAN. Current authentication in the 802.11 standard is focused more on wireless LAN connectivity (i.e., getting the LAN working) than on verifying user or station identity. For wireless LANs to grow large, i.e., to scale to hundreds or thousands of users, the current method of authentication must be replaced by an authentication framework that supports centralized user authentication. The standard is flexible enough to authenticate users for WEP (Wired Equivalent Privacy), WPA (Wi-Fi Protected Access) and 802.11i WLANs (Wireless LANs). 802.1x takes advantage of an existing authentication protocol known as the Extensible Authentication Protocol (EAP), as specified in RFC 2284. 802.1x takes EAP, which is written around PPP, and ties it to the physical medium, be it Ethernet, Token Ring or wireless LAN. EAP messages are encapsulated in 802.1x messages and referred to as EAPOL, or EAP over LAN. 802.1x authentication for wireless LANs has three main components: The supplicant (usually the client software); the authenticator (usually the access point); and the authentication server (usually a Remote Authentication Dial-In User Service server, although RADIUS is not specifically required by 802.1x).

The client tries to connect to the access point. The access point detects the client and enables the client's port. It forces the port into an unauthorized state, so only 802.1x traffic is forwarded. Traffic such as Dynamic Host Configuration Protocol, HTTP, FTP, Simple Mail Transfer Protocol and Post Office Protocol 3 is blocked. The client then sends an EAP-start message. The access point will then reply with an EAP-request identity message to obtain the client's identity. The client's EAP-response packet containing the client's identity is forwarded to the authentication server. The authentication server is configured to authenticate clients with a specific authentication algorithm. The result is an accept or reject packet from the authentication server to the access point. Upon receiving the accept packet, the access point will transition the client's port to an authorized state, and traffic will be forwarded. 802.1x for WLANs makes no mention of key distribution or management. This is left for vendor implementation. At logoff, the client will send an EAP-logoff message. This will force the access point to transition the client port to an unauthorized state.

In short, 802.1x is a June, 2001 IEEE specification for port-based network access control. 802.1x makes use of the physical access characteristics of IEEE 802 LANs to provide a mechanism for authenticating and authorizing attached devices with point-to-point connection characteristics, and of preventing access to that port should the authentication and authorization processes fail. The LAN ports can be either physical (i.e., hard-wired) or logical (i.e., wireless) in nature. 802.1x is particularly aimed at the ports of MAC bridges as specified in 802.1D, the ports used to attach servers or routers to the LAN, and the associations between wireless stations and access points in 802.11 WLANs. Applications include corporations that provide LAN access to the public in certain areas, and service providers that offer high-speed Internet access in hotels and airports. 802.11 makes use of the Extensible Authentication Protocol (EAP), as specified in RFC 2284. The EAP messages are encapsulated in 802.1x messages, thereby taking the form of EAPOL (EAP Over LAN). The authentication procedure commonly involves a supplicant (generally client software) communicating with a wireless access point that consults with an authentication server, which usually is in the form of a RADIUS (Remote Access Dial In User Service) server. See also 802.11, 802.1D, EAP, EAPOL and RADIUS.

802.2 A data link layer standard used with the IEEE 802.3, 802.4 and 802.5 standards. 802.2 is the more modern form of 802.3. Novell recommends that it be used on its NetWare networks in preference to the 802.3 (which will still work). Most Ethernet networks support both 802.2 and 802.3. For more on the 802 series, see the numbers definitions at the front of this dictionary.

802.2 SNAP Cub-Network Access Protocol. A variation on the 802.2/802.3 scheme which expands the 802.2 LLAMA header to provide sufficient space in the header to identify almost any network protocol.

802.22 IEEE 802.22 is a proposed new standard to allow for wireless regional area networks of 40 kilometers or more to work within the white space between over-the-air broadcast TV channels. The 802.22 radios would be cognitive (as in "to learn") and would sense the RF environment to determine what channel(s) may be used in a particular area (and with what operating parameters, such as power, etc.) without causing interference to other services. 802.22 will specify a cognitive air interface for fixed, point-to-multipoint, wireless regional area networks that operate on unused channels in the VHF/UHF TV bands between 54 and 862 MHz. "Signals at these frequencies can propagate 40 km or more from a well-sited base station, depending on terrain," said Carl R. Stevenson, Interim Chair of the IEEE P802.22 Working Group. "This is ideal spectrum for deploying regional networks to provide broadband service in sparsely populated areas, where vacant channels are available. Our goal is to equal or exceed the quality of DSL or cable modem services, and to be able to provide that service in areas where wireline service is economically infeasible, due to the distance between potential users. "This standard will enable the creation of interoperable IEEE 802 WRAN products. It has generated a great deal of interest from wireless internet service providers, community networking organizations, government bodies and other parties." Protocols in the standard will ensure that this new service does not cause harmful interference to the licensed incumbent services in the TV broadcast bands. The standard will provide for broadband systems that choose portions of the spectrum by sensing what frequencies are unoccupied.

802.3 A local area network protocol suite commonly known as Ethernet. Ethernet has either a 10 Mbps or 100 Mbps throughput and uses Carrier Sense Multiple Access bus with Collision Detection CSMA/CD. This method allows users to share the network cable. However, only one station can use the cable at a time. A variety of physical medium dependent protocols are supported. This is the most common local area network specification. For a much fuller explanation, see Ethernet.

802.3 1Base5 IEEE standard for baseband Ethernet at 1 Mbps over twisted pair wire to a maximum distance of 500 meters. Also called Starlan.

802.3 10Base-5 IEEE standard for baseband Ethernet at 10 Mbps over coaxial cable to a maximum distance of 500 meters.

802.3 10Base-T Also called 802.3i. 10Base-T is an IEEE standard for operating Ethernet local area networks (LANs) on twisted-pair cabling using the home run method of wiring (exactly the same as a phone system uses) and a wiring hub that contains electronics performing similar functions to a central telephone switch. The full name for the standard is IEEE 802.3 10Base-T. The 10Base-T standard, issued in the fall of 1990, defined the requirements for sending information at 10 million bits per second on ordinary unshielded twisted-pair cabling. The 10Base-T standard defines various aspects of running Ethernet on twisted-pair cabling such as:

- Connector types (typically eight-pin RJ-45),
- Pin connections (1 and 2 for transmit, 3 and 6 for receive),
- Voltage levels (2.2 volts to 2.8 volts peak), and
- Noise immunity requirements to filter outside interference from telephone lines or other electronic equipment.

Ethernet is the most popular LAN in the world. Ethernet running on loop coaxial cable – typically called thin Ethernet or thinnet – is the most popular way of running Ethernet local area networks. Loop networks suffer from the major problem that one cut in the cable can destroy the complete network. 10Base-T is a much more reliable – though more expensive – way of connecting LANs, since it requires electronics at the center of the home run. As I write this, the most common form of 10Base-T electronics is a small box joining about 12 workstations together. To get more on the LAN, you simply daisy chain the boxes together. The boxes are unbelievably reliable. They're easy to install and they often come with LAN management software, which gives you statistics on who's using the network, for how long, what the performance is, and what potential problems might crop up, etc. The cable 10Base-T networks use to connect between their central electronics and their attached workstations is typically standard twisted pair phone wiring, which is a lot easier to install than coaxial cable. 10Base-T networks are now becoming most popular and are being installed at faster rate than old-style loop coaxial wired LANs. For a fuller explanation see Ethernet.

802.3 10Broad36 This IEEE standard describes a long-distance type of Ethernet cabling with a 10-megabit-per-second signaling rate, a broadband signaling technique, and a maximum cable-segment distance of 3,600 meters.

802.3ad The IEEE 802.3ad (Link Aggregation Control Protocol-LACP) is a specification for bundling multiple Ethernet links into what appears to be one spanning tree protocol (STP) link.   Before this specification was ratified, various vendors had their own proprietary mechanisms for providing this functionality, but it would not work in mixed vendor environments. The technology allows an uplink to have 8 Gbps aggregate uplink speed in the situation where 8 Gigabit Ethernet links are used between two switches. Without the technology, seven of the links would have gone into the STP blocking state because spanning-tree protocol detected a loop in the network.

802.3ae 10 Gigabit Ethernet standard offering data speeds up to 10 billion bits per second. Built on the Ethernet technology used in most of today's LANs, 10-Gigabit Ethernet is described as a "disruptive" technology that offers a more efficient and less expensive approach to moving data on backbone connections between networks while also providing a consistent technology end-to-end. Using optical fiber, 10-Gigabit Ethernet is being positioned to replace existing networks that use ATM switches and SONET multiplexers on an OC-48 SONET ring with a simpler network of 10-Gigabit Ethernet switches and at the same time improve the data rate from 2.5 Gbps to 10 Gbps. 10-Gigabit Ethernet is expected to be used to interconnect LANs, wide area networks (WANs), and metropolitan area networks (MANs). 10-Gigabit Ethernet uses the IEEE 802.3 Ethernet media access control (MAC) protocol and its frame format and size. On multimode fiber, 10-Gigabit Ethernet will support distances up to 300 meters. On single mode fiber, it will support distances up to 40 kilometers. Smaller Gigabit Ethernet networks can feed into a 10-Gigabit Ethernet network. There are seven faces of 10 Gigabit Ethernet:

Interface	Type	PM D (nm)	PHY	Fiber type/ diameter (microns)	Range (meters)
10Gbase-SR	Serial	850	LAN	Multimode/50	65
				Multimode/62.5	26
10Gbase-LR	Serial	1,310	LAN	Single mode/9	10,000
10Gbase-ER	Serial	1,550	LAN	Single mode/9	40,000
10Gbase-LX4	WWDM	1,310	LAN	Single mode/9	10,000
				Multimode/62.5	300
10Gbase-SW	Serial	850	WAN	Multimode/50	6
				Multimode/62.5	26
10Gbase-LW	Serial	1,310	WAN	Single mode/9	10,000
10Gbase-EW	Serial	1,550	WAN	Single mode/9	40,000

802.3af This IEEE standard is designed to power network devices over Ethernet wiring. The standard defines two types of power sourcing equipment – end-span and mid-span. The major objective of this new standard is to make deploying IP telephones and wireless access points easier and reduce the cost of powering the devices. Traditionally, IP phones and wireless access points have required two connections – one to the LAN and another to AC electricity. You may wish to place your wireless access point up in the ceiling. But getting AC power up there is expensive and a pain. It's clearly easier to run a single RJ-45 cord up to the ceiling. End-span refers to an Ethernet switch with embedded Power over Ethernet technology. These new switches deliver data and power over the same wiring pairs – transmisson pairs 1/2 and 3/6. Mid-span devices resemble patch panels and typically have been six and 24 channels. They are placed betwen legacy switches and powered devices. Each of the mid-span ports has an RJ-45 data input and a data/power RJ-45 output connector. Mid-span devices tap the unused wire pairs 4/5 and 7/8 to carry power, while data runs on the other wire pairs. According to engineering specs, I have read, designers have at most 12.95 Watts of available power per the IEEE 802.3af standard. The voltage available from the PSE (Power Sourcing Equipment) ranges from 44 Volts to 57 volts, but PDs (Powered Devices)

802.3ah An IEEE standard for delivering Ethernet over copper in the first mile, dubbed "Ethernet in the First Mile". The standard is designed to use existing infrastructure (i.e. copper wire) to provide higher speeds over – speeds as fast as 100 megabits per second. The idea is that if phone companies can use wiring they already have to provide Ethernet at 10Mbps to customers instead of DSL or T1 lines, and do it at significantly lower costs, they will. This standard will support Category 3 cable (standard, voice-grade phone wires) to 750 meters. In addition, the standard will support 100Mbps and 1Gbps Ethernet over optical fiber. To keep costs down, the EFMA development alliance is also proposing a standard for Ethernet over a single strand of fiber, in addition to the dual-strand fibers currently in use. Single-strand fiber would work by using different wavelengths of light on the same strand for the two directions, instead of using the same wavelength over two separate strands of fiber, as in the current dual-strand practice. In either case, the EFMA's proposed

standard calls for fiber runs as long as 10 kilometers. The current thinking of the 802.3ah committee is that these fibers would aggregate many 10Mb copper drops – with only half as much fiber. There's also support for passive optical networks in the 802.3ah proposal. This standard would allow several users of Gigabit Ethernet to share a single fiber for as many as 20 kilometers, using passive splitters to aggregate and separate traffic.

802.3b IEEE standard for 10Broad36. Approved in 1985, it is the standard for broadband Ethernet at 10 Mbps over coaxial cable to a maximum distance of 3600 meters. It was incorporated into ISO/IEC 8802-3.

802.3c IEEE standard for 10 Mbps baseband repeaters. The standard was approved in 1985, and is incorporated into ISO/IEC 8802-3.

802.3d IEEE standard for media attachment devices and baseband media over fiber optic repeater links. Approved in 1987, it has been incorporated into ISO/IEC 8802-3.

802.3e IEEE standard for 1Base-5, baseband Ethernet at 1 Mbps over twisted pair wire to a maximum distance of 500 meters. Also called Starlan. The standard addresses physical media, physical signaling and media attachment. Approved in 1987, it is incorporated into ISO/IEC 8802-3.

802.3h Standard for layer management in CSMA/CD networks. Approved in 1990, it has been incorporated into ISO/IEC 8802-3.

802.3i The IEEE standard addressing multisegment 10 Mbps networks, and twisted-pair media for 10Base-T networks. 10Base-T is an IEEE standard for operating Ethernet local area networks (LANs) on twisted-pair cabling using the home run method of wiring (exactly the same as a phone system uses) and a wiring hub that contains electronics performing similar functions to a central telephone switch. The full name for the standard is IEEE 802.3 10Base-T. The 10Base-T standard, issued in the fall of 1990, defined the requirements for sending information at 10 million bits per second on ordinary unshielded twisted-pair cabling; the standard has been incorporated into ISO/IEC 8802-3. The 10Base-T standard defines various aspects of running Ethernet on twisted-pair cabling such as:

Connector types (typically eight-pin RJ-45),
Pin connections (1 and 2 for transmit, 3 and 6 for receive),
Voltage levels (2.2 volts to 2.8 volts peak), and
Noise immunity requirements to filter outside interference from telephone lines or other electronic equipment.

Ethernet is the most popular LAN in the world. Ethernet running on loop coaxial cable – typically called thin Ethernet or thinnet – is the most popular way of running Ethernet local area networks. Loop networks suffer from the major problem that one cut in the cable can destroy the complete network. 10Base-T is a much more reliable – though more expensive – way of connecting LANs, since it requires electronics at the center of the home run. As I write this, the most common form of 10Base-T electronics is a small box joining about 12 workstations together. To get more on the LAN, you simply daisy chain the boxes together. The boxes are unbelievably reliably. They're easy to install and they often come with LAN management software, which gives you statistics on who's using the network, for how long, what the performance is, and what potential problems might crop up, etc. The cable 10Base-T networks use to connect between their central electronics and their attached workstations is typically standard twisted pair phone wiring, which is a lot easier to install than coaxial cable. 10Base-T networks are now becoming most popular and are being installed at faster rate than old-style loop coaxial wired LANs. For a fuller explanation see Ethernet.

802.3j IEEE standard for 10Base-F, which provides for fiber optics links connecting 10 Mbps active and passive starbased baseband networks. The standard was approved in 1993 and is incorporated into ISO/IEC 8802-3.

802.3k IEEE standard for layer management for repeaters in 10 Mbps baseband networks. It was approved in 1992 and is incorporated into ISO/IEC 8802-3.

802.3l A conformance statement for the media attachment unit protocol for 10Base-T networks. The statement was approved in 1992 and is incorporated into ISO/IEC 8802-3.

802.3p IEEE standard for media attachment unit layer management for 10 Mbps baseband networks. The standard was approved in 1992 and is incorporated into ISO/IEC 8802-3.

802.3q Provides guidelines for the development of managed objects. Approved in 1993, it was incorporated into ISO/IEC 8802-3.

802.3r IEEE standard for CSMA/CD and physical media specifications for 10Base-5, which is baseband Ethernet at 10 Mbps over fat coaxial cable to a maximum distance of 500 meters. This version of the original Ethernet standard was updated in 1996.

802.3t Standard for 120-ohm cables in 10Base-T simplex links. Approved in 1995, it was incorporated into ISO/IEC 8802-3.

802.3u A supplement to 802.3 that governs Carrier Sense Multiple Access/Collision Detection (CSMA/CD) for 100 Mbps networks, i.e., 100Base-T, commonly known as Fast Ethernet. Approved in 1995, this supplement covers the specification's MAC parameters, the physical layer, and repeaters for 100Base-T4, TX, and FX.

802.3v IEEE standard for 150-ohm cables in 10Base-T link segments. Approved in 1995, it was incorporated into ISO/IEC 8802-3.

802.3z Gigabit Ethernet over fiber standard, ratified on June 29, 1998. See Gigabit Ethernet.

802.4 A physical layer standard specifying a LAN with a token-passing access method on a bus topology. Used with Manufacturing Automation Protocol (MAP) LANs. Arcnet can work this way. Typical transmission speed is 10 megabits per second.

802.5 A physical layer standard specifying a LAN with a token-passing access method on a ring topology. Used by IBM's Token Ring hardware. Typical transmission speed is 4 or 16 megabits per second. Typical topology is star.

802.6 This IEEE standard for metropolitan area networks (MANs) describes what is called a Distributed Queue Dual Bus (DQDB). The DQDB topology includes two parallel runs of cable – typically fiber-optic cable – linking each node (typically a router for a LAN segment) using signaling rates in the range of 100 megabits per second.

802.7 IEEE technical advisory group on broadband LANs.

802.8 IEEE technical advisory group for fiber-optic LANs.

802.9 IEEE technical advisory on ISLAN, which stands for Integrated Services LAN. ISLAN is Isoethernet with switched or packetized voice on an Ethernet LAN, which is 10 megabits per second of Ethernet (used for data) plus six megabits per second of ISDN B channels, which gives you 96 B ISDN channels plus a D channel. You can use the B channels for voice.

802.10 This committee deals with LAN security.

802.X The Institute of Electrical and Electronics Engineers (IEEE) committee that developed a set of standards describing the cabling, electrical topology, physical topology, and access scheme of network products; in other words, the 802.X standards define the physical and data-link layers of LAN (local area network) architectures. Or, in simple language, the set of IEEE standards for the definition of LAN protocols. IEEE 802.3 is the work of an 802 subcommittee that describes the cabling and signaling for a system nearly identical to classic Ethernet. IEEE 802.5 comes from another subcommittee and similarly describes IBM's Token-Ring architecture.

811 The telephone number some, but by no means all, telephone companies use for their business office. See also N11.

822 Short form of RFC 822. Refers to the format of Internet style e-mail as defined in RFC 822.

82596 The 82596 is an intelligent, 16-/32-bit local area network coprocessor from Intel. The 82596 implements the CSMA/CD access method and can be configured to support all existing 802.3 standards. Coupled with the 82503 Dual Serial Transceiver, the 82596 provides the optimal Ethernet connection to Intel1386 and Intel486 client PCs and servers. The board space required for an 82596/82503 motherboard implementation is less than six square inches. Provides full Ethernet bandwidth performance while allowing the CPU to work independently. An on-board four-channel DMA controller along with an intelligent micro machine automatically manages memory structures and provides command chaining and autonomous block transfers while two large independent FIFOs accommodate long bus latencies and provide programmable thresholds.

855 Service A North American toll-free area code. This fifth such area code was to be introduced in November 2000, but was delayed until at least February 2001 by FCC order. See 800 Service for a full explanation.

866 Service A North American toll-free area code. This fourth such area code was introduced in July 2000. See 800 Service for a full explanation.

877 Service A North American toll-free area code. The original such area code was 800 numbers. Introduced in 1967, the 800 area code was sufficient for about 30 years. The NANP (North American Numbering Plan) was expanded in March 1996 to include 888 numbers. Then we ran out of 888 numbers. So, the NANC (North American Numbering Council) responded by assigning 877 numbers to relieve the pressure. 877 numbers were introduced on April 4, 1998. 866 numbers were added in July 2000, and 855 in February 2001. See 800 Service for a full explanation.

880 The NANP (North American Numbering Plan) area codes 880, 881 and 882 were established to provide a means by which toll-free calling could be extended beyond the borders of the country in which the party paying for the resides. Rather and for example, the Caribbean caller pays for the international segment of the call to the U.S. gateway, and

the called party pays for the domestic U.S. segment of the call. In theory, this concept can be implemented between any countries that share the NANP. However, they primarily are used to allow Caribbean callers to call the U.S.

881 See 880.

882 See 880.

888 Service A North American toll-free area code. When North America was in danger of running out of 800 numbers, the 888 prefix was introduced in April 1996. When 888 numbers began to run out, 877 numbers were introduced in April 1998. Since that time, the 855 and 866 area codes also have been introduced. All of these toll-free area codes follow the numbering convention of 8NN, with the last two digits always the same. See 800 Service for a full explanation.

8B/10B Local Fiber 8-byte/10-byte local fiber. Fiber channel physical media that supports speeds up to 149.76 Mbps over multimode fiber.

8B6T 8 Bits 6 Tri-state symbols. A data encoding/decoding scheme that encodes eight data bits into a 6-bit transmission sequence. The six bits represented via a tri-state symbol comprising positive voltage, zero voltage, and negative voltage. This method of representing bits differs from the NRZ (Non Return to Zero) method used in 4B5B. 100Base-T4 is the only standard which uses 8B6T. Once the data are encoded into the 8B6T format, the 6T codes are demultiplexed across three wire pairs. See also 4B5B, 8B10B, 100Base-T4, and NRZ.

8B10B 8 Bits 10 Bits. A data encoding/decoding scheme that encodes eight data bits into a 10-bit transmission sequence. The eight bits of data to be transmitted over the serial (i.e., one bit at a time) line are converted into a 10-bit code group prior to transmission, with the extra two bits of "special characters" serving such signaling and control functions as indicating the start of the data frame, the end of the data frame, and the link configuration. The primary purpose for the additional bits in the transmission code is to improve the transmission characteristics of the serial link, ensuring that sufficient bit-level transmissions occur such that the receiver can recover a "clock" (i.e., can synchronize) from the data stream. This type of timing mechanism generically is known as CDR (Clock/Data Recovery). The additional bits also increase the likelihood that the receiver can detect bit errors. Further, some of the special characters are in the form of bit patterns known as "commas," which enable the receiver to accomplish word (i.e., byte) alignment with respect to the incoming bit stream. 8B10B is used in some implementations of Gigabit Ethernet (GbE, or GE), which is standardized as 802.3z, and is proposed for use in 10GbE (10 Gbps Ethernet), which being standardized as 802.3ae. See also 4B5B, 5B6B, 8B6T, 10GbE, and GE.

8NN The present convention for future toll-free numbers is 8NN, with the last two digits always the same. See 800 Service.

8th-Floor Decision Refers to the 8th floor at the Washington offices of the FCC, where the commissioner's offices and meeting rooms are located. Decisions made on the 8th floor sometimes have a profound effect on new communication services.

8YY See 8NN.

9-track A standard for 1/2" magnetic tape designed for data storage. Its nine tracks hold a byte (eight bits) plus one parity bit in a row across the tape width-wise.

900 MHz The radio spectrum used in narrowband Personal Communications Services (PCS).

900 Number Rule A rule passed by the FTC (Federal Trade Commission) and which became effective November 1, 1993. The Rule requires that advertisements for 900 numbers contain certain disclosures, including information about the cost of the call. This information also must be included in an introductory message, or preamble, at the beginning of any 900-number program where the cost of the call could exceed $2.00. Any caller must be afforded the opportunity to hang up at the conclusion of the preamble, without incurring any cost for the call. The Rule also requires that all preambles state that individuals under the age of 18 must have the permission of a parent or guardian to complete the call. The 900 Number Rule has been very instrumental in reducing the level of 900-number abuse.

900 Service A generic and common (and not trademarked) term for AT&T's, MCI's, Sprint's and other long distance companies' 900 services. All these services have "900" as their "area code." Dialing a 900-number is free to the company or person receiving the call, but costs money to the person making the call. Here's the story: 900 service was introduced as the industry's "information service" area code. You'd dial 1-900-WEATHER, for example, and punch in some touchtones in response to prompts and you could hear the weather in Sydney, Australia or Paris, France, wherever you might be planning your next vacation. For this service, you'd be charged perhaps 75 to 95 cents a minute. And you'd get the bill as part of your normal monthly phone bill. That was the original idea. Then some people got the idea that 900 would make a wonderful porn number and they started advertising "Call

900-666-3333 and speak with Diana. She really wants you." And they started charging $5 a minute. When huge 900 call bills started appearing on people's bills, there was an outcry from many subscribers who wouldn't pay the bills. Some children called on their parents' phones. Employees made calls from work and the company's accountants went nuts. So the industry retreated from 900 porn. Then someone thought – "Why not sell things through an 900 number?" We could sell a set of ginzu knives for just calling this 900 number. No messing with credit cards or checks. The bill goes straight on your phone bill. At about the same time someone thought that 900 numbers would be great for running sweepstakes. "Call up, register your name for a free trip with a racing car team to the Australian Indianapolis 500. Three lucky people will be chosen. The call will cost you only $2.75." So 900 services became a new type of gambling.

The long distance companies providing 900 services reacted predictably to some of the newer services. They clamped down on who they would sign up, which service and/or product you could, or could not sell. And, rather than charging "a piece of the action" as they did in the beginning, the long distance companies began to charge for them as if they were normal long distance calls: charge a set-up fee, a fee for carrying the call, a fee for collecting the money and a fee for the possibility of bad debts. There are variations on these themes.

The 900 service business is rife with stories of people who are alleged to have made millions overnight with innovative 900 numbers. Clearly enormous monies have been made – especially in the beginning when there was novelty to 900 calls. The prognosis is that the 900 number business will grow, that it will mature and that the North American public will wake up to its various scams and discover real value in many of its services. For example, one of the author's "genuine value" and favorite 900 services is fax-back. Dial a 900 number, punch in some touchtone digits, hang up and within seconds your fax machine begins to churn out useful information.

In the summer of 1991, AT&T issued guidelines for its EXPRESS900 service. Those guidelines included:

The predominant purpose of the calls does not include Entertainment, Children's Programming, Credit/Loan Information, Fulfillment, Political Fundraising, Games of Chance, Postcard Sweepstakes, Job Lines and Personal Lines.

Every program must have a Preamble and Caller Grace Period, with notification to callers of the opportunity to hang up before charging begins.

Sponsors may not route calls to any telecommunications equipment or arrangements which allow charging to begin before the caller realizes any value on the call, e.g., Automatic Call Distribution (ACD) with call queuing, or Caller Hold.

9001 ISO 9001 is a rigorous international quality standard covering a company's design, development, production, installation and service procedures. Compliance with the standard is of increasing significance for vendors trading in international markets, in particular in Europe where ISO 9001 registration is widely recognized as an indication of the integrity of a supplier's quality processes. ISO is the International Standards Organization in Paris.

902-928 MHz A frequency range for use by, amongst other things, cordless phones. Such frequency offers much greater range than traditional cordless phones. Some new 900 MHz phones also use spread-spectrum technology to further increase range and call security. Previous cordless phones operated within the 46-49 MHz band. The 900 contains 50 channels for cordless telephone transmission. The 46-49 MHz band contains only 10.

911 Service 911 is an emergency reporting system whereby a caller can dial a common number – 911 – for all emergency services. The caller will be answered at a common answering location which will figure the nature of the emergency and dispatch the proper response teams. The first 911 service came on line in 1968. Here are the reasons why 911 benefits a community: Only one number for all emergency services. It's an easy number

to remember. It's an easy number to dial. It's great for travelers and new residents. Calls are received by trained personnel. See also E-911, which stands for Enhanced 911 service and typically includes ANI (Automatic Number Identification) and ALI (Automatic Location Information). 911 service is sometimes called B-911 which stands for Basic 911 service. The first 911 service in the United States was in 1968 in Alabama. See also Basic 911, B-911 and E-911.

9145 A common term in the southern part of America for a customer service representative (i.e. a salesman) of the local telephone company.

92 code In 1859 Western Union Company created a code book consisting of the numbers one to 92. Beside each number and associated with it was a commonplace phrase used in everyday telegraph communications. Telegraph operators were trained to use these codes instead of verbose, plain-English phrases wherever possible, in their telegraph transmissions. On the distant end, the codes were translated back into plain language. The use of 92 codes improved telegraph operators' productivity while not cutting into Western Union's revenue, since customers were still charged per-letter or per-word, based on the wording of their original message. Examples of 92 codes were 1 (Wait), 2 (Important business), 13, (I understand), and 30 (No more, end). Some of the 92 codes are still used today by ham radio operators.

950 Local exchange used by some North American long-distance carriers to let their customers access their calling card and other services.

950-XXXX The Feature Group B Carrier Identification Code (CIC). You would dial 950-XXXX (the last four digits are the CIC) if you wanted to place a paid call from a pay phone, and use a carrier other than the preselected carrier, who might overcharge you. You also would use 950-XXXX if you wanted to dial around your preselected carrier, and use another carrier that does not offer Feature Group D access (i.e., equal access) in your area. Feature Group D, which is more expensive, allows you to dial 101XXXX to achieve the same thing. See also 101XXXX, Feature Group B, and Feature Group D.

958 Dial 958 in New York City and a computer run by Nynex (now called Bell Atlantic) will tell you the phone number you're calling from. This is very useful. Imagine having a jack on your wall. You've lost track of which phone line it's connected to. Dial 958 and bingo! You know. Other phone companies have similar services but they often have different numbers. In some parts of Pennsylvania, the phone number is 958-4100.

976 A local information, pay-per-call phone exchange. An information service that lets callers listen to recorded messages such as sports scores or adult conversations, at higher rates than normal calls. Similar to 900 service but, numbers are locally assigned. In other words, two different customers could have the same 976 number in two LEC territories. The sponsor of the 976 service splits revenue with the phone service provider.

979 See UIPRN.

999 Great Britain's equivalent of the United States emergency number 911.

9PZDD End User Port Cost Recovery Charge DID.

@ The character typically above the 2 on your keyboard. It's called the "at sign." In English, its biggest use is in two apples @ 50 cents each equals $1 total." But in computerese, its big use is in electronic mail addressing. It is used to separate the domain name and the user name in an Internet address and is pronounced "at." For example, Bill Gates' e-mail address is billg@Microsoft.com. It is pronounced "Bill G at Microsoft.com. A professor at Rome University, Giorgio Stabile, announced on July 2000 that the symbol @ has been found on a commercial trade paper of Venetian merchants, dated 1536. The symbol was a commercial trademark for an amphor, a weight unit. In Spanish, the @ is called "arroba", from an Arab word, meaning quart, that which is exactly the measuring unit "amphora", widely used in Greek and Roman tradition.

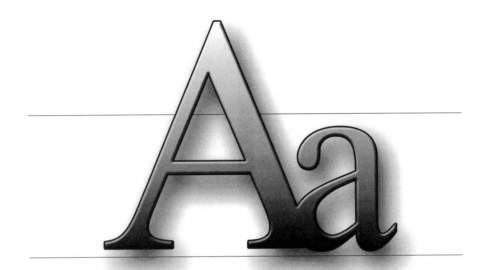

A 1. Abbreviation for AMP or AMPERE, a unit of electric current. For a longer explanation, see Ampere.

2. The non-local wireline cellular carrier. In one of its less intelligent decisions, the FCC decided to issue two cellular franchises in each city in the United States. They gave one to the local phone company (the B carrier) and one to a competitor (the A carrier). This duopoly has naturally meant little real competition. But with the issuing of PCS licences in recent years, competition has begun to heat up, i.e. prices have dropped. Meantime, the "A" carrier on your cellular phone is the non-local wireline carrier, i.e. the competitor to the local phone company and "B" is the other one, namely the local phone company's cellular company. Recently, many of the "A" carriers have been purchased by other large telecom carriers, notably AT&T.

3. Abbreviation for atto, which is ten to the minus 18th. Which means it's very very small.

A & A1 Control leads that come from 1A2 key telephone sets to operate features like flashing of lights to indicate on hold, line ringing, etc.

A & B Bits Bits used in older T-carrier channel banks for signaling and control purposes. Traditionally, T-1 frames are organized into superframes of 24 frames. Older channel banks (i.e., D1, D2, D3 and D4) use a process known as "bit robbing" for signaling and control purposes. In the sixth frame of a superframe, an "A bit" was robbed from each channel. In the 12th frame of a superframe, a "B bit" was robbed. In each case, the robbed bit was replaced with a signaling and control bit which could be used to indicate such things as "on-hook" or "off-hook" status, and to perform certain monitoring and alarming functions. This is an intrusive, in-band signaling and control technique, as the robbing of the bits violates the integrity of the bit stream and, therefore, violates the integrity of the datastream being transmitted. Newer Extended Superframe (ESF) channel banks perform their signaling and control functions without the use of this intrusive technique. See also C-Bit, Channel Bank, Extended Superframe, T-1, and T-Carrier.

A & B Leads Additional leads used typically with a channel bank two-wire E&M interface to certain types of PBXs (also used to return talk battery to the PBX).

A & B Signaling Procedure used in most T-1 transmission links where one bit, robbed from each of the 24 subchannels in every sixth frame, is used for carrying dialing and controlling information. A type of in-band signaling used in T-1 transmission. The way it works is thus. You start with 1.544 million bits per second. You knock it down by 8,000 framing bits to 1,536,000. To prove this, take eight bits for each channel and multiply by the number of channels (24) then multiply that by the sampling rate (8000 Hz) then add the framing bit multiplied with the sampling rate. Here's the arithmetic: ((8*24=192)*80 00=1,536,000 bps)+1*8000= 1,544,000 bps or 1.544 Mbps.

A Battery Another term for Talk Battery.

A Bit See A & B Bits.

A Block Cellular licenses received from the FCC with no initial association to a telephone company. Also referred to as non-wireline.

A Carrier Cellular provider in the 824-849 Mhz frequency range. In U.S. markets that have only two cellular carriers, one is designated the A carrier and the other the B carrier, which operates in the 869-894 MHz range. The A band license was generally awarded to a new entrant and this license was dubbed the non-wireline license.

a la carte cable Paying for each cable channel – channel by channel. This compares with paying for cable channels in bundles, which most of us now. The problem with bundles is we end up paying for channels we don't want. Hence, the appeal of a la carte cable.

A Law The PCM coding and companding standard used in Europe and in areas outside of North American influence. A Law Encoding is the method of encoding sampled audio waveforms used in the 2.048 Mbps, 30-channel PCM primary system known as E-carrier. See MU Law and PCM.

A Links A SS7 term. A-Links connect an end office or signal point (SP) to a mated pair of Signal Transfer Points (STPs). Two-way path diversity is recommended so that one common disaster does not isolate a signal point from the rest of the network. The other location for A-Links is between mated pairs of STPs and the SCPs. Typically, this would occur at the regional level with the A-Links assigned in a quad arrangement. See B, C and D Links.

A Number A cellular term for the number of the calling party. The originating switch analyzes the number in order to route the call to the B Number, the number of the called party.

A Port Refers to the port in an FDDI topology which connects the incoming primary ring and the outgoing secondary ring of the FDDI dual ring. This port is part of the dual attachment station or a dual attachment concentrator.

A records Address Records. All machines (i.e., host computers) that are connected directly to the Internet have IP (Internet Protocol) addresses. The DNS (Domain Name Servers) translate the "names" of the machines from URLs (Uniform Resource Locators), such as Harry_Newton@TechnologyInvestor.com, into the dotted decimal notation characteristic of the IP addressing scheme. Those machines that are directly connected to the Internet have both their URLs and their IP addresses stored in the DNS as "A Records" (Address Records). Machines that are not directly connected, have their address translation information stored as "MX Records." The MX Records point to the IP address of the mail host (i.e., the machine that is accepting mail for that target machine). See also MX Record, CNAME Records, DNS, and URL.

A,B,C, and D-Bit Signaling A signaling technique sometimes used in T-1 channel banks based on Extended SuperFrame (ESF). This approach is an in-band, or Channel-Associated Signaling (CAS), technique in which bit-robbing is used in frames 6, 12, 18, and 24 of a superframe of 24 frames. In each case, the eighth bit of each channel is "robbed," in a process known as "bit-robbing," in order to insert a signaling bit. Assuming that 16 states (i.e., values) are required for signaling purposes, each bit position in each channel can be either a "0" or a "1." (Note that 2 to the 4th power equals 16.) If 16 states are required, the signaling bits are known as A,B,C, and D-bits. If four states are required, the signaling bits are known as A,B,A, and B-bits. If only two states are required, the signaling bits are known as A,A,A, and A-bits. See also Bit-robbing, ESF, and T-1.

A-3 A voice scrambler developed by Bell Labs in the 1930s for the U.S. government and U.S. military. A-3 worked by mixing the information signal with a second signal containing noise, thereby making the original signal incomprehensible. At the distant end, the original noise was removed, thereby unhiding the message. The A-3 was a primitive security device and was eventually replaced by the more secure SIGSALY. See SIGSALY.

A-Band Carrier See A Carrier.

A-B Rolls A technique by which audio/video information is played back from two videotape machines rolled sequentially, often for the purpose of dubbing the sequential information onto a third tape, usually a composite master.

A-B Test Direct comparison of the sound/picture quality of two pieces of audio/TV equipment by playing one, then the other.

A-Carrier or A-Band Carrier Cellular provider in the 824-849 Mhz frequency range. In U.S. markets that have only two cellular carriers, one is designated the A carrier and the other the B carrier, which operates in the 869-894 MHz range. The OxECAOxEE band license was generally awarded to a new entrant and this license was dubbed the OxECnon-wirelineOxEE license.

A-Condition In a start-stop teletypewriter system, the significant condition of the signal element that immediately precedes a character signal or block signal and prepares the receiving equipment for the reception of the code elements. Contrast with start signal.

A-GPS A cell phone term. Assisted GPS, a satellite positioning system that improves the functionality and performance of GPS. Requires a special cell phone handset.

A-IMS Advanced IMS. An initiative launched by Verizon Wireless, Cisco, Lucent, Motorola, Nortel, and Qualcomm to develop a next-generation network architecture based on IMS (IP Multimedia Subsystem).

A-Interface The network (air) interface between a Mobile End System (M-ES) and the Cellular Digital Packet Data (CDPD)-based wireless packet data service provider network.

A-Key Authentication Key. An authentication mechanism of the ANSI-41 (formerly TIA IS-41C) standard for cellular intersystem inter-operability. The A-key is sent between the cellular phone and the Authentication Center (AC), and is known only to those two entities. Subsequently, the cellular phone and the AC generate a second secret key, known as a SSD (Shared Secret Data). Both the A-key and the SSD are transmitted around the ANSI-41 network to be used by switches to perform the challenge-response process of authentication. Both keys are encrypted through the CAVE (Cellular Authentication and Voice Encryption) algorithm. See CAVE.

A-Law The ITU-T companding standard used in the conversion between analog and digital signals in PCM systems. A-law is used primarily in European telephone networks and is similar to the North American mu-law standard. See also Companding and Mu law.

A-link SS7 access link. A dedicated SS7 signaling link not physically associated with any particular link carrying traffic.

A/B Switch 1. A switch that allows manual or remote switching between one input and two outputs. See A/B Switch Box.

2. A feature found on all new cellular telephones permitting the user to select either the "A" (non-wireline) carrier or the "B" (wireline) carrier when roaming away from home.

A/B Switch Box A device used to switch one input between two devices, such as printers, modems, plotters, mice, phone lines, etc. An example of how you use such a box: You plug one phone line into the "C" (for Common) jack. You plug a fax machine into the "A" jack. You plug a modem into the "B" jack. By turning the switch, you can use one phone line for either a modem or a fax. A/B switch boxes come in many flavors, including also serial and parallel port versions. There are also A/B/C switches that switch among three devices, e.g. a fax, a modem and a phone. There is also a Crossover Switch that connects two inputs and two outputs. In one position the switch might connect input A with output D and input B with output C. In the other position, it might connect input A with output C and input B with output D. See also A/B Switch.

A/B Switches Input Selector Switch. A switch used by cable customers to alternate between cable and over-the-air television reception through a cable box.

A/D Analog to Digital conversion.

A/D Converter Analog to Digital converter, or digitizer. It is a device which converts analog signals (such as sound or voice from microphone), to digital data so that the signal can be processed by a digital circuit such as a digital signal processor. See CODEC.

A/UX An alternate operating system for the Macintosh based on UNIX. A/UX has its own, unique 32-bit addressing mode.

A/V Switching A Satellite Term. A/V switching is a feature that allows users to connect one or more sources, such as a VCR, camera and/or laser videodisc player and select which source will be monitored.

A1 The A1 interface carries signaling information between the Call Control and Mobility Management functions of the MSC (Mobile Switching Center) and the call control component of the BSC (Base Station Controller).

A20 Line A control line on the Intel 80386 microprocessor that allows MS-DOS and an extended memory manager to create the High Memory Area, or HMA. Only one program can claim control over the A20 at a time.

A4 The Basic Group 3 standard that defines the scanning and printing of a page 215 mm (8.5 in) wide. An A5 page is 151 mm (5.9 in) wide, and the A6 is 107mm (4.2 in) wide.

A5 See A4.

A6 See A4.

AA 1. Automated Attendant. A device which answers callers with a digital recording, and allows callers to route themselves to an extension.

2. Auto Answer. A modem indicator light that is meant to tell you the modem is ready to pick up the phone, so long as there's a communication program running and prepared to handle the call. See also Modem.

3. In telco parlance, AAA stands for Authentication, Authorization, and Accounting, and is an essential telco OSS (Operations Support System) function.

AAA 1. Authentication, authorization, and accounting. Pronounced "triple a." A framework used for network management and security that controls access to computer resources by identifying unique users, authorizing the level of service, and tracking the usage made of resources. AAA servers typically interact with network access and gateway servers and with databases and directories that contain user information.

2. American Automobile Association. The AAA was formed in 1905 for the express purpose of providing "scouts" who would warn motorists of hidden police speed traps. Since then its functions have expanded and now it rescues broken-down automobiles and provides discount travel services.

AABS Automated Attendant Billing System. A feature which allows collect and third-number billed toll calls to be placed on an automated basis. A synthesized voice prompt guides the caller through the process, the system then seeks approval of the prospective billed party, and either completes or denies the call based on that authorization or lack thereof. AABS is automated in much the same way as calling card services have been automated, through the use of an Intelligent Peripheral (IP) device.

AAC Advanced Audio Coding. A proprietary compression codec that Apple uses for its iTunes music store. AAC also supports Apple's own DRM; purchased tracks from the iTunes Music Store are copy-protected. AAC also produces smaller file sizes than MP3 at the same quality.

AAL ATM Adaptation Layer of the ATM Protocol Reference Model, which is divided into the Convergence Sublayer (CS) and the Segmentation and Reassembly (SAR) sublayer. The AAL accomplishes conversion from the higher layer, native data format and service specifications of the user data into the ATM layer. On the originating side, the process includes segmentation of the original and larger set of data into the size and format of an ATM cell, which comprises 48 octets of data payload and 5 octets of overhead. On the termination side of the connect, the AAL accomplishes reassembly of the data. Taken together, these processes are known as Segmentation and Reassembly. AAL is defined in terms of Types

	Class A	Class B	Class C	Class D
Timing relation between source and destination	Required		Not Required	
Bit Rate	Constant	Variable		
Connection Mode	Connection-Oriented			Connectionless
Applications	Voice, Video, Circuit Emulation	Compressed Voice or Video	Frame Relay, X.25 Traffic	SMDS, LAN Traffic

AAL Service Classes, with specific attributes and example applications. Source: ITU-T Recommendation I.362 (March, 1993)

supported by the Convergence Sublayer. Each type supports certain specific types of traffic, and each offers an appropriate Quality of Service (QoS), based on traditional network references. See the next five definitions for AAL specifics.

AAL-1 ATM Adaptation Layer Type 1: AAL functions in support of Class A traffic, which is connection-oriented, Constant Bit Rate (CBR), time-dependent traffic such as uncompressed, digitized voice and video. Such traffic is isochronous, i.e., stream-oriented and highly intolerant of delay.

AAL-2 ATM Adaptation Layer Type 2: This AAL supports Class B traffic, which is connection-oriented, Variable Bit Rate (VBR), isochronous traffic requiring precise timing between source and sink. Examples include compressed voice and video.

AAL-3/4 ATM Adaptation Layer Type 3/4: AAL support of Class C and D traffic, which is Variable Bit Rate (VBR), delay-tolerant data traffic requiring some sequencing and/or error detection support, but no precise timing between source and sink. Originally two AAL types, AAL types 3 and 4, were combined in support of both connection-oriented and connectionless traffic. Examples include X.25 packet and Frame Relay traffic.

AAL-5 ATM Adaptation Layer Type 5. AAL functions in support of Class C traffic, which is of Variable Bit Rate (VBR), and which is delay-tolerant connection-oriented data traffic requiring minimal sequencing or error detection support. Such traffic involves only a single datagram in Message Mode. Examples of AAL-5 data include signaling and control data, and network management data. AAL-5 also is known as SEAL (Simple and Efficient AAL Layer). AAL-5 traffic originates in the form of a native data payload unit which is known as an IDU (Interface Data Unit). The IDU is of variable length, up to 65,536 octets. At the Convergence Layer the IDU is appended with a trailer including the UU, CPI, Length and CRC fields. The UU (User-to-User) field of one octet contains data to be transferred transparently between users. The CPI (Common Part Indicator) field of one octet aligns the trailer in the total bit stream. The Length field of two octets indicates the length of the total IDU payload. The CRC (Cyclic Redundancy Check) of four octets is used for purposes of error detection and correction in the trailer, only. When the payload user data plus the trailer data hit the ATM Adaptation Layer, the entire set of data is segmented into 48-octet payloads, with each being prepended with a 5-octet header to form a 53-octet ATM cell. See ATM.

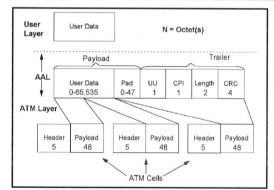

AAL Type 5 operation. User data is appended with trailer at ATM Adaptation Layer and segmented into 48-octet cells at ATM Layer.

AAL Connection An ATM term. Association established by the AAL between two or more next higher layer entities.

AALn See AAL-1, AAL-2, AAL-3/4, AAL-5.

AAN Associated Account Number. Used in wireless service.

AAR Automatic Alternate Routing.

AARP 1. AARP used to stand for American Association of Retired Persons, but it's now just AARP, since you don't have to be retired to join. You just have to be old – 50 years old, to be exact. Don't laugh. You'll be old some day, and just before your 50th birthday AARP will send you a membership invitation. AARP is a remarkably successful marketing organization. They have great discounts.

2. AppleTalk Address Resolution Protocol is a method of mapping addresses between the addresses of computers on an AppleTalk LAN and those known to Ethernet, Token Ring or some other LAN. AARP actually will map addresses at any level of protocol stacks to resolve issues of AppleTalk node address availability used by the AppleTalk DDP (Datagram Delivery Protocol), higher-layer AppleTalk protocols, and underlying data links providing

AppleTalk connectivity. AppleTalk addresses are temporary addresses that are assigned dynamically from a pool, or cache. When a host Apple Macintosh computer boots (i.e., starts up), it chooses an address and checks with the address cache to see if that address is available. If it is not, the process repeats until an available address is discovered. AARP manages this process. See also AppleTalk and DDP.

AARP Probe Packets Packets transmitted by AARP that determine if a randomly selected node ID is being used by another node in a nonextended AppleTalk network. If the node ID is not being used, the sending node uses that node ID. If the node ID is being used, the sending node chooses a different ID and sends more AARP probe packets.

AAV Alternative Access Provider. Another name for a CAP. See CAP.

ABAM A Lucent Technologies (see Western Electric) ordering code for a now-outmoded physical copper cable used in DSX (e.g., T-1 and T-2) implementations. ABAM cable is two-pair, foam insulated, twisted-pair cable of 22/24 gauge, with 100 ohms impedance at .772 MHz. ABAM is just an ordering code, which really doesn't stand for anything. ABAM has been replaced by Category/Level 2, or higher, cable.

abandon on ugly A video call center term coined by Andrew Waite. The term refers to a caller seeing the not-attractive answering agent, and hanging up instantly. Often the caller redials in the hope of reaching someone more attractive.

abandoned attempt An attempt to make a call that is aborted by the person making the call.

abandoned cable You move out of a building. You take your phone system, but you leave all the cable that connected your phone system. You've abandoned your cabling. Now a new tenant moves in. And he must decide should he use the abandoned cabling (and save some money) or should he rip the old cabling out and go for brand new cabling? Most companies opt for the new stuff, since they want "reliability." But it all depends on the individual circumstances.

abandoned call The non-technical explanation of an abandoned call is: A call that is answered, but disconnected before any conversation happens. The technical explanation is: A call which has been offered unto a communications network or telephone system, but which is terminated by the person originating the call before it is answered by the person being called. Follow this sequence for an explanation: You call an airline. You hear ringing. Their phone rings. A machine, called an Automatic Call Distributor (ACD), answers the call, plays you some dumb message like, "Please don't hang up. A real human will answer eventually. Dial 1 to order a pizza, Dial 2 for anchovies,...." You, the caller, are put on Eternity Hold. You get bored waiting and hang up before a live operator answers. You have just abandoned your phone call. Hence, the term Abandoned Call. Information about abandoned calls is highly useful for planning the number of people (also called operators, agents or telephone attendants) an owner of an automatic call distributor or other phone system should employ on what days, during what times of the day, and at what specific locations. Thus the company can organize its resources (i.e., schedule its people) to ensure that the "right" percentage of incoming calls are answered within the "right" amount of time to provide the caller (i.e., you, the customer) the service you deserve (or the service they think you deserve, or, the service they think you should deserve, or the level of service they can afford to provide to "optimize" the cost/benefit relationship).

abandoned call cost The amount of revenue lost because of abandoned calls. This is calculated based on the number of calls, your estimate of the percentage abandoning, and your estimate of the revenue per call. It's an impossible number to calculate since many callers do, in fact, call back and place their orders on another later call.

abandoned property The term that some issuers of travelers checks and telephone debit cards give to the 15% or so of their cards or checks which, for some reason, are never cashed.

abandonment rate The percentage of callers who hang up after being entered into a queue.

abandonware Any software that's no longer sold or supported by the maker.

abbreviated address calling A calling method that allows the user to employ a logical address (e.g., telephone number) involving fewer characters. The destination's assigned device addresses these characters when initiating a connection. May also be called Abbreviated Dialing when specifically used in connection with telephone systems.

abbreviated dialing A feature that permits the calling party to dial the destination telephone number with fewer than the normal digits. Abbreviated Dialing numbers must be set up before using them. Speed Dialing is a typical example of Abbreviated Dialing. See Speed Dialing.

ABC 1. Automatic Bill Calling – a method of billing for payphone calls. Changed in 1982 to Calling Card service.

2. Automated Business Connection.

3. Alternate Billed Call. A call originating on one line and charged to another line (e.g., collect call, calling card call, 3rd-number-billed call).

ABCD signaling bits These are bits robbed from bytes in each DS-0 or T-1 channel in particular subframes and used to carry in band all status information such as E&M signaling states.

ABD See Average Business Day.

abdicate To give up all hope of ever having a flat stomach.

ABDN Attendant Blocking of Directory Number.

ABEC The Alternate Billing Entity Code, which is used by long distance companies (IntereXchange Carriers – IXCs) to bill third parties' (other CICs – Carrier Identification Codes) traffic through the primary IXC's CIC code. This is used by IXCs for sending EMI billing records to LECs for bill page presentation.

ABEND ABnormal END, or ABortive END, or system "crash," and almost always very bad news. When an operating system detects a serious problem, such as a hardware or software failure, the system issues an abend (abnormal end) message. An abend recognized by Novell's NetWare, for instance, would stop the file server...and you're dead in the water. Usually caused by input or data presented to a computer which is beyond its ability to cope. If an abend happens in a single-task program (like MS-DOS), the machine will cease to take input ("lock up") and must be restarted ("re-booted"). Abends can be caused by a variety of factors, such as poorly functioning NetWare Loadable Modules (NLMs), power problems and heavy network traffic. They are the bane of a NetWare operating manager's existence. Multitasking operating systems (like UNIX) allow other programs to continue running while only stopping the one causing trouble. See VREPAIR which explains one way to repair your ABEND problem.

ABF See Air Blown Fiber.

ablation Optical memory data writing technique in which a laser burns holes (or pits) into thin metal film.

ABM 1. Asynchronous Balanced Mode. A service of the data link level (Logical Link Control) in IBM's token-passing ring. ABM operates at the Systems Network Architecture data link control level and allows devices to send data link commands at any time.

2. ABM: Anything (or Anyone) But Microsoft, i.e. Linux, Unix, etc.

ABN ABNormal alarm status.

abnormal release A condition that occurs when your cellular telephone connection is interrupted unexpectedly. An abnormal release can occur due to a switching error, a failed handoff, or any number of other reasons. The cellular providers came up with the term "abnormal release." It's a good thing that they didn't call it "unusual release," because it's not all that unusual.

ABOC Ameritech Bell Operating Company. Beth Sullivan, a reader, emails me one morning. She's desperate. She has a job to do...and there's one stupid acronym impeding her productivity – ABOC. She's convinced it's something to do with some complex technical thing she should know and will look stupid if she doesn't know. I check all my sources. No luck. She emails me the following morning. Mystery solved: Ameritech Bell Operating Company. The irony of all this? It's not even the company's name. The company's real name is SBC Ameritech Corporation.

abort To stop doing something. Often to get out of a software program. Also, to discontinue sending or receiving a message.

abort delimiter A local area network term. A signal sent by a Token Ring station indicating that the message it was sending was terminated part way through the transmission, and is known to be incomplete. The station will then increment the soft error counter for Abort Delimiter Transmitted, and send a soft error report within two seconds.

abort sequence A series of 12 to 18 1-bits appearing at the end of an AppleTalk LLAP frame. The sequence delineates the end of the frame.

above 890 decision The 1959 FCC decision which allowed companies to build their own private microwave communications systems. The decision resulted from AT&T Long Lines' reluctance to provide companies with long distance service to remote places, such as oil wells, gas pipelines, power stations and paper plants. The decision got its name because the FCC allowed privately-owned microwave systems using radio frequencies "above 890" megahertz – which are naturally called "microwave." See also Microwave and ENTELEC.

above the fold This term has several related definitions. In one context, it refers to the part of a web page that can be viewed without scrolling in a fully maximized browser window. In online advertising, "above the fold" is where an advertiser wants to see its ad placed, so as to maximize the ad's chances of being seen and clicked on. In the search engine industry, "above the fold" refers to the first few entries on a search results page. Those listings tend to get the most clicks, while results that follow those first few results get far

fewer clicks. An owner of a commercial website typically wants the listing for its website to appear "above the fold" in search results. The term "above the fold" originated in the newspaper industry. It refers to the content above the fold on the front page of a newspaper, which people see when they approach or pass by the newsstand. Newspaper editors make a point of placing the most attention-grabbing content above the fold, so as to encourage people to purchase the paper. "Above the fold" content on a newspaper is also free content, in the sense that it is what a person can see and read for free without buying the newspaper. In this context, "above the fold" signifies "free" as opposed to "for a fee."

above the line Expenses incurred by telephone company that are charged to the ratepayer by being allowed in the company's rate-base.

ABR 1. Available Bit Rate. As defined by the ATM Forum, ABR is an ATM layer service category for which the limiting ATM layer transfer characteristics provided by the network may change subsequent to connection establishment. A flow control mechanism is specified which supports several types of feedback to control the source rate in response to changing ATM layer transfer characteristics. It is expected that an end-system that adapts its traffic in accordance with the feedback will experience a low cell loss ratio and obtain a fair share of the available bandwidth according to a network specific allocation policy. Cell delay variation is not controlled in this service, although admitted cells are not delayed unnecessarily. In short, ABR provides for transport of traffic at the bit rate available at the time, and on a dynamic basis.

2. area border router. Router located on the border of one or more OSPF areas that connects those areas to the backbone network. ABRs are considered members of both the OSPF backbone and the attached areas. They therefore maintain routing tables describing both the backbone topology and the topology of the other areas.

ABRD AutoBaud Rate Detect. A process by which a receiving data device determines the speed, code level, and stop bits of incoming data by examining the first character – usually a preselected sign-on character (often a carriage return). ABR allows the receiving device to accept data from a variety of transmitting devices operating at different speeds without needing to establish data rates in advance.

abrasion resistance Ability of a wire, cable or material to resist surface wear.

abrupt close Close of a connection on a network without any attempt to prevent any loss of data.

ABS 1. See Average Busy Season.

2. alternate billing services. These are IN (Intelligent Network) services that allow subscribers to charge a call to a number or telephone other than the one they are using. For example, by using a charge card, credit card or personal identification number.

3. application bridge server. Software module that allows the ICM to share the application bridge interface from an Aspect ACD with other applications.

absent subscriber service A service offered by local telephone companies to subscribers who will be away. A live operator or a machine intercepts the calls and delivers a message. When you come back, you get your old number. But in the meantime, while you're away, you pay less money per month than you would for normal phone service. Also known as Vacation Service.

abscissa A horizontal coordinate.

absolute delay The time interval or phase difference between transmission and reception of a signal.

absolute gain In simple language, absolute gain measures how much a device improves the power of a signal. The absolute gain of an antenna, for a given direction and polarization, is the ratio of (a) the power that would be required at the input of an ideal isotropic radiator to (b) the power actually supplied to the given antenna, to produce the same radiation intensity in the far-field region. If no direction is given, the absolute gain of an antenna corresponds to the direction of maximum effective radiated power. Absolute gain is usually expressed in dB. See Isotropic Gain.

The absolute gain of a device is the ratio of (a) the signal level at the output of the device to (b) that of its input under a specified set of operating conditions. Examples of absolute gain are no-load gain, full-load gain, and small-signal gain.

absolute URL A URL that contains a scheme (for example, http) and a server address, for example www.harrynewton.com.

absolute zero A temperature about 460 degrees below zero in Fahrenheit.

absorption Attenuation (reduction in strength of a signal) caused by dissipation of energy. In the transmission of electrical, electromagnetic, or acoustic signals, absorption is the conversion of the transmitted energy into another form, usually thermal. Absorption is one cause of signal attenuation. The conversion takes place as a result of interaction between the incident energy and the material medium, at the molecular or atomic level.

absorption band A spectral region in which the absorption coefficient reaches a relative maximum, by virtue of the physical properties of the matter in which the absorption process takes place.

absorption coefficient A measure of the attenuation caused by absorption of energy that results from its passage through a medium. Absorption coefficients are usually expressed in units of reciprocal distance. The sum of the absorption coefficient and the scattering coefficient is the attenuation coefficient.

absorption index A measure of the attenuation caused by absorption of energy per unit of distance that occurs in an electromagnetic wave of given wavelength propagating in a material medium of given refractive index.

absorption loss That part of the transmission loss caused by the dissipation or conversion of electrical, electromagnetic, or acoustic energy into other forms of energy as a result of its interaction with a material medium.

absorption modulation Amplitude modulation of the output of a radio transmitter by means of a variable-impedance circuit that is caused to absorb carrier power in accordance with the modulating wave.

abstract service A mechanism to group a set of related unbound applications where some aggregator has taken the responsibility to ensure that the set of related applications work together. This is a generalization of a broadcast service to support applications not related to any broadcast TV service. A set of resident applications which an MSO has packaged together (e.g., chat, e-mail, WWW browser) could comprise one abstract service.

abstract syntax In open systems architecture, the specification of application-layer data or application-protocol control information by using notation rules that are independent of the encoding technique used to represent the information.

Abstract Syntax Notation One (ASN.1). LAN "grammar," with rules and symbols, that is used to describe and define protocols and programming languages. ASN.1 is the OSI standard language to describe data types. More formally, ASN.1 is a standard, flexible method that (a) describes data structures for representing, encoding, transmitting, and decoding data, (b) provides a set of formal rules for describing the structure of objects independent of machine-specific encoding techniques, (c) is a formal network-management Transmission Control Protocol/Internet Protocol (TCP/IP) language that uses human-readable notation and a compact, encoded representation of the same information used in communications protocols, and (d) is a precise, formal notation that removes ambiguities.

ABT Advanced Broadcast Television.

AC 1. Access Customer.

2. See AuC, Authentication Center.

3. alternating current. Typically refers to the 120 volt electricity delivered by your local power utility to the three-pin power outlet in your wall. Called "alternating current" because the polarity of the current alternates between plus and minus, 60 times a second. In an AC power system, current (AMPS) is delivered to a load through a wire called the "hot" wire and returns through a wire called the "neutral" wire. The other form of electricity is DC, or direct current, in which the polarity of the current stays constant. Direct current, for example, is what comes from batteries. Outside North America, electricity typically alternates at 50 times a second – which is neither better nor worse, just different. In North America, standard 120 volt AC may be also be referred to as 110 volts, 115 volts, 117 volts or 125 volts. Con Edison, the electricity supplier to New York City, told me that they are only obliged to deliver voltage to 120 volts plus or minus 10%. This means your outlet may deliver anywhere from 118 volts to 132 volts before your power company will get concerned. But you probably should. Telephone and computer are sensitive to voltage fluctuations – some more to high voltages; some more to low voltages. My suggestion: If your stuff is valuable, protect it with a voltage regulator. Or better, power it with stable DC current, which you've converted from fluctuating AC power. Your AC electrical circuit consists of two supply conductors – hot and neutral. There is also a "load." That is the term for the device you're running. The hot, energized or live conductor is ungrounded and delivers energy to the load. The hot conductor is connected to the fuse or circuit breaker at the main service entrance. The neutral or common conductor is grounded and completes the circuit from the load back to the utility transformer. The load is any electric or electronic appliance or gadget plugged into the AC electrical outlet. It completes the circuit from the transformer through the hot conductor, to the load, through the neutral conductor and back to the utility transformer. Standard 120 volt circuits also include an equipment ground conductor. This equipment grounding conductor provides an intended path for fault current and is never intended to be a part of the load circuit. The equipment ground serves three very important purposes:

1. It maintains metal appliance cases at zero volts, thus protecting people who touch the cases from receiving an electrical shock.

2. It provides an intentional fault path of low impedance path for current flow when the hot conductor contacts equipment cases (ground fault). This current causes the fuse or circuit breaker to open the circuit to protect people from electric shock.

3. Any electronic equipment (not electrical) uses the equipment ground as a zero volt reference for logic circuits to provide proper equipment performance.

This is a true story: Thomas Edison helped develop the electric chair in order to "prove" the "deadly dangers" of alternating current (AC) electrical systems. Edison was in direct competition with Westinghouse's brilliant Nikola Tesla, whom he detested, and whose efficient AC system was rapidly becoming the preferred method for transmission of electricity over long distances. This threatened Edison's direct current (DC) system. Realizing he was losing the war, Edison began holding demonstrations in which he electrocuted large numbers of dogs and cats by luring them onto a metal plate wired to a 1,000-volt AC generator. This display, although it attracted people, did not work to sway the public to his side and the AC system became the electricity standard. The legacy of all of Edison's efforts directly led to the development of the electric chair, which uses AC electricity.

See AC Power, Grounding, Battery and Surge Arrestor.

AC Field The Access Control field of a token. See Token.

AC Power Phone systems typically run on AC, Alternating Current. Except for very small systems, phone systems typically need their own dedicated (i.e., shared with no other device) AC power line. This line should be cleaned" with a power conditioner, or voltage regulator. It also should be protected with a surge arrestor. If possible, the phone system should also be protected by a battery-based UPS (Uninterruptible Power Supply). The most reliable battery backup is lead acid, the same technology as used in your car. Phone systems consume more power as they process more phone calls. For example, a PBX brochure says that at minimum capacity it needs less power than eight 100-watt light bulbs, but that at its maximum duplex capacity it needs the same power as 26 100-watt light bulbs – or 2600 watts. Telephone sets typically draw their power from the central telephone switch, be it a PBX or a central office. Few phones require their own AC power. ISDN devices and phones typically require a local AC power source. See AC, Ground and Grounding.

AC To DC Converter An electronic device which converts alternating current (AC) to direct current (DC). Most phone systems, computers and consumer electronic devices (from answering machines to TVs) run on DC. Most phone systems have an AC to DC converter in them. Hint: it's probably buried in the power supply.

AC-DC Ringing A type of telephone signaling that uses both AC and DC components – alternating current to operate a ringer and direct current to aid the relay action that stops the ringing when the called telephone is answered.

AC-3 The coding system used by Dolby Digital. A standard for high quality digital audio that is used for the sound portion of video stored in digital format.

AC15 AC15 is a British protocol for signaling on private wires between analog switches.

ACA 1. Automatic Circuit Assurance.

2. Australian Communications Authority. The regulatory authority for telecommunications and radiocommunications, established under the Australian Communications Authority Act 1997. See ACIF and ACCC.

3. Account Customer Address.

Academic Computing Research Facility Network ACRFNET. A network connecting various research units such as colleges and research and development laboratories in the U.S.

ACADEMMET A network within Russia which connects universities.

ACARS ARINC Communication Addressing Reporting System, airline system developed at VHF and also implemented in aeronautical satellite communications systems. See ARINC.

ACAT Additional Cooperative Acceptance Testing. A method of testing switched access service that provides a telephone company technician at the central office and a carrier's technician at its location, with suitable test equipment to perform the required tests. ACAT may, for example, consist of the following tests:

• Impulse Noise
• Phase Jitter
• Signal-to-C-Notched Noise Ratio
• Intermodulation (Nonlinear) Distortion
• Frequency Shift (Offset)
• Envelope Delay Distortion
• Dial Pulse Percent Break
• **ACB** 1. Annoyance Call Bureau.

2. Automatic Call Back.

ACC Analog Control Channel. A wireless term. The ACC is the analog signaling and control channel used in some cellular systems in support of call setup and certain features. Defined in IS-54B, the ACC is a radio frequency channel distinct from those used to support conversations in some systems using TDMA (Time Division Multiple Access). A DCCH (Digital Control CHannel) is preferred, for obvious reasons of error performance; the DCCH is specified in IS-136, the successor to the IS-54 series. See IS-54B and IS-136.

ACCC The Australian Competition and Consumer Commission is an independent statutory authority which administers Trade Practices Act 1974 and the Prices Surveillance Act 1983. It claims to be the only national government agency in Australia responsible for enforcing the competition aspects of the Trade Practices Act.

accelerated aging A series of tests performed on material or cable meant to duplicate long term environmental conditions in a short period of time. Such tests might include exposure to extreme cold, heat, variations in humidity (including "wet"), and mechanical stress. Also refers to the impact of the stress associated with revising and maintaining Newton's Telecom Dictionary, according to Ray Horak, Consulting Editor.

accelerated depreciation A method which allows greater depreciation charges in the early years of an asset's life and progressively smaller ones later on. The total amount of depreciation charged is still equal to 100% of the asset's value. By taking the charges early on in the asset's life, you get the time value of money, i.e. depreciation charged today (and tax saved today) is worth more than the same amount of depreciation charged (or tax saved) tomorrow.

Accelerated Graphics Port See AGP.

Accelerated Aging A test in which voltage, temperature, etc. are increased above normal operating values to obtain observable deterioration in a relatively short period of time. cables, generally expressed in dB per unit length. See dB.

accelerator 1. A chemical additive which hastens a chemical reaction under specific conditions. A term used in the telecommunications cable manufacturing industry.

2. In a Windows program, an accelerator is a keystroke that dispatches a message to a program, invoking one of its functions. For example, Alt-F4 tells the current Windows application of Windows itself to quit.

accelerator board A board added onto a personal computer's main board and designed to increase the PC's performance in writing to screen or disk, etc. See also Accelerator Card.

accelerator card An add-on product that upgrades the CPU of a PC to a higher speed or more powerful generation of processor. An accelerator card is usually a "daughter board" that clips onto the original CPU or is inserted into the socket that held the original CPU. See also Accelerator Board and Web Accelerator.

Accenture Accenture is the new name for Anderson Consulting. I have no idea why they changed their name. According to Accenture, the firm is reinventing itself to become the market maker, architect and builder of the new economy, bringing innovations to improve the way the world works and lives. I hope they do.

acceptable angle The biggest possible angle between a ray and the center axis. The maximum angle that a fiber optic cable accepts light for further transmission.

Acceptable Use Policy AUP. Many transit networks have policies which restrict the nature of their use or the basis on which access privileges are granted. A well known example is that of the now defunct NSFNET's (National Science Foundation NETwork) AUP which traditionally did not allow commercial use. Subsequently, of course, NSFNET did become commercialized as the Internet. AUP enforcement varies with the specific network.

acceptance 1. A formal set of criteria which you and your vendor define and agree upon that will determine whether the system he shipped you and/or installed at your place is now as you both agreed upon, and therefore, you are obligated to pay for it. See Acceptance Test.

2. Acceptance also refers to the amount of time within which a buyer has to decide whether a software or hardware element is acceptable. Different from a warranty on performance, acceptance applies to the appearance and performance of the element as initially configured and installed, and as compared to the specifications to which the seller and purchaser have agreed. Acceptance periods usually range from two to four weeks, but are determined on a case-by-case basis. It's commonly used in the secondary telecom equipment marketplace, as well. See Acceptance Test.

acceptance angle The angle over which the core of an optical fiber accepts incoming light. It's usually measured from the fiber axis.

acceptance cone In fiber optics, the cone within which optical power may be coupled into the bound modes of an optical fiber. The acceptance cone is derived by rotating the acceptance angle about the fiber axis.

acceptance pattern Of an antenna, for a given plane, a distribution plot of the off-axis power relative to the on-axis power as a function of angle or position. The acceptance pattern is the equivalent of a horizontal or vertical antenna pattern. The acceptance pattern of an optical fiber or fiber bundle is the curve of total transmitted power plotted against the launch angle.

acceptance test The final test of a new telephone system. If the system passes the test – i.e. it meets all specifications laid down in the sales contract – and is working well, then, and only then, will the customer finish paying for it. See also Acceptance and Acceptance Testing.

acceptance testing Operating and testing of a communication system, subsystem, or component, to ensure that the specified performance characteristics have been met. See Acceptance and Acceptance Test.

acceptance trial A military term. A trial carried out by nominated representatives of the eventual military users of the weapon or equipment to determine if the specified performance and characteristics have been met.

acceptor The Bluetooth device receiving an action from another Bluetooth device. The device sending the action is called the initiator. The acceptor is typically part of an established link. See Bluetooth.

access 1. A series of digits or characters which must be dialed, typed or entered in some way to get use of something. That "something" might be a PBX or KTS telephone system, a long distance carrier, an electronic mail service, a private corporate network, a mainframe computer, or a local area network. Once the user dials the main number for the service he must then enter his assigned Access Code to get permission to use the system. An Access Code becomes an Authorization Code when it is used for identifying the caller; it becomes an Account Code when used for purposes of identifying the client for billback of associated charges. Access Code may also mean the digit, or digits, a user must dial to be connected to an outgoing trunk. For example, the user picks up his phone and dials "9" for a local line, dials "8" for long distance, dials "76" for the tie line to Chicago, etc. In programming a phone system there are unique Access Codes for Startup, Configuration programming, Administration programming and other functions; technicians who have been certified at various levels are provided with the appropriate Access Codes which allow them to invoke the appropriate level of system privilege.

2. In respect to privacy, an individual's ability to view, modify, and contest the accuracy and completeness of personally identifiable information collected about him or her. Access is an element of the Fair Information Practices Act.

access attempt The process by which one or more users interact with a telecommunications system to start to transfer information. An access attempt begins with an issuance of an access request by an access originator. An access attempt ends either in successful access or in access failure.

access broadband over power line See BPL.

access bus Access Bus is correctly spelled Access.bus. It is a 100 Kbps bus currently being implemented as part of the Video Electronics Standards Association's Display Data Channel for controlling PC monitors. Access.bus allows bidirectional communication between compatible systems and displays, allowing on-the-go installation. Also supports daisychaining to reduce cable snarl. Access.bus has four pins per connector. See also USB and Firewire.

access channel capacity Cable television channel capacity dedicated to cablecasting by entities not affiliated with the cable system operator, and over which the cable system operator does not exercise editorial control. Categories of access channel capacity are:
- Public Access, dedicated for use by the general public.
- Educational Access, dedicated to local educational authorities.
- Government Access, dedicated to local government.
- Leased Access, dedicated to commercial users.

The significant distinction here is "editorial control": subject to certain narrowly-defined exceptions (obscenity, "unlawful conduct"), the cable operator is prohibited from exercising editorial control over programming carried on access channels. This distinction affects the records which must be kept in the system's PIF:
- The operator is exempt from the requirements of Section 76.225c of the FCC rules regarding limits on commercial matter in children's programming carried on access channels.
- The operator is required by Section 76.701(h) to maintain records to verify compliance with rules governing leased-access channels carrying indecent programming.

access channels Channels set aside by the cable operator for use by the public, educational institutions, municipal government, or for lease on a non-discriminatory basis.

access charge As part and parcel of the Modified Final Judgement (MFJ) which mandated the breakup of the Bell System though AT&T's divestiture of the Bell Operating Companies, the FCC declared that all end users have easy access to the long distance carrier of their choice. Further, such access to the InterExchange Carrier (IEC, or IXC) was to be provided with equal ease, i.e. on an equal 1+ basis, as traditionally enjoyed by AT&T alone. Additionally, the breakup of the Bell System invalidated the complex settlements structure which served to subsidize the Local Exchange Carriers (LECs) for providing local access to the long distance network(s). In order to replace the traditional settlements process and to compensate the LECs for the use of the vital, expensive and relatively unprofitable local access network, Access Charges were mandated, falling into two general categories.

1. The end user Access Charge, also known as Customer Access Line Charge (CALC) applies to every local loop, and is sensitive to the nature of the circuit. Subject to review and adjustment by the FCC on an annual basis, Access Charges differ for residential and business users, single lines versus trunks, leased lines versus local access circuits, and so on. A mandatory charge appearing as a separate line item on the user's bill, it applies whether or not the user ever, in fact, places a long distance call over either a wired or wireless network. This surcharge is levied per the Code of Federal Regulations, Title 47, Part 69.

2. The Carrier Access Charge (CAC) applies to all IXCs which connect to the LEC network. Paid by the IXC to the LEC, such charges are determined by special tariffs subject to regulatory approval, and are sensitive to factors including the distance between the IXC Point of Presence (POP) and the point of termination into the LEC network. Additionally, the IXC pays to the LEC a usage charge sensitive to the traffic passed to the IXC as measured by Minutes of Use. The Telecommunications Act of 1996 greatly modifies the basis on which such charges are determined and levied. Most especially, the act eliminates such charges to the extent that the IXCs provide local exchange service directly, effectively bypassing the incumbent LECs.

access code A series of digits or characters which must be dialed, typed or entered in some way to get use of something. That "something" might be the programming of a telephone system, a long distance company, an electronic mail service, a private corporate network, a mainframe computer, a local area network. Once the user dials the main number for the service he must then enter his assigned Access Code to get permission use the system. An Access Code becomes an Account Code when it is used for identifying the caller and doing the billing. Access Code may also mean the digit, or digits, a user must dial to be connected to an outgoing trunk. For example, the user picks up his phone and dials "9" for a local line, dials "8" for long distance, dials "76" for the tie line to Chicago, etc. In programming a phone system such as Nortel Networks' Norstar, there are Access Codes to begin Startup, Configuration programming, and Administration programming.

access contention In ISDN applications, synonymous with "contention." See contention.

access control The ways organizations define annd/or restrict individuals from obtaining data from, or placing data in a storage devicem typically over a network. See Access Code, access control field and list.

access control entry ACE. In Windows-based systems, an entry in an access control list containing the security identifier for a user or group and an access mask that specifies which operations by the user or group are allowed, denied, or audited.

access control field 1. A term specific to Synchronous Multimegabit Data Service (SMDS), the Access Control Field controls access to the shared DQDB (Distributed Queue Dual Bus) which, in turn, provides access to the SMDS network. It consists of a single octet which is a portion of the 5-octet header of an SMDS cell. See also SMDS.

2. A Token Ring term. A field comprising a single octet (eight bits) in the header of a Token Ring LAN frame. Three Priority (P) bits set the priority of the token, a single Token (T) bit denotes either token or a frame, a Monitor (M) bit prevents frames or high-priority tokens from continuously circling the ring, and three Priority Reservation r bits allow a device to reserve the token for network access the next time the token circles the ring. See also Token Ring.

access control list ACL. Most network security systems operate by allowing selective use of services. An Access Control List is the usual means by which access to, and denial of, services is controlled. It is simply a list of the services available, each with a list of the computers and users permitted to use the service.

access control message A message that is a user request, a resource controller response, or a request/response between resource controllers.

access control method Set of rules which determine the basis on which devices are afforded access to a shared physical element, such as a circuit or device. In a Local Area Network environment, it regulates each workstation's physical access to the transmission medium (normally cable), directs traffic around the network and determines the order in which nodes gain access so that each device is afforded an appropriate level of access. By way of example, token passing is the technique used by Token Ring, ARCnet, and FDDI. Ethernet makes use of CSMA/CD or CSMA/CA; DDS makes use of a polling technique. See Media Access Control. (MAC.)

access control system A system designed to provide secure access to services, resources, or data; for computers, telephone switches or LANs.

access controls An electronic messaging term. Controls that enable a system to restrict access to a directory entry or mailbox either inclusively or exclusively.

access coordination An MCI definition. The process of ordering, installing, and maintaining the local access channel for MCI customers.

access coupler A device placed between two fiber optic ends to allow signals to be withdrawn from or entered into one of the fibers.

Access Customer Name Abbreviation See ACNA.

access device The hardware component used in the signaling controller system: access server or mux.

access envy When I surf the web at two million bits per second (download) and you surf it at only 52,000 bits per second, you have a serious case of access envy, namely you envy my high speed.

access event Telcordia definition for information with a logical content that the functional user and the Network Access FE (Functional Entity) exchange.

access floor A system consisting of completely removable and interchangeable floor panels that are supported on adjustable pedestals or stringers (or both) to allow access to the area beneath.

access function An intelligent network term. A set of processes in a network that provide for interaction between the user and a network.

access group All terminals or phones that have identical rights to use the computer, the network, the phone system, etc.

access level Used interchangeably with Access Code. "Level" in dialing tends to mean a number.

access line A telephone line reaching from the telephone company central office to a point usually on your premises. Beyond this point the wire is considered inside wiring. See Local Loop and Access Link.

access link The local access connection between a customer's premises and a carrier's POP (Point Of Presence), which is the carrier's switching central office or closest point of local termination. That carrier might be a LEC, IXC or CAP/AAV; in a convergence scenario, the carrier might also be a CATV provider.

access list List kept by routers to control access to or from the router for a number of services (for example, to prevent packets with a certain IP address from leaving a particular interface on the router).

access manager 1. An element in some architecture implementations of a PCS infrastructure that includes functions such as subscriber registration and authentication. It may include the Home Location Register, HLR, and Visitor Location Register, VLR.

2. A means of authorization security which employs scripting.

access mask In Windows systems, a value that specifies the permissions that are allowed or denied in an access control entry of an access control list. The access mask is also used in an access request to specify the access permissions that the subject requires when accessing an object.

access method The technique or the program code in a computer operating system that provides input/output services. By concentrating the control instruction sequences in a common sub-routine, the programmer's task of producing a program is simplified. The access method typically carries with it an implied data and/or file structure with logically similar devices sharing access methods. The term was coined, along with Data Set, by IBM in the 1964 introduction of the System/360 family. It provides a logical, rather than physical, set of references. Early communications access methods were primitive; recently they have gained enough sophistication to be very useful to programmers. Communications access methods have always required large amounts of main memory. In a medium size system supporting a few dozen terminals of dissimilar types, 80K to 100K bytes of storage is not an unusual requirement. The IEEE's 802.x standards for LANs and MANs. See Access Methods.

access methods Techniques and rules for figuring which of several communications devices – e.g. computers – will be the next to use a shared transmission medium. This term relates especially to Local Area Networks (LANs). Access method is one of the main methods used to distinguish between LAN hardware. How a LAN governs users' physical (electrical or radio) access to the shared medium significantly affects its features and performance. Examples of access methods are token passing (e.g., ARCnet, Token Ring and FDDI) and Carrier Sense Multiple Access with Collision Detection (CSMA/CD) (Ethernet). See Access Method and Media Access Control.

access minutes The term Access Minutes or Access Minutes of Use is used by NECA (the National Exchange Carrier Association) and the FCC in measuring traffic between LATA service providers (CLECs or ILECs) and IXCs (IntereXchange Carriers). The formal definition is "Access Minutes or Access Minutes of Use is that usage of exchange facilities in interstate or foreign service for the purpose of calculating chargeable usage. On the originating end of an interstate or foreign call, usage is to be measured from the time the originating end user's call is delivered by the telephone company and acknowledged as received by the interexchange carrier's facilities connected with the originating exchange. On the terminating end of an interstate or foreign call, usage is to be measured from the time the call is received by the end user in the terminating exchange. Timing of usage at both the originating and terminating end of an interstate or foreign call shall terminate when the calling or called party disconnects, whichever event is recognized first in the originating or terminating end exchanges, as applicable." This comes from the FCC's 69.2 Definitions.

access network 1. Several wholesale carriers define access network as the fiber connection and associated electronic equipment that link a core network to Points of Presence (POPs) and on to Points of Interconnect (POIs) switch locations.

2. The part of the carrier network that touches the customer's premises. The access network is also referred to as the local drop, local loop, or last mile.

access node (AN) Access nodes are points on the edge of a network which provide a means for individual subscriber access to a network. At the access node, individual subscriber traffic is concentrated onto a smaller number of feeder trunks for delivery to the core of the network. Additionally, the access nodes may perform various forms of protocol conversion or adaptation (e.g. X.25, Frame Relay, and ATM). Access nodes include ATM Edge Switches, Digital Loop Carrier (DLC) systems concentrating individual voice lines to T-1 trunks, cellular antenna sites, PBXs, and Optical Network Units (ONUs).

access number The telephone number you use to dial into your local Internet Service Provider (ISP). To connect to the Internet you must first establish an account with an ISP in your area. Usually you will receive a list of telephone numbers you can use to "dial-in" to the service.

access organization An entity which originates program material for transmission over the access channel capacity of a cable television system. An access organization may be an individual, a non-profit corporation, an unincorporated non-profit association, or a for-profit corporation. However, under most cable franchises, commercial advertising is prohibited on Public, Educational, and Government Access channels.

access phase In an information-transfer transaction, the phase during which an access attempt is made. The access phase is the first phase of an information-transfer transaction.

access point AP. Basically a device that connects a computer to a network. It includes:
- A wired device that sends out wireless Ethernet signals to which you and I can connect to.
- A point where connections may be made for testing or using particular communications circuits.
- A junction point in outside plant consisting of a semipermanent splice at a junction between a branch feeder cable and distribution cables.
- A cross-box where telephone cables are cross connected.
- Network device that connects a wireless radio network to a wired LAN (local area network).

access protection Refers to the process of protecting a local loop from network outages and failures. Access protection can take many forms, such as purchasing two geographically diverse local facilities, adding protection switches to the ends of geographically diverse local loops, or buying service from a local access provider which offers a survivable ring-based architecture to automatically route around network failures.

access protocols The set of procedures which enable a user to obtain services from a network.

access provider A company, such as a telephone company, that hooks your computer up to the Internet.

access rate 1.	The maximum data rate of the access channel, typically referring to access to broadband networks and network services.

2. AR. A Frame Relay term which addresses the maximum transmission rate supported by the access link into the network, and the port speed of the device (switch or router) at the edge of the carrier network. The Access Rate defines the maximum rate for data transmission or receipt. See also Committed Information Rate.

access request A message issued by an access originator to initiate an access attempt.

access response channel Access Response CHannel. ARCH. Specified in IS-136, ARCH carries wireless system responses from the cell site to the user terminal equipment. ARCH is a logical subchannel of SPACH (SMS (Short Message Service) point-to-point messaging, Paging, and Access response CHannel), which is a logical channel of the DCCH (Digital Control CHannel), a signaling and control channel which is employed in cellular

systems based on TDMA (Time Division Multiple Access). The DCCH operates on a set of frequencies separate from those used to support cellular conversations. See also DCCH, IS-136, SPACH and TDMA.

access router An access device with built-in basic routing-protocol support, specifically designed to allow remote LAN access to corporate backbone networks. An access router is not designed to replace backbone routers or to build backbone networks.

access server An access server is a basically a special purpose computer that provides, allows and disallows access to something – typically a network or the Internet. Picture a hotel. It has decided to offer high speed Internet to its guests. It intends to charge its guests $10 a day for the Internet access. It runs lines from each room to the basement where they come into a simple hub, router or switch and eventually on a device called an access server, which also plugs into the Internet. That device will perform a number of functions:

1. If a guest plugs their laptop into their RJ-45 outlet and opens an Internet browser, the access server typically will respond by sending the guest a "Welcome" page, saying the guest needs to pay money for broadband Internet access. That money might be charged to the guest's room or charged to the guest's credit card. And the access server may send a message to a credit card company requesting authorization for the charge.

2. The access server may send the billing information to the hotel's billing computer for including on the guest's hotel bill.

3. It may also time the guest, informing him when his payment for service has or is about to expire.

4. It might also house some form of hotel Intranet site, giving the guest information on activities in the hotel (meetings), amenities, movies playing, restaurant hours, etc.

5. It might also house a firewall, preventing outsiders from mounting attacks from outside.

Today's access servers are often rack-mountable, 1U-size and housed in a rack with other communications and computing equipment.

In the old days an access server was a simpler device. It was a communications processor that connected asynchronous devices to a LAN (local area network) or WAN (wide area network) through network and terminal emulation software. It performed both synchronous and asynchronous routing of supported protocols. An access server is sometimes called a network access server. See also 1U.

access service Switched or Special Access to the network of an IXC for the purpose of originating or terminating communications.

Access Service Ordering Guidelines ASOG. Industry guidelines for issuing Access Service Requests (ASRs) as sponsored by the Ordering and Billing Forum (OBF) and the Alliance for Telecommunications Industry Solutions (ATIS). These guidelines outline the forms, data elements and the business rules necessary to create an Access Service Request.

Access Service Request ASR. A form used by a CLEC (Competitive Local Exchange Carrier) to request that the ILEC (Incumbent LEC) provide Special Access or Switched Access Services as specified in the various Access Service Tariffs. Some services that can be requested are: Feature Group A, WATS Access Line; Feature Groups B, C, D Forms; special access circuits; multipoint service legs; additional circuits; testing service; and 800 database access. The ASR has been used for many years between the RBOCs (regional Bell operating companies) to order special circuits that extend into other telcos' serving areas.

access signaling A term which Nortel Networks' Norstar telephones use to indicate their ability to access a remote system (such as a Centrex or a PBX), or dial a number on an alternate carrier by means of Access Signaling (also referred to as "End-to-End" Signaling).

access surcharge State specific usage charges applied to dedicated lines on the originating end of a circuit.

access switch Feeder node to Enterprise Network Switches that perform multiprotocol bridge/routing and support a wide range of serial-link (e.g., SDLC BSC, asynchronous) attached devices. Also known as Gateways, such devices currently are known as Routers and Encapsulating Bridges, although the differences between them are most significant.

access tandem A Local Exchange Carrier switching system that provides a concentration and distribution function for originating and/or terminating traffic between a LEC end office network and IXC POPs. In short, a distinct type of local phone company switching system specifically designed to provide access between the local exchange network and the interexchange networks for long-distance carriers in that area. The Access Tandem provides the interexchange carrier with access to multiple end offices within the LATA. More than one Access Tandem may be needed to provide access to all end offices within any given LATA. Currently, the Access Tandem function may be in the form of a physical and logical partition of a LEC Central Office switch, which also serves end users for purposes of satisfying local calling requirements. Additionally, the IXC may extend the reach of the POP through a high-speed channel extension via dedicated circuits, thereby achieving interconnection with the LEC though collocation of termination facilities in the LEC CO.

In its internal glossary, US West defines Access Tandem as the switching system that provides distribution for originating or terminating traffic between End Offices and the Interexchange Carrier's Point-of-Termination. An Access Tandem is also used to distribute originating or terminating traffic between a CLEC end office and an intraLATA toll point or an Interexchange Carrier's Point of Termination.

access time There are many definitions of access time:

1. In a telecommunications system, the elapsed time between the start of an access attempt and successful access. Note: Access time values are measured only on access attempts that result in successful access.

2. In a computer, the time interval between the instant at which an instruction control unit initiates a call for data and the instant at which delivery of the data is completed.

3. The time interval between the instant at which storage of data is requested and the instant at which storage is started.

4. In magnetic disk devices, the time for the access arm to reach the desired track and for the rotation of the disk to bring the required sector under the read-write mechanism.

5. The amount of time that lapses between a request for information from memory and the delivery of the information, usually stated in nanoseconds (ns). When accessing data from a disk, access time includes only the time the disk heads take to settle down reaching the correct track (seek time) and the time required for the correct sector to move over the head (latency). Disk access times range between 9ms (fast) and 100 ms (slow).

6. A Verizon definition: Access time (usage) is measured from the time that the originating customer's call is delivered by the Telephone Company to, and acknowledged as received by, the receiving customer's equipment when they are connected with the originating exchange. On the terminating end of an interstate or foreign call, access time is measured from the time the call is received by the end user in the terminating exchange. Access ends when the calling or called party disconnects.

access token A data structure that contains authorization information for a user or group. A system uses an access token to control access to securable objects and to control the ability of a user to perform various system-related operations on a local computer.

Access Unit 1. AU. An electronic messaging term, used for implementing value-added services such as fax, Telex, and Physical Delivery via X.400.

2. In the token ring LAN community, an access unit is a wiring concentrator. See Media Access Unit (MAU).

ACCOLC Access Overload Class. A term used in the cellular phone business to allow the cellular system some way of choosing which calls to complete based on some sort of priority. Originally, when the Federal government began designing cellular systems, the government intended to give certain emergency vehicles (such as police, ambulances, and fire departments) codes in their cellular phones that would allow them priority over other subscribers to communicate during emergencies. There is no standard in use within the United States at this time.

account On LANs or multiuser operating systems, an account is given to each user for administrative and security reasons. In online services, an account identifies a subscriber.

account code (voluntary or enforced) A code assigned to a customer, a project, a department, a division – whatever. Typically, a person dialing a long distance phone call must enter that code so the Call Accounting system can calculate and report on the cost of that call at the end of the month or designated time period. Many service companies, such as law offices, engineering firms and advertising agencies use account codes to track costs and bill their clients accordingly. Some account codes are very complicated. They include the client's number and the number of the particular project. The Account Code then includes Client and Matter number. These long codes can tax many call accounting systems, even some very sophisticated ones.

Account Executive AE. A fancy, schmancy name for a salesperson. The idea is that the customer is an "account," and the salesman is the executive running the account. Telephone companies call their salespeople account executives – especially on the equipment and non-long distance side.

account policy On networks and multiuser operating systems, account policy is the set of rules that defines whether a new user is permitted access to the system and whether an existing user is granted additional rights or expanded access to other system resources. Account policy also specifies the minimum length of passwords, the frequency with which passwords must be changed, and whether users can recycle old passwords and use them again.

accountant Someone who figures your numbers, then numbers your figures and then sends you a bill. See also Economist.

accounting management In network management, a set of functions that enables network service use to be measured and the costs for such use to be determined and includes all the resources consumed, the facilities used to collect accounting data, the facilities used to set billing parameters for the services used by customers, maintenance of the data bases used for billing purposes, and the preparation of resource usage and billing reports.

Accounting management is one of five categories of network management defined by ISO for management of OSI networks. Accounting management subsystems are responsible for collecting network data relating to resource usage. See also configuration management, fault management, performance management and security management.

accounting rate system This fading system governs the sharing of revenues received by telephone companies for international calls. Carriers would bilaterally agree on an accounting rate applicable on an international route. The accounting rate would then be split in half, with the so-called settlement rate being passed on to the carrier receiving the call. Controversy and opposition to accounting rates have arisen because they have not been cost-driven. instead, they were used to cross-subsidize cheaper domestic call rates. The realignment of rates that has accompanied deregulation is known as rebalancing. Generally, the United Kingdom, Germany, France, Scandinavia, and Switzerland have rebalanced their tariffs. Spain, Portugal, and Greece still have a way to go. This is how it used to work: If AT&T sends France Telecom one million minutes of calls, but France sends only 500,000 minutes back to AT&T, then AT&T will have to pay France Telecom for the imbalance. If the accounting rate between France and the United States is $1 per minute, then AT&T will pay France Telecom $250,000 for its work in completing the 500,000 extra calls. AT&T pays France Telecom only half the cost because AT&T does half the work itself.

accounting servers A Local Area Network costs money to set up and run. Thus it may make sense to charge for usage on it. In LANs which rely on the Novell NetWare Network Operating System, the network supervisor sets up accounting through a program called SYSCON. When this happens, the current file server automatically begins to charge for services. The supervisor can authorize other network services (print servers, database servers, or gateways) to charge for services, or can revoke a server's right to charge.

accounting traffic matrix A mobile term. A traffic matrix is a collection of information, gathered over a period of time, containing statistics on Mobile End System (M-ES) registration, de-registration, and Network Protocol Data Unit (NPDU) traffic.

ACCS Automatic Calling Card Service.

accumulator 1. A register in which one operand can be stored and subsequently replaced by the result of an arithmetic or logic operation. A term used in computing.

2. A storage register.

3. A storage battery.

accuracy Absence of error. The extent to which a transmission or mathematical computation is error-free. There are obvious ways of measuring accuracy, such as the percentage of accurate information received compared to the total transmitted.

AC/DC Ringing A common way of signaling a telephone. An alternating current (AC) rings the phone bell and a direct current (DC) is used to work a relay to stop ringing when the called person answers.

ACD See the next seven definitions, ACD, Automatic Call Distributor and ACIS.

ACD Agent A telephony end user that is a member of an inbound, outbound, skills based, or programmable Automatic Call Distribution group. ACD Agents are distinguished from other users by their ability to sign on (i.e., login) to phone systems that coordinate and distribute calls to them.

ACD Agent Identifier The identifier of an ACD agent. An agent identifier uniquely identifies an agent within an ACD group. See ACD Agent.

ACD Application Bridge Refers to the link between an ACD and a database of information resident on a user's data system. It allows the ACD to communicate with a data system and gain access to a database of call processing information such as Data Directed Call Routing.

ACD Application-Based Call Routing In addition to the traditional methods of routing and tracking calls by trunk and agent group, the latest ACDs route and track calls by application. An application is a type of call, e.g. sales vs. service. Tracking calls in this manner allows accurately reported calls especially when they are overflowed to different agent groups.

ACD Call Back Messaging This ACD capability allows callers to leave messages for agents rather than wait for a live agent. It helps to balance agent workloads between peak and off-peak hours. In specific applications, it offers callers the option of waiting on hold. A good example is someone who only wishes to receive a catalog. Rather than wait while people place extensive orders, they leave their name and address as a message for later follow-up by an agent. This makes things simpler for them and speeds up service to those wanting to place orders.

ACD Caller Directed Call Routing Sometimes referred to as an auto attendant capability within the industry, this ACD capability allows callers to direct themselves to the appropriate agent group without the intervention of an operator. The caller responds to prompts (Press 1 for sales, Press 2 for service) and is automatically routed to the designated agent group.

ACD Central Office An Automatic Call Distributor (ACD), usually located in a central office and supplied to the customer by the telephone company (telco) with tariffed pricing structures. Some data gathering equipment is often located on customer premises.

ACD Conditional Routing The ability of an ACD to monitor various parameters within the system and call center and to intelligently route calls based on that information. Parameters include volume levels of calls in queue, the number of agents available in designated overflow agent groups, or the length of the longest call. Calls are routed on a conditional basis. "If the number of calls in queue for agent group #1 exceeds 25 and there are at least 4 agents available in agent group #2, then route the call to agent group #2."

ACD Data Directed Call Routing A capability whereby an ACD can automatically process calls based on data provided by a database of information resident in a separate data system. For example, a caller inputs an account number via touch tone phone. The number is sent to a data system holding a database of information on customers. The number is identified, validated and the call is distributed automatically based on the specific account type (VIP vs. regular business subscriber, as an example).

ACD DN A Nortel term for an Automatic Call Distribution Directory Number (ACD DN), which refers to the queue where incoming calls wait until they are answered. Calls are answered in order in which they entered the queue.

ACD Group Multiple agents assigned to process incoming calls that are directed to the same dialed number. The ACD feature of the telephone switch routes the incoming to one of the agents in the ACD group based upon such properties as availability of the agent and length of time since the agent last completed an incoming call.

ACD Intelligent Call Processing The ability of the latest ACDs to intelligently route calls based on information provided by the caller, a database on callers and system parameters within the ACD such as call volumes within agent groups and number of agents available.

ACD Number The telephone number that calling devices dial to access any of the multiple agents in an ACD group. Once the incoming call arrives at the ACD number, the ACD service can then route the call to one of multiple agents in the ACD group.

ACD Skills-Based Routing See Skills-Based Routing.

ACE 1. Above ground Cable Enclosure.

2. See Access Control Entry.

ACELP Algebraic Code Excited Linear Prediction. A variation of CELP. ACELP improves on the efficiency of CELP voice compression by a factor of 2:1, thereby yielding good quality voice at only 8 Kbps, as compared to CELP at 16 Kbps, and PCM at 64 Kbps. ACELP accomplishes this minor miracle through the use of a code book which contains algebraic expressions of each set of voice samples, rather than expressions of each set as a series of numbers. ACELP has been standardized by the ITU-T in G.723.1 (Dual Rate Coder for Multimedia Communications), and in G.729 as CS-ACELP (Conjugate Structure-ACELP). Proprietary versions also exist. The Frame Relay Forum has specified ACELP in FRF.11 IA (Implementation Agreement) as one of the minimum voice compression algorithms required for network-to-network interoperability in VoFR (Voice over Frame Relay) applications. See CELP for the full background on ACELP. See also CS-ACELP, Dual Rate Speech Coder for Multimedia Communications, LD-CELP, and VoFR.

ACeS Asian Cellular Satellite, the first regional satellite communication system dedicated to mobile satcom communications and now serving users in 26 countries of the Asia-Pacific region. www.acesinternational.com.

ACF Advanced Communication Function. A family of software products used by IBM allowing its computers to communicate.

ACF/NCP Advanced Communication Function/ Network Control Program. In host-based IBM SNA networks, ACF/NCP is the control software running on a communications controller that supports the operation of the SNA backbone network.

ACF/VTAM Advanced Communication Function/Virtual Terminal Access Method. In host-based IBM SNA networks the ACF/VTAM is the control software running on a host computer that allows the host to communicate with terminals on an SNA network.

ACFG Short for AutoConFiGuration. The Plug and Play BIOS extensions, now turning up on PCs, are also known as the ACFG BIOS extensions.

ACH 1. Attempts per Circuit per Hour. This is a term you often see in call centers. It refers to the number of times someone tried to reach a circuit in one hour. In a normal phone system, ACH refers to the number of times someone tried to make a call on a circuit in one hour. Measuring ACH is useful for figuring how many inbound or outbound trunks you may need. See also CCH, which is connections per hour.

2. ACH is the automated clearing house created with the blessings of the Fed. Here's how it works: If you walk into the store, and buy something using your "check card" which is maligned by calling it a debit card, your transaction is transmitted to an ACH originating financial depository institution (OFDI) where the money appears in your merchant's account as a nearly instant "credit" (definition 2, below). The ACH transaction travels electronically to a receiving depository financial institution (RDFI) where it is "debited" (removed from) your account. This is the trick that forms the basis for all manner of financial confusionism, because most people don't know that in a check card (debit card) transaction, you get a credit into the OFDI from the RFDI (debited). Because this is pure 'effing magic to non-accountants, this is precisely where the banksters wish to lead the sheeple in order to have more shear fun with them. Taking money out of a real account is a debit and going in debt is a credit.   Whew! Unless of course, you are doing a web-based, non-recurring WEB type transaction which turns a one-way RDFI debit from your account into a named ODFI credit at the merchant's account... Need more? Ure's ACH Rules: 1: Do without rather than use recurring payments.   2: Use cash! There's a reason that everyone from the little Trinity Valley Electric Company up at our ranch place up to the biggest banks in the world want you to sign up for automatic ACH transfers: Less work, and less consumer control over their personal finances. www.nacha.org.

ACID ACID is an acronym for the four characteristics of a database transaction done correctly over a network – private or the Internet. For the transaction to be considered valid, it must be:

- Atomic; The transaction should be done or undone completely. In the event of a failure (network, or system) all operations and procedures should be undone. And all data should be able to be rolled back to its previous state.
- Consistent: A transaction should transform a system from one consistent state to another consistent state.
- Isolation: Each transaction should happen independently of the other transactions occurring at the same time.
- Durable: Completed transactions should remain permanent, even during a network or system failure.

ACIF Australian Communications Industry Forum, established in May, 1997 as a communications industry self-regulatory body. The ACIF is responsible in Australia for developing standards, codes of practice and service applications.

ACIS Automatic Customer/Caller Identification. This is a feature of many sophisticated ACD systems. ACIS allows the capture of incoming network identification digits such as DID or DNIS and interprets them to identify the call type or caller. With greater information, such as provided by ANI, this data can identify a calling subscriber number. You can also capture caller identity by using a voice response device to request inbound callers to identify themselves with a unique code. This could be a phone number, a subscriber number or some other identifying factor. This data can be used to route the call, inform the agent of the call type and even pre-stage the first data screen associated with this call type automatically. See also ANI, Caller ID and Skills-Based Routing.

ACK In data communications, ACK is a character transmitted by the receiver of data to ACKnowledge a signal, information or packet received from the sender. In the de facto standard IBM Binary Synchronous Communications (also known as BSC or Bisync) protocol, an ACK is transmitted to indicate the receipt of a block of data without any detected transmission errors. This positive acknowledgement reassures the transmitting device of that fact in order that the next block of data may be transmitted. Binary code for an ACK is 00110000. Hex is 60. See also Acknowledgment.

ACK Ahead A variation of the XMODEM protocol that speeds up file transmission across error-free links. See XMODEM.

ACK1 Bisync acknowledgment for odd-numbered message.

acknowledgment In data communications, the transmission of acknowledgment (ACK) characters from the receiving device to the sending device indicates the data sent has been received correctly.

ACL 1. Access Control List. A roster of users and groups of users, along with their rights. See Access Control List.

2. applications connectivity link. Siemens' protocol for linking its PBX to an external

computer and having that computer control the movement of calls within a Siemens PBX. See also Open Application Interface.

ACM 1. An ATM term. Address Complete Message. One of the ISUP call set-up messages. A message sent in the backward direction indicating that all the address signals required for routing the call to the called party have been received. See ATM.

2. Association for Computing Machinery. www.acm.org.

3. Automatic Call Manager. The integration of both inbound call distribution and automated outbound call placement from a list of phone contacts to be made from a database. Telemarketing and collections applications are targets for this type of system.

ACMA Australian Communications and Media Authority. The Australian Communications and Media Authority Act of 2005 established ACMA to regulate Australia's telecommunications, radio communications, broadcasting, and online content industries. ACMA replaced two earlier regulatory authorities, the Australian Communications Authority and the Australian Broadcasting Authority. Home page: www.acma.gov.au.

ACNA Access Customer Name Abbreviation. A three-digit alpha code assigned to identify carriers (including both ILECs and CLECs) for billing and other identification purposes. If your company is a CLEC applying to rent space in an ILEC's central office, you must put your ACNA number on your application. If you don't, the ILEC will hold your application up. If you don't have an ACNA number, you need to get one from Telcordia.

ACO 1. Additional Call Offering. An ISDN option that alerts a Terminal Adapter (TA) or ISDN-capable router that a second call is coming in over a given Bearer (B) channel. If the existing call is a data call, ACO provides the TA or router with the option of dropping the data call, in favor of the incoming voice call. In this scenario, it is assumed that the voice call takes priority over the data call.

2. alarm cut off. Feature that allows the manual silencing of the office audible alarm. (Subsequent new alarm conditions might reactivate the audible alarm.)

ACOM Term used in G.165, "General Characteristics of International Telephone Connections and International Telephone Circuits: Echo Cancellers." ACOM is the combined loss achieved by the echo canceller, which is the sum of the echo return loss, echo return loss enhancement, and nonlinear processing loss for the call.

ACONET A research network in Austria.

acoustic connection A connection to a device or system made by sound waves.

acoustic coupler An acoustic modem. A modem designed to transfer data to the telephone network acoustically (i.e by sound), rather than electronically (i.e. by direct wire connection). An acoustic coupler consists of a pair of rubber cups into which the user places a telephone handset. One cup picks up the telephone's sound output with a microphone which converts it to an analog signal for transmission to the computer at the other end. The other cup receives the computer's (or terminal's) output and converts it from digital to analog and then uses a speaker to convert that signal into sound which was picked up by the phone's mouthpiece. The data communications link is achieved through acoustic (sound) signals rather than through direct electrical connection. It is attached to the computer or data terminal through an RS-232-C connector. To work the acoustic coupler, start the computer's communications program, dial the distant computer on a single line telephone with a normal (e.g. old-fashioned) handset. When the distant computer answers with a higher pitched "carrier tone," you place the telephone handset in the acoustic coupler and transmit and receive data. Since the data is transmitted by sound between the handset and the acoustic coupler (and vice versa), the quality isn't always reliable. You can usually transmit up to 300 baud, which by today's standards, is painfully slow. People use acoustic couplers when they're short of time or cannot physically connect their modem electrically, e.g. they're using a payphone without an RJ-11 jack or can't get to a Wi-Fi hot space.

Acoustic Kitty During the 1960s, at the height of the Cold War, American and Soviet spies were trying to be creative to get inside information. Recently declassified documents tell the U.S. planned to use a cat named "Acoustic Kitty" to spy on the Russians. It took five years and more than $14.5 million, but the CIA surgically outfitted the cat with a transmitter, turning it into an eavesdropping platform, using the tail as an antenna. The plan was to have the cat stroll near the Kremlin, curl up on a windowsill or park bench for a nice nap and transmit private conversations by Soviet officials. All was ready, and the CIA agents brought the cat to the park, letting him out of the surveillance van so he could become America's newest spy. Technicians stood by their dials and switches, waiting to capture top secret Soviet conversations. The result? The cat, immediately upon being let out of the van, was run over.

acoustic model In automatic speech recognition, an acoustic model models acoustic behavior of words by gluing together models of smaller units, such as phonemes. (Sorry for the definition of the word model with the word model. But it's actually the best way of defining this term. HN)

acoustic noise An undesired audible disturbance in the audio frequency range.

acoustic suspension A loudspeaker system that uses an air-tight sealed enclosure.

acoustic wave A longitudinal wave that:

1. Consists of a sequence of pressure pulses or elastic displacements of the material, whether gas, liquid, or solid, in which the wave propagates.

2. In gases, consists of a sequence of compressions (dense gas) and rarefactions (less dense gas) that travel through the gas.

3. In liquids, consists of a sequence of combined elastic deformation and compression waves that travel though the liquid.

4. In solids, consists of a sequence of elastic compression and expansion waves that travel though the solid.

The speed of an acoustic wave in a material medium is determined by the temperature, pressure, and elastic properties of the medium. In air, acoustic waves propagate at 332 meters per second (1087 feet per second) at 00xAFC, at sea level. In air, acoustic wave speed increases approximately 0.6 meters per second (2 feet per second) for each kelvin above 00xAFC. Acoustic waves audible to the normal human ear are called sound waves. See Analog Wave.

acoustics That branch of science pertaining to the transmission of sound. The qualities of an enclosed space describing how sound is transmitted, e.g. its clarity. See also Sound.

acousto-optic The interactions between acoustic waves and light in a solid medium. Acoustic waves can be made to modulate, deflect, and focus light waves by causing a variation in the refractive index of the medium. See also Fiber Optics.

ACP Activity Concentration Point.

ACPI Advanced Configuration and Power Interface. A specification that enables efficient handling of power comsumption by desktop and laptop computers. With ACPI, the operating system can turn off unused peripheral devices such as CD-ROMs or displays. Users control the time a certain device will power up or down, and the level of power consumption targeted by the device when the battery reaches a certain level of discharge. ACPI also has a deep-sleep mode that allows an image file to reload without rebooting the system and launching applications when the computer is powered. See APM and WFM for a fuller explanation.

acquired taste A friend of mine, Nansi Friedman, claims that I (Harry) am an "acquired taste." She says that's a compliment. I'm not so sure. On the other hand, she is not an acquired taste. She's gorgeous and intelligent. She's also a lawyer, drat.

acquisition 1. In satellite communications, the process of locking tracking equipment on a signal from a communications satellite.

2. The process of achieving synchronization.

3. In servo systems, the process of entering the boundary conditions that will allow the loop to capture the signal and achieve lock-on. See also phase-locked loop.

4. In mobile, the process by which a Mobile End System (M-ES) locates a Radio Frequency (RF) channel carrying a channel stream, synchronizes to the data transmissions on that channel stream, and determines whether the channel stream is acceptable to the M-ES for network access.

Acquisition and Tracking Orderwire See ATOW.

acquisition time 1. In a communication system, the amount of time required to attain synchronization.

2. In satellite control communications, the time required for locking tracking equipment on a signal from a communications satellite. See also satellite.

ACR 1. Attenuation to Crosstalk Ratio. One of the factors that limits the distance a signal may be sent through a given medium. ACR is the ratio of the power of the received signal, attenuated by the media, over the power of the NEXT crosstalk from the local transmitter, usually expressed in decibels (dB). To achieve a desired bit error rate (BER), the received signal power must usually be several times larger than the NEXT power or plus several dB. Increasing a marginal ACR may decrease the bit error rate. See also Attenuation, BER, Crosstalk, dB, and NEXT.

2. Allowed Cell Rate. An ATM term. An ABR service parameter, ACR is the current rate in cells/sec at which a source is allowed to send. ACR is a parameter defined by the ATM Forum for ATM traffic management. ACR varies between the MCR and the PCR, and is dynamically controlled using congestion control mechanisms.

3. Acronym for Anonymous Call Rejection.

ACRFNET Academic Computing Research Facility Network. A network connecting various research units such as colleges and research and development laboratories in the U.S.

Acrobat A way of viewing a file without needing the associated software. For example, you run Word or QuarkXPress, make a pretty desktop published document, replete with diagrams, photos and diagrams. Now you want to send the file to someone to view it in all its glory. Simple. Convert the file to an Acrobat file (which has a .PDF extension) and send to them. The receiving person will run an Acrobat viewer program and see your beautiful work. They won't be able to change your work. But they will be able to see it. Acrobat is from Adobe, the Los Altos, CA company which produces PostScript. Acrobat has three benefits: The Acrobat viewing program is free. You can use Acrobat to view virtually any Windows or Apple software created file. Third, an Acrobat file can be up to 75% smaller than the original file in its native form, i.e. the original Word or QuarkXpress file. This saves considerable transmission time. See also PDF, PostScript.

acronym A pronounceable, artificial word formed from the initial letters or groups of letters of words in a set phrase or series of words. Examples of acronyms include WAC from Women's Army Corps, OPEC from Organization of Petroleum Exporting Countries, LORAN from LOng-RAnge Navigation, and COBOL from COmmon Business Oriented Language. That is not the end of the story. There are variations on the theme.

A recursive acronym is one created through a procedure that repeats until a specified or desired result is achieved. A hackish, and especially MIT Massachusetts Institute of Technology tradition, a recursive acronym humorously refers to itself or another acronym, or both. For example, GNU (an operating system) is a recursive acronym for "GNU's Not UNIX," EINE (an editor) for "EINE Is Not Emacs," and ZWEI (an editor) for "ZWEI Was EINE Initially." All of these recursive acronyms are products of MIT.

acrylate Acrylates are vinyl polymers that are used in a wide variety of applications, including as a coating for optical fibers. The fiber has an outside diameter of 125 microns (millionths of a meter). The acrylate coating is applied immediately, as the fiber is drawn in order to protect the glass from physical damage, and brings the outside diameter up to a total of 250 microns. In the event that a protective tight buffer is applied, the acrylate coating also serves to bind the buffer to the fiber. The buffer increases the outside diameter to 900 microns. See also tight buffer cable.

ACS 1. Asynchronous Communications Server.

2. Automatic Call Sequencer. A rudimentary automatic call distributor. See Automatic Call Sequencer.

3. Advanced Communication System. An old name for AT&T's data communications/data processing service, originally known as BDN (Bell Data Network) and later called Net 1000. ACS supported multiprotocol communications, offering protocol translation (much like X.25 packet switching), as well. In late 1986, after 10 years in birth, AT&T quietly buried ACS, which offered too little in the face of what – by then – had become cheap, powerful desktop microcomputers and 1200 bps modems priced at less than $200. ACS was depicted in AT&T presentations as a cloud – user data entered the cloud of the network on the originating end and exited the cloud on the terminating end. What went on in the cloud of the network was obscured from the user. The thinking was that the user needn't be concerned with what went on inside the cloud; rather, that was AT&T's responsibility and concern. This clever conceptual sell was never successful – it was way too obscure and offered way too little. See also Cloud.

4. ATM Circuit Steering. A means of routing ATM traffic to test facilities built in the ATM device (e.g., switch, router, or concentrator), as ATM networks have no point of entry for a test device.

ACSE 1. Association Control Service Element. An OSI application-layer protocol. The method used in OSI for establishing a connection between two applications.

2. Access Control and Signaling Equipment, a major part of a fixed Earth station that performs control and signalling functions and acts as a gateway interface between terrestrial networks and the space segment of a satcom system.

ACSR Aluminum conductor, steel-reinforced.

ACT Applied Computer Telephony is Hewlett-Packard's program that is a strategy and set of open architecture commands and interfaces for integrating voice and database technologies. The idea is that with ACT a call will arrive at the telephone simultaneously with the database record of the caller. And such call and database record can be transferred simultaneously to an expert, a supervisor, etc. ACT works on both HP 3000 and HP 9000 computers. ACT essentially controls the telephone call movement within PBXs it connects to. See also Open Application Interface.

ACTA 1. America's Carriers Telecommunications Association, an organization founded in 1985 by 15 small long distance companies wishing to create an association in which the members controlled the direction of the organization. (That's their words.) "The focus established was to provide national representation before legislative and regulatory bodies, while continuing to improve industry business relations." Perhaps ACTA's greatest claim to fame was its 1996 petition to the FCC to outlaw Voice Over the Net (VON). Sprint was among the members which broke ranks over that issue, upon which the FCC so far has declined

to act. In December, 1998, ACTA merged into CompTel (Competitive Telecommunications Association). See also CompTel. www.comptel.org.

2. Administrative Council for Terminal Attachments. The FCC privatized its Part 68 responsibilities, selecting ATIS (Alliance for Telecommunications Industry Solutions) and the TIA (Telecommunications Industry Association) as joint sponsors of the Part 68 ACTA (Administrative Council for Terminal Attachments). ACTA comprises 18 members, with two each elected from six interest segments, including Local Exchange Carriers (LECs), Interexchange Carriers (IXCs, or IECs), Terminal Equipment Manufacturers, Network Equipment Manufacturers, Testing Laboratories, and Other Interested Parties. "Invited Observers," a non-voting category, will include members approved by the council on a case-by-case basis. ACTA responsibilities include adopting and publishing technical criteria for terminal equipment submitted by ANSI-accredited standards development organizations, and operating and maintaining a database of approved terminal equipment. The first meeting of ACTA was scheduled for May 2, 2001. See also ATIS, Part 68 and TIA.

ACTAS Alliance of Computer-Based Telephony Application Suppliers was a part of the North American Telecommunications Association (NATA), which now has been fully integrated into TIA (Telecommunications Industry Alliance). See TIA.

ACTGA Attendant Control of Trunk Group Access. A complicated term for a simple concept, namely that your operator completes long distance calls. A primitive form of toll control.

action items More interesting sounding than "things to do."

actionable A legal term that's been co-opted by marketers, consultants and techies. In the legal world, it's "giving cause for legal action," such as a lawsuit. Now it's become anything you can take action on.

ACTIUS Association of Computer Telephone Integration Users and Suppliers. A British organization, ACTIUS provides an open industry forum for interchange and discussion on Computer Telephone Integration (CTI) between its members. Subscriptions are UK stlg275.00 for full membership and UK stlg150 for corresponding membership. A key aim of ACTIUS is to explain the benefits of CTI applications to the broadest possible range of users. ACTIUS also represents its members' interests on relevant regulatory and standards issues, both in the UK and the EU (European Union) Address: 11 Nicholas Road, Henley-on-Thames, OXON, United Kingdom RG9 1RB. Contact: Brian Robson, Secretary Tel: 011-44-1491-575295 Fax; 011-44-1491-410201 email brian.robson@BTInternet.com.

activated return capacity A cable TV term. The capability of transmitting signals from a subscriber or user premises to the cable headend. The typical information that can be sent back includes the ID number of the cable TV set-top box and what station you are watching. See also Cable Modem.

activation The process of enabling a subscriber device for network access and privileges on behalf of a registered account. For a cellular phone to be activated, the system must be informed of the combination of telephone number and Electronic Serial Number (ESN). Once this information has been entered, the phone can place and receive calls.

activation code Software is increasingly being bought and downloaded over the Internet. This saves time, shipping, aggravation and makes for happy customers. To get the software installed and working on your PC usually involves your putting some information into the software that identifies the software as legitimately purchased and legitimately belonging to you. That information typically includes your name as password and some random collection of numbers and digits as your passcode.

activation fee One-time fee for initial connection to the cellular system. As competition has intensified, so more and more carriers are dropping or severely reducing their activation fees. They do this in order to attract more new subscribers.

active A service flow is said to be "active" when it is permitted to forward data packets. A service flow must first be admitted before it is active. This definition from CableLabs.

active attack An attack on a computer system which either injects false information into the system, or corrupts information already present in the system. An active attack typically happens when many computers are programmed to attempt to access one computer system simultaneously and the attempts are continued relentlessly during the attack. The active attacks we read about tend to be those which happen to famous web sites. See also Denial of Service Attack, Passive Attack, Smurf Attack and Spoofing.

active branching device A device which converts an optical input into two or more optical outputs with gain or regeneration.

active call 1. A definition used in Call Centers. An active call is when the connection is in any state except Hold, Null, or Queued. In other words, any state during the establishment of the connection and/or call. This also includes the actual establishment of the connection and/or call itself.

2. A term which Hayes defines in its Hayes ISDN System adapter manual. An active call is a voice call to which you are connected that is not on hold.

active campaign A call center/marketing term. An outbound calling project that is currently running.

active channel An active channel is what Microsoft calls a Web site that has been enabled for push delivery to Internet Explorer 4.0 browsers. To create a channel, developers write and upload a CDF (channel definition format) file to their Web site. New content is delivered to users automatically when the site is updated. Developers and subscribers can control the update frequency; which channels, subchannels, and items (sections) are subscribed to; and other channel characteristics. Most Active Channels use dynamic HTML (DHTML) and other effects to spice up content and make it more interactive. See also channel definition format, DHTM and push.

active circuit Powered circuitry. MCI also defined it as circuits for which are there is a completed "install order" and a "completed date."

active components Active components, which include transistors and diodes, amplify current, convert alternating current (AC) to direct current (DC) and switch electronic signals, etc. Active components belong to a group called discrete components, which perform a single function, e.g. regulate current or switch signals. By contrast, integrated circuits combine the functions of multiple electronic components on one chip. Together, discrete components and integrated circuits are the building blocks of electronic devices. See also Passive Components.

active content Programs typically written in Java, JavaScript, or ActiveX, which are often downloaded and executed in a single step by a web browser; differs from static content, as typified by web pages built using only HTML tags, which offer a consistent, unchanging document.

active contract One you must sign. See Contract.

active coupler A fiber optic coupler that includes a receiver and one or more transmitters. It regenerates (thus "active") input signals and sends them through output fibers, instead of passively dividing input light.

active device Electronic components that require external power to manipulate or react to electronic output. These include transistors, op amps, diodes, cathode ray tubes and ICs. Passive devices include capacitors, resisters and coils (inductors).

active directory A feature of the Windows NT server and first introduced in NT 5.0, which will be called Windows 2000. Think of it as a real-time, super-fast Directory Information service – like they have when you call 411 or 555-1212 in North America. But instead of being answered by a person, it answers requests for peoples' phone numbers and addresses by sending instant messages back to the PC workstation which is asking the question. Here's a simple example. Imagine you want to call someone in your organization using your IP telephone. Instead of dialing a number as we do today, we dial a person. Our PC talks to the Windows NT server which we've logged onto and says, in essence, "I want to call Helen. Where is she?" It comes back and says this is "Helen's address. Call there." My PC says "thank you" and then dials Helen at that number. It's called "Active" Directory because its address entries change from moment to moment, as Helen moves around. See TAPI 3.0.

active discovery packet The type of packet used by PPPoE during the discovery stage.

active flip The ability to answer an incoming mobile phone call simply by opening the phone's keypad cover, and to end the call simply by closing the cover.

active hub A device used to amplify transmission signals in certain local area network topologies. You can use an active hub to add workstations to a network or to lengthen the cable distance between workstations and the file saver.

active laser medium Within a laser, active laser medium is the material that emits coherent radiation or exhibits gain as the result of electronic or molecular transitions to a lower energy state or states, from a higher energy state or states to which it had been previously stimulated. Examples of active laser media include certain crystals, gases, glasses, liquids, and semiconductors. See also Laser Medium.

active line A voice or data communications channel currently in use.

active link A logical communications circuit that is established only for the duration of communications. An active line needs a call-setup and call-clearing procedure for every connection.

active matrix liquid crystal display A technique of making liquid crystal displays for computers in which each of the screen's pixels – the tiny elements that make up a picture – is controlled by its own transistor. Active matrix LCD display technique uses a transistor for each monochrome or each red, green and blue pixel. It provides sharp contrast, speedier screen refresh and doesn't lose your cursor when you move it fast (also knowing as submarining). Some active matrix CD screens are as fast as normal glass CRTs.

active medium The material in fiber optic transmission, such as crystal, gas, glass,

liquid or semiconductor, which actually "lases." It's also called laser medium, lasing medium, or active material.

active monitor Device responsible for managing a Token Ring. A network node is selected to be the active monitor if it has the highest MAC address on the ring. The active monitor is responsible for such management tasks as ensuring that tokens are not lost, or that frames do not circulate indefinitely. See also ring monitor and standby monitor.

active open Used in TCP to request connection with another node.

active optical component Active optical components are fiber-optic bits and pieces that need power (hence the word "active") to operate and cause electrical signals within a network to create, modulate, or amplify the original light signal. Examples include source lasers, modulators, pump lasers and wavelength lockers. Active Optical Components are used to manufacture, transmit, modulate, and amplify light. Source lasers create the light; modulators (electroabsorption or lithium niobate) are used to encode the optical signal into a stream of on/off bits; and pump lasers emit light to amplify or boost the source signal.

active participation A feature in an automatic call distributor, a piece of equipment used in call centers. This feature is typically used to allow intrusion with the ability to speak and listen by a supervisor into an Agent Call. The resultant call is a conference. A Versit definition.

active passive device On a local area network, a device that supplies current for the loop is considered active. Such a device is a Token Ring MAU (Multistation Access Unit). A device, which does not supply current, is considered passive.

active pixel region On a computer display, the area of the screen used for actual display of pixel information.

active push The server on the Web interacts with the client by sending all the content to the client upon the client's request (polling), essentially the way that a client/server application might. PointCast is an example of this. See Push and Directed Push.

active satellite A satellite carrying a station intended to transmit or retransmit radio communication signals. An active satellite may perform signal processing functions such as amplification, regeneration, frequency transition and link switching, to make the signals suitable for retransmission. See Geosynchronous Satellite.

active service flow An admitted Service Flow from the Cable Modem (CM) to the Cable Modem Termination System (CMTS) which is available for packet transmission. A CableLabs definition.

Active Server Pages ASP. A specification for a dynamically created Web page with a .ASP extension that utilizes ActiveX scripting – usually VB Script or Jscript code. When a browser requests an ASP page, the Web server generates a page with HTML code and sends it back to the browser. ASP pages are similar to CGI scripts, but they enable Visual Basic programmers to work with familiar tools.

active splicing Aligning the ends of two optical fibers with the aim of minimizing the splice loss.

active star See Star Coupler and Multiport Repeater .

active tag An RFID tag that comes with a battery that is used to power the microchip's circuitry and transmit a signal to a reader. Active tags can be read from 100 feet or more away, but they're expensive – more than $20 each in the early part of 2004. They're used for tracking expensive items over long ranges. For instance, the US military uses active tags to track containers of supplies arriving in ports.

active terminator A terminator that can compensate for variations in the terminator power supplied by the host adapter through means of a built-in voltage regulator.

active video lines All video lines not occurring in the horizontal and vertical blanking intervals. In other words, the lines conveying the video and audio signals.

active vocabulary A phrase used in voice recognition to mean a group of words which a recognizer has been trained to understand and recognize. When a voice says ... PAIR, it knows the word means two, not the fruit, pear. That's one example of why voice recognition is so difficult. English, particularly, is a very complex and confusing language.

active window A Windows term. The active window is the window in which the user is currently working. An active window is typically at the top of the window order and is distinguished by the color of its title bar, typically dark blue.

active/passive device On a local area network, a device that supplies current for the loop is considered active. Such a device is a Token Ring MAU (Multistation Access Unit). A device which does not supply current is considered passive.

actives A call center/marketing term. Refers to customers who have purchased within a time period defined by the company instigating the marketing. Customers who have purchased outside the specified time period are considered inactive.

ActiveX In its simplest terms, said InfoWorld, May 19, 1997, ActiveX is an architecture that lets a program (the ActiveX control) interact with other programs over a network (such as the Internet). It's quite a different animal than Java, which is an entire new programming language plus a specification for a virtual machine. The ActiveX architecture, according to InfoWorld,uses Microsoft's Component Object Model (COM) and Distributed COM (DCOM) standards. COM allows different applications to talk to each other locally. DCOM allows them to talk over a network. ActiveX, formerly known as Object Linking and Embedding, or OLE, is an umbrella of mechanisms designed to bring sound bytes, animation and interactivity to Web documents, similar to plug-in technology for Netscape Communications Corp.'s Navigator and Sun Microsystems Inc.'s Java applets. They all provide ways to send small programs to a browser, without the involvement of any other special software on the desktop. In short, ActiveX is a software code from Microsoft which allows a developer to add move things to an otherwise static Web page. ActiveX is positioned by Microsoft as a competitive move against Java. Microsoft's ActiveX enables software components to interact with one another in a networked environment, regardless of the language in which they were created. Kind of like the Olympics. ActiveX is built on the Component Object Model (COM).

Elements of ActiveX are:

- ActiveX Controls–the interactive objects in a Web page that provide interactive and user-controllable functions.
- ActiveX Documents–enable users to view non-HTML documents, such as Microsoft Excel or Word files, through a Web browser.
- Active Scripting–controls the integrated behavior of several ActiveX controls and/or Java Applets from the browser or server.
- Java Virtual Machine–the code that enables any ActiveX-supported browser such as Internet Explorer 3.0 to run Java applets and to integrate Java applets with ActiveX controls.
- ActiveX Server Framework–provides a number of Web server-based functions such as security, database access, and others.
See also ajax.

ActiveX Controls A component that can be inserted in a page to provide functionality not directly available in HTML, such as animation sequences, credit-card transactions, real-time video sequences or spreadsheet calculations. ActiveX controls can be implemented in a variety of programming languages using C++, Visual Basic or Java. Over 1,000 ActiveX controls are available today. These include the Macromedia Shockwave for Director control and the Adobe Acrobat control.

Activity Concentration Point ACP. A location on a telecommunications network where there is high communications traffic, including voice, data, document distribution and teleconferencing. Generally, there will be some switching equipment present at the ACP.

activity costing A call center term. The costing methodology used to determine the specific cost of a given activity. (See Cost Per Phone Hour.)

activity factor A decimal fraction which represents the percentage of speech on a voice channel versus those periods of (non-talking) silence on that channel. Most voice channels carry actual speech 30% to 40% of the total available time. This represents an activity factor of 0.3 to 0.4.

activity report A report printed by a facsimile machine which lists all transmissions and receptions – their time, date, and number of documents; the remote unit type, diagnostic codes; and machine identification.

ACTL Access Customer Terminal Location. The CLLI code that identifies the location of the access customer's switch.

ACTS 1. Association of Competitive Telecommunications Suppliers. Trade association of telephone equipment dealers in Canada.

2. Automatic Coin Telephone Service includes a telephone company central office that can complete all types of payphone calls automatically without an operator. Recorded announcements are used to convey instructions to the customer.

actuator Motorized device used to position an earth-bound satellite dish for reception of programs. Actuators are built into a horizon-to-horizon mount; they look like motorized shock absorbers when attached to a polar mount.

ACU Automatic Calling Unit. Also an 801 ACU. A telephone company-provided device instructed by a computer to place a call on behalf of the computer. The call is then connected to a telephone company-provided Data Set. Anyone other than an IBM shop would simply buy a Hayes or Hayes-compatible modem, and not bother with the trouble and expense of an ACU.

ACUTA The Association for Telecommunications Professionals in Higher Education. Prior to 1998, it was the Association of College & University Telecommunications Administrators.

The new name is more meaningful, but the acronym stuck. ACUTA is an international, not-for-profit educational association serving approximately 800 institutions of higher learning. All members are director level or higher, and are responsible for data, video, communications, and all variety of networks, in addition to traditional telephony. Corporate affiliate members are welcome, as well. www.acuta.org.

AD Administrative Domain. A group of hosts, routers, and networks operated and managed by a single organization.

ad avails Advertising spots available to a cable operator to insert local advertising on a cable network.

ad click rate Sometimes referred to as "click-through." This is the percentage of ad views that resulted in an ad click.

ad clicks Number of times users click on an ad banner.

ad hoc mode A wireless network framework in which devices can communicate directly with one another without using an access point or a connection to a regular network. Contrasts with an infrastructure network, in which all devices communicate through an access point (AP).

ad hoc network A wireless network comprising only stations without access points.

ad insertion equipment Tape or solid state equipment for interrupting programming on a specific channel and putting prerecorded advertising on the channel.

ad views Number of times an ad banner is downloaded and presumably seen by visitors. If the same ad appears on multiple pages simultaneously, this statistic may understate the number of ad impressions, due to browser caching. Corresponds to net impressions in traditional media. There is currently no way of knowing if an ad was actually loaded. Most servers record an ad as served even if it was not.

ADA 1. Average Delay to Abandon. Average time callers are held in queue before they get frustrated and decide to hang up.

2. A high level computer language which the Department of Defense has been trying to foist on its suppliers and thus, make a standard. Ada is named for British mathematician Ada Lovelace, known at the time as Lady Lovelace. She was the girlfriend of Charles Babbage, the inventor of the computer.

ADACC Automatic Directory Assistance Call Completion.

ADAD Automatic Dialing and Announcing Device. Device which automatically places calls and connects them to a recording or agent. A Canadian term for an automatic dialer.

adaptable digital filtering A way of fixing twisted-pair telephone lines so they carry data more efficiently up to 12,000 feet before the need to regenerate the signal. The filter can be customized to meet the needs of a twisted pair.

adapter 1. A device used to connect a terminal to some circuit or channel so it will compatible with the system to which it is attached. An adapter converts one type of jack or plug to another, for example, from old 4-prong telephone jacks to new modular. An adapter may also combine various items, such as putting three plugs in one jack.

2. Another name for a NIC – Network Interface Card – a card which fits into a computer and joins the computer to a local area network.

adapter card A printed circuit card installed inside of a computer. It takes data from memory and transmits it over cable to connected devices such as a modem, or printer.

adapter segment A name sometimes used for the upper memory area of a PC, at hexadecimal addresses A000 through EFFF (640K to 1024K).

adaptive algorithm An algorithm that can "learn" and change its behavior by comparing the results of its actions with the goals that it is designed to achieve.

adaptive antenna array An antenna array in which the received signal is continually monitored in respect of interference (usually adjacent or co-channel). Its directional characteristics are then automatically adjusted to null out the interference. Such a concept often employs computer control of a planar type antenna.

adaptive channel allocation A method of multiplexing wherein the information-handling capacities of channels are not predetermined but are assigned on demand.

adaptive communication Any communication system, or portion thereof, that automatically uses feedback information obtained from the system itself or from the signals carried by the system to modify dynamically one or more of the system operational parameters to improve system performance or to resist degradation.

adaptive compression Data compression software that continuously analyzes and compensates its algorithm (technique), depending on the type and content of the data and the storage medium.

Adaptive Differential Pulse Code Modulation See ADPCM.

adaptive equalization An electronic technique that allows a modem to continuously analyze and compensate for variations in the quality of a telephone line.

adaptive interframe transform coding A class of compression algorithms commonly used in video codecs to reduce the data transmission rate.

Adaptive Multi-Rate Speech Codec AMR. Technology deployed into GSM networks to increase voice capacity by up to 4 times. GSM equipment vendors are deploying EDGE and AMR into GSM/GPRS networks simultaneously.

Adaptive Predictive Coding APC. Narrowband analog-to-digital conversion that uses a one-level or multilevel sampling system in which the value of the signal at each sampling instant is predicted according to a linear function of the past values of the quantized signals. APC is related to linear predictive coding (LPC) in that both use adaptive predictors. However, APC uses fewer prediction coefficients, thus requiring a higher sampling rate than LPC. See also Multipulse-excited LPC.

Adaptive Pulse Code Modulation APCM. A way of encoding analog voice signals into digital signals by adaptively predicting future encodings by looking at the immediate past. The adaptive part reduces the number of bits per second that another rival and more common method called PCM (Pulse Code Modulation) requires to encode voice. Adaptive PCM is not common because, even though it reduces the number of bits required to encode voice, the electronics to do it are expensive. See Pulse Code Modulation.

adaptive radio A radio that monitors its own performance, monitors the path quality through sounding or polling, varies operating characteristics, such as frequency, power, or data rate, and uses closed-loop action to optimize its performance by automatically selecting frequencies or channels.

adaptive retransmission algorithms Used by self-adjusting timers to determine and dynamically set timers to effectively adjust data traffic in the event the link is slower than usual due to congestion or their network conditions.

adaptive routing A method of routing packets of data or data messages in which the system's intelligence selects the best path. This path might change with altered traffic patterns or link failures.

adaptive site A web site that learns the habits of its visitors, or asks the visitors for their preferences and presents them with personalized pages each time they

adaptive speed leveling A modem technology that allows a modem to respond to changing line conditions by changing its data rate. If the line quality improves, the modem attempts to increase the data rate. If the line quality declines, the modem compensates by lowering the data rate. This is also known as adaptive equalization.

adaptive switching The adaptive switch is an intelligent node that sits in the network and evaluates the packets that are coming through the network. The switching will make the decision around what it wants to do with those packets. This platform sits on the edge of the network in the Gateway GPRS Support Node (GGSN) or acts as a replacement to the GGSN. It can only force QoS on a network-wide basis.

adaptive timeout Retry with exponential timeout: first attempt - 1 sec and the last attempt - 16 secs. A CableLabs definition.

Adaptive Transform Acoustic Coding. See ATRAC3.

ADAS Automated Directory Assistance Service. A service from Northern Telecom which automates the greeting and inquiry portion of the directory assistance call. With ADAS, directory assistance callers are greeted by the automated system and asked to state the name of the city and the listing they are seeking. They are then connected with an operator. The ADAS service knocks a few seconds off each directory assistance call.

ADB Apple Desktop Bus. A low-speed serial bus used on Apple Macintosh computers to connect input devices to the Macintosh CPU (central processing unit). Normally the ADP connects via an 8-pin round or DIN connector.

ADC 1. Analog-to-Digital Converter or Analog to Digital Conversion. A method of sampling and encoding analog signal to create a digital signal. The process is accomplished by a coded, also known as a DSP (Digital Signal Processor). See also Analog to Digital Converter, Codec and DSP.

2. Automated Data Collection. A variety of technologies that provide for automation of the function of data collection. Examples include bar code readers, OCR (Optical Character Recognition), OMR (Optical Mark Recognition), voice recognition, and smart cards.

ADCCP Advanced Data Communications Control Procedures, A bit-oriented ANSI-standard communications protocol. It is a link-layer protocol. ADCCP is ANSI's version of SDLC/HDLC.

ADCU Association of Data Communications Users.

add path request A request made by the network to add a path using the Add Path packet, which establishes a multi-hop path between two network nodes. Although the two nodes are usually the source and destination nodes of a Virtual Wavelength Path, there are cases in which other nodes might want to establish a path between them. Unlike the Restore Path request, the Add Path request is never flooded; it is instead forwarded using

information carried in the path itself (source routing). See VWP.

add/drop The process wherein a part of the information carried in a transmission system is demultiplexed (dropped) at an intermediate point and different information is multiplexed (added) for subsequent transmission. The remaining traffic passes straight through the multiplexer without additional processing.

Add/Drop multiplexer See ADM.

add-in card An expansion board that fits into the computer's slots and is used to expand the system's memory or extend the operation of another device.

add-on 1. A telephone system feature which allows connecting a third telephone to an existing conversation. This "add-on" feature is initiated by the originator of the call. The feature is also known as "Three-Way Calling."

2. Hardware, often referred to as peripheral equipment, that is added to a system to improve its performance, add memory or increase its capabilities. Voice mail, Automated Attendant and Call Detail Recording Equipment are examples of PBX add-on devices. Lucent, Nortel and some other manufacturers call them applications processors.

3. A call center/marketing term. A technique to increase the revenue of an order, for example, two dozen instead of one dozen or, two green shirts bought and sold with matching green tie.

add-on conference A PBX feature. Almost always used in conjunction with another feature called consultation hold, this feature allows an extension user to add a third person to an existing two-person conversation. The user places an existing central office call or internal call on Hold, and obtains system dial tone. The user can then call another internal extension or an outside party. After speaking with the "consulted" party, the originating phone reactivates the initiating command (typically a button push) and creates a three-party conference with the call previously placed on Hold.

add-on conference -- intercom only Allows a telephone user to add someone else to an existing intercom (within-the-same office) conversation.

add-on data module Plug-in circuit cards which allow a PBX to send and receive analog (voice) and digital (data) signals.

added bit A bit delivered to the intended destination user, to the intended user information bits and the delivered overhead bits. An added bit might be used to round out the number of bits to some error checking scheme, for example.

added block Any block, or other delimited bit group, delivered to the intended destination user in addition to intended user information bits and delivered overhead bits. See also Extra Block.

Additional Call Offering See ACO.

Additional Cooperative Acceptance Testing See ACAT.

additional period Billing periods charged after initial, first or minimum period on a call. Usually, long distance toll/DDD has a one-minute initial period at premium rate; subsequent "additional" minutes (period) are billed at a lower rate. Additional period billing increments vary by long distance company.

additive primaries By definition, three primary colors result when light is viewed directly as opposed to being reflected: red, green and blue (RGB). According to the tri-stimulus theory of color perception, all other colors can be adequately approximated by blending some mixture of these three lights together. This theory is harnessed in color television and video communications. It doesn't work so well in color printing where special colors are often printed separately.

Additive White Gaussian Noise AWGN. See White Noise.

ADDMD Administrative Directory Management Domain. A X.500 directory management domain run by a PTT (Posts, Telegraph, and Telephone administration) or other public network provider.

address An address comprises the characters identifying the recipient or originator of transmitted data. An address is the destination of a message sent through a communications system. A street address (i.e. 123 Elm Street, Normal, OK) is your physical address. A telephone number is considered the address of the called person. In computer terms, an address is a set of numbers that uniquely identifies the physical or logical location of something – a workstation on a LAN, a location in computer memory, a packet of data traveling through a network. On the Internet, addresses are based on the IP protocol, which uses a 32-bit code in the IP header to identify host addresses. Web URLs and e-mail addresses are arbitrary text addresses that correlate to IP addresses. They are maintained in directory service databases. For a longer explanation, see internet address.

Address Complete Message ACM. A CCS/SS7 signaling message that contains call-status information. This message is sent prior to the called customer going off-hook.

address field In data transmission, the sequence of bits immediately following the opening flag of a frame identifying the secondary station sending, or designated to receive, the frame.

address field extension EA. A Frame Relay term defining a 2-bit field in the address field, identifying the fact that the address structure is extended beyond the 2-octet default. Frame Relay standards provide for extension of the address field up to 60 bits, which extension will be implemented as the popularity of Frame Relay grows, placing pressure on the standard addressing convention.

address filtering A way of deciding which data packets are allowed through a device. The decision is based on the source and destination MAC (Media Access Control, the lower part of ISO layer two) addresses of the data packet.

address harvester The programs that search Web pages and/or filter newsgroup traffic looking for email addresses to unload unsolicited advertising (or worse).

address learning Each node on a network has a unique node address automatically assigned to it (embedded in the adapter card). Switches and routers "learn" this address to enable accurate transmission to and from each node.

address mapping Technique that allows different protocols to interoperate by translating addresses from one format to another. For example, when routing IP over X.25, the IP addresses must be mapped to the X.25 addresses so that the IP packets can be transmitted by the X.25 network. See also address resolution.

address mask An electronic messaging term. A bit mask used to select bits from a network address (e.g. Internet) for sub-net addressing. The mask is 32 bits long and selects the network portion of the address and one or more bits of the local portion. Sometimes called sub-net mask.

address message A message sent in the forward direction that contains address information, the signaling information required to route and connect a call to the called line, service-class information, information relating to user and network facilities and call-originator identity or call-receiver identity.

address message sequencing In common-channel signaling, address message sequencing is a procedure for ensuring that address messages are processed in the correct order when the order in which they are received is incorrect.

address munging Modifying one's e-mail address in such a way that computers can't read it but humans can.

address prefix An ATM term. A string of 0 or more bits up to a maximum of 152 bits that is the lead portion of one or more ATM addresses.

address records See A Records.

address resolution 1. The process of discovering a device's address. An internetworking term. A discovery process used when, as in LAN protocols such as TCP/IP and IBM NetBIOS, only the Network Layer address is known and the MAC address is needed to enable delivery to the correct device. The originating end station sends broadcast packets with the device's NLA to all nodes on the LAN; the end station with the specified NLA address responds with a unicast packet, addressed to the originating end station, and containing the MAC address. See Address Resolution Protocol.

2. An ATM term. Address resolution is the procedure by which a client associates a LAN destination with the ATM address of another client or the bus.

address resolution protocol ARP. The Internet protocol used to map dynamic Internet addresses to physical (hardware) addresses on local area networks. Limited to networks that support hardware broadcasts.

Address Screening A service provided by Switched Multi-megabit Data Service (SMDS). Address Screening allows the network to compare the Source Address of the transmitting party to a list of addresses for including (or excluding) end-points into (or out of) a virtual network.

address separator A character that separates the different addresses in a selection signal.

address signaling Signals either the end user's telephone or the central office switching equipment that a call is coming in.

address signals Address signals provide information concerning the desired destination of the call. This is usually the dialed digits of the called telephone number or access codes. Typical types of address signals are DP (Dial Pulse), DTMF, and MF.

address space The amount of memory a PC can use directly is called its address space. MS-DOS can directly access 1024K of memory (one megabyte). A protected mode control program like Microsoft Windows 3.x or OS/2 can directly address up to 16 megabytes of memory. Here is a definition of address space, as supplied by the Personal Computer Memory Card International Association (PCMCIA) as address space applies to PCMCIA cards: "An address space is a collection of registers and storage locations contained on a PC Card which are distinguished from each other by the value of the Address Lines applied

to the Card. There are three, separate, address spaces possible for a card. These are the Common Memory space, the Attribute Memory space and the I/O space."

address space probe An intrusion technique in which a hacker sequentially scans IP addresses, generally as the information-gathering prelude to an attack. These probes are usually attempts to map IP address space as the hacker looks for security holes that might be exploited to compromise system security.

address table A table stored in routers, bridges and switches that enables these devices to know where on the network to forward information.

addressable The characteristic of a network device enabling it to send and receive messages independently due to its unique identification code.

addressable programming A cable TV (CATV) industry term. A subscriber orders a movie or sports event. He does that calling a phone number (generally an 800 number). A computer answers, grabs the calling number, confirms the request, then hangs up. The computer passes the request onto the cable company's computer, which checks the calling phone number against its accounting records. If the subscriber has good credit, the cable company sends a coded message down its cable network to the caller's set-top cable box/converter. The message temporarily enables that particular converter to descramble the channel offering the desired program.

addressability 1. In computer graphics, the number of addressable points on a display surface or in storage.

2. In micrographics, the number of addressable points, within a specified film frame, written as follows: the number of addressable horizontal points by the number of addressable vertical points, for example, 3000 by 4000.

3. A cable TV term. The capability of controlling the operation of cable subscriber set-top converters by sending commands from a central computer. Such addressability is absolutely required for a cable system to offer pay-per-view services.

addressable point In computer graphics, any point of a device that can be addressed. See Addressability.

addressed call mode A mode that permits control signals and commands to establish and terminate calls in V.25bis. See also V.25bis.

addressee The intended recipient of a message.

addressing Refers to the way that the operating system knows where to find a specific piece of information or software in the application memory. Every memory location has an address.

ADF 1. Automatic Document Feeder.

2. Adapter Description File.

ADH 1. Average Delay to Handle. Average time a caller to an automatic call distributor waits before being connected to an agent.

2. Automatic Data Handling.

adherence A term used in telephone call centers to connote whether the people working in the center are doing what they're meant to be doing. Are they at work? Are they on break? Are they answering the phone? Are they at lunch? All these activities are scheduled by workforce management software. If they're in line, the workers are "in adherence." If not, they're "out of adherence." See Adherence Monitoring.

adherence monitoring Adherence monitoring means comparing real-time data coming out of an ACD with forecast call volumes, forecast service levels and forecast workforce employment levels. The idea is to see if the people, the calls and the system are working as forecast. This a measure of how well your forecasting works. You need to know how well it works since it's your forecasting on which you base your employment. See Adherence.

adhesive wire See tapewire.

ADI Area of Dominant Influence: A television market as delineated by the Arbitron Company.

adjacencies Service requirements that a network operator must fulfill by virtue of its being a network operator. For example, adjacencies that a telephone service provider has to deal with are E911 and CALEA support.

adjacency Relationship formed between selected neighboring routers and end nodes for the purpose of exchanging routing information. Adjacency is based upon the use of a common media segment.

adjacent cell A cellular radio term. Two cells are adjacent if it is possible for a Mobile End System (M-ES) to maintain continuous service while switching from one cell to the other.

adjacent channel Any of two TV channels are considered adjacent when their view carriers, either off-air or on a cable system, are 6 MHz apart. FM signals on a cable system, two channels apart are adjacent when their carriers are 400 to 600 kHz apart.

adjacent channel interference When two or more carrier channels are placed too close together in the frequency spectrum, they interfere with each other and mess up each other's conversations.

adjacent colocation Adjacent colocation is the same as physical colocation. See Colocation.

Adjacent MD-IS A cellular radio term. Two Mobile Data Intermediate Systems (MD-ISs) are adjacent if each MD-IS controls one of a pair of adjacent cells.

Adjacent MTA An MTA (Message Transfer Agent) that directly connects to another MTA. A Message Transfer Agent operated by a public service provider or PTT (Post, Telegraph, and Telephone administration), or a client MTA.

adjacent nodes 1. In SNA, nodes that are connected to a given node with no intervening nodes.

2. In DECnet and OSI, nodes that share a common network segment (in Ethernet, FDDI, or Token Ring networks).

adjacent channel A channel or frequency that is directly above or below a specific channel or frequency.

adjacent nodes 1. In SNA, nodes that are connected to a given node with no intervening nodes.

2. In DECnet and OSI, nodes that share a common network segment (in Ethernet, FDDI, or Token Ring networks).

adjacent signaling points Two CCS/SS7 signaling points that are directly interconnected by signaling links.

adjunct 1. Network system in the Advanced Intelligent Network Release 1 architecture that contains SLEE (Service Logic Execution Environment) functionality, and that communicates with an Advanced Intelligent Network Release 1 Switching System in processing AIN Release 1 calls. See also Adjunct Processor.

2. An auxiliary device connected to the ISDN set, such as a speakerphone, headset adapter, or an analog interface.

adjunct key system A system installed behind a PBX or a Centrex. Such a key system provides the users with several more features than the PBX or Centrex. Not a common term today.

adjunct power Power supplied to optional data or voice equipment in an equipment room, telecommunications closet, or work area, through separate power supplies.

adjunct processor 1. A computer outside a telephone switching system that "talks" to the switch and gives it switching commands. An adjunct processor might include a database of customers and their recent buying activities. If the database shows that a customer lives in Indiana, the call from the customer might be switched to the group of agents handling Indiana customers. Adjunct processors might also be concerned with energy management, building security etc.

2. An AIN (Advanced Intelligent Network) term for a decentralized SCP (Signal Control Point). An Adjunct Processor supports AIN services which are limited to one or more SSPs (Service Switching Points), which are SS7-equipped Central Office PSTN switches. Where multiple SSPs are supported, they typically comprise a regional network grouping. Adjunct Processors can include routing logic or call authorization security specific to a particular geographic area, providing switches with switching commands.

Adjunct Service Point ASP. An intelligent-network feature that resides at the Intelligent peripheral equipment and responds to service logic interpreter requests for service processing.

Adjunct Switch Application Interface. See ASAI.

adjusted ring length When a segment of Token Ring (in practice a dual ring) trunk cable fails, a function known as the Wrap connects the main path to the backup path. In the worse case – the longest path – would occur if the shortest trunk cable segment failed, so ARL is calculated during network design to ensure the network will always work.

ADK Application Definable Keys.

ADM Add/Drop Multiplexer. A multiplexer, such as a terminal multiplexer, capable of extracting and inserting lower-rate signals from a higher-rate multiplexed signal without completely demultiplexing the signal. Also a SONET/SDH term for a device which can either insert or drop DS1, DS2, and DS3 channels or SONET signals into/from a SONET bit stream. The ADM literally can reach up into the SONET pipe and extract a DS1-level signal, without going through the rigorous process of demultiplexing and remultiplexing which is required in the traditional T/E-carrier world. While the devices are much more complex than are TDMs, the process is much faster, induces no signal delay, creates no signal errors. The ADM also provides for dynamic bandwidth allocation, optical hubbing, and ring protection.

ADMD Administration Management Domain. An X.400 Message Handling System public carrier. Examples include MCImail and ATTmail in the U.S., British Telecom's Gold400mail in the U.K. The ADMDs in all countries worldwide together provide the X.400 backbone.

admin Administration.

adminisphere The rarefied organizational layers beginning just above the rank and file. Decisions that fall from the adminisphere are often profoundly inappropriate or irrelevant to the problems they were designed to solve. This definition from Wired Magazine.

administrable service provider An SCSA definition. A service provider which supports administrable services (for example, SCSA Call Router).

administration 1. The method of labeling, identifying and documenting an organization's voice/data communications cabling infrastructure.

2. A term used by the telephone industry to program features into a phone system. On a Northern Telecom Norstar system, administration includes making settings on 1. System speed dial; 2. Names on phones; 3. Time and date; 4. Restrictions; 5. Overrides; 6. Permissions; 7. Night Service and 8. Passwords.

administration by telephone The capability for the system administrator to perform most routine system administrative functions remotely from any Touch Tone pad. Such functions include mailbox maintenance (e.g. create, delete, set password, set class of service, etc.) and disk maintenance.

administration directory management domain A X.500 directory management domain run by a PTT (Posts, Telegraph, and Telephone administration) or other public network provider.

administration sub-system Part of Lucent's premises distribution system that distributes hardware components for the addition or rearrangement of circuits.

administrative alerts A Windows NT term. Administrative alerts relate to server and resource use; they warn about problems in areas such as security and access, user sessions, server shutdown because of power loss (when UPS is available), directory replication, and printing. When a computer generates an administrative alert, a message is sent to a predefined list of users and computers.

administrative assistant Admin. The PC (Politically Correct) term these days for what we used to call a secretary. Some knucklehead in HR (Human Resources), which we used to call Personnel), probably invented the term to make secretaries feel more important. Actually, admins do a lot more than secretaries used to do, and they really are more important. It still seems like a silly name to the old-timers, though.

administrative contact The individual authorized to interact with the registrar on behalf of the domain name registrant. The administrative contact should be able to answer non-technical questions about the domain name's registration and the domain name registrant.

Administrative Council for Terminal Attachments See ACTA.

administrative distance Cisco defines administrative distance as a rating of the trustworthiness of a routing information source. Administrative distance is often expressed as a numerical value between 0 and 255. The higher the value, the lower the trustworthiness rating.

Administrative Domain 1. AD. A group of hosts and networks operated and managed by a single organization. An Internet term.

2. An ATM term. A collection of managed entities grouped for administrative reasons.

administrative layer The Virtual Network Service layer that provides the customer and telco with the ability to monitor, maintain, reconfigure, and manage the network. The administrative layer consists of two components: Service View Management tools and Applications Hosts.

administrative management domain An X.400 electronic mail term: a network domain maintained by a telecommunications carrier.

administrative number Any number used by a telephone company to perform internal administrative or operational functions necessary to maintain reasonable quality of service standards.   A number that has soft dial tone is regarded as an administrative number.

Administrative Operating Company Number See AOCN.

administrative point A location at which communication circuits are administered, i.e. rearranged or rerouted, by means of cross connections, interconnections, or information outlets.

Administrative Service Logic Program ASLP. The SLP responsible for managing the feature interactions between Advanced Intelligent Network Release 1 features resident on a single SLEE (Service Logic Execution Environment).

administrative subsystem That part of a premises distribution system where circuits can be rearranged or rerouted. It includes cross connect hardware, and jacks used as information outlets.

administrative trunk groups A category of telephone company trunk groups that provide call status monitoring. The function may be one of revenue protection (i.e., coin overtime collection), telephone operator assistance (i.e., verification), protection of the message network from overloads (i.e., no-circuit announcements.), etc. Types of trunk groups in this category include: Announcement, Coin Supervisory, Coin Zone, Permanent Signal, Vacant Code and Verification.

administrative weight A value set by the network administrator to indicate the desirability of a network link. One of four link metrics exchanged by PTSPs to determine the available resources of an ATM network. See PTSP.

administrator The individual responsible for managing the local area network (LAN) and the computers on it. This person configures the network, maintains the network's shared resources and security, assigns passwords and privileges, and helps users. When Apple first came out with an Intel powered Mac, TechWeb wrote: In Windows XP, administrators have no restrictions on what they can do. They have unlimited access to the system and the Windows Registry without providing any kind of authentication. This is convenient for the user, but it means that malware has unfettered access to the system, unless anti-virus/anti-malware software stops it. It is possible to run as a more restricted user in XP, but because of the way many Windows programs work, you have to be an administrator to install or run some software. So most Windows users run as administrators all the time, leaving their systems wide open to malware. The Mac OS X version of an administrator can do a lot without restriction, but there are certain actions they can't perform and system directories they can't touch without providing their user name and password. This means users can still run programs and install software; they just have to authenticate first. It's a bit more inconvenient for the user, but it makes it harder for malware to silently infect your system. There is an unrestricted administrator account known as root in OS X, but the login for that account is disabled by default.

administrivia A silly term for administrative tasks, most often related to the maintenance of mailing lists, digests, news gateways, etc. An Internet term. On a Web site, it's the odds and ends that don't quite fit under a specific category or merit their own page. Often, it's the legal stuff about copyrights, liability, licensing, etc.

Admiration Don't accept your dog's admiration as conclusive evidence that you are wonderful. -Ann Landers

admitted A service flow is said to be "admitted" when the Cable Modem Termination System (CMTS) has reserved resources (e.g., bandwidth) for it on the Data Over Cable Service Interface Specification (DOCSIS) network.

admitted service flow A service flow, either provisioned or dynamically signaled, which is authorized and for which resources have been reserved but is not active. See admitted.

ADML Asymmetric Digital Microcell Link. A Telcordia standard for Wireless Local Loop (WLL). Using low-power, omnidirectional radio systems, ADML can be deployed to cover an area as large as 1 mile in radius. ADML supports as much as 1 Gbps aggregate bandwidth, providing individual users with bandwidth in radio channels as great as T-1 (1.544 Mbps). See also Wireless Local Loop and LMDS.

ADN Advanced Digital Network. ADN is Pacific Bell of California's low-cost leased 56 Kbps digital service. ADN is available for intraLATA calls.

ADO Auxiliary Disconnect Outlet. A device usually located within the tenant or living unit used to terminate the ADO cable or backbone cable. Source ANSI/TIA/EIA-570-A.

ADP 1. Apple Desktop Bus. A synchronous serial bus allowing connection of the Mac keyboard, mouse and other items to the CPU. A Mac keyboard or mouse is called an ADB device. Contrast with peripherals, which attach through the SCSI interface.

2. Automatic Data Processing. The same as DP, data processing.

3. The name of a company which processes my pay check.

ADPCM Adaptive Differential Pulse Code Modulation. A speech coding method which uses fewer bits than the traditional PCM (Pulse Code Modulation). ADPCM calculates the difference between two consecutive speech samples in standard PCM coded telecom voice signals. This calculation is encoded using an adaptive filter and therefore, is transmitted at a lower rate than the standard 64 Kbps technique. Typically, ADPCM allows an analog voice conversation to be carried within a 32-Kbps digital channel; 3 or 4 bits are used to describe each sample, which represents the difference between two adjacent samples. Sampling is done 8,000 times a second. ADPCM, which many voice processing makers use, allows encoding of voice signals in half the space PCM allows. In short, ADPCM is a reduced bit rate variant of PCM audio encoding. See also DPCM and PCM.

ADPE Automatic Data Processing Equipment.

ADQ Average Delay in Queue. An important measure of the customer responsiveness of a call center. See also ASA, Average Speed of Answer.

ADR See American Depositary Receipt.

ADRMP (pronounced add-rump) AutoDialing Recorded Message Player. A device that calls a bunch of telephone numbers and upon connection will play a message to the answering person. ADRMPs are used for lead solicitation and message delivery. They are often unpopular due to their indiscriminate dialing pattern and random message playing.

ADS 1. AudioGram Delivery Services.

2. The IETF working groups are grouped into areas, and managed by Area Directors, or ADs. The ADs are members of the Internet Engineering Steering Group (IESG). Providing architectural oversight is the Internet Architecture Board, (IAB). The IAB also adjudicates appeals when someone complains that the IESG has failed. The IAB and IESG are chartered by the Internet Society (ISOC) for these purposes. The General Area Director also serves as the chair of the IESG and of the IETF, and is an ex-officio member of the IAB.

ADSI 1. Analog Display Services Interface. ADSI is a Telcordia standard defining a protocol on the flow of information between something (a switch, a server, a voice mail system, a service bureau) and a subscriber's telephone, PC, data terminal or other communicating device with a screen. The simple idea of ADSI is to add words to, and therefore a modicum of simplicity of use to a system that usually uses only touchtones. Imagine a normal voice mail system. You call it. It answers with a voice menu. Push 1 to listen to your messages, 2 to erase them, 3 to store them, 4 to forward them, etc. It's confusing. You have to remember which is which. ADSI is designed to solve that. It's designed to send to your phone's screen the choices in words that you're hearing. You then have the choice of responding to what you hear or what you see. Your response is the same – a touchtone button. ADSI's signaling is DTMF and standard Bell 202 modem signals from the service to your 202-modem equipped phone. From the phone to the service it's only touchtone. With ADSI, you don't hear the modem signaling because every time the service gets ready to send you information, it first sends a "mute" tone. ADSI works on every phone line in the world. For ADSI to work visually, you'll need a special ADSI-equipped phone (Nortel has one showing 8 lines by 20 characters) or a piece of ADSI software in your PC. The nice feature of ADSI is that the standard is so flexible, it can work on cheap phones with a small display and more expensive phones with a bigger display and on a PC with a neat big display.

2. A set of Microsoft ActiveX controls that abstract the capabilities of directory services from different network providers to present a single set of directory service interfaces for accessing and managing network resources.

ADSL Asymmetric Digital Subscriber Line. One of a number of DSL technologies, and the most common one. ADSL is designed to deliver more bandwidth downstream (from the central office to the customer site) than upstream. The technical reason for this asymmetry has to do with issues of cross-coupled interference in the forms of FEXT (Far-End CrossTalk) and NEXT (Near-End CrossTalk). As it turns out, the asymmetry suits the applications perfectly, as DSL is used primarily for access to the Internet and Web, in which most people need fast downloads (music, software, presentations, etc.) but don't need high-speed uploads. They need it for e-mail and instant messaging. In ADSL, downstream rates range from 256,000 bits per second to as much as nine million bits per second, whereas upstream bandwidth ranges from 16 to 640 thousand bits per second. But these figures are changing. And these days, phone companies, which are the primary providers of ADSL service, sell their offerings in all sorts of speeds. Typically the more you pay, the faster service you get. ADSL transmissions work at distances up to 18,000 feet (5,488 meters) over a single copper twisted pair. See also HDSL, SDSL, and VDSL.

ADSL was developed by Telcordia and is now standardized by ANSI as T1.413; ETSI (European Telecommunications Standards Committee) contributed an Annex to the standards to reflect European requirements. ADSL technology splits the bandwidth of a qualifying pair to support multiple channels. An analog channel running at 4 kHz and below supports analog voice and fax. Packet data runs at 25kHz and above. Performance of ADSL lines is subject to the condition of the twisted-pair cable plant. Factors which affect performance include length of the loop, wire gauge (diameter), presence of bridge taps (better not to have any), and cross-coupled interference (NEXT and FEXT). Assuming no bridge taps and assuming 24-gauge copper,

ADSL will deliver downstream 1.5/2.0 Mbps over a distance of about 18,000 ft (5.5 km.). At 6.1 Mbps, 12,000 ft (3.7 km.) is the maximum length of the loop. Where the length of the loop exceeds those maximums, the achievable transmission rate drops precipitously due to signal attenuation and associated error performance. Error performance is addressed through FEC (Forward Error Correction), thereby maximizing throughput. Special electronics at both ends of the connection are required in order to accomplish the minor miracle of ADSL. At the carrier end of the connection is placed an ATU-C (ADSL Termination Unit-Centralized), while an ATU-R (ADSL Termination Unit–Remote) is placed at the customer premises. In order to achieve such a high data rate over UTP, relatively sophisticated compression techniques must be employed. While the standard calls for use of DMT (Discrete Multi-Tone), DMT implementations have experienced some difficulty. See also ADSL Forum, ADSL Lite, ADSL2 and DSL Filter.

ADSL Forum The ADSL Forum is an industry association formed to promote the ADSL concept and to facilitate the development of ADSL system architectures and protocols for major ADSL applications. Its name has been changed to DSL Forum. See DSL Forum.

ADSL Lite Also known as G.lite, Universal ADSL and Splitterless ADSL, a proposal of the UAWG (Universal ADSL Working Group) for a simplified version of ADSL. An interoperable extension of ANSI T-1.413 ADSL, ADSL Lite is application-specific, designed specifically for Internet access, which is unlike the original ADSL concept. ADSL Lite allows access to the Internet through a modem (either internal or external) operating at speeds of as much as 1.5 Mbps over existing twisted pair local loops of relatively short length (under three miles) and of good quality, with no loading. According to the UAWG, "By reducing the complexity of the on-site installation and the need for new wiring at the user's home, G.lite ADSL makes it possible to more cost-effectively increase bandwidth for the consumer up to 30 times the speed of the current highest-speed analog modem technology. With the ability to deliver "always-on" Internet access at higher speeds, G.lite ADSL dramatically improves the consumer's online experience." ADSL Lite subsequently was standardized by the ITU-T as G.992.4. In late March, 1999, I was lucky enough to have a G.Lite ADSL line installed at my country home. My average speed download speed (information coming at me) is 575,000 bits per second and my average upload speed is 120,000 bits per second. That about one-third as fast as my cable modem in New York City. But heck, I'm in the middle of the woods, at last two miles from my rural phone company's central office.

ADSL2 Standardized in August 2002 as ITU Recommendation G.992.3, ADSL2 improvements include line diagnostics, power management, power cutback, reduced framing and on-line configuration. ADSL2 offers improved performance over longer loops and those loops with bridged taps, improved interoperatibility between products involving different chipset manufacturers and faster start-up. ADSL2 also supports Channelized Voice over DSL (CVoDSL), which allows service providers to bundle VoDSL services to residential subscribers. ADSL2 can operate in all-digital mode, allowing for the transmission of data in the voice bandwidth, which adds 256 kbps to the upstream data rate. Finally, ADSL2 supports bonding of multiple local loops in support of Inverse Multiplexing over Asynchronous Transfer Mode (IMA). See also ADSL and ADSL2+.

ADSL2+ Standardized in January 2003 as ITU Recommendation G.992.5, ADSL2+ is an interoperable version of ADSL that increases downstream speeds to as much as 24 Mbps for spans of fewer than 5,000 feet (1.5 km). Upstream speeds are limited to 1.0 Mbps, however. Actual performance depends on line length and line and loop conditions. ADSL2+ also offers seamless bonding options, theoretically allowing two ADSL local loops to be bonded for aggregate downstream bandwidth of as much as 48 Mbps. G.992.5 Annex L, aka ADSL2+RE, extends ADSL reach up to 21,000 feet. See also ADSL and ADSL2.

ADSP Apple Datastream Protocol. A transport mechanism for interprocess communications between Apple Macintosh and Dec Vax minicomputers.

ADSTAR Automated Document STorage And Retrieval.

ADSU ATM DSU. Terminal adapter used to access an ATM network via an HSSI-compatible device. See also DSU.

ADT 1. Abstract Data Type.
2. Audio Tracking Database.

ADTF ACR Decrease Time Factor: This is the time permitted between sending RM-cells before the rate is decreased to ICR (Initial Cell Rate). The ADTF range is .01 to 10.23 sec with granularity of 10 ms.

ADTS Automated Digital Terminal System.

ADU Asynchronous Data Unit.

adultery A punishment for an adulterous wife in medieval France was to make her chase a chicken through town naked. I can offer no explanation for the logic of this punishment, nor what she was meant to do with the chicken if she ever caught it.

advance payment Payment of all or part of a charge required before start of service.

advance replacement See Advance Replacement.

advance replacement warranty A warranty service whereby the dealer sends the customer a replacement component before the customer returns the defective product. This not only accelerates the replacement time, but also helps the buyer if the component is vital. When you buy vital telecom gear, it's good to check that your equipment has an Advance Replacement Warranty or Guarantee.

Advanced Audio Coding See AAC.

Advanced Branch Exchange ABX. An uncommon term meaning a private branch exchange (PBX) with advanced features normally including the ability to handle both voice and data in some sort of integrated way.

Advanced Communications Service ACS. A large data communications network established by AT&T.

Advanced CoS Management Advanced Class of Service Management. Essential for delivering the required QoS to all applications. Cisco switches contain per-VC queuing, per-VC rate scheduling, multiple CoS queuing, and egress queuing. This enables network managers to refine connections to meet specific application needs. Formerly called FairShare and OptiClass.

Advanced Data Communications Control Procedures ADCCP. A bit-oriented, link-layer, ANSI-standard communications protocol.

Advanced Data Communications Control Protocol See AEP.

Advanced Intelligent Network AIN. The local Bell telephone companies' architecture for the 1990s and beyond. See AIN for a much fuller explanation.

Advanced Interactive Executive AIX. An IBM version of UNIX. AIX runs on PS/2 computers, IBM workstations, minicomputers, and mainframes.

Advanced Interactive Video AIV. Interactive videodisc format and system using LV-ROM, a method of storing analog video, digital audio, and digital data on a single videodisc. The system was developed by Philips UK, the British Broadcasting Corporation, Acorn Computer, and Logica Ltd. Most prominent application was the BBC's Domesday Project.

Advanced Mobile Phone Service AMPS. The analog cellular system originally developed by AT&T and currently installed throughout the United States (800 MHz) and various other regions around the world. AMPS transmits data by varying the frequency of a radio signal. The system uses Frequency Division Multiple Access (FDMA) for access control and Frequency Division Duplex (FDD) for two-way base station-to-subscriber conversation. Each AMPS cell site can accomodate 832 simultaneous calls. See AMPS.

Advanced Peer-to-Peer Networking APPN. An SNA protocol that allows network nodes to interact without using a host computer. Instead, each network device runs both client and server portions of an application.

advanced power management An industry standard for taking advantage of a computer's power saving features. Used particularly in battery-powered laptops.

advanced private line termination An AT&T/Lucent term which means the PBX user gets access to all the services of an Enhanced Private Switched Communications Services (EPCS) network. It also works when it is associated with AT&T's Common Control Switching Arrangement (CCSA) network.

Advanced Radio Data Information Service ARDIS. A network originally developed by Motorola for IBM's field service personnel. The service was established commercially in 1990 by a partnership between IBM and Motorola. In 1994, IBM sold its interest back to Motorola. ARDIS has coverage in more than 400 major metropolitan areas and more than 10,000 cities. The ARDIS network is based on a Motorola technology called Data TAC and has traditionally focused on vertical markets (i.e. specialized applications) with major customers, such as Otis Elevator, Sears, and a host of municipal public safety and emergency departments.

Advanced Research Projects Administration Network ARPANet. The precursor to the Internet. Developed in the late 1960s by the Department of Defense as an experiment in wide-area networking that could survive a nuclear attack.

advanced services This is the FCC's definition: Advanced telecommunications capability is the availability of high-speed, switched, broadband telecommunications that enables users to originate and receive high-quality voice, data, graphics, and video using any technology.

An older telephony definition of advanced services is as follows: Value-added telephony services beyond standard analog voice service. Advanced services include call waiting and call forwarding, private branch exchanges, ISDN, digital data, and other voice and data services.

Advanced Technology Attachment ATA. The standard ATA-1 standard was developed by the American National Standards Institute (ANSI) in 1994 in the X3.221 specification and is more commonly known as IDE. See IDE for a full explanation.

Advanced Television Enhancement Forum (ATVEF) A consortium of broadcast, cable and computer companies founded in 1998 that developed the ATVEF Enhanced Content Specification, an HTML and JavaScript- based format for adding content to interactive TV. ATVEF closed at the end of 1999 and turned over the specification to the ATV Forum and SMPTE.

Advanced Television System ATV. Any television technology that provides audio and video quality that is better than is provided by the current television broadcast system, or that otherwise enhances the current system. This definition, courtesy the FCC. Your taxpayer monies paid for it. See also ATV, DTV, HDTV, and SDTV.

Advanced Television System Committee (ATSC) An organization founded in 1983 to research and develop a digital TV standard for the U.S.; an international organization of 200 members that is establishing voluntary technical standards for advanced television systems.

Advanced Time Division Multiple Access See ATDMA.

Advanced Voice Busy Out AVBO.

AdvancedTCA Advanced telecom computing architecture, also called ATCA; a series of carrier-grade communications equipment specifications from PICMG. Advanced Telecom Computing Architecture*, or AdvancedTCA* is a series of industry standard specifications for the next generation of carrier grade communications equipment. As the largest specification effort in PICMG's history and with more than 100 companies participating, AdvancedTCA incorporates the latest trends in high speed interconnect technologies, next generation processors, and improved reliability, manageability and serviceability, resulting in a new blade (board) and chassis (shelf) form factor optimized for communications. AdvancedTCA provides standardized platform architecture for carrier-grade telecommunication applications, with support for carrier-grade features such as NEBS, ETSI, and 99.999% availability.

advancenet An Ethernet-based local area network from Hewlett Packard, Palo Alto, CA. See Ethernet.

adventure gaming An interactive role-playing computer game in which the player becomes a character in the narrative.

advergaming An advertising technique in which the detailed product information is embedded in a computer game played online, designed to actively engage the player with the marketing message. It's aimed at young people. These games often capture information about the players which can be used in follow-up marketing campaigns.

advertainment This is marketing-speak for television adverts that are designed to be entertaining or funny, with the placement of the product played down. Some have been created by Hollywood directors such as the Coen Brothers or Spike Lee. There is now even a video-on-demand channel in the US on which you can watch your favorites.

advertising A packet switched networking term. Advertising is a process in which routing or service updates are sent at specified intervals so that other routers on the network can maintain lists of usable routes.

Advice of Charge AOC. Basically this is fancy name for seeing on your phone what you're being charged for the call as you speak. This service comes with some cell phones and some ISDN phones.

adviser Some people think it should be spelled adviser. Some think it should be spelled advisor. The most common spelling is adviser.

advisor See Adviser.

advisory tones Signals such as dial tone, busy, ringing, fast-busy, call-waiting, camp-on and all the other tones your telephone system uses to tell you that something is happening or about to happen.

adware Also called spyware and malware. Programs that are installed via the Internet on a user's PC for the financial benefit of someone else without the user's full knowledge and consent. Such programs range from relatively harmless – telling a distant computer which web sites a user visits to capturing the user's social security number and sending it to someone who's intent on stealing the user's financial identity. There are a number of software programs – such as Ad-Aware and Spybot's Search and Destroy – which will find adware on your computer and remove it. You should run such software at least once a week. See also drive-by downloads.

AE Account Executive. A fancy, schmanzy name for a salesperson. The idea is that the customer is an "account," and the salesman is the executive running the account. Telephone companies call their salespeople account executives – especially on the equipment and non-long distance side.

AEB Analog Expansion Bus. The analog voice processing bus designed by Dialogic which allows multiple cards to route audio signals within a PC. It is used to interface DTI/124 and D/4x voice response component boards which fit in an AT-expansion slot of a PC. See also PEB and SCSA which are more modern digital expansion buses.

AEC 1. Acoustic Echo Cancellation.
2. Alternate Exchange Carrier. See CLEC.

AECN Alternate Exchange Carrier Name. A unique identifier for a CLEC. It is a 4 digit number. CLEC stands for competitive local exchange carrier. Some phone companies call the AECN an ECC, which stands for Exchange Carrier Code. See ECC and Industry Standard Codes.

AECS Plan Aeronautical Emergency Communications System Plan. The AECS Plan provides for the operation of aeronautical communications stations on a voluntary, organized basis to provide the President and the Federal Government, as well as heads of state and local governments, or their designated representatives, and the aeronautical industry, with

a means of communicating during an emergency.

AEMIS Automatic Electronic Management Information System. This was the first computerized UCD/ACD reporting system introduced by AT&T for CO UCD (Uniform Call Distribution). This package was updated to become the PRO 150/500 system for UCD management on the Dimension PBX/UCD. AEMIS was the successor to the FADS or Force Administration Data System. It was an electro-mechanical system of peg counters and different colored busy lamp fields used to note trunk and position status.

AEP AppleTalk Echo Protocol. Used to test connectivity between two AppleTalk nodes. One node sends a packet to another node and receives a duplicate, or echo, of that packet.

aerial cable Cables strung outside and overhead. They're called aerial even though they only hang from poles or buildings. Some aerial cable hangs by its own strength. Some is supported by steel wire above it. Stringing aerial cable is cheaper than burying it, though buried cable lasts longer.

aerial cross box Also called a tree stand. A cross box on a pole. Used when there's a narrow easement.

aerial distribution method A method of running cables through the air, typically pole-to-pole. The old fashioned way. Some phone companies say aerial cable is more reliable than underground. Certainly, it's cheaper to fix or add to. It just looks less appetizing.

aerial insert In a direct-buried or underground cable run, an aerial insert is a cable rise to a point above ground, followed by an overhead run, e.g., on poles, followed by a drop back into the ground. An aerial insert is used in places where it is not possible or practical to remain underground, such as might be encountered in crossing a deep ditch, canal, river, or subway line.

aerial plant Cable and other telephone paraphernalia that is suspended in the air on telephone or electric utility poles.

aerial service wire splice A device used to splice aerial service wire and attached to the aerial wire. It's also called a football or a potato. Why? Because that's its shape.

AERM SS7 MTP 2 function that provides monitoring of link alignment errors.

aeronautical advisory station An aeronautical station used for advisory and civil defense communications primarily with private aircraft.

aeronautical broadcast station An aeronautical station which makes scheduled broadcasts of meteorological information and notices to airmen. In certain instances, an aeronautical broadcast station may be placed on board a ship.

aeronautical earth station An earth station in the fixed-satellite service, or, in some cases, in the aeronautical mobile-satellite service, located at a specified fixed point on land to provide a feeder link for the aeronautical mobile-satellite service.

Aeronautical Emergency Communications System AECS Plan. The AECS Plan provides for the operation of aeronautical communications stations, on a voluntary, organized basis, to provide the President and the Federal Government, as well as heads of state and local governments, or their designated representatives, and the aeronautical industry with communications in an emergency.

aeronautical fixed service A radiocommunication service between specified fixed points provided primarily for the safety of air navigation and for the regular, efficient and economical operation of air transport.

aeronautical fixed station A station in the aeronautical fixed service.

aeronautical mobile OR (off-route) service An aeronautical mobile service intended for communications, including those relating to flight coordination, primarily outside national or international civil air routes.

aeronautical mobile R (route) service An aeronautical mobile service reserved for communications relating to safety and regularity of flight, primarily along national or international civil air routes.

aeronautical mobile satellite service A mobile satellite service in which mobile Earth stations are located on board aircraft; survival craft stations and emergency position-indicating radiobeacon stations may also participate in this service.

aeronautical mobile satellite (OR) (off-route) service An aeronautical mobile-satellite service intended for communications, including those relating to flight coordination, primarily outside national and international civil air routes.

aeronautical mobile-satellite (R) (route) service An aeronautical mobile-satellite service reserved for communications relating to safety and regularity of flight, primarily along national or international civil air routes.

aeronautical mobile service A mobile service between aeronautical stations and aircraft stations, or between aircraft stations, in which survival craft stations may participate; emergency position-indicating radiobeacon stations may also participate in this service on designated distress and emergency frequencies.

aeronautical multicom service A mobile service not open to public correspondence, used to provide communications essential to conduct activities being performed by or directed from private aircraft.

Aeronautical Radio Inc. ARINC. The organization that coordinates the design and management of telecommunications systems for the airline industry. It's one of the largest buyers of telecommunications services and equipment in the world.

aeronautical radionavigation-satellite service A radionavigation-satellite service in which Earth stations are located on board aircraft.

aeronautical radionavigation service A radionavigation service intended for the benefit and for the safe operation of aircraft.

aeronautical station A land station in the aeronautical mobile service. In certain instances, an aeronautical station may be located on board ship or on a platform at sea.

aerospace Air force publicists coined the term "aerospace" to convince everyone that space was the business of those who fly in the air. According to the Economist Magazine, the "aerospace industry" was quickly accepted into the language, perhaps because President Eisenhower's alternative, the "military industrial complex," sounded rather more sinister. After the Apollo program, which ended in 1972, the "space" in aerospace often seemed like a syllable tacked on to make building airplanes sound grander. But the growth in satellite use in the 1980s made space a respectable business in its own right. In America as of writing in the fall of 1991, the annual sales of space hardware are now bigger than those of civilian aircraft.

AES Advanced Encryption Standard. A standard for encryption which is intended to replace DES (Data Encryption Standard), a standard developed by IBM in 1977 and thought to be virtually uncrackable until 1997. The AES standard specifies a symmetric, or private key, algorithm. It is a block cipher supporting key lengths ranging from 128 to 256 bits, and variable-length blocks of data. See also Block Cipher, DES, Encryption, and Private Key.

AET Application Entity Title. The authoritative name of an OSI application entity, usually a Distinguished Name from the Directory.

AF 1. Audio Frequency. The range of frequencies which theoretically are audible to the human ear; i.e., 30 Hz - 20 KHz. Truly high fidelity audio covers the entire range. Full AF is not practical over the PSTN, as to much bandwidth is required. Most of us can't hear the full AF range, anyway. As you get older, your hearing deteriorates. See also Bandwidth.

2. Assigned Frame. Motorola definition.

AFAIK As Far As I Know.

AFCEA Armed Forces Communications and Electronics Association. It is a non-profit association of active and retired telecommunications and networking professionals, with over 30,000 members from industry, government and the military. AFCEA has over 130 chapters worldwide. Chapter activities include meetings, seminars, symposia, continung education opportunities, and other activities. AFCEA publishes SIGNAL, a monthly news magazine about communications and networking in C4ISR environments. www.afcea.org.

AFE See Analog Front End.

AFI An ATM term. Authority and Format Identifier: This identifier is part of the network level address header.

affiliate 1. This definition from the Telecommunications Act of 1996. The term 'affiliate' means a person that (directly or indirectly) owns or controls, is owned or controlled by, or is under common ownership or control with, another person. For purposes of this paragraph, the term 'own' means to own an equity interest (or the equivalent thereof) of more than 10 percent. See the Telecommunications Act Of 1996.

2. A broadcast TV station not owned by a network, but one which includes the network's programs and commercials in its programming schedule.

Affiliated Sales Agency ASA. A term for a company which resells the service of a phone company. Typically, the phone company pays the ASA a commission. Sometimes the commission is so large that it blurs the thinking of the ASA into recommending to its customers telecom products and services they would be better without.

affiliates Sites that steer user to another e-commerce site in return for a piece of the action, i.e. a percentage of any buy. Go to my site, www.harrynewton.com. There you'll find a button that suggests you buy this dictionary via ecommerce. Click on the button. It gets you to Amazon. If you buy the dictionary, I get 7% of what you paid from Amazon. I'm an Amazon sales affiliate. Please buy the book. I need the money.

AFI Authority and Format Identifier. The portion of an NSAP format ATM address that identifies the type and format of the IDI portion of an ATM address. See also IDI and NSAP.

AFIPS American Federation of Information Processing Societies. A national, highly-respected organization formed by data processing societies to keep abreast of advances in the field. AFIPS organizes one of the biggest trade shows in the data processing industry

– the NCC (National Computer Conference).

affinity Requirements of an MPLS traffic engineering tunnel on the attributes of the links it will cross. The tunnel's affinity bits and affinity mask bits of the tunnel must match the attribute bits of the various links carrying the tunnel.

AFI Authority and Format Identifier. The part of an NSAP-format ATM address that identifies the type and the format of the IDI portion of an ATM address. See also IDI and NSAP.

AFN All Fiber Network. Burlington, Vermont is building an AFN network for its municipality of 40,000 people. The AFN will first support city services. Then it will extend fast Internet service to its businesses and residences.

AFNOR Acronym for Association Francaise de Normalisation. France's national standards-setting organization.

AFP AppleTalk File Protocol. Apple's network protocol, used to provide access between file servers and clients in an AppleShare network. AFP is also used by Novell's products for the Macintosh.

AFS Andrew File System.

AFT Automatic Fine Tuning; See AFC.

After-call Wrap-up The time an employee spends completing a transaction after the call has been disconnected. Sometimes it's a few seconds. Sometimes it can be minutes. Depends on what the caller wants.

aftermarket Trading in the Initial Public Offering after its IPO offering. Trading volume in IPOs is extremely high on the first day because of flipping (immediate selling) and aftermarket orders.

AGC Automatic Gain Control. There are two electronic ways you can control the recording of something – Manual or Automatic Gain Control (AGC). AGC is an electronic circuit in tape recorders, speakerphones, and other voice devices which is used to maintain volume. AGC is not always a brilliant idea since it will attempt to produce a constant volume level, that is, it will try to equalize all sounds – the volume of your voice, and, when you stop talking, the circuit static and/or general room noise which you do not want amplified. Never record a seminar or speech using AGC. The recording will be decidedly amateurish. Manual Gain Control means there is record volume control and is thus, preferred in professional applications.

AGC Threshold The level of input current at which the AGC circuit becomes active.

AGC Time Constant The amount of time it takes to achieve the required AGC level; also the amount of time it takes to recover from AGC.

AGCOMNET US Department of Agriculture's voice and data communications network.

Age of Interruption, The The world we live in today. We're working on the great american novel. The phone rings. An urgent email comes in. Our BlackBerry beeps. Our notebook chimes. Bingo, we're interrupted. See also continuous partial attention.

aged packet A data packet which has exceeded its maximum predefined node visit count or time in the network.

agenda A list, outline, or plan of things to be considered or done. What I want out of life. My life's plan.

agent 1.The classic definition of an agent is an entity acting on behalf of another.

2. This term comes from the huge telephone call-in reservation centers which the airlines, hotels and car rental services run. An agent is the person who answers your call, takes your order or answers your question. Agents are also called Telephone Sales Representatives or Communicators. The term "agent" was first used in the airline business. It came from gate or counter ticket agent.

3. An "agent" is the person or persons you have legally authorized to order your telephone service and equipment from telephone companies.

4. In the computer programming sense of the word, an agent acts on behalf of another person or thing, with delegated authority. The agent's goals are those of the entity that created it. An agent is an active object with a mission, but agents are abstractions that can be implemented in any way, whereas an object has a formal definition. Business Week in its February 14, 1994 issue wrote, "It's what computer scientists call an 'agent' – a kind of software program that's powerful and autonomous enough to do what all good robots should: help the harried humans by carrying out tedious, time-consuming, and complex tasks. Software agents just now emerging from the research labs can scan data banks by the dozen, schedule meetings, tidy up electronic in-boxes, and handle a growing list of clerical jobs."

5. Windows defines agent slightly differently. It said that an agent was software that runs on a client computer for use by administrative software running on a server. Agents are typically used to support administrative actions, such as detecting system information or running services.

See also Bot.

agent logon/logoff A call center term. Agents begin their day by punching some buttons on their phone. This indicates to the automatic call distributor that they are now ready to take calls. Later in the day, they punch some other buttons and indicate to the ACD that they are now ready to stop working. This is called logoff.

agent sign on/sign off A feature which allows any ACD agent to occupy any position in the ACD without losing his or her personal identity. Statistics are collected and consolidated about this agent and calls are routed to this agent no matter where he sits or how many positions he may occupy at one time.

aggregate AMA record A telephone company AIN term. An AMA record generated to record multiple instances of service usage within a specified aggregation interval. It is created by formatting peg counts of AMA events.

aggregate bandwidth The total bandwidth of channel carrying a multiplexed bit stream. It includes the payload and the overhead. For example, a T-1 line has an aggregate full duplex bandwidth of 1.544 million bits per second.

aggregate rate The sum of the channel data rates for a given application.

aggregation 1. An AMA (Automatic Message Accounting) function that accumulates AMA data, resulting in a less than detailed AMA record.

2. An ATM term. Token A number assigned to an outside link by the border nodes at the ends of the outside link. The same number is associated with all uplinks and induced uplinks associated with the outside link. In the parent and all higher-level peer groups, all uplinks with the same aggregation token are aggregated.

3. Making otherwise scattered data accessible at a single location, usually via a Web page. Aggregation is necessary to make sense of the hodgepodge of otherwise useful corporate data – which can range from catalog information to pricing material, to product support, to news or marketing materials – that results when previously disparate computer systems at a company and its partners are tied together.

aggregation device A specialized ISDN terminal adapter that can aggregate, or bond, the two B channels "on the fly" into a single higher-speed connection. Some aggregation devices also include an Ethernet bridge, i.e. a connection to a local area network.

aggregator 1. A breed of long distance reseller. An aggregator is essentially a sales agent for a long distance company. Here's how it works: The aggregator goes to a long distance company and says "May I sell your long distance service at a discount?" The long distance company says Yes. The aggregator hits the street and sells cut-rate long distance service to any and everyone. The long distance provider installs the service and bills it. The aggregator makes its profit by charging a fixed monthly service fee, a percentage of savings or some other arrangement. The key to it: The end user saves some money because his calls are "aggregated" with those of ALL the customers of the aggregator and the long distance company extends a bulk savings to the aggregator. At least that's the theory. "Should you – as an end-user – consider buying your long distance from an aggregator? The simple answer is YES. Discounts are sometimes so deep it's not uncommon for a company using a major carrier to switch to billing through an aggregator and save 20% to 25% – with nothing of substance happening. They still get their bills from their normal carrier and they still place and receive calls on their existing carriers as they had been doing. No wires are touched. No routing is changed.

"What about the pitfalls? There are some: First, don't buy long distance that isn't billed directly by the long distance carrier providing the service. If the aggregator does the billing, there's too much opportunity for "mischief," says Dick Kuehn, Cleveland consultant. "There's opportunity for doing things like increasing each of your calls by 30 seconds. And because a user has no answer supervision on his call detail records, it's very hard for the user to figure his exact timing." The problem, says Dick, is there's no way for a user to verify his own bill. Dick says "Carriers are honest. Resellers (aggregators that bill) are open to question."

Dick also believes you probably shouldn't deal with an aggregator who bills you a percentage of "savings." This is also open to abuse. There are so many rates, so many changes monthly, so many options that it's virtually impossible for the user to figure out what he would have paid had he not gone with the aggregator. The calculation is too open to abuse.

The panoply of companies in the long distance business – not only aggregators – has expanded dramatically. And confusion between companies and what they did became rife. All, of course, purport to save you money on your long distance bills. And many do. Here's a simple explanation of the major categories:

FACILITIES BASED CARRIER. Owns most of its circuits. Has own sales force and possibly independent sales agents. Best examples: AT&T, WorldCom, Allnet and Sprint.

TRADITIONAL RESELLER. Rents/leases most circuits or buys bulk time from carrier. Resells under own brand name, has published prices, sends own bills. Appears to be (and for all practical purposes is) same as the carriers.

AGGREGATOR. "Sponsor" who buys carrier's (typically AT&T) multi-location 800 or outbound service; enrolls other businesses as sites; volume discounts for all based on total calling at all sites. End user is still the carrier's, not the aggregator's. The carrier typically does the billing.

REBILLER: (Also called "Switchless Reseller"). Buys service as multi-location customer from carrier. Signs up individual sites (just like aggregator). Generates own end-user bills. No switch or network, but does sales, customer service, billing for long distance calls. Sometimes the rebiller's bills are more detailed than the bills you get directly from the carrier.

SALES AGENTS: Businesses or groups who are not direct employees of carrier, but who receive sales commissions from carrier. Customers belong to carrier and carrier does billing.

OTHER THIRD-PARTY MARKETERS. Buying co-ops, user groups, long distance brokers, pyramid (legal) marketing systems, shared tenant providers, Centrex aggregators, affinity groups (like college alumni and church congregation groups).

2. In the world of blogs, an aggregator is An aggregator is software that pulls information from various web feeds that you have selected and displays any updates made to them. Aggregators make it possible to download the updated content (or an excerpt) for viewing on your computer, in your web browser, or even displayed on another website without needing to visit the original website. This definition from blogossary.com.

aggressive accounting Euphemism for any of the sleight of hand accounting practices used by companies to hide their weaknesses and artificially inflate their value. Current corporate magicians include Enron, WorldCom, Xerox, AIG, Time Warner (when they were AOL-Time Warner)– and the list continues to grow.

aggressive mode The connection mode that eliminates several steps during internet key exchange authentication negotiation) between two or more IPsec peers. Aggressive mode is faster than main mode but not as secure. See IPsec, IKE.

aging The change in properties of a material with time under specific conditions.

aging number Telephone company terminology for a disconnected phone number that is not available for assignment to another customer for a specified period of time.

agnostic An agnostic is a a freethinker. He or she is a person who holds the view that any ultimate reality, or absolute truth, (i.e., God) is unknown and probably unknowable. An agnostic is neither a believer nor a non-believer (i.e. he's not an atheist). He isn't committed either way, since, in his opinion, there's no way to know the truth. He'll work with all realities. The term generally relates to the domain of religion. In a telecommunications context, we use the term in more or less the same way. For example, if we say a service is network agnostic, we mean that it works with any network. For example, a Layer 1 (i.e., Physical Layer) protocol (e.g., T-carrier, SONET, or DWDM) neither knows nor cares what higher-layer protocols (e.g., PCM, Frame Relay, or TCP/IP) are running over it and neither knows nor cares about the native applications (e.g., voice, video, or LAN-to-LAN internetworking). The Layer 1 circuit just creates and hauls the signal. There are many other ways to use the word agnostic around the telecom and computing industries. But the meaning is always the same. An OS agnostic piece of equipment means that it will run on several operating systems – Windows, Linux, Unix, etc. But – and this is the caveat – every buzzword (like agnostic) is ultimately a marketing/sales term. And before you believe it all, you'd better check. And then check again. And again.

AGP Accelerated Graphics Port. An Intel-developed interface that enables high-speed graphics. Graphics data move between the PC's graphics controller and computer memory directly, instead of being cached in video memory. An interface specification that enables 3-D graphics to display quickly on ordinary personal computers. AGP is designed to convey 3-D images (for example, from Web sites or CD-ROMs) much more quickly and smoothly than is possible today on any computer other than an expensive graphics workstation. It is especially useful in conjuction with gaming, three-dimensional (3D) video, and sophisticated scientific/engineering graphics programs. The interface uses your computer's random access memory (RAM) to refresh the monitor image and to support the texture mapping, z-buffering, and alpha blending required for 3-D image display. AGP offers high-speed data transfer to and from RAM, optimizing the use of memory and minimizing the amount of memory necessary for high-performance graphics. The AGP main memory use is dynamic, meaning that when not being used for accelerated graphics, main memory is restored for use by the operating system or by other applications. AGP runs at several times the bus speed of conventional Peripheral Component Interconnect (PCI). Because of this, the data transfer rate using AGP is significantly greater than with PCI video cards. AGP employs eight sideband address lines, so multiple data transfers can take place concurrently. The final 8X specification was released in September 2002 yielding a throughput of 2.1 gigabytes per second. You can buy AGP graphics boards. They're very powertul. But you must have an AGP "port" on your motherboard.

AGPS Assisted Global Positioning System. Essentially GPS in a cellphone. See GPS.

AGTK Application Generator ToolKit. A set of tools that are used to implement and modify a voice-processing application. It includes software to create the script and packages for the creation and editing of prompts. See Application Generator.

AH Authentication header. An IPSec header used to verify that the contents of a packet have not been modified while the packet was in transit. See AHP.

AHP Authentication Header Protocol. A protocol used in IPSec that authenticates a packet IP header and payload (content). If a packet is modified during transmission, the recipient is notified. See IPSec.

AHD Audio High Density. System of digital audio recording on grooveless discs, employing an electronically-guided capacitance pickup.

AHFG ATM-attached Host Functional Group: The group of functions performed by an ATM-attached host that supports the ATM Forum's specification for MPOA (Multiprotocol over ATM).

AHR Abbreviation for ampere hour, measurement of battery power: how much current may be drawn for an hour. Important specification for portable computers, cellular phones, etc.

AHT 1. Average Handle Time. The amount of time an employee is occupied with an incoming call. This is the sum of talk time and after-call-work time. Contrast with Average Holding Time.

2. See Average Holding Time.

AHT Distribution Average Handle Time Distribution. A set of factors (either 48 or 96) for each day of the week that defines the typical distribution of average handle times throughout the day. Each factor measures how far AHT in the half or quarter hour deviates from the AHT for day as a whole.

Ahoy See Hello.

AI Artificial Intelligence. Perhaps the next phase of computing. The present forms of AI in computer software are called Expert or Knowledge Based systems.

Ai An ATM term. Signaling ID assigned by Exchange A.

AIA 1. An M.100/S.100 definition. Application Interface Adapter: a component providing the client side of a client server connection to an S.100 server. See M.100 and S.100.

2. Automatic Internal Administration.

3. American Institute of Architects.

AICC Automatic Incoming Call Connection. A Rolm term for connecting an incoming call to the person's phone, without requiring him/her to press any keys.

AIDS 1. Access IDentifier.

2. A Trojan Horse software program (a virus) which caused extensive damage in December 1989.

AIFF Audio Interchange File Format. This audio file format was developed by Apple Computer for storing high-quality sampled audio and musical instrument information. It is Apple's answer to Microsoft's WAV file format. It is also used by Silicon Graphics and in several professional audio packages. Played by a variety of downloadable software on both the PC and the Mac. Encoding a file in AIFF is about as uncompressed as you can get. To save some space, try encoding files in Apple's Apple Lossless format, which offers identical quality at nearly half the size. See also ADPCM, PCM, sound, TrueSpeech, VOC, WAV and waveform.

AIIM Association for Information and Image Management.

AIM 1. Amplitude Intensity Modulation.

2. Association for Interactive Media. Originally called the Interactive Television Association (ITA). The AIM is a Washington association of companies and organizations involved with interactive media. According to the AIM CEO, "ITA has long been the industry's most forceful proponents of the view that high speed Internet and interactive television development are so interrelated that, from the customer's perspective, these services will be seamless." www.interactive hq.org.

AIMS An Acronym for Auto Indexing Mass Storage. Indicates the AIMS Specification which is a standard card interface for storing large data such as image and multimedia files.

AIMUX ATM Inverse Multiplexing: A device that allows multiple T-1 or E1 communications facilities to be combined into a single broadband facility for the transmission of ATM cells.

AIN Advanced Intelligent Network. A now somewhat obsolete term. AIN was based on circuit switching, not on today's more modern packet switching. AIN was a term promoted by Bellcore (now Telcordia) and adopted by Bellcore's original owners, the regional Bell holding companies, and by AT&T and virtually every other phone company. AIN was meant to indicate the architecture of their networks for the 1990s and beyond. Much AIN architecture was introduced in varying degrees. But most of the features thought for it never happened. While every phone company had a different interpretation of what their AIN is (or was) , there seems to be two consistent threads. First, the network can affect (i.e. change) the routing of calls within it from moment to moment based on some criteria other than the normal, old-time criteria of simply finding a path through the network for the call, based on the number originally dialed. Second, the originator or the ultimate receiver of the call can somehow inject intelligence into the network and affect the flow of his call (either outbound or inbound). The concept of AIN is simple. Before calls are sent to their final destination, the network queries a database. "What should I do at this very moment with this phone call?" The disposition of the call depends on the response. That database may belong to the phone company. Or it may belong to the customer. It makes no difference, so long as they're connected. And various carriers (phone companies) have proposed and implemented various ways of joining these databases. Initial AIN services tended to be focused on inbound inbound toll-free calls. Although no two phone companies seem to have the same idea as to what an Advanced Intelligent Network is, (some call it just an Intelligent Network), it generally includes three basic elements:

1. Signal Control Points. SCPs. Computers that hold databases in which customer-specific information used by the network to route calls is stored.

2. Signal Switching Points. SSPs. Digital telephone switches, which can talk to SCPs and ask them for customer-specific instructions as to how the call should be completed.

3. Signal Transfer Points. STPs. Packet switches that shuttle messages between SSPs and SCPs.

All three communicate via out of band signaling, typically using Signaling System 7 (SS7) protocol. The AIN has increased in complexity, as carriers have added voice response equipment that can prompt callers to enter further instructions as to how they'd like their call handled. Despite the differences between AIN networks, all work fundamentally the same, according to Mark Langner at the time with TeleChoice, Verona, NJ: The SS7 identifies that a call requires intelligent network processing. The SSP creates a query to find out how this call should be handled. The query is passed via out-of-band signaling through STPs to an SCP. That interprets the query based on the criteria in its database and information provided by the SSP. Once the SCP has determined how the call is to be handled, it returns a message through STPs to the SSP. This message instructs the SSP how the call should be handled in the network. According to Langner, the number of actions that could take place at the SCP are truly infinite. The call could be translated into a different number for completion. It could be routed to a user's private network for on-net handling. It could be sent to a voice response unit in the carrier network, where a message is played to the caller. Or it could even be blocked, preventing completion of the call.

Among the IN (Intelligent Network) services, some manufacturers including Ericsson has identified:

- Enhanced number translation services functions
- Enhanced screening services, i.e. selective call diversion
- Selective forwarding of calls * Location-dependent call forwarding
- Improvements to voice announcements
- Services to support fixed and mobile integration, i.e. personal communications services, PCS and universal personal telecommunications, UPT, and
- Enhanced billing.

See AIN definitions below and NCD, SCP, SiteRP, SS7, SSP and STP.

AIN Release 0 Advanced Intelligent Network Release defined by individual Bell Operating Companies for initial deployment in 1991, or so. See AIN.

AIN Release 0.0 Advanced Intelligent Network Release based on Ameritech specifications with input from Telcordia and some vendors. Contains three trigger detection points. Deployed in 1992 (U.S.) and end of 1993 (Canada). First service for this architecture was "Switch Redirect" for Bell Atlantic (for switch or line failure.) See AIN.

AIN Release 0.1 Advanced Intelligent Network Release provides for some additional functionality and more extensions to Rel 0.0. Contains 5 trigger detection points. See AIN.

AIN Rel 1.0 Advanced Intelligent Network Release target architecture for AIN. Contains 32 trigger detection points. (Hence Rel 0.0 & 0.1). See AIN.

AIN Release 1 Logical Resources For Bell Operating Companies, the logical network resources configured and updated to provide Advanced Intelligent Network Release 1 subscriber services (e.g., SLP and trigger data). See AIN.

AIN Release 1 Switching System An access tandem, local tandem or end office that contains an ASC (Advanced Intelligent Network Release 1 Switch Capabilities) functional group. See AIN.

AIN Release 2 An Advanced Intelligent Network Release for initial deployment in 1995, evolving from AIN Release 1 and supporting an expanded range of information networking services from the Bell operating telephone companies. See AIN.

AIN Switch Capabilities ASC. A functional group residing in an Advanced Intelligent Network Release 1 Switching System that contains the Network Access, Service Switching, Information Management, Service Assistance and Operations FEs (Functional Entities). See AIN.

AINTCC Automated INTercept Call Completion. A new feature of Northern Telecom's central offices. The AINTCC feature provides options for connecting a caller automatically to an intercepted number after hearing an announcement, or connecting a caller to an intercepted number without an announcement. Not using an announcement makes the number change transparent to the caller. The called (intercepted) party then has the option of informing the caller of the number change.

AIOD Automatic Identification of Outward Dialing is the ability of the telephone system to know the specific extension placing a call. It's used as part of the process of recording the detail of each telephone call for billback and cost control purposes. See AIOD Leads and Call Accounting System.

AIOD Leads Terminal equipment leads used solely to transmit automatic identified outward dialing (AIOD) data from a PBX to the public switched telephone network or to switched service networks (e.g., EPSDS), so that a vendor can provide a detailed monthly bill identifying long-distance usage by individual PBX extensions, tie-trunks, or the attendant.

AIR An ATM term. Additive Increase Rate: An ABR service parameter, AIR controls the rate at which the cell transmission rate increases.

air coax A form of coaxial cable which uses air as a dielectric.

air core cable Cables built with no grease or jelly to surround the cable pairs. See Dry Pic.

Air Blown Fiber ABF. Small, flexible plastic microduct tubing installed prior to the installation of individual or multiple optical fibers that are "blown in" through the microduct using compressed air. Fibers can travel 300 meters (1,000 ft) or more in a single run or turn up to 300 tight corners in a matter of seconds. When network changes are needed, the installer simply "blows out" the old fiber, then blows in new fiber, with minimal disruption in an office environment.

air conditioning In the Department of Defense, air conditioning is a synonym for the term "environmental control," which is the process of simultaneously controlling the temperature, relative humidity, air cleanliness, and air motion in a space to meet the requirements of the occupants, a process, or equipment.

air gap The separation of two networks so that there is "nothing but air" between them. This security approach is used by the US military to separate its classified networks from unclassified networks.

air gap termination Nathan Oldacre, who works for ITCDeltacom, sent me this: Air gap termination is a term we have used quite a bit at our company. This phrase is used whenever something gets unplugged by accident. One day, we were working when all our systems technicians came in to install a new UPS for our LAN server. All of a sudden all 50 users in our department lost their network connections. Being friends with the technicians, I asked Marcus Flack if he was aware of our problem. He told me that the router we were connected to had experienced a brief air gap termination. Brian French looked in my eye and said, "Marcus stepped on the power cord." Since then we have had several outages due to air gap terminations of various type.

air handling plenum A designated area, closed or open, used for environmental air circulation (return air). For a larger explanation, see Plenum.

air interface Air interface is a cellular industry term. It refers to the system that ensures compatibility between equipment (cell phones) and the base stations. It involves the specification of channel frequencies and widths, modulation, power and power sensitivity levels, and data framing. The system also selects which radio channels are employed during a call. Air interface is the standard operating system of a mobile network. A four-layer protocol stack which ensures compatibility between terminal equipment and base stations, or hubs, through the development of a standard. In terms of the OSI Reference Model, the layers include the Physical (PHY), the Media Access Control (MAC) layer, the Data Link Control (DLC) layer, and the Network layer. The PHY layer specifies radio characteristics such as channel frequencies and widths, modulation schemes, power and power sensitivity levels, and data framing. The MAC layer, which cuts across the PHY and DLC layers, specifies the procedures by which the wireless terminal and the base station negotiate selection of the radio channel to be employed. The DLC layer specifies the manner in which the frames are sequenced, and the mechanism used to ensure their integrity during transmission. The Network layer specifies the mechanism used to identify and authenticate the wireless terminal to the base station. Air interfaces are specified for technologies such as AMPS (Advanced Mobile Phone System), CDMA (Code Division Multiple Access), DECT (Digital European Cordless Telecommunications), GSM (Global System for Mobile Com-

munications), PCS (Personal Telecommunications Services), and PWT (Personal Wireless Telecommunications), and TDMA (Time Division Multiple Access). See also AMPS, CDMA, DECT, GSM, OSI Reference Model, PCS, PWT, TDMA.

air PBX An IP PBX system, still in the early stages of productization, where traditional PBX desk phones are replaced by cell phones. The implementation can be via a wireless LAN or cellular or a combination of the two, the last scenario requiring the use of dual-mode Wi-Fi/cellular phones. An air PBX is designed for enterprise environments with mobile and nomadic workers.

air pressure Air pressure at sea level is roughly equal to the weight of an elephant spread over a small coffee table.

air pressure cable Telephone cable equipped with air-pressure equipment so the phone company can determine when there's a problem with the line. When a cable is cut, the pressure drops and the company is notified of the problem. Nitrogen is often used instead of air because nitrogen is noncorrosive. Nitrogen also prevents water entering the cable when there's a break.

air rights The right of a landowner to protection of access to light and air above a structure. Air rights have value in many cities and can be sold. They might be sold, for example, to an adjoining building which can then build higher and bigger.

air space coaxial cable One in which air is the essential dielectric material. A spirally wound synthetic filament of spacer may be used to center the conductor.

air time Time spent talking on a cellular network to calculate billing. See also Airtime.

Air-To-Ground See ATG.

airborne radio relay 1. Airborne equipment used to relay radio transmission from selected originating transmitters.

2. A technique employing aircraft fitted with radio relay stations for the purpose of increasing the range, flexibility, or physical security of communications systems.

airbrush A computer imaging term. A fine-mist paint tool used to create halos, fog, clouds, and similar effects. Most paint programs let you control the size and shape of the application area. Some packages provide a transparency adjustment that determines the density of the applied color.

aircraft earth station A mobile Earth station in the aeronautical mobile-satellite service located on board an aircraft.

aircraft emergency frequency An international aeronautical emergency frequency, such as 121.5 MHz (civil) and 243.0 MHz (military), for aircraft stations and stations concerned with safety and regulation of flight along national or international civil air routes and maritime mobile service stations authorized to communicate for safety purposes.

aircraft station A mobile station in the aeronautical mobile service, other than a survival craft station, located on board an aircraft.

airdrome control station An aeronautical station providing communication between an airdrome control tower and aircraft. Also called airport control station.

AIRF Additive Increase Rate Factor: Refer to AIR.

Airfone The first air-to-ground radiotelephone service. The company was founded around 1975 by Jack Goeken, who had earlier founded MCI. Goeken sold Airfone to GTE in 1986, and in 2000 the company, along with the rest of GTE, became part of Verizon. When an airline passenger places an Airfone call over North America, the call is transmitted via a radio in the aircraft's belly to the closest of 135 terrestrial radio base stations located throughout North America. From there the call travels to one of three main ground-based switching stations, and from there over a backhaul circuit for interconnection with the public telephone network. A call placed on a flight outside of North America is radioed to an orbiting satellite, and from there to a satellite earth station for subsequent transport and handoff to the PSTN.

airline control protocol Data link layer polled protocol that runs in full-duplex mode over synchronous serial (V.24) lines and uses the binary-coded decimal (BCD) character set.

airline mileage The monthly charge for many leased circuits is billed on the basis of "airline mileage" between the two points. Though it sounds as if it's the distance a crow would fly directly between the two points, in reality, it is the distance in mileage between two Rate Centers whose position is laid down according to industry standards, originally created by AT&T. The entire U.S. is divided by a vertical and horizontal grid. The coordinates – vertical and horizontal – of each rate center are defined and applied to a square root formula which yields the distance between the two points. Think back to school. There's a right-angled triangle. At the top is one Rate Center. At the side is the other Rate Center. The horizontal is the horizontal coordinate. The vertical is the vertical coordinate. The formula is simple: Square the vertical distance. Square the horizontal distance. Add the two together. Then take their square root. That will give you the distance across the hypotenuse – the side opposite the right angle in the triangle. Thus, your "airline" mileage. For sample V and H city coordinates and the formula on how to calculate airline mileage, see V & H under the letter V.

airline miles See Airline Mileage.

airline protocol Generic term that refers to the airline reservation system data and the protocols, such as P1024B (ALC), P1024C (UTS), and MATIP, that transport the data between the mainframe and the ASCUs.

Airline X.25 See AX.25.

airlink A cellular radio term. Airlink is the physical layer radio frequency channel pair used for communication between the Mobile End System (M-ES) and the Mobile Data Base Station (MDBS).

airlink interface The Cellular Digital Pack Data (CDPD)-based wireless packet data service provider's interface for providing services over the airlink to mobile subscribers.

airplane mode A mode on some cell phones whereby the phone's network connection is turned off in order to comply with FAA regulations, while letting the user continue to use the handset for games and other applications.

airtime Actual time spent talking on a cellular telephone. Most cellular carriers bill their customers based on how many minutes of airtime they use each month. Whether the calls are incoming or outgoing makes no difference, the customer is still billed. Whether the calls are going to a toll-number or a toll-free 800 number also makes no difference. The customer racks up airtime and he pays. The more minutes of time spent talking on the phone, the higher the bill. Airtime charges during peak periods of the day in North America vary from 25 to 80 cents per minute. Most carriers offer a discount on these rates for off-peak usage. Some carriers offer a discount on these rates if the customer pays a higher minimum usage charge each month.

airwave Airwave systems are transmission systems that use the "airwaves," rather than conductors, to transmit information. Airwave systems actually send information across "space," rather than through conductors. The term "airwave" comes from the fact that human speech, in its native form, is an acoustic means of communications which makes use of the physical matter in the air to conduct compression waves. From mouth (transmitter) to ear (receiver) the physical matter (e.g., molecules of oxygen, carbon dioxide and such) in the air actually carries, or conducts, the signal. Airwave transmission systems (e.g., microwave, satellite and infrared) support information transfer from transmitter to receiver through space, using electromagnetic energy in the form of radio or light signals. The presence of the physical matter which occupies the space between transmitter and receiver actually causes the signal to attenuate, or weaken. The term "airwave" persists, however. See also Free Space Communications.

AIS 1. Alarm Indication Signal. Formerly called a "Blue Alarm" or "Blue Signal." An AIS is a signal transmitted downstream informing that an upstream failure has been detected. AIS is a signal that replaces the normal traffic signal when a maintenance alarm indication has been activated. In ATM, an alarm indication signal is an all ones signal sent down or up stream by a device when it detects an error condition or receives an error condition or receives an error notification from another unit in the transmission path. See also Squelching.

2. Automatic Intercept System.

3. Automated Information System. Any equipment of an interconnected system or subsystems of equipment that is used in the automatic acquisition, storage, manipulation, control, display, transmission, or reception of data and includes software, firmware, and hardware.

AITP Association of Information Professionals. AITP began life in 1951 as the NMAA (National Machine Accountants Association), changed its name to the DPMA (Data Processing Management Association) in 1962, and then to AITP in 1996. AITP provides IT-related education, information on relevant IT issues, and forums for networking with peers. www.aitp.org.

AIW Application Implementer's Workshop. A group of vendors working with IBM to develop software and hardware consistent IBM's Advanced Peer-to-Peer Networking protocol.

AIX Advanced Interactive eXecutive: IBM's implementation of UNIX. The Open Software Foundation (OSF) based its first operating system (OSF-1) on AIX. The next revision of the OSF operation system (OSF-2) will also be based on AIX with a Mach kernel (Mach was developed by Carnegie Mellon University).

AJAX Asynchronous JavaScript and XML. AJAX is a set of technologies that can be used to make interactive Web applications run more efficiently than they would otherwise, all other things being equal. AJAX accomplishes this by updating only those parts of a web page that

need updating during a user session, rather than refreshing the entire Web page. This reduces server and network traffic (which benefits all users who who are using these network resources) and it similarly reduces demands on a user's CPU, which further improves the interactive user's experience.

AK Authorization key.

AKO Bisync acknowledgment for even-numbered message.

Aka Denwa Japanese for a red telephone. Some coin phones in Japan are red and are known as aka denwas.

Akku German for "rechargeable battery" (accumulator).

AL ATM Adaptation Layer. The third layer of the ATM Protocol Reference Model. The AAL layer comprises the Convergence Sublayer (CS) and the Segmentation and Reassembly sublayer (SAR). In total, it is at this layer that multiple applications are converted to and from the ATM cell format. ATM Adaptation Layer sits above the ATM Layer, supporting higher-layer service requirements. For data communications services, the AAL defines a segmentation/reassembly protocol for mapping large data packets into the 48-octet payload field of an ATM cell. See the next the five definitions. See also CS AND SAR.

Al-Jazeera The best known of some 150 Arab satellite channels. As of early 2005, it boasted some forty to fifty million regular viewers. It is broadcast from Qatar and basically financed by the Emir of Qatar.

ALAP AppleTalk Link Access Protocol. In an AppleTalk network, this link access-layer (or data link-layer) protocol governs packet transmission on LocalTalk.

alarm Notification that the traffic signal has degraded or failed or equipment is malfunctioning. An SNMP message notifying an operator or an administrator of a network problem. See also Event and Trap.

alarm display 1. A message notifying an operator or administrator of a network problem.

2. Attendant console indicators show the status (i.e. what's happening) in the telephone system. There are usually two types of alarms – minor and major. Minor displays may be something as "minor" as a "hung" trunk, i.e. one that didn't hang up when the person speaking on it hung up. They can often be remedied by turning the PBX off, counting to ten, and then turning it on. (Before you do, check it will load itself.) Major problems – such as a blown line card in the PBX, one console out or half the trunks out – often require a service call and are often covered under the Emergency Conditions section of telephone service contracts.

alarm holdoff See Alarm Soaking.

alarm indicating signal AIS. Also known as a "Blue Alarm Signal" or "All Ones Keep Alive," an AIS is an unframed all-ones bit pattern sent by equipment at the far end to indicate that an alarm condition exists upstream in a circuit leading to the downstream equipment. Keep-alive signals are required by the network facilities to prevent oscillation of the line repeaters which causes interference (i.e. cross-talk and bleeding) within adjacent channels. SONET defines four categories of AIS: Line AIS, STS Path AIS, VT Path AIS, DSn AIS. See the next signal.

alarm indication signal AIS. A signal that replaces the normal traffic signal when a maintenance alarm indication has been activated. See the previous definition.

alarm soaking The allowing of an error condition to persist before action is taken. Alarm holdoff is another term for it. The term "soak period" is used for the holdoff period.

ALB Analog Loop Back. A test to see if a phone line is working and how well it's working. Analog Loop Back is a common test for locating transmission problems in data transmission systems.

albatross Many Japanese golfers carry "hole-in-one" insurance. With the Japanese, who take their golf very seriously, it is traditional to share one's good luck by sending gifts to all your friends when you get an "ace." The cost of what the Japanese term an "albatross" often reaches $10,000. See Albatross Manager.

albatross manager He hangs around you constantly. You can't get rid of him. This definition courtesy Tom Henderson. See also Seagull Manager.

ALBO Automatic Line BuildOut. ALBO is a means of automatic cable equalization used in T-1 span-line interface equipment.

ALC 1. Automatic Level Control.

2. See Automatic Light Control.

3. Airline Line Control. A full-duplex, synchronous communications protocol used in airline reservations systems. ALC is a packet polling protocol which adheres to a strict master/slave relationship between the central host and the remote terminals. ALC relies on IBM's IPARS (International Airline Passenger Reservation System) character set, which comprises 6 data bits and no parity bits. Error detection is accomplished through a Cyclic Check Character (CCC), but only very limited procedures are identified for error correction. ALC was

designed for use in X.25 networks, but also works in a Frame Relay environment.

4. Assembler Language Coding.

Alcatel Alcatel is a huge French telecommunications manufacturer. Its real name is Compagnie FinanciOxE4re Alcatel. And it's based in Paris. Lucky company.

ALE 1. Approvals Liaison Engineer. This engineer acts on your behalf to assess design and component changes to your BABT approved products. BABT is the British Approvals Board for Telecommunications.

2. See Automatic Link Establishment.

3. Access Line Equipment.

alerter service A Windows 2000 term. Notifies selected users and computers of administrative alerts that occur on a computer. Used by the Server and other services. Requires the Messenger service.

alerting A signal sent to a customer, PBX or switching system to indicate an incoming call. A common form is the signal that rings a bell in the telephone set. Others signals can trigger such devices as whistles, gongs and chimes.

alerting call A call for which the subject connection is in the alerting state. This usually implies that the telephone instrument is ringing.

alerting pattern An intelligent network term. Alerting pattern is a specific pattern used to alert a subscriber (e.g. distinctive ringing, tones etc.). See Q.931.

alerting signal A ringing signal put on subscriber access lines to indicate an incoming call. Telcordia defines alerting signals more broadly; thus: "Alerting signals (for example, ringing, receiver off-hook) are transmitted over the loop to notify the customer of some activity on the line."

alerting state A state in which a device is alerting (e.g., ringing) or is being presented (offered) to a device. This indicates an attempt to connect a call to a device. The device may be a device such as a telephone station. The device may also be a routing or distribution type of device. This includes a ACD or Hunt Group device.

ALEX Software which provides Internet users with a transparent read capability of remote files at anonymous FTP sites.

Alexander Graham Bell See Bell, Alexander Graham.

algorithm A prescribed finite set of well defined rules or processes for the solution of a problem in a finite number of steps. Explained in normal English, it is the mathematical formula for an operation, such as computing the check digits on packets of data that travel via packet switched networks. Algorithm derives from the name of the ninth-century Persian mathematician al-Khomeini, who also had a lot to do with the invention of algebra. The word algebra comes from the Arabic al-jabr, which first appeared in a treatise by al-Khwarizmi.

algorithmic language An artificial language established for expressing a given class of algorithms.

ALI Automatic Location Identification or Information. Working with Automatic Number Identification, the use of a database to associate a physical location with a telephone number. ALI is a feature of E-911 (Enhanced 911) systems. ALI is provided to agents answering E-911 calls. It may include information such as name, phone number, address, nearest cross street and special pre-existing conditions (i.e. hazardous materials). On some systems it may also provide the appropriate emergency service address for the particular address. ALI is retrieved from a computer database. The database may be held on site or at a remote location and may be maintained by the local phone company (or its parent) or another agency. See also PS/ALI.

alias 1. You're called John Smith. But on computer systems, you call yourself Jack Plumppuding. "Jack Plumpudding" is your alias.

2. A shortcut that enables a user to identify a group of hosts, networks, or users under one name. Aliases are used to speed user authentication and service configuration. For example, in configuring a Firebox a user can set up the alias "Marketing" to include the IP addresses of every network user in a company's marketing department.

3. A feature of the Apple Macintosh System allowing the user to create a file that points to the original file. When you click on an alias, the original application is launched. Aliases can work across a network; so you can access a program residing on a file server or a Mac that runs file sharing.

4. Unwanted signals generated during the A-to-D (Analog to Digital) conversion process. This is typically caused by a sampling rate that is too low to faithfully represent the original analog signal in digital form. Typically, a rate that is less than half the highest frequency to be sampled.

5. A nickname for a domain or host computer on the Internet. See also Alias Email Address.

6. An alias is an e-mail address that forwards its mail to a specified mailbox, masking the true name of the mailbox in which the mail is actually received. For example, Sales@ JoesDomain.com could be an alias for JoeSmith@JoesDomain.

alias email address An Internet Service Provider (ISP) called IBM.net (now taken over by AT&T) explains their definition of alias email address as follows: "To change your email address, you need to create an alias email ID. An alias email ID allows you to create a new email address for yourself that points back to your original email address. This service is great for people that registered with IBM Internet Connection Services, but did not get the email address they wanted when registering. The email alias ID can be between 3 and 32 characters in length. This allows you to choose an email alias ID much longer than your original user ID, which must be between 5 and 7 characters in length. For example, you may have been assigned the email address smit394@ibm.net when registering. You could create an alias email ID that is more memorable, such as joes_restaurant@ibm. net."

aliasing Distortion in a video signal. It shows up in different ways depending on the type of aliasing in question. When the sampling rate interferes with the frequency of program material the aliasing takes the form of artifact frequencies that are known as sidebands. Spectral aliasing is caused by interference between two frequencies such the luminance and chrominance signals. It appears as herringbone patterns, wavy lines where straight lines should be and lack of color fidelity. Temporal aliasing is caused when information is lost between line or field scans. It appears when a video camera is focused on a CRT and the lack of scanning synchronization produces a very annoying flickering on the screen of the receiving device.

aliasing noise A distortion component that is created when frequencies present in a sampled signal are greater that one-half the sample rate. See Anti-aliasing Filter.

alien crosstalk Electromagnetic noise that can occur in a cable that runs alongside one or more other cables. The term "alien" is used because this type of crosstalk occurs between different cables rather than between individual wires within a single cable. See also crosstalk.

alien wavelength A wavelength in an optical system that was introduced from a third-party external node, such as a router, ethernet switch or transponder. See Third-party wavelength.

aligned bundle A bundle of optical fibers in which the relative spatial coordinates of each fiber are the same at the two ends of the bundle. Also called "Coherent Bundle."

alignment 1. The adjustment of components in a system for optimum performance.

2. The process of adjusting a satellite dish to receive the strongest signal.

alignment error In IEEE 802.3 networks, an error that occurs when the total number of bits of a received frame is not divisible by eight, i.e. not properly framed. Alignment errors are usually caused by frame damage due to collisions.

ALIS Access Lines In Service. See Access Line.

ALIT Automatic Line Insulation Testing. Equipment located in a Central Office which sequentially tests lines in the office for battery crosses and grounds.

ALJ An Administrative Law Judge appointed by a State Commission to review a Commission docket, such as a rate case or incentive regulation proposal, and to make recommendations to the Commissioners.

all call paging With this feature, a user can broadcast an announcement – a page – to someone through the speakers of all the telephones on the system and, possibly, any external loudspeakers. If you want instant fame, ask your secretary to call all the airports in the country and page you. Mike Todd, the movie mogul, used to have his secretary perform this wonderful task. Mr. Todd gave gigantic egos a whole new meaning.

all channel tuning Ability of a television set to receive all assigned channels: VHF and UHF, channels 2 through 83.

all dielectric cable Cable made entirely of dielectric (insulating) materials without any metal conductors.

all flash, no cash The financial version of "All Talk, No Action" and "All Hat, No Cattle."

all inputs hostile Measurement technique for troubleshooting networks, particularly for crosstalk, using worst case conditions (typically, full chroma signal on all inputs other than the one under test).

All Number Calling Once upon a time, the first two digits of telephone exchanges sort of corresponded to their location. For example, MU-8 meant Murray Hill 8 in Murray Hill, Manhattan, New York City. Then the phone company started running out of letters, so it went to All Number Calling. The All Number Calling provides a theoretical maximum of 792 central office exchange (NNX) codes per area code (NPA). This is derived on the basis of 800 NXX code combinations (8x10x10) leaving out eight special service combinations, including 411, 611, 911.

All Ones Keep Alive Alarm Indicating Signal (AIS). Also known as a "Blue Alarm Signal" or "All Ones Keep Alive," an AIS is an unframed all-ones bit pattern sent by equipment at the far end to indicate that an alarm condition exists upstream in a circuit leading to the downstream equipment. Keep-alive signals are required by the network facilities to prevent oscillation of the line repeaters which causes interference (i.e. cross-talk and bleeding) within adjacent channels. SONET defines four categories of AIS: Line AIS, STS Path AIS, VT Path AIS, DSn AIS.

All or Nothing Rule The FCC rule that requires a carrier to choose a single regulatory framework – price cap or rate of return – for all operations at the federal level. Carriers are free to be rate of return in one state and price cap (or alternative regulation) in another.

All Trunks Busy When a user tries to make an outside call through a telephone system and receives a "fast" busy signal (twice as many signals as a normal busy in the same amount of time), he is usually experiencing the joy of All Trunks Busy. No trunks are available to handle that call. The trunks are all being used at that time for other calls or are out of service. These days, many long distance companies are replacing a "fast" busy signal with a recording that might say something like, "I'm sorry. All circuits are busy. Please try your call later."

All-Silica Fiber An optical fiber composed of a silica-based core and cladding. The presence of a protective polymer overcoat does not disqualify a fiber as an all-silica fiber, nor does the presence of a tight buffer.

allan variance One half of the time average over the sum of the squares of The differences between successive readings of the frequency deviation sampled over the sampling period. The samples are taken with no dead-time between them. See Two-Sample Variance.

Allen, Robert Chairman and CEO of AT&T from April, 1988 to November 1997 when C. Michael Armstrong took over. Allen became chairman and CEO of AT&T in April 1988 following the unexpected death of then-Chairman James E. Olson. Under Allen's leadership, the company evolved from a highly regulated utility to a successful competitor in the dynamic, fiercely competitive communications industry. In 1995, Allen made the bold decision to restructure AT&T into three companies, spinning off the company's equipment arm to become the very successful Lucent Technologies and its computer arm into specialty computer maker NCR.

Allen began his AT&T career at Indiana Bell in 1957 and rose steadily to hold officer posts there, at Bell of Pennsylvania, Illinois Bell, the Chesapeake and Potomac Telephone Companies and AT&T. He was the company's president and COO prior to being named CEO.

In 1980, Bob Allen headed a task force to look into AT&T's future. The recommendation of the task force: keep equipment manufacturing at all costs. When he became chairman, he sold AT&T's manufacturing arm off. It's now called Lucent Technologies.

allocate To assign space or resources for a specific task. This is often used to refer to memory or disk space.

allocated channel A Radio Frequency (RF) channel that is configured to allow use by Cellular Packet Data (CDPD) transmissions.

allocation 1. A method by which end users not presubscribed are assigned to long distance phone companies in the same ratio as customers who selected a long distance company before an end office conversion to equal access.

2. The amount of stock in an initial public offering (IPO) granted by the underwriter to an investor. IPO allocations are determined based on a customer's commission volume, trading history, and type of investor. IPO allocations are normally communicated to investors the morning after the pricing. In the old days all technology IPOs went up on their first day of trading and, hence, if you got an allocation, you immediately made money. It was money for jam. These days, no more.

3. In radio systems, the band of radio frequencies designated by upper and lower frequency limits and specified by the ITU/WARC for use by one or more of 38 terrestrial and space radio communications services under specified conditions.

allocation of resources A reason which CEOs give for not doing what they should be doing to grow their company.

Allocations The assignments of frequencies by the FCC for various communications uses (e.g.: television, radio, land-mobile, defense, microwave, etc.) to achieve allegedly fair division of the available spectrum and minimize interference among users.

Allotment Allotment in radio communications systems, the designation of portions for specific services; for satellite communications systems, as Inmarsat, specific orbital positions of satellites may also be involved.

Allowed Cell Rate ACR. An ATM term. An ABR service parameter, ACR is the

current rate in cells/sec at which a source is allowed to send. ACR is a parameter defined by the ATM Forum for ATM traffic management. ACR varies between the MCR and the PCR, and is dynamically controlled using congestion control mechanisms.

alloy A combination of two or more metals that forms a new or different metal with specific or desirable qualities.

ALM 1. AppWare Loadable Module. A visual computer telephony applications generator that works on Novell's NetWare. An ALM works by tying into Novell NetWare's NLMs. See Appware.

2. Automated Loan Machine. Like an ATM (Automated Teller Machine), an ALM sits in the wall of a building or inside a building on a wall. However, instead of giving money you own, an ALM dispenses money in the form of an instant loan. One of the leading ALM manufacturing companies is Affinity Technologies of Columbia, South Carolina. Alan Fishman of Columbia Financial Partners contributed this definition. Mr. Fishman is a leading New York City venture capitalist, who helped Affinity get started. Mr. Fishman's company is Columbia Financial Partners.

3. Airline Miles. The method used to calculate the distance (for pricing purposes) of the point-to-point long distance lines in long distance telephone networks. See Airline Mileage.

ALO Transaction An ATP transaction (AppleTalk Transaction Protocol) in which the request is repeated until a response is received by the requester or until a maximum retry count is reached. This recovery mechanism ensures that the transaction request is executed at least once. See also ATP.

ALOHA 1. A method of data transmission in which the device transmits whenever it wants to. If it gets an acknowledgement from the device it's trying to reach, it continues to transmit. If not (as in the case of a collision with someone else trying to transmit simultaneously), it starts all over again. The ALOHA method get its name from a dying satellite that was donated to university researchers in the Pacific. It was used to transmit data by satellites among South Sea islands, especially Hawaii. The ALOHA "method" – called "transmit at will" – was invented because the users were short of funds to develop more sophisticated data transmission protocols, and they had a free satellite, which typically had more bandwidth than they had stuff to send. See Alohanet.

2. Aloha, the greeting, has multiple meanings, depending on the context. It can mean hello, goodbye, love, caring and hospitality. It can also be used at the end of a letter or email to mean "warm regards."

Aloha Suite Temporary office space, including phone, fax, email and Internet, provided to laid-off employees to help them create the appearance that they are still working while they search for a new job.

alpha pup uSED by market researchers for the coolest kid the block, as in "If the alpha pups like it, we'll sell a million of them."

Alohanet An experimental form of frequency modulation radio network developed by the University of Hawaii. Alohanet is implemented by creating transmission frames containing data, control information, and source and destination addresses which are broadcast for reception by the destination receiver and ignored by all others. Actually, Alohanet is an early version of Ethernet, the local area network technique. See Aloha.

Alpeth Aluminum-polyethylene primary covering used as the sheath for aerial cable.

alpha 1. Only alphabetic characters.

2. The first (A) version of hardware or software. It typically has so many bugs you only let your employees play with it. A beta is the next version. It's a pre-release version and selected customers (and the press, sometimes) become your guinea pigs. After beta, and when the bugs are removed, comes "general availability" or "general release." That's when the product is finally available for buying by the general public."

alpha channel The upper 8 bits of the 32-bit data path in some 24-bit graphics adapters. The alpha channel is used by some software for controlling the color information in the lower 24 bits.

alpha geek The most knowledgeable, technically proficient person in an office or work group. "Ask Harry, he's the alpha geek around here."

alpha test The first testing phase of a software version. Alpha tests are conducted in-house, before being promoted to beta test, which typically involves a real customer.

alphabet The ITU (International Telecommunications Union) phonetic alphabet is: Alpha, Bravo, Charlie, Delta, Echo, Foxtrot, Golf, Hotel, India, Juliet, Kilo, Lima, Mike, November, Oscar, Papa, Quebec, Romeo, Sierra, Tango, Uniform, Victor, Whiskey, X-Ray, Yankee and Zulu.

alphabetic Only alphabetic characters. See also Alphanumeric.

alphanumeric A set of characters that contains both letters and numbers – either individually or in combination. Numeric is 12345. Alphabetic is ABCDEF. Alphanumeric is 1A4F6HH8. American and Australian zip codes are numeric. Canadian and English postal codes are alphanumeric. No one knows why.

alphanumeric display A display on a phone or console showing calling phone number, called number, trunk number, type of call, class of service and perhaps, some other characteristics of the call. It may also contain instructions as to how to move the call around, set up a conference call, etc. The display may be liquid crystal or light emitting diode. Typically, it's liquid crystal.

alphanumeric memory A cellular radio feature that allows you to store names with auto-dial phone numbers.

ALPS circuit A communication path across a TCP connection between a host reservation system and an ASCU. When MATIP encapsulation is used on an ALPS circuit, it is equivalent to a MATIP session.

ALS 1. Active Line State, one possible state of an FDDI optical fiber.

2. Additional listing.

ALT Alternate Local Transport. Another term used for a provider of service other than the LEC. Interchangeable term with CLEC and CLSP.

Alt Newsgroups A set of Usenet newsgroups containing articles on controversial subjects often considered outside the mainstream. Alt is an abbreviation for alternative. These newsgroups were originally created to avoid the rigorous process required to create an ordinary newsgroup. Some alt newsgroups contain valuable discussions on subjects ranging from agriculture to wolves, others contains sexually explicit material, and others are just for fun. Not all ISPs and online services give access to the complete set of alt newsgroups.

Altair Ethernet Motorola's name for its wireless local area network, which transmits at the very high frequency of 18 to 18 megahertz. Altair users need to fill out a small, one-page FCC application in order to use the system.

Altazimuth Mount A mounting, e.g., for a directional antenna, in which slewing takes place in the plane tangent to the surface of the Earth or other frame of reference and elevation about (above or below) that plane. Also called an X-Y Mount.

ALTEL Association of Long distance TELephone companies. A trade association composed of alternative (to AT&T) long distance carriers and resellers of long distance services.

alternate access Has the same meaning as Local Access except that the provider of the service is an entity other than the Exchange Carrier authorized or permitted to provide such service. The charges for Alternate Access may be specified in a private agreement rather than in a published or special tariff if private agreements are permitted by applicable governmental rules.

alternate access provider A carrier providing local access and transport other than the primary local exchange carrier.

alternate answering position Usually refers to a second receptionist's desk which has a telephone switchboard or console functioning like the main one. Also refers to when the main receptionist is away from his/her desk, or is very busy taking calls, the telephone system automatically sends the calls to another console or to a phone that will be answered.

alternate buffer In a data communications device, the section of memory set aside for the transmission or receipt of data after the primary buffer is full. This helps the device control the flow of data so transmission is not interrupted due to lack of space for the incoming or outgoing data.

alternate entrance A supplementary entrance facility into a building using a different routing to provide diversity of service and for assurance of service continuity.

alternate lock code A three-digit lock code to be used with the partial lock feature in some cellular phones.

Alternate Mark Inversion See AMI.

alternate media connector A buildings network term. An optional module that plugs into a 1016, 2016, 3024 or 3124 repeater to provide an AUI, BNC, or fiber Media Expansion Port (MEP).

alternate recipient An electronic messaging term. In X.400 terms, a user or distribution list that a recipient MTA (Message Transfer Agent) delivers a message to (if allowed) when the message cannot be sent to the preferred recipient.

alternate route A second or subsequent choice path between two exchanges, usually consisting of two or more circuit groups in tandem. Sometimes called "alternative route" or "second-choice route."

alternate routing 1. AR. Redirecting a call over alternate facilities when the first choice route for that call is unavailable. AR is a mechanism that supports the use of a new path after an attempt to set up a connection along a previously selected path fails. It's a feature used in network design and also in PBXs. For example, with PBXs it's a feature used with long distance calls that permits the telephone system (typically a PBX) to send calls over different (alternate) phone lines. It might do this because of congestion of the primary

phone lines the calls would normally be sent over. Alternate routing is often confused with Least Cost Routing in which the telephone system chooses the least expensive way (available at that time) to route that call. Least Cost Routing typically works with so-called "look-up" tables in the memory of the PBX. These tables are put into the PBX by the user. The PBX does not automatically know how to route each call. It must be told by the user. That "telling" might be as simple as saying "all 312 area codes will go via the AT&T FX line." Or it might be as complex as actually listing which exchanges in the 312 area code go by which method. Least Cost Routing tells the calls to go over the lines which are perceived by the user to be the least costly way of getting the call from point A to point B. Alternate routing happens when the least cost routes get congested and alternate routes (typically more expensive) are found from the look-up tables in the PBX's memory.

2. In the US emergency services telephone network, alternate routing is the ability to route 9-1-1 calls to an alternate location if the primary public safety answering point is busy or otherwise unavailable.

Alternate Service Provider ASP. A fancy name for a new phone company which is not a traditional telephone company. Such could be a local, metropolitan or long distance phone company specializing in voice, data and video. Such a company, if it concentrated on local service, could also be known as CLEC, which stands for a Competitive Local Exchange Carrier. See CLEC.

Alternate Serving Wire Center ASWC. When a building or customer is served by two different Central Offices, the one that is not his main central office is called the ASWC.

Alternate Serving Wire Center (ASWC) One of a Carrier's primary requirements is to make certain communications will not be interrupted totally by a facility failure. Verizon answers this need with Alternate Serving Wire Center (ASWC), an optional feature available for interstate DS1 and DS3 high capacity switched and special access services.

ASWC provides both wire center diversity and loop protection by enabling you and your customers to separate multiple DS1 and DS3 circuits over two geographically diverse paths. One path leads to the serving wire center; the other path goes to an alternate serving wire center. Two separate paths can also be provided for interoffice facilities (Interoffice Circuit Diversity). If there is a fault in the loop, a wire center, or an interoffice facility on one path, the other path continues unaffected at the capacity you have pre-established.

Features:

- Available for switched and special access interstate DS1 and DS3 services
- Provides an alternate path to an alternate serving wire center and loop protection
- May be combined with the option of adding Interoffice Circuit Diversity to provide added security against facility disruptions between wire centers
- Allows you to determine what percentage of traffic (optimal is 50%) will be routed over each path
- Loop path provisioned via a self-healing SONET ring in some cases * ASWC and Interoffice Circuit Diversity available with tariff rates, where filed,
- where facilities permit (Otherwise, special construction charges may apply.)

Benefits:

- Total service will not be interrupted in the event of a facility failure.
- ASWC lets you obtain your service assurance from one point of contact, Verizon.
- This may reduce or eliminate mileage charges.

How It Works:

Alternate Serving Wire Center provides 50% survivability to customers that divide their high capacity service equally between the two different wire centers. ASWC provides both wire center diversity and loop protection by enabling separate DS1 or DS3 circuits over two geographically diverse paths. One path leads to the serving wire center and the other leads to an alternate serving wire center. Should a fault occur in the loop, a wire center, or an interoffice facility on one path, the other path continues unaffected at the capacity you pre-established. Not all products, rates, and terms are available in all areas. Please see the tariff or contact your Account Manager for availability in your specific area.

alternate use The ability to switch communications facilities from one type of service to another, i.e., voice to data, etc.

alternate voice data AVD. An older service which is a single transmission facility which can be used for either voice or data (up to 9600 bps). Arrangement includes a manually operated switch (on each end) to allow customers to alternately connect the line to their modem or PBX.

Alternating Current See AC.

Alternative Access Provider A telecommunications firm, other than the local telephone company, that provides a connection between a customer's premises to a point-of-presence of the long distance carrier. Another name for a competitive access provider

(CAP). See CAP.

alternative channel A call center/marketing term. A competitive marketing strategy to expand the means by which a company can reach its customers. Direct marketing and specifically telephone-based marketing are major examples that have emerged in the highly competitive era of the '80s and '90s.

alternative non-traffic sensitive cost-recovery plans New charges proposed by the regional Bell holding companies to supplement subscriber line charges. In short, another charge on the subscriber with an interesting, though dubious, justification. They have not been fully implemented.

alternative regulation Quasi-price cap regulatory plans at the state level that allow operators to set and raise prices and do not cap a carrier's profitability. Most independent and rural LECs participate in some state alternative regulatory plans.

Alternative Regulatory Framework ARF.

alternator A machine which generates electricity which is alternating current. See AC.

Altrac The name of the proprietary encoding system for audio which Sony used on its portable memory-based players. In December, 2004 it introduced its first hard-drive driven Walkman which handled the more popular, rival music format called MP3. See MP3.

ALTS Alternative Access Providers to the local telephone network i.e., Teleport.

ALU Arithmetic Logic Unit. The part of the CPU (Central Processing Unit) that performs the arithmetic and logical operations. See Microprocessor.

Alumina Aluminum Oxide, Al203. Alumina ceramic is used as the substrate material on which is deposited thin conductive and resistive layers for thin film microwave integrated circuits.

aluminum The first known item made from aluminum was a rattle made for Napoleon III's infant son in the 1850s. Napoleon also provided his most honored guests with knives and forks made of pure aluminum. At the time, the newly discovered metal was so rare and difficult to process, it was considered more valuable than gold. Outside America, aluminum is spelled aluminium. See aluminum wire.

aluminum wire Aluminum wire will conduct electricity and communications signals and is used in special instances. It is lighter and cheaper than copper. But it has a bigger bending radius – which means if you try and bend it sharply, it is likely to snap.

alvyn Aluminum-polyethylene, the sheath used for riser cable where a flame retardant sheath is required.

always on In the old days you needed to access the Internet via a dial-up phone line. When you wanted to get onto the Internet, you had to dial a phone number. Then along came broader band services. beginning with ISDN and moving to cable modems, DSL and now fiber optics (as in Verizon's FiOS service). These broadband connections were "always on," meaning you no longer had to dial to get onto the Internet. You were always there. And if you kept your computer on, email would flow to you constantly, virtually instantly from the time someone sent it to you. These days we don't hear the words "always on" much longer. Everything is always on. These days even cell phones are always on, able to receive text messages all the time whether you're on the phone speaking or not.

Always On/Dynamic ISDN See AO/DI.

AM See Amplitude Modulation. Also Access Module.

AM detector An electrical circuit which frequently uses diodes and filtering circuits to remove amplitude modulation from another radio frequency carrier or wave form.

AM noise Amplitude noise. The random and/or systematic variations in output power amplitude. Usually expressed in terms of dBc in a specified video bandwidth at a specified frequency removed from the carrier.

AM-PM conversion AM-PM conversion represents a shift in the phase delay of a signal when a transistor changes from small-signal to large-signal operating conditions. This parameter is specified for communications amplifiers, since AM-PM conversion results in distortion of a signal waveform.

AM/VSB Amplitude-Modulated Vestigial Sideband.

AMA Automatic Message Accounting. A fancy name for the billing of phone calls. The telephone companies call AMA the distributed network function that measures usage of the network by subscribers and produces formatted records containing this usage information. Users of this formatted information include billing systems and other Operational Support (OS) systems. Corporate telephone customers know AMA as another name for Call Detail Recording or Station Message Detail Recording (SMDR). AMA equipment records call details. In a corporation, AMA records are used to generate billing for departments, to verify phone bills from carriers, to check for unauthorized calls. See AMA Tape and Call Accounting System (for a bigger explanation).

AMA Tape A telephone company machine-readable magnetic tape which contains the

customer's long distance calling and billing data for a given month.

AMA Teleprocessing System AMATPS. The primary method for delivery of AMA data from the network to billing systems. The current AMATPS architecture consists of an AMA Transmitter (AMAT) and a collector.

AMADNS AMA Data Networking System. In OSS, the next generation Bellcore system for the collection and transport of AMA data from central office switches to a billing system. See also AMA.

amateur radio operator Also known as HAM Radio Operator. A class of non-commercial private radio operator who use interactive radio as a hobby. There are six classes of licenses that are earned by examination, which the FCC sponsors.

AMATPS AMA Teleprocessing System. In OSS, the Bellcore (now Telcordia) legacy system for collecting and transporting AMA data from central office switches to a billing system. The AMATPS consists of an AMA Transmitter and a collector. See also AMA.

amazing When Apple cult leader Steve Jobs introduced the new iBook portable in May of 2001, he did not use the word "stunning" to describe it in the media briefing. He repeatedly called it "amazing." This word "amazing" seems to be the adjective most used in the Spring of 2001 at Apple. The new OS X was described as both amazing and stunning. See Cool.

amazon-ized That sick feeling you get when you wake up one morning and find your industry being dominated by a Web-based retailer.

ambient lighting The general level of illumination throughout a room or area.

ambient noise The level of noise present all the time. There is always noise, unless you're in an anechoic chamber. When measured with a sound level meter, it is usually measured in decibels above a reference pressure level of 0.00002 pascal in SI units, or 0.00002 dyne per square centimeter in cgs units.

ambient noise level See Ambient Noise.

ambient temperature The temperature of a medium surrounding an object.

ambimousterous Able to use a mouse with either hand.

ambulance chasing See Spambulance Chasing.

AMCs Adds, Moves and Changes. Telephone-speak for adding phone extensions and trunks, moving phone extensions and changing them. In short, the AMCs covers the sort of work your telephone installer does in his monthly visit to your office. In some rapidly-changing firms, AMCs affect every phone every year. They refer to that as 100% AMC. AMCs can easily cost more than the initial cost of the phone system. Increasingly, AMCs can be done by a secretary from her PC over the LAN and connected to the phone system. The more you can do of this, the better. It will save time, money and the aggravation of new employees having to wait weeks before their new company's phone system finally recognizes that they exist.

AMD Air Moving Device. IBM-speak for a fan with a feedback sensor to let the system know if the fan has failed.

Amdahl's Law Dr. Gene Amdahl's observation in 1965 that computer system speed is governed by the speed of the slowest component. In most systems, the slowest component is the human operator, even if it's Amdahl himself.

America Online The largest North American on-line computer and Internet access service. AOL provides e-mail, instant messaging, forums, software downloads, news, weather, sports, financial information, conferences, on-line gaming, an encyclopedia, and other features, to its subscribers. www.aol.com.

American Bell, Inc. 1. The predecessor to AT&T, American Bell was formed in 1880 as a Massachusetts corporation to supersede National Bell Telephone Company, which had consolidated the interests of the original Bell Telephone and New England Telephone Company. In the face of unfriendly Massachusetts corporate law, American Bell on December 30, 1899 was folded into American Telephone and Telegraph Corporation (AT&T), its wholly owned long distance subsidiary incorporated in New York state.

2. The old name for the unregulated telephone equipment supply subsidiary of American Telephone & Telegraph. American Bell (deja vu all over again) was formed on January 1, 1983 by FCC mandate to market, sell and maintain newly manufactured equipment and enhanced services. American Bell Inc. had its name changed to AT&T Information Systems, becoming a division of AT&T. It has been reorganized many times. When it was American Bell, it was only selling telecommunications products and services to end users. When it become AT&T Information Systems, it sold AT&T phone systems and AT&T computer systems. It then merged with AT&T Long Lines, which was then called AT&T Communications and later called simply AT&T. Subsequently, it became known as AT&T Technologies, including the old Western Electric. In 1996, AT&T decided to split into three separate companies, with AT&T Technologies and Bell Telephone Laboratories becoming Lucent Technologies. Sadly, old gadgetry, knick-knacks and mementos bearing the name American Bell, Inc.

have no marketable value as antiques or examples of American folk art. See also Lucent Technologies.

American Depository Receipt ADR. A certificate issued by a U.S. bank for a share or shares of a non-U.S. company. Non-U.S. companies that wish to list on a U.S. exchange must abide by the regulatory and reporting standards of the Securities and Exchange Commission (SEC). These securities are called receipts because they represent a certain amount of the company's actual shares. Examples of ADRs are France Telecom, British Sky Broadcasting, and Equant.

American Mobile Telecommunications See AMTA.

American National Standards Institute See ANSI.

American National Standards Institute Character Set The set of characters available. The character set includes, letters, numbers, symbols and foreign language characters.

American Radio Relay League A non-profit organization that promotes interest in amateur radio in the United States and represents US radio amateurs in legislative matters. ARRL is also a member of and the International Secretariat for the International Amateur Radio Union (IARU), which is made up of similar societies around the world. ARRL was founded in 1914 in Hartford, Connecticut by Hiram Percy Maxim, an inventor and industrialist. ARRL has over 150,000 members and is the largest amateur radio organization in the United States. Website: www.arrl.org.

American Registry for Internet Numbers See ARIN.

American Standard Code For Information Interchange ASCII. The standard 7-bit code for transferring information asynchronously on local and long distance telecommunications lines. The ASCII code enables you to represent 128 separate numbers, letters, and control characters. By using an eighth bit – as in extended ASCII or IBM's EBCDIC – you can represent 256 different characters. ASCII often uses an eighth bit as a parity check or a way of encoding word processing symbols, not as a way of broadening the number of characters and symbols which it can represent. See also ASCII.

American Telephone & Telegraph See AT&T.

American Wire Gauge AWG. Standard measuring gauge for non-ferrous conductors (i.e. non-iron and non-steel). AWG covers copper, aluminum, and other conductors. Gauge is a measure of the diameter of the conductor. See AWG for a bigger explanation.

Ameritech Ameritech Corp was one of the Regional Bell operating companies formed as a result of the AT&T Divestiture. Ameritech covered five states and included the operating telephone companies of Illinois Bell, Indiana Bell, Michigan Bell, Ohio Bell, and Wisconsin Bell. On October 9, 1999, Ameritech merged with (read acquired by) SBC, which used to be known as Southwestern Bell.

AMI Alternate Mark Inversion. An older, but still common, line coding technique used in T-1 transmission systems. AMI generates alternately inverted positive and negative pulses to represent marks. In other words, pulses of alternating polarity are use to represent binary ones, with each such pulse being of the same amplitude, or signal strength. Binary zeros are represented with the generation of a null pulse, i.e., zero voltage. This technique prevents the buildup of DC voltage. As repeaters on T-1 circuits require regular pulse transitions (i.e., positive and negative voltages) to recover and regenerate timing on the circuit, long strings of zeros must be avoided. As a general rule, no more that 15 consecutive zeros can be transmitted without violating the ones density rule. B8ZS (Binary 8 with Zero Substitution), the newer line coding technique used in T-1 circuits, substitutes alternating bipolar pulses for the eighth zero in a string of eight consecutive zeros. B8ZS also supports clear channel communications of 64 Kbps per channel. See also B8ZS.

AMI Violation A "mark" that has the same polarity as the previous "mark" in the transmission of alternate mark inversion (AMI) signals. Note: In some transmission protocols, AMI violations are deliberately introduced to facilitate synchronization or to signal a special event. See AMI.

AMIS See Audio Messaging Interchange Specification. A standard for networking voice mail systems.

AML 1. Analog Microwave Link.
2. Actual Measured Loss.

AMO All things Firefox. See extensions.

amortization table Let's say you borrow $100,000 to finance your house. You'd like to know how much you have to pay each month until your loan is finally paid off. An amortization table shows you that. A good one shows you four columns. The month you have to pay. The total amount you have to pay that month. The amount of what you paid that's interest. And the amount of what you paid that's principal.

AMP See Ampere.

Amp hour AH. A rating system telling you how long a battery will last. A battery with

an amp-hour rating of 100 will supply 100 amps for one hour, 50 amps for two hours or 25 amps for four hours. It will run the four amp laptop I'm writing this on for at least 25 hours. That would be wonderful. However, the battery would weigh much more than my laptop. I probably could barely lift it. That would not be so wonderful. As usual, life is a bunch of tradeoffs.

ampacity The maximum current an insulated wire or cable can safely carry without exceeding either the insulation or jacket materials limitations. If the ampacity is exceeded, the conductor heats up due to the resistance to current flow as electromagnetic energy is converted to thermal energy. The insulation heats up, as well, perhaps to the point that it catches fire, which is not a good thing.

amperage rating The amperage which may be safely applied to a circuit, service or equipment. See also Ampere.

ampere Amp. The unit of measurement of electric current or the flow of electrons. One volt of potential across a one ohm impedance causes a current flow of one ampere. Amp is the abbreviation for ampere. It is mathematically equal to watts divided by volts. Note that in the electrical context, WATTS is spelled with two "Ts." In telecommunications, WATS, meaning Wide Area Telecommunications Service, is spelled with only one "T."

ampere-hour unit Measurement of battery capacity, determined by multiplying the current delivered by the time it is delivered for. See Ampere.

amphenol connector Amphenol is a manufacturer of electrical and electronic connectors. They make many different models, many of which are compatible with products made by other companies. Their most famous connector is the 25-pair connector used on 1A2 key telephones and for connecting cables to many electronic key systems and PBXs. The telephone companies call the 25-pair Amphenol connector used as a demarcation point the RJ-21X. The RJ-21X connector is made by other companies including 3M, AMP and TRW. People in the phone business often call non-Amphenol-made 25-pair connectors, amphenol connectors.

amplified handset An amplified handset is the best phone gadget you can buy. You use it to crank up the volume of incoming calls (and in some cases the volume of outgoing calls) and save yourself enormous amounts of money on callbacks. "We have a bad line. I'll call you back." There are three types of amplified handsets: 1. The handset with a built-in amplifier. These devices suck their power from the phone line and since the phone line doesn't have much power, you won't have much amplification. I'm not overly impressed with amplified handsets. 2. The handset with amplifying circuits built into the phone. Ditto for our comments about power. 3. The handset with the little external box amplifier which is powered by either AC or by several batteries, typically AA alkalines. Such an external amplifier will produce much greater amplification. This is the type I prefer.

amplified spontaneous emission A background noise mechanism common to all types of erbium-doped fiber amplifiers (EDFAs). It contributes to the noise figure of the EDFA which causes loss of signal-to-noise ratio (SNR). See Erbium-Doped Fiber Amplifier.

amplifier When telephone conversations travel through a medium, such as a copper wire, they encounter resistance and thus become weaker and more difficult to hear. An amplifier is an electrical device which strengthens the signal. Unfortunately, amplifiers in analog circuits also strengthen noise and other extraneous garbage on the line. Cascading amplifiers, therefore, compound, or accumulate, noise. Digital systems make use of regenerative repeaters, which regenerate (i.e., reshape, or reconstruct) the signal before amplifying it and sending it on its way. As a result, noise is much less prevalent and less likely to be amplified in digital systems; whether one or many repeaters are in place. The ultimate yield of a repeater in a digital environment is that of improved error performance, which also yields improved throughput, assuming that error correction involves retransmission. See also Repeater and Throughput.

amplitude Signal strength, or signal power. Also referred to as the wave "height." See Amplitude Modulation for better understanding.

amplitude distortion The difference between the output wave shape and the input wave shape.

amplitude equalizer A corrective network that is designed to modify the amplitude characteristics of a circuit or system over a desired frequency range. Such devices may be fixed, manually adjustable, or automatic.

Amplitude Intensity Modulation AIM. See Intensity Modulation.

Amplitude Modulation Also called AM, it's a method of adding information to an electronic signal in which the signal is varied by its height to impose information on it. "Modulation" is the term given to imposing information on an electrical signal. The information being carried causes the amplitude (height of the sine wave) to vary. In the case of LANs, the change in the signal is registered by the receiving device as a 1 or a 0. A combination of these conveys different information, such as letters, numbers, punctuation

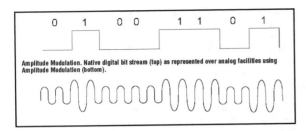

Amplitude Modulation. Native digital bit stream (top) as represented over analog facilities using Amplitude Modulation (bottom).

marks, or control characters. In the world of modems, digital bit streams can be transmitted over an analog network by amplitude modulation, with the carrier frequency being modulated to reflect a 1 bit by a high amplitude sine wave (or series of sine waves) and a 0 bit with a low amplitude sine wave or (series of sine waves). The principal forms of Amplitude Modulation are

- QDM: Double-band Amplitude Modulation
- QAM: Quadrature Amplitude Modulation
- SSB: Single-sideband Modulation
- VSB: Vestigial Sideband Modulation
- Contrast with Frequency Modulation and Phase Shift Keying.

Amplitude Modulation Equivalent AME. See Compatible Sideband Transmission.

Amplitude-vs.-Frequency Distortion Distortion in a transmission system caused by nonuniform attenuation, or gain, in the system with respect to frequency under specified operating conditions. See frequency distortion.

AMPS Advanced Mobile Phone Service. Currently known as analog mobile phone service. It's another word for the North American analog cellular phone system. The spectrum allocated to AMPS is shared by two cellular phone companies in each area or region (geographic market). This system was deployed during the 1980s in North America. Most other parts of the world deployed cellular later and went straight to digital. North America has now switched virtually entirely to digital cell phone service. See CDMA, CDMA2000 and GSM.

AMR 1. Automated Meter Reading. The automated reading of utility meters, generally power utility meters. AMR can make of wired and wireless technology, including cellular and LEOs (Low Earth Orbiting satellites). AMR involves a huge capital investment, but saves a lot of labor – i.e. the labor of having someone drive a truck to your house and physically read your meter.

2. See Adaptive Multi-Rate Speech Codec.

AMRL Adjusted Main Ring Length.

AMS 1. Account Management System.

2. Attendant Management System. An NEC term. With the NEAX2400 IMS, the AMS is an on-screen, dynamic Open Applications Interface (OAI) computer application that emulates and enhances attendant console capabilities. A typical AMS workstation combines the NEC HDAC console and headset with a color display and standard keyboard. The computer is equipped with a UNIX System V operating system, the NEC Applications Manager support platform, and a comprehensive package of software components. Communication between the HDAC and the AMS computer software is made possible by the OAI connection between the computer and the NEAX2400 IMS (ICS).

AMS-IX Amsterdam Internet Exchange. See IX.

AMTA American Mobile Telecommunications Association. An organization created to represent the interests of the U.S. commercial trunked radio industry, which generally is known as SMR (Switched Mobile Radio). In 1994, the AMTA spawned a sister organization IWTA (International Wireless Telecommunications Association), which represents similar interests on a worldwide basis. See also SMR. www.amtausa.org.

AMTS Automated Maritime Telecommunications System.

AN Access network.

ANA 1. Assigned Night Answer.

2. Automatic Network Analyzer. A computer controlled test system that measures microwave devices in terms of their small signal S-parameters. The use of this instrument by both engineering and production permits quick and accurate characterization of the input and output impedance, gain, reverse isolation of individual units and the degree of match between units.

ANAC Automatic Number Announcement Circuit. A telephone number that provides access into a telephone company system that announces the telephone number of the circuit

that you are using. It works regardless of whether the number is published or non-published. The ANAC is intended for the use of telephone company technicians when they are installing or troubleshooting your service. Hackers, crackers, phreakers, toll fraud artists and others use the ANAC to find out your number so that they can gain illegal access into your system through a modem line, for example. I could tell you what your ANAC is, or tell you how to find out, but I won't. It's for your own good. Trust me on this one.

analog Also spelled analogue (especially in Europe). The word analog comes from the word "analogous," which means "similar to." In telephone transmission, the signal being transmitted – voice, video, or image – is "analogous" to the original signal. In other words, if you were able to see the natural acoustical compression waves that transmit voice in its native form, and compare them to the representation of the corresponding electrical signals on an oscilloscope as the same voice information is transmitted on a phone line, the two signals would look essentially the same. The only basic difference is that the acoustical voice compression waves are converted into electrical signals, which also travel in smooth wave form. In consideration of such factors as crosstalk, the electrical signals are restricted along the parameters of amplitude (i.e., power level, which is used to represent volume) and frequency (which is used to represent pitch, or tone). In both acoustical and electrical forms, the voice signal is in the form of a smooth and continuous energy flow, which varies along the parameters of amplitude and frequency. In telecommunications, analog means telephone transmission and/or switching which is not digital, and which therefore is not represented in discrete terms such as voltage on/off or light pulse on/off. Outside the telecom industry, analog is often called linear and covers the physical world of time, temperature, pressure, sound, which are represented by time-variant electrical characteristics, such as frequency and voltage. See Analog Transmission, B channel and Sine Wave.

analog bridge A circuit which allows a normal two-person voice conversation to be extended to include a third person without degrading the quality of the call.

analog carrier The first carrier-loop system, which emerged during the 1970s and was used to provide improved voice-quality transmission to subscribers who were located at distances too remote to be served by the central office. In this system, multiplexing occurred at the central office, and there was little value added at the analog box deployed on the subscriber side. While analog carriers did provide an advantage over previous systems, they were difficult to install and often resulted in inconsistent quality of service.

analog cellular The original standard for cellular communications (see AMPS).

analog channel A channel which transmits in analog waveforms. See Analog.

analog channel compression A technique for squeezing more than one program into a single channel using analog processes.

analog chips Analog chips translate real-world phenomena – like motion, sound, temperature, pressure and light – into electronic or mechanical digital patterns that can be understood by their digital counterparts as for example when a song is recorded for a CD. An analog processor performs the opposite, taking the digital signals, for example off the CD, and converting them into analog sounds which we can enjoy.

analog circuit A circuit which is analog, rather than digital, in nature. As all transmission circuits make use of electromagnetic energy (i.e., electricity, radio, or light), they all start with analog. If the circuit, on the other hand, is to operate in digital mode, the analog carrier (i.e., the specific frequency that carries the signal), is varied in terms of discrete levels of voltage, amplitude, presence (i.e., on/off), or some other parameter. See also Analog.

analog computer A computer that performs its tasks by measuring continuous physical variables – pressure, voltage, flow – and manipulating these variables to produce a solution, which is then converted into a numerical equivalent. Analog computers are largely used as special purpose machines in scientific or technical applications. The earliest analog computers were purely mechanical devices with levers, cogs, cams, etc., representing the data or operator values. Modern analog computers typically employ electrical parameters such as voltage, resistance, or current to represent the quantities being manipulated. See also Analog Processor.

analog decoder A circuit that performs decoding using analog circuit techniques rather than digital circuit techniques. (For example, decoding a video signal.)

Analog Digital Converter An A/D Converter (pronounced "A to D Converter"), or ADC. A device which converts an analog signal to a digital signal.

analog driver An accessory circuit for an oscillator of filter which permits its frequency to be changed by a continuously varying signal.

analog facsimile Facsimile which can transmit and receive grey shadings – not just black and white. It is called analog because of its ability to transmit what appear to be continuous shades of grey. "Analog" facsimile is usually transmitted digitally.

analog fiber video network Implies a network in which fiber optic cable links entities together, with the video transmitted in its original analog format. Many of these systems use an "everybody on" philosophy, in which each participating entity on the network has its own channel and the other entities simply purchase a modulator (like a television tuner, for about $250) for each of the other parties, or channels with which they wish to interact. The practical limit to such a system is 16 channels, though most analog fiber networks link up no more than four entities at a time so that the participants see four monitors (TV sets) . . . one for each of the "live" participating sites. The advantage of analog fiber over digital fiber is that an expensive CODEC at each site is not required. The disadvantage is the relatively large bandwidth requirements that result in higher costs of fiber.

Analog Front End AFE. The analog front end is responsible for converting the digital signal to analog and forcing the signal onto the twisted pair line. it also the part of the fax machine that converts between the digitally modulated signal and the analog signal used on the telephone line.

analog inverter A device which inverts a wave from, i.e. turns it upside down.

analog loop-back A method of testing modems and data terminals by disconnecting the device from the telephone line and looping a signal out through the device's transmit side and in through its receive side. The test tells if the trouble is with the telephone line or with the modem.

analog microwave A microwave system in which the digital bit stream is modulated by a modem and then frequency shifted up to the appropriate microwave carrier frequency. Contrast with Digital Microwave.

analog monitor A computer screen that uses an analog signal, a smoothly varying value of current or voltage that varies continuously. VGA, SVGA and Macintosh models are examples of analog monitors. Most computer screens are analog. Most analog monitors are designed to accept input signals at a precise frequency. Higher frequencies are needed to carry higher-resolution images to the monitor. For this reason, multiscanning monitors have been developed that automatically adjust themselves to the incoming frequency. See also Analog and Digital Monitor.

analog multiplier A circuit that produces an output that is linearly proportional to the product of the two analog inputs.

analog private-line service A dedicated circuit that transmits information between two or more points. It uses analog transmission signals and is engineered for 300 to 3,000 Hz with a net maximum loss of 16 dB.

analog processor The type of processor used in an analog computer. A processor serves as the brains of a computer, taking data inputs, performing mathematical processes on them, and providing results in the form of data outputs. An analog processor works on the basis of analog, rather than digital, inputs, performs its processes on an analog basis, and provides analog outputs. While analog processors and computers may seem obsolete, they are extremely valuable in certain design applications which involve continuously variable shapes and speeds which cannot effectively be represented in discrete digital terms. Examples include automobile and aircraft design. For example, the design of an automobile carburetor, internal combustion engine, and entire drive train must consider the fact that acceleration is a relatively smooth process, rather than one that takes place in a herky-jerky mode involving discrete steps. Similarly, an automobile braking system must deal with a smooth process of deceleration. As another example, much contemporary aircraft design work focuses on the fluid and dynamic changes in the shape of aircraft wings which can reshape themselves in the event of turbulence, temperature changes, or other changes in the environment in which they must operate. See also Analog Computer.

analog recording System of recording in which music is converted into electrical impulses that form "patterns" in the grooves of phonograph record masters or in the oxide particles of master tapes representing (or analogous to) musical waveforms.

analog semiconductor Analog semiconductors are essentially the "translators" between the wave-form world of man (light, heat, pressure, and sound all move in waves) and the digital world ("ones" and "zeros") of computers. Analog semiconductors act as amplifiers in strengthening a weak signal, as converters to turn a signal from wave-form to digital and back again, and as voltage regulators, stepping down a signal from higher to lower power, also called power management.

analog signal A signal in the form of a continuous wave varying in step with the actual transmitted information; attempts to transmit an exact replica of the inputted signal down a communications channel. See Analog and all the various definitions starting with Analog.

analog switch Telephone switching equipment that switches signals without changing the analog form of the original phone call. The major form of analog switching is circuit switching. Most switching is now done digitally.

analog synchronization A synchronization control system in which the relationship between the actual phase error between clocks and the error signal device is a continuous function over a given range.

Analog to Digital Converter ADC. A converter that uniquely represents all analog input values within a specified total input range by a limited number of digital output codes, each of which exclusively represents a fractional part of the total analog input range. Note: This quantization procedure introduces inherent errors of plus or minus half LSB (least significant bit) in the representation because, within this fractional range, only one analog value can be represented free of error by a single digital output code.

analog transmission A way of sending signals – voice, video, data – in which the transmitted signal is analogous to the original signal. In other words, if you spoke into a microphone and saw your voice on an oscilloscope and you took the same voice as it was transmitted on the phone line and threw that signal onto the oscilloscope, the two signals would be essentially the same. The only difference would be that the electrically transmitted signal would be at a higher frequency. Most transmission is now done digitally.

analog video Signals represented by an infinite number of smooth transitions between video levels. TV signals are analog. By contrast, a digital video signal assigns a finite set of levels. Because computer signals are digital, analog video must be converted into a digital form before it can be shown on a computer screen.

analog wireless The dominant radio transmission standard in the United States; also called AMPS.

analogue An English/European way of spelling analog, which is the correct North American spelling. See Analog.

anamorphic Unequally scaled in vertical and horizontal dimensions.

ANC All Number Calling. The dialing plan used in telephone networks. Consisting of all numbers, ANC replaced the old U.S. system which consisted of two letters and five numbers (2L + 5N). In other words, the GR (Greenwood) exchange became 47, PA (Pennsylvania) became 72, and UL (Ulysses) became 85. Remember the Glenn Miller hit, "Pennsylvania 6-5000?" Those old exchanges were charming, often reflecting the character of the communities to which they were assigned. However, and as the number of telephone numbers grew, we ran out of alpha prefixes that included the first two letters of meaningful words. The 1 and 0 on the dial were used in the second position of the area code (prefix) in order to differentiate a local call from a long distance call because back then, there was no need to dial a 1 for a long distance call. You simply picked up the phone and start dialing either a local 7-digit number or a long distance 10-digit number – the prefix 1 was added later during the 50's. The switch (like a SxS switch or a XBar #5), looked at the second number you dialled and if it is 0 or 1, it expected you to complete a long distance call with a total of 10 digits. This is the reason at the time that all office codes were in the form [N N X] as opposed to area codes were in the form [N (0/1) X]. N = 2 to 9, X = 0 to 9, 0/1 = 0 or 1. Eventually, we had to use the 1 on the dial in the second position of the prefix; there are no letters associated with that number. Hence, All Number Calling. The reason why the numbers 1 and 0 has no letters is the following: First the number 1 had no letters because originally the customer's telephones (1900-1940) were not very reliable piece of hardware and almost every time you picked up the phone handset, you get a glitch pulse from the hook switch which looked like dialling the digit 1, hence it was decided that the first digit of a call cannot be 1, because the hook switch inside the phone was a simple metallic contact switch. Now, because you couldn't use 1 as a first digit (because of the glitching problem) and you couldn't use the 1 for the second position number either (because it was used to distinguish an area code from an office code) of any telephone number, hence you couldn't use the number 1 in either the 1st or the 2nd position in the 2-letter plan (2L + 5N), hence no letters were associated with the number 1 and the first three letters (abc) were printed on the number 2. In addition, you couldn't use 0 in either the first position (because it was dedicated to the operator – in older phones, the number 0 had the word OPERATOR engraved over it in a circular fashion or the letter "Z" was engraved standing for Zero), and also you couldn't use 0 in the second position (again, because it was used to distinguish an area code from an office code, as described above), hence the number 0 on the dial was not associated with any letters as well. That's why the numbers 1 and 0 had no letters printed over them. Since there are 26 letters in the English language and we have 8 numbers on the dial to use as the (2L + 5N) plan, AT&T decided to use 3 letters per number for a total of 24 and kept the letters Q and Z out. The letter Q was kept out of the (2L + 5N) plan because almost every time you need to use the letter Q it must be followed by the letter U, hence you only have one name associated with the 2L office (like the QUebec office). And the letter Z was used on some phones over the number 0, to indicate Zero, hence no office name included the letter Z. This is exactly why no letters exist on the numbers 1 and 0.

ANCARA Advanced Networked Cities And Regions Association. A formal association of cities and regions exploring advanced uses of information technology. ANCARA was founded in 1996 by the regions/cities of Eindhoven (The Netherlands), Kansai (Japan), Orlando (Florida), Silicon Valley (California), Singapore, and Stockholm (Sweden). The intent is to accelerate the development of the Global Information Infrastructure (GII). www.ancara. nl. See also GII.

ancestor node An ATM term. A logical group node that has a direct parent relationship to a given node (i.e., it is the parent of that node, or the parent's parent.)

anchor A hyperlinked word or group of words. An anchor is the same as a hyperlink – the underlined words or phrases you click on in World Wide Web documents to jump to another screen or page. The word anchor is used less often than hyperlink, but it maintains the seafaring theme of navigating and surfing the Net. See also Hyperlink.

anchorage accord A milestone ATM Forum document (April 12, 1998) so named because of the meeting location, the Anchorage Accord outlines which versions of ATM Forum specifications vendors should implement. ATM Forum specifications comprise approximately 60 baseline specifications for successful market entry of ATM products and services. Included are Broadband InterCarrier Interface (BICI), Interim Local Management Interface (ILMI), LAN Emulation (LANE), network management, Private Network Node Interface (PNNI), signaling, SMDS (Switched Multimegabit Data Service) and IP (Internet Protocol) over ATM, traffic management, and a number of physical interfaces. The accord also limits the conditions under which specifications are revised in order to cut down on future confusion. See also ATM Forum.

anchor system A cellular term. An anchor system is the system that maintains the connection to the PSTN (Public Switched Telephone Network) during the process of call handoff from one cell site to another. In this context, a "system" comprises all of the cellular carrier's MSCs (Mobile Switching Centers) serving a particular geographic area, and all of the cell sites supported by the MSCs. See also MSC.

ancillary charges Charges for supplementary services comprised of optional features, which may consist of both non-recurring and monthly charges.

ancillary terrestrial component ATC. In 2003 the FCC ruled that certain mobile satellite services (MSS) providers in the 2GHz, Big LEO and L-band frequency bands could install and integrate ancillary terrestrial components (ATC) in their MSS networks to enhance their ability to provide mobile services in their existing serving areas on land, in the air and over oceans without using additional spectrum resources beyond those already allocated. An ATC, for example, would better enable a MSS to provide mobile service to customers inside buildings and urban canyons.

AND Automatic Digital Network Dialing. A digital private line service that transmits voice, data, video and other digital signals.

AND Gate A digital device which outputs a high state if either of its inputs are high.

AND Logic Gate A type of logic gate which uses AND logic. The output of an AND logic gate would consider the first and second input.

And Statement The opposite of a NOT statement. See Not Statement.

anecdotal evidence Information gathered through conversations with a handful of customers, suppliers or salespeople and often used by executives to counter quantitative analyses that discredit their personal beliefs about the market and what it needs.

Anechoic Chamber A perfectly quiet room. A room in which sound or radio waves do not reflect off the walls. An anechoic chamber is the only place in which a speakerphone will work perfectly. The more a room resembles an anechoic chamber – i.e. lots of drapes, plush carpet, etc. – the better a speakerphone will work.

anemometer A device which measures wind speed and direction.

angel Investor in an early stage technology start-up. Typically an angel invests when the company is little more than an idea, a simple business plan and several management people, but rarely a full management team.

angle bracket The term for these two brackets <\<> and >. These two brackets have major use in the HTML language. See HTML.

angle modulation Modulation in which phase angle or frequency of a sine wave carrier is varied.

Angle of Arrival AOA. This is one of the technologies for finding a cell phone caller's location. This is most useful if the cell phone has called 911 (emergency). Angle of Arrival technology measures the direction of arrival of the caller's signal (generally at least three measurements are needed) at different cell sites. Each cell site receiver sends this direction information to the mobile switch where the angles are compared and the latitude and longitude of the caller is computed and sent to the PSAP. AOA works with any handset - digital, analog, TDMA, GSM, CDMA, etc. This is sometimes referred to simply as Time Difference of Arrival (TDOA). Time Difference of Arrival relies on the fact that each cell site is generally a different distance from the caller and that signals travel with constant velocity. Therefore, each signal arrives at the cell site at slightly different times. Using these properties, a signal defines a locus of points on a circle around a base station on which a mobile could be located. Then, using synchronized receivers, the times can be compared and a latitude and longitude can be computed and sent to the PSAP. At least three different receivers are needed for TDOA to work. TDOA works with any handset - digital, analog, TDMA, GSM, CDMA, etc. See also time difference of arrival and wireless location signature.

angle of deviation In fiber optics, the net angular deflection experienced by a light ray after one or more refractions or reflections. The term is generally used in reference to prisms, assuming air interfaces. The angle of deviation is then the angle between the incident ray and the emergent ray.

angle of incidence The angle between an incident ray and the normal to a reflecting or refracting surface.

angled end An optical fiber whose end is deliberately polished at an angle to reduce reflections.

angry fruit salad A terrible visual interface on a web site or a software screen that has far too many colors.

angstrom One ten-millionth of a millimeter. Angstroms are primarily used to express electromagnetic wavelengths, including and particularly optical wavelengths. It is also called an Angstrom unit. It is named after Anders Jonas Angstrom, a Swedish astronomer and physicist, who lived 1814-74. It is said that the 128-bit addressing scheme of IPv6 provides enough unique IP addresses to theoretically provide 1,500 per square angstrom of the earth's surface.

angular circumference The measurement of the amount of bend in a fiber-optic cable.

angular misalignment loss The optical power loss caused by angular deviation from the optimum alignment of source to optical fiber – fiber-to-fiber, or fiber-to-detector.

ANI Automatic Number Identification. ANI provides for the transmission through the network of the BN (Billing Number), versus the telephone number, of the originating party (i.e., the calling person, also called party in the phone business). ANI originally was intended exclusively for the use of the long distance and local phone carriers for billing purposes. ANI information is sent through the network, from the originating central office, through all intermediate tandem offices, to the terminating central office. The information originally was sent over analog trunks in the form of DTMF (Dual Tone MultiFrequency) signals, although contemporary networks usually pass the information through the digital SS7 (Signaling System 7) network. For some years, ANI has been available to end user organizations, as well. In order to gain access to ANI data, you must have a "trunk side" connection, which carries an additional charge. Much like CLID (Calling Line IDentification), ANI delivers the number of the calling party. Unlike CLID, ANI does not depend on the presence of SS7 throughout the entire network. Also unlike CLID, ANI information cannot be blocked by the calling party. So, let's pretend that you are running a large call center. A customer calls you. Before the call is even connected to your ACD (Automatic Call Distributor), ANI presents the BN of the calling party to the ACD. Your ACD captures the BN, dips into a computer database and matches that number with the profile of the caller. As your telephone agent answers the call, he gets a "screen pop" with information about the caller, and he answers the call with "Good morning, Mr. Newton. This is Ray. I read about ANI in your Dictionary. Isn't it wonderful!" Some large users say they save as much as 30 seconds on the average IN-WATS call by knowing the phone number of the person calling them and being able to use that information to access information about them in the company database. They avoid asking regular customers for routine identification information (like their address and phone number) since it is all there in the database. See also Caller ID, CLASS, Common Channel Signaling, DNIS, Flex ANI, ISDN and ISUP.

ANI Identification Want to know what number you're calling from? Call your cell phone.

ANI II ANI II information digits are two digits that are sent with the originating telephone number to identify the type of originating station. Examples include "00" for a regular line; "06" for hotel/motel; "27" for coin phone; "61," "62," and "63" for cellular phones; and "70" for private pay stations. In a SS7 network, the ANI II code is in the form of an Originating Line Indicator, which is populated in the Originating Line Information Parameter (OLIP) within a SS7 Initial Address Message (IAM).

ANI II Codes ANI II digits are two digits that are sent with the originating telephone number identifying the type of originating station (e.g., Hotel/Motel, etc.). The ANI II code is populated in the Originating Line Information Parameter (OLIP) within a SS7 Initial Address Message.

anima Someone who communicates with you telepathically.

animated ringtone A ringtone for a cell phone that is accompanied by a video or other animation.

animation The process of displaying a sequential series of still images to achieve a motion effect.

anime Anime (pronounced ah-knee-may) is an artistic and sensual type of Japanese animation that can be found on hundreds of Web sites.

anisochronous Pertaining to transmission in which the time interval separating any two significant instants in sequential signals is not necessarily related to the time interval separating any other two significant instants. Isochronous and anisochronous are characteristics, while synchronous and asynchronous are relationships.

anisotropic Pertaining to a material whose electrical or optical properties vary with the direction of propagation or with different polarizations of a traveling wave.

anisotropic filtering A graphics term. A method of filtering textures using a nonsquare area. With this method, a pixel may encompass information from many texture elements (called texel) in one direction and fewer in another. It yields sharp textures on objects that slant away from the viewing plane, such as a road running into the distance.

ankle biter A person who aspires to be a hacker/cracker but has very limited knowledge or skills related to computer systems. Usually associated with young teens who collect and use simple malicious programs obtained from the Internet. Also known as a script kiddie.

ANM ANswer Message. The fourth of the ISUP call set-up messages. A message sent in the backward direction indicating that the call has been answered. See ISUP and Common Channel Signaling.

ANN Artificial Neural Network. See Neural Network.

Anna Kournikova virus The Anna Kournikova computer virus is a viral worm that uses Visual Basic to infect Windows systems when a user unwittingly opens the attachment to an email. The attachment appears to be a graphic image of the attractive Russian tennis star Anna Kournikova. However, when the file is opened, a software enables the worm to copy itself to the Windows directory and then send the file as an attachment to all addresses listed in the computer's Microsoft Outlook e-mail address book. The virus typically arrives as e-mail with the following subject, message, and attachment:

- Subject: Here you have, ;o)
- Message body: Hi: Check This!
- Attachment: AnnaKournikova.jpg.vbs

Notice that the extension on the attachment. It's confusing. It looks like a jpeg image file, but really is a visual basic script file. And one should never open one of those coming in as an attachment to an email.

anneal The process of using heat and gradual cooling to soften a metal such as copper, making it less brittle and, therefore, less likely to break when it is flexed.

annealed wire See anneal.

Annex A The first of the frame relay standard extensions, Annex A outlines provisions for a Local Management Interface (LMI) between customer premises equipment and the frame relay network for the purpose of querying network status information.

Annex D The second frame relay standard extension dealing with the communication and signaling between customer premises equipment and frame relay network equipment for the purpose of querying network status information.

Annie A Web homepage which seems to have been abandoned for some time. Most of the links are out of date. It's been orphaned.

annotation Marking such as that done by highlighting, underlining, text, or freehand drawing. See Annotations.

annotations Notes that you can add to Web documents. These notes are stored on your local hard disk and are available each time that you access a document. This feature is found in NCSA Mosaic, but not Netscape Communicator or Microsoft's Internet Explorer.

announcement service Allows a phone user to hear a recording when he dials a certain phone number or extension. These days, announcement services are provided increasingly by totally solid-state digital announcers. These gadgets are more reliable, deliver a clearer message and last much longer than analog tape-based machines (like answering machines), which use recording tape.

announcement system An arrangement for providing information by means of recorded announcements.

annoyance call bureau The department in your local phone company which you call when you need help with annoying or harassing phone calls you are receiving. The Bureau will recommend you file a report with your local phone company. And then it may apply Trap and Trace equipment and techniques to try to locate the source of your annoying phone calls. The Annoyance Call Bureau is the stepchild of the phone industry, which means it is typically underfunded. See Trap and Trace.

annoyware You are typing away with a new shareware program that suddenly stops itself and a portal pops up from the author asking for a promise to pay in order to continue.

Annual Percentage Rate APR. A percentage calculation of the finance charge portion of a financing contact.

annular ring An indicator (or ring) around the circumference of the coaxial cable every so many feet – often 2.5 meters (8.2 feet) – to indicate a point where transceivers are to be connected. Same as transceiver attachment mark.

annunciator Original name for the indicator on magnetic switchboards which indicates the particular line that is calling the exchange. Now it is simply a light, a bell or a device that tells you something. That something might be the ringing of a phone or it might be a problem that you're having with some piece of remote equipment. A communicating annunciator is a sophisticated device that is connected to a phone line and gets on that line (dial-up or leased) to let you know that something is broken.

Anomalous Propagation AP. Abnormal propagation caused by fluctuations in the properties (such as density and refractive index) of the propagation medium. AP may result in the reception of signals well beyond the distances usually expected.

anomaly An impedance discontinuity causing an undesired signal reflection in a transmission cable.

anonymizer A device that hides the identity of a user by blocking that user's Internet Protocol (IP) address.

anonymizing technology Technology that masks a user's identity, or a server's identity, or the true originating location of Web attack.

Anonymous Call Rejection ACR. A service some local phone companies are providing their subscribers. It allows subscribers to automatically stop certain calls from ringing their phone. The calls stopped are "restricted," namely they would be displayed as "P" or "Private" on a subscriber's Caller ID device, meaning that the calling person did not send you his calling number. The person who makes such a call would hear, "We're sorry. The party you have reached is not accepting private calls. To make your call, hang up, dial *82 or 1182 on a rotary phone and re-dial." The caller will be able to reach you only by re-dialing without restricting display of his or her number.

Anonymous FTP Anonymous FTP allows a user to retrieve documents, files, programs, and other archived data from anywhere on the Internet without having to establish a userid and password. By using the special userid of anonymous, the network user bypasses local security checks and can access publicly accessible files on the remote system. See also FTP.

In short, anonymous FTP ia a way of logging in anonymously to distant hosts on the Internet and often freeware (free software) from the Internet. With an implementation of the FTP protocol, users can get public domain software from Internet sites, using the word "anonymous" for a login ID, and their userid@hostname.domain as the password. A database called Archie contains a list of what is available from anonymous FTP sites, and can be reached at "archie.mcgill.ca" and at "archie.sura.net." See FTP.

anonymous telephone number A telephone number that should not be displayed or voiced back to the called party. Such a designation is stored in switch memory and is included in signaling information sent to the terminating switch for interSPCS calls.

ANOVA ANalysis Of VAriance.

ANS Answer. (What else?) (Which is a question, the ANS to which is provided below.)

ANSA Alternate Network Service Agreement. An ISDN term. Under ANSA, customers who reside in areas where the central office switch does not support ISDN can be serviced from a neighboring central office at no additional charge. From the customer's perspective, ISDN is readily available and affordable, but the customer MUST agree to migrate to the local central office if and when service becomes available. In most cases this will involve a change in phone number. This agreement pertains to BellSouth customers only.

ANSI American National Standards Institute. A standards-setting, non-government organization founded in 1918, which develops and publishes standards for transmission codes, protocols and high-level languages for "voluntary" use in the United States. In a press release, ANSI described itself as "a private non-profit membership organization that coordinates the U.S. voluntary standards system, bringing together interests from the private and public sectors to develop voluntary standards for a wide array of U.S. industries. ANSI is the official U.S. member body to the world's leading standards bodies – the International Organization for Standardization (IOS or ISO) and the International Electronic Commission (IEC) via the U.S. National Committee. The Institute's membership includes approximately 1,300 national and international companies, 30 government agencies, 20 institutions and 250 professional, technical, trade, labor and consumer organizations." ANSI is located at 11 West 42 Street, 13th Floor, New York NY 10036 212-642-4900. ANSI puts out a biweekly newsletter called "ANSI Standards in Action. See also ANSI Character Set, CCITT, ECMA, IEEE, and ISO. www.ansi.org.

ANSI-41 A standard for transaction-based services that allow the development of applications such as Short Message Service (SMS) and access to Home Location Register/Visitor's Location Register (HLR/VLR) for wireless networks.

ANSI T1.110-1987 Signaling System 7 (SS7) - General Information.

ANSI T1.111-1988 Signaling System 7 (SS7) - Message Transfer Part (MTP).

ANSI T1.112-1988 Signaling System 7 (SS7) - Signaling Connection Control Part (SCCP).

ANSI T1.113-1988 Signaling System 7 (SS7) - Integrated Services Digital Network (ISDN) user part.

ANSI T1.114-1988 Signaling System 7 (SS7) - Transaction Capability Application Part (TCAP).

ANSI T1.206 Digital Exchanges and PBXs - Digital circuit loopback test lines.

ANSI T1.222-G The standard for the design, fabrication, and installation of telecommunication tower superstructures and foundations in the United States. The Revision G standard was issued in August 2005 with an effective date of January 1, 2006.

ANSI T1.227-1995 Telecommunications Operations Administration Maintenance and Provisioning.

ANSI T1.301 ANSI ADPCM standard.

ANSI T1.401-1988 Interface between carriers and customer installations - Analog voice grade switched access lines using loop-start and ground-start signaling.

ANSI T1.501-1988 Network performance - Tandem encoding limits for 32 Kbit/s Adaptive Differential Pulse-Code Modulation (ADPCM).

ANSI T1.601-1988 Integrated Services Digital Network (ISDN) - Basic access interface for use on metallic loops for application on the network side of the NT (Layer 1 specification).

ANSI T1.605-1989 Integrated Services Digital Network (ISDN) - Basic access interface for S and T reference points (Layer 1 specification).

ANSI T1.Q1 ANSI's standard for telecommunications network performance standards, switched exchange access network transmission performance standard exchange carrier-to-interexchange carrier standards.

ANSI TIX9.4 ANSI's SONET standard.

ANSI X3T9.5 A committee sponsored by the American National Standards Institute (ANSI) that is responsible for a variety of system interconnection standards. The committee has produced draft standards for high-speed coaxial cable bus and fiber optic ring local networks.

ANSI X3T9.5 TPDDI Twisted-Pair Distributed Data Interface (TPDDI) is a new technology that allows users to run the FDDI standard 100 Mbps transmission speed over twisted-pair wiring. Unshielded twisted-pair has been tested for distances over 50 meters (164 feet). TPDDI is designed to help users make an earlier transition to 100 Mbps at the workstation.

ANSI Character Set The American National Standards Institute 8-bit character set. It contains 256 characters.

answer back A signal or tone sent by a receiving equipment or data set to the sending station to indicate that it is ready to accept transmission. Or a signal or tone sent to acknowledge receipt of a transmission. See Answer Supervision.

answer back supervision Another word for answer supervision. See Answer Supervision.

answer call The name of a Bell Atlantic service. Here are Bell Atlantic's words. Answer Call is an answering machine without the machine. This automated messaging service answers your calls right through your touch-tone phone - 24 hours a day - even when you're on the phone. And since it's on Bell Atlantic network, there's no equipment to buy... nothing to turn to...no wires to connect... and no maintenance. By simply dialing a private passcode, you can listen to your messages, replay them or even change your greeting. What's more, Answer Call gives you the option of providing your employees (who share one line) with up to eight private "mailboxes" to receive and retrieve their own messages.

answer delay 1. The time from the beginning of ringing until the called station answers.

2. Any time delay between a "request-to-send" signal and the receipt of the first character of the response message. A three-digit number identifying one of the assigned geographic area codes in the North American direct distance dialing numbering plan.

answer detect The use of a digital signal processing technique to determine the presence of voice energy on a telephone line. It is used with call (answer) supervision, to identify an answered line. It's beginning to be used with computerized dialing equipment as it eliminates the need for a telephone representative to constantly monitor call set-up

progress on each telephone line in the event a call is answered. See Answer Supervision and Answer Signal.

Answer Message ANM. A CCS/SS7 signaling message that informs the signaling points involved in a telephone call that the call has been answered and that call charging should start.

answer mode When a modem is set by the user to receive data, it is in Answer Mode. In any conversation involving two computers, two terminals or one computer and one terminal, one side of the conversation must always be in Answer Mode. Putting a modem/computer in answer mode is sometimes done through software and sometimes through hardware, i.e. a switch on the side of the machine. You cannot run a data communications "conversation" if both sending and receiving equipment are in answer mode. Computers – mainframe and mini – which receive a lot of phone calls are typically put in answer mode." The terminals or computers calling them are typically in transmit mode.

Answer Receive Ratio ASR. A measurement of the effectiveness of a telecommunications service offering. ASR is the relationship between the number of line seizures and the number of answered (i.e., completed) calls. Did you ever get dial tone, and then a "fast busy" because the destination device is unavailable? That is not good for ASR. Did you ever get dial tone, and then an answered call because the destination device is available? That is good for ASR.

answer signal A supervisory signal, usually in the form of a closed loop, returned from the called telephone to the originating switch when the called party answers. This signal stops the ringback signal from being returned to the caller.

answer supervision Follow this scenario: I call you long distance. My central office must know when you answer your phone so my central office can start billing me for the call. It works like this: when you, the called party, answer your phone, your central office sends a signal back to my central office (the originating CO). This tells my central office to start billing me for the call. This signal is called Answer Supervision. Before the divestiture of the Bell System in early 1984, most of the nation's long distance companies – with the exception of AT&T Communications – did not receive Answer Supervision. They did not know precisely when the called party answered. So they started their billing cycle after some time – 20 or 30 seconds after the caller completed dialing. These long distance companies presumed that after this time, some one will have answered and the call will be in progress and can then be timed and billed. Without Answer Supervision, their billing of calls is inaccurate. They may bill for calls which didn't occur. And you may pay more for calls which did occur.

With the divestiture of the Bell System, and the introduction of Equal Access, the local phone companies have been told by the FCC that they must provide accurate answer supervision to all long distance phones. And with that answer supervision, the pricing of your long distance calls should be accurate. Not all long distance companies, however, choose to buy answer supervision (it costs a little more). And thus your long distance calls may still be billed inaccurately.

Check. If you are "accessing" your preferred long distance carrier by dialing a seven digit local number, then dialing your number and your account code, your carrier is probably not receiving Answer Supervision and the timing and billing of your long distance calls may be inaccurate. Check this out. Remember: just because your town has equal access doesn't mean your preferred long distance phone company has opted for it because it is expensive or for some other reason.

Virtually no hotels have answer supervision. So they start billing you arbitrarily. Some start billing you after three rings. Some after four. When you check out, carefully check your phone bill. You will, in most instances, find you have been billed for many uncompleted calls. Tell your family to pick up the phone quickly when you're out of town and may be calling them. Don't let the phone ring too many times as you're likely to be billed for the dubious pleasure of listening to ringing signals.

Answer supervision is getting better, however, as the electronics of "listening" to sounds on phone lines get better. Electronics are now available to do – to a 95% accuracy – what we as humans do – to a 100% accuracy – namely distinguish between a normal ringing sound, a fast busy sound and a person or fax machine answering the phone and saying "Hello." These electronics are getting better and less expensive, by the month. See Answer Supervision-Line Side.

Answer Supervision-Line Side Answer Supervision-Line Side is a service I first read about in a US West publication, which describes it as "providing an electrical signal that is passed back to the originating end of a switched connection. This signal indicates that the called line has gone off-hook. This service offering has applicability for record start and end, announcement start and end, dialtone reorigination prevention, call progress sequence indications, and other uses. This service offering may be used by terminal equipment (PBX, pay telephone, call diverter, etc). connected to the calling line to determine that the call has been answered."

answerback In data communications, answerback is a response programmed into a data terminal to identify itself when polled by a remote computer or terminal. This response is usually in reply to a Control-E (ASCII Character 5, Inquiry), which is known on the Telex and TWX networks as a "Who Are You?" character, or "WRU." The Answerback allows a remote computer to verify it has dialed correctly (usually on the Telex or TWX networks) by matching the Answerback received with the Answerback expected.

answering machine A machine that answers your telephone when you don't. The machine plays a message that you have recorded to greet the caller, and then allows the caller to leave you a message. Reportedly, one of the first answering machines was developed in Sweden in the 1950s; it took about three days to install the beast. Answering machines largely are consumer items. They are not the same thing as voice mail. See also Voice Mail. My favorite answering machine message is: "I am not available right now, but Thank you for caring enough to call. I am making some changes in my life. Please leave a message after the Beep.   If I do not return your call, You are one of the changes."

answering machine Detection The ability of outbound dialing systems, inside of the host ACD, the PBX, or as part of the predictive dialer product, to detect and filter out calls answered by answering machines. These systems may place the associated telephone number in a callback (see Callback) queue and may also play a recorded message.

Answering Tone The tone an asynchronous modem will transmit when it answers the phone. The tone indicates that it is willing to accept data.

ANT 1. Access Network Termination.

2. Alternate Number Translation. The ability to reroute 1-800 calls on NCP failure.

3. An ant can lift 50 times its own weight, which is equivalent to a human being pulling a 10-ton trailer.

4. When MCI first started, the conventional wisdom was that it would have as much affect on AT&T, then the dominant player, as an ant crawling up an elephant's leg with rape in mind. Forty years later neither AT&T nor MCI existed. How times changed.

ant farm Gigantic multiscreen movie theater complex with glass facade, often found near American malls. Also called multiplexes or gigaplexes.

ANTC Advanced Networking Test Center. An FDDI interoperability testing center established in 1990.

antediluvian Before the flood. Very old fashioned.

antenna 1. In basic telecommunications, an antenna is a device for transmitting, receiving or transmitting and receiving radio frequency (RF) signals. The antenna – its design, construction and placement – is the most important part of any radio system. Antennas are designed for specific and relatively tightly defined frequencies and are quite varied in design. An antenna for a 2.5 GHz (MMDS) system does not work for a 28 GHz (LMDS) design. Antennas come in all shapes and sizes. Their shape depends on the frequency of the signal they're receiving or transmitting and the use to which their communications are being put to. Antennas can broadcast signals in all directions. They're called omnidirectional. They can also broadcast signals in a fine straight line – like a flashlight. Electrical signals with frequencies higher on the spectrum, for example, are shorter and more directional. As they get higher on the spectrum, they look more like light. These must be focused and thus, require antennas which are shaped like the mirror reflector of a focusing flashlight. This parabolic shape focuses the broad beam (of the bulb or the electrical signal) into a narrow, focused beam. The weaker the received signal, the bigger the antenna must be. Antennas come in many varieties and have cute names, like parabola, caresgrain, helix, lens and horn. The plural of antenna is antennas. It used to be antennae. But then we all decided that antennae were for ants and antennas were for satellite and other telecommunications devices. See the following Antenna definitions.

2. In RFID terminology, the antenna is the conductive element that enables the tag to send and receive data. Passive tags usually have a coiled antenna that couples with the coiled antenna of the reader to form a magnetic field. The tag draws power from this field.

antenna array An assembly of antenna elements with dimensions, spacing, and illumination sequence such that the fields for the individual elements combine to produce a maximum intensity in a particular direction and minimum field intensities in other directions.

antenna beam The radio frequency energy pattern emitted by an antenna. Imagine a flashlight. Turn the head one way and the light becomes more focused, and thus more intense. Now turn it the other way and it becomes less focused. Radio and microwave antennas are designed to be less or more focused. A broadcast TV satellite will have an antenna whose beam covers the entire continental United States. A satellite like Iridium which needs

to send individual signals to individual cellular-like phones will have tightly focused radio beams.

antenna blind cone The volume of space, usually approximately conical with its vertex at the antenna, that cannot be scanned by an antenna because of limitations of the antenna radiation pattern and mount. An example of an antenna blind cone is that of an air route surveillance radar (ARSR). The horizontal radiation pattern of an ARSR antenna is very narrow. The vertical radiation pattern is fan-shaped, reaching approximately 700xAF of elevation above the horizontal plane. As the antenna is rotated about a vertical axis, it can illuminate targets only if they are 700xAF or less from the horizontal plane. Above that elevation, they are in the antenna blind cone. Also called Cone of Silence.

antenna coupler A device used to match the impedance of a transmitter and/or receiver to an antenna to provide maximum power transfer.

antenna dissipative loss A power loss resulting from changes in the measurable impedance of a practical antenna from a value theoretically calculated for a perfect antenna.

antenna diversity The use of two or more antennas to increase the odds that a usable signal will be received. For example, improved signal reception improves throughput in a wireless data environment since it means fewer packets need to be retransmitted.

antenna effective area The functionally equivalent area from which an antenna directed toward the source of the received signal gathers or absorbs the energy of an incident electromagnetic wave. Antenna effective area is usually expressed in square meters. For parabolic and horn-parabolic antennas, the antenna effective area is about 0.35 to 0.55 of the geometric area of the antenna aperture.

antenna efficiency Refers to the degree to which an antenna's power is radiated in relation to the power which is wasted in various losses.

antenna electrical beam tilt The shaping of the radiation pattern in the vertical plane of a transmitting antenna by electrical means so that maximum radiation occurs at an angle below the horizontal plane.

antenna entrance A pathway facility from the antenna to the associated equipment.

antenna feed An antenna device which collects radio signals from space and puts them onto a coaxial line. See Feedhorn.

antenna gain The ratio, usually expressed in decibels, of the power required at the input of a loss-free reference antenna to the power supplied to the input of the given antenna to produce, in a given direction, the same field strength, or the same irradiance, at the same distance. When not specified otherwise, the gain refers to the direction of maximum radiation. The gain may be considered for a specified polarization.

antenna illumination The degree to which an antenna feed pattern covers an antenna reflector.

antenna lobe A picture showing an antenna's radiation pattern. A more technical explanation: A three-dimensional radiation pattern of a directional antenna bounded by one or more cones of nulls (regions of diminished intensity).

antenna matching The process of adjusting impedance so that the input impedance of an antenna equals or approximates the characteristic impedance of its transmission line over a specified range of frequencies. The impedance of either the transmission line, or the antenna, or both, may be adjusted to effect the match.

antenna noise temperature The temperature of a hypothetical resistor at the input of an ideal noise-free receiver that would generate the same output noise power per unit bandwidth as that at the antenna output at a specified frequency. The antenna noise temperature depends on antenna coupling to all noise sources in its environment as well as on noise generated within the antenna. The antenna noise temperature is a measure of noise whose value is equal to the actual temperature of a passive device.

antenna power gain An FCC term for "the square root of the ratio of the root-mean-square free space field intensity produced at one mile in a horizontal plane in micro-volts per meter for one kilowatt antenna input power 137.6 mV/m." This ratio may be expressed in dB.

antenna power input An FCC term for the RF peak or RMS power supplied to the antenna from its associated transmission line and matching network.

antenna preamplifier A small amplifier, usually mast-mounted, for amplifying weak signals to a level sufficient to compensate for down-lead losses and to supply sufficient input to system control devices.

antenna stack Antenna tower with multiple antennas and supports.

antenna structure An FCC term which refers to the whole tower system including radiation system, supporting structure and any "surmounted appurtenances."

antenna switch A switch that enables a transmitter, receiver or transceiver to use any one of several different antennas.

antennae The plural of antenna – the type that grow on ants – is antennae. The plural of telecommunications antennas is antennas.

anthropomorphism The process of giving human qualities to inanimate objects. For example, getting a file cabinet to talk about what's inside it, or getting a modem to explain how to do communications.

anti aliasing See Antialiasing.

anti clockwise polarized wave See Left-Hand Polarized Wave.

anti curl A feature marketed by manufacturers of slimy paper fax machines (i.e. thermal paper). As the paper emerges from the fax machine, "anti-curl" simply sends the paper through a path which causes it to bend slightly in the opposite direction to which it was rolled over the roll. This bending purports to make the paper less curly when it emerges. It works to an extent. Virtually all slimy fax machines now have the anti curl "feature," though most don't advertise it, since it's like advertising that a fax machine makes faxes.

anti digit dialing league A group of people that resisted the move from named exchanges to all number dialing. The Bell System fought against the League because there was no global standardization between numbers and digits on rotary dial phones. Thus, IDDD was impossible until the advent of all digit numbers.

anti collision An RFID term. A general term used to cover methods of preventing radio waves from one device from interfering with radio waves from another. Anti-collision algorithms are also used to read more than one tag in the same reader's field.

anti dilution A private company is raising money. You buy shares. Part of the deal of buying shares is that you sign a "Shareholders Agreement" – an agreement between you and the company. One clause in that Agreement says that if the company sells shares in the future at a price lower than you bought them, you will be issued additional shares. Let's say you buy 100 shares at $10. Then the company sells shares at $5. Then it must send you an additional 100 shares for free.

anti jamming Describes features and/or mechanisms of a communications system that allow the system to continue to operate despite jamming attempts.

anti oxidant A substance which prevents or slows down oxidation of material exposed to heat.

anti reflection coating A thin, dielectric or metallic film (or several such films) applied to an optical surface to reduce its reflectance and thereby increase the transmittance of the optical fiber. The ideal value of the refractive index of a single layer film is the square root of the product of the refractive indices on either side of the film, the ideal optical thickness being one quarter of a wavelength.

anti replay Security service where the receiver can reject old or duplicate packets in order to protect itself against replay attacks. IPSec (a group of Internet security measures) provides this optional service by use of a sequence number combined with the use of data authentication.

anti siphoning FCC rules which prevent cable systems from "siphoning off" programming for pay cable channels that otherwise would be seen on conventional broadcast TV. "Anti- siphoning" rules state that only movies no older than three years and sports events not ordinarily seen on television can be cablecast.

anti static A material, such as packing material, that is treated to prevent the build-up of static electricity. The static charges gradually dissipate instead of building up a sudden discharge.

anti stuffing A mechanical flap in a coin phone which prevents the blocking by paper or other material of coin chutes. An anti-stuffing flap is meant to assure that you, the user, get your money back after you've tried to make a call but didn't get through.

anti tromboning A feature used when a call coming in a trunk is transferred back out to a trunk over the same physical link on which it arrived. A call in this state is said to be tromboned, and is consuming the expensive resource of two trunks. Transferring the call from its source to its new destination, and using no trunks is called anti-tromboning, a feature to prevent transfers from non-productively tieing up trunks. See also Tromboning.

anti viral programs Programs which scan disks looking for the tell-tale signatures of computer viruses.

antialiasing 1. A filter (normally low pass) that band limits an input signal before sampling to prevent aliasing noise. See Aliasing Noise.

2. A computer imaging term. A blending effect that smooths sharp contrasts between two regions of different colors. Properly done, this eliminates the jagged edges of text or colored objects and images appear smoother. Used in voice processing, antialiasing usually refers to the process of removing spurious frequencies from waveforms produced by converting digital signals back to analog.

anticipointment Raising people's levels of anticipation and then disappointing

them. A definition contributed by Gerald Taylor, president, of MCI.

antiphishing Before you got to a web site – manually, clickthrough or automatically – your browser checks a database on the Internet to find out if that site poses a phishing threat. If it does, you're warned or automatically denied the ability to go there. See phishing.

antistatic A material, such as packing material, that is treated to prevent the build-up of static electricity. The static charges gradually dissipate instead of building up a sudden discharge.

antitromboning See Anti tromboning.

ANVM Active Nonvolatile Memory. Memory that contains the software currently used by the network element.

ANX Automotive Network Exchange: Private network service offering that originated as a collaborated commerce solution for automakers the world over. www.anx.com.

anycast 1. A term associated with IPv6, the proposed new protocol for the Internet, Anycast refers to the ability of a device to establish a communication with the closest member of a group of devices. By way of example, a host might establish a communication with the closest member of a group of routers for purposes of updating a routing table. That router would then assume responsibility for retransmitting that update to all members of the router group on the basis of a Multicast. Through this approach, the host is relieved of the mundane task of addressing each router in the network, a task clearly best accomplished by a lesser device with lesser responsibilities. See also IPv6. Compare with Broadcast, Multicast, and Unicast.

2. In ATM, an address that can be shared by multiple end systems. An Anycast address can be used to route a request to a node that provides a particular service.

AnyLAN A high-speed local area network technology for which Hewlett-Packard Co. and International Business Machines Corp. created the original specifications. Announced in 1993, AnyLAN was intended to allow work groups operating Token Ring and Ethernet LANs to swap many more data-intensive applications at much greater speeds than possible at the time, increasing the speed and capacity of LANs sixfold to tenfold. HP and IBM announced they would give the technology of AnyLAN to any competitor free of charge, in order to establish it as a standard and expand the size of the total market. In 1996, the IEEE 802.12 committee officially recognized AnyLAN as the fast LAN standard, renaming it 100VG-AnyLAN. See 100VG-ANYLAN for a detailed explanation.

Anynet/MVS IBM product name for the ACF/VTAM feature that implements IBM's "Networking Blueprint" technology on hosts and OS/2 workstations and permits SNA LU 6.2 applications to work over TCP/IP or TCP/IP-oriented sockets applications to run over SNA.

anytime minutes A term introduced by the cell phone business to mean minutes that could be used any time. Originally cell phones charged by minutes of usage – often inbound and outbound. By the late 1990s, a bunch of bright marketing folks in the cell business had the idea to sell a bundle of minutes of usage for a flat price – e.g. 200 minutes per month. They started selling "bundles" of cell phone monthly usage. The idea was to increase the money you got out of your poor subscriber each month. The hope was the customer wouldn't use up all the minutes. Most didn't. Typically the bundles included categories of minutes – peak and non-peak. Eventually the non-peak minutes became unlimited and the peak minutes became "anytime" minutes. Anyway, the long and the short of it was that by the end of the 1990s, cell phone pricing became so complicated that few mortals (i.e. you and me) could figure out which plan was the best for them. There were some gurus who claimed this was exactly what the cell phone companies had intended all along. There are no simple ways to save money on buying cell phone service. Figure you'll always pay 25% more than what the advertised rate is. The only key I've been able to find is to get your entire family served by the same carrier. These days most carriers don't charge you to call someone on the same carrier. Don't hold me to it, however. These guys give obfuscation a whole new meaning.

anywhere fix The ability of a Global Positioning System (GPS) receiver to start position calculations without being given an approximate location and approximate time. See GPS.

AO/DI Always On/Dynamic ISDN. ISDN is a set of standards for an digital, circuit-switched network. The standards specify several types of end user access circuits, each of which involves one or more B (Bearer) channels and one D (Data, or Delta) channel. The B channels are for circuit-switched user communications (voice, data, and video), and the D channels are primarily for signaling and control (e.g., on-hook and off-hook signaling, ringing signals, performance monitoring, synchronization and error control). As the signaling and control functions are not particularly intensive, there often is bandwidth available on the D channels. AO/DI takes advantage of that bandwidth, dynamically as it is available, for end user packet data transfer (e.g., e-mail transfer) between ISDN end users. For example, as much as 9.6 Kbps is available, on average, for packet data transfer over the 16-Kbps D channel associated with an ISDN BRI. AO/DI can be set up between the BRI circuit to your home and the PRI circuit to your office. Over the D channels, you can maintain a constant (always on) virtual connection from your home to your e-mail server. (That way you can hang out by the pool in Richardson, but still get your mail from the downtown Dallas office. Your boss in New York and your clients all over the world will think you're working hard at the office. Typical text e-mail works just fine at 9.6 Kbps. When more bandwidth is required for perhaps transferring an e-mail with a graphics attachment, one or more B channels can be activated automatically. Not all service provider offer AO/DI. Those that do price it pretty attractively. See also ISDN.

AOA See Angle of Arrival.

AOC Advice Of Charge. See Advice of Charge.

AOCN Administrative Operating Company Number. Among the provisions of the Telecommunications Act of 1996 was the requirement that the RBOC's (Regional Bell Operating Companies') Code Administrators no longer work with Bellcore (now Telcordia Technologies) in the administration of NPA (Numbering Plan Administration, i.e., area codes) and NXX (Central Office prefixes) administration. To fill the void, a small number of AOCN companies were authorized to assume this responsibility. Each LEC (Local Exchange Carrier), whether ILEC (Incumbent LEC) or CLEC (Competitive LEC) must select an authorized AOCN to input various data into the various Traffic Routing Administration (TRA) tables to make sure that calls are properly routed through the PSTN (Public Switched Telephone Network), and that the traffic is properly rated. About the same time, Telcordia's role as NANPA (North American Numbering Plan Administrator) was shifted to Martin Marietta. See also CLEC, ILEC, LEC, NANPA, NNX, OCN, PSTN, and TRA.

AoE ATA over Ethernet. AoE is a proprietary Storage Area Network (SAN) protocol designed by Coraid, Inc. for accessing ATA storage devices over Ethernet networks. See ATA and SAN.

AOHell America OnLine Hell. Hacker programs that allow one to mess with AOL's software. Using these programs you can get access to, inter alia, personal electronic mail accounts.

AOL See America Online.

AOS 1. Alternate Operator Services. Today there are many Operator Services Providers not owned by the Bell Telephone Companies or AT&T. The AOS industry is dropping the descriptive term "alternate" and communicating that they be known as OSPs. AOS was coined by AT&T. See AOSP and Operator Service Providers.

2. All Options Stink. Term taken from the military, but easily applied to politics, business, etc.

AOSP Alternate Operator Service Provider. A new breed of long distance phone company. It handles operator-assisted calls, in particular Credit Card, Collect, Third Party Billed and Person to Person. Phone calls provided by OSP companies are typically far more expensive than phone calls provided by "normal" long distance companies, i.e. those which have their own long distance networks and which you see advertised on TV. You normally encounter an OSP only when you're making a phone call from a hotel or hospital phone, or a privately-owned payphone. It's a good idea to ask the operator the cost of your call before you make it.

AOSSVR Auxiliary Operator Services System Voice Response.

AOW Asia and Oceania Workshop. One of the three regional OSI Implementors' Workshops.

AP 1. See ADD-ON or Applications Processor. AP is an AT&T word for a piece of equipment which hangs off the side of their PBX and makes it do more things, like voice mail.

2. See Adjunct Processor, an AIN term for a decentralized SCP. See Adjunct Processor.

3. Access Providers.

4. Anomalous Propagation.

5. Access Point. See RF fingerprinting.

APA All Points Addressable. A method of host graphics implementation which uses vertical and horizontal pixel coordinates to create a more graphic image. An SNA definition.

Apache The web server software on about half of the world's existing web sites is Apache. Apache is UNIX freeware. Apache was originally based on code and ideas found in the most popular HTTP server of the time: NCSA httpd 1.3. It has since evolved into a far superior system that can rival – some say surpass – any other UNIX-based HTTP server in terms of functionality, efficiency, and speed. Apache includes several features not found in the free NCSA server, among which are highly configurable error messages, DBM-based authentication databases, and content negotiation. It also offers dramatically improved performance and fixes many bugs in the NCSA 1.3 code.

APAD Asynchronous Packet Assembler/Disassembler.

APC 1. Adaptive Predictive Coding. A narrowband analog-to-digital conversion technique employing a one-level or multilevel sampling system in which the value of the signal at each sample time is adaptively predicted to be a linear function of the past values of the quantized signals. APC is related to linear predictive coding (LPC) in that both use adaptive predictors. However, APC uses fewer prediction coefficients, thus requiring a higher bit rate than LPC.

2. APC finish on a fiber optic connector. See UPC.

APCC The American Public Communications Council, which is part of the North American Telecommunications Association (NATA).

APCO Association of Public-Safety Communications Officials International, Inc. An organization dedicated to the enhancement of public safety communications. APCO's more than 13,000 members come from public safety organizations including 911 centers, law enforcement agencies, emergency medical services, fire departments, public safety departments, military units, and colleges and universities. www.apcointl.org.

APCO25 APCO Project 25. A voluntary "standard" for a Common Air Interface (CAI) for SMR (Specialized Mobile Communications) public service networks in the US. APCO25 is a joint effort of APCO (Association of Public-Safety Communications Officials International Inc.) and NASTD (National Association of State Telecommunications Directors) for a uniform digital mobile radio technology. Representatives of the NTIA (National Telecommunications Industry Association), NCS (National Communications System), and DOD (Department of Defense) also sit on the steering committee. APCO25 makes use of channels of 6.24 kHz and 12.5 kHz, supporting aggregate bit rates of 9600 bps for data communications. Voice is supported at 4,400 bps. See also APCO, SMR, and TETRA.

APD Avalanche PhotoDiode. A diode that, when hit by light, increases its electrical conductivity by a multiplication effect. APDs are used in lightwave receivers because the APDs have great sensitivity to weakened light signals (i.e. those which have traveled long distances over fiber). APDs are designed to take advantage of avalanche multiplication of photocurrent.

aperiodic antenna An antenna designed to have an approximately constant input impedance over a wide range of frequencies; e.g., terminated rhombic antennas and wave antennas.

aperture For a parabolic reflector or a horn antenna, aperture is the dimension of the open mouth and represents a surface over which it is possible to calculate the radiation pattern. For a series of n stacked transmitting elements such as dipoles or slots, the vertical aperture is usually defined as n times the element spacing in wavelengths.

aperture distortion In facsimile, the distortions in resolution, density, and shape of the recorded image caused by the shape and finite size of the scanning and recording apertures or spots.

aperture grille A type of monitor screen made up of thin vertical wires. Said to be less susceptible to doming than iron shadow mask.

aphorism A truth or sentiment expressed in one sentence, for example: "Never buy a car you can't push." Now that's an aphorism.

API Application Programming Interface. Software that an application program uses to request and carry out lower-level services performed by the computer's or a telephone system's operating system. For Windows, the API also helps applications manage windows, menus, icons, and other GUI elements. In short, an API is a "hook" into software. An API is a set of standard software interrupts, calls, and data formats that application programs use to initiate contact with network services, mainframe communications programs, telephone equipment or program-to-program communications. For example, applications use APIs to call services that transport data across a network. Standardization of APIs at various layers of a communications protocol stack provides a uniform way to write applications. NetBIOS is an early example of a network API. Applications use APIs to call services that transport data across a network.

API_connection An ATM term. Native ATM Application Program Interface Connection: API_connection is a relationship between an API_endpoint and other ATM devices that has the following characteristics:

1. Data communication may occur between the API_endpoint and the other ATM devices comprising the API_connection.

2. Each API_connection may occur over a duration of time only once; the same set of communicating ATM devices may form a new connection after a prior connection is released.

3. The API_connection may be presently active (able to transfer data), or merely anticipated for the future.

APL Automatic Program Load in telecom. In data processing, it's a popular programming language.

APLT Advanced Private Line Termination. Provides the PBX user with access to all the services of an associated enhanced private switched communications services (EPSCS) network. it also functions when associated with a common control switching arrangement (CCSA) network. See Advanced Private Line Termination.

APM 1. Average Positions Manned, the average number of ACD positions manned during the reporting period for a particular group.

2. Advanced Power Management. A specification originally sponsored by Intel and Microsoft to extend the life of batteries in battery-powered computers. The idea of the specification is for the application programs, the system BIOS and the hardware to work together to reduce power consumption. An APM-compliant BIOS provides built-in power management services to the operating system of your PC. The operating system passes calls and information between the BIOS and the application programs. It also arbitrates power management calls in a multi-tasking environment (such as Windows) and identifies power-saving opportunities not apparent to applications. The application software communicates power-saving data via predefined APM interfaces. Windows 95 adopted APM to shut down the computer. It uses a special mode of the latest Intel processors – System Management Mode, or SMM. SMM lets the BIOS take control of the machine at any time and manage power to peripherals. A BIOS' APM support can't be circumvented by other software. This could cause a crash. Microsoft, Intel, Toshiba and others are now working on a new spec, called ACPI – Advanced Configuration and Power Interface. www.intel.com/IAL/power-mgm/apmovr.htm and www.ata.or/~acpi/.

APNIC Asia Pacific Network Information Center. A group formed to coordinate and promote TCP/IP based networks in the Asia-Pacific region. APNIC is responsible for management and assignment of IP (Internet Protocol) addresses in the Asia-Pacific, just as are ARIN and RIPE in the regions of the Americas and Europe, respectively. See also ARIN, IP, and RIPE.

APO Adaptive Performance Optimization. A technology used on the Texas Instruments ThunderLAN chipset, which was jointly developed by Compaq and Texas Instruments. APO dynamically adjusts critical parameters for minimum latency, minimum host CPU utilization and maximum system performance. This technology ensures that the capabilities of the PCI interface are used for automatically tuning the controller to the specific system in which it is operating.

Apocalypse, Four Horsemen Of The four horsemen of the Apocalypse were War, Plague, Famine and Death.

apogee The point on a satellite orbit that is most distant from the center of the gravitational field of the Earth. The point in an orbit at which the satellite is closest to the Earth is known as the perigee. In commercial application, the terms have most significance with respect to LEOs (Low Earth Orbiting) and MEOs (Middle Earth Orbiting) satellite constellations, which travel in elliptical orbits. See LEO and MEO.

apologize To lay the foundation for a future offense.

APON Originally specified by FSAN (Full Service Access Network) and subsequently standardized by the ITU-T as G.983.3, APON (ATM Passive Optical Network) is a local loop technology running the ATM protocol over single mode fiber. Synonymous with BPON (Broadband PON) APON runs at 155 Mbps or 622 Mbps downstream at a wavelength of 1490nm for voice and data and 1550nm for video transmission. The upstream speed is 155 Mbps at 1310nm for voice and data. The maximum logical reach of BPON is 20km, and the split ratio is 32:1. See also BPON, EPON, FSAN, GPON and PON.

APOT Additional Point Of Termination. The significance of APOT is that in the CLEC environment APOT is a requirement to submit LSR orders for collocation. These are some requirements that apply to APOT from Bell's point of view: APOT= Location "A" tie down information; CFA= Location "Z" tie down information; ACTL= Location "A" CLLI; LST= Location "Z" CLLI.

apparent power The mathematical product of the RMS current and the RMS voltage. Identical to the VA rating.

APPC Advanced Program-To-Program Communications. In SNA, the architectural component that allows sessions between peer-level application transaction programs. The LUs (Logical Units) that communicate during these sessions are known as LU type 6.2. APPC is an IBM protocol analogous to the OSI model's session layer: it sets up the necessary conditions that enable application programs to send data to each other through the network.

APPC/PC An IBM product that implements APPC on a PC.

appearance Usually refers to a private branch exchange line or extension which is on (i.e. "appears") on a multi-button key telephone. For example, extension 445 appears on three key systems.

appearance test point The point at which a circuit may be measured by test equipment.

append To add the contents of a list, or file, to those of another.

APPGEN A shortened form of the words APPlications GENerator.

Apple Computer, Inc. Cupertino, CA. Manufacturer of personal computers. Heavy penetration in the graphics/desktop publishing business and in education. Apple was formed on April Fool's Day, 1976, by Steve Wozniak and Steve Jobs, aided greatly by Mike Markkula.

Apple Desktop Bus The interface on a Mac where non-peripheral devices, such as the keyboard, attaches. A Mac keyboard or mouse is called an ADB device. Contrast with peripherals, which attach through the SCSI interface. See also USB, which is a new bus for use on PCs but fulfilling essentially the same function as the Apple Desktop Bus.

Apple Desktop Interface ADI. A set of user-interface guidelines, developed by Apple Computer and published by Addison-Wesley, intended to ensure that the appearance and operation of all Macintosh applications are similar.

Apple Menu The Apple icon in the upper left hand corner of the Apple Macintosh screen. The Apple menu contains aliases, control panels, the chooser and other desk accessories.

Apple Pie Both an American icon, and the name chosen for Apple Computer's Personal Interactive Electronics (PIE) division, chartered with extending the company into new growth areas such as Personal Digital Assistants (PDAs), e.g. the Apple Newton. The PIE division includes Apple Online Services, Newton and Telecommunications group, publishing activities, and ScriptX-based multimedia PDA development.

Apple Remote Access ARA is Apple Computer's dial-in client software for Macintosh users allowing remote access to Apple and third party servers.

Apple URP Apple Update Routing Protocol. The network routing protocol developed by Apple for use with Appletalk.

AppleShare Apple Computer's local area network. It uses AppleTalk protocols. AppleShare is Apple system software that allows sharing of files and network services via a file server in the Apple Macintosh environment. See AppleTalk.

applet Mini-programs that can be downloaded quickly and used by any computer equipped with a Java-capable browser. Applets carry their own software players. See Java.

AppleTalk Apple Computer's proprietary networking protocol for linking Macintosh computers and peripherals, especially printers. This protocol is independent of what network it is layered on. Current implementations exist for LocalTalk (230.4 Kbps) and EtherTalk (10 Mbps).

AppleTalk Address Resolution Protocol See AARP.

AppleTalk Zone and Device Filtering Provides an additional level of security for AppleTalk networks. On AppleTalk networks, network managers can selectively hide or show devices and/or zones to ARA clients. See ARA.

appliance See Edge Appliance.

appliance creep Gadget creep in an enterprise network environment. For example, over time, various groups in the enterprise, including branch offices and remote sites, may install firewalls, intrusion detection systems, load-balancing devices, various types of WAN acceleration appliances, and other network devices, each of which performs a specific, narrow function. Each of these appliances also has power, interface, and space requirements, which create network management challenges. The figurative or literal string of appliances on a network is sometimes called an appliance conga line.

appliance conga line See appliance creep.

application A software program that carries out some useful task. Database managers, spreadsheets, communications packages, graphics programs and word processors are all applications.

application acceleration The use of one or more techniques by a WAN accelerator to improve perceived application response time across a WAN. These techniques include compression and coalescing.

application based call routing In addition to the traditional methods of routing and tracking calls by trunk and agent group, the latest Automatic Call Distributors route and track calls by application. An application is a type of call, for example, sales or service. Tracking calls in this manner allows accurately reported calls, especially when they are overflowed to different agent groups. See ACD.

Application Binary Interface ABI. The rules by which software code is written to operate specific computer hardware. Application software, written to conform to an ABI, is able to be run on a wide variety of system platforms that use the computer hardware for which the ABI is designed.

application bridge Aspect Telecommunications' ACD to host computer link. Originally it ran only over R2-232 serial connections, but it now runs over Ethernet, using TCP/IP link protocol. See also Open Application Interface.

application class An SCSA term. A group of client applications that perform similar services, such as voice messaging or fax-back services.

Application Entity AE. A cellular radio term. An Application Entity provides the service desired for communication. An Application Entity may exist in an M-ES (Mobile End System) (i.e., mobile application entity) or an F-ES (Fixed End System). An Application Entity is named with an application entity title.

Application Equipment Module AEM. A Northern Telecom term for a device within the Meridian 1 Universal Equipment Module that supports Meridian Link Modules. The Meridian Link Module (MLM) is an Application Module, specially configured to support the Meridian Link interface to host computers.

Application For Service A standard telephone company order form that includes pertinent billing, technical and other descriptive information which enables the company to provide communications network service to the customer and its authorized users.

application framework This usually means a class library with a fundamental base class for defining a complete program. The framework provides at least some of the facilities through which a program interfaces with the user, such as menus and windows, in a style that is internally consistent and abstracted from the specific environment for which it has been developed.

This is an explanation I received from Borland. I don't quite understand it, yet. An application framework is an object-oriented class library that integrates user-interface building blocks, fundamental data structures, and support for object-oriented input and output. It defines an application's standard user interface and behavior so that the programmer can concentrate on implementing the specifics of the application. An application framework allows developers to reuse the abstract design of an entire application by modeling each major component of an applications as an abstract class.

application gateway A firewall that applies security mechanisms to specific applications, such as FTP and Telnet servers. An application gateway is very effective but can impose a performance degradation.

Application Generator AG. A program to generate actual programming code. An applications generator will let you produce software quickly, but it will not allow you the flexibility had you programmed it from scratch. Voice processing "applications generators," despite the name, often do not generate programming code. Instead they are self-contained environments which allow a user to define and execute applications. They are more commonly called applications generator, since one generator can define and execute many applications. See Applications Generator for a longer explanation.

application ID An emerging security concept that is being developed to deal with the proliferation of malware. An application ID is a cryptographic signature embedded in a trusted application and its supporting files. This enables a computer or network server to recognize trusted applications and block or quarantine the rest. This is easier than the impossible task of trying to keep track of an ever-growing list of malicious applications. The basic idea is that if an application isn't on a list of trusted applications and isn't authenticated by an application ID, then block or quarantine it.

application layer The topmost, visible to the user, presentation of a communications network; the user interface point in network architectures. See Open Systems Interconnection – Reference Model.

application level firewall A firewall system in which service is provided by processes that maintain complete TCP connection state and sequencing. Application level firewalls often re-address traffic so that outgoing traffic appears to have originated from the firewall, rather than the internal host.

application level proxy A firewall technology that involves examining application specific data in order to guard against certain types of improper or threatening behaviors.

application metering The process of counting the number of executions of the copies of an application in use on the network at any given time and ensuring that the number does not exceed preset limits. Application metering is usually performed by a network management application running on the file server. Most application metering software will allow only a certain number of copies (usually that number specified in the application software license) of an application to run at any one time and will send a message to any users who try to exceed this limit.

application module A Northern Telecom term for a computer that can be attached to a Northern Telecom phone system and add intelligence and programmability to the phone system. Often, the AM will be a computer conforming to open standards, such as DOS or Windows, or it may be VME-based.

Application Module Link AML. A Northern Telecom internal and proprietary link that connects the Meridian 1 (via EDSI or MSDL port) to the Meridian Link Module.

application program A computer software program designed for a specific job, such as word processing, accounting, spreadsheet, etc.

Application Program Interface API. A set of formalized software calls and routines that can be referenced by an application program to access underlying network services.

Application Programming Interface API. A set of functions and values used by one program to communicate with another program or with an operating system. See API for a better explanation.

Application Profile As SCSA term. A description of the kinds of resources and services required by a client application (or an application class). An application profile is defined once for an instance of an application; then system services such as the SCR will be able to fulfill the needs of the application without the application having to state its needs explicitly.

application server A dedicated, heavy duty PC which sits on a corporate network and contains a program which people on the network share. Such program would typically be a database – perhaps a sales force automation program, such as Goldmine, Act or Maximizer. See also Database Server.

Application Service Element ASE. A messaging term. A module or portion of a protocol in the application layer 7 of the OSI (Open Systems Interconnection) protocol stack. Several ASEs are usually combined to form a complete protocol, e.g., the X.400 P1 protocol which consists of the MTSE (Message Transfer Service Element), and the RTSE (Reliable Transfer Service Element).

Application Service Provider ASP. The definition of an ASP is evolving. Today it's a company which offers software to business users over the Internet on some sort of per-use charge. For business users, an ASP is a kind of outsourcer; users are not required to buy, own or take care of their own software. Instead of buying software, buying the heavy duty computers to run it on and the heavy duty broadband telecommunications network to get it to all their distant corporate users, the user companies (i.e. the ASP's customers) simply rent the applications from the ASPs, Examples of such applications are ERP (Enterprise Resource Planning), order entry accounting packages, software packages like Microsoft Office, etc.

application sharing Feature of many document-conferencing packages that lets two or more users on different (and usually distant) computers simultaneously use an application that resides on only one of the machines. Imagine, there are three of us and we have to jointly present a PowerPoint presentation to the "big bosses" tomorrow. Today, we have to work on the presentation. But, sadly we're in different cities. So, one of us loads application sharing software and dials the other two in a conference call. That "dialing" may be done over normal phone lines or through the Internet. See NetMeeting, the most popular application sharing.

Application Software Interface ASI. The Application Software Interface is a product of the Application Software Interface Expert Working Group of the ISDN Implementor's Workshop. The Interface focuses on the definition of a common application interface for accessing and administering ISDN services provided by hardware commonly referred to in the vendor community as Network Adapters (NAs) and responds to the applications requirements generated by the ISDN Users Workshop (IUW). The characteristics of this Application Interface shall be:

Portable across the broadest range of system architectures;

Extensible (their words, not mine)

Abstracted beyond ISDN to facilitate interworking;

Defined in terms of services and facilities consistent with OSI layer interface standards.

According to the Application Software Interface Group, the primary goal of the ASI is to provide a consistent set of application software interface services and application software interface implementation agreement(s) in order that an ISDN application may operate across a broad range of ISDN vendor products and platforms. The application software interface implementation agreements will be referenced by (and tested against) the IUW (ISDN Users Workshop) generated applications. It is anticipated that the vendor companies involved in the development of these implementation agreements will build products for the ISDN user marketplace which conform to them. ASI Implementation Agreements are likely to become a US Government Federal Information Processing Standard (FIPS).

Application Specific Integrated Circuit See ASIC.

applications engineering Applications engineering is the process of analyzing your telephone network to find products and services that will reduce your monthly bill without sacrificing network quality. It can be as simple as calling the telephone company to convert a particular service to a Rate Stabilization Plan (RSP). In many instances, the use of applications engineering concepts will increase the quality of your network. For example, putting DIDs onto a T1 will save you money and provide your network with a digital backbone. Unfortunately, most applications engineering is done by the telephone company or by their sales agents. Their main goal is not to save you money, but rather to sell telephone company products. Therefore, they are unlikely to advise you of all the hidden costs of converting to a particular service. A true application engineer will provide you with a complete cost analysis that includes all the conversion costs, and provides you with the "break-even date." The break-even date is the date that your monthly saving offsets the initial conversion cost of the service. It is often used synonymously with the term break-even point.

applications generator An applications generator (AG) is a software tool that, in response to your input, writes code a computer can understand. In simple terms, it is software that writes software. Applications generators have three major benefits: 1. They save time. You can write software faster. 2. They are perfect for quickly demonstrating an application. 3. They can often be used by non-programmers. Applications generators have two disadvantages. 1. The code they produce is often not as efficient as the code produced by a good programmer. 2. They are often limited in what they can produce. Applications generators tend to be either general purpose tools or very specific tools, providing support for specific applications, such as connecting voice response units to mainframe databases, voice messaging system development, audiotex system development, etc. There are simple AGs. There are complex AGs. There are general purpose AGs. There are specialized AGs. There are character-based AGs. There are GUI-based AGS. In researching AGs to write computer telephony and interactive voice response applications, I found three different levels of AG packages. First, there are the sort of non-generator generators. They don't really create new software, but they allow you to tweak existing application blocks. There's no compiling and they're pretty simple to use (though they often lack database and host connectivity). Then there are the pretty GUI forms-based app gens. They usually entail building a call-flow picture, using either pretty icons or easy to understand templates. When you're done filling in all the blanks, you compile it and actually "generate" new software. They're very cute. Finally, there's the script level language of a company like Parity Software, San Francisco. Real programmers dig this. They often feel it gives them a lot more power and flexibility. For very complex apps (with T-1/ISDN, ANI, host connection, speech recognition, multimedia capabilities, etc.) you'll probably need the power and flexibility of a script language. Most of the better GUI application generators let you drop down to a script-level language (and C too).

applications layer The seventh and highest layer of the Open Systems Interconnection (OSI) data communications model of the International Standards Organization (ISO). It supplies functions to applications or nodes allowing them to communicate with other applications or nodes. File transfer and electronic mail work at this layer. See OSI Model.

Applications Partner An Applications Partner is AT&T's new name for an outside company which will write software to work on AT&T phone systems, such as the Merlin, Legend and the Definitely. AT&T is setting up an Applications Partner Program to work with companies to help them develop programs and distribute their products. See also Desktop Connection.

applications processor A special purpose computer which attaches to a telephone system and allows it (and the people using it) to perform different "applications," such as voice mail, electronic mail or packet switching. We think AT&T invented the term. See also Add-On.

applique Circuit components added to an existing system to provide additional or alternate functions. Some carrier telephone equipment designed for ringdown manual operation can be modified with applique to allow for use between points having dial equipment.

APPN Advanced Peer-to-Peer Networking is, according to its creator IBM, a leading-edge distributed networking feature IBM has added to its Systems Network Architecture (SNA). It provides optimized routing of communications between devices. In addition to simplifying the addition of workstations and systems to a network and enabling users to send data and messages to each other faster, APPN is designed to support efficient and transparent sharing of applications in a distributed computing environment. Because APPN permits direct communication between users anywhere on a network, it facilitates the development of client/server computing, in which workstation users anywhere on a network can share processing power, applications and data without regard to where the information is located. Workstations on an APPN network are dynamically defined so they can be relocated easily on the network without extensive re-programming. APPN also allows remote workstations to communicate with each other, without intervention by a central computer. Also, IBM's Advanced Peer-to-Peer Networking software.

APPN End Node An APPN end node is the final destination of user data. The end node cannot function as an intermediate node in an APPN network and cannot perform routing functions. See APPN.

approved ground Grounds that meet the requirements of the NEC (National Electrical Code), such as building steel, concrete-encased electrodes, ground rings, and other devices. See AC and Grounding.

AppServer A SCSA term. AppServer defines the software environment that enables voice processing applications to run on any computing platform. AppServer sits on a PC equipped with call processing hardware and allows a remotely hosted application to control the call processing hardware.

APR Annual Percentage Rate. A percentage calculation of a finance charge portion of a financing contact.

APS Automatic Protection Switching. A means of achieving network resiliency through switching devices which automatically switch from a primary circuit to a secondary (usually geographically diverse) circuit. This switching process would take place when the primary circuit fails or when the error rate on the primary line exceeds a set threshold. There are two basic APS architectures in SONET optical fiber networks: 1+1 and 1:N. A 1+1 architecture is characterized by permanent electrical bridging to service and protection equipment, which is placed at both ends of the circuit. At the head end, or transmitting end, the same payload signal is sent over both the primary and the secondary optical circuit. The optical signal is monitored for failures at the tail end independently and identically over both optical circuits. The receiving equipment at the tail end selects either the service channel (primary circuit) or the protection channel (secondary circuit), based on pre-defined switching performance criteria. A 1:N protection switch architecture is one in which any of "N" (i.e., any Number of) service channels (primary circuits) can be bridged to a single optical protection channel (secondary circuit).

APT Asia-Pacific Telecommunity.

AQCB Automated Quote Contract Billing. System used to price non-tariffed products and services.

AR 1. Automatic Recall.

2. Access Registrar. Provides RADIUS services to DOCSIS cable modems for the deployment of high-speed data services in a one-way cable plant requiring telco-return for upstream data.

AR Coating AntiReflection coating. A thin, dielectric or metallic film applied to an optical fiber surface to reduce its reflection and thereby increase its transmitting ability.

ARA AppleTalk Remote Access. Provides an asynchronous AppleTalk connection to another Macintosh and its network services through a modem. A remote user using ARA can log on to a remote server and mount the volume on his desktop as if he were connected locally.

ARAB Attendant Release Loop. A feature of the PBX console. See Release.

Arabic Numerals Shakespeare was right when he asked, "What's in a name?" The jackrabbit is not a rabbit. It is a hare. A Jerusalem artichoke is not an artichoke; it is a sunflower. Arabic numerals are not Arabic; they were invented in India. India ink (sometimes referred to as "Chinese ink") was not known until recently in either China or India.

ARABSAT ARABSAT is the Arab Satellite Communications Organization, established in 1976 and now having 21 member Arab States. www.arabsat.com.

arachibutyrophobia Arachibutyrophobia is a fear of peanut butter sticking to the roof of your mouth.

ARAM Audio grade DRAM. DRAMS are low cost integrated circuits that are widely used in consumer electronic's products to store digital data.

aramid Aramid is a synthetic textile material which is lightweight, nonflammable, and highly impact-resistant. Dupont markets it under the trademark Kevlar. In addition to being used in construction of some fiber optic cables to provide tensile strength, aramid fibers are used in bulletproof vests, sailboat sails, and industrial-strength shoelaces. See Tight Buffer Fiber Optic Cables.

arbitrated loop Arbitrated loop is a shared, 100 MBps (200 MBps full duplex) architecture. Analogous to Token Ring, multiple devices can be attached to the same loop segment, typically via a loop hub. Up to 126 devices and one fabric (switch) attachment are allowed, although the majority of arbitrated loops are deployed with from four to 30 devices. Since the loop is a shared transport medium, devices must arbitrate for access to the loop before sending data. Fibre Channel provides a superset of commands to provide orderly access and ensure data integrity. A GBIC, or gigabit interface converter, is a removable transceiver and is commonly used in Fibre Channel switches, hubs and host bus adapters. A transceiver converts one form of signaling into another, e.g., fiber optic signals to electrical signals.

arbitrated loop topology A Fibre Channel topology that provides a (FC-AL) low-cost solution to attach multiple communicating ports in a loop. Nodes are linked together in a closed loop. Traffic is managed with a token-acquisition protocol, and only one connection can be maintained within the loop at a time. See Fibre Channel.

arbitrage The price of gold in London equals the price of gold in New York. If it didn't, traders would step in, simultaneously buy in one place and sell in another, to profit from the discrepancy. This process of buying in one market and selling in another is called arbitrage. It doesn't happen in the world of security exchanges (e.g., stock, bond, currency, and commodities exchanges) much anymore because they all are electronically linked so tightly that they balance in seconds. It does happen in the world of telecommunications. When intraLATA long distance was the sole province of the LECs (Local Exchange Carriers), for example, you could place a call to gain access to an IXC (IntereXchange Carrier) in order to place an intraLATA call at much lower rates. You weren't supposed to do it, but you could. This was known as arbitrage because you went across the LATA domain, placing an interLATA call to an IXC, and looped right back into the LATA. You profited from that practice, taking advantage of the discrepancy in rates. The IXC also profited, while the LEC lost revenue that rightfully belonged to it, and it alone. The Telecom Act of 1996 put an end to that practice, as the LATA became open to full competition and there no longer was any advantage to arbitrage. Arbitrage remains alive and well for many international long distance routes. As a hypothetical example, it may be much less expensive to call from Sao Paolo to New York via Finland, than to call New York directly. That's also arbitrage, and it exists only because of the artificially high discrepancies in the international long distance tariffs of some carriers and along some specific routes. See also Broker and Trombing.

arbitration A Fibre Channel term. The process of selecting one respondent from a collection of several candidates that request service concurrently.

ARC Attached Resource Computer, the root name of the local area networks (LAN). It was developed by Datapoint Corporation called ARCNET. It was one of the first LANs. Ethernet and Datapoint's incompetent management killed it. See ARCNET.

arcane The word "arcane" was first used in the English language about 1547. It comes from the Latin "arcanus," and means something known or knowable only to those having the key to unlock the secret. We strive to provide you with the keys to unlock the secrets of a wide range of computer and network technologies. We hope that we have been successful. If you think that we have been successful in this endeavor, buy a copy of this book for everyone you know...and for everyone each of them knows. It makes a great gift.

ARCH Access Response CHannel. Specified in IS-136, ARCH carries wireless system responses from the cell site to the user terminal equipment. ARCH is a logical subchannel of SPACH (SMS (Short Message Service) point-to-point messaging, Paging, and Access response CHannel), which is a logical channel of the DCCH (Digital Control CHannel), a signaling and control channel which is employed in cellular systems based on TDMA (Time Division Multiple Access). The DCCH operates on a set of frequencies separate from those used to support cellular conversations. See also DCCH, IS-136, SPACH and TDMA.

archie An Internet term. A corruption of "archive," Archie is a FTP search engine located on several computers around the country. It's sort of a superdirectory to the files on the Internet. If you're looking for a file or even a particular topic, Archie provides its specific location. Veronica, Jughead and WAIS (Wide Area Information Servers) are other tools for searching the huge libraries of information on the Internet. Some companies, such as Hayes, make Archie software which give you a menu driven interface that lets you browse through the various Archie servers on the Internet as though browsing through card catalogs of remote libraries.

architectural assemblies Walls, partitions, or other barriers that are not load bearing. In contrast, Architectural Structures are load bearing.

architectural freedom An AT&T term for flexibility in locating functions, such as control, storage or processing of information, at any site in or around a network, such as customer premises, central offices or regional service bureaus. Architectural freedom also means the ability to distribute functions among combinations of locations and have them interrelate through a high-throughput, low-delay, transparent network. See also Architecture.

architectural structures Walls, floors, floor/ceilings and roof/ceilings that are load bearing. In contrast, Architectural Assemblies are not load bearing.

architecture The architecture of a system refers to how it is designed and how the components of the system are connected to, and operate with, each other. It covers voice, video, data and text. Architecture also includes the ability of the system to carry narrow, medium and broadband signals. It also includes the ability of the system to grow "seamlessly" (i.e. without too many large jumps in price).

architecture police An individual or group within a company that makes sure software and hardware development follows established corporate guidelines. The architecture police tend to rein in creative development efforts.

archival Readable (and sometimes writable) media. Archival media have defined

minimum life-spans over which the information will remain stable (i.e, accurate without degradation).

archive A backup of a file. An archived file may contain backup copies of programs and files in use or data and materials no longer in use, but perhaps needed for historical or tax purposes. Archive files are kept on paper, on microfilm, on disk, on floppies, etc. They may be kept in compressed or uncompressed form. See Archiver.

archive bit A Windows NT (soon to be Windows 2000) term. Backup programs use the archive bit to mark files after backing them up, using the normal or incremental backup types.

archive server An email-based file transfer facility offered by some computers on Internet.

archiver A software program for compressing files. If you compress files, you will save on communications charges, since you will be able to transmit those files faster as they're now smaller. My favorite MS-DOS archiver, also called file compression utility is Phil Katz's PKZIP.EXE and PKUNZIP.EXE. You can cut a database by as much as 90% and a word processed file by maybe 30% by using PKZIP. How much you can cut is determined by how much fluff is in the file. PKZIP is the most widely-used archive and compression utility today. You can recognized "zipped" files because their extension is always ZIP. There are other compression programs out there which you will recognize by these extensions, ARC, AR7, ARJ, LZH, PAK and ZOO.

archiving files This is a process where the information contained in an active computer file is made ready for storing in a non-active file, perhaps in off-line or near-line storage. Typically when files are archived, they are compressed to reduce their size. To restore the file to its original size requires a process known as unarchiving. See also Archiver.

ARCNET Attached Resource Computer NETwork. One of the earliest and most popular local area networks. A 2.5M-bits-per-second LAN that uses a modified token-passing protocol. Developed by Datapoint, San Antonio, TX, Arcnet interface cards are now obsolete, having been replaced by faster Ethernet cards (IEEE 802.3) and Token Ring cards (IEEE 802.5).

ARD Automatic Ring Down. A private line connecting a telephone in one location to a distant telephone with automatic two-way signaling. The automatic two-way signaling used on these circuits causes the station instrument on one end of the circuit to ring when the station instrument on the other end goes off-hook. This circuit is sometimes called a "hotline" because urgent communications are typically associated with this service. ARD circuits are commonly used in the financial industry, but you see them at airports, where they're used to call hotels. May also have one way signaling. Station "A" rings Station "B" when Station "A" goes off hook, but Station "B" cannot ring Station "A".

ARDIS A public data communications wireless network that allows people carrying handheld devices to send and receive short data messages. Such messages might be from a sheriff standing in the street searching his department's data base for unpaid parking tickets. ARDIS network was purchased by Motient Corporation, formerly American Mobile, in 1998. The network is the largest packet data network in the US and provides packet data services using the DataTAC protocol. ARDIS was originally jointly owned by Motorola and IBM. It was an outgrowth of a network originally created for IBM service technicians.

ARE All Routes Explorer. An ATM term. A specific frame initiated by a source which is sent on all possible routes in Source Route Bridging.

Area A logical set of network segments (CLNS-, DECnet-, or OSPF-based) and their attached devices. Areas usually are connected to other areas via routers, making up a single autonomous system.

area code A three-digit code designating a "toll" center in the United States and Canada. Until January, 1995 the first digit of an area code was any number from 2 through 9. The second digit was always a "1" or "0". In January 1995, North America (i.e. the US and Canada) adopted the North American Numbering Plan (NANP) and second digits could be any number. This dramatically increased the number of possible area codes – from 152 to 792 and the number of phone numbers to more than six billion. For a full explanation, see North American Numbering Plan. For a full listing of area codes, see North American Area Codes.

area code expansion The new North American Numbering Plan (NANP) allowed basically any three numbers to become an area code. This exploded the number of area codes now possible. Some manufacturers of phone equipment, e.g. Rockwell, choose to call this happening "Area Code Expansion." They claimed that their switch would accommodate all future permutations and combinations of area codes.

area code overlay See Overlay Area Code.

area code restriction The ability of the telephone equipment (or its ancillary devices) to selectively deny calls to specific (but not all) area codes. Area code restriction is often confused with "0/1" (zero/one) restriction which denies calls to all area codes by sampling the first and second dialed digits (is it a 0 or 1?) and thus, identifying and blocking an attempt at making a toll call. For a full listing of area codes, see North American Area Codes.

area code split See Overlay Area Code.

area color code A cellular radio term. A color code that is shared by all cells controlled through a single Mobile Data Intermediate System (MD-IS). The value of the Area Color Code must be different between any two adjacent cells controlled by adjacent MD-ISs. Refer to color code.

Area Director See ADs.

area exchange Geopolitical areas set up for the administration of local telephone services. Usually a single metropolitan area or collection of towns and villages sharing a common area of community interest.

area transfer A rerouting, by splicing, of subscriber cable facilities from one Central office to another, usually within the same exchange area. An area transfer normally requires a change of telephone number for the subscribers involved and is, therefore, scheduled to occur on, or near, the Directory delivery date.

area wide centrex A centrex service, using the Intelligent Network (IN), to allow centrex service to be provided throughout a large area without dedicated facilities. See Centrex and IN.

ARF 1. A Satellite Term. Alternate Recovery Facility.
2. Alternative Regulatory Framework.

argument An argument is an addition or additions to a command that slightly change the command, either by adding options, deleting options and/or specifying filenames. For example, most MS-DOS programs will give you a list of their arguments by typing the name of the .exe and following it with /?, e.g. Type pkunzip /? and you'll get a list of all the arguments that you can follow pkunzip with.

argument of perigee To reach a geostationary orbit, the satellite is first launched on a highly elliptical transfer orbit, the perigee (point closest to the Earth) of which is approximately 200km, allowing it to reach its final altitude at the apogee (furthest point, in this case about 36,000km). The satellite then describes a transfer orbit, which causes it to pass in turn through the perigee and the apogee. The line that passes through the center of the Earth linking the perigee and apogee, known as the line of apsides, itself rotates in the orbital plane at a speed that depends on the geometry of the elliptical orbit and the inclination of the orbital plane to the equator. See Geostationary.

argument separator In spreadsheet programs and programming languages, a comma or other punctuation mark that sets off one argument from another in a command or statement. The argument separator is essential in commands that require more than one argument. Without the separator, the program can't tell one argument from another.

ARI Automatic Room Identification. In Hotel/Motel telephone system applications, the ability to display the room number on the console.

Ariane The name of a family of rockets used, amongst other things, for sending communications satellites into space. Ariane is a product of the European Space Agency, the equivalent of the U.S.'s NASA.

ARIN American Registry for Internet Numbers. A not-for-profit, voluntary, association charged with the responsibility of management of IP (Internet Protocol) addresses in the geographic areas of North America, South America, the Caribbean, and sub-Saharan Africa. ARIN membership comprises end users, including ISPs (Internet Service Providers), corporate entities, universities, and individuals. ARIN became operational on December 22, 1997 as a result of a broad-based industry agreement to separate management of IP addresses from that of URLs (Uniform Resource Locators). URL administration now is the responsibility of CORE. Both previously were the sole responsibility of InterNIC. ARIN's counterparts are APNIC (Asia Pacific Network Information Center) and RIPE (Reseaux IP Europeens). See also APNIC, CORE, InterNIC, IP, RIPE, and URL.

ARINC Aeronautical Radio INC. The collective organization that coordinates the design and management of telecommunications systems for the airline industry. It's one of the largest buyers of telecommunications services and equipment in the world. In its own words, "ARINC develops and operates communications and information processing systems for the aviation and travel industries and provides systems engineering and integration solutions to government and industry. Founded in 1929 to provide reliable and efficient radio communications for the airlines, ARINC is a $280 million company headquartered in Annapolis, MD with over 2,000 employees worldwide." www.arinc.com.

arithmetic coding A compression technique which produces code for an entire message, rather than encoding each character in a message. Arithmetic coding improves on Huffman Encoding, although it is slower. See also Compression and Huffman Encoding.

Arithmetic Logic Unit ALU. The part of the CPU (Central Processing Unit) that performs the arithmetic and logical operations. See Microprocessor.

arithmetic operation The process that results in a mathematically correct solution during the execution of an arithmetic statement or the evaluation of an arithmetic expression.

arithmetic overflow 1. In a digital computer, the condition that occurs when a calculation produces a result that is greater than a given register or storage location can store or represent.
2. In a digital computer, the amount that a calculated value is greater than a given register or storage location can store or represent. The overflow may be placed at another location. See Overflow.

arithmetic register A register (i.e. short-term storage location) that holds the operands or the results of operations such as arithmetic operations, logic operations, and shifts.

arithmetic shift A shift applied to the representation of a number in a fixed radix numeration system and in a fixed-point representation system in which only the characters representing the fixed-point part of the number are moved. An arithmetic shift is usually equivalent to multiplying the number by a positive or a negative integral power of the radix, except for the effect of any rounding; compare the logical shift with the arithmetic shift, especially in the case of floating-point representation.

arithmetic underflow In a digital computer, the condition that occurs when a calculation produces a non-zero result that is less than the smallest non-zero quantity that a given register or storage location can store or represent.

arithmetic unit The part of a computing system which contains the circuits that perform the arithmetic operations. See also ALU.

arj The extension .arj shows that a file or program has been "compressed," and must be "exploded" with the arj program before being either read or used. Groups of files may be compressed together, but this is more commonly done with the zip program. See Zip.

ARM 1. Asynchronous Response Mode. A communication mode involving one primary station and at least one secondary station, where either the primary or one of the secondaries can initiate transmission.
2. Audience Relationship Management, a business which offers publishers a single audience platform for print, online and events.

arm and a leg In George Washington's days, there were no cameras. One's image was either sculpted or painted. Some paintings of George Washington showed him standing behind a desk with one arm behind his back while others showed both legs and both arms. Prices charged by painters were not based on how many people were to be painted, but by how many limbs were to be painted. Arms and legs are "limbs," therefore painting them would cost the buyer more. Hence the expression. "Okay, but it'll cost you an arm and a leg."

arm candy Colloquial expression for a dumb, but beautiful female date. On your arm, as you arrive at the party, is a gorgeous woman. "My," say your friends, "You have great arm candy." See also Eye and Wrist Candy.

arm Supplier answered by answering machines. These systems may place the associated
A maker of hardware. A term used especially in the gaming business.

armageddon The fabled battlefield where God's heavenly forces are to defeat the demon-led forces of evil. The final battle.

Armed Forces Radio Service AFRS. A radio broadcasting service that is operated by and for the personnel of the armed services. An example of an AFRS is the radio service operated by the U.S. Army for U.S. and allied military personnel on duty in overseas areas.

ARMIS Automated Management Reporting Information System. Since 1987 the FCC has used the ARMIS system for collecting network infrastructure, financial and operating information from the largest carriers. The FCC produces 10 public reports using this system.

armor Mechanical protection usually accomplished by a metallic layer of tape, braid or served wires or by a combination of jute, steel tapes or wires applied over a cable sheath for additional protection. It is normally found only over the outer sheath. Armor is used mostly on cables lying on lake or river bottoms or on the shore ends of oceans. See Armored Cable.

armored cable 1. A stainless steel handset cord which is meant to resist vandalism. Typically used on a coin phone, most stainless steel handset cords are too short. This is said to be because they were first ordered for use in prisons, where guards wanted to be certain they would not be used by the prisoners as hanging devices. Thus, they requested Western Electric to make them too short for such a use. Whether there is any validity to this story is dubious. However, it is part of telephone industry folk history and therefore, worth preserving.

2. In outside cable an armored cable has its sheath covered with three protective layers: a vinyl jacket, a steel wrap, and another vinyl jacket. Armored cable is intended for use in direct-burial applications; the steel armor protects the sheath from damage during installation. See also Hard Cable.

army A group of frogs is called an army.

ARP Address Resolution Protocol. 1. A low-level protocol within the Transmission Control Protocol/Internet Protocol (TCP/IP) suite that "maps" IP addresses to the corresponding Ethernet addresses. In other words, ARP is used to obtain the physical address when only the logical address is known. An ARP request with the IP address is broadcast onto the network. The node on which the IP address resides responds with the hardware address in order that the packets can be transmitted. By way of example, TCP/IP requires ARP for use with Ethernet, in which case the physical address would be defined by the MAC address hard-coded on the NIC (Network Interface Card) of the target workstation. See also RARP.

2. A low-level protocol which serves to map IP addresses, or other non-ATM addresses, to the corresponding address of the target ATM device. Once the ATM address has been identified, the ARP server can stream data to the target device as long as the session is maintained.

ARP poisoning See ARP spoofing.

ARP spoofing The sending of fake Address Resolution Protocol (ARP) packets to an Ethernet LAN in order to change entries in the LAN server's cached lookup table so as to redirect traffic on the LAN from its intended destination to another destination. See ART table. See also pharming.

ARP table A table of IP addresses stored on a local computer, used to match IP addresses to their corresponding MAC addresses. See also ARP.

ARPA Advanced Research Projects Agency of the U.S. Department of Defense. (The whole DOD annual telecommunications bill exceeds $1 billion.) Much of the country's early work on packet switching was done at ARPA. At one stage it was called DARPA, which stands for Defense Advanced Research Projects Agency. ARPA was the U.S. government agency that funded research and experimentation with the ARPANET and later the Internet. The group within DARPA responsible for the ARPANET is ISTO (Information Systems Techniques Office), formerly IPTO (Information Processing Techniques Office). See also DARPA Internet. DARPA has changed its name to ARPA and back again. It's hard to keep up.

ARPANET Advanced Research Projects Agency NETwork. A Department of Defense data network, developed by ARPA, which tied together many users and computers in universities, government and businesses. ARPANET was the forerunner of many developments in commercial data communications, including packet switching, which was first tested on a large scale on this network. The predecessor of the Internet, it was started in 1969 with funds from the Defense Department's Advanced Projects Research Agency (ARPA). ARPANET was split into DARPANET (Defense ARPANET) and MILNET (MILitary NETwork) in 1983. ARPANET was officially retired in 1990.

ARPM Average Revenue Per Minute.

ARPU Average Revenue Per Unit. Pronounced R-POOH. The average revenue generated per customer unit per month. How much a customer provides a phone company – wired or wireless. This is one indicator of the financial performance of a telecom company. Is the ARPU rising or falling? How does it compare to the ARPU of competing companies? It's a metric which Wall Street uses to compare companies of different sizes. See the next definition.

ARPU erosion The decline of average revenue per customer due to price competition, substitution of one service for another (for example, use of wireless for some long-distance calls), changing customer demographics, or other factors. See also ARPU.

ARQ Automatic Retransmission reQuest, Automatic Repeat reQuest or Automatic ReQuest for ReTransmission. No one knows which one it really is. But it doesn't matter. They all mean the same thing. ARQ is the standard method of checking transmitted data, used on virtually all high-speed data communications systems. The sender encodes an error-detection field based on the contents of the message. The receiver recalculates the check field and compares it with that received. If they match, an "ACK" (acknowledgment) is transmitted to the sender. If they don't match, a "NAK" (negative acknowledgment) is returned, and the sender retransmits the message. Note: this method of error correction assumes the sender temporarily or permanently stores the data it has sent. Otherwise, it couldn't possibly retransmit the data. No error detection scheme in data transmission is foolproof. This one is no exception.

array 1. The description of a location of points by coordinates. A 2-D array is described with x,y coordinates. A 3-D array is described with x,y,z coordinates.

2. A named, ordered collection of data elements that have identical attributes; or an ordered collection of identical structures.

3. Two or more hard disks that read and write the same data. In a RAID system, the operating system treats the array as if it were a single hard disk.

4. A form of telecommunications wireless antenna. See Array Antenna and Phased Array Antenna.

array antenna Take a bunch of directional antennas. Aim them at the same transmitting source. Join them together. Presto, you now have a very powerful giant antenna. Array antennas are used for picking up weak signals. They are often used in astronomical and defense communications systems. For a bigger explanation, see Phased Array Antenna.

array connector A connector for use with ribbon fiber cable that joins 12 fibers simultaneously. A fan-out array design can be used to connect ribbon fiber cables to non-ribbon cables.

array processor A processor capable of executing instructions in which the operands may be arrays rather than data elements.

array splice ARS. A splicing device used for ribbon cable. Splices 12 fibers at once.

Array Waveguide AWG. A passive optical component used for wavelength separation (i.e. to multiplex and de-multiplex wavelengths, particularly those tightly spaced together), which becomes increasingly important in high channel count DWDM systems, which require closer channel spacing. AWGs are fabricated by depositing thin layers of glass onto silicon wafers.

Arrayed Wave Guide AWG. Chip-sized devices made of glass that combine the streams of different lasers and boost capacity. See also Array Waveguide.

arrestor A device used to protect telephone equipment from lightning, electrical storms, etc. An arrestor is typically gas filled so when lightning strikes, the gas ionizes and, bingo, a low resistance to the ground that drains the damaging high voltage elements of the lightning away.

arrival rate A call center term. The pattern in which calls arrive. Call Arrival Rates can be smooth, like outgoing telemarketing calls, or random, like incoming toll-free number calls, or peaked, where calls escalate in response to advertising.

ARRL See American Radio Relay League.

arrobe Term coined by the French (who else?) to replace the barbaric English "at" (@) in email addresses. The word is derived from arroba, the Spanish equivalent.

ARS 1. Automatic Route Selection, also called Least Cost Routing. A way that your phone system automatically chooses the least expensive way of making the call that it is presented with. That least expensive way may be a tie line or a WATS line, etc. It may even be dial-up. See Least Cost Routing and Alternate Routing.

2. See array splice.

ARSG Australian Radiocommunications Study Group.

ART Autorit0xC7 de R0xC7gulation des T0xC7l0xC7communications (French telecommunications regulator).

artefact Misinterpreted information from a JPEG or other compressed image. Colour faults or line faults that visibly impact the image negatively. The higher the level of compression, the more likely the artefacting.

article An Internet term. An article is a USENET conversation element. It is a computer file that contains a question or piece of information made available to the USENET community by posting to a newsgroup.

articulation index A measure of the intelligibility of voice signals, expressed as a percentage of speech units that are understood by the listener when heard out of context. The articulation index is affected by noise, interference, and distortion.

artifacts Distortions in a video signal. Unintended, unwanted visual aberrations in a video image. In all kinds of computer graphics, including any display on a monitor, artifacts are things you don't want to see. They fall into many categories (such as speckles in scanned pictures), but they all have one thing in common: they are chunks of stray pixels that don't belong in the image.

artificial intelligence In 1930s, Alan Turing, a British mathematician, challenged scientists to create a machine that could trick people into thinking it was one of them. The idea is that a computer will have achieved intelligence when a person chatting over a teletype is unable to tell whether a human being or a machine is at the other end of the conversation. And this for long was THE classic definition of artificial intelligence. After half a century, the prospect of passing the Turing test remains so remote that many computer scientists have abandoned it as a practical goal. The real challenge these days with artificial intelligence, now more commonly called "expert systems," is not to recreate people but to recognize the uniqueness of machine intelligence and learn to work with it in intelligent, useful ways.

artificial line interface In T-1 transmission, refers to the ability of a piece of transmission equipment to attenuate its output level to meet the required loop loss of 15-

22.5 dB normally switch selectable between 0,7.5, and dB.

Artificial Neural Network ANN. See Neural Network.

ARU Audio Response Unit. A device which gives audible information to someone calling on the phone. "Press 1 for the train timetable to Boston." The ARU reads the timetable. The caller responds to questions by punching buttons on his telephone keypad. If this sounds like Interactive Voice Response – IVR, you're 100% right because that's exactly what it is. See IVR.

AS 1. Autonomous System. An Internet term. An Autonomous System is just that – a system which is autonomous. Typically, an AS is an ISP, an Internet Service Provider. Within the ISP, routers exchange information freely – all systems are trusted, as they are under a single administration in the same domain. Therefore, such systems can run an IGP (Interior Gateway Protocol) such as IGRP (Interior Gateway Routing Protocol) or OSPF (Open Shortest Path First). As the same level of trust does not exist between ASs, they must run an EGP (Exterior Gateway Protocol) such as BGP (Border Gateway Protocol) or IDRP (InterDomain Routing Protocol). See also BGP, EGP, IDRP, IGP, IGRP and OSPF.

2. Australian Standards. Standards that have been approved by Standards Australia in response to formal requests from the community, an industry body or government departments.

as is A term used in the secondary telecom equipment business. "As is" is equipment that is bought or sold with no stated or implied warranties. You should expect any condition from good to bad, from complete to incomplete. Buy As Is equipment at your own risk.

As Is Tested or As Is Working A term used in the secondary telecom equipment business. One step up from "as is" condition. The product has been tested. It works and is complete, unless otherwise specified. Buyer should test upon receipt. There is no warranty beyond receipt. Seller is guaranteeing the product will work upon arrival. After that, the buyer is responsible for any problems.

AS&C Alarm Surveillance and Control.

AS/400 IBM's mid-range mini-computer. AS/400 stands for Application System/400. IBM has a product called CallPath/400 which allows AS/400 computers to link to PBXs from the leading manufacturers.

ASA 1. Average Speed of Answer. How long average callers have to wait before they speak to an agent. The time can vary, even over the course of one day, due to call volumes and staff levels. An important measure of service quality. ASA is used in most call centers.

2. See Affiliated Sales Agency.

ASAI Adjunct Switch Application Interface. A software message set or interface protocol on the Lucent (now called Avaya) DEFINITY PBX switch for PBX-to-file server CT (Computer Telephony) applications. ASAI supports activities such as event notification and call control. Essentially, ASAI is a Lucent (now Avaya) specification for SCAI (Switch-to-Computer Applications Interface), an early implementation of CT.

ASAM ATM subscriber access multiplexer. A telephone central office multiplexer that supports SDL ports over a wide range of network interfaces. An ASAM sends and receives subscriber data (often Internet services) over existing copper telephone lines, concentrating all traffic onto a single high-speed trunk for transport to the Internet or the enterprise intranet. This device is similar to a DSLAM (different manufacturers use different terms for similar devices).

ASBR Autonomous System Boundary Router. ABR located between an OSPF autonomous system and a non-OSPF network. ASBRs run both OSPF and another routing protocol, such as RIP. ASBRs must reside in a nonstub OSPF area. See also ABR, nonstub area, and OSPF.

ASC 1. AIN Switch Capabilities. See AIN.

2. Abnormal Station Code. Generated by the OCU because of a loss of signal from the DSU/CSU or the DSU/CSU is not attached.

3. Automatic Slope Control. A device which automatically changes the slope of an amplifier's curve to compensate for temperature changes.

Ascension Island A 35-square mile island of volcanic origin, situated in the Atlantic Ocean halfway between Africa and South America, between the Equator and the Tropic of Capricorn. From 1922 to 1964 the island was managed by the Eastern Telegraph Company (renamed Cable and Wireless in 1934), reflecting the island's heritage as a major relay point for submarine cable systems that linked the United Kingdom, Portugal and South Africa with South America and West Africa. Even today, many of the commercial and government organizations on Ascension are involved in communications.

ASCENT Association of Communications Enterprises. An association of approximately 750 entrepreneurial communications firms and their suppliers in both the wired and wireless domains. ASCENT was born in 2000 of a name change from Telecommunications Resellers Association (TRA), which was formed in 1992 through the merger of the Telecommunications Marketing Association and the Interexchange Resellers Association. TRA holds several conferences and exhibitions each year, and acts as the resale industry's lobbying group and

consumer watchdog. In 1997, TRA absorbed NWRA (National Wireless Resellers Association). www.ascent.org.

ASCI-Assisted Routing A layer 3 switch that has some of its routing functionality built within ASCIS.

ASCII Pronounced: as'-kee. American Standard Code for Information Interchange. It's the most popular coding method used by small computers for converting letters, numbers, punctuation and control codes into digital form. (Computers can only understand zeros or ones.) Once defined, ASCII characters can be recognized and understood by other computers and by communications devices. ASCII defines 128 characters, including alpha characters, numbers, punctuation marks or signals in seven on-off bits and a parity bit (used for data). A capital "C", for example, is 1000011, while a "3" is 0110011. As a seven-bit code, and since each bit can only be a "one" or a "zero," ASCII can represent 128 "things," i.e. 2 x 2 x 2 x 2 x 2 x 2 x 2 which equals 128. ASCII is the code in which virtually every personal computer in the world encodes "things," including IBM, Apple and Radio Shack/Tandy. This compatible encoding (it was developed by ANSI – the American National Standards Institute) allows virtually all personal computers to talk to each other, if they use a compatible modem, or null modem cable and transmit and receive at the same speed. There are variations of ASCII. (Nothing is totally standard anymore.) The most important variation – one originally from IBM – is called Extended ASCII. It codes characters into eight bits (or one byte) and uses those ASCII characters above 127 to represent foreign language letters, and other useful symbols, such as those to draw boxes. But at 127 and below, extended 8-bit ASCII is identical to standard 7-bit ASCII. The ITU (now called the ITU-T) calls ASCII International Telegraph Alphabet 5.

The other major method of encoding is IBM's EBCDIC (pronounced ebb'-si-dick). It's largely used on IBM and IBM-compatible mainframe computers (but not their PCs, which use ASCII and extended ASCII.) EBCDIC is an eight-bit encoding scheme, thus allowing up to 256 "things" to be encoded, i.e. 2 x 2 x 2 x 2 x 2 x 2 x 2 x 2 = 256. EBCDIC codes letters, characters and punctuation marks in a totally different way than ASCII. For ASCII files to be read by an IBM mainframe (one that reads EBCDIC), those ASCII files must be translated into EBCDIC by one of the many translation programs available. See also ASCII Editor, Baudot, EBCDIC, Extended ASCII, Extended Graphics Character Set, Morse Code and Unicode.

ASCII Editor An ASCII editor (also called a "text," "DOS" or "non-document mode" editor) does NOT use extended ASCII and printer [ESCAPE] codes, which are used by word processors to create advanced features such as bold, italic, underlining, and super/subscript printing effects; and fancy formatting such as automatic paragraph reformat, pagination, hyphenation, footers, headers, and margins. I initially wrote this dictionary using an ASCII editor called ZEdit, which is a customized version of QEdit, undoubtedly the best editor ever written. Then, the author QEdit, produced a new and more powerful editor, called The Semware Editor. And I'm now using it to write this edition. Since an ASCII editor can't do so much, why would anyone use one? Well, its strength is in the lack of those very things a word processor has, which clutter it and slow it down! Here are my benefits:

1. It's lightning fast. No word processor can match an ASCII editor's speed at loading itself, loading files, finding things in files, etc.

2. A file produced by an ASCII editor can be read and edited by any word processor (absolutely any). Thus it's the universal word processing file. A WordPerfect file typically can't be read by WordStar and vice versa. The reason is that every word processor uses different high-level codes for the same features (underlining, bolding, etc.) There is no consistency among word processors as to how they encode their text so they can tell printers to do bolding, etc.

3. An ASCII editor is better to type programming languages, such as EDLIN (for batch files), Basic, FORTRAN, Pascal, etc. If QEdit used extended ASCII and printer codes, it could not be used by these programs...for each program interprets these "high level" codes differently from another program. An ASCII editor types straight, "vanilla" text...nothing fancy about it.

ASCII File An ASCII file consists solely of ASCII 127 and below ASCII characters that are visible. You create an ASCII file using a simple editor, also called an ASCII editor. An ASCII file is also called a text file. See ASCII.

ASCII-To-Fax Conversion Allows the transfer of a word-processed file directly to your fax board so it can be faxed without being scanned from a hard copy print-out. Documents faxed with ASCII-to-Fax conversion come out much cleaner at the other end, since the scanning process always degrades the image.

ASDS Accunet Spectrum of Digital Services. AT&T's leased line (also called private line) digital service at 56 Kbps. MCI and Sprint have similar services. It is available in N x 56/64 Kbps, for N = 1, 2, 4, 6, 8, 12. The 56/64 Kbps POP-POP service (between long distance carrier central offices) costs the same as an analog line.

ASE 1. A messaging term. Application Service Element. A module or portion of a protocol in the application layer 7 of the OSI (Open Systems Interconnection) protocol stack. Several ASEs are usually combined to form a complete protocol, e.g., the X.400 P1 protocol which consists of the MTSE (Message Transfer Service Element), and the RTSE (Reliable Transfer Service Element).

2. Amplified Spontaneous Emission.

ASG Access Service Group. Generally represents the tandem or the dial tone office and associated offices subtending a tandem.

ASH Ardire-Stratigakis-Hayduk, a synchronous compression algorithm that is said to offer four times throughput on a typical synchronous channel. It can be used in bridges, routers, ISDN and modems. Transcend of Cleveland, OH said at one point that it was the exclusive licensor of ASH.

ASI 1. Alternate Space Inversion. A line coding technique used on ISDN circuits to communicate from the NI (Network Interface) device to the other CPE. It is the opposite of AMI (Alternate Mark Inversion) which is used frequently at the T-1 level. ASI assigns spaces to binary ones, and alternate polarities (at +/-750 mV) of marks to zeroes. See also AMI.

2. Advanced Services Implementation.

3. Application Software Interface. An ISDN term. See Application Software Interface.

4. Adapter Support Interface. The driver specification developed by IBM for networking over IEEE 802.5 Token-Rings.

Asia-Pacific Telecommunity APT. A telecommunications organization for the Asia Pacific region.

ASIASAT ASIASAT is the Asia Satellite Telecommunications Co. Ltd., set up in 1990 in Hong Kong and now providing coverage over Asia, northeast Africa, Australia and New Zealand. www.asiasat.com.

ASIC Application-Specific Integrated Circuit. A silicon chip that is custom-designed for a particular purpose, at least that's the pure definition. In actuality, the term is misleading because many ASICs are designed to perform multiple, generalized tasks. From the manufacturer's point of view, a microprocessor is an ASIC, though they can and are used for widely disparate purposes in the field. An ASIC requires large production volumes to be economical; long design cycles and high-priced design tools (and designers) make them expensive to create, but inexpensive to produce in high-volumes. Manufacturers use ASICs to consolidate many chips into a single package, thereby reducing system board size and power consumption. Many video boards and modems use ASICs. ASICs span programmable array logic (PAL) devices, electrically programmable logic devices (EPLDs), field programmable logic devices (FPGAs), gate arrays, standard cell-based devices, and full custom, designed-from-scratch ICs. See also ASSP, FPGA, and SOC.

ASIC Chip Application Specific Integrated Circuit Chip. A fancy name for microprocessor chips which do specific tasks. For example, an ASIC chip might be responsible for a graphics display. See ASIC.

ASL Adaptive Speed Leveling. A US Robotics term for adjusting the transmission speed of a modem up or down, depending on the conditions on the line. US Robotics says it can adjust speed in 2 or 3 seconds after detecting changed line conditions. It requires like modems on either end of the transmission.

ASME American Society of Mechanical Engineers.

ASMTP See Authenticated SMTP.

ASN 1. Abstract Syntax Notation One. LAN "grammar," with rules and symbols, that is used to describe and define protocols and programming languages. ASN.1 is the OSI standard language to describe data types. The Abstract Syntax Notation is a formal language defined by ITU X.208 and ISO 8824. Under both CMIP and SNMP, ASN.1 defines the syntax and format of communication between managed devices and management applications. See CMIP, SNMP. For a fuller definition, see Abstract Syntax Notation One.

ASOC Technology This definition from Bookham Technology: ASOC technology is the fabrication of integrated optical components from a base material of silicon. Silicon has excellent optical properties including a low optical loss at 1310nm and 1550nm – the wavelength bands at which to transmit telecommunication signals. Silicon is the world's best known manufacturing material and benefits from the mature manufacturing processes of the microelectronic industry to significantly reduce the complexity of design and manufacture. Manufacturing ASOC devices fundamentally involves the construction of low-loss, single-mode waveguides onto silicon-on-insulator wafers. Library elements, such as couplers and modulators, and hybridized active elements, such as lasers and photodiodes, are combined to provide devices with a wide range of functionality. Consequently compact and versatile integrated optical components can be produced in high volume and at low cost. The flexible nature of ASOC technology means it can also be utilized to provide application

specific, high functionality, high value devices including Arrayed Waveguide Gratings (AWG) and Variable Optical Attenuators (VOA).

ASOG See Access Service Ordering Guidelines.

ASP 1. A Northern Telecom term for Attached Support Processor.

2. Adjunct Service Point. An intelligent-network feature that resides at the intelligent peripheral equipment and responds to service logic interpreter requests for service processing. See also AIN.

3. Administrable Service Provider. A SCSA term.

4. Abstract Service Primitive. An ATM term. An implementation-independent description of an interaction between a service-user and a service-provider at a particular service boundary, as defined by Open Systems Interconnection (OSI).

5. Application Service Provider. ASPs, in reality, are nothing more than software rental agencies. They host the applications on their computers and take care of all maintenance, refinements and such. Access is generally through the Internet. They are great for some people whose mantra is "Own Nothing If Possible."

See Application Service Provider.

6. Active Server Page. See Active Server Page.

7. AppleTalk Session Protocol. A protocol that uses ATP to provide session establishment, maintenance, and teardown, as well as request sequencing. See also ATP.

8. Auxiliary signal path. In telecommunications, link between TransPaths that allows them to exchange signaling information that is incompatible with the PSTN backbone network; used to provide feature transparency.

9. Average Selling Price. You sell a bunch of microprocessors to various people at different prices. Divide the total price by the number you sold and bingo you have the average selling price.

10. Alternate Service Provider. Another name for a new phone company. See Alternate Service Provider.

11. Cleopatra's fatal attraction.

aspect What do you get when you moon a chicken? A friend named his company Aspect Communications and was mad as a hatter when I told him I knew where his company's name came from. I meant it as a joke. He took it seriously. So, Jim, you've now made history.

aspect ratio The ratio of width to height of a computer display or TV screen. The aspect ratio of NTSC and PAL TV is four units of width to every three units of height. This is expressed as 4 x 3 aspect ratio. A 35 mm frame measures 12 x 24 mm, which means it has two units of width to one unit of height. It is different in size from a TV screen. This is why the side parts of movies are chopped off on TV. For VGA and Indeo video technology, the aspect ratio is 4:3 yielding today's standard PC screen sizes in pixels of 640 x 480, 1024 x 768, 1280 x 10243, 1600 or 1200 and higher.

aspherical surface Lens surface with more than one radius of curvature, i.e. the surface does not form part of a sphere. The aspherical elements of a lens help compensate for many lens aberrations common in simpler lens designs. Aspherical elements are particularly important for wide-angle lenses, since they are prone to distortion.

ASPI Advanced SCSI Programming Interface set. In other words, software primitives and data structures which allow software using the ASPI interface to be SCSI host adapter-independent. SCSI stands for Small Computer System Interface. (Pronounced Scuzzie.) ASPI is software that acts as a liaison between SCSI device drivers (the software that drives the SCSI devices) and the interface card (also known as the host adapter). Whenever a new device is added to a computer system, a software program called a "driver" must tell the computer how to talk to the new device. Instead of forcing vendors to write drivers for every host adapter, ASPI lets them write a driver to ASPI standards, supposedly guaranteeing that the device the driver controls will work with all ASPI-compatible host adapters.

The idea behind ASPI is to create a "black box" software interface - one which allows programmers to create software without having to know anything about the details of the SCSI interface hardware used in your computer. With ASPI, it's possible to write programs that can be used with any SCSI-based device used on a computer system that supports ASPI. While things are not always 100% perfect in all cases, ASPI greatly reduces potential compatibility problems for you, the user.

How does ASPI work? Essentially, there are two parts to an ASPI implementation. First, there's the ASPI "manager" which is a device driver supplied by the hardware manufacturer, and the ASPI software application. It's important to note that without an ASPI manager, ASPI compatibility is not possible. It's the manager that creates the standard ASPI-compatibility layer between the SCSI host adapter hardware and the ASPI-compatibility application. The manager is very hardware-specific, and is almost always supplied by the manufacturer of your SCSI host adapter.

ASQ Automated Status Query.

ASR 1. Automatic Speech Recognition. See Speech Recognition.

2. Automatic Send Receive. A teletype or telex machine manufactured by the old Teletype Corporation. Such a hard-copy terminal, if left on and loaded with paper, will receive incoming messages and print them, even when nobody is present. The machine also can be commanded remotely to send the full contents of its paper tape reader. Teletype Corporation's ASR-33 was a very popular minicomputer terminal in the 1970s. They now are considered to be obsolete.

3. Access Service Request. This is a request that a telephone company gives to another telephone company for any of many kinds of interconnectivity or data sharing needs. These requests can be between local carriers or long distance carriers and can originate with either an incumbent or an alternative company. See Access Service Request for more detail.

4. Authorized Sales Representative. Many phone companies have programs which allow interconnect or other resellers companies to resell their services – from simple local lines to T-1 lines. The phone companies often pay these companies a small commission for their sales efforts.

5. Answer Seizure Ratio. A measurement of the effectiveness of a telecommunications service offering. ASR is the relationship between the number of line seizures and the number of answered (i.e., completed) calls. Did you ever get dial tone, and then a "fast busy" because the destination device is unavailable? That is not good for ASR. Did you ever get dial tone, and then an answered call because the destination device is available? That is good for ASR.

6. Average Service Rate = Percentage of calls placed that actually complete to terminating end. For instance you might hear a technician say the ASR for India is currently 15%. ASR is a term used frequently in the international arena. For instance the ASR would be much higher for calls from the USA to the United Kingdom than third world countries such as India, Zaire, Vietnam etc.

ass-vertising Advertising on bikini underwear. Worn and flashed by female models on the street, the buttborne billboards are designed to attrach young male consumers. A Wired Magazine definition.

assembler Software that translates assembly language into machine language – the code of ones and zeros (1s and and 0s) used by computers. Contrast with compiler, which is used to translate a high-level language, such as C, into assembly language first and then into machine language. Assembler code is a close approximation of machine code. It is difficult to write and different for each processor. See also Assembly language.

assembly language A computer language for writing software. It is a language which is converted by programs called compilers or interpreters into machine language programs which consist of only 1s and 0s and which a computer can understand. Even though an assembly language consists of recognizable mnemonics and meaningful words, it's not easy to program in. It is referred to as a "low-level language". Assembly language programs run faster than high-level language programs, such as Basic, COBOL or FORTRAN, which are much easier to learn and program in. Choosing a programming language is a tradeoff of ease for speed. See also Assembler.

asserted A signal is asserted when it is in the state which is indicated by the name of the signal. Opposite of Negated.

assertion language In the design of semiconductors, assertion languages let engineers check properties during simulation and formal verification of their designs. There has been some effort to standardize assertion languages, in particular in something called OpenVera.

asset allocation The practice of allocating a certain percentage of a portfolio to different types of investments (stocks, bonds, foreign stocks, cash reserves or equivalents, gold, mutual funds, futures, options, etc.).

asset sale An asset sale occurs when you sell the assets of the company, but not the company itself. There are tax advantages to both the buyer and seller from an asset sale, instead of a company sale.

assignation A secret romantic rendezvous. An invitation to an assignation doesn't work if she doesn't know the meaning of the word. Are you listening Jane Laino?

assigned cell An ATM term. Cell that provides a service to an upper layer entity or ATM Layer Management entity (ATM-entity).

assigned frequency The center of the assigned frequency band assigned to a station.

assigned frequency band The frequency band within which the emission of a station is authorized; the width of the band equals the necessary bandwidth plus twice the absolute value of the frequency tolerance. Where space stations are concerned, the assigned frequency band includes twice the maximum Doppler shift that may occur in relation

to any point of the Earth's surface.

Assigned Night Answer ANA. After business hours or when you place your phone system on "Night Answer," this feature sends calls from specified trunks to designated extensions or departments. You may use this feature to send calls directly to modems, or to emergency numbers, or even to outside home numbers.

assigned number Telephone company jargon for a telephone number that is assigned to a customer.

assigned plant concept A pair is dedicated from the central office to the subscriber home and maintained at that address, even when idle. See Reassignment.

assignment A call center term. The process of assigning individual employees to specific schedules in a Master File or Daily Workfile. Master File assignment can be done either manually or automatically (based on employee schedule preference and seniority). See Assignment Lists.

assignment lists In a non-mechanized line assignment environment in a telephone company, assignment lists of lines and numbers are prepared by the Network Administrator as a means of providing input to the Service Center for service order preparation. The lines and numbers made available for assignment are determined by the guidelines for overall loading plan and load balance objectives. The age of telephone numbers is also a consideration. Also see Intercept Interval.

Assisted Global Positioning System AGPS. See GPS.

Assisted GPS AGPS. See GPS.

assmosis The process by which some people seem to absorb success and advancement by kissing up to the boss rather than working hard or smart.

associate 1. A verb used in Windows by File Manager. You associate a three character extension with an application. This tells File Manager that, when you click twice on the file, File Manager will know which application to launch. For example, you may tell Windows that the .QXD extension is associated with QuarkXpress. When you click on a QXD file, File Manager will launch Quark and load that particular file.

2. Thanks to companies like Wal-Mart, companies no longer have "employees" – they have "associates." Of course, they're still paid like employees. In fact, some people would argue they are paid less than employees.

associated common-channel signaling A form of common-channel signaling in which the signaling channel is associated with a specific trunk group and terminates at the same pair of switches as the trunk group. The signal channel is usually transmitted by the same facilities as the trunk group.

association A relationship between two connection segments that share a common Leg 0 (i.e., a common subscriber is in control of connection segments). Definition from Bellcore.

Association Control Service Element ACSE. The International Standards Organization's Open Systems Interconnect (OSI) application layer services used, for example, in Manufacturing Automation Protocol V3.0 (MAP).

Association of Communications Enterprises See ASCENT.

Association of Information Technology Professionals See AITP.

ASSP Application Specific Standard Product. An integrated circuit that performs functions for a single application (e.g., keyboard controller). ASSPs use a standard-cell design approach to reduce chip size, costs and product development time. Think of ASSPs as a simply, cheaper way of building ASICs – Application Specific Integrated Circuits.

assurance level Probability expressed as a percent. Example: There is 90% Assurance (probability) that the mean holding time on the trunk group is between 168.5 and 191.5 seconds.

AST 1. Automatic Scheduled Testing. A method of testing switched access service (Feature Groups B, C, and D) where the customer provides remote office test lines and 105 test lines with associated responders or their functions' equivalent; consists of monthly loss and C-message noise tests and annual balance test.

2. Automatic Spanning Tree. A function that supports the automatic resolution of spanning trees in Source Route Bridging (SRB) networks, providing a single path for spanning explorer frames to traverse from a given node in the network to another. AST is based on the IEEE 802.1 standard. See also IEEE 802.1 and SRB.

ASTA Advanced Software Technology and Algorithms. Component of the U.S. Government's HPCC program High Performance Computing and Communications program (HPCC) intended to develop software and algorithms for implementation on high-performance computer and communications systems. See also HPCC.

ASTAP Asia-Pacific Telecommunity Standardization Program.

Asterisk Think of the PC on your desk. It probably runs Windows or Mac OS. Those are called computer operating systems. They tell the computer how to do elementary things like write to the screen, respond to the keyboard, search the hard drive, etc. To do useful things like writing a letter or making a spreadsheet, you need additional software, called application software – like Word or Excel – that runs on top of Windows or Mac OS. What makes a PC so useful is that each of us can make our computer useful to ourselves by simply buying and installing whatever software we need. In loose jargon, our PCs are referred to as "open systems" – though some computers are more open than others (see Linux). Telephone systems don't work this way. Each telephone system has a closed operating system and closed application software. That means you and I can't get to the telephone operating system or application software. This has one big advantage – namely a phone system is very reliable and rarely needs rebooting. You don't blue screens of death. (See BSOD.) On the other hand, a phone system can't be changed and thus made more useful to a user. There have been "open" systems in telephony. A number of manufacturers like Dialogic, Rhetorex and Natural MicroSystems made "open" phone systems in the 1990s. These systems were largely circuit boards that slide into the main bus of a PC or a server. They were largely used for specialized telephony applications, such as voice mail, automated attendants, order entry, information dispensing, etc. They weren't used as the backbone for telephone systems, since standard key systems and PBXs did a much cheaper and more reliable job. We called this peripheral industry computer telephony. And its heyday was the 1990s. The Tech Wreck of 2000-2001 hurt this industry as did the move to buying over the Internet. Never say die. Along came a company called Digium, based in Huntsville, Alabama, It created something called Asterisk software and PCI cards. Used together with a little user programming (you have to be familiar with Linux), you can created what Digium called "the industry's first Open Source PBX." The company claimed that Asterisk "offers a strategic, highly cost-effective approach to voice and data transport over TDM, switched, and Ethernet architectures." Digium claims that its solutions reduce the costs of traditional TDM and VoIP implementations through Open Source, standards-based software and next-generation gateways, media servers, and application servers. Digium claim its hardware supports traditional voice protocols, including PRI, RBS, FXS, FXO, E&M, Feature Group D, Groundstart, Loopstart, and GR-303. Data protocols include PPP, Cisco HDLC, and Frame Relay. For packet voice, Asterisk claims it supports IAX (Inter-Asterisk eXchange), (and the newer IAX2), SIP, MGCP, Skinny (SCCP), and H.323 VoIP protocols. Asterisk has a small, fanatical following. According to them, Mark Spencer's software (Asterisk) could do to the PBX market what Linus Torvalds did to the operating system market when he unleashed the first version of Linux in 1991. Linux is now the operating system for many computers that host Web sites. Mr. Spencer said the big difference between himself and Mr. Torvalds is he wants to make money from his technology. How much Asterisk catches depends on its economics. As I write this, an Asterisk "solution" doesn't seem cheap. See IAX.

asteroid event Any major news or event that pushes a company to the brink of extinction by wiping out the value of its stock almost overnight. The corporate version of what killed the dinosaurs.

ASTM American Society for Testing and Materials, a non-profit industry-wide organization which publishes standards, methods of test, recommended practices, definitions and other related material.

ASTRA ASTRA is the name of the principal series of satellites, first launched in 1988, operated by SES-ASTRA, the abbreviation of Soci0xC7t0xC7 Europ0xC7enne des Satellites – ASTRA, Europe's first private sector satellite operator, incorporated in 1985 in Luxembourg and now a leading provider of broadcast, broadband and satellite communications throughout Europe. www.ses-astra.com.

ASU Application-Specific Unit.

ASWC Alternate Serving Wire Center, when a building or customer is served by two different Central Offices. The one that is not his main central office is called the ASWC. It also means that the customer or building has two different NNXs (i.e. phone numbers). E.g. 691 (18th Street central office) and 240 (West Street central office).

asymmetric Not symmetric, i.e., unbalanced. An asymmetric telecom channel has more bandwidth (i.e. speed) in one direction than in the other. Its bandwidth is unbalanced. There are reasons for this. Take the Internet. Grabbing stuff from the Internet to your PC needs more bandwidth than sending stuff back from your PC. At least that's one theory. To accommodate this theory, for example, there's ADSL (Asymmetric Digital Subscriber Line). ADSL provides asymmetric bandwidth, as the downstream (from the network to the user premises) bandwidth of as much as 6.144 Mbps, and a return channel (from the user premises to the central office) of something like 608 Kbps. Asymmetric can also refer to the physical topology of the network. For example, a point-to-multipoint circuit might connect one device on the East Coast directly to three devices on the West Coast through the use of a bridge. For example, Miller Freeman might lease a multipoint circuit which connects its New York office to its office in San Francisco. At the San Francisco office is a bridge which has three drops, 1 for the San Francisco office and one for each of its two offices in Menlo Park. All communications between the sites take place through the multidrop bridge. The circuit is asymmetric as it lacks symmetry. There is one site connected on the East Coast and there are three sites connected on the West Coast. Multipoint circuits also are known as multi-drop circuits and fan-tail circuits, as they fan out like the tail of a fish on the distant end. See the next several definitions. See also ADSL, Full Duplex and Symmetric.

asymmetric cryptography A form of cryptography involving the use of two different (yet mathematically related) keys so that a message encrypted with one key can only be decrypted with the other key (and vice versa).

Asymmetric Digital Subscriber Line See ADSL and Asymmetric Digital Subscriber Line Transceiver.

Asymmetric Digital Subscriber Line Transceiver A microprocessor chip that is the crux of asymmetric digital subscriber line service. I found the following description of just such a chip in Motorola literature describing their MC145650 144-pin transceiver. "The MC145650 is a single integrated circuit transceiver device for ANSI (American National Standard Institute) T1.413 category 2 ADSL modems, based on the Discrete Multi-Tone (DMT) line code. The category 2 specification requires payload rates of (6.144 Mbps + 640 Kbps) downstream and 640 Kbps upstream, with crosstalk, over carrier serving area (CSA) range loops, and to achieve (1.544 Mbps + 176 Kbps) downstream and 176 Kbps upstream with crosstalk, over selected ANSI integrated services digital network (ISDN) loops. The payload makeup is flexible, thereby allowing multiple data streams to be multiplexed and demultiplexed. The MC145650 is capable of data rates up to 8 Mbps downstream and 1 Mbps bidirectionally; however, actual data rates obtained in any system are dependent on loop length, impairments, and transmitted power. The ADSL and DMT techniques are adaptive, changing system parameters based on loop characteristics in order to optimize the data route."

asymmetric encryption See Public Key Encryption.

asymmetric keys A pair of encryption keys, composed of one public key and one private key. Each key is one way, meaning that a key used to encrypt data cannot be used to decrypt the same data. However, information encrypted using the public key can be decrypted using the private key, and vice versa. This technology is commonly applied to e-mails, which are encrypted for confidentiality en route.

asymmetrical compression Techniques where the decompression process is not the reverse of the compression process. Asymmetrical compression is more computer-intensive on the compression side so that the decompression of video images can be easily performed at the desktop or in applications where sophisticated codecs are not cost effective. In short, any compression technique that requires a lot of processing on the compression end, but little processing to decompress the image. Used in CD-ROM creation, where time and costs can be incurred on the production end, but playback must be inexpensive and easy. See ASYN.

Asymmetrical Digital Subscriber Line ADSL.

asymmetrical modem A type of modem which uses most of the available bandwidth for transmission and only a small part for reception.

asymmetrical modulation A duplex transmission technique which splits the communications channel into one high speed channel and one slower channel. During a call under asymmetrical modulation, the modem with the greatest amount of data to transmit is allocated the high speed channel. The modem with less data is allocated the slow, or back channel. The modems dynamically reverse the channels during a call if the volume of data transfer changes.

asymmetrical multiprocessing A relatively simple implementation of multiprocessing in which the operating system kernel runs on one dedicated CPU and assigns tasks as they come in to other "slave processors." It is also known as "master/slave" processing. Compare to Symmetric Multiprocessing.

asymmetrical PVC Refers to a PVC (Private Virtual Circuit) which supports simplex, or asymmetrical, assignments of committed information rate in each direction of transmission. A PVC transmission path is duplex, meaning that there must be a communications path in each direction between the two points being connected. However with an asymmetrical PVC, the network capacity in each direction does not necessarily have to be equal.

asyn Greek prefix meaning "not together." See Asynchronous Transmission.

asynchronous See Asynchronous Transmission.

Asynchronous Balanced Mode ABM. Used in the IBM Token Ring's Logical Link Control (LLC), ABM operates at the SNA data link control and allows devices on a Token Ring to send data link commands at any time and to initiate responses independently.

asynchronous completion A Versit definition. A domain issues a service request and need not wait for it to complete. If the application waits for this completion, this is known as synchronous, but if it is sent off to another system entity and the domain goes on to other activities before the service request completes (and the system later sends a message to the domain announcing the service's completion), that completion is known as Asynchronous.

asynchronous device A device whose internal operations are not synchronized with the timing of any other part of the system.

asynchronous gateway A routing device used for dial-up services such as modem communications.

asynchronous mapping A SONET term. SONET optical fiber transmission systems run at a very high rate of speed, of course. In fact, SONET runs at a minimum of 51.84 Mbps, which is the foundation transmission level known as OC-1 (Optical Carrier Level 1). the OC-1 frame begins as a T-3 electrical signal at 44.736 Mbps. The native format of the incoming signals always is electrical in nature, and originates at various speeds. Examples are 64 Kbps (DS-0), 1.544 Mbps (DS-1 – specifically, T-1), 2.048 Mbps (DS-1 – specifically, E-1), or 44.736 (DS-3 – specifically, T-3). As these incoming signals of various speeds are presented to the SONET facility, they are multiplexed to form a T-3 frame and are converted from the T-3 electrical format to the OC-1 optical format. The OC-1 frames then are mapped into (presented to, accepted by, and fit into) the SONET facility in an asynchronous fashion. While the SONET transmission facility, itself, is highly synchronized, it deals with inputs on an asynchronous (start-stop) fashion. These mappings are defined for clear channel transport of digital signals that meet the standard DSX cross connect requirements, typically DS-1 and DS-3 in most practical applications, although DS-2 is also supported. See also SONET.

asynchronous network A network in which the clocks do not need to be synchronous or mesochronous. Also called a Nonsynchronous Network. See Asynchronous.

asynchronous request An SCSA term. A request where the client does not wait for completion of the request, but does intend to accept results later. Contrast with synchronous request.

asynchronous teleconferencing An interactive group communication that allows individuals to communicate as a group without being present together in time or place. Participants to join and exit the conference when it is convenient for them, leaving messages for others and receiving messages left for them. Computer conferencing is an example of asynchronous teleconferencing.

asynchronous terminal A terminal which uses asynchronous transmissions. See Asynchronous Transmission.

asynchronous time division multiplexing A multiplexing technique in which a transmission capability is organized in a priori unassigned time slots. The time slots are assigned to cells upon request of each application's instantaneous real need.

Asynchronous Transfer Mode ATM is the technology selected by the Consultative Committee on International Telephone & Telegraph (CCITT) International standards organization in 1988 (now called the ITU-T) to realize a Broadband Integrated Services Digital Network (B-ISDN). It is a fast, cell-switched technology based on a fixed-length 53-byte cell. All broadband transmissions (whether audio, data, imaging or video) are divided into a series of cells and routed across an ATM network consisting of links connected by ATM switches. Each ATM link comprises a constant stream of ATM cell slots into which transmissions are placed or left idle, if unused. The most significant benefit of ATM is its uniform handling of services, allowing one network to meet the needs of many broadband services. ATM accomplishes this because its cell-switching technology combines the best advantages of both circuit-switching (for constant bit rate services such as voice and image) and packet-switching (for variable bit rate services such as data and full motion video) technologies. The result is the bandwidth guarantee of circuit switching combined with the high efficiency of packet switching. For a longer explanation, see ATM.

asynchronous transmission Literally, not synchronous. A method of data transmission which allows characters to be sent at irregular intervals by preceding each character with a start bit, and following it with a stop bit. It is the method most small computers (especially PCs) use to communicate with each other and with mainframes today. In every form of data transmission, every letter, number or punctuation mark is transmitted digitally as "ons" or "offs." These characters are also represented as "zeros" and "ones" (See ASCII). The problem in data transmission is to define when the letter, the number or the punctuation mark begins. Without knowing when it begins, the receiving computer or terminal won't be able to figure out what the transmission means.

One way to do this is by using some form of clocking signal. At a precise time, the transmission starts, etc. This is called synchronous transmission. In asynchronous transmission there's no clocking signal. The receiving terminal or computer knows what's what because

each letter, number or punctuation mark begins with a start bit and ends with a stop bit.

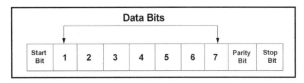

Transmission of data is called synchronous if the exact sending or receiving of each bit is determined before it is transmitted or received. It is called asynchronous if the timing of the transmission is not determined by the timing of a previous character.

Asynchronous is used in lower speed transmission and by less expensive computer transmission systems. Large systems and computer networks typically use more sophisticated methods of transmission, such as synchronous or bisynchronous, because of the large overhead penalty of 20% in asynchronous transmission. This is caused by adding one start bit and one stop bit to an eight bit word – thus 2 bits out of ten.

The second problem with large transfers is error checking. The user sitting in front of his own screen checks his asynchronous transmission by looking at the screen and re-typing his mistakes. This is impractical for transferring long files at high speed if there is not a person in attendance.

In synchronous transmission start and stop bits are not used. According to the book Understanding Data Communications, characters are sent in groups called blocks with special synchronization characters placed at the beginning of the block and within it to ensure that enough 0 to 1 or 1 to 0 transitions occur for the receiver clock to remain accurate. Error checking is done automatically on the entire block. If any errors occur, then the entire block is retransmitted. This technique also carries an overhead penalty (nothing is free), but the overhead is far less than 20% for blocks of more than a few dozen characters.

AT 1. Access Tandem.

2. Advanced Technology. Refers to a 16 bit Personal Computer architecture using the 80X86 processor family which formed the basis for the ISA Bus as found in the first IBM PC.

3. AudioTex. See AudioTex.

4. See AT Command Set.

AT Bus The electrical channel used by the IBM AT and compatible computers to connect the computer's motherboard and peripheral devices, such as memory boards, video controllers, PC card modems, bus mouse boards, hard and floppy disk controllers and serial/parallel input/output devices. The AT bus supports 16 bits of data in one slug, whereas the original IBM PC supported only 8 bits (and was called the ISA bus for Industry Standard Architecture). These days there are much faster "buses," including the EISA, MCA (Micro-Channel Architecture), Local Bus, PCI, VESA, etc.

AT Command Set Also known as the Hayes Standard AT Command Set. A language that enables PC communications software to get an asynchronous and "Hayes-compatible modem" to do what you want it to do. So called "AT" because all the commands begin with "AT," which is short for ATtention. The most common commands include ATDT (touchtone a number), ATA (manually answer the phone), ATZ (reset modem – it will answer OK), ATS0=0 (disable auto-answer), and ATH (hang up the phone). To avoid having yourself knocked off your data call by the beep that comes in on the phone company's call waiting, put the following line in your modem setup: ATS10=20. That will increase your S10 register to two seconds. This register sets the time between loss of carrier (caused by the 1.5 second call waiting signal) and internal modem disconnect. Factory default on most modems is 1.4 seconds – just perfect to be cut off by the call waiting tone! (Dumb.)

If you have to dial through several phone systems, waiting for dial tone on the way and/or going through fax/modem switches, you may consider a dial stream that looks like ATDT 1-800-433-9800 [W]212-989-4675 [W]22, where [W] means (in some software programs) "Wait for any key. When you get it, touchtone out the next digits." In other software programs – pure Hayes command – W means wait for second dialtone. If W in square brackets doesn't work for you, then change X3 in your setup line to X1; change your computer's dialed number to 9; and dial your distant computer with your phone. When you hear the modem at the other end answer, tell your computer's software to dial 9. It will dial 9, hear the modem tone at the other end and connect as though it had dialed it all by itself. X1 tells your modem to dial (or touchtone) immediately – without waiting for dial tone.

You can use several AT commands on one line. You only need AT before the first one. Some modems require commands typed in capital letters. When your dialing fails and you can't figure why, get out of your communications software program and start again. Or in total desperation, turn your computer and modem completely off and start again. The word "Hayes" comes from the manufacturer of modems called Hayes Microcomputer, Norcross, GA, the creator of the command set. Not all Hayes compatible modems are. See also AT+V and Hayes Command Set.

At Local Mode One of the command modes available on the ISDN set. It is used for compatibility with existing communications packages for analog modems or for data-only application programs. See AT Command Set.

at the glass A synonym for "on the display screen." It refers to single-window, single-sign-on access to applications, content, and processes – all of which are presented to the user in a seamless, integrated manner, so that the user doesn't realize that he is actually working with multiple applications and multiple data sources. The integration takes place "at the glass." A mashup is an example of integration at the glass. See mashup.

At Work Microsoft's office equipment architecture announced on June 9, 1993. Microsoft's idea was to put a set of software building blocks into both office machines and PC products, including desktop and network-connected printers; digital monochrome and color copiers; telephones and voice messaging systems; fax machines and PC fax products; handheld systems and hybrid combinations of the above. At Work didn't go very far. But Windows CE and then Windows mobile came out and became popular.

AT#V See AT+V below.

AT&T As of the end of 2006, AT&T now consists of the former AT&T Long Lines plus four of the seven regional Bell Operating Companies (Pacific Telesis, Southwestern Bell, Ameritech, and BellSouth) that were created by the breakup of AT&T on January 1, 1984. On December 29, 2006 the FCC approved AT&T's $86 billion buyout of BellSouth, the largest telecommunications takeover, so far, in U.S. history. Now to some history, AT&T Corporation – formerly American Telephone and Telegraph Company – was incorporated on March 3, 1885, to manage and expand the burgeoning long-distance business of American Bell Telephone Company and its licensees. It continued as the "long-distance company" until Dec. 30, 1899, when it assumed the business and property of American Bell and became the parent company of the Bell System. It remained the Bell System parent, providing the bulk of telecommunications equipment and services (local, long distance and international) in the United States, until Jan. 1, 1984, when it divested itself of the Bell operating companies that provided local exchange service. On September 20, 1995, AT&T announced that it would be splitting into three companies: a "new" AT&T, to provide long distance and international communications transmission and switching; Lucent Technologies, to design, make and sell telecommunications systems and technologies; and NCR Corp., to concentrate on transaction-intensive computing. The strategic restructuring was completed on Dec. 31, 1996. Sadly, things didn't work out for the company which retained the name AT&T. Its long distance business contracted as competition expanded and prices plummeted. And in early 2005, it agreed to be purchased by SBC, one of its former local phone companies, for $16 billion. At one stage, AT&T and the Bell System had over one million employees. By the time of the SBC acquisition announcement it had fewer than 50,000. In the fall of 2005, SBC changed its name to AT&T. I'm sure there was a logical reason. See SBC.

AT&T Consent Decree The Telecommunications Act of 1996 defined it as follows: "The term 'AT&T Consent Decree' means the order entered August 24, 1982, in the antitrust action styled United States v. Western Electric, Civil Action No. 82-0192, in the United States District Court for the District of Columbia, and includes any judgment or order with respect to such action entered on or after August 24, 1982. See Telecommunications Act of 1996." Under that consent decree which took place at the beginning of 1984, AT&T split off seven approximately the same size, local operating companies (called the Baby Bells) while it kept Western Electric, the manufacturing arm and the long distance provider. The restriction against manufacturing outside telecommunications was removed and AT&T was free to move into making and selling computers, which it did disastrously. When the history books are written, it will be shown that AT&T's boss at the time, Charlie Brown, sewed the seeds of AT&T ultimate demise by making two disastrous decisions: 1. Not securing any wirelss licenses (see the next definition) and 2. Spinning off the regional bell operating companies and keeping the long distance services and telecom manufacturing arm (Western Electric).

AT&T Wireless In 1983 Federal Communications Commission set guidelines for the rollout of cellphone service. The local incumbent (i.e. monopoly) telephone company would get one license in each market while the second would be up for grabs. At the time, AT&T was about to be split up, and it could have demanded that the new AT&T long-distance company be given the incumbent licenses. But Charles Brown, then chief executive, decided that the cellphone was largely a local business. "He felt that it was logical that the cellular business should go to the Baby Bells," says Sheldon Hochheiser, AT&T's former historian, referring to the local phone companies spun off in the 1984 breakup. A study at the time

by McKinsey & Co. predicted that by the year 2000 there would be 900,000 cellphone users in the U.S. The actual number turned out to be more than 100-fold greater than the prediction, making that prediction one of the worst ever made. The licenses that Mr. Brown had decided not to seek turned out to be worth many billions of dollars. Eventually AT&T realized it had missed the boat on cellphones. It bought a cellphone provider in 1993 but later spun it off as an independent company called AT&T Wireless. That company was acquired last year by Cingular Wireless, which is 60%-owned by SBC. Then in early 2005, SBC announced that it would buy what was left of AT&T itself. The price was a paltry $16 billion. A circle was closed.

AT+V V standards for voice. AT+V is a new ANSI standard for voice modems. It's a superset of the Hayes AT command set which worked so well in modems. AT+V combines pre-fixed Hayes AT commands with a new set of voice-related +V commands. The specification is detailed in ANSI/TIA/EIA IS-101 "Facsimile Digital Interfaces – Voice Control Interim Standard for Asynchronous DCE." The TIA TR-29.2 subcommittee details the specification in their PN-3131. Rockwell's voice modem chipset does not comply with this standard, but uses another called AT#V, which is similar. In Windows 95, the variance between these command sets is ratified by the Win 95 system registry and vendor-supplied INF files. See also Windows Telephony.

ATA 1. American Telemarketing Association. The professional industry association for telephone sales and marketing.

2. Analog Terminal Adapter. A device for a Northern Telecom Norstar phone system that lets it use analog devices, for example fax, answering machines, modems and single line phones, behind the Norstar's central telephone unit (its KSU). Before you buy the analog terminal adapter, check that its speed is fast enough for you. In mid-1995, it was constrained to 9,600 bps, or 14,400 bps if the phone line was clear.

3. AT Attachment. Refers to the interface and protocol used to access a hard disk on AT compatible computers. Disk drives adhering to the ATA protocol are commonly referred to as IDE interfaced drives for PC compatible computers. The ATA specification is fully backward compatible with the ST-506 standard it superseded. IDE drives are sometimes referred to as ATA drives or AT bus drives. The newer ATA-2 specification defines the EIDE interface, which improves upon the IDE standard. See ATA2, IDE and Enhanced IDE.

ATA Document The latest draft of the ANSI X3.T9 subcommittee Advanced Technology Attachment document. See also ATA.

ATA over Ethernet AoE is a proprietary Storage Area Network (SAN) protocol designed by Coraid, Inc. for accessing ATA storage devices over Ethernet networks. See ATA and SAN.

ATA Registers These registers are accessed by a host to implement the ATA protocol for transferring data, control and status information to and from the PC Card. They are defined in the ATA Document. These registers include the Cylinder High, Cylinder Low, Sector Number, Sector Count, DriveHead, Drive Address, Device Control, Error, Feature, Status and Data registers. The I/O and memory address decoding options for these registers are defined within this specification.

ATA2 The second generation of the Advanced Technology Attachment specification for IDE devices that defines faster transfer speeds and LBA (Logical Block Address) sector-locating methods. See ATA, IDE and Enhanced IDE.

ATRAC3 Adaptive Transform Acoustic Coding. Sony's proprietary audio format for compressing audio for use on its portable audio and MiniDisc players. ATRAC3's quality is comparable to an MP3 recorded at a bit rate of 128 Kbps.

Attack A deliberate attempt to compromise the security of a computer system or deprive others of the use of the system.

ATAPI Attachment Packet Interface specification does for CD-ROM and tape drives what ATA-2 does for hard drives. It defines device-side characteristics for an IDE-connected peripheral. The benefits of having a single interface for the most common non-disk storage device in the desktop world, the CD-ROM are obvious. For the manufacturer, there is no need to add a separate controller card for the CD-ROM. For the end-user it means no more fussing with interrupts, cards and proprietary driver software. ATAPI essentially adapts the established SCSI command set to the IDE interface.

Atari Jaguar The world's first 64-bit home console video game system. Made by IBM, the Jaguar was released in 1993, and offered high-speed action, CD-quality sound, and polygon graphics processing beyond most other machines available at the time.

ATB All Trunks Busy. One measure which your phone company or phone systems might give you of telephone traffic in and out of your office. See All Trunks Busy.

ATC See ancillary terrestrial component.

ATCA See AdvancedTCA.

ATD 1. Asynchronous Time Division.

2. ATtention Dial the phone. The first three letters in the most frequently-used command in the Hayes command set for asynchronous modems – typically those used with microcomputers.

ATDMA Advanced Time Division Multiple Access (ATDMA) is a multiplexing technique used in DOCSIS (Data-Over-Cable System Interface Specification) networks for upstream communications from the customer premises to the network headend. ATDMA is a combination of FDMA and TDMA. At the most fundamental level, DOCSIS 1.0 multiplexing uses FDMA (Frequency Division Multiple Access), which divides the available RF (Radio Frequency) channel into multiple narrower RF subchannels in order that multiple transmissions can occur simultaneously. TDMA (Time Division Multiple Access) allows transmission to occur over the same FDMA subchannel, with the individual transmissions taking place sequentially over their assigned time slots. In the context of DOCSIS, the combination of the two is known as ATDMA. The alternative technique is known as Synchronous Code Division Multiple Access (S-CDMA). See also DOCSIS and S-CDMA.

ATDT ATtention Dial the phone in touchtone mode. The first four letters in the most frequently-used command in the Hayes command set for asynchronous modems – typically those used with PCs.

ATDNet Advanced Technology Demonstration Network. A joint research effort of Bellcore, Bell Atlantic, and the U.S. Government, this network is aimed at demonstrating the efficacy of advanced technologies in the network of the future.

ATG Air-To-Ground. Communications services provided from an airplane in flight. These services have been primarily voice telephone calling services in the past, but are being extended to fax and data services with new digital Air-to-Ground (ATG) systems. ATG services in the U.S. operate in the 800-900 MHz region. In 1994, Ground-to Air services were also introduced. Air-To-Ground service is now available from some planes flying outside the United States.

atheism A non-prophet organization.

athermal effect Any effect of electromagnetic energy absorption not associated with a measurable rise in temperature.

ATIS Alliance for Telecommunications Industry Solutions, a trade group based in Washington, D.C. and open to membership of North American and World Zone 1 Caribbean telecommunications carriers, resellers, manufacturers, and providers of enhanced services. Originally called the Exchange Carriers Standards Association (ECSA), ATIS is heavily involved in standards issues including interconnection and interoperability. More recently and in connection with the privatization its Part 68 responsibilities, the FCC selected ATIS and the TIA (Telecommunications Industry Association) as joint sponsors of the Administrative Council for Terminal Attachments (ACTA). ACTA responsibilities include adopting and publishing technical criteria for terminal equipment submitted by ANSI-accredited standards development organizations, and operating and maintaining a database of approved terminal equipment. The first meeting of ACTA was scheduled for May 2, 2001. See also ACTA, Part 68, and TIA. www.atis.org

ATLAS Automated Telephone Listing Address System.

ATM 1. Automated Teller Machine. The street corner banking machine which is usually hooked up to a central computer through leased local lines and a multiplexed, secure data network. Some ATM networks work over ATM. See 2. America, I believe, is the only place in the world with drive-up ATM machines with Braille lettering.

2. Asynchronous Transfer Mode. Very high speed transmission technology. ATM is a high bandwidth, low-delay, connection-oriented, packet-like switching and multiplexing technique. Usable capacity is segmented into 53-byte fixed-size cells, consisting of header and information fields, allocated to services on demand. The term "asynchronous" applies, as each cell is presented to the network on a "start-stop" basis–in other words, asynchronously. The access devices, switches and interlinking transmission facilities, of course, are all highly synchronized.

Here's some history on ATM from the Networking Alliance: The ATM method of moving information is not completely new. Like most things it is an evolution of earlier methods. The key difference between ATM and "X.25 packet switching" and the popular "Frame Relay" technologies is that the packets of the earlier technologies varied in size. Engineers realized that as the speed was dramatically increased to be able to carry "real time" voice and video, the varied length packets would become unmanageable. During the 1980s the ITU, now the ITU-T (International Telecommunications Union-Telecommunications Services Sector), adopted ATM as the transport technology of the future. Ultimately and after a great deal of debate, the ITU-T determined that each cell would be 53 octets long. To meet current and future demands, networking technologies and protocols have evolved to optimize network performance based on traffic characteristics. ATM represents the first world-wide standard to be embraced by the computer, communications and entertainment industries.

Each ATM cell contains a 48-octet payload field, the size of which has an interesting background. Data people prefer to move data in huge blocks or frames, which are more efficient for large file transfers. Voice people, on the other hand prefer tiny blasts of data, which are more effective for moving digitized voice samples (ala PCM in a T-Carrier environment). Since ATM is positioned as the ultimate service offering in support of data , voice data, video data, image data, and multimedia data, the small payload prevailed. With that battle out of the way, the European and U.S. camps clashed, with the European Telecommunications Standards Institute (ETSI) proposing a 32-octet cell and the U.S. Exchange Carriers Standards Association (ECSA) proposing a 64-octet cell–the issue was the difference in standard PCM voice encoding techniques. After lengthy wrangling, it was decided that a 48-octet cell would be the perfect mathematical compromise. Although neither camp was perfectly pleased (such tends to be the nature of a compromise, I am told), it was a solution that all could accept.

In any event, each cell also is prepended with a 5-octet Header which identifies the Virtual Path (Virtual Circuit), Virtual Channel, payload type, and cell loss priority; as well as providing for flow control, and header error control.

The small, fixed-length cells require lower processing overhead and allow higher transmission speeds than traditional packet switching methods. ATM allocates bandwidth on demand, making it suitable for high-speed connection of voice, data, and video services. ATM services will be available at access speeds up to 622 Mbps, with the backbone carrier networks operating at speeds currently as high as 2.5 Gbps. The ATM edge and core backbone switches operate at very high speeds, and typically contain multiple buses providing aggregate bandwidth of as much as 200+ Gbps. ATM core switches currently are available with capacities of as much as one terabit per second, although none have been deployed at this level.

Here's a full explanation: Conventional networks carry data in a synchronous manner. Because empty slots are circulating even when the link is not needed, network capacity is wasted. The ATM concept which has been developed for use in broadband networks and optical fiber based systems is supported by both ITU-T (nee ITU) and ANSI standards, can also be interfaced to SONET (Synchronous Optical Network). ATM automatically adjusts the network capacity to meet the system needs and can handle data, voice, video and television signals. These are transferred in a sequence of fixed length data units called cells. Common standards definitions are provided for both private and public networks so that ATM systems can be interfaced to either or both. ATM is therefore a wideband, low delay, packet-like switching and multiplexing concept that allows flexible use of the transmission bandwidth and capable of working at data rates as high as 622.08 Mbps, with even higher rates planned. Each data packet consists of five octets of header field plus 48 octets for user data. The header contains data that identifies the related cell, a logical address that identifies the routing, header error correction bits, plus bits for priority handling and network management functions. Error correction applies only to the header as it is assumed that the network medium will not degrade the error rate below an acceptable level. All the cells of a Virtual Path (VP) follow the same path through the network that was determined during call set-up. (Note that ATM is a connection-oriented network service.) As there are no fixed time slots in the system, any user can access the transmission medium whenever an empty cell is available. ATM is capable of operating at bit rates of 155.52 and 622.08 Mbps; the cell stream is continuous and without gaps. The position of the cells associated with a particular VC is random, and depends upon the activity of the network. Cells produced by different streams to the ATM multiplexer are stored in queues awaiting cell assignment. Since a call is accepted only when the necessary bandwidth is available, there is a probability of queue overflow. Cell loss due to this forms one ATM impairment. However, this can be minimized through the use of statistical multiplexers. Bit errors in the header which are beyond the FEC capability can lead to misrouting.

While ATM was developed as a backbone WAN technology, a 25.6 Mbps version of ATM was reluctantly approved by the ATM Forum for use in a LAN workgroup environment. The Desktop ATM25 Alliance, which promoted the standard, disbanded in 1996 due to lack of interest. ATM has continued to march to the desktop, however slowly and at the higher speeds. ATM also has found its way into the LAN world through the development of cost-effective, high-performance ATM LAN backbone switches. PBX manufacturers also are working diligently to determine how best to incorporate ATM switching fabrics into voice/data/video/multimedia PBX systems, resulting in an ATM-based communications controller for premise application. See also ATM Forum, ATM Access Switch and ATM Forum UNI V3.0.

ATM25 Workgroup ATM running at 25.6 million bits per second cell-based user interface based on IBM token ring network. ATM-25 is mainly used on internal corporate local area networks. For a much fuller explanation, see ATM.

ATM Access Switch A specialized ATM switch which sits on the end user premise, providing access into a carrier ATM network. The ATM Access Switch is used for such applications as distance learning and telemedicine. It is a high-capacity, cell-based switch designed to support broadband networking. Its fully integrated access, multiplexing and switching functions provide the capability for a variety of combined data, video, imaging and voice services on a single platform. See ATM.

ATM Adaptation Layer See AAL.

ATM Address Defined in the UNI Specification as three formats, each having 20 bytes in length including country, area and end-system identifiers. See ATM.

ATM Backbone Switch A specialized ATM switch which sits in the carrier backbone network. The ATM Backbone Switch is claimed to be ideal for backbone networks supporting multiple services in corporations, telcos, cellular and Internet public service providers. Network operators can aggregate all of their traffic over a single backbone of ATM. It is ideal for service provider backbones supporting multiple services such as cell relay, permanent virtual circuits (PVCs), switched virtual circuits (SVCs) circuit emulation, LAN interconnectivity and frame relay. The Backbone Switch has throughput traffic and traffic management features needed for large-scale ATM deployment and service offerings. ATM backbone switches include internal busses providing bandwidth of as much as 200+ Gbps, and are interconnected by SONET fiber optic transmission facilities currently operating at speeds of as much as 2.5 Gbps. See ATM.

ATM Cell The ATM term for a set of data. A cell can be a segment (i.e., fragment) of a larger set of data (e.g., a file), or it can be an accumulation of smaller sets of data (e.g., analog voice samples converted into digital bytes through a codec). In its native form, the set of data to be transmitted originates at what is known as the User Layer. As that data enters the ATM network, it is formed into ATM cells by either an ATM switch or an ATM-capable router. Each cell contains a payload (i.e., text field or data field) of 48 octets, with each octet comprising eight bits. Each 48-octet payload is prepended with a header of five octets, which comprises several fields used by the ATM switches to route the data correctly, and to identify the nature of the data in order that the appropriate Quality of Service (QoS) level can be guaranteed. In total, the ATM cell comprises 53 octets. The cells travel through the ATM network in a cell stream. As the cells exit the ATM network, they are reassembled (i.e., reformed) into the native data format in a process known as Segmentation and Reassembly (SAR). See ATM for a more detailed explanation.

ATM Cell

8	7	6	5	4	3	2	1	Oct
GFC				VPI				1
VPI				VCI				2
VCI								3
VCI				PT		CLP		4
HEC								5
Information Payload (48 octets)								6 ⋮ 53

GFC:	Generic Flow Control	PT:	Payload Type
VPI:	Virtual Path Identifier	CLP:	Cell Loss Priority
VCI:	Virtual Channel Identifier	HEC:	Header Error Cor

ATM Distributed Network System ADNS. A network architecture that essentially spreads intelligence throughout an ATM network, allowing a range of services, including voice, data, and video.

ATM Edge Switch An ATM cell switch which sits at the edge of the carrier network, providing access from the end users' world to the carriers' ATM backbone network. It is analogous to a Central Office providing access to a Tandem network in the traditional, circuit-switched voice and data world. ATM Edge Switches also are known as Access Nodes and Service Nodes.

ATM Ethernet LAN Service Unit An ATM ELSU provides 12 independent virtual Ethernet bridges for running over ATM networks. ELSUs are designed for flexible deployment, either local to an ATM switch or at a remote site. ELSUs are designed for LAN internetworking services over ATM networks.

ATM Forum An industry organization with some 800 members, co-founded by N.E.T. and three other leading networking companies, which focuses on speeding the development, standardization and deployment of ATM (Asynchronous Transfer Mode) products. It has been remarkably successful. The ATM Forum is based in Mountain View, CA. Their phone number is 415-949-6700. See ATM. www.atmforum.com.

ATM Forum UNI V3.0 The ATM Forum UNI V3.0 implementation agreement is based on a subset of the ITU-TS broadband access signaling protocol standards. Additions to this subset have been made where necessary to support early deployment and interoperability of ATM equipment. The procedures and protocol defined in the agreement apply to both public and private UNIs. Moreover, since the protocol is symmetrical, it also applies in the configuration ATM-end-point to ATM-end-point. See ATM and ATM Forum.

ATM Inverse Multiplexing AIMUX. A device used to combine multiple T-1 or E-1 links into a single broadband facility, over which ATM cells can then be transmitted.

ATM Islands Local implementations of ATM equipment and applications which are not connected by ATM wide-area services.

ATM Layer ATM. The second layer of the ATM Protocol Reference Model. At this layer are included such functions as cell multiplexing, creation of headers, flow control and selection of VPIs (Virtual Path Identifiers) and VCIs (Virtual Channel Identifiers). See ATM Layer Link.

ATM Layer Link A section of an ATM Layer connection between two adjacent active ATM Layer entities (ATM-entities).

ATM Link A virtual path link (VPL) or a virtual channel link (VCL).

ATM Machine See ATM.

ATM Passive Optical Network See APON.

ATM Peer-to-Peer Connection A virtual channel connection (VCC) or a virtual path connection (VPC).

ATM Protocol Reference Model A multidimensional protocol model consisting of 4 layers and 3 planes and serving as a point of reference for understanding,

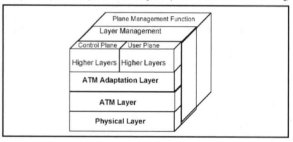

developing and implementing ATM technology. Each layer addresses a discrete set of related functions, with all layers closely interrelated. The layers include the Physical Layer, ATM Layer, ATM Adaptation Layer, and Higher Layers related to the specifics of the native user data protocol. The planes include the Control Plane, User Plane and Management Plane.

ATM Switch A networking device that forwards 53-byte ATM-standard cells between different devices. These switches can be designed for both LAN and WAN environments. An ATM switch acts as both a traffic aggregator, such as when terminating unchannelized DS-3 trunks from DSLAMs, and as a multiservice switch that is capable of forwarding traffic in different ways, depending on what has to be done with it. For example, an ATM switch forwards customer IP traffic directly to an IP network by shunting it to a router or subscriber management system, frame-relay traffic to a frame switch, voice-over-ATM traffic to a voice/data gateway, or long-distance traffic to other central offices via a SONET inter-office ring.

ATM Token Ring LAN Service Unit The ATM TLSU provides a powerful tool for offering internetworking services over ATM networks. Emulated token rings consist of up to 64 TLSU token ring ports located anywhere in the ATM network, interconnected with PVCs. These emulated token ring networks can be completely isolated from one another to ensure security and fairness among the attached LANs. The TLSUs are designed for flexible deployment, either local to an ATM switch or at a remote site. See ATM Ethernet LAN Service Unit.

ATM Traffic Descriptor A generic list of traffic parameters that can be used to capture the intrinsic traffic characteristics of a requested ATM connection.

ATM User-User Connection An association established by the ATM Layer to support communication between two or more ATM service users (i.e., between two or more next higher entities or between two or more ATM-entities). The communications over an ATM Layer connection may be either bidirectional or unidirectional. The same Virtual Channel Identifier (VCI) is issued for both directions of a connection at an interface.

ATMARP ATM Address Resolution Protocol. The means of mapping IETF classical IP addresses to ATM hardware addresses. The process works in much the same way as conventional ARP, which maps network-layer addresses to the MAC (Media Access Control) layer in a LAN.

ATNA All talk, no action. Four-letter acronym describing a person who makes promises with great fanfare, but seldom follows through.

ATOD 1. Analog-to-digital, in the context of converters.

2. "Any time of day" in the context of per-minute retail or wholesale long-distance rates.

atom Atom is a type of web feed, written in XML, which allows a user to download any updates made to the website or blog using a feed reader. See atom feed.

atom feed An atom feed, also called an RSS feed, is an XML-based document which contains content items, often summaries of stories or weblog posts with web links to longer versions. Weblogs and news websites are common sources for web feeds, but feeds are also used to deliver structured information ranging from weather data to "top ten" lists of hit tunes...

Web feeds in the form of Atom and RSS have been around for a long time and their prevalence has been steadily growing. There are millions of websites and applications that offer information in this form. In fact, chances are the majority of websites you already visit offer their content as an RSS or Atom feed.

The reason these web feeds are useful is similar to the reason a table of contents at the beginning of books and magazines are useful. Instead of needing to visit each story individually, you can see a list of short summaries and decide what is of interest to you. More importantly, you can see a list of summaries from multiple sources (say a bunch of your favorite magazines) all in one place and decide which articles from each are of interest. Imagine being able to review a summary of all the newest articles in all your favorite magazines before you buy.

But this is just part of the story. RSS and Atom feeds are also useful for abstracting the information offered by many different websites into a format that other programs can interpret. Whereas it is difficult for any given application to extract information consistently from a wide variety of differently formatted websites, it is very easy to make use of information in RSS or Atom formats. By adopting these simple information sharing standards, content publishers and application developers make their information available to other applications and content sites in ways they may never have imagined. Developers do this because it extends the exposure of their content.

atomic time See International Atomic Time. See also http://tycho.usno.navy.mil/cgi-bin/anim.

atomicity A feature of a transaction considered or guaranteed to be indivisible. Either the transaction is uninterrupted, or, if it fails, a mechanism is provided that ensures the return of the system to its state prior to initiation of the transaction.

ATP 1. AppleTalk Transaction Protocol. Transport-level protocol that provides a loss-free transaction service between sockets. This service allows exchanges between two socket clients in which one client requests the other to perform a particular task and to report the results; ATP binds the request and response together to ensure the reliable exchange of request-response pairs.

2. ALPS Tunneling Protocol. A protocol used to transport ALPS data across a TCP/IP network between an ALC/UTS router and an AX.25/EMTOX router. It consists of a set of messages (or primitives) to activate and deactivate ALPS ATP circuits and to pass data.

3. A compendium of test procedures that may be used to demonstrate compliance with certain specifications.

ATS 1. Abstract Test Suite: A set of abstract test cases for testing a particular protocol. An "executable" test suite may be derived from an abstract test suite.

2. Actual Transfer Switch – used in AC power to a telecom site. The AC power comes from the utility to the ATS which routes the power to a main AC breaker.

ATSC 1. Advanced Television Systems Committee. Formed by the Joint Committee on Inter-Society Coordination (JCIC) to establish voluntary standards for Advanced TV (ATV) systems, the ATSC focuses on digital television, interactive systems and broadband multimedia communications standards. Membership is open to American (North and South America, including the Caribbean) entities directly affected by the work of the committee. ATSC comprises 54 members including television networks, motion picture and television program producers, trade associations, television and other electronic equipment manufacturers and segments of the academic community. ATSC's proposal for ATV (Advanced TV)

was accepted by the FCC on November 28, 1995 and was formally adopted, in the most part, as the U.S. DTV (Digital TV) standard on December 24, 1996. www.atsc.org. See JCIC, HDTV and ATV.

2. Australian Telecommunication Standardisation Committee.

ATT 1. See Average Talk Time.

2. Automatic Toll Ticketing. A system which telephone companies use to automatically keep call detail records including calling number, number called, time of day and length of call. The phone company uses this information, together with the cost of phone calls, to generate an invoice to its customers.

attach A command that assigns a connection number to a workstation and attaches the workstation to the LOGIN directory on the default (or specified) file server. As many as 100 workstations can be attached to a file server running NetWare v2.2. When loaded the NetWare shell (workstation file NETx.COM) automatically attaches your workstation to the nearest file server. You can also specify in SHELL.CFG which server you prefer to attach to.

attach terminal To assign a terminal for exclusive use by the application program. Contrast with Detach Terminal.

Attached Resource Computer Network. See ARCnet.

attachment An attachment is a file in its native format – the format it is saved in by its application (e.g. Excel, PowerPoint, Word). An attachment typically accompanies a electronic message. It comes in its native format, i.e. with the extension its application recognizes – .xls (Excel), .ppt (PowerPoint) or .doc (Word). When you receive an email containing an attachment, Windows gives you two choices – open it or save it to disk. When you say "open it," Windows looks at the attachment's extension and tries to figure if it has an application on the disk that can open that file. If not, you get an error message. If you opt to save it to disk, you can then change the file's extension and open another program and view the file. One example: If I receive an attachment of a WordPerfect file, I can't open it by clicking on it. But if I save it, then open Word, I can open the file, view it and work with it. In this case I don't have to change the file's extension. Other times, changing the extension works.

Attachment Unit Interface AUI. In a LAN (Local Area Network), the interface between the medium access unit (MAU) and the data terminal equipment within a data station.

attack An attempt to circumvent the security measures in place on a network either to gain unauthorized access to the system or to force a denial of service.

attack time The time interval between the instant that a signal at the input of a device or circuit exceeds the activation threshold of the device or circuit, and the instant that the device or circuit reacts in a specified manner, or to a specified degree, to the input. The term often implies a protective action such as the provided by a clipper (peak limiter) or compressor, but may be used to describe the action of a device such as a vox (Voice Operated circuit), where the action is not protective.

attempt Trying to make a telephone call. Also defined as a call offered to a telecommunications system, regardless of whether it is completed. More technically, an attempt is a seizure of a component of equipment. Even a momentary off-hook condition of a telephone may cause a seizure (attempt).

attendant The "operator" of a phone system console. Typically, the first person to answer an incoming call. That person usually directs incoming calls to the proper person or department. That person may also assign outgoing lines or trunks to people requesting them. Few companies spend any time training their attendants. They should. There are two types of things attendants should be trained for: 1. Manners, including the correct way to keep people waiting and to screen incoming calls, and 2. The structure of the company. If a caller asks for some help, the attendant should know which department or person might be responsible for providing that help. Increasingly in North America, company phone systems are answered by devices called "automated attendants." They allow a caller to punch in the extension he wants or the department he wants. See Automated Attendant.

attendant access loop A switched circuit that provides an attendant with a manual means for call completion and control. An attendant access loop might be given a specific telephone number.

attendant busy lamp field Lamps, lights or LEDs that show whether a PBX or key system extension is busy or not. These days, many attendant busy lamp fields are being incorporated into CRT displays. We hope more will do this as many lamp-based attendant busy lamp fields are difficult to read.

attendant call waiting indication An unusual feature on a PBX console. The call waiting button on the attendant console lights to indicate a predetermined number of calls in queue. The light flashes when a second (programmable) threshold is reached.

attendant camp-on If the extension is busy, the attendant or operator can place the call in a queue behind the call already in progress. When the call is over, the "camped-on" call will automatically ring the extension.

attendant conference PBX feature that allows the attendant (or operator) to establish a conference connection between central office trunks and internal phones.

attendant console An attendant console is the larger, specialized telephone set used by the operator or attendant to answer incoming calls and send those calls to the proper extension. Consoles are becoming more sophisticated these days in several ways. Operators need to punch fewer buttons to move calls around, while the information they present to the attendant is more useful for keeping tabs on calls and letting people know what's happening. Many consoles are acquiring TV screens that report the status of each extension, who's speaking, where the call is going, and whether there are problems, such as broken lines or trunks, etc. anywhere on the system. Some of the more modern screens will allow the operator to send messages around the company that can alert someone as to who's calling before he/she picks up the phone. You can also easily program switches through consoles with CRT (also called TV) screens. In the old days you needed to punch in complex codes. Now you can respond to "Yes/No" decisions on a screen with lots of explanatory words and help menus.

attendant control of trunk group access The telephone operator or attendant controls the users' access to trunks for making local and/or long distance calls. This may reduce long distance call abuse.

attendant direct station select This feature gives an operator the ability to reach an extension by simply pushing one button. In direct station select, every extension has its own button. Direct Station Select usually comes with some form of Attendant Busy Lamp Field which shows whether the extensions are busy. Some attendants like direct station select. Others don't, preferring to simply punch in 345, instead of hunting for the button which corresponds to extension 345. The best consoles these days are using some form of easy-to-read screen prompts.

attendant exclusion A PBX feature which stops the attendant from listening in on a phone call once she or he has passed the call to the correct extension.

attendant forced release An attendant-activated (pushbutton) facility that will automatically "disconnect" all parties on a given circuit when that circuit is "entered" by the attendant.

attendant incoming call control A PBX feature which diverts incoming trunk calls automatically to a predetermined phone after a predesignated period of time or number of rings.

attendant key pad Allows the attendant to perform all functions using a standard touch tone key pad on the console or adjacent to it.

attendant locked loop operation PBX feature which allows the attendant at a console to retain supervision or recall capability of any particular call which has been processed.

attendant lockout This feature denies an attendant the ability to re-enter a phone call unless specifically recalled by that PBX extension.

attendant loop transfer Allows the attendant to transfer any call to another attendant.

attendant monitor A special attendant circuit which allows "listening in" on all circuits with the console handset/headset transmitter deactivated.

attendant override A feature that allows an attendant to enter a busy trunk connection and key the trunk number within the PBX. A warning tone will be heard by the connected parties, after which they connected parties and the attendant will be in a three-way connection.

attendant position Where a telephone operator sits to answer calls and send them on to the people in the company. This is usually in front of a telephone system with buttons, toggle switches, etc. that facilitate this process.

attendant recall When a phone call has been transferred to a telephone extension and not answered, this telephone system feature sends the call back to the attendant. Sometimes the call will return to a special part of the attendant console which will indicate to the attendant that it is a "returned" call. It's a good idea to pay attention to the speed of recall back to the operator. People hate to be extended into endless ringing. Think of calling a hotel and how aggravating it is to wait until the call comes back to an operator after she/he extended it to the room...and it rings and rings.

attendant recall on trunk hold The system will recall the attendant if a trunk placed on hold is not re-entered within a predetermined time.

attendant transfer of incoming calls A PBX and Centrex feature. A telephone extension is talking on a line but that person wants to transfer the call to someone else. The person hits his/her hookswitch a couple of times. (The hookswitch is the

toggle switch the handset depresses when you replace it.) This flashing of the hookswitch signals the attendant to join the call. The person asks the attendant "to please transfer this call." The attendant then transfers the call to the new extension. This feature is totally inefficient as it's hard to reach the attendant, who's always busy, etc. All newer phones can transfer both incoming and outgoing calls automatically by just flashing the hookswitch, dialing the extension number and hanging up.

attended A telephone system having an attendant or receptionist whose primary job is to answer all incoming calls. Many smaller systems, such as key systems, are not centrally "attended." The phone is simply answered by whoever is near. A non-attended phone system should be set up so anyone can answer an incoming call. Some systems, such as most key systems, come this way from the factory. Others, such as PBXs, have to be specially set up. Some systems can be set up so an attendant will get first shot at answering the incoming call, but then, after a couple of rings, anyone else can answer the call (perhaps a loud "night" bell will ring).

attended mode Imagine a communications situation where your computer is connected over a phone line to another user on another computer and you are uploading and downloading files. Attended mode refers to a situation where both users manually enter the commands required to send or receive a file concurrently, usually while conversing over the phone. Compare this to Unattended Mode.

attention key A key or combination of keys on a computer or terminal which signals the main computer to stop its present task and wait for a new command. The ESCape key is often the Attention Key. In Crosstalk, it's Control A.

attention management A new term covering the whole area of push technology at the desktop.

attention signal The attention signal used by AM, FM, and TV broadcast stations to actuate muted receivers for inter-station receipt of emergency cuing announcements and broadcasts involving a range of emergency contingencies posing a threat to the safety of life or property.

attenuate To decrease electrical current, voltage or power in communicating channel. Refers to audio, radio or carrier frequencies. See Attenutation.

attenuation The decrease in power of a signal, light beam, or lightwave, either absolutely or as a fraction of a reference value. The decrease usually occurs as a result of absorption, reflection, diffusion, scattering, deflection or dispersion from an original level and usually not as a result of geometric spreading, i.e., the inverse square of the distance effect. Optical fibers have been classified as high-loss (over 100 dB/km), medium-loss (20 to 100 dB/km), and low-loss (less than 20 dB/km). In other words, attenuation is the loss of volume during transmission. The received signal is lower in volume than the transmitted signal due to losses in the transmission medium (such as that caused by resistance in the cable). Attenuation is measured in decibels. It is the opposite of Gain. Some electrical components are listed as "with attenuation" which means they will compensate for irregular electrical supply (e.g. surges). See Gain.

attenuation coefficient The rate at which average power decreases with distance.

attenuation constant For a particular propagation mode in an optical fiber, the real part of the axial propagation constant.

attenuation equalizer Any device inserted in a transmission line or amplifier circuit to improve the shape of its frequency response.

attenuation limited operation The condition in a fiber optic link when operation is limited by the power of the received signal (rather than by bandwidth or distortion).

Attenuation to Crosstalk Ratio ACR. The difference between attenuation and crosstalk measured in decibels. For a longer explanation, see ACR.

attenuator A device to reduce signal amplitude by a known amount without introducing distortion. Attenuators are installed in fiber optic transmission systems to limit the power level received by the photodetector to within the limits of the optical receiver.

ATTND Attendant. What else?

atto Atto means one quintillion, which is 10 to the power of minus 18. See also FEMTO, which is 10 to the power of minus 15. See Femtosecond.

attribute The form of information items provided by the X.500 Directory Service. The directory information base consists of entries, each containing one or more attributes. Each attribute consists of a type identifier together with one or more values. See Attributes.

attributes Information about an MS-DOS or Windows file that indicates whether the file is read-only, hidden, or system, and whether it has been changed since it was last backed up. You can assign attributes to a file using the ATTRIB command. You can identify a file as read-only (meaning others can't change it, but can read it) and/or as a file you want to archive when using the BACKUP, RESTORE, and XCOPY commands. The command to make a file read only is typically

ATTRIB +R filename

By using the ATTRIB command to make a file "read only," you also make it impossible to erase the file from your disk. If you want to remove the "read only" protection, i.e. make the file "read and write," the command is

ATTRIB -R filename

ATU ADSL Transceiver Unit. The ADSL Forum uses terminology for DSL equipment based on the ADSL model for which the Forum was originally created. Thus, the DSL endpoint is known as the ATU-R and the CO unit is known as the ATU-C. These terms have since come to be used for other types of DSL services, like RADSL and SDSL. ATU generally represents xDSL services.

ATU-C ADSL Transmission Unit-Central Office. Special electronics in support of ADSL (Asymmetrical Digital Subscriber Line) and placed in the carrier's CO. The ATU-C has a matching unit on the subscriber premise in the form of an ATU-R. The two units, in combination, support a high data rate over standard UTP copper cable local loops. See ADSL and ATU-R.

ATU-R ADSL Transmission Unit-Remote. Special electronics in support of ADSL (Asymmetrical Digital Subscriber Line) and placed at the customer's premise. The ATU-R has a matching unit at the carrier's Central Office in the form of an ATU-C. The two units, in combination, support a high data rate over standard UTP copper cable local loops. See ADSL and ATU-C.

ATUG Australian Telecommunications Users Group, based in Milsons Point, Sydney, Australia.

ATV Advanced TV. Refers to any system of distributing television programming that generally results in better video and audio quality than that offered by the NTSC 525-line standard. This group of techniques is based on digital signal processing and transmission. HDTV (High Definition TV) and SDTV (Standard Definition TV) both fall under the definition of ATV. Although ATV systems are collectively considered to offer better quality than the NTSC signal, they can carry multiple pictures of lower-quality and can also support the cancellation of artifacts in ordinary NTSC signals. See also ATVEF, DTV, HDTV, and NTSC.

ATVEF Advanced Television Enhancement Forum. An industry group dedicated to creating standards for the future combination of Internet content with ordinary broadcast television, using IP, HTML and JavaScript. See ATV.

ATX Audiotex. Interactive voice response systems that deliver information or entertainment to general telephone callers, i.e. anyone with a phone. Audiotex services are typically widely advertised. They include everything from sex to the weather. See also Audiotex.

AU 1. A UNIX sound file format, i.e. filename.au. When a Sun Microsystems or other UNIX computer makes a sound, it does so in AU file format. And because the Internet is dominated by Unix boxes, you'll find a lot of AU files there. Macintosh and PC browsers are usually able to play AU files. In contrast, a sound file that originated on a PC is likely to be in WAV or MIDI format instead.

2. Administrative Unit. A SDH term. A element of a SDH frame that contains enough information for the switching and cross-connection of Virtual Channels (VCs). See also SDH and VC.

AuC AuC is also called AC. It stands for Authentication Center. A mobile term. The AuC is a piece of HLR, which is the Home Location Register – a permanent database used in GSM mobile systems (like those in Europe) to identify a subscriber and to contain subscriber data related to features and services. HLR is used to authenticate the user of mobile station equipment, i.e. a cell phone. The AuC performs secret, mathematical computations to verify the authenticity of the cell phone and thus to allow the user to make a phone call. AuC is no longer strictly coupled with the HLR. IS-41 Rev. C (now ANSI-41) defines the AuC as a stand-alone network element. See also CAVE.

auction An FCC definition. A procedure for choosing the users of spectrum space. In auction, the federal government treasury receives the profits.

Audible Audible.com's proprietary format for its audio books. Audible comes in four different formats. the first three vary in size, depending on the quality of the recording, while the fourth uses MP3 compression. The formats are also digitally protected.

audible indication control Three fancy words for the ability to turn up or down the bell or beeper on your PBX attendant console.

audible ring A sound sent from the called party's switch to inform the calling party that the called line is being rung. A long explanation for a bell or buzzer that tells you it's for you.

audible ringing tone The information tone sent back to the calling telephone subscriber as an indication that the called line is being rung.

audible sound Audible sound spans a huge range of frequencies from around 20 hertz (vibrations per second) to 20 kilohertz. See Sound.

audible tones Audible tones are the sounds provided by the network or an attached switch to inform callers of the status of the line or of an event. Audible tones most frequently encountered in computer telephony applications include: ringing, busy tones (called party busy and network busy), SIT tones (Special Information Tones), and special tones used by computer telephony systems such as the "record at the beep" (which is usually a 1,000 Hz tone).

audio Sound you hear which may be converted to electrical signals for transmission. A human being who hasn't had his ears blown by listening to a Sony Walkman or a ghetto blaster can hear sounds from about 20 to 20,000 hertz.

audio bridge In telecommunications, a device that mixes multiple audio inputs and feeds back composite audio to each station, minus that station's input. Also known as a mix-minus audio system.

audio caller ID A ringtone that is associated with a specific phone number, so that the user can identify a specific caller without having to look at the phone. Your phone rings "Here comes the bride," when your newly-engaged daughter is calling, for example.

audio crosspoint module Circuit board containing crosspoints for audio signal switching.

audio extender An arrangement consisting of a sender unit, receiver unit and a cable that connects the two, all of which enables an audio speaker to be up to 1,000 feet away from a microphone, computer, or other audio source.

audio frequencies Those frequencies which the human ear can detect (usually in the range of 20 to 20,000 hertz). Only those from 300 to 3,000 hertz are transmitted through the phone, which is why the phone doesn't sound "Hi-fi."

audio frequency The band of frequencies (approximately 20 hertz to 20,000 hertz) that can be heard by the healthy human ear.

audio indexing software Audio indexing software creates a searchable index of speech content in digitized audio and video files. It allows for quick searches and rapid access to audio content.

Audio Interchange File Format See AIFF.

audio menu Options spoken by a voice processing system. The user can choose what he wants done by simply choosing a menu option by hitting a touchtone on his phone or speaking a word or two. Computer or voice processing software can be organized in two basic ways – menu-driven and non-menu driven. Menu-driven programs are easier for users to use, but they can only present as many options as can be reasonably spoken in a few seconds. Audio menus are typically played to callers in automated attendant/voice messaging, voice response and transaction processing applications. See also Menu and Prompts.

Audio Messaging Interchange Specification AMIS. Issued in February 1990, AMIS is a series of standards aimed at addressing the problem of how voice messaging systems produced by different vendors can network or inter-network. Before AMIS, systems from different vendors could not exchange voice messages. AMIS deals only with the interaction between two systems for the purpose of exchanging voice messages. It does not describe the user interface to a voice messaging system, specify how to implement AMIS in a particular systems or limit the features a vendor may implement. AMIS is really two specifications. One, called AMIS-Digital, is based on completely digital interaction between two voice messaging systems. All the control information and the voice message itself, is conveyed between systems in digital form. By contrast, the AMIS-Analog specification calls for the use of DTMF tones to convey control information and transmission of the message itself is in analog form.

audio modulator A device which combines audio with higher frequency radio signals for transmission.

audio oscillator Hewlett-Packard's first product. They sold them at $54.40.

audio response unit A device which translates computer output into spoken voice. Let's say you dial a computer and it said "If you want the weather in Chicago, push 123," then it would give you the weather. But that weather would be "spoken" by an audio response unit. Here's a slightly more technical explanation: An audio response unit is a device that provides synthesized voice responses to dual-tone multi-frequency signaling input. These devices process calls based on the caller's input, information received from a host data base, and information carried with the incoming call (e.g., time of day). ARUs are used to increase the number of information calls handled and to provide consistent quality in information retrieval. See also Audiotex and Interactive Voice Response.

audio server An audio server is a big PC which contains digitized music and software and sits on a network, local or distance. You use your browser to talk to your audio server. Its software responds. It allows you to choose which songs you want played and where – in which rooms of your house, etc.

audio subcarrier A Satellite Term. Carrier wave that transmits audio information.

audio text The term used to describe a system that provides automated interactive telephone information, such as stock prices, sports scores and personals. Also spelled Audiotex and Audiotext.

audio track The section of a videodisc or tape which contains the sound signal that accompanies the video signal. Systems with two separate audio tracks (most videodiscs) can offer either stereo sound or two independent soundtracks.

audioconference Another term for Teleconference. Teleconferences make use of conference bridges to allow participants to join a voice conference over the PSTN. See Teleconference.

audiographic conferencing Teleconferencing that also allows participants to share and interact through graphics, figures and printed text. Hardware used during audiographic conferences includes facsimile, telewriters and film-based projectors. Transmission is via a narrowband telecommunications channel such as a telephone line or a radio subcarrier.

audiographics The technology which allows sound and visual images to be transmitted simultaneously. According to AT&T, audiographics generally refers to single frame or slow frame visual images as opposed to continuous frame image transmission (e.g. television). Audiographic transmission is often used to teach or train people in remote locations from an educational institution or business training center, saving travel and housing expense.

audiotex A generic term for interactive voice response equipment and services. Audiotex is to voice what on-line data processing is to data terminals. The idea is you call a phone number. A machine answers, presenting you with several options, "Push 1 for information on Plays, Push 2 for information on movies, Push 3 for information on Museums." If you push 2, the machine may come back, "Push 1 for movies on the south side of town, Push 2 for movies on the north side of town, etc." See also Information Center Mailbox.

Audiotext A different, and less preferred, spelling of Audiotex. See Audiotex.

Audio/visual Multimedia Services AMS is an ATM term. It specifies service requirements and defines application requirements and application program interfaces (APIs) for broadcast video, videoconferencing, and multimedia traffic. AMS is being developed by the ATM Forum's Service Aspects and Applications (SAA) working group. An important debate in the SAA concerns how MPEG-2 applications will travel over ATM (asynchronous transfer mode). Early developers chose to carry MPEG-2 over ATM adaptation layer 1 (AAL 1); others found AAL 5 a more workable solution. Recently, some have suggested coming up with a new video-only AAL using the still-undefined AAL 2.

auditory pattern recognition Auditory pattern recognition is the ability to recognize spoken words.

audit To conduct an independent review and examination of system records and activities in order to test the adequacy and effectiveness of data security and data integrity procedures, to ensure compliance with established policy and operational procedures, and to recommend any necessary changes.

audit file On some systems, each time a billing file is generated, an audit file is created to record the details of the generation process.

audit trail A record of all the events that occur when users request and use specific resources. An audit trail gives you the ability to trace who did what and who was responsible for what. An audit trail is a chronological record of system activities that is sufficient to enable the reconstruction, review, and examination of the sequence of environments and activities surrounding or leading to an operation, a procedure or an event in a transaction from its inception to final results. Audit trail may apply to information in an automated information system, to the routing of messages in a communications system, or to material exchange transactions, such as in financial audit trails. Audit trails are great for finding out what activity caused the disaster, the network to crash, etc. Audit trails are great for tracing crooks, i.e. unauthorized break-ins.

auditing 1. Checking to see if the phone bill you got from your carrier is accurate. You do this by comparing it to your own telephone system's records.

2. Tracking activities of users by record and selected types of events in the security log of a server or a workstation.

auf WiederhOxEEren Most people know that "goodbye" in German is "auf Wiedersehen," which means "until we see each other again." At the end of a phone conversation, however, "auf WiederhOxEEren" is used, which means "until we hear from each other again." "TschOxC5s" is also used, which is equivalent to saying "Bye" – or "K'bye" ("OK bye") if you're a teenager.

auger A type of drill bit typically used to make large, deep holes for passing wire or cable through wood.

augmentation of bandwidth Bandwidth augmentation is the ability to add another communications channel to an already existing communications channel.

AUI Autonomous Unit Interface or Attachment Unit Interface. Most commonly used in reference to the 15 pin D type connector and cables used to connect single and multiple channel equipment to an Ethernet transceiver.

AUI Cable Attachment Unit Interface Cable is usually a four-twisted pair cable that connects an Ethernet device to an Ethernet external receiver (XCVR).

AUP Acceptable Use Policy. The term used to refer to the restrictions placed on use of a network; usually refers to restrictions on use for commercial purposes, most commonly with respect to the Internet.

AUPC Automatic uplink power control. This is a feature of satellite modems. It allows a local modem to adjust its own output power levels in order to maintain Eb/No at the remote modem. Rain does and will affect Ku-bands links.

aural Relating to the sense of hearing.

AUR Access Usage Record.

AURP AppleTalk Update-Based Routing Protocol. Method of encapsulating AppleTalk traffic in the header of a foreign protocol, allowing the connection of two or more discontiguous AppleTalk internetworks through a foreign network (such as TCP/IP) to form an AppleTalk WAN. This connection is called an AURP tunnel. In addition to its encapsulation function, AURP maintains routing tables for the entire AppleTalk WAN by exchanging routing information between exterior routers. See also AURP Tunnel.

AURP Tunnel Connection created in an AURP WAN that functions as a single, virtual data link between AppleTalk internetworks physically separated by a foreign network (a TCP/IP network, for example). See also AURP.

AUSM ATM user service module.

AUSSAT Australian domestic satellite operator (now Optus).

Austin Wireless City Project In Austin, Texas, Less Networks has enabled the Austin Wireless City Project to network together 50 hot spots from City Hall, the Texas State Capitol, a downtown park, to dozens of coffee shops, bookstores, restaurants and bars. Even though the access is free, a common login is required providing the ISPs and the venues the security and logging data they need should someone abuse the free service. Moreover, the authentication logs provide invaluable usage statistics to the venues who foot the monthly bills for the broadband. This from the Wall Street Journal of May 27, 2004.

Australian Radiocommunications Study Group ARSG. An ITU study group which develops radiocommunications standards and regulatory procedures to allow for efficient and effective worldwide exploitation of the radiofrequency spectrum.

Australian Telecommunications Authority See AUSTEL.

Australian Telecommunication Standardisation Committee The ATSC was formerly responsible for managing Australia's input to international and regional telecommunications technical standards. ATSC functions are now performed by the ACIF.

authenticate To establish, usually by challenge and response, that a transmission attempt is authorized and valid. To verify the identity of a user, device, or other entity in a computer system, or to verify the integrity of data that have been stored, transmitted, or otherwise exposed to possible unauthorized modification. A challenge given by voice or electrical means to attest to the authenticity of a message or transmission.

Authenticated SMTP Authentication is means of verifying, or positively identifying, another party, like the sender of an e-mail message. E-mail makes use of SMTP (Simple Message Transfer Protocol), which is an application-layer extension of the TCP/IP (Transmission Control Protocol/Internet Protocol) protocol suite. TCP/IP and its various application-layer extensions form the fundamental protocols used in the Internet, which is an inherently insecure network of networks. One means of imposing a measure of security for e-mail is Authenticated SMTP, which blocks e-mail from other than trusted sources. That way you don't get a lot of spam. You also reduce the likelihood of contracting a computer virus, assuming that your trusted e-mail correspondents are as careful as you are about finding scanning for viruses and killing them. Encryption also is a good idea, as it secures your e-mails as they transverse the Internet. See also Authentication, Encryption, Internet, SMTP, TCP/IP, and Virus.

authentication The process whereby a user or information source proves they are who they claim to be. In other words, the process of determining the identify of a user attempting to access a system. Authentication codes are exchanged between two computers that use SMTP. I believed that defining authentication was pretty simple until I came across the following definition in a book on cryptography. Here it is: Authentication is any technique enabling the receiver to automatically identify and reject messages that have been altered deliberately or by channel errors. Also, can be used to provide positive identification of the sender of a message although secret (symmetric) key algorithms may be used for some authentication without the prior sharing of any secrets between the messaging parties. See Authentication Token.

Authentication Center See AuC.

Authentication Random Number A random value used in authentication procedures.

Authentication Sequence Number A sequence count that is incremented for each change of Authentication Random Number (ARN).

authentication token A portable device used for authenticating a user. Authentication tokens operate by challenge/response, time-based code sequences, or other techniques. An example is Security Dynamics Technologies Inc.'s SecurID card. See Authentication.

authority zone Associated with DNS, an authority zone is a section of the domain-name tree for which one name server is the authority. See also DNS.

auto available An ACD feature whereby the ACD is programmed to automatically put agents into Available after they finish Talk Time and disconnect calls. If they need to go into After-Call Work, they have to manually put themselves there. See Auto Wrap-up.

auto changer A jukebox-style optical media system that permits the storage and playback multiple discs. Autochangers are available for both 5-,8-, and 12-inch optical discs.

auto discovery Process by which a network device automatically searches through a range of network addresses and discovers all known types of devices present in that range. Auto discovery is what takes place when a new device is added to a network. That device (e.g., an ethernet board, a modem, a router) communicates to the network management system that it is online and defines its key characteristics.

Auto Discovery is also the process by which MPOA (MultiProtocol Over ATM) edge devices automatically find each other through the ATM network. See MPOA.

auto fill-in You're surfing the Web with your browser. You come to a form that needs being filled it. The form asks for your first name. You type H, as in Harry. A window pops up. It shows "Harry." Click on Harry. It will drop the word Harry into the line on the form that asks for your first name. Auto Fill-in is a time saving tool of new browsers. Some browsers actually do auto fill-in even more efficiently. They simply type "Harry" when they come to a form that says "first name?" Others require you to click.

auto greeting ACD Agent's pre-recorded greeting that plays automatically when a call arrives.

auto ID center A non-profit collaboration between private companies and academia that is pioneering the development of an Internet-like infrastructure for tracking goods globally through the use of RFID tags.

auto negotiation Process that allows two devices at either end of a link segment to negotiate common features, speed and functions. It's a two-part process by which a network device automatically senses the speed and duplex capability of another device.

auto partitioning A feature on some network devices that isolates a node within the workgroup when the node becomes disabled, so as not to affect the entire workgroup or network.

auto play The Microsoft AutoPlay feature of the Windows operating system. This lets your CD run an installation program or the game itself immediately upon insertion of the CD-ROM.

auto responder Auto responder is just what it sounds like. You send an email to an email address on the Internet. The person who owns the address is away. He has set up an "auto responder," which is a piece of software. That software automatically sends you a response. The most common use of auto responders is for vacation messages. "I'm away. I'll be back in the office on December 25 and will reply then. If you need help earlier, contact Joe Plumpudding on TastyMorsel@Plumpudding.com."

auto sensing Process during which a network device automatically senses the speed of another device.

Auto Spid When you install an ISDN device on an ISDN line you have to give that device the phone line's SPIDs – Service Profile IDentifier numbers. The SPID is actually a label identifier that points to a particular location in your telephone company's central office memory where the relevant details of your ISDN service are stored. Auto SPID is an industry-wide initiative of ISDN equipment makers that automatically "negotiates" a connection to the telecom provider by downloading the SPID numbers from your digital central office switch. This means that in future, when you get an ISDN line, you won't have to worry about begging SPID numbers out of the telephone company.

Auto Wrap-up An ACD feature whereby the ACD is programmed to automatically put agents into After-Call Work after they finish Talk Time and disconnect calls. When they

have completed any After-Call Work required, they put themselves back into Available. See AutoAvailable.

Auto-answer A feature of a modem that allows it to answer incoming calls automatically.

authoring Authoring is the process of using multimedia applications to create multimedia materials (including Web pages) for others to view. Multimedia authoring uses many tools, from the more familiar text editor or desktop publishing application, to tools for capturing and manipulating video images or editing audio files. Authors might include specialized creators of training, sales, or corporate applications such as insurance claims processing. Or, they might be creators of everyday business communications like voice-annotated email. Over time, everyone involved in business communications will probably have some level of multimedia authoring capability.

authoring system Software which helps developers design interactive courseware easily, without heavy computer programming. A specialized, high-level, plain-English computer language which permits non-programmers to perform the programming function of courseware development. The program logic and program content are combined. See Authoring.

authorization Think of charging things on your MasterCard, Visa, or American Express card. If the store cannot authorize the amount of your purchase, your Visa card will not allow you to make the purchase. Authorization is needed for many long distance calls, especially those made using credit cards, telephone company calling card, etc. Authorization is done by the operator's computer checking with the remote validation database service. See BVA, BVS and Validation.

authorization code A code in numbers and/or letters employed by a user to gain access to a system or service. If you are making a call out on a restricted line, the PBX will ask you for an authorization code. If you have one, your call will go through. If not, your call will be denied (i.e. not go through). Authorization codes come in various flavors. Some can be used for making long distance calls. Some can be used also for international calls, etc. See Authorized User.

authorized agent Also called Authorized Sales Agent. A term chosen by some of the Bell operating companies and many of the cellular phone companies to refer to companies which sell their network services on commission. Some of these companies have specific industry knowledge and have written specialized software. The idea is to work with businesses to arm them with the absolute best package of telecommunications hardware, software and services.

authorized bandwidth The necessary bandwidth required for transmission and reception of intelligence. This definition does not include allowance for transmitter drift or Doppler shift.

uthorized dealer See Dealer.

authorized frequency A frequency that is allocated and assigned by an authority to a specific user for a specific purpose.

authorized user A person, firm, corporation or any other legal entity authorized by the provider of the service to use the service being provided.

auto answer The capability of a phone, a terminal, a video phone, a modem or a computer to answer an incoming call and to set up a connection without anyone actually doing anything to physically answer the call. I have auto answer on my video phone. When it rings, my phone answers and I see who's calling. It's nice.

auto attendant A shortened name for an automated attendant, a device which answers a company's phones, encourages you to touchtone in the extension you want, and rings that extension. If that extension doesn't answer, it may send the call to voice mail or back to the attendant. It may also allow you to punch in digits and hear information, e.g. the company's hours of business, addresses of local branches, etc. See also Automated Attendant.

auto baud Automatic speed recognition. The ability of a device to adapt to the data rate of a companion device at the other end of the link.

auto baud detect See Auto Baud.

auto busy redial A feature of a phone or phone system where the phone has the ability to keep trying a busy number until answered. The circuit actually recognizes the busy tone, hangs up, and dials again. One of the greatest time-savers ever invented.

auto call Automatic Calling; a machine feature that allows a transmission control unit or a station to automatically initiate access to (i.e. dial) a remote system over a switched line.

auto dial A feature of phone systems and modems which allows them to dial a long phone number (usually long distance) by punching fewer buttons than there are numbers to dial. One button auto dial on electronic phones is very common these days. Most communications software programs will allow you to auto dial a string of 35 to 40 digits, which you may need if you're dialing through a complex network.

auto dial auto answer A modem feature. Auto Dial lets you dial a phone number through your modem, using your personal computer or data terminal keyboard. Auto Answer permits the modem to automatically answer the incoming call without anybody having to be there.

auto dialer See Automatic Dialer.

auto fax tone Also called CNG, or Calling Tone. This tone is the sound produced by virtually all Group 3 fax machines when they dial another fax machine. CNG is a medium pitch tone (1100 Hz) that lasts 1/2 second and repeats every 3 1/2 seconds. A fax machine will produce CNG for about 45 seconds after it dials. See also CNG.

auto line feed An instruction in a communications program which causes the program to perform a Line Feed (LF) when you hit a carriage return or the "Enter" key.

auto partition A feature of 10Base-T. When 32 consecutive collisions are sensed by a port in a hub or concentrator from its attached workstation or network segment, or when a packet that far exceeds the maximum allowable length is received, the port stops forwarding packets. The port continues to monitor traffic and will automatically begin normal packet forwarding when the first correct packet is received.

auto recognition A term used in file conversion in which your conversion software figures out by itself in what form the original file was – WordPerfect 5.0, Word 6.0, Wordstar 5.5 etc. See also Auto Styling.

auto reconfiguration The process performed by nodes within the failure domain of a Token Ring network. Nodes automatically perform diagnostics in an attempt to reconfigure the network around failed areas. See also failure domain.

auto selection tool An imaging term. A tool that selects an entire area within a specified range of color values around a selected pixel.

auto sensing See Auto Styling.

auto start A standby electrical power system that starts up when the normal supply of commercial power fails.

auto stream An AT&T ISDN term. The method of data flow in which both channels between the ISDN set and the application are in use simultaneously.

auto styling Auto styling is a term we found in a database conversion software program. What it means is that the program looks at the data in a field and determines from that data if the field is a numeric, character or memo, etc. The problem with auto styling is that it's frequently wrong. For example, it might check one field, find all numbers and decide it's a numeric field. Such a field might be a zip code, which actually is normally a character field. One reason why you might want you zip code to be a character field is that character fields are set left. Numeric fields are set right. (They line up at the decimal point.) Another name for auto styling is auto sensing.

autoanswer A feature of a telephone which automatically answers incoming calls without the user of the phone lifting a handset or otherwise answering the call. Modems and fax machines also autoanswer. In North America, it's spelled AUTO ANSWER. In Britain, it's spelled Autoanswer.

autoattendant See Automated Attendant.

autobauding The process by which the terminal software determines the line speed on a dial-up line.

AUTODIN The worldwide data communications network of the U.S. Department of Defense. Acronym for "AUTOmatic DIgital Network." AUTODIN is a labor-intensive 40-year-old Department of Defense communications network that has largely been replaced by DISN and DMS.

Autodial Button An Autodial button on a phone provides one-touch dialing of outside numbers, intercom numbers, or feature codes.

automated attendant A device which answers callers with a digital recording, and allows callers to route themselves to an extension through touch tone input, in response to a voice prompt. An automated attendant avoids the intervention of a human being in the form of a console attendant, thereby avoiding related personnel costs. Commonly implemented in Voice Processor systems and software, front-ending PBXs and ACDs.

An automated attendant is typically connected to a PBX or a Centrex service. When a call comes in, this device answers it and says something like, "Thanks for calling the ABC Company. If you know the extension number you'd like, pushbutton that extension now and you'll be transferred. If you don't know your extension, pushbutton "0" (zero) and the live operator will come on. Or, wait a few seconds and the operator will come on anyway." Sometimes the automated attendant might give you other options such as "dial 3 for a directory of last names and dial 4 for a directory of first names." Automated attendants are also connected to voice mail systems ("I'm not here. Leave a message for me."). Some

people react well to automated attendants. Others don't. A good rule to remember is before you spring an automated attendant on your people/customers/subscribers, etc., let them know. Train them a little. Ease them into it. They'll probably react more favorably than if it comes as a complete surprise. The first impression is rarely forgotten, so try to make it a good experience for the caller. See also Dial By Name.

Automated Coin Toll Service ACTS. In the old days, operators handled routine toll calls by counting the sound of coins hitting the box, checking prices, putting calls through, figuring and collecting overtime charges, etc. ACTS does all this automatically. It figures charges, tells those charges by digitized computerized voice to the customer, counts the coins as they are deposited and then sets up the call.

Automated Maritime Telecommunications System An automatic, integrated and interconnected maritime communications system serving ship stations on specified inland and coastal waters of the United States.

Automated Meter Reading See AMR.

automated radio A radio with the capability for automatically controlled operation by electronic devices that requires little or no operator intervention.

Automated Tactical Command And Control System A command and control system or part thereof which manipulates the movement of information from source to user without intervention. Automated execution of a decision without human intervention is not mandatory.

Automated Telephone Listing Address System ATLAS. Central repository for Verizon Listings data. Input data is from Verizon Retail, Wholesale, and Independent tlephone companies. ATLAS users include Verizon Directory Assistance and Directory Publishing.

automated vehicle classification A system used in commercial vehicle operations to electronically identify a vehicle's type. Automated vehicle classification decreases the amount of time required at border crossings by reducing the amount of paperwork that drivers and border officials have to prepare and process. See automate vehicle identification.

automated vehicle identification The use of a transponder on-board a motor vehicle, combined with toll-house and roadside receivers, to automatically identify a vehicle. The technology supports electronic toll collection and the tracking down of stolen vehicles. See automated vechicle classification.

Automated Voice Response Systems AVRS. Devices which automatically answer calls. They may simply inform the caller that the call is in a queue and will be answered soon; alternatively they may prompt the caller to use voice commands, or touchtones to seek more information.

automatic See Seamless.

automatic address discovery This refers to the process by which a network device can poll other network devices to discover the network addresses which each device supports. Automatic address discovery makes the set up and on-going maintenance of complex internetworks much simpler than if all address updates were performed manually.

Automatic Alternate Routing AAR. A call going through a network can get a connection via secondary routes between two locations without need for the user's intervention, i.e. automatic.

automatic button restoration When the telephone handset of a multi-line instrument (typically a 1A2 multi-line key set) is placed back in its cradle, the line button being used automatically "pops" back up. Conversely, when a user picks up the handset, he must always push down a line button to make a call. Most phones with this feature can be disabled, so the buttons stay down when the handset sits on the cradle. A twist of a single screw inside the instrument will usually solve the aggravation of the automatic button restoration. Some people like automatic button restoration because it saves a user from accidentally barging into someone else's call. This was a much greater problem with 1A2 key systems. It no longer is a problem with most electronic key systems since they usually extend the user automatic privacy once they get on a call so no one else can barge in, even if they want to.

automatic call distributor ACD. A specialized phone system originally designed simply to route an office's incoming calls to all available personnel so that calls are evenly distributed. Now increasingly used by companies also making outgoing calls. You receive and make lots of phone calls typically to customers. You need an ACD. Once used only by airlines, rent-a-car companies, mail order companies, hotels, etc., it is now used by any company that has many incoming calls (e.g. order taking, dispatching of service technicians, taxis, railroads, help desks answering technical questions, etc.). There are very few large companies today that don't have at least one ACD. Many smaller companies, like the company that publishes this dictionary, also have one.

An ACD performs four functions. 1. It will recognize and answer an incoming call. 2. It will look in its database for instructions on what to do with that call. 3. Based on these instructions, it will send the call to a recording that "somebody will be with you soon, please don't hang up!" or to a voice response unit (VRU). 4. It will send the call to an agent as soon as that operator has completed his/her previous call, and/or the caller has heard the canned message. The term Automatic Call Distributor comes from distributing the incoming calls in some logical pattern to a group of operators. That pattern might be Uniform (to distribute the work uniformly) or it may be Top-down (the same agents in the same order get the calls and are kept busy. The ones on the top are kept busier than the ones on the bottom). Or it may be Specialty Routing, where the calls are routed to answerers who are most likely to be able to help the caller the most. Distributing calls logically is the function most people associate with an ACD, though it's not the most important.

The management information which the ACD produces is much more valuable. This information is of three sorts: 1. The arrival of incoming calls (when, how many, which lines, from where, etc.) 2. How many callers were put on hold, asked to wait and didn't. This is called information on ABANDONED CALLS. This information is very important for staffing, buying lines from the phone company, figuring what level of service to provide to the customer and what different levels of service (how long for people to answer the phone) might cost. And 3. Information on the origination of the call. That information will typically include ANI (Automatic Number Identification – picking up the calling number and DNIS (Direct Number Identification Service) picking up the called number. Knowing the ANI allows the ACD and its associated computer to look up the caller's record and thus offer the caller much faster service. Knowing the DNIS may allow the ACD to route the caller to a particular agent or keep track of the success of various advertising campaigns. Ad agencies will routinely run the same ad in different towns using different 800 phone numbers. Picking up which number was called identifies which TV station the ad ran on.

The seven definitions that follow the definition "ACD" show some of the features which newer ACDs have. See also 800 Service, ACD and Automatic Call Sequencer.

Automatic Call Intercept A feature of a Rolm ACD. This feature automatically forwards calls to an attendant if the dialed number is not installed or out of order. It can also intercept an attempted trunk call that is in violation of a Class of Service restriction. Automatic Call Intercept will also recall the attendant after a predetermined period of offshoot inactivity (e.g. flash or hold).

automatic call reconnect Feature permitting automatic call rerouting away from a failed trunk line.

automatic call rescheduling When a call is unsuccessful, due to no reply or busy signal, the system will automatically dial the number again after a pre-determined time.

automatic call sequencer A device for handling incoming calls. Typically it performs three functions. 1. It answers an incoming call, gives the caller a message, and puts them on "Hold." 2. It signals the agent (the person who will answer the call) which call on which line to answer. Typically, the call which it signals to be answered is the call which has been on "hold" the longest. 3. It provides management information, such as how many abandoned calls there were, how long the longest person was kept on hold, how long the average "on hold" was, etc.

There are three types of devices which handle incoming calls. The least expensive is the Automatic Call Sequencer which is traditionally used with key systems. It differs from Uniform Call Distributors (UCDs) and Automatic Call Distributors (ACDs) in that it has no internal switching mechanism and does not affect the call in any way. It simply recommends which call should be picked up and keeps statistical information on the progress of calls. A more expensive type of device is the UCD.

The most full-featured and expensive is the ACD. Distinctions between ACDs and UCDs and/or PBXs with features called UCDs and ACDs are blurring as UCDs get more sophisticated. The main difference, as we understand it, is that a UCD offers fewer options for routing an incoming call and answering calls in any particular order. ACDs typically produce the most detailed management information reports. One company also makes something called an Electronic Call Distributor. It is essentially an automatic call distributor.

automatic callback When a caller dials another internal extension and finds it busy, the caller dials some digits on his phone or presses a special "automatic callback" button. When the person he's calling hangs up, the phone system rings his number and the number of the original caller and the phone system automatically connects the two together. This feature saves a lot of time by automatically retrying the call until the extension is free. See also CAMP ON. Wouldn't it be nice if they had this feature on long distance calls? *66 is the usual code to access Automatic Callback.

Automatic Calling Unit ACU. A device that places a telephone call on behalf of a computer.

Automatic Circuit Assurance ACA is a PBX feature that helps you find bad trunks. The PBX keeps records of calls of very short and very long duration. If these calls exceed a certain parameter, the attendant is notified. The logic is that a lot of very short calls or one very long call may suggest that a trunk is hung, broken or out of order. The attendant can then physically dial into that trunk and check it.

automatic controls Controls that are automatically started by switching systems in response to a threshold value being exceeded.

automatic cover letter In a fax transmission, an automatic cover letter allows the user to automatically attach a cover letter to the document being sent. This is especially convenient when sending material directly from your PC.

Automatic Data Speed Recognition ADSR. A function of integrated voice/data PBXs to permit data devices and terminals of various data speeds to be connected within the PBX. ADSR is frequently associated with an "automatic modem select" function in "modem pooling" arrangements or with "format and protocol conversion" capability to provide internal data calls between dissimilar devices and terminals. PBXs don't switch data calls much these days. Local area networks now do that. See Local Area Network.

automatic dialer or autodialer A device which allows the user to dial pre-programmed telephone numbers by pushing one or two buttons. Sometimes referred to as a "repertory" dialer. Dialers can be bought as a separate device and added to a phone. Today most telephone sets are outfitted with autodialers. There are four basic measures of an automatic dialer's efficiency. 1. What's the longest number it will dial automatically? This is important because using some of AT&T's long distance competitors requires dialing lots of numbers, with lots of pauses. 2. How many numbers will it dial? Some people like to have a dialer which dials hundreds of numbers. Others like a small one, just for their most frequently called numbers. 3. Will the dialer recognize dial tone? This is important because using a long distance company or dialing through a PBX requires one to recognize consecutive dial tones. 4. Can you "chain" dial? In other words, can you hit one speed dial button after another and have the machine dial through a complex network and throw in authorization codes, etc.?

Automatic Dialing See Speed Dialing.

automatic directory propagation In electronic mail, automatic directory propagation is the ability to update addresses automatically in one domain after manually entering address changes in another domain, whether on the same LAN or another LAN connected by a gateway. In general, automatic directory propagation can be peer-to-peer, where changes in any post office are sent to all other post offices, or master-to-slave, where changes in the master post office are sent to the slaves, but changes in the slave post office do not go to the master.

automatic equalization The process of compensating for distortion of data communications signals over an analog circuit.

automatic exchange A term for a central office which automatically and electronically switches calls between subscribers without using an operator. Not a common term.

automatic exclusion A telephone system function in which the first station on a line prevents access by all others on that line; identified with "bridge lifters" in local wire telephone plant; can be duplicated in digital local exchanges with software. See Automatic Privacy.

Automatic Facilities Test System AFACTS is a Rolm CBX feature. It is an automatic testing system for identifying faulty tie and central office trunks. AFACTS can pinpoint faulty trunks and generate exception and summary reports.

automatic fallback A modem's ability to negotiate an appropriate data rate with the modem on the other end of the link, depending on line quality. For example, if two 2400 baud modems can not pass data at 2400 baud, they would "fall back" to 1200 baud automatically in order to transmit data without excessive errors.

automatic forwarding A feature of many e-mail programs that automatically retransmits incoming messages to another e-mail address.

automatic frequency control A circuit in a radio receiver which automatically brings the tuning units of the set into resonance with a wave which is partially tuned in.

automatic gain This is an electronic circuit which automatically increases the volume when someone is speaking quietly and drops it when someone is speaking loudly. The idea is to keep the transmitted signal even. Most tape recorders, for example, have automatic gain circuits. This allows them to pick up voices of people in a room, even though the volume of each person's conversation arriving at the tape recorder is different. The problem with automatic gain circuits is they're always looking for something to amplify. Thus, when it is quiet (and meant to be), the automatic gain circuit will try to amplify the ambient noise in the room – to keep the sound level constant. All professionally recorded tapes are done on tape recorders with manual volume controls.

automatic hold -- station or intercom When a user is having a conversation and receives another call, he may press the button to answer that new call. The call he was on originally is automatically put on hold.

automatic identification Sometimes called automatic data capture. These are methods of collecting data and entering it directly into computer systems without human involvement. Technologies normally considered part of auto-id include bar codes, biometrics, RFID and voice recognition.

Automatic Identified Outward Dialing AIOD. The toll calls placed by all extensions on the telephone systems are automatically recorded. This information allows bills to be sent, long distance lines to be chosen, etc. Some central offices, for example, can provide an itemized breakdown of charges (including individual charges for toll calls) for calls made by each CPE telephone extension. See Call Accounting, Call Detail Recording, SMDR and AIOD.

Automatic Intercept Center AIC. A Bellcore definition: "The centrally located set of equipment that is part of an Automatic Intercept System (AIS) that automatically advises the calling customer, by means of their recorded or electronically assembled announcements, of the prevailing situation that prevents completion of connection to the called number."

automatic intercept system A type of switching system for handling intercept calls, typically used by phone companies and typically totally automated these days. When calling a number, an automatic intercept system might come on and say, "I'm sorry the number you just dialed has been changed, If you'd like to connect to the new one, please hit your star button. You will be charged 75 cents for the privilege."

Automatic Level Control ALC. A control system that adjusts the incoming signal to a predetermined level. Somewhat similar to automatic gain control. See Automatic Gain Control.

Automatic Light Control ALC. Vidicon camera control which automatically adjusts the target voltage to compensate for variations in light levels. See also Automatic Gain.

automatic line hold A PBX feature. As long as a phone does not go "on-hook," activation of various line pushbuttons will automatically place the first line on hold without the use of a special "hold" button.

Automatic Location Identification ALI. Working with Automatic Number Identification, the use of a database to associate a physical location with a telephone number.

Automatic Line Insulation Testing ALIT. Equipment located in a Central Office which sequentially tests lines in the office for battery crosses and grounds.

Automatic Link Establishment ALE. The capability of an HF radio station to contact, or initiate a circuit, between itself and another specified radio station, without operator assistance and usually under computer control. ALE techniques include automatic signaling, selective calling, and automatic handshaking. Other automatic techniques that are related to ALE are channel scanning and selection, Link Quality Analysis (LQA), polling, sounding, message store and forward, address protection, and anti-spoofing.

automatic message accounting The network functionality that measures, collects, formats, and outputs subscriber network-usage data to upstream billing OSs and other OSs (Operations Systems).

automatic message switching A technique of sending messages to their appropriate destination through information contained in the message itself – typically in its "address."

automatic network restoral Automatic network restoral is a term which reflects the ability of a network to restore service rapidly and automatically following a catastrophic failure, such as that of a network cable cut. The result is higher network availability and reliability.

Automatic Number Announcement Circuit See ANAC.

Automatic Number Identification ANI. Being able to recognize the phone number of the person calling you. You must have equipment at your office. And the network must have the ability to send the calling number to you. For a much longer explanation, see ANI, CALLER ID, CLASS, ISDN and System Signaling 7.

automatic overflow to DDD Toll calls jump to expensive direct distance dialed calls, when all lower cost FX, WATS lines, etc. are busy.

automatic phone relocation The Automatic Phone Relocation feature now available on some phone systems allows a telephone to retain its personal and system programming when it is reconnected to another physical location.

automatic privacy When someone is speaking on a phone line or on an intercom, this feature ensures no one else can accidentally or deliberately butt into that conversation. If you did, however, want somebody else to come into the conversation (for example, someone to provide some additional information), there's usually another feature called Privacy Release. By pushing this button on the phone, other people who have the same extension button or intercom button on their phones can then push their buttons and join the conversation. Or you can bring them into your conversation by dialing them in. Most modern key systems come with Automatic Privacy. Many people don't like it, especially those who live in small offices. Some newer phone systems are coming standard without it. And you have to program it in, if you want it.

automatic program load APL is a PBX feature that allows it to load its own software into RAM from a local device such as a hard disk or a floppy disk. All this takes place automatically without human intervention. APL is an important feature since it often determines how fast a PBX can get back into service after some sort of failure – usually a failure in commercial power.

automatic protection switching 1. APS. The ability of a network element to detect a failed working line and switch the service to a spare (protection) line. 1+1 APS pairs a protection line with each working line. 1:n APS provides one protection line for every n working lines.

2. Switching architecture designed for SONET to perform error protection and network management from any point on the signal path.

automatic queuing Queuing is exactly as it sounds. Something you want is being used. So you get placed in line for that device. There are two types of queuing – automatic and manual. Manual is when you're put in queue by a person, for example an operator. Automatic is when you're put in queue by a machine, for example a PBX aided by its software.

automatic recall 1. A central office feature which gives telephone subscribers the ability to automatically redial their last incoming call – without actually knowing that number. On some central offices it is now possible for the calling person to block the ability of someone they've called to automatically call them back. The service is typically known as star sixty-nine – i.e. *69.

2. A PBX feature which returns a call to the PBX attendant (or alerts the attendant) if a call extended to a telephone is not answered within a pre-set period of time. The most logical time is three rings, or 18 seconds. This feature allows the attendant to give the caller some information, take a message or connect the caller to someone else. Most hotel switches have this feature. And when the call doesn't get answered, the switch sends it back to the operator. The sad thing is that the hotel operator is usually so busy, he/she keeps you waiting another 20 or 30 seconds, irritating you.

automatic recharge You give your credit card to a long distance phone company. Instead of charging your card at the end of each month for your usage, they charge you upfront, say $25. When you've used up $20 of calls, they then charge you another $25. This is called automatic recharge. This method is also used by highways for collecting of tolls, for example EZPass in New York surrounding states.

automatic recovery Your telephone system dies – typically because its power is cut off. Once the power comes on, instructions in the machine direct it to reload its software so that within minutes the system can be back and running normally. Those "instructions" are normally not affected by power drops.

automatic redial A telephone capable of detecting a busy signal and redialing the call until a connection is obtained. Often called "Demon Dialer" after a consumer unit made for telephones. See also Automatic Recall.

automatic replication Also known as data replication. It refers to the process of automatically duplicating and updating data in multiple computers on a network. The word "automatically" in this case means that the process of duplication is handled without human intervention by the software responsible. The idea is that if one or several of the machines go down, the company using the data will still have reliable data.

automatic rerouting This refers to the process by which an intelligent voice or data network can automatically route a call, or virtual circuit, around a network failure. With frame relay, PVCs represent a fixed path through the network. However, in the event of a network failure along the primary path over which the PVC is routed, the PVC will be automatically routed to a secondary network path until the primary path is physically restored.

automatic restart 1. The process by which a mechanism automatically restarts after power fails. Restart is from the exact point of interruption.

2. System facilities that allow restart from the point of departure after system failure.

automatic repeat request. See ARQ.

automatic request for retransmission. See ARQ.

automatic retransmission request See ARQ.

Automatic Ring Down ARD. A private line connecting a station instrument in one location to a station instrument in a distant location with automatic two-way signaling. The automatic two-way signaling used on these circuits causes the station instrument on one end of the circuit to ring when the station instrument on the other end goes off-hook. This circuit is sometimes called a "hot-line" because urgent communications are typically associated with this service. ARD circuits are commonly used in the financial industry. May also have one way signaling only.

automatic ringdown tie trunk A direct path signaling facility to a distant phone. Signaling happens automatically when you lift the receiver on either phone. See also Manual Ringdown TIE Trunk.

automatic rollback A feature of the Transaction Tracking System (TTS) that returns a database on a Novell NetWare local area network to its original state. When a network running under TTS fails during a transaction, the database is "rolled back" to its most recent complete state. This prevents the database from being corrupted by the incomplete transaction.

Automatic Route Optimization 1. See Automatic Route Selection.

2. In some switching systems, may denote added ability to adjust routing depending on traffic load or time of day.

automatic route selection Your phone system automatically chooses the least costly way of sending a long distance call. See Least Cost Routing and Alternate Routing.

automatic routing management Formerly AutoRoute. The connection-oriented mechanism used in Cisco WAN switches to provide connectivity across the network. Switches perform a connection admission control (CAC) function on all types of connections in the network. Distributed network intelligence enables the CAC function to route and re-route connections automatically over optimal paths while guaranteeing the required QoS.

Automatic Scheduled Testing AST. A method of testing switched access service (Feature Groups B, C, and D) where the customer provides remote office test lines and 105 test lines with associated responders or their functions' equivalent; consists of monthly loss and C-message noise tests and annual balance test.

Automatic Secure Voice Communications Network AUTOS-EVOCOM. A worldwide, switched, secure voice network developed to fulfill Department of Defense (DoD) long-haul, secure voice needs.

Automatic Send/Receive Automatic Send Receive, or ASR. A teletype or telex machine manufactured by the Teletype Corporation. See ASR for more detail.

automatic sequential connection A service feature provided by a data service to connect automatically, in a predetermined sequence, the terminals at each of a set of specified addresses to a single terminal at a specified address.

automatic set relocation A phone system feature which allows a telephone to retain its personal and system programming when it is reconnected to another physical location.

Automatic Slope Control ASC. Circuitry that permits amplifier response compensation for varying slope, or tilt, at its input.

Automatic Spanning Tree See AST.

Automatic Speech Recognition ASR. See Speech Recognition.

automatic speed matching The ability of an asynchronous modem to auto-matically determine whether it is expected to communicate at 300, 1200 or 2400 bps.

Automatic Time-Out On Uncompleted Call A PBX feature. If a phone stays "off-hook" without dialing for a predetermined time interval, or stays connected to a busy signal longer than the predetermined time interval, the intercom switching equipment will automatically connect this phone to intercept.

automatic toll ticketing A system which makes a record of the calling phone number, the called number, the time of day, the length of the call, etc. and then generates an instant phone bill for that call. Often used in hotel/motels.

Automatic Traffic Overload Protection ATOP. A Rolm feature defined as a dynamic form of line-load control, which automatically denies a dial tone during those periods when the Rolm CBX may become overloaded. One wonders why someone would create this feature.

automatic trunk A dial connection whose destination is predetermined, thus a request for service (called a seizure) initiates dialing to the programmed destination.

Automatic Vehicle Location See AVL.

Automatic Voice Network AUTOVON, the principal long-haul, unsecure (meaning it's not secure) voice communications network within the Defense Communications System. See AUTOVON.

Automatic Voltage Regulation AVR. It can be the best friend your phone system or computer has. AVR eliminates the peaks and valleys of power surges and brown outs, providing you with clean, reliable power. Always look for a UPS with Automatic Voltage Regulation (AVR). It can be the best friend your platform has. AVR eliminates the peaks and valleys of power surges and brown outs, providing you with clean, reliable power.

automatic volume control A circuit in a radio receiver; automatically maintains various received transmissions at approximately the same volume.

automatic wakeup The capability for the user to schedule a wakeup call to a predetermined telephone number, either one time or daily.

automatic wakeup service The guest or the operator dials into a machine which records a request for a guest wakeup call the following morning. The auto wakeup machine is a glorified, programmable auto-dial answering machine. The machine is said to save hotels money and make wakeup calls more reliable, and certainly more anonymous.

Automative Gain Control AGC is used to protect a device from optical overload while maintaining bandwidth and sensitivity performance.

Automaton See Bot.

autonomic Of relating to, or controlled by the autonomic nervous system. Acting or occurring involuntarily; automatic; an autonomic reflex. See Autonomic Nervous System.

autonomic computing A term coined by IBM to describe the concept of a self-managing computing system. Imagine a computer system that can look after itself in the same way that our body's autonomic nervous system does, without requiring any conscious effort on our part. The developing technology aims to allow computer systems to self-manage and heal themselves, automatically resolving the problems systems administrators face managing performance, reliability, security, consistency and scalability with hundreds of networked computers with complex tasks and huge data sets. According to Steve Wojtowecz, director of strategy for IBM Tivoli, there are five stages of evolution in autonomic computing:

1) Basic - Systems administrators manually analyze and solve problems.

2) Managed - There are centralized tools, but actions are still taken manually.

3) Predictive - Administrators use automated tools as a cross resource and to make recommendations.

4) Adaptive - An automated tool monitors systems, correlates and takes corrective action.

5) Autonomic - the point at which administrators will simply need to set management rules for business policy and let the system take care of itself.

autonomic nervous system That part of the nervous system that governs involuntary body functions like respiration and heart rate.

autonomous confederation A group of autonomous systems that rely on their own network reachability and routing information more than they rely on that received from other autonomous systems or confederations.

Autonomous System Number ASN. A unique number assigned by the InterNIC that identifies an autonomous system in the Internet. ASNs are used by routing protocols (like BGP – Border Gateway Protocol) to uniquely define an autonomous system.

autopartitioning A feature on some network devices that isolates a node within the workgroup when the node becomes disabled, so as not to affect the entire network or group.

autoreconfiguration The process performed by nodes within the failure domain of a token ring network. Nodes automatically perform diagnostics in an attempt to reconfigure the network around the failed areas. See also Failure Domain.

autoresponder An e-mail that is automatically sent in reply to any e-mail received in a specified mailbox. Also known as a vacation message.

autoscaling A drawing feature that automatically adjusts the axis units of a graph to the minimum and maximum numerical values of a set of data.

autosearch See Recorder.

autoseed A United Kingdom definition. The process of selecting records from a database using pre-defined criteria and allocating the records to outbound or mailing campaigns.

AUTOSEVOCOM The AUTOmatic SEcure VOice COMmunications system of the U.S. Department of Defense (DoD). A worldwide, switched, secure voice network developed to fulfill DoD long-haul, secure voice needs.

AutoSPID See SPID.

autotimed recall When a user places a call on hold and forgets about it, Autotimed Recall will ring that user or the receptionist after a predetermined time. That time is usually programmable. It shouldn't be longer than 30 seconds; otherwise, your customers, sitting endlessly on your eternity hold, will go nuts and go elsewhere.

autotimed transfer This telephone system feature switches unanswered incoming calls to a backup answering position after a predetermined (usually adjustable) interval of time.

autotype protocol Hayes Microcomputer definition for a file transfer protocol which allows the user to automatically "type" a disk file, the clipboard or the contents of Smartcom Editor in either plain text (ASCII) or ANSI.SYS format to a remote computer. Pacing, send lines and await character echo options are provided. If necessary, character set mapping translates between different code pages and systems (Macintosh or Windows text files, for example).

AUTOVON AUTOmatic VOice Network. During the early years of the "Cold War," AUTOVON was built on the foundation of the U.S. Army's three-switch SCAN (Switch Communications Automatic Network). AUTOVON was an international military TTTN (Tandem Tie Trunk Network) that worked much like the PSTN (Public Switched Telephone Network). Under the control of the DOD (Department of Defense), AUTOVON was for unsecure communications between all branches of the military. The topology of the network was highly secret and many of the COs (Central Offices) were in "hardened" buildings, or placed underground, in order that the network could survive a nuclear attack. AUTOVON telephones had an extra column of four buttons on the right side of the touchtone keypad. The "FO" key was for "Flash Override," the "F" key for "Flash," the "I" key for "Immediate," and the "P" key for "Priority" calls. AUTOVON was replaced by the DSN (Defense Switch Network), which is under the control of DISA (Defense Information Systems Agency). See also TTTN.

Auxiliary Disconnect Outlet ADO. A device usually located within the tenant or living unit used to terminate the ADO cable or backbone cable. Source ANSI/TIA/EIA-570-A.

auxiliary equipment See also Peripheral Device or Applications Processor.

auxiliary equipment access The ability of a telephone system to interface with (i.e. talk to) auxiliary equipment such as a paging system or dial dictation system.

auxiliary line A telephone trunk in addition to the main number you rent from the phone company. Phone systems are often equipped for calls to hunt from a busy main number to one or more auxiliary lines (Incoming Service Group, or ISG). For example, the publisher's main office main number is 212-691-8215. But it also has 8216, 8217, 8218 and several unmarked or coded trunks. These are auxiliary lines and they don't receive their own billing or listing from the phone company. Sometimes, people have single line private lines which "appear" on their phone and no one else's. Sometimes they call these auxiliary lines. Sometimes these are called private lines. Sometimes they are also called terminal numbers.

auxiliary network address In IBM parlance, in ACF/VTM, any network address except the main network address, assigned to a logical unit which is capable of having parallel sessions.

auxiliary power An alternate source of electric power, serving as backup for the primary power at the station main bus or prescribed sub-bus. An off-line unit provides electrical isolation between the primary power and the critical technical load; an on-line unit does not. These are government definitions: A Class A power source is a primary power source; i.e., a source that assures an essentially continuous supply of power. Types of auxiliary power service include: Class B: a standby power plant to cover extended outages (days); Class C: a quick-start (10 to 60 seconds) unit to cover short-term outages (hours); Class D: an uninterruptible (no-break) unit using stored energy to provide continuous power within specified voltage and frequency tolerances.

auxiliary ringer This is a separate external telephone ringer or bell. It can be programmed to ring when a line or a telephone, or both, ring or when Night Service is turned on.

auxiliary service trunk groups A category of trunk groups that provides selected services for customer or operators and terminates on announcement systems, switchboards, or desks. Examples include Directory Assistance, Intercept, Public Announcement, Repair Service, Time, and Weather.

auxiliary storage A mass storage device capable of holding a larger amount of information than the main memory (i.e. RAM) of the computer or telephone system, but with slower access time. Examples include magnetic tape, floppy disks, etc.

auxiliary work state A call center term. An agent work state that is typically not associated with handling telephone calls. When agents are in an auxiliary mode, they will not receive inbound calls.

AUU ATM User-to-User.

availability The amount of time a computer or a telephone system is available for processing transactions or telephone calls. Here's a more technical definition: The ratio of the total time a functional unit is capable of being used during a given interval to the length of the interval; e.g., if the unit is capable of being used for 100 hours in a week, the avail-

ability is 100/168. Contrast this with the term Reliability.

In SONET, the basic Bellcore reliability criterion is an end to end two way availability of 99.98% for interoffice applications (0.02% unavailability or 105 minutes/year down time). The objective for loop transport between the central office and the customer premises is 99.99%. For interoffice transport the objective refers to a two way broadband channel, e.g. SONET OC-N, over a 250 mile path. For loop applications the objective refers to a two way narrowband channel, e.g. DS0 or equivalent. See Reliability.

availability reports Availability reports show how often and for how long nodes, links or paths were unavailable due to outages between specified dates. They can be used to monitor network reliability and to calculate rebates for users.

available In automatic call distribution language, an agent state, between calls, when an agent, having finished the previous transaction, returns to accept the next inbound caller. See also Availability.

Available Bit Rate ABR. An ATM level of service that adjusts bandwidth according to congestion levels in the network. It does not guarantee a specific amount of bandwidth, and the end station must retransmit any information that did not reach the far end.

available channel In the CDPD cellular mobile system, a Radio Frequency (RF) channel is available if it is an allocated channel that is not currently in use for either Cellular Digital Packet Data (CDPD) or non-CDPD-based wireless packet data service.

available line In voice, video, or data communications, an available line is a circuit between two points that is ready for service, but is idle. Pretty obvious.

available state According to Bellcore, an available circuit state occurs when all of the following are true:

1. The Bit Error Ratio (BER) is better than 1 in 10 to the nth power for a specific number of consecutive observation periods of fixed duration.

2. Block Error Ratio (BLER) is better than 1 in 10 to the nth power under the same conditions.

3. There are a specific number of consecutive observation periods of fixed duration without a severely errored unit time.

available time From the point of view of a user, the time during which a functional unit can be used. From the point of view of operating and maintenance personnel, the available time is the same as the uptime, i.e., the time during which a functional unit is fully operational.

avalanche multiplication A current-multiplying phenomenon that occurs in a semiconductor photodiode that is reverse-biased just below its breakdown voltage. Under such a condition, photocurrent carriers, i.e., electrons, are swept across the junction with sufficient energy to ionize additional bonds, creating additional electron-hole pairs in a regenerative action.

Avalanche Photo Diode APD. A fiber optic transmission device. A light detector that generates an output current many times the light energy striking its face. A photodiode that shows gain in its output power compared to the optical power that it receives through avalanche multiplication (signal gain) of the current that flows through a photosensitive device. This type of diode is used in receivers requiring high light sensitivity. See APD.

avalanching The process by which an electrical signal is multiplied within a device by electron impact ionization.

avatar 1. A graphical icon that represents a real person in a virtual reality system. When you enter the system, you can choose from a number of fanciful avatars. Sophisticated 3D avatars even change shape depending on what they are doing (e.g., walking, sitting, etc.).

2. A common name for the superuser account on UNIX systems. The other common name is root.

Avaya Communication Previously the Enterprise Networks Group of Lucent Technologies (originally Western Electric, the manufacturing arm of the AT&T Bell System), the spin-off of Avaya was officially announced July 3, 2000. Avaya manufactures voice, converged voice and data, customer relationship management, messaging, multi-service, networking, and structured cabling products and services. Lucent Technologies, minus Avaya, designs and delivers the systems, software, silicon and services for network service providers and large enterprises. "Avaya" is a made-up name, which is known as an "empty vessel" in the naming business. See also Lucent Technologies.

AVBO Advanced Voice Busyout. The local voice busyout feature that provides a way to busy out a voice port or a DS0 group (time slot) if a state change is detected in a monitored network interface (or interfaces). When a monitored interface changes to a specified state, to out-of-service, or to in-service, the voice port presents a seized/busyout condition to the attached PBX or other customer premises equipment (CPE). The PBX or other CPE can then attempt to select an alternate route. AVBO adds the following functionality to the local voice busyout feature:

1. For Voice over IP (VoIP), monitoring of links to remote, IP-addressable interfaces by the use of a real time reporter (RTR).

2. Configuration by voice class to simplify and speed up the configuration of voice busyout on multiple voice ports.

3. Local voice busyout is supported on analog and digital voice ports using channel-associated signalling (CAS).

AVC Automatic Volume Control. In radio it maintains constant sound level despite undesired differences in strength of incoming signal.

AVD Alternative Voice Data or Alternating Voice Data. AVD used to mean a voice circuit that would also handle data. See AVD Circuits. Now AVD means it's a normal switched analog phone line connected to which is a AVD modem, which lets you transfer bursts of fax, data and images between your voice conversation. The protocol mostly used is VoiceView. See AVD Circuits and DSVD.

AVD Circuits Alternate Voice Data Circuits. Telephone lines which have been electrically treated to handle both voice and data signals. Typically used on leased overseas circuits to save money. See AVD.

average The average and the mean are the same. What's different is the median. See Mean for a full explanation.

Average Business Day ABD. The sum of the busy hour-data (usage, peg count, or overflow) recorded for each of the five busiest days of the week and divided by the total number of days being reported during any given basic data service month. Traditionally, Monday through Friday are the five busiest days of the week. However, traffic characteristics in a particular central office may produce high loads during the weekend. This average (5 days) figure is used in selecting the busy season months and the average busy season value.

Average Busy Season ABS. Average Busy Season is the month's (normally three but not necessarily consecutive) with the highest average busy hour CCS per Network Access Line (NAL). Research suggests that these calling volumes are highly stable and thereby extremely predictable.

average call duration Divide the total number of minutes of conversation by the number of conversations. Bingo, that's your average call duration.

average customer arrival rate Represents the number of entities (humans, packets, calls, etc.) reaching a queuing system in a unit of time. This average is denoted by the Greek letter lambda. One would prefer to know, if possible, the full distribution of the calls arriving.

average delay The delay between the time a call is answered by the ACD and the time it is answered by a person. This typically includes time for an initial recorded announcement plus time spent waiting in queue. Average delay can be chosen as the criterion for measuring service quality.

Average Handle Time AHT. The period of time an employee is occupied with an incoming call. This is the sum of talk time and after-call-work time.

average holding time The sum of the lengths (in minutes or seconds) of all phone calls during the busiest hour of the day divided by the number of calls. There are two definitions. The one above refers to average speaking time (it's the more common one). There's a second definition for "average holding time." This refers to how long each call was on hold, and thus not speaking. This second definition is typically found in the automatic call distribution business (ACD). Check before you do your calculations.

average latency The time required for a disk to rotate one-half revolution.

average load The traffic load obtained by averaging a series of hourly loads. An average load may be further defined as average carried load, average offered load, etc. This term is not to be confused with load that is inherently an average of all the instantaneous loads over a basic time interval such as an hour.

average pulse density In T-1 bipolar transmissions, refers to the number of "1" pulses per "0" conditions and is usually tied to a maximum number of "0"s in a row (i.e., FCC Part 68 requires 12.5% pulse density and no more than 80 consecutive "0"s where as AT&T Pub 62411 uses a formula and no more than 15 consecutive "0"s).

Average Speed Of Answer ASA. How many seconds it takes an operator on average to answer a call.

Average 10-High Day ATHD. A mathematical average of the data generated during the 10-high day busy hour. This is the same hour for all ten days and generally will occur during the busy period. However, predictable recurring heavy traffic days which occur outside the busy period should be included.

Average Picture Level APL. In video systems, the average level of the picture signal during active scanning time integrated over a frame period. It is defined as a percentage of the range between blanking and reference white level.

average rate Average rate, in kilobits per second (kbps), at which a given virtual circuit can transmit.

average rate of transmission See Effective Transmission Rate.

Average Talk Time ATT. The average length of a complete customer transaction over the phone.

average transfer delay Average time between the arrival of a packet to a station interface and its complete delivery to the destination station.

AVI Audio Video Interleaved. File format for digital video and audio under Windows. Use the "Media Player," which comes with Windows, to play AVI files. The AVI file format is cross-platform compatible, allowing .AVI video files to be played under Windows and other operating systems. AVI is not a definition on compression but a rule on file formatting. It is primarily used with the Microsoft Windows operating systems. Its main problem is its huge file size, i.e. it encodes analog video into a huge digital file that eats up your hard disk. Therefore several companies have developed their own procedure for compressing AVI files. MJPEG, Cinepark, and Indeo from Intel are some of the popular methods used to compress AVI files. They use the "intracoding" method for compression and are widely used to edit video images frame by frame. But they are inferior to MPEG in terms of compression ratio, image quality and compatibility. MPEG is an abbreviation of Moving Picture Experts Group, one of the international standard organizations. MPEG is a rule concerning file format and compression method. Since it is an international standard, it is compatible with several operating systems including Windows. Many consumer devices such as CD player and DVD player support MPEG-1 and MPEG-2 formats. This makes MPEG easy to use. The MPEG compression ratio is very high because it uses intercoding, a superior compression technique to intracoding. See MPEG.

AVK Audio Video Kernel. DVI (Digital Video) system software designed to play motion video and audio across hardware and operating system environments.

AVL Automatic Vehicle Location – not an elegant name, but an umbrella description for the fleet management version of mobile telematics, which involves integrating wireless communications and (usually) location tracking devices (generally GPS) into automobiles. The best known example of mobile telematics is GM's OnStar system, which automatically calls for assistance if the vehicle is in an accident. These systems can also perform such functions as remote engine diagnostics, tracking stolen vehicles, provide roadside assistance, etc. www.onstar.com. Best known operators of AVL services (mobile telematics for fleets) are Qualcomm's OmniTRACS division and Teletrac. They generally involve integrating wireless communications and location sensing technology (frequently GPS) into commercial vehicles, to allow mobile communications, automated dispatching, cargo tracking, etc.

avoidable costs A wonderful concept used by the regulated telephone industry. It refers to those costs which would be avoided (i.e. not incurred) if the service were not offered. Examples of costs to avoid are maintenance, taxes, labor, and other direct costs. The concept of Avoidable Costs is to allow the phone industry the justification to price a competitive service very low.

avoidance routing The assignment of a circuit path to avoid certain critical or trouble-prone circuit nodes.

avoirdupois A pound of feathers weighs more than a pound of gold. Feathers are weighed by "avoirdupois" weight measure, which has 16 ounces to a pound, while gold is weighed in "troy" measure, which only has 12 ounces to a pound.

AVR See automatic voltage regulation.

AVRS Automated Voice Response System. See IVR and Voice Response System.

AVSS Audio-Video Support System. DVI (Digital Video) system software for DOS. It plays motion video and audio.

AW 1. Administrative Weight. The value set by the network administrator to indicate the desirability of a network link. One of four link metrics exchanged by PTSPs to determine the available resources of an ATM network.

2. Admin Workstation. A personal computer used to monitor the handling of calls in the ICM system. The AW also can be used to modify the system configuration or scripts.

AWC Area-Wide Centrex.

AWG American Wire Gauge. Originally known as the Brown and Sharpe (B&S) Wire Gauge. The U.S. standard measuring gauge for non-ferrous conductors (i.e., non-iron and non-steel), AWG covers copper, aluminum, and other conductors. Gauge is a measure of the diameter of the conductor. The AWG numbering system is retrogressive (i.e., backwards): The higher the AWG number the thinner the wire. This is due to the fact that the AWG number originally indicated the number of times the copper wire is drawn through the wire machine to reduce its diameter. For example, a 24-gauge wire was drawn through the wire machine 24 times; therefore, it is thinner than a 22-gauge wire, which was drawn through the wire machine only 22 times. A 24-gauge (AWG) wire has a diameter of .0201 in.

(.511 mm), a weight of 1.22 lbs/ft (1.82 kg/km), a maximum break strength of 12.69 lbs (5.756 kg) and D.C. resistance ohms of 25.7/1000 ft (84.2 km). For example, heavy industrial electrical wiring may be No. 2; homes are typically wired with No. 12 or No. 14. Telephone systems typically use No. 22, No. 24 or No. 26. The thicker the wire, the more current it can carry farther without creating heat and without suffering attenuation (signal or power loss) due to resistance. You need thicker phone cabling when your phones are farther away. Also, you need thicker wire when transmitting at higher frequencies such as would be the case in a data application (e.g. 10/100Base-T); high-frequency signals attenuate to a greater extent than do low-frequency signals. Some vendors save money by installing systems with thin wire. Make sure you specify.

AWM Appliance Wiring Material.

AX.25 An amateur radio implementation of the X.25 protocol. Used by some private VANs (Value Added Networks) to avoid PTT (Post, Telephone and Telegraph Administration) monopolies (and thus high prices) on X.25 transmission and switching.

axial propagation constant In an optical fiber, the propagation constant evaluated along the optical axis of the fiber in the direction of transmission. Note: The real part of the axial propagation constant is the attenuation constant. The imaginary part is the phase constant.

axial ratio Of a wave having elliptical polarization, the ratio of the major axis to the minor axis of the ellipse described by the tip of the electric field vector.

axial ray A ray that travels along the axis of an optical fiber.

axial slab interferometry Synonym for Slab Interferometry.

axis The center of an optical fiber.

AXT Abbreviation for alien crosstalk.

AZ/EL Mount Antenna mount that requires two separate adjustments to move from one satellite to another.

azimuth The horizontal angle which the radiating lobe of an antenna makes in angular degrees, in a clockwise direction, from a north-south line in the northern hemisphere. In the southern hemisphere, the reference is the south-north line. Azimuth actually involves a lot more than antennas. For example, it covers the alignment of a recording head in a tape recorder.

azimuth beam width The angular measurement of an antenna pattern as viewed along the horizon. It is sometimes called the "horizontal beam width."

B 1. A capital B stands for Byte. A small b stands for bit. Typically a byte is eight bits (but it could be more or fewer). Virtually all telecommunications transmission – data, voice or video, etc. – is stated in bits per second (bps, or b/s). Virtually all transmission inside a computer (like movement between a disk and the computer's central microprocessor is written in bytes per second. Thus Mbps would be million bits per second, while MBps would be million bytes per second. For a much longer explanation, see the introduction. See also Bit and Byte.

2. The local wireline cellular carrier. In one of its less intelligent decisions, the Federal Communications Commission decided to issue two cellular franchises in each city of the United States. They gave one to the local phone company (the B carrier) and one to a competitor (the A carrier). This duopoly has naturally meant little real price competition. At least it did until the FCC issued PCS (Personal Communications Services). licences and competition started. See PCS.

3. beta. As in beta test. A beta test is a test of a product in a real production environment with a real, live customer. Beta tests come after alpha tests, and before the general release of the product.

B 1 See B-1.

B Battery A section of a phone system power supply that provides unfiltered Direct Current for operating relays and various other components. Typically 20 volts. See A Battery.

B Bit See A & B Bits.

B carrier A cellular carrier operating in the 869-894 MHz range. In U.S. markets that have only two cellular carriers, one is designated the A carrier and the other the B carrier, which operates in the 869-894 MHz range. The " B" band license was generally awarded to the wireline telephone operator in the market and, hence, the B Carrier was historically referred to as the "wireline" license.

B channel B channel digital channels carrying PCM voice or other services in a TDM signal, such as European ISDN at 2.0489 million bits per second or American ISDN at 1.544 million per seconds. The designation "B" has no connection with any characteristics of the channel. In the early days of the development of digital telephony, a need arose to distinguish it from analog telephony. As analog telephony was oldest and the word analog begins with the letter "a," it seemed logical to use a straight alphabetical approach. The analog channel became the A Channel, with a capital "A." Consequently the new digital channel was called the B Channel, which seemed logical, as "B" could mean "Binary" or "Bits." The signaling channel, far smaller than the collected PCM channels it served, was seen as an increment. So the mathematical symbol for an increment, the Greek letter "delta," was used, and the signalling channel became the delta Channel. This was long before word processing became available, and typewriters of the time usually did not have Greek letters. So the Roman alphabet capital "D," the equivalent of the Greek capital delta

was used, and the signalling channel became the D Channel. Consequently, the European 2.048 million bits per second bearer channel often is called the 30B+D and the American 1.544 million bit per second bearer channel is called the 23B+D. See analog, Basic Rate Interface (BRI) and ISDN.

B connector A commonly-used wire-splicing device consisting of a flexible plastic sleeve over a toothed metal cylinder that bites through insulation when crimped with pliers or a special crimping tool. It is about one inch long and can hold three or four wires. A gel-filled version (water-retardant jelly) is available for installation in damp or humid areas. B connectors are also known as chiclets, beans, beanies, and rodent rubbers.

B frame Bi-directional or B frames (often called B pictures) refer to part of the MPEG video compression process whereby both past and future pictures / frames are used as references. B frames typically produce the most compression.

B interface An interface used in Cellular Digital Packet Data (CDPD) which is deployed over AMPS. The B interface connects the Mobile Data Intermediate System (MD-IS) to the Mobile Data Base System (MDBS).

B links A SS7 term. Bridge links assigned in a quad arrangement. Bridge Link. A CCS/SS7 signaling link used to connect STP (Signal Transfer Point) pairs that perform work at the same functional level. These links are arranged in sets of four (called quads). A minimum of three-way path diversity is recommended to allow three completely separate paths. The B-Links are assigned so that local Signal Transfer Points (STPs) can communicate with other local STPs or mated pairs of STPs at the same hierarchical level: local to local, regional to regional, or mated pair to mated pair.

B number A cellular term for the number of the called party. The originating switch analyzes the A Number (the number of the calling party) in order to route the call to the B Number.

B port The port which connects the outgoing primary ring and the incoming secondary ring of the FDDI dual ring. This port is part of a dual attached station or concentrator. See FDDI.

B&S Brown and Sharp Gauge. See AWG (American Wire Gauge).

B-1 An unassociated central office line or a central office line that is not in rotary hunt that is used for a business line or a business fax/modem.

B-911 Basic 911. There is also an E-911, which stands for Enhanced 911. B-911 is a centralized emergency reporting system which may have many features but which does NOT provide ALI (Automatic Location Information) to the 911 operator. In most cases, it does not provide ANI (Automatic Number Identification) either. B-911 provides a common emergency response number and relies on Emergency Hold and Forced Disconnect to maintain effective service. See 911 for a full explanation.

B-Block Cellular licenses received from the FCC with an initial association to a telephone company. Also referred to as wireline.

B-Block Carrier A 30-MHz PCS carrier serving a Major Trading Area (MTA) in the frequency block 1870-1885 MHz paired with 1950-1965 MHz.

B-CDMA Broadband Code Division Multiple Access.

B-Crypt A symmetric cryptographic algorithm designed by British Telecom.

B-DCS Broadband Digital Cross-connect System. B-DCS is a generic term for an electronic digital cross-connect system capable of cross-connecting signals at or above the DS3 rate.

B-ICI B-ISDN Inter-Carrier Interface: An ATM Forum defined specification for the interface between public ATM networks to support user services across multiple public carriers.

B-ICI SAAL B-ICI Signaling ATM Adaptation Layer: A signaling layer that permits the transfer of connection control signaling and ensures reliable delivery of the protocol message. The SAAL is divided into a Service Specific part and a Common part (AAL5).

B-ISDN Broadband ISDN. A very vague term that defines a communications channel as anything larger than a single voice channel of 64 Kbps. Under this terminology, "broadband" can be as little as two voice channels. The official ITU-T recommendation I.113 [45] defines Broadband ISDN as "a service or system requiring transmission channels capable of supporting (transmission) rates greater than the primary rate [DS1]." Thus, broadband ISDN is a new concept in information transfer, although exactly what it is isn't clear yet. There is some discussion that broadband ISDN begins at 155 Mbps (the OC-3 version of SONET/SDH). In another ITU-T recommendation (I.121 [47]), the ITU-T presents an overview of what it sees as B-ISDN capabilities:

"B-ISDN supports switched, semipermanent and permanent point-to-point and point-to-multipoint connections and provides on demand, reserved and permanent services. Connections in B-ISDN support both circuit mode and packet mode services of a mono- and/or multi-media type and of connectionless or connection-oriented nature and in a bidirectional or unidirectional configuration. A B-ISDN will contain intelligent capabilities for the purpose of providing advanced service characteristics, supporting powerful operation and maintenance tools, network control and management." The ITU-T (nee ITU) was to decide on an international standard for B-ISDN by 1996, although it has not yet been released. The ITU-T has defined two types of services in the context of B-ISDN, Interactive and Distribution. For a definition of those services, See Interactive Services and Distribution Services.

Bellcore says that "National and international standards bodies have made the Asynchronous Transfer Mode (ATM) the target solution for providing the flexibility required by B-ISDN. ATM provides a common platform capable of supporting both broadband and narrowband services ... The physical layer-transmission standard for B-ISDN is the Synchronous Optical Network (SONET), also known as the Synchronous Digital Hierarchy (SDH)." See N-ISDN and SONET.

B-ISDN Inter-Carrier Interface An ATM Forum specification for a public UNI (User Network Interface) providing the interface via PVCs between public ATM networks to support user services across multiple public carriers. See UNI.

B-ISUP Broadband Integrated Services Digital Network user part. See ISDN and ISUP.

B-LLI Broadband Lower Layer Information: This is a Q.2931 information element that identifies a layer 2 and a layer 3 protocol used by the application.

B-PCS Broadband Personal Communications Services. The FCC designated 140 MHz in the 1850-1990 MHz range for PCS services. The B-PCS auctions ended in 1995, yielding over $7 billion in revenues to the U.S. government. B-PCS is intended to support features such as "follow-me." See also 500 Service and PCS.

B-Picture Bi-directionally predictive-coded; an MPEG term for a picture that is coded using compensated prediction from a past and/or future reference picture.

B-TA Broadband Terminal Adapter. A form of DCE which provides the interface into a B-ISDN network from B-TE2, which is Terminal Equipment not compatible with the B-ISDN network.

B-TE Broadband Terminal Equipment: An equipment category for B-ISDN which includes terminal adapters and terminals.

B-WLL See Broadband Wireless Local Loop.

B/I Busy/Idle bits.

B2B Business to Business. A term that defines a company selling its wares to other businesses and buying its wares from other businesses, as against selling its wares to consumers, which is B2C. One of the strongest B2B areas is setting up systems for companies to buy their supplies from other companies, using the Internet to streamline the process. Here is a neat piece from Ariba, one of the leaders in this field: In today's economy, where the new Internet rules apply, how can companies ensure that they will be winners? The answer is to leverage the Internet to gain a closer alignment with customers and partners, to integrate supply chains, and to take advantage of new revenue opportunities. Only by embracing the change and using it for competitive advantage will the bricks-and-mortar companies of today become the dot-com companies of tomorrow. Today, eCommerce is

about using the Internet to gain competitive advantage, not only by improving the efficiency and effectiveness of existing business processes, but also by enabling constant adaptation to rapidly changing competitive landscapes.

Despite these imperatives for moving business on-line, today a single corporate purchase is still quite cumbersome and costly. AMR estimates that the cost per procurement transaction ranges from US$75 to US$175, often exceeding the cost of the items being purchased. This staggering fact results from most transactions happening via a complex, paper-intensive request, approval, and order process with high administrative costs and no economies of scale. These processes often include the re-keying of information, lengthy approval cycles, and significant involvement of financial and administrative personnel, often resulting in delays to end-users and productivity losses. Beyond the time and expense associated with manual processing costs, organizations suffer even greater costs when they cannot fully leverage procurement economies of scale. Most organizations lack the systems that enable them to monitor purchases and compile data necessary to negotiate better volume discounts with preferred suppliers. In addition, most organizations suffer from a problem known as "maverick buying," which occurs when personnel do not follow internal guidelines as to which suppliers to use for operating resource purchases. When employees do not buy from preferred suppliers, organizations pay a premium. AMR estimates that maverick buying accounts for one-third of operating resource expenditures, costing organizations a 15% to 27% premium on those purchases. In addition, low visibility into spending patterns and limited information further limit the benefits of volume procurement contracts.

Traditional procurement processes also result in missed revenue opportunities and additional costs to suppliers. When buyers are unable to channel purchases to preferred suppliers, these suppliers lose revenue. Suppliers also suffer from inefficient, error prone and labor-intensive order fulfillment processes. Many suppliers dedicate significant resources to the manual entry of information from faxed or phoned-in purchase orders and the manual processing of paper checks, invoices and ship notices. Suppliers also spend significant resources on customer acquisition and sales costs, including the production and distribution of paper catalogs. Without fully automated and integrated electronic commerce technologies, both buyers and suppliers incur substantial extraneous costs in conducting commerce.

Business-to-business eCommerce addresses these issues by providing a modern, electronic infrastructure that allows corporations to manage their internal and external procurement processes strategically. Business-to-business eCommerce leverages technology, lowering transaction costs through electronic commerce, automation, and use of decision support techniques to identify opportunities to rationalize the supply chain. By adopting business-to-business eCommerce, companies can build strategic supplier relationships and use aggregate buying to gain volume discounts from suppliers.

B2C Business to Consumer. A term that defines a company selling its wares to consumers, as against selling its wares to other businesses, which is B2B. See B2B.

B3ZS Bipolar with 3-Zero Substitution. The line coding technique used in the SONET STS-1 (Synchronous Transport Signal-Level 1) electrical signal, which is then converted to an optical signal for transmission over the SONET optical fiber transmission system. B3ZS looks for a series of 3 consecutive zeros, removes them, and replaces them with either BOV or OOV. The choice between BOV and OOV is such that the number of B pulses between consecutive V pulses is odd, where B represents the normal bipolar pulse and V represents a bipolar violation. See also B8ZS, Bipolar Violation, SONET and STS.

B7Sub A reader asked us if we knew what this term meant. Ray Horak, who's the best, emailed me, "I can find only three mentions of this anywhere, and no details. I suspect that it is an example of a "jam bit 7" process. This is an earlier, nonstandard feature that was used in DDS channels and channel banks, again, to preserve 1's density. The deal was that if the DS-0 carried a DDS formatted data channel, the least significant bit was a signaling bit and could often be 0. If the "data" were all 0's, then the entire byte could be all 0's. The third time in a row that an all-0 byte came up it would violate the 15-0's limit, so the channel bank would "jam" a 1 into the least significant bit (that is bit 7, if you count 0-7, which is the way we count them, rather than 1-8). This didn't affect data (which was 56 kilobits per second in the seven most significant bits) nor the signaling (which required every signal to be repeated multiple (7?) times before it meant anything to the receiver)."

B8ZS Bipolar with 8 Zero Substitution. A technique specified in the ITU-T G.703 recommendation, B8ZS is used in North American T-1 circuits to accommodate the ones density requirement in the public network. B8ZS inserts one of two special violation codes for strings of eight consecutive zero voltage states, with the intentional bipolar violation codes being inserted in bit positions 4 and 7 of the datastream. The bipolar violation codes are of opposite polarity, thereby preventing the buildup of excessive DC voltage levels, with the specific bipolar violation code depending on the polarity of the preceding pulse. B8ZS, which generally is used on newer T-1 and ISDN PRI circuits, offers clear channel communications of 64 Kbps per channel. AMI (Alternate Mark Inversion) the older technique, suffers from a loss of timing recovery when 8 consecutive zeros are transmitted. E-1 standards utilize HDB3 (High Density Bipolar 3) line coding, which also is specified in G.703. See also AMI and HDB3.

BA Business Address.

BAA Blanket Authorization Agreement. Signed by interconnectors guaranteeing that they have authority of the end user customer to request CPNI and place service orders on the customer's behalf. CPNI is Customer Proprietary Network Information. It's information pertaining to a customer's selected service arrangements with a bell operating company. This information is only available to competitors when written authorization from the customer is provided.

Babbage, Charles 1791 to 1871. An English researcher who contributed a great deal to the theory and practice of computing and conceived his now-famous 'analytical engine' – an early computer.

babble Just what it sounds like – crosstalk from several interfering communications circuits or channels.

babbling tributary "A station that continuously transmits meaningless messages," as defined by John McNamara, of DEC and author of "Local Area Networks, an introduction to the technology." Some people might argue this was another word for Harry Newton, the author who didn't know when to stop and wanted to make this dictionary the most comprehensive telecom dictionary ever.

BABT British Approvals Board for Telecommunications. You need their approval before you can sell telecom equipment in Great Britain.

Baby Bells At the beginning of 1984, under a Divestiture Agreement with the Federal Government, AT&T spun off seven operating phone companies of roughly equal size. The phone companies were known officially as the Regional Bell Operating Companies (RBOCs) or the Regional (Bell) Holding Companies (RHCs). Unofficially they were call the Baby Bells. The original seven Baby Bells were Ameritech (bought by SBC), Bell Atlantic (since merged with GTE to form Verizon), BellSouth, NYNEX (since acquired by Bell Atlantic and now called Verizon), Pacific Telesis (since acquired by SBC), Southwestern Bell (now called SBC) and US West (since acquired by Qwest). In early 2005, SBC announced that it would acquire the company that had spawned it, namely AT&T. And in doing so, it would become the largest communications company in the United States. See also AT&T, Divestiture, MCI, Qwest and WorldCom.

These days, Qwest and its siblings don't even like to be called "Baby Bells." They say they have evolved into modern telecommunications companies that offer fiber-optic-based Internet and television services, not just phone connections. "We don't want to be thought of as a local phone company any more," says Janis Manning, assistant general counsel in charge of trademark issues for Verizon Communications Inc.

baby bills 1. A term for the numerous companies formed by ex-employees of Microsoft. A play on the "Baby Bell," the reference is to Bill Gates, co-founder and chairman of Microsoft.

2. A term for the many companies that the Federal Government might split Microsoft into, as a result of the Microsoft antitrust case. It, of course, never happened. the government's case fizzled and Microsoft stayed intact.

babyphone Feature allowing calls to an off-hook telephone to listen to room noises, for example, to check if a baby is crying.

back board A piece of plywood mounted on a wall. Phone equipment is mounted on the plywood. It is more efficient to first mount phone equipment on plywood in the service bay, test it out while it's convenient and diagnostic tools are handy. Then take the phone equipment and the back board (which typically consists of the KSU, power supply and 66-blocks) and install them on the customer's premises. This "pre-installation" makes enormous sense – economically and reliably. Sadly, few installation companies do it.

back box Think of your half-built office or house. The rooms have been framed out. There are metal or wooden studs going from the ceiling to the floor. Soon you will attach sheetrock on either side of the studs and form your finished walls. But before you do, you need to string cables through the walls. In each room you'll want a bunch of electrical outlets. You make those electrical outlets by attaching a a metal or plastic box to the metal or wooden stud. That box is typically three inches or so deep and has an opening that is 2" x 3 1/2." In electrical and telecom industry terminology that box is called a "back box" or a single gang receptable box. By attaching various face plates to it and changing the insides, you can make your back box deliver electricity, cable TV, telephone, local area network and/or broadband service. You can also get back boxes in a double width, which

is known (surprise, surprise) as a double gang receptacle box. In a normal electrical installation a single gang box would have two 120 volt A/C outlets, and a double gang box would have four. In a telecom installation a single gang receptacle could contain as many as six telecom outlets, e.g. a couple of coaxial cables, a couple of phones and a couple of local area network connections for your computer and your networked printer.

back door A way of getting into a password-protected systems without using the password. Usually a carefully guarded secret to prevent abuse and misuse. Undocumented or unpublished access path to either a network or its components, such as routers, firewalls and so on.

back end First we have the front end and then we have the back end. Obviously if you think about either, both terms are totally meaningless. But people pretend they mean something. For example, one competitive dictionary (and a not a good one at that) defines back end as node or software program that provides services to a front end. (You figure what that means.) See also client, FRF.11, and server. Another dictionary defines back end as dDatabase server functions and procedures for manipulating data on a network.

back end processor A server is often called a back end and a workstation is often called the front end. On a LAN (Local Area Network), a back end processor runs on a server. It is responsible for preserving data integrity and handles most of the processor-intensive work, such as data storage and manipulation.

back end results A call center/marketing term. Used to describe the number of trial or risk free orders that actually pay in terms of customer orders. In a telephone sales context, a bad back-end can be the result of the reps selling the trial and not the product. See Front End Results.

back feed pull Used in tight locations where it's difficult to take large cable pulling equipment. The cable is fed in two parts from the mid-point. The first section of cable is fed in one direction. After this is fed, the remaining cable is unreeled and fed through the opposite direction to the other end point.

back haul Back haul is a verb. A communications channel is back hauling when it takes traffic beyond its destination and back. There are many reasons it might do this. The first is that it may be cheaper to go that route instead of going directly. You might, for example, have a full-time private line from New York to Dallas. You might find it cheaper to reach Nashville by going to Dallas first, then dialing back to Nashville. The economics of backhauling may change from one moment to another as the line to Dallas is empty, close to full or full. Another reason for back hauling is that you may do it to accommodate changes in your calling or staffing patterns. You may have an automatic call distributor in Omaha and one in Chicago. A call from New York may come into your Omaha ACD, but when it gets there you may discover that there are no agents available to handle the call. So it may now make sense to back haul the call to the Chicago ACD, where an agent is available. In fiber networks, back hauling is a traffic management technique used to reduce the expense of multiplexing/demultiplexing.

back hoe fade The degradation in service experienced when a backhoe cuts your buried fiber optic cable. Called fade because sometimes not all communications are cut off. Also, when they are all cut off, the term becomes a euphemism. Better to report a back hoe fade to your boss than to say, "We just lost 158,000 circuits between New York and Washington. Our customers are not pleased." See also Backhoe Fade for another variation.

back lobe In a directional antenna, there is a main lobe and there may be additional lobes, one of which extends backward from the direction of the channeled signal, called a back lobe.

back office operations Management and support tasks that can be performed away from a company's headquarters, such as telemarketing, credit card processing, data file maintenance and many clerical and accounting functions. Back office operations are an economic development opportunity for small communities that have the appropriate infrastructure (e.g. advanced telecommunications, reliable express mail services.) Back office operations are helping share a new definition of place, one in which, for example, geographic remoteness is no longer a liability – because of telecommunications and other linkages to the "outside" world.

back porch The portion of a video signal that occurs during blanking from the end of horizontal sync to the beginning of active video. The blanking signal portion which lies between the trailing edge of a horizontal sync pulse and the trailing edge of the corresponding blanking pulse. Color burst is located on the back porch.

back pressure Propagation of network congestion information upstream through an internetwork.

back projection When the projection is placed behind a screen (as it is in television and various video conferencing applications where the image is displayed on a monitor or a fabric screen) it is described as a back projection system. In these systems the viewer sees the image via the transmission of light as opposed to reflection used in front projection systems. Audiences generally prefer back projection systems since they seem brighter.

back scatter An RFID definition. A method of communication between tags and readers. RFID tags using back-scatter technology reflect back to the reader a portion of the radio waves that reach them. The reflected signal is modulated to transmit data. Tags using back scatter technology can be either passive or active, but either way, they are more expensive than tags that use inductive coupling.

back to back channel bank The connection of voice frequency and signaling leads between channel banks to allow dropping (i.e. removing) and inserting (i.e. adding) of channels.

back to back connection A connection between the output of a transmitting device and the input of an associated receiving device. When used for equipment measurements or testing purposes, this eliminates the effects of the transmission channel or medium.

back to back peering Peering is when large ISPs (Internet Service Providers) assume that the traffic is approximately equal between them, and both benefit equally from a free connectivity between them since both companies need each other's network to get people to their web sites. These ISPs allow traffic from these large ISPs to enter their network for free. Companies that are allowed into this prestigious club are called Tier 1 providers. They include Sprint, CAIS, UUNet, PSINet (the first and largest independent commercial ISP in the world), Cable and Wireless, etc. Companies that do not have large enough networks have to pay Tier 1 companies to get access to their networks.

back to square one Back to square one (or "back at square one," which was the original way of saying it) comes from football radio commentaries from the 1930s. There being no picture, these live reports would explain the position of play by dividing the football pitch into numbered grids. Square one was just in front of the goal. So, when a ball went out of play and resulted in a goal kick, the play was "back at square one." Neat.

back up server A program or device that copies files so at least two up-to-date copies always exist.

back-end system The server part of a client/server system that runs on one or more file servers and provides services to the front-end applications running on networked workstations. The back-end system accepts query requests sent from a front-end application, processes those requests, and returns the results to the workstation.

backbone The backbone is the part of the communications network which carries the heaviest traffic. The backbone is also that part of a network which joins LANs together – either inside a building or across a city or the country. LANs are connected to the backbone via bridges and/or routers and the backbone serves as a communications highway for LAN-to-LAN traffic. The backbone is one basis for design of the overall network service. The backbone may be the more permanent part of the network. A backbone in a LAN, a WAN, or a combination of both dedicated to providing connectivity between subnetworks in an enterprise-wide network.

backbone bonding conductor A copper conductor extending from the telecommunications main grounding busbar to the farthest floor telecommunications grounding busbar.

backbone cabling Cable and connecting hardware that comprise the main and intermediate cross-connects, as well as cable runs that extend between telecommunications closets, equipment rooms and entrance facilities.

backbone closet The closet in a building where the backbone cable is terminated and cross connected to either horizontal distribution cable or other backbone cable.

backbone facilities Plant and equipment used to provide transmission services to connect tributary facilities from clusters of dispersed users or devices. See Backbone.

backbone network The part of a communications facility that connects primary nodes; a primary shared communications path that serves multiple users via multiplexing at designated jumping-off points. A transmission facility, or arrangement of such facilities, designed to connect lower speed channels or clusters of dispersed users or devices.

backbone router A high-capacity system that directs and forwards information across a backbone network typically comprising leased lines. Data are directed across the network based on information contained in Layer 3 of the OSI Reference Model. The system can also be used to connect multiple LANs in a corporate campus network.

Backbone Subsystem See Riser Subsystem.

Backbone To Horizontal Cross-Connect BHC. Point of interconnection between backbone wiring and horizontal wiring.

backbone wiring The physical/electrical interconnections between telecommuni-

cations closets and equipment rooms. Cross-Connect hardware and cabling in the Main and Intermediate Cross-Connects are considered part of the backbone wiring.

backchannel A real-time, running online commentary/blog/chat by members of the audience during speeches and other activities at a symposium, conference, etc. Sometimes also spelled as two words: back channel.

backcharging A phone fraud term. Backcharging is starting the clock on a phone call at the time a customer contacts the long-distant phone service provider – not when the person being called answers the phone – which is what it should be.

backdoor A design fault, planned or accidental, that allows the apparent strength of the design to be easily avoided by those who know the trick.

backfeed pull A method used to pull cable into a conduit or a duct liner when the cable is long or when placing cable into controlled environmental vaults, central offices, or under streets. With this method, the cable pays off its reel at an intermediate manhole and is first pulled in one direction. The remaining cable is then removed from the reel, laid on the ground, and then pulled in the opposite direction.

backfile conversion The process of scanning in, indexing and storing a large backlog of paper or microform documents in preparation of an imaging system. Because of the time-consuming and specialized nature of the task, it is generally performed by a service bureau.

backfilling To designate memory on an expanded memory card and make it available for use as conventional memory.

background See Background Processing.

Background Area of Concern, Consequence and Incentive BACI. A questioning strategy used by Lucent Technologies for uncovering a customer's implied needs and converting them to clearly defined ones that may lead to a purchasing decision.

background communication Data communication, such as downloading a file from a bulletin board, that takes place in the background while the user concentrates on another application (e.g. a spreadsheet) in the foreground.

background music This feature allows music to be played through speakers in the ceiling and/or through speakers in each telephone, throughout the office, or office-by-office, or selectively. Background music is typically played through paging speakers, but it can also be played through the speakers of speakerphones. In fact, the two – paging and background music – often go hand-in-hand. When you want to page someone, the music turns off automatically and comes back on when the paging is over. The same thing happens on airplanes. Background music is said to motivate workers, often into shutting it off.

background noise The noise you hear when nothing else is being transmitted. Digital circuits are so quiet that some form of White Noise must be injected into them so as to prevent people from suspecting that the circuit they're speaking on has gone dead. See also White Noise.

background noise regeneration A technique for eliminating background noise bandwidth requirements.

background processing The automatic execution of lower priority computer programs when higher priority programs are not using the computer's resources. A higher priority task would be completing calls. A lower priority task would be running diagnostics. Some PBXs have this feature. Some insist on running their diagnostics even though they are choked with calls. The smarter ones tone down their diagnostics when they get busier, which makes sense.

background program A low priority program operating automatically when a higher priority (foreground) program is not using the computer system's resources.

background task A secondary job performed while the user is performing a primary task. For example, many network servers will carry out the duties of the network (like controlling who is talking to whom) in the background, while at the same time the user is running his own foreground application (like word processing). See also Background Processing.

background tone Music that plays in the background while you are talking on your cell phone. It's a new trend in India, where a background tone may also be a religious chant or devotional message.

backhaul See Back Haul.

backhoe An excavator whose bucket is rigidly attached to a hinged pole on the boom and is drawn backward to the machine when in operation.

backhoe fade Signal loss caused by some moron who forgot to call before he dug. Usually involves a fiber cut about the size of a backhoe blade. Typically affects a large number of very peeved customers and a small number of overworked Network Operations folks supported by the ever-encrusted Cable Cowboys called to the scene. Murphy's Law will ensure that Skillet with the Caterpillar made sure that he pulled the fiber as far out of the hole as possible so that it stretched the jacket inducing NUMEROUS breaks along it's (still-buried) length. Murphy would also state that a flood occurring before repairs are completed at the location is likely as well. Definition courtesy Todd Timlake. See Condescension Fade and Back Hoe Fade.

backoff When a device attempts to transmit data and it finds trouble, the sending device must try again. It may not try again immediately. It may "back off" for a little time so the trouble on the line can be cleared. This happens with LANs. For example, an earlier attempt to transmit may have resulted in a collision in a CSMA/CD (Carrier Sense Multiple Access/Collision Detection) Local Area Network (LAN). So the device "backs off," waits a little and then tries again. How long it waits is determined by preset protocols.

backoff algorithm The formula built into a contention local area network used after collision by the media access controller to determine when to try again to get back onto the LAN. See also Backoff.

backpack radio A radio that is carried in the field, usually on the user's back, and is operable while in this position. See also manpack radio.

backplane The physical area, usually at the rear of an electronics frame, where modules and cables plug into the system. The high-speed communications line to which individual components, especially slide-in cards, are connected. For example, all the extensions of a PBX are connected to line cards (circuit boards) which slide into the PBX's cage. At the rear of the PBX cage, there are several connectors. Each of these connectors is wired to the PBX's backplane, also called a backplane bus. This backplane bus is typically running at a very high speed, since it carries many conversations, address information and considerable signaling. These days, the backplane bus is typically a time division multiplexed line – somewhat like a train with many cars, each representing a time slice of another conversation (data, voice, video or image). The backplane's capacity determines the overall capacity of the switch. See Passive Backplane.

backplane bus See Backplane.

backpressure Propagation effects in a communications network of hop-by-hop flow control to upstream nodes.

backreflection In cases where light is launched into an optical fiber in a forward direction, backreflection refers to the light that is returned to the launch point in the reverse direction.

backronym See Acronym.

backscatter 1. In fiber optics, the return of a portion of the scattered light signal to the input end of the fiber. Conceptually, backscattering is similar to echo. See also echo.

2. Radio wave propagation in which the direction of the incident and scattered waves resolved along a reference direction (usually horizontal) are oppositely directed. A signal received by backscattering is often referred to as "backscatter."

backshell A backshell, also called a hood, is a mechanical backing that is sometimes put onto a connector. The device protects the conductors and can be assembled or injection molded. Commonly used with D-Sub connectors.

backshore Bringing an offshored job back to the United States. See backsourcing.

backside bus The bus (i.e., common electrical path) within the microprocessor that connects the CPU (Central Processing Unit) of a PC with the Level 2 cache memory. See also Cache and Frontside Bus.

backslash Also called a virgule, the backslash key achieved fame because Microsoft used it to distinguish between subdirectories in MS-DOS. This is a backslash .

backsourcing When outsourcers fail to deliver quality, service or cost effectiveness, companies will bring the job back in-house. "It's time to back-source this one, because we've lost control of what we're doing." Submitted by Bill McCormac.

backup A copy of computer data on an external storage medium, such as a floppy disk or tape. Computers and telephone systems (which are computers) are unreliable. They glitch and lose data for all sorts of unusual and impossible-to-predict reasons. Thus the necessity for backups. The theory is that when, not if, a glitch occurs the PBX's database will disappear off the face of the earth. If this happens, you have a backup. Simply retrieve it, load it up and, presto, you're back live. Only information changed (since the backup was made) is lost. Backups save time in restoring the system after a loss. Most modern PBXs work with a database and other extensive customized instructions the user loads in. Most PBX users forget to make and keep backups of their PBX data. They expect their vendor to make backups, but he rarely does. This carelessness costs weeks of aggravation, as the PBX's database and instruction set is manually (and painfully) put back together.

The method by which backups are maintained is also important. The medium should

clearly be reliable, i.e. the best quality magnetic medium. The method of backing up is also important. For example, a streaming tape backup is less reliable than a file-by-file backup. In a streaming backup, the backup medium simply captures the original data one bit after another in one long stream. In a file-by-file backup, the data moves over in logical segments – command files, data files, etc. Streaming backups will work if their data is placed back on the same precise device from which they were originally taken. But, if they are placed on a different device (even though the same model number, etc.), they may barf because the tape assumes bad sectors are in the same place. This will probably not be true. Streaming tape backup devices are less expensive to buy and much faster to use. Avoid them.

Backup Domain Controller BDC. A server in a network domain that keeps and uses a copy by a computer without interrupting its current or primary task. For Windows NT Server domains, refers to a computer that receives a copy of the domain's security policy and domain database, and authenticates network logons.

backup link A resilient (fault tolerant) link which is not used until the primary link fails.

backup ring The token ring cabling between MAUs or CAUs consists of the main ring and the backup ring. The data is normally transmitted on the main ring, but if an error occurs, the data can be transmitted on the backup ring until the main ring is repaired. In MAU networks the switching is done manually while in CAU (or CAM) networks it is done automatically.

backward channel In data transmission, a secondary channel whose direction of transmission is constrained to be opposite to that of the primary (or forward) channel. The direction of transmission in the backward channel is restricted by the control interchange circuit that controls the direction of transmission in the primary channel. The channel of a data circuit that passes data in a direction opposite to that of its associated forward channel. The backward channel is usually used for transmission of supervisory, acknowledgement, or error-control signals. The direction of flow of these signals is opposite to that in which information is being transferred. The bandwidth of this channel is usually less than that of the forward channel, i.e., the information channel.

backward compatible A general term applied to new technologies, products or services which represent a more or less graceful upgrade of those existing. As such, they can be provisioned without requiring dramatic and complete (read prohibitively expensive) changes in the current order of things. For instance, a new and enhanced PBX system technology can be implemented through an upgrade of the existing system rather than a full replacement. Although such an upgrade can be quite expensive, the investment in the current system is protected to a large extent from technical and functional obsolescence. Another example of backward compatibility is that of Frame Relay, which is based on the existing network infrastructure, including LAP-D, the link access protocol used in ISDN. Frame relay services, therefore, work over existing access links and make use of network standards already in place.

backward learning Algorithmic process used for routing traffic that surmises information by assuming symmetrical network conditions. For example, if node A receives a packet from node B through intermediate node C, the backward-learning routing algorithm assumes that A can reach B through C optimally.

backward recovery The reconstruction of an earlier version of a file by using a newer version of data recorded in a journal.

backward signal A signal sent in the direction from the called to the calling station, or from the original communications sink to the original communications source. The backward signal is usually sent in the backward channel and consists of supervisory, acknowledgement, or error control elements.

backward supervision The use of supervisory sequences from a secondary to a primary station.

BACR Billing Account Cross Reference. The reference number correlating to the account of the LEC or LECs rendering matching bills in a meet point bill environment.

bad 1. When I'm good, I'm very good, but when I'm bad, I'm better – Mae West.

2. "There is an old saying in the acquisition world: you want it bad, you get it bad," said Tom Till, who led the Amtrak Reform Council, a group created by Congress to study the railroad's problems. "That's exactly what happened with Acela," said Till. Acela is Amtrak's high-speed train which developed serious brake problems in early 2005.

bad block A defective unit on a storage medium that software cannot read or write.

bad line button A button on the console that lets the attendant remove a line from service when there's noise or static on it, or when it's just simply busted. Some bad line buttons simply mark the line to separate it from the rest so the phone company can quickly identify the problem. See Bad Line Key.

bad line key When the PBX attendant encounters a bad trunk, he/she pushes this bad line button on the console, automatically flagging the trunk for later checking and repair. See Bad Line Button.

bad line reporting Automatically reports a poor connection without interrupting the current call.

bad sectors Defective areas on a hard or soft disk. The MS-DOS FORMAT command locks out bad sectors so they are never used. Other operating systems have similar commands. See also Hot Fix.

badware Badware is the same as malware. See malware.

BAF Billing format of the 0122 structure code defined by the Bellcore Automatic Message Accounting Format (BAF) Requirements TR#030#NWT-001100. This format identifies paths according to the resource they terminate on.

baffle A partition used with a loud speaker to prevent air vibrations from the back of the diaphragm from cancelling out the vibrations from the front of the diaphragm. Particularly valuable in the reproduction of bass notes.

bag phone A slang expression for a transportable cellular phone whose characteristics are 3 WATT output, heavy weight (for a portable), and a bag with a handle. Bag phones are not designed for carrying around. They are designed to carry from one place to another and used at that place for serious conversations, including possibly faxing and modemming. Bag phones' big "plus" is that they give off more power than a handheld cellular phone. This makes them useful for semipermanent "installation" in places like construction sites, etc. Bag phones are as powerful as a car phone, which also have 3 WATT output. That compares with handheld cellular phones which are typically 0.6 WATT.

According to my friends at Motorola, the main distinction that constitutes a Bag Phone is the lack of a self-contained battery. A Bag Phone, depending on the actual manufacturer, or more often the garage-shop assembler, was a mobile transceiver and handset that was placed in a soft-sided bag and powered via a cigarette lighter connected cord. At times, the dealer would sell a camcorder battery outfitted with a female cigarette lighter receptacle that was stuffed into the bag. Often a bag phone had a clip- or suction-mounted antenna that was affixed temporarily to the vehicle.

The formal name for bag phone was "Transmobile," a phone that could be moved from car to car, thus avoiding a fixed installation. A Transportable was a distinctly different category. Transportables had their own integral battery, and antenna, and therefore could be operated anywhere, independent of 12 volt DC power supply.

bagel The Big Bagel. Zero. As in I bagelled him at tennis last night. I beat him 6-0.

bagel defense Network security that is hard on the outside, but soft on the inside. The network has a firewall to protect against intrusions from the outside, but little security to prevent employees from accessing resources that they shouldn't.

baggy pantsing A reprimand to someone who incautiously left a terminal unlocked.

Bayh-Dole Act of 1980 The Economist refers to the Bayh-Dole Act of 1980, together with the amendments in 1984 and augmentation in 1986, as "possibly the most inspired piece of legislation to be enacted in America over the past half-century." This Act basically unlocked all the inventions and discoveries that had been made in laboratories throughout the United States with the help of taxpayers' money. Before Bayh-Dole, the fruits of research supported by government agencies had belonged strictly to the federal government. Nobody could exploit such research without tedious negotiations with the federal agency concerned. Worse, companies found it nigh impossible to acquire exclusive rights to a government-owned patent. And without that, few firms were willing to invest millions more of their own money to turn a raw research idea into a marketable product. The result was that inventions and discoveries made in American universities, teaching hospitals, national laboratories and non-profit institutions sat in warehouses gathering dust. Of the 28,000 patents that the American government owned in 1980, fewer than 5% had been licensed to industry. Although taxpayers were footing the bill for 60% of all academic research, they were getting hardly anything in return. The Bayh-Dole Act, according to the Economist, did two big things at a stroke. It transferred ownership of an invention or discovery from the government agency that had helped to pay for it to the academic institution that had carried out the actual research. And it ensured that the researchers involved got a piece of the action. Overnight, universities across America became hotbeds of innovation, as entrepreneurial professors took their inventions (and graduate students) off campus to set up companies of their own. Since 1980, American universities have witnessed a tenfold increase in the patents they generate, spun off more than 2,200 firms to exploit research done in their labs, created 260,000 jobs in the process, and now contribute $40 billion annually to the American economy.

BAIC Barring of All Incoming Calls. A wireless telecommunications term. A supplementary service provided under GSM (Global System for Mobile Communications).

Bakelite Invented around 1920 by Dr. Leo Hendrick Baekeland, bakelite is a is a flour-filled (yes, wheat flour..!) phenol-formaldehyde resin, which superseded celluloid plastic. Bakelite is a "thermosetting" (versus "thermoplastic") plastic. This means it sets with heat and cannot be remolded. Bakelite is a cast material, as opposed to Catalin plastic (a phenolic plastic invented by George Catalin) which is a molded material. Most 'Bakelite' jewelry is actually Catalin plastic, which is an inferior material due to its tendency to separate and flake apart, as seen in so many, valuable old radios of this composition (e.g. the Fada Bullet and the Emerson red tabletop). Genuine bakelite is only available in two colors: Black or Brown, whereas Catalin comes in a rainbow of hues. Both Catalin and Bakelite can be identified and distinguished from their modern look-alike imitations by this simple test: Rub the sample plastic hard with your thumb until it feels rather hot, then quickly smell the heated area. The distinct smell of phenol will be evident if the piece is authentic. Frank Svoboda and Bill Layer, who works for Viking Electronics, Hudson, Wisconsin contributed this definition. b.layer@vikingelectronics.com. In the 1930s, bakelite replaced mother-of-pearl as for buttons on quality shirts.

balance To equalize load or current between parts or elements of a telephone line or circuit. Balancing helps get the best out of a phone line. In more technical terms, balancing a line is to adjust the impedance of circuits and balance networks to achieve specified return loss objectives at junctions of two-wire and four-wire circuits. See Balanced Line and Balanced Signal Transmission. See also Balanced Line.

balanced circuit Telephone circuit in which the two conductors are electrically balanced to each other and to the ground. A balanced electrical interface generally allows data to be transmitted over longer distances than does an unbalanced circuit. See Balance.

balanced configuration Point-to-point network configuration in HDLC with two combined stations.

balanced electrical interface An electrical interface on which each circuit consists of a separate pair of wires. A balanced electrical interface generally allows data to be transmitted over longer distances than does an unbalanced electrical interface.

balanced line A transmission line which has two conductors and a ground. When the voltages of the two conductors are equal in strength with respect to ground but opposite in polarity, then you have a balanced line. Twisted pair is used for balanced lines. Coaxial cables are unbalanced, as the center conductor carries a charge, but the outer conductor, or shield, is maintained at zero voltage.

balanced mode transmission Data transmission with information conveyed by differences in voltages on two circuits to minimize effects of induced voltages.

balanced modulator An amplitude modulating circuit that suppresses the carrier signal, producing an output consisting only of upper and lower sidebands.

balanced return loss A measure of the effectiveness with which a balanced network simulates the impedance of the two-wire circuit at a hybrid coil. More generally, a measure of the degree of balance between two impedances connected to two conjugate sides of a hybrid set, network, or junction.

balanced signal transmission Two voltages, equal and opposite in phase with respect to each other, across the conductors of a twisted-pair (commonly referred to as tip and ring). See also Balance.

balanced to ground In a two-conductor circuit, a balanced-to-ground condition exists where the impedance-to-ground on one wire equals the impedance-to-ground on the other. This is the preferred condition for decent data communications.

balanced transmission line A line having conductors with equal resistance per unit length and equal capacitance and inductance between each conductor and ground. Coaxial cable, for example, is configured easily as a balanced transmission system by the use of resistance-to-ground terminators.

balancing network A network used in a set ending a four-wire circuit to match the impedance of the two-wire circuit.

2. Sometimes employed as a synonym for balun.

balcony A little platform up a telephone pole where people can work or sleep safely.

Baldridge, Malcolm Served as Secretary of Commerce from 1981 until his death in 1987. Baldridge was a proponent of quality management as a key to the prosperity and long-term strength of the United States. He took a personal interest in the quality improvement act that was eventually named after him and helped draft one of the early versions. In recognition of his contributions, Congress named an award in his honor, the Malcolm Baldridge National Quality Award. Congress established the award program in 1987 to recognize U.S. organizations for their achievements in quality and business performance and to raise awareness about the importance of quality and performance excellence as a competitive edge. The award is not given for specific products or services. Three awards may be given annually in each of these categories: manufacturing, service, small business, education and health care.

ballast Device that modifies incoming voltage and current to provide the circuit conditions necessary to start and operate electric discharge lamps, e.g. fluorescent bulbs.

balloon help 1. Place your cursor on a button or a command, in a couple of seconds a balloon pops up explaining in lesser or greater detail what clicking on the button or command will do.

2. From Wired Magazine: When someone insists on explaining every obvious detail and function of an electronic device. Refers to the rarely used Balloon Help feature on Macs. "Um, I don't really need balloon help. Just give me the domain address."

balloon test In cell site construction, a balloon test is the use of a large balloon as a visual aid to demonstrate the height of a proposed cell tower.

ballot A release form that authorizes a customer's long-distance phone service to be switched to (another) long-distance carrier or reseller.

ballpark If people don't want to come to the ballpark, how are you going to stop them? - Yogi Berra.

Balun BALanced/UNbalanced. Baluns are small, passive devices that provide the physical and electrical interfaces between electrically balanced twisted pair and electrically unbalanced coaxial cable. Baluns are used often so that IBM 3270-type terminals, which traditionally require coaxial cable connection to their host computer, can run off twisted-pair. Baluns work for some types of protocols and not for others, and they often create some level of performance degradation.

bamboozle The word bamboozle, meaning to fool or to cheat, traces back to the Chinese custom of punishing swindlers by whacking them on the hands and back with bamboo poles.

BAN 1. Billing Account Number. Used by telephone companies to designate a customer or customer location that will be billed. A single customer may have multiple billing accounts.

2. Body Area Network. A   base technology for permanent monitoring and logging of vital signs, it works for supervising the health status of patients suffering from chronic diseases, such as diabetes and asthma. Another prominent area of application for long-term logging of patient data is cardiology, where 24-hour-ECGs are required for therapy control and as early indicators for impending heart attacks. The basic concept of BAN is the fusion of both ideas: a set of mobile, compact units which enable transfer of vital parameters between the patient's location and the clinic or the doctor in charge. The vital signs data flow passes a chain of BAN modules from each sensor to a main body station, which consolidates the data streams of all sensor modules attached. It transmits the data to a home base station, from where they can be forwarded via telephone line or internet.

banana A telecommunications tool. A banana is an induction probe, usually yellow, size of a banana. w/ metal clip, (ears) for clipping a test set onto.

banc Why Banks are sometimes Bancs?: My friend, the talented hedge fund manager, Dennis Mykytyn of Modern Capital Management explains: It's a law. You cannot use the word "Bank" unless you are a registered Bank. So you cannot set up a company and call it The Bank of Newton. Securities firms are not Banks, so when they start a broker, they spell Bank as Banc to get around the law. It is bizarre.

band 1. Originally referred to AT&T's WATS Bands. AT&T WATS service was organized into circles of increasing distance from the caller. Each circle or BAND (also called SERVICE AREA), cost more per minute. But within each service area, each call costs the same per minute, even though the distances the calls travel might be different. There were typically six interstate bands covering the US and several intrastate bands (depending on how large the state is) which a customer can buy. The word "band" was invented by AT&T Communications (originally known as AT&T Long Lines) when it introduced WATS service. Recently it changed the word "band" to "Service Area." Nobody knows why. See Postalized and WATS.

2. Band can also be the range of frequencies between two defined limits. For example, the band of frequencies able to be heard by the human ear ranges between 30 to 25,000 hertz. The ear can hear a band (or more correctly, a bandwidth) of about 25,000 hertz.

band elimination filter BEF. A filter that has a single continuous attenuation band, with neither the upper nor lower cut-off frequencies being zero or infinite.

Band letters Unofficial, non-standard and consequently imprecise designations of radio frequency bands in the microwave spectrum, originally devised in the USA and the UK for military secrecy during the second world war. Originally, there were six bands, assigned code letters K, X, C, S, L and P in order of ascending wavelength. An unofficial mnemonic was coined to aid remembering them: King Xerses Can Seduce Lovely Princess, and an

in-joke of the time held that the mnemonic assured secrecy, for were it to be translated into German, the order of the bands would differ and half of them would have different letters. In order of ascending frequency, the bands from L through V now are often divided into subbands, designated by suffix subscript letters. For instance, the K band used in satcom usually is divided into at least two subbands, designed Ku for frequencies in the range 10 - 14 GHz and Ka for frequencies in the range 24 - 36 GHz.

Letter	Frequency, GHz	Wavelength, cm
P	0.225-0.390	133.3-76.9
L	0.390-1.550	76.9-19.3
S	1.55-5.20	19.3-5.77
C	4.20-6.20	7.14-4.84
X	5.20-10.90	5.77-2.75
K	10.90-36.00	2.75-0.834
Q	36.0-46.0	0.834-0.652
V	46.0-56.00.	0.652-0.536
W	56.0-100.0	0.536-0.300

band marking A label placed on an insulated wire or fiber during installation or manufacture to identify it. Or, more correctnly, a continuous circumferential band applied to a conductor at regular intervals for identification.

band pass filter BPF. A device which passes a specific range of frequencies and (in theory) blocks all others.

band splitter A multiplexer designed to split the available frequency band into several smaller channels. A band splitter can use time division or frequency division multiplexing.

band stop filter BSF. A device which blocks a specific range of frequencies and (in theory) passes all others.

band, citizens One of two bands used for low power radio transmissions in the United States – either 26.965 to 27.225 megahertz or 462.55 to 469.95 megahertz. Citizens band radio is not allowed in many countries, even some civilized countries. In some countries they use different frequencies. CB radios, in the United States, are limited by FCC rule to four WATTS of power, which gives each CB radio a range of several miles. Some naughty people boost their CBs with external power. The author of this dictionary has actually spoken to Australia while driving on the Santa Monica Freeway in Los Angeles. See also CB.

band, frequency The frequencies between the upper and lower bands. See also BAND. Here is the accepted explanation of "bands:"

Below 300 Hertz	— ELF —	Extremely low frequency
300–3,000 Hertz	— ILF —	Infra Low Frequency
3–30 kHz	— VLF —	Very Low Frequency
30–300 kHz	— LF —	Low Frequency
300–3,000 kHz	— MF —	Medium Frequency
3–30 MHz	— HF —	High Frequency
30–300 MHz	— VHF —	Very High Frequency
300–3,000 MHz	— UHF —	Ultra High Frequency
3–30GHz	— SHF —	Super High Frequency
30–300GHz	— EHF —	Extremely High Frequency
300–3,000 GHz	— THF —	Tremendously High Frequency
Band	**American**	**European**
P	0.2-1.0 Ghz	0.2-0.375 Ghz
L	1-2 Ghz	0.375-1.5 Ghz
S	2-4 Ghz	1.5-3.75 Ghz
C	4-8 Ghz	3.75-6 Ghz
X	8-12.5 Ghz	6-11.5 Ghz
J	–	11.5-18 Ghz
Ku	12.5-18 Ghz	–
K	18-26.5 Ghz	18-30 Ghz
Ka	26.5-40 Ghz	–
Q	–	30-47 Ghz

banded memory In a PostScript printer, virtual printer memory is a part of memory that stores font information. The memory in PostScript printers is divided into banded memory and virtual memory. Banded memory contains graphics and page-layout information needed to print your documents. Virtual memory contains any font information that is sent to your printer either when you print a document or when you download fonts.

banded rate A price range for regulated telephone service that has a minimum floor and maximum ceiling. The minimum covers the cost of service; the maximum is the rate filed in the price list.

bandit mobile A mobile subscriber that is revealed in the toll-ticketing records as having an invalid ESN, invalid telephone number, or other problem that warrants denial of service to that mobile.

banjo Also called beaver tail. Used to connect devices to modular jack wiring for testing. See Modular Breakout Adapter.

banjo clip See Modular Breakout Adapter.

bandmarking A continuous circumferential band applied to an insulated conductor at regular intervals for identification.

bandpass The range of frequencies that a channel will transmit (i.e. pass through) without excessive attenuation.

bandpass filter A device which transmits a band of frequencies and blocks or absorbs all other frequencies not in the specified band. Often used in frequency division multiplexing to separate one conversation from many.

bandpass limiter A device that imposes hard limiting on a signal and contains a filter that suppresses the unwanted products of the limiting process.

bandwidth 1. In telecommunications, bandwidth is the width of a communications channel. In analog communications, bandwidth is typically measured in Hertz – cycles per second. In digital communications, bandwidth is typically measured in bits per second (bps). A voice conversation in analog format is typically 3,000 Hertz, carried in a 4,000 Hertz analog channel. In digital communications, encoded in PCM, it's 64,000 bits per second. Do not confuse bandwidth with band. Let's say we're running a communications device in the 12 GHz band. What's its bandwidth? That's the space it's occupying. Let's say it's occupying from 12 GHz to 12.1 GHz. This means that it's occupying the space from 12,000,000,000 Hz to 12,100,000,000 Hz. This means its bandwidth is one hundred million cycles or one hundred megahertz (100 MHz). Affiliated terms are narrowband, wideband and broadband. While these are not precise terms, narrowband generally refers to some number of 64 Kbps channels (Nx64) providing aggregate bandwidth less than 1.544 Mbps (24x64 Kbps, or T-1), wideband is 1.544 Mbps-45 Mbps (T-1 to T-3) and broadband provides 45 Mbps (T-3) or better.

2. The capacity to move information. A person who can master hardware, software, manufacturing and marketing – and plays the oboe or some other musical instrument – is "high bandwidth." The term is believed to have originated in Redmond, WA, in the headquarters of Microsoft. People there (e.g., Bill Gates) who are super-intelligent and have generally broad capabilities, are said to have "high bandwidth."

3. Microsoft jargon for schedule. For example, "I have a bandwidth problem" means that I have an overloaded schedule.

4. The combined girth of a rock band. By way of example, the band "Meatloaf" is broadband, largely due to the individual girth of the singer by the same name. On the other hand, the "Rolling Stones" are narrowband, due largely to the svelte Mick Jagger. While the "Rolling Stones" are older, they are also richer than is "Meatloaf." So, bandwidth isn't everything.

bandwidth allocation See Bandwidth Reservation.

bandwidth augmentation Bandwidth augmentation is the ability to add another communications channel to an already existing communications channel.

bandwidth broker Imagine you're a long distance carrier and suddenly you need T-3 capacity for one of your customers from New York to Washington for a week. You go to someone called a "bandwidth broker" and ask to rent the capacity. That broker has some form of connecting device, to allow you to connect your network to someone else's network. The bandwidth broker will physically do the connection and charge a fee for doing it. That connecting device is called a pooling point. That device could be anything from a patch panel to a full-fledged real-time expensive switch. Here are some words from the web site of LightTrade.com, a bandwidth broker. "The exploding telecommunications market has made inevitable a day when bandwidth will be traded as a commodity, much as electricity is today. The bandwidth market is likely to center around three areas: (1) a central Bandwidth Trading Organization ("BTO"), functioning much like a commodities exchange, (2) an independent auditor (Pooling Point Administrator or "PPA") and (3) electro-optical switches to effect the physical delivery and Quality of Service verification (Pooling Points or "PPs")."

bandwidth compression A technique to reduce the bandwidth needed to

transmit a given amount of information. Bandwidth compression is used typically in "picture type" transmissions – such as facsimile, imaging or video-conferencing. For example, early facsimile machines scanned each bit of the document to be sent and sent a YES or NO (if there was material in that spot or not). More modern machines simply skip over all the blank spaces and transmit a message to the receiving facsimile machine when to start printing dots again. A facsimile "picture" is made up of tiny dots, similar to printing photos in a magazine. Today, bandwidth compression is used to transmit voice, video and data. There are many techniques, few of which are standard. The key, of course, is that if you're going to compress a "conversation" at one end, you must "de-compress" it at the other end. Thus, in every bandwidth compressed conversation there must be two sets of equipment, one at each end. And they better be compatible.

bandwidth exchange See Bandwidth Broker.

bandwidth envy I have a dial-up connection to the Internet. You have a DSL line. You're running 20 to 30 times faster than me. I envy your good luck. I have bandwidth envy.

bandwidth glut In the United States, Europe, Japan and Australia the rapid expansion of fiber optic systems in the late 1990s resulted in a bandwidth glut, forcing many carriers into bankruptcy. But most other parts of the world suffered a bandwidth shortage.

bandwidth hog 1. A service that uses a lot of bandwidth. 2. A user who uses a lot of bandwidth.

bandwidth junkie One who worships brute speed when it comes to Internet connections. He's the type of person who has a T-1 line in his bedroom. Eric Smestad, ninfan@Limbo.Alleged.com, wrote me, "After reading your definition of "Bandwidth junkie" I started to feel rather worried, you see, I am afraid that two of my friends and I may be bandwidth junkies. We live in the same apartment building and have ethernet cable ran from apartment to apartment, a switched hub, and a T-1 line to the Internet with a Cisco router. These seem to be obvious symptoms, is there a cure?" Answer, there is no cure.

bandwidth limited operation The condition prevailing when the system bandwidth, rather than the amplitude (or power) of the received signal, limits performance. The condition is reached when the system distorts the shape of the signal waveform beyond specified limits. For linear systems, bandwidth-limited operation is equivalent to distortion-limited operation.

bandwidth on demand Just what it sounds like. You want two 56 Kbps circuits this moment for a videoconference. No problem. Use one of the newer pieces of telecommunications equipment and "dial up" the bandwidth you need. An example of such a piece of equipment is an inverse multiplexer. Uses for bandwidth on demand include video conferencing, LAN interconnection and disaster recovery. Bandwidth on demand is typically done only with digital circuits (they're easier to combine). Bandwidth on demand is typically carved out of a T-1 circuit, which is permanently connected to the customer's premises from a long distance carrier's central office, also called a POP – Point of Presence.

bandwidth reservation Bandwidth reservation is the process of assigning bandwidth to users and applications served by a network. This involves assigning priority to different flows of traffic based on how critical and delay-sensitive they are. This makes the best use of available bandwidth, and if the network becomes congested, lower-priority traffic can be dropped. This is also called Bandwidth Allocation.

bandwidth throttling A feature in some Web server software that allows the system administrator to control or alter the proportions of bandwidth available to the services that the server provides.

Bandwidth Trading Organization BTO. See Bandwidth Broker.

bang An exclamation point (!) used in a Unix-to-Unix Copy Program (UUCP) electronic mail address. People who are on AT&T Mail often give you their mail address as "Bang Their Name." My AT&T Mail address used to be Bang HarryNewton, i.e. !HarryNewton.

bang path A series of UUCP nodes mail will pass through to reach a remote user. Node names are separated by exclamation marks nicknamed "bangs." The first node in the path must be on the local system, the second node must be linked to the first, and so on. To reach user 1 on sys2 if your computer's address is sys1 you would use the following address:

sys1! sys2! sys3! user1

bank A row of similar components used as a single device, like a bank of memory. Banks must be installed or removed together. See Bank Switching.

bank switching A way of expanding memory beyond an operating system's or microprocessor's address limitations by switching rapidly between two banks of memory. In MS-DOS, a 64K bank of memory between 640K and one megabyte is set aside. When more memory is needed, the bank, or page, is switched with a 64K page of free memory.

This is repeated with additional 64K pages of memory. When the computer requires data or program instructions not in memory, expanded memory software finds the bank containing the data and switches it with the current bank of memory. Although effective, bank switching results in memory access times that are slower than true, extended memory.

bankrupt See Chapter 11 and Chapter 7.

banner An advertisement on an Internet page that is usually "hot linked" to the advertiser's site. A rectangular graphic, usually 468 x 60 pixels, which is used as an advertising element on a Web page. See Banner Ad.

banner ad You're visiting a web site, for example, www.hellodirect.com, a purveyor of telephone accessories. As I look at this site, I can see an advertisement for a product called Polycom at the top of the page. Information about Hello Direct starts below the Polycom ad. That ad is a "banner ad." It stretches across the top of a web site such as a banner would stretch across a road, e.g. "Welcome Lions' Club Members" across the main street of a small town. So the first aspect of a banner ad on a Web page is its placement across the top of a home page. The second aspect is that when you slide your cursor over it, your cursor changes to a hand and you are now able to click on the banner ad and go elsewhere. In other words, a banner ad also contains a hyperlink to some place else, typically the location on the Web of what the banner ad is advertising.

banner ad rotator Displays alternating banner ads and includes an administration area with the ability to add, edit and delete banners from the rotation list.

Bantam Bantam is an ADC trade name for the miniature phone style jacks used by the telecommunications and broadcast industries. The Bantam jack was developed by ADC in 1961 for the US military as a space saving solution for ship and submarine use. The electrical characteristics are identical (voltage rating, return loss, contact resistance, etc.) as the standard 1/4" jack that has been in use since the early 1900s. The Bantam jack proved to be so popular it has become an industry standard.

bantam connector plug A plug and jack used to connect test equipment with digital circuits such as (DS1, DS3, STS1). Wired to DSX patch panels.

BAOC Barring of All Outgoing Calls. A wireless telecommunications term. A supplementary service provided under GSM (Global System for Mobile Communications).

bar code A standard way of identifying the manufacturer and product category of a particular item. The barcode was adopted in the 1970s because the bars were easier for machines to read than optical characters. Barcodes' main drawbacks are they don't identify unique items and scanners have to have line of sight to read them. In short, a bar code is a bunch of lines of varying width printed on something. The bar code is designed to be read optically by some data capturing device. Bar codes are turning up on letters. They are read by image scanning devices in the post office and allegedly help the mails move faster. Bar codes are on most things you buy now in supermarkets. By scanning those bar codes at the checkout counter, the supermarket knows what's being sold and not being sold. And presumably the supermarket, or its computer, can order supplies to keep the supermarket stocked with goods that are selling and not re-order those which aren't. Bar codes will eventually be replaced by RFID tags. See RFID.

Bar Mitzvah Under one is you were not commanded. Under talmudic law until you hit 13 or 12 (bas mitzvah for a girl) you were not commanded to fulfill the commandments of the Torah. The origin for making a party comes from the story in the Talmud about a blind man (the blind were also free from most obligations) who said that he would make a party if he were told that he was obligated to fulfill the commandments.

Barb Wire Jim The nickname for Jim Tower, who founded the Calaveras Telephone Company in Copperopolis, California in 1895. As an enterprising 16-year-old, Tower used barbed wire fencing to string the first telephone line for what soon was to become the Calaveras Telephone Company. In 1899, Tower, only 20 years old, negotiated an agreement with Sunset Telephone Company to interconnect his telephone network with Sunset's, thereby connecting Calaveras Telephone Company with the rest of the world. The Calaveras Telephone Company, one of the last family-founded phone companies in the United States, is still owned and operated by the Tower family.

Barbarella Barbarella is a 1968 movie starring Jane Fonda. Back then I was a lowly consultant employed by an English bank to help the American president sell his stuffy old English directors on getting into cable TV business. I set up a presentation on media diversity and all the great movies you would be able to see on cable TV. As part of the "presentation" I showed clips from Barbarella. The directors loved the scene of Jane Fonda being attacked by rats and asked me to replay that scene several times for them. (They were not interested in the CATV business, but did like the rats.) The whole scene of the gargantuan boardroom table, the elderly lords, earls and barons among the directors, and the many TV monitors I'd strung around the table for their delicate viewing came back to me in May of 2001 when

I read in the New York Times that Jane Fonda was once so embarrassed by the movie that she admitted she must have had her head "up her armpit" when she agreed to do it. For a while Jane Fonda was married to Ted Turner, CNN founder, who gave a billion dollars to the United Nations. Once Ted was allegedly asked what sort of interest he got in his UN charity from audiences he spoke in front off on global issues. He answered, "Most people just want to know what it's like to sleep with Barbarella."

Barbe, Jane See Jane Barbe.

Barber Pole The modern barber pole originated in the days when bloodletting was one of the principal duties of the barber. The two spiral ribbons painted around the pole represent the two long bandages, one twisted around the arm before bleeding (blue) and the other used to bind it afterward (red from blood). Originally, when not in use, the pole had both bandages wound around it, so that both might be together when needed. The pole was hung at the door as a sign. But later, for convenience, instead of hanging out the original pole, another one was painted as an imitation and given a permanent place on the outside the shop. This was the beginning of the modern barber pole.

barcode rape When booth bunnies or other trade show fauna grab you by the name tag and swipe the barcode – and earn a commission – before even talking to you about the products they offer. See Booth Bunny.

bare board A circuit board without components is called a bare board. A bare board will typically have the correct holes drilled and the correct soldering done in order to accept the components which later will be attached.

bare wire An electrical conductor having no covering or insulation. See also Hard Cable.

barge-in Interrupting a call in progress, or interrupting a computer telephony system (say voice mail or automated attendant) while the thing is talking to you.

barge-out Leaving a call in progress without notice.

barium ferrite A type of magnetic particle used in some recording media including floptical diskettes. See also Ferrite, Hard Ferrite and Soft Ferrite.

barn raising Remember when it really meant to build a barn? Now it means to solve a difficult problem by pulling staff and resources from all parts of the company to solve a problem. "We'll need to do a little barn raising to solve this one." The term is more palatable than "multi-functional task force," some people argue.

barrel An imaging term. Distortion that swells an image in the middle and narrows it at the top and the bottom.

barrel connector This is a cylindrical (barrel-shaped) connector used to splice together two lengths of coaxial cable. In the cable TV industry, the term is often shortened to just barrel.

barrel contact A term in cabling. A barrel contact is an insulation displacement type contact consisting of a slotted tube that cuts the insulation when the wire is inserted.

barrel distortion When a screen is distorted – with the top, bottom and sides pushing outwards (like a beer barrel) – the screen is said to be suffering barrel distortion.

BARRNet Bay Area Regional Research Network. Regional network serving the San Francisco Bay Area. The BARRNet backbone is composed of four University of California campuses (Berkeley, Davis, Santa Cruz, and San Francisco), Stanford University, Lawrence Livermore National Laboratory, and NASA Ames Research Center. BARRNet is now part of BBN Planet. See also BBN Planet.

BARS A Nortel switching term meaning Basic Automatic Route Selection.

Barton Enos Barton once said he was "disgusted" when told that it would be possible to send conversation along a wire. He later co-founded (with Elisha Gray) the Western Electric Company, which became AT&T's manufacturing subsidiary and was once the largest electrical equipment manufacturer in the U.S. In addition to phones, the company made sewing machines, typewriters, movie sound equipment, radio station gear, radar systems and guided missile parts. See also Graybar.

base A component of a transistor which serves as the middle layer of the 3-layer silicon sandwich–the transistor. The base works as a faucet between the emitter and collector, controlling the current moving through the three layers.

base address The first address in a series of addresses in memory, often used to describe the beginning of a network interface card's I/O space.

base amount A call center term. One of the historical patterns; what the monthly call volume would be if there were no long-term trend or seasonal fluctuation- in other words, the average number of calls per month.

base band audio The range of frequencies from 0 to 20 kHz.

base band video The band of frequencies from 0 to 4.2 MHz.

base controlled hand-off Cell transfers managed by and initiated from the cellular radio network, typical of Advanced Mobile Phone Systems (AMPS).

Base General Premises Cabling Licence See BCL.

base load In trunk forecasting, an amount of telephone traffic measuring during a certain defined time. See Base Period.

base memory What many people refer to as the first 640 kilobytes of memory in an MS-DOS PC.

base period In trunk forecasting, a time span of consecutive study during which a base load is determined.

base schedules A call center term. A fixed set of pre-existing schedules that you can use as a starting point in scheduling. New schedules created are in addition to the base schedules.

base staff A call center term. The minimum number of people, or "bodies in chairs," required to handle the workload in a given period. The actual required number of staff is always greater than the base staff, because of various human factors such as the need for breaks and time off. Therefore, schedules need to add in extra people to accommodate breaks, absenteeism and other factors that will keep agents from the phones. See Rostered Staff Factor.

base station In cellular technology, a base station is a fixed station used for communicating with mobile stations, most commonly handsets. Each cell in a cellular network requires a base station. It is sometimes used loosely in the standards to mean any land side functionality. See also Mobile Switching Center.

base station controller In cellular communications, the Base Station Controller is a component of a base station which supervises the functioning and control of multiple Base Transceiver Stations and acts as a small switch.

base station hotel A facility that serves as the heart of a wireless service provider's distributed antenna system (DAS). It is where the system's main antenna tower and equipment are installed, and from which leased lines radiate to remote antenna nodes in dead zones and other areas that the main antenna tower cannot reliably serve. See remote antenna node.

base station transceiver In cellular communications, the Base Transceiver Station is a component of a base station which consists of all radio transmission and reception equipment, it provides coverage to a geographic area, and is controlled by a Base Station Controller.

Base Transceiver Station BTS. The electronic equipment housed in cabinets that together with antennas comprises a PCS facility or "site". The cabinets include an air-conditioning unit, heating unit, electrical supply, telephone hook-up, and back-up power supply.

Base64 A standard algorithm for encoding and decoding non-ASCII data for attachment to an e-mail message, base 64 is the foundation for MIME (Multipurpose Internet Mail Extensions). MIME was standardized by the Internet Engineering Task Force (IETF) in RFC 1521 – Appendix G – Canonical Encoding Model. Base64 uses a 65-character subset of US-ASCII to encode any file in any format. The 65 characters include all 26 characters of the English language in both upper and lower case; all 10 digits (i.e., 0-9) and the characters "+", "/" and "=". The 65th character is "=," which is used to signify special processing functions. The first 64 characters each represent a unique 6-bit value from 0 (000000) to 63 (111111). The encoding process starts with 24-bit input groups, which are formed by concatenating (linking together) three 8-bit groups. These 24-bit input groups are then treated as 4 concatenated 6-bit groups, each of which is translated into a single character in the base64 alphabet. The resulting "characters" are formed into an output stream of "lines" of no more than 76 characters each. The specific type of content (e.g., audio, image or video) is identified by a content header, which precedes the attached data, which then is transmitted across the IP network, usually the Internet. The receiving device decodes the data by reversing the process. As base64 yields a data stream which is approximately 33% greater than the original content, it is not a compression algorithm. It does, however, comprise a standard means of transmitting non-ASCII data over an IP network, and is unaffected by gateways between networks and systems. UUencode and binhex are alternative, non-standard methods of accomplishing the same thing, although their content may be affected by gateway intervention. See also binhex, MIME and UUencode.

baseband A form of modulation in which signals are pulsed directly on the transmission medium without frequency division. Local area networks as a rule, fall into two categories – broadband and baseband. The simpler, cheaper and less sophisticated of the two is baseband. In baseband LANs, the entire bandwidth (capacity) of the LAN cable is used to transmit a single digital signal. In broadband networks, the capacity of the cable is divided into many channels, which can transmit many simultaneous signals. While a base-

band channel can only transmit one signal and that signal is usually digital, a broadband LAN can transmit video, voice and data simultaneously by splitting the signals on that cable using frequency division multiplexing. The electronics of a baseband LAN are simpler than a broadband LAN. The digital signals from the sending devices are put directly onto the cable without modulation of any kind. Only one signal is transmitted at a time. Multiple "simultaneous" transmissions can be achieved by a technique called time division multiplexing (see multiplexing). In contrast, broadband networks (which typically run on coaxial cable) need more complex electronics to decipher and pick off the various signals they transmit. Attached devices on a broadband network require modems to transmit. Attached devices to baseband networks do not.

Baseband LANs typically work with one high speed channel, which all the attached devices – printers, computers, databases – share. They share it by using it in turns – for example, passing a "token" to the next device. That token entitles the device with the token to transmit. IBM's LAN is a token ring passing local area network. Another way of sharing the baseband LAN is that each device, when it is ready to transmit, simply transmits into the channel and waits for a reply. If it doesn't receive a reply, it retransmits. Thus there are two main network or baseband access control schemes – Token Ring Passing and CSMA/CD. See also CSMA/CD, Broadband, Ethernet and Local Area Networks.

baseband modem A modem which does not apply a complex modulation scheme to the data before transmission, but which applies the digital input (or a simple transformation of it) to the transmission channel. This technique can only be used when a very wide bandwidth is available. It only operates over short distances where signal amplification is not necessary. Sometimes called a limited distance or short-haul modem.

baseband signaling Transmission of a digital or analog signal at its original frequencies, i.e., a signal in its original form, not changed by modulation.

baseboard A term used in voice processing/computer telephony to mean a printed circuit board without any daughterboards attached.

baseboard raceway A floor distribution method in which metal or wood channels, containing cables, run along the baseboards of the building. The front panel of the baseboard channel is removable, and outlets may be placed at any point along the channel.

baseline 1. The line from which a graph is drawn. The base line is the X axis on vertically oriented graphs, the Y axis on horizontal bar graphs, or the line representing zero if the data contains both positive and negative numbers.

2. The imaginary line extending through a font and representing the line on which characters are aligned for printing. In conventional, alphanumeric fonts, the baseline is usually defined as the imaginary line touching the bottom of uppercase characters.

baseline report Compares two similar time ranges in a report format. A baseline time range is protected against purge action so that baseline data is available at report time. The baseline time range can be 1 to 30 days.

baseline sequential JPEG The most popular of the JPEG modes which employs the lossy DCT (Discrete Cosine Transform) to compress image data as well as lossless processes based on variations of DPCM (Differential Pulse Code Modulation). The "baseline" system represents a minimum capability that must be present in all Sequential JPEG decoder systems. In this mode image components are compressed either individually or in groups. A single scan pass completely codes a component or group of components.

Bash Bourne-again shell. Interactive UNIX shell based on the traditional Bourne shell, but with increased functionality. See also root account.

BASIC Beginners All-purpose Symbolic Instruction Code. A programming language written by mathematicians John G. Kemeny and Thomas E. Kurtz at Dartmouth College in 1963. BASIC is an easy language to learn. Not all "Basics" are the same. BASIC is an example of something called a backronym, a word that often is interpreted as an acronym, although it was not so intended. Originally, "BASIC" was just "basic." Then somebody, whose name is lost in the mists of time, decided to make an acronym out of it and coined Beginners All-purpose Symbolie Instruction Code. Designed to be used with minimum of training, BASIC was taught as part of a course required of most Dartmouth students during the 1960s. The Mathematics Department estimated that, at its peak, BASIC was understood by 80 percent of Dartmouth undergrads – at a time when computers were exotic to most Americans. BASIC laid the foundation for the software that was later used in the introduction of personal computers. See also Acronym.

Basic 911 Basic 911 is the emergency telephone system in the United States that automatically connects 911 callers to a designated answering point (e.g., public service answering point). Call routing is determined by the Central Office from which a call originates. Basic 911 may or may not support Automatic Location Identification and Automatic

Number Identification. Basic 911 is in place in most of the United States. Many locales have upgraded to Enhanced 911. See E-911.

basic budget service An inexpensive local phone service often restricted to people with limited incomes. It may or may not include any outgoing calls. It may include only a few outgoing calls.

Basic Cable Television Service A tier of cable television service which is available to all subscribers, is specifically identified as "basic" service, includes all television broadcast stations listed on the Must-Carry Station List in the system's PIF, includes all television broadcast stations which the system offers to any subscriber pursuant to any retransmission-consent agreement, includes all public, educational, and government access channels which are designated by franchise for carriage on the basic tier and may include, at the cable operator's option, any other service.

basic call A call between two users that does not require Advanced Intelligent Network Release 1 features (e.g. a POTS call). Definition from Bellcore. See AIN.

Basic Call State Model BCSM. An Advanced Intelligent Network (AIN) term described in ITU-T Standard Recommendation Q.1290. BCSM is a model which describes the call processing steps for basic call control (i.e., a non-intelligent, two-party call). The model is divided into the originating BCSM and terminating BCSM. In total, the BCSM describes the activities of a Service Switching Point (SSP) in establishing, maintaining, and clearing a basic call. This call model comprises Points In Calls (PICs), Detection Points (DPs), and triggers. PICs are the states (i.e., conditions or activities) that a call goes through, and include on-hook (i.e., idle, or available), off-hook (i.e., unavailable, origination attempt, or answer), collecting information, analyzing information, routing, and alerting (i.e., ringing). The DPs represent prospective points for the entry of services as the call progresses from PIC to PIC. Trigger Detection Points (TDPs) are the processes of the SSPs as they check to determine if there are any active triggers. Triggers cause a reaction, or the initiation of a process. If, for example, an originating device goes off-hook to initiate a call, an active trigger causes the SSP to change from a "null state" (i.e., monitoring state) to an "analyzing information state." See also AIN.

basic call vectoring A PBX feature that processes and routes incoming calls according to a set of commands that can be preprogrammed by the customer to route calls. Calls can be handled differently depending on variables such as the tie of day, the day of week, the number of calls waiting, etc.

Basic Carrier Platform Common Carrier Platform. A common carriage transmission service coupled with the means by which consumers can access any or all video provides making use of the platform. Video dialtone service at the basic platform level differs from the "channel service " that Local Exchange Carriers (LECs) currently may provide cable television operators in that the Commission will require LECs to provide sufficient transmission capacity to serve multiple video programmers.

Basic Emergency Service Provider See BESP.

Basic Encoding Rules See BER.

Basic Exchange Telecommunications Radio Service BETRS. A service that can extend telephone service to rural areas by replacing the local loop with radio communications, sharing the UHF and VHF common carrier and private radio frequencies. See BETRS for a longer explanation.

Basic Mode Link Control Control of data links by use of the control characters of the 7-bit character set for information processing interchange as given in ISO Standard 646-1983 and ITU-T Recommendation V.3-1972.

Basic Multilingual Plane See BMP.

basic rate The default rate paid by a telephone subscriber who does not elect any other calling plan. Basic rate can apply to phone service from either a local phone company or a long distance phone company. Basic rate for local service can be a flat monthly rate or a monthly rate plus charges for each call – on length of call, time of day and distance sensitive charging. A basic rate for long distance or local toll service will be length of call, time of day and distance sensitive charging.

Basic Rate Interface BRI. Also known as BRA (Basic Rate Access). There are two standard interfaces in ISDN: BRI and PRI. BRI is intended for consumer, SOHO (Small Office Home Office), and small business applications. BRI supports a total signaling rate of 144 Kbps, which is divided into two B (Bearer) channels which run at 64 Kbps, and a D (Delta, or Data) channel which runs at 16 Kbps. The B channels "bear" the actual data payload, i.e., they carry the actual information that you are sending. Such information can be PCM-encoded digital voice, digital video, digital facsimile, or whatever you can squeeze into a 64 Kbps full-duplex channel. The D channel is intended primarily for signaling and control information, including call setup, call maintenance and monitoring, call teardown,

Caller ID, and Name ID. As the signaling and control requirements actually are fairly modest, the D channel also will support packet data transfer at rates up to 9.6 Kbps, by special arrangement with the servicing telephone company and at additional cost. The preferred BRI standard is the "U" interface, which uses only two wires and makes use of the 2B1Q line coding technique in the U.S. and the 4B3T technique in Europe. Another BRI standard is the "T" interface which uses four wires. See ISDN for a much fuller explanation. See also 2B1Q, 4B3T, T Interface, and U Interface.

Basic Rate ISDN BRI. An ISDN service that offers two 64Kbps B channels used for data transfer and one 16Kbps D channel used for signaling and control information. Each B channel can carry a single digital voice call or can be used as a data channel; the B channels can also be combined into a single 128Kbps data channel. See also Basic Rate Interface.

basic service A telephone company service limited to providing local switching and transmission. Basic Service does not include equipment. The term Basic Service is unclear and varies between telephone companies and data communications service providers.

Basic Service Elements BSEs. An Open Network Architecture term. BSEs are services which value-added companies could get from their phone company. BSEs are optional basic network functions that are not required for an ESP to have a BSA, but when combined with BSEs can offer additional features and services. Most BSEs allow an ESP to offer enhanced services to their customers. BSEs fall into four general categories: Switching, where call routing, call management and processing are required; Signaling, for applications like remote alarm monitoring and meter reading; Transmission, where dedicated bandwidth or bit rate is allocated to a customer application; and Network Management, where a customer is given the ability to monitor network performance and reallocate certain capabilities. The selection of available BSEs is an ongoing process, with new arrangements being developed. ANI, Audiotext "Dial-It" Services, and Message Waiting Notification are all examples of BSEs. See AIN and Open Network Architecture.

Basic Serving Arrangement BSA. An old term defining the relationship of an enhanced service provider (value added provider) to the phone company providing the line/s. Under ONA (Open Network Architecture), a BSA is the basic interconnection access arrangement which offers a customer access to the public network (i.e. the normal switch phone service) and provides for the selection of available Basic Service Elements. It includes an ESP (Enhanced Service Provider) access link, the features and functions associated with that access link at the central office serving the ESP and/or other offices, and the transport (dedicated or switched) within the network that completes the connection from the ESP to the central office serving its customers or to capabilities associated with the customer's complementary network services. Each component may have a number of categories of network characteristics. Within these categories of network characteristics are alternatives from among which the customer must choose. Examples of BSA components are ESP access link, transport and/or usage. See Open Network Architecture.

basic trading area Geographic boundaries that segment the country for FCC licensing purposes. BTAs are based on Rand McNally's Commercial Atlas & Marketing Guide. BTA boundaries follow county lines and include the county or counties whose residents make the bulk of their shopping goods purchases in the area. The FCC has used BTAs to license a number of services including broadband and narrowband Personal Communication Services.

basic telephony The lowest level of service in Windows Telephony Services is called Basic Telephony and provides a guaranteed set of functions that corresponds to "Plain Old Telephone Service" (POTS - only make calls and receive calls). See Windows Telephony Services.

basis point A basis point is one hundredth of one percent. For example, you might read, "For the week, yields dropped eight basis points, from 4.66% to 4.58%."

BASR Buffered Automatic Send/Receive.

bat shit The bat on the Bacardi symbol is there because the soil where the sugar cane grows is fertile from the excessive guano the flying mammals contribute.

bastion host A computer placed outside a firewall to provide public services (such as World Wide Web access and FTP) to other Internet sites, hardened to withstand whatever attacks the Internet can throw at it. Hardening is accomplished by making the box as single-purpose as possible, removing all unneeded services and potential security vulnerabilities. Bastion host is sometimes inaccurately generalized to refer to any host critical to the defense of a local network. See also DMZ.

Batch 1. In data processing, the processing of data accumulated over a period of time. See Batch Processing.

2. In telecommunications, the accumulation of messages for transmission in a single group.

Batch Change Supplement 0x06BCS. A batch change supplement is a supplement to a batch change in the programmed logic associated with a general purpose computer system, or with a specialized computer or computerized system such as a PBX or router. A "batch change" refers to a change in programmed logic that is made on a batch basis (i.e., in one big bunch of changes). Sometimes, batch changes require additional changes, in the form of supplements.

batch processing There are two basic types of data processing. One is batch processing. Also called deferred time processing and off-line processing. Batch processing occurs where everything relating to one complete job – such as preparing this week's payroll – is bunched together and transmitted for processing (locally, in the same building or long distance, across the country), usually by the same computer and under the same application program. Batch processing does not permit interaction between the program and the user once the program has been read (i.e. fed into) the computer. In batch processing with telecommunications (i.e. sending the task to be done over the phone line), network response time is not critical, since no one is sitting in front of a screen waiting for a response. On the other hand, accuracy of communications is very critical, since no one is sitting in front of a screen checking entries and responses.

The second type of processing is called interactive or real time processing. Under this method of processing, a user sends in transactions and awaits a response from the distant computer before continuing. In this case, response time on the data communications facility is critical. Seconds count, especially if a customer is sitting at the other end of a voice call awaiting information on whether there's space on that airline flight, for example. See Batch File.

bathwater In England in the 1500s, most people got married in June because they took their yearly bath in May and still smelled pretty good by June. Baths consisted of a big tub filled with hot water. The man of the house had the privilege of the nice clean water, then all the other sons and men, then the women and finally the children – last of all the babies. By then the water had become so dirty you could actually lose someone in it – hence the saying, "Don't throw the baby out with the bath water."

batmobiling Batmobiling is putting up emotional shields from the retracting armor that covers the batmobile as in "she started talking marriage and he started batmobiling."

Batteries A, B and C In ancient times, when vacuum tubes were still in use, the A battery (often about 6 volts) heated the filament or cathode to boil off electrons, the B battery (usually several hundred volts), positive with respect to the cathode, sucked the negative electrons to the plate, while the C battery, a few volts negative with respect to the cathode, tended to repel the electrons back toward the cathode. By varying a small signal voltage riding on the C battery, the flow of electrons to the plate could be controlled to generate a much larger voltage.

battery Batteries are electrochemical devices that store energy in the form of chemical bonds and convert this energy directly into electricity on demand. In general, a battery consists of a cathode that is the positive pole of a cell, an anode that constitutes the negative pole, and an electrolyte that is the medium through which ions are passed and that carries current internally. A battery's performance is evaluated by many factors, the most important of which are its energy content, the maximum power it can deliver, its cyclability, and its cost. No single battery can be considered to be the "dream machine" that possesses all desirable properties. That is why there are so many different types of batteries.

All telephone systems work on DC (direct current). DC power is what you use to talk on. Often the DC power is called "talking battery." Most key systems and many PBXs plug directly into an AC on the wall, but that AC power is converted by a built-in power supply to the DC power the phone system needs. All central office switches (public exchanges) use DC power also. But they derive it from rechargeable lead acid batteries, which in turn are charged by AC, typically from the local electricity company. These batteries perform several functions: 1. They provide the necessary juice to power telephone switches. 2. They serve as a filter to smooth out fluctuations in the commercial power and remove the "noise" that power often carries. 3. They provide necessary backup power should commercial power stop, as in a "blackout" or should it get weak, as in a "brownout." To sum up, "battery" is the term used to reference the DC power source of a telephone system. Often called "Talking Battery." See also AC, AC Power, Battery Reserve, Central Office Battery and NICAD Battery.

battery backup A battery which provides power to your phone system when the main AC power fails especially during blackouts and brownouts. Hospitals, brokerage companies, airlines and hotel reservation services must have battery backup because of the integral importance of their phone systems to their business.

battery eliminator A device which has a rectifier and (hopefully) a filter. This device will convert AC power into the correct DC voltages necessary to drive a telephone system. Such a battery eliminator, or power supply, should deliver "clean" power, i.e. with little "noise" and of low impedance.

battery management An umbrella term used by various UPS manufacturers to describe a suite of functions related to charging, testing, and maximizing the life of a UPS (Uninterruptible Power Supply) battery. Battery management may include imminent battery failure diagnosis and indication (first introduced in 1989 by APC Corp.), scheduled battery testing, hot swappable user replaceable batteries, high speed battery charging, output regulation to reduce unnecessary battery usage, and/or special battery charging techniques.

battery reserve The capability of the fully charged battery cells to carry the central office power load imposed when commercial power fails and the primary power source (generators/rectifiers) is out of service. Properly described in terms of the number of hours the batteries can furnish operating power for dependent CO apparatus for a demand equal to that on the CO during its busy hour. A busy-hour reserve of eight hours is typical for a telephone office battery plant.

battle faxes A cycle of vitriolic faxes exchanged between clients and lawyers, fighting lovers, etc. "Here's the latest round of battle faxes with my record company." Wired Magazine defined this term.

BAU Business As Usual.

baud rate A measure of transmission speed over an analog phone line – i.e. a common POTS line. (POTS stands for Plain Old Telephone Service). Imagine that you want to send digital information (say from your computer) over a POTS phone line. You buy a modem. A modem is a device for converting digital on-off signals, which your computer speaks, to the analog, sine-wave signals your phone line "speaks." For your modem to put data on your phone line means it must send out an analog sine wave (called the carrier signal) and change that carrier signal in concert with the data it's sending. Baud rate measures the number of changes per second in that analog sine wave signal. According to Bell Labs, the most changes you can get out of a 3 KHz (3000 cycles per second) voice channel (which is what all voice channels are) is theoretically twice the bandwidth, or 6,000 baud.

Baud rate is often confused with bits per second, which is a transfer rate measuring exactly how many bits of computer data per second can be sent over a telephone line. You can get more data per second – i.e. more bits per second – on a voice channel than you can change the signal. You do this through the magic of coding techniques, such as phase shift keying. Advanced coding techniques mean that more than one bit can be placed on a baud, so to speak. To take a common example, a 9,600 bit per second modem is, in reality, a 2,400 baud modem with advanced coding such that four bits are impressed on each baud. The continuing development of newer and newer modems point to increasingly advanced coding techniques, bringing higher and higher bit per second speeds. My latest modem, for example, is 56,000 bits per second. Baud is named after Jean-Maurice Emile Baudot. See Baudot Code.

Baudot Code The code set used in Telex Transmission, named for French telegrapher Jean-Maurice Emile Baudot (1845-1903) who invented it. Also known by the ITU approved name, International Telegraph Alphabet 2. The Baudot code has only five bits, meaning that only 32 separate and distinct characters are possible from this code, i.e. 2 x 2 x 2 x 2 x 2 equals 32. By having one character called Letters (usually marked LTRS on the keyboard) which means "all the characters that follow are alphabetic characters," and having one other key called Figures (marked FIGS), meaning "all characters that follow are numerals or punctuation characters," the Baudot character set can represent 52 (26 x 2) printing characters. The characters "space," "carriage return," "line feed" and "blank" mean the same in either FIGS or LTRS. TDD devices (Telecommunications Devices for the Deaf) use the Baudot method of communications to communicate with distant TDD devices over phone lines. See also ASCII, EBCDIC, Morse Code and Unicode, which are other ways of encoding characters into the ones or zeros needed by computers.

bay A telephone industry term for the space between the vertical panels or mounting strips ("rails") of the rack. One rack may contain several bays. A bay is another place you put equipment.

Bayesian Inference The theoretical underpinnings can be traced back to Thomas Bayes, an 18th century English cleric whose works on mathematical probability were not published until after his death ("Philosophical Transactions of the Royal Society of London" 1763). Bayes' work centred on calculating the probabilistic relationship between multiple variables and determining the extent to which one variable impacts on another. A typical problem is to judge how relevant a document is to a given query or agent profile. Bayesian theory aids in this calculation by relating this judgement to details that we already know, such as the model of an agent. More formally, the resulting, "a posteriori," which is applicable in judging relevance can be given as a function of the known "a priori" models and likelihood theories. Adaptive probabilistic concept modelling (APCM) analyses correlation between features found in documents relevant to an agent profile, finding new concepts and documents. Concepts important to sets of documents can be determined, allowing new documents to be accurately classified. This sort of advanced mathematical concepts are used in call centers to improve customer service.

BBC World Service One of the world's most widely recognized international broadcasters of radio programming, transmitting in over 30 languages by FM radio, shortwave radio, radio relay, satellite radio, and now also by Internet radio to over 200 countries. The quality of its programming is intended, in part, to promote respect for the UK abroad. The service began in 1932 and was initially called the BBC Empire Service. BBC World Service is funded by the British Government through the Foreign and Commonwealth Office.

BBG Basic Business Group.

BBG-I ISDN Basic Business Group.

BBIN BroadBand Intelligent Network. See Broadband and IN.

BBN Communications BBN Communications was the implementer and operator of the ARPANET, the forerunner of today's Internet. It was also responsible for a number of networking firsts: the first packet switch, the first router, and the first person-to-person network email. BBN Communications also designed, built and operated the Defense Data Network.

BBS Bulletin Board System. Another term for an electronic bulletin board. Typically a PC, modem/s and communications bulletin board software attached to the end of one or more phone lines. Callers can call the BBS, read messages and download public domain software. The person who operates a BBS is called a system operator, commonly shortened to SYSOP (pronounced "sis-op"). See also Electronic Bulletin Board.

BBUL Bbumpless Build-up Layer Packaging. Intel's new technology that helps build thinner, faster processors more cheaply. The technique will debut in 2006 or 2007 and semiconductor chips based on it would be thinner than a dime.

Bc Committed Burst Size. A Frame Relay term. The maximum amount of data bits that a frame relay network agrees to transfer under normal conditions during period of time, or Measurement Interval (T). Above the Bc, the data bits can be marked DE (Discard Eligible), indicating to the various network devices that those bits are eligible to be discarded should the network suffer congestion. See also Committed Information Rate.

BCC 1. Bellcore Client Company. What Bellcore called its original owners – the seven regional Bell operating companies and their operating phone company subsidiaries. This definition is something of a historical footnote, as Bellcore was sold to SAIC and there no longer are seven RBOCs and Bellcore is now called Telcordia Technologies. See Bellcore and RBOC.

2. Block Check Character. A control character appended to blocks in character-oriented protocols and used for figuring if the block was received in error. BCC is especially used in longitudinal and cyclic redundancy checking. As a packet (or in IBM jargon, a frame) is assembled for transmission, the bits are passed through an algorithm to come up with a BCC. When the packet is received at the other end, the receiving computer also runs the same algorithm. Both machines should come up with the same BCC. If they do, the transmission is correct and the receiving computer sends an ACK – a positive acknowledgement. If they don't, an error has occurred during transmission, and they don't have the same bits in the packet. The receiver transmits a signal (a NAK, for negative acknowledgement) that an error has occurred, and the sender retransmits the packet. This process goes on until the BCC checks.

3. Blind Carbon Copy.

BCCH Broadcast Control CHannel. A wireless term for the logical channel used in certain cellular networks to broadcast signaling and control information to all cellular phones. BCCH is a logical channel of the FDCCH (Forward Digital Control CHannel), defined by IS-136 for use in digital cellular networks employing TDMA (Time Division Multiple Access). The BCCH comprises the E-BCCH, F-BCCH and S-BCCH. The E-BCCH (Extended-BCCH) contains information which is not of high priority, such as the identification of neighboring cell sites. The F-BCCH (Fast-BCCH) contains critical information which must be transmitted immediately; examples include system information and registration parameters. S-BCCH (System message-BCCH), which has not yet been fully defined, will contain messages for system broadcast. See also IS-136 and TDMA.

BCD Binary Coded Decimal. A system of binary numbers where each digit of a number is represented by four bits. See BINARY. In ATM, binary code decimal is a form of coding of each octet within a cell where each bit has one of two allowable states, 1 or 0.

BCH Bose, Chaudhuri and Hocquenghem error correction code. Named after the three guys who invented it.

BCHO Base-Controlled Hand-Off. A cellular radio term.

BCL Base General Premises Cabling Licence. In Australia, BCL is the license that allows cablers to perform building cabling. It replaced the General Premises Cabling Licence (GPCL).

BCM 1. Bit Compression Multiplexer.

2. Basic Call Model (a term from the Bellcore discussion of Advanced Intelligent Networks). See AIN.

BCOB An ATM term. Broadband Connection Oriented Bearer: Information in the SETUP message that indicates the type of service requested by the calling user.

BCOB-A Bearer Class A: Indicated by ATM end user in SETUP message for connection-oriented, constant bit rate service. The network may perform internetworking based on AAL information element (IE).

BCOB-C Bearer Class C: Indicated by ATM end user in SETUP message for connection-oriented, variable bit rate service. The network may perform internetworking based on AAL information element (IE).

BCOB-X Bearer Class X: Indicated by ATM end user in SETUP message for ATM transport service where AAL, traffic type and timing requirements are transparent to the network.

BCP See Best Current Practice.

BCS Batch Change Supplement. A batch change supplement is a supplement to a batch change in the programmed logic associated with a general purpose computer system, or with a specialized computer or computerized system such as a PBX or router. A "batch change" refers to a change in programmed logic that is made on a batch basis (i.e., in one big bunch of changes). Sometimes, batch changes require additional changes, in the form of supplements.

BCSM See Basic Call State Model.

BD See Building Distributor.

BDC See Backup Domain Controller.

BDFB Breaker Distribution Fuse Box, a distinct rack of equipment in the Switch Room.

BDLC Burroughs Data Link Control, a bit oriented protocol.

BDN Bell Data Network. The predecessor to ACS (Advanced Communication System). See ACS.

BDCS Broadband Digital Cross-Connect.

BDSL Broadband Digital Subscriber Line. Same as VDSL.

BDT Billing Data Tape.

Be Burst excess, or Excess Burst Size. A Frame Relay term. See Excess Burst Size.

beacon A token ring frame sent by an adapter indicating that it has detected a serious ring problem, such as a broken cable or a MAU. An adapter sending such frames is said to be beaconing. See Beaconing.

beacon initiative A "plan and rationale to help bring the information highway to Canadian businesses and consumers. The central theme of the initiative is an open collaborative effort with all interested players to bring enhanced interactive data, image and video services to Canadians." That's what the press release of April 5, 1994 said. Companies in the Beacon Initiative include BC Telephone, AGT Limited, SaskTel, Manitoba Telephone System, Bell Canada, NBTel, Maritime Tel & Tel, Island Telephone, Newfoundland Telephone and Stentor Communications.

beacon interval Data transmitted on your wireless network that keeps the network synchronized.

beaconing Token Ring process to recover the network when any attached station has sensed that the ring is inoperable due to a hard error. Stations can withdraw themselves from the ring if necessary. A station detecting a ring failure upstream transmits (beacons) a special MAC frame used to isolate the location of the error using beacon transmit and beacon repeat modes. See Beacon.

beam A way to exchange information between Newton users. Beaming uses the small built-in infrared unit at the top of the Newton MessagePad to send anything that's on one MessagePad to another MessagePad, or to a Sharp Wizard. This can be done across a conference table.

beam bender A repeater that is used to direct a base station's wireless signal to an area that is blocked by an obstruction from the base station antenna's coverage pattern. For example, a beam bender may be installed to reach wireless subscribers in an urban canyon or in a natural valley that is blocked from the base station by a ridgeline. At a minimum, a beam bender consists of a receive antenna and a separately mounted transmit antenna, with the two antennas being connected to each other by a cable or wireless arrangement. A beam bender is installed in a location that has line of sight to both the base station and subscribers.

beam diameter The distance between the diametrically opposed points on a plane perpendicular to the beam axis at which the irradiance is a specified fraction of the beam's peak irradiance. The term is most commonly applied to beams that are circular or nearly circular in cross-section.

beam divergence As through the air telecom signal go further, they diverge. Beam divergence measures that divergence.

beam-steering mechanism The mechanism that steers a laser in a free-space optical communication system, for example, in an optical switch.

beamsplitter A device for dividing an optical beam into two or more separate beams, often a partially reflecting mirror.

beamwidth 1. The width of the main lobe of an antenna pattern. Usually the beamwidth of an antenna's main lobe is defined as 3 dB down from the peak of the lobe.

2. The angle of signal coverage provided by a radio. Beamwidth may be decreased by a directional antenna to increase gain.

beans 1. Also called a B Connector or Plain B Wire Connector. A twisted-pair splicing connector that looks like a one-inch drinking straw. They have metal teeth inside them to pierce the vinyl insulation of the wire to make a good connection. Sometimes water-retardant jelly is placed inside. See B Connector.

2. Beans are also filled with nutrients essential to a healthy diet. My children love them because of their after-effects.

bear pit In Elizabethan England, bear baiting consists of letting a pack of crazed hounds loose on a chained bear, and watching from a safe distance while the beasts fight. Very popular. Almost as much fun as a public hanging. Even the Queen thinks this is great fun. One of the most famous of these bears is called Sackerson.

bear trap The stock market is on a long-term slide. Then suddenly it rises for 10%, 20%, or 30%. Does this mean the bear market is over or will it start up again when this little rally stops? A bear market is characterized by many "boomlets." I counted over eight boomlets in the period of 2000 through 2002. Yet the market continued to fall. When a boomlet happens many investors start talking about the beginning of a new bull market. But it's not. It's just a pause. And investors call it a bear trap.

bearer channel In its most basic definition, a bearer channel is a basic communication channel with no enhanced or value-added services included other than the bandwidth transmission capability.

In private line international telecommunications, a bearer channel can consist of any number of DS-0s between two countries that have an agreement to pass traffic between them. This type of bearer is usually an E-1 link between two countries but can be 384 Kbps or any number of 64 Kbps channels combined together. Bearer is usually referred to as a half channel between the two countries that have an agreement to pass traffic, and is referred to as the international half circuit. The middle of the ocean or satellite up-link is the demarcation point for these bearers.

In a more elevated (and more recent) definition, a bearer channel is 64,000 bits per second full duplex. It's the basic building block of the digital signaling hierarchy. An ISDN BRI channel consists of two bearer channels of 64,000 bits per second and one data signaling channel of 16,000 bits per second. It is thus called 2B (two Bearer) + 1D (one D, for data channel. An ISDN PRI (Primary Rate Interface) is 23 B + 1D. The D or Data channel for PRI is 64 Kbps. See ISDN.

BEARS Billing Exchange Account Record System. An RBOC system for ICS processed through CMDS. Similar to CATS.

beat A sum or difference frequency resulting from the mixing of two signals. See Heterodyne.

beat frequency An old radio term: The frequency resulting when an oscillation of one frequency is "beat" or heterodyned against an oscillation of a different frequency. The figure given is normally in cycles per second.

beat reception An old radio term. The resultant audible frequency when two sources of unequal undamped electrical oscillations of constant amplitude act simultaneously in the same circuit. See Beat Frequency.

beating The phenomenon in which two or more periodic quantities having slightly different frequencies produce a resultant having periodic variations in amplitude.

beauty contest The process by which the executives of a company that is considering doing an IPO interview a number of investment banks to determine which ones would do the best job of managing the offering and providing ongoing research reports once the company is public. The parade of investment bankers through a company's offices is known as a beauty contest.

beaver tail Also called Banjo. Used to connect devices to modular jack wiring for testing.

BECKON See BECN.

BECN Backward Explicit Congestion Notification. A Frame Relay term defining a 1-bit field in the frame Address Field for use in congestion management. During periods of severe congestion, the network will advise the (transmitting) device in the backward direction of this fact. This bit notifies the user that congestion-avoidance procedures should be initiated for traffic in the opposite direction of the received frame. It indicates that the frames that the user transmits on this logical connection may encounter congested resources. In other words, slow down, you move too fast. Your bits may not get through or get delayed. In reality, few devices are capable of "throttling back" based on such advice. Consider a router which connects multiple workstations transmitting across multiple LANs or LAN segments to a Frame Relay network. It is highly unlikely that the router can advise those workstations to pause or slow down the rate of transmission. It is possible that the router can buffer some small number of frames for a short period of time, but that's about it. BECN is a wonderful concept, but one which cannot effectively be implemented.

Bed of Nails Cord A description of a type of alligator clip that attaches to the end of a craft test set, also called a butt set.

Bedbug Letter Years ago, the story goes, when people still traveled in Pullman sleeping cars, a passenger found a bedbug in his berth. He immediately wrote a letter to George M. Pullman, president of the Pullman's Palace Car Company, informing him of this unhappy fact, and in reply he received a very apologetic letter from Pullman himself. The company had never heard of such a thing, Pullman wrote, and as a result of the passenger's experience, all of the sleeping cars were being pulled off the line and fumigated. The Pullman's Palace Car Company was committed to providing its customers with the highest level of service, Pullman went on, and it would spare no expense in meeting that goal. Thank you for writing, he said, and if you ever have a similar problem – or any problem – do not hesitate to write again. Enclosed with this letter, by accident, was the passenger's original letter to Pullman, across the bottom of which the president had written, "Send this S.O.B. the bedbug letter."

beeper A colloquial term for a mobile pager – one you carry on your belt or in your purse. In the very beginning, when pagers went off, they made beeping sounds. Thus, they became known as beepers. Not very exciting.

beeper sitting To take on responsibility for writing down incoming pages on behalf of a vacationing or otherwise out-of-range beeper-owning friend. "Harry is beeper-sitting for Gerry while he's in Aspen."

beepilepsy A serious disease afflicting those of us with vibrating pagers, characterized by sudden spasms, goofy facial expressions and loss of speech.

bee's wax In the 18th century, personal hygiene left much room for improvement. As a result, many women and men had developed acne scars by adulthood. The women would spread bee's wax over their facial skin to smooth out their complexions. When they were speaking to each other, if a woman began to   stare at another woman's face she was told "mind your own bee's wax." Should the woman smile, the wax would crack, hence the term "crack a smile." Also, when they sat too close to the fire, the wax would melt and hence the expression "losing face."

beets Beets reminded early cooks of a bleeding animal when they cut them open, so they started calling them "beets." This was derived from the French word bOxEOte, meaning "beast."

BEF Band Elimination Filter. A filter that has a single continuous attenuation band, with neither the upper nor lower cut-off frequencies being zero or infinite.

beggarware The author, Marius Milner, of the famous program Network Stumbler calls his program BeggarWare. He says "You may use it without any formal obligation to the author, except that you may not sell it to anybody. If you like the program and feel that you've received some benefit from it, and want to help me develop future versions, then you may pay as much as you feel the program is worth. If you are using this as part of your work, you are encouraged to help – particularly if it has alerted you to security problems. You are not obligated to pay for it."

Beidou China's navigation satellite system, which currently provides only countrywide coverage with its three geostationary satellites. As the number of satellites in Beidou's constellation increases (two more are scheduled to be launched in 2007), coverage will expand to neighboring countries. The end game is for a 35-satellite network providing global coverage.

Bel A relative measurement, denoting a factor of ten change. Rarely used in practice; most measurements are in decibels, or dB (0.1 Bel). See also dB.

Belden A major manufacturer of communications cable. Belden has set many cabling standards. Many other cable manufacturers follow their specs.

bell A bell in a telephone instrument rings when a 20 Hz signal of about 90 volts AC is applied to the subscriber loop. In contrast, the normal voltage applied to a subscriber loop and used for speaking and listening is 48 volts DC.

Bell, Alexander Graham The following biography of Mr. Bell is courtesy The Bell Homestead Museum Complex, Brantford, Ontario, Canada. www.bfree.on.ca./comdir/festivals/sesqui/bell.html. "Alexander Graham Bell was born in Scotland on March 3, 1847 and was the second son of Alexander Melville Bell and the former Eliza Grace Symonds. He was educated at the University of Edinburgh and the University of London. In the 1860s, while they were still living in Scotland, disaster struck the Bell family. Aleck's younger brother Edward Charles died of tuberculosis and, soon afterwards, his elder brother Melville James died from the same disease. Doctors warned his parents that Aleck, too, was threatened by the disease. His father promptly sacrificed his career as a noted teacher of speech and researcher into the problems of the deaf and, in July, 1870, sailed with his family for Canada in search of a better climate. Upon arrival, the Bells bought the property at Brantford now known as the Bell Homestead. Alexander Graham Bell was, at this time, 23 years old. In this wondrous new climate, Bell's health was quickly restored. Like his father, Bell was a teacher of speech, and in the spring of 1871, he accepted an invitation to teach in Boston and moved there to pursue his career. In 1872, he opened a school of his own for teachers of the deaf and, the next year, became Professor of Vocal Physiology at Boston University. The deafness of his mother no doubt provided an inspiration for Bell to follow in his father's footsteps with work in speech studies. It was this work that led Bell to both his bride, Mabel Hubbard – left totally and permanently deaf from Scarlet Fever when she was five years old - and to his ideas for the telephone. At Christmas, during the summer vacations, and at every opportunity, Bell came home to Brantford. On July 26, 1874, while visiting the Homestead, he talked far into the night while disclosing the telephone idea to his father. After young Bell returned to Boston, he began to work in earnest on his new invention and, on June 3, 1875, succeeded in transmitting speech sounds. During a visit home to Brantford in September of that same year, he wrote the specifications for the telephone. On March 10, 1876, at Boston, through the Liquid Transmitter he had designed, Alexander Graham Bell uttered the first words to be carried over a wire - "Mr. Watson, come here, I want you!" In the summer of 1876, back in Brantford again, young Bell conducted three great tests of the telephone. In the first of these three tests, Bell received the first successful telephone call carried between two communities on August 3, 1876 from Brantford to Mount Pleasant. The second great test was made on August 4, 1876, when a large dinner party at the Bell Homestead heard speech recitations, songs, and instrumental music from the telegraph office in Brantford over a line three and one-half miles long. The third great test is hailed as the first long-distance phone call in the world made on August 10, 1876 from Brantford to Paris, Ontario. At last! The invention of the telephone was complete. In his later years, Bell interested himself in the problem of mechanical flight, carrying out experiments with man-lifting kites, working with these at his summer home at Baddeck, Nova Scotia. Bell died at his beloved home at Baddeck on August 2, 1922, and is buried there."

Bell also invented the photophone, the ancestor of optical transmission systems; the audiometer, which is still used to test hearing; a device to find icebergs by listening to underwater echoes, the ancestor of SONAR; a desalinating machine for removing salt from seawater; a medical jacket that was the predecessor of the iron lung; the hydrofoil boat; and was part of a team that invented the three-wheel landing gear and wing flaps for airplanes. See also Reis, Johann Philipp, a German who may have invented the phone years before Alexander Graham Bell. See also Decibel, Photophone, and SONAR.

Bell 103 AT&T's modem and modem specification providing asynchronous originate/answer full duplex transmission at speeds up to 300 bits per second (300 baud) using FSK (Frequency Shift Keying) modulation on dialup voice phone lines. At one stage this was the most common standard for modems running with personal computers. Every dial service in the U.S. adheres to this standard, and obviously, now, a whole bunch of much faster standards. See the next few definitions and the V.xx standards. See also Acoustic Coupler.

Bell 201 An AT&T standard for synchronous 2400-bps full-duplex modems using DPSK modulation. Bell 201B was originally designed for dialup lines and later leased lines. Bell 201C was designed for half-duplex operation over dialup lines.

Bell 202 An AT&T standard for asynchronous 1800-bps full-duplex modems using DPSK modulation over four-wire leased lines as well as 1200-bps half- duplex operation over dialup lines.

Bell 208 An AT&T standard for synchronous 4800-bps modems. Bell 208A is a full-duplex modem using DPSK modulation over four-wire leased lines. Bell 208B was designed for half-duplex operation over dialup lines.

Bell 209 An AT&T standard for synchronous 9600-bps full-duplex modems using QAM modulation over four-wire leased lines or half-duplex operation over dialup lines.

Bell 212 AT&T specification for a modem providing full duplex asynchronous or synchronous data transmission at speeds up to 1,200 bits per second on the voice dial-up phone network.

Bell 43401 Bell Publication defining requirements for transmission over telco-supplied circuits that have dc continuity (i.e. circuits that are metallic).

Bell Atlantic Formed as a holding company after the AT&T Divestiture. Included Bell of Pennsylvania; C&P Telephone Companies of D.C., Maryland, Virginia, and West Virginia; Diamond State Telephone (MD); and New Jersey Bell. In April, 1997, Bell Atlantic bought NYNEX (the holding company for New York Telephone and New England Telephone) for $22.1 billion. NYNEX had customers in New York, Massachusetts, New Hampshire, Vermont, Maine and a small part of Connecticut. On June 30, 2000, Bell Atlantic merged with GTE to form Verizon Communications. See also Bell Operating Companies, and Verizon.

Bell Communications Research Bellcore. Formed at Divestiture to provide certain centralized services to the seven Regional Holding Companies (RHCs) and their operating company subsidiaries. Also serves as a coordinating point for national security and emergency preparedness communications matters of the federal government. Bellcore does not work on customer premise equipment or other areas of potential competition among its owners – the seven (now five) Regional Bell Operating companies. Bellcore also works on standardizing methods by which customers of long distance companies will reach their favorite long distance companies. At time of this writing, around 8,000 people worked at Bellcore, mostly in northern New Jersey. Bellcore had the unenviable task of trying to service the needs of the seven competitors (now five), which owned it. They agreed to sell it to SAIC in 1997. At the time, Bellcore's annual budget was around $1 billion, paid for by the seven (now five) Bell regional operating companies. Bellcore's new name is Telcordia Technologies. See also Bellcore for a longer explanation.

Bell compatible A term sometimes applied to modems. A modem is said to be "Bell compatible" if it conforms to the technical specifications set forth by AT&T for the various devices, such as Bell 212.

bell curve See Gaussian Beam.

Bell customer code A three-digit numeric code, appended to the end of the Main Billing Telephone Number. Used by Local Exchange Carriers to provide unique identification of customers.

Bell Labs The new name for Bell Telephone Laboratories. Bell Labs is the R&D (Research and Development) arm of Lucent Technologies. Bell Telephone Laboratories was organized in 1925 to consolidate the research laboratories of AT&T and Western Electric, the manufacturing arm of the Bell System. The name changed to AT&T Bell Laboratories in 1984, when the Bell System was broken up by the Modified Final Judgement (MFJ), which separated the BOCs (Bell Operating Companies) from AT&T. On January 1, 1997, AT&T voluntarily separated into AT&T Corporation, Lucent Technologies and NCR Corporation. Bell Labs became the R&D arm of Lucent, and AT&T formed its own R&D arm known as AT&T Laboratories. See also Bell System, Lucent, and MFJ.

Bell Operating Company BOC. A BOC is one the 22 regulated telephone companies of the former Bell System, which was broken apart (the Divestiture of the Bell System) at midnight on December 31, 1983. At Divestiture, the Bell operating companies were grouped into seven Regional Holding Companies (RHCs). According to the terms of the Divestiture Agreement between the Federal Courts, the Federal Government and AT&T, the divested companies must limit their activities to local telephone services, directory service, customer premise equipment, cellular radio and any other ventures as the Federal Court may approve from time to time. BOCs are specifically limited from manufacturing equipment and from providing long distance service. See also Regional Bell Operating Company.

The following definition of a Bell Operating Company comes from the Telecommunications Act of 1996: The term 'Bell operating company'

(A) means any of the following companies: Bell Telephone Company of Nevada, Illinois Bell Telephone Company, Indiana Bell Telephone Company, Incorporated, Michigan Bell Telephone Company, New England Telephone and Telegraph Company, New Jersey Bell Telephone Company, New York Telephone Company, U S West Communications Company, South Central Bell Telephone Company, Southern Bell Telephone and Telegraph Company, Southwestern Bell Telephone Company, The Bell Telephone Company of Pennsylvania, The Chesapeake and Potomac Telephone Company, The Chesapeake and Potomac Telephone Company of Maryland, The Chesapeake and Potomac Telephone Company of Virginia, The Chesapeake and Potomac Telephone Company of West Virginia, The Diamond State Telephone Company, The Ohio Bell Telephone Company, The Pacific Telephone and Telegraph

Company, or Wisconsin Telephone Company; and

(B) includes any successor or assign of any such company that provides wireline telephone exchange service; but

(C) does not include an affiliate of any such company, other than an affiliate described in subparagraph (A) or (B).

Bell South See BellSouth.

Bell Speak A term coined by Michael Marcus for insider jargon spoken by "real" telephone people – those who practiced pre-divestiture. Such old jargon is usually incomprehensible to anyone in today's telephone industry who is younger than 46.

Bell System The entire AT&T organization prior to when it was broken up – at the end of 1984. The Bell System included Bell Labs, Long Lines, Western Electric and the 23 Bell operating companies. See AT&T and Lucent.

Bell System Practices BSPs. The book that explains everything MA Bell did before divestiture in 1984. No longer used by regional Bells or AT&T. Manufacturers now issue their own instructions. Each RBOC has its own way of doing things. And many BOCs inside the RBOCs also do things their own way. This is not designed to confuse the poor business customer, who may have circuits in many states and thus deals with many of them – but just to encourage diversity, or something.

Bell System Reference Frequency BSRF. See Timing.

Bell Telephone Laboratories See Bell Labs.

belly-buttons The Web world counts eyeballs. The insurance and managed care industries count belly buttons. One person equals one belly button. So an insurance policy that covers five belly buttons actually covers five distinct individuals.

bellboy A public paging system run by local Bell phone companies. The name survived for many years and only drew criticism in the middle to late seventies with the rise of the Women's Liberation movement. It's now not used. Shucks.

Bellco Bellco = Bell company.

Bellcore Bell Communications Research. Bellcore was formed by federal mandate co-incident with the Divestiture of AT&T in 1984 to provide certain centralized research and development services for its client/owners, the seven (now five) Regional Bell Operating Companies (RBOCs), and their operating company subsidiaries, the BOCs (Bell Operating Companies). The formation of Bellcore was deemed critical at the time, as the MFJ (Modified Final Judgment) called for Bell Telephone Laboratories (Bell Labs) to remain with AT&T. This action left the RBOCs without R&D support. Bellcore was initially staffed with researchers who worked at Bell Labs and who didn't move over to AT&T. Bellcore's focus has been on standards, procedures, software development and all manner of R&D of common interest to the RBOCs, with the exception of the physical sciences. The MFJ precluded the RBOCs from equipment manufacturing, and even from close involvement in the equipment design process. Traditionally, much of Bellcore's efforts have been in the area of development and support for Operations Support Systems (OSSs) such as Centrex management, line testing, order negotiation and processing, and billing systems. Bellcore also has been a key player in the standards design and systems development work for Advanced Intelligent Networks (AINs).

Bellcore has served as a coordinating point for national security and emergency preparedness, and communications matters of the federal government; this was natural, as the RBOCs and their operating units were (and still are) the dominant carriers responsible for such implementing and supporting such things. Bellcore also was responsible for administering the North American Numbering Plan (NANP), although in 1995 this responsibility was shifted to a more impartial entity, the NANC (North American Numbering Council), in the face of a deregulated and competitive telecommunications landscape. You can acquire Bellcore documents from Bellcore – Document Registrar, 445 South Street, Room 2J-125, P.O. Box 1910, Morristown, NJ 07962-1910. Fax 201-829-5982. www.bellcore.com.

In early 1995, Bellcore's Board of directors announced that the owners (read RBOCs) intended to sell the company, as their interests largely were no longer common in anticipation of deregulation and full competition. Bellcore's 5,600 employees and $1 billion+ budget clearly were not supportable in this environment. On November 21, 1996 it was announced that SAIC (Science Applications International Corporation) had agreed to purchase Bellcore. According to the SAIC press release, "Employee-owned SAIC provides high technology products and services to government and private industry, systems integration, national security, energy, transportation, telecommunications, health care and environmental science and engineering." SAIC revenues are over $2.2 billion; the company employs over 22,000 in more than 475 locations worldwide. SAIC also owned Network Solutions, Inc. (NSI), which runs certain functions for the Internet; NSI was spun off as a public company in early 1998. www.saic.com. As of March 9, 1999, Bellcore changed its name to Telcordia

Technologies. The background to the name change? Due to the sale of Bellcore to the new parent company, SAIC, it was agreed in the sale that Bellcore (BELL COmmunications & REsearch) would not retain the name forever. The name implies they soley work for the regional Bell operating companies. Bellcore was allowed to retain the name for one year, and within six months after that, Bellcore could/can continue to use, if it chose "Formerly known as Bellcore". This was to allow Bellcore to retain market exposure using the name and allow the industry time to realize that they were the same company, with a different name and a new direction! See also BCC.

Bellcore AMA Format BAF. The standard data format for AMA (Automatic Message Accounting) data to be delivered to the Revenue Accounting Office (RAO) in Advanced Intelligent Network Release 1.

Bellcore Multi-Vendor Interactions MVI. The process for coordinating the efforts of Bellcore, the Bell Operating telephone companies and vendors to address technical issues associated with Advanced Intelligent Network.

BELLE A chess-playing computer developed by Joseph Condon, a physicist, and Ken Thompson, an electrical engineer, at Bell Labs. BELLE was orignally developed in 1973 as a software program that could run on any standard computer. It was able to evaluate around 200 moves per second. Over the next four years, Condon and Thompson developed a specialized computer for BELLE, enabling BELLE to evaluate 160,000 moves per second. Until 1983, BELLE was the most powerful chess machine in the world. In the fall of 1983, when BELLE was defeated by a chess-playing program running on a brand new Cray super-computer, BELLE held an International Chess Federation master rating of 2200.

bellhead A member of the telecom establishment, in contrast to a nethead, who believes that the PSTN is a relic, the Internet is the future of all telecommunications and networking, and Bellheads should not be allowed to control the Internet.

Bellman-Ford Algorithm The shortest path routing algorithm that figures on the number of hops in a route in order to find shortest-path spanning tree.

BellSouth Corporation The largest regional Bell holding company formed at the Divestiture of AT&T. Includes Southern Bell and South Central Bell and several other BellSouth businesses. See Regional Bell Operating Company.

bellwether One that takes the lead or initiative; an indicator of trends. The state of California, for instance, long has been a bellwether with respect to the regulation of utilities through its PUC (Public Utilities Commission). The term "bellwether" originated in the 15th century, when the medieval English began the practice of putting a bell around the neck of a male sheep which had been castrated before reaching sexual maturity. As a result of neutering, the ram was much easier to handle, and to train as the leader of the flock. The bell made it much easier to find. Not surprisingly, not every state is anxious to be known as a bellwether.

below water Because you're such a valuable employee, you have been given stock options in your company. The idea is in a year or two – whatever the rules are – you can exercise your options at $5, i.e. buy shares in your company at $5 a share. And you can then sell your $5 stock at $100 because that's what your company's stock is now selling at on the stockmarket. Bingo, you've made $95 on each of your 10,000 options – for a total of $950,000. That's the theory. But something awry happens along the way. Your company's stock craters. It's now selling at $3. Your stock options are now $2 under water.

below-the-line Expenses incurred that are charged to shareholders of regulated operating telephone companies, not ratepayers.

benchmark A standardized task to test the capabilities of various devices against each other for such measures as speed.

bend loss A form of increased attenuation caused by allowing high-order signals to radiate from the side of the fiber. The two common types of bend losses are those occurring when the fiber is curved and microbends caused by small distortions of the fiber imposed by poor cabling techniques. Bend loss is measured in decibels, i.e. dB.

bend radius The smallest bend which may be put into a cable under a stated pulling force without either causing damage to the cable or causing attenuation (i.e., signal loss). Bend radius affects the size of bends in conduits, trays or ducts. It affects pulley size. It affects the size of openings at pull boxes where loops may form. If the bend radius is too severe, the cable or fiber can crack or break, which is not a good thing. Coax and shielded copper cables can suffer cracks in the outer conductors, which also is not a good thing. Typically, the bend radius is 4X, 6X or 10X the outside diameter of the cable, dependent on specific performance characteristic limitations. See also Macrobend Loss and Microbend Loss.

bending radius See Bend Radius.

bent pipe Bent pipe refers to a conventional satellite link provisioned through a transponder, which essentially is either an amplifier (analog) or a repeater (digital). This configuration involves a transmitting earth station (i.e., terrestrial antenna) on the uplink, a satellite (i.e., non-terrestrial antenna), and a receiving earth station. The satellite transponder, which is the operational element of the satellite, does little more that receive the signal over the uplink frequency, boost the signal power, clean up the signal, shift the frequency, and retransmit the signal over the downlink frequency. The satellite link, therefore, is nothing more than a dumb RF (Radio Frequency) transmission pipe that is bent approximately 22,300 miles up and 22,300 down through a GEO (Geosynchronous Earth Orbiting) satellite. The next generations of satellites have embedded processing power, variously in the forms of circuit switches, fast packet switches, and Statistical Time Division Multiplexers (STDMs). See also GEO.

Bentogram In binary encoding, a syntax for encoding Versit's vCard. Bentograms are based on Benton, the OpenDoc standard interchange format.

BEP Back End Processor.

BER 1. Bit Error Rate. The ratio of error bits to the total number of bits transmitted. If the BER gets too high, it might be worth while to go to a slower baud rate. Otherwise, you would spend more time retransmitting bad packets than getting good ones through. The theory is that the faster the speed of data transmission the more likelihood of error. This is not always so. But if you are getting lots of errors, the first – and easiest – step is to drop the transmission speed. Bit Error Rate is thus a measure of transmission quality. It is generally shown as a negative exponent, (e.g., 10 to the minus 7 which means one out of 10,000,000 bits are in error.) See also Bit Error Rate.

2. A LAN term, which means Basic Encoding Rules. Rules for encoding data units described in ASN.1. (Abstract Syntax Notation One. LAN "grammar," with rules and symbols, that is used to describe and define protocols and programming languages. ASN.1 is the OSI standard language to describe data types.) BER is sometimes incorrectly lumped under the term ASN.1, which properly refers only to the abstract syntax description language, not the encoding technique.

Berkeley Snoop protocol This explanation is from the Daedalus Research Group at the University of California at Berkeley. The Snoop protocol is a TCP-aware link layer protocol designed to improve the performance of TCP over networks of wired and single-hop wireless links. The main problem with TCP performance in networks that have both wired and wireless links is that packet losses that occur because of bit-errors are mistaken by the TCP sender as being due to network congestion, causing it to drop its transmission window and often time out, resulting in degraded throughput. The Snoop protocol works by deploying a Snoop agent at the base station and performing retransmissions of lost segments based on duplicate TCP acknowledgments (which are a strong indicator of lost packets) and locally estimated last-hop round-trip times. The agent also suppresses duplicate acknowledgments corresponding to wireless losses from the TCP sender, thereby preventing unnecessary congestion control invocations at the sender. This combination of local retransmissions based primarily on TCP acknowledgments, and suppression of duplicate TCP acknowledgments, is the reason for classifying Snoop as a transport-aware reliable link protocol. The state maintained at the base station is soft, which does not complicate handoffs or overly increase their latency. The scheme has been demonstrated to yield 100-2000% throughput improvements for TCP limited by single-hop wireless in-building links under several circumstances.

For data transfers from the mobile host, we design a mechanism called Explicit LOss Notification and use it to implement a link-aware transport protocol. With ELN, TCP uses hints from base stations in the network and/or the receiver to decouple retransmissions from congestion control. This helps it perform congestion control only for congestion-related losses and not for losses due to corruption.

Berners-Lee English computer scientist,Tim Berners-Lee, invented the World Wide Web in 1989 while working at CERN, a physics lab on the Swiss-French border. Time Magazine wrote about Mr. Berners-Lee as follows: "It started, of all places, in the Swiss Alps. The year was 1980. Berners-Lee, doing a six-month stint as a software engineer at CERN, the European Laboratory for Particle Physics, in Geneva, was noodling around with a way to organize his far-flung notes. He had always been interested in programs that dealt with information in a "brain-like way" but that could improve upon that occasionally memory-constrained organ. So he devised a piece of software that could, as he put it, keep "track of all the random associations one comes across in real life and brains are supposed to be so good at remembering but sometimes mine wouldn't." He called it Enquire, short for Enquire Within Upon Everything, a Victorian-era encyclopedia he remembered from childhood. Building on ideas that were current in software design at the time, Berners-Lee fashioned a kind of "hypertext" notebook. Words in a document could be "linked" to other

files on Berners-Lee's computer; he could follow a link by number (there was no mouse to click back then) and automatically pull up its related document. It worked splendidly in its solipsistic, Only-On-My-Computer way. But what if he wanted to add stuff that resided on someone else's computer? First he would need that person's permission, and then he would have to do the dreary work of adding the new material to a central database. An even better solution would be to open up his document – and his computer – to everyone and allow them to link their stuff to his. He could limit access to his colleagues at CERN, but why stop there? Open it up to scientists everywhere! Let it span the networks! In Berners-Lee's scheme there would be no central manager, no central database and no scaling problems. The thing could grow like the Internet itself, open-ended and infinite. "One had to be able to jump," he later wrote, "from software documentation to a list of people to a phone book to an organizational chart to whatever. So he cobbled together a relatively easy-to-learn coding system – HTML (HyperText Mark-up Language) – that has come to be the lingua franca of the Web; it's the way Web-content creators put those little colored, underlined links in their text, add images and so on. He designed an addressing scheme that gave each Web page a unique location, or url (universal resource locator). And he hacked a set of rules that permitted these documents to be linked together on computers across the Internet. He called that set of rules HTTP (HyperText Transfer Protocol). And on the seventh day, Berners-Lee cobbled together the World Wide Web's first (but not the last) browser, which allowed users anywhere to view his creation on their computer screen. In 1991 the World Wide Web debuted, instantly bringing order and clarity to the chaos that was cyberspace. From that moment on, the Web and the Internet grew as one, often at exponential rates. Within five years, the number of Internet users jumped from 600,000 to 40 million. At one point, it was doubling every 53 days. Raised in London in the 1960s, Berners-Lee was the quintessential child of the computer age. His parents met while working on the Ferranti Mark I, the first computer sold commercially. They taught him to think unconventionally; he'd play games over the breakfast table with imaginary numbers (what's the square root of minus 4?). He made pretend computers out of cardboard boxes and five-hole paper tape and fell in love with electronics. Later, at Oxford, he built his own working electronic computer out of spare parts and a TV set. He also studied physics, which he thought would be a lovely compromise between math and electronics. "Physics was fun," he recalls. "And in fact a good preparation for creating a global system." It's hard to overstate the impact of the global system he created. It's almost Gutenbergian. He took a powerful communications system that only the elite could use and turned it into a mass medium. "If this were a traditional science, Berners-Lee would win a Nobel Prize," Eric Schmidt, CEO of Novell, once told the New York Times. "What he's done is that significant." You'd think he would have at least got rich; he had plenty of opportunities. But at every juncture, Berners-Lee chose the nonprofit road, both for himself and his creation. Marc Andreessen, who helped write the first popular Web browser, Mosaic – which, unlike the master's browser, put images and text in the same place, like pages in a magazine – went on to co-found Netscape and become one of the Web's first millionaires. Berners-Lee, by contrast, headed off in 1994 to an administrative and academic life at the Massachusetts Institute of Technology. From a sparse office at M.I.T., he directs the W3 Consortium, the standard-setting body that helps Netscape, Microsoft and anyone else agree on openly published protocols rather than hold one another back with proprietary technology. The rest of the world may be trying to cash in on the Web's phenomenal growth, but Berners-Lee is content to labor quietly in the background, ensuring that all of us can continue, well into the next century, to Enquire Within Upon Anything.

On November 7, 2002 I attended the Guggenheim Marconi International Fellowship Foundation Awards. Mr. Berners-Lee was presented the 30th Marconi Awardee. I had the privilege of sitting near him and speaking with him. He's impressive, and young.

Bernoulli Daniel Bernoulli was the 18th century Swiss mathematician who first expressed the principle of fluid dynamics – the basis of Bernoulli "boxes" also called disk cartridges. A Bernoulli box, which is a mass storage device, uses both floppy and hard disk technologies. Bernoulli disks are removable. They physically look like large floppy disks.

BERT Bit Error Rate Test, or Tester. A known pattern of bits is transmitted, and errors received are counted to figure the BER. The idea is to measure the quality of data transmission. The bit error rate is the ratio of received bits that are in error, relative to the number of bits received. Usually expressed in a power of 10. Sometimes called Block Error Rate Tester. See also BLERT.

BESP Basic Emergency Service Provider. Integral to the services a BESP provides is that it mans telephone exchanges and computer databases where the ANI (Automatic Number Identification) and ALI (Automatic Location Identifier) of a 911 emergency call are added to the message header. The header is then reconnected to the message text and directed to a communication center. The BESP maintains a database that includes the telephone number and associated ANI and ALI. A BESP is an integral component of the 911 emergency service communications network. The observed sequence of a 911 call is:

- 911 is dialed.
- The call goes to the local exchange.
- The call then to goes to the central BESP exchange.
- The central exchange simultaneously directs the call to:
- 1. A communication center where an emergency operator receives the call and
- 2. The BESP.
- The BESP adds the ANI and ALI to the call header.
- The BESP directs the header to the communication center, reconnecting with the message.
- The ANI and ALI are displayed on the terminal of the operator handling the call.
- The ANI and ALI arrive at the terminal within seconds of the voice message.

This definition provided by David Thorndike who had just completed his term as director of BRETSA, the Boulder Regional Emergency Telephone Service Authority.

bespoke Custom-made, made to order, made by engagement, requested item.

best current practice BCP. The sub-series of RFCs (Requests For Comment), which are published by the IETF (Internet Engineering Task Force). BCPs are vehicles by which the IETF community defines and ratifies the best current thinking on a statement of principle or on what is believed to the best to perform various operations of the IETF processes. See also IETF and RFC.

best effort A term for a Quality of Service (QoS) class with no specified parameters and with no assurances that the traffic will be delivered across the network to the target device. ATM's ABR (Available Bit Rate) and UBR (Unspecified Bit Rate) are both best-effort service examples.

best of breed One of the top honors at Westminster, but in the tech world it's supposedly the top software or hardware in its class.

best path An internetworking term: The optimal route between source and destination end stations through a wide area network. Determined through routing protocols such as RIP and OSPF, best path can be based on lowest delay, cost or other criteria.

best-of-need Refers to a product or service that isn't necessarily best-of-breed, but it gets the job done (i.e., meets the need) and costs a lot less than the best-of-breed solution.

Beta 1. Refers to the final stages of development and testing before a product is released to market. "Alpha" is the term used when a product is in preliminary development. "Her baby is in beta," according to Peter Lewis of the New York Times, means she is expecting soon. In the software industry, beta has been known to last a year or more. Microsoft's Bill Gates has given the word "beta" a whole new meaning by having as many as 500,000 "beta testers" for his Windows 95 operating system. At this level, beta testing is no longer testing, it's marketing. And it's positively brilliant. See Beta Test.

2. Business Equipment Trade Association (UK).

3. Informal name for Betacam, a professional color difference videotape recording format that uses the Y, R-Y, and B-Y color difference components. Also the name of a consumer videotape recording format that is completely different from the professional Betacam format.

beta site A place a beta test is conducted. See Beta and Beta Test.

beta test Typically the last step in the testing of a product before it is officially released. A beta test is often conducted with customers in their offices. Some customers pay for the equipment or software they get under a beta test; some don't. Some beta tests stay in (if they work). Some don't. Most products don't work when they're first introduced. So beta tests are a good idea. Unfortunately, most manufacturers don't do sufficient beta testing. They want to get their product to market before the competition does. This often means we now have two or three new products on the market, none of which work reliably or do exactly what they're meant to do. Our rule: always wait several months after a product is introduced before buying it. By then the major bugs will have been fixed. The test before the beta test is called the Alpha. It isn't that common. See Beta.

betacam Portable camera/recorder system using 1/2-inch tape originally developed by Sony. The name may also refer just to the recorder or the interconnect format; Betacam uses a version of the Y, R-Y, B-Y color difference signal set. Betacam is a registered trademark of the Sony Corporation.

betacam SP A superior performance version of Betacam. SP uses metal particle tape and a wider bandwidth recording system.

betamax 1. The noun. A format for video tape which Sony introduced too expensively.

VHS (Video Home System), using half-inch tape introduced by Matsushita/JVC in 1975, effectively killed Sony's attempt to make Betamax the leading video tape standard.

2. The verb. When a technology is overtaken in the market by inferior but better marketed competition as in "Microsoft betamaxed Apple right out of the market." See VHS.

3. The legal term. In 1984 the Supreme Court held that Sony, the maker of Betamax, a video recorder, couldn't be held liable for copyright infringement by Betamax customers, because Betamax had legal uses as well as illegal ones. See Grokster.

betazed A planet in the second Star Trek TV series, inhabited by Betazoids, beings with great powers of empathy and telepathy.

BETRS Basic Exchange Telecommunications Radio Service. BETRS is a digital fixed radio service, or Wireless Local Loop (WLL), that can be used to extend telephone service to rural areas by replacing the traditional wired local loop with a radio communications circuit. BETRS shares the UHF (Ultra High Frequency) and VHF (Very High Frequency) common carrier and private radio frequencies. BETRS operates in the paired 152/158 and 454/459 MHz bands, and on 10 channel blocks in the 816-820/861-865 MHz bands. As many as four simultaneous users can share a single radio channel through TDMA (Time Division Multiple Access). Licensing is restricted to state-certified carriers operating in the area where the BETRS service is to be provided, and is considered by regulators to be part of the PSTN. The technology was developed in the mid-1980s and was licensed by the FCC in 1987. See also TDMA, UHF, VHF, and WLL.

beverage antenna A very long, horizontal single-wire antenna that is strung between two poles. Picture a very long, metallic clothesline. Sometimes the center of the antenna is mounted on a mast and the two sides of the antenna slope toward the ground. Also known as a wave antenna, the Beverage antenna is named after the late Harold Beverage, holder of over 30 radio engineering patents and former chief research engineer for RCA.

bezel The metal or plastic part – in short, the frame – that surrounds a cathode ray tube – a "boob" tube.

Bezeq The name of the erstwhile-monopoly Israeli local and long distance phone company. Its full name is the Israel Telecommunications Corp. Ltd.

BEZS Bandwidth Efficient Zero Suppression. N.E.T.'s patented T-1 zero suppression technique; maintains Bell specifications for T-1 pulse density without creating errors in end-user data; uses a 32 Kbps overhead channel.

BFT Binary File Transfer. BFT is a method of routing digital files using facsimile protocols instead of traditional modem file transfer protocols. See Binary File Transfer for a fuller explanation.

BFV Bipolar violations: The digital data format consists of pulses of opposite polarity. No two consecutive pulses should be the same polarity; if two are detected in a row, the term is a violation, which is also a warning flag.

BGAN Broadband Global Area Network. Think of a satellite providing 432,000 bits per second access to the Internet anywhere in the world. That's a broadband global area network. Inmarsat is planning such a service, or was, when I wrote this definition.

BGE-I ISDN Business Group Elements.

BGID Business Group ID.

BGM Background Music. George Media of the PBX Quality Assurance Group at Panasonic Communications Company says this should be standard in the PBX industry. "At Panasonic, we use this term a lot."

BGP Border Gateway Protocol is a Gateway Protocol which routers (other non-router devices also may be involved as intermediaries) employ in order to exchange appropriate levels of routing information. In an intradomain routing environment between Autonomous Systems (ASs), IBGP (Internal BGP) is run, allowing the free exchange of information between trusted systems. IBGP is in a class of protocols known as IGPs, or Internal Gateway Protocols. In an interdomain environment, EBGP (External BGP) is run, allowing the routers to exchange only pre-specified information with other pre-specified routers in other domains in order to ensure that their integrity is maintained. EBGP is in a class known as EGPs, or External Gateway Protocols. When BGP peer routers (i.e., routers with a TCP connection for purposes of exchanging routing information) first establish contact, they exchange full routing tables, which are maintained in Routing Information Bases (RIBs). Subsequent contacts involve the transmission of incremental changes, only. Peer routers also periodically exchange "keep alive" messages. Should a router cease to receive those messages from a peer, it updates its routing table to delete the associated route, acting under the assumption that either the silent router or the interconnecting link between the two has failed. BGP uses TCP (Transmission Control Protocol) as its transport protocol, as the reliability of the datastream is of critical importance. Specifically, BGP uses TCP port 179 for establishing its connections. BGP is a Path Vector (PV) protocol, which is similar to a Distance Vector (DV) Protocol, but with a key difference. A Distance Vector protocol selects the best path between two border routers based on the hop count (i.e., number of routers transversed). A border router (BR) running a Path Vector routing protocol advertises the destinations it can reach to its neighboring border routers. Further, a path vector protocol pairs each of those destinations with the attributes of the path to it. The attributes include the number of hops (i.e., routers transversed) and the administrative "distance." The attribute of administrative distance weights routes learned from IBGP more heavily than those learned from EBGP. In other words, interior routes are weighted more heavily (and preferred) than are exterior routes, which cross network domains and which, by definition, involve multiple Autonomous Systems (ASs). Note, however, that BGP does not take into account factors such as link speed or network load. BGP-4, originally published in October 1991 in RFC 1267, and updated in RFC 1771. BGP-4 provides a set of mechanisms for supporting Classless Inter-domain Routing (CIDR), and the aggregation of routes. In combination, these mechanisms support the concept of supernetting, which allows multiple (and even noncontiguous) Class C IP network addresses to be combined into a single supernet. See also CIDR, Route Flap, and TCP.

BGP4 See BGP.

BH Bandwidth Hog. A term defined by Philip Elmer-DeWitt, technology editor of Time Magazine in 1994, who spearheaded the launch of Time Online, the first fully electronic national magazine. He defined BH "as a person who uses the online medium like a bullhorn and attracts like-minded people who then rove in a pack, filling them with up with screeds." (Screed is a long discourse or essay.)

BHANG Broadband High Layer Information: This is a Q.2931 information element that identifies an application (or session layer protocol of an application).

BHC Backbone to Horizontal Cross-connect. Point of interconnection between backbone wiring and horizontal wiring.

BHCA Busy Hour Call Attempts. A traffic engineering term. The number of call attempts made during the busiest hour of the day.

BHM Busy Hour Minutes.

BHMC Busy Hour Minutes of Capacity. For Switched Access Service-Feature Groups B and D, this term refers to the maximum amount of access minutes an Interconnector or Interexchange Carrier (IXC) expects to be handled in an End-Office switch at peak activity during any hour between 8 A.M. and 11 P.M.

bi A Latin prefix meaning twice.

bi-directional Antenna that radiates most of its power in two directions.

BIA Burned In Address. On most LAN-interface cards (also called NIC or network interface cards), the 48-bit MAC address is burned into ROM - hence the term Burned-In Address. Also called a MAC Address.

bias 1. A systemic deviation of a value from a reference value.
2. The amount by which the average of a set of values departs from a reference value.
3. An electrical, mechanical, magnetic, or other force field applied to a device to establish a reference level to operate the device.
4. Effect on telegraph signals produced by the electrical characteristics of the terminal equipment.

bias distortion Distortion affecting a two-condition (binary) coding in which all the significant intervals corresponding to one of the two significant conditions have uniformly longer or shorter durations than the corresponding theoretical durations. The magnitude of the distortion is expressed in percent of a perfect unit pulse length.

bias generator A CBX printed circuit card that generates a signal that reduces idle channel noise for all coders installed in the CBX.

bias potential The potential impressed on the grid of a vacuum tube to cause it to operate at the desired part of its characteristic curve.

bias stabilization A means of controlling the bias in a circuit so that it does not fluctuate. Heat or signal variations can throw off bias resulting in damage to components. See Bias, Bias distortion.

Bib Signaling ID assigned by Exchange B.

BIC Acronym for "butt in chair" – the designation for a technician who has to work on Saturday or Sunday, traditionally light days when not much happens. The purpose of the BIC is to be there just in case something happens.

BICC Bearer Independent Call Control.

BICEP An ATM term. Bit Interleaved Parity: A method used at the PHY layer to monitor the error performance of the link. A check bit or word is sent in the link overhead covering the previous block or frame. Bit errors in the payload will be detected and may be reported as maintenance information.

BICI See Broadband Inter-Carrier Interface. This is also the Spanish colloquial word for bicycle.

biconic A screw-on fiber optic connector with a conical shape, the biconic connector was developed by AT&T (now Lucent Technologies). It has fallen out of favor, with most fiber installers preferring SC Connectors or ST Connectors.

biconical antenna An antenna consisting of two conical conductors having a common axis and vertex. Excitation occurs at the common vertex. If one of the cones is flattened into a plane, the antenna is called a discone.

BICSI Building Industry Consulting Service International, a professional organization. For those who acquire certain requisite education and experience by BICSI, the association makes them a RCDD, Registered Communication Distribution Designer.

bicycle networking The practice among cable access TV shows of distributing programming from one local cable access station to another. They use local messengers on bicycles to transport tape.

bid sniping Bidding at the last minute of an auction - especially on an eBay auction – is called "sniping" or bid sniping. There is software and there are services that will automate this process for you. One is at http://tollfreephone.auctionstealer.com/home.cfm.

bidding credit A spectrum auction term. It is a credit given to eligible FCC auction applicants which allows them to receive a discount on their winning bids in an auction.

bidirectional Antenna that radiates most of its power in two directions.

bidirectional bus A bus that may carry information in either direction but not in both simultaneously.

bidirectional couplers Fiber optic couplers that operate in the same way regardless of the direction light passes through them.

Bidirectional Line Switched Ring Commonly referred to as BLSR, bidirectional line switched ring is a method of SONET transport in which half of the working network is sent counter-clockwise over one fiber and the other half is sent clockwise over another fiber. BLSR offers bandwidth use advantages for distributed traffic in single-ring architectures. See also Line Switched Ring, Path Switched ring and SONET.

bidirectional printing A typewriter always prints from left to right. So did the early computer printers. Today's computer printers print from left to right, drop down a line, then print from right to left. This increases the printer's speed.

bidirectional rate shaping A hardware-based technology that enforces traffic policies, tracks usage, and manages traffic concurrently by routing data packets to the logical ingress queue and processing policies in a bidirectional fashion. also referred to as bandwidth by the slice, bidirectional rate shaping enables service providers to maximize revenue and achieve scalability by dividing up available Ethernet capacity into fixed increments that their customers can buy.

BIFF Backhoe Induced Fiber Failure (BIFF). Definition is the same as Backhoe Fade, i.e. the backhoe cut the fiber and that's why it's no longer working.

bifurcated routing Routing that may split one traffic flow among multiple routes.

Big Brother "Big Brother is watching you". A phrase coined by George Orwell in his classic novel '1984', a fictional tale of a totalitarian future in which "Thought police" monitor one's every move via telescreens which broadcast propaganda and have a built in camera and microphone. The phrase is now used to describe the huge amounts of personal information collected by government and commercial organizations.

big hat, no cattle Texas expression used to dismiss a cowboy wanabee. In Lone Star IT circles, it describes a technician with a certificate or degree in computer science, but little or no field experience.

big iron A mainframe computer. Alludes to the fact that in the old days, mainframes were huge. These days, mainframes are much smaller.

big iron switch A legacy Class 4 or Class 5 switch.

Big LEO LEO stands for Low-Earth-Orbit. Big LEO is a low-earth-orbit satellite system that will offer voice and data services.

big pipe services High-speed connections or services that require high-speed connections. That being said, the definition of "big-pipe" has evolved over time. Ten years ago, T1, T3, and OC3 and above were considered big-pipes. Even today, some people still use that definition, although increasingly "big-pipe" refers to Gigabit Ethernet and OC48 and above.

big wig In Shakespeare's time, men and women took baths only twice a year – May and October. Women kept their hair covered, while men shaved their heads (because of lice and bugs) and wore wigs. Wealthy men could afford good wigs made from wool. The wigs couldn't be washed. But to enhance them, they would carve out a loaf of bread, put the wig in the shell, and bake it for 30 minutes. The heat would make the wig big and fluffy, hence the term "big wig." Hence the term, big wig, has come to mean rich and powerful.

bigamy 1. The only crime in which two rites make a wrong.

2. Bigamy is having one wife/husband too many. Monogamy is the same." – Oscar Wilde.

big-endian An architectural format for storage of binary data, big-endian is most significant in multi-byte data types. Big-endian considers the most significant bytes to be the leftmost (i.e., those with a lower address) in a multi-byte data word. Some architectures use big-endian for ordering bits within bytes. Some architectures use big-endian for ordering bits, and little-endian for ordering bytes. Little-endian considers the right-most bytes (or bits) to be most significant. IBM and many other mainframe computers use big-endian architecture, while most PCs use little-endian. The PowerPC is bi-endian, as it can understand both approaches. Conversion of data between the two data architectures is known as the NUXI problem. If the word "UNIX" were stored in two two-byte words, a big-endian system would store it as "UNIX," while a little-endian system would store it as "NUXI." The terms "big-endian" and "little-endian" are derived from "Gulliver's Travels," wherein the Lilliputians were divided politically over the issue of whether soft-boiled eggs should be opened on the big side or the little side.

bikini transmitter The bikini transmitter is a body wire developed for a special surveillance project. Law Enforcement professionals needed to secretly record a conversation between a suspect and a female agent. The suspect insisted the meeting take place at a topless beach. An audio transmitter was sewn into a string bikini with the antenna threaded through the string. The largest component, the battery, was carried, uh...internally. It is not known whether the transmitter was waterproof.

bilateral synchronization A synchronization control system between exchanges A and B in which the clock at exchange A controls the received data at exchange B and the clock at exchange B controls the received data at exchange A. Normally implemented by deriving the receive timing from the incoming bit stream.

bildschirmtext German word for interactive videotex. The German Bundespost likes this service. But the German version isn't as successful as the French because the French gave away the videotex terminals. And the Germans didn't. Also, the French really encourage videotex entrepreneurs by giving a real piece of the action – 60% of the collected revenues.

bill An itemized list or statement of charges. If you can't bill for a product or service, you're engaged in a hobby rather than a business.

bill and keep Imagine a phone call from New York to Los Angeles. It may start with the customer of a new phone company, then proceed to a local phone company (let's say New York Telephone, now called Bell Atlantic). Then it may proceed to a long distance company before ending in Los Angeles and going through another one or two local phone companies before reaching the person dialed. Under the existing rules, all the companies carrying these phone calls have to be paid in some way for their transmission and switching services. There are programs in place such that the company doing the billing and collecting the money pays over some of those monies to the other phone companies in the chain. One such program is called "reciprocal compensation." The opposite of reciprocal compensation is called "Bill and Keep." Under this program, the company billing the call gets to keep all the money. The others in the chain (or most of the others in the chain) get nothing. The concept of "bill and keep" has its roots in the international postal service where "bill and keep" has been in place for many years.

Bill Gates Years ago, Bill Gates formed a company to sell a computerized traffic counting system to cities, which made $20,000 its first year. Business dropped sharply when customers learned Gates was only 14 years old. Eventually, Microsoft grew up and Bill Gates, its founder and largest shareholder, became the world's richest man.

bill of materials A list of specific types and amounts of direct materials expected to be used to produce a given job or quantity of output.

bill scraping A market research technique in the telecommunications industry that involves capturing information from customers' telephone bills to discover what services they have, what mobile content they have downloaded, and how much they are paying for them. Bill scraping captures actual buyer behavior, rather than customers' assertions about what they buy.

bill to room A billing option associated with Operator Assisted calls that allows the calling party to bill a call to their hotel room. With this option, the carrier is required to notify the hotel, upon completion of the call, of the time and charges.

billboard Electronic sales pitches that come up on your computer screen at any time.

billboard antenna A broadside antenna array with flat reflectors.

billed number screening You (at home or your business) establish who can and cannot charge a call to your phone by making an agreement with your local telephone company to screen your calls. (e.g. refusal of all collect call requests.)

billed telephone number BTN. The primary telephone number used for billing regardless of the number of telephone lines associated with that number. Apparently, the term "billing telephone number" is more accurate than billed telephone number. For a longer explanation, see Billing Telephone Number.

billibit Someone's absolutely awful term for one billion bits. Also (and better) called a gigabit.

billing account number BAN. Used by telephone companies to designate a customer or customer location that will be billed. A single customer may have multiple billing accounts. See Billing Telephone Number.

billing company The company that will bill the customer for collect or third number billed calls. It may or may not be the same as the Earned Company.

billing increment The increments of time in which the phone company (long distance or local) bills. Some services are measured and billed in one minute increments. Others are measured and billed in six or ten second increments. The billing increment is a major competitive weapon between long distance companies. Short billing increments become important to you, as a user, when your average calls are very short – for example, if you're making a lot of very short data calls (say for credit card authorizations). Being billed for a lot of six second calls is a lot cheaper than being billed for a lot of one minute calls.

billing media converter A billing media converter, as made by the Cook division of Northern Telecom, provides a means of transporting Automatic Message Accounting (AMA) data from DMS-10 central offices to regional accounting offices with the physical transfer of magnetic tapes. The BMC is polled.

Billing Telephone Number BTN. The primary telephone number used for billing regardless of the number of telephone lines associated with that number. Multiple WTNs (Working Telephone Numbers), also known as ETNs (Earning Telephone Numbers) can be associated with a single BTN. A Billing Telephone Number is the number to which calls to given location are billed. It is the seven-digit number with the area code followed by an alphanumeric code assigned by the local telephone company (e.g. NPA-NXX-XXXX).

Billing Validation Service See BVA and BVS.

billion In North America, a billion is a thousand million. In many countries overseas, a billion is a million million.

binaries Binary, machine readable forms of programs which have been compiled or assembled. As opposed to source language forms of programs.

binary Where only two values or states are possible for a particular condition, such as "ON" or "OFF" or "One" or "Zero." Binary is the way digital computers function because they can only represent things as "ON" or "OFF." This binary system contrasts with the "normal" way we write numbers – i.e. decimal. In decimal, every time you push the number one position to the left, it means you increase it by ten. For example, 100 is ten times the number 10. Computers don't work this way. They work with binary notation. Every time you push the number one position to the right it means you double it. In binary, only two digits are used – the "0" (zero) and the "1" (the one). If you write the number 10101 in binary, and you want to figure it in decimal as we know it, here's how you do it. 1 is one thing; Zero x 2 = zero; 1 times 2 x 2 = 4; 0 x 2 x 2 x 2 = 0; 1 x 2 x 2 x 2 x 2 = 16. Therefore the total 10101 in binary = 1 + 0 + 4 + 0 + 16 = 21 in decimal.

Binary notation differs slightly from notation used in ASCII or EBCDIC. In ASCII and EBCDIC, the binary values are used for coding of individual characters or keys or symbols on keyboards or in computers. So each string of seven (as in ASCII) or eight (as in EBCDIC) ones and zeros is a unique value – but not a mathematical one.

ASCII uses a seven bit coding scheme. Thus, the maximum number of different things you can code using seven bits is 128, i.e. 2 x 2 x 2 x 2 x 2 x 2 x 2 = 128. The maximum number represented by a byte (8 bits) in the IBM EBCDIC coding system is 256. i.e. 2 x 2 x 2 x 2 x 2 x 2 x 2 x 2 = 256. See Binary Coded Decimal, Binary File and Binary Transfer.

binary code A code in which every element has only one of two possible values, which may be the presence or absence of a pulse, a one or a zero, or high or a low condition for a voltage or current.

Binary Coded Decimal BCD. A system of binary numbering that uses a 4-bit code to represent each decimal digit from 0 to 9 and multiple 4-bit patterns for higher numbers. The decimal numbers 0 to 9 are represented by the four-bit binary numbers from 0000 to 1001.

binary digit A number in the binary system of notation.

binary file A file containing information that is in machine-readable form. It is an application. Or it can only be read by an application. See Binary Transfer.

Binary File Transfer BFT. The transmission of binary files, software, documents, images, video and electronic data exchange information between communicating devices, including PCs, fax devices, etc. When binary files are transferred via telecommunications, they must not be changed in any way during the transfer; otherwise they will be destroyed. Many electronic mail services, such as CompuServe or MCI Mail, have ways of "attaching" binary files to electronic mail, such that the binary file will not be affected or changed in any way during transmission. But that only works between MCI users or between CompuServe users. The Internet uses a different system. It's called MIME, which stands for Multipurpose Internet Mail Extension. MIME uses software to encode the binary file into as ASCII file. It transmits the ASCII file. At the other end, that file is decoded into its original binary format. MIME uses many software techniques for doing this. But all have the same effect of not changing the original binary file. See MIME.

Binary Logarithmic Access Method BLAM. A proposed alternative to the IEEE 802.3 backoff algorithm.

binary notation Any notation that uses two different characters, usually the binary digits O and 1.

binary number system A number system that uses two characters (O and 1) with two as its base, just as the decimal number system uses ten characters (O through 9) with ten as its base.

Binary Phase Shift Keying See BPSK for an explanation.

Binary Runtime Environment for Wireless See BREW.

binary symmetric channel A channel designed so that the probability of changing binary bits in one direction is the same as the probability of changing them back to the correct state.

Binary Synchronous Communications BISYNC or BSC. 1. In data transmission the synchronization of the transmitted characters by timing signals. The timing elements at the sending and receiving terminal define where one character ends and another begins. There are no start or stop elements in this form of transmission. For a more detailed explanation, see BSC.

BSC Block

E O T	B C C	B C C	E O T	Data (Text) ≤ 512 B	S T X	HDR Address	S O H	S Y N	S Y N

Legend

EOT	End of Transmission
BCC	Block Check Character
EOT	End of Text
STX	Start of Text
HDR	Header
SOH	Start of Header
SYN	Synchronous Character

2. Also a uniform discipline or protocol for synchronized transmission of binary coded data using a set of control characters and control character sequences.

Binary To Decimal Conversion Conversion from base 2 to base 10. See Binary.

Binary Transfer See Binary File Transfer and MIME.

BIND Berkeley Internet Name Daemon. BIND is the software that allows us to type site names like www.yahoo.com instead of a string of numbers. BIND is the implementation of a DNS server originally developed and distributed by the University of California at Berkeley. Many Internet Hosts run BIND, and it is the ancestor of many commercial BIND implementations.

bind A request to activate a session between two logical units (LUs). See BIND.

binder A way to separate groups of 25 pairs in a twisted-pair cable with more than 25. Color plastic ribbon binds separate each group of 25. The first group (1-25) is white/blue, the second (26-50) is white/orange, the third (51-75) is white/green, the fourth (76-100) is white/brown, etc. Binders are helically applied colored thread, yarn, or plastic ribbon. They're used to confine, separate and identify groups of fibers or wires in a cable. Binders are usually used for holding assembled cable components in place.

binder group A typical phone line – business or residence, analog or digital – is a pair of wires. When the phone company installs cabling along the street – up in the air or underground – it installs a cable containing many pairs of wires. The pairs are color coded so you can find which is which. Often there are so many of them you have to do something better – like take a bunch of cable pairs and wind something (or bind something) around them. That's called a binder group. The binder is the spirally wound colored thread or plastic ribbon used to separate and identify cable pairs by means of color coding. The groups are composed of insulated twisted copper pairs that are also twisted within each binder. Typically, they are wrapped in 25-pair bundles. Normal telephone color-coding provides for only 25 pairs of wire, so binder groups allow multiple pairs of the same color wire to be in one cable. A 50-pair cable has blue and orange binder groups; a 75-pair cable has blue, orange, and green groups. Since several wire pairs have the same color markings, installers must be careful when stripping cable insulation so they do not destroy the binder threads. For example. pairs 1-25 might he in one binder group and pairs 26-50 in another. In xDSL, one often hears discussions of signal interference between adjacent pairs within a binder group. The best of all worlds is to keep a data pair separated from another data pair by assigning it to an adjacent bindergroup. If the data pairs are too close to each other they create what telephone companies call "disturbers" (i.e., crosstalk).

bindery A Novell NetWare database containing definitions for entities such as users, groups, and workgroups. The bindery contains three components: objects, properties, and property data sets. Objects represent any physical or logical entity, including users, user groups, file servers, print servers, or any other entity given a name. Properties are the characteristics of each bindery object, including passwords, account restrictions, account balances, internetwork addresses, list of authorized clients, workgroups, and group members. Property data sets are values assigned to entities' bindery properties.

binding A Windows 95 definition. Binding is a process that establishes the communication channel between a protocol driver and a network adapter driver.

binding post 1. A screw with a small nut. You take your wires and join them together on a binding post by wrapping them together around the screw and then tightening the nut on them. You'll find binding posts on huge wall-mounted things called terminal boxes. We have several in the basement of the building in Manhattan, New York, in which I live. Binding posts are numbered on the terminal box and those numbers are entered into the phone company's cable management system. That way, when the technician comes to check out a trouble report, he can quickly find the offending pairs. These days, other more modern termination devices (such as Krone Boxes, 110-connecting blocks, and even 66-blocks) are replacing binding posts. Our building's binding posts are probably 60 years old. See also House Box.

2. A post witches were attached to just before they were burned at the stake. Rude switchboard operators and ACDs with long queues should also suffer this fate.

BinHex When sending files which aren't plain ASCII across a network – dial-up, leased line or the Internet – you basically have two options. First, you can attach them as a binary file (i.e. non-ASCII file). Or second, you can encode them into ASCII characters and send the file as part of your message. The first method is preferable. But you can typically only send binary files from one account to another on the same network, or between two networks that have agreed between themselves to a method of transferring files. But that's a rarity. And it certainly doesn't work in and around the Internet. Thus something called MIME was created. MIME stands for Multipurpose Internet Mail Extensions. It is the name for encoding binary files into ASCII characters for transfer across the Internet and on-line services. Under MIME, there are a number of methods of encoding binary files, one of which is called BinHex, which is a popular encoding algorithm that uses Run-Length Encoding (RLE). To send a binary file, you encode it using BinHex encoding software. You then include the ASCII encoded file in your message. The recipient of your message then decodes the ASCII using BinHex decoding software, which may or may not be built into his electronic mail or browser package.

binomial distribution The binomial is a two-parameter distribution, the parameters being n, the number of trials, and p, the probability of a particular outcome of a single trial.

bio break Meeting-speak for "bathroom break." "Let's take a short bio break before moving on to action plans."

bio-surveillance Monitoring hospital admissions to detect increases in similar symptoms that could indicate a biological attack.

biobology The art of scrutinizing and interpreting ambiguous, lo-res imagery. Practiced by spy agencies and, more recently, neuroscientists studying brain scans.

biogorrhea Pointless or excessive filler in a weblog. As one Net diarist wrote; "I was accused recently of biogorrhea. Well, yes, that's fair. But see, here's the thing. I'm leaving town for a month..." Courtesy, Gareth Branwyn, Wired Magazine.

biohacker A hobbyist who tinkers with DNA and other aspects of genetics. Also: biohacker. Some speculate a "biohacker" – the equivalent of a computer hacker seeking thrills rather than impact – could be behind the anthrax letters. "Hacker" is usually considered a bad word.

bioinformatics Bioinformatics is the process of using computer systems for the purposes of acquisition, storage and analysis of biological data. Computer databases and algorithms are used to speed up biological research and to characterize the molecular components of living things. Bioinformatics is notably being used in the Human Genome Project, the effort to identify the 80,000 genes in human DNA. Bioinformatics includes a wide spectrum of technologies including computer architectures, knowledge management, networking and collaboration tools as well as traditional life-science equipment needed to handle biological samples.

biometrics Biometrics is the study of unique and measurable physical (or biological) characteristics such as fingerprints, retinal pattern, or handwriting, biometric authentication devices capture, encrypt and use these unique and measurable characteristics as the basis for confirming identity and determining whether to grant or deny access to physical or logical assets. When used for authentication, personal biometrics are used much like a door key, however that key cannot be duplicated or lost. The basic principles of biometric verification have been understood and practiced for thousands of years. References occur as early as ancient Egypt, where individuals were identified via unique "biometrics" such as scars or a combination of measured features such as complexion, eye color, height and so on. Biometrics as we know it today really began in the late nineteenth century when criminology researchers attempted to tie physical features and characteristics with criminal tendencies. The results were not conclusive but the idea of measuring individual physical characteristics became accepted, leading the way for fingerprinting to become the international methodology among police forces for identity verification. Today it is taken for granted that fingerprints (and foot and handprints) accurately identify not just criminals, but newborns, minors (as a means of thwarting abduction), military and police personnel, and a host of others. Once technology simplified the collection, codification and encryption of biometric data, the stage has become set for biometrics to become the security standard of the electronic world, replacing passwords, bits of paper, etc. See also biometric access control.

biometric access control Any means of controlling access through human measurements, such as fingerprinting, handwriting analysis, retinal scanning, hand geometry and voiceprinting. See Biometrics.

biometric device A device used in authenticating access to a system. A biometric device authenticates a user by measuring some hard-to-forge physical characteristic, such as a fingerprint or retinal scan. You see a lot of biometric devices in Hollywood thrillers, including Mission Impossible, a movie which was released in the summer of 1996.

bionic code A problem-solving routine for human behavior as it is exercised in the realm of networks and cyberspace. The first bionic codes were developed by Ebon Fisher based on a series of his theatrical experiments involving communication systems amongst audience members. Fisher's bionic codes have been formalized as a series of diagrams and statements which "float" in the infosphere in a variety of media.

bioinformatics Bioinformatics applies informatics to biological research. Informatics is the use of advanced computing techniques to manage and analyze data. It has become indispensable because of the large volumes of complex data pouring out of genomics, proteomics, drug screening and medicinal chemistry research. Bioinformatics is also the integration and mining (detailed searching) of the ever-expanding databases in all these areas. In short, bioinformatics concerns itself with using computer databases to help unravel the human genetic code and develop new drugs.

BIOS Basic Input/Output System of desktop computers. The BIOS contains the buffers for sending information from a program to the actual hardware device the information should go to. Every personal computer has a system basic input/output system, or BIOS, which is what takes control of your computer the moment you turn it on. The screen that you first see when you turn on your computer is called the power on self test screen, better known as the POST screen. If you purchased your computer from one of the major computer manufacturers, this screen is often hidden by the manufacturer's logo. To get rid of this logo from the screen, press the ESC button on your keyboard; you'll then see what is going on in the background. At this stage in the system boot, the BIOS is probing the hardware to test the system memory and other device connections. Once the POST is completed, the BIOS

proceeds to look for a device to boot from. Once it finds your hard drive, it will begin to load Windows. The BIOS also acts as a main system component control panel, where low-level settings for all of your hardware devices are made. The device boot order, port addresses, and feature setttings such as plug and play are all found in the BIOS setup screens. For example, if you want to change the order of the drives that your computer checks to boot from, then you will want to modify the device boot order. I have to modify this setting almost every time that I install Windows because I want my computer to boot off the CD-ROM to launch the Windows XP setup application instead booting off of the operating system on my hard drive. BIOSs on each and every PC may be made by different companies or accessed in different ways ways. Nevertheless, the most common way to access the setup screen is to press F2 or the Delete key when the POST screen is displayed. Some computers even tell you which key to push to enter setup, as my laptop does. If your PC doesn't allow you to access the setup screen in this way, consult your computer documentation or contact your computer manufacturer for instructions. See also extensible firmware interface.

BIOS Enumerator A Windows term. Bios enumerator is responsible, in a Plug and Play system, for identifying all hardware devices on the motherboard of the computer. The BIOS supports an API that allows all Plug and Play computers to be queried in a common manner. See BIOS.

biosensors Devices such as fingerprint readers and signature recognition systems.

BIP Billing Interconnection Percentage. A calculation of who owns how much of a route when there are multiple providers between two points. Used to allocate revenue in MPB arrangements.

BIP-8 Bit Interleaved Parity 8. A method of error monitoring used in SONET optical fiber transmission systems. A SONET frame of data comprises a large number of bits organized into bytes, or 8-bit values. The bytes, of course, are interleaved into a byte stream. The BIP-8 method looks across all matching bit positions for those distinct bytes in a frame, and calculates parity, which should be even, rather than odd. If an odd parity is calculated at the receiving end, a bit error is indicated. The number of such errors in frames over a period of time constitutes the Bit Error Rate (BER). The specific period of time involved is known as a "sliding time window," as it varies according to the maximum detection time, which is sensitive to the transmission rate, or level of the Optical Carrier (e.g., OC-1, OC-3, OC-24, OC-48, OC-192). The higher the OC-N level, the higher the rate of transmission, and the shorter the sliding time window. In other words, the faster the rate of transmission, the more quickly the error detection process must take place in order to establish the BER and to correct the problem.

BIP-N Bit Interleaved Parity N. A method of error monitoring. With even parity, an N bit code is generated by the transmitting equipment over a specified portion of the signal in such a manner that the first bit of the code provides even parity over the first bit of all N-bit sequences within the specified portion, etc. Even parity is generated by setting the BIP-N bits so that there are an even number of 1s in each of all N-bit sequences including the BIP-N. See BIP-8 for a concrete example.

bipolar 1, The predominant signaling method used for digital transmission services, such as DDS and T-1. The signal carrying the binary value alternates between positive and negative. Zero and one values are represented by the signal amplitude at either polarity, while no-value "spaces" are at zero amplitude.

2. Bipolar Signal: A signal that can take on two polarities, of which neither is zero.

Bipolar with 8 Zero Substitution See B8ZS.

bipolar coding The T carrier line coding system that inverts the polarity of alternate "one" bits.

bipolar signal A signal having two polarities, both of which are not zero. It must have a two-state or three-state binary coding scheme. It is usually symmetrical with respect to zero amplitude.

bipolar violation The presence of two consecutive "one" bits of the same polarity on the T carrier line. See also Bipolar Coding.

bird A satellite.

bird dog device An electronic tracking device used by law enforcement to track the physical location of a subject.

birdie A birdie is a lightweight device that you blow through underground cement pipes through which you want to pull cable. Here's how it typically works: First, you use a bore to make an underground hole. Then you fill that hole with hollow concrete cement pipes joined together to form one long underground conduit (i.e. tunnel). Then you go to one end of the tunnel and use a air compressed device to blow a very lightweight "birdie" attached to a lightweight string through the tunnel. Someone at the other end catches the birdie and pulls gently on the string. Attached to the end of the string is strong mule tape. He keeps pulling on it. Attached to the end of the mule tape is the telecommunications cable – fiber or wire – that you really want to instal in the underground conduit. The whole point of this elaborate procedure is that it's far better for the cable to lay it after the pipes are laid than it is during the installation process when the cable could be damaged.

birefringent A fiber optic term. Having a refractive index that differs for light of different polarizations.

birthdays February 11, 1847 Thomas Alva Edison born
February 16, 1982 Michael Allen Newton born
February 20, 1980 Claire Elizabeth Newton born
February 23, 1965 Michael Dell (Dell Computer) born
March 3, 1847 Alexander Graham Bell born
March 11, 1933 Ben Rosen (Compaq, SRX, etc.) born
March 15, 1949 Gerry Friesen born
April 6, 1939 John Sculley born
April 27, 1791 Samuel Morse born
May 11, 1948 International Communications Association born
June 10, 1942 Harry Newton born in Sydney, Australia
June 13, 1961 TeleCommunications Association born
June 19, 1924 Ray Noorda (Novell) born
June 27, 1968 Carterfone decision handed down by FCC
July 4, 1943 Susan Newton born in Perth, Australia
August 31, 1962 The Communications Satellite Act is born
September 2, 1936 Andy Grove (Intel)
September 12, 1948 Communications Managers Association
October 28, 1955 Bill Gates (Microsoft)
November 10, 1946 Ray Horak (Context Corporation)
November 13, 1954 Scott McNealy (Sun Microsystems)
December 10, 1928 William G. McGowan, founder of MCI

bis 1. The French term for "second" or "encore." It is used by the ITU/ITU to designate the second in a family of related standards. "Ter" designates the third in a family. See V Series.
2. Border Intermediate System
3. Bus Interface Card.

biscuit A biscuit is a regular phone jack that is frequently located in a residence.

BISDN Also spelled B-ISDN. Broadband ISDN. This is a vaguely defined term. It basically means any circuit capable of transmitting more than one Basic Rate ISDN, i.e. 144 Kbps. One definition I read recently suggested that BISDN is "a set of public network services that are delivered over ATM, including data, voice, and video. BISDN will provide services such as high-definition television (HDTV), multi-lingual TV, voice and image storage and retrieval, video conferencing. high-speed LANs, and multimedia." See B-ISDN for a longer explanation. See also ISDN.

bistable trigger circuit A trigger circuit that has two stable states.

BISUP Broadband ISDN User's Part: A SS7 protocol which defines the signaling messages to control connections and services. See Signaling System 7.

BISYNC (pronounced bye-sink). BISYNChronous Transmission. A half-duplex, character-oriented, synchronous data communications transmission method originated by IBM in 1964. See Synchronous.

bisynchronous transmission Also called BISYNC. A data character-oriented communications protocol developed by IBM for synchronous transmission of binary-coded data between two devices. BISYNC uses a specific set of control characters to synchronize the transmission of that binary coded data. See also Binary Synchronous Communication.

bit Bit is a contraction of the term BInary digiT. It is the smallest unit of information (data) a computer can process, a "1" or "0." Similarly, a bit is the basic unit in digital communications. Across a transmission facility, a bit can be represented in a variety of ways, including positive (+) and negative (-) voltage, relatively high (e.g., +3.0) volts and relatively low (e.g., +1.5) volts, positive (+) and null (0) voltage, presence and absence of light, and relatively high-amplitude and relatively low-amplitude radio signal. The term was coined by John W. Tukey, a professor at Princeton and a researcher at Bell Telephone Laboratories. Tukey also coined the term "software." See also Software and Tukey.

bit bucket Slang for throwing out bits – into a wastepaper bucket.

bit budget The number of bits that are available for storing one or more values, one or more instructions, one or more network addresses, one or more files, or network traffic.

bit buffer A section of memory capable of temporarily storing a single bit (BInary digiT) of information. Used to make data transmission accurate or consistent.

bit check A bit added to a digital signal and used for error checking, i.e. a parity bit. See also Parity.

bit count integrity A means of determining the integrity of a data packet. The counting of bits is very important. If the receiver can't confirm that it received all of the bits, it can't confirm that the integrity of the received data packet is maintained. Bit count integrity is especially important in the domain of compressed, encrypted data networking.

bit depth The number of bits used to represent the color of each pixel in a graphic file or the sound in an audio file. The higher the number, the more information included in the file and the higher the quality of the data. Common graphic bit depths are 4-bit, 8-bit, 16-bit, 24-bit and 32-bit. Common sound bit depths are 8-bit and 16-bit.

bit duration See Bit Time.

bit error A mistake in transmitting a bit. Error rate statistics play a key role in measuring the performance of a network. As errors increase, user payload (especially data) must be re-transmitted. The end effect is creation of more (non-revenue) traffic in the network. See Bit Error Rate.

bit error rate BER. The percentage of received bits in error compared to the total number of bits received. Usually expressed as a number to the power of 10. For example, 10 to the fifth power means that one in every 100,000 bits transmitted will be wrong. In transmitting data a high error rate on the transmission medium (i.e. some noise), doesn't mean there'll be lots of problems with the final transmission. It just means there'll have to be lots of re-transmissions – "until one gets it right." These re-transmissions reduce the amount of data transmitted in a unit of time and therefore, increase the time needed to send that information. If the BER gets too high, it might be worth while to go to a slower transmission rate. Otherwise, you would spend more time retransmitting bad packets than getting good ones through. The theory is that the faster the speed of data transmission the more likelihood of error. This is not always so. But if you are getting lots of errors, the first – and easiest – step is to drop the transmission speed.

bit error ratio floor A limiting of the bit-error ratio in a digital fiberoptic system as a function of received power due to the presence of signal degradation mechanism or noise.

bit flipper A person who flips bits for a living. In other words, an industrial strength member of the digiterati, which I guess would be a digiteratus. These people are way kewl (that's a NetHead term for cool, or k0Ol) when it comes to really technical data protocol stuff. See also Digiterati, Nethead and Gearhead.

bit interleaving A form of TDM for synchronous protocols, including HDLC, SDLC, BiSync and X.25. Bit interleaving retains the sequence and number of bits, so that correct synchronization is achieved between both ends. See Bit Interleaving/Multiplexing.

bit interleaving/multiplexing In multiplexing, individual bits from different lower speed channel sources are combined one bit at a time/one channel at a time into one continuous higher speed bit stream. Compare with byte interleaving/ multiplexing.

bit oriented Used to describe communications protocols in which control information may be coded in fields as small as a single bit.

Bit Oriented Protocol BOP. A data link control protocol that uses specific bit patterns to transfer controlling information. Examples are IBM's Synchronous Data Link Control (SDLC) and the ITU-T High-Level Data Link Control (HDLC). Bit-oriented protocols are normally used for synchronous transmission only. Bit-oriented protocols are code transparent (meaning they work regardless of the character encoding method used), since no encoded characters are used in the control sequence.

bit oriented transmission An efficient transmission protocol that encodes communications control information in fields of bits rather than characters or bytes.

bit parity A binary bit appended to an array of bits to make the sum of all the bits always odd or always even. See Parity.

bit pattern A group of bits arranged in specified ways to represent numbers, letters or symbols, forming a unique binary number for each character. For example, the 7-bit ASCII code produces 128 different characters, i.e. $2 \times 2 \times 2 \times 2 \times 2 \times 2 \times 2 = 128$.

bit pump A device which pumps out bits at a high rate of speed. Slang for high-speed carrier electronics such as ADSL terminating units, which can achieve speeds of multiple Mbps over standard twisted-pair local loops. "Bit pumps" possess no particular intelligence (e.g., protocol conversion, or error detection and correction); they just pump bits.

bit rate The number of bits of data transmitted over a phone line per second. You can usually figure how many characters per second you will be transmitting – in asynchronous communications – if you divide the bit rate by ten. For example, if you are transmitting at 1200 bits per second, you will be transmitting 120 characters per second. In real life, it's never this simple, however. The total bits transmitted will depend on re-transmissions, which depends on the noise of the line, etc. See BAUD RATE.

bit robbing A technique to signal in-band in digital facilities, which typically use out of band signaling, e.g. Signaling System 7. In bit robbing, we steal bits from the speech path a few line-signal bits. The remaining bits are adequate to recreate the original electrical analog signal (and ultimately, the original sound). Bit robbing typically uses the least significant bit per channel in every sixth frame for signaling. See Bit Stuffing and Robbed Bit Signaling.

bit specifications Number of colors or levels of gray that can be displayed at one time. Controlled by the amount of memory in the computer's graphics controller card. An 8-bit controller can display 256 colors or levels of gray; a 16-bit controller, 64,000 colors: a 24-bit controller, 16.8 million colors.

Bit Seven BIT7. A TR008 DS1 line code that performs zero code suppression by placing a one in bit 7 of an all-zeros byte.

bit stream A continuous flow of binary digits (bits), through some form of communications medium_e.g., fiber optics, air (wireless) or twisted-pair, with no break or separators between the characters.

bit stuffing 1. A process in some synchronous data communications protocols to ensure that the transmission is properly clocked. In HDLC (High-level Data Link Control), for example, each frame of data is both preceded and succeeded by a "flag" of six consecutive "one" bits. The flags signal the beginning and ending of the data frame, and are used by the receiving devices to synchronize on the rate of transmission. In order to avoid confusion, therefore, any set of six consecutive "one" data bits must be broken by a stuffed "zero" bit. This stuffed bit is added by the sender and stripped by the receiver. The idea of inserting the "one" bit is to avoid the data's mimicking the flag, and confusing the receiver. Stuff bits also are used in SONET (Synchronous Optical NETwork) prior to the multiplexing of DS-1s and DS-2s in order to resolve clocking differences prior to the construction of a SONET frame. See Zero Stuffing.

2. A process used in some packet data network protocols in order to fill a packet to the prescribed, fixed size. For example, X.25 requires that all packets in a specific network be of the same size (e.g., 128B, 256B, or 512B, where B = Byte of eight bits). If the amount of data to be transmitted in a packet is not sufficient, stuff bits are added to fill up the packet. Again, the stuff bits are added by the sender, which signals their presence, and are stripped away by the receiver.

Bit synchronization Bit synchronization the process by which a digital receiver comes in step with a received bit sequence, in order to extract information from it. Because the process primarily results in synchronization of the receiver clock, it is also called clock acquisition.

bit synchronous A SONET term describing the manner in which information streams are mapped into the SONET frame format for unchannelized VT (Virtual Tributary) transport. For instance, multiple VT1.5s can be mapped into a SONET frame, with each VT1.5 carrying a single T-1 signal within the STS-1 SPE (Synchronous Payload Envelope). In the LOH (Line OverHead) of the SONET are included VT pointers which identify the established portions of the SPE in which the beginning byte of each VT1.5 is located. As each VT1.5 works its way through the SONET network, its actual position may change within the various SPEs, thereby requiring that the pointer be reset; this process is very complex and expensive, and results in a small level of delay. Regardless of issues of complexity, cost and (slight) processing delay, it is essential that the timing of the various T-carrier and SONET data be maintained in order that the data arrive at an identifiable and consistent point in time; otherwise, it would be unrecognizable. As is true with many things in life, timing can be everything. Compare and contrast with Byte Synchronous. See also SONET and VT.

bit time Also known as "bit duration," the bit time is the length of time associated with a bit placed on a transmission medium. In somewhat oversimplified terms, a transmission system that runs at a signaling speed of 1 Gbps supports bits that have an individual duration of 1 billionth of a second.

Bit Torrent BitTorrent is a protocol and peer-to-peer software application that enables large files, including movie files, to be quickly downloaded over the Internet. There are multiple search engines for finding torrent files. A new meta engine called Torrent Finder, for example, enables you to search nearly three dozen Torrent sites at the same time. Like any peer-to-peer system, BitTorrent is a very use-at-your-own-risk, no-lifeguard-on-duty kind of service. Reviewers have had mixed results downloading BitTorrent movies. It's very difficult to determine the legality of some files, while other files are clearly violating copyrights. Selection is spotty at best, and it's common to spend days downloading a massive movie file, only to have the sources dry up at the last minute, leaving you with a massive, incomplete movie file on your system that you can't watch. So, despite its popularity with geeks, BitTorrent isn't the future of video-on-demand for the masses. See also Napster.

bit transfer rate The number of bits transferred per unit of time. Usually ex-

pressed in Bits Per Second (BPS).

bit twiddler A technical person. Twiddle, according to the Random House Dictionary, means "to play or trifle idly with something; fiddle." The expression bit twiddler is used thus, "I'm not a bit twiddler. You'll have to ask Joe in Engineering if you want the answer to that."

BITBLT BIT BLock Transfer. Microsoft Windows relies intensively on a type of operation called bit block transfer (BITBLT) to redraw rectangular areas of the image on the computer's screen. Generally, BITBLT operations are accomplished by software routines in the video driver, a cheap, but slow method that uses many of your CPU's clock cycles. If you add a separate video controller with a special processor to handle BITBLT, you will be able to offload video tasks from your main CPU and make your computer run faster.

BITE Built-in test equipment. Features designed into a piece of equipment that allow on-line diagnosis and testing of failures and operating status.

bitmap Representation of characters or graphics by individual pixels arranged in row (horizontal) and column (vertical) order. Each pixel can be represented by one bit (simple black and white) or up to 32 bits (high-definition color). Bitmapped images can be displayed on screens or printed. The method of storing information that maps an image pixel, bit by bit. There are many bitmapped file formats, .bmp, .pcx, .pict, .pict-2, tiff/.tif,.gif (89a), and so on. Most image files are bit mapped. This type of file gives you stair stepped edges, the 'jaggies'. When examined closely you can see the line of pixels that creates edges. Bitmap images are used by all computers. The desktop for all Windows machines uses .bmp files, while the Macintosh uses pict files. Most Internet publishing and e-mail use JPEG or .JPG and .GIF (89a) formats." See Bit Specifications.

bitmapped graphics Images which are created with matrices of pixels, or dots. Also called raster graphics. See Bitmap.

bitmask A pattern of bits for an IP address that determines how much of the IP address identifies the host and how much identifies the network. For example, if a bitmask of 24 were applied to the address 10.12.132.208, 10.12.132 identifies the network and the remainder of the address (1-254) can be used to specify individual machines on the 10.12.132 network.

BITNET Because It's Time NETwork. An academic computer network based originally on IBM mainframes connected with leased 9600 bps lines. BITNET has recently merged with CSNET, The Computer+Science Network (another academic computer network) to form CREN: The Corporation for Research and Educational Networking. The network connects more than 200 institutions and has more than 900 computational nodes.

bitnik A person who uses a coin-operated computer terminal installed in a coffee house to log into cyberspace.

bitpipe A telecommunications carrier that derives revenues only from levels 1-4 of the OSI model. If a carrier doesn't want to sit around and watch its services continue to become commoditized and its margins shrink, it will seek to become more than a "mere bitpipe" by offering higher-layer services and content.

bitorrent See BitTorrent.

bitronix Hewlett Packard's term for its bidirectional parallel port communications "standard." It introduced this standard with its 600 dps LaserJet 4 plain paper printer in the fall of 1992. It is hoping other manufacturers will adopt the standard. The big plus of the standard is that it allows a printer to tell a connected computer that it (the printer) has run out of paper, or the paper has jammed, etc. Having that communication back and forth will allow the user to clear the problem and get the printer and up and running faster and stop the computer from locking up.

BITS 1. Building Integrated Timing Supply. A single building master timing supply. BITS generally supplies DS1 and DS0 level timing throughout an office. The BITS concept minimizes the number of synchronization links entering an office, since only the BITS will receive timing from outside the office. In North America, BITS are thus the clocks that provide and distribute timing to a wireline network's lower levels. Known in the rest of the world as a SSU (Synchronization Supply Unit).

2. See bump-in-the-stack.

bits clock The bits clock provides a pulse that synchronizes the entire network. The pulse is a 1-0-1-0-1-0-1-0 stream. Used extensively in SONET network.

Bits Per Second The number of bits passing a specific point per second. See Bps.

A KILObit per second is one thousand bits per second.

A MEGAbit per second is one million bits per second (thousands of kilos).

A GIGAbit per second is one billion bits per second (thousands of millions).

A TERAbit per second is one trillion bits per second (thousands of billions).

A PETAbit per second is equal to 10 to the 15th or 1,000 terabits per second.

bits versus bytes Bits are the smallest units of information in telecom and data processing; the term bit is derived from binary digit. One byte is typically equivalent to 8 bits. Given that a bit can carry two pieces of information (represented by either 0 or 1), two bits together can represent four (or 22) pieces of information: 00, 01, 10, and II. Similarly, eight bits or one byte mean that 256 (28) pieces of information can be represented, stored, and transmitted. These pieces of information can be color codes, sound levels, or numbers and letters. See also asynchronous and synchronous. See also the Introduction to this book at the very beginning.

bitslag All the useless rubble on the Net one has to plow through to get to the rich information core.

bitspit To transmit. "Did you bitspit the file to Harry?"

BitTorrent Software that speeds downloads of large files like TV shows and movies, by grabbing bits and pieces from other users and their PCs, who have also downloaded the file.

BITW See bump-in-the-wire.

BIU Bus Interface Unit. The data circuit equipment that provides physical access to the bus.

bix A Nortel Networks' trade name for an in-building termination and cross-connect system for unshielded twisted pair cables, also called a terminal block. See also Terminal Block.

biz Slang for business, as in venture capital biz.

BL Business Line.

black ball See White List.

black body A totally absorbing body that does not reflect radiation (i.e. light). In thermal equilibrium, a black body absorbs and radiates at the same rate; the radiation will just equal absorption when thermal equilibrium is maintained.

black box 1) An electronic device that you don't want to take the time to understand. As in, "We'll put the data through a black box that will put it into X.25 format." The term has recently come also to mean PBX switches. While "Black Box" is a generic term, The Black Box Corporation of Pittsburgh, PA, has had the audacity (and brilliance) to register the term as a trademark. The Phone Phreak community has used the term black box to describe a device that's put on phone lines in electromechanical central office areas (they don't work under ESS offices). To the phone phreak community, a black box is made up of a resistor, a capacitor and a toggle switch that would "fool" the central office into thinking the phone had not been picked up when receiving a long distance call. Since the call was not "answered," the call could not be billed. Clever, eh? (Illegal, too.) See Blue Box, Red Box and White Box.

2) Cockpit voice recorder and flight data recorders that investigators use to reconstruct the events leading up to a plane crash. The box is actually bright orange.

Black Box Corporation A leading direct marketer of connectivity solutions – everything from cables to routers. It publishes and distributes the Black Box Catalog. Black Box is based in Pittsburgh. They kindly provided the pinout diagrams for the back of this dictionary. www.blackbox.com.

black box testing Network security testing where the evaluators conducting the tests are not provided with any information in advance about the client's organization and network infrastructure. Black-box testing simulates external attacks by hackers, who start off with nothing but the client's company name. In black box testing, evaluators gather as much information as they can about the company's network and business from as many sources as possible, just as a hacker would do. They use publicly available information about the company; they glean information from the company's website; they use tools of the trade such as port scanners; and they may use social engineering techniques to get information from unwitting employees. See also white box testing.

black collar workers Once it referred to miners and oil workers. Today it often refers to creative types (artists, graphic designers, video producers) who've made black attire a kind of unofficial uniform.

Black Dot Law In 1996, seven years before the Federal Trade Commission established a national Do Not Call Registry, Alaska enacted a do-not-call law commonly referred to as the "Black Dot" Law. The law prohibited telemarketers from making telemarketing calls to consumers whose phone numbers had a black dot next to them in the telephone directory. Consumers who wanted a black dot listing had to pay their local phone company anywhere from $5.00 to $35.00 for the designation. Telemarketers wanting to conduct business in Alaska had to purchase Alaskan telephone companies' black dot lists for the areas they wanted to call, and delete the black dot phone numbers from their calling lists. In 2003, the Federal Trade Commission established a national Do Not Call Registry, which is free for consumers. Telemarketers have to check the registry at least once every 31 days

for each area code they want to call and scrub their calling lists accordingly. The national Do Not Call Registry removed the need for the Black Dot Law, so the Alaska state legislature repealed it on August 16, 2006.

black facsimile transmission 1. In facsimile systems using amplitude modulation, that form of transmission in which the maximum transmitted power corresponds to the maximum density of the subject copy.

2. In facsimile systems using frequency modulation, that form of transmission in which the lowest transmitted frequency corresponds to the maximum density of the subject copy.

black hats These guys are the bad hackers. They're the ones that break into systems, steal confidential information to sell or destroy, delete documents, remove legitimate users, and above all they usually just break stuff. Sometimes they get paid to do it, sometimes they do it because of a grudge they have on their previous employer, and sometimes they do it simply for personal enjoyment.

black hole 1. Routing term for an area of the network where packets enter, but do not emerge, owing to adverse conditions or poor system configuration within a portion of the network.

2. A theorized invisible (thus perceived as dark) region in space with a small diameter in relation to its intense gravitational field that is proposed to result from the collapse of a massive star in which the escape velocity equals the speed of light. Basically an object whose gravity is so strong that not even light can escape from it.

black hole router A router (or its firewall) on a TCP/IP-based wide area network that does not send an appropriate Internet Control Message Protocol (ICMP) response when it receives a larger packet than it can handle. See also black hole.

black jack When Samuel Morse won a contract from the U.S. Congress in the 1800s to build a telegraph line from Washington, DC to Baltimore, he initially tried to install the lines underground, alongside railroad tracks. There were problems with the lines and the wires were later suspended from poles. To this day, millions of miles of communications cables are installed on utility poles. Since the wooden poles are subject to weathering, they are coated with creosote, a preservative derived from coal tar. It has come to be called black jack.

black level The lowest luminance level that can occur in video or television transmission and which, when viewed on a monitor, appears as the color 'black.'

black list See white list and gray list.

black matrix Picture tube in which the color phosphors are surrounded by black for increased contrast. See also black matrix.

black paint Henry Ford produced the model T only in black because the black paint available at the time required the least time to dry of all other colors. One can only assume he didn't like to watch paint drying. (That was intentionally bad.)

black phone Of or pertaining to the POTS business. POTS is plain old telephone service. Analog and old.

black recording 1. In facsimile systems using amplitude modulation, that form of recording in which the maximum received power corresponds to the maximum density of the record medium.

2. In a facsimile system using frequency modulation, that form of recording in which lowest received frequency corresponds to the maximum density of the record medium.

black signal 1. In facsimile, the signal resulting from the scanning of a maximum-density area of the subject copy.

2. In cryptographic systems, a signal containing only unclassified or encrypted information.

black thursday The day that began the Great Depression. It was October 24, 1929.

black wire Also called dark fiber, which is optical fiber through which no light is being transmitted and which, therefore, carries no communications signal. See also dark fiber.

BlackBerry BlackBerry is Research in Motion's popular handheld terminal. Originally it only did email. Then it also became a cell phone and you could speak on it. Now it's a software "platform." Others can write software to it and sell that software to BlackBerry users, like me. See also predictive text and thumbing.

BlackBerry dunk You're sitting on the toilet and you're using your BlackBerry to check your emails. You get a little too excited with your emails and suddenly your BlackBerry falls into the toilet. That's called a BlackBerry dunk. That's probably the end of your BlackBerry and your precious emails.

Blackbird Blackbird is a multifunction system, providing home users with a single box that can work as a game system, network computer, home broadband router, and set-top box. The primary backer behind Blackbird is Motorola. Much of Blackbird's "middleware" software was developed with partners. According to David Lammers, of EE Times, Spyglass built a browser and other networking software that will let companies customize their own offerings and get to the retail channel by next year. Another key partner is VM Labs, a Silicon Valley-based game company headed up by former Atari president Richard Miller. VM Labs worked with Motorola to create a media processor, Nuon, which, at different times, has gone under the code names of Merlin and Project X, Burgess said. Nuon is a 128-bit VLIW engine that will work in tandem with the CPU in Blackbird, a PowerPC 860 core.

blackout A total loss of commercial electric power. A blackout has a decidedly negative affect on your ability to compute and communicate, assuming that your systems are wired, rather than wireless. UPS (Uninterruptible Power Supply) systems provide battery backup protection from a blackout. Carriers use diesel power generators to keep their central offices operating during a blackout-that way, the phones still work when the lights go out.

blackout area A CATV and satellite TV definition. A pre-defined area of the country where a particular program (sports or special events) will not be available, usually because of contractual agreements.

Blacksburg Electronic Village Blacksburg is a town of 40,000 people in the mountains of Southwestern Virginia which contains the main campus of Virginia Polytechnic Institute & State University, also known as Virginia Tech, They have wired both Virginia Tech students and thousands of their town's citizens with email, internet access and bulletin boards. The official explanation is: The Blacksburg Electronic Village (BEV) is a cooperative project of Virginia Tech, Bell Atlantic of Virginia and the Town of Blacksburg. It links the town's citizens to each other and to the Internet via computer lines. Citizens gain access to the Internet from their home or office through a high-speed modem pool or by using Ethernet LANs which are available in some offices and hundreds of apartment units in town. Blacksburg residents may take advantage of a full spectrum of services including the Internet, World Wide Web, Gopher, electronic mail, electronic mailing lists and thousands of Usenet newsgroups. In addition to full Internet access, citizens enjoy the benefits of extensive online local resources. The Blacksburg Electronic Village is famous because it really is the first town to aggressively ensure that the bulk of citizenry could and would have access to electronic mail and to the various resources on the Internet. It has apparently made a major difference in how people live and communicate in Blacksburg. www.bev.net.

blade 1. A blade is any card placed into a backplane in a telephone system. Usually a blade is an additional module. The "blade" term shows the aspect of insertion of the flat plane. This definition contributed by Bob Frankenberger. A blade is also a "computer-on-a-card." Several hundred of these computers can now be crammed into a chassis that now holds 40-plus of today's popular pizza-box shaped server computers. According to the Wall Street Journal, "their small size, their ability to share Internet connections and the disk storage among the blades allow these computers to operate on a fraction of the power and air conditioning of traditional servers" – thereby promising to reduce important costs for data centers. Says the Journal, "for the companies that house and manage thousands of computers for Internet operations, such very small and cost-efficient machines allow to collect more fees per square foot of computer space." See also Blade Server.
   2. Some people in telecom also refer to a blade as software.
   3. When people get fired, they often refer to the event as "the blade fell."

blade server A Blade Server is a computer system on a motherboard, which includes processor(s), memory, a network connection and sometimes storage and packaged up in a neat box which can be rack mounted or stacked. The idea is to reduce space needs for companies who need many servers – for example a company running a popular web site. There are also storage blades – small boxes that contain multiple hard disks. Storage blades and blade servers can be rack mounted in multiple racks within a cabinet together with common cabling, redundant power supplies and cooling fans. Blades can be added as required, often as "hot pluggable" units of computing as they share a common high speed bus. Most of the Computer Manufacturers are now offering blades. See also Blade.

BLAM Binary Logarithmic Access Method. A proposed alternative to the IEEE 802.3 backoff algorithm.

blame shift To deflect responsibility by pointing the finger at someone else. "Don't blame shift," Julia Roberts warns Brad Pitt in "The Mexican." See also blamestorming.

blamestorming Blamestorming occurs when people sit around in a group and discuss why a deadline was missed or a project failed, and most importantly, who was responsible. See also blame shift.

blank A character on teletype terminals that does not punch holes in paper tape (except for feed holes to push the paper through). Also the character between words, usually called a "Space" is referred to in IBM jargon as a Blank.

blank and burst On the AMPS cellular telephone network, certain administrative messages are sent on the voice channel by blocking the voice signal (blanking) and sending a short high speed data message (burst). The blank and burst technique is one that causes a momentary dropout of the audio connection (and sometimes disconnection of cellular modem connections) when a power level message is transmitted to the cellular phone.

blank cell The hollow space of a cellular metal or cellular concrete floor unit without factory installed fittings.

blanket Covering or intended to cover a large group or class of things, conditions, situations, etc.: a blanket proposal; a blanket LOA (Letter of Agency). An LOA is a letter you give to someone whom you allow to represent you and act on your behalf. For example, a letter of agency is used when your interconnect company orders lines from your local phone company on your behalf. Letters of Agency are also used when companies switch their long distance service from one carrier to another. A blanket LOA can mean everything from a group of numbers belonging to one customer at multiple sites or multiple customers at multiple sites.

blanked Baseball term. Get shut out. No hits. No runs. NO errors.

blanking The suppression of the display of one or more display elements or display segments.

blanking interval Period during the television picture formation when the electron gun returns from right to left after each line (horizontal blanking) or from top to bottom after each field (vertical blanking) during which the picture is suppressed.

blanking pulses The process of transmitting pulses that extinguish or blank the reproducing spot during the horizontal and vertical retrace intervals.

blargon Language used on blogs.

Blast BLocked ASynchronous Transmission. A proprietary technology.

blatherer A Internet user who takes four screens to say something where four words would work a lot better.

bleaching What Window Washer software from Webroot calls its function of permanently erasing files on your PC. If you don't bleach a file, but simple "erase" it, it can easily be found by someone – especially if you haven't used your PC much and, hence, have written something new over the file you erased.

BLEC 1. Broadband Local Exchange Carrier. A service provider offering broadband services locally.

   2. Building Local Exchange Carrier. BLECs are business entities – mostly specialists, aligned with REITs (Real Estate Investment Trusts) or landlords – who invest in creating and managing the "intelligent local loop" – a combination of premise wiring, demarc systems, basement POPs filled with co-located equipment and applications servers, and, possibly higher order local aggregation points (i.e. central offices) filled with the same – and all dedicated to providing end-user customers (business and residence) with a wide range of communications services (voice, broadband data, Internet) and computer networking applications (e.g. directory, messaging, DNS, IP telecom gateway, etc.)

bleeding edge First, there was leading edge, then cutting edge. Now there's something so new that even its inventors aren't completely sure what it's going ultimately to be used for. Some make it. Others just bleed. That may also be part of the bleeding edge.

blend To have outbound and inbound phone calls answered by the same agents. See the next two definitions.

blended agent A call center person who answers both incoming and makes outgoing calls. This idea of a blended agent is a new concept in a Call Center. In the past call centers have typically kept their inbound and outbound agents separate. The reason? Management felt that the necessary skills were very different and no one could master both.

blended ARPU Average revenue per unit that is calculated by (1) combining wireless voice and wireless data or (2) by combining wireline basic local service and wireline value-added services, and possibly even combining wireline long distance.

blended call center A telephone call center whose agents both receive and make calls. In other words, a call center whose phone system acts both as an automatic call distributor and a predictive dialer.

blended floor system A combination of cellular floor units with raceway capability and other floor units with raceway capability systematically arranged in a modular pattern.

blended threat Combination of threats that attempt to wreak havoc on an Internet-connected network. For example, a worm might initially cause damage to web pages on a particular server and then proceed to perform a DDoS (Distributed Denial of Service) on the server, or vice versa.

BLER Block Error Ratio. The ratio of the blocks in error received in a specified time period to the total number of blocks received in the same period.

BLERT BLock Error Rate Test. Data transmission testing in which the error rate counted is the number of blocks containing errored bits instead of the raw number of errored bits. Many users claim this is more representative of the real throughput quality of a circuit than simple raw BER testing. See BERT.

BLES Broadband Loop Emulation Service. Developing specifications from both the ADSL Forum and the ATM Forum. BLES addresses the requirements for an end-to-end architecture to support voice and data over frame-based DSL loops, with voice supported over an emulated circuit. One BLES approach is for all traffic to be carried in ATM cells. The other approach is a mixture of IP and ATM, with data carried in IP packets, and with compressed voice carried in ATM cells using AAL 2 (ATM Adaption Layer 2) and supported by the rt-VBR (real time-Variable Bit Rate) service category. In either case, BLES extends Class 5 Central Office (CO) capabilities (e.g., custom calling and Centrex services, as well as voice trunking) over a DSL local loop for support of network-based circuit switched voice. The other approach is VoMBN (Voice over Multiservice Broadband Networks), which also supports VoDSL (Voice over DSL), but without the involvement of a Class 5 switch. Rather, VoMBN involves various gateway devices at the network edge in support of VoIP (Voice over Internet Protocol) or VoATM (Voice over ATM) packet voice traffic. BLES generally appeals to incumbent PSTN carriers while VoMBN appeals to newer carriers that have built backbone networks based on either ATM or IP, rather than traditional circuit-switching technology. See also AAL 2, ADSL, ADSL Forum, ATM, IP, LES, and rt-VBR.

BLF The Busy Lamp Field is a visual display of the status of all or some of your phones. Your BLF tells you if a phone is busy or on hold. Your Busy Lamp Field is typically attached to or part of your operator's phone. See Busy Lamp Field.

blind bore Imagine you want to lay fiber cable along the side of a highway. You know from blue staking and from the city maps that there are other utility cables along the highway you want. The first thing we know is that we can't trust the maps or the blue staking. The second thing we know is that we don't want our activities to cut someone else's cable. There are expensive implications to doing this. So what we do is we hand dig pot holes every so often along what's known as the running line – where the utilities are meant to be buried. The idea is that our pot holes will locate the existing underground cables and thus make it safer to bore our own cable. Blind boring occurs when we bore underground without digging pot holes. The reason we might do this? Some states and some cities simply don't allow pot holing. They trade the risk of hitting a utility line against creating a hole in the middle of the street. They don't want pot holes in their street, since an asphalt patch has never the same integrity as a total overlay and they don't want their streets messed up. See Blue Stake.

Blind Carbon Copy BCC. A list of recipients of an e-mail message whose names do not appear in the normal message header, so the original recipient of the message does not know that copies have been forwarded to other locations. Sometimes called blind courtesy copy.

blind dialing All modems come from the factory programmed to "listen" for a dial tone before dialing their connection. However, there are some phone lines which don't have dial tones or, more often, strange dial tones, which your modem doesn't recognize. A "strange" dial tone might be one you find in an strange place, usually not in North America. In this case, you have to tell your modem to start dialing when you want it to. This is called "blind dialing." The old way you did this was to insert an X1 in the dialing stream. The new way, in Windows 95/98, is to go into Control Panel / Modems / Properties / Connection and remove the check mark from "Wait for dial tone before dialing." Actually, you can leave this unchecked. Your modem will work just fine.

blind squirrel Old expression: Even a blind squirrel finds a nut every once in a while. The moral is keep trying.

blind survey When a company does a survey of its customers and potential customers and doesn't tell the customers who the company is, then it's called doing a blind survey.

blind testing Tests in which the brand name of the product is not disclosed during the test.

blind transfer Someone transfers a call to someone else without telling the person who's calling. Also called Unsupervised or Cold Transfer. Contrast to Screened Transfer.

bling bling Another name for ostentatious jewelry. Catrina Murphy gave me this definition. She attends the Boys and Girls Club of Boston, which my daughter was IT director of. Catrina is sweet and does not wear bling bling. But her eyes will go bling bling when she sees her name in this dictionary. Catrina loves sports and computers (?). She wants to grow up to be a professional swimmer or a lawyer. See cell phone bling.

blink An RFID technology developed under ISO 14443, which is used in most contactless credit cards. A blink card contains a small RFID microchip and a wire loop. When the card is passed within four inches of a blink terminal (an RFID-enabled point-of-sale terminal that uses blink technology), the blink terminal's magnetic field generates electric voltage in the card's wire loop through a process known as inductive coupling. This voltage powers the card's microchip, which then transmits data from the card to the terminal at a frequency of 13.56 MHz, a frequency that was selected for its suitability for inductive coupling, its resistance to environmental interference and its low absorption rate by human tissue. Firmware in the processor encrypts the credit card ID and card holder's name during transmission.

blinking An intentional periodic change in the intensity of one or more display elements or display segments.

BLIP Ericsson term for Bluetooth's local infotainment points – the 30-foot radius hot spots around wireless connectors that may, one day, feed news, entertainment and ads to the mobile masses.

blister pack A pocketed polyvinyl chloride shipping container with a snap-on cover.

blitz A call center/marketing term. Used to describe telephone sales or prospecting activity of intense, high volume accomplished in a short period of time.

blivit A problem that is intractable, a piece of hardware that can't be fixed if it breaks, an embarrassing bug that pops up during a customer demo, or anything of that ilk. An embarrassing bug at a sales pitch is also a blivit.

bloatware Ever-fatter packages of "upgraded" software that, with each upgrade, come with dozens and dozens of new features. With each upgrade, the customer has less need to look elsewhere. At least that's the theory. See also Hyperware and Vaporware.

BLOB Binary Large OBjects. When a database includes not only the traditional character, numeric, and memo fields but also pictures or other stuff that consumes a large amount of space, it is said to include these.

block In data communications, a group of bits transmitted as a unit and treated as a unit of information. Usually consists of its own starting and ending control deliminators, a header, the text to be transmitted and check characters at the end used for error correction. Sometimes called a Packet.

block character check BCC. The result of a transmission verification algorithm accumulated over a transmission block, and normally appended at the end, e.g. CRC, LRC.

block cipher A digital encryption method which ciphers long messages by segmenting them into blocks of fixed length, prior to encryption. Each block, which typically is 64 bits in length, is encrypted individually. The blocks may be sent as individual units, or they may be linked in a method known as Cipher-Block-Chaining. See also Encryption.

block diagram A graphic way to show different elements of a program or process by the use of squares, rectangles, diamonds and various shapes connected by lines to show what must be done, when it must be done and what happens if it's done this way or that. In short, it shows how all the small decision points add up to the whole process.

block down converter A device which lowers a whole band of frequencies from one band to another lower band.

Block Error Rate Test See BLERT.

Block Error Ratio See BLER.

Block IIA satellite A type of GPS satellite that was first launched in 1989, but which no longer is launched. Block IIA satellites broadcast signals on the L1 and L2 carrier frequencies, but not on the L5 carrier frequency. Block IIA satellites also lack the ability to broadcast the military's new GPS signal, M-Code, and the new civil signal, L2C. Block IIA satellites are gradually being replaced by newer satellites. See GPS.

Block IIF satellite A new type of GPS satellite that will begin to be launched into orbit in 2007. Block IIF satellites broadcast signals on the L1, L2 and L5 carrier frequencies. Block IIF satellites are able to broadcast the military's GPS signal, M-Code, on L1, and the second civil signal, L2C, on L2. A third civil signal will be broadcast on the L5 frequency. See GPS.

Block IIR satellite A type of GPS satellite that was first launched in 1997, but which no longer is launched. Block IIR satellites broadcast signals on the L1 and L2 carrier frequencies, but not on the L5 carrier frequency. Block IIR satellites also lack the ability to broadcast the military's new GPS signal, M-Code, and the new civil signal, L2C. Block IIR satellites are gradually being replaced by newer satellites. See GPS.

Block IIR-M satellite A "modernized" version of the Block IIR GPS satellite. Block IIR-M satellites were first launched in 2005. Block IIR-M satellites broadcast signals on the L1 and L2 carrier frequencies, but not on the L5 carrier frequency. Block IIR-M satellites are able to broadcast the military's new GPS signal, M-Code, on L1, and the new civil signal, L2C, on L2. See GPS.

block length Measure of the size of a transmission block in data communications stated in characters, records, language words, computer words, and bits.

block misdelivery probability The ratio of the number of misdelivered blocks to the total number of block transfer attempts during a specified period.

Block Mode Terminal Interface BMTI. A device used to create (and break down) packets to be transmitted through a ITU-T X.25 network. This device is needed if block-mode terminals (such as IBM bisync devices) are to be connected to the network without an intermediate computer.

block multiplexer channel An IBM mainframe input/output channel that allows interleaving of data blocks.

block pair BP. The telephone wires that run from the terminal box to the customer's premises.

block parity The designation of one or more bits in a block as parity bits whose purpose is to ensure a designated parity, either odd or even. Used to assist in error detection or correction, or both.

block the blocker Call Block. A feature that lets you automatically reject calls from parties that have blocked the transmission of their calling telephone number in order that you are unable to determine who is calling you. These features have meaning only if you subscribe to CLID (Calling Line ID) from your LEC, and you have call display equipment. See also Call Block and CLID.

block transfer The process of sending and receiving one or more blocks of data.

block transfer attempt A coordinated sequence of user and telecommunication system activities undertaken to effect transfer of an individual block from a source user to a destination user. A block transfer attempt begins when the first bit of the block crosses the functional interface between the source user and the telecommunication system. A block transfer attempt ends either in successful block transfer or in block transfer failure.

block transfer efficiency The average ratio of user information bits to total bits in successfully transferred blocks.

block transfer failure Failure to deliver a block successfully. Normally the principal block transfer failure outcomes are: lost block, misdelivered block, and added block.

block transfer rate The number of successful block transfers made during a period of time.

block transfer time The average value of the duration of a successful block transfer attempt. A block transfer attempt is successful if 1. The transmitted block is delivered to the intended destination user within the maximum allowable performance period and 2. The contents of the delivered block are correct.

blockage In satellite communications, blockage is a break in the signal path between a satellite and a terminal on the Earth, usually caused by an intervening object which shadows the terminal from the satellite.

blockbuster The term blockbuster was coined in the 1920s in the movie business to denote a movie whose long line of customers could not be contained on a single city block.

blocked attempt An attempt to make a call that cannot be further advanced to its destination, due to an equipment shortage or failure in the network.

blocked calls The fraction of calls failing to be served immediately are called "blocked calls." Blocking can occur in two ways: All facilities are occupied when a demand is originated, and/or a matching of idle facilities cannot be made even though certain facilities are idle in each group.

blocked calls delayed A variable in queuing theory to describe what happens when the user is held in queue because his call is blocked and he can't complete it instantly.

blocked calls held A variable in queuing theory to describe what happens when the user redials the moment he encounters blockage.

blocked calls released A variable in queuing theory to describe what happens when the user, after being blocked, waits a little while before redialing.

blocked port A security measure in which a specific port is disabled, stopping users outside the firewall from gaining access to the network through that port. The ports commonly blocked by network administrators are the ports most commonly used in attacks. See also Port.

blocked site An IP address outside the firewall, explicitly blocked so it cannot connect with hosts behind the firewall. Sites can be blocked manually and permanently, or automatically and temporarily.

blocking When a telephone call cannot be completed it is said that the call is "blocked." Blocking is a fancy way to say that the caller is "receiving a busy." There are many places a call can be blocked: at the user's own telephone switch – PBX or key system, at the user's

local central office or in the long distance network. Blocking happens because switching or transmission capacity is not available at that precise time. The number of calls you try compared to the number of times you get blocked measures "the grade of service" on that network. Blocked calls are different from calls that are not completed because the called number is busy. This is because numbers that are busy are not the fault of the telephone switching and transmission network. One might think the fewer blocked calls, the better. From the user's point of view, the answer is obviously YES, it is better. Less blockage, fewer busies and less frustration. But as one designs a switching and transmission network for less and less blocking, the network becomes more and more expensive. Logarithmically so. We keep adding extra circuits and extra equipment. Thus, in any telecommunications network design there is always a trade-off: What are you prepared to pay, compared to what can you tolerate?

Everyone designs their network with different trade-offs depending on what they and their users or customers, can tolerate and/or are willing to pay. Most companies are willing to pay more for better service if someone explains the logic of telephone design to them. Many network salesmen, however, don't believe this. They practice the sales "theory" of selling better service for less money. This doesn't work in business, and especially not in telephony. The "Grade of Service" is a measurement of blocking. It varies from almost zero (best, but most expensive case, no calls blocked) to one (worst case, all calls blocked). Grade of Service is written as P.05 (five percent blocking). "Blocking" used to be a technical term but has now become a sales tool especially among PBX manufacturers, who increasingly claim their switch to be "non-blocking." This means it will not, they claim, block a call in the switch.

There are several flaws in this logic: First, it's not logical or useful to buy a non-blocking PBX if the chances of being blocked elsewhere – the local lines, the local exchange or the long distance network – are very high. Second, a true non-blocking PBX can be very expensive, perhaps too much power and too much money for most peoples' needs. Third, most manufacturers define "non-blocking" differently. One defines it strictly in terms of switching capability and ignores the fact that his PBX might not have sufficient other "things," like devices which ring bells on phones (to indicate an incoming call) or devices which deliver dial tone to a phone (to indicate the PBX is ready to receive instructions).

blocking factor The number of records in a block; the number is computed by dividing the size of the block by the size of each record contained therein. Each record in the block must be the same size.

blocking formulas Specific probability distribution functions that closely approximate the call pattern of telephone users probable behavior in failing to find idle lines.

blocking ratio For a group of servers, the ratio of blocked attempts to total attempts within a specified time interval.

blog Also called weblog. The word blog is a combination of web and log. A blog is a fancy name for a free website containing news, information and opinion about all manner of things. A blog might be written by one person, sometimes an executive of a company, e.g. there is a popular blog written by Jonathan Schwarz, president of Sun Microsystems – http://blogs.sun.com/jonathan. Sometimes the blog might be written by someone giving out free advice and entertainment, e.g. the one I write, which you'll find at www.InSearchOfThePerfectInvestment.com. Sometimes blogs contain material contributed by the readers of the blog, e.g. www.Slashdot.com. Blogs have become popular as sources of news and information which are different to the traditional news media as we know it – network TV, daily newspapers, etc. A person who runs a blog is called a blogger. Unlike professional journalists, who often move from beat to beat, bloggers are often passionate, if not fanatical, about the subjects they cover. No one knows how many blogs there are on the Internet. I'm guessing several hundred thousand. There's nothing special or different about blogs compared to normal websites. You get to them by typing an address into your web browser. Blog has no special technical meaning. It's a website, pure and simple. See blog and blogorrhea.

blogerati People versed in operating blogs.

Blogger See blog and blogorrhea.

blogorrhea Pointless or excessive filler in a weblog. As one wrote: "I was accused recently of blogorrhea. Well, yes, that's fair. But see, here's the thing: I'm leaving town for a month and I needed to get everything said before I left town. So there."

blogosphere The world of blogs and bloggers.

bloomers Late blooming baby boomers. The term is used by marketers – especially those who sell travel and luxury goods - to describe a demographic group that's only now realizing how much money it has available to spend prior to senility and death.

blooming The bleeding of signal charge from extremely bright pixels, resulting in over-saturated pixels. Blooming compares with over-exposure in film photography. Blooming can occur in the entire image or part of an image."

blow battery When troubleshooting a line or phone with no dial tone, a technician will use a telephone or his butt set and blow into the receiver to see if he hears it in his ear. "I don't have dialtone but I have Blow Battery." This definition contributed by Keith Boe General Manager, Business Communication Services in Everett, Washington.

blower A microphone in America. A telephone in England or Australia, as in "I'm on the blower, Sheila." See also dog.

blowfish The name of the scrambling algorithm behind Philip Zimmerman's powerful encryption scheme called Pretty Good Privacy, or PGP, which lets you converse in total privacy over normal phone lines. See PGP for a fuller explanation.

blowing your buffer Losing one's train of thought. This happens when the person you're speaking with won't let you get a word in edgewise or just said something so astonishing that your brain gets derailed. "Damn, I just blew my buffer!"

blown fiber Blown fiber is an installation method where a housing with lots of smaller tubes is installed without fiber. Once the housing is checked out, they take an air compressor and use air pressure to float the fiber down the tube. This greatly lowers the rate of failures during underground burying of fiber since the housing is empty during the installation. The air flow moves in the same direction as the fiber is installed and helps lower friction and any tugging on the fiber itself. One advantage to blown fiber is that the tubes can be installed with lots of joins (as needed to accommodate the installation problems, such as tight corners), but the fiber strands are later installed with no splices required. Another is that you can upgrade your fiber when you need it.

blown fuse A broken fuse.

BLR Branch Level Revenue.

BLSR See Bidirectional Line Switched Ring.

BLT See Build-Lease-Transfer. In North America, a BLT is a bacon, lettuce and tomato sandwich, usually toasted.

Blu-ray A standard for recording onto a digital video disc (DVD). Blu-Ray boosts the disc's storage capacity from its present 4.7 gigabytes to as much as 50 gigabytes, sufficient to store a two-hour movie in high-definition video format, or over ten hours of standard television broadcast. The technique uses blue-violet laser light, which has a shorter wavelength (405 nanometres) than the red laser light traditionally used with DVD players (650 nm). This allows a greater number of bits that can fit onto a disc. A consortium of nine firms including Sony, Philips, Samsung, Sharp and Thomson Multimedia backs this format. It is in competition with another protocol backed by NEC and Toshiba, which can hold up to 20 gigabytes and is called HD-DVD. See also Blue Laser Disc and DVD.

blue alarm Used in T-1 transmission. Also known as the AIS (Alarm Indication Signal). The blue alarm is turned on when two consecutive frames have fewer than three zeros in the data bit stream. A blue alarm sends 1's (ones) in all bits of all time slots on the span. See Blue Alarm Signal and T-1.

blue alarm signal Alarm Indicating Signal (AIS). Also known as a "Blue Alarm Signal" or "All Ones Keep Alive," an AIS is an unframed all-ones bit pattern sent by equipment at the far end to indicate that an alarm condition exists upstream in a circuit leading to the downstream equipment. Keep-alive signals are required by the network facilities to prevent oscillation of the line repeaters which causes interference (i.e. cross-talk and bleeding) within adjacent channels. SONET defines four categories of AIS: Line AIS, STS Path AIS, VT Path AIS, DSn AIS.

blue books The CCITT 1988 recommendations were published in books with blue covers, hence the term "blue books." The CCITT is now called the ITU-T.

blue box A device used to steal long distance phone calls. The classic blue box was slightly larger than a cigarette container. It had a touchtone pad on the front and a single button on top. Typically, you went to a coin phone and dialed an 800 number. While the distant number was ringing, you punched the single button on the top of the blue box. That button caused the blue box's speaker to emit a 2600 Hz tone. This disconnected the ringing at the other end but left the user inside the long distance network. The user then punched in a series of digits on the touchtone pad. The phone network heard those tones and sent the call according to the instructions in the tones. The tones duplicated the tones which the touchtone pads of long distance operators emitted. They are different from those emitted by normal telephones. The first blue box was "discovered" at MIT in a small utility box that was painted blue, thus the term blue box. When they were young, Steve Jobs and Steve Wozniak, founders of Apple Computer, sold blue boxes, which Wozniak built. People who used blue boxes in their salad days included characters with adopted pseudonyms like Dr. No, The Snark and Captain Crunch, who got his name from the free 2600 Hz whistle

included as a promotion in boxes of Captain Crunch breakfast cereal. With the advent of CCIS, Common Channel Interoffice Signaling (i.e. out-of-band signaling), blue boxes no longer work.

blue chips In poker, the blue chips are worth a lot and are thus worth having. That's why high priced shares are often called blue chips. The white chips, on the other hand, are worth very little. Most professional gamblers leave the white chips as tips.

blue collar computer A colloquial term for a handheld computer which is used by "blue collar" workers for tasks such taking inventory, tracking goods, etc. Such a computer may have a pen, a large pen-sensitive or touch-sensitive screen, a bar code scanner and a modem. It may be able to capture signatures – useful for confirmation of the delivery of goods.

blue collar worker Someone who "works" for a living is a blue collar worker, which we used to call "craft" in the telephone company. In years gone by, such folks wore blue shirts with blue collars, since they didn't show dirt readily. A white collar worker may be either management or a non-management "professional," and is theoretically a step above a blue-collar worker in the company hierarchy. The term originated in the fact that, in years gone by, such folks wore white shirts with white collars as evidence of the fact that they didn't do the kind of "dirty" work that would soil a shirt.

blue glue IBM's SNA (Systems Network Architecture). Blue Glue is a "glue logic" architecture that serves to interface physical and logical units in the IBM world. See also Glue Logic and Systems Network Architecture.

blue goo Nanotechnological machines that monitor and control other machines to ensure that their replication does not get out of control.

blue grommet The rubber collar over the joint between the handset and the armored cable on a pay phone. Blue identifies a "hearing aid compatible" handset.

blue laser A blue laser is a short-wave lasers with concentration in the blue and violet part of the spectrum, as opposed to the red portion. Blue lasers have wide commercial applications. They can create high-definition DVDs that deliver HDTV quality reproduction; a single disc could contain more than 12 hours of music per side. Blue lasers can also make possible digital tape recorders that are comparable to today's VCRs, but can store high-definition images. The first blue laser with commercial possibilities was developed in January 1999 by Shuji Nakamura, a Japanese inventor. In 1993 IBM announced that it had used blue laser technology to read and write data at a rate five times faster than the infrared laser rewritable optical disk drives of the day.

blue laser disc An optical storage medium that uses shorter wavelength blue laser light instead of the red laser light used in CDs and DVDs to fit more bits onto the disc surface. Storage capacity is expected up to 50 Gb. See also Blue Laser, Blu-Ray, DVD.

blue pages Section of a phone directory commonly used for government phone numbers, as distinct from white and yellow pages.

blue screen of death BSOD. A sarcastic name for the screen displayed when Microsoft Windows encounters an error so serious that the operating system cannot continue to run. Windows then go completely blue and it may (or may not) display information about the failure and may perform a memory dump and an automatic system restart. The Macintosh equivalent is a blank screen with a text box containing a graphic of a bomb with a lit fuse. There are two typical causes of a BSOD – a bad driver or bad memory. You get a new driver from the software company whose driver was bad. The problem is figuring out which driver and which memory. A program called Alexander SPK from www.Alexander.com may help you figure out which driver. Memory is a different problem. You may never figure out which memory is bad. You got to do it by trial and error. See also Crash.

Blue Sky Laws State securities laws designed to protect individual investors. The phrase purportedly originated from a state judge who said that the securities of a particular company had all the value of a patch of blue sky. Companies and mutual funds are affected by state blue sky laws. However, the SEC and Congress have superceded most of these these state laws because they were obsolete, arbitrary, and poorly enforced.

blue stake A verb that means to mark an area. Here's how it works. Let's say you're a phone company and you want to lay a cable alongside a highway. You go to the highway authority and ask for permission to lay your cable. They give it to you. Then you call for "blue staking." This means that one organization comes out and marks on the ground using blue paint where all the stuff is located. This one organization represents all the various utilities who have stuff buried in the proposed running line (the path that you intend to lay your cable in). There is a legality here. If you then dig and hit someone's buried cable, you are absolved from legal responsibility – so long as you called for blue staking. If you hit something that was not located (i.e. not blue staked and not located on any map), it's called "off locate."

Blue Zoo The nickname for AT&T Labs' Silicon Electronic Research Laboratory (formerly known as Bell Labs' Integrated Design Circuit Capability Laboratory) in Murray Hill, NJ.

bluejack To temporarily hijack another person's cell phone by using Bluetooth wireless networking. This involves sending anonymous text messages to other phone users via Bluetooth short-range radio. Bluetooth works over a range of about 10 metres and phones fitted with it can be made to search for other handsets using it that will accept messages sent to them. The group of lanky tourists strolling through Stockholm's old town never knew what hit them. As they admired Swedish handicrafts in a storefront window, one of their cell phones chirped with an anonymous note: "Try the blue sweaters. They keep you warm in the winter." The tourist was "bluejacked" —surreptitiously surprised with a text message sent using a short-range wireless technology called Bluetooth. Already, Web sites are offering tips on bluejacking, and collections of startled reactions are popping up on the Internet, reported the Associated Press. See also Bluetooth.

Bluespamming Sending unsolicited advertising to cell phones and other Bluetooth-enabled devises. As Gizmodo.com warns, "If you're looking for a way to finally kill off Bluetooth, bluespam is probably the way to do it." See Bluetooth.

Bluetooth A short-haul wireless protocol that is used to communicate from one device to another in a small area usually less than 30 feet. The commonest use is a wireless headphone communicating with the user's cellphone or desk phone. Bluetooth uses the 2.4 ghz spectrum to communicate a one million bit per second connection between two devices for both a voice channel and a 768k data channel. The Bluetooth specification calls for different profiles such as voice and serial emulation to be used by devices to communicate. The initial effort (April, 1998) was in the form of a consortium of Intel, Microsoft, IBM, Toshiba, Nokia, Ericsson and Puma Technology. The project was code-named Bluetooth after Harald Blaatand, the 10th century Danish king who unified Denmark. The idea of Bluetooth was to create a single digital wireless protocol to address end-user problems arising from the proliferation of various mobile devices that need to keep data synchronized (i.e., consistent from one device to another). According to an article in Telecommunications Magazine, December, 1998, "The standard's proponents (Ericsson, Nokia, IBM, Intel, Toshiba) talk of a world where equipment from different vendors works seamlessly together using Bluetooth as a sort of virtual cable, where a laptop can automatically use a mobile phone to pick up e-mail or a PDA can send data wirelessly to a fax machine." Bluetooth operates using FHSS (Frequency Hopping Spread Spectrum), spreading data packets across the designated frequency range (2.45 GHz) at a rate of 1,600 hops per second to lessen interference. The nominal link range is 10 meters, and the gross data rate is one Mbps, with plans to double the data rate in the future. Bluetooth devices operate in a picocell topology in the 2.45 GHz range of the unlicensed ISM (Industrial, Scientific and Medical) spectrum. Note that this same frequency band is used by a wide variety of other devices, which fact causes some concern in terms of the potential for interference. For example, both HomeRF and IEEE 802.11, the specification for wireless Ethernet LANs, specifies this frequency range for RF (Radio Frequency) communications. Bluetooth supports both SCO (Synchronous Connection Oriented) links for voice and AC (Asynchronous Connectionless) links for packet data. Bluetooth can support 1) an asynchronous data channel in asymmetric mode of maximally 721 Kbps in either direction and 57.6 Kbps in the reverse direction; alternatively, the data channel can be supported in symmetric mode of maximally 432.6 Kbps; 2) up to three simultaneous synchronous packet voice channels; or 3) a channel which simultaneously supports both asynchronous data and synchronous voice. Full-duplex communications can be supported using TDD (Time Division Duplex) as the access technique. Voice coding uses the CVSD (Continuously Variable Slope Delta) modulation technique. Security is sort of provided through encryption and authentication, using the challenge-response mechanism, though there is talk of Bluetooth snooping by near devices. Frequency hopping, a spread spectrum technique, is used to improve performance in the unlicensed and heavily-used ISM band. www.bluetooth.com. See also 802.11, CVSD, HomeRF, ISM, Picocell, Spread Spectrum, SWAP, and TDD. For a comparison of wireless standards, see Wireless. See also the next few definitions.

Bluetooth device class A parameter that indicates the type of device and which types of services that are supported. The class is received during the discovery procedure. The parameter contains the major and minor device class fields. The term "Bluetooth device class" is used on the UI (User Interface) level. See the next definition and Bluetooth.

Bluetooth device type The term "bluetooth device type" is used on the UI (User Interface) level. This term overrides the terms "Bluetooth device class" and "Bluetooth service type" when there is a mix of information containing both Bluetooth Device Class and Bluetooth Service Types.

Bluetooth passkey The name of the PIN. The term "Bluetooth passkey" is used

in the UI. See PIN.

Bluetooth service type One or more services a device can provide to other devices. The service information is defined in the service class field of the Bluetooth device class parameter.

Bluetooth session The activity and participation of a device on a piconet.

Bluetooth special interest group An industry group comprising about 700 members of the telecommunications and computing industries promoting the development of the Bluetooth standard. According to its own promotion literature, "Bluetooth wireless technology is set to revolutionize the personal connectivity market by providing freedom from wired connections for portable handheld devices. The Bluetooth SIG is driving development of the technology and bringing it to market. The SIG is comprised of telecommunications, computing, network, and consumer electronics industry leaders and includes Promoter group companies 3Com Corporation, Ericsson Technology Licensing AB, IBM Corporation, Intel Corporation, Agere Systems, Inc, Microsoft Corporation, Motorola Inc., Nokia Corporation, Toshiba Corporation, as well as hundreds of Associate and Adopter member companies." www.bluetooth.com. See also Bluetooth.

BM Burst Modem.

Bmp 1. A Windows BitMaP format. The images you see when Windows starts up and closes, and the wallpaper that adorns the Windows desktop, are all in BMP format. BMP is the standard Windows image format on DOS and Windows-compatible computers. The BMP format supports RGB, indexed-colour, grayscale, and bitmap color modes.

2. Basic Multilingual Plane. A 16-bit coding scheme resulting from the merger of Unicode and the scheme developed by jointly by the ISO and IEC. Generically, BMP is known as a Universal Character Set (UCS). See also UCS and Unicode.

BMTI Block Mode Terminal Interface. A device used to create (and break down) packets to be transmitted through a ITU-T X.25 network. This device is needed if block-mode terminals (such as IBM bisync devices) are to be connected to the network without an intermediate computer.

BN 1. Bridge Number: A locally administered bridge ID used in Source Route Bridging to uniquely identify a route between two LANs.

2. An ATM term. BECN Cell: A Resource Management (RM) cell type indicator. A Backwards Explicit Congestion Notification (BECN) Rm-cell may be generated by the network or the destination. To do so, BN=1 is set to indicate the cell is not source-generated and DIR=1 is set to indicate the backward flow. Source generated RM-cells are initialized with BN=0.

3. Business Name.

BNA 1. Burroughs Network Architecture. Communications architecture of Burroughs, now Unisys.

2. Billing Name & Address.

BNAP A British telecommunication industry term. It stands for Broadband Network Access Point and generally refers to a site where access can be obtained to a higher bandwidth network, e.g. SDH and SONET.

BNAR Busy No Answer Reroute (or Routing). Rules that a LEC (Local Exchange Carrier) employs to redirect a call that encounters a Busy or No Answer. Often these calls are billed per call with additional message units applied to the calls that are forwarded.

BNC See BNC Connector.

BNC Barrel Connectors These connectors join two lengths of thin Ethernet coaxial cable together. See BNC.

BNC Connector A bayonet-style, twist-locking connector for slim coaxial cables, like those used in old 10Base-5 Ethernet and CATV systems. BNC is an acronym for Bayonet Niell-Concelman, which describes the bayonet style of the connector. Neill and Concelman are the inventors. BNC also sometimes is referred to as British Naval Connector, because it was originally developed by the British Navy as a trustworthy connection technique for harsh environments such as onboard ships. Researchers at the College of Engineering, California State Polytechnic University, Pomona, say it was a "Baby Nevel Connector" named after a man called Nevel who invented the large size of connector that resembles a regular BNC connector. In any event, a TNC is a threaded (T) version of a BNC.

BNC Female To N-Series Female Adapter The BNC female to N-Series female adapter is a connector which enables you to connect thin coaxial cable to thick coaxial cable. The BNC female connector attaches to the thin cable and the N-Series male connector attaches to the thick Ethernet cable.

BNC T-connectors The top of the T in a BNC T-connector functions as a barrel connector and links two lengths of thin Ethernet coaxial cable; the third end connects to the SpeedLink/PC 16.

BNC Terminators 50-ohm terminators are used to block electrical interference on a Ethernet coaxial cable network and to terminate the network at certain spots. You attach a BNC terminator to one plug on a T-connector if you will not be attaching a length of cable to that plug. You may also need to use a BNC terminator with a grounding wire to ground the network. See BNC.

BNR Bell-Northern Research. Northern Telecom's research arm. Northern Telecom is now called NorTel.

BnZS Code A bipolar line code with n zero substitution.

board 1. Short for printed circuit board. Phone systems have boards for all sorts of purposes – from boards that serve trunk lines, to boards that serve proprietary phone sets, to boards that serve T-1 lines, etc. Computers also have boards – ones for SCSI ports, for floppy and hard disks, for CD-ROMs, etc. In display/monitor terminology a board refers to the adapter (or controller) that serves as an interface between the computer and monitor.

2. An SCSA term. Any hardware module that controls its own physical interface to the SCbus or SCxbus. From a programming point of view, a board is an addressable system component that contains resources.

boat anchor 1. An old bulky computer, radio, or other piece of equipment that has outlived its usefulness and could probably be put to better use as a boat anchor.

2. A legacy system that continues to be relied on and impedes the adoption of newer, better solutions.

BOB 1. Software from Microsoft, which describes it as a "superapplication" for Windows designed for consumers intimidated by computer technology. According to Microsoft, users will use Bob to write letters, send e-mail, manage their household finances, keep addresses and dates and launch full-blown Windows applications, all under the guidance of cartoon characters. Bob hasn't done well and has effectively died.

2. BreakOut Box.

BOC 1. Bell Operating Company. The local Bell operating telephone company. These days there are 22 Bell Operating Companies. They are organized into (i.e. owned by) seven Regional Bell holding companies, also called RBOCs, pronounced "R-bocks," or RHCs. See Bell Operating Company.

2. Business Office Code. Western region FID that carries a three character alphanumeric code identifying the U S WEST business that handles the customer's service.

BoD Board of Directors.

body The main informational part of a message. Body is the information, not the address nor the addressing information. There can be single or multiple parts to a body. For example, a single part could be text, or multiple parts could include text and graphics, or voice and graphics, etc. See Body Part 14.

body belt Used to attach telephone workers to poles and structures. Also called safety belt and climbing belt. You got to be gutsy, fit and well trained to use one of these things to go up a telephone pole. It would be good if you had spikes on your boots. That way your feet could dig into the pole.

Body Part 14 BP 14. An X.400 messaging term. A non-specific, identifying body part. A binary attachment with identifying an header to explain the nature of the content such as a particular spreadsheet, word processor, etc.

body worn transmitter A body worn transmitter is an audio transmitter secretly worn by an agent for surveillance purposes. Body wires must be carefully designed to be rugged in everyday use and transmit a strong signal regardless of how the antenna and wearer are positioned.

BOE Buffer Overflow Error.

BOF 1. Business Operations Framework. A wireless telecommunications term. A document compiled to describe the operations of a telecommunications business entity in a specific area.

2. Birds of a Feather. A group of people with similar interests. If interest in a subject is strong enough, a BOF may develop into a SIG (Special Interest Group).

BOGBS Blue, Orange, Green, Brown, Slate. See WRBYV.

bogo-sort The ultimate evil algorithm, equivalent to tossing a deck of cards into the air, picking them up at random, and then testing to see whether they are in order. If a programmer sees an example of this type of algorithm, he could say, "OK, this program uses bogo-sort." Not to be confused with BogoMIPS, which is the number of million times per second a processor can do absolutely nothing. The Linux operating system actually measures BogoMIPS at startup in order to calibrate some kinds of timing loops.

bogon Something that is stupid or nonfunctional.

BOHICA Bend Over Here It Comes Again. See FUBAR and Snafu.

BOIC Barring of Outgoing International Calls. A wireless telecommunications term. A supplementary service provided under GSM (Global System for Mobile Communications).

boiler room This is an old police vice squad slang for an illegal gaming operation where bets are called in and the bookies take them down. It is now attributed largely in telemarketing circles, to describe phone room set ups where calls are made.

bollocks English Cockney slang for balls, as in dogs' bollocks. It really means "Nonsense!" As a verb, it means to make a total mess of something. The word was introduced to North America in the movie "Bend It Like Beckham."

Bollywood Slang for the Bombay, India equivalent of Hollywood, i.e. a place movies are made.

bolt from the blue On a clear day with blue skies, lightning can jump outside of its parent cloud and travel for more than five miles through clear air. This is called the "bolt from the blue" phenomenon. The study of lightning is called kerauno-pathology.

bolt pattern The pattern that the bolts on the back of the device make. The idea is that if you attach something to the device, you don't want to mess up its warranty by opening the box and putting your stuff inside. So you design your stuff to fit the bolt pattern. This does not mess up the warranty. His product is designed to fit the bolt pattern, but not to intrude into the box.

bolt-on An application software system that performs certain specific tasks, and that interfaces with another, broader system. For example, a Warehouse Management System (WMS) might bolt on to an Enterprise Resource Planning (ERP) system.

bolt-on acquisition A product or company acquisition that fits naturally with the buyer's existing business lines or strategy. There aren't many.

BOM An ATM term. Beginning of Message: An indicator contained in the first cell of an ATM segmented packet.

bomb technician If you see a bomb technician running, follow him.

bond 1. The electrical connection between two metallic surfaces established to provide a low resistance path between them.

2. See also pairing.

bond strength A cable term. Designates the amount of adhesion between surfaces.

bonding 1. In ISDN BRI transmissions, bonding refers to joining the two 64 Kbps B channels together to get one channel of 128 Kbps. Also known as dial-in channel aggregation.

2. Bonding is also the name of a group known as the Bandwidth ON Demand INteroperability Group (BONDING). The group's charter is to develop common control and synchronization standards needed to manage high speed data as it travels through the public network. This will allow equipment from vendors to interoperate over existing Switched 56 and ISDN services. Version 1.0 of the standard, approved on August 17, 1992, describes four modes of inverse multiplexer (I-Mux) interoperability. It allows inverse multiplexers from different manufacturers to subdivide a wideband signal into multiple 56- or 64-Kbps channels, pass these individual channels over a switched digital network, and recombine them into a single high-speed signal at the receiving end.

3. Bonding is the permanent joining of metallic parts to form an electrically conductive path that will assure electrical continuity and the capacity to conduct safely most any current likely to be imposed on it.

4. In microelectronics, the process of connecting wires from the leads on the package to the bonding pads on the chip. Part of the assembly process.

bonding conductor for telecommunications A conductor that connects the telecommunications bonding infrastructure to the building's service equipment power ground.

bonding jumper A bonding jumper is a reliable conductor to ensure the required electrical conductivity between metal parts required to be electrically connected.

bone Telephone in London cockney rhyming slag is dog 'n' bone, bone obviously rhyming with phone, In most cases of cockney the rhyming word is dropped leaving just dog, but bone works for phone, too.

bong A tone that long distance carriers and value added carriers use in order to signal you that they now require additional action on your part – usually dialing more digits in order to provide billing information. For example, you hear a bong (or boing) to prompt you to enter a calling card number. The bong tone consists of a short burst of the # touch tone, followed by a rapidly decaying dial tone. See also Voice Modem.

Bonjour See Zeroconf.

BONTs Broadband Optical Network Terminations.

BOO See Build-Operate-Own.

boob tube Slang for a television set.

boodle An illegal payment, as in graft.

book-to-bill ratio Manufacturing companies use this ratio to report on their financial health. The "book" means orders booked, i.e. sold. You sell something when your customer calls up and orders something. The "bill" means when the order is actually billed and customers have to pay for it. The idea is that if your customers order more than you bill, you're doing OK. Your sales are rising. The book-to-bill ratio is used by the semiconductor industry to give an overall sense of supply and demand. If the number is below 1.00, it suggests there is more supply than demand, and times are awful. If it's bigger than 1.00, times are wonderful. Orders are pouring in. The semiconductor industry's equipment Book-to-Bill ratio is one of the most important indicators of that industry's financial health. In July, 2001, its book-to-bill ratio was 0.67. It explained that a book-to-bill of 0.67 means that $67 worth of new orders were received for every $100 of product shipped for the month.

booking factor Booking factor is a percentage of the frame relay links used by frame relay paths, based on the sum of the Committed Information Rates (CIRs) of all the frame relay paths (FRPs) on the frame relay link (FRL).

bookmark A gopher or Web file that lets you quickly connect to your favorite pre-selected page. Appropriately named. The way it works: You connect to a home page. You decide you'd like to return at some other time. So you command your internet surfing software to mark this web site with a "bookmark." Next time you want to return to that web site, you simply go to your bookmarks and click on the one you want. And bingo, you're there. A bookmark is also known as a hot list. Most Web browsers have bookmarks or hot lists. Microsoft calls a bookmark a favorite. Netscape calls it a bookmark.

boolean approach A decision-making process favored by business execs in which the answer is either "yes" OR "no." Also known as a digital decision. See the next few definitions.

boolean expression An expression composed of one or more relational expressions; a string of symbols that specifies a condition that is either true or false.

Boolean Logic Boolean Logic is named after the 19th century mathematician, George Boole. Boolean logic is algebra reduced to either TRUE or FALSE, YES or NO, ON or OFF. Boolean logic is important for computer logic because computers work in binary – TRUE or FALSE, YES or NO, ON or OFF.

Boolean operators See Boolean Logic. Boolean operators are AND, OR, XOR, NOR, NOT. The result of an equation with one or more of the boolean operators is that the result will either be true or false.

Boolean valued expression An expression that will return a "true" or "false" evaluation.

boomerang The boomerang feature of some PBXs and voice processing systems comes in two types. The simplest allows you to access your voice messages on the road and return a call based on the CLID (Calling Line IDentification) information received by the system when the original caller left the message. This feature usually can be invoked either by depressing a single key on the touchtone keypad, or by spoken command, which is interpreted by a voice recognition module contained within the voice processing system. The more powerful version allows you to obtain a second dialtone within the voice mail system. You then can dial the number of the original caller, assuming that your access privileges allow you to do so based on your Class of Service (CoS). This second version is considered more powerful because CLID information often is unavailable. Under either scenario, the boomerang feature allows you to return to your original place in the voice mail queue once the return call has been terminated. See also CLID and Class of Service.

boomerang worker Retiree returning to former employer.

booster A television or FM broadcast station, operating at relatively low power that receives a distant input signal, amplifies it, and retransmits it on the same channel.

booster amplifier A radio device that is used to improve wireless signal coverage in buildings, tunnels, parking garages, and other areas where there are coverage gaps inside an FCC-licensed service contour. A booster amplifier can also be mounted on a motor vehicle or boat to improve coverage and reduce disconnects and drop-outs for cell phone users on board. A booster amplifier consists of antenna that receives signals from a cell site. The signals are amplified and sent to the cell phone. When the cell phone transmits, its signals are sent to the amplifier, which amplifies and transmits them to the cell site. Booster amplifiers fall into two categories: narrowband (Class A) and broadband (Class B). A narrowband amplifier (also called a channelized amplifier) amplifies a single 25 KHz or 12.5 KHz channel. A broadband amplifier amplifies the entire band or a large portion of the band.

boot 1. Abbreviation for the verb to bootstrap. A technique or device designed to bring itself into a desired state where it can operate on its own. For example, one type of boot is a routine whose first few instructions are sufficient to bring the rest of itself into memory from an input device. See Bootstrap and Rebooting.

2. Slang for to steal.

boot loader A Windows 2000 term. Defines the information needed for system startup, such as the location for the operating system's files. Windows NT automatically creates the correct configuration and checks this information whenever you start your system.

boot mail Military slang for email that, due to incompatible communications technologies, must be transferred to a disc and carried from the squadron field radio to each soldier's personal computer.

boot partition A Windows term. The volume, formatted for an NTFS, FAT or HPFS file system, that contains the Windows operating system and its support files. The boot partition can be (but does not have to be) the same as the system partition.

boot priority Which disk drive the computer looks to first for the files it needs to get started. Modern PCs start their boot cycle with the hard disk and then move to the floppy disk drive. Older PCs started their boot cycle with the floppy disk drive.

Boot Protocol BOOTP. The protocol used for the static assignment of IP address to devices on the network.

BootP Bootstrap Protocol. A TCP/IP protocol, which allows an internet node to discover certain startup information such as its IP address.

Boot ROM A read-only memory chip which contains instructions on how to start the computer. Typically the boot ROM starts the computer and then instructs the computer to take its information from the hard disk. Boot ROM are also used in workstations that are not PCs. In this case the boot ROM tells the workstation to talk to the file server and to take its software from the server. Workstations can thus operate on the network without having a disk drive. These are commonly called diskless PCs or diskless workstations. In Wind River's VxWorks, boot ROMs are used to download the VxWorks kernel from a host computer over the network.

booth bunny An attractive, scantily-clad woman who attends a booth at a trade show. Her job is to attract tired, bored men to the booth by displaying a pretty smile, long handsome legs or large breasts, or, preferably, all three. Booth bunnies might give away literature, imprint badges, serve coffee or just smile and direct people to a salesperson. At night, they might act as hostess for a company party. Sexual favors are not usually part of the job description, though most of the tired, bored men visiting the booth fantasize otherwise. A sure fire way of figuring whether the lady at the booth is a bunny or the CEO is to ask, "Does it work on DOS?" See also Barcode Rape.

boot sector A critical disk structure for starting your computer, located at sector 1 of each volume or floppy disk. It contains executable code and data that is required by the code, including information used by the file system to access the volume. The boot sector is created when you format the volume.

bootstrap The process of starting up a computer. Think about the following explanation in regard to your desktop PC. Usually, when you turn your computer on, it goes to a location of permanent Read Only Memory (See ROM) for instructions. These instructions, in turn, load the first instructions from the disk telling the computer what tasks to start performing. The name of this process comes from the expression "pulling oneself up by one's own bootstraps." The typical personal computer BOOT (startup) throws a message on the screen instructing the user to "insert a disk."

To confuse matters, there are WARM boots and COLD boots. Cold boots occur when the ac power switch on the computer is turned on. Warm boots occur when you hit the reset button (or Ctrl/Alt/Del) while the ac power switch stays on. A warm boot – reset – is done when you're changing disks or programs, or have done something dumb, like tried to access a drive that didn't access, or tried to print without connecting up to a printer.

You do a cold boot when the machine locks up rock hard and a warm boot doesn't work. To do a cold boot with your computer, turn the ac power off, count to ten and then turn it on. Remember: never leave disks in your computer when you're turning it on and off. The surge of electricity might destroy the disks. When modems give trouble, do a cold boot on them. In fact, when phone systems give trouble, do a cold boot on them also. See also Boot RAM, Device Driver.

bootstrap loader A computer input routine in which preset operations are placed into a computer that enable it to get started whenever a reset condition occurs. In electronic PBXs this is often called Automatic Program loading. In personal computers it is the sequence that searches predetermined disks for a Command Interpreter program, then a Configure System file, and finally an Autoexecution Batch file. See also Bootstrap.

Bootstrap Protocol BOOTP. An Internet protocol that provides network configuration information to a diskless workstation. When the workstation first boots, it sends out a BOOTP message on the network. This message is received by the server, which obtains the appropriate configuration information and returns that information to the workstation.

This information includes the workstation's IP address, the IP address of the server, the host name of the server, and the IP address of a default router.

boot volume The volume that contains the Windows operating system and its support files. The boot volume can be, but does not have to be, the same as the system volume.

BOP Bit Oriented Protocol. See Bit Oriented Transmission.

border A security perimeter formed by logical boundaries that can only be crossed at specifically defined locations known as border gateways.

border cell A cellular term. A cell that is located on the edge of the serving area of a MSC (Mobile Switching Center). As the caller moves outside the range of the MSC border cell, it must be handed off to another MSC. See also MSC.

Border Gateway Protocol See BGP.

border node An ATM term. A logical node that is in a specified peer group, and has at least one link that crosses the peer group boundary.

Boresight Boresight is the axis of a directive antenna, either 1) the electric boresight, which is the tracking axis determined by radio-frequency and electronic means, such as nulls of conical scan or simultaneous lobing antennas or the maximum of the beam pattern, or 2) the reference boresight, which is geometrical and usually determined by mechanical and/or optical means.

born digital Documents (books, manuscripts, reports,etc.) not published on paper. They are created on a computer and distributed electronically usually via the Internet.

borscht A group of functions provided in Line Circuits (LCs). It stands for:

B: Battery supply to subscriber line.

O: Overvoltage protection.

R: Ringing current supply.

S: Supervision of subscriber terminal.

C: Coder and decoder.

H: Hybrid (2 wire to 4 wire conversion).

T: Test.

Borscht is a group of functions provided to an analog line from a line circuit of a digital central office switch. An analog electronic switch can omit C and possibly H. A line circuit on a switch with a metallic matrix (SXS, Xbar, 1,2,3ESS) only detects call originations and disconnects itself.

BOS Bill Output Specifications. Provides exchange companies with generic detailed specifications to support the billing function of CABS. An industry standard CABS data format used for both paper and mechanized billing.

BOSS 1. Billing and Order Support System, or Business Office Support System. An OSS (Operations Support System) used by the RBOCs (Regional Bell Operating Companies) and other telcos in support of their internal processes, BOSS essentially is a database containing customer records and billing information. Telephone company service representatives in incoming call centers can access BOSS online for information that will assist them in issuing orders, issuing billing adjustments, investigating complaints, and answering customer questions across a wide range of service-related issues.

2. Boss. As the old joke says, "If the boss calls, get a name."

bot 1. Shortened word for Robot, from the Czech "robata," meaning "work." A bot is a program that works for you on an automated basis, perhaps running on a computer 24 hours a day, seven days a week, automating mundane tasks for the owner. Bots are used on the Internet in many ways. Most popular is their use in Internet Relay Chat (IRC) and Web search engines. IRC bots are programs that connect to the Internet and interact with the Internet in very much the same way a normal users do. In fact, IRC servers treat bots as regular users, where they typically are used for channel control. Bots have also been called automatons. In the world of Web searching, bots are also called spiders, crawlers, and agents. They explore the World Wide Web by retrieving a document and following all the hyperlinks in it. Then they generate catalogs that can be accessed by search engines. See also Robot and Spider.

2. See Build-Operate-Transfer.

bot networks Also called botnets. In September 2004, Symantec, the antivirus and security company, released its sixth semiannual Internet Security Threat Report. It says the firm found a vast increase in the number of "bot networks" that are under the control of hackers. Each network consists of thousands of machines that have been infected with Trojan horses viruses and are now controlled by criminals. During the first six months of 2004, Symantec detected a rapid growth of bot networks from fewer then 2,000 to 30,000. The number of PCs in each network is said to average around 2,000. Multiply the number of networks by the average population of controlled machines and it works out to

60 million "zombie" PCs – that we know about. Symantec found one bot network consisting of 400,000 zombies, according to an article by John Markoff in the New York Times. Each network can be used to broadcast spam, launch devastating denial-of-service attacks against Web sites the hackers don't like, and more. How much of this you believe depends on your own experiences. Of course, Symantec does sell anti-virus and security software. So it's always good to magnify, or at the very least, publicize the dangers. Suffice keep your spyware and virus-checking software up-to-date and run them at regular intervals – at least once a week. See also zombie PC.

bot storming Automated probes that search a website for unprotected or unpatched software components.

botnet herder A hacker who controls a large number of bot-infected PCs.

botnets See bot networks and botnet herder. See bot networks.

bottle A water-tight device shaped like a glass bottle which contains amplifiers, regenerators and other equipment is used at regular distances along an underwater cable.

bottlenecks Think of a Coca Cola bottle. It has a neck at the top. It's the narrowest part of the bottle. When you upend the bottle, the neck is what slows the coke from coming out. Likewise in a network the narrowest or slowest part of the network is the "bottleneck." Bottlenecks, of course, never go away. Bottlenecks just move around from one point to another.

bottleNets Wi-Fi antennas and networks built largely of recycled materials – plastic water bottles, bamboo, TV parts, etc. – and used mainly in developing nations.

bottom line A phrase that can mean net profit, the lowest possible price that someone will take or the basic meaning with all the frills and nonsense cut away.

bottom lining From Wired's Jargon Watch column. What phone and cable companies consider when picking areas for trials and early deployment of interactive services. They look for areas full of upper- and middle-class households with enough money to pay for these services and generally ignore areas with lower incomes.

bottom-of-the-bill discount A discount applied to a customer's total bill. The percentage discount may be based on the customer's total bill amount, or total spending during a certain timeframe, or some other criterion.

bounce 1. The return of an email message to the sender when the message is undeliverable. This usually means that you have gotten the address wrong, the destination address has been changed or the destination server has died. The bounce often includes information from the email system that explains the nature of the problem.

2. To reboot or restart a PC or a phone system, as in "let's bounce the server (reboot)." In some circles bouncing the server has replaced the traditional term of rebooting the server. Some people think it's more PC – politically correct. See also Bounced.

bounce board A large microwave reflector resembling an outdoor movie screen which is use to redirect (bounce) microwave telephone signals between two remote transceivers.

bounced "We bounced the system" means we rebooted or restarted the system. Bouncing the system means that you basically clean out its random access memory memory and load the operating system and programming afresh. Theoretically, after you bounce the system, it should work better. Theoretically. See Bounce.

bounced mail Mail that is returned to the originator due to an incorrect e-mail address or a downed mail server.

bouncing busy hour The daily busy hours that do not remain consistent over a number of days.

bouncing circuit K. Nathan Casassa of Bell Atlantic Global Networks tells me that it's a term he uses when a dedicated Internet or WAN circuit goes up and down, up and down, for a period of time. He says "The circuit has been bouncing all morning." A bouncing circuit is also called a flapping circuit.

bound mode In an optical fiber, a mode whose field decays monotonically in the transverse direction everywhere external to the core and which does not lose power to radiation. Except in a single-mode fiber, the power in bound modes is predominantly contained in the core of the fiber.

boundary conditions Boundary conditions are those that are found at the cusp of valid and invalid inputs and parameters. Many faults are found in a computer telephony system's ability to handle boundary conditions, especially when the computer telephony system is under load. For example, for a network that expects a switch to reset a trunk port within two seconds, the associated boundary conditions would be found at 1.9 to 2.1 seconds.

boundary function Capability in an SNA sub-area node to handle some functions that nearby peripheral nodes are not capable of handling.

boundary node In IBM's SNA, a sub-area node that can provide certain protocol support for adjacent sub-area nodes, including transforming network addresses to local addresses, and vice versa, and performing session level sequencing and flow control and less intelligent peripheral nodes.

boundary routing A 3Com proprietary name for a method of accessing remote networked locations, such as a bank branch office. Effectively a form of bridging, the idea is to reduce the need for technical expertise locally and the cost of equipment at the remote site and manage the communications from the head office.

bounding box Traditionally, computer programs have dealt with onscreen objects, such as images, by placing them in an invisible rectangle called a bounding box. You can see an example of a bounding box by clicking an image inside a word processor such as Word. The outline that appears around the image is the bounding box.

bound trap In programming, a problem in which a set of conditions exceeds a permitted range of values that causes the microprocessor to stop what it is doing and handle the situation in a separate routine.

bouquet This term is used by pay-TV operators, especially satellite TV operators, to refer to a collection of TV programs channels, and possibly additional multimedia services, such as radio programs, that are marketed as a single package. Delivering multimedia programming as bouquets also offers a technical advantage to a pay-TV operator, in the sense that the programs making up the bouquet and be multiplexed and compressed into a single data stream, thereby conserving bandwidth.

bounty A bounty is a premium or reward for doing something.

Bourne Shell A UNIX command processor developed by Steven Bourne.

boutique A firm which has few clients, but charges each of them a great deal of money. It's a "designer" company. A research boutique is a high-priced research firm that issues reports costing thousands of dollars. The word "boutique" is designed to justify the high prices. There is no correlation between the cost of the information and its truth, nor its usefulness.

BOVPN Branch Office Virtual Private Network. A type of VPN that creates a securely encrypted tunnel over an unsecured public network, either between two networks that are protected by the WatchGuard Firebox System, or between a WatchGuard Firebox and an IPSec-compliant device. BOVPN allows a user to connect two or more locations over the Internet while protecting the resources on the Trusted and Optional networks.

bow A type of distortion in which opposite sides of the screen image curve in the same direction.

bow-tie antenna A type of dipole antenna that is shaped like a bow tie. A form bow-tie antenna is made up of rods in the shape of a bow tie. A solid bow-tie antenna has a wire mesh between the rods. Like other dipole antennas, a bow-tie antenna is often used to improve TV reception.

bowl surfer A person who takes ubiquitous Wi-Fi Internet access a little too far – to the toilet. A favorite activity of my son, Michael, and his friends at Dartmouth College, Hanover, NH who thoroughly enjoy the benefits of being on the most wired campus in the world.

box 1. A box is a hip way of referring to a PC or a server, as in "We have 10 Compaq boxes" or "Attach the CD tower to the NT Box."

2. FCC definition. Electronic equipment used to process television signals in a consumer's home, usually housed in a "box" that sits atop a TV set or VCR. See also converter and descrambler.

bozo filter Imagine that you're receiving zillions of emails from MotherInLaw@aol.com. You don't want to receive. Simple. You set up a "bozo filter." This piece of software automatically deletes any incoming emails from MotherInLaw@aol.com. Bozo filters are best set up by your email provider at this site. You don't want to set them up on your machine. See Mail Bomb.

BP Block Pair. See Block Pair.

BP 14 Body Part 14, an X.400 electronic messaging term referring to a nonspecific body part, commonly used to transfer binary attachments.

BPAD Bisynchronous Packet Assembler/Disassembler.

BPDU Bridge Protocol Data Unit: A message type used by bridges to exchange management and control information.

BPI Bytes Per Inch. How many bytes are recorded per inch of recording surface. Typically used in conjunction with magnetic tape.

BPL Broadband over Power Line is the contemporary and updated version of Powerline Carrier (PLC), a means for sending and receiving voice and data over electric utility power lines. PLC has been used for many years by electric utilities for low speed data communica-

tions applications such as telemetry and control between power plants and substations. PLC also has long been used by telephone companies to provide voice service to extremely remote subscribers who have electric service but for whom the construction costs associated with telco local loops would be prohibitively expensive. BPL can be broken into two component technologies: Access BPL and In-House BPL.

Access BPL is a form of PLC that uses components of the existing electrical power grid for the delivery of broadband services. Specifically, Access BPL uses special injectors (i.e., modems and couplers) to interface the telecommunications network to medium voltage (MV) lines in the electrical distribution network. MV operates at a manageable 7,200 volts or so. The Radio Frequency (RF) carrier supporting the communications signals shares the same line with the electrical signals as they operate at different frequencies. Repeaters spaced every 300 meters or so serve to re-amplify, re-time and regenerate the signal as it travels from the utility substation toward the customer premises. The signals are removed by extractors placed just ahead of the transformers, which typically serve a number of households. Typically, the extractors bypass the transformers and bridge the communications signals between the MV lines and the low voltage (LV) (110/220 volts) drops to the premises. Alternatively, they may bridge the signal to a Wi-Fi (802.11b/g) node that serves multiple premises through a wireless hotspot. Although Access BPL services are not widely available, a small number of utilities began offering service in 2003 and 2004, and several others have announced their intentions to do so.

In-House BPL is a form of indoor PLC that operates over residential or business interior electrical cabling at 110/220 volts. Standards for in-house PLC are relatively recent, with the HomePlug standards perhaps being the most notable. HomePlug 1.0 was published in 2001 by the HomePlug Powerline Alliance (www.homeplug.org), which was founded by vendors including Cisco and Intel. Those standards are loosely based on Ethernet and support up to 16 devices communicating at speeds up to 14 Mbps over a shared electrical path. HomePlug compatible devices (e.g., include PCs, routers and bridges that use Ethernet, USB or Wi-Fi technologies) can simply plug into an electrical socket through a bridge or adapter about the size of a typical voltage adapter and, thereby, connect directly to the LV wiring. So every electrical socket effectively becomes a port into a high speed LAN. In-House BPL speeds don't compare well with more conventional Ethernet LANs, but can be an attractive solution where buildings can't easily be rewired and where a building's physical layout is not conducive to wireless LANs. The next step is the HomePlug AV standard, which is being built from the ground up to support entertainment applications such as HDTV and Home Theater. HomePlug AV will run at speeds up to 200 Mbps. Dozens of HomePlug 1.0 compatible products have been certified and many more are under development. In a home a BPL modem plugs into a wall outlet inside the house, receives those signals from the outside power lines and typically converts them into a familiar 802.11b Ethernet connection.

This technology is in its infancy. If successful, it would give the power companies the ability to compete against DSL offerings from the phone company and cable modem offerings from cable TV companies. There are variations on this theme. For example, AT&T and Pacific Gas and Electric Company are jointly doing tests in which the PGE power grid is being used to deliver broadband via wireless signal repeaters on street lampposts. Customers then use Cisco Wi-Fi phones to make VoIP phone calls and receive up to three megabits per second of wireless broadband service for their computer to work quickly over the Internet. Because electricity travels at a lower frequency than Internet signals, the two can theoretically coexist on the same line without interference. EarthLink, and other ISPs, see the nation's power grid as a perfect infrastructure alternative to leasing cable or local loop telephone facilities from competitors. Access speeds in BPL tests average between one and three megabits per second, roughly comparable to what most residential DSL and cable modem Internet hookups offer. After hearing arguments that BPL would interfere with two-way radio communications which could affect military, commercial and homeland security radio service, the FCC in the summer of 2004 unanimously affirmed the widespread deployment of BPL services and technology. There are two "big" problems to getting widespread deployment: First, the cost. It's not cheap to install all the communications equipment, put up regular repeaters, install the switching gear necessary, etc. Second, power companies are not phone companies. They're uncomfortable with the complexities of telecommunications. And rightly so. Further, many are not convinced that they should be the third supplier on the block – after the cable company and the phone company. See PLC.

BPM Business Process Management.

BPON Broadband Passive Optical Network. BPON standards were set by the ITU-T as G.983.3 for a local loop technology running the ATM protocol over single mode fiber. Synonymous with APON (ATM PON) BPON runs at 155 Mbps or 622 Mbps downstream at a wavelength of 1490nm for voice and data and 1550nm for video transmission. The upstream speed is 155 Mbps at 1310nm for voice and data. The maximum logical reach of BPON is 20km, and the split ratio is 32:1. See also APON, EPON, GPON and PON.

BPP 1. Brokered Private Peering, an evolving industry plan designed to revamp the way providers exchange traffic.

2. Bits Per Pixel. The number of bits used to represent the color value of each pixel in a digitized image.

Bps Bps is confusing. Is it bits per second or bytes per second? In telecommunications, bps always means bits per second. In computing, BPs (note the capital "B") often means bytes per second. But don't trust people to always be correct – using the correct upper or lower case "B." You have to figure what context you're working in. The "Rule of Thumb" is that outside the computer, in the telecom world – and that means from the computer to the world, on the USB, on the LAN, on the local loop, on the WAN, across the country, across the ocean – it's bits per second. Raw bits per second. In telecom you don't always get the speed you pay for. All telecom circuits require signaling and timing and that requires bits. You need to know if your signaling is "inband" or "out of band." For example, a 64 Kbps circuit might use 8 Kbps for inband signaling. This means you only get 56 Kbps (64 minus 8) for sending your precious material. On the other hand, a 64 Kbps ISDN BRI B channel circuit is actually a full 64 Kbps. The signaling for that ISDN channel is handled on a side channel of 16 Kbps, called the D channel.

Inside the computer, Bps is bytes per second. More commonly, it's KBps – kilobytes per second, or MBps – million bytes per second. Virtually all hard disk drive transfer rates (between the hard disk and the main microprocessor, or KSU) are in megabytes per second, Mbps. Sometimes the computer industry refers to KBps (kilobytes per second) when it's talking about transferring files from distant places to your machine – you see the number when you download files over the Internet. You wonder why the number is so much smaller than the alleged speed of your modem, which is measured in bits per second. You can translate between the two by knowing that the computer industry is referring to serial data communications in which each byte is actually ten bits – eight bits for the letter, number, or character of the information you're receiving and two bits for start and stop information.

Virtually all telecom transmission is full duplex and symmetrical. This means that if you read that T-1 is 1,544,000 bits per second (1,544,000 bps or 1.544 Mbps), it's full duplex (both ways simultaneously) and symmetrical (both directions the same speed). That means it's 1,544,000 bits per second in both directions simultaneously. If the circuit is not full duplex or not symmetrical, this dictionary points that out. For now, the major asymmetrical (but still full duplex) circuit is the xDSL family, starting with ADSL, which stands for asymmetric, which means unbalanced. The DSL "family" no longer starts with "A," but most of it is still asymmetrical. Our definitions point out which is which.

There's one more complication. Inside computers, they measure storage in bytes. Your hard disk contains this many bytes, let's say eight gigabytes. That's fine. But they're not bytes the way we think of them in internal or external computer transmission terms. They're different and they have to do with a way computer stores material – on hard disks or in RAM. They're what I call "storage bytes." When we talk 1 Kb of storage bytes, we really mean 1,024 bytes. Which comes from the way storage is actually handled inside a computer, and calculated thus: two raised to the power of ten, thus 2 x 2 x 2 x 2 x 2 x 2 x 2 x 2 x 2 = 1,024. Ditto for one million, two raised to the power of twenty, thus 1,048,576 bytes.

Finally, in telecom, when talking about transmission speed, the rule to be aware of is that the speed of a circuit is determined by the slowest part of the circuit. If one part of your circuit can only transmit at 9600 bps, then that's going to be the speed of your circuit – irrespective of the fact that other parts can go much faster. When measuring speed, you also have to factor in accuracy. All data communications schemes have error-checking systems, some better than others. Typically such systems force a re-transmission of data if a mistake is detected. You might have a fast, but "dirty" (i.e. lots of errors) transmission medium, which may need lots of re-transmissions. Thus, the "effective" bps (transmission speed) of that communications network is likely to be lower than what it's billed as. See also Baud and Mbps.

BPSK Binary Phase Shift Keying. Also known as BiPhase Shift Keying. A simple modulation technique that involves the phase shifting of the RF (Radio Frequency) signal 180 degrees in accordance with a digital bit stream. A "one" bit prompts a phase shift; a "zero" bit does not. BPSK is used in some implementations of IEEE 802.11a wireless LANs, and in some analog, coax-based CATV systems. See also 802.11a, CATV, FSK and PSK.

BPV BiPolar Violations.

BQM Business Quality Messaging. An initiative intended to facilitate the collaboration of vendors of e-mail and other messaging-enabled applications toward business systems that run reliably on both corporate networks and the Internet. The BQM SIG (Special Interest Group) was formed in April 1997. Founding members include AT&T, Hewlett Packard, IBM, Intel and Microsoft. www.bqm.com.

BR BackReflection.

BRA Basic Rate Access. A Canadian term for the ISDN 2B+D standard, which is called BRI in the U.S. – Basic Rate Interface. See ISDN.

Brador The first Trojan horse designed to attack Windows Mobile 2003 devices. A handset that is infected by Brador sends the handset's IP address to the attacker and opens TCP port 2989 on the handset, giving the hacker backdoor access to the device. Brador ican be easily removed. Visit your security software's website for info on how to remove it.

BRADS Bell Rating Administrative Data System. A Bellcore shared notification system for changes in NXXs. Also governs administration of RAOs.

bragg grating A process in which multiple lines are etched into a fiber optic cable to form a type of filter or reflector. These gratings act as prisms by diffracting light within the fiber. By using these gratings, an optical network adds the ability to select, divert, or focus light from within the fiber, negating the need for additional equipment to perform these functions. See Bragg Reflector.

bragg reflector A device designed to finely focus a semiconductor laser beam. Dennis Hall, a professor at the University of Rochester's Institute of Optics in New York, told the Economist Magazine in the Spring of 1993 that he and his colleague Gary Wicks have etched into the surface of his gallium-arsenide laser a grating of 600 concentric grooves, each a quarter of a millionth of a meter apart. The grating acts as what is known as a Bragg reflector. As the waves of laser light pass through each of its ridges, they are reflected by each of its ridges, a process which causes them to come together into an even, circular beam. See Bragg Grating.

braid A fibrous or metallic group of filaments interwoven cylindrically to form a covering over one or more insulated conductors.

braid angle The smaller of the two angles formed by the shielding strand and the axis of the cable being shielded.

braid carrier A spool or bobbin on a braider which holds one group of strands or filaments consisting of a specific number of ends. The carrier revolves during braiding operations.

braid ends The number of strands used to make up one carrier. The strands are wound side by side on the carrier bobbin and lie parallel in the finished braid.

brain fart A brain fart is a temporary lapse in doing or describing something which is usually self-evident to the person suffering, and suddenly realizing, the lapse. On receiving an answer, it is often followed by a slap on the head and expostulation, "Duh, I knew that!"

Brainerd, Paul S. Founder of Aldus Corporation in 1984, Mr. Brainerd is reputed to be "the father of desktop publishing." His program Aldus PageMaker allowed the average PC user to produce professional-looking documents.

branch A path in the program which is selected from two or more paths by a program instruction. "To branch" means to choose one of the available paths.

branch feeder A cable between the distribution cable and the main feeder cable to connect phone users to the central office. An outside plant term.

branch manager "Waiter, there's a twig in my Bird's Nest Soup." "Just a moment, sir, I'll call the Branch Manager."

branching filter A device placed in a waveguide to separate or combine different microwave frequency bands.

brand A set of differentiating promises that link a product to its customers. The brand assures the customer of consistent quality plus 'superior' value for which the customer is willing to give loyalty and pay a price that results in a reasonable return to the brand – presumably one above a similar commodity product. Think Coca Cola (a powerful brand) versus a supermarket fizzy cola drink (with no brand image). Coca Cola will typically cost more.

branding A term for identifying the Operator Service Provider (OSP) to the caller. Picture calling from your hotel room. You dial long distance. You have no idea which carrier you're using. But a message comes on: "Thanks for using MCI." Now you know. That's called branding.

brassiere In 1912, Otto Titzling invented a breast supporter for an amply endowed singer by the name of Swanhilda Olafsen. Eighteen years later, in 1930, Philippe de Brassiere copied the Titzling design and started a successful business selling the breast supporters. When word of Brassiere's success reached Titzling, he sued but lost because he had not thought of patenting his invention. This is why women today wear brassieres instead of titzlings.

Brazil Brazil is home to the world's largest snake (the anaconda, measuring up to 35 feet in length), largest spider, largest rodent (the capybara, a guinea-pig-lookalike the size of a police dog), and the world's largest ant.

BRB Be Right Back. Used in online chat to tell other participants in the session that you'll be away from the keyboard for awhile (and that your silence shouldn't be misinterpreted). Often used for a bathroom break.

BRCS Business and Residence Customer Services. Also known as Custom Calling Features (CCF) including call waiting, call forwarding, 3-way calling, etc., available through the central office without requiring the subscriber to use special equipment.

breadboard Writes Alex Richardson, "A breadboard is called a breadboard simply because that's what it was. You bought a ten cent wooden breadboard at the dime store, screwed down the tube sockets and heavy components such as transformers, and wired her up. Metal chassis were beyond those of us who had no sheet metal skills or tools. That came later. There were a lot of us."

break An interruption. As in "Make and Break." Make means contacts which are usually open, but which close during an operation. "Make and Break" accurately describes rotary dialing.

break-before-make Refers to a switch that is designed to break (open) the first set of contacts, thereby ending the old connection path, before engaging (closing) the new contacts to establish a new connection path. This prevents a momentary concurrent connection of the old and new signal paths. See also make-before-break.

break in The attendant can interrupt conversations and announce an emergency or an important call.

break key A break key is found on some PCs. It is used to interrupt the current task running. On PCs, two keys touched together – Ctrl and C – will sometimes stop the present task. In contrast, touching your Pause key simply stops something momentarily. Hitting the key a second time causes the task to begin again.

break optimization A call center term. The automatic adjustment of break start times for schedules in the Daily Workfile so as to more closely match staff to workload in each period of the day. The program can thus improve upon the originally scheduled break arrangement because it now has information about schedule exceptions, newly added schedules, and additional call volume in AHT (Average Handle Time) history. See Break Parameters.

break out box A testing device that permits a user to cross-connect and tie individual leads of an interface cable using jumper wires to monitor, switch, or patch the electrical output of the cable. The most common break out box in our industry is probably the RS-232 box. Some of these boxes have LEDs (Light Emitting Diodes), which allow you to see which lead is "live." See also Breakout Box.

break parameters A call center term. A group of scenario assumptions you set to govern the placement of breaks in employee scheduling. These are typically:

Earliest allowable break start time

Latest allowable break start time

Duration of the break

Whether the break is paid or unpaid

break strength A term denoting the greatest amount of weight or longitudinal stress, flexing, or bending a substance can bear without tearing apart or rupturing. See also Flex Strength and Tensile Strength.

break test access Method of disconnecting a circuit, which has been electrically bridged, to allow testing on either side of the circuit. Devices that provide break test access include: bridge clips, plug-on protection modules, and plug-on patching devices. Break test access also provides a demarcation point.

breakage In telecommunications, there are many services which are provided in a bundle or a monthly fixed fee. For example, a prepaid phone card might have 100 minutes of long distance usage on it. A cell phone might carry a $50 a month price tag to include 500 minutes. The user may only use 90 minutes of the 100 minutes of the prepaid phone card. The cell phone user might only use 450 minutes of the 500 minutes in that one month. The amount of unused minutes is called "breakage." Most telecom providers expect 10% to 15% breakage in each of their plans. And most are not disappointed.

breakdown set A device that attaches to copper telephone pairs and sends current down the pairs. The current causes the wire to heat slightly, thus slowly drying out the cable. The device is used by telephone companies to dry out pairs of cables which have become wet.

breakdown voltage The voltage at which the insulation between two conductors breaks down.

Breaker Distribution Fuse Box BDFB. A distinct rack of equipment in the Switch Room.

breaking strength The amount of force needed to break a wire or fiber.

breakout A wire or group of wires in a multi-conductor configuration which terminates somewhere other than at the end of the configuration.

breakout box A device that is plugged in between a computer terminal and its connecting cable to re-configure the way the cable is wired. When hooking up a terminal that is wired as if it were a computer itself (such as a VT-100), a break out box is used to break out, or fan out the 25 connections in the RS-232 cable. Each wire in the break out box goes through a switch that can be turned off, and a wire jumper is provided to connect each pin on one side to one or the other pin on the other side. This allows you, for example, to switch pins 2 & 3, thus fooling two computer devices into thinking one is talking to a terminal. (Now you have the essence of a null modem cable.) Break out boxes are necessary because there is no such thing as "standard" pinning on an RS-232 cable. To connect one computer to a printer one minute and to another computer the next minute, usually requires totally different wiring in the RS-232 cable, i.e. two sets of cables. This lack of standardization is why you'll always see dozens of RS-232 cables lying around where computers are used.

Brendan Smith Logic What is the dollar conclusion you want? Figure the number. Then find the numbers and the logic that you can use to justify your conclusion. Thus your arguments will be impeccable and your conclusion will be unassailable.

Brevity code According to the U.S. Army, a brevity code is the shortened form of a frequently used phrase, sentence, or group of sentences, normally consisting entirely of upper case letters; for example, COMSEC means communications security, REFRAD means release from active duty, and SIGINT means signals intelligence. When originating brevity codes: (1) Use letters that convey the meaning of the language they represent. (2) Do not represent the same word with more than one brevity code. (3) Make the first letter of the brevity code and the first letter of the phrase should be the same.

BREW Binary Runtime Environment for Wireless. BREW was developed by Qualcomm to provide a standard set of APIs for developers to quickly and easily add new features and applications to Qualcomm-based wireless hardware. The BREW development process is heavily reliant on a relationship with Qualcomm, and requires "True Brew" certification for use as an approved application. BREW is basically Qualcomm's open source application development platform for wireless devices equipped for code division multiple access (CDMA). BREW makes it possible for developers to create portable applications that will work on any handsets equipped with CDMA chipsets. Because BREW runs in between the application and the chip operating system software, the application can use the device's functionality without the developer needing to code to the system interface or even having to understand wireless applications. Users can download applications – such as text chat, enhanced e-mail, location positioning, games (both online and offline), and Internet radio – from carrier networks to any BREW-enabled phone. BREW is competing for wireless software market share with J2ME (Java 2 Micro Edition), a similar platform from Sun Microsystems.

BRI Basic Rate Interface. There are two subscriber "interfaces" in ISDN. This one and PRI (Primary Rate Interface). In BRI, you get two bearer B-channels at 64 kilobits per second and a data D-channel at 16 kilobits per second. The bearer B-channels are designed for PCM voice, slow-scan video conferencing, group 4 facsimile machines, or whatever you can squeeze into 64,000 bits per second full duplex. The data (or D) channel is for bringing in information about incoming calls and taking out information about outgoing calls. It is also for access to slow-speed data networks, like videotex, packet switched networks, etc. See Basic Rate Interface and ISDN.

brick A large hand-held cellular phone or handheld two-way radio. In more technical language, a "brick" is a station in the mobile service consisting of a hand-held radiotelephone unit licensed under a site authorization. In some circles, calling something a brick means that it doesn't work – i.e. it has the electronic functionality of a real housing brick.

bricks and mortar Slang referring to businesses such as retail stores that exist in the real world and have quaint things like retail stores and building as opposed to those that exist just on the Internet. In reality, all Internet-based businesses have some bricks and mortar.

brick-wall filter An ideal low-pass filter with a completely rectangular characteristic; unrealizable in practice, unless modified by roll-off.

bridge 1. In classic terms, a bridge is a data communications device that connects two or more network segments and forwards packets between them. Such bridges operate at Layer 1 (Physical Layer) of the OSI Reference Model. At this level, a bridge simply serves as a physical connector between segments, also amplifying the carrier signal in order to compensate for the loss of signal strength incurred as the signal is split across the bridged segments. In other words, the bridge is used to connect multiple segments of a single logical circuit. Classic bridges are relatively dumb devices, which are fast and inexpensive; they simply accept data packets, perhaps buffering them during periods of network congestion, and forward them. Bridges are protocol-specific, e.g., Ethernet or Token Ring in the LAN domain. Bridges also are used in the creation of multipoint circuits in the WAN domain, e.g., DDS (Dataphone Digital Service).

Bridges also can operate at Layer 2 (Link Layer) of the OSI Reference Model. At this level, a bridge connects disparate LANs (e.g., Ethernet and Token Ring) at the Medium Access Control (MAC) sub-layer of Layer 2. In order to accomplish this feat, the MAC Bridge may be of two types, encapsulating or translating.

Encapsulating bridges accept a data packet from one network and in its native format; they then encapsulate, or envelope, that entire packet in a format acceptable to the target network. For instance, an Ethernet frame is encapsulated in a Token Ring packet in order that the Token Ring network can deliver it to the target device, which must strip away several layers of overhead information in order to get to the data payload, or content. In order to accomplish this process, a table lookup must take place in order to change basic MAC-level addressing information.

Translating bridges go a step further. Rather than simply encapsulating the original data packet, they actually translate the data packet into the native format of the target network and attached device. While this level of translation adds a small amount of delay to the packet traffic and while the cost of such a bridge is slightly greater, the level of processing required at the workstation level is much reduced.

Bridges also can serve to reduce LAN congestion through a process of filtering. A filtering bridge reads the destination address of a data packet and performs a quick table lookup in order to determine whether it should forward that packet through a port to a particular physical LAN segment. A four-port bridge, for instance, would accept a packet from an incoming port and forward it only to the LAN segment on which the target device is connected; thereby, the traffic on the other two segments is reduced and the level of traffic on the those segments is reduced accordingly. Filtering bridges may be either programmed by the LAN administrator or may be self-learning. Self-learning bridges "learn" the addresses of the attached devices on each segment by initiating broadcast query packets, and then remembering the originating addresses of the devices which respond. Self-learning bridges perform this process at regular intervals in order to repeat the "learning" process and, thereby, to adjust to the physical relocation of devices, the replacement of NICs (Network Interface Cards), and other changes in the notoriously dynamic LAN environment.

While bridges are relatively simple devices, in the overall scheme of things, they can get quite complex as we move up the bridge food chain. (Please don't blame me. I didn't invent this stuff!) Bridges also can be classified as Spanning Tree Protocol (STP), Source Routing Protocol (SRP), and Source Routing Transparent (SRT).

Spanning Tree Protocol (STP) bridges, defined in the IEEE 802.1 standard, are self-learning, filtering bridges. Some STP bridges also have built-in security mechanisms which can deny access to certain resources on the basis of user and terminal ID. STP bridges can automatically reconfigure themselves for alternate paths should a network segment fail.

Source Routing Protocol (SRP) bridges are programmed with specific routes for each data packet. Routing considerations include physical node location and the number of hops (intermediate bridges) involved. This IBM bridge protocol provides for a maximum of 13 hops.

Source Routing Transparent (SRT) bridges, defined in IEEE 802.1, are a combination of STP and SRP. SRT bridges can act in either mode, as programmed.

2. In the context of either audioconferencing (voice) or videoconferencing, a bridge connects three or more telecommunications channels so that they can all communicate together. In either case, compensation is made for signal loss (called balancing) in order to maintain consistent quality, thus allowing all participants to hear and see each other with equal ease. In video conferencing, bridges are often called MCUs – Multipoint Conferencing Units. One feature of some video bridges is their ability to figure who's speaking and turn on the camera which is on that person and have that person's face be on everyone's screen.

3. Finally, we'll explain bridge as a verb, as in "to bridge." Imagine a phone line. It winds from your central office through the streets and over the poles to your phone. Now imagine you want to connect another phone to that line. A phone works on two wires, tip and ring (positive and negative). You simply clamp each one of the phone's wires to the cable coming in. That's called bridging. Imagine bridging as connecting a phone at a right angle. When you do that, you've made what's known as a "bridged tap." The first thing to know about bridging is that bridging causes the electrical current coming down the line to lose power. How much? That typically depends on the distance from the bridged tap

to the phone. A few feet, and there's no significant loss. But that bridged tap can also be thousands of feet. For example, the phone company could have a bridged tap on your local loop, which joined to another long-defunct subscriber. The phone company technicians simply saved a little time by not disconnecting that tap. If you want the cleanest, loudest phone line, the local loop to your phone should not be bridged. Instead it should be a direct "home run" from your central office to your phone.

Bridging can be a real problem with digital circuits. Circuits above 1 Mbps (e.g., T-1) should never, ever be bridged. Because of the power loss, they simply won't work or will work so poorly they won't be worth having. ISDN BRI channels are also digital. But they were specifically designed to work with the existing telephone cable plant, which has a huge number of bridged circuits. Telephone companies typically will install ISDN BRI circuits with up to six bridged taps and about 6,000 feet of bridged cabling. But that's a rule of thumb. And frankly, if I were getting an ISDN line, I'd ask for a line that had no taps and no bridges.

4. To conference on another party. For example, when the repair tech says "let me bridge on my Supervisor" or "who's on this bridge?". See Conference and Conference Bridge.

See Internetworking, Loading Coil, Routers, Source Routing and Transparent Routing.

bridge/router A device that can provide the functions of a bridge, router, or both concurrently. A bridge/router can route one or more protocols, such as TCP/IP and/or XNS, and bridge all other traffic.

bridge amplifier An amplifier installed on a CATV trunk cable to feed branching cables.

bridge and roll Also called a "facility roll." A telecom term that refers to the process of moving a signal from one line to another in a network. It's a two-stage process: first the "bridge" and then the "roll." In the "bridge" phase, the signal to be moved is "bridged" by running it along its original path and a second path at the same time. Then it is "rolled" to the second path and removed from the first path completely. The sequence would be:

1. signal on path A. 2. signal on path A and B. (The Bridge) 3. signal on path B; Path A cleared for new signal. (The Roll)

Bridge and roll is used mainly in network maintenance, to move traffic off a line so the line may be tested or repaired without interrupting service.

bridge battery A small supplementary battery on a laptop which holds the contents of the memory and the system status for a few minutes while you replace a drained battery. NEC uses the term on its UltraLite Versa laptops.

bridge clip A small metal clip that used to electrically connect together two sides of a 50 pair block. Removing the bridging clips breaks the circuit. You might remove the clips when you want to insert a piece of test gear and check to see which side the trouble is on.

bridge equipment Equipment which connects different LANs, allowing communication between devices. As in "to bridge" several LANs. Bridges are protocol-independent but hardware-specific. They will connect LANs with different hardware and different protocols. An example would be a device that connects an Ethernet network to a StarLAN network. With this bridge it is possible to send signals between the two networks, and only these two networks.

These signals will be understood only if the protocols used on each LAN are the same, e.g. XNS or TCP/IP, but they don't have to be the same for the bridge to do its job for the signals to move on either LAN. They just won't be understood. This differs from gateways and routers. Routers connect LANs with the same protocols but different hardware. The best examples are the file servers that accommodate different hardware LANs. Gateways connect two LANs with different protocols by translating between them, enabling them to talk to each other. The bridge does no translation. Bridges are best used to keep networks small by connecting many of them rather than making a large one. This reduces the traffic faced by individual computers and improves network performance.

bridge group Virtual LAN terminology for a group of switch interfaces assigned to a singular bridge unit and network interface. Each bridge group runs a separate Spanning Tree and is addressable using a unique IP address.

bridge lifter A device that removes, either electrically or physically, bridged telephone pairs. Relays, saturable inductors, and semiconductors are used as bridge lifters.

Bridge Protocol Data Unit BPDU. The implementation of the spanning tree protocol (STP) and rapid spanning tree protocol (RSTP) protocols allows network devices to detect and block links that could cause logical loops within a network and to manage redundant links to maintain network integrity in the event of a link failure. Bridges and switches that use the spanning tree protocol (STP) or the rapid spanning tree protocol (RSTP) use the bridge protocol data unit (BPDU) to communicate with each other and exchange information. The BPDU is a datagram that has a specific format to relay the following information about the switch that transmits it:

- Media Access Control (MAC) addresses (switch and port)
- Switch priority
- Port priority
- Port cost
- Root switch identifier
- Root port and designated port identifiers
- Path cost from port to root switch
- Spanning tree enabled devices gather the BPDUs from other devices on the network and use the information to make configuration decisions such as the election of a root device, the election of a designated switch to become a link between a subnet and the root device, the designation of root and designated ports that are used to communicate STP and RSTP information, the shortest best path between a device and the root switch, and finally the detection and removal of loops in the network.

When a change occurs in a network topology BPDUs are resent between the network devices to determine if a reconfiguration is required. For instance, if the root switch fails, BPDUs can be resent to figure out a new root switch. Also if a link between network devices fails, a previously blocked redundant link can be opened to maintain network communication. The exchange of BPDUs makes configuration and reconfiguration of the spanning tree topology possible, however, STP and RSTP BPDUs are not the same. RSTP BPDUs are optimized for quicker configuration of the network and are therefore different than traditional STP BPDUs. Steps have been taken though to ensure the compatibility between the two standards such that data exchanged between STP and RSTP devices is unhindered.

bridge static filtering The process in which a bridge maintains a filtering database consisting of static entries. Each static entry equates a MAC destination address with a port that can receive frames with this MAC destination address and a set of ports on which the frames can be transmitted. Defined in the IEEE 802.1 standard. See also IEEE 802.1.

bridge tap An undetermined length of wire attached between the normal endpoints of a circuit that introduces unwanted impedance imbalances for data transmission. Also called bridging trap or bridged tap. See Bridged Tap.

bridged jack A dual position modular female jack where all pins of one jack are permanently bridged to the other jack in the same order.

bridged ringing A system where ringers on a phone line are connected across that line.

bridged tap A bridged tap is multiple appearances of the same cable pair at several distribution points. A bridged tap is any section of a cable pair not on the direct electrical path between the central office and the user's offices. A bridged tap increases the electrical loss on the pair – because a signal traveling down the pair will split its signal between the bridges and the main pair. Since most existing telephone company cable pair is bridged, the phone company puts loading coils in the circuit. The effect of load coils is to modify the loss versus frequency response of the pair so it is nearly constant across the voice band. This works for voice. However the loss above the voice band due to load coils increases rapidly. ISDN, T-1, DSL and other digital circuits operates above the voice band. So, when the phone company installs digital circuits, it must remove the load coils. See Bridge and Loading Coil.

bridgehead server A routing server that accepts mail or other traffic from another server and then distributes it to the next server in the route. The arrangement is similar to the hub-and-spoke system used by airlines to avoid having to fly direct flights to and from every city. A bridgehead server minimizes network facilities needs and site-to-site direct-traffic flow by exchanging mail and messages between bridgehead servers. By consolidating traffic between two widely separated points, a bridgehead server enables economies of scale and efficient use of overall network bandwidth. For example, a multinational company needs to send an email with a 25 megabyte attachment from its US headquarters to managers in 12 different European locations. In a network without a bridgehead server, this would entail 300 megabyte of mail being routed between the US and Europe (25 MB x 25 destinations). However, if the organization has a bridgehead server in the UK, then only one email, with a 25 MB attachment, would go between the US and the UK. The bridgehead server in the UK would then forward the email to each of the 12 final destinations in Europe. This would mean saving 275 MB of traffic between the US and the UK.

bridgehead services Services performed by a bridgehead server.

bridger Bridger Amplifier. An amplifier which is connected directly into the main trunk of a CATV system, providing isolation between the main trunk and multiple (high level)

outputs.

bridging Bridging across a circuit is done by placing one test lead from a test set or a conductor from another circuit and placing it on one conductor of another circuit. And then doing the same thing to the second conductor. You bridge across a circuit to test the circuit by listening in on it, by dialing on it, by running tests on the line, etc. You can bridge across a circuit by going across the pair in wire, by stripping it, etc. You can bridge across a pair (also called a circuit path) by installing external devices across quick clips on a connecting block.

bridging adapter A box containing several male and female electrical connectors that allows various phones and accessories to be connected to one cable. Bridging adapters work well with 1A2 key systems and single line phones, but usually not with electronic or digital key systems and electronic or digital telephones behind PBXs.

bridging clip A small piece of metal with a U-shape cross-section which is used to connect adjacent terminals on 66-type connecting blocks.

bridging connection A parallel connection by means of which some of the signal energy in a circuit may be extracted, usually with negligible effect on the normal operation of the circuit. Most modern phone systems don't encourage bridging connections, since the negligible is nearly negligible.

bridging loss The loss at a given frequency resulting from connecting an impedance across a transmission line. Expressed as the ratio (in decibels) of the signal power delivered to that part of the system following the bridging point before bridging, to the signal power delivered to that same part after the bridging.

bridle cards Proprietary Basic Rate ISDN Dual Loop Extension that lets ISDN service be provided up to 28,000 feet away. See ISDN.

BRIDS Bellcore Rating Input Database System.

briefcase A Windows 95 feature that allows you to keep multiple versions of a file in different computers in sync with each other.

brightness An attribute of visual reception in which a source appears to emit more or less light. Since the eye is not equally sensitive to all colors, brightness cannot be a quantitative term.

bring-your-own-access Of or pertaining to an online service, often a bandwidth-intensive service, that is provisioned by an entity other than the customer's Internet service provider. Bring-your-own-access (BYOA) services are in the cross-hairs of telcos, cablecos, and satellite operators these days because these network operators are footing the bill to upgrade their networks to carry these services, without compensation from the BYOA services that are making money off of these services.

BRISC Bell-Northern Research Reduced Instruction Set Computing.

brite cards and services Basic Rate Interface Transmission Extension lets telephone companies extend service from ISDN-equipped central offices to conventional central offices. See ISDN.

British Telecommunications Act In 1981 in the U.K. this act separated telecommunications from the post office and created British Telecommunications (BT). See also Post Office Act.

brittle Easily broken without much stretching.

broadband Today's common definition of broadband is any circuit significantly faster than a dial-up phone line. That tends to be a cable modem circuit from your friendly local cable TV provider, a DSL circuit, a T-1 or an E-1 circuit from your friendly local phone company. In short, the term "broadband" can mean anything you want it to be so long as it's "fast." In short, broadband is now more a marketing than a technical term. See also the definitions following.

broadband amplifier An amplifier with a relatively wide frequency response as distinguished from a single channel or narrower band amplifier.

broadband bearer capability A bearer class field that is part of the initial address message.

Broadband Circuit See Broadband.

Broadband Integrated Services Digital Network B-ISDN.

Broadband Inter-Carrier Interface BICI. A carrier-to-carrier interface line PNNI (private network-to-network interface) that is needed because carriers do not permit their switches to share routing information or detailed network maps with their competition's equipment. NOTE: BICI supports permanent virtual circuits between carriers; however, the ATM Forum is currently addressing switched virtual circuits.

Broadband Loop Emulation Services See BLES.

broadband multimedia Broadband multimedia is the present obsession of Terry Matthews, the only man in Canada who founded two companies to reach annual sales of over $1 billion. He is now working on his third, called March Networks, which focuses on broadband multimedia. Terry's obsession in a nutshell:

- As we wire the world for broadband communcations and as the cost drops dramatically (a factor of a hundredfold over the past five years), we open the world to an entire new range of new telecommunications opportunities – those involving video, voice and data combined as a viewable, storable, retrievable record. Visiting patients electronically makes for happier nurses, happier, longer living patients. Ditto for on-line, broadband education. Shrinkage (i.e. stealing) is a $32 billion "industry" in the U.S. Cut it by 10% with extensive video surveillance tied to cash register transactions and you'll increase retail store net income by 18%. In the utility industry (pipelines, electricity, oil, etc.) security and operations managers must manage hundreds of remote installations, mitigating threats to reliable power delivery. Centralizing video and data records from remote sites allows utilities to collect valuable multimedia (graphic and useful) information that can significantly lower operations cost. Such applications include verification of alarms reported by SCADA (Supervisory Control And Data Acquisition) systems, visual equipment inspection, remote project management and monitoring of conditions at dams, rivers and other electricity generating sites.
- The telecommunications industry is about to enter a new era – selling speciality multimedia vertical industry applications. This contrasts with what we do today. We sell horizontal applications. This means that the industry's services are the same for every customer. Every customer buys bandwidth in various widths. And because my bandwidth is indistinguishable from your bandwidth, our major method of competing as telecom carriers has been to cut prices. No more.
- Selling these new broadband multimedia applications will help chew up the excess bandwidth carriers installed in recent years.
- Selling these applications as applications, not as bandwidth, will significantly boost profits.
- Selling these new applications as applications is akin to selling additional channels of television programming on one common pipe – the coaxial cable which your CATV brings to your house.

Broadband over Power Line See BPL.

Broadband Passive Optical Network See BPON.

Broadband Personal Communications Standards BPCS. Consists of 120 MHz of new spectrum available for new cellular networks. Also known as wideband PCS.

Broadband Service Provider BSP. Once upon a time, companies which provided access for you and me to the Internet were called Internet Service Providers (ISPs, for short). But they largely provided dial-up service. Then came broadband. Now an ISP that provides broadband access to the Internet is called a broadband service provider, or BSP. Such companies may provide service by cable modems over cable TV coaxial cable, DSL or ADSL, T-1 lines, wireless or satellite. Broadband service providers (BSPs) are also trying to generate both revenue and stickiness with broadband content.

Broadband Switching System See BSS.

Broadband Wireless Local Loop B-WLL is also known as local multi-point distribution service, i.e. LMDS. B-WLL is a way of getting various multimedia services such as high-speed Internet, cable TV, and VOD (video-on-demand) to subscribers. The great advantage of B-WLL is that wireless technology can be used to connect the costly last mile of high data speed networks from an operator's backbone network to individual users. The technology uses millimeter wave signals in the 28 GHz spectrum to transmit voice, video, and data signals within a three-mile to 10-mile radius.

LMDS differs from an ordinary transport system in the way a train differs from a pipeline. Both are data transport systems, but a pipeline can transport only one product from one place to another. A train, on the other hand, can transport many different products over the same infrastructure. LMDS, implemented with multi-service protocol such as AIM, can transport, among others, voice, Internet, Ethernet, video, computer files, and transaction data. It is the multipoint radio technology, combined with the appropriate protocol and access method that gives LMDS its potential tremendous potential. LMDS/B-WLL infrastructure technology can be divided into two basic multiple access technologies: FDD and TDD. FDD equipment uses separate frequencies for the up-link and down-link channels, as opposed to TDD, which uses the same frequency channel for both up-link and down-link, separating the traffic by the use of time slots. FDD equipment differs among vendors in the type of backbone network technology incorporated into the system. The two primary divisions are cable-modem-based versus telecom-network-based. With respect to the telecom-backbone-based solutions. there are two basic architectures being developed: time division multiplex (1DM) and packet-based (either ATM or IP). B-WLL has some advantages: (1) It can be engineered to provide 99.99% availability, rivaling that of the best fiber backbones.

(2) It can be deployed quickly. Once a hub is installed (a matter of days), new customers can be added in a matter of hours. (3) It is estimated that deployment of a B-WLL system is about 60% cheaper than fiber-optic cable-based networks. Physical technologies such as copper or fiber require individual rights-of-way to each building, as well as the physical placement of the transport media. (4) Wireless equipment is less vulnerable to sabotage, theft, or damage resulting from exposure to the elements. There are negatives. (1) It requires line-of-sight. You typically can't shoot it through buildings or hills. (2) Bad weather can affect it.

broadcast 1. To send information to two or more receiving devices simultaneously – over a data communications network, voice mail, electronic mail system, local TV/radio station or satellite system. Broadcast involves sending a transmission simultaneously to all members of a group. In the context of an intelligent communications network, such devices could be host computers, routers, workstations, voice mail systems, or just about anything else. In the less intelligent world of "broadcast media," a local TV or radio station might use a terrestrial antenna or a satellite system to transmit information from a single source to any TV set or radio capable of receiving the signal within the area of coverage. See also Narrowcasting and Pointcasting. Contrast with Unicast, Anycast and Multicast.

2. As the term applies to cable television, broadcasting is the process of transmitting a signal over a broadcast station pursuant to Parts 73 and 74 of the FCC rules. This definition is deliberately restrictive: it does not include satellite transmission, and it does not include point-to-multipoint transmission over a wired or fiber network. In spite of the fact that the broadcast industry and the cable television industry are forever bound together in a symbiotic relationship, they are frequently at odds over policy issues. See Broadcast Station. Compare with Cablecast.

Broadcast Channel BCCH. A wireless term for the logical channel used in certain cellular networks to broadcast signaling and control information to all cellular phones. BCCH is a logical channel of the FDCCH (Forward Digital Control CHannel), defined by IS-136 for use in digital cellular networks employing TDMA (Time Division Multiple Access). The BCCH comprises the E-BCCH, F-BCCH and S-BCCH. The E-BCCH (Extended-BCCH) contains information which is not of high priority, such as the identification of neighboring cell sites. The F-BCCH (Fast-BCCH) contains critical information which must be transmitted immediately; examples include system information and registration parameters. S-BCCH (System message-BCCH), which has not yet been fully defined, will contain messages for system broadcast. See also IS-136 and TDMA.

broadcast domain Set of all devices that receive broadcast frames originating from any device within the set. Broadcast domains typically are bounded by routers because routers do not forward broadcast frames.

broadcast list A list of two or more system users to whom messages are sent simultaneously. Master Broadcast Lists are shared by all system users and are set up by the System Administrator. Personal Lists are set up by individual subscribers.

broadcast message A message from one user sent to all users. Just like a TV station signal. On LANs, all workstations and devices receive the message. Broadcast messages are used for many reasons, including acknowledging receipt of information and locating certain devices. On voice mail systems, broadcast messages are important announcement messages from the system administrator that provide information and instructions regarding the voice processing system. Broadcast messages play before standard Voice Mail or Automated Attendant messages.

broadcast net A British Telecom turret feature that allows each trader single key access to a group of outgoing lines. This is designed primarily for sending short messages to multiple destinations. The "net" function allows the user to set up and amend his broadcast group.

broadcast quality A specific term applied to pickup tubes of any type – vidicon, plumbicon, etc. – which are without flaws and meet broadcast standards. Also an ambiguous term for equipment and programming that meets the highest technical standards of the TV industry, such as high-band recorders.

broadcast station An over-the-air radio or television station licensed by the FCC pursuant to Parts 73 or 74 of the FCC Rules, or an equivalent foreign (Canadian or Mexican) station. Cable television systems are authorized by FCC rules to retransmit broadcast stations; however, such retransmission is subject to a number of restrictions:

- The cable television operator is liable for copyright royalty fees collected by the Copyright Office.
- Under certain conditions, certain broadcast stations are eligible for mandatory carriage.
- Under certain conditions, the cable operator must obtain the permission of the

licensee of the broadcast station. This term includes satellite-delivered broadcast "superstations" such as WGN-TV and WWOR, but it does not include:

- Satellite-delivered non-broadcast programming services (HBO, ESPN, C-SPAN, QVC, etc.).
- Video services delivered by terrestrial microwave systems such as MDS, MMDS, or ITFS, unless the actual signal being delivered was originally picked up from a broadcast station.
- Cablecasting programming originated by the cable operator or an access organization.

broadcast storm A pathological condition that may occur in a TCP/IP network that can cause a large number of broadcast packets to be propagated unnecessarily across an enterprise-wide network, thereby causing network overload. Broadcast storms happen when users mix old TCP/IP routers with routers supporting the new releases of TCP/IP protocol. Routers use broadcast packets to resolve IP addressing requests from stations on LANs. If a station running an old version of TCP/IP sends such a request, TCP/IP routers in an enterprise-wide network misunderstand it and send multiple broadcasts to their brother and sister routers. In turn, these broadcasts cause each router to send more broadcasts, and so on. This chain reaction can produce so many broadcast messages that the network can shut down. It should be noted that this is extremely rare and it happens only in TCP/IP networks that use two specific TCP/IP protocol releases.

broadcast transmission A fax machine feature that allows automatic transmission of a document to several locations.

broadwing The name for the merged company comprising the old Cincinnati Bell Inc., a LEC (Local Exchange Carrier), and IXC Communications, an IXC (Interexchange Carrier) which acquired Cincinnati Bell. The merged company changed its name to Broadwing Inc. in 2000. Cincinnati Bell continues to operate as a LEC division of Broadwing. www.broadwinginc.com.

brochureware A pejorative term for what companies can pull off with a clever copy writer, some nice graphics, and a bit of an advertising budget. Ever read a brochure and compared it to the product? You get the idea. See Webware.

broken arrow An error code displayed on certain computer terminals for various kinds of protocol violations and unexpected error conditions, including connection to a down computer. In the military, it is jargon for a nuclear accident.

broken clock Todd Kingsley has an expression that even a broken clock is right twice a day. This is his kind of saying I'm really stupid, but occasionally (not often) I fluke it and get it right.

broken link A link to a file that does not exist or is not at the location indicated by the URL. In short, you click on a hyperlink on a Web page you're viewing, but nothing happens or you get an error message. Bingo, broken link. You've been sent somewhere that doesn't exist. This is neither exciting, nor good programming.

broken pipe This term is usually seen in an error message by browser programs to let the user know that the stream of information which was downloading at the time has been forcibly cut. This can occur for many reasons, most commonly because you are on a very crowded network or your access provider is experiencing heavy traffic.

broken record In the 1960s, 1970s and 1980s there was an expression that you sound "like a broken record." This meant that you were repeating yourself. The expression came from the fact that when a needle got stuck in the groove of a vinyl record, the sound simply repeated itself. Then came the compact disc and the needle never got stuck in the groove since there was no needle. As a result, college kids today have no idea what the expression "broken record" means, since most have never owned nor seen a record player.

broker A company (or person) that acts as an agent, or intermediary, in contractual relationships between a buyer and a seller. A broker takes neither ownership nor physical possession of the thing being bought and sold. Equipment brokers, for example, buy and sell equipment without taking ownership, sometimes without testing or refurbishing the equipment, and often without even seeing it. The broker finds a seller and a buyer, and has the equipment shipped directly from one to the other, relying on the seller's representation of its condition. Equipment brokers have been around since about 1968, when it became legal in the US for you to own your own equipment. Actually, they have been buying and selling gear between telephone companies for many years. See also Junk Dealer and Secondary Equipment.

Bandwidth brokers are a more recent phenomenon, having been born out of the Telecom Act of 1996, which opened the local exchange to competition and spurred the growth of competitive carriers in both the voice and data and the local and long-haul markets. Bandwidth brokers put themselves in the middle of a bandwidth transaction between buyer and seller. Both buyer and seller may be carriers, or the buyer may be an end user organization and the seller a carrier – either way, the arrangement works the same way. Most brokers act

like a commodities exchange, never taking title to the capacity, and often using a Website to post example asking prices for certain levels of bandwidth (e.g., T-1 or T-3, OC-3 or OC-24) between specific points of termination (e.g., New York City and Los Angeles) for various contract terms (e.g., 6, 12, or 24 months). If you're interested, you make an offer, and the broker works with you and the seller to put a deal together. An SLA (Service Level Agreement) may be part of the negotiated terms. You may not even know the name of the seller until after the contract is signed, if ever. Some owners of telco hotels also serve as bandwidth brokers. In addition to renting space to both incumbent and competitive carriers, they serve as bandwidth brokers, and even take responsibility for establishing the physical connection between the parties. They may even go so far as to monitor the performance of the seller's network over time, providing reports to both parties on a regular basis. In addition to acting as a broker for leased lines, some also broker frame relay, VPN (Virtual Private Network), and Internet access services. See also SLA, Carrier Hotel, Telecom Act of 1996, and VPN.

bronze Alloy of copper and tin, widely used and known since ancient times. Copper content in bronze varies between 89% and 96%.

brouter Concatenation of "bridge" and "router." Used to refer to devices which perform both bridging and routing functions. In local area networking, a brouter is a device that combines the dynamic routing capability of an internetwork router with the ability of a bridge to connect dissimilar local area networks (LANs). It has the ability to route one or more protocols, such as TCP/IP and XNS, and bridge all other traffic.

brown bag session A meeting scheduled during lunch hour in which the employee not only has to work but must bring his or her own lunch.

Brown and Sharpe Wire Gauge An older name for American Wire Gauge, the U.S. standard measuring gauge for non-ferrous conductors (i.e., non-iron and non-steel), AWG covers copper, aluminum, and other conductors. Gauge is a measure of the diameter of the conductor. See AWG for a full explanation.

Brown, Charles L. The AT&T chairman who presided over the second dumbest decision in the telecommunications industry – namely the AT&T Divestiture of 1984. The dumbest decision was in 1877 when Western Union rejected spending $100,000 to buy the patent rights to the invention of the telephone. Western Union believed the telegraph superior technology. See also McGowan, William G.

brown down In the context of DSL, a situation that occurs when a CO (or local exchange) cannot handle all of the calls attempted and even disrupts calls in progress. Also called a "brown down".

brownfield The opposite of greenfield. Brownfield is the sum of all legacy material (equipment, architectures, procedures, etc.) in any given network project. Greenfield refers to the material being developed anew. See Greenfield and Legacy.

brownout 1. When you lose all your electricity, it's called a blackout. When your voltage drops more than 10% below what it's meant to be, it's a brownout. If a brownout lasts less than about a second, it is called a SAG. Brownouts are sometimes caused by overloaded circuits and are sometimes caused intentionally by the AC utility company in order to reduce the power drawn by users during peak demand periods (like during hot summers when everyone is using their air conditioners). Studies have shown that brownouts of all durations make up the vast majority of power problems that affect telephone systems and computers.

2. In Internet terms, when a system is so overloaded by requests that it slows down to the point of near unusability, it is suffering a "brownout."

brownout reset Circuitry that forces a device to a reset state if the device's power drops below a specified voltage level.

browse list A list of computers and services available on a network.

browser 1. Also called a web browser. A browser is software that translates the digital bits that flow to your computer from the Internet into pictures and text so you can read them and look at them. A browser displays the documents, images and video you find on the Internet and the World Wide Web on your computer. A web browser is software which allows a computer user (like you and me) to "surf" the Internet. It lets us move easily from one web site to another. Every time we alight on a web page, our web Browser moves a copy of documents and images on that page to our computer. In actual fact, a web page doesn't look on a web server anything like it looks like as presented by a browser. A web page basically consists of raw ASCII (i.e. plain) text with words that say include this picture here, this headline over there, this table here. A web page on a server may consist of ten or twenty elements, each of which flow to your browser bit by bit. To receive those bits and transit responses (like "I got them"), your browser uses HTTP – the HyperText Transfer Protocol. Invisible to the user of a Web Browser, HTTP is the actual protocol used by the

Web Server and the Client Browser to communicate over the Internet. Despite the alleged standard coding of each web page, browsers often "see" web pages slightly differently. The two most famous browsers are Microsoft's Internet Explorer and Mozilla Firefox. You can often see the differences by opening the same web page in two different browsers. The last point about browsers is that the speed at which you can effectively surf the web, i.e. have browser pull up pages quickly, is determined by five factors: 1. The speed of your connection to the Internet. 2. The speed of your browser. Some work faster than others. 3. The speed of your computer's primary microprocessor. The faster the better. A browser requires significant processing power to assemble and display web pages. 4. The amount of RAM memory your computer has. 5. The number of other software programs that may be using up your computer's processing power while your browser is attempting to do its job.

See also browser sniffing, browsing, Internet and surf.

2. A developed tool used to inspect a class hierarchy in an object-oriented software system.

browser compatibility A term that compares the way a Web page looks on one WWW browser as opposed to another. Usually this is done with Microsoft Internet Explorer (MIE) and Netscape Navigator, but can also refer to cross-platform compatibility. (For example, the way a page renders or displays on a Windows system as opposed to a Mac.) The reason these incompatibilities exist is due to the way a browser interprets the Web page's code (HTML). The differences are usually very slight, but they're enough to annoy some Web designers and sometimes even their clients to the point in which great time and energy is spent in making a Web site compatible with any browser on any type of system. Browser compatibility is also used in conjunction with (and should not be confused with) the term browser support.

browser plug-in A Plug-in is a small program that you download and install to add a specific feature to your Web browser. These features can include additional multimedia such as video and sound.

browser sniffing The term "browser sniffing" refers to figuring a website visitor's browser and version number, and possibly any installed plug-ins. Typically, this is done in order to grant or deny access to certain features of that site that will work only on specific versions of specific browsers (there being few standards in this business). Usually accomplished with JavaScript, there are two common methods that a website can use to figure browser information. You can evaluate the User Agent (UA) string revealed in the navigator object properties or you can test for the existence of other specific objects. Using this information, you can make decisions as to what content / information your web site should serve up to specific visiting users. See Browser.

browser support This refers to the ability of a particular browser to even recognize and interpret certain HTML or other Web page codes. For example, Netscape Navigator 1.0 did not have the ability to render a page layout in frames. This feature did not come along until version 2.0, therefore it can be said that Navigator 1.0 did not "support" frames.

browsing We've all browsed the Internet. We've all surfed, which is the same as browsing. It means jumping from one web site to another using a piece of software called a browser. The formal definition of browsing is the act of searching through automated information system storage to locate or acquire information without necessarily knowing of the existence or the format of the information being sought. There is another less common definition, largely used by Windows – namely the process of creating and maintaining an up-to-date list of computers and resources on a network or part of a network by one or more designated computers running the Computer Browser service.

brush A computer imaging term. A paint package's most basic image-creation tool. Most packages let you select a variety of sizes and shapes. Many let you customize shapes.

brussel sprouts In his book "The Ideas That Conquered The World," Michael Mandelbaum tells a story about a young girl who is eating dinner at a friend's house and her friend's mother asks her if she likes brussels sprouts. "Yes, of course," the girl says. "I like brussels sprouts." After dinner, though, the mother notices that the girl hasn't eaten a single sprout. "I thought you liked brussels sprouts," the mother said. "I do," answered the girl, "but not enough to actually eat them."

brute force attack A cracker term. Brute force attack means hurling passwords at a system until it cracks.

BS Base Station: A fixed land station in the land mobile service that relays signals to and from mobile voice and data terminals or handsets.

BSA See both Basic Switching Arrangement and Open Network Architecture.

BSAC BSAC stands for bit sliced arithmetic coding that provides one of the forms of scalability in MPEG-4 audio. See MP-4.

BSAFE BSAFE is a cryptography engine and suite of RSA security components. BSAFE's low-level, general-purpose core crypto engine provides cryptographic base components (primitives) for securing applications or services. BSAFE uses crypto algorithms such as RSA Diffie-Hellman, DES, Triple DES, RC2, RC4, RC5 and SHA-1.

BSC 1. Binary Synchronous Communications. A set of IBM operating procedures for synchronous transmission used in teleprocessing networks, BSC has become a de facto standard protocol. BSC is a character-oriented protocol which involves the communication of data in blocks of up to 512 characters. Each block of TeXT (TXT) data is preceded by a header which includes SYNchronizing bits (SYN) in order that the receiving device might synchronize on the rate of transmission, a Start Of Header (SOH), a HeaDeR (HDR) containing application address information, and a Start of Text (STX). Each block is succeeded by a trailer which includes End Of Text (EOT), Block Checking Characters (BCC) for error detection and correction, and End Of Transmission (EOT). BSC is a polling protocol which operates in a half duplex (HDX) mode, generally over the analog PSTN (Public Switched Telecommunications Network). As each block of data is transmitted from the polled device, the receiving computer system responds with either an ACKnowledgement (ACK) indicating successful and error-free receipt of the subject data, or a Negative AcKnowledgement (NAK) indicating detection of an error created in transmission. An ACK prompts the polled device to transmit the next block of data. A NAK prompts the device to retransmit the subject block. The process continues, block-by-block, until such time as all data have been transmitted and received. By way of example, BSC is commonly used in a Call Accounting application to transfer Call Detail Recording (CDR) information to a centralized processor from a pollable buffer attached to a PBX system.

2. Base Station Controller. A wireless telecommunications term. The BSC is a device that manages radio resources in GSM (Global System for Mobile Communications), including the BTS (Base Transceiver Station), for specified cells within the PLMN (Public Land Mobile Network).

BSD Berkeley Software Distribution. Term used when describing different versions of the Berkeley UNIX software, as in "4.3BSD UNIX".

BSE 1. Basic Switching Element. See Open Network Architecture.

2. Basic Service Elements. A term used in voice processing to describe technical telephone system features such as ANI, DID trunks, call forwarding, stutter dial tone, suppressed ringing, and directory database access.

BSGL Branch Systems General Licence. A British term. A licence that must be obtained by any organization seeking to link a private network to the British PSTN (Public Switched Telephone Network). A separate licence must be held on each site.

BSI British Standards Institution. The body responsible for development of UK standards across a wide range, including telecommunications. BSI also is responsible for input to European standards bodies like CEN and CENELEC, as well as international standards bodies such as the ISO and the ITU-T. BSI claims to be the oldest of the over 100 national standards bodies in the world. www.bsi.org.uk. For the U.S. counterpart see also ANSI.

BSIC Base Station Identity Code. An attribute of a GSM (Global System for Mobile Communications) cell which is a code allowing a distinction between local cells having the same radio frequency.

BSIS Branch Sales Information System.

BSOD Blue Screen of Death. See Blue Screen of Death.

BSP 1. Bell System Practice. A very defined way of writing and presenting instruction and installation manuals. BSPs also establish standards for splicing cable, for installing phones, answering phones, collecting debts, finding phone taps, climbing poles. They are (or once were) the instruction manuals that dictated how to do everything. Divestiture has changed the rules. BSPs are not as important as they were when AT&T handed down all the BSPs.

2. Butt Sweat Protector. David Bomelyn, an engineer who works for a cellular phone out west, claims that "a BSP is the plastic or hard covering of a book or some technical manual that the phone tech keeps in his or her back pocket or by himself at all time. In most cases when this book is pulled out, the tech is in a real jam and his or her butt is most likely is puckered and is sweating and we don't want the pages in this book to get wet." David claims the definition was originally published in the Motorola Dictionary book. Apparently, "The big wigs at Motorola didn't know it was in their dictionary. Some one had slipped it in. When the powers to be found out, they chopped it out of the dictionary, never to be seen again. So maybe you do the world a favor and bring this back to life and give all of us fone techs a good laugh."

3. See Broadband Service Provider.

BSRF Bell System Reference Frequency. See Timing.

BSS 1. Base Station System. A wireless telecommunications term. A GSM (Global System for Mobile Communications) device charged with managing radio frequency resources and radio frequency transmission for a group of BTSs. See also GSM.

2. Business Support System. The system used by network operators to manage business operations such as billing, sales management, customer-service management and customer databases. A type of Operations Support System (OSS).

3. Broadband Switching System. A carrier (e.g., LEC or IXC) switch for broadband communications. Such switches are capable of switching frames (Frame Relay) or cells (SMDS and ATM) at a very high rate of speed. They contain multi-Gigabit busses which may be stacked or chained. For example, an "edge" switch, which is located at the edge of the network much as is a Class 5 Central Office in the voice world, may involve a 20 Gbps bus. "Core" switches, which are located in the core of the network and which are the equivalent of a tandem switch in the voice world, may involve eight busses of 20 Gbps each for a total capacity of 160 Gbps.

BST Base Station Transceiver. A component of a base station which consists of all radio transmission and reception equipment. A BST provides coverage to a particular geographic area and is controlled by its corresponding BSC.

BT 1. British Telecom. See 1981.

2. Burst Tolerance: BT applies to ATM connections supporting VBR services and is the limit parameter of the GCRA.

3. Bit Time. See Bit Time.

BTA 1. Basic Trading Area. A wireless telecommunications term. The United States is broken down into 493 major trading areas for economic purposes. These boundaries were used for licensing PCS wireless phone systems. Several BTAs make up Metropolitan Trading Area, an area defined by the FCC for the purpose of issuing licenses for PCS. Thus, each MTA consists of several Basic Trading Areas (BTAs).

2. Broadband Telecommunications Architecture, an architecture introduced by General Instrument's Broadband Communications Division at the Western Cable Television Show on December 1, 1993. General Instrument said the plant is built to 750 MHz and can support reduced node size and add services such as video-on-demand, telephony, interactivity, data services, etc.

3. Business Technology Association. Previously NOMDA/LANDA, BTA was formed by the merger of the National Office Machine Dealers Association and the Local Area Network Dealers Association. BTA holds several conferences a year. Members include manufacturers, distributors, retailers and consultants. www.btanet.org.

BTAM Basic Telecommunications Access Method. One of IBM's early host-based software programs for controlling remote data communications interface to host applications, supporting pre-SNA protocols. See IBM.

BTag An ATM term. Beginning Tag: A one octet field of the CPCS_PDU used in conjunction with the Etag octet to form an association between the beginning of message and end of message. See ATM.

BTB Ratio See Book-To-Bill Ratio.

BTexaCT The British Telecommunications communication technologies research group that includes the former BTRL (British Telecommunications Research Laboratories). www.labs.bt.com

BTI British Telecom International.

BTL Bell Telephone Laboratories.

BTN Billed Telephone Number. The primary telephone number used for billing, regardless of the number of phone lines associated with that number. According to Bellcore, BTN is sometimes known as a screening telephone number. The BTN is a telephone number used by the AMA process as the calling-party number for recording purposes. See Billing Telephone Number.

BTO 1. Built To Order. PC makers like Toshiba and Dell are delivering PCs and notebooks that contain the hardware specifications that you particularly want – RAM, size of hard disk, etc.

2. Bandwidth Trading Organization. See Bandwidth Broker.

3. See Build-Transfer-Operate.

BTOS A UNIX program which translates binary files into ASCII.

BTRL British Telecom Research Laboratories.

B-tree A tree structure for storing database indexes. Each node in the tree contains a sorted list of key values and links that correspond to ranges of key values between the listed values. To find a specific data record given its key value, the program reads the first node, or root, from the disk and compares the desired key with the keys in the node to select a subrange of key values to search. It repeats the process with the node indicated by the corresponding link. At the lowest level, the links indicate the data records.

Btrieve Btrieve is a key-indexed database record management system. You can retrieve, insert, update, or delete records by key value, using sequential or random access methods. First introduced in 1983, Btrieve was one of the first databases designed for LANs. Novell bought the company in the late 1980s and then later sold it. It's now called Pervasive Software.

BTS Base Transceiver Station. A wireless telecommunications term. A GSM (Global System for Mobile Communications) device used to transmit radio frequencies over the air interface.

BTSM BTS (Base Transceiver Station) Management. A wireless telecommunications term. Devices configured to manage BTS functions and equipment.

BTSOOM Beats The S*** Out Of Me. Acronym used in e-mail, during online chat sessions, and in newsgroup postings.

BTU 1. Basic Transmission Unit.

2. British Thermal Unit. A measure of thermal energy often used in designing building heating and cooling systems. The heat output of computer equipment is often specified and must be taken into account when sizing building climate control systems. Computer equipment heat output is expressed in BTU per hour. 3.7 BTU per hour is equivalent to 1 Watt of dissipation.

BTV See Business Television.

BTW By The Way. An acronym used in electronic mail on the Internet to save words or to be hip, or whatever.

BTx BTx means Base Transmit. It is the frequency on which a base station transmits and a user station receives.

Bubble Up 1. The act of letting an idea or issue rise up the organization chart to a superior. "The best ideas are the ones that bubble up from front line employees."

2. A vague Microsoft term that describes how information is produced in response to an inquiry in a database. Ask the database for how much my portfolio is worth and the program goes out and does some calculations which are transparent to me before it "bubbles up" (i.e. produces) the information. I first heard the term from Microsoft when they referred to their new improved www.moneycentral.com. It's neat word.

bubbleVision An affectionate name for CNBC, CNNfn, Bloomberg TV and other 24/7 cable / satellite channels that concentrate on financial news and "interpretation."

buck See megabuck and passing the buck.

bucket RMON terminology for a discrete sample of data. The RMON History group specifically uses buckets in its sampling functions of the different data sources. See Buckets.

bucket o' dial tone Once upon a time, every new central office technician was sent to another central office for a "bucket o' dial tone," when the CO was overloaded. There is no such thing, of course, but the old hands had a big laugh over it. It was sort of a "right of passage." It worked only once. See also Fiber Exhaust and Frequency Grease.

bucket shop Brokerage firms with dubious reputations. Many of these are fly-by-night operations, consisting of many brokers making cold phone calls to little old ladies. These shops specialize in low priced "penny stocks," which they sell to one fool and then to a greater fool. The brokers tend to hop from shop to shop, just ahead of federal regulators.

bucket truck A truck equipped with a cherry picker – used by many utility companies to access power, cable, or telephone lines.

buckets When competition in cell phones heated up in North America in the mid-1990s, some of the newer competitors pushed increased coverage patterns, small phones, and huge packages of monthly minutes so large that their salespeople started calling the deals "buckets."

buckywalter What a payphone is called in Boontling, a local dialect formerly spoken in Anderson Valley, California, and now spoken only by old-timers. The word "buckywalter" was coined by local residents because the first payphone in Boonville was owned by Levi Walter and it cost a nickel (called a "buckey" back then) to use. Variant spellings: bucky walter, Bucky Walter.

buddy beacon A cell phone feature that lets you see your location on a street map shown on your phone's screen. The phone has GPS semiconductors and software and makes use of GPS satellites. For it to work well, you have to be outside (at best) or within view of the soluthern sky if you're in northern hemisphere. Vice versa if you're in the southern hemisphere.

buddy list A list of the screen names of other users in an instant messaging environment.

budget dust Year-end money that must be spent before it is swept away by the storms of a new fiscal year. See also budget flush.

budget flush The "use it or lose it" spending spree that occurs near the end of the fourth quarter. The last-minute draining of the budget is such a common practice by IT departments that Wall Street analysts include it in their technology stock projections.

buff and puff Sometimes when a tech from the phone company or cable company makes a service call to the customer premises, he finds that there is no equipment problem or at most just a loose connection that is easily tightened. In the interest of customer service or maybe just to make the repair visit seem a smidgen more substantial than it turned out to be, the tech usually will proceed to wipe the equipment with a cloth and get rid of any lingering dust or lint with a small puff of breath. This type of repair visit is a "buff and puff."

buffer 1. In data transmission, a buffer is a temporary storage location for information being sent or received, and serves the purpose of flow control. Usually located between two different devices that have different abilities or speeds for handling the data. The buffer acts like a dam, capturing the data and then trickling it out at speeds the lower river can handle without, hopefully, flooding or overflowing the banks. Buffers can be located at the incoming ports of a switch or router, at the outgoing ports, or even internal to the switching matrix.

2. A coating material used to cover and protect the fiber. The buffer can be constructed using either a tight jacket or loose tube technique. Tight buffering is extruded directly on the fiber coating. Loose tube buffering is an extruded tube around the coated fiber to isolate the fiber from stresses in the cable system.

3. A circuit or component which isolates one electrical circuit from another.

buffer amplifier An amplifier, usually in a gain of 1 or 2, used to drive a heavy capacitive or resistive load.

buffer box A collection device connected to a host phone system (PBX) that collects, stores and reports data from the host phone system. Such a box could be used to collect information on phone calls and have that information downloaded once a month by a remote call accounting service and then assembled into reports, showing who made the most calls, the most expensive calls, etc. A buffer box would typically be attached to a PBX by a serial cable. A buffer box would have some of the ingredients of a PC – a microprocessor and a storage device (e.g. hard disk).

buffer coating Protective material applied to fibers. Increases apparent fiber size. May be more than one layer. Stated in microns. Usually thicker or multi-coated on tight-buffer cables.

buffer memory Electronic circuitry where data is kept during buffering. See Buffer.

buffer overflows Buffer overflows are sometimes called stack smashing. They are the most common form of security vulnerability in the past decade. They are also the easiest to exploit. More attacks are the result of buffer overflows than any other problem. Computers store everything – programs, data, photographs – in memory. If the computer asks a user for an 8-character password and receives a 200-character password, these extra characters may overwrite some other area in memory. They're not supposed to. That's the bug. If it is just the right area of memory, we overwrite it with the just the right characters, we can change a "deny connection" instruction to an "allow access" command or even get our own code executed. The Morris worm is probably the most famous overflow-bug exploit. It exploited a buffer overflow in the UNIX fingerd program. It's supposed to be a benign program, returning the identity of a user to whomever asks. This program accepted as input a variable that is supposed to contain the identity of the user. Unfortunately, the fingerd program never limited the size of the input and Morris wrote specific large input that allowed his rogue program to (install and run) itself. ... Over 6,000 servers crashed as a result; at that time (in 1988) that was about 10 percent of the Internet. – This explanation is from Bruce Schneier, "Secrets and Lies: Digital Security in a Networked World (2000)."

buffer overrun A network security term. An attack in which a malicious user exploits an unchecked buffer in a program and overwrites the program code with their own data. If the program code is overwritten with new executable code, the effect is to change the program's operation as dictated by the attacker. If overwritten with other data, the likely effect is to cause the program to crash.

buffer storage Electronic circuitry where data is kept during buffering. See Buffer.

buffer tubes Loose fitting covers over optical fibers used for protection and isolation.

buffered repeater A device that amplifies and regenerates signals so they can travel farther along a local area cable. This type of repeater also controls the flow of messages to prevent collisions.

buffering 1. Applying a protective material extruded directly on the fiber cladding to protect it from the environment or a secondary extrusion (usually 900 microns O.D.) over the primary buffer.

2. Extruding a tube around the coated fiber to allow isolation of the fiber from stresses

on the cable.

bug 1. A concealed microphone or listening device or other audio surveillance device.

2. To install the means for audio surveillance.

3. A semiautomatic telegraph key.

4. A problem in software or hardware. The original computer bug, a moth, is enshrined at the Washington Navy Yard. It was the cause of a hardware failure in an early computer in 1945. The story goes like this: a team of top Navy scientists was developing one of the world's first electronic computers. Suddenly, in the middle of a calculation, the computer ground to a halt, dead. Engineers poured over every wire and every inch of the massive machine. Finally, one of the technicians discovered the cause of the problem. Buried deep inside its electronic innards, crushed between two electric relays, lay the body of a moth. These days, "bugs" in telecom or computer systems are not insects. They're indescribable glitches that adversely affect smooth operations. Bugs usually originate in software. Some programmers call bugs "undocumented features." And they are, indeed. All the above is the story the navy likes to put out. In fact, the word "bug" for problem in design has been around for eons. It's mentioned in a 1910 book called, "Edison, His Life and Inventions" by Frank Lewis Dyer. The book talks about Edison harassing his employees to "get all the bugs out."

bug mix A silly term for the precise collection of bugs in a particular piece of software.

bug rate The frequency with which new bugs are found during the testing cycle is referred to as the bug rate. As these bugs are fixed, and as time passes, bugs will become more and more difficult to find, and new bugs will be found less frequently, i.e., the bug rate decreases. Usually, when the bug rate drops to zero and all major bugs (and most of the minor bugs) have been fixed, a product is ready for the next stage in its lifecycle.

bugbashing Bugbashing is Microsoft-speak for fixing a bug in software.

BUI Browser user interface. The use of a web browser as an interface to an application.

Build To Suit BTS. A system made to a customer's specific specifications.

Build-Operate-Own BOO. A project whereby a private company is awarded a concession to build a telecommunications network or services and operates it for a certain period of time before taking ownership.

Build-Operate-Transfer BOT. A project whereby a private company is awarded a concession to build a telecommunications network or service and operates it for a certain period of time before handing over ownership to the national telecommunication administration or PTO.

Build-Lease-Transfer BLT. A project whereby a private company is awarded a concession to build a telecommunications network or service and leases it for period of time before handing over ownership to the national telecommunication administration or PTO.

Build-Transfer-Operate BTO. A project whereby a private company is awarded a concession to build a telecommunications network or service, hands over ownership to the national telecommunication administration or PTO, and operates it for a certain period of time.

building core A three-dimensional space permeating one or more floors of the building and used for the extension and distribution of utility services (e.g., elevators, washrooms, stairwells, mechanical and electrical systems, and telecommunications) throughout the building.

Building Distributor BD. The international term for intermediate cross-connect. A distributor in which the building backbone cable(s) terminates and at which connections to the campus backbone cable(s) may be made.

building distribution frame Somewhere in the building where all telephony wiring for the building is. All the cables from the outside the building would first come here and be punched down on the building distribution frame. Additional cables would take the telephone circuits upstairs.

building entrance agreement A piece of paper that lets phone companies and utilities enter a building, gives them permission to build facilities and store equipment and lets them access their equipment. In great demand by Competitive Local Exchange Carriers (CLECS) who give residents a choice of providers and often force existing carriers to improve services or reduce rates. Often the CLECS are charged fees, which they agree to, to avoid paying the RBOCs' access charges.

building entrance area The area inside a building where cables enter the building and are connected to riser cables within the building and where electrical protection is provided. The network interface as well as the protectors and other distribution components for the campus subsystem may be located here. Typically this area is the end of the local telephone company's responsibility. From here on it's your responsibility. You should protect your equipment inside the building from spikes and surges and other electrical nonsenses which the phone company's cables might bring in. For the best disaster protection, it's wise to have two building entrances by which your telecommunications cables can enter. And they should enter from separate telephone central offices. Some telephone companies, e.g. New York Telephone, are now tariffing such services. See Building Entrance Agreement.

building entrance terminal Cable entrance point where typically a trunk cable between buildings is terminated.

building footing The concrete base under the foundation of a building in which copper wire may be laid to form an electrical ground.

building integrated timing supply BITS. A clock, or a clock with an adjunct, in a building that supplies DS1 and/or composite clock timing reference to all other clocks in that building. See BITS.

building module The standard selected as the dimensional coordination for the design of the building, e.g., a multiple of 100 mm (4 in), since the international standards have established a 100 mm (4 in) basic module.

building out The process of adding a combination of inductance, capacitance,and resistance to a cable pair so that its electrical length may be increased by a desired amount to control impedance characteristics.

building steel The structural steel beams that make up the frame of a building. If the steel frame is buried deep in the earth, it can be used as an electrical ground. But you'd better be careful: Unbalanced three-phase power is also probably using the frame as a ground, and you may pick up huge quantities of 60 hz hum.

building wiring A generic term wire used for light and power in permanent installations using 600 volts or less, typically not exposed to outdoor environments.

bulk billing A method of billing for long distance telephone services where no detail of calls made is provided. WATS is a bulk billed service. Therein lies the problem for the cost conscious user. There's no verification of calls made. See Call Detail Recording.

bulk encryption Simultaneous encryption of all channels of a multichannel telecommunications trunk.

bulk storage Lots of storage. Usually reels of magnetic tape or hard disks.

bulk TCP Describes a client/server application that uses TCP for transport. The term "bulk" refers to the fact that the size of the application's transactions are generally large – often hundreds of millions of bytes. Bulk TCP applications are not interactive, nor do they have the same latency and jitter requirements as real-time traffic. Bulk TCP applications include email and Web downloads.

bullet amplifier A Satellite Term. Small device used to increase signal power and offset signal loss caused by coaxial cable and splitting devices.

bulletin board system A fancy name for an electronic message system running on a microcomputer. Call up, leave messages, read messages. The system is like a physical bulletin board. That's where the name comes from. Some people call bulletin board systems electronic mail systems. See also BBS.

bulletize To highlight supposedly key information using bullet points. "To help explain my idea, I've bulletized the main points on the next slide…" Often used by people who can't explain themselves in complete sentences. Microsoft's PowerPoint seriously accelerated the trend to think in bullets – serially, one after another. PowerPoint and bullets are not the way the world works. But go tell that to all the corporate types who are in love with PowerPoints.

bulletproof Used to describe an algorithm or implementation that's thought to be capable of recovering from any imaginable exception or condition – a rare and valued quality. It implies that the programmer has covered all bases.

bump-in-the-wire Describes a network service that is implemented as a hardware device that is installed directly on the network. For example, a firewall may be implemented as a bump-in-the-wire (BITW) network device.

bump-in-the-stack Describes a network service that is implemented as software in the protocol stack. For example, a firewall may be implemented as a bump-in-the-stack (BITS) software module.

bumper beeper Radio beacon transmitter, hidden in or on a vehicle for use with radio tailing equipment.

bunch stranding A group of wires of the same diameter twisted together without a predetermined pattern.

bunched frame-alignment signal A frame-alignment signal in which the signal elements occupy consecutive digit positions.

bundle 1. A group of fibers or wires within a cable sharing a common color-code.
  2. In T-1, specifically M44 Multiplexing, a bundle consists of 12 nibbles (4 bits) and

may represent 11 channels of 32 Kbps compressed information plus a delta channel. A bundle is typically a subset of a DSI and treated as an entity with its own signaling delta channel.

  3. Some telecommunications carriers have tried to entice buyers to buy more than one service, i.e. long distance, local service, call forwarding, call waiting, Internet access, etc. It's gotten harder to sell a bundle of services as telecommunications has become more competitive in recent years and large telecom carriers have lacked marketing skills, preferring to sell on cheaper price, which is easier. the better carriers sell bundles.

bundle fodder Junk software included on CD-ROMs or packaged with peripherals such as modems and designed to bulk up the presumed value of the total package to entice an unsuspecting consumer to buy. See Shovelware.

bundled Combining several services under one telephone tariff item at a single charge. See Bundled Services.

bundled cable An assembly of two or more cables continuously bound together to form a single unit prior to installation (sometimes referred to as loomed, speed-wrap or whip cable constructions).

bundled rates See Bundled Services.

bundled services Combining several services under one telephone tariff item at a single charge. Bundled services sound like a good idea, especially since you tend to think that you're getting some sort of discount for a suite of services. Oftentimes, that's just not true. Rather, bundling is a way of getting you to subscribe to a bundle of services, some of which you may not even need and will never use, at a potentially inflated price. In a bundle of services, you never quite know the price of a given component of the bundle, so you really can't objectively compare different companies' offerings. Of course, that's part of the reason for the bundling.

bundling A marketing term used by local, long-distance and cable companies whereby several services are combined into one package. The main idea is to drive revenues higher. The secondary idea is to reduce customer churn (it works). The third idea is to derive some economies of scale – e.g. put more services on the same copper wire or coaxial cable. Bundling may allow for cross-subsidization (when allowed by the regulator). Bundling may let the carrier channel a new product into an existing stream of demand for another product. Examples of bundling are rife within telecom. Cable companies, for example, have bundled cable TV and broadband Internet service together. Some users believe they get two "bargains" – cheaper service and easier service calls (only one carrier to deal with – no fingers to point). Sometimes they're right. Usually they're not. See Bundled Services.

bunny suit A layered, hooded outfit that covers every part of your body, except your eyes. Bunny suits are worn by people who work in places where cleanliness is absolute. The human skin sheds about 30,000 particles of skin a second. If one of these particles made it into a semiconductor or a piece of optical fiber it could seriously impair the usability of the device.

burbling What my wife, Susan, calls my endless lectures about telecommunications and other subjects of great fascination to me, but not, sadly, to her.

burden test A semi-legitimate test used in regulation to determine if the offering of a new or continued service will cause consumers of other services to pay prices no higher than if the service were not offered. In other words, the question is "Who carries the Burden?" It's sometimes called the "avoidable cost test."

buried cable A cable installed under the surface of the ground in such a manner that it cannot be removed without disturbing the soil.

buried service wire splice A watertight splice filled with an encapsulant.

burn in To run new devices and printed circuits cards, often at high temperatures, in order to pinpoint early failures. The theory is all semiconductor devices show their defects – if any – in the first few weeks of operation. If they pass this "burn-in" period, they will work for a long time, so the theory goes. "Burn-in" should probably be 30-days under full power and working load. Burn-in should also take place in a room with lots of heat and at least 50% humidity, since this will simulate the poorly-ventilated places most people install telephone systems.

burn out A condition, where stress causes agents to be apathetic and lethargic, caused by intensity of calling, lack of variety and poor working conditions. It is particularly associated with outbound cold calling and inbound complaint handling, both of which are stressful for agents if not carefully managed.

burn rate Cash divided by operating expenses plus non cash charges (depreciation and amortization) in the most recent quarter. This is a measure of how long a company can survive before it runs out of cash. In short, burn rate is the speed at which a new company uses up its cash en route to developing a product before it turns cash positive.

burned-in address BIA. The hardware address on a network interface card (NIC). This address is assigned by the manufacturer of the interface card, which ensures that every card has a unique address. Also called a MAC address.

burning a pole Slang expression to describe when an installer accidentally slides down a telephone pole. This usually happens on an old pole full of gaff holes when his gaff breaks out of the pole. Burning a pole results in painful chemical burns because the installer usually winds up with a chest and legs full of splinters coated with creosote, a coal tar distillate used to preserve the wooden pole. Installers are taught to "kick out," rather than hug the pole and suffer burns.

burrus diode A surface-emitting LED with a hole etched to accommodate a light-collecting fiber. Named after its inventor, Charles Burrus.

burst 1. In data communication, a sequence of signals, noise, or interface counted as a unit in accordance with some specific criterion or measure.

2. To separate continuous-form or multipart paper into discrete sheets.

3. A small reference packet of the subcarrier sine wave, typically 8 or 9 cycles, which is sent on every line of video. Since the carrier is suppressed, this phase and frequency reference is required for synchronous demodulation of the color information in the receiver.

burst isochronous Isochronous burst transmission. See Isochronous.

burst mode A way of doing data transmission in which a continuous block of data is transferred between main memory and an input/output device without interruption until the transfer has been completed.

burst switching In a packet-switched network, a switching capability in which each network switch extracts routing instructions from an incoming packet header to establish and maintain the appropriate switch connection for the duration of the packet, following which the connection is automatically released. In concept, burst switching is similar to connectionless mode transmission, but it differs from the latter in that burst switching implies an intent to establish the switch connection in near real time so that only minimum buffering is required at the node switch.

burst traffic Burst traffic is many phone calls (usually incoming telephone calls) that simultaneously arrive at a computer telephony system. Burst traffic tests are usually performed as part of a systems load and stress testing. Burst traffic tests are particularly important for computer telephony systems as usually they must perform a lot of processing to set up to handle a call. The arrival of many calls simultaneously, such as in response to a television sales offer (or when Oprah asks you to call 1-800-xxxxxxx to vote on whether OJ is guilty or innocent), can place significant strain on a computer telephony system.

burst transmission 1. A method of transmission that combines a very high data signaling rate with very short transmission times.

2. A method of operating a data network by interrupting, at intervals, the data being transmitted. The method enables communication between data terminal equipment and a data network operating at dissimilar data signaling rates.

bursty Refers to data transmitted in short, uneven spurts. Or, discontinuous signals occurring at random intervals.

bursty information Information that flows in short bursts with relatively long silent intervals between. LAN traffic is characterized as "bursty" in nature, as devices tend to transmit substantial amounts of data at irregular intervals.

bursty seconds Bursty seconds is a measure of the amount of time spent at maximum data transfer rate.

bursty traffic Communications traffic characterized by short periods of high intensity separated by fairly long intervals of little or no utilization. Data traffic and certain kinds of video traffic are inherently bursty.

bus 1. An electrical connection which allows two or more wires or lines to be connected together. Typically, all circuit cards receive the same information that is put on the bus. Only the card the information is "addressed" to will use that data. This is convenient so that a circuit card may be plugged in "anywhere on the Bus." There are two common buses inside a PC – the older ISA bus, capable of only five megabytes per second and the newer PCI bus, capable of transmitting up to 132 megabytes per second. All computers and most telephone systems use buses of some type. Computer buses are typically open. Telephone system buses are typically closed. See also Backplane and Bus Network.

2. BUS (Broadcast and Unknown Server). An ATM term. More specifically, a LAN Emulation (LANE) term. This server handles data sent by a LAN Emulation Client (LEC) to the broadcast MAC address ('FFFFFFFFFFFF'), all multicast traffic, and initial unicast frames. The BUS works in conjunction with a LES (LAN Emulation Server), which automatically registers and resolves differences between LAN MAC addresses and ATM addresses. This is accomplished by labeling each device transmission with both addresses. See also ATM and LANE.

bus and tag cables Multi-wire copper cables used by IBM from the 1960s to the early 1990s to connect a peripheral device, such as a printer or tape drive, to a local mainframe. Bus and tag cables were always used in pairs; the "bus" carried the data, and the "tag" carried the control information.

bus and tag channel A parallel channel between an IBM mainframe and attached peripherals, created with bus and tag cables. See bus and tag cables.

bus card An expansion board that plugs into the computer's expansion bus. See also PCMCIA.

bus enumerator A Windows 95 term. A new type of driver required for each specific bus type, responsible for building ("enumerating") the hardware tree on a Plug and Play system.

bus extender A device that extends the physical distance of a bus and increases the number of expansion slots.

bus hog A device connected to a transmission bus which, after gaining access to the transmission medium, transmits a large number of messages regardless of whether other devices are waiting.

Bus Interface Unit BIU. In LANs, the device furnishing direct connection of a DTE to the LAN bus.

bus master A VME board (usually a CPU) that can contend for, seize and control the VME bus for the purpose of accessing bus resources such as voice boards or even other CPUs. See VME.

bus mastering A bus design that enables add-in boards to process independently of the CPU and access the computer's memory and peripherals on their own. Bus mastering is a way of transferring data through a bus in which the device takes over the bus and directly controls the transfer of data to the computer's memory. Bus mastering is a method of Direct Memory Access (DMA) transfer.

bus mouse A mouse that is attached to your computer's not via a serial port or a USB port but through a special card attached to the motherboard. Note that such a mouse takes up an expansion slot in a PC. Bus mice are increasingly rare.

bus network All communications devices share a common path. Typically in a bus network, a "conversation" from each device is sampled quickly and interleaved using time division multiplexing. Bus networks are very high-speed – millions of bits per second – forms of transmission and switching. They often form the major switching and transmission backbone of a modern PBX. The printed circuit cards which connect to each trunk and each line are plugged into the PBX's high-speed "backbone" – i.e. the bus network. Similarly, Broadband Switching Systems (BSSs) make use of internal buses, which run at Gbps rates.

In the LAN world, bus networks include Ethernet, which is by far the most common LAN standard. The Ethernet bus specification provides for a common electrical highway which can be shared by as many as 1,024 attached devices per physical Ethernet or Ethernet segment. The Ethernet bus runs at 10 Mbps, although throughput is typically much less. See also BUS.

bus slave A VME board (usually a subsystem or I/O board) which can only respond to VME bus accesses mapped to its address. Slaves can usually interrupt the VME bus on one of 7 levels. See VME.

bus speed 1. The speed at which the computer's CPU (central processing unit) communicates with other elements of the computer. For example, the speed at which data moves between the CPU and its various bus-attached devices, such as video controller, disk controllers, voice cards, etc. In defining bus speed, Walt Mossberg of the Wall Street Journal wrote: A "bus" is a techie term for a pathway through which bits of data travel among various components inside a PC. In recent years, computer makers have been increasing the "bus" speeds, to keep up with advances in the speeds of processors. Though these high speeds sound enticing, they have little relevance to an average computer user. As with memory speeds, faster bus speeds won't make Web surfing or word processing discernibly faster. Again, the publicizing of bus speeds is mainly a marketing trick to get unfamiliar consumers to pay more." See AGP, AT BUS and PCI.

2. The speed at which the internal bus operates in a network switch. For example, an ATM Broadband switch may provide multiple, redundant buses, each of which might operate at 20 Gbps, yielding a total rate of 160 Gbps for an 8-bus switch.

3. The speed at which a network bus operates. For instance, an Ethernet bus provides a raw transmission rate of 10 Mbps, although the rate of data throughput typically is much less.

bus topology A network topology in which nodes are connected to a single cable with terminators at each end. One form of Ethernet LANs used bus topology, joining the various PCs along one piece of coaxial cable.

bush telegraph An Australian term for the informal word-of-mouth way that news spreads in a community. It is similar to the "coconut wireless" in Hawaii.

business analytics software Software that helps businesses understand what makes their customers happy and how to keep them that way so companies can improve retention and increase sales. Business analytic applications extract the events and attributes locked away in the text fields of transactional software systems, integrating them with structured and numeric data for automated analysis. The software has several aims: customer retention, new idea testing, loyalty measuring, product and price testing, brand assessment and web analytics. According to one brochure I read on this area, "Analysis of behavior can tell a company that support call volume, customer attrition or sales dropout rates are on the rise, but only analysis of communications can explain why customers are calling, why they're dropping out of the sales pipeline, and why they're closing existing accounts."

Business and Residential Custom Services BRCS. Also known as Custom Calling Features (CCF) including call waiting, call forwarding, 3-way calling, etc., available through the central office without requiring the subscriber to use special equipment.

business audio At one stage business audio was thought to involve voice annotating spreadsheets. Now it means making phone calls through and by your PC.

business card The electronic date equivalent to a printed business card. This electronic version of the business card is treated like a file and can be exchanged between Bluetooth devices. See vCard.

Business Management Computer BMC. The computer system used at a CATV (Cable TV) or MMDS (wireless) business office to keep track of subscriber accounts.

business model The term business model came into use in the late 1990s with the advent of the investor enthusiasm for the Internet and the resulting stock market boom. The concept was the Internet was so revolutionary it would allow the creation of companies that worked differently to normal companies. For example, one popular business model was that you could give away your product and services on the Internet and pay for them by taking advertisements. Another concept was that you could sell things without direct mail catalogs, advertising and salespeople. All you needed was a web site. In fact very few of the new business models ultimately succeeded. But the term crept into the English language. Now when someone asks you "what's your business model?" they're really asking, how do you intend to get paid for whatever you intend to do?

business need A generic term generally used by upper management to explain or defend a business decision – particularly a bad one.

business PCS Advanced communications technology that adds wireless capability to create an in-building or campuswide wireless communication network. This is also called Wireless PBX or Enterprise PCS. See those definitions.

business process management Defining and documenting a set of guidelines of how an organization or an industry works. For example, from the original order to cash is the main business process for all retail companies.

business service Service used primarily for any purpose other than that of a domestic or family nature. A telephone industry definition.

Business Support System See BSS.

business technology association An association formerly known as Nomda/Landa.

Business Television BTV. Point-to-multipoint videoconferencing. Often refers to the corporate use of video for the transmission of company meetings, training and other one-to-many broadcasts. Typically uses satellite transmission methods and is migrating from analog to digital modulation techniques.

business to business Basically there are three types of ecommerce: First, where one business sells something to another. Second, where one business sells to a consumer, i.e. you or me. Third, where one consumer sells to another consumer. The medium for all this ecommerce is typically the Internet or a private Intranet, i.e. a network confined to one or several organizations. See also eCommerce.

business to consumer See Business to Business.

busker A person who entertains in public places, especially for money. When Singapore decided to legalize street performances in 1997, artists were required to audition and to donate any money collected to charity.

busy In use. "Off-hook". There are slow busies and fast busies. Slow busies are when the phone at the other end is busy or off-hook. They happen 60 times a minute. Fast busies (120 times a minute) occur when the network is congested with too many calls. Your distant party may or may not be busy, but you'll never know because you never got that far.

busy back A busy signal.

busy call forwarding When you call a busy phone extension, your call is automatically sent to another predetermined telephone extension.

busy hour The hour of the day (or the week, or the month, or the year) during which a telephone system carries the most traffic. For many offices, it is 10:30 A.M. to 11:30 A.M. The "busy hour" is perhaps the most important concept in traffic engineering – the science of figuring what telephone switching and transmission capacities one needs. Since the "busy hour" represents the most traffic carried in a hour, the idea is if you create enough capacity to carry that "busy hour" traffic, you will be able to carry all the other traffic during all the other hours. In actuality, one never designs capacity sufficient to carry 100% of the busy hour traffic. That would be too wasteful and too expensive. So, the argument then comes down to, "What percentage of my peak busy or busy hour traffic am I prepared to block?" This percentage might be as low as half of one percent or as high as 10%. Typically, it's between 2% and 5%, depending on what business you're in and the cost to you – in lost sales, etc. - of blocking calls.

PSTN busy hours during the past 20 years or so were from 2:00 to 3:00 P.M., when children arrived home from school and called their working parents to let them know that they were O.K. During the past year or so that busy hour has shifted to 7:00 to 11:00 P.M., reflecting the fact that working parents have put the children to bed and started to "surf the 'Net." See Busy Hour Usage Profile.

busy hour (average busy season) A telephone company definition. A time-consistent hour, not necessarily a clock hour, having the highest average business day load throughout the busy season. This must be the same hour for the entire busy season.

Busy Hour Call Attempts BHCA. A traffic engineering term. The number of call attempts made during the busiest hour of the day.

Busy Hour Call Completion BHCC. A measurement of telephone traffic determined by the system's busiest hour and used to determine system capacity.

busy hour usage profile The busy hour usage profile identifies how a system will normally be used (i.e., who the users are and what type of transactions they are performing) during the busy hour. Different things that users of CT systems do stress the CT system in different manners. For example, sending a broadcast message in voice mail (where one message is automatically sent to many recipients) may force the system to perform 10 times as much disk I/O as sending messages to single recipients. Or, updating an account balance in an IVR transaction forces a strain on the IVR to mainframe link. Or, setting up conference calls across a switch or network stresses the use of the network database resources. Understanding and characterizing the type and mix of calls that will take place during the busy hour is key to designing and placing a real-world load on the CT system. Load testing a system should always incorporate tests using the busy hour call profile. See Busy Hour.

busy lamp A light on a telephone showing a certain line or phone is busy. See Busy Lamp Field.

busy lamp field A device with rows of tiny lights that shows which phones in a telephone system have conversations on them, which phones are ringing, which phones are on hold. Each light corresponds to a telephone extension on the system. The busy lamp field usually sits attached to the attendant's console, telling the attendant if an extension is busy, free, on hold, etc. The benefit of having a busy lamp field is that the operator doesn't have to dial the number to find out what's happening with the extension. This saves the attendant time in handling incoming calls and gives the caller better service. A busy lamp field is often combined with DSS (Direct Station Select.) Next to each light on the busy lamp field there is a button which the operator can push which will dial the corresponding extension (i.e. directly select it) and will typically transfer the call automatically. This button is like an autodial button. This saves the time of dialing the two, three or more numbers of the extension. These days, busy lamp fields are often built into phones on a key system, and everyone, not just the operator, can have one. This gives everyone information on what's happening in the system. It makes transferring calls, etc. easier. See also Direct Station Select.

busy line verification A service whereby at the request of a local calling party, the telco (telephone company) checks to see if a local line is clear or in use, and lets the calling party know. This is not a free service; it generally costs around $10 per call, maybe even more.

busy line verification with interrupt A service whereby at the request of a local calling party, the telco checks to see whether if local line is clear or in use, and if the line is in use, the operator interrupts the call if requested to do so. This is not a free service; it generally costs around $10 per call, maybe even more.

busy out Let's say you have three lines in a hunt group. Let's say your first line dies.

This will mean that your callers will now receive an endless ring tone – with no one answering, even though you're in the office or at home. The best solution is to ask your friendly local phone company to "busy out" the broken line. This means to fool its switch that your line is busy, i.e. someone is speaking on it, and that the calls should be forwarded over to the next line. The easiest way to busy out a line is simply to short it, i.e. physically connect the two conductors – tip and ring – together. Just remember when you do get the line fixed, to unbusy out your line. Busying out lines going into a computer is useful when the computer is not available, i.e. during maintenance. This way callers do not get connected to modems with no computers to talk to. This is also known as "taking the phone off the hook." In a voice phone system with trunks that rotary (or hunt) on, sometimes busying one or more broken trunks out helps calls progress onto trunks that are still working. This way, someone doesn't end up on your third trunk with endless ringing, while your 4th, 5th, 6th etc. trunks are free. See Rollover Lines.

busy override A feature of some PBXs which allows the attendant or other high priority user to barge in on a telephone conversation. A warning tone is usually thrown into the conversation to alert the parties to an override. The feature is also called "Barge-In." Sometimes when conversations are overridden, only the person within the organization can hear the barge-in.

busy period A telephone company definition. A three to eight month period within which the three busy season months will occur. Example: November through April.

busy redial Another name for Automatic Callback. (Verizon and some other companies use this name.) See also Automatic Callback.

busy season An annual recurring and reasonably predictable period of maximum busy hour requirements – normally three months of the year, and typically the three months preceding Christmas.

busy season prior to exhaust A telephone company definition. The busy season prior to exhaust of an addition is defined as the latest busy season for which an addition will provide objective levels of service.

busy signal A signal indicating the line called is busy. The busy signal is generated by the central office. There are two types of busy signals. See Busy.

busy test A method of figuring whether something which can carry traffic is actually doing so or whether it's broken or free and available for use.

busy tone See Busy Signal.

busy verification of station lines 1. An attendant can confirm that a line is actually in use by establishing a connection (dialing in and listening) to that apparently busy line.

2. In the public switched telephone network, a switching system service feature that permits an attendant to verify the busy or idle state of station lines and to break into the conversation. An alternating tone of 440 Hz is applied to the line for 2 seconds, followed by a 0.5-second burst every 10 seconds to alert both parties that the attendant is connected to the circuit.

busy/idle flag A cellular radio term. A busy/idle flag is an indicator that is transmitted by the Mobile Data Base Station (MDBS) periodically to indicate whether the reverse channel is currently in the busy state or the idle state.

butcher Colloquial term for a telephone technician who tends to use cutters without verifying whether the cabling is active or inactive.

butt connector A nifty little tubular connector, made of plastic, which is used for splicing two wires together. The connector is narrower in the middle than on the ends, in order to bring the wires to be spliced into physical contact with each other. After the wires are inserted into the butt connector, one wire from each end of the tube, the connector is crimped using a crimping tool. Heat is then applied to the connector, which softens it and causes it to adhere to the spliced wires, forming a sealed, water-tight connection. Also called a crimp barrel or crimp terminal.

Butt in Chair BIC. Butt in Chair is the designation for a technician who has to work on Saturday or Sunday, traditionally light days when not much happens. The purpose of the BIC is to be there just in case something happens.

butt set See Buttinsky.

butt splice Connecting the ends of two wires with a "butt splice connector," from such manufacturers as AT&T and 3M.

buttinsky Or Butt Set. The one-piece telephone carried on the hips of telephone technicians. It's called buttinsky or butt set because it allows technicians to "butt in" on phone calls, not because the device is worn on their butts. Butt sets used to be essentially telephones without ringers. But now they are much more sophisticated. They will pulse out in rotary or dial out in touchtone and allow you to talk or to monitor a call. They will run com-

puterized tests on the line. Some even have the equivalent of an asynchronous computer terminal built in, which can be used to talk to a distant computer over a phone line. This distant computer could assign them their next jobs, allow them to check and assign features (touchtone, rotary, hunt), report the time spent on this job, etc. In short, the terminal and the computer could replace a raft of clerks and a deluge of paperwork.

The derivation of the term "buttinsky" has long been lost in history. Butt can also mean to attach the end of something to the end of something else. If the clips of a butt set are attached to a pair of wires, it is a "butt" connection. There are those people who also think that the term buttinsky came from some middle European language – Polish or Yiddish – and is slang for someone who butts in a lot, which is what you can do with a Butt Set. I shall keep researching this one. If you can help, let me know, please.

button caps Interchangeable plastic squares fit over the buttons of electronic telephones, and are used to label the features programmed onto each programmable button location. Button caps can be either pre-printed or have clear windows which allow features, lines, and Autodial numbers to be labelled on the button.

buttons Why do men's clothes have buttons on the right while women's clothes have buttons on the left? Answer: When buttons were invented, they were very expensive and worn primarily by the rich. Because wealthy women were dressed by maids, dressmakers put the buttons on the maid's right. Since most people are right-handed, it is easier to push buttons on the right through holes on the left. And that's where women's buttons have remained since.

butyl rubber A polymer of isobutylene with small amounts of isoprene. An ozone resistant insulation compound.

buyside Term that describes institutional investors and members of the investment community including mutual funds, hedge funds, trusts, and financial advisers. Sellside are brokerage firms.

buzz 1. To check the continuity of a cable pair by putting an audible buzzer on one end and then checking with a "buttinsky" to see if you can hear the buzz and thus identify the correct cable pair.

2. A feature of a Rolm CBX which lets the user signal one Rolm desktop product without picking up the handset. Only one buzz per extension is permitted.

buzzer An electromechanical device that makes a buzzing noise when power is applied, often used to signal someone to answer an intercom call. Battery-powered buzzers were once used to help trace phone circuits. TONE GENERATORS are more common today, but old terminology is still used, as in "buzzing out a line."

buzzer leads The wires inside a telephone intended for the connection of a buzzer, usually as part of an intercom system.

buzzword quotient The percentage of words in a speech, conversation, lecture or document that are buzzwords, i.e. words that sound great but mean little. Some words might be synergy and convergence. If one out of every ten words in a press release is a buzzword, the buzzword quotient is 10%.

BVA AT&T's exclusively-owned Billing Validation Application database. Today, BVA contains all the Regional Bell Operating Companies (RBOCs) calling cards, and other billing information such as billed number screening and payphone numbers. The RBOCs and AT&T access that database today. Prior to Divestiture three market players, the RBOCs, AT&T and most Independent Companies, dominated the "O" Operator Services business which provided alternate billing arrangements such as collect calls, bill to third number and charging calls to calling cards. The three market players still exclusively employ BVA which allows them to validate or authorize alternate billing arrangements. No other long distance carrier or a company needing access to the data for billing validation can use the system. The database is owned by AT&T and is updated daily by the RBOCs and Independent Companies with local exchange information, billing number screening and calling card information. The scenario is further complicated by the 1984 AT&T Plan of Reorganization's exclusive BVA access restrictions. In other words, the three original market players (AT&T, the RBOCs and Independent Companies) have exclusive access to BVA for a predetermined contract length. In most cases these arrangements run into the 1990's. See BVS.

BVAS Broadband value-added services. For example, streaming audio and video, VoIP, and multiplayer games.

BVS Business Validation Service. US West Service Link was the first in the nation to develop and make available a nationwide Billing Validation Service open to any company that needs to verify the legitimacy of their callers' requests to place charge calls to their local telephone calling cards. US West Service Link developed BVS in 1987 and turned up the system for "on-line" customers in early 1988. Today, the US West BVS system is a national validation source containing calling card data of customers served by the RBOCs, GTE,

Southern New England Telephone, United Telecommunications, Cincinnati Bell, Rochester Telephone and Telecomm Canada, a consortium representing all of Canada's local telephone operating companies. In all, more than 60 million records are stored. Sprint, MCI and ITI are among the carriers using BVS. BVS uses X.25 and SS7 protocol. See BVA.

BW Bandwidth: a numerical measurement of throughput of a system or network. See Bandwidth.

BWA Broadband Wireless Access. See 802.16e and WiMAX.

BWG Broadband Wireless Group.

BWM Bandwidth Management System.

BX Armored building wire, 600 volts.

BX.25 AT&T's rules for establishing the sequences of events and the specific types and forms of signals required to transfer data between computers. BX.25 includes the international rules known as X.25 and more.

by-Wire In days gone by, when you turned the steering wheel in a car or an airplane, it moved some wires which moved the wheels or the rudder. Ditto for the brakes. Nowadays, when you do something, the motion you make is detected by some electronic sensors, which is relayed to some computer chips. These chips then signal motors to turn the wheels or make the brakes work. Such a system is said to be more reliable and to allow the injection of computer "intelligence" into the process. For example, the computer might determine that you shouldn't be pushing the brakes because that will make the car skid and it will not apply the brakes.

BYOI Bring your own infrastructure.

bypass 1. A term coming from the idea of using a method to bypass the local exchange network of the Local Exchange Carrier (LEC). Also known as Facilities Bypass, this approach is employed for several reasons: 1. Because the phone company is too expensive, or 2. Because the phone company can't provide the desired levels of bandwidth, quality, or responsiveness.

Bypass means you might be transmitting between two of your offices in the same city, perhaps between your office and your factory, using private microwave or lines you lease from the local power company. You might bypass both the LEC and IXC to link two or more company sites through microwave or satellite transmission systems. You also might be bypassing the LEC to connect directly to your friendly long distance carrier, which then carries your calls to distant cities. In the last case, fiber optic networks deployed by AAVs (Alternative Access Vendors), also known as CAPs (Competitive Access Providers) typically are used, although private microwave may be used as well. Bypass vendors increasingly include CATV providers, some of which have upgraded their coaxial cable networks to support two-way, interactive, switched voice and data communications.

Bypass is a word created by the local telephone industry to sound very threatening. The theory is major users will bypass their local phone company, depriving their phone company of needed revenues. This will drive the local phone company close to imminent bankruptcy, or at the very least, to the state regulators for huge rate increases, hurting the remaining customers, who would, presumably, be customers in areas where bypass is not available (i.e. the poor and the disadvantaged). Additionally, the thought is that the Universal Service Fund would be threatened, thereby depriving users in high cost areas (read "remote and rural") of the right to affordable telephone service. The reality of this threat has not been proven. Nevertheless, the rhetoric frightens sufficient regulators to look at the evils of bypass and to outlaw it, or at least severely restrict it – as several states have done. The Telecommunications Act of 1996 changed the rules on bypass. Now it will happen big-time in the United States. See Telecommunications Act of 1996.

2. An AC power path around one or more functional units of a UPS. An automatic bypass is controlled by UPS control logic and activated when some part of the UPS malfunctions or intentionally shuts down due to an overload or other abnormal condition in order to maintain power to the protected load. A manual bypass is a user controlled switch on a UPS that allows a complete electrical bypass of the unit and may be engaged when there is a total UPS failure or when performing certain types of diagnostics or repair. A service bypass is a manual bypass that allows complete maintenance, or even removal, of the UPS without shutting off the load. A true service bypass is commonly a separate device from the UPS. This definition courtesy APC, a maker of UPSes.

bypass cabling Bypass cabling or relays are wired connections in a local area ring network that permit traffic between two nodes that are not normally wired next to each other. Such bypass cabling might be used in an emergency or while other parts of the system are being serviced. Usually such bypass relays are arranged so that any node can be removed from the ring and the two nodes on either side of the removed node can

then talk.

bypass circuitry Circuits that automatically remove a device from the data path when valid signals are dropped.

bypass relay A relay used to bypass the normal electrical route in the event of power, signal, or equipment failure.

bypass switcher An audio-follow-video switcher usually associated with a master control switcher. Used to bypass the master control switcher output during emergencies, failures, or off line maintenance.

bypass trunk group A trunk group circumvents one or more tandems in its routing ladder.

byte Abbreviated as "B" (big "B."). A set of bits (ones and zeros) of a specific length represent a value, in a computer coding scheme. A byte is to a bit what a word is to a character, which is why a byte sometimes is referred to as a "word." A byte might represent a letter, number, punctuation mark or other typographic symbol (e.g., ,, : , S, @, &, or !), or control character (e.g., carriage return, line feed, beginning/ending flag, or error check). The term generally is thought as designating a computer value consisting of eight bits, which technically is known as a "physical byte." For example, ASCII code makes use of an 8-bit byte, comprising seven information bits and one parity bit for error control. EBCDIC (IBM's invention) makes use of an 8-bit byte, with all bits being information bits. In some circles, a byte is called an octet. This is the case in the world of broadband networking, although an octet, more correctly, is a set of 8 bits, which may comprise more than one byte. A "logical byte," as opposed to a "physical byte," may comprise fewer than 8 bits, or more. A 4-bit byte is often referred to as a "nibble," proving that humor is pervasive, even in the world of computer code. A byte can consist of more than 8 bits, as is the case with Unicode, which involves 16-bit bytes. Unicode is a standard coding scheme used to accommodate complex alphabets such as Chinese and Japanese. Note that computer storage capacity is measured in bytes, while transmission capacity and speed (also called bandwidth) is measured in bits per second (bps). The exceptions to this rule are Fibre Channel, ESCON and other transmission standards which are used in Storage Area Networks (SANs); those standards measure bandwidth in Bytes per second (Bps). See also Bps and Byte Count.

Now, let's have some fun with numbers as we count bytes. The real fun starts when we get a Kilobyte (KB). The actual definition can be confusing, since there are two measurements. In the metric system, a kilobyte is 1,000 bytes, or 10 to the third power. In the computer world, things tend to be measured in binary terms – 1s and 0s. In other words, we work with powers of 2, rather than powers of 10. In binary terms, a kilobyte is 2 to the 10th power, or 1,024. It's not exactly 1,000 bytes, but it's pretty close. Following is a summary of sizes, in binary terms, noting that each step in the progression involves a multiplier of 1024, with the exception of googolbyte.

KB = Kilobyte = 1,024 bytes (2 to the 10th power)
MB = Megabyte = 1,048,576 bytes (2 to the 20th power)
GB = Gigabyte = 1,073,741,824 bytes (2 to the 30th power)
TB = Terabyte = 1,099,511,627,776 bytes (2 to the 40th power)
PB = Petabyte = 1,125,899,906,842,624 bytes (2 to the 50th power)
EB = Exabyte = 1,152,921,504,606,846,976 bytes (2 to the 60th power)
ZB = Zettabyte = approximately 1,000,000,000,000,000,000,000 bytes (2 to the 70th power)
YB = Yottabyte = 2 to the 80th power, i.e., approximately 10 with 80 zeroes
One googolbyte equals 2 to the 100th power bytes, i.e., approximately 10 with 100 zeroes bytes.

byte bonding From Wired Magazine. Byte Bonding occurs when computer users get together and discuss things that noncomputer users don't understand. When byte-bonded people start playing on a computer during a noncomputer-related social situation, they are "geeking out."

byte code The executable form of Java code that executes within the Java virtual machine (VM). Also called interpreted code, pseudo code, and p-code.

byte count The number of 8-bit bytes in a message. Since ASCII characters typically have 8 bits, the byte count is also called the character count.

byte count protocol A class of data link protocols in which each frame has a header containing a count of the total number of data characters in the body of the frame.

byte interleaving Byte interleaving is the process used in TDM (Time Division Multiplexers) in applications such as digital voice. Using PCM (Pulse Code Modulation) and T-1 TDM as an example, 24 uncompressed voice conversations can be supported over a single T-1 facility. The amplitude (i.e., volume level) of the native analog voice signals are sampled 8,000 times a second by a codec (coder/decoder). Each sample is expressed in the form of an eight-bit byte. The bytes are interleaved, beginning with a byte from conversation #1, then conversation #2, and so on through conversation #24. The process continues over and over again, always in the same order. On the receiving end, the process is reversed, and the analog voice signals are reconstructed. See also Byte, PCM, T-1, and TDM.

byte multiplexer channel An IBM mainframe input/output channel that allows for the interleaving, or multiplexing, of data in bytes. Compare with block multiplexer channel.

byte multiplexing A byte (or value, or character) from one channel is sent as a unit, with bytes from different channels following in successive time slots. In this manner the bytes are interleaved. See Multiplex.

byte stuffing The process whereby dummy bytes are inserted into a transmission stream so that the net data transmission rate will be lower than the actual channel data rate. The dummy bytes are identified by a single controlling bit within the byte.

byte synchronous A SONET term describing the manner in which channelized DS-0 level data is mapped into the SONET frame. There are two approaches: Floating Mode and Locked Mode. Floating Mode requires that each frame in the originating T-1 frame be identified in order that the constituent DS-0 bytes can be mapped into the SONET frame effectively through alignment with SONET bytes. The individual T-1 frames, which are converted into VT1.5s as they are mapped into the SONET frame, can float within the SONET frame; this approach is highly effective, but complex and expensive. Locked Mode mapping involves the direct mapping of a T-1 frame into a VT1.5 established at a fixed and inflexible location within the SONET SPE (Synchronous Payload Envelope). This latter approach is much simpler and more efficient than is Floating Mode, but not as effective in maximizing the load balance of the SONET network. In any case, the timing of incoming DS-0 signals must be maintained as they enter the SONET network, are transported through it, and exit it to re-enter the T-carrier network digital hierarchy to be presented to the target device. Loss of timing renders the data useless. Timing is everything. Compare and contrast with Bit Synchronous. See also SONET and VT.

byte time Byte time is the time it takes to transmit a byte at wireline speed. Examples of Usage: Frame Relay is much faster than X.25: frames are switched to their destination with only a few byte times delay, as opposed to several hundred milliseconds delay on X.25. Late collisions (of Ethernet packets) are detected by the transmitter after the first "slot time" of 64 byte times. They are only detected during transmissions of packets longer than 64 bytes. Its detection is exactly the same as for a normal collision; it just happens "too late."

byte timing circuit Optional X.21 circuit used to maintain byte or character synchronization.

bytecode A Java compiler creates platform-independent bytecode, an intermediate data format between the source code written by a programmer and the machine code required by the target computer. The bytecode then is morphed into machine-specific executable code by the a Java Virtual Machine (JVM), a real-time interpreter that resides on the client computer. Bytecode allows a Java applet to be executed on any machine that is running a Java Virtual Machine. See also Compiler, Interpreter, and Java Virtual Machine.

BZT Bundesamt fur Zulassungen in der Telekommunikation. The name of the German telecom approval authority. It was established in 1982 under the name of the Central Approvals Office for Telecommunications. The name was changed to BZT on March 10, 1992. It is currently based in Saabrucken.

C 1. A powerful programming language developed at Bell Labs by Brian Kernighan and Dennis Ritchie in the early 1970s. C has wide application, including central office switches and voice processing systems. C operates under Unix, MS-DOS, Windows (all flavors) and other operating systems.

2. Symbol for capacitance. See also Capacitance.

3. Symbol for Celsius, as in temperature.

C band A portion of the electromagnetic spectrum used heavily for both terrestrial microwave and non-terrestrial (i.e., satellite) microwave radio transmission. In terrestrial microwave applications, the C Band is in the range of 4-6 GHz. Satellites split the C Band into an uplink channel at approximately 6 GHz and a downlink channel at 4 GHz. Traditionally, C-Band satellites ranged in power from five to 11 watts per transponder, requiring receive antennas of five to 12 feet in diameter. The fleet of C-Band satellites is being gradually replaced by higher-powered (10-17 watt) satellites, which allow the size of the dish to be reduced to 90 inches in diameter. Traditional applications include voice communications, videoconferencing, and broadcast TV and radio. The large dish size and associated high cost of such dishes have contributed to their lack of popularity for TV reception by individuals; Ku band dishes largely have replaced them in support of DBS (Direct Broadcast Satellite) TV reception. Contrast with KU Band and KA Band. See also C-Band.

C band dish Large (6- to 10- foot) satellite dish antenna, usually motorized, used to intercept signals from C band satellites. Many big dish antennas today receive both C band and Ku band signals.

C battery A source of low potential used in the grid circuit of a vacuum tube to cause operation to take place at the desired point on the characteristic curve.

C bit Signaling and control bits used in certain T-carrier systems. Classic M13 multiplexers serve to multiplex 28 T-1s into a T-3 signal. M13 multiplexers accomplish this process by first combining four T-1s, at a signaling speed of 1.544 Mbps, into a T-2, at a signaling speed of 6.312 Mbps. At this stage, 136 Kbps of overhead is added, including justification, or bit stuffing. (Note that 1.544 Mbps x 4 = 6.176 + .136 = 6.312). At the next stage, seven T-2s, are combined into a T-3, at a total signaling speed of 44.736 Mbps, with 552 Kbps of bit stuffing. (Note: 6.312 x 7 = 44.184 + .552 = 44.736) The stuff bits added during this last stage are known as C Bits, as A & B Bits are used in older channel banks for signaling and control purposes at the T-1 level. C Bits are used for a variety of signaling and control purposes, including synchronization, and parity checking for error control. See also A & B Bits.

C BIT parity framing T-3 framing structure that uses the traditional management overhead bits (X,P,M,F), but differs in that the control bits (C bits) are used for additional functions, e.g. FID, FEAC, FEBE,TDL and CP. See C Bit for a longer explanation.

C conditioning A type of line conditioning which controls attenuation, distortion and delay so that they lie within specified limits. See Conditioning.

C connector Also called a female amp connector or 25-pair female connector. The male version is called a P connector.

C drop clamp Clamp used to fasten aerial wire to buildings.

C lead The third of three wires which make up trunk lines between central office switches. There are three wires – positive, negative, and the "c lead." The purpose of the "c lead" is to control the grounding, holding and releasing of trunks.

C links An SS7 term. Cross links used between mated pairs of Signal Transfer Points (STPs). They are primarily used for STP to STP communications or for network management messages. If there is congestion or a failure in the network, this is the link that the STPs use to communicate with each other. See A, B and D Links.

C message weighting This definition from James Harry Green, author of the excellent Dow Jones Handbook of Telecommunications. C Message Weighting is a factor in noise measurements to approximate the lesser annoying effect on the human ear of high and low-frequency noise compared to mid-range noise.

C plane The control plane within the ISDN protocol architecture; these protocols provide the transfer of information for the control of user connections and the allocation/deallocation of network resources; C Plane functions include call establishment, call termination, modifying service characteristics during the call (e.g. alternate speech/unrestricted 64 kbps data), and requesting supplementary services.

C wire C wire is what the phone company calls the last piece of its wire that comes into your house or office. It is typically the piece of underground cable that comes in from its pedestal on the street to your network interface box on the side of your house or building.

C&C Computers and Communications. An NEC slogan which focused on the deployment of computer and telephony elements to create an integrated environment. Later on, NEC changed it to "Computing and Communicating" and expanded it into a "Fusion" strategy. See Fusion.

C++ A high-level programming language developed by Bjarne Stroustrup at AT&T's Bell laboratories. Combining all the advantages of the C language with those of object-oriented programming, C++ has been adopted as the standard house programming language by several major software vendors.

C-Band Conventional Wavelength Band. The optical wavelength band from 1490 to 1570nm (nanometers). The C-Band has been used by conventional WDM (Wavelength Division Multiplexing) optical fiber systems since about 1993. Most WDM and DWDM (Dense WDM) systems, therefore, currently make use of this band. According to ITU-T standards, the C-Band will support 8 optical channels at a frequency spacing of 400 GHz, or wavelength spacing of 3.2nm; 16 channels at 200 GHz, or 1.6nm; 40 channels at 100 GHz, or 0.8nm; 80 channels at 50 GHz, or 0.4nm; and 160 at 25 GHz, or 0.25nm. See also C Band, L-Band and S-Band.

C-Block Carrier A 30-MHz PCS carrier serving a basic trading area in the frequency block 1895-1910 MHz paired with 1975-1990 MHz.

C-DTE Character mode Data Terminal Equipment. A term to describe most PCs (personal computers) and printer-terminals that use asynchronous signals for data communications.

C-Link A signaling link used to connect mated pairs of Signal Transfer Points (STPs). An Ericsson term.

C-Message Weighting A type of telephone weighting network that allows for equal attenuation of all frequencies within the voice band in the same manner as it appears to be attenuated by the media.

C-Netz An older German analog mobile phone network now limited almost exclusively to car phones. This system was scheduled to be retired at the end of 2000.

C-Notched Noise The C-message frequency-weighted noise on a channel with a holding tone that is removed at the measuring end through a notch (very narrow band) filter.

C.100 Fremont, California, June 30, 1999 - The Enterprise Computer Telephony Forum (ECTF) announced the release of its latest Interoperability Agreement, C.100. This agreement specifies that certain packages of Java Telephony API (JTAPI) 1.3 constitute a portable, object-oriented, call control API that meet all ECTF interoperability requirements. C.100 (available for downloading from the ECTF web site at www.ectf.org) serves as a formal reference to the core, call center, call control, and private data packages of JTAPI 1.3 – see Sun Microsystem's JTAPI web site at http://java.sun.com/products/jtapi/index.html. According to Sun, "JTAPI was designed to be simple to implement. Application developers must still be knowledgeable about telephony, but they will not need implementation-specific knowledge to successfully develop their applications. It can be implemented without existing telephony APIs, but it was also designed to allow layering above APIs such as TAPI, TSAPI and others." The specific JTAPI packages covered by C.100 are: JTAPI Core, Call Center, Call Control and Private Data.

C/N C/N is abbreviation for Carrier-to-Noise Ratio, a measure of the intelligibility of a radio communications channel. See also decibel.

C.A.R.E. See CARE.

C.E.R.T. See CERT.

C/A Code The standard Clear/Acquisition GPS (Global Positioning Code) – a sequence of 1023 pseudo-random, binary biphase modulations on the GPS carrier at the chip rate of 1,023 MHz. Also known as the "civilian code." See GPS.

C/N Carrier-to-Noise Ratio, a measure of the intelligibility of a communications channel.

C/R Command Response. A Frame Relay term defining a 1-bit portion of the frame Address Field. Reserved for the use of FRADs, the C/R is applied to the transport of data involving polled protocols such as SNA. Polled protocols require a command/response process for signaling and control during the communications process.

C2 Command and Control. The exercise of authority and direction by a properly designated commander over assigned forces in the accomplishment of the mission. Command and control functions are performed through an arrangement of personnel equipment, communications, facilities, and procedures employed by a commander in planning, directing, coordinating, and controlling forces and operations in the accomplishment of the mission. See C3.

C3 Command, Control and Communications. The capabilities required by military commanders to exercise command and control of their forces. See C4.

C4 Command, Control, Communications, and Computers. Once it was C2, then C3, now C4. It basically refers to all the computers and telecommunications which the U.S. military needs to run itself.

C4I Command, control, communications, computers, and intelligence.

C4ISR Command, control, communications, computers, intelligence, surveillance and reconnaissance.

C7 European equivalent of the North American System Signaling 7. C7 is not 100% compatible with North American System Signaling 7 and that's where gateway and signaling conversion switches come in. These switches convert the signaling between one and the other and do it in real time. See Signaling System 7.

CA 1. Call Appearance.

2. Canada, as in a Web address, i.e. www.Corel.ca. Do not type Corel.com, thinking the CA is a mistake. It's not. See URL and Web address.

3. Certificate Authority. See Certificate Authority.

CAAGR Compound Annual Average Growth Rate.

cabinet 1. A container that may enclose connection devices, terminations, apparatus, wiring, and equipment.

2. In telecommunications, an enclosure used for terminating telecommunications cables, wiring and connection devices that has a hinged cover, usually flush mounted in the wall.

cable 1. May refer to a number of different types of wires or groups of wires capable of carrying voice or data transmissions. The most common interior telephone cable has been two pair. It's typically called quad wiring. It consists of four separate wires each covered with plastic insulation and with all four wires wrapped in an outer plastic covering. Quad wiring is falling into disrepute as it is increasingly obvious that it does not have the capacity to carry data at high speeds. The wire and cable business is immense. The assortment of stuff it produces each year is mind-boggling. In telecommunications, there is one rule: The quality of a circuit is only as good as its weakest link. Often that "weak link" is the quality of the wiring or cabling (we used the words interchangeably) that the user himself puts in. Please put in decent quality wiring. Don't skimp. See Category of Performance.

2. To send someone a telegram, as in "I cabled John with the good news that his mother-in-law had passed away."

Cable Act Of 1984 An Act passed by Congress that deregulated most of the CATV industry including subscriber rates, required programming and fees to municipalities. The FCC was left with virtually no jurisdiction over cable television except among the following areas: (1) registration of each community system prior to the commencement of operations; (2) ensuring subscribers had access to an A-B switch to permit the receipt of off-the-air broadcasts as well as cable programs; (3) carriage of television broadcast programs in full without alteration or deletion; (4) non-duplication of network programs; (5) fines or imprisonment for carrying obscene material; and (6) licensing for receive-only earth stations for satellite-delivered via pay cable. The FCC could impose fines on CATV systems violating the rules. This Act was superseded by the Cable Reregulation Act of 1992.

cable assembly A completed cable that typically is terminated with connectors and plugs. It is ready to install.

cable bays Lots of cable arranged like bays in a harbor.

cable bend radius Cable bend radius during installation infers that the cable is experiencing a tensile load.   Free bend implies a small allowable bend radius since it is at a condition of no load.

cable binder In the telephone network, multiple insulated copper pairs are bundled together into a cable called a cable binder. Each binder group contains 25 cable pairs, which are color-coded in order to make it easier to splice and terminate them properly. That way you won't get the tip and ring reversed, or connect your boss' phone to his administrative assistant's cable.

cable-blowing machine A machine that uses compressed air to install fiber cable into a pre-installed innerduct or direct-buried duct at speeds of up to 300 feet per minute.

cable bot See Denial of Service Attack.

cable budget A local area network term. The overall length of cable allowed between the DTEs located farthest apart within a common collision domain.

cable buffer The protective material used to coat the next layer up from the fiber coating. This is usually 900 0xCAm. Also referred to as buffered fiber. This layer is applied by the cable manufacturer.

cable buoy When a submarine cable breaks, a cable ship sails to the location of the fault, lowers a grappling hook to the sea floor, pulls up one loose end of the cable, attaches a buoy to it, and then releases the cable so that the ship can freely maneuver to find the other loose cable end on the ocean floor. The cable buoy, attached by a line to the first loose cable end that was located, enables the cable ship to easily find it again after it has located, pulled up, repaired and joined it to the second loose cable end.

Cabling Business Magazine A magazine on cabling run by Steve Paulov and family in Mesquite (Dallas), TX. A great magazine. 214-270-0860.

cable card See CableCard.

cable casters A dart gun which shoots a small string across suspended ceilings.

cable channel The number assigned to a television channel carried by a cable television system. Cable channels 2 through 13 are assigned to the same frequencies as broadcast channels 2 through 13; cable channels above Channel 13 are not. Cable channel assignments are specified in EIA Interim Standard EIA/IS-132, and are incorporated by reference into the FCC's cable television rules.

cable cleaner Orange citrus based cleaner. Great for cleaning up yourself and your tools after a messy splice job.

cable consortium A group of companies that club together to finance and build an international submarine cable.

cable converter box Equipment often provided by a cable company in a subscriber's home that allows access to cable TV services.

cable cut Service outage caused by cutting or damaging a cable.

cable cutoff wavelength For a cabled single-mode optical fiber, Cable Cutoff Wavelength specifies a complex inter-relation of specified length, bend, and deployment conditions. It is the wavelength at which the fiber's second order mode is attenuated a measurable amount when compared to a multimode reference fiber or to a tightly bent single-mode fiber.

cable diameter Expressed in millimeters or inches. Affects space occupied, allowable bend radius, reel size, length on a reel and reel weight. Also affects selection of pulling grips.

cable distribution hub The interchange point between the regional fiber network and the cable plant. At the hub, the cable modem termination system (CMTS) converts data from a wide-area network (WAN) protocol, such as packet over SONET (POS), into digital signals that are modulated for transmission over HFC plant and then demodulated by the cable modem in the home or business. The CMTS provides a dedicated 27 Mbps downstream data channel that is shared by the 500 to 1,000 homes served by a fiber node, or group of nodes. Upstream bandwidth per node typically ranges from two Mbps to ten Mbps.

cable dog Slang expression. In the West, lifelong cable installer who seeks no upward mobility. In the East, worker who deals with underground cable.

cable driver An amplifier, usually in a gain of 2, suitable for driving the low resistance of a double terminated cable. Load resistance = 150 ohms for video, 100 ohms for instrumentation.

cable drop The segment of cable that typically runs from the street or a telephone pole into the home.

cable entrance facility An entry point in a central office for outside plant cables.

Cable Europe The new name, as of September 2006, of the European Cable Communications Association. Cable Europe is an association of the leading European broadband cable communications operators and their national trade associations. Cable Europe's mission is to foster co-operation and the exchange of best practices among its members, and to promote and defend the European cable industry's policies and business interests through lobbying.

cable gland A type of connector. I cannot describe it. Go to this web site. There's a photo of one type of cable gland connector. www.josef-schlemmer.com/1PRODUCT.html.

cable head The point where a marine cable connects to terrestrial facilities. See cable station.

cable headend The cable headend connects the cable network with dishes that receive both satellite and traditional broadcast TV signals, and with cable modems linking to the Internet.

Cable Information Technology Convergence Forum This forum is a organization of the cable television, telephone, computer and switching network industries. The Forum was created to promote greater communication between vendors in the information technology industry, cable television companies and CableLabs, the research and development consortium serving most of the cable operators in North America. The Convergence Forum, based in Louisville, CO, was conceived and sponsored by CableLabs. Companies that have agreed to join the Forum include Apple Computer, Bay Networks, Cisco Systems, Compaq, Digital Equipment Corporation, Fore Systems and LANCity. See also CableLabs.

cable Internet Internet access via traditional cable television networks with upstream capability supported by either telco return (in nonupgraded one-way cable plants) or RF return (in upgraded two-way cable plants).

cable jacket See Sheath.

cable landing Where a submarine cable comes ashore. See cable station.

cable landing station Just what it sounds like. The buildings where undersea cables begin and end.

cable loss The amount of radio frequency (RF) signal attenuated (lost) while it travels on a cable. There are many reasons for cable loss, including the cable's shape, its type, its size, its length and what it's made of. For coaxial cable, higher frequencies have greater loss than lower frequencies and follow a logarithmic function. Cable losses are usually calculated for the highest frequency carried on the cable. See Attenuation.

cable management Companies have oodles of telephone cables in and around their buildings. Cable management is the science and art of managing those cables. Typically, cable management covers keeping track of where the cables are (maps are useful),

what type and quality the cable is, and what is attached to either end of it. Cable management for corporations is critical, since stringing, laying and snaking of new cable can be inordinately expensive.

cable mapping Cable mapping is the task of trying to track every single pair of wire or circuit from beginning to end. You will need to know where all cables reside, not just the circuits that are in use. Cable mapping is critical for any organization – from company to university – which has a lot of cables floating around. Installing more of it – when there are plenty of spare pairs – is stupid and expensive. Thus, the need for cable mapping.

cable mile Also known as sheath mile. The measurement, in miles, of fiber optic cable that is deployed. Contrast with fiber mile and route mile.

cable modem A cable modem is a device that will let you transmit and receive computer information over your cable TV line – just as a phone modem will let you transmit and receive computer information over your dialup telephone line. A dial-up modem on your local analog phone line will provide online Internet access through the public telephone network at up 53,000 bits per second. A cable modem will give you Internet access through your cable TV network at more than one million bits per second, or about 20 times faster. When a cable modem unit is installed next to your computer, a splitter is placed between the coaxial cable coming in from CATV provider. One side of the splitter will go to your cable set-top box and the other to your cable modem. Your cable modem will connect to your computer through a standard 10Base-T Ethernet RJ-45 interface. Data is transmitted between the cable modem and computer at standard Ethernet local area network speeds of 10 million bits per second. You connect your cable modem directly to your computer using a standard Ethernet NIC (network Interface card) card – the exact same card you use to connect to your office's local area network. Here's how it works. A cable television system typically has 60 or more TV channels. Most of them are used for receiving channels like ABC, CBS, NBC, CNN, ESPN and HBO. In a typical installation, the cable company chooses one TV channel and sets it aside for data transmission. That one channel can deliver 27 million bits per second downstream and 10 mbps of upstream capacity. Typically the cable company organizes that capacity is shared by a cluster of homes or apartments. Because data traffic is bursty, several hundred cable modem users can surf roughly at the same time. If speeds begin to fall off due to heavy traffic (we're all sharing one line), the cable operator will eventually allocate more channel space. A device called a cable modem termination system (CMTS) is located at the local cable operator's network hub and controls access to cable modems on the network. Each of the cable modems have their own network numbers – just as each NIC card has its own network number. Your carrier's access software can turn off or on service to the your cable modem. Traffic is routed from the CMTS to the backbone of a cable Internet service provider (ISP), such as @Home or Road Runner, which, in turn, connects to the Internet. With newer cable modem systems, all traffic from the CMTS to the cable modem is encrypted to ensure privacy and security for users. Some cable modem ISPs, such as @Home and Road Runner use proxy and caching servers to store copies of popular Web sites closer to their subscribers. The upshot: A customer with a cable modem connection isn't forced to travel across the Internet. The problem: sometimes a cable modem subscriber can be fed an old web site. Several million cable modems have been installed in the U.S. and Canada. The hardware and software supporting those connections aren't always completely interoperable, or able to work together. If, for example, a cable company uses Motorola network equipment, at one stage only a Motorola modem can be plugged into it. To try to promote cable modem rollouts, as well as relieve technological confusion, CableLabs, an industry trade organization, drafted a standard for cable modem products in 1996 called DOCSIS, which stands for Data Over Cable Service Interface Specification. The standard was developed to ensure that cable modem equipment built by a variety of manufacturers is compatible, as dial-up modems are. CableLabs tests DOCSIS cable modems, stamping the products that pass the test "CableLabs Certified." My personal experience with cable modems has been excellent – far better than my experience with DSL lines. I've learned two things about getting cable modems to work. First, when you first install your cable modem, it's best to replace the entire cable from and within your house to the pole coming in from the cable company. The fewer splitters in that line the better and the newer the cable the better. Second, your cable modem and attached router will occasionally lock up. Before you call Repair, unplug the cable modem and the router from electricity, count to ten, then plug them back in again. 99% of the time, your system will then work fine. See DOSCIS and the next definition.

Cable Modem Termination System CMTS. To deliver data services over a cable network, one six-MHz television channel (in the 50-750 MHz range) is typically allocated for downstream traffic to homes and another channel (in the 5-42 MHz band) is used to carry upstream signals. A head end CMTS communicates through these channels with cable modems located in subscriber homes to create a virtual LAN connection. See also

Cable Modem, CableLabs Certified, and DOCSIS.

cable normal switch A mechanism incorporated into a consumer television receiver which allows the user to select the channel assignment plan. In older receivers, this mechanism is usually a physical switch; in newer receivers, it is usually incorporated as an option in the setup menu. All cable/normal switches allow two choices: "standard" (sometimes called "normal" or "off air") which tunes to the channel assignments used by broadcast stations for over-the-air transmission; and "cable" (sometimes called "CATV" or "STD") which tunes to cable channels. Many receivers also include a third option called HRC or Harmonically-Related Carriers.

cable operator As defined by the Cable Act, a cable operator is a CATV (Community Antenna TeleVision) system operator that provides video programming using closed transmission paths and using public rights-of-way. Not included are open video systems, MMDS (Multichannel Multipoint Distribution Systems), or DBS (Direct Broadcast Satellite). See also Cable Act of 1984.

cable plant A term which refers to the physical connection media (optical fiber, copper wiring, connectors, splicers, etc.) in a local area network. It is a term also used less frequently by the telephone company to mean all its outside cables – those going from the central office to the subscribers' offices.

cable programming service Any tier of cable television programming except: - The basic tier (see Basic Cable Television Service). - Any programming offered on a per-channel or per-program basis.

cable protection There are three basic types of protection in addition to standard plastic cladding:

ElectroMagnetic (EM) Shielding: Prevents passive coupling. EM shielding can be a metallic conduit or metal wrapping-with appropriate grounding-on the wires.

Penetration-Resistant Conduit: Used to secure the cable from cutting or tapping. Note, however, not all penetration-resistant conduits provide EM shielding.

Pressurized Conduit: Detects intrusion by monitoring for pressure loss. Fiber optic cable is extremely difficult to tap and if tapped, the intrusion can be detected through signal attenuation. But since fiber optic cable can be cut, penetration-resistant conduit is recommended to protect the cable.

cable racking Framework fastened to bays to support cabling between them.

cable radio Radio programming delivered over a cable TV operator's network. An example is WALN FM 92 in eastern Pennsylvania and Western New Jersey, which has been broadcasting over Service Electric Cable's cable TV network since 1974.

cable ready Label for consumer electronic devices, such as television sets and VCRs, that are designed to allow direct connection to a cable television network.

cable relief system A fancy way of saying that you have a communications system which can pump a lot of bits over some cables that weren't originally meant to carry that many bits. For example let's say you put digital subscriber line electronics on a standard phone line. That would be called a "cable relief system."

Cable Reregulation Act Of 1992 Cable Reregulation Bill 1515 passed Congress in October 1992, forcing the FCC to reregulate cable television and cable television rates (after the Cable Act of 1984 effectively de-regulated the cable TV industry). After the Act was passed, the FCC forced the industry to reduce its rates by 10% in 1993 and then again by 7% in 1994.

cable restraint system A cable organizer that allows a dozen or more cables to be organized, restrained, and secured in order to reduce clutter in a wiring closet, facilitate access to cables for maintenance and repair, and to prevent accidental power and/or data losses.

cable riser Cable running vertically in a multi-story building to serve the upper floors.

cable run Conduit used to run cables through a building. Also, path taken by a cable or group of cables.

cable scanner A device which tests coaxial, twisted-pair, and fiber-optic cable. It measures the length of a cable segment, tests for opens and shorts, and can report on the distance to the problem so the problem can be found and fixed. Many scanners also indicate if a cable segment has RFI or EMI.

cable segment A section of network cable separated by switches, routers, or bridges.

cable sheath A covering over the conductor assembly that may include one or more metallic members, strength members, or jackets. See Cable Shield.

cable shield A metallic component of the cable sheath which prevents outside electrical interference and drains off current induced by lightning.

cable ship A ship that is custom built to lay telecommunications cables on the ocean floor and haul them up for repair. At one stage the big phone companies, like AT&T, owned and ran a fleet of them. I'm guessing now the cable ships are owned and run by outside speciality companies, such as Kokusai Cable Ship Co out of Japan.

cable signal leakage Excessive levels of radio frequency (RF) energy that leak from cable television systems. Leak can cause interference to communications users, including safety service users such as aviation, police and fire departments. FCC rules specify the maximum RF leakage, and require that cable television systems be operated within certain guidelines.

cable station The on-shore facility where submarine cables arrive and terminate, and where they connect to overland backhaul circuits. A cable station is often a hardened facility, with most of its structure underground.

cable stripper 1. Tool used to strip the jackets off ALPETH and lead-jacketed telephone cable. Cable strippers include cable knives and snips.

2. A professional or amateur stripper who appears on X-rated, or community access channels. Quality varies widely. Pay is often non-existent.

cable telephony cable telephony is transmitting anything other than TV pictures over a cable TV system. That "anything" might be anything from a data connection to the Internet to simple, standard, analog voice phone calls – local, long distance and international. Typically transmitting anything other than TV over the standard coaxial cable CATV providers install at your house requires a cable modem. See Cable Modem.

cable television CATV. See CATV.

Cable Television Laboratories, Inc. See CableLabs.

Cable Television Relay Station See CARS.

cable TV See CATV.

cable type The type of cable used. Also called the media. Examples are coaxial, UTP (Unshielded Twisted Pair), STP (Shielded Twisted Pair) and fiber. Factors including cost, connectivity and bandwidth are important in determining cable type. Choosing cable is getting more and more complex. Our tip: Choose and buy well in advance of when you'll need it. The cable you want will not always be in stock.

cable vault Room under the main distribution frame in a central office building. Cables from the subscribers lines come into the building through the cable vault. From here they snake their way up to the main distribution frame. The cable vault looks like a bad B-movie portrayal of Hell, replete with thousands of dangerous black snakes. Cable vaults are prime targets for the spontaneous starting of fires. They should be protected with Halon gas, but usually aren't because some parts of the phone industry think Halon is too expensive.

cable weight Expressed in lbs per 1000 (without reel weight included). Affects sag, span and size of the messenger in aerial applications.

CableB2B CableB2B is a trademarked term for an initiative led by CableLabs to develop interoperable interface specifications to support the automation of B2B (Business-to-Business) communications between CATV operators and Internet content providers. The B2B message set is anticipated to include specifications for service availability and order management, provisioning, network management, and customer care. See also CableLabs.

CableCard CableCard is a small metal card, the size of the so-called PC card for laptops, that slides into a slot on the back of many new high-definition TV sets. The CableCard's simple mission is to eliminate your cable box. The card stores all the account information that used to be monitored by the box, like descramblers for your movie channels. Life without the large and price cable box is now simple. The coaxial cable from your wall plug directly into your new TV. You change channels using the TV's own remote control. Losing the box frees up one power outlet on your wall, one valuable input on the TV and one component's worth of space in your equipment rack or wall unit. Furthermore, if you ever move, waxes the New York Times, you won't have to learn how to use a new cable company's box. You'll operate the same TV using the same remote in the same way. Eliminating a detour through the cable box also spares your video signal an analog-to-digital conversion or two. As a result, your picture should be clearer and sharper. On top of all these advantages, it typically costs a lot less to rent a CableCard than a cable box – though the price for service is the same. When I wrote this definition in December, 2004, CableCard was only one-way. It receives information from the cable company, but it can't send information back to the cable company. This has several implications: First, you no longer receive the cable company's onscreen TV guide. Most CableCard TV sets (marketed as "Digital Cable Ready") have their own built-in channel guides, and so do hard-drive recorders like the TiVo. Second, you lose the ability to order pay-per-view movies with your remote control. You have to order them using your cable company's Web site or by calling its toll-free number. Third, today's CableCard can't handle video-on-demand services. (They're like pay-per-view movies, except that you can start a movie whenever you like, and even pause it while it plays.) By most estimates, two-way CableCards are at least two years away.

cablecast Cablecasting. Non-broadcast radio or television programming transmitted by a cable television system to its subscribers. Cablecast programming may be originated by the cable operator itself ("origination cablecasting" or "local origination") or by an access organization.

cableco Cableco stands for the cable TV company.

cablehead The point where a marine cable connects to terrestrial facilities.

CableHome CableHome is a trademarked term of CableLabs for a project to develop interface specifications to extend high quality cable-based (i.e., cable TV-based) services to network devices within the residence. The initiative addresses issues of device interoperability, QoS (Quality of Service), and network management. See also CableLabs.

CableLabs Cable Television Laboratories, Inc. A research and development consortium of cable television system operators established in 1988, CableLabs membership includes system operators in North, Central, or South America, and the Caribbean. CableLabs plans and funds research and development projects to help member companies and the cable industry take advantage of opportunities and meet challenges in the telecommunications industry. A good deal of emphasis is placed on digital cable and cable modem technologies. Current projects include CableModem (formerly known as DOCSIS), PacketCable, OpenCable, CableB2B, and CableHome. www.cablelabs.com. See also the definitions for the above projects.

CableLabs Certified CableModem CableLabs Certified CableModem is a trademarked term for a cable modem initiative previously known as DOCSIS, a project led by the MCNS (Multimedia Cable Network System Partners Limited), which consisted of leading CATV operators. The DOCSIS project, now spearheaded by CableLabs, the research and development organization supporting the CATV industry, was aimed at developing on behalf of the North American CATV industry a set of necessary communications and operations support interface specifications for cable modems and associated equipment. The project led to the deployment of HFC (Hybrid Fiber-Coax) cable television systems in support of high-speed bi-directional data transfer, as well as entertainment TV. Activities are in phases, with the interfaces to be addressed including Cable Modem to CPE Interface (CMCI), Cable Modem Termination System-Network Side Interface (CMTS-NSI), Operations Support System Interface (OSSI), Cable Modem Telco Return Interface (CMTRI), Cable Modem to RF Interface (CMRFI), Cable Modem Termination System-Downstream RF Interface (CMTS-DRFI), Cable Modem Termination System-Upstream RF Interface (CMTS-URFI), and Data Over Cable Security System (DOCSS). Overlaying the CableLabs Certified infrastructure is the PacketCable project, a set of interoperable interface specifications for delivering advanced, real-time multimedia services over two-way cable plants. PacketCable will use IP (Internet Protocol) technology in support of multimedia conferencing, IP telephone, interactive gaming, and a wide range of multimedia services. For more detail see CMCI, CMTS-NSI, OSSI, CMTRI, CMRFI, CMTS-DRFI, CMTS-URFI, and DOCSS. See also CableLabs and MCNS.

cablese Abbreviations and skeletonized language used in transmitting cables/telegrams in order to reduce transmission cost.

cablese captions A US State Department term referring to standard abbreviations that can be used in State Department cables/telegrams.

cableway An opening in a work surface that allows access to cords or cables from below, or mounting of an electrical receptacle or telephone jack. Cableways typically come with removable plastic grommets.

cabling The combination of all cables, wire, cords, and connecting hardware installed. A term used to refer collectively to the installed wiring in a given space.

Cabling Reference Panel (of the ACIF) CRP. One of the five reference panels of the Australian Communications Industry Forum, the CRP is responsible for the formulation of telecommunications cabling standards and management of the cabling industry licensing arrangements.

CABS Carrier Access Billing System. Basically, when a company has a network in a CLEC environment, other companies can use that network for their own traffic. The CABS system allows each company to bill the others for the balance of traffic which they passed.

CABS BOS Carrier Access Billing Specifications - Billing Output Specifications.

CAC 1. Carrier Access Code. The digits you must dial in North America to reach the long distance carrier of your choice. Those digits fit the following format 101X-XXX.

2. Customer Administration Center. A type of terminal used by a PBX user to maintain and troubleshoot his PBX.

3. Connection Admission Control is defined as the set of actions taken by the network during the call setup phase (or during the call re-negotiation phase) in order to determine whether a connection request can be accepted or should be rejected (or whether a request for re-allocation can be accomplished).

CACH Call Appearance Call Handling.

cache From the French "cacher," which translates "to press or hide," especially in terms of tools or provisions. In the context of computer systems and networks, information is cached by placing it closer to the user or user application in order to make it more readily and speedily accessible, and transparently so. At the same time, information which is cached places less strain on limited computer I/O (Input/Output) resources and limited network resources. Let's consider two specific definitions, the first of which relates to computer systems and the second of which relates to computer networks. Let's also consider a combination of the first two, in the context of the Internet.

1. In the context of a computer system, cache memory generally is a partition of SRAM (Static Random Access Memory). Since much of computing is highly repetitive or predictable in nature, and since solid state components (silicon chips) are much faster than mechanical disk drives, the speed of information access can be enhanced if certain information can be stored in RAM. That information typically is in the form of program information, memory addresses, or data. Thereby, the information can be stored in anticipation of your need for it, and can be presented to you faster than if the computer needed to access the hard drive through the execution of an I/O function. The cache memory sits (logically and, perhaps physically) between the CPU and the main memory (RAM). Caching works because of a phenomenon known as the locality principle which states that a von Neumann CPU (i.e., one that performs instructions and makes database calls sequentially, one after another) tends to access the same memory locations over and over again. A cache works like this. When the CPU needs data from memory, the system checks to see if the information is already in the cache. If it is, it grabs that information; this is called a cache hit. If it isn't, it's called a cache miss and the computer has to fetch the information by accessing the main memory or hard disk, which is slower. Data retrieved during a cache miss is often written into the cache in anticipation of further need for it. Let's assume that you open a CD-ROM application with hyperlinks. As the system can reasonably assume that you will exercise the hyperlink options, the information associated with them can be stored in cache memory. If you do, indeed, exercise those options, it's a cache hit and the data is there waiting for you. The cache also will hold information that you recently accessed, in anticipation of your wanting to back up, or access it again. Caching can take place through partitioned or segmented cache memory, which can be in the form of L1 (Level 1) primary cache and L2 (Level 2) secondary cache. L1 cache memory is accessed first, L2 second, the main memory (RAM) and then hard drive last. Also, one cache might hold program instructions and the other might hold data. Generally when the cache is exhausted, it is flushed and the data is written back to main memory, to be replaced with the next cache according to a replacement algorithm. The cache freshing and flushing mechanism is designed differently by different vendors. It behaves slightly different. However it mainly depends on main memory type, like write back or WB, write through or WT, write protected or WT, write combining or WC and uncached or UC. See also Cache Memory.

2. In the context of a computer network such as a LAN, or the combination of the Internet and World Wide Web, data can be cached in a server which is close to you. In anticipation of your imminent request for that data in a logical sequence of data access, it will be transmitted from the main server to the remote server. Thereby, the data is accessible to you more quickly than if it had to be transmitted across the entire network each time you had a need for it. Should you access a certain set of data frequently, it might be permanently stored on a server in proximity, and refreshed by the main server from time to time in order to ensure its currency (i.e. that it remains up to date).

3. In the context of an Internet client/server application, caching really shows its stuff. First, the network uses distributed cache servers to house the WWW information that users in your region use frequently. As you access a Web site, your speed of access and response is improved because the data is housed on a server closer to you. The data then is loaded into cache memory on your client computer workstation. As you move forward, from page to page and link to link, your client caches the information provided by the cache server, with all of this happening in anticipation of your next move. As you move backward, the same thing happens, in anticipation of that next move, as well. Just in case you don't believe the client side of this story, go to Internet Explorer or Netscape, and click on cache. (The fastest way to regain space on your hard disk is to flush the cache which these programs dump to your hard disk.) For detailed explanations of specific caching protocols, see also CARP, HTCP, ICP, NECP, Squid, WCCP, WPAD, and WREC.

Cache Array Routing Protocol See CARP.

cache coherency Managing a cache so that data is not lost or overwritten. See also Cache.

cache controller A chip, such as the Intel 82385, that manages the retrieval, storage and delivery of data to and from memory or the hard disk. Cache controllers may reside in either clients or servers. See also Cache.

cache engine A cache engine is a carrier-class, high-speed dedicated Internet appliance that performs Web content caching and retrieval. When a user accesses a Web page, the cache engine locally stores the page's graphics and HTML text. When another user later requests the same Web page, the content is pulled from the cache engine. This process improves download time for the user and reduces bandwidth use on the network. Here is a an explanation of a cache engine from Cisco, which makes one. How does the cache engine work? The cache engine communicates with a Cisco router, which redirects Web requests to the cache engine using the Web Cache Control Protocol (WCCP), a new standard feature of Cisco IOS software. The WCCP also enables load balancing of traffic across multiple cache engines and ensures fault-tolerant, fail-safe operation. What are the benefits of Web caching? By reducing the amount of traffic on WAN links and on overburdened Web servers, caching provides significant benefits to ISPs, enterprise networks and end users. Those benefits include cost savings due to a reduction on WAN usage and dramatic improvements in response times for end users. The cache engine also provides network administrators with a simple method to enforce a site-wide access policy through URL filtering. See also Cache.

cache hit When the data you want is actually in cache. Thus you don't have to access your hard disk and your computing is faster. See Cache, Cache Miss and Cache Memory.

cache memory Available RAM (Random Access Memory) or SRAM (Static RAM) that you set up to allow your computer to "remember" stuff – so the next time your computer wants that information, it can find it fast from RAM, instead of searching through a slower hard disk I/O (Input/Output) process. This high speed cache memory eliminates the CPU wait state. When the CPU reads data from main memory, a copy of this data is stored in the cache memory. The next time the CPU reads the same address, the data is transferred from the cache memory instead of from main memory. Novell's NetWare, for example, uses cache memory to improve file server access time. In NetWare, cache memory contains the directory and file caches, along with the FAT (File Allocation Table), the turbo FAT, the Hash table, and an open space for other functions. See also Cache.

cache miss When the caching software guesses wrongly and you have to read your data off your hard disk rather than reading it from the cache in memory. In short, a cache miss occurs when a processor looks for data in cache memory and finds that it is not there. As a result, it needs to look to the main memory for the missing data, slowing operation. See also Cache, Cache Hit and Cache Memory.

caching Sometimes spelled cacheing. A process by which information is stored in memory or on the server in anticipation of the next request for information. See Cache for a full explanation.

CAD 1. Computer Aided Dispatch.

2. Computer Aided Design. A computer and its related software and terminals used to design things. A CAD system might be as simple as computerized drafting tools or as complex as detailed layouts of integrated circuits. CAD systems often have terminals on peoples' desks and a central maxi-computer in the company's main computer room. CAD terminals are often run over LANs (local area networks) or through telephone systems. The terminals are often moved, thus having universal wiring and a universal switching system – a LAN or a phone system – is extremely useful.

CAD/CAM Computer Aided Design/Computer Aided Manufacturing. See CAD.

Cadabra In October 1994, Jeff Bezos wanted to name his new Web venture "Cadabra" – as in "abracadabra." But his attorney convinced him that this magical moniker sounded a bit too much like "cadaver." Reluctantly, Bezos went with his second choice: Amazon.com.

cadaver See Cadabra.

CADB Calling Area Data Base. An MCI definition. An MCI System that stores reference data for various MCI Systems and reconciles MCI Calling Areas with those of Bell.

caddie When Mary, later Queen of Scots, went to France as a young girl (for education & survival), Louis, King of France, learned that she loved the Scottish game "golf". So he had the first golf course outside of Scotland built for her enjoyment. To make sure she was properly chaperoned (and guarded) while she played, Louis hired cadets from a military school to accompany her. Mary liked this a lot and when she returned to Scotland (not a very good idea in the long run), she took the practice with her. In French the word cadet is pronounced 'ca-day' and the Scots changed it into "caddie".

caddy The shell of an optical disc. Protects it from grubby fingerprints, and includes write protection devices. AKA case.

cadence In voice processing, cadence is used to refer to the pace and rhythmic pattern of tones and silence intervals generated by a given audio signal. Examples are busy signals and ringing tones. A typical cadence pattern is the US ringing tone, which is two seconds of tone followed by four seconds of silence. Some other countries, such as the UK, use a double ring, which is two short tones within about a second, followed by a little over two seconds of silence. See also Ring Cadence and Ring Cadence Acceptance.

CADS Code Abuse Detection System.

CAE Computer Aided Engineering.

cafeteria officing When an employer gives a worker the choice of setting up an office (desk, chair, computer) at home or at a central office. See Hoteling.

cage antenna An antenna having conductors arranged cylindrically.

CAGR Compound Annual Growth Rate.

CAI 1. Computer Assisted Instruction. Commonly known as CBT (Computer Based Training). See CBT. See also CAD for a discussion on telecom needs.

2. Common Air Interface. A standard for the interface between a radio network and equipment. A CAI allows multiple vendors to develop equipment, such as radio terminal devices (e.g., cordless phones, cellular phones and PCS terminals) and base stations (e.g., cellular antenna sites), which will interoperate. The yield is a competitive (read less expensive) market for equipment. The British CT2/Telepoint system incorporated one of the first CAI standards. See also CT2.

CALA Central America / Latin America.

CALC Carrier Access Line Charge. A per minute charge paid by long distance companies to local phone companies for the use of local networks at either or both ends of a long distance call. This charge goes to pay part of the cost of local telephone poles, wires, etc. See Access Charge and Carrier Common Access Line Charge.

CALEA Communications Assistance to Law Enforcement Act. Passed in 1994, CALEA is a U.S. law granting law enforcement agencies the ability to wiretap newer digital networks. The act also requires both wireline and wireless carriers to enable such wiretapping equipment. See DCS1000.

calendar routing A call center term for directing calls according to the day of the week and time of day. See also Source/Destination Routing, Skills Based Routing and End-of-Shift Routing.

calibrate To test and reset a measuring or timing device against a standard to make sure it is functioning correctly.

caliph The caliph is a concept. The caliph is the leader of the umma, the Islamic community of believers, i.e. all religious Muslims. It is a global idea. There have been rulers who considered themselves the caliph, but no true caliphate has ever been established.

caliphate An Islamic caliphate ruled the Middle East for over six hundred years, until the thirteenth century. See caliph.

call Everyone has a different definition for "call." My definition is simplest: Two people or two machines are on a phone line speaking to each other. That's a call. This definition by Bellcore (now Telcordia Technologies) of a call: An arrangement providing for a relation between two or more simultaneously present users for the purpose of exchanging information. The ATM Forum's definition: A call is an association between two or more users or between a user and a network entity that is established by the use of network capabilities. This association may have zero or more connections. Here are some more formal definitions:

1. In communications, any demand to set up a connection.
2. A unit of traffic measurement.
3. The actions performed by a call originator.
4. The operations required to establish, maintain, and release a connection.
5. To use a connection between two stations.
6. The action of bringing a computer program, a routine, or a subroutine into effect, usually by specifying the entry conditions and the entry point.

call abandons Also called Abandoned Calls. Call Abandons are calls that are dropped by the calling party before their intended transaction is completed. The call may be dropped at various points in the process. The point in the call at which the call is abandoned will have varying impacts on a computer telephony system. Many callers upon hearing an automated system will hang up. For systems that expended significant energy in setting up to answer a call, a large percentage of call abandons can negatively impact the call capacity of the system.

call accepted signal A control signal transmitted by the called equipment to indicate that it accepts an incoming call.

call accounting system A computer, a magnetic storage device (floppy or hard disk), software and some mechanical method of attaching itself to a telephone system. A call accounting system is used to record information about telephone calls, organize that information and upon being asked, prepare reports – printed or to disk. The information which it records (or "captures") about telephone calls typically includes from which extension the call is coming, which number it is calling (local or long distance), which circuit is used for the call (WATS, MCI, etc.), when the call started, how long it lasted, for what purpose the call was made (which client? which project?). A call accounting system might also include information on incoming calls – which trunk was used, where the call came from (if ANI or interactive voice response was used), which extension took the call, if it was transferred and to where and how long it took.

There are 12 basic uses for call accounting systems:

1. Controlling Telephone Abuse. It's the 90-10 rule. 10% of your people sit on long distance calls all day to their friends and family. The others work. Some people still think WATS calls are free. Knowing who's calling where and how much they're spending is useful. Often they appreciate being told they're spending money. Big money...and they stop.

2. Controlling Telephone Misuse. I figured once you could call between two major cities for five cents a minute and $1 a minute. That's a 20-fold difference! Often you need different lines. Often a company has different lines. Sometimes the phone system makes the dialing decision. Sometimes the person makes the dialing decision. Whoever's doing it can be wrong. A call accounting system is a good check to see if you're spending money needlessly.

3. Allocating telephone calling costs among departments and divisions. Telephones – voice, data, video and imaging – are some of your biggest expenses. They're a cost that should be allocated to the products you're making, or the departments or divisions in your company. Telephone costs can determine which product is profitable. Which isn't. Item: A software company recently dropped one of its three "big" software packages because phone calls for support got too expensive.

4. Billing Clients and Projects back for telephone charges incurred on their behalf. Every lawyer, government contractor, etc. does it. Makes sense.

5. Sharing and Resale of long distance and local phone calls, as in a hotel/motel, hospital, shared condominium, etc. Someone's got to send out the bills. And it's not the phone company. In fact, with a call accounting system you can be your own phone company!

6. Motivation of Salespeople. The more phone calls they make the more they sell. This rule is as obvious as the nose on your face. You WANT salespeople to make more calls? Hang a list of all their calls on the wall. Give prizes to those who make the most! Or those who make more than last week. Or those who set a new record.

7. Customer Service Measurement. An auto dealer wanted to know how long it took after a caller left a voicemail until the call was returned, so a report was created by the nice folks at www.CallAccounting.com that captured the ANI of the caller and looked for an outgoing call with that same number. It was called the "Elapsed Time To Return Calls" report. Management can now monitor the efficiency of their reps in returning calls from prospects.

8. Personnel Evaluation. Which employees are doing better at being productive on the phone (however you define "productive"). You want them to get on and off the phone fast? Or you want them to stay on and coddle your customers? You can now correlate phone calls with income – from service or just straight sales.

9. Network Optimization. Two fancy words for figuring which is the best combination of MCI, AT&T, MCI, Sprint, Wiltel, etc. lines. And which is the best combination of all the various services each offer. A rule of thumb: There's a 20-fold difference in per minute telephone calling costs between any two major cities in the US. And – amazing – you won't hear any difference in quality, despite the huge difference in price. I think it's the biggest price difference in any product anywhere. It's amazing.

10. Phone System Diagnostics. Is the phone system working as well as it should? Are all the lines working? Are all the circuit packs (circuit cards) working? Call accounting systems can tell you which lines you're getting no traffic on. Or which line carried the 48-hour call to Germany (it's happened). Either way, you can figure quickly which lines are working and which aren't.

11. Long Distance Bill Verification. Was the bill we received from our chosen long distance phone company accurate? Mostly it isn't. In fact, there's no such thing as an accurate phone bill. That's an oxymoron. Using your call accounting systems to check your long distance gives you some peace of mind. It's cheap peace of mind. Everyone should have one.

12. Tracing Calls. True story: Every third or fourth Friday afternoon a large factory in the south received bomb threats. They'd clear the factory, search the factory and not find anything. By the time they'd checked, it was too late to start up production. One day they checked their call accounting records. The calls were coming from a phone on the factory floor. The whole thing was a ruse to get an afternoon off...And now that many phones give you the number of who's calling, call accounting systems are turning out to be great for checking the effectiveness of regional ad campaigns, figuring the profitability of direct mailings and even figuring the profitability of individual customers.

Call accounting systems often are in the form of a module of a much more comprehensive system which typically includes modules for directory/personnel management, inventory management, cable and wire (i.e., connectivity) management, traffic analysis

or network optimization, and bill reconciliation. Such systems commonly make use of a RDBMS (Relational DataBase Management System), upon which all modules rely for access to a common set of data. Thereby, a single point of data entry is provided, and all data is continuously synchronized. Note that call accounting addresses only voice calls. Network accounting software addresses the management of data traffic. See also Network Accounting.

call admission precedence An MPLS traffic engineering tunnel with a higher priority will, if necessary, preempt an MPLS traffic engineering tunnel with a lower priority. Tunnels that are harder to route are expected to have a higher priority and to be able to preempt tunnels that are easier to route. The assumption is that a lower-priority tunnel can find another path. See MPLS.

call announcement A telephone operator or person acting as a telephone operator can announce a call to the called party before putting the call through. All modern phone systems have this feature.

call answering The name for a central office based answering service, provided by your local phone company. The major advantage of this service is that if you're speaking on your line and another call comes through, it won't receive a busy, it will hear your melodic voice asking if you'd like to leave a message and it will take your message. Once the machine takes the message, it will put some sound on your phone line, which you'll hear next time you pick up. That sound will alert you to the fact that you have a message in your mailbox.

call agent See home agent and MGCP.

call attempt An attempt to make a telephone call to someone. Tally up call attempts and compare them to completions and you'll have some idea of corporate frustration and, thus, the need for more lines or more phone equipment. The measures in this in call attempts, calls answered, calls overflowed, and calls abandoned.

call back A security procedure in which a user dials in to access a system and requests service. The system then disconnects and calls the user back at a preauthorized number to establish the access connection. Same as dial-back. See also International Call Back.

call barring The ability to prevent all or certain calls from reaching to or from a phone.

call before dig A preventive maintenance measure in which signs are posted near buried cables advising people to phone before digging in the area.

call blending A phone system has a bunch of people answering and making calls. The calls are coming (say in response to an ad). The calls are going out, courtesy of a dialing machine (perhaps a predictive dialer). The idea is to keep the calls at a constant level. The idea is to blend incoming with outgoing calls. Some predictive dialers have call blending. Others don't. They need a dedicated workforce. Call blending automatically transfers staff members between outbound and inbound programs as call volumes change. Some predictive dialers let you choose which workstations will be used for call blending, to avoid training of every staff member.

call block 1. A name for an enhanced custom calling service, one of several known as CLASS services. Call Block helps you avoid unwanted calls by rejecting calls from a list of numbers you specify. Depending on the specifics of the LEC offering, the caller may get a message indicating that you are not accepting calls at the present time. Call Block does not work either for numbers outside your local calling area or for calls connected through an operator. See also Class.

2. A feature that allows the calling party to prevent the calling number from being transmitted and displayed on the Caller ID equipment of the called party. Call Block can be provided, as a matter of course, on all your outgoing calls, although you can override Call Block on a call-by-call basis in order to transmit your number. You might choose to do this in order to receive better service from an incoming call center. You can request Call Block from your LEC. You also can invoke Call Block on a call-by-call basis, typically by pressing *67 before placing an outgoing call–this feature also is known as "Cancel Calling Number Delivery."

call blocking 1. Check into a hotel. Dial a 0+ call. You're connected to an Alternate Operator Service company. But you know their rates may be high. You ask to be connected to AT&T or MCI, or whoever is the carrier of your choice. Sadly, the AOS cannot connect and neither can (nor will) your hotel's operator. This is called "Call Blocking." The FCC has barred the practice. But it continues. See also Call Splashing.

2. An AIN (Advanced Intelligent Network) service allowing the user to block calls to specific numbers or country codes. Also known as Call Control Service. Content Blocking, a variation on the theme, allows the blocking of calls to either all or specified 900 numbers. While contemporary PBXs and Electronic Key Telephone Systems have such capabilities, not all users enjoy the benefits of working behind such a system. Additionally, PBX systems typically provide the ability to block only all 900 numbers, although some such numbers might have legitimate application.

call by call A common feature setting on ISDN switching equipment made in North America. It is properly titled "Call-By-Call Service Selection". This feature enables a single ISDN-PRI trunk group to carry call traffic to more than one facility or service. Rather than dedicating an entire ISDN-PRI trunk group to a single service, Call-By-Call Service Selection allows the customer to run various multiple services, such as ACCUNET, SDDN, MEGA800, etc., over a single trunk group. This can reduce costs and lower the chances of blocked services. Note: Call-By- Call Service is only available on ISDN PRI network using Country Protocol option 1 (U.S.).

call card A British term. A paper record, used in manual telebusiness systems, to record the results of a call.

call center A place where calls are answered and calls are made. A call center will typically have lots of people (also called agents), an automatic call distributor, a computer for order-entry and lookup on customers' orders. A Call Center could also have a predictive dialer for making lots of calls quickly. The term "call center" is broadening. It now includes help desks and service lines. Alcatel wrote a piece on them: "Call centers are the heart and soul of many businesses. They find new customers, create revenue, and help fix the problems of existing customers. Call centers are divided into two types, inbound and outbound. An inbound call center receives calls from current and potential customers for various reasons such as technical support, customer service, response to special offers, and order processing. Outbound call centers are generally geared to solicit new business (and/or collect outstanding bills). However, regardless of the function of the call center, they all have one thing in common, a centralized system to manage their internal workings. The centralized system to manage a call center will contain some or all of the following services to help customers reach agents and help agents manage their calls: + Specialized routing capability known as automatic call distribution (ACD), and (specialized commands for agents) including Log-on/log-off, Idle, In conversation, Wrap-up, Unavailable, Pause, Intrusion, Help and Silent monitoring.

Some larger call centers use more advanced technology such as interactive voice response (IVR), and multimedia capabilities such as co-browsing with a customer service specialist or engaging in a chat session. Routing and presenting of customer information to an agent has also become more sophisticated because of advances in computer telephony integration (CTI). ACD technology routed and queued calls on a first in, first out (FIFO) basis. Advances in this technology now allow a call to be routed based on the caller's telephone number, account number or even the number of times the caller was put on hold.

CTI advances now allow a full view of a customer's buying patterns, past encounters and opportunities to improve service. Customer relationship management (CRM) software is now tightly integrated in today's modern call center. Advances in IVR speech recognition enable customers to engage in some level of self-service, which adds up to happier customers and lower call volumes that must be handled by an agent. Finally, a well designed reporting and business intelligence system will provide agents, supervisors, and business managers with useful customer information to provide better customer services, to understand buying patterns, and to determine scheduling and service level goals. Better tools leads to more productive agents and ultimately happier customers."

For more information on Call Centers, please read Call Center Magazine. www.CallCenterMagazine.com. See also Call Centre.

call centre A British term. An area in an organization where business is conducted by phone in a methodical and organized manner. Call centres are typically based on the integration of a computerized database and an automatic call distribution system. (Note that this definition contains British spellings.) See Call Center.

call clearing The process by which a call connection is released from use.

call clear packet An information packet that ends an X.25 communications session, performing the equivalent of hanging up the phone.

call collision 1. Contention that occurs when a terminal and a DCE simultaneously transfer a call request and an incoming call specifying the same logical channel. The DCE will proceed with the call request and cancel the incoming call.

2. The condition that arises when a trunk or channel is seized at both ends simultaneously, thereby blocking a call.

call completion rate The ratio of successfully completed calls to the total number of attempted calls. This ratio is typically expressed as either a percentage or a decimal fraction.

call completion service An AIN (Advanced Intelligent Network) service which provides the Directory Assistance operator with the ability to automatically extend the call to the listed party. In a typical landline application, there is an additional charge to the calling party for such a service. (Frankly, I'd rather dial the number myself and save the 50 cents.) Cellular telephone providers often provide the service free of charge as a value-added service, in order to minimize "driving and dialing" traffic accidents. (I take advantage of this service–It just makes sense.)

call control Call control is the term used by the telephone industry to describe the setting up, monitoring, and tearing down of telephone calls. There are two ways of doing call control. A person or a computer can do it via the desktop telephone or a computer attached to that telephone, or the computer attached to the desktop phone line (i.e. without the actual phone being there). That's called First Party Call Control. Third-party call control controls the call through a connection directly to the switch (PBX). Generally third-party call control also refers to the control of other functions that relate to the switch at large, such as ACD queuing, etc.

call control procedure Group of interactive signals required to establish, maintain and release a communication.

call control signal Any one of the entire set of interactive signals necessary to establish, maintain, and release a call.

Call Control eXtensible Markup Language See CCXML.

call data Call data refers to any data about a phone call that is passed by a switch to an attached computer telephony system. Call data is usually used by the computer telephony application to process the call more intelligently. Call data may be passed In-Band, over the same physical or logical link as the call – usually via tones, or Out-Of-Band, over a separate link – usually a serial link. Call data may also be passed as part of the data designed to control telephone networks, such as SS7 (Signaling System 7) links. In addition to information about the call, status about the call and even control over the call, can be available as part of the call data link services. Call data almost always includes what number dialed the call (ANI) and/or what number called (DNIS). More complex call data links used for "PBX integration" may also indicate why the call was presented (such as forwarded on busy), tell what trunk the call is coming in on or to pass message waiting on or off indications, and other functions. Full blown computer telephony links, such as are now being offered by many switching vendors, enhance the call data path, providing additional status information about calls. These links can even provide a level of call control to the attached computer telephony system. The above definition courtesy of Steve Gladstone, author of a great book called Testing Computer Telephony Systems, available from 212-691-8215.

call delay The delay encountered when a call reaches busy switching equipment. In normal POTS telephone service, the delay is considered OK if no more than one and a half percent of the calls are delayed by three seconds during the busy hour.

Call Detail Recording CDR. A feature of a telephone system which allows the system to collect and record information on outgoing and incoming phone calls – who made/received them, where they went/where they came from, what time of day they happened, how long they took, etc. Sometimes the data is collected by the phone system; sometimes it is pumped out of the phone system as the calls are made. Whichever way, the information must be recorded elsewhere – dumped right into a printer or into a PC with call accounting software. See also Call Accounting System.

call diverter 1. A device which when connected to a called telephone number intercepts calls to that number and connects them to a telephone operator or prerecorded message.

2. An ancillary device which is connected to a telephone line. The device will, when the called telephone rings, initiate a telephone call on another line to a different telephone number. The calling party may or may not be aware that his call has been diverted to another telephone.

call duration The time from when the call is actually begun (i.e. answered) to the instant either party hangs up. Call Duration is an important concept for traffic engineering.

call establishment The process by which a call connection is created.

call forward busy When your phone is busy, an incoming call is transferred to another number. That number might be one appearing on your phone system. It might be one at your home in the same city. It could even be in another city. Call Forward Busy can perform the same as Rollover Lines. I use Call Forward Busy to move calls from the first line of my residence to my second line, because my local phone company charges too much for Rollover Lines. (Don't ask why. They don't know either.) You can get Call Forward Busy from central offices, as well as PBXs and some key systems. See also Rollover Lines.

call forwarding A service available in many central offices, and a feature of many PBXs and some hybrid PBX/key systems, which allows an incoming call to be sent elsewhere. There are many variations on call forwarding: Call forwarding busy. Call forwarding don't answer. Call forwarding all calls, etc.

Call forwarding is a useful feature. For example, you're going to a meeting but you're expecting an important call. Pick up your phone, punch in some digits and all your calls will go to the new number – perhaps the phone outside the meeting room. The big disadvantage is that many people return to their offices but forget they forwarded their calls elsewhere. As a result, they usually miss a whole bunch of important calls. Some electronic phones now have a reminder light or message on them saying "all calls are being forwarded." Some people program their PBXs to cancel all call forwards at noon and at midnight every day. This makes sense.

Call forwarding is used to send calls to voice mail systems. For example, tell your PBX that if your phone isn't answered in four rings, send that call to your voice mail.

If you are getting call forwarding service from a central office in North America, the code to begin call forwarding is #72 and the number you want to be forwarded to. To cancel it, you punch in #73.

call frame Harris' PBX to computer link. Harris' protocol for linking its PBX to an external computer and having that computer control the movement of calls within a Harris PBX. See also Open Application Interface.

call gapping A control application that limits the rate of flow to a specific destination code or station address.

call girls One of the many colloquial names give to the early female telephone operators. Others include Hello Girls, Central, and Voice with a Smile.

call guide A paper or screen "cheat sheet" that provides bullet points or actual copy for call center agents to use while they are on the telephone making marketing calls. They provide responses to commonly asked questions or objections in the most effective way. Call guides are excellent training tools as well as monitoring aids for coaches. See Call Guide Routing.

call guide routing The process by which a call center agent navigates through a call guide. The routing may be driven from a computer-based menu or function key or automatically by the computer system based on responses entered into a field.

call handoff A cellular phone term. Call handoff happens when a wireless call is transferred to another cell site in mid-conversation.

call hold If you hang the phone up, you lose the caller. Call hold – a feature of most phone systems – allows you to "hold" the call, so the other person can't hear you. You can then return to the conversation by pushing a button on your phone, typically the button flashing which shows which line the person is sitting on hold. Call hold is useful when you have someone on another line calling you.

call identifier A network utility that is an identifying name assigned by the originating network for each established or partially established virtual call and, when used in conjunction with the calling DTE address, uniquely identifies the virtual call over a period of time.

call in absence horn alert A cellular car phone feature that sounds your car's horn when you are receiving a call.

call in absence indicator A cellular car phone feature that ensures that power to the cellular phone is not lost if the car's ignition is turned off.

call in progress override A cellular car phone feature that keeps power to the phone during a call even though you've turned off the car's ignition.

call letters Certain combinations of letters assigned to radio stations by the FCC. The group of letters assigned the U.S. by the International Radiotelegraph Convention are all three and four letter combinations beginning with N and/ or W and all combinations of KDA to KZZ inclusive.

call mix Call mix is the pattern of call types (each call type defines what the caller will do for that call) that goes into creating a busy hour call profile or other call profile. A voice mail system's busy hour call profile call mix may consist of 10% call abandons, 20% login and send one message, 30% login and listen to one message, and so on. Varying the call mix can often be useful to stress particular parts of a system. For example, a call mix of 100% call abandons is frequently used to stress a computer telephony system's ability to handle high traffic call setup scenarios. This definition courtesy of Steve Gladstone, author of the book Testing Computer Telephony Systems, available from 212-691-8215.

call model An abstraction of the call processing functionality of the architecture and the relationship that exists between the functionality of the Service Switching FE in an ASC and the Service Logic and Control FE in a SLEE (Service Logic Execution Environment). The call model consists of two components: Connection View and Basic Call State Model. Definition from Bellcore (not called Telcordia Technologies) in reference to Advanced Intelligent Network.

call not accepted signal A call control signal sent by the called terminal to indicate that it does not accept the incoming call.

call notification service See Call Pickup Service.

call packet A block of data carrying addressing and other information that is needed to establish an X.25 switched virtual circuit (SVC).

call park The phone call is not for you. Or maybe it is, but you don't want to answer it on your phone. Put it into CALL PARK, then you or anyone else can answer it from any other phone. Call Park is similar to placing a call on hold, but you retrieve the call by dialing a code, rather than by pressing a line button. The attendant may have a call for you, but you're not there. So s/he places the call in Call Park, pages you and tells you the call is in Call Park. You pick up the nearest phone, dial one or two digits (the code for grabbing the call out of Call Park) and you have the call. It's faster than looking for you, then telling you to hang up while s/he transfers the call.

call pickup A phone is ringing but it's not yours. With call pickup, you can punch in a button or two on your phone and answer that person's ringing phone. Saves time. See Call Pickup Group.

Call Pickup Group CPUG. All the phones in an area that can be answered by each other by simply punching in a couple of digits. See Call Pickup.

call pickup service An AIN (Advanced Intelligent Network) service similar to those offered by many premise-based voice processors but residing on a network-based platform. In the event of an unanswered call, the voice processor records the call and notifies the called party of the message via pager, fax or some other technique.

call processing The system and process that sets up the intended connection in a switching system. The system scans the trunk and/or station ports for any "requests" for service. Upon detecting a request, the system checks the stored instructions and look-up tables and sets the connection up accordingly.

call processing language See CPL.

call progress The status of the telephone line; ringing, busy ring/no answer, voice mail answering, telephone company intercept, etc. See Call Progress Analysis and Call Progress Tone.

call progress analysis As the call progresses several things happen. Someone dials or touchtones digits. The phone rings. There might be a busy or operator intercept. An answering machine may answer. A fax machine may answer. Call progress analysis is figuring out which is occurring as the call progresses. This analysis is critical if you're trying to build an automated system, like an interactive voice response system.

call progress signaling All telephone switches use the same three general types of signals: + Event Signaling initiates an event, such as ringing. + Call Progress Signaling denotes the progress (or state) of a call, such as a busy tone, a ringback tone, or an error tone. + Data Packet Signaling communicates certain information about a call, for example, the identity of the calling extension, or the identity of the extension being called.

call progress tone A tone sent from the telephone switch to tell the caller of the progress of the call. Examples of the common ones are dial tone, busy tone, ringback tone, error tone, re-order, etc. Some phone systems provide additional tones, such as confirmation, splash tone, or a reminder tone to indicate that a feature is in use, such as confirmation, hold reminder, hold, intercept tones.

call queuing Incoming or outgoing calls may be queued pending an answer. The idea of call queuing is to save money. See also Callback Queuing.

Call Rate CR. The number of calls within a span of time, such as within an hour, or within a day, etc. It may be confined to a narrow usage, such as the busy hour (BH) originating call rate per main station, or to a broader usage. Hence, its usage should include enough modifying words to assure that it will be properly understood. A telephone company definition.

call record The data record of a call transaction. The record is made up of event details that typically include date, time, trunk(s) used, station(s) used and duration. In an ACD, these events may also include time in queue, call route used, system disposition flag, inbound or outdialed digits and wrap-up data entered.

call reference Information element that identifies to which call a Layer 3 message pertains.

Call Reference Value CRV. A number carried in all Q.931 (I.451) messages, providing a local identifier for a given ISDN call. Also called Call Reference Number.

call release time The time it takes from sending equipment a signal to close down the call to the time a "free condition" appears and the system is ready for another call.

call reorigination Caller reorigination allows a caller with a telephone debit card account number or a telephone credit card account to make unlimited calls without hanging up and redialing their access and their account numbers. At the end of the first call, the caller remains on the line. The caller then presses the pound key (the # key) for a prescribed number of seconds and receives a confirmation tone (which sounds like a high-pitched dial tone). After receiving the tone, the caller immediately dials their next phone number. And so on. Sometimes you can hit the # button after the person you were talking to has hung up. My experience has been that it's better to hit the # before the person has hung up.

Just tell them what you're doing. All this allows the card account holder to make a series of calls without ever hanging up and redialing the often lengthy access card account numbers. This saves a lot of time for callers with a long list of calls to make. It also saves money for callers from hotels, which charge for each connection to the long distance provider, but don't charge based on the length of the call (especially if it's a local call or toll-free 800 call). The prescribed number of seconds that the user needs to hold down the pound key is configurable. Depending on the application the time may range from 1 to 5 seconds. With most carriers, it's 2 to 3 seconds. The above definition was kindly provided by Karen Shelton, Systems Engineer, IEX Corporation, Richardson, TX.

call request packet In packet data switching, a call request packet carries information, such as sender and recipient identification, that is needed to establish an X.25 circuit. In more technical terms, a call request packet is sent by the originating data terminal equipment (DTE) showing the requested network terminal number (NTN), network facilities and either X.29 control information or call user data.

call restrictor Equipment inserted in a telephone line or trunk which restricts outgoing calls in some way. Usually from making a toll call.

call return An enhanced custom calling feature included in what are known as CLASS services and which are offered by local exchange carriers, courtesy of SS7. Call Return service which allows you to automatically dial the number of the last caller, even if you did not answer the telephone. It's a great idea, assuming that you are not blindly making long distance calls to long distance salespeople. Enhanced Call Return allows the called party to access a network-based voice system which announces the date, time and telephone number of the last incoming call. Should you chose to do so, you can launch call return by pushing a button on the telephone keypad.

call rotation A feature of the automated attendant of a phone system such that when some one is busy the phone system automatically connects the incoming call to someone else. The call progression is set up in a table or otherwise defined inside the software of the phone system.

call routing A list of choices set up within an ACD (Automatic Call Distributor) where to send incoming calls.

call routing tree A graphical display of complex call routing decision logic.

call screening There are several definitions. Here are two. 1. A PBX feature that looks at the digits dialed by the caller to figure whether the call should be completed.

2. A receptionist or secretary answers the executive's phone and checks that the person calling is important enough be put through to the almighty executive whose calls are being screened.

call second A unit for measuring communications traffic. Defined as one user making one second of a phone call. One hundred call seconds are called "ccs," as in Centum call seconds. "ccs" is the U.S. standard of telephone traffic. 3600 call-seconds = 1 call hour. 3600 call-seconds per hour = 36 CCS per hour = 1 call-hour = 1 erlang = 1 traffic unit. See also Erlang and Traffic Engineering.

call selector A local phone company service which alerts the subscriber with a distinctive ring that one of the six numbers your pre-selected is calling.

call sequencer A call sequencer, also called an Automatic Call Sequencer, is a piece of equipment which attaches to a key system or a PBX. The Call Sequencer's main function is to direct incoming calls to the next available person to answer that phone. It typically does this by causing lights on telephones to flash at different rates. The light with the fastest flashing is the one whose call has been waiting longest. This call is answered first. Call Sequencers also might answer the phone, deliver a message and put the person on hold. They might keep statistical tabs of incoming calls, how fast they were answered, how long the people waited, how many people abandoned (hung up while they were on hold waiting for their call to be answered by a human being), etc. Call Sequencers are usually simple and inexpensive. Better, but much more expensive devices for answering incoming phone calls are Automatic Call Distributors. These are the devices which typically answer when you call an airline. See Automatic Call Distributor.

call setup The first six PICs (Point In Call) of the Originating BCSM (Basic Call State Model), or the first four PICs of the Terminating BCSM. Definition from Bellcore (now called Telcordia Technologies) in reference to Advanced Intelligent Network.

call setup time The amount of time it takes for a circuit-switched call to be established between two people or two data devices. Call set-up includes dialing, wait time and time to move through central offices and long distance services. You don't pay for call set-up, but you will need extra lines to take care of it. See also Answer Supervision and Traffic Engineering.

call shedding In many states, laws require that a real-life breathing person be available for a phone call being outdialed with an automated device, e.g. a predictive dialer.

The reason these laws were enacted is because a lot of times call-center managers have their predictive dialers going so crazy in search of real-life people answering the phone that, when they fluke it and actually hit a person, they don't have an agent ready. Most systems simply hang up the connection when this occurs and they mark it down for later calling. This is called "call shedding" and is illegal in many states.

call spill-over In common-channel signaling, the effect on a traffic circuit of the arrival at a switching center of an abnormally delayed call control signal relating to a previous call, while a subsequent call is being set up on the circuit.

call splashing A "splash" happens when an Alternate Operator Service (AOS) company, located in a city different to the one you're calling from, connects your call to the long distance carrier of your choice in the city the AOS operator is in. Let's say you're calling from a hotel in Chicago. You ask AT&T to handle your call. The AOS, located in Atlanta, "splashes" your call over to AT&T in Atlanta. But you're calling Los Angeles. Bingo. Your AT&T call to LA is now more expensive than it would be – if you had been connected to AT&T in Chicago.

call splitting A feature allowing a phone user to speak privately with either party of a conference call by alternating between the two. Call splitting by an attendant allows the attendant to speak to the called person privately while effectively putting the calling person on hold, or vice versa.

call stalker An AT&T PC-based product which gives the 911 attendant the phone number and address of the person calling.

call store The temporary memory used in a stored program control switch (SPC) to hold records of calls in progress. These records are then transferred to permanent memory.

call stream British Telecom's premium rate service.

call tag A term used in the secondary telecom equipment marketplace. A ticket directing a freight carrier (e.g., UPS) to pick up equipment at another site. The company issuing the ticket pays the freight charge. Normally used to return defective equipment, it ensures the dealer a quick return and an accurate tracking mechanism.

call teardown The procedure of disconnecting a call between the central office and the subscriber.

call ticket A report maintained by a manufacturer in its Technical Support Database that contains pertinent information on a single technical support issue as reported by a customer. Information typically includes: call tracking number (CTN), customer contact information, system configuration information, customer issue description, actions taken by manufacturer, actions suggested to and taken by customer, and a record of all customer contact events regarding the technical support issue.

call trace A name for local telephone company service which permits the tracing of the last call received and holds the results for later use by an authorized law enforcement agency. (Results of the trace are not available to the customer.)

Call Tracking Number CTN. Unique ticket number issued by a manufacturer of hardware or software and used for tracking the continuing status of an ongoing technical issue – typically the customer is having a problem with his hardware or software.

call transfer Allows you to transfer a call from your phone to someone else's. On some phones you do this by punching in a bunch of numbers. Some you do it by hitting the "transfer" button and then the number you want to send the call to. The fewer buttons and numbers you have to punch, the easier it will be for your people.

If you're choosing a telephone system, check how easy it is to transfer a call. It is the most commonly used (and misused) feature on a phone system. How many times have you been told, "I'll transfer you to Mr. Smith, but if we get disconnected, please call back on extension 234." If your people are saying this to your customers or prospects, you are giving the outside world the wrong impression of your business. And since 97% of your prospects' contact with your company is first through your phone system, you could be losing precious business.

call type A call center term used in Rockwell ACDs. A portion of your call center traffic corresponding to one or more ACD gates or splits. This division of the total ACD traffic is the level at which forecasting and scheduling are done. At setup time, each Call Type is defined in the ACD by a unique three-letter code and specific gate or split number(s) that identifies the corresponding ACD report data.

call user data In packet data networking technology, user information transmitted in a call request packet to the destination data terminal equipment (DTE).

call wading Also known as Call Waiting, this feature allows you accept calls from, and talk to, additional callers during a conversation without any equipment other than an ordinary telephone. It is a CLASS feature for loop start lines. So called because once you have more than one call involved it becomes nearly impossible to know who you are talking to as you try to switch between (wade through) the calls. This feature is mostly called Call

Waiting. See Call Waiting and CLASS.

Call Waiting Call Waiting is a feature of phone systems that lets you know someone is trying to call you. You're speaking on the phone. A call comes in for you. You might hear a beep in your ear or see a light on your phone turn on. Or you might hear a beep and see a message come across the screen of your phone. When you hear the beep, you can, if you wish, put the present call on hold and answer the new one. You do this typically by hitting the touchhook on your phone (the cradle that sits under your handset.) Or you can ignore the new one, hoping it will go away, and perhaps send it to your attendant/operator, or voice mail. Call Waiting can be done manually by your telephone system operator. Or it can be a service which you buy monthly from your local phone company.

A major problem with call waiting is if you're on a data call from your PC, the call waiting "beep" will often cause your modem to hang up, thus destroying your data call. There are two solutions to this, the obvious one being to turn off call waiting. Some phone systems will allow you to turn it off. The way to turn it off is to include *70 in your modem dial string before you dial the phone number. That will tell your phone company (in most cases) to turn off the call waiting sound while you're on that (and only that) phone call. Another way is modify your modem's initialization string. Here's how. In all Hayes and Hayes-compatible modems, there's a S10 register. It tells the modem how long before it hangs up after losing carrier. In Hayes modems, the S10 register is set for 1.4 seconds. The typical call waiting tone is 1.5 seconds. Solution, increase the S10 register to six seconds (to be sure). Use your communications software. Go into terminal mode, then type: ATS10=60. You must put this command in every time you power up, because the Hayes 1200 modem (and others) have volatile memory. But the Hayes 2400 and higher speed asynchronous modems have non-volatile memory. They remember the six seconds after they've been switched off. The command to write this to memory is ATS10=60&W. The "&W" means write it to memory.

callback 1. A feature of some voice and data telephone systems. You dial someone. Their phone or computer is busy. You hit a button or code for "callback." When their phone becomes free, the phone system will call you and them simultaneously. You can only use this callback feature on things internally in your phone system – calling other people, calling long distance lines (which might be busy), calling the dictation pool, etc. See Call Waiting, Callback Modem and Callback Queuing.

2. A quick way of referring to international callback, which works thus: Calling the United States from many countries abroad is far more expensive than calling those countries from the United States. A new business called International Callback has started. It works like this. You're overseas. You dial a number in the United States. You let it ring once. It won't answer. You hang up. You wait a few seconds. The number you dialed in the U.S. knows it was you calling. There is a piece of equipment on that number that "hears" it ring and knows it's you since no one else has that number. (Typically it's done with Centrex service.) That was your special signal that you want to make a call. A switch attached to that line then calls you instantly. When you answer (overseas, obviously) it conferences you with another phone line in the United States and gives you U.S. dial tone. You can then touchtone from overseas your American number, just as if you would, were you physically in the U.S. There are huge savings. U.S. international callback operators can offer as high as 50% savings on calls from South America, where international calling rates are very high. The process of international callback is being automated with software and dialing devices. International callback is also helping to bring down the high cost of calling the U.S. from overseas. In recent years, deregulation has caused the price of international calls in many countries to fall dramatically. And now international callback or just callback is being done from other countries, including and especially Israel. A company called Kallback in Seattle, WA. has received a service mark from the U.S. Patent and Trademark Office for the words "callback" and "kallback" and sends letters to and threatens law suits against companies who use "their" words. See also Refile.

3. A security procedure in which a user, dials into access a system and request service. The system then disconnects and calls the user back at a pre-authorized number to establish the access connection. This capability is often implemented into a modem, known as a Callback Modem, surprisingly enough. Same as dial-back.

callback modem A modem that calls you back. Here's how it works. You dial into a network. A modem answers. You put your password in. It accepts the password. It says "Please hang up. I will now call you back." You hang up. It calls you back. There are two reasons for doing this instead of allowing you to just go straight into the network. 1. It's better security. You have to be at a pre-determined place – an authorized phone number. 2. It may save on phone calls. The modem uses the company's communications network, which is probably cheaper than what the person calling in can use.

callback queue The queue used to hold callers who have requested a busy pool or

extension. See Callback Queuing.

callback queuing An option on a telephone system which allows outgoing calls to be put in line for one or several trunks. When a trunk becomes available, the phone system calls the user, his phone rings and then the phone system dials the distant party on the trunk it grabbed before calling the user. Phone systems typically have two types of queuing. The first is called Hold-On Queuing. With this, the user dials his long distance number, the phone system searches for the correct trunk, finds it's not available and tells the user with a beep or message. The user then elects to stay on the line and wait. The instant the trunk becomes free, the phone system connects the user to it. The second type of queuing is called Callback Queuing. The user hangs up and the phone system calls you back, as we explained above.

There are tradeoffs between the two types of queuing. Callback queuing obviously can tolerate longer queues. The longer you wait, the more chance you have of reaching a very low-cost trunk. But users don't like waiting so long for a trunk. And when the call does come, it may likely reach a phone, newly-deserted by a user who's gone to the bathroom.

In contrast, hold-on queuing is more efficient of the user's time, but less efficient of the user's trunks. The less time you wait, the less chance you have of reaching a low-cost trunk. Life is a trade-off. Queuing is no exception. See also Queuing.

callbridge Rolm (now Siemens) open architecture interface. A method of connecting a Rolm CBX (Siemens telephone system) to an outside computer, so that the computer may "talk" to the PBX and make certain things happen, e.g. moving a screen of client information around simultaneously with the phone call from the client. This feature is especially useful in customer service and customer order-entry environments – for example with direct mail order catalog companies, etc. See Open Application Interface.

Called DTE A DTE which receives a call from another DTE.

called line identification facility A service feature provided by a network (private or public), which enables a calling terminal to be notified by the network of the address to which the call has been connected. See Caller ID.

called line identification signal A sequence of characters transmitted to the calling terminal to permit identification of the called line.

called party subaddress Information element that is passed transparently by the SPCS (if certain conditions are met) and can be used to further identify the destination party.

called party camp-on A communication system service feature that enables the system to complete an access attempt in spite of issuance of a user blocking signal. Systems that provide this feature monitor the busy user until the user blocking signal ends, and then proceed to complete the requested access. This feature permits holding an incoming call until the called party is free.

callender switch A very rudimentary, early telephone switching system developed by Romaine Callender and the Lorimer brother in the late 1800s.

caller ID Your phone rings. A name pops upon on your phone's screen. It's the name and number of the person calling you. Actually, it's the originating telephone number and the name the phone company thinks is the subscriber. The originating telephone number is stored in the originating central office equipment register, which is a database. That number supports a further database lookup, which associates the directory listing, assuming that the originating number is listed (i.e., not unlisted, or "nonpub" for nonpublished). The name and number information is passed through the local and long distance networks, and appears on your Caller ID box or your display telephone between the first and second rings. The delivery of Caller ID information assumes several things. First, the entire network of switches must be supported by SS7 (Signaling System System #7). Second, the calling party must originate the call from a single-channel line, rather than a multichannel trunk (e.g., T-1). Third, the originating line/caller must not block the transmission of the information. If all of these criteria are not met, your Caller ID box will display "ANONYMOUS" or "NOT AVAILABLE." Caller ID is one of several CLASS (Custom Local Area Signaling Services) provided by your LEC (Local Exchange Carrier). There generally is both a small installation charge and a monthly charge for Caller ID. Caller ID lets you amaze your parents and scare your technophobic friends, when you answer the phone with something like "Hi, Harry! Great Dictionary!" Caller ID also helps you avoid those dinnertime calls from telemarketers. They always block their numbers. By the way, Caller ID is not the same as ANI, although they often are confused. See also ANI, Caller ID Message Format (for a very detailed explanation), and CLASS.

caller ID message format Calling Number Delivery (CND) came about as an extension of Automatic Number Identification (ANI). ANI is a method that is used by telephone companies to identify the billing account for a toll call. Although ANI is not the service that provides the information for CID, it was the first to offer caller information to authorized parties. The CID service became possible with the implementation of Signaling System 7

(SS7). The CID information is transmitted on the subscriber loop using frequency shift keyed (FSK) modem tones. These FSK modem tones are used to transmit the display message in American Standard Code for Information Interchange (ASCII) character code form. The transmission of the display message takes place between the first and second ring. The information sent includes the date, time, and calling number. The name associated with the calling number is sometimes included also. Since the time CID was first made available, it has been expanded to offer CID on Call Waiting (CIDCW) as well. With CIDCW, the call waiting tone is heard and the identification of the second call is seen. In earlier editions of my dictionary, I included the complete formatting, down to individual bits. It's too technical for this dictionary. However, if you want the entire story in all its detail, go to http://www. testmark.com/develop/tml_callerid_cnt.html and read the article on "Caller ID Basics", by Michael W. Slawson of Intertek Testing Services, TestMark Laboratories. Michael has assured me that he will leave his excellent paper on the Web forever.

caller ID spoofing Your phone has a small screen. When your phone rings, your screen displays the phone number (and perhaps the name) of the person calling you. Now imagine if the person calling you had some software or hardware which changed the number that appeared on your screen to a number that didn't match their own number. This is known as caller ID spoofing. It's a technique that is employed by callers who want their identity concealed. Why? Many phones allow you to cut off incoming calls from people you don't want to hear from – bill collectors, telemarketers, mothers-in-law, etc. Bill collectors might want to use caller id spoofing to ensure their demanding calls get through and they get their money. Caller ID spoofing can be done in many ways – the Internet, with computer telephony, etc. It first appeared in mid-2004. See Caller ID and spoofing.

caller independent voice recognition Having a voice response unit recognize the voice of a caller without having been trained on the caller's voice.

caller name An enhancement of Caller ID. Prior to sending the originating telephone number to your display, the carrier associates that number with an electronic white pages listing, thereby transmitting both the originating number and the associated directory listing. Assuming that the number is listed and that the Caller ID number is not blocked, this provides a much better indication of the identity of the caller.

calling A procedure which consists of transmitting address signals in order to establish a link between devices that want to talk to each other.

calling card A credit card issued by Bell operating companies, AT&T, MCI, Sprint and other phone companies (local and long distance) and used for charging local and long distance calls. Typically, the number on your calling card is the phone number at which you receive bills (home or business phone) plus a four digit Personal Identification Number (PIN). Increasingly often it's not. I prefer to carry a calling card with digits completely different to my phone number since this provides me with greater security. It's harder for someone to figure out my calling card. Some phone companies – local and long distance – charge more for a call made with a Calling Card. Some don't. Bell Canada claims they trademarked the term "Calling Card" in Canada. If they did, good luck protecting it, since the term "calling card" is generic. See Breakage, Debit Card and Prepaid Calling Card.

calling DTE A DTE (Data Terminal Equipment) which places a call to another DTE.

calling jack In manual switchboard systems, the jack that is used by the operator to connect the call that came in through the answering jack, to the circuit for the subscriber who will be receiving the call.

calling line ID Also called Caller ID. You are called. As the call comes in, you receive the phone number of the person calling you. See Caller ID for a full explanation. See also ANI, Calling Number Display, CLI, ISDN and Signaling System 7.

calling line identification See CLI and Calling Line ID.

calling line identification facility A service feature, provided by a network, that enables a called terminal to be notified by the network of the address from which the call has originated. See Calling Line ID.

calling list A call center term. A collection of records from a database that is used for a specific telemarketing campaign. See Calling List Penetration.

calling list penetration A call center term referring to outbound calls. The percentage of a call list for which the decision makers have been reached after a given number of attempts. See Calling List.

calling number An international term for what Americans call "ANI" – or automatic number identification. In other words, calling number simply tells you, the receiver of the call, who's calling. It tells you that by displaying the caller's number on the screen of your phone or the screen of your PC. And sometimes it might tell you who's actually calling. It can do that because some central offices (also called public exchanges) have the ability to dip into a database and replace the calling number with the name of the person who owns that phone number.

calling number display Your phone has a LCD (Liquid Crystal Display) or LED (Light Emitting Diode) display. When your phone rings, it will show which telephone number (internal or external) is calling you. Some phone systems allow you to add the person's name to the calling number display. See also ANI and Caller ID.

calling party The person who makes (originates) the phone call.

calling party camp-on A feature that enables the system to complete an access attempt in spite of temporary unavailability of transmission or switching facilities. Systems that provide this feature monitor the system facilities until the necessary facilities become available, and then proceed to complete the requested access. Such systems may or may not issue a system blocking signal to let the caller know of the access delay.

Calling Party Control CPC. Sometimes referred to as CPC Wink. A call supervision feature which provides the ability for a CO (Central Office) to signal the called party when the calling party hangs up. Some switches also provide CPC to the calling party when the called party hangs up. CPC allows the PBX, key system, or telephone answering device to reset the line so that it is ready to either accept or initiate another call. CPC is accomplished by either a loop current drop or reversal.

calling party identification A telephone company service which tells the person being called the number and sometimes the name of the person calling them. They can then decide to answer or not answer it. See ANI, which stands for Automatic Number Identification.

Calling Party Number CPN. When a call is set up over an ISDN network, SS7 sends an IAM (Initial Address Message) as part of the ISUP (ISDN User Part) protocol. Included in the IAM is the Calling Party Number subfield, which contains the number of the calling party. Also included is a two-bit Presentation Indicator (PI), which indicates whether the terminating switch (cellular or PSTN) should pass the CPN to the called party. If the PI says "yes," the originating number is passed to the called party, who can see that number displayed on the telephone set or on an adjunct display device. The end result is Caller ID, perhaps also with the Caller Name, assuming that the called party has subscribed to those features. Based on recognition of the calling party, or the lack of it, Caller ID can prompt the called party to either accept or reject the call. See also Caller ID, Caller Name, IAM, ISUP, and SS7.

calling party pays In the United States, cell phone users pay for incoming as well as outgoing calls. In Europe and in most countries elsewhere, the calling party pays. According to many people, paying for incoming calls retards the industry's growth. As a result, there are many people in the U.S. cell industry, who would like this changed to "Calling Party Pays."

calling party subdividers Information element that is passed transparently by the SPCS (if certain conditions are met) and can be used to further identify the originating party. An AIN term.

calling pattern Telecommunications managers are great at looking at phone bills, seeing patterns and smelling out calls that don't fit into those calling patterns. A simple example: zillions of calls over a weekend to Pakistan. The company doesn't do business with Pakistan. Nor does anyone work on the weekend.

calling sequence A sequence of instructions together with any associated data necessary to perform a call.

calling tone See CNG.

callPath IBM's telephone system link to IBM's computers. See Callpath Coordinator, Callpath Services Architecture, Callpath Cics and Callpath Host.

CallPath CallCoordinator CallCoordinator is IBM's integrated call management application that uses CallPath Services APIs to integrate data processing applications with telephone systems. IBM has versions of CallCoordinator for MVS CICS, OS/2 and Windows workstations. CallCoordinator provides features such as Intelligent Answering (based on ANI, DNIS, or Calling Line ID), Coordinated Voice and Data Transfer, Consultation (both voice and data), Conferencing (both voice and data), Transfer Load balancing between a single or multiple telephone systems, Outbound dialing, Event logging for Management Information Reporting, Personal Dialing Directory (Windows Only), Personal telephony facilities (answer phone, disconnect, transfer, etc.), Integration with CallPath DirectTalk/2 and CallPath DirectTalk/6000, Customizable Application Programming Interfaces. CallCoordinator integrates with existing 3270 or 5250 applications, and on the workstation versions, has the ability to communicate with existing applications via Dynamic Data Exchange or standard LAN communications protocols (such as TCP/IP). See CallPath Services Architecture.

CallPath CICS Enabling software that connects your telephone systems with your IBM 370 or 390 (i.e. the mainframe version of CallPath/400, which works on the AS/400 platform). See CallPath CallCoordinator.

CallPath Services Architecture CSA is IBM's architecture that defines the protocols for communication between computers and telephone switches. CallPath Services Architecture, announced in 1991, provides an Application Programming Interface (API) that enables a call management application to interact with telephone systems, with little regard to the protocols or communications interface provided by the telephone system. The idea is that with CallPath a call will arrive at a computer terminal simultaneously with the database record of the caller. And such call and database record can be transferred simultaneously to an expert, a supervisor, etc. CallPath has especial value in telephone call centers. As of writing, IBM provided connectivity to PBXs (Lucent Definity Generic 3, Nortel Meridian 1, Siemens/ROLM 9751 and Hicom, Bosch, Alcatel, SDX, Ericsson, Philips, Deutsche Telecom, Cortelco, and GPT), central office switches (AT&T 5ESS and Northern Telecom DMS-100), and ACDs (Aspect and Rockwell). IBM's CallPath products provide support for locally attached applications and client/server applications. IBM has CallPath APIs available for mainframes, minicomputers and workstations, in particular IBM System 390 and ES9000, AS/400, RISC System/6000, OS/2 workstations, Windows workstations, Sun Solaris, HP UX, and SCO UNIX workstations. See Open Application Interface and DirectTalk.

callpower A Rockwell ACD term. An integrated voice and data workstation for use in combining ACD capabilities with host computer database management.

CallWare CallWare is a company in Salt Lake City, UT, which makes computer telephony software that runs on the Novell NetWare operating system. CallWare software includes voice mail, autoattendant, IVR database lookup, etc.

CALLS Proposal Access charge reform adopted by the FCC in May 2000 that applies only to price cap carriers. The order intended to lower consumer rates and make implicit subsidies explicit. It (1) reduced per minute access charge rates paid by interexchange carriers (IXCs); (2) eliminated the presubscribed interexchange carrier charge (PICC) (3) increased the fixed subscriber line charge (SLC); and (4) established a $650 million interstate universal service support mechanism.

CALNET The California Network (CALNET) implemented service in September 1991 with the objective of providing cost-effective telecommunications services to state and local government in California by reducing costs through consolidation of user service requirements. In the summer of 1998, CALNET provided services to 300,000 government customers statewide. The Department of General Services' (DGS) Telecommunications Division oversees CALNET.

CAM 1. Call Applications Manager. The name of the Tandem software interface which provides the link between a call center switch telephone switch (either a PBX or an ACD) and all Tandem NonStop (fault tolerant) computers. CAM supports most major PBXs and automatic call distributors.

2. Computer-Aided Manufacture. The actual production of goods implemented and controlled by computers and robots. Often used in conjunction with CAD. Only a few factories are completely automated. Usually, there is some human intervention in the actual construction of the product, often to make sure a part is placed in the robot correctly.

3. Controlled Attachment Module. Intelligent Token-Ring hub.

4. Content Addressable Memory. It is a key component used in high performance routers. See Content Addressable Memory for a full explanation.

cam wrench Multiple use tool used for unlocking pedestals and NID's (SNI's). 99% of telephone repair people have this tool.

CAMA Centralized Automatic Message Accounting. See CAMA/LAMA.

CAMA-ONI Central Automatic Message Accounting-Operator Number Identification. An operator located at a position that is connected temporarily on a customer-dialed station-to-station call.

CAMA/LAMA Centralized Automatic Message Accounting/Local Automatic Message Accounting. Specific versions of AMA in which the ticketing of toll calls is done automatically at a central location for several COs (CAMA) or only at the local office for that office's subscribers. See CESID.

camcorder A camera and a video recording system packaged as a whole.

CAMEL Customized Application of Mobile Enhanced Logic. An ETSI standard for GSM (Global System for Mobile Communications). CAMEL enhances GSM for the provisioning of international IN (Intelligent Network) services. In order to effect CAMEL, the GSM operator installs a CSE (CAMEL Service Environment), similar to the wired IN equivalent. The CSE comprises a SSP (Service Switching Point), IPs (Intelligent Peripherals), a SCP (Service Control Point), the SCE (Service Creation Environment) and some additional SS7 (Signaling System 7) software. CAMEL supports the availability of IN services internationally, across GSM networks. Initial services will include voice mail, call waiting, call forwarding, and Freephone (toll-free) access. While only approximately 10% of GSM users currently roam internationally, that number is expected to increase significantly in the future. See also ETSI, GSM and IN.

camel A camel can lose up to 30% of its body weight in perspiration and continue to cross the desert. A human would die of heat shock after sweating away only 12% of body weight. Traveling at a rate of 2 to 3 miles per hour, camels can carry 500 to 1,000 pounds on their backs. They are able to keep up this pace for 6 or 7 hours a day. Camels will refuse to carry loads that are not properly balanced. See Camel Droppings.

camel droppings Some camel droppings are so dry that they can be set on fire as soon as they are dropped. In Idaho, there's a law that you may not fish from a camel. See also Camel and Shit.

camel toes I don't quite understand this one. But I am assured that it exists. Camel toes describes what some telephone men describe as the sight of a female telephone operator in tight polyester pants – as in a frontal view below waist level.

camgirl A young woman who broadcasts live pictures of herself over the World Wide Web. Also called cam-girl, cam girl or Webcam girl.

camp on You're calling a telephone an extension or you want to transfer a call to a phone but it's busy. This telephone system feature will allow you to lock the call you're trying to transfer onto the line that's busy. When it becomes free, the phone will ring and the "camped-on" call will be connected automatically.

camp on privacy A telephone system feature that enables a caller to camp on the line of a called person whose phone is in privacy mode. As soon as privacy mode is released, the caller is automatically connected to the called party.

campus The buildings and grounds having legal contiguous interconnection.

Campus Area Network CAN. A network that provides interconnectivity in a confined geographic area such as a campus or industrial park. Such networks operate over fairly short distances, and do not require public rights-of-way.

campus backbone Cabling between buildings that share telecommunications facilities.

Campus Distributor CD. The international term for the main cross-connect. The distributor from which the campus backbone cable emanates.

campus environment An environment in which users – voice, video and data – are spread out over a broad geographic area, as in a university, hospital, medical center, prison. There may be several telephone systems. There may be several LANs on a campus. They will be connected with bridges and/or routers communicating over telephone, microwave or fiber optic cable.

campus network A campus network is a LAN that is spread over multiple buildings. Campus Networks are typically created by small companies or divisions of larger firms.

campus subsystem The part of a premises distribution system which connects buildings together. The cable, interbuilding distribution facilities, protectors, and connectors that enable communication among multiple buildings on a premises.

CAN 1. Abbreviation for cancel. The binary code is 100001 and the HEX is 81.

2. See Campus Area Network.

3. Controller Area Network, a shared data communications link used by the automobile industry for high-speed, real-time applications such as controlling the car's engine or the transmission. See Controller Area Network.

cancel By touching the "cancel" button on a phone system you're telling the phone system to ignore the last command you gave it. That command might have been transfer, hold, park, etc. The "cancel" button is often mistakenly confused with the "release" button. The "release" button acts the same as hitting "Enter" on a computer system, i.e. it tells the system to go ahead and do what you just told it to do, no matter how stupid your command. In short, "Cancel" means kill the last command. You use it when you make a mistake. "Release" means "Enter" – Do it and do it now.

cancel call waiting On a touchtone phone in North America, you typically can cancel the feature, Call Waiting, by touchtoning *70.

cancelmoose A Newsgroup/Usenet Term. An individual who wages war against spamming.

candy bar design Describes a mobile handset whose shape resembles a candy bar.

cannibalize To devour a phone system by stripping parts from it to repair another system. A common technique for maintaining equipment whose original manufacturer no longer supplies parts. Before you cannibalize, check out the monthly publication Telecom Gear. That publication lists sources of secondary telecom equipment. Good stuff, too.

cannibalism Two men accused of eating human body parts, washed down with a bottle of wine, were freed by a Cambodian provincial court because there was no law against cannibalism. The two men, both crematorium workers, were arrested for eating fingers and toes of a body they were cremating. Police in Banteay Meanchey province,

140 miles northwest of Phnom Penh, were alerted by villagers, who said the men often ate human parts after relatives of deceased had left the crematorium. Eating human parts was common during the 1975-79 Khmer Rouge "killing fields" rule, when an estimated 1.7 million people died from torture, overwork, disease, execution and widespread famine. Besides having no law against cannibalism, the men were hungry.

canonical Conforming to a generally accepted rule or procedure, commonly reduced to the simplest or clearest schema possible. A simple matrix used for translating addresses is one example. When using Windows XP faxing service, telephone numbers must be in the canonical form in which a U.S. number would appear as +1 (626) 555-1212. If you use even a slightly different form, says Microsoft, such as (626) 555-1212 or 1-626-555-1212, the dialing rules won't be applied and the fax transmission will fail. See also Canonical Address and CNAME Records.

canonical address A method for storing unique telephone numbers. Canonical addressing is used by Windows Telephony TAPI (Telephony API) for making telephone calls from a database of numbers. A canonical address describes all possible aspects of a telephone number. You can call a telephone number using canonical addressing independent of calling location or access method. A canonical address is stored in a database and preceded by an ASCII Hex (2B) to indicate its address type. It includes delimiters and strings for Country Code, Area Code, Subscriber Number, Subaddress and Name. See also Canonical.

canopy beds In England in the 1500s, the roof was thatched. Insects and other animals lived there. There was little to stop things from falling into the house. This posed a real problem in the bedroom where bugs and other droppings could really mess up your nice clean bed. Thus came into existence a bed with big posts and a sheet hung over the top afforded some protection. Hence canopy beds.

cantenna A cantenna is a homemade Wi-Fi antenna. It is made out of a soup can or such. It is also the brand name of such a device, www.Cantenna.com.

Canuck Slang for a Canadian. Canadians call each other Canucks.

Cao's Law According to the November, 2000 Gilder Technology Report, Cao's Law tells us that the communications spectrum is virtually infinite and that wavelength division multiplexing (WDM) will follow a sort of turbo version of Moore's Law. WDM will spread across an optical fiber more and more and finer and finer channels of light each using less and less power. It will multiply these lambdas two to three times as fast as Moore multiplied transistors. Channels on a fiber will recapitulate the saga of transistors on a chip and exhibit many of the same trade-offs between power and connectivity. On optical fiber, the trade-off is between bitrate and channel count. So far, we can pump a high bitrate on each channel, or we can transmit lots of channels. But we can't do both on the same fiber. The dispersive effects of 10 and 40 Gbps systems, in which the modulated signals tend to "mush" together, can disable high channel count WDM. At the other extreme, each of Avanex's 100,000 channels – if they ever escape from the lab – will probably bear multi-gigabit signals. Nevertheless, there is today among telecom carriers a real world pattern emerging that manifests Simon Cao's law in action.

CAP 1.   Competitive Access Provider. Also known as AAV (Alternative Access Provider). CAPs provide an alternative means of establishing a connection between a user organization and an IXC (IntereXchange Carrier), completely bypassing the ILEC (Incumbent Local Exchange Carrier). CAPs typically deploy high-capacity SONET fiber optic transmission systems in a ring topology around geographic areas in which are found a high density of large businesses. Drops from the fiber optic rings are terminated at both the customer locations and the IXC POPs (Points of Presence). Thereby, end user organizations with substantial levels of interLATA voice and data traffic can bypass the ILEC facilities, which often are made up of poor quality UTP (Unshielded Twisted Pair) and who may take months to provision a T-1 circuit. In addition to providing superior performance and much reduced provisioning time, such fiber optic transmission facilities offer incredible levels of bandwidth, which quickly can be increased, and generally are provided at much lower cost than leased-line ILEC circuits. CAPs also offer the inherent advantage of loop diversity. In the event that the ILEC local loop suffers a catastrophic failure, the CAP loop likely will not be affected, unless both loops follow the same physical path and are destroyed by the same post-hole digger (or other catastrophe). In the unlikely event that the redundant CAP loop fails, the user organization can still access the IXC through the ILEC on a circuit-switched, 1+ dial-up basis. Since the Telecom Act of 1996 and various state initiatives have relaxed regulatory constraints and opened the local exchange to competition, many CAPs have become CLECs (Competitive Local Exchange Carriers). As CLECs, they are free to offer switched voice and data services within the local exchange area. Where they do not provide their own fiber optic local loops, they lease UTP local loops from the ILECs for resale, with those loops terminating in colocated termination equipment in the ILEC central offices and with the traffic then being directed to the CLEC fiber optic transmission facilities. As facilities-based

carriers, the traffic then is transported to the CLEC's own switching centers and wire centers for local and long-haul service access. See also CLEC, ILEC, IXC and SONET.

2. Cellular Array Processor.

3. Carrierless Amplitude and Phase modulation is a bandwidth-efficient line coding technique. CAP is a variant of Quadrature Amplitude Modulation (QAM), which is used in today's rate-adaptive voice band V.32/V.34 dial modems. AT&T Bell Laboratories first began development of CAP in the mid-1970's for more efficient implementation of a digital signal processor (DSP), while providing the same high level of performance. Used in conjunction with advanced error correcting codes and channel equalization, CAP modulation provides robust performance and excellent loop reach in the presence of bridge-taps, cross-talk, and other interferers. Carrierless Amplitude & Phase Modulation is now a transmission technology for implementing a Digital Subscriber Line (DSL). The transmit and receive signals are modulated into two wide-frequency bands using passband modulation techniques. CAP is bandwidth-efficient and supports ADSL, HDSL, RADSL, and SDSL line coding.

4. Client Access Protocol. See iCalendar.

5. Cap is short for capitalization. See market cap.

Cap Code Every functioning pager is assigned a unique cap code that identifies to the paging terminal what signals should be sent to the pager. No two cap codes in the same system are the same. When a beeper's special phone number is dialed, it causes only that one beeper to be signaled.

cap-and-grow A strategy for protecting legacy revenue streams, creating new revenue streams, reducing migration costs, and making change easy by introducing a next-generation switch alongside a legacy switch, using the new switch for new subscribers and new services, and using the legacy switch for existing customers who are content with legacy services.

cap-and-replace The next and final stage after the deployment of an interim cap-and-grow solution, namely, migrating existing customers to the next-generation switch and ditching the old legacy switch that was serving them.

Cap'N Crunch See Captain Crunch.

Capability Sets In 1989, the ITU-T developed the standard for Intelligent Networks (IN). The full standard is quite extensive to implement. In order to facilitate acceptance of the standard, the ITU developed a set of upwardly compatible "capability sets". The first of these (CS-1) was approved in 1992 is the standard that is implemented in most IN elements today. CS-2 was supposed to follow on in 1997 and CD-3 in 1998, however, these have not been finalized. CS-1 includes the following services: Number translation, Alternate Billing, Call Screening (based on source or destination number), Automatic callback, Conference calling, Call logging, Mass calling, Televoting and VPN. VPN

CS-2 will define standards for voice services over the internet, mobility services, broadband services etc.

capacitance The capacity of a conductor (e.g., copper wire or bus), or the dielectric insulation surrounding a conductor, to store an electrical charge. Capacitance is measured in farads. The capacitance of cable systems generally is measured in picofarads (pF). See also capacitor.

capacitance, direct The capacitance measured directly from conductor to conductor through a single insulating layer.

capacitance, mutual The capacitance between two conductors with all other conductors, including shield, short circuited to ground.

capacitive coupling The transfer of energy from one circuit to another by virtue of the mutual capacitance between the circuits. The coupling may be deliberate or inadvertent. Capacitive coupling favors transfer of higher frequency components, whereas inductive coupling favors transfer of lower frequency components.

capacitor Capacitors provide a means of storing electric charge so that the charge can be released at a specific time or rate. The simplest type of capacitor is a parallel plate capacitor and consists of two closely spaced plates of conductive material with an insulating material known as a dielectric sandwiched between them. Dielectrics are chosen for their ability to enhance a capacitor's performance. Capacitors are rated for their capacitance which is measured in farads and voltage. The rated voltage is usually the breakdown voltage, I.e. The maximum voltage the dielectric can insulate against before the voltage discharges between the two plates of the capacitor as an electric spark.

Because capacitors store charge, they can be used in electronic circuits in place of a battery. However, a battery generates electricity through a chemical reaction. A capacitor will generate an electric current only after it has been charged by another current source. When working on electronic equipment – even equipment unplugged from a power supply – make sure you are well insulated against shock. Capacitors in a circuit can hold a charge, sufficient to cause injury or death, sometimes for many hours after an appliance or device has been turned off. Capacitors in TV sets, for example, can store up to 100,000 volts.

capacity 1. The information carrying ability of a telecommunications facility. What the "facility" is determines the measurement. You might measure a data line's capacity in bits per second. You might measure a switch's capacity in the maximum number of calls it can switch in one hour, or the maximum number of calls it can keep in conversation simultaneously. You might measure a coaxial cable's capacity in bandwidth.

2. The measure of the amount of electrical energy a condenser can store up. The unit of capacity is the farad.

capacity study A local document issued at least once a year for each entity within the telephone company. The capacity study includes information relative to the network access line/trunk capacity of each item of switching equipment as well as the network access line capacity of lines and numbers.

capacity transfer control A Northern Telecom term for a feature which permits single allocation of capacity to be shared among members in a digital switched broadcast connection. For teleconferencing, for instance, a conference leader can transfer transmission capacity among the digital ports in the circuits. 95% of such transfers will take place within 10 seconds.

capacity swaps Imagine you're a long distance telecom company in need of circuits in California. You find another long distance company in need of circuits of circuits in say New York, where you have too many. So you swap (also known as barter) the use of your New York circuits for the use of their circuits in California. Typically money won't change hands. But the transaction can get more interesting. Let's say you're a phone company in need of profits – for example, to maintain your growth, please your shareholders and Wall Street. You may want to treat the swap as two separate transactions. And if your accountants are sufficiently creative, you can create "sales" and report the accompanying "profits". Of course, the "profits" won't be real... but in the heydays of the late 1990s, auditing was lax, corporate governance was non-existant and expectations were high. Anything went. Most of the companies that indulged in capacity swaps that made "profits" eventually went Chapter 11 (bankrupt).

CAPC Competitive Access Provider Capacity. The highest possible reliable transmission speed that can be carried on a channel, circuit, or piece of equipment.

capcode A capcode is a four or seven digit number on either side or rear of the casing of a pager, the type you wear on your belt. This number is a paging system necessity to know how to generate the right sequence of tones to alert the pager. Also spelled cap code.

capex A shortened way of saying capital expenditures. See also OPEX.

CAPI Cryptography Application Program Interface. The first API developed by Microsoft for encryption programs.

capout Capout means broken in bad French. In German, the word is kaput.

CAPS 1. Code Abuse Prevention System.

2. Competitive Access Providers to the local telephone network i.e., Teleport or Metropolitan Fiber System.

capsizing Downsizing gone awry. It's the process of a company repeatedly reducing head count.

capstan 1. A flangeless pulley used to control speed and motion of magnetic tape through a recorder or playback unit.

2. A rotating drum or cylinder used for pulling cables by exerting traction upon a rope or pull line passing around the drum.

CAPTAIN Character And Pattern Telephone Access Information Network System. A form of videotext developed in Japan and operated through the public switched telephone network. Displays are on a TV set. It's interactive.

Captain Crunch In the 1960s, boxes of a breakfast cereal called Cap'n Crunch had a promotion. It was a toy bosun's whistle. When you blew the whistle, it let out a nearly precise 2,600 Hz tone. If you blew that whistle into the mouthpiece of a telephone after dialing any long distance number, it terminated the call as far as the AT&T long distance phone system knew, while still allowing the long distance connection to the distant city to remain open. If you dialed an 800 number, blew the whistle and then pressed in a series of tones (called multi-frequency or MF tones) on your "Blue Box," you could make long distance and international calls for free, since the only thing the billing machine at the local telephone company central office knew was the original toll-free call to the 800 number. It assumed the call was free.

The man who discovered the whistle was John Draper and he picked up the handle of Cap'n Crunch from the Quaker Oats breakfast cereal, Cap'n Crunch. A marvelous account of the exploits of phone phreaks was published in the October 1971 issue of Esquire Magazine. That article described how the Cap'n would call himself (he needed two lines and two phones) – choosing to route the call through Tokyo, India, Greece, South Africa, South America, London, New York and California – to make his second phone next to him ring. He'd have a wonderful time talking to himself, albeit with a round-the-world delay (despite the speed of light) of as long as 20 seconds. Later, AT&T closed the loophole Cap'n Crunch had discovered. AT&T turned from in-band signaling to out-of-band signaling. Cap'n Crunch's legacy (he got put in jail four times during the 1970s) is Signaling System 7, a system of immense benefit to us all. See 2600 Tone, Multi-Frequency Signaling and Signaling System 7.

Captcha Captcha is an acronym for "completely automated public Turing test to tell computers and humans apart." It's even trademarked by Carnegie Mellon University. According to Leo Noteboom, of AskLeo! "One of the oldest problems in computer science is to build a computer or software that mimics "thinking" like a human, and does it so well that you couldn't tell the difference. You could ask it a series of questions, and you wouldn't be able to tell whether the responses came from a real human or a computer. That's referred to as a form of "Turing test", after the computer scientist Alan Turing. A Captcha is a kind of "reverse Turing test". In essence it's a way of proving that you're human. You know those images with slightly scrambled letters, where you're supposed to type in what you see? That is a Captcha. Deciphering those letters is currently beyond the ability of contemporary mass market computers and software. Since you and I can typically make out what those letters are, and then type them in correctly, we must not be computers. We've proven that we're human. Why do we care? As with so many things these days, it mostly comes back to spam. Here's one example: without a Captcha, it would be easy to write a computer program to open hundreds or thousands of Hotmail accounts, and start spamming from them. Once the accounts are blocked, the program can just as easily start creating thousands more. It really would be as easy as it sounds for a reasonably competent programmer. However, at some point along the account creation process, Hotmail presents a Captcha, saying in effect "prove to me that you are not a computer, and I'll let you create this account". The computer program is stopped dead in its tracks."

captive effect An effect associated with the reception of frequency-modulated signals in which, if two signals are received on or near the same frequency, only the stronger of the two will appear in the output. The complete suppression of the weaker carrier occurs at the receiver limiter, where it is treated as noise and rejected. Under conditions where both signals are fading randomly, the receiver may switch from one to the other.

captive portal You're at Wi-Fi hot spot. You open your laptop computer, turn on your wireless and start your browser. Bingo, you see a screen showing the Wi-Fi's page asking you to log in and/or pay them money. Once you do pay them money, or prove you've paid them money, that screen will disappear and you'll be transported to the Internet. That initial Wi-Fi screen is called a captive portal. A captive portal is basically a default web page that is downloaded to everyone that logs onto the Internet through that Wi-Fi spot, or wired spot – think a hotel room.

captive screw Let's say you have a couple of screws at the front of an industrial grade computer. You unscrew the two screws and a panel pops down. You can then get access to something inside the computer. The screw, however, doesn't leave the computer. No matter how much you unscrew it, you cannot remove the screw. It's "captive."

Capture Division Packet Access CDPA. Capture Division Packet Access is a packet-oriented cellular access architecture able to support the constant bit rate traffic and variable bandwidth on demand for multimedia traffic. CDPA integrates multiple access and channel reuse issues to achieve a high rate of spectral efficiency, and presents general advantages even if used for delay-constrained circuit-oriented traffic. Unlike CDMA and TDMA, wherein the effective data rate of each connection is typically a small fraction of the total radio channel allocated for PCN, the CDPA approach allows each user to access the entire channel, if necessary, for brief periods of time (packet access). Spectrum sharing is accomplished by exploiting the different path losses suffered by the various signals as they appear at the base stations (the capture effect), with co-channel interference abated through time diversity (colliding users do not successively retry in the same time interval). Results suggest that abating co-channel interference by random transmission may be more effective than spatial isolation at cells using the same channel, as is usual in FDMA/TDMA systems. See also Capture Effect, CDMA, FDMA, and TDMA.

capture effect An effect associated with the reception of frequency-modulated signals in which, if two signals are received on or near the same frequency, only the stronger of the two will appear in the output. The complete suppression of the weaker carrier occurs at the receiver limiter, where it is treated as noise and rejected. Under conditions where both signals are fading randomly, the receiver may switch from one to the other.

capture ratio The ability of a tuner or receiver to select the stronger of two signals at or near the same frequency. Expressed in decibels, the lower the figure, the better.

CAR Committed Access Rate. CAR is a term Cisco applies to Cisco Weighted Rate-limit, a traffic control method that uses a set of rate limits to be applied to a router interface.

CAR is a configurable method by which incoming and outgoing packets can be classified into QoS (Quality of Service) groups, and by which the input or output transmission rate can be defined.

car phone The type of cellular phone that's installed in a vehicle. There are four types of cellular phones being sold today – mobile, transportable, portable and handheld. A car phone (also called a mobile unit) is attached to the vehicle, its power comes from the vehicle's alternator (or battery if the car is not running) and the car phone has an external antenna, which works best if it's mounted in the middle of the highest point of the car and wired directly with no breaks in the wire. Many window-mounted antennas have a break in their wiring. The wiring ends at the inside. There is no electrical connection between the inside of the window and the antenna glued onto the outside of the window. The "connection" is done through signal radiation. In North America, the car phone transmits with a standard three watts of power.

carbon block A device for protecting cable from lightning strikes. The carbon block consists of two electrodes spaced so that any voltage above the design level is arced from line to ground. Carbon block protectors are used commonly in both local customer offices and central offices. They are effective, but can be destroyed if high voltage is directly applied – as in a direct strike by lightning. A more expensive, but more effective method of protection is the gas tube. These are glass capsules that are connected between the circuit and the ground. When a voltage higher than the design voltage strikes the line, the gas ionizes and conducts the excess voltage to ground. When the voltage is gone, the protector restores itself to normal. Gas tubes, however, take a tiny time to ionize. This may not be fast enough for very sensitive things, like PBX circuit cards. So gas tube protectors are often equipped with diodes, which clamp the interfering voltage to a safe level until the gas tube ionizes.

carbon fiber A strong synthetic material that is low in mass with excellent damping characteristics, used in the manufacture of tonearms.

carbon rheostat A rheostat using carbon as the resistance material. See Rheostat.

carbon transmitter The microphone of an telephone set from yesteryear which uses carbon granules and a diaphragm. The diaphragm responds to our voice and varies the pressure on the granules and hence, their resistance. If your carbon mike isn't working well, the humidity has got to it. Tap it lightly on your desk and the carbon granules will line up and it will work much better. Carbon microphones are very reliable but are being increasingly replaced with more sensitive electret microphones.

Carbon XMTR See Carbon Transmitter. The carbon XMTR is in the sending part of older telephone hand sets.

card 1. A printed circuit card, or Printed Circuit Board (PCB). See also Card Cage and Printed Circuit Board.

2. A discrete unit of data specifically designed to be read easily on the small screen of a handheld wireless device such as a cell phone, pager, or PDA (Personal Digital Assistant). A card can be for entering data, displaying data, or listing indexes or available menu options. Packet radio networks generally send a "deck of cards" in a single data packet for reasons of efficiency, as the individual cards are small in terms of byte count.

Card Authorization Center CAC. A computer directly linked to MCI switches for authorization and determination of billing center ID for MCI card calls.

card cage A frame in a telephone system or computer for mounting circuit cards, power supply, backplane and other equipment.

card dialer A device attached to a telephone which accepts a special plastic card and then automatically dials the number on the card as indicated by the holes punched in it. A card dialer is now obsolete except for unusual applications, like systems whereby you carry your card with you and use it as a security device.

Card Issuer Identifier Code CIID (pronounced "sid") A code issued with certain calling cards. AT&T's CIID cards cannot be used by other interexchange carriers but can be used by LECs, local exchange carriers.

card services The software layer above Socket Services that coordinates access to PCMCIA cards, sockets and system resources. Card Services is a software management interface that allows the allocation of system resources (such as memory and interrupts) automatically once the Socket Services detects that a PC Card has been inserted. This is called "hot swapping." The idea is that you can slide PCMCIA cards in and out of PC at will and your Socket and Card services will recognize them and respond accordingly. It's a great theory. In practice, it doesn't work because certain cards, like network cards, simply can't be connected and disconnected at will. Socket Services is a BIOS level software interface that provides a method for accessing the PCMCIA slots of a computer. Card Services is a software management interface that allows the allocation of system resources (such as memory and interrupts) automatically once the Socket Services detects that a PC Card has been inserted. Both of these specifications are contained in the PCMCIA Standards docu-

ment. You do not need either Socket or Card Services to successfully use PCMCIA cards in your desktop or laptop. You simply need the correct device drivers and the proper memory exclusions. See PCMCIA, Socket Services and Slot Sizes.

card slot A place inside a phone system or computer into which you slide a printed circuit board. See Board.

card walloper A programmer who provides batch programs that do mundane things such as payroll.

CardBus Laptops typically come with slots for what are now known as PC cards – little credit card size devices who do various things – like become a modem, become a network interface card, become a video conferencing card, become an ISDN card, become a wireless LAN, etc. These cards were originally called PCMCIA cards. (For a full explanation see PCMCIA). The original PCMCIA spec was 16-bit. The new spec, called CardBus, which combines the PCI bus, has a 32-bit interface. The CardBus specification is the significantly improved successor to the previous PC Card standard. The two standards are not compatible. You cannot run a laptop with both PCMCIA and CardBus cards. You must run them with cards of the same standard. And these days, the best standard to go with is CardBus. Here are some of CardBus; benefits:

20 times the throughput of previous 16-bit PC Card slots. The 32-bit CardBus interface can transmit data at 400-600 Mbps, compared to 16-bit PC Card's 20-30 Mbps. Users must have that higher bandwidth for linking to a 100 Mbps Fast Ethernet network, for quickly moving data to and from SCSI-2 storage devices (such as Zip drives) and for handling bandwidth-hungry applications like video conferencing.

Better systems performance. Bus mastering lets a CardBus device transfer data to computer memory directly, without intervention from the notebook's processor. This boosts overall computer performance multitasking the newer Windows systems.

Lower power consumption. CardBus devices run at 3.3 volts, instead of 16-bit PC Card's 5 volts. That means CardBus devices use less power than conventional PC Card devices, and generate less heat inside the computer. Thus batteries last longer.

Easier installation of multifunctional devices. The CardBus specifications enables sharing of multiple resources on a single card with no need for special drivers.

Optimized video performance. The CardBus Zoomed Video feature handles streamed video transmissions more efficiently by transferring the data directly to the PC's video controller over a dedicated bus. That way, video doesn't have to compete for bandwidth on the computer's PCI bus. See Card Services and PCMCIA.

cardioid pattern An antenna pattern similar to a half-hemisphere.

CARE Customer Account Record Exchange. A system developed to make easy the exchange of customer account information between the IXC (long distance phone company) and the LEC (local phone company) to make easy the provisioning of telecom services. CARE generically identifies data elements that might be exchanged between the IXC and LEC in an industry format. It is intended to provide a consistent definition and data format for the exchange of common data elements. The C.A.R.E. records (kept at the LEC) inform the customer's long-distance provider of changes in the customer's account (i.e., customer has selected Company X as its provider, or has terminated service, etc.)

CARE/ISI Customer Account Record Exchange/Industry Standard Interface. National guidelines for the formats and language used in mechanized exchanges of Equal Access-related information between Interexchange Carriers and telephone companies.

caret The symbol ^ which is found above 6 on most keyboards. Also used to indicate the "Ctrl" key in some instruction manuals. Sometimes it is used to indicate the power to be raised, as in 2^4, which equals $2 \times 2 \times 2 \times 2$.

Carnivore 1. Carnivore is the name of the FBI's Internet surveillance system - a system that supposedly can monitor e-mail for evil intentions (bomb plots, terrorists, scams, etc.) without violating the civil rights of ordinary folks. At least that's the theory.

2. See DCS1000.

CAROT Centralized Automatic Reporting On Trunks. A test and maintenance facility associated primarily with electronic toll switching systems like the AT&T Communication's #4-ESS. CAROT is a computerized system that automatically accesses and tests trunks for a maximum of fourteen offices simultaneously. It enables rapid routine testing of all trunks to ensure quick identification of faults and potential failures.

CARP Cache Array Routing Protocol. A protocol developed to route client requests to one of a cluster, or array, of proxy servers on which databases are cached from origin Web servers. CARP contains a Proxy Array Membership Table from which an HTTP client agent (i.e., proxy server or client browser) can allocate and intelligently route URL requests to any member. Microsoft has implemented CARP in its proxy servers. See also Client, HTTP, Proxy, Server, and URL.

carpal tunnel syndrome Carpal tunnel syndrome is a serious disorder of the arm caused by fast, repetitive work, such as typing without support for your wrists or with insufficient time for rest. In carpal tunnel syndrome, the tendons passing through the wrist bones swell and press on the median nerve. Surgery to take pressure off the nerve can relieve numbness and pain, but it's not always effective and many victims remain permanently disabled. The best prevention is using a wrist rest and undertaking specific exercises. A lot of "knowledge workers" have claimed that carpal tunnel syndrome is the result of working at computer keyboards all day long, day after day. There is a good book on the subject – Conquering Carpal Tunnel Syndrome by Sharon J. Butler, New Harbinger Publications, Oakland, CA. See also Computer Vision Syndrome.

carriage deals Let's say I want to start a new TV channel. Let's call it Harry's 24-Hour All Tennis Channel. I figure out how to fill 24-hours a day, seven days a week with great tennis. Super idea. Now I have my programming. All I need is to get it out there. I have to work "carriage deals" with satellite operators and cable TV operators to get my channel on their network (i.e. to carry my channel on their network) so their customers can see it. There are no "standard" carriage deals. Sometimes the cable TV operator or satellite operator might pay me a flat monthly fee, or a per subscriber fee. Sometimes I will pay them. It all depends on how "hot" my programming is.

carriage return By hitting this key, the printing head or the cursor on your screen will return to the left hand margin. Usually hitting a Carriage Return or the "Enter" key includes a line feed, i.e. the paper will move up one line or the cursor will drop down one line. "Usually" does not mean always. So check. You can usually correct the problem of not having a line feed with a carriage return by moving a dip switch on the printer, changing one of the parameters of the telecommunications software program (the part where it says something about auto linefeed) or changing the computer's operating system (by doing a "Config" or the like). In most microcomputers, a Carriage Return is equivalent to a "Control M," or ASCII 13. A line feed is a "Control J."

Carriage Service Provider CSP. A commercial entity that acquires telecommunications capacity or services form a carrier for resale to a third party.

carried load 1. A telephone industry definition. Carried load is the usage measured on a circuit group. A circuit has a potential carried load capacity of 36 CCS per hour which is rarely approached because of the idle time between calls.

2. A data networking definition. The traffic that occupies a group of servers on a LAN.

Carried Traffic The part of the traffic offered to a group of servers that successfully seizes a server on a LAN.

carrier 1. A company which provides communications circuits. Carriers are split into "private" and "common." A private carrier can refuse you service. A "common" carrier can't. Most of the carriers in our industry – your local phone company, AT&T, MCI, Sprint, etc. – are common carriers. Common carriers are regulated. Private carriers are not.

2. An electrical signal at a continuous frequency capable of being modified to carry information. For analog systems, the carrier is usually a sine wave of a particular frequency, such as 1800 Hz. It is the modifications or the changes from the carrier's basic frequency that become the information carried. Modifications are made via amplitude, frequency or phase. The process of modifying a carrier signal is called modulation. A carrier is modulated and demodulated (the signal extracted at the other end) according to fixed protocols. Some of the wideband (i.e. multi-frequency) circuits are also called "carriers." T-1, which typically has 24-channel PCM voice circuits, is known as a carrier system.

Carrier Access Billing System See CABS.

Carrier Access Code CAC. A code used in North America to reach a long distance carrier, called an Interexchange Carrier (IXC). The primary carrier of choice is reached by dialing "1" plus the area code and called party number. Secondary IXCs can be reached by dialing either 101XXXX, which is the Feature Group D (equal access) CIC, or 950-XXXX, which is the Feature Group B CIC, The XXXX CIC numbers are used to dial around the carrier presubscribed to the calling telephone number. See 101XXXX, 950-XXXX, Feature Group B, and Feature Group D. See also Carrier Identification Code.

Carrier Access Line Charge CALC. A per minute charge paid by long distance companies to local phone companies for the use of local networks at either or both ends of a long distance call. This charge goes to pay part of the cost of local telephone poles, wires, etc. See Access Charge and Carrier Common Access Line Charge.

carrier access tarriff The rate charged to all long-distance carriers, including telephone companies, for access to the local telephone network.

carrier band The range of frequencies that can be modulated to carry information on a specific transmission system. See also Carrierband.

carrier bypass A long distance phone company provides a direct link between its own switching office and a customer's office, thus bypassing the local phone company. Bypass is done to save the customer or the long distance company money. Bypass is also

done to get service faster. Sometimes the local phone company simply can't deliver fast enough.

carrier circuit A higher level circuit (DS-1, DS-3, Transmission System, etc.) that has been designed to carry lower-level circuits (DS-0, DS-1).

carrier class Carrier class means telecom switching and transmission equipment that is targeted at local, long distance and international phone companies, but not at end users. There are two implications to this. It means the equipment is (theoretically) more reliable because it's built better and stronger. And it's more expensive. As phone companies face greater budget constraints, the term "carrier class" means less and less. Most telecom gear is now carrier class. See Carrier Class IP Switch.

Carrier Class IP Switch A Carrier Class IP Switch is a high volume, high reliability hybrid device for routing IP packets. It separates out high priority packets that must all arrive together, like voice and video, and delivers them immediately. All other packets are delivered through normal routing. It adds the timing precision of a switch to the low cost, speed and efficiency of a router. See also Carrier Class.

Carrier Common Line Charge CCL. The charge which IXCs (IntereXchange Carriers) pay to LECs (Local Exchange Carriers for the privilege of connecting to the end user through LEC local loop facilities. The CCL is a charge to cover a portion of the costs associated with the local loop, which is used for origination of local, intraLATA long distance (also known as "local toll"), and interLATA long distance calls. In combination, the CCL, the CALC (Customer Access Line Charge), and the monthly tariff charge for the local loop are intended to cover the costs of provisioning and maintenance of the loop, as well as to provide the LEC with a reasonable rate of return (i.e., profit) on its investment. That they do. They also encourage bypass and may, in the long term, be self-defeating. See also Access Charge.

Carrier Detect CD. The little red LED light on most modems. When this light is on, your modem is connected to another modem or communications device.

carrier detect circuitry Electronic components which detect the presence of a carrier signal and thus determine if a transmission is about to happen. Used in modems.

carrier Ethernet Wide-area Ethernet services used for high-speed connectivity within a metropolitan, nationwide, or even internationally. A typical use of carrier Ethernet is for LAN extensions over a wide area. Unlike earlier Ethernet services, which were best-effort, carrier Ethernet supports SLA (service level agreement) reliability.

carrier extension A proposal for modifying the CSMA/CD access mechanism for Gigabit Ethernet. Under a carrier extension, when a device in the network transmits, the signal stays active for a longer time before another device can attempt to transmit. This lets an Ethernet frame travel a longer distance, and thereby increases the potential network diameter.

Carrier Failure Alarm CFA. An alarm telling you that timing has been lost in your digital transmission because there are too many zeros in the message. When this happens, all the calls are lost until the equipment regains timing.

carrier frequency The frequency of a carrier wave. The frequency of an unmodulated wave capable of being modulated or impressed with a second (information-carrying) signal. In frequency modulation, the carrier frequency is also referred to as the "center frequency."

carrier grade A term that describes network hardware or software designed for telecommunications service providers. Carrier-grade components have the performance, robustness, scalability and reliability to support continuous network availability in a service provider environment.

carrier hotel A term for a building that houses many local and long distance telephone companies and many different types of local and long distance companies. Those companies typically provide voice, data, video transmission, Internet access and perhaps switching. They may also provide Internet services, such as web site hosting, and web site caching. New York City has the classic carrier hotel. It's 111 8th Avenue in Chelsea, Manhattan, near where I live. It's the old Port Authority building. It's a huge well-constructed building, with floors that can support heavy machinery. It covers an entire square block. The new owner put in heavy duty and emergency power, heavy-duty air conditioning, and tons of duct space in and around the building and to the local manholes. Each floor in the building has its own loading dock. You can drive a two-ton truck to the loading dock on the ground floor and then lift the truck and its trailer up to your floor, back the trailer into position, unhitch and drive the truck out of the building. In this way, it's possible to roll in a complete central office and have it up and running in hours. As carrier hotels are neutral sites owned by "disinterested" third party landlords, who are only motivated by the rent, the entire process of installing, maintaining and operating a central office is much simpler, faster and less expensive than colocating equipment in an ILEC (Incumbent Local Exchange Carrier) CO (Central Office). Therefore, CLECs (Competitive LECs) and IXCs (Interchange Carriers) often prefer locating in a carrier hotel. Further, the companies that live in the carrier hotel can interconnect with each other directly over very short distances (e.g., the fifth floor to sixth floor) with cable very simply, quickly and cheaply – without having to contact the local ILEC and wait an eternity while they engineer the connection to death and delay things. Some carrier hotels also do bandwidth brokering, helping their tenants do deals with each other. My friend, who's in real estate, tells me that the act of making an old building into a modern carrier hotel added a minimum of $250 million to value of the building. See also Broker, Colocation, Peering, Peering Point, Private Peering Point and One Wilshire.

Carrier Identification Code CIC. Four digit numbers used by end-user customers to reach the services of Interexchange Carriers (IXCs). The primary carrier of choice is reached by dialing "1" plus the area code and called party number. Secondary IXCs can be reached by dialing either 101XXXX, which is the Feature Group D (equal access) CIC, or 950-XXXX, which is the Feature Group B CIC, The XXXX CIC numbers are used to dial around the carrier presubscribed to the calling telephone number. See 101XXXX, 950-XXXX, Feature Group B, and Feature Group D.

Carrier Information Parameter An SS7 parameter. See SS7.

carrier interconnection plan The plan now largely implemented for connecting local and long distance phone companies in North American. The carrier interconnection plan provides the features known as Feature Group D. Exchange access plan and equal access are two other names that have been used to refer to the features provided by this plan. See Feature Group D and 101XXXX.

carrier leak The unwanted carrier remaining after carrier suppression in a suppressed carrier transmission system.

Carrier Liaison Committee CLC. A committee formed to help industry participants work together to resolve the issues of implementing 800 Portability. CLC is sponsored by the Exchange Carriers Standards Association (ECSA) and is comprised of the LECs (local exchange carriers), long distance carriers and users of 800 service.

carrier loss In T-1, carrier loss means too many zeros. A carrier loss in T-1 is said to occur when 32 consecutive zeros appear on the network. Carrier is said to return when the next 1 is detected.

carrier neutral When the Telecommunications Act of 1996 was passed it said that local phone companies must allow other phone companies to locate their equipment on their premises. This meant that other phone companies could get access to local loops and provide customers with services such as DSL. The only problem with this business is that the new carrier became locked into dealing with one phone company – the one where their equipment was. Later, several real estate companies had the idea of renting space to all companies and bringing cables in from many other carriers. Thus you could locate your equipment there but you wouldn't be tied into dealing with one carrier. You would, in essence, now be "carrier neutral."

carrier noise level The noise level resulting from undesired variations of a carrier in the absence of any intended modulation.

carrier of last resort A designation that the California Public Utilities Commission applies to California's remaining small telephone companies, which have a regulatory requirement to provide basic service to any customer requesting such service within their franchise territory.

carrier portal A network operator's branded portal, where customers access content and services.

carrier power (of a Radio Transmitter) The average power supplied to the antenna transmission line by a transmitter during one radio frequency cycle taken under the condition of no modulation. Does not apply to pulse modulation or frequency-shift keying.

carrier provided loop A local phone line owned by a long distance company that is resold as part of a WAN service. This is generally separated from your long distance service, the same way local calls are.

carrier select keys Buttons at the bottom of a payphone used to choose a long distance carrier.

carrier selection As a result of Judge Greene's Modified Final Judgment which led to the breakup of the Bell System, most local phone companies must offer their customers (business and home) the opportunity to select which long distance company they would like to use on a "primary" basis. That means when you dial 1+ (one plus) you get that carrier. To use any other long distance company you have to dial more digits, e.g. 1-0288 (for AT&T). See NANP.

carrier sense In a local area network, a PC or workstation uses its network card to detect if another station is transmitting. See CSMA.

Carrier Sense Multiple Access CSMA. In local area networking, CSMA is a way of getting onto the LAN. Before starting to transmit, personal computers on the LAN "listen" to make sure no other PC is transmitting. Once the PC figures out that no other PC is transmitting, it sends a packet and then frees the line for other PCs to transmit. With CSMA, though stations do not transmit until the medium is clear, collisions still occur. Two alternative versions (CSMA/CA and CSMA/CD) attempt to reduce both the number of collisions and the severity of their impact. See CSMA/CA and CSMA/CD.

Carrier Sense Multiple Access/Collision Avoidance CSMA/CA. A protocol that requires the PC to sense if another PC is transmitting. If not, it begins transmitting. Under CSMA/CA, a data station that intends to transmit sends a jam signal; after waiting a sufficient time for all stations to pick up the jam signal, it sends a transmission frame; if while transmitting, it detects another station's jam signal, it stops transmitting for a designated time and then tries again. For a longer explanation see CSMA/CA and Ethernet.

carrier sense multiple access/collision detection A network control scheme. It is a contention access control scheme. It "listens" for conflicting traffic to avoid data collisions. The Ethernet LAN uses CSMA/CD, then waits a small amount of time and then tries again. For a longer explanation, see CSMA/CD and Ethernet.

Carrier Serving Area CSA. The geographic area served by a PSTN (Public Switched Telephone Network) CO (Central Office). The CSA generally is considered to have a radius of 12,000 feet. It is the geographical portion of a wire center which will be provided with customer facilities primarily via digital loop carrier systems.

carrier shift 1. A method of keying a radio carrier for transmitting binary data or teletypewriter signals, which consists of shifting the carrier frequency in one direction for a marking signal and in the opposite direction for a spacing signal.

2. In amplitude modulation, a condition resulting from imperfect modulation whereby the positive and negative excursions of the envelope pattern are unequal, thus effecting a change in the power associated with the carrier. There can be positive or negative carrier shift.

carrier signal A continuous waveform (usually electrical) whose properties are capable of being modulated or impressed with a second information-carrying signal. The carrier itself conveys no information until altered in some fashion, such as having its amplitude changed (amplitude modulation), its frequency changed (frequency modulation) or its phase changed (phase modulation). These changes convey the information.

carrier synchronization In a radio receiver, the generation of a reference carrier with a phase closely matching that of a received signal.

carrier system A system where several different signals can be combined onto one carrier by changing some feature of the signals transmitting them (modulation) and then converting the signals back to their original form (demodulation). Many information channels can be carried by one broadband carrier system. Common types of carrier systems are frequency division, in which each information channel occupies an assigned portion of the frequency spectrum; and time division, in which each information channel uses the transmission medium for periodic assigned time intervals.

carrier terminal The modulation, demodulation and multiplex equipment used to combine and separate individual channels at the ends of a transmission system.

Carrier To Noise Ratio CNR. In radio receivers, the ratio, expressed in decibels, of the level of the carrier to that of the noise in the receiver bandwidth before any nonlinear process such as amplitude limiting and detection takes place.

carrier transitions Carrier transitions appear on a serial link whenever there is an interruption in the carrier signal (such as an interface reset at the remote end of a link). Another example of a carrier transition is with frame relay. The keep-alive times are too far apart, causing the line protocol to go down but the frame pvc is stays up.

carrier wave A carrier wave is the radio frequency wave generated at a transmitting station for the purpose of carrying the modulated frequency wave. Carrier waves are a form of analog signal that is used to encode information. The coding used to impress information on the wave can be a function of the frequency (the number of waves or cycles per second) or amplitude (height) of the waves or cycles. See PCM, TDM.

Carrier-to-Interference Ratio C/I. The ratio of the amplitude of a radio frequency carrier to the amplitude of any form of interference including both noise and other undesired carriers. The C/I is a broader measure than C/N (Carrier-to-Noise Ratio) because it includes undesired radio frequency carriers.

Carrier-to-Noise Ratio C/N. The ratio of the amplitude of a radio frequency carrier to the amplitude of background noise.

carrierband Same as single-channel broadband. See also Carrier Band.

Carrierless Amplitude and Phase CAP. A transceiver technology that can be used in ADSL systems, CAP is a variation of Quadrature Amplitude Modulation (QAM).

With CAP, the POTS upstream and downstream channels are supported by splitting the frequency spectrum. CAP was the first ADSL transceiver to be commercially deployed, but Discrete Multitone (DMT) was selected as the standard.

carrier's carrier A carrier's carrier is a company that provides telecommunications services to interexchange carriers or telephone companies. A carrier's carrier does not provide service to the public and therefore is subject to fewer regulations.

CARS Cable Television Relay Service, a microwave service authorized by Part 78 of the FCC Rules for the purpose of transmitting signals intended for carriage over a cable television system. Back in the days when "CATV" stood for Community Antenna Television, "CARS" stood for Community Antenna Relay Service. Over the years, "CATV" has evolved to mean Cable Television, but CARS remains CARS.

Carson's Rule A radiocommunications term. Carson's Rule is a method of estimating the bandwidth of an FM (Frequency Modulation) subcarrier system. It is commonly used in satellite systems in order to ensure that a high-fidelity, sharp TV picture will be delivered over a subcarrier TV channel. Violation of Carson's rule results in a higher video signal-to-noise ratio, at the expense of streaking in fast-moving scenes, sharpness of picture, and loss of audio fidelity. Carson's rule states that $B = 2x(Df+fmax)(A-3)$, where B is the bandwidth, Df is the peak deviation of the carrier frequency, and fmax is the highest (maximum) frequency in the modulating subcarrier signal.

Carterfone A device for connecting a two-way mobile radio system to the telephone network invented by Thomas Carter. It was electrically connected to the base station of the mobile radio system. Its electrical parts were encased in bakelite. When someone on the radio wanted to speak on a "landline" (the phone system), the base station operator would dial the number on a separate phone then place the telephone handset on the Carterfone device. The handset was acoustically, not electrically, connected to the phone system. No more than 4,000 Carterfones were ever installed, yet the Bell System thought they were the most dangerous device ever invented. Tom Carter died in Gun Barrel, TX where he lived, in the early part of 1991. He died a poor man. See Carterfone Decision.

Carterfone Decision In the summer of 1968 the FCC determined that the Carterfone and other customer phone devices could be connected to the nation's phone network – if they were "privately beneficial, but not publicly harmful." The Carterfone decision was a landmark. It allowed the connection of non-telephone company equipment to the public telephone network. This decision marked the beginning of the telephone interconnect business as we know it today. The Carterfone decision made a lot of lawyers rich before all the rules on connection to the network got cleared up, and finally codified in something called Part 68 of the FCC's Rules. See Carterfone, NATA and Network Harm.

Carterphone Incorrect spelling for Carterfone. See Carterfone.

cartridge 1. A device which holds magnetic tape of some kind.

2. A device to translate (transduce) stylus motion to electrical energy in a phonograph. It comes in three basic types – moving magnetic coil, induced magnet and ceramic. A phono cartridge is also call a pickup. Most record players use ceramic cartridges because they have higher output than the three magnetic types and can work with a less powerful (i.e. cheaper) amplifier.

CAS 1. Centralized Attendant Service. One group of switchboard operators answers all the incoming calls for several telephone systems located throughout one city. CAS is used by customers with several locations in the same geographic area, i.e. retail stores, banks.

2. Communicating Applications Specification. A high-level API (Application Programming Interface) developed by Intel and DCA that was introduced in 1988 to define a standard software API for fax modems. CAS enables software developers to integrate fax capability and other communication functions into their applications. See CAS 2.0.

3. CAS is a generic acronym for Channel Associated Signaling or Call-path Associated Signaling. CAS is in-band signaling used to provide emergency signaling information along with a wireless 911 call to the Public Safety Answering Point (PSAP). This signaling information includes the phone number of the wireless phone and coding used to derive a general location of the caller, and meets the Enhanced 911 Phase 1 FCC requirements. This coding can be either a p-ANI or an ESRD. This in-band signal is made up of tones which pass within the voice frequency band and are carried along the same circuit as the talk/call path that is being established by the signals. NCAS is a generic acronym for Non Call-path Associated Signaling. This definition contributed by Glenda Drizos and Doug Puckett of Sprint PCS, Overland Park, Kansas.

CAS 2.0 Communicating Applications Specification. Fax standard for both fax and voice applications. Developed by Instant Information, Inc. (I3), CAS 2.0 offers a simple, yet highly flexible and scaleable model that has allowed vendors such as Brooktrout, Dialogic and FaxBack to add greater and more sophisticated functionality to their products. Originally designed and developed by Intel and Digital Communications Associates in 1988, CAS is

an API (Application Programming Interface) specification that provides support for programs sending data to other devices and computers. It is one of the world's most popular software interfaces to a fax board and is a universal standard embraced by hundreds of developers. Since its inception, more fax ports have shipped supporting CAS standards than any other communications protocol. New features in CAS 2.0 include:

Full 32-bit Windows 95, NT 3.51, NT 4.0, support.

Full support for asynchronous real-time fax applications.

Class 1 and Class 2 support.

C++ class library support.

An intuitive redesigned setup user interface.

Full Brooktrout, Gammalink and WildCard co-processed hardware support.

cascade 1. To connect the output of a device into the input of another device, which then may in turn be connected to another device. Imagine the organization of a company. At the top is the president. Reporting to him are three vice presidents. Reporting to each of these vice presidents are five directors. Reporting to each of these directors are five managers. If each of these positions were a piece of network gear, you'd have a classic cascaded topology. For example, a long-haul circuit may involve multiple, cascading repeaters; multiple, cascading virtual circuits may be involved in a Frame Relay network.

2. A Windows term. When windows cascade, they are arranged in an overlapping pattern so that the title bar of each window remains visible.

3. A series of reply posts to a USENET message, each adding a trivial or nonsense theme to the collection of previous replies. Some consider this art; there is a USENET newsgroup devoted to propagating this art form (alt.cascade).

cascade amplification Successively using two or more amplification systems. Most radio and audio products have more than one stage of amplification.

cascaded amplifier Two or more amplifiers coupled together. Most radio and audio products have more than one stage of amplification.

cascaded stars Local area network topology in which a centralized multiport repeater serves as the focal point for many other multiport repeaters.

cascaded topology See Cascade.

cascading faults Faults that cause other faults. Typically faults in a network causing other faults.

cascading menu A Windows term. A menu that is a submenu of a menu item. Also known as a hierarchical menu.

cascading notification A feature of some sophisticated voice mail systems. Let's say someone leaves a message for you in your voice mail box. Your voice mail system then automatically goes out to find you, i.e. to notify you. It may start by lighting your message light, calling your home phone number, calling your cellular phone, calling your beeper, etc. I like this feature because when I want you, I want you. And a little mechanized help is much appreciated. I first saw the feature in Macrotel's MVX voice mail series.

Cascading Style Sheets (CSS. According to CNET, cascading style sheets are a big breakthrough in Web design because they allow developers to control the style and layout of multiple Web pages all at once. Before cascading style sheets, changing an element that appeared on many pages required changing it on each individual page. Cascading style sheets work just like a template, allowing Web developers to define a style for an HTML element and then apply it to as many Web pages as they'd like. With CSS, when you want to make a change, you simply change the style, and that element is updated automatically wherever it appears within the site. Both Navigator 4.0 and Internet Explorer 4.0 support cascading style sheets. If you needed any more proof of the problem-solving nature of CSS, the World Wide Web Consortium (W3C) has recommended cascading style sheets (level 1) as an industry standard. See also: DHTML, HTML.

CASE Computer Aided Software Engineering. CASE is a new, faster, more efficient way of writing software for some applications. The idea with CASE is to sketch out relations between databases, events, and options and then have the computer write the code.

case method A traditional way of load testing computer telephony systems, the case method involves gathering many individuals together in a room full of telephones along with several cases of an appropriate libation (frequently beer), and using these individuals to simulate real users calling into (or being called by) the computer telephony system. Case method testing usually continues until all the cases have been consumed, the testing is completed, it becomes too late in the evening to continue, or the perspective of the gathered individuals becomes too subjective to be of use any longer to those conducting the test. This definition courtesy of Steve Gladstone, author of the book Testing Computer Telephony Systems.

case sensitive This means that uppercase letters must be typed in uppercase on your keyboard, and that lowercase letters must be typed in lowercase. It is important to key in your data in the exact combination of upper or lower case characters. Inputting in the wrong case could make your entry invalid for some fields (for example, password). DOS and Windows are much less case sensitive than Unix, for example.

CASE Tools These tools provide automated methods for designing and documenting software programming. Computeraided software engineering (CASE) sketches relations between databases, events, and options. It then provides a language in which the computer writes the code, letting programmers develop applications faster. It's new and seems to work in limited instances.

cassegrain antenna An antenna in which the feed radiator is mounted at or near the surface of a concave main reflector and is aimed at a convex secondary reflector slightly inside the focus of the main reflector. Energy from the feed unit illuminates the secondary, reflects it back to the main reflector, which then forms the desired forward beam. This technique is adapted from optical telescope technology and allows the feed monitor radiator to be more easily supported.

cassette tape A slow, inefficient method of storing and retrieving data which uses the same technology as audio cassettes – like the Sony Walkman. Some PBXs use cassette tape to backup their user programming and database.

castellation A series of ribs and metallized indentations that defines edge contact regions.

casual billing A billing agreement between the local and long distance company that provides for the local company to bill long distance calls for an IXC made by end users who dial 10XXX. Normally, the IXC does not have an account for the end user.

casual caller A long distance telephone user who uses a long distance call around number, e.g.; 10-1xxxx. This person usually does not went to be pic'ed (chose a Primary Interexchange Carrier). See 101xxxx.

casual calling As a result of divestiture most long distance companies in the United States have a 10-1X-XXX code. These codes are used for various purposes, one of which is long distance calling. To bypass your local exchange carrier and to bypass your selected long distance carrier, your must dial as follows: 10-1X-XXX + 1 + telephone number. You do this to save money or to make a long distance call if your long distance phone company is kaput. You may also use this service if can't be equal accessed to your preferred long distance carrier. In addition to using the 10-1X-XXX code for long distance calling, the code is also used for the following: Temporary access. While you're waiting for your account to become active, you may use the 10-1X-XXX code to temporarily access the long distance company's network. The bill for these calls typically arrives on the phone bill provided by your chosen local phone company.

You need to watch your phone bills. A class action lawsuit has been filed in the United States District Court for the District of Columbia, in Washington, D.C., on behalf of all MCI subscribers who were charged MCI's "casual calling" rates (in some cases as high as $2.87 for a one-minute call) instead of the lower rates which MCI advertises and which subscribers expected to be charged. The lawsuit alleges that this practice of MCI violates the Communications Act of 1934. MCI has one set of rates for subscriber direct dialing, including its widely promoted "Five Cents Sunday" Calling Plan. MCI also has another, much higher set of rates it charges for non-subscriber or casual calling: $0.38 per minute, plus a surcharge of $2.49 for every call. The lawsuit alleges, however, that MCI has charged many of its own subscribers the higher non-subscriber, casual calling rates.

casual day See Dress-down day.

casual encryption Want to send something to someone over a network? There are various levels of encryption – some more hard to crack than others. Remember the harder it is to crack, the harder (and more expensive) it is to administer and work with. PC Magazine describes casual encryption as encoding your documents using compression software such as WinZip and assigning a password to the file – so that only the person knowing the password could actually unzip the file and read it.

CAT 1. Shortened way of saying "Category," as in Cat 1 cabling. Say Cat 5 or CAT5 and you mean Category 5 wiring. See Category 1 through 5 and Category of Performance.

2. CAT (concatenate) - a UNIX command that dumps a file to a standard output. If you type 'cat myfile', myfile would be displayed to the screen. If you typed 'cat myfile | wc' UNIX would dump the file contents to the program 'wc' (word count) and just print the number of words in the file.

cat whisker A fine metal thread resembling the arched shape of a cat's whisker, used in early radio wave detecting crystal sets.

catastrophe theory A special branch of dynamical systems theory that studies and classifies phenomena characterized by sudden shifts in behavior arising from small changes in circumstances.

Category 1 CAT 1. An unspecified Category of Performance for inside wire and cable

systems. CAT 1 cables can be of various gauges, and are useful in support of applications requiring a carrier frequency less than 1 MHz, which roughly translates into 1 Mbps, depending on the compression scheme employed. Example applications include analog voice, ISDN BRI, and doorbells. CAT 1 is specified in ANSI/ICEA S-80-576 and S-91-661. See also Category of Performance.

Category 2 CAT 2. A Category of Performance for inside wire and cable systems. CAT 2 cables can be of either 22 or 24 gauge, and are useful in support of applications requiring a carrier frequency of up to 4 MHz; the transmission rate achievable depends on the compression scheme employed. CAT 2 requirements are based on the IBM Type 3 cabling system for low-speed data cable suitable for 4-Mbps Token Ring. Example applications include 4 Mbps Token Ring LANs (802.5), 1Base-5 Ethernet LANs (802.3), and T-1. CAT 2 is specified in ANSI/ICEA S-80-576 and S-91-661.See also Category of Performance.

Category 3 CAT 3. A Category of Performance for inside wire and cable systems. CAT 3 cables can be of either 22 or 24 gauge, and are useful in support of applications requiring a carrier frequency of up to 16 MHz; the transmission rate achievable depends on the compression scheme employed. Example applications include POTS, ISDN, T-1, 4/16 Mbps Token Ring (802.5), and 10Base-T Ethernet (802.3). Cat 3 technical specifications are defined by FCC Part 68, ANSI/EIA/TIA-568, TIA TSB-36 and TIA TSB-40. CAT 3 safety requirements are defined by UL 1459 (Telephone), UL 1863 (Wire and Jacks) and NEC 1993, Article 800-4. See also Category of Performance.

Category 4 CAT 4. A Category of Performance for inside wire and cable systems. CAT 4 cables can be of various gauges, and are useful in support of applications requiring a carrier frequency of up to 20 MHz; the transmission rate achievable depends on the compression scheme employed. Example applications include 4/16 Mbps Token Ring (802.3) and 10 Mbps Ethernet (802.3). CAT 4 technical specifications are defined by FCC Part 68, EIA/TIA-568, TIA TSB-36, and TIA TSB-40. CAT 4 safety requirements are defined by UL 1459 (Telephone), UL 1863 (Wire and Jacks) and NEC 1993, Article 800-4. See also Category of Performance.

Category 5 CAT 5. A Category of Performance for inside wire and cable systems. CAT 5 cables can be of various gauges, and are useful in support of applications requiring a carrier frequency of up to 100 MHz; the transmission rate achievable depends on the compression scheme employed. Example applications include 4/16 Mbps Token Ring, 10/100Base-T, 100VG-AnyLAN, and even 155 Mbps ATM LANs. CAT 5 is now the most common cabling being installed for LAN connectivity. Increasingly, CAT 5 cabling is being installed for both data and voice use, and is the cabling of choice for forward-looking companies. Category 5 technical specifications are defined by FCC Part 68, EIA/TIA-568, TIA TSB-36, TIA TSB-40, and ANSI/ICEA S-91-661. See also Category of Performance.

Category 5e See Category 5 Enhanced.

Category 5 Enhanced CAT 5e. CAT 5e has become the de facto minimum standard for cabling. It supports signaling rates of up to 100 MHz over distances of up to 100 meters. Specifications call for a tighter twist, electrical balancing between pairs, and fewer cable anomalies, such as inconsistencies in the core diameter. CAT 5e supports 100Base-T, ATM, and Gigabit Ethernet, which is 100 million bits per second. By way of comparison, category 3 cable is rated for any application up to 16 Mhz (Frequency) while category 5e is rated for up to 100 Mhz (Frequency) Computer networks and phone systems all operate at one or another frequency. Category 5e will accomodate both analog phones as well as any computer protocol such as Gigabit Ethernet. See also Category of Performance.

Category 6 CAT 6. Proposed Category 6 standards from the TIA (Telecommunications Industry Association), known as Class E standards at the ISO (International Organization for Standardization), describe a new performance range for unshielded and screened twisted-paid cabling. Category 6/Class E is intended to specify the best performance that UTP and ScTP cabling solutions can be designed to deliver based on current technology. Category 6e will be specified in the frequency range of 1-250 MHz. The cabling system actually will be rated at 200 MHz in consideration of positive PSACR (Power Sum Attenuation-to-Crosstalk Ratio, also known as PS-ACR), although equipment crosstalk cancellation capabilities will allow the system to be characterized to 250 MHz. The ScTP version of CAT 6 comprises four twisted pairs of 100 Ohm wire, with the entire cable being protected from ambient noise by means of a metallic screen. Most CAT 6 cables have a center filler that separates each pair within the cable jacket. The interface at the workstation will be an 8-position modular jack. According to the Telecommunications Industry Association, "Category 6 cabling is the latest addition to the structured cabling standards and has twice the bandwidth of category 5e cabling. This improved bandwidth, together with vastly improved immunity from external noise, provides the potential for category 6 to support multi-gigabit applications. This white paper provides an update on category 6 cabling and applications standards together with references for finding category 6 information and products." See also Category of

Performance and PS-ACR.

Category 6e See Category 6 Enhanced.

Category 6 Enhanced Proposed category 6/class E standards describe a new performance range for unshielded and screened twisted-paid cabling. Category 6/Class E is intended to specify the best performance that UTP and ScTP cabling solutions can be designed to deliver based on current technology. Category 6e will be specified in the frequency range of 1-250 MHz. For Category 6e, the 8-position modular jack interface will be mandatory in the work area. See Category 6.

Category 7 CAT 7. Also known as Category 7/Class F. A developing cabling standard from the TIA (Telecommunications Industry Association) for STP (Shielded Twisted Pair), intended to support signaling rates up to 600 MHz in support of ATM and Gigabit Ethernet. A corresponding standard under development by the ISO (International Organization for Standardization) is known as Class F. A typical CAT 7 cable is anticipated to comprise four 23 AWG (American Wire Gauge) twisted pairs, each enveloped within a solid metallic foil wrap. An overall braided sheath typically surrounds the four foil-wrapped pairs, and a drain wire will serve to ground the cable in order to reduce electrical noise potential. Such double-shielded cabling systems sometimes are known as SSTP (Double-Shielded Twisted Pair, as in Shielded-Shielded Twisted Pair) and PiMF (Pairs in Metal Foil). See also Category of Performance.

category of performance As we try and push more and more data faster and faster down a pair or two of wires, so the quality of the wires and the components they connect to has become increasingly important. You can't push 100 Mbps (Million bits per second) down junky phone lines. As a result, the EIA/TIA (Electronic Industries Alliance/Telecommunications Industry Association) has defined cabling and cabling component standards. The idea is that if your stuff conforms to the standard, users will be able to achieve the data rates and reliability they want, assuming that the cabling system is installed properly. Some standards specify physical characteristics, such as thickness of cable, plastic material used in the outer jacket, etc. In the main, these "Category of Performance" standards specify tests which the cabling and cabling components must pass. There were originally five categories of tests. But now there are effectively only two categories that anyone buys – Category 3 and Category 5. In simple terms, if all you want is to support voice and data to 10 Mbps (Megabits per second), i.e. standard Ethernet, then Category 3 will do the trick. If you are transmitting at 100 Mbps on your local area network, then you need Category 5. Today, Cat 5 is 20% to 30% more expensive than Cat 3, but it's well worth it. In my unhumble opinion, all cable going to the desktop today should be Cat 5, since it will support both voice and high-speed data. While Cat 5 isn't necessary for either voice or 10 Mbps data, it's easier and ultimately less expensive to install a single wiring system.

Now, let's talk about how they measure Cat 3 and Cat 5. First, you should understand that all the tests are self-certifying, which means that while there are standards, each manufacturer is itself responsible for conforming. No one will put a manufacturer in jail if his Cat 5 stuff doesn't perform to Cat 5 standards. The only thing likely to happen to the manufacturer is that the world will find out his stuff is garbage and he'll go broke.

The concept of the test is simple. The test for cable and components is a swept frequency test. Note: The standards all speak to the carrier frequency supported, which is measured in Hertz, usually in MHz, or MegaHertz. The transmission rate, which is what we ultimately are interested in, is measured in bps (bits per second). The transmission rate relies on the carrier frequency. At a given carrier frequency, the signal is modulated in various ways to allow the impression of one or more bits per Hertz. The bottom line is that the standards address Hz, not bps.

For example, Cat 3 must pass all signals up to 16 MHz. Cat 5 must pass signals from up to 100 MHz. Cat 3 and Cat 5 cables and components are designed to support any applications intended to operate over those frequencies. Interestingly, both Cat 3 and Cat 5 cabling are the same thickness, namely 24 gauge. Cat 5 conductors, however, are

Category	Cable Type	Application & Speed Supported
1	UTP	Analog Voice
2	UTP	Digital Voice, 1 Mbps Data
3	UTP/STP	16 Mbps Data
4	UTP/STP	20 Mbps Data
5	UTP/STP	100 Mbps Data
5e	UTP/STP	up to 1,000 Mbps Data
6	UTP/STP	1,000 Mbps Data
7*	UTP/STP/Fiber	1,000 Mbps Data
*Proposed standard		

manufactured according to tighter specifications, which result in fewer anomalies, such as nicks in the conductor, and variations in diameter. The twist structure on Cat 5 is tighter, and the insulation is better. Connecting hardware is definitely different. Altogether, the standards specify parameters including frequency rating, attenuation (signal power loss), impedance (resistance), crosstalk, structural return loss, and delay skew. The standards also define maximum distances, as attention renders the signal useless beyond a certain distance.

Below is Harry's quick rule of thumb for what you should buy based on what you want to transmit, as against the swept frequency test:

Harry's Rules of Thumb: If you want the most flexibly wired office – the Office of The Future – install nothing less than Cat 5e cabling to everyone's desk. And put in twice as much cabling as you ever dreamed you will need. When installing cable, never use staples; never use tie wraps (some idiot will tighten them too tightly); never untwist wire before you punch it down; (the twists should be right up to the termi0x2265ation); never strip more jacket off the wire than is needed to terminate it; never pull too hard on any cable - especially around a corner; never pull on a cable to straighten out a kink or loop (always go back and untwist it); never stuff too many cables in too small a conduit and never use less than a quad-shielded coaxial cable. See Commercial Building Telecommunications Wiring Standard.

CAT1 See Category 1 and Category of Performance.

CAT2 See Category 2 and Category of Performance.

CAT3 See Category 3 and Category of Performance.

CAT4 See Category 4 and Category of Performance.

CAT5 See Category 5 and Category of Performance.

CAT5e See Category 5 Enhanced and Category of Performance.

CAT6 See Category 6 and Category of Performance.

CAT6e See Category 6e and Category of Performance.

CAT7 See Category 7 and Category of Performance.

catenet As defined by Louis Pouzin in 1974, a catenet is "an aggregate of networks (which would) behave like a single logical network." In other words, an "internetwork" in which networks are interconnected by "gateways," which are known as "routers" in contemporary terminology. Pouzin's concept was later accepted by DARPA (Defense Advanced Research Project Agency) as the goal of the project it was supporting – that project developed into what we now know as the Internet.

CATNIP Common Architecture for Next Generation Internet Protocol. One of the three IPng candidates.

cathode The heated element which emits electrons in a vacuum tube. It may be a filament, or may be a separate element, heated by proximity to a filament. It is maintained at a negative potential in respect to the anode or plate. Cathodes have other applications, also.

cathode ray The beam of electrons emitted by a cathode. See Cathode Ray Tube.

Cathode Ray Tube CRT. A TV screen. A CRT is a tube of glass, used in television, oscilloscope and computer terminals, from which air has been removed (i.e. vacuum tube). At the back of the CRT is an electron gun which directs an electron beam to the front of the tube. The inside front of the tube has been coated with fluorescent material which reacts to and lights up once the electron beams hit. CRTs are very reliable if they are vented, since the electron gun gets hot. CRTs have a "memory." They will memorize what's been left on their screen for a while, i.e. the image is burned into the screen. And you'll see it even though the screen is turned off. In short, turn your screen off when you're not using it. Or run a "CRT-saving" program which varies the image on the screen.

cathodic protection A means of controlling corrosion of metal through use of a sacrificial metallic anode. A form of galvanic corrosion protection in which a conductor with a negative charge will repel chlorine ions, rather than attract them, as would be the case with a positive charge. Cathodic protection is the reason that twisted-pair, copper local loops use -48 volts, rather than +48 volts, for "Ring." At least it was in the olden days of uninsulated local loops. It still works. The same principle works to protect bridges, pipelines, and other metallic structures.

cathodic protection cable Cable used for direct burial dc service in cathodic protection installations for pipelines and other buried or water submerged metallic structures.

CATI Computer Assisted Telephone Interviewing, a market research term for a call center based on the use of a computerized database.

CATLAS AT&T software standing for Centralized Automatic Trouble Locating and Analysis System. CATLAS is used as a maintenance tool for locating and diagnosing problems in AT&T electronic central offices.

CATNIP Common Architecture for Next Generation Internet Protocol. One of the 3 IPng candidates.

CATS 1. Consortium for Audiographics Teleconferencing Standards, San Ramon, CA. 510-831-4760. CATS describes itself as a non-profit corporation dedicated to promoting standards for this technology, which it describes as enhancing audioconferencing by allowing people at different sites to work together in real time to create, manipulate, edit, annotate and reference still images. Now called IMTC.

2. Calling Card & Third Number Settlement. An RBOC system for ICS processed through CMDS. Similar to BEARS.

Cats and Dogs In 17th century England, houses had thatched roofs-thick straw-piled high, with no wood underneath. It was the only place for animals to get warm, so all the dogs, cats and other small animals (mice, bugs) lived in the roof.   When it rained it became slippery and sometimes the animals would slip and fall off the roof.
Hence the saying "It's raining cats and dogs."

catsup Another American word for ketchup, which people in Australia and England call tomato sauce.

CATV CAble TeleVision. This term originally stood for "community antenna television," reflecting the fact that the original cable systems carried only broadcast stations received off the air; however, as cable systems began to originate their own programming, the term evolved to mean Cable Television. CATV is a broadband transmission facility. It generally uses a 75-ohm coaxial cable which simultaneously carries many frequency-divided TV channels. Each channel is separated by guard channels. Some of the industry's first CATV pioneers were TV-set dealers who figured that cable would drive demand. See Addressable Programming and Broadband.

CAU 1. Northern Telecom term for Connection Arrangement Unit.

2. Controlled Access Unit. CAU. An intelligent hub from IBM for Token Ring networks in conjunction with IBM LAN Network Manager software.

CAVE 1. Cave Automatic Virtual Environment. A sophisticated virtual reality facility developed by the Electronic Visualization Lab at the University of Illinois at Chicago. CAVE is much like the HoloDeck of "Star Trek: The Next Generation," although a headset and a wand are required to create the illusion. CAVE allows scientists to see, touch, hear and manipulate data in order to do such things as create and test new models of various machines, or to manipulate atoms of a molecule. See also Virtual Reality.

2. Cellular Authentication and Voice Encryption. An encryption algorithm specified in ANSI-41 (formerly TIA standard IS-41C) to initiate the system of authentication challenges in order to prove the identity of a mobile phone. CAVE is used in conjunction with the A-key (Authentication key), MIN (Mobile Identification Number) and ESN (Electronic Serial Number) in order to prevent cellular fraud artists from capturing the MIN and ESN data. Such data typically is transmitted "in the clear." The A-key is intended to be known only to the phone and the AC (Authentication Center). Once the A-key is transmitted, the phone and the AC generate a second secret key, known as a SSD (Shared Secret Data), the value of which can be updated in the event that the IS-41 network suspects that it has been compromised.

cavity A volume defined by conductor-dielectric or dielectric-dielectric reflective boundaries, or a combination of both, and having dimensions designed to produce specific interference effects (constructive or destructive) when excited by an electromagnetic wave.

cavity magnetron An early British innovation in radar systems development during World War II which permitted the use of extremely short waves (microwaves) and improved the quality of the information and images possible through radar systems. See Magnetron.

cavity wall A wall built of solid masonry units arranged to provide air space within the wall.

CB Why 10-4, good buddy, that stands for Citizens Band. Also known as Children's Band, not because of Radio Shack's toy walkie talkies, but for the inane chatter that sometimes goes on in these channels. In short, CB is low-power (up to four Watts permitted) public radio. You do not need permission from the FCC to transmit or receive at these frequencies. Thus CB's great popularity. CB went through a boom (perhaps a craze?), then it ran out of radio frequencies and public enthusiasm. Its original frequencies were 26.965 to 27.225 Mhz. Now the FCC's given it new frequencies – 462.55 to 469.95 MHz. These new frequencies are much better, clearer and less congested. If you buy a CB set, make sure you get one that operates in these higher frequencies. In some countries they use different frequencies. CB radio is not allowed in many countries, even some civilized countries, though it will obviously work there.

CBA Your PBX is busted. You call the phone man. He comes. You ask him what's wrong with it? He shrugs his shoulders. "CBA," he answers. What's that? you ask. He says "Could Be Anything." Stupid, but funny.

CBAC Context Based Access Control. CBAC is a sophisticated way of managing different types of traffic on a single network. It allows an intelligent network (i.e. one carrying the different types of traffic) to recognize a given traffic and prioritize its movement over the network accordingly. In other words, voice will have greater priority over data, because voice is more sensitive to delays and dropouts, etc. See also QoS (Quality of Service) and RSVP (Resource Reservation Protocol).

CBB Circuit Breaker Bay is a distribution panel used to power equipment in a rack/row of a site.

CBC CBC Cipher Block Chaining. A technique commonly used by encryption algorithms like Data Encryption Standard (DES) - CBC, where a plain text message is broken into sequential blocks. The first block is encrypted using a given cipher, creating cipher text. That cipher text is used to encrypt the second block of plain text. This pattern continues, with each subsequent block of plain text being encrypted using the cipher text encrypted just before it.

CBDS Connectionless Broadband Data Service: A connectionless service similar to Telcordia Technologies' SMDS defined by European Telecommunications Standards Institute (ETSI). In short, the European version of SMDS.

CBEMA Computer Business Equipment Manufacturers Association. A lobbying group created to protect the interests of its members. CBEMA is the author of the AC voltage disturbance tolerance specification to which all computing and business equipment is designed. Specifies overvoltage and undervoltage events that computing equipment must withstand. This standard, for example, provides that all computing and business equipment must withstand a power loss or transfer time of 12ms.

CBF Computer Based Fax.

CBFM Cured By F...ing Magic. When the telco carrier takes the circuit for test Wednesday morning and calls you back 2 days later with no reason for the outage..the circuit problem resolution is CBFM.

CBH See Component Busy Hour.

CBI You get the phone call at 2 A.M. that XYZ has broken/gone down, and you dial into the network to troubleshoot. While looking, the problem clears, but not as the result of what you did or didn't do. It's "cleared before isolation," since you can not point to what actually fixed the outage. John Schubert, a telecom engineer, told me that he commonly guessed the technician on-site actually found the problem but was too embarrassed, proud, or whatever, to admit their mistake as to what caused the problem and what they did to fix it. See also CBA and CBFM.

CBK Change BacK.

CBQ Class-Based Queuing. A queuing algorithm used in routers to manage congestion. Through user-definable class definitions, incoming packet traffic is divided into classes. A class might include all traffic from a given interface, all traffic associated with a particular application, all traffic intended for a particular network or device destination, or all traffic of a specific priority classification. Each class of traffic is assigned to a specific FIFO (First In First Out) queue, each of which is guaranteed some portion of the total bandwidth of the router. Should some class(es) of traffic not make full use of their allocated bandwidth, CBQ portions out that available bandwidth to other class-specific FIFO queues on a proportionate basis. See also FIFO, RED, Router and WFQ.

CBR 1. Constant Bit Rate. CBR refers to processes such as voice and video that require a constant, repetitive or uniform transfer of information. The ATM Forum defines it as "an ATM service which supports a constant or guaranteed rate to transport services such as video or voice as well as circuit emulation which requires rigorous timing control and performance parameters."

2. Committed Bit Rate.

CBT Computer Based Training. Also known as CAI (Computer Assisted Instruction), CBT commonly supports self-paced learning through the use of a CD-ROM storage technology. You plug the CD-ROM into the multi-speed CD-ROM drive of your high-performance PC, and begin the course. CD-ROM storage technology provides enough memory to support great graphics, and is fast enough to provide hyperlinks so you can move around the material quickly. At the end of each section, the CBT system will offer you a quiz to test your knowledge, and will present you with your score. CBT was crude prior to the development of CD-ROM and high-performance PCs. See also CAI and CD-ROM.

CBTA Canadian Business Telecommunications Alliance. The largest organization of business telecom users in Canada. According to the CBTA's own literature, the CBTA is a national non-profit organization that has been working on behalf of business telecommunications customers for over 34 years. As a major voice in the Canadian telecommunications industry, the CBTA represents about 400 organizations from all sectors of the Canadian economy including industry, commerce, education, health care, and government. CBTA member organizations have combined annual expenditures of over $4 billion (Canadian dollars).

The CBTA strives to facilitate a competitive advantage in Canadian business (where business is understood to include all for-profit, not-for-profit, governmental, educational, and medical institutions), through the strategic application of telecommunications by proactively advocating the common interests of its members and by promoting information exchange and professional development. The main goals of the CBTA are to encourage innovation, quality, and choice in the telecommunications marketplace; to build a strong, influential, and representative national organization; to provide professional development opportunities and relevant services to its members; to enhance the Alliance's visibility as a leading industry authority. CBTA was at the Canadian Trust Tower, 161 Bay Street, Suite 3650, Toronto, Ontario, Canada M5J 2S1. 416-865-9993 and fax 416-865-0859. www.cbta.ca The CTBA closed its doors in June, 1999 due to lack of funding.

CBUD Call Before U Dig. Operational management system for protection of fiber facilities. May have electronic geographic maps of states, counties and city streets where the carrier has buried facilities, upon which reported construction activities are automatically mapped. Human technicians verify that the activities do not pose a danger to the facilities, or dispatch on-site technicians when facilities may be at risk.

CC 1. Call Control. A wireless telecommunications term. A term used to refer to circuit communications management.

2. Country Code. The portion of an international telephone number used to identify the country of the called party. That country code may be one, two or three digits. An international phone number consists of a Country Code (CC) and a NSN (National Significant Number). Until December 31, 1996, the CC can be up to three digits and the NSN up to 11 digits – for a total of no more than 12 digits. After December 31, 1996, the CC stays at up to 3 digits and the NSN goes to up to 14 digits, for a total of no more than 15 digits. See the Appendix for a full list of International Country Codes.

3. Carbon Copy. That's where it comes from. It refers to the person who received one of the copies of a memorandum or document. Now CC is an electronic mail term that means that someone will receive a copy of the electronic mail.

4. Company Code. Also called OCN (Operating Company Number). A unique four-place alphanumeric code (NNXX) assigned to all U.S. domestic telecommunications service providers by NECA (National Exchange Carrier Association). See Company Code for a complete explanation.

5. Dot cc (.cc) is a premium, international top-level domain (TLD) that works worldwide on the Internet just like .com, .net and .org. The ccTLD identifies the country in which the Internet address resides. See Domain Name for a full discussion.

CCB Common Carrier Bureau. One of the largest divisions of the FCC, this Bureau regulates interstate telephone systems – licensing them, monitoring their charges, the conditions they offer service under, etc. The Common Carrier Bureau opened long distance communications to competition in America.

CCBM Came Clear By Magic. This term created (but not contributed) by Nynex (now part of Verizon). This is pseudo-technical lingo used by Nynex when they repair something that they broke (i.e. the problem was their fault), but they are unwilling to admit it was their fault. They say this "CCBM" to interconnect companies who are trying to get their customers' phone lines fixed.

CCC 1. Clear Coded Channel. A 64 Kbps channel in which all 64 Kbps is available for data.

2. Clear Channel Capability. The bandwidth of a data transmission path available to end users after control and signaling bits are accounted for.

3. Communications Competition Coalition. Lobbying organization established to encourage competition in telecommunications in Canada.

CCCM Computer Controlled Cable Modem, a host-based device that shares PC resources such as memory and processing.

CCD Charge Coupled Device, a light sensitive chip used for image gathering. CCDs are the "eyes" of a scanner or digital camera – still or movie. They convert incoming light into bit-map images, which can be stored by the device on digital media – a flash memory card, a CD, a hard disk. CCDs are small electronic devices with arrays of light-sensitive elements. The number of these elements and their width determine the scanner's or camera's resolution. Light is bounced off the image onto the CCD, which translates the varying intensities of the reflected light into digital data. In their normal condition CCDs are grayscale devices. To create color, a color pattern is laid down on the sensor pixels, using a color mask like RG BG, (Red, Green, Blue and Green). The extra green is used to create contrast in the image. The CCD pixels gather the color from the light and pass it to the shift register for storage. CCD are analog sensors, the digitizing happens when the electrons are passed through the A to D converter. The A to D converter (analog to digital converter) converts the analog signal to a digital file or signal. CCDs are currently the device of choice for digital and consumer imaging devices like camcorders, scanners, and digital cameras.

CCDN Corporate Consolidated Data Network. It is the name for IBM's main internal data communications network. It used to be managed by IBM. It's now managed by Advantis, an IBM spin-off company, which is majority-owned by IBM and the rest by Sears.

CCE See Call Carrying Equipment.

CCF See Consumer Consultative Forum.

CCFL Cold Cathode Fluorescent Lamp. A technology several laptop computer manufacturers use to light their LCD screens.

CCH Connections per Circuit per Hour. How many phone calls one circuit was able to complete in one hour. Compare to ACH, which is call Attempts per Circuit per Hour. See ACH. CCH and ACH are terms useful in traffic engineering, i.e. figuring out how many trunks you need for your incoming and/or outgoing telephone calls.

CCIA Computer and Communications Industry Association. A trade organization of computer, data communications and specialized common carrier services companies headquartered in Arlington VA. It runs seminars, does lobbying and generally tries to take care of the common interests of its members. See CBEMA.

CCIE Cisco Certified Internetwork Expert. Cisco's equivalent of Novell's CNE or Microsoft's MCSE. Attend a class. Study. Become a CCIE. A way of giving some certification as to your ability to manage a network.

CCIR Comite Consultatif International des Radiocommunications. The agency responsible for the international use of the radio spectrum. Effective in 1993, the CCIR is now known as the International Telecommunications Union – Radio. ITU-R and ITU-T form the International Telecommunications Union.

CCIR 601 An internationally agreed-upon standard for the digital encoding of component color television that was derived from the SMPTE RP125 and the EBU 324E standards. It uses the 4:2:2 sampling scheme for Y, U and V with luminance sampled at 13.5 MHz and chrominance (U and V components) sampled at 6.75 MHz. After sampling, 8-bit digitizing is used for each channel. These frequencies are used because they work for both 525/60 (NTSC) and 625/50 (SECAM and PAL) television systems. The system specifies that 720 pixels be displayed on each line of video. The D1 digital videotape format conforms to CCIR 601. See CCIR 656. The CCIR is now known as the ITU-R. See ITU-R.

CCIR 656 The international standard defining the electrical and mechanical interfaces for digital TV operating under the CCIR 601 standard. It defines the serial and parallel interfaces in terms of connector pinouts as well as synchronization, blanking and multiplexing schemes used in these interfaces. The CCIR is now known as the ITU-R. See ITU-R.

CCIRN Coordinating Committee for Intercontinental Research Networks. A committee that provides a forum for North American and European network research organizations to cooperate and plan. The CCIR is now known as the ITU-R. See ITU-R.

CCIS Common Channel Interoffice Signaling. A way of carrying telephone signaling information along a path different from the path used to carry voice. CCIS occurs over a separate packet switched digital network. CCIS is separate from the talk path. A special version of CCIS is called Signaling System #7. SS#7 is integral to ISDN. CCIS offers basically two benefits: first, it dramatically speeds up the setting up and tearing down of phone calls. Second, it allows much more information to be carried about the phone call than what is carried on in-band (old-fashioned) signaling. That information can include the calling number, a message, etc.

Signaling for a group of voice telephone circuits is done on CCIS by encoding the information digitally on one of the voice circuits. In the previous method of signaling – the one replaced by CCIS – multi-frequency tones were sent down the same talkpath that the conversation would eventually travel. By taking the signaling information out of the talk-path, the "phone phreak" community could no longer get free calls by using so-called "blue boxes" which duplicated the multi-frequency tones used by switching machines. CCIS is a much more efficient method of signaling, since it doesn't require a full voice grade channel just to check if the called party in LA is free and whether the call coming in from New York should be put through. See also Common Channel Signaling, System Signaling 7, ISDN and Common Channel Interoffice Signaling.

CCITT Comit0xC7 consultatif international t0xC7l0xC7graphique et t0xC7l0xC7-phonique, which, in English, means the Consultative Committee on International Telegraphy and Telephony. The CCITT is now known as the ITU-T (International Telecommunications Union-Telecommunications Services Sector), based in Geneva, Switzerland. The scope of its work is now much broader than just telegraphy and telephony. It now also includes telematics, data, new services, systems and networks (like ISDN). The ITU is a United Nations Agency and all UN members may also belong to the ITU, represented by their governments. In most cases the governments give their rights on their national telecom standards to their telecommunications administrations (PTTs, or TOs). But other national bodies (in the US, for example, the State Department) may additionally authorize Recognized Private Operating Agencies (RPOAs) to participate in the work of the ITU. After approval from their relevant national governmental body, manufacturers and scientific organizations may also be admitted, as well as other international organizations. This means, says the ITU, that participants are drawn from the broad arena. The activities of the ITU-T divide into three areas:

Study Groups (at present 15) to set up standards ("recommendations") for telecommunications equipment, systems, networks and services.

Plan Committees (World Plan Committee and Regional Plan Committee) for developing general plans for a harmonized evolution of networks and services.

Specialized Autonomous Groups (GAS, at present three) to produce handbooks, strategies and case studies for support mainly of developing countries.

Each of the 15 Study Groups draws up standards for a certain area - for example, Study Group XVIII specializes in digital networks, including ISDN. Members of Study Groups are experts from administrations, RPOAs, manufacturing companies, scientific or other international organizations - at times there are as many as 500 to 600 delegates per Study Group. They develop standards which have to be agreed upon by consensus. This, says the ITU, can sometimes be rather time-consuming, yet it is a democratic process, permitting active participation from all ITU member organizations.

The long-standing term for such standards is "CCITT (ITU-T) recommendations." As the name implies, recommendations have a non-binding status and they are not treaty obligations. Therefore, everyone is free to use ITU-T recommendations without being forced to do so. However, there is increasing awareness of the fact that using such recommendations facilitates interconnection and interoperability in the interest of network providers, manufacturers and customers. This is the reason why ITU-T recommendations are now being increasingly applied – not by force, but because the advantages of standardized equipment are obvious. ISDN is a good example of this. NOTE: ISDN and other standards recommendations include options which allow for multiple "standards," in recognition of differing national and regional legacy "standards;" as a result, international standards recommendations do not necessarily yield evenly applied standards options.

The ITU-T has no power of enforcement, except moral persuasion. Sometimes, manufacturers adopt the ITU-T specs. Sometimes they don't. Mostly they do. The ITU-T standardization process runs in a four-year cycle ending in a Plenary Session. Every four years a series of standards known as Recommendations are published in the form of books. These books are color-coded to represent different four cycles. In 1980 the ITU published the Orange Books, in 1984 the Red Books and, in 1988, the Blue Books. See ITU Study Groups and ITU V.XX below.

The ITU has now been incorporated into its parent organization, the International Telecommunication Union (ITU). Telecommunication standards are now covered under Telecommunications Standards Sector (TSS). ITU-T (ITU-Telecommunications) replaces ITU. For example, the Bell 212A standard for 1200 bps communication in North America was referred to as ITU V.22. It is now referred to as ITU-T V.22. See ITU.

CCK Complementary Code Keying is a modulation technique used by IEEE 802.11b compliant wireless LANs for transmission at 5.5 and 11 Mbps. CCK is compatible with the DSSS (Direct Sequence Spread Spectrum) spreading technique specified in the IEEE 802.11 standard for wireless Ethernet LANs, and uses complementary codes to spread the data frames across the assigned spectrum. CCK codes are defined by a set of 256 8-chip code words, which work well in an indoor multipath environment and which can be demodulated efficiently. See also 802.11b, Chip, DSSS, and Spread Spectrum.

CCL 1. Configuration Control Link, works with ACD.

2. Carrier Common Line. A usage-sensitive charge billed to IXCs by the local telephone company to recover costs for the portion of the telephone company facility, normally to the end user location, that is used to complete a call.

CCMA Call Centre Management Association, an association based in the United Kingdom. The CCMA is run by a Committee made up of elected representatives who have an interest in Call Centres. The annual subscription is stlg95.00 and a joining fee of 30 English pounds sterling. One of CCMA's primary goals is to set up an educational program where members can attend courses and have those attributed to a recognized qualification.

CCMI Computer Controlled Cable Modem.

CCNA 1. Customer's Carrier Name Abbreviation. Identifies the common language code for the IXC providing the interLATA facility. This code reflects the IXC to be contacted for provisioning whereas the ACNA reflects the IXC to be billed for the service. See ACNA.

2. Cisco-Certified Network Associate.

CCNE Cisco-Certified Network Engineer.

CCR 1. Customer Control Routing, works with ACD. See ACD.

2. An ATM term. Current Cell Rate: The Current Cell Rate is an RM-cell field set by the source to its current ACR when it generates a forward RM-cell. This field may be used to facilitate the calculation of ER, and may not be changed by network elements. CCR is formatted as a rate.

CCS 1. Centi Call Seconds. One hundred call seconds or one hundred seconds of telephone conversation. One hour of telephone traffic is equal to 36 CCS (60 x 60 = 3600 divided by 100 = 36) which is equal to one erlang. CCS are used in network optimization. See also Erlang and Traffic Engineering.

2. Common Channel Signaling: A high-speed, packet switched communications network that is separate from the public packet switched and message networks. It is used to carry addressed signaling messages for individual trunk circuits and/or database-related services between signaling points in the CCS network. *A form of signaling in which a group of circuits share a signaling channel. See Signaling System 7.

3. Custom Calling Services was changed a few years ago to "Custom Calling Features". These services include speed calling, call forwarding, 3-way calling, etc.

CCS Node A network element or a network system connected to the CCS network via SS7 links.

CCS/SS7 A Telcordia Technologies term for Common Channel Signal/Signaling System 7. See Signaling System 7.

CCS7 Common Channel Signaling 7. See ISDN and Signaling System 7.

CCSA Common Control Switching Arrangement. A private network set up by AT&T for very large users and using parts of the public switched network. One important feature of a CCSA is that any user anywhere in a CCSA network can reach any other user by dialing only seven digits. Only very large customers subscribe to this service. It's expensive. AT&T has fewer than 100 customers.

CCSA ACCESS A PBX feature which allows a PBX user to get into a CCSA network. See CCSA.

CCSD Command Communications Service Designator; control communications service designator JP 1-02. It's a circuit designator. Also used as a circuit order number. A U.S. military term somewhat like USOC code in the civilian world.

CCSDS The Consultative Committee on Space Data Systems, the international organization that promulgates standards for space data systems; secretariat in Washington, D.C. www.ccsds.org.

CCT Continuity Check Tone.

ccTLD Country Code Top Level Domain. Eg: harry@newton.uk. The uk shows it's from the United Kingdom. See Top Level Domain, Country Code, Domain Name, gTLD, and TLD.

CCTA Central Computer and Telecommunications Agency, a governmental agency in the UK which compiles publicly available guidelines used by many European computer and telecommunications producers; as of 1 April 2001, integrated into the Office of Government Commerce (OGC). www.ogc.gov.uk.

CCTV Closed Circuit TV.

CCU 1. Communication Control Unit. A processor, often a minicomputer, associated with a host mainframe computer that performs a number of communications-related functions. Compare with cluster control unit.

2. Camera Control Unit.

CCXML Call Control eXtensible Markup Language. CCXML is a new software language designed to allow developers to program telephone switches and computer telephony devices. It is designed to address call management and event processing. CCXML has been designed to complement and integrate with a VoiceXML system. The two languages are separate and are not required in an implementation of either language. For example CCXML could be integrated with a more traditional IVR (Interactive Voice Response) system and VoiceXML could be integrated with some other call control system.

CCXML operates around voice network events such as making a call or attempting a conference call. In the case of working with VxML, when a call control message is sent within a VxML message, that command is handed over to CCXML as an event. All events are detected and acted upon by a CCXML interpreter, which is essentially the brains or the logic built into CCXML. Once the event happens CCXML is triggered and used for call control. CCXML provides the following functions:

- Call routing
- Call bridging
- Outbound calling
- Dialog execution
- Conferencing
- Coaching
- Call answering based on caller information

The reason VxML and CCXML are important is that they are truly platform independent, fairly easy for a developer to learn, and use very little bandwidth when compared to other forms of voice over net technologies. With both VxML and CCXML standardized it is possible for developers from many different companies to create telephony systems that will innately be able to talk to each other and interoperate fully.

According to the W3 working group, "There are a number of needed features that VoiceXML currently can't supply:

Support for multi-party conferencing, plus more advanced conference and audio control. Any large conference application requires such features.

Ability to give each active call leg its own dedicated VoiceXML interpreter. Currently, the second leg of a transferred call lacks a VoiceXML interpreter of its own, limiting the scope of possible applications.

Sophisticated multiple-call handling and control, including the ability to place outgoing calls. Multiple-party conferences mean VoiceXML needs a more effective way of handling telephony resources.

Handling for richer and more asynchronous events. Advanced telephony operations involve substantial amounts of signals, status events, and message-passing. VoiceXML does not currently have a way to integrate these asynchronous "external" events into its event-processing model.

Ability to receive events and messages from external computational entities. Interacting with an outside call queue, or placing calls on behalf of a document server, means the VoiceXML must be contacted by an outside party."

An implementation of VoiceXML is not required to support CCXML. Such implementations may choose to support proprietary methods of call control, and still be deemed compliant with the W3C VoiceXML Recommendation. An implementation of CCXML is not required to support VoiceXML. Such implementations may choose not to support interactive dialogs at all, or may do it in a proprietary way, and still be deemed compliant with the W3C CCXML Recommendation." For more information, www-voice@w3.org.

CD 1. Carrier Detect. CD is a signal generated by a dial-up modem. CD indicates its connection status. If your CD light is on, then your modem is speaking to another modem.

2. Compact Disc. A 12 centime diameter (around 4 3/4 inches) disk containing digital audio or digital computer information, which can be played back and (now recorded) on a laser-equipped player. It was introduced by Sony and Philips in 1982. Philips (the inventor) chose the diameter of a Dutch 10-cent coin for the diameter of the hole in the CD. A compact disc originally came in only one flavor – read only. And most music tapes can only be listened to, not recorded to. For music it was a major breakthrough. It recorded music digitally (that is, coded as the zeros and ones of computer-speak) instead of trying to make an electrical copy of the sound waves themselves as devices like audiocassettes and LP records had. The Economist described the CD well. It said, "Instead of using a needle, the sound was plucked from the CD's surface by a tiny beam of laser light and then processed by a microcomputer. To the ear, the leap in performance between a compact disc and a long-playing record was even greater than the difference between color and black and white TV was to the eye."

A CD can typically hold up to 650 megabytes of information. That is the equivalent of 1,500 floppies or 250,000 pages of print. Most computer CD-ROM drives can play audio CD disks – if they have the software and the speakers. Audio CD players, though, cannot play computer CD-ROM discs. But most computer CD-ROM players can play CD audio discs. As CD-ROM have become more popular and their makers have tried to do more and more with them, so CD-ROM formats have proliferated and some are not compatible with each other. See CD-R, CD-ROM, CD-I, CD-V, DVD and WORM.

3. Cell Delineation. An ATM term. See Cell Delineation.

4. See Campus Distributor.

CD I See CD-I below.

CD UDF This format defines specifications for a unified logical file format for CD-Recordable. The new format, known as CD UDF, defines a common scheme of "packet writing" to assure interchangeability of CD-R (CD Recordable) discs and to greatly increase performance and flexibility when storing large and small files on recordable disc media. As a result, CD-Recordable drives can be integrated into computer systems to behave much like other removable disk storage products. www.osta.org

CD-Audio Sometimes called "Redbook audio," is the digital sound representation used by CD-ROMs. CD audio is converted to analog sound output within the CD-ROM drive. The sampling frequency for CD audio is 44.1 KHz.

CD-I Compact Disc Interactive. Geared toward home entertainment, the drive connects to a television.

CD-Plus A format for CDs created by Sony and Philips Electronics which makes the multimedia track on CDs invisible to CD players. The problem: there are few enhanced CDs

– discs for both CD players and CD-ROM drives. The reason: the format prior to CD-Plus puts the multimedia data on track one, which listeners must skip over on their CD players. Thus the new format.

CD-R CD-Recordable. A standard and technology allowing you to write to and read from a Compact Disc, but not erase or change what you record. This technology is compatible with existing CDs, i.e. you are able to read these discs in existing CD-players, and often in both PC and Macintosh machines. See CD-ROM and Multi-Session.

CD-Recordable See CD-R.

CD-ReWritable CD-ReWritable discs are erasable and can be used again and again, such as an audio cassette.

CD-ROM Compact Disc Read Only Memory. Also called CD or CD-ROM. The familiar five inch diameter Compact Disc which you see in the audio stores, but now made for computers. These discs hold huge amounts of data – as much as 700 megabytes per disk, or 300,000 pages of ASCII text. But CD-ROMs are catching up. Most of today's newer CD-ROMs can now be used in multimedia applications and virtually all CD-ROM drives available today support the Multimedia PC or MPC standard. See CD, CD-ROM XA, C-V, CD-WO and Shovelware. See also DVD for an update on what CD-ROMs are becoming.

CD-R CD Recordable. A compact disc you can write once to but read many times from. The disc has a spiral track which is preformed during manufacture, onto which data is written during the recording process. This ensures that the recorder follows the same spiral pattern as a conventional CD, and has the same width of 0.6 microns and pitch of 1.6 microns as a conventional disc. Discs are written from the inside of the disc outward. The spiral track makes 22,188 revolutions around the CD, with roughly 600 track revolutions per millimetre. Instead of mechanically pressing a CD with indentations, a CD-R writes data to a disc by using its laser to physically burn pits into the organic dye. When heated beyond a critical temperature, the area "burned" becomes opaque (or absorptive) through a chemical reaction to the heat and subsequently reflects less light than areas that have not been heated by the laser. This system is designed to mimic the way light reflects cleanly off a "land" on a normal CD, but is scattered by a "pit," so a CD-R disc's data is represented by burned and non-burned areas, in a similar manner to how data on a normal CD is represented by its pits and lands. Consequently, a CD-R disc can generally be used in a normal CD player as if it were a normal CD. CD-R's real advantage is that the writing process is permanent. The media can't be erased and written to again. Only by leaving a session "open" - that is, not recording on the entire CD and running the risk of it not playing on all players - can data be incrementally added to a disc. The most recent solution is CD-RW. See CD-RW.

CD-RW CD Rewritable, also spelled rewriteable. It's a compact disc that allows repeated recordings. CD-RW drives can write both CD-R and CD-RW discs and can read any type of CD. Like regular CDs, CD-Rs and CD-RWs are composed of a polycarbonate plastic substrate, a thin reflective metal coating, and a protective outer coating. CD-R is a write once, read many (worm) format, in which a layer of organic polymer dye between the polycarbonate and metal layers serves as the recording medium. The composition of the dye is permanently transformed by exposure to a specific frequency of light. In a CD-RW, the dye is replaced with an alloy that can change back and forth from a crystalline form when exposed to a particular light, through a technology called optical phase change. The patterns created are less distinct than those of other CD formats, requiring a more sensitive device for playback. Only drives designated as "MultiRead" are able to read CD-RW reliably. Similar to CD-R, the CD-RW's polycarbonate substrate is preformed with a spiral groove to guide the laser. The alloy phase-change recording layer, which is commonly a mix of silver, indium, antimony, and tellurium, is sandwiched between two layers of dielectric material that draw excess heat from the recording layer. After heating to one particular temperature, the alloy will become crystalline when it is cooled; after heating to a higher temperature it will become amorphous (won't hold its shape) when it is cooled. By controlling the temperature of the laser, crystalline areas and non-crystalline areas are formed. The crystalline areas will reflect the laser, while the other areas will absorb it. The differences will register as digital data that can be unencoded for playback. To erase or write over recorded data, the higher temperature laser is used, which results in the non-crystalline form, which can then be reformed by the lower temperature laser. CD-RW discs usually hold 74 minutes (650 MB) of data, although some can hold up to 80 minutes (700 MB) and, according to some reports, can be rewritten as many as 1000 times.

CD-ROM XA Stands for Compact Disc - Read Only Memory eXtended Architecture. Microsoft's extensions to CD-ROM that let you interleave audio with data. Though it is not a video specification, limited video can be included on disc. Demand for multimedia applications is increasing use of CD-ROM XA. To use it, you must have a drive that reads the audio portions of the disc and audio card in your computer that translates the digital into sound. Not all drives can recognize the extensions. See CD-R, CD-RW, CD-WO.

CD-V Compact Disc Video. A format for putting 5 minutes of video on a 3-inch disc. This format has come and gone. Video is shifting towards CD-ROM XA.

CD-WO Compact Disc Write Once. A CD-ROM version of the WORM (Write Once Read Many) technology. For companies performing all CD-ROM publishing in-house, this format is useful for creating test discs before sending the master for duplication. CD-WO discs conform to ISO 9660 standards and can be played in CD-ROM drives.

CDA See Communications Decency Act.

CDAR Customer Dialed Account Recording.

CDC Customer Data Change.

CDCS Continuous Dynamic Channel Selection.

CDDI Copper Distributed Data Interface is a version of FDDI (Fiber Distributed Data Interface– a 100 million bit per second local area network) that runs on unshielded twisted-pair cabling rather than optical fiber.

CDE Common Desktop Environment. A graphical user interface which is common and consistent across UNIX platforms. CDE is designed to make UNIX systems easier to use, simpler to support and more cost-effective to target for application development. CDE is defined to converge the major components of the desktop (X Window System, OSF/Motif, and CDE), adding basic features such as printing APIs (Application Programming Interfaces), and providing a competitive, common approach to delivering on-line information access and on-line publishing technology. A major initiative of The Open Group, CDE sponsors include Digital Equipment Corporation, Fujitsu Limited, Hewlett-Packard, Hitachi, IBM, Novell, and Sun.

CDEV Control panel DEVice. An Apple Macintosh term.

CDF 1. An ATM term. Cutoff Decrease Factor: CDF controls the decrease in ACR (Allowed Cell Rate) associated with CRM.

2. Channel Definition Format. A term used in Internet/Intranet push technology.

3. See Combined Distribution Frame.

CDFP Centrex Data Facility Pooling.

CDFS Compact Disc File System, which controls access to the contents of CD-ROM drives in PCs.

CDG The CDMA Development Group (CDG), according to its Web site, is an industry consortium of companies who have come together to develop the products and services necessary to lead the adoption of CDMA wireless systems around the world. In working together, the 100 member companies will help ensure interoperability among systems, while expediting the availability of CDMA technology to consumers. The CDMA Development Group is committed to the definition of CDMA features, services, technical requirements and other activities that promote the availability and evolution of CDMA (IS-95 based) wireless systems worldwide. Specific objectives include:

- Leading the adoption of CDMA based systems around the world.
- Maintaining a forum to address issues impacting manufacturers and carriers actively involved in CDMA deployments.
- Developing next-generation CDMA systems.
- Minimizing the time required to implement CDMA services and features. www.cdg.org.

CDH Interface An interface once required by the Bell System to protect their phone lines from "foreign" (i.e. non-AT&T) phone equipment. CDH devices were eventually ruled a total waste of money and the phone companies refunded the money – at least to the subscribers who asked. If you still have the stuff installed, you may be due a huge refund. Watch out for the statute of limitations.

CDL Coded Digital Locator.

CDLC Cellular Data Link Control. A public domain data communications protocol used in cellular telephone systems. In other words, you can attach a data terminal to a cellular telephone and send and receive information. There are more 5,000 modems using CDLC on the Vodaphone Cellular System in the UK, where it is the de facto standard for cellular data communications. Features like improved synchronization field, forward error correction, bit interleaving, and selective retransmission make CDLC ideal for cellular transmissions, according to Millidyne who makes the CDLC modems in the US.

CDLRD Confirming Design Layout Report Date. The date a common carrier accepts the facility designed proposed by the Telco.

CDMA Code Division Multiple Access is a digital, spread spectrum, packet-based access technique generally used in RF (Radio Frequency) radio systems. Perfected and commercialized by Qualcomm, CDMA is used in certain cellular phone systems and in some WLANs (Wireless Local Area Networks). CDMA organizes the data to be transmitted into discrete packets of various lengths. Prior to being packetized, the data may be converted from analog to digital format by a vocoder, as would be the case in a voice transmission over a cellular network. Once in a digital format, the data may be compressed in order to reduce the raw number of bits to be transmitted and, therefore, make more efficient use of limited RF spectrum. In a CDMA-based cellular voice network, for example, each transmission comprises a stream of data packets, which stream is assigned a unique 10-bit code sequence known as a PN (PseudoNoise) sequence. That PN code is prepended (i.e., added to the front of) each packet in the packet stream, enabling each receiver to separate that specific transmission from both the inherent background noise and the other transmissions that share the RF channel. Thereby, multiple packets associated with multiple conversations can share the same spectrum, overlapping in both frequency and time, without mutual interference. The major benefit of CDMA is increased capacity (up to 20 times analog cell service) through more efficient use of spectrum. CDMA also provides three features that improve system quality: 1) The "soft hand-off" feature ensures that a call is connected before handoff is completed, as the cellular phone moves from cell-to-cell; this reduces the probability of a dropped call. 2) Variable rate vocoding allows speech bits to be transmitted at only the rates necessary for high quality, which conserves the battery power of the subscriber unit. 3) Multipath signal processing techniques combines power for increased signal integrity. Additional benefits to the subscriber include increased talk times for portable units, more secure transmissions and special service options such as data, integrated voice and data, fax and tiered services. See also FDMA, OCDMA, S-CDMA, Spread Spectrum, TDMA, and Vocoder.

CDMA Development Group See CDG.

CDMA2000 Code Division Multiple Access 2000. A third generation (3G) wireless system, cdma2000 is a trademark of Qualcomm, the company that commercialized CDMA. cdma2000 essentially is the CDMA approach to IMT-2000, the ITU's concept for a single, global standard for 3G wireless technology. Based on earlier CDMA versions (also known as TIA/EIA IS-95a and IS-95b), cdma2000 (also known as IS-2000) has been approved by the ITU. The initial version, known as cdma2000 1xMC (one times Multi-channel), offers 3G capabilities within a single standard 1.25 MHz channel, effectively doubling the voice capacity of the predecessor cdmaOne systems and offering data speeds up to 307 Kbps. The high-speed version is known officially as cdma2000 3xMC (Three times Multi-Carrier), as it makes use of three standard 1.25 MHz channels within a 5 MHz band to deliver data speeds up to 2 Mbps in support of integrated voice, data and video. cdma2000 runs in the 800 MHz and 1.8-2.0 GHz spectrum.

CDMA2000 IXEV-DO CDMA2000 IXEV-DO is a third-generation wireless protocol that is a stepping stone in the evolution of cdma2000. 1 XEV-DO "data optimized"is a data only overlay that uses a 1.25 MHz channel to provide a peak rate data throughput of 2.4 Mbps. UBS Warburg believes a lx-EV DO network is better suited for the enterprise. Several Korean mobile operators have rolled out I x-EV DO networks.

CDMA2000 1xEV-DO Release 0 Another term for EV-DO Release 0.

CDMA2000 1xEV-DO Revision A Another term for EV-DO Revision A.

CDMA2000 1xEV-DO Revision B Another term for EV-DO Revision B.

CDMA2000 1xEV-DO Revision C Another term for EV-DO Revision C.

cdmaOne A brand name, trademarked and reserved for the exclusive use of the member companies of the CDMA Development Group (CDG). cdmaOne describes a complete wireless system that incorporates the IS-95 CDMA air interface, the ANSI-41 network standard for switch interconnection and many other standards that make up a complete wireless system. As of the end of second quarter 2002, there were approximately 127 million total global CDMA subscribers (according to the CDMA Development Group). cdmaOne is basically a brand name. That's it. Nothing more. See also CDMA, and CDG.

CDMP Cellular Digital Messaging Protocol.

CDN 1. Control Directory Number.

2. Content Data Network. A network that bills the end user on the basis of the specific content accessed, rather than on either a flat-rate or usage-sensitive basis. The emerging concept of the CDN centers on the Internet and World Wide Web, and has application in both the wired and wireless domains.

CDO Community Dial Office. A small automatic central office switching system that is completely unattended. Routine maintenance is provided by a traveling technician once or twice each year, or as troubles develop. Such an office usually serves a small community with a few hundred lines in a rural area.

CDP 1. Customized Dial Plan.

2. See continuous data protection.

CDPA See Capture Division Packet Access.

CDPD Cellular Digital Packet Data. A radio technology that supports the transmission of packet data at speeds of up to 19.2 Kbps over the existing analog AMPS (Advanced Mobile Phone Service) cellular network, with appropriate CDPD upgrades. The data is structured in packets that are transmitted during pauses in cellular phone conversations, thereby avoiding issues of developing an overlay cellular network for data communications. Estimates suggest that as much as 20%-30% of an AMPS network is idle, even during periods of peak usage. This idle capacity is due to short pauses between the point in time at which you disconnect your circuit-switched cellular telephone conversation and the time when someone else seizes that same radio channel to place a call. Idle capacity also is created when you are "handed-off" from one cell to another as you travel through the area of coverage in your vehicle. While 19.2 Kbps transmission rates are possible, throughput commonly drops to 2.4 Kbps or so during periods of peak usage.

Connectionless protocols such as IP (Internet Protocol) and the OSI CLNP (Connectionless Network Protocol) are employed to accomplish this minor miracle. The contention method employed is DSMA/CD (Digital Sense Multiple Access/Collision Detection), which is much like CSMA/CD used in Ethernet LANs. CDPD offers a number of advantages over competing wireless mobile data technologies. Among those advantages is the fact that CDPD packets use forward error correction (FEC) to reduce the impacts of noise and interference over the air link. Additionally, CDPD incorporates authentication and encryption to yield much improved security.

CDPD uses a full 30-KHz voice channel, but it can move your connection from one channel to another to avoid congesting voice communications. The drawbacks to CDPD are that it requires you to have a CDPD modem to access your upgraded cellular network, and not all cellular service providers are willing their to upgrade their cellular equipment to CDPD. On July 21, 1993, the group of cellular carriers that supports the Cellular Digital Packet Data (CDPD) project released the complete version 1.0 of its open specification designed to enable customers to send computer data over an enhanced cellular network. The group said the packet data approach is ideally suited to those applications that require the transmission of short bursts of data, for example, authorizing a credit card number, exchanging e-mail messages or making database queries. The cellular networks deploying CDPD will enable mobile workers to use a single device to handle all of their voice and packet data needs. This version of the specification includes input from parties that reviewed the earlier release (0.8 and 0.9). The new version provides details of the CDPD architecture, airlink, external network interfaces, encryption and authentication, network support services, network applications services, network management, radio resource management and radio media access control.

CDPD was originally developed by IBM and is backed by the CDPD Forum. Carriers offering CDPD in the U.S. include Ameritech, AT&T Wireless, Bell Atlantic NYNEX Mobile, and GTE Mobilnet (now Verizon Wireless). Those carriers offer the service in over 60 markets. Limited CDPD coverage also is provided in Canada, Ecuador, Indonesia and Mexico. Pricing generally is on the basis of a monthly fee, plus a usage charge calculated on the number of kilobytes of data transmitted. Terminal equipment can be in the form of a laptop which plugs into your cellular phone through an adapter, or which has its own CDPD modem and antenna. Recently developed are CDPD terminals which look much like a cellular telephone, although with enhanced display capabilities. The original scope of the CDPD specifications was expanded in July 1995 to include CS-CDPD (Circuit Switched CDPD), which operates on a connection-oriented basis, much like a cellular voice call. This approach supports longer file transmissions and yields improved throughput, as a radio channel is seized and maintained for the duration of the transmission. CDPD has application beyond AMPS, as it can be run over TDMA and CDMA networks, as well. See CDPD Forum.

CDPD Cell Boundary The locus of points at which a Mobile End System (M-ES) should no longer access service by using the transmission of a particular cell. See CDPD.

CDPD Forum The CDPD Forum Inc. is an not-for-profit special interest group formed in 1994 to promote the development, deployment and use of CDPD. The forum comprises companies that develop, deliver or use CDPD products or services. companies with an interest in CDPD. www.cdpd.org. See CDPD.

CDPD SNDCP Cellular Digital Packet Data (CDPD) SubNetwork Department Convergence Protocol. See CDPD.

CDR 1. Call Detail Recording (as in Call Accounting) or Call Detail Record, as a record generated by customer traffic later used to bill the customer for service. See Call Accounting.

2. Clock/Data Recovery. An technique used in Local Area Networks (LANs) whereby a data octet is subdivided, scrambled, and encoded into an expanded form. The expanded expression of the data value includes bits which are used for clock recovery (i.e., synchronization) and data recovery (i.e., error detection) between the LAN hub or switch, and the attached terminal device. See also 4B5B, 5B6B, 8B6T, and 8B10B.

CDR Exclude Table A table listing local central office codes which are not monitored (i.e. ignored) by a call accounting system.

CDSA Common Data Security Architecture. A security framework for developing security and authentication application programs.

CDSL Consumer Digital Subscriber Line (also called Consumer DSL) service, introduced and trademarked by Rockwell in the fall of 1997. CDSL is a one megabit modem technology. The key difference between it and ADSL is that ADSL requires a splitter installed at each home to divide voice and data traffic onto separate lines. CDSL doesn't. According to Rockwell, customers would simply buy a CDSL modem at the local electronics store and call the phone company to have the service turned on. See ADSL.

CDV 1. Compression Labs Compressed Digital Video, a compression technique used in satellite broadcast systems. CDV is the compression technique used in CLI's SpectrumSaver system to digitize and compress a full-motion NTSC or PAL analog TV signal so that it can be transmitted via satellite in as little as 2 MHz of bandwidth. (A normal NTSC signal takes 6 Mhz.)

2. An ATM term. Cell Delay Variation: CDV is a component of cell transfer delay, induced by buffering and cell scheduling. Peak-to-peak CDV is a QoS delay parameter associated with CBR and VBR services. The peak-to-peak CDV is the ((1a) quantile of the CTD) minus the fixed CTD that could be experienced by any delivered cell on a connection during the entire connection holding time. The parameter "a" is the probability of a cell arriving late. See CDVT.

3. CD Video. A small videodisc (5" diameter) that provides five minutes of video with digital sound plus an additional 20 minutes of audio.

CDVT Cell Delay Variation Tolerance-ATM layer functions may alter the traffic characteristics of ATM connections by introducing Cell Delay Variation. When cells from two or more ATM connections are multiplexed, cells of a given ATM connection may be delayed while cells of another ATM connection are being inserted at the output of the multiplexer. Similarly, some cells may be delayed while physical layer overhead or OAM cells are inserted. Consequently, some randomness may affect the inter-arrival time between consecutive cells of a connection as monitored at the UNI. The upper bound on the "clumping" measure is the CDVT.

CE 1. An ATM term. Connection endpoint: A terminator at one end of a layer connection within a SAP.

2. Circuit emulation.

CE Mark Conformite Europeene (French) Mark; the English translation is European Conformity Mark. CE Mark is a type of pan-European equipment approval which indicates that the manufactured product complies with all legislated requirements for regulated products. Obtaining the CE Mark allows a product to be sold into 18 European countries without any further in-country testing. Several country regulatory bodies are set up to do the testing and thus the certification. These "notified bodies" comprise CEN, CENELEC and ETSI. CEN is responsible for European standardization in all fields except electrotechnical (CENELEC) and telecommunications (ETSI). The CE Mark is now a requirement for all telecommunications terminal equipment (TTE) products sold into the European Union (EU), effective January 1, 1996. Countries covered are Austria, Belgium, Denmark, Finland, France, Germany, Greece, Ireland, Italy, Luxembourg, the Netherlands, Portugal, Spain, Sweden, and the United Kingdom. CE marking confirms that a product has been tested and meets the essential requirements of the European Telecom Directive to market it throughout the EU. The European TTE Directive 91/263/EEC specifies approved products to meet appropriate telecommunications technical standards. These include personal safety, protection of public networks, interoperation with public network equipment, and electromagnetic compatibility.

Several country regulatory bodies are set up to do the testing and thus the certification. These bodies are called "notified bodies." When one company, Larscom, reported one of its NDSUs had been granted a CE Mark, it said that the European "notified body" issuing the approval for its product was the British Approval Board for Telecommunications (BABT).

CE router Customer edge router. A router that is part of a customer network and that interfaces to a provider edge (PE) router.

CeBIT is the world's largest computer and office automation show. It attracts 600,000 plus people to Hannover, Germany in March each year. It is also called the Hannover Fair. It is about five times the size of Comdex, which is North America's largest computer show. Many of the "booths" at Hannover are really small buildings, which are used year round. Many of the "booths" are three stories high, with an open air restaurant on the top floor. Space at the show is sometimes rented for four years.

CEBus Consumer Electronics Bus. EIA IS-60 (Electronics Industry Association Interim Standard-60), known as the home automation standard, includes specification of the CEBus. Essentially a Home Area Network (HAN) standard, CEBus allows connectionless, peer-to-peer communications over a common electrical bus, using standard electrical wiring rather than special voice/data cabling. CEBus employs a CSMA/CD contention protocol similar to that used in Ethernet LANs. Each CEBus has two channels. One channel is for real-time, short-packet, control-oriented functions; the other channel is dedicated to intensive data transfer. The standard includes error detection, automatic retry, end-to-end acknowledgement, and duplicate packet rejection, as well as authentication to ensure the identity of the user and encryption to provide for data security. For more information, contact the CEBus Industry Council in Indianapolis, Indiana.

CEC Canadian Electrical Code. The Canadian equivalent of the US National Electrical Code (NEC).

CED 1. CallEd station iDentification. A 2100 Hz tone with which a fax machine answers a call. See CNG.

2. Capacitance Electronic Disc. System of video recording a grooved disc, employing a groove-guided capacitance pickup.

3. Caller-Entered Digits. Digits entered by a caller on a touch-tone phone in response to prompts. Either a peripheral (ACD, PBX, or VRU) or the carrier network can prompt for CEDs.

4. CED Compression. A method of compression used in faxing.

CEI 1. Comparable Efficient Interface. The idea is that the telephone industry will let all its information providers have this interface – defined by technical specs and pricing – and, if it does, then the phone companies can themselves use this information to become information providers themselves. The concept has merit. Implementation has been agonizingly slow.

2. An ATM term. Connection endpoint Identifier: Identifier of a CE that can be used to identify the connection at a SAP.

Ceiling Distribution Systems Cable distribution systems that use the space between a suspended or false ceiling and the structural floor for running cable. Methods used in ceiling distribution systems include zone, home-run, raceway, and poke-through.

Ceiling Feed A method of routing communications and/or power cabling vertically from the ceiling/plenum to a cluster of workstations.

CEKS Centrex Electronic Key Set.

celestial Celestial a term first applied to communications with the advent of satellite communications and used to denote "via satellite" communications, to distinguish it from terrestrial communications conducted entirely on the surface of the Earth or within its atmosphere. However, because celestial has long been used in astronomy, in communications it usually appears in combining forms, such as in "celestial-based." See terrestrial.

celestrial sphere Celestial sphere an imaginary sphere convenient in astronomy and consequently in describing satellite orbits.   To an observer on Earth, the night sky appears as a hemisphere resting on the horizon. Hence the simplest descriptions of astronomical observations are in terms of positions on an immense, imaginary spherical shell formed by the sky and concentric with the centre of the Earth, with an equator that is a projection of the Earth's equator.   This is the celestial sphere.   It differs from the Earth in that it is seen from inside, whilst the Earth is viewed from its outside. See celestial.

cell 1. The basic geographic unit of a cellular system. It derived its name "cell" from the honeycomb pattern of cell site installations. Cell is the basis for the generic industry term "cellular." A city or county is divided into smaller "cells," each of which is equipped with a low-powered radio transmitter/receiver. The cells can vary in size depending upon terrain, capacity demands, etc. By controlling the transmission power, the radio frequencies assigned to one cell can be limited to the boundaries of that cell. When a cellular phone moves from one cell toward another, a computer at the Mobile Telephone Switching Office (MTSO) monitors the movement and at the proper time, transfers or hands off the phone call to the new cell and another radio frequency. The handoff is performed so quickly that it's not noticeable to the callers. For a longer explanation, see CMTS.

2. The basic unit of a battery, consisting of plates, electrolyte and a container. A chemical device that produces electricity through electrolysis.

3. A unit of transmission in ATM and SMDS. A fixed-size packet consisting of a 48-octet payload and 5 octets of control overhead in the form of a header in the case of ATM, and a header and trailer in the instance of SMDS. See also Cell Switching, ATM and SMDS.

4. The smallest component of a table. In a table, a row contains one or more cells.

cell cluster In cellular and PCS systems, a grouping of physically proximate cells in which each cell uses a different subset of the total spectrum allocated for the wireless service, but the grouping or cluster of cells represents the total allocated spectrum. For example, in AMPS cellular a frequency re-use pattern of seven is usually used. Thus, a cell cluster consists of seven proximate cells, each of which uses a different subset of the AMPS spectrum.

Cell Delineation CD. An ATM term. Cell Delineation is accomplished at the Transmission Convergence (TC) sublayer of the ATM Physical Layer (PHY), working in tight formation with the Physical Medium (PM) sublayer. It is at the TC sublayer that the re-

sponsibility is assumed for Physical Layer operations that are not medium independent. For example, it is at this sublayer that the ATM cell switch interfaces with a SONET transmission system. Cell Delineation is responsible for defining the cell boundaries at the originating endpoint (e.g., ATM cell switch or ATM workstation) in order that the receiving endpoint can reassemble (i.e., identify and recover) all cells associated with a data payload that has been segmented (i.e., cut up into ATM cells). Cell Delineation is achieved by the receiving endpoint's locking onto the 5-octet cell headers. A failure in this process is known as Loss of Cell Delineation (LCD).

cell dragging An AMPS (Advanced Mobile Phone System) phenomenon in which a mobile cellular terminal moving away from the current serving cell penetrates deeply into or passes through a neighboring cell before the signal weakens to the point of reaching a predetermined signal level threshold, thereby becoming a candidate for handoff. The phenomenon is caused by irregularities of terrain and various other factors affecting radio propagation.

cell extender A wireless repeater that is used to reach customers located beyond the periphery of a base station's coverage pattern.

cell fi handset See cell-fi phone.

cell fi phone A Wi-Fi-enabled cellular phone that enables voice and data to roam seamlessly across private Wi-Fi networks and public cellular networks.

cell header A cell header precedes payload data (user information) in an ATM or SMDS cell. The header contains various control data specific to the cell switching protocol.

Cell Interarrival Variation CIV. An ATM term. "Jitter" in common parlance, CIV measures how consistently ATM cells arrive at the receiving end-station. Cell interarrival time is specified by the source application and should vary as little as possible. For constant bit rate (CBR) traffic, the interval between cells should be the same at the destination and the source. If it remains constant, the latency of the ATM switch or the network itself (also known as cell delay) will not affect the cell interarrival interval. But if latency varies, so will the interarrival interval. Any variation could affect the quality of voice or video applications.

cell group color code A cellular radio term. A color code assigned to a set of cells. Each member of the set is adjacent to at least one other member of the set and no two members of the set are allocated the same Radio Frequency (RF) channel for Cellular Digital Packet Data (CDPD) use. Each cell is assigned exactly one Cell Group Color Code.

cell ID One of three main location-based service (LBS) technologies, Cell ID fixes the location of the user by identifying which cell in a network is carrying the user's call and translates that information into latitude and longitude. Best used in less dense, rural areas, Cell ID is less accurate than competing technologies such as E-OTD and GPS, and is not well suited to commercial applications such as location-based advertising, which requires fixing the user's exact location. See Location Services.

cell loss priority A bit in an Asynchronous Transfer Mode header, used to indicate a cell's priority level. A cell set with a CLP bit to zero has higher priority than a cell with the CLP bit set to one. In the case of congestion, cells with a CLP bit set to one may be discarded. See Cell Loss Priority Field.

Cell Loss Priority Field CLP. A single priority bit in the ATM cell header; when set, it indicates that the cell may be discarded should the network suffer congestion. Voice and video data do not tolerate such loss; therefore, such cells would not carry a set CLP bit.

cell loss ratio A negotiated Quality of Service parameter in an Asynchronous Transfer Mode network. This parameter indicates a ratio of lost cells to total transmitted cells.

cell padding The space between the contents and inside edges of a table cell.

cell phone A cellular telephone. In Austria and Germany, it's called a "Handy," in China a "da ge da" (big brother), in Finland a "kanny" (extension of the hand), in France "le portable" or "le mobile", in Greece "kinito" (movable), in Australia a "mobile," in Israel "pelephone" (wonder phone), in Italy "telefonino" (little phone), in Japan "keitai" (portable), and in Turkey "cep" (pocket). See also GSM and SIM card.

cell phone bling Bling-bling (ornaments, crystals, charms, etc.) that teenagers put on their cell phones to personalize them. Crystal-studded cell phone straps are also cell phone bling. The practice appears to have originated in Japan and has now spread worldwide. See also bling bling.

cell relay A form of packet switching using fixed length packets which results in lower processing and higher speeds. Cell relay is a generic term for a protocol based on small fixed packet sizes capable of supporting voice video and data at very high speeds. Information is handled in fixed length cells of 53 octets (bytes). A cell has 48 bytes of information and 5 bytes of address. The objective of cell relay is to develop a single high-speed network based on a switching and multiplexing scheme that works for all data types. Small cells (like 53 bytes) favor low-delay, a requirement of isochronous service. The downside to small cells is that the address information is almost 10 percent of the total packet. That equates to high overhead and raw inefficiency. See ATM and SMDS.

cell reversal The reversal of the polarity of the terminals of a battery cell as the result of discharging.

cell site A transmitter/receiver location, operated by the WSP (Wireless Service Provider), through which radio links are established between the wireless system and the wireless unit. The area served by a cell site is referred to as a "cell". A cell site consists of an antenna tower, transmission radios and radio controllers. The cell site of an analog cellular radio system handles up to 5,000 users (but not all at once).

cell site controller The cellular radio unit which manages the radio channels within a cell.

Cell Site on Wheels COW. A portable cell site positioned to fill in or increase coverage, as a backup for emergency use.

cell spacing The amount of space between cells in a table. Cell spacing is the thickness, in pixels, of the walls of each cell.

cell splitting A technique used by cellular operators to improve capacity. Involves splitting cells to create smaller cells to enable frequencies to be re-used. more often. Cell splitting in the city of London, for example, has resulted in cell sizes measuring less than 100 yards.

cell switching A term that refers to how cellular calls are switched. Cellular systems are built to accommodate moving phones – ones in cars, buses, etc. These phones are low-powered. The "moving-ness" and the low power of the phones pose major design constraints on the design of cellular switching offices, which are called Mobile Telephone Switching Offices (or MTSOs). First, you have to build many cellular switching sites. That way each phone is close to a cell site. Thus there's always a cell site which can pick up the transmission. Second, because of the closeness of the cell sites, any phone conversation may be simultaneously heard by several MTSOs. As a result, the MTSO constantly monitors signal strength of both the caller and the receiver. When signal strength begins to fade, the MTSO locates the next best cell site and re-routes the conversation to maintain the communications link. The switch from one bcell site to another takes about 300 milliseconds and is not noticeable to the user. All switching is handled by computer, with the control channels telling each cellular unit when and where to switch.

The Cellular Mobile Telephone System is a low-powered, duplex, radio/telephone which operates between 800 and 900 MHz, using multiple transceiver sites linked to a central computer for coordination. The sites, or "cells", named for their honeycomb shape, cover a range of three to six, or more, miles in each direction. Their range is limited only by certain natural or man-made objects.

The cells overlap one another and operate at different transmitting and receiving frequencies in order to eliminate cross-talk when transmitting from cell to cell. Each cell can accommodate up to 45 different voice channel transceivers. When a cellular phone is activated, it searches available channels for the strongest signal and locks onto it. While in motion, if signal strength begins to fade, the telephone will automatically switch signal frequencies or cells as necessary without operator assistance If it fails to find an acceptable signal, it will display an "out of service" or "no service" message, indicating that it has reached the limit of its range and is unable to communicate.

Each mobile telephone has a unique identification number which allows the Mobile Telephone Switching Office (MTSO) to track and coordinate all mobile phones in its service area. This ID number is known as the Electronic Security Number (ESN). The ESN and Telephone Number are NOT the same. The ESN is a permanent number engraved into a memory chip called a PROM or EPROM, located in the telephone chassis. This number cannot be changed through programming as the telephone number can, although it can be replaced. Each time the telephone is used, it transmits its ESN to the MTSO by means of DTMF tones during the dialing sequence. The MTSO can determine which ESN's are good or bad; thus individual numbers can be banned from use within the system. See Cellular Radio.

cell tax A reference to the demands that ATM (Asynchronous Transfer Mode) places on bandwidth. ATM is a cell-switching technology which segments a data stream into cells, each of which comprises 48 octets of payload (data) and 5 octets of overhead (control information). Therefore, each cell is approximately 10.4% overhead. ATM signaling adds another 5%-10% in overhead. While ATM is wasteful of bandwidth, its benefits are significant. See also ATM and Cell.

cell trace box Wireless Integration/Interface Device. Also referred to as a "Proctor" box (name of the vendor), Cell Trace Box (US West's name for it), and protocol converter.

cell transfer A cellular radio term. Cell Transfer is the procedure of changing the channel stream in use to a channel stream originating at a different cell.

cell yell Cell yell is the tendency of many cellphone users to speak into their phones more loudly than necessary. It has created many cell-yell haters. Apparently many people don't realize how loudly they talk on their cell phones. A person who cell yells is called a cell yeller.

cell-fi handset A cell-fi phone. See next definition.

cell-fi phone A Wi-Fi-enabled cellular phone that enables voice and data to roam seamlessly across private Wi-Fi networks and public cellular networks.

cellco Cellular phone company.

cellpadding The syntax used to control the "padding" or area around the contents of a table's cell. In HTML tables are used as a layout tool which allows an HTML author to render text and graphics on a Web page in columns and rows. There are many options available with tables and cellpadding is one of them. The syntax looks something like this: <\<>table border=0 cellpadding=5 cellspacing=10>. This syntax would produce a table with a cellpadding of 5 pixels in width.

celliquette Social etiquette for using your cell phone. Don't use them in theaters. Don't talk loudly in the presence of others, etc.

cellphone A British term for a cellular telephone – whether car-based or handheld.

cells on wheels Mobile cellular towers that are used temporarily until a permanent tower is operational.

Cellsite On Wheels COW. A trailer with antenna and transmitting/receiving hardware used to provide temporary cell phone service in emergencies, special events, remote testing and repair, until a normal, permanent tower can be erected. A cow comes with climate control, diesel generator, and self-supporting wind-resistant antenna mast. Both full-size COWs and mini-COWs are available. The COW is especially useful in system changeouts and to get service where a permanent building may be a while in coming and plans call for a Self Contained Cell Site (SCCS) in the future. See also COLT and COW (for a larger explanation).

cellular alarm system An alarm system for a home or business that uses a cell phone instead of a phone line to make the call when the alarm is activated. There are clear advantages to such an alarm system. For one, it's much harder for the thieves to cut the phone lines and thus disable the alarm.

Cellular Data Link Control CDLC is a public domain data communications protocol used in cellular telephone systems. In other words, you can attach a data terminal to a cellular telephone and send and receive information. There are more than 5,000 modems using CDLC on the Vodaphone Cellular System in the UK, where it is the de facto standard for cellular data communications. Features like improved synchronization field, forward error correction, bit interleaving, and selective retransmission make CDLC ideal for cellular transmissions, according to Millidyne who makes the CDLC modems in the US.

Cellular Digital Packet Data CDPD is an open standard developed by a group of cellular carriers led by McCaw. The specification provides a standard for using existing cellular networks for wireless data transmission. Packets of data are sent along channels of the cellular network. See CDPD for a much more detailed explanation.

Cellular Digital Packet Data Group In the summer of 1993, a group of cellular carriers that supports the Cellular Digital Packet Data (CDPD) project released the complete - version 1.0 - of its open specification designed to enable customers to send computer data over an enhanced cellular network. According to the group, the packet data approach is ideally suited to those applications that require the transmission of short bursts of data, for example, authorizing a credit card number, exchanging e-mail messages or making database queries. The cellular networks deploying CDPD will enable mobile workers to use a single device to handle all of their voice and packet data needs. The 1.0 specification provides network and customer equipment manufacturers the parameters for building to this nationwide approach that sends packets of data over existing cellular networks. This version of the specification includes input from parties that reviewed the earlier release (0.8 and 0.9) and provides details of the CDPD architecture, airlink, external network interfaces, encryption and authentication, network support services, network applications services, network management, radio resource management and radio media access control. Copies of the specification can be had from CDPD Project Coordinator Tom Solazzo at Pittiglio Rabin Todd & McGrath, 714-545-9400 extension 235.

cellular extension service A PBX service whereby the PBX simultaneously rings a cellular phone and a desktop phone. While an employee can be reached this way, the problem is that when an employee makes a call from his cellular phone, the call is outside of the control of the company's PBX and routes through the cellular network instead of through the least-cost network that the PBX is programmed to send calls through. Furthermore, the cellular number is displayed as the caller ID on the called party's caller ID display. A dual-mode phone that automatically uses the company's wireless LAN when the employee is on the company's premises takes care of these problems when the employee is on the premises, but the problem remains when the employee is off-site and using the phone's cellular mode.

cellular floor method A floor distribution method in which cables pass through floor cells, constructed of steel or concrete, that provide a ready-made raceway for distributing power and communication cables.

cellular geographic service area CGSA. The geographic area served by the wireless (cellular) system within which a WSP is authorized to provide service.

cellular Macarena The dance that occurs when a cellular phone rings in a public place. Everyone reaches for their coat pocket, front pants pocket, back pants pocket, etc.

cellular mobile radio See Cellular Radio.

cellular mobile telephone service See CMTS.

cellular modem A device that combines data modem and cellular telephone transceiver technologies in a single unit. This allows a user to transfer data on the cellular network without the use of a separate cellular telephone.

cellular on light truck See COLT.

cellular phone service See CMTS.

cellular polyethylene Expanded or "foam" polyethylene consisting of individual closed cells suspended in a polyethylene medium.

cellular protocols Conventions and procedures which relate to the format and timing of device communications. In data transmission communications, there are currently three major protocols, which are converging into a de facto standard: MNP, SPCL, and PEP.

cellular radio A mobile radio system. In the old days, there was one central antenna and everything homed in on that and emanated from it. With cellular radio, a city is broken up into "cells," each maybe no more than several city blocks. Every cell is handled by one transceiver (receiver/transmitter). As a cellular mobile radio moves from one cell to another, it is "handed" off to the next cell by a master computer, which determines from which signal the strength is strongest. Cellular mobile radio has several advantages:

1. You can handle many simultaneous conversations on the same frequencies. One frequency is used in one cell and then re-used in another cell. You can't do this on a normal mobile radio system.

2. Because one cellular system can accommodate many more subscribers than a normal mobile radio system, and therefore because it can achieve certain economies of scale, it has the potential of achieving much lower transmission costs.

3. Because the transceiver is always closer to the user than in a normal mobile system, and the user's radio device thus needs less power, the device can be cheaper and smaller. Cellular radios started at over $5,000 and are now well under $500. From the first portable units, weight has already dropped to under one pound. There are several units that will fit in your breast pocket and not overly stretch your suit.

The following are specific cellular radio terms, or general telecom terms that mean something special in cellular radio:

- A/B Switch Permits user to select either the wireline (B system) or the nonwireline (A system) carrier when roaming.
- Alphanumeric memory Capability to store names with phone numbers.
- Call-in-absence horn alert User-activated feature that sounds car horn upon receiving a call.
- Call-in-absence indicator Feature that displays what calls came in while user was absent.
- Call-in-progress override Insures that power to the phone is not lost if the car's ignition is turned off.
- Call restriction Security feature that limits phone's use without completely locking it. Variations might include dial from memory only, dial last number only, seven-digit dial only, no memory access, etc.
- Call timer Displays information on call duration and quantity. Variations might include present call, last call, total number of calls, or total accumulated time since last reset. Call-timer beep serves as a reminder to help keep calls brief. It might be set to go off once a minute, ten seconds before the minute, for example.
- Continuous DTMF (touch-tones) Sends DTMF (dual-tone, multi-frequency) tones – also called touchtones – allowing access to voice mail and answering machines that require long-duration tones. "Continuous" means you get the tone so long as your finger is on the button. This may seem obvious to you and me, except that some "modern" phones just give a short tone no matter how long you keep your finger on the touchtone button.
- Dual NAM Allows user to have two phone numbers with separate carriers (see multi-NAM).

- Electronic lock Provides security by completely locking phone so it can't be used by unauthorized persons.
- Hands-free operation Allows user to receive calls and converse while leaving handset in cradle (similar to office speakerphone).
- Hands-free answering Phone automatically answers incoming call after a fixed number of rings and goes to hands-free operation.
- Memory linkage Allows programming specific memory locations to dial a sequence of other memory locations.
- Multi-NAM A cellular telephone term to allow a phone to have more than two phone numbers, each of which can be on a different cellular system if desired. This lets the user register with both carriers in home city, expanding available geographic coverage.
- Mute Silences the telephone's microphone to allow private conversations without discontinuing the phone call. Audio mute turns off the car stereo automatically when the phone is in use, and turns it back on when the call is completed.
- NAM Numerical Assignment Module. Basically, your cellular phone number, although it refers specifically to the component or module in the phone where the number is stored.
- On-hook dialing Allows dialing with the handset in the cradle.
- Roaming Using any cellular system outside your home system. Roaming often incurs extra charges.
- Scratch pad Allows storage of phone numbers in temporary memory during a call. Silent scratch pads allows number entry into scratch pad without making beep tones.
- Signal strength indicator Displays strength of cellular signal to let user know if a call is likely to be dropped.
- Speed dialing Dialing phone number from memory by pressing a single button.
- Standby time Maximum time cellular phone operating on battery power can be left on to receive incoming calls.
- Talk time Maximum time cellular phone operating on battery power can transmit.
- Voice-activated dialing Your cellular phone recognizes your words and dials accordingly. You say "Dial Mom" and it dials mom.

cellular radio switching office The electronic switching office which switches calls between cellular (mobile) phones and wireline (i.e. normal wired) phones. The switch controls the "handoff" between cells and monitors usage. Different manufacturers call their equipment different things, as usual.

Cellular Switching See Cell Switching.

Cellular Telecommunications Industry Association See CTIA.

CELP Code Excited Linear Prediction. An analog-to-digital voice coding and compression scheme used in the transmission of voice over packet data networks. VoFR (Voice over Frame Relay) and VoIP (Voice over Internet Protocol) are excellent examples. CELP, and its several derivatives, depend on a "code book," which essentially is a binary (digital) description of a set of voice samples. All voice starts out as an analog signal, of course. The first step is the conversion of that analog voice to convert it to a digital (data) format. This step is accomplished in a codec (coder/decoder) through the use of a standard technique known as PCM (Pulse Code Modulation). PCM involves the sampling of the analog voice stream at very precise points in time. Specifically, PCM involves the sampling of the amplitude (volume level) of the voice stream every 125 microseconds, or every 8,000th of a second. Each sample is converted into an 8-bit byte, which describes its value in digital terms. Now, PCM was designed with digital circuit switching and TDM (Time Division Multiplexing) in mind. Circuit switches provide bandwidth (capacity) on demand, as available, and on a basis which is temporary, exclusive and continuous in nature. Digital circuits such as T-1 rely on TDM, as do digital circuit switches. TDM provides time slots (slots of time) spaced every 8,000th of a second, in consideration of the voice application for which they were developed. Therefore, from end-to-end and through all of the switches and transmission facilities, PCM-based voice is supported very effectively. But it turns out that PCM is not very efficient. That is to say that voice can be compressed to a considerably greater extent than PCM does. Further, compressed voice can be transmitted quite effectively over packet-switched networks, which certainly are more efficient than their circuit-switched counterparts.

CELP involves the gathering in a buffer of a set of 80 PCM voice samples, representing 10 ms (10 milliseconds, or 1/100th of a second) of a voice stream. Once gathered, the set of voice samples is considered as a block of data. The data block is compressed to remove silence and redundancy, the volume level is normalized, and the resulting smaller data set is compared to a set of candidate descriptions in the codebook. The data transmitted across the network are only the index number of the selected code description, and the average loudness level of the set of samples. Every 10 ms, the code is sent across the network in a block of 160 bits, yielding a data rate of 16 Kbps, which compares very favorably with PCM voice over circuit-switched TDM networks at 64 Kbps. The compression rate is 4:1. At the receiving end of the transmission, the transmitted code is compared to the code book, the PCM signal is reconstructed and, eventually, an approximation of the analog signal is reconstructed. The reproduction is not perfect, but generally is close enough to yield good perceived quality through this process of voice synthesis. The devices which perform these processes of compression and decompression are DSPs (Digital Signal Processors), which include the basic codec function. This definition is courtesy of "Communications Systems & Networks," a great book written by Ray Horak, who is my Contributing Editor. Since you already have either bought this book, or have borrowed it from a colleague, or have stolen it, let me recommend that you buy Ray's book. Together, his book and this book tell the whole story. See also Circuit Switching, PCM, and TDM. See also the following derivatives of CELP: ACELP, CS-ACELP, LD-CELP and QCELP.

celsius At a glance, the Celsius scale makes more sense than the Fahrenheit scale for temperature measuring. But its creator, Anders Celsius, was an oddball scientist. When he first developed his scale, he made freezing 100 degrees and boiling zero degrees. No one dared point this out to him, so fellow scientists waited until Anders Celsius died to change the scale.

CEM Customer Experience Management. As if buying wasn't fun enough, they aim to make it even more fun and to measure it, to boot.

CEMH Controlled Environment ManHole. Environmental control of the CEMH is maintained by a heat pump (a fancy name for an airconditioner – cooler and heater).

CEN Comite European de Normalisation (French); the English translation is European Committee for Standardization. CEN is responsible for European standardization in all fields except electrotechnical (CENELEC) and telecommunications (ETSI). Certified products are awarded the CE Mark, signifying that a company has met the applicable essential health and safety requirements and the specific conformity assessment requirements to market its product in the European Union under the "New Approach" directives. CEN membership includes all EU countries, as well as affiliate members including Turkey, Cyprus, and many countries which formerly were members of the Soviet Bloc. Technical committees address standards in the areas of medical informatics, geographic information systems, character set technology for multilingual information, infrastructure, and advanced manufacturing technologies.

CENELEC Comite European de Normalisation ELECtrotechnique (French); the English translation is European Committee for Electrotechnical Standardization. CENELEC is responsible for European standardization in the electrotechnical field, working closely with ETSI (telecommunications) and CEN (all other fields). Certified products are awarded the CE Mark, signifying that a company has met the applicable essential health and safety requirements and the specific conformity assessment requirements to market its product in the European Union under the "New Approach" directives. CENELEC is the European technical organization responsible for coordination of standards for safety and electromagnetic emissions for electrical equipment in the European Economic Community (EEC). The EEC is working toward having a uniform set of standards that will apply for all EEC countries. Membership includes all EU countries, as well as affiliate members including Turkey, Cyprus, and many countries which formerly were members of the Soviet Bloc.

center conductor The solid or stranded wire in the middle of coaxial cable. The conductor diameter is measured by the American Wire Gauge (AWG).

center wavelength In a laser, the center wavelength is the nominal value central operating wavelength. It is the wavelength defined by a peak mode measurement where the effective optical power resides. In an LED (Light Emitting Diode), it is the average of the two wavelengths measure at the half amplitude points of the power spectrum.

Central Office CO. (pronounced See-Oh). Central office is an ambiguous term in North America. It can mean a telephone company building where subscribers' lines are joined to switching equipment for connecting other subscribers to each other, locally and long distance. Sometimes, that central office means a wire center in which there might be several switching exchanges. That means there will be switches, cable distribution frames, batteries, air conditioning and heating systems, etc. But a central office is sometimes simply a single telephone switch, what Europeans call a public exchange. In short, you have to figure out by the context if central office means a building or a switch, or a collection of switches. Simple, eh?

central office battery A group of wet cells joined in series to provide 48 volts DC. Central office batteries are typically charged off the main 120 volts AC. The batteries have two basic functions. 1. To provide a constant source of DC power for eight hours or so after AC power drops, and 2. To isolate the central office from glitches on the AC line.

central office distribution frame See main distribution frame.

central office code Part of the North American Numbering Plan (NANP), the central office code also is known as the central office prefix, the NXX code, and the end office code. A ten-digit telephone number in the U.S., for example, follows the format NXX-NXX-XXXX, where N must be a number other than "0" or "1," and X can be any number. The first three digits (NXX) comprise the area code, the second three (NXX) comprise the local central office code, and the last four (XXX) comprise the "line number." Here is a definition from Bellcore (now called Telcordia Technologies): A 3-digit identification under which up to 10,000 (0000-9999) station numbers are subgrouped. Exchange area boundaries that generally have billing significance are associated with the central office code. Note that multiple central office codes may be served by a single central office. Also called NXX code or end office code. Several central office codes in North America are kept for special purposes:

    555 – Directory Assistance
    950 – Feature Group B Access
    958 – Local Plant Test
    959 – Local Plant Test
    976 – Information Delivery Service

Central Office Equipment Reports COER. A telephone company definition. A large scale computer software package which accepts Central office Engineering Data properly formatted by a Data Collection System (DCS), subjects this data to a series of validation tests, and produces final summarized reports designed to meet both administrative and engineering requirements.

central pffice override A third party may interrupt or join in your conversation.

central office trunk 1. A trunk between central offices. It may be between major switches or between a major and a minor switch.

    2. A trunk between public and private switches.

Central Processing Unit CPU. The part of a computer which performs the logic, computational and decision-making functions. It interprets and executes instructions as it receives them. Personal computers have one CPU, typically a single chip. It is the so-called "computer on a chip." That chip identifies them as an 8-bit, 16-bit or 32-bit machine.

Telephone systems, especially smaller ones, are not that different. Typically they have one main CPU – a chip – which controls the various functions in the telephone. Today's telephone systems are in reality nothing more than special purpose computers. As phone systems get bigger, the question of CPUs – central processing units – becomes harder to figure. The design of phone systems has, of late, tended away from single processor-controlled telephone systems (as in single processor controlled PCs). There are several reasons for this move. First, it's more economical for growth. Make modules of "little" switches and join little ones together to make big ones. Second, it's more reliable. It's obviously better not to rely on one big CPU, but to have several. In short, the issue of Central Processing Units – CPUs – is blurring. But the concept is still important because by understanding how your telephone switch works (its architecture), you will understand its strengths and weaknesses.

central site A central site is a location that acts as a data collection point for remote and branch offices, as well as telecommuters and travelers.

centralized attendant service Calls to remote (typically branch) locations are automatically directed to operators at a central location. Imagine four retail stores in a town. There are three branch stores and one main, downtown store, each having its own local phone numbers, which customers call. It's clearly inefficient to put operators at each of the stores – when one group is busy, the other will be free, etc. What this feature does is to direct all the calls coming into each of the stores into one bank of operators, who then send those calls back to the outlying stores.

Despite the extra schlepping of calls around town, having one large group of operators is cheaper than maintaining many small groups. Each store has its own local Listed Directory Number (LDN) Service. Special Release Link Trunk circuits connect each unattended location (each store) to the main attendant location. These trunks are only temporarily used during call processing. An incoming call to an unattended store seizes such a trunk circuit for completion of the call to the centralized attendant, who then uses the same trunk circuit to process the call to the remote location's internal extension. (After all if the caller was calling that store, they obviously want to talk to someone in that store.) The circuit is then released and is available for other calls. Since such special trunk circuits are only used during that part of a call that requires connection between locations, such trunks are more efficient than normal tie trunk circuits.

Centralized Automatic Message Accounting CAMA. The recording of toll calls at a centralized point.

Centralized Network Administration See CNA.

Centralized Ordering Group COG. An organization provided by some communications service providers (like a local phone company) to coordinate services between the companies and vendors.

Centrex Centrex is a contraction of Central Exchange. Centrex is a business telephone service offered by a local telephone company from a local central office (also called a public exchange). Centrex is basically normal single line telephone service with "bells and whistles," added. Those "bells and whistles" include intercom, call forwarding, call transfer, toll restrict, least cost routing and call hold (on single line phones).

Think about your home phone. You can often get "Custom Calling" features. These features are typically fourfold: Call forwarding, Call Waiting, Call Conferencing and Speed Calling. Centrex is basically Custom Calling, but instead of four features, it has 19 features. Like Custom Calling, Centrex features are provided by the local phone company's central office.

Phone companies peddle Centrex which is leased to businesses as a substitute for that business buying or leasing its own on-premises telephone system – its own PBX, key system or ACD. Before Divestiture in 1984, Centrex was presumed dead. AT&T was, at that time, intent on becoming a major PBX and key system supplier. Then Divestiture came, and the operating phone companies recognized they were no longer part of AT&T, no longer had factories to support, but did have a huge number of Centrex installations providing large monthly revenues. As a result, the local operating companies have injected new life into Centrex, making the service more attractive in features, price, service and attitude. Here are the main reasons businesses go with Centrex as opposed to going with a stand-alone telephone system:

1. Money. Centrex is typically cheaper to get into (the central office already exists). Installation charges can be low. Commitment can also be low, since most Centrex service is leased on a month-to-month basis. So it's perfect for companies planning an early move. There may be some economies of scale, also. Some phone companies are now offering low cost, large size packages.

2. Multiple locations. Companies with multiple locations in the same city often are cheaper with Centrex than with multiple private phone systems and tie lines, or with one private phone system and OPX lines. (An OPX line is an Off Premise eXtension, a line going from a telephone system in one place to a phone in another. It might be used for an extension to the boss' home.)

3. Growth. It's theoretically easier to grow Centrex than a standalone PBX or key system, which usually has a finite limit. With Centrex, because it's provided by a huge central office switch, it's hard, theoretically, to run out of paths, memory, intercom lines, phones, tie lines, CO lines, etc. The limit on the growth of a Centrex is your central office, which may be many thousands of lines.

4. Footprint Space Savings. You don't have to put any switching equipment in your office. All Centrex switching equipment is at the central office. All you need at your office are phones.

5. Fewer Operators because of Centrex's DID features. Having fewer operator positions saves money on people and space.

6. Give better service to your customers. With Centrex, each person has their own direct inward dial number. Many people prefer to dial whomever they want directly rather than going through a central operator. Saves time.

7. Better Reliability. When was the last time a central office crashed? Here are some of the features built into modern central offices: redundancy, load-sharing circuitry, power back-up, on-line diagnostics, 24-hour on-site personnel, mirror image architecture, 100% power failure phones, complete DC standby backup and battery power. Engineered to suffer fewer than three hours down time in every 40 years.

8. Non-blocking. Trunking constraints are largely eliminated with Centrex, since a central office is so large.

9. Minimal Service Costs. Repair is cheap. Service time is immediate. People are right next to the machine 24-hours a day. Phones and wires are the only things that require repair on the customers' premises. You can easily plug new phones in and unplug them yourself. All other equipment is in the central office. You need not hold inventory or test equipment.

10. No technological obsolescence. Renting Centrex means a user has the ultimate flexibility – ability to jump quickly into new technology. Central offices are moving quickly into new technologies, such as ISDN.

11. Ability to manage it yourself. You can now get two important features previously available only on privately-owned self-contained phone systems (like PBXs): 1. The ability for you, the user, to make changes to the programming of your own Centrex installation without having to personally call a phone company representative. 2. The ability to get call

detail accounting by extension and then have reports printed by a computer in your office. The phone company does this call accounting by installing a separate data line which carries Centrex call records back to the customer as those calls are made.

The above arguments are pro-Centrex. There are also anti-Centrex arguments. Central offices often run out of capacity. Centrex is also cable-intensive. A PBX with 20 trunks and 100 phones only needs 20 cable pairs from the user's office to the telephone company. A Centrex installation with the same configuration needs 100 pairs. Every time someone new joins your company, the phone company needs to install another cable pair from the central office to your new employee's desk. Sometimes, they have it. Sometimes, they don't. Delays can get extensive. What with the explosion of telecom demand in recent years – individual fax machines, the Internet, etc. –there just isn't enough copper in the ground, and a typical telco won't plow in the cable unless they receive a pay-off in three years.

The "big" key to Centrex traditionally comes down to price. In some cities the price of Centrex lines is lower than "normal" PBX lines. Of course, you can buy Centrex lines and attach your own PBX or key system to those Centrex lines. The big disadvantage of Centrex is that there are very few specialized Centrex phones able to take better advantage of Centrex central office features the way electronic PBX phones take advantage of PBX features.

Centrex is known by many names among operating phone companies, including Centron and Cenpac. Centrex comes in two variations – CO and CU. CO means the Centrex service is provided by the Central Office. CU means the central office is on the customer's premises. See the following Centrex definitions.

Centrex Call Management A Centrex feature that provides detailed cost and usage information on toll calls from each Centrex extension, so you can better manage your telephone expenses.

Centrex CCRS Centrex Customer Rearrangement System. Computer software from New York Telephone that allows their Centrex customers to make certain changes in their own line and features arrangements. Other phone companies have similar services under different names.

Centrex CO Indicates that all equipment except the attendant's position and station equipment is located in the central office. See Centrex.

Centrex CU Indicates that all equipment including the dial switching equipment, is located on the customer's premises. See Centrex.

Centrex Extend Service The name of a Bell Atlantic service. If you maintain offices in multiple locations – or have work-at-home employees – this service allows you to tie all your locations into one phone system. So everyone in your company can take advantage of Centrex features and services on a cost-efficient, call-by-call sharing basis. With Centrex Extended Service, you can even tie non-Centrex locations into your Centrex system.

Centrex SMDI Have you ever called someone and been forwarded to their voice mail, and then had to enter their extension again. This is because their voice mail does not know where the call originated. The voice mail system does not know who you just called, All it knows is that it just received another call. SMDI is simply a modem link back to the central office supplying your Centrex system. This modem link will feed a computer at your location the information about incoming calls as they are forwarded through your system. It feeds the originating number and why the call is transferred, so you know whether the user didn't answer their phone or that it was busy.

To indicate the health of the SMDI link, it generates a heartbeat message every few seconds. If you don't receive a heartbeat within the time window, you know there is a problem with the SMDI link and know to restart the link. The one good thing about SMDI is the option that your voice mail system can control the status of message waiting indicators on the user's phone. They can either have a message waiting lamp on their phone, or use the "stutter dial tone" which causes a broken dial tone when the user picks up their phone.

Problems to watch out for... Since SMDI is a communications link, it can be broken. If the SMDI link goes down, make sure you build in the old two step method of finding out what extension the caller was attempting to call. Make sure you also offer a user directory in case the caller does not know the extension number. When the link goes down, make sure the system does not continue to send the message waiting status commands down an inactive link. SMDI stands for Standard Message Desk Interface or Simplified Message Desk Interface. See also SMDI.

Centronics The name of the printer manufacturer whose method of data transmission between a computer and a parallel printer has become an industry standard. See Centronics Printer Standard.

Centronics Printer Standard The Centronics standard was developed by the Centronics company which makes computer printers. The Centronics standard is a 36-pin single plug/connector with eight of the 36-pins carrying their respective bits in parallel (eight bits to one character), which means it's much faster than serial transmission which sends only one bit a time. There are several types of Centronics male and female plugs and receptacles. So know which you want before you buy. The pinning – the location and function of each of the 36-individual wires – is standard from one Centronics cable to another.

The Centronics printer standard has been adopted by many printer and PC companies, including IBM. It is a narrower standard than the RS-232-C standard. The Centronics works only between a computer and a printer. It won't work over phone lines, unless conversion is done at either end. However, it is standard and has none of the dumb interface problems the RS-232-C standard does.

Centum Call Second 1/36th of an erlang. The formula for a centum call second is the number of calls per hour multiplied by their average duration in seconds, all divided by 100.

CEO Chief Executive Officer.

CEP Certificate Enrollment Protocol. Certificate management protocol jointly developed by Cisco Systems and VeriSign, Inc. CEP is an early implementation of Certificate Request Syntax (CRS), a standard proposed to the Internet Engineering Task Force (IETF). CEP specifies how a device communicates with a CA, including how to retrieve the public key of the CA, how to enroll a device with the CA, and how to retrieve a certificate revocation list (CRL). CEP uses Public Key Cryptography Standard (PKCS) 7 and PKCS 10 as key component technologies. The public key infrastructure working group (PKIX) of the IETF is working to standardize a protocol for these functions, either CRS or an equivalent.

CEPT Conf0xC7rence des administrations Europeanes des Postes et Telecommunications (European Conference of Postal and Telecommunications Administrations). Standards-setting body whose membership includes European Post, Telephone, and Telegraphy Authorities (PTTs). It in turn participates in relevant areas of the work of CEN/CENELEC. It was originally responsible for the NET standards, but these have subsequently been passed on to ETSI.

CEPT Format Defines how the bits of a PCM carrier system of the 32 channel European type T-1/E-1 will be used and in what sequence. To correctly receive the transmitted intelligence, the receiving end equipment must know exactly what each bit is used for. CEPT format uses 30 VF channels plus one channel for supervision/control (signaling) and one channel for framing (synchronizing). All 8 bits per channel are used to code the waveshape sample. For a much better explanation, see T-1.

CER An ATM term. Cell Error Ratio: The ratio of errored cells in a transmission in relation to total cells sent in a transmission. The measurement is taken over a time interval and is designed to be measured on an in-service circuit.

CERB Centralized Emergency Reporting Bureau. A Canadian term similar to PSAP – Public Safety Answering Position. See PSAP.

Cerf, Vinton Founding President of the Internet Society (ISOC) from 1992 to 1995 and co-creator of the transmission control protocol/Internet Protocol (TCP/IP), which enables computers to talk to each other over the Internet. Cerf proved that a network can reconfigure itself so that no communications are lost. He did this by breaking apart the Defense Department's Arpanet network artificially, and showing that it could be reconnected by way of flying packet radios in Strategic Air Command jets. Cerf used the airborne radios to link to ground radios which were, in turn, linked to internet gateways (today's routers) to effectively interconnect the artificially separated pieces of ARPANET artificially separated. See also Kahn, Robert.

CERN European Laboratory for Particle Physics Research in Geneva, Switzerland.

CERNET2 China Education and Research Network. The world's largest next-generation Internet network with a native IPv6 backbone now under working and under expansion in China. China is motivated to complete the migration from IPv4 (what the rest of the world typically has at present) to IPv6 as soon as possible because China was allocated only 22 million IPv4 addresses 25 years go (out of the inventory of about four billion IPv4 addresses), which it is projected to deplete next year. The United States, by contrast, was allocated around 70% of the inventory of IPv4 addresses 25 years ago. The IPv4 address allocations made sense 25 years ago, but not any more. Migrating to IPv6 everywhere will give China many orders of magnitude more IP addresses than it has in the present IPv4 environment.

CERT Computer Emergency Response Team. The CERT is a group of computer experts at Carnegie-Mellon University chartered to work with the Internet community to facilitate its response to computer security events involving Internet hosts, to take proactive steps to raise the community's awareness of computer security issues and to conduct research targeted at improving the security of existing systems. The CERT was formed by DARPA in November 1988 in response to the Internet worm incident. They maintain an archive of security-related issues on their FTP server at "cert.org." Their email address is "cert@cert.org" and their 24-hour telephone Hotline for reporting Internet security issues is 412-268-7090.

certificate A cryptography term. Also known as a digital certificate, a "certificate" is a password-protected, encrypted data file which includes the name and other data which serves to identify the transmitting entity. The certificate also includes a public key which serves to verify the digital signature of the sender, which is signed with a matching private key, unique to the sender. Through the use of keys and certificates, the entities exchanging data can authenticate each other.

Certificate Authority CA. A trusted third-party organization or company that issues digital certificates used to create digital signatures and public-private key pairs. These pairs allow all system users to verify the legitimacy of all other system users with assigned certificates. The role of the certificate authority is to guarantee that the individual granted the unique certificate is, in fact, who he or she claims to be. Usually, this means that the certificate authority has an arrangement with a financial institution, such as a credit card company, which provides it with information to confirm an individual's claimed identity. Certificate authorities are a critical component in data security and electronic commerce because they guarantee that the two parties exchanging information are really who they claim to be. See also Certificate and Digital Certificate.

Certificate of Compliance C of C. A certificate which is normally generated by a Quality Control Department, which shows that the product being shipped meets customer's specifications.

Certificate of Public Convenience and Necessity See CPCN.

certification practice statement This term originates in the American Bar Association's Digital Signature Guidelines where it is defined as a "statement of the practices which a certification authority employs in issuing certificates."

certified Several companies in the "secondary" industry test used equipment, parts and/or systems. They have various ways of testing them. Typically they test with working phones operating for extended periods at different temperatures. The idea is to check that this used equipment works the way it's meant to work – to the original manufacturer's design specification. Once these tests have been completed a secondary dealer will "certify" such equipment, usually in writing. Such certification carries the assurance that the used equipment works as it's meant to. Sometimes certified equipment is upgraded to the most current revision level of hardware and software. Sometimes it's not. You, the buyer, must check. Certified equipment typically carries a guarantee – that guarantee being as good, obviously, as the company that backs it. See also NATD, Refurbished, and Remanufactured.

certified equipment A term used in the secondary telecom equipment business. Equipment carrying the written assurance that it will perform up to the manufacturer's specifications. It qualifies for addition to existing maintenance contracts.

certified test report CTR. A report providing actual test data on a cable. Tests are normally run by a Quality Control Department, which shows that the product being shipped conforms to test specifications.

CES An ATM term. Circuit Emulation Service: The ATM Forum circuit emulation service interoperability specification specifies interoperability agreements for supporting Constant Bit Rate (CBR) traffic over ATM networks that comply with the other ATM Forum interoperability agreements. Specifically, this specification supports emulation of existing TDM circuits over ATM networks.

CESID Caller Emergency Service ID. Several states in the United States require PBXs to send a telephone extension number to a PSAP (E-911 emergency answering point). This is helpful in cases where the caller is not located near the address listed for the business' Listed Directory Number. A "campus" environment of separate buildings comes to mind. The location information displayed at the PSAP usually comes from the ALI (Automatic Location Identification) database. Customers with PBXs are normally required to provide initial telephone location information as well as updates to the ALI database maintainer (usually the LEC). The CESID information is transmitted in one of two formats: ISDN-PRI ANI (around Chicago and in some areas of New York), or CAMA. CAMA uses R2MF signaling to send the CESID to the PSAP, and, like ISDN-PRI, has to be ordered as a specific type of trunk from the LEC. Unlike ISDN-PRI, however, it appears to have no other use aside from E-911 calls. Until or unless PSAPs in the rest of the country switch over to ISDN-PRI trunks for call receipt, CAMA trunks will be around for awhile. See CAMA, E-911, ISDN PRI, LEC and PSAP.

cesium See cesium clock and International Atomic Time.

cesium clock A clock containing a cesium standard as a frequency-determining element. It's a very accurate clock. The current definition of a second is the duration of 9,192,631,770 oscillations of cesium atoms excited by microwaves. Today's cesium atomic clocks are accurate to within one million billionth of a second or 1 second in 30 million years. Cesium clocks are the most accurate clocks ever made. Cesium clocks are used in Global Positioning System (GPS) satellites and as primary reference clocks in telecommunications carrier networks. See cesium standard. See Cesium Standard and International Atomic Time.

cesium standard A primary frequency standard in which a specified hyperfine transition of cesium-133 atoms is used to control the output frequency. Its accuracy is intrinsic and achieved without calibration. See cesium clock and International Atomic Time.

CEU Commercial End User. See SU, service user.

CEV Controlled Environmental Vault. A below ground room that houses electronic and/or optical equipment under controlled temperature and humidity.

CFA 1. Carrier Facility Assignment. CAPs/CLECs give RBOCs/LECs a slot or channel assignment where their T-1s or T-3s will be connecting. A CFA is the identifier or location where an IXC, CAP, CLEC, or LEC will interconnect with the incumbent Telco. It will come in one of three forms: ACTL/CLLI, APOT, or tie/cable pair. ACTL/CLLI looks like 1001/T3/18/WASHD-CAB123/WASHDCXY789, where 1001 is the DS-3 off a SONET OC ring, while T3 signifies that connection will be made at the T-3 slot 18 (the T-1 will connect to this slot). The first location CLLI is the ACTL WASHDCCAB123. The second location CLLI is the central office CLLI WASHDCXY789. If the CFA is in APOT form, it will be like 1.9.13.23.18 (i.e. floor.aisle.bay.panel.jack). Tie/cable pair looks like 10011/t3/T1TIE/WASHDCAB123/WASHDCAB where the last 1 in 10011 signifies slot 1. The Telco tech will then do the interconnect or x-connect at that facility assignment. See CAP, CLEC and RBOC.

2. Carrier Failure Alarm. The alarm which results from an out-of-frame or loss of carrier condition and which is combined with trunk conditioning to create a CGA.

3. Connecting Facilities Arrangement. Identifies a complete communications channel between two places.

4. Connecting Facility Assignment. The facility designation of the high-capacity system that the line or trunk is using.

CFAC Call Forward All Calls.

CFAMN Call Forwarding Address Modified Notification.

CFB 1. Call Forward Busy.

2. Call Forwarding on mobile subscriber Busy. A wireless telecommunications term. A supplementary service provided under GSM (Global System for Mobile Communications).

CFCA Communications Fraud Control Association. Founded in 1985, CFCA is a not-for-profit international educational association working to help combat telecommunications fraud. CFCA seeks to promote a close association among telecom security personnel, to enhance their professional status and efficiency, and to serve as a clearinghouse of information pertaining to the fraudulent use of telecommunications services. CFCA membership is includes interexchange carriers (IXCs); local exchange carriers (LECs); competitive local exchange carriers (CLECs), incumbent local exchange carriers (ILECs), private network companies, law enforcement officers; users, e-mail providers, security product vendors, and corporations that use telecommunication services. www.@cfca.org.

CFDA Call Forward Don't Answer.

CFGDA Call Forward Group Don't Answer.

CFM 1. Cubic Feet per Minute. A measure of how much air you move through the fan of a PC.

2. Carrier Financial Management.

CFNRc Call Forwarding on mobile subscriber Not Reachable. A wireless telecommunications term. A supplementary service provided under GSM (Global System for Mobile Communications).

CFNRy Call Forwarding on No Reply. A wireless telecommunications term. A supplementary service provided under GSM (Global System for Mobile Communications).

CFO Chief Financial Officer.

CFP Channel Frame Processor.

CFR 1. Confirmation to Receive frame.

2. Code of Federal Regulations.

CFRP Carbon Fiber Reinforced Plastic. A light and durable material, which has been used (for the wings of advanced fighter jets) in the defense business and which Toshiba introduced in 1991 as casing for a line of notebook sized computer laptops.

CFUC Call Forwarding UnConditional. A wireless telecommunications term. A supplementary service provided under GSM (Global System for Mobile Communications).

CFV Call For Votes. Begins the voting period for a Usenet newsgroup. At least one (occasionally two or more) email addresses is customarily included as a repository for the votes.

CFW Call Forward.

CG Character Generator: Device that electronically displays letters and numbers on the television screen.

CGA 1. Carrier Group Alarm. A service alarm generated by a channel bank when an out-of-frame (OOF) condition exists for some predetermined length of time (generally 300 milliseconds to 2.5 seconds). The alarm causes the calls using a trunk to be dropped and trunk conditioning to be applied.

2. Color Graphics Adapter. An obsolete IBM standard for displaying material on personal computer screens. The simplest (and conventional) CGA displays 320 horizontal picture elements, known as pels or pixels, by 200 pels vertically. There is also an Enhanced CGA, which is 640 x 400, or 128,000 pixels per screen. Older portables may use CGA monochrome mode. CGA has essentially been obsoleted by VGA. See Monitor and VGA.

CGI Common Gateway Interface. An Internet term. Let's start simple. You write a form on a web site which you want visitors to fill in – let's say their email address and phone number. You write some software, also called a script, which presents the visitor with the form and ask him/her to kindly fill it out. That software is usually executed on the Web server. CGI originally referred to the pre-defined way in which these programs communicated with the Web Server but lately it has come to refer to the programs themselves. The preferred programming language for CGI scripts is Perl. Forms are usually processed by CGI (Common Gateway Interface) scripts. CGI is basically a standardized way of sending information between a server and a processing script. CGI scripts are typically written in Perl or some other programming language such as C++, Java, VBScript, or JavaScript. Before creating interactive forms on your web site, you must check with your ISP or server administrator to see if CGI scripts can run on your server. See also CGI-Bin, PERL and Servlet.

CGI-Bin A directory on a server that "houses" all of the CGI programs. When you see this as a directory in your browser's URL window, it usually means you are either running or about to run a CGI program. The "binary" part refers to when many of the files placed in that directory were binary files. More recently, many of these files are text-based. CGI-Bin is the most common name of a directory on a web server in which CGI programs are stored. The "bin" part of "cgi-bin" is a shorthand version of "binary" because once upon a time, most programs were referred to as "binaries." In real life, most programs found in cgi-bin directories are text files – scripts that are executed by binaries located elsewhere on the same machine.

CGI Joe A Wired Magazine definition. A hardcore CGI script programmer with all the social skills and charisma of a plastic action figure.

CGI Script See CGI.

CGM Computer Graphics Metafile. A standard format that allows for the interchanging of graphics images.

CGSA Cellular Geographic Service Area. The actual area in which a cellular company provides cellular service. CGSAs are usually made up of multiple counties and often cross state lines.

CGSA Restriction If you own a cellular phone, you are prevented from making calls outside your own local Cellular Geographic Service Area. This restriction is an option that is available to subscribers in most cellular cities.

chad 1. The little solid round dots of paper made when paper tape is punched with information. Also the little rectangular pieces of paper that are made when any sort of punch card is used for data input purposes. Carriers used to use a lot of paper tape for AMA (Automated Message Accounting) purposes in recording long distance calls, but that's all been replaced by magnetic tape and other electronic media. We used to use a lot of punch cards for all sorts of computer data input purposes, be that's all been replaced, as well. One of the rare exceptions is the continuing use of obsolete punch card technology is U.S. voting machines. It's interesting to note that Al Gore, who erroneously claimed to be the "Father of the Internet," also erroneously claimed to have lost the 2000 presidential election due to obsolete punch card voting machines in Florida.

2. CHAnge Display.

Chad Tape Punched tape used in telegraphy/teletypewriter operation. The perforations, called "chad," are severed from the tape, making holes representing the characters.

chadless tape 1. Punched tape that has been punched in such a way that chad is not formed.

2. A punched tape wherein only partial perforation is completed and the chad remains attached to the tape. This is a deliberate process and should not be confused with imperfect chadding. See Chad.

chain mailboxes Mailboxes that are connected together to provide a service or a number of messages (e.g. Directory, Product Information, etc.).

chaingang A group of Web homepages which merely link to each other.

chaining A programming technique linking one activity to another, as in a chain. Each link in the chain may contain a pointer to the next link, or there may be a master control or program instructing the programs to link together.

chainsaw consultant An outside "expert" brought in to reduce the employee headcount, leaving the top brass with clean hands and a clean conscience.

chairman In the late 1700s, many houses consisted of a large room with only one chair. Commonly, a long wide board was folded down from the wall and used for dining. The "head of the household" always sat in the chair while everyone else ate sitting on the floor. Once in a while, a guest (who was almost always a man) would be invited to sit in this chair during a meal. To sit in the chair meant you were important and in charge. Sitting in the chair, one was called the "chair man."

chalk talk A PowerPoint presentation.

challenge email Let's say you're sick of all the spam mail. You set up a system whereby to every email that comes your way, you send an email asking the sender of your email a question. If the person replies to your email the correct way, they are allowed to send emails to you. This method of cutting back on spam is called "challenge email."

Challenge-Handshake Authentication Protocol CHAP. An authentication method that can be used when connecting to an Internet Service Provider. CHAP allows you to log in to your provider automatically, without the need for a terminal screen. It is more secure than the Password Authentication Protocol (another widely used authentication method) since it does not send passwords in text format. An Internet term.

challenge-response A type of authentication procedure into a system in which a user must respond correctly to a challenge, usually a secret key code, to gain access.

challenged Indicating an undesirable or unappealing condition. People who are "intellectually challenged" are stupid.

chalk talk When your professor explains it on the blackboard, that's called chalk talk. This term has now been extended to anyone using foils, overheads or PowerPoint.

change + A woman marries a man expecting he will change. He doesn't. + A man marries a woman expecting that she won't change. She does. Both are disappointed.

change freeze I'm sitting in Air New Zealand's lounge in the Auckland Airport. They have wireless Internet access. But they have blocked off access to POP3 and SMTP email. I don't know why and there is no logic for this decision, since every other air lounge provides it. A man called Sean Kemball, Network and Web Operations Manager for Air New Zealand agrees with me. He says they are changing in the New Year. I ask "Why not today?" He replies, "Although I agree that the change is not technically difficult, the purpose of a change freeze is to ensure stability of our core systems during a time when we have reduced levels of technical support available, and as I'm sure you can appreciate, this means we need to apply additional scrutiny to all changes. This is a common approach taken by most large organisations, and balances a reduced level of agility against a higher level of systems availability and defined service." The good news: Two weeks later, on January 5, I revisit Air New Zealand's lounge. SMTP and POP3 email access are working perfectly. Their change freeze was temporary. They got it fixed.

change management Change management is the process of introducing controlled change during the project life cycle. The intent behind a change control process is to evaluate the risk, at the end-user level, against the urgency and importance of the change. IT organizations must establish a specific change control process for every type of change and consider a procedure specific to rapid and emergency changes. The creation of these processes must involve several groups, including users, developers, and operations. Successful change management is not only a matter of skill and expertise, it is also a question of where the team draws its support from the company hierarchy. Source Giga Information Group. See also change freeze.

channel 1. Typically what you rent from the telephone company. A voice-grade transmission facility with defined frequency response, gain and bandwidth. Also, a path of communication, either electrical or electromagnetic, between two or more points. Also called a circuit, facility, line, link or path.

2. An SCSA term. A transmission path on the SCbus or SCxbus Data Bus that transmits data between two end points.

3. A channel of a GPS (Global Positioning System) receiver consists of the circuitry necessary to tune the signal from a single GPS satellite.

4. A shortened way of saying "distribution channel." Let's say you make a product – hardware or software. You need to have some way of selling it. You can sell it yourself with your own salespeople. Or you can give it to distributors to sell. Such distributors could be wholesalers, small retailers, large retail chains, direct mail catalogs, etc. Each one of these categories is called a "channel." See also Channel Conflict, Channel Management and Channel Ready.

5. A Fibre Channel term. A point-to-point link, the main task of which is to transport data from one point to another.

Channel 1 When the FCC first allocated broadcast TV frequencies in 1945 in the United States. Later, the FCC decided that TV was taking up too much broadcast spectrum. Each channel requires a bandwidth 600 times as wide as an individual radio station does.

Thus, the Channel 1 band, 44 to 50 MHz, was reassigned for mobile radio use.

channel aggregator Also known as inverse multiplexors. Devices that allow very large amounts of data to be sent down the narrow band channels of ISDN. The aggregator effectively pulls together ISDN channels at one end to form a higher bandwidth and then re-synchronizes the information at the other end. Re-synchronization is necessary because during transmission the ISDN channels may send the information along different routes, so it arrives at its destination at fractionally different times.

Channel Associated Signaling CAS. A form of circuit state signaling in which the circuit state is indicated by one or more bits of signaling status sent repetitively and associated with that specific circuit.

channel attached Describing the attachment of devices directly to the input/output channels of a (mainframe) computer. Devices attached to a controlling unit by cables rather than by telecommunications circuits. Same as locally attached (IBM).

channel bank A multiplexer. A device which puts many slow speed voice or data conversations onto one high-speed link and controls the flow of those "conversations." Typically the device that sits between a digital circuit – say a T-1 – and a couple of dozen voice grade lines coming out of a PBX. One side of the channel bank will be connections for terminating two pairs of wires or a coaxial cable – those bringing the T-1 carrier in. On the other side are connections for terminating multiple tip and ring single line analog phone lines or several digital data streams. Sometimes you need channel banks. Sometimes, you don't. For example, if you're shipping a bundle of voice conversations from one digital PBX to another across town in a T-1 format – and both PBXs recognize the signal – then you will probably not need a channel bank. You'll need a Channel Service Unit (CSU). If one, or both, of the PBXs is analog, then you will need a channel bank at the end of the transmission path whose PBX won't take a digital signal. See Channel Service Unit and T-1.

channel blanket topology A wireless LAN architecture that involves enabling each radio channel to be used everywhere, i.e., on every wireless access point on the WLAN, to create blankets of coverage. A switch on the network keeps track of radio channel usage on all access points and controls channel allocation in real-time in order to maximize channel re-use network-wide while avoiding co-channel interference. Vendors offering channel-blanket WLAN solutions claim that a channel-blanket topology improves bandwidth utilization up to tenfold over traditional WLAN designs, which are based on meticulous RF cell-planning designed to avoid co-channel interference. A user-experience benefit of channel-blanket topology, in theory at least, is that the ubiquitous availability of every radio channel reduces handoff latency for mobile users.

channel capacity A measure of the maximum possible bit rate through a channel, subject to specified constraints.

channel capture A condition that occurs when the Ethernet MAC layer temporarily becomes biased toward one workstation on a loaded network, thereby making that one station the contention winner more times than would randomly occur.

channel checks When an analyst asks questions of a company's customers, suppliers, employees and even rivals in order to find out how well a business is really doing, he (or she) is doing what Wall Street calls "channel checks." Channel checks are meant to be an integral part of a serious Wall Street analyst's job description.

channel coding The process of adding redundant information into a transmitted bit stream before transmission in order to protect the bit stream from errors that may occur. Channel coding therefore reduces the error rate in a channel, but increases the amount of information (overhead) that must be transmitted. Typical methods of channel coding include forward error correction, error detection schemes, and interleaving of bits.

channel compression The process of fitting more than one program into a single channel. See Analog Channel Compression and Digital Channel Compression.

channel conflict Channel conflict happens when a manufacturer wants to sell over the web, but a brick-n-mortar retailer that manufacturer sells through says, "If you sell your products directly on your web site, we'll stop carrying your brand or relegate it to the bottom shelf." The physical distribution channel is in conflict/competition with the e-commerce distribution channel. Sometimes this problem is solved by the manufacturer agreeing to keep his prices at full retail on his web site. Other times it's solved by the manufacturer simply saying – Sorry, but we're going ahead anyway. Other times it is solved by the manufacturer dropping its idea of selling on the web and simply having its web site as an product information site, with links directly to its retailers – some of which may run sell over the Internet.

channel converter A device which converts signals from one channel to another channel. There are two types – heterodyne converters and those which use frequency multiplication principles. See Heterodyne Converter and Processor.

channel definition format An open standard announced by Microsoft in

March 1997 to be presented to the World Wide Web Consortium (W3C) as a suggested future open standard for "push" technology.

channel efficiency In a LAN environment, a measure of the total information that can be communicated in a channel in a unit of time, accounting for noise, collisions and other disruptions.

channel gate A device for connecting a channel to a highway, or a highway to a channel, at specified times.

channel hopping A cellular radio term. Channel Hopping is the process in CDPD of changing the Radio Frequency (RF) channel supporting a CDPD channel stream to a different RF channel on the same cell. This is typically used to avoid collisions with voice traffic use of the RF channel.

channel identification Information element that requests or identifies the channel to be used for a call. An AIN term.

channel loopback In network management systems, diagnostic test that forms a loop at the multiplexer's channel interface that returns transmitted signals to their source. See also Loopback.

channel management Since the last of the HP/Compaq merger much has been written. Now that shareholders have approved the deal, it is interesting to look at one of the most critical aspects of this and any merger - channel management. HP and Compaq each support broad product lines with unique distribution strengths. HP is known for its reseller channel, while Compaq is lauded for its direct marketing strength. The potential for conflict is obvious. While the aim of the merger it to increase sales, HP resellers are rattled by the fear of being cut out by greater adoption of Compaq's direct selling tactics. Moreover, much of the Compaq and HP reseller network find themselves with indistinct product lines. Pundits correctly predict that the dealer channel will be cut back to reduce costs. Only Value Added Resellers (VARs) offering true added value will survive, as the industry moves towards the Dell direct sales model. In this rather confusing environment, resellers naturally become nervous and some of the better VARs defect to the competition. In meeting this challenge the new HP needs to focus on two key words, clarity and communication. 0x15 Clarity - the merged company must offer a clear and consistent policy regarding a wide range of questions concerning VARs. Which product lines will be retained and which will be dropped? What happens to brands? Which products will be pushed through the expanded HP direct sales channel and which will remain exclusively in the VAR channel? How will VAR pricing change, if at all? Which VARs will be dropped? 0x15 Communication - the new HP cannot afford to confuse VARs. Ongoing communication is critical if VARs are to stay happy and continue to push products. If the messaging is badly managed, HP can expect to loose good VARs to the competition and see sales drop. The aim of all mergers is to create a larger, stronger company. With a huge VAR channel to manage, the new HP must move fast to shore up its partner network.

channel mode An AT&T term for a method of communications whereby a fixed bandwidth is established between two or more points on a network as a semi-permanent connection and is rearranged only occasionally.

channel model See Inifiniband.

channel modem That portion of multiplexing equipment required to derive a desired subscriber channel from the local facility.

channel packing A technique for maximizing the use of voice frequency channels used for data transmission by multiplexing a number of lower data rate signals into a single higher speed data stream for transmission on a single voice frequency channel.

channel queue limit Limit on the number of transmit buffers used by a station to guarantee that some receive buffers are always available.

channel rate-adaptation protocols These protocols tell the ISDN terminal adapter (TA) how to change its transmission/ reception speeds to match those of the connecting device. Europe and Japan primarily use the V.110 protocol. The U.S. uses V.120. See ISDN and Protocol.

channel ready A channel is a shortened way of saying "distribution channel." Let's say you make a product – hardware or software. You need to have some way of selling it. You can sell it yourself with your own salespeople. Or you can give it to distributors to sell. Such distributors could be wholesalers, small retailers, large retail chains, direct mail catalogs, etc. Each one of these categories is called a "channel." Channel Ready means that your product is in a form your chosen channel can handle. Typically this means that your software and/or hardware comes and can be delivered in a shrinkwrapped box and that you have set an organization which can service, support, train and otherwise satisfactorily deal with customers contacting you.

channel seized The time when a connection is established between the cellular user's mobile equipment and the mobile telephone switching office (MTSO). Channel sei-

zure occurs before the number dialed begins to ring.

Channel Service Unit CSU. A device used to connect a digital phone line (T-1 or Switched 56 line) coming in from the phone company to either a multiplexer, channel bank or directly to another device producing a digital signal, e.g. a digital PBX, a PC, or data communications device. A CSU performs certain line-conditioning, and equalization functions, and responds to loopback commands sent from the central office. A CSU regenerates digital signals. It monitors them for problems. And it provides a way of testing your digital circuit. You can buy your own CSU or rent one from your local or long distance phone company. See also CSU and DSU.

channel splitter A channel splitter transmits channels one through 12 over one wire pair and channels 13 through 24 over another wire pair.

channel stream A cellular radio term. Channel Stream is a shared digital communications channel between a Mobile Data Base Station (MDBS) and a set of Mobile End Systems (M-ESs) considered as a logical concept, separate from the frequency of the Radio Frequency (RF) channel used to implement the channel at any given time.

channel surfing Flipping channels on a TV set. A person who channel surfs is called a Mouse Potato.

channel terminal That portion of multiplexing equipment required to derive a desired subscriber channel from the bearer facility.

channel termination Central Office (CO) equipment at the telephone company side of a local loop, a channel termination is used to terminate an end user's network access circuit, channel-by-channel. For a T-1 circuit, channel termination equipment commonly is in the form of two ports on a channel bank, one for the upstream circuit and one for the downstream circuit. Dedicated, multichannel T-1 circuits typically involve a recurring monthly charge in each direction, per channel.

channel tier An AT&T term for the tier within the Universal Information Services network that partitions transmission capacity into channels and offers the channels to the nodes' higher tiers.

channel time slot A time slot starting at a particular instant in a frame and allocated to a channel for transmitting a character, in-slot signal, or other data. Where appropriate a modifier may be added.

channel translator Device used in broadband LANs to increase carrier frequency, converting upstream (toward the head-end) signals into downstream signals (away from the head-end).

channel virtual area Where Internet Relay Chat (IRC) users communicate in real time. There are thousands of channels on the Internet.

channelization The process of subdividing the bandwidth of a circuit into smaller increments called channels. Typically, each channel carries an individual transmission, e.g. a voice conversation or a data conversation – a computer-to-computer session. Multi-channel circuits always are four-wire in nature, whether physical four-wire or logical four-wire, and the process of channelization is always accomplished through some form of multiplexer (MUX), which can be in the form of either a Frequency Division Multiplexer (FDM) or a Time Division Multiplexer (TDM). The most basic type of MUX is known as a channel bank. Traditional T-1 service is a channelized service, for instance. Through a Statistical Time Division Multiplexer (STDM), a T-1 circuit of 1.544 Mbps is subdivided into 24 channels of 64 Kbps each, with an additional 8 Kbps needed for framing overhead (do the math and you'll see). Each channel (at least for contemporary multiplexers employing ESF, or Extended SuperFrame) provides a full 64 Kbps of bandwidth in support of a single "conversation" (data or voice), which requirement was established for transmission of digitized voice according to the original PCM (Pulse Code Modulation) encoding algorithm. Older multiplexers provided only 56 Kbps of usable channel capacity due to the fact that they employed an intrusive form of signaling and control. The STDM subdivides the digital bandwidth of the circuit into 24 smaller units of bandwidth known as "time slots" through a sampling process. Picture a 24 lane highway leading up to one toll booth. On the other side of the toll booth is a one lane highway. The vehicles in the 24 lanes are allowed through the toll booth (each of the 24 input devices is allotted a piece of time in which it is allowed to send data) if they are ready to pay the toll. If they do not have the toll fee available the attendant will ask the next vehicle to go through (the TDM polls the next device during the next, (fixed) time slot). In TDM devices are polled (asked) if they have anything to transmit during their preordained time slot. If they do not have anything to send, that time slot cannot be used by another device which may have something to send. This is wasted bandwidth. This is traditional TDM. In contrast, Statistical Time Division Multiplexing (STDM) asks why waste the time slot if some other device is ready to send? STDM is a non-channelized way of using the available bandwidth. It allocates time slots to those devices who are ready to send. This principle is called "Bandwidth on Demand". FDMs subdivide bandwidth through a process of frequency separation, with each conversation occupying a separate and distinct frequency within a larger range of frequencies supported by the circuit. While FDM is seldom employed in the contemporary world of electrically-based (read copper wire) circuits, it is widely used in the wireless world (e.g., cellular and PCS). FDM also is widely employed in the fiber optic world, where it is known as WDM (Wavelength Division Multiplexing) or DWDM (Dense WDM). WDM supports multiple, very high-capacity virtual circuits over a single optical fiber, with each virtual circuit being subdivided into a very large number of TDM channels. See also Bandwidth, Channel, Channel Bank, DWDM, FDM, MUX, PCM, STDM, T-1, TDM, and WDM.

channelize See Channelization.

channelthink The perceived need to reproduce in the realm of television over IP (TVoIP) the legacy concept of television channels. It is a pejorative term used by those who believe that alternative, IP-enabled service models may be better from technical, marketing, customer-service and billing perspectives.

Chaos Theory Developed by Edward Lorenz in the 1960s, chaos theory states that simple systems may produce complex behavior. It also has been proven that complex systems possess a simple underlying order. The emerging scientific discipline of chaos theory deals with systems with boundaries that are not clearly defined, with a "system" being defined as a set of things which interact. A computer system, for instance, is a set of elements including perhaps hardware, firmware, application software, CPU, hard drive, I/O devices, terminal devices, peripheral devices, drivers, and so on. Ideally, these system elements work together to consistently yield a predictable, desired result. In the context of a complex computer system, chaos theory describes a condition in which the system behaves in a nonlinear, inherently unpredictable manner. Chaos theory commonly is known as the "butterfly effect," with the analogy being that a butterfly flapping its wings in the Amazon Valley may create slight disturbances in the air which may ultimately affect worldwide weather patterns, perhaps resulting in the creation of a cyclone in Southeast Asia some years later. In this analogy, the complex weather system possesses a simple underlying order, although the interaction of the various elements clearly is nonlinear. But I digress. In the context of an information system, the addition of line of a line of computer code or the installation of a new application software package may cause the system to crash or to corrupt data residing in a seemingly unrelated application. That is chaos theory. The theory is fascinating. The practical impact is frightening, especially as systems and networks (comprised of systems) grow ever more complex.

CHAP Challenge-Handshake Authentication Protocol. An authentication method that can be used when connecting to an Internet Service Provider. CHAP allows you to log in to your provider automatically, without the need for a terminal screen. It is more secure than the Password Authentication Protocol (another widely used authentication method) since it does not send passwords in text format. An Internet term.

Chapter 7 See Chapter 11.

Chapter 11 The Chapter 11 process is started when a company files a reorganization petition with the federal Bankruptcy Court. From that moment on, creditors are prevented from suing the company, and any creditor lawsuits in process are halted, pending the outcome of the Chapter 11 reorganization. Creditors of the company file claims with the Bankruptcy Court. A creditors committee, usually made up of the seven creditors who have filed the largest claims against the company, represents the interest of all creditors. Under Chapter 11 protection, the company's management usually continues to manage the company's business, subject to judicial review. In rare circumstances, such as fraud, a party may ask the court to appoint a trustee to manage the company during reorganization. The ultimate objective of a Chapter 11 reorganization is to restructure creditors' claims so that the company can move ahead with its business. Company management includes a negotiated partial payment to creditors. The plan also can include exchanges of debt for equity, a moratorium on repayment or a combination of these actions. In some cases, more than one plan may be proposed. For example, a creditor, or group of creditors, may develop its own plan. The complex process of reaching a consensual plan entails extensive negotiations among the company, its creditors and its shareholders.

Once developed, the company's reorganization plan – or one of the competing plans – must be accepted by specified margins of creditors and shareholders. Creditors representing two-thirds of the total dollar amount of bankruptcy claims against the company and 51 percent of the total number of those voting must accept the plan, and two-thirds of the amount of shares represented by shareholders voting on the plan must approve it for a plan to be accepted. Once accepted, the Bankruptcy Court reviews the plan to ensure that it conforms to certain additional statutory requirements before confirming it. With a restructured balance sheet, the company then emerges from Chapter 11 protection to implement the plan. Some companies emerge from Chapter 11 and become normal operating companies

again. Some don't and move into Chapter 7 bankruptcy, which is complete and relatively immediate liquidation of the company (i.e. sale of all the company's assets).

character A letter, a number or a symbol. A character is sometimes described by the digit represented by the bit pattern that makes up the Character. i.e., the letter A is ASCII code 65, a carriage return is ASCII code 13.

character cell In text mode on a PC, each pel (Picture ELement) is called a character cell. Character cells are arranged in rows and columns. A typical PC will support two text modes – 80 columns by 25 rows and 40 columns by 25 rows. The default text mode on virtually all PCs is 80 x 25. See Pel.

character code One of several standard sets of binary representations for the alphabet, numerals and common symbols, such as ASCII, EBCDIC, BCD.

character distortion In telegraphy, the distortion caused by transients that, as a result of previous modulation, are present in the transmission channel. Its effects are not consistent. Its influence upon a given transition is to some degree dependent upon the remnants of transients affecting previous signal elements.

Character Generator CG. A computer used to generate text and sometimes graphics for video titles.

character impedance The impedance termination of an electrically uniform (approximately) transmission line that minimizes reflections from the end of the line.

character interleaving A form of TDM used for asynchronous protocols. A 20% saving can be obtained by omitting the start and stop bits. This can be used either with extra channels or by carrying RS-232-C control signals.

character interval The total number of unit intervals (including synchronizing, information, error checking, or control bits) required to transmit any given character in any given communication system. Extra signals that are not associated with individual characters are not included. An example of an extra signal that is excluded in the above definition is any additional time added between the end of the stop element and the beginning of the next start element as a result of a speed change, buffering, etc. This additional time is defined as a part of the intercharacter interval.

character oriented protocol A communications protocol in which the beginning of the message and the end of a block of data are flagged with special characters. A good example is IBM Corp's. Binary Synchronous Communications (BSC) protocol. Character oriented protocols are used in both synchronous and asynchronous transmission.

Character Oriented Windows Interface COW. An SAA-compatible user interface for OS/2 applications.

character printer A device which prints a single character at a time. Contrast with a line printer, which prints blocks of characters and is much faster.

character set All the letters, numbers and characters which a computer can use. The symbols used to represent data. The ASCII standard has 256 characters, each represented by a binary number from 1 to 256. This set includes all the letters in the alphabet, numbers, most punctuation marks, some mathematical symbols and some other characters typically used by computers. See ASCII.

character stuffing A technique used to ensure that transmitted control information is not misinterpreted as data by the receiver during character-based transmission. Special characters are inserted by the transmitter and then removed by the receiver.

character terminal A computer terminal that cannot show graphics, only text.

characteristic frequency A frequency that can be easily identified and measured in a given emission. A carrier frequency may, for example, be designated as the characteristic frequency.

characteristic impedance The impedance of a circuit that, when connected to the output terminals of a uniform transmission line of arbitrary length, causes the line to appear infinitely long. A line terminated in its characteristic impedance will have no standing waves, no reflections from the end, and a constant ratio of voltage to current at a given frequency at every point on the line.

Characters Per Second CPS. A data transfer rate generally estimated from the bit rate and the character length. For example, at 2400 bps, 8-bit characters with Start and Stop bits (for a total of 10 bits per character) will be transmitted at a rate of approximately 240 characters per second (cps). Some protocols, such as USR-HST and MNP, employ advanced techniques such as longer transmission frames and compression to increase characters per second.

chargeback The process of allocating network and telecommunications line, equipment, and usage costs to departments or to individuals. Companies charge back to departments or to users. Colleges and universities charge costs back to departments and to students.

Charged Coupled Device CCD. The full name of the term is Interline Transfer

Charge Coupled Device or IT CCD. CCDs are used as image sensors in an array of elements in which charges are produced by light focused on a surface. They consist of a rectangular array of hundreds of thousands of light-sensitive photo diodes. Light from a lens is focused onto the photo diodes. This frees up electrons (charges) which accumulate in the photo diodes. The charges are periodically released into vertical shift registers which move them along by charge-transfer to be amplified.

chargen Character Generation. Via TCP, a service that sends a continual stream of characters until stopped by the client. Via UDP, the server sends a random number of characters each time the client sends a datagram.

Charlie-Foxtrot Slang. Seriously beyond all hope. Very badly broken.

chat A common name for a type of messaging done over a network, involving short messages sent from one node to another. Chatting usually happens in real-time, sometimes in just short messages, replied to quickly. Sometimes, chatting software is RAM-resident, meaning it can be "popped up" inside an application program. Users are usually notified of an incoming chat by a beep and a message at the bottom of their screens.

chat room Real-time chat services offered by many Internet Information Service Providers such as America Online. Supporting a dozen or so participants, they act much like a teleconference, although on a text basis. Private rooms are those that can be entered by invitation. Public rooms allow anyone to participate.

chatty Describes an application or network protocol that makes abundant use of handshaking, request/response transactions, and/or acknowledgements. "Chattiness" increases latency and degrades performance on the network, in part because of the resources that are consumed by this chattiness, and in part due to the round trip time (RTT) associated with each request/response transaction.

chattiness Overhead caused by a network protocol's abundant use of handshaking, request/response transactions, and/or acknowledgements.

cheapernet A slang name for the thin wire coaxial cable (0.2-inch, RG58A/U 50-ohm) that uses a smaller diameter coaxial cable than standard thick Ethernet. Thin Ethernet is also called "Cheapernet" due to the lower cabling cost. Thin Ethernet systems tend to have transceivers on the network interface card, rather than in external boxes. PCs connect to the Thin Ethernet bus via a coaxial "T" connector. Thin Ethernet is now the most common Ethernet coaxial cable, though twisted pair is gaining. Thin Ethernet is also referred to as ThinNet or ThinWire. See also 10BASE-T.

cheat codes In the Spring of 2002 two Vermont teenagers were caught for killing two Dartmouth College professors. The teenagers believed that the entire world was like a giant computer game and there were "cheat codes" that could allow you to take a shortcut to success. Cheat codes are often used in computer games to lessen the time needed to learn the game and speed up your chances of winning. Such cheat codes are often found on Web sites.

check bit A bit added to a unit of data, say a byte or a word, and used for performing an accuracy check. See also Parity.

check characters Characters added to the end of a block of data which is determined by an algorithm using the data bits which are sent. The receiving device computes its own check characters. It compares them with those sent by the transmitter. If they do not match, the receiver requests the sender to send the block again. If the check characters match, then all the bits used to compute the check characters have been received properly.

check switch A procedural term used in DISH network receivers to establish a good signal connection between the receiver and the multi-sat dish switch. Running a check switch procedure will start a series of test. At the end of the test you will see a display of what satellites you can receive.

check-in mailbox The Centigram VoiceMemo II mailbox used to assign names and passcodes for guests checking into a hotel.

check-out mailbox The Centigram VoiceMemo II mailbox used to clear out guest mailboxes when the guest checks out of the hotel.

checkpoint cycle HDLC error recovery cycle formed by pairing an F bit with a previous P bit or vice versa.

checkpoint restart A managed file transfer option which, when enabled, involves sending special data packets along with the file being transferred, which inform the receiver where the file pointer is in the source file. The receiver commits the latest data received to the file system and records the sender's checkpoint and the position of the file pointer in the destination file. If a link connecting the sending and receiving machines goes down during a transfer or the transfer is not able to complete successfully for another reason, the file transfer automatically restarts at the last completed checkpoint, thereby not requiring the file transfer to restart from the beginning. When transferring very large files, checkpoint restart can save a lot of time and network bandwidth. See also managed file transfer.

checkpointing HDLC error recovery based on pairing of P and F bits and giving the equivalent of a negative acknowledgment without using either REJ or SREJ.

checksum The sum of a group of data items used for error checking. Checksum is computed by the sending computer based upon an algorithm that counts the bits going out in a packet. The check digit is then sent to the other end as the tail, or trailer of the packet. As the packet is being received, the receiving computer goes through the same algorithm, and if the check digit it comes up with is the same as the one received, all is well. Otherwise, it requests the packet be sent again.

cheese The content of a commercial site that mainly consists of pictures of the products or other equally useless information.

cheese box This is an old trick used by the boiler room operators to hide their real physical location from the vice squad. It was a call forwarding device placed in a empty room. Police would attempt to trace the location of a boiler room raid the spot where they thought the calls were being terminated at and find nothing but this device. Meanwhile a lookout for the operation would be watching and have his people shut down to avoid detection as the police got one step closer. The name for this device originated because the first time the police came across this piece of hardware it was fitted into a box originally containing cheese. See Boiler Room.

cheesing When a buffered fiber cable appears to stretch during stripping and then cheeses (creeps) back into the outer jacket of the cable, to resume its original place.

chemical rectifier A chemical device for changing alternating current to pulsating direct, usually used for charging storage batteries.

chemical stripping Soaking an optical fiber in a chemical to remove its coating.

Chemical Vapor Deposition CVD. In optical fiber manufacturing, a process in which deposits are produced by heterogeneous gas-solid and gas-liquid chemical reactions at the surface of a substrate. The CVD method is often used in fabricating optical fiber preforms by causing gaseous materials to react and deposit glass oxides. The preform may be processed further in preparation for pulling into an optical fiber.

cherophobia If you are afraid that you might die laughing you are suffering from cherophobia.

cherry picker An industrial crane arranged with a one or two person 'bucket' to raise workers to levels that cannot easily be reached by other means. These are used to access fruit trees, windows, utility poles, and other high places. See lineman.

cherry picking A call center term. Calls come in and are identified in some way – by ANI (automatic number identification), Caller ID, or caller touchtone input. The identity of the callers is known to the agents in the call center, who can now answer the callers they wish. They decide to answer those callers who they think will buy the most and presumably give them the highest commission or best reward. Thus the expression "cherry picking."

Chernobyl packet A network packet that induces a broadcast storm and/or network meltdown. Named after the April, 1986 nuclear accident at Chernobyl in Ukraine.

Cheyenne Mountain The U.S. military has built a underground base inside Cheyenne Mountain, Colorado. The base is built on giant springs and is designed to withstand a Soviet ICBM attack. For 40 years it has been the home of Norad, the North American Aerospace Defense Command –the U.S.-Canadian early warning system that scanned the globe looking for the telltale launch of an intercontinental missile. The base is a relic of the cold war, now being revived as the possible headquarters of the United States Space Force.

CHI Concentration Highway Interface, pronounced "Ki." A user-programmable, full-duplex interface in the form of a TDM (Time Division Multiplexed) bus. CHI was developed by Lucent, and is used in ISDN controllers.

Chicago Chicago's name comes from an American Indian word meaning "place that smells bad."

chicken A chicken will lay bigger and stronger eggs if the lighting is managed in such a way as to make it appear that a day is 28 hours long.

chicken feet Chicken feet are an extremely popular dim sum dish in Asia. Not surprisingly, they aren't popular with Americans. Simply prepared, chicken feet are cooked in a black-bean sauce. The proper way to eat them is to put the entire foot in one's mouth, suck off the meat, and spit out the bones.

chicken soup During the Middle Ages, chicken soup was believed to be an aphrodisiac.

chiclet 1. Another term for a B Connector. See B Connector.

2. IBM once came out with a PC that had small keys. The press said the PC had a chiclet keyboard, after the chewing gum.

Chief Information Officer CIO. The person responsible for planning, choosing, buying, installing – and ultimately taking the blame for – a company's computer and information processing operation. Originally, CIOs were called data processing managers.

Then they became Management Information System (MIS) managers. Then, CIOs. The idea of calling them CIOs was to reflect a new idea that the information they controlled was a critical corporate advantage and one that could give the company a competitive edge over its competitors – if played correctly. See also Chief Technology Officer.

Chief Technology Officer Once there was data processing managers. They grew up to become MIS managers – Management Information Systems. Then they became Chief Information Officers, i.e. CIOs. The changing title was a simple and blatant attempt to improve their status (and pay) within the corporation. The latest step "up" is to call these people "Chief Technology Officers," or CTOs. And that's fair in the self-aggrandizement "progress" – except that I believe the Chief Technology Officer is not only responsible for the company's computer and information processing but he (or she) is also responsible for figuring how to take the entire spectrum of technology and apply it to endow his/her company with one or many competitive advantages. In short, a CTO is a broader kettle of fish than the CIO. Typically, I would see the CIO reporting to the CTO.

child domain Same thing as a subdomain. See Subdomain.

child group In some systems, a new group of users created under a parent group is called a child group. Child groups sometimes have more properties than their parent groups.

child node An ATM term. A node at the next lower level of the hierarchy which is contained in the peer group represented by the logical group node currently referenced. This could be a logical group node, or a physical node.

child object An object that resides in another object. A child object implies relation. For example, a file is a child object that resides in a folder, which is the parent object.

child peer group An ATM term. A child peer group of a peer group is any one containing a child node of a logical group node in that peer group. A child peer group of a logical group node is the one containing the child node of that logical group node.

children's programming Cable Television programming originally produced and broadcast primarily for an audience of children 12 years old and younger (reference: FCC Rules, 47 CFR 76.225). This rule also requires:

- Commercial matter in children's programming carried on Origination Cablecasting channels must not exceed specified time limits.
- The cable system must maintain, in its PIF, records ("certifications") to verify compliance with this requirement. Certifications for satellite-delivered programming must be obtained from the programmer.

These requirements apply only to Origination Cablecasting. These requirements do not apply to:

- Broadcast stations.
- Access channel capacity designated by franchise for public, educational, or governmental use.

CHILL ITU HIgh Level Language. A computer language developed by the ITU for the standardization of software in telecommunications switches. Not widely adopted. C is more widely adopted.

chime An electromechanical or electronic substitute for the conventional telephone bell, that sounds like a musical chime being struck, typically in a "bing-bong" sequence.

chimney effect Picture a phone system. We have an upright, rectangular cabinet full of printed circuit cards and all getting hot. How to cool them? Simple, raise the machine a little off the ground, put holes in the bottom of the cabinet and holes in the top of the cabinet. Hot air rises. Bingo, air will rise through the top of the cabinet and cool air will get sucked in the bottom of the cabinet. And bingo, you don't need a fan. This natural cooling technique is called the Chimney Effect and many modern phone systems now use it.

chimney mount A mounting system used for attaching an antenna mast to a brick chimney.

Chinese Wall A "Chinese Wall" refers to barriers to the flow of information that are designed to prevent the misuse of material nonpublic information (that is, to prevent "insider trading" and other abuses). Brokers, dealers and investment advisers are required by law to have "Chinese Walls" in place, although that term itself does not appear in the statutes. The reason they are required to have such barriers is that they frequently have access to material nonpublic information in one part of the firm that cannot legally be used by other parts of the firm. For example, a broker dealer's investment bank may be advising Client X on a merger at the same time the research department is writing research and the trading department trades X's common stock. With an effective "Chinese Wall" in place, the broker dealer can do all three. Without it, the firm's activities must be limited.

The SEC and other regulators have provided guidance on what is required for a "Chinese Wall" in this context. These include physical barriers and security measures, review of employee and proprietary trading, memorialization and documentation of firm procedures,

substantive supervision of inter-departmental communications by the firm's Compliance/Legal Department and procedures concerning proprietary trading when the firm is in possession of material nonpublic information.

The term "Chinese Wall" is also used more generically to refer to the policies and procedures that public companies put in place to prevent the misuse of inside information about the company by its employees and agents. That aspect of "Chinese Walls" stems from securities law requirements separate from the specific provisions that apply to regulated entities (brokers, dealers, investment advisors). As with regulated entities, there are a number of ways public companies erect "Chinese Walls." These include providing access to material nonpublic information only to a relatively few employees, requiring employees to "pre-clear" trades in the company's stock, blackout periods (designated times when employees are not permitted to trade the company's securities) and physical barriers and security measures.

chintzy More than 5,000 years ago, the Chinese discovered how to make silk from silkworm cocoons. For about 3,000 years, the Chinese kept this discovery a secret. Because poor people could not afford real silk, they tried to make other cloth look silky. Women would beat on cotton with sticks to soften the fibers. Then they rubbed it against a big stone to make it shiny. The shiny cotton was called "chintz." Because chintz was a cheaper copy of silk, calling something "chintzy" means it is cheap and not of good quality.

chip 1. An integrated circuit. The physical structure upon which integrated circuits are fabricated as components of telephone systems, computers, memory systems, etc. Typically a chip refers to an integrated circuit that has been "packaged" in insulating plastic. The plastic casing traditionally has protruding metal pins that are used to connect with other chips to per-form specific functions. Examples include microprocessors and DRAMs.

2. The transition time for individual bits in the pseudo-random sequence transmitted by the GPS satellite.

3. A term used in wireless transmission. In the IEEE 802.11b standard for WLANs (Wireless Local Area Networks), for example, each bit is encoded into an 11-bit Barker code, with each resulting data object forming a "chip." The resulting chips are placed on a RF (Radio Frequency) carrier for transmission.

chip head Anyone whose education, entertainment and employment is primarily derived from computer-based devices. Also called a Bit Head.

chip jewelry A Wired Magazine definition: Chip Jewelry is a euphemism for old computers destined to be scrapped or turned into decorative ornaments. "I paid three grand for that Mac SE, and now it's nothing but chip jewelry."

chip rate Also known as the spreading rate. The rate at which radio signals are spread across a range of frequencies in a spread spectrum transmission system. See also Spread Spectrum.

Chipless RFID tag An RFID tag that doesn't depend on an integrated microchip. Instead, the tag uses materials that reflect back a portion of the radio waves beamed at them. A computer takes a snapshot of the waves beamed back and uses it like a fingerprint to identify the object with the tag. Companies are experimenting with embedding RF reflecting fibers in paper to prevent unauthorized photocopying of certain documents. But chipless tags are not useful in the supply chain, because even though they are inexpensive, they can't communicate a unique serial number that can be stored in a database.

chirping 1. A rapid change (as opposed to a long-term drift) of the wavelength of an electromagnetic wave. Chirping is most often observed in pulsed operation of a source.

2. A pulse compression technique that uses (usually linear) frequency modulation during the pulse.

choice chip Your new TV will come with electronics that will allow you to program it not to receive certain programs, e.g. violent ones, you choose not to receive. The idea is that shows will be rated. Before they start, the show will broadcast a digital signal containing its rating. The "choice" chip in your TV will recognize the rating, check it against your instructions and block it or allow it through. The "choice" chip is so named as to give parents a choice of programs they and their children will watch. The provision for a choice chip was contained in telecommunications reform legislation passed by the Senate in mid-1995.

choke An obsolete term: An inductance with either an air or iron core, designed to retard certain frequencies; as a radio frequency choke or an audio frequency choke. See also Choke Exchange.

choke exchange A telephone exchange or central office which is assigned to Radio and TV stations, Promoters, and other users which receive large numbers of simultaneous calls. The idea is to group all of these users on a single exchange so when all routes into that exchange are in use, "normal" users (on other exchanges) will not experience blocking of their incoming or outgoing calls. Trunks from other local exchanges into the choke exchange are deliberately limited to just a few paths so callers will get an "all trunks busy" instead of completely blocking their local exchange. However, when one of the choke exchange users experiences a large number of calls (as when your favorite radio station runs a contest) the other choke exchange users will be blocked because all trunks into the choke exchange will be busy due to the first user. See Blocking and Concentration.

choke coil A coil so wound as to offer a retarding or self inductance effect to an alternating current.

choke exchange A telephone exchange which is assigned to Radio and TV stations, Promoters, and other users which will be receiving large numbers of simultaneous calls. The idea is to group all of these users on a single exchange so when all routes into that exchange are in use, "normal" users (on other exchanges) will not experience blocking of incoming or outgoing calls. However, when one of the choke exchange users experiences a large number of calls (as when your favorite radio station runs a contest) the other choke exchange users will be blocked because all trunks into the choke exchange will be busy. See Blocking and Concentration.

choke packet Packet used for flow control. The node detecting congestion generates the choke packet and sends it toward the source of congestion, which is required to reduce input rate.

choke point See Infiniband.

chooser A desk accessory on the Apple Macintosh that allows a user to choose items such as a printer or file server by clicking on an icon of the device.

chopper A device for rapidly opening and closing a circuit. An ancient radio term.

choppiness This is a less-than-optimum circumstance in which a caller's words are intermittently cut off, creating gaps in the voice transmission. This is usually the result of packet loss when transmitting voice over a packet-switched data network. Choppiness makes it difficult or impossible to have a normal conversation.

Christmas Tree Lights The first electric Christmas lights were created by a telephone company PBX installer. Back in the old days, candles were used to decorate Christmas trees. This was obviously very dangerous. Telephone employees are trained to be safety conscious. The installer took the lights from an old switchboard, connected them together, strung them on the tree, and hooked them to a battery. Then he spent the next 40 years looking for the one burnt bulb...

chroma The level of saturation or intensity of a color. Name sometimes applied to color intensity control in a receiver. The color portion of a video signal "C".

chroma key Method of electronically inserting the image from one video source into the picture from another video source using color for discrimination. A selected "key color" is replaced by the background image.

chromatic dispersion Chromatic dispersion is a characteristic of all optical fibers, caused by the fact that different wavelengths of light travel at different velocities in glass. A prism is a demonstration of this phenomenon, for example. Optical fibers can be designed to control the dispersion profile versus wavelength. The amount of chromatic dispersion is a measure of the relative velocity of light (photons) in adjacent wavelengths in the fiber. High dispersion means that light (photons) at a given wavelength is traveling down the fiber at a very different speed than light in a wavelength right next door. Low dispersion means that light in adjacent wavelengths is traveling down the fiber at about the same velocity. It is important in multi-wavelength systems to have enough dispersion to break up cross wavelength interference problems but not so much dispersion that high bit rates become costly to transmit. Some fibers are designed to have nonzero-dispersion, i.e. they must have just the right amount of dispersion to balance these effects.

The uniformity of dispersion relates to how the amount of dispersion varies across a range of wavelengths. If you plotted the dispersion value at each wavelength across a range of wavelengths and they all had exactly the same amount of dispersion, the slope of the plotted line would be zero. To get optimum performance and cost in multi-wavelength transmission systems, it is important that the amount of dispersion not vary too much across the different wavelengths used by the system.

Chromatic dispersion is one of the mechanisms that limits the bandwidth of optical fibers by producing pulse spreading because of the various colors of light traveling in the fiber. Different wavelengths of light travel at different speeds. Since most optical sources emit light containing a range of wavelengths, each of these wavelengths arrive at different times and thereby cause the transmitted pulse to spread as it travels down the fiber.

Chromatic dispersion is the sum of material and waveguide dispersion. Dispersion can be positive or negative because it measures the change in the refractive index with wavelength. Thus, the total chromatic dispersion can actually be zero (really close to zero). For example, step-index single-mode fibers have zero dispersion at 1300nm, almost exactly at the same wavelength where the optical loss of the fiber is at a minimum. This is what allows single-mode fibers to have low loss and high bandwidth. See also PMD (Polarization Mode Dispersion).

chrominance The color portion of the video signal. Chrominance includes hue and saturation information but not brightness. Low chroma means the color picture looks pale or washed out; high chroma means the color is too intense, with a tendency to bleed into surrounding areas. Black, gray and white have a chrominance value of 0. Brightness is referred to as luminance.

chromium dioxide Tape whose coating is of chromium dioxide particles. Noted for its superior frequency output.

chronic service deficiency When you work a delay with a service provider – a telephone or data carrier, you need to create certain definitions of service so that you can figure penalties if such levels of service are not maintained. For example, we might define service deficiency as being a service outage lasting for more than ten seconds. We might define Repeated Service Deficiency as a service deficiency that occurs at least four times in any given 30 day period. and we might define Chronic Service Deficiency as a service deficiency that occurs more than ten times in any given 30 day period. Of course, how these terms are defined will depend on the SLA – Service Level Agreement – which you sign with your carrier.

CHS Cylinder-Head Sector. The method of identifying a given location on a hard drive used by the original PC-AT BIOS (INT 13) and original IDE specification. Differences between details of the two methods resulted in the 528 MB limit on IDE drives. Enhanced IDE-compliant BIOSes can translate between the two methods, allowing drive sizes up to 8.4 GB. See Enhanced IDE and IDE.

chuck hole Also known as Pot Hole. Slang for when your system hangs up on-line.

church A place in which gentlemen who have never been to Heaven brag about it to people who will never get there. H.L. Mencken.

churn Cellular phone and beeper users drop their monthly subscriptions often. Long distance users change their preferred carrier as often as they change their underwear. DSL customers switch to cable modem providers. The industry calls this phenomenon "churn." And it's very expensive. Churn is defined as the level of disconnects from service relative to the total subscriber base of the service. Often referred to on a percentage basis monthly, quarterly or annually. Sometimes it's as high as 2% or 3% or even 4% a month. It drives the cellular, beeper and long distance business nuts. It's very expensive to sign up a new customer. Many cell, beeper and long distance companies offer incentives to prospective customers to switch their service. Sometimes you have to stay a customer for months and months for your supplier to recoup his sign-up incentive Some users have found ways to switch their long distance service often enough so that they never pay for a long distance phone call. The only solution to "churn" is to develop a close and binding relationship with the customer. This is not easy. And most telecom companies haven't figured it.

churn rate Monthly cancellation rate of subscribers as a percentage of total subscribers. This is a metric used for service companies (such as cell phone companies, Internet service providers, and CLECs) as an indication of how successful they are at retaining customers.

chutzpah Chutzpah is a Yiddish word that means unmitigated gall. The word is typically explained by the story of the 15-year old who goes into court, having killed his father and mother, and falls on the mercy of the court because he's now an orphan. In a commentary on Jewish law, Jack Achiezer Guggenheim explained, "A federal court in the Northern District of Illinois noted in a decision a couple of years ago that chutzpah means shameless audacity; impudence; brass. Leo Rosten's The Joys of Yiddish defines chutzpah as a Yiddish idiom meaning "gall, brazen nerve, effrontery." But neither English translation can do the word justice; neither definition can fully capture the audacity simultaneously bordering on insult and humor which the word chutzpah connotes. As a federal district court in the District of D.C. noted in 1992 that chutzpah is "presumption-plus-arrogance such as no other word, and no other language can do justice to." I included the term in this dictionary because many telecommunications professionals are excessively timid about negotiating price and conditions with their telecom vendors. I wanted to explain that adding a little chutzpah to their negotiations is likely to get them better prices and better conditions.

CI 1. Customer Interface.

2. Certified Integrator.

3. An ATM term. Congestion Indicator: This is a field in a RM-cell, and is used to cause the source to decrease its ACR. The source sets CI=0 when it sends an RM-cell. Setting CI=1 is typically how destinations indicate that EFCI has been received on a previous data cell.

CIBER Cellular Intercarrier Billing Exchange Record. A billing record format used between cellular carriers.

CIC 1. See Carrier Identification Code.

2. See Circuit Identification Code.

CICS Customer Information Control System. An IBM program environment designed to allow transactions entered at remote computers to be processed concurrently by a mainframe host. Also, IBM's Customer Information Control System software.

CID 1. A generic term in Britain to identify a customer identity, client identity or contract identity. It is a single record and all the fields of information associated with it; for example, name, address, phone number, contact history and so on.

2. Compatibility ID. Motorola definition.

3. CIrcuit Designator.

4. Caller Identification or Caller ID.

CIDB Calling Line Identification Delivery Blocking. A "feature" of central offices which lets you block the sending of your phone number to the person you're calling.

CIDCW CID on Call Waiting. See Caller ID Message Format.

CIDR Classless Inter-Domain Routing. An internetworking routing protocol. It is a way of using the existing 32-bit Internet address space more efficiently commonly used by Internet Service Providers. It allows the assignment of Class C IP addresses in multiple contiguous blocks. CIDR solved a major problem with IP address assignment. Specifically, IPv4 addresses in the Class C block were limited to 254 addresses. If a user required more than 254 addresses, the next step up the IP food chain was Class B, with 65,534 addresses. Clearly, this was wasteful, as only a few more addresses required a huge chunk of precious addresses. Although this was not an issue for the first two decades of IP, the recent popularity of the Internet (and other IP networks) quickly strained the existing IPv4 addressing scheme. The backbone routers driving much of the Internet traffic in the early 1990s had to track every Class A, B and C network, at times creating routing tables that were 10,000 entries long. The maximum theoretical routing table size is roughly set at 60,000 entries. If the Internet community didn't act fast, it was estimated that the Internet would reach maximum by 1994. CIDR came to the rescue...and will continue to do so, even with the advent of IPv6. CIDR replaces Class A, B and C addresses with a "network prefix" that indicates the number of bits used for identifying the network. Prefixes range from 13 to 27 bits, instead of the eight, 16 or 24 bits of class-based addresses. This means that address blocks can be assigned in groups as small as 32 hosts or as large as over 500,000 hosts. CIDR builds on the concept of "supernetting," with more than one block of network addresses being linked together logically into a "supernet." The problem of IP address exhaustion is similar to, but much more complex than, that of 800 numbers, which was relieved with the introduction of 888, and the 877, numbers. CIDR requires the use of routing protocols that support it, examples being RIP (Routing Information Protocol) Version 2, OSPF (Open Shortest Path First) Version 2, and BGP (Border Gateway Protocol) Version 4. See also IP, IPv4, IPv6, and TCP/IP.

CIF 1. Common Intermediate Format. An option of the ITU-T's H.261/Px64 standard for videoconferencing codes. It produces a color image of 288 non-interlaced luminance lines, each containing 352 pixels to be sent at a rate of 30 frames per second. The format uses two B channels, with voice taking 32 Kbps and the rest for video. QCIF (Quarter CIF), is a variation on the theme, requiring approximately 1/4 the bandwidth of CIF and delivering approximately 1/4 the resolution. CIF works well for large-screen videoconferencing, due to its greater resolution; QCIF works well for small-screen displays, such as videophones. QCIF is mandatory for ITU-T H.261-compliant codes, while CIF is optional. See QCIF.

2. Cost, Insurance and Freight are included. That means the seller pays the freight. The opposite of CIF is FOB, which stands for Free On Board. What this means is that you buy something, F.O.B. The seller puts it on a truck or railroad, plane, i.e. some carrier. He's responsible for getting it on the carrier. You – the buyer – are responsible for paying for the cost of the freight of getting you the goods you ordered.

3. Cells In Flight: An ATM term for an ABR (Available Bit Rate) service parameter, CIF is the negotiated number of cells that the network would like to limit the source to sending during the idle startup period, before the first RM-cell returns.

4. Cells In Frames. Referring to ATM over Ethernet (or Token Ring), CIF involves the insertion of one or more ATM cells into Ethernet frames for transport over an Ethernet LAN. CIF allows the user organization to maintain the existing Ethernet wiring, NICs (Network Interface Cards) and other hardware to support ATM applications. The drawback is that CIF must be used in a switched Ethernet environment in order to maintain ATM QoS (Quality of Service) commitments. Additionally, the SAR (Segmentation and Reassembly) process must be accomplished in software at the workstation, which is slow unless the workstation is really fast. CIF is something of a band-aid approach to bridge the gap until such time as ATM really takes hold in the LAN world.

CIF-AD Cells In Frames-Attachment Device. The device which attaches the CIF-ES to the ATM network.

CIF-ES Cells In Frames-End Station. An Ethernet- or Token Ring-attached workstation which supports CIF.

CIFS See Common Internet File System.

CIG CallInG subscriber identification. A frame that gives the caller's telephone number. See Caller ID.

CIGOS Canadian Interest Group on Open Systems. Canadian organization which promotes OSI.

CIGRR Common Interest Group on Rating and Routing.

CIID Caller Issue Identifier Card. A calling card issued by an RBOC.

CIIG Canadian ISDN Interest Group. Canadian organization which promotes ISDN.

CIM 1. Computer-Integrated Manufacturing.

2. Common Information Model. CIM is the DMTF's model for describing management information to work with disparate systems.

CIN See Customer Identification Number.

cinching A tightening, or increase in pressure in which something becomes more difficult to undo, or becomes locked up. This effect is sometimes seen in reel-to-reel systems, where the pull on a reel is greater than the speed at which it unwinds so the remaining tape, or other material, slips and becomes very tightly packed.

Cincinnati Bell One of the "original" LECs (Local Exchange Carriers) that once was part of the Bell System. While AT&T owned some stock in Cincinnati Bell, it was not wholly owned. Therefore, it was not limited by the Divestiture Decree that broke up the Bell System in 1984. So, Cincinnati Bell became Cincinnati Bell Inc., which started and acquired all sorts of interesting businesses. In 1999, the company was acquired by IXC Communications, an IXC (Interexchange Carrier). The merged company changed its name to BroadWing Inc. in 2000. Cincinnati Bell continues to operate as a LEC division of BroadWing. See BroadWing.

CIO See Chief Technology Officer.

CIP 1. Carrier Identification Parameter. An SS7 term. A 3 or 4 digit code in the initial address message identifying the carrier to be used for the connection. See SS7.

2. Channel Interface Processor. A Cisco term. Channel attachment interface for Cisco 7000 series routers. The CIP is used to connect a host mainframe to a control unit, eliminating the need for an FEP for channel attachment.

3. See Common Industrial Protocol.

cipher A means of transforming, or encrypting, data in order to disguise its meaning. Block ciphers, such as DES, are encryption algorithms which encrypt specific blocks of data. Stream ciphers, such as the RC4 algorithm from RSA Data Security, encrypt a steady flow of data. See also Encryption.

cipher text The unreadable form of an original plain text message after it has been encrypted. Also spelled Ciphertext. See also Ciphertext.

ciphertext The result of processing plaintext (unencrypted information) through an encryption algorithm. Ciphertext is thus the content of an encrypted message. See Clipper Chip.

ciphony The enciphering of voice communication for transmission via radio or telephone. The word is a contraction of "ciphered telephony."

CIPTUG Cisco Internet Protocol Telephony User Group.

CIR 1. Committed Information Rate. A Frame Relay term. When you subscribe to Frame Relay service through a public Frame Relay service provider, you identify the sites that you want to interconnect. Between each of those sites and based on agreement with the Frame Relay service provider (i.e., carrier), you typically establish a PVC (Permanent Virtual Circuit), which is a permanently identified path across which all data will flow between those sites. Each PVC has associated with it a CIR (Committed Information Rate). The CIR is the level of data traffic (in bits) which the carrier agrees to handle over a period of time – not at every instant of time, but averaged over a period. The CIR can be anywhere between 0% and 100% of the speed of the access line (e.g., a 56 Kbps circuit, or a T-1 circuit at 1.544 Mbps) and the speed of the port on the device (typically a router) to which you connect at the edge of the service provider's Frame Relay network. The Offered Load to the network can burst above the CIR for a measured interval of time (T). The Offered Load can never exceed the speed of the circuit, and it can never (usually never, but that's a long story) exceed the speed of the port. Burst levels are measured as Bc and Be. Bc (Committed Burst Size) is the maximum amount of data, measured in bits, that the carrier agrees to transfer under normal circumstances. Be (Excess Burst Size) is the maximum amount of additional data, measured in bits, that the carrier will attempt to handle, assuming that congestion conditions in the network permit. All the data bits are contained within Frame Relay frames, of course. The excess frames at the Be level will be marked as DE (Discard Eligible), either by the user's Frame Relay Access Device (FRAD) or by the carrier's FRND (Frame Relay Network Device). Both the FRAD and the FRND typically are routers. The FRAD is your router. The FRND is the carrier's router that sits at the edge of the carrier network. The FRAD connects to the port of the FRND over an access circuit. Through a mechanism known as Graceful Discard (GD), the DE frames which fall within the Be will be discarded gracefully (i.e., only when absolutely necessary) by the carrier somewhere in the network during periods of congestion. If the network is not congested, the carrier will deliver the excess frames and bits. Any bits (in frames) above the Bc+Be will be discarded, usually at the point of network ingress and not very gracefully, I might add. For example, you might have a bunch of LANs at each site. At each site, the LANs connect to a FRAD, in the form of a Frame Relay-capable router. Each router connects to the public Frame Relay network over a 56 Kbps circuit which terminates in a 56 Kbps port on the carrier's closest FRND. As each PVC has a CIR of 32 Kbps, you be pretty sure that you can transfer data at 32 Kbps, at least on the average over a period of time (e.g., a month). You also can burst above the CIR to the level of the Bc (Committed Burst Size), and with reasonable assurance that the data will get to the destination site. You can burst above the Bc, although all bits in the Be (Excess Burst Size) are subject to being marked DE (Discard Eligible). The DE bits wind up on the switchroom floor in the event that the network is congested and the buffers in the network switches and routers overflow. You also can transmit above the total of the Bc+Be, but those bits in those frames may wind up on the switchroom floor at the ingress FRND. See also Committed Burst Size, Discard Eligible, Excess Burst Size, FRAD, Frame Relay, FRND, Graceful Discard, Measurement Interval, Offered Load, Permanent Virtual Circuit, and Router.

2. An ATM term. Committed Information Range: CIR is the information transfer rate which a network offering frame relay services (FRS) is committed to transfer under normal conditions. The rate is averaged over a minimum increment of time.

circling the drain When you're kept working on a project that is going nowhere but refuses to die, it's circling the drain.

circuit A circuit is a connection or line between two points. The connection can be made through various media, including copper, coaxial cable, fiber, or radiowave. A dedicated circuit is, as it sounds, a link that is dedicated to the two entities at either end. Dedication is expensive because it does not use the network as efficiently as it could. If A. B, C, and D all need to be connected with dedicated lines, then a total of six separate links are needed to connect them. Instead, a central switch to which each is connected reduces the number of links required to four. This difference in efficiency increases exponentially as the number of users increases.

circuit board Same as a Printed Circuit Board, namely a board with microprocessors, transistors and other small electronics components. Such a board slides into a slot in a telephone system or personal computer. Also called a circuit card.

circuit breaker A special type of switch arranged to open a circuit when overloaded, without injury to itself. A circuit breaker is basically a re-usable fuse. According to APC, a circuit breaker is a protective device that interrupts the flow of current when the current exceeds a specified value. Circuit breakers are calibrated when manufactured to a specific overcurrent value. Building or equipment wiring may overheat and become a fire hazard if excessive current is passed through such wiring. Circuit breakers or fuses are installed and coordinated with wiring by selecting the appropriate trip value so that if equipment malfunction or user error causes too much current to flow through a wire, the circuit breaker will trip to prevent the wire from overheating. For building wiring and power distribution, the values of circuit panel breakers are specified in America by the National Electrical Code.

circuit card Same as a Circuit Board. See Circuit Board.

circuit conditioning Modification of (most typically) analog data circuits to bring transmission parameters of the channel into narrower limits than provided by randomly-selected voice channels. Conditioning is also used to a lesser extent in certain other services. See Load Coil.

circuit emulation A connection over a virtual channel-based network providing service to the end user that is indistinguishable from a real, point-to-point, fixed bandwidth circuit.

Circuit Emulation Switching CES. Part of the ATM Forum's proposed Service Aspects and Applications (SAA) standard.

circuit end The local channel termination needed to connect the customer's location to the carrier's POP.

circuit grooming The practice of directing selected circuit-switched DS-0s (64 kbit/s channels) from many T-1 trunks into a single T-1 (typical application is voice leased lines from a T-1 access line being 'groomed' in a DACS onto a dozen or more T-1s going to other central offices where those channels may again be groomed with other circuits onto T-1 access lines at those sites). Also used to separate voice circuits from data circuits, and for combining them for delivery to service-specific switches in the CO.

Circuit Identification Code CIC. The part of a CCS/SS7 signaling message

used to identify the circuit that is being established between two signaling points (14 bits in the ISDNUP).

circuit level gateway A circuit level gateway ensures that a trusted client and an untrusted host have no direct contact. A circuit level gateway accepts a trusted client's requests for specific services and, after verifying the legitimacy of a requested session, establishes a connection with an untrusted host. After the connection is established, a circuit-level gateway copies packets back and forth-without further filtering them.

circuit level proxy A firewall technology that involves examining transmitted data for certain types of improper or threatening behaviour without taking into account the specifics of the application involved; SOCKS is a common example of a circuit-level proxy. See SOCKS.

circuit mode 1. An AT&T term for the method of communications in which a fixed bandwidth circuit is established from point to point through a network and held for the duration of a telephone call.

2. An AIN term for a type of switching that causes a one-to-one correspondence between a call and a circuit. That is, a circuit or path is assigned for a call between each switching node, and the circuit or path is not shared with other calls.

circuit noise level At any point in a transmission system, the ratio of the circuit noise at that point to some arbitrary amount of circuit noise chosen as a reference.

Circuit Order Management System COMS. An automated processing system of MCI circuit- and service-related information. Processes hardwire service circuit orders from order entry through scheduling and completion. COMS also provides circuit order data, hardwire customer data, and circuit inventory data to other MCI systems in Finance, Engineering, and Operations.

Circuit Order Record COR. Report generated by the COR Tracking System within NOBIS, indicating circuit installations, changes, and disconnects.

circuit provisioning The telephone operating company process that somehow organizes to get you a trunk or other special service circuit.

circuit segregation Differentiating between services that are maintained by separate technicians or departments. Can be accomplished through visual and/or mechanical means.

Circuit Switched Digital Capability CSDC. A service implemented by some regional Bell Operating Companies that offers users a 56 Kbps digital service on a user-switchable basis. See Circuit Switching.

circuit switched network A network that establishes a physical circuit temporarily on demand (typically when telephone or other connected device goes off hook) and keeps that circuit reserved fo the user until it receives a disconnect signal.

circuit switching Imagine making a phone call to Grandma. You pick up the phone and dial Grandma. When you finish dialing, the various telephone company switches along the way pick a path for your call and move your call along its way to Grandma. When Grandma answers, you and she are now able to speak. Both of you now have the exclusive and full use of the circuit that was set up between you. You have that circuit until you (or she) hang up, at which time it goes idle until the system of switches grabs it for another "call." That call might be voice, data and video. Circuit switching has one big advantage: You get the full circuit for the full amount of the time you're using it. And for the most part, it's full duplex. Circuit switching has one big disadvantage. Because you get the full circuit for the full amount of the time, you pay for the privilege of tying up that circuit (no one else can use it, even when there are pauses in your conversation). Which means it's expensive.

There are basically three types of switching – circuit, packet and message:

Circuit Switching, which I just explained, is like having your own railroad track for your conversation to travel on. It's yours for as long as you keep the connection open. No one else can use it. Once you hang up, the next caller gets to use that track. Virtually all voice telephone calls are circuit switched, though that won't be true in the future. All dial-up modem calls are circuit switched also.

Packet Switching is like having your own railroad cars which you're sharing with other railroad cars on a railroad track. You slice the information you want to send so it fits into the cars, which join other cars to travel on the railroad track to the other end. You get pretty well as many railroad cars as you need. They will travel on different railroad tracks until they reach the station at the other end, where they'll be assembled in the order you sent them and then dropped off at your destination. Packet switching was originally created for sending data, since it's very efficient (and therefore cheap). One railroad track gets to carry a lot of "conversations." In circuit switching, it only carries one conversation. Packet switching does have the problem that it takes a little time to break up the data "conversations" into many packets, send them on their different ways and then reassemble them at the other end. For data, that delay is barely noticeable. It can be noticeable in a voice conversation, which is why packet switching hasn't been used much for voice – until recently. (See IP Telephony.) In packet switching, the addresses on your packets are read by the switches as they approach, and are switched down the tracks. The next packet is read to throw the switches to send that packet where it needs to go. The data conversation is sent in packets. Each packet can be sent along different tracks as they are open. The packets are assembled at the other end – typically in the last switching office before the packets reach the distant computer or distant user.

Message Switching sends a message from one end to the other. But it's not interactive, as in packet or circuit switching. Of course, you can reply. But it's not like having a "conversation." In message switching, the message is typically received in one block, stored in one central place, then retrieved or sent in one clump to the other end. Message switching is like the post office, or like email. It can be slow. But it can also be cheap. Message Switching can use a combination of circuit switching and packet switching to get its message through.

circuit, Four Wire A path in which four wires are presented to the terminal equipment (phone or data), thus allowing for simultaneous transmission and reception. Two wires are used for transmission in one direction and two in the other direction.

circular antenna A horizontally polarized, half-wave dipole antenna formed into the shape of a circle except that the terminating ends do not touch to make a continuous loop.

circular extension network Permits two or more single-line phones connected to a PBX, each with its own extension, to operate like a "square" key telephone system. An incoming call directed to any non-busy phone in the group will ring at all of the non-busy phones. The first extension to answer will be connected to the incoming call. At any time, a non-busy extension can make or receive calls.

circular hunting When calling a phone, the switching system makes a complete search of all numbers within the hunting group, regardless of the location within that group of the called number. For example, if the hunt group is 231, 232, 233 and 234, the call is directed to 233. If it is busy, the equipment will search 234, 231, and 232 to find a non-busy phone or line. Essentially it goes around the ring, remembering where it last connected and then goes to the next line or phone in the circle. See also Hunt Group and Terminated Hunt Group.

circular mil The measure of sectional area of a wire. The area of a circle on mil (.001") in diameter; 7.845 x 10 to the minus 7 sq. in. Used in expressing wire cross sectional area.

circular polarization In electromagnetic wave propagation, polarization such that the tip of the electric field vector describes a circle in any fixed plane intersecting, and normal to, the direction of propagation. The magnitude of the electric field vector is constant. A circularly polarized wave may be resolved into two linearly polarized waves in phase quadrature with their planes of polarization at right angles to each other.

circling the drain What a struggling company does just before it goes down the tubes. "Jackson knew the company was circling the drain when he jumped ship."

circulator 1. In networking, a passive junction of three or more ports in which the ports can be accessed in such an order that when power is fed into any port it is transferred to the next port, the port counted as following the last in order.

2. In radar, a device that switches the antenna alternately between the transmitter and receiver.

circumnaural A type of headphone that almost totally isolates the listener from room sounds.

CIS Contact Image Sensor. A type of scanner technology in which the photodetectors come in contact with the original document.

CISC Complex Instruction Set Computing. PC Magazine defines CISC as a microprocessor architecture that favors robustness of the instruction set over the speed with which individual instructions are executed. The Intel 486 and Pentium are both examples of CISC microprocessors. See also RISC – Reduced Instruction Set Computing. See RISC.

Cisco IOS Internetwork Operating System. An OS incorporated as part of the CiscoFusion architecture designed to provide centralized integrated, automated installation and management of Internet and intranet networks.

CISPR 22 This is a European Community standard specifying the limits of radio frequency emissions which appliances and other electrical equipment are allowed. The standard indicates the maximum allowable emissions either radiated or conducted via the power cord at various frequencies. Some countries still use the older VDE 0871 emission standards, which are nearly identical. In the USA, the FCC has a similar standard.

CIT Computer Integrated Telephone is Digital Equipment Company's program, announced in October 1987, that provides a framework for functionally integrating voice and data in an applications environment so that the telephone and terminal on the desktop can be

synchronized, the call arriving as the terminal's screen on the caller arrives. CIT uses the DEC VAX line of computers. According to DEC, CIT supports both inbound and outbound telecommunications applications. In an inbound scenario, the application may recognize the caller's originating phone number through Automatic Number Identification (ANI) and/or the dialed number through Dialed Number Identification Service (DNIS), match the information to corresponding data base records and automatically deliver the call and the data to the call center agent. In an outbound application, dialing can be automated, increasing the number of connected calls. In either scenario, the telephone calls and associated data can be simultaneously transferred to alternate locations within an organization, adding a new level of customer service to call center applications. Digital made its first CIT announcements at Telecom '87 in Geneva, Switzerland. The CIT product set, consisting of client and server software implementing a variety of switch-to-computer link protocols, and providing a robust applications interface, was first shipped in 1989. The company announced its latest release, CIT Version 2.1, in January 1991. See also Open Application Interface.

Citizens Band One of two bands used for low power radio transmissions in the United States – either 26.965 to 27.225 megahertz or 462.55 to 469.95 megahertz. The Citizens Band Radio Service is an HF two-way voice communication service for use in your personal and business activities. Expect a communication range of one to five miles. Operation is authorized by rule. Citizens band radio is not allowed in many countries, even some civilized countries. In some countries they use different frequencies. CB radios, in the United States, are limited by FCC rule to four WATTS of power, which gives each CB radio a range of several miles. Some naughty people boost their CBs with external power. The author of this dictionary has actually spoken to Australia while driving on the Santa Monica Freeway in Los Angeles. See also CB and Citizens Band Radio Service.

CIV Cell Interarrival Variation. An ATM term. See Cell Interarrival Variation.

CIVDL Collaboration for Interactive Visual Distance Learning. A collaborative effort by 10 US universities that uses dial-up videoconferencing technology for the delivery of engineering programs.

civil time Civil time ordinary or "everyday" time; equal to mean solar time plus 12 hours, as the mean solar day begins at noon.

CIX 1. Commercial Internet Exchange. Pronounced "kicks." As the Internet began to be commercialized, an agreement was reached among a number of commercial network providers that allowed them to exchange traffic. The first CIX router was installed in the Wiltel equipment room in Santa Clara, California. The Santa Clara CIX and that in Herndon, Virginia remain operational, although the CIX concept later gave way to that of the NAP (Network Access Point). See also FIX, MAE and NAP.

2. Commercial Internet eXchange Association. A non-profit trade association of Public Data Internetwork service providers that promotes and encourages development of the public data communications internetworking services industry, nationally and internationally. The CIX states that it provides a "neutral forum to exchange ideas, information, and experimental projects among suppliers of internetworking services...Together, the membership may develop consensus positions on legislative and policy issues of mutual interest." www.cix.org.

CKL Circuit Location.

CKR Circuit Reference. The access service tariff customer's overall circuit name.

CKT Circuit.

CL ConnectionLess. A service which allows the transfer of information across a network without the need for the establishment of a defined path for the data to travel. IP (Internet Protocol) is a connectionless protocol, as the individual data packets travel across the network, from edge-to-edge, over the path of least resistance. IP packets work their way across a network on the basis of link-by-link forwarding, with each link selected solely on the basis of its availability. Therefore, each packet may travel a different path. TCP is a connection-oriented which ensures that a pre-defined path is set up across the network, from edge-to-edge. TCP runs on top of IP, which is the lowest common denominator protocol in the TCP/IP protocol suite. UDP is a connectionless protocol which can be used in place of TCP. Like TCP, UDP provides reliability of data transfer, with error control being the responsibility of the application program, rather than the network switches and routers. See also Connection-Oriented, IP, TCP, and UDP.

CL2 Cables intended for general purpose use within buildings in accordance with the National Electric Code Section 725-53 (e).

cladding 1. The transparent material, usually glass, that surrounds the core of an optical fiber. Cladding glass has a lower refractive index than core glass. As the light signal travels down the central core transmission path, it naturally spreads out due to a phenomenon known as "modal dispersion." The cladding causes the light to be reflected back into the central core, thereby serving to maintain the signal strength over a long distance. See

Cladding Diameter.

2. When referring to a metallic cable, a process of covering with a metal (usually achieved by pressure rolling, extruding, drawing, or swaging) until a bond is achieved.

cladding beam In fiber optics transmission, a beam that transmits within the core and cladding layers by being reflected off the edge surface of the cladding glass. See Cladding Mode.

cladding diameter The diameter of the circle that includes the cladding layer in an optical fiber.

cladding glass A type of glass or other transparent material used in fiber optic cables which has a lower refractive index than the glass used in the core.

cladding mode In an optical fiber, a transmission mode supported by the cladding; i.e., a mode in addition to the modes supported by the core material.

cladding mode stripper A device for converting optical fiber cladding modes to radiation modes; as a result, the cladding modes are removed from the fiber. Often a material such as the fiber coating or jacket having a refractive index equal to or greater than that of the fiber cladding will perform this function.

cladding ray In an optical fiber, a ray that is confined to the core and cladding by virtue of reflection from the outer surface of the cladding. Cladding rays correspond to cladding modes in the terminology of mode descriptors.

claim I got this from Verizon's dictionary. Claim is a report to Verizon that describes what the customer believes to be an incorrect charge. Claims may be made for recurring charges, non-recurring charges and usage; they may be general or individual. Verizon also has something called Claims Input File, which is a file format through which claims are submitted electronically to Verizon.

claim process A technique used to determine which station will initialize an FDDI ring.

claim token A token ring frame that initiates an election process for a new active monitor station. Claim tokening can result from expired timers, or from any other condition that causes any station to suspect a problem with the current active monitor.

Claire Harry Newton's favorite daughter. Why "favorite?" Simple. She's his only daughter. Her full name is Claire Elizabeth Newton. "Claire" is a form of "Clare," which means "clear" or "bright" from the Latin "clarus." True to her name, she's both clear and bright. Brilliant naming, if I might say so. Her mother chose the name, Claire. Frankly, I didn't want my daughter to be called Claire. My dumb idea was not to burden her with a difficult-to-spell name. As a test, I called six information operators all around the U.S. Could they spell "Claire?" I swear this is a true story. Not one spelled Claire correctly. There are, it seems, more ways to spell Claire than there are Claires in the world. But, by then the name had stuck. Her mother insisted on it. In early 2007, Claire was 27 going on 35. Four years years ago she graduated from Bowdoin College, magna cum laude and phi beta kappa. Graduation should have been a respite for me. But she promptly announced her intention to go to law school and is now in her final year at Boston College law school. And then, once graduated, she plans to marry Ted Maloney of Boston, a truly wonderful gentleman, in September 2007. As to what happens then (grandchildren for Daddy?), well, predictions don't work well with investments, why should they work well with children? At one stage she promised me she wouldn't practice law. She now will. I'm still hoping that she will go into politics and become (my dream) the first woman president of the United States. She is a born leader and people tell me she has genuine charisma. She has promised me "first sleep" in the Lincoln bedroom if I clean up after myself and – more importantly – contribute handsomely to her campaign and stay out of her decision-making. Look closely. That's me selling Newton's Telecom Dictionaries out the side door of the White House. Claire is a great kid. I couldn't be happier. See Michael for the other one. And see Susan for the mother, who made it all happen and brought them up while Daddy was doing something "important" – like writing definitions for this dictionary. Once, Claire read this "definition" and made two points in an email to me: "1. Would you really rather me clean my room than be charismatic? 2. You say that Mom raised us while you were "doing something important"... Since when was raising children less important than making money or writing definitions?" It's not possible to argue with your daughter or your wife, or both – especially when they gang up on you, as they always do. But I wouldn't have it any other way.

CLAMN Called Line Address Modification Notification.

clamper An electronic circuit which sets the level of a signal before the scanning of each line begins to insure that no spurious electronic noise is introduced into the picture signal from the electronics of the video equipment.

clamping 1. Holding within an established operating range, or baseline or midline range in a circuit, in order to maintain various processes or electrical charges at a stable or safe level.

2. In a cathode ray tube (CRT), a process which establishes a level for the picture display at the beginning of each scan line within a frame.

clamping voltage The voltage at which a surge protector begins to stop electricity from getting through. A good surge protector in a 120 volt circuit (the one common in North America) has a clamping voltage of about 135 volts. Damage to computer equipment can occur as low as 160 volts.

clarifying bar In cheap hotels – i.e. the ones I stay in – soap is called. In ritzy hotels – i.e. the ones my wife likes – a bar of soap is now called a "clarifying bar," which costs more but does nothing different.

Clark's Law This Law was created by Jim Clark, the only man who ever created three companies, each with a market capitalization of over $1 billion. His Law states that "a person's wealth grows faster than his or her ability to manage it." His companies were Silicon Graphics, Netscape and Healtheon.

Clarke, Arthur C. Arthur C. Clarke is credited with inventing the concept of satellite communications. In 1945, he published an article entitled "Extra-Terrestrial Relays" in a magazine called Wireless World. He described what we know today as GEOs (Geosynchronous Earth Orbiting) satellites, which are placed in equatorial orbits at altitudes or approximately 22,237 miles. Many years later, he published an article entitled "A Short History of Comsats, Or: How I Lost a Billion Dollars in My Spare Time." Clarke failed to patent the concept. Outside the communications domain, Clarke is best known for his science fiction works, including "2001: A Space Odyssey." He currently lives in Sri Lanka, previously called Ceylon. See also Clarke Orbit and GEO.

Clarke Belt Named after its founder Arthur C. Clarke, the Clarke Belt is an orbit used by satellites at a height of 22,250 miles, in which satellites make an orbit in 24 hours, yet remain in a fixed position relative to the earth's surface.

Clarke Orbit Named after Arthur C. Clarke, who invented GEOs (Geosynchronous Earth Orbiting) satellites, a Clarke Orbit is better known as a geosynchronous orbit, or a geostatic orbit. See also Clarke, Arthur C.; and GEO.

CLAS Centrex Line Assignment Service. A service from local phone companies, which allows Centrex phone subscribers to change their class of service by dialing in on a personal computer, reaching the phone company's computer and then changing things themselves – without phone company personnel assisting or hindering.

In short, load your PC with communications software. Dial your local central office. Change your Centrex phone numbers. Turn on, turn off features. Change pickup groups. Add numbers to speed dialing, etc. Your on-line changes are checked by the phone company's computers. If they make sense (i.e. one change doesn't conflict with another), they take effect by early the following day – at which time you can call up and get a report on which took, which didn't and who's got what. Saves calling in person. Is more accurate. And, best of all, saves money. Typically just one flat monthly charge. No charge for any of your changes.

CLASS 1. Custom Local Area Signaling Services. It is based on the availability of channel interoffice signaling. CLASS consists of number-translation services, such as call-forwarding and caller identification, available within a local exchange of a Local Access and Transport Area (LATA). CLASS is a service mark of Bellcore, now Telcordia Technologies. Some of the phone services which Telcordia promotes for CLASS are Automatic Callback, Automatic Recall, Calling Number Delivery, Customer Originated Trace, Distinctive Ringing/Call Waiting, Selective Call Forwarding and Selective Call Rejection. See also Calling Line Identification.

2. In an object-oriented programming environment, a class defines the data content of a specific type of object, the code that manipulates it, and the public and private programming interfaces to that code. See ANI and ISDN.

3. See also Class 1 and 2, below.

Class 1 Also called Class 1/EIA-578. It's an American used between facsimile application programs and facsimile modems for sending and receiving Class 1 faxes. The Class 1 interface is an extension of the EIA/TIA's (Electronics Industry Association and the Telecommunications Industry Association) specification for fax communication, known as Group III. Class 1 is a series of Hayes AT commands that can be used by software to control fax boards. In Class 1, both the T.30 (the data packet creation and decision making necessary for call setup) and ECM/BFT (error-correction mode/binary file transfer) are done by the computer. A specification being developed (fall of 1991), Class 2, will allow the modem to handle these functions in hardware. Industry analysts believe Class 2 will be the standard for the long haul, but approval is slow. Even so, some modem makers will shortly deliver data/fax modems. See also Class 1 Office.

Class 1 Office A regional toll telephone switching center. The highest level toll office in AT&T's long distance switching hierarchy. There are essentially five levels in the hierarchy, with the lowest level – Class 5 – being those central offices owned by the local telephone companies. Each of the classes can complete calls between themselves. But, if the routes are busy, then calls automatically climb the hierarchy. A Class 1 office is the office of "last resort."

Class 2 Also known as Class 2.0/EIA-592. An American standard used between facsimile application programs and facsimile modems for sending and receiving Class 2.0 faxes. This class places more of the task of establishing the fax connection onto the fax modem, while continuing to rely on the host's processor to send and receive the image data. The Class 2 standard (known as PN-2388) is still under study by the EIA's (Electronic Industries Association) TR.29 committee, with further revisions expected. See Class 1.

Class 2 Office The second level in AT&T's long distance toll switching hierarchy.

Class 3 Office The third level in AT&T's long distance toll switching hierarchy.

Class 4 Office The fourth level in AT&T's long distance toll switching hierarchy – the major switching center to which toll calls from Class 5 offices are sent. In U.S. common carrier telephony service, a toll center designated "Class 4C" is an office where assistance in completing incoming calls is provided in addition to other traffic. A toll center designated "Class 4P" is an office where operators handle only outbound calls, or where switching is performed without operator assistance.

Class 5 Office An end office. Your local central office. The lowest level in the hierarchy of local and long distance switching which AT&T set up when it was "The Bell System." A class 5 office is a local Central Office that serves as a network entry point for station loops and certain special-service lines. Also called an End Office. Classes 1, 2, 3, and 4 are toll offices in the telephone network.

Class A, B, C amplifiers Class A, B, C amplifiers are categories of amplifier operation, originally defined for vacuum tube amplifiers in terms of input signal amplitude related to the grid bias, but now more generally defined for all amplifiers. Class A: linear amplifier in which there is current in the output circuit at all times. Class B: partially-linear amplifier in which there is current in the output circuit only when the input signal is positive; a push-pull Class A amplifier usually comprises two Class B amplifiers. Class C: highly nonlinear amplifier in which there is an output signal only when the input signal exceeds a certain threshold, at which time the output circuit is driven to saturation. Class C amplifiers are more efficient than class A or B amplifiers and are less susceptible to amplitude noise disturbance.

Class A, Class B, Class C See Internet Address Class.

Class A Certification A Federal Communications Commission (FCC) certification that a given make and model of computer meets the FCC's Class A limits for radio frequency emissions, which are designed for commercial and industrial environments. See Class B Certification.

Class A Data Center See Carrier Hotel.

Class A Networks See Internet Address.

Class A Traffic A class of ATM (Asynchronous Transfer Mode) traffic defined by the ITU-T. Class A traffic is defined as being connection-oriented, Constant Bit Rate (CBR) traffic that must be carefully synchronized between transmitter and receiver. Further, such traffic is stream-oriented, and is highly intolerant of both latency (i.e., delay) and jitter (i.e., variability in latency). Class A traffic is supported by AAL (ATM Adaptation Layer) Type 1. Examples of Class A traffic include uncompressed voice and video. See also AAL 1.

class action suit Litigation undertaken on behalf of a group of alleged aggrieved people, for example owners of a product that was found defective and which killed some of its owners. A class action suit might be undertaken by shareholders of companies whose shares have declined in price. The suit might allege misstatements or omissions in the preliminary prospectus or other material communicated to the public. Shareholder law suits, are now harder to mount due to Federal legislation.

Class B Certification A Federal Communications Commission (FCC) certification that a given make and model of computer meets the FCC's Class b limits for radio frequency emissions, which are designed to protect radio and television reception to residential neighborhoods from excessive radio frequency interference (RFI) generated by computer usage. Class B computers also are shielded more efficiently from external interface. Computers used at home are more likely to be surrounded by radio and television equipment. If you plan to use your computer at home, avoid computers that have only Class A certification (that is, they failed Class B).

Class B Networks See Internet Address.

Class B Traffic A class of ATM (Asynchronous Transfer Mode) traffic defined by the ITU-T. Class B traffic is defined as being connection-oriented, Real-Time Variable Bit Rate (rt-VBR) traffic that must be carefully synchronized between transmitter and receiver. Further, such traffic is stream-oriented, and can tolerate some levels of both latency (i.e., delay) and jitter (i.e., variability in latency). Class B traffic is supported by AAL (ATM Adaptation Layer)

Type 2. Examples of Class B traffic include compressed voice and video. See also AAL 2.

Class C Networks See Internet Address.

Class C Traffic A class of ATM (Asynchronous Transfer Mode) traffic defined by the ITU-T. Class C traffic is defined as being connection-oriented, Non Real-Time Variable Bit Rate (nrt-VBR) traffic that requires no synchronization between transmitter and receiver. Class C traffic can tolerate considerable levels of both latency (i.e., delay) and jitter (i.e., variability in latency). Class C traffic is supported by AAL (ATM Adaptation Layer) Type 3/4. Examples of Class C traffic include X.25 and Frame Relay. See also AAL 3/4.

Class C IP block. A class C is a block of 256 IP address – 254 usable (0 is reserved for a broadcast for the subnet and 255 is for a loopback.)

Class D Traffic A class of ATM (Asynchronous Transfer Mode) traffic defined by the ITU-T. Class D traffic is defined as being connectionless, Non Real-Time Variable Bit Rate (nrt-VBR) traffic that requires no synchronization between transmitter and receiver. Class D traffic can tolerate considerable levels of both latency (i.e., delay) and jitter (i.e., variability in latency). Class D traffic is supported by AAL (ATM Adaptation Layer) Type 3/4. Examples of Class D traffic include LAN and SMDS. See also AAL 3/4. See Internet Address.

Class E Networks See Internet Address.

Class n Office The way a telephone company defines its switching facilities. Class 5 is an end office (local exchange), Class 4 is a toll center, Class 3 is a primary switching center, Class 2 is a sectional switching center, and Class 1 is a regional switching center. See Class 1, Class 2, Class 3, Class 4 and Class 5.

class of emission The set of characteristics of an emission, designated by standard symbols, e.g., type of modulation of the main carrier, modulating signal, type of information to be transmitted, and also if appropriate, any additional signal characteristics.

class of office A ranking assigned to switching points in the telephone network, determined by function, interfaces and transmission needs.

class of service 1. Here's the definition of Class of Service internal to a PBX: Each phone in a corporation telephone system may have a different collection of privileges and features assigned to it, such as access to long distance, international calls, 900 area code calls, 976 local calls, etc. Let's say you are afraid that your people will waste the company's money by frivolously calling some expensive numbers, you might wish to define "Class of Service" assignments in your PBX. You could have one that's called "ability to dial everywhere except 900 area code, international calls and all 976 numbers." That could be Class of Service Assignment B. When you give a phone to an employee, you could simply give that person COS B. Big bosses, on the other hand might need to call internationally, but not 900 area code or 976 calls. That could be called Class of Service Assignment A. Class of Service assignments if properly organized, can become an important tool in controlling telephone abuse.

2. Here's the definition on the public switched network: A subgrouping of telephone users for the sake of rate distinction. This may distinguish between individual and party lines, between Government lines and others, between those permitted to make unrestricted international dialed calls and others, between business or residence and coin, between flat rate and message rate, and between restricted and extended area service.

3. Here are words courtesy Cisco relating to class of service issues on a packet switched network. "Networks typically operate on a best-effort delivery basis. All traffic has equal priority and an equal chance of being delivered in a timely manner. When congestion occurs, all traffic has an equal chance of being dropped. However, network managers are increasingly presented with a variety of bandwidth-hungry applications that compete for limited bandwidth on the enterprise network. These applications have a variety of characteristics. They may be mission-critical legacy applications with a Web interface, online business-critical applications, or newer multimedia-based applications such as desktop videoconferencing, Web-based training, and voice (telephone) over IP. Some of these applications are vital to core business processes, while many are not. It is the network manager's job to ensure that mission-critical application traffic is protected from other bandwidth-hungry applications, while still enabling less critical applications such as desktop videoconferencing. Enterprises that want to deploy new bandwidth-hungry applications are judging that it is paramount to also ensure the continued success of mission-critical applications over both the LAN and WAN. This can be achieved by defining network policies, which align network resources with business objectives and are enforced by means of QoS (Quality of Service) mechanisms. Without these QoS controls, non-vital applications can quickly exhaust network resources at the expense of more important ones, such as mission-critical applications, thus compromising business processes and certainly productivity. The QoS feature on the Cisco Catalyst 6000 family of switches prioritizes network traffic with IEEE 802.1p class-of-service (CoS) values that allow network devices to recognize and deliver high-priority traffic in a predictable manner. When congestion occurs, QoS drops low-priority traffic to allow delivery of

high-priority traffic. Ports can be configured as trusted or untrusted, indicating whether or not to trust the CoS values in received frames to be consistent with network policy. On trusted ports, QoS uses received CoS values. On untrusted ports, QoS replaces received CoS values with the port CoS value."

Class X Traffic A class of ATM (Asynchronous Transfer Mode) traffic defined by the ITU-T. Class X traffic is defined as being either connection-oriented or connectionless, traffic that accepts either Unspecified Bit Rate (UBR) or Available Bit Rate (ABR) transmission, and that requires no synchronization between transmitter and receiver. Class X traffic can tolerate considerable levels of both latency (i.e., delay) and jitter (i.e., variability in latency). Class X traffic is supported by AAL (ATM Adaptation Layer) Type 5. Examples of Class X traffic include LANE (LAN Emulation) and IP. See also AAL 5.

Class-4 Switch Class-4 is a type of circuit switch used in a tandem office. In the past, Class-4 switches managed only high-speed, four-wire T-1, T-3, and OC-3 connections (used to deliver long-distance services) in contrast to two-wire local lines on Class-5 switches. All switches now support four-wire lines.

Class-5 Switch Class-5 is a type of circuit switch used in a local telephone end office. It provides end-customer services, such as call waiting and call forwarding.

Classical IP A set of specifications developed by the Internet Engineering Task Force (IETF) for the operation of LAN-to-LAN IP connectivity over an ATM network.

classified ad Log Records required by Section 76.221(f) of the FCC Rules which relate to origination cablecasts or classified advertisements sponsored by individuals. This rule provides that the sponsor of such programming need not be identified within the content of the advertisement or program itself provided that two conditions are met:

- The true sponsor must be an individual offering services which he or she personally provides (examples: yard work; babysitting).
- The system must maintain a written record of the name, address, and telephone number of the individual.

classified heading A heading in the yellow pages that describes a type of business or service, and under which businesses of that type have their directory listings and advertising.

classified section guide A special section of the yellow pages that includes specialty headings and in which businesses may purchase advertising and special listings. For example, there may be a Physicians and Surgeons Guide, an Attorneys Guide, and a Restaurants Guide. Each of these section guides contains specialty headings not found in the regular classified headings. For example, in the Attorneys Guide, the specialty headings may include Animal Protection Law, Appellate Practice, Automobile Accidents & Injuries, Aviation Law, Bankruptcy Law, Business Law, Civil Litigation, Condominium Law, Construction Law, Maritime Law, and other specialties. A classified section guide typically is filled with ads, many of which may be full-page and two-page ads.

Classless Inter-Domain Routing See CIDR.

classmark A designator used to describe the service feature privileges, restrictions, and circuit characteristics for lines or trunks accessing a switch; e.g., precedence level, conference privilege, security level, zone restriction. See Class of Service.

CLAW Common Link Access for Workstations. Data link layer protocol used by channel-attached RISC System/6000 series systems and by IBM 3172 devices running TCP/IP off-load. CLAW improves the efficiency of channel use and allows the Channel Interface Processor (CIP) to provide the functionality of a 3172 in TCP/IP environments and to support direct channel attachment. The output from TCP/IP mainframe processing is a series of IP datagrams that the router can switch without modifications. See CIP.

CLC Carrier Liaison Committee. A committee formed to help industry participants work together to resolve the issues of implementing 800 Portability. CLC is sponsored by the Exchange Carriers Standards Association (ECSA) and is comprised of the LECs (local exchange carriers), long distance carriers and users of 800 service.

CLCI Common Language Circuit Identification. An industry standard format for identifying a special access circuit by the characters used in the circuit code. This designation code is unique and in a form that is acceptable to both manual and mechanized procedures.

CLD Competitive Long Distance carrier, a European term for what people in North America call a competitive IntereXchange Carrier (IXC).

CLD Coalition Inc. Competitive Long Distance Coalition, a Washington, DC-based lobbying group.

CLE Customer Location Equipment, same as Customer Premise Equipment or CPE. CLE is used by fancy, schmantzy telecom salespeople who want to impress their dumb customers. Don't be impressed. It means any and all telecom equipment in your office, factory, hospital, hotel or home.

Clear To cause one or more storage locations to be in a prescribed state, usually that

corresponding to a zero or to the space character.

Clear Channel 1. In radio broadcasting a frequency assigned for the exclusive use of one entity. The FCC defines a clear channel as protecting radio stations designated as Class A stations from objectionable interference within their primary and secondary service areas. The secondary service areas of the stations may extend outward from a distance of up to 750 miles at night. To provide this wide area service, Class A stations operate within a power range of 10 to 50 kilowatts.

2. In networking, a signal path that provides its full bandwidth for a user's service. No control or signaling is performed (or needed) on this path.

3. In digital networking, it's a circuit where no framing or control bits (i.e. for signaling) are required, thus making the full bandwidth available for communications. For example, a 56 Kbps circuit is typically a 64 Kbps digital circuit with 8 Kbps used for signaling. Sometimes called Switched 56, DDS or ADN. Each of the carriers have their own name for clear channel service. The phone companies are obsoleting the 56 Kbps service in favor of the more modern ISDN BRI, which has two 64 Kbps circuits and one 16 Kbps packet service, part of which is used for signaling on the 64 Kbps channels.

4. An SCSA term. A channel which is used exclusively for data transmission, with no bandwidth required for administrative messages such as signaling or synchronization. All SCbus data channels are clear.

clear collision Contention that occurs when a DTE and a DCE simultaneously transfer a clear request packet and a clear indication packet specifying the same logical channel. The DCE will consider that the clearing is completed and will not transfer a DCE clear confirmation packet.

clear confirmation signal A call control signal to acknowledge reception of the DTE clear request by the DCE or the reception of the DCE clear indication by the DTE.

clear signed message A message that is digitally signed but not encrypted. See Digital Signature.

clear text Characters in a human readable form prior to encryption or after decryption. Also called plain text.

Clear To Send CTS. One of the standard attributes of a modem in which the receiving modem indicates to the calling modem that it is now ready to accept data. One of the standard pins used by the RS-232-C standard. In ITU-T V.24, the corresponding pin is called Ready For Sending.

clearinghouse A service company that collects and processes roaming and billing information from a number of carriers. It then transfers the compiled data to the proper carriers for credits and billing.

cleaver A device used to cleave, or cut, a fiber optic cable. There are a number of such devices, some of which are much more precise than others. Fusion splices require very precise cleaves at 90 degrees, as they must meet very tight tolerances. Mechanical splices and mechanical connectors are more forgiving. Mechanical splices and some mechanical connectors are pre-loaded with index-matching gel that mitigates imperfections in the cleave. Most connectors require that the fiber endface be polished after the fiber is cleaved, which also serves to compensate for imperfections in the cleave.

cleaving To cut the end of fiber at 90 degrees with as few rough edges as possible for a fusion splice. With mechanical splices the ends are hand-smoothed with a polishing puck before splicing. See cleaver.

CLEC Competitive Local Exchange Carrier. The term and concept was coined by the Telecommunications Act of 1996. Essentially the idea of the CLEC was that it would be a new local phone company that would be compete with the incumbent, i.e. existing, monopoly local phone company. The idea behind the Act was that the incumbent would be forced to lease local wired loops and other bits and pieces of its phone equipment – called unbundled network elements (UNE) – to the new phone company, i.e. the CLEC. Ultimately, the theory went, the CLEC would start building his own local phone lines and installing his own equipment and the public would benefit by better, cheaper, more innovative telecom service – especially broadband service to the Internet. The idea of leasing some of the ILEC's plant was to give the CLEC a "leg-up." This was the theory. The first problem was that the legislation was the worst-written piece of legislation ever passed by Congress. The second problem was the ILECs deeply resented the idea that they were to allow competitors to get started in business at their expense and using their equipment and their lines. So the ILECs basically did everything they could get away with to mess up the CLECs. That meant delaying CLECs orders, creating onerous, cumbersome, new rules for doing business with them and creating huge, new charges for new services. For example, SBC (the new SouthWestern Bell) came up with something called "Unbundled Local Switching" and stated in their new tariff that "The Rate Structure for ULS will be one of 2 rate structures: Stand Alone ULS or ULS-Interim Shared Transport (ULS-IST)." SBC laid out "General Principles for Stand Alone ULS: Stand

Alone Unbundled Local Switching (ULS) which included charging for a single usage sensitive component in addition to the "appropriate" non-recurring and monthly recurring rates contained in the rate table. No one, of course, knew what any of this meant but it didn't make any difference. It delayed and confused things. It was sort of like laying siege to your enemy. And when you have unlimited resources (like the ILECs) you clearly will win. The CLECs' final problem was marketing and sales. They were basically selling a service – phone or data service – that someone (their potential customers) can't see, feel, touch or smell. The only differentiating criterion falls on sound – the quality of which is totally indistinguishable between the CLEC and the ILEC or between the CLEC and any other phone company in the country. The lack of a market and sales differentiator made selling CLEC services very very difficult. You couldn't sell a better product, so you sold lower prices. But no one believes the pitch for lower telecom prices. They've heard that "cut price" story a thousand times since long distance was de-regulated in the US in the late 1950s. One CLEC, looking for a marketing magic bullet, did some market research among its potential customers and found that virtually all believed that the local ILEC was "The devil you know and the local CLEC was the devil you don't know." As a result virtually all CLECs formed in the U.S. after 1996 have essentially failed – gone bankrupt, about to go bankrupt or are only surviving because some kindly soul is pouring good money after bad and hasn't the guts to close down his disaster. This may be too harsh. There are variations on the CLEC theme that may make it, but they need to be in no way dependent on the local ILEC for anything and they need to figure some clever way to save on their horrendously high capital expenditures and come up with some clever highly-demanded, new telecom services. Right now, the CLECs compete on a selective basis for local phone service, long distance, international, broadband Internet access, and entertainment (e.g., Cable TV and Video on Demand). CLECs include cellular/PCS providers, ISPs, IXCs, CATV providers, CAPs, LMDS operators and power utilities. See Telecommunications Act of 1996 and UNE.

CLEI Codes Common Language Equipment Identifier codes that are assigned by Bellcore (now Telcordia Technologies) to provide a standard method of identifying telecommunications equipment in a uniform, feature-oriented language. It's a text/barcode label on the front of all equipment installed at RBOC facilities et. al. that facilitates inventory, maintenance, planning, investment tracking, and circuit maintenance processes. Suppliers of telecommunication equipment give Telcordia Technologies' technical data on their equipment, and Bellcore assigns a CLEI code to that specific product. Bellcore's GR-485-CORE specification contains the generic guidelines for Common Language Equipment Coding Processes and Guidelines. See also CPR.

CLEO CLEO, the Cisco router in Low Earth Orbit, is an assembly of a Cisco 3251 mobile access router and serial card running commercial off-the-shelf Cisco IOS software in orbit: IOS release 12.2(11)YQ of September 2002. CLEO has been in orbit for over two years and has been tested in orbit for over a year. CLEO has been powered up for use more than fifty times. Access to the CLEO router was demonstrated to the AFEI Net-Centric Operations 2005 conference (May 2005) and at the IEEE Milcom 2005 conference and exhibition (October 2005).

CLEOS Conference of the Lasers and Electro-Optics Society.

clergyman A ticket speculator outside the gates of Heaven. – H. L. Mencken.

CLI 1. Command line interface.
 2. Cumulative Leakage Index. As used in the FCC Rules (in Section 76.611(a)(1)), this term identifies the results of a ground-based measurement of the signal-leakage performance of a cable television distribution system. Under the procedure specified in this rule, each leak is measured on the ground and the CLI is then calculated from measurement data (this term does not include the results of airspace measurements specified in Section 76.611(a)(2)). The calculated CLI value must be reported to the FCC by July 1 of each year on FCC Form 320.
 3. See CLLI Code.
 4. Calling Line Identification. Data generated by a network that displays the calling party's number.

cliche A common thought or idea, that has lost originality, ingenuity, and impact by great overuse. "Let's have some new cliches." Samuel Goldwyn, film producer.

click A kilometer.

click farm A shady company set up in a place where the cost of labor is low, and whose employees spend their workday committing click fraud.

click fraud Click fraud is the term used to describe someone clicking on a web site advertisement with ill intent. A fraudulent clicker can exploit the way Web ads work to rack up fees for a business rival, to boost the placement of his own ads or to make money for himself. Some people even employ software that automatically clicks on ads multiple times. Click fraud is one of the serious problems plaguing the Internet, alongside spam, identity

theft and online-auction fraud. Some believe 20% of the clicks on web ads are from people not necessarily interested in the product advertised, and are therefore in the industry's view, fraudulent. Others say the problem is less severe. What's clear is that if left unchecked, click fraud could damage the credibility of Google, Yahoo and the whole search industry, and cut into their future growth.

click stream A click stream is the sequence of clicks of pages requested by a Web surfer. This is important in gauging the value of different sites and the impact of different advertising initiatives on the Web.

click through See Clickthrough.

click through rate See Clickthrough Rate.

click tones A particular progress tone injected onto the forward voice channel (mobile unit receive, base station transmit) to indicate to the subscriber that the call has not been abandoned by the system. Basically click tones indicate acknowledgment by the cellular system that the cellular system's computer is processing the call.

click-to-call Of or pertaining to a PC-to-telephone call that is initiated when a user clicks on a link or phone icon on a webpage.

clicker's remorse What a user experiences when he clicks on a link in an email or in a popup ad and unleashes a virus or an endless cascade of windows, or has his home page changed, or is victimized in some other way.

clicks and mortar A business that combines traditional retail (bricks and mortar, i.e. real buildings) and on-line ecommerce shopping. Probably the real winners of the Internet revolution.

Clicks-for-chicks Adult sites.

clickstreams The paths a user takes as he or she navigates cyberspace. Advertisers and online media providers are developing software that can accurately track user's clickstreams.

clickthrough Imagine you're surfing the Web. You arrive at a site. Scattered across the page are boxes. Such boxes beckon you with enticements of fantastic products, services or deals. In short, they're advertising something to you. Interested? Click on the box and you'll be transported to another page somewhere else on the Web. That page tries more aggressively to sell you something. Your mouseclick on that box is called a "clickthrough" in Internet parlance. How many clickthroughs an advertiser gets from a particular box is a very important measure of how successful that box of advertising is, whether he should continue buying the ad, how much he should pay for it, etc. For example, box that generates a lot of clickthroughs is a far deal for the advertiser. But how good the "deal" is depends on the cost of the ad and the quality of the clickthroughs, i.e. how many of the clickthroughs actually resulted in a product or service being bought and thus how much money you made. How well you keep track of the profitability of your Internet advertising (often called banner ads) may well determine how company well survives (or doesn't). See also Eyeballs. See Clickthrough Rate.

clickthrough rate Imagine you've bought an advertisement on someone else's web site. You want visitors to that site to see your ad and be intrigued enough to click on your ad and be transported to your web site. When someone mouseclicks on your ad, it's called a "clickthrough." Now let's say that 10% of the people who visit that site click on your ad, that means you have a "clickthrough rate" of 10%. That is one (very minor) measure of the effectiveness of your ad. A better measure would be to figure how many people who clicked through actually bought your product and services and how much money you made on that sales. See Clickthrough.

CLID Calling Line IDentification. Also called Caller ID. See Caller ID for a full explanation. See also ANI and CLASS.

client 1. Clients are devices and software that request information. Clients are objects that use the resources of another object. A client is a fancy name for a PC on a local area network. It used to be called a workstation. Now it is the "client" of the server. See also Client Server, Client Server Model, Fat Client, Mainframe Server, Media Server and Thin Client.
 2. Customer.

Client Access Protocol CAP. See iCalendar.

client application Any computer program making use of the processing resources of another program.

client operating system Operating System running on the client platform. See Client.

Client Pull See Meta Tag.

client server A computer on a local area network that you can request information or applications from. The idea is that you – the user – are the client and it – the slave – is the server. That was the original meaning of the term. Over time, client server began to refer to a computing system that splits the workload between desktop PCs (called "workstations") and one or more larger computers (called "servers") joined on a local area network (LAN). The splitting of tasks allows the use of desktop graphic user interfaces, like Microsoft's Windows or Apple Macintosh's operating system, which are easier to use (for most people) than the host/terminal world of mainframe computing, which placed a "dumb terminal" on a user's desk. That dumb terminal could only send and receive simple text-based material. And the less it sent, the faster it worked (lines were slow), so some of the "human interfaces" were very cryptic. You often were forced to spend weeks at school learning simple mainframe programs.

A good analogy of client-server computing, according to Peter Lewis of the New York Times is to think of client server as a restaurant where the waiter takes your order for a hamburger, goes to the kitchen and comes back with some raw meat and a bun. You get to cook the burger at your table and add your favorite condiments. In computerese, this is client/server, distributed computing, where some processing work is done by the customer at his or her table, instead of entirely in the kitchen (centralized computing in the old mainframe days). It sounds like more work, but it has many advantages. The service is faster. The food is cooked exactly to your liking, and the giant, expensive stove in the kitchen can be replaced by lots of cheap little grills. See Client Server Model, Downsizing, Reengineering and Server.

Client Server Computer Telephony Client server computer telephony delivers ten benefits:

1. Synchronized data screen and phone call pop. Your phone rings. The call comes with the calling number attached (via Caller ID or ANI). Your PBX or ACD passes that number (via Telephony Services) to your server, which does a quick database look up to see if it can find a name and database entry. Bingo, it finds an entry. It passes the call and the database entry simultaneously to whoever is going to answer the phone: The attendant. The boss. The sales agent. The customer service desk. The help desk. All this saves asking a lot of questions. Makes customers happier.

2. Integrated messaging. Also called Unified Messaging. Voice, fax, electronic mail, image and video. All on the one screen. Here's the scenario. You arrive in the morning. Turn on your PC. Your PC logs onto your LAN and its various servers. In seconds, it gives you a screen listing all your messages – voice mail, electronic mail, fax mail, reports, compound documents .... Anything and everything that came in for you. Each is one line. Each line tells you whom it's from. What it is. How big it is. How urgent. Skip down. Click. Your PC loads up the application. Your LAN hunts down the message. Bingo, it's on screen. If it contains voice – maybe it's a voice mail or compound document with voice in it – it rings your phone (or your headset) and plays the voice to you. Or, if you have a sound card in your PC, it can play the voice through your own PC. If it's an image, it will hunt down (also called launch) imaging software which can open the image you have received, letting you see it. Ditto, if it's a video message.

Messages are deluging us. To stop them is to stop progress. But to run your eye down the list, one line per entry. Pick the key ones. Junk the junk ones. Postpone the others. That's what integrated messaging is all about. Putting some order back into your life.

3. Database transactions. Customer look ups. There are bank account balances, ticket buys, airline reservations, catalog requests, movie times, etc. Doing business over the phone is exploding. Today, the caller inputs his request by touchtone or by recognized speech. The system responds with speech and/or fax. Today's systems are limited in size and flexibility. The voice processing application and the database typically share the same processor, often a PC. Split them. Spread the processing and database access burden. Join them on a LAN (for the data) and on new, broader voice processing "LANs," like SCSA or MVIP. You've suddenly got a computer telephony system that knows no growth constraints. You could also get the system to front-end an operator or an agent. Once the caller has punched in all his information, then the call and the screen can be simultaneously passed to the agent.

4. Telephony work groups. Sales groups. Collections groups. Help desks. R&D. We work in groups. But traditional telephony doesn't. Telephony today is BIG. Telephony today is one giant phone system for the building, for the campus. Everyone shares the same automated attendant, the same voice mail, the same ubiquitous, universal, generic telephone features. But they shouldn't. The sellers need phones that grab the caller's phone number, do a look-up on what the customer bought last and quickly route the call to the appropriate (or available) salesperson. The one who sold the customer last time. The company's help desk needs a front end voice response system that asks for the customer's serial number, some indication of the problem and tries to solve the problem by instantly sending a fax or encouraging the caller to punch his way to one of many canned solutions. "The 10 biggest problems our customers have." When all else fails, the caller can be transferred to a live human, expert at diagnosing and solving his pressing problem. A development group might need e-mails and faxes of meeting agendas sent, meeting reminder notices phoned and

scheduled video conferences set up. All automatically. The accounts receivable department needs a predictive dialer to dial all our deadbeats. The telemarketing department also needs a predictive dialer, but different programming.

5. Desktop telephony. There are two important aspects. Call control and media processing services. Call control (also called call processing) is a fancy name for using your PC to get to all your phone system's features – especially those you have difficulty getting to with the forgettable commands phone makers foist on us. *39 to transfer? Or it is *79. With attractive PC screens, you point and click to easy conferencing, transferring, listening to voice mail messages, forwarding, etc. There are enormous personal productivity benefits to running your office phone from your PC: You can dial by name, not by a number you can't remember. You can set up conference calls by clicking on names and have your PC call the participants and call you only when they're all on the phone. You can transfer easily. You can work your voice mail more easily on screen, instead of having to remember "Dial 3 for rewind," "Dial 2 to save," and other obscure commands. Here's a wonderful quote from Marshall R. Goldberg, Developer Relations Group at Microsoft. He says "Voice mail systems that could benefit through integration with the personal computer largely remain isolated, difficult to use, and inflexible. Browsing, storing messages in hierarchical folders, and integration of address books – functions just about everyone could use – are either unavailable or unusable."

The second benefit is media control. Media control is a fancy name for affecting the content of the call. You may wish to record the phone call you're on. You may wish to have all or part of your phone call clipped and sent to someone else – as you often do today with voice mail messages. You may wish to simply file your conversations away in appropriate folders. You may wish to be able to call your PC and get it to read you back any e-mails or faxes you received in the last day or so.

6. Applying intelligence. A PC is programmable. The typical office phone isn't. A PC can be programmed to act as your personal secretary, handling different calls differently. It can be programmed to include commands, such as "If Joe calls, break into my conversation and tell me." "If Robert calls, send him to voice mail." etc.

7. The Compound Document. The typed document lacks life. But add voice, image and video clips to it and it gets life. The LAN makes the compound document easier to achieve. The Compound Document gets attention.

8. Management of phone networks. Today, phone networks are very difficult to manage. Often the PBX is managed separately from the voice mail, which is managed separately from the call accounting, etc. It's a rare day in any corporate life when the whole system is up to date, with extensions, bills and voice mail mailboxes reflecting the reality of what's actually happening. The latest generations of LAN software – NetWare 4.1 and Windows NT – have solid enterprise-wide directories and far easier management tools. Integrate these LAN management tools with telecommunications management, and potentially all you need is to make one entry (for a new employee, a change, etc.) and the whole system – telecom and computing – could update itself automatically and even issue change orders to the MIS and telecom departments and vendors.

9. No dedicated hardware in the PC. With only one link – from the switch to the LAN – there's no need to open the desktop PC and place specialized telephony hardware in each PC that wants to take advantage of the new LAN-based telephony features.

10. Switch elimination. The ultimate potential advantage of LAN-based telephony is to eliminate the connection to the switch (PBX or ACD) by simply populating the LAN server (now called a telephony server) with specialized computer telephony cards and running the company's or department's phones off the telephony server directly.

client server model In most cases, the "client" is a desktop computing device or program "served" by another networked computing device. Computers are integrated over the network by an application, which provides a single system image. The server can be a minicomputer, workstation, or microcomputer with attached storage devices. A client can be served by multiple servers. See Client Server.

client software An Internet access term. Multiprotocol PPP client software allowing dial-in access to the public Internet or corporate LANs via a dialup switch or a remote access server. The client software dialer is responsible for establishing/terminating the dial-in connection. The client software PPP driver manages the traffic sent/received across the network link.

client tennis Harris Lydon is a great tennis player, but he's also my stockbroker. When he plays me in tennis, he lets me win occcasional points. That's called playing client tennis. If he weren't playing client tennis, I'd never win a point. He's that good. I'm not that bad.

Clifton Clifton Powell is the best educator at Bellcore, now Telcordia Technologies. I know because he told me so. He also said he would recommend my dictionary even more strongly to his people than he does. Powell to the people. Right on!

climbers What personnel wear to climb wooden poles. Officially called linesman's climbers, but are often known as spurs, hooks, and gaffs. They consist of a steel shanks that strap to a person's leg. The inside has a spike used to stab the pole.

climbing belt A belt that communications/power/construction personnel use to attach themselves to poles or tower structures. Also called safety belt and body belt.

CLIP 1. Calling Line Identification Presentation. Also known as CNIP (Calling Number Identification Presentation). A supplementary service provided by certain wireless "cellular" services, such as GSM (Global System for Mobile Communications) and PCS (Personal Communications Services). Assuming that your subscriber profile includes this service, the number of the calling party displays on your wireless telephone. If the calling party uses CNIR (Calling Number Identification Restriction) however, your display says "PRIVATE." You can choose to have PRIVATE CALLS automatically transferred to your voice mailbox, while retaining the option to override that instruction with a simple keypad command. This feature is known variously as Caller ID and Calling Line ID (CLID) in the wired world. Whether in the wireless or the wired world, this feature is made available courtesy of SS7. See also GSM, PCS, and SS7.

2. As in Video Clip. A small piece of video. Originally a video clip was a piece of celluloid physically cut out of a larger movie. Now it could be digital. And it just means it's a small video.

clip on toll fraud Clip on toll fraud occurs when someone connects a phone between someone else's phone (typically a coin phone) and the central office and makes unlawful toll calls. The term "clip on" comes because the telephone service thief "clips on" to the line. Clip on toll fraud is often done on COCOT (Customer Owned Coin Operated Telephone) phone lines because these lines do not enjoy the same protection from toll fraud which is afforded to coin phone lines which local telcos provide to their own coin phones.

clipboard A generic term for a place in software which holds text, pictures or images that you are copying or moving between applications. The clipboard is a temporary holding place only. When you cut or copy a new item it will replace the current clipboard contents. In other words, there's usually only one clipboard. Windows uses a clipboard and so does the Apple Macintosh.

clipbook A Windows NT term. Windows NT is soon to be known as Windows 2000. Permanent storage of information you want to save and share with others. This differs from the Clipboard which temporarily stores information. You can save the current contents of the clipboard, which temporarily stores information. You can then share that information, allowing others to connect to the Clipboard. See Clipbook Page.

clipbook page A Windows NT term. Windows NT is soon to be known as Windows 2000. A unit of information pasted into a local ClipBook. The ClipBook page is permanently saved. Information on a ClipBook page can be copied back onto the Clipboard and then pasted into documents. You can share ClipBook pages on the notework.

clipbook service A Windows NT term. Supports the Clipbook Viewer application, allowing pages to be seen by remote ClipBooks.

clipped frame Transmit packet lost at the interface because no buffer space was available to the host transmit driver for outgoing data.

clipper A circuit or device that limits the instantaneous output signal amplitude to a predetermined maximum value, regardless of the amplitude of the input signal. See Clipper Chip.

clipper chip A microprocessor chip, officially known as the MYK-78, which the Federal Government wants to add to phones and data communications equipment. The chip would ensure that conversations in Clipper-equipped communicating equipment would be private – from everybody except the Government. With a court-approved wiretap, an agency like the FBI, could listen in, since the Government would have the key to Clipper. On February 4, 1994, the Clinton White House announced its approval of the Clipper chip and the "Crypto War" broke out – with many companies and individuals urging a stop to Clipper. In late Spring, 1994 an AT&T Bell Labs researcher revealed that he had found a serious flaw in the Clipper technology. As of writing it wasn't clear what would happen to the Clipper Chip. See NSA.

clipping Clipping has two basic meanings. The first refers to the effect caused by a simplex (one way at a time) speakerphone. Here the conversation goes one way. When the other person wants to talk, the voice path has to reverse (to "flip"). While the flipping takes place, a few sounds are "clipped" from that person's conversation. This phenomenon happens on some long distance and many overseas channels. These channels are so expensive, they are simultaneously shared by many conversations. Gaps in your conversation are filled with other people's conversation. But when you start talking, the equipment has to recognize you're now talking, find some capacity for your conversation, and send it. In the

process of doing this, your first word or part of your first word might be "clipped" and the conversation will sound "broken."

The second way your voice is clipped is what happens every day on the telephone. You're squeezing your own voice which typically spans 10,000 Hertz into a voice channel which is only 3,000 Hertz. This clips the extremes of your conversation – the higher sounds. As a result, your voice sounds flatter over the phone. As you become more economical and try to squeeze your voice into smaller capacity channels, so it becomes increasingly clipped.

CLIR Calling Line Identification Restriction, A wireless telecommunications term. A supplementary service provided under GSM (Global System for Mobile Communications). See GSM.

CLLI Code Common Language Location Identifier. Pronounced silly code. An alphanumeric code of 11 characters, CLLI was developed by Bellcore (now Telcordia Technologies) as a method of identifying physical locations and equipment such as buildings, central offices, poles, and antennas. Here is information from Telcordia: How Are CLLI Codes Developed? Each CLLI code conforms to one of three basic formats (Network Entity, Network Support Site and Customer Site). Each format, in turn, determines how these six coding elements are used:

Geographical Codes (Example: DNVR = Denver) Typically assigned to cities, towns, suburbs, villages, hamlets, military installations and international airports, geographical codes can also be mapped to mountains, bodies of water and satellites in fixed-earth orbit.

Geopolitical Codes (Example: CO = Colorado) Typically assigned to countries, states and provinces, geopolitical and geographical codes can be combined to form a location identifier that is unique worldwide.

Network-Site Codes (Example: 56 = A central office on Main Street) This element is used with geographical and geopolitical codes to represent buildings, structures, enclosures or other locations at which there is a need to identify and describe one or more functional entities. This category includes central office buildings, business and commercial offices, certain microwave-radio relay buildings and earth stations, universities, hospitals, military bases and other government complexes, garages, sheds and small buildings, phone centers and controlled environmental vaults.

Network-Entity Codes (Example: DS0 = A digital switch) This element can be used with geographical, geopolitical and network-site codes to identify and describe functional categories of equipment, administrative groups or maintenance centers involved in the operations taking place at a given location.

Network Support-Site Codes (Example: P1234 = A telephone pole) This element can be used with geographical and geopolitical codes to identify and describe the location of international boundaries or crossing points, end points, fiber nodes, cable and facility junctions, manholes, poles, radio-equipment sites, repeaters and toll stations.

Customer Site Codes (Example: 1A101 = A customer) This element can be used with geographical and geopolitical codes to identify and describe customer locations associated with switched-service networks, centrex installations; trunk forecasting, cable, carrier or fiber terminations, NCTE, CPE and PBX equipment, military installations, shopping malls, universities and hospitals.

Consider the real-life example of NYCMNY18DS0. The first four characters identify the place name (NYCM is New York City Manhattan). The following two characters identify the state, region or territory (NY is New York). The remaining five characters identify the specific item at that place (18DS0 is the AT&T 5E Digital Serving Office on West 18th Street, between Seventh and Eight Avenues). Phone companies use CLLI Codes for a variety of purposes, including identifying and ordering private lines and trapping and tracing of annoying or threatening calls. See Annoyance Call Bureau, CFA, CLLI Code – Facility Identification, Trap and Trace, and Wire Tap.

CLLI Code - Facility Identification CLFI codes provide unique identification of facilities (cables and carrier systems) between any two interconnected CLLI coded locations. The CLFI code is a variable length, mnemonic code with a maximum of 38 characters. Example: 101T1LSANCA03NWRKNJAA. This example says that there is T-1 carrier connected between the Los Angeles, California Central Office to the Newark, New Jersey Central Office. See CLLI Code.

CLM Career Limiting Move. An ill-advised activity. Trashing your boss while he or she is within earshot is a serious CLM. Sending emails through the company's server which can be read by management long after you think you've deleted them is another CLM.

CLNP Connectionless Network Protocol. An OSI network layer protocol that does not require a circuit to be established before data is transmitted. The OSI protocol for providing the OSI Connectionless Network Service (data gram service). CLNP is the OSI equivalent to Internet IP, and is sometimes called ISO IP.

CLNS Connectionsless Network Service. A Network Layer methodology which does not require a receiver's immediate acknowledgment of communications. See Connectionless Network Service.

cloaking See Masking.

CLOB Central Limit Order Book. According to the New York Times, if stock market regulators in the United States have their way, stock prices will eventually be listed in one place, though they might still be traded on multiple exchanges.

clock 1. A clock is an oscillator-generated signal that provides a timing reference for a transmission link. A clock provides signals used in a transmission system to control the timing of certain functions such as the duration of signal elements or the sampling rate. It also generates periodic, accurately spaced signals used for such purposes as timing, regulation of the operations of a processor, or generation of interrupts. In short, a clock has two functions: 1. To generate periodic signals for synchronization on a transmission facility. 2. To provide a time base for the sampling of signal elements. Used in computers, a clock synchronizes certain procedures, such as communication with other devices. It simply keeps track of time, which allows computers to do the same things at the same time so they don't "bump" into each other.

2. A clock is an internal timing device in the form of a computer chip that uses a quartz crystal to generate a uniform electrical frequency from which digital pulses are created. A clock keeps track of hours, minutes, and seconds and makes this data available to computer programs.

3. A clock is a timer set to interrupt a CPU (Centralized Processing Unit) at regular intervals in order to provide equal time to all the users of the computer. A clock also maintains the uniform transmission of data between the sending and receiving terminals and computers. In short a clock has many functions. See Network Slip, Timing, and Stratum Level.

clock and data recovery In digital communication systems, binary bits of data are sent as a series of optical or electrical pulses. The transmission channel (copper, fiber, coax, radio, or infrared) distorts the signal in various ways. Clock and data recovery is the use of special circuitry to extract embedded clock information (used for synchronization) and data from an incoming data stream.

clock bias The difference between the GPS clock's indicated time and true universal time. GPS is Global Positioning System. See GPS.

clock cycle The time that elapses from one read or write operation to another in the main memory of a computer's central processing unit (CPU). The more tasks that can be accomplished per cycle, the more efficient the chip. Some chips like the i860 chip can execute two instructions and three operations per clock cycle.

clock difference A measure of the separation between the respective time marks of two clocks. Clock differences must be reported as algebraic quantities measured on the same time scale. The date of the measurement should be given.

clock doubling Refers a computer whose internal CPU clock runs twice as fast as the clock for the rest of computer. This has the effect of increasing the computer's speed without the expense of high-speed hardware.

clock slack See PPM.

clock speed The speed at which the microprocessor in your PC synchronizes and regulates its workflow, measured in megahertz. The higher the clock speed, the faster the processor can process data. Other factors such as RAM, hard disk speed, hardware and bus widths have an effect on its performance. It is also known as clock rate. Each CPU (Centralized Processing Unit) of a computer contains a special clock circuitry which is connected to a quartz crystal that is much like the one in your watch. The quartz crystal's vibrations, which are very fast, coordinate the CPU's operation, keeping everything in step, and regulating the rate at which instructions are executed. CPU clock speeds are measured in megahertz, or MHz (million cycles per second), with each cycle known as a "clock tick." While the clock speeds of contemporary PC CPUs range from a slow of 4.77 MHz (the original IBM PC) to 3 GHz and more coming, the main system clock ticks at a rate of 66 MHz for all PCs. Some PCs are "superscalar," executing multiple instructions per clock cycle. So, clock speed is a misleading term. It is only one way of measuring the speed of a computer. One other critical way is how fast you can read and write information to the hard disk. How important that is depends on whether you're running a program with lots of access to your hard drive (e.g. a database program) or running a program which uses a lot of calculations in RAM (e.g., a spreadsheet).

clock synchronization Clock acquisition is a synonym for bit synchronization, the process by which a receiver synchronizes its clock with the transmitter in order to read a received bitstream.

clock tolerance The maximum permissible departure of a clock indication from a designated time reference such as Coordinated Universal Time.

clocking In synchronous communication, a periodic signal used to synchronize transmission and reception of data and control characters.

clockwise polarized wave An elliptically or circularly polarized electromagnetic wave in which the direction of rotation of the electric vector is clockwise as seen by an observer looking in the direction of propagation of the wave.

clone A clone, as a noun, is a person or thing that duplicates, imitates, or closely resembles another in appearance, function, performance or style. A clone, as a verb, means to produce an exact copy of. There are many different "clones" in the telecommunications and PC business. A cloned PC is the same as another one. You might need to make clone PCs if you want to give "standard issue" PCs to your people. To clone a PC, you create all the software you want and the configuration you want on one PC. You connect that PC to a network. Then you use a program – a popular one is called "Ghost" – which copies the PC's entire hard disk to a server on your network. You then make a normal boot floppy disk, add "Ghost" to the floppy disk and go to another new machine, which you've attached to the network. You insert the boot disk into the floppy drive and turn on the machine. It reads the boot disk, loads MS-DOS and Ghost, goes to the server and retrieves (i.e. copies over) all the software it needs to make itself an identical clone of the original PC.

A clone can be a mobile device, say a cellular telephone, that claims to possess the same address identifier as another mobile device. See Clone Fraud.

clone fraud A way of using cellular phones to steal phone calls. In clone fraud, a legitimate serial number is programmed into an imposter's cellular telephone. This allows unauthorized calling to go on until a huge bill appears on the mailbox of the bewildered subscriber to whom the serial number actually belongs. Crooks get the numbers because the numbers are broadcast with every cellular call and can be picked up by ordinary radio scanners, which you can often buy at your local electronics store. According to the Wall Street Journal, cellular thieves take advantage of the fact that when a cellular phone call is placed, the phone's unique electronic code is transmitted over the airwaves to update the cellular network on the user's location. Service thieves wait near the busy areas - highways, financial districts - and use scanners to lift the codes from legitimate users. Thousands of such hijacked numbers can be later downloaded through a computer into "clone" phones. Thus, a clone phone lets a user place potentially hundreds of call that are billed to the legitimate owner of the original number. See Cloned Phone and Tumbling.

cloned phone A cellular phone has two basic ways it identifies itself to the cellular phone company it wants to use - its own telephone number (which can be changed) and a special secret number that's embedded into silicon inside the phone. That number is called an Electronic Serial Number, or ESN. When the phone wants to make a call, it sends those numbers and the cellular carrier uses them to check if the call is authorized. But because the information is traveling through the air, anyone with a scanner can pick up the information and retransmit it later, thus creating a "cloned phone" and pretending that he's authorized to make the call. Of course, the owner of the cloned phone ultimately gets the bill and a nasty shock.

CLONES Central Location On-Line Entry System. CLONES is the repository for CLLI codes.

cloning See disk cloning.

cloning fraud Cloning Fraud Occurs when criminals use scanners to obtain legitimate MIN/ESN/PIN combinations and then program them into illegitimate phones. Such number combinations with PINs are even more valuable to cloners. These cloners can rack up millions of dollars in losses for carriers by creating phone banks, where either illegitimate minutes are resold or cloned phones are sold.

close coupling The condition in which two coils are placed in close magnetic relation to each other, thus establishing a high degree of mutual induction.

close talk A voice recognition term. An arrangement where a microphone is fewer than four inches from the speaker's mouth.

closed architecture Proprietary design that is compatible only with hardware and software from a single vendor of a single product family. Contrast with Open Architecture.

closed captioning A service, designed for people with hearing disabilities, that provides a simultaneous visual presentation of the sound associated with a television program. Closed captioning is not visible except on a TV receiver designed to display it or by use of a specially installed decoder.

closed end The end of a Foreign Exchange – FX – line which ends on a PBX, a key system or a telephone. The closed end is the end of the circuit beyond which a call cannot progress further. The other end of the FX circuit is called the "open end," because calls can progress further.

closed loop system 1. A closed electrical circuit into which a standard signal is fed and received instantly. A measure of the difference between the input signal and the output signal is a measure of the error, and potentially what's causing it.

2. RFID tracking systems set up within a company. Since the tracked item never leaves the company's control, it does not need to worry about using technology based on open standards.

closed test Testing a cable pair by shorting the pairs together at the far end and testing for ohms readings of each wire to ground and to each other. If a cable pair or pairs passes an open and closed test, it can normally be determined that it is fit to use for dial tone service. Dial tone must be removed from the cable pair to do an open and closed test.

closed user group A group of specified users of a data network that is assigned a facility that permits them to communicate with each other but precludes communications with other users of the service or services.

closet Telecommunications closet. An enclosed space for housing telecommunications equipment, cable terminations, and cross-connect cabling that is the recognized location of the cross-connect between the backbone and horizontal facilities.

closure A cabinet, pedestal, or case used to enclose cable sheath openings necessary for splicing or terminating fibers.

cloud 1. Beginning with AT&T's BDN (Bell Data Network) and ACS (Advanced Communication System) offerings, the data network was depicted as a "cloud." The user data was presented at one side of the carrier cloud, and was delivered at the other side. What went on inside the cloud was obscured from view. The thinking behind this heavily conceptual sell was that the user needn't be concerned with what went on inside the carrier network; rather, that was the concern and responsibility of the carrier. While BDN, later known as ACS and later still as Net 1000 was unsuccessful, the concept of the cloud was a huge success.

The cloud is particularly appropriate for illustration of the Internet, as it is a network of networks of uncertain definition. In fact, the use of packet switching and the TCP/IP protocol suite are about the only things about the Internet that are certain. The nature of the physical circuits; the placement of switches and routers; the use of various tunneling protocols, if any; the use of various protocols for grade of service support, if any; and many other specifics vary widely from network to network in this network of networks. The specifics are sensitive to the physical locations of the originating and terminating devices, the applications supported, the ISPs and backbone carriers involved, the path taken by the data, and other variables.

Some of the newer high-speed data, phone company-offered services resemble a local area network. You connect to them directly. To make a call, you don't actually dial a number as you do on a circuit-switched service, you just transmit, putting an address at the front of your transmission. The service reads the address and sends it where you want. Like a LAN, everything is connected and on line. The concept is to get stuff sent from one place to another much faster than would be possible if you had to wait to dial, for the circuit to be set up, for the machine at the other end to answer, etc. In these high-speed services, the circuit is "always set up." The provider (the phone company) refers to its network as a "cloud." And when you see diagrams of these newer high-speed services, like ATM and frame relay, you see the carrier portion drawn as a cloud (like the one you see in the sky).

2. The cloud could also be the Internet. Microsoft's .net initiative revolves around creating software programs that do not reside on any one computer but instead exist in the "cloud" of computers that make up the Internet. The move from the desktop-based computing paradigm that Microsoft has controlled to an open-network approach would be a crucial one for all computer users and software programmers. Beginning in the summer of 2000, Bill Gates of Microsoft has been working to transform his company through systems like the recently announced Hailstorm project, which aims to move most of a computer user's personal information – from daily calendar to banking information – from the desktop or laptop PC and into the network cloud, where a user could have access to it from a variety of devices and locations. To hasten this grand migration, according to the New York Times, Microsoft has been courting software developers, hoping to persuade them to write for its new operating-system-in-the-sky.

3. The cloud is also marketing speak. A true story: A vendor of 802.11b wireless networking told the management of the Ritz Carlton in Jamaica that it would install a "wireless cloud" over the entire hotel such that no matter where you were in the hotel, you could be on the wireless network and access the Internet and your email, etc. The hotel believed this. When I visited, wireless service was fine in the lobby and the convention center and pretty well zero everywhere else (including your room).

See Cloud Nine.

cloud nine When someone is feeling great they're "on cloud nine." The reason for this is that clouds are numbered according to the altitudes they attain, with nine being the highest cloud. If someone is said to be on cloud nine, that person is floating well above their worldly cares.

CLP Cell Loss Priority: This bit in the ATM cell header indicates two levels of priority for ATM cells. CLP=0 cells are higher priority than CLP=1 cells. CLP=1 cells may be discarded during periods of congestion to preserve the CLR of CLP=0 cells. See also Cell-loss Priority Field.

CLR 1. Cell Loss Ratio: CLR is a negotiated QoS (Quality of Service) parameter and acceptable values are network specific. The objective is to minimize CLR provided the end-system adapts the traffic to the changing ATM layer transfer characteristics. The Cell Loss Ratio is defined for a connection as: Lost Cells/Total Transmitted Cells. The CLR parameter is the value of CLR that the network agrees to offer as an objective over the lifetime of the connection. It is expressed as an order of magnitude, having a range of 10-1 to 10-15 and unspecified.

2. Circuit Layout Record. Is an identifier of what type of service each telecommunications circuit has.

CLS Control Line Setting.

CLSP Competitive Local Service Provider. Interchangeable term with ALT and CLEC.

CLT Additional Listing.

CLTP Connectionless Transport Protocol. Provides for end-to-end Transport data addressing (via Transport selector) and error control (via checksum), but cannot guarantee delivery or provide flow control. The OSI equivalent of UDP.

CLTS ConnectionLess Transport Service.

club I don't care to belong to a club that accepts people like me as members – Groucho Marx.

cluster 1. Collection of terminals or other devices in a single location. A cluster control unit and a cluster controller in IBM 3270 systems are devices that control the input/output operations of a group (cluster) of display stations. See also Clustering.

2. Unit of storage allocation used by MS-DOS usually consisting of four or more 512-byte sectors.

3. Physical grouping of workstations that share one or more panel runs.

4. A mini-network of PCs that work in a fault-resilient manner.

5. A cluster is a group of computers and storage devices that function as a single system.

cluster analysis A statistical technique for looking at "complex" relationships between data. It takes into account relationships that are non-linear. This is the main technique used for developing a segmentation model to group individuals together who have similar characteristics. i.e. people of a similar type "cluster" together ("birds of a feather flock together"). The technique analyses the spatial distances between the data to determine the characteristics of the clusters. For example, one cluster may be made up of predominately young males over 45 years old who have several saving and investment plan insurance and tend to live in inner cities. Another cluster may be made up of similar demographic data but these people have several credit cards, loans and credit agreements.

cluster controller A device that can control input/output operations of more than one device connected to it (e.g. a terminal). An interface between several bisynchronous devices and a PAD, NC or communication facility. The cluster controller handles remote communications processing for its attached devices. Most common types are IBM 327X.

cluster size An operating function or term describing the number of sectors that the operating system allocates each time disk space is needed.

clustering A client/server term describing the collection of servers or data in a central location for reasons of increased effectiveness and efficiency of security, administration and performance. Clustering is to help servers become fault resilient. Clustering helps to overcome problems associated with the client/server paradigm in general. In the old days of "heavy iron" (mainframes), user access to applications and files was carefully controlled. The target applications and files were resident on a highly redundant, carefully administered and tightly secured mainframe computer. As client/server has taken hold, the processors tend to be distributed in order that they are located in closer proximity to users, thereby relieving the strain on network resources, improving response times, and so on. The downside is that the resources are more difficult to manage, secure and control. Clustering the servers in a centralized location places them back in the hands of centralized MIS management, relieving these problems. There are two forms of clustering. The first form involves segmenting and spreading the database across multiple servers, with each segment of the database residing on multiple servers in order to achieve some level of redundancy. The second form positions the servers in a communications role, with each providing access for a group of users to data housed in a central repository, generally in the form of a minicomputer or mainframe.

cluster remapping A recovery technique used when NTFS detects a bad sector.

NTFS dynamically replaces the cluster containing the bad sector and allocates a new cluster for the data. If the error occurs during a read, NTFS returns a read error to the calling program, and the data is lost. If the error occurs during a write, NTFS writes the data to the new cluster, and no data is lost.

CLUT An imaging term. Color Look-Up Table. The palette used in an indexed color system. Usually consists of 256 colors.

clutter Wave reflections from obstructions such as terrain and buildings, which may show up as echoes or unidentifiable blips on a radar screen, thus interfering with scanning.

CM 1. Computing Module.

2. Configuration Management. A wireless telecommunications term. The tracking, co-ordination, and administration of software and hardware related to telecommunication or information systems. Versions are controlled and tracked.

3. Cable Modem.

4. CoMmercial Grade. Cables intended for general purpose communications use within buildings in accordance with the National Electric Code Section 800-53 (c). it, as distinguished from copper-plated. See also CMP and CMR.

CM8 Pairgain system which combines eight plain old telephone lines (POTS) onto one cable pair. This is accomplished by sending out eight different frequencies. Channel units are installed in the field, while channel cards in the central office (CO) communicate to these channel units. CM8 is a very primitive type of pairgain technology , and has many troubleshooting problems, including crosstalk, and inadvertent dialing.

CMA Communications Managers Association. An independent, not-for-profit users group formed in 1948 and serving the New York/New Jersey area. CMA provides a forum for peer-to-peer discussion of common issues, evaluation of technologies and their business applications, and the fostering of constructive relationships between suppliers and users. In addition to end users, CMA welcomes non-voting "Partners" in the form of vendors, consultants and associations. www.cma.org.

CMC 1. Common Messaging Calls. A messaging standard defined by the X.400 API Association. CMC 1.0 defines a basic set of calls to inject and extract messages and files and access address information. CMC is intended to define a useful common denominator across a wide variety of messaging systems. The idea is that an electronic mail system, no matter how crude, should be able to support a CMC front end. CMC's major "competition" is MAPI – Messaging Application Programming Interface – from Microsoft, though simple MAPI is almost identical to CMC.

2. Cellular Mobile Carrier.

CMD Charge Modulated Device, an active pixel sensor for imaging derived from CCD pixel technology and CMOS transistor technology. CMD are analog sensors, the digitizing happens when the electrons are passed through the analog to digital converter. The A to D converter converts the analog signal to a digital file or signal. Like the CCD, the CMD is used as an image capture device, CMDs are noisier imaging devices.

CMDS Centralized Message Distribution System. An RBOC-owned clearinghouse for CATS billing data. The RBOCs use it to exchange billing data with each other and to determine the financial settlements between them. They also clear ITC data through CMDS and charge the ITCs for doing this.

CMEA Cellular Message Encryption Algorithm. See ECMEA.

CMCI Cable Modem to CPE Interface. An element of DOCSIS (Data Over Cable Service Interface Specification), a project intended to develop a set of specifications for high-speed data transfer over cable television systems. CMCI is the interface between the cable modem and the CPE (Customer Premise Equipment), which typically would be in the form of a PC. See also DOCSIS.

CMI Control Mode Idle: C Idle Byte CO.

CMIP Common Management Information Protocol. CMIP is the network management standard for OSI networks. It has some features that are lacking in SNMP and SNMP-2, and is more complex. CMIP has a far smaller mind share and market share than SNMP in North America, though support for this standard is sometimes mandated, especially in Europe. CMIP is an ITU-TSS standard for the message formats and procedures used to exchange management information in order to operate, administer, maintain and provision a network. In short, CMIP is the protocol used for exchanging network management information. Typically, this information is exchanged between two management stations. CMIP can, however, be used to exchange information between an application and a management station. CMIP has been designed for OSI networks, but it is transport independent. Theoretically, it could run across a variety of transports, including, for example, IBM's Systems Network Architecture. See CMIP/CMIS, MIB and SNMP.

CMIP/CMIS Common Management Information Protocol/Common Management Information Services. An OSI network management protocol/service interface created and standardized by ISO for managing heterogeneous networks.

CMISE Common Management Information Service Element. A wireless telecommunications term. The functionality provided by CMIP in transporting network management information.

CMNS Connection-Mode Network Service. Extends local X.25 switching to a variety of media (Ethernet, FDDI, Token Ring).

CMOL Short for "CMIP Over Logical Link Control". An implementation of the CMIP protocol over the second layer of the OSI protocol stack, to be proposed as a standard by 3 Com Corp. and IBM. The goal of CMOL is to create agents that require significantly less memory than CMIP implemented over OSI, or SNMP implemented over UDP.

CMOS Complementary Metal Oxide Semiconductor, a technology used in transistors that are manufactured into most of today's computer microchips. In CMOS technology, negative charge (N-type transistors) and positive charge carriers (P-type transistors) are used in a complementary way to form a current gate that creates an effective means of electrical control. CMOS transistors draw almost no power when not in use. See CMOS RAM and CMOS Setup.

CMOS RAM Complementary Metal Oxide Semiconductor Random Access Memory. Memory which contains a personal computer's configuration information. CMOS RAM must have continuous power to preserve its memory. This power is typically supplied by a lithium battery.

CMOS Setup A program which prepares the system to work. CMOS setup records your PC's hardware configuration information into CMOS RAM. It must be modified when you add, change or remove hardware.

CMOT CMIP Over TCP/IP. More correctly, Common Management Information Protocol over TCP/IP. The original CMOT was described by RFC 1095, which is now obsolete. The new RFC 1189 "defines the means for implementing the IS version of CMIS/CMIP on top of both IP-based and OSI-based Internet transport protocols...", an expanded charter. The portion that is CMIS/CMIP over TCP/IP is still referred to as "CMOT". Someone referred to the new RFC as "CMIP over RFC 1066". In short, CMOT is an Internet standard defining the use of CMIP for managing the TCP/IP-based Internet and other attached networks. While the OSI-based CMIP is viewed as the most elegant long-term network management solution for such networks, it has not received the widespread acceptance of SNMP (Simple Network Management Protocol), which is much simpler (hence the name) and much more easily implemented. See also CMIP, OSI Reference Model and SNMP.

CMP Communications Plenum Cable. A grade of cable specified in the National Electrical Code, CMP cable is the plenum version of CM (Commercial Grade). Not only can CMP cable be run through walls without being placed in a conduit, but it also can be run in fire-critical areas such as drop ceilings or raised floors, where air circulates, often as the air return for air conditioning. CMP is insulated and sheathed in a low-smoke, fire-retardant material. It also costs a lot more than normal telephone or data wiring. See also CM, CMP-50 and CMR.

CMP-50 See NFPNA 90A.

CMR 1. An ATM term. Cell Misinsertion Rate: The ratio of cells received at an endpoint that were not originally transmitted by the source end in relation to the total number of cells properly transmitted.

2. CoMercial Riser grade. A grade of cable specified in the National Electrical Code (NEC), CMR cable can be run vertically between floors of a building without being housed in a conduit. See also CM and CMP.

CMRFI Cable Modem to Radio Frequency Interface. An element of DOCSIS (Data Over Cable Service Interface Specification), a project intended to develop a set of specifications for high-speed data transfer over cable television systems. At the CPE (Customer Premise Equipment) end of the network, the CMRFI provides the interface between the cable modem and the cable system coax drop. The CMRFI specification will include all physical, link and network level aspects of the communications interface, including RF levels, modulation techniques, coding schemes, and multiplexing. See also DOCSIS.

CMRS Commercial Mobile Radio Service. One type of wireless carrier, as defined by the Federal Communications Commission. Public use licenses are issued to carriers intending to serve multiple parties as their own subscribers or customers. Licenses are exclusive and are specific to geographic areas and a specified period of time. CMRS providers can include Personal Communications Services (PCS) providers; Air-to-Ground Carriers; Radio Common Carriers, including Paging and Radio Mobile Service providers; Cellular Carriers; Enhanced Specialized Mobile Radio Carriers (ESMRs); and Personal Mobile Radio Carriers (PMR).

CMS 1. Call Management Services. Canadian term for local calling features based on CLID (Calling Line Identification).

2. Content Management System. A fancy name for web authoring software. See Content Management System.

CMS 8800 Cellular Mobile Telephone Service (North American version).

CMTRI Cable Modem Telco Return Interface. An element of DOCSIS (Data Over Cable Service Interface Specification), a project intended to develop a set of specifications for high-speed data transfer over cable television systems. At the head-end of the network, the CMTRI provides the interface between the cable modem system and PSTN. See also DOCSIS.

CMTS 1. CMTS stands for the Cellular Mobile Telephone System. The original and still, most common CMTS is a low-powered, duplex, radio/telephone which operates between 800 and 900 MHz, using multiple transceiver sites linked to a central computer for coordination. The sites, or "cells," named for their honeycomb shape, cover a range of one to six, or more, miles in each direction. The cells overlap one another and operate at different transmitting and receiving frequencies in order to eliminate crosstalk when transmitting from cell to cell. Each cell can accommodate up to 45 different voice channel transceivers. When a cellular phone is activated, it searches available channels for the strongest signal and locks onto it. While in motion, if the signal strength begins to fade, the telephone will automatically switch signal frequencies or cells as necessary without operator assistance. If it fails to find an acceptable signal, it will display an "out of service" or "no service" message, indicating that it has reached the limit of its range and is unable to communicate. Each cellular telephone has a unique identification number which allows the Mobile Telephone Switching Office (MTSO) to track and coordinate all mobile phones in its service area. This ID number is known as the Electronic Security Number (ESN). The ESN and cellular phone's telephone number are NOT the same. The ESN is a permanent number engraved into a memory chip called a PROM or EPROM, located in the telephone chassis. This number cannot be changed through programming as the telephone number can, although it can be replaced. Each time the telephone is used, it transmits its ESN to the MTSO by means of DTMF tones during the dialing sequence. The MTSO may be able to determine which ESNs are good or bad, thus individual numbers can be banned from use within the system. See also Cell and Cellular.

2. Cable Modem Termination System. An element of DOCSIS (Data Over Cable Service Interface Specification), a project intended to develop a set of specifications for high-speed data transfer over cable television systems. CMTS comprises CMTS-DRFI (CMTS-Downstream RF Interface), CMTS-NSI (CMTS-Network Side Interface), and CMTS-URFI (CMTS-Upstream RF Interface) in order to provide two-way communications. See also CMTS-DRFI, CMTS-NSI, CMTS-URFI and DOCSIS.

CMTS-DRFI Cable Modem Termination System-Downstream RF Interface. An element of DOCSIS (Data Over Cable Service Interface Specification), a project intended to develop a set of specifications for high-speed data transfer over cable television systems. At the head-end of the network, the CMTS-DRFI provides the interface between the cable modem system and the downstream RF (Radio Frequency) path, which terminates in the cable modem at the customer premise. It works in conjunction with the CMTS-URFI (CMTS-Upstream RF Interface) in order to provide two-day communications. See also CMTS-URFI and DOCSIS.

CMTS-NSI Cable Modem Termination System-Network Side Interface. An element of DOCSIS (Data Over Cable Service Interface Specification), a project intended to develop a set of specifications for high-speed data transfer over cable television systems. CMTS-NSI is the interface of the cable modem system at the head-end of the network; i.e., at the CATV provider's premise. CMTS-NSI provides the interface between the backbone cable system and the CATV provider's server complex. See also DOCSIS.

CMTS-URFI Cable Modem Termination System-Upstream RF Interface. An element of DOCSIS (Data Over Cable Service Interface Specification), a project intended to develop a set of specifications for high-speed data transfer over cable television systems. At the head-end end of the network, the CMTS-URFI provides the interface between the cable modem system and the upstream RF (Radio Frequency) path, which terminates in the cable modem at the customer premise. It works in conjunction with the CMTS-DRFI (CMTS-Downstream RF Interface) in order to provide two-day communications. See also CMTS-DRFI and DOCSIS.

CMTRI Cable Modem Telco Return Interface. An element of DOCSIS (Data Over Cable Service Interface Specification), a project intended to develop a set of specifications for high-speed data transfer over cable television systems. At the head-end of the network, the CMTRI provides the interface between the cable modem system and PSTN. See also DOCSIS.

CMY A computer imaging term. A color model used by the printing industry that is based on mixing cyan, magenta, and yellow. It's also referred to as CMYK, with the K denoting black. The K was added after printers discovered they could obtain a darker black using special black colorants rather than by combining cyan, magenta, and yellow alone. See also CMYK.

CMYK A computer imaging term. A color model used by the printing industry that is

based on mixing cyan, magenta, yellow and black (called "K.") It used to be called CMY. The K was added after printers discovered they could obtain a darker black using special black colorants (i.e. black ink) rather than by combining cyan, magenta, and yellow. CMYK is the basis of what's known now as "four-color" printing. But there is also five, six, seven and eight color printing, etc. Each of these "extra" colors are basically "colors" which are better printed as their own color rather than as a combination the basic three. Silver, copper, gold, aluminum, etc. are all printed traditionally as extra colors. They cannot be created by combining CMYK. Sometimes people are picky about the way their colors come out – i.e. Coca Cola red – so they may be printed with that color ink rather than combining CMYK. Typically a full color printing job requires passes under four printing presses – each laying down C,M,Y and K. These extra colors – silver, gold, copper, special colors – will need additional printings by additional printing presses. Thus a color print job could easily become five, six or seven color print job. The most common full color print job we hear about is four color. But you see an awful lot of jobs that contain silver and copper and may be six color print jobs.

CN Complementary network.

CNA 1. Cooperative Network Architecture.

2. Centralized Network Administration is an AMP-defined architecture that consolidates all network electronics into a single closet instead of distributing them throughout the building. According to AMP, CNA saves money over the long haul because of reduced administration costs. Centralized Network Administration can be executed with optical fiber for runs up to 300 meters, with Category 5 unshielded twisted pair (UTP) if no user is more than 90 meters from the central cross-connected and equipment room. Compare with Distributed Network Administration.

3. Customer Name and Address Bureau.

CNAM Caller ID with NAMe. See LNP (as in Local Number Portability). See SSN.

CNAME Canonical Name. The Canonical Name resource record, CNAME, specifies an alias or nickname for the official, or canonical, host name. Alias records assign an alternate hostname to a specific hostname. Both hostnames point at whatever IP address the primary hostname is assigned to. See also CNAME Records.

CNAME Records Canonical Name Records. Records stored in a DNS (Domain Name Server) to create an alias to redirect traffic from one Internet domain name (i.e., URL, or Uniform Resource Locator) to another. See also A Records, MX Records, and URL.

CNC Complementary Network Service. See Open Network Architecture.

CND 1. Calling Number Delivery. See Caller ID Message Format and Call Block.

2. Calling Number Display. See also CLID.

CNE Certified (local area) Network Engineer. When you graduate from Novell's third level class, you become a certified network engineer. CNEs are an elite group in the LAN industry. See also CCIE and MCSE.

CNET Centre National d'Etudes de Telecommunication. The French organization that approves telecommunications products for sale in France.

CNG Also called Auto Fax Tone, or Calling Tone. This tone is the sound produced by virtually all fax machines when they dial another fax machine. CNG is a medium pitch tone (1100 Hz) that lasts 1/2 second and repeats every 3 1/2 seconds. A fax machine will produce CNG for about 45 seconds after it dials. The CNG tone is useful for owners of fax/phone/modem switches. Such switches answer an incoming call. If they hear a CNG tone, they will transfer the call to a fax machine. If they don't, they'll transfer the call to a phone, answering machine or perhaps a modem. Depends on how they're set up. Some fax machines do not transmit a CNG tone with manually-dialed transmissions – i.e. where the caller picked up the handset on the fax machine, dialed and waited for a high-pitched squeal before pushing his fax machine's "start" button. A manual dialed fax transmission will "fool" fax/voice switches. See CED and Facsimile.

CNIP Calling Number Identification Presentation. See CLIP.

CNIR Calling Number Identification Restriction. A wireless "cellular" term (GSM and PCS) for Call Block, also known as Calling Number Delivery (CND), in the wired world. See Call Block.

CNIS Calling Number Identification Services.

CNM An ATM and SMDS (Switched Megabit Data Service) term. Customer Network Management. All activities that customers perform to manage their communications networks. SMDS CNM service enables customers to directly manage many aspects of the SMDS service provided by telecommunications carriers. See Customer Network Management.

CNN effect The impact that live, ongoing TV coverage of news events has on world events like military operations or government policy. CNN's 24-hour coverage and satellite technology, viewers often find out what's going on at the same time as the commanders and the politicians.

CNO Corporate Networking Officer, a term invented by William Y. O'Connor, CEO of Ascom Timeplex.

CNR 1. Telephone company term for re-scheduling a telephone installation appointment because the "Customer is Not Ready."

2. An ATM term. Complex Node Representation: A collection of nodal state parameters that provide detailed state information associated with a logical node.

CNRI The Corporation for National Research Initiatives, a Reston, VA-based not-for-profit organization that works with industry, academia, and government on national-level initiatives in information technology. It will host the initial operations of IOPS.ORG. "IOPS.ORG will play a key role in the healthy technical and operational evolution of the Internet as an increasingly important component of the economy," said CNRI President Robert Kahn. www.cnri.reston.va.us.

CNS Complementary Network Service. CNSs are basic services associated with end users' lines that make it easier for Enhanced Service Providers (ESPs) to offer them enhanced services. Some examples of CNSs include Call Forwarding Busy/Don't Answer, Three Way Calling, and Virtual Dial Tone. See Open Network Architecture.

CO Central Office. Pronounced "See-Oh". In North America, a CO is that location which houses a switch to serve local telephone subscribers. Sometimes the words "central office" are confused with the switch itself. In Europe and abroad, the words "central office" are not known. The more common words are "public exchange." But those words tend to refer more to the switch itself, rather than the site, as in North America. See also Central Office or Public Exchange.

CO Lines These are the lines connecting your office to your local telephone company's Central Office which in turn connects you to the nationwide telephone system.

CO Location See Colocation.

CO Simulator A desktop device which pretends to act like a mini-central office. The smallest version will consist of two lines and two REJ-11 jacks. Plug a phone into both jacks. Pick up one phone. You hear dial tone. Dial or touchtone two or three digits. Bingo, the second phone rings. You pick up the second phone. You can have a conversation with yourself or with a machine – like a voice processing system. Most central office simulators can simulate normal on-hook, off-hook, dialing, answering, speaking, etc. Some now can simulate caller ID features – including the number of person calling.

co-carrier status A relationship between a CLEC and an ILEC that affords each company the same access to and right on the other's network and provides access and services on an equal basis.

co-channel interference Interference between signals transmitted in a given Radio Frequency (RF) channel in a particular cell and signals transmitted on the same RF channel in a different cell. A receiver that is in a position to receive from both cannot filter out the undesired signal and consequently the noise level at the receiver increases.

co-digging See Co-trenching.

co-directional Angelo Velez found this reference to co-directional in a technical student manual on a course on baseband equipment concerning source of signals, specifically on Transmit timing (DTE source), or terminal timing, etc. "The signal out of a DTE and into a DCE on pins 24/23 travels in the same direction as a transmit data bit. Therefore, this signal is known as the 'co-directional' transmit clock. The co-directional clock is generated by the DTE and is of the same accuracy as the DTE's reference; if the DTE is accepting the counter-directional clock (pins 15/16 in) as a reference, then the co-directional clock (24/23 out) will be synchronized with it."

co-location 1. The ability of a someone who is not the local phone company to put their equipment in the phone company's offices and join their equipment to the phone company's equipment. See Colocation for a bigger explanation.

2. Imagine you're running an ecommerce application that's important to your company, i.e. it's bringing in oodles of money. It's running on one or several servers which are physically in the same place. Now let's say you're manic about reliability. You may decide to outsource the physical hosting of one, some or all of these servers to a dedicated facility to make sure your servers are always up and running. These co-location facilities offer the customer a secure place to physically house their hardware and equipment as opposed to locating it in their offices or warehouse where the potential for fire, theft or vandalism is much greater. Most co-location facilities offer high-security, including cameras, fire detection and extinguishing devices, multiple connection feeds, filtered power, backup power generators and other items to ensure high-availability which is mandatory for all Web-based, virtual businesses. Also spelled Colocation.

co-marketing You sell something I make. You make money on what you sell. In addition, I give you extra money – what we call "co-marketing" funds. Co-marketing is misnamed. In the retail trade, it's called a "spiff" – the extra money that a manufacturer pays the salesman for selling his product. See also Bounty.

co-trenching Also known as co-digging. A form of colocation that involves multiple fiber-optic carriers to coordinate their cable implementation activities so that they share a single trench, and the cost and disruption associated with digging it. Local governments increasingly are requiring this level of coordination to minimize the disruption associated with multiple carriers digging up streets and roads. See also Colocation.

coalescing A WAN acceleration technique whereby multiple small packets are aggregated into a single larger packet. Since packet headers are overhead, coalescing multiple smaller packets' payloads of user data under a single packet header reduces the amount of overhead that is put on the WAN, thereby allowing more bandwidth to be used for moving user data rather than moving overhead bits. This reduces network traffic and improves perceived performance. Packet coalescing is particularly beneficial for a network application such as VoIP, which uses small packets, and interactive enterprise applications, whose packet payloads may consist of nothing but a few keystrokes and/or mouseclicks.

COAM Customer Owned And Maintained telephone equipment. Similar to CPE, which stands for customer premise equipment, except that COAM equipment is a broader term, referring also to customer equipment outside of their immediate premises. It could refer to a customer-own microwave system, for example.

Coarse Wave Division Multiplexing See CWDM.

Coase's Law In 1931, Coase studied why organizations are formed, what guides their growth and what leads to their demise. He observed that companies will expand until "the costs of organizing an extra transaction within the firm become equal to the costs of carrying out the same transaction on the open market." That is now known as Coase's Law. It represents an Information Age reformulation of the law of diminishing returns, which applied only to capital assets.

coasters Unsolicited floppy disks or CD-ROMs, such as the ubiquitous American Online software, that arrive in one's mailbox too often.

coasting mode In timing-dependent systems, a free-running operational timing mode in which continuous or periodic measurement of timing error is not available. In some systems, operation in this mode can be enhanced for a period of time by using clock or timing error (or correction) information obtained during a prior tracking mode to estimate clock or timing corrections to be made in the free-running mode.

coated filament A vacuum tube filament coated with a metallic oxide to provide greater electron emission and longer life.

coating A protective material (usually plastic) applied to the optical fiber immediately after drawing to preserve its mechanical strength and cushion it from external forces that can induce microbending losses.

coaxial cable A cable composed of an insulated central conducting wire wrapped in another cylindrical conducting wire. The whole thing is usually wrapped in another insulating layer and an outer protective layer. A coaxial cable has capacity to carry great quantities of information. It is typically used to carry high-speed data (as in connections of 327X terminals to computer hosts) and in CATV installations.

coaxial lightning suppressor A device which grounds a coaxial cable shield and shorts surge voltages on the inner wire and the outer tubular wire.

coaxial switch A switch used for disconnecting or re-routing signals on coaxial cables. See Transfer Switches.

COB Close Of Business.

CoBAMs Consortium of Brick and Mortars. This is not a real Consortium. It's an analysts' way of saying (and I quote) ""CoBAMs in any and all industries/sectors are here to stay provided they offer to be the de facto platform on which to conduct vertical industry B2B trade. Such CoBAMs will make life easier for the suppliers (?) by agreeing to do business on a common platform, business processes and interface. (A B2C example of this is to be found in Pegasus for Hotel reservation systems used by RedRoof Inns and Marriot etc.)" He also raised the antitrust flag to the extent that blood-from-turnip-squeezing takes place."

COBOL Common Business Oriented Language. A very popular computer programming language for business applications.

COBRA A misspelling of CORBA, Common Object Request Broker Architecture. See CORBA, Object Request Broker.

cobweb site A Web Site that is so old and hasn't been updated so long that it has figuratively grown cobwebs.

COC Central Office Connection. Separately tariffed part of a T-1 circuit.

COCE Condition of Continued Employment, which means follow the process or get approval to deviate; otherwise, find another job. In other words, "take my way or the highway."

Cochannel interference C/I. Cochannel interference refers to the interference caused between two cells transmitting on the same frequency within a network. Since cochannel interference is caused by another cell transmitting the same frequency, you can't simply filter out the interference. You can only minimize the cochannel interference through proper cellular network design. A cellular network must be designed to maximize the C/I ratio. The C/I ratio is the carrier-to-cochannel interference ratio. One of the ways to maximize the C/I ratio is to increase the frequency re-use distance, i.e. increase the distance between cells using the same set of transmission frequencies. The C/I ratio in part determines the frequency re-use distance of a cellular network.

cockpit effect An acoustics phenomenon describing the difficulty we have in modulating our speech if there is background noise significant to impair our ability to hear ourselves. Under such circumstances, the lack of feedback causes us to speak loudly, and to alter our speech patterns. Also known as Lombard Speech, the cockpit effect particularly is a problem when using a cellular phone in an automobile to communicate with a voice processor employing speech recognition technology.

cockpit problem Also known as "pilot error." A problem caused by the user's inability to operate a device or system correctly. A poorly-designed user interface often contributes to cockpit problems. Often, it's just a matter of the user's not taking the time to read the user manual, which would be largely unnecessary if the user interface were as "intuitive" as the manufacturer usually claims. Cockpit problems are the most common reason for calls to the "help desk." See also Help Desk, Idiot-Proof, Intuitive, and RTFM.

cockroach stock Companies which repeatedly issue unusual bad news that can only be attributed to their own idiocy are called Cockroach stocks. The price of their stocks tends to fall and fall and fall. In their own time, WorldCom and Enron were cockroach stocks.

cockroaches 1. Entomologists have observed that cockroaches can change course as many as 25 times in one second, making them the most nimble animals known. Some people argue that politicians and lawyers can do the same thing.

2. Crushed cockroaches can be applied to a stinging wound to help relieve the pain.

3. Companies which repeatedly issue unusual bad news that can only be attributed to their own idiocy are called Cockroach stocks.

cocktail party A device for paying off obligations to people you don't want to invite to dinner." – Charles Merrill Smith

coconut wireless What the "grapevine" is called in Hawaii.

COCOT Customer Owned Coin Operated Telephone. See also Clip on Toll Fraud.

COCUS Central Office Code Utilization Survey. The COCUS is an annual survey that seeks information on the number of central office (CO or NXX) codes currently assigned to telephone companies in America, as well as a forecast of the number of additional CO codes the companies will need over the next several years. The COCUS has been used by the industry for a number of years and is intended to provide an overall view of both present and projected CO code utilization, information that is critical for area code relief planning purposes. Although in prior years the regional CO Code Administrators performed this function, this year (1999) the COCUS is being conducted by the neutral North American Numbering Plan Administrator (NANPA), Lockheed Martin CIS.

COD An ATM term. Connection Oriented Data: Data requiring sequential delivery of its component PDUs to assure correct functioning of its supported application, (e.g., voice or video).

code 1. As a verb, it means to write instructions in computer language. As a noun, it means software.

2. In telecommunications, code is the system of dots and dashes used to represent the letters of the alphabet, numerals, punctuation and other symbols.

code bit The smallest signaling element used by the Physical Layer for transmission on the fiber cable.

code bloat Computer program code that has become swollen, i.e., bigger and more resource-intensive than it needs to be. Bloated code takes up more memory (both hard drive and RAM) than necessary, which also means that it takes an excessive amount of time to install, load, execute, transmit and work with. My classic example of code bloat is using Macromedia's Dreamweaver program to write web pages. For some reason I can't figure, it closes your choice of typeface at the end of each paragraph. Then opens each paragraph with code to indicate which typeface and which size. This is an example of the code it puts in at the beginning of every paragraph. What a waste. <\<>font face="Verdana, Arial, Helvetica, sans-serif" size="-1">. These words are not necessary at the beginning of each paragraph. In true HTML, you only need this code at the beginning of the file – until your typeface or size changes. See also Easter Egg.

code blocking A switch's ability to block calls to a specified area code, central office code or phone number.

code blue A PBX feature for hospital application. If a patient is in distress, he can simply knock the telephone handset off of the cradle. After a brief period, the PBX recognizes the lack of dialing activity and sends a "code blue" alarm to the nurses station. Nurses and doctors come running. People turn blue when they can't breathe – hence the term.

code breaker See Key and Key Holder.

code call access A very useful PBX feature. It allows attendants and extension users to activate, by dialing an access code followed by a two or three digit called code, customer-provided signaling devices throughout the premises. The signaling devices then issue a series of tones or visual coded signals corresponding to the called code. The called or paged party responds by dialing a meet-me answering code from any phone and is then connected to the paging party.

code coverage Modern computer telephony systems are composed largely of software, or "code." Invariably this code has many different logic paths and options. During normal system usage, many code paths are used only infrequently, if at all, meaning that normal usage will really only test a small portion of the total system. Code coverage refers to the amount of the system code that has been accessed during the testing of the system. The greater the code coverage, the more code that has been tested. 100% code coverage means that all the code has been tested. The amount of code coverage that has been achieved is usually determined through the use of a code coverage tool. Code coverage tools are available for most computer operating systems. Code coverage is especially important for computer telephony applications because many features are only infrequently used or are turned on or off based on the user's class of service. Also, many features interact with other features and the use of one feature often turns off another. Functional anomalies frequently exist in little-used paths and feature interactions. Tests should be designed to make sure that all code paths have been accessed and are adequately exercised. This definition courtesy of Steve Gladstone, author of the book Testing Computer Telephony Systems, available from 212-691-8215.

code conversion A process which converts the codes coming in from one network into codes that can be recognized on another network, such as converting from the Baudot code in a telex network to the ASCII code on the TWX network. Usually, the hardware will convert differences in transmission speed.

Code Division Multiple Access CDMA, also called Spread Spectrum, is a name for a new form of digital cellular phone service. The idea is that each phone call is combined with a code which only one cellular phone plucks from the air. Business Week said CDMA works "by spreading all signals across the same broad frequency spectrum and assigning a unique code – the company says one of 42 billion – to each. The dispersed signals are pulled out of the background noise by a receiver which knows the code. This method, developed by a San Diego company called Qualcomm Inc. is very new. Much of the equipment to support it – like the cellular switches– has not yet been developed." CDMA is also being used by wireless PBXs. See CDMA for a longer and better explanation.

Code Excited Linear Prediction CELP. An analog-to-digital voice coding scheme.

code independent data communication Data communication mode using a link procedure associated with the character and not dependent on the set of characters or the code used by the data source.

code level Number of bits used to represent a character.

Code Of Federal Regulations CFR. CFR is a codification of the general and permanent rules published in the Federal Register. It is divided into 50 titles that represent broad areas subject to federal regulation. Title 47 of the CFR pertains to telecommunications and contains the rules covering Part 22 Common Carriers and Part 90 Private Carriers.

code violation Violation of a coding rule; for example, the AMI coding rule is corrupted by a bipolar violation.

code word When used in the context of the Reed-Solomon encoding, it refers to the 63, 6-bit symbols (378 bits) resulting from the encoding of 47 6-bit (282 bits) information symbols. This is done by appending 16 6-bit parity symbols.

CODEC Originally CODEC stood for CODer-DECoder, i.e. microprocessor chip. Now the PC industry thinks it stands for COmpression/DEcompression, i.e. an overall term for the technology used in digital video and stereo audio. The original CODEC (still in big use in today's telephony industry) converts voice signals from their analog form to digital signals acceptable to modern digital PBXs and digital transmission systems. It then converts those digital signals back to analog so that you may hear and understand what the other person is saying. In some phone systems, the CODEC is in the PBX and shared by many analog phone extensions. In other phone systems, the CODEC is actually in the phone. Thus the phone itself sends out a digital signal and can, as a result, be more easily designed to accept a digital RS-232-C signal.

CODEC Conversion The back-to-back transfer of an analog signal from one CODEC into another CODEC in order to convert from one proprietary coding scheme (for instance, that used by CLI) to one used by another CODEC manufacturer (PictureTel, VTEL, GPT, BT, NEC, etc). The analog signal, instead of being displayed to a monitor, is delivered to the dissimilar CODEC where it is redigitized, compressed and passed to the receiving end. This is obviously a bi-directional process. Conversion service is offered by carriers such as AT&T, MCI and Sprint.

coded character set A set of unambiguous rules that establish a character set and the one-to-one relationships between the characters of the set and their coded representations.

coded image A representation of an image in a form suitable for storage and processing.

Coded Orthogonal Frequency Division Multiplexing COFDM. See 802.11a.

coded trunks You buy several trunks. They hunt on. The main number is 555-3000. If the main number is busy, the call goes to the next line. There are two types of "next lines." One type can have an actual number, like 555-3001, which you can call directly. The other can be a coded trunk with no actual number and which you can't call directly. It's better to have no coded trunks because it's hard to test coded trunks. You can't dial them directly. Actual dialable numbers are better.

coder An analog-to-digital converter that changes analog voice signals to their digital equivalents. See CODEC.

codernauts Also known as programmers or coders. Codernauts are on the ultimate quest for the perfect software.

codes Programs. It's an inside term used by hackers and heavy-duty grinders. Use it instead of "programs" like the Masons do with that secret handshake and those in the know will nod knowingly with respect. The majority will just nod sleepily.

Codial Office CDO. A small central office designed for unattended operation in a distant community. Usually a community dial office is fairly small, rarely more than 10,000 lines.

coding gain Coding gain is the ratio in decibels with and without coding, of the energy per bit (Eb/No) required to attain a specific bit error rate (BER) on a channel with a fixed noise power.

coding theory The mathematical theory describing how to encode data into streams of digital symbols at the transmitter and decode it at the receiver to maximize accuracy of data presented to user.

COE 1. Central Office Equipment; as in Central Office Equipment engineer.

2. Central Office Engineer.

coefficient of variation A telephone company definition. Relates the standard deviation of a distribution to the mean of the distribution, usually as a percent. Example: If all of the busy hour, busy season loads for an office have a mean of 10,000 CCS and a standard deviation of 1000 CCS, the coefficient of variation is 10%.

COER Central Office Equipment Reports. A telephone company definition. A large scale computer software package which accepts Central office Engineering Data properly formatted by a Data Collection System (DCS), subjects these data to a series of validation tests, and produces final summarized reports designed to meet both administrative and engineering requirements.

COFA Change Of Frame Alignment.

COFDM Coded Orthogonal Frequency Division Multiplexing (COFDM). See 802.11a.

coffee 1. A person who is coughed upon.

2. According to Dr. Bruce Ames, University of California at Berkeley, there are more than 1,000 chemicals in a cup of coffee. Of these, only 26 have been tested, and half caused cancer in rats.

3. Frederick the Great of Prussia wanted to make coffee off limits to his subjects because of the huge sums of money that was going to foreign exporters. "My people must drink beer," Frederick demanded in a manifesto. Rumors flew furiously, including one that claimed coffee made people sterile. Acclaimed musician Johann Sebastian Bach disagreed vehemently with Frederick and his anti-coffee crowd. In retaliation, the composer wrote his "Coffee Cantata," published in 1732. Bach's composition told the story of a father who threatens to break off his daughter's marriage plans unless she gives up her vile coffee-drinking habit. The girl agrees, but changes her mind when her mother and grandmother reveal that they have always been passionate, although secretive, coffee drinkers (and obviously not infertile). Bach himself was the father of 20 children.

4. On average, coffee drinkers have sex more often than non-drinkers. I don't know why.

COFETEL Mexico's "COmision FEderal de Telecomunicaciones", analogous to the U.S. FCC, the regulatory governmental body that oversees deployment of networks and services,

etc. Web site is www.cft.gob.mx.

cognitive radio A cognitive radio is a radio/system that can sense the RF environment to determine what channel(s) may be used in a particular area (and with what operating parameters, such as power, etc.) without causing interference to other services. See also 802.22.

cognitive science Cognitive science is the interdisciplinary study of the acquisition and use of knowledge. During the 1960s and 70s, cognitive science emerged from the fields of linguistics, psychology, computer science, philosophy, neuroscience and anthropology, as a multidisciplinary effort centered around a number of common problems using the computer as a research tool. In cognitive science, the computer has been used to simulate cognitive processes. It's been used widely in the marketing and advertising of products, particularly in the promotion of goods to children, who are highly susceptible to fast-paced suggestive advertising messages.

cohere To come together firmly, to be cohesive, to coalesce, to hold together, join, unite, merge, especially small, discrete parts or granules.

coherence area In optical communications, the area in a plane perpendicular to the direction of propagation over which light may be considered highly coherent.

coherent communications In fiber optics, a communication system where the output of a local laser oscillator is mixed optically with a received signal, and the difference frequency is detected and amplified.

coherent interference Any form of interference which is intelligible as a television picture. Typically either an offset ghost of a strong television picture or a ghost of a completely different picture mixed with a strong picture. See Non-Coherent Interference.

coherence length The propagation distance over which a light beam may be considered coherent. See Coherent Light.

coherent light Light signals emitted from lasers and some LEDs (Light Emitting Diodes) is coherent. Coherent light is made up of light waves that all travel in the same direction (spatial coherence), and are of the same frequency and phase (temporal coherence). Since the signals are locked in phase, their peaks and troughs are all in alignment, and the waves reinforce, or amplify, each other. This results in the intense, pure light beam that is characteristic of lasers. See also Incoherent Light, Laser Diode and LED.

coil 1. In electronics, a number of turns of wire, so wound as to afford inductance.

2. In telecommunications a coil refers to a load coil. It's a voice-amplifying device for twisted-pair wire. A load coil is usually placed every 3000 feet past the CO. They are usually placed in vaults with twisted-pair splices. They should be removed for high speed data communications.

coil antenna One consisting of one or more complete turns of wire. See Loop Antenna.

coin acceptor/rejector A mechanical or electromechanical device that checks and validates the coins deposited in a coin pay phone. They measure the coin's size and weight and steel content. These coin acceptor/rejector units transmit the value of the coin deposits to the processing part of a smart payphone or they signal the information to the telephone company central office via coded tones.

coin supervisory trunk group A trunk group that lets a switchboard operator collect overtime monies due on coin phones and check for stuck coins.

coin telephone A pay telephone that takes coins. The coin telephone was invented by William Gray, an American whose previous inventions included the inflatable chest protector for baseball players. Mr. Gray's first phone lacked a dial. Its instructions read:

"Call Central in the usual manner. When told by the operator, drop coin in proper channel and push plunger down."

In today's nomenclature, Mr. Gray's original phone is known as a post-pay coin phone. See also Payphone and several entries following it.

COLD Computer Output to Laser Disk. A computer storage management term referring to hardware and software solutions which store, index and retrieve formatted computer output on various media, including optical disks. Large scale COLD systems can be used to manage and archive storage-intensive image files such as those associated with credit card bills, telephone bills, brokerage statements, or tax returns. COLD systems also are used to store compressed graphics and image data on high-powered Web servers for Internet access.

cold bend test Bending a wire or cable around a specified diameter at low temperature to determine if the insulation or jacket material will crack. reaction or galvanic action.

cold calls See "How Are You Today?".

cold clamp A device that uses technology developed by Telstra Research Laboratories and which, when used with an optical time domain reflectometer (OTDR), can detect an optical cable fault without disrupting traffic on the cable. The cold clamp gets its name from its use of liquid nitrogen, whose boiling point is a frigid -196 degrees Celsius, -321 degrees Fahrenheit.

cold docking Docking is to insert a portable computer into a base unit. Cold docking means the computer must begin from a power-off state and restart before docking. Hot docking means the computer can be docked while running at full power. See Cold Start.

Cold Flow A cabling industry term for deformation of the insulation due to mechanical force or pressure (not due to heat softening).

cold fusion Like alchemy or the perpetual motion machine, cold fusion is one of those scientific pipe dreams we wish would come true. Alas, the methods behind the claim that we could produce an almost infinite, magically efficient energy source by doing nuclear fusion at room temperature turned out to be inconsistent or fraudulent, depending on your level of skepticism. Still an underground network of cold fusion experimentation continues. Arthur C. Clarke is among the believers and funds Infinite Energy magazine. If the truth is out there, cold fusion is way out there.

cold standby See Data Center.

cold start Everything starts from scratch. The power to the computer or telephone system is turned off. Everything in the system's volatile memory is erased. A cold start may be needed on a microcomputer when something has happened to "lock up" the keyboard and the Reset button (if there is one) doesn't clear the problem completely. A Cold Start is also needed when you want to load a new operating system. When your phone system gives troubles you find hard to diagnose, turn it off, count to ten and turn it on. This cold boot to your phone system will often fix the problem, as it will typically do on a computer system.

cold transfer An incoming phone call transferred without notice or explanation from the transferring party. "Someone in customer service cold transferred the call to me. By that point the guy was ready to crawl through the wires and kill somebody." I got this definition from Wired Magazine. In the telephone industry, they call the same thing "blind transfer."

ColdFusion A Rapid Application Development (RAD) system created by the Allaire Corporation of Cambridge, Mass, ColdFusion integrates browser, server and database technologies into Web applications. Cold Fusion Web pages include tags written in ColdFusion Markup Language (CFML) that simplify integration with databases and avoid the use of more complex languages like C++ to create translating programs. ColdFusion is the industry's leading cross-platform Web application server. With ColdFusion, Web developers can quickly develop and deliver a new generation of large-volume, transaction-intensive Web applications for everything from e-commerce to business automation and more.

collaboration A multimedia term. Collaboration involves two or more people working together in real-time, or in a "store-and-forward" mode. Applications will enable a group of people to collaborate in real-time over the network using shared screens, shared whiteboards, and video conferencing. Collaboration can range from two people reviewing a slide set on line to a conference of doctors at different locations sharing patient files and discussing treatment options.

collaboration software Software that lets two or more people do a task together. See Collaboration.

collaborating Two or more people working together on a project to share information and ideas, view suggestions, and make modifications. Computers can enable users to collaborate in real-time over a network or phone line using tools such as shared documents, shared whiteboards, and video conferencing, or time-efficient workflow such as document forwarding.

collaborative filtering A new, fancy name for database mining. See database mining.

collapsed backbone The backbone network connecting all network segments is collapsed (shortened considerably), and contained within a hub, or chassis. In the case of 10Base-T (the most common local area network), the collapsed backbone is in the form of a collapsed bus (multipoint circuit) architecture. 10Base-T is a considerable improvement on traditional Ethernet, which involves a big, thick coaxial cable. 10Base-T effectively collapses a coax segment and places it in a chassis. Devices attach to the shared bus via UTP (Unshielded Twisted Pair), which is a much less expensive and much less troublesome medium than coax. Additionally, as the bus housed in a chassis, it is protected from both physical damage and sources of ambient EMI (ElectroMagnetic Interference). Think of it this way. Your original Ethernet consisted of as many as 1,024 devices (workstations, printers and servers, etc.). They connected to a big, fat, heavy, expensive coaxial cable that looked like an orange hose, and that could run as far as 2,500 meters through your reinforced ceilings and between the floors of your building. The orange hose was a big electrical bus, or multipoint circuit. Every frame of Ethernet data traveled in both directions along the hose, and passed every attached device. Only one device could transmit across the hose at a time, or data collisions occurred. The devices connected to the hose through vampire taps that looked like big, ugly alligator clips (those of you who were hippies in the 1960's will

remember those). The vampire taps had a big sharp tooth that pierced the hose to establish a mechanical and electrical connection with the center coax conductor. If you disconnected a device and removed the vampire tap, the cable picked up noise through the hole you left in the cable. You replaced the orange hose and the vampire taps with a bunch of tiny hubs about half the size of a VCR (but with a clock that doesn't blink 12:00). Each hub contains a collapsed backbone, which is a tiny electrical bus that works like a small segment of the orange hose. Each hub and each collapsed backbone support a workgroup of perhaps 8-24 individuals with a common set of interests like accounting (which isn't very interesting) or engineering (which is very interesting) or sales (which is scary). Every user's workstation connects to the collapsed bus through twisted pair and a connector that looks much like your telephone wire and plug, only a little bigger. (It's called an RJ-45). The workgroup hubs are connected together by the same kind of wires and plugs, at least if they are close together. All together, the hubs, the tiny busses, and the little wires and plugs cost a lot less to buy and connect than the orange hose and vampire taps. They also are a lot easier on the eye. Also, they pass data by only a small number of attached devices within the workgroup, unless you need to communicate with a device attached to another hub. That means that the opportunity for congestion is much less. Think about the electrical wiring in your house. Your breaker box contains a collapsed bus to which the circuit breakers connect. Thin insulated wires from the circuit breakers connect to electrical outlets. It's much less expensive and much prettier to do it that way, rather than to run a thick, expensive and dangerous (high-voltage) electrical bus all around the house so that you can tap your lamps and appliances into it. See also 10Base-T, Ethernet, and Hub.

collapsed bus See Collapsed Backbone.

collapsed ring A SONET term. SONET optical fiber systems are deployed in a ring architecture, with two or four fibers for redundancy and, therefore, network resiliency. A collapsed ring topology is one in which the ring fibers are laid in the same fiber bundle. If the fiber bundle is cut, and all fibers in the ring are cut, the ring collapses. A collapsed ring is a very bad thing. See also SONET.

collar A financial term that occurs in conjunction with takeover bids. When AT&T offered to buy Comcast in the Spring of 1999, it offered Comcast shareholders a "collar." The collar was a guarantee that if the value of AT&T's stock dropped, AT&T would cover up to a 10% decline with additional cash.

collateral duties A call center term. Non-phone tasks (e.g., data entry) that are flexible, and can be scheduled for periods when call load is slow.

collect call A telephone call in which the called person pays for the call. The person calling calls a number and asks that the call be made "collect." Sometimes collect calls are handled by live operators, sometimes by machines. In a collect call, the phone company has to get some authorization from the person receiving the call that he will pay for it. This may be done by saying "Yes" or hitting a button a touchtone phone.

collected digits The keys a caller presses in response to menu choices or other requests for information, such as account number, made by a Voice Response Unit.

collector ring Metallic ring generally on the armature of a generator in contact with brushes for completing the circuit to a rotating member.

collimate To make light rays parallel. The process of collimation is accomplished by a collimator. See Collimation.

collimation The process by which a divergent or convergent beam of electromagnetic radiation is converted into a beam with the minimum divergence or convergence possible for that system (ideally, a parallel bundle of rays). In the context of telecommunications, collimation generally is used in fiber optics. In fiber optic transmission systems, a diode laser couples to an optical fiber in order to present light signals to the clear center of the core of the fiber through a focusing lens. The focused light signals naturally criss-cross and dance around the edges of the optical fiber, which causes the individual light pulses to lose their shape. This is due to a phenomenon known as "modal dispersion," as different portions of the light signal take different paths, or "modes," as the transverse the fiber. Modal dispersion results in what is known as "pulse dispersion," which describes the overlapping of the individual light pulses at the distant, or receiving, end of the fiber. The bottom line is that the light detector may not be able to distinguish between the individual pulses, and the integrity of the datastream is compromised. Collimation mitigates this effect by lining up the highly-divergent elements of the light beam so that it travels in parallel, rather than criss-crossing. The resulting light beam is relatively uniform, which improves the integrity of the datastream over a distance. Collimation is accomplished by coupling lenses known as "collimators," which fit inside the steel barrels of optical fiber connectors. Collimators can be positioned at the light source, the coupling point between the light source (i.e., transmitter) and the optical fiber, and the coupling point between the optical fiber and the light detector (i.e., receiver). See also Fiber, Modal Dispersion, Pulse Dispersion.

collimator A collimator is an assembly that is used to straighten and make parallel diverging light as it exits a fiber. See Collimation.

Colline du TOxC7lOxC7graphe French for "Telegraph Hill." The name was given to various hills in France where optical telegraphs were located. Telegraph Hill in San Francisco was so named for the same reason.

collinear antenna A cellular car antenna which looks like a pigtail, because it has a little curlicue in the middle. The curlicue is not a spring, but a clever bit of electro-mechanical magic known as a phasing network, which allows the antenna to boost the effective power of the transmitter's signal. Typically a collinear cellular car antenna is 13 inches high.

collision The result of two workstations (or PCs) trying to use a shared transmission medium (cable) simultaneously – a local area network, for example. The electrical signals, which carry the information they are sending bump into each other. This ruins both signals and both will have to re-transmit their information. In most systems, a built in delay will make sure the collision does not occur again. The whole process takes fractions of a second. Collisions in LANs make no sound. Collisions do, however, slow a LAN down. See Aloha, Collision Detection, Collision Domain, Contention and CSMA/CD.

collision detection The process of detecting that simultaneous (and therefore damaging) transmission has taken place. Typically, each transmitting workstation that detects the collision will wait some period of time and try again. Collision detection is an essential part of the CSMA/CD access method. Workstations can tell that a collision has taken place if they do not receive an acknowledgement from the receiving station within a certain amount of time (fractions of a second). See Aloha, Collision Domain, Contention and Ethernet.

collision domain Today's most popular local area networking topology is Ethernet. To communicate, Ethernet uses the Carrier Sense Multiple Access/Collision Detect (CSMA/CD) protocol. What this means is that each computer listens to what's going on on the network when it wants to transmit. If no other computer is talking, it will start to send its message. If two computers start to send their messages at the same time, a collision of the messages they're trying to send happens and both stop for a random amount of time and then try to transmit their messages again after that period elapses. This way of controlling traffic on a local area network works well in a lightly-trafficked network. But as traffic builds up (as it always does), it's a problem. Think of a telephone party line. The more people are on, the harder for everyone to get to speak. To control this problem we can break our one-wire Ethernet network into pseudo-separate networks – what are known as collision domains. We insert some intelligent electronics at the entrance to every pseudo-network. That electronics acts as an intelligent traffic cop, keeping traffic that is meant for Joe's PC which resides on subnetwork A away from all the other subnetworks. In this way, we have fewer machines trying to use one wire (i.e. the whole Ethernet LAN), and thus there'll be fewer collisions. Clearly the equipment to do is more costly than an Ethernet hub (which is, in essence, a giant teleconference bridge), since this equipment has to be intelligent, recognizing which messages are to go where. The first equipment that was tried was Bridges. Bridges did not change the packets at all but learned what machines were on which sides and let traffic flow when it needed to. Later on, Bridges were replaced with routers. Routers are able to look at where packets are coming from and where they are going and determine which way to send them on. (Remember that traffic on any particular segment goes to the entire segment.) These routers would act as gatekeepers to allow only the traffic destined outside the segment to pass through. Multiple network cards in the network server allow the server to function as a router and perform these functions without much cost or overhead. Routers have more intelligence than bridges and can broadcast information to other routers to find the best possible route for a particular data conversation. Routers do modify your packet in terms of the addresses so it can change the path of the packet. That also adds more latency (delay) as now each packet has more processing done to it as it goes through the network.

A few years ago switches came on the market. Switches replaced some or all of the hubs and broke the network into segments. Switches are really a marketing term for bridges. They store and forward the packets without modification and divide your network into multiple segments. The big buzz right now is level 3 switches. Level 3 switching refers to the layer of the OSI model also known as the networking layer. This adds intelligence to the switch to see where the packet is coming from and going to. This happens to be the same way that routers work. In a lightly loaded network switches will slow things down because of latency in getting the packet processed. They make it more difficult to monitor your network using a protocol analyzer. They can cause timeout or expiration issues if you add too many hops into your network. See Ethernet.

collision window The time it takes for a data pulse to travel the length of the

network. During this interval, the network is vulnerable to collision.

collocation See Colocation, which is my preferred way of spelling this term.

colocation Colation occurs when a competing local phone company (often called a CLEC, Competitive Local Exchange Carrier or an Other Common Carrier) locates (i.e. puts) its switches within an incumbent local exchange carrier's (ILEC) central office. An ILEC is the dominant phone carrier within a geographic area as determined by the FCC. Section 252 of the Telecommunications Act 1996 defined Incumbent Local Exchange Carrier as a carrier that, as of the date of enactment of the Act, provided local exchange service to a specific area. The Act provided that the Commission may treat "comparable carriers as incumbents" if they either "occupy a position in the market for telephone exchange service within an area that is comparable to the position occupied by the ILEC or such a "carrier that has substantially replaced an ILEC...." or if "such treatment is consistent with the public interest..."

There are basically two types of colocation – adjacent / physical, and virtual. Adjacent and physical are the same. They mean that your equipment sits in the same building as the ILEC's switching and cable termination equipment. Typically it sits in a locked cage. Only the CLEC and its personnel have the key. Access to that equipment is negotiated between the ILEC and the CLEC through an interconnection agreement. The CLEC will have 24x7 access to it because they have to maintain it for customers who expect service 24 hours a day seven days a week. The concept of colocation – at this stage a peculiarly North American idea – came about through the Telecommunications Act of 1996. U.S. It was a federal bill signed into law on February 8, 1996 "to promote competition and reduce regulation in order to secure lower prices and higher quality services for American telecommunications consumers and encourage rapid deployment of new telecommunications technologies." The Act required local service providers in the 100 largest metropolitan areas of the United States, the Baby Bells, to implement Local Number Portability by the end of 1998. The Act also allowed the local regional Bell operating phone companies into long distance once they had met certain conditions about allowing competition in their local monopoly areas – thus the concept of colocation. That's what the word colocation means. OK, now to the real stuff – how to spell it. Several readers have complained that in previous editions of this dictionary it was spelled "colocation." They point out that their non-technical English language dictionaries spell it collocation – with two "l"s. Random House Dictionary says that back in 1505-15 the word collocation appeared and was based on the Latin collocatus, which derives from collocare. But Random House also includes a spelling from the era of 1965-1970, which it spells colocate and defines as to locate or be located in jointly or together, as two or more groups, military units, or the like; share or designate to share the same place. My preference is colocation, since it seems to me a logical shortening of co-location. But I'm not arguing. Choose which spelling you'd like. See also Carrier Hotel, CLEC, ELEC, ILEC, and Virtual Colocation.

colocation cage A cage in a central office that is erected by the ILEC and rented to a CLEC. CLEC personnel can access and maintain the equipment in the cage.

colophon Did you ever notice a paragraph at the end of a book describing the typefaces used, the production methods, and so forth? That little paragraph is called a colophon.

color See CMYK.

color code A color system for circuit identification by use of solid colors, contrasting stripes, tracers, braids, surface marking, etc.

color difference signal The first step in encoding the color television signal. The color difference signals are formed by subtracting the luminance information from each primary color: red, green or blue. Color difference conventions include the Betacam format, the SMPTE format, the EBU-N10 format and the MII format.

color model A technique for describing a color (see CMY, HSL, HSV, and RGB).

color picture signal The electrical signal which represents complete color picture information excluding all the synchronizing signals.

color space inversion A video compression technique which reduces the amount of color information in each of a series of still images. It is based on the fact that the human eye is not highly sensitive to variations in color.

color subcarrier The 3.579545 MHz subcarrier that carries the chrominance information of the television signal. This signal is superimposed on the luminance level. Amplitude of the color subcarrier represents saturation and phase angle represents hue.

color temperature Selecting the color temperature determines the overall color cast of a display. 9,300 degrees Kelvin is good in environments lit by fluorescent lights, 6,500 degrees Kelvin is preferable under incandescent light.

colorless An optical networking term meaning "wavelength-independent" or "able to work with any wavelength."

COLT Cell Site On Light Truck. A temporary, mobile wireless antenna base station installed on a light truck. The truck is parked at a location in order to temporarily increase either the coverage or the traffic capacity of the wireless network. I first heard the expression when I heard that BellSouth was using them to help expand its communications capacity in Atlanta for the 1996 Summer Olympics. See also COW (Cell Site on Wheels).

Coltan A mineral that is found in eastern Congo, Africa and Australia. Coltan is three times heavier than iron, slightly lighter than gold. Once coltan is refined it becomes tantalum, a metallic element that is a superb conductor of electricity, highly resistant to heat. Tantalum powder is a vital ingredient in the manufacture of capacitors. Capacitors made of tantalum can be found inside almost every laptop, pager, personal digital assistant and cell phone.

column A database definition: The logical equivalent of a field, a column contains an individual data item within a row or record.

COM 1. Continuation of Message: An indicator used by the ATM Adaptation Layer to indicate that a particular ATM cell is a continuation of a higher layer information packet which has been segmented.

2. Component Object Model. COM is Microsoft's cornerstone of its ActiveX platform. COM is a language independent component architecture (not a programming language). It is meant to be a general purpose, object-oriented means to encapsulate commonly used functions and services. The COM architecture provides a platform independent and distributed platform for multi-threaded applications. COM also encompasses everything previously known as OLE Automation (Object Linking and Embedding). OLE Automation was originally for letting higher level programming languages access COM objects. An object is a set of functions collected into interfaces. Each object has data associated with it. The source of the data itself is called the data object. With COM, the transfer of the data itself is separated from the transfer protocol. See ActiveX and Windows Telephony.

3. A type of Internet domain assigned to URLs which are business or commercial entities (for example, www.bidworld.com). There is also .edu, .gov, .net, and .org. See Domain.

COM Port The communications port on a PC, a workstation, server, or other DTE (Data Terminal Equipment). This port is sometimes referred to as the serial, RS-232, DB-9 or DB-25 port (depending on if it has nine or 25 pins).

Combat Net Radio CNR. A radio operating in a network, providing a half-duplex circuit employing a single radio frequency or a discrete set of radio frequencies (frequency hopping). Combat net radios are primarily used for command and control of combat, combat support, and combat service support operations between and among ground, naval, and airborne forces.

combination network A network designed to combine two or more input channels into one output channel. See Combiner.

combination system An alternative to upgrading older telephone equipment, combination systems make it possible to add network-based features to an equipment-based telephone system.

combination trunk A central office trunk circuit that supports both incoming and outgoing traffic. Traditionally, PBX trunks are directional. That is to say that they are incoming only, outgoing only, or combination. More recently, the individual channels in a T-1 PBX trunk can be defined in terms of their directional nature, with some being defined as DID (Direct Inward Dialing), some as DOD (Direct Outward Dialing), and some as combination (both incoming and outgoing). See also DID, Direct Inward System Access, and DOD.

Combined Distribution Frame CDF. A distribution frame that combines the functions of main and intermediate distribution frames. The frame contains both vertical and horizontal terminating blocks. The vertical blocks are used to terminate the permanent outside lines entering the station. Horizontal blocks are used to terminate inside plant equipment. This arrangement permits the association of any outside line with any desired terminal equipment. These connections are made with jumpers. These connections are made with twisted pair wire, normally referred to as jumper wire, or with optical fiber cables, normally referred to as jumper cables. In technical control facilities, the vertical side may be used to terminate equipment as well as outside lines. The horizontal side is then used for jackfields and battery terminations.

Combined Station MDLC station containing both a primary and a secondary and used in asynchronous balanced mode.

combiner 1. A network for putting two or more frequency bands or channels together for transmission along a single line.

2. A passive device in which optical power from several input fibers is collected at a common point.

Comcode AT&T's old numbering system for telecom equipment, replacing older KS-prefix numbers, that supplements standard industry part designations. Comcode No. 102092848 is touchtone Princess phone with a transparent plastic housing. See also KS

Number.

COMDEX COMDEX was once, the largest computer show of the year. It got its name from COMputer DEalers eXposition. It was held twice a year in the U.S. and many times overseas. Nearly three million people attended COMDEX Fall since 1979 when the show was launched with a first-year attendance of 4,000. Attendance at the Las Vegas show peaked in the late 1990s at approximately 250,000. The show was so important that IBM first showcased their new PC during COMDEX 1981. Alas, COMDEX has fallen on hard times, as have most conferences and exhibitions in the technology space. The November 2004 event was cancelled due to lack of interest. I believe Comdex continues to be held in some cities overseas.

ComForum The Network Reliability ComForum was a gathering of senior telecommunication industry executives and government officials that served on the Network Reliability Council. The ComForum used to meet to report on the Council's findings. Apparently Comforum no longer exists. See USITA.

comfort noise Comfort tone means the same as comfort tone. See comfort tone.

comfort tone Michael Boom heard this one at an engineering meeting at Ascend Corporation, where an engineer was describing digital phone lines. It turns out that many phone connections are so clean now that there is no background noise at all. The phone customer, hearing absolutely nothing, believes the connection is broken and hangs up. The solution? A "comfort tone." A comfort tone is a very low-level synthesized white noise deliberately added to a digital line to give a comforting "hiss" to the connection. It assures customers that there is indeed a connection and gives the voices on the other end that slightly distant quality the person making the telephone call expects to hear.

COMINT COMunications INTelligence.

Comité Consultatif International des Radiocommunications CCIR. The agency responsible for the international use of the radio spectrum. Effective in 1993, the CCIR is now known as the International Telecommunications Union – Radio. ITU-R and ITU-T form the International Telecommunications Union. See ITU for a much longer explanation.

comma-free code A code constructed such that any partial code word, beginning at the start of a code word but terminating prior to the end of that code word, is not a valid code word. The comma-free property permits the proper framing of transmitted code words, provided that: (a) external synchronization is provided to identify the start of the first code word in a sequence of code words, and (b) no uncorrected errors occur in the symbol stream. Huffman codes (variable length) are examples of comma-free codes.

Comma Separated Values See CSV.

command See Command Set.

Command And Control C2. The exercise of authority and direction by a properly designated commander over assigned forces in the accomplishment of the mission. Command and control functions are performed through an arrangement of personnel equipment, communications, facilities, and procedures employed by a commander in planning, directing, coordinating, and controlling forces and operations in the accomplishment of the mission.

command and control system The facilities, equipment, communications, procedures, and personnel essential to a commander for planning, directing and controlling operations of assigned forces pursuant to the missions assigned. See Command, Control and Communications.

command buffer A segment of memory used to temporarily store commands. The command buffer only holds a copy of the last command issued.

command conference system A conference calling arrangement in a Northern Telecom PBX which allows a designated phone to originate a conference to and between a group of PBX extensions. Any phone that is busy when the conference begins is automatically connected to the conference as soon as that phone becomes free.

Command, Control And Communications C3. The capabilities required by military commanders to exercise command and control of their forces.

command interpreter The operating system that controls a computer's shell. The command interpreter for MS-DOS is COMMAND.COM. The command interpreter for Windows is WIN.COM.

command line The line on the screen, in MS-DOS, where the cursor is. The command line is where you enter MS-DOS commands.

Command Line Interpreter CLI. A Rolm user interface to the CBX software and used for things like testing.

command net A communications network which connects an echelon of command with some or all of its subordinate echelons for the purpose of command control. See C2 and C3.

command path The list of path names that tells MS-DOS where to look for files that aren't in the current directory.

command port In network management systems an interface used to monitor and control the system.

command processor The MS-DOS program, COMMAND.COM, that contains all MS-DOS's internal commands, like DIR, ERASE and REName.

command prompt The MS-DOS command prompt appears on the screen as the default drive letter followed by a greater than > sign. The command prompt lets you know MS-DOS is ready to receive a command.

command pulses Pulses transmitted from a control device (such as a dialer) to either a direct control switch or to an intermediate device (e.g., a pulse register).

Command Response C/R. A Frame Relay term defining a 1-bit portion of the frame Address Field. Reserved for the use of FRADs, the C/R is applied to the transport of data involving polled protocols such as SNA. Polled protocols require a command/response process for signaling and control during the communications process.

command save A Rockwell ACD term. The introduction of a new demand command defines up to 10 commands per terminal position to enhance the productivity of both IST and non-IST supervisors.

command set In computer telephony, a command set is a collection of special software instructions that do special jobs. These software instructions are often called function calls. For example, the command M_Make_Call (plus parameters) tells Northern's Norstar phone system to have telephone set number 21 dial a phone number. Northern's Norstar and other open phone systems (those that can be commanded by an external computer which you and I can program) all have their own command sets. Each command set is made up of function calls with funny words like M_Make_Call. Typically those function calls work in C, a common software language. A function call will reach into the specialized driver that controls the phone system (an exact analogy is the driver that drives a laser printer) and get the phone system to do something. A programmer must use these function calls if he/she wants to control the phone system from software. Exactly how M_Make_Call works is typically not revealed to the programmer. That keeps the manufacturer's technology proprietary and secret. It also saves the programmer the time and expense of writing the driver.

Command.com The program that carries out MS-DOS commands. The generic term for this program is command interpreter.

COMMDesk Banker A communications software package offered by MCI International that provides all the capabilities of COMMDesk, plus a security feature essential for financial transactions.

COMMDesk Manager An MCI definition. A communications software package designed to run on an IBM PC/XT/AT or compatible that gives the user full access to all MCI International and MCI communications services.

comment In the world of blogs, a comment is a response left on the author's blog regarding one of her blog entries. When you leave a comment, you are starting a dialog by leaving a response on a blog entry.

comments Let's say you're writing some software – whether it be something as simple as an HTML web site or something as complex as C++. You'll always want to put some words inside your code so that you (and others who come after you) will understand what you were trying to do. In the HTML "language," such remarks are called "comments." (In other languages they're called other things, like "remarks." You'll recognize comment lines in HTML because they look like this:

    <\\>!– Your comment in text goes here –>

Commercenet A not-for-profit industry association that works to accelerate the development of electronic commerce. In conjunction with the World Wide Web Consortium (W3C), Commercenet developed the Joint Electronic Payments Initiative (JEPI), an E-Commerce standard. The approximately 500 members of CommerceNet include leading banks, telecommunications companies, VANs, ISPs, online service, software and services companies, as well as major end users. See also Electronic Commerce and JEPI. www.commerce.net.

Commercial Building Telecommunications Wiring Standard In 1985, the Electronic Industries Association undertook the task of developing a standard for commercial and industrial building wiring. Approved and published on July 9, 1991, the EIA/TIA-568 "Commercial Building Telecommunications Wiring Standard" defines a generic wiring system which will support a multiproduct, multivendor environment and which will have a useful life of over 10 years. The EIA/TIA standard is based on star topology in which each workstation is connected to a telecommunications closet situated within 90 meters of the work area. Backbone wiring between the communications closets and the main cross-connect is also organized in a star topology. However, direct connec-

tions between closets are allowed to accommodate bus and ring configurations. Distances between closets and the main cross-connect are dependent on backbone cable types and applications. Each workstation is provided with a minimum of two communications outlets (which may be on the same faceplate). One outlet is supported by a four-pair, 100-ohm unshielded twisted-pair (UTP) cable. The other may be supported by (a) an additional four-pair UTP cable, (b) a two-pair, 150 ohm shielded twisted pair (STP) cable or (c) a two-fiber 62.5 /125 micron fiber optic cable. For more on cabling and cabling components, see Category OF Performance.

commercial internet The part of the Internet provided by commercial services. Allows business usage of the Internet without violating the appropriate usage clause of the National Science Foundation NETwork (NSFNET), who actually runs the Internet.

Commercial Internet Exchange Association This is a non-profit trade association for public data internetworking service providers. www.cix.org/cixhome.html.

Commercial Mobile Radio Service CMRS. CMRS is an FCC designation for any carrier or licensee whose wireless network is connected to the public switched telephone network and/or is operated for profit.

Committed Access Rate See CAR.

committed burst size Bc. A Frame Relay term defining the maximum data rate that the carrier agrees to handle over a subscriber link under normal conditions. See also Offered Load, Committed Information Rate, Discard Eligibility, and Excess Burst Size.

Committed Information Rate CIR. A Frame Relay term identifying the user's commitment to a certain average maximum data transmission rate. The monthly bill a customer receives may include at least two elements, depending on the pricing algorithm employed by the carrier; those charges can include a charge for the CIR and a surcharge for usage above the CIR. Usage above the CIR may be measured in terms of the Burst Size and Burst Interval. Usage above the CIR may be subject to discard in the event of network congestion. In the early days of frame relay when few people were using the service, customers were opting for a low Committed Information Rate, thus keeping their bills low, but knowing that they could always get their transmissions through – because there were few other people on the service. As the service got more popular, many customers found they had to hike their Committed Information Rate if they wanted to get their information through. For a more technical explanation, see CIR.

commodus See Sensitivity Training.

common audible The same as Common Bell. Ringer wiring is such that ringing occurs on more than one CO or PBX line.

common audible ringer A loud ringer connected to a phone line in a noisy area. When the phone rings, the loud ringer also rings.

common battery A battery (or several batteries) that acts as a central source of energy for many pieces of equipment. A common battery provides 48 volts of power to a central office switch and to all the phones connected downstream.

common battery signaling A system in which the signaling power of a telephone is supplied by the battery at the servicing switchboard. Switchboards may be manual or automatic, and "talking power" may be supplied by common or local battery.

common bell A bell or ringer which sounds when any of the lines terminating on that phone rings. A term harking back to 1A2 key system days.

Common Business Line CBL. An option with 800 Service that has been replaced by 800 Business Line.

common carrier A phone company which sets itself up to carry communications service for any one who wants to buy that service. In the old days if you were a common carrier you had to file your prices at the various regulatory agencies and they had to pass on them. These days most telecommunications carriers don't need to file their prices and virtually all telephone companies are common carriers. So the definition is moot. The distinction applies more to companies like trucking companies.

commercial internet exchange association A company that furnishes communications services to the general public. It is typically licensed by a state or federal government agency. A common carrier cannot refuse to carry you, your information or your freight as long as you conform to the rules and regulations as filed with the state or federal authorities. See Other Common Carrier.

Common Carrier Bureau A department of the Federal Communications Commission responsible for recommending and implementing regulatory policies on interstate and international common carrier (voice, video, data) activities.

Common Channel Interoffice Signaling CCIS. A way of transmitting all signaling information for a group of trunks by encoding that information and transmitting it over a separate channel using time-division digital methods. By transmitting that signaling information over a separate channel, CCIS saves huge long distance bandwidth, which in the past was used to switch calls across the country only to find a busy signal and then come all the way back again to signal the calling party a busy. For the biggest explanation of common channel signaling, see Signaling System 7. See also MTP, SCCP, ISUP, ISDN and TCAP.

common channel signaling This is a Bellcore definition: A network architecture which uses Signaling System 7 (SS7) protocol for the exchange of information between telecommunications nodes and networks on an out-of-band basis. It performs three major functions: 1. It allows the exchange of signaling information for interoffice circuit connections. 2. It allows the exchange of additional information services and features, e.g. CLASS, database query/response, etc. 3. It provides improved operations procedures for network management and administration of the telecommunications network. For the biggest explanation of common channel signaling, see Signaling System 7. See also ISDN, ISUP, MTP, SCCP, STP and TCAP.

common channel transit exchange An intermediate exchange where networking of common channel signaling systems occurs.

common control A method of telephone switching in which the central logic system (or control equipment) is responsible for routing calls through the network. The control equipment is connected with a given call only for the period required to accomplish the routing function. In other words, the common control equipment is associated with a given call only during the periods required to accomplish the control functions. All crossbar and electronic switching systems have common control.

common control equipment An automatic switching system that makes use of common equipment to establish a connection. Once the connection is made, the common control equipment is available to establish another connection.

Common Control Switching Arrangement CCSA. An AT&T offering for very big companies. Those big companies can create their own private networks and dial anywhere on them by dialing a standard seven digit number, similar to a local phone number. The corporate subscriber rents private, dedicated lines and then shares central office switches. CCSA uses special CCSA software at the central office.

common costs Costs of the provision of some group of services that cannot be directly attributed to any one of those services.

common equipment In telephone systems common equipment are items that are used by several or all phones for processing calls. On a key system, the device that permits a light on any instrument to flash on and off may be common equipment when used to control all lights on all instruments.

Common Gateway Interface See CGI.

Common Industrial Protocol CIP. A protocol that provides real-time I/O and messaging, and peer-to-peer messaging. CIP is used in industrial automation environments, where real-time systems are required. ControlNet and DeviceNet networks use CIP.

common intermediate format A videophone ISDN standard which is part of the ITU-T's H.261. It produces a color image of 352 by 288 pixels. The format uses two B channels, with voice taking 32 Kbps and the rest for video.

Common Internet File System CIFS. Common Internet File System. A remote collaborative file sharing technology that, according to Microsoft, dramatically reduces the time it takes to open and work with remote files. According to Microsoft, the Common Internet File System (CIFS) is an enhanced version of the native file-sharing technology used in the Microsoft MS-DOS, Windows and Windows NT operating systems and IBM OS/2, and widely available on leading UNIX systems. It enables users to open and share remote files directly on the Internet, expanding the Internet's ability to support interactive network computing. CIFS technology provides reliable direct read and write access to files stored on remote computers without first requiring users to download or copy the files to a local machine, as done previously on the Internet. Because CIFS is based on existing standards and supports the Internet's Domain Name Service (DNS) for address resolution, users are able to use existing applications over the Internet as well as integrate them with browser applications designed for the World Wide Web. In addition to remote file sharing, CIFS has mechanisms to support remote printer sharing.

CIFS runs over TCP/IP and is an enhanced version of the earlier open, cross-platform protocol for distributed file sharing called Server Message Block (SMB). The SMB protocol is the native file-sharing protocol in Windows 95, Windows NT and OS/2.

For all of its advantages, CIFS operates sluggishly over a WAN, on account of its 'chattiness,' meaning a large number of back-and-forth transactions are required to complete a file-access request over a network. Furthermore, the largest block of data that CIFS can transfer in a single trip between server and client, or vice-versa, is 61 KB. Each CIFS request requires a response before the next request is sent. Therefore, in order to transfer a single 10 MB file across a WAN, the CIFS protocol would have to make 163 roundtrips between client and server. On a typical LAN this would take only seconds, but on a WAN where bit rates are much slower and latencies are much higher, the same file transfer could likely take several minutes. This is one of the reasons why WAN acceleration devices are popular among enterprise network administrators. A WAN accelerator with support for CIFS can cut file transfer times by two-thirds or even more.

common language code Codes used to ensure uniform abbreviation of equipment and facility names, place names, etc.

Common Language Location Identification Code CLLI. The CLLI code is an 11 character mnemonic code used to uniquely identify a location in the United States, Canada or other countries. These codes are known as CLLI or 'Location Codes' and may be used in either a manual or mechanized record keeping system. For a bigger explanation, see CLLI Code. See also CFA.

common line A subscriber line, PBX trunk, paystation line, or other cable facility provided under an ILEC's local general service tariffs, that is terminated on a central office switch. A common line-residence is a line or trunk provided under the residence regulations of the local general services tariffs. A common line-business is a line provided under the business regulations of the local general services tariffs.

common mail calls New APIs (Application Programming Interfaces) from Microsoft which allow you to move information around your various mail services – the ones on your LAN, on your wireless pager, etc.

common mode The potential or voltage that exists between neutral and ground. Electronic equipment requires this to be as close to 0 volts as possible or not to exceed 1/2 volt. For AC power systems, the term common mode may refer to either noise or surge voltage disturbances. Common mode disturbances are those that occur between the power neutral (white wire) and the grounding conductor (green wire) Ideally, no common mode disturbances should exist since the neutral and grounding wires are always connected at the service distribution panel in most countries. However, unwanted common mode disturbances exist as a result of noise injection into the neutral or grounding wires, wiring faults, or overloaded power circuits. Modern computers are quite immune from common mode noise. Common mode noise is frequently mistakenly confused with inter-system ground noise, a distinct problem which frequently causes computer damage and data errors. See Common Mode Interference.

common mode interference 1. Interference that appears between signal leads, or the terminals of a measuring circuit and ground.

2. A form of coherent interference that affects two or more elements of a network in a similar manner (i.e., highly coupled) as distinct from locally generated noise or interference that is statistically independent between pairs of network elements.

Common Mode Rejection Ratio CMRR. The ratio of the common mode interference voltage at the input of a circuit to the corresponding interference voltage at the output.

common mode transmission A transmission scheme where voltages appear equal in magnitude and phase across a conductor pair with respect to ground. May also be referred to as longitudinal mode.

common mode voltage 1. The voltage common to both input terminals of a device.

2. In a differential amplifier, the unwanted part of the voltage between each input connection point and ground that is added to the voltage of each original signal.

Common Object Request Broker Architecture See CORBA.

Common Open Policy Service See COPS.

common path distortion Common path distortion is the interference of return-path signaling caused by the forward path.

common peer group An ATM term. The lowest level peer group in which a set of nodes is represented. A node is represented in a peer group either directly or through one of its ancestors.

common return A return path that is common to two or more circuits and that serves to return currents to their source or to ground.

common return offset The dc common return potential difference of a line.

common sense "Why would any person want to use this ungainly and impractical device when he can send a messenger to a local telegraph office and have a clear written message sent to any large city in the United States?" Excerpt from a report to the President of Western Union written by the committee charged with investigating the potential purchase of Bell's telephone patent for $100,000. Western Union rejected the opportunity, and of course, passed on one of the greatest business opportunities in the history of business.

common trunk In telephone systems having a grading arrangement, a trunk accessible to all groups of the grading.

common user circuit A circuit designated to furnish a communication service to a number of users.

common user network A system of circuits or channels allocated to furnish communication paths between switching centers to provide communication service on a common basis to all connected stations or subscribers.

commonality 1. A quality that applies to material or systems: (a) possessing like and interchangeable characteristics enabling each to be utilized, or operated and maintained by personnel trained on the others without additional specialized training; (b) having interchangeable repair parts and/or components; (c) applying to consumable items interchangeably equivalent without adjustment.

2. A term applied to equipment or systems that have the quality of one entity possessing like and interchangeable parts with another equipment or system entity.

communicating applications A General Magic term. An application whose design presupposes the user's desire to send and receive messages. For a Personal Intelligent Communicator to be effective, it needs to be equipped with a suite of communicating applications. All Magic Cap applications are built to communicate.

communicating applications platform A General Magic term. The Cap in Magic Cap. Software on which Personal Intelligent Communicators are based. It is designed to make it easy for developers to create communicating applications and services. Magic Cap can run on dedicated devices as well as other computer operating systems.

Communicating Applications Specification A facsimile specification. See CAS 2.0.

communicating objects A term created in the fall of 1992 by Mitel's VP Tony Bawcutt for a new Mitel division which specializes in making PC printed cards and software drivers and developer tools for those cards. Those cards are designed to be the building blocks of what Mitel calls multimedia applications – but what are more properly called PC-based voice and call processing telecom developer building blocks. One of the first cards Mitel introduced was an ISDN S-access card which converts PCs into ISDN telephones, also called voice and data workstations.

communicating word processor A dedicated word processor that includes software for sending word processed files over phone lines. Communicating word processors have now largely been replaced by PCs (Personal Computers) running word processing programs and asynchronous communications software programs.

communication channel A two-way path for transmitting voice and/or data signals. See also Circuit.

communication controller Another name for a Front End Processor, a specialized computer which was common in 3270 data communications networks. The FEP acted as a data communications "traffic cop," removing the communications traffic routing and controlling burden from the mainframe computer which lay behind the FEP. In short, the FEP designates a device placed between the network and an input/output channel of a processing system (i.e. the computer).

communication endpoint An ATM term. An object associated with a set of attributes which are specified at the communication creation time.

communication holder A fancy name for a pocket on a piece of clothing that holds a cell phone. I first saw the word on a Nike Sno tech Jacket. The jacket also contains a small channel through which you can thread the wires to a compact headset you can wear. This way you can speak on the phone without touching the phone.

communication port A port on a computer that allows asynchronous communication of one byte at a time. A communication port is also called a serial port.

communication server A dedicated, standalone computer that manages communications activities for other computers.

communication service monitor A multi-functional instrument that is used in the testing and repair of radio transmitters, radio receivers, radio transceivers and radio communication systems. A communication service monitor has a signal generator, sensitive receivers, SINAD meter, frequency counter, frequency error meter, oscilloscope and other built-in instrumentation to help test radio equipment and diagnose problems with it. See SINAD.

Communication Workers Of America CWA. A national union of telephone industry employees, currently very worried about its future membership growth given the phone industry's propensity to let its surplus workers go.

Communications Act Of 1934 Federal legislation which established national telecommunications goals and created the Federal Communications Commission to regulate all interstate and international communications.

Communications Act of 1996 It is really called the Telecommunications Act of 1996. It is a federal bill signed into law on February 8, 1996 "to promote competition and reduce regulation in order to secure lower prices and higher quality services for American telecommunications consumers and encourage rapid deployment of new telecommunications technologies." The Act, amongst other things, allowed the local regional Bell operating phone companies into long distance once they had met certain conditions about allowing competition in their local monopoly areas. You can download a copy of this Act (all 391,861 bytes) from http://www.fcc.gov/Reports/tcom1996.pdf. See also CLEC, the Communications Decency Act of 1996 and the Telecommunications Act of 1996.

Communications Assistance to Law Enforcement Act CALEA. A law passed in 1994, CALEA grants law enforcement agencies the authority to wiretap digital networks and requires wireless and wireline carriers to enable eavesdropping equipment to be used in digital networks.

Communications Decency Act of 1996 CDA. An element of the Telecommunications Act of 1996, the CDA provided for penalties of as much as $250,000 and 2 years in jail for U.S. citizens who transmit indecent material that minors could access by computer. Targeted at those who would make such material available over the Internet through Web sites, this element of the act was blocked in June 1996 by a panel of federal judges on the basis of successful arguments that the provision violated the right of free speech, as guaranteed in the Constitution. The Supreme Court in mid-1997 said the Communications Decency Act was unconstitutional. The Supreme Court handed down a seven-to-two decision that upheld a lower-court ruling against the CDA. The case, Reno vs. ACLU, was the first time that the Supreme Court has dealt with issues involving the Internet. See Telecommunications Act of 1996.

communications adapter Device attached to an IBM System 3X computer or an IBM PC that allows communications over RS-232 lines.

communications control character A character intended to control or help transmission over data networks. There are ten control characters specified in ASCII which form the basis for character-oriented communications control procedures.

communications/modem server In a network, a server equipped with a bank of modems, which can be shared by users for outgoing calls.

communications overload What a person experiences when he feels "over-connected," i.e., in a state of having to constantly respond to wireline and wireless phone calls, SMS alerts, IM alerts, email, and voicemail.

communications parameters Any of several settings required to allow computers to communicate successfully. In asynchronous transmissions, commonly used in modem communications, the settings for baud rate, number of data bits, number of stop bits, and parity parameters must all be correct.

communications protocol Procedures which are employed to ensure the orderly transfer of data between devices on a communications link, over a communications network, or within a system. The major functions of a protocol are those of Handshaking and Line Discipline. Handshaking is a specific sequence of data exchange between devices over a circuit. This initial step establishes the fact that the circuit is operational, establishes the level of device compatibility, determines speed of transmission, and so on. Line discipline is a sequence of operations which includes transmission and receipt of data, error control, and sequencing of message sets (e.g., characters, blocks, packets, frames and cells). Line discipline also includes error detection and correction processes, providing for the confirmation or validation of data received, and by implication, the failure to receive data sets.

communications satellite A satellite circling the earth, usually at a distance of about 22,000 miles, with electronic equipment for relaying signals received from the earth back to other points on the earth. See Geostationary Satellite.

communications server A communications server is a machine with intelligence that does communications tasks. Exactly which tasks it does depends on your idea of what a communications server is. It's a loose term. Here are three definitions of what people do call a communications server:

1. Its earliest configuration was a device that handled communications from several PCs and allowed those PCs access to printers, modems, etc.

2. A communications server is also a type of gateway that translates the packetized signals of a LAN to asynchronous signals, usually used on telephone lines or on direct connections to minicomputers and mainframes. It handles different asynchronous protocols and allows nodes on a LAN to share modems or host connections. Usually one machine on a LAN will act as a gateway, sharing its serial ports or an RS-232 connection to a minicomputer. All devices on the LAN can use this machine to get to the modems and the minicomputer.

3. A communications server is also the new name for an UnPBX. See UnPBX.

communications settings Settings that specify how information is transferred from your computer to a device (usually a printer or modem).

communications system engineering The translation of user requirements for the exchange of information into cost-effective technical solutions of equipment and subsystems.

communications toolbox An extension of the Apple Macintosh operating system that provides protocol conversion and the drivers needed for communications tasks.

Communications Trouble and Analysis Center CTAC. A Verizon center that provides a single point of contact for wholesale customers. The CTAC is the administrative center for monitoring and dispatching trouble calls, and for resolving usage, billing and connectivity issues.

Communications Workers of America CWA. The main labor union of the RBOCs. www.cwa.org.

communications zone A military term: Rear part of theater of operations (behind but contiguous to the combat zone), which contains the lines of communications, establishments for supply and evacuation, and other agencies required for the immediate support and maintenance of the field forces.

communicator A British term. An alternative, and probably more meaningful, name for a telebusiness agent. A communicator is called a telemarketer in North America.

community For the purposes of the FCC's cable television rules, when this term is used to specify the location of a transmitter or an antenna structure, it generally includes any named, urbanized area, without regard to size.

Community Antenna Television CATV. Signals from distant TV stations are picked up by a large antenna, typically located on a hill, then amplified and piped all over the community below on coaxial cable. That's the original definition. See CATV for a more up-to-date definition.

Community Dial Office CDO. A type of central telephone switching office that is most often found in small rural communities. It is an unattended switching center that is serviced only as needed, and maintained on an occasional basis by a traveling maintenance technician.

community name Community name is a password shared by a Network Agent and the Network Management Station so their communications cannot be easily intercepted by an unauthorized workstation or device.

community of interest A grouping of telephone users that call each other with a high degree of frequency. Often several Communities of Interest exist within an organization. This phenomenon can influence design for service when new switches are planned.

community of license The community to which a broadcast station is licensed, as specified on the station license. The transmitter may or may not be located within the community (indeed, it is frequently located at some distance, and may even be in a different state). A broadcast station is required to provide a specified field intensity over the entire Community of License.

community string A password used with the SNMP protocol, SNMP community strings are used for both read only and read/write privileges. A community string is case sensitive, and may include some punctuation characters.

community unit A discrete geographic area served by a single cable television system, to which a single Community Unit Identification Number has been assigned by the FCC. At a minimum, each franchise area served by one cable operator constitutes one Community Unit; however, a single franchise area may include two or more Community Units in the following situations:

- If a cable television operator serves a community from two different headends, the portion of the community served by each headend constitutes a separate community unit.

- If two cable operators hold separate franchises for the same geographic area, each operator is assigned a separate Community Unit Identification Number.

- If the system serves discrete, unincorporated areas within a township or county pursuant to a township-wide or county-wide franchise, each area may constitute a separate Community Unit.

- If the system serves an incorporated municipality which overlaps two or more counties, the portion of the municipality within each county is considered a separate Community Unit.

Community Unit Identification Number An identification number assigned by the FCC to each Community Unit. The format of the number is: SSNNNN, where:

- SS is the U.S. Postal Service two-letter abbreviation for the state or territory in which the Community Unit is located.
- NNNN = A four-digit serial number assigned by the FCC.

commutator A device used on a dynamo to reverse the connection periodically in

order to cause the current flow in one direction, i.e., to produce direct current.

Compact Disc A standard medium for storage of digital audio data, accessible with a laser-based reader. CDs are 12 centimeters (about 4 3/4") in diameter. CDs are faster and more accurate than magnetic tape for audio. Faster, because even though data is generally written on a CD contiguously within each track, the tracks themselves are directly accessible. This means the tracks can be accessed and played back in any order. More accurate, because data is recorded directly into binary code; mag tape requires data to be translated into analog form. Also, extraneous noise (tape hiss) associated with mag tape is absent from CDs. See CD-ROM and CVD.

Compact Disc Interactive A compact disc format, developed by Philips and Sony, which provides audio, digital data, still graphics and limited motion video. See CD-I.

Compact Wireless Markup Language See CWML.

CompactPCI CompactPCI is a ruggedized variation of the PCI bus, which a bunch of industrial grade PC makers have designed for two reasons: First, to be able to put more PCI cards into one PC (8 versus 4). Second, to make the resulting PC more rugged, i.e. better able to withstand shaking, etc. The physical configuration of the hardware conforms to the Eurocard (VMR-style) standard. The cards, are identical to VME cards in size. They differ, however, in that they use a high density 2mm (contact spacing) pin-and-socket connector for interface to a passive backplane. CompactPCI typically comes in a rugged 3U or 6U Eurocard form factor and has a 32/64 data bus with transfer rates up to 528 megabytes per second. The PC makers told me that "CompactPCI is an adaptation of the Peripheral Component Interconnect (PCI) Specification for industrial and/or embedded applications requiring a more robust mechanism form factor than desktop PCI. CompactPCI uses industry standard mechanical components and high-performance connector technologies to provide a system optimized for rugged applications. CompactPCI is electrically compatible with the PCI Specification, allowing low-cost PCI chipsets to be used in a mechanical form factor suited for rugged environments."

compander See Companding.

companding The word is a contraction of the words "compressing" and "expanding." Companding is the process of compressing the amplitude range of a signal for economical transmission and then expanding it back to its original form at the receiving end.

companion virus A virus which "infects" EXE files by creating a COM file with the same name and contains the virus code. They exploit the MS-DOS property that if two programs with the same name exist, the operating system will execute a COM file in preference to an EXE file.

company code A unique four-place alphanumeric code which must be assigned by NECA (National Exchange Carrier Association) to all U.S. domestic local exchange telecommunications providers. The alphanumeric code is expressed as NNXX, where N=0-9 and X=A-Z. Company Codes are assigned for each type of service provided by a company. Additionally, separate and distinct codes are required for ILECs (Incumbent Local Exchange Carriers), certified facilities-based CLECs (Competitive LECs), local exchange resellers, and wireless carriers. In 1996, NECA assumed this responsibility. Previously, the codes were assigned unofficially from a series of numbers, sensitive to either the type of service provided or the nature of the carrier entity. See also NECA.

compartmentation A military/government term: A method employed to segregate information of different desired accessibilities from each other. It may be used for communications security purposes.

compatible A widely misused and especially vague word, "compatible" has several meanings. Hardware or software systems or components that are capable of working together in harmony (i.e., interoperating smoothly) can be characterized as compatible. The same goes for a product that is deemed by someone (usually a marketing type or a salesperson) to be equivalent to a better known product, as in "IBM-compatible." That means that that product can be used interchangeably. Hardware or software systems or components that fulfill the basic specifications defined in a formal standard are said to be compatible with that standard. In the computer world, two computers are said to be compatible when they will produce the identical result if they run identical programs. Being "compatible" doesn't always assure that the compatible thing will work. Check. See also Compliant, Compatible Sideband Transmission, Conformance Test and Interoperability Testing.

compatibility database A compatibility database is a list of equipment and software that works with whatever the compatibility database is attached to. For example, Windows XP comes with a compatibility database. It tells you if the software application you're trying to install and run on Windows XP has been certified by Microsoft to work with XP. Try to install something not on the list and you get a warning.

compatible sideband transmission That method of independent sideband transmission wherein the carrier is deliberately reinserted at a lower level after its normal suppression to permit reception by conventional AM receivers. The normal method of transmitting compatible SSB (AME) is the emission of the carrier plus the upper sideband.

compelled signaling A signaling method in which the transmission of each signal in the forward direction is inhibited until an acknowledgement of the satisfactory receipt of the previous signal has been sent back from the receiver terminal.

compelling A favorite Microsoft word to describe what it regards as irresistibly great features of a new piece of software it has just created. You typically hear the word from Microsoft presenters at trade shows or seminars.

Competitive Access Provider See CAP.

Competitive Local Exchange Carrier See CLEC.

Competitive Long Distance Coalition A Washington, D.C. lobbying group.

compilation The translation of programs written in a language understandable to programmers into instructions understandable to the computer. Think of programmers writing in every language but Greek and computers understanding only Greek. In this case, Greek is called machine language. The other languages (the programmer languages) are called things like COBOL, FORTRAN, Pascal, dBASE. A compiler is a special program that translates from all these other languages into machine language.

compile To translate a program written in a higher language into machine language so it can be executed by a computer.

compiler A program that takes the source code a programmer has written and translates it into object code the computer can understand. For example, a compiler takes instructions written in a "higher" level language such as BASIC, COBOL or ALGOL and converts them into machine language that can be read and acted upon by a computer. The translated code is in the form of an executable program, which can be run on the target computer without additional translation software. Just-in-time compilers run on the client (i.e., target) machine, translating the code "on the fly." Optimizing compilers used in contemporary high-performance computers also ensure that the translations manage memory (i.e., caching and parallelization) utilization as effectively as possible. Compilers convert large sections of code at one time, while interpreters translate commands one line at a time. See also Bytecode and Interpreter.

complement to fill up or complete. Do not confuse it with compliment, which means too be nice to someone, as in I compliment you on a job well done.

Complementary Code Keying CCK. Modulation technique used by IEE 802.11 compliant wireless LANs for transmission at 5.5 and 11 Mbps. See CCK for a full explanation.

Complementary Metal Oxide Semiconductor CMOS. See CMOS.

Complementary Network Service CNS. CNSs are basic services associated with end user's lines that make it easier for ESPs to offer them enhanced services. Some examples of CNSs include Call Forwarding Busy/Don't Answer, Three Way Calling, and Virtual Dial Tone. See Open Network Architecture.

complete document recognition The ability to perform recognition on documents, retaining as much information as possible about the features and formatting of the original, and including the ability to capture images as well as text.

completed call Careful with this one. In telephone dialect, a completed call is one that has been switched to its destination and conversation has begun but has not yet ended.

completion ratio The proportion of the number of attempted calls to the number of completed calls.

Complex Programmable Logic Device See CPLD.

compliant The term "compliant" is used to refer to a software or hardware component or system that fully conforms, or adheres, to a standard. Compliance is a higher level of conformance than is compatibility. See also Compatible and Conformance Test.

compliment To be nice to someone, as in I compliment you on a job well done. Do not confuse it with complement, which means to fill up or complete.

Complimentary Network Services CNS. The means for an enhanced-service provider's customer to connect to the network and to the enhanced service provider. Complimentary network services usually consist of the customer's local service (e.g.,business or residence line) and several associated service options, e.g., call-forwarding service.

component An element of equipment which unto itself does not form a system. Components can be semiconductors, resistors, capacitors, etc.

componentry The science of analyzing and classifying the various elements of equipment. See Component.

Component Busy Hour CBH. A telephone company definition. The busy hour of an individual component of a switching system. Often, component busy hours coincide

with the overall office busy hour. Each component or group will have its own time consistent busy hour during the busy season. While the hour may or may not vary from one busy season year to another, only one hour may be used during a busy season year. It is upon the data collected during this component busy hour that trends are established, projections made, capacities set and future requirements derived. The component busy hour is used to determine the high day (HDCBH), ten highest days (10HDCBH), average of the ten highest days (ATHD), average busy day (ABD) and average busy season (ABS) CBH values.

component busy season The busy season during which the highest levels of traffic generally occur for which components of network facilities should be engineered.

component execution environment The runtime technical infrastructure, services, and facilities required to provide the appropriate separation layer for distributed components and to enable business components to collaborate. The reason this is done is to hide low-level technical issues from the functional developer.

Component Object Model COM is Microsoft's cornerstone of its ActiveX platform. COM is a language independent component architecture (not a programming language). It is meant to be a general purpose, object-oriented means to encapsulate commonly used functions and services. The COM architecture provides a platform independent and distributed platform for multi-threaded applications. COM also encompasses everything previously known as OLE Automation (Object Linking and Embedding). OLE Automation was originally for letting higher level programming languages access COM objects. An object is a set of functions collected into interfaces. Each object has data associated with it. The source of the data itself is called the data object. With COM, the transfer of the data itself is separated from the transfer protocol.

component software Component software is software constructed from reusable components. It was popularized by Microsoft Visual Basic and its successful custom control architecture. This architecture allows third party software components to "plug" into and extend the Visual Basic development environment. Hundreds of third party components, or custom controls, exist – for everything from accessing a mainframe database to programming a computer telephony board. Component-based software development is a productive way to build software. System developers benefit from being able to tailor their development environment for a specific need. Consider the development of an IVR system that allows callers to access their account balance stored on an IBM mainframe. To build this system, Visual Basic developers extend their development environment with a custom control for telephony and another that provides access to an IBM mainframe. There's no need to learn a new and proprietary language for telephony development. Plus, every control is accessed through a common interface of actions, properties, and events.

component video A 3-plug video cable that can transmit progressive video signals in higher resolutions. Component video transmits color television in three separate channels of red, green and blue.

composite 1. Output of a multiplexer that includes all data from the multiplexed channels. Contrast with Aggregate.

2. Refers to a type of color monitor in which the color signals all come in on the same line and are separated electronically inside the monitor. Compare this type of monitor to RGB, where the colors come in on different cables.

composite cable assembly See bundled cable.

composite clock A bipolar timing signal containing 64 khz bit-clock and 8 khz byte-clock frequencies.

composite link The datastream composed of all the input channels and control and signaling information in a multiplexed circuit. The Composite Link Speed is the transmission speed of the circuit.

composite materials Composite materials consist of two or more components. They make it possible to combine the best properties of different materials; for example, the compression strength and low price of concrete with the tensile strength of reinforcing rods. Composite materials include: Reinforced concrete, fiber-reinforced plastic, fiber-reinforced metals, plywood, chipboard and ceramics. The composites mainly considered for antennas are fiber-reinforced plastics. They combine the low weight and protective properties of plastics with the stiffness and strength of fiber.

composite second order An important distortion measure of analog CATV systems. It is mainly caused by second order distortion in the transmission system.

composite sync A signal consisting of horizontal sync pulses, vertical sync pulses, and equalizing pulses only, with a no-signal reference level. - A combination of horizontal and vertical sync pulses.

composite signaling A direct current signaling system that separates the signals from the voice band by filters. Two pairs (a quad) provide talking paths and full-duplex signaling for three channels. Also called CX Signaling.

composite timing See Composite Clock.

Composite triple beat An important distortion measure of analog CATV systems. It is mainly caused by third-order distortion in the transmission system.

composite video Composite video is a mixed signal comprised of the luminance (black and white), chrominance (color), blanking pulses, sync pulses and color burst. Composite video is a television signal where the chrominance (color) signal is a sine wave that is modulated onto the luminance (black and white) signal which acts as a subcarrier. This is used in NTSC and PAL TV systems. Composite video is the visual wave form representation used in color television. Composite video is analog and must be converted to digital to be used in multimedia computing. See also composite video signal, digital video and NTSC.

composite video signal The completed video signal that is the combined result of the primary colors of red, green and blue (RGB) producing all the necessary picture information, such as in the NTSC or PAL TV formats.

composited circuit A circuit that can be used simultaneously for telephony and dc telegraphy, or signaling, separation between the two being accomplished by frequency discrimination.

compound A term used to designate an insulating or jacketing material made by mixing two or more ingredients.

compound document The simple explanation: A compound document contains information created by using more than one application. It is a document often composed of a variety of data types and formats. Each data type is linked to the application that created it. A compound document might include audio, video, images, text, and graphics. Compound documents first became possible to the world of PCs with the introduction of Windows 3.1, which included OLE (Object Linking and Embedding). OLE allows you to write a letter in your favorite Windows word processor, embed a small voice icon in your document, send your letter to someone else, have them open your letter, place their mouse on the voice icon and hear whatever comments you recorded. To make this possible, both you (the creator) and your recipient would need access to programs that could read both the text and the voice. Ideally, you would both be on a LAN (Local Area Network) and would both get access to the identical applications software, resident, presumably, on the LAN's file server. See Compound Mailbox.

compound document mail See Compound Document.

compound mailbox A mailbox for mail from all sources- fax, voice mail, e-mail, pager, etc. See Compound Document.

compressed video Television signals transmitted with much less than the usual bit rate. Full standard coding of broadcast quality television typically requires 45 to 90 megabits per second. Compressed video includes signals from 3 mb/s down to 56 Kbps. The lower bit rates typically involve some compromise in picture quality, particularly when there's rapid motion on the screen. See MPEG.

compression Compression is the art and science of squeezing out unneeded information in a picture, or a stream of pictures (a movie) or sound before sending or storing it. Without compression, you'd never get a movie onto a single DVD platter, and our cell phone network would accommodate a fraction of the people it does. The most famous compression of all involves MP3 audio files, which have actually been compressed twice. First, the analog data from the recording studio is squeezed down to a CD. To make an MP3, that sound is then compressed again, this time by as much as 90%. MP3 compression is brilliant. Most people can't tell that they've lost much of the information. In imaging, compression techniques vary by whether they remove detail and colour from the image. Lossless techniques compress image data without removing detail; lossy techniques compress images by removing detail. Compression typically reduces the bandwidth or number of bits needed to encode information or encode a signal, typically by eliminating long strings of identical bits or bits that do not change in successive sampling intervals (e.g., video frames). Compression saves transmission time or capacity. It also saves storage space on storage devices such as hard disks, tape drives and floppy disks.

compression algorithm The arithmetic formulae which convert a signal into smaller bandwidth or fewer bits.

compression artifacts Compression artifacts are introduced by filtering, conversion transformation, quantization and transmission compression. Loss of resolution, quantization noise and block errors are typically observed as a result of these processes.

compression wave An element of sound. When you speak in native mode, or acoustically, you create disturbances in the molecules in the air. Those disturbances vary in terms of frequency (i.e., pitch or tone) and amplitude (i.e., volume or power), and travel in a waveform. The wave comprises the compression phase and the rarefaction phase. The compression phase is the phase of high pressure in which the molecules are packed together more tightly than normal. The rarefaction phase, or decompression phase, is the phase in

which the high pressure is relaxed and the molecules snap back into position.

compressor See Companding.

compromise equalizer Equalizer set for best overall operation for a given range of line conditions. This is often fixed but may be manually adjustable.

COMPSURF COMPrehensive SURFace Analysis. A Novell program that checks the surface of a hard disk, marks off sections that are lousy and therefore shouldn't be written to, and then low level formats the disk. The program is slow, but thorough and rigorous. No hard disk should ever be used on a file server on a Novell local area network without being subjected to this wonderful program. Don't believe Novell when it says that you don't need to subject new disks to COMPSURF. You should submit ALL disks.

CompTel Competitive Telecommunications Association. A national U.S. organization of competitive local and long distance carriers, most of which are facilities-based, and suppliers. ACTA (America's Carriers Telecommunications Association) merged into CompTel in December 1998. www.comptel.org.

CompTIA Computing Technology Industry Association. An organization of over 7,500 computer hardware and software manufacturers, distributors, retailers, resellers, VARs, systems integrators, telecommunications, Internet companies, and others. CompTIA aims to provide a unified voice for the industry in the areas of public policy, workforce development, and electronic commerce standards. CompTIA certifies the competency of IT and service professionals. www.comptia.org.

CompuCALL Northern Telecom DMS central office link to computer interface. With Compucall, an agent can get a screen of information about a caller concurrently with receipt of a call. Callers can work with interactive voice response systems (VRUs) to deliver information to send their call to the appropriate agent.

Communications A recent creation meaning the combination of telephones, computers, television and data systems.

compulsory arbitration If something goes wrong in deal, you must go to arbitration before you sue each other. The idea is to save money (fewer lawyers) and time (no lengthy court hearings). Arbitration means the hearing and determination of a case between parties in controversy by a person or persons chosen by the parties or appointed by some organization you agreed on when you signed the original contract.

compulsory redundancies In November, 2005 Germany's Deutsche Telekom AG said 32,000 workers will leave the company over three years as part of a restructuring program that will cost 3.3 billion euros due to "massive changes in the industry." In particular, Deutsche Telekom cited the tough competitive environment in the fixed-network and broadband sector in Germany, where all the job cuts will take place. The operator said there would be no compulsory redundancies until 2008. .... I love those words: "compulsory redancies." Under German law, they clearly call a spade a spade. They're firing the workers. Face it. Sad.

CompuServe An on-line, dial-up service – one of the largest worldwide. CompuServe has everything from electronic mail to manufacturer-sponsored forums where you can download files for updated drivers, etc. CompuServe is one of the hardest on line services to find your way around. See the following definitions. CompuServe is now owned by American Online.

CompuServe Electronic Mail You can send electronic mail to CompuServe addresses. Here's the formula: All CompuServe addresses are either of the form 7xxxx,xxx or 1xxxxx,xxx. (where each "x" signifies a digit from 0 to 7). There can be from 2 to 4 digits following the comma. To send mail to such an address from the Internet, change the comma to a period and attach "@CompuServe.com" as is shown in the following examples:

74906.1610@compuserve.com or 100906.1610@compuserve.com.

CompuServe B+ File Transfer This file transfer protocol is used by the CompuServe information service and no one else. Recovery of interrupted transfers is supported. In CompuServe B+, the host initiates the transfer. In contrast, in XMODEM, the receiver initiates the transfer, i.e. tells the distant computer to begin sending the file.

CompuServe Mail Hub A facility of CompuServe which enables users on a local area network operating Novell Message Handling Service (MHS) software to exchange electronic messages with other MHS users, CompuServe Mail subscribers and users of other E-mail services that can be reached via a CompuServe gateway.

compute servers Very powerful computers that sit on networks and are dedicated to heavy mathematical calculations. Brokerage firms use them for complex yield calculations, mathematical modeling, derivatives analysis, etc. Such servers often have as much as a gigabyte in RAM.

computer This is a definition straight from AT&T Bell Laboratories. "An electronic device that accepts and processes information mathematically according to previous instructions. It provides the result of this processing via visual displays, printed summaries or in an audible form." When it works, it's wonderful. When it doesn't, it's a disaster. The major lessons every computer user should learn: Save your work regularly. Back it up regularly. back it up to many different media. The value of your work on your computer exceeds the value of your computer many, many, manyfold.

computer aided dialing A newer (and allegedly less offensive) term for predictive dialing. See predictive dialing.

Computer Aided Professional Publishing CAP. The computerization of professional publishing (as opposed to desktop operations), including true color representation of the layout on the workstation screen.

Computer And Business Equipment Manufacturers Association CBEMA. Association active before Congress and the FCC promoting the interests of the competitive terminal, computer and peripheral equipment industries.

Computer And Communications Industry Association CCIA. Organization of data processing and communications companies which promotes their interests before Congress and the FCC.

computer cases A series of three FCC inquiries beginning in 1970 that culminated in a 1990 federal court decision that established that the FCC, if it is to exercise its authority to re-think its post-divestiture policies in light of changed circumstances, must provide reasoned explanations for its policy decisions. The Computer cases involved what was then a new industry whereby providers of data processing used the transmission facilities of common carriers to deliver computer based information to customers' terminals. The FCC labeled telecomm services combining both data processing and communications components "enhanced services". The FCC had two competitive concerns in its Computer inquiries: discriminatory access (that carriers would gain an unfair advantage by discriminating in favor of their own enhanced service offerings over that of competitors because non-carrier providers of enhanced services needed access to the telecomm network through local exchange bottlenecks) and cross-subsidization (that carriers would exploit their monopoly in local exchange services by passing on costs of unregulated enhanced services business to telephone rate payers).

Computer I, 1970 - The FCC required that any telephone carrier offering enhanced services do so by means of a separate corporate subsidiary. This structural separation was originally thought not to apply to AT&T and their local exchange affiliates because those companies were under a 1956 consent decree barring them from offering data processing services. The FCC's separation requirements were intended to create an even playing field for the BOC's competitors.

Computer II, 1980 - The FCC continued to rely on structural separation to prevent cross-subsidization and discriminatory access, but it restricted the requirement to the Bell System and removed its application to all other carriers.

Computer III, - the FCC reversed course and announced its intention to relieve the BOCs of the separation requirements in favor of a plan that would allow the BOCs to integrate their basic and enhanced services upon implementation of behavioral safeguards such as engaging in open network architecture (ONA) and being monitored closely by accountants. The FCC had determined that the BOC's loss of efficiency was greater than the risk of anticompetitive behavior in a market which had changed drastically since the first Computer inquiry. State regulatory agencies brought suit, charging that it was irrational for the FCC to have abandoned its earlier position. The Court acknowledged that emergence of powerful competitors like IBM had reduced the BOCs ability to discriminate in providing quality access to the network, but it did not agree with the FCC that changed circumstances reduced the danger of cross-subsidization. In deciding that the FCC's substitution of non-structural safeguards was arbitrary and capricious, the Court established that if the FCC is to exercise its authority to re-think its post-divestiture policies in light of changed circumstances, it must provide reasoned explanations for its policy decisions. Here, the Court found that the FCC failed to meet its burden of showing that its preemption orders were necessary to avoid frustrating its regulatory goals. (California v. FCC, 905 F.2d 1217 (9th Cir 1990).

Computer Emergency Response Team CERT. A group of computer experts at Carnegie-Mellon University who are responsible for dealing with Internet security issues. The CERT is chartered to work with the Internet community to facilitate its response to computer security events involving Internet hosts, to take proactive steps to raise the community's awareness of computer security issues, and to conduct research targeted at improving the security of existing systems. The CERT was formed by DARPA in November 1988 in response to the Internet worm incident. CERT exists to facilitate Internet-wide response to computer security events involving Internet hosts and to conduct research targeted at improving the security of existing systems. They maintain an archive of security-related issues on their FTP server at "cert.org." Their email address is "cert@cert.org" and their 24-hour

telephone Hotline for reporting Internet security issues is 412-268-7090.

computer fingerprinting A concept relatively new to the computer forensics world is computer "fingerprinting" which, in effect, writes a copy of the owner's identity to all "empty" parts of the media – typically the hard drive – so that, if stolen, the media might later be recovered by law enforcement. This is also known as a variation on low level formatting. See Format.

computer fraud Deliberate misrepresentation, alteration or disclosure of computer-based data to obtain something of value.

computer inquiry A series of ongoing FCC proceedings examining the distinctions between communications and information processing to determine which services are subject to common carrier regulation. The FCC decision in 1980 resulting from the second inquiry was to limit common carrier regulation to basic services. Enhanced services and customer premises equipment are not to be regulated. This meant the Bell operating companies had to set up separate subsidiaries if they were to offer non-regulated services.

Computer Inquiry III, adopted by the FCC in May, 1986, removed the structural separation requirement between basic and enhanced services for the BOCs and for AT&T. CI III replaced that requirement with "nonstructural safeguards." This action resulted in the imposition of such concepts as "comparably efficient interconnection" (CEI) and Open Network Architecture (ONA). The FCC's jurisdiction regarding Computer Inquiry I, II and III has now been usurped by Judge Greene, who insists on fairly tight control over the non-basic telephone company activities of the Bell operating companies. Sometimes he gives dispensations (waivers). Sometimes he doesn't. His word these days is final law on what the Bell operating companies can and can't do.

computer port The interface through which the computer connects to the communications circuit. The place where the circuit is "plugged" in.

computer security service An AIN (Advanced Intelligent Network) service providing for additional computer access security to be embedded in the network. Based on Caller ID or ANI (Automatic Number Identification), plus password protection and other authorization schemes, callers would be afforded or denied access to a networked computer on a customer premise. The network security service also would maintain an audit of all access attempts; the user organization could access that audit data in order to reconfigure access privileges, plug holes in network security, identify access anomalies, and so on. In this fashion, organizations supporting such applications as remote LAN access would realize an extra measure of security through a security system physically and logically separate from the premise.

computer server farm Picture a hall full of PC servers, lined one after another. Now you have the concept of a farm.

Computer Support Telephony See CST.

computer telephony Computer telephony adds computer intelligence to the making, receiving, and managing of telephone calls. Harry Newton coined the term in 1992. Computer telephony has two basic goals: to make making and receiving phone calls easier, i.e. to enhance one's personal productivity and second, to please corporate customers who call in or who are called for information, service, help, etc. Computer telephony encompasses six broad elements:

1. Messaging.

Voice, fax and electronic mail, fax blasters, fax servers and fax routers, paging and unified messaging (also called integrated messaging) and Internet Web-vectored phones, fax and video messaging.

2. Real-time Connectivity.

Inbound and outbound call handling, "predictive" and "preview" dialing, automated attendants, LAN / screen-based call routing, one number calling / "follow me" numbers, video, audio and text-based conferencing, "PBX in a PC," collaborative computing.

3. Transaction Processing and Information Access via the Phone.

Interactive voice response, audiotex, customer access to enterprise data, "giving data a voice," fax on demand and shopping on the World Wide Web.

4. Adding Intelligence (and thus value) to Phone Calls.

Screen pops of customer records coincident with inbound and outbound phone calls, mirrored Web page "pops," smart agents, skills-based call routing, virtual (geographically distributed) call centers, computer telephony groupware, intelligent help desks and "AIN" network-based computer telephony services.

5. Core Technologies.

Voice recognition, text-to-speech, digital signal processing, applications generators (of all varieties – GUI to forms-based to script-based), VoiceView, DSVD, computer-based fax routing, USB (Universal Serial Bus), GeoPort, video and audio compression, call progress, dial pulse recognition, caller ID and ANI, digital network interfaces (T-1, E-1, ISDN BRI and

PRI, SS7, frame relay and ATM), voice modems, client-server telephony, logical modem interfaces, multi-PC telephony synchronization and coordination software, the communicating PC, the Internet, the Web and the "Intranet."

6. New Core Standards.

The ITU-T's T.120 (document conferencing) and H.320 (video conferencing), Microsoft's TAPI – an integral part of Windows 95 and NT, Novell's TSAPI – a phone switch control NLM running under NetWare. Intel's USB and InstantON. Natural MicroSystems / Mitel's MVIP and H-MVIP. Dialogic has SCSA. And the industry has ECTF.

That's today. But what really excites is the potential. It's huge. Despite the above, phone calls today are dumb, seriously bereft of common sense. Few phones have "backspace erase." 75% of business calls end in voice mail! Often in voice mail jail. Every call not completed is an irritated customer and a lost sale. Computer telephony addresses the waste. Computer telephony adds intelligence to the making and receiving of phone calls. Bingo, happier customers and more completed transactions.

The best news: We now have the technology, the resources, the computer power, the new standards and the muscle to back our hype. We also have many new players who are, thankfully, not burdened by the assumptions of yesteryear's telecommunications industry. We also have legions of developers and systems and integrators who are grabbing these computer telephony tools and are cranking out hundreds of customer-pleasing, productivity-enhancing solutions for your business. Computer telephony delivers. And fortunately, industry now wants it.

Once a year, typically in March, industry leaders meet at a trade show called Computer Telephony Conference and Exposition. There is also a monthly magazine covering the industry called Computer Telephony Magazine. See also Telephony, Telephony Services, and Windows Telephony.

Computer Telephony Integration CTI. The integration of the telephony function with computer applications, commonly used to automate call centers.

computer telephony integratiogon assistance Programs that help nonprofit organizations, small businesses and other groups conduct an assessment of their needs and select, implement and effectively use hardware and software that allows them to merge voice and data applications; and enables their computer to perform functions traditionally accomplished by the telephone and integrate them with other desktop functions. Outgoing calls, for example, can be made or forwarded by pointing to an address book entry; caller identification (if available) can used to automatically start an application or bring up a database file; and voicemail and incoming faxes can be integrated with electronic mail, all independently of telephone equipment. Voicemail and Voice over IP (Internet Protocol) are examples of computer telephony integration applications. See also Voicemail System Selection Assistance.

Computer Vision Syndrome CVS. First there was carpal tunnel syndrome, also known as repetitive stress injury, a condition in which swollen and pinched nerves in your wrist result from long hours of repetitive activity like typing on your computer keyboard. Now comes computer vision syndrome, a complex of eye problems supposedly caused by staring at your computer screen for hours on end. Symptoms of CVS include headaches, redness, contact lens discomfort, blurred vision, light sensitivity, dry or burning eyes, focusing difficulties, and even pain in the back, neck, or shoulders. Thank goodness it's a temporary problem that can be relieved by taking short breaks. It is also recommended that you position the monitor to avoid glare, position the monitor 20-28 inches away from your eyes, and remember to blink. (Good Grief!) According to a Harris poll taken in early 2000, CVS is the #1 health complaint of U.S. office workers. See also Carpal Tunnel Syndrome, but remember to blink first.

computername The name by which the LAN identifies a server or a workstation in lan Manager terminology. Each computername must be unique on the network.

COMSAT The COmmunications SATellite corporation was created by Congress as the exclusive provider to the U.S. of satellite channels for international communications. COMSAT was once the U.S. representative to Intelsat and Inmarsat, two international groups responsible for satellite and maritime communications. Comsat Corporation ceased to exist as an independent company in August 2000, when it was acquired by Lockheed Martin Corporation (LMT). LMT decided to split the company apart, selling the mobile communications division to Telenor Satellite Services Holdings, Inc. in January 2002, announcing the sale of the Intelsat line of business to Intelsat in March 2002, and planning to sell off the rest of the former Comsat assets in coming months.

COMSEC COmmunications SECurity. A U.S. federal government term for "measures and controls taken to deny security unauthorized persons information derived from telecommunications and ensure the authenticity of such telecommunications." The government takes this very seriously.

COMSTAR A domestic communications satellite system from Comsat.

CON Circuit Order Number.

Concatenated STS-1 A signal in which the STS envelope capacities from several (i.e. N) STS-1s have been combined to carry an STS-Nc Synchronous Payload Envelope (SPE). It's used to transport signals that don't fit into an STS-1 (52Mbps) payload.

Concatenated VT A virtual tributary (VT x Nc) that is composed of several (N) VTs combined. Its payload is transported as a single entity rather than separate signals.

concatenation Linking together in a series or chain. A SONET/SDH term. Concatenation is a mechanism for allocating very large amounts of bandwidth for transport of a payload associated with a "superrate service," which is a service at a transmission rate greater than the normal maximum rate of OC-1. The set of bits in the payload is treated as a single entity, as opposed to being treated as separate bits or bytes or time slots. The payload, therefore, is accepted, multiplexed, switched, transported and delivered as a single, contiguous "chunk" of payload data. Certain data protocols (e.g., ESCON and Fibre Channel) require huge chunks of bandwidth – far more than can be provided by the STS-1 data rate. In SONET, STS-1 is the electrical equivalent of OC-1 (Optical Carrier Level 1), which is 51.84 Mbps, which is T-3 at 44.736 Mbps, plus SONET overhead. Through the use of Concatenation Pointers, multiple OC-1s can be linked together, end-to-end. Hence the term "concatenation" – to provide contiguous bandwidth through the network, from end-to-end. OC-3c and OC-12c are standardized. The same approach can be used at higher SONET OC-N levels, as well. OC-48c, for example is a concatenated OC-48 comprising 48 OC-1s, and supporting a datastream operating at approximately 2.488 Gbps. Concatenation applications include bandwidth-intensive video (e.g., HDTV) and high-speed data. Concatenation has two real benefits in support of such bandwidth-intensive applications. First, the ability to treat multiple OC-1s as a single entity eliminates much of the need to involve SONET equipment such as ADMs (Add/Drop Multiplexers). Such equipment is expensive, induces a small amount of latency (delay), and may fail on occasion. In other words, concatenation is cleaner and simpler. Second, only the first OC-1 in a concatenated SONET datastream requires the nine bytes (i.e., 72 bits) of Path Overhead (POH) for signaling and control purposes. This frees up 576 Kbps of bandwidth for data payload per subsequent OC-1, as each OC-1 frame is sent 8,000 times a second. See also SONET, OC-1, OC-3c, OC-12c, and OC-N.

concentration A fundamental concept to telephony. Applies to a Switching Network (or portion of one) that has more inputs than outputs. For example, communications from a number of phones are sent out on a smaller number of outgoing lines. The theory is that, since not all the phones are being used at any one time, fewer trunks than phones are needed. Some phone system designs assume that only 5% of the phones will be in use at any one time. Some phone systems design assume 10%. In some phone-intensive industries, you can't make any assumptions about concentration. You have to assume one line per phone. No concentration. See Concentration Ratio.

concentration ratio The ratio between lines and trunks in a concentrated carrier system or line concentrator. See Concentrator.

concentrator 1. A device which allows a relatively large number of devices or circuits (typically slow speed ones) to share either a single circuit or a relatively small number of circuits. In other words, the traffic is concentrated through a process of multiplexing, in which many relatively low capacity inputs from devices or circuits are folded together in order that they might share a single and typically higher capacity circuit which connects to a device or network of a higher order. Assuming that the capacity of the shared facility is sufficient to support all the lower order inputs in a satisfactory manner (i.e., transmission time and response time are not compromised to an unreasonable extent), the benefit of the concentrator is that communications costs are typically lowered through the process of sharing. An further, underlying assumption is that not all of the devices will be active at the same time. While a concentrator is akin to a multiplexer, it is limited to a single type of information stream and it is not capable of accomplishing some of the more sophisticated processes of the latter. For instance, an ATM concentrator might simply concentrate traffic from downstream ATM switches in a backbone LAN environment in order to share a very expensive high-capacity access circuit to an ATM WAN. The ATM adaptation process would have been accomplished previously either in the ATM LAN switch or in the workstation. In the realm of xDSL (generic Digital Subscriber Line) technology, a DSLAM (DSL Access Multiplexer) can be considered as a concentrator, as it simple concentrates traffic from xDSL circuits through a multiplexing process in order that the traffic might share a high-capacity circuit to the Internet backbone. See also DSLAM, Hub and Multiplexer.

2. A Multistation Access Unit (MAU). Token Ring LANs make use of MAUs to concentrate traffic from multiple nodes (e.g., workstations) to the LAN backbone, which may consist of nothing more than multiple interconnected MAUs. Through these central points of connec-

tion, the nodes are attached in a physical star configuration, also called home run, typically using Unshielded Twisted Pair (UTP) for reasons of lower cost of connectivity. Assuming that the concentrator is able to support all attached nodes without seriously degrading their access to the larger network, MAUs offer the advantages of lower overall cost of LAN connectivity and increased manageability.

3. A LAN hub. See also Hub.

concentric stranding A central wire surrounded by one or more layers of helically wound strands in a fixed round geometric arrangement.

concentricity In a wire or cable, the measurement of the location of the center of the conductor with respect to the geometric center of the circular insulation.

concentricity error The amount by which a fiber's core is not centered in its cladding. The distance between the center of the two concentric circles specifying the cladding diameter and the center of the two concentric circles specifying the core diameter.

conclusion A conclusion is the place where you got tired of thinking.

concrete fill A minimal-depth concrete pour to encase single-level underfloor duct.

concurrency The shared use of resources by multiple interactive users or applications at the same time. Concurrency often means that a company need only buy as many licenses to a program as it has people using the program at one time – concurrent users, in other words. See Software Metering.

concurrency control A feature that allows multiple users to execute database transactions simultaneously without interfering with each other.

concurrent computing When you process the same transaction in several places simultaneously on different equipments. This idea is to provide the ultimate in moment-to-moment disaster protection.

concurrent site license Companies that buy software for multiple computers typically buy one copy of the program and a license to reproduce it up to a certain number of times. This is called a site license, though it may apply to its use throughout an organization. Site licenses vary. Some require that a copy be bought for each potential user – the only purpose being to indicate the volume discount and keep tabs. Others allow for a copy to be placed on a network server but limit the number of users who can gain simultaneous access. This is called a Concurrent Site License. And many network administrators prefer this concurrent license, since it gives them greater control. For example, if the software is customized, it need be customized only once, namely on the server.

conical scan Conical scan is a method of tracking in which the antenna beam is rotated about the axis; if the resultant output signal is constant, the antenna is pointed at the satellite; if it varies, correction signals are generated.

condescension fade Disabling Internet connectivity at a meeting or conference. Organizers do it to encourage participants to "focus." The term evolved from backhoe fade – hacker lingo for a sudden loss of connectivity.

condenser A device for storing up electrical energy and consisting of two or more conducting surfaces or electrodes separated by an insulating medium called a dialectic.

condenser antenna An antenna consisting of two capacity areas.

condenser microphone Microphone which operates through changes in capacitance caused by vibrations of its conductive diaphragm.

conditional access system A "gatekeeper" system that allows or disallows a customer to access a service or content based on the customer's account status, credit status, or some other criterion.

conditioned circuit A circuit that has conditioning equipment to obtain the desired characteristics for voice or data transmission. See Conditioning.

conditioned loop A loop that has conditioning equipment to obtain the desired line characteristics for voice or data transmission. See Conditioning.

conditioning The adjustment of the electrical characteristics of transmission lines to improve their performance for specific uses. Conditioning involves the "tuning" of the line or addition/deletion of equipment to improve its transmission characteristics. Conditioning may involve the insertion of components such as equalizers, resistors, capacitors, transformers or inductors. Long voice-grade, twisted-pair local loops, for instance, often have inductors, or "loading coils," installed every 6,000 feet or so in order to amplify the analog signal. Such "loaded" circuits, however, have a decidedly negative impact on data communications, especially at high transmission rates. Therefore, it is necessary to condition the circuit by removing all such electronics, thereby yielding what is known as a "dry copper" circuit. It often is required that the circuit be further conditioned by removing all bridged taps, ensuring that all pairs in the circuit are of consistent gauge, that all cross-connects are mechanically sound, and so on. Carriers provide two types of conditioning for leased lines. C conditioning controls attenuation, distortion, and delay distortion. D conditioning controls harmonic distortion and signal-to-noise ratio.

conditioning equipment Equipment added to a circuit for the express purposes of matching transmission levels and impedances or equalizing transmission and delay to bring circuit losses, levels, and distortion within specified limits of CCITT standards, or in U.S. practice, common carrier tariffs. See Load Coil.

conditions Busy. Voice Mail. Out of service. All the situations that a phone line is likely to find itself in.

condofiber A shared tenancy cable or shared ownership facility such as a transatlantic fiber cable. Multiple vendors such as Sprint, MCI and AT&T may all own a group of fibers with responsibility for maintaining their own operation while at the same time paying an overall "association" fee for the common maintenance of the overall cable.

conductance The opposite of resistance; a measure of the ability of a conductor to carry an electrical charge. Conductance is a ratio of the current flow to the potential difference causing the current flow. The unit of conductance is Mho (a reversed spelling of Ohm).

conducting materials Substances which offer relatively little resistance to the passage of an electric current.

conductivity A term used to describe the ability of a material to carry an electrical charge, i.e., to allow electrons to flow. Conductivity is the reciprocal of specific resistance. Usually expressed as a percentage of copper conductivity.

conductor An uninsulated wire suitable for carrying electrical current. Some atoms do not hold their electrons tightly, and in materials made of these atoms, the electrons can drift randomly from one atom to the next very easily. These materials make good electrical conductors. Most metals have electrons that can move easily this way and are generally good conductors. The best conductors are silver, copper and aluminum. Another type of good conductor is an electrolyte. An electrolyte is composed of charged ions that are free to move, carrying a charge from one location to another. One example of an electrolyte is a solution of table salt in water. The positively charged sodium ions and the negatively charged chloride ions are capable of carrying charge from one part of the solution to another.

conduit A pipe, usually metal but often plastic, that runs either from floor to floor or along a floor or ceiling to protect cables. A conduit protects the cable and prevents burning cable from spreading flames or smoke. Many fire codes in large cities thus require that cable be placed in metal conduit. In the riser subsystem when riser closets are not aligned, conduit is used to protect cable as well as to provide the means for pulling cable from floor to floor. In the horizontal wiring subsystem, conduit may be used between a riser or satellite closet and an information outlet in an office or other room. Conduit is also used for in-conduit campus distribution, where it is run underground between buildings and intermediate manholes and encased in concrete. Multiduct, clay tile conduit may also be used.

conduit caps Plastic coverings which cover conduit ends or openings to keep water and critters out of equipment and pedestals.

conduit run The path taken by a conduit or group of conduits.

conduit system Any combination of ducts, conduits, maintenance holes, handholes and vaults joined to form an integrated whole.

Coney Island Whitefish A condom. So called because used condoms are often found under the boardwalk.

conferee Participant in a conference call who is not the call controller. This definition courtesy Hayes. According to Hayes, a "controller" is the person who sets up the conference call.

conference bridge A telecommunications facility or service which permits callers from several diverse locations to be connected together for a conference call. The conference bridge contains electronics for amplifying and balancing the conference call so everyone can hear each other and speak to each other. The conference call's progress is monitored through the bridge in order to produce a high quality voice conference and to maintain decent quality as people enter or leave the conference.

conference call 1. Connecting three or more people into one phone conversation. You used to have to place conference calls through an AT&T operator (you still can). But now you can also organize conference calls with most modern phone systems or a conference bridge. If conferencing is important to you, make sure your conferencing device has amplification and balancing. If not, it will simply electrically join the various conversations together and people at either end won't be able to hear each other. There are different types of conference devices you can buy, including special teleconferencing devices that sit on conference tables and perform the function of a speakerphone, albeit a lot better. There are also dial-in devices called conference bridges. But, however, you use these devices, they will requires lines (and/or trunks). If you install one inside your phone system, be careful to have the extra spare extensions. For a conference of 10 people, you'll typically need 10 extensions connected to your conference bridge. See Conference Bridge.

2. "He's on a conference. He can't speak with you at present."

Conérence des administrations Europeanes des Postes et Telecommunications See CEPT.

Conference, Meet-Me A conference call in which each of the people wishing to join the conference simply dials a special "Meet-Me" Conference phone number, which automatically connects them into the conference. It is a feature of some PBXs and also some special Conferencing Equipment. See Conference Bridge.

conferencing Several parties can be added to a phone conversation through Conferencing.

confidence interval A confidence interval is the range of values within which the true value is assured to lie. Confidence level must be two figures.

confidencer A noise-cancelling microphone for use on a telephone in noisy places. A confidencer is not an easy device to use.

confidential reception The ability to receive a facsimile transmission directly into memory which can be printed out or viewed at a later time.

confidential transmission A facsimile message that is sent confidentially into memory or a private mailbox, to be retrieved by the receiver at a later time. It's usually retrieved by using a confidential passcode or password.

configuration 1. The hardware and software arrangements that define a computer or telecommunications system and thus determine what the system will do and how well it will do it. This information can be entered in the CMOS and EEPROM setup programs.

2. An ATM term. The phase in which the LE Client discovers the LE Service.

Configuration Databases Rolm/IBM words for those databases which represent unique user specifications relating to system and phone features. These databases can be entered on-site and are not part of the generic software which runs the phone system.

configuration file An unformatted ASCII file that stores initialization information for an application.

configuration management One of five categories of network management defined by the ISO. Configuration management is the process of adding, deleting and modifying connections, addresses and topologies within a network. See ISO.

configuration manager 1. A SCSA system service which manages configuration information and controls system startup.

2. An Intel Plug'n Play term. A driver, such as the ISA Configuration Utility, that configures devices and informs other device drivers of the resource requirements of all devices installed in a computer system. The Windows 95 Resource Kit defined configuration manager as the central component of a Plug and Play system that drives the process of locating devices, setting up their nodes in the hardware tree, and running the resource allocation process. Each of the three phases of configuration management-boot time (BIOS), real mode, and protected mode-has its own configuration manager.

configuration registry A database repository for information about a computer's configuration.

configuration tool 1. Service management tool with a Graphical User Interface (GUI). 2. Element management service tool with a GUI.

Confirming Design Layout Report Date CDLRD. The date a common carrier accepts the facility design proposed by the Telco.

conformance test A test performed by an independent body to determine if a particular piece of equipment or system satisfies the criteria of a particular standard, sometimes a contract to buy the equipment. See also Compatible and Compliant.

conforming end office Central office with the ability to provide originating and terminating feature group D local access and transport area access service.

congestion A condition that arises when a communications link, path, or network experiences an offered load (i.e., the amount of traffic offered) that exceeds its capacity. For example, consider a T-1 link connected to the outgoing port of a switch. If the switch attempts to offer a traffic load in excess of 1.544 Mbps, a congestion condition arises, and can be resolved in one of several ways. First, the switch can simply discard the excess data. Discard Eligible (DE) data applications generally will not suffer beyond their expectations unless the congestion condition becomes extreme. Second, the switch can buffer the excess data until such time as the congestion condition eases; this process is known as "congestion control," and is limited to the maximum capacity of the buffers involved. If the congestion condition persists and the switch discards no data, eventually the congestion backs up all the way to the user terminal, and the application ceases to function in an acceptable manner. See also Utilization.

congestion collapse Condition in which the retransmission of frames in an ATM network results in little or no traffic successfully arriving at the destination. Congestion collapse frequently occurs in ATM networks composed of switches that do not have adequate and effective buffering mechanisms complimented by intelligent packet discard or ABR

congestion feedback mechanisms.

congestion control The process whereby packets are discarded to clear buffer congestion in a packet-switched network.

congestion management The ability of a network to effectively deal with heavy traffic volumes; solutions include traffic scheduling and enabling output ports to control the traffic flow. See BECN and Ethernet Switch.

connect:direct A direct electronic method of delivering CLEC and Reseller usage data files and Reseller bills, and transmitting CLEC ASRs. Available in several platforms including NDM-MVS for mainframe and NDM-PC for personal computers. Also known as Network Data Mover (NDM).

connect time Measure of computer and telecommunications system usage. The interval during which the user was on-line for a session.

connectable mode In Bluetooth terminology, connectable mode means a device that responds to paging (an attempt to establish a communication link) is said to be in connectable mode. The opposite of connectable mode is non-connectable mode.

connected 1. On line.

2. A voice recognition term for words spoken clearly in succession without pauses. For recognition to occur, words or utterances must be separated by at least 50 milliseconds (1/20th of a second). Generally refers to digit recognition and sometimes used to describe fast discrete recognition.

connected state A state in which a device is actively participating in a call. This state includes logical participation in a call as well as physical participation (i.e., a Connected device cannot be on Hold).

connected time The length of time a path between two objects is active.

connected user A Windows NT term. A user accessing a computer or a resource across the network.

connecting arrangement The manner in which the facilities of a common carrier (phone company) and the customer are interconnected.

connecting block A plastic block containing metal wiring terminals to establish connections from one group of wires to another. Usually each wire can be connected to several other wires in a bus or common arrangement. A 66-type block is the most common type of connecting block. It was invented by Western Electric. Northern Telecom has one called a Bix block. There are others. These two are probably the most common. A connecting block is also called a terminal block, a punch-down block, a quick-connect block, a cross-connect block. A connecting block will include insulation displacement connections (IDCs). In other words, with a connecting block, you don't have to remove the plastic shielding from around your wire conductor before you "punch it down."

connecting hardware A device providing mechanical cable terminations.

connection 1. A path between telephones that allows the transmission of speech and other signals.

2. An electrical continuity of circuit between two wires or two units, in a piece of apparatus.

3. An SCSA term which means a TDM data path between two Resources or two Groups. It connects the inputs and outputs of the two Resources, and may be unidirectional (simplex) if either of the Resources has only an input or an output. Otherwise it is bi-directional (dual simplex). It usually has a bandwidth that is a multiple of a DS0 (64kbit) channel. Inter-group connections are made between the Primary Resource of each Resource Group.

4. An ATM connection consists of concatenation of ATM Layer links in order to provide an end-to-end information transfer capability to access points.

connection master Software from Mitel, which brings the Connection Control Standard to an even higher level for the MVIP developer. Connection Master interacts with circuit switches on multiple MVIP cards to make connections and resolve switching contention. It also interfaces between applications and makes connections in such a way that simple one-chassis applications become networked applications. Connection Master fully supports MC-MVIP, Multi-Chassis MVIP. See also MVIP.

connection number A number assigned to a workstation that attaches to a server; it may be a different number each time the workstation attaches. Connection numbers are also assigned to print servers, as well as other applications and processes that use the server connections.

connection oriented The model of interconnection in which communication proceeds through three well-defined phases: connection establishment (call setup), information transfer (call maintenance), connection release (call teardown). Connection-oriented services ensure that all data follow the same path through the network. That is to say that all data travel across the same circuits, and through the same switches and other devices. Examples include ordinary circuit-switched voice and data calls, ISDN calls, X.25, TCP, Frame

Relay, and ATM. See Connection Service and Connectionless Mode Transmission.

Connection Oriented Network Service CONS. An OSI protocol for packet-switched networks that exchange information over a virtual circuit (a logical circuit where connection methods and protocols are pre-established); address information is exchanged only once. CONS must detect a virtual circuit between the sending and receiving systems before it can send packets.

connection oriented operation A communications protocol in which a logical connection is established between communicating devices. Connection-oriented service is also referred to as virtual-circuit service.

connection orientated protocol A protocol in which a connection is established prior to initiation of data transmission, maintained during transmission, and effectively terminated on completion of transmission. All data travel exactly the same path through the network. Examples include SPX, TCP, and Frame Relay.

connection oriented transmission Data transmission technique involving setting up a connection before transmission and disconnecting it afterward. A type of service in which information always traverses the same pre-established path or link between two points. See Connectionless Service.

connection protocol A protocol in which it is not necessary to establish, maintain, and terminate a connection between source and destination prior to transmission. (Example: IPX, IP)

connection service A circuit-switching service whereby a connection is switched into place at the beginning of a session and held in place until the session is completed. Also referred to as circuit switching. The circuit switched in place may be real or virtual. See Circuit Switching.

connection speed The speed of a data communications circuit. Some circuits are symmetrical and can maintain the same speed in both directions; others are asymmetrical and use a faster speed in one direction, usually the downstream side.

connectionless The model of interconnection in which communication takes place without first establishing a connection and without immediate acknowledgment of receipt. Sometimes it is (imprecisely) called datagram. Examples: Internet IP and OSI CLNP, UDP.

connectionless communication A form of communication between applications in which all data is exchanged during a single connection.

connectionless mode transmission A mode of data transmission in which the transmitting device accesses the network and begins transmission without the establishment of a logical connection to the receiving device. In other words, the transmitter simply begins "blasting" data. Connectionless mode is very much unlike "Connection Oriented Transmission," wherein communications involves a process of call set-up, call maintenance and call teardown. Connectionless mode is limited to LAN communications and SMDS, which essentially is a MAN extension of the LAN concept.

In connectionless transmission, each packet is prepended with a header containing destination address information sufficient to permit the independent delivery of the packet. In other words, each packet within a stream of packets is independently survivable. While this approach is characteristic of connectionless mode, it also is characteristic of connection-oriented protocols such as X.25 (packet switching), Frame Relay and ATM. See also Cloud, Connectionless Network, Connectionless Service, Connection Oriented, SMDS, X.25 and ATM.

connectionless network A type of communications network in which no logical connection (i.e. no leased line or dialed-up channel) is required between sending and receiving stations. Each data unit (datagram) is sent and addressed independently, and, thereby, is independently survivable. IEEE 802 LAN standards specify connectionless networks. SMDS also is a connectionless network, as an extension of the LAN concept for broadband data communications over a metropolitan area. Connectionless networks are becoming more common in broadband city networks now increasingly offered by phone companies.

Connectionless Network Service CLNS. Packet-switched network where each packet of data is independent and contains complete address and control information; can minimize the effect of individual line failures and distribute the load more efficiently across the network.

connectionless packet A packet of data is broadcast over the network without targeting a specific recipient to receive the packet.

connectionless service A networking mode in which individual data packets in a network (local or long distance) traveling from one point to another are directed from one intermediate node to the next until they reach their ultimate destination. Because packets may take different routes, they must be reassembled at their destination. The receipt of a transmission is typically acknowledged from the ultimate destination to the point of origin.

A connectionless packet is frequently called a datagram. A connectionless service is inherently unreliable in the sense that the service provider usually cannot provide assurance against the loss, error insertion, misdelivery, duplication, or out-of-sequence delivery of a connectionless packet.

connectionless transmission Data transmission without prior establishment of a connection.

connections per circuit hour CCH. A unit of traffic measurement; the number of connections established at a switching point per hour.

connectivity A domain of connected components that adhere to a defined set of connection rules. The set of rules is termed Connectivity Architecture. Connectivity is the property of a network that allows dissimilar devices to communicate with each other. See also Connectivity Law.

connectivity junkie A highly mobile knowledge-worker who wants voice and data connectivity anywhere at anytime.

Connectivity Law The Law of Connectivity is another way to describe Metcalfe's Law. That is, the value of a network rises by at least, and probably more than, the square of the number of nodes, n, or units connected. For the Internet, the value is probably much greater than n-squared because, unlike most telephone connections, a single website can connect to more than one node at the same time. See Metcalfe's Law.

connectoid Connectoid is the icon you create for a connection in the Dial-up Networking window in Windows.

connector A device that electrically connects wires or fibers in cable to equipment, or other wires or fibers. Wire and optical connectors most often join transmission media to equipment (host computers and terminal devices) or cross connects. A Connector at the end of a telephone cable or wire is used to join that cable to another cable with a matching Connector or to some other telecommunications device. Residential telephones use the REJ-11C connector. Computer terminals with an RS-232-C interface, use the DB-25 connector. The RS-232-C standard is actually the electrical method of using the pins on a DB-25. See RS-232-C.

Connector Panel Module A module designed for use with FDC units; it contains four, five, or six connectorized fibers that are spliced to trunk cable fibers.

connector plug A male device used to terminate a cable.

connector receptacle The fixed or stationary half of a connection that is mounted on a panel/bulkhead. Receptacles mate with plugs. Receptacles are typically female.

connector variable The maximum value in dB of the difference in insertion loss between mating optical connectors (e.g., with remating, temperature cycling, etc.). Also called Optical Connector Variation.

connector variation The maximum value in dB of the difference in insertion loss between mating optical connectors (e.g. with remating, temperature cycling, etc.). Also known as Optical Connector Variation.

connectorize To put a connector on a cable. Techs (i.e. technicians) use the term all the time.

Connexion Connexion is the name of the in-flight, high-speed Internet service developed by Boeing for its own commercial jetliners. With Connexion, flyers can access their email, surf the Internet and even make international phone calls. It costs about $500,000 to outfit a Boeing plane with Connexion service. Sadly, few airlines decided it was worth the cost. In October 2006, Boeing said it would cease the service at the end of 2006 – but in the meantime, service would be free. (Go figure.) I used Connexion once – on an El Al plane going to Tel Aviv. It was great.

CONS Connection Oriented Network Service. See Connection Oriented Network Service.

conscience 1. The inner voice which warns us that someone is looking. H.L. Mencken.

2. A conscience is what hurts when all your other parts feel so good.

consensus An opinion or position no one really likes, but everyone is able to live with.

Consent Decree 1982 The agreement which split the Bell Operating Companies off from AT&T. It took effect at midnight on December 31, 1983. It is also known as the MFJ (Modified Final Judgment), as it modified the 1956 Consent Decree. Under that consent decree which took place at the beginning of 1984, AT&T split off seven approximately-the- same size, local operating companies (called the Baby Bells) while it kept Western Electric, the manufacturing arm and the long distance company. The restriction against manufacturing outside telecommunications was removed and AT&T was free to move into making and selling computers, which it did disastrously. When the history books are written, it will be shown that AT&T's boss at the time, Charlie Brown, sewed the seeds of AT&T ultimate demise by making two disastrous decisions: 1. Not securing any wirelss licenses and 2. Spinning off the regional bell operating companies and keeping the long distance services and telecom manufacturing arm (Western Electric) – when he should have spun off the manufacturing arm and the long distance companies and kept the local Bell operating companies. See also consent decree.

conservation of radiance A basic principle stating that no passive optical system can increase the quantity L/n2, where L is the radiance of a beam and n is the local refractive index. Formerly called conservation of brightness, or the brightness theorem.

console 1. A large telephone which a PBX attendant uses to answer incoming calls and transfer them around the organization. Before you buy a PBX for your company, make sure your operator has checked out its console. Some are very difficult to use. Some are easy. Some operators hate some consoles. Some consoles hate some operators. You can measure the efficiency of consoles by counting keystrokes to do simple jobs and comparing them – e.g. answer an incoming line, dial an extension and transfer the call. How many keystrokes does your PBX take?

2. The device which allows communications between a computer operator and a computer.

3. The console is the Novell NetWare name for the monitor and keyboard of the file server. Here you can view and control the file server or router activity. At the console, you can enter commands to control disk drives, send messages, set the file server or router clock, shut down the file server, and view file server information. NetWare commands you can enter only from the console (for example, MONITOR) are called console commands. Keep your file server locked up and away from prying eyes. It's clearly not just a case of changing passwords and getting in and mucking around. There have been examples of thieves simply removing the file server's hard disk, putting it in their briefcase and walking off with it.

4. In a network (local or wide area) many of the devices connected to it – such as switches, routers and hubs – are called "consoles." See the next several definitions.

console management In a network (local or wide area) many of the devices connected to it – such as switches, routers and hubs – are called "consoles." These devices need running (i.e. managing). Often they are "managed" by a device called a console server. See the next definition.

console server A console server a specialized computer-like device that lets you get to virtually every piece of equipment in the data center, figure how it's working, fix it if it's not doing what it's meant to do and re-arrange how the various devices talk to each other – if the connection wires allow for that. Console management provides a simple and flexible solution for secure local or remote management, and is widely deployed as a critical management component in many small to large data centers.

consoleless operation Some PBXs can work without a console. Some must have a console. It's good to check. Consoles are expensive. If you don't want one – because your company is small – you don't want to be forced to buy one, only to have it sit idly by.

consolidated carrier Carriers that provide connection both as interexchange carriers and international carriers.

Consolidation Point CP. A location for interconnection between horizontal cables extending from building pathways and horizontal cables extending into furniture pathways.

Constant Bit Rate CBR. A data service where the bits are conveyed regularly in time and at a constant rate, carefully timed between source (transmitter) and sink (receiver), i.e., following a timing source or clock just as members of a marching band follow the beat of the drummer. Examples include uncompressed voice and video traffic, which have to be transported at constant bit rate because they are sensitive to variable delay and, as such, have to be transported without any interruptions in the flow of data.

constant carrier Physical line specification selection indicating full duplex line in bisync network. See SNA.

constant holding time A telephone company definition. Certain devices used in dial equipment for setting up calls may well have practically constant holding times. For estimating the probabilities of congesting, the result of substituting a constant holding time equal to the average of a varying holding time seems to be of negligible moment from a theoretical standpoint. (See Holding Time).

constellation The assemblage of satellites in a LEO (Low Earth Orbiting) or MEO (Middle Earth Orbiting) system. See LEO and MEO.

Constraint-based Routed Label Distribution Protocol. See CR-LDP.

constraint-based routing Procedures and protocols that determine a route across a backbone take into account resource requirements and resource availability instead of simply using the shortest path.

construction budget A detailed plan of placement, removal, and rearrangement of facilities to modernize and expand the capacity of the facilities network. A telephone company term.

construction zone The building of the new information infrastructure by telecommunications and cable companies.

consult 1. To ask or seek the advice of another.

2. To seek another's endorsement of a decision you've already made.

consultant A person who gets paid more than you to tell your boss what you told him. See Consult and Consultant Liaison Programs.

consultant liaison programs Large users often use communications consultants to help them choose systems and long distance phone lines. In recognition of the important role consultants play, many suppliers have consultant liaison programs. Such programs typically consist of a toll-free number and somebody on the other end to answer technical and pricing questions, a three-ring binder containing information on all the company's products and services, occasional seminars and, for those extra-privileged consultants, all expense paid trips to exotic places and "something" else. With MCI that "something else" is a dial-up, toll-free, bulletin board. Dial it up with your PC, you can download MCI's latest prices and services. It's truly splendid as most of the paperwork others issue is obsolete the moment it's issued.

consultation See Consultation Hold.

consultation hold PBX feature which allows an extension to place a call on hold while speaking with another call. The idea is "consulting with" someone while you have someone else on the phone.

Consultative Committee for International Telegraph and Telephone See CCITT.

consumables You buy a $200 "photo quality" printer. The printing is fabulous. Looks just like a Kodak glossy photo. You ask yourself: How can they afford to produce such gorgeous quality when they charge so little money for the printer? Easy. They charge an enormous amount for the ink and the paper – what's known as the "consumables." They're the things we consume every time we print out another photo of Baby Jane. And they're very expensive. In the old days, this marketing strategy was known as sell the razor blade strategy. Sell the handle cheap. Sell the razors expensively. This naming concept fell into disrepute when razor handles themselves became expensive.

Consumer Consultative Forum CCF. A forum convened by the ACA (Australian Communications Authority) to consult with consumers and their representatives about communications issues.

Consumer Protection Act, Telephone TCPA. Legislation passed by Congress and signed by the president in 1991. The Telephone Consumer Protection Act of 1991 restricts specific types of unsolicited telephone calls. Among the provisions were a prohibition on calling emergency numbers or numbers for which the recipient was charged, limiting the placement of unsolicited calls to between 8 am and 9 pm, and removing people from calling lists who request that they not be called again. The Act also makes it unlawful for any person to use a computer or other electronic device, including fax machines, to send any message unless such message clearly contains in a margin at the top or bottom of each transmitted page or on the first page of the transmission, the date and time it is sent and an identification of the business or other entity, or other individual sending the messagte and the telephone number of the sending machine or such business, other entity, or individual. The telephone number provided may not be a 900 number or any other number for which charges exceed local or long-distance transmission charges.

contact A strip or piece of metal which makes an electrical contact when some electromechanical device like a relay or a magnet operates. Contacts are often plated with precious metal to prevent them from oxidizing (i.e. rusting) and thus messing up the switch. Contacts can be male (pins) or female (sockets).

contact arrangement The number, spacing and arrangement of contacts in a connector.

contact card See Smart Card or vCard.

contact center A fancy name for a call center which includes email and instant messaging and possibly a tie-in to a customer database – so that each time the phone rings the customer's record appears on a computer screen. Thus the agent is able to say "Good morning, Mr. Smith" and address his needs far more quickly. The alternative is to ask for the customer's account number, type it into a screen, wait for the computer to respond, etc. All this, while the customer sits on the phone and twiddles his thumbs. See also call center.

contact history A log of all the contacts, either by phone or letter, made with a prospect or customer. This is an important factor in building up a marketing database which can be used to accurately target prospects.

contact image sensor Uses a flat bar of light-emitting diode that directly touches the original. It eliminates the step of having the diodes move through the lens, which causes poorer resolution. This method is more sophisticated than the charged-coupled device scanning method.

contact management A business has customers and prospects. In computerese, they're called "contacts." Software to "manage" your customers and prospects is called contact management software. It has three elements: First, a screen or two of information about that contact (address, phone number, notes about your conversations, etc.) Second, the ability to print lists, and mailers, etc. And third, often a tie-in with your phone system to let your computer dial your clients and fax them stuff. With many newer phone systems, you have one extra benefit – namely when your phone rings, your contact management software will receive the calling phone number and pop up the screen or two about your contact. This way you'll be a little prepared before you answer the phone. See also ANI and CLID.

contact management software See Contact Management.

contact region The section of the jack wire inside the plug opening as shown in Subpart F of FCC rule 6B, figures 6B.500 (a) (3) and 6B.500 (b) (3).

contact resistance Resistance is basically the opposition to electron flow in an electrical circuit and connector manufacturers strive to attain the lowest amount of resistance possible for each contact. Contact resistance is the cumulative resistance value for mated contacts.

contactless smart card An awkward name for a credit card or loyalty card that contains an RFID chip to transmit information to a reader without having to be swiped through a reader. Such cards can speed checkout, providing consumers with more convenience.

CONTEL See Continental Telecom Inc.

contended access In local area networking technology it's the shared access method that allows stations to use the medium on a first-come, first-served basis.

contending port A programmable port type which can initiate a connection only to a preprogrammed port or group of ports.

content In today's information rich and hyped society, "carriage" is the new name for transmission. And "content" is the new name for what we carry. Content is a more than just phone calls, of course. It's movies, music, games, on-line books, information, etc. Content used to be called information. Now it's called content. You figure. If words, pictures, sound or video are used as part of buying or selling, they are "transactive content," says Stanley Dolberg, an analyst at Forrester Research Inc. In the computer world, according to William Safire, content means "information on a Web site." Companies who provide content are called content suppliers, or OSPs (on-line service providers). A content provider was once called an information provider. See also Content Supplier.

Content Addressable Memory CAM. Imagine you're building a machine to switch conversations on the Internet. Since the Internet uses packets switching, you have to read the packets that come flying at you, figure quickly where to switch them (based on rules and a database) and switch them quickly. A CAM does holds the rules and the database and works very quickly. Today's CAMs let you ask them millions of questions each second. CAMs are used in today's heavy duty IP (Internet Protocol) routers. Here's a technical definition:

A CAM is a semiconductor integrated circuit that allows a table of data to be stored in a memory array that incorporates special circuitry to permit a search function. In addition to read, write and special CAM specific operations (e.g., invalidate entry, move entry, etc.), the CAM storage array allows a search word or "comparand" to be applied to the device whereby the CAM performs a massively parallel compare of the comparand to all of the valid entries stored within the CAM. If a match is found, the address of the matching entry is outputted from the CAM. In the case where multiple entries match the comparand (as is frequently the case with "ternary" CAMs that permit entries to have bits with a "don't care" state that match against either a "1" or a "0"), the CAM may have a priority encoder to resolve the address of the highest priority matching entry (typically the matching entry at the address closest to 0).

CAMs are now used in data packet forwarding and classification in networking equipment, commonly called routers and switches. Here, CAMs perform look-up functions based on elements of a cell/frame/packet/datagram header to make intelligent forwarding decisions at wire speed. To satisfy the constant need for increasing bandwidth and table size, CAMs have become very fast and dense. State-of-the-art CAMs are pipelined to increase the look-up rate of the device, allowing several look-up functions to be executed in parallel, achieving look-up rates of up to 100 million look-ups per second. Ternary storage densities

of up to nine megabits is becoming possible in one device. CAMs also typically include the ability to "cascade" whereby several devices are connected together to form very large tables capable of storing well over one million entries.

Content Data Network See CDN.

content delivery On the Internet, content delivery (sometimes called content distribution or content caching) is a service that entails copying pages from a Web site to geographically dispersed servers and, when a page is requested, dynamically identifying and serving the page from the closest server to the user, enabling faster delivery. See also Content Delivery Service Providers.

content delivery network CDN. When web traffic gets heavy, performance suffers and connections can become excrutiatingly slow. A growing number of organizations are using content delivery networks (CDNs) to solve the problem. By storing data on different servers across the Internet on content delivery service providers, it is possible to ensure that content is closer – and more quickly accessible – to users. Such network architecture are devised to maximize the efficiency of distributing and delivering information across public Internet and large-scale private networks. A CDN uses intelligent switches, cache servers, traffic managers, content routers, and other technologies. See Content Delivery Service Providers.

content delivery service providers These companies own servers that reside on the Internet and basically guarantee the delivery of Web content to end-users. Customers of these content delivery providers are heavily trafficked Web sites, e-commerce sites, and push content providers such as news channels. The main idea is to place content as close to customers as possible to help speed delivery. Once a data request comes into a Web site, instead of the original Web site's server handling the request, a content delivery service provider takes over the task. These service providers all use proprietary software algorithms and probing techniques to guarantee that the content gets from their servers to the end-users quickly. Typically, the data that reside on these content delivery service provider's servers for transmission to end-users have been bandwidth-intense data flow such as streaming video or audio. Some of the early content delivery service providers were Akamai Technologies, Adero, and Sandpiper Networks (acquired by Digital Island).

content filtering The filtering of email based on the contents of the header and the body of the email message.

content management system CMS. Imagine you're running website selling household goods. You have photos, descriptions. physical specifications and pricing of thousands of products. Your customers expect detail and accuracy. But every day your suppliers introduce new products and obsolete old ones. How do you keep up? Answer: software called a content management system. Here's a formal description: A content management system is a method of managing unstructured content (i.e., information) for Web-based (i.e., Internet, Intranet, and Extranet) access via a browser. A CMS is an applications software system, commonly modular in nature, that performs a number of functions. Once the content (e.g., a white paper, product description, article, brochure, spreadsheet, drawing or schematic, or executable program) has been developed, it is checked into the content repository, which resides on a content server. Along with the content, various metadata (i.e., data that can change, and can help others find and view the content), is checked in. The content is then transformed into a Web-viewable format though a refining process, and all words in the content are indexed for future searches. The publishing module may be capable of handling dynamic content "on the fly," which feature is critical if fast-breaking news must be posted in near-realtime. Finally, the content is run through a content publisher module, which creates a file in a content folder, and establishes links to files and folders containing previously published content. The publishing process typically is template-based. The CMS also may include a security mechanism that grants or denies access to content, perhaps at the content, folder, or even page level. Such a security mechanism is particularly important if links are available to databases, as access privileges to sensitive data must be managed carefully. An embedded search engine might support searches by subject, full-text description, author, department, date range, or content type. Content subscription may be supported, allowing the subscriber to be automatically notified via e-mail when specific content changes. The CMS may also automatically reformat content for access by cellular phones, PDAs, pagers, and other Web-enabled wireless devices.

content processing Voice processing is the broad term made up of two narrower terms – call processing and content processing. Call processing consists of physically moving the call around. Think of call processing as switching. Content consists of actually doing something to the call's content, like digitizing it and storing it on a hard disk, or editing it, or recognizing it (voice recognition) for some purpose (e.g. using it as input into a computer program).

content provider 1. In the worlds of the Internet and the World Wide Web

(WWW), the Content Provider is the company which provides the material (content), rather than the network. NTT DoCoMo runs a cell phone company whose users can access thousands of special NTT pseudo-Web sites and do things – like download ring tones for their phones, find out what the price of their favorite stocks are, read their astrology forecast for the day. The people who run these sites are typically called content providers. When AOL bought Time Warner, it did so because it wanted an assured supply of content. See bitpipe and content supplier.

2. A fancy name for a writer, also called a language therapist by William Safire in the Sunday New York Times Magazine of January 28, 1996.

content service provider CSP. Another name for a content provider. See Content Provider.

content supplier Content is a new fancy name for what telecommunications facilities carry. It includes movies, music, games, on-line books, information, etc. Content suppliers are thus movie studios, publishers, and music companies.

content switching See Layer 7 Switching.

content to go Content designed for mobile devices.

contention Contention occurs when several devices (e.g., phones, PCs, or workstations) are vying for access to a line and only one of them can get it at one time. Some method is usually established for selecting the winner (e.g., first in, first out; or camp on) and accommodating the loser (e.g., giving them a busy tone or putting it in queue). When you cannot get an outside line from your PBX extension you have been in contention and lost.

Context-Based Access Control See CBAC.

context dependent soft keys Many telephones now have an LCD screen. Sometimes such screens have unmarked keys underneath them and/or at their side. What these keys do depends on the "labels" appearing on the screen. They are called "context dependent" because what those keys do depends on where the call is at that time. The first context dependent soft keys were on the Mitel SuperSet 4 phones. When the handset was resting on the phone, only three of the six context sensitive keys had meaning. One said "Program," one said "Msg" and one said "Redial." When you picked the phone up, three buttons would now be alive. One would say "Page," one would say "Redial" and one would say "Hangup." If the phone rang and you picked it up, one button would now say "trans/conf" (meaning transfer/conference. When another phone was ringing, one button would say "Pickup," letting you push that button and answer someone else's phone. And so on. The neatest implementation of context sensitive keys was probably on the Telenova (now no longer manufactured). At one point when you were in voice mail, this phone's six buttons looked exactly like a cassette recorder – record, play, fast forward, fast reverse, etc. It was brilliant. No one has ever made using voice mail so easy.

context keys Buttons on a phone or device that have a display next to them. The buttons perform different functions depending on the what the screen shows when you press the button. See Context Dependent Soft Keys.

context sensitive A term from the computer industry which means that "Help" is only a keystroke away. Hit F1 and Help information will flash on the screen. That information will be relevant to what you're doing now, i.e. that help is within the context of what's going on right this moment. See also Context Dependent Soft Keys.

context switch The act of stopping one running task in a microprocessor and starting another. Context switches are performed by the kernel. The technique with which an Intel microprocessor handles multitasking is called a context switch. The CPU performs a context switch when it transfers control from one task to another. In the process, it saves the processor state (including registers) of one task, then loads the values for the task that is taking control. Context switching is the kind of multitasking that is done in standard mode Windows, where the CPU switches from one task to another, rather than allocating time to each task in turn, as in timeslicing.

contextual ecommerce Imagine you receive a email from your friendly CD supplier. In it, he talks about the latest from Madonna. The email mentions the name of the CD. You notice its title is in blue and underlined – like a hot link to a Web site. You click on it. Instantly, you've bought the CD. You receive it the next day by Fedex. Bingo, we now have contextual ecommerce.

contiguous port Ports occurring in unbroken numeric sequence.

contiguous slotting This term refers to the process of selecting individual DS-0 circuits, within a DS-1 circuit or DS-3 circuit, which are adjacent to one another. Due to the timing difference which can result when non-adjacent channels are selected, contiguously slotted channels are preferable when the end equipment is designed to multiplex the individual low-speed channels into a single, higher speed connection.

Contiguous United States CONUS. The area within the boundaries of the District of Columbia and the 48 contiguous states as well as the offshore areas outside the boundaries of the coastal states of the 48 contiguous states, (including artificial islands, anchored vessels and fixed structures erected in such offshore areas for the purpose of exploring for, developing, removing and transporting resources therefrom) to the extent that such areas appertain to and are subject to the jurisdiction and control of the United States within the meaning of the Outer Continental Shelf Land Act, 43 U.S.C. Section 1331, et seq. CONUS is often used to refer to the footprint (i.e., area of coverage) of a satellite that covers the contiguous United States.

Continental Telecom Inc. CONTEL. A telephone company made up of more than 600 small phone companies. In 1990 it merged with GTE in a tax-free swap of shares. CONTEL was formed and grown by Charles Wohlstetter, an ex-stockbroker, who became financially comfortable (to say the least) in the process of growing CONTEL. In late 1990, CONTEL merged with GTE, which is a euphemism for GTE buying CONTEL. The folks who used to work for CONTEL call themselves "ex-cons."

Continental Morse Code See Morse Code.

continuity An uninterrupted electrical path.

continuity check A check to determine whether electrical current flows continuously throughout the length of a single wire on individual wires in a cable. See also wiremap.

Continuity Check Tone CCT. A single frequency of 2000 Hz which is transmitted by the sending exchange and looped back by the receiving exchange. Reception of the returned indicates the channel is working. See ITU-T Recommendation.271.

continuous A word used in voice recognition to mean a type of recognition that requires no pause between utterances.

continuous backup See continuous data protection.

continuous data protection The use of mirroring, replication and other backup technologies to continuously track and save copies of data to disk in real time or in near real time, so that it can be recovered from any point and at any point in time. Also called continuous backup.

continuous DTMF This is a feature of some phones (especially cellular phones) that sends touchtone sounds for as long as the key is held down, allowing access to services such as voice mail and answering machines that need long-duration tones. Some phones automatically have continuous DTMF; some don't. It's worth checking. Continuous DTMF makes a lot more sense.

continuous information environment A term for the world we live in – in which information (text, voice, video, images, etc.) is flowing at us continuously. And our job is, somehow, to manage the information. The idea is to use the new computer telephony terms to manage the information.

continuous phase modulation CPM. An efficient means of modulation for purposes of digital transmission over a radio system, such as microwave. CPM modulates the signal by changing its phase, or position, much as does Phase Shift Keying (PSK) in modems. CPM is a memory-dependent technique which requires that the receiving device compare the value of the starting phase of the transmitted signal to the value of the ending phase of the previously transmitted signal. Thereby, the value of the transmitted symbols can be determined, as long as the transmitter and receiver are carefully synchronized and the bit intervals, therefore, are consistent in time. Each value can represent one or more bits, depending on whether a compression technique is used to improve the efficiency of data transmission. See also PSK.

continuous partial attention A disease of the Internet age. Two people doing six things simultaneously, devoting only partial attention to each one. It seems that we can no longer find the off switch on our BlackBerries, our laptops and our cellphones, etc. See also Age of Interruption, the.

continuous vulcanization Simultaneous extrusion and curing of elastomeric wire coating materials. wire or cable can withstand prior to breakdown.

continuous waves CW. A series of electromagnetic waves or cycles, all of which have a constant or unvarying amplitude. Continuous wave usually refers to the output of a device (e.g., an optical fiber laser) which is turned on, but which is not modulated with a signal.

continuously variable Capable of having one of an infinite number of values, differing from each other by an arbitrarily small amount. Usually used to describe analog signals or analog transmission.

contract For the purpose of developing applications in the telecommunications industry, there are two types of contracts: Active and Passive. An active contract is one you must sign. A passive contract is the type of contract you find in a software package. By opening the shrink wrapped package, you are committing yourself to the terms of the contract inside the package – the terms of which mostly consist of not duplicating the software in an unauthorized way.

control In switching systems, the overall control of the switches. This includes monitoring to determine when action is needed, logic to determine what action is needed, and command, to initiate the actions.

control cable A multiconductor cable made for operation in control or signal circuits.

control channel A control channel is a logic channel carrying network information rather than actual voice or data messages. Within a cellular telephone system, several of the channels are assigned as 'control' channels. Instead of supporting voice communications, these channels allow the base station to broadcast information to the cellular phones in its area. Cellular phones continuously monitor this broadcast information, selecting the base station that provides the best signal.

control character A non-printing ASCII character which controls the flow of communications or a device. Control characters are entered from computer terminal keyboards by holding down the Control key (marked CTRL on most keyboards) while the letter is pressed. To ring a bell at a remote telex terminal, an operator could hold down the CTRL key, and tap the "G" key, since Control-G is the BELL character. Most computers display Control as the "^" character in front of the designated letter. For example, ^M is the Carriage Return character.

control circuit X.21 interface circuit used to send control information from DTE to DCE.

control connections A Control VCC links the LEC to the LECS. Control VCCs also link the LEC to the LES and carry LE_ARP traffic and control frames. The control VCCs never carry data frames.

control equipment 1. The central "brains" of a telephone system. That part which controls the signaling and switching to the attached telephones. Known as the KSU (or key service unit) in a key system.

2. Equipment used to transmit orders from an alarm center to remote site to enable you to do things by remote control.

control field Field in frame containing control information.

control flag A cellular phone term. A 6-bit flag transmitted in the forward channel data stream, comprised of a 5-bit busy/idle flag and one bit of the 5-bit decode status flag.

control head roam lights Indicates that the cellular phone is outside the "home" system.

control messages Signalling messages that provide the control of setup, maintenance, and teardown of L2TP sessions and tunnels. See L2TP.

control of electromagnetic radiation 1. Measures taken to minimize electromagnetic radiation emanating from a system or component, or to minimize electromagnetic interference. Such measures are taken for purposes of security and/or the reduction of interference, especially on ships and aircraft.

2. A national operational plan to minimize the use of electromagnetic radiation in the United States and its possessions and the Panama Canal Zone in the event of attack or imminent threat thereof, as an aid to the navigation of hostile aircraft, guided missiles, or other devices.

control of flow language Programming-like constructs (IF, ELSE, WHILE, GOTO, and so on) provided by Transact-SQL so that the user can control the flow of execution of SQL Server queries, stored procedures, and triggers. This definition from Microsoft SQL server.

control panel The control panel on the Apple Macintosh is for general hardware and software settings. Icons allow a user to customize the system or application, or select a particular service, such as a specific printer, set the sound level, the date and time and choose an Ethernet connection through the network control panel.

control plane The ATM protocol includes a Control Plane which addresses all aspects of network signaling and control, through all 4 layers of the model.

Control Point CP. In IBM SNA (Systems Network Architecture) networks, a Control Point is a type of NAU (Network Accessible Unit, previously known as Network Addressable

Pin	Control Signal	From	To
4	Request-To-Send (RTS)	DTE	DCE
5	Clear-To-Send (CTS)	DCE	DTE
6	Data Set Ready (DSR)	DCE	DTE
8	Carrier Detect (CD)	DCE	DTE
20	Data Terminal Ready (DTR)	DTE	DCE
22	Ring Indicator (RI)	DCE	DTE

Unit). A CP manages the network resources within its domain of control, controlling the activation and deactivation of resources and the monitoring of their status. Such resources can include physical resources such as links and nodes, and logical resources such as network addresses. As a Network Accessible Unit, a Control Point is accessible over the SNA network itself. See also SNA.

control segment A worldwide network of Global Positioning System monitoring and control installations that ensure the accuracy of satellite positions and their clocks.

control signal 1. In the public network, control signals are used for auxiliary functions in both customer loop signaling and interoffice trunk signaling. Control signals are used in the customer loop for Coin Collect and Coin Return and Party Identification. Control signals used in interoffice trunk signaling include Start Dial (Wink or Delay Dial) signals, Keypulse (KP) signals or Start Pulse (ST) signals.

2. In modem communications, control signals are modem interface signals used to announce, start, stop or modify a function. Here's a table showing common RS-232-C and ITU-T V.24 control signals

control station On a multi-access link, a station that is in charge of such functions as selection and polling.

control theory The mathematical analysis of the systems and mechanisms for achieving a desired state under changing internal and external conditions.

control tier An AT&T term for the tier within the Universal Information Services network node that provides the transport network's connection control function.

control unit An architectural component of a processor chip which orchestrates processor activity and handles timing to make sure the processor doesn't overlap functions.

controlled access When access to a system is limited to authorized programs, processes or other systems (as in a network).

controlled environment vault CEV. It is a low maintenance, water-tight concrete or fiberglass container typically buried in the ground which provides permanent housing for remote switches, remote line concentrators, pair gain and fiber transmission systems. Because it is buried, it can often be installed in utility easements or other places where local building laws may be a problem. This below ground room that houses electronic and/or optical equipment is under controlled temperature and humidity conditions.

controller 1. In the truest sense, a device which controls the operation of another piece of equipment. In its more common data communications sense, a device between a host and terminals that relays information between them. It administers their communication. Controllers may be housed in the host, can be stand-alone, or can be located in a file server. Typically one controller will be connected to several terminals. The most common controller is the IBM Cluster Controller for their 370 family of mainframes. In an automated radio, a controller is a device that commands the radio transmitter and receiver, and that performs processes, such as automatic link establishment, channel scanning and selection, link quality analysis, polling, sounding, message store and forward, address protection, and anti-spoofing.

2. Participant in a conference call who sets up the conference call.

controller area network A serial bus network, originally developed for automotive applications in the early 1980s, for use in distributed real-time control systems (e.g., auto manufacturing, factory automation and industrial machine control, building automation, elevators, and avionics). The CAN protocol was internationally standardized in 1993 as ISO 11898-1.

Controller Card Also called a hard disk/diskette drive controller. It's an add-in card which controls how data are written to and retrieved from your PC's various floppy and hard drives. Controller cards come in various flavors, including MFN and SCSI. Controller cards are the devices used to format hard drives. Controller cards are not hard drive specific (except within categories). Controller cards will format many drives. But once you have a hard drive that has been formatted by that one controller card, it tends to prefer talking to that controller card forever. If you switch your hard disk to another machine, switch the controller card along with it. If you switch your hard disk to another machine, but not the controller card, then format the hard disk. That's not a "100% Do It Or Else You'll Be Disappointed" rule. But just a "Play It Safe and Switch Them" rule.

controllerless modem A modem that shifts all the protocol management, error detection and correction, and data compression onto software running on the system's CPU. This allows the modem manufacturer to make a much cheaper modem that does not require the memory or processing power of a traditional modem. Also known as a soft modem.

conturing In digital facsimile, density step lines in received copy resulting from analog-to-digital conversion when the original image has observable gray shadings between the smallest density steps of the digital system.

CONUS A term for CONtiguous United States (lower 48 states). See Contiguous United

States.

convection cooling Design techniques used in switching system construction to permit safe heat dissipation from the equipment without the need for cooling fans.

convector The device which covers the steam heating radiator in buildings and typically sits underneath a window. Also called a weathermaster.

convector area An area allocated for heat circulation and distribution. Convector areas, typically built into a wall, can be used as a satellite location only if a more suitable area is unavailable.

convention A rule of conduct or behavior which has been reached by general agreement, commonly by a standards-making body, whether formal (e.g., the ITU) or ad hoc (e.g., Bell Telephone Laboratories) in nature. For example, the T-1 framing conventions were developed by Bell Labs for use within the Bell System network in North America, and later were formalized at the international level by the ITU-T. See also Bell Telephone Laboratories, ITU-T, and T-1.

conventional signaling The inter-machine signaling system that has been traditionally used in North America for the purpose of transmitting the called number's address digits (telephone number) from the originating end office to the switching machine that will terminate the call. In this system, all dialed digits are received by the originating switching machine, a path is selected, and the sequence of supervisory signals and outpulsed digits is initiated. No overlap outpulsing, ten-digit Automatic Number Identification (ANI), information digits, or acknowledgment wink are included in this signaling sequence.

converged network, Used to describe trends toward the bundling of services by operators. Principally found in the United States, where regulation historically has separated the local and long distance carrier functions.

convergence 1. A measure of the clarity of a color monitor. A measure of how closely the red, green and blue guns in a color monitor track each other when drawing a color image. The other measures are focus and dot pitch.

2. A routing term. The point at which all the internetworking devices share a common understanding of the routing topology. The slower the convergence time, the slower the recovery from link failure.

3. The word to describe a trend – now that most media can be represented digitally – for the traditional distinctions between industries to blur and for companies from consumer electronics, computer and telecommunications industries to form alliances, partnerships and other relationships, in order to raid each others markets.

4. The word "convergence" as a fashion word of the "new" management was set in motion in 1992 when Tele-Communications Inc. chairman John C. Malone told a cable-show audience that his vision of all-digital, fiber-optic networks would enable TCI and other cable operators to offer 500 TV channels, interactive programming, electronic mail, and telephony. According to Business Week Magazine of June 23, 1997 that picture of digital convergence was so compelling that cable, media and phone companies promptly hopped on the bandwagon. Business Week continued, "Several billion dollars later, it become clear that convergence was a bust. Cable companies, perennially strapped for cash, scaled back on their plans to upgrade their networks to handle huge amounts of interactive data. Phone companies that had hoped to offer television service – on their own wires or in joint ventures with cable companies – went back to their core businesses." However, such concerns should never let a good word die. In the May 17, 1998 issue of The New York Times, Richard C. Notebaert, chairman and chief executive of the Ameritech Corporation, wrote, "Conventional wisdom holds that convergence – the gradual blurring of telecommunications, computers and the Internet – is primarily about technology and the inevitable clash of voice and data networks. But that narrow viewpoint misses the bigger picture...Convergence is about fundamental changes in the way we work – even behave." Now the latest concept of convergence is that all communications – the Internet and the PSTN (the public switched telephone network) – shall run over one IP telephony network. As Notebaert says, "Our public voice network will become the public multimedia network and the Internet as we know it will cease to exist. With such a robust and ubiquitous network, we'll never have to go to the time and trouble of dialing into a private network when we want to surf the Web. In essence, we'll always be on line. And that will let us develop applications we can't even dream of today." No industry is immune from the convergence fever. Every year, a trade magazine or two changes its name to include the word convergence in the hope that major profits will magically appear. Ultimately the concept of broadband IP telecommunications took hold – and slowly our CATV coaxial cable is bringing us TV channels, the Internet, phoning around the world and movies on demand. The phone companies are running to catch up with their DSL service. And soon some of these services will come through the air – see WiMAX. For now, the world convergence is out of fashion.

convergence billing Also known as convergence or composite billing. This is a fancy name for one phone company – local or long distance – providing a total communications bill to the customer. That total bill would include everything the customer buys in telecommunications services – from local, long distance, Internet access, cell phones, paging, etc. In late 1996, the belief developed in the telecom industry that if you "controlled" the bill to the customer, you would be in far better shape to sell the customer more services. The concept has some validity, especially if you also believe in fairies.

Convergence Sublayer CS. An ATM term. SEE CS.

convergent Convergent billing software is software which allows telecom companies (such as local and long distance companies) to bundle services, such as long distance, cellular, paging and cable, together onto a single monthly invoice. Bundling helps service providers offer competitive rates, boost revenue per customer and reduce customer turnover. Customers love the simplicity and convenience of one bill for all their telecom services. One company calls itself a "one stop shop with an integrated bill."

conversation path The route from originating port to terminating port for a two-way call. A conversation thus typically requires two ports on most PBXs.

conversation time The time spent on a conversation from the time the person at the other end picks up to the time either of you hang up. Conversation time plus dialing, searching and ringing time equal the time your circuit will be used during a call.

conversational mode Also called chat mode. Interactive data communications carried on between data terminals in a fashion similar to speech conversation.

conversational mode telex An MCI International product providing real time exchange between Telex terminals or other compatible devices that allows instantaneous, two-way conversations in writing.

conversion In signaling, the substitution of one, two, or three digits for received digits for the purpose of directing the call through the next office.

conversion rate Conversion rate is a Web term. It is a measure of the people who log on to your site and then click through to a second page. The higher that percentage is, the more interesting the site looks to them. Basically, a web site owner wants to get people off the first page and into the areas that make you money.

converter 1. A vacuum tube which combines the functions of oscillator and mixed tube.

2. A device for changing AC to DC and vice versa. An ancient radio term.

3. An adapter, such as one that allows a modular phone to be plugged into a 4-hole jack.

4. A British term. A repeater that also converts from one media type to another, such as from fibre (British spelling) to copper. Often called a media adaptor.

5. A device used in RF distribution systems to convert from one frequency to another. May also control channel access.

6. A device used to convert from one transmission media to another (Example: Fiber/Copper Media Converter). Converters are usually externally powered as they physically "repeat" or regenerate the signal.

convolutional code Error protection code encoding data bits in a continuous stream. An error-correction code in which each m-bit information symbol to be encoded is transformed into an n-bit symbol (n>m) where the transformation is a function of the last k information symbols, and k is referred to as the constraint length of the code. Convolutional codes are often used to improve the performance of radio and satellite links.

cookie The first batch of cookies were originally cooked up as simple mechanism to help make it easier for users to get to their their favorite Web sites without having to go through the lengthy process of identifying themselves every time they visited. A cookie is now a simple text file on your hard disk that has been placed by a web site you visited. For example, some shopping cart technologies allow you to return to shopping at a later time. What this means is that a "cookie" containing your order is placed on your computer for the site's computer to retrieve when you return. A cookie is a basically a mechanism which is a feature of the HTTP (Hypertext Transport Protocol) protocol used in the Internet and WWW. In this client/server environment, a cookie allows the server side of the connection to both store and retrieve information on the client side. When connecting to a WWW computer in the form of a server, that server can store information on your client PC. This process takes place when the server returns a HTTP object such a graphic or screen to the client computer; at that point, the server also may send a set of "state" information which is stored on the client hard drive. This information is persistent, meaning that it remains in memory for subsequent use by the specified URL (Uniform Resource Locator, which is the address of the Web site) or group of URLs. As future client requests are made against the server, the cookie is automatically passed from client to server.

Cookies are used widely in the client/server environments of the WWW element of the Internet, as well as in Intranets. Their advantage is that they can automatically identify the client to the server, thereby shortening or eliminating the user identification element of the log-in process. For example, an electronic shopping application can use a cookie to identify the shopper during subsequent access sessions, storing information about shopping preferences. Further, the service provider can alter the content of the accessed Web site to appeal more to the client user based on that specific user's profile. The downside is that cookies are placed on the client computer without the knowledge of the user, giving rise to concerns about privacy through electronic trespass. Some Web browsers such as Netscape Navigator will alert the user to the desire of the server to apply a cookie, thereby providing the user with the option of rejecting that request. Cookie blocking also can be accomplished through the use of "Cookie.Cutter," developed by Phil Zimmerman, who gained fame with his development of PGP (Pretty Good Privacy) encryption software.

So, why are people so concerned about cookies? Let's consider an example. I recently was cruising the Web and tapped into the Web site of a book distributor. I had the opportunity to develop a profile of my interests. I chose not to do that for the same reason that I choose to shop for clothes without the assistance of a salesperson. Basically, I don't want to be slotted into a particular style, price range, etc. Rather, I prefer to scan the options, quickly, unassisted and without pressure. When I need the assistance of a salesperson, I'll ask for it. Further, I prefer that my identity be a private matter. I like Caller ID (when I am the calling party), but there may be times when I want to make a conscious decision to block it. I especially don't like cookies downloaded to MY computer, and I really especially don't want someone else taking the liberty of putting stuff into my computer without my OK. It's sort of like putting something in my car or in my house without my approval. It may be harmless and well-intentioned, but it's a "privacy" thing. See Client/Server, Cookie File, HTTP, URL and Caller ID.

According to Microsoft, "A cookie is a very tiny piece of text we're asking permission to place on your computer's hard drive. If you agree, then your browser adds the text in a small file. Its purpose is to let us know when you visit microsoft.com. This text, by itself, only tells us that a previous microsoft.com visitor has returned. It doesn't tell us who you are, or your email address or anything else personal. If you want to give us that information later, that's your choice.

So why do we offer cookies? Cookies help us evaluate visitors' use of our site, such as what customers want to see and what they never read. That information allows us to better focus our online product, to concentrate on information people are reading and products they are using. And guess what? A cookie can help you. If you accept a cookie, nothing affects you immediately. But you know what happens whenever you want to download software, access a premium site or even request permission to use a Microsoft logo on your Web page? You get asked questions like who you are and your email address. And that happens every time you want to download stuff. If you have accepted a cookie, however, those questions eventually will be asked just once, no matter how often you download software or how many Microsoft sites you visit. In the future, a cookie will allow you to tell us what information you prefer to read and what you don't. If you're a gamer, for example, we can advise you on content specific to games. Why are we telling you all this? Because we want you to know why we ask you to accept a cookie. We want to be sure you understand that accepting a cookie in no way gives us access to your computer or any personal information about you. Cookies are harmless, occupying just a few bytes on your hard drive. They also can be a Web site browser's very good friend.

That having been said, consider the concept of a "third party cookie," also known as a DoubleClick cookie, after DoubleClick Inc. (www.doubleclick.net, but don't touch that website unless you want to run the risk of eating a DoubleClick cookie). After being planted by a participating Web site, a third-party cookie will follow you around the Web, recording your movements and interests.

cookie file The file (usually in your browser's directory structure) where cookies are kept. The file name is cookie.txt, just in case you'd like to delete it. See Cookie.

cookie jar reserves A kind of corporate slush fund used to doctor quarterly earnings reports – typically to make earnings seem to grow more consistently. Enron created such slush funds in late 2000 and early 2001 in order to reduce its profits and so dampen the likely political firestorm that might occur, especially in California, which had suspected Enron of "price gouging."

cool When I was growing up, things that were fun were "hot" or "groovy." Now they're "cool," which means they're in fashion. I first heard the word used by some Microsoft employees in the early 1990s. Ray and Margaret Horak say I should remove this definition since they used the term in the 1950s, when they were cool and I was groovy.

CoOP Continuity in operations. Special planning that provides the continuation of critical functions during emergency conditions such as natural disasters, accidents, technology glitches, and military or terrorist attacks.

Cooper's Law An observation by Martin Cooper, founder and chairman of ArrayComm, that the number of transmissions (voice or data) that can take place over a given area of the useful radio spectrum doubles every two-and-a-half years, as it has been doing for over 100 years. The improvements in spectrum utilization are due to improvements in frequency division, modulation techniques, spatial division (i.e., confining the area used for individual conversations to smaller and smaller areas, thereby increasing overall frequency re-use), and an increase in the magnitude of the usable radio frequency spectrum.

Cooperation for Open Systems Interconnection Networking in Europe See COSINE.

cooperative content distribution See file swarming.

Cooperative Processing Mainframe and intelligent workstations dividing application code between them.

coopetition A made-up word which means that you partner with your competition. In short, the word is a blending of cooperation and competition. You might be a wholesaler of PCS services who is now partnering with one of your retailers to create a new service which might compete with you, at some point. That's coopetition. Joe Nacchio, head of Qwest, uses the term to describe the love/hate relationship of telecom carriers who compete with each other but often find themselves sharing facilities, equipment, resources, and personnel.

coordinates "Thank you for all your coordinates." A cool way of saying "Thank you for your address, phone and fax numbers and email addresses." The origin of the noun is in geometry. George Crabb in 1823 defined co-ordinates as "a term applied to the absciss and ordinates when taken in connexion," later better known as the magnitudes that determine the position of a point; geographers and navigators still later used coordinates to describe the use of longitude and latitude in locating a spot on the globe.

COP Cable Organizer Panel. A place to organize cables in a rack-mounted telecom system.

Copernicus, Nicolaus Polish astronomer (1473 - 1543) who advanced the theory that the Earth and other planets revolve around the Sun (the "heliocentric" theory). This was highly controversial at the time, since the prevailing Ptolemaic model held that the Earth was the center of the universe, and all objects, including the sun, circled it. The Ptolemaic model had been widely accepted in Europe for 1,000 years when Copernicus first proposed his new model.

copia Latin for abundance. Steve Hersee called his Wheaton, Illinois all-things-fax company, Copia International in the hope that he would become rich as a horn of plenty. See Cornucopia in your normal dictionary.

copolymer Compound resulting from the polymerization of two different monomers.

copper clad Steel with a coating of copper welded to it, as distinguished from copper-plated.

copper distribution data interconnection CDDI. CDDI is an FDDI technology that is adapted to work over copper wires rather than fiber.

copper-covered steel wire A wire having a steel core to which is fused an outer shell of copper.

coppertone Bob Metcalfe of InfoWorld Magazine coined coppertone for bare copper wire which you can rent from your local phone company. By "bare" he means that the copper you rent will contain no electronics on it anywhere, will be unloaded, unconditioned and unpowered. How might you use such copper wire? Let's say you have an office in New York City on on 12 West 21 Street and home on 215 West 19 Street. Let's say the office LAN had a T-1 connection to the Internet. Let's say you wanted to extend the LAN with its T-1 to your home. Simple, add networking equipment on your LAN and add similar equipment at your home. Bingo, your home is now on your LAN. Can it be done easily? Yes, but only if your phone company will rent you coppertone, i.e. bare copper wire. If they gave you something else – like one of their tariffed items – e.g. a 56 Kbps circuit – you would find yourself with a slower, more inconvenient, less flexible and most likely, more expensive solution. Telephone companies don't like to rent you coppertone to do this, for obvious reasons; in fact, several now refuse to lease coppertone except to "legitimate" burglar alarm companies. See also ADSL, HDSL, and Dry Copper.

The story goes that Metcalfe got his inspiration for the term coppertone because of the old Coppertone suntan oil advertisements, picturing a dog partially pulling down a child's bathing suit and exposing (hence, the term "bare") the pale butt of an otherwise very tan child. They had billboards of this all over Southern California for years.

coprocessor An additional processor which takes care of specific tasks, the objective being to reduce the load on the main CPU. Many IBM PCs and IBM clones have the capacity to install a coprocessor chip which does only arithmetic functions. This significantly speeds up your computer if you do a lot of calculations. See Math Coprocessor.

COPS Common Open Policy Service. A standard under consideration by the IETF (Internet Engineering Task Force) for exchanging policy information in order to support dynamic QoS (Quality of Service) in an IP (Internet Protocol) network. Such policy information is exchanged between PDPs (Policy Decision Points) and PEPs (Policy Enforcement Points). The PDPs generally are in the form of network-based servers that decide which types of traffic (e.g., voice and video) are afforded priority treatment. The PEPs are in the form of routers or IP switches that implement the decisions made by the PDPs. COPS-PR is a derivative used for device provisioning. COPS currently is intended to work in conjunction with RSVP (Resource Reservation Protocol), but may well take evolve to the level of independence. COPS enjoys broad vendor support, and is thought to be a likely replacement for SNMP (Simple Network Management Protocol). SNMP offers device-monitoring capabilities, but does not do an effective job of documenting device-configuration. COPS also involves active participation between PDPs and PEPs, while SNMP is a polling protocol. See also IETF, IP, Policy-Based Networking, SNMP, and QoS.

COPW Customer Owned Premises Wire. You own the telephone wiring in your office.

copy A nice new telephone system programming feature. We found it on Northern Telecom's Norstar phone. With this button, certain programmed settings can be copied from one line to another, or from one telephone to another. Line programmable settings that can be copied on the Norstar are Line Data, Restrictions, Overrides, and Night Service. Telephone settings that can be copied are Line Access, Restrictions, Overrides, and Permissions.

copyright A copyright protects the original author of a story, software program, song, movie, piece of sculpture, or other original work from direct copying. Copying may be inferred where the alleged copyist had access to the copyrighted work. The copyright notice (the copyright symbol (c in a circle), or the word "Copyright", the year of creation, and the name of the copyright owner) should be provided on each copy of the work. Copyrights may also be registered with the Library of Congress, but this is not necessary in all cases. Copyright protects the expression of an idea, not the idea itself. (In appropriate cases, patents can be used to protect the idea.) Where the idea is so simple that there is only one way to express it, the idea and its expression may merge, preventing copyrightability. This logic was used successfully in defense of several suits involving "clean room" reverse engineering of microcode: A first group hacked out the code, and prepared a complete functional specification defining the function of each instruction. A second group then wrote new code implementing these functions. Since there had been no copying, there could be no copyright infringement; the fact that both versions of the code for some instructions were identical merely showed that the idea and expression had merged. There are only so many ways to code an ADD instruction, after all. See Intellectual Property, Patent and Trade Secret.

COR Functionality that provides the capability to deny certain call attempts based on the incoming and outgoing class of restrictions provisioned on the dial peers. This functionality provides flexibility in network design, allows users to block calls (for example, to 900 numbers), and applies different restrictions to call attempts from different originators. COR specifies which incoming dial peer can use which outgoing dial peer to make a call.

CORBA Common Object Request Broker Architecture. An ORB (Object Request Broker) standard developed by the OMG (Object Management Group). CORBA provides for standard object-oriented interfaces between ORBs, as well as to external applications and application platforms. The yield is that of interoperability of object-oriented software systems residing on disparate platforms. Additionally, CORBA provides for portability of such systems across platforms. See also Object Request Broker and OMG.

cord 1. A small, flexible insulated wire.

2. The Cibernet On-Line Roaming Database. CORD is an on-line database that acts as a repository for information that wireless carriers need to exchange in order to support roaming in their territories.

cord board The earliest manual PBX. Usually an elegant wooden device consisting of lots of cords with plugs on them. These cords sat horizontally sticking up, like missiles in a silo. Each cord corresponded to an extension. Whenever the phone rang, the cord board attendant would answer it. Each incoming line was a vertical hole. When the operator had figured for whom the call was, he/she would simply plug the cord corresponding to the desired extension into the hole corresponding to the incoming trunk. The operator would reverse the process if the internal user wanted to make an external call. Either the operator would dial the call first, or simply plug in the user's extension and thus allow the user to dial the call directly. The tip of the plug and the circular ring of the plug gave the term "tip and ring" to telephony. In electronics, it's known as positive and negative. See Cord Circuit.

cord circuit A switchboard circuit, terminated in two plug-ended cords, used to establish connections manually between user lines or between trunks and user lines. A number of cord circuits are furnished as part of the manual switchboard position equipment. The cords may be referred to as front cord and rear cord or trunk cord and station cord. In modern

cordless switchboards, the cord circuit is switch operated. See Cord Board.

cord lamp The lamp associated with a cord circuit that indicates supervisory conditions for the respective part of the connection. See Cord Board.

cord switchboard A switching system consisting of positions at which trunk-to-trunk and line-to-trunk connections are established by operators using cords and plugs. In other words, the calls are connected by taking the caller's phone line and plugging it into the phone line of the person he wants to reach, or the next operator along the way. Such switching systems were called cordboards. They were useful because of the enormous level of service they provided phone users. The operators were often delightful. Cordboards, however, are very slow, and very labor intensive. Someone once estimated that, if they hadn't invented automatic people-less switching systems, everybody in the world would have needed to become a telephone switchboard operator.

cordboard See Cord Board.

cordless switchboard A telephone switchboard in which manually operated keys are used to make connections. See Cord Board.

cordless telephone A telephone with no cord between handset and base. Each piece contains a radio transmitter, receiver, and antenna. The handset contains a rechargeable battery; the base must be plugged into an AC outlet. Depending on product design, radio frequency, environmental conditions, and national law, range between handset and base can be 10 feet to several miles. Cordless phones were once all analog. Now a breed of digital ones is out. They work much better in electrically noisy environments – like the typical office.

Cordless Telephony Generation 2 CT-2. CT-2 is a European-designed, low-cost telephone system that is based on TDMA technology. The network operates with small, light, and inexpensive handsets. The CT-2 system consists of microcells that are several hundred yards wide. The service is designed to overcome the shortage of pay phones in certain areas.

Cordless Telephony Generation 2-Plus CT-2+. CT-2+ is an expansion of the CT-2 interface specification that would extend network capabilities and allow backward compatibility with CT-2 handsets.

core The central glass element of a fiber optic cable through which the light is transmitted (typically 8-12 microns in diameter for single mode fiber and 50-100 microns in diameter for multimode fiber). This light conducting portion of the fiber is defined by the high refraction index region. The core is normally in the center of the fiber, bounded by the cladding material. See also CORE.

CORE Council of REgistrars. An organization proposed to be charged with the responsibility for establishing and maintaining a new set of gTLDs (generic Top Level Domains) for the Internet. Effective March 1998, those gTLDs were to comprise the following: .arts (entities emphasizing cultural and entertainment activities); .firm (businesses, or firms); .info (entities providing information services); .nom (individual or personal nomenclature, i.e., a personal nom de plume, or pen name); .rec (entities emphasizing recreation/entertainment activities); .shop (businesses offering goods to purchase); and .web (entities emphasizing activities related to the World Wide Web).

The administration of the new gTLDs was contracted by CORE to Emergent Corporation, which was to develop, maintain, and operate the Shared Registry System (SRS). SRS is a neutral, shared, and centralized database of the new gTLGs. As many as 90 independent entities, known as "registrants," were authorized to register domain names, or URLs (Uniform Resource Locators), with each relying on the SRS. A URL, such as www.happypaintings. arts (this is not a real URL, at least not at the time of this writing), is translated into an IP address by a Domain Name Server (DNS), also known as a resolver. At the time of this writing, the proposal for new gTLDs has been forestalled. See also Core, DNS, IP Address, SRS, TLD, and URL.

core class A public class (or interface) that is a standard member of the Java Platform. The intent is that the Java core classes, at minimum, are available on all operating systems where the Java Platform runs. A 100% pure Java program relies only on core classes, meaning it can run anywhere. All core classes reside in the Java Package.

core eccentricity A measure of the displacement of the center of the core relative to the cladding center.

core ellipticity A measure of non-circularity – the departure of the core from roundness.

core gateway The primary routers in the Internet. Historically, one of a set of gateways (routers) operated by the Internet Network Operations Center at BBN. The core gateway system formed a central part of Internet routing in that all groups would advertise paths to their networks from a core gateway, using the Exterior Gateway Protocol (EGP).

core network 1. A combination of high-capacity switches and transmission facilities which form the backbone of a carrier network. End users gain access to the core of the network from the Edge Network.

2. Cellular networks essentially consist of two parts: the Radio Access Network (RAN), which controls transmission and reception of radio signals, and the Core Network, which provides switching, transport, and enhanced services for traffic emanating from and directed to the cellular network's RAN.

3. As used in some countries, core network refers to public networks for providing fixed telephone services, telex and leased lines.

core non-circularity The percent that the shape of the core's cross section deviates from a circle. Sometimes referred to as core ovality.

Core Processing Unit CPU. The card or shelf that controls the system or part of the system. It's called the CPU because all the RAM, subprocessors, buffers, clocking circuitry and ROM are included in this part of the system.

core router In a packet-switched star topology, a router that is part of the backbone and that serves as the single pipe through which all traffic from peripheral networks must pass on its way to other peripheral networks.

core services As used in some countries, core services refers to the international and national telephone, telex and leased lines.

core size Primary description of a fiber. Stated in microns. Does not include cladding. Determines end surface area which accepts and transmits light.

core switch A Broadband Switching System (BSS) which is located in the core of the network. Conceptually equivalent to a Tandem Office in the voice world, a core switch serves to interconnect "Edge Switches," which provide user access to the broadband network much as do Central Offices in the circuit switched voice world.

core wall A wall that runs between structural floor and structural ceiling to separate stairwells, elevators, etc. from the rest of the building.

cornea gumbo A visually noisy, overdesigned Photoshopped mess. "We've got to redesign that page, it's become total cornea gumbo."

corner case A problem that occurs when two or more particular conditions co-occur, or when two or more variables concurrently take on particular values, generally, but not necessarily always outside of the normal range of their individual expected values. A corner case would be something like the following made-up example: "If a network's timestamp clock is slow and the network is fast and if a packet's position is perturbed by more than three packets, then a timeout will occur." Ideally, corner cases are identified during testing, but often they are not identified until users use hardware, software, and networks in unanticipated ways. See edge case.

corner reflector 1. A device, normally consisting of three metallic surfaces or screens perpendicular to one another, designed to act as a radar target or marker.

2. In radar interpretation, an object that, by means of multiple reflections from smooth surfaces, produces a radar return of greater magnitude than might be expected from the physical size of the object.

3. A reflected electromagnetic wave to its point of origin. Such reflectors are often used as radar targets.

4. Passive optical mirror, that consists of three mutually perpendicular flat, intersecting reflecting surfaces, which returns an incident light beam in the opposite direction. 5. A reflector consisting of two mutually intersecting conducting flat surfaces.

cornet A Siemens protocol for PBX-to-PBX signaling over a Primary Rate connection.

corporate DNA A company's core values, culture, personality, etc., that supposedly gets passed along to all new employees. Corporate DNA is altered slightly every time a new person is hired. And a wholesale shift can occur by simply replacing the CEO.

Corporate GSM. This is how Opuswave defines the term on its web page: "Corporate GSM seamlessly integrates public wireless (GSM), enterprise voice (PBX) and enterprise data (LAN) communications networks and integrates fully with the legacy infrastructure of the enterprise." Personally, I think it's a made-up term, but my friend Joel Solkoff sent it and he wants it in. So it's in. Let's see if the term sticks.

corporate network Also called an internetwork or a wide area network. A network of networks (the mother of all networks) that connects most or all of a corporation's voice, data, and video resources using various methods, including the phone system, LANs, private data networks, leased telecommunications lines, and public data networks. Connections between networks are made with bridges and routers.

Corporate networks come in many shapes and sizes. Often, they will consist of networks within the same building or facility. Here, networks are combined using bridges and routers. Corporate networks may also span great distances. Such internetworks require different types of connections than single-facility internetworks, though the fundamentals are similar. Internetworks that connect remote facilities usually rely on some type of public

or leased data communications network provided by the phone company or a data network service company. Bridges and routers are still required to connect networks to the long-distance data service, whether it's an X.25 packet switched network, a T-1 line, or even a regular phone line. See also Bridge and Router.

correlation The AMA (Automatic Message Accounting) function that permits the association of AMA data generated at the same network system or at physically separate network systems. There are three levels of correlation that affect Advanced Intelligent Network Release 1: record level, service level, and customer level. Definition from Bellcore in reference to Advanced Intelligent Network.

corresponding entities Peer entities with a lower layer connection among them.

corridor cruisers The growing number of workers who spend most of their time in – or en route to – meetings. They're one of the main targets of the fledgling pocket PC industry. See corridor warrior.

corridor service A term that Verizon is using for calls to and from the New York City area to and from Northern New Jersey, or between Philadelphia and Southern New Jersey.

corridor warriors Those employees and execs, who spend their day racing from meeting to meeting, tethered to laptops so they can retrieve even the most basic of information, take notes, and remain linked to the rest of the world via e-mail. There is a second definition, namely in-house, IT service technicians who spend their days going from one office to another, fixing PC, LAN, network printer problems, etc. Corridor warriors use cell phones, BlackBerry devices or other wireless devices for messaging throughout the day as they roam the corporate corridors.

corrosion The destruction of metal by chemical or galvanic action.

COS 1. See Class of Service.

2. Compatible for Open Systems.

3. Corporation for Open Systems international. A Federal Government blessed organization which aims towards standardizing OSI and ISDN. COS members include everyone from end-users to manufacturers. COS deals with private and public networking issues.

COSINE Cooperation for Open Systems Interconnection Networking in Europe. A program sponsored by the European Commission aimed at using OSI to tie together European research networks.

cosine filtered Descriptive of a signal passed through a brick-wall filter modified by roll-off following a cosine law.

cosmic rays Atomic nuclei (mostly protons) and electrons that are observed to strike the Earth's atmosphere with exceedingly high energy.

COSPAS-SARSAT A satellite-based Search And Rescue (SAR) system that locates distress beacons transmitting at 121.5 MHz, 243 MHz and 406 MHz. It operates in liaison with two United Nations agencies, the International Maritime Organization (IMO) and the International Civil Aviation Organization (ICAO), and serves all organizations in world that are responsible for search and rescue (SAR) operations, whether the distress occurs at sea, in the air or on land. The system actually is two systems that work together. COSPAS is the English transliteration of the Russian acronym for Space System for Search of Distress Vessels. SARSAT is an acronym for Search and Rescue Satellite-Aided Tracking System. It wsa developed jointly by Canada, France and the USA, starting in the mid 1970s, after Canada successfully demonstrated the location of emergency transmitters from orbit. The two systems joined in 1979, and the first spacecraft, the Cospas-1, was launched that year. COSPAS-SARSAT serves three types of beacons: ELT (Emergency Locator Transmitter) for aircraft, EPIRB (Emergency Position Indicating Radio Beacon) for ships, and PLB (Personal Locator Beacon) for persons on land. Since 1982, more than 11,000 people have been rescued worldwide with the assistance of the system. By 2009, the 121.5 MHz and 243 MHz beacons will be phased out, in favor of the 406 MHz beacons, which, though they currently cost more, provide better alert data and can use Global Positioning System (GPS) data for accurate position fixes. www.cospas-sarsat.org.

CoSN Consortium for School Networking A non-profit organization that promotes the use of telecommunications in Kindergarten to 12th grade education to improve learning. Members represent state and local education agencies, as well as hardware and software vendors, Internet Service Providers (ISPs) and interested individuals. www.cosn.org.

COSName Identifies class of service SNA.

cost component The price of each element of telecommunications service and/or equipment that comprises a configuration.

cost of service pricing A procedure, rationale or methodology for pricing services strictly on the basis of the cost to provide those services.

cost per phone hour A call center term. Basic unit of resource measurement. Total costs (fixed, variable and semi-variable) divided by the number of workstation call hours that are projected or actually achieved.

COT 1. Continuity Check Message. The second of the ISUP call set-up messages. Indicates success or failure of continuity check if one is needed. See ISUP and Common Channel Signaling.

2. Central Office Terminal or Termination. The termination of a local loop facility at the central office. See Digital Loop Carrier.

3. Customer Originated Trace. A CLASS (Custom Local Area Signaling Services) feature that allows the customer (e.g., you) to originate a trace to track harassing callers. When you get a nuisance call, you depress the switchhook and release it quickly. Then you listen for a special stuttered dial tone. You then depress *57 on your touchtone dial pad, or dial 1157 on your rotary phone. If the last call has been successfully traced, you'll hear an announcement. The results of the successful trace are recorded by the telephone company, and are released only to law enforcement agencies, after you have signed an authorization. Your telephone company may charge you for this service, and the charge may be as high as $100. See also CLASS.

coterminated plant Plant which has an assumed retirement dependent upon the retirement of some other item of equipment or building, etc. A telephone company term.

COTS 1. COnnection Transport Service.

2. Commercial Off The Shelf.

couch commando A couch potato who insists on taking charge of what he and the rest of the couch potatoes are watching on the TV.

couch potato A person who spends their life sitting on a couch surfing TV channels with a remote control TV device. See Mouse Potato.

coulomb The quantity of electricity transferred by a current of one ampere in one second. One unit of quantity in measuring electricity.

Council of Registrars CORE. An organization charged with the responsibility for development, implementation, and maintenance of a set of new Top Level Domains (TLDs) for the Internet. See CORE for a longer explanation.

counter rotating An arrangement whereby two signal paths, one in each direction, exist in a ring topology. See counter rotating ring.

counter rotating ring An arrangement whereby two signal paths, the directions of which are opposite, exist in a physical ring topology. Such rings typically are described as "Dual Counter Rotating Rings," such as described in SONET and FDDI standards. In such a physical configuration, one or more transmission paths operate in a clockwise manner, while one or more other paths operate counter-clockwise, or anti-clockwise. Should the primary path suffer catastrophic failure, the secondary path comes on line. It does this to ensure virtually uninterrupted communications. See also FDDI and SONET.

countermeasure Action, device, procedure, or technique that reduces a threat, a vulnerability, or an attack by eliminating or preventing it, or by minimizing the harm it can cause, or by discovering and reporting it so that some corrective action can be taken. See also blended threat.

counterpoise A system of electrical conductors used to complete the antenna system in place of the usual ground connection.

country code 1. The one, two or three digit number that, in the world numbering plan, identifies each country or integrated numbering plan in the world. In short, the one, two or three digits that precede the national number in an international phone call. This code is assigned in and taken from Recommendation E.163 (Numbering Plan for International Service) adopted by the ITU-T. There's a list of country codes and key country area codes in the Appendix at the back of this book. See also www.the-acr.com and www.sprint. com/ssi/intl_codes.html.

2. In international record carrier transmissions, the country code is a two or three alpha or numeric abbreviation of the country name following the geographical place name.

3. A two-character alphabetic code suffixed to a URL (Uniform Resource Locator) for use in communications over the Internet and WWW (World Wide Web). The country code is a portion of the Top Level Domain (TLD), and is used when the domain of the target country differs from that of the country of origin of the transmission. For example, if you send an e-mail from the U.S. to South Africa, the e-mail address would be in the format "user@ userorganization.entity.za." Example country codes include .au for Australia, .jp for Japan, .sw for Sweden, .us for United States, and .za for South Africa.

county For the purposes of the FCC's cable television rules, this term includes: - Borough (in Alaska).

- - District (in District of Columbia).
- - Independent City (in Alaska, Maryland, Missouri, Nevada, and Virginia).
- - Municipio (in Puerto Rico).
- - Parish (in Louisiana).

coupled modes 1. In fiber optics, a condition wherein energy is transferred among modes. The energy share of each mode does not differ after the equilibrium length has been reached.

2. In microwave transmission, a condition where energy is transferred from the fundamental mode to higher order modes. Energy transferred to coupled modes is undesirable in usual microwave transmission in a waveguide. The frequency is kept low enough so that propagation in the waveguide is only in the fundamental mode.

coupler An optical device that combines or splits power from optical fibers.

coupling Any means by which energy is transferred from one conductive or dielectric medium (e.g., optical waveguide) to another, including fortuitous occurrences. Types of electrical coupling include capacitive (electrostatic) coupling, inductive coupling, and conductive (hard wire) coupling. Coupling may occur between optical fibers unless specific action is taken to prevent it. Coupling between fibers is very effectively prevented by the polymer overcoat, which also prevents the propagation of cladding modes, and provides some degree of physical protection. See also Inductive Coupling.

coupling efficiency Efficiency of optical power transfer from one component to another.

coupling loss The power loss suffered when coupling light from one optical device to another. The ratio/loss of optical power (expressed as a percent) from one output port to the total output power.

coupling ratio The ratio/loss of optical power (expressed as a percent) from one output port to the total output power.

coupon From Britain: A tear-off slip to encourage response to advertisements or to a promotion on packaging. The information is keyed into a telebusiness system which automatically handles the follow-up. This may be a phone call, acknowledgement letter, brochure, distribution of a lead to a distributor and so on.

courriel A dumb French word for email. The word came because the French government, in its infinite wisdom, did not like the use of email, which it regarded as a bad American word.

courseware A combination of Web pages, E-mail, threaded discussions, chat rooms, listservs and distance learning tools used to provide online educational services or supplement regular classroom instruction.

courtesy network See guest network.

COV Control Over Voice. Mitel's proprietary signaling protocol which they use between their PBX and their proprietary analog phones.

cover page The first page of a fax message. It generally includes a header, typically the sender company's logo; the recipient's name and fax telephone number; the sender's fax and voice telephone numbers; the system's date and time; a message; a footer.

coverage The percent of completeness with which a metal braid covers the underlying surface.

coverage antenna An antenna in a distributed antenna system that provides coverage within its designated area and serves as a repeater for communications between antennas on either side of it.

coverage area The geographic area served by a cellular system; that is, the area in which service is available to users of the system. Once the mobile telephone number has traveled outside the coverage area, the mobile telephone will show "NO SERVICE."

covert couture Designer one-offs modeled after off-the-rack merchandise. For example, Gucci might make you a handbag just like the one in the store but add a monogrammed lining - for say, $17,000. As The New York Times put it: "The finished item may not look so very different from the store-bought version, but it's secretly special."

covertone A cell phone term. A realtone where the song is sung by someone other than the original performing artist.

COW Cellsite On Wheels. A COW is trailer used to house transmitting/receiving equipment. A COW can be temporary, providing cell phone service during network expansions, emergencies and short duration events. When replacing an existing cellsite, the COW processes traffic while the permanent cellsite is changed out. In the case of a change out or upgrade, the COW will use existing components, such as antennas, batteries, network facilities whenever possible. Some components may be re-engineered and expanded. Antennas may require duplexers, power may have to be reconfigured, batteries may have to be upgraded, additional network facilities may be arranged. In the event, the COW is used at a "new" site or for a special event, network access, power, generators, remote testing and repair are arranged and a temporary tower may be placed until a permanent tower can be erected. The use of the COW may or may not require a cellular system re-design depending on the application. Since the COW can be used for omni or single-directional applications, each deployment of the COW will require a review of the RF Plan for the given area. COWs come with climate control and can be configured in any manner, depending on the physical restrictions of the trailer. The COW is especially useful to get service in place while the permanent network is being built because COWs can roll through a network from site to site as the Project demands. Since the COW contains the same network equipment as a permanent cellsite, it can be removed from the COW and installed in a permanent building if that is part of the network plan.

COW Interface Character-Oriented Windows Interface. An SAA-compatible user interface for OS/2 applications.

cowboy coding Cowboy coding or Cowboy Software and means shooting from the hip and not really planning how you go about a particular process.

cow India is the only country in the world that has a Bill of Rights for Cows.

CP See Control Point.

CP/M Control Program for Microcomputers. An erstwhile popular operating system for primarily 8-bit microcomputer systems based on the family of Intel 8080 family of microprocessor chips. The CP/M system was originally written by Gary Kidall a programmer and consultant who later formed a company called Intergalatic Digital Research (later just Digital Research). Sadly, that company never upgraded CP/M to 16-bit machines. Thus it left the way open for Bill Gates and the company he formed, Microsoft, to create MS-DOS, which, in its initial form, bore a remarkable resemblance to CP/M.

CPA Cost Per Action. Consider an advertisement on a Web site. There are basically four ways you can pay for such an ad. 1. You can pay a flat monthly fee for it. 2. You can pay a cost per viewer, per eyeball. 3. You can pay a cost per click. 4. You can pay a "cost per action." You pay if someone seeing your ad actually does something. CPA pricing can range from registration forms filled out, contests entered, questionnaires answered, or cost per ultimate product purchase, and lots of other variables along a continuum of steps toward the sale. Advertisers often favor such pricing strategies because they pay only for measurable results. The problem for publishers is the guy putting up the Web is that they carry all the risk - if a poorly designed or badly targeted ad (something the Web site owner has no control over) draws low activity levels. And the poor Web site gets no money for his precious real estate space.

CPC 1. See Calling Party Control.
  2. Calling Party Connected.

CPCN Certificate of Public Convenience and Necessity. A CPCN is required from the FCC (47 U.S.C. S 214(e)(2)) for the "construction, extension, acquisition, operation, or discontinuance, reduction, or impairment of service", or more generally, to provide interstate phone lines from point A to point B. For intrastate lines, a CPCN is required from the state public utilities commission. As the name implies, CPCNs are awarded based on whether there is a public need for the service. Given FCC and Congressional policy to promote competition, the requirements are not onerous and are even less so for resellers who do not need to construct facilities.

CPCS An ATM term. Common Part Convergence Sublayer: The portion of the convergence sublayer of an AAL that remains the same regardless of the traffic type.

CPCS-SDU An ATM term. Common Part Convergence Sublayer-Service Data Unit: Protocol data unit to be delivered to the receiving AAL layer by the destination CP convergence sublayer.

CPE 1. Customer Provided Equipment, or Customer Premises Equipment, or Consumers' Personal Equipment. Originally it referred to equipment on the customer's premises which had been bought from a vendor who was not the local phone company. Now it simply refers to telephone equipment – key systems, PBXs, answering machines, etc. – which live on the customer's premises. "Premises" might be anything from an office to a factory to a home. These days the phone company ends their service and their cable at a box or punchdown block which they call the demarc point – demarcation point. They're responsible for problems to the demarc point. You're responsible for problems from the demarc point on – unless you pay them to take care of your CPE wiring and/or phone equipment. GTE once used CPE to refer to "Company Provided Equipment." It doesn't any longer. What the Americans call CPE, the Europeans now call CTE, which stands for Connected Telecommunications Equipment. See CTE Directive and demarc.
  2. Chlorinated Polyethylene. Jacketing compound characterized by physical, aging, flame and oil resistant properties comparable to Neoprene and Hypalon. Lower coefficient of friction than Neoprene and Hypalon for easier installation. Its halogen content is equivalent to Hypalon - significantly lower than PVC.

CPI 1. Computer to PBX Interface. This proprietary hardware/software interface provides direct connectivity between a PBX's switching network and a host computer to allow switched access between the host computer and data terminal equipment connected with the PBX. The interface is based on the North American Standard T-Carrier specification (24

multiplexed 64 Kbps channels operating at a combined speed of 1.544 Mbps). Developed by Northern Telecom, Inc. this interface uses in-band signaling and provides bidirectional data transmission at speeds up to 56 Kbps synchronous per channel. See Open Application Interface.
  2. Cost Per Inquiry. The total advertising cost divided by the number of inquiries received. Used for analyzing the efficiency of a medium or vehicle.

CPI-C IBM SAA Common Programming Interface-Communication between SNA and OSI environments.

CPIP Short for Carrier Pigeon Internet Protocol. Using birds to send datagrams from one network node to another, where they are scanned and sent on electronically. A group of Norwegian Linux enthusiasts pulled it off. Why? Because they could.

CPL Call Processing Language. Based on XML (Extended Markup Language), CPL is used to describe and control Internet telephony services. Currently in development within the Internet Engineering Task Force (IETF), CPL is unique insofar as it is specifically geared towards service creation by end users. This makes it a complement to Session Initiation Protocol (SIP) when considering advanced Internet telephony services that integrate VoIP with existing Web and Email features.

CPLD Complex Programmable Logic Device. Also known as Complex Programmable Gate Array (CPGA). A user-configurable logic device very similar to the Field Programmable Gate Array (FPGA). The difference is that the basic logic cells are more complex and structured than those used in FPGAs; the logic cells are larger and more capable at the expense of interconnect routing resources. See also ASIC and FPGA.

CPM 1. Customer Premise Management.
  2. Critical Path Method. See also CP/M.
  3. Cable Plant Management. See Cable Management.
  4. See Continuous Phase Modulation.
  5. Cost Per Thousand. A way of comparing the price of advertising. Assume your ad is running for a month on Lycos' home page, how much will the CPM be? It will be the price of the ad divided by the number of people visiting that site for a month. For example, let's say that the ad cost $200,000 and was visited by 10,000 people. That means that the cost per thousand (the CPM) is $20.

CPN 1. Computer PBX Network.
  2. Customer premises network.
  3. Calling Party Number. See Calling Party Number.

CPNI Customer Proprietary Network Information. Information which is available to a telephone company by virtue of the telephone company's basic service customer relationship. This information may include the quantity, location, type and amount of use of local telephone service subscribed to, and information contained on telephone company bills. This is the definition of CPNI that the independent voice mail and live telephone answering industry uses.

CPODA Compression Priority Demand Assignment. Another protocol for converting voice into data bits. See also PCM.

CPP Calling Party Pays. A billing option that changes the billing of landline calls received by wireless subscribers so that the originating caller is billed for calls to the wireless subscriber. A new concept in cell phone service in the U.S. An old concept in Europe where GSM has always had it and the caller has always paid more to call a cell phone.

CPR Continuing Property Record. Assigned by Telcordia to provide a methodology for property record number assignment for retirement units and less than retirement units. These record numbers are used by telephone service providers to maintain detailed records of telephone equipment assets throughout the lifecycle of the equipment usage. CPR numbers classify equipment into hardwired, deferrable plug-in, non-deferrable plug-in, capital tool and portable test set, minor item, and expense item. See also CLEI Code.

CPS Characters per second, or cycles per second. In asynchronous communications, there are typically 10 bits per character – 8 bits for the character and one stop and one start bit.

CPSI Customer Premises Satellite Interface. The interface between a satellite, a satellite receiver on the premises, and a user computer network, the CPSI is envisioned as an open interface that support high speed Internet access through broadband satellites.

CPU See Central Processing Unit.

CPUG Call Pickup Group. All the phones in an area that can be answered by each other by simply punching in a couple of digits. See Call Pickup.

CR 1. Carriage Return. The key on a computer called Carriage Return or sometimes "ENTER." Touching this key usually signals the computer that the entry has been completed and is now ready for processing by the computer. See Carriage Return.
  2. Critical (alarm status). Indicates a failure affecting more than 96 customers. An AT&T definition.

3. Call Reference.

4. Call Register. It is a place in memory in a telephone switch that dialed digits are stored when placing a call.

CR-LDP Constraint-based Routed Label Distribution Protocol. An alternative to RSVP (Resource ReSerVation Protocol) in MPLS (MultiProtocol Label Switching) networks. RSVP, which works at the IP (Internet Protocol) level, uses IP or UDP datagrams to communicate between LSR (Label Switched Routing) peers. RSVP does not require the maintenance of TCP (Transmission Control Protocol) sessions, although RSVP must assume responsibility for error control. CR-LDP is designed to facilitate the routing of LSPs (Label Switched Paths) through TCP sessions between LSR peers through the communication of label distribution messages during the session. See also MPLS.

crackberry Another name for the Research in Motion Blackberry that refers to its addictive nature and the inability of its users to focus on anything else while they have a BlackBerry in their hands.

cracker A person who "cracks" computer and telephone systems by gaining access to passwords, or by "cracking" the copy protection of computer software. A cracker usually does illegal acts. A Cracker is a "Hacker" whose hacks are beyond the bounds of propriety, and usually beyond the law. The term "cracker" is said to derive from the word "safe-cracker." See also Hacker, Phreak, Script Kiddies, and Sneaker.

cradle cams On-line cameras attached to computers attached to the Internet that allow parents to monitor their children from their desks at their offices. Cradle cams (cameras) – also known as Kiddie cams – are often installed in daycare centers and grade schools.

CRAFT 1. Cooperative Research Action For Technology.

2. Craft. Nonmanagement telephone company staff. Many craft employees are members of the Communications Workers of America (CWA) or the International Brotherhood of Electrical Workers (IBEW)...at least they used to be.

craft terminal A PCS wireless term. A craft terminal is a device built specifically to provide a man-machine interface that is otherwise not available. The interface is customized to provide a view into a particular device's operation such as a proprietary switch or BSS, which is a Base Station Sub-system charged with managing radio frequency resources and radio frequency transmission for a group of BTSs, which is a Base Transceiver Station, used to transmit radio frequency over the air.

craft test set Also called Goat or Butt-Set. Portable telephone used to test analog phone lines.

craftsperson In the phone industry, a craftsperson has two distinct meanings. First, it is the person who toils to install phones, repair outside plant and fix problems inside central offices. This person typically carries tools and dresses in jeans. Second, craftspeople are at the bottom of the management hierarchy in most phone companies. They typically belong to a union. Craftspeople are not "in management." See Level.

cramming A practice in which customers are billed for unexpected and unauthorized telephone charges or telephone services, which the companies didn't order, authorize or use. "Cramming" refers to the fact that the charges are crammed onto the telephone bill in an inconspicuous place in order that such charges will go unnoticed by the customer. Most of us poor, dumb customers don't bother to check our telephone bills, and quickly write a check for the total. We really don't examine phone bills in great detail since most of them are incomprehensible. Cramming is a practice of only the most unethical and desperate phone companies, i.e. most of them. (That's a tasteless joke.)

crankback An ATM term. A mechanism for partially releasing a connection setup in progress which has encountered a failure. This mechanism allows PNNI to perform alternate routing.

crapplet A poorly written or totally useless Java applet. "I just wasted 30 minutes downloading this awful crapplet!"

crash A crash is the complete failure of a hardware device or a software operation. The term usually is used to mean a "fatal" crash in which the device or software must be started from a "power up" condition. The crash of a Windows machine often is accompanied by a blank blue screen known lovingly as the "blue screen of death." A Macintosh crash is accompanied by a blank screen with a small text box that contains a graphic of a bomb with a lit fuse. See also Boot.

crash cart A cart with a monitor, keyboard, mouse, power strip, and possibly a shuttle server, which can be wheeled by a network administrator or technician to the location of a server or PC that needs on-site maintenance or restoration. See shuttle server.

CRATT Cryptographic Radio TeleType equipment. See also Five By Five.

Crawler See Bot.

crazing The minute cracks that can be found on the suffce of plastic materials such as cable insulation.

CRC Cyclic Redundancy Check. A process used to check the integrity of a block of data. A CRC character is generated at the transmission end. Its value depends on the hexadecimal value of the number of ones in the data block. The transmitting device calculates the value and appends it to the data block. The receiving end makes a similar calculation and compares its results with the added character. If there is a difference, the recipient requests retransmission. CRC is a common method of establishing that data was correctly received in data communications. See CRC Character.

CRC Character A character used to check the integrity of a block of data. The character is generated at the transmission end. Its value depends on the hexadecimal value of the number of ones in the data block. And it is added to the data block. The receiving end makes a similar calculation and compares its results with the added character. If there's a difference, there's been a mistake in transmission. So, please, re-send the data.

CRD Contention Resolution Device.

cream skimming Selecting only the most profitable markets or services to sell into. Choosing the cream of the market. An erstwhile popular economic theory to deny new entrants into the telephone and telecommunications industry. The theory included quaint ideas that it was cheaper and thus more profitable to serve metropolitan areas with phone service than to service rural areas. But the theory fell part on logic like – if it's so much cheaper to serve metropolitan areas, why are rates there so much higher than in rural communiities?

creative disruption Economist Joseph A. Schumpeter's theory that entrepreneurs generate innovation by rendering their predecessors' ideas obsolete. See also See Distruptive Technologies.

credential leakage Credentials are a digital set of attributes about the credential owner. The decision to grant or deny access to an online resource is based on the requester's credentials. The problem with this scenario, however, is that when a user divulges his credentials to an online resource in order to gain access to it, doing so exposes or "leaks" sensitive information, namely, the credentials, to that entity. For example, the access control policy for an online resource may be defined in such a way that only people with a top-secret clearance can access the resource. When the user obtains access to that resource by presenting the correct credentials, the resource learns immediately that the user has a top-secret clearance credential. This is credential leakage. Another type of credential leakage occurs when credentials are presented unencrypted by a user in a wireless environment. In this scenario, the credentials can be picked up wirelessly by a nearby bad guy.

credentials A way of establishing, via a trusted third party, that you are who you claim to be.

CREDFACS Conduit, Risers, Equipment space, Ducts and FACilitieS. Collective term for pathway elements used in communications cabling.

credit card phone A pay telephone that accepts credit cards with magnetic strips on them instead of coins.

credit crunch People use the term "credit crunch" to describe any situation in which banks seem to be unwilling to lend to businesses, consumers, or both. A true credit crunch, however, is a sudden disruption in which the credit markets stop working. The term dates from a specific economic episode in 1966. Back then, Depression-era government rules (notably the Federal Reserve's Regulation Q) capped the interest rates that banks could pay on most deposits. In 1966, rates on short-term securities zoomed higher than the federal cap, and banks suffered sudden, massive withdrawals as depositors shifted funds from savings accounts into money-market securities. Banks were forced to liquidate assets at distress prices to meet the outflow, and briefly stopped lending to anyone – institution or individual – at any price. The crunch ended quickly when the Fed cut short-term rates and flooded the markets with liquidity, which probably also forestalled a recession.

Creep The dimensional change with time of a material under load.

CREN Corporation for Research and Educational Networking. An organization formed in October 1989, when Bitnet and CSNET were combined. CSNET is no longer around, but CREN still operates Bitnet.

crest factor The crest factor is the ratio of the crest (peak, maximum) value of a current to the root-mean-square (RMS) value. A square wave of current has a crest factor of 1. A sine wave has a crest factor of 1.414. The current drawn by a typical computer power supply when powered from a typical wall outlet has a crest factor of 4. The crest factor in this case results from a complex interaction between the power supply and the utility power sine wave. The crest factor of a computer or telephone system power supply is usually reduced when it is operated from a UPS. The reduction in crest factor when operating from a UPS does not adversely affect a computer or telephone power supply, and in fact actually makes it run cooler. Crest factor is always a property of the interaction between a load and a source, so it is meaningless to attribute to either a load or source independently. Factors which generally affect the ability of a UPS to supply high crest factors are: output impedance at harmonic frequencies, output distortion, and current limit. Although a high crest factor rating of a UPS has been considered to be a measure of UPS output stability and quality, differences in measurement techniques make product comparisons on this basis useless. A preferred method is to specify the output voltage response to a step load or output voltage distortion under load. Definition supplied by APC. See also Peak-to-Average Ratio (PAR).

CREX4 Custom Toll Restriction.

CRF 1. An ATM term. Cell Relay Function: This is the basic function that an ATM network performs in order to provide a cell relay service to ATM end-stations.

2. An ATM term. Connection Related Function: A term used by Traffic Management to reference a point in a network or a network element where per connection functions are occurring. This is the point where policing at the VCC or VPC level may occur.

crimp In fiber optics, crimp means to secure buffer tubes under the designated tabs on a splice tray. See also crimp die.

crimp barrel See butt connector.

crimp die These are the part of the crimp tool that actually come in contact with the connector that is being crimped. They slide into the jaws of the crimp tool. Crimp dies are interchangeable and many types are available. See Crimp Tool.

crimp terminal See butt connector.

crimp tool Crimp tools form connectors onto cables. They are used for BNC, F-Type and RJ-11, RJ-45 connectors, among others. They have a padded handle and jaws where the crimp dies are inserted. Crimps are installed by inserting the cable into the crimp, then the crimp into the crimp die, and squeezing the handles of the crimp tool. See crimp die.

crimping A means of securing an electrical contact to a wire using tools that compress the metal contact around the wire.

crippleware The term "crippleware" comes from the plaintiff in a class-action lawsuit, Melanie Tucker v. Apple Computer Inc., that was in early 2007 making its way through Federal District Court in Northern California. The suit contends that Apple unfairly restricts consumer choice because it does not load onto the iPod the software needed to play music that uses Microsoft's copy-protection standard, in addition to Apple's own. The core argument is that the absence of another company's software on the iPod constitutes "crippleware." I disagree.

CRIS 1. Cryptography & Information Security Research Laboratory.

2. Customer Record Information System. A data format used by some LECs for billing end user customers. See DUF.

3. See Customer Record Information System.

critical angle The smallest angle at which a ray will be totally reflected within a fiber.

critical mess An unstable stage in a software project's life when any single change or bug fix can result in two or more new bugs. Continued development at this stage leads to an exponential increase in the number of bugs.

critical technical load That part of the total technical power load required for synchronous communications and automatic switching equipment.

criticality assessment Criticality Assessment is the assessment of the importance of individual computing and communications resources, including computer systems and their resident databases, telecommunications systems, and voice and data networks and subnetworks. See Ray Horak essay on Disaster Recovery Planning at the beginning of this dictionary.

crittercam A miniature camera attached to a wild animal so researchers not only can track them, but see the world from the animal's point of view.

CRL Certificate Revocation List. An up-to-date list of previously issued certificates that are no longer valid. See also Revocation.

CRLF Carriage Return Line Feed.

CRM 1. An ATM term. Cell Rate Margin: This is a measure of the difference between the effective bandwidth allocation and the allocation for sustainable rate in cells per second.

2. Customer Relationship Management. A fancy name for putting software, hardware and networking in place that improves a company's dealings with its customers. In the simplest example, a customer might like to be able to access his supplier's shipping system, so he can find out if the goods were shipped, when they were shipped and where they are now. CRM includes such customer touch functions as help desk, marketing, order entry, technical information and sales automation. According to a study I saw in late 2001, the primary applications of CRM software were field sales (26%), marketing (24%), call center service (23%), call center sales (8%), distributor sales (4%), field service (2%), online sales (2%) and other (11%).

CRO Complete with Related Order. FID used on service orders identifying two or more orders that are dependent upon each other for proper completion. Orders must be completed together.

crooked nose In ancient Rome, it was considered a sign of leadership to be born with a crooked nose. There's hope for us all.

crop dusting Surreptitiously farting while passing thru a cube farm, then enjoying the sounds of dismay and disgust. Crop dusting leads to prairie dogging.

cross assembler An assembler that can run symbolic-language on one type of computer and produce machine-language output for another type of computer.

cross certification A status where two or more organizations or certificate authorities share some level of trust.

cross compiler A compiler that runs on one computer but produces object code for a different type of computer. In short, a cross compiler generates code for a different environment than where it is run. Cross compilers are used to generate software that can run on computers with a new architecture or on special-purpose devices that cannot host their own compilers. Cross-compiler: For example, normally a compiler that runs on a Motorola 68020 will generate code that runs on a Motorola 68020. A cross-compiler might run on a Motorola 68020, but generate code for an Intel 80960.

cross connect Cross connect can be written with or without a dash, i.e. cross connect or cross-connect. I've read 15 definitions of cross-connect. They're all awful. Let's try to do better. Let's imagine you have an office that you need to wire up for voice and data. So you wire every desk with a bunch of wires. You punch one end of the wires into various plugs at the desk. You punch the other onto some form of punchdown block, for example a 66-block. That punchdown block may be in a closet on the same floor or it may be down in the basement. Then you bring the wires in from your telecom suppliers. The T-1s, the ATM, the frame relay, the local lines, the analog lines, the digital lines, etc. You punch them down on another punchdown block, for example a 66-block. Now you have two sets of blocks (they can be any form of punchdown block) – one for those going to the office and those coming in from the outside world. You now have to join them in a process known as "cross-connecting" in the telecom world. You simply run wires from one 66-block (or other punchdown device) to the other one. The reason you use cross-connect wires rather than just punching down an incoming phone line, for example, directly to your phone system is that moves, adds and changes would, over time, horribly confuse things, screw connections up, and eventually become a total mess. Easier to simply have all the changes accomplished through the cross-connect wires and wiring. Follow the short wires. Easy to see what's connected to what. Easier for labeling, documentation, etc. In short, cross-connect is a connection scheme between cabling runs, subsystems, and equipment using patch cords or jumpers that attach to connecting hardware on each end. A cross-connection is the attachment of one wire to another usually by anchoring each wire to a connecting block and then placing a third wire between them so that an electrical connection is made. The TIA/EIA-568-A standard specifies that cross-connect cables (also called patch cords) are to be made out of stranded cable. See also Cross Connect Equipment, Cross Connect Field, DACS, and Patch Panel.

cross connect equipment Distribution system equipment used to terminate and administer communication circuits. In a wire cross connect, jumper wires or patch cords are used to make circuit connections. In an optical cross connect, fiber patch cords are used. The cross connect is located in an equipment room, riser closet, or satellite closet.

cross connect field Wire terminations grouped to provide cross connect capability. The groups are identified by color-coded sections of backboards mounted on the wall in equipment rooms, riser closets, or satellite closets, or by designation strips placed on the wiring block or unit. The color coding identifies the type of circuit that terminates at the field.

cross connection See Cross Connect.

cross coupling The coupling of a signal from one channel, circuit, or conductor to another, where it becomes an undesired signal. See Cross Connect.

cross extension cable When you make an REJ-11 extension cable, the wiring crosses over. Conductor 1 becomes 4. Conductor 2 becomes 3. Conductor 3 becomes 2. And conductor 4 becomes one. Next time you have an REJ-11 extension cable in your hand, hold the REJ-11s next to each other and compare them. You'll notice the cross-over of the conductors.

cross linked A term denoting intermolecular bonds between long chain thermoplastic polymers, effected by chemical or irradiation techniques.

cross modulation distortion The amount of modulation impressed on an unmodulated carrier when a signal is simultaneously applied to the RF port of a mixer under specified operating conditions. The tendency of a mixer to produce cross modulation is decreased with an increase in conversion compression point and intercept point.

cross over cable See Crossover Cable.

cross phase modulation A fiber nonlinearity caused by the nonlinear index of refraction of glass. The index of refraction varies with optical power level which causes different optical signals to interact.

cross pinned See Crossover Cable.

cross plan termination The conversion of ten-digit telephone numbers to seven digits, or vice versa.

cross polarization The relationship between two radio waves where one is polarized vertically and the other horizontally.

cross selling You buy a shirt from me. I sell you a tie. You buy a car from me. I sell you a mobile phone for your car. There is another term. It's called "up selling." That's when I sell you a more expensive shirt or a more expensive car.

cross subsidization Supporting one area of a business from revenues generated by another area. Local phone companies in the U.S. have long argued that if they are required by government or regulatory decree to provide "universal service" to households, they should be allowed to price business service higher. This way they can cross subsidize low-priced residential with high-priced business service. At least that's the theory. There are, however, many other cross-subsidies in the telephone business – people who stay longer with one phone line cross-subsidize those who move frequently; international service in most countries is priced high and the profits used to provide other services. The problem with cross-subsidies is that everyone knows they exist, but no one knows the actual financial extent of them. The problem is of allocation. A phone company runs with one plant – switches and wires. Those switches and wires provide everything from local to international calling. How to figure how to allocate how much is used for what? It's a question that has provided millions of dollars in consulting fees for thousands of economists over the years. Despite the money, there are no conclusive answers. See also Tariff Rebalancing.

cross tabulation Method for describing frequency distributions of two variables simultaneously.

cross talk A type of interference caused by signals from one circuit being coupled into adjacent circuits. See also crosstalk, which is the preferred spelling.

cross wye A cable used at the host system, or network interface equipment that changes pin/signal assignment in order to conform to a given wiring standard (USOC, AT&T PDS, DEC MMJ, etc).

cross-connect See Cross Connect.

cross-connection Non-permanent wire connections that run between terminals of a cross-connect field. See Cross Connect.

cross-platform A piece of software that runs between an application and many operating systems, such as Windows, Mac and Linux. The concept is that you run your application on the cross-platform software which then, somehow, works magically and seamlessly on all three operating systems. Programmers have been working on cross-platform software for eons, so far without success.

crossbar Xbar. A switching system that uses a centrally-controlled matrix switching network of electromagnetic switches which work with magnets and which connect horizontal and vertical paths to establish a path through the network. Crossbar switches are circuit switches, typically in the form of voice PBXs and Central Offices (COs). Crossbar was known for its reliability, at least in comparison to earlier Step-by-Step (SxS) electromechanical switches, but is now largely obsolete because it takes up a lot of space and isn't programmable. The first crossbar switch was a central office installed in Brooklyn, NY in 1937. For a bigger explanation, see Switching Fabric and Xbar.

crossbar tandem A 2-wire common-control switching system with a space-division network used as local tandem, toll tandem, and CAMA switching. While originally designed to switch trunks, some systems have been locally modified to accept loop-start or ground-start lines.

crossed pinning Configuration that allows two DTE devices or two DCE devices to communicate. See also Crossover Cable.

crosslink An X.25 link connecting two XTX NCs on the same level.

crossover cable 1. Ethernet cables have multiple wires inside them. Some are dedicated to sending; some are dedicated to receiving. A crossover cable is a special cable in which the receive and send wires cross so that the sending leads on one device can directly connect to the receiving leads on the other device. When WatchGuard, a manufacturer, encloses a crossover cable with its products, it is typically color-coded red for easy identification.

2. Another word for a null modem cable or a cross-pinned cable. Such a cable is a RS-232 cable that enables two DTE devices or two DCE devices to be connected through serial ports and transmit and receive information across the cable. The sending wire on one end is joined to the receiving wire on the other. In an RS-232 cable, this typically means that conductors 2 and 3 are reversed. See RS-232-C.

crosspinned cable See Crossover Cable.

crosspoint A single element in an array of elements that comprise a switch. It is a set of physical or logical contacts that operate together to extend the speech and signal channels in a switching network.

crosspolarized operation The use of two transmitters operating on the same frequency, with one transmitter-receiver pair being vertically polarized and the other pair horizontally polarized (orthogonal polarization).

crossposting Crossposting is putting one copy of an electronic file up on the Internet in such a way that it can be viewed from any of several newsgroups (discussion areas). Today's Internet software lets readers avoid seeing a widely crossposted article more than once. They see it in the first group they find it. Crossposting is frowned upon in the Internet when it becomes excessive and off-topic. Crossposting is a less serious offense than spamming, which is seriously frowned upon. See Spamming.

crosstalk Crosstalk occurs when you can hear someone you did not call talking on your telephone line to another person you did not call. You may also only hear half the other conversation. Just one person speaking. There are several technical causes for crosstalk. They relate to wire placement, shielding and transmission techniques. CROSSTALK is also the name of a once popular telecommunications software program for 8- and 16-bit microcomputers. See also alien crosstalk and crosstalk attenuation.

crosstalk attenuation The extent to which a communications system resists crosstalk.

crowdsourcing Using the Web to economically outsource work to individuals operating as independent contractors in a diverse, worldwide labor pool, rather than paying more to have employees do the work. Crowdsourced tasks can be rote or highly specialized, and can take anywhere from ten minutes to a year or more to complete. See also outsourcing.

CRP 1. Command repeat.

2. Cabling Reference Panel (of the Australian Communications Industry Forum).

3. Customer Routing Point. AT&T's terminology for third-party processors that accept routing requests from the CCSS7 network. Within the ICM, the Network Interface Controller (NIC) acts as a CRP.

CRS An ATM term. Cell Relay Service: A carrier service which supports the receipt and transmission of ATM cells between end users in compliance with ATM standards and implementation specifications.

CRT Cathode Ray Tube. The glass display device found in television sets and video computer terminals. See Cathode Ray Tube.

CRTC Canadian Radio-television and Telecommunications Commission. The Canadian equivalent of the FCC (Federal Communications Commission) in the US. The CRTC has its origins in the Royal Commission on Broadcasting (1928). In 1906, amendments to the Railway Act granted the Board of Railway Commissioners for Canada the power to regulate telephone and telegraph companies under federal jurisdiction. In 1936, the Canadian Radio Broadcasting Commission (CRBC) was formed. It wasn't until 1976 that the CRTC was formed to assume regulatory responsibilities for radio, television, telegraphy and telecommunications. www.crtc.gc.ca.

cryptanalysis 1. The steps and operations performed in converting encrypted messages into plain text without initial knowledge of the key employed in the encryption.

2. The study of encrypted texts. The steps or processes involved in converting encrypted text into plain text without initial knowledge of the key employed in the encryption.

crypto A term used to describe encrypted information. The use of encryption on data communications circuits lessens the chance that the information will be successfully copied by eavesdroppers.

cryptochannel A complete system of crypto-communications between two or more holders. The basic unit for naval cryptographic communication. It includes: (a) the cryptographic aids prescribed; (b) the holders thereof; (c) the indicators or other means of identification; (d) the area or areas in which effective; (e) the special purpose, if any, for which provided; and (f) pertinent notes as to distribution, usage, etc. A cryptochannel is analogous to a radio circuit.

cryptography The process of concealing the contents of a message from all except those who know the key. Cryptography is unregulated in the United States. See Clipper Chip.

crystal A piece of natural quartz or similar material that has been ground to the proper specification which determines the operating frequency of the quartz.

crystal microphone A microphone, the diaphragm of which is attached to a piezo-electric crystal, which generates electrical currents when torque is applied, due to the vibration of the diaphragm. The earliest form of microphone, now obsolete. See also

Condenser and Electret Microphone.

CS 1. Convergence Sublayer. The upper portion of BISDN Layer 3. As an ATM term, it covers the general procedures and functions that convert between ATM and non-ATM formats. It describes the functions of the upper half of the AAL layer. It is also used to describe the conversion functions between non-ATM protocols such as Frame Relay or SMDS and ATM protocols above the AAL layer. The exact functions of the CS are dictated by the particular AAL (1, 2, 3/4, or 5) in support of the specific Service Class (A, B, C, or D). SEE AAL.

2. Capability Sets. CSs. Stages of implementation of the Intelligent Network architecture (as proposed by the ITU and the ETSI). Each stage (CS) is actually an incremental subset of the full IN (Intelligent Network) architecture.

CS-1 Capability Set 1. Term used by ITU-T to refer to their initial set of Advanced Intelligent Network (AIN) standards. Contains 18 trigger detection points. Bellcore (Bell Communications Research) plans to adopt the CS-1 terminology for its own AIN.

CS-ACELP Conjugate Structure-Algebraic Code Excited Linear Prediction. Standardized by the ITU-T as G.729, CS-ACELP is used for voice compression at rates of 8 Kbps. VoFR (Voice over Frame Relay) and VoIP (Voice over Internet Protocol) both make use of CS-ACELP, among other compression options. See also ACELP, CELP and LD-CELP.

CS-CDPD Circuit Switched-Cellular Digital Packet Data. A variation on the CDPD theme. Developed in July 1995, the expanded specification provides for packet radio transmission on a circuit-switched basis over the analog AMPS cellular network. See also CDPD.

CSA 1. CallPath Services Architecture. IBM's computer host to PBX interface. It links computer and telephone systems. See Callpath Services Architecture for detail. See also Callbridge and Open Application Interface.

2. Canadian Standards Association. A non-profit, independent organization which operates a listing service for electrical and electronic materials and equipment. It is the body that establishes telephone equipment (and other) standards for use in Canada. At least in part, CSA is the Canadian counterpart of the Underwriters Laboratories. CSA also, by way of example, is heavily involved in the development of the ISO 9000 series of standards on quality and the ISO 14000 series on Environmental Management.

3. Carrier Serving Area. A concept which categorizes local loops by length, gauge and subscriber distribution in order to determine how a specific geographic area can best be served. The concept is critical when LECs evaluate the potential for deployment of services which challenge the capabilities of the embedded voice-grade, twisted-pair cable plant. Such services include xDSL, e.g., ADSL and IDSL.

CSA T527-94 Canadian guidelines for Grounding and Bonding for Telecommunications in Commercial Buildings.

CSA T528-92 Canadian Design Guidelines for Administration of Telecommunications Infrastructure in Commercial Buildings.

CSA T529-M91 Canadian Design Guidelines for Telecommunications Wiring Systems in Commercial Buildings.

CSA T530-M90 Canadian Building Facilities Design Guidelines for Telecommunications CEC Canadian Electrical Code, Part I - 1994.

CSA T-530 Canadian equivalent of EIA-569 standard. Also has Rcv Clock, and both Xmit Clocks for synchronous systems.

CSC 1. Customer Service Center.

2. Customer Service Consultant.

3. Customer Service Coordinator.

4. Customer Support Center.

5. Customer Support Consultant.

CSCD Circuit Switched Cellular Data. A developing alternative to CDPD (Cellular Digital Packet Data) for transmitting data over analog AMPS (Advanced Mobile Phone System) networks. The problems with CDPD are that it is optimized for short messages (smaller than 1KB), it is expensive (approximately $.10 per KB), and it is not universally available. CSCD is optimized for large files, typical of contemporary e-mail transmissions. It also is billed at the same rate as a voice call. Unlike CDPC, CSCD does not require a TCP/IP interface on both ends of the connection.

CSDC Circuit Switched Digital Capability. AT&T defines it as a technique for making end-to-end digital connections. Customers can place telephone calls normally, then use the same private connection to transmit high-speed data. CSDC is a circuit-switched, 56 Kbps, full-duplex data service that provides high-speed data communications over regular telephone lines.

CSFI IBM Communications Subsystem For Interconnection: networking software.

CSI 1. Called Subscriber Identification. This is an identifier whose coding format contains a number, usually a phone number from the remote terminal used in fax.

2. Capability Set I. A set of service-independent building blocks for the creation of IN services developed by the European Telecommunications Standards Institute and the ITU-T.

CSI/NIB ComputoService Inc./National Independent Billing.

CSID Calling Station ID. When you receive a fax from someone, you'll see on the top of the page the phone number of the fax machine that sent the fax to you. That's called the Calling Station ID. Most people think that that number is the phone number from which they're receiving the fax. In fact, it's not. It's the number you enter yourself into your fax machine when you first set it up. You could happily put in a completely different phone number to the number you're sending from. And no one would be any the wiser.

CSMA Carrier Sense Multiple Access. In local area networking, CSMA is a way of getting onto the LAN. Before starting to transmit, personal computers on the LAN "listen" to make sure no other PC is transmitting. Once the PC figures out that no other PC is transmitting, it sends a packet and then frees the line for other PCs to transmit. With CSMA, though stations do not transmit until the medium is clear, collisions still occur. Two alternative versions (CSMA/CA and CSMA/CD) attempt to reduce both the number of collisions and the severity of their impact. See CSMA/CA and CSMA/CD.

CSMA/CA Carrier Sense Multiple Access with Collision Avoidance. A MAC (Media Access Control) protocol designed to avoid the potential for data collisions between devices sharing a transmission medium, e.g. a local area network. There are several implementations of this approach. The IEEE 802.11a standard for wireless LANs, for example, requires that a device that desires to transmit send a RTS (Request To Send) packet. The RTS packet contains the source and destination addresses, and specifies the duration of the desired transmission. If the network is available, the destination station responds with a CTS (Clear To Send) packet. All other devices on the network recognize this ACK (positive ACKnowledgement) mechanism, and back off until the transmission is completed. If the source station doesn't receive a CTS within a specified period of time, it retransmits RTS packets until such time as a CTS reply is received. See also 802.11a and CSMA/CD.

CSMA/CD Carrier Sense Multiple Access with Collision Detection. A MAC (Medium Access Control) technique used in Ethernet LANs. CSMA/CD requires that all devices attached to the network listen for transmissions in progress (i.e., sense the carrier frequency) before starting to transmit (multiple access). If two or more begin transmitting at the same time and their transmissions crash into each other, a data collision occurs. All stations that sense the collision (i.e., collision detection) transmit a collision notification over a sub-carrier frequency, which all stations likewise monitor. The stations that were transmitting during this time backs off, and calculates a random number of milliseconds before again attempting to transmit.

If you didn't understand the above definition, try this one: CSMA/CD: Abbreviation for Carrier Sense Multiple Access with Collision Detection, a method of having multiple workstations access a single transmission medium (multiple access) by listening until no signals are detected (carrier sense), then transmitting and checking to see if more than one signal is present (collision detection). Each workstation attempts to transmit when it "believes" the network to be free. If there is a collision, each workstation attempts to retransmit after a preset delay, which is different for each workstation. It is one of the most popular access methods for PC-based LANs. Think of it as entering a highway from an access road, except that you can crash and still try again. Or think of it as two polite people who start to talk at the same time. Each politely backs off and waits a random amount of time before starting to speak again. Ethernet-based LANs use CSMA/CD. See also CSMA/CA, Ethernet and IEEE 802.3.

CSMDR Centralized Station Message Detail Recording.

CSMI Call Screening, Monitoring and Intercept.

CSO 1. Central Services Organization. An Internet service that makes it easy to find user names and addresses.

2. Corporate Security Officer.

CSP 1. Competitive Service Provider. A general term for all companies competing to deliver telecommunications service to companies and individuals. The term includes the RBOCs, the CLECs, the IXCs and the ILECs.

2. Certified Service Provider. Initially developed for the automotive industry, a CSP is an ISP which has met the mission requirements of performance, reliability, security, and manageability for the big three auto manufacturers and their trading partners (suppliers) to exchange critical transaction and planning documents over the web.

3. Commerce Service Provider. Service providers that build and host e-commerce Web sites. CSPs relieve merchants of the complexity and expense of building, maintaining, and securing their own e-commerce sites.

4. See Carriage Service Provider.

CSPDN Circuit-Switched Public Data Networks.

CSPE Chlorosulfonated Polyethylene. Insulation compound with good electrical properties and oxidation resistance to high temperatures, the environment and chemicals. Excellent resistance to ozone.

CSQP Customer/Supplier Quality Process is a program designed to help suppliers improve the quality of their products and services and strengthen customer relationships. I first heard about CSQP from Newbridge Networks, which told me that it had been nominated for the program by several regional Bell operating companies because of the volume of products it was selling to these customers. These RBOCs funded Bellcore to work with Newbridge to assist in improving product and service quality to attain CSQP registration. The CSQP program is built upon ISO9000 standards, Malcolm Baldrige National Quality criteria and additional Bellcore criteria. According to Newbridge, CSQP establishes clear, concise requirements for each element of the requirements, including the ISO elements. Evidence of compliance must be given to the CSQP Management Team for all elements or action items opened in order to track the problem area to closure.

CSR 1. Customer Station Rearrangement (as in Centrex).

2. Customer Service Representative. A customer care agent that provides direct customer support.

3. Customer Service Record. Computer printout that details the fixed monthly charges billed by your local telephone company. The CSR is composed of computer codes called USOCs, which in turn correspond to a particular tariffed service. USOCs tell the telephone company's billing system what tariff rate should be billed for a particular service. In order to ensure your telephone bill is correct you must request and review this document. No telecom manager should be without this important document.

4. Cell Switch Router. A technology which is proposed in the form of an IETF submission to fill gaps in ATM (Asynchronous Transfer Mode) standards. The objective of CSR is to provide a standard means of building enterprise backbones or carrier-level infrastructures which support high throughput, multicast, and QoS (Quality of Service) capabilities. CSR offers ATM-level connectivity from edge-to-edge of the network via cut-through paths in support of legacy networks such as Ethernet, bypassing packet-level switch processing at the intermediate and core switches.

CSS 1. Cascading Style Sheets. See Cascading Style Sheets.

2. Capability Sets. Stages of implementation of the Intelligent Network architecture (as proposed by the ITU and the ETSI). Each stage (CS) is actually an incremental subset of the full IN (Intelligent Network) architecture.

3. Cellular Subscriber Station. A cellular phone.

CST Computer Supported Telephony, a term coined by Siemens. Here is an explanation from Dr. Peter Pawlita of Siemens. "More people communicate by telephone than by any other means. The reason is simple: The telephone bridges any distance, saves travel time and can be used spontaneously and is universally available. Unfortunately telephone usage is often associated with annoying delays and frayed nerves resulting from such things as time wasted in finding a number, dialing errors, and the absence of the dialed party. Added to this the person to whom you are speaking does not have the knowledge you require, or has to spend a long time looking for documents. What could be more obvious, therefore, than to turn these problems over to the computer – to implement Computer Supported Telephony (CST). CST denotes the functional connection of a computer system to a PBX at the application level. CST applications can automatically initiate calls, receive incoming calls, and provide "just-in-time" business data, documents and notes on the screen. All this makes telephony more convenient, time-saving, efficient and largely error-free."

CSTA Computer Supported Telephony Application. A standard from the European Computer Manufacturers Association (ECMA) for linking computers to telephone systems. Basic CSTA is a set of API calls agreed upon by the ECMA. See also CST and Open Application Interface.

CSTP Customer Specific Term Plan. See Customer Specific Term Plan.

CSU 1. Channel Service Unit. Also called a Channel Service Unit/Data Service Unit or CSU/DSU because it contains a built-in DSU device. A device to terminate a digital channel on a customer's premises. It performs certain line coding, line-conditioning and equalization functions, and responds to loopback commands sent from the central office. A CSU sits between the digital line coming in from the central office and devices such as channel banks or data communications devices. A Channel Service Unit is found on every digital link and allows the transfer of data at a range greater than 56 Kbps. A 56 Kbps circuit would need a 56 Kbps DSU on both ends to transfer data from one end to the other. A CSU looks like your basic "modem," except it can pass data at rates much greater and does not permit dial-up functions (unless it has an asynch dial-backup feature).

2. Channel Sharing Unit. Line bridging device that allows several inputs to share one output. CSUs exist to handle any input/output combination of sync or asynch terminals, computer ports, or modems and thus these units are variously called modem sharing units,

digital bridges, port sharing units, digital sharing devices, modem contention units, multiple access units, control signal activated electronic switches or data-activated electronic switches.

CSU/DSU See CSU.

CSUA Canadian Satellite Users Association. Trade association of satellite users.

CSV Comma Separated Values. Commonly used no-frills text file format used for importing from and exporting to spreadsheets, HTML editors and SQL databases. What *.txt is to word processing, *.csv is to the spreadsheet industry - a simple, common format.

CT 1. Call Type.

2. Cordless Telephone.

3. Computer Telephony. See Computer Telephony.

CT Connect A computer telephony call control server software that connects a wide range of telephone switches (PBXs and ACDs) to a variety of data processing environments. By bridging the PBX and IT infrastructure, CTI applications such as screen pops and intelligent call routing are easily implemented in call centers. CT Connect runs on Windows NT and SCO UnixWare and supports standard programming interfaces such as TAPI, TSAPI and DDE. See Computer Telephony.

CT1 Cordless Telephony Generation 1. A new type of low-cost public cordless telephone system getting popular in Europe. You carry a cheap handset. You go to within several hundred yards of a local antenna and you make your phone call. You can't receive calls as you can on a cellular radio. You can't make calls unless you're close to the antenna. The service helps overcome the serious lack of street-side coin and public phones in Europe. CT1 is the analog version of the interface specification. See CT2, CT2+, CT3 and DECT.

CT2 Cordless Telephony Generation 2, interface specification for digital technology, currently in use in the U.K. for telepoint (payphone) applications. Think of telepoint phones as cellular phones but using micro-cells. By having smaller cells than normal cellular cells, CT2 phones can be smaller, cheaper and lighter. The first generation of these phones didn't do well, since they weren't smaller and lighter; there weren't many micro-cells and you couldn't receive an incoming call. See CT1.

CT2+ An expansion of the CT2 interface specification that would extend network capabilities and allow backwards compatibility with CT2 handsets. See CT1 and CT2.

CT3 Ericsson's proprietary cordless phone system.

CTA Competitive Telecommunications Association. Trade association of alternate long distance carriers (resellers) in Canada.

CTAC See Communications Trouble and Analysis Center.

CTAS A Carrier Test Access Switch is a device that sits in a carrier's telecom network and is used to test multiple copper pair local loops. Such testing device is often used to test the quality of lines that will be used for high-speed DSL data service.

CTCA Canadian Telecommunications Consultants Association. Professional organization of telecommunications consultants.

CTD 1. Continuity Tone Detector.

2. An ATM term. Cell Transfer Delay: This is defined as the elapsed time between a cell exit event at measurement point 1 (e.g., at the source UNI) and the corresponding cell entry event at measurement point 2 (e.g., the destination UNI) for a particular connection. The cell transfer delay between two measurement points is the sum of the total inter-ATM node transmission delay and the total ATM node processing delay.

CTE 1. Connected Telecommunications Equipment. The European term for what the Americans call CPE – Customer Premises Equipment. See CPE and CTE Directive.

2. Channel Translation Equipment.

3. Coefficient of Thermal Expansion.

CTE Directive CTE stands Connected Telecommunications Equipment. The European term for what the Americans call CPE – Customer Premise Equipment. The CTE Directive refers to a paper on the proposed European-wide regulation of telecommunications terminals. That paper was published in the summer of 1997 by the European Commission. The proposed title is "European Parliament and Council Directive connected telecommunications equipment and the mutual recognition of the conformity of equipment". The timetable indicated by the EC is for a common position to be agreed by the end of 1997 with formal adoption by the Parliament and Council by mid 1998 and the legislation coming into force one year later ie, July 1999. The new Directive is designed to complement other relevant "horizontal" legislation such as that on electrical safety, EMC (Electro-Magnetic Compatibility) and ONP (Open Network Provision);

Conformity assessment will be based upon the principle of manufacturers' declarations and the principle that products reaching the market which do not conform to the applicable essential requirements will be considered to be defective, with the possibility of heavy penalties – equipment using radio comms techniques is included;

CTRs and ACTE disappear with the repeal of Directive 91/263 but a new Telecommunications Conformity Assessment and Market surveillance committee (TCAM) will advise the Commission and Notified Bodies still have a role. CTRs remain applicable until replaced;

Operators of all networks will be required to publish, and regularly update, accurate and adequate technical specifications of the available network termination points and the terminal types supported.

Flexibility is achieved by means whereby the essential requirements applicable to new network termination types can be determined in a timely manner. The essential requirements are restricted to:

(a) Prevention of misuse of public network resources causing a degradation of service to third parties.

(b) Interworking via the public network(s) and Community-wide portability between ONTPs specifying a basic level of interworking, e.g. simple voice telephony but excluding supplementary services.

(c) Effective use of spectrum allocated to terrestrial/space radio communication and used for radio services recognizing that trades-off will be necessary between the quality, capacity, and availability.

For each type of Connected Terminal Equipment (CTE) formerly defined as Telecommunications Terminal Equipment, the essential requirements applicable are to be selected from a master list contained in the Directive. The technical requirements will be defined in appropriate technical specifications. These will be harmonized European standards or, in cases where such standards do not yet exist, other appropriate technical specifications. The specifications of essential requirements will take into account the following additional requirements for the common good:

(a) Protection of health, e.g. minimizing the health hazards of radio frequency radiation.

(b) Features for users with disabilities.

(c) Features for emergency and security services.

(d) Protection of individual privacy.

CTI Computer Telephone Integration. A term for connecting a computer (single workstation or file server on a local area network) to a telephone switch (a PBX or an ACD) and having the computer issue the telephone switch commands to move calls around. The classic application for CTI is in call centers. Picture this: A call comes in. That call carries some form of caller ID – either ANI or Caller ID. The switch "hears" the calling number, strips it off, sends it to the computer. The computer then does a lookup for the numbers in a database, sends the switch back instructions on what to do with the call. The switch follows orders. It might send the call to a specialized agent or maybe just to the agent the caller dealt with last time. Meantime, the agent sees a screen pop of information about the caller – such information having been pulled up out of the server's database, using the caller ID information.

CTI and CT (computer telephony) are often confused. In fact, CTI is the older and smaller term. CTI has been the dismal part of computer telephony – the difficult integration of reluctant, closed phone systems with outside computers to which they were never meant to talk. Computer telephony (or CT) is more exciting because it's building new phone systems with fantastic new features based on open standards, open hardware and open software. CTI covers integration with switches. CT covers that AND a lot more – like callback, the UnPBX (communications server), the central office in a PC, IP telephony, one number find me, predictive dialing, unified messaging, interactive voice response, fax blasting and serving, etc.

See also Computer Telephony, TAPI, TAPI 3.0, TSAPI and Windows Telephony.

CTIA Cellular Telecommunications & Internet Association, aka Cellular Telecommunications Industry Association. The name change was made to reflect the increasing focus on wireless Internet access. Based in Washington, D.C., the CTIA is a trade association representing the interests of the wireless telecommunications industry. CTIA also is the parent of CIBERNET Corp., which provides the cellular industry with inter-company billing protocols, roaming administration tools, and financial settlement programs. www.ctia.org.

CTIP Computer Telephony Interface Products. Adapters that allow telephones to work with computers. An example is the Konexx connector, which fits between the handset and the phone, and allows a connection to a PC modem or fax machine. This definition contributed by Larry Kettler of San Diego.

CTN 1. See Call Tracking Number.

2. Consumers' Telecommunications Network. A telecommunications association in Australia that represents consumers and looks after their interests: the CTN is part of the Australian Communications Authority's Consumer Consultative Forum (CCF). See Consumer Consultative Forum.

CTO See Chief Technology Officer.

c-track A cable guide mechanism manufactured of either plastic or metal used in continuous flexing applications.

Ctrl Control. The label on the control key on your computer.

CTS 1. Clear To Send. Pin 5 on the 25-conductor RS-232-C interface or an RS-232-C signal used in the exchange of data between the computer and a serial device. In short, Clear to send is one of the nine wires in a serial port used in modem communications, CTS carries a signal from the modem to the computer saying, "I'm ready to start when you are."

2. Communication Transport System. CTS is The Siemon Company's proprietary structured wiring system. It consists of the methodology and the connecting hardware products to plan, design, and implement the communications wiring infrastructure for commercial buildings (for more information see the company's CTS Design Workbook and CTS Training Videotape). The Siemon Company is based in Watertown, CT.

3. Conformance Testing Services.

CTTS Coax To The Curb. An approach that provisions a multiline remote terminal to deliver voice and data to concentrated residential applications.

CTTU Centralized Trunk Test Unit. An operational support system providing centralized trunk maintenance through a data link on a switch.

CTX Centrex.

CTY Console TeletYpe, a contraction of "Console" and "TTY" (TeleType). A terminal keyboard associated with the console of a computer system, such as a PBX. Also, the designation for the cable that connects the CTY to the computer system. See also Console and TTY.

CU-SeeMe A popular videoconferencing and videophone product that works over the Internet. CU-SeeMe was originally developed at Cornell University and is available free for the PC and the Macintosh. An enhanced commercial version that adds an electronic chalkboard is available from White Pine Software. The software is designed for personal use and for use in instruction and in business communications.

CUA Common User Access. The policy of using the same command for a given function in all software. This makes the software easier to learn and use because you only have to learn one set of commands. Windows has a set of CUA guidelines which many Windows programs follow. For example, Alt+F4 always means close this window.

cube dweller A worker who spends all or most of his/her workday working in a cubicle. See cube farm.

cube farm An office filled with cubicles. See also Cubical Myopia and Prairie Dogging.

cubicle myopia The condition in which "cube dwellers" lose sight of how their actions and decisions affect others beyond the plastic and nylon borders of their eight by ten foot world. This often chronic condition seems to be most prevalent engineers and middle management personnel. This definition came from a gentleman called Tom Blyth, who wrote me, "I am an IT (Information Technologist formerly Telephone Technician) with the United States Coast Guard stationed in Maine. We needed a term to describe the repeated bouts of short-sightedness that are constantly sent our way by the engineers and Middle management at the Boston regional offices."

cubicle vultures Officemates who circle a laid-off worker's desk then swoop in to pick it clean – appropriating prized chairs, lamps, file cabinets, staplers, etc. for their own cubicles.

CUBIS See Customized User Billing Interface System.

cuckoo-clock Telecom slang for some 6- and 10-button models of AT&T 1A2 wall phones shaped vaguely like traditional cuckoo-clocks. These were probably the first multi-line phones to come with handsets that plugged into the base with Trimline-style 5-pin plugs, before the current modular connectors were adopted. Often seen in hospitals on TV shows.

CUG Closed User Group. Selected collection of terminal users that do not accept calls from sources not in their group and that are also often restricted from sending messages outside the group.

curfew The word "curfew" is derived from an old French word that means "cover fire." In Europe during the Middle Ages, a curfew was a metal cone or shield that was used to put out the hearth fire in the evening. The word "curfew" came to mean the end of the day's activities.

Curie Point The temperature at which certain elements (usually so-called "rare earth" elements) relax their resistance to magnetic changes. In a magneto-optic disk drive the surface to be marked is heated briefly by a laser light to its Curie point. Magnetism is then applied in the proper polarity to make the spot a "1" or a "0." It cools, and is locked in that position, until it is re-heated and changed again. This is how magneto-optic drives can be erasable.

current A measure of how much electricity passes a point on a wire in a given time frame. Current is measured in amperes, or amps. The abbreviation for current is I. See Ohm's Law.

current carrying capacity The maximum current an insulated conductor can safely carry without exceeding its insulation and jacket temperature limitations.

current limit The function of a circuit or system that maintains a current within its prescribed limits. A circuit breaker terminates current flow when current exceeds the trip limit. Most UPS systems have an electrical subcycle current limit that regulates the output current to a value within the UPS design limits. This subcycle current limit may activate when a load demanding high inrush current (like a computer or phone system) is switched on. The activation of the subcycle current limit protects the UPS from damage but allows the output voltage to become distorted or even collapse momentarily. Most on-line UPS systems will have the subcycle current limit activated by computer load switching and use a bypass in order to maintain load continuity when the current limit activates. Standby and line-interactive UPS systems can draw on the utility grid directly to supply load switching current transients and therefore do not activate the subcycle current limit or need to use the automatic bypass feature. This definition from APC.

current loop Transmission technique that recognizes current flows, rather than voltage levels. It has traditionally been used in teletypewriter networks incorporating batteries as the transmission power source. In this serial transmission system, a pair of wires connecting the receiving and sending devices transmits binary 0 (zero) when no current flows and binary 1 (one) when current is flowing.

Curriculum Vitae CV. Latin for resume. CV is summary of your academic and work history. That's the traditional definition. I personally believe that your resume should include less history and more on your accomplishments – since that's what employers are interested in.

cursor A symbol on a screen indicating where the next character may be typed. Cursors may be solid, blinking, underlines, etc. Many programs, computers and phone systems allow you to reprogram the cursor to what you like. One author of this dictionary, Harry Newton, likes a non-blinking solid block, which came standard with his original CP/M version of WordStar, but doesn't any longer.

cursor submarining A liquid crystal display on a computer laptop screen doesn't write to screen very fast. When you move a cursor across your screen or move your mouse quickly across the screen, the cursor disappears. This phenomenon is known as cursor submarining. Cute.

curves and arcs A computer imaging term. Paint packages handle curves and arcs in a variety of ways. Examples include spline curves, where-in you specify a series of points and the package draws a curve that smoothly approaches those points, and "three point" curves, in which the first two points anchor the ends of the curve and the third selects the apex.

Cus Code See Customer Code.

CUSEEME An Internet videoconferencing system that enables up to eight users to see and hear each other on their computer screens. Pronounced "See You, See me."

cuspy Well-written program. Excellent work. A program that does all that it says it will, and more, is said to be cuspy.

custom calling A group of special services available from the central office switching system which the telco can offer its subscribers without the need for any special terminal equipment on their premises. Basic custom calling features now available include call waiting, 3-way calling, abbreviated dialing (speed calling), call forwarding, series completing (busy or no answer) and wake up or reminder service.

custom chip A type of microchip that is custom-made to perform a specialized job or is customized to provide a particular function or feature not found in standard microchips.

custom controls Controls are software objects that you embed in a Visual Basic or other Windows development tool. In the old days you would compile your DOS program with a "library" of some precompiled subprograms and functions. Controls take the idea a step further and give you tremendous power, all within the Windows Graphical User Interface (GUI). The original Visual Basic "custom controls" were programmed by third parties and behave identically to controls shipped with Visual Basic:

- They appear in the Visual Basic toolbox.
- You control their behavior from your software.
- They generate events that your program can respond to.
- And they have properties that your program can change.

There are hundreds of controls out there for Visual Basic for database management, multimedia presentations, imaging, host connectivity, etc. The ones that concern us do computer telephony stuff (though anything can be leveraged, like host connectivity for IVR, etc.):

Custom ISDN A version of ISDN BRI (Basic Rate Interface) provided off an AT&T 5ESS central office. It actually offers more features and is easier to install than a National ISDN-1 BRI line. We are all awaiting the specifications on National ISDN-2, which is meant to be "standard." Meantime, Custom ISDN is the most popular, most versatile and most understood ISDN service in North America. See ISDN.

Custom Local Area Signaling Services CLASS. A generic term (like WATS) describing several enhanced local service offerings such as incoming-call identification, call trace, call blocking, automatic return of the most recent incoming call, call redial, and selective forwarding and programming to permit distinctive ringing for incoming calls. See Class.

custom software loads Let's say you want your new PC to come with Microsoft Office, Macromedia Dreamweaver, Adobe Photoshop and perhaps some customer customer relationship management software which your company has written. Normally you'd have to install all the software yourself. But some PC suppliers will supply your new PC already loaded with legal certified versions the software. Your PC vendor will charge you for the software and perhaps a software load fee. Dell leads in this area.

customer access line charge CALC. Also known variously as Access Charge, EUCL (End User Line Charge), and SLC (Subscriber Line Charge). See Access Charge.

Customer Account Record Exchange An Alliance for Telecommunications Industry Solutions, ATIS industry standard for formatting exchange of subscription information. See CARE.

customer acquisition cost Customer Acquisition Cost is the average cost to a carrier of signing up an individual subscriber. Some of the factors included in this cost are handset subsidies, marketing, advertising and promotions.

Customer Aggregation Points See Leaf POP.

customer care center A term created by Alex Szlam, the president of Melita International, Norcross, GA to describe a telephone call center with three basic elements: First, the database technology and the marketing savvy to fill that database with individual customer preference information. Second, the ability to intelligently handle inbound phone calls. Third, the ability to intelligently make outbound calls. See also Customer Sensitivity Knowledge Base.

customer centric This comes from the process of taking any noun and adding "centric" to the end of it. In this case, it means a business, product or service is focused on the customer. The marketing manager says, "Our new product line was produced using a customer-centric process." What an unusual idea.

customer code Cus Code. A new Customer Code is assigned to distinguish a converted CLEC sub-account from the old Verizon end user account.

customer contact zone A term invented by Keith Dawson, editor of Call Center Magazine. It refers to all the information a customer requires which is delivered through multiple media, including manned call centers, interactive voice response machines, fax back devices, etc.

customer control An AT&T term for the ability for an end user to monitor, choose, modify, redesign and/or program the type of service received from a network.

customer convenience port A FireWire-based interface port designed as a customer interface to a motor vehicle's on-board computing system. The customer convenience port (CCP) is designed for use with PDAs, laptop computers, portable CD players, and other devices. See FireWire, IEEE 1394.

Customer Identification Number CIN. A unique number that identifies a customer. Also known as Master Customer Number.

Customer Information Manager CIM. An MCI definition. A component of the NCS which supports the creation and maintenance of customer databases for Vnet customers. Customers have remote access to and control over their portion of the NCS database via a terminal at the customer's location.

customer interaction software Customer Interaction Software is a vague term for software that handles your entire relationship with your customers. See also Contact Management.

customer intimacy A measure of the closeness of an organization's relationships with its customers. Research has shown that better, or more intimate, relationships with customers contribute to customer loyalty, increased sales, and higher profits, even when there is an unfavorable price differential. A Lucent definition.

Customer Network Management CNM. An ATM term. CNM allows users of ATM public networks to monitor and manage their portion of the carrier's circuits. Thus far, the ATM forum has agreed that the CNM interface will give users the ability to monitor physical ports, virtual paths, usage parameters, and quality of service parameters.

Customer Originated Trace See COT.

Customer Premises Equipment CPE. Terminal equipment – telephones, key systems, PBXs, modems, video conferencing devices, etc. – connected to the telephone network and residing on the customer's premises. What North America calls CPE, Europe calls CTE – for Connected Telecommunications Equipment.

Customer Premises Satellite Interface See CPSI.

Customer Proprietary Network Information CPNI. Information which is available to a telephone company by virtue of the telephone company's basic service customer relationship. This information may include the quantity, location, type and amount of use of local telephone service subscribed to, and information contained on telephone company bills. This is the definition of CPNI that the independent voice mail and live telephone answering industry uses.

customer provided loop The customer assumes responsibility for ordering, coordinating, maintaining, and billing for the local loop.

Customer Provided Terminal Equipment Or just Customer Provided Equipment (CPE). Terminal equipment connected to the telephone network which is owned by the user or leased from a supplier other than the local telephone operating company.

customer qualification Same as Customer Validation.

Customer Record Information System CRIS. A Verizon database containing end user information used for billing.

Customer Relationship Management See CRM.

customer retrial A subsequent attempt by a phone users to make a phone call within a measurement period.

Customer Routing Point See CRP.

customer sensitivity knowledge base A term created by Alex Szlam of Melita International, Norcross, GA to describe a complex database that would keep track of your customers' preferences. Such a database would be updated almost automatically based on every contact you had with the customer. The database would probably be object-oriented since the idea is to define customer preferences based on individual preferences, not on a statistical analysis of conglomerate preferences such as those typically gleaned from existing character databases.

Customer Service Center CSC. MCI organization responsible for installing, verifying, and maintaining MCI customers and customer service.

Customer Service Record CSR. Computer printout that details the fixed monthly charges billed by your local telephone company. The CSR is composed of computer codes called USOCs, which in turn correspond to a particular tariffed service. USOCs tell the telephone company's billing system what tariff rate should be billed for a particular service. In order to ensure your telephone bill is correct you must request and review this document. No telecom manager should be without this important document.

Customer Service Unit CSU. A device that provides an accessing arrangement at a user location to either switched or point-to-point, digital circuits. A CSU provides local loop equalization, transient protection, isolation, and central office loop-back testing capability. See also CSU/DSU.

customer specific term plan A customer specific term plan is an option offered by AT&T on the purchase of its 800 services whereby customers can earn additional discounts by committing to a multiyear contract. This also is one of two plans used by aggregators to resell 800 services. The other is the Revenue Volume Pricing Plan.

customer switching system A switching system that provides service for a customer, typically a business customer. Systems in this category include key telephone systems, private branch exchanges (PBXs), automatic call distributors (ACDs), and telephone answering systems.

Customized User Billing Interface System CUBIS. A system that allows InterExchange Carriers to monitor and update subscriber service orders. This assists in maintaining an accurate end-user customer database for billing purposes.

cut To transfer a service from one facility to another.

cut back technique A technique for measuring optical fiber attenuation or distortion by performing two transmission measurements. One is at the output end of the full length of the fiber. The other is within 1 to 3 meters of the input end. Without disturbing the source-to-fiber coupling, access to the short length output is accomplished by "cutting back" the test fiber.

cut down A method of securing a wire to a wiring terminal. The insulated wire is placed in the terminal groove and pushed down with a special tool. As the wire is seated, the terminal cuts through the insulation to make an electrical connection, and the tool's spring-loaded blade trims the wire flush with the terminal. Also called punch down.

cut through 1. Cut-through, in voice processing, is what stops voice prompt playback when a touchtone key is pressed. Some of the speech recognition solutions also add cut-through that will stop voice prompt playback as soon as you start talking. Only voice cards that support continuous speech recognition are able to provide cut-through. Cut-though can be a problem in some cases. Imagine yourself at the airport trying to make a call using a speech recognition system. At the start of a new prompt, the airport public address system blares out a last boarding call for a flight. If cut-through is active, it would stop playing the prompt and wait on your response. Now what do you do?

2. The act of connecting one circuit to another, or a phone to a circuit. This is when a user dials the access code for the circuit and is immediately "cut through" to the tie line. The user controls the call. It is a tie line operation.

3. See also Cut Through Switch.

cut through dialing 10 + CIC = telephone number followed by an authorization code for intraLATA calls.

cut through resistance A measure of an insulation's ability to withstand penetration by sharp edges.

cut through switch A type of switch algorithm in which the destination address of a packet is read and the packet immediately forwarded to the switch port where the destination MAC address device is attached.

Cut to clear See shotgun.

cut-out box A plastic insert box used to install a jack in an existing sheetrock or wood wall. Area is cut out with a keyhole type saw, and inserted into the wall. Self-tightening screws keep box intact against wall, and a normal face plate is attached to the front screw holes.

cut-through Technique for examining incoming packets where Ethernet switch looks only at first few bytes of packet before forwarding or filtering it. Faster than looking at whole packet but allows some bad packets to be forwarded.

cutting edge When you're ahead of the curve, on the leading edge, you're also on the cutting edge.

cutting the cord A term that refers to the phenomenon whereby landline telephone subscribers get themselves a wireless cell phone and decide they like it so much they cancel their landline phone.

cutoff attenuator A waveguide of adjustable length that varies the attenuation of signals passing through the waveguide.

cutoff frequency 1. The frequency above which, or below which, the output current in a circuit, such as a line or a filter, is reduced to a specified level.

2. The frequency below which a radio wave fails to penetrate a layer of the ionosphere at the angle of incidence required for transmission between two specified points by reflection from the layer.

cutoff mode The highest order mode that will propagate in a given waveguide at a given frequency.

cutoff wavelength In fiber optic systems, the cutoff wavelength is the shortest wavelength at which only the fundamental node on an optical waveguide is capable of propagation. For single mode fibers, the cutoff wavelength must be smaller than the wavelength of the light to be transmitted.

cutover The physical changing of lines or trunks from one phone system to another, or the installation of a new system. It's usually done over the weekend, accompanied by heavy praying that everything will go right. There are two types of cutovers – flash cuts and parallel cuts. Parallel cuts occur when the old phone system is left functioning and the new one, central switching equipment and phones, is installed around it. This means that for some period of time there are two sets of phones, two sets of wires, two switches, two sets of phone lines, etc. The parallel cut is a far more reliable method of cutting over a new switch. But it's also more expensive. A flash cut occurs in a flash. On Friday, everyone is using the old switch. When everyone comes to work on Monday, the old switch and its phones have disappeared. In its place, there's a brand new system. Sometimes it works. More often than not, there are remaining nagging problems. With any cutover, it's a good idea to set up a complaint or cutover number. Thus, if anyone's having trouble with their phone, they can call this number and get their problems taken care of. How well these problems are taken care of will determine how well the cutover went and how well the employees perceive the new switch is working. Perception, not reality, is what's at stake here. A flash cut also is known as a hot cut.

CV 1. Old Bell-Speak for single-line phone. It stands for Combined Voice. In old Bell-Speak it meant that the two parts of the phone that dealt with voices were combined into one unit (the handset). Before this, there were phones like the HH (Hand-Held) where there was a piece you spoke into and another piece you put to your ear. From CV, you get CVW (CV Wall phone) and later on, KV (Key Voice) and KVW (Key Wall phone). Later on all this crept into the USOC codes – the Universal Service Order Code numbering systems the local Bell operating phone companies used to identify products and services. See USOC.

    2. Checksum Value.

    3. Code Violation. A violation in the coding of a signal over a digital circuit. A transmission error detected by the difference between the transmitted and the locally calculated bit-interleaved parity. Also called Coding Violation.

    4. Curriculum Vitae. See Curriculum Vitae.

CVD Chemical Vapor Deposition.

CVF Compressed Volume File. A Microsoft term which refers to a file on a compressed disk. The term was first introduced in MS-DOS 6.0, which first had double-your-disk-space technology. That technology was later removed when Stac Electronics, originator of Stacker disk doubling technology, took Microsoft to court and won.

CVP 1. Certified Vertical Partner.

    2. A British term: Co-operative Voice Processing, gives the caller the ability to move seamlessly between an Interactive Voice Processing device and a live agent.

CVS Computer Vision Syndrome.

CVSD Continuously Variable Slope Delta modulation. A method for coding analog voice signals into digital signals that uses 16,000 to 64,000 bps bandwidth, depending on the sampling rate.

CW 1. Call Waiting (as in Custom Calling Service).

    2. Continuous Wave.

    3. An amateur radio term for Morse code, so named because producing it involves switching on and off a continuous radio wave.

CWA Communications Workers of America. A national union of telephone industry employees, currently very worried about its future membership growth given the phone industry's propensity to let surplus workers go. www.cwa-union.org.

CWDM Coarse wavelength division multiplexing; A form of optical wavelength division multiplexing which relies upon wider spacings between channels in order to lower component costs. The latest standards specify a 20 nm (2500 GHz) channel separation, notably wider than the 0.8 nm (100 GHz) commonly used by dense wavelength division multiplexing (DWDM). Although this yields less aggregate fiber bandwidth (typically on the order of 10-16 wavelengths), CWDM yields significant savings in laser transponder costs, power requirements, and footprints. This technology works in smaller, cost-sensitive edge/access applications such as enterprise and storage area networking. See WDM.

CWML Compact Wireless Markup Language. A stripped-down version of HTML (HyperText Markup Language), which is used between client and server in support of the World Wide Web over wired connections. CML is used in support of i-Mode, the DoCoMo technology used to support wireless Internet access from cell phones and other devices in Japan. CML is similar to WML (Wireless Markup Language), which is used in WAP (Wireless Access Protocol). See also i-Mode, WAP, and WML.

CX Signaling A direct current (DC) signaling system that separates the signal from the voice band by filters. Also called Composite Signaling.

CXR Carrier.

cXML Commerce XML, a new set of document type definitions (DTDs) for the Extensible Markup Language (XML), will be released to the public in March 1999 for an open comment period and pilot testing. cXML is an explicit meta-language to describe the characteristics of items available for sale. It enables the development of 'intelligent shopping agents' that help to do the dirty work of corporate purchasing. By programming the characteristics you're seeking into request messages and releasing them to the network, your request will return exactly what you're seeking or nothing at all – which in itself is sometimes important to know. Think of cXML loosely in terms of 'bar coding' for the Web, but with a far richer set of attributes to uniquely identify and describe products, and can be incorporated into computer programs. www.webreference.com/ecommerce/mm/column21/index.html.

cyber Five letters which can, seemingly, be attached to a word and made into a noun or a verb. The first Cyber word was Cyberspace, a term coined by science fiction writer William Gibson in his 1984 fantasy novel "Neuromancer" to describe the "world" of connected computers and the society that gathers around them. The idea of Cyberspace is that this world of computer networks can be explored with the proper addresses and codes. People who use the system for hours on end are said to be lost in cyberspace. Today, many people say that world has arrived in the form of the Internet. And, so with projections that there will be 100 million users of the Internet by the year 2000, the word Cyber has become popular. There's "The Cyberbrary of Congress" (books Congress has on on-line). According to William Safire writing in the New York Times Magazine of December 11, 1994, "Cyber is the hot combining form of our time. If you don't have cyberphobia, you are a cyberpiliac." The US News & World Report labels its election night on-line forum a cybercast. The Washington

Post wrote that "battlefield valor belongs not to the brawny soldier but to the astrophysics major who invented smart bombs," somebody who's called a cyberwonk. See all the following definitions which begin with Cyber.

Cyber Monday The National Retail Foundation has come up with a new marketing ploy for the Monday after Thanksgiving. They have positioned it as "Cyber Monday". Why Monday? Many people get back to the office with its high bandwidth Internet connections, so they can shop at the office instead of on a slow dialup. In 2006, the Foundation exepcted the number of online shoppers to rise to well over 100 million. The total of online holiday sales will reach over $30 billion this year, and that is predicted to be up 18% from 2005.

cyberbusiness A company that does most of its business on the Internet is called a Cyberbusiness.

cybercad An Internet term. The electronic equivalent of a lounge lizard.

cybercafe Establishment with both coffee and Internet access. Trendy in some places, unknown in others. Often used as a retail store to sign up customers to Internet service by a local ISP.

cybercash An electronic payment system integrated into E-Commerce (Electronic Commerce) servers, which typically make use of the Internet. Also called digital cash, the term "Cybercash" was coined by CyberCash Inc. to describe its systems for verification of credit cards and processing of payments. See also E-Commerce.

cyberchondriac A person who obsessively searches health sites on the Internet for ailments, diseases and symptoms with which to disagnose (or, often misdiagnose) themselves. Much medical information on the Internet is tainted by the need to sell something – usually drugs. Caveat emptor.

cybercrud The computer equivalent of bureaucratese. There's also a lot of it found on email systems. It looks like useless clutter, but it's important and the better interfaces hide it.

cyberia Electronic stuff for the cyberpeople. An advertisement in the November 12, 1995 issue of the New York Times (Sunday) Magazine showed Cyberia covering everything from modern chairs to laptop computers, to cellular phones to an Apple Newton PDA.

cybermall A Web site designed for online shopping, shared by two or more commercial organizations.

cybernetics A term invented in 1948 by Norbert Wiener, the automation genius, who declared "We have decided to call the entire field of control and communications theory, whether in the machine or the animal, by the same term Cybernetics." From the Greek "kybernetes," meaning "pilot, "steersman" or "governor." The science of communication and control theory which is concerned most especially with the comparative study of automatic control systems. Examples include the brain and nervous system, and mechanical/electrical/electronic communication systems.

cybernoir An Internet term. Used to describe dark, trippy, weird "cyber" films and shows like "Wild Palms," "Tank Girl," and "VR.5."

cyberpork Government money flowing to well-connected information superhighway contractors.

cyberpunk A work coined by a book called "Cyberpunk: Outlaws and Hackers on the Computer Frontier" by Katie Hafner and John Markoff. The book defines Cyberpunk as what you and I know as a computer hacker – a person who manages to get into other people's computer systems. He does this usually through telephone lines. In most cases, hackers see themselves as harmless electronic joyriders. But they occasionally steal data, inject viruses and misleading information and disrupt legitimate business and research. Sometimes they get caught.

cybersex Adult-oriented computer games, images and chat lines. A place where people can discuss their sex lives and wanton desires with total strangers in online (over phone line) forums, even falling in love without having ever met face to face.

cyberskating Browsing the Internet. See Cyberspace.

cyberslacker See Snoopware.

cyberspace A term coined by science fiction writer William Gibson in his 1984 fantasy novel "Neuromancer" to describe the "world" of connected computers and the society that gathers around them. The idea of Cyberspace is that this world of computer networks can be explored with the proper addresses and codes. People who use the system for hours on end are said to be lost in cyberspace. Today, many people say that world has arrived in the form of the Internet. John Perry Barlow, a rock-'n'-roll lyricist turned computer activist, defined cyberspace in Time magazine as "that place you are in when you are talking on the phone." Thus by Barlow's definition, just about everybody has already been to cyberspace. I prefer Gibson's definition.

cybersquatting Let's say you register www.CocaCola.com and the names of the

100 most common corporate names in America. When the company comes to you and demands the Web site, you ask for $250,000. That practice is called cybersquatting. Basically you're squatting on someone else's property. The analogy between electronic squatting and physical squatting isn't 100% accurate. But you get the idea. Stories of big profits being made got around and the Federal government enacted the Anticybersquatting Consumer Protection Act (15 U.S.C.S. 0x15 1125(d)), which provides a cause of action against a domain name registrant based on the bad faith registration of a domain name that is identical or confusingly similar to, or in the case of a famous mark, dilutive of, the trademark owner's mark or marks. Since then, a number of cases have succeeded in U.S. courts using the newly enacted legislation. Similarly, the global community is keeping busy fighting cybersquatters, including the creation of new anticybersquatting laws, litigation in a variety of countries, and arbitration before the World Intellectual Property Organization, a specialized agency of the United Nations. According to Adrian Copiz, an attorney in Washington, D.C, "although a good number of the disputed registrations may be legitimate, it looks like the days of quick profits and holding domain names hostage may be over."

cyberstalking The New York Times of April 17, 2006 carried the following, "Claire E. Miller, a 44-year-old publishing executive in Manhattan, recently stripped her nameplate from the tenant directory at the entrance to her Kips Bay apartment building, where she has lived for more than 11 years. She has also asked the landlord to disconnect the buzzer and is in the process of changing her phone number. Drastic measures, all, for an otherwise cheerful and outgoing person. But Ms. Miller has been unnerved by a sudden and, since last September, steady onslaught of unsolicited and lusty phone calls, e-mail messages and even late-night visits from strange men - typically seeking delivery on dark promises made to them online by someone, somewhere, using her name. She is being harassed - cyberstalked, by modern definition."

cybertechnology A term I first saw in the mid-December, 1997 injunction from Judge Thomas Penfield Jackson of the U.S. District Court in Washington. Judge Jackson ruled that Microsoft could not force PC makers to load a Windows operating system bundled with Microsoft's browser, Internet Explorer. In making the ruling, Judge Jackson controversially appointed a "special master", Harvard Law School Professor Lawrence Lessig to review the "complex issues of cybertechnology and contract interpretation" in the Microsoft antitrust case. Cybertechnology has come to mean the areas of technology that deal with the Internet and the World Wide Web. Including web design, web development, web marketing, web management, networking, databases, programming, scripting, graphics, and multimedia.

cybored State one quickly gets in while waiting for the screen to change on busy (or just plain slow) sites.

cyborg A contraction of CYBERnetics and ORGanism. A human being who is linked to one or more devices on which he is dependent for survival in a hostile environment. See Cybernetics.

cybrarian A person who makes a living doing online research and information retrieval (comes from cyberspace librarian). According to one definition I read, a cybrarian is a futurist librarian who swims in the electronic ocean of cyberspace. The term is alleged to have been coined by Michel Bauwens of BP Nutrition. A cybrarian is also known as a data surfer or a super searcher.

cycle One complete sequence of an event or activity. Often refers to electrical phenomena. One electrical cycle is a complete sine wave. (A complete set of one positive and one negative alternation of current.) In the battery business, a cycle is the process of one complete battery discharge and recharge. See Cycle Life.

cycle brokering The farming out of number-crunching tasks to a distributed network of consumer PCs.

cycle life In the battery business, cycle life is the useful life of a rechargeable battery, expressed as the total number of discharges and recharges.

cycle manager extraction An MCI system which selects processable calls from Distribution and forwards them to the appropriate MCI Reference System for billing.

cycle master Part of the bus management scheme used in the IEEE 1394 connection technology. The cycle master broadcasts cycle start packets, which are required for isochronous operation. An isochronous resource manager, for DV and DA applications, is also included for those nodes that support isochronous operation. Also included is an optional bus master.

cycle pools Where dial-up call records are stored in MCI's Revenue System until extracted for billing.

cycle life The number of repetitive flex motions that a wire or cable can withstand prior to breakdown.

cycle slip A discontinuity in the measured carrier beat phase resulting from a temporary loss-of-lock in the carrier tracking loop of a Global Positioning System receiver.

cycle time The time to complete a cycle. The amount of the time between one RAM access and the next.

cyclic distortion In telegraphy, distortion that is not characteristic, bias, or fortuitous, and which in general has a periodic character. Its causes are, for example, irregularities in the duration of contact time of the brushes of a transmitter distributor or interference by distributing alternating currents.

Cyclic Redundancy Check See CRC.

cylinder A hard disk drive contains a number of platters, which are divided into tracks. A cylinder is a collection of all corresponding tracks on all sides of the platters in a disk drive. Think of a hard disk consisting of dozens of concentric cylinders, each of slightly different diameters. These distinct concentric storage areas on the hard disk roughly correspond to the tracks on a floppy diskette. Generally, the more cylinders a hard disk has, the greater its storage capacity.

cyperpunk According to the New York Times, cyperpunks are a movement of American computer mavens, a largely libertarian group espousing the idea that advanced computer encryption technologies can create electronic privacy and provide liberty and freedom from potential government Big Brothers.

D The Wall Street Journal's famous D conference, which stands for digital. D was cooked up by two of the Journal's star columnists – Walt Mossberg of "Personal Technology" and Kara Swisher of "Boom Town" – back in the heady days of 1999. The premise remains unchanged: A digital revolution continues to transform the way we work and live. Or, as Swisher claims in the event guide, "The tidal wave of technological change continues to advance upon us." But there's an added caveat: Tough times are the best times to gain an edge. D could just as easily stand for deja vue – from the setting (the Four Seasons resort, where the last of the Industry Standard's briefly legendary digital summits was held) to the 400-person crowd that included billionaires mingling in khakis, conspiring to push the restart button for a tech community limping out of the desert of unrelentingly bad news. These comments came from a report on the August 2003 show.

D Bank Also called Channel Bank. It breaks down a T-1 circuit to its 24 channels.

D Block A FCC designation for Personal Communications Services (PCS) license granted to a telephone company serving a Major Trading Area (MTA). This grants permission to operate at certain FCC-specified frequencies.

D Channel In an ISDN interface, the "D" channel (the Data channel) is used to carry control signals and customer call data in a packet switched mode. In the BRI (Basic Rate Interface, i.e. the lowest ISDN service) the "D" channel runs at 16,000 bits per second, part of which will carry setup, teardown, ANI and other characteristics of the call. 9,600 bps will be free for a separate "conversation" by the user. That "conversation" will typically be data. And many phone companies are now selling it as an "on the Internet all the time" channel, allowing you to receive and send email continuously. In the PRI (Primary Rate Interface, i.e. ISDN equivalent of T-1), the "D" channel runs at 64,000 bits per second. The D channel provides the signaling information for each of the 23 voice channels (referred to as "B channels"). The actual data which travels on the D channel is much like that of a common serial port. Bytes are loaded from the network and shifted out to the customer site in a serial bit stream. The customer site responds with its serial bit stream, too. An example of a data packet sent from the network to indicate a new call has the following components:

- Customer Site ID
- Type of Channel Required (Usually a B channel)
- Call Handle (Not unlike a file handle)
- ANI and DNIS information
- Channel Number Requested
- A Request for a Response

This packet is responded to by the customer site with a format similar to:

- Network ID
- Channel Type is OK
- Call Handle

The packets change as the state of the call changes, and finally ends with one side or the other sending a disconnect notice. The important concept here is the fact the information on the D channel could actually be anything – any kind of serial data. It could just as well be sports scores! So with that in mind, consider the Channel Number Requested packet above. This is the networks' selected channel for the customer site to use. Normally, this number is between 1 and 23, but could be a higher number if needed. This is what NFAS is all about. NFAS (Non Facility Associated signaling, pronounced N-FAST without the T) allows a D channel to carry call information regarding channels which may not even exist in the same PBX or PC system. See also ISDN and DS-0.

D Conditioning A type of line conditioning which controls harmonic distortion and signal-to-noise ratio so that they lie within specified limits. See also Conditioning.

D Connector A cable connecting standard housed in a shell that resembles the letter D. D connectors, especially DB-9, DB-15, and DB-25 are used to connect computers to peripheral devices such as modems.

D Link Diagonal Link. A SS7 signaling link used to connect STP pairs that perform work at different functional levels. These links are arranged in sets of four (called quads). Connected mated pairs of Signal Transfer Points (STPs) at different hierarchical levels. For example, from a local pair of STPs to a regional pair of STPs. Like the B-Links, D-links are assigned in a quad arrangement. Once again, the recommendation is for three way path diversity. See A, B and C Links.

D Mark See Demarc.

D Region That portion of the ionosphere existing approximately 50 to 90 kilometers above the surface of the Earth. Attenuation of radio waves, caused by ionospheric free-electron density generated by cosmic rays from the sun, is pronounced during daylight hours. See also E and F region.

D Type The standard connector used for RS-232-C, RS-423 and RS-422 communications. D-type connectors are typically seen in nine, 15 and 25 pin configurations.

D-AMPS Digital Advanced Mobile Phone Service. The EIA/TIA Interim Standard 136 (IS-136) which succeeded IS-54, and which addresses US digital cellular systems employing TDMA (Time Division Multiple Access). See IS-136.

D-Bank Another name for channel bank. A device that multiplexes groups of 24 channels of digitized voice input at 64 Kbps into T-1 aggregate outputs of 1,544,000 bits per second.

D-Block Carrier A D-Block Carrier is a 10-MHz PCS carrier serving a Basic Trading Area (BTA) in the frequency block 1865-1970 MHz paired with 1945-1950 MHz.

D-Bit Also called DBIT. the delivery confirmation bit in an X.25 packet used to indicate whether or not the DTE wishes to receive an end-to-end acknowledgment of delivery. In short, a bit in the X.25 packet header that assures data integrity between the TPAD and the HPAD.

D-CLEC A CLEC which specializes in delivering only data, most typically DSL services. See CLEC.

D-Inside Wire Direct-Inside Wire. Made of 24-gauge, annealed-copper conductors with color-added PVC, which allows it to be pulled in conduit without the aid of lubricants. Generally used in the horizontal subsystem.

D-Marc See Demarc.

D-Mark See Demarc.

D-Netz Digital network in Germany used by the two competing companies T-Mobile (Deutsche Telekom, D1-Netz) and Vodafone (D2-Netz).

D.176 The ITU recommendation that defines the way that a file containing call record details for reverse charge calls is to be formatted by an international carrier on the originating end of those calls, for end-user billing by the international carrier at the terminating end of those calls. D.176 defines the file's header record, call detail records, and trailer record.

D/A Digital to Analog conversion.

D/A Converter Digital to Analog converter. A device which converts digital pulses, i.e. data, into analog signals so that the signal can be used by an analog device such as amplifier, speaker, phone, or meter.

D/I Drop and Insert. See Bit Stuffing and Bit Robbing.

D1, D1D, D2, D3, D4 and D5 T-1 framing formats developed for channel banks. All formats contain a framing bit in every 193rd bit position. The Superframe (introduced in D2 channel banks) is made up of 12 193-bit frames, with the 193rd bit sequence being repeated every 12 frames. D2 framing also introduced robbed bit signaling, where the eighth bit in frames 6 and 12 were "robbed" for signaling information (like dial pulses). D1D was introduced after D2 to allow backwards compatibility of Superframe concepts to D1 banks.

D2-MAC One of two European formats for analog HDTV.

D3 Format 24 data channels on one standard (North American standard) T-1/D3 span line. Each data channel is 8 bits wide and has a bandwidth of 8 KHz. See also DS-1.

D3/D4 Refers to compliance with AT&T TR (Technical Reference) 62411 definitions for coding, supervision and alarm support. D3/D4 compatibility ensures support of digital PBXs, M24 services, Megacom services and Mode 3 D3/D4 channel banks at a DS-1 level.

D4 In T-1 digital transmission technology, D4 is the fourth-generation channel bank. A channel bank is the interface between the T-1 carrier system and an analog premises device such as an analog PBX (private branch exchange).

D4 Channelization Refers to compliance with AT&T TR (Technical Reference) 62411 in regards to the DSl frame layout (the sequential assignment of channels and time slot numbers within the DSl).

D4 Framing First read T-1 FRAMING. The most popular framing format in the T-1 environment is D-4 framing. The name stems from the way framing is performed in the D-series of channel banks from AT&T. There are 12 separate 193-bit frames in a super-frame. The D-4 framing bit is used to identify both the channel and the signaling frame. In voice communications, signaling is an important function that is simulated and carried by all the equipment in the transmission path. In D-4 framing, signaling for voice channels is carried "in-band" by every channel, along with the encoded voice. "Robbed-bit-signaling" is a technique used in D-4 channel banks to convey signaling information. With this technique, the eighth bit (least significant bit) of each of the 24 8-bit time slots is "robbed" every sixth frame to convey voice related signaling information (on-hook, off-hook, etc.) for each voice channel. See also Extended Super-Frame Format.

D7Z Dedicated hand-off facility.

DA 1. Doesn't answer, as in "The phone rang DA."

2. Directory Assistance.

3. Demand Assignment.

4. Discontinued Availability. Meaning a circuit that was once available is now no longer.

5. Destination Address, a field in FDDI, Ethernet and Token Ring packets which identifies the unique MAC (Media Access Control, the lower part of ISO layer two) address of the recipient. A six octet value uniquely identifying an endpoint and which is sent in IEEE LAN frame headers to indicate frame destination.

6. Desk Accessory. Standard desk accessories on the Apple Macintosh include a calculator, alarm clock and the chooser. Desk accessories are available to the user regardless of the application currently in use, networked or non-networked. Desk accessories are installed in the Apple menu and accessed from there.

7. Distribution Area. Geographical area that correlates with a Service Area Interface. See also DSA.

DAA Data Access Arrangement. A device required before the FCC registration program if a customer was going to hook up CPE (Customer Provided Equipment), usually modems and other data equipment, to the telephone network. Today, equipment is FCC registered (under Part 68) meaning that the device itself is approved for connection to the phone network. DAAs can still be found in old DP (data processing) installations.

DAB 1. Dynamically Allocable Bandwidth.

2. Digital Audio Broadcasting. Radio broadcasting using digital modulation and digital source coding techniques.

3. Digital Audio Broadcast. The international term for DARS (Digital Audio Radio System), which are proposed satellite-delivered audio/radio systems. See DARS.

DAC 1. Digital to Analog Converter. A device which converts digital pulses, i.e. data, into analog signals so that the signal can be used by analog device such as amplifier, speaker, phone, or meter. In the imaging field, a DAC is a chip that converts the binary numbers that represent particular colors to analog red, green and blue signals that a color monitor displays.

2. Dual Attachment Connector. See Dual Attachment Connector.

DACC 1. Digital Access Cross-Connect.

2. Directory Assistance Call Completion.

DACD Digital Automatic Call Distributor. Another a central office-provided ACD supplied by a local phone company.

DACOMNET A packet-switched network in South Korea.

DACS Also abbreviated as DCS, DCCS and DXC, a DACS is a Digital Access and Cross-Connect System. It is a high-capacity, non-blocking electronic cross-connect device for directing and re-directing circuits and channels in T-carrier and SONET/SDH systems. In a T-carrier system, a DACS works at the DS-0, DS-1 (T-1/E-1), and DS-3 (T-3/E-3) levels. In a SONET/SDH system, a DACS works at the STS-1 level. A DACS is a simple form of channel switch. Unlike a typical voice switch, for example, it does not switch circuits and channels "on the fly" according to dialing instructions. Unlike a packet switch, it does not direct individual packets based on packet addresses. Rather, it is pre-programmed to switch specific circuits or channels from incoming port to outgoing port. In other words and for example, you might sit at a DACS console in Los Angeles and give it specific instructions to connect this T-1 line from San Diego to that T-1 line heading to San Francisco, perhaps in support of a full-motion, broadcast quality videoconference. The next day, you can redirect the San Diego T-1 to a T-1 heading to Bakersfield, if the company president wants to set up a videoconference with the staff there. A DACS, in effect, is a programmable electronic cross-connect or patch panel.

Daemon 1. An agent program which continuously operates on a UNIX server and which provides resources to client systems on the network. Daemon is a background process used for handling low-level operating system tasks. In Greek mythology, "Daemon" was a supernatural being acting as an intermediary between the gods and man.

2. Disk And Execution MONitor. A harmless UNIX program that waits in the background and runs when a request is made on the port that it is watching. It normally works out of sight of the user. On the Internet, it is most likely encountered only when e-mail is not delivered to the recipient. You'll receive your original message plus a message from a "mailer daemon."

DAF 1. Destination Address Field.

2. Decrement All Frame. Motorola definition.

daily fix I am addicted to direct mail, now Internet direct mail. My "daily fix," according to my office, is the package that arrives daily from a direct mail supplier, such as LL Bean, Eddie Bauers, NewEgg.com, Buy.com.

daily usage file See DUF.

daisy chain A method of connecting devices in a series, much as one might interweave daisies to make a lovely floral wreath, or so the story goes. Signals are passed through the chain from one device to the next. Jack 1 is connected to jack 2, which is connected to jack 3 and so on. The last jack in the chain is not connected to jack 1. A SCSI adapter, for instance, is a daisy chain, supporting a daisy chain of up to seven devices. Intel's Universal Serial Bus also is a daisy chain. Stackable hubs, switches and other devices are daisy-chained. While this approach yields the lovely advantage of scalability, interconnection of such devices in this manner also yields some less-than-lovely level of performance degradation, as each device in the chain becomes a point of contention and, therefore, a point of potential congestion. See also Scalable and Stackable.

DAL Dedicated Access Line. A private tie line from you to your long distance or local phone company. The line may be analog or digital, e.g. a T-1 circuit.

DAMA Demand Assigned Multiple Access. A way of sharing a channel's capacity by assigning capacity on demand to an idle channel or an unused time slot.

damped wave A wave consisting of a series of oscillations or cycles of current gradually decreasing amplitude.

dampen To prevent excessive route change announcements from entering a carrier's Internet network and degrading router performance. Many carriers dampen route announcements when the customer exceeds its Dampen Limit. Sprint stops dampening and renews announcing customer routes when the customer reaches its Reuse Limit. Sprint has a dampening policy to dampen Internet traffic to confine network instabilities to a localized area. Network instabilities are caused by customer route flapping. Dampening prevents network instabilities from destabilizing the Sprint Internet Network, other Sprint customer networks, and other portions of global Internet traffic. Sprint uses Cisco Router IOS BGP to dampen Internet traffic. See Dampen Limit and Reuse Limit.

dampen limit Customer penalty value at which point Sprint dampens the customer route announcements. The current Dampen Limit is 2000. See Dampen and Reuse Limit.

damping 1. The decreasing of the amplitude of oscillations caused by resistance in the circuit. 2 The progressive diminution with time of certain quantities characteristic of a phenomenon. 3. The progressive decay with time in the amplitude of the free oscillations in a circuit. 4. More generally, decreasing some dimension of a phenomenon, such as its power.

DAMPS Digital Advanced Mobile Phone Service. Originally, AMPS was used as a 900 MHz frequency modulation (FM) transmission technology with bandwidth allocated according to frequency division multiple access (FDMA) schemes. To increase capability and security, digital techniques for cellular were introduced and systems are being converted from AMPS to DAMPS. The two most prevalent means of dividing frequencies in DAMPS are time division multiple access (TDMA) and code division multiple access (CDMA). These two formats are not directly compatible. See AMPS, NAMPS.

dancing baloney Gratuitous animated GIF files and other Web special effects that are used to impress people. "This page is kinda dull. Maybe a little dancing baloney will help?" This definition courtesy Wired Magazine.

dancing frog A problem or image on your computer screen that disappears just as soon as you try to show it to someone else. The same thing seems to happen with automobiles when you take a normally troublesome car in for a checkup with the mechanic.

DAP Directory Access Protocol. The protocol used between a Directory User Agent (DUA) and Directory System Agent (DSA) in an X.500 directory system. See X.500 and LDAP.

dark Optical fiber through which no light is currently being transmitted. See Dark Fiber.

dark copper Copper over which no communications signal is flowing. In other words, just raw copper in the ground. Sometimes dark copper pairs are sold to a CLEC (competitive local exchange carrier) or to an end customer. See also dark fiber.

dark current The flow of electricity through the diode in a photodiode when no light is present. Photodiodes are often used as light-sensitive switches. When light hits them, they turn on. Here's a more technical explanation: Dark current is the induced current that exists in a reversed biased photodiode in the absence of incident optical power. It is better understood to be caused by the shunt resistance of the photodiode. A bias voltage across the diode (and the shunt resistance) causes current to flow in the absence of light. See also Dark Fiber.

dark fiber Optical fiber through which no light is transmitted and which, therefore, no signal is being carried. Generally speaking, a dark fiber is one of many fibers contained within a cable. Carriers commonly deploy a large number of fibers (432 is a common number) at any given time, since the incremental cost is quite modest compared to pulling them one at a time as the need arises. In fact, a carrier often has little choice, as the right of way may be granted once, and only once. The fibers that the carrier is using immediately are "lit," and those that currently are unused are left "dark." The dark fiber is available for future use. Sometimes dark fiber is sold by a carrier without the accompanying transmission electronics. The customer, which may be either an end user organization or another carrier, is expected to light it up with his own electronics. See also dark copper, dark current, dim fiber and lit fiber.

dark output Output from an optical device that represents a logical 0. See light output.

dark parties Matchmaking gimmick where prospective paramours eat dinner in total darkness, waited on by servers using night-vision goggles. It's extreme blind dating. I can't imagine anything more dumb.

dark space Unallocated IP address space. Spammers, phishers and other bad guys are attracted to dark space in order to conceal their own identities. A number of proposals have been made to prevent the routing of traffic to/from dark space, in part to prevent the misdeeds of the bad guys who are illicitly using this space and in part out of concern that network operators' routing of traffic to/from dark space may tacitly confer "squatter's rights" on the users of that space. See dark space alarm.

dark space alarm A message that is sent to a network administrator whenever traffic to/from the network is to an IP address that is supposedly dark (i.e., unallocated). The idea behind the alarm is that if there is traffic to/from a dark IP address, the traffic probably is associated with an illicit activity. See dark space.

dark side At Apple trade shows, people who use Windows machines are known as being on the Dark Side.

dark swap Round-trip commodity trading of unused broadband – so-called dark fiber – among providers. The technique creates the appearance of trade activity. An unscrupulous carrier can book the swapping as revenue and thus make his financials look better to investors in the stockmarket. It happened in the late 1990s and early 2000s. And by the time you read this, some of the creative executives who thought this up should be sitting firmly in jail.

dark wavelength A Dense Wavelength Division Multiplexing (DWDM) term. Dark wavelength refers to a virtual channel in a fiber optic system utilizing DWDM. Each virtual channel is supported through a specific wavelength of light, with many such channels riding over the same fiber. Once the fiber system is deployed and the DWDM equipment is activated, some of the wavelengths may be activated immediately and others may be left dark for future needs. Such a fiber system is call "Dim Fiber," as it's neither completely dark, nor fully lit. When the need arises, those dark wavelengths are lit up. See also DWDM, Fiber, Optical Fiber and SONET.

darknet Networks and web sites that let people illegally share copyrighted material (movies, songs, photos, etc.) with little or no fear of detection. Despite all the openness of the Internet, there are still places you cannot saunter into on the Web. You must be invited. These are "darknets": exclusive peer-to-peer networks in which membership is based on circles of trust, whose activities are veiled from the general public. And though people who are adept at configuring servers and comfortable with File Transfer Protocol have used such systems for years, a spate of new online services aimed at everyday users is sure to draw new attention to under-the-radar file sharing. Darknets, like their peer-to-peer predecessors Napster, Kazaa and Gnutella, allow users to browse and download digital files like movies and music from other people's computers. But while Napster and its ilk have allowed unrestricted access to files on any of the millions of connected computers, darknets are more discriminating. In a darknet, users get access only through established relationships – and only when they have been invited to join. This selectivity promises greater privacy, regardless of whether the networks are used for sharing personal or pirated media. Grouper, among the largest of the new services, hosts more than 100,000 private groups. Users can build their own darknets or request admission to thousands of publicly listed clubs whose members can browse through group folders, download files and communicate by instant messaging or group blogs. A corporate darknet offers all the security of a private in-house network, but it allows users to send encrypted messages and documents.

DARPA Defense Advanced Research Projects Agency. Formerly called ARPA, it is a US government agency that funded research and experimentation with the ARPANET and later the Internet. The group within DARPA responsible for the ARPANET is ISTO (Information Systems Techniques Office), formerly IPTO (Information Processing Techniques Office). DARPA had sponsored research in the 1960s and the 1970s to create a computer network that could survive a nuclear detonation. See also DARPA Internet, IAB, IETF and Internet.

DARPA Internet World's largest internetwork, linking together thousands of networks around the world. Sponsored by U.S. Defense Advanced Research Projects Agency. Now called DARPANET. See next definition.

DARPANET Defense ARPANET. Also known as DARPA Internet. In 1983 the ARPANET was officially split into DARPANET and MILNET. World's largest internetwork, linking together thousands of networks around world. Sponsored by U.S. Defense Advanced Research Projects Agency. DARPANET was the beginnings of the Internet. See Internet.

DARS Digital Audio Radio System. Also known as DAB (Digital Audio Broadcasting) outside the U.S. Proposed satellite-delivered audio/radio systems, similar to DBS (Direct Broadcast System) TV systems, which have been enormously successful in competition with CATV. DARS has been debated by the FCC and the ITU-R since the initial application by CD Radio Inc. in 1990. Assuming that the FCC and ITU-R eventually agree on frequency assignments (and they now have), you may want to make room for one more satellite dish on your rooftop or on your car.

DAS Tape A cellular term. The magnetic tape that is used at the MTSO to record traffic statistics and call billing information. This tape is sent to a third-party 'billing-house' where the actual billing of the subscribers is done.

DASD Direct Access Storage Device. Any on-line data storage device. Usually refers to a magnetic disk drive, because optical drives and tape are considered too slow to be direct access devices. Pronounced DAZ-dee. The term is said to have been invented by IBM.

dashboard Key indicators used to track the progress of a business project. Of course, Microsoft and others are happy to sell you a "digital" dashboard to group and monitor them for you.

DASS Direct Access Secondary Storage. Same as near-line: storage on pretty-fast storage devices (e.g., rewritable optical) that are less expensive than hard drives but faster than off-line devices.

DASS1 Digital Access Signaling. A British term. The original British Telecom (BT) ISDN signalling developed for both single line and multi-line Integrated Digital Access but used in the BT ISDN pilot service for single line IDA only.

DASS2 Digital Access Signaling System No. 2. A British Term. A message-based signalling system following the ISO-based model developed by British Telecom to provide multi-line IDA interconnection to the BT network.

DAT Digital Audio Tape used to identify a type of digital tape recorder and player as well as the tape cassette. DAT tape machines record music that is much crisper, and free of the hisses and pops that mar traditional analog recordings. The drawback with DAT tape machines is they require considerable tape to store music digitally. In a DAT machine, the music is recorded by sampling the music 48,000 times each second. Each of those samples is represented by a number that is written as a 16-digit string of zeros and ones. There are two such signals, once for each stereo channel, meaning that storing a single second of music requires about 1.5 million bits. On top of that, extra bits are added to allow the system to mathematically correct errors and help the machine automatically find a particular song on the tape. All together, according to Andrew Pollack writing in the New York Times, a single second of music on a digital audio tape requires 2.8 million bits. But compression techniques are cutting down the amount of information required to be recorded.

data This is the old AT&T Bell Labs' definition: "A representation of facts, concepts or instructions in a formalized manner, suitable for communication, interpretation or processing." Typically anything other than voice.

data abstraction A term in object-oriented programming. An object is sometimes referred to as an instance of an abstract data type or class. Abstract data types are constructed using the built-in data types supported by the underlying programming language, such as integer and date. The common characteristics (both attributes and methods) of a group of similar objects are collected to create a new data type or class. Not only is this a natural way to think about the problem domain, it is a very efficient way to write programs. Instead of individually describing several dozen instances, the programmer describes the class once. Once identified, each instance is complete with the exception of its instance variables. The instance variables are associated with each instance, i.e., each object; methods exist only with the classes. See Object Oriented Programming.

Data Access Arrangement DAA. Equipment that allows you to attach your data equipment to the nation's phone system. At one stage, DAAs were required by FCC "law." Now, their limited functions are built into directly attached devices, such as terminals, computers, etc.

Data Access Point DAP. MCI computer that holds the number translation and call-routing information for 800 and Vnet services. These computers respond to inquiries from MCI switches on how to handle these calls.

data arrangement In public switched telephone networks, a single item or group of items present at the customer's premises, including all equipment that may affect the characteristics of the interface. An obsolete term. Historically, it came from the time when the phone industry insisted on an interface between its lines and equipment provided by others.

data attribute A characteristic of a data element such as length, value, or method of representation.

data bank A collection of data in one place. The data is not necessarily logically related, nor is it necessarily consistently maintained. See Database.

data base See Database, which is our preferred spelling.

data broadcasting A method of high speed data distribution for text and graphics which uses the spare capacity in the broadcasting television, cable and satellite transmission systems.

data bubble A new organization within BellSouth to provide high-speed digital services. No one seems to know why it's called "Data Bubble," except that someone inside BellSouth clearly thinks the term is cute.

data burst Burst transmission.

data bus A bus that transmits and receives data signals throughout the computer or telephone system. See BUS.

Data Carrier Detect DCD. A hardware signal defined by the RS-232-C standard that indicates that the device, usually a modem, is online and ready for transmission.

data center A centralized location where computing resources (e.g., host computers, peripherals, applications, databases, and network access) critical to an organization are maintained in a highly controlled environment. A data center is manned by highly trained professionals, many of whom are specialized in fields such as programming, security management, system repair, and technical support. The facility is highly secure with both physical and network security. All systems are provided clean power, with backup power supplies (i.e., both battery power and diesel generators) on-line in the event of the failure of the commercial power source. Application software is carefully controlled in order to ensure that software licenses are not violated, no software conflicts arise, versions are released in a controlled fashion, and software bugs are screened out through a testing phase. System backups are accomplished on a regular basis, with copies of backed-up data stored both on-site and off-site. Admittedly, this definition lists some of the characteristics of an ideal data center, and this is not a complete list. Note that even the ideal data center should have a backup in the form of a hot, warm, or cold standby. A hot standby is the ultimate, and the most expensive. It consists of an exact, up-to-the-second duplicate of the primary data center, including not only the host computers and peripherals, but also the applications and databases, and the network configuration. A hot standby is ready to take over instantaneously in the event of a catastrophic failure. A warm standby is much the same, but takes a bit longer to take over, as it must be activated, the most recent updates data must be loaded, etc. A cold standby takes longer still, as equipment might need to be reconfigured, applications loaded, entire databases loaded, etc.

data circuit Communications channels provided specifically for the exchange of data as compared to voice. See also IP Telephony.

Data Circuit Terminating Equipment DCTE. Also known as DCE (Data Communications Equipment). See DCE.

data circuit transparency The capability of a circuit to transmit all data without changing its content or structure.

data cleansing and scrubbing A process of removing redundancies and inconsistencies in operational data.

Data CLEC Data Competitive Local Exchange Carrier, i.e. a CLEC which only sells data services – most of which are typically high speed access to the Internet. See CLEC.

data communications The transfer of data between points. This includes all manual and machine operations necessary for this transfer. In short, the movement of encoded information by means of electrical transmission systems. See Data Communications Equipment.

data communications channel A three-byte, 192 Kbps portion of the SONET signal that contains alarm, surveillance and performance information. It can be used for internally or externally generated messages, or for manufacturer specific messages.

Data Communications Equipment DCE. Also known as Data Circuit Terminating Equipment (DCTE). A device which provides the interface between a circuit and DTE (Data Terminal Equipment. See DCE and DTE.

data compression Reducing the size of a file of data by eliminating unnecessary information, such as blanks and redundant data. The idea of reducing the size is to save money on transmission or to save money on storing the data. The file or program which has been compressed is useless in its compressed form and must be "decompressed," i.e. brought back to normal before use. One method of data compression replaces a string of repeated characters by a character count. Another method uses fewer bits to represent the characters that occur more frequently. See also Compression.

Here's another definition, courtesy the US Department of Commerce: 1. The process of reducing (a) bandwidth, (b) cost, and (c) time for the generation, transmission, and storage of data by employing techniques designed to remove data redundancy. 2. The use of techniques such as null suppression, bit mapping, and pattern substitution for purposes of reducing the amount of space required for storage of textual files and data records. Some data compaction methods employ fixed tolerance bands, variable tolerance bands, slope-keypoints, sample changes, curve patterns, curve fitting, floating-point coding, variable precision coding, frequency analysis, and probability analysis. Simply squeezing noncompacted data into a smaller space, e.g., by transferring data on punched cards onto magnetic tape, is not considered data compression. See Data Compression Table.

data compression protocols All current high-speed dial-up modems also support data compression protocols. This means the sending modem will compress the data on-the-fly (as it transmits) and the receiving modem will decompress the data (as it receives it) to its original form. There are two standards for data compression protocols, MNP-5 and ITU-T V.42 bis. Some modems also use proprietary data compression protocols. A modem cannot support data compression without using an error control protocol, although it is possible to have a modem that only supports an error control protocol but not any data compression protocol. A MNP-5 modem requires MNP-4 error control protocol and a V.42 bis modem requires V.42 error control protocol. Note that although V.42 includes MNP-4, V.42 bis does not include MNP-5. However, virtually all high-speed modems that support ITU-T V.42 bis also incorporate MNP-5. The maximum compression ratio that a MNP-5 modem can achieve is 2:1. That is to say, a 9,600 bps MNP-5 modem can transfer data up to 19,200 bps. The maximum compression ratio for a V.42 bis modem is 4:1. That is why all those V.32 (9,600 bps) modem manufacturers claim that their modems provide throughput up to 38,400 bps.

Are MNP-5 and V.42 bis useful? Don't be fooled by the claim. It is extremely rare, if ever, that you will be able to transfer files at 38,400 or 57,600 bps. In fact, V.42 bis and MNP-5 are not very useful when you are downloading files from online services. Why? How well the modem compression works depends on what kind of files are being transferred. In general, you will be able to achieve twice the speed for transferring a standard text file (like the one you are reading right now). V.42 bis and MNP-5 modem cannot compress a file which is already compressed by software. In the case of MNP-5, it will even try to compress a precompressed file and actually expand it, thus slow down the file transfer! The above information courtesy modem expert, Patrick Chen.

data compression table A term from US Robotics, makers of fine modems. A data compression table is a table of values assigned for each character during a call under data compression. Default values in the table are continually altered and built during each call. The longer the table, the more efficient the throughput gained.

data compressors Also called compactors. These devices take over where high speed modems and statistical multiplexers leave off. They save phone lines by a doubling of data throughput by further compressing async or sync data streams.

data concentrator A device which permits the use of a transmission media by a number of data sources greater than the number of channels currently available for transmission.

data connections An ATM term. Data VCCs connect the LECs to each other and to the Broadcast and Unknown Server. These carry Ethernet/IEEE 802.3 or IEEE 802.5 data frames as well as flush messages.

data contamination Data corruption.

data control block A data block usually at the beginning of a file containing descriptive information about the file.

data convergence Conergence is one of the most abused working in networking. Originally it meant that you carried voice and data on the same circuit. But the voice was circuit switched and not really co-mingled with the data. It was separated physically from the data by time division multiplexing. Now it's voice over IP (VoIP) and is integrally co-mingled with the data, both being now in standard Ethernet format. In 2003 or thereabouts Microsoft started using the term data convergence to mean handling voice, data, video, songs, music, etc. on one PC or PC-like device.

data conversion Converting data from one format to another. Conversion typically falls into three basic categories. 1. To convert to a form usable by the equipment you have, e.g. you convert some data from tape to disk (because you don't have a tape drive). Or you may convert from one method of encoding data to another, say from EBCDIC to ASCII, because you don't have software which can understand IBM's EBCDIC method of coding. 3. Or you may convert from one format to another, e.g. from the dBASE method of encoding databases to the Paradox method, or from WordPerfect to Word. There are many service bureaus whose job is to convert computer data from one form to another and there are now many programs out to do the conversion. Most of the leading PC programs now contain data conversion software built into them, so it's possible to open a dBASE database file in Microsoft Access or a WordPerfect file in Word. Conversion becomes part of the process of opening the file. It wasn't always that way. See also Data Compression.

data country code A 3-digit numerical country identifier that is part of the 14-digit network terminal number plan. This prescribed numerical designation further constitutes a segment of the overall 14-digit X.121 numbering plan for a ITU-T X.25 network.

data dialtone Networking as widespread as the telephone. The Internet is widely thought to contain the beginnings of Data Dialtone. See Internet, Metcalfe's Law, and World Wide Web.

data dictionary 1. A part of a database management system that provides a centralized meaning, relationship to other data, origin, usage, and format.

2. An inventory that describes, defines, and lists all of the data elements that are stored in a database.

data diddling Unauthorized altering of data before, during or after it is input into a computer system.

data directory An inventory that specifies the source, location, ownership, usage, and destination of all of the data elements that are stored in a database.

data dump A sample or whole extract of the information contained on a database. Used to develop a better picture of the data on a database.

data element A basic unit of information having a unique meaning and subcategories (data items) of distinct units or values. Examples of data elements are military personnel grade, sex, race, geographic location, and military unit.

Data encryption Standard DES. A 56-bit, private key, symmetric cryptographic algorithm for the protection of unclassified computer data issued as Federal Information Processing Standard Publication. DES, which was developed by IBM in 1977, was promulgated by the National Institute of Standards and Technology (NIST) – formerly the National Bureau of Standards (NBS) – for public and Government use. It was thought to be uncrackable until 1997, when a nationwide network of computer users broke a DES key in 140 days. Triple DES, a later version, encodes the data three times for additional security. As the cost of computer equipment has dropped and as computer power has increased, DES no longer is considered to be totally secure. It has been said that anyone who can afford a BMW can afford a DEScracker. As a result, NIST is searching for a replacement encryption standard to be known as AES (Advanced Encryption Standard). See also Encryption and AES.

data entry Using an I/O device (input/output device), such as a keyboard on a terminal, to enter data into a computer.

data file A database typically contains multiple files of information. Each file contains multiple records. Each record is made up of one or more fields. Each field contains one or more bytes of data. The terms file, record and field find their roots in manual office filing systems.

data fill One name for the specifications of your ISDN phone lines. Ask for your ISDN Data Fill. It will give you useful information, such as how your lines are set up – voice, data, data/voice, etc.

data flow Grouping of traffic, identified by a combination of source address/mask, destination address/mask, IP next protocol field, and source and destination ports, where the protocol and port fields can have the values of any. In effect, all traffic matching a specific combination of these values is grouped logically together into a data flow. A data flow can represent a single TCP connection between two hosts, or it can represent all the traffic between two subnets. IPSec protection (A collection of IP security measures) is applied to data flows.

data frame An SCSA term. A set of time slots which are grouped together for synchronization purposes. The number of time slots in each frame depends on the SCbus or SCxbus Data Bus data rate. Each frame has a fixed period of 125us. Frames are delineated by the timing signal FSYNC.

data freight The long-haul transport of bits in bulk. Any corporation with cross-country rights-of-way can get into the datafreight biz by laying fat fiber. Qwest is in the datafreight business.

data grade circuit A circuit which is suitable for transmitting data. High speed data needs better quality phone lines than normal dial-up phone circuits. You can acquire such circuits from many telephone companies. To upgrade voice phone lines to high-speed data circuits, you must sometimes "condition" the phone line. See Conditioning.

data group A number of data lines providing access to the same resource.

data hunt group An AT&T Merlin term. A group of analog or digital data stations that share a common extension number. Calls are connected in a round-robin fashion to the first available data station in the group.

data integrity 1. The data you receive is exactly what was sent you. Typically the concept of data integrity relates to data transmission. Data integrity is also a performance measure based on the rate of undetected errors. A measure of how consistent and accurate computer data is. In data transmission, error correcting protocols, such as LAP-M, MNP and X-modem – provide methods of ensuring that data arrives at the destination in its full integrity. This is done in many ways, including retransmitting messed-up blocks. Data integrity can be threatened by hardware problems, power failures and disk crashes, but most often by application software. In a database system, data integrity can be threatened if two users are allowed to update the same item or record of data at the same time. Record locking, where only a single user at a time is allowed access to a given data record, is a method of insuring data integrity.

2. A service provided by cryptographic technology that ensures data has not been modified. In a network environment, data integrity allows the receiver of a message to verify that data has not been modified in transit. Windows 2000 and Windows XP Professional use access control mechanisms and cryptography, such as RSA public-key signing and shared symmetric key one-way hash algorithms, to ensure data integrity.

data leakage The leakage of confidential data or information from the corporate network.

data line interface The point at which a data line is connected to a telephone system.

data line monitor A measuring device that bridges a data line and looks at how clean the data is, whether the addressing is accurate, the protocol, etc. Being only a monitor, it does not in any way affect the information traveling on the line. See Data Monitor.

data line privacy Prohibits activities which would insert tones on a data station line used by a facsimile machine, a computer terminal or some other device sensitive to extraneous noise.

data link A term used to describe the communications link used for data transmission from a source to a destination. In short, a phone line for data transmission. Or, a fiber optic transmitter, cable, and receiver that transmits digital data between two points.

Data Link Control DLC. Characters used in data communications that control transmission by performing various error checking and housekeeping functions – connect, initiate, terminate, etc.

data link control protocol A Micosoft driver used to connect Windows 95 and NT workstations to IBM mainframes when TCP/IP is not available. However, the protocol is not routable and it is not designed for peer-to-peer communications between workstations. It is designed only for connectivity to mainframes and minicomputers.

data link escape The first control character of a two-character sequence used exclusively to provide supplementary line-control signals.

data link layer The second layer of the Open Systems Interconnection data communications model of the International Standards Organization. It puts messages together and coordinates their flow. A layer that packages raw bits from the physical layer into frames (logical, structured packets for data). This layer is responsible for transferring frames from one computer to another, without errors. After sending a frame, the data-link layer waits for an acknowledgment from the receiving computer. Also used to refer to a connection between two computers over a phone line. See OSI Model.

Data Local Exchange Carrier DLEC. A DLEC is a CLEC (Competitive Local Exchange Carrier) that specializes in delivering only data, most typically DSL services.

data mart A small, single-subject warehouse used by individual departments or groups of users.

data message A message included in the GPS (Global Positioning System) signal which reports the satellite's location, clock corrections and health. Included is rough information on the other satellites in the constellation.

Data Message Handler DMH. A method used in the cellular industry for exchanging non-signaling messages between service providers on a near real-time basis. DMH originally was used to facilitate call hand-offs between carriers. It was extended to serve a number of other purposes. The DMH extension for fraud detection and prevention in known as NSDP-F (Non-Signaling Data Protocol for Fraud), and the extension for the transfer of billing and settlement information is known as NSDP-B&S. DMH has been standardized by the TIA as IS-124.

data mining Data mining refers to using sophisticated data search capabilities that use statistical algorithms to discover patterns and correlations in data. A comparison to traditional gold or coal mining, data mining is defined as a way to find buried knowledge ("data nuggets" – no kidding!) in a corporate data warehouse (or information that visitors have dropped on a Web site) and to improve business users' understanding of this data. The data mining approach is complementary to other data analysis techniques such as statistics, on-line analytical processing (OLAP), spreadsheets, and basic data access. In a nutshell, data mining is another way to find meaning in data. On September 22, 1999, Gile Felton described data mining in the New York Times: "Every click tells a story. See Data Warehouse.

data monitor A device used to look at a bit stream as it travels on a circuit. It will show the user what is going down both sides of a data channel. It will show what the user at his terminal is typing, and what the computer is responding with. Extremely useful for troubleshooting data communications problems. Also called a Data Scope.

data multiplexer A device allowing several data sources to simultaneously use a common transmission medium while guaranteeing each source its own independent channel. See Multiplexer.

data nazi Jim Seymour. His writing style was folksy and irreverent. Mr. Seymour regarded the PC mainly as a tool that put computational power in the hands of individual workers, an important departure from the centralized, mainframe era of computing. He derided the mainframe diehards as "numbskulls" and "data Nazis."

Data Network Identification Code DNIC. In the ITU-T International X.121 format, the first four digits of the 14-digit international data number; the set of digits that may comprise the three digits of the data country code (DCC) and the 1-digit network code (which is called the "network digit"). See DNIC.

Data Numbering Plan Area DNPA. In the U.S. implementation of the ITU-T X.25 network, the first three digits of a network terminal number (NTN). The 10-digit NTN is the specified addressing information for an end-point terminal in an X.25 network.

data object An individually addressable unit of information, specified by a data template and its content, that can persist independently of the invocation of a service.

Data Over Cable Service Specification See DOCSIS.

Data Over Voice DOV. See DOV.

data overrun Also called UART Overrun. This definition from Derrick Moore of JDR Microdevices (1-800-538-5000). Data overrun occurs when your PC is unable to accept interrupts as fast as needed from your serial port chip. Because your PC may be busy reading from a disk or refreshing the screen, it may not handle a serial interrupt before the next byte of data overfills the receiver buffer on the chip. Early PCs used the 8250 which only had a one byte buffer. It was enough because modems were slow and PCs only did one thing at a time. Later, when modems became faster, the serial chips like the 16550 were upgraded to 16 byte buffers. Startech's 16C650 chip has a 32 byte buffer and allows more time for the operating system to service interrupts as well as reduce the number of interrupts that occur. If you run Windows 3.1, 95, NT, or OS/2, and use a high speed modem, you have probably experienced data overrun. Since most communications programs merely request that data be re-sent, the problem is masked. In short, make sure your PC has at a least a 16-byte UART. See UART and UART Overrun.

data packet Although a computer and modem can send data one character at a time, when you're surfing the Internet, downloading files, or sending email, it's much more efficient to send information in larger blocks called data packets. Modems generally send packets of around 64 characters along with some extras for error checking. When downloading files using a protocol like Xmodem, however, the packets are larger. And when using Internet protocols such as TCP/IP, the packets are larger still – around 1,500 characters. Such packets of data contain the information you're sending or receiving, an address (i.e. where it's going) and some start and stop information. See also asynchronous communication, Ethernet, TCP/IP and Xmodem.

data packet signaling All telephone switches use the same three general types of signals: + Event Signaling initiates an event, such as ringing. + Call Progress Signaling denotes the progress (or state) of a call, such as a busy tone, a ringback tone, or an error tone. + Data Packet Signaling communicates certain information about a call, for example, the identify of the calling extension, or the identity of the extension being called.

data packet switch System-common equipment that electronically distributes information among data terminal equipment connected to a data transmission network. The switch distributes information by means of information packets addressed to specific terminal devices.

data PBX A PBX for switching lots of low-speed asynchronous data. A switch that allows a user on an attached circuit to select from among other circuits, usually one at a time and on a contention basis, for the purpose of establishing a through connection. Distinguished from a PBX in that only digital transmissions, and not analog voice, are supported. Like a telecommunications PBX that makes and breaks phone connections, a data PBX makes and breaks connections between computers and peripherals. In response to dynamic demand, it establishes communications paths between devices attached to its input/output ports by receiving, transmitting and processing electrical signals. Usually, data PBXs work off PCs' serial ports rather than through cable attached to a network interface card. For that reason, they are restricted to serial speeds, topping out at about 19.2K bits per second. For switching lots of low-speed asynchronous data, a data PBX (also called a line selector) can be better than a LAN. Total throughput can actually be higher. See also Line Selector.

data phase A phase of a data call during which data signals may be transferred between DTEs that are interconnected via the network.

data port Point of access to a computer that uses trunks or lines for transmitting or receiving data.

data protection A means of ensuring that data on the network is safe. Novell's NetWare protects data primarily by maintaining duplicate file directories and by redirecting data from bad blocks to reliable blocks on the NetWare server's hard disk. A hard disk's Directory Entry Table (DET) and File Allocation Table (FAT) contain address information that tells the operating system where data can be stored or retrieved from. If the blocks containing these tables are damaged, some or all of the data may be irretrievable. NetWare reduces the possibility of losing this information by maintaining duplicate copies of the DET and FAT on separate areas of the disk. If one of the blocks in the original tables is damaged, the operating system switches to the duplicate tables to get the location data it needs. Data protection within standard NetWare also involves such features as read-after-write verification, Hot Fix, and disk mirroring or duplexing. See Disk Mirroring and Hot Fix.

data rate The rate at which a channel carries data, measured in bits per second, also known as data signaling rate. If there are restrictions on the pattern of bits, the information capacity of the channel could be less than the data rate. In short, data rate is the measurement of how quickly data is transmitted – but it may be very different (i.e. less) than what the channel is theoretically capable of.

data rate mismatch A condition that occurs when a packet's transmission frequency (data rate) does not match the local transmit frequency.

data record See Data File and Database Management System.

data scope See Data Monitor.

data scrambler A device used in digital transmission systems to convert an input digital signal into a pseudo random sequence free from long runs of marks, spaces, or other simple repetitive patterns.

data secure line A single tip and ring line off the PBX which is protected against any tones (like call waiting) or break-ins that would otherwise mess up any ongoing data transmission call.

data security The protection of data from unauthorized (accidental or intentional) modification, destruction, disclosure, or delay.

data segment A pre-defined set of data elements ordered sequentially within a set, beginning and ending with unique segment identifies and terminators. Data segments combine to form a message. Their relation to the message is specified by a Data Segment Requirement Designator and a Data Segment Sequence. Data Segment is an EDI (Electronic Data Interchange) term.

data segment requirement designator An EDI (Electronic Data Interchange) requirement designator determines that if and when the data segment will occur in a message: - MANDATORY. The segment must appear in the message. - CONDITIONAL. The segment will occur in the message depending on agreement conditions. The relevant conditions must be given as part of the message definition. - OPTIONAL. The segment may or may not occur.

data segment sequence In Electronic Data Interchanges, each data segment has a specific place within a message. The data segment sequence determines exactly where a segment will occur in a message: - HEADING AREA. A segment occurring in this area refers to the entire message. - DETAIL AREA. A segment occurring here is detail information only will override any similar specification in the header area. - SUMMARY AREA. Only segments containing total or control information may occur in this area (e.g., invoice total, etc.)

Data Service Unit DSU. Device designed to connect a DTE (Data Terminal Equipment like a PC or a LAN) to a digital phone line to allow fully-digital communications. A DSU is sort of the digital equivalent of a modem. In more technical terms, a DSU is a type of short haul, synchronous data line driver, normally installed at a user location that connects a user's synchronous equipment over a 4-wire circuit to a serving dial-central-office. This service can be for a point-to-point or multipoint operation in a digital data network. DSUs are typically used for leased lines. For switched digital services, you need a CSU/DSU (also called a DSU/CSU). See CSU/DSU and DSU/CSU.

data set In AT&T jargon, a data set is a modem, i.e. a device which performs the modulation/demodulation and control functions necessary to provide compatibility between business machines which work in digital (on-off) signals and voice telephone lines. In IBM jargon however, a data set is a collection of data, usually in a file on a disk. See also Modem.

data set ready One of the control signals on a standard RS-232-C connector. It indicates whether the data communications equipment is connected and ready to start handshaking control signals so that transmission can start. See RS-232-C and the Appendix.

data signaling rate The total of the number of bits per second in the transmission path of a data transmission system. A measurement of how quickly data is transmitted,

expressed in bps, bits-per-second.

data sink Part of a terminal in which data is received from a data link.

data slicer A circuit interface to a radio scanner which provides a means of determining the performance of a radio system (e.g., microwave) using FSK (Frequency Shift Keying) for purposes of digital transmission. FSK shifts frequencies at specific points in time in order to represent ones and zeros. The data slicer allows a technician or engineer to determine whether the radio system is working properly. It does this by logically "slicing" the radio waves in time in order to view the ones and zeros of the data stream.

data source 1. The originating device in a data communications link.

2. An object identifier in RMON that represents a particular interface.

data span Any digital service, T-1, 56K, ISDN or data carrying service.

data steward A new role of data caretaker emerging in business units. Individual takes responsibilities for the data content and quality.

data stream 1. Collection of characters and data bits transmitted through a channel.

2. An SCSA term. A continuous flow of call processing data.

data surfer A person who makes a living doing online research and information retrieval. Also known as a Cybrarian (comes from cyberspace librarian) or a super searcher. See Cybrarian.

Data Switching Exchange DSE. The equipment installed at a single location to perform switching functions such as circuit switching, message switching, and packet switching.

data synchronization The process of keeping database data timely and relevant by sending and receiving information between laptops, between desktops in the field and between bigger computers at headquarters. See also Synchronization and Replication.

data terminal A generic term for a piece of equipment in a system capable of sending and/or receiving data signals.

Data Terminal Equipment DTE. A terminal device in the data world. DTE is part of a broader grouping of equipment known as CPE (Customer Premises Equipment), which includes voice, as well as data, terminals. At the terminal end of a data transmission, DTE comprises the transmit and receive equipment. DTE can be in the form of a dumb terminal (i.e., a terminal without embedded intelligence in the form of programmed logic), a semi-intelligent terminal, or an intelligent host computer (i.e., a PC, mid-range or mainframe computer). DTE interfaces to a circuit through DCE (Data Communications Equipment). See DCE and DTE.

data terminal ready One of the control signals on a standard RS-232-C connector. It indicates if the data terminal equipment is present, connected and ready and has had handshaking signals verified. See RS-232-C and the Appendix.

data transfer rate The speed at which data is transferred from one device to another. Data transfer rates are typically expressed in megabits (a million bits) or megabytes (a million bytes) per second.

data transfer request signal A call control signal transmitted by a DCE to a DTE to indicate that a distant DTE wants to exchange data.

data transfer time The time that elapses between the initial offering of a unit of user data to a network by transmitting data terminal equipment and the complete delivery of that unit to receiving data terminal equipment.

data typing When converting a database from one format to another, several conversion programs will convert the data to a common format before converting it to the final version. During the conversion process a program may check through the data in the database to determine what it is and arbitrarily make one field numeric, one field character, one field memo, etc.

Data User Part DUP. Higher layer application protocol in SS7 for the exchange of circuit switched data; not supported by ISDNs.

data warehouse A data warehouse is a central repository for all or significant parts of the data that an enterprise's various business systems collect. This data can either be accessed quickly by users or put on an OLAP server for more thorough analysis. Data warehouses often use OLAP servers. OLAP stands for On Line Analytical Processing, also called a multidimensional database. According to PC Week, these databases can slice and dice reams of data to produce meaningful results that go far beyond what can be produced using the traditional two-dimensional query and report tools that work with most relational databases. OLAP data servers are best suited to work with data warehouses. See Data Warehousing.

Imaging Magazine (now called Transform Magazine) once wrote a story on data warehouses. The writer, Joni Blecher, found "defining a data warehouse to be puzzling at best."

She said these definitions seem to make the most sense:

1) A collection of physical data stores designed to concisely present a historical perspective of the events that occur in an enterprise. Data warehousing is a set of activities some of which are optional and some mandatory that create, operate and evolve the collection of data stores that make up the data warehouse. – Actium.

2) An extremely comprehensive solution that includes hardware, software, middleware, partner products as well as their own professional services focused on solving business problems through the enterprise level. – NCR.

3) The place where business managers can access information for managerial processes. They're built for decision making purposes. It's an elaborate process that consists of a solution made up of many products. – Oracle.

4) A group of individuals, processes, methodologies – all the things that deal with and manage data including cleansing, enhancing, standardizing, consolidating and disseminating it. – Acxiom.

5) A data store that companies build where they're storing their information assets so they can extract knowledge and understanding to the operation and performance of their business. – Logic Works.

The term was coined by W. H. Inmon. IBM sometimes uses the term "information warehouse."

data warehousing A software strategy in which data is extracted from large transactional databases and other sources and stored in smaller databases, making analysis of the data somewhat easier. See Data Warehouse.

datacasting The term is typically used in relation to information services that are made available by digital TV, such as complimentary information presented in addition to TV programming. A datacasting service is broadly defined as a service that delivers content in the form of text, data, speech, music or other sounds or visual images (or in any form or combination of forms) to persons with appropriate reception equipment, when the delivery of the service uses the broadcasting service bands. One such service is called Moviebeam, which uses a radio receiver to receive a bunch of movies each month. The Moviebeam box will be regularly refreshed with new digital movies that are delivered to it not via digital cable, satellite or the Internet but through the old-fashioned pipeline of broadcast airwaves. Manufactured by Samsung Electronics, the receiver has a small but powerful antenna that resembles a tiny propeller. Through a process known as datacasting, Moviebeam perpetually transmits movies to the device's hard drive in tiny bits of data that travel alongside the normal over the air TV broadcast stream of a local ABC or PBS station, without interfering with the regular TV broadcast. While the service currently uses the analog broadcast spectrum, it is equipped to take advantage of digital broadcasts when they become more common. I have not see Moviebeam work. See data broadcasting.

database A collection of data structured and organized in a disciplined fashion so that access is possible quickly to information of interest. There are many ways of organizing databases. Most corporate databases are not one single, huge file. They are multiple databases related to each other by some common thread, e.g. an employee identification number. Databases are made up of two elements, a record and a field. A record is one complete entry in a database, e.g. Harry Newton, 205 West 19 Street, York, NY 10011, 212-206-7140. A field would be the zip code field, namely 10011.

Databases are stored on computers in different ways. Some are comma delineated. They differentiate between their fields with commas – like Gerry's record above. A more common way of storing databases is with fixed length records. Here, all the fields and all the records are of the same length. The computer finds fields by index and by counting. For example, Gerry's first name might occupy the first 15 characters. Gerry's last name might be the next 20 characters, etc. Where Gerry's names are too short to fill the full 15 or 20 characters, their fields are "padded" with specially-chosen characters which the computer recognizes as padded characters to be ignored. The most important thing to remember about databases is that all the common database programs, like dBASE, Paradox, Rbase, etc. don't automatically make backups of their files like word processing programs do. Therefore, before you muck with a database file – sort it, index it, restructure it, etc. Please make sure you make a backup of the main database file.

database administrator 1. A person who organizes, designs, implements and runs the company's databases. Since I personally believe databases – especially of prospects and customers – are pretty well a company's most important asset, this job of database administrator is very important.

2. DBA. A computer at MCI Worldcom that maintains the master file of Vnet translation information. The master file is created when a customer begins service and can be changed at anytime through CIM. The updated copies of the database are downloaded each night to

the DAPs.

database clustering Database clustering is a technology that allows a bunch of computers connected together to work as one and serve a single relational database. The idea is that together they can create a fault-tolerant, high performance, scalable solution that's a low-cost alternative to large, expensive servers. Database clustering is also called real application clustering (RAC). When I wrote this it was new and relatively unproven. There are two common methods of clustering relational databases, known as "shared-disk" and "shared-nothing" clustering. IBM's DB2 Universal Database (DB2 UDB) uses a shared-nothing approach. In this architecture, each node in the cluster holds only one segment of the database, and the node also handles all of the computational work that corresponds to the data it stores. A master server assesses the task at hand and then parcels it out, distributing a portion of the job to each node that contains data to be processed. The task is then executed by all of the nodes in parallel, and the master server reports the result. Oracle uses shared disk clustering – a design that's structured around a single large data store (for example, a disk array). Each node on the cluster has equal access to all of the information in the data store. Only the processing work is divided amongst them, and not the data itself. The result is particularly fault tolerant database. Even if one or more servers fails, all of the application data remains available to the other nodes. By comparison, if one node on a shared-nothing database crashes, all of the data stored on that node likewise goes offline, until a failover system can recover from the fault.

database dictionary A specific type of system table that stores information about the structure of a particular database. Primarily used in relational databases to store the names and data types of the tables and columns in a database.

database dip See dip.

database lookup A software program which allows telephone users to find information on someone calling via the LCD window on their phone. This information comes to the user via CLID (Calling Line IDentification) or ANI (Automatic Number Identification). See also Class.

Database Management System DBMS. Computer software used to create, store, retrieve, change, manipulate, sort, format and print the information in a database. Database management systems are probably the fastest growing part of the computer industry. Increasingly, databases are being organized so they can be accessible from places remote to the computer they're kept on. The "classic" database management system is probably an airline reservation system.

database mining You have a database of your customers. You have information on your customers' buying habits. You slice and dice your database to find out which of your customers might be interested in purchasing new items and which items. For example, let's say you make clothes. 20% of your customers buy in "tall" mens sizes. Next time you make some tall mens' clothes, you may want to tell those customers – with an email or a direct mail catalog. That, in its simplest example, is called database mining or collaborative filtering.

database object One of the components of a database: a table, view, index, procedure, trigger, column, default, or rule.

database server A specialized computer that doles out database data to PCs on a LAN the way a file server doles out files. Where a traditional DBMS runs both a database application and the DBMS program on each PC on the LAN, a database server splits up the two processes. The application you wrote with your DBMS runs on your local PC, while the DBMS program runs on the database server computer. With a regular file server setup, all the database data has to be downloaded over the LAN to your PC, so that the DBMS can pick out what information your application wants. With a database server, the server itself does the picking, sending only the data you need over the network to your PC. So a database server means vastly less network traffic in a multi-user database system. It also provides for better data integrity since one computer handles all the record and file locking. See Server.

datablade Datablades are components for particular types of data that plug into a central database, similar to the way razor blades snap into a razor. Informix is rewriting some of its databases so other companies can produce datablades.

datagram A transmission method in which sections of a message are transmitted in scattered order and the correct order is re-established by the receiving workstation. Used on packet-switching networks. The Dow Jones Handbook of Telecommunications defines it as, "A single unacknowledged packet of information that is sent over a network as an individual packet without regard to previous or subsequent packets." Here's another definition I found. A finite-length packet with sufficient information to be independently routed from source to destination. In packet switching, a self-contained packet, independent of other packets, that carries information sufficient for routing from the originating data terminal equipment to the destination data terminal equipment, without relying on earlier exchanges between the equipment and the network. Unlike virtual call service, there are no call establishment or clearing procedures, and the network does not generally provide protection against loss, duplication, or misdelivery. Datagram transmission typically does not involve end-to-end session establishment and may or may not entail delivery confirmation acknowledgment. A datagram is the basic unit of information passed across the Internet. It contains a source and destination address along with data. Large messages are broken down into a sequence of IP datagrams. See Connectionless Mode Transmission.

datagram packet network The type of packet-switched network in which each packet is individually routed. This may result in a loss of sequence within a message because of alternate routing, or a loss of portions of a message because of packet elimination for congestion control. See Datagram.

DATAP The programmer and system house in Atlanta which provides several long distance carriers with value-added services – what Sprint calls its SCADA system manufacturer.

Datapak A packet-switched network run in Denmark, Sweden, Finland and Norway and operated by their respective governments.

Datapath A name that once was a data service that provided digital, full-duplex data transmission at speeds of 300 bps through 19.2 Bps asynchronous and 1,200 through 64 Kbps synchronous. Datapath has built in autobaud and hand-shaking protocols.

datascope A diagnostic tool for monitoring and capturing data transmissions which displays real-time transmissions of raw data in hexadecimal, binary, or character-oriented displays.

DataSPAN DataSPAN is generally characterized as a "fast packet" service and is based on frame relay standards recommended by the International Consultative Committee for Telephone & Telegraph (ITU) (now called the ITU-T) and the American National Standards Institute (ANSI). Northern Telecom has introduced DataSPAN as a new DMS SuperNode value-added, data communications service that is targeted toward connecting high-speed local area networks. Northern Telecom asserts that DataSPAN's rapid and efficient data transport assures reliable delivery and substantial performance improvement over current LAN interconnect solutions. DataSPAN switching and transmission delay is less than 3 ms per node; X.25 switching and transmission delay can be up to 50 ms per node. Using Frame Relay, wide-area packet switching can be accomplished with the same level of performance that is traditionally limited to complex, dedicated private-line networks. DataSPAN is accessed through standard DS-0 or DS-1 links.

date and time stamp Many voice mail systems will append the date and time of receipt of voice messages for their users/subscribers.

Datex P An Austrian packet-switched network in Austria and Germany and run by their respective governments.

dating format The format employed to express the time of an event. The time of an event on the UTC time scale is given in the following sequence; hour, day, month, year; e.g., 0917 UTC, 30 August 1997. The hour is designated by the 24-hour system. UTC stands for Coordinated Universal Time, believe it or not.

DATU See Direct-Access Test Unit.

DAU Dumb Ass User.A term meaning user induced error in any technical application: computing, telecom or wherever mankind meets the mechanical. It's a stupid acronym. I don't make this stuff up.

daughterboard First there is the motherboard. That's the main circuit board of a computer system. The motherboard contains edge connectors or sockets so other PC (printed circuit) boards can be plugged into it. Those PC boards are called Fatherboards. Some fatherboards have pins on them into which you can plug smaller boards. Those boards are called Daughterboards. In a voice processing system, you might have a Fatherboard to do faxing. And you might have a range of Daughterboards, which allow you to connect different types of phone connections. Different boards exist for standard analog tip and ring, digital switched 56, etc. See also Motherboard.

daughtercard Same as Daugherboard.

DAV See Distributed Authoring and Versioning.

DAVIC Digital Audio Video Council. A voluntary SIG (Special Interest Group) organized to "favour the success of emerging digital audio-visual applications and services, by the timely availability of internationally agreed upon specifications of open interfaces and protocols that maximize interoperability across countries and applications/services." The current set of DAVIC 1.0 specifications are to further the deployment of broadcast and interactive systems that support initial applications such as TV distribution, near video on demand, video on demand, and basic forms of teleshopping. DAVIC is one of the organizations examining

the developing specifications for VDSL (Very-high-bit-rate Digital Subscriber Line), which is intended for video applications. www.davic.org. See also Digital Video Broadcast.

Day A day is customarily the time of one complete rotation of the Earth about its axis. The sidereal day, as used in astronomy and in computing satellite orbits, is the elapsed time for one complete 360 degree rotation of the Earth (and the apparent 360 degree rotation of the celestial sphere). Its duration is measured relative to the fixed stars and is equal to the time elapsed between two successive crossings of any particular meridian by the same fixed star. Its length is 86,164.091 seconds (23 hours, 56 minutes, 4.091 seconds). The mean solar day is the annual average of the elapsed time between two successive crossings of any particular meridian by the Sun. Consequently, the mean solar day begins at noon civil time (in which the day begins at midnight). The Earth rotates 360.9856 degrees in a mean solar day of length 86,400 seconds (24 hours). In a solar day, a geostationary satellite completes 1.002734 revolutions of its orbit.

Day 2 support Ongoing management of a network or system.

Day-Of-Week Factors A call center term. A historical pattern consisting of seven factors, one for each day of the week, that defines the typical distribution of call arrival throughout the week. Each factor measures how far call volume on that day deviates from the average daily call volume.

day-to-day variation That component of the variance of a set of daily load measurements.

dB 1. Decibel. Decibel is one tenth of a Bel, with "Bel" referring to Alexander Graham Bell. A decibel is a unit of the measure of a signal's strength. It is usually the relationship between a transmitted signal and a standard signal source, known as a reference. The decibel was invented by the Bell System to express the gain or loss in telephone transmission systems. Loss in such systems was the result of signal attenuation in passive devices (e.g., conductor media such as copper cables, inductors and capacitors and radiated media such as the air between transmitting and receiving radio antennae). Gain in such systems is the result of active devices such as amplifiers and repeaters. The ultimate objective of measuring the loss in such a passive system is to compensate for it with measured gain accomplished by an active device in order to produce a signal at the receiver which is of the same intensity as the signal created by the transmitter.

Decibel is a logarithmic unit. The logarithmic expression of gain or loss is much more effective for several reasons. First, the human ear interprets loudness (i.e., signal strength) on a scale much closer to a logarithmic scale than a linear scale. Second, the quantities measured often exhibit huge ranges of variation that can be expressed more conveniently in logarithmic, rather than linear, terms. A 10 dB gain (i.e., increase in signal strength) describes a ten-times increase. A gain of 3 dB means that the power of the signal was doubled; loss of 3 dB (-3 dB gain) means that the power of the signal was halved.

One area of concern – the difference between dB and dBm. When we speak of dBm (decibels to a milliwatt) we are measuring absolute power. For example 10dBm equates to 10 milliwatts, watts being a power denomination we recognise. True loss of a signal or the difference in dBms, is expressed in units of decibels, commonly abbreviated as dB. Fiber attenuation and connector loss are generally expressed in units of dB. As an example, singlemode fiber loss may be expressed as .3dB/km and a connector loss may be .2dB. In fiber optics the negative sign (-.2 dB) is generally ignored.

2. Database.

3. Data Bus connector. Usually shown with a number that represents the number of wire conductors in the connector, e.g. DB-9 or DB-25, both of which are very common connectors to plug into serial ports on PCs. See DB-9, DB-15 and DB-25.

dB Loss Budget dB Loss Budget is the amount of light available to overcome the attenuation in the optical link and still maintain specifications.

DB-9 This is the standard nine-pin RS-232 serial port on all laptop PC computers and most desktop computers today. The term DB-9 is used to describe both the male and female plug. So be careful when you order. See also the Appendix in this Dictionary for more information on the pinning of RS-232-C plugs. The above is contemporary usage, i.e. the way it's used today. In fact, it's wrong historically. The "D" originally described the shape of the housing. The second letter: A, B, C, D or E originally specified the size of the housing where the "E" is used somewhat like it's used as a drawing size (i.e., smaller than a "D"). There are connectors made with, e.g., size "B" housing with other than 9, 15 or 25 pins. Sometimes coax "pins" are included, which use up several little pin locations. See the Appendix in this dictionary for more information on the pinning of RS-232 plugs.

DB-15 A standardized connector with 15 pins. It can be used in Ethernet transceivers. It can also be used for connecting VGA monitors. DB-15 is used to describe both the male and female plug. So be careful when you order. See also DB-9 for a longer explanation.

DB-25 The standard 25-pin connector used for RS-232 serial data communications. In a DB-25 there are 25-pins, with 13 pins in one row and 12 in the other row. DB-25 is used to describe both the male and female plug. So be careful when you order. See also DB-9 and the Appendix in this dictionary for more information on the pinning of RS-232 plugs.

DB-60 A connector with 60 pins in four rows (15 pins in each row).

DB2 IBM's relational database system that runs on System 370-compatible mainframes under the MVS operating system.

DBA 1. Dynamic Bandwidth Allocation.

2. Database administrator. The individual in the organization with the responsibility for the design and control of databases.

3. dBA. dBA is a measure of sound level. The ratio for determining dB in this case is the ratio between the sound level measured with a microphone and an implicit reference sound level, namely 0db, which is defined to be approximately equal to the threshold of human hearing. 45dBA is a very faint whisper, 75 dBA is typical conversation, 100 dBA is about how loud a Walkman can get, and 120dBA is a jet plane taking off at 20 feet away from the engine. The "A" represents a special filter which is used to take into account the fact that people are less sensitive to very low and very high frequencies. The dBA system is used for measuring background sounds like computer fans or office noise. A system with a different filter called dbC is used when very loud sounds are being measured. This definition courtesy APC.

dBd Decibels dipole. A measurement of signal gain used in radio antenna design. Specifically, dBd refers to signal gain in a dipole radiator, which is a basic transmitting antenna consisting of two rods with their ends slightly separated. In the most basic form, the dipole radiator comprises two parallel horizontal rods. See also dB, dBd, and Decibel.

dBc Decibel relative to a carrier level. See also dB.

dBi 1. dBi. Decibels isotropic. A measurement of signal gain used in radio antenna design. Specifically, dBi refers to signal gain in an isotropic radiator, which is a transmitting antenna that radiates a signal equally well in all directions. See also dB, dBd, and Decibel.

2. Don't Believe It. This is used in telecontrol systems for electricity transmission/distribution systems to denote an unreliable state change at a substation RTU (Remote Termination Unit) i.e. both inputs have either gone high or low. They should normally be different. An alarm legend is then generated and sent to the NMS (Network Management System) with a status of DBI instead of the normal ON or OFF i.e.

BREAKER ON BREAKER DBI 00 BREAKER OFF

See also RTU.

DBIT Also called D-BIT. The delivery confirmation bit in an X.25 packet that is used to indicate whether or not the DTE wishes to receive an end-to-end acknowledgment of delivery. In short, a bit in the X.25 packet header that assures data integrity between the TPAD and the HPAD.

dBm dBm is used to describe the relative power of a transmitter and the sensitivity of a receiver. It is defined as Decibel (dB) ratio (log10) of Watts (W) to one milliwatt (1mW). In other words, the output power of a signal referenced to an input signal of 1mW (milliWatt). Similarly, dBm0 refers to output power, expressed in dBm, with no input signal. (0 dBm = 1 milliwatt and -30 dBm = 0.001 milliwatt). See also dB and Decibel.

dBm0 Identifier meaning "decibels referred to one milliwatt and corrected to a zero dBm effective power level;" used to state the relation of a signal level on a transmission line at other than a one-milliwatt point.

dBmp An identifier meaning decibels below reference noise referred to one milliwatt using psophometric weighting, dBmp is the ITU-T method for noise measurements. dBmp has a variance of approximately 2 dB from dBrn methods. See also dBrn.

DBMS Database management system. A computer program that manages data by providing the services of centralized control, data independence, and complex physical structures. Advantages include efficient access, integrity, recovery, concurrency control, privacy, and security. A DBMS enables users to perform a variety of operations on data, including retrieving, appending, editing, updating, and generating reports.

dBmV A decibel measure in relation to one millivolt across a specific impedance. In CATV the impedance used is 75 ohms. See Cable TV.

dBrn DeciBels above Reference Noise. A ratio of power level in dB relative to a noise reference. dBrnC uses a noise reference of -90 dBrn, as measured with a noise meter, weighted by a frequency function known as C-message weighting which expresses average subjective reaction to interference as a function of frequency. dBrn is used mostly in North American telecommunications work. See also dB and dBmp.

dBrnC An identifier meaning deciBels above Reference Noise using C-message weighting. The measurement is accomplished through a filter approximating a type C voice messaging channel, and is the North American nomenclature for a DDD (Direct Distance Dialing) trunk channel. The reference is 90 dB below one milliwatt of power. See also dBrn and dBrnC0.

dBrnC0 Pronounced "de-brink-o," it is an identifier meaning deciBels above Reference Noise using a filter approximating a type C voice messaging channel adjusted for equivalence to a 0 dBm equivalent circuit point. It is the same as dBrnC, except that it is corrected to a TLP (Transmission Level Point) of 0dB. See also dBrnC and TLP.

DBS Direct Broadcast Satellite. A term for a satellite which sends relatively powerful signals to small (typically 18-inch diameter) dishes installed at homes. See C Band, 1994 and Direct Broadcast Satellite.

DBU 1. Dial Back-Up. A method of providing redundancy in the event of the failure of a leased line or even a network, dial back-up automatically re-establishes the connection through the PSTN (Public Switched Telephone Network) on a dial-up basis. For example, ISDN commonly is used for dial back-up for Frame Relay networks.

2. Decibels below 1uW. Decibels relative to microwatts. See also dB.

dBuV Decibel ratio of Volts to one microvolt. See also dB and Decibel.

dBW A decibel measure referenced to one watt without reference to any impedance.

DC 1. Direct Current. The flow of free electrons in one direction within an electrical conductor, such as a wire. The current may be constant or it may pulsate, but it always is in one direction. See also AC.

2. Delayed Call.

DC Block A device which blocks direct current but passes radio frequencies, audio frequencies, or alternating current depending upon the function of the block.

DC Power Supply See Power Supply.

DC Signaling A collection of ways of transmitting communications signals using direct current – the type of current produced by a dry cell household "D" cell battery. DC signaling is only used on cable. It's an out-of-band signal.

DCA 1. Defense Communication Agency. The U.S. government agency under the DOD (Department of Defense) that was responsible for installation and operation of Defense Data Networks, including the ARPANET and MILNET, and PSNs. DCA was folded into DISA (Defense Information Systems Agency) in 1991. See also DISA.

2. Document Content Architecture. The IBM approach to storing documents as two types of document group: draft documents and final form documents. For presentation, the draft document is transformed into a final document through an office system.

DCAS Direct Carrier Administration System.

DCC 1. Data Communications Channel. Channels contained within section and line overhead used as embedded operations channels to communicate to each network element. An AT&T SONET term.

2. An ATM term. Data Country Code: This specifies the country in which an address is registered. The codes are given in ISO 3166. The length of this field is two octets. The digits of the data country code are encoded in Binary Coded Decimal (BCD) syntax. The codes will be left justified and padded on the right with the hexadecimal value "F" to fill the two octets.

3. Digital Compact Cassette. A digital version of the familiar analog audio cassette. A DCC recorder can play and record both analog and digital cassettes. But the digital ones will sound a lot better.

4. Digital Cross-Connect.

DCCH Digital Control CHannel. A channel used in most newer digital cellular and PCS systems for signal and control purposes between the mobile terminal device and the radio base station. See also Cellular and PCS.

DCD 1. Data Carrier Detect. Signal from the DCE (modem or printer) to the DTE (typically your PC), indicating it (the modem) is receiving a carrier signal from the DCE (modem) at the other end of the telephone circuit.

2. Dynamically Configurable Device. A dynamically configurable device is a fancy name for a Plug and Play device, so-called because you don't have to reboot the system after installing one.

3. Duty Cycle Distortion. See Jitter.

DCE 1. Data Communications Equipment. Also known as DCTE (Data Circuit Terminating Equipment). The classic definition of DCE is that it resolves issues of interface between Data Terminal Equipment (DTE) and a transmission circuit. Examples include LAN Network Interface Cards (NICs), CSUs and DSUs, modems, and ISDN Terminal Adapters (TAs). DCE may accomplish such functions as changes in electrical coding schemes, electro-optical conversion, and data formatting. The physical interfaces between DTE and DCE can take a variety of forms, one of which is the RS-232 "standard" developed by the Electronic Industries Alliance (EIA). The main difference between a DCE and DTE in RS-232 is the wiring of pins two and three in the male and female 25-pin connectors. But there is, of course, no standardization. When wiring one RS-232 device to another, it's good to know which device is wired as a DCE and which as a DTE. But it's actually best to go straight to the wiring diagram in the appendix of the device's instruction manual. Then you compare the wiring diagram of the device you want to connect and build yourself a cable that takes into account the peculiar (i.e., strange) vagaries of the engineers who designed each product. In short, with an RS-232 connection, the modem is usually regarded as DCE, while the user device (terminal or computer) is DTE. In a X.25 connection, the network access and packet switching node is viewed as the DCE. DCE devices typically transmit on pin 3 and receive on pin 2. DTE (Data Terminal Equipment) devices typically transmit on pin 2 and receive on pin 3. See also DTE and RS-232. See also the Appendix for an excellent graphic representation of the RS-232 pinout.

2. Distributed Computing Environment. An industry-standard, vendor-neutral set of distributed computing technologies developed by the Open Software Foundation (OSF). According to The Open Group, successor to the OSF, DCE provides security services to protect and control access to data, name services that make it easy to find distributed resources, and a highly scalable model for organizing widely scattered users, services and data. DCE runs on all major computing platforms, supporting distributed applications in heterogeneous hardware and software environments.

DCE-RPC Distributed Computing Environment Remote Procedure Call. A Microsoft implementation of a portmapping service. A portmapper is a service that runs on a specific port, redirecting clients that send a request to that port. These initial calls typically result in a response from the trusted machine that redirects the client to a new port for the actual service the client wants. See also RPC.

DCG Dispersion Compensation Grating. DCG overcomes the distortion of optical signals as they are transmitted through a network. Instead of trying to compensate for large amounts of signal dispersion at the end of a network, DCG periodically removes the distortion where needed along the transmission line. See Solitons.

DCH D-Channel Handler.

DCLEC A CLEC which specializes in delivering only data, most typically DSL services. See CLEC.

DCM 1. Digital Circuit Multiplication. A means of increasing the effective capacity of primary rate and higher level PCM hierarchies, based upon speech coding at 64 Kbit/s. Also a Digital Carrier Module.

2. Distributed Order Management.

DCME Digital circuit multiplication equipment is a compression equipment capable of handling voice, data and facsimile signals. This equipment is utilized as a means of augmenting the capacity of digital transmission systems. It can be used in Satellite as well as cable. Some can compress as high as 20:1. See Digital Circuit Multiplication.

DCN Disconnect frame. Indicates the fax call is done. The sender transmits before hanging up. It does not wait for a response.

DCOM Distributed Component Object Model. DCOM is Microsoft's distributed version of its COM, a language-independent component architecture (not a programming language). It is meant to be a general purpose, object-oriented means to encapsulate commonly used functions and services. COM encompasses everything previously known as OLE Automation (Object Linking and Embedding). Microsoft describes DCOM as "COM with a longer wire." See also COM.

DCP Digital Communications Protocol.

DCPR Detailed Continuing Property Records. See PICS/DCPR.

DCS 1. Distributed Communications System. See Distributed Switching.

2. Digital Crossconnect System. A device for switching and rearranging private line voice, private line analog data and T-1 lines. A DCS performs all the functions of a normal "switch," except that connections are typically set up in advance of when the circuits are to be switched – not together with the call. You make those "connections" by calling an attendant who makes them manually, or by dialing in on a computer terminal – one similar to an airline agent's. See also Network Reconfiguration Service.

3. Digital Cellular System.

4. Digital Communications System.

5. DCS. A packet-switched network implemented in Belgium and operated by the Belgian government.

6. DCS is also the Avaya PBX-to-PBX private networking protocol over point-to-point or ISDN-PRI T-1s, (not unlike the Siemens' CORNET service.)

DCS 1000 See DCS1000 below.

DCS 1800 Digital Cellular System at 1800 MHz. A GSM (Global System for Mobile Communications) standard for cellular mobile telephony established by ETSI (European Telecommunications Standards Institute) for operation at 1800 MHz. In short, DCS 1800 is GSM adopted to the 1800 Mhz frequency band. This means that existing GSM phones won't be able to talk on the DCS 1800.

DCS1000 DCS stands for Digital Collection System or Data Collection System, according to unconfirmed reports. (The FBI doesn't particularly like to confirm things.) DCS1000 is a surveillance program developed by the FBI (Federal Bureau of Investigation) for the interception and analysis of e-mail. DCS1000 was originally known as Carnivore, a poorly-chosen name that created quite a controversy as the public conjured up visions of the FBI indiscriminately examining and eating their mail. DCS1000 is packet-sniffing program that is installed, on the basis of a federal warrant under U.S. wiretap laws, at an ISP site. Once installed, the program sorts through packet traffic to find the subject of e-mail packets on the basis of various criteria, including IP addresses. Privacy advocates fear that the program will be used to capture and analyze other packet traffic, indiscriminately.

DCT 1. Digital Carrier Termination.

2. Discrete Cosine Transform. A compression algorithm used in most of the current image compression systems for bit rate reduction, including the ITU-T Px64 standard for video conferencing. DCT represents a discrete signal or image as a sum of sinusoidal wave forms.

3. Digital Consumer Terminal. A set-top box that receives and converts digital video and audio signals to standard television format. The box also processes addressability transactions and stores and transmits pay-per-view (PPV) transaction information back to the MSO. Some advanced units include support for digital video recording (also called Personal Video Recorder), video-on-demand, and/or HDTV signal processing. See PVR.

DCTE Data Circuit Terminating Equipment. Also known as DCE (Data Communications Equipment). See DCE.

DCTI Desktop Computer Telephone Integration. Basically, providing a way for your computer to control your telephone set.

DCTU Digital Cordless Telephone US.

DCV Digital Compressed Video.

DD 1. Dotted Decimal Notation. See Dotted Quad.

2. DD, as in 600 DD. It represents the number of circuit pairs that are terminated, but not spliced. It seems that 600 DD is the term used in the West, while 600 XD would be used by those trained in the East.

DDA Domain Defined Attribute. A way of adding additional information to the address of your electronic mail in order to avoid confusion between people of the same or similar name.

DDB Digital Data Bank.

DDC 1. Direct Department Calling.

2. Data Display Channel is a standard that defines communication between a monitor and a host system.

DDCMP Digital Data Communications Message Protocol. A byte-oriented, link-layer protocol from Digital Equipment Corp., used to transmit messages between stations over a communications line. DDCMP supports half- or full-duplex modes, and either point-to-point or multipoint lines in a DNA (Digital Network Architecture) network.

DDD Direct Distance Dialing. The "brand name" the Bell System used to call its Message Toll Telephone Network. It used the words "Direct Distance Dialing" to convince the public to dial their own long distance calls directly without the help of an operator.

DDE See Dynamic Data Exchange.

DDEML Dynamic Data Exchange Management Library. A feature of Microsoft Windows.

DDF Digital distribution frame. See distribution frame.

DDI Direct Dialing Inward. A British term. It is a service where a call made to a DDI number arrives direct, without the intervention of an organization's operator, at an extension or, if routed via an ACD, to a group of extensions. A specific DDI number can be assigned to each campaign. When an agent answers a call made to one of these DDI numbers, the relevant script and screen are displayed. This way the agent can give an appropriate response. British Telecom, the U.K.'s biggest phone company typically allocates DDI in contiguous ranges of 1,000-10,000 numbers.

DDJ Data Dependent Jitter. See Jitter.

DDM Distributed Data Management Architecture. An IBM SNA LU 6.2 transaction providing users with facilities to locate and access data in the network. It involves two structures: DDM Source, and DDM Target. The DDM Source works with a transaction application to retrieve distributed data and transmits commands to the DDM Target program on another system where the data that has been requested is stored. The DDM Target interprets the

DDM commands, retrieves the data and sends it back to the DDM Source that originated the request.

DDN Defense Data Network. A network that provides long haul and area data communications and interconnectivity for DOD (Department of Defense) systems, and supports the DOD suite of protocols (especially TCP and IP). All equipment attached to the DDN by military subscribers must incorporate, or be compatible with, the DOD Internet and transport protocols. The packet-switched portion of DDN was split off from ARPANET to handle U.S. military needs, at which time is was renamed MILNET (MILitary NETwork). The remaining ARPANET evolved into the public Internet. DISA (Defense Information Systems Agency) is responsible for the DDN. See also DDN NIC and MILNET.

DDN NIC Defense Data Network Network Information Center. Also called the "NIC," by those in the defense domain, the DDN NIC is responsible for the assignment of Internet addresses and Autonomous System numbers, the administration of the root domain, and the provision of information and support services to the DDN. See also DDN, INTERNIC and NIC.

DDNS Dynamic Domain Name Servers. DDNS allows the hosting of a website, FTP server, or e-mail server with a fixed domain name (e.g., www.HarryNewton.com) and a dynamic IP address. See DNS.

DDoS Distributed Denial of Service. See Denial of Service Attack.

DDP 1. See Distributed Data Processing.

2. Datagram Delivery Protocol is an AppleTalk protocol that provides for datagram delivery to higher-layer protocols. DDP supports connectionless service between network sockets at the network layer (Layer 3) of the OSI Reference Model. See also AARP and AppleTalk.

DDR-SDRAM Double Data Rate-Synchronous Dynamic Random Access Memory. A next-generation SDRAM that transfers data twice per CPU (Centralized Processing Unit) clock cycle, which doubles the data transfer of SDRAM. While the resulting performance level is higher, so is the cost. See also DRAM, EDO RAM, Flash RAM, FRAM, Microprocessor, RAM, RDRAM, SDRAM, SRAM, and VRAM.

DDS 1. Dataphone Digital Service, also called Digital Data System. DDS is private line digital service, offered in point-to-point and point-to-multipoint configurations. Data rates available include 2,400 bps; 4,800 bps; 9,600 bps; 56/64 Kbps; and 1.544 Mbps (T-1). DDS is offered on an inter-LATA basis by AT&T, MCI and Sprint, and on an intraLATA basis by the ILECs (Incumbent Local Exchange Carriers). CLECs (Competitive LECs) often use DDS circuits for local loops, leasing the DDS facilities from the ILECs. DDS originally was an AT&T tariff offering, but now is considered to be a generic term.

2. Digital Data Storage. A DAT format for storing data. It is sequential – all data that is recorded to the tape falls after the previous block of data.

DDSD Delay Dial Start Dial. A start-stop protocol for dialing into a switch.

DE 1. Discard Eligible. The frame relay standard specifies that data sent across a virtual circuit in excess of that connection's Committed Information Rate (CIR) will be marked by the user as being eligible for discard in the event of network congestion. DE data is the first to be discarded by the network when congestion occurs, thus providing protection for data sent within the parameters of the CIR. It is the responsibility of the intelligent end equipment and/or protocol to recognize the discard and respond by resending the information. As a practical matter, few users are likely to volunteer to have frames discarded, and few manufacturers provide for the setting of the DE bit. See Oversubscription.

2. Designated Entities. Small businesses, businesses owned by members of minority groups and/or women, and rural telephone companies that meet size or other criteria established for specific services. Specifically, companies are either "entrepreneurs" (defined as entities, together with affiliates, having gross revenues of less than $125 million and total assets of less than $500 million) or "small businesses" (defined as entities, together with affiliates, having gross revenues of less than $40). The blocks reserved, in part, at the FCC for designated entities are the C and F blocks.

de average See de-average.

de facto Used to describe a standard reflecting current or actual practice, but not having approval or sanction by any official standards-setting organization. In other words, a de facto standard is an unofficial standard that exists because sufficient companies adhere to it or because of market acceptance. Usually created by an individual manufacturer or developer – hence, often used as a synonym for proprietary. The Hayes AT auto-dial modem command language is an example of a de facto standard. Very often, de facto standards form the basis for de jure standards. Ethernet was the basis of Institute of Electrical and Electronics Engineers (IEEE) or IEEE standard 802.3. Contrast with de jure.

De Forest, Lee Lee de Forest invented the vacuum tube in 1907. Until the invention of the transistor after World War II, all radio, long distance telephony and complicated

electronics, including electronic computers, were derived from de Forest's invention.

de jure Used to describe standards approved or sanctioned by an official standards setting organization. ITU-T Recommendation X.25 for packet data networks is an example of a de jure standard. Contrast with de facto.

de-average A telephone subscriber frequently pays a rate based on the average amount it costs a phone company to provide service to a group of subscribers, as opposed to the actual amount it costs the company to provide the particular service to that particular subscriber. As we move towards a more competitive, deregulated environment, costs are frequently being "deaveraged". When rates are "deaveraged", a subscriber's rate is based on the amount it actually costs a company to serve that subscriber, as opposed to the average cost. Of course, when you have a factory – a gigantic phone system – that serves many subscribers with many different services, fast, slow, analog, digital, long, short, etc. – it's really very difficult, if not impossible, to figure out the cost of various services. So the term "de-average" tends to mean the phone companies drop the price of competitive services and raise the price of non-competitive services.

de-emphasis The reverse of pre-emphasis. A process used in frequency modulation detectors to reverse transmission pre-emphasis. See Pre-emphasis.

de-encapsulation The process of extracting a data packet from the user data field of the encapsulating data packet.

de-install De-install means to remove a piece of equipment that has been installed, but keep it in good condition so that it can be successfully installed some place else. When much of the telecom industry overexpanded in the late 1990s, several companies came along to offer "de-installing" services. Here's a description of one: Somera Communications, Inc. provides telecommunications carriers with a broad range of infrastructure equipment and related services designed to meet their specific and changing equipment needs. The Company offers its customers a combination of new and de-installed equipment from a variety of manufacturers, allowing them to make fast multi-vendor purchases from a single cost-effective source. See deinstall.

de-regulation See Deregulation.

de-spreading The process used by a correlator to recover narrowband information from a spread-spectrum signal.

DEA Digital Exchange Access. This is a service that Bell Canada offers. It gives the customer a local number and access to their digital equipment where the number resides.

dead air 1. Dead Air refers to the silence received when placing a call to a valid telephone number. Dial tone is received, the valid number is dialed but you do not receive ring back tone, only dead silence. See also High and Dry.

2. When a telephone company worker begins work in a manhole, he or she must first open the manhole cover, and pump fresh air into the manhole for at least 20 minutes. The reason is that the air in the manhole will likely lack oxygen or contain toxic gases – possibly leaked in from adjacent natural gas pipes. Should a worker descend into the manhole without first pumping fresh air in, he could die from breathing the air. This has happened. Most phone companies have lost workers in manholes. Telephone company workers have a generic name for the bad air in manholes. They call it dead air.

dead cat bounce A dead cat bounce occurs when the price of a stock has fallen so constantly so far, that sufficient people think it can't fall any further and they buy it and the stock rises. The expression comes from the Wall Street saying: "Even a dead cat will bounce once if it is dropped from high enough."

dead link A hyperlink to a web page that no longer exists or whose URL has been changed.

dead load An antenna tower must be designed to support three types of loads: dead load, ice load, and wind load. The dead load is the weight of the tower itself and any equipment and cables installed on it. The ice load is the added weight of ice that can form and accumulate on the tower. The wind load is the added force of wind blowing on the tower and equipment and cables installed on it.

dead man feature A vigilance mechanism implemented in portable radios of police officers, firefighters and other personnel who work in dangerous environments. When the vigilance function is activated, the device alerts the user at regular intervals to press a special button or make some other required response, the doing of which signifies that the user is OK. Failure to respond to the alert triggers further alerts which, if not responded to, triggers a call for help, under the assumption that the user may be hurt.

dead man preferred Preferred stock that has no recession and no conversion rights.

dead ringer In the 1500s, local folks started running out of places to bury people. So they would dig up coffins and would take the bones to a "bone-house" and reuse the grave. When reopening these coffins, 1 out of 25 coffins were found to have scratch marks on the inside and they realized they had been burying people alive. So they thought they would tie a string on the wrist of the corpse, lead it through the coffin and up through the ground and tie it to a bell. Someone would have to sit out in the graveyard all night the ("graveyard shift") to listen for the bell; thus, someone could be "saved by the bell" or was considered a "dead ringer."

dead sector In facsimile, the elapsed time between the end of scanning of one line and the start of scanning of the following line.

dead spot A cellular radio term. It denotes an area within a cell where service is not available. A dead spot is usually caused by hilly terrain which blocks the signal to or from the cell tower. It can also occur in tunnels and indoor parking garages. Excessive foliage or electronic interference can also cause dead spots. If you encounter a dead spot, tell your cellular radio supplier. They often will do something.

dead tree edition The paper version of a publication available in both paper and electronic forms. If you are reading my dictionary on paper, you are reading the dead tree edition.

deadbug Industry slang for a component that is added and wired onto a circuit board to facilitate active circuit modifications, rather than redesign and manufacture a new board. The term comes from the way the component looks on the board: upside down, with its termination leads up in the air, like a dead bug on its back with its legs up in the air. See also tombstone.

deadline During the Civil War, captured troops were placed in a field. They were told they could move wherever they wanted in the field except beyond a line that was drawn around the field. If they did, they would be shot. Thus the origin of the word deadline.

deadlock See Deadly Embrace.

deadly embrace Stalemate that occurs when two elements in a process are each waiting for the other to respond. For example, in a network, if one user is working on file A and needs file B to continue, but another user is working on file B and needs file A to continue, each one waits for the other. Both are temporarily locked out. The software must be able to deal with this.

deal flow Venture capitalists need to invest monies in promising new companies. But where to find them? You have to hang out a shingle, have a neat web site inviting submissions for money and have oodles of contacts who know millions of people. The key is deal flow. The more potential great deals that can flow your way, the better.

dealer A dealer is simply a person who sells equipment – hardware or software – made by someone else. That dealer may be a distributor of Novell LAN software. Or that dealer may be in the secondary telecom equipment business, i.e. a company that buys and sells secondary market telephone equipment. Generally in the secondary equipment business, a dealer takes ownership of the equipment (see BROKER). He tests and refurbishes it before remarketing. An authorized dealer has a contract with the manufacturer to buy equipment at a preset price. He also has the added support of a manufacturer's warranty. An independent dealer has no formal agreement with the manufacturer. He uses a variety of sources to obtain equipment and his warranty is backed only by the company's internal resources.

dealer board British term meaning "trading turret."

dealer locator An application of ANI (Automatic Number Identification) technology that matches the caller's phone number with a database of dealer locations. The caller can then be transferred automatically or manually referred to the closest location.

dealing room British term meaning "trading room."

deallocate To return media to the available state after they have been used by an application.

death When the French diplomat Talleyrand died, his Austrian rival Prince Metternich is supposed to have mused: "I wonder what he meant by that?"

death rate of technicians David Potter runs Source, a Dallas telephone and telecom supply and maintenance house with service nationwide. When I detailed the phone system I wanted for my new house – namely an old-fashioned circuit-switched TDM system, David, the great salesman he is, commented he could find such an antiquity but we had to take into account the death rate of the few remaining ancient technicians who could fix it.

death star villages Suburbs around New Jersey where many AT&T employees live. Makes reference to the AT&T logo, which employees have dubbed "The Death Star" (from the Star Wars films).

DEBE Does Everything But Eat. DEBE (pronounced Debbie) applies to any and all all-in-one devices that seemingly does everything you could ever want. In other words, it does everything but eat.

debit card The term telephone "debit card" covers three categories of a new type of

telephone calling card, variously called "calling card" and "prepaid telephone card." The definitions are in flux. Here's the best shot at debit card: A telephone debit card is a piece of credit-card size plastic with some technology on it or embedded into it which represents the value of the money remaining on the debit card. Such money can be used to make phone calls. The technology on the card is most typically an integrated circuit, a magnetic strip, or bar codes which can be read by an optical reader. See Breakage, Calling Card and Prepaid Telephone Card.

Debitel A big provider of mobile telephone services in Germany. It is owned mainly by Daimler-Benz and Metro, a big retail group.

Debug A MS-DOS program to examine or alter memory, load and look at sectors of data from disk and create simple assembly-language programs. MS-DOS DEBUG.COM lets you write some small programs. You can use DEBUG to correct problems in some programs.

debugger A tool (program) that is used by programmers to help identify and fix problems (bugs) in a program.

DEC Digital Equipment Corporation. An erstwhile leading manufacturer of minicomputers when there was still a market for minicomputers. The Unix operating system, developed at Bell Labs, ran on DEC computers. DEC, with its headquarters in Massachusetts and having sold so many computers to Western Electric for inclusion in central office switches and toll switches, used to be referred to as "Eastern Electric." Kenneth H. Olsen founded DEC in 1957 with three employees in 8,500 square feet of leased space in a corner of an old New England woolen mill. DEC fell upon hard times, when it completely ignored the PC, and quickly developed more power than the Dumb. Compaq bought DEC in June 1998 and fired many of its people and closed much of its production. The alleged reason for the costly acquisition was the DEC salesforce and its ultra-fast Alpha microprocessor. See DECnet.

decadic dialing The term for pulse dialing in the UK. Pulse dialing is called rotary dialing in the U.S. The term decadic is derived from "dec" meaning ten (as in decade, decathlon, decimal, etc.) Dialing a phone number in pulse dialing (as against touchtone dialing) requires up to ten pulses per digit dialed.

decapsulation A process used in networking in which the receiving system looks at the header of an arriving message to determine if the message contains data. If the message does contain data, the header is removed and the data decoded.

DECnet Group of communications products (including a protocol suite) developed and supported by Digital Equipment Corporation. DECnet/OSI (also called DECnet Phase V) is the most recent iteration and supports both OSI protocols and proprietary Digital protocols. Phase IV Prime supports inherent MAC addresses that allow DECnet nodes to coexist with systems running other protocols that have MAC address restrictions. See DEC.

DEC LAT A proprietary data communications protocol developed by DEC. See DEC.

decibel 1. dB. One tenth of a Bel, with "Bel" referring to Alexander Graham Bell. A unit of measure of the power of sound or the strength of a signal, usually expressed as the relationship between a transmitted signal and a standard signal source, known as the reference. The decibel was invented by the Bell System to measure the level of gain or loss in telephone transmission systems. Loss in such systems is the result of signal attenuation (dropping in signal strength) in passive devices (e.g., resistance/impedance in media such as copper cables, and the quality of the space between the transmitting and receiving radio antenna in an airwave system). Gain in such systems is the result of active devices such as amplifiers and repeaters, which serve to boost the strength of the signal. The ultimate objective of measuring the loss in such a passive system is to compensate for it with measured gain accomplished by an active device in order to produce a signal at the receiver which is the same intensity as the signal created by the transmitter. Decibel is a logarithmic unit. The logarithmic expression of gain or loss is much more effective than a linear expression for several reasons. First, the human ear interprets loudness (i.e., signal strength) on a scale much closer to a logarithmic scale than a linear scale. Likewise, the human eye has a logarithmic response to changes in light. Second, the quantities measured often exhibit huge ranges of variation that can be expressed more conveniently in logarithmic, rather than linear, terms.

We also use dB to measure "insertion loss", which is the difference between input power and output power. We measure such loss across a section of cable, or across a splitter, or across a connector or splice, or perhaps across an entire transmission system. Figure that a loss of 3 dB translates into a loss of 50% of signal strength. Therefore, a loss of 6 dB translates into a loss in signal strength of approximately 75%, as a 50% loss of a signal that already is at 50% strength translates into a signal with 25% of the strength of the original. (50% of 50% equals 25%) Taking this a step further, a loss of 10 dB translates into a signal that has lost 90% of its power and, therefore, has only 10% of its power remaining. So, a loss of 20 dB translates into a signal with only 1% of its power remaining. (10% of 10% equals 1%)

Let's put this into human terms in the acoustics domain, courtesy of the Deafness Research Foundation: 30 dB equals a quiet library, soft whispers. 40 dB equals a living room, a refrigerator, a bedroom away from traffic. 50 dB equals light traffic, normal conversation or a quiet office. 60 dB equals an air conditioner at 20 feet or a sewing machine. 70 dB is the sound of a vacuum cleaner, a hair dryer, or a noisy restaurant. 80 dB equals the sound of average city traffic, garbage disposals, or an alarm clock at two feet. According to the Foundation, the following noises can be   under constant exposure can damage your hearing: 90 dB is a subway, motorcycle, truck traffic or lawn mower; 100 dB is a garbage truck, chain saw, or pneumatic drill; 120 dB is a rock band concert or a thunderclap; 140 dB is a gunshot blast or a jet plane taking off; 180 dB is the sound of a rocket launching pad.

Decide A deadly computer virus.

decimal Our normal numbering system. It is to the base 10.

decipher Use of a cryptological process involving a "key" to recover information which has been "enciphered." See Cipher.

decision circuit A circuit that measures the probable value of a signal element and makes an output signal decision based on the value of the input signal and a predetermined criterion or criteria.

decision distance In the signal phase representation of a digital phase shift keying demodulator, the distance between the extremes of any particular phase state to the quadrant of a neighboring phase state represents the certainty of assessing the phase correctly.

decision feedback equalization A technique used in high-speed transmission systems to reduce the likelihood of error propagation. Such systems can suffer from intersymbol interference (ISI), which is a type of interference caused when adjacent symbols (values), interfere with each other and cause errors in the data stream.

decision instant In the reception of a digital signal, the instant at which a decision is made by a receiving device as to the probable value of a signal condition.

Decision Support System DSS. Computerized systems for transforming data into useful information, such as statistical models or predictions of trends, which is used by management in the decision-making process. There are several aspects of the best decision support systems: First, they are connected to mainframe databases. Second, they are accessible by executives from their desktops. Third, there are usually lots of programs for producing graphs, charts and writing simple reports, i.e. for the executives to extract and portray information in forms that are most useful to them.

decision tree The organization of the call flow for a particular application, expressed in a tree-like logic structure.

deck of cards A "card" is a discrete unit of data specifically designed to be read easily on the small screen of a handheld wireless device such as a cell phone, pager, or PDA (Personal Digital Assistant). A card can be for entering data, displaying data, or listing indexes or available menu options. Packet radio networks generally send a "deck of cards" in a single data packet for reasons of efficiency, as the individual cards are small in terms of byte count.

decoder A device that converts information from one form to another – typically from analog to digital and vice versa. See Decoding.

decode status flag A cellular industry term. A 5-bit flag used by the CDPD Mobile Data Base Station (MDBS) in the forward channel transmission to indicate the decoding status of Reed-Solomon blocks received on the reverse channel from the CDPD Mobile End System (M-ES).

decoding Changing a digital signal into analog form or into another type of digital signal. The opposite of Encoding. See also MODEM. Decoding and coding should not be confused with deciphering and ciphering. See DES.

decollimation In optics, that effect wherein a beam of parallel light rays is caused to diverge or converge from parallelism. Any of a large number of factors may cause this effect, e.g., refractive index inhomogeneities, occlusions, scattering, deflection, diffraction, reflection, refraction.

decolumnization The process of reformatting multi-column documents into a signal column. Generally, when you are processing a document for use in a word processing program, a single column of text is preferable to multiple columns.

decommissioned state A state that indicates that media have reached their allocation maximum.

decompile To decompile is to convert executable (ready-to-run) software code (sometimes called object code) into some form of higher-level programming language so that it can be read by a human. Decompilation is sort of reverse engineering that does the opposite of what a compiler does. The tool that accomplishes this is called a decompiler. A similar tool, called a disassembler, translates object code into assembler language. There

are a number of different reasons for decompilation or disassembly, such as understanding a program, recovering the source code for purposes of archiving or updating, finding viruses, debugging programs, and translating obsolete code. Decompilation was first used in the 1960s to make easy the migration of a program from one platform to another. Decompilation doesn't always work for a number of reasons. It is not possible to decompile all programs, and data and code are difficult to separate, because both are represented similarly in most current computer systems. The meaningful names that programmers give variables and functions (to make them more easily identifiable) are not usually stored in an executable file, so they are not usually recovered in decompiling.

decompress To expand a compressed file or group of files back to their normal size so that the file or files can be opened. See also compress.

decompression Decompression is the process of expanding a compressed image or file so it can be viewed, printed, faxed or otherwise processed.

decrypt To convert encrypted text into its equivalent plain text by means of a cryptosystem. This does not include solution by cryptanalysis. The term decrypt covers the meanings of decipher and decode.

DECT Digital European Cordless Telecommunication. The pan-European wireless standard based on time division multiple access used for limited-range wireless services. Based on advanced TDMA technology, and used primarily for wireless PBX systems, telepoint and residential cordless telephony today, uses for DECT include paging and cordless LANs. DECT frequency is 1800-1900 MHz. Stop press; the old meaning of the word DECT was "Digital European Cordless Telephones" but since the DECT standard also has spread to China and South America the correct definition is now "Digital Enhanced Cordless Telephones". See also Generic Access Profile.

dedicated A facility or equipment system or subsystem set aside for the sole use of a specific customer.

dedicated access A connection between a phone or phone system (like a PBX) and an IntereXchange Carrier (IXC) through a dedicated line. All calls over the line are automatically routed to a particular IXC.

Dedicated Access Line DAL. See Dedicated Access Line Service.

dedicated access line service A type of service often used by large companies which have a direct telephone line going directly to the long distance companies' "Point of Presence" (POP), thereby bypassing the local telephone company and reducing the cost per minute. Often referred to as "T-1" service.

dedicated array processor A microprocessor on a hardware-based RAID array that controls the execution of RAID array-specific functions, such as rebuilding. See RAID.

dedicated attendant link Assures that there will always be an intercom link available for your attendant or receptionist or operator to announce incoming calls.

dedicated bypass A connection between a phone or phone system (like a PBX) and an IntereXchange Carrier (IXC) through a dedicated line that is not provided by the dominant local provider of local phone service. For example, I live in New York. I might order a leased T-1 to MCI. If that line is not provided by Verizon, it is "dedicated bypass." Such bypass circuits are often cheaper and better quality than what the dominant local carrier can provide.

dedicated channel or circuit A channel leased from a common carrier by an end user used exclusively by that end user. The channel is available for use 24 hours a day, seven days a week, 52 weeks of the year, assuming it works that efficiently.

dedicated circuit A circuit designated for exclusive use by specified users. See also Dedicated Channel.

dedicated feature buttons The imprinted feature buttons on a telephone: Conf or Conference, Drop, HFAI (Hands Free Answerer on Intercom), Hold, Mute or Microphone, Speaker or Speakerphone, Transfer, Message, and Recall.

Dedicated Inside Plant DIP. Inside plant is the portion of a LEC's (Local Exchange Carrier's) plant, or physical facilities, that are located inside its buildings. Such inside plant comprises a wide variety of equipment such as channel banks, multiplexers, switching systems, Main Distribution Frames (MDFs) and Intermediate Distribution Frames (IDFs). Dedicated Inside Plant, most commonly, is a term describing an IDF which is dedicated to the purpose of providing a CLEC (Competitive LEC) with a point of interconnection between the local loops it has leased from the ILEC (Incumbent LEC) for purposes of customer access, and the facilities which the CLEC uses to serve those customers. The dedicated IDF also is known as a Single Point of Termination (SPOT) frame. From the SPOT frame, the circuit commonly is directed to a secure enclosure in which the CLEC has collocated in the ILEC building a concentrator or multiplexer which is connected to a high-speed transmission link which hauls traffic to the CLEC's own facilities-based network. A common SPOT is a shared dedicated IDF for use by multiple CLECs which are unable to cost justify a SPOT of their own – the cost of the SPOT, plus a reasonable profit margin – is passed on to the CLEC by the ILEC. See also Collocation and Dedicated Outside Plant.

dedicated line Another name for a private leased line or dedicated channel. A dedicated line provides the ability to have a constant transmission path from point A to point B. A dedicated line may be leased or owned. It may be assigned a single purpose, such as monitoring a distant building. It may be part of a network, with the ability for many to dial into it. It may be a tie-line between your offices or it may be a line to a long distance carrier. In this case, you do not have to dial a local connection number or put in an authorization code. A WATS line is in effect a Dedicated Line to AT&T or whomever you purchased WATS from.

dedicated machine A computer designed to run only one program or do one thing. This machine cannot easily be re-programmed to do another task, as, for example, a general-purpose machine can. A general-purpose machine, however, can be dedicated to running only one task if programmed to do so.

dedicated mode When a file server or router on a local area network is set up to work only as a file server or router, it runs in dedicated mode.

Dedicated Outside Plant DOP. Outside plant is the local loop facilities which connect the customer premises to the LEC (Local Exchange Carrier) network. Typically, the connection is to the LEC central office (CO) exchange. Dedicated outside plant, at the extreme, means that each customer premises has one or more local loops which connect directly to the "wire center" in which the CO is housed. While this approach is copper-intensive, it allows local loops to be activated remotely, as it is not necessary to "roll a truck" in order to make cross connections between various trunk and feeder facilities in the outside plant network in order to effect the connection. Once the initial investment is made, therefore, the ongoing installation and maintenance expenses are much reduced. In the context of a competitive local exchange environment, the CLEC (Competitive LEC) commonly desires to lease from the ILEC (Incumbent LEC) a dedicated outside plant in the form of a "dry copper" circuit. A dry copper circuit is one which is has no electronics (e.g., load coils, repeaters or subscriber carrier systems) between the wire center and the customer premises. Such electronics interfere with the provisioning and support of most data services, including high-speed DSL (Digital Subscriber Line) services, which many of the CLECs are interested in providing. See also CLEC, Dedicated Inside Plant, Dry Copper, DSL, ILEC.

dedicated server A computer on a network that performs specialized network tasks, such as storing files. The word "dedicated" means that the computer is used exclusively as a server. It is not used as a workstation, which means no one is sitting in front of it, using it. A dedicated server sits all alone, attached to its network, working happily all by itself.

dedicated service A communication network devoted to a single purpose or group of users, e.g., AUTOVON, FTS. It may also be a subset of a larger network, e.g., AUTOVON, FTS.

dedicated trunk A trunk which bypasses the Attendant Console and rings through to a particular phone, hunt group or distribution group.

dedupe A file processing term for removing duplicate records from the base file. Duplicate information my be present in the same file or could occur through merging several files from different sources.

deep computing According to IBM, deep computing is supercomputer-scale processing that combines massive computation and very sophisticated software algorithms to attack problems previously beyond the reach of information technology.

deep packet inspection A term used to describe the capabilities of a firewall or an intrusion detection system (IDS) to look within the application payload of a packet or traffic stream and make decisions on the significance of that data based on the content of that data. The software/computer engine that drives deep packet inspection typically includes a combination of signature-matching technology along with heuristic analysis of the data to determine the impact of that communication stream. While the concept of deep packet inspection sounds appealing, it is not so simple to achieve in practice. The inspection engine must use a combination of signature-based analysis techniques as well as statistical, or anomaly analysis, techniques. Both of these are borrowed directly from intrusion detection technologies. See intrusion detection.

deep space Space at distances from the Earth approximately equal to or greater than the distance between the Earth and the Moon.

deep UV Printed circuits are made by optical lithography, by shining light through a negative, or mask. In the beginning, it was visible light, but lately it's been made with more precise ultraviolet light, which can print lines as thin as 0.15 micron wide. They call this light "deep uv" and it is ultraviolet light emitted by an exciter or pulsing laser.

deep Web Documents and other data on Web servers that lie beyond the reach of search engines' conventional search methods and which, therefore, are not indexed. For this reason, this hidden content does not show up in search engines' search results. It is estimated that the amount of content in the deep Web may be 500 times greater than that which is part of the searchable Web.

defacto Misspelling of de facto. See de facto.

default 1. The default is a factory-set hardware of software setting or configuration. It is the preset value that the program or equipment comes with. It will work with default values in the absence of any other command from the user. For example, communications software programs, such as Crosstalk, Blast, etc., have as their default settings 300 baud, 8 bit, one stop bit, no parity. If you want to run at 1,200 baud, you have to change that "default" setting.

2. When you sign a contract with someone and that contract says you will do something and you don't do it, you are technically in "default" on your contract. What that will then mean depends on the person with whom you signed the contract. For example, if you didn't pay them the money you were meant to, they might file to place your company in bankruptcy. Or perhaps you might work out a deal to skip a payment or two and take you out of default.

default carrier Generic name given to the long distance carrier which will carry the traffic of customers who haven't presubscribed to a long distance carrier.

default gateway When individual machines on a network segment send data packets, they check the packet's destination to figure out whether the destination is local (meaning, on the same network segment) or not. If the packet's destination is not local, the machine forwards it to a node on the network serving as the entrance to all other networks. This node is called the default gateway, and could be any routing device, such as a router or a firewall appliance. If you're using static IP addresses, you will need to tell your computer what the address of your default gateway is. It might be something like 10.1.1.1.

default node representation An ATM term. A single value for each nodal state parameter giving the presumed value between any entry or exit to the logical node and the nucleus.

default route 1. A routing table entry that is used to direct any data addressed to any network numbers not explicitly listed in the routing table.

2. Entry in a routing table that can redirect any frames for which the table has no definitive listing for the next hop.

Defense Data Network DDN. The Department of Defense integrated packet switching network capable of worldwide multilevel secure and non-secure data transmission.

defense in depth A term borrowed from the military used to describe defensive measures that reinforce each other, hiding the defenders activities from view and allowing the defender to respond to an attack quickly and effectively. In the network world, defense in depth describes an approach to network security that uses several forms of defense against an intruder and that does not rely on one single defensive mechanism.

Defense Information System Network Part of the Department of Defense's Global Information Grid (GIG), DISN is a worldwide network for the support of military operations.

Defense Message System A Department of Defense system for the exchange of messages electronically between organizations and individuals in the DoD. DMS uses DISN.

Defense Message System Transition Hub Legacy Messaging Network A worldwide, general purpose, record communications network formerly known as the Automatic Digital Network (AUTODIN), which provides legacy messaging service to both the Defense Special Security Communications System (DSSCS) and General Service (GENSER) communities. DMS is replacing AUTODIN as the system of record for organizational messaging.

Defense Red Switch Network A network of the Defense (Department) Information Systems Network (DISN) consisting of a global secure voice switching network whose subscribers are served by a variety of automatic switches and manual operator switchboards. All switchboards are interconnected within the network by dedicated wide-band trunk circuits and/or via the narrowband Defense Switched Network (DSN). The Defense Red Switch Network (DRSN) provides secure command and control switches that offer high-quality secure voice and conferencing capabilities to the senior decision makers and staff of the National Command Authorities (NCA), the Commanders in Chief (CINCs), Major Commands (MAJCOMs), other U.S. government departments and agencies, and allies. The DRSN also provides secure voice conferencing and access to secure strategic, tactical, airborne, and seaborne equipment and platforms.

Defense Switched Network A Department of Defense network that provides command-and-control (C2) circuit-switched service. As a C2 network, the DSN has military requirements to provide assured service and global connectivity under stress conditions. The switched voice service of the DSN allows calls originated from DISN locations to be connected to any other DISN location. The service includes long-haul switched voice, facsimile and conference calling.

deferred processing Performing operations as a group or batch, all at once. Using batch mode, you can quickly prescan your documents, capturing just the image of each page, then perform recognition on these images later, freeing your computer for interactive work.

definition A figure of merit for image quality. For video-type displays, it is normally expressed in terms of the smallest resolvable element of the reproduced received image.

definity A family of digital PBX platforms first introduced and marketed in 1990 by AT&T, then by spinoffs Lucent Technologies and currently Avaya Communication. It is a descendant of the System 75 and 85 architectures first introduced in 1983.

deflection Some automatic call distributors can be programmed to give callers a busy signal if the waiting time is past a certain threshold for callers to your 800 numbers (since you pay for all the waiting time). This is called "deflection" by Intecom and may have another name for other manufacturers. Of course the financial benefit of this must be weighed against the impression it gives your callers.

deflection routing See Hot Potato Routing.

degauss To demagnetize. To degauss a magnetic tape means to erase it. See Degaussing Coil and Degausser.

degaussing coil Degaussing is to demagnetize. A degaussing coil is a long piece of wire bent into the shape of a circle. When a CRT becomes magnetized an area of the screen becomes discolored. If you wave the degaussing coil around this area, the screen becomes demagnetized and the picture quality improves.

degausser Device to demagnetize a color picture tube for color purity. See Degaussing Coil.

degradation In communications, that condition in which one or more of the established performance parameters fall outside predetermined limits, resulting in a lower quality of service.

degraded service state The condition wherein degradation prevails in a communication link. For some applications e.g., automatic switching to a non degraded standby link, degradation must persist for a specified period of time before a degraded service state is considered to exist.

degree of coherence A measure of the coherence of a light source.

degree of isochronous distortion In data transmission, the ratio of (a) the absolute value of the maximum measured difference between the actual and the theoretical intervals separating any two significant instants of modulation (or demodulation) to (b) the unit interval. These instants are not necessarily consecutive. The degree of isochronous distortion is usually expressed as a percentage. The result of the measurement should be completed by an indication of the period, usually limited, of the observation.

degree of start-stop distortion 1. In asynchronous data transmission, the ratio of (a) the absolute value of the maximum measured difference between the actual and theoretical intervals separating any significant instant of modulation (or demodulation) from the significant instant of the start element immediately preceding it to (b) the unit interval.

2. The highest absolute value of individual distortion affecting the significant instants of a start-stop modulation. The degree of distortion of a start-stop modulation (or demodulation) is usually expressed as a percentage.

deinstallation A term used in the secondary telecom equipment business. The shutoff and disconnection of power and the disassembly of the equipment to prepare for its removal from the building. A properly executed deinstallation will include all necessary parts for the reassembly, operation, maintenance and acceptance of the switch and any of its components at the next location. See also de-install.

dejitterizer A device for reducing jitter in a digital signal, consisting essentially of an elastic buffer into which the signal is written and from which it is read at a rate determined by the average rate of the incoming signal. Such a device is largely ineffective in dealing with low-frequency impairments such as waiting-time jitter.

dejure standard A dejure standard that involve committees, such as, ISO, ANSI, IEEE, ITU to name a few. Because of the number of people involved and the nature of the standards process, a consensus must be reached, usually by a vote, therefore this type of standard takes a long time to complete. By contrast, see de facto.

DEK Data Encryption Key. The key by which one can unlock an encrypted message. There are "private" and "public" keys. See Encryption.

delay The wait time between two events, such as the time from when a signal is sent to the time it is received. There are all sorts of reasons for delays, such as propagation delays, satellite delays, the additional time introduced by the network in delivering a packet's worth of data compared to the time the same information would take on a full-period, dedicated point-to-point circuit, etc. See also Latency.

delay announcements These are pre-recorded announcements to incoming callers that they are being delayed and being placed in an ACD queue. Sample: "Please wait. All our agents are permanently busy. You are being placed on Eternity Hold. Don't go away or you'll never be allowed back." Some announcements are giving callers sales pitches and some idea of how long they'll have to stay on line until someone helps them.

delay distortion The difference, expressed in time, for signals of different frequencies to pass through a phone line. Some frequencies travel slower than others in a given transmission medium and therefore arrive at the destination at different times. Delay distortion is measured in microseconds of delay relative to the delay at 1700 Hz. Also called Envelope Delay.

delay encoding A method of encoding binary data to form a two-level signal. A binary zero causes no change of signal level unless it is followed by another zero, in which case a transition takes place at the end of the first bit period. A binary "1" causes a transition from one level to the other in the middle of the bit period. Used primarily for encoding of radio signals since the spectrum of the encoding of signal contains less low frequency energy than an NRZ signal and less high frequency energy than a biphase signal.

delay equalizer A corrective piece of electronic circuitry designed to make communications circuit delays constant over a desired frequency range. A delay equalizer is a device that adds a delay to analog signals, which will travel through a medium faster than other frequencies used to transmit portions of the same data. The objective is to create a medium that transfers the information on all the used frequencies in the same time over the same distance, thus eliminating transmission delay distortion.

Delay Length Call Factor DLCF. An intermediate factor found in an Erlang C table that defines service level as a delay factor. It measures the delay of a call compared to the average handle time (AHT) of a call. It is used to determine the number of staff necessary to meet service level goals. For example, if the desired delay time is 20 seconds and is divided by a AHT of 150 seconds, the corresponding service level to locate in the Erlang table is 0.133. The table will give a staffing level.

delay line A transmission line, or equivalent device, designed to introduce delay.

delay modulation A modulation scheme that uses different forms of delay in a signal element. Frequently used in radio, microwave and fiberoptic systems.

delay-sensitive See time-sensitive.

delay skew The difference in timing of the transmission of signals between pairs in a cable. Delay skew is an issue when multiple pairs in a cable are used to support a single transmission. Examples include 100Base-T, which uses as many as four pairs to support a connection between a LAN hub and an attached device. By splitting the transmission across multiple pairs, the distance between the hub and the attached device can be increased considerably, as each element of the split signal runs at a lower carrier frequency than would the native signal over a single pair. This enhanced performance is due to the fact that lower frequency signals suffer less from the effects of attenuation (loss of signal strength). Delay skew, however, can have significantly adverse effects on this approach. See also 100Base-T and Attenuation.

delay spread See 802.11a.

delay, absolute The time elapsed between transmission of a signal and reception of the same signal.

delayed call forwarding A phone system feature in which you have your incoming calls forwarded to another number only after several rings.

delayed calls The fraction of calls delayed longer than a given time for service are called delayed calls. A telephone company definition.

delayed delivery Hold a message for delivery later. Just as the words say.

delayed delivery facility A facility that employs storage within the data network whereby data destined for delivery to one or more addresses may be held for subsequent delivery at a later time.

delayed ring transfer An optional KTU facility that provides for automatic transfer to the ringing signal from a principal telephone set to the attendant telephone station after an adjustable number of rings.

delayed sending A feature of fax machines which allows the machine to be programmed to send its transmissions at a later time – to take advantage of lower phone rates, for example.

delayed transmission A fax machine feature that allows a document to be transmitted automatically at a specific time.

delimiter A character that separates the parts of a DOS command. For example, a backslash is the delimiter between subdirectory names. Also, the character (typically a comma or tab) that separates field items in a database.

delivered block A successfully transferred block.

delivered overhead bit A bit transferred to a destination user, but having its primary functional effect within the telecommunication system.

delivered overhead block A successfully transferred block that contains no user information bits.

delivery confirmation Information returned to the originator indicating that a given unit of information has been delivered to the intended addresses.

delivery envelope An X.400 term.

DELNI Digital Ethernet Local Network Interconnect. An industry adopted term indicating a multiport transceiver, also known as a fan-out, and generally limited to AUI connections in Ethernet. Originally, this was a product introduced by Digital Equipment Corporation as the DEC Local Network Interconnect.

Delphi Forecasting One of the silliest methods of forecasting the future. Namely, to ask a bunch of alleged experts (often academic eggheads) what they think might happen and then averaging out their opinions, sort of.

delta 1. The mathematical term for a finite increment in a variable. Once a value has been established, the delta is the difference between that value and the succeeding value. The delta can be either positive or negative. The benefit of delta modulation, for instance, is that the difference between the subject value and a previous value can be expressed in fewer bits than can the absolute subject value – it is a form of compression, in effect, yielding better utilization of expensive network bandwidth. See Delta Modulation.

2. A wiring system for distributing and utilizing three phase electrical power. In this system, three power carrying conductors are used, possibly with a fourth safety ground wire. The voltage between any two of the three power wires is the rated distribution voltage, which is most commonly 380 to 415 VAC in most countries or 208 VAC in North America. The other type of three phase power distribution is called the WYE style.

delta channel In T-carrier/ISDN communications, a delta channel/"D channel" contains signaling and status information.

delta frame Also called Difference Frame. Contains only the pixels different from the preceding Key Frame. Delta frames reduce the overall size of the video clip to be stored on disk or transmitted on phone lines.

delta modulation A method for converting analog voice to digital form for transmission. It is the second most common method of digitizing voice after Pulse Code Modulation, PCM. Sampling is done in all conversions of analog voice to digital signals. The method of sampling is what distinguishes the various methods of digitization (Delta vs.' PCM, etc.). In delta modulation, the voice signal is scanned 32,000 times a second, and a reading is taken to see if the latest value is greater or less than it was at the previous scan. If it's greater, a "1" is sent. If it's smaller, a "0" is sent.

Delta modulation's sampling rate of 32,000 times a second is four times faster than PCM. But delta records its samples as a zero (0) or a one (1), while PCM takes an 8-bit sample. Thus PCM encodes voice into 64,000 bits per second, while delta codes it into 32,000. Because delta has fewer bits, it could theoretically produce a poorer representation of the voice. In actual fact, the human ear can't hear the difference between a PCM and a Delta encoded voice conversation.

Delta modulation has much to recommend it, especially its use of fewer bits. Unfortunately no two delta modulation schemes are compatible with each other. So to get one delta-mod digital PBX to speak to another, you have to convert the voice signals back to analog. With AT&T making T-1 a de facto digital encoding scheme, PCM has become the de facto standard for digitally encoding voice. And although there are three types of PCM in general use, they can be made compatible on a direct digital basis (i.e. without having to go back to analog voice). One problem with PCM is that American manufacturers typically put twenty four 64,000 bit per second voice conversations on a channel and call it T-1. The Europeans put 30 conversations on their equivalent transmission path. Thus, you can't directly interface the American and the European systems. But there are "black boxes" available...(In this business, there are always black boxes available.)

deltas The changes.

delta sigma modulation A variant of delta modulation in which the integral of the input signal is encoded rather than the signal itself. A normal delta modulation encoder by an integrating function. See Delta Modulation.

delta technology An Internet access term. Delta technology consists of specialized remote adaptive routing protocols for optimizing bandwidth. It prevents unnecessary traffic from being sent over slow WAN connections by only sending the changes (deltas).

delurking Coming out of online "lurking mode," usually motivated by an irresistible need to flame about something. "I just had to delurk and add my two cents to that conversation about a woman's right to abortion."

deluxe queuing A feature that allows incoming calls from phone users, tie trunks and attendants to be placed in a queue when all routes for completing a particular call are busy. The queue can be either a Ringback Queue (RBQ)– the user hangs up and is called back when a trunk becomes available – or an Off-Hook Queue (OHQ) – the user waits off-hook and is connected to the next available trunk. Deluxe Queuing is a term used mainly by AT&T. Most modern PBXs have this feature. Most have simpler names, however.

Demand And Facility Chart D&F. A telephone company definition. A chart designed to: a. Record an up-to-date picture of working network access lines, actual usage rates, future gains in working network access lines, and future usage rates.

b. Record the capacity of existing equipment and the current picture of the planned capacity additions.

c. Provide a recording vehicle to report consistent data (using standardized terminology and definitions) for planning, and budget review evaluation purposes.

Demand Assigned Multiple Access See DAMA.

demand assignment A technique where users share a communications channel. A user needing to communicate with another user on the network activates the required circuit. Upon completion of the call, the circuit is deactivated and the capacity is available for other users.

demand factor The ratio of the maximum demand on a power system to the total connected load of the system.

demand load In general, the total power required by a facility. The demand load is the sum of the operational load (including any tactical load) and non-operational demand loads. It is determined by applying the proper demand factor to each of the connected loads and a diversity factor to the sum total.

demand paging The common implementation in a PC of virtual memory, where pages of data are read into memory from storage in response to page faults.

demand priority Access method providing support for time sensitive applications such as video and multimedia as part of the proposed 100BaseVG standard offering 100Mbit/s over voice grade UTP (Unshielded Twisted Pair) cable. By managing and allocating access to the network centrally, at a hub rather than from individual workstations, sufficient bandwidth for the particular application is guaranteed on demand. Users, say its proponents, can be assured of reliable, continuous transmission of information.

demand publishing The production of just the number of printed documents you need at the present time, as in "just in time." In short, the immediate production of printed documents which have been created and stored electronically.

demand service In ISDN applications, a telecommunications service that establishes an immediate communication path in response to a user request made through user-network signaling.

demand shaping I call Dell asking for a 40 gig hard drive in my new laptop. They don't have one. They don't tell me they don't have one. They tell me that I'd be better off with a 60 gig hard drive, which they happen to have a special on today. Guess what? It's the same price as a 40 gig drive. Thus Dell has shaped my demand, and achieved the sale.

demarc Pronounced D-Marc. The demarcation point is the physical point at which the separation is made between the carrier's responsibilities for the circuit and those of the end user organization. The carrier is responsible for the local loop, which connects the user organization's premises to the carrier's CO (Central Office) or POP (Point of Presence) at the edge of the network. In a residential or small business application, the demarc is at the NIU (Network Interface Unit), which typically is on the side of the house or inside the garage. In a larger business application, it is at the MPOE (Minimum Point of Entry), which is the closest practical point to where the carrier facilities cross the property line or the closest practical point to where the carrier cabling enters a building. While the MPOE typically is in the form of a physical demarc, in older installations it may simply be in form of a tag hung on the entrance cable to identify a point of logical demarcation. There are exceptions. In some older Centrex installations in some states, the demarc is at the jack for each individual voice or data terminal. In some older campus environments, there may be a demarc for each of several cables coming from various directions, and the demarcs may be well inside the property line. In either case, it is the responsibility of the carrier to install and maintain the local loop and the demarc device, which includes some form of protector against lightening and other electrical anomalies, and some form of intelligence to support loopback testing. It is the responsibility of the end user organization or building owner to install and maintain the inside cable and wire system, which typically terminates in the demarc through a plug-and-jack arrangement. A demarc for voice services might be in the form of a simple RJ-11C jack (one line or trunk) connection, an RJ-14C (two trunks), an RJ-21X (up to 25 trunks), or a 66-block. A demarc for data services typically supports an RJ-48 termination. See also Smart Jack.

Demark An incorrect spelling of Demarc. See Demarc.

demarcation point The point of a demarcation and/or interconnection between telephone company communications facilities and terminal equipment, protective apparatus, or wiring at a subscriber's premises. Carrier-installed facilities at or constituting the demarcation point consist of a wire or a jack conforming to Subpart F of Part 68 of the FCC Rules. See Demarc.

demarcation strip The terminal strip or block (typically a 66 block) which is the physical interface between the phone company's lines and the lines going directly to your own phone system. See also Demarc.

Demilitarized Zone DMZ. See DMZ and Screened Subnet.

DEMKO Denmark Elektriske MaterielKOntrol (Denmark Testing Laboratory).

democratically synchronized network A mutually synchronized network in which all clocks in the network are of equal status and exert equal amounts of control on the others.

Demod/Remod Demodulation/Remodulation. See Fax Relay.

demodulation The process of retrieving an electrical signal from a carrier signal or wave. The reverse of modulation. See Modem.

demodulation dial tone digital The process of retrieving data from a modulated signal. A tone indicating that automatic switching equipment is ready to receive dial signals. Refers to the use of digits to formulate and solve problems or to encode information.

demodulator In general, this term refers to any device which recovers the original signal after it has modulated a high frequency carrier. In television, it may refer to an instrument which takes video in its transmitted form (modulated picture carrier) and converts it to baseband.

demolition hammer Portable jack hammer. These smaller types are powered by AC Voltage power, usually supplied by the service truck generator or inverter. Great tool for digging, especially hard ground.

demon dialer See Automatic Recall.

demountable walls Metal walls that can be disassembled and moved to other locations. They contain vertical and horizontal slots through which cable can be run, also called modular panels.

demultiplex DEMUX. To separate two or more signals previously combined by compatible multiplexing equipment. As an ATM term, it is a function performed by a layer entity that identifies and separates SDUs from a single connection to more than one connection.

demultiplexer A device that pulls several streams of data out of a bigger, fatter or faster stream of data.

demultiplexing A process applied to a multiplex signal for recovering signals combined within it and for restoring the distinct individual channels of the signals.

demux Jargon for demultiplexer.

DEN Directory Enabled Network. DEN is designed to integrate network hardware with directory services, such as Novell's Directory Services or Microsoft's Active Directory. DEN is designed as an extension to the Desktop Management Task Force's (DMTF's) Common Information Model (CIM). The DEN specification is under development by the Desktop Management Task Force (DMTF). It includes a standard approach to defining schema for integrating network equipment, such as switches and routers, with a directory service.

DENet The Danish Ethernet Network which consists of many Ethernet networks in universities connected together by bridges.

Denial of Business. DoB. See Impact Analysis.

denial of service You're no longer allowed to use a service. That service might be anything from normal phone service (you didn't pay your bills) to not being allowed into the company's email because you were just fired. See also Denial of Service Attack.

denial of service attack You're a horrid person with mean intentions. You ping a website 400 or 500 times a second. Or, even worse, you create a virus and use e-mail attachments to infect a lot of networked computers, which become Zombies under your control. You command them all to go to that   website, which is then overwhelmed and no longer able to serve its legitimate customers. In other words, your attack has the effect of denying service to the customers. Hence the term Denial of Service Attack. Such an attack is very hard to prevent, since sometimes it's done from computers all over the country or the world. This is called a Distributed Denial of Service Attack, or DDoS. InfoWorld Magazine of June 11, 2001 reported "to orchestrate DDoS assaults, a hacker first installs "cable bots" on computers that have cable or other high-speed modems, but that lack adequate firewalls against intrusion. These bots are then instructed to send massive amounts of data to a victim's site. Steve Gibson (president of Gibson Research Corporation) found that bots running on just 474 Windows PCs worldwide were enough to completely overwhelm his two T-1 lines."

According to Alcatel, the prevention of Denial of Service attacks requires five key components: 1) Upgrade. Many types of DoS attacks become useless as the vulnerabilities they leverage are detected and fixed. However, if a system still uses an older version of an application, certain vulnerabilities may still be available to attackers. One of the best prevention methods is to keep computer systems updated with the latest versions of the operating system and applications. It's also important to keep track of security updates/patches and make sure they are applied to all vulnerable systems. 2) Use a firewall. Firewalls can help prevent attacks by blocking unauthorized users and filtering packets to keep unwanted traffic out of the network. 3) Intrusion detection. Denial of Service attacks can occur without an obvious sign. Intrusion detection tools help monitor the network and detect the signs of different attacks, allowing you to identify and deal with attackers and harden the network against future attacks. 4) Virus protection. Denial of Service attacks are often propagated by worms, viruses, and Trojan horses that plant small programs on your system. These programs are designed to disrupt services, crash computers, or burden network connections. Current virus scanning products can help identify and remove infected files, as well as prevent future infections. 5) Diligence. It is important to keep up to date on current alerts, warnings, and security techniques. Organizations like the CERT Coordination Center and the Computer Incident Advisory Capability (CIAC) publish regular updates advising the computing community on current security threats and new technologies and information that can help system administrators make their networks more secure. See also Active Attack, Passive Attack, Smurf Attack, Spoofing, Trojan Horse, Virus, Worm and Zombie.

Denis The Little The 6th century monk who decided that history should be split between B.C. and A.D. and inadvertently created the Y2K problem. See Y2K.

denizen A low life citizen of the Internet. Not a complimentary term.

Dense Virtual Routed Networking DVRN. This is the ultimate IP networking fantasy – a network where bandwidth is unlimited and free, speed is not even taken into consideration, IP will never again bump heads with routers and their static routing tables, and data moves through "virtual routers" which will perform tasks way beyond today's elementary "connect the dot" computations. It's almost impossible to imagine such a stress free network with practically no limitations, but advances such as Dense Wave Division Multiplexing (DWDM), which advanced fibre optic transport by creating "virtual strands" used to carry multiple wavelengths of light multiplexed together gives us a sign of hope. All the above from the November 2000 of Computer Technology Review.

Dense Wavelength Division Multiplexing See DWDM.

density 1. The number of bits (or bytes) in a defined length on a magnetic medium. Density describes the amount of data that can be stored.

2. The number of circuits that can be packed into an integrated circuit.

3. In a facsimile system, a measure of the light transmission or reflection properties of an area, expressed by the logarithm of the ratio of incident to transmitted or reflected light flux.

denwa Japanese for telephone. An aka denwa is a red telephone. Some coin phones in Japan are red and are often known as aka denwas. A denwa bango is a telephone number in Japanese.

departmental firewall The NEC PrivateNet Systems Group issued a White Paper called Connecting Safely to the Internet – A study in Proxy-Based Firewall Technology. In that White Paper, they defined a Departmental Firewall:

A departmental firewall is identical to an Internet firewall except that it controls access to and from a single department in a larger organization. It is used to protect sensitive corporate data, such as financial information and personnel records, from access by unauthorized people. A departmental firewall tends to be more generous in the access it allows, but if insecure services, such as NFS (Network File System), are allowed through a departmental firewall, the purpose of installing the firewall in the first place might be defeated. See Firewall and NFS.

departmental LAN A local-area network used by a relatively small group of people working on common tasks; it provides shared local resources, such as printers, data, and applications.

departure angle The angle between the axis of the main lobe of an antenna pattern and the horizontal plane at the transmitting antenna.

depersonalization In 1879 a flu epidemic in Lowell, MA made it likely that all four of the telephone operators would get sick simultaneously. To help substitute operators, management numbered each of the exchange's two hundred plus customers. No problem. The customers accepted the change easily.

deplaning Getting off a plane. See Detraining.

deploy The word comews the military. It means to put troops in the field. In today's technology business it simply means to place and install equipment.

depolarization 1. In electromagnetic wave propagation, that condition wherein a polarized transmission being transmitted through a nonhomogeneous medium has its polarization reduced or randomized by the effects of the medium being traversed.

2. Prevention of polarization in an electric cell or battery.

depopulate A technique to reduce the traffic load on a switch by removing devices from the shelf or cabinet. Depopulating reduces the effective device capacity of a switch but can increase switching capacity. This is a ploy used to give older PBX systems traffic capacity nearer true ACD systems.

depressed cladding fiber An optical fiber construction, usually single mode, that has double cladding, the outer cladding having an index of refraction intermediate between the core and the inner cladding.

derating factor A factor used to reduce the current carrying capacity of a wire when used in environments other than that for which the value was established.

deregistration A cellular radio term. In the CDPD network, the process of dissociating an Network Entity Identifier (NEI) from the CDPD network.

deregulation The removal of regulatory authority to control certain activities of entrenched telephone companies. An attempt by federal authorities to make the telephone industry more competitive. Deregulation is meant to benefit the consumer. Sometimes it does. Sometimes it doesn't. Often, it's a scapegoat for whatever subsequently goes wrong. But it means different things in different countries. In some it means "a new company can come in and bash the living hell out of the local supplier," according to Philip Khoo, of Miller Freeman, Singapore. In other countries, it simply means giving the newcomers 5% of the market, while the old-timer keeps 95%.

derivation equipment Equipment that produces narrowband facilities from a wider band facility. Such equipment can, for instance, derive telegraph grade lines from the unused portion of a voice circuit.

derived facilities Transmission paths created by use of multiplexing methods that provide more than one virtual path per physical facility.

Derived MAC PDU DMPDU. A Connectionless Broadband Data Service (CBDS) term that corresponds to the 1.2 PDU in Switched Multimegabit Data Service (SMDS). CBDC is the European equivalent of SMDS.

DES 1. Data Encryption Standard. A block cipher algorithm for encrypting (coding) data designed by the National Bureau of Standards so it is impossible for anyone without the decryption key to get the data back in unscrambled form. The DES standard enciphers and deciphers data using a 56-bit key specified in the Federal Information Processing Standard Publication 46, dated January 15, 1977. DES is not the most advanced system in computer security and there are possible problems with its use. Proprietary encryption schemes are also available. Some of these are more modern and more secure. The quality of your data security is typically a function of how much money you spend. See Block Cipher, Clipper Chip and NSA.

2. Destination End Station: An ATM termination point which is the destination for ATM messages of a connection and is used as a reference point for ABR services. See SES.

descenders Those parts (or tails) of the letters p, y, j and g which descend below the base line. This type style is much easier to read than one in which the tails rest on the base line. Watch out for true (i.e. real) descenders when you're buying a system. Most have them these days, but many telephone screens don't. Careful.

descrambler A device which corrects a signal (often video) that has been intentionally distorted to prevent unauthorized viewing. Used with satellite TV systems. See DES.

Descriptive Video Service Developed by WGBH TV in Boston, this service is used by the Public Broadcasting Service (PBS) in the United States to provide narrated descriptions of a television program's key visual elements without interfering with the program's regular audio. The narration describes visual elements such as actions, settings, body language and graphics. Descriptive Video Service (DVS) helps the visually impaired to better understand what is happening on the video portion of the program. DVS is available only on TV sets that have stereo sound.

deserialization The process of changing a serial stream of bits to a parallel stream of bytes.

DESI Strip A slang term for Designation Strip, the small printed piece of paper or card that slides into or attaches onto a telephone and tells you which button answers which line or which button does what in the way of features or intercom.

Design Layout Report DLR. A record containing the technical information that describes the facilities and terminations provided by a local telephone company to a long distance telephone company. The technical information is needed by the long distance carrier to design the overall service and includes such items as cable makeup (gauge, loading, length, etc.), carrier channel bank type and system mileage, signaling termination compatibility, etc. The DLR is sent to the designated carrier representative via the local telephone company's engineering department.

Designated Entities See DE.

Designated Router OSPF router that generates LSAs for a multiaccess network and has other special responsibilities in running OSPF. Each multiaccess OSPF network that has at least two attached routers has a designated router that is elected by the OSPF Hello protocol. The designated router enables a reduction in the number of adjacencies required on a multiaccess network, which in turn reduces the amount of routing protocol traffic and the size of the topological database.

designation strip Also called Desi Strip. A Designation Strip is the small printed piece of paper that slides into or attaches onto a telephone and tells you which button answers which line or which button does what in the way of features or intercom.

designing around Designing around is a legal process which deals with designing a new patent around an existing patent. This process often yields a better product than the patented device or method. Designing around is a perfectly legal operation. It has only been recognized and sanctioned by the courts for about 5 or 6 years. Most engineers and business owners are completely unaware of this option and think it sounds illegal when described to them. Designing around can be used to protect an existing patent or to file for a new one.

Deskewing An imaging term. Adjusting – straightening – an image in software to compensate for a crooked scan.

desktop The computer's working environment. The screen layout, the menu bar, and the program icons associated with the machine's operating environment. Apple's Macintosh (introduced on January 24, 1984) really started the idea that the computer's screen was a desktop. With Windows, PCs now also have desktops.

desktop collaboration Using ISDN lines or analog lines with high speed lines, you can link your desktop computers so teleworkers, suppliers and clients can share documents and work together no matter where they are.

desktop management software Desktop management software are products that perform software distribution, inventory and asset management, remote control an remote access capabilities. In other words, let's say that you're the IT manager of a big company and your says "By lunchtime tomorrow we need this new corporate software on 1800 desktops, 350 laptops and 890 PDAs. and we want everybody to have the same software." Desktop management software is what you need to give your company a consistent desktop configuration. Your boss wants a standard configuration because it saves a enormous number of tech support calls.

desktop management task force A consortium of vendors working toward a set of standards for network management software which will ease the management of desktop systems, their components, and peripherals. Activities address "standard groups" such as software, PC systems, servers, monitors, network interface cards (NICs), printers, mass storage devices, and mobile technologies. The most notable of its accomplishments is the development of the Desktop Management Interface (DMI). www.dmtf.org. See also DMI.

desktop metaphor A desktop metaphor is the conceptual way a workstation screen area is used to emulate a user's physical desktop through graphic icon images. The icon maps directly to its real life function. For example, a trash can icon will allow a user to "throw out" a document. Gives an application a "user friendly" feel. Desktop metaphors are consistent throughout all Windows and windows-like applications, like Sparc's Open Look.

desktop nazi An IT person who overcontrols the company's computing resources

– especially what you're allowed to run and not allowed on your own PC.

desktop pattern A Windows 3.1 term. A design that appears across the desktop. You can use Control Panel to create your own pattern or choose a pattern provided by Windows.

desktop replacement A full featured laptop that is powerful and has a fast microprocessor, a large display, full-sized keyboard, and a large hard drive and is considered to be a replacement for a desktop PC. Desktop replacements are typically heavier and larger than the average laptop. When you measure a desktop's performance next to a laptop's performance, you'll typically find the desktop to be much faster – even though they may have the same CPUs, same memory, etc.

desktop search tool A piece of software that indexes, theoretically, everything on your hard disk. By entering a word or phrase you're looking for, your desktop search goes into its indexed database and finds what you need, either in an Excel spreadsheet, your email or your PowerPoints, etc. There are some issues. The more comprehensive and more accurate you want your search to be, (1) the better the quality of the search engine – it has to be able to index all varieties of files on your hard disk/s, (2) the larger the index file will have to be. Often that index file can be several megabytes large and take hours to create and update – a constant drag on your computer. A number of vendors have announced free desktop search tools. The only desktop search tool I find really useful is one called dtSearch, which costs money. I hear that Apple's newest OS (operating system) has a great desktop search tool called Spotlight.

desktop video Communications that rely either on video phones or personal computers offering a video window.

desktop videoconferencing By combining ISDN technology and individual PCs, people can meet "face-to-face" without leaving their offices. It's a way to reduce costly and time-consuming travel, maybe.

desktop virtualization The hosting and centralized management of desktops on virtual machines on servers in data centers, as a way to cost-effectively secure, manage and support desktop PCs of remote and mobile workers, telecommuters, and outsourced operations.

despew To automatically generate a large amount of garbage to the 'Net, especially from an automated posting program gone wild.

despun The orientation of a satellite antenna, which keeps it pointed to the earth. Also used to refer to the communications payload section of a Hughes spinning satellite.

despun antenna Of a rotating communications satellite, an antenna, the direction of whose main beam with respect to the satellite is continually adjusted so that it illuminates a given area on the surface of the Earth, i.e., the footprint does not move.

destaging In systems that employ cache memory, destaging is the operation that reads data from magnetic disks and writes it to cache memory. "Destaging" occurs when data being removed from the cache to make room for new data has changed; it refers to the movement of changed data from cache back to disk.

destination address That part of a message which indicates for whom the message is intended. Usually a collection of characters or bits. Just like putting a destination address on an envelope. On a token ring network this is a 48-bit sequence that uniquely defines the physical name of the computer to which a LAN data packet is being sent. The IEEE assures that in the world of LANs no two devices have the same physical address. It does so by assigning certain numbers to vendors of token ring adapters, the devices that connect computers to a token ring network.

destination address filtering A feature of bridges that allows only those messages intended for the extended LAN to be forwarded.

destination code See Destination Field.

destination directory The directory (or folder) to which files are copied or moved.

destination document The document into which an object is linked or embedded via OLE. The destination document is sometimes also called the container document.

destination field 1. A telephone company definition. A combination of digits that provides a complete address to reach a destination in the message network. Most destination codes are made up of some of the following components: Access Code, Area (NPA) Code, End Office Code (NNX), Main Station Number, Service Code, Toll Center Code.

2. A networking term. The field in a message header that contains the network address of the individual for whom the message is meant and who will (with luck and good management) receive the message.

destination host A computer system on the network that is the final destination for a file transfer or for an e-mail message.

destination node Those system nodes which receive messages over the control packet network from the source or transmitting node.

Destination Point Code DPC. The part of a routing label that identifies where the SS7 signaling message should be sent. See also Point Code.

Destination Service Access Point DSAP. The logical address of the specific service entity to which data must be delivered when it arrives at its destination. This information may be built into the data field of an IEEE 802.3 transmission frame.

destuffing The controlled deletion of stuffing bits from a stuffed digital signal, to recover the original signal.

detector 1. In a radio receiver, a circuit or device which converts or rectifies high frequency oscillations into a pulsating direct current or which translates radio frequency power into a form suitable for the operation of an indicator. This is most frequently a vacuum tube, less commonly a crystal. Coherers and delicate chemical rectifiers were used in former years.

2. In an optical communications receiver, a device that converts the received optical signal to another form. Currently, this conversion is from optical to electrical power; however, optical-to-optical techniques are under development.

detem An opto-electronic transducer that combines the function of an optical detector and emitter in a single device or module. Do not confuse with DTERM, a name for one of NEC's telephones that works on its NEAX 2000 and 2400 PBXs.

detent tuner Click type of TV tuner.

DETER The Cyber Defense Technology Experimental Research Network (DETER) is a model of the Internet built specifically to test its vulnerability to hackers, viruses, worms, trojan horses, denial of service attacks and other threats to its survival.

determine A verb that is part of the the internationally accepted way of finalizing and agreeing upon standards. For a full explanation, see ADSL Lite.

deterministic load distribution A technique for distributing traffic between two bridges across a circuit group. Guarantees packet ordering between source-destination pairs and always forwards traffic for a source-destination pair on the same segment in a circuit group for a given circuit-group configuration.

detour Difficulty in gaining access to the information highway. Often involves a Highway Construction Supervisor solving your access problem.

detraining Getting off a train. An absolutely ghastly word invented by the railroad industry to keep them on a par with new, awful language invented by the airline industry. See also Deplaning, an equally awful word.

detuning The measures taken by a cellular tower operator to prevent the tower from reradiating broadcast transmissions from a nearby AM station. See reradiation.

detuning skirt A series of wires that are strung from the top of a cellular antenna and extended to the ground below, to help prevent reradiation of AM station transmissions. See also reradiation.

Deutsche Welle Germany's government-funded international radio service, whose mission is to communicate German points of view and promote Germany's reputation worldwide. Deutsche Welle was founded in 1953. It transmits in over 30 languages by FM radio, short-wave radio, radio relay, satellite radio, and now also by Internet radio to over 200 countries.

device contention 1. Occurs when more than one application is trying to use the same device, such as a modem or printer. Some of the newer operating systems do a better job handling device contention.

2. The way Windows 95 allocates access to peripheral devices, such as a modem or a printer, when more than one application is trying to use the same device.

device control A multimedia definition. Device control enables you to control different media devices over the network through software. The media devices include VCRs, laser disc players, video cameras, CD players, and so on. Control capabilities are available on the workstation through a graphical user interface. They are similar to the controls on the device itself, such as play, record, reverse, eject, and fast forward. Device control is important because it enables you to control video and audio remotely – without requiring physical access.

device discovery The mechanism to request and receive the Bluetooth address, clock, class of device, used page scan mode, and names of devices. See Bluetooth.

device driver A device driver acts as a translator between the device and software that is trying to use the device. Think of a printer attached to your PC. You type "A." You want it to print as an A. Your word processing software sends the "A" to software called a device driver, which translates it into bits the printer understands. A more formal explanation: A device driver is a special type of software (which may or may not be embedded in firmware) that controls devices attached to the computer, such as a printer, a scanner, a voice card, a diskette drive, a CD-ROM, a DVD-drive, a hard disk, monitor or mouse. A device driver is software that expands an operating system's ability to work with peripherals. A device driver controls the software routines that make the peripherals work. There are device drivers in virtually everything electronic today – including all types of Walkmen, automobile windows, etc.

2. A program that enables a specific piece of hardware (device) to communicate with Windows 95. Although a device may be installed on your computer, Windows 95 cannot recognize the device until you have installed and configured the appropriate driver.

device fingerprint Data that positively identifies a device on a network, such as a PC, server, PDA, or IP phone. Used on a network for security purposes, such as device authentication, access control, and secure content delivery.

device ID A Plug and Play term. A code in a device's Plug and Play extension that indicates the type of device it is. The device ID and the vendor ID create a unique identifier for each PnP (Plug N Play) device.

device layer The group of protocols that handles the hardware in a Bluetooth device. The device layer handles components such as the display, keypad, and RF communications.

Device Management Protocol See DMP.

device network A network whose primary role is the connecting together (i.e. interconnection) of physical devices, such as sensors, locks, lights, motors, RFID readers, bar-code readers and cameras. See also the Internet of Things.

device rendering Most sites are written for, and tested exclusively on desktop computers with large color monitors. Mobile wireless devices typically have much smaller screens, and until today, it has been a challenge to present Java and Web based enabled pages on these. Device Rendering technology intelligently reformats the today's Web sites and Windows based applications to fit inside the screen width, thereby eliminating the need for horizontal scrolling.

device tree A hierarchical tree that contains the devices configured on a computer.

DFA Doped Fiber Amplifier. An amplifier used in fiber optic systems to amplify light pulses. Such amplifiers typically are known as EDFAs (Erbium-Doped Fiber Amplifiers) as they are doped with erbium, a rare-earth element. DFAs are more effective than regenerative repeaters in many applications, as they simply amplify the light pulses through a chemo-optical process. Regenerative repeaters, on the other hand, require that the light signal be converted to an electrical signal, amplified, and then re-converted to an optical signal. Additionally, DFAs can simultaneously amplify multiple wavelengths of light in a Wavelength Division Multiplex (WDM) fiber system. See also EDFA, SONET and WDM.

DFB Distributed Feedback Laser. A type of laser used in fiber-optic transmission systems, at the distribution level of the local loop. DFBs are point-to-point lasers distributed among nodes in a geographic area such a neighborhood. They transmit and receive optical signals between the distributed nodes and the centralized node, where the signals are multiplexed over a higher-speed fiber link to the head-end (point of signal origin). DFBs can be more effective than the traditional approach of using a single laser which serves multiple nodes through a broadcast approach, as the available bandwidth can be segmented. DFBs have application in a FTTN (Fiber-To-The-Neighborhood) local loop scenario. See also FTTN and SONET.

DFI Digital Facility Interface. An 5ESS switch circuitry in a DTLU responsible for terminating a single digital facility and generating one PIDB (Peripheral Interface Data Bus).

DFS Dynamic Frequency Selection. See 802.lla.

DFT Direct Facility Termination. A telephone company trunk that terminates directly on one or more telephones.

DG Directorate General (CEC).

DGE Dynamic Gain Equalizer. See Dynamic Gain Equalizer.

DGPS Differential Global Positioning Service – a new venture of the US Coast Guard, which it hopes to have ready by 1996. It will use an existing network of radio beacons throughout the US to create a fixed grid of known reference points in order to improve the accuracy of the Defense Department's GPS signal. The Coast Guard hopes to achieve an accuracy of about 10 meters.

DGT Direccion General de Telecommunicaciones (Spanish General Directorate of Telecommunications).

DHACP Dynamic Host Automatic Configuration Protocol. See DHCP.

DHCP Dynamic Host Configuration Protocol. In the old days, all computers on networks had unique numbers. Like phone numbers, these numbers were designed so you (or another computer) could find your computer. Ted Lemon, who worked at DEC at the time, was tired of having to walk to each computer around the big campus and, every time a new computer wanted to get on the network, to manually set its number (its IP address) and other configuration information. He went looking for a solution to the problem, and found a working group at IETF, chaired by Ralph Droms, that was working on a new protocol called DHCP that was intended to solve this problem. DHCP saves Ted walking around. In DHCP, when a computer turns on, the central server or computer assigns the IP address dynamically – as it happens, as the computer switches one. What this means that you computer is assigned a different IP address every time it logs on. The computer server assigning the IP address might be on company premises or it may be distant, i.e. it could be AOL or MSN, etc. Here's a more technical explanation: DHCP is a TCP/IP protocol that enables PCs and workstations to get temporary or permanent IP addresses (out of a pool) from centrally-administered servers. The host computer runs the DHCP server, and the workstation runs the DHCP client. DHCP allows a server to dynamically assign IP addresses to nodes (workstations) on the fly. Like its predecessor, The Bootstrap Protocol (Bootp), DHCP supports manual, automatic and dynamic address assignment; provides client information including the subnetwork mask, gateway address, and DNS (Domain Address Server) addresses; and is routable. DHCP offers the advantage of automatic configuration, whereas Bootp must be configured manually. A DHCP server, generally in the form of a dedicated server, verifies the device's identity, "leases" it the IP address typically for the duration of that computing session, i.e. until the person logs off AOL or MSN, etc. See also BOOTP and IPv6.

DHCP leasing See DHCP.

Dhrystones Benchmark program for testing the speed of a computer. It tests a general mix of instructions. The results in Dhrystones per second are the number of times the program can be executed in one second. The Dhrystone benchmark program is used as a standard indicating aspects of a computer system's performance in areas other than its floating-point performance, for instance, integer processes per second, enumeration, record and pointer manipulation. Since the program does not use any floating-point operations, performs no I/O, and makes no operating system calls, it is most applicable to measuring the performance of systems programming applications. The program was developed in 1984 and was originally written in Ada, although the C and PASCAL versions became more popular by 1989. See Whetstones.

DHT Direct to Home satellite TV.

DI/DO Abbreviations for Dispatch In (inside technicians), Dispatch Out (outside field technicians).

DIA/DCA Document Interchange Architecture/Document Content Architecture. IBM promulgated architectures, part of SNA, for transmission and storage of documents over networks, whether text, data, voice or video. Becoming industry standards by default.

Diablo Wind As explained in Jim Carlton's excellent book, "Apple, the Inside Story of Intrigue, Egomani and Business Blunders:" Infrequently a heat wave bakes the entire San Francisco Bay area including Silicon Valley, when the high pressure system that is usually parked over the Pacific and is responsible for generating those delicious marine breezes shifts over to nearby Nevada and funnels hot, dry air down that same mountain range in a reverse flow. The effect is known locally as a "Diablo wind," because the range these pour down is called the Diablo. "Diablo" is a Spanish word meaning "devil," so these are devil winds and appropriately named because they can spark wicked firestorms that wreak destruction and havoc across the region.

diagnostic programs Programs run by the computer portion of a PBX to detect faults in the system. Such programs may run automatically at regular intervals or continuously. The goal of diagnostic programs is to detect faults before they become serious and to alert someone – typically the attendant – to go fix it. Some diagnostic programs stop running when the switch gets too busy. Some don't. You can dial into some diagnostic programs from afar. You can't dial in remotely on some others. Remote diagnostic programs are among the greatest boons to improved reliability of telephone systems.

diagnostics A term used in the secondary marketplace. Original Equipment's Manufacturer (OEM) prescribed test procedure whose successful completion is normally required for maintenance acceptance of a switch, cabinet, or peripheral piece of equipment. Comment: A new maintenance contract will not go into effect until the maintenance company accepts the results of the diagnostics.

dial A round face or display upon which some measurement is registered, usually in the form of graduations of values. Examples include sundials; analog watches and clocks; and analog speedometers, pressure gauges, and tachometers. In telecommunications, "dial" generally refers to dial telephones, which generally are considered to be obsolete. Dial telephones, for those of you who are relative youngsters and have never seen one, have a round dial on the face of the telephone set. The dial has finger holes which correspond to numbers on a faceplate. The numbers are 1, 2, 3, 4, 5, 6, 7, 8, 9, and 0. To dial a number, you put your finger in the hole, and turn the dial clockwise until it stops. You then remove your finger, thereby allowing the dial to return to its original position. You do this successively until you have dialed the entire number. As the dial returns to its original position in each case, the telephone set mechanically opens and closes an electrical circuit, thereby pulsing the number to the central office switch. Dial telephone are very different from touchtone telephones, which have keypads, or tonepads. Dial telephones are very slow, and they wear out your finger. The actual "dial" on rotary dial phones is called the finger wheel. It's OK to say dial, even if you make your calls by tapping buttons on a touch-tone pad. Touch-Tone was originally a trademark of AT&T, but they let the trademark lapse. A maker of cheapie phones used Touch-Tone as a brand name in the mid-80s, but they seem to have disappeared. Most phones and phone systems can be switched to produce either touch-tones or dial pulses (clicks), like old rotary dial phones, for use with central offices that don't accept touch-tones. The technical term for touch-tone is DTMF (dual-tone/multi-frequency). See also Dial Pulsing and DTMF.

DIALAN DMS Integrated Access Local Area Network from Northern Telecom.

dial a prayer A sarcastic name for the local 611 number run by the local telephone companies as their centralized number for repair.

dial around A method used by callers to purposely bypass a payphone company's local or long distance carrier services. Such methods include calling cards and alternative carrier's collect services, such as 1-800-COLLECT or 1-800-CALL-ATT. Payphone operators receive little or no revenue from such calls.

dial backup 1. A network scheme using dial-up phone lines as a replacement for failed leased data lines. In one typical case, two dial-up lines can be used. One dial-up link is used to transmit data and the other to receive data, thus giving us full-duplex data transmission.

2. A security feature that ensures people do not log into modems that they shouldn't have access to. When a connection is requested, the system checks the username presented for validity, then "dials back" the number associated with that username.

dial by name You can dial someone by spelling their name out on the touchtone pad. Typically, the system plays a recorded announcement giving directions for using the Dial by Name feature: the caller then inputs the appropriate digits/letters. When the system recognizes a match, a recorded announcement states the name of the dialed party for confirmation by the caller before automatically completing the call. If the input digits are not uniquely associated with a particular station the system may ask the caller to pick a name from a menu of choices. Dial by Name is getting cheaper. Automated attendants are being programmed to have the feature. And you shouldn't buy an auto attendant unless it has this feature.

dial call pickup A phone user on a PBX or hybrid can dial a special code and answer calls ringing on any other phone within his own predefined pickup group.

dial dictation access A service feature available with some switching systems that permits dialing a special number to access centralized dictation equipment.

dial in banner An Internet Access Term. Optional pop-up window for dial-in connections. Allows network managers to display information or warning messages when users dial into remote networks.

dial in channel aggregation The ability to use more than one communications channel per connection. By aggregating both 64 kbit/s ISDN B channels, users can take advantage of 128 kilobits per second dial-in connections. Fast 128 kbit/s data transfer rates reduce large file transfer times. The same as Bonding.

dial in tie trunk A Dial In Tie trunk is a trunk that may be accessed by dialing an access code and then seizing a dedicated transmission path to a distant PBX (or another PBX a short distance away). Once the trunk is seized in the distant PBX, the caller may then use the features of that PBX, depending on the class-of-service and restrictions assigned to the trunk.

dial it 900 service A special one-way mass calling service that allows prospects, customers and others to reach you from anywhere across the country. In contrast to 800-service, the caller pays the 900 charge, generally one charge for the first minute, with a lesser charge for each additional minute. DIAL-IT 900 Service is a great way to involve your customers and prospects in a promotion! Premium Billing lets you select a rate above standard DIAL-IT 900 rates. The long distance carriers (through their deals with local phone companies) handle the billing. You, the information provider, split the revenues with the long distance provider. International DIAL-IT 900 service is currently available from a growing number of countries.

dial it service A telecommunications service that permits simultaneous calling by a

large number of callers to a single telephone number. There is usually a fee for calls to 900 number or 976 numbers. See also Dial It 900 Service.

dial level The selection of stations or services associated with a PBX, based on the first digit(s) dialed.

dial pacing An element of Predictive Dialing. It is important to draw a distinction between the simple placement of the calls (dialing the numbers) versus the logic that goes into when calls are made and how many to make at the same time. These decisions are made automatically by pacing algorithm software. In effect, the computer telephony equipment is a slave to the pacing algorithm. The switch dutifully makes the calls that are requested under the control of this software. Pacing software is based on huge database records. These records include information on time of day, anticipated odds of the called party answering, the number of live agents that are available to take calls, the work schedule for the call center employees, etc. It is possible for a predictive dialing system to pace calls based on a canned algorithm. For example, the software may instruct the VRU to dial 25 calls a minute no matter what. This may be suitable if the need for operator transfer is not critical, or if the information to be delivered does not require operator intervention. See also Predictive Dialing.

dial peer A dial peer, also known as an addressable call endpoint, is a device that can originate or receive a call in a telephone network. In voice over IP (VoIP), addressable call endpoints can be categorized as either voice-network dial peers or POTS dial peers. Voice-network dial peers include VoIP-capable computers, routers, and gateways within a network. POTS dial peers include traditional telephone network devices such as phone sets, cell phones, and fax machines.

The term dial peer is sometimes used in reference to a program that matches a specific dialed sequence of digits to an addressable call endpoint. According to this definition, there is one dial peer for each call leg (connection between two addressable call endpoints).

To give you some context: we had a customer who, when dialing off of his PBX, across our Cisco VoIP router, would ring back to his PBX when dialing numbers to his own NPA-NXX. Our Cisco admin changed the dial peers on the Cisco box, fixing the problem.

dial pick-up PBX feature. A phone on a PBX can answer another ringing phone by dialing a few digits. Also called an access code.

dial plan The set of rules for determining whether the digits dialed are a legitimate dialing string. A dial plan describes permissible quantities and patterns of digits in a dialing string, and also describes access codes, area codes, and specialized codes. The North American public switched telephone network (PSTN) uses a 10-digit dial plan that includes rules for a 3-digit area code and a 7-digit telephone number, and rules for local and long-distance dialing strings, directory assistance and other special services. PBXs and routers on VoIP networks generally support variable-length dial plans ranging from 3 to 11 digits. Such dial plans must comply with the dial plans of telephone networks to which they connect, for example, the PSTN. A completely private voice network that is not connected to the PSTN or to any other telephone network can use a dial plan that is incompatible with other network's plans.

dial pulse DP. A type of switched access line address signaling that uses rapid loop open and loop closure signals (pulses) to indicate the digit being dialed. The digits 1 through 9 are represented by a defined number of pulses; the digit zero is represented by ten pulses.

dial pulse signaling A type of address signaling in which dial pulse is implemented to signal the distant equipment. See Dial Pulsing.

dial pulsing A means of signaling consisting of regular momentary interruptions of a direct or alternating current at the sending end in which the number of interruptions corresponds to the value of the digit or character. In short, the old style of rotary dialing. Dial the number "five" and you'll hear five "clicks." See Dial Speed, Dial Train and DTMF.

dial repeating trunks PBX tie trunks used with terminating PBX equipment capable of handling telephone signaling without attendant involvement.

dial selective signaling A multipoint network in which the called party is selected by a prearranged dialing code; typified by the Bell SS-1 Selective Signaling System.

Dial Service Assistance DSA. A service feature associated with the switching center equipment to provide operator services, such as information, intercepting, random conferencing, and precedence calling assistance.

dial speed The number of pulses a rotary dial can send in a given period of time, typically 10 per second. A Hayes modem with a communications package, like Crosstalk, can send 20 pulses per second.

dial string A dial string is the sequence of characters sent to a device which can dial a phone number. Such a device might be a modem or a voice processing card. Here are some "digits" in a dial string: ! – flashhook (TAPI standard); & – flashhook (Dialogic); T – use tone dialing; , – pause (typically of half a second to two seconds); W – wait for dial tone.

Dial String/Command String A sequence of characters and digits used for dial-in access; ATDT5107861000,,,,,,,123456, H<\<>CR> for example.

dial through A technique, applicable to access circuits, that permits an outgoing routine call to be dialed by the PBX user after the PBX attendant has established the initial connection.

dial tone The sound you hear when you pick up a telephone. Dial tone in North America is a combination of two pure signals of 350 Hz and 440 Hz that is generated by your local telephone office for about one minute (followed by a short silence, then off-hook messages and finally a loud off-hook alert tones). By hearing the dial tone, you know that your phone company is alive and ready to receive the number you dial. If you have a PBX, dial tone will typically be provided by the PBX. Dial tone does not come from God or the telephone instrument on your desk. It comes from the switch to which your phone is connected to. Outside North America, dial tone often sounds very different. And modems PCs, which are set up to "wait for dial tone" often don't recognize these unusual dial tones. The key then is a disable the modem's property that says "wait for dial tone," and have it begin dialing the second it goes off hook.

dial tone delay 1. The specific time that transpires between a subscriber's going off-hook and the receipt of dial tone from a servicing telephone central office. It's a measure of the time needed to provide dial tone to customers. Many of the local public service commissions in the United States say that 90% of customers should receive dial tone in fewer than three seconds.

2. A telephone company definition. Percent Dial Tone Delay (% DTD) over three seconds is a measurement of calls that did not receive dial tone within three seconds. The average busy season objective for an entity in the busy season of exhaust is a maximum of 1.5 percent and is the engineering objective ceiling for all types of equipment. In addition the following maximum DTD engineering ceilings for an entity to be included in the equipment design:

1. Highest Annually Recurring Day – Not over 20%. This maximum ceiling is to be applied to all types of offices.

2. Average 10 High Day – Not over 8%. This maximum ceiling is to be applied in analog ESS offices.

dial tone first coin service A type of pay phone service in which dial tone is received when the caller goes off-hook and coins must be inserted only after the call is connected.

Dial Tone Speed DTS. A telephone company definition. The length of time required for switching equipment to provide dial tone to a subscriber originating a call. Usually expressed as the percentage of attempts that delayed over three seconds.

dial train The series of pulses or tones sent from the phone that's calling and the switching system it's attached to in order to signify the call's destination.

dial up The use of a dial or push button telephone to create a telephone or data call. Dial-up calls are usually billed by time of day, duration of call and distance traveled. A connection to the Internet, or any network, where a modem and a standard telephone are used to make a connection between computers. See Dial Up Line.

dial up account You want to access the Internet. You dial your local Internet Service Provider (ISP) via a local phone number. That's called a Dial Up Account. You can also have a dedicated account, which means that physically a piece of wire (and other electronics) connects you (i.e. your computer) to your ISP 24-hours a day, seven days a week. Mostly, dedicated is more expensive than dial up (depending on rates, etc.). But there are big advantages to dedicated – including, faster access, getting your mail the instant it comes in, etc. See xDSL and Cable Modem.

dial up line A telephone line which is part of the switched nationwide telephone system. Typically a "dial up line" is a standard analog POTS line. These days, ISDN lines are dial up, also. So this definition is changing also.

dial up modem A modem that works on the public switched telephone network (PSTN) and connects to a remote computer resource for the duration of an individual call. See Modem.

dial up networking A Windows 95 and 98 definition. It is a service that provides remote networking for telecommuters, mobile workers, and system administrators who monitor and manage servers at multiple branch offices. Users with Dial-Up Networking on a computer running Windows 95 can dial in for remote access to their networks for services such as file and printer sharing, electronic mail, scheduling, and SQL database access.

dial zero phone A telephone on a Northern Telecom Norstar phone system which is

assigned to ring when someone dials 0 (zero) from another Norstar telephone.

dial-a-barn In rural telephony, a term for revertive ringing, i.e, an intercom-like service between different extensions on a single phone line.

dial-it-service See Dial It Service.

dialback security Dialback security is a telecom security feature. If a person calls in wanting remote access, the system asks for a password. Once it receives a correct password, it hangs up on the caller and dials back a pre-defined remote number, only then giving the caller access. Unless the hacker has you tied up in your living room, it makes things very secure. It can be made even more secure with multiple passwords and features like voice recognition.

dialed To have your shit together; to have all the parts on your equipment working smoothly. As in "Hey, I finally got that bike dialed." Or, for indoorsy types, "Hey, I finally got that phone dialed."

Dialed Number Identification Service See DNIS and 800 Service.

dialed number recorder Also called a Pen Register. An instrument that records telephone dial pulses as inked dashes on paper tape. A touchtone decoder performs the same thing for a touchtone telephone.

dialer A standalone device which automatically dials a telephone number. Once upon a time (circa the 1960s), mechanical card dialers were invented to allow the automated dialing of numbers. These peripheral devices connected to a telephone set and made use of little plastic punch cards. You punched out little round holes in the card, with the holes corresponding to the number you dialed frequently, and you wrote the person's name on the card. When you wanted to dial the number, you inserted the card into the dialer, you punched the "start" button, and the dialer automatically dialed the number. Those early dialers saved a lot of wear and tear on your fingers. Dialers today are electronic devices which dial via tones, rather than dial pulses and who store their numbers in random access memory on chips, just like personal computers. Before the days of equal access and intraLATA long distance competition, dialers were used to "dial around" the LEC (Local Exchange Carrier), so that you could place cheaper intraLATA long distance calls. Today, you find dialers in airports. You find a phone with no keypad. You pick up the handset and the phone dials paging. Near to the luggage collection, you'll often find phones with many buttons. Each button may be for a particular hotel in the area. See also Dial.

dialing parity Dialing parity is a technological capability that enables a telephone customer to route a call over the network of the customer's preselected local or long distance phone company without having to dial an access code of extra digits. Here's a more technical definition from the Telecommunications Act of 1996: The term "dialing parity" means that a person that is not an affiliate of a local exchange carrier is able to provide telecommunications services in such a manner that customers have the ability to route automatically, without the use of any access code, their telecommunications to the telecommunications services provider of the customer's designation from among 2 or more telecommunications services providers (including such local exchange carrier). See also Telecommunications Act of 1996.

dialing pattern Dialing pattern refers to the digits you need to dial to place local, long distance, collect calls, or other phone calls. Dialing patterns will vary due to different types of telephone carrier switching equipment, computer software and the host carrier's credit policies (e.g. automatic roaming versus credit card roaming when using a cellular phone).

dialing plan A description of the dialing arrangements for customer use on a network.

dialing report A call center term. A report that summarizes the results of all numbers dialed for one or more telemarketing campaigns over a specified period of time.

dialog box A dialog box is a temporary window which prompts you to input information or make selections necessary for a task to continue. If you leave it open, the program or another allied one may not continue and will keep prompting you to close the Dialog Box. Microsoft also calls a Dialog Box a "Child Window."

dialup line A nondedicated communication line in which a connection is established by dialing the destination code and then broken when the call is complete.

dialup switch An Internet Access Term. Category of switching equipment designed to manage the dialup connections between the PSTN and either the Internet or a corporate LAN internetwork, providing security, accounting, and service management capabilities.

diameter mismatch loss The loss of signal power at a joint that occurs when the transmitting fiber has a diameter greater than the diameter of the receiving fiber. The loss occurs when coupling light from a source to fiber, from fiber to fiber, or from fiber to detector.

diamonds I have never hated a man enough to give his diamonds back. – Zsa Zsa Gabor

diaper change One of several daily visits by a tech-support person to the desk of a particularly cranky, lazy, or technically incompetent user. "Sorry I was late, but I had to do a diaper change down in accounting." Definition courtesy Wireless Magazine.

diaphragm The thin flexible sheet which vibrates in response to sound waves (as in a microphone) or in response to electrical signals (as in the speaker or the receiver of a telephone handset).

dibit A group of two bits which can be represented by a single change of modulation of the carrier signal. On phase modulation, one of four phases in four-phase modulation is used to represent 00, 01, 10 or 11. See DPSK.

dichroic filter An optical filter designed to transmit light selectively according to wavelength (most often, a high-pass or low-pass filter).

dichroic mirror A mirror designed to reflect light selectively according to wave-length.

dictation access and control A telephone system feature which allows a user to dial a dictation machine and use that machine (giving it instructions by push button) as if it were in his office. Typically, the material on that dictation machine is taken off by one or several typists of a centralized pool and word processed into letters, reports, legal briefs, etc. Telephone suppliers usually don't supply the dictation equipment. Newer telephone dictation machinery is, in reality, a specialized application of voice processing equipment. See Voice Processing.

dictation tank A recording gadget which receives messages dictated through the telephone system. This tank contains tape which can then be transcribed into letters or documents. See Dictation Access and Control.

dictionary attack The bad guy is trying to figure out a password to your network. So they organize their computers to throw millions of possible passwords at your system, one word after another after another. This is called a dictionary attack. The simple solution to such an attack is to configure your network so that it takes three attempts at typing in the correct password and then hangs up on the caller. Dictionary attacks can also be used to send milions of spam mails. For example, a spammer sets up software to email as follows:

a@mydomain.com
aa@mydomain.com
aaa@mydoman.com
...infinite emails follow...
zzzzzzzzzz@mydomain.com

Sometimes, they make it 'intelligent' and use all first names (i.e. chris@mydomain. com), combinations of firstnames and lastnames, first initials + lastnames, first names + last initials (i.e. chrish@mydomain.com).

These dictionary attacks really work ISP mail servers to death. As one friend wrote me, "We get these 'bad' addresses and have to send out a 'bounced email' rejection. They (the SPAMers) use that to clean up or build their lists. There's just about NO WAY to stop it without false rejections of actual email." See also SPAM.

dictionary method A group of advanced compression methods that significantly reduce the number of bits transmitted across a circuit or network. There are several basic approaches. One approach looks at the input data (e.g., a sequence of characters) to see if it has been sent previously. If so, the output is in the form of a small pointer, pointing to the previously sent data, rather than the data itself. All methods which fall into this first approach are based on the LZ77 approach developed in 1977 by Abraham Lempel and Jakob Ziv. LZ77 was refined in the LZSS algorithm developed in 1982 by Storer and Szymanski. See also LZ77, LZSS, LZ78, LZJH, and LZW.

DID Direct Inward Dialing. You can dial inside a company directly without going through the attendant. This feature used to be an exclusive feature of Centrex but it can now be provided by virtually all modern PBXs and some modern hybrids, but you must connect via specially configured DID lines from your local central office. A DID (Direct Inward Dial) trunk is a trunk from the central office which passes the last two to four digits of the Listed Directory Number to the PBX or hybrid phone system, and the digits may then be used verbatim or modified by phone system programming to be the equivalent of an internal extension. Therefore, an external caller may reach an internal extension by dialing a 7-digit central office number. Notice: DID is different from a DIL (Direct-In-Line) where a standard, both-way central office trunk is programmed to always ring a specific extension or hunt group. Traditionally, DID lines could not be used for outdial operation, since there was no dialtone offered. More recently, the individual channels in a T-1 trunk can be defined in terms of their directional nature, with some being defined as DID, some as DOD (Direct Outward Dialing), and some as combination (both incoming and outgoing). See also Combination Trunk, Direct Inward System Access, and DOD.

DID director A standalone box which interprets DID (Direct Inward Dialing) data from trunks and routes incoming calls to the appropriate fax mailboxes on your fax server. The boxes strip DID, read it, then tell the server who the fax is for by sending out DTMF. In a real fax, this means the answering fax sends its capabilities in HDLC (High-Level Data-Link Control)-encoded data frames: a Digital Identification Signal (DIS) frame spouts off the standard feature set (defined by the TSS, or ITU the fax has). An NSF (Non-Standard Facilities) frame about what vendor-specific features the fax has comes next; a CSI (Called Subscriber Identification) frame gives the calling fax's telephone number. The sending fax responds with its Digital Command Signal (DCS) frames, informing the answering fax of modem speed, image width, image encoding and page length. The sender's phone number then comes across in a Transmitter Subscriber Information (TSI) frame, as well as a response to the answering fax's non-standard facilities frame.

die 1. The silicon block onto which circuits have been etched.

2. A small cube marked on each face with from one to six spots, used in games. In 18th century English gambling dens, there was an employee whose only job was to swallow the dice if there was a police raid.

DIEL Advisory committee on telecommunications for DIsabled and Elderly People (UK).

dielectric Not a conductor of DIrect ELECTRICal current. A non-conducting or insulating substance which resists passage of electric current, allowing electrostatic induction to act across it, as in the insulating medium between the plates of a condenser. Also an insulating material otherwise used (e.g. a Bakelite panel, or the cambric covering of a wire is a dielectric material). As glass is dielectric, and therefore immune to EMI (ElectroMagnetic Interference) and RFI (Radio Frequency Interference), optical fiber transmission systems offer virtually perfect error performance. See also Semiconductor.

dielectric absorption The penetration of a dielectric by the electric strain during a period of time.

dielectric breakdown Any change in the properties of a dielectric that causes it to be come conductive; normally a catastrophic failure of an insulation because of excessive voltage.

dielectric cable A nonconducting cable, such as a fiber cable, without metallic members. See Dielectric.

dielectric constant Represented by the symbol "K," the dielectric constant is the ratio of the capacity of a condenser with a given dielectric between the conductors (e.g., insulation between twisted pairs, or between the center conductor and the outer shield of a coaxial cable) to the capacity of the same condenser with air as the dielectric. Also known as Permitivity and Specific Inductive Capacity.

dielectric heating The heating of an insulating material when placed in a radio-frequency field, caused during the rapid polarization reversal of molecules in the material.

dielectric lens A lens made of dielectric material that refracts radio waves in the same manner that an optical lens refracts light waves.

dielectric process A printing process that uses a specially treated, charge-sensitive paper. Paper is roller-fed past an electrode array where an electrical charge is applied line-by-line to form a latent image, then passed through a toner. The toner adheres to the charged image and heat fuses the toner to the paper to create the printed document.

dielectric sheath or cable A sheath or cable that contains no electrically conducting materials such as metals. Dielectric cables are sometimes used in areas subject to high lightning or electro-magnetic interference. Synonym for nonmetallic cable.

dielectric strength The property of material which resists the passage of an electric current. It is measured in terms of voltage required to break down this resistance (such as volts per mil).

dielectric test A test in which a voltage higher than the rated voltage is applied for a specified time to determine the adequacy of the insulation under normal conditions.

differential backup A backup that copies files created or changed since the last normal or incremental backup. It does not mark files as having been backed up (in other words, the archive attribute is not cleared). If you are performing a combination of normal and differential backups, restoring files and folders requires that you have the last normal as well as the last differential backup.

differential encoding Digital encoding technique whereby a binary value is denoted by a signal change rather than a particular signal level.

differential gain A type of distortion in a video signal that causes the brightness information to be distorted.

differential manchester encoding A digital signaling technique in which there is a transition in the middle of each bit time to provide clocking. The encoding of a zero or one is represented by the presence (absence) of a transition at the beginning of the bit period.

differential mode For AC power systems, the term differential mode may refer to either noise or surge voltage disturbances. The terms normal mode and differential mode are interchangeable. Differential mode disturbances are those that occur between the power hot (black wire) and the neutral conductor (white wire). Most differential mode disturbances result from load switching within a building, with motor type loads being the biggest contributor. Surge voltages that come from outside of the building, such as surges caused by lightning, enter the building on the hot (black) wire and are therefore primarily differential mode in nature since the neutral (white) wire is nominally at ground voltage. Surge suppressors sometimes divert differential mode noise and surges into the neutral wire, resulting in voltages on the neutral wire called common modem noise or surge voltages. This definition courtesy APC.

differential mode transmission A transmission scheme where voltages appear equal in magnitude and opposite in phase across a twisted-pair with respect to ground. May also be referred to as balanced mode.

Differential Mode Delay See DMD.

differential mode termination A type of cable termination where a pair of wires is terminated by a resistance matching the cable impedance, but there is no termination resistance between that pair and any adjacent pairs. For low-frequency signals this is often acceptable, but for a high-frequency environment (whether due to high-speed network protocols, or due to transmission towers nearby), this allows large voltages to exist between one pair and an adjacent pair.

differential phase A type of distortion in a video signal that causes the color information to be distorted.

Differential Phase Shift Keying DPSK. Also called dibit phase shift keying. A modulation technique used to improve the efficiency with which the naturally analog electromagnetic waveform is employed to carry digital bits in a digital bitstream. DPSK is a form of "coherent demodulation," in which the phase of the incoming signal is compared to a replica of the carrier waveform. The carrier waveform (the carrier frequency "carries" the data, and the waveform is characteristic of all electromagnetic energy), is used as a reference point. With DPSK, the carrier waveform reference point serves to record changes in the binary data code. In other words, a "1" in the PSK (Phase Shift Keying) signal is denoted by no change in the DPSK signal, and a "0" is denoted by a change in the DPSK signal. DPSK works much better than PSK because so many things can foul up the "absolute" value of a signal sent over an Unshielded Twisted Pair (UTP) cable pair or over a microwave radio channel. ElectroMagnetic Interference (EMI) of all sorts can cause the "absolute" value of an originating signal to be "questionable" on the receiving end. Assuming some reasonable level of consistency in the impact of such factors from transmitter to receiver, it helps a lot to have a reference point. DPSK does that. See also Amplitude Modulation, Frequency Modulation, and Phase Shift Keying.

differential positioning Precise measurements of the relative positions of two receivers tracking the same GPS (Global Positioning System) signals.

Differential Pulse Code Modulation See DPCM.

differentiated services Diffserve, or DiffServe. A set of technologies proposed by the IETF (Internet Engineering Task Force) which would allow Internet and other IP-based network service providers to offer differentiated levels of service to individual customers and their information streams. On the basis of a DiffServe CodePoint (DSCP) marker in the header of each IP (Internet Protocol) packet, the network routers would apply differentiated grades of service to various packet streams, forwarding them according to different Per-Hop Behaviors (PHBs). In other words, for an additional charge, DiffServe would allow service providers to provide a certain user with a preferential Grade of Service (GoS) for all packet traffic with appropriate indicators in the packet headers. The preferential GoS, which can only be attempted and not guaranteed, would include a lower level of packet latency (delay), as those packets would advance to the head of a packet queue in a buffer should the network suffer congestion. RSVP (Resource ReserVation Protocol), a developing protocol, is an element of DiffServe. See also GoS, IETF, IP, Router, and RSVP.

Diffie-Hellman Key A technique of changing encryption keys on the fly. In a landmark 1976 paper, called New Directions in Cryptograph, IEEE Transactions on Information Theory, W. Diffie and M. Hellman describe a method by which a secret key can be exchanged using messages that do not need to be kept secret. This type of "public" key management provides a significant cost advantage by eliminating the need for a courier

service. In addition, security can be considerably enhanced by permitting more frequent key changes and eliminating the need for any individual to have access to the key's actual value.

diffraction The deviation of a wavefront from the path predicted by geometric optics when a wavefront is restricted by an opening or an edge. In other words, light (and other electromagnetic waveforms) bend as they pass through a narrow aperture (i.e., opening or slit). The extent to which they bend is sensitive to the wavelength of the signal. See also Index of Refraction and Refraction.

diffraction grating An array of fine, parallel, equally spaced reflecting or transmitting lines that mutually enhance the effects of diffraction at the edges of each so as to concentrate the diffracted light very close to a few directions depending on the spacing of the lines and the wavelength of the diffracted light. In other words, a diffraction grating uses to its advantage the interference of waves caused by the phenomenon of diffraction to separate wavelengths of light in an angular (i.e., directional) fashion. If you think this was fun, check out Index of Refraction. See also Interference.

diffuse The phenomenon when the molecules of two material in contact mix.

DiffServe See Differentiated Services.

digerati The elite of the Internet and the digital age. See Digiterati.

digest A collection of Internet mailing list posts collected together and sent out as a single large message rather than as a number of smaller messages. Using a digest is a good way to cut down on the number of noncritical e-mail messages you receive.

digicash A name for electronic money transmitted in and around the Internet.

digifeiter A made-up words that means a counterfeiter who uses digital technology to create forged money or documents.

digigroup 24 channels. See Digroup.

digirepeater Digital Repeater.

digiscents See Smell-O-Vision.

digit Any whole number from 0 to 9.

digit deletion It's nice to make it easy for people to dial their desired numbers. Part of making it "nice" is to keep their pattern of dialing consistent. The charm of our ten-digit numbering system in North America – the three digit area code and seven digit local number – is its consistency, making for easy use and easy remembering. Some corporate networks, however, don't use a common numbering scheme. They might use tie trunks to get to Chicago, and insist on the user dialing 69, instead of the more common 312 area code. They might insist on the user dialing 73 when he wants to go to Los Angeles. But if he wants to reach the LA office, he might dial 235. This can be awfully confusing. So some switches – central office and PBX – have the ability to insert or delete digits. That is, they will recognize the number dialed and change it as it progresses through the network. The user, however, knows nothing of this. He simply dials a normal phone number and listens as his call progresses normally. Digit insertion and digit deletion are components of a PBX feature called common number dialing.

digit grabber A digit grabber is a hand held piece of equipment manufactured by numerous companies. It is used to identify or "grab" the incoming digits. It will read to see if the digit came in with a good solid hit and what the digit is. It has two leads on it that are connected directly to the tip and ring of a connecting block.

digit insertion See Digit Deletion.

digital 1. In displays, the use of digits for direct readout.

2. In telecommunications, in recording or in computing, digital is the use of a binary code to represent information. See PCM (as in Pulse Code Modulation). Analog signals – like voice or music – are encoded digitally by sampling the voice or music analog signal many times a second and assigning a number to each sample. Recording or transmitting information digitally has two major benefits. First, the signal can be reproduced precisely. In a long telecommunications transmission circuit, the signal will progressively lose its strength and progressively pick up distortions, static and other electrical interference "noises."

In analog transmission, the signal, along with all the garbage it picked up, is simply amplified. In digital transmission, the signal is first regenerated. It's put through a little "Yes-No" question. Is this signal a "one" or a "zero?" The signal is reconstructed (i.e. squared off) to what it was identically. Then it is amplified and sent along its way. So digital transmission is much "cleaner" than analog transmission. The second major benefit of digital is that the electronic circuitry to handle digital is getting cheaper and more powerful. It's the stuff of computers. Analog transmission equipment doesn't lend itself to the technical breakthroughs of recent years in digital. See also PCM, as in Pulse Code Modulation.

3. In marketing promotion, digital is superior. For example, digital cable TV is better quality than old-time analog TV. The American public is fed this to sell them digital cable TV,

which usually costs more than old-time analog TV. Of course, sometimes digital is better. Sometimes it's not. It all depends on how you digitally encode the TV pictures and audio signal. Assign lots of bits. It will be better. Assign only a few bits. Digital will be awful. Don't be fooled. Compare the signals before you buy.

Digital Access And Cross-Connect System See DACS.

Digital AMPS See D-AMPS.

digital audio The storage and processing of audio signals digitally. At least 16 bits of linear coding are usually required to represent each digital sample.

Digital Audio Radio Service See DARS.

digital audio server A server is a computer that sits on a network, has software and a big hard disk. It's designed to deliver software, information or abilities to another machine also on the network. A digital audio server is a PC on a network – typically a home or office Ethernet LAN – loaded with digital recordings. The idea is that you "talk" to the digital audio server from your PC (by using a browser, for example) and have it play music in those parts of your house you wish. Software running on your digital audio server will allow to choose which music, in whatever order you wish, in whichever room you wish at whatever time you want.

Digital Audio Tape See DAT.

Digital Audio Video Council DAVIC. DAVIC is a European nonprofit organization that is creating an industry standard for end-to-end interoperability of broadcast and interactive digital audio-visual information and multimedia communication, and for the delivery of interactive data services to cable modems and set-top boxes.

digital cable set top box The primary function of a cable set top box - whether digital or analog - is to control access to TV channels provided by a cable TV carrier or a satellite TV carrier. Interestingly, the set top box is not totally necessary to get the channels from the cable onto the TV screen – but the carriers need them to make sure they can shut off your programming if you don't pay your bill. All boxes - analog or digital - are individually addressable. That means the carrier can signal the box and turn on or off channels - local channels, premium channels, movie channels, porn, etc. The carrier does this based on how much monies the subscriber (i.e. you) pay. Pay more money. Get more channels, etc. A digital cable set top box allows a television set to receive and decode digital television (DTV). The device may also enable the television set to become an interface to the Internet.

A typical digital set-top box contains one or more microprocessors for running the operating system, possibly Linux or Windows CE, and for parsing the MPEG transport stream. A set-top box also includes RAM, an MPEG decoder chip, and more chips for audio decoding and processing. The contents of a set-top box depend on the DTV standard used. European DVB-compliant set-top boxes contain parts to decode COFDM transmissions while ATSC-compliant set-top boxes contain parts to decode VSB transmissions. More sophisticated set-top boxes contain a hard drive for storing recorded television broadcasts, for downloaded software, and for other applications provided by DTV service providers. Set top boxes also provide advanced features to consumers such as enhanced program guides, more broadcast channels, personal video recording, in-home connectivity and access to customized content through "walled garden" and broadband Internet sites. Set top boxes also offer the ability for operators to provide customers with fee-based services such as on demand programming, movies, sports and other personalized programs.

digital cash Once there were only stores. To buy something you needed money. Then they invented checks. And storekeepers took them. Then they invented credit cards, which were sort of checks you paid later. Once we had credit cards, we could invent direct mail catalogs and 800 lines and call centers that took your orders via the 800 lines and you paid for what you bought by giving them your credit card number. Then came the Internet and vendors started to put catalogs on the Internet in the hope that somebody would buy from them. But they needed a way to get paid. Credit cards worked but many journalists wrote about how the Internet was "insecure" and anyone could steal your number and go on a spending binge. No one asked the journalists to cite instances of spending binges. Nor did they ask the journalists about the Federal Government legislation which limits credit card liability in the case of fraud to $50. But scaring people was a good story. Meantime, some entrepreneurs thought there was an opportunity to solve people's paranoia by creating "digital cash." No one exactly knows what digital cash is, yet. Lots of people are working on variations of it. But the idea is that it will be some form of encoded information transfer that contains instructions to take money from one person and pay it to another. We'll see how it evolves.

digital cellular The state of the art in cellular communications technology. Implementation will result in substantial increases in capacity (up to 15 times that of analog technology). In addition, digital will virtually eliminate three major problems encountered

by users of analog cellular: static, loss/interruption of signal when passing between cells (during handoff), and failure to get a connection because of congested relays. Specifications for TDMA digital systems have been developed in North America (D-AMPS), in Europe (GSM) and in Japan (PDC).

digital certificate A digital certificate is a password-protected, encrypted data file which includes the name and other data which serves to identify the transmitting entity. It's much like a passport, that unquestionably identifies a person to immigration authorities when crossing international borders. The certificate also includes a public key encryption mechanism, which serves to verify the digital signature of the sender; that key is signed with a matching private key, unique to the sender. Through the use of keys and certificates, the entities exchanging data can authenticate each other. A digital certificate is issued by a Certificate Authority (CA), a trusted third-party organization or company that is in business specifically to issue and manage digital certificates. See also Certificate Authority, Encryption, Private Key Encryption, and Public Key Encryption.

digital channel compression The process of fitting more than one program into single channel using digital techniques.

Digital Circuit Multiplication DCM is a variation of analog TASI – Time Assigned Speech Interpolation. In DCM, speech is encoded digitally and advanced voice band coding algorithms are applied to TASI's old speech interpolation techniques. DCM delivers a four to fivefold increase in the effective capacity of normal pulse code modulation (PCM) T-1 links operating at 1.544 megabits per second. DCM equipment is used on the TAT-8 transatlantic optical fiber submarine cable. Most DCM equipment has three operating elements: a speech activity detector, an assignment mapping and message unit, and a speech reconstitution unit.

digital coast The City of Los Angeles. In March of 1998, the mayor of Los Angeles, Richard J. Riordan, announced that Los Angeles would now be known as Digital Coast. It never stuck.

digital command signal Signal sent by a fax machine or card when the caller is transmitting, which tells the answerer how to receive the fax. Modem speed, image width, image encoding and page length are all included in this frame.

Digital Communications Manager DCM. An MCI monitoring system that maintains communications through the network with the Site Controllers, the Extended Super-frame Monitoring Units, and the I/O DXCs. The DCM issues requests for data and collects alarm and performance information, which is processed and stored in realtime for further computation and display.

Digital Compact Cassette DCC. A digital version of the familiar analog audio cassette. A DCC recorder can play and record both analog and digital cassettes. But the digital ones will sound a lot better.

digital compression An engineering technique for converting a cable television signal into a digital format (in which it can easily be stored and manipulated) which may then be processed so as to require a smaller portion of a spectrum for its transmission. It could allow many channels to be carried in the capacity currently needed for one signal.

digital convergence 1. A Microsoft term for getting all the digital devices of the office and the home together working in a seamless architecture. See At Work.

2. In the late 1990s, AT&T Corp. made a huge bet on "digital convergence." In 1998 and 1999, the long-distance phone giant poured $110 billion into a string of cable-television system acquisitions including MediaOne and Tele-Communications Inc. In the fall of 2001, AT&T's cable operations are on the auction block. AT&T no longer talks about digital convergence.

Digital Cross-Connect 1 DSX-1. Office twisted pair facilities for interconnecting 1.544 megabit digital streams.

Digital Cross-Connect 2 DSX-2. Office coaxial cable facilities for interconnecting 6.312 megabit digital streams.

Digital Cross-Connect 3 DSX-3. Office fiber optic facilities for interconnecting 44.736 megabit digital streams.

Digital Cross-Connect System DACS (Digital Automatic Cross-Connect System) or DCS (Digital Cross-Connect System). A specialized type of high-speed data channel switch. It differs from a normal voice switch, which switches transmission paths in response to dialing instructions. In a digital cross-connect system, you give it separate and specific instructions to connect this line to that. These instructions are given independently of any calls that might flow over the system. This contrasts with normal voice switching in which switching instructions and conversations go together. Commands to a digital cross-connect system can be given by an operator at a console or can be programmed to switch at certain times. For example, you might want to change the T-1 24-voice conversation circuit to

Chicago at 11 A.M. each day to allow for the president's 30 minute video conference call. See DACS.

digital dial-up bandwidth Digital dial-up bandwidth is communications channels created by signaling to the network from the caller's site the intended destination of the connection. These channels may be terminated when the caller or called party chooses. The user pays for the bandwidth only when it is used. Digital Dial-Up Bandwidth operates in a fashion similar to the dialed voice telephone network, but the resultant connections are digital and of specified bandwidths.

digital divide The poor people are at a big disadvantage to the rich people who have access to all the trappings of the new digital economy. Thus there is a digital divide. President Clinton used the term to describe the fact that households in the United States with annual incomes above $75,000 are more than 20 times as likely to have Internet access as the poorest households. In other words, according to President Clinton, there's a "digital divide" between rich and poor households. The Economist magazine wrote the following, "With information technology now claimed to be the main engine of growth over the next couple of decades, many people worry that developing economics, which have far fewer computers and Internet connections than the rich world, will get left behind. The income gap between rich and poor countries will widen further. But such fears about a 'digital divide'", according to the Economist, "seem to be based on a misunderstanding of the nature of growth as well as the nature of IT (Information Technology). If IT can boost growth in rich economies, why should it not do the same trick in emerging economies?" See also ICT.

digital echo canceller A digital echo canceller is an echo canceller as opposed to an echo suppressor. While the echo suppressors shuts off the entire signal, an echo canceller filters out unwanted echoes among incoming signals by using an analog voice switch. The Digital Echo Canceller is one application of a digital transversal filter.

digital enveloping Digital enveloping is an application in which someone "seals" a message m in such a way that no one other than the intended recipient, say "Bob," can "open" the sealed message. The typical implementation of digital enveloping involves a secret-key algorithm for encrypting the message (i.e., a content-encryption algorithm) and a public-key algorithm for encrypting the secret key (i.e., a key-encryption algorithm).

Digital Ethernet Local Network Interconnect DELNI. The product offered by Digital Equipment Corp. (now part of Compaq) that allows up to eight active devices to be connected to a single Ethernet transceiver. A similar device is manufactured by many other suppliers under various names. The DELNI can be thought of as "Ethernet in a box."

Digital European Cordless Telecommunication See DECT.

digital facilities management system A Northern Telecom software which integrates the maintenance of all types of digital facilities from T-1 to the high-bit fiber. Largely used by telephone companies.

digital facility A switching or transmission facility designed for the handling of data signals.

digital facsimile equipment Facsimile equipment that employs digital techniques to encode the image detected by the scanner. The output signal may be either digital or analog. Examples of digital facsimile equipment are ITU-T Group 3, ITU-T Group 4, STANAG 5000 Type I and STANAG 5000 Type II.

digital frequency modulation The transmission of digital data by frequency modulation of a carrier, as in binary frequency-shift keying.

Digital Group Interface See Digroup interface.

digital hierarchy The standardized increments for multiplexing digital channels. For twenty four 64 Kbps DS-O channels are multiplexed into one 1,544,000 bits per second DS-1 channel. See also SONET.

digital home The home of the future where household appliances, entertainment systems, computers, and phone systems are networked, Internet-connected and all controlled from one place, now called the home media center. Intel is pushing this idea. See www.intel.com/technology/digitalhome/.

digital immigrant See digital native.

digital ink We'll download our "books" from the Internet. We'll read the books on a handheld device the size of a PDA. Digital ink is the technology used for displaying the book's letters and words on a screen. As I wrote this, a small company called E Ink has created a method for arranging tiny black and white capsules into words and images with an electronic charge. Because no power is used unless the reader changes the page, devices with the technology could go as long as 20 books between battery charges. The text (created with digital ink) also looks just as sharp as ink on a printed page.

digital inverter A device which outputs a high state when its input is low or outputs a low state if its input is high.

Digital LAT Protocol The LAT protocol, announced by Digital in the mid '80s, is today one of the industry's most widely used protocols for supporting character terminals over Ethernet networks. See LAT.

digital line protection Many extensions behind PBXs deliver greater voltage to the desk than do normal tip and ring analog lines. This higher voltage can damage a PCMCIA modem inside a laptop. In fact, it can destroy the modem. Newer PCMCIA cards (now called PC cards) have Digital Line Protection, which protects against that higher voltage – what one manufacturer called "innovative isolation circuitry."

Digital Loop Carrier See DLC.

digital loopback A diagnostic feature on a modem, a short haul microwave or some other digital transmission equipment which allows the user to loop a signal back from one part of the system to another to test the circuit or the equipment. Digital loopbacks can be as long or as short as are necessary to isolate the problem. By looping a signal back and measuring it at both ends of the loop (at the beginning and at the end), you can see if the device carried the message cleanly and is thus, operating correctly.

digital media Digital media is a fancy, made-up term for technology that can manage, store, protect and distribute digital video, audio and images. Example: When the Wall Street Journal starts sending out an electronic version of itself that looks like its print edition, that's referred to as "digital media."

digital microwave A microwave system in which the modulation of the radio frequency carrier is digital. The carrier is still a standard microwave radio wave. The digital modulation may be frequency or phase shift, but the control of that modulation is the digital bit stream.

Digital Millennium Act Copyright Act This Act makes it a crime to circumvent antipiracy measures that are built into most commercial software. It outlaws the manufacture, sale or distribution or code-cracking devices used to copy software. It does permit the cracking of copyright protection devices, however, when necessary to conduct encryption research, assess product interoperability and test computer systems. It provides exemptions from anticircumvention provisions under certain circumstances for nonprofit organizations such as libraries, archives and educational institutions. In general, it limits Internet service providers from copyright infringement liability for simply transmitting information over the Internet. Service providers, however, are expected to remove material from users' Web sites that appears to constitute copyright infringement. Limits liability of nonprofit institutions of higher education - when they serve as online service providers and under other certain circumstances - for copyright infringement by faculty members or graduate students. It requires webcasters to pay licensing fees to record companies. It requires that the Register of Copyrights, after consultation with relevant parties to submit to Congress recommendations regarding how to promote distance education through digital technologies while "maintaining an appropriate balance between the rights of copyright owners and the needs of users." It states that "[nothing] in this section shall affect rights, remedies, limitations or defenses to copyright infringement, including fair use..."

digital modem A digital modem is the term given often to a piece of equipment that joins a digital phone line to a piece of communicating equipment, which may be a phone or a PC. Such equipment allows testing, conditioning, timing, interfacing, etc. But it does not do what a modem does – namely convert digital signals from machines into analog signals which can be carried on analog phone lines (like the ones that come into your house) and vice versa. The term digital modem, thus, is illogical. But we're stuck with it. Things do, however, seem to be changing. The first use of the term digital modem was to describe a piece of hardware that supported ISDN lines. Later it became known as an ISDN terminal adapter.

ISDN TAs are used to connect equipment which is not ISDN-compatible to a digital ISDN BRI (Basic Rate Interface) line in a residential or SOHO (Small Office Home Office) application. ISDN telephone sets are much more expensive than the typical analog type, so most of use don't even consider buying them. ISDN-compatible PCs are somewhat more expensive, as an additional chip set must be added, so most of us don't even consider buying them. Group III fax machines (the type most of us have) are digital machines with a fax modem so that they can work over inexpensive analog lines, so they are not ISDN-compatible, by definition. So, you need a digital modem in the form of a TA in order to effect compatibility.

A digital modem interfaces these devices to the ISDN local loop, and supports things like testing, conditioning, and circuit timing. But it does not do what a conventional modem does – namely convert digital signals from machines into analog signals which can be carried on analog phone lines (like the ones that come into your house) and vice versa. Rather, a digital modem modulates (changes) and demodulates (changes back) the signals from these devices so that they can be carried over a digital ISDN circuit. Specifically, a TA is used to multiplex as many as three transmissions so that they can be carried over a single twisted pair local loop in the two B (Bearer) channels of 64 Kbps each, and the single D (Data, or Delta) channel of 16 Kbps. The TA modulates and demodulates the carrier frequency (the frequency range which carries the data streams) using a technique known as 2B1Q (2 Binary, 1 Quaternary). See also 2B1Q, BRI, ISDN, Modem and TA.

digital modulation A method of decoding information for transmission. Information, e.g. a voice conversation, is turned into a series of digital bits - the 0s and 1s of computer binary language. At the receiving end, the information is reconverted into its analog form.

digital monitor Receives discrete binary signals at two levels; one level corresponds to Logic 1 (true) while the other corresponds to Logic 0 (false). Monitors generally were of this type before VGA models appeared. Digital monitors do not have as wide a range of color choices as analog types; digital EGA monitors, for example, can display just 16 colors out of a palette of 64.

Digital Multimedia Broadcasting DMB. One of several competing standards for sending digital data, TV and radio to handheld devices such as mobile phones. DMB comes in terrestrial (T-DMB) and satellite (S-DMB) versions. DMB is being used commercially or in trials in South Korea, China, Hong Kong, Singapore, Canada, France, and Germany.

digital multiplex hierarchy An ordered scheme for the combining of digital signals by the repeated application of digital multiplexing. Digital multiplexing schemes may be implemented in many different configurations depending upon the number of channels desired, the signaling system to be used, and the bit rate allowed by the communication medium. Some currently available multiplexers have been designated as D1-, DS-, or M-series, all of which operate at T-carrier rates. Extreme care must be exercised when selecting equipment for a specific system to ensure interoperability, because there are incompatibilities among manufacturers' designs (and various nations' standards).

digital multiplexed interface A ISDN PRI-like connection between a PBX and a computer, developed by AT&T.

digital multiplexer A device for combining digital signals. Usually implemented by interleaving bits, in rotation, from several digital bit streams either with or without the addition of extra framing, control, or error detection bits. In short, equipment that combines by time division multiplexing several signals into a single composite digital signal.

digital native In a 2005 speech, Rupert Murdoch called his two Internet-savvy daughters digital natives and himself a digital immigrant – i.e. a person who's come to the Internet late in life.

digital nervous system Coined by Bill Gates in 1997, the best definition of this term came from an interview between Gary Reiner, GE's chief information officer and a reporter from the Economist. According to the magazine, "Mr Reiner heads the company's most important initiative: 'digitising' as much of its business as possible. That not only means buying and selling most things online but, more importantly, setting up a digital nervous system that connects in real time anything and everything involved in the company's business: IT (Information Technology) systems, factories and employees, as well as suppliers, customers and products.

digital network A network in which the information is encoded as a series of ones and zeros rather than as a continuously varying wave – as in traditional analog networks. Digital networks have several major pluses over analog ones. First, they're "cleaner." They have far less noise, static, etc. Second, they're easier to monitor because you can measure them more easily. Third, you can typically pump more digital information down a communications line than you can analog information.

Digital Network Architecture. DNA. The data network architecture of Digital Equipment Corporation (DEC), now part of Compaq Corporation.

digital phase-locked loop A phase-locked loop in which the reference signal, the controlled signal, or the controlling signal, or any combination of these, is in digital form.

digital phase modulation The process whereby the instantaneous phase of the modulated wave is shifted between a set of predetermined discrete values in accordance with the significant conditions of the modulating digital signal.

digital plastic A fancy term for buying goods and services on-line over the Internet using your credit card, possibly in conjunction with some verification of who you are from an independent certification authority.

digital plumbers Service geeks specially trained to set up, maintain, and repair digital home network "pipes" that now connect everything from PCs to TVs to alarm systems and coffeemakers. One hopes they earn as much as old-fashioned water plumbers.

Digital Port Adapter DPA. A device which provides conversion from the RS-449/422 interface to the more common interfaces of RS-232-C, V.35, WE-306 and others.

Digital Private Network Signaling System See DPNSS.

Digital Pulse Origination DPO. Equipment that sends dialed digits consisting of tones or pulses. It may be used at the central office end of a DID service connection.

Digital Pulse Termination DPT. Equipment that receives and processes dialed digits consisting of tones or pulses. It may be used at the customer end of a DID service connection.

Digital Radio Radio

Digital Radio Broadcasting DRB. Radio transmission intended for general reception in the form of discrete, integral values.

Digital Radio Concentrator System DRCS. A digital radio system which transmits data via a device which connects a number of circuits, which are not all used at once, to a smaller group of circuits for economy.

digital recording A system of recording by conversion of musical information into a series of pulses that are translated into a binary code intelligible to computer circuits and stored on magnetic tape or magnetic discs. Also called PCM - Pulse Code Modulation.

Digital Reference Signal DRS. A digital reference signal is a sequence of bits that represents a 1004-Hz to 1020-Hz signal.

Digital rights Management DRM. Basically any technology used to protect the interests of owners of content (music, film, video, etc.) and services. Typically, authorized recipients or users must acquire a license to view or listen to the protected material, according to the rules set by the content owner. DRM's goal is to prevent illegal distribution of paid content over the Internet. DRM products were developed in response to the rapid increase in online piracy of commercially marketed material, which exploded through the widespread use of Napster and other peer-to-peer file exchange programs. Apple uses a form of DRM in its iTunes music sales over the Internet. See Fairplay.

Digital Selective Calling DSC. A synchronous system developed by the International Radio Consultative Committee (CCIR), used to establish contact with a station or group of stations automatically by radio. The operational and technical characteristics of this system are contained in CCIR Recommendation 493.

digital sequence spread spectrum A wireless term. An RF (radio frequency) modulation technique, which uses algorithms to code transmissions in sequential channels and then decode them at the other end.

Digital Service Cross-Connect DSX. A termination/patch panel that lets DS1 and DS3 circuits be monitored by test equipment. R

digital set-top box A device that hooks up to a TV and can collect, store, and display digitally compressed TV signals. See also Digital Cable Set Top Box.

digital signage A fancy name for flat-panel displays that show a constantly rotating series of advertisements, mixed with news and entertainment. The screens are placed in "high traffic" (in other words, busy) parts of supermarkets, and the advertisements they show can be updated at will via satellite, by Internet or by personnel directly in the store. This makes it possible to vary the advertisements shown depending on the time of day, season, or local factors such as demographics, weather or an item on sale (because of overstocking, for example).

digital signal A discontinuous signal. One whose state consists of discrete elements, representing very specific information. When viewed on an oscilloscope, a digital signal is "squared." This compares with an analog signal which typically looks more like a sine wave, i.e. curvy. Usually amplitude is represented at discrete time intervals with a digital value.

Digital Signal Cross-Connect DSX. Also known variously as a DACS (Digital Access Cross-Connect System) and a DCC (Digital Cross-Connect), a DSX is a device that is used to connect digital circuits together. A DSX-1 interconnects DS-1 (T-1 or E-1) circuits, as DSX-2 interconnects DS-2 (T-2 or E-2) circuits, and a DSX-3 interconnects DS-3 circuits (T-3 or E-3).

Digital Signal Level DS-n. A hierarchical arrangement of digital signals used in North America beginning with DS-0 (64 Kbps) up to DS-4 (274 Mbps).

digital signal processor A digital signal processor is a specialized semiconductor device or specialized core in a semiconductor device that processes very efficiently and in real time a stream of digital data that is sampled from analog signals ranging from voice, audio and video and from cellular and wireless to radio and television. As opposed to a general-purpose processor, a DSP is often designed to solve specific processing problems. A DSP architecture focuses on algorithmic efficiency and may use an instruction set that is more or less tailored toward the problem the DSP is solving. General purpose processors, on the other hand, may sacrifice algorithmic efficiency for general-purpose capability and push clock-speed to achieve performance. A DSP typically has much greater mathematical computational abilities than a standard microprocessor. In some applications, like wireless, PDAs and cell phones, constraints on power consumption require performance improvements other than faster clock speed. In other applications, like cellular base stations and high definition TV, where the number of channels or the high data rate require signal processing capabilities an order of magnitude greater than general purpose processors, a DSP that uses processing parallelism can provide much higher performance much more efficiently than even the fastest general-purpose processor. A DSP often performs calculations on digitized signals that were originally analog (e.g. voice or video) and then sends the results on. There are two main advantages of DSPs – first, they have powerful mathematical computational abilities, more than normal computer microprocessors. DSPs need to have heavy mathematical computation skills because manipulating analog signals requires it. The second advantage of a DSP lies in the programmability of digital microprocessors. Just as digital microprocessors have operating systems, so DSPs have their very own operating systems. DSPs are used extensively in telecommunications for tasks such as echo cancellation, call progress monitoring, voice processing and for the compression of voice and video signals as well as new telecommunications applications such as wireless LANs and next-generation cellular data and cellular Internet services . They are also used in devices from fetal monitors, to anti-skid brakes, seismic and vibration sensing gadgets, super-sensitive hearing aids, multimedia presentations and desktop fax machines. DSPs are replacing the dedicated chipsets in modems and fax machines with programmable modules – which, from one minute to another, can become a fax machine, a modem, a teleconferencing device, an answering machine, a voice digitizer and device to store voice on a hard disk, to a proprietary electronic phone. DSP chips and DSP cores in custom chips are already doing for the telecom industry what the general purpose microprocessor (e.g. Intel's Pentium) did for the personal computer industry. DSP chips are made by Analog Devices, AT&T, Motorola, NEC and Texas Instruments, among others. DSP cores are made by BOPS, DSP Group, Infineon and others.

digital signature A digital signature is the network equivalent of signing a message so that you cannot deny that you sent it and that the recipient knows it must have come from you. In short, a digital signature is an electronic signature which cannot be forged. It verifies that the document originated from the individual whose signature is attached to it and that it has not been altered since it was signed. There are two types of digital signatures. Ones you encrypt yourself and are the result of an ongoing relationship between you and the other party. Second, there are encrypted certificates issued by a company that is not affiliated with you. That company basically certifies that you are who you say you are. It does this because it's sent you a code. And it has retained a code for you, too. Join the two codes together mathematically, come up with the correct answer, and bingo, it's you. Utah has a Digital Signature Program whose goal is to develop, implement and manage a reliable means of secure electronic messaging over open, unsecured computer networks, minimize the incidence of forged digital signatures and possible fraud in electronic commerce and establish standards and develop uniform rules regarding verification and reliability of electronic messages. According to Utah, "digital signatures will enable us to determine who sent a document, identify what document was sent, and determine whether the document had been altered in route. It reasonably ensures the recipient that the message came from an identifiable sender and contains a specific, unaltered message. It may be used where there needs to be sufficient confidence in the source, content and integrity of a message." www.commerce.state.ut.us/web/commerce/digsig/domain.htm.

Digital Sound Broadcasting DSB. Sound Transmission intended for general reception in the form of discrete, integral values.

Digital Speech Interpolation DSI. A voice compression technique implemented in voice codecs. DSI allocates the silent periods in human speech to active users. At least 50% of bandwidth in a voice conversation generally is unused, as the human-to-human communications protocol is half-duplex. In other words, we take turns talking. Additionally, the voice stream of the active talker contains lots of periods of silence, since we pause from time-to-time, if only to take a breath. DSI statistically predicts these periods of silence, and takes advantage of them to interleave the voice transmissions of active talkers. Assuming that as many as 96 voice conversations are being supported, DSI offers a compression ratio as high as 8:1 (eight to one), supporting reasonable quality voice at 8 Kbps rather than the 64 Kbps required by Pulse Code Modulation (PCM). The predecessor to this technique was

TASI (Time Assigned Speech Interpolation). TASI and DSI are lousy for data because they "clip" the first little bit of every new snippet of conversation – unless you hog the channel the whole time by talking incessantly or transmitting continuously. If you pause, you'll get clipping as the system drops you and then reconnects you. Clipping can ruin the meaning of the beginning of data conversations, unless the header knows that TASI or DSI is coming up or the data transmission is following some reasonable protocol and can resend the data. Unfortunately, this slows transmission. See also Codec and PCM.

Digital Storage Media Command And Control DSMCC. Network protocols specified in Part 6 of MPEG-2 (ISO 13818) standards dealing with user-to-network and user-to-user signaling and communications.

Digital Subscriber Line DSL. A generic name for a family of digital lines (also called xDSL) being provided by CLECs and local telephone companies to their local subscribers. Such services go by different names and acronyms – ADSL (Asymmetric Digital Subscriber Line), HDSL (High Bit Rate Digital Subscriber Line) and SDSL (Single Pair Symmetrical Services). Such services propose to give the subscriber up to eight million bits per second one way, downstream to the customer and somewhat fewer bits per second upstream to the phone company. DSL lines typically operate on one pair of wires – like a normal analog phone line. See ADSL, G.990, G-Lite, HDSL, IDSL, RADSL, SDSL, Splitter, Splitterless, VDSL, xDSL, for more detailed explanations.

digital switching A connection in which binary encoded information is routed between an input and an output port by means of time division multiplexing rather than by a dedicated circuit.

digital telephony A digital telephone system transmits specific voltage values of "1" and "0" to transmit information. An analog system uses a continuous signal that uses the entire range of voltages. The human voice is an analog signal. To transmit it digitally, it must be converted into a digital signal. This is accomplished by sampling the value of the analog signal several times a second and converting each value into an binary number. These binary numbers are transmitted as a series of voltage levels representing ones and zeros. See PCM.

Digital Television See DTV.

Digital Test Unit See DTU.

digital timestamping When transmitting time-sensitive information it is essential that you be able to irrefutably document when an event occurred. A "postmark" is a legally valid physical timestamp. In e-commerce, there need to be "trusted" third parties that will timestamp sensitive documents. One drawback to having sensitive documents timestamped is relinquishing the document itself to a third party. This is where message digests come in handy.

digital to analog conversion A circuit that accepts digital signals and converts them into analog signals. A modem typically has such a circuit. It also has other circuits, such as those concerning with signaling. See Modem.

Digital Traffic Channel DTC. A digital cellular term. Defined in IS-136, the DTC is the portion of the air interface which carries the actual data transmitted. The DTC operates over frequencies separate from the DCCH (Digital Control CHannel), which is used for signaling and control purposes. See also DCCH and IS-136.

digital transmission The transmission of a digital signal between two or more points. The unusual definition applies to the manner in which the transmission carrier is modified to carry the transmitted information. For example, in digital microwave systems, the radio frequency carrier is an analog signal, but its information modulation is derived from a digital signal.

Digital Versatile Video See DVD.

digital video Digital video is video recorded and played digitally. Traditional analog video – the one we have seen in our homes for eons – is recorded and played back in analog format, i.e. in analog wave forms. Why Digital Video? Digital video has several advantages over analog video. You can edit it, store it, and transmit it easily. Digital video may be taken from analog source – such as standard over-the-air National Television Systems Committee (NTSC) analog source. Or it may be taken from an analog video camera and a VCR. To convert analog video into digital video typically requires a board inside a PC. Analog video is typically recorded on tape, such as a VCR. Digital video is typically recorded on a hard disk (magnetic or optical) or on a CD-ROM. Two common digital video technologies are Intel's Digital Video Interactive (DVI) and Microsoft's Audio/Video Interleaved (AVI). See Analog, Digital Video Interactive, INDEO Video and PVR.

Digital Video Broadcast DVB. Digital Video Broadcast and Digital Audio Video Council (DAVIC) technology is the incumbent European standard for digital set-tops and is now starting to gain momentum for cable modems. The standard describes the out-of-band

and in-band transmission options applicable to interactive set-top boxes and cable modems. respectively, enabling the deployment of interactive TV, data, and voice services over a common platform. DVB/DAVIC is challenging DOCSIS for dominance in Europe. EuroCableLabs (ECL), operating under the direction of the European Cable Communications Association (ECCA), has championed a DVB/DAVIC-based "EuroModem" as an alternative to DOCSIS. DVB cable modems meet the preference of some European operators for a standard that better fits their set-top architectures. There is also a desire to support homegrown Euro products, rather than importing solutions from American suppliers. See also DAVIC and DOCSIS.

Digital Video Compression. DVC. Process by which multiple videoservices, or channels, are broadcast on one satellite transponder. Performed by turning analog audio and video into digital (computer) information for broadcasting.

Digital Video Interactive DVI. A compression and playback technology originally developed by RCA's Sarnoff Research Institute and eventually acquired by Intel Corp. DVI is not a compression technique per se but a brand name for a set of processor chips that Intel is developing to compress video onto disk and to de-compress it for playback in real time at the U.S. standard motion video rate of 30 frames per second. The chip set includes both a pixel processor, which performs most of the decompression and also handles special video effects, and a display processor, which performs the rest of the decompression and produces the video output. DVI's greatest long-term advantage, according to Nick Arnett writing in PC Magazine, is that its microprocessors are programmable, so DVI can be adapted to a variety of compression and decompression schemes. See also DVI for a different meaning of DVI.

digital video interface See DVI.

digital video recorder DVR. Also called a Personal Video Recorder. See PVR and TiVo.

Digital Visual Interface DVI. A secure connection between a television and an external set-top box which uses encryption to prevent piracy.

digital voice coding Technology by which linear audio (voice) samples are collected and then compressed using an encoding algorithm. Typically used to store voice data for future decoding.

digital wallet A digital wallet is an application or service that assists consumers in conducting online transactions by allowing them to store billing, shipping, payment, and preference information and to use this information to automatically complete merchant forms. This greatly simplifies the check-out process and minimizes the need for a consumer to think about and complete a merchant's form every time. Digital wallets that fill forms have been successfully built into browsers, as proxy servers, as helper applications to browsers, as stand-alone applications, as browser plug-ins (including the Google Toolbar), and as server-based applications. But the proliferation of electronic wallets has been hampered by the lack of standards. ECML (Electronic Commerce Modeling Language) provides a set of simple guidelines for web merchants that will enable electronic wallets from multiple vendors to fill in their web forms. The end-result is that more consumers will find shopping on the web to be easy and compelling. (That's the theory.) Multiple wallets and multiple merchants interoperably support ECML, which is an open standard. The theory is that a merchant can adopt ECML and gain the support of these multiple Wallets by making small and simple changes to their web site. Use of ECML requires no license. See also ECML, www.ecml.org and www.ietf.org/rfc/rfc3106.txt.

digital wireless standards This chart shows in convincing terms why you can't take most U.S. cell phones overseas and expect them to work. They simply work on different protocols and usually on different radio frequencies – not shown here. Of course, there are phones that work in different places, but they usually have the equivalent of two or three phones built into them – in effect obeying several standards and working on different frequencies. Such phones usually cost more.

digital wrapper The developing ITU-T G.709 specification, "Interface for the optical transport network (OTN)," is for a digital wrapper as a means of providing network management capabilities in a DWDM (Dense Wavelength Division Multiplexing) environment. DWDM is replacing SONET (Synchronous Optical Network) and SDH (Synchronous Digital Hierarchy) in many networks due to its much higher speeds and much lower costs. DWDM, however, is pure photonics, without the robust network management capabilities offered by SONET. The digital wrapper in G.709 is intended to address that shortcoming. While the specification currently is in the development stages, it likely will include framing conventions, non-intrusive performance monitoring, error control, rate adaption, multiplexing mechanisms, and ring protection, and network restoration on a per-wavelength basis. Indicators of traffic type and destination will assist in optical switching processes. In total,

Standards	Japan	U.S.	Europe
Cordless	PHS	PACS	DECT
Cellular	PDS	IS54, IS136, IS95	GSM
PCS	PHS	IS136, IS95	DCS 1800 (GSM)
		PCS1900 (GSM)	
		PACS	
		Omnipoint/T-Mobile	
		BB CDMA	
Data	PDC	CDPD (AMPS)	GSM
	RAM	RAM, ARDIS	RAM
	PHS	IS136, IS95	DCS1800
		PACS	HyperLAN
		Omnipoint/T-Mobile	DECT
	802.11	802.11 (Wireless LAN)	802.11
		BBCDMA	

G.709 is intended to provide much of the network management capabilities of SONET/SDH, but without being nearly so overhead-intensive. See also DWDM, SDH, and SONET.

digiterati The digital version of literati. That vague collection of people who seem to be hip and know something about the digital revolution. Also spelled digerati.

digitize Converting an analog or continuous signal into a series of ones and zeros, i.e. into a digital format. See Delta Modulation and Pulse Code Modulation. See also the Appendix.

digitized voice Analog voice signals represented in digital form. There are many ways of digitizing voice. See Pulse Code Modulation for the most common.

digitizer 1. A device that converts an analog signal into a digital representation of that signal. Usually implemented by sampling the analog signal at a regular rate and encoding each sample into a numeric representation of the amplitude value of the sample.

2. A device that converts the position of a point on a surface into digital coordinate data.

digroup Two groups of 12 digital channels combined to form one single 24-channel T-1 system.

digroup interface Digital group interface. Another term for T-1. See also Digroup and T-1.

Dijkstra's Algorithm An algorithm that is sometimes used to calculate routes given a link and nodal state topology database.

dikes A wire-cutter.

DIL Direct-In-Line. A standard, both-way central office trunk that is programmed to always ring a specific extension or hunt group within the PBX. This contrasts with Direct Inward Dialing, which allows an external caller to reach an internal extension by dialing a 7-digit central office number. A DID (Direct Inward Dial) trunk is a trunk from the central office which passes the last two to four digits of the Listed Directory Number into the PBX, thus allowing the PBX to switch the call to and thus ring the correct extension.

dilution. See Anti-Dilution.

dilution of precision The multiplicative factor that modifies the ranging error. It is caused solely by the geometry between the user and his set of GPS (Global Positioning System) satellites. Known as DOP or GDOP. See GPS.

DIM Document Image Management. The electronic access to and manipulation of documents stored in image format, accomplished through the use of automated methods such as high-powered graphical workstations, sophisticated database management techniques and networking.

dim fiber A Dense Wavelength Division Multiplexing (DWDM) term. An optical fiber traditionally supported one light stream, using one frequency of light generated by the light source, which is in the form of a diode laser in the high-capacity, carrier-class systems. As it became necessary to increase the bandwidth of the transmission system, the traditional solution was to increase the speed of the components, including the diode laser light sources, the Avalanche PhotoDiode (APD) light detectors, and the intermediate optical repeaters. That was an expensive "forklift upgrade." Along came Wavelength Division Multiplexing (WDM) and it's more capable successor technology, DWDM (Dense WDM), both of which now are standardized by the ITU-T. DWDM currently supports as many as 160 wavelengths, each of which is a different frequency of light, which we would see as a different visible color in the light spectrum. The DWDM system essentially is a Frequency Division Multiplexer (FDM) operating at the optical frequency levels. The DWDM acts much like a prism, multiplexing the light frequencies created by tunable diode lasers at the transmit end of the connection to send them over the same optical fiber, and demultiplexing them at the receive end. Since the light signals are separated by wavelength (which is inversely related to frequency), they don't interfere with each other any more than do radio signals at different frequencies. Thereby and for example, one light wavelength can operate at 10 Gbps, and so can another, and so can another, and so on up to 32 x 10 Gbps = 320 Gbps, according to ITU-T standards. Up to 160 can be multiplexed at 10 Gbps on a non-standard basis... and more wavelengths are on the way. When carriers install new optical fiber transmission systems, they often install the SONET (Synchronous Optical NETwork) level of OC-192, at 10 Gbps (rounded up slightly). They also install DWDM gear, so that the system is ready for multiples of OC-192. But they usually leave some of the wavelengths "dark," because they don't need that much bandwidth immediately. They can always "light" them up when they need to do so. This is "dim fiber" – it's not completely dark and it's not fully lit. See also APD, Dark Fiber, Diode Laser, DWDM, FDM, Forklift Upgrade, Lit Fiber, SONET, and WDM.

dime See drop a dime.

dimension An analog PBX that used PAM techniques, first introduced in the late 1970s by Western Electric (now AT&T Technologies now Lucent) for AT&T. Now effectively discontinued. Thank the Lord. Some claim Archie McGill was responsible for Dimension. Others claim it was Bob Hawke. Archie claims Bob, who worked for him, was responsible. Arch later retired to Scottsdale and Vale, having made oodles of money in the wine business. In December of the year 2000, he proudly told me "at least he drank a lot."

DIMM Dual In-line Memory Module. A DIMM has a lot more bandwidth than a single in-line memory module (SIMM). It's a small circuit board filled with RAM chips, and its data path is 128 bits wide, making it up to 10 percent faster than a SIMM. DIMMs are prevalent on the Power Mac platform but are also creeping into high-performance systems.

dimmed In Windows, dimmed means unavailable, disabled, or grayed. A dimmed button or command is displayed in light gray instead of black, and it cannot be chosen.

DIMS Document Image Management System.

DIN Deutsche Institut fur Normung (German Institute for Standardization). DIN specifications are issued under the control of the German government. Some are used on a worldwide basis to specify, for example, the dimensions of cable connectors, often called DIN connectors. See also VME.

DINA Distributed Intelligence Network Architecture.

dingy A small useful gadget. You don't know its name or you can't remember its name. So you call it a dingy. A dingy is the same thing as a thingy, and or just plain "thing." But dingy is more endearing than thingy, and a lot more endearing than just plain thing.

DINK Dual Income, No Kids. A major market force in the general economy these days, many DINKs are old yuppies who just never quite got around to having children. I guess they were just dinking around. See also Yuppie.

DIOCES Distributed Interoperable and Operable Computing Environments and Systems.

diode Diodes, the simplest semiconductor devices, are devices that conduct electromagnetic energy (i.e., electricity or light) in one direction, only, much like a one-way valve. They are sometimes referred to as PN (Positive-Negative) devices because they are made of a single semiconductive crystal with a P-type region (Positive terminal, or anode) and a N-type region (Negative terminal, or cathode). The terms "diode" and "rectifier" often are used interchangeably. However, a diode typically is a small signal device with current in the milliamp range, while a rectifier is a power device conducting current from 1 to 1,000+ amps. Diodes also are distinguished from rectifiers, as diodes conduct electrons in only one direction; i.e., from anode (+) to cathode (-). In actuality, there is some level of current leakage in the reverse direction (think of it as "echo," or reflected electrical energy). Above the small average threshold current level, or "breakdown voltage," the diode shuts off, much like a light switch.

This wonderful explanation comes from George Gilder's book Microcosm: "Named from the Greek words meaning two roads, an ordinary diode is one of the simplest and most useful of tools. It is a tiny block of silicon made positive on one side and negative on the other. At each end it has a terminal or electrode (route for electrons). In the middle of the silicon block, the positive side meets the negative side in an electrically complex zone called a positive-negative, or p-n, junction. Because diode is positive on one side and negative on the other, it normally conducts current only in one direction. Thus diodes play an indispensable role as rectifiers. That is, they can take alternating current (AC) from your wall and convert it into direct current (DC) to run your computer.

"In this role, diodes demonstrate a prime law of electrons. Negatively charged, electrons flow only toward a positive voltage. They cannot flow back against the grain. Like

water pressure, which impels current only in the direction of the pressure, voltage impels electrical current only in the direction of the voltage. To attempt to run current against a voltage is a little like attaching a gushing hose to a running faucet.

"It had long been known, however, that if you apply a strong enough voltage against the grain of a diode, the p-n wall or junction will burst. Under this contrary pressure, or reverse bias, the diode will eventually suffer what is called avalanche breakdown. Negative electrons will overcome the p-n barrier by brute force of numbers and flood "uphill" from the positive side to the negative side. In erasable programmable read-only memories (EPROMs), this effect is used in programming computer chips used to store permanent software, such as the Microsoft operating system in your personal computer (MS-DOS). Avalanche breakdown is also used in Zener diodes to provide a stable source of voltage unaffected by changes in current." See also Zener Diode.

diode laser Synonymous with injection laser diode.

diode matrix ringing A method of connecting a common audible line to a system so that all stations do not ring on all lines. See also Matrix Ringing.

dip 1. The act of consulting a database for information. Much like dipping into a bucket of water to extract a drink, carrier switches must dip into centralized databases in order to access various types of information. The database is housed in one or more SCPs (Service Control Points), which are centralized in the networks in order that many switches can share access to them, generally via SS7 links. Dips are made into such databases in order to accomplish tasks like calling card verification. 800/800 number routing requires that a dip be made in order to determine the serving IXC or LEC. LNP (Local Number Portability), mandated by the Telecommunications Act of 1996, requires the deployment of SCPs in order that the call can be terminated by the serving LEC. See also LNP and Number Portability.

2. See DIP Switch.

3. Document Image Processing. A term for converting paperwork into electronic images manipulable by a computer. Components include input via scanner, storage on optical media and output via video display terminal, printer, fax, micrographics, etc.

4. See Dedicated Inside Plant.

DIP Switch Dual In-line Package. A teeny tiny switch usually attached to a printed circuit board. It may peek through an opening in a piece of equipment. It may not. It usually requires a ball point pen or small screwdriver to change. There are only two settings – on or off. Or 1 (which is on) or 0 (which is off). But printed circuit boards often have many DIP switches. They're used to configure the board in a semi-permanent way. The DIP switches are similar to integrated circuit chips which have two rows (dual) of pins in a row (in-line) that fit into holes on a printed circuit board. If something doesn't work when you first install it, check the DIP switches first. Then check the cable connecting it to something else. See also PGA, SIMM, and SIP.

DIP/DOP Dedicated Inside Plant/Dedicated Outside Plant.

diphones Speech segment beginning in the middle of one phoneme and concluding in the middle of another. See Phoneme.

diplexer A device that permits parallel feeding of one antenna from two transmitters at the same or different frequencies without the transmitters interfering with each other. Diplexers couple transmitter and receiver to the same antenna for use in mobile communications.

dipole A radio antenna fed from the center and across two rods which are in line with each other and which have the ends slightly separated. The term often is applied to "rabbit ear" antenna such as those used on TV sets. See also dBd.

DIR An ATM term. This is a field in an RM-cell which indicates the direction of the RM-cell with respect to the data flow with which it is associated. The source sets DIR=0 and the destination sets DIR=1.

Direct Access Test Unit DATU. Also called Mechanized Loop Test (MLT) added or built into a central office switch. With DATU a technician can execute tests for shorts, opens and grounds remotely. The technician gets a digital voice, enters a password and is given a series of options. The technician can get results as a digital recording or through an alpha-numeric pager. DATU units can send a locating tone as TIP, RING or a combination of both. The unit can short lines and remove battery voltage for testing.

direct bond An electrical connection using continuous metal-to-metal contact between the things being joined.

Direct Broadcast Satellite DBS. Direct broadcast satellite DBS refers to satellite television systems in which the subscribers receive signals directly from geostationary satellites via small and relatively inexpensive dish antennas typically mounted on either the roofs or sides of houses. A DBS subscriber installation consists of a dish antenna two to three feet (60 to 90 centimeters) in diameter, a conventional television set, a signal converter placed next to the television set, and a length of coaxial cable between the dish and the converter. The dish intercepts microwave signals directly from the satellite. The converter produces output that can be viewed on the television. The receiving dish is stationary, being locked in on the position of the DBS service provider with which you have a subscription agreement. As a result, it can receive only those channels broadcast by that specific provider. Telephone companies began early on using satellites for transmitting communications as terrestrial distribution methods became overloaded. The use of satellites for distribution of television programming in the U.S. began on March 1, 1978 when the Public Broadcasting Service (PBS) introduced the Public Television Satellite Service for the distribution of programming to local PBS stations. DBS satellites operate in the Ku-band spectrum, and at fairly high power levels; hence the small size of the receiving dish, which commonly is as small as 19.7 inches in diameter. DBS has virtually eliminated the old C-band satellite dishes – huge things about 3 meters (118.1 inches) across which you mounted in your back yard. The C-band systems were tunable, however; that is to say that they could be adjusted to pick up programming from just about any TV broadcast satellite which didn't encrypt its signals. Three important attributes of digital communication helped create a satellite mass market. First, digital compression allows huge programming capacity. Second, you can multiplex several digital channels onto a single carrier, and third, digital error correction allows a much lower power and a far more noise-tolerant signal over that of an analog transmission. Because you can quantify all information that you push through a digital channel, you can calculate the limits of the amount of information. In 1977, the World Administrative Radio Council allocated three regions for the DBS service: Region 1 for Europe, Region 2 (Western Hemisphere), and Region 3 (Asia/Pacific). The Council defined the standard 9˝ spacing between satellite clusters. The FCC set up Region 2 DBS satellites and transponders by granting licenses for the CONUS (Continental United States) for the following orbital locations: 61.5˝ West, 101˝ West, 110˝ West, 119˝ West, 148˝ West, 157˝ West, 166˝ West, and 175˝ West. The satellites at 101, 110, and 119˝ cover the full continental United States and are known as Full CONUS satellites, while the others are Partial CONUS. The two primary standards for DBS systems are DVB (digital video broadcasting), an "open," worldwide standard, and DirecTV, a "closed," North American standard. DVB-S (1994) specifies normal QPSK (quadrature-phase-shift-keying) modulation, and DVB-DSNG (1997) specifies advanced modulation for news and other data services (DVB is now creating a new DBS standard, DVB-S2, to use newer technological innovations in modulation and FEC.) The DirecTV standard (1994) is similar to DVB-S, but differences, such as data-packet size (188 bytes/packet for DirecTV versus 204 for DVB), and slight differences in the transport streams keep them from being functionally interchangeable. The ODU (outdoor unit) attaches to the outdoor end of a 75 ohm RF cable; the other end attaches to the satellite receiver at your TV. The parabolic dish antenna is the initial portion of the ODU. The main electronics of the ODU downconverts the signal from the satellite's 12 to 13 GHz to the receiver's 1 to 2 GHz, so you can use a convenient coaxial cable to carry the signal. See Shannon Limit and Digital Set Top Box.

direct builds A direct line to a customer from a phone company that is not the local ILEC. A "direct build" can be accomplished either via copper cable, an optical fiber or a wireless (e.g., LMDS or other microwave) technology. But it is not leased from the local phone company.

direct connect A term describing a customer hooking directly into a long-distance telephone company's switching office, bypassing the local phone company. Such "direct connect" could be via a leased copper pair, a specially-run copper pair, a fiber optic or a private microwave system. See Direct Electrical Connection, which is different. See Direct Connect Modem.

direct connect modem A modem connected to telephone lines using a modular plug or wired directly to the outside phone line. It thus transfers electrical signals directly to the phone network without any intermediary protective device. Direct connect modems must be certified by the FCC. Direct connect modems are much more reliable and more accurate than acoustically coupled modems. Virtually all modems these days are directly connected. One day pay phones will even come with RJ-11 jacks into which you can plug the modem of your portable laptop computer.

direct connection Connection of terminal equipment to the telephone network by means other than acoustic and/or inductive coupling.

direct control switching The switching path is set up directly through the network by dial pulses without the use of central control. The Telex network is an example of direct control switching. A step by step central office also uses direct control switching.

direct current DC. The flow of free electrons in one direction within an electrical conductor, such as a wire. The current may be constant or it may pulsate, but it always is in one direction. Contrast with alternating current (AC). See AC.

Direct Current Signaling DX. A method whereby the signaling circuit E & M leads use the same cable pair as the voice circuit and no filter is required to separate the control signals from the voice transmission.

Direct Department Calling DDC. A telephone service that routes incoming calls on a specific trunk or group of trunks to specific phones or groups of phones.

Direct Distance Dialing DDD. A telephone service which lets a user dial long distance calls directly to telephones outside the user's local service area without operator assistance.

direct electrical connection A metallic connection between two things. The normal electrical way of connecting two things. This dumb definition is included in this dictionary because there was a time back in the early 1970s and before, when you couldn't (i.e. weren't allowed to) directly electrically connect your own phone or phone system to the nation's phone network. Those were the "good old days" when they (the Bell System) were trying to convince the world that electrically connecting anyone else's phones could harm the network. They never did prove this, and so today we have direct electrical connection of FCC-certified phone equipment. It's certified so it won't cause any harm to the network. See Part 68.

Direct In Line See DIL.

direct in termination Incoming calls on a PBX may be programmed to route directly to preselected telephones without the attendant intervening. DIT features may be assigned to trunk circuits on a day, night or full time basis. Direct In Termination is slightly different from Direct Inward Dialing, though how different depends on whose PBX you're using.

direct interLATA connecting trunk groups Those trunk groups used for switched LATA access that interconnect Interexchange Carriers (IXCs) used to connect that Point of Presence (POPs) directly with the Bell Operating Company (BOC) end office switching system.

Direct Inward Dialing DID. The ability for a caller outside a company to call an internal extension without having to pass through an operator or attendant. In large PBX systems, the dialed digits are passed down the line from the CO (central office). The PBX then completes the call. Direct Inward Dialing is often proposed as Centrex's major feature. See also Direct Inward System Access (DISA) for another approach to DID.

Direct Inward System Access DISA. This feature of a telephone system allows an outside caller to dial directly into the telephone system and to access all the system's features and facilities. DISA is typically used for making long distance calls from home using the company's less expensive long distance lines, like WATS or tie lines. It's also used for leaving dictation for the typing pool. With DISA, you can dial individual extensions without the aid (or hindrance) of an operator. To use DISA, you must punch in from your touchtone phone a short string of numbers as a password code.

The problem with DISA is that "phone phreakers" (i.e. unauthorized people) often acquire that number or figure it out and run up expensive long distance phone calls. It's best to restrict DISA to trusted people and check the numbers called and bills generated. Changing the password code from time to time can help prevent this. DISA is acquiring a whole new life. It's becoming something called automated attendant. That automated attendant, an additional piece of equipment, is placed next to the phone system. You dial a special phone number (as you do with DISA). You're answered by a recording that says "Dial the extension you want." In DISA, the response is typically just a tone. An automated attendant is designed to save on operators and speed up outside people getting to talk to your inside people. Automated attendant is being suggested as a lower cost alternative to Centrex.

The following is excerpted from a document Northern Telecom sent to its PBX users. Read it. It's well-done:

PBX features that are vulnerable to unauthorized access include call forwarding, call prompting and call processing features. But the most common ways hackers enter a company's PBX is through Direct Inward System Access (DISA) and voice mail systems. They often search a company's trash for directories or call detail reports that contain 800 numbers and codes. They have also posed as systems administrators and conned employees into telling them PBX authorization codes. More "sophisticated" hackers use personal computers and modems to break into databases containing customer records showing phone numbers and voice mail access codes, or simply dial 800 numbers with the help of sequential number generators and computers until they find one code that gives access to a phone system. Once these thieves have the numbers and codes, they can call into the PBX and place calls out to other locations. In many cases, the PBX is only the first point of entry for such criminals. They can also use the PBX to access the company's data system. Call-sell operators can even hide their activities from law enforcement officials by using "PBX-looping" - using one PBX to place calls out through another switch in another state.

To minimize the vulnerability of the Meridian 1 system to unauthorized access through DISA, the following safeguards are suggested:

1) Assign restricted Class of Service, TGAR and NCOS to the DISA DN.

2) Require users to enter a security code upon reaching the DISA DN.

3) In addition to a security code, require users to enter an authorization code. The calling privileges provided will be associated with the specific authorization code.

4) Use Call Detail Recording (CDR) to identify calling activity associated with individual authorization codes. As a further precaution, you may choose to limit printed copies of these records.

5) Change security codes frequently.

6) Limit access to administration of authorization codes to a few, carefully selected employees.

direct line terminations The term refers to central office/PBX lines which terminate directly on telephones, and are generally common to all instruments within the system. In a square configuration on a Key Telephone System, these lines must appear at the same button location on each phone.

Direct Memory Access DMA is a technique in which an adapter bypasses a computer's CPU, and handles the transfer of data between itself and the system's memory directly.

direct modulation Direct Modulation occurs when a laser (the light source in a network) is turned on and off.

Direct Outward Dialing DOD. The ability to dial directly from an extension without having to go through an operator or attendant. In PBX and hybrid phone systems, you dial 9, listen for a dial tone, and then dial the number you want to reach. In some phone systems, you don't have to listen for the second dial tone. You can dial straight through. All phone systems now have DOD. The older ones didn't, especially cordboard PBXs. Some Club Meds and lots of cheap hotels (especially the ones Harry – the editor – stays in) do not have DOD.

Direct Pick-Up Interference DPI/DPU. Interference picked up on the shield of a coaxial cable connected to a set-top converter box.

direct routing There are two types of routing on a IP network, such as the Internet. There is direct and indirect Routing. If a packet does not need to be forwarded outside its subnet, i.e. both the source and destination addresses have the same network number, then direct routing is used. Indirect routing is used when the network numbers of the source and destination do not match. This is the case where the packet must be forwarded outside the subnet by a node that knows how to reach the destination (a router).

direct sales When you employ your own salesforce and sell to customers (i.e. without going through a wholesaler, reseller or OEM) you are said to have direct sales.

Direct Sequence Spread Spectrum DS. RF modulation technique that uses algorithms to code transmissions in sequential channels and then decode in the receiving end.

Direct Set An ATM term. A set of host interfaces which can establish direct layer two communications for unicast (not needed in MPOA).

Direct Show See TAPI 3.0.

Direct Station Select DSS. A piece of key system equipment usually attached to an operator's phone set. When the operator needs to call a particular extension he/she simply touches the corresponding button on the Direct Station Select equipment. Typically the DSS equipment/feature is part of a Busy Lamp Field (BLF), which shows with lights what's happening at each extension. Is it busy? Is it on hold? Is it ringing? See Busy Lamp Field.

Direct Station Select Intercom DSS. An interoffice caller can punch one button on his or her phone and dial his desired person, instead of dialing the full intercom number. Direct station select is like having an auto dial or speed dial button for everyone in the office. DSS saves time, but adds more buttons to the phone – one button for each extension the user wants to dial.

Direct Termination Overflow DTO. An optional MCI Vnet and 800 Service feature, which allows a call to "overflow" to shared lines for completion by the local telephone company if the dedicated line is busy.

direct to home Direct to home is a term used by the FCC to refer to the satellite TV and broadcasting industries.

direct trunk A trunk between two class 5 central offices.

direct trunk access A PBX feature. By dialing some digits, the attendant can directly access any specific trunk. You'd do this if you want to check the trunk for problems, etc.

direct trunk select Permits you, the user, the attendant, or an attached computer telephony system to access an individual outgoing trunk instead of one chosen by the PBX from a group of trunks. You may want to grab a special trunk to get access to a specially conditioned data line, for example. Direct trunk select is particularly important in testing computer telephony systems and facilities as it is usually the best method to use to address specific ports on a VRU (Voice Response Unit) or switching system.

directed call pickup A telephone system feature. An extension user on a phone system user can answer calls – ringing or holding – on any other phone by dialing a unique answer code. If the call has already been answered by the called phone, the user who dials the answer code will join the connection in conference. Some tones will alert the conversing parties to the intrusion.

directed pickup A PBX feature. Directed pickup is when you pick up a call ringing at another, specific, known extension. You dial the code for directed pickup and the other extension number and you now answer the ringing call at your own phone. Also called Directed Call Pickup. In contrast, Undirected Pickup picks up any call ringing at any extension in the pickup group in which your extension is a member. The pickup groups are pre-programmed in the switch.

directed push The server interacts with the push client only occasionally, providing directions (agents, modules, and so on) for how content should be handled or where content is located. The client then gets the information directly from a variety of services and processes it locally. Lanacom Headliner is a good example of this. See Active Push and Push.

directional antenna An antenna which impels electromagnetic waves (i.e., radio signals) with more energy in one direction than in another, or which receives electromagnetic waves more readily from one direction than from another. An omnidirectional antenna, on the other hand, transmits with equal force in all directions, or receives signals with equal ease from all directions.

directional boring A technique used to drill into the ground allowing precise guidance of the drilling operation to avoid pre-located underground obstacles and existing services, allowing the laying of conduit such as polyethylene, copper, and steel piping as well as fiber and bare cabling. This is especially useful in built-up environments, as it minimizes impact to surrounding grounds because it does not require an open trench to bury cable. Directional Boring allows the cable to be "steered" underground into the right location, accelerates construction by avoiding other utilities such as gas and telephone lines. At the end of the bore, a back reamer is attached to the drill rod and pulled back together with the materials to be installed.

directional coupler 1. A device put in a microwave system's waveguide to couple a transmitter and receiver to the same antenna.

2. A transmission coupling device for separately sampling (through a known coupling loss) either the forward (incident) or the backward (reflected) wave in a transmission line. A directional coupler may be used to sample either a forward or backward wave in a transmission line. A unidirectional coupler has available terminals or connections for sampling only one direction of transmission; a bidirectional coupler has available terminals for sampling both directions. For optical fiber applications.

directivity The degree to which an antenna focuses its power in a specific direction if the antenna were 100% efficient. Since microwave frequency antennas are highly efficient, the difference between gain and directivity is not significant at microwave frequencies.

directories Places within a hard disk volume where you can store files or subdirectories. The term subdirectory is relative. A directory is a subdirectory only in relation to the directory above it. To a directory below it, the same directory is a parent directory.

directory 1. A list of all the files on a floppy diskette or hard disk. A directory may also contain other information such as the size of the files and the amount of free space remaining.

2. A telephone directory.

Directory Assistance DA. Formerly known as "Information", but changed to "Directory Assistance" because the operators were getting too many stupid questions like "Who was the third president of the United States?" DA allows you to get telephone numbers from an operator. It comes in real handy when you don't have access to a phone book, or when the number is a new one not included in the book. DA once was provided exclusively by the local telephone company. It also once was free – even long distance DA was free. In most states of the United States, the local phone company now charges for this service, or will begin to very shortly, perhaps even from payphones.

Directory Assistance has changed a lot in the past few years. Not only have the local telephone companies begun to charge for the service, but also the service has become competitive. The local phone companies offered to sell their Directory Assistance databases to their competitors, although at a cost that was claimed to be unreasonably high. The competitors generally refused those offers, opting instead to build their own databases from whatever sources they could find. The end result is that Directory Assistance information often is grossly inaccurate. It also is provided by operators who have no knowledge of the area in question. For instance, the Verizon operators once could provide you directions to my office; the operator you talk to today may live in Omaha and may never have been to New York City. Most local phone companies will give you the person's address as well as his phone number if you ask for it.

directory caching A method of decreasing the time it takes to find a file's location on a PC's disk. The FAT (File Allocation Table) and directory entry table are written into the file server's memory. The area holding all directory entries is called the directory cache. The file server can find a file's address (from the directory cache) and the file data (from the file cache) much faster than retrieving the information from disk.

directory date A telephone company term. The issue date of the telephone directory. Improvements which will require changes in dialing procedures, directory numbers, etc., are usually timed to occur coincidental to directory issue dates.

Directory Information Base DIB. Made up of information about objects. The collective information held in the directory. An X.500 term.

Directory Information Tree DIT. Information that outlines the structure of an X.500 directory.

directory name As defined in the X.500 Recommendations, the Directory Name is an ITU term for the name of a directory entry. For example, used to retrieve an O/R (Originator/Recipient) Name for message submission.

directory number 1. The full complement of digits associated with the name of a subscriber in a telephone directory. This is a very long way of saying the obvious, namely your phone number. In North America that is typically a seven digit number. For a longer explanation, see DN.

2. A Northern Telecom definition: A unique phone number which is automatically assigned to each telephone during System Startup. The DN, also referred to as an intercom number, is often used to identify a telephone when settings are assigned during programming. A DN may be changed during programming.

directory package The process of adding, deleting and moving people attached to PBX or Centrex phones is more than simply programming in new extensions. Or should be. First, there are the changes necessary for the call accounting system. Second, there are changes necessary for "The Corporate Phone Directory". In the past, the directory bore little relation to the telephone system and it was often months, and sometimes years behind the actual phone system. Now some phone systems are incorporating various Directory Software Packages into the PBXs.

Some features included in these systems are the ability to dial someone by name – both for the attendant and for the individual phone user, i.e. dial HARRY, or HAR, instead of 3245. Some also include the ability to interface to the call accounting system. So that who's in what corporate department corresponds to which department's bill. There's also that important thing called a Telephone Directory. It would be useful if you could hook a laser printer to the telephone system and tell it to print, in neat, photo-ready columns, an alphabetical (by last name), departmental or any other sorted telephone directory. Some of the newer PBXs have "directory package" features which include some or all of the above. Most users find today's necessity of at least three different systems to be a pain in the behind. Rightly so. The three different systems are CDR, phone directory and extension dialing in the PBX. The three are often out of synch.

directory replication A Windows NT term. The copying of a master set of directories from a server (called an export server) to specified servers or workstations (called import computers) in the same or other domains. Replication simplifies the task of maintaining identical sets of directories and files on multiple computers, because only a single master copy of the data must be maintained. Files are replicated when they are added to an exported directory and every time a change is saved to the file.

directory replicator service A Windows NT term. Replicates directories, and the files in those directories, between computers. See Directory Replication.

directory service 1. A simple term for the information service which the telephone company runs on 411 or 555-1212.

2. A computer networking term. The facility within networking software that provides information on resources available on the network, including files, users, printers, data sources, applications, and so on. The directory service provides users with access to resources and information on extended networks.

directory services A service that provides information about network objects. DNS (Domain Name System) provides node address information. An X.500 Directory service provides any appropriate information an enterprise wishes to include in the X.500 directory itself.

directory synchronization The reconciliation of user directories from two electronic mail post offices. Many gateways and messaging switches have software to automate reconciling these directories.

directory tree A list of directories. A directory tree looks like an organizational chart and shows how your directories and subdirectories are related. The main directory is called the 'Root Directory'. these days directories are called folders.

DirectShow See TAPI 3.0.

DirectX Microsoft's application programming interface (API) for developing games and multimedia applications for Windows. It gives programmers access to the hardware while hiding specific hardware details. There are six components, including Direct3D for 3-D imaging, DirectDraw for 2-D and (starting with DirectX 8) DirectVoice for communicating with other players during online game play. DirectX is used by many computer games on the market today. It is made and freely distributed by Microsoft for Windows users. Having the latest version of DirectX is essential to playing the latest games. To update your version of DirectX, go to http://www.microsoft.com/directx/. Download the latest version of DirectX.

DirecWay DirecWay is the new name for the DirecPC satellite internet access system offered by Hughes Networking. The system uses geostationary satellites to provide speeds up to 400Kbps (much higher is typical but 400Kbps is the maximum advertised speed). A 3 foot dish is used along with a USB modem. Two connection types are offered. A dial-return-system (DRS) uses your existing dial-up service to provide a connection to the Internet for all out going data and the satellite is used to route all returning data via satellite. The DRS systems offer a very nice Internet browsing experience at about 1/2 the upfront and monthly cost. The satellite-return-systems (SRS) use no phone lines after the initial setup and provides a 24/7 connection. Networking is possible but not supported by Hughes customer support. DirecWay services area marketed through several partners including DirecTV, EarthLink, AOL and others.

dirigible A balloon capable of being directed by a pilot was called a dirigible in 1885. It comes from the Latin dirigere which means to direct.

dirt poor In 17th century England many houses had dirt floors. Only the wealthy had something other than dirt, hence the saying "dirt poor."

dirt-style Having hip graphic design that remixes stodgy, amateurish Web content for ironic aesthetic ends. Think: the visual equivalent of a mash-up. – from Wired Magazine.

DIRTBAGS Digitally Initiated Resale of Telecommunications Bypass Applications by Scumbags. A term created by Ron Adams of Maryhurst College in Washington. There is some confusion between this definition of DIRTBAGS – Digitally Initiated Reorigination of Telecommunications Bypass Access Generated by Scumbags. It is ascribed to Karen Corcoran of MCI's Atlanta office to describe as a group the various hackers, phreakers, and others who invade our networks and/or steal long distance service.

dirty power Dirty power typically refers to alternating current that is not a perfect sine wave and not perfectly 120 volt. There are all sorts of ways electricity can be made "dirty." It can be affected by spikes. Spikes are transient impulses (sometimes called glitches) of relatively high amplitude but very short duration. Spikes so short that a very high-speed oscilloscope is needed to observe them can often cause problems. Many spikes can occur in a fraction of a cycle.

Power can also be affected by sags and surges. Sags and surges are rapid changes in the amplitude of an AC voltage. These are generally caused by abrupt changes in the load on a power source or circuit (such as when an air conditioner starts up), and can range from a fraction of a cycle to several complete cycles. Power can also be delivered consistently beyond its rating. In New York Con Edison guarantees 120 volts plus or minus 10%. Any level below 108 volts or above 132 volts Con Edison would consider dirty power. These are called low or high average variations. And they occur when the average voltage is above or below a desired level for significant periods of time, usually measured in seconds, minutes or longer.

Other kinds of dirty power include: Blackouts or brownouts. They occur when the power is switched off or lost completely (blackout), or when the voltage feeding a load is deliberately or inadvertently reduced significantly for a sustained period (brownout). Common mode noise is a small (+1V-2V) signal that appears between a neutral line and ground (earth) where there should be no signal. High/low frequency variations occur when the instantaneous frequency of an AC power source differs from its normal frequency, e.g., 60 Hz, by 0.5 Hz or more. Phase angle variations can be observed in three-phase systems whenever the phase relationships vary from their normal 120 degrees.

DIS Draft International Standard. As specified by ISO, a development step representing near final status on a specification. Once a specification has reached DIS status, companies are encouraged to develop actual products based on it.

DISA 1. Direct Inward System Access. DISA is a way of dialing into a phone system. It can be used to access one's voice mail, often over 800 lines. It can be used to make long distance calls, often on the theory that it's cheaper for employees to dial over the company's leased line phone network than to dial direct from their homes. DISA has been the major way crooks have dialed into PBXs and stolen toll calls. See Direct Inward System Access.

2. Data Interchange Standards Association, Inc. DISA is the Secretariat and administrative arm of the Accredited Standards Committee (ASC) X12 which has responsibility for developing Electronic Data Interchange (EDI) standards. DISA OnLine is an electronic messaging and information system designed for use by DISA's member constituency. www.disa.org.

3. Defense Information Systems Agency. DISA's mission is a combat support agency responsible for planning, developing, fielding, operating, and supporting command, control, communications, and information systems that serve the needs of the President, Vice President, Secretary of Defense, the Joint Chiefs of Staff, the Combatant Commanders, and the other Department of Defense (DOD) Components under all conditions of peace and war.

disable You figured this one. It means to prevent a hardware device from working. Unplugging it is the easiest way. It also refers to a tone or other signal which you send over a phone line to disable the equipment at the other end.

disaggregation Disaggregation the splitting of a spectrum license into two or more licenses of fewer frequencies.

disarmed The state of a firewall when it is not actively protecting a network.

disaster recovery A generic term for all the tools and planning you need to bring your telecommunications facilities back to where they were before the disaster hit you. See the next definition.

disaster recovery plan The procedures that are to be executed to restore telecom, data, phone and network operations if catastrophe were to occur (i.e. a fire).

disaster routing service A tariffed service whereby incoming calls are rerouted to a different location when the primary location experiences an out-of-service condition. It is also an efficient way to redirect office communications during office relocations or other special non-disaster situations where calls cannot be taken in the office.

disc An older method of spelling DISK, as in Floppy DISK. Disc (spelled with a C) is now more commonly used to refer to optical storage devices, like CD (Compact Discs), MO (Magneto Optical) discs and DVDs. See also MS-DOS.

Discard Eligibility DE. A bit set by the user in a Frame Relay network, to indicate that a frame may be discarded in the event of congestion.

discharge block A protective device through which unwanted voltages discharge to ground.

disclaimer A provision in a license that allows a person or organization to deny responsibility for problems related to the use of misuse of the product. Some EULAs (End-User License Agreements) include a disclaimer that purports to free the manufacturer of a piece of software from liability for the product, even if the manufacturer knows of defects in the design. Where it applies, UCITA would make the law governing disclaimers uniform and limit the chance to collect damages in cases where a faulty product has caused actual harm. See EULA.

disco 1. Some phone and cable companies call orders to disconnect service "disco orders." See Disco Tech.

2. A club where annoying loud music is played, often similar to the "music on hold" on many PBXs.

disco tech Slang expression for a technician who only handles disconnections.

disconnect 1. The breaking or release of a circuit connecting two telephones or data devices.

2. Once upon a time, Teleconnect Magazine used to run comedy supplements in April called Disconnect, which poked fun at our industry. Sadly, the magazine is no more and neither are the supplements.

disconnect frame Indicates in a fax call that the call is done. The sending fax machine sends the disconnect frame before hanging up. It does not wait for a response.

disconnect signal The signal sent from one end to indicate to the other to shut down the connection.

disconnect supervision The change in electrical state from off-hook to on-hook. This indicates that the transmission connection is no longer needed.

disconnect switch In a power system, a switch used for closing, opening, or changing the connections in a circuit or system or for purposes of isolation. It has no interrupting rating and is intended to be operated only after the circuit has been opened by some other means, such as by a circuit breaker or variable transformer.

disconnect tone A cell phone term. A tone that plays when a cell phone call ends.

disconnected recordset A recordset in a client cache that no longer has a live connection to the server. If something must be done with the original data source, such as updating data, the connection will need to be re-established.

discontinuity An interruption or drop out of the optical signal.

discounted payback period The number of years in which a stream of cash flows, discounted at an organization's cost of money, repays an initial investment.

discounting The process of computing the present worth of a future cash flow by reducing it by a factor equivalent to the organization's cost of money (or some other measure of the value of money as measured by an interest rate) and the time until the cash flow occurs.

discoverable device A Bluetooth device in range that will respond to an inquiry – normally in addition to responding to page.

discoverable mode A device that can respond to an inquiry is said to be in a discoverable mode. There are two types of discoverable modes: limited discoverable mode and general discoverable mode. The opposite of discoverable mode is non-discoverable mode. See also Silent Device.

discovery mechanism A way of finding other servers on the network.

discrete In voice recognition, refers to an isolated word. A discrete word is preceded and followed by silence, hence isolated in speech. Discrete words need to be separated by about half a second of silence when spoken to a discrete recognizer.

Discrete Cosine Transform DCT. A pixel-block based process of formatting video data where it is converted from a three dimensional form to a two dimensional form suitable for further compression. In the process the average luminance of each block or tile is evaluated using the DC coefficient. Used in the ITU-T's Px64 videoconferencing compression standard and in the ISO/ITU-T's MPEG and JPEG image compression recommendations.

discrete logic device Discrete logic devices are simple devices that typically serve one function – whether it's a gate, a buffer, a transceiver, etc. The list of discrete logic devices includes products that transfer bits of information across a backplane or bus in either a single-bit fashio or all the way up to 32-bit wide devices. In the ideal world, no designer would include discrete logic components on a printed circuit board. But the world is not ideal.

Discrete Multi-Tone See DMT.

discrete semiconductor device A semiconductor device that is specified to perform an elementary electronic function and is not divisible into separate components functional in themselves. Note 1:1 Diodes, transistors, rectifiers, thyristors, and multiple versions of these devices are examples. Other semiconductor structures having the physical complexity of integrated circuits but performing elementary electronic functions (e.g., complex Darlington transistors) are usually considered to be discrete semiconductor devices. Note 2: If a semiconductor device is not considered to be an integrated circuit in both complexity and functionality, it is considered to be a discrete device.

Discrete Wavelet Multi-Tone DWMT. A transceiver technology that can be used in ADSL systems, DWMT uses a large number of carriers that are individually modulated. Unlike DMT, the carriers are not equally spaced. With DWMT, more carriers are allocated in the lower frequency spectrum.

discretionary preview dialing A single button dialing technique where the agent initiates a call with a single keystroke. Often used in association with a CRT tied to a database. Upon hitting a single button the system selects the phone number field from the screen and dials the number. Contrast this with Forced Preview Dialing. When the call ends, the computer brings up the next screen and starts dialing the call without the agent helping or hindering.

discriminating ringing See Distinctive Ringing.

discriminator A specific type of FM detector circuit which senses phase relationships.

discussion A method of confirming others in their errors.

disengage request Message with the Billing Information Token (which contains the duration of the call) sent by the gateway to the gatekeeper when a call ends.

disengagement denial Disengagement failure due to excessive delay by the telecommunication system.

disengagement failure Failure of a disengagement attempt to return a communication system to the idle state, for a given user, within a specified maximum disengagement time.

DiSEqC Abbreviation for Digital Satellite Equipment Control, a protocol for controlling a satellite receiver outdoor unit (ODU) from the indoor unit (IDU) by sending 22 kHz signals over the existing coaxial cable between the two units. Originally developed by Eutelsat, DiSEqC can be integrated into satellite facilities to provide an open standard system for switching between frequency bands, polarizations and satellites.

dish Typically a parabolic microwave antenna – used for receiving line-of-sight terrestrial signals or signals from satellites.

dish farm An area on a roof or in a field with a lot of satellite dish antennas.

disintermediation The decline of middlemen companies that today operate between the buyer and maker of goods. Pundits predict this will happen with the rise of commerce on the Internet. For example: the insurance, auto, mortgage, news delivery, and stock brokerage industries may change dramatically over the next few years. The Internet allows many industries such as these to do business directly with their customers.

disk A piece of plastic or metal upon which a coating has been applied and which can thus, record computer information magnetically. The present convention is that a "disk" with a K refers to magnetic storage, while "disc" with a C refers to optical storage. A CD is a disc. See also MS-DOS.

disk array Also called a Drive Array. Although any set of disk drives put into a common enclosure could be called an array, in terms of RAID technology disk arrays tend to be those drives which are subject to a hardware or software-based controller that makes them appear to be a single drive to the host CPU or operating system. In short, a disk array is a disk subsystem combined with management software which controls the operation of the physical disks and presents them as one or more virtual disks to the host computer.

disk cache On a PC, a disk cache is the part of RAM that is set aside to temporarily hold data read from disk. A disk cache doesn't have to hold an entire file, as a RAM disk does, but can hold parts of running application software or parts of a data file. Disk-caching software manages the process of swapping data to and from the disk cache. See Disk Caching.

disk caching A technique used to speed up processing. Each time your application retrieves data from the disk, a special program, called a disk caching program, stores data read from the disk in an area of RAM. When the application next requests more data, some of it may already be in RAM, thereby dramatically speeding the retrieval of data. See Disk Cache and Caching.

disk cloning Disk cloning is making an exact (bit by bit) copy of your hard drive on another hard drive. You clone because you want to back up your hard drive software and data alike. Let's say your hard drive dies. You want to be able to open your PC or laptop, remove the failed drive and replace it with the "new" cloned drive. Since this is my objective, I always clone my drives onto identical drives. There are many types of cloning software. Not all allows you to clone onto identical drives. I use software called E-Z-Gig Transfer Utility. It's made by Apricorn. It's saved my life on many occasions. I make clone once a week. It takes an hour.

disk controller A hardware device that controls how data is written to and retrieved from the disk drive. The disk controller sends signals to the disk drive's logic board to regulate the movement of the head as it reads data from or writes data from or writes data to the disk. Gateway Computers defines disk controller thus: "Circuitry that manages the physical activity of a disk drive, such as moving the drive heads and creating the actual signals recorded on the disks. The controller is usually a card on the underside of the drive itself. The expansion card that connects to the drive by a ribbon cable enables communication between the disk drive and the computer and is called an adapter or host adapter. Sometimes the adapter circuitry is directly on the system board. Older hard drives had two ribbon cables to the expansion card because one part of the card held controller circuitry and another part held the adapter circuitry."

disk dancers Teenagers who use the America Online CDs and floppy disks given away in magazines and via direct mail to hop from one free account to another.

disk drive A device containing motors, electronics and other gadgetry for storing (writing) and retrieving (reading) data on a disk. See Disk Duplexing, Disk Mirroring, Disk Operating System and other definitions below starting with the work "Disk." See also IDE and Enhanced IDE.

disk drive performance Three basic things affect your perception of the speed of the hard disk drive inside your PC.

1. The disk transfer rate – the speed at which data can be read from or written to the drive's media. This speed is governed mainly by drive mechanics – especially rotational speed, latency and seek time.

2. The controller transfer rate – the speed at which the drive's controller electronics can move data across the interface. This is governed by the design of the drive controller.

3. The host transfer rate – the speed at which the computer can transfer data across the interface. This is a matter of both CPU and bus speeds as well as the hardware that provides the bridge between the host bus and the disk interface.

disk duplexing A method of failsafe protection, used on file servers on local area networks. Disk duplexing involves copying data onto two duplicate hard disks, each using a separate controller and a separate disk channel. Disk duplexing protects data against the failure of a hard disk or of the hard disk channel between the disk and the file server. The hard disk "channel" includes the disk controller and interface cable. If any component on one channel fails, the other channel continues to operate normally. (You hope.) The operating system sends a warning message to the workstations and PCs on the network to indicate the failure. It's a good idea then to fix the problem fast. Very fast. See also Disk Mirroring.

disk mirroring A technique for protecting the information on your hard disk. Disk mirroring writes data simultaneously to two identical hard disks using the same hard disk controller. Here's how it works: You have a special hard disk controller card. That's the card which organizes getting information into and off your hard disk. When you come to write information to your hard disk, your hard disk controller writes first to the first hard disk, called the primary hard disk. The controller retains that information in its memory and then writes it to the second hard disk. This causes a 50% degradation in performance since it now takes twice as long to write to the disk. When the controller comes to read, it reads only from the primary disk. Thus there is no performance degradation in reading. Mirroring is designed to protect against mechanical problems with one of your hard disks. If one of the hard disks breaks, the other one will take over instantly. You will get a warning message. You will be told to repair the broken disk and you will be told to designate the other disk now as your primary disk (it may already be). That primary disk will now become your bootable disk, the one you boot your computer from. Mirroring does not protect against viruses, or corrupt data or losing data. Any idiocy you can perform on one hard disk you can now happily perform on two. Mirroring does not protect you against a lightning strike which could knock out both your hard disks. Mirroring does not protect against the loss of a controller card since you're only using one. Another protection technique called disk duplexing uses two separate controllers to drive two separate hard disks. See also Disk Duplexing.

Disk Operating System See DOS.

disk pack A series of disks mounted horizontally and arranged as a single unit. A disk pack contains more space for storing and retrieving information than one single disk.

disk sector Magnetic diskettes are typically divided into tracks, each of which contains a number of sectors. A sector typically contains a predetermined amount of data, such as 256 bytes.

disk server A device equipped with disks and a program that permits users to create and store files on the disks. Each user has access to their own section of the disk. It gives users disk space which they would not normally have at their own personal computers. The disk server is linked to the PCs via a LAN. The next level of sophistication would be a file server, which would allow users to share files.

disk spanning When you want to save a single big file to several floppy disks, it's called disk spanning. Many of the zipping programs allow you to do this.

disk striping Writing data in stripes across a volume that has been created from areas of free space on from two to 32 disks.

disk subsystem A collection of physical disks and the hardware required to connect them to a host computer or network.

disk/file server A mass storage device that can be accessed by several computers, usually through a local area network (LAN).

diskless PC Just what it says: a PC without a disk drive. Used on a LAN, a diskless PC runs by booting DOS from the file server. It does this via a read-only memory chip on its network interface card called a remote boot ROM. Diskless PCs are cheaper than PCs with disks, they're more compact and they offer better security since users can't make off with floppy disks of important and sensitive data or add their own virus-ridden programs to file servers. Diskless PCs appeal primarily to users interested in security. One system of diskless PCs allows the system operator to disable and physically lock the machine's various ports and the computer case itself. Should something go wrong with the machine, the ports can be restored to operating condition by letting the system allow a technician to attach a laptop computer and run diagnostic programs.

diskless workstation See Diskless PC.

dismount To remove a removable tape or disc from a drive.

DISN See Defense Information Systems Network.

DISOSS IBM's Distributed Office Support System.

disparity control A function of the 8B/10B transmission encoding scheme, in which the two remaining bits are used for error detection and correction. 8B/10B transmits 8 bits as a 10-bit group, thereby leaving two additional bits for disparity control.

dispatch A radio communications technique where one communicates to many through short bursts of communication. Users of dispatch services include taxis, trucking companies and service personnel. For a longer explanation, see LMR.

dispatch walkie-talkie mode Where one subscriber talks and all the subscribers listen on the same talk channel. the service is used by taxi cab and trucking companies, etc.

dispersion In the broadest sense, dispersion refers to the spreading out of a light signal. Optical fiber transmission systems variously suffer from a number of different types of dispersion. For example, chromatic dispersion describes a phenomenon in which an optical signal is distorted because the various frequency components of that signal have different propagation characteristics. (Note: Wavelength is the inverse of frequency. In the visible light spectrum, we see each wavelength as a different color, hence the term chromatic.) Now, a light source (e.g., laser diode), no matter how finely tunable, creates light pulses across a range of wavelengths. Each wavelength travels at a slightly different speed through the fiber core and, especially, through the clearer cladding that surrounds the core. As the pulse propagates through the fiber, therefore, it tends to lose its shape and overrun other pulses in a phenomenon known as pulse dispersion. Dispersion is a particularly significant problem in high-speed networks, which operate at Gbps speeds, and in long haul networks involving links exceeding several hundred kilometers. In order to defeat dispersion, the network must be optimized through the placement of regenerative repeaters at appropriate intervals, or through the use of one of several types of Dispersion Shifted Fiber (DSF). See also Chromatic Dispersion, DSF, Modal Dispersion, Polarization-Mode Dispersion, Pulse Dispersion, and Waveguide Dispersion.

Dispersion Compensation Grating DCG. DCG overcomes the distortion of optical signals as they are transmitted through a network. Instead of trying to compensate for large amounts of signal dispersion at the end of a network, DCG periodically removes the distortion where needed along the transmission line. See DCG and Solitons.

dispersion compensating fiber A fiber that has the opposite dispersion of the fiber being used in a transmission system. It is a short length of fiber spliced into a fiber optic transmission in order to nullify the dispersion caused by that fiber.

Dispersion Shifted Fiber See DSF.

dispersion management Dispersion management comprises a number of techniques and mechanisms used in the system design of a fiber optic transmission in order to cope with the various types of dispersion experienced by signals traveling over optical fiber transmission systems.

dispersion penalty Dispersion in an optical fiber causes individual pulses to lose their shape as the edges become smeared. As the edges smear, the receiver has more difficulty distinguishing between individual ones (pulses of light) and zeroes (non-pulses of light). See also Dispersion.

display The visual presentation of information, usually on a TV-like screen or an array of illuminated digits.

display driver A piece of software which translates instructions from the software you are running into thousands of colored dots, or pixels, that appear on your video monitor. A display driver is also called a Video Driver. Symptoms of a video driver giving trouble can range from colors that don't look right to horizontal flashing lines to simply a black screen. In the Macintosh world, Apple rigidly defined video drivers. Windows, in contrast, is a free-for-all. Windows 3.1 defined the lowest common denominator of displays – namely 16 colors at 640 x 480 pixels. But most multimedia programs and many games won't run with only 16 colors. They require at least 256 colors.

display phone A telephone that has a LED display. Also called an Executive Phone. Display phones are usually difficult to read because the displays are not backlit. Phone lines and PBX extension lines don't have enough power to run the display without an external power adapter and most telephone equipment makers don't want to sell phones with an external power adapter. It limits where phones can be placed and therefore how many can be sold. Display phones typically allow both "programmable keys", buttons that can be

programmed for speed dial, conference, etc. and "soft keys", buttons that change function as different features are used, and also can show the number that is being dialed, and internal callers' names and extensions.

display rotation What the computer industry refers to as the ability to change a monitor from landscape (its normal mode) to portrait (useful portrai is supported, to switch to portrait mode you'll need to select a portrait display mode in Display Properties. Changing display

displayboards Also called Readerboards or Wall Displays. Readerboards are typically found in call centers. They are electronic displays, sort of like giant TVs. They are typically hooked into the ACD or PC monitoring the machine and they throw up information about how many people are waiting in line, how long the longest person has been in line, how well the agents are doing and, often, whose birthday it is today. The idea is that all the agents in the call center can see the Readerboards and change their behavior accordingly.

disruptive technology A disruptive technology is a new technology which destroys an existing technology or an existing way of doing things by being better, faster and cheaper. The telegraph and the telephone were both disruptive technologies, as was the communications satellite, fiber optics and the Internet. Most disruptive technology is not introduced by existing companies. In his 1997 study of disruptive vs. sustaining technologies, Clayton Christensen of Harvard Business School pointed to the dilemma that all industrial leaders face. On the one hand, they have to listen carefully to their customers, who want predictable improvements in sustaining technology. Yet, on the other hand, they must also be ready for the mayhem that could ensue if, out of the blue, some radical, disruptive technology were suddenly to rewrite the rules and render their company's products irrelevant. Christensen tells of a meeting with Andy Grove of Intel in which Grove interrupted Christensen to exclaim, "I get it! It's not the technology that's disruptive, it is how the technology disrupts the business model!" Christensen has studied hundreds of technology introductions, and found that incumbent companies are expert at bringing new technology to market if it sustains their existing value proposition, but they are horrible at bringing disruptive technology to market. For example, if we were waiting for the leading mini-computer makers (DEC and Wang) to bring us PCs, we'd probably still be waiting.

distance learning A form of instruction in which video and audio technologies are used so as to allow students to attend classes in a location distant from where the course is being presented.

distance sensitive pricing Product pricing based on the distance (airline mileage) between the originating and terminating locations of a call/data transmission.

distance vector An approach, or algorithm, used by network equipment in selecting the best available network path by calculating the total distance over which a packet would travel on each alternative route. The shortest distance is usually the most preferred.

distance vector protocol A routing protocol designed to minimize the number of hops (i.e., link-level connections between routers) which a data packet must travel from the originating device to the terminating device. Such a protocol causes each router to regularly broadcast the entire contents of its routing tables to all neighboring routers. Each recipient adds its own "distance" vector by incrementing the distance value by 1, and passes the data to its neighbors, and so on. (Note that "distance" refers to the number of routers transversed, rather than the geographic distance over which the signal propagates.) This method automatically allows the involved routers to discover and rediscover any route failures and restorals, and to "converge" on consensus-level route selections. However, this method is bandwidth-intensive in large, complex networks, as large routing tables are passed around the network frequently; during such processes, the bandwidth available in support of user data is diminished significantly. Therefore, distance vector protocols are best used in relatively small router networks with relatively few inter-router connections. Network factors such as link speeds and congestion levels are not considered in distance vector protocols. Distance vector protocols include IP RIP, IPX RIP, and AppleTalk RTMP. Distance vector algorithms also are known are known as Bellman-Ford algorithms. See also Link-State Protocol, Path Vector Routing Protocol, Policy Routing Protocol, Router, and Static Routing.

distant learning A Pacific Bell term for students sitting in front of TVs and phones and participating in classes that are being held and delivered elsewhere. In one of PacBell's trials, they used a T-1 signal, so the distant lecturer could see and hear his distant students using full-color video.

distant signal An FCC definition. A television channel from another market imported and carried locally by a cable television system.

distinctive dial tones In some phone systems, dial tones sound different. An internal dial tone sounds different to an external dial tone. The logical reason for this is simply to alert the user as to whether he or she is making an intercom or an outside local or long distance call.

distinctive ringing First, distinctive ringing is a feature that offers extra numbers which cause different ringing patterns on a line. When the main number is called, the called party will receive the normal ringing pattern. If one of the extra numbers is dialed, that line would ring with a different cadence. In North America, the normal ringing pattern is a single ring every six seconds. The distinctive ring patterns are 1) two short rings every six seconds, or 2) a short-long-short ring.

Different ringing patterns are also used in conjunction with such features as busy call return to indicate a freed line. One test done by Bell Canada set up a special ringing pattern (different from any of the featured distinctive rings) to indicate an incoming long distance call.

Each telephone company has its own name for this feature: Ident-a-Call, Teen Ring, Feature Ring, etc. In any case, different ringing patterns allow for calls to certain people, or to sort out different call purposes such as for voice, fax, modem, or answering machine.

distort To change some characteristic of a signal during its transmission. See Distortion.

distortion 1. The difference in values between two measurements of a signal – for example, between the transmitted and received signal. "Distortion" typically refers to analog signals.

2. In imaging, distortion is any deformation of the on-screen image. Two common types of distortion are pincushion and barrel.

3. When used in relation to AC power distribution, this refers to deviations between the actual AC voltage waveform delivered to the user and the ideal sine wave of voltage. Total distortion is usually expressed as a percentage of desired sine wave, for example, a square wave has approximately 33% distortion. Distortion in AC power systems can also be resolved into a series of harmonics. In this case percentages for each harmonic (such as the third, fifth, seventh, etc.) are provided. The square root of the sum of the squares of the individual harmonics is equal to the total distortion. This definition courtesy APC.

distributed antenna system A network of antennas, typically inside of a building, that are connected with a wired network because antenna-to-antenna communications are not possible or feasible in the facility. A distributed antenna system (DAS) takes wireless signals and injects them onto the wireline portion of the network, which transports them throughout the facility and then reconverts them back into wireless signals for delivery to mobile wireless users. A DAS thereby provides or extends wireless coverage in a facility for mobile users.

distributed capacity The capacity in a coil due to the proximity of the turns.

distributed common control There are two elements of telephone switching: the switching itself and the control of that switching. The earliest step-by-step telephone switches had their "Control" built into them. The dialing information at the beginning of the call physically moved switches. You could say, as a result, that control was distributed throughout the switching system. Then came the 1940s and crossbar exchanges, and the economics pointed to centralizing control. Then came computerized or stored program control (SPC) switches in which large computers were used centrally to perform virtually all the functions of the erstwhile electromechanical senders, registers, markers, etc. – those things which affect the setting up and tearing down of the call. As computers got smaller and as microprocessors appeared (the so-called computer on a chip), it became economical and efficient to place inexpensive microprocessors in the telephone circuits themselves, in essence getting much of the processing done before it hits the central processing unit. Increasingly, as special microprocessors (so-called "computers on a chip") for telecommunications evolve, we will see more and more of the processing being distributed to further and further away from the central point and closer and closer to the originating telephone instrument. It will be rare in coming years for telephones to come without microprocessors. One day, each phone will have its own switch and the rest of the system will just be one gigantic loop of cable – not unlike today's local area networks.

Distributed Component Object Model See DCOM.

Distributed Computing Environment DCE. A comprehensive integrated set of services that supports the development, use, and maintenance of distributed applications. Digital Equipment Corporation's DCE is an implementation of the Open Software Foundation's DCE (OSF DCE). In response to OSF's request for distributed computing technology, Digital submitted for consideration four of Digital's established distributed computing technologies:

Remote Procedure Call (RPC), a joint effort with HP/Apollo; Threads Service, based on Digital DECthreads; Cell Directory Service (CDS), based on the Digital Distributed Name Service (DECdns); Distributed Time Service (DTS), based on the Digital Distributed Time

Service (DECdts). See DCE for more detail.

Distributed Data Processing DDP. A data processing arrangement in which the computers are decentralized – i.e. scattered in various places. Hence, processing occurs in a number of distributed locations and only semi-processed information is communicated on data communications lines from remote points to the central computers. The object of DDP is to split processing among multiple computers, to save telecommunications charges and to improve network response time.

distributed database A database managed as a single system even though it includes many clients and many servers at both local and remote sites. A distributed database requires that data redundancy is managed and controlled.

distributed denial of service attack See Denial of Service Attack.

distributed environment Refers to a network environment, or topology, in which decision making, file storage and other network functions are not centralized but instead are found through the network. This type of environment is typical for client-server applications and peer-to-peer architectures.

Distributed Feedback Laser DFB. A type of laser used in fiber-optic transmission systems, at the distribution level of the local loop. DFBs are point-to-point lasers distributed among nodes in a geographic area such a neighborhood. They transmit and receive optical signals between the distributed nodes and the centralized node, where the signals are multiplexed over a higher-speed fiber link to the head-end (point of signal origin). DFBs can be more effective than the traditional approach of using a single laser which serves multiple nodes through a broadcast approach, as the available bandwidth can be segmented. DFBs have application in a FTTN (Fiber-To-The-Neighborhood) local loop scenario. See also FTTN and SONET.

distributed file system A type of file system in which the file system itself manages and transparently locates pieces of information from remote files and distributes files across a network. It can recognize multiple servers and be accessed independently of where it physically resides on the network.

distributed management environment A compilation of technologies now being selected by the Open Software Foundation to create a unified network and systems management framework, as well as applications. Those technologies will complement OSF's own Unix implementation, OSF/1, as well as other operating systems.

distributed microprocessor common control In telephone systems, this means that the system employs many individual microprocessors to control system functions. The microprocessors may be located in central processing equipment or in the telephones themselves.

distributed name service A technique for storing network node names so that the information is stored throughout the network (either one LAN or many joined together), and can be requested from, and supplied by, any node.

distributed network service Introduced in March 1991, AT&T's Distributed Network Service was designed expressly for the switchless resale community unlike SDN. It allows resellers to purchase large volumes of services and receive progressive discounts on all direct dial domestic and international calls. Resellers may designate any number of locations to participate in the plan with the flexibility of adding locations.

distributed nodes PBX and its "slave" switches which are physically in separate buildings, in separate areas of the campus, in separate parts of the town.

distributed order management The customer order drives the supply and demand chain – touching everything from marketing and finance to production – yet most companies today lack a cohesive architecture to address this critical piece of the business. Compounded by the increasing number of channels, suppliers, and customer relationships, these businesses are struggling to manage orders across their extended enterprise. Orders must be coordinated and fulfilled across internal divisions as well as external supply networks such as contract manufacturers and third-party logistics service providers. Thus the term Distributed Order Management.

distributed processing A network of computers such that the processing of information is initiated in local computers, and the resultant data is sent to a central computer for further processing with the data from other local systems. The term also covers computing jobs "farmed out" from a central site to remote processors where faster processing or specialized databases are available. Distributed processing is often a more efficient use of computer processing power since each CPU can be devoted to a certain task. A LAN is the perfect example of distributed processing. See also Distributed Data Processing.

Distributed Queue Dual Bus DQDB. A connectionless packet-switched protocol, normally residing in the Medium-Access Control sublayer of the data link layer. Definition from Bellcore in reference to Switched Multimegabit Data Service (SMDS). See also DQDB and SMDS.

distributed storage Whether it's organizing documents, spreadsheets, music, photos, and videos or maintaining regular backup files in case of theft or a crash, taking care of data is one of the biggest hassles facing any computer user. Wouldn't it be better to store data in the nooks and crannies of the Internet, a few keystrokes away from any computer, anywhere? A budding (but not yet here) technology known as distributed storage could do just that, transforming data storage for individuals and companies by making digital files easier to maintain and access while eliminating the threat of catastrophes that obliterate information, from blackouts to hard-drive failures.

distributed switching When electronics and computers were expensive it made sense to centralize them and run individual lines out for miles to subscribers. Then the economies changed. Electronics and computers became cheaper and running phone lines for miles became very expensive. So switching companies started building small switches which they could put closer to subscribers. Thus, individual local loops would be shorter and the long lines going back to the larger central office would be more efficiently used – namely by more people. The remote, or distributed switches, are called everything from remote switches to slave switches (because they slave off the main one which is distant). Usually these remote switches are unattended.

distribution 1. The portion of a switching system in which a number of inputs is given access to an equal number of outputs.

2. Refers to the arrangement of premises wiring runs and their associated hardware required to implement the planned customer premises wiring system extending from the network interface jack to each communications outlet at the desktop.

distribution cable, inside plant Cables usually running horizontally from a closet on a given floor within a building. Distribution cables may be under carpet, simplex, duplex, quad, or higher fiber count cables.

distribution cable, outside plant The cable running from a central office or remote terminal to the side of a subscriber's lot.

distribution channel The route which the goods or title to the goods follow from the original supplier to the end user, determined by the type of trading parties, e.g. wholesaler, retailer, etc.

distribution duct A piece of rectangular metal or plastic within or just below the finished floor in your office or factory and used to extend AC power cables and telecommunications wires to a specific work area. Also called a raceway.

distribution frame Cables coming in from thousands of subscribers need to connect to the correct ports on a central office. Similarly, cables coming in from many PBX extensions need to connect to the PBX. The cables could be directly wired to the CO or to the PBX. This would be inflexible. It would make future moves and changes a nightmare. So the solution is something called a Distribution Frame. Basically it's a giant wire connecting device made of metal. There are no electronics in it whatsoever. On one side we punch down the wires coming in from the outside world. On the other side, we punch down the wires coming in from the CO or PBX. Both sides are connected with wire that's called "jumper" wire. By pulling off one end of the jumper wire and moving it to another location we can quickly change phone numbers, add or subtract cabling (one, two or three pairs for normal or electronic phones, etc.). In big central offices, distribution frames can span whole city blocks and the "jumper" wires can be several hundred yards long. Designing distribution frames and their layout in advance is critical, otherwise, it becomes a mess and tracing where jumper wires go becomes an enormously time consuming job.

distribution frequency The number of times used in the Internet for translating names of host computers into addresses. It is a network database system that provides translation between host names and addresses.

distribution group 1. A group made of phone extensions on a PBX arranged to share the load. In the Rolm PBX, each group is assigned a dummy extension number called a pilot number.

2. A group of telephone extensions on an automatic call distributor (ACD). The ACD answers the incoming calls then checks to see if any agents' phones are free. If none are free, it delivers the caller a message and then puts the caller on hold. Which line the call has come in on may determine which group of agents should handle that call. They would be called a Distribution Group. Once the call is released from hold, it may be sent to a member of that Distribution Group following some pre-determined mathematical formula – for example, so that everyone's workload is kept constant, or a group of people are kept busy.

distribution rack A device used to mount communications equipment and cables.

distribution service In ISDN applications, a telecommunications service that allows a one-way of information from one point in the network to other points in the network with or without user individual presentation control. See Distribution Services.

distribution services In the world of B-ISDN (broadband ISDN) applications, distribution services are communications services that emphasize one-way, bandwidth-intensive transfer of information from one point in the network to other point(s) in the network. There are two classes of Distribution Services defined within this context, revolving around the issue of "Presentation Control:" Services requiring no presentation control include what we normally think of as broadcast services. Such services include such data as TV, Video On Demand (VOD), audio and multicast data. The data is broadcast or multicast across the network, with no requirement that the receiving device exercise any form of control over the transmission or presentation of the data. Services requiring presentation control include Interactive TV. While engaged in an Interactive TV session, a viewer might wish to control a TV broadcast in much the same way as he would control a videotape through a VCR remote control. While watching the Super Bowl, for instance, Ray Horak might wish to rewind and replay the touchdown scored by the Dallas Cowboys' Emmit Smith against the San Francisco 49'ers. Further, the viewer might wish to view that play from multiple angles covered by cameras positioned around the stadium. While the viewer is exercising these options, the live broadcast is buffered in large-scale temporary memory in the TV set of the future. Once the viewer has sufficiently relished the play, he can play rejoin the live broadcast, begin the program where he left off, or exercise other options. Harry Newton, on the other hand, might wish to exercise the same control over a tennis match broadcast from Australia.

distribution voltage drop The voltage drop between any two defined points of interest in a power distribution system.

distributor 1. A company with a contractual relationship with a manufacturer to buy equipment at a preset price. The manufacturer provides training, advertising and warranty support. Often called an authorized dealer, although a dealer may be one step lower in the distribution chain. A distributor is often used as a generic term for any supplier. Therefore you should clarify whether a distributor is an authorized distributor.

2. See Cross-connect.

disturber A telephone company word for crosstalk. See Binder Group and Crosstalk.

DIT Directory Information Tree. The global tree of entries corresponding to information objects in the OSI X.500 Directory.

dithering Dithering is an imaging term with at least two meanings. One meaning that is the processing of an image containing more colors than a system can handle to an image containing exactly the right number of colors that the system can handle. For example, some of the color images on my laptop contain 16 million colors. But my laptop (the way I have it set up) will only handle 256 colors. If I ask my image display software to display that image, it will "dither" it to 256 colors. This means it will give its best shot guess at what the image should look like.

In another meaning, dithering is patterning black and white dots to approximate shades of grey on a scanned image.

diurnal phase shift The phase shift of electromagnetic signals associated with daily changes in the ionosphere. The major changes usually occur during the period of time when sunrise or sunset is present at critical points along the path. Significant phase shifts may occur on paths wherein a reflection area of the path is subject to a large tidal range. In cable systems, significant phase shifts can be occasioned by diurnal temperature variance. See also Diurnal Wander and Wander.

diurnal wander A loss of signal synchronization in digital cable systems caused by temperature variations over the course of 24 hours. (Diurnal means "daily cycle.") As the ambient temperature varies from the heat of the day to the cool of the night, the cable stretches and contracts, with the overall length of the cable changing, if only ever so slightly. As the length of the medium changes, the speed of signal propagation (the time it takes for the signal to transverse the cable) is affected. As a result, the number of digital pulses effectively stored in the medium changes. The end result is that the network elements (e.g., repeaters and multiplexers) can get out of synch. Diurnal wander affects all types of cable systems – twisted pair, coax cable, and optical fiber; it especially affects cables hung from poles, rather than buried, as such cables are more exposed to temperature variations and as the weight of the cable adds to the problem. The impacts of diurnal wander are particularly great in very high-speed transmission systems. See also Diurnal Phase Shift and Wander.

diverse entry You have a building with phone service. You are concerned about the reliability of your phone service. You are concerned that the wires coming in from your phone company might be cut. So you organize to have service coming in from the phone company along different routes and entering your building from opposite sides of your building. Thus the term diverse entry.

diversity 1. In microwave communications, the strength of a microwave signal can decrease for many reasons – heat, rain, fog, etc. This is not good if the objective is to get reliable communications. One solution is to simultaneously send and receive two microwave signals at slightly different frequencies. Since different frequencies respond differently to weather problems, the likelihood is that at least one will get through well. This is called diversity.

2. A means of effecting redundancy in a network, with the result being protection from catastrophic failure. Consider the typical end user – one cable entrance to one group of wire pairs housed in one cable connected to one central office provided by one local exchange carrier and connecting to one interexchange carrier. That is a catastrophe waiting to happen. Consider the alternative of full diversity. The fully redundant end user has several cable entrances into the building – this is Entry Diversity. The local loop connection is provided through multiple, non-adjacent pairs in multiple cables – this is Pair and Cable Diversity. The several cables follow different routes to the CO – this is Route Diversity. The local loops terminate in multiple COs – this is Central Office Diversity. The carrier connection is to multiple LECs and IXCs – this is Carrier Diversity. In this scenario of full diversity, no single point of failure can totally isolate the user organization from the network. Diversity is good – perhaps expensive and complex, but good.

diversity combiner A circuit or device for combining two or more signals carrying the same information received via separate paths or channels with the objective of providing a single resultant signal that is superior in quality to any of the contributing signals.

diversity receive A method commonly employed by cellular manufacturers to improve the signal strength of received signals. Uses two independent antennas that receive signals which differ in phase and amplitude resulting from the slight difference in antennas position. These two signals are either summed or the strongest is accepted by voting. The most popular methods include dual-antenna phase switching, dual-receiver audio switching and "ratio diversity" audio combining. The most effective method is ratio diversity combining.

Divestiture There have been two really dumb business decisions in the history of telecommunications. The first was in 1876 when Western Union declined to buy the Bell patents on the telephone, saying the phone was a dumb idea. The telegraph was much better. The second one was the divestiture of AT&T in 1984. The explosion of telecom technology in the 1960s and 1970s enticed many entrepreneurs to attempt to compete against AT&T, the dominant supplier at that time. AT&T didn't like this. It acted stupidly and pulled every illegal trick under the sun to keep competition at bay. It figured, wrongly, that being a regulated utility it could do whatever it wanted. Wrong! Eventually the U.S. Justice Department hit it with a huge anti-trust suit. Instead of modifying its behavior or taking the case to court, AT&T decided to settle it – not for what the Justice Department had wanted – but what it wanted. On January 8, 1982 AT&T signed a Consent Decree with the U.S. Department of Justice, stipulating that on midnight December 30, 1983, AT&T would divest itself of its 22 telephone operating companies. According to the terms of the Divestiture, those 22 operating Bell telephone companies would be formed into seven regional holding companies of roughly equal size. Terms of the Divestiture placed business restrictions on the BOCs. Those restrictions were threefold: The BOCs weren't allowed into long distance, equipment manufacturing, or information services. In exchange for the divestiture, AT&T would be allowed to manufacture and sell computers. It had been restricted previously from that business by a 1956 Consent Decree. AT&T executives agreed to the Divestiture because they had been assured by executives of Western Electric, AT&T's manufacturing arm, and executives of Bell Labs, AT&T's research arm, that the company was brimming with computing inventions that would revolutionize the industry and make AT&T rich beyond its wildest dreams. It was total nonsense. Neither Western Electric nor Bell Labs had any significant marketable computer inventions – as was to become apparent in following years. What was further depressing (i.e. made the Divestiture decision even more stupid) was the fact that AT&T agreed to give away the ONLY part of its business that had any monopolistic value – and was most unlikely to be mangled in coming years by the inexorable and rapid gains in telecommunications technology. (In fact, the local phone companies were the only part of the old AT&T to prosper in the next 20 years.) The delicious and ironic epitaph to Divestiture of 1984 is that by 2005 – just over 21 years later – AT&T itself had ceased to exist at all. During those 21 years, it had disposed of one business after another until the tiny bit remaining (the long distance bit) was finally bought (basically put out of its misery) by SBC, which, for reasons that still leave me incredulous, decided to change its name to at&t – note the lower case. See also Baby Bells and Lucent Technologies.

divorce Rod Stewart says from now on he'll just avoid the whole marriage scene. "I'm going to find a woman I don't like and just give her the house."

Divx A new DVD (Digital Video Format) format that supports encryption and timed rentals. Divx is essentially a more expensive version of DVD. The foremost difference between it and standard DVD is the disk's encryption technology. The second difference is the way the Divx system allows for an on-demand viewing experience. Divx disks require a Divx player, which will sell for about $100 more than the price of a standard DVD player, and which will also play standard DVD disks and audio CDs. Users buy a Divx disk for about $5, and can watch it as many times as they like for a two-day period; the period begins when they first hit "play" on the Divx player. After the initial two-day viewing period, customers can play the disk at a much later time by just paying a fee. Moviephiles can also buy a special password to gain unlimited playback of the movie. Future viewings are billed in two-day periods, just like the initial viewing. The player keeps track of the number of periods the consumer has used and transmits this information over a household phone line to the consumer's account; the consumer then receives a bill in the mail. This system includes the key characteristics of the video rental model, like convenience, variety and low cost, but with important advantages. For example, the user never has to return the disks so, there are never any late fees. "It basically moves DVD technology into the rental domain," said one observer at the time of the announcement of the Divx technology in the fall of 1997.

DIW Type D Inside Wire. Originated as a specific AT&T cable. Now commonly used to describe any 22, 24, or 26 gauge PVC jacketed twisted-pair cable used primarily for inside telephony wiring.

DIX Digital/Intel/Xerox. The early 1980s consortium of manufacturers that promoted the Ethernet Version 1 and Ethernet Version 2 variations of a CSMA/CD media access protocol. This DIX "standard" was then submitted to the IEEE, where after some modifications it was released as IEEE Standard 802.3. The DIX version did not include specifications for UTP or Fiber Optic cable.

DIX Connectors A local area network connector. DIX connectors on the transceiver local area network cable link it to the network; the male DIX connector plugs into the SpeedLink/PC16 and the female DIX connector attaches to an external transceiver.

DIX Ethernet The DEC, Intel, Xerox Ethernet standard, also known as Version 1 or Bluebook Ethernet. There are subtle differences between IEEE 802.3 and the DIX Ethernet.

Djam Karet Pronounced Jam Karet. In Indonesian, it literally means rubber time. Appointments in Indonesia are either scheduled on "djam karet" time – meaning they start an hour or two late – or they are scheduled on "American time", in which case they start on time.

DL 1. Distribution List.

2. Distance Learning.

DLC 1. Digital Loop Carrier. Network transmission equipment used to provide pair gain on a local loop. The digital loop carrier system derives multiple channels, typically 64 Kbps voice-grade, from a single four-wire distribution cable running from the central office to a remote site. In the traditional deployment, Central Office Termination (COT) comprises multiplexing equipment in the central office (CO). A four-wire, twisted-pair circuit is deployed from the CO to the remote location at the point of Remote Termination (RT), where it terminates in matching DLC electronics. From the remote node, the interface is joined to individual voice-grade local loops which extend to the individual customer premises. Effectively, traditional DLCs are channel banks – devices which multiplex and demultiplex multiple channels over a high-bandwidth, electrical distribution facility. Such DLCs are used in situations in which the cost of the equipment is more than offset by the savings in distribution facilities through eliminating the need for a large number of individual copper pairs. Traditional DLCs also are known as SLCs (Subscriber Loop Carrier systems).

In a more contemporary scenario, the carriers deploy high-bandwidth fiber optic facilities from the COT to the RT. The carrier electronics at each end accomplish the optoelectric conversion process, as well as that of multiplexing/demultiplexing. The final leg of the local loop remains embedded unshielded twisted pair (UTP). This type of system can be characterized as a hybrid local loop system, and the DLC is known as a ngDLC (next generation DLC). ngDLCs support FTTN (Fiber-to-the-Neighborhood), which offers clear advantages in comparison to FTTC (Fiber-To-The Curb) and, certainly, to FTTH (Fiber-To-The-Home), given the high cost of the conversion processes. Additionally, the deployment of fiber optic distribution facilities yields much greater aggregate bandwidth – typically a minimum of 51.84 Mbps (SONET OC-1), which is the optical equivalent of 45 million bits per second (T3) in the electrical world. See also Channel Bank, DSLAM, Multiplexer, ngDLC, OC-1, SONET, and SLC.

2. See Data Link Control.

3. **Direct Line Console.** An AT&T Merlin term. An answering position used by system operators to answer calls, transfer calls, make calls, set up conference calls, and monitor system operations. Calls can ring on any of the line buttons, and several calls can ring simultaneously (unlike the QCC where calls are sent to a common QCC queue and wait until a QCC is available to receive a call).

DLCI 1. Data Link Connection Indicator: The number sequence that identifies public data networks.

2. Data Link Connection Identifier. A Frame Relay term defining a 10-bit field of the Address Field. The DLCI identifies the data link and its service parameters, including frame size, Committed Information Rate (CIR), Committed Burst Size (Bc), Burst Excess Size (Be) and Committed Rate Measurement Interval (Tc). See Tuple Address.

DLE Data Link Escape. A control character used exclusively to provide supplementary line control signals, control character sequences or DLE sequences. In packet switching, Data Link Escape is a name applied to the Control P non-print character which is used to swap the PAD from the data mode to the command mode in packet switched networks.

DLEC Data Local Exchange Carrier. A company which delivers high-speed access to the Internet, and not voice.

DLL 1. Dynamic Link Library. A feature of OS/2 and Windows that allow executable code modules to be loaded on demand and linked at run time. This lets library code be field-updated – transparent to applications – and then unloaded when they are no longer needed. Unlike a standard programming library, whose functions are linked into an application when the application's code is compiled, an application that uses DLL links with those DLL functions at runtime. Hence, the term dynamic.

2. Data Link Layer driver. A driver specification developed by DEC primarily to work with DECnet PCSA for DOS. DLL is a shared driver specification, allowing multiple protocol stacks to share a single network interface card.

DLO Diesel Locomotive. Cable with bunch or rope stranded copper conductors and a thermoset insulation/jacket. Usually manufactured in compliance with AAR specification.

DLP Digital Light Processing. It's a display technology based on Texas Instruments' Digital Micromirror Device (DMD) optical semiconductor chip. The chip's reflective surface holds many microscopic mirrors, each representing a single chip. These mirrors operate as optical switches to create a high-resolution, full-color image when hit with light from the projector's lamp. When I wrote this definition, TI had a chip out that sported 1.3 million mirrors and high-end DLP projectors gave out 7,000 lumens and use three DLP color chips – one for each RGB color. Some people think that DLP projectors are better than than LCD projectors. Some people like LCD projectors. DLP-based projectors are smaller and more portable.

DLPBC Dual Loop Port Bypass Circuitry. FC-AL (Fibre Channel Arbitrated Loop) routing circuitry that makes it possible to add and remove nodes without disrupting the network.

DLPI An ATM term. UNIX International, Data Link Provider Interface (DLPI) Specification: Revision 2.0.0, OSI Work Group, August 1991.

DLR Design Layout Report. A description of how a circuit is engineered. Often used between LECs and CLECs. See LEC and CLEC.

DLS Data Link Switching. IBM's method for carrying SNA and NetBIOS over TCP/IP operating at the Data Link layer. DLS, now an open Internet spec, can be used with OSPF or PPP.

DLSE Dial Line Service Evaluation.

DLSw DataLink Switching Workgroup. This workgroup has issued a new interoperability standard for integrating SNA and NetBIOS over the TCP/IP protocol. According to Cisco, the new DLSw standard provides interoperability and functionality not currently offered by Informational RFC 1434 or existing DLSw implementations.

DLTU Digital Link Trunk Unit. An AT&T term for a device which provides the interface to digital trunks and lines such as T-1, EDSL, and remote line units.

DM Delta Modulation.

DMA 1. Direct Memory Access. A fast method of moving data from a storage device or LAN interface card directly to RAM which speeds processing. In essence, DMA is direct access to memory by a peripheral device that bypasses the CPU.

2. Direct Marketing Association.

3. Document Management Alliance.

4. Designated Market Area. This term specifies the geographic area established by Nielsen Media Research for the purpose of rating the viewership of commercial television stations. DMAs represent the geographic areas covered by groups of competing commercial television broadcast stations. The boundaries of these areas are of considerable financial importance to the stations involved because they determine the number of viewers each station can claim, and, hence, the dollar amount the station can charge for advertising time.

Section 76.55(e) of the FCC Rules provides that, for the purpose of the cable television must-carry rules, the local market area of a commercial television broadcast station is the Nielsen DMA. A map showing DMA boundaries may be obtained from:

Nielsen Media Research 1290 Avenue of the Americas New York, NY 10104-0061 212-708-7500

DMA Channel A channel for direct memory access that does not involve the microprocessor, providing data transfer directly between memory and a disk drive. See DMA.

DMB 1. Digital Multipoint Bridge.

2. See Digital Multimedia Broadcasting.

DMD Differential Mode Delay. A fiber optic term. DMD refers to the fact that multimode and monomode (single mode) fiber optic cabling systems differ considerably in their performance characteristics. Specifically and in a LAN environment, some multimode systems are not capable of supporting signals from high-performance laser diodes, which operate at very high speeds and which emit narrowly-defined pulses of light. Traditionally, LEDs (Light-Emitting Diodes), which operate at lower speeds and which emit light pulses of a broader range of frequencies, have been use in conjunction with multimode fiber systems. In a LAN environment, such a combination has proved very satisfactory, even at high speeds and over relatively long distances. The combination of LEDs and multimode fiber also is much less costly. Ethernet at 10/100 million bits per second, Token Ring at 4/16 million bits per second, and ATM at 25/155/622 million bits per second all have made use of this traditional combination. However, the emerging Gigabit Ethernet standard requires the use of laser diodes to achieve such speeds. Multimode fiber, at least in some cases, appears to underperform in this environment, even over short distances. The problem is DMD, which yields unacceptably unmanageable levels of signal delay. See Modal Dispersion.

DMH Data Message Handler.

DMI 1. Digital Multiplexed Interface. AT&T's Digital Multiplexed Interface. A PBX to computer interface that divides the T-1 trunk into 23 user channels and one signaling channel. Also used as a T-1 PBX to computer interface. See Open Application Interface.

2. Desktop Management Interface, a protocol independent management interface developed by the DMTF, the Desktop Management Task Force, a group working on improving network printing. Making network printers easier to use – more standard and easier to use – requires three basic elements: standardized definitions of printer objects, a protocol for communicating with those objects and an application interface. What is DMI? The Desktop Management Interface (DMI) is the result of a cooperative, industry-wide effort to make PC systems easier to manage, use and control. Specifically, DMI is a specification developed by the Desktop Management Task Force (DMTF), a consortium of hardware, software, and peripheral vendors. DMI describes hardware and software components in a format that can be easily accessed over a phone line by PC management applications and technical support personal. Thanks to DMI, technical support personnel and PC management applications can:

- Access an inventory of hardware and software components
- Access and change parameter values and settings
- View data generated by software agents and diagnostics routines

With DMI, technical support applications and PC management applications use a Management Interface (MI) to access data stored in Management Information Files (MIFs).

DMI-BOS Digital Multiplexed Interface-Bit Oriented Signaling. A form of signaling, which uses the 24th channel of each DS1 to carry signaling information, allowing clear channel 64 Kbps functionality.

DMO Digital Modification Order.

DMP Device Management Protocol. The session-layer communications protocol used within the ICM. Different application-level protocols might be running beneath DMP.

DMS Digital Multiplex System. Also the name of a line of digital central office switches from Northern Telecom, also called Nortel. There are DMS-10s, DMS-100s, DMS-100Fs, DMS-200s, DMS-250s, and DMS-300s and by the time you read this, probably more.

DMS Defense Message System.

DMT Discrete Multi-Tone. A new technology using digital signal processors to pump more than 6 megabits per second of video, data, image and voice signals over today's existing one pair copper wiring. DMT technology, according to Northern Telecom, provides the following:

- Four "A" channels at 1.5 million bits per second. Each "A" channel may carry a "VCR"- quality video signal, or two channels may be merged to carry a "sports"- quality real-time video signal. In the future, all four channels operating together will be able to transport an Extended Definition TV signal with significantly improved quality over anything available today. ("A" channels are asymmetric – carrying information only from the telephone company to the subscriber's residence. All other channels within ADSL are symmetric or bi-directional.)

• One ISDN "H zero" channel at 384 Kbps (kilobits per second). This channel is compatible with Northern Telecom's multi rate ISDN Dialable Wideband Service or equivalent services. This channel could also be used for fast, efficient access to corporate LANs for work-at-home applications, using Northern Telecom's DataSPAN or other frame-relay services.

• One ISDN Basic Rate channel, containing two "B" channels (64 Kbps) and one "D" channel (16 Kbps). Basic Rate access allows the home user to access the wide range of emerging ISDN services without requiring a dedicated copper pair or the expense of a dedicated NT1 unit at the home. It also permits the extension of Northern Telecom's VISIT personal video conferencing to the home at fractional-T-1 rates (Px64).

• One signaling/control channel, operating at 16 Kbps giving the home user VCR-type controls over movies and other services provided on the "A" channel including fast-forward, reverse, search, and pause.

• Embedded operations channels for internal system maintenance, audits, and telephone company administration.

• Finally, the home user can place or receive telephone calls over the same copper pair without affecting the digital transmission channels listed above. And since ADSL is passively coupled to the POTS line, the subscriber's POTS capability is unimpaired in the event of a system failure.

DMTF Distributed Management Task Force. A consortium of vendors working toward a set of standards for network management software which will ease the management of desktop systems, their components, and peripherals. Activities address "standard groups" such as software, PC systems, servers, monitors, network interface cards (NICs), printers, mass storage devices, and mobile technologies. The most notable of its accomplishments is the development of the Distributed Management Interface (DMI). According to their web site, www.dmtf.org, "We are the industry organization leading the development, adoption and unification of management standards and initiatives for desktop, enterprise and Internet environments. Working with key technology vendors and affiliated standards groups, the DMTF is enabling a more integrated, cost effective, and less crisis-driven approach to management through interoperable management solutions." See also DMI.

DMZ DeMilitarized Zone. A partially-protected zone on a network, not exposed to the full fury of the Internet, but not fully behind the firewall. Also known as a screened subnet, a DMZ is a host computer or computer network which is inserted as a neutral zone between two other computer networks, one or both of which are untrusted. Most commonly, a DMZ sits between the Internet and an internal corporate network, and generally comprises a firewall and a bastion host. The firewall is some combination of software and hardware which examines TCP/IP packet traffic, blocking that packet traffic which does not meet pre-defined security requirements. The bastion host is a gateway designed to defend against hostile attacks launched from the external networks (i.e., the Internet), and aimed at the internal network. No connections are permitted from the untrusted network to the private network. Rather, remote users accessing the corporate network (e.g., Intranet or Extranet) through the Internet connect through the DMZ, which provides all services to remote users which the corporate network chooses to make available. Thereby, damage to the internal network is prevented, or at least made more difficult to accomplish by hackers and other unsavory characters. Further, the DMZ is not equipped to establish outgoing connections to the Internet, which prevents its being used as a tool to launch cascading attacks on other company networks. The term DMZ comes from the buffer zones established to separate untrusting nations which previously had been at war. At the end of WWII (World War II), a DMZ was established along the 16th Parallel, which separates North Vietnam and South Vietnam. That DMZ no longer exists, as the North Vietnamese were successful in reuniting the country. In 1953, at the end of the Korean Conflict (a euphemism for Korean War), a DMZ was established along the 38th Parallel between North and South Korea. That DMZ still exists, as the two nations technically remain at war, since no armistice was ever signed. See also Screened Subnet.

DN Directory Number or subscriber number or telephone number entry. The Directory Number is officially the seven digit number (723-8231) listed in the directory, hence the name. The Telephone Number is officially the 10-digit number (613-723-8231) of that subscriber, hence it includes the area code. The subscriber number used to be in the old days the same as the line number, the last four digits (8231) of the directory number. Nowadays, the subscriber number is synonymous with the directory number. The international format for a phone number consists of four items:

1. A + plus, which indicates that the caller should dial whatever digits are necessary to get him internationally. In the U.S., the prefix is 011.

2. The country code. The country code for the U.S. and Canada is 1. For Australia, it's 61. For England, it's 44.

3. The routing code, which is also called an area code.

4. The local number. In North America it's 7-digits. But in other countries it varies. At one stage in Sydney, Australia, some local numbers were five digits; some were six, some were seven and some were eight.

See Directory Number.

DNA 1. Digital Network Architecture. The framework within which Digital Equipment Corporation (DEC) designs and develops all of its communications products. DNA includes many standards of the OSI Model. Some of these standards will be adopted into ISDN. Acronym also by Network Development Corporation for their network offering.

2. Dynamic Node Access. A high-speed bus invented by Dialogic to join together multiple voice processing PEB-based systems. PEB stands for PCM Expansion Bus. See PEB.

3. Distributed Network Administration is an AMP wiring architecture that decentralizes network electronics, into closets on every floor. It saves money on initial installation and works especially well in multi-tenant building and small workgroup situations where you don't share want to share facilities, according to AMP. Distributed Network Administration can be executed with optical fiber or with Category 5 UTP. Since closets will be placed throughout the building, distance is usually not an issue in the choice of the medium.

DNA Computing Deoxyribonucleic Acid Computing. (This is not a joke.) DNA computing was first tested successfully by Leonard Adleman, a professor at the University of Southern California in Los Angeles. DNA computing is technique that takes advantage of massive parallelism through the use of DNA, test tubes, enzymes, electrophoretic gels and beads, gold-plated glass slides, and E. Coli bacteria. While several prototype DNA computers exist, the development of a practical system is years away.

DNAL Dedicated Network Access Line.

DNAR Directory Number Analysis Reporting.

DNC 1. Dynamic Network Controller.

2. Do not call (from laws prohibiting certain unsolicited telemarketing calls)

DNIC Data Network Identification Code. An address to reach a host computer system residing on a different packet switched network than the one you're on. The data equivalent of a telephone number with country code and area code. Typically the DNIC is a four digit number. The first three digits of a DNIC specify a country. The fourth digit specifies a public data network within that country. See also Data Network Identification Code.

DNIS Dialed Number Identification Service. DNIS is a feature of 800 and 900 lines that provides the number the caller dialed to reach the attached computer telephony system (manual or automatic). Using DNIS capabilities, one trunk group can be used to serve multiple applications. The DNIS number can be provided in a number of ways, inband or out-of-band, ISDN or via a separate data channel. Generally, a DNIS number will be used to identify to the answering computer telephony system the "application" the caller dialed. For example, a 401K status program may be offered by a service provider to a number of different companies. The employees of each company are provided their own 800 number to call to access their account status. When the computer telephony system sees the incoming DNIS number, it will know to which company the call was directed, and can so answer the phone correctly with a customized "you have reached the 401K line for xyz company. Please enter your personal account code and password..."

Here's another application: You use one 800 phone number for testing your advertisements on TV stations in Phoenix; another number for testing your ads on TV stations in Chicago; and yet another for Milwaukee. The DNIS information can be used in a multitude of ways – from playing different messages to different people, to routing those people to different operators, to routing those people to the same operators, but flashing different messages on their screens, so the operators answer the phone differently. In Ireland, incoming toll free phone calls from the rest of Europe arrive with DNIS. As a result a phone call arriving from Germany is routed to a computer telephony system playing messages in German. The advantage of DNIS is basically economic: You simply need fewer phone lines. Without DNIS you would need at least one phone line for every different 800 or 900 number you gave out to your callers. Make sure you understand the difference between DNIS and ANI and Caller ID. DNIS tells you the number your caller called. ANI or Caller ID is the number your caller called from. See 800 Service and 900 Service.

DNP3 Protocols define the rules by which devices talk with each other. DNP3 is a protocol for transmission of data from point A to point B using serial communications. It has been used primarily by utilities like the electric companies. A typical electric company may have a centralized operations center that monitors the state of all the equipment in each of its sub-stations. In the operations center, a computer stores all of the incoming data and displays the system for the human operators. Substations have many devices that need monitoring (are circuit breakers opened or closed?), current sensors (how much current is flowing?) and voltage transducers (what is the line potential?). The operations personnel often need to switch sections of the power grid into or out of service. One or more computers are situated in the substation to collect the data for transmission to the master station in the operations center. The substation computers are also called upon to energize or de-energize the breakers and voltage regulators. DNP3 provides the rules for substation computers and master station computers to communicate data and control commands. DNP3 is a non-pro-prietary protocol that is available to anyone. This means a utility can purchase master sta-tion and substation computing equipment from any manufacturer and be assured that they will reliably talk to each other. Vendors compete based upon their computer equipment's features, costs and quality factors instead of who has the best protocol. The theory is that electric utilities are not stuck with one manufacturer after the initial sale.

DNPA Data Numbering Plan Area. In the U.S. implementation of the ITU-T X.25 network, the first three digits of a network terminal number (NTN). See also Data Numbering Plan Area.

DNPIC Directory Number Primary InterLATA Carrier.

DNPS Divisional Network Product Support.

DNR 1. Dialed Number Recorder. Also called a Pen Register. An instrument that records telephone dial pulses as inked dashes on paper tape. A touchtone decoder performs the same thing for a touchtone telephone.

2. Dynamic Network Reconfiguration. Allows IBM networks to change addresses with-out reloading and bringing the network down.

3. A Satellite Term. Dynamic Noise Reduction is a filter circuit that reduces high audio frequencies such as hiss.

DNS 1. The Domain Naming System is a mechanism used in the Internet and on private Intranets for translating names of host computers into addresses. The DNS also allows host computers not directly on the Internet to have registered names in the same style. The DNS is a distributed database system for translating computer names (like ruby.or-a.com) and vice-versa. DNS allows you to use the Internet without remembering long lists of numbers. On TCP/IP networks (like the Internet), the Domain Naming System provides IP address translation for a given computer's domain name, or URL (Uniform Resource Locator). DNS would change a computer name such as harry.newton.com to the machine's actual numeric IP address, which is in the format xxx.xxx.xxx.xxx. The DNS makes it easier to remember where you want to go. Reverse DNS does just the reverse, translating an IP address into a URL. See also DNS Name Resolution and Domain Naming System for a more detailed explanation. See also 1983 and Reverse DNS.

2. Domain Name Server. Domain Name Servers, also known as resolvers, are a system of computers which convert domain names into IP addresses, which consist of a string of four numbers up to three digits each. Each applicant for a domain name (e.g., www.harrynewton.com) must provide both a primary and a secondary DNS server; a domain name which fails to provide both primary and secondary DNS servers is known as a "lame delegation." See also the first definition above.

3. See Distributed Network Service.

DNS Flood Attack See Domain Name Server Flood Attack.

DNS Name Resolution The matching of domain name to an IP number is known as DNS name resolution. In other words, you type in www.TechnologyInvestor.com and bingo the web site http://209.35.112.118/.

DNS spoofing An attack technique where a hacker intercepts your system's requests to a DNS server in order to issue false responses as though they came from the real DNS server. Using this technique, an attacker can convince your system that an existing Web page does not exist, or respond to requests that should lead to a legitimate Web site, with the IP address of a malicious Web site. This differs from DNS cache poisoning because in DNS spoofing, the attacker does not hack a DNS server; instead, he inserts himself between you and the server and impersonates the server.

DO A word in a high-level language program which comes before a collection of things to be done, i.e. statements to be executed.

Do-Not-Call A legal requirement to remove from a calling list the telephone numbers of people who have asked that they not receive unsolicited telephone calls. See Telephone Consumer Protection Act.

Do-Not-Disturb Makes a telephone appear busy to any incoming calls. May be used on intercom-only, by extension line-only or both.

Do-While A programming statement used to perform instructions in a loop while a certain condition exists – i.e. do something while the variable Y is less than 20.

DOA Dead On Arrival. A term several manufacturers use to refer to equipment which ar-rives at the customer's premises not working. A person who receives a DOA machine will ask the company for a NPR number – New Product Return number. This allows them to return the product and have the factory replace it with another new one.

DoB Denial of Business. See Impact Analysis.

DOC 1. Department of Communications. Canadian government department. The federal Department of Communications was merged into another department, Industry Canada, in the early 1990s.

2. See Dynamic Overload Control.

dock To insert a portable computer into a base unit. Cold docking means the computer must begin from a power-off state and restart before docking. Hot docking means the computer can be docked while running at full power.

docket Formal FCC/State regulatory commission proceeding, also referred to as a case.

docket number An FCC term. A number assigned to a proceeding opened by the issuance of either a Notice of Proposed Rule Making or a Notice of Inquiry, or if an adjudica-tory or tariff proceeding has been instituted. Since January 1, 1978, the docket numbers indicate the year they were initiated and the Bureau which initiated the docket (e.g.,MM89-494 and CC 87-313).

docking station Base station for a laptop that includes a power supply, expansion slots, monitor, keyboard connectors, CD-ROM and extra hard disk connectors. A user slides his laptop into a base station and in effect, gets the equivalent of a desktop machine.

docobjects See Document Objects.

DOCSIS Data Over Cable Service Interface Specification. A North American cable modem initiative, DOCSIS is now known as CableLabs Certified CableModem, a term trademarked by CableLabs. The specifications themselves remain known as DOCSIS, and are of several versions. DOCSIS 1.0 specifications, set in the 1995-1996 timeframe, specifies the rela-tionship between customer premises equipment in the form of cable modems and the CMTS (Cable Modem Termination System) at the head end of the service provider's network, and the specifics by which two-way data channels are carved out of the coaxial cable distribution systems. DOCSIS 1.1 (1999) enhancements include QoS (Quality of Service) mechanisms, support for concatenation and fragmentation, the addition of security in the form of authentication, support for SNMPv3 (Simple Network Management Protocol version 3). DOCSIS 2.0 is a developing specification that is intended to support upstream-intensive applications such as, peer-to-peer networking, videoconferencing, Web hosting, VOD (Video-On-Demand), and on-line gaming (i.e., gambling). See also CableLabs, CableLabs Certified CableModem, CMTS, Concatenation, DOCSIS 1.1, DOCSIS 2.0, DVB (European standard) and SNMP.

DOCSIS 1.1 While DOCSIS 1.0 is successful in delivering its designed task - high-speed data transmission over the cable network - it does not provide all of the Quality of Service (QoS), latency controls, and authentication capabilities. DOCSIS 1.1 was created in 1999 to incorporate the technology enhancements in the above areas. DOCSIS 1.1 opens a technological doorway to augmented revenue streams for cable providers. Equipment built to comply with the DOCSIS 1.1 specification becomes the foundation for expanding the list of advanced cable services offered by cable providers, including home networking through the CableLabs CableHome project, and packet telephony and multimedia services through CableLabs PacketCable project. Overall, DOCSIS 1.1 enables cable operators to deliver twice the level of functionality while reducing operating costs by half. See also DOCSIS.

DOCSIS 2.0 DOCSIS 2.0 is a major future cable data initiative spearheaded by Cable-Labs to further extend the capability of existing DOCSIS 1.0 and 1.1 standards. With DOCSIS 2.0, cable operators should be able to realize higher revenues through new up-stream-intensive broadband applications, including voice-over-Internet protocol (VoIP), peer-to-peer networking, video conferencing, Web hosting, video-on-demand, on-line gaming, and application services. Through the inclusion of advanced physical layer technologies, such as S-CDMA and A-TDMA, DOCSIS 2.0 not only increases the critical upstream data bandwidth, but also reduces susceptibility to noise and interference and guarantees back-ward compatibility and coexistence with DOCSIS 1.0 and 1.1 systems. CableLabs plans to complete the 2.0 version of the specification this year, and intends to accept DOCSIS 2.0 devices for certification and qualification in 2002. See also DOCSIS.

document camera A specialized camera on a long neck that is used for taking pictures of still images – pictures, graphics, pages of text and objects which can then be sent stand alone or as part of a video conference.

document commenting A Microsoft Web-based feature in which users will be able to add comments to a document, which are then saved in a database, not as HTML. Once the document is posted on a Web site, only users on a special discussion list will be able to view the comments.

document database An organized collection of related documents.

Document Image Management DIM. The electronic access to and manipulation of documents stored in image format, accomplished through the use of automated methods such as high-powered graphical workstations, sophisticated database management techniques and networking.

document objects DocObjects for short. Microsoft Office for Windows95 has an app called Office Binder which is a "document container that enables a user to manipulate text files, spreadsheets, graphics presentations, and other documents as a single entity." DocObjects is the "core technology that makes Office Binder work." This technology can support document containers other than Office Binder and document server as well. "The DocObjects technology consists of a set of extensions to OLE Documents. One application for this technology is in Internet browsers; a user would only need one navigation tool to browse and view all documents, whether local or network-based. A document provided from a DocObjects server is essentially a full-scale, conventional document that is embedded as an object within another DocObjects container (binder, browser, etc.)."

document recognition The ability to capture all the information on a page (text and images) and perform not only character recognition, but page structure analysis as well.

document root On a Web server, a directory that contains the files, images, and data you want to present to all users who access the server with a browser.

Document Type Definition DTD defines how XML markup tags should be displayed by an application. See DTD.

documentation Written text describing the system, how it works and how to work it. In most cases of high technology products, documentation is awful. Better documentation helps sell equipment and software. Please write your instruction manuals better. Please.

DOD 1. Direct Outward Dialing. A PBX trunk that supports outgoing dialing, only, and without the intervention of an operator or attendant. In PBX and hybrid phone systems, you pick up the phone, get "soft" (i.e., internal PBX) dial tone, dial 9, listen for a "hard" (i.e., outside) dial tone, and then dial the number you want to reach. In some phone systems, you don't have to listen for the second dial tone. You can dial straight through. All phone systems now have DOD. Traditionally, DOD trunks supported outgoing calling, only, and did not support incoming calls. More recently, some manufacturers allow individual channels in a given T-1 PBX trunk to be defined as DOD, some as DID (Direct Inward Dialing), and as combination (both inward and outward). See also Combination Trunk and DID.

2. Department of Defense.

DoD Common-User Systems The portion of the GIG (Global Information Grid), both switched and dedicated, that serves the entire Department of Defense community.

DOD Master Clock The U.S. Naval Observatory master clock, which has been designated as the DoD (Department of Defense) Master Clock to which DoD time and frequency measurements are referenced. This clock is one of two standard time references for the U.S. Government in accordance with Federal Standard 1002; the other standard time reference is the National Institute for Standards and Technology (NIST) master clock.

dodgy British slang for slightly suspicious. See also Bollocks.

DOE Direct Order Entry, now moving to the Internet for many industries.

dog Telephone in London Cockney rhyming slang is dog 'n' bone, bone obviously rhyming with phone, In cockney the rhyming word is often dropped leaving just dog. In cockney, the wife is called your trouble and strife. In England and Australia, a telephone iks also called a blower.

dog food Interim software used internally for testing purposes. Developers who use their own software will learn what is needed and what needs to be fixed. Eat your own dog food is the suggestion to software developers to actually use the products they develop.

doghouse 1. The closure containing the cellular PCS equipment. Some have heaters and airconditioning units for environmental control. The doghouse is located in a hut near the antenna.

2. Doghouse is where your customers will put you if they have problems with their cellular or fixed phone service.

Dolby A system of noise/hiss reduction invented by Ray Dolby, widely used in consumer, professional and broadcast audio applications. This technology is now also used to provide High Definition Television's clear sound. See HDTV.

domain 1. In the broadest sense, a "domain" is a sphere of influence or activity. In the vernacular, domain equals turf. In the MIS world, a domain is "the part of a computer network in which the data processing resources are under common control." In the Internet, a domain is a place you can visit with your browser – i.e., a World Wide Web site. In reality, "a place you can visit" might be a single computer. It might be a group of computers masquerading as a single computer. Or it might even be a logical/physical partition of a computer or group of computers. On the Internet, the domain is the address that gets you there, and consists of a hierarchical sequence of names (labels) separated by periods (dots). Examples of the 100 million plus Internet domain addresses include harrynewton.com and technologyinvestor.com.

A Top Level Domain (TLD) is defined as the alphabetical address suffix, which identifies the nature of the organization, e.g. com, edu, net, org. Currently, the central naming registry on the Internet is administered by the Internet Assigned Numbers Authority, (IANA), which includes the National Science Foundation (NSF), InterNIC, Network Solutions Inc., and the International Ad Hoc Committee (IAHC)..IANA, in turn, assigned responsibility to ICANN (Internet Corporation for Assigned Names and Numbers) for coordination of the following globally unique identifiers required for the Internet to function: Internet domain names, Internet Protocol (IP) address numbers, and protocol parameters and port numbers.

Only a select few Top Level Domains currently can be registered. The original TLDs include .com (commercial), .edu (educational institution), .gov (government), .org (nonprofit organization), .net (network provider) and .mil (military). There was a proposal that beginning in March 1998, new TLDs would be created in order to expand the available address options. After much wrangling, the decision was made on November 16, 2000 to create the following new TLDs: .aero (aviation industry), .biz (business), .coop (business cooperatives), .info (general use), .museum (museums), .name (individual, personal use), and .pro (professionals).

Example domain names include "teleconnect.com" and "computertelephony.com." They are different sub-domains, also known as secondary domains under the Top Level Domain. The secondary domain name can be up to 67 characters in length. Allowable characters include English letters, numbers, dots (i.e., "."), and underscores (i.e., "_"). Proposals are being considered to include certain non-English alphabets, such as Arabic, Chinese, and Japanese.

What you put in front of your domain name is entirely up to you, as long as the name is globally unique, and as long as it does not violate the naming convention in terms of total length and allowable characters. For example, everything addressed to SomethingInFrontOf@harrynewton.com or Something.In.Front.Of@HarryNewton.com will be routed to the domain "harrynewton.com" for resolution (further routing, processing, etc.). Under the supervision of ICANN, the Council of Registrars (CORE) has authorized over 100 independent, for-profit companies around the world to assign TLDs. The authorized registrars assign those TLDs through the SRS (Shared Registry System), which is a neutral, shared repository of URLs (Uniform Resource Locators), i.e., domain names or Web addresses.

Preceding both the secondary and Top Level Domain, increasingly can be found tertiary domains. For instance, a large organization might have hosts "www.company.com" and "mail.company.com." The tertiary domain might be called "support.company.com", with hosts "www.support.company.com" and "ftp.support.company.com." The domain may also include a geographical suffix in the form of a country code, indicating that the target organization is physically located outside the U.S. For example, .au is for Australia, .nz for New Zealand, and .po for Poland, and .za for South Africa (from the Boer "zud," meaning "south"). Without a geographical suffix, it is assumed that the address identifies a device physically located in the U.S., where the Internet originated. Note that if a country code is used, it becomes the TLD and the alphabetical address suffix such as .com becomes the secondary domain name.

In order to route the user or the user's mail to the correct location (correct domain) a "dip" is made into a database housed on a Domain Name Server (DNS). The DNS translates the alphabetical address into an IP (Internet Protocol) address, which corresponds to a logical address which is tied to a physical device in the form of a server, which is located at physical address, which is connected to the Internet. ICANN is responsible for the stable operation of the Internet's root server system. That master DNS database is downloaded to local DNS servers on a periodic basis in order that all DNS server databases are synchronized. See also Domain Controller, Domain Defined Attribute, Domain Name, Domain Name Server, Domain Naming System, Federated States of Micronesia, gTLD, Tuvalu, and URL.

2. Electronic mail vendors define domain in various ways. PC Magazine's networking editor, Frank Derfler defines a domain as referring to a set of hosts on a single LAN that needs only one intermediary post office to move mail from one host to another. A domain

Domain Name	Country Codes		
.biz Business	Country Codes are included where national boundar-	.dk (Denmark)	.lu (Luxembourg)
.bz Business	ies are crossed, and succeed the organizational	.do (Dominican Republic)	.lv (Latvia)
.com Commercial	extension. Where used, these ccTLDs (country code	.ec (Ecuador)	.mt (Malta)
.edu Educational	Top Level Domains) become the Top Level Domain,	.ee (Estonia)	.mx (Mexico)
.gov Government	and the gTLD becomes the Second Level Domain.	.es (Spain)	.my (Malaysia)
.info I nformation providers	Examples of ccTLDs include the following:	.fi (Finland)	.nl (Netherlands)
.mil Military	.ae (United Arab Emirates)	.fr (France)	.no (Norway)
.net Network providers	.ag (Antigua and Barbuda)	.gr (Greece)	.nz (New Zealand)
.org Non-Profits	.ar (Argentina)	.gt (Guatemala)	.pe (Peru)
New ones now include	.at (Austria)	.hk (Hong Kong)	.ph (Philippines)
.aero for services and companies dealing	.au (Australia)	.hr (Croatia)	.pl (Poland)
with air travel	.be (Belgium)	.hu (Hungary)	.pt (Portugal)
.coop for co-operative organizations	.bm (Bermuda)	.id (Indonesia)	.ro (Romania)
.museum for museums, archival institutions, and	.bn (Brunei Darussalam)	.ie (Ireland)	.ru (Russian Federation)
exhibitions	.bo (Bolivia)	.il (Israel)	.se (Sweden)
.name for individuals' and personal websites	.br (Brazil)	.int (International)	.sg (Singapore)
.pro for professions such as law, medicine,	.ca (Canada)	.in (India)	.si (Slovenia)
accounting, etc.	.cc (Cocos Islands)	.is (Iceland)	.su (Former USSR)
	.ch (Switzerland)	.it (Italy)	.th (Thailand)
The reality is anyone can secure any web site	.cl (Chile)	.jm (Jamaica)	.tw (Taiwan)
they want if no one else has it. Domain extensions	.cn (China)	.jp (Japan)	.ug (Uganda)
have become meaningless, Except that everyone is	.co (Colombia)	.kr (South Korea)	.uk (United Kingdom)
familiar with .com. That means it has a marketing	.cr (Costa Rica)	.kw (Kuwait)	.us (United States)
cache.	.cy (Cyprus)	.kz (Kazakhstan)	.uy (Uruguay)
	.cz (Czech Republic)	.lk (Sri Lanka)	.ve (Venezuela)
	.de (Germany)	.lt (Lithuania)	.za (South Africa)

may consist of only one host, depending on its design and implementation.

3. In IBM's SNA, a domain is a host-based systems services control point (SSCP), the physical units (PUs), logical units (LUs), links, link stations and all the affiliated resources that the host (SSCP) can control. In Microsoft networking, a domain is a collection of computers that share a common domain database and security policy that is stored on a Windows NT/2000 Server domain controller. Each domain has a unique name.

domain controller For a Windows NT Server domain, the primary or backup domain controller that authenticates domain logons and maintains the security policy and the master database for a domain.

Domain Defined Attribute DDA. In X.400 addressing, the DDA is a special field that may be required to assist a receiving E-mail system in delivering a message to the intended recipient. Up to four DDAs are allowed per address, with each DDA address entry made up of two parts, a type and a value. For example, if I were a subscriber to MCI Mail and I wanted to send a message to Harry Newton, this is how I would address it:

TO: Harry Newton
EMS: CompuServe / 592-7515
MBX: P=CSMAIL
MBX: DDA=ID=70600,2451

If I were a subscriber to CompuServe, the only addressing information I would need would be the number 70600,2451.

domain forwarding See Domain Pointing.

domain hopping A tactic by a spammer to avoid being detected and shut down, domain hopping is when the spammer abandons a domain and starts his spamming operation again on a new domain.

domain name In the broadest sense, a "domain" is a sphere of influence or activity. In the vernacular, domain equals turf. In the MIS world, a domain is "the part of a computer network in which the data processing resources are under common control." In the Internet, a domain is a place you can visit with your browser – i.e. a World Wide Web site. In reality, "a place you can visit" might be a single computer. It might be a group of computers masquerading as a single computer. Or it might even be a logical/physical partition of a computer or group of computers. On the Internet, the domain name is the address that gets you there, and consists of a hierarchical sequence of names (labels) separated by periods (dots). Examples of the 100 million plus Internet domain addresses include computertelephony.com (Computer Telephony magazine), harrynewton.com, technologyinvestor.com,

teleconnect.com (Teleconnect magazine), or ctexpo.com (CT Expo). All domain names have extensions. The common Top Level Domain extensions include the following:

domain name extension See Domain name.

domain name hijacking An attack technique where the attacker takes over a domain by first blocking access to the victim domain's DNS server, then putting up a malicious server in its place. For example, if a hacker wanted to take over fnark.com, he would have to remove the fnark.com DNS server from operation using a Denial of Service attack to block access to fnark's DNS server. Then, he would put up his own DNS server, advertising it to everyone on the Internet as fnark.com. When an unsuspecting user went to access fnark.com, he would get the attacker's domain instead of the real one.

Domain Name Server Domain Name Server (DNS) is a computer on the Internet which contains the programs and files which make up a domain's name database. Using a name server is much like placing a call to a 800/888 voice telephone number. The 800/888 number requires a "dip" into a database (on the DNS) in order to translate the name (e.g., harry@ctexpo.com) into a telephone number (IP address), which you then use to establish connection with the person (host computer). In other words, the DNS translates the logical alphanumeric address into a logical IP address associated with an applications server (or perhaps, a logical partition of a server), which is connected to the Internet. The telephone network (Internet) addresses by telephone number (IP address), not really by the person's (domain's) name. See Domain and Domain Naming System.

Domain Name Server DNS. Flood attacks send a high volume of Internet traffic to the name servers that are responsible for a particular Web domain, rendering those servers unresponsive. During the 2003 Iraqi War, Al-Jazeera's English and Arabic language Web sites were subject to DNS Flood Attacks. These attacks came shortly after Al-Jazeera published photos of U.S. soldiers who had been taken prisoner by Iraqi forces inside Iraq.

domain name squatter A person who registers many URLs – e.g. www.radioshack.com, www.att.com, www.harrynewton.com – in his own name and then sits on them until someone expresses an interest in buying them. He then sells them for whatever he can get – from $100 to $100,000, to who knows.

Domain Naming System DNS. in Unix-based networks, of which the Internet is the largest, a domain naming system is the commonly accepted way of naming computers on the network. Domain naming system is sometimes referred to as the BIND (Berkeley Internet Name Domain) service in BSD Unix, a static, hierarchical naming service for tcp/ip hosts. A DNS server computer maintains a database for resolving host names and IP addresses, allowing users of computers configured to query the DNS to specify remote computers by host names (in words) rather than IP addresses (which are only numbers). DNS domains should not be confused with Windows NT networking domains, although Windows NT does support and can use the Internet's DNS scheme. See Domain.

domain parking Same as Domain Pointing.

domain pointing I have one domain site. It's called www.TechnologyInvestor.com. I have another called www.InSearchOfThePerfectInvestment.com. when you punch up the second one in your Internet Browser, you'll end up at the first one. That's called Domain Pointing, Forwarding or Parking.

domain tasting A practice that lets people snatch up Internet domains, pay $6, then five days later return the domain name and receive their $6 back. This way they have the site for free for those five days. During that time, they jam their web sites full of ads that will bring them money should users click on the ads. If the sites produce a little money, then the person grabbing the site pays the money and registers it. If it doesn't, then the person returns the web site, and gets their money back. Some people register sites that look as though they're affiliated with someone else and then test them. For example, someone has registered www.NewtonsTelecomDictionary.com. That site has nothing to do with me. Some register names like www.Verizontone.com. The practice of domain tasting has soared recently. In late 2004, roughly 100,000 domain names were tested on any given day. By the time I wrote this – in early 2007 – the number has ballooned to over four million, Some people estimate that fewer than 2% of the sites that are tried out for a few days are ultimately purchased by registrants. As one article put it, domain tasting is a bit like being able to get clothes from a store, wear them for five days, and then return them at no charge. The explosion in domain tasting can be traced back to the practices and policies of the Internet. The organization responsible for regulating domain names is the Internet Corporation for Assigned Names & Numbers, better known as ICANN. Long ago, it established the "Create Grace Period," a five-day stretch when a company or person can claim a domain name and then return it for a full refund of the $6 registry fee. It was originally intended to allow registrars who had mistyped a domain to return them without cost. There's also the practice of kiting, which is the practice of registering a domain name, returning it within five days without paying, then quickly grabbing it again, without ever paying for it.

domestic arc The portion of the geostationary orbit allowing a satellite to return a footprint that almost covers the continental United States.

dominant carrier The long distance service provider which dominates a particular market and is subject to tougher regulation than its competitors. An FCC term. Essentially it means AT&T.

dominant mode The mode in an optical device spectrum with the mode power.

domino server The Internet and Intranet server packaged with Lotus Notes. Domino is the server, Lotus Notes is the client.

DOMSAT Domestic Communications Satellite.

don't answer recall Allows an extension user on a PBX to automatically retry a call by dialing a special digit code.

Donald Elliptical Projection Named after Jay K. Donald of AT&T, who invented in 1956 the basis for the V&H system used to rate calls in North America and many other countries. The concept is one of flattening the Earth, as it is much easier to calculate the distance between two points using a flat plane. The more difficult process would be that of performing a trigonometric function using degrees, minutes and seconds of latitude and longitude. See also V&H.

done deal A term used in the secondary telecom equipment business. Term used between seller and buyer to signify that a sale has been agreed to and an oral contract is now in effect, binding both parties to the agreed-to-sale as if a written contract has been signed. A written contract submitted later that does not conform to the original oral agreement is not justification for dissolving the original agreement unless both parties agree to the new written contract.

dongle 1. A device to prevent copies made of software programs. A dongle is a small device supplied with software that plugs into a computer port. The software interrogates the device's serial number during execution to verify its presence. If it's not there, the software won't work. A dongle is also called a hardware key.

2. A small pigtail cable with a connector at one end that attaches to a PCMCIA card (also called a PC card) inside your laptop and has an RJ-11, RJ-45, coax, or other connector at the other end. A dongle is used to connect the laptop to a network, a telephone line, or directly to another computer. See also Xjack – which is another alternative to a PC Card modem dongle.

3. An issue of SunExpert Magazine defines dongle as a 15-pin to 13-pin adapter. It is primarily used for large color monitors that require the 13W3 13-pin adapter.

donkeypower Donkeypower is a unit of power equal to 250 watts. It is computed that one donkeypower is 1/3 of a horsepower.

donor antenna An antenna that serves as the point of interconnection between the outside world and a distributed antenna system inside of a building or tunnel.

dooced Losing your job over something you wrote online. Named after Dooce.com, a blog run by Web developer Heather Armstrong. Armstrong got canned after anonymous critiques of her coworkers wer linked to her.

door A software program that allows access to files and programs not built into an electronic bulletin board system, thus letting users run them on-line.

door hanger Note left when customers are not home. Often information on installation, troubleshooting, and contact numbers are left.

doozy The original definition is that a doozy is an extraordinary one of a kind – a good things. For example, "The deal that AT&T gave us on those circuits was a real doozy" – meaning that AT&T gave the customer a great deal. Lately the term doozy has come to mean an extraordinary one of a kind screwup. But it's still used both ways. So you have to be careful to figure which meaning you're hearing – good or bad.

DOP See Dedicated Outside Plant.

dope To add impurities. See Doped Fiber and Semiconductor.

doped fiber Fiber optical cable treated with erbium. Such cable can carry signals three times farther than untreated fiber. Doped fiber is now the fiber of choice in long distance networks. See doped fiber amplifier and gene doping.

Doped Fiber Amplifier DFA. An amplifier used in fiber optic systems to amplify light pulses. Such amplifiers typically are known as EDFAs (Erbium-Doped Fiber Amplifiers) as they are doped with erbium, a rare-earth element. DFAs are more effective than regenerative repeaters in many applications, as they simply amplify the light pulses through a chemo-optical process. Regenerative repeaters, on the other hand, require that the light signal be converted to an electrical signal, amplified, and then re-converted to an optical signal. Additionally, DFAs can simultaneously amplify multiple wavelengths of light in a Wavelength Division Multiplex (WDM) fiber system. See also EDFA, SONET and WDM.

Dopeler Effect The Washington Post's Style Invitational asked readers to take any word from the dictionary, alter it by adding, subtracting or changing one letter, and supply a new definition. This one is one of the winners. The Dopeler Effect is the tendency of stupid ideas to seem smarter when they come at you rapidly.

doping See Semiconductor and Doped Fiber.

Doppler Aiding A signal processing strategy that uses a measured doppler shift to help the GPS Global Positioning System) receiver smoothly track the GPS signal. It allows more precise velocity and position measurement.

Doppler Effect The apparent change in wavelength of sound or light caused by the motion of the source, observer or both. Waves emitted by a moving object as received by an observer will be blueshifted (compressed) if approaching, redshifted (elongated) if receding. It occurs both in sound and light. How much the frequency changes depends on how fast the object is moving toward or away from the receiver.

Doppler Shift Doppler shift In radio communications is the apparent change of received frequency due to the relative motion of a transmitter, such as one on board a satellite and/or the motion of a mobile receiver; named after Christian J. Doppler (1803-53), Austrian physicist.

DOS 1. Disk Operating System, as in MS-DOS, which stands for MicroSoft Disk Operating System. A disk operating system is software that organizes how a computer reads, writes and reacts with its disks – floppy or hard – and talks to its various input/output devices, including keyboards, screens, serial and parallel ports, printers, modems, etc. Until the introduction of Windows, the most popular operating system for PCs was MS-DOS from Microsoft, Bellevue, WA.

2. DoS is Denial of Service. See Denial of Service.

DOSA Distributed Open Signaling Architecture. AT&T's preferred IP telephony architecture. It is incompatible with MGCP. See MGCP.

dot A dot is an integral part of an email address or a web site. You say "dot," not "period," since for most people a period means the end of a sentence. The English say "full stop," which is even more specific. My email address is "harry underscore newton (my mailbox) at msn (the site domain) dot com (commercial domain)." Keyboard translation: "Harry_Newton@msn.com" Dots are now replacing dashes in phone numbers. My phone number used to be 212-206-7140. Now many people believe it's 212.206.7140. Artists prefer the dots in phone numbers since dots save space. Some people claim it's the Swiss minimalist influence. See also Dot Address.

dot address Also known as "Dotted Quad" and "Dotted Decimal Notation." A set of four numbers connected with periods that make up an Internet address; for example, 147.31.254.130.

dot addressable graphics Refers to the mode of operation on a dot matrix printer which allows you to control each element in the dot matrix printhead. With this feature, you may produce complex graphics drawings.

dot com company A company which operates its business mainly on the Internet, using ".com" URLs. Amazon.com is the best example for the dot com company. As known as a Web retailer.

dot commer A person who works for a dot com company.

dot pitch A measure of the clarity of a color monitor. Dot pitch measures the vertical distance between the centers of like-colored phosphors on your screen measured in millimeters. The smaller the distance, the sharper the monitor. Dot pitch is the major determinant in the clarity of an image on screen. You can't do anything about it. When you buy a monitor, you buy it with a certain dot pitch and you're stuck with that dot pitch. You may be able, however, to do something about improving convergence and focus – the other measures of the clarity of a color monitor.

dot zero When new software is issued, it often bears the number .0 (i.e. dot zero) as in MS-DOS 5.0. The theory among software gurus is that you should always avoid a "Dot Zero" revision, since it will likely contain bugs and that one should wait for 4.01 or 5.01 etc. This theory has some validity, although MS-DOS 5.0 came out very clean and was not revised until 6.0.

dot-com company A company which operates its business mainly on the Internet, using ".com" URLs. Amazon.com is the best example for the dot com company. As known as a Web retailer.

dotted decimal notation Syntactic representation for a 32-bit integer that consists of four 8-bit numbers written in base 10 with periods (dots) separating them. It is used to represent IP address on the Internet, as in 192.168.67.2. This is also called dotted quad notation.

dotted notation The notation used to write IP addresses as four decimal numbers separated by dots (periods), sometimes called dotted quad. Example: 123.212.12.4.

dotted quad Also known as "Dot Address" and "Dotted Decimal Notation." A set of four numbers connected with periods that make up an Internet address; for example, 147.31.254.130.

dotting sequence A cellular radio term. An alternating series of 38 bits used for the purpose of symbol (bit) timing recovery at the CDPD Mobile Data BAse Station (MDBS) for reverse channel transmissions by the CDPD Mobile End System.

double buffering The use of two buffers rather than one to temporarily hold data being moved to and from an I/O device. Double buffering increases data transfer speed because one buffer can be filled while the other is being emptied.

double camp-on indication A PBX feature. A phone attempting to camp on to another phone which is already being "camped on" shall receive a distinctive audible signal and may be denied the ability to camp-on.

double click With a Mac and an IBM-compatible PC running Windows, double-clicking carries out an action, such as beginning a new program. Press and release the mouse button twice in rapid succession to double-click. If you don't double click fast enough, it won't work. Some mouse software allow you to assign the left hand button on the mouse to a single click and the right hand button to a double click. This is very useful.

double crucible method A method of fabricating optical fiber by melting core and clad glasses into two suitably joined concentric crucibles and then drawing a fiber from the combined melted glass.

double density Refers to a diskette which can contain twice the amount of data in the same amount of space as a single-density diskette. For example, a double-density 360k diskette has a 720k storage capacity. These days double density is an obsolete term, since there are now disks that are "double double" density. Most 3 1/2 inch disks will now hold twice 720K – or 1,440,000 bytes. These are called high density disks. Toshiba has introduced a "double double double" floppy, which will hold 2,880,000 bytes. But it hasn't caught on, yet.

double ended synchronization A synchronization control system between two exchanges, in which the phase error signals used to control the clock at one exchange are derived from comparison with the phase of the internal clock at both exchanges.

double gang receptable See Back Box.

double interrupted ring Two quick rings followed by a period of silence indicating the arrival of an outside call in some systems.

double modulation Modulation of a carrier wave of one frequency by a signal wave, this carrier then being used to modulate another carrier wave of different frequency.

double pole A double pole switch is one which opens and closes both sides of the same circuit simultaneously. Most electrical circuits open and close with only one side being broken.

double pull A method for pulling cable into conduit or duct liner that is similar to backfeed pulling except that it eliminates the need to lay the backfeed cable on the ground.

Double-Shielded Twisted Pair See SSTP.

Double Sideband Carrier Transmission DSBTC. That method of transmission in which frequencies produced by the process of Amplitude Modulation (AM) are symmetrically spaced above and below the carrier. The carrier level is reduced for transmission at a fixed level below that which is provided to the modulator. Carrier is usually transmitted at a level suitable for use as a reference by the receiver except in those cases where it is reduced to the minimum practical level (suppressed carrier). See also Amplitude Modulation, DSBSC, SSB and VSB.

doublewide Two trailer homes stapled together with a modest gabled roof.

doubler A device which doubles the distance of certain types of circuits. HDSL (High bit-rate Digital Subscriber Line) is a repeaterless means of provisioning a T-1 access circuit over a standard UTP (Unshielded Twisted Pair) local loop. Signal attenuation issues limit the range, however, to 12,000 feet on 24 gauge wire, and 9,000 feet on 26 gauge wire. A doubler can be added at that point in the outside plant to double the distance range, and a second doubler can be added to triple the range. As the doublers are line-powered, no local power supply is required. See also Attenuation, HDSL, Outside Plant, Repeater, T-1 and UTP.

doubly clad fiber An optical fiber, usually single mode, that has a core surrounded by an inner cladding of lower refractive index, which is in turn surrounded by an outer cladding, which has a higher refractive index than the inner cladding. This type of construction is often employed in single-mode fibers to reduce bending losses.

DOV Data Over Voice. A technology that allows data to be transmitted along with voice over a single voice-grade local loop. As the circuit typically is analog, a modem is required. There are several approaches to DOV. One approach uses FDM (Frequency Division Multiplexing), with voice traveling at 4 KHz and below, and data at higher frequencies. This approach is used in most DSL (Digital Subscriber Line) technologies, including ADSL and G.lite. Other approaches send data during periods of silence in a voice conversation. See also ADSL, DSL, FDM, and G.lite.

down converter A device for lowering the frequency of a modulated radio frequency carrier – typically one from a satellite or a distant microwave antenna. See Downconverter.

down sampling A sampling technique used in conjunction with certain compression algorithms, such as Wavelet Transform. Down-sampling computation involves disregarding certain samples; for instance, "down-sampling by two" considers only every other sample. This approach is an operation fundamental to the Fast Pyramid Algorithm used in wavelet transform, which is commonly applied to compression of image information. Up-sampling effectively is the reverse process of decompression on the receiving end of the data transfer. See Wavelet Transform, Fast Pyramid Algorithm and Up-Sampling.

down tilt A modification of an antenna radiation pattern which focuses energy along a line tilted below the horizontal radiation line. Down tilt uses energy which would otherwise be radiated toward the sky or into the ground.

downconverter A circuit that lowers the high-frequency signal to a lower, intermediate range. The three types of downconversion are signal, dual and block downconversion. Downconversion is accomplished by an integrated assembly of components which convert microwave signals (e.g. those from a satellite) to an intermediate frequency range for further processing. A downcoverter generally consists of an input filter, local oscillator filter, IF filter, mixer and frequently an LO frequency multiplier and one or more stages of IF amplification. May also incorporate the local oscillator, AGC/gain compensation components and RF preamplifier.

downline loading A system in which programs are loaded into the memory of a computer system, such as a LAN bridge, router or server, via the same communication line(s) the system normally uses to communicate with the rest of a network. As opposed to systems in which all programs are loaded into the computer from a disk or tape associated with the computer. A PC connected to a LAN may use this type of loading when it is first turned on in the morning to get the information it needs from a file server. Diskless PCs always work this way.

downlink 1. The part of a transmission link reaching from a satellite to the ground. Some satellite transmission circuits, especially international ones, are priced and billed separately for the uplink and the downlink. This is because their transmissions are provided by different carriers.

2. In packet data communications, a downlink is a link from an NC or PAD to another NC or PAD on a different level. The defining of downlinks and uplinks depends on the network configuration of PADs, their relationships to each other and the direction of data transmission.

download 1. To receive data from another computer (often called a host computer or host system or just plain host) into your computer. It's also called to RECEIVE. The opposite is UPLOAD or TRANSMIT. You have to be very careful distinguishing between the two. Choosing the "Download" option in some communications programs automatically erases a file of the same name that was meant for transmission. If that happens, stop everything. Grab your file unerase program and use it. Don't wait. If you wait, you may write over your erased file and never get it back. See Download Fonts.

2. The Vermonter's Guide to Computer Lingo defines download as to take the firewood off the pickup.

downloaded fonts Fonts that you send to a printer either before or during the printing of a document. When you send a font to a printer, it is stored in printer memory until it is needed. Downloaded fonts are one reason for loading lots of RAM memory into your printer.

downsampling See Down Sampling.

downsizing Downsizing is what happens when companies move from large computer systems to smaller systems. There are four major reasons companies downsize from mainframe-based computer to local area network-based computing: 1. They save money. There are several reasons: a. Mainframe computers cost lots each month in maintenance. They require costly maintenance agreements with the supplier, e.g. IBM. Servers are usually bought without maintenance agreements. When they break, the managers or the workers simply replace the broken parts themselves. b. Mainframe computers cost lots to program. There are comparatively few programs available for mainframes, compared to the plethora of off-the-shelf programs available for workstations. c. Servers require a far less costly home to live. You don't need air-conditioning, special buildings with raised floors, etc. 2. Servers today have the power of mainframes 10 years ago. In fact, servers are now beginning to acquire more power than mainframes of ten years ago. And as servers increasingly acquire several processors, they will leap in power beyond what mainframes have. 3. Servers are typically manufactured from off-the-shelf, standard components that are usually available from several manufacturers. As a result, there is constant competition and constant improvement in quality and features. 4. Servers are much more flexible tools to design networks. You can start with one baby network containing one server and several workstations (a.k.a. clients) and confined to one floor of one small building and grow to a huge, complex network containing thousands of workstations, dozens of servers and spanning the globe.

To most people, downsizing is not only swapping out the "big iron" (the mainframe) and bringing in servers and local area networks. It's also a new way of thinking about the way corporations are organized. Downsizing is often accompanied by re-engineering, which is basically re-organizing for a greater focus on the customer – a focus which means responding faster to customer needs. See Servers.

downstream 1. Refers to the relative position of two stations in a local area network ring topology. A station is downstream if it receives a token after the previous station. See also Downstream Channel.

2. In a communications circuit, there are directions of transmission – coming to you and going away from you. Downstream is another term for the transmission coming towards you. Downstream is used in cable TV – for the signal flowing to you from the cable head end. Downstream is used in modem connection for information coming at you. See Upstream Channel.

downstream channel The frequency multiplexed band in a CATV channel which distributes signals from the headend to the users. Compare with Upstream Channel, the band of frequencies on a CATV channel reserved for transmission from the user to the CATV company's headend.

Downstream Physical Unit See DSPU.

downtime The total time a telephone system is not working due to some software or hardware failure. Downtime is also defined as a time interval when a system is not in use either because of equipment failure (unplanned downtime) or scheduled maintenance (planned downtime).

DP 1. Dial Pulse (as in dialing a phone) or Data Processing. Also called EDP for Electronic Data Processing. Now more commonly called Management Information Systems – or MIS. See Dial Pulse.

2. Demarcation Point. The point of a demarcation and/or interconnection between telephone company communications facilities and terminal equipment, protective apparatus, or wiring at a subscriber's premises. Carrier-installed facilities at or constituting the demarcation point consist of a wire or a jack conforming to Subpart F of Part 68 of the FCC Rules.

DPA 1. Digital Port Adapter. A Northern Telecom word.

2. Demand Protocol Architecture. A technique for loading protocol stacks dynamically as they are required. It is associated with adapter cards in workstations and servers. Only the protocol stacks that are required for a particular communications sessions are loaded. Examples of such stacks include TCP/IP, XNS, SPX/IPX and NetBios.

DPBC Double Pole Back Connected.

DPFC Double Pole Front Connected.

DPS Digital passage service. DPS allows overseas carriers to interconnect with other overseas carriers between cable heads, earth stations, or border crossing points. The DPS is based on DS3s, which are diversely routed and continuously monitored.

DPBX Digital PBX. Not a common term. Most PBXs these days are digital. And they are simply called PBXs.

DPCM Differential Pulse Code Modulation, or Delta Pulse Code Modulation. DPCM is a variant of PCM, the technique generally used for converting analog voice signals into digital signals for transmission over a digital transmission system and through a digital switch. PCM samples the amplitude (i.e., signal strength, or volume level) of the native analog voice information stream at a rate of 8,000 times per second at precise intervals of 125 microseconds, i.e., 1/8000th of a second. PCM encodes each voice sample into an 8-bit byte, which yields a bandwidth requirement of 64 Kbps (64,000 bits per second). DPCM expresses each sample as the difference, or delta, between one sample and the previous sample, thereby requiring only four bits, rather than eight. As a result, DPCM voice requires bandwidth of only 32 Kbps, rather than 64 Kbps. Unfortunately DPCM sacrifices some voice quality in the process. DPCM also is used in some video compression applications, but also suffers from loss of quality. See also ADPCM and PCM.

DPCS Digital Personal Communications Services. See PCS.

dpi Dots Per Inch. A measure of a scanner's ability to scan. The higher the number, the sharper the image and the greater the potential size of the image. See Scanner Accuracy.

DPLB Digital Private Line Billing.

DPMI An acronym for DOS Protected Mode Interface. DPMI is an industry standard that allows MS-DOS applications to execute code in the protected operating mode of the 80286 or 80386 processor. The DPMI specification is available from Intel Corporation. It is a superset of the VCPI (Virtual Control Program Interface) specification for controlling multiple programs inside a PC, as well as programs that use protected mode.

DPMS Display Power Manager Signaling is a power reduction feature that places a computer monitor in reduced power when the monitor is still on but has been idle for some time.

DPNSS Digital Private Network Signaling System. A standard in Britain which enables PBXs from different manufacturers to be tied together with E-1 lines and pass calls transparently between each – as easily as if the phones were extensions off the same PBX and were simply making intercom calls. The international version of DPNSS is called Q.SIG, which is now becoming Q.931, which is Euro-ISDN.

DPO Dedicated Pair Out (DPO) and Dedicated Pair In (DPI). If you move out of a residential apartment, the pair that you were using is dedicated to that Apt. so when the next occupant moves in, it is a flick of a switch to get phone service turned on. That's called Dedicated Pair Out. The DPI pair is the OE (Office Equipment) or the Switch pair. Now the pair in the C.O. (Central Office), when it's joined to the DPO pair, is often called a MATED PAIR. The DPI pair is mated to the DPO pair.

DPP Distributed Processing Peripheral.

DPRAM Dual port RAM.

Dprovisioning Elimination of an existing subscriber account; deprovisioning of a subscriber account includes subscriber account deregistration and device de-activation.

DPSK Differential Phase Shift Keying, also called dibit phase shift keying. A 4-level phase shift technique. See Differential Phase Shift Keying.

DPST Double-pole single throw.

DPT Dial pulse terminate.

DPU 1. Digital Processing Unit.

2. Dynamic Path Update. Allows IBM networks to add new network nodes or change backup routing paths while the front-end is still operating.

3. Director of Public Utilities, a title often found at a state regulatory commission.

DPX DataPath loop eXtension.

DQDB Distributed Queue Dual Bus. The Metropolitan Area Network (MAN) access technique defined by the IEEE 802.6 standard for Switched Multimegabit Data Service (SMDS). Based on QPSX (Queued Packet Synchronous Exchange), developed at the University of Western Australia (of which fact Harry Newton is ever so proud), DQDB operates by maintaining a queue at each station to determine when the station may access its dual buses. The dual buses provide bidirectional transmission between originating and terminating stations with excellent congestion control ensured through the network, from end-to-end. DQDB consists of two uni-directional buses connected as an open ring (i.e., each bus is not directly interconnected). It provides for full duplex operation between any two nodes. Each SMDS cell consists of 5 bytes of control information, involving a header and trailer, plus 48 bytes of payload. See also SMDS and ATM.

DQPSK Differential Quadrature Phase Shift Keying. An improvement on QPSK, this compression technique transmits only the differences between the values of the phase of the sine wave, rather than the full absolute value. QPSK makes use of two carrier signals, separated by 90 degrees. See also PSK and QPSK.

DR Digital Radio.

Draconian Draco was an Athenian lawgiver. Around 621 B.C., the Greeks became tired of the privileged few making all the decisions. So, Draco was appointed to draft a set of laws, which turned out to be especially harsh. Those in debt could be enslaved, but only if they were members of the lower class. (He was tough, but not stupid.) Draco introduced the concept of "intent" to murder. He also introduced the distinctions between justifiable, accidental and intentional homicide. When asked about the harshness of his punishments, he supposedly said that the death penalty was appropriate for even so slight a crime as stealing a cabbage. Excessively severe rules or laws have come to be known as draconian in nature. See also Trivia.

draft proposal An ISO standards document that has been registered and numbered but not yet approved.

drag Dragging is a way of moving an item on the screen using your mouse. To drag a window in Windows, for example, move the mouse pointer onto a window's title bar, then hold down the mouse button while moving the mouse across your desktop. When you release the mouse button, the window will remain in its new location. Apply this technique to drag any data object, such as icons or list box items.

drag and drop The "drag and drop" definition defines how objects from one desktop application can be "dragged" out of that application, through clicking on the object with a mouse, across the desktop and "dropped" on another application. Most of the graphics operating systems, like Windows, Apple's Macintosh and Sun Sparc use Drag and Drop.

drag line A length of rope or string used to pull wire and cable through conduit or inaccessible spaces. Drag lines are often inserted in wall and ceilings during construction to ease future wire installation.

drag queen A drag queen is a man who likes to dress as a woman. In the 1940s, California law made it illegal to dress as a member of the opposite sex. Drag queens avoided this restriction by pinning pieces of paper to their dresses which read "I'm a boy!" The courts accepted the argument that anyone wearing such a notice was technically dressed as a man, not a woman. Some people claim that DSL lines are really T-1 lines in drag, but at a much lower price.

drain wire In a cable, an uninsulated wire laid over the component or components and used as a ground connection. It is the uninsulated wire in intimate contact with a shield to provide for easier termination of such a shield to ground. See also ground wire and ground.

DRAM 1. Dynamic Random Access Memory. Pronounced "dee-ram." The readable/writable memory used to store data in older PCs. DRAM stores each bit of information in a "cell" composed of a capacitor and a transistor. Because the capacitor in a DRAM cell can hold a charge for only a few milliseconds, DRAM must be continually refreshed to retain its data. In contrast, Static RAM, or SRAM, requires no refresh and delivers better performance, but it is more expensive to manufacture. See also DDR-SDRAM, EDO RAM, Flash RAM, FRAM, Microprocessor, RAM, RDRAM, SDRAM, SRAM, and VRAM.

2. Digital Recorder Announcer Module.

drawing In the manufacture of wire, pulling the metal through a die or series of dies in order to reduce the diameter to a specified size.

drawing tools A computer imaging term. The means of creating freehand lines or basic geometric shapes. Paint packages often provide an ellipse-drawing function as a variation of the circle (or vice versa) and a square drawing function as a variation of the rectangle. Virtually all packages offer filled geometric figures, the fill item being either a solid color or a pattern.

DRB Digital Radio Broadcasting.

DRCS Digital Radio Concentrator System.

dress down day A workday when employees are allowed to dress casually. When things were better economically, there were more dress down days.

dressing cable You "dress" cable by taking multiple cables and joining them together neatly with cable ties. A nicely-dressed, clean installation is a sign of telephony professionalism. It still exists.

drift 1. When a carrier frequency changes due to a transmitter problem. It can be caused by bad connections or defective components, temperature changes or diffraction. Crystal oscillators are the most drift-reliable circuits.

2. When customers leave and go to another carrier. See also Churn.

3. A long-term and usually steadily increasing change in a parameter, such as frequency, from its nominal value.

4. The tendency of a geosynchronous satellite in orbit to move towards one of the two satellite stable points, at 75 ⁻E and 105 ⁻W.

drill down Jargon which means to learn more about a subject.

drip irrigation One thing not to do is mess your company's entire relationship with a customer in the beginning of that relationship by unveiling a massive, complicated profile or questionnaire. Instead, gather the information that is useful for the current transaction, remember it, and build your relationship's context over time, little by little. Think of the task as a kind of "drip irrigation" for the relationship. This definition from Garage.com.

drip loop Several inches of slack in a cable that prevents water from collecting on the cable or running along the surface of the cable. A drip loop between the LNB of a satellite dish and the entry point into the building also allows some free movement of the dish while adjusting it.

drive Shortened word for hard drive.

drive by download Many of us have software on our PCs that we didn't put there. Such software might be malicious – stealing our security number and sending it to a distant computer. Or it might software that's semi-benign – perhaps telling a distant computer which web sites we visit, when and how often. This unrequested software is often call adware. In some cases, adware installs via "drive-by downloads," exploiting weaknesses in Internet Explorer that allow Web sites to run programs on users' PCs without them even clicking "OK." Programs downloaded in this way, can operate like the worst traditional viruses. The downloaded programs may install keylogging software to capture user passwords, send personal information back to a central server, and more. Service Pack 2 for Windows XP closes some but not all of the security flaws in Internet Explorer.

drive by spamming You have an "office" in your truck. You find a local Wi-Fi hotspot – an area you can receive and transmit 8021.11(b) or (a) wireless transmissions. You park your truck in the area. You send out three million spam emails. Then you disappear. No one can trace who sent them (and thus stop future transmissions) because you're long gone. Next time, one hopes, the owner of the Wi-Fi hotspot you used will improve his security and you won't be able to get on. Many Wi-Fi hotspots don't have security and can easily be found with a sniffer. See Sniffer.

drive icon A Windows NT term. An icon in a directory window in File Manager that represents a disk drive on your system. Different icons depict floppy disk drives, CD-ROM drives, network drives, etc.

drive mappings A Novell NetWare term. Drive mappings provide direct access to particular locations in the directory structure. They are a "shorthand" method for accessing directories on a disk. Instead of typing in the complete path name of a directory that you want to access, you can simply enter a drive letter that has been assigned to that directory. NetWare recognizes two types of drives (physical drives and logical drives) and three types of drive mappings (local, network, and search drive mappings).

drive test Hang a cell phone antenna from a crane in its proposed spot before construction of the permanent tower. Then technicians drive around with phones to see the range of the proposed spot. This way they see if it is worth having a tower in that proposed location.

drive type A number representing a standard configuration of physical parameters (cylinders, heads and sectors) of a particular type of hard disk drive. You need to know your drive's drive type; otherwise the BIOS of your machine will not recognize your drive on boot up and your PC will not work. Normally this information of your drive's type resides in memory kept alive by a small lithium battery. However, should the lithium battery die, your PC will "forget" which drive it has and it will ask you. If you don't know, you're in big trouble. Your PC simply won't work. My recommendation: write the drive type on two labels – one to stick on the drive and the other to stick on the bottom of your machine. This way you'll be able to find it easily. You will usually find the drive type's number on paperwork sent originally with your machine.

drivebar A Windows NT term. Allows you to change drives by selecting one of the drive icons.

driver A driver (which is always software) provides instructions for reformatting or interpreting software commands for transfer to and from peripheral devices and the central processor unit (CPU). Many printed circuit boards which you drop into a PC require a software driver in order for the other parts of the computer and the software you're running to work correctly. In other words, the driver is a software module that "drives" the data out of a specific hardware port. The port in question will usually have another device connected, such as a printer or modem, and the driver will be organized in software (i.e. configured) to communicate with the device.

drivethrough The McDonald's in Ft. Huachuca was the first restaurant to have a drive through window. The drive through allowed soldiers from Ft. Huachuca to get food since uniforms were not permitted in business establishments.

driveway effect A special quality exhibited by a radio program that causes listeners to stay in their cars after they have arrived home so they can hear the end of the program.

DRM See Digital Rights Management.

drop 1. A wire or cable from a pole or cable terminal to a building.

2. That portion of a device that looks toward the internal station facilities, e.g., toward an AUTOVON 4-wire switch, toward a switchboard, or toward a switching center.

3. Single channel attachment to the horizontal wiring grid (wall plate, coupling, MOD-MOD adapter).

4. The central office side of test jacks.

5. To delete, intentionally or unintentionally, part of a signal for some reason, e.g., dropping bits.

drop a dime A slang term meaning to inform on someone. The term originated in the days when it cost ten cents to make a local call on a payphone. An informant would "drop a dime" in a payphone to make an anonymous call to the police.

drop and insert That process wherein a part of the information carried in a transmission system is demodulated (dropped) at an intermediate point and different information is entered (inserted) for subsequent transmission.

drop cable 1. The outside wire pair which connects your house or office to the transmission line coming from the phone company's central office. See also DROP WIRE, which is different.

2. In local area networks, a cable that connects a network device such as a computer to a physical medium such as an Ethernet network. Drop cable is also called transceiver cable because it runs from a network node to a transceiver (a transmit / receiver) attached to the trunk cable.

drop channel operation A type of operation where one or more channels of a multichannel system are terminated (dropped) at any intermediate point between the end terminals of the system.

drop clamp A piece of equipment used to attach aerial wire to a J hook or Ram's Horn on a building or pole.

drop loop The segment of wire from the nearest telephone pole to your home or business.

drop on ground Usually a temporary condition where a drop is laid on the ground to get service to a location. This may be due to a cut cable, cable damage, or the original drop has yet to be installed.

drop outs Drop outs are one major cause of errors in data communications circuits. The technical definition is that the signal level drops more than 12 dB (decibels) for more than 4 milliseconds. It means some of your data will not arrive. A four millisecond drop out in a transmission at 2,400 bits per second will lose about ten bits. A "drop out" is similar to a person's voice fades away in a telephone conversation. To correct the problem of drop out, we will ask the person (or computer) at the other end to repeat what they just said. "Huh?" This is called retransmission of data. In telephony, drop outs are defined as incidents when signal level unexpectedly drops at least 12 dB for more than 4 milliseconds. (Bell standard allows no more than two drop outs per 15 minute period.)

drop reel A reel used to transport and distribute drop wire.

drop repeater A repeater that is provided with the necessary equipment for local termination (dropping) of one or more channels of a multichannel system.

drop set All parts needed to complete connection from the drop (wall plate, coupling, MOD-MOD) to the terminal equipment. This would typically include a modular line cord and

interface adapter.

drop shipment Equipment shipped to a buyer from a location separate from the seller's premise. If this third location is a different company, then this third-party supplier bills the seller and the seller bills the buyer. This saves time, but the seller loses control over the equipment's condition.

drop side Defines all cabling and connectors from the terminal equipment to the patch panel or punch down block designated for terminal equipment at the distribution frame.

drop wire Wires going from your phone company to the 66 Block or protector in your building. See also Distribution Cable.

dropout A short period of time during which a transmission service looses the ability to transmit data. Bell System specifications define a dropout as any such loss which lasts for more than four milliseconds. See DROP OUTS for a longer explanation.

dropped The system will be dropped. Typically it means that a set of changes has been made operational in the telephone or computer system. This is interesting because the meaning is the opposite of what you would think.

dropped call A call in which the radio link between the cellular customer and the cell site is broken. Dropped calls can happen often, and for many reasons, including terrain, equipment problems, atmospheric interference, and traveling out of range. In short, a dropped call is a call terminated by other than the calling or called party.

DRP 1. Distribution and Replication Protocol. A proposal for an alleged improvement to HTTP. DRP is intended to improve on HTTP in two ways: by enabling multiple files to be downloaded over a single connection and by limiting the amount of data that needs to be downloaded each time a user returns to a Web site or downloads an update of an application. Currently HTTP requires a separate connection for each item being downloaded, which can slow the download process if many connections need to be opened. HTTP is also stateless, meaning that no information about a Web site is maintained on the client machine once the user moves on to another site, with the exception of a temporary cache and cookies. See HTTP. DRP is currently being studied by The World Wide Web Consortium (W3C).

2. Disaster Recovery Plan. The procedures that are to be executed to restore telecom, data, phone and network operations if catastrophe were to occur (e.g. a fire, a floor, a terrorist act, etc.).

DRS 1. Digital Reference Signal.

2. Digital Reconfiguration Service. DRS is intended for those private line customers who reconfigure their networks regularly or from disaster recovery to accommodate their applications such as video conferencing, high speed backups, etc.

DRSN Defense Red Switch Network.

DRT DS3 Redundancy and Termination.

DRTL 1. Dial repeating tie line.

2. Dial repeating trunk line.

DRTT Dial repeating tie trunk.

DRU 1. DACS Remote Unit.

2. Digital Remote Unit. An NEC term. It's a multiplexer used to distribute NEC Dterm digital telephones and analog sets throughout the user's communications network, whether that network is local or geographically dispersed. The multiplexing technology used is North American Standard T-1 and European Standard E-1.

drug runs Bus trips to Mexico and Canada chartered by US senior citizens looking to "score" cheaper prescription drugs.

drum factor In facsimile systems, the ratio of drum length to drum diameter. Where drums are not used, it is the ratio of the equivalent dimensions.

drum speed The angular speed of the facsimile transmitter or recorder drum, measured in revolutions per minute.

drunken Swede A way of describing the sound of a computer doing text-to-speech conversion. "Why, he sounds like a drunken Swede." This great definition from Stuart Segal of Phone Base Systems, Inc. in Vienna, VA. Says Stuart, "Our people think that a drunken Swede has recorded this message." It is possible to have a computer generate speech that doesn't sound like a drunken Swede if you throw sufficient horsepower (MIPS and memory) at it. Throwing sufficient horsepower, however, has been expensive, until recently. Drunken Swedes are going to get less and less common as horsepower gets cheaper and cheaper.

dry Cable with no electronics and with no telecommunications transmission on it. In short, raw copper pair. You rent raw copper pair when you want to put your own electronics on it.

dry cell A type of primary cell in which the electrolyte is in the form of a paste rather than that of a liquid.

dry contact A dry contact refers to a circuit with an energy level such that a spark is not created in a mechanical relay or switch contracts when the circuit is opened. As a result, no cleaning of the contacts takes place (sparking vaporizes contact materials thereby continually exposing fresh contact material). Sometimes general purpose contacts are gold-flashed so the relay or switch can be used on dry circuits; when used on higher energy circuits the gold coating is destroyed, but no damage is done except that the contact should not be used to carry dry circuit signal levels again. A dry contact might operate a relay which might turn something of higher power on or off. For example, a low voltage signal in a key system might cause a dry contact to close, thus causing much higher voltage to flow to a bell, a klaxon, a strobe light. And, yes, there is a "wet" contact. The term "mercury-wetted relay" refers to a relay or switch in which the movable contact of the device makes contact with a pool of mercury. In fact, before solid-state devices, this was a common technology for switching dry circuits.

dry copper pair A dry copper pair or circuit is one which is has no electronics (e.g., load coils, repeaters or subscriber carrier systems) between the wire center and the customer premises. Such electronics interfere with the provisioning and support of high-speed data lines, such as DSL (Digital Subscriber Line) services, which many CLECs are interested in selling. See also CLEC, Dedicated Inside Plant, Dry Copper, DSL, ILEC and See Dry Twisted Pair.

dry electrolytic condenser An electrolytic condenser in which the electrolyte is in the form of a paste or jelly rather than that of a liquid.

dry loop Another term for a naked DSL. See naked DSL.

dry loop powering Refers to local (not span) powering, and a transmission medium other than copper wire (microwave/fiber optic).

Dry Martini Draft An industry nickname for a draft IETF specification for using pseudowires to provide virtual private network services such as leased lines, Frame Relay, ATM, and Point-to-Point Protocol over a sub-IP network such as SDH and SONET. The Dry Martini Draft builds on the Martini Draft and Pseudowire Emulation Edge-to-Edge. The "Martini" nickname comes from Luca Martini, the primary author of an earlier IETF draft specification, which the Dry Martini Draft generalizes and extends. See Martini Draft, Pseudowire Emulation Edge-to-Edge.

dry pair A "dry pair" is a simple pair of wires with no voltage, signals or protocols, just two wires. These are sometimes used for alarm systems but can conceivably be used for other communications if the pair is "loop qualified" before implementation.

Dry T-1 A T-1 line with an unpowered interface. A T-1 line with a power is called "Wet."

dry twisted pair Also known as "dry copper," a dedicated twisted-pair circuit without loading coils or any sort of electronics, whatsoever. Dry twisted pair circuits are commonly leased from the LEC for burglar alarm circuits, with one end terminating at the customer's premise in an alarm box and with the other end terminating at the premise of the alarm company. Recently, a number of CAPs (Competitive Access Providers), CLECs (Competitive Local Exchange Carriers) and ISPs (Internet Service Providers) have been leasing dry copper circuits from telephone companies for purposes of provisioning high-speed xDSL services. Since such xDSL services compete with LEC services and since xDSL can cause crosstalk problems with adjacent pairs in the same cable system and since it deprives them of revenue, several of the LECs recently have refused to lease dry copper. See Coppertone.

DS 1. Digital Signal. See DS-, DS-0, DS-1.

2. Danske Standardiseringsrad (Danish Standards Institution).

3. An ATM term. Distributed Single Layer Test Method: An abstract test method in which the upper tester is located within the system under test and the point of control and observation (PCO) is located at the upper service boundary of the Implementation Under Test (IUT) - for testing one protocol layer. Test events are specified in terms of the abstract service primitives (ASP) at the upper tester above the IUT and ASPs and/or protocol data units (PDU) at the lower tester PCO.

See also DS-.

DS- Digital Signal (level). A hierarchy of digital signal speeds used to classify capacities of digital lines and trunks. The fundamental speed level is DS-0 (64 Kbps, i.e., 64 Kilobits per second, or 64,000 bits per second), which is a voice grade channel. That is the bandwidth you need when you use traditional PCM (Pulse Code Modulation) to sample a voice call 8,000 times a second and encode it in an 8-bit code, yielding 8 x 8,000 = 64,000 bits per second. The highest speed in the digital signal (DS) hierarchy is DS-4, at about 274 Mbps. The full DS hierarchy is as follows: DS-1, DS-1C, DS-2, DS-3, DS-4. The DS hierarchy takes different forms in different parts of the world, where different standards generally apply. In North America, it takes the form of T-carrier; in Europe, E-carrier (generally known as the international standard); and in Japan, J-carrier. In North America, therefore, DS-1 becomes T-1, DS-2 becomes T-2, and so on. The only exception is DS-0, which is known as DS-0 across all of the various standards. The full T-carrier hierarchy, for example, comprises T-1 at 1.544 Mbps, T-1C at 3.152 Mbps, T-2 at 6.312 Mbps, T-3 at 44.736 Mbps and T-4 at 274.176 Mbps. Technically speaking, the originating data stream is in the form of a DS-1, at the speed specified by the prevailing standard. When the data stream arrives at the channel bank, which is the point of demarcation, it is framed according to the prevailing standard, signaling and control bits are added, and the signal is encoded into the proper electrical format. For example, a data stream of 1.536 Mbps arrives at the channel bank, 0.008 Mbps of framing bits are added, and the signal exits the channel bank at a T-1 signaling rate of 1.544 Mbps. See the following DS-x definitions. Also, see channel bank, and T-.

DS-0 Digital Signal, level 0. A DS-0 is a voice-grade channel of 64 Kbps (i.e., 64,000 bits per second. This channel width is the worldwide standard speed for digitizing one voice conversation using PCM (Pulse Code Modulation. The analog signal is sampled 8,000 times a second, with each sample encoded into an 8-bit byte (thus 8 x 8,000 = 64,000). There are 24 DS-0 channels in a T-1, the North American version of DS-1. See also DS-, and PCM.

DS-1 Digital Signal, level 1. It is 1.544 Mbps (i.e., 1.544 million bits per second) in North America (T-1) and Japan (J-1), and 2.048 Mbps in Europe (E-1), i.e. 2,048,000 million bits per second. Those speeds are symmetrical. That means they're the same in both directions. Many people ask why are they different? The first thing you have to understand is the purpose * The reason for developing it was to increase the number of voice grade interoffice trunks that could function over a single twisted pair of wires. The Beta testing was conducted in New York's Manhattan in 1961. When the T-1 standard was developed in North America, the engineers looked at the distribution cables then in use. What they found were cables consisting of 24 AWG wire pairs with loading coils installed every 6000 feet. Bell Labs did some experiments and found that the maximum bit rate that could be achieved over each 6000 foot span was 1.544 Mbps. The loading coils were replaced with repeaters to terminate and re-generate the signals. At the same time, Ericsson was following a similar path in London. The loading coil spacing in Europe was only 4000 feet, so the maximum achievable bit rate was higher (2.048 Mpbs). Why there's no consistency is one of those wonderful, unanswered, questions. T-1 is the original standard, having been developed by Bell Telephone Laboratories in the 1950s. Subsequently, the ITU-T developed the European E-1 standard, variant, which runs at 2.048 Mbps. The Japanese developed the J-1 standard, which also runs at 1.544 Mbps, but is incompatible with T-1. The T-1 standard, at 1.544 Mbps, for example, supports 24 voice conversations, each encoded at 64 Kbps. The E-1 standard, at 2.048 Mbps, supports 30 conversations, plus two signaling and control channels, for a total of 32 channels, each of 64 Kbps. See also T-1.

DS-1C Digital Signal, level 1 Combined. The total signaling rate is 3.152 Mbps in North America and comprises two T-1s (two J-1s in Japan), at 1.544 Mbps each, which are interleaved to support 48 DS-0s. The additional 64 Kbps is overhead used to support additional signaling and control requirements. DS-1C is seldom used, outside of limited telco applications. There is no European equivalent. See DS- and DS-1.

DS-2 Digital Signal, level 2. DS-2 effectively translates to T-2 in North America, and J-2 in Japan. (There is no European equivalent.) DS-2 supports a total signaling rate of 6.312 Mbps. It is the equivalent of 4 T-1s, and supports 96 DS-0 channels of 64 Kbps, plus overhead in support of additional requirements for signaling and control functions. DS-2 is used in carrier (telco) applications, and only rarely. See DS-, DS-0, DS-1, T-1 and T-2.

DS-3 Digital Signal, level 3. In North America and Japan, DS-3 translates into T-3, which is the equivalent of 28 T-1 channels, each operating at total signaling rate of 1.544 Mbps. The 28 T-1s are multiplexed through a M13 (Multiplex 1 to 3) Multiplexer), and 188 additional signaling and control bits are added to each T-3 frame. As each frame is transmitted 8,000 times a second, the total T-3 signaling rate is 44.736 Mbps. In a channelized application, T-3 supports 672 channels, each of 64 Kbps. In the European hierarchy, a DS-3 is in the form of an E-3, which runs at a total signaling rate of 34.368 Mbps, supports 480 channels, and is the equivalent of 16 E-1s. A J-3 runs at 32.064 Mbps, supports 480 channels, and is the equivalent of 20 J-1s. If you're moving a DS-3 (or any other DS signal) across continents, the standards of the target country rule. Channels get muxed and demuxed, with signaling conventions translated, as well. Here is a question from a reader: On the US side T-1s are in multiples of 24x64 Kbps circuits and in the UK we have two-megabit circuits with 30x64 Kbps. If we were to want to interconnect to the US at the DS-3 level, would we receive 28 T-1's with 6 spare channels, or do they get muxed and demuxed into multiples of 30 when they arrive over this side of the world? Answer: They get muxed and demuxed, along with signaling conventions translated into multiples of 30 when they arrive in the U.K. DS-3 is also called T-3. See DS-, DS-0, DS-1, T-1, T-2 and T-3.

DS-4 Digital Signal, level 4. T-4 runs at a total signaling rate of 274.176 Mbps in North America in support of 168 T-1s, yielding 4032 standard voice-grade channels. E-4 runs at 139.264 Mbps in Europe in support of 64 E-1s, yielding 1920 voice-grade channels. J-4 runs at 397.200 Mbps in Japan in support of 240 J-1s, yielding 5760 channels. See DS-0, DS-1, T-1, T-2 and T-3.

DS0 Usually written DS-0 (and pronounced D S zero). Digital Signal, level Zero. DS0 is 64 Kbps. As the basic building block of the DS hierarchy, it is equal to one voice conversation digitized under PCM. Twenty-four DS-0s (24x64 Kbps) equal one DS-1, which is T-1, or 1.544 Mbps. See DS-0.

DS0-A Refers to a process where a subrate signal (2.4, 4.8, or 9.6 Kbps) is repeated 20, 10 or 5 times, respectively to make a 64 Kbps DS-0 channel.

DS0-B Refers to a process performed by a subrate multiplexer where twenty 2.4 Kbps, ten 4.8 Kbps, or five 9.6 Kbps signals are bundled into one 64 Kbps DS-0 channel.

DS1 See DS-1.

DS2 See DS-2.

DS3 See DS-3.

DS3 PLCP An ATM term. Physical Layer Convergence Protocol: An alternate method used by older T carrier equipment to locate ATM cell boundaries. This method has recently been moved to an informative appendix of the ATM DS3 specification and has been replaced by the HEC method.

DS4 See DS-4.

DSA 1. Digital Signature Algorithm. A public key digital signature algorithm proposed by the National Institute of Standards and Technology.

2. Distributed Systems Architecture, the network architecture developed by Honeywell.

3. Directory System Agent. The software that provides the X.500 Directory Service for a portion of the directory information base. Generally, each DSA is responsible for the directory information for a single organization or organizational unit.

4. Data Service Adapter.

5. Digital Service Area   (geographical area that correlates with Remote Terminal and/or Digital Loop Carrier).

DSAP Destination Service Access Point. The logical address of the specific service entity to which data must be delivered when it arrives at its destination. This information may be built into the data field of an IEEE 802.3 transmission frame.

DSAT Digital Supervisory Audio Tones. A supervisory signaling scheme used in NAMPS – a new form of digital cellular radial called Narrow-band Advanced Mobile Phone service. See also NAMP.

DSB Digital Sound Broadcasting.

DSBSC Double SideBand Suppressed Carrier. A variation on the theme of DSBTC (Double SideBand Transmitted Carrier), DSBSC uses less power since the carrier is not sent. The same amount of bandwidth is used, and carrier synchronization is lost. See also Amplitude Modulation, DSBTC, SSB and VSB.

DSBTC Double SideBand Transmitted Carrier. A form of Amplitude Modulation (AM) used to encode analog signals for transmission over a digital facility. DSBTC multiplies the carrier signal by modulating the amplitude (volume) of the carrier signal, plus adding a dc (direct current) component. The resulting output has some level of redundancy as the output is the sum of the carrier signal plus symmetric components in sidebands, which are some frequency separation from the carrier frequency. See also Amplitude Modulation, DSBSC, SSB and VSB.

DSC Digital Selective Calling, a simplex radio system used for transmitting distress alerts from ships and for transmitting the associated acknowledgements from coast stations. DSC is a synchronous system using a 10-unit, error-detecting code. Information in a call is presented as a sequence of seven-unit binary combinations. The frequencies used are 2187.5 kHz in the medium-frequency (MF) band, 4207.5, 6312.0. 8414.5, 12,577.0 and 16,804.5 kHz in the high-frequency (HF) band, and 156.525 MHz (Channel 70) in the Very-High Frequency (VHF) band. Received DSG messages trigger audible or visible alarms and are either displayed or printed out. A message normally contains an address, the identification of the transmitting station and the message itself. The transmission speed is 100 baud at MF and HF and 1200 baud at VHF. On board a ship, the DSG equipment usually consists of an add-on unit connected to the MF/HF or VHF radio transceiver. DSG is included in GMDSS. DSCS abbreviation of Defense Satellite Communication System, the operational Department of Defense (DOD) satcom system evolved in three phases over two decades starting in 1962.

DSF Dispersion Shifted Fiber (DSF) improves on Non Dispersion Shifted Fiber (NDSF) by shifting the interface between the fiber core and cladding. DSF suffers from the noise phenomena of chromatic dispersion and waveguide dispersion. Those two types of dispersion seriously impact the integrity of the optical signals at windows (i.e., wavelength ranges) other than 1310nm (nanometers). However, other wavelengths in higher windows are much better suited for long haul transmission. DSF solves the dilemma through several types of fiber: Zero Dispersion Shifted Fiber (ZDSF) and Non Zero Dispersion Shifted Fiber (NZDF). ZDSF shifts that interface to the point that chromatic dispersion and waveguide dispersion cancel each other out at 1550nm (nanometers), rather than the 1310nm window used in NDSF. DWDM (Dense Wavelength Division Multiplexing) and EDFAs (Erbium-Doped Fiber Amplifiers) work in this window. In this window, however, another noise problem surfaces in the form of four-wave mixing, in which the signal wavelengths interact to create additional wavelengths, which serve to confuse the signal. Non Zero Dispersion Shifted Fiber (NZDF) shifts the point of optimal dispersion just a bit higher than the range in which EDFAs operate, thereby essentially eliminating the problem of four-wave mixing. See also Chromatic Dispersion, Dispersion, DWDM, EDFA, Four-Wave Mixing, NDSF, Waveguide Dispersion, Wavelength and Window.

DSCP DiffServ Code Point. See also Differentiated Services.

DSCWID Call Waiting Display with Disposition.

DSDC Direct Service Dialing Capability. Network services provided by local switches interacting with remote data bases via CCIS.

DSE 1. Data Switching Equipment.

2. An ATM term. Distributed Single-Layer Embedded (Test Method): An abstract test method in which the upper tester is located within the system under test and there is a point of control and observation at the upper service boundary of the Implementation Under Test (IUT) for testing a protocol layer, or sublayer, which is part of a multi-protocol IUT.

DSI Digital Speech Interpolation. A compression technique for squeezing more voice conversations onto a line. DSI digitizes speech so it can be cut into slices, such that no bits are transmitted when no one is speaking. As soon as speech begins, bits flow again. See Digital Speech Interpolation.

DSIA Defense Systems Information Agency. A U.S. government agency under the DOD (Department of Defense). DSIA's mission is "to plan, engineer, develop, test, manage programs, acquire, implement, operate, and maintain information systems for C4I (Command, Control, Communications and Computers and Intelligence) and mission support under all conditions of peace and war." DSIA was established on May 12, 1960, and assumed all responsibilities of the DCA (Defense Communications Agency) on June 25, 1991. DSIA is responsible for all aspects of systems and networks, including the DDN (Defense Data Network) and the DSN (Defense Switch Network). See also DDN and DSN.

DSL See ADSL, CLEC and Digital Subscriber Line.

DSL Bonding Take two DSL 1.5 megabit per second lines. Join them together. Get double the speed. When I wrote this, DSL bonding was available to join two DSL lines together. But I've heard they're working on bonding eight 1.5 megabit per second lines together, enabling service to 12 megabits per second.

DSL Filter Digital Subscriber Line filter. Let's say you have one phone line. On that phone line you run a high-speed DSL broadband circuit to the Internet and several normal household analog phones. Without a DSL filter, you are likely to hear excessive hum on your phone conversations. Also your caller ID probably won't work. A DSL filter is a low-pass filter that removes the high-end frequencies that could cause interference between the analog voice signals and the high-speed packet data signals over an ADSL line. ADSL (Asymmetric Digital Subscriber Line) supports both telephone service over an analog channel operating at frequencies of 4 kHz and below, and high-speed packet data at frequencies of 25 kHz and above. The DSL filter is physically a small box, one end of which protrudes an male RJ-11 plug. That's for inserting into a phone jack. On the other side of the box are two female RJ-11 connectors – one for your phone and/or your phone-based alarm system and one for your ADSL modem. The DSL filter essentially provides a fourth-order low-pass filter circuit that allows voice and low-speed data to pass through unfiltered, while blocking the high frequencies of DSL and other potential sources of disturbance. The filter also provides a way for alarm companies to install equipment at a place where DSL service has been previously installed. You have to install DSL filters at every outlet in your home. Or ... there is an alternatives. It's called installing a splitter. Splittered ADSL involves a centralized splitter, into which the DSL filter is embedded. ADSL Lite, or G.Lite, the new version of DSL, is a splitterless version involving microfilters that sit between the individual analog telephone sets and jacks. The DSL filters do the same thing, in either case. See also ADSL and Filter.

DSL Forum The DSL Forum, formerly known as the ADSL Forum, is an industry association formed to promote the DSL concept and to facilitate the development of DSL system architectures and protocols for major DSL applications. As the forum comprises competing companies, it does not publish material that discusses line codes or basic modulation systems, or any other material that addresses individual company or product attributes. www.dslforum.org See also ADSL and VDSL.

DSLAM Digital Subscriber Line Access Multiplexer. A device used in a variety of DSL technologies, which are lumped under the category of xDSL, with "x" being the generic "whatever." A DSLAM serves as the point of interface between a number of subscriber premises and the carrier network. At each subscriber premises is a splitter or a standalone modem, depending on the specific form of xDSL involved. The DSLAM generally is positioned in the ILEC's (Incumbent Local Exchange Carrier's) Central Office (CO). Alternatively, a Mini-DSLAM may be enclosed in a ngDLC (next generation Digital Loop Carrier), essentially servicing as a remote DSLAM. The DSLAM is a packet multiplexer, serving to multiplex the data (and perhaps voice) packets from multiple customers in order to transmit them over one or more high-speed circuits. In an integrated voice/data DSL application, the voice packets are forwarded to the PSTN (Public Switched Telephone Network), or perhaps an IP voice network, and the data packets are forwarded to the Internet, often over an ATM-based data network. See also ADSL, Digital Subscriber Line, DLC, ngDLC, and xDSL.

DSMCC Digital Storage Media Command And Control. Network protocols specified in Part 6 of MPEG-2 (ISO 13818) standards dealing with user-to-network and user-to-user signaling and communications.

DSML Directory Services Markup Language, or the most promising way to generate directory services in XML, the standard language used in e-commerce. DSML makes directory services easily able to use XML, allowing a simpler way for customers and partners to link with companies regardless of their current directories at their own sites.

DSN 1. Distributed Systems Network, the network architecture developed by Hewlett-Packard.

2. Double Shelf Network.

3. Defense Switched Network. The private, long haul, switched network used by the U.S. military in support of voice communications. DSN replaced AUTOVON, which was developed during the "Cold War." DISA (Defense Information Systems Agency) is responsible for the DSN. See also AUTOVON and DISA.

DSO 1. See DS-0.

2. Days Sales Outstanding. Another way of saying accounts receivable, usually measured in days.

DSOB An AT&T Digital Data Service standard that specifies a means of multiplexing several subrate data channels within one DS0. Five 9.6, ten 4.8 or 20 2.4 Kbps subrate channels may be multiplexed within one DS0.

DSP 1. Display System Protocol.

2. Digital Signal Processor. A Digital Signal Processor is a specialized computer chip designed to perform speedy and complex operations on digitized waveforms. Useful in processing sound (like voice phone calls) and video. See Digital Signal Processor for a much better explanation.

DSP Modem Digital Signaling Processor chip set used for analog modem emulation. The software is programmable for easy upgrades.

DSPU Downstream Physical Unit. A Physical Unit that is located downstream from the host.

DSR Data Set Ready. This signal is on pin 6 of the RS-232-C connector. It means the modem (which some telephone companies call a "data set") is ready to send data from the terminal. Some modems use Data Set Ready. Some don't. Modems that are snooty enough to give you the DSR signal, are obnoxious enough to not work until they receive the DTR (Data Terminal Ready) signal from the terminal on pin 20. By bridging pins 6 and 20 on the connector at the modem, you can usually get it to work. If it doesn't, bridge in pin 8 (carrier detect) as well.

DSRR Digital Short Range Radio.

DSS 1. Direct Station Select. A piece of key telephone equipment usually attached to an operator's phone set. When the operator needs to call a particular extension he/she simply touches the corresponding button on the Direct Station Select equipment. Typically DSS equipment/feature is part of a Busy Lamp Field (BLF), which shows with lights what's happening at each extension. Is it busy? Is it on hold? Is it ringing?

2. Decision Support Systems. Computerized systems for transforming data into useful information, such as statistical models or predictions of trends, usually in a graphical format, which is used by management in the decision-making process. See Decision Support Systems.

3. Digital Signature Standard. A standard for digital signatures proposed by the National Institute of Standards and Technology.

DSS2 Setup Digital Subscriber Signaling System 2 is a signaling protocol that specifies the procedures for the establishment, maintenance and clearing of point-to-multipoint VCCs (Virtual Channel Connections) at the B-ISDN (Broadband ISDN) UNI (User Network Interface). Such a connection is a collection of associated ATM VCC unidirectional links connecting two or more endpoints. See also ATM, B-ISDN, UNI and VCC.

DSSS Direct Sequence Spread Spectrum. A technique used in spread spectrum radio transmission systems, such as Wireless LANs and some PCS cellular systems. DSSS involves the conversion of a data stream into a stream of packets, each of which is pre-pended by an ID contained in the packet header. The stream of packets then is transmitted over a wide range of frequencies, using an approach known as "scattering." A large number of other transmissions also may share the same range of frequencies at the same time, with the potential for overlapping of packets. The receiving device is able to distinguish each packet in the packet stream by reading the various IDs, treating competing signals as noise. In a wireless LAN environment, DSSS typically operates in the 2.4 GHz frequency band, which is one of the ISM (Industrial Scientific Medical) bands defined by the FCC for unlicensed use. Although some manufacturers continue to use DSSS, Frequency Hopping Spread Spectrum (FHSS) is the preferred approach. See also 802.11b, CDMA, FHSS, ISM and Spread Spectrum.

DST Daylight Savings Time. In 2007, Daylight Saving Time was extended by about four weeks. In compliance with this provision in the Energy Policy Act of 2005, DST dates in the United States and Canada will start three weeks earlier (2:00 A.M. on the second Sunday in March) and will end one week later (2:00 A.M. on the first Sunday in November). The theory behind daylight savings time is that it extends the time we are working while in daylight, thus saving heating and lighting energy. This latest extension of daylight savings allegedly saved the U.S. a couple of hundred million dollars.

DSTM See DTM.

DSTN Dual-scan Super Twisted Nematic). A laptop-display technology that uses two display layers to overcome color shifting that occurs with supertwist displays. It's a less expensive alternative to thin film transistor and active matrix displays and has a faster video scan rate than the lower-cost STN design. The screen pointer is less apt to temporarily disappear when moved rapidly, plus the screen will scroll more smoothly. Although the displays are brighter, the screens tend to look blurry when viewed at an angle.

DSU Digital Service Unit, also called Data Service Unit. Converts RS-232-C or other terminal interface to DSX-1 interface. See Data Service Unit.

DSU/CSU The devices used to access digital data channels are called DSU/CSUs (Data Service Unit/Channel Service Units). At the customer's end of the telephone connection, these devices perform much the same function for digital circuits that modems provide for analog connections. For example, DSU/CSUs take data from terminals and computers, encode it, and transmit it down the link. At the receive end, another DSU/CSU equalizes the received signal, filters it, and decodes it for interpretation by the end-user.

DSVD Digital Simultaneous Voice and Data. Technology in a modem that allows you to send and receive voice and data (fax, images, files, etc.) on the same "conversation" on one analog phone line. DSVD allows the simultaneous transmission of data and digitally-encoded voice signals over a single dial-up analog phone line. DSVD modems use V.34 modulation (up to 33.6 kilobits per second), but may also use V.32 bis modulation (14,400 kilobits per second). DSVD modems reserve eight kilobits per second for voice transmissions. The remaining bandwidth is available for data transmission. The DSVD voice coder is a modified version of an existing specification and is defined as G.729 Annex A. The DSVD voice/data multiplexing scheme is an extension of the V.42 error correction protocol widely used in modems today. DSVD also specifies fallbacks that enable DSVD modems to communicated with standard data modems (i.e. V.34, V.32 bis, V.32 and V.22). With the arrival of broadband Internet access, DSVD is now basically dead.

DSX DSX. Also known variously as a DACS (Digital Access Cross-Connect System) and a DCC (Digital Cross-Connect), a DSX is a manual bay, panel or some other device that is used to interconnect digital circuits. A DSX-1 interconnects DS-1 (T-1 or E-1) circuits, as DSX-2 interconnects DS-2 (T-2 or E-2) circuits, and a DSX-3 connects DS-3 circuits (T-3 or E-3).

DT Direct termination. DT allows long distance carriers to bypass the LEC and terminate the switched traffic over dedicated (non-LEC) facilities to the customer premises. Direct termination calls are routed using six or ten-digit screening in the terminating long distance switch.

DTC 1. Digital Trunk Controller. See also STC.

2. Digital Transmit Command.

3. Digital Traffic Channel. A digital cellular term. Defined in IS-136, the DTC is the portion of the air interface which carries the actual data transmitted. The DTC operates over

frequencies separate from the DCCH (Digital Control CHannel), which is used for signaling and control purposes. See also DCCH and IS-136.

DTD Document Type Definition often used in relationship with the Extensible Markup Language (XML). See cXML, SMGL, XML, and XML Schema.

DTE Data Terminal Equipment. A terminal device in the data world. DTE is part of a broader grouping of equipment known as CPE (Customer Premises Equipment), which includes voice, as well as data, terminals. At the terminal end of a data transmission, DTE comprises the transmit and receive equipment. DTE can be in the form of a dumb terminal (i.e., a terminal without embedded intelligence in the form of programmed logic), a semi-intelligent terminal, or an intelligent host computer (i.e., a PC, mid-range or mainframe computer). DTE interfaces to a circuit through DCE (Data Communications Equipment). See DCE and DTE.

DTE-DCE Rate Data terminal equipment/data communications equipment rate. A designation for the maximum rate at which a modem and a PC can exchange information, expressed in kilobits per second (kbps). For maximum performance, a modem must support a DTE-DCE rate in excess of its maximum theoretical throughput.

DTF DTF is the reason I hate acronyms. In telecom DTF could and does stand for digital trunk frame, digital transmission facility or dial tone first. Now you know why I avoid defining every single, made-up acronym under the sun. You and I could sit in a room and think of ten more acronyms for DTF and fill up even more space with more useless words. Enough already.

Dterm A line of proprietary electronic phones made by NEC for use with its PBXs. The Dterm terminal derives its intelligence from its own microprocessor, which detects events and accepts direction from the PBX.

DTH 1, Direct To Home. Intended as a replacement for C-band satellite systems, DTH was proposed to operate on medium-powered FSS (Fixed Satellite Systems) in the Ku-band. DTH was superseded by DBS (Direct Broadcast Satellite), which allows the use of even smaller receive antennas than possible with DTH. That's the system currently being used and which you sitting on the side of houses today and facing south. See also Direct Broadcast Satellite and KU Band.

2. Defense Message System Transition Hub.

DTI Digital Trunk Interface.

DTIM Delivery Traffic Indication Message. A message included in data packets that can increase wireless efficiency.

DTL An ATM term. Designated Transit List: A list of nodes and optional link IDs that completely specify a path across a single PNNI peer group.

DTL Originator An ATM term. The first switching system within the entire PNNI routing domain to build the initial DTL stack for a given connection.

DTL Terminator An ATM term. The last switching system within the entire PNNI routing domain to process the connection and thus the connection's DTL.

DTLBX Dial Tone Line.

DTLU Digital Trunk and Line Unit. Provides system access for T1-carrier lines used for inter office trunks or remote switching module umbilicals.

DTM According to www.netinsight.se, DTM (Dynamic synchronous Transfer Mode) is a network protocol for high speed networking developed for dynamic transport of integrated traffic. It is a transport network architecture based on circuit-switching augmented with dynamic reallocation of bandwidth. The protocol is designed to be used in integrated services networks. It supports point-to-point, multicast and broadcast communication, i.e. a DTM network will be used for both distribution and unicast communication. DTM includes switching and a signaling protocol and can thus, in contrast to say SDH/SONET, set up multi-rate channels (circuits) on demand, and the capacity of a channel can be changed according to traffic characteristics during operation. Additionally, resources can be reallocated between nodes according to the current demands. In this way, free bandwidth is allocated to nodes with highest demands, providing an autonomous and efficient dynamic infrastructure.

DTMF Dual Tone Multi-Frequency. A fancy term describing push button or Touchtone dialing. (Touchtone is a not registered trademark of AT&T, though until 1984 it was.) In DTMF, when you touch a button on a push button pad, it makes a tone, actually a combination of two tones, one high frequency and one low frequency. Thus the name Dual Tone Multi Frequency. In U.S. telephony, there are actually two types of "tone" signaling, one used on normal business or home pushbutton/touchtone phones, and one used for signaling within the telephone network itself. When you go into a central office, look for the test board. There you'll see what looks like a standard touchtone pad. Next to the pad there'll be a small toggle switch that allows you to choose the sounds the touchtone pad will make – either normal touchtone dialing (DTMF) or the network version (MF).

The eight possible tones that comprise the DTMF signaling system were specially selected to easily pass through the telephone network without attenuation and with minimum interaction with each other. Since these tones fall within the frequency range of the human voice, additional considerations were added to prevent the human voice from inadvertently imitating or "falsing" DTMF signaling digits. One way this was done was to break the tones into two groups, a high frequency group and a low frequency group. A valid DTMF tone has one tone from each group. In other words, each DTMF tone has two tones. Here is a table of the DTMF digits with their respective frequencies. One Hertz (abbreviated Hz.) is one cycle per second of frequency.

When you touchtone, each button makes a sound that is the combination of two tones. Here's how to figure out what they are. They were deliberately designed so people couldn't whistle them.

Each Touchtone button makes two tones, a high tone and a low one				
Touchtone phone button				Low Tones
1	2	3	A	697Hz
4	5	6	B	770Hz
7	8	9	C	852Hz
*	0	#	D	941Hz
High Tones 1209Hz	1336Hz	1477Hz	1633Hz	

Normal telephones (yours and mine) have 12 buttons, thus 12 combinations. Government Autovon (Automatic Voice Network) telephones have 16 combinations, the extra four (those A, B,C and D above) being used for "precedence," which in Federal government parlance is a designation assigned to a phone call by the caller to indicate to communications personnel the relative urgency (therefore the order of handling) of the call and to the called person the order in which the message is to be noted. See also long tones and the four following definitions.

DTMF Automatic Routing This is a term relating to a fax server operating on a Novell file server. In this system, the fax software assigns a four-digit number to each user. A fax sender dials the fax line, and after the fax server answers, it sends a special auto routing request signal. The sender dials the four-digit number for the correct user, and the fax is automatically sent to the user's workstation on the LAN.

DTMF Cut-Through The capability of a voice response system to receive DTMF tones while the voice synthesizer is delivering information, i.e. during speech playback. This capability of DTMF cut-through saves the user waiting until the machine has played the whole message (which typically is a menu with options). The user can simply touchtone his response anytime during the message – when he first hears his selection number, when the message first starts, etc. When the voice processor hears the touchtoned selection (i.e. the DTMF cut-through), it stops speaking and jumps to the chosen selection. For example, the machine starts to say, "If you know the person you're calling, touchtone his extension in now." But before you hear the "If you know" you push button in 230, which you know is Joe's extension. Bingo, the message stops and Joe's extension starts ringing. DTMF Cut-Through is also known as touchtone type-ahead.

DTMF Register A printed circuit card in a switch that converts the DTMF signals coming from the phone into signals which can be used by the switch's stored program control, central computer to do its switching, etc.

DTMF To Dial Pulse Conversion A PBX feature. DTMF (push button) phones are very popular. But sometimes you install a PBX with push button phones in an area which doesn't have a central office which will respond to push button tones. It's old. In this case, anyone dialing on a push button phone will find that the PBX converts that dialing to rotary pulsing when the PBX accesses a trunk which can't handle push button dialing. All this doesn't speed up the time the call takes to get through. It just speeds up the user's dialing and makes him or her feel she is dealing with a more modern phone system.

DTO 1. Dial Tone Office.

2. Direct Trunk Overflow, which has to do with the ability of a trunk group (such as a PRI (primary rate interface)), when exceeding capacity, to gracefully overflow to another telecom facility of some kind.

DTP DeskTop Publishing.

DTR Data Terminal Ready. A control signal sent from the DTE to the DCE that indicates that the DTE is powered on and ready to communicate. DTR can also be used for hardware flow control.

DTS 1. Digital Termination Systems. DTS is a digital microwave transmission technology designed for facilities bypass applications in short-hop, line-of-sight situations. DTS useful in high-volume, pure-data applications in urban settings where line costs are high. DTS requires FCC license and is referred to formally by the FCC as Digital Electronic Message Service, or DEMS.

2. Distributed Time Service. DTS is a feature of DCE (Distributed Computing Environment), a developing open standard championed by The Open Group. DTS is a mechanism that would allow distributed computing systems to synchronize their system clocks. See also DCE and The Open Group.

DTSR Dial Tone Speed Recording.

DTT Digital Trunk Testing.

DTU Digital Test Unit. A generic term for a device used to test the performance of digital circuits. While a DTU can be in the form of a standalone device, it also can be in the form of a set of programmed logic on a chipset contained under the skin of another device. For example, and in a telecommunications network context, a DTU might be termed as the capability of a DSU/CSU (Digital Service Unit/Channel Service Unit) to respond to the request of a carrier for a loopback test, which is used to test the integrity of a four-wire digital circuit such as a T-1. Similarly, it might be the capability of a NIU (Network Interface Unit) to respond to an ALTS (Automated Line Testing System) for a line test on an ISDN line. A DTU might test across a wide range of circuit integrity parameters including physical integrity (e.g., electrical opens and shorts), electrical ground, signal strength, electrical noise levels, and bit error rate (ber). See also CSU, DSU, Loopback Test and NIU.

DTV Digital TV. The generic term for television systems employing digital, rather than analog, technology. All HDTV (high definition TV) is digital. Not all digital TV is high definition. An HDTV picture is made up of 1,080 interlaced scan lines (1080i) or 720 progressive (780p) scan lines. Standard-definition digital television and other digital-TV formats aren't as sharp but still look better than standard analog TV. This is easily confused with digital cable, which does not necessarily deliver an enhanced picture, but typically delivers additional channels. Since most TV content originally starts out as analog, rather than digital, it has to be digitized through a codec (coder/decoder). Once digitized, it can be compressed for efficiency of storage, and can be more easily edited and otherwise manipulated, and reproduced. DTV transmission offers the advantages of compression for efficiency of bandwidth utilization (in other words, it uses less bandwidth), improved error performance for improved picture and sound quality, and more effective network management and control. The full benefits of DTV are best realized if the terminal device (i.e. your TV set) is digital. A digital TV set sometimes allows the program to be compressed and archived (i.e., stored) for replay, and certainly offers better quality of picture and sound. See also 16:9, ATV, HDTV and SDTV.

DTX Battery-saving feature on a cellular phone that cuts back the output power when you stop speaking.

DU Fiber Optic Connector developed by Nippon Electric Group.

DUA Directory User Agent. The software that accesses the X.500 Directory Service on behalf of the directory user. The directory user may be a person or another software element.

DUAL Distributed Update Algorithm. A routing algorithm that provides fast rerouting (convergence) with minimal consumption of resources.

Dual Attachment Concentrator DAC. A concentrator that offers two S ports for connections to the FDDI network and multiple M ports for attachment of devices such as workstations and other concentrators. A DAC takes maximum advantage of the resiliency of the FDDI dual counter-rotating ring architecture through connection to both rings, unlike a SAC (Single Attachment Concentrator). Connection of devices through the concentrator obviates the need for each device to attach directly, thereby reducing the number of optoelectric conversions and resulting in reduced cost of attachment. The concentrator also serves as a point of contention for network access. See also Single Attachment Concentrator, Single Attachment Station, Dual Attachment Station and FDDI.

Dual Attachment Station DAS. A device such as a workstation which connects directly to both FDDI dual counter-rotating rings, rather than gaining ring access through a Dual Attachment Concentrator (DAC). A DAS allows access to two separate cable systems at the same time, providing protection against cable failure or damage. See also Single Attachment Concentrator, Single Attachment Station, Dual Attachment Concentrator and FDDI.

dual band 1. Dual band describes a cellular phone handset which works on dual frequency bands. For example, both the 800 MHz (analog AMPS) and 1900 MHz (digital) bands for North America. This allows the phone to switch between the two bands. The reason you'd want this is simple: Digital service is often cheaper and better in areas you can get it. But you can't get it everywhere. If you travel you need a cell phone you can use everywhere. Thus the idea of carrying a dual band cell phone and subscribing to a service that gets you access to both. In Europe and many other countries, a dual band cell phone will work on both the 900 MHz and 1800 MHz GSM bands, supporting seamless hand-off between the two frequency bands for carriers using both. See also AMPS, Dual Band GSM Phone, GSM and Tri-Mode.

2. A Satellite that can operate on C-band and Ku-band signals.

dual band GSM phone Dual Band wireless phones are capable of operating on GSM 900 and 1900 networks simultaneously. If your local U.S. service is GSM 1900, a dual-band phone would allow roaming almost anywhere in the world. Dual band phones are also available for operation on European GSM 900 and 1800 networks.

dual-band Wi-Fi card A Wi-Fi card for a notebook computer that provides wireless connectivity both in the unlicensed 2.4 GHz band and the licensed 4.9 GHz public safety band. The card is designed for use by emergency first responders.

dual cable A two-cable system in broadband LANs in which coaxial cable provides two physical paths for transmission, one for transmit and one for receive, as against dividing the capacity of a single cable.

dual cable entrance facility A central office with two physically separated cable entrance facilities, with outside plant cables evenly distributed between the two cable entrances.   The idea behind a dual cable entrance facility is to improve network diversity and therefore reliability in the event that one of the cable entrance facilities becomes severely damaged.

dual coaxial cable Two individually insulated conductors laid parallel or twisted and placed within an overall shield and sheath.

dual coat An optical fiber coating structure consisting of a soft inner coating and a hard outer coating.

dual core microprocessors Dual core microprocessors contain two computing "engines" on a single chip. This allows computers to complete tasks faster without needing substantially more power. Many microprocessor companies, such as Intel, IBM and AMD, are pursuing dual core architecture for their microprocessors to drive desktop and laptop computers because other ways of lifting computer performance – such as increasing the number of instructions per second that a processor can handle – have reached a point of diminishing returns.

dual duplex A term sometimes used to describe HDSL (High bit rate Digital Subscriber Line), a relatively new technology which commonly is used to provide T-1 and E-1 local loops. HDSL makes use of 2 twisted pair loops, with each operating in full duplex. The traditional approach to T-1 provisioning involves 2 twisted pairs, each operating in simplex. See also HDSL.

dual feedhorn A satellite term. A feedhorn that can simultaneously receive both horizontally and vertically polarized signals.

dual fiber cable A type of optical fiber cable that has two single-fiber cables enclosed in an extruded overjacket of polyvinyl chloride with a rip cord for peeling back the overjacket to access the fibers.

dual headset Also known as an integrated headset. A special type of headset for the blind. One jack plugs into a telephone and another jack plugs into a telephone and another jack plugs into a specially configured PC. This PC provides voice synthesized output. The dual headset allows a visually impaired TSR (Telephone Sales Representative) hands-free capability. Example: The Social Security Administration has numerous blind TSRs handling incoming public calls. Dual headsets allow these blind TSRs to perform their duties with no deterioration in public service. This definition provided by Matt Gottlieb, telecommunications specialist for the Social Security Administration.

dual homed firewall A firewall that connects simultaneously to two different networks (e.g. the public Internet and a private intranet); as such it acts as a security gatekeeper that determines which traffic is allowed to pass and which traffic is blocked.

dual homed gateway A dual homed gateway is a computer that runs firewall software and has two network interface boards: One board is attached to an untrusted network, and the other board is attached to a trusted network. A dual-homed gateway relays information between the two networks and prevents any direct contact between them. Both circuit-level gateways and application-level gateways are dual homed gateways.

dual homing 1. The process of using two geographically diverse frame relay port connections, each with its own set of virtual circuits, to support a network location running critical business applications which cannot afford network down time.

2. A method of cabling FDDI concentrators and stations that permits a alternative path to the dual ring. Can be used in a tree or dual ring of trees topology.

3. Where a device is connected to the network via two independent access points (points of attachment). One access point is the primary connection; the other is a standby connection that is activated in the event the primary connection fails.

dual line registration The ability to have two cellular telephone numbers in a single cellular telephone. This allows the user to have service on two cellular systems without "roaming" in a second city (and paying higher toll charges) or to have one number on the wireline and one number on the non-wireline system. This assures the cellular user of backup. I know a salesman who lives in Los Angeles, but spends much of his week serving customers in Phoenix. He has a cellular phone with Los Angeles number and a Phoenix number.

dual line service Telephone service where two pairs of wires are connected to the premises. One or both could be in service.

dual LNB A LNB is a line noise block downconverter. It is used in a TVRO (TV Receive Only) system to allow you to receive satellite signals. A dual LNB allows you to simultaneously receive two satellite channels. You can, for example, watch live TV on one channel, while recording another channel on your recording device, e.g. VCR or TiVO.

dual mode 1. Dual mode is the cellular industry's term for a cellular phone which will work for both analog and digital cellular phone systems. The cellular phone industry is going digital. Today's analog phones won't work on tomorrow's digital systems. But some phones – dual mode – will work on both. You may need to buy them over today's analog phones. You also need to be careful that the digital mode which you get in your dual mode cellular phone will work with the digital technology of your local carrier. That's not as standard, as yet, as today's analog technology, which works universally in North America. In short, there are many variations of digital cell phones service in North America, many of which are incompatible.

2. A handset that combines cellular capability with either voice over Wi-Fi capability, or voice over wireless LAN capability, as a way of complementing or enhancing in-building coverage for voice.

dual NAM Allows a cellular phone user to have two phone numbers with the same or separate carriers. Very useful for someone who spends half his life in one place and half in another. For example, a friend of mine lives in LA, but works weekdays in Phoenix. His handheld cellular has phone numbers from LA and Phoenix carriers.

dual processing An SFT II configuration under Novell's NetWare that assigns parts of the operating system to separate processors. Because SFT II is split into two engines (the IOEngine and the MSEngine), it is possible to install each engine on a separate CPU, creating dual processing system. However, unless such a system is extremely busy, the extra CPU will not help network performance. Dual processing improves performance only when the servers are being used at near-maximum capacity.

dual rate speech coder for multimedia communication A voice compression algorithm specified as ITU-T G.723.1 (1991) and falling under the H.324 umbrella. Dual Rate compresses voice from 64 Kbps down to 6.3 and 5.3 Kbps. High-rate compression at 6.3 Kbps uses the MP-MLQ (MultiPulse-Maximum Likelihood Quantization) algorithm, and low-rate uses ACELP (Algebraic Code Excited Linear Prediction). These compression levels yield highly efficient use of bandwidth, but impact voice quality to a considerable extent. Dual rate also imposes compression delay of 30.0 ms (milliseconds, or thousandths of a second). H.324 is the ITU-T recommendation for low bit-rate multimedia communications over the analog PSTN (Public Switched Telephone Network) through V.34 modems, which are limited to 28.8 Kbps. (Nobody in his right mind would buy a V.34 modem today.) Such a high level of compression is necessary in support of multimedia, as video and other visual information streams require a lot of bandwidth to yield even poor quality. Crummy video and crummy voice quality, both in a single standard. Who could ask for anything more? See also ACELP, H.324, and MP-MLQ.

dual ring of trees topology An FDDI network topology of concentrators and nodes that cascades from concentrators on a dual ring.

dual scan display An LCD display used in laptop computers. The screen is refreshed much quicker than in standard displays and use less power.

dual stack A term that describes concurrent network support for IPv4 and IPv6 protocol stacks. A dual-stack network environment will characterize the transition that most organizations will make from a wholly IPv4 network environment to a wholly IPv6 network environment.

dual tone multi-frequency DTMF. A way of signaling consisting of a push button or touchtone dial that sends out a sound which consists of two discrete tones, picked up and interpreted by telephone switches – either PBXs or central offices. See DTMF for a bigger explanation.

dual universal asynchronous receiver transmitter DUART. A DUART provides hardware support for two serial communications ports.

dual watch A marine radio that is able to monitor two radio frequencies at one time, generally channel 16 (the "hail and distress channel") and another channel, for example, a weather information channel. See also tri-watch.

DUART Dual Universal Asynchronous Receiver Transmitter. A DUART provides hardware support for two serial communications ports.

dub dub dub The weird shortening of www for the World Wide Web, as in: "My web site is dub dub dub HarryNewton.com."

duck test If it walks, talks and quacks like a duck, it's a duck.

duct A pipe, tube or conduit through which cables or wires are passed. Duct space is always at a premium. If you ever install a duct, make sure it's twice the diameter you think you need. If you're lucky, it will last a couple of years. The cost of putting in thicker or extra ducts is peanuts compared to the cost of having to install additional ones later. Digging up places is getting very expensive, despite Ditch Witch, a company that makes the greatest backhoe trenching equipment. And also has the greatest name.

duct cycle The relationship between the time a device or facility is used and the time it is idle.

duct liner A small diameter pipe or tubing placed inside conventional underground conduit so you can install fiber optic or cables. Its main purpose is to provide a clean, continuous path with known frictional characteristics.

ductbank An arrangement of ducts, for wires or cables, arranged in tiers.

ductile Capable of being drawn out, hammered thin or being flexed or bent without failure.

ducting See Pathway.

dude The word "dude" was coined by Oscar Wilde and his friends. It is a combination of the words "duds" and "attitude." He meant it in the use of a man excessively concerned with his clothes, grooming and manners. These days dude just means a switched-on (i.e. knowledgeable) person.

DUE Dumb User Error. A term that lacks in cleverness what it makes up for in usefulness. Used by Help Desks.

due date The date an event is to occur, i.e., an installation, a move, a change or a disconnect. Some vendors quote accurate due dates. Penalty clauses work most effectively in ensuring due dates are met. See also PISD.

due diligence An in-depth examination of a proposed private placement or IPO that the potential investors (you and me), investment bankers and lawyers for the underwriters conduct as part of the process of raising money or taking a company public. They speak with management about the company's prospects, strategy, competitors, and financial statements. Information that is material to the company's prospects must be disclosed in the private placement memorandum (PPM) or IPO prospectus.

duf file Daily Usage File. This is a Qwest term. Here's Qwest's explanation: Some Qwest products generate usage records or call detail that is recorded at a Qwest Central Office Switch. This recorded usage is subsequently processed within the Customer Records and Information System (CRIS) billing system and, if identified as belonging to you, is packaged into a file, the Daily Usage File (DUF). The DUF is transmitted to you on a daily, weekly or monthly basis depending on what you requested in the New Customer Questionnaire. Qwest interconnect and resale products that generate recorded usage records can include, but are not limited to:

- Business and Residence Exchange Lines
- Centrex
- Private Branch Exchange (PBX) Service
- Public Access Line (PAL)
- Single Line Integrated Services Digital Network/Basic Rate Interface (ISDN BRI)
- Unbundled Switching (UBS)
- Unbundled Network Elements - Platform (UNE-P)

duff When you are trail-building, what you remove is called "duff".

dumb Something, or someone, lacking in intelligence, i.e., stupid. In our context, there are dumb terminals, dumb networks, dumb switches, and so on. Dumb things work, but at a very simple level and very cheaply. Any intelligence involved in an associated process is provided by something else, typically by a distant company, also called a server or mainframe. See Divestiture.

dumb business decision. See Divestiture.

dumb decision See Divestiture.

dumb network A communications network that, in contrast to the public switched telephone network (PSTN), relies on information provided by the sender to route the transmission. A network in which the intelligence is located on the periphery, e.g. Ethernet.

dumb pipes The network infrastructure, through which content providers can provide content, service providers can provide services, and to which customers can attach devices. Some incumbent carriers (also known as "pipe owners" or "pipe holders") chafe at this characterization of their networks, since they want to be regarded as more than bit-haulers.

dumb remote A dumb remote is a remote physical partition of an intelligent switch, which usually is in the form of a CO (Central Office), although it also can be a PBX. The CO, for example, contains all, or most of, the intelligence and memory required to perform the call processing functions and to provide special services. The dumb remote, which may be in the form of either a dumb switch or a line shelf (i.e., line concentrator) is connected to the centralized CO by a signaling and control link over which the call processing queries and instructions are transmitted between the two devices. See also Remote Line Concentrator.

dumb switch A slang word for a telecommunications switch that contains only basic switching software and relies on instructions sent by an outside computer. Those instructions are typically fed the "dumb" switch through a cable from the computer to one or more RS-232 serial ports which the dumb switch sports. The switch makes no demands on what type of computer it talks to, but simply insists that it be able to feed the computer questions and promptly receive responses in a form that it (the switch) can understand. Plain ASCII is OK. For example, the dumb switch might signal the computer, "A call is coming in on port 23, what do I do now?" The computer might reply "Answer it and transfer it to extension 23." Or it might say "answer it and put it on hold," or "answer it, put it on hold and play it recording number three." In essence, a dumb switch is anything but. It is in reality an empty cage containing whatever network interface cards the user has chosen. Each of these network interface cards is designed to "talk" to one type of telephone line. That line might be a T-1 line. It might be a normal tip and ring loop start line. It might be a tie trunk with E&M signaling. The card may handle one or many lines, but always of the same type. The card knows how to answer a call or pulse out a call on that particular type of line. It has all the telephony smarts. What it lacks is the intelligence of what to do with the calls. That is provided by the outside computer. Well, almost. Most "dumb" switches do contain rudimentary intelligence – a small computer and some memory. That computer is usually programmed to handle "default" calls – and to handle calls should the link to the outside computer fail, or the outside computer itself fail. Dumb switches come in flavors all the way from residing in their own cabinet to being printed circuit cards which reside in one or more of the personal computer's slots. Dumb switches are programmed to do "specialized" telecom applications, for example emergency 911, added value 800 services, cellular switching, automatic call distributors, predictive dialers, etc. They can, of course, be programmed to be "normal" PBXs. The question increasingly being asked is "If I want to program a specialized telecom application should I use a dumb switch or should I use an open PBX?" And the answer is "It depends." Depends on what you want to do. Depends on what software is available, etc. See also OAI.

dumb terminal A computer terminal with no processing or no programming capabilities. Hence, it derives all its power from the computer it is attached to – typically over a local hardwire or a phone line. A dumb terminal does not employ a data transmission protocol and only sends or receives data one character at a time, sequentially. There are many reasons for "dumb" terminals. They're cheap. They're foolproof. Operators don't have to mess with floppy disks, etc. The require minimal training. Dumb terminals are typically used for simple data entry and data retrieval tasks. Their disadvantage is that everything must come from the central computer – not only the information (data record) but also the form in which to put it. This has led to the creation of "intelligent" terminals, which have a modicum of capabilities – such as an inbuilt (with software) form, some smart function keys and perhaps, a modicum of processing power, etc.

dumb terminal access An Internet Access Term. Telnet shell command that allows remote terminals to connect to a LAN host. Basic terminal server support over the same modems and phone lines used for remote access.

dumbwaiter One who asks if the kids would care to order dessert.

dummy load A dissipative impedance-matched network, used at the end of a transmission line to absorb all incident power, usually converted to heat.

dumb pipes The network infrastructure, through which content providers can provide content, service providers can provide services, and to which customers can attach devices. Some incumbent carriers (also known as "pipe owners" or "pipe holders") chafe at this

characterization of their networks, since they want to be regarded as more than bit-haulers.

dump To copy the entire contents of something – memory, a file on a disk, a complete disk – to a printer or another magnetic storage medium. A dump is often called a "core dump," which is a bigger dump than dump.

dumpster dipping Same meaning as Dumpster Diving.

dumpster diving Also known as Dumpster Dipping, Dumpster Diving is searching for access codes or other sensitive information in the trash. In North America, large trash receptacles – those found outside buildings – are often known as dumpsters. One pieced of folk lore is that Bill Gates of Microsoft fame started off as a dumpster diver. It's now against the law.

DUN Microsoft's acronym for Dial-Up Networking. Files on PCs with the extension DUN mean that they contain instructions to dial up an ISP – Internet Service Provider – and get access to the Internet. Such DUN file includes the phone number to dial, the account name and the password. It should dial your PC's modem, handshake with the ISP's modem at the other end, handshake with the ISP's server as regards your user name and password and get you on the Internet.

DUNDi Distributed universal number discovery. DUNDi is a peer-to-peer system for locating Internet gateways to telephony services. Unlike traditional centralized services (such as the ENUM standard), DUNDi is fully-distributed with no centralized authority whatsoever. DUNDi is not itself a Voice-over IP signaling or media protocol. Instead, it publishes routes which are in turn accessed via industry standard protocols such as IAX, IAX2, SIP and H.323. DUNDi can be used within an enterprise to create a fully-federated PBX with no central point of failure, and the ability to arbitrarily add new extensions, gateways and other resources to a trusted web of communication servers, where any adds, moves, changes, failures or new routes are automatically absorbed within the cloud with no additional configuration. An Internet Draft of the DUNDi protocol can be found here: www.dundi.com/dundi.txt

duobinary signal A pseudobinary-coded signal in which a "0" ("zero") bit is represented by a zero-level electric current or voltage; a "1" ("one") bit is represented by a positive-level current or voltage if the quality of "0" bits since the last "1" bit is even. and by a negative-level current or voltage if the quantity of "0" bits since the last "1" bit is odd. Duobinary signals require less bandwidth than NRZ. Duobinary signaling also permits addition of error-checking bits.

duopoly Similar to monopoly except there are two licensed competitors per market instead of one. The cellular business in the United States is a duopoly. Each major city in the United States has two licensed cellular providers – a feat accomplished by the FCC. As a result, prices in most cities as between the carriers are identical.

DUP Data User Part. Higher-layer application protocol in SS7 for the exchange of circuit switched data; not supported by ISDNs.

duplex 1. Simultaneous two-way transmission in both directions. A data communications term.

2. Two-sided printing.

3. Two hard disks that have separate disk controllers and are mirror copies of each other. Data is written simultaneously to both. See Mirroring.

duplex cable A two-fiber cable suitable for duplex transmission. One fiber is used for communicating one-way. The other fiber is used for communicating the other way.

duplex circuit A telephone line or circuit used to transmit in both directions at the same time. Also referred to as full duplex as opposed to half duplex which allows transmission in only one direction at one time.

duplex insulated In the thermocouple industry, a combination of dissimilar metal conductors of a thermocouple or thermocouple extension wire.

duplex operation The simultaneous transmission and reception of signals in both directions.

Duplex Signaling DX. A direct current signaling system that transmits signals directly on the cable pair. Duplex signaling is a facility signaling system and range extension technique that used bridge type detection of small dc changes. Duplex signaling is typically used on long metallic trunks.

duplex transmission The simultaneous transmission of two series of signals by a single operating communicating device. A data communications term.

duplexer 1. A device which splits a higher speed source data stream into two separate streams for transmission over two data channels. Another duplexer at the other end puts the two slower speed streams back together into one higher-spread stream.

2. A waveguide device designed to allow an antenna to be used both transmission and reception simultaneously.

3. A printer which prints on both sides of the paper simultaneously.

duppies Depressed urban professionals; former white-collar workers who are now unemployed or working for minimum wage.

duration The elapsed time between answer and disconnect for any given call.

dustbuster A phone call or e-mail message you send to someone after a long silence – just to "shake off the dust" and see if the connection still works.

duty cycle The ratio of operating time to total elapsed time for a device that operates intermittently. Usually expressed as a percentage.

DV Digital Video.

DVAC line Dedicated Voltage AC – dedicated phone line installed by the phone company, usually for high end alarm monitoring systems and also used by some merchants for their POS systems. With an alarm system, it enables the central station to transmit a signal to the alarm panel about every 30 seconds to make sure the line has not been cut. Typically very expensive and used in high risk sites such as banks, jewelry stores, apartment building fire systems etc.

DVB Digital Video Broadcast and Digital Audio Video Council (DAVIC) technology is the incumbent European standard for digital set-tops and is now starting to gain momentum for cable modems. The standard describes the out-of-band and in-band transmission options applicable to interactive set-top boxes and cable modems. respectively, enabling the deployment of interactive TV, data, and voice services over a common platform. DVB/DAVIC is challenging DOCSIS for dominance in Europe. DVB started in Europe in 1993 in liaison with the European Broadcasting Union (EBU), the European Telecommunications Standards Institute (ETSI) and the European Committee for Electrotechnical Standardization (CENELEC). The DVB Project is a consortium of more than 300 organizations in 57 countries round the world. In contrast to traditional government-driven regulatory activities, the DVB Project is market-driven and works to commercial terms with the aim of promoting DVB technologies through attaining economies of scale and through making its specifications and standards open and available to all. DVB uses MPEG-2 video and audio compression and though video in name, is capable of delivering any information that can be digitized, whether it is standard-definition TV (NTSC, PAL or SECAM), high-definition TV (HDTV) or broadband data and interactive services, which are supported by return channels (RC). DVB specifications and standards now are available for conventional broadcast, microwave, cable, satellite, satellite master antenna and satellite TV, as well as for satellite news gathering. Return channel specifications and standards are available for cable, Digital Enhanced Cordless Telecommunications (DECT), Local Multipoint Distribution System (LMDS), Public Switched Telephone Network (PSTN), Global System for Mobile (GSM) and satellite return paths. In 15 countries in Europe and in Australia and New Zealand, terrestrial (conventional broadcast) digital TV is based on DVB. The satellite version, DVB-S, is almost universally adopted round the world, and several international satcom systems are implementing DVB with various configurations of the return channel. www.dvb.org.

DVC Digital Video Compression. Process by which multiple videoservices, or channels, are broadcast on one satellite transponder. Performed by turning analog audio and video into digital (computer) information for broadcasting.

DVD Digital Video Disc. Also called Digital Versatile Disc. A DVD is a 5" diameter optical disc – the exact same size as the common compact disc (CD) which is used for distributing software or music. The big difference is that a DVD holds 4.7 gigabytes, which is seven times as much as a CD, which holds 650 or 700 gigabytes. DVDs are standard animals these days and can played in all manner of players, including PCs. They are often called DVD-video and come with a movie on it. DVDs are following the path of CDs. First the playback variety came out. Then the record variety came out. In DVDs, the record variety takes all manner – DVD-R, DVD-RAM, DVD-RW, DVD+RW and DVD+R. The distinctions are actually based on how the data is written to and read from the disk. In my readings I came across an excellent explanation by Honda Shing, Chief Technology Officer for InterVideo, the company that makes WinDVD, the popular software DVD player. Honda writes, "Imagine if the entire world agreed on a single language for all written documents, say English. In the United States we would continue to create English documents and books in which one reads left to right. But another country might prefer to write and read English right to left. And yet another, top to bottom, and so on until the effect became that although all documents were created in a single language, it would be very difficult for a person from one country to read a document from another country. DVD recording is in a similar state of confusion. DVD is such a new technology that these issues have yet to work themselves out, so you have no choice other than understanding them all. One easy way to think about the formats is as five completely different kinds of DVD disks.

"First, let's divide the formats up. The first thing to note is that DVD-R and DVD+R disks can only be recorded once. You only get one chance to record your DVD movie (or

whatever) to this kind of disk It's like pouring cement, once it is done you'll need to destroy it to change it. Further, DVD-R discs come in two types: DVD-R(A), for "authoring," and DVD-R(G), for "general." Both DVD-R and DVD+R discs will play in most DVD players, even older ones. So if you put your movie on this kind of disk there is a high probability that it will play in your living room. However, DVD-R(A) drives can not record to DVD-R(G) disks, and vice versa.

"There are also DVD formats that can be recorded more than once. DVD+RW, DVD-RW and DVD-RAM disks can all be recorded thousands of times. If you don't like how your DVD movie turns out, you can record a new version right on the same disk. These disks are more like painting a wall – if you don't like the color you just put on a new coat. Each of these rewritable formats are a little different. DVD-RAM, for instance, was created for storage of computer data – like backing up your hard drive. If you want to get a DVD writer to back up computer data, DVD-RAM is a solid option. However, if you plan to make your own DVD movies, one of the other formats may be better suited for that activity. Most DVD players can't play DVD-RAM disks.

"The DVD-RW and DVD+RW formats are both good for making DVD movies but are essentially engaged in a Beta versus VHS-type battle. The consumer market will ultimately determine which format wins or if they end up combining into a single standard, but it is important to understand that neither is yet a universal standard. Another thing to note is that many DVD players won't play any kind of rewritable disk. Most of the newer players will play these kinds of disks, but if you have an older DVD player it may not. In general, the newer your DVD player, the more likely it is to play all the recordable formats. There are web sites like Apple.com, HomeMovie.com and DVDplusRW.org that list compatible players and formats, but these are not unabridged resources either. Use them as a general guide.

"That covers the basic DVD writable formats. The last point of concern is the DVD drive itself. This is the part of your computer that will actually record your data or movie onto the DVD disc. Thankfully, if you have made it this far you are almost home. The different types of DVD drives basically break down into the same formats as the DVD writable formats. Therefore, there are DVD-RAM drives, DVD-R drives, etc. on down the line. It is also increasingly likely that DVD burners will come with the ability to record to more than one format, for example a manufacturer may offer a DVD+R/RW drive, meaning that it can record both DVD+R and DVD+RW discs. When considering DVD media (the actual silver discs) and DVD burners, make sure that both the discs and drive are the same format. Step one: Decide what you want to use your DVD burner for. If you want to back up computer data, a DVD-RAM burner is a good choice. If you want to record movies and music choose another kind of burner. Step 2: Decide where you will be watching your DVDs. If you plan to send your homemade DVD movies to friends and relatives with older DVD players, you will want to make sure you get a DVD-R or DVD+R burner. If you plan to watch your DVD movies on a computer or a newer DVD player, any format will likely do. Step 3: Match 'em up. Remember the children's clothing brand Garanimals? If a shirt had a lion on it, you had to find a pair of shorts with a lion on it. If you got a Garanimal shirt with a lion on it and shorts with a fox then they would be different sizes and would not match. DVD is the same way. You need to make sure that your DVD media and DVD drive are the same."

DVD-A DVD-Audio. DVD-A is a DVD disc designed primarily to play high-quality audio. DVD-Audio can carry up to six channels of 96kHz/24-bit audio (music for 5.1-channel mobile theater systems), or two channels of ultra high-resolution 192kHz/24-bit audio. Most DVD-Audio discs also carry Dolby Digital or stereo soundtracks for playback on DVD players that lack DVD-Audio decoders. A DVD-Audio disc may also contain liner notes, lyrics, menus, and still pictures that can display on your video monitor. If you thought that CD-audio sounded pretty good, you haven't heard anything until you've heard DVD-A.

DVD-R See DVD.

DVD-RAM See DVD.

DVD-ROM See DVD.

DVD-R/W See DVD.

DVD+RW See DVD.

DVD-Video A generic term for DVD that has a movie recorded on it. The good news is that all DVD players and DVD drives can play DVD-Video. See DVD for a much fuller explanation.

DVI 1. Digital video interface is a new, clearer, faster, better connection between a computer and a video device – monitor or LCD. DVI allows uncompressed, standard and high-definition video signals to be sent to and from devices equipped with the DVI port.

2. Digital Video Interactive. A name for including still and moving video pictures in material shown on a PC's screen. DVI is part of multimedia. DVI is also Intel's old name for its scheme for digitizing and compressing video and audio for storage, editing, playback

and integration into PC applications. The name has been replaced on the software side with Indeo video technology, on the retail side with Smart Video Recorder and on the hardware side with i750 Processors. See digital video interactive for a much bigger explanation.

DVMRP The Distance Vector Multicast Routing Protocol is used for the communication and distribution of multicast routing table information. It is based on the RIP protocol used in unicast routing.

Dvorak Keyboard A keyboard, invented mainly by August Dvorak, on which letters and characters are arranged for faster and easier typing than on the standard QWERTY keyboard. The QWERTY keyboard was actually designed to be difficult to use, to slow down typists so they wouldn't jam the old typewriters' mechanisms! In that respect the QWERTY keyboard resembles the present touchtone in that it also was designed to be slow and difficult to use so that it wouldn't confuse the early and slow telephone central offices of the time.

DVR Digital Video Recorder. Another name for a Personal Video Recorder. See PVR.

DVS Digital Video Services.

DVTS Pronounced "Divitz." Stands for Desktop Video conferencing Telecommunications System.

DWDM Dense Wavelength Division Multiplexing. DWDM is the higher-capacity version of WDM (Wavelength Division Multiplexing), which is a means of increasing the capacity of fiber-optic data transmission systems through sending many wavelengths of light down a single strand of fiber. WDM systems support as many as four wavelengths. Commercially available DWDM systems support from 8 to 40 wavelengths. The capacity is steadily increasing, both by ever-expanding channel counts and faster signal rates supported by the individual wavelengths. Each wavelength operates as though it were a separate light pipe, with each currently supporting signal rates as high as OC-192 transmission (9.953 Gbps – thousand million bits per second). Generally, existing systems make a trade-off between channel count and maximum supported rate: the current maximum channel count of 40 is limited to OC-48 rates of 2.488 Gbps, thereby yielding an aggregate signal rate on one strand of fiber of almost 100 Gbps. Systems which support higher signal rates (i.e., OC-192), support fewer than half as many channels. At OC-192, a 40-channel system would yield an incredible 400 Gbps, rounded up. While such a system is not currently available, it may not be far away. Within the next 2-3 years it is reasonable to expect to see systems supporting on the order of 100 wavelengths of OC-192 each, providing almost one terabit per second transport! In fact, the next generation of DWDM, dubbed UDWDM (Ultra-Dense Wavelength Division Multiplexing), is planned to support as many as 400 wavelengths. The highest rate as of December, 2001 is 80 Gbps x 80 wavelengths (also called Lambdas) = 6.4 Tbps (Terra bits per sec) per a SINGLE fiber hair strand!

DWDM lately has challenged SONET/SDH as the broadband optical technique of choice in the carrier domain. SONET offers considerable advantages, including strong standards development, and very powerful network management capabilities. The criticisms of SONET include its TDM (Time Division Multiplexing) nature, which is considered inappropriate for IP (Internet Protocol) traffic; its bandwidth limitations, even at OC-768 rates of 40 Gbps; and its high level of overhead, which directly reduces user data payload, although it yields considerable network management capabilities advantages. Perhaps the greatest criticism is SONET's high cost, especially considering that an increase in bandwidth (e.g., OC-48 to OC-192) requires that the transmitting laser diode, the receiving light detector, and all intermediate optical repeaters be upgraded. DWDM, as we have discussed, is an optical transmission technique that allows multiple light signals operating at different wavelengths (i.e., frequencies of light) to share a single fiber. The aggregate bandwidth yielded can be much greater than that supported by SONET. DWDM also is far less expensive. On the downside, DWDM does not offer the same inherent network management capabilities and does not offer the same level of standards development, which translates into lack of interconnectivity and interoperability between network elements of disparate origin. Further, and as we mentioned above, each wavelength in a DWDM system is, in essence, a separate circuit. Therefore, all traffic riding over that wavelength is transported and switched as a single entity, from point of origin to point of termination. As a result, a wavelength must carry traffic of the same type (e.g., circuit-switched voice, packet voice, IP packet data, ATM, or Frame Relay), with the same QoS (Quality of Service) requirements, originating at the same place, and destined for the same place. All of that means that each wavelength must be filled to capacity, or that there must be enough available wavelengths that capacity can afford to be underutilized. The arguments over SONET vs. DWDM rage, and will continue to do so for many years. Either approach is correct, and even optimal, depending on the applications focus of a given carrier. In fact, SONET and DWDM can, and often will, coexist, with SONET-framed data riding over DWDM wavelengths. That's my view, at least.

See WDM (Wavelength Division Multiplexing) for much more detailed explanation. See also SONET and WWDM.

DWELL Dots-in-a-well. A manufacturing technique used to improve performance of diode lasers, which are lasers that input information into optical networks. The simplest diode laser is a sandwich of n-type and p-type semiconductors. The areas where the n-type and p-type materials meet contain tiny wells known as "quantum wells", which are layers of semiconductor so thin that they are essentially two-dimensional. The performance of diode lasers is improved when tiny islands ("dots") of a different semiconductor are embedded into these wells. This is achieved by depositing a semiconductor on top of another semiconductor of different atomic spacing. The atoms in the two materials attempt to line up, however the top layer deforms under the strain and breaks into billions of islands, the quantum dots. The dots reduce the lasers operating current because they trap charge very effectively and make the lasers resistant to changes in temperature. Another advantage is that the dots can be tuned so that they only emit light at certain wavelengths, which can be used where optical fibers are used to carry many different channels of data at different wavelengths. See also Semiconductor, Injection Laser Diode.

Dwell Time The period of time that a satellite is over the desired area of coverage. The term has commercial significance in LEO (Low Earth Orbiting) and MEO (Middle Earth Orbiting) satellite systems. In such systems, the individual satellites in a constellation travel in elliptical orbits, rather than in traditional geosynchronous orbits. See LEO and MEO.

DWI Data Warehousing Institute. A Special Interest Group dedicated to "helping organizations increase their understanding and use of business intelligence by educating decision makers and I/S professionals on the proper deployment of data warehousing strategies and technologies." DWI has over 3,000 members. www.dw-institute.com.

DWMT See Discrete Wavelet Multi-Tone.

DWS Dialable Wideband Service.

DX 1. Distant.
2. Distant exchange.
3. Distant transmission.
4. Distant station reception.
5. Duplex.

DX Century Club An honorary organization whose members are radio amateurs who are certified by the American Radio Relay League (ARRL) for having contacted other radio amateurs in 100 different countries. Criteria for DX Century Club membership are on the ARRL's website at www.arrl.org.

DX Signaling A form of DC (direct current) signaling in which the differences in voltage on two pairs of a four-wire trunk indicates the supervision information, i.e. the call's beginning, its end, etc. See Duplex Signaling.

DXC Digital Cross Connect. See DCS.

DXCC 1. Abbreviation for "DX Century Club." 2. A radio amateur who is a member of the DX Century Club.

DXer A radio amateur who likes to listen to distant radio signals. The term also refers to radio amateurs who go on DXpeditions.

DXI Data eXchange Interface. A specification developed by the SMDS (Switched Megabit (or Multi-megabit) Data Services) Interest Group to define the interaction between internetworking devices and CSUs/DSUs that are transmitting over an SMDS access line. SMDS is a way for a corporate network to dial up switched data services as fast as 45 megabits per second. The ATM Forum defines DXI as "a variable length frame-based ATM interface between a DTE and a special ATM CSU/DSU. The ATM CSU/DSU converts between the variable-length DXI frames and the fixed-length ATM cells."

DXing A hobby of ham radio operators that involves listening to distant radio signals and logging information about them, such as the call sign and location of their source, their frequency, and the date and time they were received.

DXpedition An expedition by amateur radio operators to a remote, favorably situated place for the purpose of picking up distant radio signals. The remote location could be on a seacoast within a few hours' driving time, where the right equipment can pick up radio signals from across the ocean; or the remote location could be a place halfway around the world, and for which a good deal of planning and the permission of a foreign government are needed.

dye sublimation a spectacular printing process where exactly measured temperatures control the amount of ink transferred from colored ribbons to paper. Under high temperature and pressures, the ink is not melted, but is transformed directly into gas, which hardens on the paper after passing through a porous coating. Dye sub printers create very nearly continuous tones, making them great for natural images. Because the gas makes "fuzzy" dots, dye sub is not recommended for sharp-edged "computer-y" graphics or type. But it does turn out gorgeous photo-like images.

dynamic In English, dynamic means that things are changing. In telecomese, it tends to mean that our equipment – hardware and/or software – can respond instantly to changes as they occur. For example, dynamic routing in the call center world means that we can switch incoming calls from moment to moment. We may want to do this because we want calls from the east to go to our call center in the west when our eastern call center is busy. So we may want to flip the calls over to our northern call center when both our eastern and western call centers are busy. You get the idea.

dynamic adaptive routing See Dynamic Routing.

dynamic address mapping service A service which provides a lookup function between text-based strings and IP addresses and/or telephone numbers, in which the result of the lookup can change relatively quickly over time (hence the use of the word "dynamic.")

dynamic answer This a term typically used in Automatic Call Distributors. The ability to dynamically assign the number of ring cycles (interrupt, more or less) to the queue period when agents are unavailable. The implication of being able to assign this number allows return supervision to the calling in person to be delayed and thus not allow billing on 800 INWATS lines to begin. This is a money saving feature. But it can cost you some customers if they get bored waiting for your phones to pick up.

dynamic backup A backup made while the database is active.

dynamic bandwidth allocation The capability of subdividing large, high-capacity network transmission resources among multiple applications almost instantaneously, and providing each application with only that share of the bandwidth that the application needs at that moment. Dynamic bandwidth allocation is a feature available on certain high-end T-1 multiplexers that allows the total bit rate of the multiplexer's circuits to exceed the bandwidth of the network trunk. This works because the multiplexer only assigns channels on the network trunk to circuits that are transmitting.

dynamic beam focusing When you have a curved cathode ray tube, the distance between the gun which shoots the electrons and all the parts of the screen are equal. When you have a flat screen, the distance varies slightly. Some beams have to travel further. When some have to travel not so far, Dynamic beam focusing, a term I first heard used by NEC, focuses each electron to the precise distance it must travel, thus ensuring edge-to-edge clarity on the screen.

dynamic binding Binding is converting symbolic addresses in the program to storage-related addresses. Dynamic binding occurs during program execution. The term often refers to object-oriented applications that determine, during run time, which software routines to call for particular data objects. Also called Late Binding.

dynamic capacity allocation The process of determining and changing the amount of shared communications capacity assigned to nodes in the network based on current need.

dynamic configuration registry A part of Chicago (Windows 4.0) which contains a list of all the various hardware bits and pieces that make up your computer. The dynamic configuration registry is a vital element of what Microsoft calls "Plug and Play," which is the ability to remove and add bits and pieces of hardware while the machine is running and have the machine automatically recognize those hardwares and alert applications accordingly.

dynamic data exchange DDE. A form of InterProcess Communication (IPC) in Microsoft Windows and OS/2. When two or more programs that support DDE are running simultaneously, they can exchange information, data and commands. In Windows 3.xx this capability is enhanced with Object Linking and Embedding (OLE). See OLE.

dynamic frequency selection. DFS. See 801.11a.

dynamic gain equalizer Erbium doped fiber amplifiers (EDFAs) are one of the critical enablers of dense wavelength division multiplexing (DWDM). However, EDFAs have a gain that varies across the wavelength spectrum, creating a problem in cascaded amplifier systems. If the gain varies by 1 dB between the best and worst DWDM channels and there are 10 amplifiers in a link, then the signals will vary by 10 dB by the end of the transmission line. Since the signal-to-noise performance depends on the received power, the worst channel will have much poorer transmission quality than the best channel. To avoid this problem, it is important to make the gain constant across the wavelength spectrum of the EDFA, i.e. to flatten the gain. Static approaches, such as pre-emphasis and gain flattening filters, exist to flatten the output of an EDFA. However, static approaches cannot respond to variations in the EDFA gain due to changes in channel usage, changing traffic patterns, amplifier aging or to the power transfer due to stimulated Raman scattering. These varying power differences again limit the signal-to-noise ratio that can be achieved at the receiver. Dynamic gain equalizers (DGEs), also called dynamic gain flattening filters (DGFFs), are use to control the EDFA gain even when faced with varying amplifier gain effects. By dynamically adjusting the amplifier gain, the total gain error function can be reduced, enabling new extended-reach and ultra-long-haul DWDM systems.

Commercial dynamic gain equalizer products are available based on planar lightwave circuits, MEMS, liquid crystal, and acousto-optic technology. Key parameters on the filtering performance of the device include the insertion loss, polarization dependent loss, dynamic range, residual gain ripple and spectral response function. The control algorithm is also critical for a dynamic device, and can affect the overall response time and iterations required to achieve the desired gain flattening.

While there are multiple technologies of dynamic gain equalizers, there are two main approaches used for DGEs. The first approach creates a dynamic spectral response by controlling the amplitude of multiple sinusoidal filters with differing periods. The number of Fourier elements or stages used in the device determines the resolution of the DGE and the residual gain ripple. The acousto-optic effect, either in a bulk crystal or directly in the fiber, liquid crystals and lattice-filter planar lightwave circuits all employ this approach to create the controllable response filter. With the Fourier filter approach, the control algorithm often uses a convergence method, iterating the response multiple times to arrive at the minimum error function.

The second approach with a much more predictive control algorithm is based on slicing the spectrum and controlling individual elements. These approaches employ a demultiplexing element to provide the spectral slices with the control provided by micro-electro-mechanical systems (MEMS) or thermo-optic planar circuits. To obtain a continuous response function, the filtering element can be placed inside an interferometer, where the phase of the signal rather than the amplitude is controlled. The phase control creates attenuation when combined with the unaltered path in the interferometer.

dynamic gain flattening filters DGFFs. See Dynamic Gain Equalizer.

dynamic host configuration protocol DHCP. A protocol for automatic TCP/IP configuration that provides static and dynamic address allocation and management. See DHCP for a much longer explanation.

dynamic HTML Dynamic HTML combines HTML, scripts and style sheets to bring animation to the Web. With Dynamic HTML, you can program your Web site such that a visitor surfing it alights on a button or some object and instantly a "help" or "explanation" balloon pops up. This balloon explains in greater detail what will happen if the visitor clicks on the button. Or the type may change and suddenly become bigger. Dynamic HTML is being incorporated into both Netscape and Microsoft Internet Explorer. The World Wide Web Consortium is considering the various flavors of dynamic HTML as part of its DOM (Document Object Model) specification. See also www.astound.com.

dynamic IP addressing The Internet has an address for every "thing" attached to it at that very moment. That "thing" might be your browser-equipped tiny PC, or it might be a gigantic web site server with thousands of people grabbing files off it. That address is in a standard form and is called an IP (Internet Protocol) address. It is in the following form: xxx.xxx.xxx.xxx. You need an IP address so the Internet can route files, emails and messages back and forth to your PC. There are two types of IP addresses – static and dynamic. Static is an address you have all the time. If you're on a corporate network, you'll have a static address. If you are behind a hardware firewall, you'll have a static address. A dynamic address is one that changes every time you connect to the Internet. For example, you might connect to the Internet, you might be assigned this address: 129.37.213.178. Next time you connect that might be changed to 32.101.8.205. Those numbers are real. They're actually numbers I was assigned when I connected on two occasions one after another. The reason for dynamic addresses is that it things easier for users and for ISPs – Internet Service Providers. ISPs, for example have a limited number of addresses they can assign to their users. So they assign them on the fly – the essence of dynamic IP addressing. Virtually all the 500 or so million people who regularly surf the Internet do so through dynamic IP addressing. See Dynamic IP Addressing.

dynamic IP address allocation Allows a user to be assigned an IP address which is selected on the fly from a list of available addresses. See Dynamic Host Configuration Protocol (DHCP) and IP Network Control Protocol (IPCP). See also Dynamic IP Addressing.

dynamic link library DLL. An executable code module for Microsoft Windows that can be loaded on demand and linked at run time, and then unloaded when the code is no longer needed.

dynamic load & stress testing An advanced and accurate form of load testing, dynamic load and stress testing more accurately presents the variety of stimulus of external callers, systems and networks to a computer telephony system. It will test your system using real-world user actions and a realistic call mix and busy hour usage profile. Most modern computer telephony network and systems provide a number of services to their users. Each of those services may stress the system in a different fashion. As the usage patterns of those users is varied and what each user will do can vary significantly, it is critical to load test systems with traffic patterns as dynamically and as close as possible to the way the system will be used in the real world. This definition courtesy of Steve Gladstone, author of the book Testing Computer Telephony Systems, available from 212-691-8215.

dynamic load balancing A technique where a switching system, particularly multiple connected ACDs, apportion incoming calls (the load) to balance the workload. This is done dynamically in real time.

dynamic loud speaker A loud speaker in which the diaphragm is driven by means of a small "voice coil" suspended in a powerful magnetic field.

dynamic memory The most common form of memory, used for RAM, with an access speed ranging from about 60 to 150 nanoseconds. (A nanosecond is a billionth of a second.) Dynamic memory is an inexpensive but relatively complicated form of semiconductor memory with two states: presence and absence of electrical charge. Dynamic memory requires a continuous electrical current. All data is lost when the power is cut. Frequent saving files to disk helps preserve your data.

dynamic microphone A microphone, the coil of which is moved in a strong magnetic field by vibrations striking the diaphragm to which it is attached. Electrical currents are thus generated in the moving coil.

Dynamic NAT On outgoing requests from your network, a firewall replaces all private IP source addresses with one public address (usually its own). See Network Address Translation, and IP masquerading. NAT stands for Network Address Translation.

dynamic node access A high-speed bus invented by Dialogic to join together multiple voice processing PEB-based systems. PEB stands for PCM Expansion Bus. See PEB.

dynamic overload control DOC. The feature of a switch which uses its translation tables and intelligence to allow the switch to adapt to changes in traffic loads by re-routing and blocking call attempts.

dynamic page assembly You visit a web site. It knows who you are because of your cookie. It makes a web page just for you –knowing your likes and dislikes. Also known as User Personalization.

dynamic path Fibre Channel term. A communication path in which any node can communicate with any other node (assuming no third node is already communicating with the target). See Permanent Path.

dynamic port allocation In a voice processing system running multiple applications, dynamic port allocation is automatic allocation of ports based on the traffic being used by each application.

dynamic quarantining A security approach for dealing with "zero-day" exploits, i.e., attacks that have never been seen before and therefore have no known signature. Dynamic quarantining involves programmatically looking for anomalous traffic on the network, identifying the hosts that are generating the anomalous traffic, and disconnecting those hosts from the network.

dynamic RAM RAM memory that requires data to be refreshed periodically to prevent its loss in memory.

dynamic random access memory RAM which requires electronic refresh cycles every few milliseconds to preserve its data. See also Random Access Memory.

dynamic range In a transmission system, the ratio of the overload level to the noise level of the system, usually expressed in decibels. The ratio of the specified maximum level of a parameter (e.g., power, voltage, frequency, or floating point number representation) to its minimum detectable or positive value, usually expressed in decibels.

dynamic resource allocation The assignment of network capacity to specific users and specific services as required on a moment-to-moment basis.

dynamic routing Routing that adjusts automatically to changes in network topology or traffic. Dynamic routing automatically accomplishes load balancing, therefore optimizing the performance of the network "on the fly." Static routing, on the other hand, involves the selection of a route for data traffic on the basis of predetermined routing options preset by the network administrator. Dynamic routing is more effective, but the routers are more costly and the more complex decision-making process imposes additional delays on the subject packet traffic. See also Router.

dynamic storage allocation The allocation of memory space while a program is running. The memory is released when the program is complete.

dynamic synchronous transfer mode See DTM.

dynamic T-1 A T-1 line is 1.544 million bits per second transmission in both directions. Because these 1.544 million bits are raw bits, they can be configured in many ways – as 24 voice lines or 1.544 million bit per second line to another office or the Internet. A dynamic T-1 is not a technical term. It's a marketing term. And what one vendor means by a dynamic T-1 may be very different from what another vendor means. Essentially the idea is that this T-1 is chameleon. It changes its colors from time to time. One moment it's 24 voice lines. The next moment it's 1.544 million bits of data per second to the Internet. Or somewhere in between, perhaps 12 lines of voice and 772,000 bits per second of data per second to and from the Internet. In other words it assigns bandwidth dynamically – as it's needed. How Dynamic T-1 works depends on your vendor, how they implement the service and how they assign priorities as between voice and data. The typical approach is a black box at your office which monitors your T-1 traffic and reallocates bits to voice, to data, to a bit in between. This dynamic reallocation isn't instantaneous and in fact may take minutes or even hours to occur. Typically preference is given to voice, with data taking a slightly back seat. But how dynamic T-1 is implmented varies from one vendor to another. It pays to ask.

dynamic variation A short time variation outside of steady-state conditions in the characteristics of power delivered to communication equipment.

dynamically adaptive routing An algorithm, used for route determination in packet-switched networks that automatically routes traffic around congested, damaged, or destroyed switches and trunks and allows the system to continue to function over the remaining portions of the network.

dynamically assigned ip address See IP Addressing.

dynamo An electrical machine which generates a direct current.

dynamotor A direct current machine having two windings on its armature: one acting as a motor, the other as a generator.

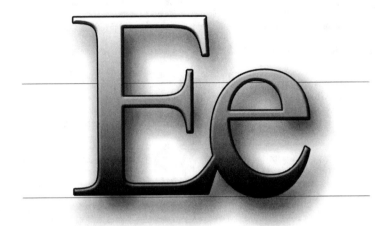

E 1. The symbol designation for electromotive force, or voltage.

2. E equals electronic, as in e-mail or e-commerce. Typically e means something over the Internet.

E region The second lowest region of the ionosphere. The E region exists only during the day. Under certain conditions, it may refract radio waves enough to return them to Earth.

EAP See extensible authentication protocol (EAP).

EAP-PEA Extensible Authentication Protocol-Protected Extensible Authentication Protocol. A mutual authentication method that uses a combination of digital certificates and another system, such as passwords.

EAP-TLS Extensible Authentication Protocol-Transport Layer Security. A mutual authentication method that uses digital certificates.

EAPOL See 802.1x.

E & M Ear & Mouth, Earth and Magneto, rEceive and transMit (take your choice). In telephony, a trunking arrangement that is generally used for two way (either side may initiate actions) switch-to-switch or switch-to-network connections. It also frequently used for computer telephony system to switch connections. See E & M Leads.

E & M Leads The pair of wires carrying signals between trunk equipment and a separate signaling equipment unit. The "M" lead transmits a ground or battery conditions to the signaling equipment. The "E" lead receives open or ground signals from the signaling equipment. These leads are also known as Ear and Mouth Leads. The Ear lead typically means to receive and the Mouth lead typically means to transmit. Changes of voltage on these leads convey such information as seizure of circuit, recognition of seizure, release of circuit, dialed digits, etc. In the old days it was the PBX operators who originated trunk calls by asking the long distance carrier for free trunks using their mouth or M lead. If the carrier had a free trunk, the PBX heard about it through its ear or E lead. See also E & M and E & M SIGNALING.

E & M Signaling In telephony, an arrangement that uses separate leads, called respectively the "E" lead and "M" lead, for signaling and supervisory purposes. The near end signals the far end by applying -48 volts dc (vdc) to the "M" lead, which results in a ground being applied to the far end's "E" lead. When -48 vdc is applied to the far end "M" lead, the near-end "E" lead is grounded. The "E" originally stood for "ear," i.e., when the near-end "E" lead was grounded, the far end was calling and "wanted your ear." The "M" originally stood for "mouth," because when the near-end wanted to call (i.e., speak to) the far end, -48 vdc was applied to that lead.

When a PBX wishes to connect to another PBX directly or to a remote PBX or extension telephone over a leased voice grade line, a channel on T-1, the PBX uses a special line interface which is quite different from that which it uses to interface to the phones it's attached directly to (i.e. with in-building wires). The basic reason for the difference between a normal extension interface and the long distance interface is that the signaling requirements differ – even if the voice signal parameters such as level and two-wire, 4-wire remain the same. When dealing with tie lines or trunks it is costly, inefficient and too slow for a PBX to do what an extension telephone would do, i.e. go off hook, wait for dial tone, dial, wait for ringing to stop, etc. The E&M tie trunk interface device is the closest thing there is to a standard that exists in the PBX, T-1 multiplexer, voice digitizer telco world. But even then it comes in at least five different flavors. E&M signaling is the most common interface signaling method used to interconnect switching signaling systems with transmission signaling systems.

E&M signaling can be either 2-wire or 4-wire. There are mainly 5 types of E&M signaling:

Type 1: 2-wire; E and M leads. Commonly used in North America

Condition	I	M lead	E lead
—I———			
On-hook	I	Ground	Open
Off-hook	I	Battery	Ground

Type II: 4-wire; E, M, SG, SB leads. Slightly less common than type 1 in North America

Condition	I	M-lead / SBE-lead / SG		
—I———				
On-hook	I	Open		Open
Off-hook	I	Battery		Ground

Type III: 4-wire; E, M, SG, SB leads. Rarely used in North America

Condition	I	M-lead / SBE-lead / SG		
—I———				
On-hook	I	Ground		Open
Off-hook	I	Loop Current		Ground

TypeIV: 4-wire; E, M, SG, SB leads. Extremely rare in North America

Condition	I	M-lead / SBE-lead / SG		
—I———				
On-hook	I	Open		Open
Off-hook	I	Ground		Ground

TypeV: 2-wire; E and M leads. Commonly used outside of North America

Condition	I	M lead	E lead
—I———			
On-hook	I	Open	Open
Off-hook	I	Ground	Ground

Type SSDC5: 2-wire; E and M leads. British Telecom Standard. This type is similar to type 1 and V, but not the same.

Condition		M lead	E lead
On-hook		Earth-Off	Earth-Off
Off-hook		Earth-On	Earth-On

See E & M and E & M Leads.

E Band The optical band, or window, specified by the ITU-T at a wavelength range between 1360nm and 1460nm (nanometers) for fiber optic transmission systems. See also C-Band, L-Band, O-Band, S-Band, and U-Band.

E Block Carrier A 10 MHz PCS carrier serving a Basic Trading Area in the frequency block 1885 – 1890 MHz paired with 1965 – 1970 MHz.

E Channel E stands for echo. It is the 16 Kbps ISDN basic rate channel echoing contents of DCEs to DTEs. Used in bidding for access to multipoint link.

E Link Extended Link. A Signaling System 7 (SS7) connection. This protocol controls all transfers between COs in North America. A SS7 signaling link used to connect a Signaling End Point (SEP) to an STP pair not considered its home STP pair.

E Mail Electronic Mail.

E Mail Gateway A LAN application that fetches messages from one electronic mail system, translates them to the format of another electronic mail system, and then sends them to the "post office" of that other system. The post office is the public entry point – the place you put mail you want the other system to receive.

E Port A expansion port on a switch. It is used to link multiple switches together into a Fibre Channel fabric. See Fibre Channel.

E Purse Electronic purse. An electronic monetary transaction card being proposed by several government agencies.

E Rate From Network World, December 15, 1997: "E Rate is the program President Clinton and Vice President Gore are referring to when they say they want every classroom connected to the Internet by the year 2000. It grants elementary and secondary schools a discount on carrier services, including not only Internet access but also a raft of other offerings."

E region The second lowest region of the ionosphere. The E region exists only during the day. Under certain conditions, it may refract radio waves enough to return them to Earth. See also D and F regions.

E-1 The European equivalent of the North American 1.544 million bits per second (Mbps) T-1,except that E-1 carries information at the rate of 2.048 million bits per second. This is the rate used by European CEPT carriers to transmit 30 64 Kbps digital channels for voice or data calls, plus a 64 kilobits per second (Kbps) channel for signaling, and a 64 Kbps channel for framing (synchronization) and maintenance. CEPT stands for the Conference of European Postal and Telecommunication Administrations. Since robbed-bit signaling is not used (as it is for T-1 in North America) all 8 bits per channel are used to code the waveshape sample. See E1, E2, E3, and T-1.

E-2 Interim data signal that carries four multiplexed E-1 signals. Effective data rate is 8.448 million bits per second (Mbps). See E1, E3 and T-1.

E-3 CEPT signal which carries 16 CEPT E-1s and overhead. Effective data rate is 34.368 million bits per second (Mbps).

E-911 See the next two definitions.

E-911 Control Office In the US emergency services telephone network, the E-911 Control Office is the central office that provides the tandem switching of 911 calls. Each E-911 public safety answering point (PSAP) connects to one or more E-911 Control Offices. The E-911 Control Offices delivers 911 voice calls, with Automatic Number Identification, to the PSAP and provides normal and emergency-specific switching functions. The specialized switch at the E-911 Control Office is known as an E-911 Tandem or Selective Router.

E-911 Service Enhanced 911 service. Dial 911 in most major cities and you'll be connected with an emergency service run typically by a combination of the local police and local fire departments. 911 service becomes enhanced 911 emergency reporting service when there is a minimum of two special features added to it. E-911 provides ANI (Automatic Number Identification) and ALI (Automatic Location Information) to the 911 operator. Picture: A call comes in. Someone is dying. The 911 operator's screen comes alive as his phone rings. The number calling is on the screen. The caller is dying and needs an ambulance. The operator punches a button or two and his screen immediately indicates the location of the ambulance dispatch center nearest the caller. The operator contacts the dispatch center, another button may dispatch a fax of a map of how to get there to the ambulance and an ambulance gets there in short order and saves a life. (Remember, this is a book, not the real world.) See also CESID and PSAP for a full explanation of how the caller's location is sent. See E-911 Control Office.

E-BCCH Extended-Broadcast Control CHannel. A logical channel element of the BCCH signaling and control channel used in digital cellular networks employing TDMA (Time Division Multiple Access), as defined by IS-136. See also BCCH, IS-136 and TDMA.

E-Band See E Band.

E-Bend A smooth change in the direction of the axis of a waveguide, throughout which the axis remains in a plane parallel to the direction of electric E-field (transverse) polarization.

E-Check An Ecommerce term for an electronic check. The E-check is a demand for payment which is sent electronically over a network from the buyer to the seller. The e-check subsequently is sent from the seller to the seller's bank, and then to the buyer's bank. See also E-Commerce.

E-Commerce Electronic Commerce. Buying and selling over the public Internet, the public Web and corporate Internets. I prefer ecommerce to e-commerce. But you see it spelled both ways. Predictions for the amount of ecommerce should not be underestimated. See the Internet.

e-discovery The efforts of a party to a lawsuit and the party's attorneys to obtain information from email and other electronic sources as part of the entire pre-trial discovery process.

E-DSS1 E-DSS1 is the European shorthand way of saying Euro-ISDN,

E-IDE Enhanced IDE. An enhancement to the original IDE disk drive found on many PCs. E-IDE raises the storage capacity limit from 504 megabytes to 8,033 megabytes and the data transfer rate from up to 3 megabytes per second to up to 16.6 megabytes per second. See also Enhanced IDE.

E-Interface The network interface between the Cellular Digital Packet Data (CDPD) networks and other external networks.

E-Mail Electronic Mail. Also spelled email, which is this dictionary's preferred spelling.

E-Netz A German digital mobile phone network started in 1994. This high density system allows cell phones to function at a low wattage of from 0.25 to one watt of power. Until October 1998 the only provider using the E-Netz was E-Plus. Since then O2 online is also on the E-Netz.

E-Nose Electronic nose, gas chromatograph detection systems. These can detect and analyze target samples for a wide variety of applications as in security control, environmental protection control, food and drug safety control.

E-OTD Enhanced Observed Time Difference is a new technology that could be used in mobile phones for location based services. The E-OTD positioning method, generally relies upon measuring the time at which signals from the Base Transceiver Station (BTS) arrive at two geographically dispersed locations – the mobile phone/station (MS) itself and a fixed measuring point known as the Location Measurement Unit (LMU) whose location is known. The position of the MS (also called a cell phone) is determined by comparing the time differences between the two sets of timing measurements. To obtain accurate triangulation, OTD measurements are needed from at least three geographically distinct BTSs. Based on the measured values, the location of the MS can be calculated either in the network or in the MS itself, if all the needed information is available in the MS. The terms "MS-assisted" applies to the former method and "MS-based" to the latter. The MS performs measurements without the need for any additional hardware. To obtain accurate triangulation, OTD measurements are needed from at least three geographically distinct BTSs. Based on the measured values, the location of the MS can be calculated either in the network or in the MS itself, if all the needed information is available in the MS. The terms "MS-assisted" applies to the former method and "MS-based" to the latter. See Location Services.

E-Rate Electronic Rate. A special discounted rate for Internet access for schools and libraries, e-rate was established as part of the Telecommunications Act of 1996. Technically known as the Schools and Libraries Universal Service Program, e-rate is funded through a portion of the special surcharges of up to five percent on every telephone bill. The carriers remit a portion of those surcharges to the Universal Service Fund, which is administered by the Universal Service Administrative Company (USAC), a not-for-profit corporation established and operated by the National Exchange Carriers Association (NECA). USAC distributes the designated funds to the Schools and Libraries Corporation (SLC) and to the Rural Health Care Corporation (RHCC), which now are divisions of USAC. The SLD (Schools and Libraries Division) get up to $2.25 billion per year, and the RHCD (Rural Health Care Division) up to $400 million per year, with the funds parceled out after the usual ton of paperwork is submitted and dissected. E-rate subsidies are for internal wiring, telecommunications services, and Internet access.

E-Signatures As it sounds, e-signatures are signatures that are recorded and transmitted digitally. They have advantages over actual signatures in that 30 physical aspects of a signature are recorded in an e-signature, making it harder to forge.

E-Stamp Electronic Stamp. Developed by E-Stamp Inc., and planned for trial by the USPS (U.S. Postal Service), the E-Stamp is a means for buying postage over the Internet. In support of a corporate Intranet post office, the system comprises PC software, a small security device which attaches to a user's printer port, and 1,024-bit encryption software for purposes of security. Think of it as a PC-based postage meter which can be refreshed over the Internet.

E-Tail A squeezing of electronic and retail. Really awful. I prefer ecommerce. See Ecommerce and E-tailer.

E-Tailer A company that does most of its retailing, consumer-to-consumer business over the Internet. Such companies include priceline.com, ftd.com, 1-800-flowers.com, ebay and expedia.com.

E-TDMA Extended Time Division Multiple Access. A proposed, new, standard for cellular. Other standards are TDMA (Time Division Multiple Access), CDMA (Code Division Multiple Access) and NAMPS (Narrow Advanced Mobile Phone Service). Refers to the extended (digital cellular) transmission technology developed by Hughes Network Systems. E-TDMA is alleged to have 15 times the capacity of today's analog cellular phone systems – in other words to allow the simultaneous use of 15 times as many cellular phones as today's analog cellular phone system.

E-Wear E-wear is a line of wearable portable audio products from Panasonic.

E-Zine Magazines that are published (i.e. made public) on the World Wide Web. Typically an e-zine is available for anyone to read who wants to visit the site the electronic magazine is located at. An e-zine is also called a Webzine.

E.164 The ITU-T recommendation for GSTN (Global Switched Telephone Network) numbering. E.164 is a 16-digit numbering scheme that provides a unique telephone number for every subscriber in the world. See also ENUM.

E.169 The ITU-T recommendation for international toll-free numbers, known as Universal International Freephone Numbers (UIFNs). Such numbers remain the same throughout the world, regardless of country or telecommunications carrier. "Freephone" is a service that permits the cost of a telephone call to be charged to the called party, rather than the calling party. In North America, "freephone" numbers are known as "toll-free" or IN-WATS, and include the 800, 888, 877 and 866 area codes. See also IN-WATS and UIFN.

E=MC^2 Energy (E) equals Mass (M) times the Velocity of Light (C) squared (^2, as in superscript 2). E equals MC squared is Albert Einstein's famous equation which shows the relationship of mass to energy. One gram of mass – half the weight of a Lifesaver candy - was converted into energy over Hiroshima in 1945.

E1 Name given to the CEPT (Conference of European Postal and Telecommunication Administration) digital telephony format devised by the ITU-T that carries data at the rate of 2.048 million bits per second (DS-1 level). It's designed to carry 32 (thirty two) 64 Kbps digital channels. E1 is the rate used by European CEPT carriers to transmit thirty 64 Kbps digital channels for voice or data calls, plus a 64 Kbps channel for signaling, and a 64 Kbps channel for framing and maintenance. Plesiochronous means "almost synchronous." In the network sense, when two networks operate with clocks of sufficiently high quality such that the signals in the two networks are nearly synchronous, the networks are plesiochronous. In the synchronization hierarchy, stratum I clocks are required for plesiochronous operation. My personal preference is to use E-1 instead of E1, but it's written both ways. See E-1 and T-1, which is the North American equivalent.

E112 Emergency call number in Europe. The European equivalent of E911. E stands for electronic.

e164.arpa The domain established by the IAB (Internet Architecture Board) for ENUM. See also E.164 and ENUM.

E2 Data signal that carries four multiplexed E-1 signals. Effective data rate is 8.448 million bits per second or 128 simultaneous conversations.

E3 Signal which carries 16 CEPT E-1 circuits and overhead. Effective data rate is 34.368 million bits per second or 512 simultaneous voice conversations. Also known as CEPT3.

E4 Signal which carries four E3 channels – or 139.264 million bits per second, or 1,920 simultaneous voice conversations.

E5 Signal which carries four E4 channels – or 565.148 million bits per second, or 7,680 simultaneous voice conversations.

E911 See Enhanced 911.

EA See Equal Access and Address Field Extension.

EACEM The European Association of Consumer Electronics Manufacturers.

EADAS Engineering and Administrative Data Acquisition System.

EADAS/NWM EADAS NetWork Management.

EAI Enterprise Application Integration. See also EPO.

EAN European Article Numbering. EAN is the universal bar coding system for product identification at point of sale, now adopted almost worldwide. There are variants, e.g. in Japan in the term is EAN/J.

EAP Extensible Authentication Protocol, as defined in IETF RFC 2284, is an authentication protocol that runs over Layer 2, the Data Link Layer (DLL), of the OSI Reference Model. As EAP does not require IP (Internet Protocol), it includes its own support for message delivery and retransmission. EAP was developed for use over PPP (Point-to-Point Protocol), although it now is in use in IEEE 802 LAN environments, including 802.3 Ethernet LANs, 802.5 Token Ring LANs, and 802.11 Ethernet WLANs (Wireless LANs), and is specified in 802.1X. In such a LAN environment, EAP takes the form of EAPOL (EAP Over LAN). EAP supports multiple authentication mechanisms. See also 802.1X, 802.3, 802.5, Authentication, IP, LAN, OSI Reference Model, PPP, and WLAN.

ear and mouth signaling See E & M Signaling.

earcon An sound version of the icon, a short burst of sound that is musical but not music. Earcons will be used by telephone companies and wireless providers as tiny branding signatures. Cynthia Sikora and Linda Roberts of Bell Labs are working on earcons. In a research paper they talk about "steer(ing) the emotional reaction of the listener in support of the desired image." Expect to be inundated by a symphony of earcons, emanating from phones, Internet enabled PDAs, wearable electronics, computers, refrigerators, medicine chests, says Faith Popcorn's "Dictionary of the Future."

early adopter When he was young and immature, Dan Good was an early adopter. That meant he would beg, borrow and steal to get the beta release of the first release of any software that would, allegedly, improve his life, his productivity, his well-bearing or his stature among his friends. Now he's older and wiser, he knows "early adopter" software is riddled with bugs. Dan now knows not to waste the time removing the buggy software and going back to his original software. He knows it is better to stick with tried and true software – preferably software that has enjoyed at least one "service release" – a fancy Microsoft word for code that fixes the bugs in the original piece of software.

Early Bird Early Bird is the nickname of Intelsat I, the first commercial geostationary communications satellite. It was launched in 1965.

early packet discard See EPD.

early termination fee ETF. A penalty fee that applies when a customer leaves a carrier before his service contract is up. Fees can often be negotiated downwards, depending on how long the customer has been with the carrier and how good his bargaining skills. ETFs are most commonly seen in wireless (cell phone) carriers, which force a customer to sign a contract for two reasons. 1. Carriers are afraid of churn – a customer jumping ship to another carrier. A long-term customer and the early termination fee penalty holds the customer. 2. Carriers "sell" the cell phone to their customer often for less than it cost them to buy the phone from the manufacturer. They need to recoup this loss. And they do it by forcing the customer into a long-term contract. Whether this makes sense or not is not my role to question.

early token release This is a method of token passing which allows for two tokens to exist on the network simultaneously. It is used primarily in 16 million bit per second (Mbps) token ring LANs. On a regular four million bit per second token ring LAN, the token is passed on only after the sending computer receives its message back from the destination computer. With early token release, the sending computer does not wait for its message to return before passing the token. This means there are two tokens on the network at the same time. This is done to take advantage of the idle time created on the faster token ring. While the message is moving to its destination and back the sending computer is idle. On the four Mbps token ring, this is not much of a problem since most of the time the message is on the ring. On the 16 Mbps token ring less of this idle time is transmission time and more is taken by copying the message to the token ring card. That is, on the faster ring, there is more of a window for a second token. Early token release is especially helpful when traffic is heavy.

EARN European Academic Research Network. A network using BITNET technology connecting universities and research labs in Europe.

earned company The company whose network carried an LD call and who is entitled to the revenue associated with the call. This may or may not be the same as the Billing Company.

earned month The month service or equipment was delivered to a customer. In usage, this is the month the phone call took place. For nonrecurring charges (NRCs), it is the month the service or equipment was delivered.

earnings per share Calculated by dividing a company's total after-tax profits by the company's number of common shares outstanding. Earnings per share can be used as an indicator of growth and profitability.

EAROM An acronym for Electrically Alterable Read-Only Memory. A type of ROM chip which can be erased and reprogrammed without having to be removed from the circuit board. An EAROM is reprogrammed electrically faster and more conveniently than an EPROM (Erasable Programmable Read-Only Memory). An EAROM chip does not lose its memory when power is turned off.

earth The term the English use for what Americans call ground. See Grounding.

earth ground The connection of an electrical system to earth. This connection is necessary to provide lightning and static protection as well as to establish the zero-voltage reference for the system. See Earth Grounding.

earth ground connection The conductor which connects directly to earth ground usually via a water pipe or possibly via a copper rod driven into the earth. This ground is different than the logic ground used in electronic circuits.

earth ground electrode The conducting body in contact with the earth. The grounding electrode may be a metallic cold water pipe when used in conjunction with a driven rod, a mat, a grid, etc. Earth should never be used as the sole equipment grounding conductor.

earth grounding The purpose of earth grounding is essentially threefold: 1. Lightning protection; 2. Static protection; and 3. Establish a zero voltage reference. See Ground and Grounding.

earth segment Also called ground segment; in satellite communications, one of the two segments of the overall services, comprising the facilities involved in communicating via a satellite, except for the satellite itself which is in the space segment.

earth station A ground-based antenna and associated equipment used to receive and/or transmit telecommunications signals via satellite. Earth stations come in all sizes, shapes and purposes. For example,M most earth stations used by cable television operators are receive-only. Other terms used synonymously with "earth station" include downlink, ground station, and TVRO ("Television Receive Only").

earth terminal Microwave radio equipment used to communicate with communications satellites. This term is most commonly associated with VSAT's, TVRO's, portable and mobile satellite operations, particularly domestic.

Earth-Moon-Earth communication See moonbounce.

earthing There are two distinct and unique categories in the broad area called grounding. One is earthing, which is designed to guard against the adverse effects of lightning, assist in the reduction of static and bring a zero-voltage reference to system components in order that logic circuits can communicate from a known reference. The other category of grounding is known as equipment grounding. This is the primary means of protecting personnel from electrocution. According to the Electric Power Research Institute, "electrical wiring and grounding defects are the source of 90% of all equipment failures." Many telephone system installer/contractors have found that checking for and repairing grounding problems can solve many telephone system problems, especially intermittent "no trouble found" problems. As electrical connections age, they loosen, corrode and become subject to thermal stress that can increase the impedance of the ground path or increase the resistance of the connection to earth. Equipment is available to test for proper grounding. One of our favorite devices is made by Ecos Electronics in Oak Park, IL. Before you attach any equipment (computer, telephone, hi-fi set, etc.) to an improperly grounded electrical outlet, you should have the problems corrected. See also Ground and Grounding.

EAS 1. Emergency Alert System. A system for radio and TV that is designed to provide warnings in the event of emergencies. The system was originally designed during the Cold War to provide the President with a means to address the American people in the event of a national emergency. It has never been used for that purpose (thank God), but it has been used to warn of natural disasters. The current use of the Emergency Alert System is voluntary and involves participation by three groups – 1. Radio, TV and cable broadcasters, 2. The National Weather Service and 3. State and local emergency management agencies. The FCC reports that 85% of the messages have dealt with weather emergencies.

2. Extended Area Service. A novel name for a larger than normal local telephone calling area. The local phone company extends its subscribers the option of paying less per month for a small calling area and paying extra per individual call outside that area (i.e. the extended area), or paying more per month flat rate but having a larger calling area (i.e. having extended area service).

3. Electronic Article Surveillance. See also RFID.

EASE A voice processing applications generator from Expert Systems, Inc. in Atlanta, GA.

Easter Egg An Easter Egg, in computer parlance, is a tiny credits screen or secret surprise (e.g., graphic or sound effect) hidden by a programmer in the program code. An Easter Egg is designed to appear only when you perform some incredibly improbable sequence of keystrokes and mouse clicks, commonly if you're another programmer disassembling or browsing the source code. Easter Eggs are intended to be amusing, rather than destructive, although they do result in code bloat (i.e., eat up memory and other resources). Easter Eggs are also used for checking software theft. By hitting the secret keystrokes, you can tell if the code was stolen. Easter Eggs are hidden in all sorts of programs, including Lotus Notes and Microsoft Excel and Word. See also Code Bloat.

eave mount A triangular antenna mounting system used for attaching a mast to the eave or gable of a peaked roof.

EAX Electronic Automatic eXchange. Term used throughout the non-Bell telephone industry to refer to an electronic central office. Similar to ESS (Electronic Switching System), the term used by AT&T and the Bell operating telephone companies.

EB Exabyte. See Exabyte.

EBCDIC (Pronounced Eb-si-dic.) Extended Binary Coded Decimal Interchange Code. It is the way IBM codes characters, letters and numbers into a digital binary stream for use in its larger computers. EBCDIC codes characters into eight bits. This gives it 256 possible characters, $2 \times 2 \times 2 \times 2 \times 2 \times 2 \times 2 \times 2 = 256$. See also Extended Character Set. EBCDIC is mainly used in IBM mainframes and minicomputers, while ASCII is used in IBM and non-IBM desktop microcomputers. EBCDIC is not compatible with ASCII, meaning that a computer which understands EBCDIC will not understand ASCII. But there are many real-time and non-real time translation programs that will convert text files back and forth. See ASCII.

EBD Effective Bill Date. This date is used when the effective date for billing differs from the completion date. In the bill section of an order, EBD applies to the total order.

EBDI Electronic Business Data Interchange. Term for EDI (Electronic Data Interchange). See EDI.

Ebbers, Bernie Aug. 27, 1941 - Bernard Ebbers is born in Edmonton, Alberta, the second of five children.

Late 1940s - Ebbers's family moves to California, where his father works as an auto mechanic.

Early 1950s - The family moves to New Mexico and Ebbers spends five years attending boarding school on a Navajo reservation, where his father works. During summers, he works for the reservation's maintenance department.

Late 1950s - Ebbers's family moves back to Edmonton, where he finishes high school before enrolling at the University of Alberta. He hopes to become a basketball coach and a school teacher, but leaves the university after one year after getting poor grades.

Early 1960s - After spending years holding down odd jobs – delivering milk, loading bread trucks for a bakery and working as a bouncer at a local bar – Ebbers gets a job coaching high-school basketball, with the help of his former coach. He is admitted to Mississippi College, a private Baptist college in Mississippi, on a basketball scholarship.

1967 - Ebbers graduates from Mississippi College, marries college sweetheart Linda Pigott, and gets a job teaching science to middle-school students.

Late 1960s to early 1970s - Ebbers moves south with his wife to Brookhaven, Miss., after she is offered a teaching position there. Ebbers takes a job managing a warehouse for a local garment factory and in the next six years moves up the ladder. He leaves the company following a disagreement with his boss.

1974 - After noticing an advertisement in the local paper, Ebbers buys a motel in Columbia, Miss., and moves with his wife into a two-bedroom trailer parked in the motel lot.

Early 1980s - Ebbers amasses a mini-empire of eight motels, including some in the Hampton Inn and Courtyard by Marriott chains.

1983 - Four entrepreneurs, including Ebbers, get together at a coffee shop in Mississippi and hatch a plan to start a long-distance company called LDDS (Long Distance Discount Service).

1985 - Ebbers becomes chief executive of LDDS.

1995 - LDDS changes its name to WorldCom, with Ebbers as chief executive, and grows through several acquisitions into one of the biggest telecommunications companies in the U.S.

1997 - Ebbers divorces Linda, his wife of 30 years, with whom he had four children – two of them adopted – named Treasure, Ave, Joy and Faith.

Oct. 13, 1997 - Business Week hails Ebbers as the "Telecom Cowboy" in a cover story on his bold bid to buy MCI.

Nov. 7, 1997 - With the help of a Friday-night phone call to MCI's chairman, Ebbers clinches a deal to acquire the much-bigger MCI for $37 billion.

March 1999 - Ebbers marries second wife, Kristie.

September 2000 - Under increasing financial strain from margin calls on his investments, Ebbers borrows $50 million from WorldCom in what will be the first of many loans from the company.

March 15, 2001 - The Wall Street Journal reports that the loans WorldCom has extended to Ebbers represent by far the largest amount any company has lent to one of its officers in recent memory.

April 29, 2002 - Ebbers is fired as chief executive of WorldCom, amid an SEC investigation into the company's accounting, unhappiness with WorldCom's sinking stock price and controversy over Mr. Ebbers's $366 million personal loan from the company.

June 30, 2002 - Ebbers appears at his Mississippi church to teach Sunday school and attend the morning worship service. At the end of the service, he walks to the front of the church and addresses the congregation. "I just want you to know you aren't going to church with a crook," he says.

March 2, 2004 - After trying for two years to build a case against Ebbers, the federal government charges him with allegedly helping to orchestrate the largest accounting fraud in U.S. history. Ebbers pleads not guilty.

March 1, 2005 - Ebbers takes the witness stand in a risky move, telling the jury he "did not know" that WorldCom's books were being manipulated under his watch.

March 15, 2005 - The jury finds Ebbers guilty on all counts.

July 13, 2005 - Ebbers sentenced to 25 years in prison for leading the largest corporate fraud in U.S. history. Even with possible time off for good behavior, Ebbers, 63, and with what his lawyers describe as serious heart problems, would remain locked up until 2027, when he would be 85.

EBITDA Earnings Before Interest, Taxes, Depreciation and Amortization. All financial accounting can mislead or enlighten. I've never been a big fan of EBIDTA. Moody's Investors Service says it's more often used as an accounting gimmick to dress up a company's financial picture to make it look rosier than it really is and deflect investor attention away from bad news. But here's the way it was explained to me: EBIDTA is a way of looking at the earnings of a firm. By ignoring taxes, depreciation and amortization, EBIDTA attempts to establish a "common ground" by which the profitability of different firms can be measured, given the likelihood that the ownership of these firms may shortly change and the new owner will have his own taxes (perhaps lower), may renegotiate the company's loans (perhaps cheaper) and may change the company's depreciation and amortization. I hear the word EBITDA more in acquisition scenarios than in ongoing companies that are not going to be acquired. EBITDA has also been used in the valuation of companies which are losing money, such as several wireless or long distance carrier telecom companies. The major factor in their losing money is often the high interest on their heavy borrowings which they used to finance their networks. Removing that interest (as in EBITDA) makes them wonderful as they're now earning money. I believe these companies should be measured on their earnings after interest, taxes and depreciation - that's commonly called net income – since that provides a better measure of the efficiency of their business, including their borrowings which are integral to their business.

EBONE European Backbone. A pan-European network backbone service.

EBPP Electronic Bill Presentation and Payment. Billing and payment over the Internet. Developing standards include Open Financial Exchange (OFX) and GOLD. See also Electronic Commerce, GOLD and OFX.

EBR If you have an "Existing Business Relationship" with someone you are legally allowed (under the DNC –Do Not Call – requirements) to call the person. EBR is defined in the law.

EBS 1. Electronic Business Set.

2. Enhanced Business Service (also known as P-Phone) is an analog Centrex offering provided by Nortel Networks. It operates over a single-pair subscriber loop., providing normal full duplex audio conversations and a secondary 8 KHz half-duplex amplitude shift-keyed signal, which is used to transmit signaling information to and from the Nortel Networks-equipped central office.

3. Emergency Broadcast System. Some local radio stations have volunteered their services to be part of a group of radio stations which would broadcast information should there be a public emergency. They operate the EBS under the aegis of the Federal Communications Commission. Such emergency would be a natural disaster, a technological disaster or a war. In 1994, Emergency Broadcast System changed. For a full explanation, see Emergency Broadcast System.

4. End System Bye packet. Part of CDPD Mobile End System (M-ES) registration procedures.

EBU European Broadcasting Union. Formed in 1950 and headquartered in Geneva, Switzerland, the EBU was formed to address technical and legal problems in Western European broadcasting. The EBU now includes former Soviet Bloc broadcasters, having merged with the former International Radio and Television Organization (OIRT) in 1993. Associate membership includes 30 countries in Africa, the Americas, and Asia. The EBU runs EUROVISION, a permanent network of 13 channels on a Eutelsat satellite, plus 5,500 kilometers of permanently leased terrestrial circuits; the network serves as a vehicle for daily news and program exchanges. www.ebu.ch.

EBXML Electronic Business XML is a set of specifications defined to enable businesses to communicate with each other, irrespective of their location and domain, over the Internet, exchanging messages in an XML format. EbXML is a project jointly initiated by UN/CE-FACT (The United Nations body for Trade Facilitation and Electronic Business) and OASIS to standardize XML business specifications. ebXML intends to develop a technical framework that will enable XML to be used in a consistent manner for the exchange of all electronic business data. It promises to replace EDI (Electronic Data Interchange). I first read about in the automotive industry which believes that in addition to simplifying electronic forms processing, promises to cut auto dealership's data communications costs in half.

EC 1. European Community, or European Economic Community (EEC). Now called the European Union. Also known as the Common Market; an organization of 15 nations in Western Europe that has its own institutional structure and decision-making framework. It was formed on November 1, 1993. The intent of the organization is to promote trade and reduce barriers. Member nations are Austria, Belgium, Denmark, Finland, France, Germany, Greece, Ireland, Italy, Luxembourg, the Netherlands, Portugal, Spain, Sweden and the United Kingdom of Great Britain and Northern Ireland

2. Electrical Carrier. OC means optical carrier.

EC-1 electrical carrier Level 1: 51.84 Mb/s. This is an STS-1 signal conditioned for transmission over a relatively short span of coax, usually between collocated network elements.

ECA Exchange Carrier Association. An organization set up under the FCC's order on access charges to perform numerous planning and management functions required to implement and operate the industrywide access charge system.

Ecash Developed by DigiCash and the Mark Twain Bank, ecash is the ability to use real money in a electronic purchasing system over the World Wide Web. The process involves you sending a check to Mark Twain Bank which in turn sends you software which gives you access to the Ecash Mint where you draw funds to your hard drive for use when purchasing goods and services on the Internet. www.marktwain.com.

ECC 1. Exchange Carrier Code. Four letter numeric originally assigned by Bellcore (now called Telcordia Technologies) to identify CLECs – competitive local exchange carriers, i.e. local telephone companies. This code is used solely by Bell legacy computer systems. Most competitive telecom companies find this code repetitive because there is already an ACNA code and an OCN (Operating Company Number) code that identifies carriers. However, the Bell Companies do not want to do away with this code because there is a field in their system for it. Some telephone companies (such as Southwestern Bell) call the ECC number an AECN, which stands for an Alternate Exchange Carrier Name. See also AECN.

2. Elliptic Curve Cryptography. The Wireless Application Protocol (WAP) Forum has added ECC into Version 1.0 of its specification for wireless security that is bandwidth, memory and power efficient. More details www.certicom.com.

3. Electronic Common Control. The generic term that applies to the third generation of automatic voice circuit switches, such as PBXs, COs (Central Offices), and toll tandems. ECC switches essentially are highly specialized computer systems with a common set of programmed logic that controls the setup, maintenance and teardown of port-to-port connections through a switching matrix comprising multiple centralized buses. A set of common logic also supports a wide array of features. ECC switches rendered obsolete electromagnetic Xbar (crossbar), which earlier rendered obsolete electromechanical SxS (Step-by-Step) systems, which had rendered obsolete manual cordboards. See also Cord Board, SxS, and Xbar.

ECC RAM Error-correcting code memory, often used in local area network servers. ECC memory tests for and corrects errors on the fly, using circuitry that generates checksums to correct errors greater than one hit.

ECCA European Cable Communications Association.

eccentric Non-circular; elliptical (applied to an orbit).

eccentricity 1. Like concentricity a measure of the center of a conductor's location with respect to the circular cross section of the insulation. Expressed as a percentage of center displacement of one circle within the other.

2. In orbit geometry, the deviation of the center of an orbit from the center of the earth.

The theoretically perfect geostationary satellite orbit has zero eccentricity.

3. Odd or whimsical behavior. In other words, behavior which deviates from and established pattern, rule, or norm. The term comes from "eccentric," which applies to rich people. Poor people who exhibit eccentric behavior are just "crazy." Rich people who are eccentric are successful. In short, there is no valid definition.

ECCKT Exchange Company Circuit. A circuit ID for a trunk line. Such term is often used between telephone companies to identify lines that one is leasing to the other. See also FOC.

ECH Enhanced Call Handling. ECH systems are those in which a telephone call is handled "intelligently" b a variety of network, human, computer and telecommunications resources. ECH systems cover those from voice mail, to interactive voice response to computer telephone integration to fax-on-demand to complex telephone networks.

echelon 1. The name of a startup company in Palo Alto that is making a microprocessor chip destined for mundane household appliances like toasters, air conditioners, ovens, etc. The idea is that chip will be used by these devices to talk to other devices and thus coordinate their coming on and going off and doing things. The chips are destined to be networked together. Early uses for the chip includes smoke detectors which call you when they detect smoke and wall switches that detect when you come into a room and turn the lights on.

2. The code name for an alleged automated global interception and relay system operated by the NSA (National Security Agency) and intelligence agencies in four other countries, including Australia, Britain and Germany. "Alleged" as the NSA refuses to either confirm or deny its existence. Echelon allegedly comprises a networked supercomputer that monitors all international voice, e-mail and fax traffic, and other electronic data transfers, including wireless communications such as cellular and pager transmissions. The Echelon supercomputers reportedly analyze as many as two million transmission per day, looking for key words such as "bomb," "assassin," "militia," "gun," "explosives," "Delta Force," "Branch Davidian," and "cocaine" and all manner of drug-related jargon. If such a key word is detected during one of your transmissions, the transmission is recorded and analyzed. What happens from that point forward has yet to be reported. Echelon allegedly comprises five spy satellites, numerous terrestrial radio antennas, Internet packet sniffers, Web search engines, and surveillance devices placed on submarine cables by divers.

ECHO 1. Sound waves, when they meet a discontinuity such as a canyon wall or a mountain, bounce; the reflected sound, when it returns to the sound source, is an "echo." Because sound travels at about 1100 feet per second, the delay between the original sound and the return of the echo can be used to measure the distance to the discontinuity.

2. A voice signal, in electrical form as in long distance telephony, can also be reflected from discontinuities, causing a speaker's words to be returned after a delay proportional to the length of the circuit. A person speaking hears speech as it is uttered; this causes no difficulties. Indeed, "sidetone" circuits are built into all telephone sets so that a speaker will be able to hear his own speech when the telephone receiver is pressed against his ear. However, hearing an echo is disturbing, and the greater the echo's delay, the more disturbing it turns out to be. Reducing echo in long-distance communication is perhaps the most difficult problem the transmission engineer has to face.

The principal discontinuity causing telephone echo is the "hybrid" circuit at each end of a connection where a 4-wire transmission facility interfaces 2-wire lines to the calling and called customers (i.e. the ones we have coming to our house). The purpose of the hybrid is to allow the nearby caller to reach the outgoing side of the trunk to the distant party, deliver the incoming signal from the distant party to the nearby caller, and block that incoming signal from the path going back to the distant party. In its simplest form, a hybrid circuit is a transformer with four ports: one for the nearby line, one for the incoming side of the trunk, one for the output side, and one for a "balance network." If the balance network exactly "matches" the nearby line, the incoming trunk signal divides between the nearby line and the balance network, and nothing is left to reach the outgoing side of the trunk.

However, a single balance network has to meet several thousand possible customer lines; it is a compromise at best, and some signal leaks through from the incoming side of the trunk to the outgoing side to produce echo at the distant end. This echo is controlled by introducing loss in trunks to reduce the echo to acceptable limits. Such loss is called "Via Net Loss" (VNL) in trunks between toll switches, and "Terminal Net Loss" in toll-connecting trunks from a toll switch to a local switch. When a switched or direct connection is so long that too much loss would be required to control the echo, echo suppressors were originally used to disconnect the outgoing side of a trunk when speech was detected on the incoming side. Today, however, echo cancelers are used to subtract the echo from the outgoing signal, permitting both parties to talk simultaneously if they so desire. Echo cancelers are also used in speakerphones where acoustic coupling from the loud speaker to the microphone would

also cause an echo at the distant end.

3. European Commission Host Organization.

4. The sound heard by a listener when holding a seashell to the ear does not come from the shell itself. It is the echo of the blood pulsing in the listener's ear.

echo attenuation In a communications circuit (4- or 2-wire) in which the two directions of transmission can be separated from each other, the attenuation of echo signals that return to the input of the circuit under consideration. Echo attenuation is expressed as the ratio of the transmitted power to the received echo power in decibels. See also Attenuation.

echo cancellation Technique that allows for the isolation and filtering of unwanted signals caused by echoes from the main transmitted signal. An echo cancellation device puts a signal on the return transmission path which is equal and opposite to the echo signal. Echo cancellation allows full duplex modems to send and receive on the same frequency. Network-based echo cancellation can interfere with modems which perform their own echo cancellation. To avoid cancellation problems, modems capable of echo cancellation (such as V.32 modems) send a unique answer tone with a phase reversal every half second. The network echo cancellers detect the phase reversal in the answer tone and disable themselves. Contrast with echo suppression. See echo canceller.

echo canceller Device that allows for the isolation and filtering of unwanted signal caused by echoes from the main transmitted signal. In data communications networks, echo cancelers are used in the same way as PADS are in the network, but some brands of echo cancelers have the ability of being disabled by a 2100 Hertz tone transmitted by the data device prior to the exchange of the data device's handshaking protocol. If the echo canceler cannot be disabled by the data device, it will block the data call from completing. See also echo cancellation.

echo check A technique for verifying data sent to another location by returning the received data (echoing it back) to the sending end.

echo modeling A mathematical process where an echo is conceptually created from an audio waveform and subtracted from that form. The process involves sampling the acoustical properties of a room and guessing what form an echo might take, then removing that information from the audio signal.

Echo Return Loss ERL. The difference between a frequency signal and the echo on that signal as it reaches its destination. For example, a way of measuring echo return loss is to say it's the frequency-weighted measure of return loss over the middle of the voiceband (approx. 560-1965 Hz) where talker echo is most annoying.

echo suppression The process of turning off reverse transmissions on a telephone line to reduce the annoying effects of echoes in telephone connections, especially on satellite circuits. An active echo suppressor impedes full-duplex data transmission. Contrast with echo cancellation.

echo suppressor Used to reduce the annoying effects of echoes in telephone connections. The worst echoes occur on satellite circuits. An echo suppressor works by turning off transmission in the reverse direction while a person is talking, thus effectively making the circuit one way. An echo suppressor obviously impedes full-duplex data – data flowing both ways simultaneously. Echo suppressors are turned off by the high-pitched tone (typically 2025 Hz) in the answering modem, which it uses to signal it's answered the phone and is ready for a data conversation.

echo suppressor disabler An echo suppressor disabler is a device which causes an echo suppressor to be disabled (i.e. turned off). Echo suppressors are turned off by the high-pitched tone (typically 2025 Hz) in the answering modem, which it uses to signal it's answered the phone and is ready for a data conversation. A disabled echo suppressor stays disabled until the circuit is disconnected and restored to its "ready" connection. Because an echo suppressor hinders full duplex transmission in data communications, it is necessary to disable the echo suppressor.

echomail A public message area or conference on a bulletin board system (BBS) that is "echoed" to other systems in a BBS network. EchoMail is organized into different groups, each with a different topic and the term normally references communications on a FidoNet network. Also a term referring to the electronic transfer of messages between bulletin board systems.

echoplex A way of checking the accuracy of data transmitted whereby the data received are returned to the sender for comparison with the original data. Somewhat time consuming. Used typically in slow speed transmissions. See Echo.

ECITC European Committee for Information Technology testing and Certification.

ECL 1. Emitter Coupled Logic.

2. EuroCableLabs.

eclipse A satellite is said to be in eclipse when it is in the earth's shadow. For a geosynchronous satellite, this occurs twice a year, over a period of 46 days, centered around and reaching a maximum (of slightly more than 70 minutes) at both the autumn equinox and the spring equinox. For this reason, satellites carry batteries to provide power when their solar panels receive no energy.

Ecliptic The intersection of the Earth's orbit plane (which contains the centers of the Earth and the Sun) and the celestial sphere. It may also be defined as the apparent annual path of the sun against the stars. Its name is from the phenomena that solar eclipses may occur when the Moon crosses it. The Sun, the Moon, all the planets and the constellations of the Zodiac lie within 8 degrees of the ecliptic.

ECM Error Correction Mode. An enhancement to Group 3 fax machines. Encapsulated data within HDLC frames providing the received with an opportunity to check for, and request retransmission of garbled data. See Facsimile and V.17.

ECMA Originally the European Computer Manufacturers Association, the name was changed to ECMA in 1994. It is an international, Europe-based industry association founded in 1961 and dedicated to the standardization of information and communications systems. ECMA addresses standards in areas such as software engineering and interfaces, APIs and languages, data presentation, character sets and coding schemes, file structures, LAN protocols, IT security, and optical disks. www.ecma.ch. See also CEPT.

ECMEA Enhanced Cellular Messaging Encryption Algorithm. The Telecommunications Industry Association's CMEA is used for confidentiality of the control channel in the most recent American digital telephony systems. It is a variable-length block cipher. A paper called "Cryptanalysis of the Cellular Message Encryption Algorithm" by Wagner, Schieier, and Kelsey) describes an attack on CMEA that "demonstrates that CMEA is deeply flawed." ECMEA (Enhanced CMEA) corrects the flaw.

ECML Electronic Commerce Modeling Language. This is an Internet commerce tool. Customers are frequently required to enter substantial amounts of information at an Internet merchant site in order to complete a purchase or other transaction, especially the first time they go there. A standard set of information fields is defined as the first version of an Electronic Commerce Modeling Language (ECML) so that this task can be more easily automated, for example by wallet software that could fill in fields. Even for the manual data entry case, customers will be less confused by varying merchant sites if a substantial number adopt these standard fields. In addition, some fields are defined for merchant to consumer communication. See also www.ietf.org/rfc/rfc3106.txt.

ECMP Equal Cost Multipath Routing. ECMP distributes traffic across multiple high-bandwidth links to increase performance. Extreme's OSPF implementation supports multiple equal-cost paths between points and divides traffic evenly among the available paths. As many as four links may be involved in an ECMP link and traffic is shared on an IP source/destination address session basis.

ECN Electronic Communications Network. A term that is applied to those communications networks set up for trading stocks and bonds using PCs and the Internet, or a dial-in circuit into a private network. ECNs are hot and are stealing trading in many of the hottest stocks from New York Stock Exchange and Nasdaq.

ECOC European Conference on Optical Communication.

economic bandwidth An AT&T term for the maximum bandwidth that a physical medium can support without a significant increase in its cost.

Economic CCS ECCS. The load that should be carried on the last trunk of a high usage trunk group to minimize the total cost of routing the offered traffic, assuming that overflow from the high usage route is offered to an alternate route engineered to meet an objective blocking probability.

economics See economist.

economist Economics is called the dismal science. The standing joke is that an economist is someone who didn't have the personality to become an accountant. There is also a theory that God invented economists to make weather forecasters and astrologers look good. Economists are incredibly prescient at predicting the past. People claim "If you laid every economist end to end, it would not be a bad thing." There are three types of economists: Those who can count, and those who can't. On the first day God created the sun; the Devil countered and created sunburn. On the second day God created sex; the Devil countered with marriage. On the third day God created an economist. This was a tough one for the Devil, but in the end and after much thought he created a second economist! For those of you who don't know (or don't care), my first degree was in Economics. See Accountant.

eCommerce Electronic Commerce. Buying and selling over the public Internet, the public Web and corporate Internets. Predictions for the amount of ecommerce should not be underestimated. According to Peter Drucker writing in the October, 1999 Atlantic Monthly Magazine, "In the new mental geography created by the railroad, humanity mastered distance. In the mental geography of ecommerce, distance has been eliminated. There is only one economy and only one market." Peter Drucker continued, "One consequence of this is that every business must become globally competitive, even if it manufactures or sells only within a local or regional market." See the Internet and the World Wide Web.

economy of scale As throughput gets bigger, so the per unit cost comes down. This is the argument used by economists to justify monopolies – namely that the per unit costs of one supplier are far lower than having two suppliers. The economy of scale argument is used to justify having only one water company in town. It makes more sense in that industry than in the telephone industry. It was once used in the telephone industry to justify one combined local phone company/long distance phone company/one supplier of terminal equipment. This argument does not really apply to telephony as technological breakthroughs have brought down the cost of getting into the telephone industry and have allowed smaller, competitive companies to become cost effective. Some large telecommunications monopolies, in fact, are experiencing diseconomies of scale. In this case, their cost of per unit business starts to rise as they get very large. Diseconomies of scale are caused by bloated bureaucracies and inertia in management decision making.

ECP 1. Enhanced Call Processing. An Octel term for an interactive customized menu in its voice mail system which provides levels of call routing. See Enhanced Call Processing.

2. Extended Capabilities (Parallel) Port. An upgrade to the original parallel port on a PC, which gives you: Transfer rates of more than two million bytes per second; bidirectional 8-bit operation)a standard parallel port has only 4 input bits); support for CD-ROM and scanner connections; 16-byte FIFO buffer; support for run length coding data compression. See also USB.

ECPA See Electronic Communications Privacy Act.

ECRIT Emergency Context Resolution with Internet Technologies. An IETF initiative to develop a standard technical solution for the routing of any IP-based e911 call to the appropriate PSAP and to provide the caller's location to the PSAP so that emergency assistance can be dispatched to the correct location.   See E911, PSAP.

ECSA Exchange Carriers Standards Association. See Exchange Carriers Standards Association.

ECTA European Competitive Telecommunications Association. ECTA aims and objectives are to assist and encourage market liberalisation and competition, represent the telecommunications industry to key government and regulatory bodies, maintain a forum for network and business development throughout Europe, assist new market entrants through pro-competitive policies, and continually reflect the dynamic nature of the telecommunications industry. www.ectaportal.com.

ECTEL The European Telecommunications and Professional Electronics Industry.

ECTF The ECTF (Enterprise Computer Telephony Forum) is an industry organization formed to foster an open, competitive market for Computer Telephony technology. Participants include industry suppliers, developers, system integrators and users working to achieve agreement on multi-vendor implementations of Computer Telephony technology based on international de facto and dejure standards.

According to its own words, "the ECTF facilities the development, implementation and acceptance of Computer Telephony (CT) solutions by bringing together suppliers, developers, systems integrators and users. The Forum discusses, develops and tests interoperability techniques for dealing with the diverse technical approaches currently available. The ECTF incorporates and augments existing industry standards and publishes CT interoperability agreements."

Principal members of the ECTF are: Aculab plc; Amarex Technology, Inc; Amtelco; Amteva, Inc.; Aspect Telecommunications; AT&T; Brite Voice Systems; Brooktrout Technology; CallScan Limited; Centigram Communications; Cintech Tele-Management; COM2001 Technologies; CSELT; Deutsche Telekom AG; Dialogic Corporation; Digital Equipment Corporation; Ericsson Business Networks AB; Excel Inc.; Fujitsu Limited; Hewlett-Packard; IBM Corporation; InterVoice; Lernout & Hauspie Speech Products; Linkon Corporation; Lucent Technologies; Mitel Corporation; Natural MicroSystems Corp.; NEC America Inc.; Northern Telecom; Periphonics; Rhetorex; Rockwell Telecommunications; Siemens AG; Sun Microsystems; Tandem Computers; Texas Instruments; Trident Data Systems; Unimax Systems Corporation and Voicetek Corporation.

The ECTF has a Web site at www.ectf.org. Membership inquiries – 510-608-5915 or ectf@ectf.org.

ECTF definitions Enterprise Computer Telephony Forum M.100 Revision 1.0 Administration Services "C" Language Application Programming Interfaces See M.100 and ECTF.

ECTUA European Council of Telecommunications Users Association.

ED 1. CallED party. The ED receives a call from the ING, or callING party, to set up a data transfer. ING and ED apply to any type of data transfer, including both voice and data.

2. Ending Delimiter.

EDA 1. Electronic Directory Assistance. A method by which companies can get access to telephone directories electronically using an X.25 packet switched connection.

2. Electronic Design Automation.

EDAC Error Detection And Correction.

EDCH Enhanced D-Channel Handler.

EDDA The European Digital Dealers Association. An European association of DEC resellers. EDDA members include VARs, systems integrators, leasing companies and service organizations.

Eddy Current Losses Losses in electrical devices using iron, due to the currents set up in it by magnetic action.

EDE Electronic distortion equalization = Adaptive compensation for optical and electrical distortions that accumulate along a transmission link.

EDF Erbium-Doped Fiber. See EDF.

EDFA Erbium-Doped Fiber Amplifier. An OOO (Optical-Optical-Optical) technique for boosting signal strength in fiber optic transmission networks. As a purely optical technique, EDFAs generally are considered to be an improvement over the earlier OEO (Optical-Electrical-Optical) technique used in regenerative repeaters. EDFAs accept an attenuated (i.e., weakened) optical signal and pass it through a section of fiber doped (i.e., infused with) with erbium, a rare-earth element. The light signal stimulates, or excites, the erbium. As the erbium atoms drop from their excited state, they emit extra energy, which serves to amplify the attenuated light signals. A single EDFA can serve to amplify multiple wavelengths in the 1550nm (nanometer) range in which DWDM (Dense Wavelength Division Multiplexing) operates. The advantage is the elimination (or at least reduction in the number) of regenerative repeaters, which are costly, power hungry, and generally problematic. EDFAs, however, cannot amplify wavelengths below 1525nm. Also, they are limited to no more than 10 spans over a distance of roughly 800km, at which point regenerative repeaters must be used to correct for accumulated noise. EDFAs increasingly are used in conjunction with Raman amplifiers. See also Dynamic Gain Equalizer, DWDM, Raman Amplification, Repeater, SONET and Wavelength.

EDGAR Electronic Data Gathering Archiving and Retrieval. A database of corporate disclosure, transaction, and financial status data maintained by the U.S. Securities and Exchange Commission (SEC).

EDGE 1. Enhanced Data for GSM Evolution. Currently being standardized by ETSI (European Telecommunications Standards Institute), EDGE is touted as the final stage (as if anything is ever "final") in the evolution of data communications within the existing GSM standards. EDGE is intended to support data transmission rates up to 384 Kbps. EDGE is also anticipated to be used with IS-136 TDMA networks in the US. See also ETSI, GSM, IS-136, and TDMA.

2. Edge in Internet usage doesn't mean the edge of the Internet. It actually means the edge that's closest to you. The idea of "edge" Internet computing or networking or service is to put a server near you and have it serve you quickly with special services you want – like voice, like software, etc. One vendor sent me their "definition" of edge. I include it for your amusement: "It's where mission-critical Internet content and applications can be assembled in real-time, customized, and delivered to end users with unfaltering performance and reliability."

edge appliance A speciality computer which does special tasks. Typically it sits on a network and performs tasks for many users. Some of the various appliances and what they do include:

- Security appliances – firewall, intrusion detection, intrusion prevention, packet inspection.
- Management appliances – network management, network operations management.
- Performance enhancement appliances – SSL accelerators, XML accelerators.
- Traffic management appliances – packet shapers, packet inspection, routers/switches.
- Load balancers – performance/traffic monitoring, traffic routing. + Storage appliances – NAS, iSCI, specialized storage management.

edge broadcasting In satellite broadcasting systems, edge broadcasting is the technique of storing video and other program material at a cache near final users. A typical implementation may entail bypassing the Internet backbone by transmitting live video and audio via satellite to a cache from which it is broadcast to users. This is called caching of streaming media. Edge broadcasting is held to have two principal advantages over conventional TV broadcasting. First, content providers may use Web facilities to track and obtain information on customers. Second, viewers may have additional control, such as recording and play-back, as well as selection of camera view in a pay-per-view event.

edge case A problem that occurs only when a single variable or condition takes on an extreme value. One objective of networking testing is to determine edge case boundaries, i.e., thresholds, beyond which problems occur. See corner case.

edge concentrator See IP Edge Concentrator.

edge connector A connector made of strips of brass or other conductive metal found at the edge of a printed circuit board. The connector plus into a socket of another circuit board to exchange electronic signals.

edge device 1. A physical device which is capable of forwarding packets between legacy interworking interfaces (e.g., Ethernet, Token Ring, etc.) and ATM interfaces based on data-link and network layer information but which does not participate in the running of any network layer routing protocol. An Edge Device obtains forwarding descriptions using the route distribution protocol.

2. A physical device which sits on the edges of the Internet under the control of an ISP (Internet Service Provider) and allows the ISP to provide its customers with ancillary services, such as voice mail, fax forwarding, video downloading, etc. For an example of an edge device maker, see www.mediagate.com.

edge effect A video term. The overemphasizing of well defined objects from the addition of black or white outlines to the vertical edges of the objects. Examples of this phenomena are, trailing white, leading black around the outline of a figure in movement within a scene.

edge router A new device for Internet Service Providers which will forward packets at high speed, do tunneling, authentication, filtering, packet accounting, traffic shaping and address translation. With such a device, service providers will be able to save money and create different services – combining features in different ways. Edge routers sit at the edge of the Internet – just at the connection by the local phone company to the Internet.

edge site A remote network site. A site at the edge of the network.

edge switch A Broadband Switching System (BSS) which is located at the edge of the network. Conceptually equivalent to a Central Office in the voice world, an edge switch is the first point of user access (and the final point of exit) for a broadband network (i.e., Frame Relay, SMDS and ATM). Edge switches are interconnected by "Core Switches," which are the functional equivalent of Tandem switches in the circuit-switched voice world.

EDH Electronic Document Handling.

EDI Electronic Data Interchange. A series of standards which provide computer-to-computer exchange of business documents between different companies' computers over phone lines and the Internet. These standards allow for the transmission of purchase orders, shipping documents, invoices, invoice payments, etc. between an enterprise and its "trading partners." A trading partner in EDI parlance is a supplier, customer, subsidiary, or any other organization with which an enterprise conducts business. EDI is used for placing orders, for billing and paying for goods and services via private electronic networks or via the Internet. According to studies, it costs about $50 to process a paper-based purchase order and about $2.50 to process the same order with EDI. Internet-based EDI can lower the cost to less than $1.25. EDI software translates fixed field or "flat" files that are extracted from applications into a standard format and hands off the translated data to communications software for transmission. EDI standards are supported (i.e. have been adopted) by virtually every computer company in the country and increasingly, by every packet switched data communications company. The formats used to convert the documents into EDI data are defined by international standards bodies and by specific industry bodies. See also Electronic Data Interchange and IES. www.edi-info-center.com/html/hotline.html.

EDI Map A Verizon definition. A template used to capture and format Order data into an EDI message. There is a different map for each Pre-Order transaction and each LSR Order form. Pre-Order EDI maps are contained in the Verizon EDI Guide for Pre-Order Local Service Requests (Pre-Order LSR).

Edison battery A type of storage battery in which the elements are nickel and iron and the electrolyte is potassium hydroxide. An old type of battery.

Edison effect A discovery attributed to Thomas Alva Edison, the Edison Effect initially was called the "Ether Effect." Technically, it is known as thermionic emission. During Edison's studies (1882) of the light bulb, William J. Hammer, one of Edison's engineers, noted that the heated filament in his incandescent lamp blackened the inside of the bulb. The phenomenon is due to the fact that the heat in a filament supplies some electrons with at least the minimal energy to overcome the attractive forces that hold them in the structure of the metal. These free electrons are thrown off, and coat the inside of the glass bulb. Quite uncharacteristically, Edison never pursued the study of thermionic emission. Some years later, he admitted that he didn't even understand Ohm's Law at the time. Note: The electron wasn't discovered until some 15 years later. The subsequent study of the Edison Effect eventually led to the development of the vacuum tube and conventional electron tubes, such as the picture tube inside your TV set. By the way, "incandescent" comes from the Latin "incandescere," which translates as "to become hot." That's exactly what happens. The filament becomes hot. Tungsten is commonly used in light bulb filaments because it conducts electricity, but also gets hot. As the electricity passes through the filament, electrons in the tungsten filament vibrate. As they do so, they convert electrical energy into thermal energy. The filament becomes hot and glows, thereby giving off light. That's why incandescent light bulbs are hot to the touch. See also Edison, Thomas Alva; Luminiferous Ether; and Ohm's Law.

Edison, Thomas Alva One of the most prolific inventors of all time, Edison (1847-1931) is best known for inventions such as the practical incandescent light bulb, the phonograph, the motion picture projector, the stock ticker, and the electric vote tabulating machine. At the age of 15, Edison was publishing the "Grand Trunk Herald" from a freight car, which also served as his laboratory. During this time, he saved the life of an executive of the rail company, was rewarded with lessons in telegraphy, and became a master telegrapher. He went on to invent a method of automatically transmitting telegraphs over a second line with the need for an operator. With the $40,000 he earned from his telegraphy invention, the stock ticker, Edison opened his legendary research laboratory in Menlo Park, New Jersey, where inventions included the microphone and the carbon telephone transmitter. In 1877, his team invented the first practical incandescent light bulb, which was based on DC (Direct Current). George Westinghouse and others improved on this work, basing their inventions on AC (Alternating Current). In 1889 Edison formed the Edison Electric Light Company, which later became General Electric. Edison died on October 18, 1931. Three days later, at the request of President Herbert Hoover, millions of Americans turned off their electric lights in tribute. See also the two previous definitions.

editing Editing is a familiar process of changing the content of files to achieve more effective communication by cutting, pasting, cropping, resizing, or copying. Multimedia editing can be done on all types of media: voice annotations, music, still images, motion video, graphics and text. Tools for editing vary from simple tools for email voice annotations to more sophisticated tools for video manipulation. See also Electronic Mail.

editor A software program used to modify programs or files while they are being prepared or after they are (allegedly) complete. You can use an editor to write in. I wrote this dictionary using an editor called The Semware Editor. This program is much faster and far more flexible than a word processing program like Microsoft Word. But it can't do all the fancy layout, bolding, underlining, etc. that Word can do. An editor produces only ASCII text. The editor that comes with Windows95 is called Notepad. I use my editor to write in, because when I'm finished with this dictionary, I send it to sophisticated desktop publishing software called QuarkXpress. And that's where the layout for this book is done. I believe that the two functions - of writing and desktop publishing - are different and shouldn't be combined in one tool.

EDLIN The MS-DOS line editor that came originally with DOS and which you can use to create and edit batch files and other small text files. It's not very good. There are far better editors around, including The Semware Editor.

EDM Electronic Document Management. EDM unites the disparate workflow, document management and imaging.

EDMS An imaging term. Engineering Document Management System.

EDO RAM Extended Data Output Random Access Memory. A much faster form of DRAM (Dynamic RAM). EDO RAM speeds access to consecutive locations in memory by assuming that the access to next memory will target an address in the same transistor row as the previous one, and latching data at the output of the chip so it can be read even as the inputs are being changed for the next memory location. EDO RAM reduces memory access times by an average of 10 percent over with standard DRAM chips and costs only a little more to manufacture. EDO RAM largely has been replaced by SDRAM (Synchronous DRAM). See also DDR-SDRAM, DRAM, Flash RAM, FRAM, Microprocessor, RAM, RDRAM, SDRAM, SRAM, and VRAM.

EDP Electronic Data Processing. Also DP, as in Data Processing. Basically, a machine (also called a computer) that receives, stores, operates on, records and outputs data. The word "electronic" was added to Data Processing when the industry moved away from tab cards – the 80 column "do not spindle," etc. – and was able to accept data electronically, instead of electromechanically as with the tab cards. People in the industry used to be called EDPers.

Now the term MIS – Management Information System – is more common.

EDR Event Data Record. EDR consist of all records that come from different sources like phone calls (CDRs), VoIP (IPDRs), data sources (DDRs), etc.

EDRAM 1. Enhanced Digital Announcement Machine.

2. Enhanced Dynamic Random Access Memory. A form of DRAM (Dynamic RAM) that boosts performance by placing a small complement of static RAM (SRAM) in each DRAM chip and using the SRAM as a cache. Also known as cached DRAM, or CDRAM.

EDS See 1962.

EDSL Extended Digital Subscriber Line. The ISDN EDSL combines 23 B-channels and one 64-Kbps D-channel on a single line. Also called the Primary Access Rate.

EDSS1 European Digital Signaling System no. 1. The European variant of the SS7 (Signaling System 7) signaling and control protocol for use over the ISDN D (Delta) channel. EDSSI was developed by ETSI (European Telecommunications Standards Institute). See also ETSI, ISDN, and SS7. The ITU-T specifications Q.921 and Q.931 DSS1 (Digital Subscriber Signaling 1) are the baseline international specifications used throughout the world. The European specification from ETSI represents a modified form of Q.931, known as EDSS1 (European Digital Subscriber Signaling). Due to the early acceptance within the industry, EDSS1 is the common variant of Q.931. See Euro-ISDN.

EDTV See Extended Definition TV.

edu An Internet address domain name for an educational organization.

EDUSAT Education Satellite. An Indian space satellite that was launched in 2005 for exclusive use by educational institutions in India. The satellite connects state-run engineering colleges, polytechnics, industrial training institutes, and other educational institutions. One benefit of EDUSAT is that it enables India to deal with a teacher shortage for some courses by having a single teacher deliver lectures via satellite to many different classrooms around the country. Databases of recorded lectures enable lectures to be viewed over and over again, throughout the country. The satellite also broadcasts educational TV shows. Currently four educational TV channels are broadcast via EDUSAT. Although EDUSAT was built for state-run educational institutions, the plan is for private engineering colleges, polytechnics and industrial training institutes to use the satellite for a fee.

edutainment The answer to the question "What do you get when you cross educational material with interactive video?" A term coined by "someone who obviously knows nothing about either education or entertainment," says Laura Buddine, president of multimedia games maker Tiger Media. But it is becoming popular in residences and it's typically played on PCs with CD-ROM players. See also MMX.

Edwards, Elwood Elwood Edwards' voice is heard more than 27 million times a day (which comes to more than 18,000 times per minute). He is the man behind those special 3 words (not "I love you") but "You've got mail!" which you hear all too often on American Online.

EE End to End signaling. Punch DTMF are sent through the lines to signal the end of a conversation. A tone code is also used to access long distance carriers, to signal your answering machine, or to access your voice mail.

EE Credit End-to-end credit. Used to manage the exchange of frames by two communicating devices, and set the maximum number of frames that may remain unacknowledged.

EEC European Economic Community. See EC.

EEHLLAPI Entry Emulator High Level Language Applications Programming Interface. An IBM API subset of HLLAPI.

EEHO Either End Hop Off. In private networks, a switch program that allows a call destined for an off-net location to be placed into the public network at either the closest switch to the origination or the closest switch to the destination. The choice is usually by time of day and is usually done to take advantage of cheaper rates.

EEL An enhanced extended link (EEL) consists of a combination of an unbundled loop, multiplexing/concentrating equipment, and dedicated transport. The EEL allows new entrants (CLECs, etc.) to serve customers without having to collocate in every central office in the incumbent telephone company's territory.) This service is very neat, and opens up a great deal of opportunity for CLECs... Basically, the EEL: - consists of a loop and interoffice transport at very attractive rates to the CLEC... - applies in those cases where CLECs reside in a different central office (CO) than their end users.

So, for example, I could serve a customer outside of my CO by ordering from the ILEC: (a) local loop from customer prem. to the nearest CO (a CO which I am NOT collocated in!) (b) an interoffice transport to a CO which I DO reside in

...this is a great service for fat pipe (DS3 and up)... I can serve large customer need OR I could mux together small customer need (T1, etc) and run the traffic over the large

pipe (i.e., DS3) from non-collocated CO to a CO in which I am collocated...
yes... serving customers outside of my "normal" wirecenter/territory.

EEMA European Electronic Messaging Association. See also Electronic Messaging Association.

EEOC Equipment Exceeds Operator Capability. A silly way of saying User Error.

EEPROM Electronically Erasable Programmable Read Only Memory. A read only memory device which can be erased and reprogrammed. Typically, it is programmed electronically. It is also erased electronically, unlike EPROM, which is erased electromagnetically with ultraviolet light. EEPROMs don't lose their memory when you lose power. EEPROM used to be often used in PBXs and were the way manufacturers of older style PBXs upgraded their software. In other words, every time they sent you a software upgrade, they'd send you a bunch of chips. You'd pull out a bunch of chips on one of the main boards in the your PBX. And you'd replace them with the new chips. When you don't have a disk drive (and in the olden days disk drives were very expensive), EEPROMs were the only way to go. EEPROM Setup in a computer allows it to recognize certain system board configurations during initialization. You, the user, can then choose options such as the type of memory chips installed and base memory size without changing jumpers on the system board. See EPROM.

EES Escrow Encryption Standard. A security system proposed by the U.S. Department of Justice for the U.S. government's data communication. It involves inserting into all new federal computers a special encryption chip whose output would be reasonably secure but could be tapped by law enforcement agencies.

EESN Expanded Electronic Serial Number. Due to the limited numerical combinations and the increasing use of mobile/portable telecommunications devices, the Telecommunicaion Industry Association has mandated that the telecommuncations industry expand the Electronic Serial Number of such communications devices from 32-bit to 56-bit. It is predicted that the growth in device numbers will exceed the space available for devices that include pagers, digital/cellular telephones, PDAs and the like. A need has been determined to modify telecom switches, operational and billing systems, and associated counterparts to expand the ESN capabilities to accomodate a larger format for these devices as their numbers grow. New 56-bit ESNs sometimes called EESNs will debut in 2005. See also ESN.

EETDN End-to-End Transit Delay Negotiation.

EF Entrance Facility (also called TEF or Telecommunications Entrance Facility). Services provided by a facility that is used to carry traffic between the access customer's POP and the exchange carrier's serving wire center.

EF&I Engineer, Furnish and Install.

eFAX eFax (as in electronic facsimile) occurs when someone faxes to a phone number, which belongs to another company, which accepts the fax, then sends the fax to your email. Many companies now provide this service for free in the hope that they'll accompany your emailed fax with advertising, which they hope they'll sell for money. So far, it seems to work. I happily use a service from www.callwave.com. There's also a company called efax, see www.eFAX.com.

EFCI Explicit Forward Congestion Indication: EFCI is an indication in the ATM cell header. A network element in an impending-congested state or a congested state may set EFCI so that this indication may be examined by the destination end-system. For example, the end-system may use this indication to implement a protocol that adaptively lowers the cell rate of the connection during congestion or impending congestion. A network element that is not in a congestion state or an impending congestion state will not modify the value of this indication. Impending congestion is the state when a network equipment is operating around its engineered capacity level.

EFD Event Forwarding Discriminator. A wireless telecommunications term. Software that contains a discriminator that determines if a notification should be forwarded on to a particular destination.

EFF See Electronic Frontier Foundation.

effective number of bits ENOB. A measure of the quality of an analog-to-digital converter. The measurement is affected by sampling rate, bandpass, and noise.

effective radiated power An FCC term for "the product of the antenna power input and the antenna power gain" expressed in watts.

effective ground-fault current path As per NEC, effective ground-fault current path is an intentionally constructed, permanent, low-impedance electrically conductive path designed and intended to carry current under ground-fault conditions from the point of a ground fault on a wiring system to the electrical supply source.

effectively grounded According to the TIA/EIA-607, effectively grounded means intentionally connected to earth through a ground connection or connections of suffi-

ciently low impedance and having sufficient current-carrying capacity to prevent the buildup of voltages that may result in undue hazard to connected equipment or to persons.

efficiency factor In data communications, the ratio of the time to transmit a text automatically and at a specified modulation rate, to the time actually required to receive the same text at a specified maximum error rate.

EFM Ethernet in the First Mile. See 802.3ae and Gigabit Ethernet.

EFMA Ethernet in the First Mile Alliance. An alliance developing a new Ethernet standard designed to use existing infrastructure (i.e. copper wire) to provide higher speeds over – speeds as fast as 100 megabits per second. The standard is being dubbed 802.3ah by the IEEE. See 802.3ah.

EFOC European Fiber Optics and Communications conference.

EFR Enhanced Full Rate. A enhanced full rate codec (coder/decoder) which improves speech quality over existing GSM cellular networks. EFR supports speech quality comparable to that of a wireline network, also offering improved tolerance for interference. See also Codec and GSM.

EFS Error Free Seconds: A unit used to specify the error performance of T carrier systems, usually expressed as EFS per hour, day, or week. This method gives a better indication of the distribution of bit errors than a simple bit error rate (BER). Also refer to SES.

EFT Electronic Funds Transfer. The moving of bits of data from one bank to another. Done in place of moving little green pieces of paper, called money.

EFTA European Free Trade Association.

EFTPOS Electronic Funds Transfer Point Of Sale.

EGA Enhanced Graphics Adapter. Second color video interface standard established for IBM PCs. Maximum resolution is 640 x 350 pixels. See MONITOR.

egghead The 80s' version of Propeller Head or Gear Head.

egocasting The consumption of on-demand music, movies, television, and other media that caters to individual, and not mass-market, tastes. Critics argue that it's leading to a culture of isolated iPod and TiVo users with no shared reference points.

egosurfing Scanning the Net, databases, print media, or research papers looking for mentions of your own name.

EGP Exterior Gateway Protocol. An Internet protocol for exchanging routing information between autonomous systems.

egress 1. The exit point. This typically refers to information being sent out of, as opposed to being sent in to a frame relay port connection or other network element. See also Egress Filtering.

2. In video it's often called "signal leakage" and it's a condition in which signals carried by the distribution system leak into the air.

3. A measure of the degree to which signals from the nominally closed coaxial cable system are transmitted through the air. Also know as signal leakage.

egress filtering A security technique in which potentially harmful traffic is blocked from leaving a network.

EHF 1. Error Hold File.

2. Extremely High Frequency. Frequencies from 30 GHz to 300 GHz.

EHz Exahertz (10 to the 18th power hertz). See also Spectrum Designation Of Frequency.

EIA Electronic Industries Alliance, previously Electronic Industries Association. A trade organization of manufacturers which sets standards for use of its member companies, conducts educational programs, and lobbies in Washington for its members' collective prosperity. Founded in 1924 as the Radio Manufacturers Association, the EIA is organized along specific electronic product and market lines. Each group or division has its own board of directors, and sets its own specific agenda, designed to enhance the competitiveness of its own business sector. Under the umbrella of EIA, and representative of the range of interests of the organization, are the Consumer Electronics Manufacturers Association (CEMA); Electronic Components, Assemblies, Equipment & Supplies Association (ECA); Electronic Industries Foundation (EIF), Electronic Information Group (EIG); Government Electronics & Information Technology Association (GEIA); JEDEC Solid State Products Technology Division; and Telecommunications Industry Association (TIA). Membership is open to companies and individuals. www.eia.org See EIA Interface.

EIA 232-D New version of RS-232-C physical layer interface adopted in 1987. See also RS-232.

EIA 530 Interface using DB-25 connector, but for higher speeds than EIA-232. Has balanced signals (like EIA-422) except for three maintenance signals which are EIA-423. See also RS-530.

EIA 561 EIA-232E interface on DIN-8 connector (like Macs use). See also RS-561.

EIA 568 An EIA standard for commercial building wiring. The standard covers four general areas; the medium, the topology of he medium, terminations and connections, and general administration. See EIA/TIA 568 below.

EIA 569 Commercial Building Standard for Telecommunications Pathways & Spaces - EIA/TIA, 1991. Lays out guidelines for sizing telecom closets, equipment rooms, conduit, etc . Every architect doing commercial buildings should have to memorize it. Few have even heard of it!

EIA 574 EIA 232E interface on DE-9 connector. (OK, DE is a bit pedantic. Most call it a DB-9, though you'll never find one in a parts book).

EIA 606 Telecommunications Administration Standard for Commercial Buildings. Guidelines covering design and identification of two- level backbone cabling for individual buildings and for campuses.

EIA Interface A set of signal characteristics (time, duration, voltage and current) set up by the Electronic Industries Association to standardize the transfer of information between different electronic devices, like computers, modems, printers, etc. The most famous EIA interface is the RS-232-C (now called the RS-232-E.) EIA-232 specifies three things: the functions of the interchange circuits, the electrical characteristics AND the connector (EIA-232-E includes two different connectors).

In contrast, the ITU-T's V.24 specifies ONLY the interchange circuit FUNCTIONS. V.28 specifies electrical characteristics compatible with EIA 232. ISO 2110 is the internal standard that defines the 25-pole D-shell connector compatible with EIA-232. Following the merger of EIA and the ITG part of EIA, all formed EIA telecommunications standards are now EIA/TIA publications and the standard referred to is now known as EIA/TIE-232-D, edition D, being the most recent. See EIA, EIA/TIA-RS-232-E and RS-232-E.

EIA/IS-132 An EIA Standard, "Cable Television Channel Identification Plan", which specifies the cable-television channel-numbering plan proposed by the EIA and the NCTA. The cable television channel assignments specified in this document are incorporated by reference into the FCC Rules.

EIA/TIA-232-E The latest version of the familiar RS-232-C serial data transfer standard for communicating between Data Terminal Equipment (DTE) and Data Circuit Terminating equipment employing serial binary data interchange. (EIA/TIE's exact words.) This standard defines the serial ports on computers, which communicate with such things as external modems, serial printers, data PBXs, etc. See also EIA and RS-232.

EIA/TIA 568 EIA/TIA 568 Commercial Building Wiring Standard. This telecommunications standard in early 1991 was out for industry review under draft specification SP-1907B. Its purpose is to define a generic telecommunications wiring system for commercial buildings that will support a multi-product, multi-vendor environment. It covers:

- Recognized Media
- Topology
- Cable Lengths/Performance
- Interface Standards
- Wiring Practices
- Hardware Practices
- Administration

EID Equipment Identifier. A cellular radio term.

EIDE Enhanced IDE. See IDE and Enhanced IDE.

EIDQ European International Directory Enquiries (EIDQ) Group. EIDQ consists of 21 European telecommunications organizations with a direct involvement in the provision of international directory information – work formerly performed under the auspices of CEPT. Membership in EIDQ is voluntary. The primary objective of EIDQ is to work towards the development of enhanced international directory information services both via traditional operator assisted methods and by new media, such as on-line services. www.eidq.org.

EIEIO The chorus from "Old McDonald Had A Farm." It's not an acronym, but it's fun to sing. When all of these acronyms start to make your head spin, just sing "Old McDonald Had A Farm." You'll feel better.

EIF See Electronic Interface Format.

Eiffel Tower The famous Paris landmark built in 1889 for the 1890 World's Fair. On October 21, 1915 the first wireless transAtlantic telephone call took place between H.R. Shreeve, a Bell Telephone engineer listening at the Eiffel Tower and using borrowed French equipment, and B.B. Webb in Arlington, Virginia. The transmission was " Hello, Shreeve! Hello, Shreeve! And now, Shreeve, good night!" During the period following W.W.I, it was often proposed that the Eiffel Tower be dismantled and sold for scrap. Literally the only thing that stalled those plans was the usefulness of the Eiffel Tower as the world's largest radio transmitter/receiver. Because the sun's heat expands the metal unevenly, the Eiffel Tower

always leans away from the sun.

eight hundred service 800-Service. A generic and common (and not trademarked) term for AT&T's, MCI's, Sprint's and the Bell operating companies' IN-WATS service. All these IN-WATS services have "800," "888" or "887" as their "area code." Dialing an 800, 888 or 887-number is free to the person making the call. The call is billed to the person or company being called.

800 Service works like this: You're somewhere in North America. You dial 1-800 and seven digits. Your local central office sees the "1" and recognizes the call as long distance. It ships that call to a bigger central office (or perhaps processes the call itself). At that central office it's processed, a machine will recognize the 800 "area code" and examine the next three digits. Those three digits will tell which long distance carrier to ship the call to.

Until 800 portability, happened in May, 1993 each 800 provider (local and long distance company) was assigned specific 800 three digit "exchanges." For example, MCI had the exchange 999. AT&T had the exchange 542. If you wanted a phone number beginning with 800-999, then you had to subscribe to MCI 800 service. If you wanted a phone number beginning with 800-542, you had to subscribe to AT&T 800 service. With 800 Portability that is no longer the case.

Here is a history of what the phone industry calls 800 Data Base Access Service. It comes courtesy Bellcore:

"After divestiture (1984), the seven regional telecommunications companies began to provide limited 800 Service on their own as well as in conjunction with interexchange carriers. The regional companies transported 800 calls only within their own calling areas. The 800 number – containing 10 digits in accordance with the North American Numbering Plan (NANP) – was routed onto the long distance carrier's networks."

"The Common Carrier Bureau of the Federal Communications Commission (FCC) endorsed an incremental approach that would ultimately give the seven companies the right to create their own 800 Service architecture and eliminate reliance on the only existing signaling system (AT&T's). That approach involved assigning to 800 service providers one or more special numbers from the NANP. These numbers, known as "NXX codes," allowed carriers (MCI, Sprint, NY Telephone, etc.) to identify their own 800 numbers and offer their customers 800 numbers."

"Bellcore – Bell Communications Research – began to develop a new network architecture that would allow an 800 Service subscriber to change to another carrier without changing their existing 800 number (full number portability), in accordance with a September 1991 FCC order. That order declared that 800 data base service should be implemented by March 1993 (later extended to May, 1993) and that the old NXX plan be eliminated as long as access times met certain FCC standards."

"In September 1991, the FCC endorsed the plan initially set forth by the Bell operating companies, which provided that the administration of the Number Administration and Service Center (NASC) be transferred from Bellcore to an independent third party outside the telecommunications industry. Lockheed Information Management Systems Company (IMS) was selected by competitive bid to succeed Bellcore as NASC administrator. NANPA (North American Numbering Plan Administration) since has been shifted to NANC (North American Numbering Council.)"

"How 800 data base service works: The telecommunications network architecture that supports 800 Data Base Access Service is considered "intelligent" because data bases within the network supplement the call processing function performed by network switches. The Service uses a Common Channel Signaling (CCS) network and a collection of computers that accept message queries and provide responses. When a caller dials an 800 number, a Service Switching Point (SSP) recognizes from the digits "8-0-0" that the call requires special treatment and processes that call according to routing instructions it receives from a centralized database. This database, called the Service Control Point (SCP), can store millions of customer records."

"Although each regional company maintains whatever number of SCPs it needs to provide 800 Data Bases Access Service, information about how an 800 call should be handled is entered into the SCP through the off-line Bellcore support system called the Service Management System (SMS). SMS is a national computer system which administers assignment of 800 numbers to 800 service providers. It is located in Kansas City, maintained by Southwestern Bell Telephone Company, and administered by Bellcore with information received from 800 Number Administration and Service Center (NASC). The NASC provides user support and system administration for all 800 Service providers who access the SMS/800."

Because 800 service is essentially a data base lookup service, there are a endless "800 services" you can create. Here are the variables that can be used to influence how an 800 phone call is handled and where an 800 phone call ultimately gets sent:

- The number calling. Virtually all 800 calls in North America (excepting Mexico) now come with the information as to from which number the call came.
- The number being called.
- The time of day, week, month etc.
- The instructions given at that particular moment. A computer might say "Sorry, our phone system is busy. We can't take any more calls in New York. Please send this one and all subsequent ones – until informed otherwise – to our phone system in Kansas City.

Here are a few examples of the services 800 providers have created using the above variables:

- TIME OF DAY ROUTING: Allows you to route incoming calls to alternate, predetermined locations at specified days of the week and times of the day.
- PERCENTAGE ALLOCATION ROUTING: Allows you to route pre-selected percentages of calls from each Originating Routing Group (ORG) to two or more answering locations. Allocation percentages can be defined for each ORG (typically an area code), for each day type and for each time slot.
- SINGLE NUMBER: The same 800 number is used for intrastate and interstate calling.
- CALL BLOCKAGE: You can block calling areas by state or area code. The caller from a blocked area hears the message: "Your 800 call cannot be completed as dialed. Please check the number and dial again or call 1- 800-XXX-XXXX for assistance." (You may want to block callers from areas which didn't see your special commercial, for example.)
- POINT OF CALL ROUTING: Allows a customer to route calls made to a single 800 number to different terminating locations based on the call's point of origin (state or area code.) You establish Originating Routing Groups (ORGs) and designate a specific answering location for each ORG's call.
- CALL ATTEMPT PROFILE: A special service that allows subscribers to purchase a record of the number of attempts that are made to an 800 number. The attempts are captured at the Network Control Point, and from this data a report is produced for the subscriber.
- ALTERNATE ROUTING: Allows a customer to create alternate routing plans that can be activated by the 800 carrier upon command in the event of an emergency. Several alternate plans can be set up using any features previously subscribed to in the main 800 routing plan. Each alternate plan must specify termination in a location previously set up during the order entry process.
- DIALED NUMBER IDENTIFICATION SERVICE: DNIS. Allows a customer to terminate two or more 800 numbers to a single service group and to receive pulsed digits to identify the specific 800 number called. DNIS is only available on dedicated access lines with four-wire E & M type signaling or a digital interface. The customer's equipment must be configured to process the DNIS digits.
- ANI: The carrier will deliver to you the incoming 800 call plus the phone number of the calling party. See also ANI, Common Channel Interoffice Signaling and ISDN.
- COMMAND ROUTING: Allows the customer to route calls differently on command at any time his business requires it.
- FOLLOW ME 800: Allows the customer to change his call routing whenever he wants to.

Now to the question of how to complete an 800 call. There are essentially two ways to terminate an 800 call. You can end the call on your normal phone line – business or residence. This is the phone line you use for normal in and out calling. That's called not having a dedicated local loop. Or you can end the call on a dedicated phone line. By "dedicated," we mean there's a leased line between your office and the local office of your 800 provider, local or long distance carrier. There are several ways this dedicated "line" might be installed. It could be part of a T-1 circuit. It could be one circuit on one single copper pair. It could even be a phone number dedicated to your 800 number – a phone number you can't make an outgoing call on.

There is one major problem with 800 lines. They're hard to test. You may have bought an 800 number to cover the country, but you may be unreachable from certain parts of the country for weeks on end and not know it. That's part of the problem of a service which uses multiple databases lookup tables and relies on many exchanges to carry the calls. Many companies – like the airlines – recruit their distant employees to call their 800 number regularly. The only part of your 800 IN-WATS line you can test is your local loop (assuming you have one) from the local central office to your office. If you have a dedicated, leased

line, you may have local Plant Test Numbers – standard seven digit numbers. You can call these numbers. If they work, you know that the end parts of your lines are working. One of our WATS lines is 1-800-LIBRARY. When it had a dedicated local phone number, it had a plant test number of 212-206-6870. So we could call this and all the subsequent hunt-on numbers every day first thing. Just to check. And when we go traveling, we call our own numbers. Just to check. Now we don't have any dedicated local phone numbers, we rely on prayer. The most common problem we have with our 800 numbers is at our local central office. Seems that it crashes every so often for very short amounts of time. When it starts up, it's meant to load all the tables to give us the features we're paying for – like hunting. Sadly, it doesn't always do this. We then report the trouble to our local phone company. It's usually fixed within an hour or so. Depends on how busy they are. See also 800 at the front of this dictionary. 800 Service now includes 888 (April 1996) numbers and 877 (April 1998) numbers.

EIGRP Enhanced Interior Gateway Routing Protocol; also known as Enhanced Internet Gateway Routing Protocol. Cisco System's newest version of its proprietary routing algorithm, IGRP, EIGRP provides link-to-link protocol-level security to avoid unauthorized access to routing tables.

eight way server A motherboard with up to eight processors, e.g. eight Intel Pentium chips. Eight-way servers were introduced in 1997 to bring more power to the Windows NT operating system.

EIM Ethernet Inverse Mapper.

EIP Early Implementers Program. A term Novell coined to refer to those companies who had early on committed to adopt its Telephony Services architecture. See Telephony Services.

EIR Equipment Identity Register. A database repository used to verify the validity of equipment used in mobile telephone service. It can provide security features such as blocking calls from stolen mobile stations and preventing unauthorized access to the network. Black-listed equipment prevents call completion for a user.

EIRENE European Integrated Railway Radio Enhanced Network. See GSM-R.

EIRP Equivalent Isotropically Radiated Power. A wireless term. The product of the power supplied to the transmitting antenna and the antenna gain in a given direction relative to an isotropic antenna radiator. This product may be expressed in watts or dB above one watt (dBW). See also Isotropic and Isotropic Antenna.

EIRPAC A ITU-T X.25 packet-switched network operating in Ireland under the control of the Irish government.

EISA In a computer a "bus" is an electrical channel for getting information and commands in and around the computer. It is the way the central microprocessor running the computer gets its information and commands to the various peripheral devices or device controllers, such as video controllers, hard disk controllers, etc. The original IBM PC was "balanced" in that the microprocessor matched the speed of the bus that came in the machine. But the microprocessor got faster and more powerful and the bus lagged behind. So there has been much effort to speed the bus up, including EISA, which stands for Extended Industry Standard Architecture. EISA is the independent computer industry's alternate to IBM's Micro-Channel data bus architecture which IBM uses in some of its high end PS/2 line of desktop computers. EISA, like Micro-Channel (also called MCA), is a 32-bit channel. But, unlike IBM's Micro-Channel, plug-in boards which work inside the XT and AT-series of IBM and IBM clone desktop computers will work within EISA machines. They won't work in Micro-Channel machines. EISA expands the 16-bit ISA (Industry Standard Architecture) to 32-bit. EISA technology is useful in computing environments where multiple high performance peripherals are operating in parallel. The intelligent bus master can share the burden on the main CPU by performing direct data transfers into and out of memory. EISA capabilities are valuable when the system is being used as a server on a local area network or is running a multi-user operating system such as UNIX or OS/2. As of writing, over 200 manufacturers had endorsed EISA. Broader, wider buses than EISA are now available. 64-bit is not uncommon, especially among servers. See Local Bus, PCI and VESA for examples of newer, faster buses.

Either End Hop Off EEHO. In private networks, a switch program that allows a call destined for an off-net location to be placed into the public network at either the closest switch to the origination or the closest switch to the destination. The choice is usually by time of day and is usually done to take advantage of cheaper rates.

either way operation Same as half-duplex.

EIU Ethernet Interface Unit.

EKE Electronic key exchange.

EKTS Electronic Key Telephone System.

ELAN Emulated Local Area Network: A logical network initiated by using the mechanisms defined by LANE (LAN Emulation). This could include ATM and legacy attached end stations. See also LANE.

elapsed carrier connect time The time used to bill IXCs. Timing begins on originating traffic when the end office receives the wink-start signal from the IXC. Timing ends when the calling party disconnects or the IXC returns disconnect supervision. Timing begins on terminating traffic when the end office returns the wink-start signal and ends when the called party disconnects or the IXC returns disconnect supervision.

elapsed time Conversation time or the time from answer to disconnect, in MM:SS:T format.

elastic buffer A variable storage device having adjustable capacity and/or delay, in which a signal can be temporarily stored.

elasticity of demand The relationship between price and the quantity sold. The theory is the lower the price, the more you'll sell. In telecommunications, this has traditionally been true, though sometimes it has taken time for demand to catch up with dramatic price cuts.

elastomer A class of long-chain polymers capable of being crosslinked to produce elastic compounds, e.g. polychloroprene and ethylene propylene rubber.

elastomeric One type of mechanical fiber splice.

elastomeric firestop A firestopping material resembling rubber. See also firestopping.

ELDAP The beginnings of an Internet directory protocol.

ELEC Enterprise Local Exchange Carrier. Generally, a larger corporation operating as their own LEC as a means of obtaining better carrier rates for themselves, possibly selling services to others for a profit to enhance revenue in the process. ELECs are a new breed of LEC. The ILECs (Incumbent Local Exchange Carriers) for around 100 years have had the exclusive right and responsibility for providing local telephone service. During the recent past, many state PUCs (Public Utility Commissions) have allowed competition for local exchange service; hence, the origin of the CLECs (Competitive LECs). ELECs actually are a subset of the CLEC concept, although they actually preceded it by a few years. It works like this:

Let's say that you are the telecommunications manager for a large college or university, or for a large corporation with theme parks and hotels around the country. In other words, your enterprise owns a piece of property on which sit a lot of buildings. You provide a wide range of voice, data and video services to management, staff, and guests in dorms or hotels. In a very significant sense, you are providing communications services to what, in effect, is a self-contained town or small city. You go to the state PUC and file for certification as a LEC. You also file local exchange tariffs, defining available services, the terms under which they are offered, and all associated costs. You arrange to interconnect your PBX with an IXC, just like an ILEC would connect a CO to an IXC POP. The IXC pays you access charges for all interLATA traffic you hand off to it, just as they would pay an ILEC for that traffic. You have just become the manager of a telephone company – perhaps a very profitable telephone company. You also have become the manager of a facilities-based long distance resale company.

Take this scenario one step further. Perhaps you become your own facilities-based IXC, with leased-line connections between your properties. Take it still one step further. In addition to hauling your own interLATA traffic, you market your long distance service and your Internet access service to other companies close to your ELEC properties, connecting those companies to your switch via dedicated circuits leased from the LEC. Once built, you can run this ELEC as a separate profit center, or you can sell it to a traditional LEC or CLEC. See also CLEC and LEC.

electret microphone Electret is a combination of ELECTRicity and magnET. An electret microphone operates on the basis of a dielectric (non-electrically conducting) material in which a permanent state of polarization, or electrical bias, has been established. The dielectric material is then spread over a conductive metal backplate. A back-electret microphone involves a diaphragm of dielectric material which is not electrically biased, but which is spread over a conductive metal back-plate which is electrically biased. Variants on standard condenser microphones, electret microphones are very inexpensive and require very little electrical energy to operate. Electret microphones are preferred for today's telephone handsets, since they are more sensitive (and cheaper) than the older carbon microphones, which many old-fashioned people (like me) still prefer because (a) I think it sounds better, (b) It's far less susceptible to interference from external sources, such as neon signs and (c) It works reliably with hearing aids (i.e. people who wear hearing aids have no trouble with carbon mics.).

electric banana Telecom installers' slang for tone probe. A testing device used to detect signals from a tone generator to identify phone circuits, often the size of a fat pencil or skinny banana. Some models contain speakers; others must be used with a headset or a butt set. See also Tone Generator.

electric lock A cellular phone feature that provides security by locking a cellular phone so it can't be used by unauthorized persons.

electrical closet Floor-serving facility for housing electrical equipment, panelboards, and controls.

electrical service equipment That portion of the electrical power installation, the service enclosure or its equivalent, up to and including the point at which the supply authority makes connection.

Electrically Erasable Programmable Read-Only Memory EEPROM. An EEPROM is a memory chip that holds its content without power. It can be erased, either within the computer or externally and has a lifespan of 10,000-100,000 write cycles.

electrically powered telephone A telephone in which the operating power is obtained either from batteries located at the telephone (local battery) or from a telephone central office (common battery).

electrician's scissors Used to cut cables. They have flat blades and look like very heavy scissors except they have notches on the side of one blade that are used to strip cable.

electricity Electricity is the flow of electric charge. Normally this is thought of as electrons flowing through wire but it can also be protons or electrically charged ions flowing through a fluid.

electro optical transducer A device used to convert electrical signals into light signals and vice versa. It is used at the ends of fiber optic transmission systems.

electrodeposition The deposition of a conductive material from a plating solution by the application of electric current.

electroluminescence The direct conversion of electrical energy into light.

electrolysis The production of chemical changes by passage of current through an electrolyte.

electrolyte A chemical solution used in batteries, chemical rectifiers, and certain types of fixed condensers.

electrolytic process A printing process where paper is treated with an electrolyte and a stylus passes the signal current through the paper to produce an image. Paper is roll-fed past the stylus and changes color depending on the intensity of current passing through the paper.

electromagnetic compatibility EMC. The ability of equipment or systems to be used in their intended environment within designed efficiency levels without causing or receiving degradation due to unintentional EMI (Electromagnetic Interference). EMI is reduced by, amongst other things, copper shielding.

electromagnetic emission control The control of electromagnetic emissions. e.g., radio, radar, and sonar transmissions, for the purpose of preventing or minimizing their use by unintended recipients. A military term. Electromagnetic emission is reduced by, amongst other things, copper shielding.

electromagnetic energy The transmission of energy in the form of waves having an electronic and magnetic component.

electromagnetic force EMF. Also called Voltage.

electromagnetic interference EMI. Interference in signal transmission or reception caused by the radiation of electrical and magnetic fields. That's the easy explanation. Here's a more comprehensive explanation: Any electrical or electromagnetic phenomenon, manmade or natural, either radiated or conducted, that results in unintentional and undesirable responses from, or performance degradation or malfunction of, electronic equipment.

electromagnetic lines of force The lines of force existing about an electromagnet or a current carrying conductor.

electromagnetic radiation EMR. The combined electric and magnetic field components of a radio wave. Electromagnetic radiation can be harmful to you and me. But it is way beyond the scope of this definition to discuss all its ramifications. Suffice: our society is swarming with radio waves. There are waves from radio and TV towers, from microwaves and cell phones, cell sites, mobile phones, cordless phones, and microwave ovens. The EMR EMR we are exposed to has been rising significantly by factors of thousands since the Second World War. There is strong evidence that excessive electromagnetic radiation damages human cells in a way that is potentially cancer causing. And that's why governments all over the world regulate and limit electromagnetic radiation of devices such as cell phones, radar, portable phones, microwave ovens, etc. Being personally careful is critical.

electromagnetic spectrum The entire range of wavelengths (the inverse of frequency) of electromagnetic waves extending from cosmic and gamma rays down through visible light and heat to every form of radio communications signal. The electromagnetic (EM) spectrum is basically just a name that scientists give a bunch of types of radiation when they want to talk about them as a group. Radiation is energy that travels and spreads out as it goes – visible light that comes from a lamp in your house or radio waves that come from a radio station are two types of electromagnetic radiation. Other examples of EM radiation are microwaves, infrared and ultraviolet light, X-rays and gamma-rays. Hotter, more energetic objects and events create higher energy radiation than cool objects. Only extremely hot objects or particles moving at very high velocities can create high-energy radiation like X-rays and gamma-rays.

electromagnetic wave The electric wave propagated by an electrostatic and magnetic field of varying intensity. Its velocity is 186,300 miles per second.

electromechanical ringing The traditional bell or buzzer in a telephone which announces incoming calls.

electromigration A phenomenon in which metal migrates along a current path in current rails, which are power-carrying circuits. Eventually, this phenomenon causes an open in the circuit or a shorting of an adjacent circuit. It is caused as the metal ions move in the direction of the current flow through metal wires comprising the circuit. Electromigration is particularly likely to affect the very thin and very tightly-spaced power distribution lines of sub-micron chip designs. As the phenomenon occurs only after months or years of use, it cannot be prevented by production testing, but only through careful circuit design. (Just in case you don't have enough things to worry about, Newton's Telecom Dictionary is pleased to bring you this source of consternation.)

electromotive force EMF. That force which determines the flow of current; a difference of electric potential. Another way of saying voltage.

electron An electron is a light, subatomic particle that carries a negative charge. Electrons are found in atoms where they balance out the positive charge of the protons in the nucleus. Electrons in an atom are arranged in layers or shells around the nucleus. All atoms follow the pattern where the shells fill in from the inside to the outside. Each layer must fill to capacity before the next layer can be started. See electricity.

electron gun Device in a television picture tube from which electrons are emitted toward screen.

electron tube rectifier A device for rectifying an alternating current by utilizing the flow of electrons between a hot cathode and a relatively cold anode.

electromagnetic compatibility EMC. The ability of a system or product to function properly in environment where other electromagnetic devices are used and not be a source itself of electromagnetic interference. See also RFID.

Electronic article surveillance EAS. Simple electronic RFID tags that can be turned on or off. When an item is purchased (or borrowed from a library), the tag is turned off. When someone passes a gate area holding an item with a tag that hasn't been turned off, an alarm sounds. EAS tags are embedded in the packaging of most pharmaceuticals.

electronic bidding A process by which bidders in an auction use computers to place their bids.

electronic blackboard This is a teleconferencing tool. At one end there's a large "whiteboard." Write on this board and electronics behind the board pick up your writing and transmit it over phone lines to a remote TV set. The idea is that remote viewers can hear your voice on the phone and see the presentation on the electronic blackboard. The product has not done well because it is expensive – typically several hundred dollars a month just for rent, plus extra hundreds for transmission costs. In Japan, there are similar boards called OABoards – Office Automation Boards. They do one thing differently – they will print a copy on normal letter-size paper of what's written on the board. This takes about 20 seconds. Some of these Japanese OABoards will also transmit their contents over phone lines. So far, neither the OABoards nor the electronic blackboards have found a sizable market in the United States.

electronic bonding EB. A term for the exchange of information between carriers' Operations Support Systems (OSSs). Through secure gateways, the carriers can exchange information such as trouble tickets, which is very important in a multivendor network. The specific technique generally is either EDI (Electronic Data Interchange) or Telecommunications Management Network (TMN). See also EDI and TMN.

electronic bracelet A device attached to the legs of criminals who have been sentenced to confinement in their homes. The device allows them to move around a confined area. In most iterations, the device emits a regular signal to a nearby receiving station. If the criminal leaves the permitted area or tampers with the device, the in-home receiving device will dial local police authorities and effectively say, "the crim has flown his coop." An electronic bracelet is not the same as an electronic leash, which is a beeper.

electronic bulletin board A computer, a modem, a phone line and a piece of software. Load communications software in your computer, dial the distant electronic bulletin board. The system will answer and present you with a menu of options. Typically those options will include leave messages, pick up messages, find out information, fill in a survey and upload and download a file.

electronic business card Also known as vCard. See Versit and VersitCard.

electronic call distribution Another term for Automatic Call Distribution. See Automatic Call Distributor.

electronic cash E-Cash. A term referring to money that is exchanged over the Internet or some other form of electronic network.

electronic commerce Using electronic information technologies to conduct business between trading partners, using or not using EDI (Electronic Data Interchange), using or not using the Internet. See EDI and OFX.

Electronic Commerce Services ECS. A set of e-mail authentication and certification services announced by the U.S. Postal Service in October 1996 and scheduled for early 1997 rollout. The service will provide an electronic postmark aimed at making e-mail authentic and traceable and, therefore, legally binding in support of electronic commerce. E-mail fraud effectively will become mail fraud, as a result. The service initially will be priced at 22 cents. ECS seems like a great idea! Hopefully, it will be more successful than the USPS' failed 1980s attempt at offering e-mail services. Critical to its success is the support of leading Internet e-mail vendors such as Microsoft and Netscape, both of whom are planning such support.

Electronic Common Control See ECC.

Electronic Communications Privacy Act ECPA. A federal act passed in 1986, the ECPA grants employers the right to review stored communications (e.g., e-mail, voice mail, and other computer data) stored on company systems. Supporters of the act mainly are businesses that insist that they have the right to monitor the use of company resources. Opponents mainly are individuals who insist that they have the inherent right to privacy. Check your privacy at the front door, my friend. See also Call Accounting, Keystroke Monitoring, and Network Accounting.

electronic custom telephone service Provides deluxe key telephone features and simplified access to certain AT&T Dimension PBX phones.

electronic data interchange EDI. The process whereby standardized forms of electronic commerce documents are transferred between computer systems often run by different companies and without human intervention. EDI is used for placing orders, for billing and paying for goods and services via private electronic networks or via the Internet. According to studies, it costs about $50 a process paper-based purchase order and about $2.50 to process the same order with EDI. Internet-based EDI can lower the cost to less than $1.25. The form and format of EDI documents may be defined by vendor specifications, ITU-T standards, the ANSI X.12 standard, or the United Nations EDIFACT standard. See EDI for a fuller explanation. See also the next definition. See also http://www.edi-info-center.com/html/hotline.html and http://www.ecworld.org/Members/edi-uk.html.

Electronic Data Interchange Association EDIA. An organization which works to provide a common platform to communicate global EDI activity, bypassing language conventions and national boundaries.

Electronic Data Processing See EDP.

Electronic Frontier Foundation EFF. A foundation established in July 1990 by Mitch Kapor, founder of Lotus, to "ensure that the principles embodied in the Constitution and Bill of Rights are protected as new communications technologies emerge." The EFF addresses a wide range social and legal issues arising from the impact on society of the increasingly pervasive use of computers as the means of communication and information distribution. Efforts include working to defeat the Communications Decency Act, in order to protect the right to free speech over the Internet. EFF also works to support both legal and technical means of enhancing privacy in communications, specifically focusing on the unfettered use of encryption algorithms. See appendix for address. www.eff.org.

Electronic Funds Transfer EFT. A system which transfers money electronically between accounts or organizations without moving the actual money.

electronic hug My friend Jennifer Durst carries a Blackberry. She is starting a company. It is hard work. Every so often I send her an electronic hug. My message to her Blackberry simply reads: "Electronic Hug." She says this makes her feel good and doesn't cost her any time or money since, under her Blackberry plan, she is entitled to unlimited messages. When I sent her this definition, she blackberried back, "I think you should mention something about what an efficient means of showing support and affection this is for start-up entrepreneurs who don't have time for the usual means."

electronic image mail The transmission of slow scan TV or facsimile via "Store and Forward." Not a common term.

electronic ink Want to read a book on screen? Most people have difficulty. They prefer the printed word on paper. They find the paper page easier on their eyes. For eons learned computer inventors have been trying to figure better ways of viewing books on screen. One of the latest is something called "electronic ink." The technology, developed by E Ink of Cambridge, Massachusetts, forms images by electronically pulling around microscopic particles of black and white pigment that float in tiny capsules inside the screen. The result is a display that uses very little power and looks almost identical to black print on white paper. For reading, it's a vast improvement over the liquid-crystal displays common in notebook computers, PDAs and cellphones.

Electronic Interface Format EIF. A standardized file format required by Verizon to communicate with DCAS.

electronic key exchange A security procedure by which two entities establish secret keys used to encrypt and decrypt data exchanged between them. The procedure used in CDPD is based on a form of public key cryptology developed by Diffe and Hellman.

electronic key system A key telephone system in which the electromechanical relays and switches have been replaced by electronic devices – often in the phone and in the central cabinet. The innards of the central cabinet of an electronic key system more resemble a computer than a conventional key system. These days, virtually all key systems are electronic. Most manufacturers have stopped making electromechanical key systems (such as 1A2).

electronic leash Pagers or beepers are often called "electronic leashes" because they allow your boss to contact you, to control you, to keep you on a leash. At one stage, beepers were carried by doctors and technicians who could never be "out of touch." As a result, beepers got a bad reputation and "real" people, i.e. bosses, wouldn't carry them. But that's getting better now and real people are now carrying them. See also ELECTRONIC BRACELET.

electronic lock Lets you lock your cellular phone so no one can use it. If you use Electronic Lock, you'll have to punch in some extra digits – like a password – to unlock the lock.

electronic mail A term which usually means Electronic Text Mail, as opposed to Electronic Voice Mail or Electronic Image Mail. Sometimes electronic mail is written as E-Mail. Sometimes as email. These days electronic mail is everything from simple messages flowing over a local area network from one cubicle to another, to messages flowing across the globe on an X.400 network. Such messages may be simple text messages containing only ASCII or they may be complex messages containing embedded voice messages, spreadsheets and images. See Electronic Text Mail, Electronic Voice Mail, Electronic Image Mail and Windows, Windows Telephony.

electronic mail gateways A collection of hardware and software that allows users on an E-mail system to communicate and exchange messages with other mail systems that use a different protocol.

electronic mall A virtual shopping mall where you can browse and buy products and services online.

electronic message registration A system to detect and count a phone user's completed local calls and then tell the central office the number of message units used. Also used in hotels.

Electronic Messaging Association EMA. The trade association for electronic messaging and information exchange. Formerly known as the Electronic Mail Association, EMA is a membership forum that seeks to enable users to work in partnership with providers of the technologies. Vendor members offer a wide range of services, including electronic mail, network, directories, computer facsimile, electronic data interchange (EDI), paging, groupware, and voice mail. EMAs technology programs aim to remove barriers to global interconnectivity and interoperability through assisting in the definition, endorsement, development, demonstration, and implementation of all messaging standards, operating conventions, and practices for use in electronic commerce. EMA lobbies governments, standards bodies, and consumer groups in advocacy of favorable public policies. www.ema.org.

Electronic Order Exchange EOE. Inter-company transactions between buyers and sellers handled electronically via standard data communications protocols. EOE can be employed to send purchase orders, price and product listings and order-related information.

electronic perception Electronic perception is a method of bringing primitive 3D vision to a variety of less expensive devices. The approach uses infrared light and a sensor chip like those in digital cameras. Light is beamed at the target, and the tiny differences in the time it takes rays to return to specific points on the chip are measured. Gauging distance based on the time that signals travel to and from a target is the principle behind radar. Electronic perception uses the time differences to compute a three-dimensional relief map of the target.

electronic phone General description for most phones designed after about 1980, where many mechanical and electrical parts are replaced by smaller, lighter, and cheaper electronic parts. Features such as mute, redial and memory became popular with these phones, which range in price from $5 to hundreds of dollars.

Electronic Product Code EPC. A 96-bit code, created by the Auto-ID Center, that will one day replace barcodes. The EPC has digits to identify the manufacturer, product category and the individual item. It is backed by the United Code Council and EAN International, the two main bodies that oversee barcode standards. See also Auto ID Center.

electronic publishing Electronic Publishing is synonymous with Desktop Publishing. Electronic Publishing software packages give the user the ability to perform page composition, insert images and manipulate text on the computer screen and display the document on the screen exactly as it will look when it is printed.

electronic receptionist A fancy name for a voice processing automated attendant, except that in addition to all the normal auto attendant features, it also sends messages to personal PCs on LANs telling the owner who's calling and giving the owner (the called party) the choice of doing something with the call – like answering it or putting it into voice mail.

electronic redlining A term for disenfranchising people and institutions because of their lack of telecommunications services and apparatus. In December, 1993, Vice President of the United States, Al Gore, told the National Press Club, When it comes to ensuring universal service, our schools are the most impoverished institution in society. Only 14% of our public schools used educational networks in even one classroom last year. Only 22% possess even one modem. Video-on-demand will be a great thing. It will be a far greater thing to demand that our efforts give every child access to the educational riches we have in such abundance. The recent article in the Washington Post on the proposed video communication network in the D.C. area is a wake-up call to all of us concerned about "electronic redlining."

electronic ringer A substitute for the conventional telephone bell, that uses music synthesizer circuitry to generate an attention-getting signal played through a speaker. Typical sounds include warbles, chirps, beeps, squawks, and chimes. The writer of this entry, Michael Marcus, once installed a phone with a chirp sound. A few days later, the customer complained that she had not been receiving any calls, and the birds in her yard were chirping much more than usual.

Electronic Serial Number ESN. A 32-bit binary number which uniquely identifies each cellular phone. The ESN consists of three parts: the manufacturer code, a reserved area, and a manufacturer-assigned serial number. The ESN, which represents the terminal, is hard-coded, fixed and supposedly cannot be changed. Paired with a MIN (Mobile Identification Number), the ESN and MIN are automatically transmitted to the mobile base station every time a cellular call is placed. The Mobile Telephone Switching office checks the ESN/MIN to make sure the pair are valid, that the phone has not been reported stolen, that the user's monthly bill has been paid, etc., before permitting the call to go through. At least that's the theory. It doesn't always work this way on calls made from roaming cellular phones. And some cellular phones have been known to have their ESNs tampered with (it's called fraud) which tends to mess up the billing mechanisms. See MIN.

Electronic Signature Act Also known as the E-Sign Act, the Electronic Signatures in Global and National Commerce Act took effect on October 1, 2000. The Act defines an electronic signature as "an electronic sound, symbol or process, attached to or logically associated with a contract or other record and executed or adopted by a person with the intent to sign the record." (I'll bet you my next paycheck that a whole gaggle of attorneys wrote that definition.) The Act provides that no contract, signature, or record shall be denied legal effect solely because it is in electronic form. It also provides that most electronic contracts and records are legally enforceable only if they are in a form that is capable of being retained and accurately reproduced for later reference by relevant parties. The Act will go a long way towards ensuring the viability of E-commerce, since electronic signatures now have the same effect as do paper and wet ink contracts. Any requirement that a stamp, seal, or other embossing device be used to authenticate a signature of document is eliminated. The Act will also speed up the contractual process, as documents can be sent electronically over the Internet, rather than by fax, courier, and postal service. The Act applies equally to transactions in interstate and foreign commerce. However, the Act does not apply to contracts or records relating to wills, adoption, divorce, or other family law matters; much of the Uniform Commercial Code (UCC); court orders; notices of cancellation of utility services; recall notices for products endangering health or public safety; and a whole host of other such things.

electronic sweep Variation in the frequency of a signal over a whole band as a means of checking the response of equipment under the test.

electronic switching system A telephone switch which uses electronics or computers to control the switching of calls, their billing and other functions. The term is now vaguely defined, with each manufacturer defining it as something somewhat different. In fact, every telephone switch sold today is electronic. The term originally came about because early telephone switches were entirely electro-mechanical. The switch consisted entirely of a moving switch. Devices like relays physically moved in order to send the call through the exchange and on its way. These things moved in direct response to the digits dialed by the telephone subscriber. These switches contained no "intelligence" – i.e. no ability to deviate from a set number of very simple tasks which could be accomplished by electromechanical relays.

Then someone said: it would be more efficient if the "instruction part" of the process were divorced from the switching mechanism. This lead to the creation of the "electronic" switch in which the "brains" of the switch are separated from the switching mechanism itself. Thus the "brains" can do simple things like collect the dialed digits as they are slowly dialed and pulse them out quickly to the switch – as fast as it can handle them. Now, the "brains" are typically a digital computer.

electronic tandem network 1. Two or more switching systems operating in parallel as part of providing network services (usually voice) to large users.

2. A telephone company switching device used to connect telephone company toll offices located in the same geographic area.

electronic telephone directory service A PBX feature which stores and produces, on demand, a directory of all extension phone numbers. The directory may include all users in a network. A CRT with keyboard and/or printer is usually required for input and retrieval. In some systems, the CRT or another type of alphanumeric display is part of the Attendant Console. In some systems, the directory may also include names and telephone numbers of frequently called outside people, especially those in the speed calling system. The directory may also be enhanced to include SMDR data such as client codes, account codes and client telephone numbers.

electronic text mail A "Store and Forward" service for the transmission of textual messages transmitted in machine readable form from a computer terminal or computer system. A message sent from one computer user to another is stored in the recipient's "mailbox" until that person next logs onto the system. The system then can deliver the message. Telex, in which a machine readable form of message transmission takes place, is also considered an Electronic Text Mail medium, albeit a very slow one. For an example of electronic mail, please dial our electronic mail system on 212-989-4675. It's free. Parameters are 300, 1200 or 2400 baud, 8 data bits, one stop bit, and no parity.

electronic voice mail A system which stores messages usually spoken over a telephone. These messages can be retrieved by the intended recipient when that person next calls into the system. Also called Voice Mail, it operates just like a touch-tone controlled answering machine.

electronic wallet See Digital Wallet.

electronic warfare See EW.

electrophotographic printing A printing method that uses light to modify electrostatic charges on a photoconductive substrate.

Electrostatic Pertaining to static electricity or electricity at rest. A constant intensity electric charge.

electrostatic charge An electric charge at rest.

electrostatic discharge ESD. Let's say you;re a maker of components that go into a cell phone., also called a wireless or mobile phone. Every time someone picks the phone to use it, they transfer a static electricity charge to the phone. Think what happens when you walk across a nylon carpet in the winter and go to shake someone's hand. A spark flies. You get a shock. Well, you're giving your phone also a shock. Though you mightn't feel it, your phone sure does. If the components don't have good ESD, they will get damaged and your phone will no longer work. All this is why manufacturers talk a lot about their ESD and how good their components are in resisting and dissipating static electricity charges. It's clearly important. ESD's formal definition is "Discharge of a static

charge on a surface or body through a conductive path to ground. Can be damaging to integrated circuits."

electrostatic printing A method of printing, very common in photocopying, in which charges are beamed onto the surface of paper. The charges attract particles of a very fine (typically black powder) which sticks to the charges. The black powder is fused permanently on the paper by great heat. "XEROXing" is electrostatic printing. In xeroxing, the black powder is called toner.

elegant An elegant program is one that is efficiently written to use the smallest possible amount of main memory and the fewest instructions.

Elektrosvyaz A Russian phone company.

element 1. Any single piece of data. For example, a user name is an element of a login string, and a BORI is an element of a header field.

2. Network Element (NE). A constituent part of a network. An element might be in the form of a modem, a multiplexer, a switch, or some other basic unit of a network.

3. The structural building blocks of HTML documents. Blocks of text in HTML documents are contained in elements, according to their function in the document, for example, headings, lists, paragraphs of text and links are all surrounded by specific elements. See HTML.

4. An XML element contains data that is transferred between a client and a server, or between two servers. See also XML.

Element Management Layer See EML.

elephant cage In layperson's terms... An elephant cage is an antenna array where you have two concentric circles of antennas set up with cabling the same length going back to the radio control house. By having the antenna cables the same length and using signal strength measurements you can determine the location of a radio transmitter without the need for an external remote site. The size of the antenna masts and the general shape of the antenna field looks like a cage for elephants. In more technical terms...Wullenweber – A circular disposed antenna array, used for direction finding invented in World War II. From a distance, one sees this huge, round collection of poles and wires, the low-frequency end of the array. Closer up, the rows of smaller wires and spikes come into view. From the air, one sees something that looks very much like Stonehenge, and in fact Wullenwebers are among the easiest man-made objects to spot from space. The antenna is sometimes called an "elephant cage," though the scale more resembles a Godzilla cage. These are very popular with the government and military for High Priority Intercept, real spook stuff, and the classic one is at Imperial Beach, CA, where the US Navy trains crypto types. It replaced a rhombic farm, four of them in Adcock configuration, but actually it's not all that much smaller. It's visible 10 miles away. Directional bearings are obtained by comparing signals as they enter the different segments of the circle.

elevation The angular distance (up and down) of a satellite above the horizon. During installation of your direct broadcast system (DBS) system, you (or the installer) can punch your zip code or latitude and longitude into the DBS receiver's setup screen and get precise elevation and azimuth angles for your location. You need this information to make sure that your dish is accurately aimed at the satellite.

elevation beam width The vertical measure (usually in degrees) of an antenna pattern.

elevator eyes A term used in sexual harassment to mean viewing someone up and down.

elevator pitch What you can tell about your company and your shiny new thing in 30 seconds – the average length of an elevator ride.. Someone once told me the longest scene in Sesame Street was 26 seconds – before the camera switched to a new angle. I figure that 26 seconds is the maximum time any of us (specially me) can concentrate on something – hence the need to capture everyone's attention quickly. You're selling someone on your ideas. They only want to hear a summary. They ask you for an "elevator pitch" – i.e. presentation that lasts no more than an elevator ride. Clearly, that's too short. The idea is that if you grab their attention in 30 seconds, they'll ask you to stick around and pitch some more.

elevator seeking Organizes the way data is read from hard disks and logically organizes disk operations as they arrive at the Novell NetWare local area network server for processing. A queue is maintained for each disk driver operating within the server. As disk read and write requests are queued for a specific drive, the operating system sorts incoming requests into a priority based on the drive's current head position. As the disk driver services the queue, subsequent requests are located either in the vicinity of the last request or in the opposite direction. Thus, the drive heads operate in a sweeping fashion, from the outside to the inside of the disk. Elevator seeking improves disk channel performance by significantly reducing disk head thrashing (rapid back-and-forth movements of the disk head) and by minimizing head seek times. Imagine how inefficient an elevator would be if the people using it had to get off the elevator in the order they got on.

ELF Extremely Low Frequency. Frequencies from 30 Hz to 300 Hz.

ELFEXT Equal Level Far End Crosstalk. Not a measurement, but a calculated result, that is derived by subtracting the insertion loss of the disturbing pair from the FEXT this pair induces in an adjacent pair. See PESELFELT.

ELIU Electrical Line Interface Unit.

elliptic curve Elliptic curves are created using mathematical expressions from number theory and algebraic geometry. Elliptic curve cryptosystems replace conventional modular discrete logarithm cryptosystems with the elliptic curve operations. There are currently no specialized attacks, which means that shorter key sizes for elliptic cryptosystems give the same security as larger keys in other cryptosystems.

elite hacker One of a reasonably small number of hackers who possess great skill and imagination, Elite hackers are able to devise novel attacks which are technically sophisticated and ingenious.

elmer A seasoned amateur radio (i.e., ham radio) operator who generously teaches others about amateur radio.

elmer list A list of elmers and their contact information.

elmering A catch-all term referring to the things that elmers do to teach others about amateur radio.

elongation A cable term. The fractional increase in length of a material stressed in tension.

ELOT Hellenic Organization for Standardization (Greece).

ELSU Ethernet LAN Service Unit. An ELSU provides 12 independent virtual Ethernet bridges for running over ATM networks. ELSUs are designed for flexible deployment, either local to an ATM switch or at a remote site. ELSUs are designed for LAN internetworking services over ATM networks.

ELT Emergency Locator Transmitter carried on board aircraft; in case of a crash, it transmits a 406 MHz locating signal to COSPAS-SART satellites. A story now legend in aircraft rescue illustrates the worth of having an ELT on board. In July 1982, an airplane crashed in the mountainous forests of British Columbia, Canada. Two months later, in September, a single-engine light airplane suffered a similar fate while searching for the first airplane. However, the second airplane was easily located, and its three survivors happily saved. The difference in survival was due to the second airplane having an ELT on board. www.icao.org.

Elvis Year The peak year of something's popularity.

EM 1. Element Manager. Software and hardware used to manage and monitor components of a telecommunications network at their lowest level.

2. Abbreviation for End of Medium. The binary code is 1001001, the Hex is 91.

3. Electromagnetic. See Electromagnetic Spectrum.

EMA See Electronic Messaging Association.

EMACS A standard Unix text editor preferred by Unix types that beginners tend to hate.

EMAG ETSI MIS Advisory Group.

email A colloquial term for electronic mail. See Email address.

email address The UUCP or domain-based address by which a user is referred to. My email address is HARRYNEWTON@MCIMAIL.COM.

email client The name of the software that runs your email. Most likely, it's Microsoft's Outlook, which is the most popular email client. But there are thousands of others.

email gateway An email gateway is typically a PC on LAN. The PC has one or more modem and/or fax/modem cards. Its job is to send and receive e-mails and/or send and receive faxes for everyone on the LAN. To pick up emails, it might dial once an hour into various mail systems, like MCI Mail, CompuServe, and download all the messages for all the people on the LAN. Once it has those messages, it brings them onto its hard disk and then alerts the recipients that they now have an e-mail. See Server.

email hygiene Principles or practices that reduce spam and protect a computer from viruses and other threats embedded in or attached to e-mail messages.

email reflector An Internet electronic mail address which automatically sends you back a reply (i.e. reflects mail to you) if you include certain key words in your message to it. Such key words might be "subscribe" or "lists help."

email retention policy An organization's policy for saving, purging, and managing email. Email retention policy may be determined, in part by industry regulations. For example, the SEC requires brokers' email to be retained for three years, and HIPAA requires hospitals and doctors to retain email discussions of a patient for the lifetime of the patient. of course, this incents hospitals and doctors wrongly. But no one said government regulations always make sense.

email server See Email Gateway.

email shorthand Acronyms for commonly used phrases that one would otherwise type. Some of the most popular ones are: IMHO: In My Humble Opinion; BTW: By The Way; RTM: Read The Manual; LOL: Laughing Out Loud; FWIW: For What It's Worth; and ROFL: Rolling On The Floor Laughing.

email threads A fancy way of saying on-going correspondence by electronic mail. I first heard this term from Sean Purcell, a smart fellow working on the excellent product called Outlook from Microsoft.

EMBARC Motorola's company which does wireless electronic mail to people carrying laptops and palmtops. EMBARC, according to Motorola, stands for Electronic Mail Broadcast to A Roaming Computer. Actually EMBARC does more than mail. It also broadcasts snippets of news.

embed To make something an integral part of some larger, like a fossilized insect is embedded in a rock. In our context, one example is to insert information (an object) that was created in one document into another document (most often the two documents were created with different applications). The embedded object can be edited directly from within the document. To embed under Windows 3.1, you must be using applications that support OLE (Object Linking and Embedding). Another example is to embed a network interface card in a laptop.

embedded base equipment All customer-premises equipment that has been provided by the Bell Operating Companies (BOCs) prior to January 1, 1984, that was ordered transferred from the BOCs to AT&T by court order.

Embedded Code Formatting ECF. A NetWare definition. This is something of a programming language, in which faxing commands or other program that automatically generates information, formats it, and faxes it without user intervention.

embedded customer-premises equipment Telephone-company-provided premises equipment in use or in inventory of a regulated telephone utility as at December 31, 1983.

embedded hyperlink A hyperlink that is in a line of text. A hotspot is the place in a document that contains an embedded hyperlink.

embedded network interface See ENI.

embedded object A Windows term. An embedded object is information in a document that is a copy of information created in another application. By choosing an embedded object, you can start the application that was used to create it, while remaining in the document you're working in.

Embedded Operations Channel See EOC.

Embedded SCSI A hard disk that has a SCSI (Small Computer System Interface) and a hard disk controller built into the hard disk unit. See also SCSI.

Embedded SQL SQL statements embedded within a source program and prepared before the program is executed.

embedded system An embedded system is a combination of computer hardware and software, either fixed in capability or programmable; that is specifically designed for a particular kind of application device. Industrial machines, automobiles, medical equipment, cameras, household appliances, airplanes, vending machines, and toys (as well as cellular phones and PDAs) are among the possible hosts of an embedded system. Embedded systems that are programmable are provided with a programming interface. Some operating systems or language platforms are tailored for the embedded market; however, some low-end consumer products use inexpensive microprocessors and limited storage. In these devices, the application and operating system are both part of a single program written permanently into the system's memory, rather than being loaded into RAM and changeable by the user. See Embedded Windows.

embedded system processors National Semiconductor's line of high-performance microprocessors used in dedicated systems, such as fax machines and laser printers.

embedded windows When you start your PC you have to wait while it loads Windows operating system off its hard disk into memory and configures itself to start accepting your humble commands. This process is known as bootup. Imagine a device which has Windows OS built into it, so that Windows pops up ready to go immediately when you switch it on. Developers increasingly use the embedded version of Microsoft's Windows in devices such as cell phones, PDAs, industrial machines, automobiles, medical equipment, cameras, household appliances, airplanes, vending machines, and even toys. The rationale is that the devices embedded with the Windows operating system have a familiar feel, as well as compatibility with PCs. Microsoft began pursuing small embedded systems in 1996 with its release of Windows CE 1.0, a from-scratch code base targeting handheld device applications with a graphical interface (GUI) that had the look and feel of Windows 95. Embedded devices have shifted from using relatively simple 8-bit processors to faster and more powerful 16 and 32-bit units as users' wanted their gadgets to be more like their PCs. See Embedded System and Palm.

embossing A means of marker identification by thermal indentation leaving raised lettering on a cable's sheath material.

EMC ElectroMagnetic Compatibility.

EMD See equilibrium mode distribution.

EME 1. Electromagnetic Energy.

2. See Earth-Moon-Earth communication.

EMEA Europe, Middle East, and Africa.

Emergency Access An alarm system built into some PBXs. In an emergency it rings all phones.

Emergency Alert System See Emergency Broadcast System and EAS.

Emergency Broadcast System EBS. The EBS is composed of AM, FM, and TV broadcast stations; low-power TV stations; and non-Government industry entities operating on a voluntary, organized basis during emergencies at national, state, or operational (local) area levels. "This is a test of the Emergency Broadcast System – this is only a test." That warning, a remnant of the cold war, is about to disappear. The high-pitched tone is to be replaced by a few short buzzes, and the "this is a test" warning may be dropped altogether. The buzzes are generated by new computer technology. The new system, approved by the FCC in 1994, is expected to be fully operational by 1998 as the Emergency Alert System. The current test lasts ca. 35 or 40 seconds; the new test will be shorter, although the duration is not yet decided upon. The system has never been used for a nuclear emergency, but is used regularly for civil emergencies and severe weather alerts. The current emergency broadcast system is serial, that is it works on a daisy chain where one station receives the warning and sends it on to the next. That means that, if one station's equipment fails, the warning may not get further down the line. The new system looks more like a 'web' in which a station does not rely on one sole source for the signal, but will receive digital signals that will activate computers at broadcast facilities and download emergency messages.

emergency dialing A variation on speed calling to call numbers for police, fire department, ambulance, etc. Typically found as special buttons on an electronic phone.

emergency hold "Emergi-hold" allows a 911 caller's line to be held open in the event that a caller attempts to hang up. This gives the PSAP (Public Service Answering Position) agent full control of the call. It will not be released until the agent finishes the call.

Emergency Information Services Interface A standard (ATIS-PP-0500006.200X) that defines how IP-based information is delivered to the native 911 network.

Emergency Services Messaging Interface A standard (ATIS.PP.0500002.200X) that defines how the call-handling equipment at a public safety answering point (PSAP) will receive IP-based information.

emergency power A stand-alone secondary electrical supply source not dependent upon the primary electrical source.

emergency ringback This feature enables the 911 PSAP (Public Service Answering Position) attendant to signal a caller who has either hung up or left the phone off hook. Emergency Ringback enables the PSAP agent to ring a phone which has been hung up or issue a loud "howling" sound from the customer's phone if it has been left off hook.

Emergency Service Number ESN. In the US emergency services telephone network, a three to five digit number used to represent an Emergency Service Zone. The ESN is used for 9-1-1 call routing to and between public safety answering points. The ESN for each call is derived from the call's Automatic Location Identification. See ESN.

Emergency Service Zone ESZ. In the US emergency services telephone network, a geographic area served by a single public safety answering point (PSAP). Each ESZ contains a unique combination of emergency service agencies (police, fire, medical); multiple ESZs may be served by one PSAP. Each ESZ is represented by a unique Emergency Service Number.

emergency stand alone service A feature of a central office switch which allows it to keep working – switching and transferring calls – even though some of its connections to other central offices switches have been broken.

emergency telephone A single line telephone that becomes active when there is no commercial AC power to the Key Service Unit.

EMF ElectroMotive Force, or ElectroMagnetic Force, a synonym for voltage. See also Ohm's Law and Voltage.

EMI 1. Electromagnetic Interference, (EMI) happens when one device leaks so much energy that it adversely affects the operation of another device. EMI is reduced by copper shielding. National and international regulatory agencies (FCC, CISPR, etc.) set limits for these emissions. Class A is for industrial use and Class B is for residential use.

Here's a definition from APC: EMI usually refers to unwanted electrical noise present on a power line. This noise may "leak" from the power lines and affect equipment that is not even connected to the power line. Such "leakage" is called a magnetic field. Magnetic fields are formed when unwanted noise voltages give rise to noise currents. Such noise signals may adversely affect electronic equipment and cause intermittent data problems. EMI protection is provided by noise filters placed on the AC power line. The filter reduces the noise voltage on the protected line, and by doing so also eliminates the magnetic fields of noise generated by the protected line. Noise signals that act over a significant distance are called RFI (Radio Frequency Interference). Equipment power cords and building wiring often act as antennas to receive RFI and convert it to EMI.

2. Exchange Message Interface.

EMI Segregation Isolation of the telecommunications signal from electromagnetic interference.

EMI/EMR See Exchange Message Interface/Exchange Message Record.

EMI/RFI Filter A circuit or device containing series inductive (load bearing) and parallel capacitive (non-load bearing) components, which provide a low impedance path for high-frequency noise around a protected circuit.

emission 1. Electromagnetic energy propagated from a source by radiation or conduction. The energy thus propagated may be either desired or undesired and may occur anywhere in the electromagnetic spectrum.

2. Radiation produced, or the production of radiation, by a radio transmitting station. For example, the energy radiated by the local oscillator of a radio receiver would not be an emission but a radiation.

emissivity Ratio of flux radiated by a substance to the flux radiated by black body at the same temperature. Emissivity is usually a function of wavelength.

emitter The source of optical power.

EML 1. Element Management Layer. A layer representing the management and monitoring of components, at their lowest level, in a telecommunications network. In short, an abstraction of the functions provided by systems that manage each network element on an individual basis.

2. Expected Measured Loss.

EMM 1. Entitlement Management Message Stream. Entitlement Management Messages define access rights for each individual decoder. The EMM stream is processed with the access control device, but the user processor is responsible for buffering EMMs and feeding them via an interface to the access control device.

2. See external meter modem.

emotags Mock HTML tags (<♦>smile>, <♦>smirk>) used in WWW-related e-mail and newsgroups in place of ASCII emoticons, for example: "<♦>flames> Someone tell that jerk to shut up, I'm sick of his vapid whining! <♦>/flame>." Definition from Wired Magazine. See Emoticon.

emoticon From Emotional Icon, one of a growing number of typographical cartoons used on BBSs (Bulletin Board Systems) to portray the mood of the sender, or indicate physical appearance. They are meant to be looked at sideways.

See also Smiley Face.

EMP A large and fast-moving electromagnetic pulse caused by lightning.

emphasis In FM transmission, the intentional alteration of the amplitude-versus-frequency characteristics of the signal to reduce adverse effects of noise in a communication system. The higher frequency signals are emphasized to produce a more equal modulation index for the transmitted frequency spectrum, and therefore a better signal-to-noise ratio for the entire frequency range.

employee tailgating You check into your company with your ID badge. Because you're chivalrous you hold the door open for the employee following you, who thus doesn't have to slide his ID badge through the system. This is called tailgating. Most companies frown on it.

EmPower EmPower is a standard for a plug which many airlines have adopted to allow their flyers adapter for aircraft. EmPowerT is the number one choice of airlines around the world. Adopted by almost 40 airlines, there are over 86,000 seats on almost 1600 aircraft using this system.

empty slot ring In LAN technology an empty slot ring is a ring LAN in which a free packet circulates through every workstation. A bit in the packet's header indicates whether it contains any messages for the workstation. If it contains messages, it also contains source and destination addresses.

empty suit You don't need to pass an IQ test to buy a suit. Dumb executives who wear nice suits are often called "empty suits" around telephone companies, especially by craft workers who never wear suits, except to funerals, marriages and bar mitzvahs. See Suit.

EMR 1. Exchange Message Record. Bellcore standard format of messages used for the interchange of telecommunications message information among telephone companies. Telephone companies use EMR to exchange billable, non-billable, sample, settlement and study data. EMR formatted data is provided to all interdepartmental applications and to large customers (users) who request reproduced message records for control and allocation of their communication costs. Bellcore BR-010-200-010 Issue 15, Oct 96. In November of 1998, I heard that EMR was being replaced by something called EMI so that it applies to IXCs as well as LECs.

2. Electromagnetic Radiation.

EMS 1. Enhanced Messaging Service. An enhanced version of Short Messaging Service (SMS) is comprised of several text messages that are clustered together. EMS provides capabilities for more rich messaging features such as sending/receiving ring tones and other melodies/sounds, pictures and animations, and modified (formatted) text.

1. Enterprise Messaging Server. A Microsoft concept which allows users to transparently access the messaging engine from within desktop applications to route messages, share files, or retrieve reference data. According to Microsoft, corporate developers will be able to add capabilities using Visual Basic and access EMS by writing either to the X.400 Application Program Interface Association's (XAPIA's) Common Mail Calls (CMC) or to Microsoft's Messaging API (MAPI). See MAPI.

2. Electronics Manufacturing Services. A fancy name for outsourcing.

EMT Electrical Metal Tubing. In many towns you must run your electrical AC wire inside metal tubing. In other towns you can run normal plastic insulated wiring. Theoretically, EMT is a safer fire hazard. What you are allowed to run depends on local laws and regulations. Tip: Dimmers for incandescent lights raise havoc with LAN data. Solution: Put the plastic electrical wires inside EMT (Electrical Metal Tubing) and ground the conduit.

EMTA Embedded Multimedia Terminal Adapter. Connects a cable TV customer's traditional telephone lines to the cable operator's IP network. Provides all the features of a data cable modem, including high-speed data Internet access. An EMTA provides the interface to the broadband network and performs voice compression, packetization, security and call signaling for IP communications services.

emulate To duplicate one system or network element with another. For instance, to imitate a computer or computer operating system by a combination of hardware and software that allows programs written for one computer or terminal to run on another. For example, at one stage the most common data terminal was a DEC VT-100. Our communications program, Crosstalk, allowed us to "emulate" a DEC-VT100 on our IBM PCs and PC clones. At another stage several companies were trying to write "emulator" software which would allow software written for Windows to successfully run on a computer running the Linus operating system. There are also emulator programs which allow Windows software to run on Apple machines. Emulator programs usually extract their toll in a loss of efficiency versus running the software directly in "native implementation," i.e. running the software on the operating system it was originally written for.

Circuit emulation, an ATM term, refers to the ability of an ATM network to emulate a circuit over a channel in a T-carrier electrical environment or the over a Virtual Channel in a SONET/SDH fiber optic transmission system. LANE (LAN Emulation) allows an ATM network to emulate a LAN, offering LAN functionality over an ATM network. See LANE for more detail.

emulation What happens one gadget emulates another. See Emulate.

emulation mode Function of a network control point (NCP) that enables it to perform activities equivalent to those performed by a transmission control unit. See NCP.

emulator A device or computer program which can act as if it is a different device or program, that is Emulate (i.e. pretend to be) another device. Certain computer terminals are necessary in specific systems and a terminal that is not that type may be able to act as if it was. If it can, it is an Emulator. This is not a common term. See also the verb Emulate.

en bloc See En-bloc.

en-bloc Enbloc is considered a buffer that holds on to all dialed digits and sends all digits to the central office at the same time. This must be enabled for PRI. It is an ISDN term for a process of call establishment. En-bloc (from French, meaning "in a group") places all of the necessary information in a block of data which is part of the call setup message sent to the

network in order to request the establishment of a connection. Such information includes originating number, dialed number, and type of call (e.g., voice, data, or video). En-bloc is used in ISDN PRI (Primary Rate Interface) implementations, as the device (e.g., PBX or router) is sophisticated enough to implement this approach. It also is used in BRI (Basic Rate Interface) implementations, assuming that an intelligent TA (Terminal Adapter) or ISDN router is involved. En-bloc signaling is highly efficient in comparison to "overlap sending," where each dialed digit is sent individually, as it is dialed. The most common example of en-bloc dialing is what we do when we dial on a cell phone. We punch out our digits. They stay in the phone until we hit "Send." The phone then sends our complete dial stream into the network. En-bloc dialing at its best. See also En-bloc signaling.

En-bloc Signaling Signaling in which address digits are transmitted in one or more blocks, each block containing sufficient address information to enable switching centers to carry out progressive onward routing. See en-bloc.

EN50-091 A European test standard for UPS system safety. Supercedes and is a superset of the IEC950 standard formerly used for UPS testing. In addition to the typical safety tests found in the IEC950 standard, this standard includes special sections on batteries and other safety concerns specific to UPS systems. UPS products are normally certified to this standard by VDE, TUV, SEMKO or other authorized certification body.

ENA Enterprise Network Accounting. "Enterprise Network Accounting is software that allows end users to collect call data from routers and generate communications management reports. ENA software tracks and allocates the costs of using a corporate network or the Internet, which allows network administrators to bill users for time spent on the network. ENA software also generates traffic statistics reports that show traffic patterns, potential misuse/abuse, and network inefficiencies. As voice traffic moves to the net, communications managers need tools to track and account for network usage. ENA represents the next phase in call accounting products. "Network World" coined the term concerning an announcement by Cisco Systems regarding a partnership with Telco Research, which is developing ENA products for Cisco Systems.

enable To make something happen. Or, in more complex language, to set various hardware and software parameters so that the central computer will recognize those parameters and start doing what you want.

enabler An "enabler" is a strange name for a piece of software.

enabling signal A signal that permits the occurrence of an event.

enbloc See En-bloc.

Encapsulated Postscript File EPS. A file that prints at the highest possible resolution for your printer. An EPS file may print faster than other graphical representations. Some Windows NT and non-Windows NT graphical applications can import EPS files.

encapsulating bridge A LAN/WAN term. A special bridge type usually associated with backbone/subnetwork architectures. Encapsulating bridges place forwarded packets in a backbone-specific envelope – FDDI, for example - and send them out onto the backbone LAN as broadcast packets. The receiving bridges remove the envelope, check the destination address and, if it is local, send the packet to the destination device. For a much longer explanation, see Bridge.

encapsulating security payload ESP The portion of the IPSec virtual private networking protocol which is used predominantly to provide data privacy.

encapsulation 1. Encasing a splice or closure in a protective material to make it watertight.

2. In object-oriented programming, the grouping of data and the code that manipulates it into a single entity or object. Encapsulation refers to the hiding of most of the details of the object. Both the attributes (data structure) and the methods (procedures) are hidden. Associated with the object is a set of operations that it can perform. These are not hidden. They constitute a well-defined interface – that aspect of the object that is externally visible. The point of encapsulation is to isolate the internal workings of the object so that, if they must be modified, those changes will also be isolated and not affect any part of the program. See Object-Oriented Programming.

3. Component lingo. Encapsulation is the isolation of a component's attributes and behaviors from surrounding structures. The technique protects components from outside interference and protects other components from relying on information that may change over time. Components are often encapsulated.

4. An electronic messaging term. The technique used by layered protocols in which a layer adds header information to the PDU (Protocol Data Unit) form the layer above. As an example, in Internet terminology, a packet would contain a header from the physical layer, followed by a header from the network layer (IP), followed by a header from the transport layer (TCP), followed by the application protocol data.

5. A networking term. It means carrying frames of one protocol as the data in another. Often the encapsulating protocol will be TCP/IP.

6. See also Encapsulation Bridging.

encapsulation bridging Method of bringing dissimilar networks where the entire frame from one network is simply enclosed in the header used by the link-layer protocol of the other network.

encipher Use of a cipher process to conceal some form of communicated intelligence.

enclosure Usually refers to a "headend enclose," relay equipment enclosure hut, or a small weather proof equipment box. In short, a place to put your precious equipment and protect it from the weather and from vandals.

enclosure reverberation A phenomenon of acoustics in which sound is reflected, or echoed, within an enclosure. This effect is particularly troublesome in automobile hands-free cell phone applications, as the coupling of the speakerphone, cell phone microphone, and the enclosed automobile chamber can degrade the quality of the transmitted signal.

encoding The process of converting data into code or analog voice into a digital signal. See also PCM and ADPCM.

encrippling Encrippling is the name of a technology which Hyperlock Technologies (www.hyperlock.com) has created which allows CD owners to unlock premium content stored on music compact discs. According to Hyperlock, instead of typical encryption approaches that wrap the equivalent of a digital security envelope around a complete piece of content, Hyperlock's system removes key pieces of data from content stored on the compact disc. The content can only be played them, by retrieving the missing data from a preselected Web site, e.g. the publisher of the compact disc.

encryption A fancy term for scrambling a message so that no one can read it except for the person for whom it's intended. In more formal terms, encryption is the transformation of data into a form unreadable by anyone without a secret decryption key. Its purpose is to ensure privacy by keeping the information hidden from anyone for whom it is not intended. In security, encryption is the ciphering of data by applying an algorithm to plain text to convert it to ciphertext. Symmetric encryption uses the same key to both encrypt and decrypt the message. Asymmetric encryption, also known as Public Key Encryption, equips each user with two keys – a private key and a public key, both of which are provided by a trusted third party known as a Certificate Authority (CA). The public key, which is known by everyone, is used to encrypt the message. The private key, which is known only to the intended recipient, is used to decrypt the message. Each public key and private key are linked in a manner such that only the public key can be used to encrypt messages and only the private key held by each individual recipient can be used to decrypt them. See also Public Key Encryption for a longer explanation.

Here is a definition courtesy, Alcatel: Encryption is the use of an algorithm to hide the meaning of a piece of information so that it cannot be read and understood. Encryption techniques have been around since the Roman Empire in the form of simple replacement techniques. One such technique is shown in the table below where the top row represents a character from the original message and the bottom row will be the replacement for the original character.

ABCDEFGHIJKLMNOPQRSTUVWXYZ
EFGHIJKLMNOPQRSTUVWXYZABCD

If we wanted to disguise the word trumpet we could use our table to replace the letter t with x, r with v and so on until trumpet becomes xvyqtix. Such a simple type of encryption is great for word games, but electronic commerce and sensitive data require a much more sophisticated encryption algorithm. Current strong types of encryption are mathematical in nature and fall into three general classes: symmetric, asymmetric, and cryptographic hash algorithms.

A symmetric algorithm uses a single key to encrypt and decrypt the data. The keys in an encryption algorithm are a binary number with a specific length measured in bits. For instance a four-bit key is a binary number with values from 0000 to 1111 totaling 16 possible combinations. Different encryption algorithms use keys with different lengths. An algorithm that uses keys with 128 bits and higher is typically considered to be strong enough for electronic commerce, because a key created with 128 bits has 2^{128} possible combinations, which is very difficult for an attacker to crack using brute force methods. There are several types of strong encryption that use symmetric algorithms. A few of the more popular ones are: Advanced Encryption Standard (AES), the International Data Encryption Algorithm (IDEA), and finally the Digital Encryption Standard (DES).

An asymmetric algorithm, which is used in a public-key infrastructure, requires two different yet related keys to encrypt and decrypt the data. One key is a private key and the other is a public key. The private key is never given out while the public key is readily available for use. Each key can decrypt information encrypted by the other key, but cannot decrypt information encrypted by itself. Therefore, the public key is given freely but the private key is needed to decrypt anything encrypted with the public key. When two people using a public-key infrastructure exchange public keys, it is very difficult for anyone to decrypt the transmission because of the length of the keys used and the fact that the private keys are always kept secret.

Hash functions are mathematical techniques that create a binary number (called a hash value) that has a fixed-length. This number cannot be run through the algorithm in reverse to figure out the original message. The hash value is used to authenticate a message in case it is intercepted and altered in transit.

How it works is the sender will compute the hash value of the original message and then encrypt both the message and the hash value and send it to the intended receiver. Once the message is received, it is decrypted and the hash value is recomputed using the sent message. If the hash value on the receiving end is the same as the one included with the message then the message is considered to be secure. The Secure Hash Algorithm and Pretty Good Privacy (PGP) are both examples of hash functions in use today..

Encryption key A unique, secret password, table or data block used to encrypt (i.e., encode) or decrypt (i.e., decode) data. See also Encryption and Public Key Encryption.

End Access End Office EAEO. An end office that provides Feature Group D.

End Delimiter ED. Sequence of bits used by IEEE 802 MAC to indicate the end of a frame. Used in token bus and ring networks, with nondata bits making ED easy to recognize.

end distortion In start-stop teletypewriter operations, the shifting of the end of all marking pulses except the stop pulse from their proper positions in relation to the beginning of the next pulse. Shifting of the end of the stop pulse would constitute a deviation in character time and rate rather than being an end distortion. Spacing end distortion is the termination of marking pulses before the proper time. Marking end distortion is the continuation of marking pulses past the proper time. Magnitude of the distortion is expressed in percent of a perfect unit pulse length.

end finish Surface condition at the optical fiber face.

end instrument A communication device that is connected to the terminals of a circuit.

end node A node such as a PC that can only send and receive information for its own use. It cannot route and forward information to another node.

end of data pattern A unique pattern of bits, matched against an identical pattern stored in a receiver, to signify the end of a message.

End Of File EOF. A control character or byte used in data communications that indicates the last character of the last record of a file has been read.

End of Medium EM. A control character used to denote the end of the used (or useful) portion of a storage medium.

End Of Message EOM. A control character used in data communications to indicate the end of a message.

end of shift routing A call center term for a process that calls won't be left in limbo when a shift ends. See also Source/Destination Routing, Skills-Based Routing and Calendar Routing.

End Of Text Message ETX. A control character used in data communications to indicate the end of a text message. See ETX.

end of transmission block A communications control character indicating the end of a block of Bisync data for communication purposes.

end of transmission block character A control character used in data communications to indicate the end of a block where data are divided into blocks for transmission purposes.

end office A central office to which a telephone subscriber is connected. Frequently referred to as a Class 5 office. The last central office before the subscriber's phone equipment. The central office which actually delivers dial tone to the subscriber. It establishes line to line, line to trunk, and trunk to line connections. See End Office Code.

end office code That part of a destination code consisting of the first three digits of a customer's seven digit directory number. It is usually expressed as an "NXX Code" where N represents digits 2 through 9 and X represents digits zero through 9.

end office conversion When an end office offers "equal access." See Carrier Identification Code and 101XXXX.

end point A network element (component) at the end of the network. In other words, a transmitter or receiver, or an originating or terminating device.

end-point security Security pertaining to users and devices logging onto the network. End-point security involves setting and enforcing criteria for trust and identity management.

end span See 802.3af.

end station An ATM term. These devices (e.g., hosts or PCs) enable the communication between ATM end stations and end stations on "legacy" LAN or among ATM end stations.

end system A host computer, in the context of the Internet.

end to end communications Data delivered between a source and destination endpoint.

end to end confidentiality The provision of data confidentiality between the sender and receiver of a communication.

end to end connection Connections between the source system and the destination system.

end to end loss The loss of an installed transmission path. The loss consists of the loss of the transmission cable or fiber, splices and connectors.

end to end service Service that enables the end user to pass information from one point to another. The Telephone Company provides access service to the carrier and local exchange service to the end user. The carrier provides inter-exchange transmission.

end to end signaling A signaling system capable of generating and transmitting signals directly from the originating station to the terminating end after the connection is established, without disturbing the connection. Touchtone dialing is such a system, allowing the user to send tones to a remote computer for data or other access. See Point To Point.

end to end testing Refers to the testing (with assistance from the telephone company) of a Common Carrier-provided facility and access services provided by the telephone company.

end user A highfalutin' term for a user. Any individual, association, corporation, government agency or entity other than an IXC that subscribes to interstate service provided by an Exchange Carrier and does not resell it to others. Telcordia's definition: A user who uses a loop-start, ground-start, or ISDN access signaling arrangement. In the past, "end user" meant the person placing or receiving the call. But the explosion of information technology has sparked robust disagreement over where a telecommunications transmission begins and ends.

end user access line The facility between the EO (End Office) and the Network Interface (NI) at the end user's premises. The end user access line includes certain non-traffic sensitive central office equipment, the outside plant facilities, the Network Channel Terminating Equipment (NCTE), when necessary, and the NI located on the end user's premises.

endpoint See End Point.

endpoint security In a layered approach to security, endpoint security enforces enterprise security policies on an end-point, for example, any network-connected device with which a user accesses the network. End-point security includes end-user authentication and device security compliance. The latter involves enforcement of policies regarding anti-virus software and virus definitions, personal firewalls, content filtering, desktop software patch levels, and other endpoint compliance requirements.

endurability The property of a system, subsystem, equipment, or process that enables it to continue to function within specified performance limits for an extended period of time, usually months, despite a potentially severe natural or man-made disturbance, e.g., nuclear attack, and a subsequent loss of external logistic or utility support.

Energy Communications EC. A PBX feature which communicates with energy consuming and monitoring devices and perform functions like dimming the lights or turning down the heat in a vacant hotel room. See also Energy Control.

energy control Indicates that phone system has software and hardware necessary to control and regulate the energy consuming devices in a user's facility (heating, ventilating, air conditioning, electrical machinery etc.). The system's processor transmits control signals, over existing telephone wiring where possible, to control units at each power-consuming device. This feature always includes user reconfiguration of the system's control parameters in response to operational and/or environmental changes. At one stage, AT&T and some other telephone equipment manufacturers sold energy control as a integral feature of their phone systems. The idea didn't take off for a lot of reasons.

energy density A beam's energy per unit area, expressed in joules per square meter. Equivalent to the radiometric term "irradiance."

energy star A U.S. government program that mandates strict limits on power consumption on electronic equipment, like computers and monitors, to the Federal Government. Products that comply often carry the symbol of a green star.

ENET 1. A silly way of saying Ethernet.
2. Enhanced Network.

ENFIA Exchange Network Facilities for Interstate Access. A tariff providing a series of options for connecting long distance carriers with local exchange facilities of the local telephone company.

Engineer Furnish and Install EF&I. A way to buy a product. If you buy a PBX (or anything else) the company will ask you if you want to buy the equipment and install it yourself, or get them to engineer, furnish and install it.

engineered capacity A telephone company term. The highest possible load level for a trunk group or a switching system at which service objectives are met. In general, for a switching-system, carried-load is equal to offered-load below engineered capacity, but is less than offered load above engineered capacity. Engineered capacity does not include equipment provided for maintenance or service protection.

Engineering Administration Data Acquisition System A telephone company term. EAD. The system is composed of traffic measuring and indicating devices, data converters, data accumulators, an EADAS central control unit (CCU), and a general purpose computer. The downstream general purpose computer provides data to the data management system which in turn provides the raw data, properly formatted and for the measurement intervals requested, to other downstream programs.

engineering judgment A telephone company definition. A term used by Network Engineering Managers and in various system publications to describe a behavior; expected of engineers when factual data and calculations are unavailable to justify engineering decisions.

Engineering Orderwire EOW. A communication path for voice or data, or both, that is provided to facilitate the installation, maintenance, restoral, or deactivation of segments of a communication system by equipment operators, attendants, and controllers.

engineering period A telephone company definition. Usually a one to four year period starting with the required service date of a new office or addition and concluding at the planned exhaust date of the switching equipment.

enhanced 800 services A name MCI uses for a family of 800 services with additional features added to them. It includes time of day and day of week routing.

enhanced 911 1. Landline Enhanced 911 is an advanced form of 911 service. With E-911, the telephone number of the caller is transmitted to the Public Safety Answering Point (PSAP) where it is cross-referenced with an address database to determine the caller's location. That information is then displayed on a video-monitor for the emergency dispatcher to direct public safety personnel responding to the emergency. This enables police, fire departments and ambulances to find callers who cannot orally provide their precise location.

Here's an E-911 example. An emergency call is placed from a subscriber's home; the ANI (calling number) accompanies the call through the network. The dialed digits "9-1-1" identify the call as emergency, which allows the telephone network to route the call to the specialty Tandem Switch, or Selective Router, at the E-911 Central Office. Here, the Tandem uses the ANI to look up the caller's ALI (address), and uses the ALI to derive the ESN (call routing number) from the Master Address Street Guide. The ESN determines the PSAP to which the call is delivered, still carrying the ANI. If desired, the PSAP can use the ANI to again derive the ALI (address), for example, to aid in dispatching emergency personnel.

2. Cell phone Enhanced 911. By December 31, 2005, your cell-phone company will always know exactly where you are. That's when the FCC will complete Phase II of its Enhanced 911 (E911) program, requiring all U.S. wireless carriers to provide the location --within about 165 to 330 feet in most cases – of anyone dialing 911 from a cell phone.

enhanced call processing An Octel term for the interactive voice response option in its voice mail system. Here's how Octel defines the term: "Companies and departments that receive a heavy volume of calls can use ECP to create menus that are presented to callers. When the system answers a call, a recorded voice instructs the caller how to use a touch-tone telephone to send call routing instructions to the system. Depending on which option is chosen, ECP's customized call routing feature allows a caller to press a single key to reach a predetermined extension, a voice messaging mailbox where he can leave a message, an Information Center Mailbox where he can listen to a series of recordings giving frequently requested information or additional levels of ECP menus. ECP menus are easily custom-built by the customer to meet its specific needs. Each menu can offer as many as ten options."

enhanced call routing An AIN (Advanced Intelligent Network) service which is an enhancement to 800 / 888 services. The calling party is voice prompted through a set of menu options which serve to define the specifics of the request and the particular needs of the caller. Based on that input, the caller is directed to the most appropriate incoming call center and agent. By way of example, language preference might be a cause for changing call routing.

enhanced dialing Features allow for speed dialing, preview dialing, and manual dialing from a host or workstation application.

Enhanced DNIS Enhanced DNIS is a combination of ANI and DNIS delivered before the first ring on a T-1 span. The number of digits delivered is configurable on a per span basis.

Enhanced IDE An improved interface to the IDE hard disk interface. Enhanced IDE allows you to attach hard disks of larger than 528 megabytes (the largest normal IDE will handle) up to a maximum of 8.4 gigabytes. Enhanced IDE has a data transfer rate of between 11 and 13 megabytes per second, compared to the 2 to 3 megabytes per second, which normal IDE drives sport. See IDE.

Enhanced Parallel Port EPP. A new hardware and software innovation (and now a standard) which allows computers so equipped to send data out their parallel port at twice the speed of present parallel ports. There's no difference in the shape of the plug or the number of conductor. See EPP for a fuller explanation.

Enhanced Private Switched Communications Service EPSCS (pronounced EP-SIS). A private line networking offering from AT&T which provides functions similar to CCSA. Big companies are its customers.

Enhanced Serial Interface ESI. Now totally obsolete. It was a broader serial interface announced by Hayes Microcomputer Products, Norcross, GA, and placed in the public domain. The ESI is an extension of the familiar COM card used in personal computers. ESI includes the definition of I/O, control registers, buffer control, Direct Memory Access (DMA) to the system and interaction with attached modem devices. ESI specification is available from Hayes Customer Service at no charge. Combined with Hayes' announcement of ESI was their announcement of new Enhanced Serial Port hardware products for the IBM microchannel and IBM XT/AT or EISA bus personal computers. According to Hayes, the ESI spec and the supporting ESP hardware provide a "cost-effective" communication coprocessor to manager the flow of data between an external high speed modem and PC. This technology prevents loss of data resulting from buffer overflow errors and provides maximum data throughput for high speed modems. Hayes said that the combination of ESP and ESI would allow through-the-phone modem speeds of up to 38.4 Kbps.

enhanced serial port See Enhanced Serial Interface.

Enhanced Service Provider ESP. An ESP is a company that provides enhanced or value-added services to end users. An ESP typically adds value to telephone lines using his own software and hardware. Also called an IP, or Information Provider. An example of an ESP is a public voice mail box provider or a database provider, for example, one giving the latest airline fares. An ESP is an American term, unknown in Europe, where they're most called VANs, or Value Added Networks. See also Open Network Architecture and Information Provider.

enhanced services Services offered over transmission facilities which may be provided without filing a tariff. These services usually involve some computer related feature such as formatting data or restructuring the information. Most Bell operating companies (BOCs) are prohibited from offering enhanced services at present. But the restrictions are disappearing.

The FCC defines enhanced services as "services offered over common carrier transmission facilities used in interstate communications, which employ computer processing applications that act on the format, content, code, protocol or similar aspects of the subscriber's transmitted information; provide the subscriber additional, different or restructured information; or involve subscriber interaction with stored information." In other words, an enhanced service is a computer processing application that messes in some way with the information transmitted over the phone lines. Value-Added Networks, Transaction Services, Videotex, Alarm Monitoring and Telemetry, Voice Mail Services and E-Mail are all examples of enhanced services.

enhanced small device interface An interface which improves the rate of data transfer for hard disk drives and increases the drive's storage capacity.

Enhanced Unshielded Twisted Pair EUTP. UTP (Unshielded Twisted Pair) cables that have enhanced transmission characteristics. Cables that fall under this classification include Category 4 and above.

Enhanced Variable Rate Vocoder See EVRC.

ENI Embedded Network Interface. An ENI might be in the form of an applications program that includes network controller logic for a wide range of printers, for example. Or it might be in the form of network interface logic such as PPP and TCP/IP embedded in a chip that is part of a modem card that fits inside a laptop.

ENIAC Electronic Numerical Integrator and Computor (spelled with an O). Early computer, built in 1944.

ENOB See effective number of bits.

ENOS Enterprise Network Operating Systems. A Sun Microsystems term. Part of Sun's Networking Solutions, ENOS provides the foundation for Sun's networking environment. ENOS combines NFS (Network File System) and the TCP/IP protocol suite into its WebNFS. NFS is a Sun system that has become the de facto standard for global file sharing.

ENQ ENQuiry character. A control character (Control E in ASCII) used as a request to obtain identification or status. Abbreviation for enquiry. The binary code is 00000101 and the hex is 05.

ENQ/ACK Protocol Hewlett-Packard communications protocol in which the HP3000 computer follows each transmission block with ENQ to determine if the destination terminal is ready to receive more data. The terminal indicates its readiness by responding with ACK.

enquire See Berners-Lee.

enriched services providers Those third-party service providers (other than Network Providers) who provide value-added services that are accessed through telecommunications networks.

ENS Emergency Number Services.

ENSO ETSI National Standardization Organizations (ETSI).

ENTELEC ENergy TELECommunications and electrical association, the oldest nationwide user group in telecommunications. It is an association of communications managers and engineers in the oil, gas, pipeline and utility industries. ENTELEC played an important role in the early opening of competition in the telecommunications industry, including the famous "Above 890" decision, which allowed private companies to build their own long distance microwave system. The decision was called "Above 890" because electromagnetic waves in the radio frequency spectrum above 890 Megahertz (million cycles per second) and below 20 Gigahertz (billion cycles per second) are typically called microwave. Microwave used to be a common method of transmitting telephone conversations and was used by common carriers as well as by private networks. Now fiber is far more common. Microwave signals only travel in straight lines. In terrestrial microwave systems, a single transmission is typically good for 30 miles, at which point you need another repeater tower. Microwave is the frequency for communicating to and from satellites. ENTELEC was formerly known as the Petroleum Industry Electrical Association.

entering distribution A call center term. In this mode of the alerting state, a call is being presented to an ACD group or hunt group in preparation for distribution to a device associated with that group. This mode is indicated by a Delivered event with a cause code of Entering Distribution.

enterprise Enterprise means the whole corporation. It tends to refer to corporations with more than one location. See Enterprise Computing.

enterprise calendaring See iCalendar.

enterprise computing Enterprise means the whole corporation. Enterprise computing refers to the computing applications on which a company's life depends: order entry, accounts receivable, payroll, inventory, etc. It is also known by the phrase "mission critical." See also Enterprise Network.

enterprise network The word Enterprise was invented by IBM. It means the whole corporation. An enterprise-wide network is one covering the whole corporation. Local PBXs. Local area networks. Internetworking bridges. Wide area networks, etc, etc. See also Corporate Network and Enterprise Computing.

enterprise number A service provided by AT&T and the Bell operating companies (a.k.a. the Bell System) years ago which allowed people to make collect calls and have their calls automatically accepted by the company at the other end. It was very expensive. It has largely been replaced with 800 IN-WATS service, which is much more successful.

enterprise peering The direct interconnection of enterprise networks for the twin purposes of exchanging voice and data traffic and doing so without touching the PSTN, thereby avoiding carrier charges.

Enterprise Resource Management ERM. Also known as Enterprise Resource Planning (ERP). See ERP for a full definition.

Enterprise Resource Planning ERP. A concept developed by The Gartner Group to describe the next generation of manufacturing business systems and MRP (Materials Resource Planning) software. See ERP for a full definition.

Enterprise RMON A proprietary extension of RMON and RMON-2, Enterprise RMON was developed by NetScout Systems (formerly Frontier Software Development) and is supported by several other vendors, including Cisco Systems. Enterprise RMON's extensions monitor FDDI and switched LANS. See RMON, RMON-2.

enterprise server 1. A Sun Microsystems term, Part of Solaris' Server Suite. Used to develop and deploy mission critical applications on large server systems. Provides distributed computing. Comes with Solstice DiskSuite and Networker products for on-line backup and recovery.

2. A waiter on Star Trek. – from comedian Don McMillan.

enterprise solution Software that enables individuals and groups (either within an organization or part of a virtual organization beyond one company) to use computers in a networked environment to access information from a wide range of sources, collaborate on projects, and communicate easily with text, graphics, video, or sound.

entertainment bypass The providing of television entertainment and other video entertainment content directly to consumers over the Internet by content companies, rather than using intermediaries such as cable TV, local broadcast TV, and telco TV.

entity 1. An active element within an OSI layer or sublayer.

2. A telephone company definition. A group of lines served by common originating equipment.

entity coordination management The portion of connection management which controls bypass relays and signals connection management that the medium is available.

entity nongrowth A telephone company term. Also referred to in some areas as 'capped' or 'floating.' The term non-growth entity will be used to identify those entities where we do not intend to add capacity. However, we must always insure that these entities continue to provide objective levels of service.

entrance and exit ramps The companies who control access to the internet and other networks of the information superhighway, whatever that is.

entrance bridge A terminal strip that is an optional component in a network interface device and is provided for the connection of ADO cable.

entrance facility EF. An entrance to a building for both public and private network service cables (including antennas) including the entrance point at the building wall and continuing to the entrance room or space. Entrance facilities are often used to house electrical protection equipment and connecting hardware for the transition between outdoor and indoor cable. The Entrance Facility includes overvoltage protection and connecting hardware for the transition between outdoor and indoor cable.

entrance point/telecommunications The point of emergence of tele-communications conductors through an exterior wall, a concrete floor slab, or from a rigid metal conduit or intermediate metal conduit.

entrance room/telecommunications A space in which the joining of inter or intra building telecommunications backbone facilities takes place.

entrapment The deliberate planting of apparent flaws in a system for the purpose of detecting attempted penetrations.

entrenched transactors Banking industry jargon for people who refuse to use cost-saving ATMs, preferring to deal only with more expensive human bank tellers.

entropic coding Entropic coding is lossless compression. It exploits information-theoretic redundancy in the signal. This redundancy originates in the fact that not all sequences of bits are equally likely in the PCM data; some sequences occur more often than others. Entropic coding uses a lossless compression scheme, such as Lempel-Ziv or Huffman coding, to use fewer bits to represent the sequences that occur most often, and more bits to represent the sequences that occur less often. In this way, the average sequence length may be compressed by a factor which depends on the classical information content of the bitstring. Entropic coding is used in MP4 audio.

entropy coding A category of compression and coding algorithms that preserves all source information (i.e. it's lossless) so that it can be reconstructed with no loss of information. Compression is achieved by a more efficient coding which reduces the entropy (or "disorder", roughly). Entropy encoding is most successful in images where there is a great deal of redundancy (e.g. solid background, text foreground). See Entropic Coding, which is the preferred way of spelling.

entry border node An ATM term. The node which receives a call over an outside link. This is the first node within a peer group to see this call.

ENUM ENUM, in short, is a proposal to map all phone numbers to IP addresses. ENUM isn't an acronym, but it could have been short for Electronic NUMber, or something of the sort. ENUM is a proposed standard (RFC 2916) from the IETF (Internet Engineering Task Force) for a DNS-based (Domain Name Server) method for mapping telephone numbers to URLs (Uniform Resource Locators, i.e., Web addresses) and, ultimately, to IP addresses. The format for telephone numbers is specified in the ITU-T E.164 standard, and the formats for URLs, IPv4 and IPv6 are standardized by the IETF. So, the translation between the two is relatively straightforward. Missing is a standardized method, set of protocols, database responsibility, and various other specifics for doing so. ENUM addresses those issues through the DNS-based approach for number registration, directory mapping, translation processes, and various other specifics that must be resolved for the concept to work on a global and seamless basis. A global domain, e164.arpa, has been set aside by the IAB (Internet Architecture Board) for the system. Here's how ENUM is proposed to work:

You dial a telephone number for a company connected to the Internet. The telephone number is reversed, with the last digit becoming the first, and the first digit becoming the last. This reversal is necessary for the telephone number order to match the logic used in URLs, where the TLD (Top Level Domain) – ".com" or ".us," for example – must be considered first in order to drive the remainder of the address search to the proper DNS administered under the proper registration authority. Traditional telephone numbers, of course, take the reverse approach, with the access code (e.g., 011 for international long distance, or 1 for domestic long distance) coming first, followed by the country code, the area code, the CO prefix, and the line number. This numbering scheme must be reversed in order for the E.164 number to be resolved by the DNS and translated into a URL, which then is translated into an IP address. Now, if the number dialed is not in the ENUM DNS database, the call is connected over the PSTN. If, however, it is in the database, the available services are identified, and the call is completed over the Internet or other IP-based network. As a result, ENUM has the potential to become one of the basic underpinnings of a convergence between the PSTN and the Internet.

ENUM proposes to employ the E.164 telephone number as a global identifier that can be used to direct a message to any device or application connected to the Internet, assuming that the device or application has registered the availability of that particular service. Any necessary conversions in protocols would have to be made, of course, either by the devices or through gateways. For example, a fax machine could send a message directly to an e-mail address, and an e-mail could be sent directly to an IP-enabled fax machine. Similarly, instant messaging and unified messaging could be supported across device types and networks. Voice over IP (VoIP) could be supported more easily. Voice mail over the Internet would be enabled through the use of the existing VPIM (Voice Profile for Internet Mail) standard.

enumerator A Windows term. A Plug and Play device driver that detects devices below its own device node, creates unique device IDs, and reports to Configuration Manager during startup. For example, a SCSI adapter provides a SCSI enumerator that detects devices on the SCSI bus.

envelope 1. In mathematics, the outer boundary of a family of curves obtained by varying a parameter of a wave has been known since mid-19th century as the envelope. This is the strict mathematical definition. From this we have the expression "pushing the envelope," which today means pushing something's performance past what it was designed for. Actually this expression should really be "pushing the outside of the envelope." Aeronautical engineers have since applied this word to the limits of aircraft operation. When a test pilot presses against those outer limits (of speed, gust, maneuver and flight) he pushes the outside of the envelope. Tom Wolfe popularized the phrase in his 1979 book "The Right Stuff" about astronauts.

2. The part of messaging that varies in composition from one transmittal step to another. It identifies the message originator and potential recipients, documents its past, directs its subsequent movement by the MTS (Message Transfer System) and characterizes its content.

3. The first envelopes with gummed flaps were produced in 1844 in Britain. They were not immediately popular because it was thought to be a serious insult to send a person's saliva to someone else. A duel was fought because the person receiving the letter suspected that the sender had sealed the envelope with his tongue.

envelope capacity There are two common usages of this term:

1. The number of bytes the payload envelope of a single frame can carry. The SONET STS payload envelope is the 783 bytes of the STS-1 frame available to carry a signal. Each virtual tributary (VT) has an envelope capacity defined as the number of bytes in the VT less the bytes used by VT overhead.

2. The bandwidth allocated within each SONET STS-1 channel to carry information end-to-end. Also known as information payload. 50.112 Mb/s.

envelope delay The difference, expressed in time, for signals of different frequencies to pass through a phone line. Some frequencies travel slower than others in a given transmission medium and therefore arrive at the destination at different times. Delay distortion is measured in microseconds of delay relative to the delay at 1700 Hz. Also called

Delay Distortion.

envelope delay distortion The distortion that results when the rate of change of phase shift with frequency over the bandwidth of interest is not constant. It is usually stated as one-half the difference between the delays of the two frequency extremes of the band of interest. See Envelope Delay.

envelope distortion Distortion of the transmitted signal which results from the different transmission speed characteristics of different frequency components to the signal. Mathematically it is the derivative of the phase shift with respect to frequency.

environment The place your telephone system's main cabinet and main electronics live. While most PBX vendors will specify the room's characteristics, the ultimate responsibility for the room is yours, the user. Not designing your telephone system's environment correctly is tantamount to jinxing your telephone system from the start.

Here are some things to watch out for (your vendor has a more comprehensive list): 1. Sufficient air conditioning? Telephone systems give off heat. You need some way of getting rid of the heat. If you don't, you will blow some of your phone system's delicate electronic circuitry. 2. Sufficient space? Is there room for technicians to get in and around your telephone system so they can repair it? Will you have room for additional cabinets when you need to grow your phone system? 3. Sufficient and correct power? Will you have sufficient clean commercial AC power? Will you require isolation regulators? Or you will require extensive wet cell batteries? Will you have space? 4. Will you have a solid electrical ground? Can you find somewhere solid to ground your telephone system to – other than the third wire on the AC power, which is not suitable for most telephone systems? Beware of cold water pipes which end in PVC plastic pipes.

environment variable Originally a UNIX term, now also used in Windows. It means a variable that is set in the shell in such a way that it is available to all child processes (programs, other shells, etc.) of the shell. (It's what you get when you "export" a variable.) Windows defines it as a string consisting of environment information, such as a drive, path, or filename, associated with a symbolic name that can be used by Windows. You use the System option in Control Panel or the set command from the Windows NT command prompt to define environment variables.

envoy 1. A palmtop communicator introduced by Motorola in March of 1994. The device lets its users receive and transmit messages via Ardis, a network owned by Motorola and IBM. Envoy contains software from General Magic.

2. Spectrum Envoy is a DSP-based PC-board used for "telephone management" from a company called Spectrum Signal Processing, Burnaby, BC. Telephone management includes voice mail, contact manager, upgradable fax/modem, business audio, etc.

EO 1. End Office. Typically your own telephone company central office – the one that gives you dial tone and through which you make your local and long distance phone calls.

2. Erasable Optical drive. EO drives act like hard drives yet offer virtually unlimited storage because their cartridges are removable. Each cartridge sports at least 650 MB. Some sport 1 gigabyte.

3. EO was a startup in Mountain View, CA which did wireless data. It made a device called EO Personal Communicator 440 and 880. It uses GO's PenPoint operating system and the Hobbit microprocessor made by AT&T, which is "optimized" for telecommunications. In fall of 1994, AT&T closed EO down and stopped the sale of EO devices. It was too expensive and wasn't selling. An excellent book was written about EO. It is called "Startup; A Silicon Valley Adventure Story." It was written by Jerry Kaplan, one of EO's founders. The book is published by Houghton Mifflin.

EOA End Of Address. A header code.

EOB End Of Block. A control character or code that marks the end of a block of data.

EOC Embedded Operations Channel. An operations channel for purposes of network management purposes (e.g., circuit monitoring and testing) which is embedded in a communications protocol. An EOC is a dedicated channel for such purposes, ensuring that network management functions can always be accomplished on a non-intrusive basis. In other words, the management of the performance characteristics of the circuit will not intrude on the ability of that circuit to support the transmission of the data that supports the end user's applications. For example, ISDN BRI (Basic Rate Interface) provides for two B (Bearer) channels of 64 Kbps each, and a D (Delta, or Data) channel of 16 Kbps, for a total of 144 Kbps. The 16-Kbps D channel always is available, on a priority basis, for network management (i.e., signaling and control) purposes, and without affecting the circuit's ability to support the end user's applications running over the two B channels. See also BRI.

EOD 1. End Of Day, a favorite Microsoft expression.

2. See everything on demand.

EOE See Electronic Order Exchange.

EOF The abbreviation for End Of File. MS-DOS files and some programs often mark the end of their files with a Ctrl Z – or ASCII 26.

EOM End of Message (indicator). In ATM network, EOM is an indicator used in the AAL that identifies the last ATM cell containing information from a data packet that has been segmented.

EOP End of Procedure frame. A frame indicating that the sender wants to end the call.

EOT End of Transmission, End of Tape.

EOTC European Organization for Testing and Certification.

EOW Engineered OrderWire.

EP BRAN See HiperLAN.

EPA Energy Star Monitors that comply with this standard consume less electricity by powering down when not in use.

EPABX Electronic Private Automatic Branch eXchange. A fancy name for a modern PBX. Other fancy names include CBX, Computerized Branch Exchange.

EPC Electronic Product Code. See also RFID.

EPD Early Packet Discard. A technique used in ATM networks for congestion control in support of both Classical IP over ATM and Local Area Network Emulation (LANE). Such data is transmitted in the form of packets and frames, respectively, each of which typically is a subset of a much larger set of data such as a file. In the case of Classical IP over ATM, each data packet can be variable in size, up to a maximum of 65,536 octets (e.g., bytes). As the IP data packet enters the ATM switch on the ingress side of the ATM network, it is stored in a buffer until such time as the ATM switch can segment it into cells, each with a payload of 48 octets – there can be a great many such cells for each packet – and act to set up a path and circuit to forward the stream of cells which comprise the original packet. If a given cell is dropped for some reason (e.g., there is not enough buffer space at either the incoming or the outgoing buffer within the switch, the integrity of the original packet is lost through this phenomenon known as "packet shredding." Early implementations of Classical IP over ATM simply forwarded the remainder of the cells associated with that packet. So, some cells made it to the ATM switch at the egress edge of the network, and some cells didn't. When the cells were reassembled into the packet as they exited the ATM network, the result was an incomplete packet. The higher layer protocols then requested a retransmission of the entire packet. If the ATM network was highly congested, this occurrence was repeated many times, thereby contributing to further congestion. Partial Packet Discard (PPD) involves numbering each cell associated with a segmented packet as it enters the ATM domain through the inbound buffer of the ingress switch. If any cell is dropped, the entire stream of cells associated with the packet is dropped. PPP enhances the performance of the ATM network by dropping those cells, which serve no purpose as the entire packet will be transmitted in either case. PPP is an earlier, and less sophisticated, technique that largely has been replaced by Early Packet Discard (EPD), which acts to discard the entire cell stream associated with that packet if there is not enough buffer space at either the incoming or the outgoing buffer within the switch, with that determination being a function of a programmable threshold. Discarded packets are detected as missing by the higher layer protocols, and retransmissions are requested. See also ATM, Classical IP over ATM, LANE, and PPD.

ephemeral key A public key or a private key that is relatively short-lived.

ephemeris The predictions of current satellite position that are transmitted to the user in the data message of a GPS (Global Positioning System) satellite message.

EPIRB Emergency Position-Indicating Radio Beacon carried by ships at sea to provide an alert in a disaster situation. EIPRBs transmit at 406 MHz, the COSPAS-SARSAT frequency, or at 1.6 GHz, the Inmarsat frequency. They are designed to transmit for at least 48 hours and may be fully automatic and float free in case a ship sinks, or may be manually activated, as those located in or close to survival craft. EIPRBs are included in GMDSS. See COSPAS-SARSAT.

epitaxy Actually, it's molecular beam epitaxy. A fabrication process for growing silicon wafers of exceptional quality. The process involves heating an element, or compound, in an effusion oven to a temperature sufficient to release some of the atoms. (It's not as extreme as vaporization, but the idea is much the same.) Some of the atoms, or molecules, are drawn in a linear beam into an intense vacuum chamber, where they are deposited on a substrate (i.e., foundation) silicon wafer, one atomic, or molecular, layer at a time. The yield is a wafer comprising films that can be measured in atomic, or molecular, levels of thickness, with each film being identical in structure to the substrate wafer. Molecular beam epitaxy was perfected by A.Y. Cho of Bell Telephone Laboratories. The fabrication process was invented in 1960 by Messrs. Kleimack, Load, Ross, and Theurer of Bell Labs as the demand developed for layered semiconductors and semi-insulators of precise film thickness. Epitaxy has made possible the manufacture of high-speed transistors packed by the millions on silicon chips. It also is used in the manufacture of optoelectronics and high-speed magnetic storage devices.

EPLANS Engineering, PLanning and ANalysis Systems. Software offered by Western Electric (now called AT&T Technologies) to help operating telephone company people run their business better.

EPO Enterprise Profit Optimization. This is a new acronym. Companies previously doing EAI, CRM and ERM are now doing EPO. There's no reason to figure out what EPO means. By the time you do, they'll have another acronym. And you won't be able to figure out what that one does, either.

EPOC EPOC is an operating system developed by Psion and now owned by Symbian, the joint venture between Psion, Nokia, Ericsson, Motorola, and Panasonic. It is designed for small, portable computer-telephones with wireless access to the Internet and other information services. EPOC is an alternative to Microsoft's Windows CE for smartphones, PDAs, etc.

EPON Ethernet Passive Optical Network. A PON running the Ethernet protocol over single mode fiber in the local loop. EPON standards were established by the IEEE as 802.3ah and incorporate fiber specifications 1000Base-PX10 for distances up to 10km and 1000Base-PX20 for distances up to 20km. EPON runs at 1.25 Gbps symmetrical, i.e., the same speed upstream and downstream. EPONs typically support a 16:1 split ratio, although 32:1 is permitted. See also APON, BPON, GPON and PON.

epoxy A liquid material that solidifies upon heat curing, ultraviolet light curing, or mixing with another material. Epoxy is sometimes used for fastening fibers to other fibers or for fastening fibers to joining hardware.

EPN Expansion Port Network, which contains line and trunk ports of proprietary Avaya systems.

EPP Enhanced Parallel Port. A new hardware and software innovation (and now a standard) which allows computers so equipped to send data out their parallel port at twice the speed of older parallel ports, i.e. those that came on the original IBM PC. The EPP conforms to the EPP standard developed by the IEEE (Institute of Electrical and Electronics Engineers) 1284 standards committee. The EPP specification transforms a parallel port into an expansion bus that theoretically can handle up to 64 disk drives, tape drives, CD-ROM drives, and other mass-storage devices. EPPs are rapidly gaining acceptance as inexpensive means to connect portable drives to notebook computers. There's no difference in the shape of the ordinary, 25-pin D-connector plug/connector or the number of conductors. The Enhanced Parallel Port (EPP) was developed by Intel Corp., Xircom Inc., Zenith, and other companies that planned to exploit two-way communications to external devices. Many laptops built since mid-1991 have EPP ports. See also ECP.

EPROM Erasable Programmable Read Only Memory. A read only memory device which can be erased and reprogrammed. Typically, it is programmed electronically, but it is erased electromagnetically with ultraviolet light. EPROMS are typically returned to the vendor or factory for reprogramming. An EPROM on a graphics card might contain the default or ROM character set. EPROM chips normally contain UV-permeable quartz windows exposing the chips' internals. See also ROM and EEPROM.

EPS Encapsulated PostScript. An extension of the PostScript graphics file format developed by Adobe Systems. EPS lets PostScript graphics files be incorporated into other documents. FrontPage supports importing EPS files.

EPSCS (Pronounced Ep-Sis.) Enhanced Private Switched Communications Service. An AT&T offering for large businesses with offices scattered all over the country. This service allows such businesses to rent space on AT&T electronic switches and join that switching capacity to leased lines. EPSCS customers get a network control center in their offices which gives them information on the continuing operation of their network and allows them some limited options for changing their services.

EPSN Enhanced Private Switched Network.

EQ See Equalization, Equalizer.

equal access All long distance carriers must be accessible by dialing 1 – and not a string of long dialing codes. This is laid down in Judge Green's Modified Final Judgment (MFJ), which spelled out the terms of the Divestiture of the Bell Operating phone Companies (BOCs) from their parent, AT&T. Under the terms of this Divestiture, all long distance common carriers must have Equal Access for their long distance caller customers. City by city telephone subscribers are being asked to choose their primary carrier who they will reach by dialing 1 before their long distance number. All other carriers (including AT&T, if not chosen as primary) can be reached by dialing a five digit code (10XXX), thus providing Equal Access for all carriers. Not all long distance companies will opt for full equal access since this involves considerable expense to the local phone companies. See also Feature

Group A, B, C and D.

equal access end office A central office capable of providing equal access. See also EQUAL ACCESS.

Equal Cost Multipath Routing See ECMP.

equal gain combiner A diversity combiner in which the signals on each channel are added together. The channel gains are all equal and can be made to vary equally so that the resultant signal is approximately constant.

Equal Level Far End Cross Talk ELFEXT. A Calculation of the FEXT between two pairs corrected for length. This calculation is made taking the measured FEXT (Far End Cross Talk) of a cable or system and subtracting the attenuation of the cable or system. Measured in units of dB. The higher the magnitude of the ELFEXT the better.

equality Always try to keep the number of landings you make equal to the number of take offs you make. – US Air Force Training Manual.

equalizer A device inserted in a transmission line or amplifier circuit to improve its frequency response. An equalizer adds loss or delay to specific frequencies to produce a flat frequency response. The signal may then be amplified to restore its original form.

equalization The process of reducing distortion over transmission paths by putting in compensating devices. The telephone network is equalized by the spacing and operation of amplifiers along the way. In recording, equalization is frequency manipulation to meet the requirements of recording; also the inverse manipulation in playback to achieve uniform or "flat" response. Also called Compensation. See equalizer and equalization circuit.

equalization circuit A compensation circuit designed into modems to counteract certain distortions introduced by the telephone channel. Two types are used: fixed (compromise) equalizers and those that adapt to channel conditions. U.S. Robotics high speed modems use adaptive equalization.

equalizing network A device which is connected to a transmission path to alter the characteristics of that path in a specified way. It is often used to equalize the frequency response characteristics of a circuit for data transmission.

equatorial coordinate system One of the four common celestial coordinate systems and the one used in expressing the motions of satellites. It resembles the Earth's coordinate system, with the same origin and same positive z axis, respectively at the centre of the Earth and from the centre through the geographic North Pole. However, the positive x axis is not at the Greenwich meridian, which rotates with the Earth, but is at the first point of Aries (the vernal equinox). Angles measured from the positive x axis, in the west-to-east direction of rotation of the Earth are termed right ascension and are equivalent to longitude in the Earth's coordinate system. Angles measured from the equator of the celestial sphere are termed declination and are exactly equivalent to latitude in the Earth's coordinate system. The system is also known as the inertial coordinate system, as within it Newton's laws hold.

equatorial orbit An orbit with a zero degree inclination angle, i.e. the orbital plane and the Earths' equatorial plane are coincident.

equatorial plane A geometrical plane passing through the earth at its equator. By definition, the orbit of a geostationary satellite is in the equatorial plane.

Equilibrium Mode Distribution EMD. The condition in a multimode optical fiber in which the relative power distribution among the propagation modes is independent of length.

equipment cabinet The metal box which houses relays, circuit boards or other phone apparatus. Usually also contains the power supply, which converts the 120 volt AC current into the low voltage direct current necessary to run the telephone system.

equipment cable A cable or cable assembly used to connect telecommunications equipment to horizontal or backbone cabling systems in the telecommunications closet and equipment room. Equipment cables are considered to be outside the scope of cabling standards.

equipment compatibility One computer system will successfully do the same thing that another computer will do with the same data. There are many levels of "equipment compatibility." The only true compatibility, however, is identical machinery. And identical means "identical" down to the very last chip and very last integrated circuit. We have found that some computers – even those consecutively numbered – do not always perform the same. We have empirically proven this for both IBM and AT&T computers.

Equipment Identity Register See EIR.

Equipment Room ER. A centralized space for telecommunications equipment that serves the occupants of the building or multiple buildings in a campus environment. An equipment room is considered distinct from a telecommunications closet because it is considered to be a building or campus serving (as opposed to floor serving) facility and because of the nature or complexity of the equipment that it contains.

equipment wiring subsystem The cable and distribution components in an equipment room that interconnect system-common equipment, other associated equipment, and cross connects.

equipped for capacity The maximum number of lines and trunks that can be supported by the available hardware. It is not a totally effective measure of the size of a PBX. See Wired-For-Capacity.

equivalent four-wire system Transmission using frequency division to get full duplex transmission over only one pair of wires. Normally two pairs are needed for full duplex.

equivalent network 1. A network that may replace another network without altering the performance of that portion of the system external to the network.

2. A theoretical representation of an actual network.

equivalent PCM noise Through comparative tests, the amount of thermal noise power on an FDM or wire channel necessary to approximate the same judgment of speech quality created by quantizing noise in a PCM channel.

ER 1. Explicit Rate. The current mechanism for flow control in ATM networks. ATM RM (Resource Management) cells are circulated by the transmitting device, indicating both the current and the desired rates of transmission. Assuming that the receiving device is able to accommodate that desired rate without overflowing its buffers, the request is granted and is honored by all intermediate switches in the network.

2. See Equipment Room.

Erasable Programmable Read-Only Memory See EPROM.

erasable storage A storage device whose contents can be changed, i.e. random access memory, or RAM. Compare with read-only storage.

erase head On a magnetic tape recorder – voice or video – this is the "head" which erases the tape by demagnetizing it immediately before a new recording is placed on the tape by the adjacent record head.

erbium A rare earth element that when added to fiber optic cabling could obviate the need for repeaters every 20 miles on undersea cables and expand fiber optic cabling to capacities of trillions of bits a second. See Erbium-Doped Fiber Amplifier.

Erbium-Doped Fiber Amplifier See EDFA.

ERC Easily Recognizable Code. ERCs are U.S. area codes with the same number in the second and third positions. Examples include toll free numbers (e.g., 800, 888, 877 and 866).

erector set telecom In North America, there's a children's game of building blocks called Lego. The game comes with hundreds of small plastic blocks, which can be assembled into all sorts of wonderful designs, from castles to gas stations. In England, Lego sets are also called Mecano sets. The generic term for Lego and Mecano sets is erector sets. The term "erector set telecom" is a concept created by Harry Newton as a way of explaining "the new open" telecommunications equipment, namely that you build your own computer telephony system from freely-available, non-proprietary hardware and software components. In short, a telecom industry along the same open hardware and software lines as the PC industry. History: A.C. Gilbert introduced the Erector Set in 1913 in the U.S. He went on to win 150 patents for a variety of inventions. His company, the A.C. Gilbert Company, created chemistry sets, microscope sets, and magic sets. Legos (from Denmark) came to the U.S. much later. The international LEGO Group was established in 1932 and is now one of the world's largest toy manufacturers, employing about 10,000 people in 50 companies in 30 countries. See Computer Telephony.

ergonomics The science of determining proper relations between mechanical and computerized devices and personal comfort and convenience; e.g., how a telephone handset should be shaped, how a keyboard should be laid out.

Erlang 1. A measurement of telephone traffic. One Erlang is equal to one full hour of use (e.g. conversation), or 60 x 60 = 3,600 seconds of phone conversation. You convert CCS (hundred call seconds) into Erlangs by multiplying by 100 and then dividing by 3,600 (i.e. dividing by 36). Numerically, traffic on a trunk group, when measured in erlangs, is equal to the average number of trunks in use during the hour in question. Thus, if a group of trunks carries 12.35 erlangs during an hour, a little more than 12 trunks were busy, on the average.

Erlang gets its name from the father of queuing theory, A. K. Erlang, a Danish telephone engineer, who, in 1908, began to study congestion in the telephone service of the Copenhagen Telephone Company. A few years later he arrived at a mathematical approach to assist in designing the size of telephone switches. Central to queuing theory are basic facts of queuing life. First, traffic varies widely. Second, anyone who designs a telephone

switch to completely handle all peak traffic will find the switch idle for most of the time. He will also find he's built a very expensive switch. Third, it is possible, with varying degrees of certainty to predict upcoming "busy" periods. See also Erlang, A.K., Erlang B, Erlang C and Poisson.

2. Erlang is also a programming language designed at the Ericsson Computer Science Laboratory. Open-source Erlang is being released to help encourage the spread of Erlang outside Ericsson. Ericsson has released, free of charge, a. The entire source code of the current Erlang system. b. The entire source code for Mnesia a distributed Database Management System, appropriate for telecommunications applications and other Erlang applications with need of continuous operation and soft real-time properties. c. Extensive libraries of code for building robust fault-tolerant distributed applications. All with documentation. All the above software has been battle tested in a number of Ericsson products, for example the new Ericsson ATM switch. See www.erlang.org.

Erlang, A. K. In 1918, A. K. Erlang, a Danish telephone engineer, published his work on blocking in "The Post Office Electrical Engineers' Journal," a British publication. Like E.C. Molina, an AT&T engineer, Erlang assumed a Poisson distribution of calls arriving in a given time. Molina had assumed a constant holding time for all calls, whereas Erlang assumed an exponential distribution for holding times. That means that longer calls occur less frequently than shorter calls. Erlang assumed that blocked calls are immediately cleared and lost and do not return. A formula that Erlang worked out based on these assumptions (Erlang B) is still in use in telephone engineering. See Erlang, Erlang B, Erlang C and Poisson.

Erlang B A probability distribution developed by A.K. Erlang to estimate the number of telephone trunks needed to carry a given amount of traffic. Erlang B assumes that, when a call arriving at random finds all trunks busy, it vanishes (the blocked calls cleared condition). Erlang B is also known as "Lost Calls Cleared." Erlang B is used when traffic is random and there is no queuing. Calls which cannot get through, go away and do not return. This is the primary assumption behind Erlang B. Erlang B is easier to program than Poisson or Erlang C. This convenience is one of its main recommendations. Using Erlang B will produce a phone network with fewer trunks than one using Poisson formulae. See also Erlang, Erlang A. K., Erlang C, and Traffic Engineering.

Erlang C A formula for designing telephone traffic handling for PBXs and networks. Used when traffic is random and there is queuing. It assumes that all callers will wait indefinitely to get through. Therefore offered traffic (see ERLANG) cannot be bigger than the number of trunks available (if it is, more traffic will come in than goes out, and queue delay will become infinite). Erlang C is not a perfect traffic engineering formula. There are none that are.

Erlang Formula A mathematical way of making predictions about randomly arriving work-load (such as telephone calls) based on known information (such as average call duration). Although traditionally used in telephone traffic engineering (to determine the required number of trunks), Erlang formulas have applications in call center staffing as well. See Erlang.

ERM Enterprise Resource Management. Also known as Enterprise Resource Planning (ERP). See ERP for a full definition.

ERMES 1. European Radio MEssaging System.

2. One of the communications protocols used between paging towers and the mobile pagers/receivers/beepers themselves. Other protocols are POCSAG, ERMES, FLEX, GOLAY and REFLEX. The same paging tower equipment can transmit messages one moment in GOLAY and the next moment in ERMES, or any of the other protocols.

ERP 1. Effective Radiated Power.

2. Enterprise Resource Planning. A concept developed by The Gartner Group to describe the next generation of manufacturing business systems and MRP (Manufacturing Resource Planning) software. ERP software links together back-office computer systems such as manufacturing, financial, human resources, sales force automation, supply-chain management, data warehousing, document management, and after-sales service and support. Such systems typically run on networks of PCs, replacing older mainframe-based systems. ERP software typically makes heavy use of telecommunications. ERP is also known as ERM (Enterprise Resource Management).

error burst A sequence of transmitted signals containing one or more errors but regarded as a unit in error in accordance with a predefined measure. Enough consecutive transmitted bits in error to cause a loss of synchronization between sending and receiving stations and to necessitate resynchronization.

error checking and correction Error checking is the process of checking a "packet" being transmitted over a network to determine if the package, or the data content within the package, has been damaged. If checked and found wanting, damaged packets are discarded. Error correction is the process of correcting the damage by resending a copy of the original packet. In public frame relay services, the network performs the function of error checking, but not error correction. That function is left to the intelligent end equipment (at the user's site).

error control Various techniques which check the reliability and accuracy of characters (parity) or blocks of data sent over telecommunications lines. V.42, MNP and HST error control protocols (three common dial-up phone line modem protocols) use error detection (CRC) and retransmission of errored frames (ARQ). See Error Control Protocols.

error control protocols Besides high-speed modulation protocols, all current models of high-speed dial-up modems also support error control and data compression protocols. There are two standards for error control protocols: MNP-4 and V.42. The Microcom Networking Protocol, MNP, was developed by Microcom. MNP 2 to 4 are error correction protocols. V.42 was established by ITU-T. V.42 actually incorporates two error control schemes. V.42 uses LAP-M (Link Access Procedure for Modems) as the primary scheme and includes MNP-4 as the alternate scheme. V.42 and MNP-4 can provide error-free connections. Modems without error control protocols, such as most 2400 bps Hayes-compatible modems, cannot provide error-free data communications. The noise and other phone line anomalies are beyond the capabilities of any standard modem to deliver error-free data. V.42 (and MNP 2-4) copes with phone line impairments by filtering out the line noise and automatically retransmitting corrupted data. The filtering process used by V.42 (and MNP 2-4) is similar to the error correction scheme used by file transfer protocols (such as XMO-DEM). The two modems use a sophisticated algorithm to make sure that the data received match with the data sent. If there is a discrepancy, the data is re-sent.

What is the difference between error control protocols (such as V.42) and file transfer protocols (such as XMODEM)? For one thing, file transfer protocols provide error detection and correction only during file transfers. File transfer protocols do not provide any error control when you are reading e-mail messages or chatting on line. Even though an error control protocol is "on" all the time, we still need file transfer protocols when two modems establish a reliable link. A modem works with bit streams, timing and tones. It does not understand what a file is. When you download or upload a file, your communications software needs to take care of the details related to the file: the filename, file size, etc. This is handled by the file transfer protocol which does more than error-checking.

The other benefit of V.42 (or MNP-4) is that it can improve throughput. Before sending the data to a remote system, a modem with V.42 (or MNP-4) assembles the data into packets and during that process it is able to reduce the size of the data by stripping out the start and stop bits. A character typically takes up 1 start bit, 8 data bits and 1 stop bit for a total of 10 bits. When two modems establish a reliable link using V.42 or MNP-4, the sending modem strips the start and stop bits (which subtracts 20% of the data) and sends the data to the other end. The receiving modem then reinserts the start and stop bits and passes the data to the computer.

Therefore, even without compressing the data you can expect to see as much as 1150 characters per second on a 9600 bps connection. Although the modem subtracts 20% of the data, the speed increase is less than 20% due to the overhead incurred by the error control protocol. This definition, with great thanks to modem expert Patrick Chen.

error correcting code A code stored on an RFID tag to enable the reader to figure out the value of missing or garbled bits of data. It's needed because a reader might misinterpret some data from the tag and think a Rolex watch is actually a pair of socks. See the next definition.

error correcting mode A mode of data transmission between the tag and reader in which errors or missing data is automatically corrected.

error correcting protocol 1. A method of transmitting bit streams in a mathematical way such that the receiving computer verifies to the sending computer that all bits have been received properly. SNA and XMODEM protocols, in the mainframe and microcomputer environments respectively, are Error Correcting Protocols. See Error Control Protocol.

2. A set of rules used by readers to interpret data correctly from the tag.

error correction code In computers, rules of code construction that facilitate reconstruction of part or all of a message received with errors.

error correction mode A method of transmitting and receiving data that eliminates errors.

error free second A Bellcore (now Telcordia Technologies) definition. An error free second is, surprise, surprise, a one second time interval of digital signal transmission during which no error occurs. That's it.

error level A numeric value set by some programs that you can test with the errorlevel option of the "If" batch command. It works as follows. Some programs set the DOS

errorlevel to a certain number depending on a certain input or response to an event. Let's say when you type the letter "Y" in response to a question the errorlevel is set to 32. Once this is done, you may condition other events based upon this number using an If command in a batch file. You can say "IF ERRORLEVEL = 32 THEN GOTO END." That way, when you type "Y" you will get whatever is at END. This can be very helpful in batch files and other programs for providing "branching" from one event to another based on certain inputs.

error logical An error in the binary content of a signal, for example, bit error.

error rate In data transmission, the ratio of the number of incorrect elements transmitted to the total number of elements transmitted.

error suspense An MCI definition. An automated process which allows billable MCI calls on switch tapes to be processed for billing, while calls with errors are held in the Error Suspense File (a separate file for each switch).

error trapping In software programming, an exception is an interruption to the normal flow of a program. Common exceptions are division-by-zero, stack overflow, disk full errors and I/O (input/output) problems with a file that isn't open. The quality of a software program depends on how completely it checks for possible errors and deals with them. Code used for trapping errors can be excessive. Some programming languages have error trapping built in. Others don't and you have to program it in.

erstwhile An English word meaning previous. I define this word because I use it in this dictionary and lots of readers have told me they don't know it.

ES 1. Errored-Second. A count of the number of seconds in which at least one code violation (CV) was detected on a digital circuit. See also Code Violation.

2. End System: A system where an ATM connection is terminated or initiated. An originating end system initiates the ATM connection, and terminating end system terminates the ATM connection. OAM cells may be generated and received.

ES-IS End system to Intermediate systems protocol. The OSI protocol used for router detection and address resolution.

ES/9000 IBM Enterprise System/9000: mainframe computer family.

ESA 1. Emergency Stand Alone.

2. European Space Agency, the international organization that manages all aspects of European joint space programs; headquarters in Paris, France. www.esa.int.

ESC The ESC key on the keyboard. Often used to leave (escape) a program. Appears on the upper left of some keyboards on the IBM or compatibles but moves around with IBM's latest keyboard redesign whim. See also ESCAPE.

escalation A formal word for taking your trouble up through the levels of management at the vendor – until you get your problem resolved. Some users have formal Escalation Charts, which detail action to be taken depending on how many hours the problem persists, etc. Escalation sometimes works and sometimes doesn't, depending on the vendor. Usually it does. The rule in telecommunications (and we guess most other industries) is that "the squeaky wheel gets the most attention." Escalation works well with honey, flowers, plants and chocolates.

escape 1. The button on many computer keyboards which allows you to "escape" the present program. ESCape is the ASCII control character – code 27. It is often used to mark the beginning of a series of characters that represent a command rather than data. So called "ESCAPE" because it escapes from the usual meaning of the ASCII code and allows commands to be interspersed in a file of data, especially for data transmission to peripheral devices such as printers and modems. See Escape Sequence.

2. A means of aborting the task currently in progress.

3. A code used to force a smart modem back to the command state from the on-line state.

escape guard time An idle period of time before and after the escape code sent to a smart modem, which distinguishes between data and escapes that are intended as a command to the modem.

escape sequence A series of characters, usually beginning with the escape character, that is to be interpreted as a command, not as data. Escape sequences are used with ANSI.SYS to change the color of a screen. They are mostly used to send print commands to printers. The name Escape is due to the fact that it "escapes" from the usual meaning of the ASCII code, letting characters be commands instead of data, yet interspersed with data in a transmission.

ESCON Enterprise Systems CONnectivity. A high-speed fiber optic channel for linking IBM mainframe computers to disk drives and other mainframes. ESCON initially ran at 200 Mbps, but runs at much higher speeds now. ESCON uses the 8B10B signal encoding technique for Clock/Data Recovery (CDR). Competing standards are 10GbE (10 Gigabit Ethernet) and Fibre Channel. See also 8B10B, 10GbE, and Fibre Channel.

escrow bucket A hopper at the outlet of a coin phone's acceptor/rejector that is tipped electrically to return money through the Coin Return or to send the money to the Cash Box as a collection for a completed call.

ESD 1. Electrostatic Discharge. See Electrostatic Discharge.

2. Electronic Software Delivery. A technique whereby software (both initial installations and upgrades) on computers can be accomplished electronically without the need for floppy disks or CDs. One ESD scenario might be for a bunch of PCs attached to a LAN to be upgraded by a file server on the LAN. Another ESD scenario might be for a single user being able to update his software by downloading upgrades directly from the software vendor via the Internet.

ESD Protection Electro-Static Discharge Protection. Procedures followed and devices worn by anyone handling manufacturer hardware, in order to protect delicate electronic components from damage due to static electricity. See Electrostatic Discharge.

ESDI 1. Enhanced Small Device Interface. An interface which improves the rate of data transfer for hard disk drives and increases the drive's storage capacity.

2. Nortel Networks term for Enhanced Serial Data Interface.

ESF Extended Super Frame or Extended Superframe Format. A T-1 format that uses the framing bit for non-intrusive signaling and control. A T-1 frame is sent 8,000 times a second, with each frame consisting of a payload of 192 bits, and with each frame preceded by a framing bit. Therefore, there are 8,000 framing bits per second. Previous generations of channel banks (D1, D2, D3, and D4) used the framing bit exclusively for network synchronization purposes. As ESF requires only 2,000 framing bits for synchronization, the remaining 6,000 framing bits can be used for error detection, using cyclic redundancy checking, and data link monitoring and maintenance. As a result, the channel banks need not rob bits from the data payload for such purposes. The ultimate yield is that the payload is not compromised, and the full 1.536 Mbps is available for user data. In a channelized T-1 application, such as traditional PCM-encoded voice, each channel is a reliable 64 Kbps, rather than the 56 Kbps realized through use of older channel banks. While voice is unaffected by the process of bit robbing, data suffers greatly. The impact of intrusive signaling and control is the reason that most carriers limit data communications to 56 Kbps. They have a lot of old channel banks still in use. Note that channel banks must be matched throughout the entire network. In other words, buying an ESF channel bank won't do you any good unless the carrier also has them in place. In networking, as in life in general, the lowest common denominator rules. See also Bit Robbing, Channel, Channel Bank and T-1.

ESH End System Hello packet. Part of CDPD Mobile End System (M-ES) registration sequence.

ESI 1. See Enhanced Serial Interface.

2. End System Identifier: This identifier distinguishes multiple nodes at the same level in case the lower level peer group is partitioned.

ESM Extended subscriber module.

ESMA Expanded Subscriber Module-100A.

ESMR Enhanced Specialized Mobile Radio. An enhancement of SMR technology, allowing two-way radio service with the capability to provide wireless voice telephone service to compete against cellular. It uses TDMA technology to put six voice conversations into one 25 kilohertz UHF radio channel in the 806-821 MHz band. ESMR can be deployed on a cellular basis, and supports hand-off, like cellular radio. ESMR was developed by Nextel and Geotek Communications. Nextel acquired a large number of SMR frequencies, applied ESMR technology, and began to offer what is planned to be a nation-wide radio service in direct competition with cellular and PCS. ESMR will support low-speed data, as well as voice, with digital technology yielding both improved efficiency of bandwidth utilization and improved security. The network remains to be fully deployed. See CDMA.

ESMTP Extended Simple Mail Transport Protocol (see SMTP), described in RFC 1651. Seldom used outside the Unix community.

ESMU Expanded Subscriber Carrier Module-100 URBAN.

ESN 1. Emergency Service Number. An ESN is a "list" of emergency numbers that corresponds to a particular ESZ (Emergency Service Zone). This list has to do with 911 service. Usually this ESN list is unique and contains a listing of the corresponding police, fire and ambulance dispatch centers for the caller's area. This "list" is used for selective routing and one button transfer to secondary PSAPs – Public Safety Answering Positions. The ESN/ESZ concept is especially useful in fringe areas.

2. Electronic Serial Number. A 32-bit binary number which uniquely identifies each cellular phone. The ESN consists of three parts: the manufacturer code, a reserved area, and a manufacturer-assigned serial number. The ESN, which represents the terminal, is hard-coded, fixed and supposedly cannot be changed. Paired with a MIN (Mobile Identification Number), the ESN and MIN are automatically transmitted to the mobile base station every time a cellular call is placed. The Mobile Telephone Switching office checks the ESN/MIN to make sure the pair are valid, that the phone has not been reported stolen, that the user's monthly bill has been paid, etc., before permitting the call to go through. At least that's the theory. It doesn't always work this way on calls made from roaming cellular phones. And some cellular phones have been known to have their ESNs tampered with (it's called fraud) which tends to mess up the billing mechanisms. See also EESN and MIN.

3. Electronic Switched Network.

ESO Equipment Superior to Operator. When closing help desk tickets, it describes situations where the problem was user stupidity, such as the power cord not plugged in, the monitor unplugged, the keyboard not attached, etc. Also called SUT, for Stupid User Tricks.

ESOs European Standardization Organizations.

ESP 1. Enhanced Serial Port. The Hayes Enhanced Serial Port (ESP) adapter, introduced in late 1990, replaces and extends the traditional COM1/COM2 serial port adapter. The ESP combines dual 16550 UARTS with an on-board communications coprocessor. The ESP has two distinct modes of operation to provide both old and new standards in the same package: Compatibility Mode and Enhanced Mode. Each ESP port can be independently operated in either mode. Default modes are configured via DIP switches and can be modified by ESP commands. The MCA-bus version of the ESP uses Programmable Option Selection (POS) rather than DIP switches. See Enhanced Serial Interface.

2. Enhanced Service Provider – a vendor who adds value to telephone lines using his own software and hardware. Also called an IP, or Information Provider. An example of an ESP is a public voice mail box provider or a database provider, say one giving the latest airline fares. An ESP is an American term, unknown in Europe, where they're most called VANs, or Value Added Networks. See also Open Network Architecture and Information Provider.

3. Ethernet Service Provider.

4. EncapSulated Postscript File.

5. Encapsulating Security Payload. The portion of the IPSec virtual private networking protocol which is used predominantly to provide data privacy.

ESPA European Selective Paging Association.

ESPRIT European Strategic Program for Research and development in Information Technology. A $1.7 billion research and development program funded by the European Community.

ESQ End System Query packet. Part of CDPD Mobile End System (M-ES) registration procedures.

ESS 1. Electronic Switching System. ESS was originally a designation for the switching equipment in Bell System central offices but has slightly more general use now. In the independent telephone company industry, the abbreviation for the same thing is EAX. The first 1ESS switch went into service in May 1965 in Succasunna, New Jersey.

2. European Standardization System.

ESS No 4 AT&T's large toll telephone switch. It will handle over 100,000 trunks and over 500,000 attempts at making a call each hour. It's large and sophisticated and can probably be configured to be the largest telephone switch in the world.

ESS No 5 AT&T's Class 5 digital central office. See also End Office.

essential lines A telephone company definition. In order to guarantee to certain customers the ability to make outgoing calls during an emergency, the telephone company's Customer Services Department designates these customers as "essential." Examples of essential lines are:. police and fire departments, ambulance companies, hospitals, etc. Whenever Line Load Control is activated, outgoing service may be selectively denied to nonessential customers in order to preserve originating calling capacity for those customers having a documented priority. Also see - Class A lines.

essential service 1. A service provided by a telecommunications provider, such as an operating telephone company or a carrier, for delivery of priority dial tone. Generally, only up to 10 percent of the customers may request this type of service. See Essential Lines.

2. A service that is recommended for use in conjunction with NS/EP (national emergency) telecommunications services.

ESSX ESSX (pronounced essex) is some local phone companies' name for Centrex. See Centrex.

established connection A telephone company term. A connection on which all necessary switching or operating steps have been taken to connect the calling and called lines. Generally speaking, it is somewhat broader than the term "completed call," in that it includes established connections to tones or announcements, as well as completed calls. A completed call is a connection between two telephones.

estimated position error A term used by Airbus Industries to measure the estimated navigational performance of a GPS device.

ESZ Emergency Service Zone. This term is used in conjunction with 911 emergency service. An ESZ is a geographic area that is served by a unique mix of emergency services. Each ESZ has a corresponding ESN (a list of Emergency Service Numbers) which enables 911 service to properly route incoming calls.

ET Exchange Termination. Refers to the central office link with the ISDN user.

ETACS Extended TACS. The cellular technology used in the United Kingdom and other countries. It is developed from the U.S. AMPS technology. See also AMPS, TACS, NTACS and NAMPS.

etailer A retailer who conducts his business by electronic commerce – i.e. over the Internet and the world wide web.

ETB Abbreviation for end-of-transmission block. The binary code is 0111001, the hex is 71.

ETC 1. Enhanced Throughput Cellular is an error correction cellular communications protocol, which helps prevent disruptive signal fading and thus reduces the number of dropped calls.

2. Eligible telecommunications carrier. An eligible telecommunications carrier (ETC) is a common carrier that has been designated to receive Lifeline and other universal service support in the area for which the carrier is designated an ETC. To be designated an ETC, a carrier must file an application with the state's utility regulator and meet all Federal and state ETC criteria.   States have primary responsibility for designating telecommunications providers as ETCs. Carriers should contact their state utility regulator to find out how to initiate the ETC designation process. The procedures differ for carriers seeking designation on non-tribal lands and carriers seeking designation on tribal lands. Eligibility for federal universal support is covered by Section 214(e) of the Telecommunications Act of 1996.

etched Antiglare treatment that prevents glare but also reduces screen sharpness and clarity on monitors. Generally considered an obsolete technology.

eternity hold My own creation for what happens when an unthinking corporation has a machine which answers your phone call and then says something like "All our agents are currently busy. Our customers (i.e. you) are important to us. (Some actually mean it.) And we'll connect you with the next available agent." And then they leave you waiting forever – or an eternity. Governmental agencies, airlines and police departments (especially when you need them) tend to be firm believers in placing their callers on Eternity Hold. A new enhancement to Eternity Hold is Conference Hold. Here everyone on Eternity Hold can speak to everyone else on hold. You can imagine the conversation. "Let's all talk about why we hate this airline..." I made this service up. It doesn't exist, but I love the idea. Imagine the conversations....

ETF See early termination fee.

ether An undefined path. See Ethernet and Luminiferous Ether.

EtherCAT Ethernet for control automation technology, developed by Beckhoff Automation LLC, and which is now an International Electrotechnical Commission (IEC) specification. EtherCAT is designed for high-speed, real-time industrial automation environments, supporting speeds greater than conventional fieldbuses.

ethical hacking A means for identifying vulnerabilities by having an authorized individual(s) attempt to break into a computer system or network and report their findings to the people who own and/or run the computer system or network.

ethical lapse Oops.

EtherLEC ELEC). Ethernet Local Exchange Carrier. Facilities-based service providers employing Optical Ethernet technology. They typically build metro Ethernet networks by leasing dark fiber and installing Optical Ethernet equipment in metro POPs and collocation centers. The "EtherLEC" business model is straightforward: provide low-cost Internet access, metro network and WAN services based on "native Ethernet." Subscribers benefit from a lower cost of ownership combined with lower transport costs than traditional TDM-based private lines.

EtherRing ADC Telecommunications' name for a "revolutionary new idea allowing transport of native mode Ethernet and Fast Ethernet (100 million bits per second) data packets over a wide area network (WAN). Unlike typical Ethernet transport solutions," according to ADC Telecommunications (www.adc.com), "EtherRing has no distance limitations."

ethernet A Local Area Network (LAN) standard officially known as IEEE 802.3 (1980)/Ethernet, and other LAN technologies are used for connecting computers, printers, workstations, terminals, servers, etc., within the same building or campus. Ethernet operates over twisted wire and over coaxial cable at speeds beginning at 10 Mbps. For LAN

interconnection, Ethernet is a physical link and data link protocol reflecting the two lowest layers of the OSI Reference Model. The theoretical limit of 10-Mbps Ethernet, measured in the smallest 64-byte packets, is 14,800 pps (packets per second). By comparison, Token Ring is 30,000 and FDDI is 170,000.

Ethernet specifies a CSMA/CD (Carrier Sense Multiple Access with Collision Detection) MAC (Media Access Control) mechanism. CSMA/CD is a technique of sharing a common medium (e.g., twisted pair, or coaxial cable) among several devices. As Byte Magazine explained in its January, 1991 issue, Ethernet is based on the same etiquette that makes for a polite conversation: "Listen before talking." Of course, even when people are trying not to interrupt each other, there are those embarrassing moments when two people accidentally start talking at the same time. This is essentially what happens in Ethernet networks, where such a situation is called a "collision." If a node on the network detects a collision, it alerts the other nodes by jamming the network with a collision "notification." Then, after a random pause, the sending nodes try again. The messages are called frames (see the diagram).

AN ETHERNET FRAME					
Preamble	Destination Address	Source Address	Type	Data up to 1500	Frame Check sequence
8 bytes	6 bytes	6 bytes	2 bytes	bytes	4 bytes

The first personal computer Ethernet LAN adapter was shipped by 3Com on September 29, 1982 using the first Ethernet silicon chip from SEEQ Technology. Bob Metcalfe created the original Ethernet specification at Xerox PARC and later went on to found 3Com. In the October 31, 1994 issue of the magazine InfoWorld, Bob Metcalfe explained that Ethernet got its name "when I was writing a memo at the Xerox Palo Alto Research Center on May 22, 1973. Until then I had been calling our proposed multimegabit LAN the Alto Aloha Network. The purpose of the Alto Aloha Network was to connect experimental personal computers called Altos. And it used randomized retransmission ideas from the University of Hawaii's Aloha System packet radio network, circa 1970. The word ether came from luminiferous ether – the omnipresent passive medium once theorized to carry electromagnetic waves through space, in particular light from the Sun to the Earth. Around the time of Einstein's Theory of Relativity, the light-bearing ether was proven not to exist. So, in naming our LAN's omnipresent passive medium, then a coaxial cable, which would propagate electromagnetic waves, namely data packets, I chose to recycle ether. Hence, Ethernet."

According to Metcalfe, "Ethernet has been renamed repeatedly since 1973. In 1976, when Xerox began turning Ethernet into a product at 20 million bits per second (Mbps), we called it The Xerox Wire. When Digital, Intel, and Xerox decided in 1979 to make it a LAN standard at 10 Mbps, they went back to Ethernet. IEEE tried calling its Ethernet standard 802.3 CSMA/CD – carrier sense multiple access with collision detection. And as the 802.3 standard evolved, it picked up such names as Thick Ethernet (IEEE 10Base-5), Thin Ethernet (10Base-2), Twisted Ethernet (10Base-T), and now Fast Ethernet (100Base-T)."

Ethernet PC cards now come in a couple of basic varieties – for connecting to an Ethernet LAN via coaxial cable or via two twisted pairs of phone wires, called 10Base-T. See also 10Base-T, Collision Domain, Ethernet Controller, Ethernet Identification Number, Ethernet Switch, Ethertalk, Frame, Gigabit Ethernet, Luminiferous Ether, Thinnet and Token Ring.

The May 22, 2003 issue of my favorite magazine, the Economist, had an article on Ethernet. I excerpt it because it's so good: "WHEN Ethernet, now by far the most popular way of distributing data around local networks, was devised by Bob Metcalfe in a memo on May 22, 1973, at Xerox's celebrated Palo Alto Research Centre (PARC), it was designed to send data at about three megabits per second. Today, one gigabit per second Ethernet is common and speeds of 100 gigabits per second are being developed. The vast majority of the Internet's traffic begins and ends its journey on Ethernet networks, which are found in nearly every office network and home broadband connection. It was not supposed to be this way. Few imagined that this particular networking protocol would last as long as it has. Indeed, the landscape is littered with better-financed, better-backed rival protocols that failed against Ethernet. IBM's Token Ring system is one famous casualty. Asynchronous Transfer Mode, supported by the telephone industry, is another. So the case of Ethernet is worth examining: the reasons for its longevity may offer lessons to the information-technology industry. Keep it simple, stupid. The first reason is simplicity. Ethernet never presupposed what sort of medium the data would travel over, be it coaxial cable or radio waves (hence the term "ether" to describe some undefined path). That made it flexible, able to incorporate improvements without challenging its fundamental design.

Second, it rapidly became an open standard at a time when most data-networking protocols were proprietary. That openness has made for a better business model. It enabled a horde of engineers from around the world to improve the technology as they competed to build inter-operable products. That competition lowered the price. What is more, the open standard meant that engineers in different organisations had to agree with each other on revised specifications, in order to avoid being cut out of the game. This ensured that the technology never became too complex or over-designed. As Charles Spurgeon, author of "Ethernet: The Definitive Guide" puts it, "It always stayed close to the ground. It addressed problems customers came up against, not problems that networking specialists thought needed to be addressed." That, coupled with the economies of scale that come from being the entrenched technology, meant that Ethernet was faster, less expensive and less complicated to deploy than rival systems.

Third, Ethernet is based on decentralisation. It lets smart "end-devices", such as PCs, do the work of plucking the data out of the ether, rather than relying on a central unit to control the way those data are routed. In this way, Ethernet evolved in tandem with improvements in computing power-a factor that was largely overlooked by both critics and proponents when Ethernet was being pooh-poohed in the 1980s and early 1990s.

Beyond the technology, there is even a lesson for companies investing in research, albeit one learned through tears rather than triumph. Xerox failed to commercialise Ethernet, as it similarly missed exploiting other inventions created at PARC, such as the mouse and the graphical user interface. To develop Ethernet fully, Dr Metcalfe had to leave PARC and found 3Com, now a big telecommunications-component firm. The lesson may have sunk in. In January 2002 PARC was carved out as an independent subsidiary of Xerox. That allows it to explore partnerships, spin-offs and licensing agreements without having to get its parent's permission.

And the future? Many geeks believe the fashionable wireless standard 802.11b, better known as Wi-Fi, is poised to become the next iteration of the technology, once concerns about its security have been resolved. Wi-Fi is able to deliver cheap internet access at a rate of many megabits-per-second, and it can do so at a range of up to 100 metres. Dr Metcalfe, who based Ethernet's early designs on a radio network called AlohaNet, that linked the Hawaiian islands, sees Wi-Fi as "Ethernet coming home".

That raises the question of whether a technology that outlived even its inventor's expectations will ever be supplanted, or whether it will continue to be upgraded indefinitely. As Dr Metcalfe quips, "When something rises up to defeat Ethernet, it's very likely that they're going to call it Ethernet." Now that, indeed, is adaptability.

ethernet address The address assigned to a network interface card by the original manufacturer or by the network administrator if the card is configurable. This address identifies the local device address to the rest of the network and allows messages to reach the correct destination. Also known as the media access control (MAC) or hardware address.

ethernet controller The unit that connects a device to the Ethernet cable. An Ethernet controller typically consists of part of the physical layer and much or all of the data link layer and the appropriate electronics.

ethernet extender Normal Ethernet – the kind you have in your office – only works a few hundred feet. But Ethernet extenders literally extend that several miles. For example, one extended whose specifications I'm looking at shows it can achieve 16 Mbps over two km, 4.6 Mbps over 3 km, 2.3 Mbps over 5 Km and 128 Kbps over 8 Km. you can place an extender on

ethernet identification number This is a unique, hexadecimal Ethernet number that identifies a device, such as a PC/AT with a SpeedLink/PC16 network interface card installed, on an Ethernet network.

ethernet II (DIX) Defined by Digital, Intel and Xerox. The frame format for Ethernet II differs from that of IEEE 802.3 in that the header specifies a packet type instead of the packet length.

ethernet meltdown An event that causes saturation on an Ethernet-based system, often the result of illegal or misdirected packets. An Ethernet meltdown usually lasts for only a short period of time.

ethernet packet A variable-length unit in which information is transmitted on an Ethernet network. An Ethernet packet consists of a synchronization preamble, a destination address, a source address, a field that contains a type code indicator, a data field that can vary from 46 to 1500 bytes, and a cyclical redundancy check (CRC) that provides a statistically derived value used to confirm the accuracy of the data.

Ethernet Passive Optical Network See EPON.

ethernet switch Ethernet is the most common local area network (LAN) standard

in the world today. Ethernet is very cheap, but its approach to congestion management is on the ugly side. There is a solution and it's called an Ethernet Switch. Here's an explanation: In typical Ethernet LAN, transmission is over a shared bus. When an Ethernet-attached device (e.g. a PC) puts information onto the network (called data frames or data packets), those frames move in both directions, and pass by all devices on the network. Every attached workstation, printer and host sees all the information on the Ethernet LAN. At some point, a device will see data that's intended for it. It will recognize the address in the data frame, as it passes by, and take that information off the network. But, since every bit of information must pass by every device, Ethernet networks can become very heavily congested very quickly. While the theoretical 10Base-T Ethernet data rate is 10Mbps (ten million bits per second), they often effectively run at a fraction of that speed – as slow as 500,000 bits per second. One way of reducing Ethernet congestion is through the use of hubs. Hubs allow the Ethernet to be divided into physical segments at the workgroup level, with each workgroup representing a "community of interest." This segmentation process serves to reduce congestion by confining traffic to all those Ethernet stations on the shared hub, passing it on to other hubs only if necessary. In other words, the single logical LAN is subdivided into multiple physical and logical segments. However, the hub still presents the data to all connected devices over a common, shared bus. Thus, hubs really don't do anything internally to control congestion. A much better, but more expensive solution is something called an Ethernet Switch. What it is does is simple. It breaks out the stuff for each device and sends it directly to and from only that device. It's wired directly to each device. Think phone system. You get the idea. Let's consider an Ethernet switch at the workgroup-level. Workgroup-level switches serve a community of interest, such as the marketing department or the accounting department. An Ethernet workgroup switch is used to interconnect Ethernet-attached devices (e.g., workstations, printers, and servers), each of which connects to a port on the switch via UTP (Unshielded Twisted Pair). When the transmitting workstation, for instance, sends an Ethernet frame of data to the switch, that frame is captured in a buffer and is fragmented into smaller data units. The switch then flows those fragments over a common, shared bus inside the switch, from the input port of the originating device to the output port associated with the destination device. The common bus also can support other transmissions, as it is carved into time slots through a process akin to TDM (Time Division Multiplexing). More complex workgroup-level switches may involve a switching matrix comprising multiple, interconnected busses, and yielding multiple possible transmission paths between ports. In either case, the switch directs the data only to the target device, rather than to all attached devices. Therefore, and unlike hubs, switches do a lot to reduce Ethernet LAN congestion. What this means to you is that a 36-port Ethernet switch can allow you and all 35 of your co-workers to communicate simultaneously. You can all pass data between each other at the same time, or you can all access the same e-mail server at the same time. Depending on the nature, design and quality of the internal switching matrix, you each may get 10Mbps or more (switches come in 100 megs also). Your traffic will be confined to your own workgroup unless it needs to go somewhere else. Workgroup switches can be interconnected directly. For example, the accounting department might have its own switch on the 10th floor, and the sales and marketing department might have its own switch on the first floor. Connecting from switch to switch, the top salesman can check his commission statement, assuming that the accounting department allows him access to that database (highly unlikely). Again, the data remain confined to your workgroup switch unless they need to go elsewhere. Workgroup switches also can be interconnected through either a backbone switch or a router, which is a highly intelligent (read much more capable, but much slower and much more expensive) switch. In either case, the data are forwarded from one workgroup switch to another only if the destination address requires. A backbone switch has a higher-speed internal switching matrix, and connects to workgroup-level hubs and switches through high-speed links, commonly in the form of fiber optics. Backbone switches also can be connected to other backbone switches. Again, the data go only where they need to go, rather than being blasted all over the company, and slowing down the whole network in the process. See also Ethernet, Gigabit Ethernet, Hub, Router and Switch.

ethernet transceiver A device used in an Ethernet local area network that couples data terminal equipment to other transmission media.

EtherTalk An Ethernet protocol used by Apple computers. AppleTalk protocol governing Ethernet local area network transmissions. Also the Apple Computer Ethernet adapter and drivers. Apple's implementation of Ethernet is compliant with IEEE specification 802.3.

Ethertype A two-byte code indicating protocol type in an Ethernet local area network packet.

ethylene propylene rubber EPR. An ozone-resistant rubber consisting primarily of ethylene propylene copolymer (EPM) or ethylene propylene diene terpolymer

(EDPM).

ETI Electronic Telephone Interface.

ETISALAT Emirates Telecommunications Corporation is the sole provider of telecommunications services throughout the United Arab Emirates. The head office is in Abu Dhabi.

ETM Electronic Ticketing Machine. A machine that looks like a banking Automated Teller Machine (ATM), except that it will dispense airline tickets and possibly, hotel reservations, car rental agreements, etc.

ETN 1. Electronic Tandem Network. An ETN is a large private network which comprises dedicated leased lines interconnecting electronic tandem switches. ETNs were deployed in the 1970s and 1980s by very large user organizations such as state and federal governments.

2. Earning Telephone Number. A billing term which is synonymous with WTN (Working Telephone Number). Multiple ETNs can be associated with a single BTN (Billing Telephone Number).

ETNO The European Public Telecommunications Network Operations Association (ETNO).

eTOM enhanced Telecom Operations Map. See Telecom Operations Map.

ETR 1. Estimated Time to Repair. "The DSL line is down. The ETR is two hours."

2. ETSI (European Telecommunications Standards Institute) Technical Report. See also ETSI.

ETS 1. ETSI (European Telecommunications Standards Institute) Technical Specification. A standard defined by the European Telecommunications Standards Institute (ETSI). See also ETSI.

2. Electronic Tandem Switching. See Electronic Tandem Switching.

3. Ethernet Terminal Server. A cost-effective and flexible way of connecting serial-based asynchronous terminals and peripherals to one or more host computers across an Ethernet LAN. See Terminal Server.

ETS 300 211 Metropolitan Area Network (MAN) Principles and Architecture.

ERS 300 212 Metropolitan Area Network (MAN) Media Access Control Layer and Physical Layer Specification.

ETS 300 217 Connectionless Broadband Data Service (CBDS).

ETS Set A Nortel Networks term for an electronic Telephone Set.

ETSI European Telecommunications Standards Institute, is the European counterpart to ANSI, the American National Standards Institute. ETSI is based in Sophia-Antipolis, near Nice, France. ETSI's task is to pave the way for telecommunications integration in the European community as part of the single European market program. ETSI was founded in 1988 as a result of an in initiative of the European Commission. It was established to produce telecommunications standards by democratic means, for users, manufacturers, suppliers, administrations, and PTTs. ETSI's main aim is the unrestricted communication between all the member states by the provision of essential European standards. It is now an independent, self-funding organization, which works closely with both CEPT (European Conference of Posts and Telecommunications Administrations) and EBU (European Broadcasting Union).

ETSI also works closely with CENELEC, which is responsible for electrotechnical standards, and CEN, which is responsible for European standardization in all remaining fields. Certified products are awarded the CE Mark, signifying that a company has met the applicable essential health and safety requirements and the specific conformity assessment requirements to market its product in the European Union under the "New Approach" directives. www.etsi.org. See also ANSI, CE Mark, CEN, CENELEC, CEPT, and EBU.

ETSI V5 European Telecommunications Standards Institute's open standard interface between an Access Node (AN) and a Local Exchange (LE) for supporting PSTN (Public Switched Telephone Network) and ISDN (Integrated Services Digital Network). Examples of Access Nodes include Digital Loop Carrier (DLC) systems, wireless loop carrier system, and Hybrid Fiber Coax (HFC) systems.

ETX End of Text. Indicates the end of a message. If multiple transmission blocks are contained in a message in Bisynch systems, ETX terminates the last block of the message. ETB is used to terminate preceding blocks. The block check character is sent immediately following ETX. ETX requires a reply indicating the receiving station's status. The binary code is 0011000, the hex is 30. See Packet.

EUC EUC stands for Extended Unix Code. It is a standard way of representing extended character sets such as Japanese.

EUCL End User Common Line charge. A FCC tariff term defined in FCC Rules 69.104 as follows: "A charge that is expressed in dollars and cents per line per month shall be assessed upon end users that subscribe to local exchange telephone service, Centrex or semi-public coin telephone service to the extent they do not pay carrier common line charges. Such charge EUCL shall be assessed for each line between the premises of an end user and a Class 5 office that is or may be used for local exchange service transmissions. Each Single Line Service is charged one CALC or EUCL. The amount varies by state." The intent is that the EUCL, in combination with the CCL (Carrier Common Line Charge) compensate the LEC for the use of the local loop for the purposes of originating/terminating interLATA long distance calls. The LEC receives no direct benefit from such traffic, although its investment in, and maintenance cost associated with, the local loop is considerable. The EUCL is known variously as the Access Charge, CALC, and SLC. See also Access Charge and CCL.

Eudora An electronic mail sending and receiving software program that runs on Macs and under Windows and is probably the most common e-mail service used by people on the Internet. It is manufactured and distributed by Qualcomm Enterprises, San Diego, California. Eudora Pro is the commercial version (i.e. the one that costs money). It includes features that are not in Eudora Light, the freeware version and one in heaviest use on the Internet.

EUI-48 Extended Unique Identifier-48 (bits). The IEEE (Institute of Electrical and Electronics Engineers) administers the addressing scheme for all LANs (Local Area Networks) adhering to the Project 802 standards, including 802.3 Ethernet and 802.5 Token Ring. The addressing scheme comprises a Company ID and an Extension ID, each of 24 bits. The Company ID, which is administered directly by the IEEE RAC (Registration Authority Committee), identifies the manufacturer of the NIC (Network Interface Card) or other interface hardware. The Extension ID, also known informally as the Board ID, is assigned by the manufacturer. Theoretically, the EUI-48 address is unique, although duplications occasionally are reported, most likely due to manufacturing errors. The EUI-48 comprises what we normally refer to as the MAC (Medium Access Control) address. The IEEE refers to it as the Equipment Identifier, as the 48-bit address effectively is a serial number that identifies the manufacturer of the hardware, the manufacturing run, etc. It generally resides in ROM (Read-Only Memory) on the NIC. See also EUI-64 and MAC Address.

EUI-64 Extended Unique Identifier-64 (bits). The IEEE expanded the 48-bit EUI-48 addressing scheme for LANs to a 64-bit global identifier known as EUI-64. This Equipment Identifier is what we normally think of as a MAC (Medium Access Control) address. It is a concatenation of the 24-bit Company ID assigned by the IEEE RAC (Registration Authority Committee), and an expanded 40-bit Extension Identifier assigned by the organization with that Company ID assignment, i.e., the manufacturer of the NIC (Network Interface Card). This 40-bit Extension Identifier replaces the 24-bit EI used in the predecessor EUI-48. This new scheme increases the available number of unique addresses, which typically are hard-coded in ROM (Read-Only Memory) on a chipset in a NIC. Thereby, more devices can have globally unique addresses – about 1 trillion per manufacturer. The move to EUI-64 is not entirely due to the increasing popularity of LANs, at least not traditional LANs. Rather, it is due to the increasing popularity of the both Internet and the IP (Internet Protocol) protocol. IPv6 (IP version 6), which ultimately will replace the current IPv4, addresses often include an Interface ID field, which is defined according to the EUI-64 format. Ultimately, you might be running the IPv6 protocol on your LAN-attached workstation, and essentially have a single address that you would use for seamless access to both the LAN-attached resources and to the public Internet. You really wouldn't want that (due to security, privacy, and other concerns), but you certainly could. In any event, the 40-bit EI address would be the same length across both the LAN and the IPv6 Internet, which makes necessary address translations a lot easier. See also EUI-48, IPv6, and MAC Address.

EULA End User License Agreement. An agreement between you, the user, and the company which allows you to use its software. The Agreement usually says you don't own the software but can use it. And don't even think about re-selling it. See Disclaimer.

EUnet European UNIX Network. (Original name). Now a major European Internet Service provider.

Eur Another way of writing Euro, the European currency. Since Euro.

EURESCOM EUropean institute for REsearch and Strategic studies in TeleCOMmunication.

Euro New European single currency introduced on January 1, 1999 by 11 European countries, who agreed to lock their exchange rates to the euro, which was, at the time of its birth, worth around $1.17. But is no longer.

Euro ISDN See Euro-ISDN.

Euro-ISDN European ISDN defined by European Telecommunications Standards Institute (ETSI). Also known as NET5.2 and V5.2. It differs from North American National ISDN-1 in that Euro-ISDN is very limited in the options it offers. In the United States, ISDN comes with many options, including two call appearances, conference calling, call forwarding variable, call forwarding – busy, call forwarding – no answer, voice mail with indicator, two secondary directory numbers, etc. That makes North American ISDN more full-featured, but much less easy to order. Users, can however, call from the United States to Europe and complete ISDN calls. They can not carry their end-user ISDN equipment from the United States and use it in most places in Europe. Other subtle differences between Euro ISDN and its North American counterpart. Euro ISDN does not require the user to send a channel ID information element, giving network side full control of channel selection and eliminating glare. Channels are always enabled. Overlap dialing is allowed in the user to network direction, eliminating outbound number configuration on the user side. Provides a sending complete information element that indicates the called number is complete or there is no called number. Does not allow NFAS (non-facility associated signaling) and hence eliminates the need for backup D-channel signaling. Euro ISDN sends calling name using the Display IE whereas North American NI-2 did not support calling name and NI-3 uses the Facility IE.

EUROBIT European Association of Business Machines Manufacturers and Information Technology Industry.

EuroCableLabs An autonomous department of Cable Europe, established to coordinate technical activities, such as research, standards development, and certification, related to broadband cable technologies for TV and IP services.

Euromodem See Digital Video Broadcast.

European Article Numbering EAN. The bar code standard used throughout Europe, Asia and South America. It is administered by EAN International.

European Cable Communications Association See Cable Europe.

European Commission The administrative body of the European Union, and a central source of policy, legislation and funding for pan-European research and development in ICT applications.

European Competitive Telecommunications Association See ECTA.

European Computer Manufacturers Association See ECMA.

European Digital Signaling System no. 1. EDSS1. The European variant of the SS7 (Signaling System 7) signaling and control protocol for use over the ISDN D (Delta) channel. EDSSI was developed by ETSI (European Telecommunications Standards Institute). See also ETSI, ISDN, and SS7.

European Space Agency The European equivalent of the U.S.'s NASA – National Aeronautics and Space Administration.

EUROSINET-EUROTOP International ISDN pilot project for travel agents.

EUROTELDEV EUROpean TELecommunications DEVelopment. An organization involved in telecommunications standardization.

EUTELSAT EUropean TELecommunications SATellite organization. Inter-governmental organization that aims to provide and operate a communications satellite for public intra-European international telecommunications services. The segment is also used to meet domestic needs by offering leased capacity, primarily for television. U.K. and France are the largest shareholders, with about 44 member countries in total. www.eutelsat.org.

EUTP Enhanced Unshielded Twisted Pair. UTP Cables that have enhanced transmission characteristics. Cables that fall under this classification include Category 4 and above.

EV European Videotelephony.

EV-DO EV-DO (Evolution Data Only or Evolution Data Optimized) is a wireless telecommunications technology that provides wireless data connections to the Internet that are 300,000 to 600,000 bits per second – as much as 10 times as fast as a regular dial-up modem – but slower than a cable modem, a DSL line or WiFi. EV-DO's biggest advantage is that you can pop an EV-DO card into your laptop (or buy a laptop with EV-DO built in) and get on the Internet in most places in the U.S. Frankly, I love my Verizon EV-DO card. There are some neat things you can do with your EV-DO laptop card. You can pop it into a little box and make it be a router and serve several computers, laptops and desktops. You can also buy EV-DO signal boosters, which will give you better reliability and possibly higher speed. The carriers with EV-DO – Verizon, Nextel and Sprint – are constantly working to speed their EV-DO and thus attract more subscribers, including those presently subscribing to landline technologies such as cable modems and DSL. EV-DO is a CDMA 2000 technology based on the 1xRTT standard. It is also called / spelled EVDO, EvDO, 1xEV-DO and 1xEvDO. The most common spelling is EV-DO. See also www.EVDOInfo.com and EV-DO Release 0, Revision A, B and C.

EV-DO Release 0 A generation of EV-DO that offers a theoretical peak downlink speed of 2.4 Mbps and a theoretical peak uplink speed of 153 Kbps. The actual average downlink throughput is considerably lower than the theoretical maximum – more in the range of 300-600 Kbps – in order to avoid network congestion and accommodate network overhead. EV-DO Release 0 supports data applications such as audio and video downloads, TV broadcasts, and video conferencing. Its first commercial deployment was in

Korea in 2002. In countries with limited wireline infrastructure it has been deployed as a DSL substitute. In the United States, EV-DO Release 0 has been deployed by Sprint Nextel and Verizon Wireless.

EV-DO Revision A A generation of EV-DO that offers a theoretical peak downlink speed of 3.1 Mbps and a theoretical peak uplink speed of 1.8 Mbps. The actual average downlink throughput is in the range of 450-800 Kbps, and the actual uplink throughput averages 300-400 Kbps, in order to avoid network congestion and accommodate network overhead. EV-DO Release A was released commercially in October 2006, and was deployed by Sprint Nextel in 20 markets by the end of 2006, with plans to deploy it in the rest of its markets by the end of 2007. Verizon Wireless announced plans to begin rolling out EV-DO Revision A in its serving areas in late 2006 or early 2007. EV-DO Revision A supports quality of service tiers and time-sensitive applications such as VoIP and video telephony.

EV-DO Revision B A generation of EV-DO that is expected to become commercially available in 2008. The Revision B standard was published by the Third Generation Partnership Project 2 (3GPP2) as 3GPP2 C.S0024-B and by the Telecommunications Industry Association (TIA) and Electronics Industry Association (EIA) as TIA/EIA/IS-856-B. EV-DO Revision B involves aggregating multiple EV-DO Revision A channels to provide higher bandwidth and lower latency for multimedia delivery, two-way data transmissions, VoIP-based concurrent services, and dynamically scalable bandwidth. Peak data rates are directly proportional to the number of channels that are aggregated. For example, when 2 channels are combined, Revision B delivers a peak downlink rate of 6.2 Mbps (i.e., 2 x 3.1 Mbps) and a peak uplink rate of 3.6 Mbps (i.e., 2 x 1.8 Mbps).

EV-DO Revision C In October 2006 this was rebranded. It is now called Ultra Mobile Broadband (UMB).

EV8 Busy Line Transfer.

EVA Economic Value Added. A financial measure of whether you're making more money with your plant, factory, assets, etc. in your present business than you would be if you sold everything and stuck the proceeds in a investment. The common assumption is that you can get a 10% or 11% return. If you earn more than that, you're EVA positive. EVA is a term you hear a lot around AT&T. You can become EVA positive in any ways – writing down the value of your capital is one way. Sacking people works too. And, so does selling people (presumably, at a decent price).

Evanescent Wave Light guided in the inner part of an optical fiber's cladding rather than in the core.

EvDO See EV-DO, which is a more common spelling.

even parity In data communications there's something called a parity bit that's used for error checking. The transmitting device adds that parity bit to a data word to make the sum of all the "1" ("one") bits either odd or even. If the sum is odd, the result is called ODD parity. If it's even, it's called EVEN PARITY. See also Parity.

event An unsolicited communication from a hardware device to a computer operating system, application, or driver. Events are generally attention-getting messages, allowing a process to know when a task is complete or when an external event occurs.

event code A code that an agent in a call center enters at the conclusion of a call. Event codes can trigger a variety of follow-up activities such as an acknowledgement letter, or inclusion in a list for a subsequent campaign.

event driven A style of programming under which programs wait for messages to be sent to them and react to those messages. See Event Driven Alarms/Triggers/Ticklers.

event driven alarms/triggers/ticklers In a parallel process, an event trigger can be set to move the processing forward when a set of criteria is met (ex. the last piece of documentation is added to the file). Alarms can also be time-driven, as when a folder is automatically routed to exception processing if no action is taken within a specified time frame. This term is often found in workflow management.

event history A history of the activities that have been carried out on a record, (customer or prospect). For example, phone calls and mailers, offers and so on.

event mask The set of events that the SLEE (Service Logic Execution Environment) designates the ASC (AIN Switch Capabilities) to report for a particular connection segment, and an indication for each event if the ASC should suspend processing events for that connection segment until the SLEE sends a message back. Definition from Bellcore in reference to its concept of the Advanced Intelligent Network.

event message A message provided by the switching domain to the computing domain to indicate one of the following: 1) A change in the state of a telephone call by reporting state transitions of each connection in the call (Call Control Events); 2) A physical or logical device change that has taken place (such as Do Not Disturb being set or a device's microphone being muted) at a device physical and Logical Device Feature Events); 3) A change in call-associated information (Call Associated Feature Events); 4) A change in switching domain specific information that is associated either a device or call (Private Data / Information Events). See Call-Progress Event Message.

event report Synonymous with Event Message.

event signaling All telephone switches use the same three general types of signals: + Event Signaling initiates an event, such as ringing. + Call Progress Signaling denotes the progress (or state) of a call, such as a busy tone, a ringback tone, or an error tone. + Data Packet Signaling communicates certain information about a call, for example, the identity of the calling extension, or the identity of the extension being called.

everything on demand EOD. A content model that includes any type of video content (for example, video-on-demand and subscription video on demand), at anytime, and excluding none.

evil When choosing between two evils I always like to take the one I've never tried before – Mae West.

EVO Call forwarding, busy line.

EVRC Enhanced Variable Rate Vocoder. A vocoder (voice coder) used in CDMA (Code Division Multiple Access) cellular systems. EVRC digitizes and compresses analog voice signals through an EVRC DSP (Digital Signal Processor), improving CDMA bandwidth utilization and, therefore, enhancing system capacity by supporting voice transmission at a rate of approximately 8 Kbps. This bit rate is comparable to that of 13-Kbps QCELP (Qualcomm Code Excited Linear Prediction) which can run at either 13 Kbps or 8 Kbps. It is a significant improvement over the 64 Kbps required to support toll-quality voice in the wireline PSTN (Public Switched Telephone Network). See also QCELP.

EW-20 A specific size of rigid waveguide with an elliptical cross sectional shape.

EW Electronic Warfare. The military use of radar, electronic counter measures and electronic counter-counter measures to keep an enemy from finding invading forces, on land or in the air. It covers such methods as sending out planes equipped with equipment which transmit thousands of signals purporting to be signals that an enemy radar might see on locating an incoming plane. By sending thousands of such signals, the enemy's radar becomes a myriad of "radar" signals, of bright spots. Thus it's impossible for the enemy to read any intelligent information. There are also anti-radiation missiles which home in on and destroy air-defense radar facilities. The only defense against such anti-radiation missiles is to turn off the radar. Electronic warfare also covers such techniques as jamming radio frequencies, anti-jamming.

EWOS European Workshop for Open Systems.

EWP Electronic White Pages.

EWSD Generic name for digital central office switches from Siemens. EWSD actually stands for Elektronisches WahlSystem Digital. In German, the common meaning of wahlen is to select, as in "I'll select blue for the new color for my house." In the phone business, it was a simple jump to selecting a phone number, to its present meaning today – to dial. So if you translate EWSD into English, it really means electronic digital dialing system. You figure.

ex parte presentation A Federal Communications Commission definition. Any communication addressing the merits or outcome of a particular proceeding made to decision-making personnel (or in some proceedings, from the decision-making personnel), which, (1) if written, is not served on the parties to the proceeding, or (2) if oral, is made without opportunity for the parties to the proceeding to be present. A simpler defintion: Ex parte refers to statements, meetings or filings that are made outside of an official comment-and-replay period. They must be reported and a summary of them made available in the public record.

Exabyte EB. A combination of the Greek "hex," meaning "six," and the English "bite," meaning "a small amount of food." A unit of measurement for physical data storage on some form of storage device – hard disk, optical disk, RAM memory etc. and equal to two raised to the 60th power, i.e. 1,152,921,504,606,800,000 bytes.

KB = Kilobyte (2 to the 10th power)
MB = Megabyte (2 to the 20th power)
GB = Gigabyte (2 to the 30th power)
TB = Terabyte (2 to the 40th power)
PB = Petabyte (2 to the 50th power)
EB = Exabyte (2 to the 60th power)
ZB = Zettabyte (2 to the 70th power)
YB = Yottabyte (2 to the 80th power)
One googolbyte equals 2 to the 100th power.

exalted carrier reception A method of receiving either amplitude- or phase-modulated signals in which the carrier is separated from the sidebands, filtered and amplified, and then combined with the sidebands again at a higher level prior to demodulation.

EXCA Exchangeable Card Architecture. ExCA is a hardware and software architectural implementation of PCMCIA 2.0 from Intel that allows card interoperability and exchange-ability from system to system, regardless of manufacturer. See PCMCIA.

exception 1. In telecom, when something happens that's "unusual," it's an exception. The key is to define what's "unusual." For example, you might define that every phone call of longer than 15 minutes is an "exception." Now you have defined an "exception," the question is how to use that information. You might ask the phone system to print out each "exception" call on a printer next to your desk immediately after the call is over. Or you might ask the machine to print "Exceptions" reports at the end of the month listing all the calls over 15 minutes. These reports might be by perpetrator. Or in chronological order, or order of phone number called, etc. In short, any event you define by certain strict parameters can be an "exception." Management reports printed in full are almost useless because they contain so much information, so much paper. Management reports which list only previously-defined "exceptions" are more useful. They show you where to focus your attention so as to improve your or your company's performance.

2. In software programming, an exception is an interruption to the normal flow of a program. Common exceptions are division-by-zero, stack overflow, disk full errors and I/O (input/output) problems with a file that isn't open. The quality of a software program depends on how completely it checks for possible errors and deals with them. Code used for trapping errors can be excessive. Some programming languages have error trapping built in. Others don't and you have to program it in.

exception condition In data transmission, the condition assumed by a device when it receives a command that it cannot execute.

exception reports Reports generated by "exceptions," often detailing extra long calls or indications of bad circuits. See Exception.

excess burst size Be, or Burst excess. A Frame Relay term defining the maximum data rate that your carrier network will attempt to transport over a specified period of time, known as the Time Interval, known as Tc (Time committed). Any data rate above the CIR (Committed Information Rate) is above the commitment between the carrier and you, the user organization. Excess data frames, therefore, may be marked as DE (Discard Eligible). See also Committed Burst Size, and Committed Information Rate, Discard Eligible, Measurement Interval, and Time Interval.

excess insertion loss In a optical fiber coupler, the optical loss associated with that portion of the light which does not emerge from the operational ports of the device.

Excess Rate In ATM, traffic in excess of the insured rate for a given connection. Specifically, the excess rate equals the maximum rate minus the insured rate. Excess traffic is delivered only if network resources are available and can be discarded during periods of congestion.

excessive zeros More consecutive zeros received than are permitted for the selected coding scheme. For AMI-encoded T-1 signals, 16 or more zeros are excessive. For B8ZS encoded serial data, 8 or more zeros are excessive.

exchange 1. Sometimes used to refer to a telephone switching center – a physical room or building. Outside North America, telephone central offices are called "Public Exchanges."

2. A geographic area established by a common communications carrier for the administration and pricing of telecommunications services in a specific area that usually includes a city, town or village. An exchange consists of one or more central offices and their associated facilities. An exchange is not the same as a LATA. A LATA consists of several adjacent exchanges.

3. A term that refers to one of the Fibre Channel "building blocks," composed of one or more non-concurrent sequences for a single operation. See Fibre Channel.

exchange access In the telephone networks, the provision of exchange services for the purpose of originating or terminating interexchange telecommunications. Such services are provided by facilities in an exchange area for the transmission, switching, or routing of interexchange telecommunications originating or terminating within the exchange area. The Telecommunications Act of 1996 defined exchange access as follows: the offering of access to telephone exchange services or facilities for the purpose of the origination or termination of telephone toll services. See also the Telecommunications Act of 1996.

exchange area Geographic area in which telephone services and prices are the same. The concept of exchange is based on geography and regulation, not equipment. An exchange might have one or several central offices. Anyone in that exchange area could get service from any one of those central offices. It's good to ask which central offices could serve your home or office and take service from the most modern. There will be no difference in price between being served by a one-year old central office, or a 50-year old step-by-step central office.

exchange carrier Any individual, partnership, association, joint-stock company, trust, government entity or corporation engaged in the provision of local exchange telephone service.

exchange carrier code ECC. See company code.

exchange carriers association An organization of long distance telephone companies with specific administrative duties relative to tariffs, access charges and payments. See Exchange Carriers Standards Association.

Exchange Carriers Standards Association ECSA. According to their literature ECSA is "the national problem-solving and standards-setting organization where local exchange carriers, interexchange carriers, manufacturers, vendors and users rationally resolve significant operating and technical issues such as network interconnection standards and 800 database trouble reporting guidelines. The Association was created in 1983. The major committees sponsored by ECSA are The Carrier Liaison Committee (to coordinate and resolve national issues related to provision of exchange access); the Telecommunications Industry Forum (TCIF) (to respond to the growing need for voluntary guidelines to facilitate the use of new technology that offers cost savings throughout the telecommunications industry – e.g. EDI, bar coding, automatic number identification); and the Information Industry Liaison Committee (IILC) (an inter industry forum for discussion and voluntary resolution of industry wide concerns about the provision of Open Network Architecture (ONA) services and related matters and Committee T1-Telecommunications (an accredited standards group under ANSI to develop technical standards and reports for US telecommunications networks. In October, 1993, The Exchange Carriers Standards Association changed its name to the Alliance for Telecommunications Industry Solutions (ATIS). It is based in Washington, D.C. See ATIS. www.atis.org.

exchange facilities Those facilities included within a local access and transport area.

Exchange Network Facilities For Interstate Access See EN-FIA.

Exchange Message Record/Exchange Message Interface EMR/EMI. The standard format used for exchange of telecommunications message information among Local Exchange Carriers for billable, non-billable, sample, settlement and study data. EMR format is contained in Telcordia Technologies (formerly Bellcore) Publication BR-010-200-010 CRIS Exchange Messaging.

Exchange Message Record EMR. Bellcore standard format of messages used for the interchange of telecommunications message information among telephone companies. Telephone companies use EMR to exchange billable, non-billable, sample, settlement and study data. EMR formatted data is provided to all interdepartmental applications and to large customers (users) who request reproduced message records for control and allocation of their communication costs.

Exchange, Private Automatic Branch EPABX. A private telephone exchange which transmits calls internally and to and from the public telephone network.

exchange service All basic access line services, or any other services offered to customers which provide customers with a telephone connection to, and a unique telephone number address on, the PSTN, and which enable these customers to place and/or receive calls to all other stations on the PSTN.

Exchange Termination ET. In Integrated Services Digital Network (ISDN) nomenclature, ET refers to the central office link with the end user.

excise tax The long-distance federal excise tax was 3% on long-distance calls. It was originally established in 1898 to finance the Spanish-American War. On May 25, 2006 the U.S. Treasury Department announced that this tax would be discontinued. The Department of Justice will no longer pursue litigation and the Internal Revenue Service (IRS) will issue refunds of tax on long-distance service for the past three years. Taxpayers will be able to apply for refunds on their 2006 tax forms, to be filed in 2007.

excite An RFID term. The reader is said to "excite" a passive RFID tag when the reader transmits RF energy to wake up the tag and enable it to transmit back.

exciting The most boring and the most over-used word in the whole high-tech world. If I read another press release describing their shiny new product as "exciting," I'll puke. If it excites, I'll get excited. But I don't need (or want) to be told I'm about to be excited.

exclude A memory management command-line option that tells the memory manager in an MS-DOS machine not to use a certain segment of memory. For example, you may exclude upper memory locations D200 through D800 (hexadecimal) because your network

adapter card uses that space. The reciprocal term – include – specifically directs the memory manager to use an area of memory.

exclusion A PBX feature that prevents the attendant from silently monitoring a call once he/she has extended it.

exclusion zone A geographical area where radio transmissions within certain frequency bands are forbidden, in order to avoid interference with sensitive and critical operations, such as U.S. Coast Guard and coastal maritime radio stations and radio observatories.

exclusive hold Only the telephone putting the call on hold can take it off. This feature assures that the call on hold will not be picked up by someone at another telephone who can then listen to your call.

exclusive hold recall When a call is placed on "exclusive hold" and is not picked up after a predetermined amount of time, you will hear a beeping at that phone, which indicates the call is still on hold.

exclusive or private unit A circuit card installed in each key telephone set sharing the same line or intercom path that causes the first caller on the line to lock out (exclude) all other stations from using or listening in, until the line is released (or privacy feature is defeated by the active caller).

Execunet An intercity switched telephone service introduced by MCI in 1975. Execunet was the first dial-up switched service introduced by a long distance phone company in competition with AT&T. At that time, all of AT&T's competitors, including MCI, were selling full-time private lines and shared private lines. The service was named by Carl Vorder-Bruegge, MCI's VP marketing at that time and introduced and made successful by Jerry Taylor, who was MCI's regional manager in Texas and is now president of MCI. The service was the forerunner of what is today a $20 billion plus per year industry – the non-AT&T provided switched long distance business. MCI no longer uses the word Execunet to describe its switched long distance service. It's just plain long distance. Jerry Taylor started Execunet using a 104-port Action WATSBOX in Dallas, Texas. He deserves a place in the history books, not just a dictionary.

executable file A computer program that is ready to run. Application programs, such as spreadsheets and word processors, are examples of executable files. Such files in PCs running MS-DOS and Windows usually end with the BAT, COM or EXE extension.

execute To begin a task.

execution time The time needed to complete a task.

executive barge-in See Executive Override.

executive busy override See Executive Override.

executive camp on A feature for use by executives or other privileged people. When they call a someone lowly, that low person hears a special distinctive tone or sees a special light or sees a special signal that their phone has been camped on by someone significant. These days many PBXs let you know who's calling – even though you're on the phone. So executive camp-on is not that useful.

executive coach A new "profession." One of them told me this: An Executive Coach is the person who helps coach you to your goals and keep you on track on a daily basis. An Executive Coach knows everything about the company. They could step in and run it in your absence, and therefore they can help you to focus on the most important aspects of building your business!

executive override A feature of some telephone systems which permits certain users to intrude on conversations on other extensions. In some systems, executive barging-in will not be heard by the person outside the office, only the one inside the office. In some systems, such as the Mitel SX series with the Mitel Superset 4 phones, this feature activates the hands-free speakerphone of the called party, who is using his other line to speak on a normal phone conversation.

executive priority A telephone system feature that lets an executive or other designated party complete a call to a phone extension that is in privacy mode or interrupt a phone conversation on a phone that is in use.

exhaust A telephone company definition. Equipment is said to exhaust when it has reached its most limiting network access line capacity level.

exhaust date A telephone company definition. The exhaust date for an entity(s) refers to the calendar date on which the entity(s) will have reached the most limiting network access line capacity level.

exit border node An ATM term. The node that will progress a call over an outside link. This is the last node within a peer group to see this call.

exit event An event occurring in an ASC (AIN Switch Capabilities) that causes call processing to leave a PIC (Point in Call).

exit strategy In 1995, the New Yorker magazine had a cartoon showing a bride-to-be and asking her suitor on bended knee, "OK, but what's our exit strategy?" That may have been the first time the term made it into the general language. Previously, it had been confined to the venture capital business. With every new startup, they laid out a plan from the day you put your money in to the day you took your money and the mega-profits out. How you did it was the "Exit Strategy." There were and are typically two types of Exit Strategies. First, your shiny company goes public. Second, it gets bought (i.e. taken over) by someone else.

EXM Exit Message. The seventh ISUP message. It's a message sent in the backward direction from the access tandem to the end office indicating that call setup information has successfully proceeded to the adjacent network. See ISUP and Common Channel Signaling.

exoneree A person who has been convicted of a crime

EXOS Abbreviation for EXtension OutSide; a phone connected to a key system based in another building. The wiring belongs to the telephone company, even though the phone equipment may not. Unlike an OPX, the circuit between the two locations does not pass through a central office.

exosphere This region lies beyond an altitude of about 400km from the surface of the earth. The density is such that an air molecule moving directly outwards has an even chance of colliding with another molecule or escaping into space.

exotic media network A network made up of exotic communication media such as satellites, deep-space spacecraft with propagation delays measured in seconds or minutes, communication devices that use acoustic modulation in air or water, and various types of optical communications devices. Exotic media networks may have high latencies, and predictable or unpredictable interruptions lasting seconds, minutes or hours.

expanded diameter Diameter of shrink tubing as supplied. When heated, the tubing will shrink to its extruded diameter.

expanded spectrum A cellular telephone term for having the full 832-channel analog cellular spectrum currently available to you, the user of the cellular phone.

expander That device in a transmission facility which expands the amplitude of received compressed signals to their approximate normal range. The receiving side of a compandor.

expandor See Expander.

expansion The switching of a number of input channels, such as telephone lines onto a larger number of output channels.

expansion carrier An AT&T Merlin term. A carrier added to the control unit when the basic carrier cannot house all the modules needed. An expansion carrier houses a power supply module and up to six additional modules.

expansion slots In a computer there are card slots for adding accessories such as internal modems, extra drivers, hard disks, monitor adapters, hard disk drivers, etc. Most modern PBXs are actually cabinets with nothing but expansion slots inside. Into these slots we fit trunk cards, line cards, console cards, etc. Some phone systems have "universal" slots, meaning you can put any card in any slot. Some phone systems have dedicated expansion slots, meaning that they expect only a certain card in that slot. In the PC industry, many manufacturers make cards for IBM and IBM compatible slots. In the phone industry, nobody makes cards for expansion slots in anyone else's phone system. A reader once asked me why? Here's the answer I gave him: The reason no one makes expansion cards for phone systems is because the phone industry doesn't trust anyone else. They don't issue open standards and the manufacturers are NOT committed to keeping their cages and their buses the same from one model of phone system to another. Manufacturers are afraid that outsider manufacturers would screw up the much-vaunted reliability of phone systems. And they have a point - people don't want to reboot phone systems like they reboot PCs. See also EISA and MCA.

expansive controls Control applications that reroute traffic from congested portions of the network to trunk routes and switching systems having available capacity.

exparte presentation An FCC definition. Any communication addressing the merits or outcome of a particular proceeding made to decision-making personnel (or in some proceedings, from the decision-making personnel), which (1) if written, is not served on the parties to the proceeding, or (2) if oral, is made without opportunity for the parties to the proceeding to be present.

expedited delivery Option set by a specific protocol layer telling other protocol layers (or the same protocol layer in another network device) to handle specific data more rapidly.

experiment See Shannon's Law.

expert What we all become when we're away from the office. See Expert System.

expert system A very sophisticated computer program consisting of three parts. 1. A stock of rules or general statements, e.g. Some long distance phone calls are free. These rules are generally based on the collective wisdom of human "experts" who are interviewed. 2. A set of particular facts, e.g. Three companies provide the bulk of long distance service in the United States. 3. Most importantly, a "logical engine" which can apply facts to rules to reach all the conclusions that can be drawn from them – one of which might be "Three companies give away long distance phone calls." (Which would be wrong.) The idea of expert systems is to help people solve problems. For example, Compaq is trying to improve its customer service by installing automated assistants that work on the principle that reasoning is often just a matter of remembering the best precedent. The simplest expert systems, according to the Economist Magazine, assume that their rules and facts tell them everything there is to know. Any statement that cannot be deduced from the system's rules and facts is assumed to be false. This can lead machines to answer "YES" or "NO," when they should say "I don't know." Slowly we are beginning to find ways of dealing with the inflexibility of machines. One such gadget is a "truth maintenance machine" invented by Dr. Jon Doyle of MIT. As each fact is fed into the system, Dr. Doyle's program checks to see if it (or the deductions derived from it) contradict any of the facts or deductions already in the system. If there is a contradiction, the machine works backward along its chain of reasoning to find the source and dispose of that troublesome fact or deduction. So the system maintains one consistent set of beliefs.

expire interval For DNS, the number of seconds that DNS servers operating as secondary masters for a zone will use to determine if zone data should be expired when the zone is not refreshed and renewed.

explicit access In LAN Technology, explicit access is a shared access method that allows workstations to use the transmission medium individually for a specific time period. Every workstation is guaranteed a turn, but every station must also wait for its turn. Contrast with Contended Access. See also CSMA.

exponential back-off delay The back-off algorithm used in IEEE 802.3 systems by which the delay before retransmission is increased as an exponential function of the number of attempts to transmit a specific frame.

exponential holding time A telephone company term. A great number of calls are assumed to have a relatively short holding time while decreasing numbers of calls are assumed to have longer and longer holding times out to the point where a very small number of calls exhibit exceedingly long holding times.

export Imagine you have a software program, like a spreadsheet or a database. And you have information in that program. Let's say it's Microsoft Word or Lotus 123. And you want to get it into a different program, say to give it to a workmate who uses WordPerfect or Excel. You have to convert it from one format to another. From Word to WordPerfect or from Lotus to Excel. That process is typically called "exporting." And you'll typically see the word "EXPORT" as a choice on one of your menus. The opposite is called importing. See Import.

export script First read my definition of EXPORT. An export script is a series of specifications which control the export process. It contains the fields to be sent, which records to be sent, the name of file to send as well as the name of the import script (if there is one) located at the receiver's end which will control the merge. See Export.

export server A Windows NT term. In directory replication, a server from which a master set of directories is exported to specified servers or workstations (called import computers) in the same or other domains.

express call completion Someone calls an information operator. "What is the name?" the operator answers. "Here is the number. Would you like me to get that number for you now? If so, please hit 1." Express Call Completion lets the operator complete the call for you while you're on line. Express Call Completion was begun in September of 1990 by Pacific Bell using a Northern Telecom central office. Express Call Completion is part of Nortel Networks' Automated Directory Assistance Call Completion (ADACC) software and Traffic Operator Position Systems Multipurpose (TOPS MP).

ExpressCard Once upon a time there was a PCMCIA card that fit into laptops and was used for everything from tieing the laptop into a local area network and later for tieing it into a wireless network, or powering external monitors. Think of the PCMCIA card as serving the same function as a card you plug into a bus inside a desktop PC. The card was originally called a PCMCIA card, then it became known as a PC Card. Then the specs got improved – i.e. got faster – and it became known as a CardBus. But the physical dimensions were typically the same, i.e. 85.6 mm long by 54 mm wide. Then, the PCMCIA group came out with something called ExpressCard, which is faster, better and, of course, smaller. The ExpressCard comes in two sizes – ExpressCard54 – which is 75 mm long by 54 mm wide (with a small 22mm square cutout at the top and ExpressCard34 which is rectangular and measures 34 mm by 75 mm. ExpressCard will support both the USB 2.0 and PCI Express interfaces. New ExpressCards in the works in include those for WIMAX, digital TV tuners, flash memory, Bluetooth connectivity, Gigabit Ethernet connectivity, etc.

express client installation An Internet Access Term. A client installation scripting utility that enables network managers to establish defined defaults that make client installation and deployment easier.

express orderwire A permanently connected voice circuit between selected stations for technical control purposes.

EXT See Extension.

extend A verb used by the phone industry to describe an operator transferring a call to a telephone extension. The word is used thus: The operator extended the call to Mr. Smith on extension 200. "Putting a call through" is a clearer way of saying "extending" a call. The word "extend" probably comes from the old days when the operator extended her arm to plug you in on her cordboard.

extended addressing In many bit-oriented protocols, extended addressing is a facility allowing larger addresses than normal to be used. In IBM's SNA, the addition of two high-order bits to the basic addressing scheme.

extended area service A geographic area beyond the local service area to which traffic is classified as local for selected customers, i.e., telephone service that allows subscribers in one exchange to call subscribers of another exchange without a toll charge. Sometimes subscribers may be given the option of paying more for the privilege of calling these more distant phone companies. Sometimes, they have no option. The local public service commission deems EAS a "good idea." And, typically, everyone's monthly rate for basic telephone service goes up.

Extended ASCII ASCII is a seven bit code. Extended ASCII adds an eighth bit. The eighth Bit is called a parity bit and is used for error checking, not error correction. See ASCII.

Extended Binary Coded Decimal Interchange Code EBCDIC. (Pronounced Eb-Si-Dick.) An IBM standard of coding characters. It's an 8-bit code and can represent up to 256 characters. A ninth bit is used as a parity bit. See Parity and EBCDIC.

extended BIOS data area In PCs, extended BIOS data area is 1KB of RAM located at 639KB. It is used to support extended BIOS functions including support for PS/2.

extended call management A Nortel Networks term for a collection of features being added to its DMS Meridian central office Automatic Call Distribution (ACD) service. Using Switch-to-Computer Applications Interface (SCAI), ECM will work with user-provided computer equipment to integrate call processing, voice processing (recorded announcements, voice mail and voice response) and data processing. For example, ECM will allow an outboard computer device to coordinate the presentation of customer data on the ACD agent's computer screen with an incoming call. The D channel of an ISDN Basic Rate Interface (BRI) serves as the transport mechanism from the DMS-100 central office switch to an outboard computing device. Communication is peer-to-peer, meaning that neither the switch or the computer is in a "slave" relationship to the other. The application layer messaging – i.e. layer 7 messaging as defined by the Open Systems Interconnection (OSI) reference model – is in the Q.932 format and is designed to conform to the T1S1 SCAI message protocol.

extended character set The characters assigned to ASCII codes 128 through 255 on IBM and IBM-compatible microcomputers. These characters are not defined by the ASCII standard and are therefore not "standard." See Extended Graphics Character Set.

Extended Definition Television EDTV. Television that includes improvements to the standard NTSC television system, which improvements are receiver-compatible with the NTSC standard, but modify the NTSC emission standards. Such improvements may include (a) a wider aspect ratio, (b) higher picture definition than distribution-quality definition but lower than HDTV, and/or (c)any of the improvements used in improved-definition television. When EDTV is transmitted in the 4:3 aspect ratio, it is referred to simply as "EDTV." When transmitted in a wider aspect ratio, it is referred to as "EDTV-Wide."

extended digital subscriber line The ISDN EDSL combines 24 B-channels and one 64-Kbps D-channel on a single line, ISDN primary rate interface.

extended graphics character set The characters assigned to ASCII codes 128 through 255 on IBM and IBM-compatible microcomputers. These characters are not defined by the ASCII standard and are therefore not "standard." The original ASCII code used a seven bit one-or-zero code. There are two to the seventh power, or 128 possible combinations. The IBM PC uses a 16-bit CPU with an eight bit data bus and thus transmits data internally in eight bit bytes. Instead of using the seven bit ASCII code, the PC uses the equivalent eight bit code, by simply making the left most digit, a zero. In seven bit code,

an R is 1010010. In 8-bit, it's 01010010. The only difference between the first 128 characters and the second 128 characters is that in the second, the first bit is a 1.

extended hamming code See Hamming Code.

extended Internet The extended Internet (X Internet) consists of technologies for connecting the digital world to the physical world, e.g. computers to connect vehicles, pallets of consumer products, smart devices, mobile devices and the human body. This is the all about technologies like RFID, telematics, sensor networks, and the network protocols and bandwidth that link them together. The idea is to link machines together. The X Internet also encompasses the sorting, sifting, and analysis of data gathered by networks of intelligent devices.

extended LAN A term used to describe a network that consists of a series of LANs connected by bridges.

extended life battery Toshiba's wonderful name for a battery for a laptop that's physically larger than the normal one, costs more than the normal one and lasts longer than the normal one. Extended life batteries make lightweight laptops into heavyweight laptops.

extended key code The two digit code that represents pressing a key outside the typewriter portion of the keyboard, such as a function key, cursor-control key or combinations of CTRL (control) and ALT keys with another key. The first number is always 0 (zero) and is separated from the second number by a semicolon.

extended LAN A collection of local area networks connected by protocol independent devices such as bridges or routers.

extended memory Memory beyond 1 megabyte in 80286, 80386, 80486 and Pentium computers. Windows uses extended memory to manage and run applications. Extended memory can be used for RAM disks, disk caches, or Microsoft Windows, but requires the processor to operate in a special mode (protected mode or virtual real mode). With a special driver, you can use extended memory to create expanded memory. Extended memory typically is not available to non-Windows applications or MS-DOS. See also Expanded Memory.

Extended Superframe Format ESF. A new T-1 framing standard used in Wide Area Networks (WANs). With this format 24 frames – instead of 12 – are grouped together. In this grouping, the 8,000 bps frame is redefined as follows:

2,000 bps for framing and signaling to provide the functions generally defined in the D-4 format.

2,000 bps are CRC-6 (Cyclic Redundancy Check-code 6) to detect logic errors caused by line equipment, noise, lightning and other interference. Performance checking is done by both the carrier and the customer without causing any interference with the T-1 traffic.

4,000 bps are used as a data link. This link is to perform functions such as enhanced end-to-end diagnostics, networking reporting and control, channel or equipment switching, and/or optional functions or services. See also T-1 Framing and D-4 Framing.

extended superframe monitoring unit ESFMU. An MCI definition. Placed on customer data circuits to provide performance monitoring throughout MCI's Digital Data Network.

extended telephony level The lowest level of service in Windows Telephony Services is called Basic Telephony and provides a guaranteed set of functions that corresponds to "Plain Old Telephone Service" (POTS - only make calls and receive calls). The next service level is Supplementary Telephone Service providing advanced switch features such as hold, transfer, etc. All supplementary services are optional. Finally, there is the Extended Telephony level. This API level provides numerous and well-defined API extension mechanisms that enable application developers to access service provider-specific functions not directly defined by the Telephony API. See Windows Telephony Services.

extended text mode Standard text mode is 80 columns wide. So-called extended text mode is 132 columns wide. This mode allows you to view more text on-screen when using such applications as Lotus 1-2-3.

extensible In strictest terms, the word means "capable of being extended." When Microsoft introduced its At Work operating system on June 9, 1993 it said that one of the operating system's key features was that it was "extensible." Microsoft's explanation: The software is designed to allow both manufacturers and customers to add new features. For example, local area network connectivity will be able to be added easily by installing an optional LAN hardware module and a software driver. Additional memory will be able to be added to the system, and the system will automatically make use of this memory. New image-processing software and communications protocols will be able to be added on the premises, and it will even be able to be done over the phone line, allowing manufacturers to create basic models that can be enhanced in many different ways to fit the needs of different user groups.

extensible authentication protocol See EAP.

extensible firmware interface The EFI is a tiny, secure operating system that sits between the hardware of a PC – or any computing device – and the high-level operating system (like Windows or Linux) that humans (like you and me) normally interact with. Although the EFI can emulate a traditional BIOS, it also can do much more. For example, it can provide a full mouse-driven graphical interface for controlling the low-level hardware functions that today can only be controlled by hitting a special key at startup and entering a limited, arcane, and text-only "BIOS Setup" routine.

But that's only the beginning, because EFI is really a kind of blank slate that will allow a total rethinking of how computers start up. For example, a traditional BIOS is space-limited, so most are programmed in compact, low-level "machine language," which is notoriously difficult to do well–in fact, very few engineers are proficient in machine language. In contrast, EFI is written in C, the world's most popular high-end programming language, and EFI isn't space-constrained because its data resides in a special reserved area of the hard drive. This means that far more engineers will be able to do more creative things with PC hardware than is now possible. There's no telling where EFI will lead, but it almost surely will initially result in new forms of system maintenance, repair, and recovery tools; new ways to install operating systems and peripherals; and more.

For example, at last year's Intel Developer Forum, Intel's Yosi Govezensky reeled off several likely EFI near-term spin-offs:

Portable, operating-system-neutral disk-management and boot-management tools

Remote configuration and installation options

Platform management utilities outside the operating system, such as a bootable flash update CD without DOS, customer-support utilities, and country (regionalization) kits.

Intel's EFI project began in 2000, and many companies, including Adaptec, AMI, ATI, Hewlett-Packard, LSI, Microsoft, and PowerQuest, are working with Intel to make EFI a production reality. EFI is steadily gathering momentum and is definitely a technology to watch. More info is available at http://www.intel.com/ technology/efi/efi.htm .

extensibility This means it's easy to add new technologies without reinventing the wheel.

extensions 1. Additional telephones connected to a line. Allows two or more locations to be served by the same telephone line or line group. May also refer to an intercom phone number in an office.

2. Extensions are small add-ons that add new functionality to popular software. An example is the Firefox browser. Firefox extensions can add anything from a toolbar button to a completely new feature. They allow the application to be customized to fit the personal needs of each user if they need additional features, while keeping Firefox small. There are around a thousand extensions for Firefox, fewer for Internet Explorer. See www.addons.mozilla.org.

3. The optional second part of an PC computer filename. Extensions begin with a period and contain from one to three characters. Most application programs supply extensions for files they create. Checking a file's extension often tells you what the file does or contains. For example, most BASIC files use a filename extension of .BAS. Most backup files have an extension of .BAK. MS-DOS programs have .EXE or .COM. dBASE database files have the extension .DBF and .DBT. Paradox files have the extension .DB. Files of sounds have their own extensions. Here are the typical extensions on sound files of various computers:

Microsoft Windows – .wav

Apple – .aif

NeXT – .snd

MIDI – .mid and .nni

Sound Blaster – .voc

Here are the typical extensions on graphics formats: .TIFF, .EPS, .CGM, .PCX, .DRW, .WMF, and .BMP

Extensions are also short for domain name extensions.

extension cord A multi-conductor, male/female modular line cord generally used to permit greater separation between the Communications Outlet and the telephone equipment. Available in various lengths up to 25 feet. May be of tinsel or stranded wire construction.

extensions to UNIX Extensions to UNIX are additional features or functions not found in the standard UNIX implementation. The extension are classified either as "open extensions" or "proprietary extensions." An "open extension" usually consists of a surface addition, such as a driver for peripheral or a software patch for a new mode of I/O. The "open extension" is transparent to standard UNIX and its application programs. The "pro-

prietary extension" is for the implementation of custom hardware or software. It results in a version of UNIX that is not transparent to UNIX and its applications.

exterior An ATM term. Denotes that an item (e.g., link, node, or reachable address) is outside of a PNNI routing domain.

exterior link An ATM term. A link which crosses the boundary of the PNNI routing domain. The PNNI protocol does not run over an exterior link.

exterior reachable address An ATM term. An address that can be reached through a PNNI routing domain, but which is not located in that PNNI routing domain.

exterior route An ATM term. A route which traverses an exterior link.

External Data Representation See XDR.

External F-ES A Fixed End System (F-ES) connected to the CDPD network outside the administrative domain of the service provider.

external interface A cellular radio term. The Cellular Digital Packet Data (CDPD)-based wireless packet data service provider's interface to existing external networks. The external application service providers communicate with CDPD subscribers through this external interface.

external interference The effects of electrical waves or fields which cause spurious signals other than the desired intelligence.

external memory Storage devices, such as magnetic disks, drums or tapes which are outside (externally attached) to the main telephone or computer system.

external meter modem A radio modem that is mounted on the outside of an electric, gas or water meter on the premises of a large commercial or industrial customer, and which interfaces with the meter through an external data port. The external meter modem transmits meter data via radio to a hub, sometimes via one or more relays, for forwarding to a host computer system at the utility company.

external modem A modem external to the computer, it sits in its own little box connected to a computer through the computer's serial port. Compare with an internal modem, which typically comes on one printed circuit card and is placed into one of the computer's expansion slots and thus connects to the computer through the computer's "backplane." Internal modems cost less because they don't need any external housing and separate power supply. But because they're mounted inside the computer, it's harder to see what they're doing. You can't see the various status lights, like OH (for Off-Hook) and CD (for Carrier Detect). They also take up valuable slots instead of a serial port.

external photoeffect In fiber optics, an external photoeffect consists of photon-excited electrons that are emitted after overcoming the energy barrier at the surface of a photo-emissive surface.

external storage See also External Memory.

external timing reference A timing reference obtained from a source external to the communications system such as one of the navigation systems. Many of which are referenced to Coordinated Universal Time (UTC).

external viewer This is the program that is launched or used by Web browsers for presenting graphics, audio, video, VRML, and other multimedia found on the Internet. Sometimes referred to as helper applications. Usually when you initially set up your browser you configure what external viewers you want to use by associating a program with a file type or extension. This way the browser knows what to do when these files are "clicked on" by the user.

EXTN Extension.

extractor A combination of a special modem and a coupler used to extract voice, data and video signals from an electrical power line. The term is used in Broadband over Power Line (BPL) technology. See BPL.

extranet An extranet, coined by Bob Metcalfe in the April 8, 1996 issue of InfoWorld, is a Internet-like network which a company runs to conduct business with its employees, its customers and/or its suppliers. Extranets typically include Web sites that provide information to internal employees and also have secure areas to provide information to customers and external partners like suppliers, manufacturers and distributors. A company might place a call for product on its extranet's web site. Its suppliers will check the site regularly and bid on the product. It is called an extranet because it typically uses the technology of the public Internet (TCP/IP and browsers) and customers and suppliers often access the extranet through the Internet via their local ISP – Internet Service Provider. But an Extranet is not a public entity. You typically need accounts and passwords, typically issued by the firm running the Extranet. The word Extranet, however, is a term in evolution. In the October 21, 1996 issue of InfoWorld, Bob Metcalfe defined Extranets as IP networks through which companies run Web applications for external use by customers. He explained that ISPs have deployed private networks (i.e. extranets) which operate on the same principles and make use of the same network technologies as the Internet, but are external to it. Access to the Internet can be gained through an Extranet. Extranets, he said, were for electronic commerce. See also Internet and Intranet.

Extremely High Frequency EHF. Frequencies from 30 GHz to 300 GHz.

Extremely Low Frequency ELF. Frequencies from 30 Hz to 300 Hz.

extrinsic joint loss For an optical fiber, that portion of a joint loss that is not intrinsic to the fibers, e.g., loss caused by end separation, angular misalignment, or lateral misalignment.

extrusion Method of continuously forcing plastic, rubber or elastometer material through an orifice to apply insulation or jacketing over a conductor or cable core. In short, the mechanical process of coating a wire or group of wires with insulating material.

extrusion detection The inverse of intrusion detection. Extrusion detection involves monitoring outbound traffic from a network so as to detect spam, viruses, denial of service attacks, and other network abuses that are originating from inside the network.

eye candy 1. A term used in information technology for visual elements displayed on computer monitors that are aesthetically pleasing or attention-getting.

2. A popular plug-in for Adobe Photoshop that provides extra "filters" for simulating various effects.

3. Donald Trump's girlfriends. They're all gorgeous. That's their function. See also Arm Candy and Wrist Candy.

eye pattern An oscilloscope display used to visually determine the quality of an equalized transmission line signal being received. So called because portions of the pattern appearing on the scope resemble the elliptical shape of the human eye.

eye phone Several researchers are studying something they call "virtual reality." One version of it, a system developed by a company called VPL Research, Redwood City, CA, is based around three things: a three-dimensional display worn on the head (called an Eye Phone), an electronic glove (the Data Glove) and a high-speed computer. The whole system cost $250,000 in the fall of 1990, which may be one reason we haven't seen or heard much of it since.

eyeball A viewing audience. "There are plenty of new eyeballs available in this time slot." Also refers to the number of people visiting your Web site.

eyeball shot Also called microwave link. The link is made by two radio transceivers equipped with parabolic dish antennas pointed at each other. The transmissions can be carried on many bandwidths including DS1, DS2, DS3, STS1 and OC1. The range varies from 0-50 miles depending on the dish size, weather and transmitter power.

EZTV Software developed under direction of Fabrice Florin and Peter Maresca of Apple Computer's Discovery Studio. EZTV was an interactive system intended to enable consumers to order movies, go shopping and play games on their TV sets.

F The symbol designation for frequency is "f."

F connector A 75 ohm coaxial cable connector commonly found on consumer television and video equipment.

F link Fully Associated Link. A link used to connect two SS7 signaling points when there is a high community of interest between them and it is economical to link them. Also called associated signaling.

F port Fabric port. A Fibre Channel term, referring to the port residing on the fabric (switch) side of the link. It attaches to a N Port (Node Port) at the connected device, across a link. See Fibre Channel.

F region The highest region of the ionosphere. It is made up of two subregions: F1 and F2. The F region refracts radio waves and returns them to Earth. The height of the F region varies by time of year, time of day, and sunspot activity.

F type connector A low cost connector used by the TV industry to connect coaxial cable to equipment. See also F-Type Connector (which is the same thing, except spelled with a dash.)

F/A 1. Fault alarm.
2. Foreign administration.

F-BCCH Fast-Broadcast Control CHannel. A logical channel element of the BCCH signaling and control channel used in digital cellular networks employing TDMA (Time Division Multiple Access), as defined by IS-136. See also BCCH, IS-136 and TDMA.

F-Block Carrier A 10 MHz PCS carrier serving a Basic Trading Area in the frequency block 1890 - 1895 MHz paired with 1970 - 1975 MHz.

F-ES Fixed End System.

F-Type Connector These are used to terminate coaxial cable. This connector is mostly used for video applications. It's a male single-conductor connector and screws into the female jack.

F2F 1. Face to Face. When you actually meet someone with whom you have been corresponding electronically, perhaps through a chat room over the Internet. F2F often is quite a surprise, as your "pen pal" may not be anything like he said he was. F2F also can be very dangerous. Never, ever meet someone F2F unless you have a companion with you and you meet in a well-lit public place. Never, ever give the other person your real name, address or telephone number until you have met him F2F and are confident that he is who he says he is. Tell your children to never, never, ever agree to meet someone F2F unless you approve in advance and you are with them at the meeting. This is a very, very dangerous world full of very, very dangerous people who prey on the unsuspecting.
2. Friend to friend network. A variation on darknets. See darknets.

FAA Federal Aviation Administration. If you build a tower of over 200 feet in height you must install FAA approved warning lights on the top. Anything under 200 feet does not need

lighting. You can paint your towers all different colors but the most common (for cell phone towers) is orange and white in 20 foot sections.

faber Factory that makes ("fabricates") IC chips.

fabless Fabless is a term used to describe a company that designs and sells semiconductor chips but doesn't own a semiconductor manufacturing factory, also called a fabrication plant. Such fabless manufacturer has others make their chips under contract. There are arguments for owning and not owning factory to make semiconductor chips.

fabric 1. A descriptive term referring to the physical structure of a switch or network. Much like a piece of cloth, physical/logical communications channels (threads) are interwoven from port-to-port (end-to-end). Ideally, data are transferred through this switch or network on a seamless basis. In ATM and Fibre Channel, the switching fabric generally is non-blocking, or virtually so, from port-to-port. In the Internet, data works its way through a complex, and even unpredictable, interwoven network of networks comprising transmission facilities, packet switches and multiple carriers. See also Fabric Application Platform.
2. Multiple Fibre Channel switches interconnected and using Fibre Channel methodology for linking nodes and routing frames in a Fibre Channel network. See Fibre Channel.

fabric application platform What Brocade, a maker of FC (Fibre Channel) switches, calls a technique for running storage software in the network rather than in the host or the FC array. See also Fabric.

fabric blindness An inability to look deeply into the guts of a storage area network to identify the root cause of a performance problem.

Fabry-Perot Laser A low cost, long wavelength laser used in short and intermediate reach transmitters. Because of a wide spectral width, these devices are not typically used in WDM systems. Lasers are available in either 1310nm or 1550nm wavelength windows. Most Gigabit Ethernet 1000Base-LX modules use 1310nm Fabry-Perot lasers.

FAC See Forced Authorization Code.

face I never forget a face, but in your case I'll be glad to make an exception – Groucho Marx.

face time Time in front of your boss. Time in front of someone important. Time where you show your face as against working from home or working on a trip. Time where you pitch your ideas to a venture capitalist, as against sending emails or business plans. Time when you show your face, or see someone else's face in the flesh. A fairly stupid term.

faceplate A cover that fits around the pushbuttons or rotary dial of a telephone. Hotels and motels put instructions on them. More businesses should also.

facilitate An overly pompous way to say "to make easy" or "to make it happen." Ditto for utilize when you mean use.

facilities A stupid, imprecisely defined word that means anything and everything. To me it sounds like toilets. But it's not. It can mean the equipment and services which make up a telecom system. It can mean offices, factories, and/or building. It can be anywhere you choose to put telecom things. Oops, I nearly said telecom facilities. So "facilities" means practically anything you want it to mean so long as it covers a sufficiently broad variety of "things" which you haven't got a convenient name for. "Facilities" sounds better than "things", especially if you want to sound pompous and erudite.

Facilities Administration And Control A PBX feature which allows you, the subscriber, to assign to your users features and privileges like authorization codes, restriction levels and calling privileges.

facilities assurance reports This feature allows a subscriber to get an audit trail of the referrals produced by the automatic circuit assurance feature of some PBXs. The audit trail will identify the trunk circuit, the time of referral, the nature of the problem and if a test was performed, the outcome of the test.

facilities based carrier A telecommunications carrier which owns most of its own facilities (i.e., stuff), such as switching equipment and transmission lines. A non-facilities based carrier is one which leases most of its switching and lines from others. There are probably no 100% facilities based carriers in the world, since even the old government monopoly phone companies leased international lines and all today's competitive carriers lease circuits from each other. There are three benefits to being a mostly-facilities based carrier:

1. In the long run it tends to be cheaper to own your own plant.

2. You can roll out new features faster. You don't have to rely on someone else.

3. You can set your own standards of service and thus achieve better network integrity than if you had to rely on bits and pieces of other peoples' networks.

ILECs (Incumbent Local Exchange Carriers) such as Bell Atlantic, BellSouth, and SBC and IXCs (IntereXchange Carriers) such as AT&T and MCI Worldcom are facilities based carriers under this definition. These IXCs have switching offices, or POPs (Points Of Presence) in all service areas of the United States and provide both originating and terminating service nationwide. Major facilities based carriers sell their services to business and residential users and to other carriers which resell those services.

Non facilities based long distance carriers are known as switchless resellers. To be recognized as a CLEC (Competitive Local Exchange Carrier) by most local regulatory authorities in the United States and to receive reciprocal compensation from the local ILEC, you must, at minimum, own a central office switch; thus you must be a facilities based carrier to some degree. There's probably not one single carrier – local, long distance or international – in the entire North America that is 100% facilities based these days. Everyone seems to be renting someone else's lines. The most facilities based would be the ILECs. The least facilities based would be the CLECs (Competitive LECs). They tend to resell local loops from the local ILEC which they terminate in their own switching centers. See also CLEC, ILEC, IXC and POP.

facilities data link FDL. An Extended SuperFrame (ESF) term. ESF extends the superframe from 12 to 24 consecutive and repetitive frames of information. The framing overhead of 8 Kbps in previous T-1 versions was used exclusively for purposes of synchronization. ESF takes advantage of newer channel banks and multiplexers which can accomplish this process of synchronization using only 2 Kbps of the framing bits, with the framing bit of only every fourth frame being used for this purpose. As a result, 6 Kbps is freed up for other purposes. This allows 2 Kbps to be used for continuous error checking using a CRC-6 (Cyclic Redundancy Check-6), and 4 Kbps to be used for a FDL which supports the communication of various network information in the form of in-service monitoring and diagnostics. ESF, through the FDL, supports non-intrusive signaling and control, thereby offering the user "clear channel" communications of a full 64 Kbps per channel, rather than the 56 Kbps offered through older versions as a result of "bit robbing." FDL implementations can vary, and the 4Kbps data bandwidth can be allocated among different functions, such as managing line side and equipment side operations. For example, ADC just introduced a new HDSL service to compete with T-1. ADC allocates 2 Kbps of the FDL channel for managing the HDSL interface and 2K bps for remotely managing attached channel banks and data terminals. As you can see, FDL has great potential for reducing the cost of onsite provisioning and maintenance. See also ESF and T-1.

facilities management Also called Outsourcing, facilities management is having someone else run your computers or your telecommunications system. The concept is that you're a great bank and you should concentrate on being in the banking business. Your outside facilities manager should concentrate on running your computers or telecom systems. He can do it cheaper, allegedly. Ross Perot's Electronic Data Systems (EDS) probably started facilities management. Mr. Perot incorporated EDS on June 27, 1962. At that time he was a leading IBM salesman. See also Outsourcing.

facilities restriction level Which types of calls a PBX user is entitled to make.

facility A telephone industry term for a phone or data line. Sometimes (but rarely) used to describe equipment. See Facilities for a far better description.

facility compensation Compensation for the use of direct trunk transport, based on relative usage. Relative usage should reflect the percentage of use for all carriers involved. Traffic studies to determine relative usage are typically done only on local terminating minutes of use during the busy hour.

facility grounding system The electrically interconnected system of conductors and conductive elements that provides multiple current paths to the earth electrode subsystem. The facility grounding system consists of the earth electrode subsystem, the lightning protection subsystem, the signal reference subsystem, and the fault protection subsystem. Faulty grounding causes more phone and computer problems than any other single factor.

facility work order An order to a phone company to rearrange things.

FACS 1. How they abbreviate the word facsimile in Bermuda.

2. Facilities Access Control Systems. A collection of dozens of interrelated computer applications developed by the former AT&T Bell Operating Companies which manage the local loops connecting customers to the Public Switched Telephone Network.

facsimile equipment FAX. Equipment which allows hard copy (written, typed or drawn material) to be sent through the switched telephone system and printed out elsewhere. Think of a fax machine as essentially two machines – one for transmitting and one for receiving. The sending fax machines typically consists of a scanner for converting material to be faxed into digital bits, a digital signal processor (a single chip specialized microprocessor) for reducing those bits (encoding white space into a formula and not an endless series of bits representing white), and a modem for converting the bits into an analog signal for transmission over analog dial-up phone lines. The receiving fax consists of a modem and a printer which converts the incoming bits into black and white images on paper. More modern and more expensive machines also have memory – such that if the machine runs out of paper, it will still continue to receive incoming faxes, storing those faxes into memory until someone fills the machine with paper and it prints the faxes out.

There are six internationally accepted specifications for facsimile equipment. Group 1, Group 2, Group 3, Group 3 Enhanced, Super GE and Group 4. Only 1, 2, 3 and 3 Enhanced will work on "normal" analog dial-up phone lines. Group 4 is designed for digital lines running at 56/64 Kbps, e.g. ISDN lines. Among the analog line fax machines, Group 2 is faster than Group 1. Group 3 is faster than Group 2, etc. Virtually all machines sold today are Group 3, though an increasing percentage are Group 3 enhanced, which has speeded up Group 3's transmission speed from 9,600 bps to 14,400 bps and improved its error correction. Group 3 faxes send an 8-1/2 x 11 inch page over a normal phone line in about 20 seconds. How much time it actually takes depends on how much stuff is actually on the paper. Unlike older machines, Group 3 machines are "intelligent." They only transmit the information that's on the paper. They do not transmit white space, as earlier machines did. Super G3 is a new "standard" for higher speed fax machines, which contain a 33.6 Kbps V.34 modem, V.8 handshaking and the new ITU-T T.85 JBIG image compression. On most phone lines such a machine should get close to double the speed of the highest speed Group 3 fax machines – 14.4 Kbps. But, the JBIG image compression will speed faxing of gray scale images by as much as five to six times. In short, these machines will send faxes much faster – if they send to a Super G3 machine at the other end. Super G3 is compatible with and can communicate with older fax machines, Group 1, 2, 3 and 3 Enhanced.

When a fax machine calls a phone line, it emits a standard ITU-T-defined, "CNG tone" (calling tone) – 1100 Hz tone every three seconds. When the receiving fax machine hears this tone, it knows it's an incoming fax call and it can automatically connect. With this tone it is possible to insert a "fax switch," which would "listen" for an incoming fax call and switch it to a fax machine if it heard the CNG tones or to something else – like a phone or answering machine – if it didn't. It is not possible to do this with a modem. A calling modem does not issue any tones whatsoever. A modem works backwards – when the receiving modem answers the phone, it emits a tone.

Typically, a Group 3 machine can speak to a Group 2 and a Group 1 machine. A Group 2 can speak to a Group 1. Speaking down means slowing down. Fax machines are dropping in price. "Personal" fax machines are emerging. Most fax machines today at Group 3 or Group 3 enhanced.

Group 4 machines are 100% digital, transmit at 64,000 bps and directly attach to the B (bearer channel) of a digital ISDN line. They will transmit a sheet of 8 1/2 x 11 paper in under six seconds. The author of this dictionary has seen a working Group 4 fax machine. It's mighty impressive.

The latest ITU standards include T.37 and T.38. T.37 is a new ITU standard for transferring of facsimile messages via store and forward over packet-switched IP networks – the Internet, corporate Intranets, etc. T.38 is a new ITU-T standard for sending real-time facsimile messages over packet-switched IP networks – the Internet, corporate Intranets, etc.

Some warnings on fax machines:

1. All analog Group 3 and enhanced Group 3 fax machines pose a security risk. Anyone can attach a normal audio cassette recorder to a phone line, record the incoming or outgoing fax "tones" of an analog fax machine. By playing back to another fax machine at a later time, you'll get a perfect reproduction of the fax. There are now fax encryption devices which make the fax transmission unintelligible to any machine other than the one it's intended for – i.e. the one that has a similar un-encryption device.

2. Some plain paper fax machines present a different security risk. Some (not all) use a carbon ribbon the width of their paper. As a result, if you want to read what came in, you simply read the carbon ribbon, which you open like a scroll, which the cleaning lady finds in the trash. These machines are increasingly less common, as plain paper fax machines acquire laser printing engines.

3. Most fax machines record all the digits dialed into them which were used to set up a fax call. If a fax machine is sitting behind a PBX (as many are these days) it will capture all the confidential authorization codes of all the company's employees. To get those codes all you need do is ask the machine to print out a report. There is no easy solution to this problem as at the time of writing this dictionary, except that some fax makers have told me they intend to obscure these numbers on their reports, at some stage. Some may, by the time you read this.

4. Slimy paper fades. How long it takes to fade depends on a bunch of factors – from what's sitting on top of the fax, to the temperature in the room, to whether it's exposed to sunlight, etc. Recommendation: If you want to retain a slimy fax, make a photo copy of it the moment you get it and throw out the original.

5. Poor quality slimy fax paper can abrade the fax machine's drum and cause a costly repair. Don't buy cheap slimy fax paper.

6. Plain paper fax machines cost more to buy, but less to run. You can buy a second tray for some plain paper fax machines which will hold 8 1/2" wide x 14" long paper, which is useful for receiving faxes from outside the US where they use longer paper. This way you save a sheet of paper.

7. It makes sense to have banks of fax machines attached to phones which roll over – also called "hunt." It makes absolutely no sense to have multiple fax machines on separate phone lines that don't hunt, i.e. one for everybody in the office. Two fax machines in rotary can receive and transmit more than twice the number of faxes that two machines on separate, non-hunting phone lines can send and receive. "Personal" fax machines should be out. Banks of fax machines should be in. Egos, though, usually prevail over logic.

8. The paper feed mechanism on plain fax machines has a tendency to jam. Slimy paper fax machines don't jam because their paper typically comes in rolls. And roll paper doesn't jam. The feed mechanism is much simpler.

9. Plain paper fax machines, like laser printers (which many are) use supplies, like toner, which run out. When the supplies run out, such machines usually accept incoming faxes into memory – until that runs out. Then they just ring and ring and ring. Which means that incoming faxes don't get through and don't roll over to the next machine. There is no simple solution since the FCC (Federal Communications Commission) has ruled that fax machines must not return a busy signal to the central office if it runs out of supplies or paper. We have a separate machine that automatically busies out a line if it failed to answer on the fifth ring. But so far, the device is not commercially available. I don't know the answer to this problem except to make sure your fax machine is always stuffed with supplies. Especially check every Friday night. A final note: If your plain paper fax machine is missing supplies, but stuffed with incoming messages in memory, don't turn it off, since you'll lose the messages. Simply replace the supplies and pray your messages will emerge.

10. Some slimy fax paper rolls are coated on the inside of the paper. Others are coated on the outside. When you put one in a fax machine and images don't appear on the paper, then turn the roll over and feed it from underneath. In short, ignore what the instruction book says.

11. Fax modem switches only work when they're called automatically by a fax machine – not by a person using a fax machine manually and is waiting the sound of the distant fax prior to pushing the "Send" button. Make sure you warn your senders. It's remarkable how many people manually dial their faxes and thus penetrate fax modem switches.

11. Think about putting your fax machine on "fine." You'll transmit better quality faxes and may only cost yourself 10% more in transmission time. But that savings depends on the quality of the fax machine at the other end. If it's an older machine, it may cost you as much as double the transmission time. Here are the numbers: Standard is 203 x 98 dpi. Fine is 203 x 196 dpi. "Fine" faxes obviously look much better.

12. Printed circuit cards which slide into slots of PCs and allow you to transmit and receive faxes work well – when transmitting faxes. They work far less well when receiving faxes – largely because of the difficulty of reading faxes. Faxes conform to one type of digital encoding and PC screens conform to another. Moreover a PC screen is landscape (i.e. horizontal), while a fax message is portrait (i.e. vertical). Viewing vertical images on horizontal screens is difficult. Here is a comparison of how fax machines and how personal computer screens encode their images. Obviously, the more digits or pixels, the clearer the end picture. Notice that the encodings are completely dissimilar:

FAX ENCODING

Standard, Group III	203 x 98
Fine, Group III	203 x 196
Superfine, Group III	203 x 391
Standard, Group IV	400 x 400

PC SCREEN ENCODING

CGA	320 x 200
Enhanced CGA	640 x 400
EGA	640 x 350
Hercules	720 x 348
VGA	640 x 480
Super VGA	800 x 600
8514/A (also called XGA)	1,024 x 768
WXGA (wide XGA)	1,366 x 768

See also 1966, 1978, 1980, demodulation, facsimile converter, Facsimile Recorder, Facsimile Signal Level, Facsimile Switch, Fax, Fax At Work, Fax Back, Fax Board, Fax Data Modem, Fax Demodulation, Fax Mailbox, Fax Modem, Fax Publishing, Fax Server, Fax Switch, Faxbios, Group 1, 2, 3, 3 BIS and 4, Phase A thru E, T.37 and T.38 and Windows Telephony.

facsimile converter A facsimile device that changes the type of modulation from frequency shift to amplitude and vice versa.

facsimile data Denotes alphabetic, numeric, or graphic information which can be transmitted and received by facsimile machines.

facsimile recorder That part of the facsimile receiver that performs the final conversion of the facsimile picture signal to an image of the original subject copy on the record medium.

facsimile signal level The facsimile signal power or voltage measured at any point in a facsimile system. It is used to establish the operating levels in a facsimile system, and may be expressed in decibels with respect to some standard value such as 1 milliwatt.

facsimile switch A new breed of "black box." Its purpose is to avoid having to lease a separate phone line for your facsimile machine, for your phone and for your modem. You buy this box, connect it to an incoming line, connect it to your fax machine, your phone and, possibly, your modem. When a call comes in, the fax switch answers the call, listens if the call coming in is from a fax machine (it can hear the fax machine's CNG calling tone) and switches the call to the fax machine, or switches the call to your modem if a computer is calling. It knows if a computer is calling because the calling computer will, when it hears the fax switch answer, send out some ASCII characters – e.g. 22. (You must put those numbers in your modem dialing stream.) And it knows if a person is calling because it hears neither a CNG tone from a fax machine nor touchtones from the dialing stream of a modem.

The above are the basics of how fax switches work. There are variations on this theme. Some fax switches work automatically. Some work by the incoming caller punching in digits. Some allow you to switch from fax machine to modem to phone and back again. And some fax switches will answer and connect to three modems and one fax or other combinations. The major problem with fax switches is that they typically send a DC ringing tone to whatever device they're trying to connect you (the incoming caller with). Sometimes some devices – for example, high-speed 9600 baud and higher modems – have difficulties responding to low power, DC ringing signals. And they just sit there not answering. Better to buy one that sends standard telephone company AC ringing signals. In short, before you buy a fax/modem/phone switch, test it on your favorite 9,600 or 14,400 bps modem. The more expensive switches tend to work better.

FACTL Facility Access Customer Terminal Location.

factoid Factoids are paragraph size pieces of "Gee Whiz" information. They were originally made famous in the newspaper, USA Today.

factory programming An RFID definition. Some read-only have to have their identification number written into the silicon microchip at the time the chip is made. The process of writing the number into the chip is called factory programming.

factory refurbished A term used in the secondary telecom equipment business. Equipment that has been returned to the factory and the factory has replaced plastic, repaired, upgraded boards, or otherwise reconditioned.

fade A reduction in a received signal which is caused by reflecting, refraction or absorption. See also Fading.

fade margin The depth of fade, expressed in dB, that a microwave receiver can tolerate while still maintaining acceptable circuit quality.

fading 1. The reduction in signal intensity of one or all of the components of a radio signal.

2. A video term. A progressive deterioration of picture quality due to increasing losses in an electromagnetic (radio) propagation path. The term "fading" may be illustrated by the following sequence: (a) Noise appears on the porches and tip of the sync pulses. (b) Noise appears in the picture. (c) Loss of picture due to loss of synchronization which in turn is caused by distortion of the sync pulse by noise.

FADS Force Administration Data System. A system which takes basic statistics on telephone traffic and gives hints as to how many operators should be employed to answer the incoming calls and when they should be present.

fail safe A specially designed system that continues working after a failure of some component or piece of the system. There are precious few, genuinely fail safe systems. To be genuinely fail safe, a system needs to be completed duplicated. It is prohibitively expensive for most commercial users to duplicate every part of their system. But you can duplicate selectively and bring yourself closer to "fail safe." The extent of the duplications you choose (and thus the cost of your telephone equipment and transmission system) depends on how important it is that your system function as close to 100% as possible. The idea is to identify those things most likely to break and to duplicate them. Power is clearly the first area to focus on. These days, the words "FAIL SAFE" are increasingly being replaced with "FAULT TOLERANT." Given the number of times your local, friendly airline has told you that its "computer is down," you can understand the reason for the wording change.

fail-to-wire A fail-safe mechanism in a WAN accelerator that ensures that traffic continues to flow through the device unimpeded even if the device's value-added functionality (i.e., its performance acceleration capability) fails.

failover When one individual computer fails, another automatically takes over its request load. The transition is invisible to the user. Failover involves switching off the failed redundant component and switching on the backup unit. A disk subsystem is running in failover mode when it switches to a hot spare or begins to use the backup disk in a mirrored pair.

failure I never thought it necessary to define this obvious word. But then I'm reading an IPO prospectus for a new switching company and bingo, it defines failure as, "A termination of the ability of an item to perform a required function. A failure is caused by the persistence of a defect." In case you're wondering, the IPO succeeded.

failure domain Area in which a failure occurred in a Token Ring, defined by the information contained in a beacon. When a station detects a serious problem with the network (such as a cable break), it sends a beacon frame that includes the station reporting the failure, its nearest active upstream neighbor (NAUN), and everything in between. Beaconing in turn initiates a process called autoreconfiguration. See NAUN.

failure rate The number of failures of a device per unit of time.

fair condition A term used in the secondary telecom equipment business. One step up from "as is" condition. Equipment may have been tested; i.e., product is in working order but looks semi-awful.

Fair Market Value See FMV.

fairness An ATM term. As related to Generic Flow Control (GFC), fairness is defined as meeting all the agreed quality of service (QOS) requirements, by controlling the order of service for all active connections.

FairPlay Apple's copy-protection software that runs in iPods. When you buy songs at the iTunes Music Store, you can play them on one – and only one – line of portable player, the iPod. And when you buy an iPod, you can play copy-protected songs bought from one – and only one – online music store, the iTunes Music Store. The only legal way around this built-in limitation is to strip out the copy protection by burning a CD with the tracks, then uploading the music back to the computer which feeds your iPod. If you're willing to go to this trouble, you can play the music where and how you choose - the equivalent to rights

that would have been granted automatically at the cash register if you had bought the same music on a CD in the first place. The name for the umbrella category for copy-protection software is itself an indefensible euphemism, according to the New York Times: Digital Rights Management. As consumers, the "rights" enjoyed are few. As some wags have said, the initials D.R.M. should really stand for "Digital Restrictions Management." There are online music stores, e.g. eMusic, which sell music in pure MP3 format (i.e. without DRM) that can be played on all MP3 players, including Apple's iPod. At eMusic, says the New York Times, there is no copy protection, no customer lock-in, no restrictions on what kind of music player or media center a customer chooses to use – the MP3 standard is accommodated by all players.

fairydust Per PacBell lore, this is the tinkle tinkle sound you get when you use some calling cards. Per the lore, the sound was recorded from Tinkerbell in the musical Peter Pan and is the intro to the calling card recording, i.e. they play Fairydust then the talent comes on and says "thank you for using xxxxxx long distance, please enter your card number etc."

faith Faith is believing in something in spite of the fact that you have no hard evidence that it exists.

fake blog A blog that an organization creates and maintains but which really isn't a blog at all. Instead, the blog is simply repackaged marketing messages. Sometimes an organization fakes a blog at multiple levels, first by using the blog simply as another outlet for marketing messages, and second, by hiding the fact that the organization is behind the blog. In the latter scenario, the identity of the blog's author is faked. A fake blog is sometimes called a flog, and the practice of fake-blogging is sometimes called flogging.

fake root A subdirectory on the file server of a local area network that functions as a root directory, where you can safely assign rights to users. Fake roots only work with NetWare shells included with NetWare v2.2 and above. If you use older versions of the workstation shell, you will not be able to create fake roots.

FAL Foreign Listing.

fall time Also called turn-off time. The length of time required for a pulse to decrease from 90 to 10 percent of its maximum amplitude. Also, the length of time required for a component to achieve this result.

Fall.com An early virus which made the characters on a screen fall to the bottom.

fallback Mechanism used by ATM networks when rigorous path selection does not generate an acceptable path. The fallback mechanism attempts to determine a path by selectively relaxing certain attributes, such as delay, in order to find a path that meets some minimal set of desired attributes.

fallback rate A modem speed that is lower than its normal (that is, maximum) speed of operation. May be used when communicating with a slower, compatible modem, or to help transmission over a line that is too noisy for full speed operation.

false negative When used in the context of biometrics this term refers to the case where an authentication system erroneously denies access to an authorized person. See the next definition.

false positive When used in the context of biometrics this term refers to the case where an authentication system erroneously grants access to an unauthorized entity. See the previous definition.

false ringing False ringing is a recording of a telephone ringing signal (two seconds on, four seconds off, which is played while a call is transferred or while a switching device listens for modem for facsimile CNG (calling) tones.

falsing In telecom signaling, DTMF tones are created using specific combinations of frequencies to prevent the possibility of "falsing." Falsing is the condition where a DTMF detector incorrectly believes a DTMF is present when in fact it is actually a combination of voice, noise and/or music.

family planning The art of spacing your children the proper distance apart to keep you on the edge of financial disaster.

family radio service According to the FCC,   Family Radio Service (FRS) is one of the Citizens Band Radio Services. It is used by family, friends and associates to communicate within a neighborhood and while on group outings and has a communications range of less than one mile. You can not make a telephone call with an FRS unit. You may use your FRS unit for business-related communications. License documents are neither needed nor issued. You are provided authority to operate a FRS unit in places where the FCC regulates radio communications as long as you use only an unmodified FCC certified FRS unit. An FCC certified FRS unit has an identifying label placed on it by the manufacturer. There is no age or citizenship requirement. You may operate your FRS unit within the territorial limits of the fifty United States, the District of Columbia, and the Caribbean and Pacific

Insular areas ("U.S."). You may also operate your FRS unit on or over any other area of the world, except within the territorial limits of areas where radio-communications are regulated by another agency of the U.S. or within the territorial limits of any foreign government. Some manufacturers have received approval to market radios that are certified for use in both the Family Radio Service (FRS) and the General Mobile Radio Service (GMRS). Other manufacturers have received approval of their radios under the GMRS rules, but market them as FRS/GMRS radios on the basis that:

Some channels are authorized to both services, or * A user of the radio may communicate with stations in the other service.

Radios marketed as "FRS/GMRS" or "dual-service radios" are available from many manufacturers and many retail or discount stores. The manual that comes with the radio, or the label placed on it by the manufacturer, should indicate the service the unit is certified for. If you cannot determine what service the unit may be used in, contact the manufacturer. If you operate a radio that has been approved exclusively under the rules that apply to FRS, you are not required to have a license. FRS radios have a maximum power of half a watt (500 milliwatt) effective radiated power and integral (non-detachable) antennas. If you operate a radio under the rules that apply to GMRS, you must have a GMRS license. GMRS radios generally transmit at higher power levels (1 to 5 watts is typical) and may have detachable antennas.

fan antenna An aerial consisting of a number of wires radiating upwards from a common terminal to points on a supporting wire.

fan out Equipment that breaks down DS1 or DS3 service to the size demanded by the customer. On a DS3 line it breaks out the 28 DS1 channels. On a DS1 line it breaks them into 24 DS0 channels.

fan out cable Multi-fiber cable constructed in a tight buffered design: for ease of connectorization and applications for intra-or interbuilding requirements.

fanatic Someone who's overly enthusiastic about something in which you have zero interest.

FAP Formats And Protocols. The set of rules that specifies the format, timing, sequencing and/or error checking for communication between clients and servers.

FAQ Either Frequently Asked Question, or a list of frequently asked questions and their answers. Many Internet USENET news groups, and some non-USENET mailing lists, maintain FAQ lists (FAQs) so that participants won't spend lots of time answering the same set of questions.

far end Opposite end of a cable you are testing. Usually the end you open or close in an open test and closed test. Often, this is the end of a cable or cable pair, since you usually want to test a cable pair in its entire length.

Far End Block Error See FEBE.

far end crosstalk Crosstalk which travels along a circuit in the same direction as the signals in the circuit. The terminals of the disturbed channel at which the far-end crosstalk is present and the energized terminals of the disturbing channel are usually remote from each other.

far field pattern Synonym for Far-Field Radiation Pattern.

far talk In voice recognition, far talk is an arrangement where a microphone is more than four inches from the speaker's mouth. The opposite is CLOSE TALK, where the microphone is closer than four inches.

farad The practical unit of capacity. A capacitor which retains a charge of one coulomb with a potential difference of one volt. See Faraday and Faraday Cage.

faraday As a faraday shield: refers to the protection a material or container provides to electronic devices to keep them from exposure to electrostatic fields. Named after M. Faraday, the English physicist.

faraday cage A structure designed to isolate a sensitive electronic system or device from outside interference, usually constructed of metal screens. Named for 19th century inventor Michael Faraday, whose name also gave us the FARAD, the unit of measuring capacitance.

faraday effect Also known as the Magneto-Optic Effect. A phenomenon that causes some materials to rotate the polarization of light signals in the presence of a magnetic field that is parallel to the direction of the propagation of the light signal. Also called magneto-optic effect.

Farland Farland is an alliance of European telecom carriers operators whose common network unites the infrastructures of its members' national networks to one pan-European fiber optic network. Farland launched its operations in March 1999. Farland is owned by BT (British Telecom).

farm Picture a hall full of computers (PC servers and/or mainframes), lined one after another. Now you have the concept of a server farm. Picture a field full of satellite antennas. Now you have a satellite farm. These days there are many "farms."

farmer's line In the late 1800s and early 1900s the only way that farmers and homesteaders in rural areas could get telephone service was to set it up themselves. A farmer's line was a small telephone network that was built, operated, and maintained by farmers and homesteaders themselves. Sometimes the wife of one of the farmers served as a daytime switchboard operator. In other cases, the farmer's line didn't have a switchboard and was set up simply as one big party line. In addition to connecting farmers with each other, in many cases via telephone wires strung along fence posts and in some cases using fences' own barb wire as telephone lines, a farmer's line was often extended to the nearest town, where it connected with the telephone network of the phone company that served the town.

FAS Frame Alignment Signal or Frame Alignment Sequence. See Frame Alignment Signal.

fashion Toward the end of the 15th century, men's shoes had a square tip, like a duck's beak, a fashion launched by Charles VIII of France to hide the fact that one of his feet had six toes.

Fast Broadcast Control Channel F-BCCH. A logical channel element of the BCCH signaling and control channel used in digital cellular networks employing TDMA (Time Division Multiple Access), as defined by IS-136. See also BCCH, IS-136 and TDMA.

fast busy A busy signal which sounds at twice the normal rate (120 interruptions per minute vs. 60 a minute). A "fast busy" signal indicates all trunks are busy.

fast clear down A call center term. A caller who hangs up immediately when they hear a delay announcement.

fast ethernet 100BaseT. Ethernet at 100 Mbps, a tenfold improvement over the original Ethernet speed of 10 Mbps. Fast Ethernet is in the form of an Ethernet hub with an internal bus that runs at 100 Mbps. The interface to the hub is through a port which generally is selectable (i.e., programmable) to run at either 10 Mbps or 100 Mbps, depending on the requirement of the attached device. Connection between the hub and the attached workstation or other device is over data-grade UTP (Unshielded Twisted Pair) in the form of Cat (Category) 5, at a minimum, and over distances of up to 100 meters, at a maximum. The attached device connects to the UTP connection via a 10/100 Mbps NIC (Network Interface Card). 100BaseT hubs interconnect over fiber optic facilities, which can support 100 Mbps over relatively long distances with no loss of performance. Fast Ethernet is no longer all that fast- Gigabit Ethernet switches were standardized in 1998. See also 10BaseT, 100BaseT, Cat 5, Ethernet, Gigabit Ethernet, NIC and UTP.

fast ethernet alliance A group of vendors that participated in writing the 100Base-X technical hub and wiring specifications, which would allow fast Ethernet (100 megabits a second) to run over Category 5, data-grade unshielded twisted pair wiring.

Fast File Transfer FFT. An ISDN term referring to the fact that file transfers can be accomplished "fast." Reason #1: Two B channels at 64 Kbps each are available to be bonded to provide as much as 128 Kbps. Reason #2: Data transfer is accomplished in an "optimistic" streaming mode, rather than a "pessimistic" packet mode. Therefore, there is no delay associated with acknowledgments. This is possible due to the excellent level of error performance inherent in digital services. The end result is FFT.

Fast Fourier Transform FFT. A signal processing term for a common computer implementation of Fourier Transforms. The FFT, as a practical implementation, will always result in a finite series of sine and cosine waves as an extremely close approximation of the possibly infinite series described by the purely mathematical application of the Fourier Transform. See Fourier's Theorem and FFT.

fast IR A four million bits per second extension to the Serial Infrared Data Link Standard that provides wireless data transmission between IrDA-compliant devices.

fast network An AT&T term for a network with low delay relative to the needs of the application.

fast packet multiplexing Multiplexing, from Latin "multi" and "plex" translates as "manyfold." In other words, folding many "conversations" onto a single circuit. You can do this in either of two ways – by splitting the channels sideways into subchannels of narrower frequency. This is called Frequency Division Multiplexing (FDM), which is used in analog networks. Or you could split it by time, through a process of Time Division Multiplexing (TDM), which is used in digital networks. TDM is much like a railroad train. The first car carries "Conversation 1." The second carries "Conversation 2." And split them apart at the other end.

Fast packet multiplexing is a combination of three techniques – time division multiplexing, packetizing of voice and other analog signals, and computer intelligence. Here are the main advantages fast packet multiplexing has over today's industry standard time division multiplexing:

1. Fast packet multiplexing doesn't blindly slot in "information" from devices if there's no information to send. Most other multiplexing techniques, including the most common – time division and frequency division – slot in capacity, whether the device is "talking" or not.

2. The fast packet multiplexer can start sending a packet before it has completely received the packet. This is accomplished by reading the destination address, which is contained in the header portion of the packet. This speed of movement is critical to voice, for example, which must move ultra-fast. Delays are devastating. (No one can afford to replace the phone instruments broken in anger.)

3. Fast packet multiplexing can interrupt the delivery of one packet in favor of sending another. It's OK to delay a packet of data by several milliseconds. It's not OK to delay a packet of voice or video.

fast packet services 1. An umbrella term for ATM, Frame Relay, and SMDS service offerings, all of which operate at broadband speeds and all of which make use of Fast Packet Switches and Multiplexers.

2. A Verizon definition. Refers to the following high-speed data offerings: Frame Relay Service, Switched Multimegabit Data Service (SMDS) and FDDI Network Services. Fast packet applications include real-time inventory control, credit verification, gathering marketing data, and sending or receiving customer information.

fast packet switching A wide area networking technology capable of switching data at a very high rate of speed in the context of a broadband network service such as Frame Relay, SMDS or ATM. The term "packet" is generic, referring to the manner in which data is formatted. "Data" is also generic in this context, referring to voice data, video data, and image data, as well as data data. Should the data be analog in its native form, it is digitized and packetized before being presented to the network for transport and switching. The packets are in the form of short (53 octets), fixed length cells in the case of SMDS and ATM. The packets are in the form of variable (0-4,096 octets) frames in the case of Frame Relay.

The underlying switching technology is based on the statistical multiplexing of data contained within the cells or frames. While any of these packets could carry digital voice, video, data or image information, only ATM is specifically intended for other than data use. All the packets travel at Level Two of the OSI Model, and routing is performed on the basis of the Level Two addressing. Fast packet is claimed to be very effective way of make best use of available bandwidth. It is claimed to offer the benefits of conventional multiplexing techniques and circuit switching techniques because of the way it operates. It is one of the transmission technologies being developed for use with B-ISDN (Broadband ISDN), which is based on ATM. The switch used to route packets in a fast packet network is termed a fast packet switch. See ATM, Fast Packet Multiplexing, Frame Relay, and SMDS.

Fast Pyramid Algorithm Pyramid algorithms are used in image compression to compute the wavelet transform. The algorithm implements a complex mathematical procedure, using far fewer calculations than are nominally required, thereby yielding an approximation of the original data. The algorithm involves a series of linear filtering operations, in combination with down-sampling by two of the output. Up-sampling is employed to reconstruct a highly satisfactory approximation of the original data. See Down-Sampling, Up-Sampling and Wavelet Transform.

fast scan receiver A cellular term. A piece of equipment that scans all 1,300 channels in an entire cellular network. It is a quick way to determine channel usage and signal strength.

fast select In packet switched networks, a calling method which allows the user to send a limited amount of information along with a "call req packet" rather than after the packet. A more technical explanation: An optional user facility in the virtual call service of ITU-T X.25 protocol that allows the inclusion of user data in the call request/connected and clear indication packets. An essential feature of the ITU-T X.25 (1984) protocol.

Fast Stat MUX MICOM's advanced statistical multiplexer that uses data compression, priority echoplex handling and fast packet technology to improve throughput.

fast switching channel A single channel on a GPS (Global Positioning System) which rapidly samples a number of satellite ranges. "Fast" means that the switching time is sufficiently fast (2 to 5 milliseconds) to recover the data message.

Fast-20 A type of SCSI, introduced in the SCSI-3 specification, in which the data rate is quadrupled to 20 MBytes per second for narrow SCSI or 40 MBytes per second for wide SCSI. Also known as Fast-20 or Double Speed SCSI.

Fast-40 A type of SCSI in which the data rate is increased to 40 MBytes per second for narrow SCSI or 80 MBytes per second for wide SCSI. Also known as Fast-40.

FAT 1. File Allocation Table. The FAT is an integral part of the MS-DOS and Windows operating systems. It is like a roadmap (or index) of a hard, floppy disk, optical, zip drive, etc. It keeps track of where the various pieces of each file on a disk are stored. A hard disk's directory and file allocation tables are extremely important because they contain the address and mapping information the operating system needs to figure where to store and where to retrieve our precious data. If any of the data storage blocks containing these tables is damaged, it will be very hard, if not impossible to find the data on the hard disk. As a result, all operating systems keep multiple updated copies of their file allocation tables on your computer's hard disk. See FAT32 and the newer system, NTFS.

2. Wide or large, as in fat bandwidth.

3. Factory acceptance testing.

fat AP Refers to a wireless LAN architecture where all of the WLAN (wireless local area network) intelligence is concentrated in the access point. The AP handles the network's wireless communications, authentication of users, encryption of communications, secure roaming, routing, and network management. See thin AP, fit AP.

fat client A client is a fancy name for a computer, terminal or workstation on a network, local or long distance. Now it is the "client" of the server. Clients come in two varieties – fat and thin. A fat client is typically a device with a microprocessor, some memory and software. A thin client has less. A fat client used to cost a lot of money. Network designers often preferred thin clients because they had less "stuff" and cost less money. These days most clients are fat. See also client, client server, client server model, mainframe server, media server and thin client.

FAT File System A FAT is a file allocation table. It is a roadmap to what's on your hard disk, your floppy, your magneto optical disk, your CD – any disk associated with your PC. The FAT file system is a fancy way of saying the whole procedure the FAT uses to organize information on your disks. Various computer operating systems use different FAT file systems.

fat finger dialing The telephone equivalent of typing www.whitehouse.com when you meant www.whitehouse.gov. For example if you misdial AT&T's collect collecting number, 1-800-COLLECT (1-800-265-5328), you're likely to hit another carrier, like OPticom, which has the following number, 1-800-265-5329 and which may or may not charge a higher fee than AT&T. The Today Show on April 4, 2002 talked about someone called, Joel Drizen who noticed something didn't look right on his phone bill, a four-minute collect call placed to him by his brother three miles away using what he thought was 1-800-COLLECT. The cost: over $16. A charge almost three times higher than he expected from a company he'd never heard of before, ASC Telecom. After investigation, they realized that he dialed 1-800-COLLLECT with three L's as opposed to two L's. According to Robert Tolchin, an attorney, ASC was basing its entire business model on what they call fat fingers dialing. That means they're going to take advantage of people who have fat fingers and mis-dial the telephone.

fat pipe A fiber-optic cable used on a network backbone for high-speed communications. It has a wide bandwidth for baseband and broadband high-capacity communications.

fat server In a client/server architecture, a server computer that performs most of the processing, with little or none performed by the client.

FAT32 A version of the file allocation table (FAT) available in Windows. FAT32 increases the number of bits used to address clusters and also reduces the size of each cluster. The result is that it can support larger hard disks of up to 2 terabytes. A terabyte is one thousand billion. See also NTFS, which is a more modern verion of FAT32.

FATbits A computer imaging term. Extreme magnification of individual pixels to allow easy pixel-by-pixel editing of images.

fatherboard First there is the motherboard. That's the main circuit board of a computer system. The motherboard contains edge connectors or sockets so other PC (printed circuit) boards can be plugged into it. Those PC boards are called Fatherboards. Some fatherboards have pins on them into which you can plug smaller boards. Those boards are called Daughterboards. In a voice processing system, you might have a Fatherboard to do faxing. And you might have a range of Daughterboards, which allow you to connect different types of phone connections. Different boards exist for standard analog tip and ring, digital switched 56, t-1, etc.

Father's Day Father's Day has more collect calls than any other day of the year.

fatigue resistance A cable term. Resistance to metal crystallization which leads to conductors breaking from flexing.

fatwa Fatwas are legal opinions proclaimed by Islamic scholars. They have proliferated in the Muslim world since the 1980s, driven by rising literacy rates and the explosion of the Internet. The growth in fatwas – some of them contradictory – has led to a debate over who can legitimately issue them and has alarmed governments in the Middle East, since the decrees sometimes challenge state-sanctioned interpretations of Islam. The New York Times wrote a piece on them and quoted a fatwa dealing with how soccer was meant now to be played.

FAU Fixed Access Unit. A fixed, wireless telephone placed in a user's home or business using cellular or PCS (Personal Communications Service). A new, lower powered, higher-frequency technology. The device provides local telephony service circumventing existing LEC (Local Exchange Carrier) transmission equipment using wired connections.

fault A hard failure or a performance degradation so serious as to destroy the ability of a network element to function effectively. Opens, short circuits and breaks are examples of common cable faults.

fault current The current that can flow in a circuit as a result of a undesired short circuit.

fault domain A fault domain defines the boundaries of an isolating soft error on a Token Ring network. The fault domain limits the problem to two stations, their connecting cables, and any equipment (a MAU, for example) between the two stations. The two stations involved are the station reporting the error and its Nearest Active Upstream Neighbor (NAUN).

fault isolation The process of determining where a network problem, or fault located.

fault management Detects, isolates and corrects network faults. It is also one of five categories of network management defined by the ISO (International Standards Organization). See also Fault.

fault resilient A fault resilient computer tends to means that it must be relied on to run 99% of the time. In contrast, a fault tolerant machine must run 100% of the time, which typically means that the design must duplicate the CPU microprocessor. A fault resilient machine will typically be less expensive than a fault tolerant machine. Which you buy depends on what your needs are and the possible cost of losing transactions should your machine go down. See Fault Tolerant.

fault tolerant A method of making a computer or network system resistant to software errors and hardware problems. A fault tolerant LAN system tries to ensure that even in the event of a power failure, a disk crash or a major user error, data isn't lost and the system can keep running. In fact, the general concept is that a fault tolerant machine must be designed with sufficient duplicated parts that it can be relied upon to run 100%. Cabling systems can also be fault tolerant, using redundant wiring so that even if a cable is cut, the system can keep running. True fault tolerance is very difficult to achieve. See Fault Resilient.

faults Conditions that degrade or destroy a cable's ability to transmit data. Opens, short circuits and breaks are examples of common cable faults.

Faustian bargain Faust, in the legend, traded his soul to the devil in exchange for knowledge. To "strike a Faustian bargain" is to be willing to sacrifice anything to satisfy a limitless desire for knowledge or power.

fauxcellarm See phantom ring tones.

favorite Microsoft's term for what Netscape and others call a bookmark. See also Bookmark.

fax An abbreviation for facsimile. See Facsimile and Facsimile Switches.

fax adapter Hook one up between your printer and your phone. Bingo, your printer now becomes a fax machine.

fax at work Fax at work was a subset of Microsoft's office equipment architecture called At Work which was it announced on June 9, 1993. Microsoft's idea was to put a set of software building blocks into both office machines and PC products, including:

- Desktop and network-connected printers.
- Digital monochrome and color copiers.
- Telephones and voice messaging systems.
- Fax machines and PC fax products.
- Handheld systems.
- Hybrid combinations of the above.

Microsoft Fax At Work has not been adopted by the fax and telecommunications industry, since (the story goes) Microsoft has wanted too much in the way of royalties and Fax at Work is now effectively dead. Newer fax standards have been issued by the ITU – including T.37 and T.38 – which have interested the industry far more. They cover sending faxes over packet switched networks, such as the Internet and corporate Intranets.

fax back You dial a computer using the handset of your fax machine. The distant computer answers. "What documents would you like? Here's a menu." You touchtone in 123. It says "Touch your Start button." You do. Seconds later your fax machine disgorges the document you wanted. Fax-back is the generic term for the process of ordering fax documents from remote machines. Fax-back uses a combination of fax and voice processing technology. Fax-back is also called fax on demand.

fax board A specialized synchronous modem for designed to transmit and receive facsimile documents. Many fax boards also allow for binary synchronous file transfer and V.22 bis communication. See also Fax Server.

fax branding The Telephone Consumer Protection Act of 1991 makes it unlawful for any person to use a computer or other electronic device, including fax machines, to send any message unless such message clearly contains in a margin at the top or bottom of each transmitted page or on the first page of the transmission, the date and time it is sent and an identification of the business or other entity, or other individual sending the messagte and the telephone number of the sending machine or such business, other entity, or individual. The telephone number provided may not be a 900 number or any other number for which charges exceed local or long-distance transmission charges.

fax broadcasting Automatically distributes faxes to preselected destinations.

fax data modem See Fax Modem.

fax demodulation A technique for taking a Group III fax signal and converting it back to its original 9.6 Kbps. It works like this: When a sheet of paper is inserted into a fax machine, the fax machine scans that paper into digital bits – a stream of 9600 bps. Then, for transmission over phone lines, that 9.6 Kbps is converted into an analog signal. But if you wish to transmit the fax signal over a digital line, then it makes sense to convert it back to its original 9.6 Kbps. That means you can put several fax transmissions on one 56 Kbps or 64 Kbps line – the capacity you'd normally need if you transmitted one voice conversation, or one erstwhile analog fax transmission. See Fax/Data Modem.

fax enhancer A standalone add-on device that connects to legacy fax machines for the purpose of boosting fax transmission speeds and enabling faxes to be sent over any network (Internet, PSTN and private networks).

fax jack A device that connects to a phone line with two jacks, one for a phone and one for a fax machine and one for a phone line. When a call comes in the fax jack answers the call and waits for the mechanical tone from the other fax machine. If it doesn't hear the tone, it rings the phone. The downside is that they block caller-ID (ANI) signals.

fax mailbox Imagine a phone number attached to a PC which accepts your incoming faxes and stores them on your PC's hard disk. Two things could happen. First, you could dial in and retrieve your fax, just as you do with your voice mail messages. Hence the name fax mailbox. Or second, the PC could simply send you your fax as an attachment to an email. I find the second service more useful. It allows me to receive faxes wherever I am in the world. Most of the time I'm NOT standing next to the fax machine in my office. My favorite company providing this service is called www.CallWave.com. Others, like www.efax.com, also do it. Most do it for free.

fax mode The mode in which the fax modem is capable of sending and receiving files in a facsimile format. See Fax Modem.

fax modem A combination facsimile machine/modem. A device which lets you send documents from a computer to a fax machine. It comes in many shapes and sizes. It may come as a card which you slip into a vacant slot in your desktop PC (called an internal fax/modem). It may come as a PCMCIA card which you slip into your laptop. It may come as a small box which you connect by a cable to your computer's serial port. It may also come as a small self-contained package about the size of a cigarette package. It may also come as part of your motherboard, which it increasingly does these days in laptops. The technology of "fax modems" is changing radically. Originally they contained dedicated fax/modem chipsets, i.e. microprocessors designed as fax modems and good for nothing else. Increasingly, fax modems are now coming with powerful, general purpose digital signal processors (DSPs), instead of dedicated fax modem chipsets. These DSP devices become fax modems when you load the appropriate software. When you load other software they can also become the equivalent of sound blaster cards.

There are big advantages to sending faxes from a fax modem, as compared to sending it from a fax machine. First, faxes sent are cleaner because they're not scanned but computer generated. Second, sending faxes directly from your computer is faster than printing the document, then sliding it in a fax machine, dialing and sending it. Third, a fax modem is typically cheaper than a fax machine. Fourth, because a fax modem uses computer software it may have some neat features, like the ability to send faxes when phone costs are low, like running the fax software in the background while you're doing something else.

There is one main disadvantages: Keeping copies of the faxes you send on your hard disk is consuming of hard disk space. A typical one page fax can easily use between 40,000 and 50,000 bytes. Twenty pages and you've used up a megabyte. See also Fax Demodulation, Fax Server and Fax Switch.

fax on demand You dial a computer using the handset of your fax machine. The distant computer answers. "What documents would you like? Here's a menu." You touchtone in 123. It says "Touch your Start button." You do. Seconds later your fax machine disgorges the document you wanted. Fax on demand is one term for the process of ordering fax documents from remote machines. Fax on demand uses a combination of fax and voice processing technology. Fax on demand is also called fax-back. See Fax Server for a more complete explanation.

Fax Over IP See FoIP.

fax publishing Fax publishing allows a caller to have electronically stored information automatically faxed to them via a touchtone telephone. By pressing touchtone keys, callers can have timely information, including product brochures, business forms and benefits information, automatically faxed to them anytime, anywhere. See also Fax Server.

fax relay Also known as "demod/remod," fax relay is one of the methods for IP fax transmission, as defined in ITU-T Standards Recommendation T.38. Fax relay defines the specification for the demodulation of standard analog fax transmission from originating machines equipped with modems, and their remodulation for presentation to a matching destination device, with the long-haul portion of the transmission being supported over an IP-based network. Fax relay depends on a low-latency (i.e., one second or less) IP network in order that the session between the fax machines does not time out. See also Fax Spoofing and T.38.

fax server A fax server sits on a local area network and literally serves faxes to those people using it. Those people may be on the LAN physically, i.e. joined by wires to the server. Or they may be outside, reaching the LAN over phone lines. Basically anything that a live person can do with a fax machine, a fax server can do. It can receive faxes and distribute them to people they're addressed to. It can send faxes for people who are typically sending those faxes from their PCs. It can send faxes to people who call it on the phone and request certain faxes – those stored on its hard disk. A fax is typically a relatively high-powered computer which has one or more PC fax boards in its slots. It can receive faxes. It can send faxes. It can store faxes. It can forward them. If it doesn't know for whom the faxes are meant, it may send the faxes to a printer or alert a supervisor to manually check the incoming faxes and distribute them – electronically or on paper. The fax server also accepts from PCs on the LAN, stores them and gets them ready for sending out over phone lines. It might send the faxes immediately or wait until later, when phone calls are cheaper. It might send the same fax to thousands of people. It might send a personalized fax to thousands of people, grabbing the names from a database on it or on another computer. A fax server can also be an interactive voice response system which you call. When you call it, it answers, reads you a menu of options – including various documents it can send you. You choose which documents you want by touchtoning in numbers. Then you designate to which fax machine you want the documents sent. The fax machine you designate might be the one you're calling from (i.e. you dialed using your fax machine's handset).

There are two types of interactive voice response fax servers. One is a one-call machine. The caller calls from his own fax machine. When he's chosen his faxes and he's ready to receive a fax, he simply hits the "Start" button on his fax machine and his machine receives the chosen faxes. There is also a two-call machine. The caller will call from a phone and touchtone in the phone number of a fax machine he wants the fax of his desired documents sent. One-call IVR fax servers are the newer breed, harder to build than the older two-call machines. There are obvious advantages to both. The one call machine – in which the user pays the phone bill – will, I suspect, become the more popular type. See also Fax Publishing and other Fax definitions.

fax spoofing Fax spoofing is included in ITU-T recommendation T.38 as the method for fax transmission over IP (Internet Protocol) networks characterized by relatively long and unpredictable levels of packet latency, which could cause conventional fax machines to time out. Fax spoofing fools the machines by padding the line with occasional "keep alive" packets to keep the session active. Thereby, the receiving machine thinks that the incoming transmission is taking place over a realtime, timed circuit-switched network. Delays up to 5 seconds per packet can be tolerated using this approach. See also Fax Relay and T.38.

fax switch A device which allows you to share one phone line with a fax machine, a phone and a modem. Here's how it works. A call comes in. The device answers the call. The switch listens for the distinctive CNG (Calling) tone which a calling fax machine emits (the "cry" of the fax machine). When it hears this sound, it switches the call to the fax machine. If it doesn't and hears nothing (or at least nothing it can recognize) it switches the call to the phone. If it hears some touchtones – e.g. 44, or *6 – it will switch the call to the modem (and therefore the attached computer) or whatever other device you've designated,

including a modem-equipped cash register, etc. Some fax switches allow you to have a data conversation with one device (the cash register), then switch to another device (the second cash register) and another, etc. – all on the one conversation. The advantage of a fax switch is that it saves having to buy several phone lines. Phone lines are expensive compared to fax switches. There are disadvantages to a fax switch – it typically must hear an incoming CNG tone to switch the call to the fax machine. This means if your friend wanting to send you a fax is dialing manually (i.e. not letting his fax machine do it), your fax switch may not ever send the call to your fax machine. Some fax switches don't send the "right" ringing signal to their attached devices. Some 9,600 baud and 14,400 baud modems, for example, are very sensitive and won't answer certain fax switches' ringing signals, especially if the fax switch's ringing signal is a DC square wave, not an AC sine wave. All this can be solved, however, with intelligence, checking and proper programming. I used to use a fax modem switch every day. It saved me money and was convenient. But now I send and receive faxes on my PC. See CNG.

FaxBios The FaxBios Association is an organization of fax printed circuit card manufacturers who have formed an association in order to promulgate a standard applications programming interface (API) which they are calling FaxBios. Phone 801-225-1850; 2625 Alcatraz Avenue, Berkeley CA 94705.

faxed The past tense of the new verb "to fax," as in "I faxed the document to him."

FB Framing bit.

FBG See Fiber Bragg Grating.

FBT Fused Biconic Tape.

FBU 1. Functional Business Unit. A fancy name for a group of workers inside a company. An FBU might be your sales department, your accounting department, etc.

2. BellSouth term for Failed Before Utilization. It means that BellSouth has turned up a circuit to a LEC to be "cut" for a customer and the circuit does not work. It won't pass traffic, does not pass go and does not collect $200. When it's FBU, BellSouth writes up a FUBI ticket. The term is also TUB (Turned Up Broke).

FBus Frame Transport Bus.

FBWA Fixed Broadband Wireless Access.

FC 1. Fiber optic Connector (developed by NTT).

2. Frame Control. On Token Ring networks, this data supplies the frame type.

3. Feedback Control: Feedback controls are defined as the set of actions taken by the network and by the end-systems to regulate the traffic submitted on ATM connections according to the state of network elements.

4. Fibre Channel.

FC adapter Fibre Channel adapter or host adapter. A Fibre Channel I/O (Input/Output) bus adapter board that operates at 100MBps in half-duplex mode, 200MBps in full-duplex mode.

FC and PC Face Contact and Point Contact. Designations for fiber optic connectors designed by Nippon Telegraph and Telephone which feature a movable anti-rotation key allowing good repeatable performance despite numerous matings.

FC connector A simplex fiber optic connector developed in the 1980s by NTT (Nippon Telegraph and Telephone). The FC connector is a keyed, all-metal connector with screw-on mechanics. The D4 connector is a derivative with a hood over the end of the connector to prevent damage to the fiber. See also SC Connector, SFF Connector, and ST Connector.

FC switch Fibre Channel switch. Intelligently manages connections between ports, routing frames dynamically. A nonblocking topology, it allows multiple exchanges of information to occur at the same time between ports. A switch offers better system throughput than a hub, but at greater expense. Some switches have a special FL-port to link arbitrated loops and other devices on the switch. By cascading multiple switches, more than 16 million devices can be connected together. See Cascade.

FC/IP Fibre Channel over Internet Protocol. See FCIP.

FC-0 Lowest level of the Fibre Channel Physical standard, covering the physical characteristics of the interface and media. FC-0 defines the physical point-to-point portion of the fiber channel. This includes the fiber, the connectors, and the optical parameters for a variety of data rates. (A serial coaxial version is also defined for limited-distance applications). The following signaling rates are defined: 132.813 Mbaud, 265.625 Mbaud, 531.25 Mbaud, and 1.0625 Gbaud. These signaling rates correspond to the data transfer rates of 12.5 Megabytes per second (100 million bits per second), 25 Mbytes/s (200M bps), 50 Mbytes/s (400M bps), and 100 Mbytes/s (800M bps). FC-0 operates with a BER (Bit Error Rate) of less than ten to the minus twelve.

FC-1 A Fibre Channel term. Middle level of the FC-PH standard, defining the 8B/10B encoding/decoding and transmission protocol.

FC-2 A Fibre Channel term. Highest level of FC-PH, defining the rules for signaling protocol and describing transfer of the frame, sequence, and exchanges.

FC-3 A Fibre Channel term. The hierarchical level in the Fibre Channel standard that provides common services, such as striping definition.

FC-4 A Fibre Channel term. The hierarchical level in the Fibre Channel standard that specifies the mapping of upper-layer protocols (ULPs) to levels below.

FC-AL Fibre Channel Arbitrated Loop. A Fibre Channel network topology that allows as many as 126 nodes to share a loop. The physical topology is usually a star, logically configured as a double loop employing a Fibre Channel hub. The physical star is advantageous, as the network administrator can isolate a failed node, or either add or disconnect a node, without disrupting the entire network, as would be the case with a physical ring. See also Fibre Channel.

FC-EL Fibre Channel-Enhanced Loop.

FC-PGA Flip Chip Pin Grid Array. A package of silicon chips that form certain Intel Pentium processors, the FC-PGA uses chips that have been turned upside down (i.e., flipped) and attached to the board, which then connects to a Socket 370 motherboard socket via an array of pins, or leads. See also Socket 370.

FC-PH Fibre Channel Physical standard, consisting of the three lower levels, FC-0, FC-1, and FC-2.

FCAPS Fault, Configuration, Accounting, Performance, and Security. FCAPS is the ISO (International Organization for Standardization) framework for network management. Here's an explanation

In the early days of network management there was little consistency in the way administrators maintained their networks - each organization was left to their own devices. In addition, each vendor of networking equipment also created their own way of managing their equipment. For a network containing switches and routers from various companies there may have been several different network management platforms to learn and operate. In an effort to standardize management practices, the International Organization for Standardization (ISO) studied how networks are managed and identified five important areas.

- Fault management, configuration management, accounting management, performance management and security management. The ISO then created a standardized model for the management of each area and called it FCAPS, using the first letter from each of the five important areas of management.

- Fault Management The ability to monitor and fix problems in the network is in the fault management domain. Typically, fault management provides the ability to collect alarms and traps coming from network devices, filter these alarms to isolate problems, acknowledge alarms, and in advanced cases, predict when problems will occur in the network based on historical information.

- Configuration Management Network configuration is very important for smooth operation. If one device is not set up properly the effects can be felt throughout the network. Configuration management records the configuration of network components and stores them so if something goes wrong, it is easy to go back to a previous working configuration.

- Accounting Management The ability to provide information on network usage is the goal of accounting management. Typically used in the telephony area, accounting management is responsible for collecting call detail records (CDRs) containing information regarding a particular phone call (i.e., length of call, end points, etc.), which are used in creating billing records.

- Performance Management Performance management is responsible for monitoring and configuring changes in the network to insure the best possible performance. QoS might be considered a performance feature since it allocates bandwidth based on application needs. The ability to monitor and configure QoS in this regard is considered to be a performance management function.

- Security Providing security by defining how users can access the network is one of the most important aspects of network management. Implementation of authentication mechanisms as well as other types of authorization procedures must be carefully designed, implemented, and maintained to ensure the integrity of the network. The five functional areas of FCAPS are the major components of any network management implementation. They are used as a reference for the development of network management products and services. Implementations of FCAPS services are determined by the various standards such as SNMP and CMIP as well as particular vendors' implementations. See Network Management for a detailed explanation.

FCB File Control Block. FCBs are used by older MS-DOS application programs to create, open, delete, read, and write files. One FCB is set up for each file you open.

FCC Federal Communications Commission. See Federal Communications Commission.

FCC Number Sequential number assigned by the Secretary's Office (Agenda Branch) to all documents approved by the Commission. This number is assigned after the item has been adopted by the Commission. Example: FCC 96-123. The first two digits reflect the year.

FCC Record A bi-weekly comprehensive compilation of decisions, reports, public notices and other documents released by the Federal Communications Commission. The FCC Record replaced the FCC Reports in October 1986. The Record is available for a fee. The ordering address: Superintendent of Documents, PO Box 371954, Pittsburgh, PA 15250-7954; by phone: 202-512-8200.

FCC Registration Number A number assigned to specific telephone equipment registered with the FCC, as set forth in FCC docket 19528, part 68. The presence of this number affixed to a device indicates that the FCC has approved it as being a compatible device for direct connection to telephone line facilities.

FCC Tariff #9 The FCC tariff for private line services including Accunet T-1.5, DDS, Voice Grade circuits, and Accunet T45.

FCC Tariff #11 AT&T's tariff file at the FCC for local private line services.

FCC Tariff #12 AT&T's tariff filed at the FCC tariff for custom-designed integrated services. A special tariff that allows AT&T to develop custom network solutions, including allowing customers to install their networking multiplexers in AT&T central offices and letting AT&T manage the network.

FCC Tariff #15 AT&T's FCC tariff filed at the FCC that allows AT&T to lower rates after all bids are placed to be competitive with other carriers.

FCD Final Committee Draft. I don't get. But they use the term as they create standards. See VC-1.

FCFS A silly abbreviation for First Come First Served. See FIFO (First In, First Out).

FCIP Fibre Channel over Internet Protocol (FCIP or FC/IP) is also known as Fibre Channel tunneling and storage tunneling. FCIP is a Storage Area Network (SAN) technology developed by the Internet Engineering Task Force (IETF) to enable the transmission of Fibre Channel (FC) data through tunneling between SANs over IP networks. See also Fibre Channel and SAN.

FCKT Facility Circuit ID.

FCN Abbreviation for Function. This button enables your cellular phone or fax machine or other telecom device to access special features, like switching from one cellular phone company to another. See also Dual NAM.

FCOS Fully programmable classes of service that control user (Feature Class of Service) access to mailbox features, operations and options. Feature Classes of Service (FCOS) are entirely independent of Limits Classes of Service (LCOS).

FCOT See Fiber Control Office Terminal.

FCS 1. Frame Check Sequence. Any mathematical formula which derives a numeric value based on the bit pattern of a transmitted block of information and uses that value at the receiving end to determine the existence of any transmission errors. In bit-oriented protocols, a frame check sequence is typically a 16-bit field that contains transmission error checking information, usually appended to the end of the frame. See Frame Check Error and Frame Check Sequence.

2. Federation of Communications Services.

3. An MCI term for Fraud Control System.

FCS Error A Frame Check Sequence error occurs when a packet is involved in a collision or is corrupted by noise.

FCSI Fibre Channel Systems Initiative formed by Hewlett-Packard, IBM and Sun in 1993.

FD See Floor Distributor.

FDA Food and Drug Administration. Organization responsible for laser safety.

FDCCH Forward Digital Control CHannel. A digital cellular term defined by IS-136, which addresses cellular standards for networks employing TDMA (Time Division Multiple Access). The FDCCH includes all signaling and control information passed downstream from the cell site to the user terminal equipment. The FDCCH acts in conjunction with the RDCCH (Reverse Digital Control CHannel), which includes all such information sent upstream from the user terminal equipment to the cell site. The FDCCH includes the BCCH, SCF and SPACH. See also BCCH, IS-136, SCF, SPACH and TDMA.

FDD 1. Floppy Disk Drive. A Hard Disk Drive is a HDD.

2. Frequency Division Duplex. A method used to achieve full duplex communications in wireless systems. The principle is that forward and reverse directions each use a separate and equally large frequency band. FDD is appropriate for symmetrical services such as voice and bidirectional data transfers. See also Long Range Ethernet and TDD. See also UMTS

TD-CDMA.

FDDI Fiber Distributed Data Interface. FDDI is an ANSI standard (X3T12) for a 100-Mbps fiber optic LAN employing a dual counter-rotating ring topology, using a token-passing technique for media access control, and using the 4B5B encoding scheme for Clock/Data Recovery (CDR). FDDI rings may use up to 200 km of optical fiber, or may employ twisted copper pairs for short hops, including terminal connections. FDDI generally is deployed in backbone applications, where it is used to join file servers, routers, switches, hubs, and other significant computing resources. The theoretical limit of Ethernet, measured in 64 byte packets, is 14,800 packets per second (pps). By comparison, Token Ring is 30,000 and FDDI is 170,000 pps. See also 4B5B, CDDI, FDDI TERMS, and FDDI-II.

FDDI Follow-On LAN A faster FDDI. Said to operate at up to 2.4 gigabits per second.

FDDI acronyms DAC Dual Attachment Concentrator

DAS Dual Attachment Station
ECF Echo Frames
ESF Extended Service Frames
LER Link Error Rate
LLC Logical Link Control
MAC Media Access Control
MIC Media Interface Connector
NIF Neighborhood Information Frame
NSA Next Station Addressing
PDU Protocol Data Unit
PHY Physical Protocol
PMD Physical Media Department
PMF Parameter Management Frames
RAF Resource Allocation Frames
RDF Request Denied Frames
SAC Single Attachment Concentrator
SAS Single Attachment Station
SDU Service Data Unit
SIF Station Information Frames
SMT Station Management
SRF Status Report Frame
THT Token Holding Timer
TRT Token Rotation Timer
TTRT Target Token Rotation Timer
TVX Valid Transmission Timer
UNA Upstream Neighbor Address

FDDI-II Fiber Distributed Data Interface-II is a recently standardized enhancement to FDDI. It still runs at 100 million bits per second on fiber or on twisted copper pairs, but in addition to transporting conventional packet data like other LANs, FDDI-II allows portions of the 100 Mbps bandwidth to carry low delay, constant bit rate, isochronous data like 64 Kbps telephone channels. This means the same LAN that carries computer packet data can carry live voice or live video calls. Some additional terms used with FDDI-II are: I-MAC which stands for Isochronous Media Access Control; P-MAC which stands for Packet Media Access Control; and WBC which stands for Wide Band Channel. See FDDI, FDDI Terms, Isochronous and Isoethernet.

FDL Facilities Data Link. A T-1 term, specifically relating to Extended SuperFrame (ESF). ESF extends the superframe from 12 to 24 consecutive and repetitive frames of information. The framing overhead of 8 Kbps in previous T-1 versions was used exclusively for purposes of synchronization. ESF takes advantage of newer channel banks and multiplexers which can accomplish this process of synchronization using only 2 Kbps of the framing bits, with the framing bit of only every fourth frame being used for this purpose. As a result, 6 Kbps is freed up for other purposes. This allows 2 Kbps to be used for continuous error checking using a CRC-6 (Cyclic Redundancy Check-6), and 4 Kbps to be used for a FDL which supports the communication of various network information in the form of in-service monitoring and diagnostics. ESF, through the FDL, supports non-intrusive signaling and control, thereby offering the user "clear channel" communications of a full 64 Kbps per channel, rather than the 56 Kbps offered through older versions as a result of "bit robbing." See also ESF and T-1. The FDL is embedded in the framing bits, using half the bits or 4000 bit/s. It is over this channel that two schemes operate:

1. In the original Bell System scheme, the repair station in the CO queries the CSU at the customer site, which responds with error statistics for the last 24 hours (in 15-min

increments). The repairman uses this info to diagnose line condition.

2. The more modern ANSI method has the CSU broadcast the error statistics for the last three seconds, every second (with overlap). Automatic monitoring equipment in the CO can tell when the line is going bad. Both systems can co-exist and operate on the same link, but that's unlikely in reality.

FDM Frequency Division Multiplexing. A technique in which the available transmission bandwidth of a circuit is divided by frequency into narrower bands, each used for a separate voice or data transmission channel. This means you can carry many conversations on one circuit. The conversations are separated by "guard channels." At one point, FDM was the most used method of multiplexing long haul conversations when they were transmitted in analog microwave signals. No more. Fiber optic transmission (today's preferred method) uses TDM – Time Division Multiplexing.

FDMA Frequency Division Multiple Access. One of several technologies used to separate multiple transmissions over a finite frequency allocation. FDMA refers to the method of allocating a discrete amount of frequency bandwidth to each user to permit many simultaneous conversations. In cellular telephony, for example, each caller occupies approximately 25 kHz of frequency spectrum. The cellular telephone frequency band, allocated from 824 MHz to 849 MHz and 869 MHz, consists of 416 total channels, or frequency slots, available for conversations. Within each cell, approximately 48 channels are available for mobile users. Different channels are allocated for neighboring cell sites, allowing for re-use of frequencies with a minimum of interference. This technique of assigning individual frequency slots, and re-using these frequency slots throughout the system, is known as FDMA. See CDMA, TDMA.

FDP Fiber Optic Distribution Panel.

FDS Frequency Division Switching. Seldom used for voice switching. Primarily used for radio and TV broadcasting.

FDX See Full Duplex.

FE Extended Framing ("F sub E"). An old name for ESF, also known as Extended Super-Frame, a T-1 carrier framing format that provides a 64 Kbps clear channel, error checking, 16 state signaling and some other nice data transmission features.

FE D4 Superframe Extended Another designation for AT&T's ESF (Extended SuperFrame).

feather An imaging term. An effect in which the edges of a pasted selection or paint tool fade progressively at the edges for a seamless blend with the background.

feather in your cap The term came from the American Indian tradition of obtaining feathers for headdresses. Birds were captured, some feathers plucked, and the birds were released. Each feather represented an act of bravery. The fashion of decorating hats with feathers declined in the twentieth century because too many birds were being slaughtered for their feathers.

feathers A pound of feathers weighs more than a pound of gold. Feathers are weighed by "avoirdupois" weight measure, which has 16 ounces to a pound, while gold is weighed in "troy" measure, which only has 12 ounces to a pound.

feature/function access code A code in the form *XX or XX and currently used by end users for control of and access to custom calling services (such as activation and deactivation of call forwarding and making changes to speed calling lists). Such features might also include "dial the last number that called me," "dial the last number I just called," "conference two phone conversations together."

feature-rich When a vendor can't tell you in simple words what his equipment does, he says his equipment (or software) is "feature-rich."

feature A feature is one of the many tasks a piece of equipment can accomplish. In the old days of selling telephone systems, there used to be an expression among salespeople that "The one with the biggest list of features wins." As a result many salespeople used to inflate their list of features by calling one feature by many names. Do not confuse a feature with the word "function," which is a much higher level word. The function of a telephone system is to be a phone system. While a phone system can have many features, it can have only one function.

feature boards Modular system cards that perform specific functionality - video or modem cards, for example.

feature buttons Think of a feature button on a telephone as a collection of numbers stored in a bin. When you hit the button, the bin quickly disgorges all the numbers one after another. Feature buttons are fast ways of doing things. You have a feature button labelled "Conference." Hit the button, set up a conference call. Without a feature button, you'd probably have to hit the switch hook and some numbers on your touchtone pad. In computer terms, a feature button on a phone is the same as a macro – an easy way of doing something. On most phones with feature buttons, the feature buttons are "programmable." This means you can assign different features to different buttons, i.e. the ones you want. For example, I always assign "Last Number Redial," "Saved Number Redial" and "Conference Call" to the buttons of any phone I'm programming. Some phones have many feature buttons. Some don't.

feature cartridge A replaceable software cartridge containing software features. The Feature Cartridge is inserted into the central cabinet, or Key Service Unit (if it's a key system). Several small phone systems (under 100 lines) use cartridges to upgrade their software. The manufacturers find cartridges are cheaper than equipping their phone systems with a floppy drive and the associated electronics.

feature code This is a number that is used to activate a particular feature on a phone system.

feature creep Occupational hazard The enemy of the good is the better. A term to show how features tend to get added to telecom equipment as time passes and new models appear. The term "feature creep" makes no judgments about whether the new features are actually useful. In book called "Startup; A Silicon Valley Adventure Story," Jerry Kaplan, the author, describes "Feature creep as the irresistible temptation for engineers to load a product down with their favorite special features."

feature function testing Feature/function testing is designed to assure that everything a system is supposed to do is done correctly, e.g. calls are switched to the correct destination, messages are left and deleted, billing records collected accurately, and so on. Feature/function testing is the most detailed portion of the test process. The people who perform functional testing must be extremely detail oriented and have the discipline to test every feature to their written functional requirement. No function of the system should be overlooked. Definition courtesy Steve Gladstone, from his book "Testing Computer Telephony Systems."

Feature Group A, B, C, D FGA, FGB, FGC, FGD, are four separate switching arrangements available from local exchange carrier (LEC) end central offices to interexchange (long distance) carriers. These switching arrangements allow the LECs' end-users to make toll calls via their favorite long distance carrier. Feature groups are described in a tariff filed by the National Exchange Carrier Association with the FCC. The feature group used by each IX (IntereXchange, also called long distance) carrier together with any special access surcharge determines the service they can provide their customers and the carrier common line access fee they will pay to the local exchange carrier involved. The most common Feature Group now is D. See the next four definitions. See Feature Group A, Feature Group B, Feature Group C, Feature Group D.

Feature Group A Offers access to the local exchange carrier's network through a subscriber-type line connection rather than a trunk. It is a continuation of the ENFIA arrangement used in the early days of OCCs, until equal access using an access tandem central office is available. Remember, without equal access the IX carrier had to require its customers to dial a local number to reach their long distance facilities, then dial an identification number, then dial long distance numbers of the called party desired. This service handicap, compared to AT&T's superior connections, qualifies the OCC for a discount off the FGA rate until access is equal. The IX carrier is billed by the LEC based upon actual monthly use rather than the ENFIA method of projected "minutes of use" rate.

Feature Group B Is similar to FGA, but provides a higher quality trunk line connection from end CO to the IX carrier's facilities, instead of the subscriber-type line. The IX customer can originate a call from anywhere within the LATA, while FGA requires customers to initiate the call from within the local exchange of the exchange carrier connecting to the IXC. FGB billings to the IX are on a flat usage basis, and a discount is applicable. To access a long distance carrier with Feature Group B capability, you dial 950-XXXX (XXXX is the Carrier Identification Code, or CIC), and then 1+ the number. Feature Group D is better. See also 950-XXXX and CIC.

Feature Group C Is the traditional toll service arrangement offered by LECs to AT&T prior to breakup of the Bell System. Quality is superior, and the service includes automatic number identification of the calling party, answerback, and disconnection supervision, and the subscribers can use either a dial or touchtone pad. This FGC service is offered only to AT&T without a discount.

Feature Group D FGD. The class of service associated with equal access arrangements. All facilities based IXCs (IntereXchange Carriers) and resellers of significance pay extra for Feature Group D terminations (connections), which is a trunk-side connection provided by the ILECs (Incumbent Local Exchange Carriers). Feature Group D is required for equal access, which allows phone users in the United States to pick up the telephone and dial 1+ to place a long distance call, with the call being handled by the IXC they have preselected. Without FGD, the user must first dial a 7- or 10-digit number, a calling card

number and PIN number, and then the desired telephone number. FGD also is required for an end user organization desiring ANI (Automatic Number Identification) information. Feature Group D also lets you dial around your preselected IXC to use another of your choice by dialing 101XXXX. See also 1+, 101XXXX, ANI, Equal Access, ILEC and IXC.

feature keys Same as FEATURE BUTTONS. A key is to a telephone man what a switch is to an electrical man.

feature phone A generic name for a telephone that has extra features (often speed dial buttons) designed to simplify and speed making and receiving phone calls.

feature transparency A PBX to PBX signaling exchange which trigger additional PBX services after a connection have been established, e.g. to display calling party name and number, call back when busy, call forwarding, executive override etc. Feature transparency is implemented by all major PBX vendors in their proprietory signaling systems and it will work between like phone systems made by the same manufacturer (though I personally wouldn't trust it, unless I saw it working and I got a guarantee in writing with a penalty). It is also supported (to an extent) in such open signaling systems as Q.SIG and DPNSS.

feature/function testing Feature/function testing is designed to assure that everything a system is supposed to do is done correctly, e.g. calls are switched to the correct destination, messages are left and deleted, billing records collected accurately, and so on. Feature/function testing is the most detailed portion of the test process. The people who perform functional testing must be extremely detail oriented and have the discipline to test every feature to their written functional requirement. No function of the system should be overlooked. Definition courtesy Steve Gladstone, from his book "Testing Computer Telephony Systems."

FEBE Far End Block Error. A maintenance signal transmitted in the PHY (Physical) overhead, indicating that a bit error(s) has been detected at the PHY layer at the far end of the link. The PHY is the Physical Layer (Layer 1) of the OSI Reference Model, and refers to the transmission facility, such as T-3 or SONET link. A block is a data block, such as a T-3 frame or an OC-1 SONET frame. The "far end" is the end of the physical circuit farthest from the edge of the network, i.e., the premises end. Errors are detected through an error detection algorithm such as block parity. When the Circuit Terminating Equipment (CTE), such as an ATM switch or a router, at the customer premises detects an error in a data block in a downstream transmission, it sends the CTE at the near end an error message. Based on the FEBE, both CTE increment the error count by one, in order to monitor and maintain a record of the bit error performance of the link. That information is used by a network management system to generate alarms and historical reports of network performance. NEBE (Near End Block Error) is essentially the same thing, only at the "near end," i.e., the network end, of the link.

February On the Roman holiday Lupercal (February 14) goats were sacrificed and the blood was smeared on two specially chosen youths. The youths would then run all around Rome with strips of goat hide in their hands. Women would strive to be beaten with these strips, known as februa (purifiers). Hence, February gets its name as the month of purification.

FEC 1. Forward Error Correction. A technique used by a receiver for correcting errors incurred in transmission over a communications channel without requiring retransmission of any information by the transmitter. Typically involves a convolution of the transmitter using a common algorithm and embedding sufficient redundant information in the data block to allow the receiver to correct. While this technique is processor-intensive, it improves the efficiency with which the network is used. See Forward Error Correction.

2. Forwarding Equivalence Class. A MPLS (Multiprotocol Label Switching) term. See MPLS.

FECN Forward Explicit Congestion Notification. A Frame Relay term. This bit contained within the Address Field notifies the receiving device that the network is experiencing congestion. Thereby, the target device is advised that frames may be delayed, discarded, or damaged in transit. It is the responsibility of the target device to adjust to that condition. In conjunction with BECN, devices in both the forward and backward directions are advised. See BECN.

FED 1. Field Emission Display. A new way of making TV and computer screen displays. FED screens are flat and potentially cheap. Like conventional glass screens, they emit light. LCDs, by comparison, don't. A typical FED screen packs millions of tiny individual emitters between two ultra-thin glass layers. Each emitter fires electrons simultaneously across a minuscule vacuum gap onto a phosphor coating very much like a CRT's. See also Field Emission Displays.

2. Fire Emitting Diode. Diodes not installed properly can become SEDs (Smoke Emitting Diodes), which then can become FEDs. See also Diode.

FED-STD A system of standards numbered FED-STD-1001 to 1008 which set modulation specifications for data transmission.

Federal Aviation Administration FAA. The federal regulatory agency responsible for air safety. Establishes antenna tower marking requirements.

Federal Communications Commission FCC. The federal organization in Washington D.C. set up by the Communications Act of 1934. It has the authority to regulate all interstate (but not intrastate) communications originating in the United States. The FCC is the U.S. federal regulatory agency responsible for the regulation of interstate and international communications by radio, television, wire, satellite and cable. Established by the Communications Act of 1934, it is responsible directly to Congress and is directed by five Commissioners appointed by the President and confirmed by the Senate for 5-year terms. The President designates one of the Commissioners to serve as Chairman. The Chairman's tenure is at the pleasure of the President. No more than three Commissioners may be members of the same political party. None can have a financial interest in any Commission-related business.

Stripped of all the extensive regulatory and legal mumbo jumbo, the FCC does three things: 1. It sets the prices for interstate phone, data and video service. 2. It determines who can or cannot get into the business of providing telecommunications service or equipment in the United States. 3. It determines the electrical and physical standards for telecommunications equipment and services. The FCC's powers, although strong, are tempered (limited) by the Federal Courts. Anyone who disagrees with FCC rulings can appeal them to a Federal Court. The FCC's power and rulings are also affected by the Justice Department (The Justice Department changed the industry with Divestiture), Congress and The 50 state public service commissions. The FCC changed with the passage of the Telecommunications Act of 1996 – the first telecom act passed by Congress since the Communications Act of 1934.

How is the FCC organized? Most items considered by the Commission are developed by one of seven operating bureaus and offices organized by substantive area:

- The Common Carrier Bureau handles domestic wireline telephony.
- The Mass Media Bureau regulates television and radio broadcasts.
- The Wireless Bureau oversees wireless services such as private radio, cellular telephone, personal communications service (PCS), and pagers.
- The Cable Services Bureau regulates cable television and related services.
- The International Bureau regulates international and satellite communications.
- The Compliance & Information Bureau investigates violations and answers questions.
- The Office of Engineering & Technology evaluates technologies and equipment.
- In addition, the FCC includes the following other offices:
- The Office of Plans and Policy develops and analyzes policy proposals.
- The Office of the General Counsel reviews legal issues and defends FCC actions in court.
- The Office of the Secretary oversees the filing of documents in FCC proceedings.
- The Office of Public Affairs distributes information to the public and the media.
- The Office of the Managing Director manages the internal administration of the FCC.
- The Office of Legislative and Intergovernmental Affairs coordinates FCC activities with other branches of government.
- The Office of the Inspector General reviews FCC activities.
- The Office of Communications Business Opportunities provides assistance to small businesses in the communications industry.
- The Office of Administrative Law Judges adjudicates disputes.
- The Office of Workplace Diversity ensures equal employment opportunities within the FCC.

Federal excise tax The long-distance federal excise tax was 3% on long-distance calls. It was originally established in 1898 to finance the Spanish-American War. On May 25, 2006 the U.S. Treasury Department announced that this tax would be discontinued. The Department of Justice will no longer pursue litigation and the Internal Revenue Service (IRS) will issue refunds of tax on long-distance service for the past three years. Taxpayers will be able to apply for refunds on their 2006 tax forms, to be filed in 2007.

Federal Information Processing Standards FIPS. The identifier attached to standards developed to support the U.S. government computer standardization program. The FIPS effort is carried out by the U.S. Department of Commerce, Springfield, VA.

Federal Open Market Committee See FOMC.

Federal Reserve Board The Fed. The federal government agency that sets interest rates and monitors, regulates, and exerts influence over the nation's monetary supply and banking system. It often does this by buying or selling government securities and taking other regulatory actions. The Federal Reserve, the central bank of the United States, was founded by Congress in 1913 to provide the nation with a safer, more flexible, and more stable monetary and financial system. Today the Federal Reserve's duties fall into four general areas: (1) conducting the nation's monetary policy; (2) supervising and regulating banking institutions and protecting the credit rights of consumers; (3) maintaining the stability of the financial system; and (4) providing certain financial services to the U.S. government, the public, financial institutions, and foreign official institutions. www.federalreserve.gov.

Federal Standard 1037C A U.S. Federal Standard entitled "Telecommunications: Glossary of Telecommunications Terms." It is prepared by the National Telecommunication System Technology & Standards Division and published by the General Services Administration (GSA) Information Technology Service pursuant to the Federal Property and Administrative Services Act of 1949, as amended. Federal Standard 1037C is a good source of official definitions used by the U.S. Federal Government. A number of definitions in this book are at least partially based on 1037C.

Federal Telecommunications Standards Committee FTSC. A U.S. government agency established in 1973 to promote standardization of communications and network interfaces. FTSC standards are identified by the designator FED-STD. The FTSC's address is General Services Administration, Specification Service Administration, Bldg 197, Washington Navy Yard, Washington DC 20407.

Federal Telecommunications System FTS. The private network used primarily by the civilian agencies of the federal government to call other government locations and to place calls to phones connected to the PSTN (Public Switched Telephone Network). FTS is a TTTN (Tandem Tie Trunk Network). A TTTN is a large, complex, private switched network which generally involves dedicated COs (Central Offices), as well as dedicated transmission facilities. See also AUTOVON and TTTN.

Federal-State Joint Board An organization with representatives from the FCC and the state public service commissions which tries to resolve Federal and State conflicts on telecommunications regulatory issues. Sometimes successfully and sometimes not successfully.

federated network Two or more autonomous networks that are linked together in a trusted relationship to facilitate communications and collaboration among users across the linked networks. Organizations whose networks are federated may have a close commercial relationship or a shared strong aversion to high carrier charges, or they have other common objectives or shared purposes of a continuing nature.

federated services Services offered by a network operator or other service provider to help an enterprise federate its network with networks of other enterprises. Federated services include helping the enterprise to develop a peering policy and helping it to implement that policy. This is an example of the principle, "If you are going to be bypassed, you might as well be part of the bypass solution," which telcos have been practicing since the mid-1980s. See federated network.

Federated States of Micronesia This is an interesting story about the commercialization of the Internet. The Federated States of Micronesia is a developing island nation in the Western Pacific Ocean, with a population of about 100,000. The country has little need for its Internet TLD (Top Level Domain) of .fm. So, the government of FMS announced on October 8, 1998 a joint agreement with BRS Media to begin registering and marketing the .fm TLD to the broadcasting industry. A number of FM radio stations have paid a lot of money to use the .fm TLD. How's that for e-commerce? See also TLD and Tuvalu.

feed A television signal source.

feedback 1. The return of part of an output signal back to the input side of the device. Think of the high-pitched squeal you hear when someone brings a microphone too close to the loudspeaker. Not all feedback is as obvious or as irritating. Some feedback is good. See Sidetone, which is what happens when you hear a little in the receiver of you're saying in the transmitter of a phone.

2. The inevitable result when your baby doesn't appreciate the strained carrots.

feeder cable A group of wires, usually 25-pair or multiples of 25-pair, that supports multiple phones in a single cable sheath. These cables may or may not be terminated with a connector on one or both ends. Feeder cable typically connects an intermediate distribution frame (IDF) to a main distribution frame (MDF). But the term "feeder cable" is also used in backbone wiring. And Bellcore defines the term slightly differently: A large pair-size loop cable emanating from a central office and usually placed in an underground conduit system with access available at periodically place manholes.

feedback control A process by which output or behavior of a machine or system is used to change its operation in order to constantly reduce the difference between the output and a target value. A simple example is a thermostat that cycles a furnace or air conditioner on and off to try and maintain a fixed temperature.

feeder line The cable running between bridges, line extenders and taps.

feeder route A network of loop cable extending from a wire center into a segment of the area served by the wire center.

feeder section A segment of a feeder route that is uniform throughout its length with respect to facility requirements and facilities in place.

feedholes Holes punched in paper or papertape which allow the paper or paper tape to be driven by sprocket wheels.

feedhorn The feedhorn is the focal point of a dish antenna. The feedhorn collects the signals reflected by the dish. The feedhorn is the device used for receiving or radiating microwave signals to or from a parabolic dish reflector.

feedware Software designed to get demand for a product or a new market segment started. Feedware is typically a less-full featured piece of software than the software you're really trying to sell. Feedware typically costs very little. It may even be free. See also Seedware.

FEFO First Ended, First Out. A rule for dealing with things in a queue. For example, higher priority messages will be sent before lower priority messages.

FEI Front end interface.

female amp connector Also called a C Connector or 25-pair female connector. The male version is called a P connector.

femtocell A small cellular base station designed for use in both residential and organizational environments. Femtocells are a recently emerged technology, to address the need for better in-building wireless coverage, greater efficiency on wireless carriers' networks (e.g., to reduce tromboning and to reduce OPEX on backhaul facilities), and to support fixed-mobile convergence.

femtosecond One-millionth of a billionth of a second. Femtoseconds are used in laser transmission and in other measures of very small happenings. It's 10 to the minus 15. There are as many femtoseconds in one second as there are seconds in thirty million years. There are 1,000,000,000,000,000 femtoseconds in one second. How small is a femtosecond? In a little more than a second, light can travel from the moon to the earth, but in a femtosecond it only travels one hundredth the width of a human hair.

FEP 1. Front End Processor. The "traffic cop" of the mainframe data communications world. Typically sits in front of a mainframe computer and is designed to handle the telecommunications burden, so the mainframe computer can concentrate on handling the processing burden. Here's a more technical definition: A dedicated communications system that intercepts and handles activity for the host. Can perform line control, message handling, code conversion, error control, and such applications functions as control and operation of special-purpose terminals. Designed to offload from the host computer all or most of its data communications functions. Front end processors are not used in the client/server world.

2. Fluorinated Ethylene Propylene. Also known by the trade name Teflon, a registered trademark of Dupont, and NeoFlon, a registered trademark of Daikin. FEP is the insulation of choice for high performance cable and wire systems installed in return air plenums. As FEP is really slick, it makes the wire really easy to pull through conduits, around corners, and so on – the same property that makes it so wonderful in the kitchen. FEP also is fire retardant and produces little visible smoke. FEP is used CAT 5 (Category 5) and higher-rated inside wire and cable systems. FEP's high cost has led many manufacturers to use PVC as an insulating compound, and to coat it with flame retardant compounds. Many countries require the use fire-retardant, low-smoke cable jackets, particularly in plenum ceilings. See also CAT 5, Plenum, and PVC.

FER Frame Error Rate. A computation based on the number of frames received with errors compared to frames received without errors. See also Frame Error Rate.

FERF 1. Far-End Receive Failure. A yellow alarm. A message from a remote network element that is having trouble receiving a signal.

2. Far-End Remote Failure. An alarm indicating a failure at the far end of an ATM network, identifying the specific circuit in a failure condition.

fernsprechvermittlungsstelle A central office in German. In Europe, they call a central office a "public exchange," or just plain "exchange." They look at you kinda strange when you say the North American word, namely "central office."

ferreed assembly A glass enclosed reed relay switch in which the reeds are made of some metal which can be opened or closed by an external magnetic field.

ferri chrome A coating used on tape comprising a layer of ferric oxide particles and a layer of chromium dioxide particles and combining the attributes of both.

ferric oxide A coating used on tape comprised of red iron oxide, the original material used for magnetic recording tapes.

ferrite A type of ceramic material having magnetic properties and consisting of a crystalline structure of ferric oxide and one or more metallic oxides, such as those of nickel or zinc. See Barium Ferrite, Hard Ferrite and Soft Ferrite.

ferroresonant transformer A special transformer which puts out regulated AC voltage even when the input voltage is variable. A ferroresonant transformer may be used by itself to correct brownouts or it may be built into a UPS. A ferroresonant transformer has an undesirable characteristic called "high output impedance" which can prevent protective devices such as circuit breakers on equipment plugged into it from functioning, resulting in a possible safety hazard. Another problem is that computer loads applied to ferro based line conditioners or UPS systems cause the voltage waveform applied to the computer to be very distorted, which may result in undervoltage conditions within the computer. Ferro based UPS systems are becoming obsolete because they can become unstable and oscillate when supplying modern power factor corrected power supplies. This definition courtesy APC.

ferrule A component of a fiber optic connection that holds a fiber in place and aids in its alignment.

Fessenden, Reginald A Canadian-born (1866-1932) inventor who was the first to surmise and demonstrate, in a series of experiments from 1900-1906, that voice, music and other audio content could be transmitted wirelessly. This laid the groundwork for broadcast radio and later, broadcast TV. Guiglielmo Marconi, the father of wireless communications, who was Fessenden's contemporary, hadn't believed it possible before Fessenden demonstrated otherwise that sound could be transmitted on radio waves. Marconi's transmissions up to that point had been Morse code. Fessenden is sometimes called the father of radio broadcasting.

festoon cable A cable, generally flat, that is used to deliver power and control aerially on jib and gantry cranes, hoists, plating and cleaning lines, and other industrial and manufacturing automation systems involving mobile/retractable equipment. Festoon cable is often mounted on C-track, T-track and I-beam trolley systems, and its hanging undulations resemble a garland, hence the name festoon cable. Since 2000, festoon cable that is manufactured for outdoor use must be enclosed in an outer rubber jacket.

festoon network A submarine (i.e., undersea) network built to link coastal cities instead of using a terrestrial network – microwave or optical. Festoon submarine cable systems are used because they're faster and cheaper to install and because disasters like hurricanes, tsunamis, volcano eruptions and don't hurt them. Many nations such as Italy, Thailand, Japan and South Africa already use or are planning to use festoon networks. Corning defines festoon as "a string or garland suspended in a loop or curve between two points." Merriam Webster defines it as "a decorative chain (as of flowers or leaves) hanging typically in a curve between two points." The cable goes out to sea, circles around and comes back to the next port. Hence the name festoon.

FET Field Effect Transistor. Very thin and small transistors are used to control pixels in a TFT (Thin Film Transistor) display.

FEXT Far-End CrossTalk. A type of crosstalk which occurs when signals on one twisted pair are coupled to another pair as they arrive at the far end of multi-pair cable system. FEXT is an issue on short loops supporting high-bandwidth services such as VDSL (Very-high-bit-rate Digital Subscriber Line), given the relatively high carrier frequencies involved. Services such as ADSL and HDSL, are not affected to the same extent, as the loops are longer and as such interference tends to be attenuated (weakened) on longer loops. See also Crosstalk. Compare with NEXT.

Fewer A smaller number. The word "fewer" is always confused with the word "less." According to the Oxford American Dictionary, the word "less" is used of things that are measured by amount (for example, eat less butter, use less fuel). Its use with things measured by numbers is regarded as incorrect (for example in "we need less workers"; correct usage is "fewer workers").

FF Form Feed. A printer function used to skip to the top of the next page or form.

FFA Field Force Automation. Refers to information technology solutions designed to help companies improve communication with employees in the field and yield increased productivity. FFA solutions typically comprise integrated software, hardware and networking components.

FFDI Fast Fiber Data Interface. A proprietary 100 megabit per second local area network that uses fiber optic, coax, shielded twisted pair or unshielded twisted pair. It is manufactured by PlusNet, Phoenix, Arizona.

FFOL Fiber Follow On LAN. Emerging LAN technology.

FFT 1. Fast File Transfer. An ISDN term referring to the fact that file transfers can be accomplished "fast." Reason #1: Two B channels at 64 Kbps each and a D channel at up to 16 Kbps are available to be bonded to provide as much as 144 Kbps. Reason #2: Data transfer is accomplished in an "optimistic" streaming mode, rather than a "pessimistic" packet mode. Therefore, there is no delay associated with acknowledgements. This is possible due to the excellent level of error performance inherent in digital services. The end result is FFT.

2. Fast Fourier Transform. A signal processing term for a common computer implementation of Fourier Transforms. The FFT, as a practical implementation, will always result in a finite series of sine and cosine waves as an extremely close approximation of the possibly infinite series described by the purely mathematical application of the Fourier Transform. See Fourier's Theorem.

FTD Originally called "Florists' Telegraph Delivery" and renamed "Florists Transworld Delivery" in 1965, FTD was the world's first flowers-by-wire service. FTD was established in 1910 by 15 American retail florists who agreed to exchange orders for out-of-town flower deliveries. In 1965 FTD expanded internationally and today connects florists in over 150 countries.

FTD Mercury Network An electronic network used by major wire services in the floriculture industry to process wire orders and messages. The network carries approximately 15 million orders and messages annually. It has been dubbed the Floral Information Superhighway.

FTTCS Fiber to the cell site.

FG An ATM term. Functional Group: A collection of functions related in such a way that they will be provided by a single logical component. Examples include the Route Server Functional Group (RSFG), the IASG (Internetwork Address Sub-Group), Coordination Functional Group (CFG), the Edge Device Functional Group (EDFG) and the ATM attached host Behavior Functional Group (AHFG).

FGA Feature Group A. Characterized by a seven-digit local telephone number access code and line side termination to the customer at the first point of switching.

FGB See Feature Group B.

FGC See Feature Group C.

FGC-EA See Feature Group C and Equal Access.

FGD See Feature Group D.

FGD-EA Feature Group D - Equal Access. See Feature Group D, also see Equal Access.

FHSS Frequency Hopping Spread Spectrum. A technique used in spread spectrum radio transmission systems, such as Wireless LANs and some PCS cellular systems. FHSS involves the conversion of a data stream into a stream of packets, each of which is prepended (prepend means added to the front of) by an ID contained in the packet header. Short bursts of packets then are transmitted over a range of 75 or more frequencies, with the transmitter and receiver hopping from one frequency to another in a carefully choreographed "hop sequence." FCC regulations specify that each transmission can dwell on a particular frequency no more than 400 milliseconds. A large number of other transmissions also may share the same range of frequencies at the same time, with each using a different hop sequence. The potential remains, however, for the overlapping of packets. The receiving device is able to distinguish each packet in a packet stream by reading the various IDs, treating competing signals as noise. In a wireless LAN environment, DSSS typically operates in the 2.4 GHz frequency band, which is one of the ISM (Industrial Scientific Medical) bands defined by the FCC for unlicensed use. Although most manufacturers prefer FHSS, Direct Sequence Spread Spectrum (DSSS) is used by some. See also CDMA, DSSS, ISM and Spread Spectrum.

fiasco The word "fiasco," meaning a failure, is derived from the ancient Italian art of glass blowing. If a Venetian glass blower made a mistake while creating a fine, delicate bottle, the imperfect vessel was turned into an ordinary drinking flask which, in Italian, is called a "fiasco." A fiasco is a disaster. My favorite use of the word comes from John Doerr, superstar of the Silicon Valley venture capital powerhouse, Kleiner Perkins Caufield & Byers. Referring to the skill of picking startups, he said, "You're only as good as your last fiasco."

fiber A shortened way of saying "fiber optic." Fiber is made of very pure glass. In Bill Gates' book called "the Road Ahead," he says that optical fiber is so clear and pure that if you looked through a wall of it 70 miles thick, you'd be able to see a candle burning on the other side. Digital signals, in the form of modulated light, travel on strands of fiber for long distances. The big advantage that fiber has over copper is that it can carry far, far more information over much, much longer distances. The short history of fiber optics for communications is that scientists keep discovering more and more ways of putting more and

more information down one single strand of fiber. Based on my own personal researches, no one has any idea what the eventual capacity limit of a strand of fiber optic might be. I have personally asked many scientists (including one Nobel Physics prize winner) and all seem to think there must be a theoretical limit. But they don't know what it is. And they believe we have many, many years of breakthroughs in fiber still to go. As of the time of this writing, SONET OC-192 (Synchronous Optical NETwork Optical Carrier Level 192) systems are being deployed fairly routinely by a number of major long distance carriers. Each OC-192 strand supports approximately 10 Gbps. With DWDM (Dense Wavelength Division Multiplexing), as many as 32 "windows," or wavelengths of light, can be overlaid into a single strand at OC-192, yielding a total of approximately 320 Gbps. Fiber is the American spelling. The spelling in England, Europe, Canada, Australia and New Zealand is fibre. See also the following definitions beginning with fiber. See Optical Fiber for an essay on the advantages of fiber as a communications medium. See also Chromatic Dispersion, OC-192 and SONET.

fiber axis In an optical fiber, the line connecting the centers of line circles that circumscribe the core, as defined under "tolerance field."

Fiber Bragg Grating FBG. An FBG is a narrowband reflection filter permanently written into the core of single-mode optical fiber that enables the type of wavelength precision that is necessary for WDM systems to successfully combine, control, and route multiple "colors" of light within a single optical fiber. FBG act as fine optical filters separating and filtering multiple wavelengths of light propagating in the same fiber that can select a wavelength with a precision of plus or minus 0.02 nanometer. These gratings are typically used in signal monitoring and gain flattening applications.

fiber buffer The material surrounding and immediately adjacent to an optical fiber that provides mechanical isolation and protection. Buffers are generally softer than jackets.

fiber bundle An assembly of parallel unbuffered optical fibers, in intimate contact with one another and secured, usually with an epoxy or other adhesive. Each endface of the bundle is typically finished to a flat or other optical surface, usually at right angles to the axis of the bundle. Such bundles are used to transmit optical power or images. Bundles used to transmit images must maintain spatial coherence amongst the relative positions of the respective fibers at each end (aligned bundles). There is no requirement for this if the bundle is used to transmit optical power only. Fiber bundles were employed in early, short-distance communication applications, but have become obsolete in modern telecommunications.

fiber channel There is no such thing as Fiber Channel. See Fibre Channel.

fiber coating The protective layer of material above the fiber cladding. Usually 250Êm diameter, however, it can be 500Êm or 900Êm. This layer is applied by the manufacturer.

fiber control office terminal FCOT is a generic term for a fiber terminal that can be configured for full digital, full analog or mixed communications.

fiber darkening The opacification of an optical fiber due to radiation, aging, or some other process. What this means in simple language is the fiber won't transmit as well as it did when it was new and first installed. See opacification.

Fiber Distributed Data Interface FDDI. A set of ANSI/ISO standards that, when taken together, define a 100 million bits per second (Mbps), timed-token protocol, Local Area Network that uses fiber optic cable as the transmission medium. The standards define Physical Layer Medium Dependent, Physical Layer, Media Access Control, and Station Management entities. The standard specifies: multi-mode fiber, 50/125, 62.5/125, or 85/125 core-cladding specification; and LED or laser light source; and 2 kilometers for unrepeatered data transmission at 40 million bits per second (Mbps).

fiber exhaust A fiber optic term. Fiber exhaust comprises the noxious light emissions from a fiber optic engine, the device which generates the light signals signals in an optical fiber transmission system. Much as an internal combustion engine creates noxious emissions, fiber optic engines (e.g., LEDs and Laser Diodes) create noxious light emissions, which can be extremely hazardous to your health. Actually, the preceding is a joke, like "frequency grease," which is used to overcome static noise in a radio system. Now for the truth: Fiber exhaust simply means that the capacity of a fiber optic transmission has been exhausted. The solution is 1) to lay more fiber, 2) to increase the speed of the system through an upgrade of light sources and detectors, or 3) to use WDM (Wavelength Division Multiplexing) or DWDM (Dense WDM) to increase capacity through the support of multiple wavelengths of light. See also Bucket o' Dial Tone, DWDM, Frequency Grease, SONET and WDM.

fiber grating A fiber optics term. Fiber gratings are sections of optical fiber that have periodic changes in their refractive index "written" into the core with ultraviolet light. The fiber grating creates nearly arbitrary combinations of signal reflection and transmission spectra. Fiber gratings can be used as optical filters and taps. Fiber gratings also can be used as lasers (believe it or not) through the use of resonant cavities defined by highly-reflective regions in the grating. EDFAs (Erbium Doped Fiber Amplifiers) use fiber grating to flatten the gain (i.e., power amplification level) across a range of wavelengths amplified, thereby reducing issues of crosstalk between closely-spaced wavelengths. EDFAs also use fiber grating to create filters that separate wavelengths and group them into bands, allowing certain wavelengths to be transmitted or reflected selectively. Fiber gratings are used in OADMs (Optical Add/Drop Multiplexers), in order to add and drop individual light wavelengths, which can be separated by as little as fractions of a nm (nanometer, or billionths of a meter). See also Diffraction Grating, EDFA, and OADM.

fiber identifier A test instrument that can differentiate between live and dead fibers in a working cable and can identify a preselected fiber to which a special transmitter has been attached.

Fiber In The Loop See FITL.

Fiber Loop Carrier See FLC.

fiber loss The attenuation (deterioration) of the light signal in optical fiber transmission.

fiber mile Let's say that you have two sheaths of fiber, each of which contains ten fibers and runs for one mile. That is one route mile (total distance of all fibers), two sheath miles (two sheaths running one mile), and twenty fiber miles (20 fibers running one mile).

fiber optic See Fiber and Fiber Optics.

fiber optic amplifier As light, like electricity or any other form of electromagnetic energy, travels through a physical medium, it attenuates, or loses intensity. At some point in a communications transmission system, you must take your increasingly weak signal and boost it back to its original strength. In analog systems, you simply amplify the weak incoming signal and send it on its way. In addition to amplifying the information signal, any accumulated noise is also amplified. In digital signals, you first regenerate the signal, then amplify it, then send it on its way, with no recognition of or regard for any noise present. Most fiber transmission systems accomplish this process by converting the original light signal on the fiber to electrical impulses, regenerate the signal, then amplify it, then convert it back to light pulses, then send it on its way. This takes significant energy and equipment – not altogether convenient for an underwater cable of several thousand miles. With new fiber optic amplifiers you no longer need to convert the light signal to electrical impulses. A fiber optic amplifier uses special fiber doped with erbium to act as the amplifier. Light comes into this special fiber, is pumped with the correct frequency laser and is amplified with extremely high gain and very low noise through a process of chemical light amplification. It's truly amazing technology. See Erbium-Doped Fiber Amplifier for a more technical explanation.

fiber optic attenuator A small device with two connectors. It reduces the amount of light passing through it, similar to the way sunglasses reduce the amount of light entering your eyes so you can see better.

fiber optic buffer Plastic coating on individual fibers. There are 12 colors to distinguish them from each other.

fiber optic cable See Fiber.

fiber optic connector There are three types: SC, ST and FC. The FC connector can be considered an earlier vintage connector but it was pre-dated by the biconic and SMA. FC connectors were specifically designed for telecommunications applications and initially came into prominence overseas. It came into its own in the 1980s and was the earliest design to incorporate the now standard 2.5mm ceramic ferrule. It also provided non-optical disconnect performance. Another advantage was this connector offered tunable keying which allowed it to be adjusted in order to minimize loss. The FC connector incorporated a threaded coupling nut which provided a secure connection, however as with the biconic and SMA, it did not allow for quick connect and disconnect as the coupling nut had to be rotated many times to thread or unthread the connector. Technicians also found it difficult to work with the threaded connectors particularly if the technician had large hands. The ST connector can be considered as the first of the contemporary connector designs. It utilizes the 2.5mm ferrule which is available in ceramic, zirconia, stainless steel, or plastic materials and is produced as a singlemode or multimode version. A variety of physical contact ceramic and zirconia ferrule styles allow the ST to achieve optimal performance. It is also available as an epoxyless version. The ST connector is easily recognizable because of its quick release bayonet-style locking mechanism. Unlike the threaded connectors, the ST only requires a quarter turn to complete a mating or demating cycle. In the opinion of some people in the industry, one of the early drawbacks to the ST connector was that it was not designed to be pull proof which meant that it could optically disconnect when the patchcord was pulled. The SC connector began to see widespread deployment in the early 1990's particularly among the RBOCs (Regional Bell Operating Companies) and several Independents. It is available in both singlemode and multimode designs. The SC is easily distinguishable by its square body style and utilizes a push - pull feature for mating and demating. The push - pull feature is preferred by many customers because when the ferrules come into contact in the mating sleeve of the adapter there are no rotational forces which could potentially damage a fiber or score the ferrules. This feature also eliminates a potential buildup of particulate in the mating sleeve which could possibly degrade optical performance. The SC is also designed as a pull-proof connector. A pull-proof connector is designed so that the ferrule is decoupled from the connector housing to which the fiber cable is attached. Therefore the connector maintains optical contact when fiber cable is pulled outward or sideways, thus eliminating the possibility of a service outage. In a non-pull-proof connector such as the earlier ST's, the connection can be broken when cable is pulled with a force that is greater than the spring force, which is around 2 - 3 pounds. Since the SC mating and demating is accomplished by push - pull force rather than rotational force, it readily lends itself to duplex connection configurations because it requires less space to engage and disengage. See also fiber optic coupler.

fiber optic coupler A device to join fibers together.

fiber optic distribution panel A termination device and organizer. It houses splice trays where connector plugs (called pigtails) are attached to the ends of fiber-optic cables.

fiber optic gyroscope A coil of optical fiber that can detect rotation about its axis.

fiber optic inter repeater link An 802.3 Ethernet standard for connecting two repeater devices at 10 million bits per second.

fiber optic link A transmitter, receiver, and cable assembly that can transmit information between two points.

fiber optic modem A modem that connects a copper cable or coax network with a fiber optic network.

Fiber Optic Test Procedure See FOTP.

fiber optic tracer A device typically based on a flashlight used to test the continuity of optic fibers and to trace multimode fibers from proper connections.

fiber optic transmission system See Fiber Optics, SDH and SONET.

fiber optic waveguide A relatively long thin strand of transparent substance, usually glass, capable of conducting an electromagnetic wave of optical wavelength (visible region of the frequency spectrum) with some ability to confine longitudinally directed, or near-longitudinally directed, lightwaves to its interior by means of internal reflection.

fiber optics A technology in which light is used to transport information from one point to another. More specifically, fiber optics are thin filaments of glass through which light beams are transmitted over long distances carrying enormous amounts of data. Modulating light on thin strands of glass produces major benefits in high bandwidth, relatively low cost, low power consumption, small space needs, total insensitivity to electromagnetic interference and great insensitivity to being bugged. All these benefits have great attraction to anyone who needs vast, clean transmission capacity, to the military and to anyone who runs a factory with lots of electronic machinery. The first field trial of an AT&T lightwave system took place in Chicago in 1977. There has been a rapid improvement in cost effectiveness of fiber systems, expressed as cost per bit per kilometer. A one hundredfold increase in cost performance in one five-year period – from 1980 to 1985. Some versions of fiber optics now carry 40 Gbps (forth billion bits per second) in support of more than 600,000 uncompressed voice conversations, and that's over a single wavelength. Some systems run as many as 40 wavelengths through a technique known as Dense Wavelength Division Multiplexing (DWDM). See also DWDM and SONET.

fiber pair Optical fiber transmission systems often, but not always, are deployed in pairs. First, you have to understand that light can only travel in one direction through a strand of fiber. It can go fast. But it can only travel in one direction. Now, let's consider the LAN (Local Area Network), where fiber is installed with a single fiber or a dual fiber. In the LAN, fiber often is deployed on the basis of a single fiber because it's cheaper. In a single fiber LAN, the devices (e.g., a switch or hub, and a server) transmit in half-duplex (i.e. one way at a time). Then they switch quickly to other direction. This is the way a typical local phone line works, i.e. the phone line from your telephone company's central office to the phone on your desk. A fiber communications system, like a LAN, may also be installed with two fibers, or fiber pairs, with the devices using one fiber to transmit and one to receive in full-duplex (i.e., simultaneous two-way) mode. While this approach requires twice the fiber and twice the port interfaces at the device level, the speed of transmission is improved by a factor of at least two. In a FDDI (Fiber Distributed Data Interface) LAN backbone network, each of two fibers continuously is active, with one transmitting in the clockwise direction and one in the counter-clockwise direction. This dual counter-rotating ring configuration is for purposes of redundancy and resiliency, as a break across both fibers at a single physical point will not isolate any two devices. In other words, any device can communicate with any other device on the FDDI network, in one direction or the other. Now let's consider the WAN (Wide Area Network) domain. In the WAN, fiber typically is deployed in a pair configuration. In some implementations, one fiber is used to transmit and one to receive. In a SONET/SDH implementation, fiber networks are deployed in a dual counter-rotating ring configuration, much like that of FDDI networks, and for the same reasons of redundancy and resiliency. See also FDDI, Fiber Optic, and SONET.

fiber pigtail A short length of optical fiber, permanently fixed to a component, used to couple power between the component and the transmission fiber.

fiber remote Fiber Remote extends what Northern Telecom calls Intelligent Peripheral Equipment using dark single or multimode fiber cable. Fiber Remote operates over a range of typical campus distances – from thousands of feet to several miles. Distance is site specific, determined by variables such as the type and quality of the fiber installed and the number of connectors and splices. The signal attenuation between the local PBX system and the remote IPE shelves should not exceed a 13 dB loss. Fiber Remote is used when users prefer a single switch; when users have a campus with right of way for running fiber where the distance between the local and remote site is within 6 miles; where there's limited riser space in a large high rise building; where a user needs to alleviate switch room congestion and where security is of utmost importance (e.g. in military bases).

fiber ring You want reliable communications. You want communications that will never go out. The traditional way is to duplicate everything, from phones to lines. That way, if something breaks, you can switch to the backup. Instead of having two fiber lines into your business, some carriers will connect you to a two-way fiber ring. The theory is that if one connection breaks, you can happily transit and receive the other way. Once you get the idea of a fiber ring's reliability, you can get creative and have multiple rings all joined together. For example, one I read about recently, was an eight-node optical fiber ring network, using four wavelength division multiplexed (WDM) channels and eight bidirectional add-drop multiplexers (BADM) to route analog or digital data between individual nodes. Specific wavelength channels are routed bi-directionally through the eight nodes, minimizing the number of hops a message must take to reach its destination. Information on multiple wavelengths propagates through the BADMs in both directions. Each BADM adds (drops) two wavelengths to (from) the ring network. The BADMs are constructed using thin film filters, an all-optical technology.

fiber seeking backhoe The telecommunications corollary to the heat seeking missile.

fiber spudger Shaped like a pencil, it's a gadget technicians use to move around and find their way through fiber optic cables on their hunt for one single fiber optic.

fiber swap Winstar also was one of the first companies to engage in the type of fiber "swap" transactions that have become so controversial in 2001-2002 in the wake of Global Crossing Ltd.'s collapse. Starting in 1998, for instance, Winstar signed a series of contracts with Williams Communications Group Inc., in which Williams agreed to pay Winstar $400 million over four years for access to its portions of its telecom network. Winstar then turned around and agreed to pay Williams $644 million over seven years. This transaction gave both companies "profits" they could book and thus make their shares look more attractive to investors. Winstar eventually went broke.

Fiber Switch Cross-Connect FXC. A type of Optical Cross-Connect (OXC). See OXC.

fiber to the cabinet A network architecture in which an optical fiber connects the telephone switch to a streetside cabinet where the signal is converted to feed the subscriber over a twisted copper pair. See also Fiber to the Curb.

fiber to the curb FTTC refers to the installation and use of optical fiber cable directly to the curbs near homes or any business environment as a replacement for "plain old telephone service" (POTS). Think of removing all the telephone lines you see in your neighborhood and replacing them with optical fiber lines. Such wiring would give us extremely high bandwidth and make possible movies-on-demand and online multimedia presentations arriving without noticeable delay. The term "fiber to the curb" recognizes that optical fiber is already used for most of the long-distance part of your telephone calls and Internet use. Unfortunately, the last part - installing fiber to the curb - is the most expensive. For this reason, fiber to the curb is proceeding very slowly. Meanwhile, other less costly alternatives, such as ADSL on regular phone lines and satellite delivery, are likely to arrive much sooner in most homes. Fiber to the curb implies that coaxial cable or another medium might

carry the signals the very short distance between the curb and the user inside the home or business. "Fiber to the building" (FTTB) refers to installing optical fiber from the telephone company central office to a specific building such as a business or apartment house. "Fiber to the neighborhood" (FTTN) refers to installing it generally to all curbs or buildings in a neighborhood. Hybrid Fiber Coax (HFC) is an example of a distribution concept in which optical fiber is used as the backbone medium in a given environment and coaxial cable is used between the backbone and individual users (such as those in a small corporation or a college environment).

fiber to the premise See FTTP.

fiberoptichead Slang expression to describe a customer who thinks he knows everything about cable. Usage: "That fiberoptichead wouldn't know a drop from a fish job."

fiberphone A battery-powered device that connects to both ends of a fiber optic cable allowing people (typically craftspeople) to talk over the cable. The complete device (two ends) costs $1,000 to $2,000 and often comes with a headset.

Fiberworld In simultaneous media events in Washington and Montreal on October 12, 1989, Northern Telecom and Bell Northern Research (BNR) unveiled "FiberWorld," which they referred to "as a vision and commitment to deliver the world's first completely family of fiber-optic access, transport, and switching products."

fibre The European, Australian, Canadian, British and New Zealand spelling. The American spelling is fiber. Whoops, except in the case of Fibre Channel, which is correct, even in American English. See Fiber, Fiber Optics and Fibre Channel.

Fibre Channel Servers and networks are the mainstays of today's businesses. The dependence on data exchange between computers and between devices is pushing the limits of current computing architectures, which must evolve to support high-speed connections. To address this need, the American National Standards Institute (ANSI) developed several standards to define high-speed connections between devices that later became known as fibre (or fiber) channel. The potential in the fibre channel standard encouraged several major companies and organizations to back the movement, and in 1994 it was ratified as an official ANSI standard. Fibre channels are divided into five layers (listed below). Much like the Open Systems Interconnection (OSI) model, each layer defines a particular function of the system.

FC-0: Layer 0 defines the physical aspects of a fibre channel including the connection types, transmission media, signaling, and other optical and electrical parameters.

FC-1: Layer 1 outlines the encoding and decoding of information on a channel and error correction routines.

FC-2: Layer 2 is the meat of the fibre channel architecture in that it defines how to break down a data stream into frames and reassemble them again. This layer also negotiates flow control between devices, as well as other advanced aspects of transmitting and receiving data.

FC-3: Layer 3 allows access to advanced fibre channel functions, such as the ability to use multiple fibre channel ports working in parallel to act as a larger virtual connection and the ability to send a single transmission to multiple destinations.

FC-4: Layer 4 is the interface between the fibre channel hardware and the high level protocols that use the hardware to communicate. Because a fibre channel is a combination of the best of channel and networking architectures, it can handle protocols from both equally well and even allow them to work at the same time over the same interface.

Because of the structure of the architecture, fibre channels can operate at speeds from 133 Mbps to over 1,000 Mbps and can transmit and receive at the same time. In addition, high-level application interfaces like IPI, SCSI, IP, and ATM can operate flawlessly over the fibre channel architecture while using the same physical interfaces. The fibre channel topology is also flexible. It can be configured in a ring, point-to-point, or star topology. The adaptability and flexibility of a fibre channel make it ideal for storage area networks and network attached storage systems because it can provide a high-speed hardware connection between computers, servers, devices, and even displays and can be configured to drop into nearly any existing network configuration.

FC's speed of data transmission is due not only to the fundamental nature of the transmission system, but also to the fact that FC is a serial link technology. In other words, FC is an I/O (Input/Output) interface over which data is streamed in serial fashion across an established link. FC provides a channel connection for dedicated or switched point-to-point connection between devices. "Channel connections" are hardware-intensive, low in overhead, and high in speed; "network connections" typically are software-intensive, high in overhead, and therefore slower. The downside is that channel connections tend to be limited to a relatively small number of devices with pre-defined addresses. Actually, FC will support, through separate ports, both channel and network connections. It also will support not only its own protocol, but also higher level protocols such as FDDI, SCSI, HIPPI and IPI. The physical topology of Fibre Channel can be point-to-point, ring, or star. In a star configuration, the interconnecting switching device is known as a "Fabric," which can be a circuit switch (star), an active hub (star) or a loop (ring).

Fibre Channel supports the transfer of data in frames, with a payload of 2,048 bytes. A CRC (Cyclic Redundancy Check) mechanism is employed for purposes of detection and correction of transmission errors. Flow control is supported through switch buffers, in a Fabric implementation. Three service classes are supported. Class 1 provides the equivalent of a dedicated physical connection; it is the highest quality of service, and is most effective for very high-speed transfers of large amounts of data. Class 2 is a connectionless grade of service, making use of multiplexed frame switching, with multiple sources sharing the same channels; Class 2 supports confirmation of frame delivery. Class 3 is identical to Class 2, minus confirmation of frame delivery.

The applications for Fibre Channel initially were for high-speed data and image transfer in NAS (Network-Attached Storage) and SAN (Storage Area Network) applications. In the recent past, much attention has been focused on the real-time transfer of audio and video, as well. As a result, Fibre Channel is being implemented for video file transfer and video playback applications in post-production digital video and movie studios, and in the broadcast backbone. Fibre Channel over IP (FCIP) has been developed for tunneling between Fibre Channel SANs over IP networks. Competing standards include 10GbE and ESCON. See also 10GbE, CRC, ESCON, FCIP, FDDI, HIPPI, IPI, and SCSI. www.FibreChannel.com

Fibre Channel -- Arbitrated Loop FC-AL. A Fibre Channel application for Storage Area Networks (SANs), FC-AL supports high-speed access to storage arrays over loops as long as 10 kilometers over a single Fibre Channel link, non-amplified, and at data rates as high as 100 MBps (MegaBytes per second). In a dual loop architecture, data rates are doubled to as much as 200 MBps; rates of 400 MBps are anticipated in the near future. Logically, FC-AL operates as a full-duplex, point-to-point, serial data channel. As many as 126 hosts can be connected to a given storage device; intermediate and cascading FC-AL hubs and concentrators can serve to improve costs, although there are corresponding performance degradations. FC-AL earns the tag "arbitrated" by virtue of the fact that access to the storage system is arbitrated on the basis of level of privilege, with fractional bandwidth services supported. The next generation of FC-AL, designated FC-EL (Fibre Channel-Enhanced Loop) is under development at ANSI. See also Fibre Channel and SAN.

Fibre Channel -- Enhanced Loop FC-EL. The next generation of FC-AL (Fibre Channel-Arbitrated Loop), FC-EL is under development at ANSI. See also FC-AL and Fibre Channel.

Fibre Channel Fabric A fabric is one or more Fibre Channel switches in a single configuration. Fabric switches provide 100 MBps per port, so adding devices to a switch actually increases the aggregate bandwidth. Fabric switches also provide enhanced services to allow registration of devices and the discovery of targets (disks) by initiators (servers). Fabric switches may be connected together to form extended storage networks for large enterprises. The addressing scheme used for fabrics supports up to 15 1/2 million devices, making fabrics a very scalable architecture.

Fibre Channel GLM GLM stands for Gigabit Linking Module, a generic Fibre Channel transceiver unit that integrates the key functions necessary for installation of a Fibre Channel media interface on most systems. They provide complete Fibre Channel FC-0 functionality on easy to install daughter cards. A specification for this standardized universal module was developed in Fibre Channel's early days to help streamline the development process for various Fibre Channel products and their multiple versions. For example, optical and Copper GLMs can be swapped on many vendors' Fibre Channel adapter cards. Included are transmit and receive optics, drivers, clock and data recovery, serializer, deserializer and laser safety features.

Fibre Channel Industry Association Formed in January 1993, as the Fibre Channel Association, the FCA worked to encourage the utilization of Fibre Channel, complementing the standards development efforts of the ANSI T11 committee. The mission of the FCA is "to provide a support structure for system integrators, peripheral manufacturers, software developers, component manufacturers, communications companies and computer service providers." On August 18, 1999, the Fibre Channel Association (FCA) and the Fibre Channel Community (FCC), two organizations for the advancement of Fibre Channel technology and Storage Area Network solutions, merged into one association, the Fibre Channel Industry Association (FCIA). www.fibrechannel-europe.com.

Fibre Channel Node Device connected to a Fibre Channel fabric, possibly a PC, disk drive, or RAID array, as well as an FC-AL.

fibre jack Fibre jack is a small-form full-duplex (i.e. two fibres) fibre connecter devel-

oped by Panduit. It has been approved for publication as a standard under TIA/EIR-604.6 (FOCIS-6).

FICON Fiber Connectivity. A high-speed fiber optic channel for connecting IBM mainframes to each other and/or to attached peripherals such as printers and storage devices. FICON channels are up to eight times as efficient as ESCON channels, IBM's previous fiber optic channel standard. FICON is IBM's implementation of the S/390 ESCON storage protocols over a Fibre Channel switch fabric.

FID Field Identifier. A USOC and ISDN SPID term. FID is widely used in the Telco business related to USOCs. The purpose of USOCs (Universal Service Ordering Codes) is to define a customer's service and equipment. FIDs (field identifiers) are used to describe more detailed and specific attributes of those USOCs. For example, the USOC for Remote Call Forwarding - RCD- will have a FID associated with it specifying the telephone number that calls will be forwarded to. See USOC.

fidelity Fidelity is the quality or state of being faithful. In the domain of electronic devices such as TV sets, radios, and record players (Lest I risk dating myself, systems that play tapes and CDs are lumped under my category of record players.), fidelity refers to the extent to which the device faithfully (i.e., accurately) reproduces the original audio or visual signal. Hi-Fi systems are noted for their high level of fidelity. In the domain of WLANs (Wireless Local Area Networks), Wi-Fi (Wireless Fidelity) refers to 802.11b wireless Ethernet. In the domain of marriage, fidelity refers to remaining faithful (i.e., not cheating on) your spouse. In the domain of the U.S. Marine Corps, "Semper Fidelis" translates from Latin as "Always Faithful" (to the Corps). See also 802.11.

FidoNet An electronic bulletin board technology for transfer and receipt of messages. According to PC Magazine, the origins of FidoNet date back to the early 1980s, when the two authors of the BBS software Fido, who lived on opposite coasts, needed an easy way to exchange modifications they made to the source code. They designed a system where, as a nightly event, the board would shut down and run utilities that automatically transferred the changed files between the author's BBSs. The logical next step was to permit the exchange of private mail messages called NetMail, between the sysops. The author found these capabilities so useful that they include them as part of the Fido BBS (Bulletin Board Software) package. It didn't take long for an informal network of Fido nodes to come into existence, all running the Fido software and exchanging various utility and program files and NetMail among sysops. Like other BBSs, the FidoNet BBSs had their own SIGs, or Special Interest Groups, where users with similar interests could exchange messages in a way similar to what on-line services call conferences or forums. By 1986 a Fido sysop had extended the NetMail concept to allow SIGs to share public messages among the BBSs, and EchoMail was born. In the years since, BBS authors and FidoNet users and sysops extended these capabilities to other BBS packages, and FidoNet grew. It currently has over 11,000 nodes covering most of the world. Many of the existing public and private networks go through FidoNet gateways into the Internet Mail system, which carries e-mail over a group of interconnected networks to universities, government agencies, military branches, and corporations. FidoNet technology uses store-and-forward messaging and is fbased on point-to-point communications between nodes.

field 1. One half (every other line) of a complete television picture "frame", consisting of every other analog scan line. There are 60 fields per second in American television.

2. A place with no phones or other communications capability where an important person inevitably is when you need some vital information, service or device that only he or she can provide. "I'm sorry, the chief technician is in the field today, and can't be reached." Few "fields" are actually fields. They're usually downtown office buildings.

3. The specific location of data within a record. In the jargon of database management systems, many fields make up one record Many records make up one file. A field is one of the basic subdivisions of a data record. The record on you in your company's database might include your name, your address, your salary, etc. A field is simply one of these – e.g. your salary, your last name, or your street address. All the records of all the employees in your company make up a file, also called a database.

4. The name given to that part of an electrical system in which electromagnetic lines of force are established.

5. In Windows, the field is the empty line in a dialog box where you enter data.

6. In call center jargon. A field is a single piece of data, such as an employee ID, stored in a record. The fields are organized under column headings.

Field, Cyrus W. Cyrus Field first conceived the idea of laying a transatlantic telecommunications cable. His first and second attempts were in 1858, and were unsuccessful. His fourth attempt was a success, even though it took one hour to transmit one word. The cable went dead within a month thereafter. An electrician attempted to correct the problem by injecting 2000-Volt impulses, which fried the cable. Field tried again several more times. In 1866, he was finally successful. That cable landed at Heart's Content, Newfoundland, and was capable of transmitting telegraph messages at the rate of eight words per minute. We've come a long way since 1866.

Field Effect Transistor A field effect transistor (FET) is composed of a single piece or channel of either P or N-type semiconductor surrounded by a ring or collar of opposite semiconductor. The collar is called the gate and the ends of the channel are called the source and the drain. As the voltage across the gate and source is varied, the resistance between the source and drain changes. A large current between the source and drain can be controlled by a small gate-source voltage.

Field Emission Displays FED. Another way of making thin, flat, lightweight computer displays for laptops, planes, etc. The other way is called "active matrix liquid crystal display." In field emission displays, a tiny color cathode ray tube sits behind each of the many pixels in the screen. This results in a brighter picture that uses less energy than the active matrix LCD displays. See also FED.

Field Identifier See FID.

field intensity The irradiance of an electromagnetic beam under specified conditions. Usually specified in terms of power per unit area, e.g., watts per square meter, milliwatts per square centimeter.

field interlacing In television, field interlacing is the process of creating a complete video frame by dividing the picture into two halves with one containing the odd lines and the other containing the even lines. This is done to eliminate flicker.

field measurement Refers to both signal strength and qualitative field tests of wireless networks.

field programmable The ability of a system to have changes made in its program while it is being installed – without having to be returned to the factory. RFID tags that use EEPROM, or non-volatile memory, can be programmed after they are shipped from the factory.

field programmable gate array FPGA. A user-configurable logic device in the form of a microprocessor. FPGAs comprise a mind-boggling variety of devices which contain memory that holds user-defined logic constructs and interconnects. Memory technologies include EEPROM, EPROM, FLASH EPROM, SRAM, fuse elements (mainly in lower density devices), anti-fuse elements, and laser-etched metal. The memory type defines whether the configuration is maintained when power is removed (EEPROM, EPROM, FLASH, fuse, antifuse, and etched metal versions) or whether the configuration must be reloaded on during power-on (SRAM versions). Fundamental design trade-offs include the complexity of the basic cells (usually a 3-5 input look-up table and flipflop) and the richness of routing resources available to connect between cells. Modern FPGAs also include various types of signal compatibility (differential and/or low voltage inputs and outputs) and user-memory elements (either distributed or provided as array blocks). Modern designs allow the placement of preconfigured logic (microcontrollers, FIFOs, RISC processors, UARTS) for complete SOCs (systems-on-chips). FPGAS are available with huge amounts of logic (up to 5 million gates announced) and pin counts (24-500+ pins). This area of technology is very dynamic and exciting. Thanks to Ken Coffman for this definition. Ken literally wrote the book on "Real World FPGA Design with Verilog." Also see ASIC, CPLD and SOC.

field repairable A characteristic of an unfortunately-decreasing number of electronic devices, that allows users or technicians to fix them where they are used ("in the field"), instead of having to send them to a centralized repair facility where esoteric parts and tools are available.

field replaceable unit Hardware component that can be removed and replaced on-site. Typical field-replaceable units include cards, power supplies, and chassis components.

field rheostat A variable resistance device. The field current and consequently the strength of the electromagnetic field regulate the speed or power of the motor, or the output of the generator.

field sequential system Field sequential system was the first broadcast color television system, approved by the FCC in 1950. It was later changed to the NTSC standard for color broadcasting.

field strength The intensity of an electric, magnetic, or electromagnetic field at a given point. Normally used to refer to the rms value of the electric field, expressed in volts per meter, or of the magnetic field, expressed in amperes per meter.

field strength meter Electronic instrument that measures the intensity of the magnetic field.

field upgradable A desirable characteristic of telecom equipment, computers, etc.,

that allows new features to be added and other improvements to be made, where the device is used, rather than having to return it to the manufacturer or a repair facility.

field wire A flexible insulated wire used in field telephone and telegraph systems. WD-1 and WF-16 are types of field wire. Usually contains a strength member. See Field Wiring.

field wiring An electrical connection intended to be made at the time of installation, in the field, as opposed to factory wired.

fieldbus A local area network in factory environment. It connects electronic measurement and control devices in real-time, not end-users who would congest the network with big print jobs, downloads from websites, and large email attachments. A fieldbus provides the real-time connectivity needed for advanced process control in an automated manufacturing environment.

FIFO First In, First Out. All telephone networks are a trade-off. It's simply too expensive to build a phone network which will be ready to give everyone dial tone and a circuit – if everyone picked up the phone simultaneously and tried to make a call. There are basically two ways of handling calls which cannot be sent on their way – i.e. for which there's no present available capacity. First, you can "block" the call. This means giving the caller a busy or a "nothing" (also called "high and dry"). Second, you can put the call into a queue. Now you have people waiting in queue, how do you handle them? The most equitable – the way most queues work – is to handle the calls on the basis of First In, First Out. (First call to come in is handled first.) There are other ways of handling calls in a queue – including First In, Last Out, by priority (e.g. which line you came in on and how much it cost, or how high you are in the corporation, etc.)

FIFO queuing also is used in some routers, although it has disadvantages in TCP/IP application. For example, if the buffer memory is full and data packets are dropped from the tail of the queue, the originating devices will assume that they are sending too rapidly and will slow down their rate of transmission. As multiple devices subsequently probe to seek the capacity of the network by sending data at higher rates, they can create another congestion condition.

FIFO also is a term used in data communications. It is a buffering scheme in which the first byte of data that enters the buffer is also the first byte retrieved by the CPU. This scheme is used in the 16550 (the UART chip which controls the serial port on most PCs and most other serial-buffering designs), because it closely mimics the way serial data is actually transmitted; that is, one bit at a time.

fifth generation Fifth generation computers and telephone systems will be based on artificial intelligence. A fifth generation phone system may make far more sophisticated decisions about routing calls across networks. Those decisions may be made on how many calls have already happened so far that month, the choice of carrier by the likely quality of his connection, etc.

FIGS FIGure Shift. A physical shift in a terminal using Baudot Code that enables the printing of numbers, symbols and upper-case letters.

figure-of-eight cable Another term for Siamese cable.

filament 1. An electrically heated wire in an evacuated glass bulb, forming one element (the cathode) of a vacuum tube.

2. The part of an incandescent light bulb that heats and lights. Filaments are often made from Tungsten.

file 1. A set of similarly structured data records (such as personnel records using a standardized form). See Field.

2. A call center term. A logical division of the data stored on a disk or diskette; for example, employee information vs. supervisor information vs. call volume history. Files generally consist of one or more records of a certain structure.

File Allocation Table A file allocation table is essentially a road map of the location of files on a hard disk. See FAT.

file caching A Novell local area network NetWare file server can service requests from workstations up to 100 times faster when it reads from and writes to the file server's cache memory (in RAM) rather than executing direct reads from and writes to the file server's hard disks.

file extensions Under Windows and MS-DOS, a file extension consists of a period and up to four characters at the end of computer a file name. The extension identifies the type of file, and often helps a computer know what to do with the file. For example, if a file is named PaintingProgram.exe, the .exe tells a Windows computer that the glossary file is executable and it should run the program. If the extension is .doc, that tells your computer that it should open the file with Microsoft Word, and so on. While most extensions are arbitrarily assigned by users or manufacturers of software, some extensions are reserved by Windows for special purposes, including exe, com and bat. Windows has a built in program called Explorer. When you click on that file, Explorer uses the file extension to launch an application it has associated with that extension. When you install a new application, it usually tells Explorer which file extensions it creates. Thus when you click on a file with that extension, it will launch the correct application.

Here are some common extensions:

ASP Active Server Page for use on the Internet
EXE executable file
BAS Basic language file
BAT DOS executable batch file
COM DOS executable command file
DOC Word document
GIF Graphic Interchange Format
HLP help screens which typically appear by pressing F1
HTM and HTML HyperText Markup Language file for use on the Internet
INI initialization file
JPG jpeg graphics file
OVL overlay file
PS Postscript file
SYS operating system file
TXT plain text file
XLS Excel spreadsheet
ZIP compressed file

For a list of virtually every file extension, go here: http://www.ace.net.nz/tech/Tech-FileFormat.html

file format The way in which data is stored. The file's format is indicated by the three or four letter extension after its name. For example, Word documents end in .doc and Excel documents in .xls. See also http://www.ace.net.nz/tech/TechFileFormat.html and File Extensions.

file gap A short length of blank tape used to separate files stored on linear magnetic tape.

file locking Picture a cabinet of file folders. Now I remove a folder to work on it. I make a photocopy of the folder in the cabinet and leave the original. You come along and remove the original because you want to work on it. You make changes and replace the changed copy in the cabinet. Ten minutes later I pull your file out and replace it with mine. Bingo, all your changes are lost. But let's say when I remove the file to work on it, I staple the remaining folder shut. That's a message to anyone else – including you that you shouldn't mess with the file. When I return, I unstaple the file, and add my changes. Now it's ready for you to do your thing. File locking ensures that a file will be updated correctly before another user, applications, or process will be allowed to write to the file. When a file is locked, no one else can write to it. Without file locking, one user could overwrite the file update of another user. In contrast to file locking, record locking allows many users to access the same file at once, but have only one access the record. See also ATTRIB and Record Locking.

file maintenance The job of keeping your data base files up to date by adding, changing or deleting data.

file management The system of rules and policies for maintaining a set of files – including how files can be created, accessed, retrieved and deleted.

file server A file server is a device on a local area which "serves" files to everyone on that local area network (LAN). It allows everyone on the network to get to files in a single place, on one computer. It typically is a combination computer, data management software, and large hard disk drive. A file server directs all movement of files and data on a multi-user communications network, namely the LAN. It allows the user to store information, leave electronic mail messages for other users on the system and access application software on the file server – e.g. word processing, spreadsheet. In computer telephony applications, potentially many users or voice channels need to access data on a file server. The file server may therefore present a significant bottleneck for computer telephony especially if it is used to store large files, such as voice prompts. The ability for the file server to handle the transaction load planned for your computer telephony application is therefore a key design consideration and issue to test.

file server console operator A user or a member of a group to whom a Novell NetWare SUPERVISOR delegates certain rights in managing the file server. A file server console operator has rights to use FCONSOLE to broadcast messages to users, to change file servers, to access connection information, to monitor file/lock activity, to check LAN driver configurations and to purge all salvageable files.

File Service Protocol See FSP.

file sharing A topology-independent feature of Apple Macintosh's System 7 operating system which allows users to share files and folders on their disks with other users across the LAN. File sharing is slow but acceptable for sharing small numbers of files among small groups. For larger networking, the user must consider AppleShare, Netware, Vines, etc.

file spanning Creating one compressed file that contains many files, typically retrieved from many removable media, e.g. floppy disks.

filed trail Also known as Beta Test.

file swarming This is a file delivery technique whereby chunks of a file that have been replicated on cooperating, geographically dispersed servers are downloaded to the requesting computer. Spreading the workload like this prevents a single server and/or network link from becoming overburdened, thereby avoiding bottlenecks and improving throughput, even though there is overhead associated with the replication of file chunks and scheduling their transmission in response to a request. A variety of file-swarming techniques have been developed and research into improved techniques continues. File swarming is also known by the more technical terms cooperative content distribution and peer-assisted file delivery. BitTorrent is a well-known example of a file-swarming application.

file transfer protocol There are two meanings to a file transfer protocol. First, the generic – software that enables file is basically

Second, the specific name of a software. For this explanation, see FTP.

1. See FTP.

Sending any information from point in

FTP. A service that enables file transfer between local and remote computers using many media, including the Internet. FTP supports several commands that allow bidirectional transfer of binary and ASCII files between computers. The FTP client is installed with the TCP/IP connectivity utilities. See also FSP, FTP, and File Transfer Protocols. One problem with transmitting information over phone lines is the noise on the phone line. One way to overcome the problem of noise is a file transfer protocol. The idea is simple: send your information in bundles (called packets). Accompany those packets with a special number derived in some way from the information in the packet. Send it all to the other end. Have the computer at the other check the number and see if corresponds to the packet. If not, send a signal back, saying "Something went wrong. Please send the packet of information back again."

Most asynchronous file transfer protocols use some form of error detection, typically checksum or cyclic redundancy check (CRC). Both the checksum and the CRC are values derived from the data being sent (or received) according to mathematical algorithms. The protocol sends the value long with the information (the bits) in the packet. The receiving program compares with the check values with the values it calculates. If the check values do not match, the receiver asks the sending computer to retransmit the packet. Older protocols required a positive acknowledgement (an ACK) before they sent another packet. But newer protocols allow transmission of several packets before they receive an acknowledgement. This is particularly useful for protocols with long delays, especially satellites. See also XModem.

file transfer software Software to transmit files between computers, over phone lines or over a direct cable connection between the two computers.

fill Bit Stuffing. See Fill Bits.

fill bits Also known as Stuff Bits. Fill bits are used to fill up a data packet or a frame, which must be of a certain minimum size, or perhaps of a certain specific size. If the transmitting device doesn't have enough bits to fill the packet, it stuffs in some nonsense bits which the receiving device tosses away. This process is known as "bit stuffing." "Keep alive bits" are fill bits used to keep a session alive between two devices across a circuit. Without the "keep alive bits," the receiving device would "time out," thereby aborting the data communication session.

filler 1. A material used in multiconductor cables to occupy large interstices formed by the assembled conductors. 2. An inert substance added to a compound to improve properties or decrease cost. measurement devices.

fills A computer imaging term. Designated areas that are flooded with a particular color. Most paint packages let you create geometric shapes in filled form. All packages also let you fill irregular closed regions. Two types of such fills exist: A seed fill floods all connected regions with the color specified by the mouse or stylus pointer; a boundary fill floods a color until the algorithm encounters a specified boundary color.

FILO First In, Last Out.

filter 1. A device which transmits a selected range of energy. An electrical filter transmits a selected range of frequencies, while stopping (attenuating) all others. It is used to suppress unwanted frequencies or noise, or to separate channels in communications circuits.

Such a filter might be called a BANDPASS filter. You can also use a filter to remove certain characters you might be receiving over a data communications channel, for example control characters or higher-order nonstandard ASCII bits.

2. An operating parameter used in LAN bridges and routers that when set will cause these devices to block the transfer of packets from one LAN to another. Filters can be set to prevent the internetworking of several types of messages. They may be set to block all packets originating from a specific destination, called source address filtering, or all packets heading for a particular destination, called destination address filtering. Filters may also be set to exclude packet of a particular protocol or any particular filed in a LAN packet.

3. Generally, a process or device that screens network traffic for certain characteristics, such as source address, destination address, or protocol, and determines whether to forward or discard that traffic based on the established criteria.

4. See TAPI 3.0.

filtering 1. A process used in both analog and digital processing to pass one frequency band while blocking others or visa-versa. Filters can be designed to remove information content such as high or low frequencies, for example, or, in image processing, to average adjacent pixels, creating a new value from two or more pixels. "Tap" refers to the number of adjacent lines or pixels considered in this process. MPEG, for instance, makes use of a 7-tap filter.

2. Bridges can reduce LAN congestion through a process of filtering. A filtering bridge reads the destination address of a data packet and performs a quick table lookup in order to determine whether it should forward that packet through a port to a particular physical LAN segment. A four-port bridge, for instance, would accept a packet from an incoming port and forward it only to the LAN segment on which the target device is connected; thereby, the traffic on the other two segments is reduced and the level of traffic on the those segments is reduced accordingly. Filtering bridges may be either programmed by the LAN administrator or may be self-learning. Self-learning bridges "learn" the addresses of the attached devices on each segment by initiating broadcast query packets, and then remembering the originating addresses of the devices which respond. Self-learning bridges perform this process at regular intervals in order to repeat the "learning" process and, thereby, to adjust to the physical relocation of devices, the replacement of NICs (Network Interface Cards), and other changes in the notoriously dynamic LAN environment.

filtering agent A new form of smart agent whose basic job is to keep away all the stuff you don't want and find the stuff you do want – such as information gleaned from the Internet. See also V-Chip.

filtering bridge See Bridge.

filtering router Internetwork router that selectively prevents the passage of data packets according to a security policy.

filtering software Filtering software seeks to identify spam automatically by certain characteristics and then block it or segregate it.

filtering traffic This is the process of selecting which traffic will be allowed into a certain portion of a network, such as the wide area network. It is also the process of determining which traffic is transmitted first, then next, and so on. The traffic is compared to a filter, or a set of specifications, to determine if it can pass through or not.

FIN Field inspection notice.

final draft Lisa Kiell's strange idea of completing a transaction. See Oxymoron.

final trunk group A last-choice trunk group that receives overflow traffic and which may receive first-route traffic for which there is no alternate route.

find me service An AIN version of call forwarding, allowing the forward numbers to be programmed or re-programmed from any location. Additionally, priority access can be extended to specific callers based on password privilege. For instance, only highly privileged callers would be forwarded to your cell phone, in consideration of the high cost of airtime.

finder The user interface portion of the Apple Macintosh operating system. Unlike running Windows on top of DOS, tight integration of the finder and system requires both to be running.

finger 1. A standard protocol specified in RFC-742. A program implementing this protocol lists who is currently logged in on another host. In short, finger is a computer command that displays information about people using a particular computer, such as their names and their identification numbers.

2. Also known as a tine. An individual digital channel of a wireless rake receiver. A rake receiver can support a number of tines, which can be combined to form a stronger received signal.

3. A piece of software that tells you the name associated with an email address.

finger wheel The actual "dial" on a rotary dial phone is called the finger wheel.

fingerprinting See RF fingerprinting.

finite state machine A computer system with a defined set of possible states and defined transitions form state to state. Given the same inputs, two identical state machines will change states identically.

FiOS FiOS is the name for a Verizon service that brings fiber optic into your house or office and delivers very fast speeds accessing the Internet. I don't normally write about about proprietary services. But I'm very impressed with its speed – it easily beats cable modems and traditional telephone company DSL. If you can get it, get.

FIPS Federal Information Processing Standard. See also FIPS PUBS nn.

FIPS PUBS nn Various standards for data communications.

FIPS 140-2 A US federal government standard that specifies encryption requirements for products that are used by the federal government and its contractors to protect sensitive but unclassified information. The standard applies to all federal agencies that use crypto-graphic-based security systems to protect sensitive information in computer and telecommunication systems (including voice systems) as defined in Section 5131 of the Information Technology Management Reform Act of 1996, Public Law 104-106.

FIPS-60 Bus and tag channel interface for IBM 360/370 mainframes (multidrop, two copper cables) one for data, one for control information, 1.5-4.5 Megabytes per second with a maximum distance of 121 meters.

FIR Fast Infrared. This infrared standard from IrDA supports synchronous, wireless communications at 4Mbps at a distance of up to one meter.

fire To discharge someone. In Scotland during medieval times, if your clan wanted to get rid of you, but not kill you, they would set fire to your house. Hence, the origin of the expression, "to get fired." The story goes that in the early part of the 20th century, if an NCR salesman lost an order, when he returned to his office, they put his desk out on the front lawn and burned the desk. Then they "fired" the salesman.

fire break A material, device, or assembly of parts installed along a cable, other than at a cable penetration of a fire barrier, to prevent the spread of fire along a cable.

firecrackers Used by the ancient Chinese to frighten away demons.

fireplug In the late 1700s the larger American cities, such as Philadelphia, laid pipes to bring in water. This was not for drinking, but for firefighting purposes. The pipes were made out of hollowed out logs placed end to end and buried under the streets. When there was a fire, the firemen would punch through to the pipe to pump the water out, and once finished would plug the hole using a wooden stake of the proper size. Hence the name, "fireplug."

fireproof A property in material such as masonry, block, brick, concrete or gypsum board that does not support combustion even under accelerated conditions, i.e. really pushing the fire at the product.

firestop 1. A material, device, or assembly of parts installed after penetration of a fire-rated wall, ceiling or floor area to prevent passage of flame, smoke or gases through the rated barrier.

2. The use of special devices and materials to prevent the outbreak of fire within telecommunications utility spaces and to block the spread of fire, smoke, toxic gases and fluids through openings, cable apertures and along cable pathways. The techniques used are often mandated by local building codes.

Classifications are available under the rating criteria of ASTM:

Rating/Achievement F: Withstands the fire test for the rating period without: Permitting flames to pass through the firestop flame occurring on any element of the unexposed side of the firestop (auto-ignition) developing any opening in the firestop that permits a projection of water beyond the unexposed side during the hose strength test.

Rating/Achievement T: Meets the criteria of an "F" rating and prevents the transmission of heart during the heating period so that the temperature rise is not more than 325 degrees Fahrenheit on any exposed surface, thermocouple or penetrating item.

firestop system A specific construction consisting of the material(s) (firestop penetration seals) that fill the opening in the wall or floor assembly and any items that penetrate the wall or floor, such as cables, cable trays, conduit, ducts, pipes, and any termination devices, such as electrical outlet boxes, along with their means of support.

firestopping The process of installing specialty materials into penetrations in fire-rated barriers to reestablish the integrity of the barrier.

firestop zoning A unique group of architectural structures or assemblies that prevents the passage of fire or toxicity from one contained area to another, thus reducing the possible spread of combustion through the fire barrier. Refer to NFPA specifications for the intended application. See also Firestop.

firewall A firewall is a piece of hardware or software or hardware and software that prevents unauthorized people from gaining access to your computer network. That's its main function. Its secondary function is to check on information to and from your PC and/or your network to make sure the information is kosher, i.e. non-damaging. Basically, a firewall will protect against two threats from outside the firewall:

1. Denial of service. Someone bad is attempting to bring down a network-connected computer by bombarding it with traffic. This computer might be a much-visited web site or a computer at the first point of entry into an internal corporate network.

2. Intrusion. Someone bad is trying to get user IDs and passwords in order to gain access to a network and then grab data from inside the network. That data might be credit cards. It might also be trade secrets on how to make Coca Cola. The idea is to get into the network and grab the information without anyone knowing.

Firewalls – whether hardware or software, or both – do not protect against computer viruses and spam emails. For that you need other protection. In short, a firewall is a system or combination of systems that enforce a boundary between two or more networks, one of which is likely to be the Internet and the other your own computer or your local area network. There are several types of firewalls – packet filter, circuit gateway, application gateway or trusted gateway. A network-level firewall, or packet filter, examines traffic at the network protocol packet level. An application-level firewall examines traffic at the application level – for example, FTP, E-mail, or Telnet. An application-level firewall also often readdresses outgoing traffic so it appears to have originated from the firewall rather than the internal host. The New York Times asked the question: "How does a firewall program know if someone or some program is trying to deliberately break into a computer?" And it answered:

A. Firewalls – named after the flame-blocking structures designed to keep fires from spreading – are programmed to analyze the network traffic flowing between your computer and the Internet. The firewall compares the information it monitors with a set of rules in its database. If it sees something not allowed, say, another computer trying to connect to one of the machines on your network, the firewall can block and prevent the action. Most firewall programs let you adjust the rules to allow certain types of data to flow freely back and forth without interference, in cases where you want to do things like stream music or share files with another computer on your network. Both Windows XP and Mac OS X include basic firewall programs as part of the operating system. There are also more complex third-party firewall programs available from independent software companies. Makers of antivirus software often have a combination package that includes a firewall program and other security software as well, which can help protect your computer from a variety of Internet threats. Firewall programs are especially important if you have a high-speed Internet service through a cable modem or D.S.L., because your computer is typically connected to the Internet for longer periods of time, increasing the exposure to malicious people or programs looking to infect and infiltrate unprotected systems. And it doesn't take long. A recent study by a security company found that an unprotected Windows computer was infected within 12 minutes of connecting to the Internet.

NEC PrivateNet Systems Group issued a White Paper called "Connecting Safely to the Internet – A study in Proxy-Based Firewall Technology." In that White Paper, they defined an Internet firewall: "The primary purpose of an Internet firewall is to provide a single point of entry where a defense can be implemented, allowing access to resources on the Internet from within the organization, and providing controlled access from the Internet to hosts inside the organization's internal networks. The firewall must provide a method for a security or system administrator to configure access control lists to establish the rules for access according to local security policies. All access should be logged to ensure adequate information for detailed security audit. A traditional firewall is implemented through a combination of hosts and routers. A router can control traffic at the packet level, allowing or denying packets based on the source/destination address of the port number. This technique is called packet filtering. A host, on the other hand, can control traffic at the application level, allowing access control based on a more detailed and protocol-dependent examination of the traffic. The process that examines and forwards packet traffic is known as a proxy. A firewall based on packet filtering must permit at least some level of direct packet traffic between the Internet and the hosts on the protected networks. A firewall based on proxy technology does not have this characteristic and can therefore provide a higher level of security, albeit at the cost of somewhat lower performance and the need for a dedicated proxy for each type of connectivity. Each organization needs to choose one of these basic types of technologies. The right choice depends on the organization's access and protection requirements."

See also Firewall friendly, packet filtering, port services, spam and viruses.

firewall friendly Refers to an application's ability to work across a firewall. This is achieved by designing an application so that it does things that firewalls regard as safe

and does not do things that firewalls regard as unsafe. Examples of activities that a firewall regards as safe and allows to take place include initiating a connection only with a port that the firewall regards as safe, such as the port used by HTTP or SMTP, and an establishing just a single connection for the entire session. Activities that a firewall doesn't like, and which the firewall will block, include trying to initiate a connection with an unknown port on with a port within a certain range, or dynamically changing connections during a session.

FireWire FireWire is the IEEE 1394 standard high performance serial bus. For a longer explanation, see IEEE 1394 and USB.

FireWire 400 The new name for the original FireWire/IEEE-1394/IEEE-1394a interface specification. It provides a speed of 400 Mbps. See FireWire 800.

FireWire 800 The latest version of FireWire, defined in the IEEE-1394b interface specification. Operating at 800 Mbps, FireWire 800 is twice the speed of the original FireWire. It also supports a larger variety of transmission media, each offering different speed/distance capabilities. A FireWire 800 cable can connect devices up to 15 feet away, while a FireWire 800 optical repeater can connect devices up to 1000 meters away. See also FireWire 400.

Firm Order Confirmation FOC. The form a local phone company submits to another phone company indicating the date when the circuits ordered by the other company will be installed. See FOC for a longer explanation.

Firmware Software kept in semipermanent memory. Firmware is used in conjunction with hardware and software. It also shares the characteristics of both. Firmware is usually stored on PROMS (Programmable Read Only Memory) or EPROMS (Electrical PROMS). Firmware contains software which is so constantly called upon by a computer or phone system that it is "burned" into a chip, thereby becoming firmware. The computer program is written into the PROM electrically at higher than usual voltage, causing the bits to "retain" the pattern as it is "burned in." Firmware is nonvolatile. It will not be "forgotten" when the power is shut off. Handheld calculators contain firmware with the instructions for doing their various mathematical operations. Firmware programs can be altered. An EPROM is typically erased using intense ultraviolet light.

firmware over-the-air FOTA. A mobile phone's intelligence resides in embedded software called firmware. And like any software, firmware may need to be updated from time to time to fix bugs and/or add new features. If customers were to bring their cell phones into a service facility whenever a firmware update was needed, this would be a huge expense for the cell phone manufacturer and/or network provider, and a hassle for customers; and there would be no guarantee that all customers would bring their cell phones, which would be a problem if some updates were to fix security problems. To address this challenge and at the same time eliminate expensive equipment recalls, technology known as "firmware over-the-air" (FOTA) has been developed to enable mobile device manufacturers and network operators to remotely and wirelessly transmit firmware updates to mobile handsets while the handsets are powered on.

First In, First Out See FIFO.

first mile See Last Mile.

first night I am enclosing two tickets to the first night of my new play, bring a friend... If you have one." - George Bernard Shaw to Winston Churchill

"Cannot possibly attend first night, will attend second... If there is one." - Winston Churchill, in reply

first office application The first office to have the guts to try a new system in a real, live production mode. The same thing as a beta test. See also Beta Test.

first party call control A call comes into your desktop phone. You can transfer that call. When the phone call has left your desk, you can no longer control it. That is called First Party Call Control. If you were still able to control the call (and let's say, switch it elsewhere) that would be called Third Party Call Control. First party call control is mostly done at your desk with your telephone or with a card in your PC, which emulates a telephone. Third party call control is usually done via a computer (often a server on a LAN) attached to a special link directly into your PBX. There are some evolving standards in call control – chiefly Microsoft's Windows Telephony and Novell's TSAPI (Telephony Services API). There is no such animal as Second Party Call Control. See Call Control, Telephony Services and Windows Telephony.

First point of Aries The reference point on the celestial sphere used in celestial navigation and in describing satellite orbits. The direction of the vernal equinox, which defines the positive x axis used in the equatorial coordinate system. The equinoxes are the two points at which the ecliptic cuts the equator of the celestial sphere, and are the times of the year when the Sun crosses the celestial (and apparently the Earth's) equator, once when moving from south to north, about March 21 (vernal, or spring equinox), and once when moving from north to south, about September 22 (autumnal equinox). The first point of Aries was once literally that, the first point on the constellation Aries, which is the first constellation of the Zodiac. However, it has now shifted westward, due to the precession of the equinoxes (due to the movement of the Earth's axis with a period of 26,000 years), so it lies in the constellation of Pices. The remnant evidence of its original position is that in astrology, Aires (the Ram) is the first sign of the zodiac and designates a period starting at 21 March, the vernal equinox. Opposite to the first point of Aries is the first point of Libra (the Balance), at the autumnal equinox, which has now shifted to Virgo (the Virgin). The first point of Aries is most widely used in celestial navigation and is tabulated in the Air Almanac published jointly by the US Naval Observatory and HM Nautical Office of the UK. aa.usno.navy.mil.

first point of switching The first exchange carrier location at which switching occurs on the terminating path of a call proceeding from the IXC terminal location to the terminating end office, or the last exchange carrier location at which switching occurs on the originating path of a call proceeding from the originating end office to the IXC terminal location.

first responder radios See Project 25.

first ring suppression Caller ID in North America comes in just after the first ring. You don't want to answer the call before the second ring otherwise you will mess up your receiving of Caller ID information. You can simply not answer until the second ring. Or you can get a trunk-based gadget which will turn off the first ring so you or your voice processing equipment won't hear it.

FIS Forms Interchange Standard.

fish 1. To push a stiff steel wire or tape through a conduit or interior wall. Pull through wires, cable or a heavier pulling-in is then attached to one end of the steel wire. The other end is then pulled until the wire or cable appears.

2. First In Still Here. A non-standard term used in inventory accounting. Roughly equivalent to FILO (First In Last Out), but suggesting that the inventory is not moving because you aren't selling any of it.

fish food 1. Webmasters who want to draw attention to new on-line content call the tidbits "fish food." The morsels are posted in "What's new" buttons or "Click here" icons on the home page, so, like the flakes that feed your guppies, they float at the top and attract hungry users. Contributed by Judy Ehrenreich.

2. Fish food refers to sales leads among a group of salespeople. Basically sales people throw a lead out, like fishfood and all the others come up to see if they can sell it. Contributed by Dino Guglietta.

fish job Running cables inside walls. Usage: "That fish job is too tough for a rookie."

fish paper R. Michael Seng, III, senior Sales Engineer at Broadwing Communications, writes me that he grew up using "Fish Paper" to separate two mildly energized , or to provide a layer of resistance for, pieces of metals. It would also be used to mitigate the electrolysis between dissimilar metals. It was a dark (light black / dark grey) somewhat brittle paper that was as thick as three to five sheets of typing paper. Basically it was a sheet of insulator that could be cut to size ad hoc.

fish tape Non-conductive tape with a reinforced fiberglass core and slippery outer nylon coating which slides easily through conduit without jamming. The idea is to push the tape through, attach it to the cable and pull the cable back. You also might use wire pulling lubricants to make the job even easier. They come in various formulations – for use in different temperatures, for pulling different cable, etc.

fishing expedition An investigation that has no clearly defined objective.The goal is to find something immoral, incriminating, unethical or illegal.

FISK Fax a dISK. Method of sending information on 3 1/2" disks painlessly across phone lines. Plug your disk into a fisk machine, choose which files you want to send, dial the number you want to send your files to and walk away.

fit AP A variation of the thin AP wireless LAN architecture. In the fit AP architecture the access point handles the network's wireless communications and the encryption of those communications, while the network switches and centralized management controller handle end-user authentication, secure roaming, quality of service, and network management. See fat AP, thin AP.

FITC Fiber To The Curb. See FTTP.

FITH FIber To The Home. (I kid you not. That's what it stands for.) See FTTP.

fist A term telegraphers use to describe the unique signature made by someone tapping the key of a wireless transmitter.

FITL Fiber In The Loop. An outside plant architecture, deployed by some telephone companies for providing broadband services to subscribers. In the FITL architecture, SONET fiber runs from the telephone company central office to an optical networking unit (ONU) at the curb, which serves 8064 subscribers. Subscribers are served from the ONU in a star topology, each home with a drop of coax, twisted pair, or composite coax-and -twisted-pair. With FITL, an individual video or voice signal is switched to each subscriber. Therein lies the major difference between FITL and the alternative broadband architecture, a hybrid fiber/coax bus. In the bus architecture, all signals serving dozens of subscribers are multiplexed and sent to all subscribers. See also Fiber Optics, FTTC, FTTH, Local Loop and POP.

five by five The term 5x5 describes the transmit and receive quality of teletype communications (on a scale of 1 to 5, with 5 being the best) between two stations. Example, 1x5 means transmitting poorly but receiving excellent. The source of this info comes from Alan DiGilio's training, as a former Naval Flight Officer, in the use of cryptograhic radio teletype equipment. ("cratt" as we called it.) Amongst my comrades in arms, when we asked each other how transmissions went, we would slip into slang by saying "Fivers," to mean 5x5.

five nines 99.999% uptime. Five nines typically refers to the reliability of a system (computer, telephone system, etc.) that works 99.999% of the time. This is far better than a recent common standard of three nines, i.e. 99.9%.

fivers See Five by Five.

FIX Federal Internet Exchange. A connection point largely serving to interconnect network traffic from MILNET (MILitary NETwork), NASA Science Net and other federal government networks, as well as providing those network users with access to the Internet. FIX-EAST is located at the University of Maryland in College Park, Maryland; FIX-WEST is located at the NASA Ames Research Center at Moffet Field between Sunnyvale and Mountain View, California. See also CIX, MAE and NAP.

fixed Attached and permanent. The opposite of mobile. See Fixed Wireless.

fixed condenser A condenser, the plates of which are stationary and the capacity of which cannot be changed.

fixed disk Old name for a hard disk.

Fixed End System F-ES. A non-mobile end system. A host system that supports or provides access to data and applications.

fixed format A way of communicating in which everything to be sent follows a predetermined sequence, i.e. it fits into a specific length and format. The idea is to allow you to predict message length, the location of the message, where the control characters are, etc.

Fixed IP Address See IP Addressing.

fixed length records A set of data records all having the same number of characters in them. Think of a database of name, address, city, state, zip. Clearly, not everyone's record will be the same length. In order to make a fixed length record, the computer will pad the record with "padding characters" which the computer will ignore when it reads the record. But by including the padding characters it has effectively given everyone the same fixed length record.

fixed line A wired phone line. The one you have in your house, as compared to your wireless phone line, the one you have in your pocket.

fixed link A communications link between two fixed points. Such links may be uni-directional (e.g carrying television program material to a transmitter) or bi-directional (e.g. carrying telephone traffic), and may be point-to-point or point-to-multipoint.

fixed loop A services feature available in some switching systems that permits an attendant on an assisted call to retain connection through the attendant position for the duration of the call. The attendant will normally receive a disconnect signal when the call has been completed.

Fixed Priority-Oriented-Demand Assignment FPODA. Medium access technique in which one station acts as master and controls channel based on requests from stations.

fixed rate A fixed monthly price. See also Flat Rate.

fixed satellite service A radiocommunication service between Earth stations as specified fixed points when one or more satellites are used; in some cases this service includes satellite-to-satellite links, which may also be effected in the inter-satellite service, the fixed-satellite service may also include feeder linker for other space radiocommunication services.

Fixed Satellite System FSS. A system of Geosynchronous Earth Orbiting (GEO) satellites. GEOs are positioned in equatorial orbits approximately 22,300 miles above the Earth's surface. Positioned in this manner, they are synchronized with the rotation of the Earth. Therefore, they are always (more or less) fixed in the same physical location relative to the Earth's surface. This allows satellite antennas on the ground to be fixed and not have to move to follow the movement of the satellites. Actually, fixed orbit satellites do slide out of their orbit slightly. Their onboard rocket engines are then started and they are brought back into geosynchronous orbit. See also GEO. Contrast with LEO and MEO.

Fixed to mobile convergence See Fixed-to-mobile convergence.

fixed wireless See Fixed Wireless Local Loop.

Fixed Wireless Local Loop FWLL. Imagine a community of 100 people spread out in a huge area in one of the Western states in the United States – e.g. Montana or Wyoming. Imagine a city of people eager to get faster Internet access and better, cheaper phone service than their local phone company can provide. The local loop is best described as the "last mile" of phone service. It's the distance between you, the customer, and the switching office down the street or across the county that's owned by the local telephone company. In most cases, local loop service uses old-fashioned twisted copper wire installed and provided by the ILEC – the incumbent local exchange company (your local phone company). However, several phone companies and several of their competitors are installing coaxial cable, fiber optics, their own cable and now fixed wireless, also called "Wireless Fiber." Such systems operate at the 38 GHz portion of the spectrum. They generally consist of a pair of digital radio transmitters placed on rooftops – one at one end at the central office and the other end at the customers' offices. It's called "fixed" to contrast it with "mobile," e.g. cellular. In order to attract customers, some fixed wireless providers are offering higher data transmission rates than wire. In 1998, for instance, one firm announced a new, fixed wireless network that would carry 128 Kbps of digital transmission right into most households. Based on the 10 MHz spectrum, the new system would connect a home to a digital switching center via a neighborhood antenna mounted on a utility pole or other structure. A single antenna would serve up to 2,000 homes. Meanwhile, the customer would only need to secure a transceiver to the side of their house.

Fixed-to-mobile convergence Companies making soft switches – switches with lots of software – are promoting the term. It suggests their switches have great flexibility merging, switching and transmitting various different types of networks – from wireless cellular to Voice Over IP to Ethernet to standard circuit switched long distance service.

fixie A bicycle with one fixed gear and no brakes. To ride it, you pedal and keep pedaling. To stop, you stop pedaling. If you stop pedaling and the bike continues moving, you will go flying over the handlebars. People buy fixies because they're "cool." Fixies are illegal in many states.

FL Fault Locating.

FL-port Fibre Channel term. Port that connects an FC-AL to a fabric.

FLAC Free Lossless Audio Codec. Files compressed using the open-source FLAC suffer absolutely no quality loss-perceptible or otherwise-from the original recording. FLAC is not compatible with many players out there, though iAudio and Rio make compatible devices.

flag 1. A variable in a program to inform the program later on that a condition has been met.

2. In synchronous transmission, a flag is a specific bit pattern (usually 01111110) used to mark the beginning and end of a "frame" of data. Frame Relay and lots of other protocol use this approach in order to delineate one frame from another, and to allow the devices in the network to synchronize on the rate of transmission for purposes of improved bandwidth efficiency. See Frame Relay, Synchronous and Zero Stuffing.

3. Fiberoptic Link Around The Globe. The longest man-made structure in the world, Flag is a 17,000 mile long fiber optic cable made of four strands of fiber, carrying 10 gigabits of data per second. The cable is laid mostly undersea, linking Japan to Britain, making land in China, Thailand, the United Arab Emirates, Italy and six other countries. Companies in most of those countries own a share of Flag's bandwidth, however the United States Verizon is the biggest owner by far with a 38% share, which it leases out. Each strand of Flag's fiber is unidirectional, i.e. you need two to make a conversation, one for going and one for coming.

flag fall Also spelled flagfall. Older taxis had a metering system which had a large metal lever facing vertically up. It was called a flag because it sort of looked like one. When a customer got into the taxi, the driver pulled the metal flag down. This action started the meter. All taxis typically charge a fixed money the moment you get into the taxi and then so much a mile and/or a minute after that. In some countries, that initial money became known as flag fall. In some countries, when you first make a phone call, they charge you a fixed amount however long the call is and then a certain amount of money based on how long you talk and how far you talk. That initial call setup charge has become known as a flag fall in some countries, including Australia.

flag sequence HDLC, SDLC, ADCCP, Frame Relay. The unique sequence of eight bits (01111110) employed to delimit the opening and closing of a frame.

flagfall See Flag Fall.

flame An outpouring of verbal abuse that network users write about other users who break the rules. A wonderful term for getting mad via electronic mail. People who frequently write flames are known as "flamers." You can flame by simply sending messages ALL IN CAPS!!!!!!!!! See Flame Fest, Flame War and Mail.

flame bait An intentionally inflammatory posting in a newsgroup or discussion group designed to elicit a strong reaction thereby creating a flame war.

flame fest Massive flaming. See Flame.

flame mail Slang term for rude electronic mail. Bill Gates, Microsoft chairman, is said to be famous for the flame mail he sends to employees who don't perform according to his likings. Mr. Gates is famous for flame mail sent by him between midnight and 2:00 AM.

flame resistant Insulated wire which has been chemically treated so it will not aid the spread of flames.

flame retardant Constructed or treated so as not to be able to convey flame.

flame war What happens when people send too much flame mail at each other. The online discussion degenerates into a series of personal attacks against the debaters, rather than discussion of their positions. A heated exchange.

flaming To send an insulting message, usually in the form of a tirade, sent via online postings but also as personal. Flaming is the verb. Flame is the noun. And too much of flaming can cause a nasty flamewar.

flamingo, pink Don Featherstone, of Massachusetts, is the father of the pink flamingo plastic lawn ornament. He graduated from art school and went to work as a designer for Union Products, a Leominster, Mass. company that manufactured flat plastic lawn ornaments. He designed the pink flamingo in 1957 as a follow-up project to his plastic duck. Today, Featherstone is president and part owner of the company, which sells an average of 250,000 to 500,000 plastic pink flamingos a year. www.flamingomania.com/outdoor1.html.

flammability Measure of a material's ability to support combustion.

flapping Flapping occurs when a routing table entry changes too often in a relatively short time. It is sometimes caused by a link going up and down or by receipt of conflicting routing updates. See also Bouncing Circuit.

flash 1. Quickly depressing and releasing the plunger in or the actual handset-cradle to create a signal to a PBX or Centrex that special instructions will follow such as transferring the call to another extension.

2. A binary file that provides flash-memory-based firmware's functionality. See also reflash.

flash chips Flash chips are memory semiconductor chips, which retain data when electrical current is switched off. They are used in products such as cellular phones, pocket computers and digital cameras.

flash crowds This refers to a published event (i.e. Victoria Secret's Fashion Parade) that receives an overwhelming amount of interest leading to network congestion. The term dates back to a 1973 Larry Niven science fiction story in which the development of teleportation causes thousands of people to suddenly teleport to a location where something interesting is happening.

flash button A button on a phone which performs the same thing as quickly pressing the switch hook on a phone. See Flash, Flash Hook, Flasphone.

flash cut The conversion from an old to a new phone system occurs instantly as one is removed from the circuit and the other is brought in. There are advantages and disadvantages to Flash Cuts. For one, they're likely to be much more dangerous than the opposite view, known as a Parallel Cut, in which the two phone systems run side by side for a month or so. Also known as Cutover and Hot Cut.

flash EPROM A type of EPROM that can be electronically erased. It differs from EEPROM in that generally the entire memory must be erased at once.

flash hook Another name for Switch Hook. The little button on the telephone that you place your receiver into. It obviously hangs the phone up, releasing that line to receive another call. If you push the flash hook quickly, you can signal the switch at the other end (central office or PBX) to do something, such as place a call on hold and switch to the incoming one (call waiting), or transfer the call to another phone. See Flash and Flash Button above.

flash memory Flash memory is nonvolatile storage – i.e. storage that can retain information without electricity – but which can be electrically erased and reprogrammed. Flash memory occupies little space and doesn't need continuous power to retain its memory. Some laptop companies, like Toshiba, are using flash memory as nonvolatile storage for the BIOS (Basic Input/Output System) and the instructions that start the computer (the bootstrap loader). Flash memory is also known as NAND. See also Flash Rom, Memory Cards and NAND.

Flash-OFDM Computerworld describes Flash-OFDM as a proprietary cellular broadband technology that network operators can deploy either for notebook computers of mobile users or serve as a fixed wireless access system, bridging the "last mile" to connect computers in homes and small offices. Key features include an all-IP architecture and fast speeds. The technology is capable of letting users traveling at 250 kilometers per hour download data at speeds up to 1.5 Mbit/sec. or upload at speeds up to 500 Kbit/sec. Orthogonal Frequency Division Multiplexing (OFDM) works by splitting radio signals into smaller low-speed signals that are transmitted in parallel, reducing crosstalk and using bandwidth efficiently, but decreasing range.

flash RAM Flash Random Access Memory. A very fast type of RAM that can quickly be erased and rewritten, Flash RAM also retains data when powered off. Typical applications include modems and removable storage media for devices such as PDAs and laptops. See also DDR-SDRAM, DRAM, EDO RAM, FRAM, Microprocessor, RAM, RDRAM, SDRAM, SRAM, and VRAM.

flash ROM Flash Read Only Memory. Read Only Memory that can be erased and reprogrammed, but stays on when power to your computer is turned off. Flash ROM is used in modems, for example, to hold software known as firmware. When a later software release comes out, you dial a distant computer which downloads new software into your Flash ROM, updating it. See Flash Memory.

flasher This is a true story of hard times in telecom. A lady drives up to a fired telecom worker who is manning a toll booth. She opens her shirt and exposes her breasts, then keeps driving, without paying the toll. The lady has never been reported, stopped or arrested. Toll collectors, allegedly, only call State Troopers on men who flash and don't pay the toll.

flashlights Flashlights got their name because early batteries had such a short life, they were flashed on and off to conserve energy. In Australia and England, flashlights are called torches.

FlashPix FlashPix is a new file format designed to optimize the electronic display, manipulation, and distribution of high-resolution images. Developed cooperatively by Eastman Kodak Company, Hewlett-Packard, Live Picture, and Microsoft Corporation, FlashPix has gained recognition as an enabling technology for companies seeking to sell and license images for reuse over the World Wide Web.

FlashROM See Flash ROM.

flat base mount A system for mounting a mast on a flat roof using a ground mount plate. See Ground Mount.

flat cable Multiconductor cable arranged in a parallel type configuration manufactured with controlled tolerance spacing. I've never been a big fan of flat cable. But I'm guessing newer flat cable with twisted wiring works better for data than it did in the past.

flat color A smooth expanse of color in Web design with no blends or interruptions.

flat file The simplest database structure. Data are arranged in a single table, with all records in identical formats (i.e., identical data field structures), and stacked together in rows and columns to form a large table similar to a spreadsheet. The term "flat file" comes from the fact that the file is two-dimensional, and that all data relevant to a given object are contained in a single file. See Relational Database.

flat monitor Monitors are screens for PCs. Some people call flat monitors "flat panels." Others call them "flat screens." These two similar-sounding terms are yet another way to befuddle consumers. If you're looking for a sleek, thin, bright computer monitor, you're looking for a flat-panel screen. And you should be looking for a digital one. These flat panels are liquid-crystal display (LCD) monitors like those found on laptops, and bear no resemblance to television-type monitors. Makers of traditional monitors, which are losing ground to flat panels, have started using the term "flat screen" to describe boxy monitors with flat rather than curved glass on the front. Frankly, I still prefer the glass ones.

flat network A term used primarily in the LAN (Local Area Network) domain to describe a network employing bridges, hubs, or OSI Layer 2 switches. As all such devices are protocol-specific, they are relatively inexpensive and are very fast. Such devices read the address of the target device, which address is contained in the packet header, and forward the packet. Some such devices also can filter packets or encapsulate them to resolve protocol differences (e.g., between Ethernet and Token Ring), although this latter function is accomplished at a relatively low level in order not to compromise speed of packet transfer. Flat networks are fairly easy to configure, but tend to be limited in terms of scalability (i.e.size they can grow to); they also can suffer from congestion caused by broadcast traffic. Flat networks are distinguished from hierarchical networks, which employ more complex router technology, operating at OSI Layer 3. See also Hierarchical Network, Bridge, Hub, and Router.

flat panel display A very thin display screen used in portable computers. These screens usually use LCD (liquid crystal display) technologies, which are backlit to make them more readable even in bright light.

flat-rate calling plan A long-distance calling plan where the customer pays the same per-minute rate any time of day or night, 7 days a week. See also flat rate service.

flat rate service FR or FRS. A fast-disappearing method of pricing local phone calls in North America. The concept was that for a fixed amount of money – say $10 a month – you received a plain old desk telephone and an unlimited number of local calls. For years, most residential and most business phones were on a flat rate service. The first thing to go was the phone instrument. You had to pay a dollar or so a month to continue renting it, or you could send it back and buy your own. Second to go was the size of the local calling area you could call. It got smaller. Third to go were the phone calls themselves. This happened first with businesses and now increasingly with residential service. Under this new "pay-per-call" you get charged a "message unit" for each local calls. A message unit is typically eight to ten cents. But psychologically, "message units" sound better than dimes. Fourth to go was the definition of local calls. What was now a "local" call got smaller, i.e. you could call less far for the price of a local call. And what was now a "local" long distance call changed. Calls which, years ago, were free (i.e. on flat rate service) have now become long distance calls. You can witness this phenomenon of changing local pricing in California, New York and Jersey. In other states, it's taking a little longer. There are cities where flat rate service still exists. Treasure them. They're disappearing, too.

flat screen See Flat Monitor.

flat top antenna An aerial consisting of one or more parallel horizontal wires supported between masts. The "T" type and the inverted "L" type belong in this class.

flat topping Flat topping is where the frame relay carrier limits the ability of the customer to burst above the CIR (committed information rate), thereby flat topping the customer to only the CIR. This is especially important when the customer expects to be able to burst above the CIR. In practice, many customers burst all the time above the CIR to their port speed.

Flatiron Building The Flatiron Building, 175 Fifth Avenue, New York, NY, was completed in 1902 for the George A. Fuller construction company, one of the most notable in NYC history. This 22 storey building became immediately a Midtown landmark with a unique triangular shape at the diagonal crossing of Broadway and Fifth Avenue.

The building has a steel frame which is covered with a non-load-bearing limestone and terra-cotta facade built to resemble a classical column with protruding and ornamented base and top. The mid-facade undulates slightly with a vertical wave pattern of decorative protrusions. Above the arcaded top floors a continuous, triangular cornice runs around the whole building.

The building's lobby is located in the middle of the long facades, with entrances from both sides. Along with the elevator banks, the first floor is divided into retail space. When opened, the building was equipped with an electric generator to provide it with its own electricity and heating.

Originally called the Fuller Building, after the Fuller Construction Co. that built it and originally occupied the building (later the firm moved to the Fuller Building in Midtown), the triangular shape gave the building its nickname "Flatiron" and subsequently the name stuck.

It must be noted that contrary to a popular belief (including mine for a long time), the Flatiron has never been the world's tallest building.

flatpack In general microwave usage, a miniature hermetic package for MIC components, designed for a minimum height, with pins for RF and DC connections exiting through the sides (narrowest dimensions), and designed to be surface mounted or "dropped in" to a cutout in a micro-strip printed circuit board. The leads and the largest surface of the package are in parallel planes.

flatten a network Streamline a network by removing network elements from it, such as switches and trunks, and by consolidating various types of functionality performed by separate network devices into a single multifunctional device. Flattening a network reduces capital expenditures (CAPEX) and operational expenses (OPEX), and reduces network latency, theoretically.

flatter giggle What Rona Peligal does out of politeness. Rona is married to Aaron Brenner, who needs the ego boost, according to Rona.

flattery The art of telling someone exactly what he thinks of himself. Flattery is the most powerful sales tool.

flattopping See Flat Topping.

flavor A slang expression meaning type or kind, as in "Unix comes in a variety of flavors."

FLC Fiber Loop Carrier. A FITL (Fiber In The Loop) system. FLC comes in several flavors. FLC-A is the Analog version. FLC-B is for a Building, as in an MTU (Multi-Tenant Unit). FLC-C is to the Curb, as in FTTC (Fiber To The Curb). FLC-N apparently is the Next version. There also are FLC-D, FLC-S, FLC-X, and probably some others. There's not a lot of information available on this, except in Korean.

fleas If humans could jump like fleas, they'd be able to leap over a 100-story building in a single bound.

FLEC Forward Looking Economic Cost. The general pricing methodology adopted by the Federal Communications Commission in its implementation of the Local Competition Provisions in the Telecommunications Act of 1996, August 8, 1996, FCC 96-325. FLEC consists of two parts: the total element long-run incremental cost of the network element and an allocation of common costs. The total element long-run incremental cost is the cost of providing the total quantity demanded in the future of a network element based upon an efficient network configuration, projected values of the cost of capital and economic depreciation rates. The allocation of common costs is required to be forward-looking as well. Appropriate common costs are those that are realized by an efficient company that cannot be attributed directly to a network element or set of network elements. FLEC has been contested in Federal Court. The U.S. Court of Appeals, Eighth Circuit, vacated the specific FCC rule requiring that state commissions use FLEC in arbitration proceedings. However, state commissions have adopted versions of FLEC as an acceptable methodology to be used in arbitration proceedings. This definition courtesy Douglas Meredith.

fleeting alarm An alarm condition that is sporadic and/or irregular in occurrence. See also Alarm Soaking.

flex One of the communications protocols used between paging towers and the mobile pagers/receivers/beepers themselves. Other protocols are POCSAG, ERMES, GOLAY and REFLEX. The same paging tower equipment can transmit messages one moment in GOLAY and the next moment in ERMES, or any of the other protocols. In mid-February, 1997 Motorola announced tht its Products Sector was now shipping its 68175 FLEX chip paging protocol IC (integrated circuit) in volume to customers worldwide. FLEX protocol, an open paging standard developed by Motorola, offers product developers, according to Motorola, a common set of rules that ensure applications work across different service providers' equipment. Currently the FLEX protocol has been adopted by service providers around the world, including providers in China, Southern Asia, India, Japan and the Middle East, along with North America and Latin America. According to Motorola, the FLEX chip IC processes information that has been received and demodulated from a FLEX radio paging channel, selects messages addressed to the paging device, and communicates the message information to the host. In a press release dated August 3, 2000, Motorola, Inc. the developer of the FLEX protocol, said that FLEX was "the worldwide de facto standard for high-speed wireless messaging."

flex ANI Flexible Automatic Number Identification. Additional two-digit ANI identifiers for PSPs (Payphone Service Providers). Flex ANI provides a means of identifying the specific class of service associated with the originating telephone number in order that the PSP can be compensated properly for originating long distance calls. Flex ANI also provides for enhanced routing, call screening. Carriers can order Flex ANI to identify calls originating from 1) dumb payphones with switch-generated coin signaling, 2) smart payphones with coin signaling resident in the phone, and 3) inmate/detention facility payphones. A FCC mandate (March 1998) requires that facilities based LECs (Local Exchange Carriers) deploy Flex ANI. Flex ANI is provided per end office (central office) on a CIC (Carrier Identification Code) basis; FGD (Feature Group D) is required. See also ANI and CIC.

flex life The ability of a cable to bend many times before breaking (stranded). See also Flex Strength.

flex strength The ability of a wire or cable to withstand bending and twisting. See also Flex Life.

flexible dialing pattern A PBX dialing pattern that allows you to set your PBX so it can have one, two, three or four digit numbers for its extensions. See also Flexible Numbering of Stations.

flexible drill bit A long drill bit that bends and is used for pulling cable and wire through walls in one operation. This means you drill the hole and then reverse the drill and it pulls the cable through the hole, while the drill bit is still inserted. Diversified Manufacturing of Graham, NC, makes such a marvelous product.

flexible intercept Allows you to assign "operator intercept" service to those extensions you wish for whatever reason, unassigned number, temporary disconnect, etc.

flexible line ringing A PBX feature which allows different phones to have different ringing for incoming calls from inside the building and from outside. Different ringing for intercom calls, different for inter-net calls, different ringing for outside calls, etc.

flexible numbering of stations A PBX feature which allows you some flexibility in the way you number the extensions off your PBX. How much flexibility depends on the particular PBX and the number of extensions you have. Hotels like giving their hotel phones the same number as the rooms. Makes sense. See Flexible Station Numbering.

flexible pricing tariffs A regulatory procedure which permits rates for certain services to be changed quickly to meet market conditions, i.e. competition.

flexible release The ability of the switching system to release a connection when either party hangs up.

flexible ringing Also called Distinctive Ringing. A PBX feature that lets phones ring differently. Useful to separate inside and outside calls. Also good when phones are close to each other because people can recognize their phone instantly.

flexible routing The ability to choose different physical paths through a network for different calls as circumstances warrant.

Flexible Service Logic FSL. The concept of supplementing application program logic through the use of non-executable code (i.e., data). In FSL, the logic the data represents is stored in a decision graph. The decision graph once loaded into the SCP or other network element can be interpreted by the decision graph traversal capability. See SCP and SS7.

flexible station numbering A feature that allows telephone extensions to be numbered according to their physical location or departmental location, etc. No rewiring is required for in-place telephones. It's all done in software. See Flexible Numbering of Stations.

flexibility The quality of a cable or cable component which allows for bending under the influence of outside force, as opposed to limpness which is bending due to the cable's own weight. Also refers to the ability of a cable to bend in a short radius.

FlexRay FlexRay is an ultra high-speed link being developed for automotive systems that need to interact, such as having the anti-lock braking system communicate with the engine so they can work together to slow the car.

flicker The wavering or unsteady image sometimes seen on monitors. A major cause is a refresh rate that's too low. Above 60 Hz, flicker disappears completely. See Monitor.

flight mode See airplane mode.

FLINK A FLash and a wINK makes a flink signal.

Flip Chip Pin Grid Array See FC-PGA.

flip flop A device or circuit which can assume either of two stable states. Flip flop devices are used to store one bit of information.

flip phone Slang name given to mainly Motorola brand wireless phones with a flip down lid or cover to expose the keypad on the phone.

flipping You're chummy with a company who's about to go public, i.e. do an initial public offering. It's a hot company, whose shares are likely to rise significantly above what the shares are sold to the public. It would be a real coup if you could get some shares before it went public. You do. And, bingo, the shares, which you paid $10 for, are now selling on the market, on their first day of trading at $20. You sell your shares instantly. That's called "flipping" in the United States and "stagging" in Australia and Great Britain. Flipping often pits the flipper against the brokerage firm/underwriter. The broker wants to control the trading in the IPO immediately after it goes public, i.e. keep the price up. Brokers try to curb flipping by individual investors by imposing waiting periods and fees on sellers. However, the largest institutional investors and mutual funds continue to flip with impunity because of their size and influence. See also Spinning.

float charging The battery charging technique for which sealed lead acid batteries are designed. A float charger maintains a voltage on the battery known as the "float voltage". The float voltage is the ideal maintenance voltage for the battery which maximizes battery life. When the float voltage is applied to a battery a current known as the "float current" flows into the battery, exactly cancelling the batteries' own internal self discharge current. Sealed lead acid batteries require float charging at least occasionally or they will become permanently degraded by a process called "sulfation". Maximum lifetime is obtained when a sealed lead acid battery is permanently float charged. This definition courtesy APC.

floating batteries The normal technique for powering telephone equipment in which batteries are simultaneously charged from a commercial source or generator and discharged to operate the telephone equipment.

floating point Using an exponent with numbers to indicate the location of the decimal point. It's more precise than integer but slower. See Floating Point Arithmetic.

floating point arithmetic Calculations performed on floating point, or exponential numbers. These numbers have two parts, a mantissa and an exponent. The mantissa designates the digits in the number, and exponent designates the position of the decimal point. Essentially floating point arithmetic allows the representation of very large numbers using a small number of bits. The speed of scientific computers is often rated in the Millions of FLoating Operations Per Second (MFLOPS) they can perform.

floating selection An imaging term. A selected area that is conceptually floating above the image, allowing it to be manipulated without affecting the background (for example, the contents of the Clipboard).

Floating Virtual Connection FVC. The ability to resume an on-demand connection on a port other than the port on which the original on-demand connection was established.

floating virtual tributary mode The timing of the virtual tributary is not locked (in frequency or phase) to the timing of the STS-1, but is allowed to float. A VT pointer is used to locate the VT Synchronous Payload Envelope (VT SPE).

flog Another word for a fake blog. See fake blog.

flogging See fake blog.

flood attack See Domain Name Server Flood Attack.

flood projection In facsimile, the optical method of scanning in which the original is floodlighted and the scanning spot is defined by a masked portion of the illuminated area.

flood search routing A routing method that employs an algorithm that determines the optimum route for traffic within a network, avoiding failed and congested links.

floodgaters From Wired's Jargon Watch column. Individuals who send you email inquiries and, after receiving only a slightly favorable response, begin flooding you with multiple messages of little or no interest to you.

flooding A packet-switched network routing method whereby identical packets are sent in all directions to ensure that they reach their intended destination.

Floor Distributor FD. The international term for horizontal cross-connect. The distributor used to connect between the horizontal cable and other cabling subsystems or equipment.

floor feed An access point in a raised or cellular floor used for the exit of communications or power cables. Floor feeds can be fixed or drilled as required, depending upon the floor type.

flop FLoating point Operation. Performing an operation on a floating point number. One measure of microprocessor speed is FLOPs per second, or MFLOPs (million flops per second).

floppy disk A thin, flexible plastic disk resembling a phonograph record upon which computer data is stored magnetically. Called a floppy disk because it is flexible and can (and will) flop inside a drive as it is being turned. And it may sound as though it is flopping. Floppy disks were never designed as the permanent storage many people are using them for at present. Floppy disks were designed by IBM as a way of having its sellers and engineers carry programs and program updates to its customers. Floppy disks were lighter and less cumbersome than carrying heavy spools of magnetic tape. IBM designed its floppy disks to be thrown away once their information was loaded into the mainframe computer. The moral of this story is that floppy disks are NOT permanent reliable storage. Anything stored on floppy disks should be backed up at least once and, if possible, twice. Floppy disks come in three standard sizes – 3 1/2, 5 1/4 and 8 inches. Floppy disks can be now safely put through X-ray machines at US airports.

floppy mini A floppy disk smaller than the traditional 5 1/4 inch diameter floppy disk. Now most commonly the 3 1/2 inch size invented by Sony, and used by the Apple Macintosh, among others. All MS-DOS laptop computers have 3 1/2 inch disks.

floptical technology The combination of optical servo track positioning and magnetic read-and-write technologies used in 3 1/2-inch Very High Density floppy disk drives. Floptical is a registered trademark of Insite Peripherals.

flow control The hardware, software and procedure for controlling the transfer of messages or characters between two points in a data network – such as between a protocol converter and a printer – to prevent loss of data when the receiving device's buffer begins to reach its capacity. In flow control, you can also deny access to additional traffic that would further add to congestion. (Think about flow control and the airlines.) See Flow Control Procedure, Rate-Based Flow Control, QFC and ER.

flow control parameter facility X.25 facility that allows the negotiation of packet and window sizes in both directions of transmission.

flow control procedure The procedure for controlling the rate of transfer of data among elements of a network, e.g., between a DTE and a data switching exchange network, to prevent overload.

flow through provisioning The ability to turn up a new circuit or increase bandwidth for a customer in real time, at the time the order is placed, rather than taking the order and waiting as long as two or three weeks for the circuit or increased bandwidth to be provisioned.

flowchart A graphic or diagram which shows how a complex operation, such as programming, takes place. The flowchart breaks that operation down into its smallest, and easiest-to-understand events.

flower Term for scotchlok bundle nicely spliced together forming a beautiful telecom flower of colors.

fluidic self-assembly A manufacturing process for RFID tags, patented by Alien Technology. It involves flowing tiny microchips in a special fluid over a base with holes shaped to catch the chips.

fluoride glasses Materials that have the amorphous structure of glass but are made of fluoride compounds (e.g., zirconium fluoride) rather than oxide (e.g., silica). Suitable for very long wavelength transmission.

Fluorinated Ethylene Propylene FEP. Also known by the trade name Teflon, a registered trademark of Dupont. FEP is the insulation of choice for high performance cable and wire systems installed in return air plenums. As FEP is really slick, it makes the wire really easy to pull through conduits, around corners, and so on – the same property that makes it so wonderful in the kitchen. FEP's fire retardant properties have led many countries to require its use, particularly in plenum ceilings. See Plenum.

flush jack 1. A telephone or data-connection jack mounted on and recessed in a wall. Each flush jack can have up to six connections on its face.

2. A toilet in a casino that works even after you have lost all your money (you still have to tip the attendant).

flush protocol An ATM term. The flush protocol is provided to ensure the correct order of delivery of unicast data frames.

flush-to-grade Describes an underground equipment vault whose top cover is flush with the surface of the ground around it.

flushing out the queue A call center term. Changing system thresholds so that calls waiting for an agent group are redirected to another group with a shorter queue or available agents.

flushofone A cordless headset with a noise cancelling microphone so you can't hear a toilet flush when on the phone in the bathroom.

flutter A rapid change in an electrical signal. The change may be in strength, frequency or phase. Distortion due to variation in loss resulting from the simultaneous transmission of a signal to another frequency.

flux 1. In soldering, a substance used to remove oxides from metal so the metal can be wet with molten solder for soldering.

2. A magnetic field that develops around a conductor of electricity.

fly-by-wire In traditional airplanes, the controls pilots moved were attached to heavy cables and hydraulic systems which themselves physically moved the rudder or the flaps, etc. Fly-by-wire replaced these wires and the hydraulic systems with computers and thin electrical wires. There are two main advantages to fly-by-wire. The computers can continuously adjust the aircraft's controls without the input of the pilot, trimming control surfaces so that the plane slides through the air with a minimum of air drag. Second, by eliminating heavy control cables and cutting down on hydraulic lines you can cut several hundred pounds off the weight of the plane, thus saving huge amounts of fuel over the life of the plane.

flying lead A grounding lead that exits the back of the connector hook on the outside of the cable jacket. It's normally attached to the drain wire or shield and then connected to the chassis of the switch, modem, etc.

flywheel A flywheel is a large heavy wheel used in electrical power generation. It's connected to an electrical power generator and will keep the generator spinning after the power source (a waterfall, or whatever) is unavailable.

FM 1. Fault Management. A network management function designed to receive fault information into a centralized management function. The faults are monitored, tracked, and resolved.

2. Frequency Modulation. See Frequency Modulation.

3. Freaking (or other similar word) magic. When something dead suddenly starts to works, it's "FM." In other words, nobody knows how or why it suddenly started to work.

But we're all pleased.

FM blanketing That form of interference to the reception of other broadcast stations, which is caused by the presence of an FM broadcast signal of 115 dBu (562 mV/m) or greater signal strength in the area adjacent to the antenna of the transmitting station. The 115-dBu contour is referred to as the "blanking area."

FM cable service The offering of FM radio signals over a cable system for a fee. A cable is connected to the subscriber's FM stereo receiver for service. Such a service hasn't been especially popular. More popular has been cable modem broadband service.

FM capture A cellular radio term. In cases of extreme co-channel interference, a receiver may experience "FM capture", which is a co-channel interference condition where is selected. Cellular users often experience FM capture as a momentary burst of someone else's conversation.

FM stereo separation A measure of a radio tuner's ability to separate the left and right hand channels of a stereo broadcast. The higher the number, the greater the separation. The unit of measure is the Decibel (dB), a logarithmic unit which expresses the ratio between two voltage, current or power levels, usually relating to a standard reference level, or a background noise level.

FM subcarrier One-way data transmission using the modulation of an unwanted portion of an FM broadcast station's frequency band.

FM threshold That point at which the input signal power is just strong enough to enable the receiver demodulator circuitry successfully to detect and recover a good quality television picture from the incoming video carrier. Using threshold extension techniques, a typical satellite TV receiver will successfully provide good pictures with an incoming carrier noise ratio of 7db. Below the threshold a type of random noise called "sparkles" begins to appear in the video picture. In a digital transmission, however, signal is sudden and dramatically lost when performance drops under the threshold.

FMAS Facility Maintenance and Administration System.

FMC Fixed Mobile Convergence. This is an evolving concept. In its simplest terms it means you have one phone, which is portable and you carry wherever you go. You use that phone when you're at home or in your office or on the road. When you're at home or in your office that phone would connect to your wireless Wi-Fi network and dial through the Internet as a VoIP call. When you're outside of your home or office, you would use that phone as a cell phone. The theory of FMC is that you can be talking on your phone in your house then walk outside and be seamlessly switched to the local cellular network. At least that's theory. Of course, I've found few cell phones or VoIP phones to be that reliable or that good quality or their service so cheap that I'd give up my wired phones. See also IP multimedia subsystem.

FMIC Flexible MVIP Interface Circuit. The FMIC provides a complete MVIP compliant interface between the MVIP bus and a variety of processors, telephony interfaces and other circuits. A built-in digital time slot switch provides Enhanced-Compliant MVIP switching between the full MVIP bus and any combination of up to 128 full duplex local channels of 64 kbps each. An 8-bit microprocessor port allows real-time control of switching and programmable device configuration. On board clock circuitry, including both analog and digital phase-locked loops, supports all MVIP clock modes. The local interface supports ST_BUS (Mitel), PCM Highway (Siemens), CHI (AT&T) signal formats at programmable rates of 2.048 MHz, 4.096 MHz, and 8.192 MHz as well as parallel DMA through the microprocessor port. See MVIP.

FMOD FM modem or channel unit. These modems handle voice, data, and fax calls. We use them on a standard A system, which is a first generation Inmarsat Maritime telecommunication system. It has mobile to fixed, and fixed to mobile, or in other words, shore to ship or ship to shore services) voice, telex, and Data. It runs at 56 or 64 kilobites per second, i.e. not very fast in today's terms.

FMP Forced Management Plan. It's AT&T's euphemism for firing its people.

FMV Fair Market Value. A special lease for IRS purposes. Be careful. With Fair Market Value (FMV) leases, there is a catch to having the lessor guarantee the dollar amount or the percentage of your buyout. In order to be a FMV lease there must be a risk. That is why you are paying a lower rate of interest. If you agree on an amount up front, make sure it is not in writing, otherwise it does not meet the IRS test for a FMV lease. With a FMV lease, the lessor owned the asset and depreciates it; lessee expenses monthly payments and deducts them for tax purposes. If the buyout is determined in writing and the IRS can prove it, then it is a financing lease and lessee owns asset and depreciates it. Beware of this. This advice from Jane A Blank, telecom consultant, Westerville OH.

FNA A Brussels-based strategic alliance, which exists to facilitate global communications connections for companies in the financial services sector. The 12 founding FNA companies are Stentor of Canada, AOTC of Australia, RTT-Belgacom of Belgium, France Telecom, Deutsche Bundespost Telekom of Germany, Hong Kong Telecom, Italcable of Italy, KDD of Japan, Singapore Telecom, Telefonica of Spain, Mercury Communications of the United Kingdom, and MCI of the United States.

FNC Federal Networking Council. The body responsible for coordinating networking needs among U.S. Federal Agencies. A US group of representatives from those federal agencies involved in the development and use of federal networking, especially those networks using TCP/IP, and the connected Internet. The FNC coordinates research and engineering. Current members include representatives from the DoD, DOE, DARPA, NSF, NASA and HHS.

FNG F..ing New Guy. A term for the new, ignorant guy on the team. Originated in the military. It was used in the movie Forrest Gump, when Forrest and Bubba first meet Lieutenant Dan.

FNPA Foreign numbering plan area.

FNPRM Further Notice of Proposed Rulemaking. FCC jargon. The FCC (Federal Communications Commission) makes rules and regulations that govern the conduct of federal telecommunications business in the United States. When an issue is identified, the FCC conducts research, holds hearings to gather information, solicits opinions, and develops a position. Before they make a ruling, they publish a Notice of Proposed RuleMaking (NPRM). Opinions are offered, and the process begins anew. The NPRM then becomes a FNPRM. Sometimes this goes on for years. It can be a sloooooooooooooooow process.

FNR Fixed Network Reconfiguration.

FNS Fiber Network Systems.

FO Fiber Optics.

FOA 1. First Office Application. A telephone company term for what you and I know as beta testing.

2. Fiber Optic Amplifier. See Fiber Optice Amplifier for a full definition.

foamed plastics Insulations having a cellular structure.

FOB Free On Board. Term indicating where the seller's responsibility ends and the buyer's begins. You buy something, F.O.B. The seller puts it on a truck or railroad, plane, i.e. some carrier. He's responsible for getting it on the carrier. It's FOB, the truck. You – the buyer – are responsible for paying for the cost of the freight of getting you the goods you ordered. The opposite of F.O.B. is C.I.F. That stands for Cost, Insurance and Freight are included. That means the seller pays the freight. See FOB, FOB Destination, FOB Place of Delivery and FOB Shipping Point.

FOB Destination Seller retains ownership until delivered to buyer. See FOB.

FOB Place of Delivery Seller retains ownership until delivered to buyer. See FOB.

FOB Shipping Point Seller responsibility ends when item is turned over to carrier. The buyer is responsible for payment if goods are damaged in transit. The buyer also handles any insurance claim. See FOB.

FOC Firm Order Confirmation. An FOC is a confirmation that a telephone company received a order from a customer, has processed it, and has provided a due date back. For most practical applications, the Due Date from the FOC is "firm", but not always set in stone. For instance in between when the FOC is issued and the Due Date, a backhoe cuts the fiber in the ground or a rainstorm floods the basement of an office building. The due date is going to change. Therefore the date is no longer firm.

focus A measure of the clarity of a color monitor, CRT or LCD. Focus is the harpness of a pixel or series of pixels on the CRT face plate. Also measured as the spot size. Focus relates to the sharpness of a monitor's electron beam as it paints the face of a Cathode Ray Tube (CRT). The other measures are convergence and dot pitch.

FOD Fax On Demand. See Fax Back.

FODU Fiber Optic Distribution Unit. A FODU is the fiber equivalent of a DSX panel, which is a Digital System Cross-connect frame. A DSX is a manual bay or panel to which T-1 lines and DS1 circuit packs are wired. A DSX permits cross-connections by patch cords and plugs, i.e. by hand.

FOG Fiber Optic Ground.

Fogging In computer graphics, fogging is (surprise, surprise) simulating the effects of fog, smoke and haze. Fogging takes considerable processing power.

FOI Act Freedom of Information Act 1982. Australian legislation dealing with access by the general public to information gathered and held by Commonwealth Government agencies.

foil A slang term for an overhead transparency. The expression "he gives good foil" reflects an executive's ability to make great presentations using overhead transparencies. In the 1970s and early 1980s, so many managers at IBM made presentations that some senior executives actually got overhead projectors built into their desks.

foilware Foilware is a slang term for an overhead transparency. There are various iterations in the development of a product. One of the first is a description of the product on overhead transparencies. Such overheads are often used to convince investors to put money into the company or to convince distributors to sell the product. This often happens long before the product actually exists. Sometimes the company will pretend with its foilware that its products actually exist. In this case, the products then become the foilware. See also Foil.

FoIP Fax Over IP (Internet Telephony). A FoIP server is a VoIP endpoint used specifically for receiving and sending faxes. FoIP Servers don't use any specialized fax, instead they use VoIP gateways to send and receive faxes. This has several advantages:

Share and leverage existing hardware deployed for voice application (and vice versa);

Consolidate and reduce the cost of phone lines by sharing them between voice and fax applications; and

IT specialists can now manage data/voice/fax network with a single set of skills.

Just like VoIP-based phone calls, the FoIP server uses standard signaling protocol like SIP and H.323 to establish calls. At that point there is absolutely no difference between a FoIP and a VoIP call. The difference lies in what is transported during the call. Voip devices use voice codecs like G.711 and G.723 to transport the voices of the calling parties. A FoIP server uses a T.38 fax codec instead. While G.711 and G.723 are specialized to transport human voice, T.38 is specialized to transport fax machine language (i.e., T.30 messages exchanged between fax terminals). Both T.30 and T.38 are ITU standards, T.30 defines how fax machines communicate together over analog lines and T.38 defines how T.30 messages can be transported over a packet network.

A FoIP server can handle several hundred simultaneous fax calls. For each one, a virtual T.30 engine simulates a fax terminal. From that point FoIP servers behave like regular LAN fax servers, all faxes are queued until they are routed/delivered and they are archived for future retrieval.

Sine both the telephony and fax services share the same VoIP infrastructure, it is possible to provide a single number for fax and voice communications. If a fax call is detected by the telephony service, the call is transferred automatically to the FoIP server.

FOIRL Fiber Optic Inter Repeater Link. Defined in IEEE 802.3 and implemented over two fiber links, transmit and receive, this medium may be up to one kilometer in length, depending on the number of repeaters in the network. A FOIRL is the perfect transmission medium to join a local area network on the eleventh floor to the fourth floor of the same building.

folder A subdirectory or a file folder. The Apple Macintosh was the first to use them. Microsoft picked up on the idea when it introduced Windows 95.

follow me 800 service Basically, Follow Me 800 Service is call forwarding of your personal 800 line. MCI announced this service in the Spring of 1991. It differs from local call forwarding in that you can dial into MCI from anywhere in the world and change the number your 800 line will send its calls to. Your 800 number always stays the same. What changes is the number it calls. A simple explanation: We buy a personal 800 line from MCI. The number is 800-555-6534. When someone calls that number, MCI looks up a database, checks where to send the number and sends it to my office at 212-691-8215. However, one day I go traveling. So I call another MCI 800 number, punch in my identification number and then give it the new number I will be at – namely 212-206-6660. From then on, MCI will send all my calls to that number – until I call and change the number again.

follow me call forwarding Progressive Call Forwarding. Allows a previously forwarded call to be forwarded from that to another phone extension.

follow me roaming The ability for the cellular system to automatically forward calls to a roaming mobile that has left it's primary service area. Without this feature, the calling party must know the location of the roamer and place a call to that area.

follow me services Also called One Number Services. Follow me systems and services are based on the premise that people are mobile (e.g., they move around a lot in and out of the office), and have many phone numbers or places they might be. A person could have an office number, a cellular number, a voice mail number, a home number, and a pager number. Which phone number will a caller be at? Follow me systems will "trackdown" the user being called no matter where they are and connect the caller to the user. The caller need only dial a single phone number. Usually, network or local switch provided call data (or data gathered by a voice response unit) is used to identify each call as being intended for a specific user. Based on options the user has selected, the caller will hear an answering prompt customized to that user, and the one number system will then automatically attempt to locate the user at one of several locations. The tracking-down process varies considerably between follow me systems. Some systems try multiple locations at once,

others will try the possible destination locations sequentially. Almost all follow me services provide the caller an exit to voice mail at various points of the call.

follow the stream Identify the packets of a specific network session in a logfile created by a protocol analyzer so as to follow the conversation that took place. This might be done, for example, to reconstruct and view HTML content offline, or to pick off usernames and passwords from an insecure session. A utility such as Ethereal or Microsoft's Network Monitor simplifies following a stream by letting you filter and display only traffic pertaining to a particular network session.

follow the sun dialing A technique used in call centers whereby the agents call those parts of the country where it's convenient to call and move the calling across the country as the sun moves. Our agents might call New York households between 6 P.M. and 9 P.M. When the time hits 9 P.M., the agents stop calling New York households and start focusing calls on households in the central time zone. To accomplish Follow The Sun Dialing, a call center needs software which knows in which phone numbers are in which time zones.

Foma 1. Freedom of Mobile Multimedia Access. Foma is the brand name used in Japan for NIT DoCoMo's 3G (WCDMA) service. Launched in October of 2001. As of late summer, 2004, Foma had 5.3 million subscribers. Foma was one of the world's first third generation mobile services

FOMC Federal Open Market Committee. The Federal Open Market Committee consists of twelve members: the seven members of the Board of Governors of the Federal Reserve System; the president of the Federal Reserve Bank of New York; and, for the remaining four memberships, which carry a one-year term, a rotating selection of the presidents of the eleven other Reserve Banks. The FOMC's main job is to set interest rates. It's the group primarily responsible for monetary policy in the United States. It holds eight regularly scheduled meetings per year.

Fonda, Jane See Barbarella.

font Alphanumeric and other printable or displayable text characters of a single distinctive style and size. The actual representation of a single character is known as a "glyph." The basic design of the style or "face" of the font is known as the "typeface." The size of the type and fonts is measured in archaic units known as "points," with one point being approximately 1/72 of an inch. The point size does not measure the size of an individual character, by the way. Rather, it measures the distance from the highest part of a letter that reaches the highest (e.g., lower case "f") to the lowest part of a letter that reaches the lowest (e.g., lower case "g") Common fonts are Arial, Helvetica, and Times New Roman.

font family A group designation that describes the general look of a font. Familiar font families are Helvetica and Times Roman.

font size See Point Size.

foo An Internet term. A place-holder for nearly anything – a variable, function, procedure, or even person. "A given user foo has the address 'foo@bar.com'."

foo foo dust Magic dust when sprinkled on equipment makes the equipment work perfectly. Foo foo dust doesn't exist, of course. It's just a term for the machine suddenly coming to life when someone "talks" to it or "touches" it. In other words, a fluke situation. Also spelled fu fu dust. See also Foo Foo School.

foo foo school This one was sent to be me in 2003 by a disappointed graduate. A foo foo school is a telecommunications school that offers assistance in job placement upon graduation. My reader commented "What better magic to put to work what you learned." See also Foo Foo Dust.

foolproof No product can never be made to be foolproof, since they will always make a bigger fool.

football Also called Aerial Service Wire Splice. A device used to splice aerial service wire shaped a bit like a football when it's installed.

footprint 1. The area on the earth's surface where the signals from a specific satellite can be received. A footprint is shown as a series of concentric contour lines that show the area covered and the decreasing power of the signal as it spreads out from the center.

2. The area on a desk a device occupies, i.e. the computer's footprint.

forbearance Forbearance is the power of a regulator not to regulate a service or market if it believes the market is "workably competitive."

Force Administration Data System See FADS.

force feed An arrangement in an outbound telebusiness unit where agents are force fed with a new call which is automatically dialed, a pre-determined time after finishing the previous call.

Force Majeure You'll find a force majeure clause in most contracts. It says essentially that the supplier agrees to do what he says he agrees to do and you can penalize if he doesn't – unless something untoward happens. Namely force majeure, which is defined (and the following is excerpted from an actual contract) as causes beyond the Company's control, including but not limited to: acts of God, fire, flood, explosion or other catastrophes; any law, order, regulation, direction, action or request of the United States Government, or of any other government, including state and local governments having or claiming jurisdiction over the Company, or of any department, agency, commission, bureau, corporation, or other instrumentality of any one or more of these federal, state, or local governments, or of any civil or military authority; national emergencies; insurrection; riots; wars; unavailability of rights-of-way or materials; or strikes, lock-outs, work stoppages, fraudulent acts of a third party, or other labor difficulties.

forced account code billing A telephone feature which prevents call from being completed if the user does not pushbutton in a billing code. That billing code may correspond to the department within the company. Or it may conform to the client and to the client's matter number the call must be billed to.

Forced Authorization Code FAC. A PBX feature which requires all or certain users to enter a code before dialing an outside number.

forced hop A channel hop made by the Mobile Data Base Station (MDSB) because non-Cellular Digital Packet Data (CDPD) activity is detected on the channel that is currently in use.

forced on net Calls that originate via switched access are forced to terminate to an on-net location.

forced perfect terminator A type of terminator containing a sophisticated circuit that can compensate for variations in the power supplied by the host adapter, as well as variations in bus impedance of complex SCSI systems.

forced release/disconnect The switching center's automatic hang-up if the calling party fails to do so at the end of a conversation.

forced route override Allows a PBX user to automatically redirect an outgoing call to a different trunk if the first trunk is busy or the connection is poor.

Ford Ford makes jet engines under license from General Electric, Ford built thousands of jet engines over the years for the USAF KC-135 fleet. When an engine failed, it is noted in the log as a FORD failure (Found On Runway Dead).

forecasting Taking historical data (what happened in the past) from your ACD and using that information to predict what might happen in the future. Has your call volume always doubled on Tuesday? It will probably double next Tuesday too. A very important function of call center management software.

foreground processing Automatic execution of computer programs designed to preempt the use of the computing facilities. Usually a real time, urgent program. Contrast this with Background Processing, which might be something less urgent, for example, diagnostics of the system.

foreign address An ATM term. An address that does not match any of a given node's summary addresses.

foreign agent A Mobile IP term. A service which enables mobile nodes associated with nomadic users to register their presence at a remote location. The foreign agent communicates with the home agent in order that data packets can be forwarded to the remote subnet. See also Mobile IP.

foreign aid Foreign aid might be defined as a transfer of money from poor people in rich countries to rich people in poor countries. - Douglas Casey

foreign area translation Translating the office codes of a distant (foreign) area to codes that make sense to a PBX which has more than one way of completing the call to that area.

foreign central office service Getting telephone service in a multi-office exchange from a central office other than the one you are normally served by. Not a common term any longer. Foreign central office service is the same price as normal local telephone service. It typically just involves asking for service off another central office. For example, our main number in New York City is 212-691-8215. Our 691- central office is in the 18th Street Exchange, a tall building on 18th Street. There is another central office in the same building. It is 206- and it is a more modern central office. When we ordered additional lines, we ordered them from this central office. You can now also call us on 212-206-6660. Don't trust my definition, however. Ask your local telephone company. See also Foreign Exchange Service.

foreign EMF Any unwanted voltage on a telecommunications circuit.

foreign exchange service FX. Provides local telephone service from a central office which is outside (foreign to) the subscriber's exchange area. In its simplest form, a user picks up the phone in one city and receives a dial tone in the foreign city. He will also

receive calls dialed to the phone in the foreign city. This means that people located in the foreign city can place a local call to get the user. The airlines use a lot of foreign exchange service. Many times, the seven digit local phone number for the airline you just called will be answered in another city, hundreds of miles away. See also Foreign Central Office Service and Foreign Exchange Trunk.

foreign exchange trunk A Foreign EXchange (FEX) trunk provides a direct connection between a PBX switch and a remote central office other than the central office that serves the location of the PBX.

foreign numbering plan area FNPA. Any other NPA (Numbering Plan Area) outside the geographic NPA where the customer's number is located.

foreign prefix service Getting dial tone in a multi wire center exchange from a foreign wire center other than the one you are normally served by. Similar to Foreign Central Office Service, except that you may get charged extra for Foreign Prefix Service. Don't trust my definition, however. Ask your local telephone company. See also Foreign Central Office.

foreploy The Washington Post's Style Invitational asked readers to take any word from the dictionary, alter it by adding, subtracting or changing one letter, and supply a new definition. This one is one of the winners. Foreploy means any misrepresentation about yourself for the purpose of obtaining sex.

Forex FOReign currency eXchange. Definitely not an area you want to invest your money in.

fork When you come to a fork in the road, take it. This profound piece of advice from the great American philosopher, Yogi Berra.

forking The splitting of an incoming call so that it rings at more than one telephone number. Whichever target telephone answers first establishes the connection, and the other telephone drops out of the equation.

forklift upgrades A forklift is a self-propelled machine used to lift and transport heavy objects by means of steel fingers inserted under the load. An upgrade is an improvement or advancement in size or functionality. A forklift upgrade has its roots in the days of "heavy iron," when it literally took a forklift to upgrade the PBX or mainframe computer technology. One drove a forklift into the switchroom or computer room, picked up the system, transported it out the door, and brought in an improved system. The old system, which had little use, was made into a boat anchor or artificial reef, or so the story goes. Such upgrades typically cost an arm and a leg.

Such upgrades are increasingly uncommon today. Rather, much gear is upgradable by simply changing the generic software load, possibly inserting a card or microchip or two, and perhaps swapping out a power supply. Most switch manufacturers are trying to figure ways to avoid forcing their customers into forklift upgrades when their requirements outgrow the existing system. Examples include stackable hubs and distributed switches, which are scalable to one degree or another. See Scalable and Stackable.

form A HTML formatted document containing blank fields that users can fill in with data. Fields include: text input, radio buttons, drop-down menus, and check-boxes. The process is usually completed by a Submit function, where the results are electronically entered into a database for further analysis. These are very common on the World Wide Web, where the form appears on the user's computer monitor, and can be filled in by selecting options with a pointing device, or typing in text directly from the computer keyboard.

Form 230 Form 730 Application Guide is a collection of literature you'll need to register your telephone/telecom equipment under Part 68 of Title 47 at the Federal Communications Commissions. To get this material (it's free) drop a line or call the Federal Communications Commission, Washington DC 20554. As I write this edition, the person at the FCC in charge is William H. Von Alven, who also puts out a newsletter for Part 68 applicants. See PART 68 for a much larger explanation.

Form Effectors FEs. Control characters intended for the layout and format of data on an output device such as a printer or CRT. Examples are CR (carriage return) and LF (line feed).

form factor A fancy way of referring to the shape and size (width, depth, height) of some device like a computer, a telephone system, a circuit board, a processor or a computer chip.

Formal Call Centre A British term. A telebusiness unit in which all of the staff are dedicated to telephone based work. See also Informal Call Centre.

formal standards Specifications which are approved by vendor-independent standards bodies, such as ANSI (American National Standards Institute), ISO (International Standards Organization), IEEE (Institute of Electrical and Electronic Engineers) and NIST (National Institute of Standards and Technology).

formant A point of excitation, or high energy, in a speech waveform caused by resonance in the human vocal tract. Formants are responsible for the unique timbre of each individual's voice.

formant synthesis A form of synthesized speech in which the computer creates the voice. The result is smooth but sometimes artificial-sounding. Formant synthesis is used in text-to-speech (TTS) technology in which the computer "reads" text as voice. Another technology used in TTS is called concatenation synthesis, which uses actual samples of human voice, chopped up and put back together. Concatenation synthesis sounds choppy.

format 1. Arrangement of bits or characters within a group, such as a word, message, or language.

2. Shape, size and general makeup of a document. As a verb, its most common usage is in "to format this disk."

3. As a verb in "to format." In the computing world, the common meaning of to format a hard disk is to completely clean the hard disk of old material so it will accept new material. Formatting a hard disk is typically done when it's new or when the disk has become so cluttered with bad files and junk that it's simply better to start all over again with a clean disk. There are typically two format commands. One is a quick format which simply destroys and rebuilds the file allocation table on the media, making it appear blank and unused to the computer's file system. But the material on the hard disk stays intact until overwritten by another file. Then there is a full format, also called low level format, which destroys and rebuilds the file allocation table and actually removes all trace of data on the media by overwriting it with all zeros. Full format takes longer. See also Computer Fingerprinting.

format shift To transcode multimedia content so that it can be played on a different device and/or over a different network. For example, a video designed for TV may be format-shifted for delivery to and display on the smaller screen of a cell phone. See transcoding.

Fortezza A cryptology mechanism developed by Mykotronx, Inc., a subsidiary of Rainbow Technologies, in conjunction with the NSA (the National Security Agency), which holds the registered trademark. The family of FORTEZZA security products includes PCMCIA-based client cards, and server boards; compatible implementations are available variously in hardware and software. All FORTEZZA Crypto implementations support data privacy, user ID authentication, data integrity, non-repudiation, and timestamping. FORTEZZA is the crypto token chosen to secure the Defense Messaging System (DMS), including both the MIL-NET and the Internet. Applications include e-mail, voice communications and file transfer. Depending on the application, the encryption keys are either 80 or 160 bits in length, thereby providing excellent security for "Sensitive But Unclassified" (SBU) government data, as well as for commercial applications. FORTEZZA opponents suggest that the NSA is attempting to force the mechanism on the private sector as a replacement for the rejected Clipper Chip technology. The fear is that the NSA holds the keys to the secret encryption algorithm, and that the agency, therefore, can gain access to your data even more easily that it could have through the "backdoor" built into the Clipper Chip. See also Clipper Chip and MISSI. www.nsa.gov.

FORTRAN FORmula TRANslation. A computer programming language developed in 1954 for scientific applications. It is still used by scientists and engineers.

fortuitous conductor Any conductor that may provide an unintended path for intelligible signals, e.g., water pipes, wire or cable, metal structural members.

fortune cookie 1. An inane/witty/profound comment that can be found around the Internet.

2. The fortune cookie was invented in 1916 by George Jung, a German Los Angeles noodlemaker. Clearly it was a successful invention. He obviously used his noodle.

forum A section within an online service (such as CompuServe, America Online, etc.) where you can find out information on a specific subject – computers made by Toshiba or printers made by Hewlett Packard. Forums may include a library from which you can download various files (programs, bug fixes, printer drivers, text, press releases of new products and so on). Many forums also include one or more "conference rooms" which users may "enter" for conversations (on-line or off-line) with representatives of companies or the person running the forum, who is typically called the "sysop," as in system operator. Most manufacturers run forums as a relatively cheap and painless way of getting help information to their customers.

forward A switch feature that temporarily redirects incoming calls. The incoming calls are redirected from the forwarding telephone to another destination by the person associated with the telephone or by the computing domain. The other destination has previously been defined to the switch by the device associated with the telephone.

forward busying That feature of a telecommunications system wherein supervisory signals are forwarded in advance of address signals to seize assets of the system before attempting to establish a call.

forward channel The communications path carrying data or voice from the person who made the call. The Forward Channel is the opposite of the Reverse Channel.

forward direction 1. The forward direction of data away from the head-end in a broadband LAN.

2. Forward link communications direction from a fixed earth station via a satellite to a mobile terminal.

forward disconnect Disconnecting of a call path as a result on the called party hanging up. Prior to forward disconnect the called party could not initiate a disconnect of a call. This was the first denial of service attack, call someone that you don't like and then don't hang up and the call is never torn down and their phone is useless.

forward echo An echo propagating in the same direction as the original wave in a transmission line, and formed by energy reflected back from one irregularity and then onward again by a second. Forward echoes can occur at all irregularities in a length of cable, and, when they add systematically, can impair its performance as a transmission medium.

Forward Error Correction FEC. A technique of error detection and correction in which the transmitting host computer includes some number of redundant bits in the payload (data field) of a block or frame of data. The receiving device uses those bits to detect, isolate and correct any errors created in transmission. The idea of forward error correction is to avoid having to retransmit information which incurred errors in network transit. The additional bits add a small amount of overhead to the block or frame. Therefore, they create some level of inefficiency in transmission. The alternative is retransmission of the block or frame of data, which can be much more inefficient where large numbers of errors occur during transmission. This inefficiency is compounded when the retransmitted block or frame is errored, as well. From the standpoint of network throughput, FEC can be much more effective, particularly when bandwidth is expensive or limited. On the other hand, FEC is processor-intensive, as it places a load on the computational capabilities of the receiving computer. The simple idea of forward error correction is to avoid having to retransmit information sent incorrectly. The technique is consuming of bandwidth and can make the transmission take longer.

I asked Ray Horak, how can a few redundant bits of information significantly reduce error rates on transmission. Here's his reply: The process is extraordinarily complex. Explaining it would take pages and pages and would do no one but a mathematician any good. Essentially, a few redundant data bits are added at strategic places in the data field (for example, just suppose that every 50th bit were repeated – the exact repeated bits vary according to the specific algorithm used). The very few redundant bits significantly lower the potential for an individual data bit to be transmitted in error and go undetected, given the complex sampling technique and complex algorithms used to develop a description of the data field. The receiving host computer is intelligent enough and has enough computational horsepower at its disposal to figure it out, unlike most of us real human types. The issue and the tradeoff is one of the cost of processing power vs. the cost of retransmission across the network. As the cost of computers comes down and the cost of bandwidth comes down, the best solution remains specific to the specifics of the user and the application.

forward prediction A technique used in video compression, specifically compression techniques based on motion compensation, where a compressed frame of video is reconstructed by working with the differences between successive video frames.

forwarding description An ATM term. The resolved mapping of an MPOA Target to a set of parameters used to set up an ATM connection on which to forward packets.

forwarding equivalency class See MPLS.

FOSI Formatting Output Specification Instance. An old SGML DTD standard for document management in the US military, replaced by the ISO standard DSSSL.

FOSSIL Fido/Opus/Seadog Standard Interface Layer. This is the interface used as an add-on to mailer software packages to connect them to PCs that are not 100% IBM-compatible.

FOTA See firmware over-the-air.

FOTP Fiber Optic Test Procedure. Standards developed and published by the EIA (Electronics Industries Alliance) as EIA-RS-455.

FOTS Fiber Optic Transmission System. A generic term. SONET and SDH are the standards-based FOTS used by the carriers. See also SDH and SONET.

foundation graphics A set of graphics libraries or imaging models that form the lowest level graphics programmer's interface in Sun's OpenWindows. Examples: a graphics sub-routine library that a program could call to draw graphics primitives like arcs, circles, rectangles, etc.

four horsemen of the apocalypse War, Plague, Famine and Death.

four pair UTP cable There are four pairs of conductors in this cable for a total of eight conductors. The cable jacket (also called the cable sheath) holds all four pairs together. Many manufacturers also include a ripcord, used for cutting the cable jacket. Pull on it, the sheath opens, allowing you to get to your conductors to attach them to things. Sometimes, the ripcord works. Sometimes, it doesn't. See UTP Cable.

four wave mixing See Four-Wave Mixing.

Four Wavelength Wave Division Multiplexing 4WL-WDM, also called Quad-WDM. MCI announced this technology in the Spring of 1996 as a method of allowing a single fiber to accommodate four light signals instead of one, by routing them at different wavelengths through the use of narrow-band wave division multiplexing equipment. The technology allowed MCI to transmit four times the amount of traffic along existing fiber. At that time MCI's backbone network operated at 2.5 Gbps (2.5 billion bits per second) over a single strand of fiber optic glass. Using Quad-WDM the same fiber's capacity rose to 10 gigabits – enough capacity to carry approximately 130,000 simultaneous voice transmissions over one single strand of fiber. Since then, a number of carriers have deployed OC-192 (Optical Carrier Level 192) fiber, running at 10 Gbps. They are opening four "windows," or wavelengths, each running at 10 Gbps. That's 40 Gbps over a single strand through DWDM (Dense Wavelength Division Multiplexing). They are pulling as many as 620 strands at a time. While most carriers have elected to implement WDM/DWDM in their networks by purchasing equipment from vendors, MCI has concentrated on developing their own WDM capability internally. Generally speaking, the intense competition among the various DWDM vendors is pushing capacity upwards faster than MCI's own internal development. See also WDM and DWDM.

four-wave mixing A source of noise in some fiber optic transmission systems, four-wave mixing is a phenomenon by which some wavelengths interact to create additional wavelengths. The problem occurs in the 1550nm (nanometer) window (wavelength band), which is the band in which DWDM (Dense Wavelength Division Multiplexing) and EDFAs (Erbium-Doped Fiber Amplifiers) work. Non Zero Dispersion Shifted Fiber (NZDF), a type of Single Mode Fiber (SMF), addresses the problem by shifting the optimal dispersion point just above the range at which EDFAs operate. The optimal dispersion point is the point at which chromatic dispersion and waveguide dispersion cancel each out. See also Chromatic Dispersion, DSF, DWDM, EDFA, SMF, Waveguide Dispersion, and Wavelength.

four-wire See Four-wire Circuit.

four-wire adapter A device which allows the connection of two-wire telephone equipment to a four-wire line. See Four-Wire Circuit.

four-wire circuit A high-performance circuit, which offers lots of bandwidth and which is capable of multi-channel communications. Four-wire circuits are of two types: physical and logical. Physical four-wire was the original approach. In other words, they all comprised four wires, which were organized into two copper pairs of UTP (Unshielded Twisted Pair). These original four-wire circuits were analog, and used amplifiers to overcome the effects of signal attenuation, which is a significant problem at high frequencies. As the amplifiers worked in only one direction, two pairs of wires and two sets of amplifiers were needed: one for transmission in one direction and another in the reverse direction. A lot of physical four-wire circuits remain in use, and more are being deployed every day. Even though such circuits mostly are digital today, it generally still requires four physical UTP wires to provide four-wire service such as T-1.

Logical four-wire performs like physical four-wire, but with fewer wires. ISDN BRI (Basic Rate Interface) is an example of logical four-wire, as it usually uses only two wires to achieve relatively lots of bandwidth (144 Kbps), and multiple channels (2B+D, or 2 Bearer channels plus 1 Data channel). HDSL2 (High bit-rate Digital Subscriber Line, version 2), an emerging local loop technology, provides T-1 service over only 2 UTP wires. SONET fiber optic technology provides incredible amounts of bandwidth and supports hundreds of thousands of channels using only 2, or even 1, physical wires (glass fibers). Microwave, satellite and infrared transmission systems support four-wire service without any wires at all. See also ISDN, T-1 and SONET.

four-wire repeater See Four-wire Circuit.

four-wire terminating set An electrical device which takes a four-wire circuit – one pair coming and one pair going – and turns it into the "normal" tip and ring circuit you need for a typical telephone, key system or PBX. See Four-Wire Circuit.

Fourier's Theorem In the early 1800s, the French mathematician Emile Fourier proved that a repeating, time-varying function may be expressed as the sum of a (possibly infinite) series of sine and cosine waves. Digital data is a bit stream, which can be sent as a sequence of square waves. Fourier's Theorem shows that to send a square wave (digital signal), a series of sine waves (analog signals) are actually summed together. If 1,000 square waves are to be sent every second, for example, the frequency components of the sine waves that are summed together are 1 kHz, 3 kHz, 5 kHz, 7 kHz, etc. The point of this analysis is to show that high frequency signals are required to form a stable, recognizable square wave.

As the bit rate increases, the square wave frequency increases and the width of the square waves decrease. Thus, narrower square waves require sine waves of even higher frequencies to form the digital signal. Note, then, that there is insufficient bandwidth in the 3 kHz voiceband to send square waves due to the absence of frequency components above 3,300 Hz. Even low frequency square waves cannot be sent because sine waves below 300 Hz are also absent. Thus, the local loop, according to Fourier's Theorem, cannot be used for the transmission of digital signals! The last paragraph is, in fact, no longer totally correct, as the increasingly successful ISDN trials are proving.

fourth estate The press. In May 1789, Louis XVI, King of France, summoned to Versailles a full meeting of the "Estates General." The First Estate consisted of 300 nobles; the Second Estate, 300 clergy; the Third Estate, 600 commoners. Some years later, and well after the French Revolution, Edmund Burke, looking up at the press gallery of the British House of Commons, said "Yonder sits the Fourth Estate, and they (i.e. the press) are more important than them all."

fourth utility The non-vendor specific communications premise wiring system which you use for integrated information distribution (voice, data, video, etc.) Leviton in Bothell, Washington has trademarked the term Fourth Utility. They make a broad range of premise wiring products.

fox message A standard sentence for testing teletypewriter circuits because it uses most of the letters on the keyboard. That sentence is "The quick brown fox jumped over the lazy sleeping dog, 1234567890".

FP 1. Feature Package. A software release for a telephone system. Originated with AT&T's Dimension PBX, now manufacturer discontinued.

2. File Processor.

FPC Flexible Printed Circuit.

FPD Flat Panel Display.

FPDL Foreign Processor Data Link. A link from a Rockwell ACD to an external computer.

FPG Feature Planning Guide.

FPGA See Field Programmable Gate Array.

FPI Formal Public Identifier. A string expression that represents a public identifier for an object. FPI syntax is defined by ISO 9070.

FPL Power Limited Fire Protection; general use, NEC 760 UL Subject 1424.

FPLMTS Future Public Land Mobile Telecommunication System. A concept developed by the ITU for a family of technological solutions which would enable the development and implementation of a 3G (3rd Generation) wireless network in support of voice and high-speed data communications. FPLMTS was abandoned in favor of IMT-2000. See also IMT-2000.

FPLP Plenum rated Power Limited Fire Protection; NEC 760 UL Subject 1424.

FPLR Plenum rate cable for use between floors (Riser) Power Limited Fire Protection; NEC 760 UL Subject 1424. See also FPLP.

FPM DRAM Fast Page Mode Dynamic Random Access Memory.

FPP Fiber Optic Patch Panel.

FPS 1. Fast Packet Switching.

2. Frames Per Second. A measure of the quality of a video signal. NTSC TV – the standard in North America – uses 30 fps. Film is 24 FPS. PAL/SECAM (European) is 25 FPS.

FPT Forced Perfect Terminator. A high-quality type of single-ended SCSI terminator, developed by IBM, with special circuitry that compensates not only for variations in terminator power but also for variations in bus impedance. See also Active Terminator and Passive Terminator.

FPU Floating Point Unit. A formal term for the math coprocessors (also called numeric data processors, or NDPs) found in many PCs. The Intel 80387 is an example of an FPU. FPUs perform certain calculations faster than CPUs because they specialize in floating-point math, whereas CPUs are geared for integer math. Today, most FPUs are integrated with the CPU rather than sold separately. See also CPU and DSP.

FQDN Fully Qualified Domain Name. An Internet term. The FQDN is the full site name of an Internet computer system, rather than just its hostname. A fully qualified domain name consists of a host and domain name, including top-level domain. For example, www.jabber.com is a fully qualified domain name where www is the host, jabber is the second-level domain, and .com is the top-level domain. A FQDN is used to locate a machine on a network.

FR See Flat Rate Service.

FR-1 A flammability rating established by Underwriters Laboratories for wires and cables that pass a specially designed vertical flame test. This designation has been replaced by VW-1.

FRA Fixed Radio Access. Fixed radio access provides customers with a connection to a network via a radio link between their premises and a fixed antenna. Also known as WLL. See WLL.

fractal A word coined in 1975 by Benoit B. Mandlebrot from the Latin "fractus," meaning "to break." Along with raster and vector graphics, fractals are a way of defining graphics in a computer. Fractal graphics translate the natural curves of an object into mathematical formulas, from which the image can later be constructed. One fractal creator called fractals a shape with the property of "self-similarity." See also Fractal Compression.

fractal compression An compression technique commonly used for color image files, fractal compression is well suited for images of natural objects. Fractal compression shrinks an image into extremely small resolution-independent files by storing it as a mathematical equation as opposed to storing it as pixels. The process starts with the identification of patterns within an image and results in a collection of shapes that resemble each other but that have different sizes and locations within an image. Each shape-pattern is summarized and reproduced by a formula that starts with the largest shape and repeatedly displaces and shrinks it. These patterns are stored as equations and the image is reconstructed by iterating the mathematical model. Depending on the specifics of the image file involved, fractal compression offers compression ratios as high as 100:1. As a result, it can store as many as 60,000 images on one CD-ROM. One disadvantage of fractal compression is that it is time consuming, taking as long as four minutes to convert a 1.3 MB TIFF file to a 228 KB file. As an asymmetric compression technique, however, it takes much less time to decompress the image than it does to compress it. As a lossy compression technique, it's not perfect – some data is lost in the process. See also Fractal.

fractal geometry The underlying mathematics behind fractal image compression, discovered by two Georgia Tech mathematicians, Michael Barneley and Alan Sloan.

Fractal Image Format FIF. A compression technique that uses on-board ASIC chips to look for patterns. Exact matches are rare and the process works on finding close matches using a function known as an affine map.

fractals Along with raster and vector graphics, fractals are a way of defining graphics in a computer. Fractal graphics translate the natural curves of an object into mathematical formulas, from which the image can later be constructed. See Fractal and Fractal Compression for a longer explanation.

fractional services A British term. Bandwidth available from carriers in increments of 64Kbit/s such as Mercury's Switchband. See Fractional T-1 for the North American definition.

Fractional T-1 FT-1. Fractional T-1 refers to any data transmission rate between 56/64 Kbps (DSO rate) and 1.544 Mbps (T-1). Fractional T-1 is a four-wire (two copper pairs) digital circuit that's not as fast as a T-1. Fractional T-1 is popular because it's typically provided by a LEC (Local Exchange Carrier) or IXC (IntereXchange Carrier) at less cost than a full T-1, and in support of applications that don't require the level of bandwidth provided by a full T-1. While FT-1 is less costly than a full T-1, it is more costly on a channel-by-channel basis, as you would expect. Users love FT-1, but carriers hate it. FT-1 costs the carriers just as much to provision as does as full T-1, they just turn down some of the channels. FT-1 is typically used for LAN interconnection, videoconferencing, high-speed mainframe connection and computer imaging.

Fractional T-3 A telephone company service in which portions of a T-3 (44.7364 Mbps) transmission service are leased to provide a service similar to a T-1 (1.544 Mbps) or T-2 (3.152 Mbps) channel, but normally at a lower cost.

FRAD Frame Relay Access Device, also sometimes referred to as a Frame Relay Assembler/Disassembler. Analogous to a PAD (Packet Assembler/Disassembler) in the X.25 world, a FRAD is responsible for framing data with header and trailer information prior to presentation of the frame to a Frame Relay switch. On the receiving end of the communication, the FRAD serves to strip away the Frame Relay control information in order that the target device is presented with the data packaged in its original form. On the receiving end, the FRAD also generally is responsible for detecting errors in the payload data created during the process of network switching and transmission; error correction generally is accomplished through a process of retransmission. A FRAD may be a standalone device, although the function generally is embedded in a router. In a Voice over Frame Relay (VoFR) implementation, the VFRADs (Voice FRADs) also are responsible for compression and decompression processes. See also Frame Relay, VoFR, and X.25.

frag Shooting or killing one of your opponents in a first-person shooter game, typically one conducted online.

fragging The term "fragging" was in widespread use during the Vietnam war and referred to the killing of officers by their own men using a fragmentation bomb such as a hand grenade.

fragment The pieces of a frame left on an FDDI ring, caused by a station stripping a frame from the ring.

fragmentation 1. In messaging it is the process in which an IP (Internet Protocol) datagram is broken into smaller pieces to fit the requirements of a given physical network. The reverse process is termed "reassembly."

2. ATM and SMDS networks routinely perform a process of Segmentation and Reassembly (SAR), segmenting the native PDU into 48-octet payloads which are carried in 53-octet cells. The process is reversed on the receiving end.

3. A condition that affects data stored on a hard disk. Adding and deleting records in a file, creates what is sometimes called the Swiss cheese effect. The operating system stores the data for an individual file in many different physical locations on the disk, leaving large holes between records. Fragmented files slow system performance because it takes the computer time to locate all parts of a file. Windows comes with software that defragments hard disks. It's not perfect software but it works. I use it regularly and notice a small increase in speed.

FRAM Ferro-electric Random Access Memory. A type of RAM that uses very little power, FRAM also retains data when powered off. Typical applications include small devices such as PDAs, phones, and smart cards. See also DDR-SDRAM, DRAM, EDO RAM, Flash RAM, Microprocessor, RAM, RDRAM, SDRAM, SRAM, and VRAM.

frame 1. A frame is a packet. It's a generic term specific to a number of data communications protocols. A frame of data is a logical unit of data, which commonly is a fragment of a much larger set of data, such as a file of text or image information. As the larger file is prepared for transmission, it is fragmented into smaller data units. Each fragment of data is packaged into a frame format, which comprises a header, payload, and trailer. The header prepends (prepend means added to the front of) the payload and includes a beginning flag, or set of framing bits, which are used for purposes of both frame delineation (beginning of the frame) and synchronization of the receiving device with the speed of transmission across the transmission link. Also included in the header are control information (frame number), and address information (e.g., originating and terminating addresses). Following the header is the payload, which is the data unit (fragment) being transmitted. Appending the payload is the trailer, which comprises data bits used for error detection and correction, and a final set of framing bits, or ending flag, for purposes of frame delineation (ending of the frame). This frame format, in the broader generic sense, also is known as a data packet. Frame, therefore, is a term specific to certain bit-oriented data transmission protocols such as SDLC (Synchronous Data Link Control) and HDLC (High-level Data Link Control), with the latter being a generic derivative of SDLC. In the case of SDLC, a frame is very similar to a block, which would be employed in a character-oriented protocol such as IBM's BSC (Binary Synchronous Communications), also known as Bisync. See also BSC, HDLC, Packet, and SDLC.

2. In TV video, a frame is a single, complete picture in video or film recording. A video frame consists of two interlaced fields of either 525 lines (NTSC) or 625 lines (PAL/SECAM), running at 30 frames per second (NTSC) or 25 frames per second (PAL/SEACAM). 24 frames are sent in moving picture films and a variable number, typically between 8 and 30, sent in videoconferencing systems, depending on the transmission bandwidth available. Up to about 12 frames a second looks "jerky."

3. One complete cycle of events in time division multiplexing. The frame usually includes a sequence of time slots for the various sub channels as well as extra bits for control, calibration, etc. T-Carrier makes use of such a framing convention for packaging data. Channelized T-1, for instance, frames 24 time slots with a framing bit which precedes each set of sampled data.

4. A unit of data in a Frame Relay environment. The frame includes a payload of variable length, plus header and trailer information specific to the operation of a Frame Relay network service.

5. A metal framework, such as a relay rack, on which equipment is mounted. A distribution frame. A rectangular steel bar framework having "verticals and horizontals" which is used to place semipermanent wire cross connections to permanent equipment. Found in telephone rooms and central offices. See Distribution Frame.

frame alignment The extent to which the frame of the receiving equipment is correctly phased (synchronized) with respect to that of the received signal.

frame alignment errors A frame alignment error occurs when a packet is received but not properly framed (that is, not a multiple of 8 bits).

frame alignment sequence See Frame Alignment Signal.

Frame Alignment Signal FAS. Frame Alignment Signal or Frame Alignment Sequence.

The distinctive signal inserted in every frame or once in n frames that always occupies the same relative position within the frame and is used to establish and maintain frame alignment, i.e. synchronization. See Frame Alignment Errors.

frame buffer A section of memory used to store an image to be displayed on screen as well as parts of the image that lie outside the limits of the display. Some systems have frame buffers that will hold several frames, in which case they should be called "frames buffers." But they're not.

frame check sequence Bits added to the end of a frame for error detection. Similar to a block check character (BCC). In bit-oriented protocols, a frame check sequence is a 16-bit field added to the end of a frame that contains transmission error-checking information. In a token ring LAN, the FCS is a 32-bit field which follows the data field in every token ring packet. This field contains a value which is calculated by the source computer. The receiving computer performs the same calculation. If the receiving computer's calculation does not match the result sent by the source computer, the packet is judged corrupt and discarded. An FCS calculation is made for each packet. This calculation is done by plugging the numbers (1's and 0's) from three fields in the packet (destination address, source address, and data) into a polynomial equation. The result is a 32-bit number (again 1's and 0's) that can be checked at the destination computer. This corruption detection method is accurate to one packet in 4 billion. See Frame Check Sequence Errors.

frame check sequence errors Errors that occur when a packet is involved in a collision or is corrupted by noise.

frame dropping The process of dropping video frames to accommodate the transmission speed available.

frame duration The sum of all the unit time intervals of a frame. The time from the start of one frame until the start of the next frame.

Frame DS1 The DS1 frame comprises 193 bit positions. The first bit is the frame overhead bit, while the remaining 192 bits are available for data (payload) and are divided into 24 blocks (channels) of 8 bits each.

frame error An invalid frame identified by the Frame Check Sum (FCS). See also Frame Errors.

Frame Error Rate FER. The ratio of errored data frames to the total number of frames transmitted. If the FER gets too high, it might be worth while stepping down to slower baud rate. Otherwise, you would spend more time retransmitting bad frames than getting good ones through. In other words, throughput would suffer. The theory is that the faster the speed of data transmission the more likelihood of error. This is not always so. But if you are getting lots of errors, the first – and easiest – step is to drop the transmission speed. Frame Error Rate is thus a measure of transmission quality. It is generally shown as a negative exponent, (e.g., 10 to the minus 2 power (10^{-2}) means one out of 100 frames are in error.) The FER is directly related to the Bit Error Rate (BER). See also Bit Error Rate.

frame errors In the 12-bit, D4 frame word, an error is counted when the 12-bit frame word received does not conform to the standard 12-bit frame word pattern.

frame flag sequence The unique bit pattern "01111110" used as the opening and closing delimiter for the link layer frames.

frame frequency A video term. The number of times per second a frame is scanned.

frame grab To capture a video frame and temporarily store it for later manipulation by a graphics input device.

frame grabber A PC board used to capture and digitize a single frame of NTSC video and store it on a hard disk. Also known as Frame Storer. See Video Capture Board.

frame ground FGD. Frame Ground is connected to the equipment chassis and thus provides a protective ground. Frame Ground is usually connected to an external ground such as the ground pin of an AC power plug.

frame header Address information required for transmission of a packet across a communications link.

frame length X.25 packets are fixed in length. ATM cells are fixed in length. Frame Relay frames (packets in the generic sense) are variable in length, which is due to their intended use for LAN internetworking. LAN frames (packets in the generic sense) are variable in length.

frame multiplexing The process of handling traffic from multiple simultaneous inputs by sending the frames out one at a time in accordance with a specific set of rules. Instead of multiplexing traffic from a lower-speed connection into a higher speed connection based on a specific time duration for each low-speed channel, frame multiplexing using the length of a given frame as the measurement.

frame rate The number of images displayed per second in a video or animation file. The Frame Rate is highly significant in determining the quality of the image, with a high frame rate creating the illusion of full fluidity of motion. 30 frames per second (30 fps) is considered to be full-motion, broadcast quality. On the other end of the scale, 2fps is most annoying. At 30 fps, the brain processes the images, filling in the blanks due to the "Phi Phenomenon." See PHI Phenomenon.

frame relay Frame relay, technically speaking, is an access standard defined by the ITU-T in the I.122 recommendation, "Framework for Providing Additional Packet Mode Bearer Services." Frame relay services, as delivered by the telecommunications carriers, employ a form of packet switching analogous to a streamlined version of X.25 networks. The packets are in the form of "frames," which are variable in length, with the payload being anywhere between 0 and 4,096 octets. The key advantage to this approach is that a frame relay network can accommodate data packets of various sizes associated with virtually any native data protocol. In other words, a X.25 packet of 128 bytes or 256 bytes can be switched and transported over the network just as can an Ethernet frame of 1,500 bytes. The native Protocol Data Unit (PDU) is encapsulated in a Frame Relay frame, which involves header and trailer information specific to the operation of the Frame Relay network.

Further, a Frame Relay network is completely protocol independent. Not only can any set of data be accepted, switched and transported across the network, but the specific control data associated with the payload is undisturbed in the process of encapsulation. Additionally, and unlike an X.25 network, a Frame Relay network assumes no responsibility for protocol conversion; rather, such conversions are the responsibility of the user. While this may seem like a step down from X.25, the data neither require segmentation into fixed-length packets nor does the network have to undertake processor-intensive and time-consuming protocol conversion. The yield is faster and less expensive switching.

A Frame Relay network also assumes no responsibility for errors created in the processes of transport and switching. Rather, the user also must accept full responsibility for the detection and correction of such errors. The user also must accept responsibility for the detection of lost packets (frames), as well for the recovery of them through retransmission. Again, this may seem like a step down from X.25 networks, which correct for errors at each network node, and which detect and recover from lost packets. Once again, however, the yield is faster and less expensive switching. In fact, it is unlikely that frames will be damaged, as the switches and transmission facilities are fully digital and offer excellent error performance.

Much like X.25, Frame Relay employs the concept of a shared network. In other words, the network switches accept frames of data, buffer them as required, read the target address and forward them one-by-one as the next transmission link becomes available. In this fashion, the efficiency of transmission bandwidth is maximized, yielding much improved cost of service. The downside is that some level of congestion is ensured during times of peak usage. The level of congestion will vary from time-to-time and frame-to-frame, resulting in latency (delay) which is unpredictable and variable in length. This is especially true in a Frame Relay network, as the length of the frames is variable – the switches never quite know what to expect.

Access to a Frame Relay is over a dedicated, digital circuit which typically is 56/64 Kbps, Nx56/64 Kbps, T-1 or T-3. The device which interfaces the user to the network is in the form of a Frame Relay Access Device (FRAD) which serves to encapsulate the native PDU before presenting it to the network. The FRAD at the destination address unframes the data before presenting it to the target device, with the two FRADs working together much as do PADs in a X.25 environment. Further, it generally is the responsibility of the FRAD to accomplish the error detection and correction process, although this responsibility may be that of the eventual target device. Across the digital local loop, the FRADs connect functionally to Frame Relay Network Devices (FRNDs, pronounced "friends"), proving once again that the carriers want to be your friends (especially as Frame Relay users tend to be large organizations with lots of $$$ to spend).

Frame Relay is intended for data communications applications, most especially LAN-to-LAN internetworking, which is bursty in nature. Frame Relay is very good at efficiently handling high-speed, bursty data over wide area networks. It offers lower costs and higher performance for those applications in contrast to the traditional point-to-point services (leased lines). Additionally, Frame Relay offers a highly cost-effective alternative to meshed private line networks. As the Frame Relay network is a shared, switched network, there is no need for dedicated private lines, although special-purpose local loops connect each customer location to a frame switch.

Frame Relay is a connection-oriented protocol, as transmission of frames between the user sites generally is on the basis of Permanent Virtual Circuits (PVCs), which are pre-determined paths specifically defined in the Frame Relay routing logic. All frames transmitted between any two sites always follow the same PVC path, ensuring that the frames will not arrive out of sequence. Backup PVCs, generally offered by the carrier at trivial cost, provide redundancy and, therefore, network resiliency in the event of a catastrophic network failure. Switched Virtual Circuits (SVCs) are VCs selected on a call-by-call basis. SVCs are advantageous as they result in automatic load balancing, which improves overall service. As SVC logic is more costly, however, not all carriers offer it; those that do impose a surcharge on SVC users.

With Frame Relay, a pool of bandwidth is made instantly available to any of the concurrent data sessions sharing the access circuit whenever a burst of data occurs. An addressed frame is sent into the network, which in turn interprets the address and sends the information to its destination over broadband facilities. Those facilities may be as "slow" as 45 Mbps, but more often are SONET fiber optics in nature and operating at much higher speeds. Like traditional X.25 packet networks, frame relay networks use bandwidth only when there is traffic to send.

Frame Relay, while intended for data communications, also supports compressed and packetized voice and video. While such isochronous data is highly sensitive to the variable latency characteristic of packet networks, improved voice compression algorithms such as ACELP provide quite acceptable support for Voice over Frame Relay (VoFR), subject to the level of congestion in the network. For voice to be supported satisfactorily in a packet network, the receiving end compensates for delay and delay variation.

In addition to public network services, Frame Relay can also be implemented in a private network environment consisting of unchannelized T-Carrier circuits. Such an implementation offers exceptional data communications performance over an existing leased line network. Additionally, framed voice and video can ride over such a network, essentially for "free" when the circuits are not being used for data communications purposes. Thereby, the usage of the circuits is maximized, with little concern for poor quality due to network congestion.

A Frame Relay frame consists of a header, information field, and trailer. The header comprises a Flag denoting the beginning of the frame, and an Address Field used for routing of the frame, as well as for purposes of congestion notification. The Information Field is of variable length, from 0 to 4,096 Bytes. The trailer consists of a Frame Check Sequence (FCS) for detection and correction of errors in the Address Field, and an ending Flag denoting the end of the frame.

The American National Standards Institute (ANSI) describes frame relay service in the following documents:

- ANSI T1.602 – Telecommunications – ISDN – Data Link Layer Signaling Specification for Application at the User Network Interface.
- ANSI T1.606 – Frame Relaying Bearer Service – Architectural Framework and Service Description.
- ANSI T1S1/90 - 175 - Addendum to T1.606 - Frame Relaying Bearer Service – Architectural Framework and Service Description.
- T1.607-1990 ISDN Layer 3 Signaling Specification for Circuit-Switched Bearer Service for DSS-1.
- T1.618 DSS-1 Core aspects of Frame Protocol for use with frame relay bearer service, ANSI, 1991.
- ANSI T1.617a, Signaling specification for Frame Relay bearer service for DSS-1, 1994.

Frame relay access makes use of the LAP-D signaling protocol developed for ISDN. Frame relay, technically speaking again, does not address the operation of the network switches, multiplexers or other elements. Both the ITU-T and ANSI were highly active in the development of Frame Relay standards, as was ETSI in Europe. See the next three definitions and also LAP-D, FRAD, FRND, PVC, SVC, VoFR, and X.25.

Frame Relay Access Device See FRAD.

frame relay forum The frame relay forum is an organization of manufacturers, carriers, end users, and consultants committed to the implementation of Frame Relay in accordance with both national and international standards. The forum's technical committee begins with formal standards, from which it creates Implementation Agreements (IAs). IAs are formal agreements of all members with respect to the specific manner in which the standards will be applied, thereby helping to ensure interoperability of systems and networks.

The forum began in 1990 as the Frame Relay Working Group, which was formed by the "Gang of Four," comprising Cisco, Digital Equipment Corporation, Nortel, and Stratacom (since acquired by Cisco). In 1991, the Frame Relay Forum was officially formed, and has grown to over 300 members. See also Frame Relay. www.frforum.com.

frame relay implementors forum See Frame Relay Forum.

frame relay modem A data communications device which connects to a PC's COM (serial) port and emulates a dial tone while actually establishing a dedicated 56Kbps frame relay connection.

frame slip That condition in a TDM network under which a receiver of a digital signal experiences starvation or overflow in its receive buffer due to a small difference in the speeds of clocks and the clock (transmission rate) at the transmitter. The receiver will drop or repeat of a full TDM frame (193 bits on a T-1 line) in order to maintain synchronization.

frame store A system capable of storing complete frames of video information in digital form. This system is used for television standards conversion, computer applications incorporating graphics, video walls and video production and editing systems.

frame switch A device similar to a bridge that forwards frames based on the frames' layer 2 address. Frame switches are generally of two basic forms, cut-through switch (on-the-fly-switching) or store and forward switch. LAN switches such as Ethernet, Token Ring, and FDDI switches are all examples of frame switches.

frame synchronization The process whereby a given digital channel (time slot) at the receiving end is aligned with the corresponding channel (time slot) of the transmitting end as it occurs in the received signal. Usually extra bits (frame synchronization bits) are inserted at regular intervals to indicate the beginning of a frame and for use in frame synchronization.

Frame UNI Frame-based User-Network Interface, a frame format for access to ATM networks. Defined by the Frame Relay Forum, Frame UNI is a derivative of the DXI standard. For low-speed access application, it provides for a router to send frames (much like Frame Relay frames) to an ATM Edge Switch, where the conversion to cell format takes place.

frames A term used to describe a viewing and layout style of a World Wide Web site, it refers to the simultaneous loading of two or more web pages at the same time within the same screen. Originally developed by Netscape and implemented in their Navigator browser, today many other popular Web browsers support this feature. Some Web sites come in two versions; a "frames" and "no frames" version. The frames version usually takes a little longer to load and may contain other "enhanced" features such as Java and Animation. Frames are now falling into disrepute among web site designers and are being replaced with extensive use of tables, which are a much easier way to design web pages and also load a lot faster.

frames received OK The number of frames received without error. See Frames Received Too Long.

frames too long An Ethernet statistic that indicates the number of frames that are longer than the maximum length of a proper Ethernet frame, but not as long as frames resulting from jabbering.

framework A Taligent definition. A set of prefabricated software building blocks that programmers can use, extend, or customize for specific computing solutions. With frameworks, software developers don't have to start from scratch each time they write an application. Frameworks are built from a collection of objects, so both the design and code of a framework may be reused.

framing An error control procedure with multiplexed digital channels, such as T-1, where bits are inserted so that the receiver can identify the time slots that are allocated to each subchannel. Framing bits may also carry alarm signals indicating specific alarms. In TDM reception, framing is the process of adjusting the timing of the receiver to coincide with that of the received framing signals. In video reception, the process of adjusting the timing of the receiving to coincide with the received video sync pulse. In facsimile the adjustment of the facsimile picture to a desired position in the direction of line progression.

framing bit 1. A bit used for frame synchronization purposes. A bit at a specific interval in a bit stream used in determining the beginning or end of a frame. Framing bits are non-information-carrying bits used to make possible the separation of characters in a bit stream into lines, paragraphs, pages, channels etc. Framing in a digital signal is usually repetitive.

framing error An error occurring when a receiver improperly interprets the set of bits within a frame.

franchise The exclusive right to operate telephone service in a community. This right – also called the franchise – is granted by some government agency. Some phone companies existed before the appropriate regulatory authority, so they're "grandfathered" in their exclusivity. Some phone companies have an exclusive area to serve more because of their presence than because of the legal right conferred on them. The question of who has a franchise to serve what community with what service is becoming increasingly unclear as competition penetrates all aspects of the phone industry.

franchise authority The franchise authority is the local body, usually a local government entity, that enters into a contractual agreement with a cable company. A franchise agreement defines the rights and responsibilities of each in the construction and operation of a cable system within a specified geographical area. The reason for the form agreement is that the cable company is typically a monopoly provider of cable TV. In return for grating that monopoly, the local government body typically extracts a "price" for granting that monopoly. It might be the existent of several community channels, available for free to the community or it may be free broadband circuits linking up the town's police, fire, schools and city offices.

franchising authority See franchise authority.

FRD Fire RetarDant. A rating used for cable within duPont's Teflon or equivalent fluorpolymer material. FRD cable is used when local fire codes call for low flame and low smoke cable. FRD cable is typically run in forced air plenums as an alternative to metal conduits.

FRED A system for searching the international X.500 user directory.

free address office Office arrangement in which all personal spaces are eliminated in favor of employees picking up their supplies at a front desk upon arriving and then choosing a temporary work area each day.

Free Lossless Audio Codec See FLAC.

Free On Board FOB. Board. Term indicating where the seller's responsibility ends and the buyer's begins. You buy something, F.O.B. The seller puts it on a truck or railroad, plane, i.e. some carrier. He's responsible for getting it on the carrier. It's FOB, the truck. You – the buyer – are responsible for paying for the cost of the freight of getting you the goods you ordered. The term FOB is typically enhanced (or made clear), thus:

FOB Destination: Seller retains ownership until delivered to buyer.

FOB Place of Delivery. Seller retains ownership until delivered to buyer.

FOB Shipping Point. Seller responsibility ends when item is turned over to carrier. The buyer is responsible for payment if goods are damaged in transit. The buyer also handles any insurance claim.

The opposite of F.O.B. is C.I.F. That stands for Cost, Insurance and Freight are included. That means the seller pays the freight.

free range workspaces Office space available to anyone in the company on a first-come, first-served basis – no registration required.

free space communications Any form of telecommunications that doesn't use a conductor (e.g., copper wire, or glass or plastic fiber). In other words, free space communications is accomplished using "space," rather than a conductor, as a medium. In simple terms, the system works through the air. Radio (e.g., radio, microwave, satellite, wireless LANs, cellular) and optical (i.e., infrared) transmission systems communicate through space, rather than through a conductor. See also Airwave and Free Space Loss.

free space loss This is simply the power loss of the signal as a result of the signal spreading out as it travels through space. As a wave travels, it spreads out its power over space, i.e. as the wave front spreads, so does its power. Sometimes called "spreading loss" to distinguish it from other losses which occur when radio waves pass through various gasses or material.

free space optics Free space optics is fancy way of saying that you send your telecom signal through the air using infrared (Ir) frequencies. Such transmission is line of sight and can run at speeds of, typically, up to 10 gigabits per second, depending on the encoding scheme chosen. You can send and receive voice, video, images and data. Free space optical systems are often installed where wired systems are not easily installed – e.g. shooting from one side of the highway to the other, or across a lake. Free space data transceivers:

Requires no RF spectrum licensing from the FCC or local municipal authority.

Support equipment from a variety of vendors, which helps enterprises and service providers protect their investment in embedded telecommunications infrastructures.

Need no security software upgrades.

Are pretty well immune to radio frequency interference or saturation. (They're operating on a different frequency.)

Can be deployed behind windows, eliminating the need for costly rooftop rights.

The primary challenge to free space optics is physical objects (e.g. birds) or dense fog. Rain and snow have little effect. See also Infrared, Laser, and WLL.

free spectral range A measurement used to ensure components in optical systems such as connectors, splices, fiber ends and bulk optic interfaces do not reflect too much light back to the transmitter, causing the modulation characteristics and spectrum of the laser transmitter to change. Free spectral range is the period variation of the return loss of a device versus wavelength. See Return Loss.

freedom Freedom is when the last kid graduates from college and the dog dies.

freedom link Freedom Link is a type of cellular service that can be purchased by organizations to allow employees to communicate with one another via a closed-circuit system, similar to a wireless PBX. Nextel's Direct Connect is one such service.

Freedom of Mobile Multimedia Access See FOMA.

FREENET An organization to provide free Internet access to people in a certain area, usually through public libraries.

freephone A service which permits the cost of the call to be charged to the called party, rather than the calling party. Freephone was pioneered in the US in 1966, where it is known generically as In-WATS (Incoming Wide Area Telecommunications Service), and makes use of the 800, 877 and 888 area codes, listed in order of introduction. Freephone service also is known as Greenphone or Freecall in some countries; outside the US, the dialing prefix can be 0800, 0500, or some similar string of preceding digits.

Freephone has proven particularly popular with business subscribers, who are often willing to bear the cost of a telephone call in order to promote their services or to encourage customers to order their products by phone. Prior to the relatively recent approvals by the ITU-T of Recommendations E.169 and E.152, however, companies have been restricted to using their Freephone number in one country as the dialing patterns differ from country-to-country. Therefore, organizations wishing to offer products or services to customers on an international basis have had no choice but to register a separate International Toll Free Service (IFTS) number in each country, which has proved unwieldy and often inefficient. The new standard for Universal International Freephone Numbers (UIFNs) will encourage the development of international Freephone service through a standard dialing plan which can work across international borders. This will greatly free up companies' abilities to operate across international markets, benefiting consumers by allowing them to obtain information or to shop around for goods and services at no personal expense. It is hoped the new standard might also stimulate the market for Freephone services in Europe and Asia-Pacific, regions that until now have been slow to take up the service.

The potential market for the UIFN service is expected to be considerable. The globalization of markets via new technologies such as the Internet means that many companies are now able to offer their products and services to users in different countries, and will benefit from being able to advertise a single toll-free number to potential customers all over the world. Calls to the new global market number can also be routed to different destinations, allowing companies to direct their incoming calls to the most appropriate location for efficient handling. See also 800 Service, InWATS, International Toll Free Service, and Universal International Freephone Number.

free space communications Radio communications including microwave, satellite and cellular.

freeware Software that doesn't cost anything, but may work just as well as the software you pay for.

freeze 1. No more price hikes for three years. That a three year freeze on prices.

2. In digital picture manipulators, the ability to stop or hold a frame of video so that the picture is frozen like a snapshot.

freeze frame The transmission of discrete video picture frames at a data rate which is too slow to provide the perception of natural motion, referred to as "full-motion." An uncompressed, digitized full-motion video signal is typically transmitted at many millions of bits per second. Freeze frame can be carried on anything from a simple voice grade phone line running at 9.6 Kbps (the same speed as a Group 3 facsimile machine).

french braid A French braid shield is a type of shield used in coaxial cable systems comprising two serves (foil shields wound around the inner conductor) braided along one axis. It looks much like a French braid hairdo made of metal.

French, the Regardless of what has been written, Napoleon Bonaparte was not short. For his time, he was of average height. Interestingly enough, he liked to surround himself with tall men, and he imposed a minimum height limit for recruits for the French Army. After the military disasters in Russia and at Waterloo, so many tall Frenchmen were killed that the genetic signature for tall- ness almost disappeared from France's gene pool. For well over the following century the average height of the French male was below that of the average European.

freq A term or abbreviation for a "file request" for a file from another node in a network. In FidoNet a node user usually freq's a file through mailer software which sends an appropriate request to a distant node that has the desired file. Freqing is the ability in FidoNet to transfer files back and forth between BBSs (bulletin board systems) automatically. Equivalent to file transfer in PCRelay.

frequency Frequency defines number of events during a time period. Frequency is also the rate at which an electromagnetic waveform (e.g., electrical current) alternates, usually measured in Hertz (HZ). Hertz is a unit of measure which means "cycles per second." So, frequency equals the number of complete cycles of energy (e.g., current) occurring in one second. See Bandwidth, Frequency and Hertz.

frequency agile modem A modem used on some broadband LANs (Local Area Networks. A frequency agile model can search the frequencies on the LAN to find one available in order to communicate with other attached devices.

frequency agility The ability of a cellular mobile telephone system to shift automatically between frequencies.

frequency band The portion of the electromagnetic spectrum within a specified upper- and lower-frequency limit. Also known as Frequency Range. See also Band, Frequency for a complete list of all the frequencies.

frequency combiner A diplexer or frequency splitter used backwards or in reverse.

frequency deviation The extent to which an FM carrier moves from its center frequency during modulation.

frequency diversity A way of protecting a radio signal by providing a second, continuously operating radio signal on a different frequency, which will assume the load when the regular channel fails. Here's another way of saying the same thing: Frequency diversity is a any method of diversity transmission and reception wherein the same information signal is transmitted and received simultaneously on two or more independently fading carrier frequencies.

frequency division multiple access A technique for sharing a single transmission channel (such as a satellite transponder) among two or more users by assigning each to an exclusive frequency band within the channel.

Frequency Division Multiplexing FDM. An older technique in which the available transmission bandwidth of a circuit is divided by frequency into narrower bands, each used for a separate voice or data transmission channel. This means you can carry many conversations on one circuit.

frequency domain Waveforms, such as speech signals, are typically viewed in the time domain, i.e. as power levels or voltages varying over time. The 19th century French mathematician Fourier demonstrated an algorithm called "Fast Fourier Transform", or FFT, which can express any complex waveform over a fixed interval as the sum of a series of sine waves of different energy levels. Analyzing signals in the frequency domain has proven an extremely powerful technique with diverse applications, including filtering, recognition and speech modeling.

frequency drift The extent to which the unmodulated center frequency of an AM or FM carrier deviates from an FCC assigned frequency.

frequency frogging The interchanging of the frequency allocations of carrier channels to prevent singing, reduce crosstalk, and to correct for a transmission line frequency-response slope. It is accomplished by having the modulators in a repeater translate a low-frequency group to a high-frequency group, and vice versa. Because of this frequency inversion process, a channel will appear in the low group for one repeater section and will then be translated to the high group for the next section. This results in nearly constant attenuation with frequency over two successive repeater sections, and eliminates the need for large slope equalization and adjustment. Also, singing and crosstalk are minimized because the high-level output of a repeater is at a different frequency from the low-level input to other repeaters.

frequency grease A special kind of radio lubricant that is used to overcome problems of static in radio transmissions. Actually, there is no such thing, but every new radio technician falls prey to the joke. It's much like a "pot stretcher." Ray Horak, my Contributing Editor, was a Mess Sergeant in the US Army. He would send the privates on KP (Kitchen Patrol) to another mess hall to get a pot stretcher if the pot was too small, or to get a screen door for the refrigerator during the summer. His buddies in Communications would send the new radio technicians to get some radio grease. It was a lot of fun during the Vietnam War, which was not a lot of fun. It worked only one time per private (usually). See also Bucket o' Dial Tone.

frequency hopping Another name for spread spectrum transmission. A technique developed by Hedy Lamarr, the actress, in the early part of the second world war to prevent the enemy from jamming or eavesdropping on conversations and on commands to steer torpedoes, etc. The idea is to hop from one frequency to another in split-second intervals as you transmit information. Attempts to jam the signal succeed only in knocking out a few small bits of it. So effective is the concept that it is now the principal antijamming device in the US military. Ms. Lamarr never got paid for the invention. But it was definitely hers. She invented it because of her patriotism for the United States. She had fled Austria in 1937. She received a U.S. patent in 1940.

Frequency hopping is used by cell phones for example to make it difficult to listen in to calls. The voice signal jumps around (hops) between frequencies in a given range at some set rate. This rate is, in theory, too rapid to tune in to before the next hop occurs. The range in which your piece of equipment operates, for example, is 2400-2438.5 MHz. These discrete ranges are assigned and controlled by the FCC so that things like your TV reception are not interfered with when the phone rings. A 20 db bandwidth implies that all the useful signal is contained in that area. In other words envision a standard bell shaped curve like that used in grading. Pretend there is a value on either side where the area under the curve is so small that you do not want to consider it and call those points the +20 db and -20 db points. The 20 db represents a ratio of signal that has become so weak so as not to be useful. If the distance from the +20 db point to the -20 db point exceeds 1 MHz then you are allowed to channel hop. With the caveat that at least 15 of the channels that you have designated as you hopping channels do not overlap. If they did, signal tracking is easier and interference becomes more probable. See also Frequency Hopping Spread Spectrum and Spread Spectrum.

Frequency Hopping Spread Spectrum FHSS technology uses a narrow band carrier to broadcast packets that "hop" from frequency to frequency in a pattern synchronized between the transmitter and the receiver, thus maintaining a single logical channel. In an FHSS system, the total frequency band is divided into a number of fractional hop channels, spreading the broadcast traffic across the entire spectrum band, serving to reduce interference and to increase security. Complex algorithms determine the order in which data is sent and received. See also Frequency Hopping.

frequency modulation A modulation technique in which the carrier frequency is shifted by an amount proportional to the value of the modulating signal. The amplitude of the carrier signals remains constant. The deviation of the carrier frequency determines the signal content of the message. Commercial TV and FM radio use this technique, which is much less sensitive to noise and interference than is amplitude modulation (AM). In the world of modems, digital bit streams can be transmitted over analog facilities through this same technique, whereby a 0 bit might be represented by a high-frequency sine wave (or set of sine waves) and a 1 bit by a low-frequency sine wave (or set of sine waves). Contrast with Amplitude Modulation and Phase Shift Keying.

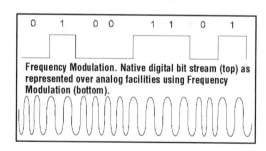

Frequency Modulation. Native digital bit stream (top) as represented over analog facilities using Frequency Modulation (bottom).

frequency offset Non-linear distortion that causes a shift in the frequency of a received signal.

frequency response The variation (dB) in relative strength between frequencies in a given frequency band, usually the voice frequency band of an analog telephone line.

frequency reuse The ability to use the same frequencies repeatedly within a single system, made possible by the basic design approach used in cellular. Since each cell is designed to use radio frequencies only within its boundaries, the same frequencies can be reused in other cells not far away with little potential for interference. The reuse of frequencies is what allows a cellular system to handle a huge number of calls with a limited number of channels.

Frequency Shift Keying FSK. A modulation technique for data transmission. See FSK for a full explanation.

frequency splitter A device which separates an input band into two output bands.

Devices with more than two outputs are also possible. A frequency splitter is a frequency combiner or diplexer used backwards or in reverse. It takes one band and divides it into two independent bands while suppressing alternate bands.

frequency standard Generally a special receiver which receives synchronized signals transmitted by the National Bureau of Standards.

frequency tolerance The maximum permissible departure by the center frequency of the band occupied by an emission from the assigned frequency or by the characteristic frequency of an emission from the reference frequency. By international agreement, frequency tolerance is expressed in parts per 10 (6) or in hertz. This includes both the initial setting tolerance and excursions related to short- and long-term instability and aging. In the United States, frequency tolerance is expressed in parts per 10(n), in hertz, or in percentages.

frequency translator In a split broadband cable system, a frequency translator is an analog device at the headend that converts a block of inbound frequencies to a block of outbound frequencies.

Fresnel loss The loss at a joint that is caused by a portion of the light being reflected.

Fresnel reflection In optical physics, fresnel reflection is the reflection of a portion of incident light at a planar interface between two homogeneous media having different reflective indices. Fresnel reflection occurs at the air-glass interfaces at the entrance and exit ends of an optical fiber. Resultant transmission losses (on the order of 4 percent per interface) can be virtually eliminated by using antireflection coatings or index-matching materials. Fresnel reflection depends upon the index difference and the angle of incidence. In optical elements, a thin transparent film is sometimes used to give an additional Fresnel reflection that cancels the original one by interference. This is called an antireflection coating.

Fresnel reflective losses For optical fiber communication, the losses incurred at the terminus interface that are due to refractive index differences. See previous definition.

Fresnel region In radio communications, the region between the near field of an antenna and the Fraunhofer region. The boundary between the two is generally considered to be at a radius equal to twice the square of antenna length divided by wavelength.

Fresnel zone Fresnel zone is the line-of-sight path between two microwave antennas. It is an elliptical zone between the two antennas where the total path distance varies by more than half of the operating wavelength. The concept is extended to describe the distance by which the direct wave clears any intervening obstacle such as a mountain peak. If the total path distance between transmitter, peak and receiver, is 1 wavelength greater than the direct distance, then the clearance is said to be two Fresnel zones.

FRF11 Basically, there are two main standards regarding voice transmission over data networks: H.323 and "Voice Over Frame Relay Implementation Agreement" (FRF.11). Both specify that the following coders should be used: G.711, G.728, G.729, and G.723.1. The H.323 adds the G.722, and the VoFR (Voice over Frame Relay) adds the G.726/7 coders. The G.711 is a PCM coder that uses 64ks/s and two companding techniques: A-law and Mu-law. Recommendation G.722 describes 7 kHz audio-coding within 64 kbit/s. The G.723.1 describes a Dual rate speech coder for multimedia communications transmitting at 5.3 and 6.3 kbit/s and is based on Multi Pulse Maximum Likelihood Quantizer (MP-MLQ) (Voice frame duration of 30mSec). The G.726 describes 40, 32, 24, 16 kbit/s Adaptive Differential Pulse Code Modulation (ADPCM). The G.727 describes 5-, 4-, 3- and 2-bits sample embedded adaptive differential pulse code modulation (ADPCM). The G.728 describes Coding of speech at 16 kbit/s using Low-Delay Code Excited Linear Prediction (LD CELP). The G.729 describes Coding of speech at 8 kbit/s using Conjugate-Structure Algebraic-Code- Excited Linear-Prediction (CS-ACELP) (Voice frame duration of 10mSec).

FRF92.02 Multiprotocol Interconnect over Frame Relay.

FRF92.07 Frame Relay Multicast Draft Service Description.

FRF92.08 Frame Relay Network-to-Network Interface Implementation Agreement Draft.

frictional electricity Static electricity produced by friction (e.g., by rubbing a hard rubber rod with a silk cloth.)

Friday In the 19th century, the British Navy attempted to dispel the superstition that Friday is an unlucky day to embark on a ship. The keel of a new ship was laid on a Friday, she was christened HMS Friday, commanded by a Captain Friday, and finally went to sea on a Friday. Neither ship or crew were ever heard of again.

friendly name A name, typically identifying a network user or a device, intended to be familiar, meaningful, and easily identifiable. A friendly name for a printer might indicate the printer's physical location (e.g. "Sales Department Printer").

friendly users When a network operator tests a new service, it will often test the service initially with "friendly users," i.e., employees and/or existing customers who can be counted on to be patient and helpful during the test.

Friesen, Gerry One of the smartest people you'll meet in a long time. Gerry was, and remains my partner in most everything business I do. Originally, he was half the operation which published this dictionary (I was the other half). Now CMP Freeman publishes the dictionary, though I still write it. And Gerry has retired to his own personal paradise. On January 1, 1999 he sent me the following email: HAPPY NEW YEAR PARTNER!!! It's been a pleasure working with you over the years. I hope 1999 brings you everything you hope for. My New Year's resolution is to do something fun every day. Hope yours is as rewarding, whatever it is, as I plan mine to be."

fringe benefit Generally this term refers to a reward received in addition to one's wage, such as a car, travel allowance, laptop computer. However this term can also refer to strips of toilet paper the ever-altruistic editor of this fine dictionary uses to bookmark reference material with and bestows upon lowly copy clerk, aspiring lexicographers visiting from Australia, such as the said Gavin Wedell.

fritterware At the dawn of the personal computer age, the writer Stephen Manes introduced the term "fritterware" to describe programs and systems that let you feel very busy – adjusting fonts and settings, tweaking color schemes or screen-saver graphics – without being productive in any normal sense.

FRL Facility Restriction Level. A term created by AT&T for its Dimension PBX. These levels define the calling privileges associated with a line; for example, intragroup calling only in the warehouse, but unrestricted calling from the boardroom.

FRND Frame Relay Network Device. Pronounced "friend." A device that sits at the edge of a public frame relay network. The FRND is the point of ingress into the cloud of the network on the inbound side of the network and the point of egress on the outbound side. The FRND connects to the user's FRAD (Frame Relay Access Device), which often is in the form of a router, over a digital access link, which usually is some form of T-carrier circuit (e.g., T-3, T-1, or Fractional T-1). A FRND can be in the form of either a switch or a router, although it usually is a router. See Frame Relay, FRAD and T-1.

frog 1. Frog means to switch pairs. e.g., there are two trunks connected to your PBX (trunks "A" and "B") Static is reported on trunk "B". A technician would "frog" the cable pairs to see if the problem was a PBX issue or a telephone company issue. If trunk "B" still has static when being connected to the PBX port that "A" was originally connected to, the problem would be isolated to the telephone company. If the static on trunk "B" goes away after being connected to the PBX port that trunk "A" was originally connected to, the problem would be the PBX and trunk "A" (which would now be connected to the PBX port that trunk "B" was originally connected to) would have static. Basically, it's a way of troubleshooting to see if the problem is a PBX problem or a telco problem.

2. Regardless of what Budweiser would have you believe, frogs never drink. They absorb water from their surroundings by osmosis.

frogging Frogging is the process of inverting line frequencies of a carrier system so that incoming high-frequency channels leave at low frequencies and vice versa. Frogging equalizes the transmission loss between high and low frequency channels.

front-end networking Method for national and regional paging in which a network intercepts paging messages before they reach facilities of an RCC (Radio Common Carrier) or other conventional paging service provider. The network then routes messages to the appropriate transmitting facilities, either locally or in various distant locations.

front-to-back ratio A measure to compare the main lobe of an antenna in relation to the next strongest lobe of an antenna.

front end The client part of a client/server application that requests services across a network from a server, which is known sometimes as the back end. It typically provides an interactive interface to the user, for example, a data entry front end, allowing database to be entered into a database server. The term "front end" is, of course, a contradiction in terms. But, who said language had to be consistent, or logical? See back End, Front End Controller, Front End Development Tools and Front End Processor.

front end controller See Front End Processor.

front end development tools These tools let a programmer control the design and manipulation of applications using visual techniques. Front-end development creates a graphical user interface, providing more flexibility and making it easier to link users with data accessed from database servers.

front end equipment The equipment positioned between a computer and the communications line(s). Its purpose is to organize data being sent and received.

front end mailer A program that operates on a bulletin board system and determines if a caller is another computer that wants to exchange mail or a human that wants to exchange mail or a human that wants to access the BBS resources. Usually the mailer transmits the prompt "Press ESC" and upon receiving an ESC character, or the passing of a timeout period, considers the caller to be human and gives it the resources of the BBS. Also known as a mailer.

Front End Processor FEP. An FEP is a computer under the control of another, larger computer (typically a mainframe) in a network. The FEP does simple, basic "house-keeping" operations on the data streams as they arrive to be processed by the bigger computer. The FEP acts as a sort of intelligent traffic cop. It relieve the bigger, host computer of some of its telecommunications Input/Output burden, so that the host computer can concentrate on handling the processing burden. Depending on its sophistication, the front end processor might also perform serial to parallel conversion, protocol conversion, block or message assembly, etc. Here's a more technical definition: A dedicated communications system that intercepts and handles activity for the host. Can perform line control, message handling, code conversion, error control, and such applications functions as control and operation of special-purpose terminals. Designed to offload from the host computer all or most of its data communications functions. IBM 3705, 3725 and 3745 are Front End Processors. Front end processors are not used in client/server networks.

front end results A call center/marketing term. Used to describe the rate of expected or tentative, rather than actual, orders generated, usually as a result of a trial or free offer.

front page extensions See FrontPage Extensions.

front porch The blanking signal portion which lies between the end of the active picture information and the leading edge of horizontal sync.

front run A prohibited practice whereby an employee of a stock brokerage firm, prior to executing a large customer order, places the same order for his/her personal account, thus "front running" the customer. The idea is that the customer's large order will push the price of the stock up, at which point the employee can sell his shares at a handsome profit.

front side bus See Frontside Bus.

FrontPage Extensions Microsoft FrontPage Extensions is software which runs on a computer attached permanently to the Internet, i.e. a server acting as a web site. This software allows the "prepackaged functionality" of web sites built in Microsoft FrontPage to run. In English this means that if you use Front Page to write your web site you can skip some complex coding. If you build a web site using FrontPage, you (or your web site hosting company) will need to have FrontPage Extensions on the web server in order to properly display and process your web pages. Advanced web developers and webmasters do not use FrontPage Extensions because they choose to code fancy server side programming themselves by hand. This gives them, they say, greater control, security and flexibility. But it "costs." Most web sites don't run Windows which FrontPage Extensions must have to run. For the novice webmaster, according to Jim Koretz who worked on this definition, "You can't beat the built in functionality of FrontPage Extensions" Personally, I don't use Front-Page Extensions on either of my web sites – www.HarryNewton.com or www.InSearchOf-ThePerfectInvestment.com. I prefer Macromedia's Dreamweaver to Microsoft's FrontPage.

Frontside Bus Also known as the Memory Bus or System Bus, the Frontside Bus is the bus (i.e., common electrical path) within the microprocessor that connects the CPU (Central Processing Unit) of a PC with the main memory, the chipset, the L2 cache and other computer components. All data traverses the front side bus. See also Backside Bus.

FRMR Frame Reject.

FRS 1. Frame-Relay Service: A connection oriented service that is capable of carrying up to 4096 bytes per frame. See Frame Relay.

2. Family Radio Service. A very low-power, short-range two-way radio service. See Family Radio Service for a much longer explanation.

3. Field Routing System.

FRTT Fixed Round-Trip Time: This is the sum of the fixed and propagation delays from the source to the furthest destination and back.

FRU Field Replaceable or Replacement Unit.

FS Failed Signal.

FSA Foreign Serving Arrangement.

FSAN Full Service Access Network. FSAN is a set of specifications for an APON (ATM-based Passive Optical Network) scheme developed by an international consortium of vendors, and ratified by the ITU-T within the G.983.1 standard (October 1998) . An FSAN network comprises three primary network elements: Optical Line Terminal, Optical Network Unit, and passive splitter. The Optical Line Terminal (OLT) is located in the carrier's Central Office

(CO) or headend, where it serves to terminate the local loop that connects the subscriber premises to the edge of the network. The Optical Network Unit (ONU), also known as Optical Network Terminal (ONT), terminates the circuit 1) at the customer premises, 2) at the curb, or 3) at a centralized location in the neighborhood. The passive optical splitter sits in the local loop between the OLT and the ONUs. The splitter divides the downstream signal from the OLT at the network edge into multiple, identical signals that travel to the ONUs at the premises. Each ONU is responsible for figuring out which data are intended for it, and for ignoring all others. Upstream signals are supported by a time-division multiple access scheme, with the transmitters in the ONUs operating in burst mode. Current FSAN implementations run at speeds of 155/622 Mbps downstream over a wavelength running at approximately 1500nm (nanometers), and 155 Mbps upstream running at about 1310nm. FSAN trunk lengths can be up to 12 miles, and as many as 32 users and 64 endpoints can be supported per trunk at the current speeds and with the current splitter technologies. (Note: As a passive network, FSAN does not amplify the signal. Therefore, trunk lengths and signal splits are limited.) Encryption is built in for security. See also APON, FTTB, FTTC, FTTH, HFC, and PON.

FSB Front Side Bus.

FSK Frequency Shift Keying. A modulation technique for transmitting data in digital format over an analog carrier, FSK involves shifting the frequency level of the carrier. There are two fairly common variations on the FSK theme. 2FSK (Two-level FSK) shifts the carrier signal between two frequencies, with one used to represent 1's and the other to represent 0's. 4FSK (Four-level FSK) shifts the carrier between four frequencies. HomeRF SWAP (Shared Access Wireless Protocol) specification provide for both modulation methods, with 2FSK supporting a maximum data of 0.8 Mbps and 4FSK supporting a maximum data rate of 1.6 Mbps. See also SWAP.

FSL See Flexible Service Logic.

FSO 1. Foreign Switching Office. Identifies the switching entity for WATS circuits when the customer is not being served out of its normal serving wire center.

2. Free Space Optics. See Free Space Optics.

FSP 1. File Service Protocol. FSP is a non-standard protocol that is similar to FTP (File Transfer Protocol). FSP is an application layer (Layer 7 of the OSI Reference Model) connectionless protocol that makes use of UDP (User Datagram Protocol) sockets, thereby placing the responsibility for error control on the receiving application program. While FSP is somewhat slower than FTP, it is a lightweight protocol that creates less of a system load than FTP. FSP is designed specifically for anonymous (i.e., no user name or password is required) file transfers, and does not require root access in order to install it on a system. FSP can run on a wide variety of platforms, including UNIX, Windows 95/NT/2000, OS/2, Macintosh, VMS, and MS-DOS. FSP originally was written for UNIX by Wen-King Su. See also Connectionless, FTP, ODP and OSI Reference Model.

2. Fiber Optic Splice Panel.

FSR See Free Spectral Range.

FSS See Fixed Satellite System.

FSX FSX is the same as a DSX except it is used for optical cross connects only. A DSX is a Digital System Cross-connect frame. It is a manual bay or panel to which T-1 lines and DS1 circuit packs are wired. It permits cross-connections by patch cords and plugs, i.e. by hand. A DSX panel is used in small office applications where only a few digital trunks are installed. See also DACS.

FT-1 Fractional T-1. Any part of a T-1 circuit that's smaller than a full T-1 circuit. Fractional T-1 circuits are cheaper than full T-1 circuits. That's their reason for existing. See Fractional T-1 for a bigger explanation.

FT-3 Fractional T-3. typically fractional T-3 delivers between four megabits per second, all the way to the full T-3 capacity of 45 megabits per second. Sometimes also Fiber T-3. See T-3.

FT1 1. Fractional T-1. See FT-1 and T-1.

2. Flammability rating established by Canadian Standards Association for a vertical flame test of wire and cable.

FT3 Fractional T-3 or Fiber T3. See FT-3 and T-3.

FT4 Flammability rating established by Canadian Standards Association for a vertical flame test of cables in cable trays.

FT6 Flammability rating established by Canadian Standards Association for horizontal flame test and smoke testing of cables.

FTAM File Transfer and Access Management. The OSI (Open Systems Interconnection) standard for file transfer (i.e., the communication of an entire file between systems), file access (i.e, the ability to remotely access one or more records in a file) and management (e.g., the ability to create/delete, name/rename a file). FTAM is also an international standard.

FTE Full Time Equivalent. A call center term. A scenario assumption used in budget forecasting and scheduling that defines the number of hours per week full-time employees are normally suppose to work. As a measure of staffing level, an FTE is equivalent to a full-time position, even though the hours may actually be filled by part-time schedules.

FTIP Fiber Transport Inside Plant.

FTP 1. File Transfer Protocol. FTP lets users quickly transfer text and binary files to and from a distant or local PC, list directories, delete and rename files on the foreign machine, and perform wildcard transfers between machines. That distant or local PC (also called an FTP host) might be on your local area network, or a phone line across the world or connected to the Internet. It might also be your web site. I use FTP to transfer files to and from my various web sites, such as www.HarryNewton.com and www.TechnologyInvestor.com. Most versions of Windows come with elementary ftp software. Frankly, I prefer software called WS_FTP Pro., which comes from a company called Ipswitch. See www.Ipswitch.com.

FTP is an application layer (Layer 7 of the OSI Reference Model) extension of the TCP/IP protocol suite, and was originally developed for transfers of large files of 50 KB (Kilobytes) or more. FTP is best known as an Internet tool for accessing file archives around the world that are linked to the Internet. If you have a modem with a terminal emulation program or are running Windows, you have all the software you need to visit ftp sites. Here's what it's like to visit a the ftp site. The site address may be ftp.pht.com. Most public ftp sites accept anonymous login. That is, when the site asks who you are, you answer "anonymous." Then when they ask for your password, you give them your Internet email address. There are two ways an FTP data connection is made. In active mode, the FTP server establishes the data connection. In passive mode, the client establishes the connection. In general, FTP user agents use active mode and Web user agents use passive mode. See also Anonymous FTP, FSP, OSI Reference Model, and TCP/IP.

2. Foil Twisted Pair. A type of STP (Shielded Twisted Pair) cable which is employed to protect the signal-carrying conductors from EMI (ElectroMagnetic Interference). FTP uses a thin metallic foil; ScTP (Screened Twisted Pair) uses a heavy braided mesh for this purpose. See also STP.

3. Financial Transaction Processing. A Wall Street term.

FTP Mail Server A server which permits the retrieval of files via e-mail. See FTP.

FTS Federal Telecommunications System is a private telephone network sometimes shared enthusiastically by all federal government agencies. And sometimes not. See FTS2000.

FTS2000 The U.S. General Services Administration in Washington, D.C. describes the FTS2000 as "the state of the art, digital, long distance telecommunications program that provides voice, data and video transmission services to federal government agencies."

FTTB Fiber To The Building. See FTTC, FTTN, FTTP and HFC.

FTTC Fiber To The Curb. A hybrid transmission system which involves fiber optics to the curb, and either twisted pair or coaxial cable to the premises. FTTC is less extreme than FTTH, but more so than FTTN. See also FTTH, FTTN and HFC.

FTTCab Fiber To The Cabinet. Also known as FTTN (Fiber To The Neighborhood). A Hybrid Fiber Coax (HFC) network architecture involving an optical fiber which terminates in either a street-side or neighborhood cabinet which converts the signal from optical to electrical. The subscriber connection is over either UTP (Unshielded Twisted Pair) or coaxial cable. FTTCab can be either FTTC or FTTN. See HFC.

FTTD FTTD = fiber to the desktop.

FTTE Fiber to the Telecom Enclosure. The telecom enclosure is not the network interface which is often on the outside of the building. It's the box inside which the user company uses as its central telecom closet. Often it contains one or more large and small Ethernet switches and a horizontal crossconnect.

FTTH Fiber to the Home. The fiber deployment architecture in which optical fiber is carried all the way to the customer's home (or premises). This is the ideal architecture and also the most costly to implement. See FTTN and FTTP.

FTTM Fiber to the MDU (multidwelling unit).

FTTN Fiber To The Neighborhood. Also known as FTTCab (Fiber To The Cabinet). A hybrid network architecture involving optical fiber from the carrier network, terminating in a neighborhood cabinet which converts the signal from optical to electrical. The connection from the cabinet to the user premises is over UTP (Unshielded Twisted Pair) or coaxial cable. ILECs (Incumbent Local Exchange Carriers), i.e., local telephone companies, use the embedded UTP for this purpose; CATV providers use the embedded coaxial cable. The advantages of the fiber include incredible levels of bandwidth, outstanding error performance, and transmission over long distances without the requirement for expensive and troublesome repeaters.

The advantage of the UTP and coax is simply it is already there. The advantage of the neighborhood cabinet is that the expensive optoelectric conversion process takes place at a single location per neighborhood of perhaps 100 or 200 users. The cabinet also serves as a sophisticated multiplexer, allowing all the users to share the single, high-capacity fiber optic system for connection to the carrier network. FTTN is the preferred local loop architecture in a full convergence scenario, which involves the delivery of voice, Internet access, and entertainment TV over the same hybrid cable plant. While FTTH (Fiber To The House) is preferable in terms of overall performance, it currently is too expensive for serious consideration. Further, ADSL (Asymmetric Digital Subscriber Line) technologies will support the necessary levels of bandwidth over considerable distances, assuming the UTP is of good quality. See also ADSL, Fiber Optics, FTTH, FTTN, FTTP, HFC and SONET.

FTTP Fiber To The Premise. A telecommunications carrier can bring service to you in several ways – twisted pair cable, coaxial cable (as in cable TV), wireless, satellite and optical fiber. Of all these methods, fiber offers potentially the fastest service. Fiber itself has basically an unlimied bandwidth – but how big it actually is depends on the electronics on either end. The early implementations are promising speeds of 25 to 50 million bits per second or greater, as compared with a maximum of about 5 to 6 million bits per second for other types of broadband services. FTTP also supports fully symmetric services. There are several architectures for FTTP systems, but all start with an optical line termination (OLT) device. An OLT device acts as a switch, and interfaces with the Internet and other systems via standard interconnections such as Ethernet or ATM. Service providers place one OLT device in a central office and another in the field closer to subscribers to provide a longer reach between the central office and customers. Most FTTP systems employ passive splitting in the field. An optical splitter divides the FTTP signal among multiple homes and businesses. At the subscriber site – a business or home – an optical network unit (ONU) converts optical signals to standard forms that can be used by customers. These generally include Ethernet (10/100/1000Base-T), plain old telephone service lines for voice and cable TV-like signals for video. Using FTTP, service providers can deliver analog and digital video that is compatible with what cable TV systems provide. Voice quality is as good or better than what traditional phone companies provide. This is a result of QoS features and the fact that voice signal is digitized closer to subscribers, eliminating signal degradation. With FTTP, QoS mechanisms control the data rate provided to each subscriber, and rates can be set in increments as low as 64K bit/sec. FTTP also lets service providers offer different bandwidth to different applications used by the same subscriber. This is enabled by QoS techniques that identify traffic based on source, destination, application or Differentiated Services. Network operators also can use QoS classification to create premium services. For example, one service might prioritize corporate VPN traffic for telecommuters or prioritize traffic to a business Web site over email exchanges. Layer 3 bandwidth management lets FTTP networks support other services as well, such as IP Centrex, IP phones, conferencing, Web-based call management and Web-based subscriber self-care. As IP over Gigabit Ethernet has reached maturity and all network architectures continue to migrate toward IP, more FTTP platforms are incorporating Ethernet technology to take advantage of various capital and operational savings. Ethernet components are less expensive than ATM counterparts and offer substantial savings in terms of fiber deployment and management – two more factors that will help fuel the growth of FTTP networks. Much of this information is from Jim Farmer of Wave7 Optics. He can be reached at jim.farmer@w7optics.com. See also FTTN.

Fu Fu Dust See Foo Foo Dust.

FUB See FBU.

FUBAR F...d Up Beyond All Recognition. In short, a mess. A term often used in electronic mail messages. See also BOHICA, SNAFU and TARFU.

FUD Fear, Uncertainty, Doubt. A marketing tactic which a dominant player (once IBM) has used to discourage its customers from buying from its competitors.

fuel-cell mobile phone A mobile phone that is powered by a fuel-cell instead of a battery. A fuel cell uses a catalyst like platinum to start an electrochemical process to generate electricity from the reaction of oxygen and a fuel such as hydrogen, ethanol or methanol. Japanese carrier KDDI, and consumer electronics companies Toshiba and Hitachi, have been jointly developing fuel cells for mobile devices since 2004. In 2005, the companies demonstrated a prototype mobile phone powered by a fuel cell. The system used a hybrid fuel-cell/battery arrangement, with power supplied by a methanol-fed fuel cell and a lithium ion battery. Mass production of fuel-cell-powered mobile phones is still years away.

fugitive glue Glue used by printers to affix stuff into magazines. The glue is designed to stick until the magazine is delivered. At that point, the stuck-in thing becomes easier to remove and/or falls into your lap. This definition contributed by Rich Kubik.

fugitive odor A smell that leaks out of a composting plant or landfill.

full availability Idealized condition which exists when your phone system can provide connections for every telephone connected to it. Also called Non-Blocking.

full deck In the 18th century in England, common entertainment included playing cards. However, there was a tax levied when purchasing playing cards but only applicable to the "ace of Spades." To avoid paying the tax, people would purchase 51 cards. Yet, since most games require 52 cards, these people were thought to be stupid or dumb because they weren't "playing with a full deck."

Full Duplex FDX. A transmission mode which supports transmission in two directions simultaneously, or, more technically, bidirectional, simultaneous two-way communications. Simplex mode supports transmission in one direction, only. Half-duplex (HDX) mode supports transmission in both directions, but in only one direction at a time. The best two-direction phone conversations take place on four-wire circuits, which traditionally are physical four-wire. That is to say that there are four physical wires, with two (i.e., one pair) for transmission in one direction and two (i.e., one pair) for transmission in the other. (Note: There also are logical four-wire circuits, which comprise two wires, or even one wire, but which behave as though there were four wires.) All long distance circuits are four-wire – they have to be in order to support multichannel communications and, thereby, to haul large numbers of separate conversations. Most local loops are two wire, which also can support full-duplex communications, but for only one conversation at a time. It's important to contrast full duplex with symmetrical, or symmetric, and asymmetrical, or asymmetric. Again, full duplex means simultaneous transmission in both directions. Asymmetric means more bandwidth (or speed) in one direction than in the other. Symmetric means the same bandwidth (or speed) in both directions. For example, V.90 (56 Kbps) modems operate in full-duplex mode on an asymmetrical basis, while older V.34 (28.8 Kbps and 33.6 Kbps) modems operate in full-duplex mode on a symmetrical basis. T-1 operates in full-duplex mode on an asymmetric basis, as does ISDN. All DSL (Digital Subscriber Line) technologies run in full-duplex mode, and most (e.g., ADSL and G.lite) operate on an asymmetrical basis. Most speakerphones (except the newer more expensive ones, and the really old toadstool-looking ones made by Western Electric) are half-duplex, meaning they only transmit in one direction at one time. The speakerphone flips its direction based on who's talking, or, more precisely, who's talking the loudest. Full duplex speakerphones are the best. See ADSL, Asymmetric, DSL, Four-Wire, Half-Duplex, Simplex, Speakerphones, Symmetric, T-1, Two-Wire, V.34, and V.90.

full duplex audio Audio that allows remote sites to speak simultaneously without losing audio contact (two-way simultaneous audio). See Full Duplex.

full duplex transmission The process of operating a circuit so that each end can transmit and receive simultaneously. See Full Duplex for bigger explanation.

full echo suppressor An echo suppressor in which the speech signals on each path are used to control the suppression loss in the other path of a 4-wire circuit. Used for long-distance communications. Compare with split echo compressor.

full mesh Term describing a network in which devices are organized in a mesh topology, with each network node having either a physical circuit or a virtual circuit connecting it to every other network node. A full mesh provides a great deal of redundancy but because it can be prohibitively expensive to implement, it usually is reserved for network backbones. See also Mesh.

full motion video Television transmission where images are sent and displayed in real-time and motion is continuous. Video reproduction at 30 frames per second (NTSC-original signals) or 25 frames per second. (PAL-original signals). Compare with freeze frame. See Freeze Frame Video.

full name What you call your child when you're mad at him or her.

full period In private line telephone and telegraph service, a circuit rented for the exclusive use of a single customer on a month-by-month basis. Archaic in the sense that part-time circuits have not been rented for years.

full service When a pager screen reads Full Service, this indicates you are in prime coverage range and will receive all new messages as well as any undelivered, stored messages.

Full Service Access Network See FSAN.

full system battery backup This means there's sufficient battery power backing the phone system so that during a power outage, the telephone system will continue to work, i.e. you won't even know the commercial power has gone out. All programming will be intact. Calls will get through, etc. Full System Battery Backup is critical to many businesses, especially those in the "life or death" business, such as hospitals, police, fire departments, etc. Other businesses who depend heavily on the phone for their revenues – airlines, brokerage companies, hotel/motels, etc. – often also use full system battery

backup.

full time equivalent FTE. A call center term. A scenario assumption used in budget forecasting and scheduling that defines the number of hours per week full-time employees are normally suppose to work. As a measure of staffing level, an FTE is equivalent to a full-time position, even though the hours may actually be filled by part-time schedules.

fully connected network A network topology in which each node is directly connected by branches to all other nodes. This architecture becomes impractical as the number of nodes in the network increases in complexity. Such networks normally go to distributed nodes.

fully perforated Paper tape on which information is represented by the holes punched through the paper.

fully qualified domain name FQDN. In Internet terms, the full name of a system, rather than just a host name. For example, if the host name is Harry, the full name is harry.company.com).

fully restricted stations In a PBX, fully restricted stations (also called phones) can't place any outside calls. They can make intercom calls as well as receive incoming calls.

fumble What a cell site does when it drops your call.

function What it does. The action carried out by a piece of equipment or software program. See also Feature.

function key 1. One of up to 12 keys on a PC keyboard labeled with the letter F followed by a number. The effect (if any) of pressing a particular function key depends on which program you are running at the time.

2. An undefined key on a computer or telephone that can be defined to perform one function, which would normally require the user hitting one or several keys in succession.

Functional Entity FE. A set of functions that provides one or more specified capabilities. Seven FEs have been identified for the Advanced Intelligent Network Release 1 architecture: Network Access, Service Switching, Service Logic and Control, Information Management, Service Assistance, Automatic Message Accounting and Operations. Definition from Bellcore in reference to its concept of the Advanced Intelligent Network. See AIN.

functional group A collection of FEs (Functional Entities) that reside together in a system.

functional management layer A communications layer in SNA that formats presentations.

functional profile A defined stack of ISO OSI-Layer elements, such as GOSIP, MAP or TOP. Functional profiles were developed in order to ensure that, when defined, ISO OSL stacks could interoperate. Due to the number of different protocol elements at each OSI layer, it was possible to define stacks that were syntactically correct, but would not be able to exchange information due to differences at particular layers. A functional profile that has been defined as a standard is termed a standardized profile. Likewise, an International Standard Profile is an ISO OSI functional profile.

functional resource An abstraction of physical entities (e.g., voice synthesizers) that the Service Assistance FE (Functional Entities) can manipulate.

functional signaling In an ISDN circuit, function signaling provides messages with unambiguous, defined meanings known to both the sender and receiver of the messages. Signaling is generated by the terminal.

functional specification A description of a system from a working point of view. It differs from a precise technical description which includes each piece of equipment precisely spelled out. A system can often work the same using different hardware and software configurations. By functionally describing a system, a user allows sellers to use their imagination to solve the problem in the most creative, cost-effective way. Most sellers prefer functional descriptions.

functional split A division within an automatic call distributor (ACD) which allows incoming calls to be directed from a specific group of trunks to a specific group of agents.

functional test A test carried out under normal working conditions to verify that a circuit or particular part of the equipment works properly.

functional transparency The ability of a network to carry any user information regardless of its form, so that user applications can operate through the network.

functional user An entity external to the functional architecture that uses the functional architecture capabilities to exchange information with other functional users. Definition from Bellcore in reference to its concept of the Advanced Intelligent Network.

funeral "I didn't attend the funeral, but I sent a nice letter saying I approved of it." – Mark Twain

"The reason many people showed up at Louis B. Mayer's funeral was because they wanted to make sure he was dead." – Samuel Goldwyn. See Goldwyn, Samuel.

FUNI Frame-based User-Network Interface, a frame format for access to ATM networks, very much like Frame Relay but with a few additional bits reserved for mapping into the ATM control bits in the cell format. The Frame Relay format and FUNI both pass through a frame switch. See Frame Relay and Frame UNI.

Further Notice of Proposed Rule Making FNPRM. An Federal Communications Commission term. A Further Notice of Proposed Rule Making is issued by the Commission to further clarify and seek more information and public comment on the Commission's proposed changes. See Notice of Proposed Rulemaking.

fuse Verb: To blend together through melting. Noun: An electrical device typically consisting of a wire or strip of fusible metal that melts to interrupt an electrical circuit when current exceeds the rated level of the fuse. The idea is that in any electrical circuit, the fuse should be the weakest point – thus the point that heats up when things go wrong and melts. Better the fuse melts than your expensive PBX. See also Circuit Breaker.

Fuse Alarm Panel FAP. A distribution panel at the top of the rack. Each device gets its power from the rack. To protect the rectifier from an over-current condition, each device has its own fuse.

fused coupler A method of making a mulmode or single-mode coupler by wrapping optical fibers together, heating them, and pulling them to form a central unified mass so that light on any input fiber is coupled to all the output fibers.

fused fiber A bundle of fibers fused together so they maintain a fixed alignment with respect to each other in a rigid rod. See also Fused Coupler.

Fused IT An environment in which the customer, IT consultant and technology partners work together throughout the selection process. Partners will demonstrate not only the benefits of their own products but explain how they work together with components from other manufacturers to provide a complete solution. They hold conferences on Fused IT. I don't believe this is a generic term. But the vendor who made it up insists that it is.

fused quartz The precise term for glass made by melting natural quartz crystals.

fusible links Short lengths (about 25 feet) of fine-gauge wire pairs inside metallic sheath cable that melt to interrupt an electrical circuit and to prevent overheating in building wiring and equipment.

fusing In fiber optics, fusing is the actual operation of joining fibers together by fusion or melting. See also the next three definitions.

fusion The process in which atomic nuclei collide so fast that they stick together and emit a large amount of energy. In the center of most stars, hydrogen fuses into helium. The energy emitted by fusion supports the star's enormous mass from collapsing in on itself, and causes the star to glow.

fusion splice A permanent splice accomplished by the application of localized heat sufficient to fuse or melt the ends of two lengths of optical fiber. See also fusion splicing.

fusion splicing Fusion splicing is a fancy way of saying we're going to join fiber optic cables together. It's done by butting the two optical fibers together, forming an interface between them, and then removing the common surfaces so that there is no interface between them. In plain English, this means that the individual fibers are melded, i.e., fused, together to form what essentially is a single fiber. This is accomplished by firing a precise electrical arc at 2000 degrees Centigrade. This technique is so good that no reflection occurs at the splice point, and insertion loss is in the range of 0.00-0.05 dB. In fact, the splice, if done by skilled technicians, often is so good that it cannot even be detected.

FUT Fiber Under Test.

futility computing See Utility Computing.

future 1. The future isn't what it used to be.
   2. Economists are incredibly accurate at forecasting the past.
   3. "Never prophesize about anything, especially the future." – Samuel Goldwyn.

Future Public Land Mobile Telecommunication System See FPLMTS.

future shock "Man has a limited biological capacity for change. When this capacity is overwhelmed, he is in future shock," defined by Alvin Toffler in his book of the same name published in 1970.

futureproof When an adjective, a term used to describe a phone system (or any technology) that supposedly won't become technologically outdated (at least anytime

soon). There's no such thing. When a verb, as in to futureproof, it means installing fiber to each outlet in your grand new house, even though you have zero idea of what you'll do with the fiber.

fuzzing A slang term for robustness testing, also known as functional protocol testing in a network environment. In application testing, fuzzing involves sending an application problematic input, such as values that have the wrong data type or are unexpectedly large or small values, in order to discover vulnerabilities in the application. In functional protocol testing, fuzzing involves sending malformed packets and unexpected messages over the network, sometimes randomly, thereby generating conditions never anticipated by the protocol developers and network device designers.

fuzzy logic Fuzzy logic is the newest wrinkle in the ancient science of controlling processes that involve constantly changing variables. Contrary to its name, fuzzy logic is a very precise sub discipline in mathematics. It was invented in the 1960s by University of California at Berkeley's Russian-born Iranian computer science professor Lotfi Zadeh. It enables mathematicians and engineers to simulate human thinking by quantifying concepts such as hot, cold, very far, pretty close, quite true, most usually, almost impossible, etc. It does this by recognizing that measurements are much more useful when they are characterized in linguistic terms that when taken to the fourth decimal point. Fuzzy logic reduces a spectrum of numbers into a few categories called membership groups. Within five years virtually all consumer goods will come with fuzzy logic. Already fuzzy logic is inside video camcorders (to reduce the motion of the camera), in washing machines (to figure the optimum mix of washing conditions for that weight and filth).

FVO Field Verification Office.

FVR Flexible Vocabulary Recognition.

FWA Fixed Wireless Access. Fixed wireless consists of a radio link to the home or the office from a cell site or base station. This "fixed" wireless link replaces the traditional wireless local loop. According to Northern Telecom, FWA is the solution of choice in sparsely-developed areas where potential subscribers have been on lengthy waiting lists, in dense urban areas where rapid expansion is desirable and in suburban settings where new neighborhood developments can be provisioned quickly with FWA.

FWIW Abbreviation for "For What It's Worth;" commonly used on E-mail and BBSs (Bulletin Board Systems).

FWLL See Fixed Wireless Local Loop.

FX Foreign Exchange. A Central Office trunk which has access to a distant central office. Dial Tone is returned from that distant Central Office, and a location can be reached in the area of the foreign Central Office by dialing a local number. This will provide easier access for customers in that area and calls may be made anywhere in the foreign exchange area for a flat rate. In short, a FX line is a special access circuit that provides service between a telephone at a customer location and a distant central office. A foreign exchange area is any area other than that serving the geographic area in which the telephone is located. See also Foreign Exchange and FXO.

FXC Fiber Switch Cross-Connect. A type of Optical Cross-Connect (OXC). See OXC.

FXO Foreign Exchange Office. Foreign exchange (FX) service is a service that can be ordered from the telephone company that provides local telephone service from a central office which is outside (foreign) to the subscriber's exchange area. In its simplest form, a user can pick up the phone in one city and receive a dial tone in the foreign city. This type of connection is provided by a type of trunk called foreign exchange (FX) trunks. FX trunk signaling can be provided over analog or T-1 links. Connecting POTS telephones to a computer telephony system via T-1 links requires a channel bank configured with FX type connections. To generate a call from the POTS set to the computer telephony system, you will need a FXO (foreign exchange office) connection configured. To generate a call from the computer telephony system to the POTS set, you will need a FXS connection configured. See FX.

FXS Foreign Exchange Station. See FXO.

FYI An Internet term. An abbreviation for the phrase "for your information." There is also a series of RFCs put only by the Network Information Center called FYIs. They address common questions of new users and many other useful things.

FZA Fernmeldetechnisches Zentralamt. Telecom approval authority Austria, literally translated "long distance communications technical central office." All that in two words. Not bad.

G 1. G stands for giga, which means a billion or one thousand million. In telecommunications, a gig is actually 1,000,000,000. In computers it is ten to the ninth power, which is actually 1,073,741,824. One thousand gigas are a tera. One thousand teras are one peta, which is equal to 10 to the 15th.

2. Abbreviation of "Grin," commonly typed within pointy brackets as <\<>G>, at the end of an item uploaded to a BBS (Bulletin Board System), where the sender wants to make sure that readers realize that the message was meant to be humorous or sarcastic, and not to be taken literally. Example: "If my wife makes meat loaf one more time, I'm going to cut her fingers off <\<>G>." Usage is similar to appending Wayne's-World usage of "Not" to reverse the meaning of a sentence.

3. Refers to the status of an employee no longer employed. They are "G" (Gone).

4. A multiconductor mining cable with a grounding conductor in the valleys between the phase conductors.

G recommendations A series of standards defined by the ITU-T covering transmission facilities. Namely: G.703 transmission facilities running at 2.048 megabits/second (E1) and 64 kilobits per second. G.703 is the ITU-T standard 1984 current version for the physical and logical traits of transmission over digital circuits. G.703 now includes specifications for the US 1.544 megabits per second as well as the European 2.048 megabits/second, and circuits with larger bandwidths on both continents. G.703 is still generally used to refer to the standard for 2.048 megabits per second; G.821 is the ITU-T Recommendation that specifies performance criteria for digital circuits for ISDN. The G.990 series covers xDSL technology. See G.990.

G2 Intelligence. The term G2 is a term used by the military (chiefly the U.S. Army) for intelligence. If someone asks, "What's the G2 on Microsoft?" they're asking for the latest information on the company, preferably information which is not public.

G3 Third Generation Mobile System. An ITU-T discussion over a proposed worldwide worldwide cellular phone GSM standard. Usually called 3G. See 3G.

G-Lite Spelled G.Litge. A ITU standard for xDSL high-speed local loop access to the Internet. For a full explanation, see G.Lite.

G-style handset A G-style handset is a standard round screw-in, screw-out handset, as compared to the K-style handset, which is the newer square handset with the two screws in the middle.

G.5 See G5.

G.703 ITU-T Recommendation G.703, "Physical/Electrical Characteristics of Hierarchical Digital Interfaces". See G Recommendations. 64 Kbps PCM and used for E-1 and T-1.

G.704 ITU-T Recommendation G.704, "Synchronous Frame Structures Used at Primary and Secondary Hierarchy Levels".

G.707 The ITU-T standards recommendation (1990) for Synchronous Digital Standard (SDH) bit rates. See SONET for a full listing of current bit-rate levels.

G.708 The ITU-T standards recommendation (1990) for Synchronous Digital Hierarchy (SDH) Network Node Interface (NNI). See SDH and SONET.

G.709 The ITU-T standards recommendation (1990) for Synchronous Digital Hierarchy (SDH) multiplexing structure. Development work on G.709 continues on "Interface for the optical transport network (OTN)," which focuses on the creation of a much less overhead-intensive "digital wrapper" for network management purposes in a carrier network based on pure optics such as DWDM (Dense Wavelength Division Multiplexing). See also Digital Wrapper, SDH, and SONET.

G.711 ITU-T Recommendation for an algorithm designed to transmit and receive mu-law PCM voice (for North America) and A-law (for the rest of the world) at a digital bit rate of 64 Kbps. It is used for digital telephone sets on digital PBX and ISDN channels. Support for this algorithm is required for ITU-T compliant videoconferencing (the H.320/H.323 standard).

G.721 Old ADPCM speech coding standard at 32 kbps from ITU-T, but it has now been replaced by G.726.

G.722 ITU-T Recommendation: This algorithm produces digital audio through a wideband speech coder operating at 64, 56 and 48 kbps. The sampling rate is 16 kHz. All the other ITU-T speech coding standards use a sampling rate of 8 kHz.

G.723 Old ADPCM speech coding standard at 40, 32, and 24 kbps from ITU-T, but it has now been replaced by G.726.

G.723.1 ITU-T Recommendation: Speech encoding/decoding with a low bit rate, 5.3 kbps or 6.3 Kbps output quality. This is what the Voice over IP would use over the Internet. This is the default encoder required for H.323 compliance.

G.726 G.726 defines ADPCM voice coder operating at 40, 32, 24, and 16 kbps. This has replaced the old ADPCM standards of G.721 and G.723.

G.727 G.727 defines embedded ADPCM with voice encoded at 40, 32, 24 and 16 kbps.

G.728 ITU-T Recommendation: Encoding/decoding of speech at 16 kbps using low-delay code excited linear predictive (LD-CELP) methods. Like G.722, it is optional for H.320 compliance.

G.729 The International Telecommunications Union's standard voice algorithm – CS-ACELP (Conjugate Structure Algebraic Code Excited Linear Predictive) voice algorithm for the coding of encoding/decoding of speech at 8 Kbps using conjugate-structure, algebraic-code excited linear predictive methods. G.729 is supported by, inter alia, AT&T, France Telecom and Japan's NTT. G.729

See V.70, the specification for DSVD, which uses G.729.

G.729A A simplified G.729 for DSVD applications. The bit stream of G.729A is compatible with G.729.

G.804 ITU-T Recommendation G.804, "ATM Cell Mapping into Plesiochronous Digital Hierarchy (PDH)".

G.990 On October 29. 1998, the International Telecommunications Union issued the following press release. (My edits in brackets.) Geneva - The International Telecommunication Union today closed a vital link in the high capacity Information Highway by reaching agreement on a set of new technical system specifications for Multi-Megabit/s network access, and initiating the formal approval process. The new specifications, designated as the G.990 series of Recommendations, specify several techniques to provide megabit per second network access on existing telephone subscriber lines (i.e. copper local loops) simultaneously with the regular voice communication. Main applications are high-speed Internet access, video and other on-line data communications such as electronic commerce, home office, distance learning.

"These new specifications for multi-megabit network access link well into the already existing ITU-T fiber- and coax-based standards on Gigabit/s transport systems for the core network, enabling network providers to offer on-demand, high capacity digital services over the last mile – another major step towards building the information society", said Peter Wery, Chairman of ITU-T Study Group 15.

The new access systems are industry's response to the yearning of subscribers for quicker network access without long waiting times and at high bit rates. Commercially very important, industry analysts foresee a market potential of several billion dollars world-wide. The new access network specifications provide for:

Symmetrical bi-directional access at bit rates of up to 2 million bits per second (New Recommendation G.991.1). Asymmetrical bi-directional access bit rates of up to 640 kilobits per second in the upstream (subscriber to network) and up to 6 million bits per second in the downstream (network to subscriber) direction, depending on the subscriber line length (Draft new Recommendation G.992.1). Splitterless, asymmetrical bi-directional access (Draft new Recommendation G.992.2, previously known as G.lite). This is a simpler, splitterless asymmetrical system which can be installed by the user. Depending on the subscriber line length, the system provides upstream access up to 512 kilobits per second and enables the subscriber to download data and video at speeds of up to 1.5 million bits per second. The standard eliminates the need for a piece of equipment called "splitter" at the consumer's premises. New G.992.2 compliant modems will simply plug into the back of the PC as current modems do. Industry analysts expect that the adoption of the standard will speed up the rollout of high-speed Internet access to consumers over existing phone lines. It is also expected that this type of Megabit per second system to become a 'best seller' in the network access arena, with transmission speed of Internet data 25 times faster than today's 56k analog modems and close to speeds achieved on cable modems. Today's agreement on a single open standard also means that consumers can choose freely from any supplier providing G.992.2-compliant products as all DSL modems will be able to interoperate. "One of the keys to the mass deployment is standardization, which allows a situation where an end-subscriber can comfortably buy a modem and be reasonably assured that they can move to a different location and have it work.

In addition to the system specifications above, a number of complementary technical specifications have also been agreed upon, addressing test procedures, system management, and 'handshaking' procedures.

The ITU, a United Nations agency, coordinates the development of global communications standards. Study Group 15 of the ITU Telecommunication Standardization Sector (ITU-T), where the work on these specifications has been carried out, is responsible for the standards development in the area of transport networks, systems and equipment.

Recommendations G.991.1 and G.992.1 have been approved and have taken effect. In respect of Recommendation G.992.2 (previously G-Lite), the Study Group has agreed to apply the approval procedure under which the draft text is circulated to all ITU-T members to determine whether the Study Group is to be assigned the authority to give it final approval ("decision") at its next meeting. After unanimous approval by the Study Group, the standard takes effect. For the G.992.2 draft standard, the "decision" step is scheduled for end of June 1999 (June 21 to July 2). The agreement by the Study Group covers the key technical specifications, thus providing the technical stability required by manufacturers and service providers to bring compatible products to the market. The next step in June/July is the formal approval of the standard before it can take effect.

G.991 See G.990.

G.992 See G.990.

G.992.2 This spec was previously known as G.Lite. See ADSL Lite and G.990.

G.DSVD Voice digitizer for V.DSVD.

G.Lite See ADSL Lite.

G.O.O.D. Job A "Get-Out-Of-Debt" job. A well-paying position you take to pay off your debts, and one you'll quit as soon as you're solvent.

G.shdsl See HDSL2, SDSL and SHDSL.

G-GC A multiconductor mining cable with two grounding conductors and an insulated pilot wire.

G/T A figure of merit expressing the efficiency of a receiver, pertains to the combination of an antenna and the receiver; G is the overall gain in decibels and T is the noise temperature, expressed in decibels relative to 10 Kelvin. Hence G/T is expressed in decibels, with the appended symbol dB/K to indicate the Kelvin scale of temperature.

G5 messaging An idea that apparently never happened. The idea was to create a 5th generation electronic messaging service, to include file negotiation, directory referencing, electronic invoicing.

GA 1. Generally Available or General Availability. A vague term manufacturers use to refer to when their new product will be generally available.

2. Abbreviation for "Go Ahead," used in real-time computer communications to indicate that you have finished a sentence and are awaiting a reply.

GAAP Pronounced gap. Generally Accepted Accounting Principles. American companies traditionally report financials according to GAAP. The problem with GAAP is that it's neither "generally accepted," nor are its principles real principles – as in the sense of principles (or laws) of physics. How you report financials under GAAP is open to much interpretation and depends on such factors as your accountant, your auditor and/or the rules and regulations of the Internal Revenue Service. See Operating Income.

GAB Group Access Bridging. A service for bridging of multiple calls to create a conference call.

gable mount See Eave Mount.

Gabriel See Saint Gabriel.

gaff Equipment worn by telecommunications and power company staff when they climb poles. The official name is linesman's climbers. They consist of a steel shank that can be strapped to a person's leg. The inside of the shank has a spike that can stab the pole. It's the spike that most people mean when they refer to a gaff.

gain 1. The increase in signaling power that occurs as the signal is boosted by an electronic device. It's measured in decibels (dB).

2. A radio term. Formally, and according to Bell Telephone Laboratories, "gain" is the ratio of the maximum radiation intensity in a given direction to the maximum radiation intensity in the same direction from an isotropic radiator (an antenna radiating equally in all directions). In other words, "gain" is a measure of the relative efficiency of a directional (focused) radio antenna systems, as compared to an omnidirectional (broadcast) system.

gain hits A cause of errors in data transmission over phone lines. Usually the signal surges more than 3dB and lasts for more than four milliseconds. AT&T's standard calls for eight or fewer gain hits in a 15-minute period.

GAIT GSM ANSI-136 Interoperability Team is a technology that enables GSM and TDMA networks to interoperate. Special handsets must be manufactured (often called "GAIT phones") and used in conjunction with GAIT networking.

GAIT Phones Named after its standards body, the GSM ANSI Interoperability Team, GAIT phones operate on both GSM and TDMA networks. See GAIT.

Galileo Galileo is the proposed European version of the US Global Positioning System (GPS). Galileo, called the European Satellite Navigation System, will have 30 satellites. GPS has 24. Sometimes not all of GPS's 24 are working. Why is Galileo necessary? According to the Europeans, "Galileo will ensure European economies' independence from other states' systems, which could deny access to civil users at any time, and to enhance safety and reliability. The only systems currently in existence are the United States Global Positioning Service (GPS) and the Russian GLONASS system, both military but made available to civil users without any guarantee for continuity." The Europeans also say that "important macro-economic benefits will be derived from Galileo, in particular through achieving a European share in the equipment market, efficiency savings for industry as well as social benefits e.g. through cheaper transport, reduced congestion and less pollution. Above that, with its open service at least offering the same performances as GPS by the time of Galileo's deployment, Galileo will offer also value added services with integrity provision and, in some cases, service guarantees, based on a certifiable system." Why should I pay for Galileo when GPS is free? Answers the Europeans, "Like GPS, Galileo will be free of charge to basic users (open service). Some applications will have to be paid for – those requiring a quality of service which GPS is unable to provide. The GPS of the future could perhaps offer such services too, but there is no guarantee that they will be free, least of all if GPS would hold a monopoly. In any case, GPS will remain a system conceived primarily for military applications." Galileo will be fully operational by 2008, according to the Directorate-General for Energy and Transport, European Commission, Belgium, Brussels. See GPS.

gallium arsenide A substance from which microprocessor and memory chips are made. Compared with silicon, GaAs is three to ten times faster or, depending on its speed, uses as little as one-tenth the power; GaAs can detect, emit and convert light into electrical signals, opening the possibility of providing optoelectronic properties on a single chip; GaAs can resist up to 10,000 times the radiation; GaAs can withstand operating temperatures of 200 degrees Centigrade, and GaAs have a higher electron mobility. Gallium arsenide chips run six to seven times faster than those made from silicon. The newer indium phosphide chips will run four times faster still.

galvanic isolation A characteristic of a UPS or transformer in which the output is completely electrically disconnected from the input. Power is coupled from input to output by magnetic fields in a transformer within the UPS. A galvanically isolated output is considered to be a separately derived source according to the US National Electrical Code and is required to be grounded, that is, the output grounding wire must be directly bonded to the input grounding wire. It is commonly but falsely believed that galvanic isolation eliminates ground loops. This definition courtesy American Power Conversion Corp.

galvanometer A delicate instrument used for measuring minute currents.

Galvo Man Telco-talk for Galvanometer Man, a technician who uses a galvanometer to find and repair circuit faults. It's common for the phone company rep to tell you that your new lines can't be installed when promised because the galvo man hasn't finished his work.

gambler You got to know when to hold them, know when to fold them, know when to walk away, know when to run.......- Kenny Rogers, "The Gambler"

game theory Game theory is a branch of economics introduced by in the 1940s. It is the mathematical analysis of interdependent decisions – finding hidden patterns or puzzles. For example, whether a union calls a strike depends on whether the union leaders think the management will respond with a better pay offer; whether I bluff at poker depends on whether I think you are likely to call the bluff. Game theory and the atomic bomb arrived at the same time with the help of the same mathematician, John von Neumann, and the early game theorists tried to use the theory to understand nuclear war, according to an October 12 article in the Financial Times. But their analysis was weak. Von Neumann told Life magazine: "If you say why not bomb them tomorrow, I say why not today?" Mr. Thomas Schelling's 1960 book, The Strategy of Conflict, revolutionized both strategic thinking and game theory. Mr. Schelling ditched the mathematics of his peers and applied the rigorous thinking of game theory to a richer world in which the superpowers tried to understand the tacit signals behind each others' threats and promises. He showed that even the deadliest wars involved significant elements of common interest and co-operation between foes. Indeed, the striking fact that the cold war never became a hot one is the co-operative feat of the century. In accepting the Nobel Prize for Economics in the fall of 2005, Mr. Thomas Schelling said the most important even of the second half of the 20th century is one that didn't happen – namely nuclear war.

gamer A gamer is a person who likes to play computer games, but who typically doesn't see them as a "game." He sees them a something far more serious, even a way of life.

gaming A euphemism for gambling. In our context, the term most typically is used with respect to gambling over the Internet, which sometimes is known as "cybergambling." A number of companies have set up gaming sites on the World Wide Web (WWW). The companies are headquartered and the Web sites are located in "friendly" countries abroad (e.g., countries in the Bahamas), where they are outside the reach of US federal and state laws against it. Some of the Web sites are run by US Native American (a euphemism for what we used to call Indian) tribes from their reservations, where they largely are outside the reach of state and federal laws. The way that cybergambling works is that you open an account in an offshore bank in the same country where your gaming company is located. You then place bets on the gaming Web site. Your losses automatically are debited to your account. Your winnings, if any, are automatically credited to your account. At least that's the way it's supposed to work. Lots of people have complained that their winnings never were credited. You just need to make sure that you gamble with honorable people. Good Luck! While the specifics are the laws are somewhat unclear, there continue to be legislative efforts to tighten both the laws and the associated regulations against gaming. As the FCC and WTO have declared a multi-year moratorium on regulation of the Internet, it is unlikely that we will see any aggressive moves in the near future to curtail this sort of activity.

GAMMA Distribution A telephone company term. A particular type of right-skewed probability distribution which closely resembles COE load distribution. GAMMA extends farther to the right which means that "for a given average busy season load" the GAMMA will predict higher peak day loads than will the normal distribution.

gamma ray The highest energy, shortest wavelength electromagnetic radiations. Usually, they are thought of as any photons having energies greater than about 100 keV.

gammic ferric oxide The type of magnetic particle used in conventional floppy disks.

GAN Global Area Network. Companies like Global Crossing and Globalstar are considered global area network companies. I hope they're still around by the time you read this.

gang of four The Frame Relay Forum began in 1990 as the Frame Relay Working Group, which was formed by The "Gang of Four," comprising Cisco, Digital Equipment Corporation, Nortel, and Stratacom (since acquired by Cisco). In 1991, the Frame Relay Forum was officially formed, and has grown to over 300 members. See also Frame Relay Forum.

ganged antennas See stacked antennas.

Gap 1. GAP. Generic Access Profile. A wireless term. See Generic Access Profile.

2. Gap. An open space in a circuit through which a condenser discharges for producing electric oscillations.

See Gapp Loss.

gap loss That optical power loss caused by a space between axially aligned fibers. For waveguide-to-waveguide coupling, it is commonly called "longitudinal offset loss."

garage Silicon Valley, according to contemporary lore, started in a garage in Palo Alto in 1939. In that year Bill Hewlett and Dave Packard started Hewlett Packard with $538. Hewlett was the inventor and Packard the manager. Their first product, an audio oscillator, was an immediate success. Walt Disney used it in making Fantasia. Microsoft also was started by Bill Gates and Paul Allen in a garage. Even more recently, Excite, the Web browser company, was started in a garage. Excite no longer works out of a garage, but there is a conference room in their new headquarters, that looks just like a garage, from the outside. East coast companies are often started in basements. West coast companies are started in garages because there no basements in West coast houses.

garage sale See Yard Sale.

garbage band A pejorative term for the ISM (Industrial Scientific Medical) radio frequency bands, also known as Part 15.247 of FCC regulations. ISM operates in the 902-928 MHz, 2.4-2.483 GHz, and 5.725-5.875 GHz ranges. Traditionally used for in-building and system applications such as bar code scanners, industrial microwave ovens and wireless monitoring of patient sensors, ISM also is used in many Wireless LANs. As the ISM band is unlicensed, anyone can use it for anything, anywhere in the U.S. Some garage door openers use it-hopefully not the garage door openers at the same hospital that's using it to monitor your pulse rate in the ICU. It's a catch-all, hence the term "garbage band."

garbage can What Australians call a garbage can, Americans call a trash can.

garbage collection A software program or routine that is used to solve "memory leaks." Garbage collection is the process of searching memory for program segments or data that are no longer active, in order to reclaim that memory space for other computer programs. See also Memory Leak.

Garbage In, Garbage Out GIGO. If the input data is wrong or inaccurate, the output data will be inaccurate or wrong. GIGO is problem with data entered by hand into computer systems. Ask yourself how many times you've received "junk" mail with the wrong spelling of your name? That's called Garbage In, Garbage Out.

garbitrage Sending garbage from one city to another, usually organized by garbitrageurs on the phone.

GARP 1. Growth At a Reasonable Price. An investment philosophy that focuses on picking stocks that provide growth, but without taking significant risk. The term means different things to different people.

2. Generic Attribute Registration Protocol. An IEEE standard for a generic method by which various devices (e.g., clients, servers, and bridges) can automatically disseminate attribute information across a bridged LAN. GARP is a Layer 2 (i.e., Data Link Layer) protocol used extensively in VLANs (Virtual LANs). A GARP participant consists of a GARP application software component, and a GARP Information Declaration (GID) component which is associated with each port of the bridge. GARP participants in a given application disseminate their attribute information through the use of the GARP Information Propagation (GIP) component.

Relying on GARP services is GMRP (GARP Multicast Registration Protocol), which provides a mechanism by which bridges and end stations can automatically and dynamically register their membership in a group with the MAC bridges by which a physical LAN segment attaches to the larger logical LAN. Once the bridges receive that registration information, they propagate it to all other bridges that support extended filtering services. The GARP VLAN Registration Protocol (GVRP) is a GARP application that provides registration services in a VLAN context. See also VLAN.

gas The word "gas," coined by the chemist J.B. van Helmont, is taken from the word "chaos," which means "unformed" in Greek.

gas carbon Used for lightning protection by phone companies. In the telephone's early days lightning often struck telephone lines, electrocuted people or burned their houses down. Early lightning protectors were made of carbon. When hit they took phone out of action and needed to be replaced by a technician. Newer lightning protectors are made with a gas. When hit by lightning they temporarily short, then re-enable the phone line. This invention has greatly reduced the number of bad lines a phone company has after a storm. Despite their name, there is no carbon in them. Gas carbons are the same size and shape as the older carbon protectors so they fit easily into the old slots.

gas pressurization A method for preventing water from entering openings in splice closures or cable sheaths by keeping the cables under pressure with dry gas.

gas tube A method of protecting phone lines and phone equipment from high voltage caused by lightning strikes. See CARBON BLOCK (another protection technology) for a more detailed explanation. Here is a definition from American Power Conversion Corp. Gas tube is a surge suppression device that clamps a surge voltage to a limited value. Also called a "spark gap", a gas tube is simply two electrodes that are held at a close distance so that high voltages between the electrodes simply arc through the air or other gas within the tube, thereby effectively clamping the voltage. Gas tubes are very slow, but can handle very large surges. The main problem with the use of gas tubes in AC power circuits is that when they clamp the surge they momentarily short out the utility line which usually trips the circuit breaker feeding the circuit which the tube is connected to. In this case the operation of surge clamping leads directly to power interruption. They are well suited to use in data line surge suppression, but have protective clamping voltages that are too high to provide effective protection for most modems or computer ports.

gaseous conductors The gases which, when ionized by an electric field, permit the passage of an electric current.

gasoline curtain The distance that separates a person's home from his office, and both of those places from "third places." Telecommunications advances are dissolving that curtain. Here's a quote from Gartner, the consultants: "As the 'gasoline curtain' between the office, home, vacation and all locations in transit dissolves, end users who are radically reachable will respond by demanding unified and flexible offerings, not discrete, 'one price fits all' packages.'"

gate 1. This term is typically used in Automatic Call Distributors, devices used for handling many incoming telephone calls. Gate refers to a telephone trunk or business transaction grouping that may be handled by one group of telephone answerers (called attendants, operators, agents or telemarketers). That one group of telephone answerers is called "the gate." All calls coming into that gate can, theoretically, be handled by any of the telephone answerers. A telephone call is homogeneous throughout the gate. An automatic call distributor may have one gate – all calls coming in can be handled by everyone. Or it may have many gates, each one consisting of the line (or lines) bringing the call in – e.g. Band 5 WATS, New York City foreign exchange line. Or it may have two gates – one for orders and one for service. ACDs with multiple gates will establish rules for moving the calls between the gates, should one gate become overloaded.

2. A circuit on a silicon chip. See Gate Array.

gate array A circuit consisting of an array of logic gates aligned on a substrate (a piece of silicon) in a regular pattern.

gate assignments Used in context of ACD (Automatic Call Distribution) equipment. Gates are made up of trunks that require similar agent processing. Individual agents can be reassigned from one gate to another gate by the customer via the supervisory control and display station. Also called splits.

Gate D Gateway Daemon. A popular routing software package which supports multiple routing protocols. Developed and maintained by the GateDaemon Consortium at Cornell University.

gate rape Frequent random searches of flight crews by airport security. Coined by pilots, who claim they're easy targets because they can be punished for objecting to it.

gatekeeper In the classic sense of the word, a gatekeeper is someone who is in charge of a gate. His or her job is to identify, control, count, supervise the traffic or flow through it. A network gatekeeper provides the same functions, including terminal and gateway registration, address resolution, bandwidth control, admission control, etc. A gatekeeper is a fancy name for a network administrator. A server that uses a directory to perform name-to-IP address translation, admission control, and call management services in H.323 conferencing is also called a

gateway 1. A gateway is what it sounds like. It's an entrance and exit into a communications network. That "communications network" may be huge, for example, at the point where AT&T Communications ends and Comsat begins – for taking my satellite call overseas. Gateways may be small – between one LAN and another LAN. Technically, a gateway is an electronic repeater device that intercepts and steers electrical signals from one network to another. Generally, the gateway includes a signal conditioner which filters out

unwanted noise and controls characters. In data networks, gateways are typically a node on both two networks that connects two otherwise incompatible networks. For example, PC users on a local area network may need a gateway to gain access to a mainframe computer since the mainframe does not speak the same language (protocols) as the PCs on the LAN. Thus, gateways on data networks often perform code and protocol conversion processes. Gateways also eliminate duplicate wiring by giving all users on the network access to the mainframe without each having a direct, hard-wired connection. Gateways also connect compatible networks owned by different entities, such as X.25 networks linked by X.75 gateways. Gateways are commonly used to connect people on one network, say a token ring network, with those on a long distance network. According to the OSI model, a gateway is a device that provides mapping at all seven layers of the model. A gateway may be used to interface between two incompatible electronic mail systems or for transferring data files from one system to another. Electronic mail systems that sit on local area networks often have gateways into bigger e-mail systems, like Internet or MCI Mail. For example, I might use MCI Mail to send a e-mail to someone's internal LAN e-mail. It might travel from MCI Mail to Internet via a gateway and then from Internet via another gateway to the company's e-mail on its own LAN.

2. A Gateway is an optional element in an H.323 conference. Gateways bridge H.323 conferences to other networks, communications protocols, and multimedia formats. Gateways are not required if connections to other networks or non-H.323 compliant terminals are not needed. Gatekeepers perform two important functions which help maintain the robustness of the network – address translation and bandwidth management. Gatekeepers map LAN aliases to IP addresses and provide address lookups when needed. Gatekeepers also exercise call control functions to limit the number of H.323 connections, and the total bandwidth used by these connections, in an H.323 "zone." A Gatekeeper is not required in an H.323 system-however, if a Gatekeeper is present, terminals must make use of its services. See TAPI 3.0.

gateway city A city where international calls must be routed. New York, Washington, DC, Miami, New Orleans, and San Francisco are the five gateway cities in the United States.

Gateway Protocol Converter GPC. An application-specific node that connects otherwise incompatible networks or networked devices. Converts data codes and transmission protocols to enable interoperability. Routers are capable of running gateway protocols – we used to call routers "gateways." Contrast to Bridge.

gateway server A communications server that provides access between networks that use different access protocols.

gating 1. Enabling or disabling a signal through applied logic. If it's turned on, the signal gets through. If not, the signal doesn't get through.

2. The process of selecting only those portions of a wave between specified time intervals or between specified amplitude limits.

gatored To be gatored means that while surfing the Internet, you're bombarded by pop-up ads. The term, according to Wired Magazine, comes from Gator, the ad-feeding app that's increasingly bundled with popular file sharing programs.

gauge A term for specifying the thickness (diameter) of cables. Thicker cables have a lower number in the American Wire Gauge (AWG) scale. Thicker gauge cables can carry phone conversations further and more cleanly than thinner gauge cable. But thicker cables cost more and take up more room, especially when you bundle them together and put them in a duct. When buying a phone system it is good to specify the thickness of the cables that will be installed – especially if some of your extensions will be a great distance from the central telephone switch, if you intend to carry high-speed data on them or you intend to live with your cabling scheme for more than a few months. You should, of course, not only specify the cable's thickness, but also whether it's stranded or solid core, coax, etc. Gauge is but one part of a cable description. See AWG for a fuller explanation.

gauge, wire The method of specifying the thickness and size of wire. The two important American gauges are the American Wire Gauge (AWG), previously known as Brown & Sharpe, and the Steel Wire Gauge. See AWG for a fuller explanation.

gauss The unit of magnetic field intensity in terms of the lines of force per square centimeter.

Gaussian beam A beam pattern used to approximate the distribution of energy in a fiber core. It can also be used to describe emission patterns from surface-emitting LEDs. Most people would recognize it as the bell curve.

Gaussian channel A radio channel in which the noise that detracts from the signal is characterized as Gaussian noise. Most radio communications channels are Gaussian channels. The exceptions are those subject to both Gaussian noise and multipath fading; see Rician channel and Gaussian noise.

Gaussian noise Gaussian noise, more correctly, is "average white Gaussian noise," also known as "white noise" and "thermal noise." It is the natural noise which occurs when electricity is passed through a conductor, and is due to the random vibration of electrons in the conductor. Gaussian noise is uniform across the entire range of frequencies involved. Gaussian noise is named after Karl Friedrich Gauss (1777-1855), the German mathematician who is generally recognized as the father of the mathematical theory of electricity. Gauss also invented the "Gaussian Distribution," or "bell curve," which is the frequency distribution of many natural phenomena. See White Noise for more detail.

gazillion An extremely large, indeterminate amount. See Gigabyte.

GB Gigabyte. See Gigabyte.

GbE See Gigabit Ethernet.

GBH Group Busy Hour.

GBIC Gigabit Interface Connector. The physical connection to Gigabit Ethernet media. A removable optical interface transceiver module designed to carry Gigabit Ethernet or Fibre Channel traffic. Used as a physical-layer transport interface on Gigabit Ethernet and fibre Channel Fabric switches.

Gbps Gigabits per second. Gig is one thousand million bits per second.

GCAC An ATM term. Generic Connection Admission Control: This is a process to determine if a link has potentially enough resources to support a connection.

GCCS Global Common and Control System. Military talk for an umbrella system that tracks every friendly tnak, plane, ship and soldier in real time, plotting their positions as they move on a digital map. It can also show enemy locations gleaned from intelligence. GCCS was used with great succcess in the second Iraq War in 2003.

GCI A TDM (Time Division Multiplexed) bus technology developed by Siemens.

GCRA An ATM term. Generic Cell Rate Algorithm: The GCRA is used to define conformance with respect to the traffic contract of the connection. For each cell arrival the GCRA determines whether the cell conforms to the traffic contract. The UPC function may implement the GCRA, or one or more equivalent algorithms to enforce conformance. The GCRA is defined with two parameters: the Increment (I) and the Limit (L).

GCS Global Communications Service.

GCT Greenwich Civil Time.

GD Graceful Discard. A Frame Term. See Committed Information Rate and Graceful Discard.

GDDM An SNA definition: Graphical Data Display Manager (GDDM) system software used for graphics display and printer devices and performs the same functions as QuickDraw in Macintosh computers.

GDDR-I A slower version of GDDR-II. See GDDR-II.

GDDR-II Graphics double data rate. GDDR-II is a new, higher data transfer "bus" between a CPU (central processing and a video card. One version of GDDR-II runs at 500 MHz and transfers data the one gigabit per second rate. That's one thousand million bits second. This incredible speed is allegely necessary for 3D graphics in high-speed gaming.

GDF Group Distribution Frame.

GDI Graphics Device Interface. The part of Windows that allows applications to draw on screens, printers, and other output devices. The GDI provides hundreds of convenient functions for drawing lines, circles, and polygons; rendering fonts; querying devices for their output capabilities; and more.

GDMO Guidelines for the Definition of Managed Objects.

GDOP See Geometric Dilution of Precision.

GE Gigabit Ethernet. See Gigabit Ethernet.

GEANT Gigabit European Academic NeTwork. A high-speed optical fiber network proposed to cover 30 European countries at speeds of 2.5 Gbps in 2001 and 100 Gbps by 2004. GEANT is the European version of Internet2. See also Internet2.

gearhead A geek who particularly loves new hardware. See Geek.

GEDCOM GEnealogical Data COMunication. GEDCOM is the accepted Genealogical Data Exchange format that allows users of different genealogy programs to exchange data. It was first developed by the Mormon Family History Library in conjunction with the PAF (Personal Ancestry File program). PAF may have been the basis of many of the commercial and shareware programs available today. Genealogy research via the Internet is pursued by millions of people. The major sites are www.ROOTSWEB.com, which supposedly has 50 million hits a month. There is also www.JewishGen.org, which has many special interest groups for various regions (e.g GerSig for Germany). Major commercial sites include www. Ancestry.com, which seems to be buying up many of the earlier programs and companies.

geek A computer enthusiast who doesn't have a life beyond computers and the Internet. Also called a Techno-Geek. Coined in the early 1940s, a geek was a carnival performer usually billed as a wild man whose act often consisted of biting the heads off live chickens

or snakes. "Geek" has its roots in the Greek "geck," meaning "fool." See also Geek Gab and Geekspeak.

geek gab "Variety" is a weekly magazine that covers the Hollywood entertainment business. It coined the word "geek gab," which it refers to as the proliferation of Web sites claiming to put forward the latest hot news on films, studios and networks. The "news," however, is often unsubstantiated rumor, can be vicious and can destroy a film.

geek testosterone When Microsoft turned over internal company materials to the court in Washington that was hearing its anti-trust case, several people believed that they revealed a company running on "geek testosterone." A geek is a computer enthusiast who doesn't have a life beyond computers. Testosterone is the sex hormone, C19H2802, secreted by the testes, that stimulates the development of male sex organs, secondary sexual traits, and sperm.

geekosphere A definition courtesy Wired Magazine: The area surrounding one's computer where trinkets, personal mementos, toys and "monitor pets" are displayed. A place where computer geeks show their colors.

geekspeak Geekspeak is the language geeks speak. A geek is a computer enthusiast who doesn't have a life beyond computers and the Internet.

geezer glut The large number of seniors that will result as the baby boom generation ages.

gefilte fish Translated literally, it means fish that's filled with something. In this case it's a mix of congealed fish parts and transparent slime jelly. It's considered a Jewish delicacy. But like all ethnic delicacies, it came about because the Jews were poor and could not afford to throw out any part of the expensive fish. So they took the truly revolting bits, cooked them and mixed them with jelly. Bingo a new delicacy. Any ethnic cooking that's basically mush was created when the ethnic people were dirt poor, though creative.

gelbe seiten Yellow Pages, in German-speaking countries.

gender Connectors, plugs and receptacles are assigned a gender to describe their physical type. Ones with pins are male, and those with holes into which the male pins slide are female. See Gender Bender.

gender bender A device which changes the gender of a connector, plug or receptacle. A gender bender is typically a small plug with all male pins on one side and all male pins on the other. By plugging a female connector into one side of a gender bender, you've effectively changed the female gender of the cable to male. Alternatively, a gender bender could be female on either side. But a gender bender must be the same on both sides.

gender changer Another name for a gender bender. See Gender Bender.

genderless connector Also called data connector or hermaphroditic connector. Invented by IBM. The connector doesn't require male and female plugs to make a connection. It was designed for token-ring applications. It was too big and clunky for my taste.

gene doping Tweaking an athlete's genes to produce more hormones or other performance-boosting substances. See doped fiber.

general availability How a product gets to market varies from one company to another. But typically, along the way, there's something called an alpha – the first version of hardware or software. It typically has so many bugs you only let your employees play with it. A beta is the next version. It's a pre-release version and selected customers (and the press) become your guinea pigs. They give you feedback. After beta, and when the bugs are removed and the features have been fine-honed, comes "general availability." That's when the product is finally available for buying by the general public.

general call The letters CQ in the international code and used as a general inquiry call.

General Claims A Verizon definition: Claims that cover more than one working telephone number.

General Packet Radio Service GPRS. General Packet Radio Service is the data service enhancement for GSM, the European standard digital cellular service. GPRS, a packet-switched service which will support the X.25 and TCP/IP packet protocols, is widely expected to be the next major step forward in the evolution of GSM technology. GPRS, an important component in the GSM evolution entitled GSM+, enables high-speed mobile datacom usage. It is most useful for "bursty" data applications such as mobile Internet browsing, e-mail and push technologies. These easily can be supported at speeds of 56 Kbps. GPRS has been demonstrated as fast as 115 Kbps, which also makes it suitable for high-speed file transfers. Notably, GPRS will support simultaneous voice and data communications over the same wireless link; voice, of course, always takes precedence. GPRS-compatible terminal equipment will include cell phones with microbrowsers and displays supporting WAP (Wireless Application Protocol), smart phones with full-screen display capability, and mobile credit card readers. Laptops can be equipped with GPRS modems, or can connect to a GPRS-capable cell phone. See also HSCSD.

General Pact Radio Service An enhancement for GSM cell phone service, based on packet switching technology enabling high-speed data transmission (115Kbit/sec).

general premises cabling licence In Australia, a previous cabling license that was replaced on July, 1, 1996, by the Base General Premises Cabling Licence (BCL). See BCL.

General Protection Fault GPF. A General Protection Fault is an indication that Windows 3.xx has tried to assign two or more programs to the same area in memory. Obviously that's not possible, since two things can't occupy the same area in memory. As a result, your screen stops and says "General Protection Fault." If you can save what you're doing, do it. If you have other programs open, try and save the material in them. Close Windows and then do a cold reboot. Do not continue to work after you have received a General Protection Fault. You must reboot. Better do a cold reboot, too.

general purpose network An AT&T term for a network suitable for carrying many forms of communication – voice and data, circuit and packet, image, sensor or signaling, for example.

general release When software is finally finished and ready to be sold to the general public, it said to be in "General Release." Before it is in general release, it is still in beta. Beta software is not alleged to be bug-free. General release software is meant to be bug-free. Sometimes it is.

General Telemetry Processor GTP. A device that receives and processes telecommunications equipment alarming protocols such as TBOS (Telemetry Bit-Oriented Serial).

General Trunk Forecast A telephone company term. GTF. A forecast of future trunk circuit requirements. This forecast covers the current year and the future four years.

Generation D Generation Digital refers all the people who have grown up with computers and digital technology. There are people now who have never known a world without PCs, MP3 players, touchtone phones, CDs and DVDs. My children look askance at me when I ask what the name of the record playing on the radio? I am definitely not a member of Generation D.

generational loss The reduction in picture quality resulting in the copying of analog images for editing and distribution.

generations, computer As computers have improved, so the industry's pundits have assigned "generations" to those improvements. The concept of generations is not perfect nor finite. Here's our best shot on generations in computers:

First generation: 1951-1958, core memory 8 Kbytes to 32 Kbytes.

Second generation: 1958-1964, transistor technology, memory 32 Kbytes to 64 Kbytes.

Third generation: 1964-1975, integrated circuitry.

Fourth generation: 1975-date, non procedural languages, software driven.

Fifth generation: into the 1990s, natural language programming, parallel processing and super computing.

Sixth generation: in the 2000, will process knowledge rather than data.

generations, PBX As PBXs have improved, so the industry's pundits have assigned "generations" to those improvements, as they did in computers. The concept of generations is not perfect nor finite. Here's our best shot on generations in PBXs:

First generation: 1920s to the late 1960s. Step-by-step mechanical equipment. The first and the last of the step-by-step Bell PBXs switches was called a 701. Lee Goeller says the 701 "was the best PBX ever built. It was infinitely flexible. It was just too BIG. In fact, it was usually bigger than the office it served. This era of the stepper will be remembered as the era the Bell System was intact and had the gaul to rent operator chairs."

Second Generation: Late 1960s: Bell 801 reed relay switch. Stromberg Carlson 800 series reed relay switch. GTE had a series, also. Reed relay switches were not very popular.

Third Generation: 1974 and 1975: Rolm introduces its first CBX, an electronic, solid-state PBX. AT&T introduces Dimension. Digital Telephone Systems introduced the D1200. Northern introduced its first stored-program controlled SL-1. Some of these PBXs switched voice digitally, though they used different techniques, including PCM, PAM, and Delta Modulation. The codecs were in the switch, not in the instruments.

Fourth generation: early 1980s. Distributed processing. Northern, Rolm and NEC and others introduced remote modules – slaves to the master switch at headquarters. These switches also added the capability of handling data without using modems. Switches like Lexar and InteCom were designed from the beginning to handle data without modems, thus requiring digital capability out to the set.

Fifth generation: CXC, Anderson Jacobson and Ztel and others called themselves "fourth generation." When they started to fail, some people called them the "fifth generation." It wasn't clear exactly what that generation was. But they all got lots of publicity and the PBXs from CXC, Anderson Jacobson and Ztel ultimately failed.

Sixth generation: Networked PBXs. Sit in New York. Operate your national network as if it were in the same building. Bingo, you can transfer calls across the country. All your messaging is the same wherever you sit on the network. Lee Goeller, however, says you can network stepper PBXs. In fact, in 1971 he says he managed one of the biggest integrated voice and data networks in the US using step-by-step electromechanical PBXs (701s made by AT&T). It was the world's largest dial tandem network. He had 63 different locations and three hubs.

Seventh generation: Open Architecture. You can now program your own PBX. For more, see the NORSTAR Command Set.

Eighth generation: Dumb switches. You can now buy completely dumb phone systems which are just basically switches. To get them to do anything they require an external computer (and software programming that computer) to drive the dumb switch. Often the "driving" is done through one or more serial ports.

Ninth generation: PC-based switches. Picture a PC server running Windows NT or 2000. The server contains some telephony boards that interface with the telephone network and switch calls around. The server also contains a NIC card (Network Interface Card) board which allows the server to connect to the corporate LAN and thus to the Internet. This PC-based switch thus does standard telephony things – take calls, switch them, conference them, etc. But because it's a PC, it adds "intelligence" so it can read emails and recognize simple speech ("Call my mother"). Because it's networked and joined to the Internet, it can also let you pick up voice mail messages over the Internet — from anywhere in the world, for free.

In reality, the concept of generations amongst PBXs is very flimsy. But it's the stuff dictionaries are made of.

generator A machine which converts mechanical energy, such as the power from a piston engine into electrical energy.

generator polynomial A polynomial, G, used in the Cyclic Redundancy Check (CRC); the message signal M is multiplied by G before transmission; upon reception the signal is divided by the same G; a residue indicates error(s).

generic access profile A wireless term defining signaling standards between a base station and a DECT wireless phone. There is no assumption in the DECT set of standards that handsets from one maker will work with base stations from another. GAP compliant handsets will work with GAP compliant base stations. Each manufacturer has added features to their DECT equipment which are outside the GAP specification, so a "foreign" handset may still not perform as well as the base station manufacturer's own. DECT stands for Digital European Cordless Telecommunication. The pan-European wireless standard based on time division multiple access used for limited-range wireless services. Based on advanced TDMA technology, and used primarily for wireless PBX systems, telepoint and residential cordless telephony today, uses for DECT include paging and cordless LANs. DECT frequency is 1800-1900 MHz.

Generic Attribute Registration Protocol See GARP.

generic cell rate algorithm An ATM function that is carried out at the user-to-network interface (UNI) level. It guarantees that traffic matches the negotiated connection that has been established between the user and the network.

Generic Flow Control Field GFC. A 4-bit value in an ATM header for purposes of flow control between the user equipment and the carrier ATM Edge Switch across the User Network Interface (UNI). The GFC field tells the target end-station that the switch may implement some form of congestion control.

Generic Framing Procedure See GFP.

generic program A set of instructions for an ESS central office or electronic PBX that is the same for all installations of that particular equipment. Detailed differences for each individual installation are listed in a separate parameter table. Here's a more formal definition, from Telcordia. A generic program is a set of instructions for an electronic switching system or operations system that is the same for all central offices using that exact type of system. Detailed differences for each individual office are usually listed in a separated parameter table.

Generic Requirements GR. See GR.

Generic Services Framework GSF. Generic Services Framework is a set of software designs being implemented by Bell Northern Research (a subsidiary of Northern Telecom) to accelerate the development and testing of new features and provide a platform for the later stages of DMS SuperNode interworking with Advanced Intelligent Networking (AIN). GSF applies the principles of object-oriented programming to DMS SuperNode software, and delivers enhancements that simplify feature development and reduce testing needs. The GSF has three distinct elements:

1. Call Separation - The new architecture uses separate software data and processes to handle the two "halves" of a call (originating and terminating).

2. GSF Agent Interworking Protocol (AIP) - The AIP uses a standardized set of instructions to allows the two call halves to communicate.

3. Event-Driven Call Processing (EDCP) - Just as the AIP mediates communications between call halves, EDCP handles communications within each call half. EDCP also uses a standardized set of instructions to simplify messaging.

Generic Top Level Domain see gTLD and Domain.

generica Features of the American landscape that are exactly the same no matter where one is, such as fast food joints, strip malls, subdivisions.

genlock Circuitry that synchronizes video signals for mixing. The video circuitry determines the exact moment at which a video frame begins. Genlock allows multiple devices (video recorders cameras, etc.,) to be used together with precise timing so that they capture a scene in unison. A genlock display adapter converts screen output into an NTSC video signal, which it synchronizes with an external video source.

GEO Geosynchronous Earth Orbit. A term for a satellite which is placed in a geosynchronous, or geostationary, orbital slot. Such orbits are always equatorial; the satellites are placed at altitudes of approximately 22,300 miles (or 35,888 kilometers). As a result, they are synchronized with the rotation of the earth, maintaining their relative position to the earth's surface. In other words, they always appear to be in the same spot. See also Geostationary Satellite. Contrast with LEO and MEO.

geofencing Imagine a tractor trailer or truck with GPS (Global Positioning System) service. The truck driver knows instantly where he is. Now figure that the truck contains a computer that also receives the GPS signal and compares the truck's indicated position to where it should be or, most importantly, where it shouldn't be. If it's somewhere it shouldn't be, the computer might dial a number on its cell phone and alert someone or something somewhere that the truck is presently somewhere it shouldn't be. The person or computer alerted might check with the driver of the truck – "Why are you there? – and, if the answer is not satisfactory, it might dispatch the police or security force. Thus the idea of geofencing – fencing the truck in with GPS, also called allowing fleet managers to erect invisible boundaries using satellite tracking. Geofencing also works with door sensors – telling whether trailer doors are open.

Geographic Information Systems GIS. Computer applications involving the storage and manipulation of electronic maps and related data. Applications include resource planning, commercial development, military mapping, etc. Imagine a satellite capable of photographing the distance from London to Paris with the ability to identify objects within the width of a car headlight. Or, imagine the Los Angeles Police Department possessing a "gang tracking" program that can be used to record and predict gang movements throughout the city. Welcome to the changing technological world of cartography. Researchers at the University of Missouri-Columbia believe this new form of computer-based cartography, Geographic Information Systems, can create maps that are adaptable, open-ended and smart. These maps will be able to make decisions, predict the future and effectively persuade an audience. "In today's world of new borders and increased technology, the era of the Rand McNally road map is gradually becoming superseded by an era of hi-tech mapping," said Karen Piper, assistant English professor at MU. "GIS can track and map movement – from wildfires to gangs to migrating animals – in a way paper maps could not."

geographic interface First there was the ASCII interface - a screen containing nothing but ASCII letters. You saw them on airline reservation terminals. Just plain boring green type against a dark, unlit, black background. Then you got the GUI - the Graphical User Interface. Windows is the most famous GUI. It used icons to represent actions. An eraser to indicate you could remove something. A calendar to indicate that you could enter your day's schedules, etc. Now there is the "Geographic Interface." It attempts to depict objects from the real world – or at least some circumscribed part of it, like your office. The first of these products was Magic Cap from General Magic. Another was the opening screen for Apple's eWorld on-line service. Another is Novell's Corsair technology. Click on a filing cabinet to get to its database contents. Click on a bloodhound find something. One idea behind these geographic interfaces is that there should be a libraries of objects. A set useful for an auto mechanic. Another set useful for a bond trader. The idea is that everyone gets to choose the bits and pieces of the geographic interface that he or she is most comfortable with. Some of these new geographic interfaces are endearing. Some are tiresome. We'll see if they endure.

geographic north North based upon some geographic grid system which may or may not correspond to "true north."

geographical portability The ability to take your New York phone number to Boston, including the area code. In short, geographical portability is the ability to take all your phone number (i.e. all its 10-digits) with you wherever you go – anywhere in North America. No timetable has been set for geographical portability in North America. See LNP (Local Number Portability) and 800 Service.

geographical redundancy A level of redundancy that involves more than one system's serving a given geographical area. Geographical redundancy is a step above physical redundancy. Physical redundancy involves a backup system (e.g., router) in the same physical location as the primary system. In the event of a failure in the primary system, the backup system automatically activates. Geographical redundancy provides an additional measure of protection through another system, or set of primary and backup systems, in another physical location. While each set of systems has primary responsibility for a given geographical area, each can assume responsibility for others in the event of a catastrophic failure affecting a given physical location. The term most commonly applies to Internet access.

geolocation technology Also known as Position Determination Technology (PDT). A technology used to determine the geographic coordinates of a radio-equipped mobile device, e.g., a cellular handset. See Position Determination Technology.

Geometric Dilution of Precision GDOP or DOP (Dilution of Precision). In systems employing position determination technology such as Enhanced 911, GDOP refers to the phenomenon that causes the calculated position's accuracy to be a factor of the relative geometry of the involved units. For example, in an angle of arrival system, two fixed units measure the angle to a mobile unit. The resulting lines intersect in a point defining the mobile's position. If each measured angle has some uncertainty, the intersection is no longer a point, but say, a quadrilateral. In this case, the mobile is known only to be within the area of the quadrilateral. Depending on where the mobile is in relation to the fixed measurement units, the area and shape of the quadrilateral (and therefore the accuracy of the calculated position) may be larger or smaller. GDOP also affects time difference of arrival systems.

geometric optics A field of physics that deals with light as if it were composed of rays diverging in various directions from a source.

geostationary arc The part of the geostationary orbit that can be "seen" by, and consequently the orbital positions that can be accessed by an observer at any particular latitude and longitude on Earth.

geostationary orbit Also called Geostationary Earth Orbit (GEO); a particular type of Geosynchronous Orbit that nominally is circular and lies within the plane of the Earth's equator, with a radius of 26,200 miles (22,237 miles above the Equator). Satellites are launched into in orbits to move in the same direction as the Earth revolves to have a period equal to the sidereal day of 23 hours, 56 minutes and 4.091 seconds. Ideally, as seen from Earth, the satellite remains fixed at an Orbital Position and consequently seems to be stationary in space, which gives the orbit its name. However, in practice, there are no purely geostationary orbits, principally because a satellite in orbit is subject to three forces that disturb its orbit. First, the Earth is an imperfect sphere, so its gravitational field is asymmetrical. This causes longitudinal drift of the orbital position, toward positions where the drift is zero, at 102 degrees West and 76 degrees East latitude. Second, the sun and the moon exert gravitational force on a satellite, which tip its orbit at an inclination to the equatorial plane at a rate of about 0.8 degrees per year. Finally, the sun exerts radiation pressure on the satellite, accelerating it during one half of its orbit and decelerating it during the other half, with the result that the orbit becomes elliptical. If left to itself, the satellite would be moved by these forces and would gradually depart from its orbit, most likely to one of the libration points between the Earth and the Moon where forces balance and gravity is zero. Hence, keeping a satellite in geostationary orbit requires exerting some force, which in turn requires energy, which is both expensive and must be stored on board. So orbits are designed to minimize the use of energy stored on board satellites by routinely accepting deviations from the geostationary ideal. The satellite is not "stationary", but moves about in a defined box, or window determined by Station Keeping. See Geostationary Satellite.

geostationary satellite A satellite with a period of revolution of one sidereal day. A satellite in geostationary orbit, also called geosynchronous orbit. A satellite placed in an geosynchronous orbit – 22,300 miles (or 35,888 kilometers) directly over the earth's equator – will appear to be stationary in the sky, turning synchronously with the earth. This means you can plant a satellite receiving/transmitting antenna on the ground, and point it at that one place in the sky to receive signals from and transmit signals to "the bird", as satellites are sometimes called. Most communications satellites are in geostationary or geosynchronous orbit. The Russians have some satellites that orbit the earth and require antennas which move. These satellites are used to transmit to far northern communities which are difficult to reach with normal geosynchronous satellites.

Geostationary Satellite Orbit GSO. A satellite orbit 23,000 miles

(35,888 kilometers) over the equator with an orbit time exactly 24 hours. Thus a satellite in a Geostationary Satellite Orbit appears motionless to an earth station which can receive it with a stationary antenna. One drawback is that a two-way communication channel through a geostationary satellite incurs a one-way delay, due to the finite speed of light (186,000 miles per sec) or about one-quarter second. This can be annoying because, after a pause in the conversation, the two users may start talking at about the same time and "collide" with each other, due to the fact that each started talking within a quarter second of each other.

Geosynchronous Orbit Synchronous with the Earth. An orbit 22,300 miles (or 35,888 kilometers) above the earth's equator where satellites circle at the same rate as the earth's rotation, thereby appearing stationary to an earth-bound observer. See also Geostationary Orbit.

Geotech Geotechnical Report. Soil boring to determine soil classification (rock, clay, etc) prior to installing the foundation for a site. See also SHPO, State Historic Preservation Officer.

GET The standard method by which a Web browser gets information from a server. See Browser.

get a life What your kids say to you when you start talking too much about the information superhighway, or the Internet, of fiber optic, or something that doesn't interest you.

get someone's goat The expression, "To get someone's goat," meaning to anger or irritate, originated in 19th century racing stables in England. High-strung race horses were kept calm by having them share the stable with a goat. Evidently, the company calmed them and allowed them to rest and relax. Unprincipled touts would sometimes sneak into the stable at night and remove the goat. The horse get upset, would not have a good rest, and lose the race the next morning.

GETS Government Emergency Telecommunications Service. An integration of commercial networks to create a telecommunication backbone in case of a national disaster. GETS is a set of switch-based and Advanced Intelligent Network features that provide a high probability of completion for critical users such as the military and government before, during and after a national security emergency. GETS provides the ability to complete more authorized calls across public, defense, and federal networks in times of emergencies that include: natural disasters, military attacks, technological emergencies, or other emergencies that degrade or threaten the security of the United States. The system was developed in response to a White House executive order, and satisfies the requirements of the National Information Infrastructure (NII) and Defense Information Infrastructure (DII). GETS is paid for by the United States government. Many service providers have entered into GETS contracts with the government as this service helps facilitate additional call completions, thereby generating the potential for additional revenue.

GFC Generic Flow Control. GFC is a field in the ATM header which can be used to provide local functions (e.g., flow control). It has local significance only and the value encoded in the field is not carried end-to-end.

GFCI Ground Fault Circuit Interrupter. A device intended to interrupt the electrical circuit when the fault current to ground exceeds a predetermined value (usually 4 to 6 milliamps) that is less than required to operate the overcurrent protection (fuse or breaker) for the circuit. This device is intended to protect people against electrocution. It does not protect against fire from circuit overload. GFCI outlets are typically installed in bathrooms, kitchens and garages because the presence of water in these area increases the possibility of electric shock. Sometime GFCI circuits are incorrectly wired. The way to find out if your GFCI is wired correctly is to press the test button on its face. This should shut off power to the GFCI outlet and to those outlets connected to it.

GFF Gain flattening filter.

GFI Group Format Identifier. In packet switching, refers to the first four bits in a packet header. Contains the Q bit, D bit and modulus value.

GFLOPS One billion FLoating point Operations Per Second. (G stands for GIGA, meaning billion). Today's fastest supercomputers are able to maintain a sustained throughput of over one billion floating point operations per second (GFLOPS) while performing real-world applications. By contrast, a 25-MHz 486 personal computer can sustain about one million floating operations per second (one MFLOP), or about one-thousandth the throughput of a supercomputer. See also G.

GFP Generic Framing Procedure. GFC is a draft ANSI (American National Standards Institute) standard that defines a generic procedure for encapsulating and delineating variable-length payloads from higher-level signals for transport over a SONET (Synchronous Optical Network) or OTN (Optical Transport Network). The higher-level client payloads may be native data client PDUs (Protocol Data Units) such as Frame Relay frames, Ethernet frames, PPP (Point-to-Point Protocol) frames, or IP (Internet Protocol) packets; block-code oriented data such as Fibre Channel or ESCON; constant rate bit streams; or data character

sets. GFP provides a mechanism by which that higher-level traffic is adapted to run over an octet-oriented synchronous transport network.

GFRC Gravel Formed or Glass Fiber Reinforced Concrete building. A GRRC building is a small shelter housing telephone equipment. It's strong. One of the first users of GFRC buildings was the U.S. Military. One of the unique properties of the material used is its ability to accept multiple bullet rounds at close distances without penetration.

GFSK The modulation technique Gaussian Frequency Shift Keying.

GFXO Ground start FXO.

GFXS Ground start FXS.

GGP Gateway to Gateway Protocol. The protocol that core gateways use to exchange routing information, GGP implements a distributed shortest path routing computation.

GGSN Gateway GPRS Support Node. An SGSN is a component sold by Ericsson, Lucent, Nokia that fit in the Mobile Switching Center of a wireless network. See GPRS.

GH Effect Gordon-Haus Effect. The arrival time of a post-pulse soliton is a random variable governed by the medium composition/delivery, a random variable whose variance is proportional to the cube of the propagation distance.

GHA See Greenwich Hour Angle.

Ghost 1. A secondary image resulting from echo or envelope delay distortion. It can be a false radar return caused by weather effects.

2. Ghost is also the name of Symantec software used backup and clone your PC's hard drive and easily transfer it to other PC's. It is used in large PC deployment and standard PC image creation.

3. A term coined by Fluke to mean energy (noise) detected on the cable that bears similarities to a real frame, but does not include a valid start frame delimiter. To qualify for this category of error, the event must be a minimum 72 octets. Ghosts are a strong indication of a physical problem on the local segment.

ghost work The unfinished projects left behind by laid-off employees. Survivors are often haunted by them.

GHz One billion, or one thousand million, hertz, or cycles per second. See also Bandwidth.

Gi Interface Reference point between a GPRS network and an external packet data network. See GPRS.

GID GARP (Generic Attribute Registration Protocol) Information Declaration. See GARP for more information.

GIF See Graphics Interchange Format.

GIG See Global Information Grid.

Gig-E Gigabit Ethernet. See Gigabit Ethernet.

giga From the Greek "gigas," meaning "giant." Prefix meaning one billion, which is one thousand million. 1,000,000,000. Giga is the reciprocal of nano. See also Gigabit, Gigabyte, GFLOP.

gigabit 1. In transmission terms, exactly one billion bits, or one thousand million bits. In the world of transmission systems, we speak of the number of bits which can be transmitted in a period of time - -specifically, one second. Hence, 1 Gbps is one billion bits per second. Let me illustrate. In eight seconds at a transmission rate of 1 Gbps, I could send you 200 copies of my dictionary.

2. In computer terms, a Gigabit is 1024 times mega; in other words, actually 1,073,741,824. One thousand gigas are a tera. One thousand teras one peta, which is equal to 10 to the 15th.

Think of it another way, just to put it in personal perspective. If you are a Gigasecond (one billion seconds) old, you have lived to the ripe old age of 31 years, 8 months, 18 days, 18 hours, 50 minutes and 24 seconds. (Feel free to check my math.) At a rate of 1 Gbps, your entire life could flash across your network in a single second's time. Think about SONET fiber optic transmission facilities, which can operate at 2.5 Gbps or more.

See also GigaSTaR.

Gigabit Ethernet Gigabit Ethernet (GE) uses the same framing as Ethernet and Fast Ethernet, but has a much higher clock speed (one billion bits per second). There are slower Ethernets: 10Base-T Ethernet (the kind on our desktop) runs at 10 million bits per second , while Fast Ethernet runs at a clock speeed of one hundred million bits per second (100BaseT) – the kind we are increasingly seeing on our desktop. The Gigabit Ethernet over fiber optic cable standard was finalized and formally approved on June 29, 1998, as IEEE 802.3z. The GE over Category 5 cable standard was ratified as IEEE 802.3ab. The IEEE 802.3ab standard uses all four pairs (8 wires) of cable in the Category 5 cable for transmission. This was a departure from previous Ethernet and Fast Ethernet copper standards which only used two pairs (4 wires) in the IEEE 802.3 and 802.3u standards. Although GE is available in both shared and switched varieties, the pricing difference between the two has become so marginal that most manufacturers are only producing the switched (and thus faster-feeling) variety. While GE is much like traditional Ethernet, differences include frame size options. The clock speed of GE is ten orders of magnitude greater than its predecessors and the previous frame size of 1,518 bytes was a bottleneck in the transmission of information. The maximum frame size has been increased from 1,518 bytes to a jumbo frame size of 9K (9,216 bytes). The larger frame size improves the frame throughput of a GE switch as each frame requires switch processing of only header information (cut-through switching). The fewer frames presented to the switch, the more data the switch can process, switch and deliver in a given period of time. Multi-mode fiber will support Gigabit Ethernet transmission at distances up to 550 meters, and single mode fiber up to 40 kilometers. The distance that really can be transmitted is dependent upon the optics used to transmit and receive the signal. In each case, there is a minimum distance of 2 meters dues to issues of signal reflection (echo). GE switches adhering to the IEEE 802.3ab standard, offer auto-negotiating 10/100/1000 Mbps ports. Both half-duplex and full-duplex interfaces are supported, with full-duplex offering the advantage of the need for the CSMA/CD protocol because data collisions are impossible with a full-duplex mechanism. QoS (Quality of Service) guarantees are not currently an inherent element of GE like ATM has, but Ethernet has a prioritization mechanism revolving around the three IEEE 802.1p tagging bits written about in the IEEE 802.1q standards. Gigabit Ethernet has recently been extended to 10 gigabit speeds with the ratification of the IEEE 802.3ae. Service providers have started to adopt 10 GE as an alternative to SONET with the advent of the IEEE 802.ah Ethernet in the First Mile (EFM) standardization. See also 64b/66b, 802.3ah, 802.3ab, 802.3ae, and 10 Gigabit Ethernet.

Gigabit Ethernet Alliance A multi-vendor forum comprised of 86 members committed to driving the industry's adoption of networking standards at up to 1,000 megabits per second. The forum supports the CSMA/CD (Carrier Sense Multiple Access with Collision Detection) protocol of the original Ethernet standard. According to the Alliance, gigabit Ethernet will initially be deployed in backbone environments as the preferred interconnection for switches which support multiple transmission speeds between Ethernet segments (10 Mbps and 100 Mbps). Technical proposals developed by Alliance members have been submitted to the IEEE 802.3z standards committee, furthering efforts to standardize 1,000 Mbps Ethernet technology. See Gigabit Ethernet Alliance Interoperability Consortium. www. gigabit-ethernet.org.

Gigabit Ethernet Alliance Interoperability Consortium An organization of 15 vendor members formed within the Gigabit Ethernet Alliance for the purpose of testing the interoperability of 1Gbps products. The group will work with the University of New Hampshire Lab to conduct its interoperability tests.

Gigabit Passive Optical Network. See GPON.

Gigabits One thousand million bits. In the U.S., that's the same as one billion bits.

Gigabyte GB. A combination of the Greek "gigas," meaning "giant," and the English "bite," meaning "a small amount of food." A unit of measurement for physical data storage on some form of storage device-hard disk, optical disk, RAM memory etc. and equal to two raised to the 30th power, i.e., 1,073,741,824 bytes. This is the progression:

    KB = Kilobyte (2 to the 10th power)
    MB = Megabyte (2 to the 20th power)
    GB = Gigabyte (2 to the 30th power)
    TB = Terabyte (2 to the 40th power)
    PB = Petabyte (2 to the 50th power)
    EB = Exabyte (2 to the 60th power)
    ZB = Zettabyte (2 to the 70th power)
    YB = Yottabyte (2 to the 80th power)
    One googolbyte equals 2 to the 100th power.

GigaE Gigabit Ethernet. Ethernet running at the speed of one billion bits per second, which is the same as 1,000 million bits per second. See Ethernet.

gigaflop A unit of microprocessor processing speed equal to one billion flops, or floating point operations per second.

gigahertz GHz. A measurement of the frequency of a signal equivalent to one billion cycles per second, or one thousand million cycles per second.

gigaplane A Sun Microsystems term, Center plane bus used in Sun's Ultra Enterprise Server line. Uses separate paths of address, data, and control lines. Communicates with several subsystems concurrently. With a 167MHz UltraSPARC CPU, the Gigaplane can do rates of 2.5 Gbytes per second.

GIGAPOP A POP (Point Of Presence) with a throughput in the range of a billion (giga) packets per second. GIGAPOPs are being implemented in support of Internet2, a high-speed Internet supported by the National Science Foundation and a project of the University

Corporation for Advanced Internet Development (UCAID). Internet2 uses the MCI vBNS (very high-speed Backbone Network Services) fiber optic network for transport between the GIGAPOPs, and for access to the GIGAPOPs from member universities and NSF-funded supercomputing centers. A GIGIPOP differs from a NAP (Network Access Point) in that it is a value adding, OSI layer 3 (Network Layer) meet point between customers and network providers. A NAP is a neutral, OSI layer 2 (Link Layer) meet point for ISPs (Internet Service Providers) to exchange traffic and routes.

GigaSTaR Gigabit/s Serial Transmit and Receive. GigaSTaR is a short-haul (up to 50 meters), point-to-point communication link that transmits data with a bandwidth up to 1.32 gigabits per second over one pair of ordinary shielded twisted copper cables over 50 meters (e.g. CAT5/7) and up to 500 meters with conventional multimode fibre cable and standard optical modules. The data rate can be scaled in multiples of 1.32 gigabits per second. The link consists of two devices, the transmitter and the Receiver which convert any parallel data word with up to 36 bits and a max 33 MHz (= 148.5 megabytes per second) to a serial data stream and back to the original parallel word. Line coding reduces the overhead to 10 percent, thus 90% of the bandwidth (= 1.188 gigabit per second payload data) is available for data transmission. Applications for GigaSTaR include high-Speed scanning, printing, mass storage, security, tomography, high-speed sensoring, high-resolution camera links.

GigE Gigabit Ethernet. See Gigabit Ethernet.

GIGO Garbage In, Garbage Out. Regardless of the capability of the computer, bad data yields bad results.

GII Global Information Infrastructure, a term first advocated in March, 1994 and since used as a concept around which to form many international committees. For more, see the IEEE Communications Magazine, June 1996.

Gilder's law Bandwidth will rise at a rate of three times the rate at which processing power is increasing, or three times the rate of Moore's Law. In other words, with processing power doubling every 18 months, bandwidth will double every six months. At the moment processing power is doubling every year and bandwidth every four months. If Kurzweil's Law holds, this rate will continue to accelerate. This law from George Gilder.

Gilder's paradigm The reversal of the calculus of abundances and scarcities that governed the previous era. While applied power and silicon surface area (spread over computer backplanes and motherboards and daughter cards) were abundant in the old era, they are scarce in the new era of single chip systems and mobile devices. And, while bandwidth was scarce in the old era, it is abundant in the new. While the old rule was to waste power, transistors, and silicon area to compensate for inadequate bandwidth, the new rule is to waste bandwidth to compensate for inadequate silicon area and power. This law from George Gilder.

Gillette units Back in the 1950s, when the laser was invented and began being experimented with, it was half-jokingly suggested that lasing power be measured in "Gillette" units, because often a laser was tested by having it burn through a razor blade.

GIM Group Identification Mark. The Group ID mark is a two digit number used by cellular sites other than your home system to determine if your cellular phone should be allowed to make phone calls, i.e. access on "roam" status. This feature is not yet fully implemented. As cellular systems are upgraded, the GIM will be on line real time, requiring all NAM information, including the Mobile Identification Number (MIN), to be validated before a subscriber is allowed to call outside of their home area.

Gimlet Sir T.O. Gimlette, a British navel surgeon, believed that drinking straight gin was unhealthy and impaired the efficiency of navel officers, so he began to dilute it with lime juice, hence, the Gimlet.

GIP 1. See Global Internet Project.
　　2. GARP (Generic Attribute Registration Protocol) Information Protocol. See GARP.

giraffiti The Washington Post's Style Invitational asked readers to take any word from the dictionary, alter it by adding, subtracting or changing one letter, and supply a new definition. This one is one of the winners. Giraffiti is vandalism spray-painted very high.

GIS Geographic Information Services. Computer applications involving the storage and manipulation of electronic maps and related data. Applications include resource planning, commercial development, military mapping, etc. Raw input comes often from satellite photographs.

GIX Global Internet eXchange. A common routing exchange point which allows pairs of networks to implement agreed-upon routing policies. The GIX is intended to allow maximum connectivity to the Internet for networks all over the world.

GK A totally silly acronym for Gatekeeper. See Gatekeeper.

GL Graphics Library.

gladiator Person engaged in a fight to the death for public entertainment in ancient Rome. See also Sensitivity Training.

glare Glare occurs when both ends of a telephone line or trunk are seized at the same time for different purposes or by different users. Most embarrassing – glaringly so, in fact. Blame Ray Horak for this awful pun. See Glare Hold and Glare Release and Glare Resolution. See also Ring Splash.

glare hold and glare release A method of glare resolution. Glare occurs when both the local and distant end of a trunk are seized at the same instant; this usually results in deadlock of the trunk. To prevent this, one end of the trunk is assigned a glare hold status and the other a glare release status. In the event of glare, the glare hold end holds the trunk and the glare release end releases the trunk and attempts to seize another. This approach is used in cellular systems between the MTSO (Mobile Traffic Switching Office) and the connecting cell sites. See also Ring Splash.

glare resolution Ability of a system to ensure that if a trunk is seized by both ends simultaneously, one caller is given priority and the other is switched to another trunk. See Glare, Glare Hold and Glare Release.

glare window The period of time in a trunk is susceptible to glare, a situation in which a trunk simultaneously is seized by the switches (e.g., a CO and either a PBX or ACD) at both ends. The size of the window can be reduced through ring splash and other techniques. See also Glare Hold and Glare Release, and Ring Splash.

glass clinking In simpler times, it was not uncommon for someone to try to kill an enemy by offering him a poisoned drink. To prove to a guest that a drink was safe, it became customary for a host to hold out his own glass and allow the guest to pour a small amount of his drink into it. Then both would drink simultaneously. When a guest wished to show that he trusted his host, he would not pour the liquid, but simply touch the host's glass with his own. The clinking of glasses before a toast is what remains of this ritual.

glass house 1. A colloquial word for a mainframe computer. It derives from the fact that all mainframe computers were once housed in a separate, locked room, with glass windows. You typically needed to pass through heavy security to gain admittance to the glass room.

2. A room, closet, department, floor, or entire building in which special equipment and/or procedures are implemented which allow the data processing within to proceed without interruption even if power or other services are cut off. Creating a "glass house" environment may be a simple as installing a UPS in a wiring closet or may involve providing backup power, heat, light, and telecommunications services for an entire building.

glass insulators Glass insulators were widely used in the 1800s to fasten wire to telephone poles and to protect insulator pins from moisture so they couldn't conduct electricity. This was a technique developed by the telegraph industry over a 40-year period of experimentation and was one of the few basic telegraph practices carried over into telephone line construction. Insulators are found in a variety of different shapes and colors depending on the time period they were developed and on their application. Most have a greenish color from traces of iron oxide from the sand used to make the glass. When insulator design was in its heyday in the mid-to-late 1800s, hundreds of patents to improve the product were issued. For example, the double petticoat, a second lip on the bottom of the insulator, was added to reduce the amount of moisture that could travel up the inside of the cap. The above explanation courtesy Tellabs of Lisle, IL.

glass mount antenna A type of car phone antenna used in cellular service. A glass mount antenna is glued to a car's rear window. Many window-mounted glass antennas have a break in their wiring. The wiring ends at the inside. There is no electrical connection between the inside of the window and the antenna glued onto the outside of the window. The "connection" is done through signal radiation. This type of antenna is not as efficient as one in which the wire goes unbroken from the radio to the antenna.

glass terminal A keyboard and screen that conveys data generated by the user directly to a computer or network without buffering or otherwise acting upon the data, and also returns data unchanged from the computer to the user. This terminal type does not provide for cursor addressing or escape sequences.

glazing Corporate-speak for sleeping with your eyes open. A popular pastime at conferences and early-morning meetings. "Didn't he notice that half the room was glazing by the second session?"

glibido The Washington Post's Style Invitational asked readers to take any word from the dictionary, alter it by adding, subtracting or changing one letter, and supply a new definition. This one is one of the winners. Glibido means all talk and no action.

GLines GSF Lines.

glitch A jargon term used in data communications to describe an extraneous bit that has been introduced into a bit stream usually by a noise source. It can also be a problem or a delay. Can be a noun. "What's the glitch?" Or a verb: "Who glitched this thing up?" Glitch is also a momentary interruption in electrical power.

global Universal. An adjective meaning the whole world. See Global Search.

global 800 International toll-free numbers. See UIFN for more detail.

global access A new service of MCI Mail. It allows you, an MCI Mail user, to use your computer and its modem to dial a local number in a foreign (i.e. non-North American city) and reach a port of a packet switched operation called InfoNet. When you reach InfoNet you will then punch in a few letters and reach MCI Mail in the U.S. You can then leave MCI Mail messages, send telexes, send faxes and send paper mail, i.e. do all the normal services MCI Mail allows you to do. The advantage of Global Access is that you don't have to dial back to the U.S. (which usually doesn't work because of all the garbage on the line) or subscribe to a foreign packet switched operation (they have them in all industrialized countries). Sadly, it usually takes weeks to subscribe to a foreign packet switched operator.

global beam An antenna down-link pattern used by the Intelsat satellites, which effectively covers one-third of the globe. Global beams are aimed at the center of the Atlantic, Pacific and Indian Oceans by the respective Intelsat satellites, enabling all nations on each side of the ocean to receive the signal. Because they transmit to such a wide area, global beam transponders have significantly lower EIRP outputs at the surface of the Earth as compared to a US domestic satellite system which covers just the continental United States. Therefore, earth stations receiving global beam signals need antennas much larger in size (typically 10 meters and above (i.e.30 feet and up).

Global Common and Control System. See GCCS.

global directory Imagine a bunch of local area networks all connected together. Today you log onto one server on one LAN, tell it who you are and what your password is. You want to connect to another server? You have to tell it who you are and what your password is, which may be different from the first time. Global Directory, a feature of NetWare 4.x, gives you a central directory. You establish your user name once and associated with your name is your authorized service on all the connected LANs. This way you don't have to sign on again for another server. If telephone systems can attach to NetWare (see Computer Telephony), then they should also be able to benefit from this central directory – for phone bill allocation, people location, phone moves and changes, etc. The reason NetWare never had a central directory is historical. Novell introduced NetWare in 1983 as software to allow a handful of personal computers to share a single hard disk, which at that stage was a costly and scarce resource. As hard disks became more available, the product evolved to allow the sharing of printers and file servers. It was always designed as a departmental computing solution. It's only recently, with more powerful desktop machines, that the Client-Server LAN concept has become more corporate-wide in concept.

global directory service Allows desktop clients to transparently access data on servers across a network.

global gateway A web page that a user lands on the first time that he visits an international organization's web site, where the user sets his country of preference for subsequent visits. The country preference is stored in a cookie on the user's PC.

global group A security or distribution group that can contain users, groups, and computers from its own domain as members. Global security groups can be granted rights and permissions on resources in any domain in its forest. Global groups cannot be created or maintained on computers running Windows XP Professional. However, for Windows XP Professional-based computers that participate in a domain, domain global groups can be granted rights and permissions at those workstations and can become members of local groups at those workstations.

Global Information Grid A globally interconnected, end-to-end set of information capabilities, associated processes, and personnel for collecting, processing, storing, disseminating, and managing information on demand to the US military, government policymakers, and support personnel. The GIG includes all owned and leased communications and computing systems and services, software, data security services, and other associated services necessary to achieve information superiority. It also includes National Security Systems (NSS) as defined in section 5124 of the Clinger-Cohen Act of 1996. The GIG supports all Department of Defense, National Security, and related Intelligence Community missions and functions (strategic, operational, tactical, and business) in war and in peace. The GIG provides capabilities from all operating locations (bases, posts, camps, stations, facilities, mobile platforms, and deployed sites). The GIG provides interfaces to coalition, allied, and non-DoD users and systems.

global internet exchange See GIX.

global internet project GIP. Comprised from a group of senior executives, representing sixteen leading Internet software, telecommunications and digital commerce companies worldwide. Its purpose is to promote the growth of the Internet across geographic boundaries worldwide. Explores present and future impact of the Internet upon commerce and society.

global login A mechanism that permits users to log on to the network, rather than repeatedly logging on to individual servers. A global logon can provide access to all network resources.

Global Maritime Distress and Safety System See GMDSS.

Global Mobile Personal Communications Services GMPCS. A term coined by the ITU-T to refer to satellite telephony to be provided by the proposed Big LEO (Low Earth-Orbiting Satellite) systems such as Teledisc and Globalstar and MEO (Middle Earth-Orbiting Satellite) systems such as ICO and Odyssey.

global network An international network that spans all departments, offices, and subsidiaries of the corporation. Global networks bring their own set of problems, including those of different time zones, languages, established standards, and PTT (Postal Telephone and Telegraph) companies.

Global Network Navigator GNN. An application developed at CERN in Switzerland which provides information about new services available on the Internet, articles about existing services, and an online version of Internet related books. The GNN is a World Wide Web (WWW) based information service.

global one A joint venture of Deutsche Telekom, France Telecom, and Sprint. Launched in January 1996, Global One provides Virtual Network Services (VNS) in more than 65 countries. See also VNS.

Global Positioning System See GPS.

global roaming When you go traveling, you want access to the Internet, preferably by making a local phone call. This capability is called global roaming.

global search A word processing term meaning to automatically find a character or group of characters wherever they appear in a document.

global search and replace A word processing term meaning to automatically find a character or group of characters wherever they appear in a document and replace them with something else.

global security service Provides networkwide security functions, including single log-in to multiple systems.

Global Service Application See UIPRN.

Global Subscriber Number GSN. See UIFN.

Global Switched Telephone Network See GSTN.

Global Title GT. An address such as customer-dialed digits that does not explicitly contain information that would allow routing in the SS7 signaling network, that is, the GTT (Global Title Translation) function is required. See Global Title Translation.

Global Title Translation GTT. The process of translating a Global Title from dialed digits to a point code (network node) address and application address (subsystem number). This process is accomplished by the STP (Signal Transfer Point) in the SS7 network. GTT is defined in IS-41B. See also IS-41, Global Title, SS7 and STP.

Global Transaction Network See GTN.

global village A term coined in the 1960s by Marshall McLuhan, who wrote a number of very popular books, including "The Medium is the Massage," "War and Peace in the Global Village" and "The Gutenberg Galaxy; The Making of Typographic Man." McLuhan coined the term to foreshadow a world of personal computers joined together over a global network. In short, he was talking about the Internet, though he didn't know about it at the time. It was just starting.

Globally Unique Identifier GUID. Also known as Universally Unique IDentifier (UUID). An identifier (ID) is a numeric or alphanumeric string of characters or bits. A Globally Unique Identifier is unique in all the world, or is highly unlikely to be duplicated. A GLID generally is understood by all systems, rather than being proprietary in nature, and often is controlled at some level by a centralized registration authority. URLs (Uniform Resource Locators), which we use to locate resources on the World Wide Web (WWW) are GUIDs; www.telecombooks.com is an example. ISBNs (International Standard Book Numbers) and SSNs (Social Security Numbers) also are examples of GUIDs.

In the context of computing and computer communications, GUIDs are 128-bit identifiers that are unique in both space and time. GUIDs identify objects (e.g., your computer, a piece of software code, or an application software program) to the network. GUIDs originally were developed Apollo as part of its NCS (Network Computing System) for making RPCs (Remote Procedure Calls) from a terminal to a host. Subsequently, the GUID concept was adopted by the OSF (Open Software Foundation), of which Apollo was a founding member. Generally speaking, GUIDs are based on the IEEE 802.1 LAN (Local Area Network) addressing scheme, with the IEEE serving as a central registration authority for numbering LAN nodes through hard-coded IDs stamped on LAN NICs (Network Interface Cards). If an 802.1 NIC is not in place, some other mechanism is used either to generate a random number identifier, or make use of some other system attribute or characteristic. The

GUID also contains several timestamp fields, which rely on timing mechanisms (i.e., clocks) embedded in the host systems and represented by UTC (Universal Time Coordinated).

Microsoft got a lot of attention through its use of GUIDs associated with Windows 98. According to Microsoft, that particular GUID was used to track your progress through the www.microsoft.com website, much like a cookie. The GUID also embedded hidden code in Word and Excel files, thereby allowing the original author of a document to be identified through a hardware identification number. This became a huge privacy issue. See also Cookie, NIC, OSF, RPC, URL, and UTC.

GlobalNet A free, electronic amateur bulletin board system network which operates based upon FidoNet technical standards. GlobalNet nodes are located in North America and Europe.

Globalstar Imagine a hand-held, light, low-cost telephone that looks like a cell phone, but works by talking to a satellite. Several companies have proposed a collection of low-orbiting satellites. The idea is that the closer the satellite, the stronger the signal on the ground, and the smaller the size of the telephone. The low earth orbit also reduces propagation delay, which plagues communications using GEOs (Geosynchronous Earth Orbiting Satellites), which are in equatorial orbits at altitudes of approximately 22,300 miles (35,888 kilometers). Companies proposing such systems include Teledesic and Globalstar.

Globalstar is a Low Earth-Orbiting (LEO) satellite-based digital telecommunications system that will offer wireless voice, fax, low-speed data, messaging and position location services worldwide. Globalstar service will be delivered through a 48-satellite LEO constellation (plus 4 backup satellites) at an orbital altitude of 1,414 kilometers (877 miles). In total, the constellation will provide wireless telephone service in virtually every populated area of the world where Globalstar service is authorized by the local telecommunications regulatory authorities. (Globalstar service providers will be required to obtain such approvals before beginning to offer Globalstar service in their territories.) The system will work on a "bent-pipe" signal relay scheme in which the call is launched from a terrestrial wireless device (e.g., wireless phone or data terminal) directly to the satellite. The satellite, with minimal processing, will relay the call to a gateway groundstation (i.e., authorized service provider's satellite dish) for connection to the destination device over the existing local terrestrial wireline or wireless network.

Users of Globalstar will make or receive calls using hand-held or vehicle-mounted terminals similar to today's cellular telephones; fixed wireless terminals (e.g., wireless pay phones) also will be supported. Because Globalstar will be fully integrated with existing fixed and cellular telephone networks, Globalstar's dual-mode handsets units will be able to switch from conventional cellular telephony to satellite telephony as required. In remote areas with little or no existing wireline telephony, users will make or receive calls through fixed-site telephones, similar either to phone booths or ordinary wireline telephones.

Globalstar planned to begin launching satellites in the second half of 1997, and to commence initial commercial operations via a 24-satellite constellation in 1998. That didn't happen. The first eight satellites were launched in 1998. The second launch of 12 satellites was a failure. An additional 24 satellites are expected to be launched by May 1999, and commercial service will be initialized in Fall 1999. Full 48-satellite coverage will occur sometime in the future. Based in San Jose, California, Globalstar is a limited partnership founded by Loral Corporation of New York City, and QUALCOMM Inc., of San Diego, California. Strategic partners represent the world's leading telecommunications service providers and equipment manufacturers. See also GEO, LEO and Propagation Delay.

Globus A collaborative academic project centered at Argonne National Laboratory focused on enabling the application of grid concepts to computing.

GLONASS The Russian Global Navigation Satellite System is similar in operation and may prove complimentary to the American NAVSTAR system. Launched in 1996, it is a 24 satellite constellation 19,100 Km above the earth in three orbital planes.

glovebox user After the September 11, 2001 World Trade Center disaster, there was a huge boom in the purchase of cell phones. Many of these phones were for using only in emergencies, for keeping in the glovebox of the car. Hence the name of the person who uses them, a glovebox user.

glue logic A generic term for program logic or a protocol that interconnects physical or logical units. For example, glue logic is used to interface microcontrollers with external memory. Glue logic is used by system designers to connect VLSI (Very Large Scale Integrated) circuits. "Blue Glue" is IBM's SNA (Systems Network Architecture). See also Glueware and Systems Network Architecture.

glueman A glued-shut Sony discman that contains a pre-release CD. As The New York Times noted, the device is designed to "keep writers from converting the music to MP3s that can then be traded over the Net."

glueware The trend of joining software applications to physical networks through the deal AT&T and Novell have struck to adapt Novell local area networking software to communicate over AT&T's long-distance network. Intel and Microsoft are considering similar arrangements, according to The Wall Street Journal.

GMD Gesellschaft fur Mathematik und Datenverarbeitung: a German government computer science research institute.

GMDSS The Global Maritime Distress and Safety System, a worldwide radio system for distress and safety at sea. Under the provisions of the Safety Of Life At Sea (SOLAS) convention and the International Maritime Organization (IMO), after 1 February 1999, GMDSS is mandatory for all cargo ships over 300 tons and for all passenger ships on international voyages, regardless of size. GMDSS overcomes the weaknesses of limited range, limited communication and high probability of human error of the traditional Morse telegraphy and radiotelephony used round the world up to February 1, 1992 when GMDSS first became operational. GMDSS covers the globe in four defined sea areas. A1: coastal areas within radiotelephone coverage of at least one VHF coast radio station providing continuous digital selective calling (DSG), usually no more than 30 nautical miles from shore and mostly serving coastal shipping. A2: beyond A1 but within radiotelephone coverage of at least one medium-frequency (MF) coast radio station providing DSG, usually no more than 150 nautical miles from shore and similar to traditional ship radio services. A3: beyond A1 and A2 and within the coverage of geostationary satellites providing continuous service round the globe from 70˜S to 70˜N latitude for 360˜ azimuth free sight at 5˜ ship satellite antenna elevation and provided by the Inmarsat-A, B, C and E systems. A4: beyond A1, A2 and A3 and comprising the polar regions beyond 70˜ latitude and served by Arctic and Antarctic radio stations as well as by satellites in polar orbits, including COSPAS-SARSAT. For further info, visit www.imo.org and/or see COSPAS-SARSAT, DSG, ELT, EPIRB, NAVAREA, NAVTEX and SART.

GMLC Gateway Mobile Location Center. The Gateway Mobile Location Centre contains functionality required to support LCS (LoCation Services). In one PLMN (Public Land Mobile Network), there may be more than one GMLC. The GMLC is the first node an external LCS client accesses in a GSM or UMTS network. The GMLC may request routing information from the HLR (Home Location register) or HSS (Home Subscriber Server). After performing registration authorization, it sends positioning requests to either the VMSC (Visited Mobile Switching Centre), SGSN (Serving GPRS Support Node) or MSC (Mobile Switching Centre) Server and receives final location estimates from the corresponding entity. See also Location Services.

GMP Global Managed Platform.

GMPCS Global Mobile Personal Communications Services. A term coined by the ITU-T to refer to satellite telephony to be provided by the proposed Big LEO (Low Earth-Orbiting Satellite) systems such as Globalstar, and MEO (Middle Earth-Orbiting Satellite) systems such as ICO and Odyssey. See Gobalstar.

GMPLS Generalized Multiprotocol Label (Lambda) Switching. From the www.mplsrc. com web page: GMPLS represents a natural extension of MPLS to allow MPLS to be used as the control mechanism for configuring not only packet-based paths, but also paths in non-packet based devices such as optical switches, TDM muxes, and SONET/ADMs." In short, all of the mechanisms that allow MPLS to be used as a viable solution for Traffic Engineering.

GMRP GARP (Generic Attribute Registration Protocol) Multicast Registration Protocol is a mechanism by which bridges and end stations can automatically and dynamically register their membership in a group with the MAC bridges by which a physical LAN segment attaches to the larger logical LAN. See GARP for more information.

GMRS General Mobile Radio Service. See Personal Radio Service for a bigger explanation.

GMSC Gateway Mobile services Switching Center. A wireless telecommunications term. A means to route a mobile station call to the MSC (Mobile Switching Center) containing the called party's HLR (Home Location Register).

GMSK Gaussian Minimum Shift Keying. A wireless telecommunications term. A means of radio wave modulation used specifically in the GSM (Global System for Mobile Communications) air interface.

GMT Greenwich Mean Time, sometimes known as Greenwich Meridian Time and World Time. Local time used to be good enough. International shipping changed all of that many years ago, creating the need for a global standard and reference point. GMT was created in October 1884 to satisfy that need, with the reference being the average (i.e., mean) time it takes the earth to rotate from noon-to-noon as measured from the Greenwich Meridian Line, or prime meridian, which runs through the principal Transit Instrument at the Royal Observatory in Greenwich, England. The Greenwich Meridian Line is used as it is at zero degrees longitude, thereby serving as the reference point for all measurements of longitude. Although GMT technically has been replaced by Coordinated Universal Time (UTC), it still is widely regarded as the correct time for every international time zone. See also UTC (now called Coordinated Universal Time) and Zulu Time.

GNCT Generalized No Circuit Treatment.

GND Ground.

GNE Gateway Network Element. A SONET Network Element (NE) that provides a direct OS/NE interface. The GNE provides an indirect OS/NE interface. The GNE provides an indirect OS/NE interface for other NEs in its own network subnetwork.

GNMC Global Network Management Center.

GNN See Global Network Navigator.

GNOME GNU Object Model Environment. Pronounced "guh-NOME." A graphical, Windows-like desktop environment designed to be used on multiple GNU (GNU's Not UNIX) platforms. Like GNU, GNOME is distributed without cost (i.e., free) on the basis of an open license. GNU, by the way, is a recursive acronym. See also Acronym and GNU.

GNU A recursive acronym for GNU's Not UNIX. Pronounced "ghu-NEW." Note: A recursion is a procedure of mathematics or grammar that can repeat itself indefinitely until a specified or desired condition is met. The GNU Project of the Free Software Foundation was launched in 1984 at MIT (The Massachusetts Institute of Technology) by Richard Stallman to develop a freeware UNIX-like operating system. Variants of GNU that use the Linux kernel sometimes are known as GNU/UNIX or Linux/GNU. See also Acronym.

GNX Development Tools A set of fax software development tools offered by National Semiconductor.

go back N protocol See go-back-N protocol. **go bedouin** Downsize a business by eliminating all but the core assets: employees and the communications links between them. A company that has gone bedouin lacks a physical location, operating simply as a network of engineering, sales, and support staff connected 24/7 by the Internet and cell phones. Definition from Wired magazine.

go gold Finish that particular release of a software program and release it to manufacturing, i.e. send it the company who's going to duplicate the disk, print the manuals, bundle the whole package and shrinkwrap it all.

go local A command typically given in a asynchronous data communications program to tell the computer that it will connect to something without a modem over a null modem cable. The command "Go Local" also refers to modem connections and can tell one modem to overlook some of the handshaking and assume it's already taken place.

go-back-N protocol Any protocol that requires the sender to go back and resend the most recent N number of packets or frames, when the receiving node determines that it has received a bad packet or frame. The reason why N is not simply 1 in the event of the receipt of a single bad packet is because, for overall network performance reasons, the receiving node cannot afford to keep well-formed, out-of-sequence packets buffered while waiting for its go-back-N message to reach the sender and get processed by it. The buffer space can be profitably reused in the time it takes for the N packets to be requested and resent.

GO-MVIP Global Organization for MVIP. GO-MVIP is a non-profit trade association established in 1993 to move the Multi-Vendor Integration Protocol standards forward. The stated goals of GO-MVIP are to 1) Develop and establish design specifications for further enhancements of MVIP, 2) Drive MVIP to an official industry standard, 3) Establish a testing laboratory and quality assurance program for current and future MVIP products, and 4) Ensure the continued growth and long-term success of MVIP. As of July 1996, GO-MVIP comprises 170 members; over 300 MVIP-compliant products exist. www.mvip.org. See MVIP.

goat 1. Also called Craft test Set or Butt Set. A telephone used to test analog phone lines.

2. A hairy animal that can become nasty when upset, like telephone customers who have a problem with their phone lines.

3. People in our population whose voices cannot – under any circumstances – be recognized by voice recognition machines. In biometric verification, a goat is a system end-user who is refused access to the system because their biometric data pattern is outside the range recognized by the system. The term comes from a research paper on speech recognition published in 1998 by George L. Doddington. The paper, "Sheep, Goats, Lambs and Wolves - An Analysis of Individual Differences in Speaker Recognition Performance" used a menagerie analogy to explain the differences in speech recognition. Sheep were speakers whose voice patterns were easily accepted by the system, goats were speakers who were exceptionally unsuccessful at being accepted, lambs were speakers who were exceptionally vulnerable to impersonation, and wolves were speakers who were exceptionally successful at impersonation. Because false rejection rates are often high when testing a biometric

verification system, goats are probably better known than the other animals in Doddington's menagerie.

god In 1970, an Arizona Lawyer named Russel H. Tansie filed a $100,000 damage suit against God. The suit was filed on behalf of his secretary, Betty Penrose, who accused God of negligence in His power over the weather when He allowed a lightning bolt to strike her home. Ms. Penrose won the case when the Defendant failed to appear in court. Whether or not she collected has not been recorded.

god games Computer games, also called simulators, in which the player manages a complex system of interacting variables, inputs and agents. There are games that let you simulate railroad empires, manage city government, enlarge tropical dictatorships, expand golf resorts and countless other worlds.

Goeken, Jack Founder of MCI (1963), FTD Mercury Network (1974), Airfone (1976), and In-Flight Phone Corporation (1989). Goeken is generally regarded as the father of air-to-ground telephone communication. Goeken's interest in telecommunications began early in life. By the time he was a senior in high school he was repairing radios and TVs. After high school he joined the Army and served in the Signal Corps. Goeken now runs Goeken Group, an incubator company that helps and takes an ownership stake in startup companies.

GOF Glass Optical Fiber. This may seem obvious, but there also is something called Plastic Optical Fiber (POF). See Fiber Optics and POF.

going bedouin Downsizing a business by eliminating all but the core assets; employees and communications links between them. A company that has gone bedouin lacks a physical location, operating simply as a network of engineering, sales and support staff connected 24/7 by Internet and cell phone.

going cyrillic Going cyrillic is when a graphical display (LED panel, bit-mapped text and graphics) starts to display garbage. "The thing just went cyrillic on me."

going postal Euphemism for being totally stressed out, for losing it. Makes reference to the unfortunate track record of postal employees who have snapped and gone on shooting rampages.

GOLAY One of the communications protocols used between paging towers and the mobile pagers/receivers/beepers themselves. Other protocols are POCSAG, ERMES, FLEX and REFLEX. The same paging tower equipment can transmit messages one moment in GOLAY and the next moment in ERMES, or any of the other protocols.

GOLD An EBPP (Electronic Bill Presentation and Payment) specification developed for billing and payment over the Internet. GOLD was developed by Integrion Financial Network, which is owned by VISA USA, IBM and a number of banks. GOLD was designed to support the display and manipulation of financial data such as bank account information and stock holding, and funds transfer. GOLD also supports transactional Web sites. The competing specification is OFX (Open Financial Exchange). See also Electronic Commerce and OFX.

gold bar Alok Das of the Pentagon's Space Vehicle's Directorate claims "it costs a bar of gold to launch a can of Coke" into a satellite into space.

gold codes Named after Robert Gold, Gold codes are used in direct-sequence spread spectrum transmission. Each transmitted signal is assigned to unique Gold code, which correlates the original information signal into a pseudo random sequence. This sequence is then modulated and transmitted as a spread spectrum signal. The receiver, which uses the same Gold code, is able to de-correlate the spread spectrum signal and recover the original information. Gold codes possess two very desirable qualities which are important in a high quality communications system. The first quality is called "auto-correlation." When a receiver is subjected to several spread spectrum signals, it must extract the desired information and reject the remainder. Auto-correlation allows for an excellent signal recovery when the transmitted code matches the reference code in the receiver.

The second quality is called "cross-correlation." Cross-correlation simply means that an undesired transmitted code cannot produce a false match at the receiver. The advantage of Gold codes is that they consistently exhibit superior cross-correlation performance, which is critical in an environment with multiple transmitters, each representing a potential interfering source. See Spread Spectrum.

gold disk You have finally finished your new software. And you're now ready to go into production, to have your software reproduced onto disks you can sell. That final, completed version of software, from which you reproduce commercial production disks, is called your "Gold Disk." From what I can see, it's a "Gold Disk," though there may be more than one disk in your package.

gold farming Using low-wage laborers to accumulate virtual scrip in online multiplayer games, like World of Warcraft. The currency is then sold for real-world cash.

gold number Also called vanity number or golden number. It's a phone number that's easy for your customers to remember, e.g. 555-LIMO. But occasionally hard for them

to dial. Tip: If you buy a vanity number, make sure it doesn't have numbers in it. 555-LIMO is harder to remember than CAR-RENT.

Golden Boy The 40-ton, 24-foot statue that since 1916 graced the top of AT&T Corporation's headquarters at 195 Broadway, New York City and then graced several other AT&T headquarters, including Basking Ridge and Bedminster, New Jersey. The statue, which is officially known as The Spirit of Communications, is covered in more than 40,000 pieces of gold leaf. The statue was commissioned in 1914 by Western Electric, then In 1914, Western Electric, then AT&T's manufacturing equipment arm, now known as Lucent, commissioned the statue and called it The Genius of Electricity. Artist Evelyn Beatrice Longman sculpted the statue in 1916. Golden Boy was then hoisted 465 feet above street level to the top of AT&T's headquarters at 195 Broadway in New York. And that's where he stood for 65 years. During the 1930s and '40s Golden Boy appeared on every telephone directory sent to homes across the country. His common appearance in homes throughout the U.S. inspired former Poet Laureate Robert Pinskey to write a poem called "A Phonebook Cover Hermes of the Nineteen-forties." Pinskey calls him "pure, the merciless messenger," in his poem, which is now embossed in bronze and sits at the foot of Golden Boy's 5-foot-high granite cylinder stand. The statue has a strange history. While it was being moved, the company was surprised to find that Golden Boy, a nude, was anatomically correct. It's been said, according to an article in Network World, that the CEO at the time, John de Butts, requested that the statue find some modesty, and Golden Boy's manhood essentially was removed in the restoration and re-gilding process. After a 10-year stint indoors, Golden Boy longed for the great outdoors once again. In 1992 AT&T moved its headquarters to Basking Ridge, N.J., and the statue followed. Golden Boy won a prominent place on a pedestal in front of the building. After another 10 years, Golden Boy moved again to his current home – outside AT&T's new world headquarters in Bedminster, NJ.

golden egg "That's the way with these directors. They're always biting the hand that lays the golden egg." Samuel Goldwyn, film producer.

golden gate bridge The first year the Golden Gate Bridge in San Francisco was open, it carried 3.5 million vehicles. Today, it carries close to 40 million. It cost $35 million to build with the construction bonds being paid off in 1971. The replacement cost is estimated to be about $1.3 billion. Bridges are not paid investments. May I sell you one?

golden hello For the potential chief executive of a large corporation, it is the one thing – more than the corporate jet or any other perk – that must be paid before the executive will move to run another company. Typically this involves bonuses, stock options, restricted stock and pension benefits that the new potential CEO would have abandoned by leaving his previous employer. Such golden hello payments are intended to make the executive "whole" – in essence to treat the executive as if his career was one smooth ascent with no costly interruptions.See also golden parachute.

golden parachute When a senior executive gets fired from his company, he is often given lots of money (thus golden) by the board to make his landing outside the company soft (thus parachute). See also Golden Rolodex.

golden rolodex, the The small handful of experts who are always quoted in news stories and asked to be guests on TV discussion shows. Example: Henry Kissinger appears to be in The Golden Rolodex under foreign policy.

Goldwyn, Samuel Famous film producer. He was born Shmuel Gelbfisz in Warsaw, Poland. He made great movies and great aphorisms, known affectionately as Goldwynisms: "A verbal contract isn't worth the paper it's written on." "A hospital is no place to be sick in." "Anyone who goes to a psychiatrist ought to have his head examined." "I had a great idea this morning, but I didn't like it." "Gentlemen, include me out." "A wide screen just makes a bad film twice as bad." "I can give you a definite perhaps." "I can tell you in two words: im possible." "I paid too much for it, but it's worth it." "If I could drop dead right now, I'd be the happiest man alive." On the film set of a tenement: Goldwyn: Why is everything so dirty here? Director: Because it's supposed to be a slum. Goldwyn: Well, this slum sure cost a lot of money. It should look better than an ordinary slum." "Pictures are for entertainment. Messages should be delivered by Western Union." "I don't want yes-men around me. I want everyone to tell me the truth, even if it costs them their jobs." "If I look confused, it's because I'm thinking." "Why did you name him Sam? Every Tom, Dick and Harry is named Sam." "I don't think anyone should write his autobiography until after he's dead." "You fail to overlook the crucial point."

golf Many years ago, in Scotland, a new game was invented. It was ruled "Gentlemen Only...Ladies Forbidden"....and thus the word GOLF entered into the English language. If you have ever wondered why you play 18 holes and not 20, or 10 or an even dozen in golf, the reason is simple – and logical. During a discussion among the club's membership board at St. Andrews in 1858, a member pointed out that it takes exactly 18 shots to polish off a fifth of Scotch. By limiting himself to only one shot of Scotch per hole, the Scot figured a round of golf was finished when the Scotch ran out.

gonk On-line jargon. It means To prevaricate or to embellish the truth beyond any reasonable recognition. "You're gonking me. That story you just told me is a bunch of gonk."

Good, Dan Former Wall Street executive who was singed, not burned by the High Tech Bubble. He knew the bubble would burst but was naive enough to think he could safely traverse the profit wave and land safety in calm water on the other side. The IPO surfboard he was riding got swept under like all the other amateur surfers as the wave came crashing down. The little high-tech knowledge he acquired proved dangerous, but fortunately not lethal. Today, older but wiser, he is an evangelist of cash and capital preservation. He is most well known for uttering the now famous phrase, "Sue the bastards". I am proud to be his friend. Dan is teaching me about investing. The goal is to make me dangerous in my ignorance, just like him. Most days, he spends acting as contributing and copy editor to www. InSearchOfThePerfectInvestment.com. In this job, he is both creative and eagle-eyed, a lethal combination. As his latest money-making project, he has just opened a public storage facility where useless stuff can be stored for a year at an exorbitant fee. His first customer is me. And Mr. Good has zero idea of just how useless the stuff he's storing really is.

good condition A term used in the secondary telecom equipment business. One step up from fair condition. Product is in working condition and looks good.

goodwill Goodwill is a term you'll find on balance sheets. It represents the difference between what a company paid to buy other companies and what those other companies' assets were valued at on their books. If a company pays a lot of money to acquire another company, then its balance sheet will have a high number for "goodwill." That number is carried as an asset on its balance sheet. But is it really an asset or not? That can only be answered by guessing whether the asset can be sold at at least what the company originally paid for it. The issue of goodwill became very important in the telecom industry after AT&T paid $50 billion to buy a bunch of cable TV companies in the hope that they would provide a local loop to the consumer. As time passed, AT&T's $50 billion looked like an increasingly stupid decision. And the value of the goodwill on AT&T's balance sheet looked increasingly valueless. (You know what I mean.)

gooey interface A slang way of saying graphical user interface. See GUI.

Google Google is arguably the world's best search engine. It was founded by two Stanford graduates, Larry Page and Sergey Brin. They understand the true secret of sustainable success. When asked at a Wall Street Journal D Conference, "How do you get better than you already are?" by a devoted fan in the audience, Page replies, "You may think using Google's great, but I still think it's terrible." In fact, Page won't rest until Google can produce "the exact right answer instantly about everything in the world."

Google hacking The illicit accessing of website information stored in search engine indexes, usually for malicious purposes, for example, to map out a Web application's components and internal structure.

Google juice Google juice refers to how high a Web site ranks in Google's search results – the higher the ranking, the more juice. Google juice is all about links. Google ranks Web sites based largely on the quantity and quality of other sites linking to it.

Googleplex Googleplex is the common name for Google's headquarters in Mountain View, Calif., where the cafeteria is lavish, the bathrooms feature high-tech toilets with warm seats and nozzles for squirting water at delicate spots, and a glassed-in laundry room offers employees a chance to take a personal hygiene break. Google has outgrown the Googleplex and occupies several nearby buildings.

googol A term coined in 1929 by Milton Sirotta, an American mathematician, a googol is a number written as 1 followed by 100 zeroes, i.e., 10 to the 100th power. See Googolbyte.

googolbyte Two raised to the power 100. See Byte.

googolplex A googolplex is 10 to the power of a googol, which is 1 followed by 10 to the power of 100 zeros. Frank Philhofer has determined that, given Moore's Law (which is that computer processor power doubles about every 18 months or so), it would make no sense to try to print out a googolplex for another 524 years – since all earliest attempts to print a googolplex out would be overtaken by the faster processor. I'm not too sure what all this means, but it sure sounds incredible. See googol.

goodput The amount of good data that is put through a network. Goodput commonly is used to describe throughput in ATM and other packet networks. Throughput is the end result of a data call, as measured by the relationship of what went in one end and what came out the other. Throughput, therefore, is a measure of the efficiency of that communications network. Throughput is a function of bandwidth, error performance, congestion, and other factors. In an ATM network, goodput excludes duplicate cells and packets. Goodput also excludes ATM cells that cannot be reassembled at the receiving node into a complete and meaningful data packet, which was segmented at the transmitting node. See also

Throughput.

GOP Grand Old Party. Another way of saying the Republican Party. In December of 2002, the Wall Street Journal stopped referring to the Republican Party as the GOP. Now they're just the Republican Party.

gopher Programmers at the University of Minnesota – home of the Golden Gophers – developed a kind of menu to "go for" items on Internet, bypassing complicated addresses and commands. If you want to connect to the State Library in Albany you select that option off the menu. Time Magazine once described Gopher as a tool used for "tunneling quickly from one place on the Internet to another." Hence the term Gopher. See also Gopherspace.

gopherspace The vast number of servers and areas of interest accessible through the Internet gopher. See Gopher.

GORIZONT The Russian geostationary telecommunications satellite.

GOS Grade of Service. Telecom traffic term. The probability that a random call will be delayed, or receive a busy signal, under a given traffic load.

GOSIP Government Open Systems Interconnection Profile. The U.S. government's version of the OSI protocols. GOSIP compliance is typically a requirement in government networking purchases. GOSIP addresses communication and inter operation among end systems and intermediate systems. It provides specific peer-level, process-to-process and terminal access functionality between computer system users within and across government agencies.

gossip Early politicians required feedback from the public to determine what was considered important to the people. Since there were no telephones, TVs or radios, the politicians sent their assistants to local taverns, pubs, and bars who were told to "go sip some ale" and listen to people's conversations and political concerns. Many assistants were dispatched at different times. "You go sip here" and "You go sip there." The two words "go sip" were eventually combined when referring to the local opinion and, thus we have the term "gossip."

Gotcha Law, The This law comes from "Got You." It's my favorite law. Think about life. You buy a beautiful laptop. You fall in love with it. You want to buy one for your wife. When you go back to the store all hot to do your wife a wonderful favor, they tell you, "Sorry, it's been discontinued." Gotcha! Compare with Murphy's Law.

Government Emergency Telecommunications Service A PSTN-based service that provides nationwide priority voice and low-speed data service during a national emergency or crisis situation, when the PSTN is congested and the probability of completing a call via normal means has significantly decreased. GETS is limited to national security and emergency preparedness (NS/EP) personnel.

government radio publications Publications on radio subjects by the Bureau of Standards and Signal Corps and sold by the superintendent of Documents, Government Printing Office, Washington D.C.

Goy A non-Jew. Goyim is the plural. The word is not pejorative, but simply descriptive. My wife is a Goy. She thinks of herself as Catholic.

GPA General Purpose Adapter. An AT&T Merlin device that connects an analog multi line telephone to optional equipment such as an answering machine or a FAX machine.

GPB Grand Pooh-Bah. Pooh-Bah is a character in Gilbert and Sullivan's opera "The Mikado." Pooh-Bah held the title "Lord-High-Everything-Else."

GPC Gateway Protocol Converter. An application-specific node that connects otherwise incompatible networks. Converts data codes and transmission protocols to enable interoperability. Contrast to Bridge.

GPCL General Premises Cabling Licence.

GPF General Protection Fault. A problem that happens too often under Windows 3.xx. A General Protection Fault is an indication that Windows 3.xx has tried to assign two or more programs to the same area in memory. Obviously that's not possible, since two things can't occupy the same area in memory. As a result, your screen stops and says "General Protection Fault." If you can save what you're doing, do it. If you have other programs open, try and save the material in them. Close Windows and then do a cold reboot. Do not continue to work after you have received a General Protection Fault. You must reboot. Better do a cold reboot, too.

GPI GammaFax Programmers Interface. C-level programming language. Real-time applications for fax switched and gateways.

GPIB An interconnection bus and protocol that allows connection of multiple instruments in a network under the direction of a controller. Also known as the IEEE 488 bus, it allows test engineers to configure complete systems from off-the-shelf instruments and control those systems with a single, proven interface. GPIB was originally called HPIB because it was developed by Hewlett Packard.

GPON Gigabit Passive Optical Network. Standardized by the ITU-T as G.984.2, GPON is a local loop technology that runs either the ATM or Ethernet protocol over single mode fiber. Downstream transmission rates are 2.488 Gbps or 1.244 Gbps. Upstream rates are 2.488 Gbps, 1.244 Gbps, 622 Mbps or 155 Mbps. The maximum specified logical reach is 60km, but the maximum practical distance is typically 20km. The maximum split ratio is 64:1, although 32:1 is more typical. See also APON, BPON, EPON and PON.

GPOP Global Point of Presence.

GPRS Want to connect your laptop to your cell phone and surf the web or send emails? GPRS is for you. GPRS stands for General Packet Radio Service. And it's the always-on packet data service for GSM, which is the cell phone standard which most countries of the world use, including Europe, Australia, America (not all carriers) and some parts of Asia. The idea is that you'll connect your GPRS-equipped (also called 2.5G) cell phone to your laptop with a cable or insert a small GPRS -equipped PCMCIA card into your laptop and transmit. GPRS will be most useful for "bursty" data applications such as mobile Internet browsing and e-mail. GPRS has been demonstrated as fast as 115 Kbps. And in theory it can go that fast. But the reality is that you'll get between 20Kbps and 50Kbps throughput – about the speed you get from your dial-up home landline. One big advantage of GPRS is that it's "always on," just like your DSL line, your cable modem or your office network. To send a data message you won't have to waste a minute dialing a number and listening to the modems go through their interminable screeching/connect dance. GPRS is the primary feature of what has become known as 2.5G – the upgrade to today's 2G cell phone network. 1XRTT is the CDMA equivalent of GPRS. See GSM and HSCSD.

GPRS introduces two new network nodes in the GSM Public Land Mobile Network (PLMN): The Serving GPRS Support Node (SGSN), which is at the same hierarchical level as the mobile services switching center (MSC), keeps track of the location of the individual mobile stations (MSs) and performs security functions and access control. The SGSN is connected to the base-station system via Frame Relay. The GGSN provides interworking with external packet-data networks, and is connected with SGSNs via an IP-based GPRS backbone network. Also (from TCP): RTO Retransmission Timeout – a value that triggers packet resend in TCP-IP if no acknowledgement is received, because the protocol assumes the packet has been lost. And RTT -Round Trip Time – is time required from emission (i.e. initial sending) of a packet until an acknowledgement is received. This helps determine the adaptive characteristics of the TCP protocol. See also Bluetooth.

GPS 1. Global Positioning System. A constellation of 24 orbiting satellites system which allow all of us to figure out precisely where we are anywhere on earth to within one meter's accuracy, both height and longitude/latitude. The GPS consists of a constellation of 24 satellites orbiting the earth at 10,900 miles. Each satellite orbits the earth twice a day. Think of them as "man-made stars" to replace the stars that we've traditionally used for navigation. The U.S. Government has over $10 billion to build and maintain the system. Each GPS satellite transmits radio waves on two frequencies: 1575.42 MHz referred to as L1, and 1227.60 MHz referred to as L2. It transmits a host of somewhat complicated data occupying about 20 MHz of the spectrum on each channel. But basically, it boils down to three items. The satellite transmits its own position, its time, and a long pseudo random noise code (PRN). The noise code is used by the receiver to calculate range. If we know precisely where a satellite is located and our precise distance from it, and if we can obtain similar readings from other satellites, we can calculate precisely our own location and altitude by triangulation. Satellite position and time are derived from on-board celestial navigation equipment and atomic clocks accurate to one second in 300,000 years. But the ranging is the heart of GPS. Both in the receiver, and in the satellite, a very long sequence of apparently random bits are generated. By comparing internal stream of bits in the receiver to the precisely duplicate received bits from the satellite, and "aligning" the two streams, a shift error or displacement can be calculated representing the precise travel time from satellite to receiver. Since the receiver also knows the precise position of the satellite, and its range from the receiver, a simple triangulation calculation can give two-dimensional position (lat/long) from three satellites and additional elevation information from a fourth.

There are actually two PRN strings transmitted: a coarse acquisition code (C/A code) and a precision code (P code). The coarse code sequence consists of 1,023 bits repeated every 266 days. But each satellite transmits a seven day segment re-initialized at midnight Saturday/Sunday of each week. By using both codes a very accurate position can be calculated. By transmitting them at different frequencies, even the signal-attenuating effects of the ionosphere can to some degree be factored out. At the present time, civilian users are only authorized to use the coarse acquisition code, which is the basis for the GPS Standard Positioning Service (SPS), and which yields a best accuracy of about one meter. Military users use both the coarse acquisition code and the precision code in what is referred to as the Precise Positioning Service (PPS), which yields accuracy to within centimeters. The Department of Defense (DOD) can at any time encrypt the precision code with another

secret code. This process of "anti-spoofing" ensures that no hostile military forces can also use the GPS service at that time. The DOD can also purposely degrade the accuracy of the coarse acquisition code, in a process known as "selective availability," to degrade accuracy to about 100 meters. But other than during brief test periods and national emergencies, the service is generally available to all. New techniques now make the civilian use of GPS almost as accurate as the military use. The accuracy of any given GPS receiver at any point in time depends on how many GPS satellite signals it can acquire. Some low-cost receivers can pick up five satellites and are thus accurate to within one meter.

That's not the end of the story, however. Regardless of the sophistication of a given GPS receiver, satellite signals can be difficult, if not impossible, to acquire in certain places such as canyons (either natural or of the man-made urban sort), indoors, in vehicles, in shaded areas such as under trees or canopies, or in other areas where the signals are too noisy, attenuated (i.e., weakened), or distorted by multipath fading. If, however, the GPS receiver is integrated into a cell phone, pager, PDA, or other device running over a terrestrial wireless network, that network can assist in the GPS process by providing additional data. This technique of Assisted GPS (AGPS), also known as Wireless Assisted GPS (WAGPS), can involve several alternative approaches. One such approach involves a special chipset built into the wireless device. That chipset accepts signals from both the GPS satellites and the base stations of the cellular network, correlates the information, and thereby informs the network of its location with an accuracy of several meters. Additional intelligence in network-based servers can further refine that location data. An alternative approach is entirely network-based, and does not involve additional intelligence embedded in the handset. This latter approach uses signal strength information derived from multiple network base stations to triangulate on the location of the terminal device. That information is correlated with the specific GPS data that pinpoints the location of the base stations, themselves. AGPS, which is relatively new and certainly not universally available, is critical to the delivery of location-based services. Such services allow the network to pinpoint your location in order to provide you with information such as a list of nearby Thai restaurants, and directions to the one you select. More importantly is the ability of AGPS to pinpoint your location for emergency purposes. E911 (Enhanced 911) service providers increasingly will depend on AGPS in order to render all sorts of emergency services more effectively and more quickly. The FCC has rules that in October of 2001, at least half the mobile phones being sold in the U.S. must be 911 location service equipped. See also Galileo, the proposed European GPS, WAAS, M-code, Y-code and the various definitions for Block II.

2. Navstart Global Positioning Satellite. Used by networks for synchronization.

GPS phone A GPS-enabled mobile phone. There are huge advantages in

gpsOne A solution combining Global Positioning System satellite and wireless network infrastructure to provide position location services. The gpsOne solution enhances location services availability, expands terrain coverage, accelerates the location determination process and provides better accuracy for callers, whether during emergency situations or while using GPS-enabled commercial applications. Or at least that's what its advocates say.

GPT A disk-partitioning scheme that is used by the Extensible Firmware Interface (EFI) in Itanium-based computers. GPT offers more advantages than master boot record (MBR) partitioning because it allows up to 128 partitions per disk, provides support for volumes up to 18 exabytes in size, allows primary and backup partition tables for redundancy, and supports unique disk and partition IDs (GUIDs).

GR Generic Requirement. A Bellcore (now Telcordia) document type replacing the Framework Technical Advisory (FA), Technical Advisory (TA) and Technical Reference (TR) document types. FA, TA and TR documents previously reflected the maturity level of the proposed requirements. In contrast, a Generic Requirements (GR) is a living document that represents Telcordia's preliminary and current view of a technology, equipment, service or interface. It does not necessarily reflect the views of any other company. See GR-303.

GR-1209 Telcordia generic requirements for fiber optic branching components.

GR-1221 Telcordia generic reliability assurance requirements for fiber optic branching.

GR-196 Telcordia generic requirements for optical time domain reflectometer (OTDR) type equipment. See GR.

GR-20 Telcordia generic requirements for optical fiber and fiber optic cable. See GR.

GR-303 GR-303 is a set of technical specifications from Telcordia to help define what the next generation of the world's telecommunications network (i.e. the new PSTN) might look like. The following words are from Bellcore (now Telcordia):

"What is GR-303? Network providers are looking to deploy Next Generation Integrated Digital Loop Carrier (NG-IDLC) systems that take advantage of leading edge technology to help reduce operating and capital equipment costs while delivering a full range of telecommunications services. Telcordia's GR-303 family of requirements specifies a set of NG-IDLC generic criteria that creates an Integrated Access System, supporting multiple distribution technologies and architectures (e.g., xDSL, HFC, Fiber-to-the-Curb, etc.), and a wide range of services (narrowband and broadband) on a single access platform. The GR-303 family of generic criteria defines a set of requirements for Integrated Access Systems that includes open interfaces for mix-and-match of (1) Local Digital Switches (LDSs) with Remote Digital Terminals (RDTs) as well as (2) RDTs and Element Management Systems (EMSs). Facilities connecting to the narrowband digital switch (i.e., LDS) are efficiently assigned and managed through the Time-Slot Management Channel (TMC), with remote operations functions supported over the Embedded Operations Channel (EOC) of the GR-303-based Integrated Access System. GR-303-based Integrated Access Systems promote increased network architecture flexibility by providing a consistent approach to deploying a wide range of access system technologies in a consistent manner. Many vendors are developing NG-IDLC products that, although they use different distribution technologies and architectures (e.g., hybrid fiber coax and fiber in the loop), meet the open interfaces described in the GR-303 requirements. This allows network providers to tailor the access system technology deployed area-by-area while utilizing core network features such as the LDS interface and Telecommunications Management Network (TMN) operations capabilities. GR-303-based Integrated Access Systems are intended to reduce capital costs through supplier competition and operating costs through a standards-based, Telecommunications Management Network (TMN) compatible operations environment that provides remote operations capabilities. Integrated Access System products will help to reduce capital costs by enabling mix-and-match among LDS, RDT and EMS products from a wide variety of vendors. The open interfaces described in the GR-303 requirements will help enable the network providers to pursue competitive bids from multiple suppliers for Integrated Access Systems products, thereby potentially obtaining better prices." For more: www.telcordia.com/resources/genericreq/gr303/index.html

According to a company called Zarak Systems Corporation, a maker of bulk call simulator GR-303 test equipment, "GR-303 is a specification for a digital loop carrier system (DLC) that operates on T-1 circuits. The GR-303 specification encompasses all aspects of the functionality of the DLC system. Thus, the term GR-303 is commonly used to describe a system or the framing on a set of T-1 circuits. GR-303 is used by telephone operating companies to concentrate telephone traffic and provide better maintainability. The system provides:

a. T-1 circuits exiting a switch (referred to as the IDT), and going directly to the remote digital terminal (RDT) equipment, without the need for additional equipment in the central office (CO)

b. concentration from 1:1 to 44:1.

c. a timeslot management channel (TMC) data link that uses messages for call setup and tear down.

d. the use of signalling bits to indicate call control.

e. a separate embedded operations channel (EOC) data link.

f. redundancy on the circuits that carry the data links.

g. expandability from two to 28 T-1 circuits that can carry up to 668 channels simultaneously.

h. expandability from 1 to 2048 subscriber channels.

i. ability to handle ISDN circuits (both BRI and PRI) for the subscriber.

j. multiple interface groups (IGs), so that the remote equipment can simultaneously interface to multiple switches.

The T-1 circuits are configured for ESF framing, and usually have B8ZS enabled. The first two T-1 circuits each carry the TMC and EOC for redundancy. The EOC is carried in timeslots 12 of the first and second T-1 circuits, and the TMC is carried in timeslots 24 of the first and second T-1 circuits.

According to Zarak, GR-303 has its foundation on SLC-96 mode 2. The two specifications differ in many aspects:

a. GR-303 is expandable, whereas SLC-96 is fixed at 2 T-1 circuits and 96 subscriber channels.

b. GR-303 has continual redundancy, whereas SLC-96 has an optional back up scheme.

c. the GR-303 protocols emanate directly from the switch, whereas SLC-96 requires equipment in the CO that is separate from the switch.

d. GR-303 has a comprehensive EOC which allows an operating company to do OAM&P remotely, whereas SLC-96 is limited in its capabilities.

Zarak highlights these disadvantages of GR-303:

1. The EOC is enormously complex in its implementation. This had led many manufacturers to implement a minimum number of its features, known as "EOC-light."

2. There is a combination of hybrid signalling, using messages to set up and tear

down the allocation of a timeslot, and then robbed bit signalling is used to indicate the call control.

3. There is only 56 kb/s data path because of the robbed bit signalling.

4. The TMC uses messages based on ISDN and Q.931 in particular. However, because the objectives are different, the messages are not standard.

5. To add BRI, a separate channel must be allocated for the D-channel (call set up and tear down). Four BRI D-channels can be merged into one GR-303 channel, called a QDS0 (quad DS0).

6. There is no scheme to handle concentrated PRI, and a whole T-1 circuit must be permanently dedicated. www.zarak.com.

GR-485-CORE GR-485-CORE specification contains the generic guidelines for Common Language Equipment Coding Processes and Guidelines (CLEI codes)m which are a standard method of identifying telecommunications equipment in a uniform, feature-oriented language. It's a text/barcode label on the front of all equipment installed at RBOC facilities et. al. that facilitates inventory, maintenance, planning, investment tracking, and circuit maintenance processes. Suppliers of telecommunication equipment give Bellcore technical data on their equipment, and Bellcore assigns a CLEI code to that specific product. See CLEI.

grace login Allows a user to finish logging on using an expired password without changing it. You can set the number of grace logons a user is allowed.

grace period 1. Long ago, the ICANN established the "Create Grace Period," a five-day stretch when a company or person can claim a domain name and then return it for a full refund of the $6 registry fee. It was originally intended to allow registrars who had mistyped a domain to return them without cost. But

1. The explosion in domain tasting can be traced back to the practices and policies of the Internet. The organization responsible for regulating domain names is the Internet Corporation for Assigned Names & Numbers, better known as ICANN. 2. This is the amount of time that the FCC will allow to elapse following the expiration of an amateur radio license before the radio amateur will have to retake the exam to renew the license. If the radio amateur renews his license during the grace period, he can do so without retaking the licensing exam. The holder of an expired license may not operate an amateur station until the license is renewed.

graceful close Method terminating a connection at the transport layer with no loss of data.

graceful degradation A condition in which a system continues to operate, providing services in a degraded mode rather than failing completely. See ABEND.

Graceful Discard GD. A Frame Relay term. A congestion management mechanism in Frame Relay networks. Based on agreement between the user organization and the Frame Relay service provider (i.e., carrier), associated with every PVC (Permanent Virtual Circuit) is a CIR (Committed Information Rate). The CIR is the level of data traffic (in bits) which the carrier agrees to handle over a period of time. The Offered Load to the network can burst above the CIR for a measured interval of time (T). Burst levels are measured as Bc (Committed Burst Size), which is the maximum amount of data that the carrier agrees to transfer under normal circumstances, and Be (Excess Burst Size), which is the maximum amount of additional data that the carrier will attempt to handle, assuming that congestion conditions in the network permit. Data bits at the Be level are contained within frames, of course, as are all data bits in the Frame Relay world. The excess frames at the Be level will be marked as DE (Discard Eligible), either by the user's Frame Relay Access Device (FRAD) or by the carrier's FRND (Frame Relay Network Device), which typically is a router that sits at the edge of the carrier network. The DE frames which fall within the Be will be discarded gracefully (i.e., only when absolutely necessary) by the carrier during periods of congestion. If the network is not congested, the carrier will deliver the excess frames and bits. Typically, at the point of network ingress, the carrier will discard any frames above the Bc+Be-and not very gracefully, I might add. See also Committed Burst Size, Committed Information Rate, Discard Eligible, Excess Burst Size, FRAD, Frame Relay, FRND, Measurement Interval, Offered Load, Permanent Virtual Circuit, and Router.

Grade 1 Cable Twisted pair cables specifically designed for analog voice circuits and data transmissions up to 1 Mbps. Applications – Key systems, analog and digital PBX, low speed data, RS-232, etc.

Grade 2 Cable Twisted pair cables designed to meet the IBM Type 3 specification. These cables are capable of data transmissions at 4 Mbps, IBM 3270, STAR-LAN I, IBM PC Network, ISDN, etc.

Grade 3 Cable Twisted pair grade 3 LAN cables have performance characteristics that permit data transmissions at 10 Mbps. Each have been tested to insure they meet the EIA/TIA 568 emerging standard. Applications – 802.3 10BASE-T at 10 Mbps, STARLAN 10 and 802.5 token ring at 4 Mbps.

Grade 4 Cable The highest quality twisted pair cables available. Super grade cables have been tested up to speeds of 20 Mbps, 802.5 token ring at 4 Mbps and 802.3 10Base-T at 10 Mbps.

Grade 5 Cable These are the IBM-type individually shielded 2 pair twisted data cables. They're currently being tested for data rates at 100 Mbps. Applications – IBM Cabling System, 802.5 token ring at 16 Mbps and FDDI at 100 Mbps. Grade 5 cable is not the same as CAT 5 cable. See Category of Performance.

Grade Of Service GOS. A term associated with telephone service indicating the probability that a call attempted will receive a busy signal, expressed as a decimal fraction. Grade of service may be applied to the busy hour or to some other specified period. A P.01 Grade of Service means the user has a 1% chance of reaching a busy signal. See Traffic Engineering.

graded-index fiber Graded-index fiber is a type of MultiMode Fiber (MMF) optical cable that is improved over earlier step-index fiber. Graded-index fiber has a gradual, rather than an abrupt, change in refractive index from the inner core to the outer core, at which point the refractive index matches that of the cladding. The gradual change in the refractive index takes into account the fact that different light rays that make up a digital pulse travel in different modes (i.e., physical paths), some of which are longer than others. The change in refractive index supports higher speed signal propagation for errant light rays that travel out towards the cladding, while those rays that travel down the center of the core travel at slower speeds. The yield is that graded-index fiber eliminates modal dispersion. As graded-index fiber can support transmission at 100 Mbps over distances of 2km or so, and speeds of 622 Mbps over shorter distances, it is widely used in LAN backbones to interconnect high speed switches, routers, servers, etc. See also Modal Dispersion, MMF, Refractive Index and Step-Index Fiber.

gradient In graphics, having an area smoothly blend from one color to another, or from black to white, or vice versa. See Gradient Fill.

gradient fill A computer imaging term. A fill composed of a smooth blend from a starting color to an ending color. There are many variations on this theme. Most programs let you apply textures, and others have "smart" gradient fill routines that lend a three-dimensional appearance.

gradium glass Gradium glass is capable of reducing optical aberrations inherent in conventional lenses and performing, with a single lens, tasks traditionally performed by multi-element conventional lens systems. By reducing optical aberrations and the number of lenses in an optical system, Gradium glass may provide more efficient light transmission and greater brightness, lower production costs, and a simpler, smaller product. Gradium glass is used in collimation in fiber optics. See Collimation.

grafitti The name of the handwriting recognition software used by Palm and other hand-held devices.

gralloch The little-known verb "gralloch" means to disembowel a deer. This is irrelevant, of course, except that you need to know how many times – five, and counting – I've hit a deer driving between my city home and my country home. And each time I hit one, it costs several thousand dollars

grand alliance Also known as HDTV Grand Alliance. Comprises AT&T, General Instrument Corporation, Massachusetts Institute of Technology, Philips Electronics North America Corporation, Thomson Consumer Electronics, The David Sarnoff Research Center, and Zenith Electronics Corporation. These organizations had developed and promoted competing digital standards for HDTV. In May 1993, and under pressure from the FCC, they joined together in a "Grand Alliance" to develop a final digital standard for HDTV, which then became known as ATV (Advanced TV). The resulting single standard was documented in the ATSC (Advanced Television Systems Committee) DTV (Digital TV) Standard, which was accepted in large part by the FCC in December 1996. See ATV and HDTV.

Grand Pooh-Bah See GPB.

grandfather clause See Grandfathered.

grandfather tape The first backup of a program or a data record, saved so that you can always go back to step one if something goes wrong.

grandfathered A piece of equipment or service that has a right to exist at the price it is being charged and in the configuration it is by reason of it being that thing before laws or rules about it were introduced and to change it all. The term goes back to the Civil War. Grandfathering was a provision in several southern state constitutions designed to enfranchise poor whites and disfranchise blacks by waiving voting requirements for descendants of men who voted before 1867. The word derives from a "grandfather clause." Grandfather clauses stated that the right to vote was only available to those Americans whose grandfathers had been eligible to vote. These clauses were used, primarily in the South, to discriminate against

blacks and immigrants shortly after Lincoln's issuance of the Emancipation Proclamation and congressional ratification of the Fourteenth Amendment. As a result, "grandfather" has come to mean something allowable because it was allowable before prohibitive legislation. See also Grandfathered Equipment.

grandfathered equipment Non-FCC registered telephone equipment that was directly connected to the telecommunications network without a phone company-provided protective connecting arrangement (PCA) prior to the formalized FCC registration program. See Grandfathered.

grandparents The people who think your children are wonderful even though they're sure you're not raising them right.

grand slam bundle Another name for a quadruple play pricing bundle – one that includes mobile phone service, landline phone service, Internet access and cable television.

granularity 1. Microsoft jargon for complexity. For example, when you "achieve granularity," you grasp the complexity of the issue or problem.

2. Scalable in the most agreeable terms. A granular technology is scalable in very small increments, like grains of sand. In other words, it can be upscaled in small increments, matching the small, incremental requirements of the user while avoiding disproportionately large increases in cost. It's an overused, overly optimistic, and misleading term which finds its application primarily in sales presentations and brochures. See Brochureware.

graphic character A character, other than a character representing a control function (like Ctrl G being, in WordStar and dBASE nomenclature, to delete the character on the right) that has a visual representation normally handwritten, printed, or displayed, and that has a coded representation consisting of one or more bit combinations.

graphic equalizer A device which adjusts the tone by changing specific frequencies. The tone control on a radio is a type of equalizer. A radio transmitter may amplify low-end signals better than high-end signals. An equalizer can reduce or increase the amplification of the broadcast for an even and accurate reproduction of the input.

graphic violator Picture the home page of a typical Web site. Somewhere on the page is a moving graphic – perhaps an animated GIF – that screams at you and violates the visual integrity and consistency of the page. Most often, such graphic is designed to deliberately violate the integrity of the page. It is often a paid-for advertisement. And the advertiser wants to draw your attention to his graphic ad. After all, he paid big money for the graphic.

graphical browser A graphical browser is another, more commonly used, term for a World Wide Web (WWW) client program. A graphical browser can display inline graphics and allows the user to choose hyperlinks to move between hypertext documents. All browsers are graphical these days. The two leading browsers are Netscape and Microsoft Internet Explorer.

Graphical User Interface GUI. A fancy name probably originated by Microsoft which lets users get into and out of programs and manipulate the commands in those programs by using a pointing device (often a mouse). Microsoft's own definition is more elaborate. Namely that GUI puts visual metaphor that uses icons representing actual desktop objects that the user can access and manipulate with a pointing device.

graphics coprocessor A programmable chip that speeds video performance by carrying out graphics processing independently of the computer's CPU. Among the coprocessor's common abilities are drawing graphics primitives and converting vectors to bitmaps.

graphics engine The print component that provides WYSIWYG (What You See Is What You Get) support across devices.

graphics file In terms of the World Wide Web (WWW), a graphics file is a file in graphics format that can be retrieved through a Web browser. The Web browser may need an add-on or file viewer in order to be able to display the file.

Graphics Interchange Format Graphics Interchange Format (GIF) is pronounced "Jiff." GIF is a format for encoding pictures (pictures, drawings, etc.) into bits so that a computer can "read" the GIF file and throw the picture up on a computer screen. The advantage of GIF files of images is that they're small, i.e. few bytes. It is a format for encoding images (pictures, drawings, logos, etc.) into bits so that a computer can "read" the GIF file and display the picture up on a computer screen. GIF can only handle 256 colors. The major advantage of GIF is that it compresses the image to a very small size, thus making it faster to transmit across phone lines. As a result, the GIF format has now become the standard for putting images up on web sites. Web page authors also often use GIF images because GIFs can be interlaced, which produces a melting effect on the client screen as the image is loading. CompuServe, the on-line service, invented GIF in 1987. GIF is based on a mathematical algorithm called LZW, which is a set of mathematical formulae used to compress images into GIF files. Unisys has a patent on LZW. GIF was originally used in the electronic bulletin board world and is used primarily to carry photographs of women in semi-clad and naked poses. According to Jack Rickard, ex-publisher of Boardwatch Magazine, "some of the photographs are reasonably good, but most feature strikingly plain women rather artlessly photographed by those whose higher calling is probably more aptly found in the building trades or automotive repair." Some argue that GIF really stands for Girls in Files. (That's a joke.) See also Internet.

graphics mode PCs work in two modes – text and graphics. In graphics mode, the pixels are individually addressable. In text mode, the graphics card inside your PC throws a type font on your screen. And that's it. You can't change it. In graphics mode, you can. Graphics and text modes are mutually exclusive.

graphics pipeline The conceptual framework for 3-D graphics processing. The pipeline consists of the application/scene, gemoety, triangle setup and rendering/ rasterization.

grasshopper fuse A fuse that indicates that it has been blown by the movement of a piece of springy metal.

grating See Fiber Grating.

grating lobes Secondary main lobes.

grave dancing Rejoicing in someone else's disaster.

graveyard orbit An orbit located 300 km (186.5 miles) above the geostationary orbit and into which Inter-Agency Space Debris Coordination Committee (IADC) guidelines state that dead satellites should be boosted to prevent their interfering with or colliding with working geostationary satellites. However, not all satellite owners park their end-of-life satellites there, either because an unexpected fuel shortage or control malfunction may cause loss of satellite control, or because the fuel used to boost a satellite to the orbit could keep it operational and consequently earning income for another three months in geostationary orbit. See IADC.

graveyard shift In the 1500s, local folks started running out of places to bury people. So they would dig up coffins and would take the bones to a "bone-house" and reuse the grave. When reopening these coffins, 1 out of 25 coffins were found to have scratch marks on the inside and they realized they had been burying people alive. So they thought they would tie a string on the wrist of the corpse, lead it through the coffin and up through the ground and tie it to a bell. Someone would have to sit out in the graveyard all night the ("graveyard shift") to listen for the bell; thus, someone could be "saved by the bell" or was considered a "dead ringer."

gravity cell A closed circuit cell used where a continuous flow of current is desired. This type consists of copper and zinc electrodes with copper sulfate and zinc sulfate electrolyte. These are separated because of difference in their specific gravity.

Gray 1. Elisha Gray was an inventor who filed for a patent on his own telephone design a few hours after Alexander Graham Bell. Gray was involved in a number of lawsuits with the young AT&T, and ultimately co-founded (with Enos BARTON) electrical equipment maker Western Electric, which was later sold to American Bell, which became AT&T. In 1984, Western Electric was renamed AT&T Technologies, and was spun off from AT&T in 1996 as part of Lucent Technologies. Stay tuned for another 10 years or so to see what becomes of Mr. Gray's company.

2. In America, it's gray. In Britain, Australia, New Zealand and South Africa, it's grey.

gray code A binary coding scheme first used in mechanical shaft encoders and then in digital radio communications to improve resistance to noise by minimizing the number of bit errors produced by a symbol error. Gray code is named for Stanford University professor

Decimal	Hexadecimal	Binary	Gray
0	0	0000	0000
1	1	0001	0001
2	2	0010	0011
3	3	0011	0010
4	4	0100	0110
5	5	0101	0111
6	6	0110	0101
7	7	0111	0100
8	8	1000	1100
9	9	1001	1101
10	A	1010	1111
11	B	1011	1110
12	C	1100	1010
13	D	1101	1011
14	E	1110	1001
15	F	1111	1000

Robert M. Gray who first proposed it in the early 1970s. The basic principle is that every transition from one value to the next involves only one bit change, which simplifies the detection of errors.

gray lady An endearing term for the New York Times newspaper.

gray list A cross between a white list and black list. A message transfer agent (MTA) that has implemented a grey list temporarily rejects email from an unwhite listed sender if the MTA has not previously encountered any of the following three items together (called a "triplet") in an email: (1) the IP address of the connecting host, (2) the sender's email ID address, and (3) the recipient's email ID. The temporary rejection is a crucial feature of a grey list. The first time that an email from an unwhitelisted source has an unrecognized triplet the email is rejected and the triplet is added to the grey list. On subsequent occasions when a received email contains the exact same triplet it is allowed to pass through. Grey listing is based on the assumption – correct or not – that a spammer's bulk-mailing process does not have a mechanism for handling temporary bounces, while a legitimate mailserver will attempt to reconnect and deliver initially rejected email.

gray scale The spectrum, or range, of shades of black that an image has. An optical pattern consisting of discrete steps or shades of gray between black and white. Early facsimile machines could only receive black and white images and print them in black or white. Now they can print 16 shades of gray. This way if they receive a photo, it will look like a photo.

gray whale One of the first non-Bell key systems sold in North America. It was a 1A2 electromechanical key system from TIE/communications. TIE was a shortening of the words Telephone Interconnect Equipment. It was a wonderfully reliable phone system. A gray whale is now a prized possession.

Graybar Probably the oldest distributor of phone equipment, as well as various electrical products. Named for Elisha GRAY and Enos BARTON, who formed Western Electric. The company is headquartered in St. Louis, though there is a famous Graybar building in Manhattan, attached to Grand Central Terminal. A plaque in the terminal near stairs leading into the Graybar building shows Gray and Barton (if it hasn't been defaced or stolen).

graybeards Older, more mature executives and scientists who are typically more conservative and more skeptical than the younger workers on the team. Consent of the graybeards is typically required for any wild-eyed R&D idea in the military or bigger companies.

grayscale monitor Any monitor capable of showing levels of gray and not just black or white.

GRE Generic Routing Encapsulation. An Internet term. GRE is one of the basic operations performed by tunnel servers when tunneling through the Internet in order to provide a secure VPN (Virtual Private Network). GRE simply provides for the encapsulation of one data packet inside another data packet. The original packet becomes the payload (i.e., data field, or content) for the final packet, which also includes a new header and trailer. Tunnel servers also encrypt the payload, and continuously authenticate the identity of the communicating machines on a packet-by-packet basis. GRE for IPX tunneling is defined by the IETF (Internet Engineering Task Force) in RFC 1701, and for IP-in-IP or bridge tunneling in RFC 1702. See also IETF, IP, IPX, Packet, and VPN.

grease monkey The word "grease monkey" comes from the person (usually a young boy) that would crawl up in the rafters to grease all of the pulleys and belts that ran all of the equipment in a blacksmith shop or machine shop.

great circle A circle defined by the intersection of the surface of the Earth and any plane that passes through the center of the Earth. The shortest distance, over the idealized surface of the Earth, between two points, lies along a great circle.

Great Wall of China A vast Chinese defensive fortification begun in the 3rd century B.C. and running along the northern border of the country for 2,400 kilometers or 1,500 miles.

greek prefixes Remember the word "chronous?" It's used to mean the process of adjusting intervals or events of two signals to get the desired relationship between them. Here are the Greek prefixes that describe different timing conditions:

  asyn - not with
  hetero - different
  homo - the same
  iso - equal
  meso - middle
  piesio - near
  syn - together

green Green is a term being applied by manufacturers to mean their equipment uses less electricity than other equipment. Classic PC equipment include screens that shut themselves almost down if they haven't been accessed for a minute or two. Much of the technology of "green" has already been used in laptops. When it moves to the desktop, there are occasionally compromises – such as moving from a 3/4rds off to full-on might take a moment or two. Are you willing to live with it?

green bean A consultant who has no practical experience, i.e. most of them.

green screen A monochrome display terminal, usually associated with mainframes in the 1970s and 1980s, that displayed fixed-size characters on a green phosphor screen.

Green Eggs and Ham Dr. Seuss wrote "Green Eggs and Ham" after his editor dared him to write a book using fewer than 50 different words. He never made it as a lexicographer, like me. Lucky man.

green energy See Green.

greenfield Also spelled green field. Greenfield is a fancy way of saying new. A Greenfield Internet opportunity is a brand new idea for a brand new business that's never been tried before. Greenfield is the opposite of "legacy," also called brownfield. A greenfield telecom network is one that is being designed and built from scratch, with no need to accommodate legacy (i.e., old) equipment or architectures. Usage of greenfield extends to almost any aspect of describing a new venture: greenfield companies, greenfield evaluation, greenfield opportunities, greenfield factors, etc. An existing network being expanded has both greenfield and "brownfield" components. A variant is to describe a greenfield effort as a "greenstart." See brownfield, greenfield operator and legacy.

greenfield operator A greenfield operator is at the opposite end of the spectrum from an incumbent operator. A greenfield operator has no legacy network assets, no legacy service focus, and no customers. It is attempting to build a network, service strategy, and market presence all at the same time.

greenwasher A business that uses the fact that it recycles to promote itself.

Greenwich Hour Angle Greenwich Hour Angle (GHA) is the angle measured westward from the Greenwich meridian to the meridian of an observed celestial object on the celestial sphere. In celestial navigation, where tabulations of GHA for the Sun, the Moon and selected planets and stars are used in computations, GHA is usually expressed in hours and minutes. In computations of satellite orbits it is usually expressed in degrees. The latitude of an object is equal to its right ascension minus the GHA of Aries at the instant of observation.

Greenwich Mean Time GMT. Also called Zulu Time (Z) and Universal Time Coordinated (UTC). Greenwich is a borough in South East London, England. It is located on the prime meridian, which is zero degrees geographic longitude. Greenwich was formerly the site of the Greenwich Observatory. For historic reasons, Greenwich is the place from which world time starts. When it is midnight there, it is 00:00:00:0000, or "all balls." For example, Greenwich Mean Time (GMT) is five hours later than United States Eastern Standard Time – i.e. the time in the northern hemisphere summer. Local time – what it was at your place – used to be good enough for business and government International shipping and commerce changed all of that many years ago, creating the need for a global standard and reference point. GMT was created in October 1884 to satisfy that need, with the reference being the average (i.e., mean) time it takes the earth to rotate from noon-to-noon as measured from the Greenwich Meridian Line, or prime meridian, which runs through the principal Transit Instrument at the Royal Observatory in Greenwich, England. The Greenwich Meridian Line is used as it is at zero degrees longitude, thereby serving as the reference point for all measurements of longitude. Although GMT technically has been replaced by Coordinated Universal Time (UTC), it still is widely regarded as the correct time for every international time zone. Communication network switches are typically coordinated on UTC, which is based on GMT, but is controlled by highly sensitive atomic clocks which keep accurate time measured in microseconds or better. See also UTC and Zulu Time.

greetings only mailboxes Mailboxes that deliver a message to incoming callers but do not allow a message to be left. The Greeting Only Mailbox may transfer a caller to a designated telephone number.

gregorian Dual-reflector antenna system employing a paraboloidal main reflector and a concave ellipsoidal subreflector.

GREP Generalized Regular Expression Parser. A really powerful UNIX utility which can search a text file or program, finding and displaying or printing lines of computer code which contain specific character strings.

greylist See graylist.

gribble Random binary data rendered as unreadable text. For example, noise characters in a data stream are displayed as gribble. Dumping a binary file (a file containing more than plain text, such as images, etc., which must be converted to ASCII–usually automatically–before being sent as email) is a great way to create great gribble. ASCII is, of course, a plain text file containing letters normally found on an, um, ordinary typewriter.

GRIC Global Reach Internet Connection. An alliance of ISPs (Internet Service Providers) and IAPs (Internet Access Providers) to provide roaming capabilities for travelers. Based on proprietary standards, roamers are authenticated before being afforded Internet access. Usage is cross-billed through the GRIC clearinghouse, with fees being set by each ISP for use of its facilities by roamers. GRIC includes over 100 member ISPs in approximately 30 countries, and includes over 1,000 POPs (Points of Presence). Members include Prodigy in the U.S., and Telstra in Australia. GRIC competes with I-PASS. The IETF's Roamops working group is developing a standard for roaming, as well. See also I-PASS and Roamops.

grid 1. That element in a vacuum tube having the appearance of a grid and which controls the flow of electrons from the filament to the plate. "Grid" generally refers to the control grid.

2. Global Resource Information Database: part of the United Nations environment program.

3. A nickname for Internet2. See Internet2.

grid computing The promise of grid computing is immense. The ability to manage many independent systems joined together by a powerful network as if it were a very powerful single system. Grid computing can join computers in the same room or across the globe. The idea is that the users get a unified view and have access to the power of multiple computing resources such as CPU cycles, disk space, software and files. Grid computing includes what is often called "peer-to-peer" computing. Open Grid Services Architecture is a standard set of protocols developed for grid computing by the Open Grid Services Architecture Working Group (OGSA WG) of the Global Grid Forum (GGF). More information is available at www.gridforum.org.

MIT's Technology Review writes: "In the 1980s internetworking protocols allowed us to link any two computers, and a vast network of networks called the Internet exploded around the globe. In the 1990s the "hypertext transfer protocol" allowed us to link any two documents, and a vast, online library-cum-shoppingmall called the World Wide Web exploded across the Internet. Now, fast emerging "grid protocols" might allow us to link almost anything else: databases, simulation and visualization tools, even the number-crunching power of the computers themselves. And we might soon find ourselves in the midst of the biggest explosion yet. "We're moving into a future in which the location of [computational] resources doesn't really matter," says Argonne National Laboratory's Ian Foster. Foster and Carl Kesselman of the University of Southern California's Information Sciences Institute pioneered this concept, which they call grid computing in analogy to the electric grid, and built a community to support it. Foster and Kesselman, along with Argonne's Steven Tuecke, have led development of the Globus Toolkit, an open-source implementation of grid protocols that has become the de facto standard. Such protocols promise to give home and office machines the ability to reach into cyberspace, find resources wherever they may be, and assemble them on the fly into whatever applications are needed. Imagine, says Kesselman, that you're the head of an emergency response team that's trying to deal with a major chemical spill. "You'll probably want to know things like, What chemicals are involved? What's the weather forecast, and how will that affect the pattern of dispersal? What's the current traffic situation, and how will that affect the evacuation routes?" If you tried to find answers on today's Internet, says Kesselman, you'd get bogged down in arcane log-in procedures and incompatible software. But with grid computing it would be easy: the grid protocols provide standard mechanisms for discovering, accessing, and invoking just about any online resource, simultaneously building in all the requisite safeguards for security and authentication. Construction is under way on dozens of distributed grid computers around the world – virtually all of them employing Globus Toolkit. They'll have unprecedented computing power and applications ranging from genetics to particle physics to earthquake engineering. The $88 million TeraGrid of the U.S. National Science Foundation will be one of the largest. When it's completed later this year (2003) the general-purpose, distributed supercomputer will be capable of some 21 trillion floating-point operations per second, making it one of the fastest computational systems on Earth. And grid computing is experiencing an upsurge of support from industry heavyweights such as IBM, Sun Microsystems, and Microsoft. IBM, which is a primary partner in the TeraGrid and several other grid projects, is beginning to market an enhanced commercial version of the Globus Toolkit. Out of Foster and Kesselman's work on protocols and standards, which began in 1995, "this entire grid movement emerged," says Larry Smarr, director of the California Institute for Telecommunications and Information Technology. What's more, Smarr and others say, Foster and Kesselman have been instrumental in building a community around grid computing and in advocating its integration with two related approaches: peer-to-peer computing, which brings to bear the power of idle desktop computers on big problems in the manner made famous by SETI@home, and Web services, in which access to far-flung computational resources is provided through enhancements to the Web's hypertext protocol. By helping to merge these three powerful movements, Foster and Kesselman are bringing the grid revolution much closer to reality. And that could mean seamless and ubiquitous access to unfathomable computer power."

2. Global Resource Information Database: part of the United Nations environment program.

GRIN Gradient index. Generally refers to the SELFOC lens often used in fiber optics.

gritch A computer complaint.

grok Grok comes from the 1961 sci-fi classic novel "Stranger in a Strange Land" by Robert A. Heinlein. Grok was a Martian word meaning "to drink" and metaphorically "to be one with." In contemporary usage among the techie crowd, you grok something if you understand it so completely that know it so intimately and exhaustively that you are "one with it," i.e., it is part of your very being. If you read this dictionary thoroughly, you will grok the language of telecom. See also Zen.

groom/fill In telephony, terms associated with more efficient use of T-1 trunks by combining partially filled input T-1 trunks into fully filled outgoing T-1 trunks.

grooming Consolidating or segregating traffic for efficiency. Managing bandwidth on a wide area, public or private network to use the long haul transmission facilities as effectively as possible. The basic concept is much like that of grooming your hair with a comb or brush, which is a matter of getting things in nice, neat little bundles which you then organize by direction through a part or a mechanism like a beret or a clip. Grooming of network traffic takes place at various points in the network, like COs (Central Offices), tandem offices, POPs (Points Of Presence), or peering points. At those physical locations, incoming traffic is organized by attributes such as type (e.g., voice, Frame Relay, and ATM), destination, and QoS (Quality of Service) requirement. Once organized, traffic with similar attributes is aggregated (i.e. joined together) and sent over outgoing circuits. Different types of traffic headed for the same destination, or intermediate point in the network, can share the same circuit, as long as each traffic type is separated by some sort of framing convention so that one can be distinguished from another.

gross additions A cellular industry term. The amount of new subscribers signing up for the service before adjusting for disconnects (churn).

gross margin A financial term, gross margin is the percentage of sales remaining after the costs of production are subtracted. To get to net profit from gross margin, you deduct marketing, sales and overhead costs.

ground Any zero-voltage point. Earth is considered a zero voltage grounding point. Grounding is connecting equipment by some conductor (wire) to a route that winds up in the earth (ground), or some suitable alternative, for electrical purposes. A "hard ground" is a direct connection between a device to ground through a wire or other conductor that has a negligible resistance to ground. One purpose of a "ground" wire is to carry spurious voltage (e.g. lightning strikes) away from the electrical and electronic circuits to which it can cause harm. Incorrect grounding is probably the major cause of telephone systems problems. See Grounding (the major explanation), Ground Return and Ground Start.

ground absorption The loss of energy in transmission of radio waves due to dissipation in the ground.

ground block A device used to ground the shield of a coaxial cable usually at the point where the cable enters a building.

ground bulkhead A special grounded metal entrance plate to which the shields of up-tower transmission lines, waveguides, or other up-tower cables are bonded.

ground button A button needed on phones used for power failure transfer behind a PBX. You need the button because many trunks behind a PBX are ground start (as compared to loop start).

ground clamp A clamp or strap used to provide make a secure connection to a water pipe or grounding rod. It connects a wire to earth ground.

ground constants The electrical constants of the earth, such as conductivity and dielectric constant. The values vary with frequency, and also with local moisture content and chemical composition of the earth.

ground fault In AC electricity, a ground fault is any unintended connection between a supply conductor and ground (i.e.: hot conductor in contact with the metal case of a piece of equipment). A ground fault will cause a high current flow and should operate the overcurrent protection (fuse or breaker provided such devices are functionally adequate) only if the ground path impedance is sufficiently low – but under no circumstances greater than two ohms.

Ground Fault Circuit Interrupter GFCI or GFI. A device intended to interrupt the electrical circuit when the fault current to ground exceeds a predetermined value (usually 4 to 6 milliamps) that is less than required to operate the overcurrent protection (fuse or breaker) for the circuit. This device is intended to protect people against electrocu-

tion. It does not protect against fire from circuit overload.

ground-fault current path An electrically conductive path from the point of a ground fault on a wiring system through normally non-current-carrying conductors, equipment, or the earth to the electrical supply source.

Ground Fault Protector GFP. A device designed to protect electrical service equipment from arcing ground faults. A GFP does not provide protection for people.

ground kit A kit of the parts necessary to ground the shield of a coaxial cable or body of a waveguide to the structure of an antenna tower at periodic intervals.

ground lead The conductor leading to the ground. Connection.

ground loop This occurs when a circuit is grounded at one or more points. It can cause telephone system problems. Here's an explanation from American Power Conversion Corp.: Common wiring conditions where a ground current may take more than one path to return to the grounding electrode at the service panel. AC powered computers all connected to each other through the ground wire in common building wiring. Computers may also be connected by data communications cables. Computers are therefore frequently connected to each other through more than one path. When a multi-path connection between computer circuits exists, the resulting arrangement is known as a "ground loop". Whenever a ground loop exists, there is a potential for damage from inter system ground noise.

ground mount A flat plate used for mounting antenna masts on the ground or sometimes on flat roofs.

ground noise Thermal Noise emanating from either an electrical ground or from the earth itself.

ground plane The surface existing or provided, that serves as the near-field reflection point for an antenna.

ground potential The electrical potential of the earth with respect to another body or region. The ground potential of the earth will vary with locality and also as a function of certain phenomena such as meteorological disturbances.

ground radials Horizontal copper rods buried at the base of an antenna tower to establish an earth ground either for lightning protection or for a propagation plane depending upon the type of antenna being used.

ground return If a battery is connected to a closed electrical circuit, an electric current will flow in the circuit. In the early days of the telegraph, the circuit consisted of a long wire, the telegraph key, the electromagnet of a telegraph sounder and a return path through the ground, which served as a conductor. Thus the current flowed from one terminal of the battery through the wire, through the electromagnet, to a metal stake driven into the ground (a "ground" electrode), back through hundreds of miles of earth to the distant stake at the distant telegraph office and then to the other terminal of the battery. In later telecommunications, the ground return path was replaced by a second wire.

ground return circuit A circuit in which the earth serves as one conductor. A circuit in which there is a common return path, whether or not connected to earth ground.

ground rod Usually a six or eight foot copper rod driven into the earth to establish an earth ground.

ground start A way of signaling on subscriber trunks in which one side of the two wire trunk (typically the "Ring" conductor of the Tip and Ring) is momentarily grounded (often to a cold water pipe) to get dialtone. There are two types of switched trunks typically for lease by a local phone company – ground start and loop start. PBXs used to work best on ground start trunks. Now most work on loop start. Normal single line phones and key systems typically work on loop start lines. You must be careful to order the correct type of trunk from your local phone company and correctly install your telephone system at your end – so that they both match. In technical language, a ground start trunk initiates an outgoing trunk seizure by applying a maximum local resistance of 550 ohms to the tip conductor. See Loop Start.

ground start supervision Telephone circuitry developed to prevent Glare.

ground start trunk A phone line that uses a ground instead of a short to signal the CO for a dial tone. required by some PBXs. See Ground Start.

ground station A cluster of communications equipment, usually including signal generator, transmitter, receiver and antenna that receives and/or transmits to and from a communications satellite. Also called a satellite earth station.

ground strap A wide copper strap used to ground equipment.

Ground to Air Paging GAP. The ability to deliver a message (also called a page) from the ground to a user in an airplane.

ground wave In radio transmission, a surface wave that propagates close to the surface of the Earth. The Earth has one refractive index and the atmosphere has another, thus constituting an interface. These refractive indices are subject to spatial and temporal changes. Ground waves do not include ionospheric and tropospheric waves.

ground wire Also called a drain wire. It is an extra conductor (usually a bare wire) added to a cable for connection of the grounding path. See ground.

grounding There are two distinct and unique categories in this broad area called grounding. One is earth grounding. The purpose of an earth grounding system is essentially threefold: 1. To guard against the adverse effects of lightning, 2. To assist in the reduction of static and 3. To bring a zero-voltage reference to system components in order that logic circuits can communicate from a known reference. The other category of grounding is known as equipment grounding. The purpose of equipment grounding is threefold: 1. To maintain "zero volts" on all metal enclosures under normal operating conditions. This provides protection from shock or electrocution to personnel in contact with the enclosure. This is the safety aspect. This is the primary means of protecting personnel from electrocution. 2. To provide an intentional path of high current carrying capacity and low impedance to carry fault current under ground fault conditions; and 3. To establish a zero voltage reference for the reliable operation of sensitive electronic equipment. Effective equipment grounding is defined in the National Electrical Code, Article 250-51 and the Canadian Electrical Code Article 10-500. These Codes read almost identically. They say: The path to ground from circuits, equipment and metallic enclosures for conductors shall;

1. Be permanent and continuous.

2. Have the capacity to conduct safely any fault current likely to be imposed on it.

3. Have sufficiently low impedance to limit the voltage to ground and to facilitate the operation of the circuit protective devices in the circuit.

The Earth shall not be used as the sole equipment grounding conductor.

According to the Electric Power Research Institute, "electrical wiring and grounding defects are the source of 90% of all equipment failures." Many telephone system installer/contractors have found that checking for and repairing grounding problems can solve many telephone system problems, especially intermittent "no trouble found" problems. As electrical connections age, they loosen, corrode and become subject to thermal stress that can increase the impedance of the ground path or increase the resistance of the connection to earth. Equipment is available to test for proper grounding. (Ecos Electronics of Oak Park, IL makes some.) Before you install power conditioning equipment such as voltage regulators, surge arresters, etc. you should test for and correct any problems you have with grounding and wiring.

This story is related by Pat Routledge of Winnepeg, Ontario, Canada about an unusual telephone service call he handled while he was living in England. It is common practice in England to signal a telephone subscriber by signaling with 90 volts across one side of the two wire circuit and ground (earth in England). When the subscriber answers the phone, it switches to the two wire circuit for the conversation. This method allows two parties on the same line to be signalled without disturbing each other. This particular subscriber, an elderly lady with several pets called to say that her telephone failed to ring when her friends called and that on the few occasions when it did manage to ring, her dog always barked first. Torn between curiosity to see this psychic dog and a realization that standard service techniques might not suffice in this case, Pat proceeded to the scene. Climbing a nearby telephone pole and hooking in his test set, he dialed the subscriber's house. The phone didn't ring. He tried again. The dog barked loudly, followed by a ringing telephone. Climbing down from the pole, Pat found: several things. First, the dog was tied to the telephone system's ground post via an iron chain and collar; Second, every time a call came in, the dog received 90 volts of signalling current. Third, after several 90 volt jolts, the dog began to urinate on the ground. Lastly, the wet ground now conducted the 90 volt signaling current, which caused the phone rang. See also AC, AC Power and Battery.

grounding electrode conductor The conductor used to connect the grounding electrode to the equipment grounding conductor, to the grounded conductor of the circuit at the service equipment, or at the source of a separately derived system.

grounding field Grounding rods placed in the ground and connected together around an antenna site or central office site. This provides the best possible ground for electronic equipment.

grounding strap A device worn on the wrist or on the shoe when handling a static-sensitive component to prevent static shocks (sparks) which could damage the component. Don't even think of touching a printed circuit card without wearing a grounding strap. See Grounding.

group 1. In call centers or in automatic call distributors, a group is the same as GATE or SPLIT. A group is an ACD routing division that allows calls arriving on certain telephone trunks or calls of certain transaction types to be answered by specific groups of employees.

2. A group is a collection of voice channels, typically 12. In AT&T jargon, a group is 12 channels. A supergroup is 60 channels. A mastergroup is 10 supergroups or 600 voice channels.

3. A collection of users, computers, contacts, and other groups. Groups can be used as security or as e-mail distribution collections. Distribution groups are used only for e-mail. Security groups are used both to grant access to resources and as e-mail distribution lists.

4. An SCSA definition. A group is an associated set of one or more Resource Objects. Groups encapsulate the functionality of the Resource Objects that are associated with them. Resource Objects within a Group have defined connectivity. The Group provides three services to the application: implicit management of connectivity between group members; representation of a single entity to the applications (group ID); and reservation of all physical resources (CPU, memory, time slots) required to provide the application with exclusive use of configured resources.

Group 1, 2, 3, 3 bis & 4 These relate to the facsimile machine business. They are essentially standards of speed and sophistication. They were created by the ITU-T in Geneva, Switzerland to make sure facsimile machines from one maker could speak to facsimile machines of another maker.

Group 1 transmits an 8 1/2 by 11-inch page in around six minutes. It conforms to ITU-T Recommendation T.2.

Group 2 transmits an 8 1/2 by 11-inch page in around three minutes. It conforms to ITU-T Recommendation T.3.

Group 3 – the most common fax in the world today – transmits an 8 1/2 by 11-inch page (also called A4) in as little as 20 seconds. It is a digital machine and includes a 9,600 baud modem. It transmits over dial up phone lines. Group 3 standards for facsimile devices were developed by ITU-T adopted in 1980 and modified in 1984 and 1988. Group 3 defines a resolution of 203 x 98 dots per inch and 203 x 196 for "fine." Group 3 uses modified Huffman code to compress fax data for transmission. For example, a white line with no text, called a run, extending across an 8.5" page equals 1728 bits. Modified Huffman Code compresses the 1728 bits into a 17-bit code word. The lengths for all possible white runs are grouped together into 92 binary codes that will handle any white run length from 0 to 1728.

Group 3 bis. This is an update to Group 3. It includes an image resolution of 406 x 196 dpi and a transfer rate of 14,400 bits per second. Fax machines that are Group 3 bis can transmit 50% faster to fax machines that are also Group 3 bis, which is a big speed improvement. Group 3 bis can drop to Group 3 if there's a Group 3 on the other end. Most of the modern plain paper fax machines and most of the today's computer fax modems are Group 3 bis. That means they transmit and receive at up to 14,400 bps.

Group 4 Fax. The latest and fastest international standard for facsimile machines. It specifies a machine which operates at 64 Kbps, which can only work on a digital channel and which takes six seconds to transmit a 8 1/2 x 11 inch page. The Group 4 standard was promulgated in January, 1987. Group 4 fax machines are designed to use one of the 64,000 bit per second B (Bearer) channels on ISDN. The main difference between Group 3 and Group 4 fax machines is that Group 4 fax machines do not convert the scanned information into an analog format before transmitting it down phone lines. Group 4 fax machines simply send the digitally scanned information down ISDN lines. The advantages of Group 4 fax are that quality is much higher, and call costs are much lower due to the increase in speed of transmission.

Most Group 3 machines will transmit and receive from Group 1, Group 2 and Group 3 machines (but at their slower speeds). Group 1, 2, 3 & 4 are international standards. Group 3 is now by far and away the most common. All Group 3 machines will transmit and receive from each other. Some manufacturers have improved on the standards by offering Group 3 "fine," for example. These "fines" can talk to the same machines. But often can't talk to other "fines." If you're buying a facsimile machine and want super-quality transmission, check its compatibility with other machines. Or, easier, buy all identical machines. See Facsimile.

group address A single address that refers to multiple network devices. Synonymous with multicast address.

group addressing In transmission, the use of an address that is common to two or more stations. On a multipoint line, where all stations recognize addressing characters but only one station responds.

Group Busy Hour GBH. The busy hour offered to a given trunk group.

group call A special type of station (i.e. extension) hunting that requires a special access number to permit a call to the special access number and ring the first available phone in that group.

group channel Twelve voice-grade channels. This is one of the levels of the transmission hierarchy.

Group Distribution Frame GDF. In frequency-division multiplexing, a distribution frame that provides terminating and interconnecting facilities for the modulator output and demodulator input circuits of the channel transmitting equipment and modular input and demodulator output circuits for the group translating equipment operating in the basic spectrum of 60 kHz to 108 kHz.

group hug A group hug is Microsoft's Strategic relationship.

group hunting Automatically finds free telephones in a designated group. See Hunt.

group scheduling software Software designed to coordinate and manage both worker schedules and office resources such as equipment and conference rooms.

group velocity 1. The velocity of propagation of an envelope produced when an electromagnetic wave is modulated by, or mixed with, other waves of different frequencies. The group velocity is the velocity of information propagation.

2. In optical fiber transmission, for a particular mode, the reciprocal of the rate of change of the phase constant with respect to angular frequency.

grouping A facsimile term for periodic error in the spacing of recorded lines.

GRoupIPC Ohio, January 8, 1997, GRoupIPC - North America (GRoupIPC-NA) has announced their official incorporation as a non-profit organization. Susan M. Chicoine, of Systran Corp., Dayton, Ohio, who served as a trustee during the incorporation process, was elected president of the group. The parent organization, GRoupIPC, was established in Europe in 1994 to provide a worldwide forum for the exchange of information on Industry Pack (IP) and PCI Mezzanine Card (PMC) technologies. GRoupIPC-NA promotes the embedded system industry's trend towards open, internationally recognized, standards solutions reinforcing the movement away from sole-vendor, proprietary solutions that limit the flexibility and upgradeability of embedded system designs. Member organizations support open-system solutions based upon the internationally recognized PMC and IP mezzanine board standards, and will promote their use to bring the benefits of stable, multi-vendor standards to embedded designs. GRoupIPC-NA will: (1) Promote market acceptance and the use of PMC and IP mezzanine board technology, (2) Disseminate information about products, applications and technical requirements, using or affecting IP and PMC mezzanine board technology and (3) Provide market and technical support to users, distributors, and manufacturers of IP and PMC technology and products. GRoupIPC-NA is headquartered in Dayton, Ohio. www.GRoupIPC.com, email GRoupIPCNA@aol.com or phone 937-427-9735.

groupware A term for software which runs on a local area network and which allows people on the network (typically a team) to participate in a joint (often complex) project. According to Fortune Magazine, March 23, 1992, using groupware, "Boeing has cut the time needed to complete a wide range of team projects by an average of 91%, or to one-tenth of what similar work took in the past." Groupware can be used in a meeting, with everyone sitting around a conference table and typing their ideas into the PC in front of them. Groupware can also be used off-line, with members of the "team" in different cities adding their comments. The "bellwether" of groupware software is Lotus' program called Notes.

growth addition A telephone company definition. Any equipment addition that increases the limiting capacity of an entity. Hence, a trunk relay addition will not generally be considered a growth addition since trunk relay equipments are not considered as limiting. However, if a trunk relay addition requires other equipment, such as trunk frames, or other common control equipment, then the addition should be considered a growth addition.

growth entity A telephone company definition. The growth entity in a multi-entity wire center is that entity where all future network access line growth is engineered to take place. Non growth entities are normally loaded at or near capacity and excess demand is served via the growth entity. This distribution of demand is accomplished via the loading plan.

growth factor A telephone company definition. A ratio derived by trending network access lines, traffic or loads and relating the future levels to current levels. Growth factors may be combined to develop a projection ratio or used individually as a projection ratio.

gruntle A pig's snout is called a "gruntle."

GRSU Generic Remote Switch Unit.

GS 1. Ground Separator.

2. Gateway Server: A station on the local area network that has devices necessary to provide system interoperability between one or more network users.

GSR Gigabit Switch Router.

GSS Group Switch Selector.

GS Trunk Ground Start Trunk. A trunk on which the communications system, after verifying that the trunk is idle (no ground on tip), transmits a request for service (puts ground on ring) to a telephone company. The other and more common type of trunk is called a Loop Start Trunk.

GSA General Services Administration.

GSDN See Global Software Defined Network.

GSF See Generic Services Framework.

GSM GSM originally stood for Groupe Speciale Mobile. Now it's known as Global System for Mobile Communications. It is the standard digital cellular (also called mobile) phone service you will find in Europe, Japan, Australia and elsewhere – a total of 85 countries. Most countries decided to pick a single, standard wireless phone technology years ago, and they settled on GSM. The U.S. refused to settle on a standard and that has resulted in a patchwork of multiple, incompatible technologies. GSM exists in the U.S., and is gaining ground in the U.S. though it uses a different frequency than the system used in Europe adn elsewhere. In the U.S. used by companies including VoiceStream, Cingular and AT&T, which was, at the time of writing, is in the process of converting its network to GSM.

GSM actually is a set of ETSI standards specifying the infrastructure for a digital cellular service. ETSI has trademarked ETSI. To ensure interoperability between countries, these ETSI standards address much of the network wireless infrastructure, including the radio interface (900 MHz), switching, signaling, and intelligent network. An 1,800 MHz version, DCS1800, has been defined to facilitate implementation in some countries, particularly the UK. Since GSM is limited to technical standards, an association of GSM operators called the Memorandum of Understanding (MoU) ensures service interoperability, allowing subscribers to roam across networks. GSM has gained widespread acceptance. As of mid-1999, there were operational GSM networks in 133 countries, with over 170 million customers – more than all the Internet subscribers in the world. GSM accounted for 65 percent of the world's cellular users.

GSM subscriber data is carried on a Subscriber Identity Module (SIM) or "smartcard" which is inserted into the phone to get it going. As a result, the subscriber potentially has the option of either SIM card mobility or terminal mobility across multiple networks.

There are now four frequency flavors of GSM: 450 MHz (upgrade of old NMT systems in Scandinavia), 900 MHz (original flavor everywhere except North America and most countries in South America), 1800 MHz (new flavor everywhere except North and South America – brought in to add capacity and competition), and 1900 MHz (North America and much of South America, i.e. the "PCS" frequencies. IN the United States, the GSM frequencies are 1850-1910 MHz and 1930-1990 MHz.

Here's the technology of GSM: Access method: mixed TDMA & FDMA with optional frequency hopping. Security: Optional radio interface encryption. Carrier frequency division: 200 KHz. Users per carrier frequency: 8. Speech bit rate (transfer rate): full rate (13 kbps) or half rate. Total bit rate: 21 Kbps. Bandwidth per channel: 25 KHz. The audio encoding subset of the GSM standard is best known to computer users because its data compression and decompression techniques are also being user for Web phone communication and encoding .wav and .aiff files.

The best book on GSM is "The GSM System for Mobile Communications" by Michael Mouly and Marie-Bernadette Pautet, both of France. The authors contributed to the development of GSM. See also BSS, which stands for Base Station System. See also SIM Card.

GSM-900 Global System for Mobile Communications at 900 MHz. A wireless telecommunications term. A GSM (Global System for Mobile Communications) standard for cellular phone systems operating at 900 Megahertz. See GSM.

GSM-R The European Standard for Railway Communication Technology. The European railways have developed a communication network based on GSM, called GSM-R. GSM-R will integrate all existing mobile radio applications for railway use and provide a platform for new services and applications for future evolution. In particular, GSM-R will provide a communication platform to enable the interoperability of railway traffic and provide a bearer service for the European Train Control System. European Council Directives require a fully interoperable Control and Command System on the international high-speed lines of European Union railways, and GSM-R will enable this. GSM-R brings together in a single system most, if not all, of the applications needed by the railways, including: digital technology, the integration of services based on a standardized open system, interworking with railway and public mobile and fixed communication systems, Europe-wide roaming and mobility management appropriate to railway specific services and performance. Applications for railway operation include :

- Logistics (tracing cargo coaches, containers, goods etc, polling of status information on goods).
- Enhanced operational services for railway staff (Intranet access to operational databases, customized news services for time scheduling and tariffs, automatic seat reservation with display).
- Telematic applications for rolling stock and fixed equipment.
- Optimized freight load and on-line sale of free capacity.
- On-line passenger information systems on trains and platform.

- Internet terminals for reliable on-train WWW services.
- Train journey specific car rental, taxi ordering, hotel reservations, etc.

Background: In 1993, the Union Internationale des Chemins de Fer (UIC) chose the GSM standard as a basis for its future digital mobile system and established the European Integrated Railway Radio Enhanced Network project (EIRENE) to co-ordinate user requirements, establish the high level specifications and to co-ordinate the related standardization and pre-operation activities. The EIRENE project produced Specifications for the interoperability of mobile communications. To ensure that Member States complied with these Specifications, they were converted into European Standards by ETSI (the European Telecommunications Standards Institute) and CENELEC (the European Committee for Electrotechnical Standardization. See also GSM.

GSN 1. Gigabyte System Network. See HIPPI.

2. Global Subscriber Number. See UIFN.

GSO GeoStationary Orbit. The path described by a satellite that always remains fixed with respect to all points on a rotating orbited body, is circular, lies in a plane and has points that revolve about the orbited body in the same direction and with the same period as the orbited body rotation. See Geostationary Satellite and NGSO Satellite.

GSTN Global Switched Telephone Network. The GSTN essentially is the international version of the PSTN (Public Switched Telephone Network). See also PSTN.

GT 1. Global Title. An address such as customer-dialed digits that does not explicitly contain information that would allow routing in the SS7 signaling network. The GTT translation function is required. See GTT.

2. Gain Transfer.

GTA 1. Government Telecommunications Association. An association of local, state and federal telecommunications professionals in Washington, D.C.

2. The former Canadian Government Telecommunications Agency is now called GTIS (pronounced "GEE-tiss", Government Telecommunications and Informatics Service) and is part of the department called PWGSC (Public Works and Government Services Canada).

GTAG Global Tag. A standardization initiative of the Uniform Code Council (UCC) and the European Article Numbering Association (EAN) for asset tracking and logistics based on radio frequency identification (RFID). The GTAG initiative is supported by Philips Semiconductors, Intermec, and Gemplus, three major RFID tag makers.

GTE General Telephone and Electronics. GTE was a major telecommunications company, whose main business was owning and operating independent (i.e., non Bell) local telephone companies, which were known as GTOCs (General Telephone Operating Companies). GTE also used to own part of Sprint, the long distance company, but sold its interest to United Telecom. On July 28, 1999, Bell Atlantic announced that it was merging with (i.e., acquiring) GTE. In 2000, Bell Atlantic changed its name to Verizon. See also GTE Sprint and Verizon.

GTE Sprint A long distance service once provided by GTE Sprint, then a 50-50 joint venture of GTE and United Telecom. In 1989, United Telecom acquired the majority interest (80.1%), and in 1991, bought the remaining 19.9%. During this time, the company became Sprint Corporation. See also Sprint.

gTLD Generic Top Level Domain. An term for Internet and World Wide Web naming. The idea of gTLD is to allow web sites with creative, descriptive names such as www.TechnologyInvestor.com, which is a lot easier to remember than an IP address. The original TLDs include .com (commercial), .edu (educational institution), .gov (government), .org (non-profit organization), .net (network provider) and .mil (military). There was a proposal that beginning in March 1998, new TLDs would be created in order to expand the available address options. After much wrangling, the decision was made on November 16, 2000 to create the following new TLDs: .aero (aviation industry), .biz (business), .coop (business cooperatives), .info (general use), .museum (museums), .name (individual, personal use), and .pro (professionals). The administration of the gTLDs has now been contracted by CORE (Council Of REgistrars) to Emergent Corporation, which developed, and is to maintain and operate the Shared Registry System (SRS). The SRS is a neutral, shared, and centralized database of the all gTLDs. A large number of independent entities, known as "registrants," were authorized to register domain names, or URLs (Uniform Resource Locators), with each relying on the SRS. A URL, such as www.HarryNewtonsGhastlyPaintings.museum (this is not a real URL, at least not at the time of this writing), is translated into an IP address by a Domain Name Server (DNS), also known as a "resolver." See also CORE, DNS, IP Address, SRS, TLD, and URL.

GTN Global Transaction Network. An AT&T service which adds smarts to the routing of inbound 800 calls. It offers six call processing services: Next available agent routing, call recognition routing, transfer connect service (allows agents to transfer calls to distant ACDs), network queuing, 800 select again service and multiple number database (allows

multiple 800 numbers to be assigned to a single routing plan in the network, rather than each 800 number having its own unique routing plan).

GTOC General Telephone Operating Company. An obsolete term for a local telephone company owned by the GTE (General Telephone and Electronics) system. GTE was acquired by Bell Atlantic, now known as Verizon. See also GTE and Verizon.

GTP 1. General Telemetry Processor.

2. GPRS Tunneling Protocol. GTP handles the flow of user packet data and signaling information between the SGSN and GGSN in a GPRS network. GTP is defined on both the Gn and Gp interfaces of a GPRS network.

GTP Tunnel Used to communicate between an external packet data network and a mobile station in a GPRS network. A GTP tunnel is referenced by an identifier called a TID and is defined by two associated PDP contexts residing in different GSNs. A tunnel is created whenever an SGSN sends a Create PDP Context Request in a GPRS network.

GTSS Global Technical Services Solutions.

GTT Global Title Translation. The process of translating a Global Title from dialed digits to a point code (network node) address and application address (subsystem number). This process is accomplished by the STP (Signal Transfer Point) in the SS7 network. GTT is defined in IS-41B. See also Global Title Translation, IS-41, SS7, and STP.

guard band See Guardband.

guardband A narrow bandwidth between adjacent channels which serves to reduce interference between those adjacent channels. That interference might be crosstalk. Guardbands are typically used in frequency division multiplexing. They are not used in time division multiplexing, because the technology is completely different.

guardian agent A Guardian Agent is similar to an Intelligent Agent (which hunts for and grabs information off of the web that you specify) only the Guardian Agent prevents certain sites, such as pornographic pages, gambling sites, or other areas you don't want a child to see, from being accessed.

guardian chip GPS-equipped microcircuit that can be embedded under the skin and used to ID and track lost kids, Alzheimer's patients, or people buried in collapsed buildings. The technology has civil liberties groups concerned about possible abuses.

guarding The process of holding a circuit busy for a certain interval after its release to assure that a necessary minimum disconnect interval will occur between calls.

gubment Slang for government.

guest book You own a Web site. You've spent megabucks attracting people to your Web site. You want to keep in touch with these visitors so you can sell them something. So you politely ask them to fill their names and email addresses in your "guest book," and your Web server captures the names into a emailing database.

guest mailbox A mailbox used by a hotel or motel to set up temporary mailboxes for their guests. At least that was the original definition. Now it seems every voice mail system comes with guest mailboxes that could be used for visitors, employees from out of town, etc. Same application as a hotel – temporary use.

guest network A network at a hotel, restaurant, convention center, gas station, truck stop, office lobby, airport lounge, or at some other private or semi-private establishment that is made available, generally for free, to guests, visitors, or customers so that they can access their email or the Web. The definition of "guest," the use of authentication, and the degree of access varies by establishment. Also called a courtesy network.

GUI Graphical User Interface. A generic name for any computer interface that substitutes graphics for characters. GUIs usually done with a mouse or trackball. Microsoft's Windows is the most famous GUI. Second most famous is the Apple operating system. GUI is pronounced "GOO-ey." See Graphical User Interface.

GUID Globally Unique IDentifier. GUIDs are small pieces of information which some software programs embed into documents and files which you create, file and email to others. They allow others to know who created the document and when. You can find them on most Microsoft Office products, like Word, Excel, PowerPoint, etc. I find them useful. Others find them a little more sinister. According to the New York Times Magazine of April 30, 2000 GUIDs "are making it possible to to link every document you create, message you email and chat you post with your real-world identity. GUIDs are a kind of serial number that can be linked with your name and email mail address when you register online for a product or service. IN short, a GUID is a number that is known to be unique and which is assigned to a session or user in order to identify them. See also Globally Unique IDentifier.

guided ray In an optical waveguide, a ray that is completely confined to the core. See also DWDM.

guided wave A wave whose energy is concentrated near a boundary or between substantially parallel boundaries separating materials of different properties and whose direction of propagation is effectively parallel to these boundaries.

guru In Hinduism, a guru is a personal religious teacher and spiritual guide. In contemporary techie lingo, a guru is an expert who is not only a wizard at something, but also a highly knowledgeable resource of information for others.

Gutenberg See 1453, up front in this dictionary.

gutta-percha A latex substance (now called rubber) first discovered in 1847 and derived from the sap of Malayan evergreen trees. It first found use circa 1851 as the insulation in the first international telegraph cable which ran between England and France. Gutta-Percha was also the first insulator to survive in underseas applications (particularly for submarine cables) and was still the insulator material of choice for golf balls and telephone receivers until about 1947, when polyethylene finally began to gain acceptance.

guy hook A hook bolted to telephone poles and used to attach guy wires.

guy thimble A device used to attach a guy wire to a bolt which is attached to an anchor in the ground.

guy wire A wire used to support radio mast, microwave antennas, telephone poles – in short anything that is tall and thin that needs supporting – so it doesn't fall over. Guy wires attach at various heights to what you're trying to support and are then attached to devices hammered into the ground some distance from the thing they're trying to support. Imagine a telephone pole standing in some exposed place. Let's say we attach three guy hooks equidistant around the pole, and three guy wires to each other. Bingo, the pole is supported firmly and shouldn't fall. See also Guy Hook.

guyed A type of wireless transmission tower that is supported by thin guy wires. In the early days, MCI used to pride itself that most of its microwave towers were guyed towers, as against AT&T's which were all self-standing and thus substantially more expensive. Ironically, neither AT&T nor MCI survived as separate entities, despite their tower philosophy differences.

GVRP The GARP VLAN Registration Protocol (GVRP) is a GARP (Generic Attribute Registration Protocol) application that provides registration services in a VLAN context. See GARP for more information.

Gzip A free compression software program available for Unix and MS-DOS. Appends either a .z or a .gz to the file name. Compressing a file makes it smaller and therefore it takes less time to transmit. The most popular programs are Winzip and PKZIP.

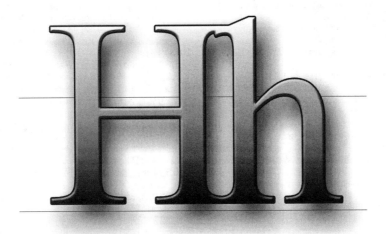

H The symbol designation for magnetic inductance.

H PAD Host Packet Assembler/Dissembler. See HPAD.

H Schedule A separate and distinct list in the translations area of program store in No. 1 or 1A ESS central office. The H (hourly) Schedule is normally used for collecting counts on items required for day-to-day administration of the central office equipment and for the engineering of general growth jobs. They include such items as call processing registers, service circuits and miscellaneous trunks, intraoffice trunks and junctors networks.

H-1B Visa A visa issued by the Immigration and Naturalization Service to skilled workers coming from overseas to work in American companies. Typically those visas are issued to to workers who have scarce skills.

H-Channel The packet-switched channel on an ISDN PRI (Primary Rate Interface) which is designed to carry user bandwidth-intensive videoconferencing information streams at varying rates, depending on type: HO – 384 Kbps; H11 – 1,536 Kbps; and H12 – 1,920 Kbps. In short, H-channels are ISDN bearer services that have pre-defined speeds, starting and stopping at locations on an ISDN PRI circuit. They are contiguously transported from one ISDN PRI site through networks to another ISDN PRI site. H-channels are accomplished by aggregation of multiple individual 64Kbps B channels; in the carrier domain, this aggregation is accomplished through a process known as "bonding." See ISDN and PRI.

H-MVIP The original MVIP standard, now called MVIP-90, was developed in 1989-90 and first deployed in 1990. MVIP-90 supports up to 512 telephony channels of 64 Kbps each between circuit boards within a single computer chassis. MVIP-90 has been widely adopted for voice, FAX, data and video services as well as for telephony switching applications. Other MVIP standards have since been developed to address multi-chassis MVIP systems and higher level software APIs. Beginning in 1993, there was interest in higher capacities within individual single-chassis MVIP nodes. Specific applications include large audio conferences, multi-media servers, and the termination and switching of all traffic on a dual FDDI-II fiber ring (as used in MC2 standard Multi-Chassis MVIP) as well as the termination of and switching of traffic from T-3/E-3 telephone trunks and SONET/SDH links at OC-3 (155 Mbps) rates (as used in MC3 standard Multi-Chassis MVIP). H-MVIP addresses this need for higher telephony traffic capacity in individual computer chassis.

H-MVIP defines three major items that together make a useful digital telephony transport and switching environment. These are the H-MVIP digital telephony bus with up to 3072 "time-slots" of 64 Kbps each; a bus interface with digital switching that allows a group of H-MVIP interfaced circuit boards to provide distributed telephony switching and a logical device driver model and standard software interface to that logical model. The bus definition includes the mechanical, electrical and timing requirements for a high capacity telephony bus that is a super-set of the existing MVIP-90 standard for computer-based telephony. Several levels of capacity expansion are defined to support a range of system implementations. Among them, H-MVIP "24/2" is a wider version of MVIP-90, while the full H-MVIP bus is both wider and faster. See ECTF, SCSA and TAO.

H.100 H.100 is a hardware specification that provides all the necessary information to implement a CT Bus interface at the physical layer for the PCI computer chassis card slot independent of software applications. It is the first card-level definition of the overall CT Bus single-communications bus specification. CT Bus defines a single isochronous communications bus, often called a mezzanine bus, across newer PC chassis card slots (PCI, and the emerging compact PCI). H.100 CT Bus will be compatible with the most popular existing implementations, SCBus, HMVIP and MVIP (as well as ANSI VITA 6) implemented in ISA/EISA card slots. A CT Bus specification for compact PCI, H.110, is also under development for a later release.

Adoption of the single-bus specification, CT Bus, will allow a fluid inter-operation of components to provide an unprecedented level of flexibility for product design and operation. CT Bus provides more capacity to allow development of a new class of applications as well as to increase the capabilities of existing applications. Its addition of greater fault-tolerance will increase the reliability of applications, and its provision for implementing a subset of the specification will provide for many lower cost applications. H.100 offers the following features:

- A PCI card slot form factor to accommodate the growing popularity of PCI slots in computer chassis.
- 4,096 bi-directional time slots (permitting up to 2,048 full duplex calls) for larger communications capacity.
- An eight megabit data rate and 128 channels per stream for greater bandwidth.
- Redundant clocking scheme for increased fault tolerance.
- Backwards compatibility and interoperability with SCBus, HMVIP and MVIP.

The H.100 is part of a complementary suite of Interoperability Agreements sponsored by the ECTF (Enterprise Computer Telephony Forum). Each specification is fully self contained, yet designed to be complementary with all of the others in the suite. Other ratified specifications include S.100 and S.200. Software developers are creating applications with the S.100 and S.200 specifications, and hardware manufacturers will introduce H.100-based communications cards.

S.100, published in March, 1996 specifies a set of software interfaces that provide an effective way to develop CT applications in an open environment, independent of underlying hardware. It defines a client-server model in which applications use a collection of services to allocate, configure and operate hardware resources. S.100 enables multiple vendors' applications to operate on any S.100-compliant platform.

H.110 H.110 defines H.100 on the CompactPCI (cPCI) Bus. The biggest difference between H.100 and H.110 is that H.110 supports CompactPCI Hot Swap (the removal and insertion of cards in a live system). www.ectf.org/ectf/home.html. See also H.100.

H.221 A framing recommendation which is part of the ITU-T's H.320 family of video interoperability recommendations. The recommendation specifies synchronous operation where the coder and decoder handshake and agree upon timing. Synchronization is arranged for individual B channels or bonded 384 Kbps (HO) connections.

H.222.0 Defines the general form of elementary stream multiplexing as the Moving Picture Experts Group 2 (MPEG-2) system part. See H.222.1.

H.222.1 Specifies the parameters of H.222.0 for communication use.

H.223 Multiplexing and control protocol for H.324.

H.225.0 This is H.323 protocol based on RAS and ISDN's Q.931. It is used for establishing calls.

H.230 A multiplexing recommendation which is part of the ITU-T's H.320 family of video interoperability recommendations. The recommendation specifies how individual frames of audiovisual information are to be multiplexed onto a digital channel.

H.231 A recommendation, formally added to the ITU-T's H.320 family of recommendations in March, 1993, which specifies the multipoint control unit used to bridge three or more H.320-compliant codecs together in a multipoint conference.

H.233 A recommendation, part of the ITU-T's H.320 family, which specifies the encryption method to be used for protecting the confidentiality of video data in H.320-compliant exchanges. Also called H.KEY.

H.234 ITU-T standard for delivering encryption key management and authentication system for audiovisual services.

H.235 An ITU-T standard (February, 1998) for securing H.323 voice and videoconference information streams over IP networks (e.g., the Internet, Intranets and LANs). H.235 provides authentication, integrity and privacy services. Authentication serves to establish as genuine the identity of all endpoints in the conference in order that unauthorized users or machines cannot participate. Integrity validates the payload of data packets, thereby ensuring that the data was neither corrupted nor altered in transit. Privacy, accomplished through an encryption mechanism, ensures that the data payload cannot be read by users or machines not authorized. Non-repudiation, planned for inclusion in future releases of H.235, protects against an endpoint's denial of participation in the conference. See also H.323.

H.242 Part of the ITU-T's H.320 family of video interoperability recommendations. This recommendation specifies the protocol for establishing an audio session and taking it down after the communication has terminated.

H.244 Recommendation on a channel aggregation method for audiovisual communications. This enables several ISDN B-channels to behave as a single higher-rate channel.

H.245 H.245 specifies the signaling protocol necessary to actually establish a call, determine capabilities, and issue the commands necessary to open and close the media channels. The H.245 control channel is responsible for control messages governing operation of the H.323 terminal, including capability exchanges, commands, and indications. H.245 does not use in-band signaling, but uses data packets sent a data network. See H.323 and TAPI 3.0.

H.245 H.245 specifies a control protocol for multimedia communication. It is responsible for establishing informational channel for voice packets exchange between two terminals just after the call is established. H.245 specifies which IP-addresses are to be used for transport, which codecs are supported by terminals and so on. It also includes some messages for MCU control.

H.246 See H.245.

H.248 The ITU-T recommendation for MGCP (Media Gateway Control Protocol), which is an open, standards-based signaling and control protocol for use between circuit-switched PSTNs (Public Switched Telephone Networks) and VoIP (Voice over Internet Protocol) networks. H.248 is compliant with Signaling System R2, which is an international signaling system within international regions, for integrating international/national signaling. See also MGCP and R2.

H.261 The ITU-T's H.261 is the standards watershed in videoconferencing. Also known as p x 64, H.261 specifies the video coding algorithms, the picture format, and forward error correction techniques to make it possible for video codecs from different manufacturers to successfully communicate. H.261 is an ITU-standard video codec designed to transmit compressed video at a rate of 64 Kbps and at a resolution of 176 x 44 pixels (QCIF). Announced in November 1990, it relates to the decoding process used when decompressing video conferencing pictures, providing a uniform process for codecs to read the incoming signals. Any H.323 client is guaranteed to support the following standards: H.261 and G.711. Other important standards are H.221: communications framing; H.230 control and indication signals and H.242d: call set-up and disconnect. Encryption, still-frame graphics coding and data transmission standards are still being developed. See H.320.

H.263 H.263 is an ITU-standard video compression protocol that improves video streaming over a corporation's LAN or WAN. It is based on and compatible with H.261. It offers improved compression over H.261 and transmits video at a resolution of 176 x 44 pixels (QCIF). See also H.264.

H.264 An ITU-standard for video compression, said to achieve more and better compression than H.263. Also called MPEG-4/AVC (for Advanced Video Coding). H.264 is the emerging standard for videoconferencing and video over the Internet. Apple video iPod (now just called the iPod) uses the H.264 MPEG-4 format, which is not compatible with the way other vendors encode their movies. You need a conversion program. They exist. Check the Internet.

H.310 Recommendations for a videoconferencing terminal in an ATM environment.

H.320 The most common family of ITU-T videoconferencing standards. These standards allow ISDN BRI videoconferencing systems and videophones to communicate with each other. I've personally had several H.320 compatible videoconferencing systems and videophones on my desk and have received from and made videoconferencing calls to many different H.320 compatible video phones. The quality is not brilliant. But you can recognize the person at the other end. And they can recognize you. Most H.320 systems allow you to bond together the two B channels of a 2B+D ISDN BRI channel and thus get better video. See all the H.2NN and H.3NN explanations above and below. See also G.711 and V.80.

H.321 The adaptation to the ATM environment of H.320 videoconferencing standards. See H.320 and V.80.

H.322 The adaption to a guaranteed quality of service LAN of H.320 terminals. See H.320 and V.80.

H.323 H.323, a standard from the International Telecommunications Union (ITU-T), which serves as the "umbrella" for a set of standards defining real-time multimedia communications for packet-based networks – what are now called IP telephony. Much of the excitement surrounding the H.323 standards involves the use of H.323 entities to communicate over the Internet or privately-managed and privately-owned Internet Protocol (IP) networks. The standards under the H.323 umbrella define how components that are built in compliance with H.323 can set up calls, exchange compressed audio and/or video, participate in conferences, and interoperate with non-H.323 endpoints.

H.323 is an ITU-T standard, which defines a set of call control, channel setup and codec specifications for transmitting real-time voice and video over networks that don't offer guaranteed service or quality of service – such as packet networks, and in particular Internet, LANs, WANs and Intranets. This ITU-T standard (ratified initially in March of 1996) defines the negotiation and adaptation layer for video and audio over packet switched networks. "Negotiation" means that this layer defines the way the devices on either end of the data conversation will figure out what is the fastest speed they can accommodate. H.323 doesn't mean you get good videoconferencing over lousy circuits. But it does mean that you should get some videoconferencing. H.323 is comprised of the following standards:

- H.225: Middleware which specifies a message set for call signaling registration and admissions, supporting call negotiations – i.e. synchronization.
- H.245: Adds the ability to open and close logical channels on the network, i.e. transmission control.
- G.711: Pulse Code Modulation (PCM) (64 Kbps) encoder/decoder specification for voice.
- G.722: 7 kHz audio-coding.
- G.723.1: Speech encoding/decoding with a low bit rate, high output quality. This is the default encoder required for H.323 compliance.
- G.728: Encoding/decoding of speech at 16 kbps using low-delay code excited linear predictive methods.
- G.729: Encoding/decoding of speech at 8 kbps using conjugate-structure, algebraic-code excited linear predictive methods.

The H.323 standard has the endorsement of several key client vendors such as Netscape, for use within their Cool Talk application; Microsoft, for use in NetMeeting, now part of Internet Explorer; and Intel, for their Internet Phone product. With Netscape and Microsoft representing approximately 95% of the Internet browser market, and Intel and Microsoft dominating the current platforms, this collective support makes H.323 a de facto standard. See H.320, H.324 and V.80.

In September, 1997, Microsoft issued a white paper on "IP Telephony with TAPI." In that paper, Microsoft said:

H.323 is a comprehensive International Telecommunications Union (ITU) standard for multimedia communications (voice, video, and data) over connectionless networks that do not provide a guaranteed quality of service, such as IP-based networks and the Internet. It provides for call control, multimedia management, and bandwidth management for point-to-point and multipoint conferences. H.323 mandates support for standard audio and video codecs and supports data sharing via the T.120 standard. Furthermore, the H.323 standard is network, platform and application independent, allowing any H.323 compliant terminal to interoperate with any other. H.323 allows multimedia streaming over current packet-switched networks. To counter the effects of LAN latency, H.323 uses as a transport the Real-time Transport Protocol (RTP), an IETF standard designed to handle the requirements of streaming real-time audio and video over the Internet.

The H.323 standard specifies three command and control protocols: + H.245 for call control

Q.931 for call signaling

The RAS (Registration, Admissions, and Status) signaling function

The H.245 control channel is responsible for control messages governing operation of the H.323 terminal, including capability exchanges, commands, and indications. Q.931 is used to set up a connection between two terminals, while RAS governs registration, admission, and bandwidth functions between endpoints and gatekeepers (RAS is not used if a gatekeeper is not present). See below for more information on gatekeepers.

H.323 defines four major components for an H.323-based communications system: + Terminals + Gateways + Gatekeepers + Multipoint Control Units (MCUs).

Terminals are the client endpoints on the network. All terminals must support voice communications; video and data support is optional. A Gateway is an optional element in an H.323 conference. Gateways bridge H.323 conferences to other networks, communications protocols, and multimedia formats. Gateways are not required if connections to other networks or non-H.323 compliant terminals are not needed. Gatekeepers perform two important functions which help maintain the robustness of the network - address translation and bandwidth management. Gatekeepers map LAN aliases to IP addresses and provide address lookups when needed. Gatekeepers also exercise call control functions to limit the number of H.323 connections, and the total bandwidth used by these connections, in an H.323 "zone." A Gatekeeper is not required in an H.323 system-however, if a Gatekeeper is present, terminals must make use of its services.

H.323 Components: Multipoint Control Units (MCU) support conferences between three or more endpoints. An MCU consists of a required Multipoint Controller (MC) and zero or more Multipoint Processors (MPs). The MC performs H.245 negotiations between all terminals to determine common audio and video processing capabilities, while the Multipoint Processor (MP) routes audio, video, and data streams between terminal endpoints. Any H.323 client is guaranteed to support the following standards: H.261 and G.711. H.261 is an ITU-standard video codec designed to transmit compressed video at a rate of 64 Kbps and at a resolution of 176x44 pixels (QCIF). G.711 is an ITU-standard audio codec designed to transmit A-law and 0xCA-law PCM audio at bit rates of 48, 56, and 64 Kbps. Optionally, an H.323 client may support additional codecs: H.263 and G.723. H.263 is an ITU-standard video codec based on and compatible with H.261. It offers improved compression over H.261 and transmits video at a resolution of 176 x 44 pixels (QCIF). G.723 is an ITU-standard audio codec designed to operate at very low bit rates. The TAPI 3.0 H.323 Telephony Service Provider The H.323 Telephony Service Provider (along with its associated Media Stream Provider) allows TAPI-enabled applications to engage in multimedia sessions with any H.323-compliant terminal on the local area network. Specifically, the H.323 Telephony Service Provider (TSP) implements the H.323 signaling stack. The TSP accepts a number of different address formats, including name, machine name, and e-mail address. The H.323 MSP is responsible for constructing the DirectShow filter graph for an H.323 connection (including the RTP, RTP payload handler, codec, sink, and renderer filters).

H.323 telephony is complicated by the reality that a user's network address (in this case, a user's IP address) is highly volatile and cannot be counted on to remain unchanged between H.323 sessions. The TAPI H.323 TSP uses the services of the Windows NT Active Directory to perform user-to-IP address resolution. Specifically, user-to-IP mapping information is stored and continually refreshed using the Internet Locator Service (ILS) Dynamic Directory, a real-time server component of the Active Directory. See TAPI 3.0.

The official ITU definition of H.323 is as follows: H.323 describes terminals, equipment and services for multimedia communication over Local Area Networks (LAN) which do not provide a guaranteed quality of service. H.323 terminals and equipment may carry real-time voice, data and video, or any combination, including videotelephony. The LAN over which H.323 terminals communicate, may be a single segment or ring, or it may be multiple segments with complex topologies. It should be noted that operation of H.323 terminals over the multiple LAN segments (including the Internet) may result in poor performance. The possible means by which quality of service might be assured on such types of LANs/internetworks is beyond the scope of this Recommendation. H.323 terminals may be integrated into personal computers or implemented in stand-alone devices such as videotelephones. Support for voice is mandatory, while data and video are optional, but if supported, the ability to use a specified common mode of operation is required, so that all

terminals supporting that media type can interwork. This Recommendation allows more than one channel of each type to be in use. Other Recommendations in the H.323-Series include H.225.0 packet and synchronization, H.245 control, H.261 and H.263 video codecs, G.711, G.722, G.728, G.729, and G.723 audio codecs, and the T.120-Series of multimedia communications protocols.This Recommendation makes use of the logical channel signalling procedures of Recommendation H.245, in which the content of each logical channel is described when the channel is opened. Procedures are provided for expression of receiver and transmitter capabilities, so transmissions are limited to what receivers can decode, and so that receivers may request a particular desired mode from transmitters. Since the procedures of Recommendation H.245 are also used by Recommendation H.310 for ATM networks, Recommendation H.324 for GSTN, and V.70, interworking with these systems should not require H.242 to H.245 translation as would be the case for H.320 systems.H.323 terminals may be used in multipoint configurations, and may interwork with H.310 terminals on B-ISDN, H.320 terminals on N-ISDN, H.321 terminals on B-ISDN, H.322 terminals on Guaranteed Quality of Service LANs, H.324 terminals on GSTN and wireless networks, and V.70 terminals on GSTN. See TAPI 3.0.

H.324 Standard for analog POTS telephone line based videoconferencing via modems. H.324 is an interoperability standard, meaning that if a vendor's videoconferencing product conforms to H.324 it should communicate with all the other vendors who say their products conform to H.324. H.324 contains several standards for videoconferencing. They are H.263 for real time video compression/decompression. G.723 for real time audio compression/decompression, H245/H.223 control protocol and multiplexing and V.80 application interface for modems. See H.320 and V.80.

H.450 Supplementary services for H.323 (since 2nd version).

H.gcp A proposed new ITU-T standard being added to the H.323 family of ITU-T recommendations, which have been widely adopted by industry as the main standards for multimedia communications over the Internet. H.gcp will permit control of gateway devices that pass voice, video, facsimile and data traffic between conventional telephony networks, i.e. the Public Switched Telephone Network and packet based data networks such as the Internet. Connections through such gateway devices allow callers from a normal telephone to make long distance voice calls over the Internet. According to the ITU-T, the H.323 family of standards already provides an extensive framework for the provision of new services. The new recommendation (i.e. H.gcp) will permit low-cost Internet gateway devices for the first time to be interfaced in a standard way with the signaling systems found in conventional telephony networks.

H.R.nnnn A proposed law introduced into the House of Representatives by a Congressman. Typically, four digits follow the H.R., signifying the proposed bill's number. The reason for including this definition in this dictionary is that every few months since divestiture some Congressman has attempted to introduce a bill into the House of Representatives changing the Communications Act of 1934. Such a bill is generally supported by a bevy of Bell telephone companies trying to use the proposed bill to remove those restrictions placed on them by Divestiture – manufacturing, creating information content and getting into long distance.

HO Channel An H zero channel is a 384 kbps channel that consists of six contiguous DSOs (64 kbps) of a T-1 line.

H10 Channel The H ten channel is the North American 1,472 Kbps channel from a T-1 or primary rate carrier. It is equivalent to twenty-three (23) 64 kbps channels.

H11 Channel The North American primary rate used as a single 1,536 Kbps channel. This channel uses 24 contiguous DSOs (DS zeros) or the entire T-1 line except for the 8 Kbps framing pattern.

H12 The European primary rate used as a single 1,920 kbps channel (30 64 kbps channels) or the entire E-1 line except for the 64 kbps framing and maintenance channel.

HA 1. Horn Alert. A cellular car phone feature that automatically blows the car's horn if a call is coming in.

2. Home Automation. See Home Automation.

HA Linux High Availability Linux. The idea of HA linux project is to provide a high-availability (clustering) solution for Linux which promotes reliability, availability, and serviceability. The basic functions required by any HA system include starting and stopping resources, monitoring the availability of the systems in the cluster, and transferring ownership of a shared IP address between nodes in the cluster.

hack The output of a hacker. Usually good programs, but sometimes just something clever of no discernible use. Just a "good hack", or something done for the "hack value."

hackademy A school – recently opened in Paris and now under police surveillance – that teaches students how to break into computer systems.

hacker The word "hacker" is widely misused. Among hackers themselves, wrote The

Economist, "it refers to someone who enjoys tinkering with technology, exploring its boundaries and getting it to do unexpected or unintended tricks." In general use the word refers to individuals who break into computers – often over the Internet – for nefarious ends (for whom hackers themselves prefer the terms "malicious hacker" or "cracker"). A hacker is not necessarily bad and is not necessarily just someone who messes around with computers. Thomas Edison was arguably a hacker, back in the 19th century. Today's technological tinkerers, however, have a far wider range of household gizmos to play with and modify, from cars to cameras. Getting them to do new things, and not merely what the manufacturer had in mind, is an increasingly popular pastime. It even has its own magazine, MAKE, which is filled with projects for the technologically intrepid. The word hacker has gone through many meanings. In the late 1950s MIT students who loved to tinker with the university's gigantic early computers started calling themselves "hackers." At one stage being a hacker was a badge of honor conferred on an elite programmer or computer hardware designer. But in 1983, the movie "War Games" presented another view of the hacker mentality – someone who tries to break into computer systems for fun and sport. Today the term tends to have positive meanings, while the word "cracker" is reserved for individuals who willfully break into computer systems seeking to wreak damage. See also Cracker, Hacker Ethic, Hacker Tourism, Phreak, Script Kiddies, and Sneaker.

hacker ethic A set of moral principles common to the first generation hacker community. According to hacker ethic, all technical information should, in principle, be freely available to all. However, destroying, altering, or moving data in a way that could cause injury or expense to others is always unethical.

hacker tourism From Wired's Jargon Watch column. Travel to exotic locations in search of sights and sensations that only a technogeek could love. The term was coined by Neal Stephenson in his colossal article for Wired on FLAG, a fiber-optic cable now being built from England to Japan.

hackle Multiple surface defects on the end of a fiber. Hackle increases connector loss possibly to the point of not transmitting light.

hacktivist Politically motivated hacker who breaks into computer systems for premeditated socially-minded purposes.

Hail Mary When photographers are in crowds, they often hold the camera above their heads and shoot. They call this method the Hail Mary photograph because they cannot see what they're photographing. They're taking the photo on a leap of faith.

hailstorm See Cloud, definition 2.

hairpinning Also known as tromboning in some applications, hairpinning is a term for data going into a network element (NE), making a hairpin turn, and going right back out. The term is fairly descriptive, if you visualize a hairpin and its u-shaped bend. Let's consider some examples of hairpinning. Specifically, let's consider hairpinning in the context of SONET fiber optics rings and routers.

A SONET OC-192 backbone metropolitan ring in Manhattan (NY) running at 10 Gbps might support multiple tributary subtending (i.e., lower level) OC-48 rings running at 2.5 Gbps, with each subtending ring serving a smaller geographic area of Manhattan. Each of the subtending rings is connected to the metro ring by a device in the form of an integrated ADM/OCX (Add/Drop Multiplexer and Optical Cross-Connect), with perhaps one such device serving to interconnect all of the rings. Traffic originating on the West Side Manhattan ring and terminating on the West Side ring is presented to the ADM/OCX and is hairpinned right back to the West Side ring without ever riding over either the internal bus of either the ADM/OCX or the transmission facility of the metro ring. Similarly, traffic originating on one West Side ring and intended for another West Side ring (assuming that there is more than one) can be hairpinned on a port-to-port basis. This low speed-to-low speed, port-to-port direct interconnection is a very commonsense approach that eliminates congestion on the metro ring. Expand this example to include metro rings connected over long-haul rings, and the advantage becomes even clearer and more compelling.

A VoIP (Voice over Internet Protocol) gateway router also can take advantage of hairpinning. Using the Manhattan example again, you are calling from your West Side apartment to another address on the West Side. To access your IP-based CLEC (Competitive Local Exchange Carrier), who also is your ISP (Internet Service Provider), you use an ILEC (Incumbent LEC) local loop that connects to the CLEC through a traditional PSTN (Public Switched Telephone Network) circuit-switched CO (Central Office). When your call hits the CLEC gateway, a router converts your voice call to an IP packet format, the router hairpins the call right back to the ILEC PSTN, without going through the process of conversion to packet format and reconversion to PCM (Pulse Code Modulation) format. There really is no point in going through the process, and congesting the gateway router in the process. See also ADN, CLEC, CO, Gateway, ILEC, IP, ISP, OXC, PCM, PSTN, Router, SONET, Subtend, Tromboning, and VoIP.

HAL The computer from the movie 2001: A Space Odyssey. HAL is an acronym for "Heuristically programmed ALgorithmic computer." The one-letter-shift transposition to "IBM" (The I became H. The B became A. The M became L) was noted shortly after the film's release and was widely accepted as a subtle Kubrick/Clarke joke directed at the computer giant. Clarke himself has pointed out that when he named HAL, he didn't catch the one-letter-shift bit, and if he had, he would have changed the name. IBM had been a huge supporter of the film project, and he wouldn't have dreamed of poking fun at them, however subtly. (Incidentally, according to Clarke, the one intentional joke in the film was the scene where a knuckle-chewing Heywood Floyd read the instructions for the Zero-Gravity Toilet.)

HALE Abbreviation for High Altitude, Long Endurance platform, an unmanned aeronautical vehicle, based on either airplane or balloon technologies, designed to remain in one geographical location by remaining at one place or tightly circling at an altitude of 10 to 20 miles, for several weeks or months, to serve as a radio communications relay. Compared to a communications satellite, a HALE communications vehicle would cost far less, introduce less delay in transmission and provide smaller coverage areas on Earth, as useful for covering high population-density areas. The US National Aeronautics and Space Administration and the European Space Agency have HALE programs, and many leading aircraft manufacturers are developing HALE vehicles. Hot Bird the name of the principal series of direct-broadcasting, geostationary-orbit satellites, the first launched in 1995, owned and operated by Eutelsat. There now are five Hot Bird satellites at the 13 degree East orbital position, ideal for covering Europe. Two more satellites, Hot Bird 7A and Hot Bird 8, are being built for the same orbital position and are scheduled for launch in late 2005. www.eutelsat.com.

Half T-1 It's exactly half a T-1, i.e., a Fractional T-1 providing 12 channels at 64 Kbps, or 768 Kbps. It's provisioned exactly the same way as a T-1. In reality, it is a T-1 with half the channels turned off. Priced at a little over half a T-1. See DS-1, Fractional T-1, T-1 and T Carrier.

half-bridge Apple Computer term for a device linking LANS over a low-speed link such as a telephone line or X.25 link. It is termed a half-bridge as one is required at each end of the link.

half-circuit When you buy an international circuit – say one from America to France, you typically pay two fees. You pay one to your American telecom carrier for half the distance and you pay one fee to the French carrier for the second half of the distance. Think of a satellite circuit. You'd pay one fee for the uplink from the U.S. to the satellite and one for downlink from the satellite to France. Those fees may be the same; then they may not be.

half-duplex A circuit designed for data transmission in both directions, but not at the same time. Telex is an example of a half duplex system, as is speaking on with most speakerphones. (The best speakerphones are full duplex. They're rare and expensive.)

half-duplex M-ES A cellular radio term. A Mobile End System (M-ES) that can either transmit or receive, but cannot do both simultaneously, for example, an M-ES that has a single transceiver (radio).

half-duplex Transmission A technique of operating a communications circuit so that each end can transmit or receive, but not simultaneously. Normal operation is alternate, one-way-at-a-time transmission. See Half Duplex.

half-life In science, the time it takes for half the radioactivity of a substance to disappear. Among techies, it is a gauge of an individual's usefulness. "He may have a short half-life here."

half-repeater A device which extends the distance a LAN can cover by joining two lengths of cable over another communication medium.

half-tapping The action of making an analog trunk appear in two places for simultaneous service. Half-tapping refers to the duplication of service on the customer side of the demarcation point and back-tapping is the description used by the telephone company when the duplicate service originates from their side of the demarcation point. Half-tapping is useful when new telephone equipment is being installed in the same location as the current equipment because the new system can be tested while the old system is still in use.

half-tone Any photomechanical printing surface or the impression therefrom in which detail and tone values are represented by a series of evenly spaced dots in varying size and shape, varying in direct proportion to the intensity of tones they represent.

half-transponder A method of transmitting two TV signals through a single transponder through the reduction of each TV signal's deviation and power level. Half-transponder TV carriers each operate typically 4 dB to 7 dB below single-carrier saturation power.

half-wave antenna An antenna which is half as long as the wave being received.

Halo Effect, The Websters defines a halo as a conventional, geometric shape, usually in the form of a disk, circle, ring or rayed structure representing a radiant light around or above the head of a divine or sacred personage. A company or person acquiring "the

Halo Effect" suggests that the company is doing that most outside observers really view as "right." In short, the company or person is on a major roll.

halogen The halogens comprise five non-metallic elements in group 7 of the periodic table of elements. The name "halogen" comes from the Greek "hals" (salt) and "genes" (born), and translates into "salt former," referring to the fact that each of the halogens can form with sodium a salt similar to common table salt (sodium chloride). The halogens exist, at room temperature, in all three states of matter: iodine and astatine in solid form, bromine in liquid form, and fluorine and chlorine in gas form. As they all are highly toxic, have a strong, unpleasant odor and will burn flesh, they are not desirable in many applications. Therefore, inside cable buffering and jacketing materials must be low smoke and zero halogen.

HAM The traditional meaning of HAM is that it derives from Home AMateur radio. And a person who operates a HAM radio is called a HAM operator. Another (perhaps more reliable) explanation is that the term "ham" comes from the fact that telegraph operators who lacked proficiency were referred to as "plugs" or "hams," as they were hamfisted. Until very recently, amateur radio service were required to know Morse code and to demonstrate proficiency in radiotelegraphy. Hence, the term "ham" was applied to a person who operates an amateur radio service. If you still don't believe, note that the first documented mention of the term was in "The Telegraph Instructor," written by G.M. Dodge and published in 1903. You can find more information about ham radio at www.ham.org and www.fcc.gov.

HAM Operator A person who operates a HAM radio. See HAM.

hamfest An expo for and by ham radio operators and vendors.

hammer 1. Don't try to drive a nail with this baby. A 64-bit core of a microprocessor is called the hammer.

2. When a military contractor refers to a "manually powered fastener-driving impact device," he is talking about a conventional carpenter's hammer.

Hamming Code An code used in Forward Error Correction (FEC). Named after R. W. Hamming of Bell Labs, the basic code has four information bits and three check bits per character. Extended Hamming Code involves a set of data viewed as a two-dimensional data block which, for example, might be 11 bits horizontal (wide) and 11 bits vertical (deep). Along both the horizontal and vertical planes in this example, "(16,11) Extended Hamming Code" takes 11 information bits, computes 5 parity bits, appends the parity bits to the information bits, and sends the entire data block across the transmission facility. In total, the result is a (256,121) code, as 16x16=256 and 11x11=121. (The number of information bits can be greater in the vertical plane. This would be advantageous if the size of the data blocks is large, as the level of overhead would be less, relative to the data bits to be transmitted.) The receiving (i.e., Forward) device re-computes the parity bits, and compares the resulting bit sequences to those sent along with the data. If the two sets of parity bits do not match exactly, there almost certainly was an error created as the data transited the network. If errors are detected, the receiving device almost always has enough information to isolate the error, and to correct it without the requirement for a retransmission. See also Forward Error Correction and SISO.

HAN Home Area Network. A residential network for data communication and control based on the same concepts as a LAN, but using standard electrical wiring. For more detail, see CEBus.

han characters Han characters are Chinese language symbols which are used to represent whole words or concepts in Chinese, Japanese and Korean. See Unicode.

hand off See Handoff.

hand-off See Handoff.

handhole A buried box whose lid is even with the surface of the ground. It provides a space for splicing and terminating cables.

handle 1. In the Windows 95 user interface, an interface added to the object that facilitates moving, sizing, reshaping, or other functions pertaining to an object.

2. In programming, a pointer to a pointer that is, a token that lets a program access a resource identified in another variable. See also object handle.

3. The name you use in an online computer service. It's typically not your own name. You adopt this name to give yourself anonymity for whatever reasons you find convenient. A handle is called a "Nom de ligne" in some circles.

handled call A call center term. A call that is answered by an employee, as opposed to being blocked or abandoned.

handoff 1. All the mobile phones we happily use each day are called cell phones. Our cell phone carrier divides a city up into many small areas, called cells. When you or I are physically in one of the cells, our mobile phone is "talking" to the wireless transmitter/receiver that's typically in the center of the cell and serves that cell. When we walk to another

cell, our phone will want to speak to the wireless transmitter/receiver that serves that cell. It will want to do that because it will get a better, clearer, stronger signal and the calls will sound much better. As I walk or travel, I clearly want to stay on my call. I don't want to get interrupted or dropped. The way this happens is that my call is handed off by the cell phone system from one wireless transmitter/receiver in one cell to the one in the next cell. Handoff is, hence, the process by which the Mobile Telephone Switching Office (MTSO) passes a cellular phone conversation from one cell to another. The MTSO's controller monitors the radio signal strength of your and mine cell phone and, when the signal weakens, switches it to a base station with a stronger signal. It then switches the communications link from the former base station to the new one and signals the terminal to begin radio communication on the new channel. Interference could be caused, for example, by other portable terminals in the same cell or an adjacent cell, or by external influences, such as nearby traffic or people moving partitions in an office. In such cases, the base station redirects the call rapidly to a less noisy channel in the same cell or an adjacent cell. There are two forms of handoff: hard and soft. A hard handoff is performed on a "break and make" basis, requiring the connection to be broken in the original cell before it is made in the successor cell. Hard handoffs are required in cellular systems using FDMA (Frequency Division Multiple Access), such as the analog AMPS (Advanced Mobile Phone System), and those using TDMA (Time Division Multiple Access), such as GSM (Global System for Mobile Communications). As AMPS and GSM employ different frequencies in adjacent cells, hard handoffs are required – you don't notice the difference in a voice conversation, as the process takes only 250 milliseconds or so, but data communications can affected adversely. A soft handoff, on the other hand, employs a "make and break" handoff algorithm. Some PCS (Personal Communications Services) systems employ CDMA (Code Division Multiple Access), which does not require the use of different frequencies in adjacent cells. Those systems, therefore, take advantage of soft handoffs. If the mobile user is moving quickly through cell sites (e.g. traveling in a car), handoff becomes harder and the user may get dropped. See also AMPS, CDMA, FDMA, GSM, MTSO, PCS, and TDMA.

2. To connect a phone call or service from one telephone company to another. These usually occur in a place called co-location.

3. The process of transferring a subscriber call from one satellite to another.

4. An SCSA definition. The change of ownership of a Group (and therefore, typically, a call) from one session to another. For example, if a call center application discovers that a caller wishes to access a technical support audiotex database, it hands off the call to an application servicing that database.

handover word The word in the GPS (Global Positioning System) message that contains synchronization information for the transfer of tracking from C/A to P code.

handset The part of a phone held in the hand to speak and listen, it contains a transmitter and receiver. In the old days, the transmitter was a carbon mike. Now it's mostly electronic. Some electronic mikes are awful. Some phone makers are going back to carbon mikes. There are two basic types of telephone handsets in North America: the G-style handset, which has round, screw-in ear and mouthpiece, and the new K-style handset, which has square ear and mouthpiece and has the two screws in the middle. I prefer the older G-style one. I think it's sturdier. It also has the advantage that you can unscrew it and quickly remove the transmitter – very useful if you want someone to listen in on your conversation, but you don't want the other party to hear his breathing and coughing.

handset management Imagine you have a phone attached to your computer through a telephony board inside your computer. Now imagine that you pick up the phone and dial a number. If the company knows you have dialed a number and knows which number you have dialed, that feature is called handset management. It is the ability of the computer to be aware of every button pushed on the phone. The advantage of this is obvious: You really want the PC to collect those digits, so it can, for example, add a price to each call and use them for monthly billing (lawyer, accountant, etc.). You also want to be able re-dial those numbers by simply clicking on the number one you want, hitting Enter and bingo, you're redialing that number, without having to key it in again. This term, handset management, has now been replaced by a more meaningful term which we're now calling "Telephone Set Management."

handsfree This a term with different meanings in the telephone business. It can mean that you have a telephone with a speakerphone and thus you are able to talk on it "handsfree," i.e. without your hands touching the handset, but you still must dial manually. In the car phone business, it means the same thing, i.e. the ability to use your phone without lifting or holding the handset to your ear, plus it may mean that your car phone has voice recognition and you can also dial handsfree by talking to your car phone, e.g. "Call mother." See also Handsfree Dialing.

handsfree answerback This feature, when activated, automatically turns ON the microphone at a telephone receiving a call so that the person receiving the call can respond without lifting the receiver. Handsfree answerback is typically used on intercom calls.

handsfree dialing A telephone feature which allows the user to place outside calls and listen to the progress of those calls without lifting the handset of his telephone. This feature is unbelievably useful when calling airlines which inevitably put you on "eternity hold." (We made that term up.)

handsfree monitoring You can dial an outside call and hear the call's progress without having to lift your handset. Similar to hands-free dialing. With hands-free monitoring, you can only listen. To speak, you must pick up the handset. To be able to speak, you need a full speakerphone. Be careful of the distinction. Many people have been caught.

handsfree telephone Could be another word for a speakerphone or for a phone that does hands-free dialing.

handshake Two modems trying to connect. Two modems trying to agree on how to transfer data. The series of signals between a computer and another peripheral device (for example, a modem) that establishes the parameters required for passing data.

Handshake In HIN. A general purpose control signal sent from the DTE to DCE in a Newbridge Networks RS-232-C connection. HIN can be used in place of Request to Send (RTS), Carrier Detect (CD) or Ring Indicator (RI).

Handshake Out HOUT. A general purpose control signal sent from the DCE to the DTE. For example, in the case of a Newbridge Networks Mainstreet Data Controller with ports configured as DCE, HOUT is sent from the Data Controller to an attached device. HOUT can be used in place of Clear to Send (CTS), Carrier Detect (CD) or Ring Indicator (RI).

handshaking The initial exchange between two data communications systems prior to and during data transmission to ensure proper data transmission. Handshaking and Line Discipline are the two most basic elements of a communications protocol. A serial (asynchronous) transmission protocol might include the handshake method (XON/XOFF), baud rate, parity setting, number of data bits and number of stop bits. Just as people shake hands, and go through a perfunctory "Hi, how are ya?", computers must go through a procedure of greeting the opposite party, verifying the identity of the other party, determining the maximum speed at which the devices can intercommunicate over a circuit, and other functions that can be described by this "humanizing term." As with human contacts, once the handshaking is complete, the business of communications begins, which is where line discipline comes in. As is always the case, the least capable device determines the specifics of the communication. In other words, the lowest common denominator rules. See also Line Discipline and Protocol.

handy A German term for a cell phone, also called a mobile phone. In German, the Handy is das Handy.

handwriting recognition A system for taking handwritten text generated with a stylus on a computer pad or directly onto the computer screen, and converting then into machine-readable text.

hang up Hang up lets you disconnect from an ISDN call. To hang up from the phone set you must depress and hold the receiver button for a specified amount of time. By default, the time is set for 0.8 of a second.

hank of wire A coil, loop, or coiled bundle of wire. The word "hank" is an old term that originally was and still is used as a measure of yarn, thread or some other textile or cordage. For example, a hank of wool yarn is 560 yards and a hank of cotton thread is 840 yards,

happiness What's the use of happiness? It can't buy you money. – Henny Youngman

happy slapping A British fad in which hooligans attack innocent people on the street, record the assults on videophones, and then share the footage with friends or upload it to the Internet.

hard cable Coaxial cable commonly used in the cable television industry for trunk and feeder. At a minimum, hard cable consists of a copper (or copper-clad aluminum) center conductor, a plastic foam dielectric and a solid-aluminum sheath. The solid aluminum sheath is quite stiff; hence the name. Hard cable is available in several configurations. The most common are:

- - Bare: There is no protective cover over the aluminum sheath. Bare cable is used in aerial installations where it's lashed to a steel supporting strand; the bare aluminum sheath is readily visible from the ground because of its dull silver color. See Strand.
- - Jacketed: The sheath is covered with a vinyl jacket, typically black. Jacketed cable is often used in aerial installations located in areas subject to corrosion from industrial pollution or salt; however, many cable companies use jacketed cable in all aerial locations.
- - Armored: The sheath is covered with three protective layers: a vinyl jacket, a steel wrap, and another vinyl jacket. Armored cable is intended for use in direct-burial

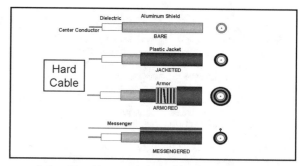

applications; the steel armor protects the sheath from damage during installation.

- - Messengered: Similar to jacketed cable, but includes an integral steel "messenger wire" to provide mechanical support. Messengered cable is intended for aerial installation without strand.

Jacketed and armored cables are also available in "flooded" configurations. Flooded cable contains a sticky, viscous substance called "flooding compound" between the sheath and the jacket. Like a self-sealing tire, flooding compound seals microscopic holes in the jacket to prevent water intrusion. Flooded cables are intended for underground use, but they are not recommended for aerial use (flooding compound is a sticky mess – literally and politically – if it drips onto a parked car). Hard cables are identified by "trade size"; i.e., the outside diameter of the sheath. The standard trade sizes are 0.412", 0.500" ("half-inch cable"), 0.625", 0.750", 0.875", and 1.000" ("one inch cable").

hard copy Anything on paper. It is all well and good to have information flash by on your CRT or video display terminal, but there are times when you want to take a Hard Copy with you. This dictionary was written on a computer screen. Now you have a hard copy in your hands. In this case, that's a lot more useful than having a disk.

hard decision See SISO.

hard disk assembly A sealed mass storage unit used for storing large amounts of data. Now available on personal computers.

hard disk imaging Today, hard drives are about the only thing that moves constantly on PCs or laptops. As a result, they wear out. i figure it takes me 11 hours from the moment I get a new hard drive to have it running the way I like. The easiest way to repair a busted hard disk is to have a cloned disk, i.e. identical copy of my disk, and pop that hard disk as a replacement for the busted hard disk. When I say "identical" I mean identical. The drive should be the same size and the software on it should be bit-by-bit the same. You produce such a replacement hard disk by using hard disk imaging software. The way it works is typically you turn off your PC, you insert or attach a second hard drive and a insert hard disk imaging or cloning software into your floppy disk drive. You now turn on your PC or laptop. Your machine boots off the floppy. It asks, "Would you like to copy all the material from drive 1 (your main drive) to drive 2 (your backup drive)?" You say Yes and off it goes, copying bit by bit (not file by file) from drive 1 to drive 2. To do this for me takes 35 minutes using a piece of software called EZ-Gig Transfer Utility. How often do I clone my main, working hard disk? Not often enough is my simple answer. The best backup procedure is to clone the disk once a week and back up the data files twice a day – before lunch and when you're finished for the evening.

hard drawn copper wire Copper wire that has been drawn to size and not annealed.

hard drive 1. A sealed hard disk. Originally the hard drive was called the Winchester magnetic storage device. It was pioneered by IBM for use in its 3030 disk system. It was called Winchester because "Winchester" was IBM's code name for the secret research project that led to its invention. A Winchester hard disk drive consists of several "platters" of metal stacked on top of each other. Each of the platter's surfaces is coated with magnetic material and is "read" to and "written" from by "heads" which float across (but don't touch) the surface. The whole system works roughly like the old-style Wurlitzer jukebox. See Winchester.

2. The Vermonter's Guide to Computer Lingo defines hard drive as getting home during mud season, i.e. it was a hard drive.

hard drive imaging See Hard Disk Imaging.

hard ferrite Ferrite that remains permanently magnetized. Used to make magnets.

hard ground A direct connection between a device to ground through a wire or other conductor that has a negligible resistance to ground. See also Ground, Grounding (the major explanation), Ground Return and Ground Start.

hard handoff A cellular radio term. A hard handoff is a handoff between cell sites that involves first breaking the connection with the previous cell site before making the connection with the new cell site. A hard handoff, or "break and make" handoff, is not noticeable in a voice conversation, but has disastrous impact on a data communication. Here's a longer explanation: What happens during a Hand-Off Sequence? Hand-off occurs when a call has to be handed from one cell to another as the user moves between cells. In traditional hard hand-off, the connection to the current cell is broken and then the connection to the new cell is made. In CDMA technology, however, it is possible to make the connection to the new cell before leaving the current cell since all cells in CDMA use the same frequency. This is known as a "make-before-break," or "soft hand-off." Soft hand-off requires less power, which reduces interference and increases capacity. See also Handoff.

hard plug loopback Also known as Hard Loopback. A loopback is a type of diagnostic test in which the transmitted signal is returned to the sending device after passing through a data communications link or network. This allows a technician (or built-in diagnostic circuit) to compare the returned signal with the transmitted signal and get some sense of what's wrong or, more correctly, what's not wrong. Loopbacks are often done by excluding one piece of equipment after another by isolating the equipment from the circuit, or the circuit from the equipment. Loopback tests usually are automated procedures accomplished through a software command which instructs a contact in a CSU/DSU to close a contact in order that the signal "loop back" across the circuit. In effect, this process isolates the equipment from the circuit, allowing the technician or automated test system to speak to the performance of the circuit without any confusion that might be imposed by the equipment, itself. It's a process of elimination. A hard loopback is test that performs the same function on a manual basis, with the technician simply twisting wires together on an RJ-45 or other cable, in order to achieve a hard-wired cross-connection. See also Loopback and Loopback Test.

hard problem A type of calculation which is easy to perform in one direction, but difficult and even impractical to perform in the other direction. In the context of cryptography, hard problems provide extreme levels of security for encrypted data. See also DES and AES.

hard RAM Carve some memory out of a computer's RAM; power it continuously and bingo you have Hard RAM, also called a virtual disk. Setting up a RAM disk lets you use your computer's conventional, extended or expanded memory to simulate a disk drive (or drives). The primary advantages of a RAM disk are its very fast access speed and its battery power-saving properties. It has no mechanical element to slow it down or to use additional power.

hard rubber A hard insulating material made of rubber, and having a dielectric constant of from two to four.

hard sectoring Physically marking the sector boundaries of a magnetic disk by punching holes in the disk where there's space available to store data. Hard sectored disks are not very common these days. Most disks – like those used on the IBM PC – are soft-sectored.

hard tubes Vacuum tubes having a high vacuum.

hard wired 1. Describes a circuit designed to do one task (e.g. a leased line).

2. A person with a very narrow and rigid view of his or her job. "That security guard is really hard-wired."

hardened Resistant to disaster. Facilities with protective features that have been designed to withstand an explosion, a natural disaster, or ionizing radiation.

hardware The actual physical computing machinery, as opposed to software which is the list of instructions to operate the hardware, or the firmware which is combination hardware/software that is "burned into" a Programmable Read Only Memory chip or chips. See Firmware and Software.

hardware address Also called physical address or MAC-layer address, a data-link layer address associated with a particular network device. Contrasts with network or protocol address, which is a network layer address.

hardware flow control Hardware flow control is the method used by the UART chip (that chip controls the serial port) to modulate the flow of data. It does this by controlling the Clear to Send/Ready to Send (CTS/RTS) lines of the serial port's interface. For example, it can turn off or re-enable the flow of data. Most high-speed sessions require hardware flow control due to their need for precise, instantaneous control over the flow of incoming and outgoing data.

hardware interrupt See Interrupt.

hardware tree A Windows 95 term. A record in RAM of the current system configuration information for all devices in the hardware branch of the Registry. The hardware tree is created each time the computer is started or whenever a dynamic change occurs to the system configuration.

hardwire To permanently connect by wire two or more devices rather than to connect them temporarily through connectors or switches. Hardwire is a term also used to represent a leased line.

hardwire services An MCI definition. Services providing intercity communications facilities dedicated to the use of a specific customer, and provided through a dedicated access line from the customer to the MCI switch.

hardwire terminating city City of circuit termination for hardwire services.

harm See Network Harm.

harmful interference An FCC term for "any radiation or induction which endangers the functioning of a radio navigation service or of a safety service or obstructs or repeatedly interrupts a radio service."

harmonic A frequency which is an exact multiple of a fundamental frequency.

harmonic distortion A problem caused when the nonlinearities in communication channels cause the harmonics of the input frequencies to appear in the output channel.

harmonic ringing A way of stopping users on a party line from hearing other than their own ring. We do this by tuning the ringer in their phone to a given ringing frequency, so it only rings when their frequency comes down the line.

harmonic signals Signals which are coherently related to the output frequency. In general, these signals are integer multiples of the output frequency.

harmonica 1. A device attached to the end of a connectorized feeder cable that converts the 25 pair into individual 4, 6 or 8 wire modular channels. I have no idea why it's called a harmonica since it bears no relation to that other harmonica, also called a mouth organ, which is small rectangular wind instrument with free reeds recessed in air slots. Tones are created by inhaling and exhaling on the air slots. See also Harmonica Adapter.

harmonica adapter An adapter that connects a 25-pair cable plug into 12 four-conductor RJ11 plugs or two 24 connector RJ11 plugs. These are often used as an alternative connectivity to hardwired blocks or temporary installations of key of PBX phone systems.

harmonica bug See Infinity Transmitter.

harmonics This definition courtesy American Power Conversion Corp. In an AC power system, distortion of voltage or current waveforms may be expressed as a series of harmonics. Harmonics are voltage or current signals that are not at the desired 50 or 60 Hz fundamental frequency, but rather at some multiple frequency. For example, the fifth harmonic of 60 Hz is 300Hz. It is a characteristic of AC signals that any distortion will have components only at integer multiples of the fundamental frequency. In AC power distribution, these distortion components only occur at odd multiples of the fundamental frequency. The third harmonic voltage distortion at a typical wall outlet in the U.S. is about 3%. Harmonic voltages have virtually no effect on modern computers, but can cause overheating in some equipment.

Harness An arrangement of wires and cables, usually with many breakouts, which have been tied together or pulled into a rubber or plastic sheath, used to interconnect electric circuits.

harry-proof To make something crash-proof, bug-proof, tamper-proof and idiot-proof. This term has been generously contributed by Steve Schone, of Danbury, CT. Micro Solutions, Inc., which has sold me Toshiba laptops and accessories since time immemorial. Steve coined the term, Harry-Proof, after he sent me a special "backup" package of hard disk and complex software. He claimed that he had constructed the whole system to be "Harry-Proof," by which he meant that someone of my awesome intellect couldn't destroy it or the hard disk I was trying to copy (as against erasing it) – though he retained the right to be wrong. He was.

harvester Professional spammers constantly scan the Web using high-speed programs known as harvesters to capture visible e-mail addresses.

hash 1. Total Adding up one or more information fields in order to provide a check number for error control. The addition is not intended to have any meaning other than for checking.

2. In some English speaking countries, like Australia, hash is the same as the pound key, also called octothorpe, also called the number key, the tic-tack-doe

1. The character on the bottom right of your touchtone keypad, which is also typically above the 3 on your computer keyboard. The # sign is correctly called an "octothorpe," but sometimes it is also spelled without the "e" on the end. There is even an International Society of Octothorpians who maintains a web page at http://www.nynews.com/octothorpe/home.htm. The octothorpe is also called the pound sign, it's also called the number sign, the crosshatch sign, the pound key, the tic-tac-toe sign, the enter key, the octothorpe (also spelled octathorp) and the hash. Musicians call the # sign a "sharp."

hash busting A spamming strategm designed to defeat filters. Hash buster programs spew random words or phrases to throw off filters that generate hash values – shorthand index numbers that represent a specific piece of junk mail.

hash mark stripe A non-continuous helical stripe applied to a conductor for identification.

hash value A small amount of binary data, typically around 160 bits, derived from a message by using a hashing algorithm. The hashing procedure is one-way. There is no feasible way of deriving the original message, or even any of its properties, from the hash value, even given the hashing algorithm. The same message will always produce the same hash value when passed through the same hashing algorithm. Messages differing by even one character can produce very different hash values. See Hash and Hashing.

hashing A cryptographic term for a small mathematical summary or digest of an original clear-text data file or message. The hash function takes a variable-length message input, and produces a fixed-length hash output which commonly is in the form of a 128-bit "fingerprint" or "message digest." For example, "Newton's Telecom Dictionary" might be hashed as "123456789123456." The resulting hashes are stored in "hash buckets" in a "hash table," with the table indexed in such a way as to speed the process of sorting through them to find a specific hash. A hash algorithm ensures data integrity through the detection of changes to the data caused either by communications errors occurring in transit, or by tampering. In combination, hashing and the use of a digital signature (digital certificate) prevent the forging of an altered message. See also Digital Certificate, Encryption, and MD5.

hashing function An algorithm that takes as input an original message and produces a fixed-length summary of that message. In order for a hashing function to be effective, the result (a.k.a. message digest) must be unique to the original message (within an acceptable range of certainty) Hashing functions are sometimes thought of as one-way encryption schemes because they cannot be easily reversed (i.e. decrypted).

HASP Houston Automatic Spooling Priority system. HASP is a spooling method used to permit many processes to perform operations such as printing at the same time, without actually "owning" the device. Time was, a job that wanted to output a line of print had to be physically attached to the printer. SPOOLing creates virtual devices and places the output on disk while it manages I/O to the real devices.

haul back trunk Haul back trunk is a term used by AT&T when referring to trunks that route calls off the AT&T network.

HAVi/Jini The Home Audio-Video interoperability architecture defines protocols and interfaces to allow digital electronics to be connected and shared on a home entertainment network, in conjunction with IEEE 1394 (also called Firewire). Philips, Sony and Sun Microsystems announced plans to bridge the HAVi architecture to Sun's Jini technology, which allows devices to be recognized and operated over a network. See IEEE 1394.

Hawthorne Effect The Hawthorne effect is the most famous psychological study in the history of the telecommunications industry. It refers to a study from 1927 to 1933 of factory workers at Western Electric's Hawthorne Plant in Illinois. It showed that regardless of the changes made in working conditions – more breaks, longer breaks or fewer and shorter ones – productivity increased. These changes apparently had nothing to do with the workers' responses. The workers, or so the story goes, produced more because they saw themselves as special, participants in an experiment, and their inter-relationships improved.

Sounds very compelling, wrote the New York Times on December 6, 1998. "The results of this experiment, or rather the human relations interpretation offered by the researchers who summarized the results, soon became gospel for introductory textbooks in both psychology and management science," said Dr. Lee Ross, a psychology professor at Stanford University. But only five workers took part in the study, Ross said, and two were replaced partway through for gross insubordination and low output.

A psychology professor at the University of Michigan, Dr. Richard Nisbett called the Hawthorne effect "a glorified anecdote."

"Once you've got the anecdote," he said, "you can throw away the data," reported the New York Times.

Hayes AT Command Set Before 1981, the modem was a dumb device. It had no memory or ability to recognize commands. It simply modulated and demodulated signals between the telephone line and the computer or terminal. In 1981, Hayes Microcomputer Products, Inc. in Norcross, GA produced the first "smart" modem, appropriately named the Smartmodem 300. It was "smart" because it understood commands, such as "ATD" which means "ATtention, Dial the phone." The Hayes Standard AT Command Set (its full name) – a language for modems – has been accepted as a standard by the modem industry. And now many modems claim to be 100% Hayes compatible, which may mean they are and may mean they aren't. As in all cases of claimed compatibility, one should check. You'll find the complete Hayes AT Command Set spelled out in virtually every manual of every modem which purports to be "100% Hayes Compatible." See also Class 1.

HBA Host Bus Adapter. A printed circuit board that acts as an interface between the host microprocessor and the disk controller. The HBA relieves the host microprocessor of data storage and retrieval tasks, usually increasing the computer's performance time. A host bus adapter (or host adapter) and its disk subsystems make up a disk channel.

HBFG Host Behavior Functional Group: The group of functions performed by an ATM-attached host that is participating in the MPOA service.

HBS Home Base Station. A wireless PCS term. Supports the PCS 1900 air interface in combination with the PCS 1900 handset.

HCI Host Command Interface. Mitel SX-2000 PBX to computer link. HCI is designed to work with Digital Equipment Corporation computers. See Open Application Interface.

HCL Hardware Compatibility List. See HCL.

HCO Hearing Carry Over. A reduced form of Telecommunication Relay Service (TRS) where the person with a speech disability is able to listen to the other end user and, in reply, the Communications Assistant speaks the text as typed by the person with the speech disability. The Communications Assistant does not type any conversation.

HCS 1. Hundred Call Seconds. One hundred seconds of telephone conversation. See CCS.

2. Hard Clad Silica.

HCT Hardware Compatibility Test. Microsoft came up with this definition and concept when it found several computers didn't run its software as well as they should. Basically the Microsoft Hardware Compatibility Test (HCT) is a series of tests for verifying the compatibility of hardware systems with Windows NT. The TCT Test Manager is an application that provides a way to launch the tests, keep track of test results and return the results to Microsoft. Microsoft maintains a Windows NT Hardware Compatibility List (HCL). If you send back the test results and your stuff passes, your hardware will be included on Microsoft's list of hardware that works with NT.

HD 1. Half Duplex circuit.

2. High definition.

3. Hard drive.

HD Radio See High Definition Radio.

HD Ready To receive and display high-definition programming, a TV set needs two basic features. One is a display capable of rendering the high-definition picture. The other is a tuner, or receiver, capable of receiving the high-definition signal, either over the air, or from a cable or satellite service. When a TV set is described as "HD-ready," it usually means the set can display high-definition pictures, but lacks the special tuner needed to receive them. It may have no tuner at all built in, or it may have just a standard tuner. With this type of TV, you must buy a separate high-definition over-the-air tuner, or obtain a high-definition cable box or satellite receiver, to get high-definition programming. A high definition tuner, cable box or satellite receiver will cost you more money than a standard unit. But, if you can get some decent high-definition TV programming, it's worth it.

HD-DVD See Blu-Ray.

HDB3 High Density Bipolar 3. Specified in the ITU-T G.703 recommendation, HDB3 is a line coding technique used in E-1 circuits. Similar to the B8ZS line coding technique used in North American T-1 circuits, HDB3 substitutes strings of four zeros with one of four bipolar violation codes. The first violation code is a single pulse in bit position 4, and matches the polarity of the last preceding pulse. Subsequent violation codes also always are in bit position 4, and are of the opposite polarity of the previous violation code. HDB3 supports clear channel communications of 64 Kbps per channel, as does B8ZS in the T-1 environment. See also B8ZS.

HDD Hard Disk Drive.

HDLC High level Data Link Control. A link layer protocol (Layer 2 of the OSI Reference Model) standard for point-to-point and point-to-multipoint communications. HDLC was based on IBM's SDLC (Synchronous Data Link Control) and ANSI's ADCCP (Advanced Data Communication Procedure). HDLC encapsulates packet data in a frame (i.e., yet another packet), with the frame header and trailer including various control information such as an error control mechanism. Variants on HDLC include Frame Relay, LAP-B (Link Access Procedure-Balanced), Link Access Procedure-Data channel), PPP (Point-To-Point Protocol), and SDLC (Synchronous Data Link Control).

HDMAC Another potential high definition TV standard. HDMAC was spawned by Britain's Independent Broadcasting Authority. Unlike Japan's Hi-Vision, HDMAC has the attraction of being compatible with existing TV sets, i.e. those in Europe.

HDMI High-definition multimedia interface, this spec combines audio and video into a single digital interface for use with digital components. HDMI supports standard, enhanced, or high definition video plus standard to multichannel surround-sound audio. See HDMI cable.

HDMI cable A cable that accommodates up to 5 Gbps of bandwidth, used for carrying high-definition digital video and audio signals without compression. HDMI cable is used to connect a set-top box or DVD player to an HDTV, and to connect an HDTV to home-theater receivers. HDMI cables are the latest rage for connecting components of a home media system. See also HDMI.

HDML Handheld Device Markup Language, which is Unwired Planet of Redwood Shores, CA's modification of standard HTML for use on mobile phones. HDML is a text-based markup language which uses HyperText Transfer Protocol (HTTP) and is compatible with all Web servers. HDML is designed to display on a smaller screen such as one might find on a cellular phone, PDA, pager, or PCS device. The basic structural unit for HDML is a "card," while that of HTML is a "page." HDML allows the mobile user to access the Internet, and send, receive and redirect e-mail. As PCS devices are graphics-challenged, a Web site must be HDML-enabled in order to allow access by such devices. www.wapforum.org. See also WAP.

HDPE High density polyethylene.

HDSL High-bit-rate Digital Subscriber Line, also known as HDSL1. The most mature of the xDSL technologies, HDSL allows the provisioning of T-1/E-1 local loop circuits much more quickly and at much lower cost than through conventional means. In the U.S., HDSL delivers T-1 at 1.536 Mbps over a four-wire loop of two pairs. E-1 capacity of 2.048 Mbps originally required three pairs, but now also requires only two pairs. Unlike ADSL, HDSL bandwidth is symmetric, as equal bandwidth is provided in each direction.

The traditional approach of provisioning T-1/E-1 access loops on copper wires requires specially-conditioned UTP (Unshielded Twisted Pair), with repeaters spaced every 6,000 feet in order to compensate for signal attenuation at the high carrier frequencies required. Each pair supports simplex (one-way) transmission at 1.544 Mbps, of which 1.536 Mbps is usable for data transmission, with the remaining .008 Mbps being used for signaling and control purposes. In combination, the two simplex circuits yield a full-duplex circuit.

HDSL, which involves special electronics at both the CO and the customer premise, delivers the same transmission capacity over standard UTP at distances up to 12,000 feet on 24 AWG (American Wire Gauge) wire and up to 9,000 feet on 22 AWG wire, without the requirement for repeaters. The UTP loop may be bridged, although loading coils are not tolerated. This is accomplished through full-duplex transmission at 784 Kbps over each pair of the four-wire circuit; 768 Kbps is usable for data transmission, with the remaining 16 Kbps being required for signaling and control. In the aggregate, the yield is 1.536 Mbps (T-1). The lower transmission rate on each pair implies a much lower carrier frequency. As lower frequency signals can travel much longer distances without experiencing unacceptable levels of attenuation (loss of signal strength), the requirement for repeaters is obviated for distances up to 12,000 feet. Note: T-1 requires 1.5 MHz over each pair, one supporting upstream transmission and the other supporting downstream transmission. While several options exist for electrical encoding, all of them are unibit schemes, impressing a single bit on each baud. HDSL operates at frequencies ranging from 80 KHz to 240 KHz, depending on the specific techniques employed. In order to improve the data transmission speed at a given carrier frequency, HDSL uses the 2B1Q electrical encoding scheme, which also is known as 4 PAM (Pulse Amplitude Modulation); this scheme is a dibit coding scheme, which allows two bits to be impressed on each baud. Echo cancellation allows each pair to support full-duplex (simultaneous two-way) transmission over each pair.

HDSL has been deployed aggressively by LECs for some years. Well over 500,000 systems reportedly are in service. Reportedly, as many as 70% of all U.S. T-1 circuits employ HDSL. Although both the COT (Central Office Termination) and the RT (Remote Termination) require the placement of HDSL electronics, the overall carrier costs of provisioning are much reduced. No special circuit engineering, no physical inspection of cable plant, and no repeater acquisition and placement is required. Additionally, the circuit can be provisioned much more quickly, which fact results in much happier customers and much faster revenue generation. In fact, several LECs have lowered their T-1 rates in consideration of the lower costs.

At the time of this writing, a proposal for a new variation on the HDSL theme recently was proposed as a standard. HDSL2, based on technology from Adtran Inc., provides the same capability over a single pair, although the local loop length is limited to about 10,000 feet. This technology also is known as SDSL (Single line DSL). S-HDSL (Single-line HDSL) is a variation on this non-standard variation (It gets confusing, doesn't it? Remember that this is an emerging technology.) run at speeds of 768 and 384 Kbps for loop lengths of 12,000 feet and 18,000 feet, respectively. See also DSL, ADSL, HDSL2, IDSL, RADSL, SDSL, T-1, T1E1.4 and VDSL. www.adtran.com and www.adsl.com.

HDSL2 High-bit-rate Digital Subscriber Line version 2. Also known as G.shdsl (G. is the ITU-T Recommendation series under which HDSL2 is being considered) and SDSL (Single line DSL). On January 6, 1998, Level One Communications, ADC Telecommunications,

Adtran, PairGain Technologies and the Siemens Semiconductor Group today announced agreement within the American National Standards Institute (ANSI) TIE1.4 committee on the basis on an HDSL2 standard. A provisional agreement, T-1/E-1 contribution number 41/97-471, has been approved marking a milestone within the ANSI HDSL2 standards effort. The elements agreed upon were line code in the form of advanced Trellis code, a precoding mechanism, spectral shaping, equalization circuits, and forward error correction. These elements make up the core of the HDSL2 standard. The agreement reached is expected to accelerate the development HDSL2 technology and promote industry interoperability. The HDSL2 standard proposal will enable service providers to deliver full T-1 (1.544 Mbps), and potentially E-1 (2.048 Mbps) performance over a single twisted pair cable, and over a longer reach than HDSL, now also known as HDSL1. Over 24 AWG (American Wire Gauge)wire, HDSL2 is intended to support a data rate of 1.536 Mbps (T-1) over a distance of up to 13,200 feet, 768 Kbps up to 17,700 feet, and 384 Kbps up to 22,500 feet. Over 25 AWG wire, the distances are 9,200 feet, 12,400 feet, and 15,500 feet, respectively. The modulation technique originally approved by the ANSI T1E1.4 committee is 16 PAM (Pulse Amplitude Modulation), an improvement on the 2B1Q (also known as 4 PAM) technique used in HDSL1. Subsequently, OPTIS (Overlapped Phase Trellis-code Interlocked Spectrum) also was approved. See also HDSL, SDSL, and T1E1.4.

HDT Host Digital Terminal. A cable telephony term. A HDT is located at the head end (i.e., the point of signal origin) of the traditional CATV network. In telephony terms, the HDT is at the edge of the carrier network. The HDT interfaces T-carrier (e.g., T-1 and T-3) circuits from the Central Office (CO) switch in support of voice services over the coaxial cable network. The HDT modulates those voice signals over the broadband coax cable, interfacing with a Remote Service Unit (RSU) at the subscriber premises, where the process is reversed. See RSU for more detail.

HDTP 1. Handheld Device Transport Protocol – a wireless-optimized protocol sitting between the UP.Browser's HDML interpreter and a datagram transport (typically UDP). HDTP provides security and reliability for the transport in a way that is significantly more efficient (optimized for wireless and communication with a minimal number of IP addresses) than is TCP. HDTP is the protocol in use on UP's UP.Link Platform V2.x. Version 3 sees UP migrating towards WAP - the new Wireless Application Protocol, from the WAP Forum. www.wapforum.org.

2. Hoofddirectie Telecommunicatie en Post. Directorate for Telecommunications and Posts, The Netherlands.

HDTV High Definition TeleVision. HDTV is a new TV standard producing a better quality picture and better quality sound. The hallmark of HDTV is the high resolution of display and the wide rectangular screen. HDTV produces a picture quality approaching 35mm film and sound approaching compact disc quality. HDTV screens use a width to height ratio of 16:9 (also referred to as 1.78:1). An analog TV (i.e. today's version) has a 4:3 ratio (or 1.33:1). HDTV should not be confused with Digital Television (DTV). DTV is an umbrella term for the transmission of pure digital television signals, along with the reception and display of those signals on a digital TV set. This is designed to replace the old analog signals. The digital signals might be broadcast over the air or transmitted by a cable or satellite system to your home. In your home, a decoder receives the signal and uses it, in digital form, to directly drive your digital TV set. HDTV is a type of digital television. It is high-resolution digital television combined with Dolby Digital surround sound (AC-3). Standard Definition Digital Television (SDTV) is also a type of DTV, without the superior resolution that HDTV offers. Standards for HDTV are still somewhat up in the air. There are two main standards for HDTV:

1) 1080 vertical lines by 1920 horizontal pixels, which is favored by many mainstream broadcasters and television networks. It produces high-quality pictures, but leaves less bandwidth for other channels and data services. This is called 1080i and is the most common in North America. The i stands for interlacing which means that every 60th of a second, all the odd-numbered lines are drawn, then in the next 60th of a second, the even lines. The p stands for progressive scan.

2) 720 vertical lines by 1280 horizontal pixels, a format favored by cable TV companies and the PC industry, as it leaves more bandwidth available.

- If you decide to buy an HDTV, here are a few things to look out for:
- Integrated HDTV. This is a high resolution set with the HD-Tuner built-in. It comes with a higher price tag, and could become obsolete if standards change. But I'm dubious they will if your TV has both 720p and 1080i.
- 'HDTV-Capable' or 'HDTV-Ready' television sets. Generally more affordable, these TV sets (also called HDTV Monitors) can display both the existing NTSC analog signals and the new ATSC digital signals. However, they require an external, High Definition tuner (set top box) to receive and display HDTV programs.

- 'Digital Ready' does not necessarily mean the TV will receive and display digital High Definition television programs. Verify the set you are considering will display true HDTV.
- Resolution. The ATSC Standard for High Definition Television requires a resolution of 1080 interlaced lines, or 720 progressive scan lines. Verify the set you choose is capable of working with both signals - i.e. converting (up/down) all signals to their native resolution.
- Audio. The Standard for HDTV is "Dolby A3" – 5.1 Channel Surround Sound. 5.1 means five channels of audio - one in the middle and two at front side - and two channels at the back, on either side. The .1 means that there is a separate channel to drive a subwoofer. Beware, some manufacturers have their own proprietary audio systems. While many of these produce a good audio experience, to get the best sound, choose DolbyA3 Surround Audio, or better.
- Connections. Choose a model that offers the most ports with the widest selection of connector types. Also check for additional front ports - these are convenient. You should look for composite, S-video and component video as a minimum set of analog jacks so you can use your existing analog audio equipment with the new set.
- Compatibility. If you are purchasing a separate set and tuner, be sure you verify that the tuner is compatible with the set you are purchasing, as well as the satellite/cable you expect to use, and is also capable of receiving over-the-air broadcasts. Some HD-tuners require an add-on module (8VSB) to receive OTA (Over The Air) Broadcasts. Also make sure the HDTV set will work with all your other components: VCR, DVD player, video game consoles, sound system, etc.

Other questions to consider:

- Are all cables included and compatible with your audio-video components?
- Do you have a surge protector to protect your investment? If a storm approaches, we recommend not relying on the surge arrestor. Unplug the system and wait for the storm to pass.
- What type of external antenna (if any) do you need for OTA (over the air) broadcasts in your area?
- Is delivery and set up included? This is especially important for the 'big-screen' projection sets, which can be heavy.
- Check for in-home maintenance/tune up contract - this may be an important consideration for large projector systems that require periodic adjustments.

HDTV does not, of course, affect a movie's plot. Samuel Goldwyn, film producer, once said, "A wide screen just makes a bad film twice as bad." See also ATV, DTV, Goldwyn, Samuel, NTSC and SDTV.

HDTV-Ready A television set capable of displaying a full high definition picture because of its high screen resolution. See HDTV.

HDX Half DupleX.

HDWDM Hyper Dense Wavelength Division Multiplexing. – 128 wavelengths and higher. See DWDM.

HE See Head End.

head 1. A device that reads, writes, or erases data on a storage medium. The device which comes in contact with or comes very close to the magnetic storage device (disk, diskette, drum, tape) and reads and/or writes to the medium. In computer devices, it performs the same function as the head on a home cassette tape recorder. Hard disks use one head for each side of each platter. The heads are attached to a common head-movement area, so that all heads move in unison. The heads are always positioned over the same logical track on each side of each platter. A head crash occurs when the head physically hits the media.

2. A sub-component accessory which serves a translator type function for signal being input to some larger system or instrument. Many pieces of test equipment have various available plug-in "sampling heads" which allow the equipment to be used with a variety of different signal types. The most common type is probably optical-to-electrical converters which convert an optical signal for display on an oscilloscope.

3. A term for a toilet. The "head" was so named because it was in the ship's bow (the front) by the catheads. On sailing ships all smelly activities, i.e., the head, the galley, and the crew's quarters were all placed in the ship's bow so the wind could blow the smell away. That way the officers and high-quality passengers who berthed in the stern were not offended.

head crash See head.

head end 1. The originating point of a signal in cable TV systems. At the head end, you'll often find large satellite receiving antennas. Now increasingly spelled headend.

2. A central control device required within some LAN/MAN systems to provide such centralized functions as remodulation, re-timing, message accountability, contention control, diagnostic control, and access.

Head End Hop Off HEHO. A method of traffic engineering whereby calls are completed by using long distance facilities directly off the switch that serves that location.

head fake Suckers rally in share prices on Wall Street. See Suckers Rally.

head landing zone In older hard drives, the head landing zone is an area of the hard disk set aside for take off and landing of the heads when the drive is turned on and off. In newer drives, the heads are retracted.

head on collision Descriptive of a condition in message telephony when two switching exchanges seize a both ways trunk at the same instant and attempt to send outbound call instructions to each other.

head slap Similar to head crash but occurs while the drive is turned off. It usually occurs during mishandling or shipping. Head slap can cause permanent damage to a hard disk drive.

head thrashing A term for rapid back and forth movements of the disk head of a hard drive.

headend The originating point of a signal in cable TV systems. At the head end, you'll often find large satellite receiving antennas.

header 1. Protocol control information located at the beginning of a protocol data unit.

2. The portion of a message that contains information that will guide the message to the correct destination. This information contains such things as the sender's and receiver's addresses, precedence level, routing instructions, and synchronization pulses.

header area The area containing preliminary information for the entire document, such as the data, company name, address, purchase order, terms, etc. An EDI (Electronic Data Interchange) term.

Header Error Control HEC. An 8-bit CRC code contained within the header of an ATM data cell. The HEC is used for checking the integrity of the cell header at the various cell switches.

header information William Safire defines it as data at the top of a credit report which the Federal Trade Commission says may be disclosed by any credit bureau with no restrictions. This includes your name, address, phone number (listed or not), social security number and mother's maiden name.

headless server A server computer with no monitor attached.

headroom A term used in the structured cabling industry to indicate additional clearance or signal margin room above the specification. This is measured in Decibels and is related to the Attenuation to Cross-Talk Ratio (ACR).

headset A telephone transmitter and receiver assembly worn on the head. Headsets are now very light and very comfortable and are no longer worn only by switchboard attendants and airline clerks. They are worn by telemarketers, customer service reps. stock brokers, order entry reps, financial service professionals, and some executives who spend a lot of time on the phone.

headset jack A place on a phone or console into which you can plug a headset.

heads up Back in the mid-twentieth century, when someone called out "heads up!" listeners knew to lift up their heads and watch out for something dangerous. Nowadays, heads ups means that you're given a briefing.

heaps Australian for lots of.

hearing aid compatible A hearing aid compatible phone may be used with inductively coupled hearing aid devices. You can find hearing aid compatible coin phones by looking for the blue grommet between the handset and the cord.

Hearing Carry Over HCO. A form of TRS (Telecommunication Relay Service) where a person with a speech disability is able to listen to the other end user and, in reply, a Communications Assistant speaks the text as typed by the person with the speech disability. See TRS.

Hearing Designation Order HDO. A Federal Communications Commission term, a Hearing Designation Order institutes a comparative or other adjudicatory hearing proceeding, usually before an Administrative Law Judge.

heartbeat Ethernet-defined SQE signal quality test function, defined in IEEE 802.3. Heartbeat is created by a circuit (normally part of the transceiver) that generates a collision signal at the end of a transmission. This signal is used by the controller interface for self-testing.

heartbeat support A function that generates a frame periodically, even if no data is sent, for network management purposes.

heat Electromagnetic waves of a frequency between that of light waves and radio waves. A form of energy.

heat distortion Distortion of a material due to the effects of heat.

heat coil An electrical protection device used to prevent equipment from overheating as a result of foreign voltages that do not trigger voltage limiting devices. It typically consists of a coil of fine wire around a brass tube that encloses a pin soldered with a low-melting alloy. When abnormal currents occur, the coil heats the brass to soften the solder, allowing the spring-loaded pin to move against a ground plate directing currents to ground.

heat shock A test to determine stability of a material by sudden exposure to a high temperature for a short period of time.

heatseeker The person who can be depended on to purchase the latest version of any existing software product as soon as it comes on the market.

heaven Everybody wants to go to Heaven, but nobody wants to die - ancient Chinese fortune cookie.

heavy iron Hardware, really BIG hardware. Contemporary hardware hardly qualifies. For example, the ENIAC (Electronic Numerical Integrator and Computor), built in 1946 at a cost of about $400,000, was the first large-scale electronic digital computer built. The ENIAC contained about 18,000 vacuum tubes, weighed 30 tons and occupied a footprint of 30x50 feet. In 1949, Popular Mechanics magazine forecast that "Computers in the future may...perhaps only weigh 1.5 tons." Technology marched on. Your laptop provides more horsepower than the ENIAC, has more memory, and more functionality.

HEC Header Error Control - a CRC code located in the last byte of an ATM (Asynchronous Transfer Mode) cell header used for checking integrity only. Using the fifth octet in the ATM cell header, ATM equipment may check for an error and correct the contents of the header. The check character is calculated using a CRC algorithm allowing a single bit error in the header to be corrected or multiple errors to be detected.

hedge fund A hedge fund is a pool of monies from rich people managed by an investment "professional" that uses more aggressive investment strategies than mutual funds. Such aggressive strategies include: the hedge can sell short, can use leverage and can buy or sell derivatives (like options, futures, etc.), Only "accredited investors" may invest in hedge funds. Such investors are high net worth individuals and institutions who allegedly know what they're doing and don't need the protection and regulation of government entities – in contrast to mutual funds which are regulated and are typically not allowed to sell short or do other risky strategies – unless specifically set up for that purpose. Hedge funds are basically not regulated by any government agency.

Hedy Lamarr See Lamarr, Hedy.

HEHO Head-End Hop Off. You have a private network. You overflow a long distance call to WATS or DDD at the originating end (the end the call is coming from). This HEHO (Head-End Hop Off) is done because it's usually cheaper than carrying the call part way through the network, then jumping off the network at that point (because the network is busy or it won't reach the end point). The opposite of HEHO is TEHO – Tail-End Hop Off. In TEHO, you carry the call as far as possible through the network, then pass it off to WATS or DDD as close to its destination as possible. The decision to go HEHO or TEHO has to do with economics, primarily which is cheaper.

heifer A young cow. A motorist was killed near Vacaville Calif., in March 1999, when an airborne cow smashed through the windshield of his pickup truck. According to the California Highway Patrol, the 700-pound heifer wandered onto a road during stormy weather, and a 1983 Mercedes Benz traveling north on the road hit it, sending it hurtling through the air into the path of the pickup, which was traveling south. The cow smashed through the windshield, killing the driver. After hitting the pickup, the heifer was batted back onto the southbound lane where it was hit by another pickup. "Vacaville" means "Cow Town" in Spanish.

Heisenberg's Realization The mere act of observing affects what is being observed.

held call A held call is a call to which you are connected but which is on hold.

held orders A telephone company term for requests for telephone lines which the phone company cannot fill. Thus it is "holding" the orders. The reasons for holding customer orders might range from lack of capacity at the serving central office to a lack of local cable plant.

helical antenna An antenna that has the form of a helix. When the helix circumference is much smaller than one wavelength, the antenna radiates at right angles to the axis of the helix. When the helix circumference is one wavelength, maximum radiation is along the helix axis.

helical scan Storage method that increases media capacity by laying data out in diagonal strips. Used in video tape recorders, etc.

helical strand A process of twisting conductors of a cable together in a helix, or spiral fashion, in order to improve the break strength of the conductors. See Helix, Stranded Copper and Stranded Fiber.

helical stripe A continuous, colored, spiral stripe applied over the outer perimeter of an insulated conductor for circuit identification purposes.

heliograph A visual signaling device that uses reflected sunlight to communicate messages over a distance. The concept goes back at least as far as the ancient Greeks, who used polished shields to signal in battle. Heliographs have been used by armies and navies around the world ever since, although their use greatly diminished in the 1900s as modern communications technologies took over. A heliograph can be as simple as a handheld mirror, or it can be more elaborate, for example, a tripod with multiple mirrors and shutters. The distance that heliograph signals can be seen depends on weather conditions, other environmental conditions, mirror size, and whether binoculars/telescopes are used for viewing the messages. See also Telegraph Hill.

helix A spiral. The shape of screw.

hello When the phone was first invented, no one was sure how to begin the conversation. Thomas Edison saw the telephone as being used by businesses with permanently open lines. How would anyone know that the other party wanted to speak? A letter was found from Thomas Edison, dated August 15, 1877 to the president of the Central District and Printing Telegraph Co. in Pittsburgh. "Friend David, I don't think we need a call bell, as Hello! can be heard 10 to 20 feet away. What do you think, Edison?" At that time Alexander Graham Bell insisted on answering the telephone with "Ahoy." Hello! became the standard as the first telephone exchanges were set up across the country. Hello first appeared in the Oxford English Dictionary in 1883. In September of 1880, the first National Convention of Telephone Companies was held in Niagara Falls. "Hello" was used on everyone's name tag for the first time. Besides electricity, the phonograph and hundreds of other inventions, we can thank Edison for the "Hello" greeting. The above from New Pueblo Communications in Tucson, AZ.

hello packet A type of PNNI Routing packet that is exchanged between neighboring logical nodes.

help desk A centralized location where queries about product usage, installation, problems or services are answered. Sometimes help-desks are provided by the manufacturer of the product. Sometimes help desks are provided by outside companies – systems integrators, independent software developers and third party companies.

helper applications Programs that can be linked to various file types and commands. Helper apps will launch automatically when linked files are accessed through a browser.

henry The inductance in a circuit in which the electromotive force induced is one volt when the inducing current varies at the rate of one ampere per second. It is 1,000,000.000 electromagnetic units, and is the unit of inductance.

HEPA High Efficiency Particulate Arrester. These high-efficiency, high-priced vacuum cleaner filters are now becoming standard equipment on some electronic and telecommunications hardware. Hepa filters were first used when they performed atmospheric testing of nuclear warheads. They used HEPA filters in the fallout shelters because the degree of filtration was so fine that it could trap radioactive particles, due to the honeycomb design. HEPA files were later used in in medical environments, soon followed by home air filtration. Finally, they are being used in vacuum cleaners. They are not cheap, but do work well, as any person allergic to dust will tell you. Trouble is, it is not clear how often they need to be replaced, unless it looks sort of obvious.

HEPNET An Internet term. A non-USENET set of newsgroups devoted to discussing the topic of high-energy nuclear physics.

HERF High Energy Radio Frequency gun. Shoots a high-powered radio signal at an electronic target (such as a computer) and puts it out of commission.

hercules graphics Hercules graphics adheres to the Hercules standard of monochrome graphics on a monochrome PC monitor. That standard is 720 x 348 pixel resolution and 64K screen memory. This encoding was never adopted as a color standard and is now pretty well obsolete. See also Monitor and Facsimile.

HERF gun A High Energy Radio Frequency gun capable of destroying magnetic data storage.

hermaphroditic connector A loopback or self-shorting connector typically used with Type 1 (STP) Token Ring cable.

hermaticity test A fine and gross leak test of a hermetically sealed IC to see if there are any leaks in the seal. The gross leak test uses a fluorocarbon fluid, and the fine leak test uses a light gas such as helium.

hermetic coating A coating applied over the cladding of a fiber that retards the permeation of moisture and hydrogen into the fiber.

hero experiment An experiment that attempts to push the capabilities of a technology to its limits. It is used most often to refer to experiments that set world records for transmission distance and/or rate of speed. The experiment itself often has no immediate commercial value but does provide valuable, basic research that may prove valuable over the longer term.

heroinware The game called Doom played on corporate networks in often called Heroinware, because the game is so addictive.

herringbone Moving or stationary lines superimposed on a television picture.

hertz Abbreviated Hz. A measurement of frequency in cycles per second. A hertz is one cycle per second, and is the basic measurement for bandwidth in analog terms. "Hertz" is named after Heinrich Rudolf Hertz, the physicist who discovered the presence of electromagnetic radio waves. Cats purr at 26 cycles per second, the same as an idling diesel engine. Electricity cycles at 50 or 60 times per second. See Analog and PCM.

Hertz, Heinrich A German physics professor, (1857 - 1894), who did the first experiments with generating and receiving electromagnetic waves, in particular radio waves. In his honor, the units associated with measuring the cycles per second of the waves (or the number of times the tip-tops of the waves pass a fixed point in space in 1 second of time) is called the hertz. See Hertz.

Hertzian Wave A name sometimes given to electromagnetic waves.

hetero The Greek prefix meaning different.

heterodyne To generate new frequencies by mixing two or more signals in a nonlinear device such as a vacuum tube, transistor, or diode mixer. A superheterodyne receiver converts any selected incoming frequency by heterodyne action to a common intermediate frequency where amplification and selectivity (filtering) are provided. A frequency produced by mixing two or more signals in a nonlinear device. See Heterodyning.

heterodyne channel converter A converter which uses heterodyne principles to down convert signals on one channel and then re-convert the signals to a new channel. Generally heterodyne converters produce higher quality results than those which use frequency multiplication principles. Occasionally the program may be demodulated to a base band frequency (in order to interface with the base band encryption equipment) and then re-modulated. Converters often have amplifiers, filters, or demodulators, and modulators to "boost" and "clean-up" signals. See Channel Converter, Converter, Heterodyne and Heterodyne Converter.

heterodyne converter A device consisting of an oscillator and a mixer which beats two signals together for the purpose of down converting or up-converting a modulated carrier.

heterodyne processor A device which converts a TV channel at one frequency (e.g. over the air) to another frequency (e.g. for carrying on a cable TV network).

heterodyne repeater A repeater for a radio system in which the received signals are converted to an intermediate frequency, amplified, and reconverted to a new frequency band for transmission over the next repeater section.

heterodyning Here is an explanation from James Harry Green's book, the Dow Jones Handbook of Telecommunications: Analog microwave repeaters use either of two techniques to amplify the received signal for retransmission: Heterodyning or Baseband. In a baseband repeater, the signal is demodulated to the multiplex (or video) signal at every repeater point. In heterodyne repeaters the signal is demodulated to an intermediate frequency, typically 70 MHz, and modulated or heterodyned to the transmitter output frequency. Heterodyne radio is reduced to baseband only at main repeater stations where the baseband signal is required to drop off voice channels. The primary advantage of baseband radio is that some carrier channel groups can be dropped off at repeater stations. Heterodyne radio has the advantage of avoiding the distortions caused by multiple modulation/demodulation and amplification of a baseband signal. Therefore, heterodyne radio is employed for transcontinental use with drop-off points only at major junctions.

heterogeneous networks Networks composed of hardware and software from multiple vendors usually implementing multiple protocols.

heterojunction A junction between semiconductors that differ in their doping level conductivities, and also in their atomic or alloy compositions.

heuristic Using much trial and error to arrive at a solution to a problem.

hexadecimal Abbreviated as Hex. "Hex" is from the Greek "hex," meaning "six." "Decimal" is from the Latin "decem," meaning "ten." Hexadecimal is a numbering system of 16 characters, ten digits and six letters. It is used to condense the long strings of zeros and ones in large binary numbers. This base-16 numeric notation system is frequently used to specify addresses in computer memory. As a sort of programming shorthand, hex makes life simpler for programmers by allowing eight-bit binary values, or bytes, to be expressed with only two hexadecimal characters (though, obviously longer numbers do have more hexadecimal characters). In hexadecimal notation, the decimal numbers 0 through 15 are represented by the decimal digits 0 through 9 and the alphabet "digits" A through F (A=decimal 10, B=decimal 11, and so forth). For example, a binary (base 2) "11111111" equates to a decimal (base 10) 255, and is expressed as "FF" in hex (base

16). There are many conventions for distinguishing hexadecimal numbers from decimal or other bases in software. In C for example, the prefix "0x" is used, e.g. 0x694A11. In Web design, hexadecimal is the alphanumeric system used to specify colors in HTML. For example, the hexadecimal equivalent of white is FFFFFF, while black is 000000. See also Byte and Nibble.

HF 1. Hands Free.

2. High Frequency. Portion of the electromagnetic spectrum, typically used in short-wave radio applications; frequencies approximately in the 3MHz to 30 MHz range.

HFAI Hands Free Answer on Intercom. A desirable feature of several phone systems.

HFC Hybrid Fiber Coax. An outside plant distribution cabling concept employing both fiber optic and coaxial cable. Fiber is deployed as the backbone distribution medium, terminating in a remote unit where optoelectric conversion takes place. At that remote unit, the signal then is passed on to coax cables which carry the data the last leg to the individual business, residence, dormitory room, etc. HFC systems provide substantial bandwidth at lower cost than a system based exclusively on fiber. Given the embedded base of coaxial cable in college and university campuses, HFC is an effective means of delivering combined voice, data, video and CATV to dormitory rooms and classrooms. HFC also is used extensively in upgraded CATV networks for the same reasons. See also FSAN.

HFPL High Frequency Portion of the Loop. The frequencies above the voice band on a copper local loop used to support traditional analog POTS (Plain Old Telephone Service) voice transmission. ADSL (Asymmetric Digital Subscriber Line) for example, runs voice in traditional analog format at 0-4 kHz, and data and even video in the HFPL at 25 kHz and above.

HFSX Apple's OS X's native file format is called HFSX.

HFU Hands Free Unit.

HG Home Gateway. A device located on a corporate LAN that accepts authorized user tunnels over the Internet.

HGC Hercules Graphic Card; long the standard monochrome graphics adapter for PCs and compatibles, now well and truly obsolete. Maximum resolution is 720 x 348 pixels.

hijack a standard Refers to a vendor's gaining control over an industry standard. There are several ways to do this. One way is for a powerful vendor to exert influence on who participates in the standards committee, and the vendor's own representative on the committee may exert a powerful influence on the decisions made by the committee. Another way a vendor may hijack a standard is to release pre-standard products that the marketplace ends up embracing, thereby causing a favorable bias for the vendor's specifications. A standard can also be hijacked after the standard is established, through a process known as "embracing and extending" the standard. What this involves is a vendor's embracing the standard and adding useful extensions to it in its own products, thereby setting the stage for its extensions to be incorporated in the next version of the standard.

Hi-Cap High Capacity. These facilities are defined as special access circuits that fall into certain specific classes of service. This definition is deliberately vague because the term "hi-cap" is vague and can apply to all sorts of circuits.

Hi-Lo Tariff A long distance private line tariff filed by AT&T whereby private lines between major cities were priced lower than private lines between smaller cities. In effect, those "larger" cities were those MCI operated in and those "smaller" cities were those MCI didn't operate it. Eventually the tariff was thrown out by the FCC and it figured in anti-trust suits by MCI and the Federal Government against AT&T.

HI8 Video The high-quality extension of the Video 8 (or 8mm) format, which features higher luminance resolution.

hibernation The concept is simple. You turn your computer off in the middle of a program. It goes to sleep. When you turn it on, it returns to exactly where you left it, without the need for rebooting. These days most laptops have some form of hibernation or suspend software. There are basically two methods. First, "Suspend" makes a copy the entire contents your RAM memory and what files you have open at that time. It dumps that information into RAM memory and keeps that memory alive with a special battery inside your laptop. Most laptops have three batteries – a big one that you can change, a small one for suspend and a third one for running the clock. When you plug your laptop into AC, it charges all three batteries. Suspend will last as long as there's charge in the battery keeping your RAM memory alive. That's typically a day. Second, "hibernation" is the same concept except that it writes the RAM's contents to a special part of your hard disk. When you start your computer next time, "hibernation" tells it to read that part first. Hibernation can last as long as 45 days. Hibernation is scarey because if your computer messes up and can't find that precious file, it screws up. Your computer might be able not start a normal Windows boot and your computer may never start. My computer technician tells me never to use "hibernation." Suspend, he says, is fine. The reason I include this definition is that you cannot use suspend or hibernation if your laptop is attached to a network. It simply won't work. If you're attached to a network, the best routine is to leave your machine on permanently or close it down completely, i.e. by turning off the power.

Hidden Markov Method HMM. A common algorithm in voice recognition which uses probabilistic techniques for recognizing discrete and continuous speech.

HIDS Host-based intrusion-detection system (HIDS). Operates by detecting attacks occurring on a host on which it is installed. It works by intercepting OS (operating system) and application calls, securing the OS and application configurations, validating incoming service requests, and analyzing local log files for after-the-fact suspicious activity.

hierarchal file system A system of arranging files in directories and subdirectories to maintain hierarchical relationships (one file ranked above the other) between the files and make them easier to find and retrieve. See Relationship Database.

hierarchical network 1. A network that includes two or more different classes of switching systems in a defined homing arrangement, meaning to home in on the telephone you wish to be connected to. This is a fancy Bell System (Oops, I mean AT&T - Old habits die hard) term meaning that when direct circuits between two switches are busy or too far apart to be directly connected, the machinery will seek a higher level of switches to route the call through.

2. A LAN (Local Area Network) term describing a network employing OSI Layer 3 routers. Flat networks, on the other hand, make use of Layer 2 bridges, hubs and switches for LAN interconnection and segmentation. Routers make intelligent decisions about routing data packets, with such decisions taking into consideration the condition of the entire network; Layer 2 devices forward packets on a link basis. Hierarchical router networks are more complex and expensive, and are slower. However, they also can add value through protocol conversion and flow control. Routers, by the way, typically serve to interconnect bridges, hubs and switches, as well as to provide access to the WAN (Wide Area Network). See also Flat Network, Bridge, Hub, Switch and Router.

hierarchical routing The process of establishing a network data path to a destination based on addresses with some kind of hierarchical addressing system. Internet Protocol routing algorithms on the Internet use IP addresses, for example, which contain network numbers, host numbers, and frequently, subnet numbers.

hierarchically complete source route An ATM term. A stack of DTLs representing a route across a PNNI routing domain such that a DTL is included for each hierarchical level between and including the current level and the lowest visible level in which the source and destination are reachable.

hierarchy A hierarchy is a group of things arranged in order of rank. It is a set of transmission speeds arranged to multiplex successively higher numbers of circuits. See also Hierarchical Network. It is also an Internet USENET newsgroup hierarchy which refers to the set of all news-groups contained within a specific broad subject category.

HIF Host Interface Node. Provides the interface between a host and the Concert Packet Service network.

high and dry 1. What happens when you dial into a long distance network and nothing happens. You don't hear anything. Your call doesn't go anywhere. You're simply left High and Dry. Also called Dead Air.

2. A local central office term. The phone has been off-hook for an extended period of time. It is open in the loop. Compare with High and Wet which means the line, battery and ground, are shorted.

high and wet The local central office term. It means the line, battery and ground are shorted. Compare with High and Dry which means the phone has been off-hook for an extended period of time. It is open in the loop.

high ascii ASCII characters whose values exceed 127. In most bulletin board networks, The use of high-ASCII in messages is prohibited since some types of personal computers cannot correctly interpret those characters. High ASCII is now supported by HTML.

high bandwidth A person who is super intelligent is said to have "high bandwidth." The term is believed to have originated at Microsoft in Seattle, Washington. See also Bandwidth.

high burst High Burst is a term used by carriers. They use it to describe a specific type of customer call demand. A typical customer who exhibits high-burst characteristics is a telemarketer. Telemarketers use what are called predictive dialers, which offers business the ability to automatically call out using algorithms to speed up or slow down dialing. Once the system has detected a live person, the system transfers the call to an available agent. Triple the productivity of your agents and watch your contacts and sales skyrocket!"

high capacity service Generally refers to tariffed, digital-data transmission service equal to, or in excess of T-1 data rates (1.544 Mbits.)

high cost and low income division A division of the Universal Service

Administrative Company (USAC), a not-for-profit corporation established by the Telecommunication Act of 1996, and operating under the supervision of the National Exchange Carriers Association (NECA). This division is responsible for disbursing funds to subsidize the cost of providing basic telephone to those who live in High Cost Areas. See also High Cost Area, Universal Service, and Universal Service Administrative Company.

high cost area A term describing a serving area of a LEC (Local Exchange Carrier) in which the cost of providing local telephone service is at least 115% of the national average. Through the "settlements" process, administered by NECA (National Exchange Carrier Association) under the direction of the FCC (Federal Communications Commission), high cost LECs are compensated for this extraordinary cost through the Universal Service Fund. See also NECA, Separations and Settlements, and Universal Service Fund.

high definition See HDTV.

high definition radio HD radio. A technology from privately held Ibiquity Digital, HD radio got its official consumer launch at the Consumer Electronics Show, January 2004, with companies like Onkyo and Kenwood announcing plans to offer radios for the new standard. The digital standard, under development for years, provides higher quality audio for both AM and FM radio broadcasts, and also allows a range of special services – identification of songs and artists on the radio display, for instance. Don't hold your breath. Some broadcasters which started broadcasting in high definition (after spending large amounts of money to upgrade their broadcast equipment) have since stopped. Few HD radios have been sold.

high definition tv 1. HDTV. A system standard for transmitting a TV signal with far greater resolution than specified by the current NTSC standard in North America or PAL overseas. For a full explanation, see HDTV.

2. "A wide screen just makes a bad film twice as bad." Samuel Goldwyn, film producer.

High Density Bipolar 3 See HDB3.

high dome Synonym for "egghead." A scientist.

High Energy Radio Frequency Gun HERF. Shoots a high-powered radio signal at an electronic target (such as a computer) and puts it out of commission.

high fidelity Systems of radio transmission and reception which permit a wide band of audio frequencies to be transmitted and/or reproduced.

high frequency noise A signal frequency more than 1,000 times the normal AC power line frequency of 60 cycles. The frequency will lie between 3 and 30MHz.

high frequency tags RFID are high frequency which typically operate at 13.56 MHz. They can be read from about 10 feet away and transmit data faster. But they are consume more power than low-frequency tags.

high level combining A processing of combining channels or carriers at the transmitter outputs.

High Level Data Link Control HDLC. A communications protocol that is bit oriented in which control codes differ according to their bit positions and patterns. For a much bigger explanation, see HDLC.

high level diplexing A processing of combining two radio frequency carriers (usually audio and video) at the final transmitter outputs.

high level languages Essentially any of the computer languages whose code is not unique to the hardware or architecture of a particular computer. High level languages are more like human language than the machine language which computers talk. High level languages translate human instructions into the machine language computers can understand, but which humans don't have to (in order to tell the computer what to do). Computer languages such as Basic, FORTRAN, COBOL and Pascal are high level languages. They are a number of levels (at a High Level) away from the actual bit manipulation (machine language, also called "bit twiddling" by the Hackers). Compare with Low Level.

high level modulation Modulation at the last amplifier stage of a transmitter.

high low tariff A tariff in which two prices are given for something – a high price and a low price. The first high/low tariff from AT&T was for leased voice lines where a lower charge was made per mile for connections between routes that have much traffic (High Density) and greater charges per mile are made for all other (Low Density) routes. The High/Low tariff was significant because it was AT&T's response to competition from long distance carriers like MCI and it was one of the first moves away from nationwide rate averaging, which was the way things were done under monopoly.

High Memory Area HMA. High Memory Area is the first 64KB of extended memory. If you're using MS-DOS 5.0 or 6.0, you can save some conventional memory (i.e. below 640K memory) by loading the operating system into HMA. Add the line DOS=HIGH to your CONFIG.SYS to use HMA for the operating system.

high order bit Hobbit. Also known as an "alt bit," "high bit," and "meta bit." The most significant bit of a byte, a high-order bit generally is the first bit in a byte. Since the hobbit is the first bit in a byte, it is the first bit that a device sees, and therefore the first bit on which action is taken. The high-order bit can be used for a wide variety of purposes in a data communications environment, all of which identify to the receiving device something of significance relative to the handling of the associated data. For example, the hobbit in the header of a packet can be used by a device to indicate the priority level of the packet data packet transfer. The hobbit also can be used to indicate the highest level of addressing, in order that the network can route the data properly.

high pass filter A filter which passes frequencies above a certain frequency and stops (attenuates) those below.

High Performance Computing Act An Act passed by Congress in 1991 to foster the creation of computer "superhighways" linking computers at universities, national laboratories and industrial organizations. One objective of the High Performance Computing Program is the establishment of a gigabit/second National Research and Education Network (NREN) that will link the government, industrial and higher education communities involved in general research activities. Such a gigabit network would provide a significant increase in bandwidth compared with the existing National Science Foundation network, which is evolving from a 1.5 megabit per second (T-1) backbone to 45 megabit per second (T-3).

High Performance Computing and Communications See HPCC.

High Performance Parallel Interface HIPPI. A high-speed multi- signal interface-analogous to an RS-232 interface but for high-speed computers, etc. HIPPI provides 800 (or1600) Mb/s interconnections using 32 (or 64) bit wide parallel data paths for distances up to 25 meters (or longer if use fiber). Standardization activity is in ANSI X3T9.3.

High Performance Routing HPR. A local area networking term. HPR is the next-generation APPN – referred to in the past as APPN+ – that adds IP-like dynamic networking – e.g., dynamic alternate routing in the event of path failure – features to APPN, and uses a routing mechanism that works at Layer 2 using a RIF concept similar to that found in SRB.

High Power Amplifier HPA. A device which provides the high power needed to shoot signals 22,000 miles plus from an earth station to a satellite.

high rejection The ability of a voice recognition system containing active vocabulary words to reject those sounds that do not match closely the words in its vocabulary.

high resolution TV Television with over 1,000 lines per screen, about double the resolution of present systems. Sometimes called HDTV, for high-definition television. We're still awaiting standards for high resolution TV. See HDTV for a bigger explanation.

high sierra format A standard format for placing files and directories on CD-ROM, revised and adopted by the International Standards Organization as ISO 9660.

High Speed Digital Subscriber Loop See HDSL.

High Speed Local Network HSLN. A local network designed to provide high throughput between expensive, high-speed devices, such as mainframes and mass storage devices.

high speed printer Any printer which can print at over 100 lines a minute. Like many definitions, this one is arbitrary. Some people claim a dot matrix is "high speed" and a letter quality, daisy wheel is a "low speed" printer. Laser printers could be classed as high speed printers, maybe.

high speed register set Registers are storage locations within the CPU that are used to hold both the data to be operated on and the instructions to accomplish the operations.

high speed signal An AT&T definition for a signal traveling at the DS-3 rate of 44.736 Mbps (million bits per second) or at either 90 Mbps or at 180 Mbps (Optical mode).

high split 1. A broadband cable system in which the bandwidth used to send toward the head-end (reverse direction) is approximately 6 MHz to 180 MHz, and the bandwidth used to send away from head-end (forward direction) is approximately 200 MHz to 400 MHz. The guard band between the forward and reverse directions (180 MHz to 220 MHz) provides isolation from interference. High split requires a frequency translator which transfers the originating signals to other frequency ranges at the head-end, in either direction. Historically, CATV systems used the spectrum below Channel 2 for inbound transmissions from the user premise to the head-end; that frequency range is 5-30/40 MHz.

2. A term used in radio communications, including paging and cellular, for several ranges of frequency used to connect a remote site to a main site. For instance, the low-split might be 806.0125 MHz and the high-split 851.0125-869.9875 MHz. Frequency translators are used to transfer the signal to another frequency range from that point forward.

high tech A high-falutin' (i.e. overly pretentious) way of saying technology. I exorcised the term out of this dictionary out of disgust.

high tier A PCS cell phone service for users moving in a high-speed automobile. High-tier PCS systems are often straightforward evolutions of current digital cellular systems. In contrast, a low-tier is a PCS cell phone service for pedestrians or slow moving vehicles (no more than 30 to 40 mph). An evolution of cordless systems originally intended for in-building applications. Systems use small cells, so they can be designed with low-power transmitters and experience fewer handoffs than high-tier PCS systems (with high-speed, mobile users). Systems provide lower cost and higher-quality services, for low-speed users only.

high usage groups Trunk groups established between two central office switching machines to serve as the first choice path between the machines and thus, handle the bulk of the traffic. See High Usage Trunk Group.

high usage trunk group A Bellcore definition. A trunk group that is designed to overflow a portion of its offered traffic to an alternate route.

high water mark A financial term. Let's say you give a money manager $100,000 of your money to manage. You agree to pay him 20% profit-sharing of all your gains. And you agree to do this annually. Let's say one year your manager loses 20% of your money. But the next year he earns 15%. He doesn't receive any profit-sharing of your 15% until he has earned back what he lost and is above the high water mark – the place you started. For a more formal definition, here's one from www.hedgeworks.com. High-water mark is an investor's capital basis in a given year used to determine the minimum value to which a manager's performance fee is measured. For example, a manager may only charge an investor a performance fee for any gains achieved over the investor's capital basis or the gains achieved since the last performance fee was charged.

highway 1. Another word for BUS. A common path or set of paths over which many channels of information are transmitted. The channels of the highway are separated by some electrical technique.

2. The Information Superhighway. In 1995, a consulting firm called Ovum defined the superhighway as a mechanism for providing access to electronic information and content held on network servers. It has four key features, according to Ovum: A. It supports two way communications. B. It offers more than just simple voice telephony. C. It is interactive and provides real-time, cooperative communications, and D. It supports electronic screen-based applications.

highway construction supervisor A consultant to provide assistance in specification, installation and/ or operation of systems and software for accessing the information highway.

highway patrol A slang term for the U.S. Congress.

hijacking Attack whereby a person, or persons, gains unlawful control of a communication, typically between two parties, and successfully assumes the identity of one of the parties.

Hindi Although Hindi is the official language of India, there are 14 regional languages that are officially recognized for conducting national affairs. In addition, there are about 170 other languages and over 500 dialects.

HIPAA Health Insurance Portability and Accountability Act. Enacted in 1996, this Act presents standards for the maintenance and transmission of personal information regarding individuals. It is incumbent upon organizations in possession of, or potentially in possession of, information to be able to protect the security and confidentiality of electronic personal information. The Act ensures that the interchange of said electronic data is standardized, and that privacy is ensured.

HiperAccess High Performance Radio Access (HiperAccess) is a broadband wireless access specification from the Broadband Radio Access Networks (BRAN) project of the European Telecommunications Standards Institute (ETSI) and is considered complementary to HiperLAN/2.This long range variant is intended for point-to-multipoint, high speed access by residential and small business users to a wide variety of networks including the UMTS core networks, ATM networks and IP networks. HiperAccess operates in the 40.5 GHz and 43.5 GHz spectrum. See also ETSI, HiperLAN, HiperLink and HiperMAN.

HiperLAN High performance radio local area network. Developed by the European Telecommunications Standards Institute (ETSI), HiperLAN is a set of WLAN communication standards used chiefly in European countries. HiperLAN is similar to the IEEE 802.11 WLAN standards used in the U.S. There are two types of HiperLAN:

HiperLAN/1: provides communications at up to 20 Mbps in the 5 GHz band. * HiperLAN/2: provides communications at up to 54 Mbps in the 5 GHz band.

According to ETSI, HiperLAN/2 marks a significant milestone in the development of a combined technology for broadband cellular short-range communications and wireless Local Area Networks (LANs) which will provide performance comparable with that of wired LANs.

Since the 5 GHz band to be exploited by the HiperLAN/2 standard is allocated to wireless LANs world-wide, HiperLAN/2 has the potential to enable the success of wireless LANs on a global basis. In collaboration with ETSI Partnership Project 3GPPTM, ETSI BRAN will draw up specifications for the access interface to UMTS (Universal Mobile Telecommunications System). This interface could also serve as a basis for the definition of interfaces to the other members of the IMT-2000 family of 3rd generation mobile systems. ETSI Project BRAN also turns its attention to the development of conformance test specifications for the core HIPERLAN/2 standards, to assure the interoperability of devices and products produced by different vendors. The test specifications includes both radio and protocol testing. In developing these test specifications, ETSI is supported by the HiperLAN2 Global Forum (H2GF), a consortium of communications and information technology companies who have joined together to ensure the completion of the HIPERLAN/2 standard and to promote it worldwide. EP BRAN has worked closely with IEEE-SA (Working Group 802.11) and with MMAC in Japan (Working Group High Speed Wireless Access Networks) to harmonize the various HiperLAN systems.

HiperLAN/1 See HiperLAN.

HiperLAN/2 A high-speed variation of the original HiperLAN (High Performance Radio Local Area Network) standard. HiperLAN2 is intended for broadband wireless LAN applications and was approved by the European Telecommunications Standards Institute (ETSI) in February 2000. HiperLAN/2 consists of three profiles for the corporate, public and home environments. HiperLAN2 operates in the 5-GHz band and supports aggregate bandwidth up to 54 Mbps. HiperLAN and HiperLAN/2 essentially are the European equivalents of 802.11a, the IEEE specification also known as Wi-Fi5. See HiperLAN, 802.11a and Wi-Fi5.

HiperLink High Performance Radio Link (HiperLink) is a planned broadband wireless specification from the Broadband Radio Access Networks (BRAN) project of the European Telecommunications Standards Institute (ETSI). HiperLink will be a very high speed method for interconnecting HiperLANs and HiperAccess over very short distances and is intended to run in the 17 GHz range. See also ETSI, HiperAccess, HiperLAN and HiperMAN.

HiperMAN High Performance Radio Metropolitan Area Network (HiperMAN) is a broadband wireless access specification from the Broadband Radio Access Networks (BRAN) project of the European Telecommunications Standards Institute (ETSI) and is considered complementary to HiperLAN/2. HiperMAN operates at frequencies between 2 GHz and 11 GHz. It is designed for point-to-multipoint configurations, but may include mesh topologies in the future. Although HiperMAN differs from the IEEE 802.16 specification, both are addressed in WiMAX specifications. See also 802.16, ETSI, HiperAccess, HiperLAN/2, HiperLink and WiMAX.

HiPot A test designed to determine the highest voltage that can be applied to a conductor without electrically breaking down the insulation.

HIPPI HIgh Performance Parallel Interface. In 1989, researchers at the Los Alamos National Laboratories began work on a standard for high-speed, point-to-point data transport between supercomputers. The result of that effort was HIPPI, which later became known as HIPPI-800 and which was standardized by the ANSI X3T9.3 committee as X.3.183-1991. HIPPI also is used to move data between supercomputers or high-end workstations and peripherals (e.g., disk arrays and frame buffers) through high-capacity, non-blocking crossbar-type circuit switches. HIPPI provides for transfer rates of 800 Mbps over 32 shielded twisted pair (STP) copper wires (single HIPPI) and 1600 Mbps over 64 pairs (double HIPPI). HIPPI connections are limited to 25 meters over STP and 10 kilometers over fiber. HIPPI is currently the most common interface in supercomputing environments. Work is in progress on HIPPI-6400, which supports transmission rates as high as 6400 Mbps, or 800 MBps (MegaBytes per second), in each direction, an 8-fold increase over the original version. Also known as SuperHIPPI and GSN (Gigabyte System Network), HIPPI-6400 is compatible with HIPPI-800. Distances of 50 meters can be bridged with parallel copper cables, and 200 meters with parallel fiber-optic cables. The connection is devised in four virtual circuits in each direction, capable of supporting various combinations of traffic such as 10Base-T, 100Base-T, Fibre Channel, ATM and HIPPI-800.

histogram 1. An imaging term. A display plotting the density of the various colors and/or values in an image.

2. A graph of contiguous vertical bars representing a frequency distribution in which the groups or classes of items are marked at equal intervals in ascending order on the x axis, and the number of items in each class is indicated by a horizontal line segment drawn above the x axis at a height equal to the number of items in the class.

history 1. When you reply to my email and say "yes," you're sending me a reply. When you reply to my email and include my original email you're sending me an email "with history." In email clients like Lotus Notes, you can select "Reply" or "Reply with History". In some, like Microsoft's Outlook, your reply typically gets sent with the history.

In AOL email, you have to work hard to send the email's history. To my tiny brain, it makes sense to include history. If you answer "Yes," it would be nice to know which request you answered "Yes" to. The original logic on not including history had something to with preserving bandwidth and server size. And I'm sure that's the reason AOL makes it so incredibly difficult to send history and why virtually all AOL users don't.

2. "She's history." The response I get from son when I ask about his latest girlfriend.

HISU Home Integrated Services Unit.

HIT 1. Electrical interference that causes the loss or introduction of spurious bits into a data stream.

2. Hit is the unit of measure of popularity most commonly cited by companies that have set up shop on the Internet's World Wide Web. It is the most commonly misunderstood measure. A hit is one file opening and transfer from a Web site. It is not a measure of how many people have visited your Web site. Since one file is needed for every chunk of text and every graphic element on a Web page, one mouse click by a viewer to your site may count as a dozen or more hits, depending on the complexity of your page. Also, because most Web sites contain more than one page, hits can rapidly multiply. So, using "hits" as a measure of the popularity of a Web site or the number of people who visit a web site is not accurate. There is software around that purports to measure the number of "unique visitors" or "eyeballs" to a site. That software tries to measure the number of unique IP addresses of computers visiting the site. But that software is confused by the single fact of America On Line (the largest ISP in the world) which often runs its subscribers' access to the Web from behind massive proxy servers which can even change IP addresses mid-session. See also HITS.

hit my clip When a teenager says "Hit my clip," he means "Page me."

hitless software upgrade A software upgrade, such as an operating system upgrade, that allows a carrier to add features or install a patch without scheduling network downtime for the upgrade.

HITS Hits is a measure of file openings done on a Web site. In the language of the Internet, hits has two meanings. The first and more common, is the number of times your Web site or a file within your Web site is accessed by people visiting it. "Hits" is often used as a measure of how popular your site is. But the measurement of "hits" is not very scientific. For example, if someone visits a home page, then jumps to another page, then comes back to the home page, that is registered as two "hits." Still, people who are selling advertising on Web pages use "hits" as a measure of how many people visit the site and therefore how much to charge for advertising. The second meaning of "hits" is how many matches you might find in a search; e.g., a Veronica search for the word "NASA" will return a long list of hits for your query.

HIVR Host Interactive Voice Response. Tying a voice response unit into a mainframe computer which has lots of data. Applications which can be produced include bank-by-phone, reservations-by-phone, etc. See Interactive Voice Response.

hizzoner His Honor. Hizzoner is the way the New York City tabloids refer to the mayor of New York. They call him hizzoner. Say it fast.

HKIX Hong Kong Internet Exchange. See IX.

HKSW Abbreviation for HOOK SWITCH, the actual electrical switch inside a phone that is controlled by the motion of the Switch Hook.

HLC High Level Committee of ITU (International Telecommunication Union).

HLD 1. High Level Domain. See Web Address.

2. High Level Design.

HLF High Level Function.

HLLAPI High Level Language Applications Programming Interface. An IBM API.

HLR Home Location Register. A database that holds subscription information about every subscriber in a mobile (i.e. cellphone) network. A HLR is permanent SS7 database used in cellular networks, including AMPS (Advanced Mobile Phone System), GSM (Global System for Mobile Communications), and PCS. The HLR is located on the SCP (Signal Control Point) of the cellular provider of record, and is used to identify/verify a subscriber; it also contains subscriber data related to features and services. The HLR is used not only when you are making a call within the area of coverage supported by your cellular provider of record. It also is used to verify your legitimacy and to support the features to which you subscribe when you are roaming outside that home area. In a roaming scenario, the local service provider queries the HLR via a SS7 link. Once verified, your data is transferred via SS7 to the VLR (Visitor Location Register), where it is maintained during your period of roaming activity within the coverage area of that provider. HLR is a key element of IS-41, the predominant wireless standard in North America. See also AMPS, GSM, IS-41, PCS, SCP, SS7 and VLR.

HMA See High Memory Area.

HMG Hyper Master Group.

HMI 1. Novell's Hub Management Interface. See HMI Driver.

2. Human-to-Machine Interface.

HMI Driver A Hub Management Interface (HMI) driver is an ODI driver running on a NetWare server that is compliant with the Novell HMI specification. A node may emulate an HMI driver by supporting the Novell NWHUB.MIB and IPX autodiscovery.

HMM Hidden Markov Method. A common algorithm in voice recognition which uses probabilistic techniques for recognizing discrete and continuous speech.

HMP Host media processing. Software that uses server CPUs rather than specialized DSPs to process signals.

HMS High-Performance Management System.

HMWPE High Molecular Weight Polyethylene.

HNDS Hybrid Network Design System.

HNPA Home Numbering Plan Area.

HO Tone A cellular term. Handoff Tone. 50ms of signaling tone sent by the mobile phone on the REVC to indicate leaving the source cell site during handoff.

hoax virus A warning for a computer virus that does not actually exist. Often these appear in the form of email messages which have been forwarded many times and contain dire warnings of impending doom and an urgent plea for the receiver to forward the warning on to others.

Hobbit 1. A microprocessor chip developed by AT&T's Bell Labs and used in the EO handheld devices. EO was closed down for lack of sales. The EO was just plain awful.

2. See High-Order Bit.

HOBIC HOtel Billing Information Center. In the days before call accounting systems, hotels relied on AT&T's HOBIC system for the call detail records (CDRs) and associated charges for long-distance calls made from hotel guest rooms. This real-time reporting was essential so that a hotel guest's bill could be updated right away, even if the guest made a long-distance call from his room right before checking out. These days, HOBIC also refers to the record layout that AT&T used with its HOBIC system for call details and associated charges. HOBIC format is one of the supported CDR (call detail report) formats used by modern call accounting systems.

HOBIC format The record layout that AT&T used with its HOBIC system for recording call details and associated charges for long-distance calls made from hotel guest rooms. Modern call accounting systems continue to support the HOBIC format, making it one of the available data formats for call records that the call accounting system sends to a hotel's property management system for customer billing.

HOBIC line A telephone line that connected AT&T's HOBIC operator or TSPS system with a hotel billing center, or with a teleprinter in the hotel's billing center, so that a HOBIC record containing call details and associated charges could be provided to the hotel immediately after a long-distance call was made from a hotel guest room.

HOBIC operator An operator at AT&T in the old days who handled long-distance calls from hotel guest rooms and reported the details of such a call to the hotel immediately afterwards.

HOBIS HOtel Billing Information System.

hobo The word hobo comes from the time after the Civil War, when men who had lost their farms and homesteads would ride the rails looking for temporary work at the various farms that were along the railroad tracks. Because they usually carried their own hoes, the farmers called them "hoe boys."

hold To temporarily leave a phone call without disconnecting it. You can return to the call at any time, sometimes from other extensions. There are several types of "HOLD" on a telephone system. How they work and what lamping they put on instruments varies from phone system to phone system.

Exclusive Hold: Prevents every other telephone from picking up the call. Only the telephone instrument that put the call on hold can retrieve it.

I-Hold: Effectively the same as Exclusive Hold.

Line Hold: The call is on hold. Anyone with a phone with the held line appearing on it can pick up the phone.

hold recall A telephone system feature which reminds you periodically that you've put someone on hold.

holding tank A queue in which a call is held until it can either use its assigned route or overflow into the next available route.

holding time The total time from the instant you pick up the handset, to dialing a call, to waiting for it to answer, to speaking on the phone, to hanging up and replacing the handset in its cradle. You are never billed for holding time. You are always billed for conversation time which is shorter than holding time. But holding time is an important figure to know when you're trying to determine how many circuits you need. For you will need sufficient circuits to take care of dialing, etc. – even though you're not being billed for that time.

holdup time The amount of time that a power supply can continue to supply the load after input power is terminated. The duration of a blackout or transfer time that a power supply can accept without any disturbance of the output. Holdup time is specified by CBEMA to be a minimum of 8 milliseconds for business and computer equipment. The typical value specified for commercial computer power supplies is 25ms. Holdup time is increased when a power supply is lightly loaded. Therefore typical computers have holdup times in the range of 100ms. This definition courtesy American Power Conversion Corp.

holey fiber Optical fiber with very tiny, closely spaced air holes that go through the entire length of the fiber. The number, size, and arrangement of air holes lead to optical fibers with a variety properties not found in standard, unperforated optical fiber. Also called photonic crystal fiber. One use of holey fiber is to guide wavelengths that can travel through the fiber's air tubes, but which can't travel through the solid portion of the optical fiber itself because it isn't transparent for those wavelengths.

holiday factor A call center term. A historical factor associated with a specific date and multiplied by the forecast call volume for that date in order to take into account an expected increase or decrease in the call volume. For example, if on a given day only half the usual number of calls occur for that day of the week and that time of year, the holiday factor for that date would be .5.

hollerith card A punched-hole 80 column card used for storing information for input into a computer. Remember the cards you got telling you "not to fold, bend, punch, spindle, etc."? They were Hollerith Cards. They're now falling into disfavor as other, less tamper-proof methods appear.

hollerith code Twelve level punched card code.

hollow pipeline Jargon for a broad bandwidth circuit that has no framing. A private out-of-band signaled (CCC, Clear Coded Channel) DS1. There is no timing, framing or error connection. You input a bit stream into one end and it comes out the other end in the same order. The maximum speed for DS-1 is 1.536 Mbps. This is 1.544 Mbps less than the framing overhead of 8 Kbps.

hollow state A jocular descriptive term for vacuum tube technology, now largely obsolete – except in high-end sound systems. See solid state.

hollowing out the corporation Hollow means to remove the innards, or the guts. The only time I heard it referred to a corporation was to AT&T. And it is the classic case. When I first started in this industry in the late 1960s, AT&T was "The Bell System" (capital letters, exclamation point) and employed a total of one million people. By the time I wrote this definition in late 2004, the hollowed-out company had dropped down to 50,000 people. That was a 95% hollowing out. When I came into the industry, AT&T was one of the most profitable companies that existed. Thirty five years it was losing money. That was serious hollowing out.

hologram A three-dimensional image produced by a system that uses lasers instead of lenses.

holographic storage A technology still in the labs. It uses lasers and crystals rather than magnetic or optical medium to store bits of data in holograms. Holographic data storage portends big reduction in physical storage space and much quicker seek time. Images of data are stored on holographic crystals. More than a million bits of data can be stored on a single hologram, and thousands of holograms can fit on a square that is just a single centimeter. Holographic storage has no moving parts. Experts predict holographic storage will emerge as a commercially available technology by 2005, and will, once again, change the economics of data storage.

holy war Arguments that involve basic tenets of faith, about which one cannot disagree without setting one of these off. For example: PCs are superior to Macintoshes.

home The beginning place of a cursor on a CRT screen. Usually it's the top left hand corner. The function key on an IBM PC or clone marked "Home" will take the cursor to the home position, namely the top left hand corner.

Homebrew Computer Club In March, 1975 a peace activist, called Frederick Moore Jr. rode his bicycle around Palo Alto and Menlo Park, tacking up small 3-by-5-inch notices on bulletin boards and telephone poles. The tiny flier read in part: "Are you building your own computer? Terminal? TV Typewriter? I/O device? Or some other digital black magic box? Or are you buying time on a time-sharing service? If so, you might like to come to a gathering of people with like-minded interests. Exchange information, swap ideas, help work on a project, whatever . . ." Thirty-two people showed up at the first meeting. Over time, the group got progressively bigger as people like Steve Jobs and Steve Wozniak (founders of Apple) and Bill Gates and Paul Allen (founders of Microsoft) visited the club to make presentations, to swap equipment and to learn. The group ultimately settled for many years at the Stanford Linear Accelerator auditorium, where as many as 750 computer enthusiasts would gather twice a month. Neither graphics nor sound were common computer features in those days, so when a hobbyist, Steve Dompier, demonstrated at the fourth meeting that he had laboriously programmed his new MITS Altair computer kit to play the Beatles' "Fool on the Hill," the assembly gave him a wild standing ovation. The typical Homebrew meeting began with a "mapping" session. The moderator – for many years Lee Felsenstein – would point to people around the room, and each would describe what they were working on. A formal presentation would follow. The meeting would conclude with a "random access" session in which people with similar interests would cluster to share information. It was at one such meeting, in the fall of 1981, that the transition from the hobbyist era of computing to today's booming world of corporate computing took place. Mr. Felsenstein pried open the case of the first IBM PC, examined its motherboard and pointed with glee to three "blue lines" – wires that were evidence, he proclaimed, of a hastily patched-together computer that had been rushed to market. See also Morrow, George.

homepage See Home Page.

HomePlug HomePlug is an industry alliance, including over 50 major industry participants such as Cisco, Motorola, Panasonic, Philips, Compaq, Conexant, Sharp, Texas Instruments, Radio Shack, and others, and was organized to promote the use of powerline as a medium for communications and connectivity. Powerline is sending telecom over electricity lines. See Powerline Communications.

HomePNA Home Phoneline Networking Alliance. An association of companies working toward the adoption of a single, unified phoneline networking standard and bringing to market a range of interoperable home networking solutions using in-place phone wiring. HomePNA solutions are intended to be plug-and-play for networking of multiple PCs, peripherals (e.g., printers, scanners and video cameras), multi-player network games, home automation devices (e.g., environmental control and security systems), digital televisions and digital telephones. An all-purpose Home Area Network (HAN) using existing telephone wiring, the HomePNA solution also is intended as a means of shared access to IP voice and video networks, the IP-based Internet, and the conventional circuit-switched Wide Area Network (WAN). Network access technologies are intended to include analog, ISDN and xDSL local loops. Initial efforts are directed at a technology that will support spatial separation of nodes by as much as 500 feet, which represents a home of up to 10,000 square feet (which is bigger than my home, and probably bigger than yours, unless you are Bill Gates and live in a monstrosity of a castle, in which case you probably already have an ATM-based LAN with SONET fiber optics pipes running at 10 Gbps, but I digress), and running at data rates of 1 Mbps. Frequency Division Multiplexing (FDM) is intended to support simultaneous voice and data traffic; frequency ranges are intended to avoid interference from devices (e.g., refrigerators and air conditioners) found in the home. HomePNA solutions are based on an Ethernet derivative, running at 1 Mbps at frequencies above 2 MHz using a proprietary compression technique from Tut Systems, and using the CSMA/CD protocol native to Ethernet; speeds of 10 Mbps are planned into the future, with the theoretical potential being as much as 100 Mbps. Members include 2Com, AT&T, Compaq, Hewlett-Packard, IBM, Intel, Lucent and Tut Systems. www. homepna.org. See also Ethernet, FDM, ISDN, SONET and xDSL. www.homepna.org.

Home Phoneline Networking Alliance The Alliance is a consortium of more than 90 companies from the PC, consumer electronics, and network equipment manufacturing industries that are delivering easy-to-use, affordable, high-speed networking solutions over existing loop telephone wires. Its 10 Mbps home networking standard, called 2.0, has been finalized. Home networking allows multiple members of a household to simultaneously use PCs or consumer electronic devices to access files from local servers and from the the Internet, print documents, and play video games from external Internet connections including cable TV, Digital Subscriber Line (DSL), satellite or conventional analog modems. See the previous definition. www.homepna.org.

homeland What Americans mean by homeland, the Russians use the word motherland and the Germans use the word fatherland. Essentially, homeland to Americans means the United States of America. Homeland came into use in late 2001 after the September 11 attack on New York City's World Trade Center. Shortly after the attacks, President George W. Bush formed The Office of Homeland Security, whose job is "coordinating national strategy to strengthen protections against terrorist threats or attacks in the United States."

Homeland Security Information Network When completed, this communications network will provide secure, real-time interactive connectivity among federal, state, local, tribal governments, and with designated private sector entities, to collect and process information, and disseminate situational awareness reports in order to detect, prevent, and respond to terrorist actions.

HomeRF Home Radio Frequency. A wireless networking specification from the HomeRF

Working Group for interoperable voice and data communications in the unlicensed 2.4 GHz band, known as the ISM (Industrial, Scientific and Medical) band. The ISM band is shared with garage door openers, cordless telephones, RF bar code scanners, microwave ovens, and a wide variety of other devices, all of which can (and do) interfere with each other. The ISM band also is shared with IEEE 802.11b WLANs (Wireless Local Area Networks). HomeRF makes use of the SWAP (Shared Wireless Access Protocol). See also 802.11b and ISM.

HomeRF Working Group Home Radio Frequency Working Group, or HRFWG. An organization "formed to provide the foundation for a broad range of interoperable consumer devices by establishing an open industry specification for wireless digital communication between PCs and consumer electronic devices anywhere in and around the home." The Shared Wireless Access Protocol (SWAP) specification is intended to enable interoperability of electronic devices from a large number of manufacturers, while providing the flexibility and mobility of a wireless solution. SWAP is expected to yield a wireless home network to share voice and data between devices such as PCs, peripherals, PC-enhanced cordless phones, and devices yet to be developed. According to its Web site, the HRFWG "was formed to provide the foundation for a broad range of interoperable consumer devices by establishing an open industry specification for wireless digital communication between PCs and consumer electronic devices anywhere in and around the home." The Shared Wireless Access Protocol (SWAP) specification is intended to enable interoperability of electronic devices from a large number of manufacturers, while providing the flexibility and mobility of a wireless solution. SWAP is expected to yield a wireless home network to share voice and data between devices such as PCs, peripherals, PC-enhanced cordless phones, and devices yet to be developed. www.homerf.org. See also HomeRF.

home agent 1. A Mobile IP (Internet Protocol) term. Mobile nodes associated with nomadic users register their presence at a remote location through a foreign agent. The foreign agent communicates with the home agent in order that data packets can be forwarded to the remote subnet. See also Mobile Agent and Mobile IP.

2. A call center agent who works from home. There are basically two ways this happens. First, the agent could have two analog phone lines – one for speaking to customers and one for a computer terminal for entering orders, checking the status of orders, etc. The second way is a broadband Internet connection, broken into two data streams, one for IP voice and one for broadband data connection back to the main computer. Airlines such as JetBlue extensively use home agents. JetBlue finds it can hire better quality agents, schedule them at short notice and save on the overhead of accommodating them inside a giant building.

Home Automation HA. Home Automation is basically anything that gives us remote or automatic control of things around the home – like lights, sound systems, security systems, thermostats, irrigation, door openers, etc. You control your remote gadgetry typically via low voltage cabling, via wireless connections (typically 802.11b) or with powerline systems, such as X10.

Home Carrier The cellular operating company which a subscriber is registered with and pays the monthly service charge and usage charges to.

Home Location Register HLR. A wireless telecommunications term. The knowledge center of the network, helping to route calls and ensure security. A HLR has a built-in Authentication Center (AuC), which is a database of subscriber information, including access rights and services subscribed to. A permanent SS7 database used in cellular networks, including AMPS (Advanced Mobile Phone System), GSM (Global System for Mobile Communications), and PCS. The HLR is located on the SCP (Signal Control Point) of the cellular provider of record, and is used to identify/verify a subscriber; it also contains subscriber data related to features and services. The HLR is used not only when you are making a call within the area of coverage supported by your cellular provider of record. It also is used to verify your legitimacy and to support the features to which you subscribe when you are roaming outside that home area. In a roaming scenario, the local service provider queries the HLR via a SS7 link. Once verified, your data is transferred via SS7 to the VLR (Visitor Location Register), where it is maintained during your period of roaming activity within the coverage area of that provider. HLR is a key element of IS-41, the predominant wireless standard in North America. See also AMPS, GSM, IS-41, PCS, SCP, SS7 and VLR.

home office The most common form of telecommuting, in which employees work at home one or more days per week. See also VPN.

home page The classic definition: The front page of an "online brochure" about an individual or organization. The Internet definition: The first page browsers see of the information you have posted on your computer attached to the World Wide Web is your "home page." It's a "welcome" page. It says "Welcome to my site, my home." It typically contains some sort of table of contents to more information which a visitor (browser, surfer, etc.) will find at your site by clicking onto hypertext links you've created. In a Web site, a home page is usually called index.htm, index.html or index.asp. The biggest mistake made by people creating Web sites is that they fail to call their home page index.* (that star depends on the operating system which the Web hoster is using). See HTML, Internet, Streaming and World Wide Web.

home run Phone system wiring where the individual cables run from each phone directly back to the central switching equipment. Home run cabling can be thought of as "star" cabling. Every cable radiates out from the central equipment. All PBXs and virtually all key systems work on home run cabling. Some local area networks work on home run wiring. See Loop Through.

home run cabling There are basically two ways you can install phone cabling in a home. The old fashioned way is to instal loop wiring – one or two pairs of wiring loop through the home, from one outlet to another. With loop wiring you install single line or two line phones around the house. Typically anyone picking up the phone can answer an incoming call or can hear the conversation. The second way to install wiring to home run cabling – in which individual cables are run directly from a central location, which may be a telephone switch or a cross-connect panel. This configuration is also known as star topology. The advantage of home run or star cabling is typically you have privacy – on one can pick up the phone while you're talking and most importantly, you can have an intercom. You can dial from one phone to other.

home tandem A tandem of a higher office class to which another tandem or an end office has a final trunk group. Home tandems may exist for all or defined subsets of tandem switched traffic.

home zone Cell phone service that is configured by the cell phone carrier to offer dual pricing – one price for calling from your phone (i.e. cheap) and another, more expensive price for calling from outside your home – like a normal cell phone. At home you may plug your cell phone into a home phone system. The idea of home zone is for your cell phone company to sell more phones, by competing with landlines.

HomePNA Home Phoneline Networking Alliance. An association of companies working toward the adoption of a single, unified phoneline networking standard and bringing to market a range of interoperable home networking solutions using in-place phone wiring. HomePNA solutions are intended to be plug-and-play for networking of multiple PCs, peripherals (e.g., printers, scanners and video cameras), multi-player network games, home automation devices (e.g., environmental control and security systems), digital televisions and digital telephones. An all-purpose Home Area Network (HAN) using existing telephone wiring, the HomePNA solution also is intended as a means of shared access to IP voice and video networks, the IP-based Internet, and the conventional circuit-switched Wide Area Network (WAN). Network access technologies are intended to include analog, ISDN and xDSL local loops. Initial efforts are directed at a technology that will support spatial separation of nodes by as much as 500 feet, which represents a home of up to 10,000 square feet (which is bigger than my home, and probably bigger than yours, unless you are Bill Gates and live in a monstrosity of a castle, in which case you probably already have an ATM-based LAN with SONET fiber optics pipes running at 10 Gbps, but I digress), and running at data rates of 1 Mbps. Frequency Division Multiplexing (FDM) is intended to support simultaneous voice and data traffic; frequency ranges are intended to avoid interference from devices (e.g., refrigerators and air conditioners) found in the home. HomePNA solutions are based on an Ethernet derivative, running at 1 Mbps at frequencies above 2 MHz using a proprietary compression technique from Tut Systems, and using the CSMA/CD protocol native to Ethernet; speeds of 10 Mbps are planned into the future, with the theoretical potential being as much as 100 Mbps. Members include 2Com, AT&T, Compaq, Hewlett-Packard, IBM, Intel, Lucent and Tut Systems. www. homepna.org. See also Ethernet, FDM, ISDN, SONET and xDSL. www.homepna.org

homeostasis A physiological constancy or equilibrium maintained by self-regulating mechanisms. In short, the state of a system in which the input and output are exactly balanced, so there is no change.

homes passed An expressed of the number of dwellings that a CATV provider's distribution facilities pass by in a given cable service area and an expression of the market potential of the area.

homing 1. When you dial a long distance number, your central office will choose a special set of trunks to send your call onto the next switching center for movement through the nationwide toll system. Those trunks are said to be the homing trunks for your central office. In other words, your central office is said to home on these trunks. If you're consistently encountering lousy long distance lines (and so are others on your central office), then ask your telephone company to check these trunks out.

2. Returning to the starting position, as in a rotary stepping switch when its connection is released.

homo 1. The Greek prefix meaning the same.

2. Home Office Mobile Office. See also SOHO, which stands for Small Office Home Office.

homogeneous networks Composed of similar hardware from the same manufacturer.

homologation Conformity of a product or specification to international telephony connection standards. What this means in simple language is that you have submitted your product to a regulatory agency or a government testing agency in a foreign country and they have said that your product is OK for use and sale in that country and is allowed to be connected to the local phone system. In short, your product has now been homologated in that country.

honey do My friend Gerry Friesen had to get off the phone with me the other day because he hadn't completed his "Honey Do" list. His wife, Paula, gives me a list of chores to do and pressages each with, "Honey Do this, Honey Do that....." I include this definition in honor or Paula Friesen, a dear friend.

honey pot A honey pot is a decoy server attached to the Internet designed to attract hackers' attention. The honey pot gives the owner of a targeted server the chance to analyze the attack, develop a strategy to thwart the attack and to block access without causing any damage to the main server.

honeycomb coil A type of inductance in which the turns do not lie adjacent to each other.

honeymoon It was the accepted practice in Babylon 4,000 years ago that for a month after the wedding, the bride's father would supply his son-in-law with all the mead he could drink. Mead is a honey beer, and because their calendar was lunar based, this period was called the "honey month" or what we know today as the honeymoon.

hook-up wire A wire used for low current, low voltage (under 1000 volts) applications within enclosed electronic equipment.

hookemware Free software that contains a limited number of features designed to entice the user into purchasing the more comprehensive version. See also Hyperware, Meatware, Shovelware, Slideware and Vaporware.

hooker The term "hooker," meaning a prostitute, honors U.S. Army General Joseph Hooker, whose penchant for war was matched only by his predilection for paid female companionship. In New Orleans during the Civil War, Hooker spent so much time frolicking with ladies of the evening that the women came to be called "Hooker's Division." Eventually, these specialized "troops" became known simply as "hookers."

hookflash Momentarily depressing (up to eight tenths of a second) the hookswitch of a telephone instrument can initiate various services such as calling the attendant, conferencing calls, transferring calls or answering a call coming in on a line equipped with call waiting. In ISDN, a hookflash signals the System Adapter to perform an operation, such as placing a call on hold. To hookflash, simply depress and release the receiver button. By default, the Hayes ISDN System Adapter recognizes a hookflash when the receiver button is depressed less than 0.8 of a second. You can change the default. See Hookswitch.

hooking signal An on-hook signal of 0.1 to 0.2 second duration used to indicate that a subscriber intends to initiate a new process such as "add-on."

hookswitch Also called switchhook or switch hook. The place on your telephone instrument where you lay your handset. A hookswitch was originally an electrical "switch" connected to the "hook" on which the handset (or receiver) was placed when the telephone was not in use. The hookswitch is now the little plunger at the top of most telephones which is pushed down when the handset is resting in its cradle (on-hook). When the handset is raised, the plunger pops up and the phone goes off-hook. Momentarily depressing the hookswitch (up to 0.8 of a second) can signal various services such as calling the attendant, conferencing or transferring calls. See Hookflash and Hooking Signal.

hookswitch dialing You can make phone calls by depressing the hookswitch carefully. If you push it five times, you dial five. Push it ten times you dial 0. Some coin phones discourage hookswitch dialing. Some don't.

Hoot'n'Holler Hoot'n'Holler are special 24-hour a day phone circuits which stay open 24-hours a day, seven days a week. Anyone picking up a phone on the circuit can listen and talk to whoever's on the line, or whoever might be within earshot. The idea is that to get someone to speak to you, you "hoot and holler," i.e. make a noise. A hoot'n'holler circuit is also called a Junkyard Circuit, Holler Down, Shout Down, Open Speech Circuit, Squawk Box System, FP or Full Period (as in FP 123456 circuit #). Hoot'n'Holler is a circuit consisting of 4 wires (2 pair – a transmit pair and a receive pair). Technically, this is how it works: Audio energy present on the transmit pair at any location will appear on the receive pair of all the other locations, usually a multipoint circuit. The transmitted audio will not return on the receive pair of the originating location. Hoot'n'Holler circuits are voice conferencing oriented party lines and are non private by nature. There is no signaling on a hoot and holler circuit except when one "shouts down" to open speakers and "listens" at the distant out points. A Hoot'n'Holler circuit is a dedicated full time voice network. Individual four wire "drops" are connected via various bridging mechanisms. These "bridges" can be analog or digital and act as mix minus devices (mixes everyone else minus yourself). They provide all drops with connectivity with each other.

hop 1. One segment of the path between routers on a geographically dispersed network. A hop is one "leg" of a journey that includes intervening stops between the starting point and the destination. The distance between each of those stops (routers) is a communications hop.

2. A change of Radio Frequency (RF) channel used to carry the Cellular Digital Packet Data (CDPD) data for a channel stream.

hop by hop route A route that is created by having each switch along the path use its own routing knowledge to determine the next hop of the route, with the expectation that all switches will choose consistent hops such that the call will reach the desired destination. PNNI (Private Network-Network Interface) does not use hop-by-hop routing.

hop channel A Radio Frequency (RF) channel that has been declared a candidate for carrying a Cellular Digital Packet Data (CDPD) channel stream after a channel hop.

hop count The number of hops it will take for a packet to make it from a source to a destination. In short, the number of nodes (routers or other devices) between a source and a destination. In TCP/IP networks, hop count is recorded in a special field in the IP packet header and packets are discarded when the hop count reaches a specified maximum value.

hop off When you make a phone call on the Internet you can call from one phone attached to the Internet to another phone attached to the Internet or you can call from one phone attached to the Internet and, at the other end, go into a PC stuffed with voice and switching cards and which is attached to local phone lines. The process of leaving the Internet is called "Hop Off."

hop sequence The carefully coordinated sequence by which radio transmitters and receivers hop from on frequency to another, hop sequence is used in FHSS (Frequency Hopping Spread Spectrum) systems. FHSS is used extensively in Wireless LANs and certain PCS (Personal Communications Systems) cellular systems. See also FHSS.

hops Term describing the number of times a message traverses different nodes.

HOPS Hardwire Order Processing System.

Horizontal 1. H. In television signals, the horizontal line of video information which is controlled by a horizontal synch pulse.

2. Descriptive of the side of a North American wire Distributing Frame on which terminal blocks are mounted horizontally; this is the side equipment is terminated on, as opposed to the "vertical side," on which cables terminate in vertically mounted blocks. In some locations, frames of horizontal blocks only are called Horizontal Intermediate Distributing Frame, or HIDF. Compare to VIDF.

horizontal beamwidth See Azimuth Beamwidth.

horizontal blanking interval The period of time during which an electron gun shuts off to return from the right side of a monitor or TV screen to the left side in order to start painting a new line of video.

horizontal cable Defines the cable used to link the communications closet / room with individual end user devices. Horizontal cabling stays typically on one floor. That's why it's called horizontal.

horizontal cross-connect A cross-connect in the telecommunications closet or equipment room to the horizontal distribution cabling.

horizontal distribution frame Located on the floor of a building. Consists of the active, passive, and support components that provide the connection between inter-building cabling (i.e. cabling coming from outside the building) and the intra-building cabling for a building.

horizontal interval The sum of Horizontal Retrace.

horizontal link An ATM term. A link between two logical nodes that belong to the same peer group.

horizontal output The amplifier that amplifies the horizontal output sync signal in a TV or monitor. The output runs through a deflection yolk. This creates magnetic fields that control tracing of the CRT beam sideways. A vertical amplifier does the same for the up and down tracing of the CRT beam. A TV's horizontal output frequency is 15.73425 kHz. On some TVs you can hear the high-pitched sound of the horizontal output circuitry when you turn on the equipment.

horizontal pathways The portion of the wiring system extending from the workstation (telecommunications outlet) to the BHC (backbone to horizontal cross-connect) in the telecommunications closet. The outlet and cross-connect facilities in the telecommunications closet are considered part of the horizontal wiring.

horizontal retrace A video term. The return of the electron beam from the right to the left side of the raster after the scanning of one line.

horizontal resolution Detail expressed in pixels that provide chrominance and luminance information across a line of video information.

horizontal scan rate The frequency in Hz (hertz) at which the monitor is scanned in a horizontal direction; high horizontal scan rates produce higher resolution and less flicker. Thus, the EGA horizontal scan rate is 21.5Hz, while the VGA standard scan rate is 31.4Hz. Some displays now offer even higher scan rates, as much as 70 Khz. See Monitor.

horizontal sync A video term. Horizontal sync is the -40 IRE pulse occurring at the beginning of each line. This pulse signals the picture monitor to back to the left side of the screen and trace another horizontal line of picture information. See Interlace.

horizontal wiring The portion of the wiring system extending from the workstation's outlet to the BHC (Backbone to Horizontal Cross-Connect) in the telecommunications closet. The outlet and cross-connect facilities in the telecommunications closet are considered part of the horizontal wiring. See Horizontal Wiring Subsystem.

horizontal wiring subsystem The part of a premises distribution system installed on one floor that includes the cabling and distribution components connecting the riser subsystem and equipment wiring subsystem to the information outlet via cross connects, components of the administration subsystem.

horn In radio transmitting, a waveguide section of increasing cross-sectional area used to radiate directly in the desired direction or to feed into a reflector that forms the desired beam.

horn alert HA. A cellular car phone feature that automatically blows the car's horn if a call is coming in.

horse race At horse race tracks, the favorite wins fewer than 30% of the time.

horsepower A horespower typically measures the power of an internal cumbustion engine. A unit of power equivalent to 550 foot pounds per second or to 746 watts. James Watt coined the term after the part-time miner observed that the average horse raised 330 pounds of coal 100 feet in one minute. Thus 33,000 foot pounds became one horsepower.

Horton A software program which provides an automatic method for creating a directory of e-mail addresses. Users can look up electronic addresses via a search key which can be a fragment of a person's name.

hose And close A pattern of behavior exhibited by phone tech-support people who spout a bunch of jargon you don't understand, ask you to perform a bunch of procedures you don't follow, and then abruptly hang up. This definition from Wired Magazine.

hosed The system is hosed means that it no longer works. The expression comes from what happens to someone who has has received the full force of a fireman's water hose, as in a street demonstration.

host 1. An intelligent device attached to a network.

2. A mainframe computer.

3. A computer with full two-way access to other computers on the Internet. A host can use virtually any Internet tool, such as WAIS, Mosaic and Netscape.

host/remote Describes a scenario where the remote is a serving end office that does not have recording capabilities and, therefore, allows a host end office to record it.

host apparent address A set of internetwork layer addresses which a host will directly resolve to lower layer addresses.

host based firewall A hostbased firewall is a firewall system that includes a bastion host (a general-purpose computer running firewall software). A host based firewall usually includes a circuit-level gateway, an application level gateway, a hybrid of both gateways, or a stateful inspection firewall.

host bus adapter HBA. A printed circuit board that acts as an interface between the host microprocessor and the disk controller. The HBA relieves the host microprocessor of data storage and retrieval tasks, usually increasing the computer's performance time. A host bus adapter (or host adapter) and its disk subsystems make up a disk channel.

host carrier The cellular operating company a subscriber from another cellular system would be billed roamer charges.

host computer 1. A host computer is one that provides services and information to a series of other devices. Host computers were traditionally mainframes that provided large and complex services to a network of smaller terminals. For this reason, host computers are often referred to as large servers. For example, airline booking from individual terminals would link into a central host to process the various transactions. IBM is still the dominant supplier of host computers for transaction processing. On the Internet, the term ~host~ means any computer that has full two-way access to other computers on the Internet. A host has a specific local or host number that, together with the network number, forms its unique Internet protocol address.

2. In the context of the Internet. this is a computer that has access to other computers on the World Wide Web.

Host Digital Terminal See HDT.

host interactive voice response A voice response system that can communicate with a host computer, typically a mainframe. Applications which can be produced include bank-by-phone, reservations-by-phone, etc.

host name The name given to a mainframe computer.

host name resolution A mechanism that provides static and dynamic mechanisms for resolving host names into numeric addresses. The Internet Name Server Protocol accesses an Internet name server that provides dynamic name-to-number translation (this process is specified in IEN 116). The Domain Name Protocol accesses a Domain Name Server that provides dynamic name-to-number translation (this process is specified in RFC-1034 and RFC-1035). A static local host table can also be accessed for name-to-number translation.

host number The part of an internet address that designates which node on the (sub)network is being addressed. See Domain.

host processor Same as Host Computer.

host server A device which connects to a LAN and then allows a computer, which cannot directly support the LAN protocols, to connect to it, providing all necessary LAN support.

host site In the transfer of files, the host site is the location receiving a file. When two individuals are exchanging files, the one who receives the file first would be the host, the other would be considered the remote.

host switch A central office switching system which provides certain functions to a smaller switch located remotely.

host table 1. A list of TCP/IP hosts along with their IP addresses.

2. In Windows 95, the host table is HOSTS or LMHOST file that contains lists of known IP addresses mapped to host names or NetBIOS computer names.

3. An ASCII text file where each line is an entry consisting of one numeric address and one or more names associated with that address. Host tables are used to resolve hostnames into numeric addresses.

hosted See Hosting.

Hosted PBX Someone else owns my PBX and rents me space, time and telecommunications services on it. That PBX is typically away from my office or offices. It's joined to me by various types of phone lines, including T-1. Essentially hosted PBX is a fancy new name for what we used to call Centrex, when the phone company owned the PBX and rented me space, time and telephone services on it. See also Centrex and Hosting.

hosting I have a web site. It's called www.HarryNewton.com. For all my friends to be able to read it, my web site has to be connected full-time to the Internet. I can do that myself if I have a full-time high-speed connection to the Internet and if I have the computing equipment and if I the full-time people to make sure it's working 100% of the time and, and, and. But most of us can't afford that. A whole bunch of businesses have sprung up to run web sites for other people, like me. These businesses are called web site hosters, web hosters, or just hosters. And the service they provide is called hosting. They typically offer four types of hosting:

1. Shared Hosting Web sites hosted on shared servers. This is what I have. I'm one web site on a server which also hosts other peoples' web sites. Clearly, this is an economical solution for people like me who have simple or moderately accessed sites. Mine is only two pages and not many people visit it. There's not much to see.

2. Dedicated Hosting Let's say my web site was larger and accessed more frequently by more people. Then clearly I'd want a server dedicated to me. This way I could have more people coming to my site and they'd get faster service. With dedicated hosting, I get my own server dedicated to me. This solution provides greater server and network resources than shared hosting and allows for hardware configuration to optimize my site's performance.

3. Co-located Hosting Let's say that I don't want the computer my hoster provides. Let's say that I want to use my to use own special computer (for whatever reason). But I want my computer on the web hoster's site for two reasons: First, the hoster can connect my computer (i.e. my Web site server) to much faster telecom lines than I can. Most often, hosters site themselves at major Internet switching sites on T-3 rings. See MAE. Second, the hoster will have people on duty 24-hours a day, 365 days a year. So, if my server goes down, he can have someone look at it quickly – even in the middle of the night. People who like this service tend to have "sophisticated mission-critical applications," for example ecommerce.

4. Application Hosting In the three services above, the hoster has no responsibility for what the customer has actually put on his web site. Ecommerce or porn, he doesn't care. With application hosting, the web hoster cares. He takes your application and works with you, sort of a combination web hoster / consultant/ webmaster. He might help you get an ecommerce site up, or a corporate email system, or a Lotus Notes collaboration system.

See also hosting service provider.

hosting service provider A third-party data center that provides, operates and maintains infrastructure services (servers, connectivity, backup, security, performance monitoring, etc.) on an outsourced basis. See hosting.

hot Live. It's on and has power flowing. Live wire. A conductor carrying a signal is said to be a hot conductor, i.e. the wire carrying the signal or the ground as opposed to the neutral or ground wire.

hot and ground reversed In AC electrical power, the correct connection of the Hot and Ground wires is reversed. This is an extremely dangerous condition because the GROUND path will rise to 129 volts and can present a lethal shock hazard to anyone in contact with equipment powered from this outlet or any outlet using the same ground path. See Hot and Neutral Reversed.

hot and neutral reversed Also called reversed polarity. A symptom of poor AC electrical wiring. In this case the correct connection of the Hot and Neutral conductors is reversed. Dangers include increased leakage current, and damage to electronic equipment or motors and appliances requiring correct polarity. See Hot And Ground Reversed.

hot attach/detach See USB.

hot chat An Internet term. Sex talk, in real time, online, usually between two or more consenting people (through not necessarily adults).

hot cut The conversion from an old to a new phone system which occurs instantly as one is removed from the circuit and the other is brought in. There are advantages and disadvantages to Hot Cuts. For one, they're likely to be much more dangerous than a Parallel Cut, in which the two phone systems run side by side for a month or so. Also known as Flash Cut. See Cutover for a longer explanation.

hot desk An employee of a company no longer has a permanent office. He works out of his home, visits customers and communicates with the office through fax and electronic mail. Occasionally that person finds it necessary to visit an office of the company. He is allocated a desk and perhaps an office for his stay. That stay might be as short as an hour or as long as several weeks. Once he checks out, someone else gets the desk. This arrangement is called a "Hot Desk."

hot desking Open office spaces with easily movable furniture and partitions that support on-the-spot group meetings or quickly assembled individual work areas. The term hot desking also refers to the ability for an employee to log on to any workstation connected to the company network and access his desktop environment, and to log on to any phone connected to the company network and access his office phone functions.

hot docking Docking is to insert a portable computer into a base unit. Cold docking means the computer must begin from a power-off state and restart before docking. Hot docking means the computer can be docked while running at full power.

hot dog The archetype of American food. The phrase 'hot dog' originated in the nineteenth century. It was a running joke that sausages were often made not from beef or pork, as claimed by those who made them, but from whatever stray animals were at hand, especially stray dogs. Students at Yale University in the 1890s referred to sausages as dogs, and the lunch wagon where they were sold as a dog wagon. Served hot, in a bun, the sausages were thus hot dogs.

hot drop Live telephone service connected to another location. For example, you have service at your house. But someone across the road has your service. Maybe they tapped in illegally. Maybe the phone company installed the line incorrectly. Maybe they forgot. Who knows what? But the person who's got the hot drop (but shouldn't have it) will often use it to make illegal long distance and international phone calls, i.e. they'll commit toll fraud.

hot fix A feature of Novell's NetWare LAN (local area network) operating system in which a small portion of the hard disk's storage area is set aside as a "Hot Fix Redirection Area." This area is set up as a table to hold data that are "redirected" there from faulty blocks in the main storage area of the disk. It's a safety feature.

hot insertion An application feature that allows users to modify or update scripts "on the fly," without closing and restarting the application.

hot key combination A combination of keys on the keyboard that are pressed down simultaneously to make the computer perform a function. For example, the Ctrl, Alt, Del hot-key combination will warm boot an MS-DOS computer.

hot line A private line dedicated between two phones. When you pick up either phone or do some act of signaling (like push a button), the other phone rings instantly. Hot lines are useful in emergencies and other areas where time is of the essence – e.g. trading currencies. The telephone hot line between the White House and the Kremlin was established in 1984. Previously, a teletypewriter hot line was used. The problem with a hot line is that it's one circuit. Cut it and you're out of business. The White House/Kremlin circuit apparently used to pass through Finland, where he has been accidentally cut by farmers. I'm guessing that the circuit now has several backups. If one goes, they can switch to another.

    1.   a direct telecommunications link, as a telephone line or Teletype circuit, enabling immediate communication between heads of state in an international crisis: the hot line between Washington and Moscow. 2. a telephone service enabling people to talk confidentially with someone about a personal problem or crisis. 3.   a telephone line providing customers or clients with direct access to a company or professional service. Also, hotline.

hot line service When you pick up the phone, you're automatically connected with a phone number. Such Hot-Line Service on a PBX typically gets you emergency service, etc. See also Hot-Line.

hot links A methodology that references and can connect information from one document to another, regardless of the type of application used.

hot list A gopher or Web file that lets you quickly connect to your favorite pre-selected page. Appropriately named. The way it works: You connect to a home page. You decide you'd like to return at some other time. So you command your internet surfing software to mark this web site on your hot list (also called marking it with a "bookmark." Next time you want to return to that web site, you simply go to your hot list or your bookmarks, click on which one you want. And bingo, you're there. A hot list is also known as a bookmark. Most Web browsers have book marks or hot lists.

hot on neutral, hot unwired In AC electrical power, the HOT wire is connected to the NEUTRAL terminal of the outlet and the HOT terminal in UNWIRED. Dangers include shock hazard from excessive leakage current and fire hazard. Depending on other conditions, equipment may or may not operate.

hot plug When a system component (e.g., a computer disk drive) fails, it may be replaced without turning the system off. During this period, the system's activity is suspended, however. Also known as a Warm Swap, it is unlike a Hot Swap, during which the system remains active. See Hot Swap.

hot plugging The ability to add and remove devices to a computer while the computer is running and have the operating system automatically recognize the change. Two new external bus standards – Universal Serial Bus (USB) and FireWire support hot plugging. This is also a theoretical feature of PCMCIA cards.

hot potato routing In Hot Potato Routing, or deflection routing, the nodes of a network have no buffer to store packets in before they are moved on to their final predetermined destination. In normal routing situations, when multiple packets contend for a single outgoing channel, packets that are not buffered are dropped to avoid congestion. But in hot potato routing, each packet that is routed is constantly transferred until it reaches its final destination because the individual communication links cannot support more than one packet at a time. The packet is bounced around like a "hot potato," sometimes moving further away from its destination because it has to keep moving through the network.

hot racking A Navy term referring to the practice in submarines of having sailors sleep in the same bunk at different times. This occurs because of the shortage of bunks on submarines. Hot racking is the reason the US Navy gave in May of 1995 for vetoing the idea of having women serve on crowded submarines.

hot redundancy A term used in conjunction with very critical telecom and computing systems, such as 911 service. With Hot Redundancy, the component or the system runs in parallel with an identical "twin." Should one twin fail, the other is already running and provides full service without interruption.

hot restart Imagine a corporate telephone system, perhaps a PBX or an ACD. It's handling phone calls to and from customers every second of the day. It's mission critical. You can't allow it to crash for even a second a day. But phone systems are nothing more than specialized computers with specialized software. They will crash or lock up just like your PC, though perhaps not as often. But they will crash and lock up. What happens when your PC crashes and locks up? You will lose data. You will probably then reboot your PC, probably from its hard disk. This will cost you anywhere from 30 to 60 seconds. You can tolerate this on your PC. You can't tolerate this on your phone system. An integral part of most phone systems' software is a feature called "Hot Restart." When the phone system's software crashes or locks up, there is "Hot Restart" software that enables the phone system to restart itself (another word for reboot) without losing the phone calls in progress or without taking 30 to 60 seconds to load a new operating system from hard disk or tape. How this exactly is done seems to vary from one telephone system maker to another. Some may

keep an operating system in RAM. Some may keep two identical processors chugging away simultaneously and switch from one to another, when things go awry. How manufacturers do "Hot Restarts" is something of a trade secret.

hot sparing A technique whereby the disk subsystem, when a disk's recoverable error rates excede some predetermined threshold, begins to copy data from the failing disk to a spare one. If the copy operation succeeds, before the disk fails for good, the subsystem switched to the spare and marks the failing disk unusable.

hot spot A hot spot (also spelled hotspot) is a small geographic area of several thousand square feet in which you can get access to a 802.11b wireless local area network (also called Wi-Fi) which in turn is connected to the Internet and, thus, the world wide web. You'll need a PC or a laptop that has a 802.11b wireless card or built in wireless capability. Hot spots exist in city squares, airport lounges, libraries, coffee shops, boardrooms, businesses, etc. You may need to pay for access to the Internet via a hot spot. You may not. How do you know if you're in a hot spot? turn on your PC. See if it "sees" a network. Open your browser. See if you can get on the network. There are also small devices called hot spot or network wireless sniffers that will pick up on wireless networks and let you know what's around. There may actually be several. The classic hot spot in one running in Bryant Park, New York City. The merchants on the surrounds on the park jointly pay the costs of their wireless network and the line to the Internet. People sitting the park can surf the Internet, pick up their email and send presentations from their laptops. The merchants are hoping that this free service will lure more people into the park and thus more customers into their shop.

hot standby Backup equipment kept turned on and running in case some equipment fails. Also known as a Hot Spare. See Data Center.

hot swap The process of replacing a failed component – e.g. a RAID drive – while the rest of the system (in this case, the disks) continues working and continues to provide function normally, i.e. providing data to the network users and providing a place for them to store their data. See Hot Swappable and Raid.

hot swappable The ability of a component (such as a redundant power supply) to be added to or removed from a device (for instance, a repeater) without powering down the device, thus providing a maximum uptime.

hot wire The ability to connect power to peripherals. Technologies capable of this include PCMCIA, USB and IEEE 1394.

hotel See Carrier Hotel.

hotel/motel console A specialized PBX console or a normal console programmed to work specifically in hotels and motels. The console will often show room status information.

hoteling Corporations get rid of permanent offices and instead assign full-time workers to a new office or cubicle each day, depending on who's in the office. Workers also have permanent lockers where they can store their files. Workers typically carry their laptops, cell phones and pagers with them. Hoteling works with companies whose employees spend a great deal of time outside the office visiting clients, customers and suppliers and working from home. Hoteling's advantage is that it prevents a company from having to keep lots of empty offices when so many people work at home or on flexible schedules. See Telecommuting.

Hotfix A Novell program that dynamically marks defective blocks on the hard disk so they will not be used. See Hot Fix.

Hotjava Java is a programming language from Sun Microsystems designed primarily for writing software to leave on World Wide Web sites and downloadable over the Internet to a PC owned by you or me. HotJava, its brother, is another piece of software installed on a Web browser at your desktop. HotJava enables Java programs delivered over the Web to run on your desktop PC. In short, Java is the programming language the programs on the Web are written in. HotJava is the software that will sit on your PC. Java is basically a new virtual machine and interpretive dynamic language and environment. It abstracts the data on bytecodes so that when you develop an app, the same code runs on whatever operating system you choose to port the Java compiler/interpreter to. What's a Java application? According to Wired Magazine, point to Ford Motor Company's website today, for instance, and all you'll get are words and pictures of the latest cars and trucks. Using Java, however, Ford could relay a small application (called an applet) to a customer's computer (the one on your desk which are using the surf the Internet). The customer could then customize options on an F-series pickup while calculating the monthly tab on various loan rates offered by a finance company or local bank. Add animation to these applications and you could get to "drive" the truck.

hotline There are several definitions. In the cellular business, a hotline is a system restriction that allows a cellular customer to call only one prearranged number. If a cellular customer hasn't paid his bill to the point where his account has been sent to collections, the cellular operator may "hotline" the customer's phone so that if he tries to make a call, he will only be connected to the carrier's collections department. No other outgoing calls will be able to be made until the bill is paid.In the landline phone business, a hotline is often a dedicated line. But it may also be a phone which only dials one number. In this case it's often call a virtual private line. See Hotline Virtual Private Line Service. For a longer explanation, see Hot Line.

hotshot dialer A piece of equipment used to create a hotline or ring-down circuit. A hotline is when one phone rings another without dialing (think of the Batphone to Commissioner Gordon's office, or the hotline between Washington and Moscow). The dialer automatically dials the number when the handset is lifted.

hotspot 1. See hot spot.

2. An embedded hyperlink is a hyperlink that is in a line of text. A hotspot is the place in a document that contains an embedded hyperlink. A hotspot is a graphically defined area in an image that contains a hyperlink. An image with hotspots is called an image map. In browsers, hotspots are invisible. Users can tell that a hotspot is present by the changing appearance of the pointer.

house cable Communication cable within a building or a complex of buildings and owned by the local phone company. House cable comes from the terminal box in the basement or the nearby outside pedestal box and goes straight to the apartment or house. Often it's not terminated. Thus, a technician installing a phone line will often have to break into a house cable and search around for an unused cable pair. In a multi-story building, a house cable is called a riser cable. Thanks to John Arias of Bell Atlantic for help on this definition. House cable owned before divestiture by the Bell System and after divestiture by the Regional Bell Operating Companies will eventually be fully depreciated and will then belong to the customer. See also Binding Post, Block Cable, Block Pair, Feeder Box, House Box, Krone Block, Riser Cable and Terminal Box.

housekeeping 1. In data processing, an operation contributing nothing to the throughput or output, but is necessary to maintaining operation.

2. In communications, housekeeping is more commonly called "overhead."

3. When technical conferences are held, the lead speaker will often start the conference with "housekeeping" messages. That messages will touch on such important information as where the bathrooms are, when lunch will held, where it will be, where the phones are, what the expected dress for the evening's party is, etc.

4. Housekeeping is a for a communications satellite in orbit, all its support systems and the power they consume; i.e., everything on board the satellite except for the communications equipment fulfilling its primary purpose.

How are you today? "Mr. Newton, how are you today?" is the question that telemarketers most often open a sales-pitch when they make cold calls, i.e. calling people they've never talked to before. I have many responses, but my favorite is, "Thank you for asking. I've just returned from the bankruptcy court having declared personal bankruptcy as a result of my double amputation which happened when the family car got sideswiped by a truck, killing my wife and two children. Now what can I do for you?" If the telemarketer still tries to sell me something (which some actually do), I hang up and pray they don't call back.

howler A device which produces a loud sound to a subscriber's phone or private branch exchange (PBX) extension to indicate that the handset is off-hook and it ought to be put back on hook.

howler tone A tone which gets increasingly louder over a short period of time. It is used to notify a user that his phone handset is off its hook.

howling Howling is typically heard in a speakerphone or conferencing unit when there is "Acoustic coupling" between the microphone and the speaker. This is due to putting the microphone too near the speaker. New circuits called acoustic echo cancelers allow you to operate the microphone and the speaker simultaneously and much closer to each other.

HP Hewlett-Packard Company. HP was formed in 1939 by Bill Hewlett and David Packard. David lost the toss to Bill, which is why the company is called Hewlett-Packard, not Packard-Hewlett. They started the company in a small garage in Addison Avenue in Palo Alto, CA. Their first sale was an audio oscillator used in a Disney film, "Fantasia." See also MBWA.

HP Openview Hewlett-Packard's Openview network management products allow network administrators to monitor and control network devices from an MS-DOS PC or UNIX workstation.

HPA 1. High Performance Addressing. A passive-matrix LCD display technology that is faster and higher contrast than regular LCD displays. HPA monitors are cheaper to make, and popular with laptop makers looking to build low-cost, high-quality products.

2. See High Power Amplifier.

HPAD Host Packet Assembler/Disassembler. The HPAD can link to a host or FEP with native protocol data, or if the host can accept it, with X.25 input. The 4400 PAD functions as either an HPAD or a TPAD See TPAD.

HPC HPCs, or Handheld PCs, are a new category of mobile companion devices for Windows-based PCs based on Microsoft's Windows CE platform. HPCs have been designed to provide the millions of mobile professionals using personal computers running Windows an affordable, easy-to-use mobile PC companion to carry their most important information when they are away from their personal computer. HPCs can access the Internet to send and receive e-mail or browse the Web. The HPC is a mobile PC companion and is designed to complement, not replace, desktop, laptop, and notebook computers. Companion applications developed for the HPC are not intended to replace the functionality of personal computer applications. The HPC is not a PDA, but rather an affordable handheld PC companion. PDAs can generally be classified as stand-alone, keyboardless devices with proprietary user interfaces and applications that require pen-based entry and navigation. See PDA.

HPCC High-Performance Computing and Communications. U.S. government-funded program advocating advances in computing, communications, and related fields. The HPCC is designed to ensure U.S. leadership in these fields through education, research and development, industry collaboration, and implementation of high-performance technology.

HPFS High-performance file system (HPFS); primarily used with the OS/2 operating system version 1.2 or later. It supports long filenames but does not provide security. OS/2 can use any file system it wants, thanks to its installable file system (IFS) architecture. Two choices available are the FAT file system, used by MS-DOS, and the High Performance File System (HPFS). You can mix and match each and select one at boot time, thanks to OS/2's Dual Boot option. IBM, which created OS/2, claims HPFS is much more efficient than FAT. It tries to store all files on disk contiguously and uses its own built-in cache. However, HPFS' most notable attribute is the long 254-character file names and case preservation. OS/2 remembers file names as upper and lower case (though it's not case-sensitive to commands).

HPO High Performance Option. A way of improving equipment transmission characteristics. For instance, the upgrading of a voice-grade line to meet standards for data transmission.

HSDPA High Speed Data Packet Access. HSDPA is a 3G enhancement to UMTS (Universal Mobile Telecommunications Systems) that provides increased data download throughput. It boosts speed and reduces latency. It operates in the 5MHz spectrum and will initially provide real-world speeds of 400,000 bits per second to 600,000 bits per second, with theoretical peak speeds of 14.4 million bits per second.

HSIN See Homeland Security Information Network.

HPT Host Processing Time.

HRC 1. See Cable Normal Switch.

2. Harmonically Related Carrier. A method of establishing picture carriers on a cable plant such that all carriers are harmonics of a single fundamental frequency close to 6mhz.

HS 601 Satellite DIRECTV uses two Hughes HS 601 body-stabilized satellites in a geosynchronous orbit at 1018 West Longitude. Each satellite contains 16 120-watt transponders. DIRECTV operates 27 of the 32 licensed DBS frequencies at this central orbit location, while USSB broadcasts on the other 5. Employing circular polarization, the satellites provide coverage over the contiguous United States.

HSA High-Speed Access.

HSB Hot StandBy.

HSC Hierarchical Storage Controller.

HSCSD High Speed Circuit Switched Data. A wireless term. A pre-third-generation standard (2.5G) for adding faster data transmission to existing GSM networks by upgrading the network software. Data speed is 14,400 bits per second to 57,600 bits per second. HSCSD is circuit switched, not packet, so not a particularly efficient use of network resources.

HSCS High Speed Circuit Switched.

HSCSD High Speed Circuit Switched Data. HSCSD has relevance in the existing GSM cellular world, where data communications currently is limited to 9.6 Kbps in support of applications such as e-mail, fax, PC file transfer, and short message service. HSCSD will enable the transmission of data at speeds up to 57.6 Kbps through the concatenation (linking) of as many as four GSM time slots of 14.4 Kbps, each. As GSM provides a circuit-switched, rather than a packet-switched, connection, HSCSD is more suited for connection-oriented applications such as video and multimedia. E-mail and other bursty data communications applications are served more cost-effectively by packet data network protocols. See also GPRS, GSM, and UMTS.

According to Ericsson, "Today's data transfer rate of 9,600 bits per second (supporting fax, e-mail, voice/fax mail, PC file transfer and short message service) will be expanded to 19.2, 28.8, and even 64 kbit/s in the near future. The first step will be to introduce high-speed circuit-switched data (HSCSD) solutions which enable users to access two time slots instead of one – thus doubling the data capability. The second step will be to introduce bandwidth-on-demand (as a built-in capability of HSCSD). By dynamically allocating up to eight time slots for each single data call (64 kbit/s; the full PCS bandwidth), new services can be offered, such as high-speed multimedia access, videoconferencing and CD-quality sound. With the HSCSD high-speed data capacity, graphics-heavy World Wide Web pages can in principle be downloaded as easily and quickly as via a terrestrial connection. See www.ericsson.se/Review/ According to www.telecoms-mag.com, "GSM already meets many of the requirements for UMTS (Universal Mobile Telecommunications Systems), with the key exception of wideband radio access. However, two new service classes under development for GSM will expand the current user data rate of 9.6 kbps to 100 kbps and beyond: high-speed circuit switched data (HSCSD) and general packet radio service (GPRS). Both techniques are designed to integrate with current GSM infrastructure. HSCSD bearer services up to 64 kbps in GSM using multi-slot transmission have already been demonstrated. This technique bundles up to eight TDMA slots within the 200 kHz GSM carrier to create a higher bandwidth channel. HSCSD is also being developed to provide bandwidth-on-demand at variable data rates.

HSD 1. Home Satellite Dish.

2. High Speed Data. Usually refers to broadband Internet access through a cable modem. Speeds are about 100 times faster than a regular POTS modem.

3. High-Speed Synchronous Data.

HSDA High Speed Data Access.

HSDL High-speed Subscriber Data Line. A Bellcore idea for a two pair phone line coming into a house or business that is a full-duplex T-1 line. See also ADSL.

HSDPA in W-CDMA High Speed Downlink Packet Access (HSDPA) is a packet-based data service in W-CDMA downlink with data transmission up to 8-10 Mbps (and 20 Mbps for MIMO systems) over a 5MHz bandwidth in WCDMA downlink. HSDPA implementations includes Adaptive Modulation and Coding (AMC), Multiple-Input Multiple-Output (MIMO), Hybrid Automatic Request (HARQ), fast cell search, and advanced receiver design. In 3rd generation partnership project (3GPP) standards, Release 4 specifications provide efficient IP support enabling provision of services through an all-IP core network and Release 5 specifications focus on HSDPA to provide data rates up to approximately 10 Mbps to support packet-based multimedia services. MIMO systems are the work item in Release 6 specifications, which will support even higher data transmission rates up to 20 Mbps. HSDPA is evolved from and backward compatible with Release 99 WCDMA systems. Currently (2002) 3GPP is undertaking a feasibility study on high-speed downlink packet access. See also MIMO.

HSDU High Speed Data Unit.

HSL A computer imaging term. A color model based on hue, saturation, and luminance. Hue is the attribute that gives a color its name (e.g., red, blue, yellow, or green). In this model, saturation refers to the strength, or purity, of the color. If you mix watercolors, saturation would specify how much pigment you added to a given amount of water. Luminance identifies the brightness of a color. For example, full luminance yields white, while no luminance yields black. See also HSV.

HSLN High-Speed Local Network. A local network designed to provide high throughput between expensive, high-speed devices, such as mainframes and mass storage devices.

HSPD High-Speed Packet Data. A term for putting data onto a digital cell phone service, largely CDMA. The latest version of HSPD services transfer packet data at speeds of up to 64 kilobits per second (kbps). The latest version of HSPD includes voice enhancements, full compliance with IS-95B, improved handoff, call setup, roaming indication, incoming call forwarding, neighbor searching and pilot reporting. See the next definition.

HSPDA High-Speed Downlink Packet Access. HSDPA supports wireless data rates as high as 19-20 Mbps. HSPDA uses shared channel transmission, individual scheduling buffers, and separate priority queues, and it can take service characteristics into account. The last three features are crucial for handling VoIP traffic, since it is desirable to give high priority to VoIP traffic so as to avoid conversation-degrading latency.

HSR Harmonic Suppression Reactor. A specially designed inductor that is inserted in series between a utility's power factor correction capacitor bank and ground. This "detunes" or breaks up a power line resonant condition (usually at 540 Hz or the 9th harmonic) that is causing telephone line power influence and noise problems.

HSRP Hot Standby Routing Protocol, a proprietary routing protocol from Cisco for fault-tolerant IP routing. According to Cisco, "HSRP enables a set of routers to work together to present the appearance of a single virtual router or default gateway to the hosts on a LAN.

HSRP is particularly useful in environments where critical applications are running and fault-tolerant networks have been designed. By sharing an IP address and a MAC address two or more routers acting as one virtual router are able to seamlessly assume the routing responsibility in the case of a defined event or the unexpected failure. This enables hosts on a LAN to continue to forward IP packets to a consistent IP and MAC address enabling the changeover of devices doing the routing to be transparent to them and their sessions." HSRP performs a function very similar to that of both IP Standby Protocol, a proprietary protocol from DEC, and VRRP (Virtual Router Redundancy Protocol), an IETF standard. See also VRRP.

HSSI High Speed Serial Interface. A serial data communication interface optimized for high speeds up to 52 Mbps. Used for connecting an ATM switch to a T-3 DSU/CSU, for example.

HSSP High Speed Switched Port.

HST High Speed Technology, a U.S. Robotics proprietary signaling scheme, design and error control protocol for high-speed modems. HST incorporates trellis-coded modulation, for greater immunity from variable phone line conditions, and asymmetrical modulation for more efficient use of the phone channel at speeds of 4,800 bps and above. HST also incorporates MNP-compatible error control procedures adapted to asymmetrical modulation.

HSTR High Speed Token Ring. Proposals have been made (1997) by IBM to the IEEE 802.5 working group for high-speed versions of the Token Ring LAN standard. HSTR proposals specify operating speeds of 100, 128 and 155 Mbps. Current and traditional versions of Token Ring operate at 4 and 16 Mbps, putting Token Ring at a decided disadvantage in comparison to Fast Ethernet and ATM. IBM plans to move forward with product development while the standards process works its magic, i.e. takes its long slow time. See also ATM, Fast Ethernet and Token Ring.

HSV 1. A computer imaging term. A color model based on hue, saturation, and value. Hue specifies the color, as in the HSL model. In this model, saturation specifies the amount of black pigment added to or subtracted from the hue. Value identifies the addition or subtraction of white pigment from the hue.

2. Hosted Software Vendor. Another term for an ASP – Application Service Provider. See Application Service Provider.

HTCP HyperText Caching Protocol. A protocol used in the Internet for discovering HTTP caching proxies and cached data. HTCP includes HTTP headers, unlike ICP (Internet Caching Protocol), which are vital to caching proxies. See also Cache, HTTP, ICP, and Proxy.

HTG Hunt Group.

HTML HyperText Markup Language. This is the authoring software language used on the Internet's World Wide Web. HTML is used for creating World Wide Web pages. HTML is basically ASCII text surrounded by HTML commands in angle brackets, which your browser interprets whichever way it feels. That means (and I reiterate this) different browsers will display the exact same HTML code differently. Here is an example of a simple line of HTML code:

<\>H2>Call Center Magazine Editorial Calendar<\>/H2>

This line says to the browser: Display those words as a type 2 headline. You can also include an image in an HTML page.

<\>img src="photos/harry.gif">

Your browser would go find the picture "harry.gif" in the subdirectory called "photos" on the computer which had the URL and the Web page you were visiting. An HTML document has three types of content: tags (which define type styles, like the one above), comments (words to tell the HTML author what he's doing, but which aren't displayed) and text (i.e. the words Call Center Magazine Editorial Calendar. Tags also let you do hyperlinking, which lets a person browsing click your HTML document, click on that hyperlink and go elsewhere – either to another page which you wrote, or to another Web page (i.e. URL) across the world.

Here's an example of text and tags you might put in your HTML document:

<\>h2>For more information on computer telephony, please visit<\>/h2> <\>a href="http://www.computertelephony.com"><\>/a>

When someone browsing your page places their mouse on the words "www.computertelephony.com," their cursor changes to a hand. They click on it. A few seconds later, they see the computer telephony home page begin to download. See HTML 1.0, HTML 2.0, HTML 3.0 and HTML 3.2.

HTML 1.0 This original specification was drafted in 1990. It was designed primarily for publishing scientific papers to the Web. The spec contained features such as six levels of headings, simple character attributes, quotations, source code listings, list, and hyperlinks to other documents and images. It is no longer used. See HTML.

HTML 2.0 This revised specification for HTML arrived in 1994. It added forms and eliminated many seldom-used tags from the original spec. It also included support for pop-up and pull-down menus, buttons, and text-entry boxes that could be used for filling in forms. As of this writing (Winter of 1996-97), this is the specification that all products support in full. See HTML.

HTML 3.0 The proposal for this HTML specification was published in 1995. The spec called for the inclusion of coding for tables, text flow around figures, and mathematical equations. The spec was too progressive and garnered only piecemeal product support. It has since been dropped in favor of HTML 3.2. See HTML.

HTML 3.2 This new specification for HTML was developed by vendors that include IBM, Microsoft, Netscape, Novell, SoftQuad, Spyglass, and Sun Microsystems. It proposed the inclusion of support for tables, applets, text flow around images, superscripts, and subscripts. Both Microsoft and Netscape have added extensions to HTML that in some cases have been included in the subsequent standard. Netscape first introduced Java, tables, and frames, and Microsoft introduced ActiveX controls. See HTML.

HTML Editor Software program that allow one to easily convert text documents to HTML code, without special training or programming skills.

HTML Tag A symbol used in HTML to identify a page element's type, format, and structure. The FrontPage Editor automatically creates HTML tags to represent each element on the page.

HTR Hard-To-Reach.

HTTP HyperText Transfer Protocol. HTTP is the standard way of transferring information across the Internet and the World Wide Web. It supports a variety of media and file formats across a variety of platforms. Invisible to the user, HTTP is the actual protocol used by the Web Server and the Client Browser to communicate over the "wire". In short, the protocol used for moving documents around the Internet. http://www is the standard prefix for the Internet of a site on the world wide web. See Domain, DRP, Internet, Surf, URL, Web Address and Web Browser.

HTTPS Hypertext Transfer Protocol Secure. A type of server software which provides the ability for "secure" transactions to take place on the World Wide Web. If a Web site is running off a HTTPS server you can type in HTTPS instead of HTTP in the URL section of your browser to enter into the "secured mode". Windows NT HTTPS and Netscape Commerce server software support this protocol.

hub The point on a network where circuits are connected. In local area networks, a hub is the core of a physical star configuration, as in ARCNET, StarLAN, Ethernet, and Token Ring. Hub hardware can be either active or passive. Wiring hubs are useful for their centralized management capabilities and for their ability to isolate nodes from disruption. Hubs work at Layers 1 (Physical) and 2 (Data Link) of the OSI Reference Model, with emphasis on Layer 1. Hubs aren't switches, as they have very little intelligence, if any, and don't set up transmission paths. Rather, hubs comprise a physical bus and a bunch of ports, to which are connected a bunch of wires, to which are connected individual terminal devices. As hubs are protocol-specific (e.g., Ethernet) and are not intelligent, they are very fast and very cheap. 10Base-T and 100Base-T, for instance, are Ethernet hub technologies. The 10/100Base-T hub is an inexpensive means of allowing LAN-attached devices to share a common, collapsed bus contained within a hub chassis. The connections are via UTP (Unshielded Twisted Pair), which is much less expensive than are the classic connections through coaxial cable. Unlike switches, hubs do nothing internally to control congestion. However, they typically are workgroup-level solutions which allow a large, logical Ethernet to be subdivided into multiple physical segments. For example, you could even use a small five-port hub on your desk to connect a couple of laptops and a desktop PC. Hubs can be interconnected directly, or through switches or routers, with the traffic being forwarded from the originating hub only if the destination address of the data packet indicates that it is necessary to do so. Therefore, hubs do reduce congestion through the control of interhub traffic. See also 10Base-T, Ethernet, Router and Switch.

HUB expansion port An older local area network term for two ports located to the right of a 2008, 2016 or 2116 repeater used to interconnect these repeaters in a stack. The interconnect cable is standard Category 5 UTP cable. These ports are now called Repeater Expansion Ports (REP).

hub junction box A box used to connect a hub interface when a node is placed at a remote location.

hub management interface A network management protocol developed by Novell to allow network managers to manage hubs anywhere on a NetWare LAN.

hub polling A polling system in which a polled station sends its traffic and passes the polling message (after it's sent its message) to the next station.

hub site The location(s) on a network where many circuits are brought together to be multiplexed into a single higher speed connection.

Hubble, Edwin P. American astronomer, (1889 - 1953), whose observations proved that galaxies are "island universes", not nebulae inside our own galaxy. His greatest discovery, called "Hubble's Law", was the linear relationship between a galaxy's distance and the speed with which it is moving. The Hubble Space Telescope is named in his honor.

hublet A mini-hub, submitted by Tracy Meyer, Network Technician, Pacific Bell Mobile Services. The term was created by Mark Alexander an Engineer at Pacific Bell Mobile Services who came up with it. "It is growing in popularity here," according to Tracy.

huddle The "huddle" in football originated because of a deaf player who used sign language to communicate, and his team did not want the opposition to see the signals he used so they huddled around him.

huddle spaces Areas such as cybercafes, which are designed for informal meetings, change encounters and work breaks to foster idea exchange and communications.

hue The attribute by which a color may be identified within the visible spectrum. Hue refers to the spectral colors of red, orange, yellow, green, blue and violet.

Huffman Coding Developed by D.A. Huffman, a popular lossless data compression algorithm that creates shorter, variable-length codewords for input symbols, i.e., the letters that comprise a phrase. Huffman Coding replaces frequently occurring characters and data strings with shorter codes. For example, Huffman Coding examines the string of symbols that comprise a phrase. The symbols then are ranked according to the frequency of their occurrence, in a top-down approach. (Shannon-Fano Coding is similar, although it is a bottom-up approach that generally is considered to be less efficient.) The two symbols with the lowest probabilities are represented by one new, composite symbol with an assigned probability equal to the sum of the probabilities of the two symbols. This process continues successively up the probability rank, building a binary tree, with each node or branch being the probability of all nodes or branches beneath it. A path can be followed to an individual leaf, which is in the form of a compressed representation of an individual letter of an alphabet. Dynamic Huffman Coding reads the information twice, the first time to determine the frequency with which each data character appears in the text and the second time to accomplish the actual encoding process. Huffman Coding suffers from the fact that it does not recognize characters between 0 and 1; for instance, 4.6 would be rounded to either 4 or 5. Huffman encoding is often used in image compression. It also is used in PKZIP, along with other compression algorithms. Modified Huffman (MH) is used in many fax machines, although it supports relatively slow transmission at 9.6 Kbps, or about 30 seconds per page. See also Compression and Shannon-Fano Coding.

huge pipes From Wired's Jargon Watch column. A high-bandwidth Internet connection. "CU-SeeMe doesn't look half-bad – if you've got huge pipes."

hum 1. Hum on phone lines sounds awful and can severely cut your data throughput. Hum on a phone line may have many sources – grounded carbon, lightning-damaged protection, left in drops or jumpers, "half-tap," or a wet cable. A wet cable pair "usually" manifests itself with a "frying" sound or crosstalk with other pairs. The solution to hum on the line is to replace the pair or remove the offending section of the pair from your circuit.

2. Noise that is present in some communications equipment, hum usually is induced either by coupling to a 60-cycle electrical source or defective filtering of 120-cycle output of a rectifier. See Hum Bucket.

hum bucker A circuit (often a coil) that introduces a small amount of voltage at power line frequency into the video path to cancel unwanted AC hum.

hum modulation Noise which unintentionally modulates a desired signal.

human interface device A fancy name for a mouse or a keyboard, or something a human talks to a computer.

human resource forensics The idea is to check that your employees aren't doing anything that might harm your company. So now there's a growing band of specialists in a field called human-resource forensics. These guys are using the latest technology to record everything from the Web sites employees visit to the files they delete to the data they download. There are obvious privacy and morale issues here. Though some forms of surveillance are perfectly proper, using technology to spy on employees can damage workplace morale and, if taken to extremes, test the boundaries of what is ethical or legal. These people (also called "experts") urge companies to think hard before turning their workplaces into areas where no one feels trusted.

humidification The process of adding moisture to the air within a critical space. Without humidity, you get static electricity. With static electricity comes strong shocks to computers equipment. With strong shocks comes loss of data. Lack of humidity is particularly bad in the middle of winter.

humility If I only had a little humility, I'd be perfect. – Ted Turner, founder of TV superstations, founder of CNN and founder of a zillion other innovations in media and telecom.

humint Humint stands for human intelligence, as contrasted to sigint, which is signals intelligence.

hundred call seconds Known by the initials CCS where C is the roman numeral for Hundred. One CCS is 36 times the traffic expressed in Erlangs. See CCS.

hunt Refers to the progress of a call reaching a group of lines. The call will try the first line of the group. If that line is busy, it will try the second line, then it will hunt to the third, etc. See also Hunt Group.

hunt group A series of telephone lines organized in such a way that if the first line is busy the next line is hunted and so on until a free line is found. Often this arrangement is used on a group of incoming lines. Hunt groups may start with one trunk and hunt downwards. They may start randomly and hunt in clockwise circles. They may start randomly and hunt in counter-clockwise circles. Inter-Tel uses the terms "Linear, Distributed and Terminal" to refer to different types of hunt groups. In data communications, a hunt group is a set of links which provides a common resource and which is assigned a single hunt group designation. A user requesting that designation may then be connected to any member of the hunt group. Hunt group members may also receive calls by station address. See also Terminal Hunt Group.

hunt group helpers Most intelligent businesses run their faxes in hunt groups. Five fax machines in a hunt group can process as many calls as 45 faxes on individual lines. The problem with hunt groups is that when the first fax machine runs out of paper, it typically won't answer the phone. This means calls won't roll over. They just keep landing on the "sleeping" fax machine. A hunt-group helper, which is a piece of hardware installed before the fax machine – between it and the central office phone line – listens to the number of rings on each line of a hunt-group set and rolls the call over to the next line if a given machine lets the line ring too many times (i.e. it's probably out of paper). A simple fix, works great.

hunting See Rollover Lines.

HUT 1. High Usage Trunk.

2. Telecommunications equipment that house loop concentrators or multiplexer, a housing device.

HVAC Heating, Ventilating, and Air-Conditioning systems. High voltage stuff. Keep your telecommunications cables away from the motors in HVAC systems, please.

HVQ Hierarchical Vector Quantization – a method of video compression introduced by PictureTel in 1988 which reduced the bandwidth necessary to transmit acceptable color video picture quality to 112 Kbps.

HW Email abbreviation for hardware.

hybrid A device used for converting a conversation coming in on two pairs (one pair for each direction of the conversation) onto one pair and vice versa. This is necessary because all long distance circuits are two pairs, while most local circuits are one pair. Here is a longer explanation from "Signals, The Science of Telecommunications," by John Pierce and Mike Noll:

The telephone instrument in your home is connected to a single pair of wires called the subscriber loop or local loop, which carries both the outgoing voice signal and the incoming one. This pair of wires creates an electrical circuit for each of the two signals. A device in your phone called a hybrid or hybrid coil keeps the two signals separate, more or less, so that what you say into your phone's transmitter doesn't blast into your ear from the receiver. In contrast, all multiplex systems provide separate talking paths in two directions. Separate paths are necessary because the amplifiers placed along the lines between terminals amplify signals traveling in one direction only. When two people talk between New York and San Francisco the call goes from one phone through a local two-wire voice circuit to a multiplex terminal. There the call is transferred to a four-wire long distance circuit that consists of two separate one-way circuits. At the end of the system, a hybrid reconverts each four-wire circuit into a two-wire circuit. See also echo.

hybrid backbone Two or more types of facilities in a corporate telecom WAN – e.g. ATM and SONET.

hybrid cable A communication cable that contains two or more types of conductors that bear electrical signals, a mixture of signal-bearing electrical conductors and optical fibers, and/or two or more different types of optical fibers. A communication cable containing signal-bearing media and electric power conductors.

hybrid CDPD Circuit-switched CDPD (Cellular Digital Packet Data), a technology known as hybrid CDPD, is the system architecture developed by the CDPD Forum for interconnecting circuit-switched data, including cellular and land-line, with the CDPD network.

hybrid coil A transformer-like device which is designed to provide the interface between a two-wire and a four-wire circuit. The device has four ports designed such that a signal input to one port will be split evenly to two adjacent ports, with no signal coupled to the opposite port. One port is connected to the two-wire line. The two adjacent ports are connected to the four-wire line. The opposite port is connected to a balance network to

cancel any stray signals. See Hybrid for a longer explanation.

hybrid communication network A communication system that uses a combination of trunks, loops, or links, some of which are capable of transmitting (and receiving) only analog or quasi-analog signals and some of which are capable of transmitting (and receiving) only digital signals.

hybrid connector A connector containing both optical fiber and electrical conductors.

hybrid coupler In antenna work, a hybrid junction forming a directional coupler. The coupling factor is normally 3 dB.

hybrid disk 1. A CD-ROM term. Under the Orange Book standard for recordable CD, a hybrid disc is a recordable disc on which one or more sessions are already recorded, but the disc is not closed, leaving space open for future recording. However, in popular use the term "hybrid" often refers to a disc containing both DOS/Windows platform is seen as a ISO 9660 disc, while on a Mac it appears as an HFS disc.

2. A CD-ROM disc which works in both a Macintosh and an MS-DOS/Windows. Most these days do.

Hybrid Fiber/Coax (HFC) At its simplest, Hybrid Fiber/Coax describes any network architecture for cable or telephone networks which employs some combination of fiber and coaxial cable. But the term "Hybrid Fiber/Coax" also describes a specific network architecture employed by most cable operators and some telephone companies. This HFC architecture is an evolution of the traditional cable distribution network, which employed a bus from the cable operator's headend, down each street, with taps outside each subscriber's home. When the cable from the headend is fiber, and an optical-to-electrical conversion is performed to place the signal on coax before we reach the subscriber's tap, that network is described as a Hybrid Fiber/Coax bus.

hybrid firewall A network protection device that includes various firewalling features (e.g. packet filtering, circuit-level proxies, application-level proxies, etc.) in order to guard against multiple forms of attack.

hybrid integrated circuits The combination of thin-film or thick-film circuitry deposited on substrates with chip transistors, capacitors and other components. Thin-film construction is used for microwave integrated circuits (MICs).

hybrid junction A waveguide or transmission line arrangement having four ports that, when terminated in their characteristic impedance, have the property that energy entering any one port is transferred (usually equally) to two of the remaining three ports. Widely used as a mixing or dividing device.

hybrid key system Term used to describe a system which has attributes of both Key Telephone Systems and PBXs. The one distinguishing feature these days is that a hybrid key system can use normal single line phones in addition to the normal electronic key phones. A single line phone behind a hybrid works very much like a single line phone behind a PBX. The second distinguishing feature of a hybrid is that it's "non-squared." This means that not every trunk appears as a button on every phone in the system – as occurs on virtually every electronic key system manufactured today.

hybrid local network An integrated local network consisting of more than one type of local network (e.g. LAN, HSLN, digital PBX).

hybrid mode A mode possessing components of both electrical and magnetic field vectors in the direction of propagation.

hybrid network 1. A communications network which has some links capable of sending and receiving only analog signals and other links capable of handling only digital signals. The current public switched telephone network is Hybrid. A Hybrid Network is also a network with a combination of dissimilar network services, such as frame relay, private lines and/or X.25.

2. An amalgam of public and private network transmission facilities.

hybrid satellite A satellite that carries two or more different communications payloads (i.e., C-band and Ku-band).

hybrid set Two or more transformers interconnected to form a network having four pairs of accessible terminals to which may be connected four impedances so that the branches connecting them may be made interchangeable.

hydra A 25-pair cable that at one end has an Amphenol connector (typical of what 1A2 phone systems were connected with) and at the other has many individual 2, 4, 6 and 8 wire connectors, typically male RJ-11s. A hydra cable is named for a mythological multi-headed monster. It's more commonly called an octopus cable. The reason it's called an octopus is that it looks a bit like an octopus – one body and many arms.

hydrogen loss Increases in optical fiber attenuation that occur when hydrogen diffuses into the glass matrix and absorbs some light.

hydrometer Instrument for determining the density of liquids. Formerly in wide use for testing radio storage "A" batteries.

hygroscopic Capable of absorbing and retaining moisture from the air.

Hypalon DuPont's trade name for their chlorosulfonated polyethylene, an ozone-resistant synthetic rubber.

hyperband A cellular wireless term meaning the ability of a radio to handle calls originating in different frequencies. On Feb. 10, 1997, Ericsson conducted the first public hyperband hand-off between the 1900 MHz and 850 MHz systems at the Universal Wireless Communications (UWC) conference in Orlando. The hyperband call, which used the TDMA IS- 136 digital wireless technology, was connected and carried across the United States through AT&T Wireless Services' network in five cities. This demonstrates the ability for operators to establish networks in either band, and connect them in a multi-vendor environment, enabling nationwide and international roaming. Advanced digital wireless applications deploying public and private networks were demonstrated including: Location and Caller ID, Short Message, Message Waiting Indicator, Voice Mail, four digit extension dialing and intelligent roaming. These advanced capabilities were demonstrated using Ericsson's 1900/850 MHz dual-band phones. These phones are equipped with the new Enhanced Full Rate ACELP vocoder delivering enhanced digital voice quality.

hypercard The first desktop program that allowed hypertext creation. It ran on the Mac. Not long after its introduction, Hypercard "stacks" became available, especially in the artistic and educational communities. Stacks are a collection of documents within one package through which the user can jump using hypertext links. Stacks were often made available free through online services.

hyperchannel An SCSA term. A data path on the SCbus or SCxbus Data Bus made up of more than one time slot. By bundling time slots into a hyperchannel, data paths with a bandwidth greater than 64 Kbps can be created.

hyperdata networking Hyperdata networking is expanding the capabilities of the LAN from simple data transfer to the delivery of complex applications through the network to the end user. I think this definition is a "stretch," but Richard Herod claims it's for real.

hyperfiction According to the New York Times, hyperfiction is a new narrative art form, readable only on the computer, and made possible by the developing technology of hypertext and hypermedia. Not all adults have familiarized themselves with hypertext, but most children have, for it is the basis of many of their computer games and is fast becoming the dominant pedagogical tool of our digitalized times. See Hypertext.

hyperlink A link from one part of a page on the Internet to another page, either on the same site or a distant site. For example, a restaurant's home page may have a hyperlink or link to its menu. A retailer of laptops might have a link to the site of the laptop's manufacturer. A hyperlink is a way to connect two Internet resources via a simple word or phrase on which a user can click to start the connection. A user can access a Web site and exercise the option to hyperlink to another, related Web site by clicking on that option. You'll recognize hyperlinks on Web pages because the links look different. Typically they're in blue and underlined. Sometimes, you'll see a button saying "For more," " "Full specs," or "Our biography," etc. During the linking process, the user remains connected in a Web session through a process known as spoofing. A hyperlink is also called an anchor.

hypermedia A way of delivering information that provides multiple connected pathways through a body of information. Hypermedia allows the user to jump easily from one topic to related or supplementary material found in various forms, such as text, graphics, audio or video.

Another definition we found is: Non-linear media, of which multimedia can be a form. Just as hypertext is a non-sequential, random-access arrangement of text, hypermedia is a non-sequential, random-access arrangement of multiple media such as video, sound and computer data.

A third definition: Hypermedia is a type of authoring and playback software through which you can access multiple layers of multimedia information related to a specific topic. The information can be in the form of text, graphics, images, audio, or video. For example, suppose you received a hypermedia document about the Sun file system. You could click on a hotspot (such as the words file system) and then read a description. You could then click an icon to see an illustration of a file structure, and then click the file icon to see and hear information in a video explaining the file system.

hyperpartnering Also known as any-to-any (A2A) collaborative commerce. Designed to help companies jump on B2B market opportunities as they appear. For example, when a distributor reports a sudden surge in demand for a product, a manufacturer could almost simultaneously forge a partnership needed to double the supply of raw materials necessary to produce the product.

hyperspectral imaging Hyperspectral is a new form of imaging much in vogue in space weaponry and space technology. Hyperspectral cameras can take a picture of an ecosystem and discern conifer from deciduous trees. Hyperspectral imaging distinguishes subtle "light signatures" that separate a field of oats from barley and tell you the precise species of oats. And then, according to the New York Times, whether the field is infested with insects or damaged by nitrogen depletion. But the ultimate idea of hyperspectral imaging is to be able to tell if tanks are sitting under trees, or tanks are covered with camouflage or tanks pained with a paint meant to make them not look like tanks. In short, by mounting hyperspectral cameras on satellite, you are able to tell very precisely what's on the battlefield that might harm you.

hypertext The term "hypertext" was coined by Theodore (Ted) Holm Nelson. His middle name comes from his mother, actress Celeste Holm. He developed the concept of hypertext in a paper delivered to the Association for Computing Machinery at its national conference in 1965. Nelson envisioned a nonsequential writing tool which included a feature he called "zippered lists," which allowed textual elements in documents to be linked. We currently associate "hypertext" with the World Wide Web and the HTML language. Imagine you're reading something. You come to a word that's in a different color or perhaps is underlined. You click on the word with your mouse. Suddenly you're transported to another sentence, to another paragraph, to another section somewhere else. That new sentence, paragraph or section may explain the original word. It may take you to another thought. It may take you to another part of the story. The New York Times defines Hypertext as "nonsequential writing made up of text blocks that can be linked by the readers in multiple ways." Hypertext is not only the words and the links. It's also the software that allows users to explore and create their own paths through written, visual, and audio information. Capabilities include being able to jump from topic to topic at any time and follow cross-references easily. Hypertext is often used for Help files. It's being used for "hyperfiction," a new narrative art form, readable only on the computer, made possible by the developing technology of hypertext and hypermedia. See HTTP and Hypertext Transfer Protocol.

HyperText Caching Protocol See HTCP.

Hypertext Transfer Protocol HTTP is the transport protocol in transmitting hypertext documents around the Internet. See Hypertext.

hyperthreading Hyperthreading is technology from Intel which allows a single physical microprocessor to execute two separate code streams (called threads) concurrently. Architecturally, each 1A-32 processor with hyperthreading consists of two logical processors, each of which has its own 1A-32 architecture. After power up and initialization, each logical processor can be halted, interrupted or directed to execute a specified thread. See Thread.

hyperware New hardware that has been announced and perhaps even publicly demonstrated, but is not being shipped to commercial customers. Vaporware is software which has been announced, but is not yet shipping to commercial customers. Years can pass between public announcement and actual commercial shipment. Be wary.

hysteresis In pure physics, hysteresis is the lag in response exhibited by a body in reacting to changes in the forces, esp. magnetic forces, affecting it. In our industry it has come to acquire a couple of meanings. See also Hysteresis Loss.

1. Hysteresis is an uninterruptible power supply (UPS) definition. The voltage output from the wall will continually shift within a certain range, causing some UPSs to constantly switch back and forth from AC to battery power.

2. Hysteresis is also a buffering approach used in digital cellular networks to prevent the "ping pong" effect which occurs when a telephone repeatedly reselects two cell sites of approximately equal strength. Here's a explanation: The retardation of the accomplishment of a process caused by changing values, especially relating to the lagging of values of resulting magnetization in a magnetic material (e.g., iron) due to a changing magnetic force. From the Greek "hysterisis," translating as "shortcoming." In a physical system, hysterisis is dependent on history. For example, you can push on something, like a sponge, or you can bend something, like a piece of metal. The object doesn't spring back to its exact original form. It remembers its previous position. In other words, the shape of sponge and the piece of metal each depend on their history. In cellular radio terminology, hysterisis is a buffer area, or offset, that prevents a cell phone from bouncing back and forth between two cell sites with roughly equal levels of signal strength. Such a process of repeated reselection causes the "ping-pong" effect that can be so aggravating in digital cellular networks.

hysteresis loss A physics term. The loss of energy by conversion to heat in a system exhibiting hysteresis.

Hytelnet A hypertext system which contains information about the Internet, such as accessible library catalogs, Freenets, Gophers, bulletin boards, etc.

Hz See Hertz.

I Used on switches to mean "ON." The "OFF" setting is "O," (the letter coming after N.)

I&M Abbreviation for Installation and Maintenance.

I&R Installation and Repair. The telephone company department responsible for these jobs. I&R refers to a person's job area, a department, or tools and test gear made for I&R.

i-appli An i-mode service provided by a downloaded and installed application program written in the Java language for Japanese i-mode-capable mobile phones. The term also refers to any such application program. See i-mode, osaifu-keitai.

I-CF ISDN Call Forwarding.

I-CFDA ISDN Call Forwarding Don't Answer.

I-CFDAIO ISDN Call Forwarding Don't Answer Incoming Only.

I-CFIB ISDN Call Forwarding Interface Busy.

I-CFIBIO ISDN Call Forwarding Interface Busy Incoming Only.

I-CFIG ISDN Call Forwarding IntraGroup only.

I-CFIO ISDN Call Forwarding Incoming Only.

I-CFPF ISDN Call Forwarding over Private Facilities.

I-CFV ISDN Call Forwarding Variable.

I-CFVCG ISDN Call Forwarding Variable facilities for Customer Groups.

I-CNIS ISDN Calling Number Information Services.

I-Commerce Internet Commerce. Same as e-commerce. Basically it refers to people buying stuff via the Internet or the World Wide Web.

I-EDI Internet-based Electronic Data Interchange.

I-ETS Interim European Telecommunications Standard.

I-HC ISDN Hold Capability.

I-Hold Indication A telephone system feature. If I put someone on hold at my phone, all the other phones which have the same line appearing on them will start flashing – indicating that the call is on hold.

I-MAC Isochronous Media Access Control. An FDDI-II term. See FDDI-II.

i-Mode i-Mode (meaning Internet-Mode) is a proprietary cell phone service from NTT DoCoMo in Japan that was launched on February 22, 1999 and lets its users access over 40,000 information and Internet services from their cell phones. Services include mobile banking, email, news, stock updates, telephone directory, downloadable ring tones, restaurant guides, ticket reservations and a wide variety of entertainment offerings, including games and downloadable cartoons. i-Mode services are linked directly to the DoCoMo i-Mode portal Web site and can be accessed by pushing the cell phone's dedicated i-Mode portal Web site and can be accessed by pushing the cell phone's dedicated i-Mode button. Several things contribute to i-Mode's incredible success (its users number in the tens of millions) – the breadth of the offerings, the fact that users are only charged for the amount of information they retrieve, and not for how long they are online, and, most importantly, an enlightened approach by NTT DoCoMo to revenue sharing and collection. i-Mode is based on packet data transmission technology. NTT DoCoMo's i-mode network structure not only provides access to i-mode and i-mode-compatible content through phones, through the Internet, and access through dedicated leased-line circuit for added security. i-Modem web sites are written in CWML (Compact Wireless Markup Language), a stripped-down version of HTML (HyperText Markup Language) that is similar to WML (Wireless Markup Language) used in WAP (Wireless Access Protocol). Transmission between the i-Mode cell site and the cell phone or other device is via packet mode, with the packets being 128 octets (i.e., bytes) in length. See also WAP and WML. In America, AT&T Wireless, in which NTT DoCoMo owns shares, introduced a i-Mode knockoff, which it calls mMode (without the hyphen). www.NTTDoCoMo.com.

I-MUX Inverse Multiplexer. See Inverse Multiplexer.

I-Order Installation Order.

I-PASS An alliance of ISPs (Internet Service Providers) and IAPs (Internet Access Providers) to provide roaming capabilities for travelers. Based on proprietary standards, roamers are authenticated before being afforded Internet access. Usage is cross-billed through the I-PASS clearinghouse, with fees being set by each ISP for use of its facilities by roamers. I-PASS includes over 100 member ISPs in approximately 150 countries, and includes over 1,000 POPs (Points of Presence). I-PASS competes with GRIC (Global Reach Internet Connection). The IETF's Roamops working group is developing a standard for roaming, as well. See also GRIC and Roamops.

I-Series Recommendations ITU-T recommendations on standards for ISDN services, ISDN networks, user-network interfaces, and internetwork and maintenance principals.

I-TV Interactive TV.

I-Use Shows a user which line the phone is connected to when the receiver is off-hook. It does this by illuminating a small light below that line button. Most key sets and PBX sets have I-Use buttons. Most of the newer two-line phones do also.

I-Way An acronym for the Information SuperHighway, that nebulous concept which refers to interconnected telecommunications channels snaking their way into every household, every company, every college, every university in the world. Essentially, the Information SuperHighway is a fancy term for the nation's phone network overlaid with heavy data communications ability.

I.122 ITU-T description of the general "bearer" services offered by ISDN networks, including both packet-switched and frame relay data services.

I.356 ITU-T Specifications for Traffic Measurement.

I.361 B-ISDN ATM Layer Specification.

I.362 B-ISDN ATM Layer (AAL) Functional Description.

I.363 B-ISDN ATM Layer (AAL) Specification.

I.430 Basic rate physical layer interface defined for ISDN. The ITU-T Layer 1 specification for the ISDN BRI S/T-interface, which consists of four wires. I.430 specifies ASI line coding. The beginning and end of each frame is marked with deliberate bi-polar violations. Each BRI frame is forty-eight bits in length including the bi-polar violations; and repeated 4,000 times per second for a total line rate of 192 Kb/s in each direction. The point-to-point limit is one kilometer. the passive bus is limited to about 10 meters.

I.431 Primary rate physical layer interface defined for ISDN. The ITU-T Recommendation for Layer 1 of the ISDN PRI. Specifies operation on North American T-1 at 1.544 megabits per second (23B+D) or European E-1 at 2.048 megabits per second (30B+D).

Primary Rate Access, which provides thirty 64 kbit/s traffic-bearing lines, plus 64 kbit/s lines for call-management and signalling information.

I.432 ITU-T Recommendation for B-ISDN User-network Interface.

I.440 The ITU-T specification, commonly known as Q.920, which describes the general network aspects of the LAPD protocol (also known as DSS1).

I.450 (Q.930) The ITU-T specification describing the general network aspects of the ISDN D channel Layer 3 protocol.

I.451 See Q.931.

I.452 See Q.932.

i.Link See FireWire.

I.R.D.A. Infrared Data Association. See Infrared.

I/G Bit Bit in IEEE 802 MAC address field distinguishing between individual and group addresses.

I/O Input/Output. See the following definitions and, most importantly, InfiniBand.

I/O Bound When a computer systems spend much of its time waiting for peripherals like the hard disk or video display, it is said to be I/O bound. If your computer is I/O Bound, going to a faster CPU (like a 386 or 486) might make little perceived difference. What you need is a faster hard disk or faster video card, etc.

I/O Channel Equipment forming part of the input/output system of a computer.

I/O Controller Provides communications between the central processor and the I/O devices.

I/O Device An input/output device, which is a piece of hardware used for providing information to and receiving information from the computer, for example, a disk drive, which transfers information in one of two directions, depending on the situation. Some input devices, such as keyboards, can be used only for input; some output devices (such as a printer or a monitor) can be used for output. Most of these devices require installation of device drivers.

I/O Request Packet IRP. Data structures that drivers use to communicate with each other.

I2 See Internet2.

I2C Inter-Integrated Circuit, with I2C actually meaning "I^2C" as in "I Squared C." A bidirectional, two-wire, serial bus specification developed in the early 1980's by Philips Semiconductors to provide a communications link between integrated circuits for audio and video (e.g., TV sets and VCRs) equipment. I2C now extends to a wide variety of computer peripherals such as keyboards, mice, printers, and monitors. I2C is a multi-master bus, meaning that multiple masters can initiate data transfers over the shared bus, with a arbitration mechanism determining which master has priority at any given time.

i386SL One version of Intel's '386 family of microprocessors. The i386SL's special feature is that it can be slowed to 0 megahertz and still maintain register integrity (memory) practically indefinitely. This results in significant power savings for computers (especially laptops) that advantage of this feature.

i750 Name of the programmable video processor family from Intel.

IA Internet Appliance. See Internet Appliance.

IA5 International Alphabet No. 5.

IAB Internet Architecture Board, formed in 1981 by the Defense Department's Advanced Research Projects Agency (DARPA). A policy setting and decision-review board for the TCP/IP-based Internet. The IAB supervises the Internet Engineering Task Force (IETF) and Internet Research Task Force (IRTF), and serves as the technology advisory group to the Internet Society (ISOC). The IAB included researchers such as Vint Cerf and Robert Kahn, who had created the TCP/IP protocol that became the universal language of the Internet. See also DARPA, Internet Architecture Board, IETF, IRTF, ISOC, and TCP/IP.

IAD 1. Integrated Access Device. A device which supports voice, data and video information streams over a single, high-capacity (i.e. broadband) circuit. A piece of equipment used to combine multiple services (ISDN, frame relay, Internet Protocol, DSL, ATM, T-1/E-1, FT-1/E-1). By combining services, a single line can replace multiple access lines. IADs allow service providers and end-users to scale their services as they need. According to literature on them, broadband integrated access devices enable you to connect to the Internet and

access multiple phone lines over a single wire. Small and medium-sized businesses already use IADs on T-1 and SDSL lines. Homes will use IADs with ADSL to access the Internet and provide advanced phone service from competing providers without busy signals, and without running more wires. Some companies are actually making "Integrated Software on Silicon" which combines a DSL modem, a digital signal processor for voice telephone, and microprocessors for network protocols and higher-level software into a single chip, along with the network protocols and routing software that runs on those chips.

2. Internet Addiction Disorder. I found this on the Internet (where else?): "A growing number of men are losing friends, family, and jobs, and sometimes all touch with reality through an addiction to the Internet, an Italian psychiatrist warned recently. Professor Tonino Cantelmi, University of Rome, told a conference he has studied 24 cases of certifiable "Internet Addiction Disorder" (IAD), a condition with symptoms of spending up to 10 hours online and a physical fallout of uncontrollable shaking hands and memory loss." Only 10 hours?

IADC Inter-Agency Space Debris Coordination Committee, the international panel of experts on space debris monitoring, mitigation, protection and risk assessment, principally concerned with preserving the commercially valuable space around the geostationary orbit to prevent its being degraded to a ring of orbiting garbage useless for satellite communications. IADC was founded in the late 1980s by ESA; NASA, ROSA VIAKOSMOS (Russian Aviation and Space Agency) and Japan, and subsequently was joined by ASI of Italy, BNSC of the UK, CNES of France, CNSA of China, DLR of Germany, ISRO of India and NSAU of the Ukraine. See also graveyard orbit.

IAHC See International Ad Hoc Committee.

IAL 1. Intel Architecture Labs, home of the ISA Bus, Plug and Play, Universal Serial Bus and other PC "advances" designed to sell more Intel products.

2. See Immediate Action Limit.

IAM Initial Address Message. In SS7 networks, a message sent in the forward direction as part of the ISUP (ISDN User Part) call set-up protocol. The IAM is a mandatory message which initiates seizure of an outgoing circuit and which transmits address and other information relating to the routing and handling of a call. Included in the IAM is Calling Number Identification (CNI), also known as Calling Line Identification (CLI). CNI is the telephone number of the calling party, which is sent to the called party for identification purposes. Many carriers also support Caller Name, which transmits the name of the calling party along with the originating telephone number. It's interesting that part of the IAM is my identification, as in "I am. See also CNI, Common Channel Signaling, ISDN, ISUP, and SS7.

IANA Internet Assigned Numbers Authority. IANA is responsible for assignment of unique Internet parameters (e.g., TCP port numbers, and ARP hardware types), and managing domain names. It also was responsible for administration and assignment of IP (Internet Protocol) numbers within the geographic areas of North America, the Caribbean and sub-Saharan Africa; on December 22, 1997, that responsibility was shifted to ARIN (American Registry for Internet Numbers). For full details see Internet Assigned Numbers Authority. See also ARIN.

IAO IntrAOffice SONET Signal. Standard SONET signal used within an Operating Company central office, remote site, or similar location.

IAP Internet Access Provider. An IAP provides companies and individuals with a way to get onto the Internet and thus the World Wide Web, or the Web for short. You typically access that IAP by having your computer dial them on a phone number (local or long distance) or access them on a dedicated, full-time connection, such as a DSL or a T-1 line. IAPs come in two flavors. There are the OLS – the online services such as American Online, Prodigy and the Microsoft Network, which offer access to the Internet, various services (such as instant messaging) and their own proprietary content – such as stock quotes, personal financial tools, advice for expectant mothers, etc. And there are the Internet Service Providers which offer few services apart from access to the Internet and email. Typically they offer very little in the way of content.

IARU See International Amateur Radio Union.

IASG Internetwork Address Sub-Group: A range of internetwork layer addresses summarized in an internetwork layer routing protocol.

IAX Inter-Asterisk Exchange – protocol for two Asterisk PBXes to talk with each other. IAX has been replaced with IAX2. See Asterisk.

IBC Initial Billing Company. The company that bills the IXC in a meet point billing arrangement. Opposite of SBC.

IBDN The Integrated Building Distribution Network (IBDN) is an unshielded twisted pair/fiber optic based structured wiring system based on the EIATIA 568 wiring standard. IBDN is a creation of Northern Telecom. IBDN is an open wiring system meaning that it can support any standards based data or voice application available today on unshielded twisted pair horizontal wiring.

IBERPAK A ITU-T X.25 packet-switched network operated in Spain by the Spanish government.

IBL Initial Binary Load.

IBM International Business Machines. Also known affectionately as I've Been Moved, International Big Mother, Itty Bitty Machines, It's Better Manually, along with others not suitable for printing (but used in the back of better computer rooms all across the North American continent).

IBM 8514/A Graphics standard introduced by IBM with 1,024 x 768 resolution. Many current monitors are 8514/A-compatible. 8514/A is also called XGA, or eXtended Graphics Array, which is IBM's high-resolution extension to its VGA adapter. It provides a resolution of 1,024 horizontally x 768 vertically, yielding 786,432 possible bits of information on one screen, more than two and a half times what is possible with VGA.

IBM and the BUNCH A term that was used in the 1960s to refer to the major players in the computer industry, namely, IBM and its smaller competitors: Burroughs, Univac, NCR, Control Data Corporation, and Honeywell. See IBM and the Seven Dwarves.

IBM and the Seven Dwarves Similar to IBM and the BUNCH, with the addition of General Electric and Xerox.

IBM Cabling System IBM's specification for the kind of cable to be used in connection its products.

IBM PC The IBM PC was first introduced in the summer of 1981. It came with a 16-bit Intel 8088 processor and no hard drive.

IBM token ring A local area network using star wiring architecture of two pair cabling to each location – one pair from the hub to the workstation and one pair from the workstation back to the hub to continue the ring. The IBM 8228 Multiple Access Unit (MAU) will support communications for eight PCs (workstations). Up to 33 IBM 8228 MAUs may be connected together into a single ring, supporting up to 260 data devices. MAU to MAU connection is accomplished with data connectors equipped with Type 1 cables from a MAU's RO (Ring Out) connection to the next MAU's RI (Ring In). The final MAU's RO connects back to the initial MAU's RI to complete the ring. See also Token Passing and Token Ring.

IBMR Internet-based market research (IBMR) is allegedly faster than traditional telephone and mall based surveys, and more cost effective. It also enables better survey design and improved targeting.

IBND Interim Billed Number Database.

IBOC In-Band On-Channel is the terrestrial digital radio format from a company called iBiquity Digital. This radio format has a novel technology for replacing traditional AM and FM radio broadcasting with more robust digital signals. The Federal Communications Commission (FCC) has officially endorsed IBOC. It has agreed to let local radio operators begin broadcasting in the new IBOC format. As the Economist wrote, "It seemed only a matter of time before car owners, equipped with an IBOC radio, would be free to enjoy stunning CD-quality local programming as they cruised America's highways and byways. Meanwhile, radio broadcasters across the country would offer ancillary data (such as promotions for movie and concert tickets, or personalised share-price tickers) by piggybacking the information on the IBOC signal. iBiquity had even started to persuade radio stations to spend tens of thousands of dollars to upgrade their broadcasting equipment for the new digital signals. Then along came Motorola, with a digital-radio chipset called Symphony, to spoil the fun. Both devices turn crackly analog broadcasting into pristine digital radio. But the similarities end there. The iBiquity scheme involves a tuner built to receive a digital bitstream from a broadcaster transmitting a signal using the IBOC format. In contrast, the Symphony chipset takes an ordinary analog AM or FM signal and pumps it through its powerful digital processor to enhance the sound and boost reception significantly. Symphony costs the broadcaster nothing, and consumers next to - and the average listener can barely hear the difference between the two digital forms of AM or FM radio reception. Symphony has some big advantages. To create CD-quality music, it uses not only hardware (its 24-bit signal processor is similar to those found in home-theatre equipment with fancy surround-sound features) but also a software engine. The built-in software, which lets users upgrade or customise the radio with third-party applications, can generate noise-cancelling signals to eliminate engine hum and other stray sounds. It can also pick up neighbouring stations more accurately than conventional tuners and thus avoid interference. The 'spectrum buffer' required by the FCC to stop adjacent stations interfering with each other could be cut in half, says John Hansen of Motorola. That could double the number of possible stations in the AM/FM bands. To confuse matters, or to enhance our radio listening pleasures, there are now two companies broadcasting radio from satellite. One is called XM Radio and the other is called Sirius."

IBR Bellcore spec 54019, which covers specs on delivering fractional T-1.

IBS 1. Intelligent Battery System. A conventional battery system interfaces to its host product and charger through a power and perhaps a simple sensor port. An intelligent battery system has state sequential intelligence, memory, and a data communications protocol to the conventional battery sensor package. Thee IBS additions allow sensor data, events and memory access to take place between the battery pack and the host device and charger.

2. International Business Systems.

3. INTELSAT Business Service.

IBSU ISBU is a two-wire digital UNE capable of ISDN BRI, i.e. basic rate at 144kbps. IBSU is the circuit identifier for a 2-wire designed ISDN Capable Loop. It typically looks like 09.IBSU.150634, where 09 is the state and 150634 is simply the circuit's number, i.e. the next available incremented number. Verizon and others use this. See ISDN.

iBTS Internet Base Transmitter Site. Think of this three-foot cubic box as a remote central office for cell phone sites. Think of it as the electronics which the cell phone tower in your backyard needs to receive and send signals to and from the air and to and from whatever landlines it's connected to – which typically include E-1 and T-1 lines. Such an iBTS can be screwed onto a pole, attached to a wall in the corner of a basement, up on the roof, etc. It's typically a digital device and is designed to handle relatively high speed connections from mobile phones to the Internet. See UMTS.

IBX Integrated Business eXchange. Another name for a PBX. This is also the name InteCom uses for their PBX family. InteCom is now owned by Matra, a French company.

IC 1. Intercom, as in speaking between two phones inside a business.

2. Integrated Circuit.

3. Intermediate Cross-connect. An interconnect point within backbone wiring. for example, the interconnection between the main cross-connect and telecommunications closet or between the building entrance facility and the main cross-connect.

4. Interexchange Carrier. Also (and more commonly) called IXC.

IC DRAM Integrated Circuit Dynamic Random Access Memory.

IC Transit Connection A Verizon definition. The connection between a Verizon access tandem and another service provider's switching entity/Point of Interface for the purpose of carrying the provider's traffic to and from an InterExchange Carrier.

ICA International Communications Association. ICA used to be the biggest trade association of the largest corporate telecommunications users – the people whose companies spent the most. ICA was founded in 1949 as the National Committee of Communications Supervisors. On behalf of its members, the ICA worked to influence the FCC, Congress, and other regulatory and law-making bodies on issues of national telecom and information distribution issues. Their former web site, www.icanet.com, is now available for sale at $1,688.

iCal See iCalendar.

iCalendar iCal. Internet Calendaring and Scheduling Core Object Specification is a specification from the Internet Engineering Task Force designed to allow people to share and coordinate their appointment calendar over the Internet. At the heart of iCalendar is the Time Zone Calendar Component, a protocol that accounts for different time zones. As a result, users of a typical day planner can schedule appointments with users in other time zones and with other calendaring programs. The iCalendar spec is the foundation for four new specifications. iTIP is the iCalendar Transport Independent Interoperability Protocol, which details how calendaring systems use iCalendar objects to interoperate and defines a message protocol for finding free time or searching to-do lists. iMIP is the iCalendar Message Based Interoperability Protocol which addresses defines interoperability among calendaring systems piggybacked on Internet email using MIME (Multipurpose Internet Mail Extension). iRIP is the iCalendar Real Time Interoperability Protocol that addresses the how diverse scheduling systems query each other in real time. Client Access Protocol (CAP) allows any calendaring client to access information from heterogeneous back-end systems. iCalendar is replacing an older specification called vCalendar. See vCalendar. www.ietf.org.

ICANN The Internet Corporation for Assigned Names and Numbers (www.icann.org) is a non-profit, international organization established by the United States government in 1998 to oversee various technical coordination issues for the Internnet globally. ICANN has broad authority to reform and administer the present system of issuing Internet addresses, including adding new Top Level Domains, like biz, as in www.mycompany.biz. Officially, ICANN coordinates the assignment of globally unique identifiers that allow the Internet to function. Specifically, those identifiers include Internet domain names, Internet Protocol (IP) address numbers, and protocol parameters and port numbers. ICANN also coordinates the stable operation of the Internet's root server system. Previously and by an accident of history (in that the ARPANet was the precursor to today's Internet) a U.S. company held what amounted to a sanctioned monopoly on the business of assigning these names. Network Solutions of Herndon, Va., maintained the master databases that map Internet names to their real, numeric IP addresses. (For example, www.informationweek.com is mapped to the real address of 192.215.17.45.) To correct that situation, the Clinton Administration authorized a nonprofit body "to take over responsibility for the IP address space allocation, protocol parameter assignment, domain name system management, and root server system management functions now performed under U.S. government contract." The idea was to open up registration to multiple competing companies; to get the Internet, as a new kind of industry, to start maintaining and policing itself; and to provide some redundancy that's absent in today's all-the-domain-names-in-one-company's-basket approach. The nonprofit body set up to do this was the Internet Corporation for Assigned Names and Numbers (ICANN). About 100 registrars have been certified, thereby ending the monopoly. Coordination domain name assignment is through the Shared Registry System (SRS), which ensures that duplicate domain names are not assigned. In the event of a dispute over a domain name, the matter is referred to the CPR Institute for Dispute Resolution under the Uniform Dispute Resolution Policy (UDRP). See also Internet and TLD.

ICAO International Civil Aviation Organization, the international aviation regulatory body that also publishes standards relevant to installations on board aircraft and at or near airports. www.icao.org.

ICAP Internet Content Adaptation Protocol. Point-to-point protocol between cache servers and network-based applications.

ICAPI International Call Control API.

ICB Individual Case Basis. A service arrangement in which the regulations, rates and charges are developed based on the specific circumstances of the customer's situation. General tariffs which apply to the vast unwashed customer base do not apply in this case. The price is negotiated between the company (i.e. the phone company) and the customer.

ICC See Intercept Call Completion.

ICCF Interexchange Carrier Compatibility Forum.

ICD International Code Designator. A 2-byte field of the 20-byte NSAP (Network Service Access Point) address, the ICD is used to identify an international organization. NSAP and ICD are used in a variety of networks, including ATM. The British Standards Institution is the registration authority for the International Code Designator. See also BSI.

ICE 1. Information Content and Exchange. An emerging protocol in the form of a XML (eXtensible Markup Language) application, ICE is designed to facilitate the automation of content syndication on the World Wide Web. The ICE architecture is intended to define business rules that would allow data (e.g., user profile) exchange among partner sites. Such data can then be processed, loaded into a repository and resold under the user interface of the licensee. For example, ICE would allow a review of a book, movie or theater production to be licensed to multiple Web site hosts. A kill date would automatically kill the content across all sites on a specified date.

2. Interactive Connectivity Establishment ICE is an emerging standard from the Internet Engineering Task Force (IETF) for a framework that would allow Voice over Internet Protocol (VoIP) traffic to be exchanged between devices on NAT-enabled (Network Address Translator) networks. NATs offer a considerable measure of network intrusion protection by allowing connection to internal network servers only if the connection requests originate within the network, refusing those that originate externally. Unfortunately, NATs also block incoming VoIP calls. ICE combines a number of network access protocols (e.g., Simple Traversal of UDP through NAT, Traversal Using Relay NAT and Realm Specific IP) to determine how devices are connecting. Thereby, an incoming VoIP call can find its way through the NATs. See also IETF, NAT and VoIP.

ice load See dead load.

ICEA Insulated Cable Engineers Association. The ICEA is a professional organization dedicated to developing cable standards for the electric power, control and telecommunications industries. Formed in 1925, the ICEA's objective has been to ensure safe, economical, and efficient cable systems utilizing proven state-of-the-art materials and concepts. See the following representative definitions.

ICEA-P-46-426 Power cable ampacities for copper and aluminum conductor cable.

ICEA- P-53-426 Ampacities 15 KV through 35 KV copper and aluminum conductor.

ICEA-S-19-81 Rubber-insulated wire and cable.

ICEA-S-61-402 Thermoplastic-insulated wire and cable.

ICEA-S-66-524 Cross-linked thermosetting polyethylene-insulated wire and cable.

ICEA-S-68-516 Ethylene propylene rubber insulated wire and cable.

ICF Internet Connection Firewall. A firewall is a security system – hardware and/or software – that acts as a protective boundary between a network (which may be your PC only or it may be your PC and your families PCs on a local area network – and the outside world. Windows XP includes something called Internet Connection Firewall software Microsoft says you can use to restrict what information is communicated between the Internet and you and/or your network. You may be connected to the Internet via a cable modem, a DSL modem, or a dial-up modem. You should not enable Internet Connection Firewall on virtual

private networking (VPN) connections, which are typically used to securely log in to a corporate network. You should not enable ICF on client computers that are part of a large company or school network with a server-client structure. ICF will interfere with file and printer sharing in these scenarios. It's very easy to enable the Microsoft's software ICF. Poking around on Network/Properties/Advanced tab will lead you to a screen that talks about the ICF. Simply select Protect my computer or network, and then click OK. The Windows XP firewall is now enabled. My personal preference is for separate firewall hardware, not just a software solution. But I'm probably an old foggy.

ICFA International Computer Facsimile Association. The mission of this new organization is to create awareness of the benefits and uses of computer fax to increase worldwide market size.

ICI 1. Interexchange Carrier Interface. The interface between carrier networks that support SMDS.

ICIS See MAG Plan.

ICIT International Center for Information Technologies. A part of MCI.

Icky PIC Sticky Plastic Insulated Conductor. A gel-filled cable which is the only type recommended for direct burial, i.e., directly in the ground without any protective conduit system. As the gel is very sticky, working with "icky PIC" is a somewhat unpleasant experience.

icky pick Industry slang for the messy gel in loose-tube fiber optic cables, whose function is to prevent water penetration and to protect the fiber from temperature fluctuations. See icky PIC.

ICM See Integrated Call Management and/or Intelligent Call Management.

ICMP Internet Control Message Protocol. ICMP is a layer 3 protocol that is tightly integrated with the TCP/IP protocol suite. It allows routers to send error and control messages about packet processing on IP networks. For example, if a packet cannot reach its destination, an ICMP message is sent to the packet's source. Functions of ICMP include:

• Reporting errors. An error message is sent when a packet cannot reach its destination.

• Reporting congestion. ICMP messages report congestion when a router's buffer is full and is unable to properly forward packets. A source quench message is returned to the data source to slow down packet transmission. This is done sparingly to reduce the chance that ICMP messages will contribute to the congestion problem.

• Relaying troubleshooting information. Troubleshooting information is relayed through ICMP's echo feature. The ping utility is based on this ability to send a packet roundtrip between two hosts.

• Reporting time outs. ICMP reports timeouts when a packet's time-to-live (TTL) counter reaches zero and the packet is discarded.

• Some common ICMP messages include:

• Echo - used by the ping utility to test the connection between two devices

• Echo reply - reply to a ping

• Destination unreachable - several reasons cause a destination to be unreachable (unavailable port, unknown network, etc.)

• Source quench - tells the data source to reduce its transmission rate

• Redirect - informs the source of a better route to the destination

• Time exceeded - sent when a packet's time-to-live reaches zero

• Parameter problem - sent when there is a problem with the IP header

• Timestamp - requests the last time a host touched a particular packet to determine delay

Timestamp reply - reply to a time stamp request

iCOMP Intel Comparative Microprocessor Performance Index. A test for measuring how fast microprocessors are.

icon An icon is a picture or symbol representing an object, task, command or choice you can select from a piece of software (e.g. a trash can for a deletion command).

iconography The science of icons. A fancy name for talking about icons – those little visual representations of objects, tasks or commands you find in Windows and other GUI (Graphical User Interface) programs and operating systems.

ICP 1. Intelligent Call Processing. The ability of the latest ACDs to intelligently route calls based on information provided by the caller, a database on callers and system parameters within the ACD such as volumes within agent groups and number of agents available.

2. Instituto das Communicacoes de Portugal (The Portuguese Institute of Communications).

3. Intelligent Call Processor. The name of an AT&T service. ICP allows users to directly link their customer premise equipment to its network for individual call processing based on customer-specific information.

4. Independent Communications Provider. An ICP is a switchless CLEC (Competitive Local Exchange Carrier). Multi-service, integrated access.

5. Integrated Communications Provider. A fancy name for a competitive local exchange carrier (CLEC) that has passed beyond being an "ordinary" CLEC, which tends to just rent incumbent LEC (ILEC) circuits, into one who builds his own circuits and has a more comprehensive sales and marketing strategy. I first saw this term in a marketing presentation by a Lucent executive. I'm not too sure he knew what the term meant, except that he felt better if he sold equipment to an ICP, not a CLEC. A company called Jetstream wrote about this:

Since the Telecommunications Act of 1996 opened the $100 billion local market to competition, more than 500 companies have registered to offer local voice and data telecommunications services in the U.S. These newly enfranchised Integrated Communications Providers (ICPs) have collectively raised more than $30 billion to fund construction of the local networks they'll need to operate as carriers, and the total continues to rise at an unprecedented rate.

To survive in this competitive arena, CLECs, DLECs, ILECs and IXCs are aggressively looking for new technologies that will let them serve their customers profitably with a differentiated range of communication services.

6. Integrated Communications Platform. A remotely managed telephone, messaging and data networking platform. The benefit of this system is it provides corporate users with a single point of contact for managing an office's entire communications infrastructure, including local and long distance voice, data, and Internet service.

7. Internet Cache Protocol. A protocol used by caches on proxy servers to query other caches on other proxy servers about Web objects in cached databases. ICP uses User Datagram Protocol (UDP), an application-layer extension of the TCP/IP protocol suite. See also Cache, Proxy, and UDP.

ICQ ICQ "I Seek You". A erstwhile popular program that lets you find your friends and associates online on the Internet in real time. You send them messages, chat with them and send files in real time. The software and its user base was bought by AOL in 1998 for $287 million. www.icq.com. Since then, instant messenging has exploded and companies like AOL, MSN and Yahoo! pretty well dominate the field. See also Instant Messaging and Skype.

ICR 1. Initial Cell Rate: An ABR service parameter, in cells/sec, that is the rate at which a source should send initially and after an idle period.

2. Internet Call Routing node. A Bellcore proposed device that would communicate with both the voice and data networks through the Signaling System 7. After a local telephone company's central office detects an incoming data call, the ICR would instruct the voice switch to reroute the connection to a data network through remote access gear, which would then send the data to the Internet.

ICS 1. See Interactive Call Setup.

2. Integrated Communications System. A Northern Telecom definition: A telecommunications based platform with advanced processing power and capacity that enables integration and orchestration of typical business equipment (telephones, fax, etc.) through open architecture interfaces, as in the Norstar-PLUS Modular ICS.

3. Intercompany Settlements. The financial settlements made between the LECs for collect type calls and third number calls. Processed by CATS and BEARS. ICSs are determined through CMDS and ITORP.

ICSA International Computer Security Association. Formerly the National Computer Security Association, founded in 1989. An independent organization which strives to improve security and confidence in global computing through awareness and certification of products, systems and people. www.ncsa.com.

ICSC Interexchange Customer Service Center.

ICT 1. The Information and Communications Technologies Standards Board.

2. Information and Communication Technologies. The availability of information and communication technologies (ICTs) is concerns policymakers, academics and non-governmental organizations. Such technology, it is generally agreed, boosts productivity, though how quickly and by how much is the subject of much debate. The far wider availability of ICTs in rich countries, goes the argument, will enable the rich to get richer, while the poor are left behind. In short, not only is there a worrying "digital divide" between rich and poor, the divide is widening-with ominous consequences. These beliefs are widely held. But a paper by two economists at the World Bank, Carsten Fink and Charles Kenny, questions the logic of this argument and highlights the woolly thinking that pervades the digital-divide discussion. The authors conclude that the divide's size and importance have been overstated, and that current trends suggest that it is actually shrinking, not growing, which means policies designed to "bridge the digital divide" may need rethinking. For a start, Fink and Kenny observe, the term "digital divide" came to prominence "more for its alliterative potential than for its inherent terminological exactitude". It is used in at least four distinct ways, and two in particular: to describe the gap in access to ICTs between rich and poor countries, and the resulting gap in usage. The digital divide is almost always described in terms of the difference in the number of telephones, internet users or computers per head in rich and poor countries. For example, there are more telephones than people in most developed countries, compared with around three telephones per 100 people in the developing world. While the gap as defined using these per-head measures looks enormous, growth rates, according to the Economist, tell a different story. Over the past 25 years, telephone penetration has been increasing faster in low and middle-income countries than in high-income countries. The same is also true of Internet usage, which grew by around 50% per year in high-income countries in the late 1990s, compared with 100% per year in low and middle-income countries. The rich are ahead, but the poor are catching up fast. See also digital divide.

ICV Integrity Check Value is a digest of a message which provides a high level of assurance that the message has not been tampered with. Also referred to as Message Authentication Code.

ID Identifier.

ID Codes ID codes are used to restrict a caller's access type, access location, or calling privileges. ID codes can be either global or access/location specific.

ID10T Pronounced ID-TEN-T. Do you ever get any of those weird error codes when an application fails? I do. Sometimes it's your fault. It's what the geeks refer to as an ID10T error. It's a joke. Look at it again. Makes you feel stupid, huh? See also Cockpit Problem and Idiot-Proof.

IDA Integrated Data Access or Integrated Digital Access.

iDAB A new Digital Audio Broadcast technology that provides enhanced sound quality, improved reception and new data services for AM and FM radio.

IDAL International Dedicated Access Line.

IDC Insulation Displacement Connection. A type of wire connection device in which the wire is "punched down" into a double metal holder and as it is the metal holders strip the insulation away from the wire, thus causing the electrical connection to be made. The alternate method of connecting wires is with a screw-down post. There are advantages and disadvantages to both systems. The IDC system, obviously, is faster and uses less space. But it requires a special tool. The screw system takes more time, but may produce a longer-lasting and stronger, more thorough (more of the wire exposed) electrical connection. The most common IDC wiring scheme is the 66-block, originally invented by Western Electric. See Punchdown Tool.

IDC Clip IDC Clips are a method of jack termination. They look like a modular jack with a mini 66-block attached to the back. They are usually more expensive than 110-type blocks but are easier to install. See IDC and Punch-Down Tool.

IDCMA Independent Data Communications Manufacturers Association, a lobbying and education group based near Washington, DC.

IDCS Integrated Digital Communications System. A computer located in Sacramento, CA that connects IDTS controllers over a X.25 Packet Switched Network.

IDDD International Direct Distance Dialing. The capability to directly dial telephones in foreign countries from your own home or office telephone.

IDDS Installable Device Driver Server. A Dialogic term.

IDE 1. Integrated Drive Electronics. IDE is the common name for the Advanced Technology Attachment (ATA) standard ATA-1 developed by the American National Standards Institute (ANSI) in 1994 in the X3.221 specification. IDE is a hard disk drive standard interface for PCs. It appeared in 1989 as a low-cost answer to two other standard hard disk interfaces, ESDI and SCSI. The distinguishing feature of the IDE interface is that it incorporates the drive controller functions right on the drive. Instead of connecting to a controller card, an IDE drive attaches directly to the motherboard with a 40-pin connector. IDE drives offer a data transfer rate of three megabytes per second, which is not very fast. Several methods of data encryption can be used with the IDE interface, including MFM and RLL. Many laptops use IDE drives. IDE has a limit of 528 megabytes. Enhanced IDE drives, which appeared around 1994 to solve the problem that computers had gotten much faster and IDE wasn't keeping up, have a data transfer rate of between 11 and 13 megabytes per second and can handle drives of up 8.4 gigabyte. See also Enhanced IDE.

2. Integrated Development Environment. A term for products such as Microsoft's Visual C++ and Borland's Delphi that combine a program editor, a compiler, a debugger, and other software development tools into one integrated software package. The first of the IDEs, Borland's Turbo Pascal changed the way programmers write code by allowing programs to be edited and compiled within the same application.

IDEA 1. International Data Encryptions Algorithm. A secret key encryption algorithm developed by Dr. X. Lai and Professor J. Massey in Switzerland to replace DES.

2. Internet Development & Exchange Association, formed in 1995 and developed as part of a capstone MBA strategy project at West Virginia University to address the ever increasing competitive nature of the ISP market. Today, IDEA claims to be the largest trade association of independent Internet Service Providers (ISPs) in the world. www.auidea. org.

iDEN Integrated Dispatch Enhanced Network. A wireless technology developed by Motorola, iDEN operates in the 800 MHz, 900 MHz and 1.5 GHz radio bands; the 900 MHz development is aimed at operators of digital Commercial Mobile Radio Service (CMRS), also known as ESMR (Enhanced Specialized Mobile Radio). iDEN is a digital technology using a combination of VSELP (Vector Sum Excited Linear Prediction) and 16QAM (Quadrature Amplitude Modulation) for compression, and TDMA (Time Division Multiple Access) over frequency channels of 25 kHz. Through a single proprietary handset, iDEN supports voice in the form of both dispatch radio and PSTN interconnection, numeric paging, SMS (Short Message Service) for text, data, and fax transmission. The iDEN system, developed by Motorola and used by Nextel, also adds a walkie-talkie feature to phones. See also ESMR, QAM, SMS, TDMA, and VSELP.

identification failure Automatic Number Identification (ANI) equipment in the originating office failed to identify the calling number. See ANI.

identified outward dialing Same as AIOD. It's a PBX feature which provides identification of the PBX extension making the outward toll calls. This identification may be provided by automatic equipment or by attendant identification of the extension.

identifier The name of a database object (table, view, index, procedure, trigger, column, default, or rule). An identifier can be from 1 to 30 characters long.

identity chaos The situation that exists when an organization's users have different userids and passwords for different networks, systems, and applications. The problem is further complicated by different systems' having different rules for password length, permissible and insisted-upon characters and numerics in userids and passwords, and different password expiration dates. The resulting chaos takes two forms: the burden on users, who have to remember multiple userids, passwords and rules; and the burden on network and systems administrators, who have to assist users when they inevitably forget their logonids and passwords. There are two "solutions:" The first is to use the browser, Firefox, which remembers some (not all) userids and passwords and fills them in automatically for you. The other is a piece of software called eWallet, which does a fairly rigorous job of remembering, managing and filling in passwords, etc. My son loves it.

identity fraud The wholesale lifting of someone's financial persona to secure bank loans, credit card and mortgages in that person's name. The key to succesful identify theft is the victim's social security number. It is not a good idea to give your number to anyone – especially online. Even when the identity theft is discovered, it can take months, sometimes years, for innocent victims to restore their tattered credit histories. Identity theft is accomplished in many creative ways – from dumpster diving to phishing. For more, see both those terms.

IDEO locator The "you are here arrow" on a map is called the IDEO locator.

IDF Intermediate Distribution Frame.

idiot-proof A device, system, or software application that has a highly intuitive user interface can be so easy to operate correctly that even a person ignorant of its intricacies can do it. That's "idiot-proof." See also Cockpit Problem.

IDLC See Integrated Digital Loop Carrier.

idle 1. Not being used but ready.

2. An SCSA term. A state of the SCbus or SCxbus Message Bus where no information is being transmitted and the bus line is pulled high.

idle cell An ATM cell used for cell stuffing where rate adaption is required. As Physical Layer cells, idle cells are required and cannot, therefore, be replaced by assigned cells during the process of cell multiplexing; this is unlike Unassigned Cells, which are not necessary at a network level and which can be replaced, therefore. See also ATM Reference Model, Physical Layer, Rate Adaption and Unassigned Cell.

idle channel code A repetitive pattern (code) that identifies an idle channel.

idle channel noise Noise which exists in a communications channel when no signals are present.

idle line termination An electronic network which is switch controlled to maintain a desired impedance at a trunk or line terminal when that terminal is in an idle state.

idling signal Any signal that indicates no data is being sent.

IDN Integrated Digital Network.

IDNX 1. International Data Exchange Network.

2. Integrated Digital Network eXchange.

IDP InterDigital Pause.

IDR Intermediate Data Rate.

IDS 1. Internal Data Services.

2. Intrusion Detection Service. See deep packet inspection.

IDSL A xDSL variant that uses ISDN BRI (Basic Rate Interface) technology to deliver transmission speeds of 128 Kbps on copper loops as long as 18,000 feet. IDSL is symmetric, i.e., equal bandwidth is provided in both directions. IDSL is a dedicated service for data communications applications, only. In that respect, IDSL differs from ISDN, which fundamentally is a circuit-switched service technology for voice, data, video and multimedia applications. IDSL terminates at the user premise on a standard ISDN TA (Terminal Adapter). At the LEC CO, the loop terminates in collocated ISP electronics in the form of either an IDSL access switch or a IDSL modem bank connected to a router. The connection is then made to the ISP POP via a high-bandwidth dedicated circuit. IDSL is used by LECs to deliver relatively low speed DSL services in geographic areas where ISDN technology is in place, but ADSL technology is not. See also xDSL, ADSL, BRI, ISDN, HDSL, RADSL, SDSL, Terminal Adapter, and VDSL.

IDT 1. Inter-DXC Trunk.

2. Inter-Machine Digital Trunk.

IDTS Integrated Data Test System. A software package that allows the user to test Analog, Digital, Fractional T-1, and T-1 Circuits, remotely, at the various MCI terminal locations.

IDTV Improved Definition TeleVision. See Improved Definition Television.

IDU 1. Interface Data Unit: The unit of information transferred to/from the upper layer in a single interaction across the SAP. Each IDU contains interface control information and may also contain the whole or part of the SDU.

2. Indefeasible Right of Use. A right to use something, which right cannot be taken away from you, i.e. voided or undone. A term that generally applies, in contemporary terms, to the purchase of an optical fiber within a sheath of fibers owned by another company. Companies that lay optical fiber generally lay much more than they need. They do so because so much of the cost is associated with securing the right-of-way, trenching, burying conduit, etc. So they "lay a lot of pipe," much of which initially is left "dark" (i.e., inactive). They then are in a position to either lease or sell the extra fibers, and the bandwidth they represent. If you buy, rather than lease, a fiber, you have an IRU.

IE Incoming Exclusion.

IEC 1. InterExchange Carrier. Also called an IXC (as in IntereXchange Carrier). In practice, an IEC or IXC is any common carrier authorized by the FCC to carry customer transmissions between LATAs. In practice this means anyone and his brother who print up stationery, rent a few lines and proclaim themselves to be in the long distance phone business. Except for AT&T, regulation of long distance carriers by the FCC is perfunctory. It is less perfunctory by the local state authorities, some of whom still think competition in telecommunications is a mild form of insanity.

2. International Electrotechnical Commission. The international standards and conformity assessment body for all fields of electrotechnology, including electricity and electronics. The IEC publishes a number of international standards and technical reports on a wide variety of subjects including telecommunications (LANs, MANs and WANs), video cameras, electrical cables, communications protocols (e.g., HDLC), Open Systems Interconnection (OSI), optical fiber cables and connectors, and diagnostic X-ray imaging equipment. The IEC was founded in 1906 by British scientist, Lord Kelvin. www.iec.ch.

3. International Engineering Consortium - The IEC was founded in 1944 as a nonprofit organization sponsored by universities and engineering societies, and is dedicated to supporting continuing education for the U.S. electronics industry. www.iec.org

IEEE Institute of Electrical and Electronics Engineers, Inc. IEEE, founded in 1884, says it's the world's largest technical professional society, consisting of IEEE has more than 360,000 members in approximately 175 countries. Through its members, the organization is a leading authority on areas ranging from aerospace, computers and telecommunications to biomedicine, electric power and consumer electronics. The IEEE produces nearly 30 percent of the world's literature in the electrical and electronics engineering, computing and control technology fields. This nonprofit organization also sponsors or cosponsors more than 300 technical conferences each year. The IEEE sponsors technical symposia, conferences and local meetings, and publishes technical papers. It also is a significant standards-making body responsible for many telecom and computing standards, including those standards used in local area networks (LANs) - e.g. the 802 series. See also www.ieee.org.

IEEE 1394 Also called Firewire. An IEEE data transport bus that supports up to 63 nodes per bus, and up to 1023 buses. The bus can be tree, daisy chained or any combination. It supports both asynchronous and isochronous data. 1394 is a complementary technology with higher bandwidth (and associated cost) than Universal Serial Bus. For a bigger explanation, see 1394 and USB.

IEEE 1588 IEEE 1588 is a precision clock synchronization protocol for networked measurement and control systems. The protocol enables precise synchronization of clocks in networked, local computing, and distributed-object environments. The protocol enables heterogeneous systems that have clocks of various degrees of precision, resolution and stability to synchronize, and to do so with sub-microsecond accuracy while consuming minimal network and local clock computing resources to achieve that accuracy.

IEEE 488 See 488.

IEEE 802 The main IEEE standard for local area networking (LAN) and metropolitan area networking (MAN), including an overview of networking architecture. It was approved in 1990. For a much larger explanation, see 802.

IEEE 802 Standards See 802 and all the definitions that follow, including IEEE 802.1, 0x06IEEE 802.11a, IEEE 802.11b, IEEE 802.11g, 0x06IEEE 802.11i, 0x06IEEE 802.11j, 0x06IEEE 802.11n, 0x06IEEE 802.12, 0x06IEEE 802.15.1, etc.

IEEE Standards Association See IEEE-SA.

IEEE-SA The IEEE Standards Association is a standards body, which develops consensus standards through an open process that brings diverse parts of an industry together. It has a portfolio of more than 870 completed standards and more than 400 in development. IEEE-SA promotes the engineering process by creating, developing, integrating, sharing and applying knowledge about information technologies and sciences for the benefit of humanity and the profession. www.standards.ieee.org.

IEN Internet Experimental Note. A standards document similar to an RFC, and is available from the Network Information Center (NIC). IENs contain suggestions and proposals for Internet implementations or specifications.

IEPG Internet Engineering Planning Group. A group primarily composed of Internet service operators. Its goal is to promote a globally coordinated Internet operating environment. Membership is open to all.

IES 1. Inter-Enterprise Systems is EDI (Electronic Data Interchange) and inter-company electronic mail, fax, electronic funds transfer, videotex/online databases and the exchange of CAD/CAM graphics. See also EDI. 2. Information Exchange Services.

IESS INTELSAT Earth Station Standard.

IETF Internet Engineering Task Force. Formed in 1986 when the Internet was evolving from a Defense Department experiment into an academic network, the IETF is one of two technical working bodies of the Internet Activities Board. Comprised entirely of volunteers, the IETF meets three times a year to set the technical standards that run the Internet. The actual technical work is done by IETF working groups, which are grouped into areas of interest such as routing, transport, and security. The working groups seek the advice of the Internet community through draft RFCs (Requests For Comment), and then submit their recommendations to the IETF for final approval. The final standards are then published as standard RFCs. Of late, the IETF has been the forum where engineers and programmers have cooperated to solve the succession of crises caused by the Internet's phenomenal growth. Examples of IETF standards include DHCP (Dynamic Host Configuration Protocol), IPv6 (Internet Protocol version 6), LDAP (Lightweight Directory Access Protocol), and MPLS (MultiProtocol Label Service). The IETF working groups are grouped into areas, and managed by Area Directors, or ADs. The ADs are members of the Internet Engineering Steering Group (IESG). Providing architectural oversight is the Internet Architecture Board, (IAB). The IAB also adjudicates appeals when someone complains that the IESG has failed. The IAB and IESG are chartered by the Internet Society (ISOC) for these purposes. The General Area Director also serves as the chair of the IESG and of the IETF, and is an ex-officio member of the IAB. The Internet Assigned Numbers Authority (IANA) is the central coordinator for the assignment of unique parameter values for Internet protocols. The IANA is chartered by the Internet Society (ISOC) to act as the clearinghouse to assign and coordinate the use of numerous Internet protocol parameters. First-time attendees might find it helpful to read The Tao of the IETF. www.ietf.org.

IF See Intermediate Frequency.

IFC Information From Controller.

IFCC Internet Fraud Complaint Center. A partnership of the FBI and the NW3C, the National White Collar Crime Center. IFCC's mission is to address fraud committed over the Internet. For victims of Internet fraud, IFCC provides a convenient and easy-to-use reporting mechanism that alerts authorities of a suspected criminal or civil violation. For law enforcement and regulatory agencies at all levels, IFCC offers a central repository for complaints related to Internet fraud, works to quantify fraud patterns, and provides timely statistical data of current fraud trends. www.ifccfbi.gov.

IFFT Inverse Fast Fourier Transform. See Fast Fourier Transform.

IFG The minimum idle time between the end of one frame transmission and the beginning

of another. On Ethernet 802.3 LANs the minimum interframe gap is 9.6 micro-seconds.

IFIP International Federation for Information Processing. A research organization that performs substantive pre-standardization work for OSI. IFIP is noted for having formalized the original Message Handling System (MHS) model. www.dit.upm.es/~cdk/lflp.html.

IFITL Integrated Fiber in the Loop.

IFP Internet Fax Protocol is specified in ITU-T recommendation T.38 as the method for supporting facsimile transmissions over IP (Internet Protocol) networks. See also T.37 and T.38.

IFRB International Frequency Registration Board.

IFS International Freephone Service. See Freephone and IFTS.

IG Isolated Ground. In AC electricity, an isolated ground is a type of outlet characterized by the following features and uses:

- It may be orange and must have a Greek "delta" on the front of the outlet. (A delta looks like a triangle.)
- It must be grounded by an insulated green wire.
- It must have insulation between the ground terminal and the mounting bracket.
- It is used primarily to power electronic equipment because it reduces the incidence of electrical "noise" on the ground path.

IGMP Internet Group Management Protocol. A protocol used by IP hosts and gateways to report their multicast group memberships. When used in concert with a multicast protocol, the IP-based network can support multicasting. See Multicast.

Ignition Key A rod arranged to strike the arc in an arc generator of high frequency currents.

IGP 1. Interior Gateway Protocol. The protocol used to exchange routing information between collaborating routers in the Internet. RIP and OSPF are examples of IGPs.

2. Integrated Graphics Processor.

IGRP Interior Gateway Routing Protocol. A distance-vector routing protocol developed by Cisco Systems for use in large, heterogeneous networks. It learns the best routes through a LAN or the Internet.

IGT Ispettorato Generale delle Telecomunicazioni (General Inspectorate of Telecommunications, Italy).

IGW International Gateway Switch.

III Information Industry Index.

IILC Information Industry Liaison Committee (part of ATIS). See Exchange Carriers Standards Association.

IINREN Interagency Interim National Research and Education Network. Evolving operating network system. Near term research and development activities will provide for the smooth evolution of this networking infrastructure into the future gigabit NREN. See NREN.

IIR Interactive Information Response.

IIS Microsoft Windows Internet Information Server, which is similar to Netscape's Webserver. IIS lets you set up a web site and control and manage it remotely though the Internet, assuming you have a necessary privileges and a Web browser.

IISP 1. The Information Infrastructure Standards Panel formed by ANSI in July 1994 to accelerate the development and acceptance of standards critical to the establishment and deployment of the information superhighway. See ANSI.

2. Interim Interface Signaling Protocol. A call routing scheme used in STM networks. Formerly known as PNNI Phase 0. IISP is an interim technology meant to be used pending completion of PNNI Phase 1. IISP uses static routing tables established by the network administrator to route connections around link failures.

IITF Information Infrastructure Task Force. This task force of high level representatives from federal agencies was formed by the Clinton Administration to identify and address the issues of creating a National Information Infrastructure. The Task Force relies on the members of the Industry Advisory Council as it assesses the requirements of individuals and businesses that will shape future networks. See also National Information Infrastructure.

IIW The ISDN Implementors Workshop, a group within the North American ISDN Users Forum.

IKE Internet Key Exchange. IKE is an IPsec (Internet Protocol security) mechanism that is used to create SAs (Security Associations) between two entities in an IP-based VPN (Virtual Private Network) application. IKE sets up a secure tunnel between the entities, authenticating their identities, negotiating the SAs, and exchanging shared key material between them in order that data can be encrypted and decrypted by those with privileged data access. As a dynamic key exchange mechanism, IKE relieves the users from the drudgery of manually configuring the SAs, which process is difficult and labor-intensive. See also Authentication, IPsec, Tunneling, and VPN.

IKP Internet Keyed Payments Protocol. An architecture for secure payments over the

Internet in the general context of Electronic Commerce. iKP is a public-key cryptography which defines transactions of a credit card nature, where a buyer and seller interact with a third party, such as a credit card company, in order to authorize transactions on a secure basis. SEPP (Secure Electronic Payment Protocol) is a standard implementation of iKP. iKP specifies RSA as the public-key encryption and signature algorithm. See also Electronic Commerce, RSA and SEPP.

ILA Intermediate Light Amplification. Several backbone telecommunications carriers use this term to refer to "repeater" points along their fiber network; points from which shunts to other networks can easily be derived.

ILAN A protocol independent router for token ring and Ethernet networks from Cross-Comm Corporation, Marlboro, MA.

ILCR International Least Cost Route.

ILD Injection Laser Diode.

ILEC An Incumbent Local Exchange Carrier is the dominant phone carrier within a geographic area as determined by the FCC. Section 252 of the Telecommunications Act 1996 defines Incumbent Local Exchange Carrier as a carrier that, as of the date of enactment of the Act, provided local exchange service to a specific area. The Act provided that the Commission may treat "comparable carriers as incumbents" if they either "occupy a position in the market for telephone exchange service within an area that is comparable to the position occupied by the ILEC or such a "carrier that has substantially replaced an ILEC...." or if "such treatment is consistent with the public interest..." See also CLEC, Colocation and ELEC.

ILLP Inter Link-to-Link Protocol.

illuminated output Another term for light output. See light output.

ILMI Interim Link Management Interface: An ATM Forum-defined interim specification for network management functions between an end user and a public or private network and between a public network and a private network. ILMI is based on a limited subset of SNMP capabilities.

ILNP Interim Line Number Portability.

ILS 1. Input buffer Limiting Scheme. A flow control scheme used in data communications that blocks overload by limiting the number of blocks arriving at a buffer.

2. Instrument Landing System.

3. Internet Locator Service. In a September 1997 White Paper on "IP Telephony," Microsoft wrote, "User-to-IP mapping information is stored and continually refreshed using the Internet Locator Service (ILS) Dynamic Directory, a real-time server component of the Active Directory, which is part of Windows NT."

ILSR IPX Link State Router. Novell's improvement on its RIP distance vector-based routing protocol.

ILT Idle Line Termination.

IM Instant Messaging.

IMA 1. Interactive Multimedia Association. Formed in 1991 (rooted in IVIA, Interactive Video Industry Association), an industry association chartered with creating and maintaining standard specifications for multimedia systems. www.ima.org.

2. Inverse Multiplexing over ATM. You have a high speed data stream. But not a high speed transmission link. You have several low speed links. Inverse multiplexing lets you join several slow speed links together and pretend that they're one high speed link. Here's a formal definition. IMA is an access specification approved in 1997 by the ATM Forum. This User Network Interface (UNI) standard allows a single ATM cell stream to be split across multiple access circuits from the user site to the edge of the carrier's ATM network. In an ATM LAN application, for instance, the ATM switch deployed in the enterprise backbone typically operates at 155 million bits per second or 622 Mbps. In this example, ATM traffic from the enterprise to the public ATM carrier-based network, requires 6 Mbps – well more than the 1.544 million bits per second provided by a T-1, but less than a full T-3 (which is 45 million bits per second). Rather than subscribing to a T-3, which requires a fiber optic access circuit and is very expensive, IMA is used. Thereby, the ATM data stream is split across four T-1 circuits by an access concentrator which possesses IMA capability. The IMA process works in a round-robin fashion, with cell number 1 traveling over T-1 number 1, cell #2 traveling over T-1 #2, and so on. Each of the four T-1 circuits is relatively inexpensive, can be provisioned over twisted-pair, and is readily available. At the edge of the carrier network, the ATM switch receives each of the four separate data streams, and reverses the IMA process to put the original datastream back together, which is then switched and transported through the network to the far edge. At that far edge, the IMA process may take place again, from the edge of the carrier network, over four T-1s, and to the IMA-capable ATM concentrator on the user premises. IMA is specified in the ATM forum specification AF-PHY-0086.000 Inverse Multiplexing for ATM Specification Version 1.0 and dated July 1997. Go to ftp://ftp.atmforum.com/pub/approved-specs/af-phy-0086.000.pdf to read and download a copy

of the 140-page document.

The definition of IMA is not easy. One reader, Rosario Brinquis of Madrid Spain, has contributed the following definition, which he thinks really captivates the essence of IMA: A methodology is described which provides a modular bandwidth for user access to ATM networks and for connection between ATM network elements at rates between the traditional order multiplex levels, for instance, between the DS1/E1 and DS3/E3 levels in the asynchronous digital hierarchies. DS3/E3 links are not necessarily readily available throughout a given network and therefore the introduction of ATM Inverse Multiplexers provides an effective method of combining the transport bandwidths of multiple links (e.g., DS1/E1 links) grouped to collectively provide higher intermediate rates.

See also ATM, Inverse Multiplexer and TDM.

image antenna A hypothetical, mirror-image antenna considered to be located as far below ground as the actual antenna is above ground.

image map A Web term. An image containing one or more invisible regions, called hotspots, which are assigned hyperlinks. Typically, an image map gives users visual cues about the information made available by clicking on each part of the image. For example, a geographical map could be made into an image map by assigning hotspots to each region of interest on the map.

image resolution The fineness or coarseness of an image as it was digitized, measured in Dots Per Inch (dpi), typically from 200 to 400 dpi.

image spam Email spam whose message content consists of an embedded graphical image instead of text, so as to evade text-recognition algorithms in anti-spam software.

IMAP Internet Message Access Protocol (1993), originally Interactive Mail Access Protocol (1986). As specified by the IETF (Internet Engineering Task Force) in RFC 1730, IMAP is an e-mail protocol that is likely to eventually replace POP3 (Post Office Protocol revision 3) for e-mail servers. IMAP allows users to create and manage remotely-held mail folders over the internet. Users can scan message headers, download selected messages, change and delete folders, and search for keywords. IMAP combines the functionality of locally held email with the convenience and connectivity of remotely-held, internet-based mail. My brilliant son Michael (he gets it from his mother), uses IMAP at school. Dartmouth College has its own propriety, server-based email system (BlitzMail) designed specifically for the campus. While BlitzMail is perfect for campus communication, it lacks some features, such as HTML support, that are necessary to communicate effectively with people outside of Dartmouth. Using an IMAP managed account through Outlook 2003, he is able to get the best of both worlds. Needing to email with me for example, Michael accesses his email account with Outlook. When communicating on campus, he uses BlitzMail. Because IMAP is totally dynamic and server based, Michael's account is always current and can be accessed from anywhere. If he was using a different protocol, such as POP3, this synchronicity would be impossible. PIMAP4, the current version, is specified in RFC 2060. See also MIME and POP3.

IMAP4 Internet Message Access Protocol revision 4. IMAP4 is an emerging Internet e-mail standard that vendors promise will make electronic messaging management easier and safer than IMAP. IMAP4 is specified in IETF (Internet Engineering Task Force) RFC 2060. See also IMAP.

IMAS Intelligent Maintenance Administration System. Northern Telecom software which is a menu-driven PC-based program that provides enhanced maintenance and administrative capabilities for DMS-10 central offices.

IMASS Intelligent Multiple Access Spectrum Sharing. A method of automatically determining the presence of existing private operational fixed microwave (OFM) systems in areas near base stations, and avoiding the use of frequencies for the PCS or cellular base station which might cause unacceptable interference. Instead the PCS or cellular systems will use frequencies in each area, which are not being used by nearby OFM (operational fixed microwave) systems. Techniques such as are helpful to PCS service providers coexisting with the incumbent OFM systems, until they can be relocated to different frequencies according to the FCC rules.

iMIP iCalendar Message based Interoperability Protocol. See iCalendar.

IMC See Internet Mail Consortium.

IMEI International Mobile station Equipment Identity. A wireless telecommunications term. An equipment identification number, similar to a serial number, used to uniquely identify a mobile phone.

IMHO Abbreviation for "In My Humble Opinion;" commonly used on E-mail, on the Internet and BBSs (Bulletin Board Systems).

Immediate Action Limit IAL. A Verizon definition: The bound of acceptable performance and the threshold beyond which Verizon will accept an end user trouble report and take corrective action.

immediate ringing A PBX feature which makes the called telephone begin ringing the instant the phone has been dialed. Normally there's a small wait between dialing the number and having the phone ring.

immunity from suit A term I first saw in licensing agreements with Microsoft. The provision says that the company signing the agreement with Microsoft agrees not to sue Microsoft or Microsoft's customers and OEMs for infringement of said company's own patents. Some observers are claiming that signing an agreement with this provision would give Microsoft a royalty-free license to an outside company's patents.

IMNSHO In My Not So Humble Opinion. An acronym used in electronic mail on the Internet to save words or to be hip, or whatever. See IMHO.

IMO Abbreviation for "In My Opinion;" commonly used on E-mail and BBSs (Bulletin Board Systems). See IMHO.

iMode See i-Mode.

IMP September 2, 1969 Professor Leonard Kleinrock births the Internet, with the installation of ARPAnet's first Interface Message Processor (IMP) in his lab at the University of California at Los Angeles. IMPs were packet-switching minicomputers, pre-Cisco routers, developed at Bolt, Bernanek and Newman (BBN) in Cambridge, Massachusetts. BBN was later merged into GTE, which then got merged into Bell Atlantic. Professor Kleinrock was the lead architect of the ARPAnet – the Defense Department's Advanced Research Projects Agency Network.

impact analysis How will a disaster impact your IT infrastructure – your computers, your networks, your telecommunications systems? That's what impact analysis is all about. PC Magazine talks about four important "exposure parameters."

- The relative value of the information or infrastructure component. For example, product plans, accounting systems, customer databases and so on, usually have a high value. A company phone list has a lower value.
- The possible publicity fallout. A defaced Web site (or a downed customer service center), means embarrassment for your company. This can translate to lost confidence in your company's products and services.
- The denial of business (DoB) potential. Will an attack affect your ability to do business? Being inconvenienced is one thing. Losing your ability to run your business is another.
- The ease of attack. The easier a component is to attack, the most often it will be struck. Components closest to the public Internet are more accessible.

The idea behind impact analysis is that you assign values to the relative disaster impact on each system based on each of the four criteria. Add the values up and assign the most resources to protecting the key systems. Impact analysis is a formalized way of figuring what's logical.

impact strength A test for determining the mechanical punishment a cable can withstand without physical or electrical breakdown by impacting with a given weight, dropped a given distance, in a controlled environment.

impact tool Also called a "punch down" tool. See Punch Down Tool.

impaired Condition that occurs when an individual circuit exceeds the transmission limits of its signaling function (e.g., seizure, disconnect, ANI) and failures occurs.

impairment standard The standard used to determine whether an ILEC is required to unbundle a specific network element. An element is deemed to be impaired if an ILEC's failure to provide access to such network elements would "impair" the ability of competitive carriers to provide services they seek to offer.

IMPDU Initial MAC Protocol Data Unit. A Connectionless Broadband Data Service (CBDS) term that corresponds to the L3 PDU in Switched Multimegabit Data Service (SMDS). CBDS is the European equivalent of SMDS.

impedance The total opposition, or resistance to flow, of electrical current in a circuit. Impedance is the term used in non-direct current (i.e., Alternating Current, or AC) applications, while resistance is used in DC (Direct Current) applications. The unit of measurement of impedance is ohms. The lower the ohmic value, the better the quality of the conductor in terms of dimensions as gauge (i.e., thickness of the conductor), and anomalies (e.g., consistency of gauge and nicks in the conductor). Low impedance will help provide safety and fire protection and a reduction in the severity of common and normal mode electrical noise and transient voltages. For telecommunications, impedance varies at different frequencies. Ohm's law says that voltage equals the product of current and impedance at any single frequency. See also Resistance.

impedance matching The connection of additional impedance to existing impedance one in order to improve the performance of an electrical circuit. Impedance Matching is done to minimize distortion, especially to data circuits. See also load coils.

impedance matching transformer See load coils.

impending event I'm selling you something on sale. The sale ends on Friday That's an "impending event." An impending event is something that causes you to make your mind up faster about buying something. The classic "impending event" is the end of the month or the end of the quarter.

impersonation attack An attack in which a hostile computer system masquerades as a trusted computer.

implantable computer A computing device that is surgically (or otherwise) implanted into a living being. Implantable medical devices for example, heart pacemaker, may contain computing components.

implementors' agreement An agreement about the specifics of implementing as a standard, reached by vendors who are developing products for the standard. Compare with De Facto Standard and De Jure.

implied acknowledgment Implied acknowledgment is a process whereby negative acknowledgment of a specific packet of information implies that all previously transmitted packets have been received correctly. See also Pipelining.

import Imagine you have a software program, like a spreadsheet or a database. And you have information in that program. Let's say it's Microsoft Word. And you want to get it into a different program, say to give it to a workmate who uses WordPerfect or Excel. You have to convert it from one format to another. From Word to WordPerfect or from Lotus to Excel. That process is typically called "exporting" and the process of your workmate getting it into his computer is called "importing." And you'll typically see the words "EXPORT" and "IMPORT" as choices on one of your menus.

import computers A Windows NT term. In directory replication, the servers or workstations that receive copies of the master set of directories from an export server.

import script First read my definition of IMPORT. An import script is a series of specifications which control the merging processes. It contains a series of merge rules which specify how the fields are to be merged and a record precedence rule which governs which records to merge of the ones received.

important call waiting Notifies you with a special ring that someone you want to hear from is calling you.

Improved Definition Television IDTV. Television that includes improvements to the standard NTSC television system, which improvements remain within the general parameters of NTSC television emission standards. These improvements may be made at the transmitter and/or receiver and may include enhancements in parameters such as encoding, digital filtering, scan interpolation, interlaced scan lines, and ghost cancellation. Such improvements must permit the signal to be transmitted and received in the historical 4:3 aspect ratio.

Improved Mobile Telephone Service IMTS. In the beginning, there was dispatch mobile service. The base operator broadcast a message to you. Everyone could hear it. You responded. Then they had mobile telephone service. You picked up the phone in your car, the operator responded. You asked for the number you wanted and she/he dialed it and connected you. You had the channel to yourself but others could still tune in. Then came Improved Mobile Telephone Service (IMTS). Now you could dial from your car without using an operator with some assurance of privacy. IMTS was the pre-cellular mobile telephone service enhancement introduced in 1965, which permitted full duplex mobile radio communications, as well as other enhancements. The original Mobile Telephone Service was introduced in 1946. See Cellular.

impulse A surge of electrical energy usually of short duration, of a non repetitive nature. See also ultrawideband.

impulse hits Errors in telephone line data transmission are caused by voltage surges lasting from 1/3 to 4 milliseconds and at a level within 6 dB of the normal signal level (Bell standard allows no more than 15 impulse hits per 15 minute period).

impulse noise High level, short duration noise that comes on a circuit. You can get impulse noise from electromechanical relays. These noise "spikes" have little effect on voice transmission but can be devastating to data. You can get a piece of test equipment called an impulse noise measuring set. Such a machine establishes a threshold and counts the number of impulses (hits) above that threshold.

impulse pay per view A pay per view system which allows the viewer to use a hand held control to select a movie for immediate viewing without making a telephone call to order. The fee is charged to the subscriber's phone or credit card bill.

impulse response The time-domain response of a network to an input impulse.

impurity level An energy level outside the normal energy band of the material, caused by he presence of impurity atoms. Such levels are capable of making an insulator semiconductor.

imputation The practice of "filling in" missing data with plausible values in order to arrive at a logical solution to a problem. Imputation is much like solving a problem for an "unknown" value through a process of elimination and inference, perhaps working backwards from a known solution in order to determine the value of the missing data (i.e., the cause of the problem). "Computation," on the other hand, is the practice of solving a problem in order to arrive at a solution based on known values. You have to be careful with imputation, as a naive or unprincipled imputation method may create more problems than it solves, thereby, distorting estimates and hypotheses, and leading to misdirected "finger-pointing."

IMS IP Multimedia Subsystem is an open, standardised, "operator friendly", Next Generation Networking (NGN) multi-media architecture for mobile and fixed IP services. It's a VoIP implementation based on a 3GPP variant of SIP, and runs over the standard Internet protocol (IP). It's used by telecom operators in NGN networks (which combine voice and data in a single packet switched network), to offer network controlled multimedia services. IMS's aim is not only to provide new services but to provide all the services, current and future, that the Internet provides. In addition, users have to be able to do everything they want when they are roaming as well as from their home networks. To achieve these goals the IMS uses open standard IP protocols, defined by the IETF. So, a multi-media session between two IMS users, between an IMS user and a user on the Internet, and between 2 users on the Internet is established using exactly the same protocol. Moreover, the interfaces for service developers are also based in IP protocols. This is why the IMS truly merges the Internet with the cellular world; it uses cellular technologies to provide ubiquitous access and Internet technologies to provide appealing services. IMS was originally defined by the 3rd Generation Partnership Project (3GPP), as part of their standardisation work for third generation (3G) mobile phone systems for W-CDMA networks. But the service model has been used by 3GPP2 (a different organisation for CDMA2000 networks) and TISPAN (fixed networks) too. It is a key technology for Fixed/Mobile Convergence (FMC).

- IMS first appeared in release 5 of the evolution from 2G to 3G networks for W-CDMA networks (UMTS), when SIP-based multimedia domain was added to NGN networks. Support for older GSM and GPRS networks is also provided.
- In 3GPP release 6, interworking with WLAN was added.
- 3GPP2 (a different organisation) based their CDMA2000 Multimedia Domain (MMD) on 3GPP IMS, adding support for CDMA2000.
- 3GPP release 7 adds support for fixed networks, together with TISPAN R1.
- "Early IMS" is defined for IPv4 networks, and provides a migration path to IPv6 Much of this definition originally appeared in www.wikipedia.

IMS/VS Information Management System/Virtual System. A common IBM large computer operating arrangement, usually using the MVS (Multiple Virtual System) operating system oriented toward batch data processing and data communications-based transaction applications.

IMSI International Mobile Subscriber Identity. An ITU-T specification used to uniquely identify a subscriber to mobile telephone service. It is used internally to a GSM (Global System for Mobile Communications) network, and has been adopted for future use in all cellular networks. The IMSI is a 50-bit field which identifies the phone's home country and carrier.

IMT 1. InterMachine Trunk. A circuit which connects two automatic switching centers, both owned by the same company.

2. International Mobile Telecommunications.

IMT-2000 International Mobile Telecommunications for the year 2000, IMT-2000 replaced FPLMTS (Future Public Land Mobile Telecommunications System) as the vision for a single global standard for wireless networks. IMT-2000 is an ITU initiative for a 21st century wireless network architecture. IMT-2000 is a G3 (Generation 3) concept (or 3G for 3rd Generation, depending on your persuasion) and set of specifications for a next-generation wireless network. IMT-2000 specifications include 128 Kbps for high mobility and ISDN applications, 384 Kbps for pedestrian speed and full-motion compressed video, and 2 Mbps for fixed E-1/T-1 access and wireless LANs. IMT-2000 also was intended operate in the 2 GHz band. Note that "2000" has several meanings: the year 2000, bandwidth of up to 2000 Kbps, and frequency range 2000 MHz. As it turns out, the year 2000 has been pushed back to at least 2002, bandwidth of 2 Mbps is only for limited in-building applications, and the 2 GHz spectrum has been allocated for other purposes in North America.

The concept of a single global standard fell apart in October 1999, when representatives from the various countries agreed to adopt "federal standards" under the IMT-2000 umbrella. Two of the three modes are based on CDMA (Code Division Multiple Access) and one on TDMA (Time Division Multiple Access). cdmaOne, also known as TIA IS-95a Telecommunications Industry Association Interim Standard-95a), was the first CDMA-based 3G approach to be introduced, and is popular with CDMA-based cellular operators in North American and Asia. cdma2000 3XMC, the high-speed version operating at 2 Mbps, was developed by Qualcomm and has been approved by the ITU. EDGE (Enhanced Data Service for GSM Evolution) is the TDMA variant of IMT-2000. GPRS (General Packet Radio Service), an interim step towards EDGE, is the choice of cellular operators with networks based on GSM (Global System for Mobile Communications) and D-AMPS (Digital Advanced Mobile Phone Service). See also D-AMPS, CDMA, cdma2000 3XMC, EDGE, FPLMTS, GPRS, GSM, and TDMA.

IMTC International Multimedia Teleconferencing Consortium. A non-profit corporation with the mission of promoting, encouraging and facilitating the development and implementation of interoperable multimedia teleconferencing solutions based on open international standards. Emphasis is on ITU-T standards such as T.120, H.320, H.323, and H.324. IMTC sponsors and conducts interoperability test sessions between suppliers of conferencing products and services based on those standards. It also focuses on market education. IMTC comprises over 140 members, including 3Com, Alcatel, BellSouth, Cisco, Compaq, Dialogic, IBM and Motorola – manufacturers, carriers, end users and others committed to open standards are welcome. www.imtc.org.

IMTS 1. Improved Mobile Telephone Service. The pre-cellular mobile telephone service enhancement introduced in 1965, which permitted full duplex mobile radio communications, as well as other enhancements. The original Mobile Telephone Service was introduced in 1946.

2. International Message Telecommunications Service, i.e., international long-distance service.

IMUIMG ISDN Memorandum of Understanding Implementation Management Group. Formed in 1992, the IMUIMG is intended to ease ISDN implementation in Europe. The organization's stated goal is to ensure consistency when ordering or using ISDN services, regardless of provider or country. Carriers in the U.S., Canada and the Asia-Pacific have been invited to join.

IMUX Inverse Multiplexer. See also IMUX DAL.

IMUX DAL A type of dedicated access line provided by ISDN PRI lines that have been tested for interconnection to Inverse Multiplexor equipment. IMUX access supports applications like video conferencing and high-speed data transfer.

IN Another name for the Advanced Intelligent Network. See IN.

In Band See In-Band.

in-territory In-territory equals in-franchise which equals in-region.

in the clear A cellular term referring to the fact that certain signaling and control information is transmitted between a cell phone and a cell site in an insecure manner. A cellular phone is equipped with 2 identification numbers, a MIN and an ESN. The Mobile Identification Number (MIN) is a changeable number assigned to the terminal by the retailer activating the service. The Electronic Serial Number (ESN) is hard-coded into each terminal at the time of manufacture. In combination, the MIN and ESN are intended to identify both the terminal and subscriber to the network for purposes of authentication and billing. Not only are the MIN and ESN transmitted as the user seeks access to a channel for purposes of initiating a call, they also are transmitted frequently by the terminal to the cell sites in order that the cellular network can keep track of the terminal for purposes of terminating incoming calls. Further, and especially in the case of analog networks, those numbers are transmitted "in the clear"; in other words, they are not encrypted. Even most digital cellular standards do not provide for encryption of such numbers. As a result, it is relatively easy for criminals to gain access to the numbers through the use of a low-cost radio scanner. Standing on a freeway overpass, for instance, it is a simple matter for the criminal to capture a number of MIN/ESN numbers and to clone them into a number of other terminals. In fact, multiple terminals may be cloned with the same MIN/ESN numbers in different cities across the nation, with the information being posted to BBSs accessible through the Internet. This definition courtesy of an article on Voice Network Fraud that Ray Horak wrote for Datapro Information Services. See also Clone.

In-band Control Control information that is provided in the same channel as data.

In-Band On-Channel See IBOC.

in-band signaling Signaling made up of tones which pass within the voice frequency band and are carried along the same circuit as the talk path that is being established by the signals. Virtually all signaling – request for service, dialing, disconnect, etc. – is in-band signaling. Most of that signaling is MF – multi-frequency dialing. The more modern form of signaling is out-of-band. Several local and long distance companies provide ANI (Automatic Number Identification) via in-band signaling. Some long distance companies provide it out-of-band, using the D-channel in a PRI ISDN loop. In cellular networks, In-band Signaling is known as CAS (Callpath Associated Signaling). See also CAS, ISDN, Out of Band Signaling and SS7 (ITU Signaling System Number 7).

in-collect A CLEC term for the process of collecting long distance calling records from

IXCs for purposes of subscriber billing. See CLEC.

In-house Broadband over Power Line See BPL.

in-line device Hardware that is physically attached between two communications lines. Also known as bump-in-the-wire.

in-the-clear See In the clear above.

in-vehicle signing On-board display of roadside sign information. The information can be received either by short-range transmission from roadside beacons or from on-board data that is retrieved based on real-time GPS. The idea is to improve the information that a driver has when driving at night and during inclement weather, thereby improving traffic flow and roadway safety.

in-WATS Incoming Wide Area Telecommunications Service. AT&T's term for toll-free service long-distance service provided by a phone company. In North America, all these In-WATS services have 800, 888 and 877 as their "area code." In other countries, the same service is available, though the area code and the name varies. Often it's called "Freephone," which is the standard name used by the ITU-T. Other names include "Freecall" and "Green Number." Dialing one of these numbers is free to the person making the call, as the charges are billed to the called party. See also 800 Service, International Freephone Service and InWATS.

INA Information Networking Architecture. Bellcore developed INA to facilitate the interoperation of proprietary software components through open interfaces based on voluntary international standards. INAsoft is the set of guidelines used to design interoperable, vendor-independent solutions, allowing rapid and successful product development.

inactivity time-outs Dial-in users can be disconnected after specific periods of inactivity. By eliminating idle connections, you reduce the number of ports required for remote access to a network.

INAP Intelligent Network Application Part. Also Intelligent Network Application Protocol. An Intelligent Network (IN) term for a mechanism, or interface protocol, for communication over a SS7 network between physical Network Elements (NEs). Such NEs might include a Service Control Point (SCP) and a Service Switching Point (SSP). INAP is defined by the ITU-T. ETSI has defined a European version known as Core INAP. See also AIN, IN, SCP, SCP, and SS7.

INAsoft See INA.

INAT Internode Alarm Transport.

In-Band On-Channel See IBOC.

inbound path On a broadband LAN, the transmission path used by stations to transmit packets toward the headend.

inbox For most people, an inbox is the place where all their electronic mail wends its way to. For many of us, it's our Outlook inbox or our Yahoo or Hotmail inbox. Many of us have many "inboxes," since many of us subscribe to multiple email services and work for one or several companies. There are two ways you can typically get to your inbox. You can jump onto the Internet and go to an Internet address and read your email on line. That's typically the way Yahoo or Hotmail does it. The second way of reading the mail in your inbox is to have your computer software visit the place your mail is sitting, pick your mail up and bring it back to your computer, attachments and all. In this case, your mail is sitting at a place called a POP3 server. That's typically a distant computer belonging to your company or the ISP (Internet Service Provider) who provides you email service. The easiest way to visualize a POP3 server is to think of it as your local post office at which you have rented a post box. Every so often you go to the post office and empty your box of mail and magazines. That's exactly what you with your POP3 server. You visit it with your computer over the Internet. When you computer gets there, it copies its entire contents to your computer's hard disk, then it empties the box, ready to be filled up with more catalogs and bills you don't want. For a longer explanation, see POP3.

Inbox Repair Tool Microsoft's popular Outlook email software was conceived as an afterthought. One year Microsoft needed "something" to add value to yet another release of its popular Office product, which had not been significantly improved. It dragged Outlook out of the back closet and bingo a star was born. There was one major problem, however. Outlook was a badly written piece of software. It pretends to be a database, but it's actually a conglomeration of databases – from inboxes to calendars. The sad part is they're not stored in multiple databases and there's no automatic process to clean them. They're stored in one gigantic "database" called a personal data store, carrying the extension "pst." The sad part is that Microsoft never figured how popular the program would become and how many people would come to rely on it and run their lives by it – as I do. As a result, our one Outlook "database" – typically called Outlook.pst – grows larger and larger. And because it is big and used frequently, it becomes corrupted. That's the word Microsoft uses. I don't know exactly what it means except that the file gets more corrupted

and eventually you can't find anything and Outlook locks. At the first sign of Outlook slowing down, you must use Microsoft's Inbox Repair Tool and uncorrupt (i.e. fix) your pst file. Search your hard disk for a file call scanpst.exe. When you find it, set up a shortcut to it. Then run the shortcut. It will start the Inbox Repair Tool. It will ask you for the name of the Outlook pst you're using and its location. Hint: It's usually the biggest and youngest file with a pst extension. Make sure you close Outlook. Give the Inbox Repair Tool the name of the file and let the Inbox Repair Tool run. It will fix your pst file. I personally run the Inbox Repair Tool once a week. It finds problems about once every two weeks. You will find a little information on the Inbox Repair Tool by searching Microsoft Office Online. Hit F1 and search for Inbox Repair Tool.

incAlliance incAlliance stands for the Isochronous Network Communication Alliance. It was announced publicly on June 13, 1995. It was formed by several high-technology businesses, including several Fortune 500 companies, to promote the use of isochronous Ethernet (also called isoEthernet) to provide interactive multimedia applications to the enterprise desktop over existing cable infrastructures. IsoEthernet never took off. It's been replaced by switched Ethernet.

incandescent The light bulb in most homes are incandescent, little changed from the bulb invented in 1879 by Thomas Edison. The bulb is produced when hot glass is blown into molds and then cooled and often, coated with a diffusing material. Placed inside the bulb is a very thin and very fragile coiled tungsten filament about 20 thousands of an inch thick. For the bulb to produce light an electric current is passed through the filament which is heated to a point where it gives off light. Incandescent bulbs are not very efficient users of energy. Only 10% of the energy is used to make visible light. The rest is wasted in heat. New light sources – such as LEDs, or light emitting diodes, are much more efficient producers of light.

INCC Internal Network Control Center.

incentive regulation Prices of services provided by the local regulated phone company are fixed or capped but incentives are provided to improve earnings through cost savings. Earning levels are flexible, within a range of rates, allowing opportunity for earnings improvement.

incestuous amplification When all the generals (i.e. bosses) listen only to those who are already in lock-step agreement with their ideas and plans. This reinforces set beliefs and creating a situation ripe for miscalculation." The term comes from warfare. See also Promiscuous Mode, which has nothing to do with this definition – except that it sounds as though it should.

incident angle The angle between an incident ray and a line perpendicular to an optical surface.

inclination The angle between the orbital plane of a satellite and the equatorial plane of the earth.

inclination angle The angle at which a satellite orbit is tilted relative to the equator. Globalstar satellites are at a 52 degree inclination. the only reason I mention Globalstar is that I don't know the inclination of other satellites. They're all different. When I find out the other tilts, I'll include them in an upcoming edition of this dictionary. See also Inclined Orbit.

inclined orbit Any nonequatorial orbit of a satellite. Inclined orbits may be circular or elliptical and may be synchronous or nonsynchronous. Inclined orbits are used for many reasons – for photographing, for reaching places in the extreme north and south which normal geosynchronous satellites can't reach.

inclusion indicator A code that identifies whether an itemized call or credit card call is included in a Calling Plan, which should not be included in the billing total.

incoherent light A random form of light whereby the phase of the light signal is unpredictable. LEDs (Light Emitting Diodes), for example, are incoherent light sources. See Coherent Light for a better understanding. See also LED and OCDMA.

incollect In the international toll business, an incollect is a collect call that is received from another country. It is also known as a received collect. The international long-distance carrier whose customer received the collect call bills its customer for the call. The international long-distance carrier whose customer initiated the collect call receives its share of the revenue for the call through the international toll settlements process.

incoming call identification ICI. Some way of telling the user who's calling. It might be the caller's extension number on an LCD screen or even the caller's name spelled out, e.g. "KATE BRODIE-DAVID CALLING." Today, most "incoming call identification" is done totally within one PBX. However, the days of ISDN and ITU Signaling System Number 7 are arriving. They promise to deliver to us the phone number of everyone calling us – from within the PBX or key system and from the outside world.

incoming calls barred An interface configuration option that blocks call delivery attempts. Only outgoing calls are allowed.

incoming first failure to match IFFM. A telephone company term. The multi-line recycle feature permits a second incoming attempt to complete to hunting lines. This feature may substantially lower the total office incoming matching loss in offices with high multiline development. At the same time, individual lines may be incurring a high %IML. IFFM registers will count all incoming first failures to match. The system objective for this measurement is 2.3%.

incoming matching loss IML. A telephone company term. Percent Incoming Matching Loss (% IML) is a measurement of incoming calls unable to complete to a line equipment because of the lack of an available path between the incoming trunk (or junctor) and the called line. The engineering objective ceiling to be included in the equipment design is 2% in the busy season of exhaust.

incoming register Equipment used to receive call completion information on an interoffice call in which the completing office uses a crossbar system.

incoming server group See ISG.

incoming trunk 1. A trunk coming into a central office. 2. A PBX trunk arranged to receive calls from the central office.

incoming WATS An incoming WATS (INWATS) trunk is used exclusively for received incoming calls from a defined geographical area to a customer's PBX. An incoming WATS trunk can only be used to receive calls via the dial-up telephone network. Originally WATS lines came in only lines that could receive calls or only lines that could make calls. Now, you can buy a WATS line that handles both incoming and outgoing lines. See WATS.

increment A small change in the value of a quantity.

incremental cursor control The user-controlled function that moves the focus in increments dictated by the application. In character-based text editing, the increment is typically one character in the horizontal direction and one line in the vertical direction.

incremental zone transfer In the Domain Name Server system which operates on the Internet and in private intranets, an incremental zone transfer is a transfer of only the changed resource records in a zone.

incubator An organization that nurtures start-up companies (especially those involved with the Internet) by providing them with office space, managerial support and financing. In return, an incubator takes an equity in the company and helps it go public. The incubator hopes to make money when it sells the shares it owns – typically after the lockup period expires, six months after the stock begins first trading publicly. Some incubators do well. Most don't. See also Industry Accelerator.

incumbent One who holds position or power. See the next definition.

Incumbent Local Exchange Carrier ILEC. A term coined from the Telecommunications Act of 1996 to describe the existing (not the new) local telephone company, such as Verizon or Bell South, that was established before the Telecommunications Reform Act of 1996. All RBOCs are ILECs but not all ILECs are RBOCs. ILECs are in competition with competitive local exchange carriers (CLECs).

indecency During the late 1800s, postage rates around the world dropped, and the obscene St. Valentine's Day card became popular, despite the Victorian era being otherwise very prudish. As the numbers of racy valentines grew, several countries banned the practice of exchanging Valentine's Days cards. During this period, Chicago's post office rejected more than 25,000 cards on the grounds that they were so indecent, they were not fit to be carried through the U.S. mail.

indefeasible right of use IRU. (or INdefeasible Right of User). A term used in the underseas cable and fiber optic carrier business. Someone owning an IRU means he has the right to use the circuit for the time and bandwidth the IRU applies to. An IRU is to a submarine or fiber optic cable what a lease is to a building.

independence If two or more events occur in nature with no influence on each other they are said to be independent. The probability of these two or more events occurring simultaneously is the product of their individual probabilities. This definition of great relevance to the network planning in the telephone industry.

independent How we want our children to be as long as they do everything we say.

independent clocks A communication network timing subsystem using precise free running clocks at the nodes for synchronization purposes. Variable storage buffers installed to accommodate variations in transmission delay between nodes are made large enough to accommodate small time (phase) departures among the nodal clocks that control transmission. Traffic is occasionally interrupted to reset the buffers.

independent operating carrier IOC.

independent sideband transmission ISB. That method of double sideband transmission in which the information carried by each sideband is different. The carrier

may be suppressed.

independent software vendor ISV. Typically a company which writes and sells software, but not hardware. Manufacturers of hardware and operating systems, i.e. IBM or Northern Telecom, often contract with ISVs to produce specialized software to make their hardware and operating system more attractive.

independent telephone company A telephone company not affiliated with one of the "Bell" telephone companies. There are about 1,400 independent phone companies. They serve more than half the geographic area of the United States, but only around 15% of its telephones. The independent phone companies used to be represented by the United States Independent Telephone Association (USITA). But once Divestiture happened, the association dropped the word "Independent" from its name, accepted membership of the Bell operating companies (but not AT&T) and became USTA, which now stands for United States Telecom Association.

index 1. Think of a filing cabinet. It contains oodles of information. Think of a computer hard disk. Same thing as a filing cabinet. Oodles of information. Now think of putting everything in the filing cabinet into filing folders and putting them in alphabetical order. Makes finding things a lot easier. Now think of a computer. You ask it to find you the name of a file folder. Nothing sophisticated here. Except it's dumber than you. It starts at the top and searches down. Of course, it searches fast. But it still searches from the top down. The fastest way for it to search is to give it less stuff to search through. Thus you make an index. Just as you do in a book. Only, compared to you and me, a computer is willing to do more stupid work. It will index in alphabetical order. It will index in date order. It will index in order of how much you sold the guy recently. It will index in any order you ask it to. And many database software programs will let you keep several indexes concurrently, thus allowing you to find things quickly. The rules of database are simple: The more indexes you keep concurrently, the more time your computer will take to update its indexes every time you enter a new record or update an old record. See also Index.htm.

2. A number, as a ratio, that describes the value of something in comparison to a baseline value. The Dow Jones Industrial Index and the Index of Refraction (IOR) are examples. See also Index of Refraction.

index dip In an optical fiber, a decrease in the refractive index at the center of the core, caused by certain manufacturing techniques.

index field The field to be used when indexing a database.

index file An (optional) file used for indexing the data in a database. Index files are usually given extensions which identify them as index files. For example, when using dBASE III+, the index files are given the NDX extension.

index matching material In fiber optics, a material (liquid, gel, or cement), the refractive index of which is nearly equal to the fiber core index. It is used in mechanical splices and connectors to reduce Fresnel reflections from a fiber endface.

index of cooperation In facsimile, the product of the total line length in millimeters times the lines per millimeter divided by r. For rotating devices, the index of cooperation is the product of the drum diameter times the number of lines per unit length.

index of refraction IOR. Refraction is the deflection from a straight path by a light ray or other energy wave as it passes obliquely (i.e., at an angle that is neither perpendicular nor parallel) from one medium (e.g., glass) to another (e.g. air), in which its velocity is different. For example, when light travels through water it refracts and makes everything appear wavy and distorted, due to the difference in the speeds with which light travels through air and water. The Index of Refraction is calculated as "$n = c/v$," "n" is the "number" for which we are solving the equation, where "c" is the "constant" speed of light in a vacuum and "v" is the "variable" speed of light in the subject medium. Snell's Law of Refraction describes the refraction in terms of the angle of incidence, in consideration of the differences in the velocity of light between the two materials, and is expressed as "$n \sin(0) = n' \sin(0')$." For example, a typical air/glass boundary, where "air $n = 1$" and "glass $n' = 1.5$," a light ray that enters the glass at 30 degrees from the "normal" (i.e., perpendicular), travels through the glass at 10.5 degrees, and straightens back out to 30 degrees when it exits the glass and re-enters the air. This shift in angle through the glass medium explains why things appear to be slightly out of place when looking through a glass windowpane at an angle, but otherwise are not distorted, assuming that the glass is pure and of consistent thickness. Note: The thicker the glass, the farther the distance over which the image (which comprises reflected light rays) travels, and the greater the shift in perceived position. As the index of refraction also is dependent on wavelength, blue light refracts more than red. This phenomenon causes rainbows, and allows prisms to separate white light (which is a combination of all wavelengths) into its constituent wavelengths, each of which we humans perceive as a color of light (e.g., red or blue). Different types of fiber optic cable have different types of glass with different refractive indexes. As light spreads from the inner core to the outer edge, it's bent back inward by the surrounding cladding, which has a higher refractive index. Thereby, the total optical signal is forced to travel through the center of the core, rather than through the edge, resulting in total internal reflection. The end result is a more coherent signal at the receiver. This optical fiber example assumes that the light signal strikes the cladding at less than a "critical angle," at which the light signal passes through the cladding, rather than being refracted, or reflected at an angle. The critical angle is sensitive to the refractive index, by the way. See also Diffraction.

index profile In an optical fiber, the refractive index as a function of radial distance from the optical axis.

index.htm All home pages on the Internet are typically called Index.htm. When you send your browser to a web site, such as www.TechnologyInvestor.com, you will reach www.TechnologyInvestor.com/index.htm, though you won't see the index.htm on your address line. If you're writing a web site, you'll need to make sure that your home page is called index.htm. Stop press: Some web sites now use default.htm or default.asp as their home page.

indexed database A database indexed on a key field. Indexing allows for rapid retrieval of records through an index field. Microsoft's Outlook is not indexed, which is why it is so slow.

indication circuit X.21 circuit used to send control information from DCE to DTE.

indication Of Lights, bells and buzzers indicating that something has or is about to happen. For example, indication of camp-on to a station: short bursts of tone are periodically transmitted to the busy phone to indicate that another call is camped on and waiting.

indirect control In digital data transmission, the use of a clock at a higher standard modulation rate, e.g., 4, 8, 128 times the modulation rate, rather than twice the data modulation rate, as is done in direct control.

indirect routing See Direct Routing.

indirect tapping A current in a conductor gives rise to a magnetic field around the conductor. When the current varies, the magnetic field changes. Conversely, if a conductor is immersed in a magnetic field, changes in the magnetic field will induce currents in the conductor. A coil of wire attached to a telephone or clamped to a telephone line can pick up conversations on the telephone. This type of wiretap, where there is no physical connection between the tap and the target line, is called Indirect Tapping.

indium gallium arsenide InGaAs, a semiconductor material used in lasers, LEDs, and detectors. See Gallium Arsenide.

indium phosphide A new type of semiconductor chip. Indium phosphide chips run much faster than normal silicon chips. See Gallium Arsenide.

individual load cycling feature This is one feature of AT&T's Dimension Energy Communications Service Adjunct. Individual Load Cycling reduces energy consumption by turning devices on and off (e.g. Air-conditioning) on an hourly basis.

individual speed calling A key system or PBX feature by which a user can dial a longer number by punching one or two buttons on his phone. Sometimes this speed dial ability is programmed into the phone. Sometimes it's programmed into the system. Whichever it is, each user has a bunch of numbers he/she can speed dial. These are his/her own. No one else can speed dial them.

induced uplink An ATM term. An uplink "A" that is created due to the existence of an uplink "B" in the child peer group represented by the node that created uplink "A". Both "A" and "B" share the same upnode, which is higher in the PNNI hierarchy than the peer group in which nk "A" is seen.

induced voltage A lightning strike or changing process in a power system that results is a surge of voltage.

inductance is the property of an electric force field built up around a conductor. Inductance allows a circuit to store up electrical energy in electromagnetic form. When current flows through a wire, lines of force are built up around the wire. The field created by DC current is steady. When AC flows through a wire, the lines of force are constantly building and collapsing. An inductor is formed by winding a conductor into a coil. In long local loops, conversation gets difficult because the long wires encounter capacitive resistance. To counter this, inductors known as load coils are connected in series, increasing the inductance. When load coils, or inductors are connected in parallel, they reduce the inductance. Inductance is measure in henrys. I kid you not. See also Inductive Connection.

induction Electromagnetic transfer of energy from one coil to another.

induction coil A coil having a high turn ratio used for raising the voltage. A step-up transformer.

induction neutralizing transformer See INT.

inductive amplifier An inductive amplified is a handheld device used by telephone installers which amplifies inductive signals and plays them over a speaker built into

one end. The other end of the tool has a metal probe to touch the wire or connector to pick up tones coming from a tone generator. The two tools are used together to locate and / or test cable. You attach tone generator at one end. You go hunting for it with an Inductive Amplifier tool. If the tone comes through to a connection point (see IDF, Termination Block, Patch Panel), the cable is OK to there, there are no breaks. Tone Generators and Inductive Amplifiers are adequate for testing for voice quality circuits, but this testing method doesn't tell you if the cable is good enough for data. See Inductive Connection.

inductive connection A connection between a telephone instrument and another device by means of the electromagnetic field generated by the telephone instrument. No direct electrical connection is established between the two. See INDUCTANCE.

inductive coupling The transfer of energy from one circuit to another by means of the mutual inductance between the circuits, i.e. energy jumping from one circuit to another without actually touching it copper wire to copper wire. The coupling may be deliberate and desired as in an antenna coupler or may be undesired as in powerline inductive coupling into telephone lines. See also Inductance.

inductive pickup A coil used to tap phone lines without direct connection.

inductive tap Wiretap that is not physically connected to the telephone wires. A voltage proportional to the varying line current is induced into a coil.

inductively coupled receiver A radio receiver in which the energy in the antenna circuit is transferred to the secondary circuit by induction.

inductivity A term sometimes used to denote the dielectric constant or the specific inductive capacity.

inductor See Inductance and Inductive Connection.

indulgence Indulgences are rewards to avoid punishment in the hereafter for sins on earth today – no matter how awful those sins. They were originally given out by Popes in the the eleventh century to European warriors who participated in the Crusades. Later some Vatican official figured that indulgences could be sold to the general population of non-warriors (i.e. you and me). The money poured in for centuries and helped finance beautiful buildings in Rome. At one point, Europe was swarming with "pardoners" – people licensed by the Church to sell indulgences, sending the money back to the Vatican after, of course, taking their commission. The beginning of the end of indulgences came in 1517, when Martin Luther nailed his 95 theses criticizing papal "indulgences" to a church door. The Reformation followed. A variation on indulgences occurred in the United States in the early part of the year 2001 in the last remaining weeks and days of the Clinton dynasty. During that time presidential pardons were put out for sale and offered to miscreants. The going price ranged from a $1 million "contribution" to the Clinton presidential library to $200,000 to the president's wife's brother, who was a lawyer for some of the miscreants. When these modern indulgences were discovered by the press, there was an uproar. But no new religion was created.

industrial revolution The Industrial Revolution happened from about 1760 to 1820 in Britain. It replaced handcrafting with machinery and brought the factory and mill system. See also the Railway Revolution.

industry accelerator First there was an incubator which was an organization that nurtures start-up companies (especially those involved with the Internet) by providing them with office space, managerial support and financing. In return, an incubator takes an equity in the company and helps it go public. The incubator hopes to make money when it sells the shares it owns – typically after the lockup period expires, six months after the stock begins first trading publicly. Some incubators do well. Most don't. Then came an industry accelerator, which was both an incubator and a venture capital company and provided marketing, legal, financial and "industry building" skills – whatever they are. Industry Accelerator.

industry standard codes There are a number of industry standard codes that Verizon requires a CLEC (Competitive Local Exchange Carrier) to obtain in order to become a CLEC in the Verizon region. These include:

1. Operating Company Number (OCN) assigned by the National Exchange Carrier Association (NECA).

2. Access Carrier Name Abbreviation (ACNA) (known as AECN within Verizon-South) assigned by Telcordia Technologies.

3. Revenue Accounting Code (RAO) (required for operation in Verizon-South only) assigned by Telcordia Technologies.

4. Exchange Carrier Code (ECC) (called AECN within Verizon-North) assigned by Telcordia Technologies.

5. Common Language Location Identifier (CLLI) assigned by Telcordia Technologies.

INE Intelligent Network Element. Network equipment which contains autonomously intelligent computing capabilities.

INET Institutional Network. Generally dedicated to linking government and other public buildings for such uses as training, meetings, data and voice. Such INETs are provided by cable TV operators pursuant to 47 U.S.C. 531 for public, educational and governmental use. It is not available for consumers.

INF File A file that provides Windows 95 Setup with the information required to set up a device, such as a list of valid logical configurations for the device, the names of driver files associated with the device, and so on. An INF file is typically provided by the device manufacturer on a disk.

inference engine The AI (Artificial Intelligence) heart of a knowledge base system. The inference engine is the technology which directs the reasoning process. The inference engine contains the general problem-solving knowledge such as how to interact with the user and how to make the best use of the domain information.

InfiniBand In all networks – whether electronic, plumbing or highway – the gating factor on the network's speed is the choke point. That may the accident up ahead, the tiny pipe in your bathroom or the PCI bus in your PC. That PCI bus only transmits at 132 megabytes per second. It is now slower than many of the microprocessors that use it. InfiniBand is an important new standard designed to speed up servers – inside them, and from them to other servers. The InfiniBand Trade Association says initially InfiniBand Technology will be used to connect servers with remote storage and networking devices, and other servers. It will also be used inside servers for inter-processor communication (IPC) in parallel clusters. Customers requiring dense server deployments, such as ISPs (Internet Server Providers) and companies with many servers (for example, those serving a popular web site), will also benefit from the small form factors being proposed." The Economist Magazine of December 6, 2001 explained it thus: "Despite all the work that goes into making microprocessors and network connections ever faster, the biggest problem in computing today is neither crunching numbers nor moving data. It is getting the data into and out of the machines themselves. Long ignored in the bowels of networks, the input/output (I/O) function in computers and servers has emerged as their most significant bottleneck-sapping the performance of high-speed chips and adding myriad complications in networks, no matter how quickly chips and fibre-optics move the data. A new I/O standard called InfiniBand aims to fix all that, promising to boost I/O speeds and to make server connections as easy as plugging a toaster into an outlet. InfiniBand in effect turns the server inside out, eliminating the bottleneck by removing the I/O function from the server entirely, and allowing disparate server components to be networked together as if they were all part of a single unit. At the heart of the problem is lagging innovation in computer I/O infrastructure. Almost all computers and servers today rely on the decade-old Peripheral Component Interconnect (PCI) bus design to move information from the microprocessor to peripheral devices and out to the network. PCI connects the peripherals directly over individual copper wires, translating from the language spoken inside the machine to the languages spoken by the peripherals. For years, PCI proved more than adequate for most computing needs, handling the workflow from databases and the relatively slow network traffic. Yet PCI development has lagged far behind progress on other components of servers and computers. As processor speeds doubled every 18 months, following the venerable rule of Gordon Moore, one of the founders of Intel, and networking speeds also accelerated at breakneck pace, growth of I/O bandwidth trudged along, doubling only once every three years or so. Today, as processor speeds cross the 2 gigahertz barrier and network speeds approach 10 gigabits per second, typical PCI connections have still reached only 133 megahertz.

"Nowhere has this proved more troublesome than in data-centers – the bunker-like facilities that house corporate information in hundreds of racks of high-speed servers. As engineers work to squeeze every scrap of power from their servers to meet the growing demand for services, the additional milliseconds taken up by PCI are proving a real drag along the information highway. Worse, the general inflexibility of PCI when increasing the capacity of a data-center has made adding new servers almost a daily chore... InfiniBand seeks to solve most of these problems by eschewing direct copper connections in favor of signals that are processed and handled by logic circuitry. In short, it adds intelligence to the communication that takes place between the various peripherals and components, forming what amounts to an I/O network among them. The new standard borrows from the design of mainframe computers, which use a so-called "channel-model" to permit components in different machines to share data simultaneously with each other in a special network channel. In much the same way, InfiniBand uses channels of data to create a "network fabric" in which all components are connected into a weaving of pathways capable of supporting multiple channels of data simultaneously. The concept is not unlike that of the USB (universal serial bus) connector found on the back of most new personal computers, which allows all manner of devices to be connected simultaneously with a simple, high speed connection. The initial I/O function into the server rests in the hands of a dedicated switch

at the gateway, which carries the brunt of the work done in translating the standards used on the network (such as Internet Protocol) into the control language used internally by the server. Within the system created by InfiniBand, special adapters connect logic and memory to devices such as storage and network controllers. In theory, that allows a microprocessor on one side of a room to connect to a hard-drive and CD-ROM on the other side at blazing speeds. The net result, say InfiniBand backers, will be a dramatic increase in performance – at least a two- to four-fold increase over the speed of PCI. Better still, the new design promises a more reliable and more easily expandable system, which is also simpler to manage. Users can add an unlimited number of devices and servers to an InfiniBand network, simply by plugging them in, with no effect on performance. And because InfiniBand servers can be made to function as foot-soldiers, doing only what they are commanded to do, they can be made without bulkier peripherals – allowing more to be stuffed into smaller areas such as server racks and so saving money." www.infinibandta.org.

infinite dial In the days before the World Wide Web, an AM radio station was limited to delivering its programming on its assigned frequency in the 525-1715 KHz band (only up to 1615 KHz outside of North America), and an FM radio station was limited to delivering its programming on its assigned frequency in the 88-108 MHz band. The FCC's rules and restrictions still apply today, however with the advent of streaming audio and audio on demand on the World Wide Web, radio stations and consumers of radio programming now have more options for delivering and receiving that audio programming. This Web-enabled capability is known as the "infinite dial."

infinite loop A state in which specific steps of a program are executed repeatedly, not allowing the program execution to advance further.

infinity transmitter Also called Harmonic Bug. Infinity transmitters got their name because the original manufacturers claimed they could pick up conversations from an infinite distance. They are sometimes called harmonica bugs because original versions were activated by a 440 Hz tone created by a harmonica. An infinity transmitter is a room listening device that uses the telephone lines to send audio back to the surveillance operator. An infinity transmitter is attached to a telephone line as if it were an extension or it can be installed on a telephone instrument. To use an infinity transmitter, the target telephone is dialed and the tone or signal is sent down the telephone line before the target phone has a chance to ring. The infinity transmitter closes the telephone circuit and activates its microphone to pick up room conversations and transmit the audio down the telephone line. The disadvantage of the infinity transmitter is that it is easily detected. Telephone calls cannot be made by the target phone when the device is operating. Some telephone systems may not send audio to the target telephone line until after the phone is picked up.

InFLEXion A voice pager technology used in the transmission and storage of voice messages. It uses the ReFLEX protocol providing privacy and guaranteed message delivery with message receipt acknowledgement.

inflight packet See Mobile IP.

info banner A satellite TV and cable TV term. An info banner is a display that appears on a channel when a TV program is selected, or when the Info button is pressed on the remote control. The display indicates channel name and number, current date and time, program name, time remaining in the program, the start and stop times of the program and a brief description of the program's content.

infobahn A new term for the Information Superhighway.

infomediary Contraction of Information Intermediary. John Hagel III and Marc Singer invented the term infomediary in the 1999 book, "Net Worth," subtitled, "The emerging role of the infomediary in the race for customer information." In the book, the authors write, "Consumers won't have the time, the patience, or the ability to work out the best deals with information buyers on their own (nor will vendors have time to haggle, customer by customer). In order for consumers to strike the best bargain with vendors, they'll need a trusted third party – a kind of personal agent, information intermediary, or infomediary – to aggregate their information with that of other consumers and to use the combined market power to negotiate with vendors on their behalf. In this book, we argue that companies playing the infomediary role will become the custodians, agents, and brokers of customer information, marketing it to businesses (and providing them with access to it) on consumers' behalf, while at the same time protecting their privacy. These new entities will emerge from combinations of companies that provide unique brand franchises, strong relationships with their customers, and radically new strategies. They will become the catalyst for people to begin demanding value in exchange for data about themselves. By offering a variety of agent and targeted marketing services, they will help consumers reduce the 'interaction' cost of searching for goods at favorable prices in an environment of proliferating increasingly complex products." In short, a lot of a big words for an expansion of services that already exist in our economy – from personal shoppers, to private banks, to shopping clubs to the

AARP (American Association of Retired Persons). The book sold well, however.

infomercial A short segment shown on the video system of a plane purporting to be informational/newsy and educational. In fact, the segment is a commercial paid for by the company whose products and/or services are featured. According to research, infomercials create three types of phone calls – order calls, inquiry/incomplete calls and customer service calls.

informatics The sciences of collecting, organizing and reporting scientific data.

information appliance A poorly-defined term for a device which transmits and receives emails and has a browser, allowing it to browse the Internet and surf the Web. the theory is that an information appliance is cheaper than a full-blown PC. That's the theory. But PCs have come down in price and information appliances have become more complex and gone up in price. Some people are now calling smart cellphones information appliances because they include a personal digital assistant (calendar, email, phone book and note taking)

information at your fingertips At Fall Comdex 1990 Bill Gates, Microsoft chairman, suggested the idea. With Information at your fingertips, he said, PC users can easily access company wide information "anywhere at anytime" through an icon-based graphical user interface. In the speech, Gates demonstrated applications that used Object Linking and Embedding (OLE), Dynamic Data Exchange (DDE), handwriting recognition, cellular communications and multimedia. See OLE.

information broker Technique used to optimize performance on large metadirectories by storing some information locally in participating directories. The metadirectory will search for and locate the information in the local directory if necessary. See also Metadirectory and Web Services.

information center mailboxes An Octel term for a voice bulletin board on a voice mail system. Here's their explanation: Multiple callers can access, directly or indirectly, recorded announcements containing information that would otherwise have been given live by employees. Callers are frequently "outside" users of the system. One type of "listen only" mailbox simply plays the messages to the callers. This technology, sometimes known as audiotex, makes it possible to create a verbal database so callers can select which information they want to hear. Another type of Information Center Mailbox prompts callers to reply to announcements. Callers wanting further information can be given the opportunity to leave their names and phone numbers after listening to a product description. They can also be transferred to a designated employee who can immediately take an order. If desired, a password can be required before confidential or controlled access information can be heard.

information digits CDR call type options. Two digit codes which precede the 7- or-10 digit destination number and inform exchange carriers and IECs about the type of line that originated the call, any special characteristics of the billing number, or certain service classes. These codes plus the destination number are part of the signalling protocol of equal access offices. These codes are defined by Bellcore. Examples: 00 - POTS, 01 - Multiparty, 02 - ANI Failure, 06 - Hotel/Motel, 07 - Special Operator Handling, 20 - AIOD, 24 - 800, 27 - Coin, 30 - Unassigned DN, 31- Trouble/Busy, 32 - Recent change or disconnect, 34 - Telco Operator, 52 - Outward WATS, 61- Cellular 1, 62 - Cellular 2, 63 - Roaming, 70 - Private Pay Phone, 93 - Private virtual Network.

information element The name for the data fields within an ISDN Layer 3 message.

information engineering Coined by James Martin, an erstwhile prolific writer in data processing, the term refers to systems within data processing and their impact on giving the corporation a greater competitive edge. In short, a fancy term for Management Information Systems (MIS), which itself was a fancy term for DP, namely Data Processing.

information field A Frame Relay term referring to the variable length field of data, which can include either user payload or internetwork control data to be passed between routers or other intelligent end user devices. The information field may be 0-4,096B, although ANSI recommendations are that the field be 1600B, which accommodates most LAN packets.

information frame Frame in HDLC, DDCMP, or related protocols containing user data.

information highway A term coined by Al Gore. This fact affirmed by Dan Lynch, the man who started the trade show, InterOp and who was very heavily involved with Internet from the very beginning. As the term got developed and people got turned on by the idea, it became known as The Information Superhighway. See also Information Superhighway.

Information Outlet IO. Sort of like an AC power outlet, but a little more cerebral. A connecting device designed for a fixed location (usually a wall in the office) on which horizontal wiring subsystem cable pairs terminate and which receives an inserted plug; it is an administration point located between the horizontal wiring subsystem and work location wiring subsystem. Although such devices are also referred to as jacks, the term information outlets encompasses the integration of voice, data, and other communication services that can be supported via a premises distribution system.

information packet A bundle of data sent over a network. The protocol used determines the size and makeup of the packet.

information page mapping See ADSI.

information payload 50.112 Mb/s of bandwidth allocated within each SONET STS-1 channel to carry information end-to-end. Also known as STS- 1 envelope capacity.

information processing Data to achieve a desired objective. Also called data processing.

information provider A business or person providing information to the public for money. The information is typically selected by the caller through touchtones, delivered using voice processing equipment and transmitted over tariffed phone lines, e.g., 900, 976, 970. Typically, billing for information providers' services is done by a local or long distance phone company. Sometimes the revenues for the service are split by the information provider and the phone company. Sometimes the phone company simply bills a per minute or flat charge. A typical "information provider" is American Express, which provides a service – 1-900-WEATHER. By dialing that number you can touchtone in city names and find out temperatures, weather forecasts, etc. Calling 1-900-WEATHER costs several dollars a minute.

information scent The visual and linguistic cues that enable a searcher to figure if a source, a Web site, has the information they seek.

information service The Telecommunications Act of 1996 defined Information Service as the offering of a capability for generating, acquiring, storing, transforming, processing, retrieving, utilizing, or making available information via telecommunications, and includes electronic publishing, but does not include any use of any such capability for the management, control, or operation of a telecommunications system or the management of a telecommunications service. The basic distinction between information services and normal voice calls is that the former are taxed. The latter aren't. According to some people Voice over IP (VoIP) is an information service. According to others, it's not.

information signals A Bellcore definition. Information signals inform the customer or operator about the progress of the call. They are generally in the form of universally understood audible tone (for example, dial tone, busy, ringing) or recorded announcements (for example, intercept, all circuits busy).

information superhighway A very vague concept which Senator Al Gore created in the early 1990s and which gained great popularity when he became vice president and the Clinton/Gore administration started pushing the concept. The Information Superhighway is a term sufficiently vague that it can mean anything to anyone. It can mean a gigantic Internet reaching everybody in North America, or the planet (if you're that expansive). It could just as easily mean a combination 500-channel interactive cable TV system with full video on demand to every household in North America. Somewhere in all this is the idea that easy access to large amounts of information will enrich our lives immeasurably. Who's going to get first access to it all, what the precise technical details will be, and who's going to pay for it are, naturally, minor details to be worked out. We can be assured that the details will be worked out, since the idea originated in Washington, DC., home of so many practical ideas.

Information Technology IT. A fancy name for data processing, which became management information systems (MIS), which became information technology. All the same thing, essentially. See also IT.

Information Technology District The Information Technology District (ITD) is New York City's fastest growing totally-wired community. Anchored by the New York Information Technology Center @ 55 Broad Street and sharing the Downtown Business Improvement District's boundaries of City Hall to the southern tip of Manhattan, the ITD serves as the headquarters for Silicon Alley. The ITD is home to more than 250 IT companies, from web page developers to financial modeling firms. According to promotion from ITD, these companies are quickly emerging as the City's prime economic generators, creating jobs and innovative products, and serving as pioneers in the ongoing revitalization of Downtown into a 24-hour, 21st Century global community.

Information Technology Services I got this email from Ferrell Mallory, Managing Director, IT Operations Brigham Young University, Salt Lake City. "Remember all that preaching you did years back about telecom operations getting swallowed up by MIS types? Well, I'm there again. I was in and out of our campus MIS operation 14 years ago when they decided my telecom operation was a "major misfit" until the campus became so dependent upon the data network we installed - mostly from telecom revenues - AND everyone blamed the network if anything vaguely related to the network was inoperative, e.g., network infrastructure, servers, clients, keyboards, electrical power, air conditioning, etc., etc. So, we reorganized again, this time along functional lines instead of technology and I'm back in what we call "Information Technology Services". I still have the telephone system and data communication system but have added servers, and all campus instructional media systems and services.

information to go A term coined by Digital Equipment Corporation to refer to the transmission of data over airwaves instead of fixed wires.

information warehouse See Data Warehouse.

INFOSEC A military term for information systems security. See NSA.

InfoSpace A service that helps surfers locate listings of people, businesses, government offices, toll-free numbers, fax numbers, e-mail addresses, maps and URLs, all on one Web site. InfoSpace has developed a patent pending technology that integrates all of these services. www.infospace.com.

Infostrada SpA A new phone company which started service in July 1998, competing against Italy's erstwhile monopoly, Telecom Italia. Infostrada is controlled by Olivetti SpA and the German company, Mannesmann AG.

infrared IR. The band of electromagnetic wavelengths between the extreme of the visible part of the spectrum (about 0.75 um) and the shortest microwaves (about 100 um). This portion of the electromagnetic spectrum is used in some fiber-optic transmission systems, but more commonly for communications through the air. In such free-space optics applications, the system typically consists of a two transmitter/receivers. The infrared light signal is transmitted through a focused lens to a collecting lens in the receiving device. Transmission rates of as much as 622 Mbps can be achieved over distances of as much as several miles, and systems running in the Gbps range are being developed for WLL (Wireless Local Loop) application. Typically deployed in campus environments or other very short-haul applications where cabled systems are not possible or practical, infrared offers the advantage of no FCC licensing requirements, thereby sometimes making it preferable to microwave. Infrared also is commonly used for short haul (up to 20 feet) through-the-air data transmission. With the adoption of infrared standards at a meeting of over 50 manufacturers in June 1994, many PC device manufacturers began development of the Infrared Serial Data Link (IRDL) with speeds up to 1.5 Mbps. This standard is designed to ensure that products sporting this link will work together and interchangeably. See also IrDA and WLL.

Infrared Data Association See IrDA.

infrared fiber Optical fibers with best transmission at wavelengths of 2 um or longer, made of materials other than silica glass.

infrared serial data link As a result of a meeting at Microsoft of over 50 manufacturers in June 1994, many PC devices will begin sporting something called the "Infrared Serial Data Link," an infrared through-the-air (up to 20 feet) link with speeds up to 1.5 million bytes per second. This standard is designed to insure that products sporting this link will work together and interchangeably. There is now an organization called I.R.D.A., the Infrared Data Association, representing over 80 manufacturers.

infrared technology IR. Infrared communications systems use very high frequencies, just below visible light in the electromagnetic spectrum to carry data. Like light, IR cannot penetrate opaque objects; it is either directed (line of sight) or diffuse technology Inexpensive directed systems provide very limited range (three feet) and typically are used for personal area networks but occasionallY used in specific wireless LAN applications. High performance directed IR is impractical for mobile users and is therefore used only to implement fixed subnetworks. Diffuse (or reflective) IR wireless LAN systems do not require line of sight, but cells are limited to individual rooms. See also Infrared.

infrastructure A collection of those telecommunications components, excluding equipment, that together provide the basic support for the distribution of all information within a building or campus.

ING CallING party. The ING calls the ED, or callED party, to set up a data transfer. ING and ED apply to any type of data transfer, including both voice and data.

ingredient technology See Indeo Video.

ingress 1. Ingress is a cable TV term. Ingress occurs when strong outside signals leak into a CATV coaxial cable and interfere with the signal quality inside the home and nearby homes. Picture a car driving along outside a house. The car has a strong CB radio. It sends the signal out. It is picked up by the coaxial CATV cable in the house, which then sends it to nearby houses. The primary cause of ingress is cheap wiring and/or loose connectors. But the interfering signal is caused by radio transmitters of all types (including short wave transmitters), electrical appliances, motors with brushes, light dimmers or speed controls on toys. Leakage is really a shielding problem. The number of houses that can be affected

by ingress depends on the strength of the signal and the number of service areas around a CATV node, which could be as many as 1,000. Companies like Trilithic in Indianapolis are expert in measuring ingress. See also Leakage.

2. The act of entering a network. It derives from the classic definition of the word meaning the act of going in or entering.

ingress policing A mechanism on a packet-switched network to regulate the processing and forwarding of incoming packet traffic, in order to ensure that all inbound traffic flows receive fair service. The goal of ingress policing is to reduce a negative impact on forwarding performance for all incoming traffic flows when one of those flows is excessive. Ingress policing involves discarding packets from the excessive inbound traffic flow. When packets are lost, for whatever the reason, the node or station that sent the packets assumes that it is the result of network congestion and responds by reducing packet flow to the node where the packets were lost. The sending station or node continues to reduce the packet flow rate until packet loss disappears, at which point it begins to slowly increase the packet flow rate to as high a level that it can without packet loss.

inheritance A term from object oriented programming. Data abstraction can be carried up several levels. Classes can have super-classes and subclasses. In moving to a level of greater specificity, the application developer has the option to retain some attributes and methods of the super-class, while dropping or adding new attributes or methods. This allows greater flexibility in class definition. It is even possible in some languages to inherit from more than one parent. This is referred to as multiple inheritance. See Object Oriented Programming.

INIC ISDN Network Identification Code.

INIM ISDN Network Interface Module (INIM) is both hardware and software. It does the job of an NT-1, so the physical network interface is ISDN-U. When calls arrive, the INIM collects the number dialed and Caller ID. This data is passed on to the Call Processing Module, which does the actual call handling. When you go off hook to place an outbound call, the INIM assigns an available ISDN B-channel to the call. For 'Find Me' scenarios, the INIM lets Front Desk place multiple, simultaneous out-bound calls. During data calls, the INIM constantly monitors the data transmission rate on the ISDN line. The INIM will automatically build up or tear down the second ISDN B channel from a data call to match bandwidth requirements. Because telcos charge for usage per B-channel, the INIM uses both B channels only when necessary. If both B channels are doing data when a new voice or fax call arrives, the INIM instantly tears down one of the B channels to let the new call through. Ditto for when you make an outbound voice or fax call.

INIT An INIT is the Macintosh System 7 equivalent of a terminate and stay resident (TSR) program . An init might load to initialize a fax modem, screen saver, etc. Similar to the DOS environment, some inits conflict. When troubleshooting operating system problems, remove inits first.

initial/additional indicator A Verizon definition. An indicator that identifies whether Local Calling Plan usage charges are billed at the Initial or Additional rate.

Initial Address Message IAM. A SS7 signaling message that contains the address and routing information required to establish a point-to-point telephone connection.

initial answer Initial answer refers to the point in time at which a computer telephony system answers an incoming call. Many computer telephony systems require significant processing to set up to answer incoming calls. For example, the system may examine the incoming ANI, DNIS, or PBX integration data to determine how to answer (which prompt to use), or where to switch the call. This can involve significant database access and processing time. Therefore, the ability to handle large number of incoming calls (especially in burst mode) may delay the initial answer. The delay from when a call reaches a computer telephony system until the computer telephony system answers the call (the initial answer) is usually a key response time to understand when testing a computer telephony system.

Initial MAC Protocol Data Unit IMPDU. A Connectionless Broadband Data Service (CBDS) term that corresponds to the L3 PDU in Switched Multimegabit Data Service (SMDS). CBDS is the European equivalent of SMDS.

initial period The minimum billing period on a call. For interstate or inter-LATA AT&T calls, the initial period is one-minute. Some non-AT&T long distance companies have initial periods under one-minute. This also applies to local calls in Measured areas.

initial program load The initial loading of generic and/or configuration software into a PBX or other phone system. The Initial Program Load is a pain in the rear end. But an even bigger pain is what happens when you lose your programming and you've forgotten to back it all up.

initial public offering IPO. The initial offering of shares in a company to the public to raise capital for any number of reasons (i.e., reduce debt, research and development, expansion). Shares are sold to investment banks, who then sell them to the public via retail brokerage firms.

initial sequence number ISN. Generated at each end of TCP connection to help to uniquely identify that connection.

initialization string A group of commands sent to the modem by a communications program at start-up – before the number has been dialed. Such a string tells the modem to set itself up in a way that will make it easy to correctly communicate with a distant modem.

initialize 1. Setting all counters, switches, addresses or contents of storage to zero at the beginning of, or at prescribed points in the operation of a computer routine or a communications transfer; a major function of "rebooting" a computer, giving everything a "reset".

2. In disk management, the process of detecting a disk or volume and assigning it a status (for example, healthy) and a type (for example, dynamic).

initializing terminals An ISDN term. These devices, sometimes called self-initializing terminals, are basically ISDN terminals that can generate their own terminal identification number. This makes it easier for the network and the terminal to agree on the number to use.

initiating event An event that causes the ASC (AIN Switch Capabilities) to assign an ID to a connection segment for a certain user ID and to communicate with a SLEE (Service Logic Execution Environment) for the first time with respect to the combination of the specific user ID and connection segment ID.

initiating switch The switch which initiates the Query on Release capability by identifying a portable NPA-NXX and populating a QoR indicator in the Initial Address Message. In the context of Local Number Portability (LNP), a dialed TN is routed to the destination switch as usual, and then a dialed Query on Release (QoR) is accomplished by the donor (destination) switch sending a release cause code back to the originating switch within the IAM (Initial Address Message), and at that point the originating switch would recognize the number as ported and initiate an LNP data base lookup. Latency is an issue with this method of doing LNP.

initiator 1. A SCSI device, usually a host system, that requests an 1/0 process to be performed by another SCSI device.

2. The Bluetooth device initiating an action to another Bluetooth device. The device receiving the action is called the acceptor. The initiator is typically part of an established link.

injection laser Another name for a semiconductor or diode laser. See Injection Laser Diode.

Injection Laser Diode ILD. A solid-state device that works on the laser principle to produce a light source for optical fiber.

Injection Molding The process used to inject molten polymer into a mold. Connector backshells are often injection molded.

injector A combination of a special modem and a coupler used to interface a communications circuit to an electrical power line in order that voice, data and video signals can be sent over the power grid. The term is used in Broadband over Power Line (BPL) technology. See BPL.

INL See Internode Link.

inline ads A format of Web-based advertising in which advertising content is placed on the same layer as site content. They are either graphical, HTML-based, or a mixture of the two. They generally reside within the flow of a page's content, though they may be placed outside the margin of a page. These ads make up the bulk of impressions served. The Internet Advertising Bureau (IAB) defines a number of ad sizes in the inline category; the dominant 468 x 60 pixel banner ad, the 120 x 60 pixel "button #2" ad, and the large 120 x 600 pixel "skyscraper" ad. See also Pop-Up Ad.

inline image A built-in graphic that is displayed by a Web browser as part of an HTML document and is retrieved along with it.

inline plug-in An application that, when inserted into a Web browser that supports it (through an installation procedure), allows greater functionality and flexibility of the browser to view multimedia that would otherwise require an outside application.

inline power In the olden days when you connected things with wires, the wires carried only the information – with one notable exception, the standard phone line with its 48 volts direct current, which was enough power to ring the phone and let you talk on it. Eventually phones got more complicated – they became multi-line PBX and key system phones and they became answering machines and speakerphones and they needed more power for them to do their magic. Hence, phones started sprouting a second set of power wires (usually carrying DC power) and, often, a power transformer that converted high voltage (110 or higher) to DC. Then came along local area networks, such as Ethernet, firewire and USB. Devices at the end of such connections – whether IP telephones, digital cameras,

PDAs, iPods, or cell phones – all needed power to run themselves. So the idea arose, let's add a pair of wires for the low voltage DC power. Hence, the concept of inline power. It's a neat, useful idea. It means each device can be run without the need for an external power supply. It also means that I can charge my BlackBerry with a thin USB cable from my laptop. Some Ethernet cables work the same way. The maker adds a pair for inline cable and bingo, the device at the other end now receives enough power to run it and/or charge it batteries. With Ethernet, there are basically two ways of delivering inline power: first, on the signal pairs (i.e., pins 1, 2, 3, 6); or second, over the spare pairs (i.e., pins 4, 5, 7, 8). In the first scenario, data and power share the same pairs. In this case the electrical DC power is sent over different frequencies to the data transmission. In the second scenario, power uses the spare pairs (pins 4, 5, 7, 8) and data goes on the signal pairs (pins 1, 2, 5, 6) as usual. An Ethernet device with PoE ports is said to have "inline power."

ILNP Interim Local Number Portability. See LNP and Number Portability.

INMAC International Network Management Center.

INMARSAT The INternational MARitime SATellite service that has satellites and provides mobile communications to ships at sea, aircraft in flight, vehicles on the road and to small stationary satellite antennas which people carry with them. Inmarsat provides dial-up telephone, telex, fax, electronic mail, Internet access, data connections and fleet management. Typically you pay a per minute charge for the use of Inmarsat's communications services. Inmarsat describes itself as the international mobile satellite organization. It is based in London. Formed as a maritime-focused intergovernmental organization, Inmarsat has been a limited company since 1999. Inmarsat Ltd is a subsidiary of the Inmarsat Ventures plc holding company. It operates a constellation of geostationary satellites designed to extend phone, fax and data communications all over the world. The constellation comprises five third-generation satellites backed up by four earlier spacecraft. Today's Inmarsat system is used by independent service providers to offer a range of voice and multimedia communications. Users include ship owners and managers, journalists and broadcasters, health and disaster-relief workers, land transport fleet operators, airlines, airline passengers and air traffic controllers, government workers, national emergency and civil defence agencies, and peacekeeping forces. Keystone of the strategy is the new Inmarsat I-4 satellite system, which from 2004 will support the Inmarsat Broadband Global Area Network (B-GAN) – mobile data communications at up to 432kbit/s for Internet access, mobile multimedia and many other advanced applications. www.inmarsat.com.

inmate call management A typical inmate call management application package serves the special requirements of correctional institutions. By assigning Personal identification Numbers (PINs), inmates may be allowed debit calling only, collect calling only, or a combination of both. Some packages can also access external commissary databases, allowing immediate debit of an inmate's account in real time. They may also do:

- billed number screening, reducing billing fraud.
- blocking of calls after commissary funds are exhausted, minimizing lost revenue on commissary funds.
- identifying announcement to called party.
- time-of-day restrictions to specified destination numbers.
- temporary call prohibition for specific PINs.
- restricted destination numbers for specific PINs.
- system-wide deny numbers.

inmate calling Dennis Squires from Bell Atlantic says, "Prison inmates can get very creative when they have a lot of time on their hands, like 20 years." So, to avoid stealing of phone calls by prisoners, some phone companies have a service called "Inmate Calling." The inmate enters his authorization code, which allows him to make phone calls – but only to authorized phone numbers, i.e. mothers, lawyers, bail bondsmen, etc.

INMC International Network Management Center.

INMLM Integrated Network Manager Link Module.

INMS Integrated Network Management Services.

INN See InterNode Network.

inner coder The second of two concatenated error-correcting coders in transmission and the first of two in reception used in the digital signal processing (DSP) of bitstreams in radio communications systems.

inner duct A flexible plastic conduit that's placed in larger conduits. It's used for two reasons: when different companies lease space in the same conduit and/or to provide yet another layer of protection for precious, fragile optical fibres.

inner wires For a standard four wires connection, the wires fastened to the inner two pin locations in a jack or connector. For example, in a six position connector, the inner wires are pins 3 and 4.

innerduct A nonmetallic raceway, usually circular, placed within a larger raceway.

inoculatte The Washington Post's Style Invitational asked readers to take any word from the dictionary, alter it by adding, subtracting or changing one letter, and supply a new definition. This one is one of the winners. Inoculatte means to take coffee intravenously.

INode Integrated Node (SSP and STP).

INOS Intelligent Network Operations System. An operations system designated to manage, monitor, and control elements of an intelligent telecommunications network.

iNOW! iNOW! is a standards-based, multi-vendor initiative announced in December 1998 to provide interoperability among IP telephony platforms. Ascend, Cisco, Clarent, Dialogic, Natural MicroSystems and Siemens will be working with the iNOW! Profile to make their gateways and gatekeepers interoperable with each other's products and with those from Lucent and VocalTec. The specs on the standard are due to be published in January 1998. According to the January, 1998 release, the iNOW! Interoperability Profile will detail how to achieve interoperability between gateways from different vendors and interoperability between gatekeepers from different vendors. Up until recently, carriers and callers were limited in the destinations they could reach since calls had to be terminated on the exact same platform from which they originated. Internet telephony service providers had to choose between dependence on a single vendor or operating multiple parallel networks of incompatible gateways. According to the release, Lucent and VocalTec are responsible for the development of the programming and engineering for the iNoW! Interoperability Profile. The interoperability guidelines are based on the International Telecommunications Union (ITU) H.323 standard, and the upcoming H.225.0 Annex G standard.

INP 1. Interim Number Portability. Interim method of porting a LEC owned phone number to a CLEC via remote call forwarding when a LEC switch does not fully support LNP functionality. See Number Portability.

2. International Network Provisioning.

INPA Interchangeable Numbering Plan Area. An area code that looks like an office code. There is no particular name for an office code that looks like an area code, according to Lee Goeller.

INPS Intelligent Network Product Support.

INPUT A signal fed into a circuit.

input buffer limiting Buffering strategy that divides buffer at a mode into two classes, both available to transit packets but only one available to packets input at the node.

input circuit The grid circuit of an electron tube.

Input/Output I/O. Input and output are two of the three functions that computers perform (the other is processing). Input/Output describes the interrelated tasks of providing information to the computer and providing the results of processing to the user. I/O devices include keyboards (input) and printers (output). A disk drive is both an input and an output device, since it can both provide information to the computer and receive information from the computer.

Inquiry A request for specific information.

INS 1500 A term for a digital T-1 line transmitting at 1.544 million bits per second.

INS 64 A term for a digital ISDN BRI line transmitting at 144,000 bits per second.

insanely great When Steve Jobs of Apple and Pixar fame evaluates something it's either "insanely great" or an expletive that's unpublishable here. Jobs is a talented fellow.

insanity Doing the same thing again and again, but expecting a different result the next time.

insatiable The demand for telecommunications services was insatiable, according to common myth around the late 1990s – just before the telecom bubble burst.

insertion gain The gain resulting from the insertion of a device in a transmission line, expressed as the ratio of the power delivered to that part of the line following the device to the power delivered to that same part before insertion. If more than one component is involved in the input or output, the particular component used must be specified. If the resulting number is negative, an "insertion loss" is indicated. This ratio is usually expressed in decibels. See Insertion Loss.

insertion loss Also called attenuation. The difference in the amount of power received before and after something is inserted into the circuit (viz. another telephone instrument) or a call is connected. In an optical fiber, insertion loss is the optical power loss due to all causes, usually expressed as decibel/kilometer. Causes of insertion loss may be absorption, scattering, diffusion, dispersion, microbending, or methods of coupling power outside the fiber. In lightwave transmission systems, the power lost at the entrance to a waveguide due to causes, such as fresnel reflection, packing fraction, limited numerical aperture, axial misalignment, lateral displacement, initial scattering, or diffusion. See also Attenuation and Insertion Gain.

inside dial tone The dial tone provided by a PBX. This lets you dial an internal num-

ber. When you dial 9 (internal extensions never start with 9), you get the dial tone provided by your local RBOC. Outside North America, you often dial 0 to get an outside line.

inside link An ATM term. Synonymous with horizontal link.

inside plant Everything inside a telephone company central office. Thus, electronic equipment in buildings. Includes central office switches, PBX switches, broadband equipment, distribution frame, power supply equipment, etc. It doesn't include telephone poles, cable, terminals, cross boxes, cable vaults or equipment found outdoors. See also Inside Wiring and Plant.

inside telephone wiring Telephone Wiring: For a residence, the wiring usually starts where the line enters the house. It's called the demarcation line. For businesses, the location of the demarcation point varies. See also Inside Wiring.

inside wiring That telephone wiring located inside your premises or building. Inside Wiring starts at the telephone company's Demarcation Point and extends to the individual phone extensions. Traditionally, Inside Wiring was installed and owned by the telephone company. But now you can install your own wiring. And most companies installing new phone systems are installing their own new wiring because of potential problems with reusing the old telephone company cable. See also Inside Plant.

insider attack Attack originating from inside protected network.

insider trading Insider trading occurs when a person sells or buys shares based on information that had a material effect on the company or its shares but was not public knowledge. The typical "insider" is a director or officer of the company. But it can also be the company's investment banker or lawyer, or friend of the CEO. For the insider to be prosecuted he or she typically must have benefited financially by the information.

INSIG In-signaling.

installation The physical hook-up and diagnostic testing of a PBX switch, cabinet, or peripheral item prior to a cutover and maintenance acceptance by the maintaining vendor.

installed base How many of whatever are in and working. Installed base is often confused with annual shipments. They're very different. Shipments is what goes out the factory. Installed base is what's out there. The equation is: Installed base at beginning of year plus annual shipments less equipment taken out of service during the year is equal to the installed base at the end of the year.

installer's tone Also called test tone. A small box that runs on batteries and puts an RF tone on a pair of wires. If the technician can't find a pair of wires by color or binding post, they attach a tone at one end and use an inductive amplifier (also called a banana or probe) at the other end to find a beeping tone.

instance ID An ATM term. A subset of an object's attributes which serve to uniquely identify a MIB instance.

Instant Trademark for MICOM's family of local data distribution and data private automatic branch exchange (PABX) products.

instant messaging I'm logged into the Internet. I load some software. It shows me that you're also logged into the Internet. I type you a message. You see it on your screen the moment I hit "send." You type your reply and send it. I see it. Bingo, an Internet service that has come to be called "instant messaging." Instant messaging is essentially real-time, on-line electronic mail. Popular Instant Messaging started in November 1996 with software called ICQ, created by an Israeli firm called Mirabilis. ICQ had over 850,000 users in six months and was bought up by America Online in 1998. AOL introduced its hugely-popular variation (AOL Instant Messenger). Then Microsoft introduced its software called MSN Messenger. Then Yahoo! introduced theirs. I like Yahoo! because you can attach a web cam and your bored daughter can watch you work, while she plays or otherwise goofs off. As of writing, none of these instant messaging softwares is compatible with the other. But there is talk in the trade press of eventual standards. There are serious reasons we need standards. Instant messaging is evolving into much more than a tool for sending typed messages to buddies online. Just as the original Web browsers revolutionized the way average users connect to Internet content, today's instant message screens are evolving into easy-to-use connections for linking people at any given moment on the Internet via text, voice and video. All the new capabilities will be built on a single critical assumption: knowing that a person is online. That, in turn, makes it possible for electronic merchants and providers of online services to reach Internet users with information or incentives – at the precise time they are able to react, namely when they are online in front of their screen, an easy target. Tools are being integrated into instant messaging software that permit the immediate delivery of an increasing array of data that does not come from friends or family. America Online has unveiled a version of its instant messaging software that automatically delivers tailored news headlines and stock quotes. Instant Messaging began life as a humble notification system in MIT's computer science department in the late 1980s. In 1983, MIT's Laboratory for Computer Science began Project Athena, a network of workstations and servers for undergraduate use. By 1987, there were thousands of workstations and several servers, it had become difficult to quickly get users vital messages, such as warnings of that a building's network was about to go down. Administrators needed something faster than e-mail, which could take hours to deliver. A team led by C. Anthony DellaFera began work on "Zephyr", the first widely used instant-messaging system, which was deployed in 1988. Though the system was intended to send system status notifications and alerts to users, students began to use it to pass messages amongst themselves. When a user logged on, notifications went out and friends could send that user messages which would pop up in separate windows on the screen.

instant on Buy a PC (Personal Computer). Turn it on. Bingo, it's already loaded with Windows or OS/2. Instant On is a new term for preloading software onto hard disks of new computers and shipping those computers already pre-loaded with that software.

instantaneous override energy function IOEF. A feature of the AT&T PBX Dimension Energy Communications Service Adjunct (ECSA), which allows the user to turn all the ECSA energy functions ON or OFF. IOEF is most often used for periodic maintenance, or to adjust to sudden changes in weather. Dimension is no longer made. I'm guessing that Lucent, which is the spun-off part of AT&T which now makes PBXs, supports this feature on their newer PBXs.

Institute for Telecommunications Sciences ITS is the research and engineering branch of the National Telecommunications and Information Administration (NTIA), which is part of the U.S. Department of Commerce (DoC). www.ntia.doc.gov. See NTIA.

instruction register The register which contains the instruction to be executed and functions as the source for the subsequent operations of the arithmetic unit.

Instructional Television Fixed Service ITFS. A service provided by one or more fixed microwave stations operated by an educational organization and used mainly to transmit instructional, cultural and other educational information to fixed receiving stations.

insulated wire Wire which has a nonconducting covering.

insulating materials Those substances which oppose the passage of an electric current through them.

insulation A material with very high resistivity used to protect conductors. Insulation is usually extruded over the wire or conductor after the drawing process. In short, insulation is a material which does not conduct electricity but is suitable for surrounding conductors to prevent the loss of current they're carrying.

Insulation Displacement Contact (or Connection) IDC. The IDC has replaced wire wrap and solder and screw post terminations as the way for connecting conductors (i.e. wires carrying telecom) to jacks, patch panels and blocks. Insulation Displacement Connections are typically two sharp pieces of metal in a slight V. As the plastic-covered wire is pushed into these metal teeth, the teeth pierce the plastic jacket (the insulation) and make connection with the inside metal conductor. This saves the installer having to strip off the conductor's insulation. This saves time. Since IDCs are very small, they can be placed very close together. This reduces the size of jacks, patch panels and blocks. IDCs are the best termination for high speed data cabling since a gas-tight, uniform connection is made. The alternate method of connecting wires is with a screw-down post. There are advantages and disadvantages to both systems. The IDC system, obviously, is faster and uses less space. But it requires a special tool. The screw system takes more time, but may produce a longer-lasting and stronger, more thorough (more of the wire exposed) electrical connection. The most common IDC wiring scheme is the 66-block, invented by Western Electric, now Lucent. But there are other systems – from other telecom manufacturers. See Punchdown Tool.

insulation resistance I.R. That property of an insulating material which resists electrical current flow through the insulating material when a potential difference is applied.

insulators Some atoms hold onto their electrons tightly. Since electrons cannot move freely these material can't easily conduct electricity and are know as non-conductors or insulators. Common insulators include glass, ceramic, plastics, paper and air. Insulators are also called dielectrics.

insurance When insurance on ships and their cargoes was introduced in 14th-century Europe, it met opposition on the grounds that it was an attempt to defeat financial disasters willed by God.

insured burst In an ATM network, the largest burst of data above the insured rate that temporarily is allowed on a PVC and not tagged by the traffic policing function for dropping in the case of network congestion. The insured burst is specified in bytes or cells.

insured rate Long-term data throughput, in bits or cells per second, that an ATM network commits to support under normal network conditions. The insured rate is 100 percent allocated; the entire amount is deducted from the total trunk bandwidth along the path of the circuit. Compare with excess rate and maximum rate.

insured traffic Traffic within the insured rate specified for an ATM PVC. This traffic should not be dropped by the network under normal conditions.

INT Induction Neutralizing Transformer. A specially designed multipair longitudinal inductor that is spliced into a wireline facility to substantially reduce low frequency steady-state or surge induced voltages and currents that may be causing noise, equipment malfunctions and/or damages or creating a personnel safety hazard. See TEN.

intaxication The Washington Post's Style Invitational asked readers to take any word from the dictionary, alter it by adding, subtracting or changing one letter, and supply a new definition. This one is one of the winners. Intaxication is the euphoria at getting a refund from the Internal Revenue Service (IRS), which lasts until you realize it was your money to start with.

INT14 A software interrupt designed to communicate with the com (serial) port in a PC. Communications programs use interrupt 14h to talk to a modem physically attached to another computer on the network.

integer A computing procedure for solving or finding the optimum solution for complex problems in which the variables are based on integers. Integers include all the natural numbers, the negatives of these numbers, or zeros.

integrated access An AT&T term for the provision of access for multiple services such as voice and data through a single system built on common principles and providing similar service features for the different classes of services.

Integrated Circuit IC. After the transistor and other solid state devices were invented, electronic circuits were designed that were more complex than ever. It became a real problem wiring all the components together. In 1958-1959, Jack Kilby and Robert Noyce independently invented the integrated circuit. An integrated circuit is a piece of silicon or other semiconductor called a chip on which is etched or imprinted a network of electronic components such as transistors, diodes, resistors, etc. and their interconnections.

integrated circuit, hybrid semiconductor A semiconductor integrated circuit in which the main parts of the circuit elements are produced as semiconductor circuit elements and that is completed by mounting additional components in the package.

Integrated Development Environment IDE. A Windows program within which a developer may perform all the essential tasks of development including editing, compiling and debugging.

Integrated Digital Loop Carrier IDLC. Access equipment which extends Central Office services; it connects to a SONET ring on the network side while providing telephony services on the subscriber side (POTS, ISDN, leased lines, etc.).

Integrated Dispatch Enhanced Network. iDEN. A wireless technology developed by Motorola, iDEN operates in the 800 MHz, 900 MHz and 1.5 GHz radio bands; the 900 MHz development is aimed at operators of digital Commercial Mobile Radio Service (CMRS), also known as ESMR (Enhanced Specialized Mobile Radio). iDEN is a digital technology using M16QAM (Quadrature Amplitude Modulation) for compression, and TDMA (Time Division Multiple Access). Through a single proprietary handset, iDEN supports voice in the form of both dispatch radio and PSTN interconnection, numeric paging, SMS (Short Message Service) for text, data, and fax transmission. See also ESMR, QAM, SMS and TDMA.

integrated EDI A term applied to the direct entry of information received electronically into the recipient's computer system. It eliminates the manual checking that is still frequently done by many recipients of EDI information, saving time and costs. It requires the sender to adhere strictly to standard and pre-agreed formats.

Integrated IS-IS Formerly Dual IS-IS. Routing protocol based on the OSI routing protocol IS-IS, but with support for IP or other networks. Integrated IS-IS implementations send only one set of routing updates, regardless of protocol type, making it more efficient than two separate implementations.

integrated messaging Also called Unified Messaging. Integrated messaging is one of many benefits of running your telephony via a local area network. Here's the scenario: Voice, fax, electronic mail, image and video. All on the one screen. You arrive in the morning. Turn on your PC. It logs onto your LAN and its various servers. In seconds, it gives you a screen listing all your messages – voice mail, electronic mail, fax mail, reports, compound documents .... Anything and everything that came in for you. Each is one line. Each line tells you whom it's from. What it is. How big it is. How urgent. Skip down. Click. Your PC loads up the application. Your LAN hunts down the message. Bingo, it's on screen. If it contains voice – maybe it's a voice mail or compound document with voice in it – it rings your phone and plays the voice to you. Or, if you have a sound card, it can play the voice through your own PC. If it's an image it may hunt down (also called launch) an imaging application which can open the image you have received, letting you see it. Ditto, if it's a video message.

Messages are deluging us. To stop them is to stop progress. Run your eye down the list, one line per entry. Pick the key ones. Junk the junk ones. Postpone the others.

It gets better. You're out. Dial in on a gateway with your laptop. Skim your messages. Dial in on a phone. Punch in some buttons. Hear your voice mail messages. Or if you're not on laptop, have your e-mail read to you. Better, have your fax server OCR your faxes and image mail and have it read them to you. A LAN server is the perfect repository for messages. It can search for them, assemble them, process them, store them, convert them, compress them, shape them, shuffle them, interpret them. Integrated messaging essentially applies intelligence and order to the messages deluging you each day. See Unified Messaging and Telephony Services.

Integrated Network Management Services 2.0 Also called INMS. INMS is customer-premises based network management platform, which allows users to monitor and manage their circuits on the MCI network. The INMS Platform is made up user workstations and a communications/database server. The server interfaces with MCI's INMS Host, which collects and forwards to the INMS server all of the customer-network related data. The INMS Host also interfaced with the CSM for service inquiry management.

Integrated Personal Computer Interface IPCI. A ROLM-designed communications printed circuit card designed to provide an IBM PC with asynchronous data transmission over two-strand wiring to and from a Rolm CBX PBX.

integrated photonics Integrated photonics are devices that include optical waveguides embedded in a semiconductor or ferroelectric substrate, and which perform some type of signal processing function under electrical control. These functions include: routing of light signals in different directions, filtering out one or several wavelengths, emitting light or modulating the intensity and/or phase of an incoming light signals. An optical waveguide consists of a region in which the refractive index is higher than in the surrounding material so that a light signal can propagate without spreading (diffraction). In any applications, only single mode waveguides are useful. This definition courtesy Ericsson.

Integrated Public Number Database IPND. A database of information about customers of telecommunications services in Australia, arranged by number, for all carriers and carriage service providers.

Integrated Services Digital Network See ISDN and Signaling System 7.

Integrated Services Digital Network User Part ISDN-UP The part of SS7 (Signaling System Number 7) that encompasses the signaling functions required to provide voice and non-voice services in ISDN and pre-ISDN architectures. The basic service offered by the ISDN-UP is the control of circuit switched connections between subscriber line exchange terminations. Definition from Bellcore in reference to its concept of the Advanced Intelligent Network.

integrated voice data There are many different meanings to this concept. The most common (we'll get arguments on this) is that a workstation or a combination telephone/personal computer on a desk can combine voice and data signals over a single communications channel. That channel might be carried digitally on one pair of wires. That is "the most integrated" voice/data. Less integrated is when you carry voice and data digitally on two pairs – one pair for transmitting and one pair for receiving. Even less integrated are some systems which use three pairs of cabling set up as one voice analog pair, one digital data pair and one power/signaling pair. In short, "integrated voice/data" means different things to different people and depends on the technology. See also ISDN.

Integrated Voice Data Workstation See ISDN, IVDT and Integrated Voice/Data.

Integrated Wireless Network A collaborative effort under way by the Departments of Justice, Homeland Security, and the Treasury to replace their legacy incompatible, standalone, land mobile radio (LMR) systems with a consolidated nationwide wireless network to support law enforcement, first responder, and homeland security requirements. When completed, IWN will provide secure voice, data, and multimedia communications among 80,000 federal agents and provide links to state, local, and tribal public safety and homeland security entities. An estimated 2,500 radio sites are being installed nationwide for IWN coverage in major metropolitan areas, on major highways, along U.S. land and sea borders, and at ports of entry.

integration ban An FCC regulation, that goes into effect on July 1, 2007, that prohibits cable operators from deploying set-top boxes with proprietary embedded security. Instead, after July 1, 2007, new digital TVs and set-top boxes will be required to have a slot installed for a platform-agnostic PCMCIA card, provided by the cable company, that provides the proprietary security. The basic idea behind the ruling is to foster consumer choices in devices that are attached to cable companies' networks like the choices that resulted after the Carterfone ruling allowed consumers to purchase third-party telephones for use with telephone companies' networks. The integration ban prohibits a cable company from providing a device such as a set-top box or digital video recorder that relies on a security mechanism that is inseparably built into the device. The thinking is that by banning integrated security, and by opening up digital TVs and set-top boxes to platform-agnostic, insertable/removable security cards, and by allowing consumer electronics manufacturers unaffiliated with cable companies to manufacture and sell to consumers set-top boxes and other devices that connect to cable companies' networks, consumers will benefit from ensuing innovation and price competition. The cable industry is now hoping that the FCC will approve, as an alternative to the PCMCIA card, a downloadable software-based security mechanism. A software-based solution will have the advantage of simplifying and lowering the cost of new-installs and the subsequent deployment of security patches, upgrades, and enhancements.

integration software If your business is like Technology Investor Magazine, it has different software programs for each business task – accounting, sales automation, order entry, inventory, etc. If you could get those pieces of software to talk to each other, and to talk sense to each other, you could save time, lower labor costs, improve your products and provide better customer service. Better yet, if you could get your internal software programs talking to software at your suppliers and customers, you could save even more money, labor and time. That's what integration software does. Every business of any size can use it to improve how their business works.

integration testing Integration (or single thread) testing is the phase in the computer telephony lifecycle that begins as individual modules are pulled together to make a complete system. Testing in this phase is related to making sure the interfaces between the various modules function correctly, and is oriented to functional issues. Inter-module functions are be checked for load stability by exposing them to a variety of real-world stimuli. Definition courtesy Steve Gladstone, from his book "Testing Computer Telephony Systems."

Integrity The decision you make when nobody is watching. Definition courtesy Alocholics Anonymous.

intel 1. Military for intelligence.

2. Also as Intel, the world's largest semiconductor manufacturer.

Intel blue Specifications required to provision the ISDN line to meet the needs of Intel's ISDN-based products. When ordering your ISDN phone line and you want to use it for data or video, tell them it's "Intel Blue." That should tell your local phone company the correct technical specifications for your line. And when you come to plug in your ISDN equipment (assuming your chosen manufacturer has made it compatible with Intel Blue), it should work. This is not a guarantee, but a probability. See ISDN.

intellectual property Intellectual property is produced by effort of the mind, as distinct from real or personal property. Intellectual property may or may not enjoy the benefit of legal protection. In the November 4, 2002 issue of Information Week, Tony Kontzer wrote that intellectual property generally takes one of four forms: inventions, ideas, trade secrets, and goodwill. Each has its own method of protection.

- A patent issued by the U.S. Patent and Trademark Office grants an inventor exclusive right to an invention for 20 years from the date of application. According to the U.S. Patent and Trademark Office's Web site, a patent can be obtained by anyone who "invents or discovers any new and useful process, machine, manufacture, or composition of matter, or any new and useful improvement thereof." In addition to being new and useful, an invention also must meet one other condition before a patent can be issued: It must not be obvious.

- A copyright registered with the Copyright Office of the Library of Congress gives authors the exclusive right to reproduce, adapt, distribute copies of, perform, or display literary, dramatic, musical, artistic, and certain other intellectual works. While the bulk of copyrights are issued for works in the arts, they're also granted to business ideas, such as source code and mission statements.

- Trade secrets fall under state law and are defined as confidential information that provides indisputable economic value. A business owner can turn to trade-secret laws if such information is improperly disclosed – by a former employee, for instance – or is otherwise illegally acquired by a competitor.

- A trademark registered with the Patent and Trademark Office grants ownership of a word, name, symbol, or device that indicates the source of traded goods and distinguishes those goods from the goods of others. The owner of a trademark can prevent others from using a confusingly similar mark, protecting the goodwill that a brand carries with it. But the trademark can't be used to prevent them from selling the same goods under a clearly different mark.

See Copyright, Patent, Trademark, Trade Secret, and WIPO.

intelligence The part of a computer which performs the arithmetic and logic functions. Also, the information impressed or modulated on a transmission carrier – either voice or data.

intelligent agent Software that has been taught something of your desires or preferences and acts on your behalf to do things for you. It might, for example, search through incoming material on networks (e-mail and news) and find what you're interested in or looking for. It might, for example, monitor your TV viewing habits, accept general instructions about your preferences and then, on its own, browse through huge databases of available videos and make recommendations about programs you might be interested in viewing.

intelligent answering A Rolm term, explained thus: "When your customers call – or you call them – the Rolm 9751 CBX system can use automatic number identification (ANI) or dialed number identification service (DNIS) to identify the caller and the reason for that call."

intelligent assistance A concept Apple is pushing for its Newton PDA. Newton can anticipate what you want to do and provide a bit of help. This is how Fortune Magazine explained it: For example, scrawl "lunch with John Thursday." My Newton would assume that Thursday means next Thursday and that John is the John I've been meeting with a lot lately, John Sculley, and that I want to eat at 12:30, my usual lunch hour. Newton updates my calendar, and presto, displays the entry for my approval. I can okay it or change it.

intelligent battery system See IBS.

Intelligent business process routing If your business is like Technology Investor Magazine, it has different software programs for each business task – accounting, sales automation, order entry, inventory, etc. If you could get those pieces of software to talk to each other, and to talk sense to each other, you could save time, lower labor costs, improve your products and provide better customer service. Better yet, if you could get your internal software programs talking to software at your suppliers and customers, you could save even more money, labor and time. That's what integration software does. Every business of any size can use it to improve how their business works. There are three types of integration software: enterprise application, business-to-business, and business-to-community. What's the difference? All integration software lets two or more software applications – e.g. accounting and inventory – exchange (transport) and understand (transformation) each other's data. That's why it's often called plumbing software. Enterprise application integration (EAI) software links a company's "inside" applications – the software only its employees use. It's what used to be called middleware, but with better management and more features. Plus, it connects all applications – a universal translator of sorts. Middleware usually connects just two specific applications. EAI also does intelligent business process routing – telling each piece of software where to send its data to complete all the necessary business processes. Think of what customer relationship management software should do after a salesperson enters an order: notify accounting to register the income, advise accounts receivable to issue a bill, tell inventory to see if the product is in stock, let shipping know to print a packing slip, and tell logistics to schedule delivery, etc.

Intelligent Call Management ICM. A generic name for a system that distributes phone calls across geographically distributed call centers. The ICM system provides pre-routing, post-routing, and performance monitoring.

intelligent concentrator A concentrator which receives signals from a device on one port and retransmits them to devices on other ports. An intelligent concentrator is one that has software and therefore has programming capabilities.

intelligent error concealment Approaches to error-handling on packet networks (such as the Internet) that do not impair a user's online experience. For example, packet errors and packet loss with time-sensitive services, such as voice or video, can seriously degrade the user experience even if the errors are corrected, since latencies associated with determining that a packet is damaged or lost and having the packet resent can be as disruptive to the user experience as the actual error itself. Intelligent error concealment involves using fast error-masking algorithms instead of more time-consuming packet-resends. This is done by recreating downstream at the audio or video decoder, those parts of the audio or video packet that are damaged or lost. There are two general approaches to do this: temporal interpolation and spatial interpolation. Temporal error concealment methods use information from adjacent error-free packets to estimate and reconstruct data values for the damaged or lost packet. Spatial methods use information from the same packet to estimate and reconstruct data values for the damaged packet.

intelligent hub A hub that performs bridging and routing functions in a collapsed backbone environment. In short, it functions both as a bridge and multiprotocol router.

Intelligent Multiple Access Spectrum Sharing IMASS. A method of automatically determining the presence of existing private operational fixed microwave (OFM) systems in the areas near base stations, and avoiding the use of frequencies for the PCS or cellular base station which might cause unacceptable interference. Instead the PCS or cellular systems will use frequencies in each area, which are not being used by nearly OFM systems. Techniques such as this will be helpful to PCS service providers coexisting with the incumbent OFM systems, until they can be relocated to difference frequencies according to the FCC rules.

Intelligent Network IN. A network that allows functionality to be distributed flexibly at a variety of nodes on and off the network and allows the architecture to be modified to control the services. The most familiar intelligent network is the Public Switched Telephone Network (PSTN). In North America, the Intelligent Network is an advanced network concept that is envisioned to offer such things as (a) distributed call-processing capabilities across multiple network modules, (b) real-time authorization code verification, (c) one-number services, and (d) flexible private network services (including (1) reconfiguration by subscriber, (2) traffic analyses, (3) service restrictions, (4) routing control, and (5) data on call histories). Levels of IN development are:

- IN/1. A protocol intelligent network targeted toward services that allow increased customer control and that can be provided by centralized switching vehicles serving a large customer base.
- IN1+. A protocol intelligent network targeted toward services that can be provided by centralized switching vehicles, e.g., access tandems, serving a large customer base.
- IN/2. A proposed, advanced intelligent-network concept that extends the distributed IN/1 architecture to accommodate the concept called the "service independence." Traditionally, service logic has been localized at individual switching systems. The IN/2 architecture provides flexibility in the placement of service logic, requiring the use of advanced techniques to manage the distribution of both network data and service logic across multiple IN/2 modules. See AIN, which stands for Advanced Intelligent Network.

See also Dumb Network.

Intelligent Peripheral IP. A network system in the Advanced Intelligent Network Release 1 architecture containing an Resource containing an Resource Control Execution Environment (RCEE) functional group that enables flexible information interactions between a user and the network.

intelligent phone When the Bell operating companies get bored they occasionally fantasize about applications for the networks they provide. Here are some of their ideas for what intelligent phones could, if motivated, do: Select entertainment on demand (movies, music, video); Order groceries or other services or products; Record customized news and sports programming; Enroll and participate in education programs from the convenience of subscribers' living rooms; Find up-to-minute medical, legal and encyclopedic information; Pay bills and manage finances; Make airline, rental car and hotel reservations and buy sports and entertainment tickets.

intelligent premises equipment This refers to modern equipment, such as routers and intelligent switches. These devices are often capable of taking on roles traditionally performed by the network service, such as error correction.

intelligent routing A voice call comes in. Your voice mail machine recognizes it as being urgent, so it gives the caller a message, "Please hold. Harry is away from his desk. I'll find Harry for you." Meantime, it dials several numbers looking for me. It also beeps me. Eventually I call in. It tells me, "John Smith is calling for you. You want him?" Yes, I say and we're connected. This is a simple form of a broad concept that many are beginning to call intelligent routing. See also At Work and Windows Telephony.

intelligent terminal A terminal is an input/output device to a distant computer. The terminal may communicate with the computer over a dedicated collection of wires or over phone lines. In the early days, terminals contained no processing power. They simply reflected what the user typed in and what the distant computer responded. As computers became cheaper and with the advent of the "computer on a chip," so it was economically possible to put computing power into a terminal. This reduced the load on the main computer and cut down on communications costs. There are levels of "intelligence" in terminals. An intelligent terminal might perform simple arithmetic functions or it might check the accuracy of input data (does the zip code match the state?). It may perform far more comprehensive processing – as doing virtually all the local processing, and only transmitting summary results to corporate headquarters once a day. A personal computer can be used and act as an Intelligent Terminal. Many personal computer communications software can emulate terminals, the most common being the DEC VT-100.

intelligent token A hardware device which generates one-time passwords. In turn, the passwords are verified by a secure server, yielding additional security.

intelligible crosstalk Crosstalk from which information can be derived.

INTELSAT INternational TELecommunications SATellite organization. At its formation, Intelsat was a worldwide consortium of national satellite communications organizations. Intelsat was originally is owned by 138 governments and Intelsat itself owned 24 satellites worldwide. At one stage, INTELSAT owned and operated the world's most extensive global communications satellite system. In June 2001, Intelsat, Ltd. was formed as a result of the privatization of the former intergovernmental organization INTELSAT is now privately owned by an international group of over 200 shareholders; major owners include Lockheed Martin Corporation (beneficial owner), Videsh Sanchar Nigam Limited, France Telecom, Telenor Broadband Services A.S., and British Telecommunications plc. The U.S. Open-Market Reorganization for the Betterment of International Telecommunications Act ("ORBIT") required that Intelsat conduct an initial public offering ("IPO") of its equity securities no later than December 31, 2002.

Intense Strong or large.

intensity modulation IM. In optical communication, a form of modulation in which the optical power output of a source is varied in accordance with some characteristic of the modulating signal. In intensity modulation, there are no discrete upper and lower sidebands in the usually understood sense of these terms, because present optical sources lack sufficient coherence to produce them. The envelope of the modulated optical signal is an analog of the modulating signal in the sense that the instantaneous power of the envelope is an analog of the characteristic of interest in the modulating signal. Recovery of the modulating signal is by direct detection, not heterodyning.

inter- 1. Means between two things, as opposed to intra, which means inside one thing. Interstate means phone calls and communications between states. Intrastate means communications inside one state. Calling between New York and California is interstate. Calling from Los Angeles to San Francisco is intrastate.

2. There is some argument about whether words should be spelled intra-state or inter-frame. In this dictionary, I spell them all without the dash. It seems to make more sense, since the word intra or inter has become an integral part of so many words.

Inter-exchange Mileage IXC mileage. The airline mileage between two cites. Synonym of Long Haul Mileage and Airline Mileage. See Airline Mileage.

interaction recording Interaction recording is the automatic recording of discussions between customers and call centers, whether those contacts occur via phone, web chat, or email. Interaction recording also includes automatically recording what how a browsing customer floats around a web site. Once all this information is captured, the job is then to analyze the information for effectiveness – how many and what type of sales? Did the customer find the answers to his questions?

interactive The ability of a person or device to talk to or communicate with another device (typically a computer) in real time, i.e. no delays. The term generally is applied in the context of interaction with a computer over a network in a conversational mode. Interactive processing is very time-dependent since a user is sitting there, waiting for the computer to ask him/her questions. The opposite of Interactive processing is batch processing. See Batch and Real Time.

interactive CATV Interactive CATV is a two-way cable system from which subscribers can receive and send signals. They will probably do this by punching buttons on their cable TV's remote control, which may look more like a computer keyboard than a traditional cable TV handheld remote signaling device.

Interactive Connectivity Establishment See ICE.

interactive data transaction A single (one-way) message, transmitted via a data channel to which a reply is required for work to proceed logically.

interactive kiosk Interactive kiosks represent a powerful new product delivery vehicle for "non-store" marketers, to increase sales by offering their products and services in high traffic areas, such as airports. According to research analyst Warren Hersch, financial services kiosks are not confined to full-service bank locations; they are found in supermarkets, shopping malls and auto dealerships. Self-service terminals offering government services reside in discount stores, libraries, outdoor pavilions and subway stations. Kiosks purveying travel-related services occupy office complexes, colleges and universities, pharmacies and other retail outlets.

interactive logon Logging onto a machine via the keyboard, in contrast to a network logon.

interactive pillows At Wired Magazine's NextFest 2005, an organization called Interactive Institute demonstrated two cushions. Each cushion was wired with electroluminescent wire and linked wirelessly to the Web. Touch, hug or lean on your pillow and its counterpart glows, no matter where it's located – so long as it's on a network somewhere.

interactive processing A computer operation in which each item is processed as it is entered (on a time scale of seconds). An example is automatic teller machines (ATMs).

interactive services A B-ISDN term referring to two-way communications in support of three types of services. "Conversational Services" include interactive voice, video and data communications. "Messaging Services" include video mail and compound mail. "Retrieval Services" include retrieval of data, image, video, and compound mail documents. Most of these services are highly bandwidth-intensive, hence their inclusion in the concept of Broadband ISDN.

Interactive Television Association ITV has now changed its name to the Association for Interactive Media (AIM). See AIM.

interactive video The fusion of video an computer technology. A video program and a computer program running in tandem under the control of the user. In interactive video, the user's actions, choices, and decisions affect the way in which the program unfolds. See Indeo Video.

Interactive Video and Data Service IVDS. See 218-219MHz.

Interactive Voice Response IVR. Think of Interactive Voice Response as a voice computer. Where a computer has a keyboard for entering information, an IVR uses remote touchtone telephones. Where a computer has a screen for showing the results, an IVR uses a prerecorded human voice that is stored (digitized) on a hard drive. In addition it can use a synthesized voice (computerized voice) for read back information that is constantly changing. (The synthesized voice is commonly referred to as Text-to-Speech.) Whatever a computer can do, an IVR can, from looking up train timetables to moving calls around an automatic call distributor (ACD). The only limitation on an IVR is that you can't present as many alternatives on a phone as you can on a screen. The caller's brain simply won't remember more than a few. With IVR, you have to present the menus in smaller chunks. See IVR.

Interagency Interim National Research and Education Network See IINREN.

Interagency Radio Advisory Committee IRAC. A government committee that advises the Commerce Secretary on the government's spectrum needs.

interarea cell transfer A cellular radio term. A cell transfer between two cells that are controlled by different serving Mobile Data Intermediate Systems (MD-ISs).

interaxial spacing Center to center conductor spacing between any two wires.

interbreak interval A call center term. A Scenario scheduling assumption specifying the minimum amount of time that must elapse between the end of one break and the beginning of another.

interbuilding backbone Telecommunications cable(s) that are part of the campus subsystem that connect one building to another.

interebuilding Cable The communications cable that is part of the campus subsystem and runs between buildings. There are four methods of installing interbuilding cable: in-conduit (in underground conduit), direct-buried (in trenches), aerial (on poles), and in-tunnel (in steam tunnels).

interbuilding cable entrance The point at which campus subsystem cables enter a building.

interbuilding wiring Consists of underground or aerial telephone wire/cables used on the premises to connect structures remote from the primary building to the premises telephone system.

intercarrier compensation What service providers charge each other to carry and terminate switched telephone traffic on each other's networks. See the next definition.

intercarrier compensation rate The rate charged by Carrier A to Carrier B to carry Carrier B's switched telephone traffic on Carrier A's network, and to terminate Carrier's B switched telephone traffic on Carrier A's network. From MCI's website: "The intercarrier compensation regime establishes the prices that carriers pay to other carriers for originating and/or terminating switched telephone calls. The regime, which is an artifact of the divestiture of AT&T and the subsequent passage of the 1996 Telecom Act, includes access charges, universal service and reciprocal compensation, among other charges. Intercarrier compensation rates vary widely, despite the fact that the functionality provided is virtually identical. For example, under the current system, the rates charged and paid depend on whether the calls in question are intrastate, interstate, local or ISP-bound. The intercarrier compensation regime is collapsing as technologies evolve and the widening use of services such as voice over Internet protocol will hasten the demise of the regime." Intercarrier compensation rates vary from a fraction of a penny to as much as 34 cents per minute, depending upon a wide range of factors, including the geographic regions from where the call originated and ended.

intercast New plug-in cards from Intel which will allow your PC to simultaneously receive TV pictures, and in the blank spaces of TV signals, Internet Web pages and text.

intercept Calls which cannot reach their destination may be intercepted and diverted to a station attendant, a recording or some other place. See Inercept Recording and Intercept Service.

intercept call completion ICC. This service automatically connects a caller who has dialed a customer's disconnected number to the subscriber's new phone number. Some LECs (local exchange carriers) may play the "you have dialed a number that is no longer in service" intercept message before connecting the caller to the subscriber's new number. This service is available only to subscribers who move within the LEC's serving area. There is a flat monthly charge for this service. The caller pays applicable usage charges, if any, for the portion of a call from its point of origin to the intercepted disconnected number; the ICC subscriber pays all applicable intraLATA toll charges, if any, for the portion of the call between the intercepted disconnected number and the new number. Third-number-billed and collect calls cannot be billed to an intercepted disconnected number. This service is useful for businesses that move to another central office in the LEC's serving area.

intercept interval A telephone company term. The intercept interval is the amount of time a changed or disconnected telephone number must remain unassigned in order to insure that after reassignment the new customer does not receive calls intended for the previous subscriber. Intercept intervals vary by customer class of service and are established by the utilities commissions and/or telephone companies.

intercept recording You make a phone call. It doesn't go through. The phone company intercepts that call and sends it somewhere. Intercept Recording is a recording telling you that your call cannot be completed and has been intercepted on its way to the destination number for some reason that will be explained by the recording. The most common voice you hear on intercept announcements is Jane Barbie's. See Barbie, Jane and Intercept Service.

intercept operator A person who provides intercept service at an intercept position of a switchboard or at an auxiliary services position of a centralized intercept bureau.

intercept service A service of the local phone in which a phone call is redirected by an operator or a recording to another phone number or a message.

Interchange Carrier IC. A common carrier that provides services to the public local exchanges on an intra or interLATA basis in compliance with local or Federal regulatory requirements and that is not an end user of the services provided.

interchangeability The ability to substitute a device from one manufacturer for another manufacturer's device on a fieldbus network without loss of functionality or interoperability with other devices on the network.

interchangeable NPA Code Code in the NXX format used as a central office code (NNX format), but that can also be used as an NPA code. Interchangeable NPA codes will be introduced on or after January 1, 1995.

intercom Intercommunication. An internal communication system which allows you to dial another phone in your building, office complex, factory or home. There are three types of intercom: 1. Dial: It allows you to dial or pushbutton another extension; 2. Automatic: One phone goes off hook and automatically dials another; and 3. Manual: The user can manually signal another phone by pushing a button for that phone. An example is a buzzer between a boss and a secretary.

intercom blocking A PBX feature by which phones with a particular Class Of Service (COS) are blocked from calling certain phones. A rare feature.

interconnect A circuit administration point, other than a cross connect or an information outlet, that provides capability for routing and re-routing circuits. It does not use patch cords or jumper wires, and typically is a jack-and-plug device used in smaller distribution arrangements or that connects circuits in large cables to those in smaller cables. See also Interconnect Companies.

interconnect agreement An agreement between an established local phone company and a new local phone company for both companies to allow their subscribers to dial each other. Such agreement covers issues such as sharing of revenues and if a subscriber, who changes local phone companies, can keep his phone number.

interconnect companies Companies which sell, install and maintain telephone systems for end users, typically businesses. AT&T coined the word "interconnect" as a pejorative word – to indicate that these companies "interconnected" to AT&T's telephone network – but didn't really belong there and, if they were there, they were probably unreliable and caused harm. These "interconnect" companies contrasted with true-blue companies belonging to AT&T which did a sterling job. Anyway, despite the changes in the industry, the term stuck and the nasty associations have pretty well gone away. Now the irony is that the independent (i.e. non-Bell, non-AT&T) interconnect companies often deliver better service at a lower price. The industry is looking for a better word. TELECONNECT Magazine once started a campaign to make "TELECONNECT" a replacement for interconnect. But TELECONNECT's lawyers and the lawyers for a manufacturing/interconnect company called Teleconnect told us to lay off and stop trying to make the word generic. Since then we rather like the terms "Telecommunications Systems Integrator," "Telecommunications VAR" or "Telecom Developer." They seem to be catching on. By the way, TELECONNECT, the magazine, is no more, but the term telecom systems integrator lives on.

interconnecting cable The wiring between modules, between units, or the larger portions of a system. Pretty obvious?

interconnection A term generally used to describe the connection, with or without a protective connecting arrangement, of customer- or phone company-provided communications equipment to facilities of the local phone companies.

interdiction In cable television networks, one method of denying a subscriber the reception of those (premium) channels that are not ordered and paid for. As such, interdiction is one option in a conditional access strategy. Special filters or scramblers on a subscriber drop usually perform interdiction, which block or garble the premium channel or channels.

interdigital pause This is the time seperation between the SS7 signaling tones so the tones (i.e. signaled digits) won't overlap or run togeather.

interdomain trust relationships With Windows NT, Unix and some other operating systems, the user accounts and global groups from one domain can be used in another domain. In the MIS world, a domain is the part of a computer network in which the data processing resources are under common control. In the Internet, a domain is a place you can visit with your browser – i.e. a World Wide Web site. When a domain is configured to allow accounts from another domain to have access to its resources, it effectively trusts the other domain. The trusted domain has made its accounts available to be used in the trusting domain. These trusted accounts are available on Windows NT Server computers and Windows NT Workstation computers participating in the trusting domain.

Hint: By using trust relationships in your multidomain network, you reduce the need for duplicate user account information and reduce the risk of problems caused by unsynchronized account information.

The trust relationship is the link between two domains that enables a user with an account in one domain to have access to resources on another domain. The trusting domain is allowing the trusted domain to return to the trusting domain a list of global groups and other information about users who are authenticated in the trusted domain. There is an implicit trust relationship between a Windows NT Workstation participating in a domain and its PDC.

In this example, the following statements are true because the London domain trusts the Topeka domain:

Users defined in the Topeka domain can access resources in the London domain without creating an account within that domain. Topeka appears in the From box at the initial logon screen of Windows NT computers in the London domain. Thus, a user from the Topeka domain can log on at a computer in the London domain. When trust relationships are defined, user accounts and global groups can be given rights and permissions in domains other than the domain where these accounts are located. Administration is then much easier, because you need to create each user account only once on your entire network, and then the user account can be given access to any computer on your network (provided you set up domains and trust relationships to allow it).

Note Trust relationships can be configured only between two Windows NT Server domains. Workgroups and LAN Manager 2.x domains cannot be configured to use trust relationships.

interdrive The name of the FTP Software client implementation of the Sun NFS protocol.

interend office trunk groups A category of trunk groups that interconnects end offices.

interenterprise communications Communications exchanged between multiple organizations, e.g., between business trading partners, collaborators, affiliates or a business and its customers.

interest groups In IEEE parlance, Interest groups are the first step in the creation of a standard. For an example, see WPAN.

Interexchange Carrier IXC. At one stage an IXC was a telephone company that was allowed to provide long-distance telephone service between LATAs but not within any one LATA. Then some states in the United States started allowing intra-LATA competition. Now an IXC is best defined as a telephone company that is allowed to provide long-distance telephone service between LATAs. See IXC and LATA. Contrast with LEC.

Interexchange Channel IXC. A communications channel or path between two or more telephone exchanges.

Interexchange Customer Service Center ICSC. The Telephone Company's primary point of contact for handling the service needs of all long distance carriers.

interexchange plant The facilities between one switching center and another switching center; sometimes including line-to-line, no-user-switching centers called "tandem exchanges."

interface 1. A mechanical or electrical link connecting two or more pieces of equipment together.

2. A shared boundary. A physical point of demarcation between two devices where the electrical signals, connectors, timing and handshaking are defined. The procedures, codes and protocols that enable two entities to interact for a meaningful exchange of information.

3. To bring two things or people together to allow them to talk, either in English or in some technical way.

4. A poorly-defined word often used when the speaker is incapable of figuring precisely what he means. No one would ever invite a pretty girl out to lunch asking her to "interface" with you. See also Interface Device.

5. According to Steven Johnson's book, "Interface Culture – How new technology transforms the way we create and communicate," the word interface "refers to software that shapes the interaction between user and computer: The interface serves as a kind of translator, mediating between the two parties, making one sensible to the other. In other words, the relationship governed by the interface is a semantic one, characterized by meaning and expression rather than physical force."

interface device A device which meets a standard electrical interface on one side and meets some other nonstandard interface on the other. The purpose of the device is to allow a device with a nonstandard interface to connect to a device with a standard interface. See also Interface.

interface functionality The characteristic of interfaces that allows them to support transmission, switching, and signaling functions identical to those used in the enhanced services provided by the carrier. As part of its comparably efficient interconnection (CEI) offering, the carrier must make available standardized hardware and software interfaces that are able to support transmission, switching, and signaling functions identical to those used in the enhanced services provided by the carrier.

interface IC Interface ICs (integrated circuits) refer to busses and switches that connect and exchange data between different devices according to standards set by various industry standard bodies such as Telecommunications Industry Association or Institute of Electrical and Electronics Engineers (IEEE).

Interface Manager The original name for Microsoft's Windows. Later called Windows, and finally shipped in its first version in November 1985.

Interface Message Processor IMP. A processor-controlled switch used in packet-switched networks to route packets to their proper destination.

interface nodes Network nodes used to move data on and off the network.

interface overhead the interface overhead is the remaining portion of the bit stream after deducting the information payload. The interface overhead may be essential (e.g. framing for an interface shared by users) or ancillary (e.g. performance monitoring).

interface payload The portion of the bit stream which can be used for telecommunications services. Any signaling is included in the interface payload. See also Interface Overhead.

interface shelves Shelves in a Rolm PBX cabinet containing the printed circuit card groups that connect telephones, terminals, lines and trunks to CBX interface channels. These shelves also contain shared electronics cards.

interference Electromagnetic energy (i.e., electricity, radio, and light) are what we use for transmission purposes in telecommunications and data communications networks. Electromagnetic energy travels in waves, and each transmitted signal is defined as operating at a given frequency, which we call a wavelength in the optical (i.e., light) world, or range of frequencies. If another signal from another source creates a wave front that exactly overlaps your signal in phase (i.e., matches the rise and fall of the wave form), the amplitude (i.e., magnitude, or power level) of the combined signal increases. If another signal from another source creates a wave front that is exactly out of phase, they cancel each other out.

Here's a simpler explanation: Energy you receive with a signal. You don't want the energy. You want the signal. Getting rid of the interference may be a pain. The interference may be man made (e.g., electrical motors, fluorescent light boxes, or radio transmitters) or it may be GOD made (e.g., lightning, sunspots, or static electricity). Some conducted transmission media (fancy term for cabling) are more immune to interference than others. Commonly used wired transmission media, in order of immunity to interference, are:

1. Optical fiber
2. Coax
3. Shielded twisted pair
4. Unshielded twisted pair
5. Unshielded untwisted pair.

interference emission Emission that results in an electrical signal being propagated into and interfering with the proper operation of electrical or electronic equipment. The frequency range of such interference may be taken to include the entire electromagnetic spectrum.

interferometer An instrument that employs the interference of light waves for measurement.

interflow The ability to establish a connection to a second ACD and overflow a call from one ACD to the other. This provides a greater level of service to the caller.

interframe See Interframe Coding.

interframe coding A video term. It's a technique to cut down the size of the video to save on transmission costs. It's a way of source coding where the temporal correlation of moving pictures is used for data reduction. Interframe coding use compression techniques which track the differences between frames of video and eliminates redundant information between frames. Interframe coding stores only once those pixels that don't change. Then multiple frames access those pixels during decompression. This results in more compression over a range of frames than intraframe coding, which compresses information within a single frame.

interframe encoding A way of video compression that transmits only changed information between successive frames. This saves bandwidth. See interframe coding.

Interframe Gap IFG. The minimum idle time between the end of one frame transmission and the beginning of another. On Ethernet 802.3 LANs the minimum interframe gap is 9.6 micro-seconds.

Interim Interswitch Signal Protocol IISO. A call routing scheme used in STM networks. Formerly known as PNNI Phase 0. IISP is an interim technology meant to be used pending completion of PNNI Phase 1. IISP uses static routing tables established by the network administrator to route connections around link failures.

interim local number portability See LNP and Number Portability.

interim number portability See LNP and Number Portability.

Interim Operating Authority IOA. Authority granted by the FCC for a company to operate a cellular system during the interim between that company's application for a cellular license and the FCC's granting of such a license. IOAs usually apply in situations where litigation (i.e., lawsuits) threatens to extend the hearings and licensing process for long periods of time. They also sometimes are granted when license requests are unopposed, as the normal licensing process can take many months, and fast-developing or transient needs of the cellular community might be otherwise left unsatisfied. See also STA.

interior An ATM term. Denotes that an item (e.g., link, node, or reachable address) is inside of a PNNI routing domain.

interlace In TV, each video frame is divided into two fields with one field composed of the odd- numbered horizontal scan lines and the other composed of the even-numbered horizontal scan lines. Each field is displayed on an alternating basis. This is called interlacing. It is done to avoid flicker. See also Interlacing.

interlaced GIF When you're downloading a Web page, which contains images, interlaced GIF images appear first with poor resolution and then improve in resolution until the entire image has arrived, as opposed to arriving linearly from the top row to the bottom row. This lets users get a quick idea of what the entire image will look like while waiting for the rest to load. Your Web browser has to support progressive display. Non-progressive-display Web browsers will still display interlaced GIFs, but only after they have arrived in their entirety.

interlaced image A Web term. A GIF image that is displayed full-sized at low resolution while it is being loaded, and at increasingly higher resolutions until it is fully loaded and has a normal appearance. See Interlaced GIFs.

interlaced scanning mode A scheme that takes two passes to paint an on screen image, painting every other line on the first pass and sequentially filling in the rest of the lines on the second pass. This scheme usually causes flicker.

interlacing Regular TV signals are interlaced. In the US there are 525 scanning lines on the regular TV screen. This is the NTSC standard. Interlaced means the signal refreshes every second line 60 times a second and then jumps to the top and refreshes the other set of lines also 60 times a second. Non-interlaced signals, which are used in the computer industry, means each line on the entire screen is refreshed X times. X times depends on what the video card is outputting to the color monitor. The more expensive the card and the monitor, the more often the monitor will be refreshed. The more it's refreshed, the better and more stable it looks – the less perceived flicker. For example, text on an NTSC United States TV set tends to "flicker." It doesn't on a non-interlaced monitor. Typical non-interlaced computer monitors refresh at 60 to 72 times a second. But good ones refresh at higher rates. Generally, anything over 70 Hz (i.e. 70 times a second) is considered to be flicker-free and therefore preferred, if you can afford it. In short, buy an non-interlaced monitor. You'll like it better.

InterLATA Services, traffic or facilities that originate in one LATA, crossing over and terminating in another Local Access and Transport Area. (LATA). This can be either Interstate or Intrastate service, traffic or facilities. Under provisions of Divestiture, the Bell operating companies cannot provide Inter-LATA service, but can provide Intra-LATA service. Some LATAs are very large. So some "local" phone companies provide the equivalent of long distance service. And some of these phone companies have different pricing packages. Some of these packages are cheap, but not highly-publicized. See also InterLATA Service and LATA.

InterLATA Call A call that is placed within one LATA (Local Access Transport Area) and received in a different LATA. These calls are currently carried by a long distance company.

InterLATA Carrier IC. Any carrier that provides telecommunications services between a point in a LATA and a point in another LATA or outside a LATA.

InterLATA Competition Originally long distance telephone companies in the United States were not allowed to provide InterLATA telecom services. Later, many Public Utility Commissions (PUC) in ARF (Alternative Regulatory Framework) Phase III started to consider it. And many state agencies started to allow long distance phone companies to compete with local monopolies to carry intraLATA toll calls.

InterLATA Service As defined by the Telecommunciations Act of 1996, the term 'interLATA service' means telecommunications between a point located in a local access and transport area (LATA) and a point located outside such area. The term LATA means a contiguous geographic area–

(A) established before the date of enactment of the Telecommunications Act of 1996 by a Bell operating company such that no exchange area includes points within more than 1 metropolitan statistical area, consolidated metropolitan statistical area, or State, except as expressly permitted under the AT&T Consent Decree; or

(B) established or modified by a Bell operating company after such date of enactment and approved by the Commission.

interleave 1. The transmission of pulses from two or more digital sources in time-division sequence over a single path.

2. A data communication technique, used in conjunction with error-correcting codes, to reduce the number of undetected error bursts. In the interleaving process, code symbols are reordered before transmission in such a manner that any two successive code symbols are separated by I-1 symbols in the transmitted sequence, where I is called the degree of interleaving. Upon reception, the interleaved code symbols are reordered into their original sequence, thus effectively spreading or randomizing the errors (in time) to enable more complete correction by a random error-correcting code.

3. Interleaving also refers to the way a computer writes to and reads from a hard disk. Understanding interleaving is critical if you want to get your hard disk to work at its maximum speed (without in any way damaging the disk). Let's look at the way MS-DOS reads information from a hard disk. All hard disks are controlled by a special card called a hard disk controller card. Let's say your computer wants a file. It tells the hard disk controller card it wants the file. The controller searches the disk for the first sector of the requested file, reads that sector (usually 512 bytes) to your computer's RAM and then transfers the information to the CPU to be processed. When this is complete, the controller goes back to the hard disk and searches for the file's second sector. The process continues in this way until the file is completely read.

The problem is that while the controller and the CPU are doing their things, the hard disk itself is spinning 3,600 times a minute. By the time the controller reads one sector and it is ready to return to the disk, the next consecutive file sector has spun past the read/write head. If the file is stored in contiguous sectors, the controller must wait for the disk to complete its revolution before it can read the next file sector. To solve this problem, hard drive makers developed a concept called interleave setting, which tells the hard drive controller to skip a certain number of sectors when it writes a file to disk. Thus, when the file is later read back, the appropriate file sectors should fall under the read/write heads at the appropriate time.

If the controller reads or writes one sector and then skips a sector, the interleave is 2 (every other sector is used to store logically consecutive blocks). The interleave is sometimes written as 2:1. If the controller writes to one sector and then skips two, the interleave is 3 or 3:1. The interleave factor is usually established by the manufacturer or reseller of

the hard disk/controller combination. If someone else assembles the hard disk/controller combination, that person may need to experiment to determine the correct interleave factor – i.e. the one that works fastest without messing up.

Setting the "correct" interleave settings on your hard disk is critical to getting maximum performance out of your hard disk. Here's a test that a writer for PC Resource Magazine did. He copied the same files from one part of his hard disk to another part using different interleave settings:

Setting	Time to copy file
3	1 min 15 seconds
4	1 min 17 seconds
5	1 min 1 second
6	35 seconds
7	41 seconds
8	1 min 10 seconds

Clearly his best interleave setting is six. There are two ways of choosing the correct interleave setting. You can do it by trial and error as the writer did. His test took three hours. Or buy a program and do it in seconds. The best program is called Disk Technician. It's from a company called Prime Solutions in San Diego. Sadly, the program doesn't work on certain laptops and on certain controller card/hard disk combinations.

interleaved memory An option on some system boards that increases processing speed by assigning memory locations on an alternating basis to two banks of RAM. The computer has to wait one cycle between accesses to a single bank of memory, but it can access a different bank without having to wait.

intermediate assist A method for pulling cables into conduits or duct liners in which manual labor or machines are used to assist the pulling at intermediate manholes.

intermediate cross-connect an interconnect point within backbone wiring. for example, the interconnection between the main cross-connect and telecommunications closet or between the building entrance facility and the main cross-connect.

intermediate distribution frame IDF. A metal rack designed to connect cables and located in an equipment room or closet. Consists of bits and pieces that provide the connection between inter-building cabling and the intra-building cabling, i.e. between the Main Distribution Frame (MDF) and individual phone wiring. There's usually a permanent big, fat cable running between the MDF and IDF. The changes in wiring are done at the IDF. This saves confusion in wiring. See also Feeder Cable and Connecting Block.

intermediate frequency IF. A microwave frequency that has been reduced from its native frequency, so that it can be processed. This is necessary because a microwave cannot be fully processed in its native frequency.

intermediate frequency transformer A transformer designed to amplify the intermediate frequencies generated in a superheterodyne radio receiver. These are normally sharply tuned to a single frequency band.

intermediate high-usage trunk group A Bellcore definition. A high-usage trunk group that receives route-advanced overflow traffic and may receive first-route traffic and/or switched-overflow traffic.

intermediate reach Intermediate reach refers to optical sections from a few kilometers (km) to approximately 15 km. An AT&T SONET term.

intermediate system An OSI term which refers to a system that originates and terminates traffic, as well as forwarding traffic to other systems.

intermittent problems Intermittent problems are issues or bugs that come to light only after systems have been running for some time, or certain infrequently performed sequences of events are performed. Often many thousands of calls need to be put through before they are discovered. And bugs may only be seen occasionally, perhaps one of every 100 times something is done. Intermittent problems are among the hardest to find and duplicate.

intermix A mode of service defined by Fibre Channel that reserves the full Fibre Channel bandwidth for a dedicated (Class 1) connection, but also allows connectionless (Class 2) traffic to share the link if the bandwidth is available.

intermodulation IM. The production, in a nonlinear element of a system, of frequencies corresponding to the sum and difference frequencies of the fundamentals and integral multiples (harmonics) of the component frequencies that are transmitted through the element.

intermodulation distortion IMD. Nonlinear distortion characterized by the appearance of frequencies in the output, equal to the sum and difference frequencies of integral multiples (harmonics) of the component frequencies present in the input. Harmonic

International Calling Codes

Albania	355
Algeria	213
American Samoa	684
Andorra	376
Angola	244
Anguilla	264
Antarctica	672
Antigua and Barbuda	1
Argentina	54
Armenia	374
Aruba	297
Australia	61
Austria	43
Azerbaijan	994
Bahamas	1
Bahrain	973
Bangladesh	880
Barbados	1
Belarus	375
Belgium	32
Belize	501
Benin	229
Bermuda	1
Bhutan	975
Bolivia	591
Bosnia and Herzegovina	387
Botswana	267
Brazil	55
Brunei Darussalam	673
Bulgaria	359
Burkina Faso	226
Burundi	257
Cambodia	855
Cameroon	237
Canada	1
Cape Verde	238
Cayman Islands	1
Central African Republic	236
Chad	235
Chile	56
China	86
Christmas Island	672
Cocos (Keeling) Islands	672
Colombia	57
Comoros	269
Congo	242
Cook Islands	682
Costa Rica	506
Cote D'Ivoire (Ivory Coast)	225
Croatia (Hrvatska)	385
Cuba	53
Cyprus	357
Czech Republic	420
Democratic Republic of Congo	243
Denmark	45
Diego Garcia	246
Djibouti	253
Dominica	1
Dominican Republic	1
East Timor	670
Ecuador	593
Egypt	20
El Salvador	503
Equatorial Guinea	240
Eritrea	291
Estonia	372
Ethiopia	251
Falkland Islands (Malvinas)	500
Faroe Islands	298
Fiji	679
Finland	358
France	33
French Antilles	596
France, Metropolitan	33
French Guiana	594
French Polynesia	689
Gabon	241
Gambia	220
Georgia	995
Germany	49
Ghana	233
Gibraltar	350
Great Britain (UK)	44
Greece	30
Greenland	299
Grenada	1
Guadeloupe	590
Guam	671
Guatemala	502
Guinea	224
Guinea-Bissau	245
Guyana	592
Haiti	509
Heard and McDonald Islands	692
Herzegovina	387
Honduras	504
Hong Kong	852
Hungary	36
Iceland	354
India	91
Inmarsat	
East Atlantic Ocean	871
Indian Ocean	873
Pacific Ocean	872
West Atlantic Ocean	874
Indonesia	62
Iran	98
Iraq	964
Ireland	353
Israel	972
Italy	39
Ivory Coast	225
Jamaica	1
Japan	81
Jordan	962
Kazakhstan	7
Kenya	254
Kiribati	686
Korea North	850
Korea South	82
Kuwait	965
Kyrgyzstan	996
Laos	856
Latvia	371
Lebanon	961

components also present in the output are usually not included as part of the intermodulation distortion.

internal bus See Local Bus.

internal F-ES A cellular radio term. A fixed End System within the administrative domain of the CDPD service provider. Typically provides value-added support services such as network management, accounting, directory, and authentication services.

internal modem A modem on a printed circuit card which is inserted into one of the slots on a PC (personal computer). The other type of modem for a PC is an external modem – essentially a modem with the same circuitry as an internal modem but with a metal or plastic case. An internal modem costs slightly less than an external one. Internal modems are good if you're short of desk space and afraid your external modem will be stolen. External modems have lights so it's easier to tell what's going on. Everybody has their theories on which type of modem is best. We prefer the external ones – largely for their lights and ease of moving around.

internal reachable address An ATM term. An address of a destination that is directly attached to the logical node advertising the address.

internal reflection As light rays enter a fiber optic link at an angle, they hit a barrier known as the cladding/core barrier. If the angle is less than a certain "critical angle" all of those rays are reflected back into the core of the fiber and continue propagating through the fiber to the end of the link. See Index of Refraction.

internal switch interface system ISIS7. The operating system used in Concert Packet Services Engines to schedule the node's tasks, allocate hardware resources among users, transfer information between devices, and run software programs.

international 800 service You can now have your customers overseas call you for free on an 800 line, just as your domestic customers do. The service is available from countries including Australia, Brazil, France, Hong Kong, Israel, Italy, Japan, Sweden, Switzerland and the United Kingdom. Overseas 800 service is often known as "Freephone."

International Ad Hoc Committee IAHC. One of the organizations which parcels out Internet domain names. The IAHC has proposed to add a number of TLD (Top Level Domain) names to the existing list of .com (commercial), .edu (education), .gov (government), .mil (military) and .org (not-for-profit organizations). Those proposed TLDs include .firm (businesses), .store (stores), .web (entities emphasizing cultural and entertainment activities), .rec (entities emphasizing recreation/entertainment activities), .info (entities providing information services) and .nom (those wishing individual or personal nomenclature). See Domain, Domain Name Server, and Domain Naming Service.

International Alphabet No. 5 IA5. Internationally standardized alphanumeric code with national options. ASCII is United States version.

International Amateur Radio Union The IARU is an international organization founded in Paris, France in 1925, whose two-fold mission is (1) the protection, promotion, and advancement of amateur radio communications and amateur radio communications via satellite within the framework of regulations established by the Interna-

Lesotho	266	Norway	47	Svalbard and Jan Mayen Islands	378
Liberia	231	Oman	968	Swaziland	268
Libya	218	Pakistan	92	Sweden	46
Liechtenstein	41	Palau	680	Switzerland	41
Lithuania	370	Palestinian Authority	970	Syria	963
Luxembourg	352	Panama	507	Taiwan	886
Macau	853	Papua New Guinea	675	Tajikistan	992
Macedonia	389	Paraguay	595	Tanzania	255
Madagascar	261	Peru	51	Thailand	66
Malawi	265	Philippines	63	Togo	228
Malaysia	60	Pitcairn	872	Tokelau	690
Maldives	960	Poland	48	Tonga	676
Mali	223	Portugal	351	Trinidad and Tobago	1
Malta	356	Puerto Rico	1	Tunisia	216
Mariana Islands	670	Qatar	974	Turkey	90
Marshall Islands	692	Republika Srpska	387	Turkmenistan	993
Martinique	596	Reunion Island	262	Turks and Caicos Islands	649
Mauritania	222	Romania	40	Tuvalu	688
Mauritius	230	Russian Federation	7	US Minor Outlying Islands	1
Mayotte	269	Rwanda	250	USSR (former)	7
Mexico	52	Saint Kitts and Nevis	1	Uganda	256
Micronesia	691	Saint Lucia	1	Ukraine	380
Moldova	373	Saint Vincent & the Grenadines	1	United Arab Emirates	971
Monaco	377	Samoa	685	United Kingdom	44
Mongolia	976	San Marino	378	United States	1
Montserrat	1	Sao Tome and Principe	239	Uruguay	598
Morocco	212	Saudi Arabia	966	Uzbekistan	998
Mozambique	258	Senegal	221	Vanuatu	678
Myanmar	95	Seychelles	248	Vatican City State	396
Namibia	264	Sierra Leone	232	Venezuela	58
Nauru	674	Singapore	65	Viet Nam	84
Nepal	977	Slovak Republic	421	Virgin Islands (British)	1
Netherlands	31	Slovenia	386	Virgin Islands (U.S.)	1
Netherlands Antilles	599	Solomon Islands	677	Wallis and Futuna Islands	681
New Caledonia	687	Somalia	252	Western Sahara	34
New Zealand	64	South Africa	27	Western Samoa (now called Samoa)	685
Nicaragua	505	Spain	34	Yemen	967
Niger	227	Sri Lanka	94	Yugoslavia	381
Nigeria	234	St. Helena	290	Zaire	243
Niue	683	St. Pierre and Miquelon	508	Zambia	260
Norfolk Island	672	Sudan	249	Zimbabwe	263
Northern Mariana Islands	670	Suriname	597		

tional Telecommunication Union (ITU), and (2) to provide support for member amateur radio societies in the pursuit of these objectives at the national level, with respect to the following: (a) representation of the interests of amateur radio at conferences and meetings of international telecommunications organizations; (b) encouragement of agreements between national amateur radio societies on matters of common interest; (c) enhancement of amateur radio as a means of technical self-training for young people; (d) promotion of technical and scientific investigations in the field of radiocommunication; (e) promotion of amateur radio as a means of providing relief in the event of natural disasters; (f) encouragement of international goodwill and friendship; (g) support of member amateur radio societies in developing amateur radio as a valuable national resource, particularly in developing countries; and (h) development of amateur radio in countries not represented by member amateur radio societies.

international ampere The current which will in one second deposit 0.001118 gram of silver from a neutral solution of silver nitrate.

international atomic time inTernational Atomic tIme (TAI) is measured in the SI second, defined in terms of vibrations of a cesium atom. It is therefore not explicitly tied to the Earth's rotation, although that was of course the motivation for the original definition of the second. An SI second is a unit of time equal to 1/60 of a minute or 1/3600 of an hour. The international definition was originally 1/86400th of a mean solar day in 1900, but is now instead the time required for 9,192,631,770 vibrations of a Cesium atom. See optical clocks.

international callback Calling the United States from many countries abroad is far more expensive than calling those countries from the United States. A new business called International Callback has started. It works like this. You're overseas. You dial a number in the United States. You let it ring once. It won't answer. You hang up. You wait a few seconds. The number you dialed in the U.S. knows it was you calling. There is a piece of equipment on that number that "hears" it ring and knows it's you since no one else has that number. (Typically it's done with Centrex service.) That was your special signal that you want to make a call. A switch attached to that line then calls you instantly. When you answer (overseas, obviously) it conferences you with another phone line in the United States and gives you U.S. dial tone. You can then touchtone from overseas your American number, just as if you would, were you physically in the U.S. There are huge savings. U.S. international callback operators can offer as high as 50% savings on calls from South America, where international calling rates are very high. The process of international callback is being automated with software and dialing devices. International callback is also helping to bring down the high cost of calling the U.S. from overseas. In recent years, deregulation has caused the price of international calls in many countries to fall dramatically. And now international callback or just callback is being done from other countries, including and especially Israel. A company called Kallback in Seattle, WA. has received a service mark from the U.S. Patent and Trademark Office for the words "callback" and "kallback" and sends letters to and threatens law suits against companies who use "their" words. See Callback.

international carrier A carrier that generally provides connections between a customer located in World Zone 1 and a customer located outside of World Zone 1, but with option of providing service to World Zone 1 points in North American Numbering Plan area codes outside the U.S.

international center for information technologies A Washington "think tank" whose mission is to bring together discussion on new telecommunications and computer technologies. Targeted at senior executives who are looking for ways to apply new technologies to gaining competitive edges for their company.

International Computer Security Association See ICSA.

international denial An optional restriction on your cellular phone that prevents the cellular number from marketing international calls. Some carriers place this restriction on all subscribers using their service.

international direct distance dialing IDDD. Being able to automatically dial international long distance calls from your own phone. The direct calling by the originating customer to the distant (international) called customer via automatic switching. IDDD is synonymous with the phrases international direct dialing and international subscriber dialing.

International Engineering Consortium The International Engineering Consortium, established in 1944, is a non-profit organization dedicated to catalyzing positive change in the information industry and its university communities. The Consortium provides educational opportunities for today's information industry professionals and conducts a variety of industry-university programs. The IEC also conducts research and provides publications addressing major opportunities and challenges of the information age. More than 70 leading, high technology universities are currently affiliated with the Consortium. www@iec.org.

International Freephone Service IFS. The ITU-T term for international toll-free service. IFS works on the basis of Universal International Freephone Numbers (UIFNs). A UIFN consists of a dialing pattern comprising an international prefix (e.g., 011), a three-digit country code (e.g., 800) for global service application, and an 8-digit Global Subscriber Number (GSN). See UIFN for much more detail.

international gateways The switches in the various domestic long distance networks (e.g. MCI, AT&T and Sprint) which interface their networks with International telecommunications networks. All US International calls are routed through an international gateway.

International Morse Code See Morse Code.

International Organization For Standardization ISO. An organization established to create standards. See ISO.

international prefix The combination of customer-dialed digits prior to dialing of the country code required to access the automatic outgoing international equipment in the originating country.

International Radiocommunications Advisory Committee See IRAC.

international record carrier IRC. One of a group of carriers that, until recently, was part of a monopoly of U.S. common carriers certified to carry data and text to locations outside the U.S. In recent years, regulation of this type of service has been markedly relaxed. Most of the IRCs got bought by MCI.

international private line circuits IPLC. The circuit lines used to connect nodes on the network. See also Trunk Circuit.

international shortwave broadcast station A station that sends programs overseas either for direct reception by listeners abroad or for intermediate reception by overseas relay stations that rebroadcast the programs on shortwave or medium wave stations to nearby audiences.

International Standards Organization ISO. See ISO.

international switching carrier ISC. An exchange whose function is to switch telecommunications traffic between national network and the networks of other countries. Also known as an international gateway office.

International Telecommunications Organization ITO. Foreign government agencies responsible for regulating communications. Formerly known as Postal Telephone and Telegraph (PTT).

International Telecommunications Union ITU. Anglicization of the proper French name of the Union Internationale des Telecommunications (UIT), resident in Geneva, Switzerland. ITU is the for agreements on telecommunications technical and operating standards and is a constituent body of the United Nations, engaging also in international development and education concerning telecommunications. ITU. Its most successful work is done in the establishment (but not enforcement of) standards and the allocation of radio frequencies worldwide – including satellites, etc. For a much bigger explanation, see ITU and ITU-T.

International Telecommunications Union-Radiocommunication Sector ITU-R Formerly the International Radio Consultative Committee (CCIR). The technical study branch of the International Telecommunication Union responsible for the study of technical and operating questions relating specifically to radio communications. See also ITU-T.

International Telegraph Alphabet #1 ITA 1. World-standard ITU-T version of the manual telegrapher's code. Colloquial name: International Morse Code.

International Telegraph Alphabet #2 ITA 2. World-standard ITU-T version of the 5-unit (also called 7.5 unit) teleprinter code used for Telex, international telegrams and most general telegraphy by wire lines; Colloquial name: Baudot code.

International Telegraph Alphabet #3 ITA 3. World-standard ITU-T version of a 6-unit extended set of ITA 2 to include characters needed for automatic typesetting directly from telegraph circuits. Colloquial names: Teletypesetter code, press code, extended Baudot code, and others.

International Telegraph Alphabet #4 ITA 4. World-standard ITU-T version of a 7-unit code in which only the combinations using 4 marking bits are valid; receiving any character with more or less than 4 marking bits is its error checking feature. Colloquial names: 7-unit ARQ Code, Moore ARQ Code, Moore Code, RCA Code, and others.

International Telegraph Alphabet #5 ITA 5. World-standard ITU-T version of a 7-unit teleprinter code with an 8th parity bit also used for asynchronous data terminals such as minicomputers or PCs. Colloquial name: ASCII code.

international telephone address A four-part code specifying a unique address for any telephone company in the world.

international toll free service See IFTS.

International Wireless Telecommunications Association See IWTA.

internesia The tendency to find wonderful things on the Internet and then forget where they were.

Internet It is very hard to define the Internet in a way that is either meaningful nor easy to grasp. To say the Internet is the world's largest and most complex computer and communications network is to trivialize it. But it is. To say it's the computer network for everybody in the world is to trivialize it. But it is. To say that it is fast becoming the world's global shopping mall (for buying and selling) is to trivialize it. To say that it's becoming the network for the world's corporations to communicate is to trivialize it. But it is. To say that it's replacing both physical mail and electronic faxes is to trivialize it. But it is. It is clearly the most important happening in the computing, communications and telecommunications industries since the invention of the computer or the transistor. The Internet is both a transport network – moving every form of data around the world (voice, video, data and images) – and a network of computers which allow you (and them) to access, retrieve, process and store all manner of information. No one really has a clear idea of the Internet's size, its growth or its capacity. We know that it grows daily. The most important part of the Internet for all of us normal people is something called the World Wide Web, as characterized by all those web sites starting with www. The World Wide Web is a subset of the Internet. Web Wid Web sites sit on computers worldwide joined by telecommunications links. Each web site has documents conforming to a specific Internet protocol called HTTP, which stands for HyperText Transfer Protocol. HTTP is the standard way of transferring information across the Internet and the World Wide Web. The reason we can read all the documents on all the Web sites is that the pages are written in HTML or Hypertext Markup Language, which tells your Web browser (e.g. Internet Explorer or Netscape) how to display the page and its elements (photos, videos, music, etc.) The defining feature of the Web is its ability to connect pages to one another – as well as to audio, video, and image files – with hyperlinks. Just click a link, and suddenly you're at a Web site on the other side of the world. How does the Internet find my web site www.InSearchOfThePerfectInvestment. com? Any computer attached at that moment to the Internet has an address – just like a phone number, except that it's typically a 16-digit address. When you type in www. InSearchOfThePerfectInvestment.com, your browser sends a command out to a database on the Internet in effect asking, "What is www.InSearchOfThePerfectInvestment.com?" Back comes the answer that it's 64.226.86.128. Your browser connects you and there you. An IP address on the web is a 4- to 12-digit number. The digits are organized in four groups of numbers (which can range from 0 to 255) separated by periods. If your computer is a web site permanently attached to the Internet, it will have a permanent address, i.e. it will always be the same. When you connect your personal computer to the Internet, your ISP (Internet Service Provider) will likely assign you a new address every time you connect. For a full explanation, see DHCP.

The first question everyone asks is, "Who runs the Internet." The simple answer is "everyone and no one." Think of the Internet as two parts. The first are technical standards – how everyone connects to the Internet. These standards are set by various committee under the direction of something called the IETF. The second are the communications circuits which carry the Internet's traffic. There are hundreds of companies – including basically every traditional local and long distance phone company in the world – that interconnect. How they get paid and pay each other other depends on arrangements they have amongst themselves.

The Internet's roots are in a U.S. Defense Department network called Advanced Research Project Agency NETwork (ARPAnet), established in 1969. ARPAnet tied universities and research and development organizations to their military customers, and provided connectivity to a small number of supercomputer centers to support timesharing applications. Quickly, the biggest application among its users became email. Much of the funding was provided by NSFNET (National Science Foundation NETwork). In the mid-1990s, the Internet was "commercialized", extending its use to anyone with a PC, a modem, a telephone line and an access provider – a special company known as an Internet Service Provider or Internet Access Provider. The Internet has become a major new publishing, research and commerce medium. I believe that its invention is as important to the dissemination of knowledge, to peoples' life styles and to the way we'll be conducting business in coming years as the invention of the Gutenberg Press was in 1453.

At its heart, the Internet is many large computer networks joined together over high-speed backbone data links ranging from 56 Kbps (now rare) to T-1, T-3, OC-1, OC-3 and higher. The Internet now reaches worldwide. Depending on the whim of the local government (which typically controls the local phone company and thus access to the Internet for its citizenry) you can pretty well get onto the Internet and roam it unchecked. The governments of Singapore, the People's Republic of China, Burma, Saudi Arabia and a few others limit their peoples' access to the Internet. The topology of the Internet and its subnetworks changes daily, as do its providers and its content. The bottom line is that the makeup of the Internet – i.e. how it works – is not all that important. It is the applications and information available on it that are important – the most significant of which are e-mail (electronic mail) and the World Wide Web. Commercial networks from AT&T, SPRINT, Worldcom and many others now carry the bulk of the traffic. As NSFNET (i.e. the U.S. Government) no longer funds the Internet, it has been commercialized, with money changing hands in complex ways between users, companies with Web sites, Internet Access Providers, long distance providers, government, universities and others. Increasingly, businesses are joining their computers to the Internet. There are now over 125 million Internet sites you can visit, running on over 75 million computers. There are probably 500 million computers worldwide equipped to reach the Internet – from the laptop I'm writing this on, to the desktop at your office.

The Internet's networking technology is very smart. Every time someone hooks a new computer to the Internet, the Internet adopts that hookup as its own and begins to route Internet traffic over that hookup and through that new computer. Thus as more computers are hooked to the Internet, its network (and its value) grows exponentially. The Internet is basically a packet switched network based on a family of protocols called TCP/IP, which stands for Transmission Control Protocol/Internet Protocol (TCP/IP), a family of networking protocols providing communication across interconnected networks, between computers with diverse hardware architectures and between various computer operating systems. Most PCs, including Windows-based machines and Macintoshes, will happily communicate using TCP/IP.

How TCP Works: TCP is a reliable, connection-oriented protocol. Connection-oriented implies that TCP first establishes a connection between the two computer systems that intend to exchange data (e.g. your PC and the host computer you're trying to reach, which may be thousands of miles away). Since most networks are built on shared media (for example, several systems sharing the same cabling), it is necessary to break chunks of data into manageable pieces so that no two communicating computers monopolize the network. These pieces are called packets. When an application sends a message to TCP for transmission, TCP breaks the message into packets, sized appropriately for the network, and sends them over the network. Because a single message is often broken into many packets, TCP marks these packets with sequence numbers before sending them. The sequence numbers allow the receiving system to properly reassemble the packets into the original original order, i.e. the original message. TCP checks for errors. And finally, TCP uses port IDs to specify which application running on the system is sending or receiving the data. The port ID, checksum, and sequence number are inserted into the TCP packet in a special section called the header. The header is at the beginning of the packet containing this and other "control" information for TCP.

How IP Works: IP is the messenger protocol of TCP/IP. The IP protocol, much simpler than TCP, basically addresses and sends packets. IP relies on three pieces of information, which you provide, to receive and deliver packets successfully: IP address, subnet mask, and default gateway. The IP address identifies your system on the TCP/IP network. IP addresses are 32-bit addresses that are globally unique on a network. There's much more on TCP/IP in my definition on TCP/IP and on Internet Addresses in that definition.

Here's how the Internet is used: As a computer network joining two (or more) computers together in a session, it is basically transparent to what it carries. It doesn't care if it carries electronic mail, research material, shopping requests, video, images, voice phone calls, requests for information, faxes ... or anything that can be digitized, placed in a packet of information and sent. A packet-switched network like the Internet injects short delays into its communications as it disassembles and assembles the packets of information it sends. And while these short delays are not a problem for non-real time communications, like email, they present a problem for "real-time" information such as voice and video. The Internet can inject a delay of as much as half a second between speaking and being heard at the other end. This makes conversation difficult. Internet telephony, as it's called when it runs on the Internet, is getting better, however, as the Internet improves and voice coding and compression techniques improve. I've enjoyed some relatively decent conversations to distant places.

Probably the most famous quote about the Internet is one from John Doerr, one of Silicon Valley's most famous venture capitalists. He said, "The Internet is the greatest legal creation of wealth in the history of the planet." Later, after the dot com bust he came to regret his words. By hyping wealth rather than invention, he has confessed he has distracted the industry from pursuing revolutionary technologies.

Now for a little history on the Internet. In the early 1990s the Internet was run by and for the United States government. There was no public use of the Internet. There were no commercial applications. In fact it wasn't even clear to the Federal Government what the Internet actually was. So an organization called the Federal Networking Council (FNC), which actually managed networking for the Federal Government, on October 24, 1995, unanimously passed a resolution defining the term Internet. This definition was developed in consultation with the leadership of the Internet and Intellectual Property Rights (IPR) Communities. RESOLUTION:

"The Federal Networking Council (FNC) agrees that the following language reflects our definition of the term "Internet". "Internet" refers to the global information system that –

(i) is logically linked together by a globally unique address space based on the Internet Protocol (IP) or its subsequent extensions/follow-ons;

(ii) is able to support communications using the Transmission Control Protocol/Internet Protocol (TCP/IP) suite or its subsequent extensions/follow-ons, and/or other IP-compatible protocols; and

(iii) provides, uses or makes accessible, either publicly or privately, high level services layered on the communications and related infrastructure described herein."

MCI Mail was the first commercial application attached to the Internet. Once it got one, all the other email services wanted on...and the rest is history. See various Internet definitions following. See also Berners-Lee, Domain, Domain Naming System, Grid Computing, gTLD, ICANN, Internet2, Internet Appliance, Internet Protocol, Internet Telephony, Intranet, IP Telephony, Surf, TCP/IP, Web Browser and Web Services.

Internet access The method by which users connect to the Internet, usually through the service of an Internet Service Provider (ISP).

Internet access provider See IAP.

Internet address When you travel the Internet or its World Wide Web area, you need an address to get to where you want to go – just like you need an address on a letter you mail or a phone number you wish to reach. All Internet addresses are expressed in dotted decimal notation of four fields of eight bits. In binary code, each bit has two possible values, 0 or 1. Therefore, each 8-bit field yield two to the eighth power, or 256 possible combinations. Since one of the possible combinations is 000, which means nothing, it is not used, thereby leaving 255 possible numbers in each field. IP addresses are written as XXX.XXX.XXX.XXX, where X is any number between 0 and 9, and where each 3-digit field has a value between 001 (i.e., 1) and 256. Internet addresses currently are based on the IPv4 (Internet Protocol version 4 protocol), which uses a 32-bit code in the 20-octet IP header to identify host addresses. A 32-bit address field yields 2 to the 32nd power possible addresses – that's 4,294,967,296 addresses. that seems like a lot of addresses, but it's not enough in the context of the commercialized Internet. Note that IPv6 has been standardized by the IETF (Internet Engineering Task Force), but has yet to be widely implemented, as equipment upgrades generally are required. Among the advantages of IPv6 is an address field expanded to 128 bits. A 128-bit address field yields 2 to the 128 power addresses – that's

340,282,366,920,939,463,463,374,607,431,768,211,456 distinct addresses. That's enough for approximately 32 addresses for every square inch of dry land on the Earth's surface, which should be enough for a while. No one wants to remember all those numbers when they go checking out their favorite site. So they came up with a neat idea of naming sites and having a bunch of computers do the translation, very similar to what happens with 800 toll-free numbers in North America. As a result Web URLs (Uniform Resource Locators) and e-mail addresses (such as www.harrynewton.com and harry@harrynewton.com) are textual addresses that are translated into correlating IP addresses through DNSs (Domain Name Servers, i.e. dedicated translation computers), which maintain tables of both domain names and IP addresses. For example, if you wish to reach www.Javanet.com, you can type www.Javanet.com in your browser or you can simply type 209.94.128.8. But www.Javanet.com is easier to remember. Internet addresses are organized into hierarchical "classes," as follows:

Class A Addresses: Begin with a "0" bit. Of a possible 128 Class A networks, only 51 networks exist. Examples include General Electric Company, IBM Corporation, AT&T, Hewlett-Packard Company, Ford Motor Company, and the Defense Information Systems Agency. They all are huge organizations, and require the highest possible categorization.

Class B Addresses: Begin with a "10" bit sequence. Of a possible 65,536 Class B networks, only about 12,000 exist.

Class C Addresses: Begin with a "110" binary bit sequence. Most applicants are assigned Class C addresses in blocks of 255 IP addresses. As of January 1998, about 800,000 Class C addresses were assigned.

Class D Addresses: Begin with a "1110" bit sequence. They are intended for multicast purposes.

Class E Addresses: Begin with a "1111" bit sequence. They are reserved for future use.

Now, the term "Internet Address" can be a bit misleading. As we have seen, it actually refers to an "IP Address," unless it's a URL, of course. Even if it's a URL, it's translated into an IP address. IP addresses often are used in the LAN (Local Area Network), as well as in the Internet and other public packet data networks. In such a case, one IP address often is used internal to the LAN domain, and another in the Internet domain, in order to mask the internal IP subnet address from the outside world. Masking the internal IP address essentially "masks," or hides, the true IP address of your workstation from the outside world. You may do this for one simple reason – you don't want the outside world to be able to get to your PC. The internal IP address might be either IPv4 or IPv6, while the Internet "outside world" address currently is always IPv4. In either event, the IP addresses are translated, one to the other, through a process of NAT (Network Address Translation), which is accomplished in an access router. On the outbound side, your true IP address is translated into an Internet IP address associated with the router. Responses to your transmissions are addressed to the router, which then translates them back into your true IP address for successful delivery. This translation and masking process secures and protects your identity. See NAT for a full explanation of this process. See also Subnet Mask.

Internet appliance A sub-$500 machine specially designed for Internet browsing and first proposed in the late Fall of 1995 by Larry Ellison, head of database software company Oracle. Part of its appeal to people outside Microsoft and Intel is that the Internet Appliance would not have to be based on standard PC technology. It need have an Intel chip and need not run Windows. This device is also called an Internet Terminal, a Network Computer or an IPC, an Interpersonal computer. The original description of the Internet Appliance was that it would come with 4mb of RAM, 4mb of flash memory, processor, monitor, keyboard and mouse – all for under $500.

Internet Architecture Board The Internet Architecture Board (IAB) is a technical advisory group of the Internet Society. Its responsibilities include:

IESG Selection: The IAB appoints a new IETF chair and all other IESG candidates, from a list provided by the IETF nominating committee.

Architectural Oversight: The IAB provides oversight of the architecture for the protocols and procedures used by the Internet.

Standards Process Oversight and Appeal: The IAB provides oversight of the process used to create Internet Standards. The IAB serves as an appeal board for complaints of improper execution of the standards process.

RFC Series and IANA: The IAB is responsible for editorial management and publication of the Request for Comments (RFC) document series, and for administration of the various Internet assigned numbers.

External Liaison: The IAB acts as representative of the interests of the Internet Society in liaison relationships with other organizations concerned with standards and other technical and organizational issues relevant to the world-wide Internet.

Advice to ISOC: The IAB acts as a source of advice and guidance to the Board of Trustees and Officers of the Internet Society concerning technical, architectural, procedural, and (where appropriate) policy matters pertaining to the Internet and its enabling technologies.

Internet Assigned Numbers Authority IANA. This group is responsible for the assignment of unique Internet parameters (e.g., TCP port numbers, and ARP hardware types), and managing domain names. It also was responsible for administration and assignment of IP (Internet Protocol) numbers within the geographic areas of North America, South America, the Caribbean and sub-Saharan Africa; on December 22, 1997, that responsibility was shifted to ARIN (American Registry for Internet Numbers). www. arin.net. The IANA has well-established working relationships with the US Government, the Internet Society (ISOC), and the InterNIC. ISOC provides coordination of IANA activities with the Internet Engineering Task Force (IETF) through the participation of IANA in the Internet Architecture Board (IAB). IANA responsibility was assigned by DARPA (Defense Advanced Research Project Agency) to the Information Sciences Institute (ISI) of the University of Southern California. ISI has discretionary authority to delegate portions of its functions to an Internet Registry (IR), previously performed by SRI International and currently performed by Network Solutions Inc. (NSI), a subsidiary of SAIC. Beginning March 1998, that function is shared with the Council of Registrars (CORE). CORE contracted (November 1997) with Emergent Corporation to build and operate the new Internet Name Shared Registry System (SRS), which is a neutral, shared database repository that coordinates registrations from CORE and propagates those names to the global Internet Domain Name System (DNS). www.isi.edu/div7/iana/ See also ARIN, CORE, DNS, Internet, InterNIC, and SRS.

Internet backbone This super-fast network spanning the world from one major metropolitan area to another is provided by a handful of national Internet Backbone Providers (IBPs). These companies and organizations use connections running at on T-3 lines and above linked up at specified interconnection points called national access points. Local ISPs connect to this backbone through routers so that data can be carried though the backbone to its final destination. The largest backbone operators include AT&T and WorldCom's UUNET Interconnection between these backbone operators is done through peering arrangements. Tele2 (based in Sweden) is one of few Pan-European fixed line operators that re-sell voice and Internet services to residential users, without owning a backbone network (but leases it instead). See also Internet backbone provider.

Internet backbone provider Courtesy the FCC: "As the market has evolved, Internet backbone providers fall into one of two tiers. The first tier consists of core Internet backbone providers that own and control their own networks; maintain nodes with default-free routers; exchange traffic with all other core backbone providers on a settlements-free basis (essentially a "bill-and-keep" system); interconnect at a minimum of five major national access points (NAPs) and on a private bilateral basis with other backbone providers and ISPs; and offer high-speed transmission facilities that connect their nodes and that transmit high volumes of Internet traffic both nationwide and globally. Under the so-called "peering arrangements" among core Internet backbone providers, these providers will only deliver traffic to each other that is destined for the core provider's end users or ISPs' customers. Access to any one of the core backbone providers offers ubiquitous Internet connectivity. The second tier of backbone providers also maintains nodes with default-free routers and offers transmission facilities – albeit at lower speeds than those of core providers – connecting their nodes. However, they typically rely on facilities obtained from core backbone providers to transmit traffic throughout the United States and to other countries. Because the core backbone providers offer services to them that are costly to provide, the second tier providers must pay for interconnection to the core providers' networks." See also Internet.

Internet Base Transmitter Site See ITBS.

Internet cable access A general term used to describe accessing the Internet using the cable TV coaxial cable for inbound Internet access (i.e. downstream) and the phone line for up sending commands and requests (i.e. upstream information). The cable TV is very fast – as much as six million bits per second. The phone is relatively slow – no more than fifty thousand bits per second. But it works because most information from the Internet flows at you, not away from you. The cable and telecom industry is working on standards to make disparate cable systems and TV set-top boxes work with each other. The industry has developed Data Over Cable Service Interface Specification (DOCSIS), which sets standards for both two-way and cable-plus-phone specifications. See DOCSIS.

Internet cache protocol See ICP.

Internet call waiting Imagine you have one phone line at your house. You're presently using that one line to surf the Internet. Someone calls you. You have installed call forwarding. Their incoming call gets forwarded to another phone line, which is answered by a service provider who is providing the Internet call waiting service. Their machine answers: "The number you called is presently surfing the Internet. Would you like to tell them you're calling? Do you have a message? Record now." The machine picks up the incoming phone line from callerID, records the message and then sends an email to you, saying this phone number is trying to reach you. And here's their message." It then plays the message.

Internet Content Adaptation Protocol See ICAP.

Internet content provider ICP. A company that will design and deliver content for your Web site.

Internet Control Message Protocol ICMP. An integral part of IP (Internet Protocol), ICMP is a mechanism by which IP software on a host or gateway can communicate with its peers on other machines to report errors and pass other information (e.g., time stamps) relevant to IP packet processing. As specified in RFC-792, ICMP also provides a number of diagnostic functions. One of the most frequently used ICMP messages is the Echo Request, commonly known as the Ping utility, which allows a device to test the communication path to another device. The ICMP header comprises 12 octets. See also Ping.

Internet Engineering Planning Group See IEPG.

Internet Engineering Steering Group IESG. The executive committee of the IETF (Internet Engineering Task Force).

Internet Engineering Task Force IETF. One of two technical working bodies of the Internet Activities Board. The IETF is the primary working body developing new TCP/IP (Transmission Control Protocol/Internet Protocol) standards for the Internet. It has more than one thousand active participants. www.ietf.org.

Internet fax Internet fax is, as it sounds, sending faxes over the Internet. There are a whole bunch of manual ways to send faxes over the Internet – most of which are akin to sending a fax over the PSTN, as we do it today. Dial up, etc. There are movements, however, to automate this process and get Internet faxing more along the lines of Internet email. Internet Fax is coming in two parts. The first is a store and forward model that is essentially based on the MIME attachment of TIFF files to standard E-Mail messages delivered by SMTP. The standards for this model are found in the IETF - ITU agreements of January 1998. The second part is an Internet draft that extends SMTP itself. The draft turns a fax machine into a virtual SMTP server so that transmission of the fax from point-to-point happens in real time. The protocol would extend SMTP beyond its function of a simple mail transport protocol to the point where, when a transport session is established, the user can exchange capabilities between devices - something that cannot be done with store and forward mail. Implementing these will be a series of hybrid "stupid-smart" devices that bridge faxes between the PSTN and the Internet. The Panasonic FO-770I, which is already on the market, is one such device with almost all the capabilities of the new standard . Load your fax, toggle "send" in one direction to transmit via the PSTN, toggle "send" in the other direction to go via the Internet. Other manufacturers are working on the introduction of inexpensive "black boxes" to connect standard G3 faxes in small-office, home-office (SOHO) environments directly to one's PC and from there to the Internet.

Internet fax protocol IFP is specified in ITU-T recommendation T.38 as the method for supporting facsimile transmissions over IP (Internet Protocol) networks. See also T.37 and T.38.

Internet firewall See Firewall.

Internet Freedom and Broadband Deployment Act A proposed 2002 act of the House of Representatives also called the Tauzin-Dingell Bill. The bill pushes the Regional Bell Operating Companies (RBOCs) to offer broadband Internet services over long-distance lines without opening up their local phone service monopolies to outside competition. The Tauzin-Dingell bill, or more formally H.R. 1542, the Internet Freedom and Broadband Deployment Act, is legislation written by Representatives John D. Dingell (D-MA) and Billy Tauzin (R-LA).

Internet gateway Internet gateways are devices which typically sit on a local area network and handle all the translations between IPX traffic on your LAN (IPX is the NetWare protocol) and the TCP/IP traffic on the Internet. TCP/IP is the protocol used on the Internet. See also Internet Servers and other definitions beginning with Internet.

Internet group name In Microsoft networking, a name registered by the domain controller that contains a list of the specific addresses of computers that have registered the name. The name has a 16th character ending in 0x1C.

Internet imposters A term created by Gretchen Morgenson of the New York Times for overheated stocks posing as growth companies.

Internet in the pocket A term that refers to an always-on, always-connected-to-the-Internet, mobile broadband experience, the user's interface to which is by means of a mobile device that supports both voice and data.

Internet integrator A fancy name for a consulting firm which specializes in helping its clients do stuff with the Internet, including transaction processing; supply chain infrastructure integration; wireless integration; business to business (B2B) and application

monitoring/management.

Internet Intellectual Infrastructure Fund A fund created in 1995 to offset government funding for the preservation and enhancement of the intellectual infrastructure of the Internet. The fund was funded by 30% of the Internet domain registration fee, which was set at $50 per year at that time. On March 16, 1998, the funding for the Intellectual Infrastructure was completed, and the InterNIC ceased to collect that portion of the annual fee, thereby reducing it to $35 for new registrations. Proceeds of the fund are to be used to build Internet2, which will be a separate Internet for institutions of higher learning. See also CORE, DNS and InterNIC.

Internet Key Exchange See IKE.

Internet Mail Consortium IMC. A technical trade association which pursues cooperative promotion and enhancement of electronic mail and messaging on the Internet. Activities cover promotion of Internet mail and the products and services which serve to implement it. IMC is involved in formative efforts for IETF (Internet Engineering Task Force) mail standards, with a focus on implementation guidelines. www.imc.org.

Internet land grab Think real estate. Now think of the Internet in real estate terms. There's a limited number of names. Once someone has HarryNewton.com. No one else can get it. Ditto for t-shirt.com, sex.com, google.com, etc. The concept is that these domain numbers have value, like owning the land at the corner of Fifth Avenue and 57th Street in New York City. Land typically derives its value from the rent it commands

Internet Message Access Protocol See IMAP.

Internet MIB subtree A tree-shaped data structure in which network devices on a local area network and their attributes can be identified within the confines of a network management scheme. The name of an object or attribute is derived from its location on this tree.

For example, an object in MIB-I might be named 1.2.1.1.1.0. the first 1 indicates the object is on the Internet. The 2 denotes that it falls within the Management category. The second 1 shows the object is part of the first fully defined MIB, known as MIB-I. The third 1 indicates which of the eight object groups is being referenced. And the fourth 1 is a textual description of the network component. The 0 indicates there is only one object instance. An object instance links a particular object to a specific node on the network. The numbering system is infinitely extendible to accommodate additions to this base identification scheme. This common naming structure permits equipment from a variety of vendors to be managed by a single management station that uses SNMP. The four main categories of the tree are Directory, Management, Experimental and Private/Enterprises.

Internet number The dotted-quad address used to specify a certain system. The Internet number for cs.widener.edu is 147.31.170.2. A resolver is used to translate between hostnames and their numeric Internet addresses. See gTLD and Internet.

Internet Numbers Registry IR. The officially designated organization responsible for the assignment of IP addresses, the IR assigns unique URLs (Uniform Resource Locators), which are translated into IP addresses through a resolver. IR is a responsibility of the IANA (Internet Assigned Numbers Authority), a function assigned to the Information Sciences Institute (ISI) of the University of Southern California. In accordance with its discretionary authority, ISI initially delegated that responsibility to SRI International and, subsequently, to Network Solutions Inc. (NSI). Beginning March 1998, NSI shares that responsibility with CORE and Emergent Corporation, which administers the Shared Numbers Registry (SRS). See also CORE, IANA, SRS, and URL.

Internet of Things, The The total interconnected collection of device networks. The phrase is borrowed from the title of a series of ITU Internet Reports, "The Internet of Things," published in 2005. Written by a team of analysts from the Strategy and Policy Unit (SPU) of the ITU, the report takes a look at the next step in "always on" communications, in which new technologies like RFID and smart computing promise a world of networked and interconnected devices that provide relevant content and information whatever the location of the user. Everything from tires to toothbrushes will be in communications range, heralding the dawn of a new era, one in which today's Internet (of data and people) gives way to tomorrow's Internet of Things. The report costs 100 Swiss Francs or around $82.

Internet offloading Internet offloading is a term used to describe Internet data traffic from a carrier's telephone voice switch and process it on a separate data switch or other equipment, in order to get rid of the Internet-bound traffic and handle that data more cheaply. Here's the logic: A normal central office telephone switch is designed to handle voice phone calls, each averaging three minutes. Pricing is done to accommodate this pattern. After the Internet became very popular, the manager of my local phone company complained to me that his average phone call had now risen to one hour (from three minutes) and this was "killing him." He had been forced to put in more capacity, even though he wasn't getting paid any more money.

Internet open trading protocol IOTP. An interoperable framework for Internet commerce, IOTP was developed by the Open Trading Protocol Consortium, and has been accepted by the Internet Engineering Task Force (IETF) for standards development. IOTP is intended to be independent of any underlying electronic payment systems, such as SET (Secure Electronic Transaction), Mondex, CyberCash, and DigiCash. According to the IETF, a "fundamental ideal of the IOTP effort is to produce a definition of these trading events in such a way that no matter where produced, two unfamiliar parties using electronic commerce capabilities to buy and sell that conform to the IOTP specifications will be able to complete the business safely and successfully."

Internet Packet Exchange IPX. Novell NetWare's native LAN communications protocol, used to move data between server and/or workstation programs running on different network nodes.

Internet peering See Peering.

Internet plumbers Internet plumbers are the companies who make the equipment that makes up the Internet's infrastructure, such as the routers and the servers. They include Cisco, IBM, Lucent and Sun.

Internet print server An Internet print server allows anyone to print a document on any printer on the Internet with the same ease as printing on a printer attached to the PC. Internet Printing is just like e-mail, Internet faxing, Internet Telephone and Video. It is a new method of data streaming through the Internet, that will further open up the Internet for new applications. Black Ice Software, was the first company to announce an Internet print server. www.blackice.com.

Internet Printing Protocol IPP. The protocol that uses the Hypertext Transfer Protocol (HTTP) to send print jobs to printers throughout the world. Windows 2000 and Windows XP Professional support Internet Printing Protocol (IPP) version 1.0.

Internet Protocol IP. Part of the TCP/IP family of protocols describing software that tracks the Internet address of nodes, routes outgoing messages, and recognizes incoming messages. Used in gateways to connect networks at OSI network Level 3 and above. See Internet, Internet Protocol Address and TCP/IP.

Internet Protocol Data Record See IPDR.

Internet protocol address Also called IP Address. It's a unique, 32-bit number for a specific TCP/IP host on the Internet. IP addresses are normally printed in dotted decimal form, such as 128.127.50.224. Once your domain is assigned a group of numbers by the Internet's central registry, it can house one or several domains and/or hosts, i.e. computertelephony.com and teleconnect.com. People looking for those domains will be pointed to that server where they will find all information in the domain – perhaps a home page, or a place to leave e-mail, etc. There are three classes of IP address A, B, and C – the most common of which is a class "C" address block. A class "C" address block can address about 256 hosts (e.g., 128.10.10.*). a class "B" address block can contain about 256*256 (e.g., 128.10.*.*) hosts. Some ip addresses are reserved for broadcasts in respective domains. See Domain and Internet.

Internet protocol datagram The fundamental unit of information passed across the Internet. Contains source and destination addresses along with data and a number of fields which define such things as the length of the datagram, the header checksum, and flags to say whether the datagram can be (or has been) fragmented, This is a self-contained packet, independent of other packets.

Internet protocol suite The TCP/IP suite of network protocols which were mandated for use in the Internet in 1983. The suite includes the following protocols (and the Layer at which each functions in the context of the OSI Reference Model): IP, or Internet Protocol (Layer 3); TCP, or Transmission Control Protocol (Layer 4); UDP, or User Datagram Protocol (Layer 7); FTP, or File Transfer Protocol (Layer 7); TELNET, or TELecommunications NETwork (Layer 7); SMTP, or Simple Mail Transfer Protocol (Layer 7); and SNMP, or Simple Network Management Protocol (Layer 7). See the definitions of these terms for much more detail.

Internet Radio As I write this, I am wearing headphones plugged into my laptop listening to Klassik Radio from Hamburg, Germany. Find it and others at http://windows-media.com/radiotuner/default.asp. The music is wonderful. The clarity is perfect. How does this work? The "radio station" simply sets up a web server attached to the Internet. Instead of putting up text and pictures like other web sites do, it puts up its "radio station." That radio station may be programming its broadcasting over the air in Hamburg, or it may be programming by just specially putting the material up on he web site. That station might also give you its program. How do I listen to Internet radio broadcasts? Most internet radio broadcasts are easily accessible using your web browser and either RealPlayer (from www.Real.com) or Windows Media Player, which comes with Microsoft's browser Internet Explorer. Both of these softwares come with preloaded addresses for interesting Internet

radio stations. But you can find thousands more by searching on Web for "Internet Radio." There are four neat things about Internet radio: The quality is perfectly fine on a dial connection; there's a huge variety of available radio stations (at least several thousand), enough to satisfy anyone's taste; it's all free and fourth, when you log onto many of the stations, a screen may pop up giving you the Radio's program guide. Internet Radio is one of the Internet's enormous benefits.

Internet Registry Activities involved in the administration of generic Top Level Domains (gTLDs) in the CORE (Council or REgistrars) Domain System. Such activities comprise all the services needed for assignment and maintenance of Internet domain names. As many as 90 registrars will be authorized by CORE as registrars to administer and maintain the new gTLDs: .arts, .firm, .info, . nom, .rec, .shop and .web. InterNIC historically has been primarily responsible for the assignment, administration and maintenance of a subset of the traditional gTLDs, specifically, .com, .edu and .org. Future responsibility for those traditional gTLDs is uncertain. See also CORE, DNS, gTLD and InterNIC.

Internet Relay Chat IRC. Sort of like CB radio, but run on the Internet, and far more confusing than CB radio. See IRC for a real definition.

Internet Research Task Force IRTF. An Internet organization that creates long- and short-term research groups concentrating on protocols, architecture, and technology issues. For more information on IRTF, see www.irtf.org.

Internet router see Router.

Internet security Information traveling on the Internet usually takes a circuitous route through several intermediary computers to reach any destination computer. The actual route your information takes to reach its destination is not under your control. As your information travels on Internet computers, any intermediary computer has the potential to eavesdrop and make copies. An intermediary computer could even deceive you and exchange information with you by misrepresenting itself as your intended destination. These possibilities make the transfer of confidential information such as passwords or credit card numbers susceptible to abuse. This is where Internet security comes in and why it has become a rapidly growing concern for all who use the Internet. See the Internet and Secure Channel.

Internet server 1. An Internet server is a device which users on the Internet access to get services. Such services might be electronic mail, news, a Web page, etc. A company will have one or more Internet servers attached to the Internet when it wants to deliver services to people on the Internet. Such Internet servers could be called e-mail servers, FTP servers, News servers and World Wide Web servers. Internet servers most commonly run on Unix. But Microsoft Windows NT is increasingly gaining popularity.

2. A Sun Microsystems term, Part of Solaris' Server Suite. Provides secure, scalable workgroup-based Internet computing.

Internet Server API See ISAPI.

Internet Service Provider ISP. A vendor who provides access for customers (companies and private individuals) to the Internet and the World Wide Web. The ISP also typically provides a core group of internet utilities and services like E-mail, News Group Readers and sometimes weather reports and local restaurant reviews. The user typically reaches his ISP by either dialing-up with their own computer, modem and phone line, or over a dedicated line installed by a telephone company. An ISP is also called a TSP, for Telecommunications Service Provider, and a ITSP, for Internet Telephony Service Provider.

Internet Small Computer Systems Interface See iSCSI.

Internet Society See ISOC.

Internet telephony In the very beginning, Internet telephony simply meant the technology and the techniques to let you make voice phone calls – local, long distance and international – over the Internet using your PC. To make these calls, both people on the phone need appropriate hardware and software. The hardware is typically a sound card or voice modem in a PC. There are almost as many ways of making phone calls on the Internet as there are software packages. The key is to figure a way that your PC can dial and reach someone else's distant PC – which must be turned on, plugged in and connected to some place that my PC can find you at. In short, making voice phone calls was the first definition of Internet telephony. But then people started thinking of other things Internet telephony could become. For example, Internet telephony could let you talk to someone while the two of you worked on making perfect a document that was on both your screens. If the Internet could send email, people started thinking of sending fax, voice, video and imaging mail/messages. And maybe, as you cruise the Internet and find a product you'd like to buy, you might see a button that says "I'd like to know more. Have an operator call me." So you click the button, and 15 seconds later your phone rings. The operator is calling, wanting to know how he can help? In short, the definition of Internet telephony is broadening day by day to include all forms of media (voice, video, image), all forms of messaging and all variations of speed from real-time to time-delayed. See Gold, Packet Switching, Tier 1 and, for the best explanation, TAPI 3.0.

Internet terminal A sub-$500 machine specially designed for Internet browsing and first proposed in the late Fall of 1995 by Larry Ellison, head of database software company Oracle. Part of its appeal to people outside Microsoft and Intel is that the Internet Appliance would not have to be based on standard PC technology. It need have an Intel chip and need not run Windows. This device is also called an Internet Terminal, a Network Computer or an IPC, an Interpersonal computer. The original description of the Internet Appliance was that it would come with 4mb of RAM, 4mb of flash memory, processor, monitor, keyboard and mouse – all for under $500. Also called a NC, or Network Computer.

Internet time This term came into the English language around 1994 and became canonical during the late 1990s. It was used to describe the accelerated pace at which, in an Internet/Web enabled world, all business was supposedly going to be conducted. Internet time and its cousins, Web Time and Warp Speed, did well as titles of books, which didn't sell well, but little else. See also New Economy.

Internet TV Just as it sounds – moving TV transmitted over the Internet. It's "the next big thing" for the Internet. The problem is twofold: First, the bandwidth required is huge. Remember TV is 30 pictures per second. Think of how long it often takes to download one picture, let alone 30 every second. Second, storing movies on servers is expensive and consuming of vast space. I've looked into this "business" on several occasions and always been impressed at how difficult and how expensive it is to organize.

Internet worm This software program caused a major part of the Internet network to crash by replicating and generating spurious data.

Internet2 Internet2 is a high-speed network created by a consortium of U.S. universities called the University Corporation for Advanced Internet Development (UCAID). It transmits high-quality audio and video with almost no delay 34 U.S. universities announced the formation of Internet2 in October 1996. This second version of the Internet is a collaboration of the National Science Foundation (NSF), the U.S. Department of Energy, 206 research universities, and a small number of private businesses. Each participating university has committed at least $500,000 to fund the project. Intended to serve as a private Internet for the exclusive use of its member organizations, it will be separate from the traditional Internet. The network eventually will operate over fiber optic transmission facilities at speeds of up to 10 Gbps. Internet2 will connect through gigiPOPs, switches with throughput in the range of billions of packets per second, and will run the IPv6 protocol. Internet2 isn't apparently just about fast networks. The consortium has also made progress developing innovative software and services. One application, called Shibboleth, is a piece of open source software that enables users to share restricted online resources. Without the software, if a college or business wanted to subscribe to some kind of online database, they'd have to create hundreds or thousands of accounts, one for each individual user. That's a huge administrative burden, and particularly complicated in schools where new students are enrolling and old ones are graduating. But Shibboleth handles all the identification and authentication of users in between the school and the database, thus reducing the complexity of management and protecting the privacy of individual users. Pennsylvania State University is using the program to allow its students to access music download service Napster to give them a legal alternative to file sharing. Development of the software was supported with funding from several public and private universities and the National Science Foundation. See also www.internet2.edu, also IPv6 and Internet.

internetwork See Internetworking.

internetwork management A generic term used to describe the actions that help maintain, a complex network.

Internetwork Operating System IOS. Cisco's massive operating system that runs most routers on the Internet.

Internetwork Packet Exchange IPX. A network protocol native to NetWare that controls addressing and routing of packets within and between LANs. IPX does not guarantee that a message will be complete (no lost packets).

internetwork router In local area networking technology, an internetwork router is a device used for communications between networks. Messages for the connected network are addressed to the internetwork router, which chooses the best path to the selected destination via dynamic routing. Internetwork routers function at the network layer of the Open Systems Interconnection (OSI) model. Also known as a network router or simply as a router.

internetworking Communication between two networks or two types of networks or end equipment. This may or may not involve a difference in signaling or protocol elements supported. And, in the narrower sense – to join local area networks together. This way users can get access to other files, databases and applications. Bridges and routers are the devices which typically accomplish the task of joining LANs. Internetworking may be done with cables – joining LANs together in the same building, for example. Or it may be done with telecommunications circuits – joining LANs together across the globe.

InterNIC Internet Network Information Center. The InterNIC registry is where you always used to go to register your domain name. Registration was free until 1995; then it changed to $50 a year; now it's $35 a year for domain names with anniversary dates on or after April 1, 1998. The InterNIC Registration Services Host computer contains information on Internet Networks, ASNs, Domains, and POCs (Points of Contact, the person or persons identified in a record). The InterNIC was established in January 1993 as a collaborative project between AT&T, General Atomics (no longer involved) and Network Solutions, Inc., and was supported through a 5-year cooperative agreement with the NSF (National Science Foundation). InterNIC participates in Internet forums to promote Internet services, explore new tools and technologies, and contribute to the Internet community. InterNIC currently is operated by Network Solutions Inc., a subsidiary of SAIC (Science Applications International Corporation), the private company which also has acquired Bellcore. On December 22, 1997, responsibility for assignment of IP numbers was shifted to ARIN (American Registry of Internet Numbers) for the geographic areas of North American, South America, the Caribbean and sub-Saharan Africa. About the same time, the decision was made to shift responsibility for domain name registration to the Council of Registrars (CORE). CORE is empowered to authorize as many as 90 independent registrars, including InterNIC, to register domain names, or URLs (Uniform Resource Locators). www.internic.net. See also ARIN, CORE, Domain, Domain Name and POC.

internode Communication paths which originate in one node and terminate in another.

internode link A data line for high-bandwidth connections between PBXs.

interoffice Between two telephone company switching offices.

interoffice channel A portion of a communications circuit between central offices of a common carrier which serves customers located in different central office areas. When associated with foreign exchange service, the term denotes the channel which interconnects a primary wire center to a different wire center.

interoffice trunk A trunk circuit connecting two local telephone company central offices.

interoperate The ability of equipment from several vendors to work together using a common set of protocols. We're not talking about identical products, but ones that conform to the same protocol and theoretically can talk to each other. For example, two laptops from different manufacturers should be able to talk to each other using infra-red communications because both conform (theoretically at least) to the same infra-red communications protocol. Interoperability is not Yes or No. There are levels of it. See Interoperability Testing.

interoperability See Interoperate.

interoperability testing In our industry there are several levels of testing of new products and services. Clearly, the manufacturer does one level of testing. His job is to ensure that his product meets the claims that he promotes for it. Then there is a interoperability testing. This is testing to ensure that his product works with other products that allegedly conform to the same standard/s. Clearly it is one thing for a bunch of engineers to create a new standard on paper. And it is another thing to have engineers build product to that standard and have their products communicate with other products from other manufacturers who also allegedly conform to the standard. The most public example of interoperability testing in the telecom industry is that which occurred so successfully in the ATM Forum. That organization was established by manufacturers who make ATM products. Their objective was not to establish new standards (they already existed), but to make sure that their products worked successfully with each other without hardware modification. Plug the equipment together. Configure the software. Bingo it would work. See Interoperate.

interoperator Modular hardware or software that implements part of the OSI model and can work with components implementing the other parts of the model.

interpacket gap IPG. A delay or time gap between CSMA/CD packets intended to provide interface recovery time for other CSMA/CD sublayers and for the Physical Medium. For 10Base-T, the IPG is 9.6 us (96 bit times). For 100Base-T, the IPG is 0.96 us (96 bit times).

InterPBX Calls coming into one PBX can be transferred to extensions on another PBX using direct tie lines between the two PBXs.

interpersonal message IPM. The term used in the 1984 X.400 recommendations to refer to a message in the Interpersonal Messaging System. The 1988 X.400 recommendations use the term "interpersonal message" (IPM).

Interpersonal Message (IPM) User Agent A class of cooperating user Agents capable of processing Interpersonal (IP) messages. An x.400 term.

interpersonal messaging IPM. Electronic exchange of information between two or more persons.

interpolation 1. The process of estimating values of (a function) between two known values.

2. A video technique used in motion compensation where a current frame of video is reconstructed by using the differences between it and past and future frames. This technique is also known as forward and backward prediction. Intel (originator of Indeo Video) defines interpolation slightly differently, namely: The process of averaging pixel information when scaling an image. When the size of an image is reduced, pixels are averaged to create a single new pixel; when an image is scaled up in size, additional pixels are created by averaging pixels of the smaller image.

interposition calling One operator in a multi-position system calling another.

interposition transfer Transfer of a call from one operator to another.

interposition trunk 1. A connection between two positions of a large switchboard so that a line on one position can be connected to a line on another position.

2. Connections terminated at test positions for testing and patching between testboards and patch bays within a technical control facility.

Interpositioning An equipment configuration in which carrier-provided terminal equipment accesses exchange carrier facilities through customer-provided terminal equipment. An old term.

interpret Interpret means that the computer will translate a stored program expressed in pseudocode into machine language and will perform the indicated operations as they are translated. See Interpreter for a longer explanation.

interpreter Much like a compiler, an interpreter translates source code written by a programmer into machine code the computer can understand. For example, a compiler takes instructions written in a "higher" level language such as BASIC, COBOL or ALGOL and converts them into machine language that can be read and acted upon by a computer. The translated code is in the form of an executable program, which can be run on the target computer without additional translation software. Compilers convert large sections of code at one time, usually translating the code so that it can be run at a later time on the target computer. Interpreters translate commands one line at a time while the application is running. Therefore, the machine running the application must also be running the interpreter. Interpreters are useful in the testing of new or modified code, and for teaching programming. See also Bytecode and Compiler.

interprocess communications The ability of programs to share information. At the most basic level, it consists of cutting and pasting information between two programs. Above that ranks the "live" paste, in which information shared between two documents is updated whenever one of the documents is modified. This is referred to as Dynamic Data Exchange (DDE). In advanced DDE, programs can send messages as well as data to other programs running locally or remotely. Beyond DDE is Object Linking and Embedding (OLE), which lets one program borrow the specialized capabilities of another program loaded on the machine (say, advanced chart creation) rather than having to implement that capability redundantly.

InterRepeater Link IRL. A networking term. A mechanism for connecting two and only two repeater sets.

interrogate To determine the state of a device or unit.

interrogator See RFID reader.

interrupt A temporary suspension of a process caused by an event outside of that process. More specifically, an interrupt is a signal or call to a specific routine. An interrupt setting allows the hardware in a file server, router, workstation or PC to send an interrupt signal to the processor. The interrupt signal temporarily suspends the other station tasks while the processor performs the task requested by the interrupting device. After the routine is completed, the processor then continues with the original tasks. Each piece of hardware (serial and parallel ports and network boards) installed in the same computer needs a unique interrupt. Interrupts are divided into two general types, hardware and software. A hardware interrupt is caused by a signal from a hardware device, such as a printer. A software interrupt is created by instructions from within a software program.

TIP: When you slide a new card into one of the empty slots on your PC and things go awry, check that the new card's interrupt is not the same as one of the other cards in your bus. An interrupt is also called a hardware interrupt or an InterRupt reQuest (IRQ). For a listing of normal IRQs see IRQs. See also Interrupt Requests and Polling.

Interrupt Driven Someone who moves through a workday responding to interruptions rather than the work goals as originally set.

interrupt flag In a PC, there is a configuration control (addressed as bit IF of the processor flag register). This control process manages the CPU's ability to receive and process interrupt requests. The flag is often set to zero (which means interrupts disabled) by device drivers or other I/O privilege-level code that needs exclusive access to the CPU during critical operations. See also Interrupt, Interrupt Handling Routine and UART.

interrupt handling routine This program, which is often part of a device driver, handles all requests from a particular interrupt line. Interrupt-handling routines are defined in the CPU's Interrupt Descriptor Table (IDT). When the CPU (the Central Processing Unit of your PC) receives an interrupt request, it looks up the matching interrupt-handling routine in the IDT, then transfers control to the routine until it (the CPU) gives an interrupt return call (IRET), indicating the task is complete.

interrupt latency The delay in servicing an interrupt request is known as interrupt latency. It is not a problem with devices that are not sensitive to timing inconsistencies (such as hard-disk controllers or video boards). But it is a problem with high-speed, asynchronous communications (9,600 bps and above), which are highly time-sensitive operations.

interrupt overhead The cumulative demand on your computer's central microprocessor by peripheral devices that generate interrupt requests is referred to as interrupt overhead. Such devices include hard-disk controllers, network interface cards, parallel and serial ports.

interrupt request IRQ. This is the communications channel through which devices issue interrupts to the interrupt handler of an IBM PC or IBM compatible PC's microprocessor. It's the channel through which these devices get the microprocessor's attention. Different IRQs are assigned to different devices. This assignment pattern differs from PC to PC. Many LAN interface cards use an IRQ to get to the microprocessor. You must be sure that your LAN interface card is not trying to use the IRQ assigned to another peripheral, like the hard disk controller or EGA card. See also IRQ for a different and longer explanation.

interrupt request lines Hardware lines over which devices can send signals to get the attention of the processor when the device is ready to accept or send information. Typically, each device connected to the computer uses a separate IRQ.

interrupter An automatically operated electromechanical device used to turn lights, bells or other signals on and off in timed sequences. An interrupter makes lights wink on and off on a key system. Or did, when everything was electromechanical. It was used on 1A key telephone systems.

interrupting equipment Motor-driven mechanical devices used to break the ringing generator's output into ringing and silent periods, creating the busy and ringback tone pulses.

intersatellite link A message transmission circuit between two communication satellites, as opposed to a circuit between a single satellite and the earth.

InterSpan The full name is InterSpan Frame Relay and it's AT&T's frame relay data communications service, announced in the late fall of 1991.

Intersputnik A Russian satellite system similar in concept to the West's Intelsat, except that it's set up by Russia and the Eastern bloc countries. Two US carriers, AT&T and IDB Communications, once used Intersputnik to alleviate their shortage of US-Russia circuits. See INTELSAT.

interstate Literally, between states (crossing a state line). Services, traffic or facilities that originate in one state, crossing over and terminating in another.

interstate highway system When the U.S. Interstate Highway System was begun during the Einsenhower era, it was required that one mile in every five must be straight. These straight sections would be usable as airstrips in times of war or other emergencies.

interstices In cable construction, the spaces, valleys or voids between or around the cable's components.

interswitch trunk A circuit between two switching machines.

intersymbol interference ISI. Intersymbol interference is a source of noise in baseband signaling that occurs when the signal pulses or symbols spread into adjacent pulses or symbols. This spreading effect occurs when the signal varies with frequency or when portions of the signal are delayed due to multipath fading. See also Baseband and Multipath Fading.

intertandem trunk groups A category of trunk groups that interconnects tandems.

intertoll trunks Trunks connecting Class 4 and higher switching machines in the AT&T long distance network.

interval Time. Pulse interval, for example, means the time from the start of one pulse to the start of the next.

interworking The ability to seamlessly communicate between devices supporting dissimilar protocols, such as frame relay and ATM, by translating between the protocols, not through encapsulation. Many carriers are planning to implement the necessary equipment and conversion algorithms to allow the network itself to transparently convert from frame relay to ATM, and vice versa.

INTFC Interface.

intra Intra means inside. Intrastate means inside the state. Interstate means between states.

intra-area cell transfer A cellular radio term. A cell transfer between two cells that are controlled by the same serving Mobile Data Intermediate System (MD-IS).

intrabuilding backbone Telecommunications cable(s) that are part of the building subsystem that connect one closet to another.

intracalling This is an outside plant term. Intracalling refers to the ability of a remote line concentrator to interconnect users served by the same concentrator without providing two trunks directly back to the central office.

intraday distribution A call center term. A historical pattern consisting of factors for each intra-day period of the week that define the typical distribution of call arrival or average handle time throughout each day. Each factor measures how far call volume or average handle time in that half hour or quarter hour deviates from the average half-hourly or quarter-hourly figure for that day. This information enables the program to forecast intra-day call volumes and staffing requirements.

IntraEnterprise Communications Communications that are exchanged within a single organization (including multiple sites of the organization).

intraexchange Any traffic or service within an exchange area or serving area.

intraflow This is an automatic call distribution term. It refers to the ability to select a second or subsequent group of agents to backup the primary agent group. This is designed to allow the caller to be serviced more efficiently and less expensively.

intraframe coding A way of video compression that compresses information within a single frame. Compare to Interframe Coding.

IntraLATA Telecommunications services that originate and terminate in the same Local Access and Transport Area. See also Local Access and Transport Area. This can be either Interstate or Intrastate service, traffic or facilities.

intramodal distortion In an optical fiber, the distortion resulting from dispersion of group velocity of a propagating mode. It is the only form of multi mode distortion occurring in single-mode fibers.

intranet A private network that uses Internet software and Internet standards. In essence, an Intranet is a private Internet reserved for use by people who have been given the authority and passwords necessary to use that network. Those people are typically employees and often customers of a company. An Intranet might use circuits also used by the Internet or it might not. Companies are increasingly using Intranets – internal Web servers – to give their employees easy access to corporate information. According to my friends at Strategic Networks Consulting, Boiled down to its simplest, an Intranet is a private network environment built around Internet technologies and standards – predominantly the World Wide Web. The primary user interface, called a Web browser, accesses Web servers located locally, remotely or on the Internet. The Web server is the heart of an Intranet, making selection of Web server software a crucial decision, even though much fanfare has focused on browsers (Netscape's Navigator vs. Microsoft's Explorer).

At its core, a Web server handles two arcane languages (HTML and CGI) that are the meat and potatoes of generating Web pages dynamically, making connections and responding to user requests. But in the rush to dominate the potentially lucrative Intranet market, these simple Web functions are being bundled into operating systems and vendors are now touting pricey "Intranet suites" which encompass everything from database and application interfaces, to e-mail and newsgroups, to the kitchen sink. Most medium- or larger-sized companies will need more than just a handful of simple Web servers to deploy a reasonably robust Intranet. To help a company post current job openings, or make up-to-date product specs and available inventory accessible by traveling sales reps, an Intranet needs the following capabilities:

- Database access. Getting at critical data housed in corporate databases can be accomplished via generic, universal ODBC linking or based on "native" links directly to Sybase, Oracle et al. allowing use of all the database's features.
- Application hooks. Used by developers, a standard programming interface (API) allows outside applications like Lotus Notes to interact with Web data and vice versa. In addition, proprietary APIs exist – most notably Microsoft's ISAPI (for "Internet Server API") which lets developers link directly to Microsoft applications.
- User publishing. In addition to dialogues via chat/newsgroup/bulletin board features, users will want to post their own content on Web servers without having to attain Webmaster status.
- Search vehicles. How does an engineer find the current specs on Project #686-2

among thousands of pages spread across a bunch of Web servers? The answer: an indexing and search engine that creates an internal Yahoo! for your own Web sites.
- Admin/management. A catch-all for loads of important, but still ill-conceived features for managing access, users, content and the servers themselves. Intranet administrators are currently fascinated with analyzing Web server logs which contain data of some sort, including user connections and page activity.

According to a white paper released by Sun Microsystems in the summer of 1996, the basic infrastructure for an intranet consists of an internal TCP/IP network connecting servers and desktops, which may or may not be connected to the Internet through a firewall. The intranet provides services to desktops via standard open Internet protocols. In addition to TCP/IP for basic network communication, these also include protocols for:

Browsing	HTTP
File Service	NFS
Mail Service	IMAP4/SMTP/POP3
Naming Service	DNS/NIS+
Directory Services	DNS/LDAP
Booting Services	Bootp/DHCP
Network Administration	SNMP
Object Services	IIOP (CORBA)

See also Extranet and Intranet.

intranodal service A feature of "intelligent remotes" (i.e., intelligent remote switching nodes), intranodal service refers to the ability of the node to continue to switch calls within its own geographic domain, even if the signaling and control link to the CO (Central Office) fails. This capability usually is limited to the basic switching of voice calls. More complicated processes, such as the support of custom calling services or other complex features, are interrupted until such time as the signaling and control link can be restored to the CO, where the majority of programmed logic and memory reside.

intranode Communications path which originates and terminates in the same node. See Intranodal Service for a fuller explanation.

intraoffice call A call involving only one switching system.

intraoffice trunk A telephone channel between two pieces of equipment within the same central office.

intrapreneur An entrepreneur who works inside a big company. Hence, intra, as in inside. It's hard to imagine it actually happening. But the word has become popular as a way for large companies to motivate their employees to take personal career risks and introduce new products.

intrastate Services, traffic or facilities that originate and terminate within the same state. Therefore, if related to telephone, falling under the jurisdiction of that state's telephone regulatory procedures.

infrastructure A term coined by "Data Communications" Magazine and referring to the software, hardware, and Internet services underlying a corporate Intranet.

intrinsic joint loss That loss in optical power transmission, intrinsic to the optical fiber, caused by fiber parameters, e.g., dimensions, profile parameter, mode field diameter, mismatches when two non identical fibers are joined.

intrinsics Intrinsics are a component of many windows toolkits. The windows toolkit intrinsics definition has been developed by the MIT X Consortium. The intrinsics define the function of specific graphical user interface and window objects. They do not define any particular look or feel, just the function. Example: A pull down menu intrinsic would define the function of a pull down menu within a toolkit but not the appearance of it.

intrusion detection A technology that gathers and analyzes information across gateways, servers, and desktops to identify possible security breaches that can occur from within or outside an organization. An intrusion detection system (IDS) can detect attacks on the network through the use of statistical analysis of network traffic as well as by monitoring reports and log files to detect abnormal network activity. Once illicit activity is detected, the IDS alerts administrators. The first concern is that someone must repond to the alarm. If nobody is present, or there is a delay, damage is being done to the network and money lost. Another problem is a false alarm. An IDS must be tuned to recognize valid traffic and usage patterns and updated to detect the most up-to-date types of attacks to reduce the number of false alarms. Intrusion prevention systems (IPS) integrate with an IDS to automate network responses to many network attacks. An IPS provides two types of prevention: reactive and proactive. Reactive prevention simply acts once an attack has actually been detected. For instance, if the IDS detects abnormal traffic coming from several computers on a network that is consistent with virus activity, the IPS could communicate with a firewall or switch to block the infected system from accessing other parts of the network through rules or access control lists (ACL). Proactive prevention occurs before any attacks occur. For instance, if an IDS is scanning the network and notices several open ports that are not normally used by the average user, but could be exploited by an attacker. The prevention system could communicate with a firewall or switch and automatically block access to those ports using rules or ACLs to eliminate avenues of attack and catalog the systems and relay the information to a system administrator. IDS and IPS are often indistinguishable because most security products have both IDS and IPS features. Combined, IDS/IPS systems give network administrators the ability to detect and prevent attacks. They can also provide a detailed log of what happened during an attack or after the attack is over and the network recovered. A related area of intrusion detection is vulnerability assessment, which uses scanning technology to determine misconfigurations or other security vulnerabilities contained within a computer system or network. See deep packet inspection.

intrusion prevention system See intrusion defection above.

Intrusive Scan See Port Services.

Intrusive Test Breaking a circuit in order to test its functionality. Testing intrusively will drop service on the circuit.

Intserv Internet Services. IETF's model to transport audio, video, real-time and traditional data within a single architecture.

INTUG International Telecommunications Users Group.

intuitive The cognition (i.e., knowledge) or understanding of something without evident rational thought or inference. User interfaces to devices, systems and software applications often are touted as being "highly intuitive." That means that the manufacturer claims that anyone of normal intelligence can operate the thing without having to read the manual. Such claims are always inflated, to put it politely. See also Cockpit Problem, Help Desk, Idiot-Proof, and RTFM.

intumescent firestop A firestopping material that expands under the influence of heat.

invar shadow mask A special type of shadow mask that is made out of Invar, an alloy that is able to withstand the high temperature generated by an electron beam. An Invar shadow mask allows the CRT to generate a brighter image than a conventional shadow mask. An Advanced Invar Shadow Mask improves brightness by 40% over a standard shadow mask.

inventory workdown In 1999 and 2000, the telecom industry shipped too much equipment. When demand dropped off, a lot of it languished in warehouses and factories. As demand picked up, inventories sitting around began to drop. That phenomenon became known as inventory workdown. The theory was when the inventory was worked down to a normal level (whatever that is), factory shipments would pick up. It sort of started to happen at the end of 2001.

Inverse ARPA See Reverse DNS.

Inverse Fourier Transform Inversion of Fourier transform to convert frequency representation of signal to time representation.

inverse multiplexer I-Mux. An inverse multiplexer performs the inverse function of a multiplexer. "Multiplexer" translates to "many fold." For example, a TDM Mux (Time Division Multiplexing Multiplexer) accepts many (typically 24) low-capacity inputs in the form of information streams, and folds them together through a process know as byte interleaving in order to send them over a single, high-capacity, shared digital circuit. In this example, each of the 24 voice-grade channels supports a transmission of 64 Kbps; the total capacity of the T-1 circuit is 1.536 Mbps. The advantage of this approach is that of economy of scale – a single, high-capacity T-1 circuit is far less expensive that are 24 individual voice-grade circuits. An inverse multiplexer does just the inverse. In other words, it accepts a single, high-capacity information stream and splits it up into multiple information streams, each of which is sent over a separate and lower-capacity circuit; the process is reversed on the receiving end. Videoconferencing, for example, may make use of inverse multiplexers. A full-motion videoconference requires a full T-1. While a user organization may have multiple T-1s at a given site, a full T-1 may not be available at the moment it is required. Therefore, an inverse mux might split that video datastream into four data streams of 384 Kbps, and send each over six channels of four separate T-1s. At the receiving end of the video datastream, the four data streams are received, demultiplexed and resynchronized in order to reconstitute the original datastream. Resynchronization is critical, as each of the four circuits may impose different levels of propagation delay on the signal due to reasons such as differing route lengths. Synchronization prevents your head (which traveled over T-1 #1) from appearing at your knees (which traveled over T-1 #3) on the receiving TV monitor. Inverse Multiplexing over ATM (IMA) is an access specification approved in 1997 by the ATM Forum. This User-Network Interface (UNI) standard allows a single ATM cell stream

to be split across multiple access circuits from the user site to the edge of the carrier's ATM network. In an ATM LAN application, for instance, the ATM switch deployed in the enterprise backbone typically operates at 155 Mbps or 622 Mbps. In this example, ATM traffic from the enterprise to the public ATM carrier-based network, requires 6 Mbps-well more than the level supported by a T-1, but less than a full T-3. Rather than subscribing to a T-3, which requires a fiber optic access circuit and which is very expensive, IMA is used. Thereby, the ATM data stream is split across four T-1 circuits by an access concentrator which possesses IMA capability. The IMA process works in a round-robin fashion, with cell #1 traveling over T-1 #1, cell #2 traveling over T-1 #2, and so on. Each of the four T-1 circuits is relatively inexpensive, can be provisioned over twisted-pair, and is readily available. At the edge of the carrier network, the ATM switch receives each of the four separate data streams, and reverses the IMA process in order to reconstitute the original datastream, which then is switched and transported through the network to the far edge. At that far edge, the IMA process may take place again, from the edge of the carrier network, over four T-1s, and to the IMA-capable ATM concentrator on the user premises. See also ATM and TDM.

inverse multiplexing The process of splitting a single high-speed channel into multiple signals, transmitting each of the multiple signals over a separate facility operating at a lower rate than the original signal, and then recombining the separately-transmitted portions into the original signal at the original rate. See Inverse Multiplexer for a full description of the process. Inverse multiplexing is also called loop bonding. See inverse multiplexer above and loop bonding.

inverse telecine A system for retoring the frame rate of film footage to improve appearance and reduce storage space.

inverted backbone A network architecture in which the wiring hub and routers become the center of the network; all the network segments attach to this hub.

inverted rotary converter A motor driven device that produced dial tones, busy tones, ringing signals, etc. and was used in conjunction with 701 type step-by-step electromechanical switches.

inverter 1. A device which converts direct current electricity to alternating current electricity, often used to power AC devices in a car.

2. See Analog Inverter and Digital Inverter.

invitation to send A character or sequence of characters which calls for a station to begin transmission. Usually this is part of a polling arrangement.

inward operator trunk group A trunk group used to provide distant operators with the means of obtaining an inward operator's assistance for the completion of calls. A typical application commonly employs a tandem to cord switchboard type trunk group.

inward restriction A Centrex service feature which stops Centrex lines from receiving certain incoming calls.

inward trunk Used only for incoming calls, these trunks cannot dial out. "800" lines, for example, can only be used to receive calls.

INWATS INward Wide Area Telephone Service. A service of interexchange carriers (e.g., AT&T, MCI, and Sprint), local exchange carriers, the Bell operating companies and the independent phone companies and long distance resellers in North America which allows subscribers to receive calls from specified areas (depending on the rate band chosen) with no charge to the person who's calling. Rather, the charges are billed to the called party. See 800 Service for a much bigger explanation.

IOA See Interim Operating Authority.

IOC 1. Inter-Office Channel.

2. Independent Operating Carrier or Independent Operating Company. See Independent Operating Carrier.

IOD Identified Outward Dialing. See also AIOD and Call Accounting System.

IOEngine Input/Output Engine. The part of the Novell SFT III operating system that handles physical processes, such as network and disk I/O, hardware interrupts, device drivers, timing, and routing. SFT III is split into two parts: the IOEngine and the MSEngine (Mirrored Server Engine). The IOEngine routes packets between the network and the MSEngine. To network workstations (i.e. PCs on the LAN), the IOEngine appears as a standard NetWare router or bridge. The primary server and the secondary server each have an IOEngine, but they share the same MSEngine. Because the IOEngines are not mirrored, NetWare Loadable Modules (NLMs) and applications that directly interface with hardware, such as backup NLMs may be installed in the IOEngines on both the primary and the secondary server. See MSENGINE.

IOF Inter Office Facility.

ion exchange technique A method of fabricating a graded-index optical fiber by an ion exchange process.

ionization The process of breaking up molecules into positively and negatively charged carriers of electricity called ions.

ionizing A type of energy, in the upper end of the electromagnetic spectrum, that has the ability to strip electrons from molecules, thus forming a new, distinct ion or compound.

IONL Internal Organization of the Network Layer. The OSI standard for the detailed architecture of the Network layer. Basically, it partitions the Network layer into subnetworks interconnected by convergence protocols (equivalent to internet working protocols), creating what the Internet community calls a catenet or internet.

ionosphere That part of the atmosphere in which reflection and/or refraction of electromagnetic waves occurs. It extends from about 70 to 500 kilometers. At that point, ions and free electrons exist in sufficient quantities to reflect electromagnetic waves.

ionospheric absorption Attenuation of the energy in a radio wave due to the interaction between it and gas molecules. Deviative absorption describes the appreciable bending that occurs in an ionospheric layer at close to critical frequency. Non-deviative absorption describes the condition where little or no bending occurs as the wave passes through an ionized layer. See Ionosphere.

ionospheric cross modulation Nonlinearities within the medium can produce nonlinear absorption. This can lead to the modulation on a strong signal being transferred to a weaker carrier. Sometimes described as the Luxembourg effect.

ionospheric disturbance An increase in the ionization of the D region of the ionosphere, caused by solar activity, which results in greatly increased radio wave absorption. See Ionosphere.

ionospheric focusing A variation in the curvature of the ionospheric layers can give rise to a focusing/defocusing effect at a receiving antenna. This may produce either an enhancement or attenuation in the received field strength due to signal phase variations.

ionospheric refraction The change in the propagation speed of a signal as it passes through the ionosphere.

IOP 1. Input/Output Processor.

2. Interoperability: The ability of equipment from different manufacturers (or different implementations) to operate together.

IOPS Internet OPerators Group. On May 20, 1997 Nine of the nation's major Internet service providers announced the formation of IOPS.ORG, a group of Internet service providers (ISPs) dedicated to making the commercial Internet more robust and reliable. IOPS. ORG will focus primarily on resolving and preventing network integrity problems, addressing issues that require technical coordination and technical information-sharing across and among ISPs. These issues include joint problem resolution, technology assessment, and global Internet scaling and integrity. IOPS.ORG will provide a point-of-contact for these industry-wide technical issues.

The founding members of IOPS.ORG are ANS Communications, AT&T, BBN Corporation, EarthLink Network, GTE, MCI, NETCOM, PSINet, and UUNET, and it is expected that additional national and international Internet operators will join. IOPS.ORG will work with other Internet organizations, with Internet equipment vendors, and with businesses that rely on the Internet. IOPS.ORG members individually will continue to support other Internet organizations such as the Internet Engineering Task Force (IETF), the North American Network Operators Group (NANOG), and the Internet Society.

The Corporation for National Research Initiatives (CNRI), a Reston, VA-based not-for-profit organization which works with industry, academia, and government on national-level initiatives in information technology, will host the initial operations of IOPS.ORG. "IOPS. ORG will play a key role in the healthy technical and operational evolution of the Internet as an increasingly important component of the economy," said CNRI President Robert Kahn. www.iops.org.

IOR Index of Refraction. The ratio of light velocity in a vacuum to its velocity in a given transmission medium, such as an optical fiber.

IOS 1. Internetwork Operating System from Cisco. This operating system runs the vast majority of routers now deployed in the core of the Internet. The IOS operating system is now being replaced by the newer IOS XR. See also Junos Code.

2. See ISO.

IOS XR See IOS.

IOTA Iota, written as i or sometimes j in electrical engineering is the square root of -1. The concept of iota was introduced by Leonhard Euler, Swiss Mathematician. The important property of i (or j) is that it maps any kind of differential, whether application in differential amplifier of differential dense wave multiplexing or differential time division multiplexing. To understand i (or j) more deeply, one has to attempt a conjecture of number theory....... or simply apply intuition.

IOTA DM IP-based over-the-air device management. See also IOTA.

IOTP See Internet Open Trading Protocol.

IP 1. The Internet Protocol. IP is the most important of the protocols on which the Internet is based. The IP Protocol is a standard describing software that keeps track of the Internet's addresses for different nodes, routes outgoing messages, and recognizes incoming messages. It allows a packet to traverse multiple networks on the way to its final destination. Originally developed by the Department of Defense to support interworking of dissimilar computers across a network. While its roots are in the ARPAnet development, IP was first standardized in RFC 791, published in 1981, and updated in RFC 1349. This protocol works in conjunction with TCP and is usually identified as TCP/IP. It is a connectionless protocol that operates at the network layer (layer 3) of the OSI model. See IP Address, IPv4, IPv5, IPv6, the Internet, and TAPI 3.0.

2. Intelligent Peripheral. A device in an IN (Intelligent Network) or AIN (Advanced IN) that provides capabilities such as voice announcements, voice recognition, voice printing and help guidance. By way of example, MCI's 1-800-COLLECT makes use of IPs, which are specialized voice processing systems. The IP prompts the caller to enter the target telephone number and speak his or her name. The system then instructs the network to connect the call. Based on a spoken acceptance of the call by the called party, the system authorizes call completion.

3. Information Provider. A customer that offers recorded information on its listed numbers.

4. Intellectual Property. A legal term that refers to original creative work (a book, a movie, software code, etc.) manifested in a tangible form that can be legally protected, for example, by a patent, trademark, or copyright.

5. Illustrative Paragraph. See Stupid.

IP Address See Internet Address.

IP Address Confirmation Also called IP Confirmation. Here's how it works. In order to send electronic mail over the Internet, you need what's called a SMTP (Simple Mail Transfer Protocol) server, a specialized computer to send your mail. If you have your own domain name, like www.technologyinvestor.com, then it's easy to set up your own mail server. Most of us, however, don't. We need to use someone else's. In the good old days you could pretty well anyone's SMTP server. Simply tell your email client. These days many owners of SMTP mail servers got unhappy with people using their servers to spam (i.e. send zillions of unwanted email). So they set up a system to check the IP address from where you were coming before they allowed your email through. If your IP Address was one of the company's addresses, e.g. one you'd been assigned temporarily while you were on line, then they let your email through. If it wasn't, your email would be stopped. This process is called IP Address Confirmation.

IP Address Mask Internet Protocol address mask. A range of IP addresses defined so that only machines with IP addresses within the range are allowed access to an Internet service. To mask a portion of the IP address, replace it with the asterisk wild card character (*), For example, 192.44.*.* represents every computer on the Internet with an IP address beginning with 192.44. See IP Addressing.

IP addressing A networking term. IP (Internet Protocol) addressing is a system for assigning numbers to network subdivisions, domains, and nodes in TCP/IP networks. IP addresses are figured as 32-bit (four-byte) numbers. The high bytes constitute the "Class A" and "Class B" portions of the address, which denote network and subnetwork. The low bytes ("Class C" address segments) identify unique nodes – individual machines or (in the case of multi-addressing) individual node processes. The Class C address segment (two bytes) can represent 65,536 unique values – enough so that in most conventional TCP/IP LANs, sufficient values are available to afford each machine its own "fixed" IP address. In public internet-access, however, the number of fixed addresses available to a provider may not be sufficient to provide each dialup client with a permanent IP address. In such scenarios, available Class C addresses can be assigned dynamically, as machines log into network access ports – on the presumption that no more than N clients will attempt to log on, simultaneously (where N denotes the number of absolute addresses in the pool). Thus:

"Fixed" or "Static" IP address: a four-byte TCP/IP network address permanently assigned to an individual machine or account.

"Dynamically-assigned" IP address: a four-byte TCP/IP network address assigned to a machine or account for the duration of a single session.

IP communications Data, voice and video communications over a single, converged IP network.

IP Confirmation Also called IP Address Confirmation. See IP Address Confirmation.

IP datagram The fundamental unit of information passed across the Internet. Contains source and destination addresses along with data and a number of fields which define such things as the length of the datagram, the header checksum, and flags to say whether the datagram can be (or has been) fragmented.

IP device control IPDC. See Simple Gateway Control Protocol.

IP edge concentrator An IP Edge Concentrator as defined by Seranoa Networks, is a subscriber aggregation device attached to an edge router at the POP of an IP service provider. It provides a high-capacity, cost-effective alternative to vendor-specific WAN line cards installed in edge routers. It operates as a layer 2 switch to forward low-speed subscriber access circuits (e.g. T-1) onto a high speed trunk (e.g., GigE) to the edge router. It may include a network processsor to offload various IP services from the edge router such PPP, Multilink PPP, IP DiffServ QOS, and traffic policing and rate limiting. This edge architecture enables IP service providers to reduce the cost of adding additional subscribers by 75%, increase edge capacity, improve performance and availability, enable auto recovery through router redundancy, and offer support for valuable multi-link services, according to Seranoa.

IP masquerade Also known as MASQ. A LINUX networking function similar to the NAT (Network Address Translation) function found in many network routers and firewalls. IP Masquerade allows multiple client computers internal to a premises-based network to share a single connection to the Internet or other IP-based network through a centralized LINUX MASQ server. The individual computers (e.g., client workstations) typically are attached to the server via a LAN (typically Ethernet), and may or or may not be running the TCP/IP protocol suite. As a workstation connects to the Internet through the MASQ server, the server translates the internal network address it uses into a registered IP address, which is required for access to the public Internet. As the destination computer responds to the public IP address, the server re-translates that address to the internal network address in order to deliver the data packets to the device that originated the data session. See also LINUX and NAT.

IP multicasting IP multicasting is a form of networking where one computer sends information out to a group of other computers at the same time. The originator achieves this by sending packets through a special multicast address, which delivers to a group of hosts, rather than a single (unicast) host. The number of receivers in a multicast session is not limited to bandwidth of the originator (as in the case of unicast). The technology has proven to be more efficient than sending a copy of the stream to all nodes since not all may want it and users are limited to a particular subnet (as in the case of broadcast). For IP multicast only one copy of the stream will pass over any link in the network – thereby conserving bandwidth.

IP multimedia subsystem IMS. IP Multimedia subsystem is an open Next Generation Networking (NGN) multi-media architecture for mobile and fixed IP services. It's a VoIP implementation based on a 3GPP variant of SIP (Session Initiation Protocol) and runs over the standard Internet protocol (IP). It's used by telecom operators in NGN (so called Next Generation Networks), which combine voice and data in a single packet switched network, to offer network-controlled multimedia services. IMS's aim is not only to provide new services but to provide all the services, current and future, that the Internet provides. In addition, users have to be able to do everything they want when they are roaming as well as from their home networks. The interfaces for service developers are also based on IP protocols. The idea is to bring the Internet to the cellular world. IMS was originally defined by the 3rd Generation Partnership Project (3GPP), as part of their standardisation work for third generation (3G) mobile phone systems for W-CDMA networks. But the service model has been used by 3GPP2 (a different organization for rival CDMA2000 networks) and TISPAN (fixed networks) too. It is a key technology for Fixed/Mobile Convergence (FMC).

IP PBX Internet Protocol PBX. An IP PBX connects its phones via an Ethernet LAN and sends its voice conversations in IP packets. There are pros and cons to IP PBXs. Moves and changes with the phones are easier. Wiring is easy. Voice quality and management controls vary between systems. The IP PBX is an evolving animal. See IP Telephony and TCP/IP. See also IP-enabled PBX.

IP router A computer connected to a multiple physical TCP/IP networks that can route or deliver IP packets between networks. See also Gateway.

IP security See IPsec.

IP service edge switch A IP switch platform for network services, such as secure Internet access, private intranets and extranets with quality of service, and managed network firewalls. This type of switch allows network operators to add services quickly through a graphical application set which is synchronized with the switch's command line interface. See IP Switching.

IP shuffling IP Shuffling is the ability of an IP phone to directly exchange RTP (real time transfer protocol) voice streams with another IP phone.

IP hijacking Attack when active, established, session is intercepted and co-opted by attacker. May occur after authentication has been made, permitting attacker to assume role of authorized user. Primary protections rely on encryption at session or network layer.

IP spoofing The use of a forged IP source address to circumvent a network firewall.

The illicit packet seeking to gain access appears to have come from inside the protected network and to be eligible for forwarding into the network. In short, IP spoofing is an attack whereby a system attempts to illicitly impersonate another system by using its IP network address. See Firewall, IP, IP Address and IP Router.

IP subnet All devices which share the same network address. Routers are boundaries between subnets so each connection to a router has a different network address.

IP switching A term coined by Ipsilon Networks to describe a new class of switch it developed, combining intelligent IP routing with high-speed ATM switching hardware in a single, scalable platform. The IP switch implements the IP protocol stack on ATM hardware, allowing the device to dynamically shift between store-and-forward and cut-through switching based on the flow requirements of the traffic as defined in the packet header. Data flows of long duration can be optimized by cut-through switching, with the balance of the traffic afforded the default ATM treatment, which is hop-by-hop, store-and-forward routing. Ipsilon suggests that first-generation IP Switches can achieve rates of up to 5.3 million PPS (packets per second) by avoiding ATM cell segmentation and reassembly, ATM overhead, and ATM switch processing of each cell header. One of the advantages of IP Switching is the use of IP (Internet Protocol), which protocol is mature, well-understood, and widely deployed across a wide range of networks. Contrast with Tag Switching. See IP Service Edge Switch.

IP telephony See VoIP for the best explanation. Here is Microsoft's definition, excerpted from their white paper on TAPI 3.0: IP Telephony is an emerging set of technologies that enables voice, data, and video collaboration over existing IP-based LANs, WANs, and the Internet. Specifically, IP Telephony uses open IETF and ITU standards to move multimedia traffic over any network that uses IP (the Internet Protocol). This offers users both flexibility in physical media (for example, POTS lines, ADSL, ISDN, leased lines, coaxial cable, satellite, and twisted pair) and flexibility of physical location. As a result, the same ubiquitous networks that carry Web, e-mail and data traffic can be used to connect to individuals, businesses, schools and governments worldwide. What are the benefits of IP Telephony? IP Telephony allows organizations and individuals to lower the costs of existing services, such as voice and broadcast video, while at the same time broadening their means of communication to include modern video conferencing, application sharing, and whiteboarding tools. In the past, organizations have deployed separate networks to handle traditional voice, data, and video traffic. Each with different transport requirements, these networks were expensive to install, maintain, and reconfigure. Furthermore, since these networks were physically distinct, integration was difficult if not impossible, limiting their potential usefulness. IP Telephony blends voice, video and data by specifying a common transport, IP, for each, effectively collapsing three networks into one. The result is increased manageability, lower support costs, a new breed of collaboration tools, and increased productivity. Possible applications for IP Telephony include telecommuting, real-time document collaboration, distance learning, employee training, video conferencing, video mail, and video on demand. See the Internet, IP Telephony algorithms, TAPI, TAPI 3.0, TCP/IP and most importantly, VoIP.

IP telephony algorithms The major IP Telephony Algorithms in the market today (fall of 1997), according to a white paper, called "IP Telephony powered by Fusion" from Natural MicroSystems (www.nmss.com), include:

- MS-GSM: This algorithm, marketed by Microsoft, runs at 13kbps and is a derivative of the ITU (International Telecommunications Union) standard GSM work. GSM is used in 85 countries around the world as the standard for digital cellular communications. Microsoft's implementation varies from the standard in several ways including how the encoded data is represented and what aspects of the encoder are supported. Natural MicroSystems provides an MS-GSM encoder that is compatible with Microsoft's Win95/WinNT embedded product.

- ITU G.723.1: This algorithm runs at 6.3 or 5.4 kbps and uses linear predictive coding and dictionaries which help provide smoothing. The smoothing process is CPU-intensive, however (30Mips on an Intel Pentium), so don't expect a PC-based implementation to work well for lots of real-time activity.

- VoxWare: This is a proprietary encoder that has been bundled by Netscape with their Browser. It delivers 53:1 compression and very low jitter. VoxWare presents very low network bandwidth requirements; however, it also has lower speech quality.

Most speech encoder algorithms have a set of rules concerning packet delivery and disposition management. This is often called jitter buffer management. "Jitter" in this case refers to when the signal is put into frames. The decoding algorithm must decompress and sequence data and make "smoothing" decisions (when to discard packets versus waiting for an out-of-sequence packet to arrive). Given the real-time nature of a live connection, jitter buffer management policies have a large affect on voice quality. Actual sound losses range from a syllable to a word, depending on how much data is in a given packet. The first buffer size is often a quarter-second, large enough to be a piece of a word or a short word – similar to drop-outs on a cellular connection in a poor coverage area.

IP Telephony Gateway ITG. Also called a Voice over IP (VoIP) Gateway. A bridge between traditional circuit-switched telephony and the Internet that extends the advantages of IP telephony to the standard telephone by digitizing the standard telephone signal (if it isn't already digital), significantly compressing it, packetizing it for the Internet using Internet Protocol (IP), and routing it to a destination over the Internet.

IP tunneling IP Tunneling means carrying a foreign protocol with a TCP/IP packet. For example, IPX can be encapsulated and transmitted via TCP/IP.

IP VPN Internet Protocol Virtual Private Network. An IP VPN is essentially your own private Internet. An IP VPN is a private network for a corporation or an institution connecting any number of end points (offices, factories, employee homes) using a combination of private and public circuits. A public circuit is usually the Internet. An IP VPN IP traffic – voice, video and data – in basically the same format as the public Internet, except that the traffic is often encrypted and thus protected from prying ears. The main advantages of an IP VPN are:

1. Cheaper than a traditional circuit switched network. I explain this in my definition for Internet and packet switching.

2. Guaranteed reliability and quality. Since the network is yours, you can choose to install as many circuits as you wish and go at whatever speed you desire in order to achieve whatever quality of service (QoS) you wish. If a line between two places crashes, the smarts in the network will find another way for the conversation to take place. This is called self-healing by some vendors and it depends on which circuits you have in your network and where they are. For example, I wouldn't expect self-healing to work if the last mile line broke and there was no other line, e. g. one line to your hotel room. (Silly example, but you get the message.)

3. Easier to maintain. At least that's the theory. IP devices are computers. They have unique identification numbers and they identify themselves when they log on. So, if they log on to the IP VPN network from a different place – say a different hotel room in a different city, they can be accommodated.

IP-enabled PBX A PBX that has both traditional TDM switching and IP switching, the former handling call processing for traditional telephones, fax machines, and modems, and the latter handling call processing for IP phones. An IP-enabled PBX usually starts out in life as a traditional PBX that has been retrofitted with a processor card or attached server to handle VoIP traffic.

IPA Intellectual Property Attorney.

IPackage Installation Package.

IPARS The International Passenger Airline Reservation System. An IBM-originated term.

IPASS See I-PASS.

IPC 1. Interprocess Communications. A system that lets threads and processes transfer data and messages among themselves; used to offer services to and receive services from other programs. Supported IPC mechanisms under MS OS/2 are semaphores, signals, pipes, queues, shared memory, and dynamic data exchange.

2. Interprocessor Communication. See InfiniBand.

3. Instructions per clock. A measure of the speed of a microprocessor. The other major measure is clock speed.

IPCEA Abbreviation for Insulated Power Cable Engineers Association.

IPCH Initial Paging CHannel is the channel number used by your cellular provider to "page" the phones on the system. The term "paging" refers to notifying a particular phone that it has an incoming call. All idle, turned-on phones on a system monitor the data stream on the IPCH. Non-wireline cellular carriers use channel 0333 as the IPCH, while wireline providers (those operated by a telephone company use channel 0334).

IPCI See Integrated Personal Computer Interface.

ipconfig Ipconfig is a useful networking diagnostic tool under Windows. You run ipconfig by doing this: (1) click on the Start button, (2) select Run, (3) type CMD in the text box, (4) press ENTER, (5) type ipconfig at the command prompt, and (6) press ENTER. You will your computer's present subnet mask, IP address (in numeric and binary format) and the gateway your computer's networking is passing though – assuming you have a router (as most of us do). If you don't see much when you run ipconfig, it means you're having networking problems.

IPCP IP Control Protocol; protocol for transporting IP traffic over a PPP connection.

IPD Intrusion protection or prevention device. Typically a piece of hardware that sits between your server and the Internet and filters out DOS (denial of service) attacks and attempts to drop viruses into your server. An IPD operates in-line: traffic passes through the

box transparently. That is, the IPD itself has no IP addresses on its ports; a pair of ports (1a 1b) passes packets, without buffering, while examining the flow for identifying marks of an attack (or of a worm, virus, phishing scam, etc.). The two adjacent devices on the LAN segment (often a router and a switch) don't see the IPD. When the IPD recognizes a packet as malicious, it stops forwarding it. Only the beginning of the packet, a fragment, leaves the IPD box. Routers and switches automatically discard fragments, so servers and workstations never see the attack. For a much bigger explanation see intrusion detection.

IPDC Internet Protocol Device Control. See MGCP and Simple Gateway Control Protocol.

IPDR Internet Protocol Detail Record. A developing standard for an open billing method for IP (Internet Protocol) traffic. IPDR is intended to specify a common format for billing records and billing procedures in order that IP-based carriers can exchange billing information easily and, thereby, to enable the internetwork provisioning and billing of a wide variety of value-added IP-based services and applications. IPDR is an initiative of IPDR.org.

IPDS Intelligent Printer Data Stream. It's IBM's host-to-printer page description protocol for printing. You can now buy kits which let you use your present printer to emulate an IBM printer.

IPE Intelligent Peripheral Equipment. Northern Telecom's term for being able to extend all the features of its PBX over distances longer than a normal extension in a building. See Fiber Remote.

IPEI International Portable Equipment Identities. A wireless term.

IPEM If the Product Ever Materalizes.

IPIC IntraLATA Primary Interexchange Carrier.

IPL Initial Program Load.

IPLC International Private Leased or Line Circuit.

IPM Interruptions Per Minute or Impulses Per Minute.

IPMM Integrated Power Monitor Module.

IPND Integrated Public Number Database.

IPNG IPng. IP Next Generation. Collective term used to describe the efforts of the Internet Engineering Task force to define the next generation of the Internet Protocol (IP) which includes security measures, as well as larger IP addresses to cope with the explosive growth of the Internet. The were three candidate protocols for IPng (CATNIP, TUBA and SIPP), were blended into IPv6, which is in trial stages at the time of this writing. See IPv6.

IPNS International Private Network Service. It actually international private line service and it's typically a circuit from 9.6 Kbit/sec up to T-1 or E-1. Domestically you would simply call it "Private line data service."

IPP IPP is the Internet Print Protocol, a collection of IETF standards developed through the Printer Work Group, www.pwg.org, that will make it as easy to print over the Internet as it is to print from your PC. IPP uses the HTTP protocols to "POST" a supported MIME Page Description Language file to a printer. Printers are given Internet addresses such as, www.mydomain.com/ipp/my_printer, so they can be located on the Internet. IPP has the support of all the major printer companies including, Xerox, HP, Lexmark, IBM as well as Novell and Microsoft. Since fax, at a sufficient level of abstraction, is "remote printing," work is under way to create a Fax Profile for IPP as well, so that IPP can duplicate the legal as well as common practices of fax transmissions. Richard Shockey. Rshockey@ix.netcom. com contributed this definition. Thank you.

IPO Initial Public Offering. Start a company. Some years later, take it public. Come out at $12. A week later, your stock is at $24. You're a success, and rich. IPOs are critical in saying "Thank You" to all your hardworking employees. See also Initial Public Offering.

iPOD IP (Internet Protocol) Phone over Data. There tend to be two variations – emulation and driving. The emulation iPOD connects directly to digital station ports on a PBX and emulates a digital PBX feature phone. The emulation iPOD also enables the new PC IP PBX vendors to interoperate with enterprise PBXs. The driving iPOD drives digital PBX phones in the same fashion as if the phone were connected directly to a PBX station circuit card. The driving iPOD can enable the new PC PBX vendors to use existing desk sets in the enterprise. Both versions of the iPOD provide a TCP/IP interface for the purpose of transporting the voice and call control signaling associated with a PBX digital station call over a packet network. Protocols, DSP algorithms, densities and different form factors all constitute possible platform variations.

IPR Intellectual Property Rights.

IPRS Internet Protocol Routing Service. Defined by Bell Atlantic as "a low-cost access service for ISPs. This service supports basic dial, ISDN, and dedicated requirements for transparent connectivity from the end-user to the ISP."

IPS 1. Internet Protocol Suite. See also IPS7.

2. Inches Per Second. A measurement of the speed of tape movement. Industry standards are as follows: 15132 ips, 15116 ips, 1-718 ips, 3-314 ips, 7-112 ips, 15 ips.

3. Intrusion Protection System. See NIP.

IPS7 Internet Protocol Signaling 7. A Nortel initiative for a standardized signaling and control protocol between PSTN networks and multimedia IP-based networks. SS7 (Signaling System 7) is the international standard for signaling and control in the circuit-switched PSTN (Public Switched Telephone Network). There is no standard protocol for VoIP (Voice over Internet Protocol), and no standard protocol for the interexchange of signaling and control messages between VoIP networks and the PSTNs. Such interexchange is required in order that support for CLASS (Custom Local Access Signaling Services) services (e.g., Call Forwarding, Call Waiting, and Calling Line ID) can be supported seamlessly across the PSTN and VoIP networks. A standard, open architecture for IPS7 has been submitted by Nortel to the IETF (Internet Engineering Task Force) for its consideration. See also IP, PSTN, SS7, and VoIP.

IPsec A collection of IP security measures that comprise an optional tunneling protocol for IPv6. IPsec supports authentication through an "authentication header" which is used to verify the validity of the originating address in the header of every packet of a packet stream. An "encapsulating security payload" header encrypts the entire datagram, based on the encryption algorithm chosen by the implementer. See also Authentication, Encryption, IPv6, and Tunneling.

iPSTAR The principal satellite communication system initiated in 1997 by Shin Satellite of Thailand to provide broadband via satellite through 18 gateways in 17 countries in the Asia-Pacific region. www.thaicom.net/ipstar/ipstar.html.

IPT IP Telephony.

IPT Gateway IP Telephony Gateway. Imagine you and I work for a company which has a PBX – a telephone system. You dial 234 to reach Harry. You dial 9 and a long distance number to dial your biggest client in Los Angeles. Now imagine you want to call your company's branch office in London. You dial 22. You hear a dial tone. You then punch in 689. You hear another dial tone. Then you punch 123. Bingo, the boss of the London office answers. Here's what all those numbers mean. Dialing 22 dials you into a PC called the IP Telephony Gateway, which, on the one side, is connected to your PBX and on the other side is connected to a data line your company has between your office and your London office. Dialing 689 is you telling the IPT Gateway that you want to speak to the PBX in your London office. Dialing 123 tells the London PBX to dial extension 123.

That connection between your PBX and your London office's PBX might be anything from a dedicated private data line (e.g. part of your company's Intranet), to a virtual circuit on a Virtual Private Network (VPN) or it might be the public Internet. The IPT Gateway's major function is to convert the analog voice coming out of your PBX into VoIP (voice over Internet Protocol) and then send it on a packet switched data circuit which conforms to the IP. In short, an IPT Gateway allows users to use the Internet (or most likely an Intranet or Virtual Private Network) to talk with remote sites using (Voice over Internet Protocol).

IPTC On April 30, 1998, Ericsson Inc. released a press release which contained, inter alia, "Ericsson Inc. has developed a new IP telephony platform called Internet Telephony Solution for Carriers (IPTC) that raises the standard for IP telephony systems. IPTC offers phone-to-phone, fax-to-fax and PC-to-phone services over a TCP/IP network. It provides a superior operations and management (O&M) facility that moves IP telephony to a true carrier-class communications system. IPTC works by taking phone and fax calls that originate in the public switched telephony network (PSTN) and passing them to the IPTC platform, which carries them over the TCP/IP network to their destination where they are fed back to the PSTN network. PC-to-phone calls are taken directly from the TCP/IP network and carried to their destination in the same way...IPTC software runs on industry standard platforms that are based on Intel Pentium processors and Microsoft Windows NT...IPTC uses a Web-based management program to update and control multiple gateways. No longer is it necessary to change the parameters in individual gateways when IPTC can update all gateways within a network through one "netkeeper" applications program. The call and traffic control for individual gateways in a network is handled by sitekeepers. The sitekeepers connect to the netkeeper, which acts as a single point of control for the O&M functions of the whole IPTC platform. The netkeeper is not involved in the processing of calls but stores the platform topology information, routing configuration and alarm information. Other features included in the IPTC platform are least-cost routing, dynamic route allocation, multiple IP networks support, and the ability to handle validated and un-validated traffic. Real-time billing with fraud prevention and call duration advice with integrated voice response software is also provided."

IPTV Internet Protocol TeleVision. IPTV describes the technique of sending digital TV programs over a broadband network running the IP protocol. IPTV requires that the TV programming be digitized and then compressed before being fragmented into IP packets before transmission over a broadband network that supports Quality of Service (QoS).

The standard compression mechanism is MPEG-4, although proprietary mechanisms also can be used, with Microsoft's Windows Media A/V technology being one example. At the subscriber premises, the signal is decompressed in a set-top box (which could also be a PC). The advantage of IPTV is that the common denominator of the IP protocol supports not only video-oriented features such as High-Definition TV (HDTV), Video on Demand (VOD), Digital Video Recording (DVR), Instant Channel Changing (ICC), multiple Picture-in-Picture (PIP) and Interactive TV (ITV), but also media services for telephones, PCs and other consumer devices. Further, the same network can support Voice over IP (VoIP) and broadband Internet access. This combination of voice, data and video is what the Local Exchange Carriers refer to as the "triple play," which generally is linked to Passive Optical Network (PON), the broadband fiber optic local loop technology. IPTV and the complete "triple play" also can run over CATV networks and, at least theoretically, over Digital Subscriber Line (DSL), WiMAX or any other broadband local loop technology. Internet TV (iTV) is much more problematic, as bandwidth and QoS issues abound over the open, public Internet. In addition to delivery over a closed, QoS-enabled IP network IPTV also, once again at least theoretically, can be delivered directly to the consumer over the public airwaves (i.e., broadcast TV) or via satellite. See also DSL, MPEG-4, PON, and WiMAX.

IPTV News defines the term as "Internet protocol television, or IPTV, uses a two-way digital broadcast signal that is sent through a switched telephone or cable network by way of a broadband connection, along with a set top box programmed with software that can handle viewer requests to access media sources. A television is connected to the set top box that handles the task of decoding the IP video and converts it into standard television signals. The Switched Video Service (SVS) system allows viewers to access broadcast network channels, subscription services, and movies on demand." In other words, IPTV sends video over the Internet, as today we send email. You can, of course, send anything that's digitized, i.e. made into digital form. The problem with sending TV over the Internet is twofold. First, A decent TV signal contains a lot of information – millions of bits per second. Second, it has to delivered quickly in real-time – in contrast to everything else on the Internet. If your connection to the Internet is slow and/or your computer receiving and processing the incoming signal is slow, the picture you receive will be blurry and, most likely, small. IPTV is also a Microsoft project. Microsoft's technology is designed to let telecommunications and cable companies offer new subscriber services that use their two-way broadband networks. Planned features for Microsoft IPTV include instant channel changing, interactive programming guides with integrated video and multiple picture-in-picture capability on standard TV sets. Microsoft said the technology will support high-definition television and "next-generation" (whatever that is) digital video recording and video-on-demand.

IPU Intelligent Processing Unit. Another way of saying CPU. See CPU. Also Intelligent Peripheral Unit, the hardware associated with an intelligent peripheral. Also Alcatel's parlance for an actual workstation that's associated mostly with one of Alcatel's applications called the local applications platform or LAP and a software applications package called the monitor reset controller-2 or MRC-2. In short, everyone is using IPU to mean whatever cool thing they want it to mean. Certainly sounds cool.

IPv4 Internet Protocol Version 4. The current version of the Internet Protocol, which is the fundamental protocol on which the Internet is based. Although its roots are in the initial development work for ARPAnet, IPv4 was first formalized as a standard in 1981. Since that time, it has been widely deployed in all variety of data networks, including LANs and LAN internetworks. While IPv4 served its purpose for some 25 years, it has lately proved to be inadequate, largely in terms of security and limitations of the address field. The address field is limited to 32 bits; although 2 to the 32nd power is a very large number, we are running out of IP addresses just as we have run out of 800 numbers and traditional area codes. Hence, the development of IPv6. See IP.

IPv5 Internet Protocol Version 5. IPv5 is not exactly a missing link, although it might appear so. Rather, IPv5 was assigned to ST2, Internet Stream Protocol Version 2, which is documented in RFC 1819. ST2 is an experimental protocol developed as an adjunct to IP for support of real-time transport of multimedia data. See IP and IPv6.

IPv6 Internet Protocol Version 6. This new Internet Protocol designed to replace and enhance the present protocol which is called TCP/IP, or officially IPv4. IPv6 has 128-bit addressing, auto configuration, new security features and supports real-time communications and multicasting. IPv6 is described in RFC 1752, The Recommendation for IP Next Generation Protocol, including the strengths and weaknesses of each of the proposed protocols which were blended to form the final proposed solution. At the time of this writing, IPv6 is standardized, but not widely deployed. It requires upgrades that are expensive. They will be fork-lift upgrades in many cases. Therefore, IPv6 is being deployed pretty much only in the NextGen carrier networks, which are being built from the ground up. IPv6 offers 128-bit addressing, auto configuration, new security features and supports real-time communications and multicasting. The 128-bit addressing scheme will relieve pressure on the current 32-bit scheme, which is nearly exhausted due to the widespread use of IP in the Internet and a wide variety of LAN, MAN and WAN networks. Clearly, 2 to the 128th power is a huge number, yielding a staggering number of IP addresses. According to Mark Miller of Diginet Corporation, it equates to approximately 1,500 addresses per square angstrom, with an angstrom being one ten-millionth of a millimeter. Another way of looking at this is that IPv6 yields about 32 addresses per square inch of dry land on the earth's surface – in other words, we are not likely to run out of IPv6 addresses. (Don't be surprised to see your telephone assigned an IP address in the future.)

Autoconfiguration Protocol, an intrinsic part of IPv6, allows a device to assign itself a unique IP address without the intervention of a server. The self-assigned address is based in part on the unique LAN MAC (Media Access Control) address of the device, which might be in the form of laptop computer. This feature allows the user the same full IPv6 capability when on the road as he might enjoy in the office when the laptop is inserted into a LAN-attached docking station. IPv6 security is provided in several ways. Data integrity and user authentication are provided by any of a number of authentication schemes. Second, the Encapsulating Security Payload feature provides for confidentiality of data through encryption algorithms such as DES (Data Encryption Standard). Several different types of IPv6 addresses support various types of communications. Unicast supports point-to-point transmission, Anycast allows communications with the closest member of a device group, and Multicast supports communications with multiple members of a device group.

IPX Internet Packet eXchange. Novell NetWare's native LAN communications protocol, used to move data between server and/or workstation programs running on different network nodes. IPX packets are encapsulated and carried by the packets used in Ethernet and the similar frames used in Token-Ring networks. IPX supports packet sizes up to 64 bytes. Novell's NCP and SPX both use IPX. See also IPX.COM.

IPX Autodiscovery The ability of a network manager to discover the node address and functionality of network devices.

IPX.COM The Novell IPX/SPX (Internetwork Packet eXchange/Sequenced Packet eXchange) communication protocol that creates, maintains, and terminates connections between network devices (workstations, file servers, routers, etc.). IPX.COM uses a LAN driver routine to control the station's network board and address and to route outgoing data packets for delivery on the network. IPX/SPX reads the assigned addresses of returning data and directs the data to the proper area within a workstation's shell or the file server's operating system. See also Netware.

IPX/SPX Internetwork Packet Exchange/Sequenced Packet Exchange. Two network protocols. IPX is NetWare protocol for moving information across the network; SPX works on top of IPX and adds extra commands. In the OSI model, IPX conforms to the network layer and SPX is the transport layer.

IPXCP IPX Control Protocol; protocol for transporting IPX traffic over a PPP connection.

IPXWAN A Novell specification describing the protocol to be used for exchanging router-to-router information to enable the transmission of Novell IPX data traffic across WAN (Wide Area Network) links.

IR 1. Infrared. The band of electromagnetic wavelengths between the extreme of the visible part of the spectrum (about 0.75 um) and the shortest microwaves (about 100 um). See free space optics, infrared and infrared technology.

2. Internet Registry. See also Internet Assigned Numbers Authority.

3. Investor Relations. That part of the company which handles investors – private or institutions. IR answers questions, sends out materials and organizes the president and CFO (chief financial officer) to speak at conferences in front of institutional and private investors.

4. Intermediate Reach. The distance specification for optical systems that operate effectively from 3 to 20 km (1.8 to 12.5 mi).

IRAC International Radiocommunications Advisory Committee. A committee established to provide advice to the Australian Communications Authority regarding international radiocommunications matters.

IRAD Internal Remote Access Device. It is the method by which a service representative authorized by the customer (password required) can dial up to and connect to a Norstar Modular ICS 7.1 and Compact ICS 7.1 phone system unit for diagnostics and troubleshooting.

IRAM Intelligent RAM. The idea is to put a microprocessor into a memory chip – a move that dramatically improve computer performance.

IRC 1. International Record (i.e. non-voice) Carrier. One of a group of common carriers that, until a few years ago, exclusively carried data and text traffic from gateway cities in the U.S. to other countries. The distinction between international companies providing

"record" and data has eroded and now both types of companies provide voice and data services internationally.

2. Internet Relay Chat. IRC is another Internet-based technology, like FTP, Telnet, Gopher, and the Web. Described in RFC 1459, IRC is live text communication between two or groups of people that uses special IRC software and ASCII commands. Each IRC is delegated to a single channel and each channel is dedicated to a different area of interest. Users enter the IRC channel on the basis of a "nick" (nickname). IRC requires special software, use of complicated ASCII-based commands and it doesn't have a graphical interface, so people more generally use World Wide Web-based chat rooms instead. See also Internet.

3. Interference Rejection Combining. A cellular term.

IRD Integrated Receiver/Descrambler. A receiver for satellite signals that also decodes encrypted or scrambled signals. Especially used in the cable TV business.

IrDA 1. A suite of protocols for infrared (IR) exchange of data between two devices, up to one meter apart (20 to 30 cm for low-power devices). IrDA devices typically have throughput of up to either 115.2 Kbps or 4 Mbps. IrDA protocols are implemented in some cell phones, PDAs, printers and laptop computers. Specific standards have been set for Serial Infrared Link (SIR), Infrared Link Access Protocol (IrLAP), and Infrared Link Management Protocol (IrLMP). IrLAP explains how link initialization, device address discovery, connection start-up (including link data rate negotiation), information exchange, disconnection, link shutdown, and device address conflict resolution occur on an IR connection. IrLAP implements the high-level data-link control (HDLC) communications protocol for infrared environments; the rules for discovery and address-conflict resolution are IrLAP's most significant departure from HDLC. Transmission speeds included in the specifications range from 1.152 Mbps to 4.0 Mbps. Imagine that you're carrying around a small portable laptop, PDA or other device and you want to exchange data with your desktop, you simply aim the device at your desktop PC and transmit information back and forth. IrDA works like a charm. I've used it many times. I simply aim the back of my laptop at the back of another laptop. All of a sudden, one laptop's taskbar will pop up with a message "I smell another IrDA port. Want to transfer something." It pops up a screen asking you which file you want to send or receive....It couldn't be easier. frankly, I was surprised. It's a great tool for casual file transfer – like the time I had given a PowerPoint presentation to some students at MIT and one asked for a copy of the presentation. I simply aimed my laptop at his...and bingo, my file was his. See also Infrared, IrLAP and IrLMP.

2. InfraRed Data Association. A not-for-profit organization formed in 1993 to set and support hardware and software standards for infrared data transmission, IrDA membership now exceeds 160 corporations worldwide. Standards activities are across hardware, software, systems, components, peripherals, communications, and consumer markets.

IRE Institute of Radio Engineers.

IREQ The Interrupt Request signal between a PCMCIA Card and a socket when the I/O interface is active.

Iridium The name for Motorola's original and incredibly ambitious satellite project "to bring personal communications to every square inch of the earth." The idea was that you could use an Iridium phone pretty well anywhere – so long as you were outside (you had to be outside) and could "see" one of the Iridium satellites. According to Motorola, "for the first time, anyone, anywhere, at any time can communicate via voice, fax, or data." Iridium used the 1610 to 1626.5 MHz band. Motorola originally estimated the service costing $3 a minute. The idea is that we all carry an Iridium handset – a device larger than today's cellular phone – and that we talk directly from the phone to one of 66 (or so) Iridium LEO satellites circling the Earth at 480 miles up and then down to the satellite closest to the called person, then down to an Iridium phone on the ground or to a satellite dish, through landlines to the phone of the called person. The big benefit is that the system knows who you are and where you are the moment you turn on your phone – like a cell phone. This way it can always complete calls from someone calling you who doesn't know – or doesn't need to know – where you precisely are. It was called Iridium after the element called iridium, which has 77 electrons, which used to be the number of satellites needed. In 1994, the number got cut to 66. But the name stuck. In November of 1992, Business Week estimated that putting up the full Iridium system would cost $3.4 billion. Iridium started launching satellites in November of 1997 and started service sometime in 1999, after many delays and several launches of new satellites to replace those that had failed. After months and months of trying, I finally got to test an Iridium phone. I was at a trade show in Australia. "Can I test the phone, please?" I asked the nice man in the Iridium booth. "Sure," he answered, He handed me a brick-sized phone and said step outside. "Why outside?" I asked. "'Cause it doesn't work inside," he answered. I stepped outside with him. It took the phone a couple of minutes to find the satellites it needed, and another minute or so to dial my friend in London. When I finally got through, the quality of the conversation was O.K. But, oh what a cumbersome process to make a phone call! On Friday, August 13, 1999 (note the date), Iridium L.L.C. filed for Chapter 11 bankruptcy protection. The company had been delinquent on a $90 million interest payment and in default on more than $1.5 billion in bank loans. The entire Iridium network was scheduled to be de-commissioned in 2000. The plan was to de-orbit the satellite constellation, and for the satellites to burn up in the Earth's atmosphere. At the last minute, an entrepreneur stepped in and paid a reputed $25 million for the whole Iridium. The entrepreneur then turned around and signed a contract with the US Government to use Iridium for government and military purposes. You can subscribe to Iridium service and use it for data, voice an paging. Iridium owners are selling service to parts of the world with inadequate landline service. According to Iridium, 86% of the earth's landmass and of its oceans (da!) are in areas with inadequate landline service. See also LEO and Teledesic. www.iridium.com.

iRIP See iCalendar.

irish corrosion Nasty English slang for a fiber optic cut.

IRL 1. Inter-Repeater Link. A networking term. A mechanism for connecting two and only two repeater sets.

2. In Real Life. An Internet term (sort of) – and (sort of) the opposite of URL (Uniform Resource Locator). People who meet "In Real Life" don't meet in chat rooms over the Internet. Real life is good! The Internet is good, too. It's just not "real life."

IrLAP InfraRed Link Access Protocol from IRDA, the InfraRed Data Association. IrLAP defines a link protocol for serial infrared links. IrLAP explains how link initialization, device address discovery, connection start-up (including link data rate negotiation), information exchange, disconnection, link shutdown, and device address conflict resolution occur on an IR (Infrared) connection. IRLAP implements the high-level data-link control (HDLC) communications protocol for infrared environments and adds procedures for infrared-based link initialization and shutdown plus connection start-up, disconnection, and information transfer. The rules for discovery and address-conflict resolution are IrLAP's most significant departure from HDLC. Until recently, IRDA's standards characterized infrared ports as serial links operating at speeds up to 115 Kbps. IRDA's latest standards allow transmission rates as high as 4 Mbps and provide for LAN access via a new IRLAN protocol. The 4 Mbps mode uses pulse-position-modulation data encoding with four possible chip or time-slice positions per data symbol. The system can recognize and prevent interference with UART-based systems by including a Serial Infrared physical-layer link Interaction Pulse (SIP) at least every 500 milliseconds. IRDA has developed APIs for accessing the infrared port. The first, IRCOMM, emulates existing communications device drivers to handle legacy serial and parallel-port connectivity. There's also a native API that infrared-aware programs can use to locate and communicate with each other.

IrLMP InfraRed Link Management Protocol. See IrDA.

IROB In Range of Building. An Underwriters Laboratories term to define where the protection of UL 1459 will apply. See UL 1459.

iron Hardware, as opposed to firmware and software. See also HEAVY IRON.

iron cord A colloquial term for a telegraph line in the mid-to-late 1800s.

IRP I/O Request Packet. Data structures that drivers use to communicate with each other. See Interrupt and IRQs.

IRQs Interrupt ReQuests. IRQs are found in PCs. IRQs are also called hardware interrupts. They are the way a device signals the data bus and the CPU that it needs attention. In more technical terms, an IRQ is a signal sent to the central processing unit (CPU) to temporarily suspend normal processing and transfer control to an interrupt handling routine. Interrupts may be generated by conditions such as completion of an I/O process, detection of hardware failure, power failures, etc. Devices that use hardware interrupts include the serial and parallel ports, mouse interface cards, modems, game ports, and even the hard disk on XTs. The original IBM PC and PC-XT had only seven hardware interrupts. The bigger AT bus extended that to 15. Until the advent of the 32-bit PS/2 micro-channel and the 32-bit EISA buses, hardware interrupts could not be shared by two or more devices within the PC. Thus if one device had a specific hardware interrupt, even though you weren't using it that time, nothing else could use it. When you start filling your PC with devices (and remember most PCs still use the old AT bus) – like serial ports, modems and mice, you may suddenly find your modem no longer works. There are two solutions – change the interrupts (either in software or using jumpers), making sure no two devices are trying to share the same interrupt – or simply remove one of the printed circuit devices you're not using from the bus. (That's typically my solution.) These are the "normal" IRQs used by current hardware devices in PCs. Below is a list of 16-bit IRQs as they have become used.
See also Interrupt and Interrupt Request.

irradiation In insulations, the exposure of the material to high energy emissions for the purpose of favorably altering the molecular structure by crosslinking.

	8 Bit XT Bus	16 Bit AT Bus
IRQ0	TIMER SERVICES	IRQ0 TIMER SERVICES
IRQ1	KEYBOARD	IRQ1 KEYBOARD
IRQ2	UNUSED	IRQ2 SLAVE INTERRUPT
IRQ3	COM2 & COM4	IRQ3 COM2 & COM4 *
IRQ4	COM1 & COM3	IRQ4 COM1 & COM3
IRQ5	LPT2	IRQ5 LPT2 *
IRQ6	FLOPPY DISK	IRQ6 FLOPPY DISK
IRQ7	LPT1	IRQ7 LPT1
		IRQ8 REAL TIME CLOCK
		IRQ9 IRQ2 VECTOR
		IRQ10 Available *
		IRQ11 Available *
		IRQ12 Available *
		IRQ13 MATH COPROCESSOR
		IRQ14 HARD DISK CONTROLLER
		IRQ15 Available *

*Available for assigning to new devices, such as network and video cards.

IRQs in alphabetical order*	
Device	**IRQ**
ARCnet card	2
Bus Mouse	2
Cascade	2
CD-ROM drive	5
COM1	4
COM2	3
COM3	4
COM4	3
Diskette Controller	6
Ethernet card (old ones)	5
Ethernet card (new ones)	10 or 15
Floppy Drive	6
Hard disk drive	14
Keyboard	1
LPT1 (PARALLEL)	7
LPT2 (PARALLEL)	5
Math Coprocessor	13
PC Timer	0
Printer 1	7
Printer 2	5
PS/2 Mouse	12
Real time	8
Scanner	7
Sound Card	7
Tape Backup	5

*This list is not set in concrete. These are suggestions and ideas. Experimentation is the best solution.

irrefragable Impossible to refute; incontestable; undeniable, as an irrefragable argument; irrefragable evidence.

irritainment Annoying entertainment and media spectacles you're unable to stop watching. The O.J. trial is a prime example.

IRSG Internet Research Steering Group. See IRTF.

IRTF Internet Research Task Force. The IRTF is a community of network researchers, generally with an Internet focus. The work of the IRTF, which is governed by its Internet Research Steering Group (IRSG), focuses on the areas of Internet protocols, applications architecture and technology. The IRSG is supervised by the Internet Architecture Board (IAB). Guidelines and procedures for IRTF Research Groups are described in RFC (Request For Comment) 2014. www.irtf.org.

IRTU Integrated Remote Test Unit.

IRU Indefeasible Right of Use (or User). A term used in the underseas cable and fiber optic carrier business. Someone owning an IRU means he has the right to use the circuit for the time and bandwidth the IRU applies to. An IRU is to a submarine or fiber optic cable what a lease is to a building.

IS 1. Information Separator. A type of control character used to separate and qualify data logically. Its specific meaning has to be defined for each application.

2. Interim Standard. EIA/TIA terminology for a "standard" before it becomes a standard. See EIA and TIA.

3. Intermediate System: OSI terminology for a router, which functionally sits between devices on the originating and terminating ends of a session. Such a system provides forwarding functions or relaying functions or both for a specific service such as Frame Relay or ATM.

IS-124 The EIA/TIA standard for Data Message Handler (DMH), a method used in the cellular industry for exchanging non-signaling messages between service providers on a near real-time basis. DMH originally was used to facilitate call hand-offs between carriers. It was extended to serve a number of other purposes. The DMH extension for fraud detection and prevention in known as NSDP-F (Non-Signaling Data Protocol for Fraud), and the extension for the transfer of billing and settlement information is known as NSDP-B&S.

IS-136 Also known as Digital AMPS (D-AMPS). The EIA/TIA Interim Standard which succeeded IS-54, and which addresses digital cellular systems employing TDMA (Time Division Multiple Access). IS-136 also specifies a DCCH (Digital Control CHannel) in support of new features controlled by a signaling and control channel between the cell site and the terminal equipment. IS-136 also allows analog AMPS (Advanced Mobile Phone System) to coexist with North American TDMA on the same cellular network, sharing frequency bands and channels, which supports a smooth transition from analog to digital cellular. IS-136 gave rise to a high-tier standard for PCS (Personal Communications Services), developed by a Joint Technical Committee (JTC) comprising representatives from ATIS and the TIA. High-Tier PCS supports fast-moving vehicular traffic, much like traditional cellular. See also DCCH, PCS and TDMA.

IS-2000 Interim Standard 2000. The EIA/TIA Interim Standard for cdma2000, the 3G (3rd Generation) wireless mobile standard for cellular networks based on CDMA. See cdma2000 for a full explanation.

IS-41 Interim Standard 41. A signaling protocol used in the North American standard cellular system. IS-41 defines the processes by which cellular providers accomplish signaling between MSCs (Mobile Switching Centers) and other devices for purposes of intersystem handoff and automatic roaming. IS-41 includes pre-call validation of the ESN/MIN combination in order to ensure the legitimacy of the originating device. The signaling protocol has been effective in countering "Tumbling" fraud. The IS-41 messaging language is supported by ISDN and X.25 networks. IS-41B defines Global Title Translation (GTT), which translates dialed digits into a point code (network node) address and application address (subsystem number). IS-41C defines the formats and procedures for Short Message Service (SMS). IS-41D includes feature and service support for Calling Number ID (CNID), E911 (Enhanced 911), law enforcement intercept (wire-tapping). See also CNID, ESN, GTT, MIN, MSC, SMS, and Tumbling.

IS-410 Interim Standard 41 Zero. The initial version of IS-41, which was released in February 1988. IS-410 defined the process for intersystem call hand-off.

IS-41A Interim Standard 41a. A version of IS-41 which supports automatic roaming.

IS-41B Interim Standard 41B. A version of IS-41 which defined Global Title Translation (GTT), the process of translating Global Titles (telephone numbers) into Point Codes. Point Codes are unique addresses of SS7 network nodes. See also Global Title, Global Title Translation, Point Code, and SS7.

IS-41C Interim Standard 41C. A version of IS-41 which supports PCS SMS (Short Messaging Service), defining message formats and authentication standards. See also IS-41, SMS and Authentication.

IS-41D Interim Standard 41D. A version of IS-41 which addresses Calling Number ID (also known as Calling Line ID, or CLID), Enhanced 911, and Law Enforcement Intercept. See also IS-41.

IS-54 Interim Standard 54. It is the dual mode (analog and digital) standard for cellular phone service in North America. In its analog form, it conforms to the AMPS standard. IS-54 is an EIA/TIA, developed with the involvement of the CTIA. Since 1995, IS-54 enhancements fall under IS-136. See IS-54B.

IS-54B Interim Standard 54B, the second version of IS-54. IS-54B defined TDMA (Time Division Multiple Access), an access technique used in digital cellular networks. See also IS-136 and IS-54.

IS-55 Interim Standard 55. Standard for TDMA digital cellular service, which is three times the capacity of today's analog cellular service. IS-55 is a fully digital cellular system.

IS-634 The Interim Standard for the interface between cellular base stations and Mobile

Traffic Switching Offices (MTSOs). Issued by the TIA subcommittee TR45.4, IS-634 standardizes the functionality of the A-interface associated with the handling of call processing in order that terminal equipment and MTSOs of disparate origin can interoperate in a predictable fashion. The interim standard is intended to support AMPS, N-AMPS, CDMA and TDMA. The first release of IS-634 employs SS7 and 64-Kbps PCM encoding.

IS-661 The Interim Standard for a hybrid CDMA/TDMA wireless system.

IS-95 Interim Standard 95. IS-95 is a TIA standard (1993) for North American cellular systems based on CDMA (Code Division Multiple Access), and is widely deployed in North America and Asia. IS-95a defines what generally is known as cdmaOne, which supports voice and 14.4 Kbps data rates. IS-95b supports data rates up to 115 Kbps. See also cdmaOne and IS-2000.

IS-IS Intermediate System to Intermediate System. OSI link-state hierarchical routing protocol, based on DECnet Phase V routing, whereby intermediate systems (routers) exchange routing information based on a single metric to determine network topology.

ISA 1. Interactive Services Association.

2. Industry Standard Architecture. The most common bus architecture on the motherboard of MS-DOS computers. The ISA bus was originally pioneered by IBM on its PC, then its XT and then its AT. ISA is also called classic bus. It comes in an 8-bit and 16-bit version. Most references to ISA mean the 16-bit version (which carries data at up to 5 megabytes per second). Many machines claiming ISA compatibility will have both 8- and 16-bit connectors on the motherboard. In 1987 IBM introduced a 32-bit bus which it called MCA for Micro Channel Architecture, which is the internal bus inside some of IBM's line of PS/2 MCA machines. But MCA isn't popular because it is incompatible with ISA, so the industry (excluding IBM) invented a 32-bit bus called EISA which stands for Extended Industry Standard Architecture, which is compatible with ISA. EISA, however, suffered from some of the same problems as the MCA bus, namely it was complicated to program to get the card's full benefit. As a result, other buses have been invented, including the VL bus from VESA (the Video Electronics Standards Association) and Intel's PCI bus. PCI stands for Peripheral Component Interconnect. Both the VL and PCI buses claim to be more than 20 times faster than the ISA bus. The PCI bus, according to Intel, can transfer data at up to 132 megabytes per second. Until the advent of complex Windows graphics and imaging programs, the speed of the ISA was not a gating factor in a computer. As of writing, a VL bus could only handle three drop-in printed cards, while a PCI bus could handle 10. See also EISA and Microchannel.

ISA Configuration Utility The Intel Architecture Lab has been co-developing the Plug and Play specifications with industry partners to ensure long-term compatibility across cards, systems and software. IAL has openly licensed the necessary BIOS software to PC manufacturers so they can add Plug and Play capabilities to their systems. Intel Architecture Labs also designed the ISA Configuration Utility for system and add-in card manufacturers to include with their products. This software utility makes it easier for users to install existing ISA cards in their PCs. The software tells the user which resources are available, but configuration is still done manually. The utility also allows the user to optimize the way resources are assigned, which is particularly important for memory addresses.

ISAKMP Internet Security Association and Key Management Protocol. A protocol that provides a method for authentication of communications between peers over the Internet. ISAKMP provides a framework for the management of security keys, and support for the negotiation of security attributes between associated peers. ISAKMP/Oakley is a hybrid protocol that uses the ISAKMP framework to support the Oakley key determination protocol, which defines a method for establishing security keys for use on a session basis by Internet hosts and routers.

ISAM Indexed Sequential Access Method. It is a procedure for storing and retrieving data from a disk file. When the programmer designs the format of the file, a set of indexes is created which describes where the records of the file are located on the disk. This provides a quick method of retrieving the data, and eliminates the need to read all the data from the beginning to find the desired information. The indexes can be stored as part of the data file or in a separate index file.

ISAPI Internet Server Application Program Interface. This API was created by Process Software Corp. and Microsoft and announced by Microsoft at the 1995 Fall Interop show. It is tailored to Internet servers and uses Windows' dynamic link libraries (DLLs) to make processes faster than under other APIs. It allows Internet browsers supporting it to access remote server applications from Microsoft and others. See ISAPI Filter.

ISAPI Filter A World Wide Web term. An ISAPI Filter A replaceable DLL which the server calls whenever there is an HTTP request. When the filter is first loaded, it communicates to the server what sort of notifications will be accepted. After that, whenever a selected event occurs, the filter is called to process the event. Example applications of

ISAPI filters include custom authentication schemes, compression, encryption, logging, traffic analysis or other request analyses.

isarithmic flow control Approach to flow control in which transmission permits circulate throughout network. Node wishing to transmit must first capture permit and destroy it, then recreate permit after transmission finished.

ISB InterShelf Bus.

ISC International Softswitch Consortium. The ISC describes itself thus: "The International Softswitch Consortium is the premiere public advocate for the worldwide advancement of softswitch network interoperability, promoting the growth of Internet-based multimedia communications and applications. The ISC, which has more than 140 members, was launched in May 1999 to support rapid advancement of application development for the evolving Internet protocol networks. Internet protocol networks are built on distributed servers generally called call agents, media gateway controllers, softswitches, application servers and media gateways. To further the development of open network architectures and standard interfaces, the ISC organizes interoperability test events, holds educational conferences, and establishes work groups to address current and emerging issues vital to the industry." www.softswitch.org. See also Softswitch.

iSCSI Internet Small Computer Systems Interface. iSCSI is an improved protocol for SANs (Storage Area Networks) or NAS ((Networked Attached Storage) to allow the transfer of large blocks of data between information repositories and hosts (i.e., file servers and clients) using an IP (Internet Protocol) infrastructure. iSCSI essentially is a transport protocol (Layer 4 in terms of the OSI Reference Model) for SCSI, riding on top of TCP/IP. In the past, such storage devices have used proprietary protocols or have appeared on the network as servers. But the devices did not appear as available storage on servers. Furthermore, because they were simply objects on the network, they suffered network traffic congestion along with everything else. iSCSI devices are different in that they are accessed via an iSCSI HBA (host bus adapter). To the computer, the HBA looks just like any other SCSI HBA: It appears to be a storage device that you'd access via the server, just as you would with a directly attached storage device, such as an internal disk or a SAN. To the network, the iSCSI HBA appears to be a NIC. It has an IP address and communicates using standard IP Ethernet network packets. The difference is, when the server needs to move some data to storage, it transfers the data to its HBA, where it becomes standard SCSI-3 data. The data is then enclosed in an IP packet and is sent out via the Ethernet network. Once it gets to the iSCSI storage device, the IP packet information is stripped off, and the data is moved to the device's internal SCSI controller, which in turn transfers it to disk. One advantage of iSCSI is that it's completely transparent. The server software sees only what appears to be a SCSI controller; the network only sees IP traffic. To the IT staff, it means that there's little new to learn. iSCSI uses standard Ethernet infrastructure and standard SCSI provisions in the server software. Because servers are talking to the network through their HBAs, implementing a separate storage network is easy and relatively cheap. Furthermore, because you're talking to your storage through Ethernet rather than Fibre Channel, you gain flexibility. For example, a standard SAN using Fibre Channel is limited to speeds of 2Gbps and fiber lengths of 30 kilometers. The iSCSI is standard TCP/IP – thus of basically infinite length. Further if you run iSCSI over 10GbE (10 Gigabit Ethernet), you can run at five times as fast as the fastest SAN, and cover more distance. The management of an iSCSI storage network might also be easier than a traditional SAN, which often use proprietary management software. See also Fibre Channel, IP, NIC, SAN, SCSI, and VPN.

ISCP Integrated Services Control Point. Bellcore's ISCP software system manages and distributes the information needed to run intelligent networks that provide inexpensive, rapidly deployed customized service, according to Bellcore.

ISD 1. Incremental Service Delivery.

2. Information Systems Department.

ISDL Integrated Services Digital Line. Part of the family of xDSL technologies, ISDL is a means of provisioning ISDN BRI capacity of 128 Kbps to the premise without the need for upgrade of existing cable pairs. Such existing copper pairs are often in such poor condition as to deny the provisioning of conventional ISDN. See ISDN, xDSL, and ADSL.

ISDN Integrated Services Digital Network. ISDN is a set of international standards set by the ITU-T (International Telecommunications Union-Telecommunications Services Sector) for a circuit-switched digital network that supports access to any type of service (e.g., voice, data, and video) over a single, integrated local loop from the customer premises to the network edge. ISDN requires that all network elements (e.g., local loops, PBXs, and COs) be ISDN-compatible, and that the SS7 (Signaling System 7) be in place throughout the entire network. ISDN also specifies two standard interfaces – BRI and PRI.

BRI (Basic Rate Interface) is the North American term for the low speed version known internationally as BRA (Basic Rate Access). It delivers a total of 144,000 bits per second

and designed for the desktop. BRI also generically is known as 2B+D. The two B (Bearer) channels are information-bearing; that is to say that they support end user data (and voice) transfer. The B channels support "clear channel" communications at 64 Kbps each. The D (Data or Delta) channel is intended primarily for signaling and control (e.g., on-hook and off-hook signaling, performance monitoring, synchronization, and error control) at 16 Kbps. The D channel also will support end user packet data transfer at speeds up to 9.6 Kbps, as the signaling and control requirements are not so bandwidth-intensive as to require a full 16 Kbps. BRI is intended primarily for consumer and small business applications. As ISDN-compatible terminal equipment generally is too expensive, most end users opt for an inexpensive Terminal Adapter (TA) that serves as the interface between the ISDN local loop and the non-ISDN terminal equipment.

PRI (Primary Rate Interface) is the North American term for an ISDN T-1 circuit. PRI runs at a total signaling speed of 1.544 Mbps in support of 24 channels. Also known as 23B+D, PRI supports 23 Bearer channels and one D channel. Multiple PRIs can be linked to share a single D channel, as the signaling and control bandwidth requirements are relatively light; however, a backup D channel is recommended in such implementations in order to ensure that the PRI links continue to function should a D channel fail. ITU-T specifications allow as many as five PRIs to be so linked, although some manufacturers support as many as eight. The European/International version is PRA (Primary Rate Access). Also known as 30B+D, PRA supports 30 B channels and one D channel, and is the ISDN equivalent of E-1. PRI and PRA are intended for application in connecting PBXs, ACDs, and data switches, routers and concentrators to the network. All of these various switching and concentrating devices must be ISDN-compatible. A PRI/T-1 is often installed on two unloaded copper pairs, or more commonly, on fiber.

As ISDN essentially is a highly sophisticated enhancement of the traditional circuit-switched PSTN (Public Switched Telephone Network), it offers the advantage of flexibility. As long as IDSN is supported by all network elements at all end user locations and throughout the service provider networks, ISDN capabilities are considerable. First, a single local loop connecting to a single service provider can support any mix of voice, data and video – channel-by-channel. Second, multiple channels can be linked together in what is known as "bonding" or "rate adaption." For example, rate adaption allows you to link together two 64-Kbps B channels to form a 128 Kbps chunk of bandwidth for a videoconference or, perhaps, a single symmetric (i.e., equal speed in each direction) Internet access experience at 128 Kbps. Third, ISDN is standardized worldwide, so connectivity generally is not an issue. That is not to say that there are not differences from country to country or region to region, but most of those differences are relatively inconsequential at the basic level.

ISDN BRI is useful if you can't get DSL or cable modem service because it can give you videoconferencing, and faster data communications. But it is not an easy service to get up and running. Cost and availability, however, often are issues. The best advice I can give you is: 1. Figure out what you want to do with your ISDN. 2. Find which equipment you're going to need that will do the best job for you. 3. Call the manufacturer of that equipment, tell him where you're located and ask him which ISDN service to order. 4. After he tells you, order your ISDN service from your local phone company. 5. Then buy the equipment. 6. Allow yourself at least two months to get up and running. 7. Any ISDN equipment you install in a PC will cause major interrupt problems. Make sure you know which "interrupts" your PC is using for what. See also IRQs. ISDN has "enjoyed" many "meanings," including "I Still Don't kNow what it means," referring to the fact that ISDN was not well explained by the service providers; "It Still Does Nothing," referring to the fact that ISDN does relatively little of significance (It took so long for the service provider to make it available that technologies like Frame Relay and DSL made it obsolete, or so some would say.); and "I Smell Dollars Now," referring to the fact that the service providers traditionally charged a lot for ISDN service. ISDN was designed originally to be a totally new concept of what the world's telephone system would eventually become. (Remember this "Vision" came long before the Internet.) According to AT&T, today's public switched phone network has the following limitations: 1. Each voice line is only 4 KHz, which is very narrow, which limits also the speed you can send data across. 2. Most signaling is in-band signaling, which is very consuming of bandwidth (i.e. it's expensive and inefficient). 3. The little out-of-band signaling that exists today runs on lines separate to the network. This includes signaling for PBX attendants, hotel/motel, Centrex and PBX calling information. 4. Most users have separate voice and data networks, which is inefficient, expensive and limiting. 5. Premises telephone and data equipment must be separately administered from the network it runs on. 6. There is a wide and growing variety of voice, data and digital interface standards, many of which are incompatible. ISDN's "vision" was to overcome these deficiencies in four ways: 1. By providing an internationally accepted standard for voice, data and signaling. That standard has pretty well achieved, though don't try and take North American ISDN equipment to Europe. 2. By making all transmission circuits end-to-end digital. 3. By adopting a standard out-of-band signaling system. 4. By bringing significantly more bandwidth to the desktop. One of the best features of ISDN is the speed of dialing. Instead of 20 seconds for a call to go through on today's still partially analog network, with ISDN it takes less than a second. The speed is truly beautiful. Here are some sample ISDN services (some of which are now available on non-ISDN phone lines): Call waiting: A line is busy. A call comes in. The user knows who is calling. He can then accept, reject, ignore, transfer the call. Citywide Centrex: A myriad of services: Specialized numbering and dialing plans. Central management of all ISDN terminals, including PBXs, key systems, etc. Credit card calling: Automatic billing of certain or all calls into accounts independent of the calling line/s. Calling line identification presentation: Provides the calling party the ISDN "phone" number, possibly with additional address information, of the called party. Such information may flash across the screen of an ISDN phone or be announced by a synthesized voice. The called party can then accept, reject or transfer the call. If the called party is not there, then his/her phone will automatically record the incoming call's phone number and allow automatic callbacks when he/she returns or calls back in from elsewhere. Calling line identification restriction: Restricts presentation of the calling party's ISDN "phone" number, possibly with additional address information, to the called party. Closed user group: Restricts conversations to or among a select group of phone numbers, local, long distance or international. Collaborative Computing. Work on the same document or drawing or design with someone 10,000 miles away. With ISDN, it doesn't really matter where members of the design team live. Desktop videoconferencing. I have an ISDN desktop videoconferencing device on my desk. It's wonderful to see the person at the other end. It makes for a far more meaningful conversation. EMail (a.k.a. Personal mailbox): ISDN can carry information to and from unattended phones as long as they're equipped with proper hardware and software. Internet Access: It's much nicer to browse the Internet at 128 Kbps than at 53.3 Kbps which is the fastest you can get on a dial-up phone line. But it is much nicer to browse even faster – see Cable Modem and DSL.) Shared Screen – Switched data services provided via ISDN lets two people in remote locations, both equipped with a computer terminal, view the same information on their screens and discuss its contents while making changes – all over one telephone line. Simultaneous Data Calls: Two users can talk and exchange information over the D packet and/or the B circuit or packet switched channel. There are two major problems to the widespread acceptance of ISDN: First, the cost of ISDN terminal equipment is too high. Second, the cost of upgrading central office hardware and software to ISDN is too high. Both costs are coming down. Integral to ISDN's ability to produce new customer services is ITU Signaling System 7. This is a ITU-T recommendation which does two basic things: First, it removes all phone signaling from the present network onto a separate packet switched data network, thus providing enormous economies of bandwidth. Second, it broadens the information that is generated by a call, or call attempt. This information – like the phone number of the person who's calling – will significantly broaden the number of useful new services the ISDN telephone network of tomorrow will be able to deliver. For more on ISDN, see also AO/DI, Euro-ISDN, Intel Blue, ISDN 2, ISDN 30, ISDN Standards, ISDN Telephone, ISUP, NT1, NT2, Q.931, Robbed Bit Signaling, S Interface, SS7, SPID, T Interface, TCAP, Terminal Adapter, and U Interface.

ISDN 2 What the Americans call ISDN BRI, the British call ISDN 2, which is ISDN with two BRI channels and one D channel.

ISDN 2e Euro-ISDN BRI. See ISDN 2 and ISDN.

ISDN 30 The name of an ISDN service which delivers 30 ISDN BRI lines over a single line. ISDN 30 is a fancy name for ISDN on an E-1 line. You find it in countries outside North America, especially Europe. ISDN 2, in the UK, is their name for what Americans call ISDN BRI. See ISDN.

ISDN Basic Link Facility The ISDN Basic Link Facility consists of a local transmission facility terminated in the local central office and in a suitable network interface device that is capable of supporting ANSI standard ISDN Basic Rate 2 Binary 1 Quaternary U interface line coding scheme. The standard ISDN Basic Link Facility is 18,000 cable feet or less from the central office termination or served via appropriate electronic equipment, to the Network Termination One device located on the customer's premises.

ISDN BRI Service 2B+D - Two bearer channels and one D channel to your desktop. There are many varieties of ISDN BRI service. The three most common are National ISDN-1 compliant, AT&T 5ESS Custom (an older form of ISDN BRI) and Northern Telecom DMS 100 ISDN. With ISDN BRI, you choose your ISDN equipment first, figure out what it needs, check whether what it needs is available from your local telephone company, then get your line installed, then buy your equipment. Changing your line specs later is expensive and slow. ISDN is still a very first generation product, though when you get it working, it usually works thereafter relatively flawlessly. See ISDN for a much longer explanation.

ISDN Forum See Vendors ISDN Association.

ISDN Integrated Access Bridge/Router A remote access device that connects the computer to an ISDN line and which performs bridging and/or routing as well as a supporting analog devices such as phones or faxes.

ISDN Modem ISDN Modem is just another name for a Terminal Adapter (TA). See Terminal Adapter.

ISDN Multirate ISDN Multirate is a network-based ISDN service that allows users' network access equipment to add B channels as needed, depending on bandwidth demands, in increments if 64 Kbps, up to 1.536 Mbps. Access to ISDN Multirate service is obtained over ISDN PRI lines.

ISDN Network Termination Device You can't plug your ISDN phone directly into an ISDN line like you can with today's analog lines. You need a black box, called a Network Termination device, called an NT1, as in Network Termination 1. In North America you can pick one of these devices up for under $250. The NT1 provides an interface between the ISDN loop and an S or T interface terminal, such as an ISDN phone, or the PCTA (Personal Computer Terminal Adapter). The PCTA is the device which turns a PC into an ISDN terminal/phone. The NT1 is the classic ISDN "black box." It sits on the subscriber's premises at the end of the subscriber loop coming in from the phone company. It talks to the ISDN central office. And, in turn, all ISDN terminals, phones and other devices on the subscriber premises are plugged into this black box. The basic NT1 functions are:

    Line transmission termination.

    Layer 1 line maintenance functions and performance monitoring.

    Layer 1 multiplexing, and

    Interface termination, including multi drop termination employing layer 1 contention resolution.

    Some ISDN devices – such as LAN hubs – now come with NT1s built in. See ISDN.

ISDN Overflow/Diversion A feature of Rockwell Galaxy ACDs. ISDN Overflow/Diversion allows Galaxy ACD users to overflow calls between multiple switches using PRI D-channels and B-channels through the public network. This gives the user a virtual private network without the cost of dedicated trunks. By using ISDN messages to overflow a call, specific information associated with the call can be passed to the destination switch, such as ANI, DNIS, and delay time in queue at the originating switch.

ISDN PRI PRI stands for Primary Rate Interface. In North America, ISDN PRI can be thought of as "enhanced T-1." And some long distance carriers are only delivering T-1 in this format. In this ISDN-PRI format, it has major benefits – chiefly the extra bandwidth and benefits derived from the much richer and much faster out-of-band signaling. For example, ANI (Automatic Number Identification, DNIS (Dialed Number Identification Service), etc. are delivered much better this way. ISDN PRI is 24 B (bearer) channels, each of which is a full 64,000 bits per second. One of these channels is typically used to carry signaling information for the 23 other channels. If you're running voice on a single ISDN-PRI, you get 23 voice channels, compared to 24 if you run voice on a T-1 line. However, if you get multiple ISDN-PRI lines, you can often carry the signaling for those voice lines on that one B channel and thus get 24 on each of the others. For example, some voice processing cards will let you support up to eight ISDN PRI channels on one B channel – i.e. the first PRI gives you 23 voice channels. All seven others will give you 24 channels. In Europe, ISDN PRI is 30 bearer channels of 64 Kbps and two signaling channels, each of 64 KBps.

ISDN Repeaters ISDN repeaters let telephone companies extend ISDN lines up to 48,000 feet from their central office. This means they can send ISDN to people further away.

ISDN Router You have a local area network in your office. You want to join that LAN to the Internet and/or your company's private Intranet. You can do the joining with any number of communications lines, from slow dial-up analog lines to high-speed T-3 and everything in between. One of those in-betweens is an ISDN BRI line. An ISDN router is a device which joins a LAN to an ISDN line. It combines the functionality of an ISDN Terminal Adapter (TA) and a TCP/IP router for access to an ISDN network. The TA function accomplishes the interface between the ISDN network and devices which are not ISDN compatible. The router function allows multiple workstations to access the ISDN network, as well each other, through what essentially is a LAN hub with TCP/IP routing capabilities. See also ISDN Terminal Adapter.

ISDN standards The path of a call in an ISDN network is based on standards:

    1. ISDN User A signals the public network over a standard interface. The 2B1Q protocol is used by the terminal as well as the line card in the telephone company central office. The 2B1Q arranges the bits of a digital ISDN signal in a standard manner over the twisted pair connecting the user and the central office. Bellcore TR268 establishes protocols for signaling between the caller and the network.

2. TR444 and TR448 define the standard protocols that allow ISDN services to be carried by the SS7 network.

3. TR317 defines the protocols for standard SS7 networking between the LEC's intra-LATA switches that, when combined with TR444 and TR448, can deliver ISDN services.

4. TR394 defines the protocols for standard SS7 networking of the interLATA switches that, when combined with TR444 and TR448, provide the interface for ISDN services and the interexchange network.

5. ISDN UserB, equipped with the standard network interface and using the 2B1Q and TR268 protocols, is prepared to receive the ISDN call from User A. The network delivers the ISDN call information carried over SS7 from User A to User B in the call set-up message.

ISDN teladapter A Nortel term for a device which connects a national ISDN 1 telephone and a Macintosh to a Nortel switch via an ISDN line card.

ISDN telephone An ISDN phone can attach to an ISDN basic rate interface. It typically has one digital voice (at 64 kbps) channel and two data options – one for packet switched services (up to 9600 bps) and another for circuit switched data (up to 64 kbps). It will also have an RS-232-C connector on its back and a two line, 48-character LCD adjustable display. It will also have a bunch of dedicated buttons for standard stuff – last number redial, speed dial, on-hook dialing, listen-on-hold, etc. Some ISDN phones work behind most ISDN central offices – e.g. the Telrad phones. Most don't. They have to work behind the central office they were designed for. Or did have to. In February, 1991, Bellcore issued a technical specification for a standard ISDN phone line. The idea of National ISDN-1 is that it be a set of standards which every manufacturer can conform to. A consumer can buy an ISDN phone (one conforming to National ISDN-1) at his local Radio Shack (or other store) take it home, plug it in and know it will work, irrespective of whose central office he's connected to. At time of writing most ISDN phones cost over $600. Some cost nearly $1,000. Some cost more than a personal computer. In late fall of 1992 there was increasing talk that PCs will soon come with telecom ports – able to accept the ISDN signal directly from the central office, without the need for a separate (and expensive ISDN) phone instrument. The PC and the software within it, will then become the phone (presumably with a handset, headset or earset attached to the back of the PC). See ISDN.

ISDN terminal adapter ISDN terminal adapters are devices that typically allow analog devices to speak on digital ISDN lines. Terminal adapters are essentially similar to modems, with the following difference. Modems connect terminals to the traditional analog network, and terminal adapters connect those analog-network (non-ISDN) terminals to the digital ISDN network.

ISE Integrated Switching Element.

ISG Incoming Service Group or Grouping. A fancy name for hunting or rollover. You receive many incoming calls. You don't want to miss a call, so you ask your phone company to set your phone lines up to roll over, also called hunt, also called ISG (Incoming Service Group) in telephonese. You order five lines in hunt. The calls come into the first. If the first one is busy, the second rings. If it's busy, the third rings. If they're all busy, then the caller receives a busy. The commonest types of hunting are sequential and circular hunting. Sequential hunting starts at the number dialed, keeps trying one number after another in number order and ends at the last number in the group. It's typically descending. For example, it starts at 691-8215, goes to 691-8216, then 691-8217, etc. But it can also be ascending – from 691-8217 up. Circular hunting hunts all the lines in the hunting group, regardless of the starting point. Circular hunting, according to our understanding, circles only once (though your phone company may be able to program it circle a couple of times). The differences between sequential and circular are subtle. Circular seems to work better for large groups of numbers. You don't need consecutive phone numbers to do rollovers. Nowadays you can roll lines forwards, backwards and jump around, for example most idle, least idle. Rollovers are now done in software. This also has its downside, since software fails. For example, theoretically if a rollover strikes a dead trunk, it should bounce to the next live trunk. But sometimes it hangs on the dead trunk and many of your incoming calls never get answered. They might ring and ring. They might hit a busy. My recommendation: Test your rollovers at least twice a day. In particular, test that your callers ultimately get a busy if all your lines are busy. Nothing worse your customer should receive a ring-no-answer or a constant busy when calling your company. See also Terminal Number.

ISI See Intersymbol Interference.

ISIS7 Internal Switch Interface System.

ISIS Slot Area that contains the software programs used for each Engine. Because slots are set up as a logical section of the computer's memory, each node may have a combination of program slots.

ISL InterSatellite Link. A relatively new development in satellite technology, ISLs allow LEOs (Low Earth Orbiting satellites) and MEOs (Middle Earth Orbiting satellites) to communicate directly, rather than through earth stations. ISL functions include selection of the shortest circuit path between originating and terminating device, selection of lowest cost terrestrial route, and complete bypass of terrestrial carriers. In the case of this last function, the following scenario best explains: A user of the Iridium LEO system, for instance, places a call from his satellite phone to another user of a satellite phone on the same system. The caller connects directly to a LEO, which finds the other user through querying a GEO (Geosynchronous Earth Orbiting satellite) which then queries the other 65 LEOs in the satellite constellation over ISL signaling links. As the target user answers the call, the connection is established and maintained over another ISL, perhaps established directly between the LEOs in best positions to communicate with the users. The terrestrial networks, both local and long-haul, are bypassed completely, and any charges associated with those terrestrial networks are avoided. As the satellites communicate directly, no earth stations are involved, and precious bandwidth is conserved. Also, issues of propagation delay are minimized, as only one uplink/downlink combination is required, as are the number of systems and processes that must act on the call. This process effectively is the same as in a cellular network, although it is much more complex – the satellites are whizzing around, the users may be mobile and, therefore, multiple handoffs may be required from satellite to satellite on both ends of the connection. While cellular users also typically are mobile, at least the cell sites are stationary. While at the time of this writing there are no ISL standards, either radio or laser light frequencies are appropriate.

ISM band Industrial, Scientific and Medical Band. A term describing several bands in the RF (Radio Frequency) spectrum, also referred to as Part 15.247 of FCC regulations. Specifically, ISM bands include 902-928 MHz, 2.4-2.483 GHz, and 5.725-5.850 GHz. ISM frequencies are unlicensed. In other words, they can be used for any variety of applications without the requirement for FCC permission. Traditionally used for in-building and system applications such as bar code scanners, industrial microwave ovens, and wireless monitoring of patient sensors, ISM also is used in many WLANs (Wireless Local Area Networks), including HomeRF and 802.11b. As there is no licensing requirement, there exists the potential for interference from other applications in close physical proximity. Therefore, spread spectrum technology is often used to protect the data transmission integrity of a Wireless LAN. See also 802.11b, HomeRF, Spread Spectrum, and WLAN.

ISN 1. Intelligent Services Node.

2. AT&T's Information Systems Network.

ISNAP Intelligent Services Network Applications Processor.

ISO 1. The Greek prefix which means equal or symmetrical, as in isometric. See also Isochronous.

2. Most people believe that ISO stands for The International Standards Organization in Paris. Actually the organization is strictly called the International Organization for Standardization (IOS) and is based in Geneva. ISO is a voluntary, non-treaty organization chartered by the United Nations. It began to function officially on February 23, 1947 and in 1951 published its first standard, entitled "Standard reference temperature for industrial length measurement." Its role is to define international standards covering all fields other than electrical and electronic engineering, which is the responsibility of the IEC (International Electrotechnical Commission). In the world of communications, the ISO is best known for the 7-layer OSI (Open Systems Interconnection) Reference Model. The U.S. representative to the ISO is ANSI. If you go to the ISO's home page, www.iso.ch, you will find the following explanation: The International Organization for Standardization (ISO) is a worldwide federation of national standards bodies from some 130 countries, one from each country. ISO is a non-governmental organization established in 1947. The mission of ISO is to promote the development of standardization and related activities in the world with a view to facilitating the international exchange of goods and services, and to developing cooperation in the spheres of intellectual, scientific, technological and economic activity. ISO's work results in international agreements which are published as International Standards. Many people will have noticed a seeming lack of correspondence between the official title when used in full, International Organization for Standardization, and the short form, ISO. Shouldn't the acronym be "IOS"? Yes, if it were an acronym - which it is not. In fact, "ISO" is a word, derived from the Greek isos, meaning "equal", which is the root of the prefix "iso-" that occurs in a host of terms, such as "isometric" (of equal measure or dimensions) and "isonomy" (equality of laws, or of people before the law). From "equal" to "standard", the line of thinking that led to the choice of "ISO" as the name of the organization is easy to follow. In addition, the name ISO is used around the world to denote the organization, thus avoiding the plethora of acronyms resulting from the translation of "International Organization for Standardization" into the different national languages of members, e.g. IOS in English, OIN in French (from Organisation internationale de normalisation). Whatever the country, the short form of the Organization's name is always ISO. See ANSI and IEC.

3. ISO also stands for Independent Service Organizations or Independent Sales Organizations in the computer sales community.

4. Independent Service Organizer. What Visa and MasterCard call their vendors who sell their credit card processing services to vendors like local retail stores.

convergenece of teh new paradigm.

ISO 8073-DAD2 (TP4) Layer 4 (transport) Connection-Mode Protocol. TP4 (Transport Protocol Class 4) provides guaranteed delivery and sequencing end-to-end. ISO's version of Transmission Control Protocol (TCP). See ISO.

ISO 8473 Layer 3 (Network) Connectionless Network Protocol (CLNP). "Datagram" routing. No guaranteed delivery nor sequencing. ISO's version of Internet Protocol. See ISO.

ISO 11172 ISO 11172 MPEG-1 and ISO 13818 MPEG-2 Specifications define audio compression algorithms at bit rates from 32 kbps to 384 kbps. The MPEG-1 define three similar compression techniques which are referred to as layer I, II and III. In progressing from layer I to layer III, improvements in compression efficiency are achieved at the expense of additional complexity and algorithmic delay. All layers support two audio channels at sample rates of 32, 44.1 or 48kHz at bit-rats from 16 to 384 kbps. The MPEG-2 standard extends the number of audio channels to five plus a low frequency effects channel. MPEG-2 also provides the additional sample rate options of 16, 22.05 and 24kHz. See ISO and MPEG.

ISO 13818 See ISO and ISO-11172 above.

ISO 8877 Information Processing Systems Q Interface Connector and Contact Assignment for ISDN Basic access interface located at reference points S and T - International Organization for Standardization. Part of this standard describes pin/pair assignments for 8-line modular connectors. The assignments are the same as EIA's T-568A. See ISO.

ISO 9000 Series The ISO 9000 series, published in 1987, outlines the requirements for the quality system of an organization. It is a set of generic standards that provide quality assurance requirements and quality management guidance. It is now evolving into a mandatory requirement, especially for manufacturers of regulated products such as medical and telecommunications equipment. ISO 9001, the most comprehensive of three compliance standards – 9001, 9002, 9003 – is a model for quality assurance for companies involved with designing, testing, manufacturing, delivering and servicing of products. ISO 9002 covers manufacturing and installation only. ISO 9003 covers product testing and final inspection of standards. See ISO and ISO 9001.

ISO 9001 ISO 9001 is a rigorous international quality standard covering a company's research and development, design, production, installation and service procedures. Compliance with the standard is of increasing significance for vendors trading in international markets, in particular in Europe where ISO 9001 registration is widely recognized as an indication of the integrity of a supplier's quality processes. ISO 9001 is the most rigorous of the three standards. See ISO and ISO 9000 Series.

ISO 9002 ISO 9002 covers manufacturing and installation only. See ISO and ISO 9000 Series.

ISO 9003 ISO 9003 covers testing and final inspections of manufactured products. See ISO and ISO 9000 Series.

ISO 9660 The CD-ROM logical file format standard adopted by ISO in 1987. Describes a table of contents but not the format of the actual data. This has led to incompatibilities between different computers. Based on a specification developed by the High Sierra Group (HSG) which included Apple, Microsoft, 3M, Philips, Hitachi, DEC. Also know as Yellow Book and High Sierra. See ISO.

ISO/IEC The International Organization for Standardization (IOS) and the International ElectroTechnical Commission. www.iso.ch.

ISOC A non-profit organization that fosters the voluntary interconnection of computer networks into a global communications and information infrastructure. According to ISOC, the society provides leadership in addressing issues that confront the future of the Internet. ISOC is the umbrella organization for the IAB (Internet Architecture Board), IETF (Internet Engineering Task Force) and IRTF (Internet Research Task Force). The ISOC has approximately 150 organizational and 6,000 individual members in more than 100 countries. www.isoc.org. See also IOPS and NANOG.

isochronous Isochronous transmission means "two-way without delay." Normal everyday voice conversations are isochronous. They have always been isochronous. We could not tolerate delays. We just never called them isochronous. The word isochronous appeared when we started digitizing voice, then joining it with data on a single channel. The data guys suddenly woke up to the fact that users would only tolerate joining voice and data – if the voice went through without delays. So they came up with this fancy new term "isochronous." By accepting this realization, they then could design buses – e.g. Universal

Serial Bus – where other flows of data (e.g. printing, keyboard entry, data communications from the Internet) could be delayed minute amounts of time, while voice went through without delay.

Isochronous comes from the Greek "iso" (equal) and "chronous" (time). Isochronous transmission is used to move stuff which must get to its destination with absolutely no delays. Voice and video need isochronous transmission. Let me explain. In the beginning, the phone network switched a call from A to B. It kept the circuit open. Whatever you said at one end went to the other end at the speed of electricity or light, effectively instantly. Then they invented other methods of transmission, where the circuit isn't open 100% from end to end during the "conversation." One example is packet switching, used widely for sending data. If you're sending an electronic mail, it clearly doesn't matter if your electronic letter arrives half a second faster or slower. It does matter with voice and video. See Isoethernet and Universal Serial Bus.

Isochronous transmission needs to be defined technically. "Instantaneous" is not a technical description. It must be delivered within certain time constraints. For example, in multimedia transmission you have to make sure that the video arrives at the same time as the audio. Isochronous can be contrasted with asynchronous, which refers to data streams broken by random intervals, and synchronous processes, in which data streams can be delivered only at specific intervals. Isochronous service is not as rigid as synchronous service, but not as lenient as asynchronous service. Certain types of networks, such as ATM, are said to be isochronous because they can guarantee a specified throughput. Likewise, new buses, such as IEEE 1394, support isochronous delivery. See also Asynchronous, Isochronous Ethernet, and Synchronous.

isochronous ethernet A 10 Mbps LAN topology that sets aside 96 ISDN channel to carry voice, data and video. See Isochronous and Isoethernet.

ISODE ISO Development Environment. An implementation of OSI's upper layers on a TPC/IP protocol stack. Pronounced "eye-so-dee-eee".

ISODE Consortium X.500 directories. www.isode.com.

IsoENET Another word for IsoEthernet. See IsoEthernet.

IsoEthernet Isochronous Ethernet Integrated Services. An extension to the Ethernet LAN standard proposed by IBM and National Semiconductor and first demonstrated at Fall Comdex 1992, and standardized as IEEE 802.9a. IsoEthernet adds six megabits per second of capacity to regular Ethernet, specifically to carry low delay, constant bit rate (CBR), isochronous data, especially voice and video. This isochronous capacity appears as up to 97 telephony channels of 64 Kbps each – 96 for transmission of information (voice, video, data, etc.) and one (called the D channel) for signaling. Like FDDI-II, IsoEthernet has the potential to carry both live voice or video calls together with LAN packet data on the same cable. IsoEthernet is limited to a single-workgroup solution in support of collaborative communications as videoconferencing and whiteboarding. It never caught on, and has been rendered obsolete by Switched Ethernet, ATM, and other high-speed LAN technologies. See also Ethernet, Isochronous, FDDH-I.

isolated ground IG. In AC electricity, an isolated ground is a type of outlet characterized by the following features and uses:

- It may be orange and must have a Greek "delta" on the front of the outlet. (A delta looks like a triangle.)
- It must be grounded by an insulated green wire.
- It must have an insulator between the ground terminal and the mounting bracket.
- It is used primarily to power electronic equipment because it reduces the incidence of electrical "noise" on the ground path.

isolation When they're close, electrical can interfere with each other. When they're close, wireless signals interfere with each other. Isolation means doing something to keep the signals apart. You can do this in many ways. With wires you can insulate the wires. You can insulate and twist each two-wire circuit around. You can also then sheath the wires in a metal foil.   With wireless – for example microwaves transmission – you can keep the signals separated. In formal terms, isolation with wireless transmission is the separation needed between a donor antenna and a coverage antenna so that their signal patterns don't interfere with each other. See also Power Conditioning.

isolator A device that permits microwave energy to pass in one direction while providing high isolation to reflected energy in the reverse direction. Used primarily at the input of communications-band microwave amplifiers to provide good reverse isolation and minimize VSWR. Consists of microwave circulator with one port (port 3) terminated in the characteristic impedance. See Optical Isolator.

isotropic Exhibiting properties of the same values when measured along axes in all directions.

isotropic antenna A hypothetical omni-directional point-source antenna that serves as an engineering reference for the measurement of antenna gain. See also EIRP and Isotropic Radiator.

isotropic radiator A completely omni-directional antenna, i.e., one which radiates equally well in all directions. This antenna exists only as a mathematical concept and is used as a known reference to measure antenna gain expressed as dBi. See also dBi.

ISP 1. Internet Service Provider. A vendor who provides access for customers (companies and private individuals) to the Internet and the World Wide Web. The ISP also typically provides a core group of internet utilities and services like E-mail, News Group Readers and sometimes weather reports and local restaurant reviews. The user typically reaches his ISP by either dialing-up with their own computer, modem and phone line, or over a dedicated line installed by a the Internet Service Provider, a CLEC, or a local or long distance telephone company. An ISP makes money from as many as four ways – monthly subscriptions for connecting customers to the Internet and providing them email, extra services (such as additional email addresses), a cut of inbound telephone revenues, advertising and commission fees. An ISP is also known as a Telecommunications Service Provider or an ITSP, for Internet Telephony Service Provider. See also IAP.

2. Information Service Provider. A company which provides information over the phone in response to touchtones punched in by a subscriber. That information may be weather, stock prices, etc. Often it is provided over 900 number, in which a phone company (local or long distance) bills the end user for calls to the ISP (Information Service Provider), paying part of the revenue collected to the ISP and keeping some of it as its collection billing and network fee. In many states, ISPs also use special local numbers, like those in New York beginning with 970 and 976.

3. Integrated Service Provider. This is a term for a new (or relatively new) company that delivers all of the different data and voice services including: Voice over IP, dialtone and Internet service.

4. ISDN Signal Processor.

5. Information Services Platform.

ISPBX Integrated Services Private Branch eXchange.

ISPC International Signaling Point Code.

ISPT Instituto Superiore delle Poste e delle Telcomunicazioni. (Superior Institute for Posts and Telecommunications, Italy).

ISR International Simple Resale. A system which allows international carriers to buy transmission capacity in bulk, to plug it into the public network at each end, and to resell it, one call at a time. This eliminates the need for settlements between international carriers.

ISS Intelligent Services Switch.

ISSI InterSwitching Interface. An interface between two SMDS switching systems within a LATA.

ISSN Integrated Special Services Network.

Issue A euphemism for "problem." Margaret Horak, wonderful wife Ray Horak, my Contributing Editor, is a top-notch consultant involved in customer service and certain related OSSs (Operations Support Systems). Seems as though her clients lately refer to "problems" as "issues." It's less scary that way.

IST Initial Service Term.

ISTF Integrated Services Test Facility.

ISUP Integrated Services Digital Network User Part. The call control part of the SS7 protocol. ISUP determines the procedures for setting up, coordinating, and taking down trunk calls on the SS7 network. ISUP is defined by ITU-T recommendations Q.761 and Q.764. ISUP also provides:

Calling party number information (including privacy indicator).

• Call status checking, to keep trunks in consistent states at both ends.

• Trunk management, and

• Grabbing and releasing of trunks and the application of tones and/or announcements in the originating switch upon encountering error, blockage or busy conditions.

There are seven ISUP Messages: Initial Address Message (IAM), Continuity Check Message (COT), Address Complete Message (ACM), Answer Message (ANM), Release Message (REL), Release Complete Message (RLC) and Exit Message (EXM). For you to benefit from these capabilities, your phone equipment must first be able to access the CCS7 network. One suggested way (but not the only way) is through the ISDN primary rate access (PRA) standard, which supports Q.931 protocol. See IAM, ISDN and Common Channel Signaling.

ISV Independent Software Vendor. Typically a company which writes and sells software, but not hardware. Manufacturers of hardware and operating systems, i.e. IBM or Northern Telecom, often contract with ISVs to produce specialized software to make their hardware and operating system more attractive.

IT 1. Information Technology. A fancy name for data processing (DP), which became management information systems (MIS), which became information technology. All mean the same thing – computers, software and networking. IT (pronounced eye tee) may have come from Europe. I heard it first from Siemens and Nixdorf who merged in 1989. IT means all the equipment, processes, procedures and systems used to provide and support information systems (computerized and manual) within an organization and those reaching out to customers and suppliers. These days virtually all IT is networked, i.e. it travels over phone lines of various sizes and speeds. Thus IT today includes control over data telecom – but typically not voice telecom. As data and voice merge onto a common transmission and switching path (it's called convergence), then IT will increasingly take over telecom management in the corporation. But right now, they're often separate.

2. Inter-Toll Trunk.

IT&T Information Technology and Telecommunications. Not a common term. IT is more common. See IT.

It's A shortened form of "It is." Not to be confused with Its, which is the possessive of it. Correct usage: Its house. It's a house.

ITA Integrated Trunk Access.

ITAA Information Technology Association of America. ITAA was founded in 1961 as ADAPSO (Automated Data Processing and Services Association). ITAA says its 9,000+ members are IT companies who create and market products and services associated with computers, communications and data. ITAA divisions are Software, IT Services, Information Services and E-Commerce, and Enterprise Solutions. www.itaa.org.

ITAR International Traffic in Arms Regulations. A U.S. government document which established the rules for import and export of goods and services which have significance in terms of national security. Such goods and services are assigned to the U.S. Munitions List, and cannot be exported. The most complex, (read "effective") encryption technologies are included in this list. For example, PGP and DES are on the list. Don't take your laptop out of the country if you have these encryption algorithms loaded on it, unless you want to spend a few years in a federal prison.

ITB Intermediate Block Character. A transmission control character that terminates an intermediate block. A Block Check Character (BCC) usually follows. Using ITBs allows for error checking of smaller blocks in data communications.

ITC 1. Information To Controller.

2. International Teletraffic Congress.

3. Japan's Telecommunications Technology Committee.

4. Independent Telephone Company. Monopoly local service provider in a given area not serviced by an RBOC.

ITCA International TeleConferencing Association. A professional association organized to promote the use of teleconferencing, including audio, videographics, video, business TV, and distance education. Membership is open to service and product providers, consultants and users. www.itca.org

ITCO An ITCO is an Independent Telephone COmpany. "Independent" as in independent from the original bell system, An ITCO is kind of like a cross between a CLEC and an RBOC, or any other ILEC. Usually they're collections of small mom-and-pop rural local phone companies that have been around for a while. ITCOs are used to being the incumbent telecom provider (thus like an ILEC or RBOC) in some small corner of the country, and they're used to providing basic local voice service. With telecom deregulation of 1996, they can move into other pockets of the country to sell more voice services. Now, however, they see the aggressive big national CLECs popping up in their back yards, advertising single-sourced local voice and data, and they're scared. They only know voice, and they're nowhere near as big powerful or marketing savvy. So they know they have to do something NOW to keep and attract customers... but many don't know what, or how. See Telecommunications Act of 1996, U.S.

iterative development An approach to application development in which prototypes are continually refined into increasingly complete and correct systems. Similar to prototyping.

iterative process The process of repeatedly processing a bunch of instructions. Each repetition, theoretically, comes progressively closer to the desired result, the "correct" answer, etc.

ITESF Internet Traffic Engineering Solutions Forum. An initiative of Bellcore formed in 2Q 1997 at the request of several Incumbent LECs (ILECs) to address common issues of Internet congestion of the Public Switched Telephone Network (PSTN). The ITESF seeks to develop generic requirements for products and features designed to off-load Internet traffic from the PSTN-such traffic is characterized by very long holding times during which relatively bursty traffic is supported. In other words, sometimes the circuit-switched network is used to full capacity while, at other times, little or no data is transmitted-regardless, the network is committed to supporting the traffic, whether or not it is present and whether or not the capacity of the network is required. The ITESF's solution, in abbreviated form, is to recognize Internet traffic for what it is by virtue of the dialed number, and to shunt it off to an IP data network. While this seems very obvious and very simple to do, it does require that some entity recognize the issue, take charge, and do something about it (i.e., set standards). Therefore, the ITESF is forming. See also Bellcore.

ITFS 1. Instructional Television Fixed Service. A service provided by one or more fixed microwave stations operated by an educational organization and used mainly to transmit instructional, cultural and other educational information to fixed receiving stations.

2. International Toll Free Service. A service which allows callers to dial a Freephone (i.e., toll-free) number from one country, with the call terminating in another country. The traditional approach to providing IFTS is to use the toll-free dialing pattern of the originating country. For example "1-800-XXX-XXXX" in the U.S., "0044-22-XXXXXX" or 0066-33-X-XXXXX in Japan, and 989-9-XX-XXXX in Columbia. This approach works, but requires that a multinational company have at least one different Freephone number in each country. A much better approach is UIFN (Universal International Freephone Number), which involves a universal, standardized numbering scheme in the format "+ 800 XXXX XXXX." See also UIFN.

ITG AT&T's Integrated Telemarketing Gateway. This is a set of specs for hooking up an outside computer to an AT&T switch. Under ITG, information travels in both directions – from the switch to the host computer and from the host computer to the switch. See also IG (one-directional link), ASAI and Open Application Interface.

ITI 1. Idle Trunk Indicator.

2. Information Technology Industry Council. ITI, according to ITI, represents the leading U.S. providers of information technology products and services. Its members had worldwide revenues of $323 billion in 1994. They employ more than one million people in the United States. 202-626-5725.

ITIC Information Technology Industry Council. www.itic.org.

iTIP iCalendar Transport Independent Interoperability Protocol. See iCalendar.

ITM See Information Technology Management.

ITORP IntraLATA Toll Originating Responsibility Plan. Performs a similar function to CMDS, but for IntraLATA toll calls within an RBOC's geographic territory. A settlement plan used between ITCs and the RBOC.

ITS 1. Institute for Telecommunications Sciences. The research and engineering branch of the National Telecommunications and Information Administration (NTIA), which is part of the U.S. Department of Commerce (DoC). www.its.bldrdoc.gov.

2. Intelligent Transportation System. A concept for a transportation system using IT (Information Technologies) to reduce highway transit time, provide necessary emergency services and traffic advisories, reduce traffic congestion, and improve travel safety. The concept is being translated into reality through the development of a number of wireless applications. For example, vehicle navigation systems relying on GPS (Global Positioning System) satellites can track your location, with directions to your destination offered through graphic maps displayed on a monitor. The same terminal can be used to display alternate routes as traffic congestion develops. The same terminal also can display emergency messages. Trucks no longer need to stop at weigh-in stations, as the gross weight of the truck can be transmitted on a wireless basis. The Intelligent Transportation Society of America (ITSA) is heavily involved in the promotion of the concept, as well as the underlying technologies and applications. www.itsa.org.

ITSEC The European Information Technology Security classification and evaluation initiative.

ITSP Internet Telephony Service Provider. See ISP.

ITT International Telephone and Telegraph. A company that once was the largest manufacturer of telecommunications equipment outside the U.S.

ITU The ITU, an organization based in Geneva, Switzerland, is the most important telecom standards-setting body in the world. In actual fact, it has no power to set standards. But if its members agree on a standard, it effectively becomes a world standard. Why fight city hall? "ITU" stands for "International Telecommunication Union," a name that the organization adopted in 1934. The ITU presently consists of three major sectors that were established in 1992: the Radiocommunication Sector (ITU-R), the Telecommunication Development Sector (ITU-D), and the Telecommunication Standardization Sector (ITU-T). When it was created, the ITU-T took over the work formerly done by the CCITT (Comite Consultatif Internationale de Telegraphique et Telephonique or, in English, International Telegraph and Telephone Consultative Committee, which until 1992 had been the most influential telecom standards organization in the world. For some reason, most reference works that mention the ITU and the ITU-T consistently use "Telecommunications" in the full form of the acronyms. Even one of the ITU's own acronyms databases lists "Telecommunications Standardization Sector" as the meaning of the "-T" in "ITU-T." However, the official names

of the organizations are "International Telecommunication Union" and "International Telecommunication Union-Telecommunication Standardization Sector."

The scope of the ITU's work is now much broader than just telegraphy and telephony. It now also includes IP voice, telematics, data, new services, systems and networks (like ISDN). The ITU is a United Nations Agency and all UN members may also belong to the ITU, represented by their governments. In most cases, the governments hand their rights on their national telecom standards to their telecommunications administrations (PTTs). But other national bodies (in the US, for example, the State Department) may additionally authorize Recognized Private Operating Agencies (RPOAs) to participate in the work of the ITU. After approval from their relevant national governmental body, manufacturers and scientific organizations may also be admitted, as well as other international organizations. This means, says the ITU, that participants are drawn from the broad arena. The activities of the ITU-T divide into three areas: Study Groups (at present 15) to set up standard ("recommendations") for telecommunications equipment, systems, networks and services. Plan Committees (World Plan Committee and Regional Plan Committee) for developing general plans for a harmonized evolution of networks and services. Specialized Autonomous Groups (GAS, at present three) to produce handbooks, strategies and case studies for support mainly of developing countries. Each of the 15 Study Groups draws up standards for a certain area – for example, Study Group XVIII specializes in digital networks, including ISDN. Members of Study Groups are experts from administrations, RPOAs, manufacturing companies, scientific or other international organizations - at times there are as many as 500 to 600 delegates per Study Group. They develop standards which have to be agreed upon by consensus. This, says the ITU, can sometimes be rather time-consuming, yet it is a democratic process, permitting active participation from all ITU member organizations. The long-standing term for such standards is "ITU (ITU-T) recommendations." As the name implies, recommendations have a non-binding status and they are not treaty obligations. Therefore, everyone is free to use ITU-T recommendations without being forced to do so. However, there is increasing awareness of the fact that using such recommendations facilitates interconnection and interoperability which is in the interest of network providers, manufacturers and customers. This is the reason why ITU-T recommendations are now being increasingly applied – not by force, but because the advantages of standardized equipment are obvious. ISDN is a good example of this. ISDN and other standards recommendations include options which allow for multiple "standards," in recognition of differing national and regional legacy "standards;" as a result, international standards recommendations do not necessarily yield evenly applied standards options. The ITU-T has no power of enforcement, except moral persuasion. Sometimes, manufacturers adopt the ITU-T specs. Sometimes they don't. Mostly they do, as for example with modem specifications, including V.90, H.XXX standards. The ITU-T standardization process runs in a four-year cycle ending in a Plenary Session. Every four years a series of standards known as Recommendations are published in the form of books. These books are color-coded to represent different four cycles. In 1980 the ITU published the Orange Books, in 1984 the Red Books and, in 1988, the Blue Books. See ITU STUDY GROUPS and ITU V.XX below. The ITU has now been incorporated into its parent organization, the International Telecommunication Union (ITU). Telecommunication standards are now covered under Telecommunications Standards Sector (TSS). ITU-T (ITU-Telecommunications) replaces ITU. For example, the Bell 212A standard for 1200 bps communication in North America was referred to as ITU V.22. It is now referred to as ITU-T V.22.

ITU itself says that it specializes in three main activities – defining and adopting telecommunications standards, regulating the use of the radio frequency spectrum and furthering telecommunications development around the world, particularly in the developing countries. It also holds a major trade show in Geneva every four years. As satellites have become more important as a method of long distance communications, so the ITU's allocation of scarce satellite frequencies among countries has become a hot bed of controversy. There are many who believe the ITU to be the most important telecommunications organization in the world. The organization owes its origins to Union Telegraphique which was formed in 1865, with the specific aim of developing standards for the telegraph industry. In 1947, under a United Nations charter, it was reformed as the ITU. This body has three main aims:

a. To maintain and extend, international cooperation for the improvement and interconnectivity of equipment and systems, through the establishment of technical standards.

b. To promote the development of the technical and natural facilities (the spectrum) for most efficient applications.

c. To harmonize the actions of national standards bodies to attain these common aims. In particular, to encourage the growth of communications facilities in developing countries.

Due to the rapid growth of the telecommunications industry, it was necessary to set up the International Consultative Committees (ITU, CCIR and IFRB) within the ITU's jurisdiction in order to adequately manage this expansion. The aims are achieved by organizing international conferences and meetings, by sponsoring technical cooperation, and by publishing information and promoting world exhibitions. Currently, the ITU has about 170 member nations. I.T.U., Place des Nations, CH-1211 Geneve 20, Switzerland. Tel +41 22 99 51 11. Fax +41 22 33 72 56. International Telecommunication Union. See the following definitions. ITU can be contacted at www.itu.org, www.itu.ch (where CH stands for Switzerland. Do not type com.)

ITU calls its recommendations for standards "pre-published." According to the ITU, this means the text of the recommendation, as approved, is supplied by the responsible "author" (normally the chairman of the Study Group which approved it). It is however not yet edited. Therefore there may be changes (mostly of an editorial nature). Here are the categories of ITU pre-published Recommendations:

Series E. Overall network operation, telephone service, service operation and human factors.
Series F. Non-telephone telecommunication services.
Series G. Transmission systems and media, digitals systems and networks.
Series H. Audiovisual and multimedia systems.
Series I. Integrated services digital network (ISDN).
Series J. Transmission of television, sound programme and other multimedia signals.
Series M. TMN and network maintenance: international transmission systems, telephone circuits, telegraphy, facsimile and leased circuits.
Series O. Specifications of measuring equipment.
Series P. Telephone transmission quality telephone installations, local line networks.
Series Q. Switching and signaling.
Series T. Terminals for telematic services.
Series V. Data communication over the telephone network.
Series X. Data networks and open systems communication.
Series Y. Global information infrastructure.
Series Z. Languages and general software aspects for telecommunication systems.

ITU E.169 An ITU-T Recommendation allowing International Freephone customers to be allocated a unique Universal Freephone Number (UIFN) which will remain the same throughout the world, regardless of country or telecommunications carrier.

ITU H.222 An ITU-T Study Group 15 standard that addresses the multiplexing of multimedia data on an ATM network. See H.222 and other H.XXX entries.

ITU Q.XXX See all the definitions beginning with Q.

ITU Study Groups The ITU operates as a series of groups considering specialist areas. Key study groups applicable to networking and communications are: Study Group VII responsible for terminal equipment for telematic services, including fax and higher level OSI standards; Study Group X covering Languages and methods for telecommunications applications; Study Group XI covering ISDN, telephone network including V-series Recommendations; Study Group XVIII covering digital networks including ISDN. See ITU above and ITU-T.

ITU V.XX A set of evolving telecom standards. For more on those standards, see under the letter "V."

ITU-T The Telecommunications Standards Section (TSS) is one of four organs of the ITU. Any specification with an ITU-T or ITU-TSS designation refers to the TSS organ.

ITU-T Recommendations series A Series – Organization of the work of ITU-T

B Series – Means of expression: definitions, symbols, classification
C Series – General telecommunication statistics
D Series – General tariff principles
E Series – Overall network operation, telephone service, service operation and human factors
K Series – Protection against interference
L Series – Construction, installation and protection of cables and other elements of outside plant
N Series – Maintenance: international sound programme and television transmission circuits
R Series – Telegraph transmission
S Series – Telegraph services terminal equipment
U Series – Telegraph switching

ITUSA Information Technology Users' Standards Association.

ITV Internet TV or Interactive TV. Also abbreviated to I-TV. See also IPTV.

IU Interface Unit.

IUT Implementation Under Test: The particular portion of equipment which is to be studied for testing. The implementation may include one or more protocols.

IUW The ISDN Users Workshop.

IVCP Installation Verification Certification Program.

IVD Incremental Volume Discount.

IVDM Integrated Voice and Data Multiplexer. A device that Northern Telecom uses to provide DIALAN, a central office provided local area network offering completely digital, full duplex data transmission at speeds of 300 bps through 19.2 asynchronous and 1,200 bps through 64 kbps synchronous. DIALAN users use existing telephone sets and an Integrated Voice and Data Multiplexer (IVDM) that plugs into a telephone jack.

IVDS Interactive Video and Data Service. See 218-219 MHz.

IVDT Integrated Voice/Data Terminal. A device with a terminal keyboard/display and a voice telephone with or without its own processing power. See Integrated Voice/Data Terminal.

IVR Interactive Voice Response. Think of IVR as a voice computer. Where a computer has a keyboard for entering information, an IVR uses remote touchtone telephones. Where a computer has a screen for showing the results, an IVR uses snippets of recordings of human voice or a synthesized voice (computerized voice). Recordings are used for repetitive messages, "Thanks for calling ABC Company. Push one for our sales department. Push two for our service department." Synthesized voice (also called Text-To-Speech) is used for reading information from files which contain information that can't be put into neat "sound bites," like numbers and dates, e.g. reading my incoming email. Whatever a computer can do, an IVR can too – from looking up train timetables to moving calls around an automatic call distributor (ACD). The only limitation on an IVR is that you can't present as many alternatives on a phone as you can on a screen. The caller's brain simply won't remember more than a few. With IVR, you have to present the menus in smaller, cascading chunks.

The benefits of Interactive Voice Response are obvious. By automating the retrieval and processing of information by phone, you can "give data a voice" and "add intelligence to the phone call." By doing that, you can:

Put information to work. The classic IVR "killer app" takes an existing database (e.g., a magazine's article archives, a freight company's package-tracking system) and makes it available by phone (or other media, such as fax, e-mail, or DSVD – Digital Simultaneous Voice and Data). You can automate telephone-based tasks. From "bank by phone" to "find my package" to "sell me an airline ticket," to "validate my new credit card," IVR gives access to and takes in information; performs record-keeping, and makes sales, 24 hours a day – supplementing or standing in for human personnel. IVR can add value to communications. Any call-handling phone system (e.g. an automatic call distributor used by airlines) can profit from IVR. Used as a front-end for an ACD, an IVR system can ask questions (e.g., "what's your product serial code?") that help routing and enable more intelligent and informed call processing (by people or automatic systems). IVR can add interactive value to what would otherwise be wait-time. The IVR can be used to distribute info, make callers aware of specials – even provide entertainment. The result: fewer callers drop off queue; you make more sales.

Periphonics, a early and leading IVR company, explains it thus:

IVR systems allow individuals to access information in an organization's computer database and to receive that information either verbally, using an ordinary touch-tone phone, or on a PC, via the Internet. In addition, these systems enable customers to execute certain transactions on-line without the intervention of customer service personnel. Typically, 30-60% of the repetitive and/or routine inbound calls are automated, which can maximize the effectiveness of the current customer service staff.

Benefits of IVR:
1. Reduces costs.
 2. Improves access to information (24 hours a day - 7 days a week).
 3. Enhances customer service.
 4. Improves competitive position with increased customer retention.
 5. Streamlines operations.
 6. Generates new revenues.
 7. Better utilization of telephone and computer systems capabilities.
 8. Improves productivity of customer support staff and reduces the need to increase staff for peak periods.
 9. Provides more services in less time and at lower costs.
 10. Reduces errors in data capture/input, with feedback for valid entries.
 11. Gives a typical ROI of six to nine months.

IVS Interactive Voice Service.

IW 1. Interworking.
 2. Information Warfare.
 3. Inside Wire.

iWarp iWarp is a form of remote direct memory access (RDMA) that will be independent of switch fabrics and applications protocols. The goal of one of the people working on this new standard "is to unify IP and InfiniBand." According to a news magazine report, Cisco Systems hopes to use iWarp to power its efforts at delivering storage network that link servers and large storage arrays over IP instead of the fast but more costly Fibre Channel links often used today."

IWN See Integrated Wireless Network.

IWS Intelligent Workstation.

IWTA International Wireless Telecommunications Association. An organization created to represent the interests of the worldwide commercial trunked radio industry, which generally is known as SMR (Switched Mobile Radio) in the U.S. IWTA grew out of its sister organization AMTA (American Mobile Telecommunications Association. See also SMR. www.iwta.org.

IWV German for pulse dialing. The German is ImpulseWahlFerfahren.

IX 1. IntereXchange carrier, i.e. a long distance phone company. More commonly known as IXC (IntereXchange Carrier) or IEC (InterExchange Carrier). Since the Modified Final Judgement (MFJ) took effect in the U.S. in 1984, the IXs have enjoyed the virtually exclusive right to haul traffic across LATA (Local Access and Transport Area) boundaries. The ILECs (Incumbent Local Exchange Carriers), on the other hand, are limited to providing local service and short-haul long distance within the LATA boundaries. See also IEC, ILEC, IXC, LATA, and MFJ.

2. Internet eXchange. A commercial peering point at which ISPs (Internet Service Providers) outside the U.S. exchange traffic directly on a regional basis, avoiding the costs of exchanging traffic through the U.S. NAPs (Network Access Points). IXs include the Amsterdam Internet Exchange (AMS-IX), Hong Kong Internet Exchange (HKIX), Japan Internet Exchange (JPIX), London Internet Exchange (LINX), and Service for French Internet Exchange (SFINX). See also MAE, NAP, and Peering Point.

IXC IntereXchange Carrier. Also known as IEC (InterExchange Carrier) and IC. Long-haul long distance carriers, IXCs include all facilities-based inter-LATA carriers. The largest IXCs are AT&T, MCI, Sprint and Worldcom; a huge number of smaller, regional companies also fit this definition. The term generally applies to voice and data carriers, but not to Internet carriers. IXC is in contrast to LEC (Local Exchange Carrier), a term applied to traditional telephone companies which provide local service and intraLATA toll service. IXCs also provide intraLATA toll service and operate as CLECs (Competitive Local Exchange Carriers) in many states. Once upon a time, the non-AT&T IXCs were called OCCs (Other Common Carriers), a status which they resented for understandable reasons.

IXM Inter-Exchange Mileage.

IXN Interconnection revenue. Only issued in reciprocal compensation states and only paid out to wholesale from the ILECs to the CLECs.

IXO A one-way protocol for ASCII-based communications between the Internet and wireless pagers, IXO was named for the company that allegedly invented it. Others suggest that it was invented by Robert Edwards, the paging pioneer who owned Radiofone Corp. In any event, it was renamed APE (Alphanumeric Paging Entry), and then TAP (Telocator Alphanumeric Protocol). Motorola calls it PET (Personal Entry Protocol). It largely has been replaced by SNPP (Simple Network Paging Protocol). See also SNPP and TAP.

iXXX The little i stands for Intel and the numbers that follow refer to the particular microprocessor chip. For example, i386 refers to the 80386 chip. Then, they started calling them by name, i.e. Pentium. Then they joined them with wireless stuff and started with meaningless numbers. All this happened because they couldn't make them faster because of too much heat. So, obfuscation became a marketing technique. Intel is not the first company to use the technique. They probably learned it in one of my marketing seminars.

J Box See Junction Box.

J-Carrier The Japanese version of the T Carrier system of North America. It's different to the North American one in more ways (e.g. signaling and such) than are apparent in the two tables below. Mbps equals million bits (not bytes) per second. A voice channel is a 64,000 PCM encoded channel.

See also PCM, T Carrier.

NORTH AMERICAN HIERARCHY (T-CARRIER)		
T-1	1.544 Mbps	24 voice channels
T-1C	3.152 Mbps	48 voice channels
T-2	6.312 Mbps	96 voice channels
T-3	44.736 Mbps	672 voice channels
T-4	274.176 Mbps	4,032 voice channels
JAPANESE HIERARCHY (J-CARRIER)		
J-1	1.544 Mbps	24 voice channels
J-2	6.312 Mbps	96 voice channels
J-3	32.064 Mbps	480 voice channels
J-4	97.728 Mbps	1,440 voice channels
J-5	397.000 Mbps	5,760 voice channels

J-Hook 1. In a microwave or satellite antenna, a j-hook is the name for a length of waveguide with one end turned through 180 degrees. This passes through the reflector vertex to illuminate the reflector's surface from an electronics unit mounted behind the microwave or satellite structure. It looks like a J-Hook.

2. J-hooks are also pieces of J-shaped pieces of bent metal used to hold cables in an equipment rack.

J2ME Java 2 Platform, Micro Edition (J2ME). The edition of the Java platform that is targeted at small, standalone or connectable consumer and embedded devices. The J2ME technology consists of a virtual machine and a set of APIs suitable for tailored runtime environments for these devices. The J2ME technology has two primary kinds of components - configurations and profiles.

jabber To jabber. In local area networking technology, continuously sending random data (garbage). Normally used to describe the action of a station (whose circuitry or logic has failed) that locks up the network with its incessant transmission.

jack Common term for communications terminals found at the end of cables, usually the female receptacle usually found on equipment.

jack contacts Metallic elements of telephone jacks that carry the central office currents/voltages to the CPE plus contacts.

jack header A raceway similar to a header duct, usually provided in short lengths to connect a quantity of distribution ducts together.

jack pins See Jack Contacts.

jack positions A numbering scheme to permit consistent identification of the Jack Contact(s) position. Position identification helps assure compatibility between the wiring system and the associated terminal equipment.

jack type Different types of jacks (RJ-11, RJ-45, or RJ-48) are used on telephone lines in North America – analog or digital. The RJ-11 is the most common in the world and is most often used for analog phones, modems, and fax machines. The RJ-11 can be wired with two conductors, four conductors and six conductors. If it has two pairs for two separate phone lines, it's called an RJ-14. One of the lines is the "normal" RJ-11 line – the red and green conductors in the center. The second line is the second set of conductors – black and yellow – on the outside. The RJ-14C is surface or flushmounted for use with desk telephone sets while the RJ-14W is for wallmounted telephone sets. The RJ-48 and RJ-45 are slightly bigger jacks and are both virtually the same, but they both have an 8-pin configuration and are often used for high-speed LANs or T-1 lines. An RJ-11 jack can fit into an RJ-45/RJ-48 connector, however, an RJ-45/RJ-48 jack cannot fit into an RJ-11 connector. See RJ for a complete listing of all available jacks.

jacket The protective and insulating housing of a cable. Not part of the fiber or the fiber buffer. See also Hard Cable.

jacket material The material used as the outer insulator of a cable. See also Hard Cable.

jackfield See patch panel.

jackrabbit Shakespeare was right when he asked, "What's in a name?" The jackrabbit is not a rabbit. It is a hare. A Jerusalem artichoke is not an artichoke; it is a sunflower. Arabic numerals are not Arabic; they were invented in India. India ink (sometimes referred to as "Chinese ink") was not known until recently in either China or India.

jacks A receptacle used in conjunction with a plug to make electrical contact between communication circuits. Jacks and their associated plugs are used in a variety of connecting hardware applications including cross connects, interconnects, information outlets, and equipment connections. Jacks are used to connect cords or lines to telephone systems. A jack can be female or male.

JAE See Java Application Environment.

Jahangir Jahangir was a 17th-century Indian Mughal ruler. He had 5,000 women in his harem and 1,000 young boys. He also owned 12,000 elephants.

jake Slang term for cheap four conductor wire used for inside wiring. Don't use it.

jam In an IEEE 802.3 network, the jam signal, which is normally produced by fixing the minimum number of data bytes that must be transmitted, is used to ensure that if a collision is produced, all devices on the network will detect it. See Jam Signal.

jam signal A signal generated by a printed circuit card to ensure that other cards know that a packet collision on a local area network has taken place. See Jam.

jamming 1. The interference with through-the-air radio transmission, the object being to hinder the receiver's ability to pick up and understand the signal. An example is the Russians' jamming Radio Free Europe.

2. The illegal interference by a monopoly local phone company of its customer's ability to choose a competing long distance carrier. Also known as an unauthorized PIC freeze.

Jane Barbe On July 27, 2003, CBSNews.com ran the following story, "One of the most recognizable voices in the country was silenced Tuesday. Jane Barbe, who recorded messages used by telephone companies across the country, died of complications from cancer at the age of 74. Over the past 40 years, if you didn't get through to the party you wanted, you probably still got through to Barbe. She was the "telephone lady" that delivered the message, "We're sorry, your call cannot be completed as dialed." A drama major at the University of Georgia, Barbe started recording the announcements in 1963. Twenty years later, she was making even the most disjointed of messages sound smooth as silk. Messages such as, "The number you have reached has been changed. The new number is ." Although she largely masked her Georgia accent on her recordings, in person she was the model of Southern hospitality, taking her odd brand of anonymous fame in stride. "I don't think anybody even knows who I am until somebody says I'm the lady on the phone," she said in a past interview. "Then the others say, 'Oh, really?'" Barbe said she always did her best not to sound like a machine. Instead, she tried to address her telephone audience one caller at a time. She said it could be overwhelming if she started to think she was talking to 22 million people a day. Trouble on the line is never fun, but for most of us Barbe's soothing telephone manner made it just a little bit easier to bear."

JANET Joint Academic Network. A British network covering universities, the UK Research Council and the UK Further Education Sector. In short the U.K> education and research community. It is managed by UKERNA and funded by the Joint Information Systems Committee (JISC). JANET is a trademark of the Higher Education Funding Councils for England, Scotland and Wales.The network's highspeed backbone is called SuperJANET.

Japander Actors who pander to the Japanese love of Western celebrities by appearing in ads that run only in Japan. Among those who have played: Madonna, Ben Stiller, Mel Gibson, and Brad Pitt.

Japanese TACS An analog system operating in the 900-MHz band in Japan.

JATE The Japanese equivalent of the U.S. FCC part 68 certification for equipment to be attached to the Japanese telephone network. It stands for Japan Approvals Institute for Telecommunications Equipment. Getting JATE approval is expensive, complex and immensely time consuming. At least that's what I wrote in the ninth edition of my edition. JATE wrote me and suggested that I correct my definition as follows:

1. Getting JATE approval is not expensive. It is not expensive compared with the same approval in U.S.A. We don't require to give detailed descriptions on a testing machine nor environmental tests, while FCC requires those. Therefore it is easier to get approval in JATE than in FCC. In some cases, FCC registration costs more than JATE approval, with fee for application and for a test laboratory.

2. Getting JATE approval is not complex. Japanese technical conditions only require not to harm network, like condition in U.S.A. does so. It's just simple.

3. Getting JATE approval is not time consuming. The average period is 23.4 days for documentation examination process, from the date an application is received by JATE to the date it is completed. And 90% were completed within 38.4 days.

Java A programming language from Sun Microsystems designed primarily for writing software to leave on Internet Web sites and downloadable over the Internet to a PC owned by you or me. Java grew out of "The Green Project," a top-secret effort headed by James Gosling. The initial excitement over Java was over its ability to bring motion to static Web pages – to make animated figures dance and stock tickers flash. But Java has a larger potential. In the past, software programs had always been written for particular computers and had resided on one machine. Java theoretically enables software to run on any machine (to "write once, run anywhere," as Sun puts it), Java would allow programs to reside anywhere on the Web, flowing across the wires of the Internet and working equally well wherever they land, thus rendering Windows irrelevant, if not obsolete (Sun's hope). In a Java-fuelled computing world, the reign of the Wintel PC (Windows/Intel) machine would be challenged by cheap, bare-bones devices known as "thin clients" – the most prominent of which is the network computer, or NC – a stripped down PC that stores and accesses files and programs on a network rather than a hard drive. The reality of Java since 1995 is that, like all new languages and computer "breakthroughs," getting it implemented into the real world of day-to-day programming has proven sticky. Java is basically a new virtual machine

and interpretive dynamic language and environment. It abstracts the data on bytecodes so that when you develop an app, the same code runs on whatever operating system you choose to port the Java compiler/interpreter to. What's a Java application? According to Wired Magazine, point to Ford Motor Company's website today, for instance, and all you'll get are words and pictures of the latest cars and trucks. Using Java, however, Ford could relay a small application (called an applet) to a customer's computer (the one on your desk which you are using the surf the Internet). The customer could then customize options on an F-series pickup while calculating the monthly tab on various loan rates offered by a finance company or local bank. Add animation to these applications and you could get to "drive" the truck. See the following definitions. http://java.sun.com.

Java Application Environment JAE. The source code release of the Java Development Kit.

Java Blend A product that enables developers to simplify data application development by mapping database records to Java objects and Java objects to databases.

Java Card API An ISO 7816-4 compliant application environment focused on smart cards.

Java Computing A computing architecture using standard network protocols that exploit the universal availability of networks, data, and Java applications to dynamically deliver services to a wide variety of Java-enabled devices. See Java.

Java Electronic Commerce Framework JECF. A structured architecture for the development of electronic commerce applications in Java.

Java Phone Mostly spelled JavaPhone. A Java API specification controlling contacts, power management, call control, and phonebook management, intended specifically for the programmability requirements of mobile phones.

Java Telephony API JTAPI. A set of modularly-designed, application programming interfaces for Java-based computer telephony applications. JTAPI is designed to serve a broad audience, from call control centers to Web-page designers. JTAPI offers telephony interface extensions grouped into building-block "packages." JTAPI consists of one Core package and several extension packages. JTAPI applications are portable across platforms without modification. Applications written to JTAPI are independent of platform or phone system. There are two configurations for JTAPI: Desktop Computer and Network Computer. In a desktop configuration, the JTAPI application or Java applet runs on the same workstation that houses the telephony resources. In a network configuration, the JTAPI application or Java applet runs on a remote workstation. This workstation can be a network computer with only a display, keyboard, processor, and some memory. It accesses resources off of the network making use of a centralized server that manages telephony resources. JTAPI communicates with this server via a remote access mechanism, such as Java Remote Method Invocation (RMI), JOE, or a telephony protocol. JTAPI interfaces for other computer telephony applications, such as SunXTL, TAPI, TSAPI and IBM Call Path are being produced.

Java Virtual Machine JVM. Java is a programming language. Instead of being compiled for a specific operating system (e.g., DOS or Unix) as most software is, Java is translated into bytecode, which is an intermediate code between the source code written by a programmer and the executable machine code run by the target computer. To run bytecode on your computer, you must install a program called a Java Virtual Machine, a real-time interpreter that creates executable bytecode as the Java applet is running. See also Bytecode and Java.

JavaMail API The JavaMail API provides a set of abstract classes that model a mail system. The API provides a platform independent and protocol independent framework to build Java based mail and messaging applications. The JavaMail API is implemented as a Java standard extension. Sun provides a royalty-free reference implementation, in binary form, that developers will be able to use and ship. See Java.

JavaPhone A Java API specification controlling contacts, power management, call control, and phonebook management, intended specifically for the programmability requirements of mobile phones.

JavaScript A scripting language for Web pages. Scripts written with JavaScript can be embedded into HTML documents. With JavaScript you have many possibilities for enhancing your Web page with interesting elements. It makes it easy to respond to use initiated events (such as form input). Some effects that are now possible with JavaScript were once only possible with CGI. Some computer languages are compiled, which means that you run your program through a compiler, which performs a one-time translation of the human- readable program into a binary that the computer can execute. JavaScript is an interpreted language, which means that the computer must evaluate the program each time it is run. Java and JavaScript are not the same thing. JavaScript was designed to resemble Java, which in turn looks a lot like C and C++. The difference is that Java was built as a general purpose object language, while JavaScript is intended to provide a quicker and simpler language for enhancing Web pages and servers.

JavaTel Java Technology Toolkit, or JavaTel, a cross-platform product designed to link any telephone, appliance or networked computer to any Java-based application. In October, 1996, IBM, Intel, Lucent Technologies, Nortel and Novell said they'll support the standard. JavaTel will offer software developers and device manufacturers a uniform interface for driving basic telephony functions, such as call setup, disconnect, hold and call transfer. A series of JavaTel Extension Packages will deliver interfaces such as advanced call control, media services, terminal management, call center management and mobile services. See JavaTel API.

JavaTel API Java Telephony Application Programming Interface. One of many Java Media APIs developed by Sun Microsystems with help from Lucent Technologies. Provides for call set-up, tear-down and media stream control. JavaTel can run on top of the Sun XTL Teleservices architecture. See JavaTel.

JB7 Jam Bit 7. It is the same zero suppression format found on T-1s as AMI, which is Alternate Mark Inversion. See AMI.

JBIG Joint Bitonal Image Group. Standard for black and white, and grayscale image representation.

JBIG Alliance In late September, 1996 12 companies announced the formation of the JBIG Alliance. The JBIG Alliance is an industry group formed to create a public forum for the dissemination of information encouraging the adoption and use of ISO/IEC Standard 11544:1993 (JBIG compression) for storage and transmission of bitonal and grayscale image data. JBIG (Joint Bitonal Image Group) is an advanced compression scheme originally developed, like the ITU/ITU Group IV standard that it is intended to replace, as an improved facsimile transmission standard. JBIG's exceptional compression is the result of an advanced compression technique known as arithmetic coding. For bitonal images of standard business documents, JBIG provides file size reductions of 20 to 60 percent with the existing Group IV standard (see table). According to the JBIG Alliance, users can use JBIG's efficient compression for either reducing storage and transmission costs, or for substituting higher resolution images without incurring substantially higher storage or transmission costs. JBIG can also store many grayscale images of equal or better quality in less space than required

	GroupIII	GroupIV	JBIG
Invoices	254,187	287,419	122,813
Line Art/text	306,256	166,098	119,060
Photo/magazines	274,883	241,742	119,066

for JPEG compressed files.

Here is a chart showing average file sizes in bytes of identical quality Group III, Group IV, and JBIG Files

JBOD Just a Bunch Of Disks. A very simple and inexpensive storage technology used in storage-intensive applications, such as imaging. JBOD is much simpler than even RAID (Redundant Array of Inexpensive Disks). See also RAID.

JCL Job Control Language.

JDC Japanese Digital Cellular. See PHS (Personal Handyphone System).

JECF Java Electronic Commerce Framework. A structured architecture for the development of electronic commerce applications in Java.

JEDEC Joint Electronic Devices Engineering Council. An organization of the U.S. Semiconductor manufacturers and users that sets package outline dimension standards for packages made in the U.S.

jeep The name Jeep came from the abbreviation used in the army for the "General Purpose" vehicle, G.P. At least that's one theory. According to Major E.P. Hogan, who wrote a history of the development of the Jeep for the Army's Quartermaster Review in 1941, the word jeep is an old Army greasemonkey term that dates back to World War I and was used by shop mechanics in referring to any new motor vehicle received for a test. The word also found later use as a less than complimentary term for new recruits. See Jeep Fishing.

jeep fishing When you use your Hummer automobile to pull Jeeps and other four-wheeled, all-terrain vehicles out of the mud on weekends. This definition contributed by Jack Rickard, ex-editor rotundas of Boardwatch Magazine, who obviously owns a Hummer, which he would love, were it more reliable.

jello Ellis Island immigrants were often served a bowl of Jello as a "Welcome to America."

jello on springs To help create her signature sexy walk –once described as "Jello on springs," – actress Marilyn Monroe sawed off part of the heel of one of her shoes.

JEMA Japan Electronic Messaging Association.

jeopardy A wonderful AT&T word meaning anything occurring during the course of accomplishing scheduled work which might cause the scheduled completion date to slip.

JEPI Joint Electronics Payments Initiative. A specification from the World Wide Web Consortium (W3C) and CommerceNet for a universal payment platform to allow merchants and consumers to transact E-Commerce (Electronic Commerce) over the Internet. JEPI comprises a standard mechanism for web clients and servers to negotiate payment instrument, protocol and transport with one another. JEPI consists of two parts: Protocol Extensions Protocol (PEP) is an extension layer that sits on top of HTTP (HyperText Transfer Protocol), and Universal Payment Preamble (UPP) is the negotiation protocol that identifies appropriate payment methodology. These protocols are intended to make payment negotiations automatic for end users, happening at the moment of purchase, based on browser configurations. See also Electronic Commerce.

JES Job Entry Subsystem. Control protocol and procedure for directing host processing of a task in an IBM host environment. Also the specific IBM software release, host-based, that performs job control functions.

Jewish Jewish refers to the religion, not the language. The language is called Yiddish.

Jewish holidays Jewish holidays are defined by three sentences: 1. They tried to kill us. 2. They failed. 3. So, let's eat.

Jewish telegram "Start worrying. Details to follow."

JFK Jackfield.

JHTML JHTML is an in-line scripting language that can be used to create dynamic Web pages. JHTML is composed of both Java and HTML code, with Java providing the data management logic, and HTML providing the Web page presentation. A Web server parses and evaluates the code for dynamic info (database, or any other real time data source) along with the plain HTML that is sent to a browser. JHTML coding can be done with a Java WebServer and a Web editor. Special Java "tags"–<\<>java> and <\<>/java>–are placed in the code by programmers to indicate the beginning and end of the Java commands. JHTML pages can send and receive cookies to a user's browser to store information. Cookies are often used to track sessions. Microsoft had developed its Active Server Pages (ASPs) to enable server-side processing, so that the user doesn't need a return trip to the server, and as a response, Sun developed JHTML and Java Server Pages (JSPs). Sun claims that the advantage of JHTML over ASP is that Java is a much richer language that those that Microsoft uses. Other examples of in-line scripting languages are Mivascript, PHP, and iHTML.

Jini A new software initiative by Sun Microsystems. Jini is 48,000 bytes of Java software code designed to let Java-enabled devices (cell phones, handheld computers, digital cameras, printers, etc) communicate simply without network headaches of drivers, complex installations. According to Interactive Week, Java was designed to allow any program to "run anywhere," Jini is Sun's attempt to "connect anything" to anything else, at any time. A Java-based program that runs from a large computer server is used to track all the Jini devices. "Spontaneous networking" is the phrase Sun uses to describe Jini. Jini (pronounced jee-nee) is named for the supernatural being that inhabited Aladdin's lamp. Sun officials intentionally misspelled the word. It's usually spelled "genie" or "jinni." But Sun's intent was to couple it with Java. So Jini has the same number of letters.

JIP Jurisdiction Information Parameter is defined in the SS7 IAM (Initial Address Message)and is used to provide location information of a roaming end user.
The JIP is a very important part of the billing record and is necessary in order that proper billing can done.

Jiro Sun's initiative to give developers a standard environment in which to create interoperable storage management components. Originally known as Project StoreX, the goal of Jiro is to let any storage management tool, applet or application intgerface with other management tools.

JIS Japanese Industrial Standards.

JIT Just In Time.

jithead An international transportation term used to describe people who order goods on a "just in time" basis and then freak out when told they didn't order early enough. Definition courtesy Wireless Magazine.

jitter Jitter is variability in latency, or delay. If a network provides varying levels of latency (i.e. different waiting times) for different packets or cells, it introduces jitter, which is particularly disruptive to audio communications because it can cause audible pops and clicks. Jitter is also the tendency towards lack of synchronization caused by mechanical or electrical changes. Technically, jitter is the phase shift of digital pulses over a transmission medium. Three forms of jitter exist: Data Dependent Jitter (DDJ), Duty Cycle Distortion (DCD), and Random Jitter (RJ). Data Dependent Jitter is caused by limited bandwidth characteristics and imperfections in the optical channel components as it relates to the transmitted symbol sequence, according to Information Gatekeepers. This jitter results from less than ideal individual pulse responses and from variation in the average value of the encoded pulse sequence which may cause baseline wander and may change the sampling threshold level in the receiver. DCD Jitter is caused by propagation delay differences between low-to-high and high-to-low transitions. DCD is manifested as a pulse width distortion of the nominal baud time. RJ is the result of thermal noise.

jitter buffer management Most speech encoder algorithms have a set of rules concerning packet delivery and disposition management. This is often called jitter buffer management. "Jitter" in this case refers to when the signal is put into frames. The decoding algorithm must decompress and sequence data and make "smoothing" decisions (when to discard packets versus waiting for an out-of-sequence packet to arrive). Given the real-time nature of a live connection, jitter buffer management policies have a large affect on voice quality. Actual sound losses range from a syllable to a word, depending on how much data is in a given packet. The first buffer size is often a quarter-second, large enough to be a piece of a word or a short word – similar to drop-outs on a cellular connection in a poor coverage area. See also IP Telephony Algorithms.

jitterati What the digital generation becomes after tanking up on too much coffee. This definition courtesy Wired Magazine.

job A file, typically sent in batch mode. Specifically a set of data, including programs, files and instructions to a computer, that together amount to a unit of work to be done by a computer.

joe job Spam that maliciously has another party's return email address, unbeknownst to the owner of that email address. A joe job is done with the intent of causing inconvenience or discomfort to the owner of the return email address, which will come about when recipients of the spam mistakenly assume that the third party behind the return address is responsible for sending the spam. The spam recipients may go further and send the third party irate or rude email responses, or even report the third party to the party's ISP.

Joel Vietnam was no preparation for Harry and the summer of 1997. Fortunately the green was bucks, not beret. Bucks are mitigating. Joel is a dear friend of mine, who's totally wonderful. So, if you don't understand this definition, don't worry. He will.

Johnson noise Thermal noise generated in conductors and semiconductors, named after Swedish-born American physicist John Bertrand Johnson (1887-1970) who first observed and reported it in 1927 at the Bell Telephone Laboratories (now part of Lucent Technologies).

join 1. A basic operation in a relational system that links the rows in two or more tables by comparing the values in specified columns.
    2. A service/feature which allows a device to join an existing call, i.e., Conference.

joining An ATM term. The phase in which the LE Client establishes its control connections to the LE Server.

joint 1. The place at which two separate things parts are brought together.
    2. A disreputable gathering place, or a dive. The term is derived from fact that the pipes used in opium were made of bamboo, and had many joints.
    3. A marijuana cigarette. Also known as a reefer or a doobie. See #2 for the likely origin of the term.

joint costs A regulatory concept. Joint costs are essentially overhead costs. They cover the costs of providing more than one service. Most costs in the telecommunications industry are joint. And being "joint" they give regulators enormous pleasure trying to allocate those costs to various services and therefore trying to figure what prices for those services should be.

joint pole A utility telephone pole which supports the facilities of two or more companies. A typical joint pole supports three: electric power, cable television and telephone. In many places, joint poles also support other devices such as street lights, municipal communications systems, signs, traffic signals, seasonal decorations, fire and police call boxes, and alarm signal wiring. The figure illustrates the typical allocation of space on utility poles in the United States. The allocation is similar in Canada except that cable television and telephone are sometimes lashed to the same strand. Starting at the top and working down, facilities on this pole are:

- Static wire: a grounded wire at the very top of the pole intended to protect lower conductors from lightning.
- Transmission: three uninsulated conductors which carry 3-phase high voltage (typically 69 to 200 kilovolts) circuits among substations.
- - MGN (multi-grounded neutral): a single uninsulated grounded conductor. The rrents in the three phases of the transmission line are never quite equal; the MGN carries the residual unbalanced current. At many poles, the MGN is physically grounded to a ground-rod at the base of the pole.

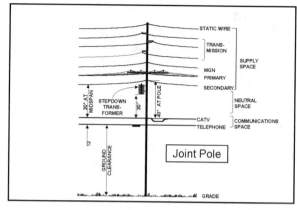

Joint Pole

- Primary: one to four uninsulated conductors, frequently supported on a crossarm, which carry power from substations to pole-mounted stepdown transformers. Primary circuits may be single-phase or three-phase, and operate at 4 to 15 kilovolts.
- - Secondary: one or two insulated conductors, accompanied by an uninsulated grounded neutral conductor. The secondary circuit (so named because it is fed from the secondary winding of the stepdown transformer) provides the standard 3-wire 115/230-volt electric service for residential and small commercial customers. Secondary conductors are usually twisted together in a bundle called "triplex," although older secondary lines may consist of three separate conductors.
- Stepdown transformer: an oil-cooled transformer which converts the primary voltage to the secondary voltage. Most stepdown transformers are designed for single-phase operation; if a three-phase secondary circuit is required, three physical transformers are sometimes mounted on the same pole.
- Neutral Space: an unused space which separates electric power facilities from communications facilities. This space is specified by the National Electrical Safety Code for safety reasons.
- CATV: cable television facilities supported by steel strand. An expansion loop at each pole absorbs expansion and contraction caused by temperature variations.
- Telephone: copper telephone cables supported by steel strand. Each telephone cable contains several individual wire pairs; a large cable may contain as many as several hundred pair. These days you may find fiber optic cable also.

joint procurement consortium An organization formed by Ameritech, Bell-South, Pacific Bell and SBC Communications to help them buy things.

joint trench A company wants to lay cable under a busy street. It applies to the city for permission. The city wants to limit the number of times a trench will be dug under that street. So the city announces that this will be a "joint trench" and it tells you that you must contact all the other utilities in town (typically giving them 30 days to respond) and check out those who might wish to locate their cable in that trench. Once you've determined who wants to participate in the joint trench, the trench will probably change in depth and width. Something called the "Western Formula" will be applied. Each of the utilities will pay then less than what they would have, had they built it by themselves.

joint user A person, firm or corporation designated by the customer as a user of access facilities furnished to the customer by the company, and to whom a portion of the charges for such facilities are billed under a joint use arrangement.

joint user service An arrangement whereby a corporation, association, partnership or individual whose telecommunications needs do not warrant the provision of separate leased service, is permitted to use the service of another customer by mutual agreement. The primary objective of joint user service is to save money by buying circuits in bulk.

Jose A common Spanish name pronounced Ho-zeh. There is an old story of the Spanish fireman who had two sons. He named the first one Jose. And the second? Why Hose B, of course! See also Hosed.

joule The unit of work or energy. The energy expended when a current of one ampere flows through a resistance of one ohm for one second. Joule's Law says the heat produced in a circuit in joules is proportional to the resistance, to the square of the current and to the time.

journal printers These are special purpose printers which provide hard copy output for audit trail and demand printing functions associated with hotel/motel management features.

journaling The logging of changes to files, tables, queues, etc. to a local file in order to support fast restarts, failover, high-availability, and/or other purposes. See remote journaling.

journalism The last refuge of the vaguely talented.

joy clicker One who nervously fiddles with a mouse.

joystick A pointing device for a computer whose upright level is used to manipulate a pointer on a screen. Named after a similar shaped control in airplanes. Joysticks are often used in computer gaming.

JPEG Joint Photographic Experts Group. So called as it was developed jointly by the International Standards Organization (ISO) and the ITU-T, it formally is known as ISO 10918-1 Recommendation T.81. JPEG is a compression technique used primarily in the editing of still images, and in color fax, desktop publishing, graphic arts and medical imaging. JPEG is symmetrical in nature, requiring equal processing power, time and expense on both the transmitting side (compression) and the receiving side (decompression). Its complexity renders it ineffective for real-time video; imaging applications are not so delay-sensitive.

The JPEG compression standard works by converting a color image into rows of pixels (picture elements), which are dots of color image, each with a numerical value representing levels of brightness and color. The picture is then broken down into blocks, each 16 pixels x 16 pixels, and then reduced to 8 pixels by 8 pixels by subtracting every other pixel. The software uses a formula that computes an average value for each block, permitting it to be represented with less data. Further steps subtract even more information from the image. To retrieve the data and thus decompress the image, the process is reversed. A specialized chip decompresses the images hundreds of times faster than is possible on a standard desktop computer. JPEG is a lossy image-compression algorithm that reduces the size of bitmapped images by a factor of 20:1 to 30:1 which compromises the absolute quality of the image in terms of resolution and color fidelity; JPEG can be pushed to yield a 40:1 compression ratio, although the loss in quality is noticeable at this level. JPEG compression works by filtering out an image's high-frequency information to reduce the volume of data and then compressing the resulting data with a lossless compression algorithm. Low-frequency information does more to define the characteristics of an image than does high-frequency information which serves to define sharp edges–losing some high-frequency information doesn't necessarily affect the image quality. In complex images, however, JPEG suffers from an effect known as "tiling," yielding a mosaic-like effect due to the block-oriented compression technique. When you see an image with the .JPG extension, that means it's JPEG image. See also JPEG 2000, JPEG ++, Motion JPEG, and MPEG.

JPEG 2000 A newer form of JPEG encoding for photographs and illustrations. JPEG 2000 uses wavelet compression, which encodes and sends an image in a continuous stream, yielding a higher resolution (than normal JPEG) as the file opens. Users with web browsers equpped to handle the JPEG 2000 images can choose to download only what is appropriate for their screen size or Internet connection. Someone surfing the web on a wireless handheld device will get just the photo or illustration's core, while someone using a 21 inch monitor and a high-speed connectikon will get everything.

JPEG++ Storm Technology's proprietary extension of the JPEG algorithm. It lets users determine the degree of compression that the foreground and background of an image receive; for example, in a portrait, you could compress the face in the foreground only slightly, while you could compress it in the background to a much higher degree. See JPEG.

JPG See JPEG.

JPIX Japan Internet Exchange. See IX.

JSP Java Server Page. See JHTML.

JT-2 6.312 Mb/s data rate. Same as T-2. Signal compatible with ITU-T document G.704 signal specification.

JTAG Joint Test Action Group. A consortium of North American manufacturers of ASICs (Application-Specific Integrated Circuits) which was formed in 1988 to explore the idea of building test capabilities into the chips for automated testing on a software-controlled basis. Previously, all chips had to be tested on a "bed-of-nails" approach, which involved gaining access to a great many hardware test points on each chip, and the use of expensive in-circuit test equipment. This approach was cumbersome, time-consuming and expensive, particularly with the advent of ASICs, which are densely packed with circuits. In 1990, the IEEE refined the concept, which was standardized as IEEE 1149.1, which commonly is known as the Standard Test Access Port and Boundary Scan Architecture. Boundary scan allows observability and controllability of the test procedure through Test Access Ports (TAPs) in the form of the input and output boundary pins (pins at the edge of the ASIC) on a software-controlled basis.

JTAPI See Java Telephony API and the definitions below.

JTAPI Address Object Part of the JTAPI Core call model. The Address object represents a telephone number. It is an abstraction for the logical endpoint of a phone call. This is distinct from a physical endpoint. In fact, one address may correspond to several physical devices.

JTAPI Call Model The JTAPI Core call model is defined in the Core API package. A call model describes a set of software objects that correspond to physical and conceptual entities in the telephony world. These objects fit together in a specified way to represent a telephone call. The Core API objects are: Provider Object, Call Object, Connection Object, TerminalConnection Object, Terminal Object and Address Object. In the physical view, each Core object represents a tangible property or telephony equipment. From a logical view, the call model represents an abstraction of telephony software entities or the functional properties of the objects. In describing these objects, it is difficult to separate the objects' physical representation from their logical properties, therefore, the description of these objects changes perspective frequently.

JTAPI Call Object Part of the JTAPI Core call model. The Call object represents a telephone call, the information flowing between the service provider and the call participants. A telephone call comprises a Call object and zero or more connections. In a two-party call scenario, a telephone call has one Call object and two connections. A conference call is three or more connections associated with one Call object.

JTAPI Connection Object Part of the JTAPI Core call model. A Connection object models the communication link between a Call object and an Address object. Relationships between Call and Address like connected, disconnected, and alerting are modeled by the Connection object as states. The Connection object also serves as a container for zero or more TerminalConnection objects. Connection objects model the logical aspects of a call connection.

JTAPI Core Package All JTAPI implementations make use of the Core package. Many application developers will only need basic telephony, in which case they will only need to use the Core API package. The Core API package provides basic telephony: placing calls, answering calls, and dropping calls. It defines the basic call model that the extension packages follow in design.

JTAPI Provider Object Part of the JTAPI Core call model. The Provider object is an abstraction of telephony service provider software. The provider might manage a PBX connected to a server, a telephony/fax card in a desktop machine, or a computer networking technology, such as IP. A Provider hides the service-specific aspects of the telephony subsystem and enables Java applications and applets to interact with the telephony subsystem in a device-independent manner.

JTAPI Standard Extension Packages The JTAPI specification defines standard extension packages. The core telephony package, Call Control, Call Center, Private Data and Terminal Set Management extension packages are at version 1.0. The specifications for Media, and Capabilities extension packages are at version 0.3. Mobile, and Synchronous are still under consideration. There are currently eight standard extension packages: Call Control, Call Center, Private Data, Terminal Set Management, Capabilities, Media Services, Mobile Phones and Synchronous.

JTAPI Terminal Object Part of the JTAPI Core call model. The Terminal object represents a physical device like a telephone and its associated properties. Each Terminal object may have one or more Address Objects (telephone numbers) associated with it, as in the case of some office phones capable of managing multiple call appearances. Additionally, Terminal objects that have more than one phone line may share a telephone number and a single Address object with another Terminal in an adjacent office. However, each Terminal has a unique TerminalConnection associated with a Call even though it may share an Address.

JTAPI TerminalConnection Object Part of the JTAPI Core call model. TerminalConnection objects model the relationship between a call and physical entities, represented by the Terminal object. The TerminalConnection object signals a Terminal when there is an incoming call and monitors the Terminal's activity during the process of a call. This object also closely communicates with the Connection object to receive information on a Call's change in state and to send information on the Terminal's state change. It models the physical aspects of a call connection.

JTC Joint technical Committee.

JTWROS Joint Tenants With Right of Survivorship. When two or more people maintain a Joint Account with a brokerage firm or a bank, upon the death of one account holder, ownership of the account assets passes to the remaining account holder(s). This transfer of assets escapes probate, but estate taxes may be due, depending on the amount of assets transferred.

JUCA Joint Use Collate Access is a bullpen environment where multiple customers are located at a site. JUCA is sold by the rack, not by the square footage.

Judge Harold Greene Judge Greene presided over the 1982 AT&T Antitrust settlement, enforcing its provisions and making decisions about requests from the participants to modify or reinterpret the provisions of the settlement. As long as he doesn't allow AT&T to be completely free of regulation, Judge Greene will probably always be involved in figuring the future of the telecommunications industry.

juggling on-demand connection An Internet Access Term. The ability to have more suspended on-demand connections than there are ports on the Dialup Switch.

jughead Jonzy's Universal Gopher Hierarchy Excavation And Display. A database of Gopher links which accepts word searches and allows search results to be used on many remote Gophers. See Archie.

judgment I asked four members of my extended family how to spell judgment in America. They said it's spelled without the "e." Australia, Canada and Great Britain commonly spell it with the "e". Most American dictionaries and American newspapers spell it without the "e". And that's the way I'm trying to spell it in this dictionary.

juice a brick From Wired's Jargon Watch column. To recharge the big and heavy NiCad batteries used in portable video cameras. "You better start juicin' those bricks, we got a long shoot tomorrow."

jukebox A jukebox is a piece of hardware that holds storage media, such as optical disks or cartridge tapes. Jukeboxes are typically designed to hold as few as five and as many as 120 devices. Like old-fashioned record playing jukeboxes, media is moved by a robot-like device from the storage slot to the drive reading it. This lets the user share one drive among several cartridges or disk. Jukeboxes are typically used for secondary and archival storage. Access to information is not fast. See Jukebox Management.

jukebox management In a network, tasks like retrieval and writes to a jukebox come randomly from all the users. These tasks vary in urgency – retrievals are higher priority than writes, for example. Jukebox management software sorts out requests from the network by priority. Management also enhances the performance of a jukebox, by intelligently reordering requests. For example, if there are three requests for images on platter 1 and two from platter 2 and the another from platter 1, jukebox management means the requests from platter 1 will get handled together, then go to platter two. Sometimes it's called "elevator sorting" – responding to requests in logical order, not in the order in which they were made.

Julian Date Not as often supposed a date according to the Julian calendar introduced by Julius Caesar, but rather a date in the Julian Period chronological system, now used chiefly in astronomy and hence in computations of satellite orbits. The Julian Period was devised by Joseph Justus Scaliger (1540-1609), a French historian of Italian extraction, principally to permit comparison of the differing computations of time made by ancient civilizations. Scaliger named the system after his father, Julius Caesar Scaligeri, and chose its period as 28 X 19 X 15 = 7,980 years, where 28 is the number of years in the solar cycle of the Julian calendar, 19 is the number of phases in the lunar cycle in which the phases of the Moon recur on a particular day in the solar year, and 15 was the cycle of indication, which originally an ancient Roman governmental tax schedule. He chose a starting point of 4713 BC, which was the nearest past year in which the three cycles began together. In astronomical computations, such as locating the Greenwich meridian with respect to a satellite orbit, Julian days are referenced to 12.00 noon (Universal Time, equivalent to GMT) on "0 January" (actually 31 December) of the year 2000, which is Julian day 2,451,545.0. Zero days of the month are used to simplify adding. The decimal place is carried to permit writing 0.5 to indicate the start of a day.

Julie Julie is the computer-generated "voice of Amtrak" who helps callers navigate the railroad's electronic answering system, also called automated voice response system. Julie's real voice belongs to Julie Stinneford, a professional voice talent.

jumbo frame In networking language, a frame is a packet of information. When information is sent across telecom lines A bigger A Gigabit Ethernet (GE) term. Gigabit Ethernet standards have adjusted the standard Ethernet frame size. The minimum size has been increased from 64 bytes to 512 bytes. The maximum frame size has been increased from 1,518 bytes to a "jumbo frame" of 9,000 bytes. This larger frame size reduces the number of frames that must be processed by a Gigabit Ethernet switch in the process of switching a large data set. As each frame must be processed by the switch, the fewer frames involved, the less the processing demands on the switch, and the less the delay in doing so. Jumbo frames, therefore, increase the throughput of the switch. However, jumbo frames require that multiple standard Ethernet frames be consolidated through a relatively minor process of protocol conversion. See also Gigabit Ethernet.

jumbo group A 3,600 channel band of frequencies formed from the inputs of six master groups. See Mastergroup and Supergroup.

jump hunting See Nonconsecutive Hunting.

jump scrolling Characteristic of a terminal with vertical motions of a whole line of characters at a time in discrete steps of one line, much as a teleprinter terminal might do. Contrast with "smooth scrolling" as done by graphics terminals.

jumper 1. A wire used to connect equipment and cable on a distributing frame.

2. Single twisted pairs used for cross connecting between 66, 110, or Krone blocks.

3. A patch cable or wire used to establish a circuit, often temporarily, for testing or diagnostics.

4. Jumpers are pairs or sets of small prongs on adapters and motherboards. Jumpers allow the user to instruct the computer to select one of its available operation options. When two pins are covered with a plug, an electrical circuit is completed, When the jumper is uncovered the connection is not made. The computer interprets these electrical connections as configuration information. 4. When errors are found on printed circuit boards, a Jumper cable is sometimes soldered in to correct the problem.

jumper cable a short length of conductor or cable used to make a connection between terminals or around a break in a circuit, or around an instrument. is often a temporary connection.

jumper wire A short length of wire used to route a circuit by linking two cross connect points. See also Jumper.

jumperless No jumpers on the hardware. Settings are accomplished with software – but not by setting jumpers. See Jumper.

junction box A metal or plastic box used as an access for cable or wire (coax, fiber, UTP, STP). When companies build a network in a building, building management usually require the J box to be located close to the building's entry point.

junctor A connection or circuit between inlets and outlets of the same or different switching networks.

June In the 1500s in England, most people got married in June because they took their annual bath in May. By June, when then weather improved, they were starting to smell, so brides carried a bouquet of flowers to hide their body odor. Hence the custom today of the bride carrying a bouquet of flowers.

JUNET Japan UNIX Network.

junk bands Refers to the 2.4GHz and 5.8 GHz bands, i.e., the industrial, scientific and medical (ISM) bands. They are sometimes called "junk bands" because of their being open to any type of compliant wireless device, including cordless phones, car alarm remotes, microwave ovens, remote-control toys and WiFi. In some environments the airwaves in these bands are filled with emissions from all these kinds of devices, and there's lots of interference.

junk bonds Junk bonds, or high-yield bonds, are rated below investment grade because they are allegedly riskier. As a result, they carry a higher yield than investment grade bonds. I included this definition because in April, 1998, Level 3 Communications, Inc., one of the newest telecommunications transmissions companies sold $2 billion in junk bonds, equaling the largest junk bond deal up to that point in the 1990s. According to the rating agencies, junk bonds are rated BB+ and below. Investment grade bonds are rated BBB- or above. Bonds from a record 60 companies were downgraded from investment grade to junk in the year 2001.

junk dealer A pejorative term for a broker of used telecom equipment in what's known as the "secondary market." See also Broker.

Junos Code The operating system which Juniper Networks has created for its router. According to Juniper, Junos allows high-speed forwarding across ever-more complex sets of paths. Junos will compete with IOS, the Cisco Internetwork Operating System (IOS) that runs the vast majority of routers now deployed in the core of the Internet.

jurisdiction A geographic area associated with and bounded by state and/or LATA boundaries. The following are the four categories: Interstate/InterLATA, Interstate/IntraLATA, Intrastate/InterLATA, Intrastate/IntraLATA. and an area in which a common carrier is authorized to provide service.

Jurisdiction Information Parameter See JIP.

JV Joint Venture.

JVM See Java Virtual Machine.

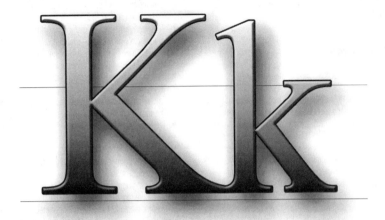

K 1. See K-Style Handset below.

2. In metric terms it means one thousand (1,000) times, taken from the Greek word kilo. It's often appended to a measurement such as kiloHertz or kHz, which means 1,000 Hertz. In data communications, a kilobit means a thousand bits per second (kbps). In computer memory terms, it means 1,024, which is the figure for two raised to the 10th power, i. e. 2 x 2 x 2 x 2 x 2 x 2 x 2 x 2 x 2 x 2. See also Byte.

3. Scientifically, K (Capital Letter) should be used ONLY for Kelvin (Absolute Temperature). k (lower case) means a thousand. See Kelvin.

K band That portion of the electromagnetic spectrum in the high microwave/millimeter range – from 10.9 GHz to 36 GHz. See also Ka BAND and Ku Band.

K plans Also called keysheets. When designing a phone system, you need to assign features and line assignments to each extension. A keysheet or a K Plan is an organized way of figuring and keeping track of those features and assignments for system design and programming. Typically it's one page per extension. These days, keysheets are often done on computer.

K plant Old Bell System lingo for equipment used in key systems.

K style handset A K-style handset is the newer, square telephone handset. The older, round handset is called the G-style handset.

K10 Old Bell-Speak for 10-button (9-lines) key telephone.

K20 Old Bell-Speak for 20-button (19 lines) key telephone.

K30 Old Bell-Speak for 30-button (29 lines) key telephone.

K56flex K56flex was one of two pre-standard modem solutions for running data over dial-up phone lines at up to 56 Kbps one way and up to 33.6 Kbps the other way. The standard was developed for use on the Internet, with the 56 Kbps channel running downstream (i.e., flowing to you) and the 33.6 Kbps channel running upstream (i.e., from you). The logic is that at 56 Kbps, Web pages fill a lot faster on your screen. 56Kflex was developed by Rockwell Semiconductor and Lucent Technologies, two of the world's leading manufacturers of modem chips. More than 700 modem makers, PC manufacturers, including Compaq and Toshiba, and ISPs like Microsoft Network, supported this de facto standard. The competing 56 Kbps "standard" was called x2, and was developed by US Robotics. x2 was not compatible with 56Kbps. In other words, a 56flex modem cannot talk at 56Kbps to a x2 modem. On February 6, 1998, the ITU-T ratified a new standard called V.90 for Pulse Code Modulation (PCM) modems running at speeds to 56 Kbps. The idea of this V.90 was to create an international standard so that all 56 Kbps modems could talk to each other. Previous to the finalization of the standard, V.90 had been referred to informally and variously as V.PCM and V.fast. It allows speeds of up to 56 Kbps in one direction only, from the central site equipment to the end user. The "back channel" upstream from end user to the central site remains limited to 33.6 Kbps (V.34 speeds). Actually, in North American use, the mo-

dem is actually limited to only 53.3 Kbps. The reason for this? The FCC determined that running the modem at 56 Kbps, which it's perfectly capable of, would entail pumping out too much power, which might interfere with adjacent telephone circuit pairs in the same bundle. Some 56Kflex modems can be software upgraded to V.90. Others can't. See 56 Kbps Modem (for a longer technical explanation), V.91, V.92, and V.PCM.

Ka Band That portion of the electromagnetic spectrum in the high microwave / millimeter range – approximately 33 GHz to 36 GHz. Ka Band is used primarily in satellites operating at 30 GHz uplink and 20 GHz downlink and is intended for applications such as mobile voice. In short, Ka Band are a A band of frequencies in the 18 to 31 GHz range that are available for global satellite use. See Ku Band.

KA9Q A popular implementation of TCP/IP (Transmission Control Protocol/Internet Protocol) and associated protocols for amateur packet radio systems.

Ka-Band See Ka Band.

Kahn, Robert Dr. Dr Robert Kahn and Dr. Vinton Cerf are considered to the fathers of the Internet. While working at the Advanced Research Projects Agency (ARPA), they published in 1967 a plan for a key forerunner of the Internet, something called ARPANET. The intent was to create a tool to link geographically dispersed research-center computers. The design incorporated the packet switching concept. Through packet switching, a message is divided into multiple packets of data that are transmitted individually and can follow different routes to their destination, where they are reassembled in their original order. ARPANET carried its first message in 1969. To transcend the network-specific boundaries of ARPANET, Kahn championed the idea of open-architecture networking, which would allow for networks of different designs to connect by means of a communications protocol. Kahn teamed up with Cerf to co-invent TCP/IP, which stands for Transmission Control Protocol/Internet Protocol. See Internet and TCP/IP.

kangaroo When England first took over Australia, colonists were astounded by the variety of previously unknown animals. One in particular puzzled them. When they asked an aborigine what it was, he replied "I don't know," which in his particular dialect is, "kan ga roo."

Kaput In German, kaput means broken. In bad French, it's capout.

Kate Hamill The world's best marathon runner and a dear friend of my daughter, Claire. I know why she's the world's best marathon runner. She told me so after she ran the Boston Marathon and I helped bring her tired and shattered body back to her home in New York City. I didn't ask for proof on "the world's best claim." The fact that anyone could complete 26 grueling miles running up and down dale without stopping is good enough for my awestruck brain.

Kazaa Kazaa software, one of the most popular downloads on the Internet, is used for sharing music. Because Kazaa's file sharing relies on routing requests through individual users' computers instead of central servers (as in the old Napster), the record industry has been unable to shut down the service in court – not for lack of trying. See also Skype, which rhymes with hype. www.Kazaa.com.

Kb Kb is Kilobit, which is one thousand bits. KB is kilobyte, which is 1,024 bytes. Kilo is one thousand. See also Kilobyte. See Byte and Bps for much more detailed explanations.

KBHCA Thousand Busy Hour Call Attempts. See BHCA.

KBIC Keyed Biconic.

KBps KBps is kilobytes per second. Kbps is kilobits per second. In short, one thousand bits or bytes per second. For a much better explanation, see Bps.

KCHAR Kilocharacter.

Kearney System An AT&T numbering scheme for telecom parts. See KS Number.

Keep Alive Bits See Fill Bits.

keep alive signal A generic term for a signal transmitted when a DTE detects a loss of input from the customer's equipment for a specified period of time (sometimes called a blue signal or AIS). T-carrier systems, for instance, transmit a keep-alive signal during periods of circuit idleness which exceed 150 msec. The purpose of the keep-alive signal is to maintain the circuit during periods of idleness; otherwise, the circuit would time out and the logical connection between devices would be terminated.

keg of nails Putting thousands of metal shards into a 16,000 mile per hour counter-orbit against the U.S.'s low-orbit satellites. Such keg of nails would be designed to destroy low orbiting satellites.

Keiretsu A Japanese term describing a group of affiliated corporations with broad power and reach. In Japan, six giant keiretsu – Mitsubishi, Mitsui, Dai Ichi, Kangyo, Sumitomo, Sanwa, and Fuyo – dominate much of the country's economic activity.

keitai A mobile phone, a cell phone, a portable phone in Japanese. NTT DoCoMo's phones are called Keitais.

kelvin A unit of absolute temperature, equal to 1/273.16 of the absolute temperature of the triple point of water, equivalent to one degree Celsius. A temperature in kelvin may be converted to Celsius by subtracting 273.16. Named after the British physicist William Thomson, first Baron Kelvin.

Kerberos A security system for client/server computing. Kerberos is a scheme, developed at MIT in the 1980s, to enable secure multiple system access to a client/server computing environment. Named for the three-headed dog, Cerberus, who guarded the gates of Hades in Greek mythology. Kerberos is a UNIX-based distributed database used for user authentication.

Kermit An asynchronous file transfer protocol originally developed at Columbia University in New York City in 1981. The protocol has become popular because of its flexibility. One of the clearest advantages of Kermit is its ability to be tailored for virtually any equipment. File transfer protocols like Kermit break a file into equal parts called blocks or packets, with the data (also known as text or payload) preceded and succeeded by specific control data. The receiving computer checks each arriving packet and sends back either an acknowledgement (ACK) or a negative acknowledgement (NAK) to the sending computer, explicitly indicating the arrival of the packet and its arrival condition. Because modems use phone lines to transfer data, noise or interference on the line will often mess up the block of data. When a block is damaged in transit, an error occurs. The purpose of a protocol is to set up a mathematical way of measuring if the block came through accurately. And if it didn't, ask the distant computer to re-transmit the block until it gets it right. Kermit believers say that Kermit is robust, platform-independent, medium-independent, extensible, and highly configurable. XMODEM, YMODEM, and ZMODEM are the file transfer protocols most commonly compared with Kermit, and which are found in numerous shareware and commercial communication software packages. XMODEM and YMODEM are stop-and-wait protocols; XMODEM uses short blocks (128 data bytes), YMODEM uses longer ones (1024 data bytes). ZMODEM is a streaming protocol. In the results of tests of file transfers shown to me, ZMODEM and Kermit are closest in terms of speed and efficiency of transfer, with Kermit edging out ZMODEM for first place. The Kermit Project can be found at www.columbia.edu/kermit/. The Kermit newsgroups are comp.protocols.kermit.misc and comp.protocols.kermit.announce.

Here are words excerpted from The Kermit Project at www.columbia.edu/kermit/: Kermit - What is it? Kermit is a file transfer protocol first developed at Columbia University in New York City in 1981 for the specific purpose of transferring text and binary files without errors between diverse types of computers over potentially hostile communication links, and it is a suite of communications software programs from the Kermit Project at Columbia University. Over the years, the Kermit Project has grown into a worldwide cooperative nonprofit software development effort, headquartered at and coordinated from Columbia University.

The Kermit protocol was named after Kermit the Frog, star of the television series, The Muppet Show, used by permission of Henson Associates, Inc. Since its inception in 1981, the Kermit protocol has developed into a sophisticated and powerful tool for file transfer and management, incorporating, among other things:

File group transmission; File attribute transmission (size, date, permissions, etc); File name, record-format, and character-set conversion; File collision options, including an "update" feature; File transfer recovery Auto upload and download; Client/Server operations; Recursive directory-tree transfer, even between unlike platforms; Uniform services on serial and network connections; and An Internet Kermit Service Daemon. Kermit software has been written for hundreds of different computers and operating systems, some of it by volunteer programmers all over the world, some of it by the Kermit Project staff. The major features of the most popular Kermit programs are:

- Connection establishment and maintenance for a wide variety of connection methods (dialup, TCP/IP, X.25, LAN, etc).
- Terminal emulation.
- Error-free file transfer.
- Numeric and alphanumeric paging.
- Character-set translation during both terminal emulation and file transfer – a unique feature of Kermit software.
- Script programming to automate complicated or repetitive tasks.

The one feature that distinguishes Kermit protocol from most others is its wide range of settings to allow adaptation to any kind of connection between any two kinds of computers.

kernel The level of an operating system or networking system that contains the system-level commands or all of the functions hidden from the user. In Unix, the kernel is a program that contains the device drivers, the memory management routines, the scheduler, and system calls. This program is always running while the system is operating. See Kernel-Based Window System.

kernel based window system Kernel-based window systems are those in which the software application executes and displays in the same physical machine. Examples include personal computers and Macintoshes. The advantage is speed. The disadvantage is that applications are closely tied to the system environment and are therefore not portable. Kernel-based window systems also do not allow users/developers to use the network as a means of sharing computer resources.

kernel driver A Windows NT term. A driver that accesses hardware.

Kerr Effect When polarized light is shone onto a magnetized surface, the light is reflected back at an angle and in a different direction, depending on the polarity of the magnetism. This quirk of nature is called the Kerr Effect and it is the basis of magneto-optical (erasable) discs. The Kerr Effect also affects optical fiber transmission systems. This phenomenon is manifested where the index of refraction of a fiber optic cable varies with the intensity of the transmitted light signal. This nonlinear phenomenon occurs in systems with milliwatt transmitters and very long span lengths, resulting in self phase modulation of the signal, which is not a good thing.

Kevlar An aramid fiber that belongs to the nylon family, Kevlar often is used in strength members surrounding optical fiber cables. Aramid fibers most commonly are used in riser cables (i.e., cables which rise vertically through a building), increasing the tensile strength of the cable so that it doesn't break under its own weight. Kevlar is a trademark of the Dupont Company. Kevlar is also used in bulletproof vests worn by police. See also Aramid.

key 1. One or more characters or perhaps a field within a data record used to identify the data and perhaps control its use.

2. The physical button on a telephone set. What normal people call a "Switch", telephone people call a "Key."

3. The physical button on a key telephone set. In a KTS (Key Telephone System environment, the user selects an outside line or intercom line by depressing the appropriate key. The term "Key" originated in the manual switchboard (cordboard) systems, with the operator flipping a key (switch) to set up a talk path. What normal people call a "Switch," telephone people call a "Key."

4. In encryption, a key is a data string which, when combined with the source data according to an algorithm, produces output that is unreadable until decrypted. A key can also be used to decrypt a data string. See Key Holder.

5. The device which unlocks your front door or perhaps your terminal or computer, assuming that you haven't lost it. I have been told that all of my lost keys will be waiting for me on my desk in my next life.

Key Escrow A Key Escrow is a system that keeps copies of encryption keys so that those keys can be accessed by an authorized agent to decrypt any messages created by those escrowed keys.

key exchange A procedure by which the value of a key is shared between two or more parties. See also Encryption.

key generation The process of creating a key.

key holder In encryption, a key is a data string which, when combined with the source data according to an algorithm, produces output that is unreadable until decrypted. A key can also be used to decrypt a data string. In the mid-1990s in the United States, there was great controversy about software that could encrypt electronic communications. Under U.S. laws, such software could not be exported. There was a movement to change the law and create organizations called "key holders." These would be organizations that would be given copies of an individual's decryption key or codebreaker. Such organizations, the theory went, would, under court order, give an individual's decryption key to a law enforcement agency. This might happen with or without the individual's consent.

key illumination A lamp under a button (called a "key" in telephony) which flashes at different rates to signal an incoming call, a steady busy and "wink" (fast) hold.

key management Digital cryptography systems are based on the use of keys. Before secure transmissions can take place, the appropriate keys must be obtained for use by the sender and the receiver. The total operations and services related to the use and distribution of cryptographic keys is known as key management.

key map A MIDI patch-map entry that translates key values for certain MIDI messages, for example, the keys used to play the appropriate percussion instrument or a melodic instrument in the appropriate octave.

key pad The touchtone dial pad on a pushbutton phone. Contrary to popular belief, touchtone is not a registered trademark of AT&T. See Touchtone.

key pad state An AT&T enhanced fax term. The KEYPAD state can be either NULL or NON-NULL. NULL: The keypad is in use by a feature and is not available for use on a call. NON-NULL: The keypad is available for use in originating a call or for sending DTMF tones on an existing call.

key pair Set of mathematically related keys-a public key and a private key-that are used for asymmetric cryptography and are generated in a way that makes it computationally infeasible to derive the private key from knowledge of the public key.

key pulse In multi-frequency (MF) tone signaling, a signal used to prepare the distant equipment to receive digits.

key pulse dialing The transmitting of telephone address signals in which digits are transmitted using pushbuttons. Each button generates a unique set of tones.

key pulsing A pulsing system in which digits are transmitted using pushbuttons. Each button corresponds to a digit and generates a unique set of tones.

key punch card A method of storing programs and other data on heavy, stiff, stock paper, measuring 3-1/4" high by 7-3/8" wide. Each card represents one line of program code, and has a pattern of 80 columns and 12 rows, where each column is typically used to represent a single piece of data such as a character. The top row is called the "12" or "Y" row; the second row from the top is called the "11" or "X" row; and the remaining rows are called the "0" to "9" rows (indicated by the numbers printed on the cards). Each position is either "punched" out, or left solid. A typewriter like machine, called a key punch machine, is used to keyboard input the punched holes into cards by means of dies. The punched cards data are fed into a computer via a card reader, which scans each card and can read each punch/hole. Punch cards are rarely used today. Where keypunchers punched data into cards, bar codes and point-of-sale machines now enter data directly into computer files. Where sorters sequenced cards, sort programs now do it electronically. Where collators merged, matched, and selected cards, programs now do it with electricity. Where tabulators accumulated and printed information from cards, programs now send electronic information to laser printers and graphical color output. There is nothing new in the methods, procedures, and processesof personnel and material accounting-only the technology used to perform the task.

key recovery 1. Process for learning the value of a cryptographic key that previously was used to perform some cryptographic operation. 2. Techniques that provide an intentional, alternate (that is, secondary) means to access the key used for data confidentiality service in an encrypted association.

key sequence number An identifier associated with a key that allows one value of a key to be distinguished from an older or newer value of the key.

key service panel An old 1A2 key telephone term. Wired or unwired connector panel for modular expansion of key system service by allowing the installation of additional Line Cards and/or other KTUs. Typically, Key Service Panels are available in different jack configurations to accommodate 18-, 20-, 36- and 40-pin KTUs. Commonly abbreviated as KSP. Most KSPs are supplied as rack-mount equipment.

Key Service Unit KSU. This is a small metal cabinet which contains all the electronics of a business key telephone system. The KSU fits between the lines coming in from the central office and the lines going to the individual phones. Be careful where you place the Key Service Unit. That place should be well-ventilated as the KSU gets hot. It should be near a power outlet (it needs one). Unless it's a very small phone system, it should be plugged into a power outlet dedicated to it (other devices, such as typewriters, computers, TV sets, and vacuum cleaners, plugged into the same electrical circuit could affect it). And the power outlet should be above the reach of the mops and brooms of the local cleaning people. Otherwise the plug will get knocked out and the phone system won't work the next day. See Key Telephone System.

key set Also called Key Telephone Set. A telephone set having several buttons which can be used for call holding, line pickup, auto-dialing, intercom and other features. Ericsson calls a keyset a touchtone telephone.

key station line The circuit which extends from the key set to the key system common equipment. May be two or four wire.

key strip The row(s) of buttons on key telephone sets used for line or extension access and for features like call hold and intercom.

key system The equipment utilized to provide features associated with key sets, including keysets, multipair cable (rapidly disappearing from the scene), key service unit, distribution blocks, and miscellaneous devices.

key system power supply The local source for all DC voltages required for talking and lamp signaling within the Key Telephone System. The power supply may or may not also provide an AC voltage output for ringing. If it does, a separate Ringing Generator will normally not be required.

Key System Unit See KSU.

Key Telephone System KTS. A system in which the telephones have multiple buttons permitting (requiring) the user to directly select central office phone lines and intercom lines. According to strict, traditional definition, a KTS is not a switch. A PBX switch allows the sharing of pooled trunks (outside lines), to which the user typically gains access by dialing "9," with software in the switch managing contention for the pooled lines, selecting an available line, and setting up the connection. A KTS system, on the other hand, requires that the user make the selection of an available outside line through the use of "grayware" (brain power).

KTS systems generally and traditionally find most appropriate application in relatively small business environments, typically in the range of 50 telephones and requiring relatively unsophisticated functionality and feature content. PBX systems generally are applied to larger and more demanding situations. Contemporary Electronic Key Telephone Systems (EKTSs), however, often cross the line into the PBX world, providing switching capabilities, as well as impressive functionality and feature content.

Key Telephone Unit A modular 1A2 Key Telephone System building block that plugs into a KSU or KSP. Commonly abbreviated as KTU. Typical KTU examples include 4000 Series Line Cards, 4448 Delayed Ring Transfer Card, 6606 Interrupter, etc.

key to disk A method of entering data whereby it's sent directly from the keyboard to a disk, usually a hard disk.

keyboard 1. A series of switches, arranged somewhat like a standard "QWERTY" typewriter that allows you to send information to a computer. There is no such animal as a "standard" computer keyboard. For speed typists, this is a terrible pity. If you are buying multiple computers for your office, check out your peoples' preference for keyboards. Getting the right one can make a big difference. You can often buy PCs without keyboards and buy third party differently designed keyboards.

2. According to the Vermonter's Guide to Computer Lingo, a keyboard is where you hang your keys. Joke.

keyboard buffer A temporary storage area in memory that keeps track of keys that you typed, even if the computer did not immediately respond to the keys when you typed them.

keyboard call setup Allows you to set up a data call using the buttons of a telephone, or it allows you to set up a voice call using the keyboard of your PC. Which definition you choose – setting up a voice or a data call – depends on which manufacturer you're working with.

keyboard plaque The disgusting buildup of dirt and crud found on computer keyboards. "Are there any other terminals I can use? This one has a bad case of keyboard plaque."

keyboarding A really stupid word for typing.

keyed A term used in data communications whereby the RJ-45 male plug has a small, square bump on its end and the female RJ-45 plug is shaped to accommodate the plug. A keyed RJ-45 plug will not fit into a female, non-keyed (i.e. normal) RJ-45. The purpose of keying a plug is to differentiate it from a "normal" non-keyed plug. Keyed RJ-45 plugs are typically used for data communications. See also RJ-11, RJ-22 and RJ-45.

keying Modulation of a carrier signal, usually by frequency or phase, to encode binary (digital) information, (as in FSK or Frequency Shift Keying).

keylog See Magic Lantern.

keylogger See keylogging software.

keylogging software Keylogging software, also caller a keylogger, is a piece of software that sits on your PC and records every keystroke you type – including passwords, email messages and which Internet sites you visit. Then it secretly sends out those records via the Internet to whoever planted the keylogger. Such software could be acceptable. Key logging software could be planted by a parent to see if a child visits pornographic sites. It could also be installed by someone trying to rip off your social security number and your various banking passwords. That someone might be trying to steal from you. Keylogging software could be installed when you surf a corrupt Web site and/or download applications such as file-sharing programs. Such keylogging software could easily be programmed to send its stolen information to remote databases offshore, where thieves sift through it for passwords, user names and account numbers. Essentially keylogging is automated identify theft via telecommunications. To protect yourself, don't download any stuff from web sites you're not familiar with, and:

- Keep your Windows up to date with Microsoft's latest patches.
- Run an antivirus program regularly. That doesn't mean run it constantly. I find running it constantly slows my PC down too much.
- Install a software or hardware firewall. I prefer a hardware firewall. I use Watch-guard.
- Tweak your browser. www.cert.org/tech_tips/securing_browser.
  See also adware.

keypad See Key Pad.

keypad lock This feature lets a user disable the keys on a mobile phone's keypad so the user won't accidentally dial an expensive (or cheap) number.

keypunch An old method of data entry in which a keyboard was used to activate a keypunch machine, which punched holes or notches in cards.The cards were fed into a computer to input data; A keyboard-actuated punch that punches holes in a data medium. See also Key Punch.

keystroke monitoring An application software program that can be installed on a workstation to monitor your keystrokes. Such programs are used to track your activities during a data session, usually a session on the Internet and Web (World Wide Web, or WWW). Some programs take snapshots of your screen at definable intervals so that those who are monitoring your activities can see what you see, so watch out. Debates rage over keystroke monitoring. Proponents mainly are companies that maintain that they right to monitor your use of company resources. Opponents mainly are individuals who maintain that they have an inherent right to privacy. PC programs are commercially available so that husbands can monitor their wives activities and (more commonly) so that wives can monitor their husbands. See also Electronic Communication Privacy Act and Network Accounting.

keystone A monitor distortion where one end of the screen – either side to side, or top to bottom – is larger than the other end.

keystone module A universal modular network-wiring component that can fit many different vendors 'faceplates, modular furniture systems adapters or surface mount assemblies. For example, a keystone module from Siemon would be interchangeable with one made by Avaya.

KF-E FK-E is a suffix to an FCC registration number, indicating the function of a system in accordance with its FCC registration. A KF-E is a Fully Protected Key System, with "KF" denoting "Key Function," and "E" denoting that the system will accept both rotary and tone signaling. See also Key Telephone System, MF-E, PF-E and Registration Number.

KHz KiloHertz. One thousand hertz. Typically written as kHz. See Hertz and K.

kiddie cams On-line cameras attached to computers attached to the Internet that allow parents to monitor their children from their desks at their offices. Kiddie cams (cameras) are often installed in daycare centers and grade schools. Also called Cradle Cams. See also Web Cam.

Kilby, Jack. S. A co-inventor of the computer chip. Working at Texas Instruments in 1958, Mr. Kilby thought up ways to put all the parts of an electric circuit – resistors, transistors, diodes and capacitors – onto a single sliver of material. It was a major step in miniaturization, and badly needed. Until then, discrete transistors and other parts required for electronic products meant gadgets had to be pretty big. Without knowing about Mr. Kilby's invention, Robert Noyce and other engineers at Fairchild Semiconductor in California came up with the same idea a few months later. The Noyce concept could be produced more efficiently than Mr. Kilby's, so the two companies eventually shared patent rights and the men became known as co-inventors. Kilby receive the Nobel Prize in physics in 2000. Kilby dies in 2005. Noyce died in 1990.

kill file A file that lets you filter USENET postings to your account to some extent. It excludes messages on certain topics or from certain people.

kill message A recorded message played at the beginning of a call to a 900 (or other pay-per-call) number that warns the caller of the charges and gives him the option to hang up before it starts.

killer app Killer application. The high-tech industry's lifelong dream. That dream is to discover a new application that is so useful and so persuasive that millions of customers will rush in, and throw money at you to buy your killer app. The term derives from the PC industry where a killer app was so powerful that it alone justifies the purchase of a computer. The first PBX killer app was probably being able to dial out without an operator (i.e. dial 9). The second was probably least cost routing. The third was probably call accounting. The next may be some form of hookup to PCs, with desktop and LAN connectivity. In the PC industry, killer apps have been spreadsheets, word processing and databases. Finding that one killer app that will make them wealthy beyond their wildest dreams is what drives many software programmers and entrepreneurs. See Windows Telephony.

kilo One thousand (1,000), as in Kilobits per second (kbps), or KiloHertz (kHz). One thousand twenty-four (1,024), as in KiloByte. See KiloByte.

Kilobyte KB. From the Greek "chilioi," meaning "thousand." A unit of measurement for physical data storage on some form of storage device – hard disk, optical disk, RAM memory, etc. The actual definition can be confusing, since there are two measurements. In the metric system, a kilobyte is 1,000 bytes, or 10 to the third power. In the computer world, things tend to be measured in binary terms – 1s and 0s. In binary terms, a kilobyte is 2 to the 10th power, or 1,024. Here is the progression:
Following is a summary of sizes, in binary terms: KB = Kilobyte (2 to the 10th power)

MB = Megabyte (2 to the 20th power)
GB = Gigabyte (2 to the 30th power)
TB = Terabyte (2 to the 40th power)
PB = Petabyte (2 to the 50th power)
EB = Exabyte (2 to the 60th power)
ZB = Zettabyte (2 to the 70th power)
YB = Yottabyte (2 to the 80th power)
One googolbyte equals 2 to the 100th power.

kilocharacter One thousand characters. Used as a measure of billing for data communications by some overseas phone companies. See also Kilosegment.

kilohertz 1000 hertz, or 1000 cycles per second. Written as kHz. See Hertz.

kilosegment 64,000 characters. Used as a measure of billing for data communications by some overseas phone companies. See also Kilocharacter.

King Kong King Kong was Hitler's favorite movie.

Kingsbury Commitment December 13, 1913: A letter from Nathan C. Kingsbury, VP AT&T, to the Attorney General of the United States committed AT&T to dispose of its stock in Western Union Telegraph Company. It also promised to provide long distance connection of Bell System lines to independent phone companies (where there was no local competition) and further agreed not to purchase any more independent telephone companies, except as approved by the Interstate Commerce Commission which regulated the phone industry at that time. See also Divestiture.

Kingsley See Todd.

kiosk In Britain, a kiosk is a telephone booth. In America, a kiosk is a small structure with one or more sides open, used as a newsstand, refreshment stand, or a place for surfing the web, or a place for receiving and sending faxes, etc. As an Internet or fax telecommunications tool, it will typically have a phone line attached and it will typically be located somewhere there's lots of foot traffic, e.g. a suburban mall.

KISS Keep It Simple Stupid. A philosophy of management that says simple is better. In the 1960s, during the heat of the space race, NASA decided it needed a ballpoint pen capable of writing in the zero-gravity confines of a space capsule, where gravity does not help the ink on its way to the tip. After considerable R&D, the Astronaut Pen was developed at a cost of about $1 million. The pen, which uses a pressurized container of ink, worked perfectly and has enjoyed modest success as a novelty item since. Faced with the same problem, the Soviets opted for pencils.

kissup A person or organization who tells their boss what he/she wants to hear.

kitting "We do equipment kitting," say some vendors. Kitting comes from the word kit. It means that a vendor will assemble all the equipment necessary to solve your problem (i.e. assemble it all into a "kit") and probably install it for you as well. Here's a description of the benefits from dealing with a telecom secondary equipment assembler and installer, Somera. "Why deal with multiple vendors when you can save time and work with one? Somera can free your resources and reduce your costs by taking over your inventory aggregation (i.e. kitting) and shipping requirements. Let Somera handle your kitting and staging for complex, multi-vendor equipment. We'll locate and acquire all the pieces from our own warehouse or from suppliers worldwide. Our staging process insures all needed components are present and installed in racks. We'll also coordinate shipping to insure that all components arrive when needed for smooth installation. The benefits to you:

- Single point of contact for multi-vendor purchases
- One purchase order
- Consolidated lead times
- Maximized capital expenditure efficiencies
- Fully-managed inventory receiving, storage and warehousing
- Equipment shipped to single or multiple locations
- Same day, next day, and pre-scheduled deliveries
- Internet-based accounting and tracking

Klingon The office that treats mental health patients in metropolitan Multnomah County, in Portland, Oregon, is looking for people fluent in Klingon, to serve as interpreters. According to the county Department of Human Services, which serves about 60,000 mental health patients, some of the people in their care will only speak in this language, created for the Star Trek series. County officials are therefore obligated to respond with Klingon-English interpreters, as they would with any other language, such as Spanish or Vietnamese. Klingon, an artificial language like Esperanto, was designed by a linguist to have a consistent grammar, syntax, and vocabulary.

kludge A hardware solution that has been improvised from various mismatched parts. A slang word meaning makeshift. A kludge can also be in software. It may not be elegant and is probably only a temporary fix. As in, "That patch to the software is a real kludge."

kluge Another way of spelling kludge. We think spelling it kludge is correct. See Kludge.

klutz Yiddish term for an incompetent. Not be confused with yutz which is Yiddish for dumb guy. See Yiddish.

Klystron An electron tube, used for converting a stream of electrons into ultra high-frequency waves which transmits as a pencil-like radio beam. A traveling-wave tube (TWT) is a specialized vacuum tube used in wireless communications, especially in satellite systems. The TWT can amplify or generate microwave signals. Two common types of TWT include the Klystron and the magnetron.

In the Klystron, a negatively charged cathode emits a beam of high-speed, high-energy electrons that travel through the cylindrical tube in straight lines to a positively charged anode. A coil is wound around the tube. When the coil is energized with a radio-frequency (RF) signal, the electrons in the beam alternately bunch up and spread out. In the magnetron, the electrons move in circles rather than in straight lines. The circular motion, produced by magnets at either end of the tube, allows the electrons to pick up energy over a greater distance.

Inside the TWT, the regions of high and low electron concentration move along or around the tube in waves. When the tube is properly operating, some of the energy from the electrons is imparted to the signal in the coil. The result is amplification of the signal.

A TWT can be made to function as an oscillator by coupling some of the output back into the input. This configuration is called a backward-wave oscillator, because the feedback is applied opposite to the direction of movement of the electrons inside the tube. Such an oscillator can generate up to approximately 0.1 watt of signal power in the microwave range.

A parametric amplifier is a TWT amplifier that operates from a high-frequency alternating current (AC) power source, rather than the usual direct current (DC) source. Some characteristic of the circuit, such as its impedance, is made to vary with time at the power-supply frequency. Parametric amplifiers are useful because they generate very little internal noise. This makes it possible to obtain excellent sensitivity in receiving systems, minimizing data-transfer errors.

KM Knowledge Management. See Knowledge Management.

knee mail I saw this first on the billboard of the Tabernacle Seventh-Day Adventist Church in Portland, Oregon. The entire slogan was "God answers knee mail." I bet he also gets a lot of spam.

KNET Kangaroo Network Hardware/software product (Spartacus/Fibronics) that lets IBM mainframes communicate over networks using the TCP/IP protocol.

knowbots Intelligent computer programs that automate the search and gathering of data from distributed databases. The creation of knowbots is part of a research project headed up by the Corporation for National Research Initiatives, Reston, VA. Two knowbot-based databases for the medical field are expected to be available in 1991. Knowbots could become more widespread for general use, according to networking experts.

knowledge base system In its most simple term, it means knowledge that is known by the system. Software in which application specific information is programmed into something called the "knowledge base" in the form of rules. The system uses artificial intelligence (AI) procedures to mimic human problem solving. It applies the rules stored in the knowledge base and the facts supplied to the system to solve a particular business problem.

knowledge management The big buzzword at the 1998 March Internet World in Los Angeles was "knowledge management." Defining precisely what it means is difficult. Some people define knowledge management as the ability to get the right information to the right people at the right time. Robert Buckman, one "father" of the term, says about knowledge management, "You have to think about using knowledge to accelerate speed to reduce costs in this environment where you have no pricing ability. You have to get better teamwork function, collaborate better, no matter what tools you use to get there." Yun Wang, wrote in InfoWorld Magazine, "In a way, this is what IT should have been about from the beginning…What has kept these networks from serving as true knowledge management systems in the past is the haphazard way in which they tend to grow. New technology routinely is added on to old without systemwide reengineering; new data sources are hooked up to the network without recategorization of the entire information base."

When planning your knowledge management system, says Wang, you should start by thinking about which information is most critical to move to which people, and begin planning accordingly. Thinking about the purpose of each kind of information can help you devise a scheme for information distribution and storage that best fits your company's business needs. The difference between a typical company network and a knowledge management system has to do with the deliberate engineering of an information structure. If you want to empower your employees to use your company's formal and informal information base to its full potential, you have to begin thinking of disparate data sources, applications, and interfaces as parts to a greater whole. From this vantage point, you can begin engineering an information retrieval system that makes the most sense for your company's purposes – the integration of multiple sources and interfaces in a logical fashion is what causes "information" to magically transform into "knowledge." See also Knowledge Worker.

knowledge worker In its simplest use, a knowledge worker is a person who uses a computer. But that's not the end of it. Some people take this term real seriously. It's as though they've discovered a new religion. A reader sent me this definition: "An organizational employee who, whenever he/she performs knowledge work, adds intellectual value to the organization's memory. A knowledge worker is an empowered person who both knows (has access to) and affects (measurably change) the organizational memory in a profitable sense. Profitable sense assumes the business-process being aligned to organization strategy, and the value (outcome) of the individual's work effort being measurable." John Perry Barlow, who is a cattle rancher, computer hacker, poet, and a lyricist for the rock band, The Grateful Dead, thinks the expression was created by the "droids" who run Microsoft and Apple.

KPI Key Performance Indicator.

KPSI A cable term. Tensile strength in thousands of pounds per square inch.

Kruegerware Browser-hijacking software that, like Freddie Krueger, just won't die. Some of these evil pop-up apps get their hooks into your Windows registry, so even if you delete everything associated with them and reinstall your browser, they still come back to haunt you.

KS Number Abbreviation for Kearney System number. AT&T's Western Electric division had a major manufacturing and distribution facility in Kearney, New Jersey. Thousands of items were assigned part numbers with KS prefixes. KS numbers still appear on certain basic telecom hardware items made by various manufacturers. AT&T now uses both a Code number which reflects standard industry numbering and a different "Comcode" number on most products. The Code No. 259C modular-to-Amphenol adapter is Comcode No. 103339396, and KS No. 21997L15. Very confusing. Kearney, by the way, is pronounced "carny."

KSR Keyboard Send Receive. A combination teleprinter/transmitter/receiver with transmission capability from the keyboard only, i.e. there is no punch paper tape device and no magnetic memory device, such as a floppy disk.

KSU Key Service Unit. The heart (or brain) of a key telephone system is its KSU. Some telecom newbies say Key System Unit. Computer guys often call it a Central Processing Unit, or CPU. Old telecom guys call it a switch. Cardiologists call it a heart. Neurosurgeons call it a brain. The key service unit is basically the main cabinet containing all the equipment, switching and electronics necessary to run a key telephone system. See also Key Service Unit.

KSU-Less Phone System KSU stands for Key Service Unit. It's a funny term for the main cabinet which contains all the equipment and electronics necessary to run a key telephone system. When you pick up your key telephone's handset and punch a button for an outside side, the KSU connects you to an outside line. When you pick up the handset and punch a button to make an intercom call, the KSU gives you intercom dial tone and receives your dialing instructions, rings the correct extension and gives you a talk path. When you have a KSU key system, you plug outside phone lines into the KSU and run lines in a star configuration to each phone. A KSU-less phone system, on the other hand, has all its electronics in the phone sets themselves. You run the outside phone lines into each KSU-less phone and typically one pair of wires looping for intercom to each one of the KSU-less phones. A KSU-less phone system is typically a very small system, consisting of usually no more than six phones. A KSU-less phone can be very easy to install. And that's its primary charm. It may also be cheaper than having a small KSU phone system. I always think of a KSU-less phones as multiple single line phones joined together in one plastic multi-line phone with one intercom path to other KSU-less phones. See also KEY. What normal people call a "Switch," telephone people call a "Key." That's were the term Key Service Unit gets it derivation from.

KTA Key Telephone Adapter. A Rolm multiplexing unit which connects a standard 1A2 key telephone to a three-pair cable coming in from the Rolm CBX.

KTI Key Telephone Interface.

KTILA Development Centre for Telecommunications (Greece).

KTS Key Telephone System, often just called a key system. A key system has multi-line phones with keys that you press to get dial tone on a specific line from the phone company's Central Office (CO). In smaller key systems, incoming calls usually ring at several – or all – phones. In bigger key systems, calls usually go to the receptionist or attendant, who will then tell someone that he or she has a call on a particular line, often using the intercom to call one phone, or by making a paging announcement to several people, or throughout a large area. With a PBX ("Private Branch Exchange"), you usually use a single-line phone and have to dial 9 to get dial tone. Incoming calls usually go to a receptionist, attendant or operator, who transfers the call to the appropriate person.

KTU Key Telephone Unit. The circuit cards found in a KSU that control telephone sets and their features in a key system.

Ku Band Portion of the electromagnetic spectrum in the 12 GHz to 14 GHz range. Used for satellites, employing 14 GHz on the uplink and 11 GHz on the downlink in support of such applications as broadcast TV for man-on-the-street interviews and other situations requiring a small, portable dish. Ku also is used in Direct Broadcast Satellite (DBS) systems, also know as Direct Satellite System (DSS), such as DirecTV.

Ku-Band See Ku Band.

Kurzweil Reading Machine Regarded by many as the most significant technological advance for the blind since the introduction of Braille in 1829, the Kurzweil Reading Machine, introduced in 1976, was the first computer to translate text into computer-generated speech, allowing visually imparied people to read almost any printed material.

KV Old Bell-Speak for key telephone. K stands for Key; V stands for Voice. CV stood for Combined Voice, a single line, simple telephone. All this had to do with letters included in USOC (Universal Service Order Code) codes – the alphanumeric naming convention the local Bell operating phone companies used to identify products and services. See USOC.

KVLY A TV station in Fargo, North Dakota whose antenna mast, which towers 2,063 feet into the air, is the highest in the country and set the standard for the maximum allowed by the Federal Aviation Administration.

KVM Switch Keyboard/Video/Mouse Switch. A switchbox used to control multiple computers from a single keyboard, monitor and mouse. Some KVMs can control multiple computers over LANs, WANs and the Internet. These KVMs are great for administration and management of remote computers. The problem according to a August 18, 2005 article in Network Computing Magazine is they're expensive and don't work as well as testdrivers for the magazine would like. The magazine recommended "inexpensive" remote access software, especially if the company is running a mainly homogeneous shop.

KVW Old Bell-Speak for a key telephone that's designed specifically for wall mounting. Most modern key telephones can be used on a desk or wall.

KWH KiloWatt Hour. One thousand WATTS of electricity used for one hour.

Kynar Pennwalt trade name for polyvinylidene fluoride, a fluorocarbon material typically used as insulation for wire wrap wire.

L band 1. A band of frequencies in the 0.5 to 2 GHz range that are used primarily for voice communications.

2. Long Wavelength Band. The optical band, or window, specified by the ITU-T at a wavelength range between 1565nm and 1625nm (nanometers) for fiber optic transmission systems. See also C-Band, E-Band, O-Band, S-Band, and U-Band.

L card Analog Line Card.

L multiplex A system of analog multiplexers built up through groups, supergroups, master groups and jumbo groups of circuits. See L Carrier.

L-Band See L Band.

L-to-T Connector A device that mates two FDM (Frequency Division Multiplexed) groups with one TDM (Time Division Multiplexed) digigroup to allow 24 voice conversations in analog form to talk to (tie into) a DS-1 line – a T-1 line.

L2 Cache Level 2 Cache A type of cache, normally external in nature. But some manufacturers are building internal caches to speed up computer processing. L2 cache speeds up processing an average of 25% over an external L2 cache

L2F Layer 2 Forwarding: VPN protocol used to establish connectivity between Host / Service provider in an Overlay or Peer-to-Peer VPN model. Similar to L2TP. (Layer 2 Tunnel Protocol.)

L2TP Layer 2 Tunneling Protocol. An IETF (Internet Engineering Task Force) standard tunneling protocol for VPNs (Virtual Private Networks). L2TP evolved from a combination of PPTP (Point-to-Point Tunneling Protocol) and the proprietary Layer 2 Forwarding (L2F) protocol from Cisco. L2TP is implemented by ISPs (Internet Service Providers) to provide secure, node-to-node communications in support of multiple, simultaneous tunnels in the core of the Internet or other IP-based network. End user access to the ISP is on an insecure basis. The ISP assumes responsibility for encryption at the network edge. See also IETF, L2F, PPTP, Tunneling, and VPN.

L3 An acronym for layer 3 switching. See Layer 3 and Layer 3 Switching.

L33tspeak Prnounced "leet speak." Sort of an "in" way of writing in simple code. The first rule is to change certain letters to similar-looking Internet slang that at one time identified the writer as a proficient hacker and now identifies anyone who uses it seriously as a hopeless wannabe.

L8R Later. Shorthand added to an email.

LA Listed Address.

label A set of symbols used to identify or describe an item, record, message or file. It can also be the same as the address in storage.

Label Forwarding Information Base See LFIB.

label swapping Routing algorithm used by APPN in which each router that a message passes through on its way to its destination independently determines the best path to the next router.

Label Switch Router LSR. A device located in the core of the network that switches labeled packets according to precomputed switching rules. This device can be switch or a router. Also see Multiprotocol Label Switching (MPLS).

labeling algorithm Algorithm for shortest path routing or similar problems which labels individual nodes, updating labels as appropriate to reach a solution.

LACP See Link Aggregation Control Protocol.

LAD LATA Architecture Database.

LADT 1. Local Area Data Transport. One of a number of similar names for LEC service offerings of in-city data transport using baseband (short haul) modems rented from the LEC, with names such as LADS (Local Area Data Service) being typical. See also WAN.

2. AT&T's product name for providing simultaneous voice and data (see SAD) by Frequency Division Multiplexing 4800 bps data above speech signals on a two-wire voice telephone exchange line; when used in combination with Centrex and a data PBX in the LEC's exchange building, the resultant data operation is called Datakit.

LAI Location Area Identity. A wireless term. A LAI is part of GSM (Global System for Mobile Communications), used in the radio interface. An LAI is part of the GSM Temporary Mobile Subscriber Identity (TMSI). The TMSI is allocated by the GSM network on a location basis. More simply put, it identifies the cell that a mobile telephone user is in.

Lake Wobegon Effect The tendency to treat all or most members of a group as above average, especially with respect to test scores or executive salaries; in a survey, you can see the Lake Wobegon Effect in the tendency for most people to describe themselves or their abilities as above average.

LAMA Local Automatic Message Accounting. A process using equipment in the central office which records the information necessary to bill your local phone calls by your local phone company.

Lamarr, Hedy Silver Screen actress Hedy Lamarr (born 1914) enjoyed one of the more memorable careers in Hollywood. Her name still ranks among the brightest lights in the history of movies. But what many people may not know is that she helped the United States win World War II. On June 10, 1941, Lamarr and composer George Antheil received Patent No. 2,292,387 for their invention of a classified communication system that was especially useful for submarines. The system was a stroke of genius. It was based on radio frequencies changed at irregular periods that were synchronized between the transmitter and receiver. While a message was being sent, both the transmitter and the receiver would simultaneously change radio frequencies according to a special code. At each end of the transmission, identical slotted paper rolls, similar to those used on player pianos, dictated the code according to their pattern of slots. Just as a player piano holds and changes notes at different intervals to make a melody, their invention held and changed radio frequencies to make an unbreakable code. Signals could be transmitted without being detected, deciphered or jammed. See Frequency Hopping.

lambda The 11th letter of the Greek alphabet. Lambda is used as the symbol for a wavelength in lightwave systems. A single strand of fiber typically carries many different colored wavelengths of light. The process is called WDM (Wavelength Division Multiplexing) or DWDM (Dense WDM), with each range of wavelengths appearing in a "window," roughly corresponding to a color in the visible light spectrum. Light wavelengths are measured in nanometers, with a nanometer being one billionth of a meter. Each wavelength is now called a lambda. When you hear a telecom carrier selling lambdas, it means he's selling the capacity of one or more wavelengths of information-carrying capacity. See also DWDM and Lambda Switch.

lambda miles The measurement, in miles, of individual wavelengths transmitted within a single fiber (i.e., fiber miles multiplied by Wavelength Division Multiplexing channels).

lambda switch A type of switch which is capable of switching light signals. Such a switch is capable of identifying different wavelengths (frequencies) of light, which roughly correspond to visible colors in the light spectrum. In a fiber optic transmission systems employing DWDM (Dense Wavelength Division Multiplexing), the lambda switch would identify those separate wavelengths of light in an incoming fiber and perhaps switch each over a separate fiber on the outgoing side. Or perhaps a lambda switch would identify various data streams in multiple wavelength light streams, select those intended to travel in a particular direction, pluck them out and redirect them to a particular fiber going in the right direction, multiplex them, and shift them to a different wavelength. If this definition seems a bit fuzzy, it's because there is no such thing as a lambda switch at the time of this writing...at least not one that's commercially available. But, that is expected to change by 2005 or so. See also Lambda, DWDM, SONET, and WDM.

lambdasphere A word created by George Gilder to celebrate the idea of pumping multiple wavelengths of light down one thin strand of fiber. George writes, "On the Forum, in the Gilder Technology Reviews, in my books, and on the road, for the last decade I have been celebrating the lightwave or "lambda" network – a circuit switched system as simple and robust and enduring for multimedia communications as the public switched telephone network has been for voice. Essential to fulfill the dreams and business plans of Internet entrepreneurs, such a broadband bonanza can spur the economy out of its current doldrums. Enabled by Wavelength Division Multiplexing (WDM) many colors of infrared light on each fiber thread, this new lambdasphere can both fuel and fund a multi-trillion dollar agenda for thousands of vendors of optical equipment over the next decade."

lamda The correct spelling is lambda. See lambda.

lame A user who behaves in a stupid or uneducated manner.

lame bug A fancy way of saying that a piece of software contains a bug, but the bug is so bleeding obvious and so minor it should have fixed eons ago by the programmers. I first heard this term from Sean Purcell, a smart fellow working on the excellent product called Outlook from Microsoft.

lame delegation On the Internet, there are Domain Name Servers, also known as resolvers, which are a system of computers which convert domain names (like www. HarryNewton.com) into IP addresses, which consist of a string of four numbers up to three digits each. Each applicant for a domain name must provide both a primary and a secondary DNS server. A domain name which fails to provide both primary and secondary DNS servers and thuse whose Name Server Record points to an incorrect server is known as a "lame delegation." This can be caused when a zone is delegated to a server that has not been properly configured to be authoritative (i.e. to work) for the zone. A server that is authoritative for the zone has an NS record that points to another that is not authoritative for the zone. This will cause resolvers to direct queries to servers that will not respond authoritatively, if at all. This causes unnecessary network traffic and extra work for servers. ONe quarter of all zones allegedly have lame delegations. See also DNS.

laminate The whole structure in the mold, consisting of several piles. A layer in a composite is called a "ply."

laminations Thin sheets of steel used as the magnetic core in electrical apparatus, (e.g., the core of an audio frequency transformer is normally composed of laminations).

lamp The technically correct term for light bulb, which nontechnical folks put into their lamps.

lamp battery A steady (unpulsing) 10 volt AC source of power to operate the lamps in key telephone sets; usually one of the outputs of the local Key System Power Supply.

lamp flash A pulsed 10 VAC source of lamp power sent to a key telephone set to indicate a CO or PBX line is ringing in. Pulse repetition rate is normally 60 Hertz with a duty cycle of .5 sec on and .5 sec off. This signal is usually provided by the local Interrupter KTU.

lamp leads Lamp and Lamp Ground (L&LG) wires connected to all lamps in the key telephone set over which steady and pulsed 10 VAC signals from the Line Card KTU are sent.

lamp steady A steady (unpulsed) 10 VAC source of lamp power sent to a key telephone set to indicate that the line is in use. See also Lamp Battery.

lamp wink A pulsed 10 VAC source of lamp power sent to a key telephone set to indicate that the line is on Hold status; pulse repetition rate is normally 120 Hertz with a duty cycle of .4 sec on and .1 sec off. This signal is usually provided by the local Interrupter KTU.

LAMs Line Adapter Modules.

LAN Local Area Network. A fancy name for a communications network connecting personal computers, workstations, printers, file servers and other devices inside a building or a campus. Devices on a LAN can transmit between each other. They can see each other email. One PC can send another a file. One PC can print to any number of printers connected to the LAN – high-speed ones, expensive color printers, etc. Devices on one LAN can often transmit to the outside world if the LAN is connected to a telecommunications link to somewhere. That "somewhere" might be a link to the Internet. It may be a link into the corporate network, which would let this LAN connect to a LAN across the country. That connection is called a "WAN," which stands for wide area network. LANs come in many flavors. But by far the most popular is Ethernet. LANs have come a long way. LAN software used to be difficult to install and expensive to buy. Now all Windows software and Apple computers have built-in LAN software. To install a LAN in your office or home, you'll need LAN network adapter cards for each PC (some already come with them), sufficient wiring and devices known as "hubs," which do what hubs for the railroads – move trains from one track to another. For a more detailed explanation, see Hub, Ethernet and Local Area Network. See also AppleTalk, Ethernet, LocalTalk and token ring.

LAN Adapter Also called a NIC card. A LAN adapter is a a PC-compatible circuit card that provides the PC-to-LAN hardware connection. In addition, LAN software drivers and LAN operating systems need to be run on the PC for it to function as a LAN station. See LAN.

LAN Aware Applications that have file and record locking for use on a network.

LAN Emulation Also known as LAN-E, it is a set of specifications developed by the ATM Forum for the operation of LAN-to-LAN bridged connectivity over an ATM network, allowing ATM to be deployed on a legacy LAN or with legacy LAN applications.

LAN ignorant Applications written for single users only. These are not recommended for use on LANs (local area networks).

LAN intrinsic Applications written for client-server networks.

LAN manager 1. A person who manages a LAN. Duties can includes adding new users, installing new hardware and software, diagnosing network problems, helping users, performing backup and setting up a security system. Unlike MIS managers, LAN managers are rarely formally trained in LAN management. Sometimes they're called LAN Network Managers.

2. The multi-user network operating system co-developed by Microsoft and 3Com. LAN Manager offers a wide range of network-management and control capabilities. It has been superseded by Windows NT Advanced Server.

LAN network manager An IBM-developed network management tool. It is a software program that runs under OS/2 and which provides management and diagnostics tools needed to manage a Token Ring LAN. A PS/2 running LAN Network Manager collects vital statistics and special management data packets on the ring to which it is connected. When multiple rings are involved, the LAN Network Manager relies on the token ring bridges and routers to help in managing those token ring LANs that are not directly connected to the LAN Network Manager station. IBM has installed software in its bridges called the LAN Network Manager Agent. The agent software acts as the eyes and ears for the LAN Network Manager station so that the station can manage the remote rings as if it were connected directly to them. If there were no such agents, managers of networks would be blind to what's going on these LANs. Remote management with LAN Network Manager includes the ability to perform ring testing, analyze traffic and error statistics, and force adapters off the network.

LAN party When a group of geeks choose to get together in their free time to network their computers in order to play games rather than engage in normal social activities.

LAN server IBM's implementation of LAN manager, now largely superseded by OS/2 2.1.

LAN telephony A made-up term that refers to the technology of using LAN data networks rather than traditional telephone lines and a PBX to carry voice traffic along with the data traffic that is typical of a LAN system. A LAN telephony system requires both a device to process incoming and outgoing calls to determine what address on the LAN a call will be sent to and also a voice gateway that connects the LAN telephony system to the PSTN. Each telephone on the LAN has its own network address and converts audio signals to digitized packets for transmission across the LAN. When a call is received from outside the LAN, the call processor determines which network address receives the call and forwards the packets to that address. When a call originates within the LAN, the call processor determines if the packet is to be forwarded to another network address or to the PSTN, in which case the packets must travel through the voice gateway.

Land attack A land attack occurs when a malicious person sends instructions over a network to a server attempting to crash the server. Officially known as land.c code, Land Attack works by tricking the targeted server into trying to set up a TCP session with itself. If the machine falls for this form of IP spoofing, it goes into a TCP closed loop and has to be physically rebooted. A number of security experts, including Chris Klaus, chief technology officer at Internet Security Systems, Inc., agree there is no reason a machine would want to talk to itself like this. Systems should be designed to prevent such attacks.

land line See Landline, the preferred spelling.

Land Mobile Radio See LMR over IP.

LANDA Local Area Network Dealers Association. It runs a number of excellent trade shows each year. Its members are LAN resellers, distributors, manufacturers, and consultants. In 1993 it merged with NOMDA, the National Office Machine Dealers Association. And shortly, thereafter, NOMDA/LANDA changed its name to the Business Technology Association, headquartered in Kansas City. www.btanet.org. See BTA.

landline Also referred to as Land Line. A terrestrial circuit, whether wired (i.e., twisted pair, coax, or fiber) or wireless (i.e., microwave or some other form or radio, or free space optics), or some combination. A landline is different to a satellite link, which is not terrestrial in nature.

landscape Most computer screens are horizontal, i.e. they are wider than they are high. In the new language of computer screens, such screens are called "landscape." When a computer screen is higher than it is wide, it's called "portrait." Some computer screens can actually work both ways. Some even have a small mercury switch in them that determines which way the screen is standing (portrait or landscape) and will adjust their image accordingly.

LANE An ATM term. LAN Emulation: The set of services, functional groups and protocols which provide for the emulation of Ethernet and Token Ring LANs over an ATM backbone. Operating at the Link Layer (Layer 2 of the OSI Reference Model), LAN Emulation takes over the MAC (Medium Access Control) layer function found on Ethernet and Token Ring NICs (Network Interface Cards). LANE supports connectionless service in either a broadcast or multicast mode. The network addresses of the LECs (LAN Emulation Clients are resolved through a LES (LAN Emulation Server), by virtue of a LUNI (LAN emulation User-to-Network Interface), as defined by the ATM Forum. The LES maintains a table of MAC-to-ATM addresses in order that the native MAC addresses of the LAN-attached devices (e.g., workstations) can be mapped into ATM addresses, with the process being reversed on the destination end of the transmission. Another server, known as a BUS (Broadcast and Unknown Server) handles data addressed to the MAC broadcast address, all multicast and unicast traffic sent by a LEC prior to the establishment of an ATM address for the destination LEC. As a Layer 2 ATM service, LANE functions only within a single ELAN (Emulated LAN) environment, but offers significant advantage in the establishment of a VLAN (Virtual LAN) consisting of multiple physical LAN segments interconnected over the WAN via an ATM VC (Virtual Channel). LANE is not simple and is prone to bottlenecks, but is a cool way to internetwork LANs over ATM. MPOA (MultiProtocol Over ATM) is even more cool, as it operates at the Network Layer (Layer 3 of the OSI Reference Model). Thereby, MPOA overcomes the limitations of LANE by supporting multiple network protocols such as IP, IPX, and AppleTalk. See also MPOA.

lane A single interactive multimedia network.

language Computer software that allows you to write programs.

language interpreter Any processor, assembler or software that accepts statements in one software language and then produces equivalent statements in another language.

LANI Local Automatic Number Identification.

LAP Link Access Procedure.

LAP-B LAPB. Link Access Procedure Balanced, the most common data-link control protocol used to interface X.25 DTEs with X.25 also specifies a LAP or link access procedure (not balanced). Both LAP and LAP-B are full-duplex, point-to-point bit-synchronous protocols. The unit of data transmission is called a frame. Frames may contain one or more X.25 packets. LAP-B is the data link level of X.25 in a packet switched network. Same as a subset of the asynchronous balanced mode of HDLC. It is the link initialization procedure that establishes and maintains communications between the data terminal equipment (DTE) and data communications equipment (DCE). All public packet data networks (PDNs) support LAPB.

LAP-D Or LAPD. Link Access Procedure-D. Also called Link Access Protocol for the D channel. Link-level protocol devised for ISDN connections, differing from LAPB (LAP-Balanced) in its framing sequence. Likely to be used as basis for LAPM, the proposed ITU-T modem error-control standard.

lapjacking The increasingly common practice, especially in airports, of stealing laptop computers. Definition courtesy Wireless Magazine.

Laplink Laplink wsa originally a DOS program that transferred files between laptop and desktop computers. It came with a special cable that attached from one computer's serial port to the other. You plugged the cable in, loaded the LapLink software and transferred files back and forth. You could do the transferring from either machine. It was a nifty program, the brainchild of Mark Eppley and he formed a company to sell it, called Traveling Software in Bothell, WA. Eventually the program became so successful that the word "to laplink" became a common verb to connote the transferring of files between computers, as in "I'll laplink these files over to Mary's machine." Eppley's program is now in Windows format. You can transfer across a parallel port cable (which is faster than the serial port). You can also transfer files across phone lines and across the Internet. You don't even have to be in the same city to transfer files. And Eppley has a new feature called SpeedSync that cuts transfer time down by sending only the changed parts of the file. We used to use LapLink every day. But when it became too complex, we switched to FileSync for backing up files. We still use the old Laplink III occasionally.

LAPM Link Access Procedure for Modems. A type of error control used in V.42 and V.42bis modems. LAPM uses the Automatic Repeat Request (ARR) method, whereby a request for retransmission of an errored data frame is automatically requested by the receiving device. As I was writing this entry, a reader asked "if I can transmit HDLC format data thru the GSM network. The data sheet for one of the GSM engines states it can handle V42 bis data transfer." According to Ray Horak, LAPM and both V.42 and V.42bis are specifically linked. If GSM supports v.42, then it supports LAPM, by definition. Such modems also use HDLC frames, which and are synchronous modems, by definition. Therefore, he should be able to accomplish all of this via GSM, as best I can determine.

laptop computer A portable computer you can use on your lap. Also called a notebook. Usually a laptop weighs fewer than than ten pounds. Mine weighs eight pounds. Laptops weighing fewer than four pounds are called subnotebooks. To get a laptop to weigh that little, you typically have to sacrifice something – the floppy disk drive and the CD-ROM drive, for example. Laptops are probably the most useful gadget to come along in years. I wrote much of this dictionary on a laptop in planes, trains, airports, etc. My laptop has a data modem and an Ethernet network card. I don't own a desktop. I do all my work on my laptop.

LARG LIDB Access Routing Guide. The LARG is a single source for Line Information Database (LIDB) access routing data used for administering and maintaining STP Global Title Translations (GTTs). It identifies the responsible signaling network control center, the intra-network LIDB data, the inter-network route effective date, and the capability code or pseudo point code. This provides routing information necessary to route an inquiry to the correct LIDB for the validation of calling cards and other Alternate Billing Services. The LARG is available on a one-time and monthly basis and is provided on disc, can be delivered via email, and is web-downloadable from the TRA Product Distribution System (PDS). See also LIDB.

large squaring capability A feature on some key systems which permits all lines to appear on all telephone sets.

laser An acronym for Light Amplification by Stimulated Emission of Radiation. It is a device which produces light. Tunable lasers can produce light of a single frequency, or visible color, in human terms. By turning the laser light signal on and off quickly, you can transmit the ones and zeros of a digital communications channel. Lasers carried through glass fiber are ideal for telecommunications transmission for two major reasons. 1. Glass fiber of such purity has now been developed that only a very minute portion of the laser light traveling through it is lost. In telecom terms, this means very little of the laser signal is attenuated, or loses power. The signal maintains its strength and thus reduces the need for frequent and expensive repeaters (the digital word for "amplifiers"). Laser transmission systems can now carry many thousands of voice conversations for hundreds of miles without repeaters on two fibers no thicker than a human hair. (You need two fibers – one for transmission in each direction) 2. The glass fibers in laser fiber optic telecom systems are totally immune to electromagnetic interference of any kind. There's no humming from electrical motors. You can't pick up the local TV station in the background. You can't pick up any interference from adjacent cables. In recent years, laser fiber optic transmission has been getting cheaper, more reliable and more powerful at roughly the same rate as computers, and like computers, nobody believes there is an end in sight. As we were writing the first edition of this dictionary, Russell Dewitt of Contel (since acquired by GTE, since acquired by Bell Atlantic, now renamed Verizon) delivered a paper entitled Evolution of fiber optics in rural telephone networks. In it, he talked about "Fiber Optics Progress" and said: "In 1860, the Pony Express could deliver a letter from St. Louis to San Francisco in ten days. For three typed pages the data transmission rate was about three bits per minute. In comparison, today in Contel we are transmitting at the rate of 565 Mbps (million bits per second) over single mode fiber. This is a capacity of 8,064 voice channels. Gigabit per second systems will be available for use this year (a gigabit is a thousand million bits) and a 20 Gbps system has been demonstrated in the laboratory. For the future, the ultimate potential of a single mode fiber has been estimated. It is about 25,000 Gbps (25,000,000,000,000 bits per second). At that rate you could transmit all the knowledge recorded since the beginning of time in 20 seconds."

That sounded pretty incredible a few years ago, and it was. SONET (Synchronous Optical NETwork) optical fiber systems currently run at speeds up to 40 Gbps on a single lambda (i.e., wavelength of light). DWDM (Dense Wavelength Division Multiplexing) hundreds of lambdas can be multiplexed to yield aggregate speeds in the range of Tbps, at least in the labs. Free-space optics systems are used in WLL (Wireless Local Loop) applications, routinely running at speeds up to 622 Mbps, with some running at up to 1 Tbps. Free-space systems run over distances of up to several miles through the air, with no associated requirement for trenching or planting poles, splicing fibers, etc. See also DWDM, Laser Diode, Laser Fax, LCD, LED, Fiber Optics, Free-Space Optics, Single Mode Fiber, and SONET.

Laser Diode Conceptually similar to LEDs, Laser Diodes are the light sources in high-speed fiber optic systems. While LEDs are limited to transmission rates of 500 Mbps or so, Laser Diodes operate at speeds of many Gbps.

laser fax A conventional laser printer that is also capable of being used as a FAX machine when combined with an optional plug-in cartridge and used with a personal computer.

laser optical System of recording on grooveless discs using a laser-optical-tracking pickup. Originally, the technology was WORM – Write Once (i.e. not erasable) Read Many. It's now erasable.

laser printer A high speed non impact dot matrix printer which uses a laser beam to electrostatically form characters on paper. The printer then heats the paper which melts a metallic dust attracted to the electrostatic areas which form the inked images on the paper. Laser printers are fast and the quality of their printing beautiful, rivaling that produced by conventional photo typeset (the way this book was produced).

lashing Attachment of a cable to a support strand by wrapping steel wire or dielectric filament around the cable.

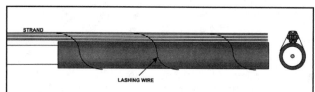

LASIK Surgery Laser in Situ Keratomileusis. Eye surgery using a laser which corrects vision problems by sculpting the cornea.

LASS See Local Area Signaling Services.

LAST Local Area Systems Technology. A Digital (DEC) protocol.

last 10 feet The distance between the customer's broadband modem and the TV, which is sometimes expressed as the distance between the customer's PC and the TV, or between the customer's home office and living room. See the next definition.

last 10 yards problem The challenge of delivering a high-quality radio signal indoors from a municipal Wi-Fi network, without the user's having to spend additional money for a repeater. The "last ten yards problem" is rooted in Wi-Fi's use of unlicensed spectrum, its required use of low-power transmitters, and the limited ability of Wi-Fi signals to bounce – all of which limit the ability of Wi-Fi signals to penetrate walls and reach inner rooms of a residence or office building. See the previous definition.

last digit dialed signal Allows the use of the # sign on Touchtone telephones to indicate that the last digit has been dialed on outgoing calls. This signal enables the PBX

to process calls more rapidly, since some PBXs count the time after a digit was dialed. If nothing else is dialed within a certain time, it assumes that the dialing is complete and then pulses out the call.

last extension called Same as Last Number Redial.

Last In First Out LIFO. The last phone call (or data) arriving is the first call (or data) to leave – to be processed, to be saved, whatever. The term LIFO comes from accounting. It's one of several ways to value an inventory. See also FIFO.

last known good configuration In Microsoft Windows NT, the last configuration that was used to boot the computer successfully. Windows NT saves this configuration and offers it as a startup option during the boot process.

last mile "Last mile" is an imprecise term that typically means the link between an end-user and the telephone company central office – local, long distance or Internet. Of course, it doesn't mean a "mile," since that "mile" could be less than a mile or several miles. The term has entered the language referring to the problems of your telecommunications making it that last mile at a price you can afford and using a modern technology. Let's take price first. Competition in telecom has been heavily concentrated on long distance and international hauls. Carriers in long distance and international have relied on local phone to deliver their customers' communications. Since there was little competition, the price has often been very high. Second, there's the issue of the technology and quality of that last mile. Often that last mile runs over old, limited bandwidth copper wire that has been in the ground for eons and is supplied by a sleepy phone company who doesn't have any competition and not much incentive to perform and hasn't improved the quality of the cable in the loop. This changing slowly. In early 2003, Motorola introduced radio equipment to solve the last mile problem (i.e. you could use their equipment to bypass the local phone company). They changed the name of the "last mile" and called it the "first mile." I thought this was an innovative use of the language.

last number redial Most modern electronic phones have a button on them called "Last Number Redial." When you touch this button, your phone will automatically dial the last number you dialed. If you also have speed dial numbers on your phone, any number you dialed with a Speed Dial button will not appear in Last Number Redial. Most Last Number Redial buttons on electronic phones attached to a PBX will only recognize completely-dialed numbers. Last Number Redial buttons are useful. Also useful – but less common – is a stored number dial button. Dial a number, punch in "save," then that number will be saved to that button, ready to be dialed later, even though you might dial some other numbers in the meantime.

last piece of unfinished business What Joel Novak's mother used to call him before he finally succumbed and got married.

LAT 1. Local Area Transport. A proprietary communications protocol developed by DEC for terminal-to-host communications. LAT allows terminal emulators to access VAX and VMS systems over Ethernet. See LAT Protocol.

2. A proprietary protocol used in Digital Equipment Corp. terminal servers, providing communication for terminals across an Ethernet LAN. See LAT Protocol.

LAT protocol The LAT protocol, announced by Digital Equipment Corporation in the mid-80s, is today one of the industry's most widely used protocols for supporting character terminals over Ethernet networks. LAT is currently licensed by more than 40 third party hardware and software developers, and is compatible with the products of more than 30 major system vendors, from Apollo, and Apple to IBM, Tandem and Wang. The basic function of the LAT protocol is to permit a terminal server to connect multiple asynchronous devices – video display terminals, printers or plotters– to a host timeshare computer. To do this, LAT (or any other terminal server protocol) puts data into packets that can be understood by both the asynchronous device and the host. Essentially, a terminal server protocol is responsible for establishing lower level communications connections, and for routing appropriate transmissions to their destinations.

LATA Local Access and Transport Area, also called Service Areas by some telephone companies. One of 196 local geographical areas in the US within which a local telephone company may offer telecommunications services – local or long distance. At one stage, AT&T was expressly prohibited from offering intraLATA calls by the terms of the Divestiture. But it is now allowed to offer intraLATA phone calls. Other competitors, such as MCI and Sprint, though rules vary by state, have always been allowed to offer intraLATA phone calls and do so in many states. LATAs serve basically two purposes. First, they provide a method for delineating the area within which the Bell Operating Companies may offer service. Second, they provided a basis for determining how the assets of the former Bell System were to be divided between the Bell Operating Companies and AT&T. While writing this edition of the dictionary, Ray Horak and I got into an argument about LATAs. I thought there were fewer than 196 LATAs. But Ray researched the subject to death and affirms 196 is the correct number. In the midst of his research, he sent me the following memo:

Dear Harry, I've got the story on LATAs...and it is very strange, indeed! There originally were 161 LATAs established by the MFJ. Those LATAs were identified by three-digit codes, as follows: 1xx designated NYNEX (now Verizon) LATAs; 2xx, Bell Atlantic (now Verizon);. 3xx, Ameritech; 4xx, BellSouth; 5xx Southwestern Bell; 6xx, US West now Qwest); and 7xx, Pacific Telesis (now SBC). 8xx was assigned by Bellcore (now Telcordia Technologies) to areas such as the Commonwealth of North Mariana Islands, Midway/Wake, Guam, and other Caribbean islands. 9xx was assigned by Bellcore to areas covered by Southern New England Telephone (SNET), Cincinnati Bell, and the Navajo Nation (one LATA in Arizona and one in Utah). Since the initial designation of LATAs, a number of subLATAs have been identified, for a variety of reasons.

There are 17 subLATAs in Florida, mandated in 1984 by the Florida Public Utilities Commission for equal access purposes. There are 23 "900" LATAs set aside for places like SNET (CT), Cincinnati Bell (OH), and the Navajo nation (1 in AZ and 1 in UT). Add to that the "800" LATAs for Puerto Rico, the Virgin Islands, the Bahamas, Jamaica, AK, HI, and the various other Caribbean, Atlantic and Pacific Islands and it gets very strange. Most of these are pseudo-LATAs set up by Bellcore for toll routing purposes, especially since many of the island nations were only recently added to the NANP. Equally interesting, if not more so, is the story on the 8xx area codes. There are 8 NPAs in the 8xx range–with 1,298 NXXs– set aside for "non-dialable toll points." These are remote areas served by a cordboard. The subscribers are reached from the cordboard via ring-down circuits.

late collision A networking term. Late collision is an Ethernet collision that takes place on the local segment after 64 bytes of a frame have been placed on the network by the originating device. Late collisions are usually detected only on coax networks, because the 10Base-T monitor station would have to be transmitting at the same time in order to detect a late collision. Late collisions may also be inferred by detecting the presence of a "jam" signal at the end of a frame that is larger than 64 bytes. Note that traditional Ethernet (versus Gigabit Ethernet) specifies a frame size minimum of 64 octets (bytes) and a maximum of 1,514 octets. Also note that a single logical Ethernet may comprise multiple physical segments, with the segments being connected by bridges, hubs, switches or routers. If all of this seems a trite confusing, it's because it is. At some level, however, it's really pretty simple. First, it takes a certain amount of time for a data bit, and certainly a frame of data bits, to propagate (move) across a wire and through all of the intermediate devices that might be involved. The original Ethernet standard specifies big, thick coaxial cable that will support LAN (Local Area Network) communications over a maximum reach of 2.5 kilometers, from one extreme end of the cable to the other. As many as 1,024 devices may be attached, each with a minimum spatial separation of one meter (due to issues of echo, or signal reflection) and with a maximum spatial separation of 500 meters (due to issues of signal attenuation, or power loss). In the most extreme case, therefore, as many as 1,022 devices might be positioned between transmitting device and receiving device. Each device must read the incoming frame of data, determine if it is intended for it, and, if not, pass it on. This process takes some time, and it takes some time for the frame of data to work its way across the wire to the next device, where the process is repeated. Second, Medium Access Control (MAC), or collision control, technique used in Ethernet is CSMA/CD (Carrier Sense Multiple Access with Collision Detection). CSMA/CD allows multiple devices to sense the status of the carrier frequency to determine whether it is "clear" to send a frame of data. Assuming that they sense that it is "clear," they can access the wire at their own option, and at their own risk. If multiple devices access the wire at about the same time, a collision is likely. All attached devices constantly monitor the wire. If a collision is detected, they broadcast a "jam" signal, or collision detection, over a separate subcarrier frequency. All attached devices also monitor the subcarrier frequency, and adjust as necessary, backing off and re-accessing the wire when it becomes available again. Third, a transmitting device is assumed to be transmitting a series of frames of data. The series of frames is assumed to be associated with the transmission of a set of information, which is organized into frames of certain minimum and maximum sizes, 46 bytes and 1,500 bytes, respectively. Small sets of information, such as a query, are very small. Some sets of information, such as file transfers, are potentially very large, and are fragmented into data frames of as much as 1,500 bytes, plus overhead. The maximum size of 1,500 is set so that no single transmitting device can lay claim to all of the capacity of the network, thereby giving other devices a chance. The minimum size is mandated so that a device can be advised of a collision, and have a chance to adjust in that event, and before it assumes that the first frame was received without collision. When you extend the traditional Ethernet with intermediate hubs and switches, you mess with the original concept, and with the underlying physics of signal

propagation, which is tuned to network length and link length and number of attached devices and the processes, all of which is tuned to minimum and maximum frame sizes, in consideration of assumptions about the supported applications. It's all very confusing at some level, but relatively simple at another. Read this lucid explanation by Ray Horak several times. You'll get it. See also Ethernet and CSMA/CD.

LATA Access Any activity or function performed by a local phone company in connection with the origination or termination of interLATA telecommunications for an IC. This includes, but is not limited to, the provision of network control signaling, answer supervision, automatic calling number identification, carrier access codes, directory services, testing and maintenance of facilities, and the provision of information necessary to bill customers. See LATA.

LATA Tandem LT. An LEC switching system that provides an intraLATA traffic concentration/distribution point for end office switching systems or other tandems within a LATA.

late target channel keyup A cellular term. A condition when the target cell does not receive the execute target order in time for the arriving mobile, caused by link delays between MTSO and target cell site. After the mobile retunes to the target cell, noise will be heard on the downlink audio from the target cell, as the assigned voice channel is not on the air (yet). This results in noise during the handoff.

latency A fancy term for waiting time or time delay. The time it takes to get information through a network. Real-time, interactive applications such as voice and desktop conferencing are sensitive to accumulated delay, which is referred to as latency. For example, telephone networks are engineered to provide less than 400 milliseconds (ms) round-trip latency. You can get latency in several ways:

1. From propagation delay – the length of time it takes information to travel the distance of the line. This period is mostly determined by the speed of light; therefore, the propagation delay factor is not affected by the networking technology in use.

2. From transmission delay – the length of time it takes to send the packet across the given media. Transmission delay is determined by the speed of the media and the size of the packet.

3. From processing delay - the time required by a networking device for route lookup, changing the header, and other switching tasks. In some cases, the packet also must be manipulated; for instance, changing the encapsulation type, changing the hop count, and so on. Each of these steps can contribute to the processing delay.

4. Rotation delay. The delay in accessing data which comes from waiting for a disk to rotate to the currant location.

In a bridge or a router, latency is the amount of time elapsed between receiving and retransmitting the LAN packet. The length of time the packet is stuck in a bridge or router. See Interrupt Latency.

latency sensitive Describes a service such as IPTV or VoIP, or an application such as an online game, which becomes severely degraded when there is network latency, i.e. when there are delays sending packets. See latency.

latent cooling capacity An air conditioner's capability to remove moisture from the air.

lateral A fancy word for a trench. In the parlance of digging up roads and the countryside and laying cable, trenching (i.e. building a trench in which to lay cable) is often called "building laterals." As one wag put it, it's where the rubber meets the road as far as enterprise customers go.

latitude The distance, expressed in degrees, from the Earth's equator to points North and South. The equator is assigned the value of 0 degrees; North and South poles are 90 degrees.

lattice tower A lattice tower is a self supported tower without guyed wires. There are generally three types of telecommunications towers: Monopole - Straight solid pole, Guyed Tower – Tower with angled guyed wires attached to support it and Lattice – which has an upside down V look to it.

LATTIS Local Area Transport Tariff Information System.

launch A new term for starting a program from within another. Typically what might happen is you're working in a messaging program, which has individual lines showing you've just received several faxes, voice mails, electronic mail documents. You click on one of the lines. Your program recognizes that it's an electronic mail message and says "quickly open the electronic messaging software and get it to read the message." So it "launches" the messaging software.

launch fiber An optical fiber used to couple and condition light form an optical source into an optical fiber. Often the launch fiber is used to create an equilibrium mode distribution in multimode fiber. Also called launching fiber.

Laurus Unit A rack for T-1 interface units.

LAVC Local Area VAX Cluster.

law of unintended consequences this "Law" states that for every action, there is an excellent chance of producing an opposite and totally disproportionate reaction.

lawn sale See Yard Sale.

LAWRS Limited Airport Weather Station.

lawsuit Action at law, attorney's apparel. In 1970, an Arizona lawyer named Russell H. Tansie filed a $100,000 damage suit against God. The action was filed on behalf of his secretary, Betty Penrose, who accused God of negligence in His power over the weather when He allowed a lightning bolt to strike her home. Ms. Penrose won the case when the Defendant failed to appear in court. It is not recorded whether or not she collected.

lawyers They're like nuclear warheads. They have theirs. I have mine. Once you use them, everything gets messed up. No one gains. – Danny DeVito Other People's Money.

A town that will not support one lawyer will easily support two. – Mark Twain.

lay The length measured along the axis of a wire or cable required for a single strand (in stranded wire) or conductor (in cable) to make one complete turn about the axis of the conductor or cable. The term originally referred to the process of placing and twisting or braiding fibers to make ropes or hawsers. See lay length.

lay length Twist Length. The distance between twists in a twisted pair cable. For example, a cable with a lay length of 3 inches has 4 twists per foot (TPF). See also Twisted Pair and Twists Per Foot.

layer See Layering.

layer 0 layer zero. The physical right-of-way.

Layer 1 The OSI (Open Systems Interconnection) Reference Model, organizes the communications process into seven separate and distinct, interrelated categories in a layered sequence. Layer 1 is the Physical Layer (PHY). It deals with the physical means of sending data over lines (i.e., the electrical, mechanical and functional control of data circuits). T-carrier and SONET are examples of Layer 1 protocols. See also OSI Reference Model.

Layer 2 The OSI (Open Systems Interconnection) Reference Model, organizes the communications process into seven separate and distinct, interrelated categories in a layered sequence. Layer 2 is the Data Link Layer (DLL). It is concerned with procedures and protocols for operating the communications lines, including the detection and correction of message errors. X.25 and Frame Relay are examples of Layer 2 protocols. See also OSI Reference Model.

Layer 2 Switching Switches run at Layers 1 & 2 of the OSI Reference Model, so Layer 2 Switching is just plain old switching. Layer 3, 4, and 7 Switching always involve routers, which are highly intelligent switches capable of running at higher layers. Layer 2 switches are simple, compared to routers. Since they run only at the Physical Layer (Layer 1) and the Data Link Layer (Layer 2), they don't make complex decisions. Rather, they simply receive incoming traffic, set up a path from incoming port to outgoing port through a switching matrix, and send it on its way. Layer 2 switches operate independently, although usually under the control of a centralized signaling and control system that coordinates their actions. Examples of pure Layer 2 switches include circuit switching and ATM. See also ATM, Circuit Switching, Layer 3 Switching, Layer 4 Switching, Layer 7 Switching, and OSI Reference Model.

Layer 2 Tunneling Protocol L2TP. A networking protocol standard that can be used to route non-IP traffic over an IP network and to authenticate sends and receivers; L2TP can be combined with IPSec to provide greater security (e.g. data privacy) if needed. See L2TP.

Layer 3 In the widely-adopted OSI (Open Standards Interconnection) model, there are seven levels defined of interconnection. Layer 3 is the Network layer. It determines how data is transferred between computers. It also addresses routing within and between individual networks. See next definitions.

Layer 3 Switching Layer 3 switching is a combination of Layer 2 switching and Layer 3 routing for use in large and complex internetworks, such as the Internet. Cisco generally is credited with inventing the concept through its development of the proprietary TAG Switching product line. TAG Switching, and competing product specifications, formed the basis for the IETF's specifications for MPLS (MultiProtocol Label Switching). Layer 3 switching typically makes use of both pure switches, which operate at Layers 1 and 2 of the OSI Reference Model, and routers, which are highly intelligent switches operating at least at Layers 1-3, and often capable of operating at all seven layers. Most commonly the routers are positioned at the edges of the networks, where they perform the processes of header analysis and packet priority assignment. As such processes can be quite complex, they can be fairly time-consuming, which translates into latency-inducing. The edge routers attach

an abbreviated tag address to the front of the packet, and each packet associated with a given packet stream. In the core of the network, Layer 2 switches read the abbreviated tag address, and set up an appropriate link to the next switch, which repeats the process, and so on until the destination edge router is reached. The advantage of Layer 3 switching is that the complex (read slow and expensive) processes are performed only at the edges of the network by routers, thereby enabling simple (read fast and inexpensive) switches to perform their tasks with much greater speed. Overall, Layer 3 Switching integrates routing and switching to provide high-speed performance without the drawbacks of a of a flat Layer 2 network, including broadcast storms, address limitations and spanning-tree loops. Additionally, Layer 3 Switching can include traffic prioritization, security, and bandwidth allocation mechanisms. Layer 3 Switching can control larger network segments than Layer 2 Switching, thereby eliminating the need to create and isolate subnets. See also Layer 2 Switching, Layer 4 Switching, Layer 7 Switching, MPLS, and OSI Reference Model.

Layer 4 The OSI (Open Systems Interconnection) Reference Model, organizes the communications process into seven separate and distinct, interrelated categories in a layered sequence. Layer 4 is the Transport Layer, which defines the rules for information exchange and manages end-to-end delivery of information within and between networks, including error recovery and flow control. TCP (Transmission Control Protocol) is an example of a Layer 4 Protocol. See also OSI Reference Model.

Layer 4 Switching Layer 4 switching refers to hardware-based routing that considers the application being processed prior to processing. In transmission control protocol (TCP) or user datagram protocol (UDP) flows, the application is encoded as a port number in the packet header. The switch then uses this information to make decisions on the flow of data. For example, the following applications are associated with the following port numbers:

Application:	Port #
FTP	20
Telnet	23
HTTP	80

TCP provides transport functions that ensure the total amount of bytes sent is received correctly at the other end. UDP is an alternate transport that does not guarantee delivery. It is widely used for real-time voice and video transmissions where erroneous packets are not retransmitted.

Layer 5 The OSI (Open Systems Interconnection) Reference Model, organizes the communications process into seven separate and distinct, interrelated categories in a layered sequence. Layer 5, the Session Layer, is concerned with dialog management. Layer 5 controls the use of the basic communications facility provided by Layer 4, the Transport Layer. Specifically, Layer 5 deals with the establishment, maintenance, and termination of a session between computing nodes. See also OSI Reference Model.

Layer 6 The OSI (Open Systems Interconnection) Reference Model, organizes the communications process into seven separate and distinct, interrelated categories in a layered sequence. Layer 6, the Presentation Layer, deals with data formatting, code conversion (e.g., conversion between character coding schemes such as ASCII and EBCDIC), and compression and decompression. See also OSI Reference Model.

Layer 7 The OSI (Open Systems Interconnection) Reference Model, organizes the communications process into seven separate and distinct, interrelated categories in a layered sequence. Layer 7 is the Applications layer. It addresses functions associated with particular applications services, such as file transfer, remote file access and virtual terminals. File Transfer Protocol (FTP), Telecommunications Network (TELNET), and User Datagram Protocol (UDP) are all Layer 7 extensions of the TCP/IP protocol suite, which runs at Layers 3 and 4. See also OSI Reference Model.

Layer 7 Switching A packet switching approach that differentiates between packet datastreams based on Layer 7 (Application Layer) information. In other words, information identifying the general nature of the application is used by the routers to set up a path that most closely matches the expectations of the application. Such information, which generally is buried deep in the packet header, includes H.323 or SIP (Session Initiation Protocol) information. Layer Switching also is known as Content Switching. See also H.323, Layer 2 Switching, Layer 4 Switching, Layer 7 Switching, MPLS, and OSI Reference Model, and SIP.

layer entity An active element within a layer. See also OSI Reference Model.

layer function A part of the activity of the layer entities. See also OSI Reference Model.

layer management An ATM term. As described in the ATM Protocol Reference Model, Layer Management is an element of the Management Plane. Layer Management acts on the management of the resources at each of the various specific layers of the model. For example, Operation, Administration and Maintenance (OA&M) information is exchanged between the layers. See ATM Protocol Reference Model for a graphic representation of the three-dimensional model.

layer service A capability of a layer and the layers beneath it that is provided to the upper layer entities at the boundary between that layer and the next higher layer.

layer service provider Each layer of the OSI Reference Model is a Layer Service Provider. Examples of layers are Presentation, Session, Transport, Network, and Link. The Layer Service is made available through Service Access Points. See also OSI Reference Model.

layer user data Data transferred between corresponding entities on behalf of the upper layer or layer management entities for which they are providing services.

layered arrays A large array that combines arrays instead of individual hard drives. This can improve performance and reliability but requires additional hard disks.

layered network architectures Currently the basis of all telecommunication network architecture standards, with functions allocated to different layers and standardized interfaces between layers. The OSI Reference Model and SNA (Systems Network Architecture) are examples of layered network architectures. See also Layering, OSI Reference Model, and SNA.

layering Layering is a technique to write complex software faster and more easily. Layering is often used with public, open software. The idea is to have layers of software on top of other layers. Each performs a specific task, yet each is interrelated to layers above and below it. The idea is that if your software works at one layer (i.e., conforms to the rules of that layer), it should be compatible with the layers of software above and below it. The most famous layered software model is the seven-layer OSI (Open Systems Interconnection) Reference Model. It breaks each step of a transmission between two devices into a discrete set of functions. These functions are grouped within a layer according to what they are meant to accomplish. Layer 2, the Data Link Layer, for example, is concerned with the transmission of frames of data between devices and covers protocols that are aimed at packaging raw data characters into frames, detecting and correcting errors when frames get lost or mutilated, arranging for retransmission and adding flags and headers so that DTE (Data Terminal Equipment) can recognize the beginning and end of a frame. Other layers serve other purposes. Each layer communicates with its counterpart through header records. The flexibility offered through the layering approach allows products and services to evolve. Accommodating changes are made at the layer level rather than having to rework the entire OSI model. Another layered software architecture is Microsoft/Intel's Windows Telephony. It has three layers. At the lowest is SAPI, which is the Service providers' API. In the center is the actual Windows Telephony code. At the top is TAPI – Telephony applications API. See also OSI Reference Model and Windows Telephony.

laying pipe Deploying transmission facilities, most especially optical fiber. Carriers are laying a lot of pipe these days – big pipes. If you were thinking of another definition, you should pray for forgiveness.

lays The twists in twisted pair cable. Two single wires are twisted together to form a pair; by varying the length of the twists, or lays, the potential for signal interference between pairs is reduced.

LB Leaky Bucket. An ATM term. Leaky Bucket is the term used as an analogous description of the algorithm used for conformance checking of cell flows from a user or network. The "leaking hole in the bucket" applies to the sustained rate at which cells can be accommodated, while the "bucket depth" applies to the tolerance to cell bursting above the rate of the ATM "contract" negotiated between the user premises equipment and the network over a given time period. ATM cells which exceed that limit are discarded at the network edge. In other words, they "leak" out of the bucket and spill out on the switchroom floor. See GCRA, UPC and NPC.

LBA An abbreviation for Logical Block Address. See Logical Block Address.

LBO Electrical Line Build Out.

LBRV Low Bit Rate Voice. Digitized voice that requires a bandwidth of fewer than 32 Kbps. LBRV digitizing techniques include packetized voice, APV, DSI, and LPC.

LBS Location Based Services. See Location Services.

LC connector The LC connector is a small form-factor fiber optic connector that resembles a scaled down SC connector.

LCA Local Calling Area.

LCAS Link Capacity Adjustment Scheme. LCAS is a non-standard technology proposed by a number of equipment manufacturers to improve the efficiency with which SONET (Synchronous Optical Network) transmission systems handle data traffic such as Ethernet. SONET supports payloads from 1.544 Mbps (T-1) or 2.048 Mbps (E-1), but only in frames of 51.84 Mbps (OC-1), concatenated frames of 155 Mbps (OC-3), or larger. In other words, SONET is not highly granular, to say the least. An end user organization desiring to transmit Ethernet at a rate of 100 Mbps, for example, therefore must lease an OC-3, of which approximately 50 Mbps of bandwidth is wasted, even under full load. LCAS intends to resolve these issues of transmission rate mismatch by shaping data into SONET VT1.5 payloads (i.e., Virtual Tributary 1.544 Mbps, which is T-1 speed), which can be concatenated (i.e., linked together). Further, the VT1.5 payloads can be dynamically incremented or decremented as necessary in support of LAN traffic, which is inherently bursty in nature. Some manufacturers claim to be able to break LCAS payloads down into increments as small as 0.5 Mbps. See also Concatenation, Ethernet, SONET, T-1, and VT1.5.

LCD 1. Liquid Crystal Display. An alphanumeric display using liquid crystal sealed between two pieces of glass. The display is divided into hundreds or thousands of individual dots, which are charged or not charged, reflecting or not reflecting external light to form characters, letters and numbers. LCD displays have certain advantages. They use little electricity and react reasonably quickly – though not nearly as quickly as a glass cathode ray tube or a gas plasma screen. They are reasonably legible. They need external light to reflect their information to the user. The newer so-called "supertwist" LCDs are much more readable. You see LCDs on computer laptops and telephone screens. The reason computer laptop LCD screens are brighter than phone screens is that laptops use fluorescent or other light sources to illuminate the LCD (typically from the back or the side). Phones typically don't have the power. The only way for a phone to get the power is to plug the phone and its screen into an AC outlet. Most users, however, don't want to have to plug their phone into both AC and phone outlets. It's cumbersome. In newer LCDs, called active matrix displays, the circuit board contains individual transistors for each pixel, or dot on the screen. The enables the crystals to shift quickly, resulting in a higher quality image and the ability to display full-motion video. Active matrix displays in color are hard to manufacture. Low production yields are common, though improving.

2. Loss of Cell Delineation. An ATM term. See Loss of Cell Delineation for a full explanation.

LCE Line Conditioning Equipment.

LCF-PMD The ANSI X3T9.5 standard which defines the requirements for the transmission of data over low cost fiber in an FDDI topology. Also refers to the ANSI working group responsible for the development and perpetuation of the standard.

LCM 1. Line Control Module.

2. Line Concentrating Module is a cabinetized peripheral which contains two duplicated Enhanced Line Concentrating Modules (LCME) to interface analog circuits.

LCN Logical Channel Number. An ISDN term which applies when multiple users share a single B or D channel for X.25 data communications. Each user is assigned a LCN. For instance, as many as eight users can share a single access to a single B channel through an 8-port Terminal Adapter in a BRI application. Effectively eight logical virtual channels are derived from a single B channel.

LCP See Link Control Protocol.

LCR 1. Least Cost Routing. A software-based telephone system feature that automatically chooses the lowest cost phone line to the destination. What actually is the "lowest cost" is determined by algorithms, equations and decision trees programmed into the PBX. Least Cost Routing typically works with "look-up" tables in the PBX's memory. These tables are put into the PBX by the user. The PBX does not automatically know how to route each call. It must be told so by the user. That "telling" might be as simple as saying "all 312 area codes will go via the AT&T FX (foreign exchange) line." Or it might be as complex as actually listing which exchanges in the 312 area code go by which method. Least cost routing tells the calls to go over the lines which are perceived by the user to be the lowest cost way of getting the call from point A to point B. There are typically two types of "least cost routing" translation – that which examines the first three digits of the phone number (i.e. just the area code) and the first six digits of the phone number (i.e. the area code and the three digits of the local central office). Six digit translation is preferred because it allows you more flexibility in routing – particularly to big area codes, like 213 in LA, where there are long distance calls within the area code. These days, least cost routing can virtually be "no cost" routing. For example, a company might choose to ship its voice calls via VoIP over its own internal data network. If there's unused capacity on that network the voice calls are essentially going for free. Least cost routing can also be done in stages. For a New York to LA call, a company might choose to use the company's data network as a VoIP call and then to hop off the end as a local call. Least cost routing was a huge advance in telephony when coast to coast calls in the U.S. were 35 cents a minute. Now they're one cent a minute for big companies, least cost routing is less significant. See also Automatic Route Selection, Alternate Routing and Six Digit Translation.

2. Line Concentration Ratio. A CO (Central Office) design and engineering term for the optimization of system capacity through the classic act of balancing cost and performance. LCR is used to determine central office switching equipment quantities and configurations based upon line usage. It is neither necessary nor economical to equip every outside line with a dedicated path through the switch. With the possible exceptions of Mother's Day or when George Strait concert tickets go on sale, not every line goes off hook at the same time. The higher ratios have been predominantly used in rural areas while the lower, more equipment intensive ratios have been the standard in urban and suburban areas. An LCR of xx:1 indicates that there exists one predetermined dial tone path for any given number (xx) of lines at a given time; in other words, an LCR of 6:1 means that six lines would have access to one dial tone path through the switch.

While the specific architecture of the CO switch manufacturer impacts the LCR calculations, in the AT&T 5ESS switch, the following calculations show the maximum allowable lines per line unit when assuming an average holding time. Exceeding the maximum allowed lines would require load balancing of line units, deloading line units, or extra line unit equipment. The following is based on the relatively short holding times characteristic of voice communications.

4:1 concentration ratio, 256 lines/unit, 5.74 ccs/line, 1470 ccs/unit
6:1 concentration ratio, 384 lines/unit, 4.30 ccs/line, 1650 ccs/unit
8:1 concentration ratio, 512 lines/unit, 3.32 ccs/line, 1700 ccs/unit

Note that there are 64 voice-grade channels of 64 Kbps per unit. Also note that "ccs" means "centum call seconds," which translates to "100 call seconds." A voice-grade channel supports up to 36 ccs, which is 3600 seconds (60 seconds x 60 minutes = 3600 seconds per hour) of traffic capacity.

The Internet has forced the re-evaluation of the LCR and equipment capacities, as circuit-switched connections through the CO to your ISP tend to last a very long time compared to voice conversations. Holding time for internet usage commonly assumes an average of 10 ccs/attempt at 9pm busy hour, with 20% of Internet customers online during busy hour, at least according to GTE internet. This definition courtesy of Kevin Knox, ADSL DSA Project Manager for Pac Bell.

LCS Live Call Screening. This is a wonderful phone/voice mail feature, at this stage confined to some Panasonic phone systems and a very few other makers'. Here's how it works. Someone calls you and ends up in your voice mail. You can hit a LCS button on your phone (or it will happen automatically) and you will hear the message being left in your voice mail. If you choose to speak to this person (based on what you're hearing) you press a button or simply pick up your handset. There are various options with this feature. You can listen through your speakerphone or through your handset. You can have it happen on every voice mail call or just the ones you choose. You'll know if someone is leaving you a voice mail because a special light lights on your phone.

LD 1. Long Distance.
2. Loop Disconnect.

LD-CELP Low Delay-Code Excited Linear Prediction. Standardized by the ITU-T as G.728, LD-CELP is variation of CELP. LD-CELP compresses voice at the same rate of 16 Kbps as does CELP, but it offers the advantage of lower delay levels, which yields improved voice quality with a lesser demand on the computational processes accomplished by the codecs. LD-CELP accomplishes this lower level of delay by gathering only 5 PCM samples in a buffer, rather than the 80 gathered by CELP. Therefore, each transmitted data block represents only 0.625 milliseconds of a voice stream. The concept behind the "Low Delay" is really very simple. If you stop to gather only 5 things, it takes a lot less time than if you stopped to gather 80 things. So LD-CELP gathers 5 voice samples, sends them through the network, gathers 5 more, and so on. See CELP for the full background on LD-CELP. See also ACELP.

LDA Long Distance Alerting.

LDAP Lightweight Directory Access Protocol. LDAP grew out of the X.500 directory standards efforts of the past decade. As Tim Howes and a small group of other students at The University of Michigan were searching for a way to run the X.500 protocol on their PCs, they created a lighter method of access. The resulting directory protocol, which they named DIXIE, was the precursor for LDAP. LDAP defines a standard manner of organizing directory hierarchies and a standard interface for clients to access directory servers. LDAP does not require vendors to mess around with the internal workings of their proprietary directories ,and they can add an LDAP to existing directories without too much trouble. This is why virtu-

ally all directory vendors support LDAP. LDAP is being touted as an Internet-based solution to the intricacies of DAP, the predecessor protocol. See DAP and X.500.

LDBS Local Data Base Services.

LDCLEP See LD-CLEP.

LDM See Limited Distance Modem.

LDN Listed Directory Number. Your main phone number. The one you list in the telephone directory and Directory Assistance.

LDS Local Digital Services. LDS is a term used by long distance companies to describe "last mile" services provided by local phone companies. LDS is generic word to describe any digital services, including T-1, T-3, OC-12, Frame Relay and dedicated Internet services.

LDT Line appearance on a Digital Trunk.

LDU Line Director Unit.

LDTV Low Definition TeleVision (e.g. VHS).

LE LAN Emulation. Refer to LANE.

LE_ARP An ATM term. LAN Emulation Address Resolution Protocol: A message issued by a LE client to solicit the ATM address of another function.

Lead Agent The first agent in an ACD group. See also Automatic Call Distributor.

lead cable Before plastic (polyethylene) was invented telephone cable was insulated with paper and covered in lead. Much of this cable is still used by the RBOCs. This cable is heavy and nonbuoyant in underwater applications. It is now being removed due to the effects of lead on the environment.

lead in 1. Wire or cable from antenna to TV set.
2. The conductor from the antenna to the radio receiver.

lead covered cable also called lead sheathed cable. a cable provided with a sheath of lead for the purpose of excluding moisture and affording mechanical protection.

leader The section at the beginning of a roll of magnetic tape which holds no data and often is not even magnetic tape. A leader is used to feed the magnetic tape through the tape mechanism and secure it onto the roll.

leadership priority An ATM term. The priority with which a logical node wishes to be elected peer group leader of its peer group. Generally, of all nodes in a peer group, the one with the highest leadership priority will be elected as peer group leader.

leading current The phrase difference in a capacitive alternating current where the current leads the E. M. F.

LEAF 1. Law Enforcement Access Field. See Clipper Chip and the NSA.
2. Large Effective Area Fiber. A high capacity fiber that can carry much data over long distances without amplification. www.corning.com.
3. See Leaf POP.

leaf internetwork In a star topology, an internetwork whose sole access to other internetworks in the star is through a core router.

Leaf POP The word leaf is used to describe POPs (Point of Presences) that are on the fringe of the network, like a leaf on a tree. Leaf POPs do not communicate with one another. They follow the chain up to get their information. They have a many-to-one relationship with their region-cores. The Leaf POPs in turn have a many-to-one relationship with their users since they are the connection between the user and the Internet. Leaf POPs can also be called customer aggregation points. See POP.

leakage 1. A cable TV term. Leakage occurs when certain radio frequencies ooze out of the CATV coaxial cable in such strength that they are evident outside the home. They might be sufficiently strong to interfere with aircraft navigation. Leakage is really a shielding problem. See Ingress.
2. See Leaking Memory.
3. A condition that occurs when a carrier cannot bill for a call. Leakage can occur due to system errors, bypass, or fraud. Leakage is especially a problem for the carriers when the call originates from a PBX system supporting hotel/motel, prison, dormitory, and campus environments. See also Leaky PBX.

leakage current The undesirable flow of current through or over the surface of an insulation.

leaking memory Under Windows, when you close a program, Windows sometimes fails to release all the memory that it's used to run that program. This is called "leakage." It is a cumulative problem. The more programs you open and close, the more memory you lose. This can eventually create problems, like slowing down and eventually locking up. The simplest solution is to get out of Windows regularly and reboot your computer.

leaky bridge A type of LAN bridge that forwards packets from one LAN to another even though the packet should not be forwarded. Usually due to poor engineering.

leaky bucket See LB and the next definition.

leaky bucket algorithm An algorithm designed to monitor the flow of cells to verify they conform to the stated traffic contract for the associated connection.

leaky coax A device to assist wireless transmission. A leaky coax is a coaxial cable that has the tops of the corrugated shield milled of to make a series of holes on one side of the cable. Instead of preventing signal loss, the cable will now leak the signal the entire length. This provides a much easier method of evenly covering tunnels, underpasses, stairwells, elevator shafts and basements. Any building constructed with steel, or re-enforced concrete will have dead areas (called nulls) that are in the shadows of traditional transmitters. Instead of boosting the power output and overpowering nearby receivers, a length of leaky coax run high in the ceiling will cover the same areas evenly. The outer insulation is applied after the holes have been cut to provide a weather tight cable. You must orient the cable so the holes are pointed where you want the coverage. Many cellular carriers are working with subways to install leaky coax in subway stations and tunnels. The English Channel Tunnel has leaky coax installed in its tunnels so people can make cell phone calls while they're traveling on that overpriced high-speed train that runs between London and Paris.

leaky PBX One of those really silly terms for which the phone industry is famous. Picture this. You dial into your company's PBX, accessing it through a DISA (Direct Inward System Access) port. You get dial tone, enter an authorization code, get a second dial tone, and place an outgoing call on a trunk-to-trunk basis. The PBX is referred to as "leaky." If you are an authorized user, the advantage is that you can place a low-cost call through the corporate network rather than placing that same business call from your home or your hotel room. The disadvantage is that toll fraud artists can do the same thing if they can hack your PBX's authorization codes. All PBXs are, of course, "leaky," or at least capable of being made leaky. DISA ports are very dangerous. Do not install them. If they are installed, disable them permanently.

Another definition for "leaky PBX" builds on the first. Picture this. You are at work at your office in Seattle and you want to call your brother in Richardson, Texas, a suburb of Dallas. Or you are home in Bellevue, Washington, a suburb of Seattle, and you dial a local number to access the PBX in Seattle for the same reason. You dial your brother's telephone number. The LCR (Least Cost Routing) software in the PBX routes the call over the corporate leased-line network to the Dallas PBX, since that is a free call, and the capacity is available. At the Dallas PBX, the call is connected to the PSTN on a trunk-to-trunk basis, in what is known as Tail-End Hop Off (TEHO). At both ends, the call was local in nature, and no charges applied. The long-haul portion of the call traveled over the corporate network, and no charges applied since the IXC (IntereXchange Carrier) was bypassed. Technically speaking, your company just became an IXC. The PBX is "leaky," since the call is not billable by any authorized IXC. See also DISA and Leakage.

leaky feeder radio See leaky coax.

leaky reply A message to an unintended recipient, often caused by hitting the Reply To All option on an email program. On some programs you actually can send your replies to people who have been blind copied. The whole experience can be embarrassing.

leaky roof syndrome You don't notice that your roof is leaky when the sun is shining. When it rains and the roof leaks, it's too late to fix it.

LEAP Lightweight Extensible Authentication Protocol. A mutual authentication method that uses a username and password system.

leapfrog technology A new technology that enables a technologically-lagging country or community to leapfrog into the technological present, often before more mature, technologically-advanced countries or communities can deploy the same technology, due to the latter's investments in legacy technologies that still must be paid for. An example is the fact that many countries have adopted fibre as a method of long distance communications, bypassing satellite and microwave altogether.

LEAPS Long-term Equipment AnticiPation Securities. Introduced in 1990, LEAPS are simply longer-term options with expirations of up to three years. LEAPs can be very profitable or they can be very expensive. www.888options.com.

LEAS LATA Equal Access System.

lease See DHCP.

leased circuit Same as Leased Line or Private Line.

leased line Same as a leased or dedicated circuit, private line, leased channel. A leased line is a telephone line rented for exclusive use of the customer 24-hour a day, seven days a week from a telephone company – a local phone company (like Bell Atlantic) or a long distance company like AT&T or MCI Worldcom.

Least Cost Routing LCR. A telephone system feature which automatically chooses the "least cost" long distance line to send out a local, regional or long distance call. The user typically dials "9" and then his 10-digit long distance number which is routed over the least costly service. For a longer explanation, see also LCR.

least privilege administration A recommended security practice in which every user is provided with only the minimum privileges needed to accomplish the tasks they are authorized to perform, and no others.

leatherneck The first U.S. Marines wore high leather collars to protect their necks from sabres. Hence the name "leathernecks."

LEC 1. Local Exchange Carrier. The local phone companies, which can be either a Bell Operating Company (BOC) or an independent (e.g., GTE) which traditionally had the exclusive, franchised right and responsibility to provide local transmission and switching services. Prior to divestiture, the LECs were called telephone companies or telcos. With the advent of deregulation and competition, LECs now are known as ILECs (Incumbent LECs). This terminology delineates them from CLECs (Competitive LECs).

2.LAN Emulation Client: An ATM term for a router capable of supporting LANE (LAN Emulation). It works like this: A LAN-attached device, typically in the form of a PC, addresses another LAN-attached device-of course, the originating device hasn't a clue where the other device is physically located. When the router receives the data packets, it exercises its LEC capability, establishing a connection to an edge ATM switch, mapping the native LAN MAC addresses to ATM addresses. Through the ATM network, a matching connection is established to a matching LEC. MPOA (MultiProtocol Over ATM) is much better, but less mature. See also LANE and MPOA.

LECID An ATM term. LAN Emulation Client Identifier: This identifier, contained in the LAN Emulation header, indicates the ID of the ATM host or ATM-LAN bridge. It is unique for every ATM Client.

LECS An ATM term. LAN Emulation Configuration Server: This implements the policy controlled assignment of individual LE clients to different emulated LANs by providing the LES ATM addresses.

LED Light Emitting Diode. A semiconductor diode which emits light when a current is passed through it, the intensity of the light varying with the amount of current. A diode only allows current to flow in one direction. You need two conductive materials sandwiched together to made a diode. When electricity is passed through the diode the atoms in one material (in the semiconductor chip) are excited to a higher energy level. The atoms in that first material have too much energy and need to release that energy. The energy is then released as the atoms shed electron to the other material within the chip. During this energy release, light is created. The color of the light from an LED is a function of the ingredients (materials) and recipes (processes) that make up the chip. In lightwave transmission systems, light emitting diodes or lasers are used as sources of light. These devices are fabricated from multi layered structures of compound semiconductors epitaxially grown on a single-crystal substrate. LEDs are used as sources for optical data link applications in which the data rates are less than about 500 megabits per second and the transmission distances do not exceed a few kilometers. LEDs are also used in alphanumeric displays on calculators and computer devices. LEDs use less power than normal incandescent light bulbs, but more power than LCDs (Liquid Crystal Displays). Contrast with Laser Diodes.

leech Person who pulls items off bulletin boards and consume knowledge without ever making a contribution. An old term.

leg My definition: A segment of a multipoint circuit which lies between any two of the points. Bellcore's definition: An object within a connection view that represents a communication path toward some addressable entity.

leg iron 1. Also called spurs and climbers. What personnel wear to climb wooden poles. Officially linesman's climbers. They consist of a steel shanks that strap to a person's leg. The inside has a spike used to stab the pole.

2. Worn by prisoners to prevent them running away. Many customers want their telephone technicians to wear them until their system is up-and-running 100%.

legacy 1. Noun. All the stuff you have on hand – equipment, software, files and paperwork. In short, everything of a data processing/telecommunications nature in your business today. The use of the word "legacy" suggests that you've inherited all this stuff from previous generations of obsolete management. Most English dictionaries define the word legacy as "Anything handed down from the past. The idea is that you're forced to update it, without junking it altogether – which is expensive and potentially problematical. All you have is the legacy of previous generations. Preserve it, because no one can afford to junk it. Or so the theory goes.

2. Microsoft defines legacy in its Windows 95 Resource Kit as hardware and devices cards that don't conform to its Plug and Play standard.

3. As an adjective, you'll find the word in legacy system, legacy media, legacy bank. In this case, it's often referred to as pre-Internet way of the doing things.

legacy technology Outdated stuff that is basically obsolete but still too expensive to trash. Also called "heritage system."

legacy wiring Preinstalled wiring that may or may not be suitable for use with a network.

legal holiday A call center term. Any holiday for which special wages are paid to employees who work on that day.

Legion of Doom See Master of Deception.

Lego Lego is a Danish-created child's game of many standard plastic pieces which fit together to make wonderful objects. Lego was invented by Gottfried Christiansen in Denmark. In 1947 he bought the first plastic moulding-injection machine and in 1949 began making plastic "bricks." I include this term in this dictionary because many people see open, standards-based telecommunications as "lego" telecommunications. They see open telecom as comprising standard building blocks – hardware and software – which fit together to make wonderful objects, just like Lego. See Legoware Telecom.

Legoware Telecom The British term for what North Americans know as Lego Set or the British term Erector Set Telecom.

lemon A lemon is a product (often a car or a computer) that is so unreliable it leaves a sour taste in your mouth. This definition courtesy Sara Hartman who once bought a lemon car.

Lempel-Ziv-Welsh LZW. The Lempel-Ziv-Welsh compression algorithm is way of reducing the number of bits to transfer. See LZW for a full explanation.

LEN Line Equipment Number.

length The number of bits or bytes in a computer word, a field, a record, etc.

lenses Lenses were named during the 13th century for their vague resemblance in shape to lentils – from the Italian word lenticchie for "lentils," which was later changed to the Italian lente for "lens." For more than 300 years, lenses were called "glass lentils."

LEO Low Earth Orbit satellites. Also called LEOS, as in Low Earth Orbiting Satellites. LEO satellites orbit 400 to 1,600 miles (644 to 2,575 kilometers) above the earth's surface. 48 to 66 LEOs are needed to cover the entire earth. Low Earth Orbit satellites move around the earth in various orbits like electrons whizzing around the nucleus of an atom (remember your high school physics). A group of such satellites is known as a "constellation." To establish a connection, you gain access to one of these satellites much as you gain access to a cell site in close proximity in the case of cellular telephony. When that satellite moves out of range, you are handed off to another satellite which has come into view. During a lengthy conversation, this process may take place many times. LEOs are being promoted for functions as diverse as worldwide paging with acknowledgement, worldwide handheld telephone service and tracking cargo (with the truck sending up a continuous stream of info about its whereabouts). A primary advantage of LEOs is that the transmitting terminal – the one on earth – doesn't have to be very powerful, because the LEO satellite is so much closer than traditional geostationary satellites, which are satellites placed in an geosynchronous orbit – 22,300 miles (or 42,164 kilometers) directly over the earth's equator. The close proximity of the satellites also minimizes propagation delay, thereby avoiding that aggravating CB-radio like problem of conversation delay and clipping. LEOs are divided into two groups, Little LEOs and Big LEOs, with each group having been assigned specific radio frequencies by international agreement. Little LEOs support data services, while Big LEOs support both voice and data communications. See Iridium and GLOBALSTAR. Contrast with GEO and MEO.

LEOS Low Earth Orbiting Satellites. Iridium and Globalstar are LEOS. See LEO and Iridium.

LEP Large Electron Positron Collider.

LER Label Edge Router: Converts IP packets into MPLS packets, and MPLS packets into IP packets. On the ingress side, the LER examines the incoming packet to determine whether the packet should be labeled. A special database in the LER matches the destination address to the label. An MPLS shim header is attached and the packet is sent on its way. The LER adds and/or removes (pops or pushes) labels. See MPLS.

LERG Local Exchange Routing Guide. A Bellcore document which lists all North American Class 5 offices (Central Offices, or end offices) and which describes their relationships to Class 4 offices (Tandem Offices). Carriers use the LERG in the network design process. See also Bellcore, Central Office and Tandem Office.

LES 1. An ATM term. LAN Emulation Server: This implements the control coordination function for the emulated LAN, in order to perform LAN Emulation (LANE) over an ATM network. Examples are enabling a LEC or IXC to extend a LAN, resolving issues of addressing between LAN MAC addresses and ATM addresses. The LES works in conjunction with a BUS (Broadcast and Unknown Server), which distributes the broadcast and multicast packets.

See Emulation and LANE.

2. Label Edge Switch. A switch at the edge of an MPLS (MultiProtocol Label Switching) network. See also MPLS.

3. Loop Emulation Service. A developing specification from the ATM Forum, LES is a service designed to emulate a conventional voice local loop over a DSL loop equipped to support VoATM (Voice over ATM). LES employs AAL2 (ATM Adaptation Layer type 2) for support of rt-VBR (realtime Variable Bit Rate) compressed voice. The DSL loop supports VoATM using AAL2 between a premises-based ATM IAD (Integrated Access Device) and a network-based ATM switch or DSLAM (DSL Access Multiplexer). Simultaneously, LES supports high-speed packet data over the bandwidth not required for voice traffic, which always takes precedence in such a convergence scenario. For example, a DSL loop running at T-1 speed might support 16 voice channels using AAL2 running compressed voice in ADPCM (Adaptive Differential Pulse Code Modulation) and using silence suppression, and claiming bandwidth of only approximately 350 Kbps. This compares with the 1.024 Mbps required to carry 16 voice channels in traditional uncompressed PCM format. This LES approach leaves bandwidth of over 1 Mbps available for packet data traffic destined for the Internet. Where LES is employed in a circuit-switched PSTN scenario, the extended concept is known as BLES (Broadband LES). Where LES is employed in a scenario where the voice data terminates in a packet-switched network based on VoIP (Voice over Internet Protocol) or VoATM (Voice over ATM), the extended concept is known as VoMBN (Voice over Multiservice Broadband Network). See also AAL2, ADPCM, BLES, DSL, DSLAM, IAD, PCM, rt-VBR, VoATM, VoIP, and VoMBN.

less than Of smaller quantity. Of less importance. The word "less" is always confused with the word "fewer." According to the Oxford American Dictionary, the word "less" is used of things that are measured by amount (for example, eat less butter, use less fuel). Its use with things measured by numbers is regarded as incorrect (for example in "we need less workers;" correct usage is fewer workers.)

let's roll Todd Beamer was a passenger on the doomed Flight 93, taken over on September 11, 2001 by terrorists who intended to use the aircraft as a missile to destroy the White House or the Capitol. He had a telephone line open to an operator in Chicago, who reported hearing him recite the Lord's Prayer before leading a group of heroic passengers to rush the suicidal hijackers. Then Beamer said: "Are you guys ready? Let's roll."

Letter Of Agency A letter sent by an end user to a telephone company – local or long distance – authorizing the end user's equipment vendor to deal on the end user's behalf with the phone company. A letter of agency is actually a specialized Power of Attorney.

Letter of Intent See LOI.

letterbox When a program or movie which has originally been created for theatre viewing on a 16 by 9 aspect screen is shown on a 4 by 3 aspect television screen there is a black area above and below the picture. This is done to preserve the entire original picture. In short, when you take a Hollywood movie and put it on a TV screen, you can run it with letterboxes and get the whole Hollywood image. Or you can run it full TV screen and lose parts of the left and right hand side of the picture. See Letterboxing. Contrast with Pan-and-Scan.

letterboxing A TV term referring to the technique in which the aspect ratio (width: height) of an original film is preserved by blacking out portions of the TV screen, typically at the top and bottom. No material is cut out, however. You've noticed this when viewing classic films like Ben Hur. Contrast with PAN-AND-SCAN. See Letterbox.

letters shift A physical shift in a teletypewriter, specifically Telex, which enables the printing of alphabetic characters.

level 1. A fancy telecom word for volume. The power, or amplitude, of a signal measured at a certain point in the circuit. Specifically, the point of measurement is known as a Transmission Level Point (TLP). See also Loss and Pad.

2. Your management position (i.e. "level" in the management structure) in a telephone company. In AT&T and members of the operating Bell telephone companies, employees are identified by their "Levels." At the bottom of the totem pole are crafts people, the installers, the repair people, the trench diggers, etc. They do not have a level. They are often unionized. Management begins one level above the union. They are called first level. They are often called supervisors. Above them are second level managers. Above them are third level managers. They are called district managers. Above them are fourth level managers. They are division managers. Fifth level managers are assistant vice presidents. Sixth level managers are vice presidents. Above vice presidents, levels get fairly vague. Salary is contingent upon level. There are several levels with different salary levels within each level. It is not uncommon for AT&T or for a Bell operating company to have as many as 16 different management levels. At one stage, there was talk about eliminating the fourth level

altogether.

3. As wiring got to carry faster and faster data flows, so the quality of wiring has become increasingly important. Thus more and more companies have started specifying cabling standards. Here is a series of standards, which Anixter has promoted:

- Level 1 VOICE

Level 1 cables are MADE to meet minimum telecommunication cable requirements. Typical uses include analog and digital voice plus low speed data (20 Kbps). Plenum constructions are available in shielded and unshielded designs while PVC constructions are available in shielded designs only.

- Level 2 ISDN & LOW SPEED DATA

Cables support the IBM Type 3 Media requirement. Most uses are defined through the IBM Cabling System guidelines. This specification defines electrical requirements through 1 MHz. These products are available in both plenum and PVC UTP (unshielded twisted pair) constructions. There are no shielded options in Level 2.

- Level 3 LAN & MEDIUM SPEED DATA

These products support the ANSI/EIA/TIA-568 Commercial Building Telecommunications Wiring Standard specification horizontal cable (also known as Category 3). This standard defines cable performance through 16 MHz and thus supports high speed LAN applications. Shielded constructions are available.

- Level 4 EXTENDED DISTANCE LAN

Level 4 identifies the first 100 ohm premises cables specifically designed for LAN applications. Most UTP LANs require a higher degree of performance than the standard telecommunications design offers. Level 4 cables require performance testing through 20 MHz and provide outstanding crosstalk isolation and attenuation. They are ideal for extended distance 10Base-T and 16 Mbps Token Ring. The specification for Level 4 is referenced from TIA TR41.8.1 Category 4 and NEMA "Low Loss."

- Level 5 HIGH SPEED LAN 100 OHM

This level requires the ultimate design for 100 ohm UTP cable. TIA TR4 and the NEMA Premises Wiring Task Force have recently defined this new specification for 100 ohm cable tested through 100 MHz. These cables are intended to be used up to and including 100 Mbps CDDI applications.

- HIGH SPEED LAN 150 OHM - DGM

The 150 ohm shielded twisted pair (STP) data grade media is the cornerstone of the IBM Cabling System. In addition to the many IBM applications, this cable is now supported by a consortium of five system vendors for 100 Mbps twisted pair transmission until the ANSI X3T9.5 standard is complete.

- Level 6

Increases UTP cable performance by requiring 10dB of ACR (Attentuation-to-Crosstalk Ratio) at 155 MHz. Level 6 cable also must meet more stringent four-pair NEXT (power sum) requirements than must Level 5.

- Level 7

Meets at least twice the Category 5 (Level 5) bandwidth requirement. Level 7 UTP achieves 10 dB ACR at 200 MHz, and is power sum-tested to higher NEXT values than is Level 6. Level 7 can support multiple applications at different frequencies with a single cable jacket, and will support Gigabit Ethernet at distances up to 100 meters.

Level 1 Relay Another name for a repeater. Level 1 indicates that the device operates at the lowest layer (physical layer), as defined by the Open Systems Interconnect (OSI) architecture.

Level 2 Relay Another name for a bridge. Level 2 indicates that the device operates at the second layer (data link layer), as defined by the Open Systems Interconnect (OSI) architecture.

Level 3 Relay Another name for a router. Level 3 indicates that the device operates at the third layer (network layer), as defined by the Open Systems Interconnect (OSI) architecture.

Level 7 Relay Another name for a gateway. Level 7 indicates that the device operates at the seventh layer (application layer) of the Open Systems Interconnect (OSI) architecture.

There are other standards. See also Category of Performance.

level playing field An area of business competition where all the players enjoy the same rights and privileges. None has special privileges, such as conferred by government regulation. The term has special meaning in the telecom industry where regulation is so pervasive. At one point, the aspiring competitors to the long distance carriers argued for a level playing field. Then when these new competitors got rights and privileges and the older long distance carriers still had the remnants of regulation, the older ones complained. They now wanted a "level playing field." There really is no definition of "level playing field." Everyone defines it the way they want.

level mode interrupt A method of transmitting an Interrupt Request from a PCMCIA Card to a socket using the IREQ signal. In this mode, the IREQ signal is asserted when the Card initiates an interrupt and is negated when the Host acknowledges to the PC Card that the interrupt has been serviced. The method of acknowledgment is specific to devices on the PCMCIA Card.

level sensitive interrupt A host system interrupt which causes repeated interrupts as long as the interrupt request signal is in the asserted state and the interrupt request is not disabled. Used in Micro Channel Architecture bus hosts and available in EISA hosts.

level sensitive interrupt trigger These adjustable triggers are the key to the operation of the new UART chip, called the 16550. They determine both the amount of data (in bytes) that the UART can receive before generating an interrupt request and the remaining buffer space available to store additional, incoming data. See 16550 and UART.

leverage The word "leverage" used as a verb in techno-babble. Really! I'm always hearing companies saying stuff like, "We're going to leverage our blah-blah application off our blah-blah software capabilities...." and trash of a similar nature. It appears they mean, enhance, exploit, capitalize on or some such. Can you enlighten? Francine Brevetti, Business Writer, The Oakland Tribune. Francine, you're right. In the same way, I'm also leveraging my fingers to write this dictionary. It's techno-babble – big words being substituted for small words plus a big leap in logic. Yuch.

lexicographer >From the Greek "lexikographos," translating as "grapher of words." In other words, the author of a dictionary. When they want to be cute, my children write down their father's profession on school forms as "lexicographer." None of their classmates or anyone at the school seems to know what the word means. This pleases my children who have inherited their father's perverse sense of humor.

LF 1. Line Feed. ASCII character 10. This character is now identified in the ASCII code set as New Line. See New Line.

2. Low Frequency.

LFIB Label Forwarding Information Base. A data structure and way of managing forwarding in which destinations and incoming labels are associated with outgoing interfaces and labels.

LFSR Linear Feedback Shift Register. Mechanism for generating a sequence of binary bits. The register consists of a series of cells that are set by an initialization vector that is, most often, the secret key. The behavior of the register is regulated by a clock. At each clocking instant, the contents of the cells of the register are shifted right by one position, and the exclusive-or of a subset of the cell contents is placed in the leftmost cell. One bit of output usually is derived during this update procedure.

LFSS Link Failed Signal State.

LG Line Group.

LGBC Lightguide Building Cable. Alternative term used for fiber cable in which individual optical fibers are stranded around central members. LGBC cable is used inside a building.

LGC Line Group Controller.

LGCI ISDN Line Group Controller.

LGN An ATM term. Logical Group Node: LGN is a single node that represents the lowest level peer groups in the respective higher level peer group.

LGX Lightguide Cross-Connect distribution system. A component of fiber optic connecting hardware. This component accommodates 24 to 216 fiber terminations. Also referred to asn n LGX shelf or frame.

LH Long Haul.

LHMC Long Haul Mileage Calculation.

liberty cabbage Sauerkraut. During the First World War Americans rechristened sauerkraut "liberty cabbage." In their denunciation of all things German, some patriots stomped dachshunds to death in the streets. During the Second World War the mutts were patriotically renamed "wiener" dogs (a patently American word) and allowed to participate in the war effort.

liberty crack The first known (non-Greek) Trojan Horse targeted for the Palm operating system. It arrived one summer masquerading as an application called "Liberty."

libor London Interbank Offered Rate. Some companies pay Libor plus XX% on their bank loans. It's a fancy way of saying you agree to pay a floating interest rate on your loan.

library A file that stores related modules of compiled code.

library of violators Picture the home page of a typical Web site. Somewhere on the page is a moving graphic – perhaps an animated GIF – that screams at you and violates the visual integrity and consistency of the page. Most often, such graphic is designed to deliberately violate the integrity of the page. It is often a paid-for advertisement. And the advertiser wants to draw your attention to his graphic ad. After all, he paid big money for the ad. Such graphic is a called a graphic violator. A collection of graphics violators is called a library.

licensed bands There are two types of wireless communications devices. Those that require a licence from the Federal Communications Commission. And those that don't. Those that require a license run in a licensed communications band, a specific frequency. Those that don't require run in unlicensed communications bands can be plugged in and run – so long as they meet FCC rules for that communications band, i.e. that frequency. The FCC's rules loosely prohibit "harmful interference" of unlicensed devices, but devices that run in an unlicensed band are not guaranteed protection from interference.

Licklider, Joseph Carl Robnett Joseph Carl Robnett"Lick" Licklider was the first person to conceive of a worldwide computer network. He didn't have any idea how to build it, but he was smart and knew it was doable. Licklider's vision inspired the developers of the ARPANET.

LIDB Line Information Data Base developed by Telcordia for the Regional Bell Operating Companies and all the local phone companies includes such services as Originating Line number Screening, Calling Card Validation, Billing Number Screening, Calling Card Fraud and Public Telephone Check. The LIDB systems contain all valid telephone and calling card numbers in their regions, and have the necessary information to perform billing validation. A national system connecting them all together started working at the beginning of 1992. See also LARG.

LIF Location Interoperability Forum, formed by Ericsson, Motorola and Nokia in September 2000 with the goal to define, develop and promote common interfaces allowing user appliances and Internet-based applications to obtain location information from wireless networks, independent of positioning methods.

Life After Death LAD. A satellite term. With the usage of the PMs inside the encoder, the encoders can remain in the scrambled mode for two program epochs' (current and next programs) worth of time per channel after the connection with the UCS (Uplink Control System) has been severed.

life cycle A test performed on a material or configuration to determine the length of time before failure in a controlled, usually accelerated, environment.

life questions Questions of a personal nature that a user answers at the time his user id is established, and which are later used to authenticate the user's identity when he calls the help desk for security-related assistance, such as a password reset. Life questions ask for such things as a user's date or place of birth, mother's maiden name, favorite color, make or color of first car, and high school. Most free magazines ask at least one life question. It's the way the auditors can check if the free subscription is real and the subscriber has been verified, or simply entered by the publisher in order to boost his magazine's circulation.

lifeline service A minimal telephone service designed for the poor and elderly to assure they can be reached by phone and have a "Lifeline" to the world in case of emergency. Typically, Lifeline Service entitles you to a phone line, a listing in the directory and a minimal number of outgoing local calls, e.g. 10. Some people who are neither poor nor elderly, subscribe to Lifeline Service and use it for incoming calls – for an answering machine or a computer electronic mail or bulletin boards. There's no difference in the quality of service provided by Lifeline Rates and normal phone lines. The cost of providing lifeline service is subsidized at the national level through the settlements pool administered by NECA (National Exchange Carrier Association) under the supervision of the FCC (Federal Communications Commission). NECA Lifeline Assistance Programs include SLC Waiver, which waives either the entire Subscriber Line Charge up to $3.50, or a portion of it; and Link-Up America, which offsets half the initial installation fee, up to $30, and defrays interest expenses. See also NECA, Separations and Settlements, and Universal Service.

lifestyle MVNO An MVNO such as Amp'd Mobile, Disney Mobile, Helio, Virgin Mobile and Mobile ESPN, that pulls in wireless subscribers by offering value-added services and content that appeal to a certain demographic – usually 15-24 year olds. See also MVNO.

LIFO Last In First Out. A method of organizing queues. See Last In First Out and FIFO.

liftoff A term referring to the moment a spacecraft first rises from the ground after its launch. "Five, four, three, two, one, liftoff," the now famous rocketry countdown, was invented by the German director Fritz Lang as a suspense builder for his 1928 science-fiction movie, "Die Frau im Mond," or "The Woman in the Moon," (a.k.a. "By Rocket to the Moon.") A young engineering student, Wernher von Braun, was impressed by the movie, and when he began work on on the V-1 and later the V-2 rocket, used it. Later, after the second World War ended and he and some of his colleagues were brought to the U.S. to

continue their work here, the countdown became a familiar part of the space program.

Ligne d'abonne Numerique French name for DSL. See DSL.

light Technically, light is electromagnetic radiation visible to the human eye. The term is also applied to electromagnetic radiation with properties similar to visible light, including the invisible near-infrared "light" (or more technically correct, radiation) that carries signals in most fiber optic communication systems. Light consists of electromagnetic waves ordinarily applied to those having a wave length of from .000075 cm. (the red ray) to .000038 cm. (the violet ray).

Light Amplification of Stimulated Emission of Radiation LASER. A device which transmits a narrow beam of electromagnetic energy in the visible light spectrum. The light waves are in phase with one another, or coherent, rather than jumbled as in normal light. See Laser.

light guide A light guide is a transmission channel that contains a number of optical fibers packaged together.

Light Emitting Diode See LED.

light pen A video terminal input device which is a light sensitive stylus connected by a cable to the video terminal. The user brings it to the desired point on the screen surface and presses a button. A light pen is used to select options from a menu on the screen or to draw images by dragging the cursor around the screen on a graphics terminal.

light piping Use of optical fibers to illuminate.

light year The distance that light travels in a pure vacuum (e.g., outer space) during a year. It's a big number. Do the math: 186,000 miles per second x 60 seconds per minute x 60 minutes per hour x 24 hours per day x 365 days per year = 5,865,696,000,000 miles (that's almost six trillion miles). All wave forms in the electromagnetic spectrum propagate at roughly this speed, assuming they are unimpaired by physical matter such as copper wires, earth's atmosphere or glass fibers. Such physical matter not only slows the rate of travel as a result of resistance, but also creates distortion in the signal. See Fiber Optics and Loss.

lightbulbs At the turn of the century, most lightbulbs were handblown, and the cost of one was equivalent to half a day's pay for the average U.S. worker. Nowadays the retail cost of a light

light output Output from an optical device that represents a logical 1. See dark output.

light trail See lighttrail.

lightbulbs At the turn of the century, most lightbulbs were handblown and all were incandescent, and the cost of one was equivalent to half a day's pay for the average U.S. worker. Nowadays the retail cost of an incandescent lightbulb is largely the packaging and freight of getting it from the factory to the retail store.

lighted fiber See Lit Fiber.

lighting fiber See Lit Fiber.

lightning arrester See Lightning Suppressor.

lightning line A colloquial term for a telegraph line in the mid-to-late 1800s.

lightning rods Ladies in Europe took to wearing lightning rods on their hats and trailing a ground wire – a fad that began after Benjamin Franklin published instructions on how to make them in his almanac, Poor Richard Improved, in 1753.

lightning suppressor A device which grounds out surge voltages in order to protect equipment from lightning.

lightpath An end-to-end optical circuit.

lighttrail A set of linearly connected nodes, each of which can insert or remove data. Multiple optical transmissions, originating at any or all of these nodes, can travel across these nodes in the same direction in an interleaved (i.e., nonoverlapping) manner, thereby improving utilization of the wavelength. The bus-like nature of this architecture enables dynamic insertion or removal of data at any node along the path; it enables dynamic bandwidth provisioning; and it improves utilization of the wavelength. The interleaving of separate transmissions is accomplished through the nodes' "drop-and-continue" architecture, i.e., a node can drop (i.e., put on hold) one transmission in order to make way for another transmission. It can then resume the paused transmission when the coast is clear. An out-of-band protocol orchestrates the "drop-and-continue" behavior.

lightwave communications Fiber Optic communications using light to carry information.

lightwave transmission This term now means laser communications systems shot through the air (as opposed to glass fiber). Also called "free space lightwave communications." Typically, a signal is radiated directly from a light transmitter to a receiver less than a mile away. Advantages to lightwave transmission: easy to install, no digging of cables, wide bandwidth, reliable, cheap, no FCC frequency clearance approvals required and the receiving and transmitting equipment occupy little space. Disadvantages: only works for a mile or so and is subject to attenuation (fading) from fog and dust. It's perfect for between downtown buildings, where installing cables is too expensive, too cumbersome, too slow, etc. See Laser.

Lightweight Directory Access Protocol. See LDAP.

Lightweight Extensible Authentication Protocol LEAP. A mutual authentication method that uses a username and password system.

LIJP An ATM term. Leaf Initiated Joint Parameter: Root screening options and Information Element (IE) instructions carried in SETUP message.

like new A term used in the secondary telecom equipment business. It means in excellent condition. Under normal conditions, the like new equipment could pass as new (i.e., not used, but not necessarily in the O.E.M. packaging). See Like New Repair Update.

Like New Repair And Update LNRU. A term in the industry which repairs telecom equipment. It means all equipment is repaired and updated to the current manufacturer's specifications. New plastic is used to refurbish to a "like new" status. Also added are a new coil cord, line cord and address tray. Included is a full diagnostic test with a burn-in (if required) and an operational system test. Definition courtesy Nitsuko America. See also Repair and Quick Clean and Repair, Update and Refurbish.

LILO Linux Loader. LILO is a versatile boot loader for Linux. It does not depend on a specific file system. It can boot Linux kernel images from floppy disks and hard disks, and can even boot other operating systems. One of up to sixteen different images can be selected at boot time. Various parameters, such as the root device, can be set independently for each kernel. LILO can even be used as the master boot record.

lily livered The practice of calling cowards "lily-livered" originated in the Middle Ages. Then, it was believed everything was made of four elements: Earth, Air, Fire, and Water. It was also believed that in humans, the Four Contraries: hot, cold, dry, and moist, would combine to form the Four Humors: cholor, blood, melancholy, and phlegm. If someone had demonstrated that they were cowardly, they would be thought of as not having any cholor, or yellow bile, the humor thought to control courage. Since at the time people thought an absence of cholor in the liver would leave it white, the color of the lily, cowards would be called "lily-livered."

LIM Link Interface Module.

LIM-EMS The abbreviation for Lotus Intel Microsoft-Expanded Memory Specification. A software technique that allows MS-DOS to access memory beyond one megabyte by mapping the memory into a window in an area that MS-DOS can access. LIM-EMS is one of the greatest techniques for speeding up getting in and out of programs. For example, when my calendar program called Maxi-Calendar is not running, it occupies only 7K of normal RAM and 350K of expanded RAM. When I need it, it swaps itself quickly out of expanded RAM into normal RAM, taking less than half a second. If I didn't have expanded memory, it would take as long as 15 seconds to swap the program onto and off my hard disk, which is the other alternative. LIM stands for Lotus/Intel/Microsoft, the founding organizations that developed the Expanded Memory Specification. AST Research was also part of the driving force behind EMS, though its name doesn't appear in the acronym.

limelight People in the public eye are said to be "in the limelight". Invented in 1825, limelight was used in lighthouses and stage lighting by burning a cylinder of lime which produced a brilliant light. In the theater, performers on stage "in the limelight" were seen by the audience to be the center of attention.

limey What they call English people in Australia. Captain Cook, who "discovered" Australia for the British, lost 41 of his 98-man crew to scurvy (a deficiency of vitamin C) on his first voyage to the South Pacific in 1768. By 1795, the importance of eating citrus was realized, and lime juice became a standard issue on all British Navy ships. Hence the term "limey" when referring to a Brit.

limit order When you place a "limit order," you instruct your broker or brokerage service to buy or sell this stock at the price I specify.

Limited Distance Modem LDM. A special purpose conversion device designed to connect two DTEs (data communications devices) over a relatively short distance, typically up to several miles. An LDM is not really a modem since it does not perform a digital-to-analog conversion, but transmits a special type of digital signal to the other LDM on the circuit. Also called a line driver, local dataset or short-haul modem.

limiter A circuit which shapes a signal sent through it to conform to certain preset tolerances, used in both audio and video to regulate signal flow and prevent overloading, which would lead to distortion and the introduction of spurious noise.

limiting amplifier Relating to analog signals and their processing. Also refers to

the operating range of an amplifier where little or no distortion occurs.

limits of error the maximum deviation (in degrees or percent) of a thermocouple or thermocouple extension wire from standard emf-temperature to be measured. integral part of the cable, or exterior to it.

limner In Elizabethan England you got your portrait painted by someone called a limner.

LIN Local Interconnect Network. A low-speed, local area network designed for automobiles. It will be used in applications such as turning on lights, fans or locking doors. The automobile industry is also developing a high-speed network called FlexRay. See also CAN, FlexRay, MOST and Safebywire.

LINCS Leased Interfaculty National Airspace Communications System.

line The word line is confusing. In traditional telecom, a line is an electrical path (two wires) between a phone company central office and a subscriber, usually with an individual phone number that can be used for incoming and outgoing calls. A line, in this definition, is the most common type of loop. In carrier systems, a line is the portion of a transmission system that extends between two terminal locations. The line includes the transmission media and associated line repeaters. A line is also used to indicate the side of a piece of central office equipment that connects to or toward the outside plant. The other side of the equipment is called the drop side. And finally, a line is a family of equipment or apparatus designed to provide a variety of styles, a range of sizes, or a choice of service features. The confusion over the word line starts with an office phone system. Some people believe a line to be the same animal as a trunk – i.e. the line coming in from the central office to the PBX. Other people think a line is an extension, i.e. the line from the PBX to the phone on the user's desk.

line adapter In communications, a device that converts a signal into a form suitable for transmission over a communications channel. A modem is a specific type of line adapter used to convert the computer's digital signals into analog form so that they can be transmitted over a telephone line.

line analyzer Any device that monitors and displays information about a transmission on a communications channel. A line analyzer is used for troubleshooting and load monitoring.

line build out Because T-1 circuits require the last span to lose 15-22.5 dB, a selectable out put attenuation is generally required of DTE equipment (typical selections include 0.0, 7.5, and 15 dB of loss at 772 KHz).

line capacity A telephone company definition. The maximum number of network access lines that can be working on installed lines at the entity's derived objective percent line fill.

line card 1. A plug-in electronic Printed Circuit (PC) card that operates lamps, ringing, holding and other features associated with one or several telephone lines or telephones in a telephone system.

2. A device that transmits and receives optical data and converts optical signals to and from electrical signals. They also transmit multiple data streams to and from other line cards. Line cards plug into switches, cross connects, multiplexers and routers that form the building blocks of optical networks. Each line card connects to a tributary of a long-haul transmission line.

line circuit The sensor in the CO which detects and advises the switching system that one of its subscribers has gone off-hook and wishes to make a call. One line circuit is dedicated to each line of each subscriber.

line coding Line coding can be D4/AMI or B8ZS. See either of these.

line concentration ratio LCR. Used to determine central office switching equipment quantities and configurations based upon line usage. It is not necessary, nor economical to equip every outside line with a dedicated path through the switch. With the exception of Mother's Day or when George Strait concert tickets go on sale, not every line goes offhook at the same time. The higher ratios have been predominately used in rural areas while the lower, more equipment intensive ratios have been the standard in urban and suburban areas. An LCR of xx:1 indicates there exists 1 predetermined dial tone path for any (xx) lines at a given time; in other words, an LCR of 6:1 means 6 lines would have access to 1 dial tone path through the switch. See LCR for a very detailed explanation.

line conditioning 1. A service offered by telephone companies to reduce envelope delay, noise and amplitude distortion. By doing this, you allow for transmission of higher speed data than over a traditional dial-up phone line.

2. The removal by the Incumbent Local Exchange Carrier (ILEC) from the local loop of any devices that may diminish the capability of the loop to deliver high-speed wireline telecommunications capability, including xDSL service. Such devices include bridge taps, low

pass filters, and range extenders (e.g., amplifiers and line doublers).

line control unit LCU. A data communications term used by some vendors for hardware that controls polling and access by remote terminals; most commonly found in minicomputer networks, LCU capability ranges from simple hardware to processor-based devices with a history dating to telegraph networks.

line cord The connecting cord between the phone and the jack in the wall. In North America, it's also known as a mounting cord.

line current A telephone's average off-hook current is about 35 milliamps (mA) or 0.035A. Line current is electrical current measured on an idle telephone line. Typical range is 20 - 100 mA DC, with 40 - 50 mA considered optimum for proper operation of the phone.

line discipline The two most basic elements of a communications protocol are Handshaking and Line Discipline. Line Discipline is the sequence of network operations which actually transmits and receives the data, controls errors in transmission, deals with the sequencing of message sets (e.g., packets, blocks, frames and cells), and provides for confirmation or validation of data received. This definition courtesy of "Communications Systems & Networks," which is Ray Horak's best-selling book. See also Handshaking and Protocol.

line disturbance analyzer A tool used in analyzing problems in a facility's incoming power. The line disturbance analyzer is connected at the power input to measure and record incoming power, then left in place long enough to gather data typical of the site.

line driver A short haul communications device used when cable lengths between RS-232 devices begin to the alleged 50-foot RS-232 limit. A line driver is a signal converter that conditions the digital signal transmitted by an RS-232 interface to ensure reliable transmission beyond the 50-foot RS-232 limit and often up to several miles; it is a baseband transmission device. Also called baseband modem, limited distance modem, or short-haul modem. See also Limited Distance Modem.

line equipment Equipment in a central office which is there to serve a phone line. That line equipment includes a line relay or equivalent which starts to work when the customer's telephone goes off-hook.

line equipment number A line equipment number identifies the physical central office line equipment for each subscriber.

line error A Token Ring error reported by any ring station that detects an FCS failure, or some type of protocol code violation in a received frame.

line feed The act of moving a cursor or the head or a printer or telex machine down one line. These days, on the keyboards of most equipment – Personal Computers, etc. – there's no single key that says "Line Feed." There's usually a key that's labelled "Enter" or "Return." This key does two functions – a line feed and a carriage return (i.e. sending the print head or cursor to the left hand side of the carriage or the screen). In many (but not all) programs, a line feed is control J or ASCII character 10. The name for this character has been changed to New Line. See New Line.

line finder The first switching element of a step-by-step phone system which recognizes a calling party is waiting for dial tone to make a call, identifies the party, and connects that party to the switching system so that the processing of the call may begin. Normally line finders serve 100 or 200 subscriber lines.

line hit Electrical interference that causes a hit, that is, a loss or introduction of spurious bits into a data stream.

line hold Provides a winking, blinking flash on the line lamp at every telephone which has the line appearing on it.

line information data base See LIDB.

line insulation test LIT. A test performed from the central office, which measures resistance and voltages on local lines to find faults.

line interface unit See LIU.

line link frame LLF. An arrangement that permits a crossbar office to transmit dial pulse information over a line to a PBX for switching Direct Inward Dial (DID) calls to the indicated phone. See Crossbar Switching.

line load control A control application that limits the number of customers that can obtain dial tone.

line lockout When a phone stays off-hook for longer than a predetermined time, line lockout provides then some loud noise and then puts the phone line out of service – until someone puts the phone back on hook again.

line loop back LLB. A troubleshooting function of CSU/DSU equipment. The receive pair of a circuit is connected directly back to the transmitter so the line can be tested. If a clear signal is "looped back" to it the line is OK. If there's a problem, it's inside or beyond the receiving equipment.

line loss What the power utility (electricity) business calls theft of service, i.e. when you steal and they can't figure out how to bill.

line noise Spurious signals introduced into a line by static, or other imbalances in the circuit. Line Noise is the most common cause of "Hits" or problems in data calls.

line number The last four digits of a telephone number are called the "line number." A ten-digit telephone number in the U.S., for example, follows the format NXX-NXX-XXXX, where N must be a number other than "0" or "1," and X can be any number. The first three digits are the area code, the second three are the central office prefix, and the last four are the line number.

line of sight Some through the air transmission media – such as microwave, infrared, and laser – operate at a frequency which transmit in a perfectly straight line. Or in "line of sight." In other words, the area between a transmitter and a receiver must be clear of obstructions.

line pool A Line Pool is a specific group of lines in certain key systems used for making outside calls. In Northern Telecom's Norstar, three Line Pools give phone access to outside lines without taking up too many Line buttons on each phone instrument.

line powered Telephone equipment that is powered solely by the CO talk battery supplied in a standard phone line.

line powered voice Short for line-powered voice services. Voice services (for example, VoIP) that are provided over an existing phone line with equipment (and associated power for it) in the central office, without the need for new equipment in the subscriber's home.

line preference User selects the line to be used simply by pressing the button associated with that line.

line printer A type of printer which prints an entire line of text at one time. This printer is obviously a high speed printer. It is used, for example, to print TELECONNECT Magazine's monthly mailing list.

line printer daemon LPD. A service on the print server that receives documents (print jobs) from Line Printer Remote (LPR) tools running on client systems.

line protocol Rules for controlling transmission on a synchronous data transmission line. Includes rules for bidding for the line, for positive and negative acknowledgements, requests for retransmission, and transmitter time-outs.

line sharing The practice by which a CLEC (Competitive Local Exchange Carrier) and an ILEC (Incumbent LEC) share a local loop. It works like this: A DSL provider cuts a deal with an ILEC to run DSL service for Internet access purposes over the same local loop that the ILEC, or voice CLEC uses for voice service. The data CLEC providing the Internet access uses the high-frequency portion (i.e., above 25 KHz) of the local loop for digital packet data transfer, and the ILEC uses the low-frequency portion (i.e., 4 KHz and below) for analog voice. On December 9, 1999, the FCC made line sharing official policy. The FCC order applies only to voice-compatible forms of DSL, such as ADSL and RADSL. HDSL, SDSL and other DSL versions which require use of the entire local loop are not covered by the order, as they are not voice-compatible and, therefore, do not allow the line to be shared. The FCC order also requires the ILECs to provide to a DLC (Digital Loop Carrier) or any other accessible terminal in the outside loop plant. The FCC order further stated that the cost of line sharing arrangements should be on a cost basis. As the ILECs charge themselves zero for such arrangement, the CLEC argument is that they should be charges zero, as well. Several ILECs have agreed to such zero-cost deals. See also ADSL, DLC, HDSL, RADSL, SDSL, and xDSL.

line signals 1. In the ITU-T sense, equivalent to "supervisory signaling" in Bell network language.

2. In North American colloquial use, any of a wide variety of communications signals as found transmitting a physical transmission medium.

line queuing Dial an outside line (typically a long distance line). It's busy. Your phone system will put you in queue for that line. The queue might involve your waiting a few seconds on hold; or it might involve your hanging up and having the phone system call you back. There are thus two types of Line Queuing – hold-on and callback queuing.

line relay A telephone company term. Relay in a subscriber's line which operates on the calling-in signal.

line ringing Provides the user with an audible indication of a call on a specific line that appears on his/her telephone.

line signals 1. In the ITU-T sense, equivalent to "supervisory signaling" in Bell network language.

2. In North American colloquial use, any of a wide variety of communications signals as found transmitting on a physical transmission medium.

line side connection A carrier term. A local loop, which connects the customer premise to the carrier network. The carrier community uses this term to describe the customer side of the network, regardless of whether it is specifically in the form of a line or a trunk. In this context, the term "trunk" refers to a local loop which connects a network switch (e.g., a central office circuit switch, a frame relay switch or router, or an ATM switch) to a customer switch (e.g., a PBX circuit switch, a frame relay router, or an ATM switch). Also in this context, a "line" connects a network switch to a non-switch (e.g., a telephone set, a computer modem, or a traditional key system). In other words, a trunk connects one switching device to another switching device, while a line connects a non-switching device to another device, which can be in the form of either a switch or a non-switch. Compare with Trunk Side Connection.

line-side termination An end office switch connection that provides transmission, switching and optional features suitable for customer connection to the public switched network, including loop start supervision, ground start supervision, and signaling for BRI-ISDN service.

line speed The maximum number of bits you can transmit over a line in a certain defined time, say one second.

line status indication Provides a visual indication on an ECTS (electronic telephone set) telephone of the idle, busy, ringing or held state for each line appearing on the telephone.

line switched ring A technique for providing redundancy in a SONET network. Line-switched rings use either 2 or 4 fibers per ring. The primary ring transmits in one direction (e.g., clockwise), with the other transmitting in the reverse direction. Through this technique, a failure in a SONET ring will not prevent devices from communicating, as they can transmit and receive at least one direction, assuming that there is no more than 1 break in a fiber. See Path-Switched Ring.

line switching Another term for circuit switching. See Circuit Switching and Switching.

line termination 1. Defines the local loop at the telephony company side of a digital connection – DSL or ISDN. The classic line termination devices is the NTI at the user side of the interface.

2. A Verizon definition. Equipment that terminates a BRI or Centrex BRI digital subscriber line on the network side of the network to the end user (or CLEC) interface. Alternatively, electronics at the ISDN network side of the user-network interface that complement the electronics equipment.

line transfers A telephone company definition. Line transfers consist of physically removing a customer line from one line equipment and moving it to another. Line transfers are used as an important corrective tool to improve load balance.

line turnaround time The delay in a circuit as the direction of communications changes, usually in half duplex communications. When one side of the communications stops sending, there is a delay before the other party stars sending in return. This is the Line Turnaround delay.

line up Plant technician colloquialism for testing and adjusting transmission performance of a circuit to specified levels and losses at various points along the circuit.

line voltage Voltage measured on a telephone circuit; typically 48 volts DC when phone is idle. Voltage may be lower at great distance from the central office, or when carrier equipment is used to multiplex several phone lines on one pair of wires. Line voltage on PBX systems is typically 24 or 48 volts.

linea de abonado digital Spanish name for DSL. See DSL.

linear distortion Amplitude distortion wherein the output signal envelope is not proportional to the input signal envelope. This distortion is often caused by part of the signal being bounced off something, while part arrives free and clear. Thus the receiver hears the same signal but bits of it arrive earlier and later than other bits, causing distortion. See Linearity for a better explanation.

linear feedback shift register See LFSR.

linear power amplifier LPA.

linear predictive coding A speech coding method that analyzes a speech waveform to produce a time-varying filter as a model of the human vocal tract. See also Digital Signal Processing.

linear programming Techniques in Operations Research (OR) to find an optimum solution to a linear function, given certain restrictions and typically expressed in many equations. A typical linear programming problem might be to find the least expensive, most efficient route between various pick-up and drop-off points in a transportation route.

linear tape open architecture LTO technology is an open architecture, high-capacity tape format created by manufacturers, Certance, HP, and IBM. LTO is largely used for backing up huge amounts of data in the midrange to enterprise-class server environments. Imagine a strip of wide tape – not unlike tape used for audio or video recordings. Now, instead of writing data across the entire tape, LTO splits the tape horizontally into four bands. It writes to one band and checks it. If that's OK, it writes to another band. It writes backwards and forwards and until all the data is written and checked and duplicated. According to the web site, www.ultrium.com, "LTO's ultra-high capacity of tape storage products is designed to deliver outstanding performance, capacity and reliability combining the advantages of linear multi-channel, bi-directional formats with enhancements in servo technology, data compression, track layout, and error correction."

linearity Think of a cheap sound system. The higher you turn up the volume, the more ugly (i.e. more distorted) your music sounds. Think of an expensive sound system. The higher you turn up the volume, the better your music sounds to you (though not to your wife). Linearity is a term for measuring the amount of distortion you get as your crank up the "volume" – which in transmission terms – might also be called the throughput or the bandwidth. Let's take an example we all understand: To increase their revenues, cable TV operators are offering more and more services to their customers – including high-speed Internet access. They call these "bandwidth-intensive services." Operators are thus faced with a big problem – how to push more information down the existing cable plant. While there are many techniques, the inevitable outcome of expanding the bandwidth (pumping more information down an existing pipe) is that a higher performance distribution system is required. That means the system has to be upgraded with better electronics, better components, better cable, etc. Increasingly, the linearity (or distortion performance) of the delivery system begins to play a critical role in the overall cost per bandwidth (how much your customer will pay) and quality of service provided to the customer. Because you lose signal power (and thus information) as your information travels down a transmission medium (copper or glass), amplifiers are needed. But amplifiers (like those in your sound system) are the key contributor to system non-linearity (lousy sound). Amplifiers with lower distortion allow delivery costs to be reduced and system performance to be improved. Hence, there is strong demand for higher output power, ultra-linear (i.e. non-distorting) amplifiers that operate on a fixed budget of AC-line or battery power -. While it's easy to increase linearity by - consuming more power, this is usually undesirable in applications running from or being backed-up by batteries. And when more AC-line or battery power is consumed, more heat is generated, which decreases operating life expectancy. Thus you see a lot of R&D work happening in this important telecom area. In more technical language, linearity is the consistency of gain   as input level is increased. If the gain response (magnitude and phase) of an amplifier changes with an increase of input level, the change indicates harmonic distortion is being generated. . In an A/D (Analog to Digital) or D/A (Digital to Analog), linearity measures the precision with which the digital output/input tracks the analog input/output.

linearly polarized A mode of operation of fiber optics for which the field components in the direction of propagation are small compared to components perpendicular to that direction.

lineman A person who fixes the telephone company's outside aerial plant – typically the wires hanging from poles dotted across the country-side. See also Lineman's Climbers.

lineman's climbers Telephone pole climbing irons which are strapped to the telephone lineman's legs, allowing him or her to climb a wooden telephone line. You can tell when a pole has been climbed by the holes left in it by the lineman's climbers.

lines A computer imaging term. The line tool draws straight lines, typically from point to point. Most paint packages let you continue lines in a fashion that permits rapid creation of polygons.

lines of force The directional lines of magnetic or static field which represent the stresses.

lines per minute The way of measuring the speed of a line printer. Like any measure of speed, the speed you will get from your printer may be different from what the manufacturer says. Your speed will depend on how fast you feed the printer from your computer – a function of how fast you're transmitting, what software you're running, how fast that software can get the information to be printed off your disk, etc.

line side See Lineside.

link 1. Another name for a communications channel or circuit. The ATM Forum defines link as an entity that defines a topological relationship (including available transport capacity) between two nodes in different subnetworks. Multiple links may exist between a pair of subnetworks. Synonymous with logical link.

2. A Windows command that takes several programs and subprograms that were meant to be used together, but were written separately, and combines them into one. Usu-

ally used to create an executable program out of modules that were not themselves directly executable.

3. An element in an HTML document that points to a document or to a specific location in a document, using a URL. When the document is displayed in a browser, clicking on a link causes the browser to display the document and/or location that it points to. Links usually appear on-screen as underlined text and are usually in blue, although Web page designers can change how they look.

4. An optical cable with connectors attached to a transmitter and receiver (source and detector).

link access protocol A version of HDLC in which the communication line has no single controller and either of the two connected stations may initiate a data transfer operation.

link aggregation The grouping of multiple network links into one logical high bandwidth link. For example, you can group four separate leased T-1 links, each offering bandwidth of 1.544 Mbps, into a single logical link offering aggregate bandwidth of 6 Mbps. That approach might well be preferable to upgrading to the next level in the T series, namely a T-3, which not only would be much more expensive and which might be oversized at 28 T-1s or 44.736 Mbps, but which also might not be available through your local service providers. Similarly, you might group four GbE (Gigabit Ethernet) links for a total of 4 Gbps, which would be less expensive than a 10GbE (10 Gbps Ethernet) link, which also might not be available or which might be way too expensive. To aggregate links you need equipment. See Inverse Multiplexing.

Link Aggregation Control Protocol LACP, also known as IEEE 802.3ad. The Link Aggregation Control Protocol-LACP is a specification for bundling multiple Ethernet links into what appears to be one spanning tree protocol (STP) link.   Before   this specification was ratified, various vendors had their own proprietary mechanisms for providing this functionality, but it would not work in mixed vendor environments. The technology allows an uplink to have 8 Gbps aggregate uplink speed in the situation where 8 Gigabit Ethernet links are used between two switches. Without the technology, seven of the links would have gone into the STP blocking state because spanning-tree protocol detected a loop in the network.

Link Aggregation Token See Aggregation Token.

link attached Describing devices that are connected to a network, a communications data link, or telecommunications circuit; compare with channel-attached.

link attribute A link state parameter that is considered individually to determine whether a given link is acceptable and/or desirable for carrying a given connection.

link budget A list of radio frequency powers, gains, and carrier-to-noise interference radios along the entire transmission bath between two stations.

Link Capacity Adjustment Scheme See LCAS.

link connection An ATM term. A link connection (e.g., at the VP-level) is a connection capable of transferring information transparently across a link without adding any overhead, such as cells for purposes for monitoring. It is delineated by connection points at the boundary of the subnetwork.

link constraint A restriction on the use of links for path selection for a specific connection.

link control facility A Fibre Channel hardware facility which attaches to the end of a link and manages the transmission and reception of data. The LCF is contained within each F Port (Fabric Port, i.e., switch port) and N Port (Node Port, i.e., device port). See Fibre Channel.

Link Control Protocol LCP. A Link Control Protocol is a protocol operating at Layer 2, the Data Link Layer, of the OSI Reference Model. Such a protocol is employed by circuit terminating equipment at each end of the link in order to communicate across it. The specifics of an LCP include packet format, packet size, compression mechanisms, link performance monitoring, handshake and authentication mechanisms, and error detection and correction. LCP examples include HDLC (High-level Data Link Control) and PPP (Point-to-Point Protocol). See also HDLC and PPP.

link converter A device for an InteCom S/80 which connects distributed switching modules to the centralized switching equipment through a coaxial cable or a fiber optic cable.

Link Encapsulation LE. A function of the HiPPI Framing Protocol (FP) layer. LE encapsulates IEEE 802.2 Logical Link PDUs (Protocol Data Units) inside of HiPPI packets, thereby allowing IP traffic to travel over a HiPPI connection.

link farm A web page that consists of lots of links to other web pages, and nothing else.

link layer The logical second layer of the OSI Reference Model for Open Systems Interconnection, located between the Physical and Network layers. See OSI.

link layer access method The algorithm that determines when any given network interface in a PC/TCP local area network is allowed to transmit. It is also known as the access method. CSMA/CD is the access method for the Ethernet.

link love Spreading link love is to post links to a site or blog, usually unsolicited, that you enjoy, admire, or find useful.

link metric An ATM term. A link parameter that requires the values of the parameter for all links along a given path to be combined to determine whether the path is acceptable and/or desirable for carrying a given connection.

link optimization ISDN feature that prevents administration packet from opening the communications link and allows only user data to open the line. Link optimization ensures that remote connections are not kept open unnecessarily, which saves usage costs.

link protocol The set of rules by which a logical data link is set up and by which data transfers across the link. It includes formatting of the data.

link pulse A communication mechanism used in 10Base-T networks to indicate link status and, in auto-negotiation equipped devices, to communicate information about abilities and negotiate communication methods. 10Base-T uses Normal Link Pulses (NLPs) which indicate link status only. These are transmitted periodically while not transmitting packets. 10Base-T and 100Base-T nodes equipped with auto-negotiation exchange information using a Fast Link Pulse (FLP) mechanism which is compatible with NLP.

link rot The process by which links on a Web page become obsolete because of changes in location or expiration of the sites to which they are connected. Link rot happens quickly.

link set A group of signaling links directly connecting two signaling points. Signaling data links are grouped into link sets. All links in a link set must connect to a single point code. Up to 16 links can be assigned to a single link set.

link state protocol A type of routing algorithm in which updates to routing tables are exchanged between neighboring routers only when modifications need to be made. Such modifications would be required, for example, if a new link or router were added to the network, or if a link between two routers suffered a catastrophic failure. Some link state protocols provide means to assess the performance characteristics of the various available links, thereby supporting a bias toward the link performing best. Distance Vector Protocols, on the other hand, exchange routing data on a highly regular and predetermined basis, regardless of whether updates are required. Therefore, link state protocols consume less networking resources and reduce network congestion by providing updates only when needed, and sending only the changes. Periodically, the entire route table will be sent as a precautionary procedure. Examples of link state protocols include OSPF (Open Shortest Path First), ISO's IS-IS (Intermediate System-to-Intermediate System), and Novell's NLSP (NetWare Link Services Protocol). See also Distance Vector Protocol, Path Vector Routing Protocol, Policy Routing Protocol, Router, and Static Routing.

link state parameter Information that captures an aspect or property of a link.

Link Status Signal Unit LSSU. A packet sent between MTPs (Message Transfer Part of the SS7 Protocol) to provide SS7 information about the sending node and its links. This information is sent during the initial alignment of the links, when there is an associated processor outrage, or when link congestion is detected.

link test A test that is performed by the hardware to ensure the integrity of the cable in a local area network. The link test can be disabled to allow old style NICs incapable of performing a link test to connect to the repeater.

link time This is a specific time delay that allows access to PBX or Centrex features through a telephone system. Link Time is also referred to as a Hookswitch Flash or Recall.

linked object A representation or place holder for an object that is inserted into a destination document. The object still exists in the source file and, when it is changed, the linked object is updated to reflect these changes.

links 1. The transmission portion of the local loop.

2. An affectionate name for Apple's electronic mail system, called AppleLink.

LINX London Internet Exchange. See IX.

Linux Linux (officially pronounced LINN-ucks) is an increasingly popular, open source computer operating system. Open source means that Linux is basically free. Many companies "sell" it but what they really sell is the help they offer in getting Linux up and running on your computer. Today you'll find Linux running mainly on servers. Every time you run a Google search or place a bid at eBay, for example, you're tapping into databases spread across thousands of Linux servers. Linux is moving to the desktop. Wal-Mart sells simple PCs that run Linux and will do email, surf the net and do basic word processing and spreadsheets. Linux owes its creation to Linus Torvalds, who started the Linux kernel as a research project at the University of Helsinki, Finland in 1991, allegedly in his undergraduate dorm room at the University of Helsinki. He built Linux on ideas borrowed from AT&T's Unix operating system and on the work of the GNU open-source project. His goal was to create something faster and more streamlined than either Unix or Windows. Torvalds invited other programmers to copy, use, and improve his offering, as long as they agreed to share any changes they might make, and he has been the movement's unofficial regent ever since, approving every new line of code. While Linux started out as a program written "for hackers by hackers," in Torvalds's words, that era is long past. Torvalds himself is now paid by an industry consortium, the Open Source Development Labs in Portland, Oregon, to oversee Linux's evolution. And the typical open-source programmer, apparently, is no longer a passionate hobbyist but a full-time professional at a company that either publishes or uses open-source software. At IBM, for example, 7,500 programmers are making the company's business software run on Linux, which many customers see as more reliable and less virus prone than Microsoft's Windows products.

While it is true that Linux is very closely modeled after Unix, and in most cases programs that run on Unix will run on Linux, Linux is not really Unix. Linux is among the most powerful and feature-packed operating systems available for the PC, offering a large base of hardware peripheral compatibility. Linux hosts an impressive array of compilers and development environments, including C/C++, Perl, Pascal, SmallTalk, and complete X-windows system that rivals many commercial offerings. For all its life, Linux has been free. But it is not shareware. Major parts of it are copyrighted. Linux has many script languages and parsers such as Awk, Sed, Yacc as well as all popular shells (Borne, Korn, C, BASH, etc.). The Linux kernel was originally written only for the Intel 80x86 microprocessor, i.e. the 386/486/Pentium family of chips. But it has been ported to Alphas, Sparcs, 68k, and PowerPC. The source code for Linux is available on the Internet to anyone who wants it." See OpenOffice, www.linuxresources.org/what.html, www.redhat.com and www.rtems.com.

Linux Phone Standards Forum November 14, 2005: A consortium of telecommunications and technology companies Monday established a trade body dedicated to driving the adoption of the Linux operating system in Mobile phones. Linux is currently used in a small number of handsets. The open-code system is an alternative to other mobile phone operating systems developed by Microsoft Corp. (MSFT) with its Windows Mobile software and Symbian PLC (SYN.YY) - majority owned by Nokia Corp. (NOK) – which develops systems for high-end mobile phones with computer-like applications. Founding members of the Linux Phone Standards Forum include ARM Holdings PLC (ARMHY), Cellon, Esmertec AG (ESMN.EB), France Telecom SA's (FTE) Orange unit, FSM Labs, Huawei Technologies Co. (HWI.YY), Jaluna, MIZI Research, Montavista Software, Open-Plug and PalmSource Inc. (PSRC). The forum will look to define standards and programming interfaces for Linux-based system services to support the development and deployment of applications and user-level services. www.lipsforum.org

Lilon Lithium Ion. The Lithium Ion battery is lightweight and does not suffer from memory effect. It also delivers a higher run time average and about 80% more power per ounce. Similar to NiMH technology, Lilon batteries have a life expectancy of 500 charge and discharge cycles. Lilon batteries are typically used in mid- to high-priced portables.

LIP Loop Initialization Protocol. Part of the FC-AL (Fibre Channel Arbitrated Loop) standard for reconfiguring the system when nodes are added or removed from the loop.

lipstick Over her lifetime, the average American woman will swallow four pounds of lipstick. The good news is that that's a lot less than the Coca Cola I drink.

lipstick indicator The tendency for lipstick sales to increase prior to and during a recession.

Liquid Crystal Display LCD. A low power display that aligns material suspended in a liquid under the influence of a low voltage so it reflects ambient light and displays alphanumeric characters. LCD displays are finding great use as methods of displaying information on new electronic telephones, especially those positioned behind PBXs. The advantage of putting such displays on telephones is that the power to drive the display is very small. The display can be line powered – i.e. powered by the one or two pairs coming from the PBX. This avoids the necessity and cost of a transformer/rectifier – the little black box you plug into the wall to run your answering machine or to power up your rechargeable calculator/laptop computer. Such LCD displays on electronic phones can perform many functions. The most useful is that of "walking" the user through the phone call – showing him/her how to transfer a call, to make a conference call, to split a conference, etc. An LCD can also alert you as to who's calling you.

LIRC Long Run Incremental Cost.

LIS 1. Link Interface Shelf.

2. Local Interconnection Service. A local interconnection service arrangement with an ILEC (Incumbent Local Exchange Carrier) to provide trunking, E911, SS7 requirements for LNP (Local Number Portability).

Lisp Lisp is the LISt Processor language– is "the greatest single programming language ever designed," according to computer scientist Alan Kay. It was born in 1958 because John McCarthy, then an assistant professor at MIT, working on new tools for artificial intelligence research, wanted a language in which one could write programs that would make logical inferences and deductions. Previous languages, including – Fortran – were numeric, which made for powerful number-crunching. But Lisp made use of symbolic expressions, which treated both data (such as numbers) and code as objects that could be manipulated and evaluated. This enabled programmers to create conditional expressions. Lisp made possible the now-familiar "if-then-else" structure – and today Lisp is used as a "macro" language, allowing users of software such as Emacs to create their own mini-applications that can automate tasks. This definition from Daniel Turner.

list box A Windows term. In a dialog box, a box that lists available choices, for example, a list of all files in a directory. If all the choices do not fit in the list box, there is a scroll bar.

list host A host computer, in the form of a server, that is used to support e-mail list services, usually on an outsourced basis. The service provider will place your e-mail list on its host computer, and associate it with an e-mail that you want to send to many e-mail addresses. You just send the e-mail to a special target mailbox, and the list host server forwards it to everyone on your list. If people want to be removed from the list, they so indicate with a return e-mail, and the list host server takes care of it. At least that's the theory.

list server An automated mailing list distribution system.

Listed Directory Number LDN. Incoming exchange network calls to the PBX via assigned listed local telephone directory number are directed to the attendant.

listen before talk LBT. Same as carrier sense multiple access (CSMA). Compare with Listen While Talk.

listen only A conference call mode wherein a participant can only listen to the call, not speak. The listen-only mode can be toggled on/off by the conference moderator or operator. Listen-only mode is useful for blocking out background noise during press announcements, investor relations presentations, and other conference calls with a large number of conferees. Listen-only mode is also useful in these scenarios where only a designated subset of conferees are authorized to have a speaking role.

listen while talk LWT. Same as carrier sense multiple access with collision detection (CSMA/CD). Compare with Listen Before Talk.

LIT See Line Insulation Test.

lit fiber Let's start with optical fiber. When a carrier initially installs an optical fiber in the ground (sub-terrestrial), under the ocean (submarine), or through the air (hung on poles), it's called "dark" fiber. That means he hasn't put any electronics on it, so he's not sending any light down the fiber and thus he's not transmitting any information. Usually a dark fiber is one of many dark fibers in a cable containing a great many fibers, commonly 432 in sub-terrestrial cabling systems. The carrier deploys such a large cabling system because the incremental cost of pulling one large cable is much less than the incremental cost of pulling fibers one at a time, as needed. The dark fibers are designated for future use, or for backup purposes in the event that a fiber fails for some reason. Sometimes dark fiber is sold or leased by a carrier without the accompanying transmission service, e.g., SONET. The customer is expected to put his own electronics and photonics on the fiber and thus be able to make transmissions. This process is calling "lighting" the fiber, since light now moves down the fiber. And the fiber is now called "lit" since that's the past participle of the verb "to light." If a given optical fiber system supports WDM (Wavelength Division Multiplexing) or DWDM (Dense WDM), multiple wavelengths of light can be transmitted across the system simultaneously. If not all of the available wavelengths are active (e.g., only 10 of a possible 32), the fiber is said to be "dim," i.e., neither dark nor fully lit. See also Dark Fiber, Dim Fiber, DWDM, SONET, and WDM.

lithium ion Type of highly efficient rechargeable battery, often used in computer laptops and cellular portables. Here's an explanation from 1-800-BATTERIES, a seller of rechargeable batteries:

NiCad: Nickel Cadmium is the most popular and durable type of rechargeable battery. It is quick to charge, lasts about 700 charge and discharge cycles, and works well in extreme temperatures. Unfortunately, NiCads suffer from "memory effect" if they are not completely discharged during each cycle. The memory effect reduces the overall capacity and run time of the battery.

Nickel Metal Hydride (NiMH) batteries do not suffer from memory effect. Compared to a NiCad battery of equal size, NiMH batteries run for 30% longer on each charge. They are also made from non-toxic metals so they are environmentally friendly. The downside to NiMH technology is overall battery life. These batteries last for 400 charge and discharge cycles.

Lithium Ion (LiON) is the latest development in portable battery technology. These batteries do not suffer from memory effect. Compared to a NiMH of equal size, a LiON will deliver twice the run time from each charge. Unfortunately, these batteries are only available for a limited number of models and are expensive. Similar to NiMH technology, LiON batteries have a life expectancy of 400 charge and discharge cycles.

little endian A format for storage or transmission of binary data in which the least significant byte (bit) comes first. See Little-endian.

little LEO Relatively small and inexpensive low earth orbiting satellites that provide low-cost, low-data rate, two-way digital communications, and location positioning to small handheld terminals. The frequency allocations are in the VHF band below 400 MHz. Systems include Leosat, Orbcomm, Starnet, and Vitasat. For example, the Orbcomm system requires 34 satellites for reliable full-world coverage. See Big LEO.

little-endian See Big-endian.

LIU Line Interface Unit. Essentially a digital transceiver, an LIU is a generic term for a type of Circuit Terminating Equipment (CTE) used to terminate a T-1/E-1 circuit or ISDN PRI (Primary Rate Interface) circuit. In an optical environment, an OLIU (Optical Line Interface Unit) serves the same function for interfacing with a SONET/SDH OC-1, or better, optical circuit. Generally speaking, an LIU provides CSU (Channel Service Unit) and DSU (Digital Service Unit) functionality, supports a variety of framing formats, provides loopback test capability, and offers performance monitoring and diagnostics. The term LIU also sometimes is used to describe a PBX line card, which serves to interface PBX extensions to the switch, as opposed to trunk cards, which serve to interface the PBX switch to a CO (Central Office) switch. See also CSU, DSU, E-1, Loopback Test, SONET, SDH, and T-1.

LIU7 Line Interface Unit for CCS7.

live bug Colloquialism used to refer to a leaded integrated circuit package when the leads are down, like a bug that is alive and standing upright.

Live Call Screening See LCS.

livelock A request for an exclusive lock that is repeatedly denied because a series of overlapping shared locks in a shared database keeps interfering. A SQL server will detect the situation after several denials, and refuse further shared locks.

liver The human liver stretches across almost the width of the body, occupying a space about the size of a football. It weighs more than 3lbs. I have no idea why anyone other than me would be interested in this trivia.

liveware People.

LL Long Lines.

LLB Line LoopBack. A maintenance and/or diagnostic mode of operation whereby a CSU regenerates a signal received from a span line and retransmits that signal back onto the span towards its point of origin.

LLC Logical Link Control. A protocol developed by the IEEE 802.2 committee for data-link-level transmission control. It is the upper sublayer of the IEEE Layer 2 (OSI) protocol that complements the MAC protocol. IEEE standard 802.2 includes end-system addressing and error checking. It also provides a common access control standard and governs the assembly of data packets and their exchange between data stations independent of how the packets are transmitted on the LAN. See 802 Standards.

LLC2 Logical Link Control 2. The frame format used to carry 3270 traffic on Token Ring LANs.

LLDP Local Loop Demarcation Point. See MPOE.

LLDPE Linear Low Density Polyethylene.

LLF 1. Line Link Frame. See Crossbar Switching.

2. Low Layer Functions.

LLPOFYNILTATW Liar, Liar Pants On Fire, Your Nose Is Longer Than A Telephone Wire.

LLU Local Loop Unbundling. This term is used in England where British Telecom (BT), the pre-eminent British telecom company, encourages other telecom companies to rent the copper wire in its local loops for applications such as broadband Internet access, e.g. DSL. LLU happened in the U.S. Because the main renters tended to be companies who wanted to use the phone companies' own cables to compete against the phone companies, the risk of success was small. And predictably, most CLECs, as they were called in the U.S. (competitive local exchange carriers), failed. In the melee that followed, the good news was that the price of DSL lines fell worldwide.

LLWAS Low Level Wind Shear Alert System.

LM Long distance Marketer.

LMB See LNB.

LMCS Local Multipoint Communication Service. The Canadian equivalent of LMDS (Local Multipoint Communications Service). Using frequencies above 25 GHz, LMCS produces a wireless broadband digital network capable of delivering high-bandwidth signals over the air. Industry Canada has allocated the frequency band 25.35 to 28.35 GHz for LMCS networks. Some observers believe LMCS may represent a form of fiber to the curb. See also LMDS.

LMDS Local Multipoint Distribution System. Developed by Bellcore (now called Telcordia Technologies) for Wireless Local Loop (WLL) applications, LMDS systems initially were trialed commercially in New York for point-to-multipoint broadcast TV, on the basis of an experimental FCC license granted to CellularVision. In that application, broadcast microwave signals operating at 28 GHz transmit to small receiver dishes, typically installed on the top of apartment buildings. Each of 12 transmitters in the boroughs of Manhattan, Brooklyn, Queens and portions of The Bronx covers an area of 28 square miles. At that high frequency, line-of-sight is required for maximum signal performance. This necessity for Line-of-Sight (LOS) is the reason it wasn't installed in Manhattan (too many tall buildings). The received LMDS signal is often then distributed through the building's central CATV system. It also can be used to broadcast directly to a subscriber's home via an 18" flat antenna sitting in the subscriber's window. There are actually all sorts of variations on the LMDS theme. In one trial, the service was used for high-speed Internet downloads to LMDS subscribers – the Internet downloads coming from LMDS, the command to initiate those downloads being sent from the subscriber's PC over his local phone line. There is R&D going on at present to enable LMDS to carry two-way voice conversations. In Brazil, CellularVision uses LMDS technology, transmitting in the FM range, which means the signal has the ability to bounce, and to reflect off virtually any surface, thereby avoiding issues of line-of-sight and increasing the coverage area significantly. Two-way or interactive communication may be inserted between video channels for transmission back on the opposite polarity. This reverse polarization, or interweaving, theoretically allows simultaneous use of signals at the same frequency for two applications. In March 1997, the FCC set aside total LMDS bandwidth of 1.15 GHz in the 28-GHz, 30-GHz and 31-GHz frequency bands, with the intent that LMDS would be used for its original intended purpose of WLL. LMDS also was viewed as competitive with conventional cable-based CATV. LMDS generally has been regarded as a commercial failure due to its high cost, performance issues and requirement for LOS. LMDS has been superseded by the IEEE 802.16 specification, which also is known as WiMAX. See also 802.16, ADML, Broadband Wireless Local Loop, LMCS, MMDL, MMDS, WiMAX and Wireless Local Loop (WLL).

LMEI Layer Management Entity Identifier.

LMHost LMHost is a text file which contains the NetBIOS name and IP addresses of other computers on a network. Microsoft Windows Network is a NetBIOS-based network where each computer is given a unique name, the NetBIOS name. In a traditional NetBIOS name, each machine sends a NetBIOS broadcast that announces its name as it boots. If another host already exists with that name, it will send a message to the new client saying that that name is in use. If it doesn't get a message back, the client assumes that the name is available. One of the ways to get around this broadcast problem is to create a text file named LMHost on every computer. Not a recommended course of action.

LMI 1. Local Management Interface. A specification for the use of frame-relay products that define a method of exchanging status information between devices such as routers.

2. Logical Modem Interface. The core of the Microsoft Fax interface. LMI lets third-party licensed vendors write plug-in modules to provide instant and transparent access to diverse underlying systems. An easy analogy for the LMI is to consider the Windows print manager. To the user, simply installing the printer driver suited to their printer is all that is required. According to Microsoft, "The LMI interface provides a similar layer between the internal fax components of Windows 95 and the fax hardware or, in our case, the fax server."

LML Lineup Maintenance Level.

LMOS Loop Maintenance Operations System.

LMP A Bluetooth term. Link Manager Protocol. The LMP is used for peer-to-peer communication.

LMP Authentication A Bluetooth term. An LMP level procedure for verifying the identity of a remote device. The procedure is based on a challenge-response mechanism using a random number, a secret key and the BD_ADDR of the non-initiating device. The secret key used can be a previously exchanged link key or an initialization key created based on a PIN (as used when pairing).

LMP Pairing A Bluetooth term. A LMP procedure that authenticates two devices based on a PIN and subsequently creates a common link key that can be used as a basis for a trusted relationship or a (single) secure connection. The procedure consists of the steps: creation of an initialization key (based on a random number and a PIN), lmp authentication based on the initialization key and creation of a common link key.

LMR This is Cisco's explanation of LMR over IP. It's a good explanation of LMR and LMR over IP. A Land Mobile Radio (LMR) system is a collection of portable and stationary radio units designed to communicate with each other over predefined frequencies. They are deployed wherever organizations need to have instant communication between geographically dispersed and mobile personnel. Typical LMR system users include public safety organizations such as police departments, fire departments, and medical personnel. However, LMR systems also find use in the private sector for activities like construction, building maintenance, and site security. In typical LMR systems, a central dispatch console or base station controls communications to the disparate handheld or mobile units in the field. The systems might also employ repeaters to extend the range of communications for the mobile users. LMR systems can be as simple as two handheld units communicating between themselves and a base station over preset channels. Or, they can be quite complex, consisting of hundreds of remote units, multiple dispatch consoles, dynamic channel allocation, and other elements. LMR systems have proven a very useful tool to many types of organizations. However, recent events have exposed limitations in the ability of LMR systems to fulfill certain communications needs, particularly system interoperability. By combining LMR systems with the connectivity of IP networks, we can solve many limitation problems. Within an organization, the radio systems tend to be homogenous, with most elements typically purchased from the same manufacturer. Although the electromagnetic spectrum is rather vendor agnostic, signaling mechanisms and other control aspects of individual radio systems can be quite proprietary. This proprietary factor means that adding equipment generally means purchasing from the same manufacturer or finding compatible equipment, assuming that it still manufactures that particular model of radio. If organizations merge or need to consolidate operations that were previously using different LMR systems, issues with interoperability could require workarounds to bridge the existing systems or ultimately require the purchase of all new equipment. Interoperability issues within an organization are one aspect of the problem. Consider the situation in which multiple public safety organizations are involved with the same incident. Organizations enjoy the autonomy of using their own radio systems with their own channels. But autonomy implies that the radios for one group will not be able to communicate with radios used by other groups. So, coordinating the activities of the field personnel from these different groups at one site requires some sort of workaround, either redeploying radios, or some sort of custom cross-patching at dispatch consoles to bring parties together. Closely associated with interoperability issue is the ability to extend the command and control function of radio systems. Generally, providing someone with the ability to participate in a radio talk group means giving that person a radio. However, if the radio user is out of range of the radio system or is an infrequent user of this capability, that solution might be physically or economically unfeasible. Today, radio systems can be linked through leased lines or over the public telephone network to extend their reach. These lines can be expensive and are often in addition to the communication services run for data purposes. With the LMR over IP service, standards-based VoIP technology voice gateways are used in combination with additional LMR specific features to address interoperability, extending command and control, and other issues. Base stations, repeaters, and dispatch consoles generally possess a wired interface that can be used to monitor audio received from their air interface, and as input for audio to be transmitted on their air interface. Although this wired interface may contain other control capabilities as well, as long as it has some sort of speaker output and microphone input, it can be connected to a voice port on a router. The audio received on the voice port is encoded with a standard audio codec, such as G.711 or G.729. Those audio samples are packaged in standards-based Real-Time Transport Protocol (RTP) packets suitable for transport on an IP network. At this point, the communication element is abstracted from the distinctive characteristics of each radio system, thus providing a solution for the interoperability problem. Now, these audio packets can be sent across the network to other LMR gateways with different brands of radio systems either individually (unicast) or as a group (multicast). The recipient of the audio packets need not be another LMR gateway. It can be any device capable of receiving and decoding the RTP stream, such as an IP telephone or PC with appropriate software. The IP network and IP-enabled devices can be used to allow users to monitor or transmit on a particular radio channel from a desk without issuing another radio. This can be done locally, nationally, or internationally, assuming the IP network has been properly designed.

LMR over IP See LMR.

LMS 1. Local Message Switch.
2. Loop Monitoring System.

LMSS Land Mobile Satellite Service.

LMU Line Monitor Unit.

LNA Low Noise Amplifier.

LNB Low Noise Blocker, also called Low Noise Block converter. Imagine a satellite dish for receiving signals from a distant satellite. The dish is parabolic (not a sphere). As the Microwaves come in from afar. The dish catches those microwaves and focuses them on a point in front of the antenna. At that point there's an LNB. The LNB is the component located at the end of the arm projecting from the satellite dish. It receives the signals sent by the satellite (e.g. "Ku-band") and converts them to a lower frequency (e.g. 3.7-4.2GHz) that can be accepted by a satellite receiver. LNBs send this converted signal to the satellite receiver via RG-6 coaxial cable. DirecTV, for example, currently has satellites in three orbital positions, and a separate LNB is needed to access each satellite position and to access the programming on these satellites, e.g. high definition. See also Triple LNB dish.

LNC Low Noise Converter.

LND Last Number dialed.

LNNI LANE Network-to-Network Interface. An ATM term for the standardized interface protocol between LANE (LAN Emulation) servers (LES-LES, BUS-BUS, LECS-LECS and LECS-LES). See also LANE and LUNI.

LNP Local Number Portability. Similar in concept to 800/888 and other toll-free number portability, LNP was mandated by the Telecommunications Act of 1996 to level the playing field between the ILECs (Incumbent Local Exchange Carriers) and the CLECs (Competitive Local Exchange Carriers). In July 1996, the FCC issued a ruling that LNP must be in place nationwide by January 1, 1998. Since each state is responsible for implementation of LNP, timetables vary; the specifics of the implementations vary, as well.

In some states, the implementation approach is exactly like that for 800 (toll-free) number portability. In other words, the originating central office "dips" into a centralized database of numbers via an signaling system 7 (SS7) data link. The database, known as a SCP (Service Control Point) in IN (Intelligent Network) terms, identifies the LEC (local exchange carrier) providing service to the target telephone number in order that the originating carrier can hand the call off to the terminating carrier.

In other states, such as Illinois, which is the first to implement LNP, a totally different approach is taken. This implementation involves the use of a new 10-digit telephone number, known as a LRN (Local Routing Number). When the originating CO switch consults the SCP, the new 10-digit number is provided along with the identification of the CLEC to which the service has been ported. The originating carrier then hands off the call to the CLEC. While this approach is claimed to be faster, clearly two telephone numbers are required, thereby placing additional pressure on the North American Numbering Plan (NANP).

To implement LNP, the FCC has mandated a system of regional databases, which will store master copies of all porting information. These databases will be maintained by regional Number Portability Administration Centers (NPACs) that will serve as number portability clearinghouses for all local operators. Originally, the deal was that Lockheed Martin would maintain the databases in four regions and Perot Systems in three regions. After Perot had some problems getting going on time, Lockheed Martin was selected to run all seven. That responsibility now rests with NeuStar, which was an independent business unit within Lockheed Martin before being spun off.

In either case, LNP will require the SCPs be established by LECs, CLECs, IXCs (inter-exchange carriers) and wireless carriers. Further, the SCPs must be synchronized in order that the databases are consistent across them all. The concept is simple, but its implementation is complex and expensive. MNP (Mobile Number Portability) is the LNP version intended for eventual use in certain mobile networks. See also Number Portability and Trigger. www. NeuStar.com

LNPA Local Number Portability Administration. See also LNP and NANC.

LNRU Like New Repair and Update. A term in the industry which repairs telecom equipment. It means all equipment is repaired and updated to the current manufacturer's specifications. New plastic is used to refurbish to a "like new" status. Also added are a new coil cord, line cord and address tray. Included is a full diagnostic test with a burn-in (if required) and an operational system test. Definition courtesy Nitsuko America. See also Repair And Quick Clean and Repair, Update And Refurbish.

LO Local Operator. In the PCS sense, a functional entity providing local wireless service to customers in a geographical region. The LO (Local Operator) is serviced by the National Services Organization in the PCN (Personal Communications Network) for long-distance communications and for marketing/sales.

LOA Letter Of Agency. A letter that you give to someone whom you allow to represent you and act on your behalf. For example, a letter of agency is used when your interconnect company orders lines from your local phone company on your behalf. Letters of Agency are also used when companies switch their long distance service from one carrier to another. A blanket LOA can mean everything from a group of numbers belonging to one customer at multiple sites or multiple customers at multiple sites.

load 1. The act of taking a program or data from external storage – a cassette, a floppy or hard disk, etc. and storing it in the computer's main RAM memory.

2. The load is any electric or electronic appliance or gadget plugged into an AC electrical outlet. It completes the circuit from the transformer through the hot conductor, to the load, through the neutral conductor and back to the utility transformer. See AC, AC Power, Ground and Grounding.

load balance A telephone company term. Load Balance is the even distribution of customer traffic volume across all loading units in a switching entity. Load Balance is not related to the absolute level of load, but only to how well the existing load is distributed. See also Load Balancing.

Load Balance Index LBI. A telephone company term. Indicates trends, identifies superior performances and points up opportunities for improvement in load balance administration of dial Central Office line equipment.

load balancer In server farms, a load balancer accepts IP packets and then distributes them among identical web servers. This enables the manager to add web servers as loads grow, or to take a server out of service and not have the clients notice. This definition contributed by Alan Simmons. See also Load Balancing.

load balancing The practice of splitting communication into two (or more) routes. By balancing the traffic on each route, communication is made faster and more reliable. In telephone systems, you can change phone and trunk terminations in order to even out traffic on the network. An example: You have a PBX of three separate cabinets, each of which are joined by tie lines. Instead of having each cabinet serve anyone in the building, you might figure which groups talk to each other the most and concentrate them into specific cabinets. The objective is to maximize the number of calls that can be handled inside each cabinet and reduce the number of calls that need to travel between the cabinets. This makes the calls go faster and reduces the need for inter- cabinet lines.

In data internetworking, bridges and routers perform load balancing by splitting LAN-to-LAN traffic among two or more WAN links. This allows for the combination of several lower speed lines to transmit higher speed LAN data simultaneously. In local area networking, load balancing is a function performed by token ring routers. In data networking, load balancing can also be a form of inverse multiplexing where data packets are alternated over all available circuits. At the receiving end, the packets are reassembled in their proper order.

In disk arrays, load balancing means using multiple power supplies within a disk array so that power usage is spread equally across all the power supplies. The failure of one supply will not cause the entire array to fail. See also Load Balancer.

load coil Load coils are also known as impedance matching transformers. Load coils are used by the telephone companies on long analog POTS (Plain Old Telephone Service) lines to filter out frequencies above 4 kHz, using the energy of the higher frequency elements of the signal to improve the quality of the lower frequencies in the 4 kHz voice range. Load coils are great for analog voice grade local loops, but must be removed for digital circuits to function. Load coils must be removed for DSL loops, as the frequencies required are well above 4 kHz. Today many phone companes offer broadband service, but often tell their customers that they can't get the service because "you live too far from the telephone company's office." Tell the company to remove the loading coils and any bridging taps on your local loop and it will work. If they baulk, offer to pay for the removal of the loading coils. If that doesn't work, offer to buy commercial ADSL service.

load coil detector A device use to detect unseen load coils on a wire pair. See Load Coil.

load factor Ratio of the Peak to Average ratio over a designated time period; has meaning in both traffic engineering and in transmission technology, particularly for data transmission.

load leveling Load can apply to telecommunications traffic or electricity (AC or DC). Load leveling in telecom typically means distributing traffic over more than one route. Load leveling in electricity typically means

load number Load number is the Canadian equivalent of the U.S. concept of Ringer Equivalence. The idea is that each phone or "phone thing" you buy (e.g. answering machine) comes with a number. You add the numbers together and if you get above a certain

number, you are drawing too much current and none of the bells on the phones will ring. In Canada, single line phones are typically rated at 10 for the newer ones with electronic "bells" or 20 for the older electro-mechanical ones with real metal bells. In Canada, the rule is not more than 100 points on a line. In the U.S., phones are typically one and the rule is not more than five points on a line.

load service curves The output from load and stress testing on a computer telephony system is a set of load service curves. Load service curves identify how individual areas of the system respond under various load amounts. Traffic is provided to the computer telephony system at defined steps (perhaps at 1,000 call per hour increments) until the system design threshold is reached. Measurements are taken at each step, and usually shown graphically, in a "curve". Most computer telephony systems are designed to handle up to a specified number of busy hour calls with specific response times. For example, the time that passes between the point in time a caller enters a DTMF digit and the point the computer telephony system speaks a response should usually be no more than 1 second 97% of the time, nor more than 3 seconds 99% of the time. A load service curve would be used to illustrate the response time at each step of increasing load. When the load curve shows the response time is slower than the above parameters, the system has reached its capacity. Of course, the load placed on the system must accurately mimic the real world load the system will experience or it is largely meaningless. This definition from Steve Gladstone, president, Hammer Technologies, makers of fine computer telephony testing systems, 508-694-9959.

load sharing In data processing, load sharing is the technique of using two computers to balance the processing normally assigned to one of them. In local area networking, load sharing is performed by token ring routers when connecting remote LANs. It allows combining Ethernet and Token Ring traffic over a common WAN (Wide Area Network) link such as T-1 or 56 Kbps circuit. Loads sharing eliminates the need for duplicate WAN links (and bridges or routers) each serving a different type of LAN.

load testing Also known as stress testing, the goal of load testing is to make sure the system will meet or exceed its busy hour load capacity objectives under all operating conditions. This requires stressing the system in incremental steps until it breaks and understanding what happens when the system is operating under its full rated transaction load? Beyond its load? Does it slow down? How? Does it fail? Where? How is service restored after an outage? Is service restoration graceful or must the system reboot? Is restart manual or will the system reset itself? Individual load tests may be performed to understand the impact of load on specific system bottlenecks. Most significant architectural problems will come to light under load testing. It is critical that any load placed on a computer telephony system be dynamic, and mimic the load characteristics the system will experience under real-world usage and conditions. See also Dynamic Load Testing and Load Service Curves.

loaded cable Twisted wire pair into which inductors have been inserted at periodic intervals to approximate the optimum ratios of the primary cable constants for minimum loss. A loaded cable acts like a lowpass filter. Transmission loss below the cutoff frequency is reduced below that for the nonloaded cable and is nearly flat. Above the cutoff frequency, loss increases very rapidly. See Loading.

loaded line A telephone line equipped with loading coils to add inductance in order to minimize amplitude distortion. See Loading and Loading Coil.

loaded loop Also called a Loaded Pair (loaded twisted pair); a loop that contains series inductors, typically spaced every 6000 feet for the purpose of improving the voiceband performance of long loops. However, high bandwidth DSL operation over loaded loops is not possible because of excessive loss at higher frequencies. See Loading.

loading A method of improving the voice quality of a phone line. Telephone companies put load coils on local lines. What this loading does is to insert inductance in a local loop circuit to offset the effect of capacitance in the cable. Loading "tunes" the circuit to the voice frequency band (500 to 2500 Hz) and thus improves the quality at the expense of overall bandwidth. You usually have to ask that the loading coils be removed if you're planning to transmit high-speed data exclusively on that circuit. See Loading Coil.

loading coil See Load Coil.

loading division A telephone company term. A group of the same type of equipment designed to be loaded similarly by both usage and classes of service.

loading high A memory management verb for loading a device driver or TSR (Terminate and Stay Resident) program into upper memory, out of conventional memory. Under DOS, the loading high commands are DEVICEHIGH for device drivers and LOADHIGH (or LH) for TSRs. Third party memory managers use their own routines to load high, though they can sometimes borrow DOS commands.

loading plan A telephone company term. A Loading Plan is a systematic scheme for fully utilizing all existing capacity in a given switching entity; Utilizing and coordinating the capabilities and capacity limitations of various entities in a multi-entity wire center and maintaining objective service levels at all times. A Loading Plan is the basis for achieving and retaining good Load Balance.

LOC An ATM term. Loss of Cell Delineation: A condition at the receiver or a maintenance signal transmitted in the PHY overhead indicating that the receiving equipment has lost cell delineation. Used to monitor the performance of the PHY layer.

local Pertaining to a system or device that resides within a subject device's switching domain.

Localmon.dll The standard print monitor for use with printers connected directly to your computer. If you add a printer to your computer using a serial or parallel port (such as COM1 or LPT1), this is the monitor that is used.

local access The connection between a customer's premises and a point of presence of the Exchange Carrier.

Local Access and Transport Area LATA. The MFJ (Modified Final Judgment), which broke up the Bell System, also defined 196 distinct geographical areas known as LATAs. The LATA boundaries generally were drawn in consideration of SMSAs (Standard Metropolitan Statistical Areas), which were defined by the Census Bureau to identify "communities of interest" in economic terms. Generally speaking, the LATA boundaries also were coterminous with state lines and existing area code boundaries, and generally included the territory served by only a single RBOC. The basic purpose of the LATA concept was to delineate the serving areas reserved for LEC (Local Exchange Carrier) activity. In other words, IntraLATA traffic (i.e., local and local long distance) became the sole right and responsibility of the LECs. InterLATA traffic, on the other hand, became the sole right and responsibility of the IXCs. Over time, a number of state PUCs allowed the IXCs to compete for IntraLATA long distance; they also allowed CAPs (Competitive Access Providers) to provided limited local service in competition with the LECs. The Telecommunications Act of 1996 (The Act) opened the floodgates for competition with the LATA boundaries. The Act also allows the RBOCs to provide InterLATA service outside the states in which they provide local service. Additionally, The Act contains provisions for the RBOCs to offer InterLATA service within the state in which they provide local service, once they have satisfied a 14-point checklist, the most significant conditions of which relate to significant, demonstrated levels of competition within their respective local exchange serving areas. California is divided into 11 LATAs. Sparsely populated states such as South Dakota comprise only a single LATA.

local airtime detail This cellular telephone carrier option (which means it costs money) provides a line-itemized, detailed billing of all calls, including call attempts and incoming calls to the mobile. What you get for free is generally a non-detailed, total summary of all calls.

Local Area And Transport Area See LATA.

Local Area Data Transport LADT. A service of your local phone company which provides you, the user, with synchronous data communications.

Local Area Network LAN. A short distance data communications network (typically within a building or campus) used to link computers and peripheral devices (such as printers, CD-ROMs, modems) under some form of standard control. Older data communications networks used dumb terminals (devices with no computing power) to talk to distant computers. But the economics of computing changed with the invention of the personal computer which had "intelligence" and which was cheap. LANs were invented as an afterthought – after PCs – and were originally designed to let cheap PCs share peripherals – like laser printers – which were too expensive to dedicate to individual PCs. And as time went on, what LANs were used for got broader and broader. Today, LANs have four main advantages: 1. Anyone on the LAN can use any of the peripheral devices connected to the LAN. 2. Anyone on the LAN can access databases and programs running on client servers (super powerful PCs) attached to the LAN; and 3. Anyone on the LAN can send messages to and work jointly with others on the LAN. 4. While a LAN does not use common carrier circuits, it may have gateways and/or bridges to public telecommunications networks. See LAN Manager, Token Ring and Ethernet.

Local Area Signaling Services LASS is a group of central office features provided now by virtually all central office switch makers that uses existing customer lines to provide some extra features to the end user (typically a business user). They are based on delivery of calling party number via the local signaling network. LASS can be implemented on a standalone single central office basis for intra office calls or on a multiple central office grouping in a LATA (what the local phone companies are allowed to serve) for interoffice calls. Local CCS7 (Common Channel Signaling Seven) is required for all configurations. The following features typically make up LASS:

Automatic Callback: Lets the customer automatically call the last incoming call directory number associated with the customer's phone when both phones become idle. This feature gives the customer the ability to camp-on to a line.

Automatic Recall: Lets the customer automatically call the last outgoing call currently associated with the customer's station when both stations become idle. This feature gives the customer the ability to camp-on to a line.

Customer-Originated Trace: Lets the terminating party request an automatic trace of the last call received. The trace includes the calling line directory number and time and date of the call. This information is transmitted via an AM IOP channel to a designated agency, such as the telephone company or law enforcement agency.

Individual Calling Line Identification: Consists of two distinct features: 1. Calling Number Delivery which transmits data on an incoming call to the terminating phone. 1. Directory Number Privacy which prevents delivery of the directory number to the terminating phone.

Also, LASS has some selective features:

Selective Call Acceptance: Allows users to restrict which incoming voice calls can terminate, based on the identity attribute of the calling party. Only calls from parties identified on a screening lists are allowed to terminate. Calls from parties not specified on a screening list are rerouted to an appropriate announcement or forwarded to an alternate directory number.

Selective Call Forwarding: Allows a customer to pre-select which calls are forwarded based on the identity attribute of the calling party.

Selective Call Rejection: Allows a customer to reject incoming voice calls from identity attributes which are on the customer's rejection list. Call attempts from parties specified on the rejection list are prevented from terminating to the customer and are routed to an announcement which informs the caller that his/her call is not presently being accepted by the called party.

Selective Distinctive Alert: Allows a customer to pre-select which voice calls are to be provided distinctive alerting treatment based on the identify attributes of the calling party.

Users can, at their convenience, activate or modify any of these features by sending commands to the central switch from their existing touchtone telephones.

local attack A network security term. An attack that targets the machine on which the attacker is interactively logged on.

Local Automatic Message Accounting LAMA. A combination of automatic message accounting equipment and automatic number identification equipment in your telephone company's central office and used by them to bill your local phone calls.

local battery Having "local battery" means the telecom equipment – the telephone, the PBX, the key system, etc. – has its own source of power and does not draw from the power coming down the phone line. The term came from telegraphy and was used to distinguish the battery which provided power to the telegraphic station as against the power that went to drive the line and the signal traveling down it. See Battery.

local bridge A bridge between two or more similar networks on a local site (within same building).

local bus A microprocessor inside a PC must communicate with certain integral devices, including memory, video controllers, hard disks. This is typically called an internal bus. That is to distinguish it from the "external" bus, such as the AT, ISA, EISA, MCA buses, which define the communications between the motherboard and the various peripheral devices, such as the I/O cards like those handling modems and LAN connections. As microprocessors have gotten faster, so they have begun to outpace the speed of their computer's internal bus, which has tended to narrow the stream of data in and out of the CPU, slowing the computer. A Local Bus is a new type of internal bus. It is a faster bus. The idea is to get a broader path between your critical components – memory, video and disk controller – and your microprocessor. The idea is to get the data in and out of the microprocessor at the same speed as the microprocessor's system clock. Local Bus is an emerging standard. See also EISA, PCI and VESA.

local call Any call within the local service area of the calling phone. Individual local calls may or may not cost money. In many parts of the US, the phone company bills its local service as a "flat" monthly fee. This means you can make as many local calls per month as you wish and not pay extra. Increasingly this luxury is dying and local calls are costing money.

local call accounting Computes the dollar amount for local calls based on the total message units stored for each phone.

local call billing Computes the dollar amount for local calls placed by guests based on total message units.

local central office Switching office in which a subscriber's lines terminate.

local channel controller An AT&T name for its family of 3270 compatible cluster controllers.

local composite loopback In network management systems. Composite loopback test that forms the loop at the output of the local multiplexer that returns transmitted signals to their source. See loopback.

local dataset Signal converter that conditions the digital signal transmitted by an RS-232 interface to ensure reliable transmission over a dc continuous metallic circuit without interfering with adjacent pairs in the same telephone cable. Normally conforms with Bell 43401. Also called baseband modem, limited distance modem, local modem, or short-haul modem. See line driver.

local digital services LDS. LDS is a term used by long distance companies to describe "last mile" services provided by local phone companies. LDS is generic word to describe any digital services, including T-1, T-3, OC-12, Frame Relay and dedicated Internet services.

local distribution frame LDF. Another word for an Intermediate Distribution Frame. It's a device for cross connecting cables – from one thing to another. On one side of the LDF are the pairs from individual phones in that part of the building or area. On the other side are trunks coming in from a central office or cables coming in from the central, larger PBX. LDFs typically help with the organization of cables in a building or area. See also Intermediate Distribution Frame.

local echo A modem feature that enables the modem to send copies of keyboard commands and transmitted data to the screen. When the modem is in Command mode (not online to another system) the local echo is invoked through the ATE1 command. The command causes the modem to display your typed commands. When the modem is online to another system, the local echo is invoked through the ATFO command. This command causes the modem to display the data it transmits to the remote system.

local exchange The telephone company exchange where subscribers lines are terminated. Also called an "End Office."

local exchange carrier A local phone company. See also LEC. As defined by the Telecommunications Act of 1996, a local exchange carrier means any person that is engaged in the provision of telephone exchange service or exchange access. Such term does not include a person insofar as such person is engaged in the provision of a commercial mobile service under section 332(c), except to the extent that the Commission (the Federal Communications Commission) finds that such service should be included in the definition of such term.

local explorer packet Packet generated by an end system in an SRB network to find a host connected to the local ring. If the local explorer packet fails to find a local host, the end system produces either a spanning explorer packet or an all-routes explorer packet.

local heap A memory storage area limited to 64K in size.

Local Interconnect Network See LIN.

Local IP A telephone company AIN term. The Internet Protocol (IP) indicated when an SCP or Adjunct requests a local AIN switch to make a connection to an IP to which the SSP or ASC switch has a direct ISDN connection.

local long distance IntraLATA long distance. A marketing term invented by the LECs (local exchange carriers) to distinguish intraLATA from interLATA toll calling. Specifically, the term was invented by the RHCs (Regional Holding Companies), which currently are limited to providing "long distance" calls only within the intraLATA toll market in their home states.

local loop The physical connection from the subscriber's premise to the carrier's POP (Point of Presence). The local loop can be provided over any suitable transmission medium, including twisted pair, fiber optic, coax, or microwave. Traditionally and most commonly, the local loop comprises twisted pair or pairs between the telephone set, PBX or key telephone system, and the LEC (Local Exchange Carrier) CO (Central Office). As a result of the deregulation of inside wire and cable in the United States, the local loop typically goes from the demarc (demarcation point) in the phone room closet, in the basement or garage, or on the outside of the house, to the CO. The subscriber or building owner is responsible for extending the connection from the demarc to the phone, PBX, key system, router, or other CPE device. See also Demarc and Subloop.

local management interface LMI. The specification for a polling protocol for use in Frame Relay networks between the user equipment in the form of a FRAD (Frame Relay Access Device) and the network equipment in the form of a FRND (Frame Relay Network Device). The LMI verifies the existence of the UNI (User Network Interface) and the Permanent Virtual Circuit (PVC).

local measured service LMS. Years ago virtually all phone lines in the United States were FLAT RATE. That meant that for a fixed amount of money each month, you,

the customer (a.k.a. subscriber) were allowed to make as many local calls as you wanted. For many reasons, the U.S. phone industry has progressively moved to LOCAL MEASURED SERVICE for local calls. Typically this means that for a fixed amount of money each month, you, the customer, can receive as many calls as you want and can make a finite number of outgoing local calls – typically 50. Each additional call beyond the 50 (or whatever the number is) costs extra. How much that call costs depends on the distant the call travels, the time of day, the day of the week, and the local company's tariffs.

local multipoint communications service See LMCS.

local multipoint distribution system See LMDS.

local net The broadband architecture used in Sytek's work. Also the product name of their network. Sytek is in Sunnyvale, CA.

local number portability LNP. Imagine a town in which there are many local phone companies. You have service from one company. But another comes along offering better service, lower price and more features. You want to switch. But you don't want to change your phone number. That's what LNP is all about – the ability to change your phone company and still keep your phone number.

Regardless of the local provider selected, consumers will continue to have access to Emergency 911 service; operator and directory assistance services; advanced services such as voice mail, Caller ID and Call Forwarding; and other customized local area and signaling capabilities, including equal access to all 800 and 888 toll-free telephone numbers.

On November 15, 1997 I received a press release from Lockheed Martin saying that they had successfully developed and tested Local Number Portability (LNP) in the Midwest, the FCC's mandated national test region. The implementation was mandated by the Federal government as part of the Telecommunications Act of 1996, which required this process be completed by October, 1997. In the press release, Lockheed Martin explain that in 1994, amidst concern over the need for fair, open telecommunications markets, the MCI Telecommunications Corporation initiated a Gallup Poll to investigate the demand for a Local Number Portability system. The poll randomly surveyed approximately 2000 businesses and consumers across the country in September and October 1994. The study assessed whether business or consumers would switch local telephone service providers under various market scenarios. The results of the poll indicated that 83 percent of business customers and 80 percent of residential customers would not change local service providers if changing service providers meant changing phone numbers. The study helped articulate the need for LNP in order to facilitate fair and open market competition in the local telecommunications industry. The LNP system developed in response to that survey and completed by Lockheed Martin in October, 1997, represents a substantial benefit to consumers: They now can choose local service from a variety of providers without changing their phone numbers. The system works for voice, data, and video lines, residential and business lines.

Here is what Lockheed wrote about local number portability technology in that press release: The database that facilitates Local Number Portability is a technological marvel in its complexity and the speed with which it operates. Each phone number has a network address. The LNP database keeps track of these addresses. When a customer places a call, the database records the caller's network address, locates the dialed number's network address, and notifies all telecommunications companies involved where to route the call and which companies to credit for the call.

Simplified, the process works like this:
- Call placed
- Network address of caller identified
- Network address of call recipient identified
- Telecommunications companies told how to route the call and which companies to credit.

All of this happens within nano-seconds of the customer placing the call – an imperceptible lapse of time. The LNP system also records the appropriate information whenever a customer changes local carriers, updating account information and ensuring that no interruption in service occurs.

In developing LNP, Lockheed Martin IMS drew on two types of existing technology. In 1993, Lockheed Martin IMS developed a portability system for 800 numbers that was used by 140 telecommunications companies in the United States at the request of Bellcore, the research and engineering division of the Regional Bell Operating Companies. This database allowed customers to handpick 800 numbers (such as 1-800-FLOWERS) and keep those numbers regardless of the long-distance carrier used. That experience laid the groundwork for developing LNP.

Lockheed Martin IMS also used existing infrastructure to run LNP. Twenty years ago, each local phone company installed computerized databases for their own internal use. LNP is an incremental application of this network, the Advanced Intelligent Network (AIN), and was made possible by the investment that local service providers made years ago. See also LNP and Location Portability.

local order wire A communications circuit between a technical control center and selected terminal or repeater locations.

local oscillator LO. A device which generates a specific single wave frequency. A local oscillators is used to reduce a high microwave frequency to a low intermediate frequency, so that it can be processed.

local phone A phone attached to your computer. See also Handset Management.

local phone service When I dial the pizza store on the corner, I'm making a local phone call. But when I'm calling the pizza store 50 blocks away, is that a local phone call? Answer, it could be. It depends. What's local phone service? What do you charge for it? Once upon a time, most Americans didn't pay for local phone service. They paid a flat monthly fee. Then the phone companies needed money, so they started charging for local service. A few cents per call. Then the phone companies timed the call and charged more the longer you talked. Then they started charging for longer local calls – maybe for calls of ten miles and further. In short, the definition and pricing of local phone calls is changing. Now local calls are looking increasingly like long distance calls – charged by time, distance, and day of the week.

local printer A printer that is directly connected to one of the ports on your computer.

local redirector A local redirector is a shim that redirects HTTP requests to a local proxy server. A local redirector is also known as a load balancer, a local redirector is a piece of software that receives a server request (e.g., HTTP, FTP or NFS) and reroutes it to one of a cluster of Web servers to be actioned. The distribution function may be based on which machine in the cluster has the lowest current level of utilization, the proximity of the server to the client, or which machine has the resources necessary to carry out the request. This definition courtesy of Mark Gibbs.

local service area The geographic area that telephones may call without incurring toll charges. A flat rate calling area. Increasingly rare.

Local Service Management System See LSMS.

Local Service Ordering Guidelines LSOG. A manual that describes and provides examples of the LSR (Local Service Request) forms that are used by a CLEC to communicate with an ILEC for the purpose of ordering changes, additions, deletions, or enhancements in service. For example, LSRs are used to order various types of local loops, to port telephone numbers. The LSOG is printed and distributed by Telcordia Technologies (nee Bellcore) under the auspices of the Ordering and Billing Forum (OBF) of the Carrier Liaison Committee of the Alliance for Telecommunications Industry Solutions (ATIS). See also ATIS, CLEC, ILEC, LSR, OBF, and Telcordia Technologies.

local service request A form used by a CLEC (Competitive Local Exchange Carrier) to request local service from an ILEC (Incumbent LEC). See LSR for a full explanation.

local switch The term local switch refers to the switch (PBX, ACD, dumb) to which the computer telephony system is directly connected. Usually, the local switch will provide better integration with the computer telephony system (more comprehensive call data) than connections that take place over the network. Additionally, the local switch will have both line and trunk side connections and will also support connection to whatever agents or desktop users that may use the computer telephony application.

local switching center The switching center where telephone exchange service customer station channels are terminated for purposes of interconnection to each other and to interoffice trunks.

local tandem A central office, usually in large metropolitan areas, serving as a transit switch between noncontiguous class 5 exchanges. It connects end office trunks.

local test desk A testing system that is used to test local loops and central office subscriber line equipment from a central point, typically a central office.

local traffic Telephone calling traffic that is classified as "local" in the landline tariff on file with the appropriate state regulatory body. The term includes single and multimessage unit traffic. Local traffic does not necessarily conform to city or county boundaries. Local traffic might include your town and the two towns adjacent to it. Or it may be just your town. Or maybe just half your town. It all depends on the size of your town and what the local state government regulator allowed your phone company to get away with. Originally all local phone calls in North America involved dialing four or seven digits. Now you sometimes have to dial ten digits, including a 1 and the three digit area code. In New York City, your neighbor in the next apartment may actually have an area code that's different to yours. Hence to call him, you have to dial ten digits. See Toll Traffic.

local trunk Trunks between Class 5 local central offices, also called switching centers.

LocDev Local Device. A Bluetooth device which initiates a SDP procedure. A Local Device is typically a master device on the piconet. However, a Local Device may not always have a master connection relationship to other devices. See also RemDev.

locality A measure of how close commonly-accessed files are to one another on a hard disk. "High locality" means the files reside on sectors or tracks which are close to each other. When this is the case, seek times during operation are shorter than average.

localize Make a language- and culture-specific version of a web site or application software.

LocalTalk Apple Computer's proprietary Local Area Network (LAN) for linking Macintosh computers and peripherals, especially LaserWriter printers is called Appletalk. AppleShare is the company's networking software, and the LAN hardware is called LocalTalk. Appletalk is a CSMA/CA network that runs at 230.4 Kbps and is therefore incompatible with any other LAN. It is also a lot slower than the present top speeds of Ethernet (100 Mbps) and Token Ring (16 Mbps). Outside manufacturers, however, make gateways which will connect an Appletalk LAN to other local area and telecommunications networks – LANs, WANs and MANs. See also AppleTalk, Ethernet, and CSMA/CA.

location One definition is the place where a telephone jack is located. This location is given a number. The wire going to that location is given a number. All this in hope of being organized for installation, moves, chan4ges and maintenance.

location based services Location based services enable personalised (customised) services to be offered based on a person's (item's) location. Services include areas of security, fleet and resource management, location based information, vehicle tracking, person-to-person location and messaging applications. To enable these services, there are a number of different technology layers that need to be co-ordinated on a network. These technologies include Applications, Middleware, Determination technologies and associated Silicon and Intellectual Property (IP). See Location Services.

location capable Able to provide a caller's location to a public service answering point (PSAP).

Location ID A feature of the IS-136 standards for digital cellular networks employing TDMA (Time Division Multiple Access). A capable telephone set will display the name of the cellular carrier providing service. In a wireless office system application, the phone can display the name of the company. When you are at home, connected to your PBS (Personal Base Station), the phone can display "cordless."

Location Interoperability Forum LIF. Open standards forum announced in September of 2000 to develop and promote a simple, ubiquitous interoperable location service solution necessary for mass consumer acceptance of mobile location services. Motorola, Nokia, and Ericsson were the charter members. See also LIF and Location Services.

location portability The ability of an end user to retain the same geographic or non-geographic telephone number (NANP numbers) as he/she moves from one permanent physical location to another. Location Portability will involve either of the following scenarios: 1) new location is within the same central office area, or 2) new location is within a different central office area. See also Local Number Portability.

location services Cellphone carriers will soon be able to figure out, within 100 meters where you, a cellphone user, are. The first application of this technology is called Emergency Location Services (or E911). What this means is that if you dial 911 in the United States on your cellphone, the operator will know where you are and be able to send help. There are two basic technologies currently being adopted; E-OTD (Enhanced Observed Time Difference) uses a software-enabled cellphone handset and cell sites to calculate your location. GPS (Global Positioning Systems) relies on a chip being installed in the cellular phone and orbiting satellites to determine position. From October 1, 2001, the FCC mandated 50% of all the new cellphone activations in the United States should be equipped with location services. See E-OTD, GPS and Location Based Services.

Within the two broad categories of cellphone location technology, there are a number of technologies available for determining a caller's location after initiating a 911 call from a mobile handset, but the state of their development differs. Here is a partial list. The first group fits into what is known as network solutions:

• Angle of Arrival (AOA). Angle of Arrival technology measures the direction of arrival of the caller's signal (generally at least three measurements are needed) at different cell sites. Each cell site receiver sends this direction information to the mobile switch where the angles are compared and the latitude and longitude of the caller is computed and sent to the PSAP. AOA works with any handset – digital, analog, TDMA, GSM, CDMA, etc.

• Uplink Time Difference of Arrival (U-TDOA). This is sometimes referred to simply as Time Difference of Arrival (TDOA). Time Difference of Arrival relies on the fact that each cell site is generally a different distance from the caller and that signals travel with constant velocity. Therefore, each signal arrives at the cell site at slightly different times. Using these properties, a signal defines a locus of points on a circle around a base station on which a mobile could be located. Then, using synchronized receivers, the times can be compared and a latitude and longitude can be computed and sent to the PSAP. At least three different receivers are needed for TDOA to work. TDOA works with any handset – digital, analog, TDMA, GSM, CDMA, etc.

• Wireless Location Signatures. Wireless Location Signature methods compare the radio signal received to a database of standard signal characteristics, such as reflections and echoes. Using this information from several cell site receivers, the caller location can be computed and sent to the PSAP. This technique works best in urban environments where lots of structures exist to provide the needed reflections. It works with any phone – digital, analog, TDMA, GSM, CDMA, etc.

• Location Pattern Matching (LPM). The wireless phone's signal is received at various antenna sites equipped with special gear. The receivers send the caller's voice call to the mobile switch, where sophisticated equipment analyzes the acoustic radio signal, and then compares it to a database of standard signal characteristics. These characteristics include signal reflections (multipath), echoes and other signal "anomalies." According to U.S. Wireless, the only supplier of gear for this technique, when a computerized match is made, the location of the caller can be determined within the FCC's requirements. The technique is effective in urban environments that include tall buildings and other obstructions, where other techniques might not succeed. The caller's voice call and the latitude and longitude are then sent to the PSAP for use by the dispatcher. LPM works with any phone – digital, analog, TDMA, GSM, CDMA, etc.

• Multi-path Fingerprinting (MP). This technology uses features of the physical environment to locate mobile handsets. A wireless signal bounces off a variety of solid objects on the way to its destination (either a base station antenna or a handset), causing what's called multipath interference. Essentially, the same signal is received multiple times due to the delay caused by bouncing off objects and taking longer paths to the destination. Multipath Fingerprinting takes advantage of this characteristic (which is normally a nuisance) to characterize signals that are received from certain locations. For instance, a base station can record what a handset signal looks like transmitted from a certain intersection of highways. A block away, the multipath signal "fingerprint" will look different, since the location of buildings, trees, and other obstructions has changed. To employ this system, an operator must send test units around to various locations so the base stations can record the fingerprints and create a database for comparison later on. Of course, if new construction occurs in an area the fingerprint will change and must be re-recorded.

• Enhanced Cell Identity (E-CID). Enhanced Cell Identity uses a combination of angular information (the cell sector receiving the signal) and timing information to approximate the location of the handset.

Then there are various handset solutions:

• GPS. GPS techniques use handsets equipped with GPS receivers. The GPS receiver determines the caller's latitude and longitude which is sent to the provider's receivers and relayed to the PSAP.

• Assisted GPS (A-GPS). Assisted GPS uses techniques and advanced chipsets designed to allow reception of GPS signals indoors. Assisted GPS can be supplemented with an advanced forward link trilateration (A-FLT) system. A-FLT is a network-based location technology that takes measurements of signals from nearby base stations and reports time and distance readings back to the network, which uses them to triangulate an approximate location of the handset.

• Wireless Assisted GPS. Wireless Assisted GPS generally uses advanced chipsets capable of acquiring very weak GPS signals and integrating the signals very quickly to determine location.

• TV-GPS. TV-GPS uses synchronization signals from television stations to determine handset location when indoors. Because of the frequency and power of TV signals, they often can be received at indoor locations where GPS signals cannot.

• Advanced Forward Link Trilateration (A-FLT). Advanced Forward Link Trilateration is a handset-based position location technology that works by using measurements, taken by the handset, of signals from nearby base stations, and reporting the time/distance readings back to the network, which are then used to triangulate an approximate location of the handset. In general, at least three surrounding base stations are required to obtain an optimal position fix.

• Timing Advance/Network Measurement Report (TA/NMR). This method relies in part on timing advance, which is the maximum amount of time that a TDMA mobile

station uses to compensate for propagation delay in order to avoid user time slot overlap when the mobile is far away from the base station. It also employs information from the Network Management Report, which is the measurement done either at the handset or base stations to improve communication flow on the air interface. Various events – such as handoff, power control, and candidates list – use the Network Management Report.

• Enhanced Observed Time Difference (E-OTD). Enhanced Observed Time Difference measures the differences in time that signals from the base stations take to reach both the handset and a fixed point in the network. This information is then sent from the handset and the fixed point to a mobile location center where a latitude and longitude are computed and sent to the PSAP.

See GMLC.

location tracking Vehicle Location Tracking Devices (VLD) are products targeted at the mobile fleet/ vehicle management space. Weighing less than eight ounces, they can be installed in almost any vehicle including: automobiles; construction equipment; trucks; buses; motorcycles or even boats. When used with a carrier-grade server, it allows users to track the locations of specific vehicles equipped with these devices, via the Internet. I first heard about this from a company called Paradigm Advanced Technologies, Inc., which provides wireless location-based electronic commerce (L-Commerce) solutions. Paradigm has licensing rights to a wireless location patent (US Patent #B1 5,043,736) covering the apparatus and method of transmitting position information from satellite navigational signals (like GPS) over cellular systems to a base unit, and displaying the location of a person or object so equipped. Paradigm owns PowerLOC Technologies Inc. , which anticipates providing a comprehensive range of L-Commerce and L-Business products and services including a family of proprietary wireless-location devices for this industry and for location-based service providers (LSPs) in particular.

location transparency More professionals are working from home, customer sites or from the road. Location transparency means that your communications system – faxing, email, voice mail, etc – works as well for you, the user, whether you're in the office or in the field, or where in the field you are.

locator A term used in the secondary telecom equipment business. A locator is a company that assists both a buyer and seller to quickly find each other. A locator contracts with dealers to provide them with daily lists of potential customers. The list develops from phone calls to an 800-number asking for a specific component.

locator service In a distributed system, a feature that allows a client to find a shared resource or server without providing an address or full name. Generally associated with Active Directory, which provides a locator service.

lock and load Originally a military term. Then it became software speak for freezing code on a program in development. Then it became "Let's make a decision and get on with it."

lock code The lock code locks a cellular telephone to prevent unauthorized use. The lock code is programmed into the NAM (Numerical Assignment Module) and is frequently factory set to either 1234 or 00004.

lock on The process by which an earth station initially acquires the signal from a satellite.

lock out In a satellite telephone circuit controlled by an echo suppressor, one or both subscribers can't get through because of excessive noise at one end. You get this also with speakerphones. The person with the speakerphone can simply hog the conversation because his speakerphone keeps transmitting his voice. There are weird variations on this. Sometimes you might call someone on your speakerphone and wait for them to pick up. They do. They shout into the phone "I'm here." All they can hear is you at the other end talking or typing. The sound at your end is hogging the channel and thus locking out the person at the other end. The solution? Turn the "mute" button on your speakerphone. This will stop your end transmitting and allow the other end to say "Hello, I'm here." See also Lockout.

locked mode In SONET networks, a mode of operation for a Virtual Tributary (VT) group. A VT group can function in either locked or floating mode. While floating mode minimizes delays in distributed VT switching, locked mode is used to enhance the efficiency of the network devices performing the switching.

locked resources An Intel Plug and Play term. Resources that must be used by the same card each time the system is booted. The configuration manager cannot assign these resources to any other card.

locked virtual tributary mode The timing of the virtual tributary is locked (in frequency and phase) to the timing of the STS-1. While offering easy visibility and access to DS0 bytes, it adds extra delay, and buffering and reformatting complexity. Intended for switch-to-switch transmission where both are already synchronized. Not expected to have much usage.

locker Telco-speak for a storage area, often an urban storefront, where phone company installers and repairman can pick up and drop off tools and installation material such as phones, wire and hardware. These places are prime targets for burglary and robbery, and often have no identification to show their valuable contents.

locking Preventing several people getting to and changing the same data in a shared database simultaneously. Locks may be permanent and prevent access completely, or they may be "advisory." A user is warned the data is being used by someone else and that the data is not presently available. Locks prevent the destruction of data that can occur if two people access a file at the same time. In any data base or other computer system, there are typically two types of "locks" – record and file locks. A record lock occurs when an airline agent pulls up your travel plans. No other travel agent can access those records at that time. A file lock occurs when the whole file is locked up. This might occur in a centralized word processing program. The whole document will be locked when it is being used by someone.

lockout A PBX feature. Denies the attendant the ability to re-enter an incoming central office connection directly terminated or held on her position, unless specifically recalled by the phone user.

lockup A company goes public, i.e. does an IPO – Initial Public Offering. Its stock is listed on the stock exchange and starts trading hands. Shares owned by insiders (management, directors, etc.) cannot be sold because they are subject to a "lock up." This is a time period that has been negotiated with them by the underwriter as part of his commitment to take the company public. Under the agreement they cannot sell their shares in a company for (typically) six to 12 months right after the initial public offering.

locust It has been documented that locusts have formed swarms measuring up to one mile wide, 100 feet deep, and 50 miles long. They may travel more than 2,000 miles. A swarm this enormous has been known to contain as many as 40 billion locusts.

locutorio In Buenos Aires, they have places (normally privately owned) where there are a lot of telephone booths, fax machines, that you walk in, make a call, pay and walk out. The word in Spanish in Buenos Aires is "locutorio," which was the name given to a place in convents or prisons where nuns or prisoners could speak to visitors from behind bars (in both cases). The word comes from Latin and is mostly used in Spain (Telefonica de Espana purchase part of the old Argentine telco). This definition contributed by Jorge E. Corbalan, head translator in the Arthur Andersen office in Buenos Aires.

LOD Letter of Disconnect (you don't pay,.. you don't play).

lodestone A magnetic ore of iron, used in the making of early compasses.

LOF Loss of Frame. LOF is a generic term with specific variations of meaning, depending on the signal standards domain in which it is being used. In the OSI/ATM world for instance, LOF is a condition at the receiver or a maintenance signal transmitted in the PHY overhead indicating that the receiving equipment has lost frame delineation. This is used to monitor the performance of the PHY layer. In the SONET world also, LOF is a condition detected in the signal overhead at the receiver, indicating that a valid framing pattern could not be obtained. There is however no "LOF indication" per se transmitted in the SONET overhead: SONET uses the AIS-L overhead indication to inform downstream equipment that the receiving equipment upstream has experienced a failure. A hierarchy of LOF defect and alarm notifications is implemented at the LOF-detecting equipment based on the duration the condition persists. See also AIS and SEF.

log See Logarithm.

logarithm Log. A mathematical function. The exponent that indicates the power to which a number is raised to produce a given number. For example, the equivalent of 100 to base 10 is 2. In other words, 10 squared = 100.

log in The process of identifying and authenticating oneself to a computer system. Used to control access to computer systems. See Login Script.

log file Also spelled logfile, which is very efficient in terms of keystrokes but which is very poor spelling. A log file is simply a file that tracks access activity for a host resource. For instance, a log file might contain information relative to those who access your Web site. Such information might identify user name, user domain, the length of time spent on each page, and which links they exercise. See also Log In and Log Off.

log off 1. To type in the needed keystrokes for ending a session that's on-line with a computer. Often those keystrokes are "Logoff." Usually it's very easy to Log Off. It's more difficult to Log On.

2. Employees wear their photo-ID badges in little plastic holders attached by a clip to their clothing. It's a good idea to "log off" when you're in public. This means to turn your badge around so nobody can see whom you work for.

log on To enter the needed keystrokes to start an on-line session with a computer. "Logging On" may be done with a computer that's local or one that's long distance and your work is done over communications lines.

According to the Vermonter's Guide to Computer Lingo, log on is making the wood stove hotter.

Logfile See Log File.

logic Logic is the application of mathematical analysis and deductive reasoning to propositions that may or may not be true or false. The logic we're interested in owes much to the work of George Boole in the mid-19th century. He formulated a system that could be applied to the relationships between propositions to which only a binary choice of truth existed, i.e. yes or no. The first application of the Boolean Algebra that derived from this was Shannon's research into the analysis of relay switching circuits in 1938. Logic is not common sense. See common sense.

logic bomb Program routine that destroys data. For example, a logic bomb may reformat the hard disk or insert random bits into data files. It may be brought into a personal computer by downloading a corrupt public domain program. Once executed, it does its damage right away and then stops, whereas a virus keeps on destroying. Another definition of a logic bomb is that it is a resident computer program that lies dormant for a period, and then triggers an unauthorized act when a certain event, such as a date, occurs.

logical In networking, logical means the way it works, the way the software sees it. In contrast, physical means how it's physically connected, which is often very different. For example, a physical addresses is translated from a logical address. Allow me to illustrate. When someone dials your telephone number, they are dialing a logical address; in other words, the series of numbers means nothing until they are translated into a physical address. The physical address is the port to which your local loop is connected to which your telephone is connected. Similarly, your postal address is a logical address. It has meaning only when translated by the post office into the plot of earth on which your house sits. A logical address, on the other hand and just to confuse you, may have no fixed physical address. For example, your e-mail address has no fixed physical address. Rather, it is translated into an IP (Internet Protocol) address which is associated with your e-mail server, which can be moved from place to place. Ultimately, your e-mail address actually is associated with you, and you and your computer can move all over the world without losing access to your e-mail. Rather, you gain access to your e-mail by going on a network connected to the Internet – e.g. dialing a telephone number (logical address) which connects you to your e-mail server which has a physical address which can change as the server is moved from one location to another.

logical address See Physical Address.

logical block address A logical block address is a sequential address for accessing blocks on storage media. The first block of the media is addressed as block 0 and succeeding blocks are numbered sequentially until the last block is encountered. This is the traditional method for accessing peripherals on a SCSI bus.

logical layer The CVNS layer used to define the virtual networks and provision the services within the physical layer. There are two components that make up the logical layer: Data Access Points (DAPs), and Service Control Manager (SCM).

Logical Terminal Group LTGRP. An ISDN term for a group of logical terminals connected to an ISDN switch, each of which terminals has a unique Logical Terminal Identifier (LTID). The LTID comprises the LTGRP, which is followed by the Logical Terminal Number (LTNUM).

Logical Terminal Identifier LTID. An ISDN term. See Logical Terminal Group.

Logical Terminal Number LTNUM. An ISDN term. See Logical Terminal Group.

logical topology The logical layout of a network, as opposed to the Physical Topology. In the LAN world, for instance, a 10Base-T network is a Star from the standpoint of Physical Topology (the way it looks). Yet the network operates as a logical Bus. In other words, the devices arrayed around the 10Base-T hub connect through ports into the hub chassis which houses a collapsed bus which supports communications over the shared physical path just as does a traditional Ethernet bus network, which is a bus in both physical and logical terms.

logical bus A LAN topology, such as Ethernet, which shares a common communications channel.

logical channel 1. A software based connection through which data is sent. The channel is assigned by the switch. In X.25 talk, a logical channel refers to a virtual connection operated over a physical connection that can support one or more virtual connections simultaneously.

2. A temporary allocation of telecommunication system resources to transmit a message; usually comprises a time slot in a multiplexed digital transmission channel, and may be viewed as a "promise" from the system that a message delivered using it will be transmitted to its destination.

logical channel number Virtual circuit identified at the packet level of X.25. See Logical Channel.

logical client Refers to one component in a pair of communicating components which is obtaining access to CTI functionality through the other component. This term is used to differentiate between two components which are communicating across an inter-component boundary. See Logical Server.

logical drive A disk drive recognized by the operating system. A computer's logical drives may differ from its physical drives. For example, a single hard disk drive may be partitioned into two or more logical drives. Before MS-DOS 5.0, the biggest logical "drive" that DOS could address was 32 megabytes. If you had a larger hard disk, you partitioned the bigger driver into logical drives of 32 megabytes.

logical formatting The third step in structuring a data medium so that data may be written to it. Logical formatting must follow physical formatting (also called low-level formatting) and partitioning (figuring into how many drives you wanted to slice the one drive into).

logical group node An ATM term. A logical node that represents a lower level peer group as a single point for purposes of operating at one level of the PNNI routing hierarchy.

logical ID An AT&T Merlin term. A numbering sequence used to identify station and trunk locations on the communications system control unit.

logical layer The CVNS layer used to define the virtual networks and provision the services within the physical layer. There are two components that make up the logical layer: Data Access Points (DAPs), and Service Control Manager (SCM).

logical link An abstract representation of the connectivity between two logical nodes. This includes individual physical links, individual virtual path connections, and parallel physical links and/or virtual path connections.

Logical Link Control LLC; A protocol developed by the IEEE 802 committee, common to all of its LAN standards, for data link-level transmission control; the upper sublayer of the IEEE Layer 2 (OSI) protocol that complements the MAC protocol; IEEE standard 802.2; includes end-system addressing and error checking.

Logical Modem Interfaces LMIs are to Microsoft's MAPI what SPIs are to TAPI. LMIs serve network-based fax servers and multi-port fax boards. MAPI sits on top of Microsoft's Exchange and other MAPI-compliant messaging systems.

logical node An abstract representation of a peer group or a switching system as a single point.

logical node ID A string of bits that unambiguously identifies a logical node within a routing domain.

logical port Lport. The logical interface between an endpoint(i.e., end process or program) and a communications or transmission facility. Multiple logical ports can be associated with a single physical port that connects to a transmission circuit that is capable of being fractionalized, or channelized. For example, a T-1 circuit commonly is channelized into 24 voice-grade channels, each of which is, in effect, a separate logical circuit contained within a single physical circuit. Each channel is associated with a single logical port that terminates the transmission is a given set of logic and, ultimately, in a single terminal device. A T-3 circuit typically comprises 28 T-1s, each of which is a logical circuit (typically comprises 24 logical circuits of its own) that terminates in a logical port. Logical ports also are used to terminate Frame Relay VCs (Virtual Circuits). See also Port.

logical provisioning An AT&T term for the establishment of network services by changing software controls, rather than by physically installing or rearranging hardware.

logical ring A network which is treated logically as a ring even though it maybe cabled as a physical star topology.

logical server Refers to one component in a pair of communicating components which is providing CTI functionality to the other component. This term is used to differentiate between two components which are communicating across an inter-component boundary. See Logical Client.

Logical Unit Interface See LU 6.2.

logiciel French for software.

login The process whereby a user gains access to a computer or network.

login script When users log into a local area network, they may wish to do many

things or the network supervisor may wish them to do several things. These commands are part of something called a "login script." In computerize (Novell's words), a login script contains commands that initialize environmental variables, map network drives, and/or control the user's program execution. Login scripts are similar to batch files. The familiar AUTOEXEC.BAT can be thought of as an MS-DOS login script.

login string A means of gaining access to the network. The string typically consists of three fields: username, destination, and password. The destination and password are optional depending on the network configuration.

logout The process whereby a user exits a computer or network.

LOH Line Overhead. 18 octets in a SONET frame for purposes of controlling the reliable transport of payload data between SONET network elements such as repeaters. LOH and SOH comprise Transport Overhead (TOH).

LOI Letter of Intent. You sell stuff. You want to sell your stuff to someone. But that someone won't give you a legally-binding Purchase Order. But you need something to show your bank in order to convince them to give you some money. So you ask your customer to give you a "Letter of Intent" which says they'll buy whatever the Letter says they'll buy, perhaps. IN short, a Letter of Intent means nothing, except that your "customer" loves you enough to write a letter on his letterhead. And the Letter is worth the paper it's written on – however you can use it to convince your banker to give you a loan.

LONAL Local Off Network Access Line. Similar to Foreign Exchange Service, this line terminates local calls from an end user's private network.

long distance Any telephone call to a location outside the local service area. Also called toll call or trunk call.

long haul communications That type of phone call which reaches outside a local exchange or serving area.

long haul data Special conditioning of the inter-office channel in order to meet performance specifications needed for processing data over the circuit. The long haul circuit will have the same type of conditioning as the local loop at each end.

long haul modem A modem or other communications device that can transmit information over long distances.

long key A long key is a character held down for a prescribed period of time. The time period is generally longer than the time for other keys. In call reorigination, the # sound is defined as a "long key." This definition was kindly provided by Karen Shelton, Systems Engineer, IEX Corporation, Richardson, TX.

long lines AT&T Long Lines. The department of AT&T which operates long distance toll service. It is no longer called Long Lines. It is called AT&T Communications.

long reach Long reach refers to optical sections of approximately 25 kilometers or more in length and is applicable to all SONET rates. See Long Reach Ethernet.

long reach Ethernet LRE. Long-Reach Ethernet is a proprietary Cisco extension to the IEEE 802.3 Ethernet LAN (Local Area Network) standard. LRE extends fully symmetric, full-duplex Ethernet connectivity at speeds of 5-15 million bits per second (Mbps) over distances of up to 5,000 (1,524 meters) feet over unconditioned Cat 1/2/3 (Category 1, 2 or 3) UTP (Unshielded Twisted Pair). While the exact transmission rate supported depends on the type and quality of the UTP, the exact distances, and numerous other factors, the most optimistic specifications are speeds of 5 Mbps over distances up to 5,000 feet, 10 Mbps up to 4,000 feet and 15 Mbps up to 3,500 feet. Note: Traditional Ethernet runs at 10 Mbps (also at 100 Mbps and 1 Gbps), over distances of up to 100 meters over Cat 3 (or better) UTP. Frequency Division Duplex (FDD) allows LRE transmissions to coexist with POTS (Plain Old Telephone Service), PBX, or ISDN signaling services over the same pairs. LRE also can be provisioned in the same bundle (i.e., binder group) as ADSL (Asymmetric Digital Subscriber Line). LRE makes use of Quadrature Amplitude Modulation (QAM), a modulation technique that uses shifts in both signal amplitude and phase to define each symbol. The system administrator or telecom manager may choose profiles that use different modulation options (QAM-256, QAM-128, QAM-64, QAM-32, QAM-16, QAM-8, and QAM-4) and frequency plans according to the line specification and rate definition. LRE is designed for application in hotels, office buildings, and other multi-tenant and multi-dwelling units with old POTS inside wire and cable systems that can't be upgraded easily, if at all. LRE provides an ISP with the ability to take high-speed DSL to the premises, connect to a whole bunch of Cisco gear (i.e., splitters, switches, servers, and the like) and software, and provide high-speed Internet access over old, beat-up telephone wire. It's a great idea. See also ADSL, Ethernet, FDD, and QAM.

long tail In 2004, Wired magazine popularized the phrase "the long tail" to refer to the large number of specialized offerings that in themselves appeal to a small number of people, but cumulatively represent a large market that can be easily aggregated on the

Internet. Plotted on a graph along with best sellers, these specialized products trail off like a long tail that never reaches zero.

long tones First, we invented touchtone, also called DTMF, Dual Tone Multi Frequency tones. You'd punch your number with tones, instead of dialing them. Then someone thought you could control telephone response gadgets, like voice mail, interactive voice response, etc. with touch tones. For these gadgets to work, they had to "hear" the tones you sent. No one really set standards as to the minimum length tone they would hear. But it was generally conceded that they were to be 120 milliseconds. So some manufacturers of telephone equipment started to make phone equipment that, if you pushed a touchtone button, the machine would only sent a touchtone of 120 millisecond duration. That was called a short tone. It wasn't very useful because the manufacturers quickly discovered that many pieces of equipment couldn't respond that quickly. And the manufacturers got complains that their customers couldn't call their voice mail, their bank, etc. As a result, some manufacturers of equipment brought out new hardware (replacing the old) to allow you to send "long tones," which are now defined as touchtones that last for as long as you hold down the button – just as it is (and has always been) on a normal single line, non-electronic, non-digital telephone. Isn't progress wonderful? See also DTMF for a much longer explanation of tone dialing.

long wavelength Light whose wavelength is greater than about 1 micrometer.

long wavelength band. The optical band, or window, specified by the ITU-T at a wavelength range between 1565nm and 1625nm (nanometers) for fiber optic transmission systems. See also C-Band, E-Band, O-Band, S-Band, and U-Band.

longest available This is a method of distributing incoming calls to a bunch of people. This method selects an agent based on the amount of time that each agent has been on the phone. This allows for an equitable distribution of calls to each agent. See also Top Down and Round Robin.

longest prefix matching Longest prefix matching is an address searching method in CIDR (Classless Interdomain routing), where a group of hosts whose addresses have a common prefix are grouped as a subnet. This common prefix is the network address of this subnet. To find out whether a host belongs to a particular subnet, a router simply masks off the lowest x number of bits in the address. Conversely, to route a packet the router simply compare the destination address with the address of all the subnets and find the subnet whose naddress has the best (longest) match. This look-up is most efficiently done with tenary CAMs. For detailed discussion, see page 164-165 in High-Performance Communication Networks by J. Walrand and P. Varaiya.

longevity testing When you're building a computer telephony system, many problems do not come to light until it's been running under high traffic for a long term. According to Steve Gladstone, author of the book "Testing Computer Telephony Systems," (available from 212-691-8215), there can be slow memory leaks, counter overflows, or disk fragmentation which slows down data access, system resets that cause calls to be dropped inadvertently, and so on. Frequently these issues are discovered only after thousands of calls are placed or after some random call pattern has occurred. Longevity tests, according to Gladstone, provide a high load to a system over an extended period of time. The most effective way to do this is to run a load test based on your busy hour usage profile over several days and track failures over the entire testing period. If failure rates increase, or specific failures occur, the events leading up to the fault can be relatively easily duplicated. If the system generating the load provides good error tracking and event logging, it may be possible to immediately identify the sequence that caused the fault, the corresponding reactions of the computer telephony system, and even immediately duplicate the fault causing scenario.

longhorn At one stage, Longhorn ws the name of the next major release of today's 32-bit Windows XP operating system. Longhorn eventually became known as Windows Vista, and was, predictably late.

longitude The distance in degrees from one Meridian to any other. Longitude is usually measured from the prime meridian (Greenwich, England).

longitudinal balance A measure of the electrical balance between the two conductors (tip and ring) of a telephone circuit; specifically, the difference between the tip-to-ground and ring-to-ground AC signal voltages, expressed in decibels.

longitudinal conversion loss LCL. A measure (in dB) of the differential voltage induced on a conductor pair as a result of subjecting that pair to longitudinal voltage. LCL is considered to be a measure of circuit balance.

longitudinal mode An optical waveguide mode with boundary condition determined along the length of the optical cavity.

Longitudinal Redundancy Check LRC. An error checking technique based on an accumulated collection of transmitted characters. An LRC character is accumulated at both the sending and receiving stations during the transmission of a block of data. This accumulation is called the Block Check Character (BCC) and is transmitted in the last character in the block. The transmitted BCC is compared with the accumulated BCC character at the receiving station for an equal condition. When they're equal, you know your transmission of that block has been fine. LRC commonly is used in combination with VRC (Vertical Redundancy Checking) to improve the reliability or error control in asynchronous transmission and in support of the ASCII coding scheme.

With VRC, a check bit, or parity bit added to each ASCII character in a message such that the number of bits in each character, including the parity bit, is odd (odd parity), or even (even parity). The term comes from the fact that the bits representing each character of data logically is viewed in a vertical fashion. When LRC is used in combination with VRC, the parity of a block (set) of data characters is checked for parity longitudinally (along the horizontal plane) of characters, as though they were laid out logically in a matrix format. For instance, the word "CONTEXT" consists of 7 letters, each of which consists of 7 bits, viewed in a block matrix format as follows:

The transmitting machine sums the bit values for each character, beginning with "nothing," which is an even value in mathematical terms. In the case of the letter "C," for

BIT/VALUE	CONTEXT
1*	11001000
2*	11100000
3*	01111010
4*	01100100
5*	00010110
6*	00000001
7*	11111110
8**	0010000
*INFORMATION BIT	** PARITY BIT

instance, the next bit is a "1" bit, which creates an odd value. The next bit is a "1" bit, which creates an even value. The next four bits are "0" bits, which do not change the even value. The seventh bit is a "1" bit, which creates an odd value, once again. Assuming that the device is set for odd parity, which is the default, it will insert a "0" bit in the eighth bit position, retaining the odd value. (Should the device be set for even parity, a "1" bit would have been inserted in the eighth bit position.) Should the value of the 7 information bits be an even value, the device appends a "1" bit in order to create an odd value. Across the longitudinal plane, the transmitting device accomplishes exactly the same process, appending "0" to retain odd values or "1" bits in order to create odd values.

After the data, character-by-character, has been formatted in this fashion, each bit sequence is transmitted across the network to the target device, which also is set for odd (or even) parity. The receiving device goes through exactly the same process, examining each character for parity. If the parity does not match the expectation of the receiving device, the subject character is flagged as errored, although no remedial action is taken. As VRC exposes the transmission to reasonable likelihood that two bits in a given character can be errored in the process of transmission of each character, that the parity of the character therefore would not be affected, and that the receiving device would not detect the fact that the character was errored, this technique is known as "send and pray." LRC substantially improves the likelihood that errors in a block of data will be detected, although there remains potential for compensating errors to affect the data, without detection. Any remedial action must be accomplished on a man-to-machine basis. See also Vertical Redundancy Checking and Parity.

longitudinal shield A tape shield, flat or corrugated, applied parallel to the axis of the core being shielded.

longitudinal shrinkage a term generally applied to shrink products denoting the discrete axial length lost through heating in order to obtain the recovered diameter.

Longitudinal Transmission Check LTC. An even or odd parity check at fixed intervals during data transmission.

longitudinal wrap A tape applied longitudinally with the axis of the core being covered, a opposed to a helical, or spiral, tape wrapped core.

look ahead routing A Common Channel Signaling System 7 (SS7) technique that determines the availability of a communications channel before the call is sent over the network. The technique maximizes efficiency and use of the public switched network. See Signaling System 7.

look angles The coordinates to which an earth station must point to "see" the satellite; azimuth and elevation are the most common specifications. However, other pairs exist, as the right ascension and declination used in radio astronomy.

look-up table 1. A translation table. You dial a certain number. But the number is meaningless. For the phone system to complete the call, it needs routing instructions. It gets that by "looking up" that number in a table, which translates that number to another number that is now meaningful to the switching network. There are lots of applications for Look-up Tables. Least Cost Routing tables are essentially look up tables. IN-WATS dialing works by looking up the 800 number and finding its real ten digit normal number. Most private networks use look up tables to translate the number dialed by the internal user into a number that the network can recognize.

2. A set of addresses (source and destination) used by a bridge or router to determine what should be done with a packet. As the packet comes in, its address information is read and compared with the information in the look-up table. Depending on the information, the bridge may forward the packet, or discard it, leaving it for the local LAN. Many bridges and routers can build their look-up tables as they operate. See also 800 Service, Eighthundred Service, Bridge, Personal 800 Number and Router.

Look@Me A real-time Internet collaboration freeware tool that gives the user the ability to view another user's screen that is equipped with the program. It can be used to edit material, review graphics or provide immediate training and support.

loomed speed-wrap See Bundled Cable.

loop 1. Typically a complete electrical circuit.

2. The loop is the pair of wires that winds its way from the phone company's central office to the telephone set or system at the customer's office, home or factory, i.e. "premises" in telephones.

3. In computer software. A loop repeats a series of instructions many times until some prestated event has happened or until some test has been passed. An infinite loop (i.e., a loop that never stops) causes a computer to crash. See also Crash.

loop antenna An antenna consisting of one or more complete turns of wire, both ends of which are to be connected to the input circuit of the radio receiver.

loop back A diagnostic test in which a signal is transmitted across a medium while the sending device waits for its return. See Loopback and Loopback Test.

loop bonding Loop bonding is also sometimes referred to as "inverse multiplexing" and is conceptually the same as what is called "trunking" in Ethernet parlance. It refers to combining several parallel data paths so that they appear as one logical bigger transmission channel capable of handling more bandwidth than just a single path. The specifics of how this all work depend on each vendor's implementation.

loop checking A method of checking the accuracy of transmission of data in which the received data are returned to the sending end for comparison with the original data.

loop circuit Generally refers to the circuit connecting the subscriber's set with the local switching equipment.

loop current detection When a modem, telephone or fax card (etc.) seizes the line (i.e. completes the connection between tip and ring terminals of the telephone cable) current flows from the positive battery supply in the telephone central office, through the twisted pair in the loop, through the card (or phone) and back to the central office negative terminal where it is detected, showing that this telephone or telephone device is off hook. The fax card or modem can detect problems such as disconnects, shutting down the connection or a busy signal.

loop emulation service See LES.

loop extender Device in the central office that supplies augmented voltage out to subscribers who are at considerable distances. It provides satisfactory signaling and speech for such subscribers.

loop plant Telco-talk for all the wires and hardware and poles and manholes used to connect their central offices to their customers.

loop qualification Test done by the phone company to make sure the customer is within the maximum distance of 18,000 feet from the central office that services that customer. 18,000 is the maximum distance an ISDN-BRI phone line will work.

loop resistance the total resistance of two conductors measured round trip form one end.

loop reverse-battery A method of signaling over interoffice trunks in which changes associated with battery reversal are used for supervisory states. This technique provides 2-way signaling on 2-wire trunks; however, a trunk can be seized at only one end. It cannot be seized at the office at which battery is applied. It is also called reverse-battery signaling.

loop signaling A method of signaling over circuit paths that uses the metallic loop formed by the line or trunk conductors and terminating circuits.

loop signaling systems Any of three types of signaling which transmit signaling information over the metallic loop formed by the trunk conductors and the terminating equipment bridges.

loop start LS. You "start" (seize) a phone line or trunk by giving it a supervisory signal. That signal is typically taking your phone off hook. There are two ways you can do that – ground start or loop start. With loop start, you seize a line by bridging through a resistance the tip and ring (both wires) of your telephone line. The Loop Start trunk is the most common type of trunk found in residential installations. The ring lead is connected to -48V and the tip lead is connected to 0V (ground). To initiate a call, you form a "loop" ring through the telephone to the tip. Your central office rings a telephone by sending an AC voltage to the ringer within the telephone. When the telephone goes off-hook, the DC loop is formed. The central office detects the loop and the fact that it is drawing DC current and stops sending the ringing voltage. In ground start trunks, ground Starting is a handshaking routine that is performed by the central office and the PBX prior to making a phone call. The central office and the PBX agree to dedicate a path so incoming and outgoing calls cannot conflict, so "glare" cannot occur. See GLARE. Here are two questions that help in understanding:

How does a PBX check to see if a CO Ground Start trunk has been dedicated?

To see if the trunk has been dedicated, the PBX checks to see if the TIP lead is grounded. An undedicated Ground Start Trunk has an open relay between 0V (ground) and the TIP lead connected to the PBX. If the trunk has been dedicated the CO will close the relay and ground the TIP lead.

How does a PBX indicate to the CO that it requires the trunk?

A CO ground start trunk is called by the PBX CO Caller circuit. This circuit briefly grounds the ring lead causing DC current to flow. The CO detects the current flow and interprets it as a request for service from the PBX.

See also POTS.

loop test A way of testing a circuit to find a fault in it by completing a loop and sending a signal around that loop. See Loopback.

loop through A type of phone system wiring that allows phones to connect to one cable in parallel going to the common central switching equipment. The most common type of Loop Through wiring is that which you have in your home. You have one cable with two conductors – a red and a green – winding through your home. Whenever you want to connect a phone, you simply attach it to the red and green conductors. The other way of connecting phones is called HOME RUN. In that system, every phone has its own one, two or three pairs of conductors which wind their lonely way back to the central PBX or key system cabinet. In Loop Through wiring, many phones share one set of cables. In Home Run Cabling, only one phone sits on that line.

loop timing A way of synchronizing a circuit that works by taking a synchronizing clock signal from incoming digital pulses.

loop up/loop down In T-1, there are generally two loopback types, LLB (line loopback) and TLB or DLB (terminal or DTE loopback). Loop Up refers to activating one of these loop backs, where as Loop Down refers to deactivating one of these loopbacks.

loop-in The output of a "loop through."

loop-out The output input of a "loop through."

loopback Type of diagnostic test in which the transmitted signal is returned to the sending device after passing through a data communications link or network. This allows a technician (or built-in diagnostic circuit) to compare the returned signal with the transmitted signal and get some sense of what's wrong. Loopbacks are often done by excluding one piece of equipment after another. This allows you to figure out logically what's wrong. (It's called Sherlock Holmes deductive reasoning.) See Loopback Test.

loopback test A test typically run on a four-wire circuit. You take the two transmit leads and join them to the two receive leads. Then you put a signal around the loop and see what happens. Measuring differences between the sent and the received signal is the essence of a loopback test. See Loopback.

looping Problem encountered in distributed datagram routing in which packets return to a previously visited node.

looping plug A device used to provide a physical loop on a T-1 or 56/64K DDS circuit. The plug is inserted into the demarcation jack and allows LECs (local exchange carriers) and CLECs (competitive local exchange carriers) to test circuits for continuity. A T-1 looping plug is pinned 1-4, 2-5 using a RJ-45 plug. A 56/64K DDS looping plug is pinned 1-7, 2-8 using an RJ-45 plug.

loopstart circuit The standard world-wide telephone circuit. For the phone to signal the phone system that it wants to make a call, it applies a DC termination across the phone line. See Loop Start for a longer explanation.

loose tube cable A cable construction in which the optical fiber is placed in a plastic tube having an inner diameter much larger than the fiber itself. The loose tube isolates the fiber from the exterior mechanical forces acting on the cable. The space between the tube and the fiber is often filled with a gel to cushion the fiber. In other words, the cable can slip and slide inside the jacket. This is particularly important in outdoor applications, since the jacket and armor will expand and contract more than the conductors (especially optical fiber) as the ambient temperature changes during the course of a year or even during the course of a day. If the cable were of tight buffered rather than loose tube construction, these differences in expansion/contraction rates would cause microbends in the conductors, which would affect performance negatively and which ultimately might cause cracks and breaks in the conductors. See also tight buffer.

loosely coupled A computer system architecture consisting of multiple computer systems, each with its own dedicated memory and its own copy of the operating system, connected over a communications link. See also Tightly Coupled.

LOP Loss of Pointer. LOP is a generic term with specific variations of meaning, depending on the signal standards domain in which it is being used. In the OSI/ATM world for instance, LOP is a condition at the receiver or a maintenance signal transmitted in the PHY overhead indicating that the receiving equipment has lost the pointer to the start of cell in the payload. This is used to monitor the performance of the PHY layer. In the SONET world also, LOP is a condition detected in the signal overhead at the receiver, indicating that a payload position pointer could not be obtained. There is however no "LOP indication" per se transmitted in the SONET overhead: SONET uses the AIS-L overhead indication to inform downstream equipment that the receiving equipment upstream has experienced a failure. A hierarchy of LOP defect and alarm notifications is implemented at the LOP-detecting equipment based on the duration the condition persists. See also AIS.

LORAN LOng Range Aid to Navigation. A radio-navigation system which helps you find where you are. It works by timing the difference in reception of pulses from one or more fixed transmitters, usually on land. It's a radio based systems that's pretty good for coastal waters where there are LORAN transmitters. The maximum range is about 1,400 miles at night (about a half that during the day), in virtually any sort of weather. LORAN doesn't cover much of the rest of the earth and its accuracy varies depending on electronic interference and geographic variations. The first LORAN transmitters were put into operation by the U.S. Navy in 1944. LORAN range limitations were overcome in 1973, when the U.S. Navy first installed the "Transit" SatNav (Satellite Navigation) system, which set the stage for a much better system known as GPS (Global Positioning Satellite System). Over time and as costs have decreased, even small pleasure craft have replaced their LORAN receivers with GPS. See also GPS.

lord of the rings While known as a painter, sculptor, architect, and engineer, Leonard da Vinci was the first to record that the number of rings in the cross section of a tree trunk revealed its age. He also discovered that the width between the rings indicated the annual moisture.

LORG Microsoft jargon for large size organizations. They are Microsoft's main customers.

LOS 1. Loss of Signal. LOS is a generic term with specific variations of meaning, depending on the signal standards domain in which it is being used. In the OSI/ATM world for instance, LOS is a condition at the receiver or a maintenance signal transmitted in the PHY overhead indicating that the receiving equipment has lost the received signal. This is used to monitor the performance of the PHY layer. In the SONET world also, LOS is a condition directly detected at the physical level (photonic or electronic) at the receiver. There is however no "LOS indication" per se transmitted in the SONET overhead: SONET uses the AIS-L overhead indication to inform downstream equipment that the receiving equipment upstream has experienced a failure. A hierarchy of LOS defect and alarm notifications is implemented at the LOS-detecting equipment based on the duration the condition persists. See also AIS.

2. Line of Sight. Another silly acronym is NLOS, which stands, predictably, for Non Line of Sight.

Loschmidt number The number of molecules in one cubic centimetre of a gas at one bar pressure and zero degree centigrade temperature, about 27 million trillion – the actual figure is 2.687 x 10 raised to the 19 – first calculated by and named for Austrian chemist Joseph Loschmidt (1821-1895); used in theoretical calculations of radio-frequency signal attenuation in the atmosphere.

losing face See Bee's Wax.

loss The drop in signal level between two points on a network. It is important to distinguish between "loss" and "level". Level is the amplitude, or signal strength, and is measured at finite points known as Transmission Level Points (TLPs). Loss is the difference between levels over a circuit, between two TLPs, and is measured in decibels (dB). Loss occurs constantly throughout telephony – from long distance circuits to switches, as switches set up connections via internal circuits. Loss is cumulative. Add two circuits each with a loss of 10 dB. You will have 20 dB loss in the total circuit. The human ear can detect a 3 dB loss. In figuring losses on fiber optic cables, you add up the individual losses from the connectors, the splices and the cable itself.

loss assessment Loss Assessment is the process of assessing the monetary cost to the business of the total failure of a specific resource or function. See Ray Horak essay on Disaster Recovery Planning in Introduction section of this dictionary.

loss budget A loss budget is the maximum amount of signal degradation a data communications network can withstand before it becomes susceptible to errors and/or loss of data. The idea is a establish a "loss budget" by consulting your equipment vendors for recommended wire types and maximum allocable lengths of cable before you build the network.

loss deviation Denotes the variation of the actual loss from the designed value.

Loss of Cell Delineation LCD. An ATM term. Cell Delineation is accomplished at the Transmission Convergence (TC) sublayer of the ATM Physical Layer (PHY), working in tight formation with the Physical Medium (PM) sublayer. It is at the TC sublayer that the responsibility is assumed for Physical Layer operations that are not medium independent. For example, it is at this sublayer that the ATM cell switch interfaces with a SONET transmission system. Cell Delineation is responsible for defining the cell boundaries at the originating endpoint (e.g., ATM cell switch or ATM workstation) in order that the receiving endpoint can reassemble (i.e., identify and recover) all cells associated with a data payload that has been segmented (i.e., cut up into ATM cells). Cell Delineation is achieved by the receiving endpoint's locking onto the 5-octet cell headers. A failure in this process is known as Loss of Cell Delineation (LCD).

Loss of Frame See LOF.

Loss of Pointer See LOP.

lossless compression Image- and data-compression applications and algorithms, such as Huffman Encoding, that reduce the number of bits a file would normally take up without losing any data. In this way, no information is lost or altered in the compression and/or transmission process. Lossless compression is particularly important for alphanumeric text files, as the integrity of the data is affected dramatically if even one bit is lost. PKZip and WinZip are examples of products that use lossless compression techniques. See also Lossy Compression.

lossy When you convert information from digital to analog or vice versa or compress, or change an image in some way – e.g. make it small for the web – you inevitably lose some of the data. How important that loss of information is to you depends on what you're going to do with the information. If you're simply going to view the image small on the Internet, then it doesn't matter that you made your image "lossy." If you want to print your photo poster size, the more information you don't loose, the better. Your picture will look better blown up. Different imaging formats – JPEG, GIF etc. – produce different results. An image which looks good in JPEG might look less good in GIF. You can play around with the concept of "lossy" with your digital camera. Take a picture of a vase with some fruit on a tripod. Take a picture of a vase with some fruit at the largest image size your camera will produce. My camera's largest size produces an image of 19 megabtyes. Now take photos of the same bowl of fruit and ask your camera to produce the smallest file. My camera's smallest is around 650 kilobytes. Open both images as large as you can on your computer and check out the differences. You'll find plenty.

lossy compression Compression techniques that result in the some loss of data are known as lossy. While the data file is not completely accurate, once compressed and decompressed, the savings in storage memory and transmission bandwidth/time are considerable. Audio, video and image files often are compressed via a lossy technique, as the savings in storage and transmission bandwidth are so significant, while the level of loss is often not noticeable. JPEG, MPEG and Wavelet are all examples of lossy compression techniques. See also GIF, JPEG, lossless compression, lossy, MPEG, and Wavelet.

lost call attempt A call attempt that cannot be further advanced to its destination due to an equipment shortage or failure in the network.

lost calls cleared Traffic engineering assumption used in Erlang C that calls not satisfied on the first attempt are held (delayed) in the system until satisfied.

lost calls held Traffic engineering assumption used in Poisson that calls not satisfied on the first attempt are held in the phone system for a period not exceeding the average holding time of all calls.

lost wages An endearing term for a town called Las Vegas whose claim to fame is gambling. Hence the term "lost wages."

lottery A method, authorized by the Congress, designed to provide the FCC with an alternative or option to comparative hearings for allocating spectrum space to competing applicants in various services.

Lotus Express An e-mail communications software program for PC users of MCI Mail that allows the user to automatically send and receive messages, as well as binary files, such as spreadsheets or documents.

loudspeaker See Sound.

loudspeaker paging access Interface to customer-provided paging equipment.

love Zero scores in tennis are called "love." In France, where tennis first became popular, a big, round zero on scoreboard looked like an egg and was called "l'oeuf," which is French for "egg". When tennis was introduced in the US, Americans pronounced it "love." I can continue further with bad yokes...

Lovejob Graphics service bureau slang for a file that an art director obsessively wants to output in every possible variation. "Yeah, I know we're ripping it to the Iris proofer for the ninth time. This one's a lovejob." Definition courtesy Wireless Magazine.

Lovelace, Ada Augusta Ada Augusta Lovelace was the daughter of Lord Byron, the English poet. Miss Lovelace is regarded as the first computer programmer because she worked for the computer pioneer Charles Babbage. A computer language was named after her. That language is called ADA.

low battery cutoff An uninterruptible power supply (UPS) definition. This UPS feature automatically switches off battery power before the batteries discharge beyond safe limits. Without this feature, batteries can be taken into deep discharge, making them useless.

low entry networking LEN. A peer-oriented extension to SNA, first implemented on IBM's System/36, that allows networks to be more easily built and managed by such techniques as topology database exchange and dynamic route selection.

low frequency The band of frequencies between 30 and 300 kilohertz.

low frequency RFID tags They typically operate at 125 KHz. The main disadvantages of low frequency tags are they have to be read from three feet closer and the rate of data transfer is slow. But they are less expensive and less subject to interference than high frequency tags.

low level combining A process frequently used in multichannel transmitters for combining channels or carriers at some processing stage prior to the transmitter output.

low level diplexing A processing of combining two radio frequency carriers (usually audio and video) at some processing stage prior to the final transmitter output.

low level formatting The first step in preparing a drive to store information after physical installation in complete. The process sets up the "handshake" between the drive and the controller. Most drivers are now low level formatted at the factory. See Physical Formatting.

low level language A programming language that uses symbols – one step away from the machine language of a computer. Low level computer languages, such as Assembler and C, actually manipulate the bits in computer registers. Higher level languages such as Basic and Fortran will take care of the piddling details of doing specific functions when you give it a broad command like "PRINT". In a lower level language, you must provide all the details of instruction necessary in the code (program) to perform the operation. It is possible to do this by calling standard routines, but still takes up the programmers' time in deciding which routines, and keeping the registers straight as he designs the program.

low loss dielectric An insulating material that has a relatively low dielectric loss, such as polyethylene or teflon.

low smoke zero halogen See LSZH.

Low Noise Amplifier LNA. Typically a parametric amplifier in a satellite earth station.

Low Noise Blocker See LNB.

Low Noise Converter LNC. A pre-amplification device consisting of a low noise amplifier and down converter built into one unit.

low order bit Least Significant Bit (LSB). See also High-Order Bit, and Least Significant Bit.

low pass filter A device that cuts frequencies off above a certain point and allow all other frequencies to pass. Opposite of high pass.

Low Power FM Radio Station See LPFM.

Low Power Television Service LPTV. A broadcast service that permits program origination or subscription service or both via low powered television translators. LPTV operates secondarily to regular television stations. Transmitter output is limited to a 1000 watts for a UHF station, 10 watts for a VHF station, except when VHF operation is on an allocated channel when 100 watts may be used.

low speed loopback A closed circuit feature useful for maintenance or testing.

low speed signal Signal traveling at the DS1 rate of 1,544 Mb/s or at the DS1C rate of 3,152 Mb/s.

low tier See High Tier.

low voltage A low voltage condition exist when fewer than 105 volts AC is present at a 120 VAC outlet or HOT conductor. This figure was chosen by many manufacturers of electronic and telephone equipment. It is also the test of "low voltage" tested by a wonderful AC electric outlet-testing product called the Accu-Test II made by Ecos Electronics Corporations of Oak Park, Illinois. Below 105 volts, motors deteriorate and electronic circuits overheat. Long-term damage can occur to most gadgets plugged into an electrical outlet which consistently delivers below 105 volts. Also 105 volts is below the stated tolerance levels of all North American power utilities who state that their acceptable power is 120 volts plus or minus 10 percent. If your power is consistently below 105 volts, you should contact your local power utility.

low voltage differential signaling LVDS: Low Voltage Differential Signaling is a new data interface standard that is defined in the TIA/EIA-644 and the IEEE 1596.3 standards. It is essentially a signaling method used for high-speed transmission of binary data over copper. It uses a 350 mV voltage swing between two wires rather than a signal referenced to the ground. This low voltage differential is what delivers high data transmission speeds and inherently greater bandwidth at lower power consumption with low electromagnetic interference (EMI) The receiver for LVDS utilizes the differential signal to determine the state and rejects the common noise for better noise immunity.

low water peak fiber The presence of moisture in optical fibers results in significant levels of attenuation (i.e., power loss) at 1383 nm (nanometers). Depending on the magnitude of that "water peak" loss, wavelengths within +/- 50 nm also can be affected. This water peak is caused by residual moisture deposited during the manufacturing process. Specifically, it largely is caused by hydrogen atoms in the water molecules. The hydrogen atoms readily diffuse through the glass matrix of an optical fiber and are trapped at points of defect in the glass structure. The process of manufacturing low water peak fibers involves reducing water levels in the fiber from 10 ppb (parts-per-billion) to less than 1 ppb. This is accomplished by exposing the fiber to deuterium gas or deuterium blends under elevated temperature and pressure. The deuterium atoms displace the hydrogen atoms at the defect sites, which lowers the water peak and increases the capacity of the fibers by supporting a wider range of spectrum. See also Rain Attenuation.

lower case Upper- and lower-case letters are named "upper" and "lower," because in the time when printers set pages using loose type, the "upper-case" letters were stored in the case on top of the case that stored the smaller, "lower-case" letters.

loyalty management Verizon Wireless has a department called "Loyalty Management." Their main job is handling cancellations from customers.

loyalty market Giving awards to customers for shopping with you. The airlines have the best loyalty program around – frequent flyer miles.

LPA Linear Power Amplier.

LPC 1. Linear Predictive Coding. Low bit rate voice (LBRV) digitizing technique that requires a bandwidth of only 2.4 or 4.8 Kbps. This technique may result in poor quality voice signals.

2. Late Payment Charge.

LPD Line Printer Daemon, a process on Berkeley spooler implementations that provides LPR systems.

LPDA-2 IBM's protocol under NetView for monitoring of dial-up modems for error correction.

LPF Low Pass Filter: In an MPEG-2 clock recovery circuit, it is a technique for smoothing or averaging changes to the system clock.

LPFM On January 20, 2000, the FCC adopted rules creating a new, low power non-commercial FM radio (LPFM). A normal FM station transmits at thousands of watts of power.

This level of power means that the station will need lots of expensive transmitting equipment. costing at least a million dollars. But for that money it will be able to cover an entire urban area. The LPFM station is designed to let individuals and small organizations own and operate radio stations for a wide variety of not-for-profit reasons and transmit only to small groups of people within limited geography. In this sense, the FCC is trying to bring station creation closer to normal people – in the same way that anyone can create a web site. A LPFM station is a 10 or 100 watt transmitter. This level of power gives the station a range of up to 3.5 miles. A transmitter this size and its antenna might cost $2,000 to $5,000. In a city, the range of an LPFM transmitter could cover a small neighborhood. The owner-operators of low power FM radio stations could include religious groups, churches, colleges, PTA-sponsored school stations, foreign language stations, race track pit and parking areas and be an extension to a school or college public address system.

LPI Lines Per Inch. The number of lines both horizontal and vertical, that a facsimile machine will print in a square inch.

LPIC 1. IntraLATA Primary Interexchange Carrier. A Primary Interexchange Carrier for IntraLATA long distance traffic, the LPIC may be different than the PIC (Primary Interexchange Carrier). While the PIC typically is your carrier for all long distance, it may be designated only for InterLATA traffic if you also have designated a LPIC. Traffic to both the LPIC and PIC is automatically routed from a given location when dialing 1+ in equal access areas. The LPIC is identified by a LPIC Code (LPIC Code) and the PIC by a PIC Code (PICC) which is assigned by the local telephone company to the telephone numbers of all the subscribers to those carriers to ensure the calls are routed over the correct network. When a subscriber switches long distance carriers, it often is referred to as a PIC change.

2. Local Presubscribed Interexchange Carrier.

LPM Lines Per Minute. A reference to printer speeds.

Lport See Logical Port.

LPP Link Peripheral Processor or Link Peripheral Processing.

LPR The LPR command is used to queue print jobs on Berkeley queuing systems.

LPRS Low Power Radio Service. See Personal Radio Services.

LPS 1. Line Profile System.
2. Lightning Protection Subsystem.

LPSW Line Protection Switch.

LPT Port A logical designation for a series of I/O (Input/Output) addresses that allows the computer to communicate with a parallel printer.

LPT1 The first or primary parallel printer port on the IBM PC or clone. LPT2 is the second parallel port. COM1 is the first serial port. LPT1 is usually the default printer port, i.e. the one your computer will print to, if you don't tell it something else.

LPTV Low Power Television Service. A broadcast service that permits program origination or subscription service or both via low powered television translators. LPTV operates secondarily to regular television stations. Transmitter output is limited to a 1000 watts for a UHF station, 10 watts for a VHF station, except when VHF operation is on an allocated channel when 100 watts may be used.

LQA Link Quality Analysis. See Automatic Link Establishment.

LRC 1. Longitudinal Redundancy Check. A system of error control based on transmission of a block check character based on preset rules. The check character formation rule is applied in the same manner to each character on a bit by bit basis. See Longitudinal Redundancy Check.

2. Loop Resiliency Circuit. Hub circuitry that allows devices to be inserted into or removed from an active FC-AL (Fibre Channel Arbitrated Loop) loop.

LRE 1. Line Regenerating Equipment.
2. Lightwave Repeating Equipment.
3. Long Reach Ethernet. See Long Reach Ethernet.

LRE-DSI Low Rate Encoding-Digital Speech Interpolation.

LRIC Long Run Incremental Cost.

LRN Local Routing Number. A 10-digit telephone number. The term is used in the context of LNP (Local Number Portability). See LNP.

LRR Long Range Radar.

LRU Least Recently Used. Refers to an algorithm that sorts items according to time last accessed and then discards the oldest items in the list to free up needed space.

LS Trunk Loop-Start Trunk. A trunk on which a closure between the tip and ring leads is used to originate or answer a call. High-voltage 20-Hz AC ringing from the telephone company signals an incoming call. See LOOP START.

LS1A-A Single Mode, Single-Fiber Interconnection Cable. An AT&T definition.

LSA Link State Advertisements: OSPF link state information comprised of four variables: Router Link Advertisements, Network Link Advertisements, Summary Link Advertisements and Autonomous System (AS) Link Advertisements. All but the AS Link Advertisements are flooded through a single OSPF area only. The AS Link advertisement s are flooded throughout the entire OSPF network.

LSAP An ATM term. Link Service Access Point: Logical address of boundary between layer 3 and LLC sublayer 2.

LSAS Line Side Answer Supervision.

LSB Least Significant Bit and Least Significant Byte. That portion of a number, address or field which occurs right most when its value is written as a single number in conventional hexadecimal or binary notation. The portion of the number having the least weight in a mathematical calculation using the value.

LSCIE Lightguide Stranded-Cable Interconnect Equipment.

LSCIM Lightguide Stranded-Cable Interconnect Module.

LSCIT Lightguide Stranded-Cable Interconnect Terminal.

LSDU Link layer Service Data Unit.

LSE Local Support Element.

LSI Large Scale Integration. Refers to micro electronic components which combine many hundreds of transistors on an integrated circuit. See CHIP.

LSL Link Support Layer. A layer within the Novell Open Data-Link Driver specification. This layer lets multiple protocol stacks access a network card simultaneously.

LSMS Local Service Management System. The database from which each LEC gets updates from the NPAC for portability requests. See NPAC.

LSO Local Service Office. Defined for North America as a six digit number consisting of the area code and the first three digits of the exchange code (i.e. 410-638 (area code for northern Maryland, with exchange code for Belair Maryland)). Used to identify a geographical area and local service provider for that circuit. An LSO is important when ordering circuits from a telephone company, especially a long distance one.

LSOA Local Service Order Administration system. An OSS (Operations Support System) used by LECs (Local Exchange Carriers) to administer service orders such as orders for new service, service rearrangements, and changes of carrier. See also LNP, NPAC and OSS.

LSOG See Local Service Ordering Guidelines.

LSOR Local Service Order Record.

LSP 1. Local Service Provider. Another term for a Local Exchange Carrier (LEC).
2. Label Switched Path. See MPLS (Multiprotocol Label Switching).
3. Local Service Provider. See Location tracking.

LSR 1. An ATM term. Leaf Setup Request: A setup message type used when a leaf node requests connection to existing point-to-multipoint connection or requests creation of a new multipoint connection.

2. Line Service Request. Document in the LEC (local exchange carrier) world which is used for porting requests, directory listing changes, trunk ordering etc. It has lots of slots for hand written letters, and uses lots of trees (often a full sheet for each number regardless of the volume of the request) also available in proprietary electronic systems from larger LECs, such as US West, Southwestern Bell and GTE.

3. Local Service Request. A term that was spawned by the Telecommunications Act of 1996, which formalized competition in the local exchange domain. A LSR is a form, or series of forms, used by the CLEC (Competitive Local Exchange Carrier) to request "local service" from the ILEC (Incumbent LEC). Specifically, the LSR is used by the CLEC to request a local loop from the ILEC, meaning that the CLEC wants the ILEC to provide a local loop from the customer's premises to the CLEC's termination equipment, which typically is housed in the ILEC's (CO) Central Office. A LSR also is used to request that the ILEC port the subscriber's telephone number to the CLEC. In full form, porting the telephone number means that the subscriber's telephone number remains the same – it's just ported to the CLEC through full Local Number Portability (LNP). If the CLEC and/or the ILEC do not have the necessary software in place to accomplish such a level of number portability, the subscriber may wind up with two telephone numbers – one number is the new CLEC number, and the other is the old ILEC telephone number. Effectively, in this latter scenario, incoming calls are forwarded to the new number in an arrangement known as Interim Number Portability (INP). The LSR may be in either paper or electronic form, and contains all information required for administrative, billing, and contact details. See also CLEC, ILEC, INP, and LNP.

4. Label Switch Router See Label Switch Router and MPLS (MultiProtocol Label Switching).

LSSGR LATA Switching System Generic Requirements.

LSV Line Supervisory Equipment.

LSZH Low Smoke Zero Halogen. LSZH stands for Low Smoke Zero Halogen and describes a cable jacket material that that is non-halogenated and flame retardant. This type of jacket material has excellent fire safety characteristics of low smoke, low toxicity and low corrosion. This type of cable jacket material is used in applications such as Central Offices, Mass Transit Rail Systems, Nuclear Plants and Oil Refineries or in any other application where the protection of people and equipment from toxic and corrosive gasses is absolutely critical. LSHZ cable is required for plenum applications, as halogen-laden smoke is particularly deadly. See also halogen and plenum cable.

LT 1. Line Terminator.

2. Logical Terminal.

3. An ATM term. Lower Tester: The representation in ISO/IEC 9646 of the means of providing, during test execution, indirect control and observation of the lower service boundary of the IUT using the underlying service provider.

LTB Last Trunk Busy.

LTC Line Trunk Controller.

LTCI ISDN Line Trunk Controller.

LTCLASS Logical Terminal Class classifies each ISDN terminal based upon the type of messaging that is exchanged between the terminal and circuit switch part of the ISDN switch. There are 3 classes: BRAKS, BRAMFT, and BRAFS. The newer (BCS-31 and above) ISDN sets should be defined with BRAMFT or BRAFS. The decision to use BRAMFT or BRAFS depends on what kind of service is being provided, for example, MADN groups should be defined on a BRAFS terminal. See also LTID.

LTC A satellite term. Longitudinal/Linear Time Code.

LTCS Label Traffic Control System. LTCS is a prepackaged implementation of the Multiprotocol Label Switching (MPLS) standard from a Boston, U.S.-based company called Harris & Jeffries. www.hjinc.com. According to Interactive Week of May 3, 1999, MPLS has emerged as a compelling technology because it offers service providers a mechanism for controlling the flow of traffic across their networks, which is a crucial requirement for offering advanced applications such as voice services and virtual private networks. In addition, MPLS has been embraced by makers or Asynchronous Transfer Mode gear because it provides a layer of translation between IP and ATM."

LTDP Line Terminator Type DP.

LTE SONET Lite Terminating Equipment: ATM equipment terminating a communications facility using a SONET Lite Transmission Convergence (TC) layer. This is usually reserved for end user or LAN equipment. The SONET Lite TC does not implement some of the maintenance functions used in long haul networks such as termination of path, line and section overhead. In short, line terminating equipment includes network elements which originate and/or terminate line (OC-N) signals. LTEs originate, access, modify, and/or terminate the transport overhead.

LTGRP Logical Terminal Group. See Logical Terminal Group.

LTID Logical Terminal Identifier. LTIDs are required for each logical terminal connected to an ISDN switch. Each LTID allocated must be unique. An LYID consists of a logical terminal group name (LTGRP), followed by a logical terminal number (LTNUM). i.e. ISDN 99 and LCME1 100, where ISDN and LCME1 represent the LTGRP, and 99 and 100 represent the LTNUM within the group.

LTN Local Transport Network. A term for fiber ring.

LTNUM Logical Terminal Number.

LTO See Linear Tape Open Architecture.

LTP 1. Line Test Position.

2. List Transmission Path.

LTPA Line Terminator Type PA.

LTPB Line Terminator Type PB.

LTR Local Transport Restructuring. An FCC ordered tariff restructuring by the LECs on access charges for IXCs.

LTRS Letters Shift. 1. Physical shift in a terminal using Baudot Code that enables the printing of alphabetic characters. 2. Character that causes the shift.

LTS Loop Testing System.

LTTB Line Terminator Test Board.

LU 1. Line Unit.

2. Logical Unit, access port for users in SNA. In a bisync network, a port through which the user gains access to the network services. A LU can support sessions with the host-based System Services Control Point (SSCP) and other LUs.

3. Local Use flag. Occasionally used to initialized approval for local cellular calls. The Cellular carrier insures that local users are registered with a local system.

LU 6.2 Logical Unit Interface. Version 6.2. An IBM SNA protocol that allows for peer-to-peer or program-to-program communications. The LU 6.2 protocol standard frees application programs from network specific details. On an IBM PC, a LU 6.2 program accepts commands and passes them on to an SDLC card to communicate directly with the mainframe or a token ring handler. LU 6.2 enables users to develop applications programs for peer-to-peer communications between PC's and IBM host systems. It increases the processing power of the PC user without the constraints of mainframe-based slave devices, i.e. 3274/3276 controllers. It creates a transparent environment for application-to-application communications, regardless of the types of systems used or their relative locations. Also referred to as Advanced Program-to-Program Communication (APPC).

LU Type 1 LU 1 is the SNA protocol that describes generic input/output devices (e.g. line printer).

LU Type 3 LU 3 is the SNA protocol that describes a print output device that uses 3270 data streams.

Lucent Technologies On September 30, 1996, AT&T spun off AT&T Technologies (previously Western Electric), its manufacturing operations and its Bell Labs into a separate company called Lucent Technologies. Lucent then comprised Bell Laboratories (R&D), Network Systems (development and manufacture of switches, and related systems and software for the carrier market), Business Communications Systems (development, manufacture, marketing and servicing of advanced communications products for business customers), Microelectronics Group (design and manufacture of high-performance integrated circuits, optoelectronic components, and power systems), and Consumer Products (design, manufacture, sales, servicing and leasing of both wired and wireless communications products for consumers and small businesses). One reason for the split was that AT&T Technologies' major customers (the RBOCs) were viewed as unlikely to purchase equipment and services from their strongest likely competitor (i.e. AT&T) in a deregulated and competitive environment. In March 2000, Lucent sold its small and mid-sized business sales group to Expanets, a division of NorthWestern Corporation. In July 2000, Lucent spun off Enterprise Networks Group, as Avaya Communication. Avaya manufactures voice, converged voice and data, customer relationship management, messaging, multi-service, networking, and structured cabling products and services. Lucent Technologies, minus Avaya, designs and delivers the systems, software, silicon and services for network service providers and large enterprises. See also AT&T and Avaya. www.lucent.com.

Lucky Dog Phone Company Once upon a time AT&T was an expensive long distance phone company. There were many other phone companies offering cheaper long distance. Instead of dropping its prices, AT&T came up with a "brilliant" strategy – have another "company" provide cheap phone service. AT&T called that "company" the Lucky Dog Phone Company. It was, of course, really AT&T providing the service under a different name.

Ludd, Ned See Luddites.

Luddites The common myth: Luddites were people who hated industrial progress and went around destroying machinery. Wrong: They were people who had no problem with progress. They simply wanted to share in the fruits, i.e. be paid more. In the Sunday New York Times of December 6, 1998, William Safire explains all. Here's an excerpt: In Leicestershire in 1779, a man named Ned Ludd broke into a house, and in what was reported to be 'a fit of insane rage,' destroyed two machines used for knitting hosiery. The breaking of such knitting frames – machinery invented two centuries before – had been going on for nearly a century. Ludd, however, did it with such gusto and flair that subsequently, whenever machines of any sort were found smashed, the excuse was given that 'King Ludd must have been here.'

What was Ludd's motive? Was he a lover of hand-knitted hosiery? Did he prefer going barefoot? Or was he making some sort of unadorned social protest of deeper significance? Revisionist historians say that Ludd and other frame-wreckers were protesting poor working conditions and low wages at the beginning of the Industrial Revolution. However, between 1811 and 1816, organized bands of masked men swore allegiance to 'King Ludd' rather than the British sovereign, and waged a war against the serflike conditions spawned by the users of textile machinery. 'If the workmen dislike certain machines,' explained The Nottingham Review in 1811, 'it was because of the use to which they were being put, not because they were machines or because they were new.' That living-condition claim was swept aside by commercial interests and officialdom, which hung the label Luddite on protesters not for demanding a living wage but for obstructing the march of technological progress. The historical revisionists argue that others attributed the anti-machinery 'cause' to the Luddites. Intellectuals and romantics like the poets Blake, Byron, Shelley and Wordsworth picked up that anti-technology theme, but identified with its other side. In the "dark Satanic mills" of industry, they saw the human spirit being stifled. Lord Byron wrote an inflammatory 'Song for the Luddites' in 1816. Its first stanza: 'As the Liberty lads o'er the sea/Bought their freedom, and cheaply, with blood,/So we, boys, we/Will die fighting, or live free,/And down with all kings but King Ludd!' Mary Shelley, daughter of the feminist Mary Wollstonecraft and wife of the poet, gave the Luddite theme dramatic power in her 1818 novel, 'Frankenstein.' The danger of rampant technology is expressed by the monster, who says to Dr. Victor Frankenstein, "You are my creator, but I am your master.' Between the sweatshop operators and the romantic poets, the meaning of Luddite became fixed as "radical opponent of technological or scientific progress.'

LUDs On TV shows like "Law & Order," cops who want info on the bad guys' phone calls, pull their LUDs. LUDs is an acronym for Local Usage Details – a record of incoming and outgoing calls for a particular phone number. The cops get this list from the local phone number. These lists are one reason the bad guys like to use coin phones.

lug Something which sticks out and onto which a wire may be connected by wrapping or soldering.

luma The brightness signal in a video transmission, i.e. Y. See luminance.

luminance The measurable, luminous intensity of a video signal. Differentiated from brightness in that the latter is nonmeasurable and sensory. The color video picture information contains two components: luminance (brightness and contrast) and chrominance (hue and saturation). The photometric quantity of light radiation. Luminance is that part of the video signal which carries the information on how bright the TV signal is to be.

Luminiferous Ether Heinrich Rudolf Hertz discovered the phenomenon of the propagation of electromagnetic energy through space. He assumed that there was some conducting matter in the air that supported the arcing of a spark from a positive to a negative terminal. To test the theory, he tried to arc a spark through a vacuum tube, and succeeded. He assumed that the magical conducting matter was so small that it could not be identified and could not be voided from the tube. He named this non-existent matter "luminiferous ether." Bob Metcalfe and his associates later named their magical LAN (Local Area Network) technology Ethernet. See also Edison Effect.

LUNI LANE User Network Interface. Pronounced "looney," as in "Sometimes all these acronyms make me looney." An ATM term for standardized network interface protocol between a LANE (Local Area Network Emulation) client and a LANE Server. See also LANE and LNNI.

LUNS Logical Unit Numbers. An identification number given to devices connected to a SCSI adapter. Each SCSI ID can have eight LUNs. Normally, there is only one device with LUN 0. See Daisy Chain and SCSI.

Lurker A person who "hangs around" online bulletin boards and forums, browsing through the messages and, if moved, replying to some of them. That's one definition. Here's another: A visitor to a newsgroup or online service who only reads other people's posts but never posts his or her own messages, thus remaining anonymous. See Lurking.

Lurking The practice of reading an Internet mailing list or Usenet newsgroup without posting anything yourself. Everywhere you look there's a Slim Shady lurking. He could be working at Burger King, spitting on your onion rings. Or so I'm told. In the online world, lurking is not considered particularly antisocial; in fact, it is a good idea to lurk for a while when you first subscribe so that you can get a feel for the tone of the discussions in the group and come up to speed on recent history. See Lurker.

Lusha A talented lady with endless patience. She checked and spell-checked every definition until her eyes fell out of her head. She said, politely, she'd had more exciting jobs. Certainly better paying ones. If you find a mistake in this dictionary, please email Lusha. It's all her fault. LDing@Amherst.edu. When she becomes a world-famous concert pianist, you'll be able to tell all your friends, you knew her when she was a lowly lexicographer.

Lux A contraction of luminance and flux and a basic unit for measuring light intensity. A Lux is approximately 10 foot candles.

LUXPAC A ITU-T X.25 packet switched network operated in Luxembourg by the Luxembourg government.

LVDM A SANs definition. Low-Voltage Differential Multipoint (for low-power bus applications requiring multi-point interconnections, where LVDS is not feasible.)

LVDS Low Voltage Differential Signaling is a new data interface standard that is defined in the TIA/EIA-644 and the IEEE 1596.3 standards. It is essentially a signaling method used for high-speed transmission of binary data over copper. It uses a 350 mV voltage swing between two wires rather than a signal referenced to the ground. This low voltage

differential is what delivers high data transmission speeds and inherently greater bandwidth at lower power consumption with low electromagnetic interference (EMI.) The receiver for LVDS uses the differential signal to determine the state and rejects the common noise for better noise immunity.

LVTTL Low Voltage TTL.

LXSMN Telephone Assistance Plan Surcharge.

Lymph To walk with a lisp.

Lynx A World Wide Web (WWW) browser developed at the University of Kansas for students and faculty who use dumb terminals (e.g., VT100) for connection to mainframe and midrange computers running either the UNIX or the MVS operating system. Lynx was built on an early version of the Common Code Library developed by the CERN WWW project headed by Tim Berners-Lee. Lynx is keyboard-oriented and text-only, rather than being fully featured, graphically-oriented, mouse-driven, audio- and video-capable. Given its origins and orientation (dumb terminals and big computers), that's understandable. In the context of high-performance PCs and Graphical User Interfaces, however, Lynx is not the browser of choice for most of us.

LZ77 A compression approach falling into the general category of "dictionary methods," LZ77 was developed in 1977 by Abraham Lempel and Jakob Ziv. LZ77 involves an implicit dictionary in which data is represented by previously processed data. In other words, a given set of data which is exactly the same as a previously transmitted of data need not be restated in full. Rather, a pointer refers to the previous data set. The result is that far fewer bits need be sent to communicate the same information. That yields more efficient use of bandwidth, which always is a limited resource. LZ77 is an asymmetric compression method, as the coding process is relatively time-consuming, while the decoding process is much less so. LZ77 was refined in the LZSS algorithm developed in 1982 by Storer and Szymanski. See also Compression, Dictionary Method, LZSS, LZ78, LZJH, and LZW.

LZ78 A refinement of the LZ77 compression method. A "dictionary method," LZ78 creates a dictionary of data phrases that occur in the input data. When the compression software encounters an input phrase that previously has been entered into the dictionary, the output is in the much-abbreviated form of the index number of that phrase. The result is that far fewer bits need be sent to communicate the same information. That yields more efficient use of bandwidth, which always is a limited resource. LZ78 was further refined in the LZW algorithm developed by Terry Welch in 1984. See also Dictionary Method, LZSS, LZ77, LZJH, and LZW.

LZJH Lempel-Ziv-Jeff.Heath. A compression algorithm developed by Jeff Heath of Hughes Network Systems for use over satellite links, LZJH is based on earlier work by Abraham Lempel and Jakob Ziv.. LZJH is the basis for the V.44 compression standard finalized on June 30, 2000 by the ITU-T. V.44, which supplants the earlier V.42bis modem compression technology, provides 6:1 (i.e., 6 to 1) compression performance. V.44 is intended for use in V.92 modems, the successor to V.90 modems. In combination with V.92 modems, V.44 will have a significant effect on the speed of data transmission, as did the earlier move from V.34 to V.90 modems. See also LZ77, LZ78, V.42bis, V.44, V.90, and V.92.

LZS Lempel-Ziv-Stac. A data compression algorithm developed by Stac Electronics and sometimes used by routers. LZS is based on the LZ77 algorithm developed by Abraham Lempel and Jakob Ziv in 1977. See also LZ77.

LZSS Lempel-Ziv-Storer-Szymanski. A data compression method developed in 1982 by Storer and Szymanski, and based on earlier work by Abraham Lempel and Jakob Ziv. LZSS offers a better compression ratio than LZ77, and the decoding process is simpler and faster. Popular archivers such as PKZip make use of LZSS.

LZW Lempel-Ziv-Welsh. A data compression method developed in 1984 by Terry Welch, based on LZ78 by Abraham Lempel and Jakob Ziv. LZW was developed for hardware implementation in high-performance disk controllers. For example, Nortel uses LZW in its Distributed Processing Peripheral (DPP), which is the Automatic Message Accounting Transmitter (AMAT) for the DMS-100 family of central office switches. Nortel selected this non-proprietary protocol and helped promote it as an industry standard. The nominal compression ration is 2.8:1, without considering field suppression. Transmitting data compressed at a ratio of 2.8:1 at 9600 bps is equivalent to transmitting non-compressed data at 27 Kbps. Compatible compression collectors can poll the DPP in compressed or non-compressed mode. DPPs equipped with the Data Compression feature can transmit in either compressed or non-compressed mode, based on the collector's polling request for a specific polling session. To preserve data integrity, AMA data are still stored in non-compressed form on the DPP disks. LZW is a lossless data-compression algorithm patented by Unisys. See Lossless, Lossy, and LZ78.

m 1. (small letter) Milli. One-thousandth. M (big letter) Mega. One million, e.g. Mbps or Mbit/s, one million bits per second. But m or M is confusing. There are some places it will mean a thousand and other places it will mean a million. You need to check the context and often ask. My brokerage statement, for example, uses a big letter M to mean thousand. In short, be careful.

2. Meter. The fundamental metric unit of length, a meter is equivalent to 39.37 inches. See Meter for more detail.

M Bit The More Data mark in an X.25 packet that allows the DTE or DCE to indicate a sequence of more than one packet.

M hop The transmission of satellite signals through an uplink to a satellite that downlinks to a receiving station half-way or between the final receiving station. The intermediate receiving station uplinks the signal again to a satellite for delivery to the final, targeted receiving station. As an example, data from a company in the UK for a business in Japan may be sent to a satellite covering North America and Europe. The signal is downlinked to an earth station in Colorado. The signal is then uplinked to a satellite covering Asia and North America for final delivery to an earth station in Japan. The signal flow forms an 'M', thus the name.

M patch bay A patching facility designed for patching and monitoring of digital data circuits at rates from 1 Mbps to 3 Mbps.

M port The port in an FDDI topology which connects a concentrator to a single attachment station, dual attachment station, or implemented in a concentrator. This port is only implemented in a concentrator.

M VTS Marconi Video Telephone Standard sends color pictures over regular phone lines at up to 10 frames per second at 14.4 Kbps. The MCI Video Phone, which conforms to this standard, has a resolution of 128 by 96 pixels.

M&C Monitor and Control.

M-code A modernized military GPS signal designed in the late 1990s to replace the military's Y-Code GPS signal. M-Code was first transmitted by Block IIR-M satellites in 2005 and can only be broadcast by those satellites and the newer, soon-to-be-launched Block IIF satellites. The M-code signal differs from the military's older Y-Code signal in terms of its data structure, security architecture, and modulation technology. See also GPS.

m-coupon A coupon delivered wirelessly to a person's mobile phone.

M-ES Mobile End System.

m-government Mobile government. There are two sides to this. The government authority uses mobile phones, PDAs and WiFi devices to provide better services to its constituents. This includes everything from scheduling garbage trucks to diverting traffic based on real-time conditions to reporting potholes, to issuing paperless parking tickets. The other side to this is the constituents' use of government services – everything from finding library opening times, to checking traffic conditions and to paying parking tickets.

M/W Microwave.

M.100 M.100, a specific API for S.100 Server Configuration and Control, now available from the Enterprise Computer Telephony Forum (ECTF). By managing the configuration, startup and shutdown of S.100-conforming computer telephony servers, M.100 allows Call Centers to provide more consistent service through better-configured, more stable servers and orderly shutdown when problems occur. In its July issue, Computer Telephony Magazine said M.100 "addresses (a) configuration management, (b) performance management, (c) statistic management and (d) fault management across the system/server and application level. M.100 was announced on April 22, 1998. At that time, the ECTF said M.100 allows users to ensure that all of the resources and other elements of a server are brought up in an orderly manner during startup to produce a stable environment. If problems occur during operation, M.100 allows the server to be shut down without loss of data, a critically important benefit to large Call Centers. During a shutdown, M.100 allows all current calls to be handled and prevents additional calls to be taken or created until shutdown has been completed. This enables Call Centers to continue servicing customers without catastrophic interruptions even when problems occur. M.100 employs session and event management, symbols and data types from S.100, and it defines functions that allow administrators and developers to create customized administration applications. These include management of configuration data, management of services, startup and shutdown, information about service providers and handling of generic administration commands. M.100 also contains the infrastructure hooks that enable inclusion of vendor-supplied diagnostic tools. This feature allows users to easily configure a system with those diagnostic routines needed to create a highly stable environment. Because M.100 is a specific API for S.100 servers, it addresses several key areas not covered by the network management APIs. First, the startup and shutdown of an S.100 server is specific to its operating system environment. If the server were running under Windows NT, it would be started as an NT service, and M.100 would control the startup and the shutdown of the various services. M.100 also supports making persistent information available on the server that can be accessed whenever new applications or services are installed or removed. This data is referred to as a profile that is similar to an INI file. An S.100 server contains multiple profiles that describe all the components that make up the server. M.100 allows for manipulation of the profiles making for much cleaner server configurations and re-configurations. Also, M.100 makes it much easier to handle KVSets, the preferred mechanism for using data within an S.100 server. M.100 is available from the ECTF Web Site at www.ectf.org.

M.500 A set of Management Information Base (MIB) objects for use in managing CT servers and resources through SNMP. The ECTF which fathered the M.500 calls it a Management Information Base (MIB) Interoperability Agreement that specifies the standardized administration of Computer Telephony (CT) servers. M.500 allows developers of CT solutions to demonstrate a more complete solution to their customers by including practical visibility into their products. This comprehensive management capability, according to the ECTF, allows users to deploy a rich set of mixed media applications using off-the-shelf hardware and component software. It also allows them to fine-tune performance and get the most out of the advanced technology inherent in today's CT solutions. The standards-based management offered by M.500 provides a very practical way for users to realize a long-term reduction in cost of ownership by providing greater interoperability options with new CT components that extend the life of the overall system. M.500 contains object specifications for both CT server vendors and resource providers. Vendors support the portion of the MIB relevant to their offering, and the user is then able to view and manage the entire CT server as a whole. Additionally, M.500 is leading the industry by releasing the first MIB for fax resources.

By using the new M.500 MIB, IT managers can confirm whether or not important resources, such as fax server ports, are active. They can also monitor fax call statistics, such as the number of fax calls placed and connection attempts and failures for a specific fax destination. Data of this type, according to the ECTF, can help IT managers make full use of available fax resources and allow them to make adjustments to imbalances between business requirements and the supporting fax resources. According to the ECTF, M.500 follows the same industry standards as the other industry MIBs that manage hubs, bridges, routers, etc. It can be implemented using Simple Network Management Protocol (SNMP) allowing customers to use off-the-shelf tools, such as HP OpenView (TM) or IBM NetView (TM), to manage their CT systems. M.500 is fully extensible allowing the addition of objects to manage unique aspects of a particular product. M.500 is available for downloading at the ECTF Web Site at www.ectf.org.

M1 1. Multiplexer in the U.S. digital signal hierarchy. See M1, which is listed as if it were M-ONE in this dictionary.

2. Management Interface 1: The management of ATM end devices.

M2 Management Interface 2: The management of Private ATM networks or switches.

M2M Machine-to-machine.

M2PA MTP2-User Peer-to-Peer Adaptation Layer. The M2PA protocol supports the transport of Signaling System Number 7 (SS7) Message Transfer Part (MTP) Level 3 signaling messages over Internet Protocol (IP) using the services of the Stream Control Transmission Protocol (SCTP). M2PA is also used between SS7 Signaling Points using the MTP Level 3 protocol. The SS7 Signaling Points may also use standard SS7 links using the SS7 MTP Level 2 to provide transport of MTP Level 3 signaling messages.

M3 Management Interface 3: The management of links between public and private networks.

M3UA SIGTRAN M3UA signaling gateways run the lower levels of the SS7 stack (MTP3 and MTP2), and application servers the higher levels (ISUP, SCCP). M3UA is used for carrying MTP3 traffic between the signaling gateway and application post. It allows reliable, resilient and flexible system architectures to be developed while offering interoperability between vendors.

M4 Management Interface 4: The management of public ATM networks.

M5 Management Interface 5: The management of links between two public networks.

M12 Multiplex 1-to-2. A designation for a multiplexer which takes four DS-1 (T-1) signal inputs at 1.544 Mbps and interleaves them into a single DS-2 (T-2) output at 6.312 Mbps, adding additional signaling and control bits. See also M13, Multiplex, T-1, and T-2.

M13 Multiplex 1-to-3. In the U.S. digital hierarchy, multiplexers are called by the digital signal levels they interface with. For example, a multiplexer, which joins DS-1 (T-1) channels to DS-3 (T-3) is called a M13 Mux. A M13 takes 28 T-1 inputs at 1.544 Mbps and interleaves them into a single T-3 output at 44.736 Mbps, adding additional signaling and control bits. See also M12, Multiplex, T-1, and T-3.

M23 Multiplex 2-to-3. A DS-3 (T-3) signal format that combines seven DS-2s (T-2s) to form a DS-3 (T-3). When T-3 technology first was developed, a M24 multiplexer (MUX) was used to combine 24 DS-0 channels at 64 Kbps into a single T-1 format at a total signaling rate of 1.544 Mbps, adding essential signaling and control bits in the process. Then a M12 combined four T-1s into a T-2 at a total signaling rate of 6.312 Mbps, adding more signaling and control bits. Then a M23 combined seven T-2s into a T-3 at a total signaling rate of 44.736 Mbps, adding even more signaling and control bits. Contemporary multiplexing generally combines these last two stages in a single M13 device. See also DS-0, M13, M24, T-1, and T-3.

M24 A T-1 service that allows a user to multiplex up to 24 voice or data channels into a single T-1 link, compatible with AT&T central office based channel banks (M24 compatibility generally refers to compliance with the channelization and coding techniques specified by AT&T TR62411).

M28 The telephone company multiplexing scheme that multiplexes 28 T-1 data streams

onto a single carrier system, the T-3. See also M13.

M34 A designation for a multiplexer which interfaces between six DS-3s (T-3s) and one DS-4 (T-4) circuit.

M44 A T-1 service that allows up to 44 voice channels (48 without signaling) to operate over a single T-1 link by using ADPCM, and is compatible with AT&T central office based equipment. MJ44 compatibility generally refers to the ability to accept the 44 channel T-1 aggregate and break out one of the channels individually for routing purposes.

MA Abbreviation for Milliamp or Milliamperes, unit of electric current.

Ma Bell A term used to refer affectionately to the old AT&T and the old Bell System. Several Women's Lib organizations objected to it some years ago on the basis that there were no women in the higher corporate structure of AT&T, and that women, as over supervised operators, were the downtrodden majority within the Bell System. There was a movement afoot to change it to Pa Bell. But then came Divestiture and the breakup of the Bell System. The term Ma Bell now largely belongs in the history books. And there are now a handful of women in senior management in the Bell operating companies.

MAC 1. Every 802.11 computer or communicating device that wants to be on a wireless or wired Ethernet network must have its own specific MAC (Media Access Control) address hard-coded into it. This unique identifier can be used to provide security for wireless networks, for example. When a network uses a MAC table, only the 802.11 radios that have had their MAC addresses added to that network's MAC table are able to get onto the network. Ditto for wired networks. See also MAC address.

2. Moves, Adds and Changes. When you first install a phone system it will cost money to run wires and install phones all over the building. Very quickly you will notice that you'll need to move people and their phones, add phones for new people and change phones around. This will cost money, often lots of it. How much it will cost you depends on the arrangement you have negotiated with the vendor of your phone system. It's a good idea to get a good deal on Moves, Adds and Changes later on BEFORE you sign your original deal to buy the phone system.

3. Mobile Advisory Council.

4. Multiply Accumulate. Computing functions accomplished by processors. Computers can multiply values, and can store both the multipliers (the original values) and the products (i.e., the end results of the calculations) in registers (temporary memory) for subsequent use. MAC speed and memory are embedded in silicon chip sets, including DSPs (Digital Signal Processors) used in telecommunications. See also DSP.

5. Shortened way of saying Macintosh, a line of computers from Apple Computer.

MAC address Medium Access Control Address. A MAC address is the hardware address of a device that is designed to be connected to a shared network medium, e.g. a LAN and/or the Internet. A MAC address is a unique address. It doesn't change. no two devices have the same MAC address. A MAC address (also called MAC name) traditionally is in the form of a 48-bit number, formally known as an EUI-48 (Extended Unique Identifier-48), which is unique to each LAN (Local Area Network) NIC (Network Interface Card). The MAC Address is programmed into the card, usually at the time of manufacture. The IEEE Registration Authority administers the MAC addresses scheme for all LANs that conform to the IEEE Project 802 series of standards, with such LANs including both Ethernet and Token Ring. The MAC address comprises two distinct identifiers (IDs), which typically are programmed into ROM (Read Only Memory) and which, therefore, cannot be changed. The first address is a unique 24-bit Company ID, also known as a Manufacturer's ID, which is assigned by the IEEE to the manufacturer of the NIC. The second address is a 24-bit Extension ID, assigned by the manufacturer. Also known as a Board ID, the Extension ID identifies the specific NIC, which is in the form of a printed circuit board. The Extension ID is intended to be unique, although occasional duplications occur, creating havoc. Unlike Network Layer Addresses, MAC names are location-independent, i.e. they stay with the card, which fits inside the device (e.g., workstation, printer, or server), which can move around. Destination and source MAC names are contained in the header of the LAN packet and are used by various devices (e.g., bridges, hubs, and switches) to filter and forward packets. The IEEE recently has enhanced the addressing scheme, expanding the Extension ID from 24 to 40 bits. This new addressing scheme is known as EIU-64, and is intended to provide NIC a globally unique identifier, which can be encapsulated in an IPv6 address. See also EUI-48, EUI-64, and Network Layer Address.

MAC Bridge See Bridge.

MAC layer That layer of a distributed communications system concerned with the control of access to a medium that is shared between two or more entities.

MAC name A MAC name (also called a MAC address) is a 48-bit number, unique to each local area network card, that is programmed into the card, usually at the time of manufacture. Unlike Network Layer Addresses, MAC names are location-independent.

Destination and source MAC names are contained in the LAN packet and are used by bridges to filter and forward packets. See also Network Layer Address.

MAC OS The operating system powering the Macintosh line of computers from Apple Computer, Inc.

MAC protocol The procedures used to control access to a medium that is shared between two or more entities.

MACs Moves, Adds and Changes. See MAC.

machine code Same as Machine Language. See Machine Language.

machine dependent Software which will only run (i.e. is dependent) on a certain computer.

machine language A computer language composed of machine instructions that can be executed directly by a computer without further compilation. Instructions and data coded in binary code. Machine language is the native language of computer hardware. Machine language is the only language recognized by the microprocessor that controls all the operations in your PC. All programs and all data to be processed by your computer (PC, mini or mainframe) have to be translated into machine language at some stage.

machinima Machine crossed with "cinema. According to the Economist magazine, machinima could be on the verge of revolutionising animation. Around the world, growing legions of would-be digital Disneys are using the powerful graphical capabilities of popular video games such as Quake, Half-Life and Unreal Tournament to create films at a fraction of the cost of Shrek or Finding Nemo. There is an annual machinima film festival in New York, and the genre has seen its first full-length feature, Anachronox. Spike TV, a cable channel, hired machinima artists to create shorts for its 2003 video game awards, and Steven Spielberg used the technique to storyboard parts of his film A.I. At www.machinima. com, hobbyists have posted short animated films with dialogue, music and special effects. Some creative players are making machinima movies enitrely using movie sequences from inside video games.

Macintosh A line of computers made by Apple.

Macro 1. MAChine ROutine. An instruction in a source language (e.g., FORTRAN) that is equivalent to a specified sequence of machine or assembler instructions. See Macro Language and Machine and Assembler.

2. A statement in software code that is expanded into one or more identifiers or statements and can consist also of one or more parameters. When invoked, a macro causes a list of conditions to occur.

3. Software which lets you alter the definitions of what the keys on your computer keyboard are. With a "macro" software program, you could change the letter "M" on your keyboard to type "Michael" every time you hit it. But this would be stupid. Better to hit a combination of letters to get "Michael." Most personal computers have extra non-alphabetic keys, like Control and Alternate. For example, type Ctrl-Alt M and bingo, the machine types "Michael Newton is a good son." Sometimes the origin of terms fades into the mist; like the software guru who recently admitted that he didn't know why it was called a "macro". That's easy, because it isn't a "micro".

4. Macro means very big. Random House's dictionary says it means "very large in scale, scope, or capability." Macro also refers to macroeconomics – the study of bigger things in economics, like the factors that affect the wealth and growth of countries. In contrast, microeconomics focuses on factors that affect the wealth and growth of organizations, such as corporations.

macro language A collection of instructions by which any kind of information in the system can be located and manipulated and by which new information types can be added to the system.

macro virus A computer virus implemented in a scripting language such as Microsoft's Visual Basic Script (VBS) that is run when a user opens an infected document in Excel or Word, or especially one attached to an email received by Outlook.

macrobend loss Loss in optical power created when a lightguide (i.e., optical fiber) is bent in such a manner that it can be seen with the naked eye. See also Microbend Loss.

macrobending In an optical fiber, all macroscopic deviations of the fiber's axis from a straight line; distinguished from microbending.

macroblock The four 8 by 8 blocks of luminance data and the tow (for 4:2:0 chroma format), four (for 4:2:2 chroma format) or eight (for 4:4:4 chroma format) corresponding 8 by 8 blocks of chrominance data coming from a 16 by 16 section of the luminance component of the TV picture. Macroblock is sometimes used to refer to the pel data and sometimes to the coded representation of the pel values and other data elements defined in the macroblock header. See also pixelation.

macroblocking See pixelation.

macrocell A large radio cell. SMR (Specialized Mobile Radio), also known as TMR (Trunk Mobile Radio), systems use macrocells. They work on the basis of a large omnidirectional (i.e., all directions) antenna placed on the highest spot in an area, in order to maximize the direct line-of-sight and, therefore, the quality of the signal. SMR systems generally cover a radius of 50 miles or so. That's a macrocell. A microcell is smaller. A picocell is smaller, still. See also microcell and picocell.

A macrocell is a new word for what we used to call a "cell," as in cellular radio. This is what some researchers at Bell Northern Research (BNR) wrote: "Today's cellular networks employ macrocells and are optimized to serve users in automobiles, moving at relatively high speeds. Yet, a growing proportion of cellular traffic is originating from users who are not driving in vehicles, but are on foot... If a portion of the radio frequencies in a geographic area were transferred from macrocell (optimized for cars) to microcell technology (optimized for pedestrians), cellular traffic would increase up to a hundredfold. Microcell networks will entail the deployment of many more transceivers than today's macrocell systems. However, microcell equipment will be less costly because it is low power, simpler and smaller, say the researchers at Bell Northern Research.

macros Software which lets you alter the definitions of what the keys on your computer keyboard are. With a "macro" software program, you could change the letter "M" on your keyboard to type "Michael" every time you hit it. But this would be stupid. Better to hit a combination of letters to get "Michael." Most personal computers have extra non-alphabetic keys, like Control and Alternate. For example, type Ctrl-Alt M and bingo, the machine types "Michael Newton is a good son."

macrovision Macrovision is an artificial video error injected into tape signals, a poor implementation of copy protection.

Macstar An old AT&T 3B2 computer-based software system that interfaces with the Remote Memory Administration System (RMAS) to effect customer moves and rearrangements on the 5ESS switch. The user needs only a terminal and printer.

mad as a hatter In the 19th century, workmen who used mercury, a poison, to cure beaver skins for top hats over time developed nervous twitches, drooled and spoke incoherently. Thus the expression, "mad as a hatter."

made-for-mobile Describes content and services that have been specifically made for mobile phones and other mobile devices.

MADI Multichannel Audio Digital Interface. MADI is an interface standard described by the Audio Engineering Society (AES) standards AES-10 and AES-10id. It was developed by Neve, Sony, and SSL as an easy way to interface digital multitrack tape recorders to mixing consoles.

MADN Multiple Appearance Directory Number. An ISDN term. A telephone number that appears on multiple ISDN telephone sets.

MAE A Network Access Point (NAP), or public peering point, where Internet Service Providers (ISPs) interconnect to exchange traffic at the national backbone level. The original MAEs were MERIT Access Exchanges, established by MERIT Access Exchange, which joined with IBM and MCI to establish the original public Internet backbone that replaced the NS-FNET (National Science Foundation NETwork). MERIT subsequently was acquired by MFS, which then was acquired by MCI, which then was acquired by Worldcom to become MCI Worldcom. MAE East, the original MAE, is located in Vienna, Virginia, a suburb of Washington, D.C. MAE-West later was established in San Jose, CA. The term MAE then was used to denote either Metropolitan Area Exchange, or Metropolitan Area Ethernet – it's your choice, as both Exchange and Ethernet are correct, despite ongoing Internet industry arguments over which is more appropriate. Actually, it's a moot point, since MCI Worldcom calls it just a plain MAE, which term it has trademarked.

Physically, a MAE is a building with zillions of wires, gigantic switches and computers with routing tables containing the locations of ISPs and how to get to them. MAEs and other NAPs are run by networking companies and carriers, to which ISPs pay a fee for the privilege of locating equipment there and exchanging traffic through the switches. Each MAE comprises a Gbps ATM switch, which supplements legacy FDDI switches. The ISPs connect to the MAE switch through various means, including Ethernet, FDDI and ATM. There also exist CIXs (Commercial Internet Exchanges), also called IXs (Internet Exchanges), which likewise function as public peering points. In order to minimize the cost of using these public peering points, consortia of ISPs have established regional private peering points, where they exchange traffic directly. See also CIX and NAP. For more on MAEs and charts of the traffic they carry, see www.nap.net/where/w_mae-east.html and www.mae.net/east.html. See also MAE-East.

MAE-East MAE presently stands for Metropolitan Area Exchange or Ethernet. Both Exchange and Ethernet are correct. MAE is a huge interconnection point for Internet Service Providers (ISPs). They use MAEs for routing their internet traffic from customers on their network to Internet sites on other peoples' Internet networks. There are two MAEs – MAE-East in Vienna, Virginia and MAE-West in San Jose, California. Physically a MAE is a building with zillions of wires, gigantic switches and computers with routing tables containing the location of Internet sites and how to get to them. MAEs are run by networking companies (MAE-East and MAE-West are run by MFS) and ISPs pay a fee to these networking companies to locate equipment there and for the transmission, switching and interconnect services. See also MAE (for a longer explanation) and Merit.

MAE-East++ An expansion of MAE-East, located in a nearby building. i++ (plus plus) gets its term from the C programming language where it means "use the current variable i and add 1." Plus plus has come to mean an expansion, an improvement, an upgrading, etc. See MAE and MAE-East.

MAE-West See MAE.

MAE-West++ An expansion of MAE-West, located in a nearby building. i++ (plus plus) gets its term from the C programming language where it means "use the current variable i and add whatever one to it." Plus plus has come to mean an expansion, an improvement, an upgrading, etc. See MAE-East.

MAG plan Access charge reform adopted by the FCC in October 2001 that applies only to rate of return regulated carriers. It (1) increased the subscriber line charge (SLC): (2) phased out the carrier common line (CCL) charge; and (3) created an explicit and uncapped Universal Support mechanism, the Interstate Common Line Support (ICIS).

magalog A mail order catalog disguised as a magazine in the hopes of confusing its recipients.

Magazine 1. Hardware unit, mounted in a frame, containing printed board assemblies.

2. From the Arabic "makhzan," a magazine is a "storehouse." The term is applied to some periodical publications that contain miscellaneous pieces such as articles, stories and poems, some of which may be illustrated in various ways. The first magazine was published In England in the 1730s, and contained nothing but short pieces snipped out of daily newspapers. See the Appendix for a list of magazines and other publications in computer and telecom.

mageiricophobia Mageiricophobia is the intense fear of having to cook.

MAGIC Multidimensional Applications and Gigabit Internetwork Consortium. One of the Information Superhighway projects funded by the Information Technology Office (ITO) of the Defense Advanced Research Projects Agency (DARPA) and the National Science Foundation (NSF). MAGIC is a gigabit-per-second ATM-based network connecting various high-tech research and development sites in Minneapolis (MN), Sioux Falls (SD), Lawrence (KS), Kansas City (KS) and Ft. Leavenworth (KS).

magic lantern The is the name for the FBI's new magic software code which the FBI sends to a distant PC in an email. The email installs so called "keylogging" software that purports to encryption-key and password-sniff everything going on with that computer. Eventually the idea is to somehow grab the material produced by magic lantern and use it to figure out that computer.

magic wand One of the devices used by con men when pretending to do TSCM (Technical Surveillance CounterMeasures). A magic wand is typically a field strength meter or box with many fancy lights.

maggots Telco slang for B-connectors, which are about the size and shape of fly larvae.

magnetic bubble A device in which information is stored in a magnetic film as a pattern of oppositely directed magnetic fields. Magnetic bubble devices hold their memory even if you lose power.

magnetic core Material used to store data in main memory.

magnetic disk A computer storage device that records data bits as tiny spots on magnetic-coated disk platters. Hard disks and floppy disks are variations on magnetic disks. See also Magneto Optical and Holographic Data Storage.

magnetic field strength The magnitude of the magnetic field vector, expressed in units of amperes per meter (A/m).

magnetic ink An ink that contains particles of a magnetic substance whose presence can be detected by magnetic sensors. Typical is the ink on your checks, which carry your name, your account number and the check's number.

magnetic field the region within which a body or current experiences magnetic forces.

magnetic flux The rate of flow of magnetic energy across or through a surface (real or imaginary).

magnetic medium Any data-storage medium and related technology including diskettes and tapes, in which different patterns of magnetization are used to represent the values of stored bits and bytes.

magnetic noise caused by change in current level, e.g. ac powerline (creates magnetic field around that cable) this magnetic field causes the magnetic noise.

Magnetic North North based upon the magnetic compass or the earth's magnetic field.

magnetic storage Any medium (generally tape or disk) upon which information is encoded as variations in magnetic polarity. The hard disk on your computer is magnetic storage.

magnetic stripe A strip of magnetic material, usually tape, attached to a credit card containing data relating to the card holder. You have a magnetic stripe now on the back of most of your credit cards. That stripe tells a computer who you are, what your account number is, etc.

magnetic tape A tape made of magnetic material upon which data may be stored for later retrieval by a computer.

magneto A small hand-cranked AC generator which uses permanent field magnets and can make electricity to ring telephone bells. See also Magneto Phone.

magneto optic Relating to the change in a material's refractive index under the influence of a magnetic field. Magneto-optic materials generally are used to rotate the plane of the polarization. This phenomenon is how magneto optical disk drives work. The Magneto-Optic Effect also is known as the Faraday Effect.

magneto optical drive MO. A popular way to back up files on a personal computer. As the term implies, a magneto optical drive employs both magnetic and optical technologies to obtain ultra-high data density. A typical cartridge is slightly larger than a conventional 3.5-inch magnetic diskette, and looks similar. Data is written using magnetism (in the form of a magnetic field called the bias field) and light (a laser beam) to a disk that resembles a CD-ROM disk. A magneto optical drive holds huge amounts of information - as much as several gigabytes on a single disk. Introduced in 1988, the magneto optical drive (also spelled with a dash between the magneto and optical), the drive provides the convenience of the removability of floppies and the Bernoulli Box, the random access convenience of hard disk, the reliability of CD-ROMs and the promise of DAT-like capacity. But, according to PC Magazine, before you rush out and buy this ultimate storage solution, note that they're expensive, can't provide as much storage on one side of a disk as the largest hard disks and they're slower than today's hard disks. An explanation of how magneto optical drives work, courtesy PC Magazine: The recording layer on the disk stores the equivalent of binary 1s and 0s on the magnetic domains. The disk is designed so that the bias field by itself is to too weak to change the polarity of the magnetic domains. But when a spot on the disk is heated by a high-powered laser beam, its resistance to changing polarity drops. The bias field can now change the disk area's polarity. To read the disk, the drive uses a laser beam that is not hot enough to allow the bias field to change the disk area's polarity.

magneto phone A magneto phone is a hand-cranked, self-powered phone, in contrast to the more modern "Central Battery" phone, which we typically have in our home and which derives its electricity from the distant telephone company central office, also called a telephone exchange.. A magneto phone has a hand-cranked magneto to generate ring current, and a local battery to power voice signals. Thus two magneto phones can be connected together via a simple pair of wires, without the need for a Central Exchange battery feed. More technically, a magneto phone contains a small hand-cranked AC generator which uses permanent field magnets to make electricity to alert the central office or another phone that it wants to make a phone call. A magneto phone has a crank handle on the side and a dial pad. It is a very old phone – typically used around the turn of the 20th century. It is totally manual, self-powered. You crank the handle which turns a bunch of permanent magnets on the inside which in turns generates electricity which in turn is used to ring a bell at the central office. When that bell rings, an operator answers, asks which number the person wants to reach, dials and connects the call.

magnetron Alex Beam in the New York Times Book Review of June 16, 2002, writes, "In 1940, with the Battle of Britain hanging in the balance, Winston Churchill's top scientists came to America bearing a gift, and a plea. The gift was a palm-size copper disk called a magnetron, which could broadcast microwave radar beams farther and more precisely than any system in the world. The plea: Help us refine this invention, which could not only target enemy bombers but could also detect a U-boat periscope poking through the waves and even direct artillery fire against fixed positions, a precursor of today's much-vaunted "smart" weapons. The British scientists literally unpacked their bags for an audience of astonished American scientists at Alfred Loomis's Tuxedo Par New York estate, and the magnetron became, at once, the focus of Loomis's whole being." The book reviewed is called "Tuxedo Park, a Wall Street Tycoon and the Secret Palace of Science That Changed the Course of World Wall II" by Jennet Conant. Alex Beam continues, "Loomis immediately convened an emergency meeting of the Establishment: his cousin Henry Stimson, the secretary of war; and Stimson's aides John McCloy, Robert Patterson and Robert Lovett. All Yale men, all Harvard Law, all Wall Streeters, all Republicans and all instrumental in starting M.I.T.'s wondrous Radiation Laboratory, the Rad Lab, which would not only develop the radar that helped win the war but spawned the microwave oven to boot."

MAHO Mobile Assisted Hand-Off. A process in which the wireless mobile station assists the base station in assigning a voice channel by reporting its surrounding F signal strengths to the base station.

MAHR Abbreviation for MILLIAMPERES HOUR, 1/1000th of an ampere hour. Term is commonly used with small rechargeable battery packs, such as those used by portable phones and laptop PCs.

mail bomb 1. A network security term. An attack in which a malicious user mails a dangerous program to an unsuspecting recipient. When the recipient runs the program, it performs some malicious action on their computer.

2. The flooding of an e-mail address with frequent messages, often done as an act of protest or harassment. See also Bozo Filter.

mail bridge Mail gateway that forwards e-mail between two or more networks while ensuring that the messages it forwards meet certain administrative criteria. A mail bridge is simply a specialized form of mail gateway that enforces an administrative policy with regard to what mail it forwards.

mail enable applications Applications that use mail as a way of addressing and transporting information to and from users on a network.

mail exchange records See MX Records.

mail exploder Part of an electronic mail delivery system which allows a message to be delivered to a list of addresses. Mail exploders are used to implement mailing lists. Users send messages to a single address (e.g., smith@somehost.edu) and the mail exploder at somehost.edu takes care of delivery to the individual mailboxes on the lists, including Smith.

mail filter A piece of software which lets a user sort his/her email messages according to information in the header.

mail flash A cellular mail filter. Software that allows e-mail messages to be sent to a cell phone, first allowing the user to determine which messages are most important. Mail flash makes use of the SMS (Short Message Service) or WAP (Wireless Application Protocol) protocols. See also Mail Filter, SMS, and WAP.

mail gateway A machine that connects two or more electronic mail systems (especially dissimilar mail systems on two different networks) and transfers messages between them. Sometimes the mapping and translation can be quite complex, and generally it requires a store-and-forward scheme whereby the message is received from one system completely before it is transmitted to the next system after suitable translations.

mail path A series of machine names used to direct electronic mail from one user to another.

mail reader Software which enables a user to select unread electronic mail and unread conferences messages and have them downloaded for reading off-line. Most mail readers also permit users to create responses off-line and upload them at their convenience.

mail reflector An Internet term. A special mail address; electronic mail sent to this address is automatically forwarded to a set of other addresses. Typically, used to implement a mail discussion group.

mail relaying A network security term. A practice in which an attacker "bounces" e-mail off another system's e-mail server in order to use its resources and/or make it appear that the mail originated from the other system.

mail server Mail Server is the "post office" of a messaging network. Mail server is a computer host and its associated software that offer electronic mail reception and (optionally) forwarding service. Users may send messages to, and receive messages from, any other user in the system.

mailbot An email server that automatically responds to requests for information and sends that information by return email.

mailbox A directory or file on a computer somewhere that stores messages for a single user. A mailbox may include email, voice mail and/or facsimile documents. See also Unified Messaging.

mailer A piece of software that sends single or multiple emails.

Mailer Daemon The Mailer-Daemon is not a person, but rather a computer program on Internet which runs automatically to perform the service of telling you why it could not deliver your email message. See Daemon.

mailgram An overnight electronic mail service of Western Union. The letter is phoned in or sent by computer to a central Western Union computer, from where it is sent to teleprinter machines located in post offices in major cities. When it's printed at the post office, the Mailgram is placed in an envelope and hand delivered by your friendly, local mailman in next day's mail.

mailing list A group of people to whom users can refer by a common name (for

example, a mailing list called Marketing). When users address a message to a mailing list, all members of the mailing list receive the same message.

mailto A piece of HTML code that sits on a Web site and, through your browser, displays an email address underlined and in blue. When you click on this email address, this mailto HTML code automatically launches your email program and automatically inserts the email address. Mailto HTML code looks like this <\<>A HREF=mailto:Harry@HarryNewton.com>Harry@HarryNewton.com<\<>/A>. In this line, the second email address is what the person browsing your Web site will set in blue and underlined. When he clicks on it, the first email address will be what's dropped into the email as the address to send to. Typically the two are the same. But you could easily write it as: <\<>A HREF=mailto:Harry@HarryNewton.com>email to Harry Newton<\<>/A>. In which case, "email to Harry Newton" in blue and underlined would appear when viewed through a browser by someone visiting my Web site.

main PBX or Centrex switch into which other PBXs or remote concentration of switching modules are homed. A PBX or Centrex connected directly to an electronic tandem switch (ETS). Also, a power source.

main cross-connect The interconnect point where wiring from the entrance facility and from the workstation is connected to telecom equipment.

Main Distribution Frame MDF. A wiring arrangement which connects the telephone lines coming from outside on one side and the internal lines on the other. A main distribution frame may also carry protective devices as well as function as a central testing point. See Main Distribution Frame Fill, Distribution Frame and Frame.

main distribution frame fill The central office mainframe is the termination point for outside plant cables. The "fill" is the percentage of pairs used by customers of the total number of pairs on the frame. Optimum fills vary based on the size of the central office and the amount of growth in the area. A low fill means idle lines and wasted investment in outside plant. A high fill, plus unexpected growth, forces budget busting and crisis construction projects.

main feeder Feeder cable that transports pairs from the central office to branching or taper points.

mainframe 1. One of the telephony synonyms for Main Distributing Frame.

2. Data processing term descriptive of very large computers.

main lobe The main lobe is the area with the maximum intensity in the pattern of radiation produced by an antenna. One presumes it's called "lobe" because the pattern in a microwave signal of the main lobe typically looks like a ear lobe.

main memory The principal random storage area inside the computer. Used for storing data and programs and under the direct control of the CPU – the main processor. Also called RAM memory.

main network address In IBM's SNA, the logical unit (LU) network address within ACF/VTAM used for SSCP-to-LU sessions for certain LU-to-LU sessions. Compare with auxiliary network address.

main PBX A main PBX is one which has a Directory Number (DN) and can connect PBX stations to the public network for both incoming and outgoing calls. A main PBX can have an associated satellite PBX, and can be part of a tandem tie trunk network (TTTN). If the main PBX provides tandem switching for tie trunks, it is called a tandem PBX. In the context of ESN (Electronic Switched Network), a main PBX has tie trunks to only one node. See PBX.

main satellite service A PBX feature that allows multi-location customers to concentrate their attendant positions at one location referred to as the Main. Other unattended locations are referred to as Satellites.

main service entrance In AC electricity, the main service entrance is the necessary equipment, usually consisting of main circuit breakers or fuses, a switch and branch circuit breakers or fuses, in a grounded enclosure (panel) connected directly to earth. Located in the building at the point of entrance of the supply conductors from the power utility. Other panels in the building are referred to as branch, service or supply panels.

main station A subscriber's telephone instrument, terminal or workstation used to originate and receive calls. Very often if two instruments have the same extension number (are bridged), one becomes the Main Station and the other is a bridged station for inventory purposes. See Main Station to Line Ratio.

main station to line ratio A telephone company term. The ratio of main stations to lines. This ratio will normally be greater than 1.0 because of 2 and 4 party service, etc.

main terminal room The location of the cross-connect point between the incoming cables from the telecommunications external network and the premises cabling system.

mainframe A powerful computer, almost always linked to a large set of peripheral devices (disk storage, printers, and so forth), and used in a multipurpose environment at the corporate or major divisional level. A mainframe is a large-scale computer typically containing hundreds of megabytes of main memory and hundreds of gigabytes of disk storage. It is capable of "serving" thousands of "on-line" terminals. The term – main frame – derives from the racks that typically hold a large computer and its memory.

mainframe chiller system Water-cooled mainframe computers rely on mainframe chillers for a continuous supply of liquid coolant to maintain processor temperature within a specified range. Exceeding the temperature specifications or an interruption of coolant flow can cause a sudden shut-down, interrupting of computer operations, and possible hardware damage, requiring costly repairs.

mainframe gateway A hardware/software system that allows PCs on a LAN (Local Area Network) to communicate with a mainframe. A single, usually dedicated, PC acts as the gateway. PCs on the LAN share its hardware and its communication link, communicating with it over the LAN cable. The most common mainframe gateway is an SNA gateway, which hooks a LAN into an IBM mainframe.

mainframe server Clients are devices and software that request information. Client is a fancy name for a PC on a local area network. It used to be called a workstation. Now it is the "client" of the server. A mainframe server is a large computer that stores lots of information and manages libraries of information. Here's a definition of Thin Client, courtesy of Oracle Corporation, writing in early 1994: "Mainframe systems store lots of data, but they're expensive, slow and difficult to use. Because all the processing happens on one large computer, they can't move large amounts of multimedia information to large numbers of users. Example, the IBM ES/9000, Amdahl's 5995-1400 or any plug compatible mainframe." See also Client, Client Server, Client Server Model, FAT Client and Media Server.

mains Some countries call their normal commercial power outlets – "mains." In Europe the frequency of commercial power is 50 Hz. In the United States, its frequency is 60 Hz. It's hard to convert the frequency of commercial power. It's easier to convert voltage. In Europe and Australia, normal household voltage is 240 volts. In the U.S., it's 120 volts.

mains modem A modem which is part of a system called remote metering which monitors electricity usage and allows electric companies to offer such services as electronic mail, burglar alarms and energy management. The idea of energy management is that if the electric companies could turn off unnecessary appliances for a few hours during peak times, they might not have to build expensive new power stations. In exchange for that favor, they undoubtedly would be prepared to offer their customers price reductions.

maintenance 1. All work needed to keep the telephone system operating properly, including periodic testing, repairs, etc. See Preventive Maintenance.

2. All work needed to keep a software program operating properly, operating on new machinery and operating with new management needs. Often, software maintenance means substantially rewriting the original software program. Most of the work done by data processing departments in large companies involves maintaining old programs. This is not a put-down.

maintenance acceptance A term used in the secondary telecom equipment business. The point at which a maintenance company has tested a system, component, or peripheral device and determined that it meets manufacturer's specifications. The product can now be added to a maintenance contract. Once under contract, the maintenance company is responsible for repairing or replacing any defective components.

maintenance contract Contract guaranteeing the repair of a PBX switch to support it at operational levels for a predetermined fixed term and fixed price.

maintenance control center MCC. A central place in a stored program control central office from which system configuration and trouble testing are controlled.

maintenance control circuit MCC. A voice circuit used by maintenance personnel over microwave links for coordination. This is not available to operations or technical control personnel.

maintenance hole A vault located in the ground or earth as part of an underground duct system and used to facilitate placing, connectorization, and maintenance of cables as well as the placing of associated equipment, in which it is expected that a person will enter to perform work. Also called a manhole.

maintenance limit A Verizon definition. The maximum margin, value, or deviation associated with normal in-service performance.

maintenance of service charge MSC. A Verizon definition. A charge to the CLEC or Reseller when the field technician is dispatched to remedy a problem found in the customer premise equipment (at the end user's premises).

maintenance release An euphemism for a new piece of software that fixes a buggy piece of old software. When software is released, it's usually buggy. A maintenance release attempts to fix the bugs. A more correct term for the new software would be a "bug fix." A maintenance release often carries the number one as the second digit after the decimal point, e.g. 3.51 or 4.01.

maintenance services In IBM's SNA, network services performed between a host SSCP and remote physical units (PUs) that test links and collect and record error information. Related facilities include configuration services, management services and session services.

maintenance termination unit MTU. A MTU is an electronic circuit that is owned and deployed by a telephone company to aid in fault sectionalization and is installed at the network interface. The MTU should meet the requirements of Bellcore Technical Advisory TSY-000324 and be testable with the Mechanized Loop Test System (MLT). The MTU is designed to work on single line residence or business service. Bellcore is now called Telcordia Technologies.

maintenance update A euphemism for a piece of software which fixes bugs in a previously-released version of the software. A maintenance update rarely has any new features and rarely costs anything. Software companies send them because they find bad bugs in their software and want to fix those bugs asap, or because they can't stand the heat from complaining customers.

maintenance usage A telephone company term. The amount of time, measured in CCS that equipment components are removed from service. Can be caused by equipment malfunction, routine maintenance, transitions, etc.

major trading area Major Trading Area. The U.S. and its territories are divided into 51 areas, called MTAs, which are based on Rand McNally's analysis in identifying areas of economic integration and documented in Rand McNally's (1-800-284-6565) "Commercial Atlas and Marketing Guide." (Note: the FCC modified Rand McNally's proposed 47 MTAs to 51 to handle Alaska, Puerto Rico, etc.) An MTA consists of an integer number of BTAs; BTAs are never split between MTAs. For example, the Chicago MTA consists of 18 BTAs (including the Chicago BTA) and 84 counties. Each MTA consists of one or more Basic Trading Areas (BTAs). For instance, the Chicago MTA includes 18 BTAs such as Rockford BTA, Springfield BTA, Ft. Wayne BTA, etc. Milwaukee is another MTA. See also local traffic.

Majordomo Translated from Latin as "master of the house". Majordomo is a freeware e-mail list server program that maintains multiple mailing lists and automatically re-distributes e-mails to all members of the list. It automatically interprets (i.e. understands) commands in e-mails sent to it by individuals who wish to subscribe to, unsubscribe from, receive periodic summaries of, or otherwise become associated with a mailing list. It also handles the mass mailing of messages to members of the list. Written in the PERL language, Majordomo originated in the UNIX environment, although it in can run under any operating language with a PERL interpreter. You find Majordomo all over the Internet. For an example of how Majordomo mailing lists work, here's a note on subscription information for an electronic newsletter called WinNews published by Microsoft. "If you know someone who might be interested in WinNews, please instruct them to:

1. Send Internet e-mail to: ENEWS99@MICROSOFT.NWNET.COM. 2. Send the message from the account that you wish to subscribe (some people use more than one e-mail account). 3. Subject line should be blank. 4. Body of message should ONLY have in the text: SUBSCRIBE WINNEWS If you wish to stop receiving WinNews, send mail to enews@microsoft.nwnet.com with a blank subject line and the body of the message should only save in the text: UNSUBSCRIBE WINNEWS."

make-before-break Refers to a switch that is designed to engage the new contacts, thereby establishing a new connection path, before breaking (opening) the previous contacts to end the old connection path. This design prevents there ever being an open circuit.

make busy To make a communication circuit unavailable for connection. The technical term for taking the phone off the hook and leaving it off hook.

malicious call tracing An ISDN service which enables to User to Network message to be sent while the call is in progress, ensuring that origination details are captured at the local exchange.

mall walker A couple of hundred come every day to the Mall of America in Minnesota and walk. They come because it's free, the weather is always clement, never too hot or too cold, and above all, it's safe, without danger, under surveillance 24/7. I learned this from an article which Bernard-Henry Levy wrote for the May, 2005 of The Atlantic Monthly.

Maloney, Ted See Ted Maloney.

malware Malware is a generic term for software that sits on your PC which you did not install and would not have installed. Some malware replicates itself, infecting computers hooked up to the Internet. Sometimes these programs cause damage, and sometimes they don't. See adware and virus.

MAM Major Account Manager.

MAN Metropolitan Area Network. A high-speed data intra-city network that links multiple locations within a campus, city, or LATA. Typically extends as far as 50-kilometers, operates at speeds from 1 Mbit/s to 200 Mbps and provides an integrated set of services for real-time data, voice and image transmission. The IEEE 802.6 standard defines MAN standards and SMDS is the MAN service offered by local phone companies. Private and public MANs may use the ANSI FDDI standard.

man-in-the-middle attack A form of active wiretapping attack in which the attacker intercepts and selectively modifies communicated data to masquerade as one or more of the entities involved in a communication association.

man machine interface A term coined by James Martin to designate the ease (or lack of ease) of a person working with a computer.

man page Manual page. On-line documentation that commonly comes bundled with computers running Unix.

managed circuit A managed circuit is a fancy name for a phone line which the supplier of the circuit, i.e. your phone company carrier, provides you. This is probably one of the dumbest definitions in this dictionary.

managed earnings Public companies can change what they report as earnings and profits by delaying and accelerating expenses and sales. They can also schedule write-downs at the right and wrong time. In short, they can manage their earnings, making them smaller or larger, depending on their needs for their shareholders and the public.

managed file transfer A file transfer application that provides more than just basic file transfer; it provides features that provide more control over the transfer, such as scheduling, checkpoint restart, and security. See also checkpoint restart.

managed modem service A managed service is a carrier service offering, which essentially involves the carrier's owning the requisite equipment and managing certain aspects of the service offering. It's a form of outsourcing. Consider the traditional means by which an ISP supports dial-up access: At the ISP premises are one or more modem banks installed in one or more access routers, which are connected to the edge of the PSTN by one or more dial-up access circuits. The circuits are in the form of either individual single-channel analog circuits or, more commonly, one or more channelized T-carrier circuits, with each voice-grade channel supporting a single connection at a maximum rate of 56/64 Kbps. Seldom does a given connection actually make effective use of the channel capacity, at least over a period of time. All traffic from all ISP customers is directed to the ISP location; traffic intended for the Internet is then rerouted to the Internet over an unchannelized T-carrier circuit. Now consider a managed modem service provided by a PSTN carrier: The carrier places one or more modem banks in its Central Office. The ISP's dial-up customers' calls connect to the network-based modem banks, where they are converted into a TCP/IP packet format by the carrier's network-based router. Traffic intended for the ISP is (e.g., outgoing or incoming e-mail) is sent over an unchannelized T-carrier circuit to the corresponding router at the ISP location. Traffic from multiple dial-up users is aggregated at the edge of the carrier network, optimizing the usage of the T-carrier access circuit. Traffic intended for the Internet or World Wide Web is routed directly in that direction, without looping through the ISP. Authentication is provided by the ISP's authentication server, which commonly would be in the form of a RADIUS (Remote Access Dial-In Authentication Server) server. The carrier provides technical support for the modem bank, the unchannelized T-carrier circuit, and all routers involved, all of which network elements it owns and leases to the ISP. The carrier supports a variety of access protocols, including V.34/V.34+/V.90 modems and ISDN. See also Authentication, ISDN, ISP, Modem, RADIUS, Router, Server, and T-carrier.

managed object A telephone company AIN term. If you understand this definition, you're a better person than I am. Given a Common Management Information Services Element (CMISE) interface between an Operations System and a network element or network system, a managed object is an abstract representation of a physical or logical network element or network system resource. The managed object constitutes an Operations Systems' view of that resource from the CMISE operations system and a network element or network system interfere. It can also be called a Managed Object Instance to emphasize the distinction between Managed Objects and Managed Object Classes.

managed object class A telephone company AIN term. A group of managed objects that all have the same types of attributes, same permissible ranges of attribute values, same semantics and pragmatics for interpretation of Common Management Information Services Element (CMISE) requests, and the same capabilities for issuing event reports.

managed service provider A definition courtesy the Yankee Group: "Managed service providers (MSP) vary considerably in the kinds of services they offer, yet they all call themselves the same thing....The term "managed service provider" simply means an organization that provides some degree of management of the services their clients need to do business. This can mean anything from hosting a web site to off-loading the entire LAN (local area network) function to the MSP location. To confuse the landscape further, MSPs sometimes refer to their services as outtasking, which is probably a better way to describe their offerings. More commonly, an MSP allows businesses to off-load certain IT functions – such as monitoring of the WAN/LAN, gateway and some even offer LAN equipment monitoring-providing an excellent value proposition for end users to gain expertise and a level of service in these areas that they could never obtain in-house... Outtasking MSPs monitor WAN/LAN ports and connections, with some managing servers and other network components, such as routers or the gateway to the internet, VPN, IP telephony, servers and security. Typically, customers run their office equipment on-site. MSPs will monitor the ports and connections on your network that you outtasked to them. The MSP monitors the devices to look for outages; starts remediating issues immediately as soon as it discovers a problem; and proactively addresses the issue before the client even knows there is a problem (VARs are reactive).

When the MSP detects problem, it can work with carriers, telcos or vendors to:
- Open the trouble ticket for the client
- Dispatch the vendor for break/fix
- Swap out hardware
- FTP new software operating system to router or other device that is out and drive the problem to resolution

MSPs offer 24x7x365 support, which in certain high availability verticals, such as healthcare and financial, is critical. The resounding problem with outtasking MSPs is support for the remaining components a company will have to maintain itself or via a relationship with a local VAR."

managed system An entity that is managed by one or more management systems, which can be either Element Management Systems, Subnetwork or Network Management Systems, or any other management systems.

management domain MD. An X.400 term describing a set of messaging systems. At least one system contains, or realizes, an MTA (Messaging Transfer Agent) managed by a single organization. It is a primary building block in the organizational construction of an MHS (Messaging Handling System), referring to an organizational area for the provision of messaging services. MD is used in a similar manner in X.500, consisting of at least one DSA (Directory System Agent). A management domain may or may not be identical with a geographical area. See Domain.

management extranet A management extranet uses the Web to electronically join entire supply chains, so the IS department can tie itself via the Web to its suppliers and service providers. See also Management Intranet.

management information base See MIB.

management information system MIS. Management information provided by computer data processing. Once upon a time called data processing.

MANET See Mobile ad hoc network.

mantelpiece The shelf above, or the finish around, a fireplace is called a "mantelpiece" because at one time people hung their coats (or "mantles") over the fireplace to dry them.

management intranet A management intranet uses World Wide Web technologies and techniques to integrate disparate management tools and databases, provide universal access to documentation and promote distributed collaboration among far-flung IS support people. See also Management Extranet.

management plane An element of the ATM Protocol Reference Model, the Management Plane addresses the management of the ATM switches and hubs, cutting through all 4 layers of the model. Included in Management Plane functions are Operation, Administration and Maintenance (OA&M) functions.

management services In IBM's SNA, network services performed between a host SSCP and remote physical units (PUs) that include the request and retrieval of network statistics.

management system An entity that manages a set of managed systems, which can be either NEs, subnetworks or other management systems.

manchester encoding A digital encoding technique in which each bit period is divided into two complementary halves. A negative-to-positive (voltage) transition in the middle of the bit period designates a binary "1" while a positive-to-negative transition represents a "0". This encoding technique is self-clocking (i.e., the receiving device can recover transmitted clock from the data stream). See also CDR.

mandatory dialing When permissive dialing is over after an area code change, it becomes mandatory dialing. See Permissive Dialing.

mandrel A fiber wrapping device used to cause attenuation within a fiber cable. See mandrel wrapping.

mandrel wrapping A mandrel is a cylindrical shaft or bar. In machining, a mandrel is inserted into the piece you are working on. This holds it in place and maintains its shape during machining. In multimode fiber optics, mandrel wrapping is a technique used to modify the modal distribution of a propagating optical signal. Basically, you wrap a specified number turns of fiber on a mandrel of specified size, depending on the fiber characteristics and the desired modal distribution. It has application in optical transmission performance tests, to simulate, i.e., establish, equilibrium mode distribution in a launch fiber (a fiber used to inject a test signal in another fiber that is under test). If the launch fiber is fully filled ahead of the mandrel wrap, the higher-order modes will be stripped off, leaving only lower-order modes. If the launch fiber is underfilled, e.g., as a consequence of being energized by a laser diode or edge-emitting LED, there will be a redistribution to higher-order modes until modal equilibrium is reached. Note: High-order modes are those that propagate towards the edge of the core; low-order modes are those that remain in the center of the core. See also mode.

manhole Now also called a "person hole". An underground concrete vault in which cables may be spliced, and transmission equipment (repeaters, etc.) may be located. A manhole is used in conjunction with an underground cable running in conduits. A manhole cover is normally made of heavy steel lid of various geometric shapes. They are designed to rest flush with the outside surface. Two most common shapes are the circle and the square. However, rectangles and triangles are also used. The equilateral triangle and the circle share a similar property. They will not fall through the smaller opening they sit on and damage equipment inside (no one replaces a lid with someone still inside the vault). The square cover is often hinged and sometimes bolted closed (high voltage danger is an example). Engineers continue to advance the manhole and cover. It often controls waterflow by location such as being in the highest or lowest position on the street. It is almost always connected to a storm drain since the lid and vault are usually not watertight. Round lids are the images most people first envision. Larger round lids with smaller singleman lids are used to facilitate equipment movement. Hole are built into lids to allow workers to use a grab bar for access and movement. Rarely does a worker lift and carry or even roll a lid anywhere, rather they drag it. Ramped skids that radiate outwards from center on the bottomside are being built into newer round lids to help the round lid do what it does best and that is to allow the worker to easily reinstall the cover and close the job. I have a rubbing done from a beautiful manhole cover in Hanover, NH.

manpack radio A radio carried in the field by one or more persons (for example, military personnel or field scientists), and must be set up for operation. It is not normally operable while being carried. See also backpack radio.

manual exclusion A PBX extension user, by entering a certain code, can block all other phones on that line from entering the call. Assures privacy on the line.

manual gain control MGC. There are two electronic ways you can control the recording of something – Manual or Automatic Gain Control (AGC). AGC is an electronic circuit in tape recorders, speakerphones and other voice devices which is used to maintain volume. AGC is not always a brilliant idea since it attempts to produce a constant volume level. This means it will try to equalize all sounds – the volume of your voice and, when you stop talking, the circuit static and/or general room noise which you undoubtedly do not want amplified. Sometimes it's better to have quiet, when you want quiet. Manual Gain Control is preferred in professional applications. Manual Gain Control is simply an elegant way of saying there's a record volume control. Never record a seminar or speech using AGC. The end result will be decidedly amateurish.

manual hold The method of placing a line circuit on 'hold' by activating a non-locking "hold" button on the phone, usually one colored red.

manual intercom A crude, single-path communications link between telephones without the ability to signal the receiving party.

manual modem adapter An external device for the Merlin key system from AT&T. It allows connection of single line accessories to any Merlin telephone. The device, in effect, draws a standard tip and ring line out of the Merlin proprietary cabling/signaling scheme. Some other key systems have similar devices. Comdial calls theirs a "data port" and their phones contain extra RJ-11 jacks.

manual originating line service The attendant must complete all outgoing calls. All other calls are blocked. This "feature" is used to cut down on long distance phone abuse. There's a wonderful story. When many of the PBXs in Europe went from manual originating line service to automatic dial "9" long distance, the number of long distance calls doubled within two months. Some of these calls were legitimate. Some were not. How much abuse there was varied from company to company. Typically, those compa-

nies with employees who were more bored suffered (or enjoyed?) more abuse.

manual PBXs Refers to PBXs which are not automatic and which require that all calls, including intercom calls, be placed through the attendant. Such PBXs are still used today, though in limited applications. You can still find manual PBXs in vacation hotels, nursing homes and in the data communications departments of some firms, who use manual PBXs as manual dataPBXs. These are especially useful in places where long data calls and sold metal-to-metal connections are an advantage.

manual ring down line Two phones connected by a pair of wires and a battery. Signaling is performed manually by flipping a switch on and off which connects and disconnects the battery. This causes a weak ringing. It's used by rescue teams in caves and mines because radio range is often limited.

manual ring down tie trunk A direct talk path between two distant phones. Signaling must be done manually from either phone. Contrast this with Automatic Ringdown Tie Trunk, in which the signaling occurs the moment one of the phones is lifted off hook.

manual signaling Pushing a button on a telephone sends an audible signal to a predetermined phone. Manual signaling can be used for secretary/boss communications.

manual telephone A telephone without a dial. Taking the receiver off hook automatically rings a predetermined number. A courtesy phone.

manual terminating line service Provides extension lines that require all calls to be completed by the attendant. For a better explanation see Manual Originating Line Service.

manual tie line A tie line which requires the assistance of an attendant at both ends of the circuit in order to complete a connection; a manual tie lie is incapable of passing dialing down the link. Also called Manual Trunk, Ringdown Trunk, Ringdown Tie Line.

Manufacturing Automation Protocol MAP. A protocol initially developed as an internal specification for its own factory floor equipment and now championed by General Motors as the industry standard to facilitate communications among the diverse automation devices found in production environments. AT&T, IBM and DEC have endorsed this standard and have already or will introduce MAP-compatible products. TOP (Technical Office Protocol) was initiated by Boeing Computer Services (one of the nine companies that helped form the MAP Users Group in 1984) and is designed for use in the engineering and office environment and to move information from the factory floor to other parts of the company. Implementation of these protocols would lead to GM's factory of the future concept.

manufacturing message format standard An Application Layer protocol developed as a part of MAP to provide a syntax for exchanging messages in the manufacturing environment.

Manufacturing Message Specification MMS. An International Standards Organization (ISO) application layer protocol that defines the framework for distributing manufacturing messages within a network. This specification is used in MAP 3.n.

MAP 1. A new term for multiplexing, implying more visibility inside the resultant multiplexed bit stream than available with conventional asynchronous techniques.

2. Mobile Application Part. As defined by IS-41 (Interim Standard 41) a User Part of the SS7 protocol used in wireless mobile telephony. MAP standards address registration of roamers and intersystem hand-off procedures. As a query-and-response procedure, MAP makes use of TCAP (Transaction Capabilities Application Part) over the SS7 network. See also IS-41, SS7 and TCAP.

3. Maintenance and Administration Position. See MAP/MAAP below.

4. Manufacturing Automation Protocol. See that definition.

map book Book of maps compiled of any of the following: Street names, lot numbers, APN's, telephone cable locations, drop locations, equipment locations, cable counts. Used by field technicians, this is a vital tool in the Outside Plant Operations.

MAP/MAAP Maintenance and Administration Panel. A device attached to a PBX to allow you to maintain and administer the system – to change phone features, etc.

MAP/TOP Manufacturing Automation Protocol/Technical Office Protocol.

MAPI Microsoft's Windows Messaging Application Programming Interface, which is part of WOSA (Windows Open Services Architecture) and thus part of all Windows operating systems. MAPI is a set of API functions and as OLE interface that lets messaging clients, such as Microsoft Exchange, interact with various message service providers, such as Microsoft Mail, Microsoft Exchange Server, Microsoft Fax and various computer telephony servers running under Windows NT server. Overall, MAPI helps Exchange manage stored messages and defines the purpose and content of messages – with the objective that most end users will never know or care about it. A friend of mine, who's a great programmer, Pete MacLean, explained MAPI as: MAPI is Microsoft's new foundation for a modular mail system. You can pick and choose among various email clients, address books, message stores (foldering systems), and transports (the message-service specific pieces) and build your own custom mail system. See also At Work, Microsoft Exchange, Windows 95, Windows Telephony and WOSA. The biggest explanation of MAPI is in the definition for Windows 95.

mapping 1. In network operations, the logical association of one set of values, such as addresses on one network, with quantities or values of another set, such as devices on another network (e.g. name-address mapping, internetwork-route mapping).

2. A Novell NetWare term. To assign a drive letter to a chosen directory path on a particular volume of a particular file server. For example, if you map drive F to the directory SYS:ACCTSRECEIVE, you will access that directory every time you enter "F:" at the DOS prompt. See also Drive Mappings.

3. In EDI (Electronic Data Interchange), mapping defines the translation between a company's unique data layout and an EDI formal structure.

4. The process by which the structure of a file is provided after the file has been compiled. A "map file" lists all of the program variables, and their memory addresses.

5. Translation from one programming language to another through the use of a compiler. For example, as translation from C to machine language is relatively easily accomplished, C is said to "map well." See also Compiler.

MAR Major Account Representative.

marriage 1. In Pennsylvania, Ministers are forbidden from performing marriages when either the bride or groom is drunk.

2. A second marriage is a triumph of hope over experience. – Samuel Johnson, 1770.

3. Marriage is a relationship in which one person is always right, and the other is a husband. – anon.

4. Next time I'll simply find a woman I don't like and give her the house. – anon.

MARS Multicast Address Resolution Server. Mechanism for supporting IP multicast. A MARS serves a group of nodes (known as a cluster); each node in the cluster is configured with the ATM address of the MARS. The MARS supports multicast through multicast messages of overlaid point-to-multipoint connections or through multicast servers.

marathon A family of products that are combination fast packet multiplexer, data compression, voice compression and fax de-modulation devices that fit many, voice, data, fax and LAN "conversations" onto one leased circuit – analog or digital. The idea of Marathon is to save money on long distance telecommunications charges. The Marathon family of products is made by Micom Communications Corporation, Simi Valley CA, now a subsidiary of Northern Telecom (Nortel).

Marconi, Guglielmo Guglielmo Marconi, born in Bologna, Italy in 1874, was on a holiday when he read of the electromagnetic wave experiments of Hertz. This article established the thought in Guglielmo's mind that electromagnetic waves could free telegraphy from the wires and submarine cables, which at that time constrained its use. Finding out if electromagnetic waves could be used to communicate at a distance became an obsession for Marconi. His mother allowed him to use two large rooms on the top floor of their house as a laboratory. She also helped persuade Guglielmo's father to provide (albeit grudgingly) the money necessary for the batteries, wire and other equipment Guglielmo needed. Marconi started by repeating Hertz's experiments. His oscillator was an induction coil equipped with four spheres for the spark discharge. The frequency of the oscillations was in what we, today, call the VHF range. The detector he used with his receiving coil was a Branly coherer, similar to that used by Oliver Lodge. The coherer provided much greater sensitivity than the spark-gap equipped loop of wire Hertz had used. Marconi placed a curved metal detector behind his oscillator to direct the waves toward the detecting circuit. Soon, Marconi was able to cause a bell, located thirty feet away, to ring when the oscillator was keyed. Through trial-and-error experimentation, he was able to increase the sensitivity of the coherer significantly over what others had achieved. The following spring, Marconi took his experiments outdoors. Connecting metal plates to the oscillator's spark gap lowered the frequency and strengthened the intensity of the oscillations produced. Similar plates were connected to each side of the coherer. By chance, Marconi found that if one of the metal plates was elevated high in the air and the other was laid on the ground, the range at which oscillations could be detected increased to over one-half mile. Soon, the elevated plates at the oscillator and detector were replaced by long vertical wires. The plates which had lain on top of the ground now were buried. This arrangement increased the distance at which signals could be received to one and one-quarter miles. An intervening hill was found to be no barrier to the reception of the signals. The combination using lower-frequency oscillations and using the Earth as an element in his antenna system were crucially important achievements. Another demonstration was held in March of 1897. This time longer wavelengths were used in conjunction with wire antennas raised some 120 feet above the ground by means of kites and balloons. This arrangement resulted in signals being received over a distance of four and one-half miles. In May of 1887, Marconi demonstrated that wireless signals could span significant lengths across water by sending signals between the shore and an island in

the Bristol Channel, a distance of 8.7 miles. This was a crucial test because the submarine cable that normally provided communications to the island had failed several times in recent months. Repairing the cable was costly both in time and in money so Marconi's system must have appeared as an excellent alternative. Marconi established the Wireless Telegraph and Signa. Ltd. in July of 1897. In 1899, he changed the name of his company to The Marconi Wireless Telegraph Co. Ltd. A major goal Marconi had in mind was to show the value of wireless for communicating with ships. In 1897, he returned home to Italy to convincingly demonstrate that wireless could communicate between naval warships. The Italian Navy soon adopted the Marconi wireless system. In 1896, in England, Marconi obtained the first patent on the wireless. In 1901 he succeeded in transmitting signals across the Atlantic. In 1909 he received jointly with C. F. Braun the Nobel Prize in Physics. Marconi was made a Marchese and a member of the Italian senate. He died in Rome on July 20, 1937. See also TESLA, Nikola.

margin account A brokerage trading account that allows you to use borrowed money from your brokerage firm when buying stocks. Not always a good idea, since the stock may decline to where the value of your stock is what you owe your broker and your broker will sell your stock to allow him to get his money back. Meantime, you can lose your entire holdings. Or worse, if the stock falls fast, you may end up owing your broker money which you no longer have.

marginal cost The cost of supplying an extra unit of output. The telecommunications transport business is the only one in the world where the marginal cost of providing an extra unit of product (i.e. a phone call) is zero. This makes for wonderful economics once your network is in place.

marginalized When they stop your salary, but give you a bigger title (also called being "kicked upstairs"), you have, in essence, been marginalized. You have been placed in a position of marginal importance. I included this definition, because one of my dearest friends just got marginalized. And he's too young to retire.

marine telephone Marine telephones operate on assigned radiotelephone frequencies much as a radio broadcast does. Marine telephones can be used to contact other marine telephones or to reach land-based telephones through an operator.

marinized terrestrial cable A type of underwater fiber optic cable that has a conventional terrestrial fiber optic cable core and additional protection to withstand a shallow water environment.

MARISAT A satellite for marine use. Conversations on MARISAT are crystal clear. Call Comsat and ask them for a demo call to a ship somewhere in the world. It's very exciting.

marine broadcast station A coast station which makes scheduled broadcasts of time, meteorological, and hydrographic information.

marine utility station A station in the maritime mobile service consisting of one or more hand-held radiotelephone units licensed under a single authorization. Each unit is capable of operation while being hand-carried by an individual.

maritime air communications Communications systems, procedures, operations, and equipment that are used for message traffic between aircraft stations and ship stations in the maritime service. Commercial, private, naval, and other ships are included in maritime air communications.

maritime broadcast communications net A communications net that is used for international distress calling, including international lifeboat, lifecraft, and survival-craft high-frequency (HF); aeronautical emergency very high-frequency (VHF); survival ultra high-frequency (UHF); international calling and safety very high-frequency (VHF); combined scene-of-search-and-rescue; and other similar and related purposes. Basic international distress calling is performed at either medium frequency (MF) or at high frequency (HF).

maritime frequencies

- channel 16 - In maritime radio communications, channel 16 is reserved for emergency distress calls and safety purposes, when there is a grave and imminent threat to life or property. Channel 16 is known as the "hail and distress channel." After initial emergency contact is made using channel 16, communication is switched to a regular channel for further exchange of information, in order to free up channel 16 for other emergency communications. Maritime radio users are required to monitor channel 16. Frequency: 156.8 MHz.

- channel 9 - In maritime radio communications, channel 9 is used as an alternate to channel 16, when the latter is already in use. Initial emergency contact is made using channel 9, and then communication is switched to a regular channel for further exchange of information. It is advisable for maritime radio users to monitor channel 9. Frequency: 156.5 MHz.

- channels 5, 12, 14, 20, 65, 66, 73, 74, 77 - In maritime radio communications, these channels are reserved for port operations, i.e., for ship-to-ship and ship-to-shore communications related to the handling, movement and safety of ships in or near ports, locks or waterways. Channel 77 is reserved for communications to and from commercial boat pilots in regards to the movement and docking of ships. Channels 11-14 are reserved for port operations on the Great Lakes, St. Lawrence Seaway and designated major ports.

- channel 6 - In maritime radio communications, Channel 6 is used for ship-to-ship communications regarding navigational and weather warnings, and for communicating with the US Coast Guard during search-and-rescue operations. Frequency: 156.3 MHz.
- channels 7, 8, 9, 10, 11, 18, 19, 67, 79, 80, 88 - In maritime radio communications, these channels are used for normal ship-to-ship communications and limited ship-to-shore communications related to a ship's mission and purpose. Recreational boats may not use these channels. Channels 8, 67 and 88 may not be used for ship-to-shore communications. Channel 88 is not available on the Great Lakes and St. Lawrence Seaway.
- channels 9, 68, 69, 71, 72, 78 - In maritime radio communications, these channels are for non-commercial boat operations, i.e., these channels are used by recreational boaters and others not engaged in commercial maritime transport. Channel 72 may not be used for ship-to-shore communications.
- channel 13 - In maritime radio communications, this channel may be used by any vessel for safety communications related to the maneuvering or directing of vessels. Channel 13 is known as the "bridge-to-bridge channel." Frequency: 156.65 MHz.
- channel 15 - In maritime radio communications, this channel is a one-way shore-to-ship channel for ships to receive broadcast information regarding environmental conditions in which ships operate, such as weather, sea conditions, time signals and hazards. Frequency: 156.75 MHz.
- channel 17 - In maritime radio communications, this channel is reserved for state and local government vessels for operational, regulatory and safety purposes. Frequency: 156.85 MHz.
- channel 22 - In maritime radio communications, this channel is used for communications between a ship and a US Coast guard ship or aircraft after initial contact has been made using channel 16. Frequency: 157.1 MHz.
- channels 24, 25, 26, 27, 28, 84, 85, 86, 87 - In maritime radio communications, these channels are used for radiotelephone calls from the ship to any place in the world. These channels can also be used for communications between a ship and another ship outside of transmitting range. Calls on these channels require the assistance of a marine operator (possibly a telco operator) on the channel assigned to the navigation area of the ship initiating the call.

maritime mobile satellite service A mobile satellite service in which mobile earth stations are located on board ships; survival craft stations and emergency position-indicating radiobeacon stations may also participate in this service.

maritime mobile service A mobile service between coast stations and ship stations, or between ship stations, or between associated on-board communication stations; survival craft stations and emergency position-indicating radiobeacon stations may also participate in this service.

Maritime Mobile Ship Identity MMSI. Maritime Mobile Ship Identity is a nine-digit number that uniquely identifies a radio station on board a ship, according to the International Maritime Organization safety requirements. The first three digits are the Maritime Identity Digit (MID), whilst the last six comprise identification of the station. MID are according to the listing of ITU Radio Regulations Appendix 34.

maritime radio navigation satellite service A radionavigation-satellite service in which earth stations are located on board ships.

maritime radio navigation service A radio navigation service intended for the benefit and for the safe operation of ships.

mark 1. A term that originated with the telegraph. It currently indicates the binary digit "1" (one) in most coding schemes. A space is zero in most coding schemes.

2. A call center term. To flag a record in a browse listing for some special purpose. Typically you mark a record that you want to copy information from. As long as the record is marked, you can continue copying the information to other records.

marker The logic circuitry in a crossbar central office that controls call processing functions. See Xbar.

marker beacon A transmitter in the aeronautical radio navigation service which radiates vertically a distinctive pattern for providing position information to aircraft.

marker tape A tape laid parallel to the conductors under the sheath in a cable imprinted with the manufacturer's name and the specification to which the cable is made.

Other information such as date of manufacture may also be included.

marker thread A colored thread lain parallel and adjacent to the strands of an insulated conductor which identifies the cable manufacturer. It may also denote a temperature rating or the specification to which the cable is made.

market capitalization Market cap. In the grand scheme of investing, few measures are more important than market capitalization, which is a simple value of the company, calculated by multiplying the total number of shares issued by their price on the stockmarket. Using this measure, we can easily compare the value of one company against another and make some judgments (along with other financial information like earnings and growth) about buying or selling shares in the company.

market order When you place a "market order," you instruct your broker or brokerage service to buy or sell this stock at whatever price they get. Buying or selling on "market order" has a reputation among the investment professionals for being an extremely stupid way to do business. These professionals believe that brokers will buy higher and sell lower if you place a market order. The other, more professional, way of placing orders is via limit orders. Specify the price you want to buy or sell.

market price Prices set at market rates but, in most cases, are not permitted to be less than cost.

market research Something the phone industry doesn't do – as indicated by the design stupidities of most phones, including – an upside-down keyboard, no backspace erase (except on cell phones), diminutive unreadable screens, voice mail jail, no dial by name (on most phones), dumb handsets that hurt your shoulder, no shoulder rests (on most phones), handset cords that are too short, etc. Don't get me started.

marketing There are two distinct meanings to the term marketing and they're very different. The traditional meaning is to all the things necessary to get your potential customer to buy from you. Marketing includes everything from direct mail to emails to telemarketing – calling your potential customers on the phone. But marketing doesn't include selling. The second meaning of marketing is less common meaning. It means to go shopping. Housewives in various parts of the English-speaking world talk about marketing for the family's groceries.

marking bias The uniform lengthening of all marking signal pulses at the expense of all spacing pulses.

mark-hold The normal no-traffic line condition where a steady mark is transmitted.

markup Special codes in a document that specify how parts of it are to be processed by an application. In a word-processor file, markup specifies how the text is to be formatted; in an HTML document, the markup specifies the text's structural function (heading, title, paragraph, and so on.)

marquee 1. A region on a Web home page that displays a horizontally scrolling message.

2. In graphics software, a sizable and movable frame that identifies a selected portion of a bit-mapped image. The marquee frame can be rectangular in shape or, in some cases, irregular. A lasso tool, for example, enables you to select all contiguous portions of an image that share the same color.

marquee client Webspeak for a customer whose name is recognizable to most Americans – Coca Cola, McDonald's, Avis, etc. A company wishing to raise lots of money should have lots of marquee clients. It gives the company some credibility.

martian Packets that turn up unexpectedly on the wrong network because of bogus routing entries. Also used as a name for a packet which has an altogether bogus (non-registered or ill-formed) Internet address.

martian mail An email message that arrives months after it was sent (as if it had been routed via Mars).

martian packet Strange fragments (data packets) of electronic mail that turn up unexpectedly on the wrong computer network because of bogus routing. Also used for a fragment that has an altogether unregistered or ill-formed Internet address.

Martini Draft An industry nickname for a draft IETF specification for encapsulating and transporting Ethernet across MPLS networks using tunneling pseudowires through label-switched paths. MPLS' support for persistent paths across connectionless packet networks is a crucial enabling technology for this specification. The "Martini" nickname comes from Luca Martini, the primary author of the draft specification. See MPLS, pseudowire.

MAS 1. Minimum Average Surcharge.

2. Mobile Application Subsystem. An MAS is application software that is independent of the Cellular Digital Packet Data (CDPD) network. A cellular radio term.

3. Multiple Address System. A microwave point-to-multipoint communications system, either one-way or two-way, serving a minimum of four remote stations. The private radio MAS channels are not suitable for providing a communications service to a larger sector of

the general public, such as channels the commission has allocated for cellular paging or specialized mobile radio services. (SMR).

MASC 1. Major Accounts Service Center.

2. Major Account Support Consultant.

MASER Microwave Amplification by Stimulated Emission of Radiation. A device that generates electromagnetic signals in the microwave range, known for relatively low noise.

mashup A mashup is a web site or web application that combines content from more than one source. Examples include housing for sale listings imposed on a map of the area, movies playing imposed on a map showing where the movie houses are.

mask 1. A field made up of letters or numbers and wildcard characters, used to filter data. For example, a mask 800xxxxxxx may be applied to the dialed digits field of a call record to identify toll-free calls.

2. A computer imaging term. The electronic equivalent of placing transparent tape over selected regions of an image, a mask marks pixels that remain unchanged by subsequent painting operations. For example, you might mask out a mountain range and add background clouds to the sky. In the final image, the clouds will appear between the peaks.

masking Also known as cloaking. It allows you to have your domain name appear in the browser instead of your destination URL. Masking comes in handy both when you have a long URL and/or if you don't want to change a URL you've been using for years.

maskable interrupt An interrupt on a computer that can be interrupted by another interrupt.

MASP Mediated Attribute Store Protocol, an XML-based protocol for a new attribute store service.

MASQ See IP Masquerade and Masquerade.

masquerade 1. To pretend to be someone else by using another person's password or Token. It's not a nice thing to do.

2. See IP Masquerade.

MASS Major Account Support Specialist.

mass splicing Simultaneous splicing of many fibers in a cable.

MASS860 An organization of computer system vendors formed to promote open system standards and the writing of applications software for the Intel i860 microprocessor. Members include Intel, Oki Electric Industry Co., Ltd., IBM, Stratus, Olivetti, Alliant, Samsung and Stardent Computer.

massively parallel systems Tightly coupled multi processing computers that house 100 or more CPUs, each with its own memory.

mast A pipe or pole used for mounting antennas.

mast clamp A piece of equipment used to attach a ram hook to a pole. Aerial service wire is attached with a ram hook (sometimes called a ram horn) with a drop clamp.

master Term applied to the data communications equipment at one end of a synchronous digital transmission network that supplies the clock timing signal that determines the rate of transmission in both directions.

Master Address Street Guide MSAG. In the emergency services telephone network in the United States, the Master Address Street Guide is a database containing the mapping of street addresses to Emergency Service Numbers within a given community. This allows the derivation of call routing information from a call's Automatic Location Identification. See Automatic Location Identification.

master boot record MBR. The first sector on a hard disk, which begins the process of starting the computer. The MBR contains the partition table for the disk and a small amount of executable code called the master boot code.

master clock An electronic timing circuit which synchronizes the entire data communications network. The source of timing signals, or the signals themselves, that all network stations use for synchronization.

master control program A part of the Burroughs operating system that monitors the host's operation.

Master Control Unit MCU. An InteCom word for the device which controls the main operating functions of the system.

Master Customer Number MCN. A Verizon definition. A unique number that identifies a customer. Also known as Customer Identification Number.

master frequency generator In FDM, equipment used to provide system end-to-end carrier frequency synchronization and frequency accuracy of tones over the system.

Master Group MG. In frequency division multiplexing (the old way of putting many voice conversations onto on communications line) a master group consists of 300 voice-grade (4 kHz) channels.

master number hunting When a call is directed to the pilot number of a hunt group, it will hunt to the first non-busy station in that group. If a call is directed to a

specific station in that hunt group it will go directly to that station and not hunt to another station in the group.

master slave switching system A configuration consisting of a central switch and one or more remote switches. The master switch typically controls all I/O (input/output) information. The slave system performs tasks as directed by the master, including switching calls between phones attached to that remote module – without sending those calls back to the central switch. There are enormous savings in wiring since not every remote phone has to have a pair back to the central switch.

master slave timing In a communication system, a timing subsystem wherein one station or node supplies the timing reference for all other interconnected stations or nodes.

master station 1. The main phone or station in a group. The one controlling the transmission of the others.

2. The unit which controls all the workstations on a LAN, usually through some type of polling. The master station on a token-passing ring allows recovery from error conditions, such as lost, busy or duplicate tokens, usually by generating a new token. Sometimes servers are referred to as master stations.

3. In navigation systems employing precise time dissemination, a station whose clock is used to synchronize the clocks of subordinate stations.

4. In basic mode link control, a data station that has accepted an invitation to ensure a data transfer to one or more slave stations. At a given instant, there can be only one master station on a data link.

Mastergroup MG. 10 supergroups each comprised of 5 groups of 12 channels summing up to 600 circuits transported as a unit in an analog FDM carrier system. First used in Bell's type L1 carrier systems on intercity coaxial cable. Six mastergroups are equal to one jumbogroup.

Masters of Deception A group of young people who gained notoriety of sorts by hacking and cracking the U.S.'s telephone system, securing free phone calls and gaining admission to networked computer systems. There were two rival groups, the Masters of Deception and The Legion of Doom. They attempted to outdo each other with greater and greater feats. The leader of the MOD was a fellow who called himself Phiber Optik. Michele Slatalla and Joshua Quittner wrote a book chronicling the whole story. It's called Masters of Deception; The gang that rule cyberspace.

mastic A cabling term. A meltable coating used on the inside of some shrink products which, when heated, flows to encapsulate the interstitial air voids.

MAT Meridian Administration Tools.

MATC Major Account Technical Consultant.

matched junction A waveguide component having four or more ports, and so arranged that if all ports except one are terminated in the correct impedance, there will be no reflection of energy from the junction.

matching loss ML. The inability to find an idle path between two idle equipment components. Usually expressed as a percentage. Example: Incoming Matching Loss, Originating Matching Loss, etc.

mated pair 1. A pair of devices which are perfectly matched. In other words, they perform identical functions. Modems, for instance, are perfectly mated as they perform the same functions of modulating and demodulating signals, depending on the direction of the transmission. Otherwise, they can't communicate. Like most things in network technology (and most things in life) communications is best accomplished between entities that are balanced and symmetrical. By the way, two mated pairs of things are known as "quads."

2. If you move out of a residential apartment, the pair that you were using is dedicated to that Apt. so when the next occupant moves in, it is a flick of a switch to get phone service turned on. That's called Dedicated Pair Out (DPO). The DPI pair is the OE (Office Equipment) or the Switch pair. Now the pair in the C.O. (Central Office), when it's joined to the DPO pair, is often called a mated pair. The DPI pair is mated to the DPO pair.

MATEL A Multiplex Automatic TELephone system. Picture one long, up to two miles wire (any decent quality works). You roll it out, then you clip phones anywhere into the wire. Then you have, in effect, a seven channel bus PBX. You have two digit extension dialing between the phones (up to 60). You also have conferencing, broadcast, call back, DID and connection to one central office line and one radio channel. The uses? String it around a rioting prison, a siege, an airport hijack, an emergency in New York City subway, etc. Three advantages: instant communications, communications where radio is bad and radio silence – you keep the press and the bad guys in the dark. This definition from John McCann, general manager of Racal Acoustics Limited, Frederick, MD, which makes the MATEL.

material dispersion Material dispersion occurs because of pulse of light in a fiber includes more than one wavelength. Because of the refractive index of a material varies with wavelength (check out a prism in the sun!) different wavelengths travel down the fiber at different paths. See also Chromatic Dispersion.

material scattering In an optical fiber, that part of the total scattering attributable to the properties of the materials used for fiber fabrication.

math coprocessor A coprocessor is a special purpose microprocessor which assists the computer's main microprocessor in doing special tasks. A math coprocessor performs mathematical calculations, especially floating point operations. Math coprocessors are also called numeric and floating point coprocessors. If you do a lot of mathematical tasks on your PC, like recalculating large spreadsheets, then installing a math coprocessor makes huge sense. Intel included a math coprocessor with its 486DX chip, but removed it for the 486SX. No other Intel chip has a math coprocessor built in. When you buy a math coprocessor make sure it's the same speed as your existing processor.

mathematical theory of communications Theory evolved by Claude Shannon and Warren Weaver; first published in 1949. Shows that all communications can be explained using a model that consists of six functions, in sequence: 1) source, 2) encoder, 3) message, 4) channel, 5) decoder, and 6) receiver. See Shannon, Claude Elwood.

MATR Minimum Average Time Requirement.

matrix 1. A switch. A device for moving calls from one input to the desired output. There are many types of switching matrices – from simple step-by-step matrices to complex digital pulse code modulated matrices. Most switching matrices are "blocking." They do not have sufficient capacity to switch every call. There are some switches that are "non-blocking." These have the ability to switch every call simultaneously. By definition, non-blocking matrices are more expensive. They are only needed in special situations of high traffic.

2. The encompassing material in the composite (i.e. the plastic).

matrix ringing Two key system phones picking up the same extensions with different lines ringing on different phones. These days, with electronic phones, matrix ringing is easy. In the old days, with 1A2 phones, you needed to do considerable wiring.

matrix switch Device that allows multiple channels connected via serial interfaces (typically RS-232C) to connect under operator control to designated remote or local analog circuits or other serial interfaces.

Matthews, Gordon February, 23, 2002. AUSTIN - Gordon Matthews, the inventor of voice mail, died Saturday from complications related to a stroke. He was 65. Matthews was the holder of more than 35 U.S. and foreign patents – including the one for voice mail. Matthews first conceived the technology known as voice mail in the late 1970s and patented it in 1982. The success of his "Voice Message Exchange" in a digital format and manipulating electronic messages gave birth to a new industry. "I'm not really pleased with some of the things I see voice mail being used for today," Matthews once said. "We didn't design this technology to annoy people, but rather make their lives easier." Matthews attended the University of Tulsa, where he studied engineering physics. After graduating in 1959, Matthews joined the Marines as an aviator, in honor of his uncle and his father, who also was a Marine. He discovered his desire to invent when a friend was killed in a midair collision. It was believed that he had to momentarily take his hands off the controls to adjust his radio frequency. Five years later, he traded in his uniform to work in a suit at IBM Inc., where he developed a system that allowed pilots to control cockpit functions with their voice. "Every one of my inventions came about when something bothered me," he once said in an interview. He believed that if a solution can be envisioned, it also can be created. He moved to Dallas in 1966 to work for Texas Instruments. He specialized in using computers to automate telephone systems of large corporations with multiple lines and then launched a series of his own businesses specializing in computers and telecommunications. In 1979, Matthews formed a new company, VMX Inc. of Dallas, which stands for Voice Message Express. Matthews applied for a patent in 1979 for his voice mail invention and sold the first system to the 3M Corp. in Minnesota in 1980. His wife, Monika, recorded the first greeting on this first commercial voice mail. By 1989, his company was earning royalties of what he said were "tens of millions of dollars." But he knew it was time to sell. "The market was growing faster than I could grow the company," Matthews once said. He sold VMX Inc. and retired to Austin after 13 years with his Dallas-based voice mail company.

MATV Master Antenna System, such as used in apartment buildings and motels. A combination of components providing multiple television receiver operations from one antenna or group of antennas; normally on a single building.

Matzo ball soup This may be apocryphal. But it's a touching Passover story: Marilyn Monroe is attending her first Seder. Arthur Miller asks her how she likes the Matzo ball soup. Marilyn answers, "I wish they could have used a different part of the Matzo."

MAU 1. Math Acceleration Unit.

2. Multistation Access Unit. A MAU is a wiring concentrator used in Local Area Networks.

In token ring networks it's called a MultiStation Access Unit (MSAU). In Ethernet networks it's called a Medium Attachment Unit. Basically a MAU is a standalone device that contains multiple NICs (Network Interface Cards), thereby allowing multiple terminals, PCs, printers, and other devices to be share a single point of connection to a Token Ring or Ethernet LAN. Each computer is wired directly to the MAU which then provides the connection to the LAN. MAUs themselves can be connected to expand the network. The MAU is a small box with eight or sixteen connectors and an arrangement of relays that function as bypass switches. When only one MAU is used, its relays and internal wiring arrange themselves so that the MAU and the connected computers form a complete electrical ring. MAUs can be cascaded to create bigger rings. The MAU listens for the "I'm here" signal sent by a computer when the computer is attached to the MAU. If the token ring adapter card in a computer is not working properly or the computer is turned off, the MAU no longer hears the "I'm here" signal and automatically disconnects the computer from the ring using the bypass relay.

maverick buying See B2B.

MaxCR Maximum Cell Rate: This is the maximum capacity usable by connections belonging to the specified service category.

maximize button The maximize button in Windows is the up-arrow button at the far right of a title bar in the Windows operating system. Click on the maximize button to enlarge the IMARA Lite window to full size. See also, Minimize button and Restore button.

maximum access time Maximum allowable waiting time between initiation of an access attempt and successful access.

maximum bandwidth zero suppression MBZS is modeled after CCITT Recommendation G.922. MBZS forces the 31st bit in a data stream to 1 if there are 30 preceding zeros. This method uses no bandwidth, but may introduce errors into the data stream. The average Bit Error Rate (BER) for the MBZS ones insertion method is 4.7×10^{-10} (10 to the negative 10). Trunk modules at both ends of a link must use the same zero suppression method. MBZS is used when the trunk does not interface with a telephone company. An example would be an intra-building connection between floors where a cable facility is used with a CSU that has been programmed not to insert ones when the TRK-2 I/F sends excessive zeros. MBZS is only used on telephone company facilities when there is a clear-channel facility that uses B8ZS to suppress excessive zeros sent out by the MBZS option.

maximum block transfer time Maximum allowable waiting time initiation of a block transfer attempt and completion of a successful block transfer.

maximum burst Specifies the largest burst of data above the insured rate that will be allowed temporarily on an ATMPVC but will not be dropped at the edge by the traffic policing function, even if it exceeds the maximum rate. This amount of traffic will be allowed only temporarily; on average, the traffic source needs to be within the maximum rate. Specified in bytes or cells. See Maximum Burst Size.

Maximum Burst Size MBS. ATM network performance parameter that defines the duration of transmission at a peak rate that would be accepted on a given ATM virtual circuit. See Maximum Burst.

maximum calling area Geographic calling limits permitted to a particular access line based on requirements for a particular line.

maximum keying frequency In facsimile systems, the frequency in hertz numerically equal to the spot speed divided by twice the X-dimension of the scanning spot.

maximum modulating frequency The highest picture frequency required for a given facsimile transmission system. The maximum modulating frequency and the maximum keying frequency are not necessarily equal.

maximum permissible exposure MPE. The maximum RF energy to which a person can be exposed. MPE is measured in milliwatts per square centimeter and is frequency-dependent. There are two MPE limits, one for the general population, and one for RF-knowledgeable workers who have controlled access to electromagnetic emissions environments.

maximum power level Maximum power output limit for Mobile end System (M-ES).

maximum rate Maximum total data throughput allowed on a given virtual circuit, equal to the sum of the insured and uninsured traffic from the traffic source. The uninsured data might be dropped if the network becomes congested. The maximum rate, which cannot exceed the media rate, represents the highest data throughput the virtual circuit will ever deliver, measured in bits or cells per second.

maximum stuffing rate The maximum rate at which bits can be inserted or deleted.

Maximum Transmission Unit MTU. The largest possible unit of data that can be sent on a given physical medium. Example: The MTU of Ethernet is 1500 bytes.

Maximum Usable Frequency MUF. The upper limit of the frequencies that can be used at a specified time for radio transmission between two points and involving propagation by reflection from the regular ionized layers of the ionosphere. MUF is a median frequency applicable to 50 percent of the days of the month, as opposed to 90 percent cited for the lowest usable high frequency (LUF) and the optimum traffic frequency (OTF).

maximum user signaling rate The maximum rate, in bits per second at which binary information can be transferred (in a given direction) between users over the telecommunication system facilities dedicated to a particular information transfer transaction, under conditions of continuous transmission and no overhead information.

maximum viewing area In a glass cathod ray tube monitor, the maximum viewing area is the actual maximum viewing area and it is dependent upon the size of the plastic or bezel around the CRT.

maybe "I'll give you a definite maybe." Samuel Goldwyn, film producer. See Goldwyn.

Mayday A radio distress call, signifying an imminent, life-threatening emergency. It is from the French "m'aidez" (help me).

MAYPAC A X.25 packet-switched network operated in Malaysia by the Malaysian government.

MB Megabyte. A unit of measurement for physical data storage on some form of storage device – hard disk, optical disk, RAM memory etc. and equal to two raised to the 20th, i.e. 1,048,576 bytes. Here is a summary of sizes:

    MB = Megabyte (2 to the 20th power)
    GB = Gigabyte (2 to the 30th power)
    TB = Terabyte (2 to the 40th power)
    PB = Petabyte (2 to the 50th power)
    EB = Exabyte (2 to the 60th power)
    ZB = Zettabyte (2 to the 70th power)
    YB = Yottabyte (2 to the 80th power)
    One googolbyte equals 2 to the 100th power.

Mbaud One million bits of information per second. Also referred to as Mbps or Mb/s.

MBG Multilocation Business Group.

Mbits Million bits. See Mbps.

MBONE Multicast Backbone. A collection of Internet routers that support IP multicasting. The MBONE is used as a "broadcast (actually multicast) channel" on which various public and private audio and video programs are sent. Circa 1992 IETF (Internet Engineering Task Force) effort. Came out of earlier ARPA DARTnet experiments. Supports multicast audio and video across the Internet. Provides one-to-many and many-to-many network delivery services for apps like videoconferencing and audio. Supports simultaneous communication between several hosts. At present, the Internet MBONE is the largest demonstration of the capabilities of IP Multicast. The MBONE is an experimental, global, volunteer effort, and topographically is layered on top of portions of the physical Internet. (IP multicast packet routing is not supported by many installed production routers.) The network is linked by virtual point-to-point links called "tunnels". The tunnel endpoints are typically workstation-class machines having operating system support for IP multicast and run the "mrouted" multicast routing daemon. It presently carries IETF meetings, NASA space shuttle launches, music, concerts, and many other live meetings and performances. www.mbone.com.

Mbps This one is confusing. When you see Mbps as the speed of a telecommunications, networking or local area networking transmission facility (i.e. something that moves information), Mbps means million bits per second – exactly one million. No more. No less.

When you see Mbps or MBps referred to the context of computing, it means million bytes per second, which is the same as one million bytes per second. How many bits that is depends on how many bits there are in a byte. Typically it's eight (but it could be more or fewer). So, in this case, one Mbps would be eight million bits per second.

To be correct, Mbps is million bits per second and MBps is million bytes per second. You will also see it written as Mb/s. That usually means million bits per second. For a much longer explanation, see Bps.

MBS 1. Maximum Burst Size. An ATM term for a traffic parameter which specifies the maximum number of cells which can be transmitted at the Peak Cell Rate (PCR). In the signaling message used for call setup, the Burst Tolerance (BT) is conveyed through the MBS. The BT, together with the SCR and the GCRA, determine the MBS that may be transmitted at the peak rate and still be in conformance with the GCRA.

  2. Meridian Business Set.

MBus A Sun Microsystems definition: An open specification for connecting multiple CPUs (such as those in SPARC modules) with a 64-bit, 320-MB/second data path. Designed by

Sun Microsystems; available from SPARC International.

MBWA Management By Walking Around. A technique pioneered by David Packard, one of the two founders of Hewlett-Packard which he and Bill Hewlett started in 1939 in a small garage in Addison Avenue in Palo Alto, CA.

MBZS See Maximum Bandwidth Zero Suppression.

MC 1. Main Cross-connect. The interconnect point where wiring from the entrance facility and from the workstation is connected to telecom equipment.

  2. Matrix Controller.

  3. Multi-Carrier.

  4. Metal Clad. A UL classification indicating a metal clad cable. An assembly of insulated conductors with a metal cladding applied over the core and with grounding conductor(s) if the cladding is interlocked armor.

MC1 Cable The inter-PC chassis MVIP bus cable, which can support up to 20 PCs. It allows a developer to distribute MVIP's resources across all the connected computers. MC-MVIP type MC1 media provides 1536 x 64 Kbps of inter-chassis connectivity using twisted-pair copper cables. See MC2 and MVIP.

MC2 Cable The advanced inter-PC chassis MVIP bus cable, which leverages FDDI-II to provide up to 3072 x 64 Kbps of inter-chassis connectivity on fiber or copper. MC3 leverages SONET/SDH fiber technology at 155 Mbps to provide 2400/4800 x 64 Kbps of inter-chassis telephony. See MC1 and MVIP.

MC-MVIP Multi-Chassis MVIP. See MVIP.

MCA Micro Channel Architecture. The internal 32-bit bus inside some of IBM's PS/2 machines. It was originally introduced by IBM as a proprietary bus which manufacturers of IBM clone PCs would have to pay IBM large royalties if they wanted to include the bus in their machines. Sadly, this strategy backfired and few people wanted IBM PCs with the MCA bus. ISA remains the most popular PC bus. See ISA and EISA.

MCC 1. Mobile Country Code. A portion of the LAI and the IMSI (International Mobile Subscriber Identity).

  2. Mobile Control Channel. See Control Channel.

MCCS Mechanized Calling Card Service was formerly known as ABC Service. MCCS is a CO switch facility that automatically bills credit card calls made on DDD without the involvement of an operator.

MCDV An ATM term. Maximum Cell Delay Variance: This is the maximum two-point CDV objective across a link or node for the specified service category.

MCF Message Confirmation Frame. Confirmation by the receiver in a fax transmission that the receiver is ready to receive the next page.

McGill, Archie See Dimension.

McGowan, William G. Few men ever tilt at windmills. Few ever bring the windmills tumbling down. William G. McGowan was one of those men. He was also my mentor and my very dear friend. He turned my onto telecommunications. It was the summer of 1969. I had just graduated from the Harvard Business School and had become editor of a small New York City based newsletter called "The Knowledge Industry Report." (We were way ahead of our time. There was no such thing as the knowledge industry.) That summer the FCC announced the historic MCI Decision in which they allowed a small company called MCI Communications Corporation to offer interstate long distance private line (i.e. leased line) service in competition with the major (and pretty well only) provider at the time, American Telephone and Telegraph. AT&T was a gigantic company employing over one million people. It was the largest private employer in the U.S. I called McGowan, who was MCI CEO and chairman, to interview him for the newsletter. I wrote a story. He invited me down to Washington to meet him and the five other employees of MCI. (That's all there were.) I went down and met a man with a twinkle in his eye. Uncle Bill and I became good friends. I stayed at his house. I went on trips with him. I even took him to the Washington National Airport on the back of my motorcycle. I invested a little in MCI. I helped him with business ideas for MCI and I helped on new product launches, including switched long distance phone service which we started in Dallas with Jerry Taylor and Ray Miller – against the wishes of all the MCI executives at MCI's Washington headquarters – except Uncle Bill. There are so many memories I have of "Uncle Bill," as many of us called him. They are flooding back as I tearfully write this. I remember he called me in early 1982, the day AT&T announced they were divesting themselves of their local operating companies in favor of keeping the long distance company, Bell Labs and Western Electric, the manufacturing arm. McGowan couldn't get over the dumbness of the AT&T's decision. Said Uncle Bill, "Charlie Brown [then AT&T's chairman] is giving up the only bits of their business that has a competitive edge. Heck," He continued, "They have a monopoly on local service. That's where the value is. They have no competitive advantage in long distance. All the new technology is going against them in long distance, making it cheaper and cheaper for competitors like us to provide long distance service and compete against them. They are keeping Bell Labs and Western Electric because Charlie Brown has been told they have great untapped computer discoveries in their labs which they'll now be allowed to exploit." [The Divestiture Decree removed the 1956 Consent Decree restriction on AT&T against making anything except communications equipment.] "But," continued McGowan, "When Charlie Brown and the AT&T management finally go visit see Bell Labs, they'll find the Labs bare of saleable, computer inventions." He was right. AT&T lost billions trying to sell computers. It even bought NCR (for $7.4 billion) and later gave it to its shareholders, got out of the computer business and was even forced to spin off its manufacturing arm. From 1969 to end-2004, AT&T shrunk 95% – from one million people to around 50,000. Few large companies have become so small so quickly. For that, Uncle Bill deserves a small part of the credit (or blame) – though he and I would agree that AT&T shot themselves many more times in the foot more than Uncle Bill shot them.

McGowan did not start MCI. Jack Goeken did. Bill was introduced to MCI by Mike Bader, a Washington lawyer who was doing some work for Jack and thought Jack needed some help. (He did.) William G. McGowan was born and raised in the coal country of Pennsylvania, the son of a second generation Irish railroad engineer and a school teacher. Throughout his schooling, both in high school and in college, he worked on the railroad that ran through his home town of Wilkes-Barre. He did his undergraduate work at a college that was just then starting – Kings College, where there is now the McGowan School of Business, which Bill funded just before his death. As graduation from Kings College approached, Mr. McGowan decided he wanted to go to the Harvard Business School. He had only enough money saved to carry him through the first year. He wasn't sure how he would pay for his second year – until he heard about the school's "Baker Scholars" program – a Harvard financial aid program available to the top business graduate students entering their final year. So, even before he started his first year, his "operating plan" was to pay his first year out of his railroad savings and win the Baker scholarship to complete HBS. He was that kind of a man – and he did just that. McGowan carried that same determined spirit and long-term planning into his business life. He was always a meticulous planner. He taught me many things. But two I remember to this day: First thing, every day, write down the five most important things you want to accomplish that day. Do the hardest first. That way, with an early accomplishment, you begin the day on a roll. Second, don't plan small or meek. One of Uncle Bill's favorite statements was "The meek shall inherit the earth, but they won't gain market share." To this day, Larry Harris, once MCI's lawyer still carries that statement with him every day, neatly folded in his wallet.

Upon graduation from HBS, McGowan began working on Wall Street, where he helped with movie financing. He is actually listed as Associate Producer on the movie, Oklahoma. I can still remember his stories on how they made the movie in the middle of winter and how much it cost to fly the cornfields into cold Oklahoma from warm Mexico. Remember the song with the words, "the corn is as high as an elephant's eye." Well it had to be. After leaving Wall Street, he became a self-employed management consultant, concentrating on helping people fix businesses that had fallen on bad times. I remember him telling me how he would walk into potential clients' offices and ask them what their problems were? They would lay out a marketing problem, an engineering problem, or a finance problem. Whatever the problem, McGowan's sales pitch would always be, "You're lucky. I happen to be expert in that area, having recently worked on solving several similar problems." He usually got the job and solved the problem. His brain was extensive, though his experience wasn't. He often got paid in kind. His clients were, after all, in trouble. One client gave him a share of valueless building on Fourth Avenue in New York City. Several years later, the City changed the street's name to Park Avenue South. The building's value skyrocketed and McGowan made his first big financial hit.

In the mid-1960s, when Uncle Bill first got into telecommunications, AT&T had a virtual monopoly on all telephone communications in the United States. AT&T owned all the local phone companies in all major cities. It owned the long distance business and it made virtually all the telecommunications equipment used to supply telephone service, including the equipment that phone users had in their homes and their businesses. It was the world's largest phone company and one of this country's strongest business empires. But Bill McGowan looked at it – and at the law – and concluded that there was nothing that gave the Bell System a legal-given or God-given right to all of the long distance business. I could spend pages detailing the stumbling blocks AT&T threw in his way – as well as the service, regulatory and legal battles he had to win to even gain the right to go into the long distance business. And, once given that right by the FCC, no one could have foreseen the struggles he went through to raise money, build a national long distance system and convince the American people to use his service. At that time, Ma Bell was next to Godliness in the American public's mind. AT&T was the archetypal widows and orphans stock. Getting the American public to switch was heresy.

McGowan was the kind of man who persevered in the face of challenges – and he always won the big ones. If you knew him, you never had any doubt. He did, however. He confided to me that MCI had nearly missed paying its payroll several times. And he worried and worried. He smoked too much. He drank too much. He ate too much. He exercised too little. All of which affected his health. In 1986, Uncle Bill suffered a heart attack that led to him becoming the highest-ranking corporate executive at that time to have a heart transplant. After the transplant operation, his University of Pittsburgh Hospital surgeons told him they kept him alive for 1 1/2 hours on a machine while they awaited the private plane delivering his replacement heart. The plane was 1 1/2 hours late. He had been technically dead for 1 1/2 hours. When he finally awoke, he asked the surgeons if they were worried about the plane's delay? They answered, "Yes." He smiled his pixie smile, grateful to have survived that one. During his recuperation, he became interested in the developments taking place in modern medicine. He died in June, 1992, becoming one of the longest-living heart transplant patients at that time.

To build MCI's national network, he raised money from local businessmen all over the country. And on June 2, 1972 Blyth & Co. issued a red herring prospectus for MCI, offering three million shares at $10 a share, giving MCI a market capitalization of $30 million. The company began trading on Nasdaq on June 23, 1972. According to the prospectus, Bill McGowan owned 1,466,550 shares. or 17.5% of the new public company. In 1998 WorldCom bought MCI for $40 billion, giving the original shareholders of MCI a fantastic gain – or at least those who had held. The sad irony is that later WorldCom went bankrupt and all the shares it issued to pay for MCI turned out to be valueless. There are many of us who believe that, had Uncle Bill lived, MCI would never have sold out to WorldCom and the history of telecommunications would have been very different.

McGowan's charitable activities are extensive. Prior to his death he served on the Board of Directors of Georgetown University and its Medical Center. McGowan established The McGowan Charitable Fund which gives money to youngsters seeking an education that may lead to entrepreneurial successes and business leadership in his mold. He also established the McGowan Center for Artificial Organ Development, now the McGowan Institute for Regenerative Medicine, at the University of Pittsburgh. The Fund continues its support of medical research through grants in the fields of cancer, diabetes, Alzheimer disease, Osteogenesis Imperfecta, paralysis, hearing, sickle cell, eyesight, hearing, organ development and regeneration and heart attack survival. For more on the McGowan Fund and/or to apply for a grant, go to www.mcgowanfund.org/index.html. The site has a stunning picture of Uncle Bill just as I remember him. Check out the smile.

By the way, there are many people, including Pete Howley, who credit McGowan with ushering in the era of cheap long distance communications, encouraging the use of innovations such as microwave and fiber optics in telecommunications, and thus enabling the Internet to begin, flourish and impact our lives as it does today. I won't argue with Pete. If it weren't for McGowan, it's possible that AT&T could still be renting us phones, charging us outrageous amounts for long distance calling and we might never have heard of cell phones (which AT&T thought had no market potential), packet switching (on which the Internet is based) and voice over IP (the present wave in phone calling).

MCHG Mass Change.

MCHO Mobile Controlled Hand-off.

MCI 1. Once upon a time it was called Microwave Communications Inc. Then it became just MCI, which stands for nothing. MCI was the largest long distance phone company in the US after AT&T. In MCI's early days, the initials were said to stand for "Money Coming In." MCI was a full-service long distance company offering every service from switched single channel voice to leased T-1. In 1996, it announced that it had accepted a takeover offer from British Telecom (BT), the leading phone company in England. BT had held a 20% stake in MCI for a number of years, and viewed MCI as a vehicle to gain a significant position in the highly lucrative U.S. market. Then GTE offered more. Then upstart Worldcom offered even more. The MCI/Worldcom merger was completed on September 20, 1998 for approximately $40 billion (not cash, but shares). MCI Worldcom, at the time of the merger, boasted annual revenues of $30 billion, and a presence in over 65 countries. In order to gain regulatory approvals in the U.S. and Europe, MCI sold its Internet backbone to Cable & Wireless for $1.75 billion. See also Ebbers, Bernie and WorldCom.

2. Media Control Interface. A standard control interface for multimedia devices and files. Using MCI, a multimedia application can control a variety of multimedia devices and files. Windows provides two MCI drivers; one controls the MIDI sequencer, and one controls sound for .WAV files.

3. Message Center Interface. An interface in some PBXs which allows you to connect an external PC and do voice mail/IVR (interactive voice response).

MCI Worldcom The company formed on September 20, 1998, when Worldcom merged with (read acquired) MCI for approximately $40 billion. MCI Worldcom, at the time of the merger, boasted annual revenues of $30 billion, and a presence in over 65 countries. In order to gain regulatory approvals in the U.S. and Europe, MCI sold its Internet backbone to Cable & Wireless for $1.75 billion. Now it's simply called Worldcom. See also MCI and Worldcom.

MCL Mercury Communications Limited (UK). The second long distance company in England. It is competitor of British Telecom, the erstwhile monopoly local and long distance company in the U.K.

MCLD Modifying Calling Line Disconnect. This is what CPC (Calling Party Control) is referred to on the Lucent 5ESS switch. See Calling Party Control.

McKinsey & Company McKinsey is a very large management consulting firm advising leading companies on issues of strategy, organization, technology, and operations. Around the time AT&T was being broken up – the early 1980s – a study by McKinsey & Co. predicted that by the year 2000 there would be 900,000 cellphone users in the U.S. The actual number, of course, turned out to be more than 100-fold greater. Some believe that Charlie Brown, AT&T's head at the time, used the McKinsey study to justify not asking the FCC to be granted wireless licenses. Had he done so, AT&T would have had assets worth many billions of dollars. Eventually AT&T realized it had missed the boat on cellphones. It bought a cellphone provider in 1993 but later spun it off as an independent company called AT&T Wireless. That company was acquired last year by Cingular Wireless, which is 60%-owned by SBC, and the AT&T Wireless name was retired.

MCLR An ATM term. Maximum Cell Loss Ratio: This is the maximum ratio of the number of cells that do not make it across the link or node to the total number of cells arriving at the link or node.

MCM One thousand circular mills.

MCN See Master Customer Number.

MCNC Microelectronic Center of North Carolina.

MCNS Multimedia Cable Network System Partners Ltd. An organization which is leading the development of DOCSIS (Data Over Cable Service Interface Specification), which is an industry specification that defines the technical equipment and interface specifications for high-speed cable modem and headend equipment in order to deliver high-speed data services over cable television systems. MCNS consists of Comcast Cable Communications Inc., Cox Communications, Tele-Communications Inc. (TCI), and Time Warner Cable. MCNS has partnered with a number of other companies in this project, as well. See also DOCSIS.

McNutt, Emma M. The first female telephone operator, Ms. Emma M. McNutt was hired in September 1878 by the New England Telephone company in Boston, Massachusetts. Her hiring caused quite a stir, as "proper ladies" didn't work outside the home in those days. Previous to Ms. McNutt's appearance on the scene, all of the operators were males, who proved themselves to be unruly. Within a few short years male operators were extinct – for about 100 years, at least. Ms. McNutt worked for the Bell System until her retirement in 1911.

MCP Master Control Program. A part of the Burroughs operating system that monitors the host's operation.

MCR An ATM term. See Minimum Cell Rate.

MCRIS See Message Customer Record Information System.

MCSE Microsoft Certified Systems Engineer.

MCSP Microsoft Certified Solutions Provider.

MCT Algorithm A compression algorithm introduced in 1986 by PictureTel. MCT reduced the bandwidth necessary to transmit acceptable picture quality from 768 kbps to 224 kbps making two-way videoconferencing convenient and economical at relatively low data rates (for those times).

MCTD An ATM term. Maximum Cell Transfer Delay: This is the sum of the fixed delay component across the link or node and MCDV.

MCU Multipoint Control Unit. A bridging or switching device used in support of multipoint videoconferencing and supporting as many as 28 conferenced sites. The devices may be in the form of CPE or may be embedded in the WAN in support of carrier-based videoconferencing services. MCU standards are defined in ITU-T H.231, with T.120 describing generic conference control functions.

MCVF MultiChannel Voice Frequency.

MD 1. Mediation Device. A SONET device that performs mediation functions between network elements and OSs. Potential mediation functions include protocol conversion, concentration of NE to OS links, conversion of languages, and message processing.

2. Manufacturer Discontinued. A product that the manufacturer no longer makes is called "manufacturer discontinued." Some people think it's a nice way of saying obsolete.

But there are many "obsolete" products that do just fine, often for less money. And I personally find that every time I fall in love with a product and want to buy another one, it's "manufacturer discontinued."

3. Message Display.

4. Message Digest. Refers to a message digest function, or algorithm, used for digital signature applications in the Internet. See also Digital Signature and MD5.

MD5 Algorithm Message-Digest (version) 5 Algorithm. As defined in RFC 1321 from the IETF (Internet Engineering Task Force), MD5 is an algorithm which takes an input message of arbitrary length, and produces an output in the form of a 128-bit "fingerprint" or "message digest." The algorithm is intended for digital signature applications, where a large file must be compressed in a secure manner before being encrypted with a private key under a public-key algorithm such as RSA. It is conjectured that it is computationally infeasible to duplicate an MD5 message, or to produce any pre-specified MD5 message. The Message-Digest function also is known as a one-way hash function. See also Compression, Digital Signature, Encryption, Hashing, Private Key, Public Key, and RSA.

MD-IS Mobile Data Intermediate System.

MD-IS Serving Area A cellular radio term. The set of cells controlled by a single serving Mobile Data Intermediate System.

MD5 Message Digest 5. Algorithm used for message authentication. MD5 verifies the integrity of the communication, authenticates the origin, and checks for timeliness.

MDA Monochrome Display Adapter.

MDBS Mobile Data Base Station.

MDC Meridian Digital Centrex. A Northern Telecom abbreviation.

MDDB MultiDrop Data Bridge. MDDB. A technique for combining data circuits in a computer environment in which the host machine polls other equipment.

MDF Main Distribution Frame. See Main Distribution Frame.

MDI Medium Dependent Interface.

MDK Modem Developer's Kit. Definition invented, I believe, by Microsoft.

MDLP Mobile Data Link Protocol. The Link Layer protocol defined in CDPD networks.

MDMF Multiple Data Message Format. See Caller ID Message Format.

MDN Mobile Dialing Number. The originating telephone number of the cellular caller.

MDPE Medium-Density PolyEthelyne. A type of plastic material used to manufacture jacketing for cable systems.

MDQ Market Driven Quality. An IBM term of the mid-1980s.

MDRAM Multibank Dynamic Random Access Memory. Memory normally used in video boards that boasts extended performance with high bandwidth and short access times. The MDRAM chip can access several memory banks at a time.

MDS 1. An FCC term for a fixed station operating between 2.15 and 2.162 GHz.

2. Multipoint Distribution Service.

MDSI Message Delivery Service Interoffice. Data link using a 9600 baud modem asynchronous 10-bit character transmission. An obsolete term.

MDT Mean Down Time.

MDU 1. Message Display Unit. See Readerboard.

2. Multiple Dwelling Unit. Any housing structure that is broken into more than one living area to accommodate multiple "family" units (apartment buildings, condominiums, duplexes, etc.). It's also called MDU/MTU which stands for multiple dwelling unit/multitenant unit.

MDUs Multiple Dwelling Units. Telephony jargon for high-rise apartment buildings. See also Planned Communities.

MDVC Mobile Digital Voice Channel. The channel between a mobile phone and a cell site antenna in a digital cellular or PCS environment. The MDVC supports both voice and data transmission, although the allocated bandwidth is designed primarily to support voice. Signaling and control functions take place over separate channels set aside specifically for that purpose.

MDWDM Metropolitan dense wave division multiplexing. It is DWDM technology which boosts the carrying capacity of fiber optic telecommunication networks. See DWDM.

ME Mechanical Engineer.

MEA Metropolitan Economic Area. See Metropolitan Statistical Area and MSA.

Meaconing A system for receiving radio beacon signals and retransmitting them on the same frequency to confuse navigation and cause inaccurate bearings to be obtained by aircraft or ground stations.

Mean The sum of all items divided by the number of items, e.g., for the five numbers 7, 7, 8, 10, and 11, the mean is $(7 + 7 + 8 + 10 + 11)/5 = 43/5 = 8.6$. The average (also the arithmetic average) and the mean are the same. What's different is the median. The median of a set is the number that divides the set in half, so that as many numbers are larger than the median as are smaller. The median of 7, 7, 8, 10, and 11 is 8. If the set has an even number of elements, the median is the number halfway between the middle pair. the median is widely used as a measure of central tendency. See several Mean definitions below.

Mean Busy Hour For a telephone line or group of lines or a switch the Mean Busy hour is the 60 minute period where traffic is the greatest.

Mean Deviation An average of all deviations, plus or minus from the mean. It is occasionally used as a measure of dispersion.

Mean Launched Power The average power forma continuous valid symbol sequence coupled into a fiber.

Mean Opinion Score See MOS.

mean power of a radio transmitter The average power supplied to the antenna transmission line by a transmitter during an interval of time sufficiently long compared with the lowest frequency encountered in the modulation taken under normal operating conditions. Normally, a time of 0.1 second, during which the mean power is greatest, will be selected.

mean solar day The yearly average of the elapsed time between two successive crossings of any particular meridian by the Sun.

mean surface temperature The mean temperature of a heavenly body that determines the noise temperature as seen by a satellite antenna pointed at it; typical temperatures: the Earth: 22 degrees centigrade; the Sun: 6,000 degrees centigrate.

mean time between failure MTBF. The average time a manufacturer estimates before a failure occurs in a component, a printed circuit board or a complete telephone system. One must check, since MTBFs are cumulative. MTBF was developed and administered by the U.S. military for purposes of estimating maintenance levels required by various devices and systems. Since accurate statistics require a basis of "failures per million hours of operation," an MTBF estimate on a single device is not very accurate; it would take 114 years to see if the device really had that many failures! Similarly, since the MTBF is an estimate of averages, half of the devices can be expected to fail before then, and half after. MTBF cannot be used as a guarantee. Telecommunications systems operate on the principle of "Availability," for which there is a body of CCITT Recommendations.

mean time between outages MTBO. The mean time between equipment failures or significant outages which essentially render transmission useless. See Mean Time Between Failure.

mean time to repair MTTR. The vendor's estimated average time required to do repairs on equipment.

mean time to service restoral MTSR. The mean time to restore service following system failures that result in a service outage. The time to restore includes all time from the occurrence of the failure until the restoral of service.

measured load The load that is indicated by the average number of busy servers in a group over a given time interval, usually determined with a scanning device.

measured rate A message rate structure in which the monthly phone line rental includes a specified number of calls within a defined area, plus a charge for additional calls. See Local Measured Service.

measured service Also known as USAGE SENSITIVE PRICING (USP). A local phone company method of pricing used to bill local phone calls. Measured service is often charged on the number of calls, the time of day, the distance traveled and the length of the call. See Local Measured Service.

measurement interval Tc (Time committed). A Frame Relay term defining the interval of time which the carrier uses to measure data rates that burst above the CIR (Committed Information Rate). See also Committed Information Rate.

meat puppet An online identity created solely to stuff a wiki or usenet ballot box. The term is an extension of sock puppet, a pseudonym created by a usenet member to second one's own opinion. See also sock puppet.

meatwagon A slang term for an ambulance.

Meatware People. See also Vaporware.

MEC Mobile ECommerce. See 3G.

MECAB Multiple Exchange Carrier Access Billing.

MECAL Master Event Calendar.

mehrfrequenzverfahren German for tone dialing.

MELCAS Mercury Exchange Limited Channel Associated Signaling. A family of signaling techniques used by Mercury Communications Limited (MCL), a service provider in the U.K. MELCAS is used for signaling over E&M and FX trunks, for example. See also CAS, E&M, FX, and MCL.

MECCA Multiplex Engineering Control Center Activity.

mechanic A programmer.

mechanical equipment room A room serving the space needs for HVAC and other building systems other than telecommunications equipment. These are often special-purpose rooms.

mechanical hold A very basic line-holding mechanism used on simple two- and three-line phones that operated by placing a short circuit or a resistor across one phone line while talking on another. Chief disadvantage was that a call put on hold at one phone could not be taken off hold at another phone. Inexpensive multi-line phones with electronic holds largely replaced mechanical holds in the 1980s.

Mechanical Loop Test MLT. Also called Direct-Access Test Unit (DATU) added or built into a central office switch. With MLT a technician can execute tests for shorts, opens and grounds remotely. The technician gets a digital voice, enters a password and is given a series of options. The technician can get results as a digital recording or through an alphanumeric pager. MLT units can send a locating tone as TIP, RING or a combination of both. The unit can short lines and remove battery voltage for testing.

mechanical splice A splice in which conductors are spliced together mechanically. Copper conductors are mechanically spliced using connectors or clips that are crimped to join them. Glass optical fibers are mechanically spliced by being inserted into a plastic or metal splice loaded with index matching gel and then being crimped in place. Glass optical fibers also can be fusion spliced, i.e., melded together, which yields much improved performance. See also fusion splicing and index matching material.

mechanical strength Mechanical strength refers to the capacity of a network element to endure various physical forces. The term commonly is used with respect to transmission media such as coaxial cable, twisted pair, and optical fiber. Mechanical strength includes flex strength, tensile strength, break strength, and bend radius. See those terms for more detail.

mechanical stripping Removing the coating from a fiber using a tool similar to those used for removing insulation from wires.

Mechanized Calling Card Service MCCS was formerly known as ABC Service. MCCS is a central office switch feature that automatically bills credit card calls made on DDD (direct distance dial) rates without the involvement of an operator.

mechanized loop testing MLT. The system provides computer control of accurate and extensive loop testing functions in the customer contact, screening, testing, dispatch and closeout phases of trouble report handling. It also provides full diagnostic outputs instead of just pass/fail indications.

mechanized loop test MLT. A test system that tests the end user's loop, which is comprised of the wires and equipment used to provide dial tone/calling service to that end user.

mechatronics From MIT's Technology Review, "To improve everything from fuel economy to performance, automotive researchers are turning to "mechatronics," the integration of familiar mechanical systems with new electronic components and intelligent-software control. Take brakes. In the next five to 10 years, electromechanical actuators will replace hydraulic cylinders; wires will replace brake fluid lines; and software will mediate between the driver's foot and the action that slows the car. And because lives will depend on such mechatronic systems, Rolf Isermann, an engineer at Darmstadt University of Technology in Darmstadt, Germany, is using software that can identify and correct for flaws in real time to make sure the technology functions impeccably. "There is a German word for it: grOxC5ndlich," he says. "It means you do it really right." In order to do mechatronic braking right, Isermann's group is developing software that tracks data from three sensors: one detects the flow of electrical current to the brake actuator; a second tracks the actuator's position; and the third measures its clamping force. Isermann's software analyzes those numbers to detect faults – such as an increase in friction - and flashes a dashboard warning light, so the driver can get the car serviced before the fault leads to failure. "Everybody initially was worried about the safety of electronic devices. I think people are now becoming aware they are safer than mechanical ones," says Karl Hedrick, a mechanical engineer at the University of California, Berkeley. "A large part of the reason they are safer is you can build in fault diagnoses and fault tolerance."

MECOD Multiple Exchange Carriers Ordering and Design.

medradio A new service for advanced medical radio communication ("MedRadio") devices in the 401-406 MHz band. In establishing MedRadio, the FCC noted that an ever-increasing number of medical devices are coming to rely upon radio transmissions for critical aspects of their functionality. In its Notice of Proposed Rule Making, the FCC proposed designating an additional two-megahertz of spectrum for these devices, at 401-402 MHz and 405-406 MHz, adjacent to the existing Medical Implant Communications Service (MICS) band at 402-405 MHz, for a total of 5 megahertz specifically designated for medical device radiocommunications. Underscoring the flexibility and scope of potential uses under this new service, the FCC proposed to revise its nomenclature and designate the entire 401-406 MHz band as MedRadio service. To accommodate a wider variety of devices than the current MICS service, which is limited to use of implant devices, the FCC proposed allowing the use of body-worn transmitting devices in the MedRadio service. The FCC also proposed increased flexibility for the newly designated 401-402 MHz and 405-406 MHz bands to allow the use of low power, low duty cycle MedRadio devices without requiring the frequency agility capability required by the current MICS rules.

media In the context of telecommunications, media is most often the conduit or link that carries transmissions. Transport media include coaxial cable, copper wire, radio waves, waveguide and fiber.

media access control MAC. The real term is Medium Access Control. But some naughty people call it, incorrectly, Media Access Control. See Medium Access Control for a full explanation.

media access control convergence functions MCF. Media Access Control Convergence performs functions or processes which map information received from IEEE 802.2 Layer 2 LLC into a format acceptable to the lower layer medium.

media compatible Here's an old definition: Usually used to refer to floppy disk media. Even though two different computers (e.g. an AT&T PC 6300 and an Apple IIe) both use 5 1/4 inch floppy disks, the information recorded on them is recorded in a different format and thus, they are not media compatible. You can put one disk in another's machine. But it won't work. You'll get a dumb error message. Updated, the newer DVD formats of Blu-Ray and HD-DVD look physically the same but are definitely not media compatible.

media converter A device used to convert from one transmission media to another. (Examples: copper-to-fiber media converter, 10BaseT-to-10BaseFL media converter.) Converters are usually externally powered as they physically "repeat" or regenerate the signal.

media filter A device used to convert the output signal from a token-ring adapter board to work with a specific type of wiring. For example, a media filter can link 16Mbps token-ring network interface cards with unshielded twisted-pair (UTP) wiring, thus saving the expense of additional cable runs.

media gateway A media gateway (also called media gateway controller) is a fancy name for the new "central office" of the new IP-based telecom industry. Think of it as third generation central office. The first generation was when all phone calls were analog and central offices were electromechanical. The central office's job was to connect one phone conversation through the network to the party at the other end. They called it circuit switched. And it was all very mechanical. Every conversation had a circuit dedicated to it while the conversation took place. The second generation occurred when the industry turned digital. At some point in its progress, each phone conversation was converted to a stream of 64,000 bits per second. The central office's function then became to draw out one digital conversation from a stream of many and connect that one conversation to the person at the other end. In this generation, the signals that controlled each conversation – told it where to go, whom to charge, etc. – were now carried in a separate network, called SS7 (Signaling System 7). In the 1990s, the world changed again. The Internet appeared, first as a way of joining University computers, then as an crude email system, then the world wide web took hold and everything Internet exploded. And the Internet got bigger than the traditional Public Switched Telephone Network (PSTN). The Internet is a very different animal to the phone network. It's packet switched. That makes it orders of magnitudes more efficient than the dedicated-circuit PSTN. As things you could exploded on the Internet and as the price of doing them plummeted, the thought occurred to the traditional phone industry that it would make enormous sense if somehow the PSTN morphed into the Internet, gaining both the Internet's efficiencies and the PSTN's ubiquity. But to merge the two (or at least bits and pieces of the two), you needed a third generation "switch" that will handle all the various telecom streams – from packet switched IP (Internet Protocol) to traditional TDM (Time Division Multiplexed) streams and all those in between. The media gateway controlled is that switch. It's also called media gateway, a call agent or a softswitch, to reflect the fact that its switching is done in software, not in hardware as in previous switching generations.

Telecom carriers of all ilk use softswitches to support converged communications services by integrating SS7 telephone signaling with packet networks. Using network processors at its core, softswitches can support IP, DSL, ATM and frame relay in the same unit. According to the International Softswitch Consortium, a softswitch should be able to (1) control connection services for a media gateway and/or native IP endpoints, (2) select processes that can be applied to a call, (3) provide routing for a call within the network based on signaling and customer database information, (4) transfer control of the call to another network element, and (5) interface to and support management functions such as provisioning,

fault, billing, etc. The switching technology in a softswitch is in software (hence its name) rather than in the hardware as with traditional switching center technology. This software programmability allows it to support existing and future IP telephony protocols (H.323, SIP, MEGACO, etc.). See also media gateway control protocol and softswitch.

Media Gateway Control Protocol First please read the definition for media gateway, then return to this definition. The media gateway control protocol (MGCP) is a protocol for the control of Voice over IP (VoIP) calls by external call-control elements known as media gateway controllers (MGCs), Softswitches, or call agents (CAs). Media Gateway Control Protocol (MGCP), also known as H.248 and Megaco, is a standard protocol for handling the signaling and session management needed for managing telecom "conversations" in different formats. The protocol defines a means of communication between a media gateway, which converts data from the format required for a circuit-switched network to that required for a packet-switched network. MGCP can be used to set up, maintain, and terminate calls between multiple endpoints. Megaco and H.248 refer to an enhanced version of MGCP. The MGCP standard is endorsed by the Internet Engineering Task Force (IETF) as Megaco (RFC 3015) and by the Telecommunication Standardization Sector of the International Telecommunications Union (ITU-T) as Recommendation H.248. H.323, an earlier UTI-T protocol, was used for local area networks (LANs), but was not capable of scaling to larger public networks. The MCGP and Megaco/H.248 model removes the signaling control from the gateway and puts it in a media gateway controller, which can then control multiple gateways. MGCP was itself created from two other protocols, Internet Protocol Device Control (IPDC) and Simple Gateway Control Protocol (SGCP). Defined in RFC 2705, the MGCP specifies a protocol at the Application layer level that uses a master-slave model, in which the media gateway controller is the master. MGCP makes it possible for the controller to determine the location of each communication endpoint and its media capabilities so that a level of service can be chosen for all participants. The later Megaco/H.248 version of MGCP supports more ports per gateway, as well as multiple gateways, and support for time-division multiplexing (TDM) and asynchronous transfer mode (ATM) communication. See also DOSA, Media Gateway, Megaco, Simple Gateway Control Protocol. and softswitch.

media hub A fancy name for a server which doles out audio, video and photographs to various devices around the house. The idea is the media hub will be connected to the Internet, so it can download movies, songs, photographs, magazines and newspapers for viewing, reading and listening to all around the house.

media independent interface MII. A part of the Fast Ethernet specification. The MII replaces 10Base-T Ethernet's Attachment Unit Interface (AUI), and is used to connect the MAC layer to the physical layer. The MII establishes a single interface for the three 100Base-T media specifications (100Base-TX, 100Base-T4, and 100Base-FX).

media interface connector An optical fiber connector which links the fiber media to the FDDI node or another cable.

media path Same as wire run. The means by which telephone signals are conveyed from the Network Interface Jack to the Communications Outlet.

media processing The processing of transactions during a telephone call; these transaction may include fax operations, speech recognition and synthesis, Touch Tone recognition, voice and fax store-and-forward messaging, and the conversion of messages from one format to another (such as from text to voice, or fax to text).

The DM/V-A multifunction and resource series boards provide up to 120 channels of continuous speech processing, conferencing, or other media processing features such as:

- silence compressed record
- message storage using G.7110xCA-law or A-law pulse code modulation (PCM) or OKI adaptive differential pulse code modulation (ADPCM)
- compressed recordings using True Speech, GSM, and G.726 low-bit rate coders
- automatic gain control (AGC) to automatically adjust the signal level of incoming calls
- application's ability to dynamically switch sampling rate and coding method to optimize data storage and voice quality
- sampling rates and coding methods that are selectable on a channel-by-channel basis
- dynamic adjustment of playback volume
- detect dual-tone multifrequency (DTMF) to control record and play functions
- local echo cancellation techniques to improve DTMF cut-through and talk off/play off
- voice player and recorder resources are linked with the DTMF
- 240 channels of basic voice and up to 1200 ports in the system

media processor A special microprocessor whose job is to perform processing for multimedia devices, e.g. videophones, audio, computer telephony devices, voice recognition and 3-D, while the computer's main microprocessor (e.g. a Pentium) handled the basic processing and input/output (I/O) processing. Such media processors might have all the characteristics of digital signal processors and then some. See also MMX.

media server 1. A media server is a fancy name for a file server on a network which contains files containing voice, images, pictures, video, etc. In short, a media server is a repository for media of all types. Basically they're called media servers because they serve up media to anyone who asks for it, is on the network and is authorized to get it. Media servers are also called file servers.

2. The IP telephony community refers to a media server as a slave IP call processing and handling device. Several companies make a combination hardware/software tool which, among other things, collects digits, plays announcements, and establishes IVRs (interactive voice response).

media service instance A logical server providing access to media services (e.g., accessing a datastream, sending and receiving faxes, playing & recording sounds, engaging other VRU services) that can be associated with a call through a Media Access Device. See MediaStreamID.

media stream The information content carried on a call-that is, what actually is transmitted and received over the line, and can, with the necessary hardware, be read and written by a media stream API.

media StreamID Allows an association to be established between a given call and Media Services available on a Media Service Instance that can be associated with the call through a Media Access Device. See Media Service Instance.

media type A call's media type describes what type of information the call is carrying, such as data or voice. A computing domain can use this information, for example, to route the call to a more appropriate computing domain, such as a data computing domain for an incoming data call.

median The average and the mean are the same. What's different is the median. If you say the median price earnings ratio of the S&P 500 was 15.9, that means that half of all stocks traded below that valuation and the other half traded above it. See Mean for a full explanation.

mediated access A Verizon definition. Mediated access allows CLECs to use Verizon network facilities, switches, and operating systems while prohibiting access to proprietary Verizon databases.

mediation system A wireless telecommunications term. A mediation system provides for three functions for transporting data from one device to another. These include protocol conversion, message routing, and store-and-forward processing.

Medical Implant Communications Service MICS. See Personal Radio Services.

medium 1. The material on which data is recorded; for example, magnetic type, diskette.

2. Any material substance that is, or can be, used for the propagation of signals, usually in the form of modulated radio, light, or acoustic waves, from one point to another, such as optical fiber, cable, wire, dielectric slab, water, air, or free space.

medium access control MAC. The IEEE sublayer in a LAN (Local Area Network) which controls access to the shared medium by LAN-attached devices. In the context of the OSI Reference Model, the MAC layer extends above to the Data Link Layer (Layer 2), and below to the Physical Layer (Layer 1). Within the MAC sublayer are defined Data Link Layer options which specify the basis on which devices access the shared medium, and the basis on which congestion control is exercised. Defined at the Physical Layer are media options such as UTP (Unshielded Twisted Pair) CAT 3, 4, and 5 (Categories of wire); STP (Shielded Twisted Pair); fiber optic cable; and wireless radio and infrared. Specific IEEE MAC standards are defined for LANs such as CSMA/CD (802.3), Token Passing Bus (802.4), Token Passing Ring (802.5), Metropolitan Area Networks (802.6), and Wireless LANs (802.11). The MAC sublayer works in conjunction with the Logical Link Control Layer of the IEEE model; at this higher level are defined specific LLC conventions such as frame format and addressing.

medium attachment unit A device used in a data station to couple the data terminal equipment (DTE) to the transmission medium.

medium dependent interface MDI-X. The physical components of a network interface which handle the electrical or optical connection to a cable. This includes the connector, transceivers, and other physical layer components. MDI-X refers to a physical connection which includes an internal crossover of the transmit and receive signals. All standard repeater ports are MDI-X and are often marked with just an X by the port. Some repeater ports are changeable to a DTE port. In this case, the port is changed to a MDI port for connection to a MDI-X port on another repeater. An example of a DTE port is the connection on a NIC.

medium earth orbit MEO. Medium (or Middle) Earth Orbit satellites orbit the earth at distances between LEOs and GEOs (Low Earth Orbit satellites and Geosynchronous

Earth Orbit satellites). Because MEOs operate at heights greater than LEOs, they have larger footprints, or areas of coverage, so that fewer satellites are needed to provide complete coverage over the earth. The planned Odyssey system, which will have 12 satellites, is one MEO with complete world coverage. See also LEO, MEO, GEO and Geosynchronous.

medium frequency MF. Radio frequencies from 300 KHz to 3000 KHz.

medium interface connector MIC. In LAN/MAN systems, the connector at the interface point between the bus interface unit and the terminal, termed the medium interface point.

MEE Multiplex Equipment Engineering.

meet-me conference A teleconferencing term. Meet-Me Conferencing is an arrangement by which you can dial a specific, pre-determined telephone number and security access code to join a conference with other participants. You are automatically connected to the conference through a conference bridge. Conference participants may call in at a preset time or may be directed to do so by a conference coordinator. Meet-Me Conferences may be set up through a teleconferencing service provider, generally with the capability to conference thousands of participants. It also can be provided through a phone system, such as a PBX, key system or hybrid. Some phone systems restrict this to intercom circuits only. In almost all phone systems there is a maximum number of parties that can be connected in such conference at one time.

meet-me intercom conference Dial a special number ("access code") and any telephone can join an intercom conference call.

meet-me page A feature which allows a person to answer an intercom page from any phone in the system.

meet point A location at which the facilities of two carriers connect.

meet-point billing MPB. The process whereby two or more local exchange carriers jointly provide to a third party the transport element of a Switched Exchange Access Service to a local exchange carrier end office switch. Under this process, each local exchange carrier receives an appropriate share of the transport element revenues, as defined by their effective Exchange Access tariffs.

meet-point billing traffic Refers to traffic that is subject to an effective Meet Point arrangement. Also see Meet-Point Billing.

meg A motionless electromagnetic generator is a device patented by Thomas Bearden (see www.cheniere.org) which extracts energy from the vacuum and converts it to electricity. This technology will be very important in the telecom industry for allowing devices to operate without using the national electric grid.

mega A prefix meaning one million, also represented as an M. A megabit equals one million bits, Megahertz equals one million cycles per second, but Megabyte equals 1,048,576 bytes. See also Megabyte.

megabit One million bits. See Bit.

megabits per second A measurement of speed indicating that one million bits of information travel past one point in the circuit in one second.

megabuck Channel surfing television one Saturday morning, I came across a program that purported to teach me how to hunt large male deer with large horns. The announcer was waxing enthusiastically on a piece of property in central Illinois which had many large male deer. He referred to it as "megabuck country." He showed a video of a large male deer and called it "the buck of a lifetime." The hunter, fortunately, missed his arrow shot at the deer, which lived to see another day and to start in another TV show. See also buck and pump and dump.

megabyte MB. A combination of the Greek "mega," meaning "large," and the English "bite," meaning "a small amount of food." A unit of measurement for physical data storage on some storage device – hard disk, optical disc, RAM memory, etc. The actual definition can be confusing, since there are two measurements. In the metric system, a megabyte is 1,000,000 bytes. In the computer world, things are measured in binary terms – 1s and 0s. In binary terms, a megabyte is 2 to the 20th power, i.e., 1,048,576 bytes, which is the closest power of two to one million, .i.e 1024 x 1024. Another way of stating Megabyte is Kilobyte (KB) multiplied by 1,024. A megabyte can be either a decimal (metric) megabyte or a binary megabyte, depending on the context. A decimal megabyte also called a "millionsbyte" or a "miobyte," and is used to describe capacity in newer ROM BIOS drives. Binary megabytes are used to describe capacity in DOS FSISK, Windows 3.x File Manager, and CMOS setup in older ROM BIOS drives.

Here is a summary of sizes:

MB = Megabyte (2 to the 20th power)
GB = Gigabyte (2 to the 30th power)
TB = Terabyte (2 to the 40th power)
PB = Petabyte (2 to the 50th power)

EB = Exabyte (2 to the 60th power)
ZB = Zettabyte (2 to the 70th power)
YB = Yottabyte (2 to the 80th power)
One googolbyte equals 2 to the 100th power.

Megaco MEdia GAteway COntrol. Designated by the ITU-T as H.248, Megaco is a low-level device protocol for interfacing the circuit-switched PSTN (Public Switched Telephone Network) with various packet networks, primarily in support of packet voice. Megaco is an evolution of MGCP (Media Gateway Control Protocol) that expands the range of packet network options to include ATM, in addition to IP-based networks, and that includes simplified provisions for signaling and control for conferencing applications, including voice conferencing and whiteboarding. Megaco links the Media Gateway (MG) and Media Gateway Controller (MGC) for intradomain remote control of connection-aware or session-aware devices, or endpoints. See also MGCP.

megaflops Million Floating point Operations Per Second. A measure of computing power usually associated with large computers. Mega means million. Also known as MFLOPS.

megahertz 1. MHz. A unit of frequency denoting one million Hz or one million cycles per second. See Bandwidth and Hertz.

2. What the Vermonters' Guide to Computer Lingo says you get when you're not careful downloading, which it defines as taking firewood off the pickup truck.

megalink Name for BellSouth's leased T-1 service.

megaplex Also known as a multiplex. A collection of different movie theaters in the same building, many playing different movies.

megaserver A large server. Possibly a Microsoft-created term. The idea is that people may keep much of their personal and professional information on large servers. The concept is that a person will be able tap into a large central database via the Web to get e-mail, personal schedules, news, weather updates and other information anywhere, anytime. See Server.

megastream British Telecom's brand name for a service of 30 64-Kbps channels (i.e. E-1).

mego Mine eyes glazeth over. It happens when the instructor gets very boring in his lecture and his audience is just about to fall asleep.

megohm A resistance of 1,000,000 ohms. See Ohm.

meme A meme is a type of online chain letter where bloggers answer questions or participate in a quiz designed to give a quick overview of the author's personality. Once the author completes the meme, it is customary to tag other bloggers to participate. This definition from blogossary.com.

memenuked Sudden and overwhelming media attention generated when an idea takes off globally. After he coined the term warchalking, London Web designer Matt Jones apologized for not keeping up with his blog "due to festivities arranged before my life got memenuked." This definition came from Wired Magazine.

memex In the 1930s, Vannevar Bush first wrote of the memex – a pocket device that sounds very much like a wireless Palm – in the 1930s. It didn't gain much attention until his seminal 1945 article, "As We May Think," published in the Atlantic Monthly. Here are his words from the article.

"Consider a future device for individual use, which is a sort of mechanized private file and library. It needs a name, and, to coin one at random, "memex" will do. A memex is a device in which an individual stores all his books, records, and communications, and which is mechanized so that it may be consulted with exceeding speed and flexibility. It is an enlarged intimate supplement to his memory. It consists of a desk, and while it can presumably be operated from a distance, it is primarily the piece of furniture at which he works. On the top are slanting translucent screens, on which material can be projected for convenient reading. There is a keyboard, and sets of buttons and levers. Otherwise it looks like an ordinary desk. In one end is the stored material. The matter of bulk is well taken care of by improved microfilm. Only a small part of the interior of the memex is devoted to storage, the rest to mechanism. Yet if the user inserted 5000 pages of material a day it would take him hundreds of years to fill the repository, so he can be profligate and enter material freely. Most of the memex contents are purchased on microfilm ready for insertion. Books of all sorts, pictures, current periodicals, newspapers, are thus obtained and dropped into place. Business correspondence takes the same path. And there is provision for direct entry. On the top of the memex is a transparent platen. On this are placed longhand notes, photographs, memoranda, all sorts of things. When one is in place, the depression of a lever causes it to be photographed onto the next blank space in a section of the memex film, dry photography being employed. There is, of course, provision for consultation of the record by the usual scheme of indexing. If the user wishes to consult a certain book, he

taps its code on the keyboard, and the title page of the book promptly appears before him, projected onto one of his viewing positions. Frequently-used codes are mnemonic, so that he seldom consults his code book; but when he does, a single tap of a key projects it for his use. Moreover, he has supplemental levers. On deflecting one of these levers to the right he runs through the book before him, each page in turn being projected at a speed which just allows a recognizing glance at each. If he deflects it further to the right, he steps through the book 10 pages at a time; still further at 100 pages at a time. Deflection to the left gives him the same control backwards." See also Internet.

memo 1. A telephone feature that enables the user to store a phone number for calling in the future. For example, while speaking to a Directory Assistance operator, you can put the number she gives you into memory, and then call that number by pushing one or two buttons.

2. A call center term. A free form field used to store descriptive text or comments. The information in a memo field can be of any length and type.

Memorandum of Understanding See MOU.

memory The part of a computer or sophisticated phone system which stores information or instructions for use. Memory comes in many variations. There is memory which is lost when the power is switched off. There is memory which is retained when power is turned off.

memory administration MA. A set of functions that provide network system database updates, network system database integrity, network system database security and network system database backup and restoration. Definition from Bellcore (now Telcordia) in reference to its concept of the Advanced Intelligent Network.

memory board An add-on board designed to increase a computer's RAM.

memory caching A technology for increasing hardware performance by storing frequently used sequences of instructions in a memory cache separate from the computer's main memory where they can be more quickly accessed by the CPU.

memory call service A family of central office based voice messaging services from BellSouth.

memory cards The memory card is a bunch of memory chips crammed into a small plastic cartridge about the size of a credit card and about three times the thickness. It is used in several palmtop computers. As this dictionary was being written, we were awaiting the release of a 16 megabyte memory card. In contrast to flash memory, a memory card requires small batteries, typically the same ones as used in watches. See Smart Card.

memory effect The gradual shorting of a battery's useful life, caused by recharging before the battery is completely discharged. This is a real problem with nickel cadmium batteries, less of a problem with Nickel Hydride and even less of a problem with Lithium ion batteries.

memory interface A PCMCIA definition. The memory interface is the default interface after power up, PCMCIA Hard Reset and PCMCIA Soft Reset for both PCMCIA cards and sockets. This interface supports memory operations as defined in PCMCIA Release 1.0 and later and is used by both Memory Cards and I/O Cards.

memory leak In order to run, a computer program requests chunks of memory from the operating system for itself and for the data it needs. As the program runs, it may make additional requests for memory. If the program is well-written and, therefore, well-behaved, that memory will be explicitly released when the program is closed. A badly written program or a badly written operating system may not release that memory, leaving it in an unusable condition. This may be especially true if the program crashes (i.e., ends unexpectedly), which is common in poorly-written software. This phenomenon is known as a "memory leak," and it's not a good thing. The longer the software runs on your computer, the more and more of your computer memory it uses, and places in an unusable state. Eventually it uses up so much of your RAM memory, it starts using space on your hard drive. Soon, it runs out of hard disk space. Eventually your whole machine freezes. You are forced to reboot, losing all unsaved data in the process. There are two solutions: The first and most obvious is to reboot your computer, i.e. turn it off, wait ten seconds, then turn it on again. Some operating systems handle memory leak much better than others. Windows 2000 and Windows XP handle it much than Windows 95 and 98. There is another solution. See Garbage Collection.

memory map An indication of what type of data is stored where in a computer's RAM memory.

memory moments See Senior Moment.

memory protection The structuring of memory resources in Novell's NetWare 4.0 that guards the NetWare server memory from corruption by NLMs. Memory protection allows you to run NLMs in a separate memory domain called the OS PROTECTED domain. Once you determine the NLM to be safe, you can load it into the OS domain, where it can run most efficiently.

memory reserve power The operating voltage, generally provided by a battery, which supplies power to the memory modules when your commercial power fails. You should check your memory reserve power before it's too late. You should test it even when you don't need it.

memory technology driver A PCMCIA card definition. A memory technology driver is a memory device specific software that interfaces to Card Services to mask the details of accessing different memory technologies.

MEMS See MicroElectroMechanical System.

MEN See Metro Ethernet Network.

menu Options displayed on a computer terminal screen or spoken by a voice processing system. The user can choose what he wants done by simply choosing a menu option – either typing it on the computer keyboard, hitting a touchtone on his phone or speaking a word or two. There are basically two ways of organizing computer or voice processing software – menu-driven and non-menu driven. Menu-driven programs are easier to use but they can only present as many options as can be reasonably crammed on a screen or spoken in a few seconds. Non-menu driven screens allow more alternatives but are much more complex and frightening. It's the difference between receiving a bland "A" or "C" prompt on the screen – as in MS-DOS and receiving a menu of "Press A if you want Word Processing," "Press B if you want Spread Sheet," etc. It's very easy to write menus in MS-DOS using BATch files. See also Audio Menus and Prompts.

menu bar call center term. The part of the menu system visible as a single row across the top of the display.

MEO Middle (or Medium) Earth Orbiting satellite. MEOs operate much like LEOs (Low Earth Orbiting satellites systems), although in slightly higher orbits. MEOs generally operate at an altitude of around 10,000 kilometers. Continuous global communications services can be achieved with six to 12 satellites. MEOs are capable of supporting both voice and data services. Contrast with GEO, Geosynchronous and LEO. See also Medium Earth Orbit.

MEP Multiplex Equipment Provisioning.

merced Merced is a 64-bit multichip module jointly developed by Intel with Hewlett-Packard Co. It extends the Intel architecture in both raw speed (one version will reach 600-MHz) and overall performance. Merced will be followed by the two-chip processor called Flagstaff in the year 2000. Flagstaff chips will be the first built using a process that creates much smaller 0.18-micron-wide circuits, which will enable Intel to build smaller and faster chips in greater volume. There will be two versions of Flagstaff, with a choice of 4 Mbytes or 8 Mbytes of secondary cache, according to writer, Tom Davey.

merchant silicon Merchant Silicon is a special microprocessor chip, which is a non-ASIC, commercially available semiconductor chip.

mercury-wetted relay A relay or switch in which the movable contact of the device makes contact with a pool of mercury. Before solid-state devices, this was common technology for switching dry circuits. See also Dry Contact.

merging traffic The telecommunications, cable, consumer electronic and media conglomerates all vying for access to the same markets. See Siliwood.

meridian A Northern Telecom (now called Nortel) name for a family of PBXs.

Meridian Link Meridian Link from Northern Telecom enables Meridian 1 Communication Systems (i.e. PBXs and ACDs) to exchange information with host computers and the application software that resides on those computers. This means that the application software can use information such as Automatic Number Identification (ANI) and Dialed Number Identification System (DNIS) to automatically perform a series of routines, such as record look-up.

meridians Lines circling the earth from pole to pole which are used to measure distance (longitude) around the globe.

meridional ray In fiber optics, a ray that passes through the optical axis of an optical fiber. This contrasts with a skew ray, which does not.

MERIT The successor to NSFNET, MERIT originally was a statewide IP network operated by the University of Michigan. It also was a substantial regional subnetwork (subnet) of the NSFNET and the Internet. MERIT provides access into the Internet through MAEs (Merit Access Exchanges) located in San Jose (MAE West) and Vienna, Virginia (MAE East); those points of access actually are provided in partnership with MFS Datanet. See MAE.

MERS Most Economical Route Selection. A term used by GTE and some other PBX manufacturers to mean Least Cost Routing. See Least Cost Routing.

MES Mobile Earth Station, the generic satcom term for a mobile terminal.

mesh Network architecture in which each node has a dedicated connection to all other nodes.

mesh connectivity A Wide Area Network (WAN) term for connectivity over a mesh network, wherein each site is directly connected with every other site. See Mesh Network for a much fuller explanation.

mesh network Imagine a bunch of network nodes – where users get on and off a network. The nodes are perhaps buildings, perhaps towers on mountain tops, perhaps antennas on telephone poles, or perhaps just wireless devices on various floors of a building. Imagine now that basically each of the nodes are joined together by telecom "lines" – wireless or wired. Now you have a mesh network. The advantages of mesh networks include: 1. They're more reliable. One link can crash and the "conversation" (data or voice) can find its way through and around. 2. They're more available, which is another way of saying they're more reliable.

The disadvantages of a full mesh network include high costs, difficulty of configuration and reconfiguration, and lead times associated with carrier provisioning. Mesh networks are relatively easy and inexpensive to configure where four or fewer sites must be interconnected; the cost and complexity increase significantly where more than four sites are involved. As a result, large organizations increasingly tend to favor alternative solutions for voice and data. Such alternatives as fiber ring networks, Virtual Private Networks (VPNs), Frame Relay, SMDS and ATM.

In a March, 2006 issue, the Economist wrote about city-wide Wi-Fi networks. It talked about a company called NeoReach who had installed a wireless network in Tempe, Arizona for use by city government agencies – police, fire, etc. – and also potentially by students and others in the town. The Economist wrote: "As a result, when the city began considering ways to extend broadband access to more residents – including nearly 60,000 students, staff and faculty members at the main Arizona State University campus in the center of town – wireless made a lot of sense, says Mr Heck. "We don't have a lot of competition for broadband in Tempe," he says. "There's just not a lot there." A local provider, NeoReach, won the contract to build a Wi-Fi "mesh" network to provide broadband throughout Tempe, using equipment made by Strix Systems. NeoReach is paying for the network's construction, and will collect access fees from subscribers; the city's administration, police, fire and emergency services will also pay to use the network, which will cost $2.3 million to build. Mesh networking allows large areas to be blanketed with wireless coverage quickly and inexpensively. As its name suggests, a mesh network consists of an array of wireless access points, only a few of which are actually connected back to the Internet via high-speed links (known as backhaul connections). The trick is that all of the access points double as relays, passing packets of data to and from their neighbors. This connects up the mesh, so that users can access the Internet at high speed at any of the access points. If the nearest access point does not have a backhaul connection, the packets of data that users send and receive simply make one or more hops across the mesh. As well as being cheap and fast to set up – partly because many of the access points can be attached to utility poles – mesh networks have several other merits. They can provide coverage in areas, such as sprawling suburbs, where fast copper or fiber-optic connections are hard to come by. "When you get out in the residential areas, there's no fiber," says Chuck Haas of MetroFi, whose company has installed mesh Wi-Fi networks in three of the San Francisco Bay Area's largest suburbs. Mesh networks are reliable, since the failure of one or more access points does not bring down the whole network, and they can also route data around obstacles, such as large buildings, which might otherwise block coverage." See also sensor mote.

meso The Greek prefix meaning the middle.

mesochronous The relationship between two signals such that their corresponding significant instants occur at the same average rate.

message 1. A sequence of characters used to convey information or data. In data communications, messages are usually in an agreed format with a heading which establishes the address to which the message will be sent and the text which is the actual message and maybe some information to signify the end of the message. A Northern Telecom Norstar definition: A message, which appears on the telephone display that informs the recipient to call the person who sent the message. Messages can only be sent within the Norstar system.

2. The Layer 3 information in the OSI model that is passed between the CPE and SPCS for signaling.

3. A SCSA definition. The transport container for SCSA requests, replies and events. Assumes a set of conventions for directing the delivery of the message to the proper entity, either a client or service provider. See also SCSA Message Protocol.

4. A completed call, i.e., a communication in which conversation or exchange of information took place between the calling and called parties.

message alert A cellular phone term, also called "call-in-absence" indicator. A light or other indicator announcing that a phone call came in, an especially important feature if the cellular subscriber has VOICE MAIL.

message alignment indicator In a signal message, data transmitted between the user part and the message transfer part to identify the boundaries of the signal message.

message backbone A single format message transport system designed for the electronic mail and messaging needs of an entire corporation or enterprise.

message center A centralized place within the corporation where messages are taken and (occasionally) delivered. Message centers are good if they are staffed with competent, motivated people and the various phones in the place have message waiting lights (like they do in hotels). If staffed by talented people, message centers work a lot better than the amateur message takers called secretaries and their part-time short-term replacements.

message circuit A long distance telephone phone line used in furnishing regular long distance or toll service to the general public. The term is used to differentiate message circuits from circuits used for private line service.

Message Customer Record Information System MCRIS. A system used by Verizon to receive and interpret central office switch usage records.

message detail recording See Call Detail Recording and Call Accounting Systems.

Message digest One-way hashing algorithm that produces a hash. Both MD5 and Secure Hash Algorithm (SHM) are hashing techniques that enhance security of data transmission.

message format The rules for placing information necessary for an electronic message. The format includes where the heading is and how long it will be as well as other control information.

message frame A SCSA term. A data link layer frame the encapsulates control and signaling data transmitted on the SCbus or SCxbus Message Bus. The form of a Message Bus frame is fully compliant with ISO HDLC UI (Unnumbered Information) Frame specifications. See SCSA.

message handling service 1. MHS. Software whose primary function is the movement of messages between application programs. MHS was developed by Action Technologies. Novell acquired full marketing and development rights to MHS for NetWare based LANs (local area networks). Under MHS, each application sends messages to the server's mhsmailsnd directory. MHS delivers messages to an application by placing them in the application's assigned directory. MHS is most commonly used for e-mail (electronic mail). Various e-mail programs use the MHS format for exchanging e-mail messages. MHS was superseded by GroupWise.

2. MHS also means the standard defined by ITU-T as X.400 and by ISO as Message-Oriented Text Interchange Standard (MOTIS). MHS is the X.400 family of services and protocols that provides the functions for global electronic-mail transfer among local mail systems.

message header The header before a string of data containing information regarding the destination of the data, usually in a packet in X.25 format.

Message Identifier MID. When the ATM Adaptation Layer segments higher layer information for transport in a sequence of ATM cells, the MID is a 10-bit field and is used to associate all the cells that carry segments from the same higher layer packet.

message integrity The notion that a message has not been altered and is, therefore, true to its original intent (a.k.a. data integrity).

message management A new term for managing all your voice mail, fax mail and electronic mail. The major concept is to join the control of the devices that produce voice, fax and e-mail through your desktop PC connected over your LAN. For example, you come into work in the morning, turn your PC on, and immediately see a screenful of messages – one line per message. By clicking on that line with your mouse, your PC would pull up the application that will then let you read or hear your message. The benefit is that you can handle your messages faster and be more discriminating about which ones you pay most and least attention to.

message packet A unit of information used in network communication. Messages sent between network devices (workstations, file servers, etc.) are formed into packets at the source device. The packets are reassembled, if necessary, into complete messages when they reach their destination. A message packet might contain a request for service, information on how to handle the request, and the data that will be serviced. An individual packet consists of headers and a data portion. Additional headers are appended to the data portion as the packet travels through the layers of the communication protocol. Any message that exceeds the maximum size is partitioned and carried as several packets. When the packet arrives at its destination, the headers are stripped off, the message delivered and the request serviced.

message rate A method of billing local phone calls that varies from one place to another. Phone calls are billed as "message units." Message units are a combination of length of call and distance of the call. In a city like New York you might buy "basic" phone service and be entitled to 50 "message units." That may mean you can call the local pizza house for under 5 minutes 50 times. Or it may mean that you can call from Manhattan to the Bronx 25 times – assuming each call is two message units.

message register leads Terminal equipment leads at the interface used solely for receiving dc message register pulses from a central office at a PBX so that message unit information normally recorded at the central office only is also recorded at the PBX.

message registration A phone system feature that records the number of message units incurred by each phone. Useful in hotels which bill local calls by message units.

message retrieval The ability of a fax machine to store material already transmitted so it can be retransmitted.

message signal unit Signal unit of CCS that carries a message corresponding to the information part or packet of the HDLC frame plus a message transfer part corresponding to the HDLC frame header.

message stick An ancient form of communication among aboriginal tribes in Australia. Village elders who wanted to send a message to another tribe painted, carved, or burnt it on a wooden stick. A village messenger then carried the message stick by hand to the tribe or tribes that the message was intended for. By tradition, a messenger carrying a message stick was granted entry into and safe passage through the other tribe's territory. Anyone in a tribe who came across an outside messenger on the tribe's land had an obligation to safely escort the messenger to the tribal elders. After receiving the message, the elders then had an obligation to ensure that the messenger was granted safe passage through the tribe's land, either to return to his own people or to move on to another aboriginal tribe to spread the message further. The a message stick ensured that the same message was communicated consistently to each tribe. Typical messages were announcements of ceremonies, meetings, or other events, or they may contain a warning or grievance.

message store An X.400 electronic mail term: A staging point, similar to a post office, in which messages are temporarily held for later transmission to one or more recipients.

message switch A message switch is another term for an electronic mail gateway, which is a LAN application that fetches messages from one electronic mail system, translates them to the format of another electronic mail system, and then sends them to the "post office" of that other system. The post office is the public entry point – the place you put mail you want the other system to receive.

message switching A technique for receiving a message, storing it for a while and then sending it on. Message switching is normally used when the desired recipient is not there. The message switch will keep attempting delivery, freeing the calling party to handle other work. Unlike voice phone calls no direct connection is made in message switching between the incoming and outgoing messages. Each message, like a Western Union telegram, contains a destination address and is recipient through intermediate nodes. Each node along the way receives the message, stores it briefly, and then passes it on to the next node. Message switching is a forerunner of what is now called packet switching. Only difference is that message switching had no fixed message length.

message switching network A public data communications network over which subscribers send primarily text messages to one another.

Message Telecommunications Service MTS. The regulatory term for long distance or message toll voice service. Misnamed. Actually there's nothing "messagey" about this service. This is a 100% switched telephone service (which can, obviously, be used for voice, data, video or fax).

Message Telephone Service MTS. Official designation for tariffed long-distance, or toll, telephone service. See Message Telecommunications Service.

message toll voice service See Message Telecommunications Service.

Message Transfer MT. The carriage of information between parties using computers as intermediaries. It is one aspect of message handling.

Message Transfer Part MTP. Provides physical, data link, and network layer functions. MTP transports information from the upper layers (including the user parts and SS7 applications) across the SS7 network and includes the network management procedures to reconfigure message routing in response to network failures. Refers to level 1 through 3 in the SS7 protocol stack (MTP1-MTP3).

message transfer agent An X.400 electronic mail term: Software usually residing in a LAN server or host computer that moves messages between senders and recipients.

Message Transfer Part MTP. The part of SS7 signaling node that is used to place formatted signaling messages into packets, strip formatted signaling messages from packets, and send or receive packets.

message unit 1. The charge for one unit of local telephone service. How many message units you will find on your bill is a function of how many calls you made, how far the calls traveled, what time of the day or night you called and for how long you talked.

2. In IBM Corp's Systems Network Architecture, the portion of the data within a message that is passed to, and processed by, a specific software layer.

Message Unit Detail MUD. A service offered by local telephone companies in which they give you a report listing the phone number of all local calls made from each of your billing numbers. The billing number may be the main number for a PBX or the individual extensions if Centrex service is used. MUD reports are usually available to a telephone company customer at additional cost. MUD reports generally have to be requested in advance. Some telephone company central offices cannot generate MUD reports. When available, MUD reports may not be in machine processable form.

message waiting A light on the phone or some letters or characters on the phone's display indicating there's a message waiting somewhere for the owner of the phone. That message might be with a special message center (as in a hotel or a larger company), with the operator or with a computer attached to the phone system or with someone else in the company. Message waiting lights are incredibly useful at hotels. It's amazing more companies don't use them also.

message-to-slave directory propagation In electronic mail on a local area network, message-to-slave directory propagation is a way of updating user addresses where changes in the master post office are sent to the slaves, but changes in the slave post offices are not sent to the master.

messages roses French term for pay telephone pornographic messages. What the English call phone sex lines.

messaging One-to-one communication. Messaging comes in several flavors (not all implemented as yet), including text messaging, audio messaging and video messaging. See also Instant Messaging.

Messaging Application Programming Interface MAPI. A set of calls used to add mail-enabled features to other Windows-based applications. See MAPI.

Messaging Enabled Applications MEA. This term defined by the Electronic Messaging Association, Arlington, VA. Applications that directly access the messaging service as a way of addressing and transporting information to and from objects on a network. Messaging Enabled Applications differ from Mail Enabled Application because they do not require electronic mail to successfully navigate the network. They may or may not have their own directory and also have the ability to directly access Directory Services. For example: A student requests a class via electronic mail to the education system. The education system automatically schedules the student (this an electronic mail enabled application).

messaging middleware Lets applications communicate and exchange data via asynchronous messages and queues.

messenger 1. A piece of heavy metal cabling attached to a pole line to support aerial phone cable. Linear supporting member, usually a high strength steel wire, used as the supporting element of a suspended aerial cable. The messenger may be an integral part of the cable, or exterior to it.

See also Hard Cable and Messengered Cable.

2. See Windows XP Real Time Communications.

messenger service In villages in developing countries around the world, individual residential phone service is rare. Instead, there may be a communal village telephone. In such a scenario, the phone company typically offers messenger service, whereby a messenger runs to the home of the called party to deliver a message to the person or to bring him/her to the village telephone to take the call.

messengered cable Messengered cable is similar to jacketed cable, but includes an integral steel "messenger wire" to provide mechanical support. Messengered cable is intended for aerial installation without strand. See Hard Cable for a much bigger explanation.

MET Multibutton Electronic Telephone for an old phone system called AT&T Horizon.

meta Change, or transformation. From Greek.

meta-signaling In broadband ISDN, some future User-to-Network interfaces may be point-to-multipoint configurations. The devices on this interface may each generate signaling messages directed to the broadband (ATM) switch. In point-to-multipoint configurations, a separate, independent, virtual channel for signaling is required between the network and each of the devices on the multipoint interface, so the switch will always know who sent the message. The meta-signal channel is a pre-established management channel which will be used to establish the point-to-point signaling virtual circuits between the network and

each of the individual devices on a multipoint interface. (narrowband ISDN's Basic Interface uses a similar process on the D-channel called the TEI assignment.) The meta-signaling channel will operate on virtual circuit 1 within each virtual path. Meta- signaling is not required on point-to-point interfaces, because the switch is only communicating with one device! In those cases, the signaling protocol (q.93B) will operate on virtual 5.

Meta Tag A meta tag is an optional HTML tag that is used to specify information about a Web document to search engines crawling the Internet. The Meta Tag appears in the <\<>head> portion of a Web page. Search engines use "spiders" to index Web pages by reading the information contained within the meta tag's code. An HTML or Web page author uses these tags to help his or her page get noticed or "come up" when an Internet surfer queries a search engine for a particular key word or topic. The meta tag can also be used to specify an HTTP or URL address for the page to "jump" to after a certain amount of time. This is known as Client-Pull. This means a Web page author can control the amount of time a Web page is up on the screen as well as where the browser will go next. Here's the HTML syntax for search engine indexing:

```
<HTML>
<HEAD>
<TITLE>Technology Investor Magazine</TITLE>
<META NAME="keywords" CONTENT="Technology, eCommerce, Investing, Investor, Stocks,">
<META NAME="description" CONTENT="This is a magazine about investing in technology stocks.">
</HEAD>
<BODY></BODY>
</HTML>
```

Here's the HTML syntax for Client-Pull:

```
<HTML>
<HEAD>
<TITLE>stopandgo.com</TITLE>
<META HTTP-EQUIV="REFRESH" CONTENT="30;URL=gothere.html">
/HEAD>
<BODY></BODY>
</HTML>
```

This code makes the Web page "refresh" or change to the URL specified in 30 seconds.

metadata Courtesy the New York Times: Metadata, a term created by the fusion of an ancient Greek prefix with a Latin word, has come to mean ''information about information" when used in technology and database contexts. The Greek meta means behind, hidden or after, and refers to something in the background or not obviously visible, yet still present. Data, the Latin term, is factual information used for calculating, reasoning or measuring. A vast amount of information (factual or otherwise) exists on the Internet, in databases, libraries and other repositories. Weeding through all of it to find what you're looking for can be maddening, but using metadata can help. For example, many Web pages have ''metatags" embedded in their underlying code that help Internet search engines home in on keywords taken from the page's content. For a Pennsylvania sports-themed site, a metatag could contain keywords like "football," "Pittsburgh Steelers" or "Terrible Towels." A search engine can scan the Web for these keywords in the page's metatag, and possibly find results more quickly and accurately from the billions of pages available online. You can also see metadata used in other everyday ways. Digital photos can contain exposure information tucked inside the file. Songs encoded as digital audio files with software like Windows Media Player or iTunes can embed text like the artist, album and song title alongside the music data, which lets you see what's playing when you look at the screen of your audio player.

metajacking Stealing the contents of another site's HTML meta tag and using it on your web site because that will place your site higher in a search engine's results. In other words, when people go to find look for something relating to what you do, they will find your site faster than your competitor's.

metatag See meta tag.

Metal Oxide Varistor Metal Oxide Varistor. A voltage dependent resistor which absorbs voltage and current surges and spikes. This low-cost, effective device can sustain large surges and switch in 1 to 5 nanoseconds. It is used as a surge protector and suppressor. It often the first electronic component that electrons coming in on an incoming phone line hit. Many trunk boards inside PBX are protected by MOVs. If the voltage or current is high, it will blow the MOV, thus protecting the remaining the far more valuable devices on the board.

metal rubber Developed by NanoSonic of Blacksburg, Virginia, the mouse-pad-like substance is infused with metal on the nano level. So while it can be stretched and twisted like a rubber band, it retains nearly all the conductivity of metal. Metal rubber is being developed for use within artificial muscles, "smart clothes," and flexibile electronics. In flight, birds can alter the shape of their wings and bodies in order to fly better in varying conditions. What if a plane could do that, too? Such a plane would need to be made from a material that could bend and stretch like rubber, and direct changes in shape by conducting electrical signals, the way metal can.

metal tape Recording tape coated with iron particles and noted for its wide dynamic range and wide frequency response.

metallic circuit A circuit completely provided by metallic wire conductors, and not containing any carrier, radio or fiber and in which the ground or earth forms no part.

metallic voltage A potential difference between metallic conductors, as opposed to a potential difference between metallic conductor and ground.

metallization Metallization is necessary if composites are to be used in antenna applications. Antennas are mainly metallized with copper or gold. Composite systems that lend themselves to metallization are those epoxy and thermoplastic composites with Kevlar, glass or carbon fiber reinforcement.

metallized fiber Basically metallized fiber is a fiber strand with a thin coating of metal all around it. The idea of metallized fiber is to cut down on the leakage of the light signal or signals being carried inside the fiber.

metaphor See Desktop Metaphor.

metasignaling ATM Layer Management (LM) process that manages different types of signaling and possibly semipermanent virtual channels (VCs), including the assignment, removal and checking of VCs.

metasignaling VCs An ATM term. The standardized VCs that convey metasignaling information across a User-Network Interface (UNI).

Metcalfe's Law Named after Robert N. Metcalfe. Metcalfe's Law states that the value of a network – defined as its utility to a population – is V=A*N*N+B*N+C where V is the value and N the number of users of a network. The value of a network grows with the square of the number of its users. Often, for small N, the COST of a network exceeds its VALUE. This means there is a critical mass phenomenon in networks. Therefore, small pilots of some networks might fail where a larger operational network might succeed wildly. It's tough getting them started, but then BAM! off they go. A simple example: One telephone is useless. Two telephones are better, but not much. Only when a good proportion of the population has a telephone does the network gain the power to change the society. Some people estimate that a totalitarian government cannot survive even one telephone per hundred people. Metcalfe's Law pertains most especially to computer networks, as Bob Metcalfe is the co-inventor of Ethernet, the first LAN (Local Area Network), which was standardized as 802.3. According to many observers, the Internet owes its extraordinary growth and impact to its ability to harness both Metcalfe's Law and Moore's Law (which says that computing power and capacity double every 18 months, while costs drop by half). See also Connectivity Law, Data Dialtone, Ethernet, and Moore's Law.

meteor burst communications Communications by radio signals reflected by ionized meteor trails.

meteor scatter propagation The refraction of radio signals off of trails of plasma left in the wake of meteors burning up in the ionosphere. The frequencies that this works best for are between 30 and 100 MHz, however with the development of new software and transmission methods, this technique has worked with some higher frequencies.

meter 1. The metric unit of length, equivalent to 39.37 inches. An instrument for measuring quantities of length. According to the NIST (National Institute of Standards and Technology), the origins of the meter go back to the 18th century, when there were two definitions. Christian Huygens, an astronomer, proposed that a meter would be the length of a pendulum having a period of one second. That definition failed, as the force of gravity varies slightly over the surface of the earth, which affects the period of the pendulum. The successful definition was that of one ten-millionth of the length of the earth's meridian along a quadrant (one-fourth of the circumference of the earth). The French Academy of Sciences made this decision in 1791.

The French, with typical French arrogance (unlike Americans, of course), therefore defined the meter as one ten-millionth of the length of the meridian from pole to equator, through Paris. Despite the fact that this measurement was off by a whopping 0.2 millimeters because of the flattening of the earth due to its rotation, this definition stuck. Over the years, the definition was refined considerably. According to the International System of Units (SI), the current definition of a meter is the length of the path traveled by light in a vacuum during a time interval of 1/299,792,458th of a second.

2. A device which measures network traffic along some parameter, which typically is temporal (e.g., the duration of a connection or session), quantitative (e.g., number of data packets), or qualitative (e.g., the level of conformance of a packet or packet stream to a pre-defined traffic profile) in nature. The meter commonly will retain in memory traffic statistics based on those measurements, and upload them to a centralized network management or other support system on command.

Metered T-1 See Fractional T-1.

metering pulses In virtually all foreign countries, periodic pulses are returned from the distant exchange to the exchange (central office) of the calling number. These pulses determine the cost of the call, local or long distance. Typically, all pulses cost the same. However, the farther you call, the quicker the pulses come. This system contrasts to the North American long distance billing scheme which typically charges a certain amount for the first minute – no matter how much of the minute is actually used in conversation. After the first minute in the U.S., conversations are billed in one minute increments. Overseas pulses can be as short as three or four seconds (especially for international calls).

method An SCSA definition. The specific implementation of an operation for a class; code that can be executed in response to a request.

metric A measurement. A benchmark. You might measure your marketing program by applying the metric of how many mailing pieces you send out each day. You might measure the value of your web advertising by how many "click throughs" you get each day. You need metrics to see if you're succeeding. Metrics originally meant the application of mathematical analysis to a field of study – for example, econometrics. In its current voguish sense, it is a hifalutin word for measurement. See also Service Level Agreement.

Metro Ethernet Network MEN. Metro Ethernet Network is a way to connect buildings on the Internet like desktops within a building. Its advantages include relatively simple scalability, due to its packet-based technology. Standards compliant interfaces are available for data communication/telecommunication devices at line rates of 10/100/1000 Mbps, and the draft-standard for 10 Gbps has been ratified. An Ethernet-based Metropolitan Area Network is generally termed a Metro Ethernet Network. Some European service providers have also introduced MEN-like technology for Wide Area Networks. In enterprise networks, Metro Ethernet is used primarily for two purposes: connectivity to the public Internet and connectivity between geographically separate corporate sites - an application that extends the functionality and reach ability of corporate networks. See Metropolitan Area Network.

metro network See Metropolitan Area Networks.

Metropolitan Area Network MAN. A loosely defined term generally understood to describe a network covering an area larger than a local area network (LAN), but less than a wide area network (WAN). A MAN typically interconnects two or more local area networks, may operate at a higher speed, may cross administrative boundaries, and may use multiple access methods. While MAN is a data term, a MAN may carry data, voice, video, image, and multimedia data. The only true MAN technology is SMDS, which, in fact, is limited to the MAN.

metropolitan dial The common rotary dial or touchtone pad that contains both numbers and letters. Dials and pads are also available without the letters. Presumably metropolitan areas required the letters because of multiple central office exchanges, but rural areas with few subscribers and only one CO, required just a few digits and no letters.

metropolitan fiber ring A metropolitan fiber ring is an advanced, high-speed local network that can also be used to connect businesses and residences directly to a long distance carrier's network, and provide alternatives to the local telecommunications services they have today. This definition, courtesy MCI.

metropolitan network See Metropolitan Area Network.

metropolitan service provider See MSP.

Metropolitan Statistical Area MSA. Sometimes known as SMSA (as in Standard MSA), MSAs are areas based on countries as defined by the U.S. Census Bureau that contain cities of 50,000 or more population and the surrounding countries. Using data from the 1980 census, the FCC allocated two cellular licenses to each of the 305 MSAs in the United States.

MExE Mobile Station Application Execution Environment (GSM 02.57): a framework to ensure a predictable environment for third-party applications in GSM or UMTS handsets (ie the Mobile Station). MExE does this by defining different technology requirements called "classmarks". MExE classmark 1 is based on WAP, classmark 2 on PersonalJava and JavaPhone, and classmark 3 on J2ME CLDC and MIDP. Other classmarks may be defined in the future. MExE specifies additional requirements for all classmarks, for instance a security environment, capability and content negotiation, a user profile, user interface personalization, management of services and virtual home environment. A handset can support any number of classmarks.

mezzanine board A printed circuit board that plugs directly into a another plug-in card. For example, a mezzanine card might plug into a VMEbus, CompactPCI or PCI card, which may be peripheral controller or CPU or just adapter card that connects the mezzanine card to the target bus. Mezzanine boards are designed to be rugged so that they do not have to be bolted down.

MF 1. MultiFrame.

2. MultiFrequency.

MF-E MK-E is a suffix to an FCC registration number, indicating the function of a system in accordance with its FCC registration. An MF-E is a Fully Protected Multi-Function (Hybrid) Key System, with "MF" denoting "Multi-Function," and "E" denoting that the system will accept both rotary and tone signaling. See also Hybrid Key System, KF-E, PF-E and Registration Number.

MF-er Another name for a blue box. MF stands for multi-frequency. A blue box reproduced the telephone industry's multifrequency tones, which it used at that time for inband signaling. Hard-core phreakers called them MF-ers – for multifrequency transmitters. The acronym was also understood to stand for "motherf***ers,"because they were used to f*** around with Ma Bell. These blue boxes no longer work because the phone industry has gone to out of band signaling. See SS7.

MFCR2 MFCR2 protocol is a signaling and control technique which is an alternative to ISUP (ISDN User Part) for call set-up. It's used in South America.

MFD Abbreviation for Microfarad; one thousandth of a farad, the unit of measuring capacitance. The capacitor is a common electrical device that can store electric charges, and pass AC but not DC. Most phones use capacitors to disconnect the bell during conversations.

MFJ The Modified Final Judgment is the federal court ruling that set up the rules and regulations concerning deregulation and divestiture of AT&T and the Bell system. See Modified Final Judgment for a bigger explanation.

MFLOPS Million Floating point Operations Per Second. A measure of computing power usually associated with large computers. Also known as Megaflops.

MFM Modified Frequency Modulation. An encoding scheme used to record data on the magnetic surfaces of hard disks. It is the oldest and slowest of the Winchester hard disk interface standards. RLL (Run Length Limited encoding) is a newer standard, for example.

MFOS MultiFunction Operations System. An AT&T term.

MFP See Multi-Function Peripheral.

MFS Multifunction Peripherals. A gadget you connect to your computer that can print, photocopy, fax and scan.

MFSK Multiple Frequency Shift Keying.

MFV German for tone dialing. Stands for Mehrfrequenzverfahren.

MG Master Group.

MGB Master Ground Bar, also called Main Grounding Busbar.

MGC See media gateway controller

MGCP See media gateway control protocol.

MGN Multi Grounded Neutral. A single uninsulated grounded conductor. The currents in the three phases of the transmission line are never quite equal; the MGN carries the residual unbalance current. At many poles, the MGN is physically grounded to a groundrod at the base of the pole. See Joint Pole.

MGTS Message Generator Traffic Simulator.

MH Modified Huffman data compression method.

MHF Mobile Home Function.

MHO The unit of conductivity.

MHS Message Handling Service. A program developed by Action Technologies (and others) and marketed by those firms and Novell to exchange files with other programs and send files out through gateways to other computers and mail networks. It is used particularly to link dissimilar electronic-mail systems. A company running e-mail on their internal LAN will dedicate one computer on the network to be a MHS machine. Every hour or so it will call MCI Mail, CompuServe, etc. and download e-mail messages for people and upload messages from people on the network. Once it has the messages downloaded it will distribute them to the people on the LAN the messages are destined for. See MHS Message Handling System.

MHS Enterprise A messaging installation either on a local or corporate-wide level that uses MHS as its backbone between several messaging applications such as E-mail, scheduling, fax, workflow and more. Gateways are used to connect to X.400 systems, public carriers and mainframe systems.

MHS Message Handling System An ISO standard Application Layer protocol that defines a framework for distributing data from one network to several others. It transfers relatively small messages in a store-and-forward manner (defined by ITU-T as X.400 and by ISO as MOTIS/Message-Oriented Text Interchange Standard). See MHS.

MHTML MIME encapsulation of aggregate HTML documents is a proposed standard that would, if deployed on the Internet, allow the easy attachment of complex Web pages – or entire sites – to an email viewer. According to InfoWorld, that means that all components of a Web site could be sent as attachments on a single e-mail and then reassembled with full integrity to produce a functional site for the end viewer, even if that viewer does not have Web access. For example, companies could "push" via email Web or intranet content – as well as applications and software downloads– to employees without giving those employees carte blanche Web access.

MHz An abbreviation for Megahertz. One million Hertz. One million cycles per second. Used to measure band and bandwidth. See Band and Bandwidth. Megahertz is also used by the computer industry to mean millions of clock cycles per second, a measure usually applied to the computer's main microprocessor. Everything that happens in a computer is timed according to a clock which ticks millions of times every second. Higher MHz computers work faster than lower MHz computers. But megahertz is not an accurate measure of a microprocessor's speed. Other factors, such as wider data paths and the ability to execute more than one instruction per clock cycle, affect the actual speed of a microprocessor. Which is why a 100 MHz Pentium chip outpaces a 100 MHz Intel DX4 chip. When comparing the speed of one PC to another, there are other factors also, such as the amount and speed of the system's random access memory (RAM).

MI/MIC Mode Indicate/Mode Indicate Common, also called Forced or Manual Originate. Provided for installations where other equipment, rather than the modem, does the dialing. In such installations, the modem operates in Dumb mode (no Auto Dial capability), yet must go off hook in Originate mode to connect with answering modems.

MI5 Britain's domestic intelligence agency.

MIA Media Interface Adapter. Converts electrical signals into optical signals and vice versa.

MIB Management Information Base. MIB is a database of network performance information that is stored on a Network Agent for access by a Network Management Station. MIB consists of a repository of characteristics and parameters managed in a network device such as a NIC, hub, switch, or router. Each managed device knows how to respond to standard queries issued by network management protocols. To be compatible with CMIP, SNMP, SNMP-2, RMON, or RMON-2, devices gather statistics and respond to queries in the manner specified by those specific standards. Within the Internet MIB employed for SNMP (Simple Network Management Protocol)-based management, ASN.1 (Abstract Syntax Notation One) is used to describe network management variables. These variables, which include such information as error counts or on/off status of a device, are assigned a place on a tree data structure. MIB is used in X.400 electronic mail. Many managed devices also have "private" MIB extensions. These extensions make it possible to report additional information to a particular vendor's proprietary management software or to other management software that's aware of the extensions. See CMIP, Internet MIB Tree, MIB-2, RMON, RMON-2, SNMP.

MIB Attribute An ATM term. A single piece of configuration, management, or statistical information which pertains to a specific part of the PNNI protocol operation.

MIB-1 The initial collection of objects and attributes defined by the TCP/IP (Transmission Control Protocol/Internet Protocol) standards community. MIB-I was elevated to Internet standard status in May 1990.

MIB-2 The expression MIB refers to the original SNMB MIB definition in IETF RFC 1157. The broader MIB-2 (RFC 1213) adds to the number of monitoring objects supported and is included in SNMP-2's MIB. However, SNMP-2s MIB (RFC 1907) is a superset of MIB-2.

MIB Instance An ATM term. An incarnation of a MIS object that applies to a specific part, piece, or aspect of the PNNI protocol's operation.

MIB Object An ATM term. A collection of attributes that can be used to configure, manage, or analyze an aspect of the PNNI protocol's operation.

MIC 1. Microphone.

2. Medium Interface Connector. FDDI de facto standard connector.

3. A designation for a multiplex which interfaces between two DSIs and one DSIC circuit.

4. Microwave Integrated Circuit - In the microwave industry, a hybrid using thin- or thick-film conductors and passive components on a ceramic substrate combined with chip-form active and passive components.

Michael Michael is my favorite son, for an obvious reason. He is my only son. Once Michael could not read what was in this dictionary. Now he disagrees with my definitions. Were he not so correct and so smart, I'd be annoyed. At 25, Michael is now happily running a $600 million dollar private equity fund, which has already displayed early signs of great.

He has graduated from Dartmouth College where he received a degree in government – a subject that gave impracticality a whole new meaning but didn't stop him getting the prestigious private equity fund job. As for concrete achievements, Michael got himself elected Student Body Vice-President – the first time a freshmen had been elected in the long history of Dartmouth College. I was very proud. He also bought himself a $130,000 apartment as a sophomore and sold the apartment on his graduation for $172,500 – his first real estate transaction. And one of many more I'm hoping to watch with awe and respect in coming years. What I'm most proud of is Michael's remarkable maturity. I shudder to remember the disaster I was at 25. Trust me, it was not pretty. Michael is different. He's a pleasure to be with. He seems to actually like his parents, especially if there's food in the refrigerator. He has an insatiable appetite that fortunately has been funded by Whole Foods' rising stock price and encouraged by his parents' "open refrigerator" policy. There is no "generation gap" between us (the parents) and our kids, as there was in my time. Michael and I are close. He actually enjoys spending time with Susan, his mother, and me. I know this. He told me so. We have developed a great friendship based on mutual respect. Michael respects my ability to provide him with expensive toys, including a car, an apartment, furniture, babysitting services for his dog (Winnie) and infinite food. I respect his ability to find expensive toys to acquire and to consume huge amounts of food. I'm hopeful our friendship, which, fortunately, is based on more than toys and food, can continue through the remainder of my life. I'm now 64. Though I don't want to live the rest of my life through my kids, it wouldn't be hard. Michael and his sister, Claire are both destined for great achievements. And now there's Ted. See Claire, Susan and Ted.

Mickey Unit of space that a mouse moves, measured at 1/200th of an inch. The speed of a computer mouse, that is, the distance the cursor moves across the screen, as it relates to the movement of the mouse across the mouse pad, is officially rated in "mickeys."

MICR Magnetic Ink Character Recognition. A process of character recognition where printed characters containing particles of magnetic material, are read by a scanner and converted into a computer-readable digital format.

micro One-millionth.

micro cassette Miniaturized version of the standard audio cassette.

micro channel A proprietary bus developed by IBM for its PS/2 family of computers' internal expansion cards. Also offered by NCR, Tandy and other vendors. See ISA and EISA.

micro components Miniaturized audio components that provide the benefits of traditional sized components in far less space.

micro farad One millionth of a farad. This is the common unit for designation capacitance in electronics and communications.

micro to mainframe link The telecommunications path over which data between a microcomputer and a mainframe computer travels.

micro to mainframe software Software which provides the logic by which data can be transferred back and forth between a microcomputer and a mainframe computer.

micro trenching See microtrenching.

microarchitecture Microarchitecture is what semiconductor chip designers call their chip's inner workings. Microarchitecture performs five basic functions – data access, execution, instruction decode and write back results. Microprcessors are typically measured in speed, i.e. cycles per second, which are called hertz. But the reality is that microachitecture also causes a chip to work faster or slower.

microbend loss Loss in optical power created when a lightguide (i.e., optical fiber) is bent microscopically, i.e., so slightly that it cannot be seen with the naked eye. Mark Mauriello at Lucent described to me a microbend less as a small diameter bend that is equivalent to the bending of a fiber around a grain of sand. It causes loss at 1550 nm but not 1310 nm. In contrast, a macrobend loss is a large diameter bend equivalent to the bending of a fiber around your finger. It causes loss at both 1550 nm and 1310 nm. Microbends can be caused during the installation process, during packaging, or by mechanical stress due to water in the cable during repeated freeze and thaw cycles. See also microbending and macrobend loss.

Microbending Curvatures of the fiber which involve axial displacements of a few micrometers and spatial wavelengths of a few millimeters. Microbends cause loss of light and inrease the attenuation of the fiber. See microbend loss.

microcap A small company with a small market capitalization, which is defined as the share price times the number of shares outstanding. How small a microcap is depends on whom you're talking with.

microcell A cellular radio cell that's smaller than a macrocell, and larger than a picocell. See Macrocell and Picocell.

microchannel A proprietary bus developed by IBM for its PS/2 family of computers,

which it introduced in 1987. The bus was one of IBM's "weapons" against the disturbing increase in clone PCs (disturbing to IBM). IBM figured that introducing a proprietary bus like the microchannel (called MCA for the MicroChannel Architecture), it could slay the clones or control them by charging large amounts of money to license the microchannel architecture. The microchannel, however, engendered a big yawn from consumers, who didn't at the time need its speed, nor its greater complexity, nor its higher cost. As the need to process large amounts of Windows' video information became more urgent in the late 1980s and early 1990s, so faster buses have become more critical. And two new buses, VL and PCI, have been created. For a fuller explanation, see ISA, EISA, PCI and VESA.

microchips What's left in the bag when the big chips are gone.

microchunk As a verb, take a product or service that is traditionally sold as a monolithic whole and chop it into smaller pieces for sale on an 0xD6 la carte basis. As a noun, a microchunk is one of these smaller pieces. Mobile content increasingly is being developed and purchased in microchunks.

microcircuit A microelectronic device that has a high circuit-element and/or component density and that is considered to be a single unit. See also Iintegrated Circuit.

microcircuit module An assembly of microcircuits, or an assembly of microcircuits and discrete parts, designed to perform one or more electronic circuit functions and so constructed that, for the purpose of specification, testing, commerce, and maintenance, it is considered to be indivisible.

microcode 1. Programmed instructions that typically are unalterable. Usually synonymous with firmware and programmable read-only-memory (PROM).

2. Translation layer between machine instructions and the elementary operations of a computer. Microcode is stored in ROM and allows the addition of new machine instructions without requiring that they be designed into electronic circuits when new instructions are needed.

microcom networking protocol The Microcom Networking Protocol, MNP, is a de facto standard protocol that provides error correction and data compression in dial-up modems. The protocol's design allows for a broad range of services to be implemented, while maintaining compatibility among modems with different levels of MNP capabilities. For example, a modem capable of MNP Class 5 and Class 7 data compression can talk to a modem that lacks MNP data compression.

According to Microcom, MNP is an error correction protocol accepted by international standards authorities (ITU-T Rec. V.42). MNP offers a reliable and widely accepted method of correcting errors in transmissions over dial-up communications lines. MNP incorporates three different data compression methods, including the ITU-T recommendation, V.42bis.

Since its original definition, MNP has evolved through nine classes of enhancements. Of those nine classes, the first four provide error control and are in the public domain. Classes 5 through 7 may be licensed from Microcom. Currently MNP error control (Classes 2,3 and 4) has been adopted, along with the LAPM protocol, as mandatory elements of the Consultative Committee on International Telegraphy and Telephony (ITU-T) V.42 recommendation for modem error control.

An Overview Of MNP Service Classes

Class 1

This is the first level of MNP performance. MNP Class 1 uses an asynchronous byte-oriented half-duplex method of exchanging data. MNP Class 1 implementations make minimal demands of processor speeds and memory storage. MNP Class 1 makes it possible for devices with few hardware resources to communicate error-free. Class 1 implementations are no longer included in modems.

Class 2

MNP Class 2 uses asynchronous byte-oriented full-duplex data exchange (i.e., data goes in both directions at once). All microprocessor-based modems are capable of supporting MNP Class 2 performance.

Class 3

This class uses synchronous bit-oriented full-duplex data exchange, eliminating the overhead of start and stop bits used in byte-oriented asynchronous communications. The user still sends data asynchronously to the modem while communications between modems is synchronous.

Class 4

This class introduces two new concepts Adaptive Packet Assembly and Data Phase Optimization, both of which further enhance performance. Adaptive Packet Assembly means that the size of the packets in which data is sent between modems is altered according to the quality of the physical link. The higher the line quality, the larger the packets. Larger packets, while more efficient (the ratio of user data to control data is higher), are also more susceptible to errors. Data Phase Optimization means that repetitive control information is removed from the data stream to make packets more efficient. Both techniques, when combined with Class 3, yield a protocol efficiency of about 120 percent (A V.22bis 2400 bps modem will realize approximately a 2900 bps throughput).

Class 5

This class implements MNP basic data compression to realize a net throughput efficiency of 200 percent on average. (A 2400 bps modem will realize 4800 bps). Class 5 uses a real-time adaptive algorithm to compress data. The real-time aspects of the algorithm allow the data compression to operate on interactive terminal data as well as file transfer data. The adaptive nature of the algorithm means data compression is always optimized for the user's data. The compression algorithm continuously analyzes the user data and adjusts the compression parameters to maximize data throughput.

Class 6

This class implements Universal Link Negotiation and Statistical Duplexing. The first feature allows a single modem to operate at a full range of speeds between 300 and 9600 bps, depending on the maximum speed of the modem on the other end of the link. Modems begin operation at a common slower speed and negotiate the use of an alternative high speed modulation technique. The Microcom AX/9624c modem is an example of a modem that uses Universal Link Negotiation, starting with 2400 bps V.22bis technology and shifting to 9600 bps V.29 fast train technology, if the other modem has that technology too. Statistical Duplexing allows the modem to simulate full-duplex service on the half-duplex V.29 modem connection.

Class 7

This class implements a more efficient data compression method than the one used in Class 5. The difference between the two classes is that Class 5 realizes an average 200 percent speed improvement over a non-MNP modem, versus an average 300 percent improvement for Class 7. Class 7 data compression uses Huffman encoding with a predictive algorithm to represent user data in the shortest possible Huffman codes. In addition to Class 5 and Class 7 data compression, MNP also supports V.42bis data compression. Based on the Lempel-Ziv-Welsh data compression model, V.42bis supports an average 400 percent efficiency improvement.

Class 8

Not defined.

Class 9

This class reduces the amount of time required for the modem to perform two frequently occurring administrative activities: to acknowledge that a message was received and to retransmit information following an error. Message acknowledgment is streamlined by "piggy-backing" the acknowledgment in its own dedicated packet. Retransmission is streamlined by indicating in the error or Negative Acknowledgment Packet (NAK) the order sequence number of each of the failed messages. Rather than sending all the messages over again (even the good ones) from the point of the error, as is usually done with error correcting protocols, only the failed messages are resent.

Class 10

MNP Class 10 consists of Adverse Channel Enhancements that optimize performance in environments with poor or varying line quality, such as cellular telephones, international telephone calls, and rural telephone service. These enhancements fall into four categories:

1. Multiple aggressive attempts at link setup 2. Adapting packet size to accommodate varying levels of interference 3. Negotiating transmission speed shifts to achieve the maximum acceptable line speed 4. Dynamically shifting to the modem speed most suitable to transmission line conditions

See Error Control Protocols and LZW.

Microcomputer The combination of CPU (Central Processing Unit) and other peripherals (I/O, memory, etc.) that form a basic computer system. See Microprocessor.

microcosm The traditional definition: A little world; a miniature universe. A smaller, representative unity having analogies to a larger unity. A Jimmy Stewart movie, Magic Town, features "Grandview," a small town in the Midwest that is a perfect statistical microcosm of the United States, a place where the citizens' opinions match perfectly with Gallup polls of the entire nation. George Gilder defines it as the domains of technology unleashed by discovery of the inner structure of matter in quantum theory early in the 20th century. The epitome of the microcosm is the microchip.

microdisplay Microdisplays are a sector of the flat panel display industry used in small optically-viewed devices such as video headsets, camcorders, viewfinders, and other portable devices. Microdisplays are typically of such high resolution that they are only practically viewed with optics (a fancy word for a specially-designed magnifying glass). Although the displays are typically sized less than two inches, many can provide a magnified viewing area similar to that of a full size computer screen. For example, when viewed through a lens, a high-resolution 3/4-inch diagonal display can be made comparable to viewing a 21-inch diagonal computer screen or a large TV screen.

MicroElectroMechanical System MEMS. Semiconductor chips that have a top layer of mechanical devices, such as mirrors or fluid sensors. MEMS devices are used to make pressure, temperature, chemical and vibration sensors, light reflectors, and switches as well as accelerometers for air bags, vehicle control, pace makers and games. They are also used in the construction of microactuators for data storage and read/write heads, and are a key component for photonic switches, which can be used in cross-connect, add/drop multiplexer, dispersion compensation, and gain-equalization applications. MEMS also are used in optical switching systems, which offer clear advantages in purely optical systems such as DWDM (Dense Wavelength Division Multiplexing) combined with EDFA (Erbium-Doped Fiber Amplification) and Raman amplification. There are two types of MEMS switches: mechanical and microfluidic. Mechanical MEMS switches involve vast arrays (as many as hundreds of thousands) of micromachined mirrors on a silicon chip. Control signals (electromagnetic or thermal) adjust the mirrors to switch optical signals between incoming ports and outgoing ports associated with optical fibers. Microfluidic MEMS switches involve the movement of fluids contained in tiny channels etched into the silicon chips in the form of an intersecting grid configuration. In default mode, the fluids allow the light signal to pass through. If the signal is to be switched, tiny bubbles are injected into and removed from the fluid hundreds of times per second, thereby reflecting the signal to the proper output port.

microelectronic assembly An assembly of unpackaged (uncased) microcircuits and/or packaged microcircuits, which may also include discrete devices, so constructed on a packaging interconnect structure that for the purpose of specification, testing, commerce, and maintenance, the package is considered to be an indivisible component. (Ref. JESD30-B.) Note 1: The passive and/or active discrete and microelectronic devices may be mounted on either one or two sides of the packaging interconnect structure, and the external terminals usually exit from one side of the assembly. Note 2: Many package sizes, shapes, and external terminal forms are possible.

microfiche A rectangular sheet of transparent film that contains multiple rows of greatly reduced page images of report, catalogs, rate books, etc.

microfilm A small roll of photographic film which can hold several thousand document pages which, when projected onto a screen, produces a legible copy of the item or form photographed.

microfloppies The latest generation of floppy disks at 3 1/2 inches diameter, invented by Sony. The microfloppy is used in the Apple Macintosh and most MS-DOS laptop computers. Used in an MS-DOS machine, a 3 1/2 inch microfloppy diskette will currently format to carry 1.44 million bytes of data – equivalent to about 500 pages of double spaced text.

microfluidic optical fibers The blazing-fast Internet access of the future - - imagine downloading movies in seconds – might just depend on a little plumbing in the network. In the February 2004 issue of MIT Technology Review, the editors wrote, "Tiny droplets of fluid inside fiber-optic channels could improve the flow of data-carrying photons, speeding transmission and improving reliability. Realizing this radical idea is the goal of University of Illinois physicist John Rogers, whose prototype devices, called microfluidic optical fibers, may be the key to superfast delivery of everything from e-mail to Web-based computer programs, once "bandwidth" again becomes the mantra. Rogers began exploring fluid-filled fibers more than two years ago as a researcher at Lucent Technologies' Bell Labs. While the optical fibers that carry today's phone and data transmissions consist of glass tubing that is flexible but solid, Rogers employs fibers bored through with microscopic channels, ranging from one to 300 micrometers in diameter, depending on their use. While Rogers didn't invent the fibers, he and his team showed that pumping tiny amounts of various fluids into them - and then controlling the expansion, contraction, and movement of these liquid "plugs" - causes the optical properties of the fibers to change. Structures such as tiny heating coils printed directly on the fiber precisely control the size, shape, and position of the plugs. Modifying the plugs' properties enables them to perform critical functions, such as correcting error-causing distortions and directing data flows more efficiently, thus boosting bandwidth far more cheaply than is possible today." See also Mirofluidics.

microfluidics Microfluidics is the scaling down of laboratory fluid tests to miniature sizes. Lab experiments can be performed by manipulating tiny amounts of chemicals and biological samples on chips containing tiny tubes and vessels. Amounts of fluids in quantities as small as microliters, nanoliters or even picoliters can be handled. This technique saves the use of large amounts of expensive chemicals and precious samples, and may help speed up new discoveries by enabling scientists to automate thousands of experiment a day with great accuracy. Microfluidics hardware requires construction and design that differs from conventional hardware, as it is not generally possible to scale conventional devices down to miniature size and expect them to work. When the dimensions of a device or system reach a certain size as the scale becomes smaller, the particles of fluid, or particles suspended in the fluid, become comparable in size with the apparatus itself. This dramatically alters system behavior, as capillary action changes the way in which fluids pass through microscale-diameter tubes, as compared with macroscale channels. See also Microfluidic Optical Fibers.

microform Microform means Microfiche and Microfilm.

micrographics Conversion of information into or from microfilm or microfiche.

micron One thousandth of a millimeter. Or one millionth of a meter. A unit of measurement corresponding to 1/25,000 of an inch or 40 millionths of an inch.. A micron can be used to specify the core diameter of fiber-optic network cabling. This diameter should match your hardware vendor's requirements; but if you install fiber before you buy the equipment, specify the 62.5-micron size.

micropayment An on-line payment of a dime or less. Touted as the key catalyst for Internet commerce, micropayments were conceived as a means of generating revenues which would be significant for vendors, in the aggregate, while being so trivial to the individual users that they would not hesitate make micropayments freely. While still rhetorical, micropayments were to apply to such services as custom newsfeeds, processing applets and data queries. The term micropayment has come to have its own significance since most payments over the Internet are made by credit card. The fees charged on credit cards have been too high to make charging micropayments worthwhile. A whole bunch of vendors have come along trying to create a framework for enabling micropayments but making a profit simultaneously. We're still awaiting "the solution."

microphone A transducer that changes the air pressure of sound waves into an electrical signal that can be recorded, amplified and/or transmitted to another location.

microprocessor An electronic circuit, usually on a single chip, which performs arithmetic, logic and control operations, with the assistance of internal memory. The microprocessor is the fabled "computer on a chip," the "brains" behind all desktop personal computers. Typically, the microprocessor contains read only memory – ROM – (permanently stored instructions), read and write memory – RAM, and a control decoder for breaking down the instructions stored in ROM into detailed steps for action by the arithmetic logic unit – ALU – which actually carries out the numerical calculations. There's also a clock circuitry which connects the chip to an exterior quartz crystal whose vibrations coordinate the chip's operations, keeping everything in step. And finally, the input/output section directs communications with devices on the outside of the chip, such as the keyboard, the screen and the various disk drives.

The Fortune Magazine issue of May 6, 1991 contained a very good explanation of chips and microprocessors (usually used interchangeably). Here is the article, slightly condensed:

Chips today can store and retrieve data, perform a simple mathematical calculation, or compare two numbers or words in a few billionths of a second. And they can carry out tens of thousands of such tasks in the blink of an eye. Today's chips contain millions of transistors, capacitors, diodes, and other electronic components, all connected by metallic threads a fraction of the diameter of a human hair. A single chip the size of a fingernail can store dozens of pages of text or combine circuits that can perform scores of tasks simultaneously.

Most chips fall into one of two categories - memory chips and logic chips. Memory chips have the easier job: They merely store information that will be manipulated by the logic chips, the ones with the smarts. Today's biggest-selling memory chip (mid-1991) is the one-megabit dynamic random access memory, or DRAM. Each DRAM is a slice of silicon embedded with a lattice of 1,000 vertical and 1,000 horizontal aluminum wires that circumscribe one million data cells. The densest DRAM designed so far has 64 million cells.

Think of those wires as streets and those cells as blocks. Each block contains a transistor that can be turned on or off – to signify 1 or 0 – and that can be identified by it's unique "address" in the wire grid, much like a house in a suburban subdivision. Each digit, letter, or punctuation mark is represented by 1's or 0's stored in eight-cell strings. (See ASCII.) The word "chip" takes up 32 cells in a memory chip. Most PCs sold today have at least eight one-megabit DRAMs.

It's the job of the logic chips to turn those transistors in the DRAMs on or off, and to retrieve and manipulate that information once it's stored. The most important and complex logic chips are microprocessors like Intel's 80386DX, the brains of the more powerful IBM-compatible PCs sold today. If the structure of a memory chip is a suburban subdivision, the layout of a microprocessor is more like an entire metropolitan area, with distinct neighborhoods devoted to different activities. A typical microprocessor contains among other things:

A timing system that synchronizes the flow of information to and from memory and

throughout the rest of the chip.

An address directory that keeps track of where data and program instructions are stored in the DRAMs.

An arithmetic logic unit with all the circuits needed to crunch numbers.

On-board instructions that control the sequence of microprocessor operations.

Other logic chips in a computer take their cues from the microprocessor millions of times each second to draw images on the screen, to feed instructions from a spreadsheet program, say, out of the disk drives into DRAMs, or to dispatch data to a modem or a printer. Perhaps most amazing of all, memory and logic chips can accomplish all this with just a trickle of electricity - far less than it takes to light a flashlight bulb.

Ted Hoff at Intel invented the microprocessor in 1971. See also 1971 in the beginning of this dictionary.

microprocessor controls A control system that uses computer logic to operate and monitor an air conditioning system. Microprocessor controls are commonly used on modem precision air conditioning systems to maintain precise control of temperature and humidity and to monitor the unit's operation.

micropundit An Internet blogger who gains a reputation as a reliable critic or authority on a particular subject of typically small consequence.

microsatellites Unlike traditional satellites, which can weigh tons, microsatellites are the size of a suitcase and weigh about 220 lbs. Since it costs "a bar of gold to launch a can of Coke," according to the New York Times, lightweight microsatellites will be much cheaper to launch than their obese precursors. The U.S. military's goal is to send microsatellites into space in flocks. In this cluster, they would be reprogrammable, able to switch to new tasks when the Pentagon required it.

microsecond One millionth of a second. A microsecond is ten to the minus six. One microsecond – a millionth of a second – is the duration of the light from a camera's electronic flash. Light that short freezes motion, making a pitched ball or a bullet appear stationary. See Atto, Nanosecond, Femto and Pico.

microsegmenting The process of configuring Ethernet and other LANs with a single workstation per segment. The objective is to remove contention from Ethernet segments. With each segment having access to a full 10 Mbps of Ethernet bandwidth, users can do things involving significant bandwidth, such as imaging, video and multimedia.

microsegmentation Division of a network into smaller segments, usually with the intention of increasing aggregate bandwidth to devices.

microslot The time between two consecutive busy/idle flags (60 bits, or 3.125 milliseconds at 19.2 kbps). It is used in CDPD only. A cellular radio term.

Microsoft Founded in 1975 by Bill Gates and Paul Allen as Micro-soft (now called Microsoft) it is (or was at the time of writing this edition of this dictionary) one of the largest software companies in the world. See the next few definitions.

Microsoft At Work A new architecture announced by Microsoft on June 9, 1993 and then put into retirement a couple of years later. Many of its features and ideas surfaced in Windows 95. It consisted of a set of software building blocks that will sit in both office machines and PC products, including:

- Desktop and network-connected printers.
- Digital monochrome and color copiers.
- Telephones and voice messaging systems.
- Fax machines and PC fax products.
- Handheld systems.
- Hybrid combinations of the above.

According to Microsoft, the Microsoft At Work architecture focuses on creating digital connections between machines (i.e. the ones above) to allow information to flow freely throughout the workplace. The Microsoft At Work software architecture consists of several technology components that serve as building blocks to enable these connections. Only one of the components, desktop software, will reside on PCs. The rest will be incorporated into other types of office devices (the ones above), making these products easier to use, compatible with one another and compatible with Microsoft Windows-based PCs. The components, according to Microsoft, are:

- Microsoft At Work operating system. A real-time, preemptive, multi tasking operating system that is designed to specifically address the requirements of the office automation and communication industries. The new operating system supports Windows compatible application programming interfaces (APIs) where appropriate for the device.
- Microsoft At Work communications. Will provide the connectivity between Microsoft At Work-based devices and PCs. It will support the secure transmission of original digital documents, and it is compatible with the Windows Messaging API and the

Windows Telephony API of the Windows Open Services Architecture (WOSA).

- Microsoft At Work rendering. Will make the transmission of digital documents, with formatting and fonts intact, very fast and, consequently, cost-effective; will ensure that a document sent to any of these devices will produce high-quality output, referred to as "What You Print Is What You Fax Is What You Copy Is What You See."

- Microsoft At Work graphical user interface. Will make all devices very easy to use and will make sophisticated features accessible; will provide useful feedback to users. Leveraging Microsoft's experience in the Windows user interface, Microsoft At Work-based products will use very simple graphical user interfaces designed for people who are not computer users.

- Microsoft At Work desktop software for Windows-based PCs. Will provide Windows-based PC applications the ability to control, access and exchange information with any product based on Microsoft At Work. Desktop software is the one piece of the Microsoft At Work architecture that will reside on PCs.

See also Fax At Work, Voice Server, Windows, Windows CE, Windows 95, Windows Telephony and WOSA.

Microsoft Exchange Email messaging software that retrieves messages into one inbox from many kinds of messaging service providers.

Microsoft SQL Server A Microsoft retail product that provides distributed database management. Multiple workstations manipulate data stored on a server, where the server coordinates operations and performs resource intensive calculations.

Microsoft TAPI See TAPI and Windows Telephony.

Microspeak A term coined by James Gleick in The New York Times Magazine of June 18, 1997 to refer to the language of euphemisms Microsoft Corporation often indulges in. For example, Mr. Gleick referred to Microsoft's seeming unwillingness to use the word "bug" and use words such as "known issue," "intermittent issue", "design side effect," "undocumented behavior," or "technical glitch."

microstrip antenna See patch antenna.

microtransaction A small electronic ecommerce transaction – under ten cents. The significance of the term "microtransaction" is that there is a real need for such small transactions on the Internet – but no one has figured a way to bill economically in such small amounts. Ultimately, someone will. You might need a microtransaction, for example, if you were renting software for small amounts of time.

microtrenching How to get broadband fiber or coax to the last mile? You can't dig the big trenches we use for long-haul. Those trenches are too expensive and easily cost $100 a foot. A new digging technique known as micro trenching to the rescue. Micro trenching involves the creation of a shallow trench in the sidewalk or street asphalt, which is typically one-quarter of an inch wide and two to six inches deep. The layer of road base is not even touched. Using this method, a crew can lay as much as a thousand feet of fiber per day. Moreover the cost is typically one-fifth to a half of what normal trenching costs. Micro trenching also eliminates red tape and saves time. Surveying and permitting may take 30 days for a micro trenching job, as opposed to 60 or more days for a traditional trenching project. The difference is even more glaring when comparing the trenching work itself. A build may take as little as two days for micro trenching, but 30 or more days for conventional trenching work. From planning to completion, the build time frame for micro trenching may be 50 to 55 days, whereas the traditional approach can take 160 days, or more. Time is money. Much of this definition comes from Bill Morrow is vice chairman and CEO of Grande Communications, a Texas telecom company.

microwave Electromagnetic waves in the radio frequency spectrum above 890 Megahertz (million cycles per second) and below 20 Gigahertz (billion cycles per second). (Some people say microwave refers to frequencies between 1 GHz and 30 GHz.) Microwave is a common form of transmitting telephone, facsimile, video and data conversations used by common carriers as well as by private networks. Microwave signals only travel in straight lines. In terrestrial microwave systems, they're typically good for 30 miles, at which point you need another repeater tower. Microwave is the frequency for communicating to and from satellites.

microwave absorber A tuned device used for blocking and preventing or reducing the re-emission or reflection of microwave frequency energy.

microwave band Loosely defined as those frequencies from about 1 gigahertz upward, Services that use microwave frequencies for point-to-point and point-to-multipoint communications include common carrier, cable TV operators broadcasters, and private operational fixed users.

microwave diffraction The tendency of microwave radiation to "bend" around objects.

microwave ducting Under certain conditions the earth and the upper atmosphere can sometimes behave as a wave guide with respect to microwave signals carrying them long distances.

Microwave Multi-Point Distribution System MMDS. Microwave Multi-point Distribution System. A means of distributing cable television signals, through microwave, from a single transmission point to multiple receiving points. Often used as an alternative to cable-based cable TV. According to an April, 1995 press release from Pacific Telesis, which was starting an MMDS service, "in digital form, it will provide more than 100 channels to a radius of approximately 40 miles from the transmitter. The MMDS transmitter delivers video to homes that are in its 'line of sight.' MMDS transmissions are limited by the terrain and foliage of a given market. The microwave signal is received by an antenna on the subscriber's home, then sent down coaxial cable to a box atop the customer's TV set. The box decodes and decompresses the digital signal."

microwave oven The following "urban legend" seems to be quite prevalent on the internet: "The microwave oven was invented after a researcher walked by a radar tube and a chocolate bar melted in his pocket." Actually, there is a kernel of truth in this, but it is, alas, inaccurate. The device was actually invented after researchers working on radar at Bell Labs noticed that a chocolate bar that had been left near a magnetron melted after a few hours of exposure. If a researcher had a chocolate bar melt in his pocket as he walked by radar equipment, he would most probably would have melted with it.

microwave pill In space weaponry, one idea is to place a microwave gun on board a microsatellite, have the microsatellite sidle alongside your enemy's satellite, then emit a pulse of microwaves and fry your enemy's electronics. This application, in U.S. military parlance, is called a "high-power microwave pill."

Microwave Pulse Generator MPG. A device that generates pulses at microwave frequencies.

microwave reflection The tendency of microwave radiation to bounce off objects and behave according to the physical principles of light.

microwave tags RFID radio frequency tags that operate at 5.8 GHz. They have very high transfer rates and can be read from as far as 30 feet away, but they use a lot of power and are expensive.

MICS Medical Implant Communications Service. See Personal Radio Services.

MID An ATM term. Message Identifier. The message identifier is used to associate ATM cells that carry segments from the same higher layer packet.

Mid Span See 802.3af.

mid-air passenger exchange Grim air traffic controller speak for a head-on collision. Midair passenger exchanges are quickly followed by "aluminum rain." Definition from Wired Magazine.

mid-span A phone service that runs from a pole to a hook attached to a cable strand before it reaches the building.

mid-span device See 802.3af.

mid-span meet Sonet's ability to mix the terminal, multiplexing and cross-connect equipment from different vendors. A major accomplishment for standardization. Wish more telecom systems could meet mid-span. Sadly, most can't.

mid-split A broadband cable system in which the cable bandwidth is divided between transmit and receive functions. A cable television system in which the guard band (or split) between the upstream and downstream signals is at 114-150 MHz. Defined in IEEE 802.7. Compare to the definitions of Sub-Split and High-Split.

midband Microsoft's word for telecom speeds that are faster than phone lines but slower than broadband networks. ISDN BRI is midband.

middle mile The connection between an ISP and the Internet Backbone. The middle mile consists of one or more carrier networks that move traffic from the access ISP to the "final" ISP. This is where path diversity takes place; indeed traffic typically takes two different paths through this middle mile - traffic flowing from the client to the content provider takes one route, while traffic from the content provider back to the client takes a different path. This is due to the carrier's policies (e.g. hot-potato routing) or to the decisions the Border Gateway Protocol (BGP) makes from each perspective. The middle mile is the section of the link where path diversity is possible.

Middleware Middleware is software which sits between layers of software to make the layers below and on the side work with each other. Essentially, middleware serves as a translation mechanism, gluing together application software across a network, and on a transparent basis. On that broad definition, middleware could be almost any software in a layered software stack. Let's consider several examples: In computer telephony, middleware tends to be software that sits right above that part of the operating system dealing with telephony – TSAPI in NetWare or TAPI in Windows – but below the computer telephony application above it which the user sees on their desktop. Database middleware supports one-way access from clients to a server-based database. Messaging middleware is similar to an e-mail system, allowing systems to exchange data.

Middlewave Communications software that acts as a universal translator between diverse radio frequencies and protocols. The software physically resides on a remote client and on a communications server located between the client and the applications server.

MIDI Musical Instrument Digital Interface, a standard for connecting musical instruments, synthesizers and computers. The MIDI standard provides a way to translate music into computer data, and vice versa. A file with a MIDI extension means that it's a file containing music. In contrast, a file with the extension WAV typically means it's a sound file, which means it may contain a voice recording.

MIDP Mobile Information Device Profile. Set of Java APIs that is generally implemented on the Connected Limited Device Configuration (CLDC). It provides a basic J2ME application runtime environment targeted at mobile information devices, such as mobile phones and two-way pagers. The MIDP specification addresses issues such as user interface, persistent storage, networking, and application model.

midrange system Medium-scale computer that functions as a workstation or as a multiuser system handling several hundred terminals.

midspan repeater A device that amplifies the signal coming or going to the central office. This device is necessary for ISDN service if you are outside the 18,000 feet distance requirement from the central office.

midsplit A broadband cable system in which the cable bandwidth is divided between transmit and receive frequencies. The bandwidth used to send toward the head-end (reverse direction) is about 5 MHz to 116 Mhz, and the bandwidth used to send away from the head-end (forward direction) is about 168 MHz to 400 Mhz. The guard band between the forward and reverse directions (100 MHz to 160 MHz) provides isolation from interference. Requires a frequency translator.

MIF 1. Management Information File. MIF is a file format for DMI that describes components within a PC. See WFM.

2. Minimum Internetworking Functionality. A general principle within the ISO that calls for minimum local area network station complexity when interconnecting with resources outside the local area network.

migration An AT&T marketing strategy designed to encourage all phone equipment month-to-month renters into long-term contracts for AT&T's "flagship products", i.e., System 75, 85 and Definity PBXs.

MII See Media Independent Interface.

mike Microphone.

MIL A unit used in measuring diameter of a wire or thickness of insulation over a conductor. One onethousandth of an inch, i.e. 0.001".

MIL-STD-188 A set of U.S. military standards relating to telecommunications.

mileage See Airline Mileage.

mileage range indicator A code that denotes the mileage range being billed.

mileage sensitive rates Mileage sensitive rates (also called "banded rates") are rates that increase with physical distance. For example, if you live in New York City, a call to (or from) New Jersey will cost less than a call to (or from) California. Things are changing, however. Most calls now in the United States are distance insensitive, and are only time sensitive. They charge by time, not for how far you call.

military-industrial complex In his farewell remarks as president in 1961, Dwight Eisenhower famously warned about a "military-industrial complex" made up of the "conjunction of an immense military establishment and a large arms industry."

millennials People born between 1980 and 2000. These are a group for whom personal computers, the Internet and various digital communications (cell phones, instant messaging, etc.) have been a persistent and integral part of their lives. Their worst nightmare: their broadband access to the Internet drops conntection for a day or two.

millennium bug Y2K, Year 2000. See Y2K.

milli One thousandth. Millisecond equals one thousandth of a second.

millimeter wireless Operating at microwave frequencies with a separation of millimeters between peaks and troughs of the radio wave, millimeter wireless includes LMDS (Local Multipoint Distribution Services) and MMDS (Multichannel Multipoint Distribution Services). See also LMDS and MMDS.

millisecond Millisecond equals one thousand of a second.

millions of floating instructions per second A measure of computing power. Processors are assigned a value in terms of their ability to perform complex mathematical operations (multiplications, division, addition, and subtraction) with "floating point" (non-integer) numbers.

millions of instructions per second A measure of computing power

measured in terms of the number of instructions in can execute in a second. Its merit is highly dependent, of course, on how powerful its instructions actually are. That is, two computers with the same MIPS ratings might have significantly different computing power if one of the computes has instructions that can accomplish twice as much per instruction as the other on the average.

milliwatt One thousandth of a WATT. Used as a reference point for signal levels at a given point in a circuit.

milliwatt (102 Type) test line An arrangement in an end office that provides a 2004 Hz tone at 0 dBm0 for one-way transmission measurements towards the customer's premises from the local exchange carrier's end office.

MILNET MILitary NETwork. Along with DARPANET, MILNET was created in 1983 as successor to the ARPANET. One of the DDN networks that make up the Internet; devoted to non-classified U.S. military communications. SEE ARPANET.

milspec Military Specification. Milspecs are very demanding.

MIM Metal-Insulator-Metal. A display technology which uses active matrix technology that uses diodes behind each pixel to produce images. It is an improvement on passive displays but a step behind TFT (Thin Film Transistor) technology. See also LCD.

MIMD Multiple Instruction Multiple Data is a type of parallel processing computer, which includes dozens of processors. Each processor can run different parts of the same program and execute those instructions on different data. This makes it more flexible, though more expensive than a computer running SIMD – single instruction multiple data.

MIME Multipurpose Internet Mail Extensions. Developed and adopted by the Internet Engineering Task Force (IETF) in RFC 1521 and 1522, and updated in RFCs 2045-2048, MIME is the standard format for including non-text information in Internet mail, thereby supporting the transmission of mixed-media messages across TCP/IP networks. The MIME protocol, which is actually an extension to SMTP, covers binary, audio and video data. MIME also is the standard for transmitting foreign language text (e.g., Russian or Chinese) which cannot be represented in plain ASCII code. Here's an explanation: When sending files which aren't plain US-ASCII across a network – dial-up, leased or the Internet – you basically have two options. First, you can attach them as a binary file (i.e., non-ASCII file). Or second, you can encode them into ASCII characters and send the file as part of your message. The first method is preferable. But you can typically only send binary files from one account to another on the same network, or between two networks that have agreed between themselves to a method of transferring files. That's a rarity. And it certainly doesn't work in and around the Internet. Therefore, MIME was created, employing a base64 coding scheme, which involves relatively simple encoding and decoding algorithms. See also base64, Bin-Hex, Multipurpose Internet Mail Extensions and UUencode.

MIMJ Modified Modular Jack. These are the 6-pin connectors used to connect serial terminal lines to terminal devices. MIMJ jacks can be distinguished from the familiar RJ11 jacks by having a side-looking tab, rather than a center-mounted one.

MIMO Multiple Input/Multiple Output (MIMO) is a wireless technology that employs multiple intelligent radio antennas to improve transmission speed and quality of wireless networks over longer distances than otherwise possible. The technique involves Space Division Multiplexing (SDM), with arrays of multiple spatially separated transmit and receive antennas. The idea is to push 802.11 (also known as Wi-Fi) WLAN (Wireless Local Area Network) transmission rates from around 20 Mbps to at least 100 Mbps and perhaps as high as 540 Mbps. MIMO also will extend 802.11 distances from 100 meters to as long as 200 meters. MIMO is the basis for the developing IEEE 802.11n standard. See also 802.11n, OFDM, SDM and Wi-Fi.

MIN Mobile Identification Number. A 24-bit number corresponding to the actual 7-digit telephone number assigned by the cellular carrier exclusively to your phone, used for both billing and for receiving calls. The MIN is meant to be changeable, as the ownership of the device may change hands, the owner may change telephone numbers, or the owner may change cities. The MIN is paired with the Electronic Serial Number (ESN). Theoretically, both numbers are verified, and in combination, every time a call is placed in order to verify the legitimacy of the device and the call. See MIN2 and ESN.

MIN Mark Can be 0 or 1. Your home cellular station sends extended address data upon origination and page response. See MIN1.

MIN2 The area code of your cellular phone number.

mini-COW See COW.

mini-floppy A floppy disk that is 3 1/2 inches in diameter. Also called a microfloppy. See Microfloppy.

mini-MAP Mini-Manufacturing Automation Protocol. A version of MAP consisting of only physical, link and application layers intended for lower-cost process-control networks. With mini-MAP, a device with a token can request a response from an addressed device.

Unlike a standard MAP protocol, the addressed Mini-MAP device need not wait for the token to respond.

mini-POP See MTU and POP.

minicomputer A computer smaller than a mainframe, but bigger than a PC. Minicomputer is a loose term for describing any general-purpose digital computer in a low-to-moderate price range. Minicomputers used to be primarily used for the processing of a single application or the processing of a number of small applications. For example, in a distributed processing network, a minicomputer could perform a specific operation and send the volume processing through communication lines to a large mainframe. Today the tasks that were performed by minicomputers are now largely performed by industrial grade PCs working singly or together on a local area network.

mining cable A flame retardant heavy duty portable power cable for use with portable power supply systems and on mobile mining equipment.

minimal regulation Regulated local telephone company under limited state regulation has the ability to file price lists for services and those price lists are usually effective on 10-days' notice to local state commission and customers. This means there are no extensive hearings.

minimize button The minimize button is the down-arrow button at the immediate right of a title bar in Windows software. Clicking the minimize button will shrink a window to its icon.

minimum cell rate MCR. ATM performance parameter that specifies the minimum rate for cell transmission that a network must guarantee to a user on a given virtual circuit. Also, a field in an RM cell specifying the smallest value to which the ER field can be set.

Minimum Internetworking Functionality MIF. A general principle within the ISO that calls for minimum local network station complexity when interconnecting with resources outside the local network.

minimum point of entry MPOE. The closest practical point to where the carrier facilities cross the property line or the closest practical point to where the carrier cabling enters a multi-unit building or buildings. The MPOE typically is in the form of a physical demarcation point (demarc), although it may simply be in form of a tag hung on the entrance cable to identify a point of logical demarcation. In either event, the MPOE establishes the point at which the carrier's responsibility ends, and that of the end user organization or building owner begins. The MPOE commonly is a point on the entrance cable which is 12 inches from the inside wall. In a campus environment, with multiple entrance cables to multiple buildings, the MPOE is the most logical point of demarcation, in the carrier's opinion. See also Demarc.

minimum point of presence See Minimum Point of Entry.

minister In Middle English the word "minister" meant "lowly person." It was originally adopted as a term of humility for men of the church.

minitel French name for videotex. See Videotex.

MIP Mapping A texturing technique used for 3D animation in games and in CAD imaging. When scenery contains acutely angled polygons that disappear into the distance, MIP mapping mixes low and high resolution versions of the same texture to reduce the jagged effect.

MIPS Millions of Instructions Per Second. A measure of computer speed that refers to the average number of machine language instructions performed by the CPU in one second. A typical Intel 80386-based PC is a 3 to 5 MIPS machine, whereas an IBM System 370 mainframe typically delivers between 5 and 40 MIPS. MIPS measures raw CPU performance, but not overall system performance.

MIR Maximum Information Rate: See also PCR.

Miramax Miramax, the movie making and distribution operation founded by Harvey and Bob Weinstein in 1979, got its name from their parents, Miriam and Max.

mIRC An Internet Relay Chat program that runs under Windows.

mirror A term used to reference Internet FTP sites that copy files from other archives every day or so. By accessing a mirror site close to your location you reduce transmission over the Internet.

mirror image Your main web site is in California. You have a server in New York, which contains exactly what's on your site in California. Your New York site is thus the mirror image of your California site. The reason you have a second site is so your east coast customers can connect to your New York site faster and more efficiently (i.e. less data loss) than connecting to your California site.

mirror server Imagine you're a big international software company with branches in every country and in every town in the entire universe. You have zillions of users. Every day your users need to download software from you – fixes, upgrades, etc. If you have only one server (big computer to serve multiple users), that server will bog down in heavy traffic. Also, all your customers overseas will receive their software very very slowly because they're farther away. So the solution is to install a bunch of "mirror servers" all around the world. Each server would be a mirror image of the main one – the one at your headquarters. That means its content and structure would be identical. The only difference is that servers all around the world would be closer to the users. And when the users come first on-line to your main server and find software they need, they would then be given the choice of having their material downloaded to them from a server that's closer to them. Go to www.microsoft.com. They often allow you to download some new software. As you are about to download, they ask you which server in which country you'd like.

mirror set A fully redundant or shadow copy of data. Mirror sets provide an identical twin for a selected disk; all data written to the primary disk is also written to the shadow or mirror disk. The user can then have instant access to another disk with a redundant copy of the information on the failed disk. Mirror sets provide fault tolerance. See also Fault Tolerance.

mirror site A duplicate Web site. A mirror site contains the same information as the original Web site and reduces traffic on that site by providing a local or regional alternative.

Mirrored Server Link MSL. A dedicated, high-speed, point-to-point connection between the primary server and the secondary server. The MSL can either be a coaxial or fiber-optic cable (with maximum distances of 100 feet and 2.5 miles, respectively). See MSL.

mirrored web-page screen pop In computer telephony, the ability for a Web-Enabled Call Center agent to assist callers by "pushing" Web pages to their computer while they are on the phone together. The agent's screen mirrors the caller's screen. The agent can do the clicking for the customer so he or she sees the information they're intended to see.

mirroring A fault tolerance method in which a backup data storage device maintains data identical to that on the primary device and can replace the primary if it fails. Mirroring will typically cost you a 50% performance degradation when your write to disk and 0% performance degradation when you are reading. For a full explanation, see Disk Mirroring.

MIS 1. Management Information System. A fancy name for Data Processing. MIS are also the first three letters of the word MISanthrope and MISguided and lots of other MIS words. MIS departments are taking over corporate telecommunications departments which is why there's little love lost between the two.

2. IBM-speak for Management Initiated Separation. Translation: You're fired.

miscellaneous common carrier A communications common carrier (typically one using microwave) which is not offering switched service to the public or to companies. A miscellaneous common carrier usually provides video and radio leased line transmission services to TV and radio networks.

miscellaneous trunk restrictions Denies preselected lines access to preselected trunk groups (e.g., FX or WATS trunks). A call attempt over a restricted group routes to an intercept tone.

mismatch A termination having a different impedance than that for which a circuit or cable is designed.

MISSI Multilevel Information Systems Security Initiative. Developed under the leadership of the NSA (National Security Administration), MISSI is a framework for the development and evolution of interoperable, complementary security products intended to provide flexible, modular security for networked information systems. MISSI encompasses a suite of security technologies developed to support national defense operations across the Defense Information Infrastructure (DII) and the National Information Infrastructure (NII), and with application in secure corporate environments. Included in the MISSI Security Solutions suite are a set of best security practices, as well as endorsement of compliant products which implement elements of the architecture. A fully compliant MISSI architecture controls system access by "level" (e.g., unclassified, sensitive, confidential, secret and top secret) and "compartment" (i.e., topical area of interest). Authentication and encryption are provided courtesy of the Fortezza family of chips developed by Mykotronx Inc. See also Fortezza.

missile mail On June 8, 1959, a postal official heralded a move by the US Post Office as being "of historic significance to the peoples of the entire world." On that date, which will live forever in the arcane annals of the USPS (United States Postal Service), the Navy submarine U.S.S. Barbero fired a guided missile carrying 3,000 letters at the Naval Auxiliary Air Station in Mayport, Florida. The official went on to say that "Before man reaches the moon, mail will be delivered within hours from New York to California, to Britain, to India or Australia by guided missiles." See also Snail Mail.

mission critical operation An operation that an organization considers key to its success and survival, such as sales, marketing, and customer service. An organization will probably be more willing to invest resources in order to fix a problem or make improvements to a "mission critical" operation than to other operations.

mission critical systems Systems on which the future success of an organization depends.

MIT Media Laboratory Founded by Nicholas Negroponte, author of the book "Being Digital," the MIT Media Lab opened its doors in 1985 at the Massachusetts Institute of Technology in the Wiesner Building designed by I.M. Pei. According to its web site, "in its first decade, much of the Laboratory's activity centered around abstracting electronic content from its traditional physical representations, helping to create now-familiar areas such as digital video and multimedia. The success of this agenda is now leading to a growing focus on how electronic information overlaps with the every day physical world. The Laboratory pioneered collaboration between academia and industry, and provides a unique environment to explore basic research and applications, without regard to traditional divisions among disciplines." During the booming 1990s, the Media Lab attracted huge donations from cash-rich technology companies whose executives wanted to be associated with the MIT Media Lab "stars." Donations ebbed when some of the sponsors hit hard times in their main technology businesses and some started to wonder what the MIT Media Lab had actually accomplished, if anything. The last I heard: Negroponte was leading an effort to bring $100 computers to the third world.

Mitel Mitel is a IP telephony PBX and computer telephony component maker. The word Mitel actually stands for MIke and TErry's electric Lawn mower. Here's how it happened. In the early 1970s, Mike Cowpland and Terry Matthews were engineers working in the semi-conductor factory of Northern Telecom in Kanata, Ontario, Canada. They decided they wanted to go out on their own. Their first idea was cordless, electric lawn mowers. Apparently they thought there was a demand for such devices. They formed Mitel and ordered electric lawn mowers from a manufacturer in England. The mowers arrived after the onset of the Canadian winter in November, i.e. too late. No one wanted lawn mowers. So Terry and Mike decided to make telephone systems. That went well. Mitel went public. Most of the company was later sold to British Telecom, who lost their shirt on it and eventually sold it on a fire sale to a venture capital company. Terry moved on and founded the eminently successful Newbridge Networks, which he later sold to Alcatel. Mike got involved with Corel, which he eventually screwed up and ultimately got kicked out of. In early 2001, Terry Matthews bought back the trademark and the telecommunications division of Mitel. He and Don Smith are now running Mitel.

mitzvah Most people think mitzvah is a Jewish word meaning joy – often good deed done without expecting anything in return. This is not correct. In modern Hebrew, mitzvah follows the traditional standard and is used to mean command or duty. "Ani metzaveh lach" means "I command you." In short, mitzvah is a duty. The traditional term for a good deed without hope of repayment was 'chesed shel emmes' or plain 'Chesed'. Joy was 'simcha'. Mitzvah was never used in the way it is commmonly used today – as a good deed or the joy of good fortune. See also Bar Mitzvah.

MIX 1. Multinational Internet eXchange.

2. Multicasting International eXchange.

mixed cable Cables have characteristic impedances which vary according to the cables' physical parameters. When cables of different characteristic impedance are mixed, an impedance imbalance will occur. It is thus bad practice to mix different wire gauges, and twisted or non-twisted pair cables. However, it is accepted practice to combine long runs of twisted pair cable with short lengths of modular patch cords and line cords. Baluns are an exception to this rule.

mixed mode An imprecise term which suggests that one digital bit stream can carry voice, data, facsimile and video signals.

mixed mode night service After-hours answering of incoming calls in which Assigned Night Answer (ANA) is specified for some trunks, and Universal Night Answer (UNA) specified for others.

mixed signal processor MSP. A multi-functional chipset that can do both digital-to-analog (D/A) and analog-to-digital (A/D) signal conversions and processing both analog and digital signals. Therefore and for example, a MSP can serve as a codec, converting analog voice into digital signals for transmission over a digital circuit such as a T-1. It also can serve as a modem, converting digital computer signals into analog format for transmission over an analog circuit. On the terminating end of the circuit, the MSP reverses the processes in order to convert the signal back into its native format.

mixed station dialing A telephone system feature which allows you to install both rotary and pushbutton phones on your phone system. Most modern PBXs have this feature.

MJ Modular Jack. A jack used for connecting voice cables to a faceplate, as for a telephone. See RJ (Registered Jack).

MJU Multi-Junction Unit.

MLHG MultiLine Hunt Group.

MLID Multiple Link Interface Driver. A layer of the Novell Open DataLink Interface specification. The MLID layer controls a specific network interface, and works below the Link Support Layer.

MLL Monthly Leased Lines.

MLM Meridian Link Module. A Northern Telecom term for an Application Module that provides a link to a host processor through the Meridian Link interface.

MLPP See Multilevel Precedence Preemption.

MLS Microwave Landing System.

MLSC Multiple Low Speed CSU.

MLST Minimum Line Scan Time. The minimum amount of time it takes a transmitting fax machine to scan each of the 1,143 fax lines of a document. The MLST of older machines can be as long as 40 milliseconds (ms), or thousandths of a second. The MLST of newer machines can be 20 ms or fewer.

MLT See Mechanized Loop Test.

MLT-3 Multi Level Transmit - 3 Levels. The ANSI approved modulation scheme used for the transmission of data on an FDDI network over shielded and unshielded copper twisted pair media (see TP-PMD).

MM Mobility Management. A wireless industry term.

MMAC Multimedia Mobile Access Communication. MMAC is a wireless standard used largely in Japan. Like IEEE 802.11a and HiperLAN/2, which are both used in North America and Europe, MMAC uses OFDM (Orthogonal Frequency Division Multiplexing) as the physical layer specification and uses data rates ranging from 6 to 54 Mbps in the 5-GHz band. See OFDM. According to the Multimedia Mobile Access Communication Systems Promotion Council (MMAC-PC), an MMAC is a mobile communication system that can "transmit ultra high speed, high quality Multimedia Information 'anytime and anywhere' with seamless connections to optical fiber networks." The frequency bands and associated applications intended to be supported include 3-60 GHz for mobile video, 30-300 GHz for WLAN (Wireless LAN) and high quality TV, 5 GHz for WLAN and ATM access, and 3-60 GHz for Wireless Home-Link. The MMAC Systems Promotion Council includes well over 100 vendors, most of which are Japanese companies. See also HiperLAN.

MMCF Multimedia Communications Forum. Formed in June 1993, the MMCF is an non-profit research and development organization of telecommunications service providers, multimedia application and equipment developers, and end users who realize the revolutionary potential of multimedia communications. Forum members are dedicated to accelerating market acceptance and multivendor interoperability of multimedia communications worldwide. MMCF acts as a central clearinghouse for all multimedia communications-related standards, specifications and recommendations.

MMDS Microwave Multi-point Distribution System or Multipoint Multichannel Distribution Service. Irrespective of what the acronym means, the definition is the same. MMDS is a way of distributing cable television signals, through microwave, from a single transmission point to multiple receiving points. Often used as an alternative to cable-based cable TV. According to an April, 1995 press release from Pacific Telesis, which was starting an MMDS service, "in digital form, it will provide more than 100 channels to a radius of approximately 40 miles from the transmitter. The MMDS transmitter delivers video to homes that are in its 'line of sight.' MMDS transmissions are limited by the terrain and foliage of a given market. The microwave signal is received by an antenna on the subscriber's home, then sent down coaxial cable to a box atop the customer's TV set. The box decodes and decompresses the digital signal." MMDS is increasingly being called "Wireless Cable." See Wireless Cable.

MME Mobility Management Entity.

MMF Multi-Mode Fiber. Due to its relatively thick inner core (25-200 microns) MMF allows light signals to travel in many modes (i.e., physical paths). Some portions of light pulses travel more or less down the center of the inner core; some disperse (i.e., spread out) and strike the edges of the core, at which point they are reflected back into the core by the cladding; and some actually enter the cladding and travel through it for a distance before they re-enter the core. Some paths are longer than others. The longer the path, the longer the time it takes to travel it. So, some portions of a light pulse arrive at the receiver before some other portions. This phenomenon of "modal dispersion" results in "pulse dispersion," which is the distortion, or smearing, of the individual pulses. There are two types of MMF: step-index fiber and graded-index fiber. MMF usage largely is restricted to application in LANs and FDDI (Fiber Distributed Data Interface), both of which are characterized by relatively short distances and relatively low speeds. Single Mode fiber (SMF), which has a thinner core, is used on long haul, high speed applications. See also FDDI, Graded-Index Fiber, SMF and Step-Index Fiber.

MMFD Abbreviation for micromicrofarad; one millionth of a farad, the unit of measuring capacitance.

MMI 1. Man Machine Interface.

2. Machine Machine Interface. The former is more common.

MMIC Microwave Monolithic Integrated Circuit.

MMITS Modular Multifunction Information Transfer System. Hiding behind something call the MMITS Forums is a group of people working to define software programmable radios. There are two areas of emphasis:

Handheld – working with being able to download software into cellular handsets so they can work with a variety of air interfaces.

Mobile – concentrating on military requirements initially. but looking at the needs of public safety (Police, Fire, etc.) for the future.

The basic idea is to bring PC concepts to the radio world by moving the Digital/Analog – Analog/Digital function very close to the antenna, and do all of the tuning, spread/de-spread, modulation/demodulation, etc with DSPs. www.mmitsforum.org.

MMJ A six wire modular jack with the locking tab shifted off to the right hand side. Used in the DEC wiring system.

MML Man Machine Language.

MMoIP 1. Multimedia over IP. 2. Multimedia mail over IP.

MMR Modified Modified Read data compression method used in newer Group 3 facsimile machines.

MMS 1.Material Management System.

2. Multimedia Messaging Service. A service that allows cell phone users to send pictures, movie clips, cartoons and other graphic materials from one cell phone to another. According to a press release from Singapore's SingTel's announcement of MMS: "With MMS, SingTel Mobile's postpaid customers can send photos and pictures with integrated text and voice clips from their MMS mobile phones to another mobile phone. Recipients will get an MMS message, if they are using MMS phones, or an SMS notification to retrieve the MMS message via the Internet or email. MMS messages can also be sent directly to email addresses."

MMSI See Maritime Mobile Ship Identity.

MMSU Modular Metallic Service Unit.

MMTA MultiMedia Telecommunications Association. The successor organization to NATA. MMTA was orignally organized around five divisions – computer telephony integration, conferencing/collaboration and messaging, LAN/WAN internetworking, Voice/Multimedia and Wireless Communications. MMTA is on 202-296-9800. In November, 1996 MMTA announced its intent to merge with TIA, another Washington organization called Telecommunications Industry Association. On December 15, 1997 MMTA announced that it had been officially combined into the Telecommunications Industry Association. The combination of all these organizations reflected the fact that the telecommunications industry had become competitive, thus depriving many of Washington organizations of their reason for existing. See also www.tiaonline.org. See ACTAS and NATA.

MMU Memory Management Unit. Circuitry that manages the swapping of blocks of memory.

MMUSIC Multiparty Multimedia Session Control. A Working Group (WG) of the IETF (Internet Engineering Task Force) , the MMUSIC is chartered to develop Internet standards track protocols to support Internet teleconferencing sessions. MMUSIC's current focus is on supporting loosely controlled conferences on the MBone, although the protocol are ensured to be general enough to be used in managing tightly-controlled conferences on any IP-based network. Among its accomplishments are the RFCs for Real-Time Stream Protocol (RTSP), Session Announcement Protocol (SAP), Session Description Protocol (SDP), Session Initiation Protocol (SIP), and Simple Conference Control Protocol (SCCP). See also SIP.

MMVF A format of rewritable DVD disc proposed by NEC. It stands for Multi Media Video File. For a longer explanation, see DVD-RAM and DVD.

MNA Multimedia Network Applications.

MNA7 Multiple CCS7 Network Addresses.

MNC Mobile Network Code. A part of the IMSI (International Mobile Subscriber Identity) or LAI.

mnemonic From the Greek mnemonikos, a shorthand label or term that is easy to remember. A mnemonic is a symbolic representation of an address (e.g., ATL for Atlanta, or DLS for Dallas) or operation code (e.g., JMP for jump). Acronyms are a type of mnemonic; LASER, for instance, is shorthand for Light Amplification by Stimulated Emission of Radiation. See also Acronym.

mnemonic dial plan Pronounced "nemonic." A way of dialing using characters typed on the keyboard of a terminal. The word Mnemonic comes from the same roots as

memory. It's a memory jogging way of remembering something, like a way to dial. See Mnemonic Prompts.

mnemonic prompts System commands represented by the appropriate alphabet letter rather than by a number, (for example, "P" to "Play", "A" to Answer"). See Mnemonic Dial Plan.

MNLP Mobile Network Location Protocol.

MNP 1. Microcom Networking Protocol.

2. Mobile Number Portability.

MNP10-EC Error correction protocol for awful communications environments, like cellular networks. Use of MNP10-EC helps prevent disruptive signal fading and reduces the number of dropped calls that occur when you're trying to send data over cell networks. See Microcom Networking Protocol for a greater explanation.

MNRP Mobile Network Registration Protocol.

MO See Magneto Optic Drive.

Mobcasting Mobile audio podcasting using a phone-in blogging service, such as AudioLink. See podcasting.

mobi A new top-level domain for websites whose content is customized for the small screens of mobile devices, like cellphones.

mobile In North america, the original term for a telephone that worked with cellular service was cell phone. In the rest of the world it was called a "mobile." Eventually America caught up and the industry is now preferring the term mobile to describe a cell phone. The technical term – i.e. what you find in the industry literature – is Mobile Station.

mobile ad hoc network MANET. I haven't seen much in the way of actual products. But a 1999 IETF paper by S. Corson of the University of Maryland and J. Macker of the Naval Research Laboratory contained the following words: "The vision of mobile ad hoc networking is to support robust and efficient operation in mobile wireless networks by incorporating routing functionality into mobile nodes. Such networks are envisioned to have dynamic, sometimes rapidly-changing, random, multihop topologies which are likely composed of relatively bandwidth-constrained wireless links. Within the Internet community, routing support for mobile hosts is presently being formulated as "mobile IP" technology. This is a technology to support nomadic host "roaming", where a roaming host may be connected through various means to the Internet other than its well known fixed-address domain space. The host may be directly physically connected to the fixed network on a foreign subnet, or be connected via a wireless link, dial-up line, etc. Supporting this form of host mobility (or nomadicity) requires address management, protocol interoperability enhancements and the like, but core network functions such as hop-by-hop routing still presently rely upon pre- existing routing protocols operating within the fixed network. In contrast, the goal of mobile ad hoc networking is to extend mobility into the realm of autonomous, mobile, wireless domains, where a set of nodes–which may be combined routers and hosts--themselves form the network routing infrastructure in an ad hoc fashion. The technology of Mobile Ad hoc Networking is somewhat synonymous with Mobile Packet Radio Networking (a term coined via during early military research in the 70's and 80's), Mobile Mesh Networking (a term that appeared in an article in The Economist magazine regarding the structure of future military networks) and Mobile, Multihop, Wireless Networking (perhaps the most accurate term, although a bit cumbersome). There is current and future need for dynamic ad hoc networking technology. The emerging field of mobile and nomadic computing, with its current emphasis on mobile IP operation, should gradually broaden and require highly-adaptive mobile networking technology to effectively manage multihop, ad hoc network clusters which can operate autonomously or, more than likely, be attached at some point(s) to the fixed Internet. Some applications of MANET technology could include industrial and commercial applications involving cooperative mobile data exchange. In addition, mesh-based mobile networks can be operated as robust, inexpensive alternatives or enhancements to cell-based mobile network infrastructures. There are also existing and future military networking requirements for robust, IP-compliant data services within mobile wireless communication networks [1]–many of these networks consist of highly-dynamic autonomous topology segments. Also, the developing technologies of "wearable" computing and communications may provide applications for MANET technology. When properly combined with satellite-based information delivery, MANET technology can provide an extremely flexible method for establishing communications for fire/safety/rescue operations or other scenarios requiring rapidly-deployable communications with survivable, efficient dynamic networking. There are likely other applications for MANET technology which are not presently realized or envisioned by the authors. It is, simply put, improved IP-based networking technology for dynamic, autonomous wireless networks."

Mobile Application Subsystems MAS. That portion of a Mobile End Systems (M-ES) concerned with the provision of application services. The MAS contains the applications software that is independent of the CDPD network. In most cases, the includes network software.

mobile attenuation The power of the mobile phone can be adjusted (or attenuated) dynamically to one of seven discrete power levels (analog cellular). This is done so that when a mobile comes closer to a base receiver its power is reduced to prevent the chance of interfering with other mobiles operating on the same voice channel in another cell (co-channel interference). Additionally, this is even more important to portable units to keep the transmit power at a minimum to increase the talk usage time before the batteries expire.

mobile banking Think of your phone as your bank. Punch some buttons on it: You can pay for a haircut you've just had, you can pay for your groceries, you can borrow money, you can send money to your relatives in another country, you can pay your gas bills, you can check how much money you have in the bank – in fact do everything a wallet stuffed with bills, a bank account and an Internet connection could do for you. Imagine that you're poor, working in a foreign country (say a Pakistani in Kuwait) and want to send money to your relatives back home. Today you visit an informal "bank", pay an outrageous commission and your money is paid out at the other end by another member of that informal "bank." If both you and your relative own a mobile phone, you can transfer the money instantly and inform him of the transaction all in one fell swoop – at minimal expense. Mobile banking is happening today in places such as South Africa, Kenya, Zambia and Congo.

mobile browser A Web browser designed specifically for a mobile handheld communications device.

mobile cellular phone See Mobile.

mobile cellular service A fancy name for cellular phone service or cell phone service. See Cell, Cell Switching, Cellular Radio and IN.

mobile communications Quite simply, the ability to communicate while moving. Wireless technologies like cordless telephony and cellular allow you to communicate on the go. Wired technologies, like copper wire and fiber optics, don't work if you might break the wire, or run out of wire. One of the great stories about mobile communications involves wires. According to William E. Kennard, FCC Chairman, and as included in a speech he made in 1998, the U.S. military faced a problem in 1907. The cavalry needed to communicate when on patrol. So, one horse soldier would ride behind the troops, unreeling wire along the ground. When a message needed to be sent back to the base camp, the soldier would jump off his horse, plant a metal stake in the ground, and send a message via Morse code. This process slowed the progress of the troop. Military engineers came up with a solution. They put a copper patch on the horse's skin. Since a horse always has one hoof on the ground, the circuit was complete. Therefore, the scouts could send a message while riding. (No mention is made of the effect on the horse.)

mobile controlled handoff This means the decision to initiate a transfer or handoff from one cell to another cell is under the control of the mobile device. Used in CDPD.

mobile data Mobile data is a generic term used to describe data communications through the air from and to field workers – from package deliverers, to car rental companies (to track cars), to field service personnel, to law enforcement officials checking license plates.

Mobile Data Base Station MDSB. Component of the CDPD network that provides data link relay functions for a set of radio channels serving a cell. An MDBS is located in each cell site, and its primary role is to relay data between Mobile End System (M-ES) and the Mobile Data Intermediate System (MD-IS). it is the stationary network component responsible for managing interactions across the airlink interface.

Mobile Data Intermediate System MD-IS. The CDPD network element that performs routing functions based in knowledge of the current location of the M-ES. Responsible for CDPD mobility management. A cellular radio term.

Mobile Data Link Protocol MDLP. The Link Layer protocol used in Cellular Digital Packet Data (CDPD). Provides Temporary Equipment Identifier (TEI) management, multiple frame operation, unidata transfers, exception condition detection with selective reject recovery, etc.

Mobile Digital Voice Channel MDVC. The channel between a mobile phone and a cell site antenna in a digital cellular or PCS environment. The MDVC supports both voice and data transmission, although the allocated bandwidth is designed primarily to support voice. Signaling and control functions take place over separate channels set aside specifically for that purpose.

Mobile End System M-ES. An end system that accesses the CDPD network through the airlink interface. The device that allows mobile users to work in an untethered fashion while remaining connected to a data network. The system's physical position may change during data transmission. A cellular radio term.

mobile entrophy My friend is a klutz. He claims that wherever he goes chaos follows.

mobile home function A Mobile Data Intermediate System, that (1) maintains an information database of the current serving area of each of its homed Mobile End Systems (M-ESs), and (2) operates a packet forwarding service for its homed M-ESs. A cellular radio term.

mobile identification number When the "SEND" key on a cellular phone is pressed, the phone transmits an origination message to the base station. This message includes the dialed digits and the identity of the calling cellular phone. The calling cellular phone is identified by its Mobile Identification Number (MIN), which is usually the same as its ten-digit phone number. See also ESN.

Mobile IP An emerging set of extensions to the Internet Protocol for packet data transmission, Mobile IP is intended to serve nomadic users connecting on a wireline, rather than a wireless, basis. Mobile IP is being developed by the IETF (Internet Engineering Task Force) to operate much like a highly secure and dynamic packet data communications version of a postal service forwarding address. The benefit is that the nomadic user will not have to continually change IP addresses and reinitialize sessions. The IETF (Internet Engineering Task Force) defines Mobile IP as enabling an IP device to roam across networks and geographies, while remaining constantly connected as if always attached to its home location _using_the_same_IP_address_ (emphasis supplied). It will work like this:

The mobile node will have one permanent address and another for location purposes and another for identifying it to other network nodes. Data will be transmitted to the permanent address, associated with the "home agent." When the nomadic node is traveling, the "home agent" will forward the data in care of the "foreign agent," the IP server serving the foreign subnet, through a process of encapsulating that data with another IP address contained in a data header preceding the original packet. Once the data packets are received by the foreign agent, the additional header will be removed through a process known as decapsulation. Should the node relocate yet another time, both the "home agent" and the previous foreign agent will be advised of that fact; thereby, inflight packets can be forwarded by the previous foreign agent to the new foreign agent through a process known as "smooth handoff." While there currently is no Mobile IP standard being developed for wireless mobility, Mobile IP promises to make life easier for users that roam from location to location within a multisite corporate enterprise. See also IP.

mobile malware Malicious software such as viruses, worms and Trojan horses that is specifically designed to infect wireless handsets, i.e. cell phones and PDAs that work as phone also.

mobile marketing The delivery of marketing messages to mobile handheld devices.

Mobile Mesh Networking See mobile ad hoc networking.

Mobile Multimedia Access Communication See MMAC.

Mobile, Multihop, Wireless Networking See mobile ad hoc networking.

mobile mounting kit An optional cellular phone accessory that allows a transportable or portable to be connected to a vehicle's power supply and antenna lead, thereby boosting power and improving reception. Sometimes referred to as a car kit or car mounting kit. Some of these kits are very expensive. Check the price of the kit before you buy your phone.

Mobile Network Location Protocol MNLP. A cellular radio term. In the CDPD network, the MNLP is the protocol used between the Home Mobile Data Intermediate System (MD-IS) and the Serving MD-IS and it used to keep the Home MD-IS updated on the location of a Mobile End System (M-ES) (i.e. the location of the cell phone).

Mobile Network Registration Protocol MNRP. In the CDPD cellular radio network, protocol used between the Mobile End System (M-ES) and the Serving Mobile Data Intermediate System (MD-IS) to announce the M-ES's Network Entity Identifier (NEI) and to confirm the service provider's willingness to provide service.

Mobile Number Portability MNP is a set of processes by which you eventually might be able to change your cellular or other wireless service provider, and retain your telephone number. MNP essentially is the mobile wireless version of LNP (Local Number Portability). See LNP for lots more detail.

Mobile Packet Radio Networking See mobile ad hoc network.

mobile phone Another name for a cellular phone. There are four main types of cellular phones – mobile (also called car phone), transportable, portable and personal. A mobile phone is attached to the vehicle, the vehicle's battery and has an external antenna. The mobile phone (the car phone) transmits with a standard three watts of power. Mobile telephone service is provided from a broadcast point located within range of the moving vehicle. That range is called a "cell." The broadcast point in turn is connected to the public network so that calls can be completed to or from any stationary telephone, i.e. one connected to a land line. See Cellular and Car Phone.

mobile phone jammer A device which transmits radio waves at cell phone frequencies causing it to jam communications between cell phones and their base station. The main purpose of a mobile phone jammer is to stop cell phones from receiving phone calls. The first mobile phone jammer came out of Israel in the summer of 1999. Costing around $1000, the device allegedly shut off all phone calls coming into cell phones in a room. Customers for such devices include recording studios, cinemas and concert halls. There is some speculation that the device may be illegal. www.cguard.com.

Mobile Phone-Throwing World Championship An annual competition in Savonlinna, Finland, started in 2000, to see who can throw a mobile phone the farthest or with the best form. There are a variety of throwing events, by age group. Distance thrown determines the winner in some events.   Style and aesthetics determine the winner in other events.

Mobile Serving Function MSF. A Mobile Data Intermediate System (MD-IS) function that (1) maintains an information database of the Mobile End Systems (M-ES) currently registered in the serving area, and (2) de-encapsulates forwarded packets from the MHF and routes them to the correct channel stream in a cell where the destined M-ES is located.

Mobile Switching Center MSC. The location of the Digital Access and Cross-connect System (DACS) in a cellular telephone network.

mobile station See Mobile.

Mobile Station Roaming Number See MSRN.

Mobile Telecommunications Sourcing Act This law, signed by President Clinton on July 28, 2000 enables states to source cellular calls for tax purposes, only at the customers place of "primary use". For residential customers the place of primary use is their home address, while for business customers it is their primary business street address. The effect of the law is that if a cellular user has a home residence in State A , their place of primary use, but originates an interstate call in State B, only State A can assess sales tax on the call.

mobile telematics Sometimes just called telematics. It involves integrating wireless communications and (usually) location tracking devices (generally GPS) into automobiles. The best known example is GM's OnStar system, which automatically calls for assistance if the vehicle is in an accident. These systems can also perform such functions as remote engine diagnostics, tracking stolen vehicles, provide roadside assistance, etc. www.onstar.com.

Mobile Telephone Exchange MTX. The Northern Telecom term for MTSO (Mobile Telephone Switching Office).

mobile terminated The term used to describe a call where the destination of the call is a mobile (i.e. cellular) telephone.

mobile unit The cellular telephone equipment installed in a vehicle. It consists of a transceiver, control head, handset and antenna.

Mobile Virtual Network Operator See MVNO and MVNE.

Mobile Virtual Network Operator Enabler See MVNE and MVNO.

mobileco Another name for a cell phone provider, also called cellco.

mobilink A "unified" cellular phone service covering 83% of North America's population. It is a consortium of six Bell cellular operators and some Canadian cellular operators. The idea is simple. Anyone calling a subscriber of one of these companies would have his call automatically routed to the subscriber, no matter where in Mobilink that subscriber was. Before Mobilink, you had to know where the person was you wanted and then dial a bunch of complex codes to get to him.

mobisode Original programming made specifically for mobile phones and devices with small video screens, such as Apple's iPod. Mobisodes are sometimes used as advertisements for television shows and movies.

Mobitex Mobitex is a packetized narrow-band data service, originally conceived by Swedish Telecom and further developed by Eritel, a joint venture of Swedish Telecom and Ericsson. The service is being offered in the United States by RAM Mobile Data/ Bell South. Base stations, which typically cover 5-15 mile radii, are arranged in a cellular-like fashion. The technical details of Mobitex, collectively referred to as the Mobitex Interface Specification or MIS, can be obtained from any operator of a public Mobitex network or from Ericsson directly. Mobitex is an open, non-proprietary system but the specification is copyrighted and is made available under a royalty-free license. The MIS is not in the public domain. All Mobitex networks operate in one of three frequency families: 80 MHz, 400 MHz, or 900

MHz. Frequencies and channels are generally assigned by a national government authority; each local operator can tell you the specific channel assignments for its country. There are 29 Mobitex networks around the world, in 22 countries on six continents. Some of the networks are operated publicly (that is, you may buy network service from the local operator) while others are owned and operated by companies for their own use and benefit.

moblogging The art of snapping a cell phone camera picture and instantly sending it to your blog for posting on the Internet.

mobo Slang for MOtherBOard.

MOC Mobile Originating Call. A cellular phone term. MOC refers to the central office where the outbound leg of a call begins. See also MTC.

MoCA Multimedia over Coax Alliance. An alliance of cable companies, telephone companies, networking gear manufacturers, and other industry players whose mission is to develop and promote specifications and certify interoperable products that enable distribution of multimedia content within the home using existing in-home coaxial cabling. MoCA also refers to the specifications developed by the alliance. MoCA-compliant products and services are now in early deployments and field trials.

mocio-economics See mociology.

mociology The study of how people adapt and use wireless technologies, from buying concert tickets to organizing political rallies. The field gets its name from mobile and sociology – and has already spawned an offshoot, mocio-economics, the study of how companies can capitalize on all this mocio-activity.

Mockapetris, Paul Inventor of the Domain Name System (DNS), and operator of the Internet's original root servers. He is also well known for his wry, often-repeated assertion that "the future of the Internet is ahead of it."

moco Mobile content.

mod Modification.

modal dispersion Also known as Differential Mode Delay (DMD), modal dispersion can be thought of as the blurring of the input signal in a fiber by the several modes (paths) that the signal may take as it propagates through (transverses) a fiber. Each light pulse in a digital optical fiber system begins life as a distinct pulse created by a light source, such as a Light-Emitting Diode (LED) or a Laser Diode, with laser diodes being used in the high-speed, long-haul systems. As each pulse propagates through the thin and very pure inner core of the fiber, it naturally spreads out. The inner core is surrounded by a layer of "cladding," glass of a slightly different refractive index which serves to bend the light signal back inward. As some portions of the light pulse travel more or less through the center of the core while other portions of the pulse bounce around, they travel different distances from point to point. Therefore, some portions of the light pulse arrive sooner than others. The result of Modal Dispersion is known as "Pulse Dispersion," which is a blurring or smearing of the individual pulses. In other words, the pulses overrun each other, losing their individual identity. The thicker the inner core, the worse the effect. The longer the cable, the worse the effect. The faster the transmission speed, the short the interval between the pulses, and the worse the effect. As Single Mode Fiber (SMF) has a thinner inner core than MultiMode Fiber (MMF), the effect of modal dispersion is less with SMF. See also Collimation, MMF and SMF.

Modal dispersion can be thought of as the blurring of the input signal in a fiber by the several modes that may propagate down a fiber. Each mode may take a separate path down the fiber, and thus that signal may arrive at slightly different times. The dispersion depends on the fiber's internal characteristics and its length. In short, modal dispersion is pulse rounding in lightwave communications that takes place because of the slightly different paths followed by the laser light rays as they arrive at the detector slightly out of phase.

modal distribution In an optical fiber operating at a single wavelength, the number of modes supported by the fiber, and their propagation time differences. In an optical fiber operating at multiple wavelengths simultaneously, the separation in wavelengths among the modes being supported by the fiber.

modal loss In an open waveguide, such as an optical fiber, a loss of energy on the part of an electromagnetic wave due to obstacles outside the waveguide, abrupt changes in direction of the waveguide, or other anomalies, that cause changes in the propagation mode of the wave in the waveguide.

modal noise A source of signal distortion in fiber optic networks, modal noise is power fluctuations in the receivers, and is a result of interactions between the fiber and the connectors when high-power lasers are used to increase transmission distances.

Modbus protocol Protocols define the rules by which devices talk with each other. Modbus is an application layer messaging protocol, positioned at level 7 of the OSI model, that provides client/server communication between devices connected on different types of buses or networks. MODBUS is a request/reply protocol and offers services specified by function codes. MODBUS function codes are elements of MODBUS request/reply PDUs. Modbus protocol was developed by Modicon in 1979 and is used in industrial manufacturing to establish master-slave/client-server communication between intelligent devices.

mode 1. In fiber optics, a mode is a physical path taken by a light ray. MultiMode Fiber (MMF) is characterized by a relatively thick core (usually either 50 or 62.5 microns), which allows the light rays that make up each digital pulse to seek many modes, depending on several factors. First, the light source can inject the light rays into the core at many angles. Second, the light rays naturally disperse, i.e., spread out, as they propagate through the fiber. The modes that are relatively straight through the center of the fiber core are known as low-order modes. Those that are less direct and involve light rays bouncing around from edge-to-edge of the core/cladding interface are known as high-order modes. This combination of low-order and high-order modes results in a phenomenon known as modal dispersion, which limits the distances and transmissions rate of MMF systems. Single Mode Fiber (SMF) has a much smaller inner core (8-10 microns), which essentially limits the number of possible modes to a single path. See Modal Dispersion.

2. Mode is essentially a switch inside a computer that makes it run like another computer, usually an older one.

mode coupling The transfer of energy between modes. In a fiber, mode coupling occurs until equilibrium mode distribution, EMD, is reached.

mode evolution A fiber optic term. The dynamic process a multilongitudinal laser undergoes whereby the changing distribution of power among the modes creates a continuously changing envelope of the laser's spectrum.

mode field diameter A fiber optic term. A measure of distribution of optical power intensity across the end face of a single-mode fiber.

mode filter A fiber optic term. A device that removes higher-order modes to simulate equilibrium mode distribution.

mode S Discrete addressable secondary radar system with data link.

mode scrambler A fiber optic term. A device that removes cladding modes. A device that induces mode coupling for uniform power distribution.

Model T Ford Henry Ford produced the model T only in black because the black paint available at the time dried the fastest.

modem 1. Acronym for MOdulator/DEModulator. Conventional modems comprise equipment which converts digital signals to analog signals and vice versa. Modems are used to send digital data signals over the analog PSTN (Public Switched Telephone Network). Although the carrier switches (e.g., central offices and tandem offices) are typically digital, as is the backbone transmission network (e.g., T-carrier), the local loop always is analog unless the user orders a more costly digital loop (e.g., ISDN or T-1). Therefore, the PSTN is analog as far as most people are concerned.

Conventional modems work like this. Your PC outputs data in the form of "1's" and "0's" which are represented by varying levels of voltage. The modem converts the digital data signal into variations of the analog sine wave so the data can be transmitted over the PSTN. A matching modem on the other end reverses the process in order to present the target device with a digital bit stream. The modulation techniques include some combination of Amplitude Modulation (AM), Frequency Modulation (FM) and Phase Modulation (PM), also known as Phase Shift Keying (PSK). Used in combination, these techniques allow multiple bits to be represented with a single (or single set) of sine waves. In this fashion, compression is accomplished, which allows more data to be transmitted in the same period of time and which therefore reduces the connect time and the associated cost of the data transfer. Contemporary, conventional modems are standardized by the ITU-T as part of the "V" series of standards. Such modems are characterized by error detection and correction mechanisms, adaptive equalization, internal dialing, and numerous other sophisticated capabilities. 56 Kbps modems are the latest development in the world of conventional modems; they remain to be standardized. The term "modem" also is applied (and correctly so, in the purely technical sense) to ISDN TAs (Terminal Adapters), ADSL TUs (Terminating Units), line drivers and short-haul modems. The last two, in fact, are voltage converters.

See also Line Driver, Modem Eliminator, Modem Pool, Modem Standards, Modulation Protocols, Serial Port, Short-Haul Modem and 56 Kbps Modem.

2. According to the Vermonters' Guide to Computer Lingo, modem is what landscapers do to dem lawns. (This is a joke.)

modem bonding A term which describes the bonding, or linking, of two 56 Kbps modems over two phone lines to double the performance. This process is accomplished through matching devices, one on each end of the connection; each modem operates at its maximum achievable rate, with the aggregate rate being roughly double that of each individual modem. Theoretically, modem bonding can yield speeds of as much as 128 Kbps

downstream and 67.2 Kbps upstream, although the FCC currently limits maximum downstream performance through each modem to 53 Kbps, for a total of 106 Kbps. Modem bonding technologies are proprietary. See 56 Kbps Modem.

modem cowboy A slang term for someone who typically lives in the mountain states of Western America and does most of his work via modem.

modem eliminator A wiring device designed to replace two modems; it connects equipment over a distance of up to several hundred feet. In asynchronous systems, this is a simple cable. Here is a specific application using a Modem Eliminator: You can connect a PC to a printer, or a PC to another printer using a cable. But you can only go a certain distance – maybe 100 feet. After that, the traditional solution has been to use a modem and go over traditional phone lines. Instead, you can connect the two devices directly by wire using a Modem Eliminator. There are two advantages of a Modem Eliminator over a normal modem. The eliminator is cheaper and it can often transmit faster. According to Glasgal Communications, there are many cases where it is either unnecessary, cumbersome or too expensive to interconnect terminals using modems or line drivers in an experimental or a very short-haul environment. A modem eliminator functionally resembles two modems back-to-back on a leased line and therefore saves the cost of two modems and a line in many situations.

modem fallback When the telephone line quality is not good enough to accommodate the top rated speed of a modem – for example 14,400 bps – the modem drops down to lower speeds – initially to 9,600 bps, then if necessary, to 4,800 bps, or even down to 2,400 bps.

modem on hold See V.92.

modem pair delay The one way delay added by a pair of modems to the propagation delay on a communications line.

modem pool A collection of modems which a user can dial up from his terminal, access one and use that one to make a data call over the switched telephone network. Modem pools are obviously designed to allow many users to share few modems, thus saving on modems. Now that modems have become less expensive, the advantages of modem pooling are no longer as great. There are also some advantages in having a modem right next to your terminal or computer – namely you can see how it is functioning. And modems have lights to indicate what they're doing. One of the most useful lights on most modems indicates whether the line is "off hook" or not. It is possible for your computer to instruct your modem to hang up the line and for your modem to forget to do it, leaving you with a huge phone bill. One problem with giving people their own modems is they (the modems) have a tendency to get pinched. It's hard to screw down modems. Harder, anyway, than computers and disk drives.

modem server A networked computer with a modem or group of modems attached to it that allows network users to share the modems for outbound calls.

Modem Sharing Unit MSU. A hardware device, most often a simple contention switch operating on RTS (Request to Send) leads to DTE interfaces, that permits only one terminal at a time to use the modem.

modem standards Definitions of electrical and telecommunications characteristics which enable modems from dissimilar manufacturers to speak to each other. Bell 103... US standard for 300 bps; ITU-T V.21...International standard for 300 bps; Bell 212A...US standard for 1200 bps; ITU-T V.22...International standard for 1200 bps; ITU-T V.22 bis... US and international standard for 2400 bps; ITU-T V.23...International videotex standard (1200/75 bps or 75/1200 bps). See also Hayes Command Set and V.Series recommendations, i.e. V.34.

modem turnaround time The time needed for a half-duplex modem (an old-fashioned one) to reverse its transmission direction.

modem-on-hold See V.92.

moderated mailing list A mailing list where messages are first sent to the list owner before they are distributed to the subscribers.

moderator A moderator is a person who controls what gets posted to a particular Internet newsgroup. A moderator is used to ensure that a newsgroup's article stick to the agreed upon subject matter. A newsgroup may or may not have a moderator.

Modified Final Judgment MFJ. The agreement reached on January 8, 1982 between the United States Department of Justice and AT&T approved by the courts on August 24, 1982 that settled the 1974 antitrust case of the United States versus AT&T. The MFJ divested AT&T of the local regulated exchange business and created seven regional holding companies – Ameritech, Bell Atlantic, Bell South, NYNEX, Pacific Telesis, Southwestern Bell and US West. The MFJ placed restrictions on the local exchange carriers, namely that they couldn't get into long distance communications. The Modified Final Judgment did not prohibit AT&T from providing local telecommunications, it prohibited AT&T from purchasing the stock of the divested RBOCs. For a full explanation, see Divestiture.

modified finite queue A traffic engineering term. Erlang C assumes an infinite queue, that is, callers will wait indefinitely to have their call answered. Since this is obviously not the case, some parties have suggested that a different algorithm should be used in order to produce more accurate forecast. In practice, while Erlang C will produce some degree of overstaffing based on its assumption of infinite queuing, alternatives that assume finite queues result in understaffing. Since most call centers would prefer slight overstaffing, and a greater likelihood of meeting grade of service, to understaffing, with a greater degree of customer dissatisfaction, Erlang C continues to be the preferred and recommended algorithm.

Modified Frequency Modulation MFM. An encoding scheme used to record data on the magnetic surfaces of hard disks. It is the oldest and slowest of the Winchester hard disk interface standards. RLL (Run Length Limited encoding) is a newer standard, for example.

Modified Huffman MH. A one-dimensional data-compression scheme that compresses data in a horizontal direction only through a process known as "run-length encoding. All Group 3 fax machines support MH, which is the lowest common denominator compression algorithm. MH scans each line of a document, looking for redundant data, i.e., the same color repeated. A document typically includes lots of white space, which MH interprets as redundant data. Rather than sending 1,728 bits of "nothing" across the circuit, MH sends a 9-bit code value, thereby compressing the data by a factor of 192. While the actual level of compression achieved, of course, depends on the amount of redundant data in a document, MH compression reduces transmission time and saves money. MH supports relatively slow fax transmission at 9.6 Kbps, or about 30 seconds per page. Group 3 standards also include Modified Read (MR) and Modified Modified Read (MMR), which are successively more efficient. All Group 3 fax machines support MH, which is the lowest common denominator compression technique. See also Compression, Huffman, Modified READ, and Modified Modified READ.

Modified Read MR. Modified READ (Modified Relative Element Address Differentiation) is a two-dimensional coding scheme for facsimile machines that works both horizontally and vertically, using the previous line as a reference. In other words, Modified READ scans each line, compressing out redundant data, much as does Modified Huffman. Modified READ then compares each line to the previous line, further compressing out redundant data. Some Group 3 fax machines use MR, which improves on Modified Huffman, assuming that a document has data continuity up and down, as well as from left to right. MR supports "business letter" quality transmission to about 20 seconds per page. See also Compression, Modified Huffman, and Modified Modified READ.

Modified Modified READ MMR. An improvement on Modified READ (MR), MMR supports the "business letter" quality transmission time of a Group 3 fax machine to about 10 seconds per page. All fax machines which operate at 14.4 Kbps use MMR, as do some high-volume machines which operate at 9.6 Kbps. See also Compression, Modified Huffman, and Modified READ.

modifier keys Keys on your PC which, when used with other keys, modify the behavior that key. Common modifier keys on your PC are the Ctrl (control) key and the Alt (Alternate) key.

modular Equipment is said to be modular when it is made of "plug-in units" which can be added together to make the system larger, improve its capabilities or expand its size. There are very few phone systems that are truly modular.

modular breakout adapter Allows the technician to access each individual conductor of a cable. Sometimes called a "banjo clip." It's a rectangular plastic box, with conductors on the sides and a modular plug at the long end.

modular cord A cord containing four twisted pairs of wires with a modular plug on one or both ends.

modular jack A device that conforms to the Code of Federations, Title 47, part 68, which defines size and configuration of all units that are permitted for connection to the public telephone network.

modular plug Connecting devices adopted by the FCC as the standard interface for telephone and data equipment to the public network. These are the plastic "ends" you see on cables. They come in two conductor, four, eight and six. Two, four and six conductor plugs are the same physical size, and are usually used for telephone voice and low speed data communications. Eight conductor (four pair) plugs are wider, and most often used for data, e.g. Ethernet LAN connections. There are several wiring configurations for modular plugs. The most common are T568A and T568B. See UTP Cable.

It's important to match the modular plugs to the type of cable you are using. Plugs made for stranded cable will not work with solid conductor wire because they're designed to pierce the cable in-between the strands. Used with a solid conductor cable, they don't pierce the cable and just get smashed. Plugs made for solid conductors usually work with stranded cable. But I wouldn't recommend trying.

modulated output spectrum The frequency spectrum of a modulated carrier that is due to modulation impressed on the carrier.

modulated waves Alternating current waves which have their amplitude varied periodically. The signals transmitted by a radio station are examples of a modulated wave.

modulation The process of varying some characteristic of the electrical carrier wave as the information to be transmitted on that carrier wave varies. Three types of modulation are commonly used for communications, Amplitude Modulation, Frequency Modulation and Phase Modulation. And there are variations on these themes called Phase Shift Keying (PSK) and Quadrature Amplitude Modulation (QAM).

modulation index In angle modulation, the ratio of the frequency deviation of the modulated signal to the frequency of a sinusoidal modulating signal. The modulation index is numerically equal to the phase deviation in radians.

modulation protocols Modem stands for MOdulator/DEModulator. A modem converts digital signals generated by the computer into analog signals which can be transmitted over an analog telephone line. It also transforms incoming analog signals into their digital signals for inputting into a computer. The specific techniques used to encode the digital bits into analog signals are called modulation protocols. The various modulation

INTERNATIONALLY ACCEPTED MODEM MODULATION PROTOCOLS					
Standard	Speed	Modulation	Duplex	Symbol Rate	Bits per symbol
Bell 103	300	FSK	Full	300	1
v.21	300	FSK	Full	300	1
v.22	1200	DPSK	Full	600	2
v.23	1200/75	FSK	Half	1200	1
Bell 202	1200/75	FSK	Half	1200	1
Bell 212A	1200	DPSK	Full	600	2
v.22bis	2400	QAM	Full	600	4
v.32	9600	QAM	Full (EC)	2400	4
v.32bis	14400	TCM	Full (EC)	2400	6
v.32ter	19200	TCM	Full (EC)	2400	8

protocols define the exact methods of encoding and the data transfer speed. In fact, you cannot have a modem without modulation protocols. A modem typically supports more than one modulation protocol. The raw speed (the speed without data compression) of a modem is determined by the modulation protocols. Here are the main internationally accepted modulation protocols:

Two modems can establish a connection only when they support the same modulation protocol. A modem with a proprietary modulation protocol can only establish a connection with another modem which also supports that modulation protocol. That protocol is typically from the same manufacturer, or from one of several manufacturers that say they are supporting it. For example, there once was a modulation protocol called V.FAST, which delivered 28.8 Kbps over normal analog phone lines. Several manufacturers supported it. Later, the ITU-T came out with V.34. Every modem maker adopted it and V.FAST went away, leaving some modem owners with modems that only worked at 28.8 Kbps with other proprietary modems.

modulation rate The reciprocal of the measure of the shortest nominal time interval between successive significant instants of the modulated signal.

modulation suppression In the reception of an amplitude-modulated signal, an apparent reduction in the depth of modulation of a wanted signal, caused by presence, at the detector, of a stronger unwanted signal.

modulator A device which converts a voice or data signal into a form that can be transmitted.

modulo Term used to express the maximum number of states for a counter. Used to describe several packet-switched network parameters, such as packet number (usually set to modulo 8 – counted from 0 to 7). When the maximum count is exceeded, the counter is reset to 0.

modulo N In communications, refers to a quantity, such as the number of frames or packets to be counted before the counter resets to zero. Relates to the number of frames or packets that can be outstanding from a transmitter before an acknowledgement is required from the receiver. Also indicates the maximum number of frames or packets stored, in case a retransmission is required (i.e., Modulo 8 or Modulo 128).

modulus 10 A modulus is a constant number or coefficient that expresses in numerical terms the extent to which a property is possessed by a substance or body. For example the LUHN formula, a form of Modulus 10 (also known as Mod 10), is a method of validating the number on a credit card or debit card when you swipe it through a card reader. Before the credit card transaction is sent over the network for authorization, the card reader performs an algorithm (i.e., calculation) on the number contained on the magnetic stripe. The result of the calculation is compared to a check digit, which is the last number on the card. If the process results in a match, the transaction is forwarded over the network for validation. If the numbers don't match, there is no need to forward the data. Modulus 10 is used in Canada to validate a Social Insurance Number (SIN). There are lots of other applications for this concept.

Here is the way that credit card numbers work: The first set of digits (1, 2, 3, or 4 digits), from left to right, are the credit card type (e.g., American Express, MasterCard, or Visa). The middle numbers are identify the issuing bank and the customer. The last digit is the check digit.

Here is the way the LUHN formula works for credit card validation:

1. Starting with the second to the last digit and moving right-to-left, multiply every other digit by two. If any resulting number is two digits, add the two digits together. For example, 7x2=14, and 1+4=5. 2. Add all of the digits skipped in the first step. 3. Add together the results of the first two steps. 4. The results of the third step must be a number that ends in zero.

modulus of elasticity The ratio of stress to strain in an elastic material.

MOH Music On Hold.

MoIP Modem over IP, i.e, data traffic of V-series voice-band modems over a mix of legacy switched networks and new IP networks. MoIP standards are defined in ITU-T V.150 and V.150.1.

moire In a video image, a wavy pattern caused by the combination of excessively high frequency signals. Mixing of these signals causes a visible low frequency that looks a bit like French watered silk, after which it is named.

moire pattern Wavy distortions, most obvious in image areas filled with solid color, that result from interference between the screen's phosphor layer and image signal.

moisture absorption The amount of moisture, in percentage, that a material will absorb under specified conditions.

moisture barrier bag MBB. A three-ply bag with characteristics that allow minimal moisture transmission, thereby preserving plastic surface-mount packages, which are packed into the bag, in a dry state.

moisture resistance The ability of a material to resist absorbing moisture from the air or when immersed in water.

MOJO The Legend of Mojo, from a web site called, citycelebrations.org. is told: Generations of Southerners have told stories of a quiet, well-dressed man wandering into town whistling an unidentifiable tune. His quick smile and endless supply of charm were legendary. Some stories say that he was aristocratic, others that he was a self-educated farmer. Wherever he went, the well-dressed stranger always carried a small leather bag filled with golden coins called Mojo. Legend has it that the golden Mojo came from the Fountain of Youth in what is now known as Florida. The mysterious golden Mojo had the reputation of imparting the owner boundless charm, wit and sophistication. In his travels, the well-dressed wanderer was known to encounter poor souls run down by life and at the end of their ropes. He would stop, carefully remove the leather pouch from his pocket and take out a single golden Mojo and place it face up in the palms of their hands. Some of history's most memorable characters were said to have "Mojo in their pockets" to explain how they became great leaders, social figures and influential artists. As the years went by, the term "Mojo" was simply believed to mean that the person possessed a great deal of personal charm and drive to succeed. Until recently that is... A small golden coin was found buried deep in the mud at the bottom of the canal this past year. Collectors and historians were unable to identify the origins of the metal or the design of the coin. It took the eyes of a very old woman spotting the coin among the collected caps, spoons and buttons found in the canal. "Mojo?" she asked. Yes, Mojo. It has been in Richmond's canal all these years. And now you can experience Mojo for yourself.

Thus MOJO defined,

1. A magic charm or spell.
2. An amulet, often in a small flannel bag containing magic items.
3. Personal magnetism; charm.

molding raceway method A cable-distribution method in which hollow metal or wood moldings support cables. Small sleeves of pipe are placed in the wall behind the molding to allow cable to pass through the wall.

molecular beam epitaxy See Epitaxy.

MOM Message Oriented Middleware. See MOMA.

MOMA 1. Museum of Modern Art. www.moma.org.

2. Message Oriented Middleware Association. According to MOMA, an international not-for-profit association of vendors, users and consultants focusing on the promotion of the use of messaging middleware to provide multi-platform, multi-tier message passing and queuing services for distributed computing architectures. MOMA serves as a point of interchange for experiences and ideas related to the development of MOM, as well as a point of concentration for interoperability and technology requirements toward influencing appropriate standards bodies. MOMA also directs its attention toward promotion of functional interoperability between applications built using disparate message-passing tools and mechanisms. See Middleware and MOM. http://198.93.24.24 (I'm sure that they would have preferred www.moma.org, but the Museum of Modern Art got there first. Hence the use of an IP address for MOMA's Web site, rather than a URL.) See IP and URL.

monaural sound Sound reproduction in which only one channel of sound is used. Compare with stereo sound in which two channels of sound are used and heard. The two channels may be different – to simulate the full depth you might hear in a concert hall. In a monaural headset, you hear out of one ear. In a biaural telephone headset, however, you hear out of two ears. Those channels will, however, be the same, since telephones transmit only one channel of sound.

Monday According to statistics, Monday is the favorite day of the week to commit suicide.

MONET Multiwavelength Optical Network. A high-end fiber optic testbed network on the US East Coast. The $100 million project, funded by ARPA (Advanced Research Projects Agency), is intended to test DWDM (Dense Wavelength Division Multiplexing), a means of increasing the capacity of SONET fiber optic transmission systems through the multiplexing of multiple wavelengths of light. MONET participants include AT&T, Bell Atlantic, Bellcore, BellSouth, Lucent Technologies, NRL (Naval Research Lab), NSA (National Security Agency), Pacific Telesis, and Southwestern Bell. See also DWDM, WDM, and SONET.

monetize Figure out how to make money from an online service or content that customers are used to getting for free. This is a problem facing many web sites, especially once the web sites' owners add up all the costs of their web sites and freak out.

money "I'm not interested in money," Marilyn Monroe once said, "I just want to be wonderful."

money laundering According to the New York Times, money laundering is a legal catch phrase that refers to the criminal practice of taking ill-gotten gains and moving them through a sequence of bank accounts so they ultimately look like legitimate profits from legal businesses. The money is then withdrawn and used for further criminal activity. The reason I have included this definition is that the monies being laundered typically make extensive use of telecommunications – especially moving monies electronically.

money market funds Funds that invest in Treasury bills (T-bills) and the highest grade government obligations.

money suck "Net Guide" Magazine before CMP closed it down was referred to internally as a "money suck," i.e. it was consuming vastly more money than it was bringing in.

monitor 1. To listen in on a conversation for the purpose of determining the quality of the attendant's or agent's response and politeness to customers.

2. Video monitor. Computer or TV screen and surrounding electronics. As computers have become more powerful so the quality of the monitor they can drive has become

PC SCREEN ENCODING	
CGA	320 x 200
Enhanced CGA	640 x 400
EGA	640 x 350
Hercules	720 x 348
VGA	640 x 480
Super VGA	800 x 600
8514/A (also called XGA)	1,024 x 768
I don't know the name	1,280 x 1,024
I don't know the name	1,600 x 1,200
I don't know the name	1,800 x 1,440
I don't know the name	2,040 x 1,664

more crammed with more information, also called pixels. My own experience with ever newer computers pretty well follows the path of this chart, showing the increasing quality of computer screens and the chips that drive them. I am currently writing this edition on a laptop which is displaying 1,280 by 1,024 pixels on its screen. It's pretty awesome. The quality of viewing photos and artwork is much improved.

monitor interface Another term for the destination interface in a SPAN session.

monitor on hold A telephone feature. If the person you're speaking with puts you "on hold," you can turn your speaker on your phone and hang up your handset but keep listening until the other person comes back to the phone.

monitoring device Records data on calls placed through a company's telephone system. The monitoring device will record the numbers called, the length of the calls, the number of calls abandoned.

monochromatic Consisting of one color or wavelength. Although light in practice is never perfectly monochromatic, it can have a narrow wavelength.

monochrome monitor A monitor with 720 x 348 pixel resolution in a single color. Most monochrome monitors display in paper white, green or amber.

monomode fiber Another term for Single Mode Fiber (SMF). See SMF.

monopole A slender self-supporting tower on which wireless antennas can be and are placed. See Monopole Colocation.

monopole colocation Monopole colocation is having many antennas share one monopole – a slender self-supporting tower on which wireless antennas are placed. As you add antennas to one slender pole, you often need to shore the pole with extra stuff – like structural steel and enhancing the monopole's foundation. If you don't, the wind will blow the monopole over and cell phone service will get even worse.

monopoly leveraging Monopoly leveraging is one of the main charges brought against Microsoft by Department of Justice in its 1998 antitrust suit. Monopoly leveraging is using a monopoly in one area to gain a monopoly in another. According to the Justice Department, Microsoft was using its 90% or so market share in desktop and laptop operating systems to gain a monopoly in the browser market. One wonders if the Federal Government is so uncreative it can't think of anything better to do with its time or money.

monopulse Originally a system designed to enable military radars to resolve both elevation and azimuth from a single pulse, through dividing the antenna beam into separate lobes and resolving information from the nulls between lobes. Used in tracking and also called simultaneous lobing.

monospaced font A font in which all characters have uniform widths. See also Proportional Font.

month to month The standard way of paying for telephone service. Some services now come in "rate stability" packages, which means if you commit to keeping the service for a while – typically three or five years – it's cheaper each month.

Monthly Factors A call center term. A historical pattern consisting of 12 factors, one each month, that tells the program how much that much that month can be expected to deviate from the average monthly traffic year after year. For example, a monthly factor of .75 means that the month will be 25% slower than average, while a factor of 1.15 means that the month will be 15% busier than average.

MOO Mud, Object Oriented. One of several kinds of multi-user role-playing environments.

mood message A simple little text message that a Skype user can send to members of his contact list. The message can indicate the mood the user is in, or it can be a witty comment, quote, information about the user's plans or whereabouts, or anything else that the user would like to share with his contacts.

MOON 1. Magneto Optical On Network.

2. To drop your trousers, to bend over and to expose your bare backside to the person you're mooning. In the technology business, mooning can be used to tell your competitors in no uncertain terms how much better you're doing." It has become something of an maritime tradition for Larry Ellison's boating rivals to drop their trousers and moon the billionaire when they pass him at sea. Larry Ellison is founder and head of Oracle, the relationship database maker.

moonbounce Using the Moon as a passive reflector for wireless communication. Also called Earth-Moon-Earth (EME) communication.

Moore's Law In the October 10, 2000 issue of the magazine, Business 2.0, the reporter asked George Moore, a co-founders of Intel and the inventor of Moore's Law, "What is Moore's Law, according to you?" George answered, "Moore's Law originally got its name from a paper I published in 1965 where we were looking at the complexity of integrated circuits, and I made what seemed like a wild extrapolation from about 60 components to 60,000 over the next 10 years. It turned out to be amazingly correct, the number of components doubling every year over the decade. And then in 1975, I updated it and said the slope was going to double about every two years, looking forward. Those were surprisingly accurate predictions. But it's gotten to the point now that anything that changes exponentially is called Moore's Law, and I'm happy to take the credit for all of it. When I

wrote that paper, the message I was trying to get across was that this was going to be the inexpensive way to do electronics. I didn't have any real feeling that my extrapolation was going to be very precise."

His forecast, which implies a similar increase in processing power and reduction in price, has proved broadly accurate: Between 1971 and 2001, transistor density has doubled every 1.96 years. Yet this pace of development is not dictated by any law of physics. Instead, according to the Economist, "it has turned out to be the industry's natural rhythm, and has become a self-fulfilling prophecy of sorts. IT (Information Technology) firms and their customers wanted the prediction to come true and were willing to put up the money to make it happen. Even more importantly, Moore's law provided the IT industry with a solid foundation for its optimism. In high-tech, the mantra goes, everything grows exponentially. This sort of thinking reached its peak during the internet boom of the late 1990s. Suddenly, everything seemed to be doubling in ever-shorter time periods: eyeballs, share prices, venture capital, bandwidth, network connections. The internet mania began to look like a global religious movement. Ubiquitous cyber-gurus, framed by colourful PowerPoint presentations reminiscent of stained glass, prophesied a digital land in which growth would be limitless, commerce frictionless and democracy direct. Sceptics were derided as bozos "who just don't get it". Today, everybody is older and wiser." See also Cao's Law.

MOP 1. Method of Procedure. A formal, written procedure detailing a job that will take place. Required for any work on the network. Includes detailed instructions for completing the work and for backing out of trouble or mistakes that may occur. For example, when network engineers extend a network by adding a new node to a SONET ring (or by doing something else) they write a MOP. The MOP tells technicians which circuits to reroute, which cards to swap and when to turn on the new node. These MOPs are essential communication tools for engineers and technicians.

2. Maintenance Operations Protocol, a DEC protocol used for remote communications between hosts and servers.

MOPS Millions of operations per second. Refers to a processor's performance. In the case of DVI (Digital Video Interactive) technology, more MOPS translates to better video quality. Intel's video processor can perform multiple video operations per instruction, thus the MOPS rating is usually greater than the MIPS rating.

MOR Multi-wavelength Optical Repeater. An optical amplifier which increases the distance between network elements in a transmission network. The MOR is used to overcome loss in a network and is often used instead of one or more regenerators, which perform a similar function. Unlike the regenerator which receives a signal and then re-transmits it, the MOR increases the power of a signal by amplifying it. This gives the MOR the ability to operate with more than one wavelength at a time, and with different bit rates (such as OC-48 and OC-192). The following is from Nortel OC-192 System documentation: MOR amplifiers are designed for a bidirectional network architecture which uses the gain region of erbium doped fiber amplifiers (1528.40 nm - 1561.00 nm). Each bidirectional channel consists of counter-propagating wavelengths, one direction in the RED Band (1547.5 -1561.0 nm) and the other direction in the BLUE Band (1528.4 - 1542.5 nm). The MOR amplifier can be used either as a post/pre-amplifier or as a line amplifier. The MOR supports eight channel wavelength over a single fiber, with four channels amplified in each direction or transmission. Four wavelengths are assigned to co-propagate in the Blue Band, and four wavelengths are assigned to counter-propagate in the Red Band. This enables a flexible and self-paced network evolution for a total aggregate capacity of up to 80 Gb/s (gigabits per second) on a single fiber.

moratorium. A moratorium in the telecommunications world is a period of time that no new service-effecting projects are begun. This includes new installations, reconfigurations, etc. Only emergency maintenance is performed during this period. A moratorium might be called by a phone company, for example, if there has been a natural disaster such as a hurricane.

morph Computer animation technique that allows figures to change from one shape to another in increments you choose.

morphology A cellular radio term. Morphology describes population density. Higher population densities cover more POPs per cell, leading to economies of scale. Lower densities imply improved propagation characteristics and a greater coverage area.

Morrow, George George Morrow, a mathematician and programmer, was a member of a group of unorthodox hobbyists who were instrumental in creating the personal computer industry. Mr. Morrow was born in Detroit. He dropped out of high school, but at the age of 28 decided to return to school and received a bachelor's degree in physics from Stanford University and a master's degree in mathematics from the University of Oklahoma. He entered a Ph.D. program in mathematics at the University of California at Berkeley, but was sidetracked by his passion for computers. He started working as a programmer in the computer laboratory at Berkeley in the early 1970's and began attending meetings of the Homebrew Computer Club, an informal group of engineers, programmers, experimenters and entrepreneurs that ultimately spun off dozens of companies that formed the core of the personal computer industry in the 1970's. Initially, most personal computers were sold as kits. Mr. Morrow formed Microstuf, a company in Berkeley, Calif., to sell expansion cards and other computer add-on products to the first generation of personal computer enthusiasts. He would later change the name of the company, first to Thinker Toys and later to Morrow Designs. A self-taught computer designer, Mr. Morrow was involved in the efforts to create and standardize the S100 bus, a hardware design that made it possible for early PC makers to share expansion cards. Morrow Designs thrived when the personal computer became an important tool for small businesses. The first machines ran the Digital Research CP/M operating system. Later, Mr. Morrow introduced a portable computer intended to compete head-to-head with the popular Osborne 1 computer. The Morrow machine matched the Osborne's $1,795 price but offered more bundled software. When I.B.M. began to dominate the PC market, Mr. Morrow was forced to shift to the industry standard. In 1985, his company introduced a popular portable design known as the Pivot and sold the design to Zenith Data Systems. But with the industry becoming increasingly dominated by large electronics companies, Morrow Designs filed for bankruptcy in 1986. See Homebrew Computer Club.

Morse Code There are (or were) two Morse Codes. One called American Morse Code and one called Continental or International Morse Code. The first one, American Morse code was invented by Samuel F. B. Morse, an American born in Boston, in 1837 for use on the electric telegraph. When the electric telegraph was adopted later in Europe, the code Morse invented was not used. The so-called "Continental" code was devised. This code, among other changes, eliminated the spaced dots for C, O, R, Y and Z and the long dash for L. These changes were needed for the satisfactory use of the early visual "needle telegraph" instruments in Europe wherein a needle swung slowly between right and left positions to indicate dots and dashes. The Continental (or International) Code was adopted as a worldwide standard at the Telegraph Conference in Berlin in 1851. American landline telegraphists, however, steadfastly refused to abandon their Morse Code. According to the history books, there was more to their refusal than just plain American stubbornness. When skillfully handled, the American Morse Code actually transmits information somewhat faster because of spaced dots being used in place of dash combinations for some letters. However, careless sending of American Morse Code will produce more errors. Morse code is referred to as International Telegraph Alphabet 1. Mr. Morse also invented the telegraph and first demonstrated it in 1844. Morse Code was used in landlines and in radio telegraphy to ships at sea. The United States Coast Guard abandoned Morse Code in 1996 and member nations of the International Maritime Organization agreed to officially stop its use by February 1, 1999. The French maritime radio authorities sent their last Morse code message on February 1, 1996. Governments are abandoning Morse code in favor of faster, better radio and satellite voice and data communications. International Morse Code is still widely used by U.S. and foreign amateur (Ham) radio operators.

Morse Code represents letters by combinations of long and short signals. Morse Code can be written in dots and dashes or signaled with flashlights and radio bleeps, or taped and scratched between cells in prison. In 1912, the easily-memorized letters SOS were chosen as the international distress signal. "Save Our Souls" was the catch phrase devised later. You'll notice that in Morse code, more commonly-used letters, such as vowels, have fewer dots and dashes. You'll also notice that some letters are represented by one dot or dash and some by as many as five. This was not a data transmission code to which you can easily apply error checking and correction. It relied heavily on the skill of the operators for error-checking.

In the American Morse Code, a space was used in five letters, namely, C, O, R, Y and Z. No such spaces are used in the modern Continental or International Morse Code. Also, the letter L was an oddball since it was defined as a long dash almost equal to three standard dashes. In fact, In the International Morse Code, the short pulse is called a dot (only when written), but often referred to as sounding a "dit", and the long pulse is called a dash (only when written), but sounded a "dah". Timing is very crucial in sending Morse Code. A "dit" timing is defined as one time unit, a "dah" must take exactly three time units, a pause between elements is exactly equal to one time unit, a pause between characters is exactly three time units, and a pause between words is exactly seven time units for precise Morse Code generation. Hence, sending the word "PARIS " including a space at the end requires 50 time units, i.e., P=11 units, A=5 units, R=7 units, I=3 units, S=5 units, there are 4 characters spaces worth 3 units each=12 units, and a word space at the end worth=7 units, hence the total of 50 time units for the whole word. This would sound like this: didahdahdit

didah didahdit didit dididit.

The Capitol Records building in Los Angeles is built to resemble a stack of records. A red airplane-warning light atop the structure flashes out the word "Hollywood" in Morse code every 20 seconds or so. All airline pilots are expected to learn Morse Code. The . (dot or dit) is short. The _ (dash) is long.

American Morse Code

A ._	B _...	C .. .
D _..	E .	F ._.
G _ _.	H	I ..
J _._.	K _._	L ___
M _ _	N _.	O . .
P	Q ._.	R . ..
S ...	T _	U .._
V ..._	W ._ _	X ._..
Y .. .. (dit dit space dit dit)	Z	

International Morse Code

A ._	B _...	C _._.
D _..	E .	F ._.
G _ _.	H	I ..
J ._ _ _	K _._	L ._..
M _ _	N _.	O _ _ _
P ._ _.	Q _ _._	R ._.
S ...	T _	U .._
V ..._	W ._ _	X _.._
Y _._ _		

See also Inmarsat, Marisat and Morse, Samuel.

Morse, Samuel Samuel Finley Breese Morse was born in Charlestown (now a part of Boston) on April 27, 1791. He entered Yale University at the age of 14 and graduated in 1810. Although he attended lectures on electricity while at Yale, after graduation he went to England to study art. He returned to the United States in 1815 and became a well-known painter. In 1832, while returning from Europe, he and a fellow passenger discussed the electromagnet and Morse conceived the idea of his telegraph. He made a working model about 1835, filed for a patent in 1838 and in 1844 inaugurated public service between Washington and Baltimore with his famous message, "What hath God wrought?" Morse made such an impression that in 1839 he became the first U.S. citizen to be photographed. He died in 1872. See Morse Code.

MORT AT&T's database of its dead employees.

MOS 1. Metal Oxide Semiconductor. Technology describing a transistor composed of a semiconductor layer including "source" and "drain" regions separated by a channel. Above the channel is a thin layer of oxide and over that a metal electrode called a gate. A voltage applied to this gate controls the current between the source and drain regions, or in another format, stops a flow between the two areas.

2. Mean Opinion Score. If you want to measure the quality of a VoIP call, there are methods, including the Median Opinion Score (MOS) test, endorsed by the International Telecommunications Union (ITU). MOS involves gathering people into a room to listen to calls, after which group members rate quality on a scale from 1 to 5. Voice-quality testing tools based on our algorithms also are available from vendors such as Agilent and Empirix. MOS is defined in the ITU-T P.800 specification "Methods for Subjective Determination of Voice Quality." P.800 involves the subjective evaluation of preselected voice samples of voice encoding and compression algorithms. The evaluation is conducted by a panel of "expert listeners" comprising a mixed group of men and women under controlled conditions. The result of the evaluation is a Mean Opinion Score (MOS) in a range from 1 to 5, with 1 being "bad" and 5 being "excellent". The components of the MOS are as follows: Opinion Scale: Conversation Test (bad to excellent), Difficulty Scale (yes or no), Opinion Scale: Listening Test (bad to excellent), Listening: Effort Scale ("No meaning understood with any reasonable effort" to "Complete relaxation possible, no effort required"), and Loudness: Preference Scale ("Much quieter than preferred" to "Much louder than preferred"). A MOS of 4.0 is considered to be "toll quality." PCM (Pulse Code Modulation), the traditional encoding algorithm (ITU-T G.711) used in the circuit-switched PSTN (Public Switched Telephone Network) rates an MOS of 4.4. Following are MOS scores for other widely used speech algorithms: Dual Rate Speed Coder for Multimedia Communication (G.723), 3.5-3.98; Adaptive Differential Pulse Code Modulation (ADPCM, G.726), 4.2; Low Delay-Code Excited Linear Prediction (LD-CELP, G.728), 4.2; and Conjugate Structure-Algebraic Code Excited Linear Prediction (CS-ACELP, G.729), 4.2 See also P.800, P.861 and TELR.

mosaic The first graphical Web browser, developed by National Center for Supercomputing Applications, which greatly popularized the Web in the last few years, and by extension the Internet, as it made the multimedia capabilities of the Net accessible via mouse clicks. Mosaic let you surf the Internet's Worldwide Web. Mosaic lets you see hypertext documents with embedded graphics and occasionally sound, movie clips and animation. Mosaic was the first popular software that allowed people to browse around the Web by pointing and clicking. In short, Mosaic is an interface to the WWW (World Wide Web) distributed-information system. Like Gopher, the WWW is functionally split into two parts, the server and the client. Using a GUI (Graphical User Interface) interface (like Mosaic), it has the ability to display:

- Hypertext and hypermedia (sounds, movies, extended character sets, and interactive graphics) documents.
- Electronic text in an enormous variety of fonts.
- Text in bold and italic.
- Layout elements such as paragraphs, bulleted lists, and quoted paragraphs. It is mostly distinguishable by its support of multiple hardware platforms, and the WWW HTML (Hypertext Markup Language) document format. Mosaic used to be the most popular Web browser. At the time I wrote this, the most popular one was Netscape and browsers based on Netscape. See Netscape.

Moshi-moshi What the Japanese say when they answer the phone. It is equivalent in meaning and tone to "Hello" in English.

MOSS MIME Object Security Services. An Internet mail security standard which was introduced in 1995 as the successor to PEM (Privacy Enhanced Mail). PEM didn't address MIME (Multipurpose Internet Mail Extension) attachments. MOSS failed to secure widespread support. S/MIME (Secure/MIME), introduced in 1996 has become the de facto standard. See also MIME, PEM and S/MIME.

MOST Media Oriented System Transfer, a fiber optic network optimized for automotive applications. See also LIN.

Most Economical Route Selection MERS. Used by several phone companies and several manufacturers to mean Least Cost Routing – the feature of a telephone system which automatically chooses the least cost route for a long distance call. See Least Cost Routing. Synonym: Least Cost Routing.

most favored nation clause A clause added to a purchase contract with a vendor saying that for a certain period after signing the contract, if the buyer finds out that the product has been bought for less, then the seller will refund the difference. The idea is to give the purchaser the assurance of the least expensive price.

most limiting capacity A telephone company term. The arithmetic minimum of (1) Line Capacity (2) Number Capacity or (3) Switching Equipment Capacity.

MOTD Message Of The Day.

mother in law booth. You visit a trade show, walk the aisles. You see all the products. One moment, you have a brainwave, "I can do better!" You quit your job, form a company, raise a little money. Twelve months later your shiny new product is ready to exhibit at next year's trade show. But you've run out of money. You need to pay for the booth before the show opens. You've run out of money. You're desperate. You borrow the booth money from your mother-in-law. It comes with a price – her. She stands in the booth and "helps" you sell your shiny new product. Her salespitch is: "My son-in-law, the genius, has this new thing. Buy it. He needs the money."

Motes From MIT's Technology Review, "Great Duck Island, a 90-hectare expanse of rock and grass off the coast of Maine, is home to one of the world's largest breeding colonies of Leach's storm petrels-and to one of the world's most advanced experiments in wireless networking. Last summer, researchers bugged dozens of the petrels' nesting burrows with small monitoring devices called motes. Each is about the size of its power source-a pair of AA batteries-and is equipped with a processor, a tiny amount of computer memory, and sensors that monitor light, humidity, pressure, and heat. There's also a radio transceiver just powerful enough to broadcast snippets of data to nearby motes and pass on information received from other neighbors, bucket brigade-style. This is more than the latest in avian intelligence gathering. The motes preview a future pervaded by networks of wireless battery-powered sensors that monitor our environment, our machines, and even us. It's a future that David Culler, a computer scientist at the University of California, Berkeley, has been working toward for the last four years. "It's one of the big opportunities" in information technology, says Culler. "Low-power wireless sensor networks are spearheading what the future of computing is going to look like." In late March, 2006 the Wall Street Journal wrote a piece headlined, "New Sensor Line Inspires Start-Ups." It read, in part: A race to popularize a new breed of wireless sensors is spawning a slew of start-up companies, including a venture by two scientists who helped pioneer the technology. The sensors, sometimes called "motes," are designed to monitor the environment and relay data that they gather using radios that consume little electrical power. Motes can be equipped to detect, say, motion or temperature, and potential uses include industrial controls, home automation, building security and tracking shipping containers. Much of the work in motes has been carried out at the University of California at Berkeley, partly funded by Intel Corp. and the Defense Department's research arm. But commercial prospects have spurred the formation of at least six sensor-related start-ups with Berkeley connections. Arch Rock is starting with $5 million in funding from Intel and two venture-capital firms, New Enterprise Associates and Shasta Ventures. It plans to develop additional software to help gather and analyze data generated from hundreds or thousands of motes. Roland Acra, a former Cisco Systems Inc. executive who is Arch Rock's chief executive, predicts that billions of motes will be deployed by the end of the decade – dwarfing the number of computers and cellphones attached to the Internet. For example, sensors deployed on a bridge could detect movements that could be a sign of structural fatigue. Motes could detect the presence of hazardous chemicals or the movement of people or sensitive equipment. Some companies are promoting a related networking technology, dubbed ZigBee, for home-automation chores such as remotely turning on lights or security systems using a cellphone. Some start-ups are working on specific mote applications. Streetline Networks Inc., based in San Francisco, has been working to equip parking meters with wireless sensors that could help drivers find unoccupied spaces with less driving around. Sensys Networks Inc., based in Berkeley, is developing sensors that could be embedded in pavement to measure the number and speed of passing cars for traffic control."

mother of The largest, greatest, grandest, of something. An expression coined in 1990 by Saddam Hussein, Iraqi dictator. The Mother Of all telephone switches would be the largest, most powerful, most elaborate etc. Such a device doesn't exist, as yet – as, in fact, Saddam Hussein's Mother of all Battles (the one against the Allies in 1990-1991) didn't exist. He lost in the Mother Of all defeats.

motherboard The main circuit board of an electronic telephone or computer. The motherboard contains edge connectors or sockets so other PC (printed circuit) boards can be plugged into it. Boards which you stick into motherboards (that's why they're called motherboards) are typically called Fatherboards, because they plug into the Motherboard. Motherboards are also called planar boards. Motherboards are common in key systems and hybrid key/PBXs. They are not common in PBXs, where all the electronics are typically on printed circuit cards which slide into the PBX's cage and which attach to a backplane, which is typically a wiring scheme connecting the PBX's printed circuit cards. See also Daughterboard.

motif Motif is the name given to the Open Software Foundation's (OSF) toolkit (Application Programming Interface) and look and feel. Standardized as IEEE 1295, OSF/Motif has become the major GUI (Graphical User Interface) for open computer systems, as defined by The Open Group, a consolidation of the OSP and X/Open industry consortia. Now in Version 2.0, Motif is the basis of the Common Desktop Environment (CDE) developed jointly by HP, IBM, Novell and SunSoft.

motion JPEG JPEG stands for the Joint Photographic Experts Group standard, a standard for storing and compressing digital images. Motion JPEG extends this standard by supporting videos. In motion JPEG, each frame in the video is stored using the JPEG format. See JPEG.

MOTIS Message Oriented Text Interchange System. Original name for the ISO (International Organization for Standardization) standard now being changed to MHS (Messaging Handling System).

MOTO Mail Order Telephone Order. A credit/debit card classification by the banking and finance industry reflecting what the banking industry thinks are its highest risk transaction type.

Motorola According to the family which founded Motorola in the 1920s, the name Motorola was chosen to mean "Music in motion" to signify one of the company's first products – a car radio.

MOU 1. Memorandum Of Understanding. In the old days, we called MOUs "Letters of Understanding." But we've become more fancy. We now call them "Memorandum of Understandings." Whatever you call them, they're basically bits of paper which detail out the essence of a legal agreement – what you will do and what I will do. An MOU is not a legally-binding contract or agreement. Think of it as a document which details what the two of us will do. Once we've agreed and detailed our agreement in a MOU, such document typically makes it to a lawyer, who quadruples its size and its complexity, and makes it into a legal Agreement. Don't tell the lawyers, however: If two of you had signed the MOU and changed its name to an "Agreement," it would be a legally binding agreement. And half the world's lawyers would be out of a job. What a joyous thought!

2. Minutes Of Use.

mount The method in NFS and other networks by which modes access network resources. The word "mount" is often used as a verb, as in my workstation "mounts" the file server, called DALLAS2.

mounting cord The connecting cord between the phone and the jack in the wall. In Europe and North America, it's called a line cord. In North America, it's also known as a mounting cord.

mouse A device that generates the coordinates of a cursor or position indicator on your computer screen (e.g. a hand, an arrow) as you move it around on a flat surface, generally in the form of a "mouse pad." The term "mouse" comes from the appearance of the device, as it generally is connected to the mouse input port by a wire which is reminiscent of the tail of a mouse (there also now are wireless mice.) The body of the mouse has one or more buttons which allow you to select objects, icons or text for the performance of certain functions, depending on the application running at the time. This "point-and-click" mode is a critical element of a Graphical User Interface (GUI), such as Windows and its variations; such GUIs were first popularized by Apple Computer.

On the bottom of the mechanical mouse is a ball that rolls on the surface of the mouse pad. In contrast, a trackball is a stationary device with a ball that you move with your finger – essentially an upside-down mechanical mouse. The mechanical mouse, which is what most of us use, was invented by Douglas Engelbart of Stanford Research Center in 1963; it was commercialized by Xerox in the 1970s.

An optical mouse makes use of a laser to detect the movement of the mouse in relationship to a grid on a special mouse pad. While optical mice are very precise, the also are relatively expensive. Optomechanical mice combine the technologies, without the requirement for the grid pad.

mouse blur Move your mouse quickly across your screen and if you're running an LCD (for example on a laptop), the mouse's pointer will blur – due to the screen's inability to change as fast as you can move the mouse. Another term for mouse blur is Cursor Submarining.

mouse over See Mouseover.

mouse potato A person who uses his mouse to view educational or entertainment on his computer. Museums are afraid, for example, that if they sell the electronic rights of the art hanging on their walls, every one will stay at home, become mouse potatoes and never visit the museums. The concept of a mouse potato derives from a couch potato – namely someone who sits on his couch and changes channels on his TV set using a remote device.

mouse trap Ralph Waldo Emerson once said, "Build a better mousetrap and the world will beat a path to your door." Over 300 people in the last 20 years have taken his advice literally by registering improved mousetrap designs with the US Patent and Trade Office, making the lowly rodent-catcher the most re-invented machine in America. The classic, cheap spring-mounted wooden trap, the Victor is the perennial winner, however. One of the problems with many of the "better mousetraps" was that they were so beautiful and so expensive, you had to open them up, remove the dead mouse, clean the trap and re-bait it. All this was offensive. Easier to pick up the dead mouse by the tail and throw it and the cheap mousetrap into the garbage. (Though some would question the morality of it all.)

mouseover You're on a Web site. You slide your mouse over a drawing, an illustration, some small object or even a blank spot. Suddenly something happens. You see a new diagram. You see a pop-up. You see some words, e.g. "click here for a great deal." That's called a mouseover. And to see the words, you typically need a Javascript enabled browser. The major ones all are. A mouseover is also called a Rollover.

MOV 1. Metal Oxide Varistor. A voltage dependent resistor which absorbs voltage and current surges and spikes. This low-cost, effective device can sustain large surges and switch in 1 to 5 nanoseconds. It is used as a surge protector and suppressor. It often the first electronic component that electrons coming in on an incoming phone line hit. Many trunk boards inside PBX are protected by MOVs. If the voltage or current is high, it will blow the MOV, thus protecting the remaining the far more valuable devices on the board.

2. A Macintosh-based audio/video (multimedia) file. A MOV file has a file extension of .mov and can be played on a Windows operating system if you have the QuickTime Movie Player application installed.

move up the stack This describes the trend among network operators to look for revenue opportunities in the areas of content and services, away from less profitable physical networking. The use of the term "stack" here refers to the OSI Reference Model, the lowest layer of which is the physical layer, and the highest layer of which is the application layer.

moves, adds and changes MACs. Any of the above ancillary work performed on a PBX switch, cabinet, or peripheral item after installation. See MAC for a fuller explanation.

Mozilla In early 1998 Netscape Communications made the source code to its browser publicly available and created the Mozilla project, in which Netscape programmers and volunteers would use the code as the basis for what they hoped would be the ultimate Internet browser. The latest version of Mozilla is now called Firefox. It's out and there are many people (including me) who really like it and use it every day. It is available for free from www.Mozilla.org. Mozilla is now available in 65 languages, with 34 more to follow.

MP-MLQ MultiPulse-Maximum Likelihood Quantization. A voice compression technique specified in ITU-T G.723.1 (Dual Rate Speech Coder for Multimedia Communications) and Frame Relay Forum FRF.11 (Voice over Frame Relay Implementation Agreement). MP-MLQ compresses voice to 6.3 Kbps. See also Dual Rate Coder for Multimedia Communications and VoFR.

MP1 MPEG Layer-1. An extension of the MPEG (Moving Picture Experts Group) standards for compressed digital video, Layer 1 supports CD-quality audio using 4:1 (4-to-1) compression, which reduces the required bandwidth from approximately 1.411 Mbps to 384 Kbps. See MP3 for more detail.

MP2 MPEG Layer-2. An extension of the MPEG (Moving Picture Experts Group) standards for compressed digital video, Layer 2 supports CD-quality audio using compression of 6:1 to 8:1, thereby reducing the bandwidth requirement from 1.411 Mbps to 256-192 Kbps. See MP3 for more detail.

MP3 MPEG Layer-3. MP3 is the most popular audio-compression format on the Internet. MP3 provides an efficient audio-coding scheme, which allows compression of audio files by a factor of up to 12, with little loss in quality from the original CD. For example, a five minute CD song takes about 50 megabytes of storage space on your computer's hard drive. In MP3 format the same song occupies only about 5 megabytes. One of the ways MP3 compression program works is by eliminating tones and frequencies the human ear cannot commonly hear. MP3 is also an extension of the MPEG (Moving Picture Experts Group) standards for compressed digital video. To play MP3 music, you'll need the music. You download it from the Internet or convert it on your PC from your favorite CD. You play your MP3 music on your PC using a software player you downloaded (for free or for pay) from the Internet. Or you can play MP3 music on portable devices, some of which have no moving parts and some of which now sport hard disks. The first MP3 player was Diamond Multimedia's Rio.

MP4 MPEG Layer-4. Rob Koenen, president of the MPEG-4 Industry Forum, which aims to increase adoption of the format, says MPEG-4 (commonly called MP4) is really "a toolbox that may be extended as need be," not a static format. Specifically, the standard consists of eight parts, some of which are still in development. Each part handles different tasks, such as video and audio representation, file format selection and format transfer. Implementation of the standard is left to companies and groups that want to create software or hardware that uses digital video and audio. "The major parts of the standard were [established] a couple of years ago," Koenen said. "Some stuff is still being added ... but there are chips available right now, [and] a number of players already." According to the MPEG-4 Industry Forum's Web site, the standard became usable in 1999, and the parts that have been added since that time do not break the standard. So, although MP4 does not get the kind of press that prior standards have received, it is alive and well. Apple's QuickTime 6, probably the most popular software media player, supports the MP4 file format, and many other companies also are delivering software and hardware products that use the standard. However, some patent and licensing issues remain to be ironed out before the specification is finalized. See Entropic Coding.

MPB See Meet Point Billing.

MPC 1. MPOA Client. An ATM term. A protocol entity that implements the client side of the MPOA (MultiProtocol Over ATM) architecture. A MPOA client, typically in the form of a host computer, establishes a VCC (Virtual Channel Connection) with a MPOA server in order either to forward data packets to a destination MPOA client, or to request information so that the originating MPOA client can establish a more direct path on the basis of a cut-through SVC (Switched Virtual Circuit). In this latter case, the server acts as a virtual router. The MPOA client implements the Next Hop Client (NHC) functionality of the Next Hop Resolution Protocol (NHRP). See MPOA.

2. Multimedia PC. See the following definitions for MPC1, MPC2 and MPC3.

3. Musepack. A lesser-known lossy audio compression codec. MPC offers superb quality when encoding audio at higher bit rates. Very few if any portable audio players support it, however. It's designed mainly for listening to music on a PC.

MPC1 Published in 1991, this original Multimedia PC (MPC) specification was adopted worldwide as the basic multimedia extension of the PC standard. MPC standards are established by the MPC Working Group of the SPA (Software Publishers Association). In 1993, MPC1 was followed by MPC2. MPC3, the latest, does not replace MPC2, but takes it one step further. See MPC2 and MPC3.

MPC2 Published in 1993 as the successor to MPC1, MPC2 standards specify elements including: 1. 4MB RAM; 2. 485 SX or equivalent microprocessor of 25 MHz; 3. hard drive of 160 MB; 4. CD-ROM drive supporting a sustained transfer rate of 300 KBps, and a maximum average seek time of 400 ms; and 5. Windows 3.0 plus multimedia extensions, or binary compatibility. No video playback standards were included. See MPC3 for the latest standards.

MPC3 MPC3 is the latest specification (Release 1.3, February 26, 1996) for multimedia PCs as defined by the Multimedia PC Working Group, an independent special interest group of the Software Publishers Association (SPA). Minimum requirements for MPC3 machines include: 1. 8 MB RAM; 2. CPU which can pass the MPC Test Suite, which is benchmarked on a 75 MHz Pentium; 3. hard drive of 540 MB; 4. CD-ROM Drive supporting a sustained transfer rate of 600 KBps and an average access time of 250 ms; and 5. Windows 3.11 and DOS 6.0 or binary compatibility. MPC3 also adds the requirement for video playback capability compatible with MPEG-1 (hardware or software). See also MPEG.

MPEG MPEG is commonly known as a series of hardware and software standards designed to reduce the storage requirements of digital video, i.e. video recorded digitally or converted into digital bits. MPEG is most commonly known as an compression scheme for full motion video. The word MPEG is actually the acronym for the Moving Pictures Experts Group, a joint committee of the International Standards Organization (ISO) and the International Electrotechnical Commission (EG). The first MPEG specification, known as MPEG-1, was introduced by this committee in 1991. The common goal of all MPEG compression is to convert the equivalent of about 7.7 meg down to under 150 Kb, which represents a compression ratio of about 52 to one. The two requirements of MPEG-1 are 30 frames per second of Standard Image Format (SIG) of 352 pixels x 240 pixels and CD-quality sound at 44.1 Khz, 16 bit stereo. MPEG image scheme offers more compression than the other poplar JPEG image compression scheme, which is largely for still images. MPEG takes advantage of the fact that full motion video is made up of many successive frames consisting of large areas that are not changed – like blue sky background. While JPEG compresses each still frame in a video sequence as much as possible, MPEG performs "differencing," noting differences between consecutive frames. If two consecutive frames are identical, the second can be stored in remarkably few bits. MPEG condenses moving images about three times more tightly than JPEG. See also JPEG.

There are two types of MPEG Playback: Software and Hardware. Software MPEG playback is the decompression of MPEG video and audio files using the processing power of the CPU. Hardware MPEG Playback uses an add-in card to deliver full-screen, full-motion, full-color video and CD-quality audio at the full NTSC video frame rate of 30 frames per second, with no dropped frames. The card plays the video from a computer file that has been compressed using the MPEG video standard. Hardware playback is typically much better quality than software playback.

There are actually two MPEG standards: MPEG-1 and MPEG-2. A third, MPEG-4, is currently under development. MPEG -1 is a small-picture mode of MPEG geared to a resolution of 352 by 240 pixels at 30 frames per second (U.S.), with full CD-quality audio. MPEG-1 was originally designed to handle much larger picture sizes than 352 by 240 through interpolation or scaling, but MPEG-2 is more efficient. MPEG-2 offers a "main profile at main level" resolution of 720 by 480 pixels at 30 frames per second (U.S.), with full CD-quality audio. This picture size enables full-screen playback on PCs or TVs. MPEG-2 can incorporate a range of compression ratios, which trade off economies of storage and transmission bandwidth against picture quality. At compression ratios of 30:1 and smaller, MPEG-2 offers the perception of broadcast-quality TV. For greater economy, MPEG-2 supports up to 200:1 compression. MPEG-2 decodes such as the IBM decoder chip can also recognize and decode MPEG-1 bitstreams, enabling the IBM chip to support both compression standards.

MPEG-3 has been dropped. It was focused on HDTV with sampling dimensions up to 1,920 by 1,080 at 30 frames per second. The standard was to address bit rates between 20 and 40 Mbit/sec. Nevertheless, it was discovered that with a little tweaking, MPEG-2 and MPEG-1 work extremely well at the HDTV rate. HDTV is now part of the MPEG-2 High-1440 Level specification.

MPEG-4 is an ISO/IEC standard developed by MPEG (Moving Picture Experts Group), the committee that also developed the Emmy Award winning standards known as MPEG-1 and MPEG-2. These standards made interactive video on CD-ROM, DVD and Digital Television possible. MPEG-4 is the result of another international effort involving hundreds of researchers and engineers from all over the world. MPEG-4 was finalized in October 1998 and became an International Standard in the first months of 1999. The fully backward compatible extensions under the title of MPEG-4 Version 2 were frozen at the end of 1999, to acquire the formal International Standard Status early in 2000. Several extensions were added since and work on some specific work-items work is still in progress. MPEG-4 builds on the proven success of three fields: 0x07 Digital television; 0x07 Interactive graphics applications (synthetic content); 0x07 Interactive multimedia (World Wide Web, distribution of and access to content) MPEG-4 provides the standardized technological elements enabling the integration of the production, distribution and content access paradigms of the three fields. More information about MPEG-4 can be found at MPEG's home page (case sensitive): http://mpeg.telecomitalialab.com This web page contains links to a wealth of information about MPEG, including much about MPEG-4, many publicly available documents, several lists of 'Frequently Asked Questions' and links to other MPEG-4 web pages. The standard can be bought from ISO, send mail to sales@iso.ch. Notably, the complete software for MPEG-4 version 1 can be bought on a CD ROM, for 56 Swiss Francs. It can also be downloaded for free from ISO's website: www.iso.ch/ittf - look under publicly available standards and then for "14496-5". This software is free of copyright restrictions when used for implementing MPEG- 4 compliant technology. (This does not mean that the software is fee of patents). As well, much information is available from the MPEG-4 Industry Forum, M4IF, http://www.m4if.org. See section 7, The MPEG-4 Industry Forum. This document gives an overview of the MPEG-4 standard, explaining which pieces of technology it includes and what sort of applications are supported by this technology.

A variation on MPEG-4 called MPEG-4 AAC-plug SBR (Spectal Band Replication) is being developed by some firms in order to get "CD-quality" sound at only 48 Kbps into cell phones.

MPEG-7, formally known as "Multimedia Content Description Standard" is unlike MPEG-1, MPEG-2 and MPEG-4, as it does not deal with encoding of moving pictures and audio. Rather, MPEG-7 provides a set of tools to describe multimedia content such as still pictures, graphics, 3D models, audio, speech, video and composition information about how these elements are combined in a multimedia presentation, regardless of how the content is coded or stored. So, MPEG-7 descriptions of analogue or digital movies can be printed on a piece of paper, as well as expressed in some digital form. An MPEG-7 description, which might include such characteristics as color, shape and texture, is associated with, rather than collocated with, the content itself to allow fast and efficient searching for and filtering of material of interest to the user. MPEG-7 includes an XML-based Description Definition Language (DDL) for extending the library of descriptions and description schemes.

MPEG-21 is an open standards-based framework for multimedia delivery and consumption. It aims to enable the use of multimedia resources across a wide range of networks and devices while protecting the copyrights of multimedia assets. MPEG-21 does not deal with encoding or compression.

MPEG-1 See MPEG.

MPEG-2 MPEG-2 is one of the most important standards developed by the Moving Pictures Expert Group the standardization of coded representations of video and audio signals. MPEG-2 has been chosen as a leading digital video compression for a broad range of future video and broadcast applications. The digital signal compression standard used by digital satellite systems. Compressing the audio and video signals allows more channels to be broadcast over the same bandwidth. MPEG stands for Moving Picture Experts Group. See MPEG and MPEG-2 Audio.

MPEG-2 Audio MPEG-2 audio is a compatible extension of the MPEG-1 audio coding which enables the transfer of mono, stereo, or multichannel audio in a single bitstream. It can operate at data rates from 32 kbps up to more than 1 Mbps, and supports sampling rates of 32, 44.1 and 48 kHz. For stereo, a typical application would operate at an average data rate of 128-256 kbps. A multichannel movie soundtrack requires an average bit rate of 320-640 kbps, depending on the number of channels (5 to 7, plus a sub woofer channel) and the complexity of the encoded audio.

MPEG-3 See MP3.

MPEG-4 See MPEG.

MPEG-7 See MPEG.

MPEG-21 See MPEG.

MPG Microwave Pulse Generator. A device that generates electrical pulses at microwave frequencies.

MPI 1. MultiPath Interface. Between a transmitter and receiver, the radio wave can take a direct path and one or more reflected paths. The direct radio wave always arrives prior to the reflected waves. If the reflected waves are of sufficient amplitude, they will interfere with the direct wave. The relationship of the amplitude and time delay between the direct and reflected waves create peaks and nulls at the receiver, causing momentary signal fading or loss. In a digital system, this can result in very significant degradation, as the receiver loses signal acquisition and frame synchronization during each fade. The net effect is an increase in the residual bit error rate.

2. Media Platform Interface libraries. Part of Sun Microsystems' XTL Teleservices architecture. MPIs provide a layer of abstraction between details of the system services, applications and providers. The system services include a message passing "server", a data stream multiplexor streams driver, a provider configuration database and a database administration tool.

MPLS MultiProtocol Label Switching. A family of IETF standards in which Internet Protocol networks can make forwarding decisions based on a pre-allocated label to setup a Label Switched Path (LSP). MPLS grew out of Cisco's proprietary TAG Switching protocol. MPLS has faster forwarding performance than IPv4 networks due to its ability to make decisions based on the pre-allocation of a 20-bit label through the IPv4 routing protocols. MPLS works like this: As an IP data stream enters the edge of the network, the ingress Label Edge Router (LER) reads the destination address of the first data packet and attaches a 32-bit shim header "label" in between the layer 2 and layer 3 headers of the packet The label is mapped to a Forwarding Equivalency Class (FEC) based on the destination network and the MPLS EXP value which signified the QoS level. The Label Switch Router (LSR) in the core of the network examine the 20-bit label, and switch es the packet with greater speed than if the device had to interrogate the IP routing table of the device. The router swaps the label with the new label that the next router needs to assist in the completion of the LSP. There are two flavors of MPLS available: Frame-based and cell-based. Cell-based MPLS is used in ATM networks, while frame-based MPLS is used in packet-based networks like Ethernet and Frame-Relay. Although MPLS offers slight performance increases, the richness of MPLS comes from the MPLS Applications. The two MPLS applications most widely deployed are MPLS VPNs (IETF RFC2547) and MPLS Traffic Engineering.

MPLS and Frame Relay Alliance The MPLS Forum and Frame Relay Forum (FRF) merged in April 2003 to form the MPLS and Frame Relay Alliance. The organization initially comprised 60 member dedicated to advancing the recognition and acceptance of MPLS (MultiProtocol Label Switching) and frame relay, and promoting their interoperability. The Frame Relay Forum was formed in May 1991 as an association of vendors, carriers and consultants committed to the education, promotion and implementation of frame relay in accordance with international standards. The FRF released 21 implementation agreements in its history. The MPLS Forum was established in 2000 to drive worldwide deployment of multi-vendor MPLS networks. See also Frame Relay and MPLS. www.frforum.com and www.mplsforum.org

MPM Marketing Product Management.

MPN Manufacturer's Part Number.

MPOA MultiProtocol Over Asynchronous Transfer Mode. A developing set of architectural specifications defined by the ATM Forum. Working at Layer 3 (Network Layer) MPOA specifies standards for Layer 2 (Link Layer) switching through a Layer 3 router – i.e., switched routing – over an ATM fabric. MPOA allows companies to build scalable, enterprise-wide LAN internetworks that seamlessly interwork ATM with LAN protocols such as Ethernet, Token Ring, FDDI and Fast Ethernet. In effect, MPOA provides for inter-LAN cut-through, for the deployment of a WAN VLAN (Virtual Local Area Network) over an ATM backbone. MPOA accomplishes this by separating the route calculation function from the Network Layer forwarding function. In support of Network Layer packets such as IP and IPX, the edge routers will recognize the beginning of a data transfer and respond with an ATM network destination address. At that point, the router network will establish a cut-through SVC (Switched Virtual Circuit) which will eliminate router-by-router delays, thereby considerably increasing the speed of associated data transfer. This is accomplished by distributing the connection intelligence through the network to the edge devices; the traditional approach involves each router's acting independently on each packet in an effort coordinated by a centralized router, which can become overloaded. MPOA draws on existing standards, including the Layer 2 LANE (Local Area Network Emulation) from the ATM Forum, and the Layer 3 NHRP (Next Hop Resolution Protocol) from the IETF. MPOA also draws on IP extensions such as RSVP (Resource ReSerVation Protocol), which is used in support of isochronous data such as streaming video over IP networks. See the following four definitions. See also Classical IP over ATM, IP, LANE, NHRP, RSVP and VLAN.

MPOA Client MPC. An ATM term. A protocol entity that implements the client side of the MPOA architecture. An MPOA client implements the Next Hop Client (NHC) functionality of the Next Hop Resolution Protocol (NHRP). See MPOA.

MPOA Server MPS. An ATM term. A protocol entity that implements the server side of the MPOA architecture. An MPOA Server implements Next Hop Server (NHS) functionality of the NHRP. See MPOA.

MPOA Service Area An ATM term. The collection of server functions and their

clients. A collection of physical devices consisting of an MPOA server plus the set of clients served by that server. See the three definitions above and one below.

MPOA Target An ATM term. A set of protocol address, path attributes, (e.g., internetwork layer QoS, other information derivable from received packet) describing the intended destination and its path attributes. See the four definitions immediately above.

MPOE Minimum Point Of Entry, pronounced em-poe. Also known as MPOP (Minimum Point Of Presence), as defined by the FCC, and the LLDP (Local Loop Demarcation Point). The MPOE is the main point of physical and logical demarcation between the LEC (Local Exchange Carrier) and the customer premises. Up to the point of the MPOE, the telco is fully responsible for deployment and maintenance of the local loop connection. Beyond the MPOE, the user organization or building owner is responsible for the extension of the connection to the PBX, Centrex telephone sets, etc. In a campus environment comprising multiple buildings, there may be multiple points of demarcation, in which case one is designated by mutual agreement as the MPOE. Here's a working explanation from Ty Osborn, who works for the best CLEC in California (he says), tosborn@email.pacwest.com, "I was first introduced to MPOE when I had a (telco) tech out on prem (an apartment building) and could not find the DS-1 Bell had delivered earlier that day. He called and asked "where the heck is the MPOE?" (pronounced em-poe). In short, a MPOE might be a phone room with some punchdown blocks. An MPOE might be a box on the outside of the building. An MPOE is the last piece of equipment a phone company installs in a building for a customer it is providing service to. The MPOE is basically the phone company's last point of responsibility for that circuit. After that it's the customer's or the equipment or service company who is servicing the customer with on-premise equipment – for example, a PBX.

MPOP 1. Minimum Point Of Presence. See MPOE.

2. Metropolitan Point Of Presence. The point of presence of a carrier within a metropolitan area. See also POP.

MPP Massively Parallel Processing. A computer in which there are many processors and each has its own RAM (memory), which contains a copy of the operating system, a copy of the application code, and its own slice of the data, on which that processor works independently of the others.

MPPP Multilink Point-to-Point Protocol. Also known as MP, MLP, and Multilink. A variant of PPP, MPPP is a method by which packet data traffic can be spread across multiple serial WAN (Wide Area Network) links in order to increase transmission speed. MPPP provides for packet fragmentation and reassembly, sequencing, and load balancing for both inbound and outbound traffic. MPPP can be configured for ISDN BRI and PRI channel aggregation (i.e., bonding), and both dialup modem and non-dialup asynchronous serial interfaces. See also Bonding and PPP.

MPRII A Swedish PC monitor standard that specified a limit of 2.5 milligauss of ELF emissions when measured at about 20 inches from the screen.

MPS 1. Multi Page Signal. A frame sent in fax transmission if the sender has more pages to transmit.

2. Mobile Positioning Service. This technology uses the idle time of a GSM cellular system to figure where you (or more precisely, your cell phone) is. An alternative to GPS. Such MPS is designed for applications like emergencies, in which the cell phone can transmit to emergency authorities – fire, police – where the person in trouble actually is at that moment.

MPSK Mobile switching centre.

MPTN Multi Protocol Transport Networking. IBM scheme addressing multi protocol network support including TCP over SNA and SNA over IP.

MPU 1. Main Processor Unit.

2. Message Processor Unit.

MPTy Multiparty. A wireless telecommunications term. A supplementary service provided under GSM (Global System for Mobile Communications).

MPX Multiplex.

MQA Multiple Queue Assignment. lets ACD system agents log into multiple queues.

MR 1. Modem Ready. An ASCII signal and visible "on" light that tells you that your modem is on and ready.

2. Modified Read. Relative element address differentiation code. A two-dimensional compression technique for fax machines that handles the data compression of the vertical line and that concentrates on space between the lines and within given characters. See MMR.

MRC 1. MF Receiver Card.

2. Monthly Recurring Charge.

MRFR Master Reference Frequency Rack.

Mrm An ATM term. An ABR service parameter that controls allocation of bandwidth between forward RM-cells, backward RM-cells, and data cells.

MRP Materials Resource Planning. MRP and MRPII are computerized systems for planning the use of resources in manufacturing processes. Such resources include the scheduling of raw materials, production equipment, and processes. MRP now is known by the broader term Enterprise Resource Planning (ERP). See ERP for a full definition.

MRR Monthly Recurring Revenue.

MRS Menu Routing System.

MRSCC MRS Central Controller.

MS 1. Mobile Station. A wireless telephone allowing mobility so that calls may be placed locally or in another geographic region. In short, another name for a cell phone.

2. Message Switch.

3. Microsoft.

4. Microprocessor System.

MS-DOS MicroSoft Disk Operating System, now buried under Windows 95 and Windows 98, and fortunately eradicated from Windows NT and Windows 2000, making these operating systems far more reliable.

MS-SPRing Multiplex- Section- Shared Protection Ring. See Self Restorable Rings.

MSA 1. Metropolitan Statistical Area. Sometimes known as SMSAs (Standard Metropolitan Statistical Areas) and MEAs (Metropolitan Economic Areas) and as defined by the U.S. Census Bureau, MSAs are geographic areas that contain cities of 50,000 or more population, and which include the surrounding counties. Such areas are characterized by the "community of interests." Using data from the 1980 census, the FCC allocated two cellular licenses to each of the 305 MSAs in the United States. The FCC developed LATA boundaries based largely on the SMSA concept.

2. Message Service Application.

3. Multi-source Agreement Industry Group. (See SFP).

MSAG See Master Address Street Guide.

MSAT Mobile SATellite. Technology for transmitting and receiving satellite transmissions in moving vehicles.

MSAU MultiStation Access Unit. See MAU.

MSB Abbreviation for Most Significant Bit and Most Significant Byte. That portion of a number, address or field which occurs leftmost when its value is written as a single number in conventional hexadecimal or binary notation. The portion of the number having the most weight in a mathematical calculation using the value.

MSC 1. Mobile Switching Center. A switch providing services and coordination between mobile users in a network and external networks. See A1.

2. Malaysian Multimedia Super Corridor. The MSC is an ambitious development concept of the Malaysia government to develop an area of Malaysia with high-tech (telecom and computer) companies. The area is scheduled to house government offices, moving them out of traffic-congested Kuala Lumpur, the capital of the country. The government is installing a 4,300 kilometer fiber optic network in the MSC. If a company agrees to invest in the area, the Malayasian government extends tax breaks and other incentives. Overseeing the MSC is something called The Multimedia Development Corporation.

3. See Maintenance of Service Charge.

MSEngine Mirrored Server Engine. The part of the Novell SFT III operating system that handles nonphysical processes, such as the NetWare file system, queue management, and the bindery. SFT III is split into two parts: the IOEngine (Input/Output Engine) and the MSEngine and the MSEngine. The primary server and the secondary server each have a separate IOEngine, but they share the same MSEngine. The file system, name buffers, and queue management system all reside in the MSEngine. Applications and NetWare Loadable Modules (NLMs) that do not address hardware directly can be mirrored by loading them in the MSEngine. If one server fails, applications and NLMs in the MSEngine continue to run. The MSEngine keeps track of active network processes; it provides uninterrupted network service when the primary server fails and the secondary server takes over. See IOEngine.

MSF 1. Mobile Serving Function.

2. Multiservice Switching Forum. A group devoted to open carrier standards especially those oriented to carrying all types of traffic, from voice to data, and focusing on using ATM as a primary switching and transport vehicle. The goal is also to support IP and frame relay services. www.msforum.org.

MSI Modular Station Interface. A Dialogic board that interfaces analog phones to SCbus and PEB-based products.

MSISDN Mobile Station ISDN number. An ISDN number provisioned to a mobile station subscriber and used to place a call.

MSIX Metered Services Information Exchange. This protocol, first announced in June of 1997, provides a common interface for Internet applications to easily exchange detailed usage information with network billing and information systems. This specification includes a way for Internet Service Providers (ISPs) to effectively meter usage and charge for Internet services. Here are words from the June press release. Compaq and NetCentric announced a New Internet Metering Technology called MSIX, which is intended to accelerate the growth of new billable Internet services such as telephony, fax, video conferencing, content distribution and gaming. The commercial deployment of such applications has been hindered by the immaturity of the management and billing tools inside the Internet infrastructure. MSIX addresses many of these issues by providing a common mechanism to effectively meter application usage by a subscriber. It provides software developers and ISPs a common accounting framework that significantly simplifies network and billing integration. MSIX will make value-added services available to a much larger audience for the first time. MSIX complements other emerging protocols such as Reservation Protocol (RSVP) that allow network resources to be reserved and different quality-of-service levels to be offered. See www.compaq.com and www.netcentric.com.

MSL Mirror Server Link. A dedicated, high-speed connection between Novell NetWare SFT III primary and secondary servers. The mirrored server link is essentially a bus extension from the primary to the secondary server. It requires similar boards in each server, directly connected by fiber-optic or other cables.

MSMQ MicroSoft Message Queueing. MSMQ provides fault-tolerant support for distributed applications which work in conjunction with Message Transaction Server, MTS, which is a component manager for Windows NT, soon to be known as Windows 2000.

MSN 1. The Microsoft Network. Microsoft's version of American On Line (AOL). It has three main features. You can use it to access the Internet. You can send and receive email. You can use some of its own information services, none of which have impressed me sufficiently to use them. I use MSN as my primary email vendor because they do a good, reliable job with email and they have high-speed (56 Kbps) local phone numbers in all the domestic places in the United States I visit. They are weak overseas. IBM.Net is much better overseas.

2. Mobile Station Number. Also known as MIN (Mobile Identification Number) The telephone number of a cellular or PCS telephone. The unique MSN is paired with a unique ESN (Electronic Serial Number) for reasons of security. See also MIN and ESN.

3. Multi Subscriber Numbering. This British (UK) feature allows you to have two to ten numbers on your ISDN BRI line, which are distinguished to your ISPBX to handle individually. One could be a published fax number, one a personal direct number, one a priority customer number etc. If you have ISDN2 (British Telecom's obsolete BRI service), they must be contiguous but not necessarily if you have ISDN2e (Euro-ISDN BRI).

MSNF Multisystem Networking Facility. An optional feature of certain IBM telecommunications access methods that allows more than one host running ACF/TCAM or VTAM to jointly control an ACF/NCP program.

MSO Multiple System Operator. A company that operates more than one cable TV system, often in different places.

MSP 1. See Mixed Signal Processor.

2. Metropolitan Service Provider. A MAN (Metropolitan Area Network) service provider. The term generally is applied to providers of GbE (Gigabit Ethernet) services.

3. See managed service provider.

MSPP MultiService Provisioning Platform. A marketing term for a network of devices (e.g., switches and multiplexers) that can support multiple protocols (e.g., ATM, Ethernet and IP) and multiple services such as voice, data and video in a variety of formats.

MSS Mobile Satellite System. MSSs are satellite systems which support mobile voice and/or data services. Constellations of such satellites are launched in various non-equatorial paths so that they whiz around the earth much like electrons whiz around the nucleus of an atom. As a result, one or more of the satellites always is in "view" of a small, low-power terminal. With orbital paths over all major land masses, one can conduct a cellular-like voice or data conversation from the jungles of New Guinea to another person in the Sahara desert.

MSRN Mobile Station Roaming Number. This number is generated by the VLR (Visitors' Location Register) of the terminating MSC (Mobile Switching Center). A switch providing services and coordination between mobile users in a network and external networks. It contains the location area identification code of the called subscriber. this number is used by the MSC for the routing of the call. for call routing MSC sends the called MSISDN number to the HLR which sends the corresponding IMSI number to the VLR which in turn generates the MSRN number and route it back to the MSC via HLR (Home Location Register).

MSX Mobile Switching eXchange. In a cellular environment, a MSX is akin to a central office (CO) in the wired world. More commonly, a MSX is known as a MSC (Mobile Switching Center) or a MTSO (Mobile Telephone Switching Office). See MTSO.

MT An ATM term. Message Type: Message type is the field containing the bit flags of a

RM-cell. These flags are as follows: DIR = 0 for forward RM-cells = 1 for backward; RM-cells BN = 1 for Non-Source Generated (BECN), RM-cells = 0 for Source Generated RM-cells CI = 1 to indicate congestion = 0 otherwise NI = 1 to indicate no additive increase allowed = 0 otherwise RA - Not used for ATM Forum ABR.

MT-RJ An emerging de facto industry standard for optical fiber connectors used in premises networks, MT-RJ features a plug and jack technology, similar to that used in copper cabling and favored by the installers as more user-friendly than SC, ST and other fiber connectors. The MT-RJ connector is much smaller than predecessor fiber connectors, offering the advantage of increased port density. MT-RJ also is a drop-in replacement for RJ-45 copper-based cabling systems. The MT-RJ connector was defined through the cooperative efforts of AMP, Fujikura, and USConec, and comprises three primary components: the connector, the connector hardware and the transceiver. See also RJ, RJ-45, SC Connector, and ST Connector.

MTA 1. Message Transfer Agent. An OSI application process used to store and forward messages in the X.400 Message Handling System. Equivalent to Internet mail agent.

2. Apple Computer's Macintosh Telephony Architecture, now effectively dead.

3. Metropolitan Trading Area. An area defined by the FCC for the purpose of issuing licenses for PCS. Each MTA consists of several Basic Trading Areas (BTAs). The United States is broken down into 51 metropolitan trading areas for economic purposes. These boundaries were used for licensing PCS. The MTA was defined by the Federal Communications Commission's (FCC) revisions of the Rand McNally 1992 Commercial Atlas and Marketing Guide, based on certain economic and industrial criteria, in order to allocate areas of service. Each MTA is typically made up of several BTAs, and is named for the city that those BTAs would be most likely to use for means of commerce. There are presently 51 MTAs in the US.

4. Multimedia Terminal Adapter. See Multimedia Terminal Adapter and CMTS.

MTBF Mean Time Between Failure. The length of time a user may reasonably expect a device or system to work before an incapacitating fault occurs. The MBTF statistical method was developed and administered by the U.S. military for purposes of estimating maintenance levels required by various devices. Since accurate statistics (i.e. lots of numbers) require a basis of "failures per million hours of operation," an MTBF estimate on a single device is not statistically very accurate. It would take over 100 years to see if the device really had that many failures. Similarly, since the MTBF is an estimate of averages, half of the devices can be expected to fail before then, and half after. Thus, MTBF cannot be used as a guarantee of how long something might or might not work for. Telecommunications systems operate on the principle of "availability," for which there is a body of CCITT Recommendations. See Availability.

MTC Mobile Terminating Call. Mobile phone receiving the inbound leg of a call. See also MTO.

MTD Memory Technology Drive.

MTIE Maximum Time Interval Error.

MTM 1. Maintenance Trunk Monitor.

2. Mean Time Maintenance.

MTNM Version 3 Multi-Technology Network Management. On October 28, 2003, The TeleManagement Forum announced the release of the Multi Technology Network Management (MTNM) Version 3. About 40 of the world's top telecommunications software vendors and network equipment suppliers have incorporated the MTNM interface into their products, making it a de facto industry standard for network management of optical technologies. Version 3 eases the integration of multi-vendor products in an operator's network, and features a host of enhancements to support new and existing technologies, such as ATM, frame relay, SONET/SDH, DSL and Ethernet. New features in MTNM Version 3, according to the Forum, include:

- Support for new transport technologies such as Digital Subscriber Line (DSL), microwave radio, and management of Ethernet tributary interfaces
- Enhanced support for existing technologies such as ATM and DWDM and generalization of Multiplex Section Protection (MSP)
- Enhancements not specific to transport such as connection, configuration, equipment, performance and fault management.
- The Forum likely will submit MTNM version 3 to the ITU (International Telecommunications Union) for acceptance as a standard.

MTP Message Transfer Part of the SS7 Protocol. It provides functions for basic routing of signaling messages between signaling points. It is Level 1 through 3 protocols of the SS7 protocol stack. MTP 3 (Level 3) is used to support BISUP. See M2PA and Message Transfer Part.

MTS 1. Message Telecommunications Service. AT&T's name for standard switched telephone service. Also called DDD, for Direct Distance Dial.

2. Member of the Technical Staff. A common term at AT&T Bell Labs, Bellcore and other R&D labs.

3. Microsoft Terminal Server. Microsoft's answer to a dumb terminal.

4. Measured Toll Service.

5. Material Transfer System.

MTSO Mobile Telephone Switching Office. This central office houses the field monitoring and relay stations for switching calls between the cellular and wire-based (land-line) central office. The MTSO controls the entire operation of a cellular system. It is a sophisticated computer that monitors all cellular calls, keeps track of the location of all cellular-equipped vehicles traveling in the system, arranges handoffs, keeps track of billing information, etc.

MTTR Mean Time to Repair. The average time required to return a failed device or system to service.

MTU 1. Maximum Transmission Unit. The largest possible unit of data that can be sent on a given physical medium. Example: The MTU of Ethernet is 1500 bytes.

2. Multi Tenant Unit or Multiple Tenant Unit. A fancy name for a building or group of buildings that house many tenants – businesses and perhaps also residences. Such MTU could be an office building, office park or corporate campus, medical facility, hotel or college dormitory. The reason for this term is that the new newer carriers (such as the CLECs) or the building's owner talk about providing telecom service to the tenants. They talk about placing a DSL access router (also called a DSLAM) into the basement of a MTU and give themselves access to the building's existing copper wiring (the copper was installed when the building was built and/or renovated to support telephone lines). DSL allows high-speed data transmission (typically for access to the Internet) over that existing wiring to and from all customers inside the MTU. According to a recent study on MTUs, MTU owners deploy broadband in order to attract and retain tenants and add more revenue streams. A new class of broadband provider is moving the Internet access point of presence, or POP, into the actual buildings where tenants reside. These Mini-POPs use scaled-down versions of aggregators used in telecommunication's companies central offices, or even enterprise office switches, allowing tenants to share the cost of an expensive T-1 or other broadband Internet link. Most MTU broadband providers are offering, or looking to offer, value-added services such as voice, video, and application services as a way for customers to build new revenue streams. Many also offer remote and on-site network management, as well as cable installation services.

MU Monitoring Unit. A wireless telecommunications term. Devices added to circuit configurations that use sophisticated trending rules with fault and topology information to determine potential outages.

Mu Law The PCM voice coding and companding standard used in Japan and North America. A PCM encoding algorithm where the analog voice signal is sampled eight thousand times per second, with each sample being represented by an eight bit value, thus yielding a raw 64 Kbps transmission rate. A sample consists of a sign bit, a three bit segment specifying a logarithmic range, and a four bit step offset into the range. All bits of the sample are inverted before transmission. See A Law and PCM.

MUA An acronym for Mail User Agent, is the end user's mail program, like Eudora.

MUD 1. Multi-User Dungeons. A term that Time Magazine in its 9/13/1993 issue called "the latest twist in the already somewhat twisted world of computer communications." Time called it "a sort of poor man's virtual reality" – created by using words, not expensive head-mounted displays. The first MUD apparently was invented in 1979 as a way for British university students to play the fantasy game Dungeons & Dragons by networked computers. MUD are basically now online games environments that use a great deal of network bandwidth.

2. Multi-User Dimension. This term refers to users who connect to each other via a host computer.

3. In the mud means low volume.

mudbox An unsheltered item of equipment that is sufficiently rugged to withstand adverse environments. It is expected to work perfectly though it sits outdoors in good and bad weather.

MUDS Multi-User Dungeons. A cyberspace term. MUDS are elaborate fictional gathering places that users create one room at a time. All these "spaces" have one thing in common, according to cyberspace wisdom, they are egalitarian. Anybody can enter the rooms (provided he has the correct equipment) and everybody is afforded the same level of respect. A significant feature of most MUDs is that users can create interactive objects that remain in the program after they leave. MUD worlds can be built gradually and collectively. See also USENET.

MULDEM A contraction for Multiplexer Demultiplexer, referring to a piece of equipment which performs both functions and generally operates between two of the AT&T digital hierarchy rates (i.e., DS1 to DS3).

mule tape Mule tape is very strong, flat tape which is used to pull cable through underground conduit. Here's how it typically works: First, you use a bore to make an underground hole. Then you fill that hole with hollow concrete cement pipes joined together to form one long underground conduit (i.e. tunnel). Then you go to one end of the tunnel and use a air compressed device to blow a very lightweight "birdie" attached to a lightweight string through the tunnel. Someone at the other end catches the birdie and pulls gently on the string. Attached to the end of the string is strong mule tape. He keeps pulling on it. Attached to the end of the mule tape is the telecommunications cable – fiber or wire – that you really want to instal in the underground conduit. The whole point of this elaborate procedure is that it's far better for the cable to lay it after the pipes are laid than it is during the installation process when the cable could be damaged.

mull A verb meaning to think it over. When faced with a BIG decision, my partner, Gerry Friesen, often says he'd like "to mull it over." Often he takes a day or two. It's a good strategy since it also gives you the time to do a little more due diligence, also call research.

multi-access The ability of several users to communicate with a computer at the same time with each working independently on their own job.

multi-address calling facility A system service feature that permits a user to nominate more than one addressee for the same data. The network may accomplish this sequentially or simultaneously.

multi-alternating routing Alternate routing with provision for advancing a call to more than one alternate route, each of which is tested in sequence in the process of seeking an idle path.

multi-carrier modulation MCM. A technique of transmitting data by dividing the data into several interleaved bit streams and using these to modulate several carriers. MCM is a form of frequency division multiplexing.

multi-cast Also spelled multicast. The broadcast of messages to a selected group of workstations on a LAN, WAN or the Internet. Multicast is communication between a single device and multiple members of a device group. For example, an IPv6 router might address a series of packets associated with a routing table update to a number of other routers in a LAN internetwork. Similarly, a LAN-attached workstation might address a transmission to a number of other LAN-attached devices. Companies are discovering they can distribute material to large numbers of employees and others on their intranets more efficiently using multicast than they can by sending such material in separate bursts to each user. In multicast mode, routers distribute a given file to all hosts that have signaled they want to receive the material, using the Class D addresses of the IP addressing hierarchy. See also Multicast and Multi-Cast Packets and IPV6. Contrast with Unicast, Anycast and Broadcast.

multi-cast packets Multi-cast packets are addressed to multiple devices within a group of devices. For example, LAN stations use multi-cast packets to deliver information to a specific set of devices such as routers, file servers, and hosts. See Multi-Cast.

multi-cast user message A user message generated at the source node and distributed to two or more destination nodes.

multi-casting The ability of one network node to send identical data to a number of end points - known as broadcast in other circles; one example is if new software or addressing updates need to be distributed to all users; also, a point-to-multipoint video transmission is a multi-cast operation.

multi-channel The use of a common channel to make two or more channels either by splitting the frequency band of the common channel into several narrower bands (called frequency division multiplexing) or by allocating time slots in the entire channel (time division multiplexing).

multi-channel aggregation A feature under some versions of Windows which gives remote users the option of using two phone lines for the same remote session. This way you double bandwidth, thus making their session go twice as fast. I've never used this feature and I'm doubtful that it works as advertised.

Multi-Channel Microwave Distribution Service MMDS. An FCC name for a service (operating in the frequency range 2150-2162 MHz and 2500-2686 MHz) where multiple NTSC video channels are broadcast within a limited geographic area (typically 25 mile radius from single omnidirectional antenna). Also called multi-channel multipoint distribution service or "wireless cable" service.

Multi-Channel Multipoint Distribution Service MMDS. An FCC name for a service (operating in the frequency range 2150-2162 MHz and 2500-2686 MHz) where multiple NTSC video channels are broadcast within a limited geographic area (typically 25 mile radius from single omnidirectional antenna). Also called "wireless cable" service. MMDS was a commercial failure due to limited bandwidth, performance issues and

Line-of-Sight requirements. MMDS has been superseded by the IEEE 802.16a specification. See also 802.16 and 802.16a.

multi-channel transmitter A transmitter using low level combining techniques to process many channels at the same time.

multi-conductor More than one conductor within a single cable complex.

multi-domain network In IBM Systems Network Architecture technology, a network that contains more than one host based System Services Control Point (SSCP).

multi-drop line A multi-drop private line or data line is a communications path between two or more locations requiring two or more LECs, but there are multiple 'drops' per LEC. For example, a hospital in Detroit has a data line going to NY, NY. But in New York, NY there are four hospitals in a several block area. Therefore, one data line with four drops. Then you can have a multipoint - multidrop line, which is a combination of both.

multi-drop line A communications channel that services many data terminals at different geographical locations and in which a computer (node) controls utilization of the channel by polling one distant terminal after another and asking it, in effect, "Do you have anything for me?"

multi-entity buildings In large metropolitan areas, it is common to find one local telephone exchange building housing more than one local switch, and for each switch to handle five or more exchanges.

multi-fiber A fiber that supports propagation of more than one of a given wavelength. See Multi-Mode.

multi-frame In PCM systems, a set of consecutive frames in which the position of each frame can be identified by reference to a multi-frame alignment signal. The multi-frame alignment signal does not necessarily occur, in whole or in part, in each multi-frame.

multi-frequency monitors Also known as multisync or multiscan monitors. They can show images in several resolution standards. Such versatility makes them more expensive than single-resolution monitors (e.g. a standard VGA) but also less prone to instant obsolescence. A multisync monitor showing a VGA may or may not look better than VGA monitor showing a VGA image. That depends on the screen's other attributes.

multi-frequency pulsing An in-band address signaling method in which ten decimal digits (the numbers on the touchtone pad) and five auxiliary signals are each represented by selecting two frequencies and combining them into one "musical" sound.. The frequencies are selected from six separate frequencies – 700, 900, 1100, 1300, 1500 and 1700 Hz. See also Captain Crunch.

multi-frequency signaling MF. A signaling code (utilizing pairs of frequencies in the 700-1700 Hz range) for communications between network switches. Code includes 10 digits and special auxiliary signals.

multi-function card A PC card that incorporates multiple peripherals such as a network adapter and modem.

multi-function peripherals MFS. These are devices which take on two or more functions generally associated with individual peripherals and combine these into one product or in a linker series of modules. A multi-function peripheral might combine a fax machine with a photocopier with a computer printer with a scanner. The term is not very precise, but it tends increasingly to mean a computer device that will print, photocopy, fax and/or scan.

multi-haul A network that is not designed solely for traditional access traffic, or for metro traffic, or for regional traffic, or for long-haul traffic, but instead is designed to carry all of those types of traffic.

multi-homed computer A computer that has multiple network adapters or that has been configured with multiple IP addresses for a single network adapter.

multi-homed Multihomed means connected to multiple networks simultaneously for redundancy and to handle load. All major telecommunications carriers are multihomed with connections to several other carriers at all major interconnection points on their networks. A multi-homed host is a computer connected to more than one physical datalink. The data links may or may not be attached to the same network.

multi-hop An example of a single hop system is a microwave system between one building (let's say downtown San Francisco) and another across town (let's say uptown San Francisco). Each with one microwave antenna on its roof. Let's say we wanted to extend that system to Oakland. We'd put a second antenna on the uptown San Francisco building and shoot across to an antenna in Oakland. That building would now have a multi-hop transmission system.

multi-hosting The ability of a Web server to support more than one Internet address and more than one home page on a single server. Also called Multi-Homing.

multi-leaving In communications, the transmission (usually via bisync facilities and using bisync protocols) of a variable number of data streams between user devices and a computer.

multi-level precedence preemption MLPP. A system in which selected customers may exercise preemption capabilities to seize facilities being used for calls with lower precedence levels.

multi-line hunt The ability of switching equipment to connect calls to another phone in the group when other numbers in the group are busy.

multi-line telephone Any telephone set with buttons which can answer or originate calls on one or more central office lines or trunks. Originally all multi-line telephones were 1A2 and they came in sizes of 2, 3, 5, 9, 11, 17, 19, 29, and 60 lines. Now, skinny wire electronic key systems come in all sizes. See Key Telephone System.

multi-line terminating system Premises switching equipment and key telephone type systems which are capable of terminating more than one local exchange service line, WATS access line, FX circuit, etc.

multi-link Multiple links. Several services can make use of multiple links in order to increase bandwidth. Examples include Multilink Frame Relay and Multilink PPP. Multilink Frame Relay is used to combine together multiple T-1s (usually two-four) in order to provide access to a Frame Relay network at speeds greater than the 1.544 Mbps provided by T-1. The alternative would be Fractional T-3, which generally is not available. Multilink Frame Relay uses an inverse multiplexing process to spread a high-speed Frame Relay data stream across multiple T-1s, recombining the original data stream at the edge of the Frame Relay network. Multilink PPP commonly makes use of two ISDN B (Bearer) channels, each at 64 Kbps, to increase Internet access speed to 128 Kbps. Multilink PPP requires that the two B channels be bonded together, in order that the data stream be multiplexed and switched as a single entity.

Multi-link PPP See MPPP.

multi-link procedure A procedure (defined in ITU-T Recommendations X.25 and X.75) that permits multiple links connecting a single pair of nodes to provide service to the network layer on a shared basis. Such sharing provides greater effective throughput capacity and availability than a single link.

multi-location billing Multilocation Billing is an option whereby a long distance carrier bills separate locations and applies volume discounts pro-rated to each site based on usage, or fixed percentage with pro-rated discounts to sites based on usage.

multi-location extension dialing An AIN (Advanced Intelligent Network) service providing network-based dialing between multiple company locations on the basis of an abbreviated dialing plan. Working much like coordinated dialing plans in the PBX world, the user need only dial an access code (e.g., "7") and a 4-digit extension number. The network connects the call to the target extension at the correct location.

multi-longitudinal mode laser An injection laser diode which has a number of longitudinal modes.

multi-master An inherent mode of the VME bus which allows the controlled sharing of the bus by multiple CPUs under a flexible priority structure.

multi-media Usually spelled multimedia these days. Multimedia is the combination of multiple forms of media in the communication of information. Multimedia enables people to communicate using integrated media: audio, video, text, graphics, fax, and telephony. The benefit is more powerful communication. The combination of several media often provides richer, more effective communication of information or ideas than a single media such as traditional text-based communication can accomplish. Multi-media communication formats vary, but they usually include voice communications (vocoding, speech recognition, speaker verification and text-to-speech), audio processing (music synthesis, CD-ROMs), data communications, image processing and telecommunications using LANs, MANs and WANs in ISDN and POTS networks. Multimedia technology will ultimately take the disparate technologies of the computer, the telephone, the fax machine, the CD player, and the video camera and combine them into one powerful communication center. Technologies that were once analog – video, audio, telephony – are now digital. The power of multimedia is the integration of these digital technologies. To many people, "multimedia" (as defined above) is a disparate collection of technologies in search of a purpose. And it's true: most of the merger of media (as above) is taking place in business communications in the moving around of compound documents. Meantime, multimedia has moved into training and in the home for education and entertainment. See Authoring, Compound Documents, Hypermedia, Ole, Shared Screens, Shared Whiteboards and Synchronization.

multi-media capabilities The ability to run simultaneous voice, image, data and video applications on a computer. A technology that requires enormous bandwidth and processing power. See Multi-Media.

multi-media network A network capable of carrying multiple forms of user information such as voice text, sounds, etc.

multi-media PC The Multimedia PC Council now defines a multimedia PC as a PC having a minimum of two megabytes of memory, a 30 megabyte hard drive, a CD ROM drive, digital sound support and Microsoft's Multimedia Extensions for Windows. See Multi-Media.

multi-media protocol A protocol suitable for handling multiple forms of information such as voice, text, pictures, numbers, etc.

multi-modal A term used to describe the ability of a contact center, or of a contact center agent, to process customer communications in several modalities (e.g., phone calls, fax, email, IM – instant messaging). Multi-modal agents are sometimes called blended agents.

multi-modal browsing An interface architecture where users interact with a system by using a variety of input/output mechanisms such as voice capabilities, touch screens, video and graphics.

multi-mode fiber See MMF.

multi-mode In fiber-optics, an optical fiber designed to allow light to carry multiple carrier signals distinguished by frequency or phase at the same time (contrasts with single-mode). (Also spelled multimode.)

multi-mode fiber an optical waveguide which allows more than one mode to propagate. step index and graded index fibers are multi-mode.

multi-mode distortion In an optical fiber, a result of different values of the group delay for each individual mode at a single wavelength. It isn't the same as "multi-mode dispersion."

multi-mode dispersion Dispersion fattens or smears the transmitted signal, making it difficult to identify the original information sent. Dispersion limits the amount of information that can be transmitted. Dispersion is much worse on the information carried on multi-mode fibers, because there's simple more going on. As a result, dispersion limits the rate and how you can send the information. Single mode is the preferred mode for long distance.

multi-mode fiber See MMF.

multi-NAM A cellular telephone term to allow a cellular phone to have two (or more) phone numbers, each of which can be on a different cellular system. This lets you get service from many cellular phone companies. For example, you could subscribe to one carrier in your home city and another in a distant city – perhaps one you travel to often. Once upon a time this saved you paying high roaming charges. But new cell phone prices make multi-NAM an anachronism.

multi-path Multiple routes taken by RF energy between the transmitter and the receiver. Signal can cancel or reinforce. Varying multipath (at sunset or sunrise) causes varying signal strength that sounds like a train's steam engine starting up.

multi-path error Errors caused by the interference of a signal that has reached the receiver antenna by two or more different paths. Usually caused by one path being bounced or reflected.

multi-plex See Multiplex.

multi-point A configuration or topology, designed to transmit data between a central site and a number of remote terminals on the same circuit. Individual terminals will generally be able to transmit to the central site but not to each other.

multi-point circuit A circuit connecting three or more locations. It is often called a multidrop circuit. See also Multi-Point and Multi-Drop Line.

multi-point distribution service A one-way domestic public radio service rendered on microwave frequencies from a fixed station transmitting (usually in an omnidirectional pattern) to multiple receiving facilities located at fixed points.

multi-point grounding system A system of equipment bonded together and also bonded to the facility ground at the nearest location of the facility ground.

multi-point private line 1. A multi-point private line or data line is a communications path between two or more locations requiring two or more LECs. For example.... A location in Detroit, MI has a data line going to NY, NY, then down to Washington, DC. See also Multi-drop private line. 2. A single communications link for two or more devices shared by one computer and or more computers or terminals. Use of this line requires a polling mechanism. It is also called a multidrop line.

multi-processing A type of computing characterized by systems that use more than one CPU to execute applications. Multi-processing is not multi-tasking, which is the ability to have more one application running on a system at the same time. The technique is not associated with multi-processing, nor does it require multi processing to take place. Multi-tasking typically uses a computer with one CPU (e.g. your desktop or laptop). Multi-processing uses a computer with several CPUs, often a server. See Multi-Tasking.

multi-processor kernel Software that enables a computer operating system to use more than one CPU chip simultaneously. Such software is typically used on servers. It is used to speed up a server.

multi-programming Computer system operation whereby a number of independent jobs are processed together.

multi-protocol message routers A device which converts different electronic mail formats. Such a router would be used to move electronic mail from a cc mail equipped-LAN to a Davinci e-mail LAN to a Wang mini-computer based system.

multi-protocol message routers A device which converts different electronic mail formats. Such a router would be used to move electronic mail from a cc mail equipped-LAN to a Davinci e-mail LAN to a Wang mini-computer based system.

Multi-Purpose Internet Mail Extension See MIME.

multi-rate ISDN "Nx64" or switched-fractional T-1 lets users combine multiple ISDN PRI "B" channels on a call-by-call basis. Since each "B" channel operates at 64 kbps, users could get from 128 kbps to 1.544 Mbps of bandwidth in 64 kbps increments.

multi-server network A single local area network (one cabling system) that has two or more file servers attached. Network addresses assigned to the LAN drivers are the same in each file server because the network boards are attached to the same cabling system. On a multi-server network, users can access files from any file server to which they are attached (if they have access rights). A multi-server network should not be confused with an inter network (two or more networks linked together through an internal or external router).

multi-session When pictures are placed on Kodak Photo CDs the first time, that's called the first session. And the result is a single session CD. The next time more photos are put on the disk, the disk is now called a multi-session Photo CD. To read such a disk, you need a CD-ROM player, which is specifically called a multi-session CD-ROM player.

multi-stage dialing The device needs to break the dialing sequence up into a number of stages in order to execute and complete the call. This is referred to as "multi-stage" or "incremental" dialing. This type of dialing is needed in cases where the switching domain prompts the device for more digits (by sending dial tone again or some other tone).

multi-stage queuing This a term typically used in Automatic Call Distributors. It is the ability to array a number of agent groups in a routing table. The notion of multiple agent groups being addressed may mean that the system may be able to "look-back" and "look forward" as it searches for a free agent in the right group to take the call presently holding.

multi-station Any network of stations capable of communicating with each other on one circuit or through a switching center.

multi-switcher A device that enables one microphone to switch between or connect to multiple radios and/or speakers, or which enables a satellite dish antenna to switch between or connect to multiple satellite receivers.

multi-tasking The concurrent management of two or more distinct tasks by a computer. Although a computer with a single processing unit (as virtually all PCs are) can only execute one application's code at a given moment, a multi tasking operating system can load and manage the execution of multiple applications, allocating processing cycles to each in sequence. Because of the processing speed of computers, the apparent result is the simultaneous processing of multiple tasks. Standard mode Windows performs multi tasking only in the form of context switching. 386 enhanced mode allows multi tasking in the form of time slicing. See Multi-Processing.

multi-tenant sharing The capability of a PBX to serve more than one tenant in a building. This process is a new option for building owners. They now become the Telephone Company for tenants in the building. There is money in this business, but chiefly on the resale of long distance phone calls.

multi-tester Usually an alternate name for VOLT-OHM-MILLIMETER, but may also apply to other MULTI-FUNCTION testing devices.

multi-threading See Thread.

multi-tier tariffs A way of paying for something (i.e. equipment) from your local phone company. The idea is that one tier of your monthly payments is to pay off the equipment, and after a finite period, this tier payment drops to zero. The next tier is to pay for your monthly service and it is ongoing. Other tiers are for other reasons. As this technique was practiced by the Bell System, it was called "two tier." You will no longer find two tier tariffs in common use.

multi-user PC A microcomputer that has several terminals attached to it, so that multiple users can simultaneously use its resources. Multi-user PCs can either slice up the time of a single microprocessor or can give each terminal-based user his own microprocessor. Multi-user PCs are an alternative to LANs and are typically used in specialized, one-ap-

plication solutions, such as a doctor's office billing system.

multi-user software An application designed for simultaneous access by two or more network nodes, i.e. two or more users on a network. It typically employs file and/or record locking. It is not associated with multi processing, nor does it require multi processing to implement.

multi-user telecommunications outlet assembly A grouping in one location of several telecommunications outlets/connectors.

Multi-Vendor Integration Protocol See MVIP.

Multi-Wavelength Optical Repeater See MOR.

multi-way communication A multimedia definition. Multi-way communication goes between two people, or between groups of people in all directions. Multi-way communication can be in real-time, or in store-and-forward mode. Examples of multi-way communication include a video conference, where one individual is giving a presentation to a group of people who listen and ask questions from their workstations; and group conferencing, where several people collaborate, supported by audio, video, and graphics on their workstation screens.

multicast 1. Sending data to many places. Also spelled multi-cast. Multicast allows messages to be sent to a selected group of workstations on a LAN, WAN or the Internet. Multicast is communication between a single device and multiple members of a device group. For example, an IPv6 router might address a series of packets associated with a routing table update to a number of other routers in a LAN internetwork. Similarly, a LAN-attached workstation might address a transmission to a number of other LAN-attached devices. Companies are discovering they can distribute material to large numbers of employees and others on their intranets more efficiently using multicast than they can by sending such material in separate bursts to each user. In multicast mode, routers distribute a given file to all hosts that have signaled they want to receive the material, using the Class D addresses of the IP addressing hierarchy. The message is sent from the transmitter down the tree only to those nodes that need to receive it in order to further distribute it to downstream nodes, from those nodes only to those that need to receive it in order to further distribute it to downstream nodes, and so on. At each of those points, the message is replicated and retransmitted only as necessary. Multicasting is much more efficient in many applications than is either broadcasting or unicasting. See also Multicast and Multi-Cast Packets and IPV6. Contrast with Unicast, Anycast and Broadcast.

2. A TV term that simply means more channels will be available for viewers. It's often used to refer to the explosion of special interest channels that will be around when digital TV hits the scene and uses the CATV network to transmit its programs. I suspect that the explosion of channels will lend new meaning to Johnny Carson quips about channels for "one-eyed, one-legged transvestites."

Multicast Address Resolution Server See MARS.

multicast backbone Mbone. A method of transmitting digital video over the Internet in real time. The TCP/IP protocols used for Internet transmissions are unsuitable for real-time audio or video; they were designed to deliver text and other files reliably, but with some delay. MBONE requires the creation of another backbone service with special hardware and software to accommodate video and audio transmissions; the existing Internet hardware cannot manage time-critical transmissions.

multicasting Sending data to a group of destinations at once. See multicast.

multichannel transmitter A transmitter using low level combining techniques to process many channels at the same time.

multichannel The use of a common channel to make two or more channels. Frequency Division Multiplexing (FDM) accomplishes this by splitting the frequency band of the common channel into several narrower bands. Time Division Multiplexing (TDM) divides the entire channel into time slots.

multichannel aggregation A feature under Windows NT which gives remote users the option of using two phone lines for the same remote session. This way you double bandwidth, thus making their session go twice as fast.

Multichannel Audio Digital Interface See MADI.

multichannel transmitter A transmitter using low-level combining techniques to process many channels at the same time.

MULTICS MULTiplexed Information and Computing System. In the mid-1960's, MULTICS was the focus of a huge development effort sponsored by MIT (Massachusetts Institute of Technology), Bell Telephone Laboratories and General Electric. The idea was to develop a "computer utility" that would provide computing resources to the population of an entire city or area – in contemporary terms, we would describe this as a multi-user time-share system. MULTICS was a failure. When Bell Labs pulled out of the project, Ken Thompson, Dennis Ritchie and a few other Bell Labs programmers developed UNICS, a single-user variant of MULTICS. UNICs later became known as UNIX. See UNIX.

multidrop Also known as a point-to-multipoint circuit, a multidrop circuit has one point of termination on one end, and multiple points of termination on the other end. For example, a hospital in Detroit leases a private data circuit going to NYC (New York City) to connect the Detroit primary data center with the smaller host computers in each of four NYC hospitals. Therefore, the "head end" in Detroit connects to NYC, where there are four drops. See also Drop.

multifiber cable A fiber-optic cable having two or more fibers, each of which can be an independent optical transmission channel.

multiframe In PCM systems, a set of consecutive frames in which the position of each frame can be identified by reference to a multiframe alignment signal. The multiframe alignment signal does not necessarily occur, in whole or in part, in each multiframe.

multifrequency monitors Also known as multisync or multiscan monitors. They can show images in several resolution standards. Such versatility makes them more expensive than single-resolution monitors (e.g., a standard VGA) but also less prone to instant obsolescence. A multisync monitor showing a VGA may or may not look better than VGA monitor showing a VGA image. That depends on the screen's other attributes.

multifrequency pulsing More commonly known as Dual Tone Multifrequency (DTMF). See DTMF.

multifrequency signaling More commonly known as Dual Tone Multifrequency (DTMF). See DTMF.

multifunctional mobile phone A mobile phone with non-telephony functions, such as a camera, web browser, email sender and receiver, alarm clock, stop watch, and SMS.

multifunction peripherals MFS. These are devices which take on two or more functions generally associated with individual peripherals and combine these into one product or in a linker series of modules. A multi-function peripheral might combine a fax machine with a photocopier with a computer printer with a scanner. The term is not very precise, but it tends increasingly to mean a computer device that will print, photocopy, fax and/or scan.

multihaul A network that is not designed solely for traditional access traffic, or for metro traffic, or for regional traffic, or for long-haul traffic, but instead is designed to carry all of those types of traffic.

multihomed computer Any computer system that contains multiple network interface cards and is attached to several physically separate networks; also know as a multihomed host. The term can also be applied to a computer configured with multiple IP addresses for a single network interface card.

multihomed host A host which has a connection to more than one physical network. The host may send and receive data over any of the links but will not route traffic for other nodes. See also host; router.

multihoming The practice of linking to multiple-access ISPs over physically discrete lines to the Internet. This is done to combat downtime caused by a malfunction of one particular line.

Multilevel Precedence Preemption MLPP. A system in which selected customers may exercise preemption capabilities to seize facilities being used for calls with lower precedence levels.

multilink See Multi-link.

multiline hunt The ability of switching equipment to connect calls to another phone in the group when other numbers in the group are busy.

multiline telephone Any telephone set with buttons which can answer or originate calls on one or more central office lines or trunks. Originally all multiline telephones were 1A2 and they came in sizes of 2, 3, 5, 9, 11, 17, 19, 29, and 60 lines. Now, skinny wire electronic key systems come in all sizes. See Key Telephone System.

multiline terminating system Premises switching equipment and key telephone type systems which are capable of terminating more than one local exchange service line, WATS access line, FX circuit, etc.

multimedia See Multi-Media.

multimedia ACD Also called multimedia queueing. Automated Contact Distribution system that is enabled to process multiple media types, such as voice calls, e-mails, incoming fax documents, web chat requests and Internet voice/video interactions using queuing/hold strategies. For example, a sales queue could be configured and staffed by sales associates that could evenly distribute all sorts of incoming customer sales requests (fax, phone calls, emails, etc.) to sales associates by blending computer and phone capabilities into a common sales response system. This makes sales associates more efficient and also provides a more consistent service level to all customers. See also Multimedia Queuing.

Multimedia Communications Forum See MMCF.

Multimedia Extensions See MMX.

Multimedia Messaging Service A service that allows cell phone users to send pictures, movie clips, cartoons and other graphic materials from one cell phone to another. See MMS.

Multimedia Mobile Access Communication See MMAC.

multimedia queuing The term was originally invented by Interactive Intelligence of Indianapolis, who makes a communications server – a telephone and messaging system based in a PC. The idea is that the communications server should do way more than servicing mere phone calls. It should handle all media streams – email, fax mail, chats, etc. – with the same discipline as phone systems (particularly automatic call distributors) handle plain ordinary phone calls. All "calls" of whatever media should be treated by the system as of equal importance. See also Multimedia ACD.

Multimedia Super Corridor The MSC is an ambitious development concept of the Malaysia government to develop an area of Malaysia with high-tech (telecom and computer) companies. The area is scheduled to house government offices, moving them out of traffic-congested Kuala Lumpur, the capital of the country. The government is installing a 4,300 kilometer fiber optic network in the MSC. If a company agrees to invest in the area, the Malaysian government extends tax breaks and other incentives. Overseeing the MSC is something called The Multimedia Development Corporation.

multimode See Multi-Mode above.

multimode dispersion Dispersion fattens or smears the transmitted signal, making it difficult to identify the original information sent. Dispersion limits the amount of information that can be transmitted. Dispersion is much worse on the information carried on multimode fibers, because there's simple more going on. As a result, dispersion limits the rate and how you can send the information. Single mode is the preferred mode for long distance.

multimode distortion Multimode fiber allows many modes of light to propagate down the fiber-optic path. Multimode fibers are generally used for short-distance data links, as they provide limited bandwidth due to modal dispersion. The relatively large core of a multimode fiber allows good coupling from inexpensive LEDs, and the use of inexpensive couplers and connectors. Multi-mode fiber typically has a core diameter of 25 to 200 microns. The core is much larger than single-mode fiber and allows several modes of light to be passed through it. Multimode fiber was the original medium specified for FDDI. See also Single Mode Fiber.

multimode fiber A fiber-optic cable with a wide core that provides multiple routes for light waves to travel. Its wider diameter of between 25 to 200 microns prevents multimode fiber from carrying signals as far as single-mode fiber due to modal dispersion.

multiparty line A single telephone line which serves two or more subscribers (network access lines). The term usually means either a "two party" or a "four party" line. Lines serving more than four parties are called "rural lines".

multipath Multiple routes taken by RF energy between the transmitter and the receiver. Signal can cancel or reinforce. Varying multipath (at sunset or sunrise) causes varying signal strength that sounds like a train's steam engine starting up.

multipath error Errors caused by the interference of a signal that has reached the receiver antenna by two or more different paths. Usually caused by one path being bounced or reflected.

multipath fading The signal degradation in a cellular radio system that occurs when multiple copies of the same radio signal arrive at the receiver through different reflected paths. The interference of these signals, each having traveled a different distance, result in phase and amplitude variations. The radio signal processing in both the base station and mobile units have to be designed to tolerate a certain level of multipath fading. See Rayleigh Fading for a longer explanation.

multipoint A configuration or topology, designed to transmit data between a central site and a number of remote terminals on the same circuit. Individual terminals will generally be able to transmit to the central site but not to each other.

multipoint circuit A circuit connecting three or more locations. It is often called a multidrop circuit. See also Multipoint and Multidrop Line.

multipoint grounding system A system of equipment bonded together and also bonded to the facility ground at the nearest location of the facility ground.

multipoint private line See Multidrop Line.

MultiPulse-Maximum Likelihood Quantization See MP-MLQ.

multipulse-excited LPC A voice codec algorithm that involves splitting the incoming speech into frames to derive predictor parameters, which are quantized and transmitted; more complex than Adaptive Predictive Coding, but offers lower error rate and fuller support of telefax functions.

multi-player gaming See Telegaming.

multiplay The carriers – phone and cable – have figured that bundles work with consumers. Bundle up phone, cable TV and the Internet in one $XXX package and the consumer (their new customer) thinks he's getting a bargain. Moreover the consumer prefers to deal with only one company when things go awry. The whole marketing ploy seems to work. The carriers are revelling in their newfound marketing expertise. Multiplay simply means a bundle of two or more of the following services: local service, long-distance service, broadband Internet access, TV, and perhaps wireless phone service. Examples: triple-play and quadruple-play.

multiple access The ability of several personal computers connected to a Local Area Network to access one another through a common addressing scheme and protocol.

multiple access schemes Methods of increasing the amount of data that can be transmitted wirelessly within the same frequency spectrum. RFID readers use Time Division Multiple Access, or TDMA, meaning they read tags at different times to avoid interfering with one another.

multiple address message A message to be delivered to more than one destination.

Multiple Address Systems MAS. A microwave point-to-multipoint communications system, either one-way or two-way, serving a minimum of four remote stations. The private radio MAS channels are not suitable for providing a communications service to a larger sector of the general public, such as channels the commission has allocated for cellular paging or specialized mobile radio services. (SMR).

multiple console operation A phone system with this feature can use more than one attendant console. It's good to know the maximum number of consoles your chosen PBX can use.

multiple customer group operation A PBX shared by several different companies, each having separate consoles and trunks.

multiple domains A set of domains on a single LAN, each of which has its own domain-wide post office. Hosts within each domain can exchange mail by going through one domain post office. Hosts in different domains must generally send mail though two intermediary post offices: the sender's domain post office and the receiver's domain post office.

Multiple Frequency-Shift Keying MFSK. A form of frequency-shift keying in which multiple codes are used in the transmission of digital signals. The coding systems may use multiple frequencies transmitted concurrently or sequentially.

multiple homing Connecting your phone so it can be served by one or several switching centers. This service may use a single directory number. It may also use several directory numbers (another term for phone numbers). It all depends on how you set the service up with your local phone company. The idea is to give you more ways of reaching the switched network – in case one or more of your local loops breaks.

multiple listed directory number service Permits more than one listed directory number to be associated with a single PBX.

Multiple Low Speed CSU MLSC. Specialized channel service unit (CSU) that combines several low speed lines onto one 64 kbps channel. 72. MULTIPLE NETWORK IDs: A feature that allows customers to implement multiple dialing plans with each one having its own network ID. This feature benefits extremely large corporations that want different divisions to have their own CVNS networks. Customers still only receive one bill.

multiple master domain Consists of two or more Single Master Domains connected through two-way trust relationships. See Trust Relationship.

multiple name spaces The association of several names or other pieces of information with the same file. This allows renaming files and designating them for dissimilar computer systems such as the PC and the Mac.

multiple parallel processing A method of fault tolerance used with host computers. Several CPUs cooperate to process data. It one CPU fails, its processing tasks are automatically assigned to other processors.

multiple protocol router A communications device designed to make decisions about which path a packet of information will take. The packets are routed according to address information contained within, and can route across different protocols.

multiple routing The process of sending a message to more than one recipient, usually when all destinations are specified in the header of the message.

multiple spot scanning In facsimile systems, the method in which scanning is carried on simultaneously by two or more scanning spots, each one analyzing its fraction of the total scanned area of the subject copy.

multiple token operation Variant of token passing for rings in which a free token on a LAN is transmitted immediately after the last bit of the data packet, allowing multiple tokens on ring (but only one free token) simultaneously.

multiple tuned antenna An antenna with connections through inductances to ground at more than one point and so determined that the total reactances in parallel are equal to those necessary to give the antenna the desired natural frequency.

multiplex 1. To transmit two or more signals over a single channel.

2. In the world of CATV and the explosion of choices that digital TV is bringing, to multiplex means to offer subscribers a choice of various starting times for movies and events.

3. A name for a collection of movie theaters under one roof.

multiplex aggregate bit rate The bit rate in a time division multiplexer that is equal to the sum of the input channel data signaling rates available to the user plus the rate of the overhead bits required.

multiplex baseband In frequency division multiplexing, the frequency band occupied by the aggregate of the signals in the line interconnecting the multiplexing and radio or line equipment.

multiplex hierarchy In the U.S. frequency division multiplex hierarchy,

12 channels = 1 group 5 groups (60 channels) = 1 supergroup 10 supergroups (600 channels) = 1 mastergroup 6 mastergroups = 1 jumbo group

In contrast, the ITU-T standard says 5 supergroups (i.e. 300 channels) = 1 mastergroup.

multiplexed channel A communications channel capable of carrying the telecommunications transmissions of a number of devices or users at one time.

multiplexed packet switching A network in which each node assembles/disassembles a packet containing data from different users. Data within the packet can have a different source and destination than other data in the packet. The one commonality is that the virtual circuits (logical pathways from the source to destination) all include the same internode link.

multiplexer 1. Electronic equipment which allows two or more signals to pass over one communications circuit. That "circuit" may be a phone line, a microwave circuit, a through-the-air TV signal. That circuit may be analog or digital. There are many multiplexing techniques. When spelled "multiplexor," it refers only to optical units. See Frequency Division Multiplexing, Time Division Multiplexing and Pulse Code Modulation.

2. An RFID definition. An electronic device that allows a reader to have more than one antenna. Each antenna scans the field in a preset order.

multiplexing efficiency Figure of merit for multiplexers. The ratio of the aggregate channel input data rate to the composite output data rate. Many statistical multiplexers achieve a multiplexing efficiency of 8 or more.

multipling Connecting together identical purpose terminals of different switches to provide a singular input or output (usually called parallel connecting outside the telephone industry).

multipoint access User access in which more than one terminal equipment (TE) is supported by a single network termination.

multipoint distribution service An FCC term for a fixed station operating between 2.15 and 2.162 GHz.

multipoint grounding system A system of equipment bonded together and also bonded to the facility ground.

multipoint line A single communications line to which more than one terminal is attached. Also called multidrop.

multipoint to multipoint connection A Multipoint to Multipoint Connection is a collection of associated ATM VC or VP links, and their associated nodes, with the following properties:

1. All Nodes in the connection, called endpoints, serve as a Root Node in a Point to Multipoint connection to all of the (N-1) remaining endpoints.

2. Each of the endpoints on the connection can send information directly to any other endpoint, but the receiving endpoint cannot distinguish which of the endpoints is sending information without additional (e.g., higher layer) information.

multipoint/multidrop An arrangement of communications facilities ranging from across a floor to across the globe in which one common facility is shared by all; one station at a time, usually under central polling control, sometimes contention for the channel, as in many LANs, has a history running deep into telegraphic communications. Contrast: Point to Point.

multiport card A circuit board with two or more ports for modems or other devices. Useful for enabling one PC to handle multiple incoming or outgoing calls at one time.

multiport repeater A repeater, either standalone or connected to standard Ethernet cable, for interconnecting up to eight ThinWire Ethernet segments.

multiport switch A local area network term. A device which allows packets to switch from one cable to another.

multiprocessing A type of computing characterized by systems that use more than one CPU to execute applications. Multi processing is not multi tasking, which is the ability to have more one application running on a system at the same time. The technique is not associated with multi processing, nor does it require multi processing to take place. Multi tasking typically uses a computer with one CPU (e.g. your desktop or laptop). Multi processing uses a computer with several CPUs, often a server. See Multi-Processing, Multi-Tasking and Multi-Threaded.

Multiprotocol Over ATM MPOA. A proposed ATM Forum spec that defines how ATM traffic is routed from one virtual LAN to another. MPOA is key to making LAN emulation, Classical IP over ATM, and proprietary virtual LAN schemes interoperate in a multiprotocol environment. At this point, its unclear how MPOA will deal with conventional routers, distributed ATM edge routers (which shunt LAN traffic across an ATM cloud, while also performing conventional routing functions between non-ATM networks), and route servers (which centralize lookup tables on a dedicated network server in a switched LAN).

multiprotocol message routers A device which converts different electronic mail formats. Such a router would be used to move electronic mail from a cc mail equipped-LAN to a Davinci e-mail LAN to a Wang mini-computer based system.

Multipurpose Internet Mail Extension MIME. An extension to electronic mail (Internet and other e-mail systems) which provides the ability to transfer non-ASCII data, such as graphics, software, audio, video, binary and fax. MIME was developed and adopted by the Internet Engineering Task Force. It was designed for transmitting mixed-media files across TCP/IP networks. The MIME protocol, which is actually an extension to SMTP, covers binary, audio and video data. Essentially what happens with MIME is that you pick a binary file (any file that isn't ASCII) to send along with your e-mail. Your e-mail software then converts your binary file to ASCII text which can easily be transmitted across one or more e-mail systems. All e-mail networks will transmit ASCII (i.e. ASCII 128 and below). The technique that makes MIME encoding possible is called UUencoding. Here is an example. Notice that every line begins with an M. Every line is the same length. At one end you UUencode the file (and it looks as below). When you receive the file, you must UUdecode it. Mostly the coding and encoding is done automatically by your e-mail program. Sometimes you must decode it manually. Don't try and UUdecode the following file. It won't work. I chopped the middle of the file out. I'm running it below purely as an example. See UUEncode and MIME.

Multiprotocol Label Switching See MPLS.

MultiPulse-Maximum Likelihood Quantization See MP-MLQ.

Multipurpose Internet Mail Extension See MIME and the next definition.

Multipurpose Internet Mail Extensions Mapping MIME mapping. A way of configuring browsers to view files that are in multiple formats. An extension of the Internet mail protocol that enables sending 8-bit based e-mail messages, which are used to support extended character sets, voice mail, facsimile images, and so on.

multiserver network A network that uses two or more file servers.

multisession An incrementally updated Kodak Photo CD. See Multi-Session for a fuller explanation.

multislacking When an employee has two browser windows open, a non-work-related site on top of a productive one, and quickly clicks on the legitimate site whenever the boss is nearby. Definition courtesy Wireless Magazine.

multistage dialing The device needs to break the dialing sequence up into a number of stages in order to execute and complete the call. This is referred to as "multi-stage" or "incremental" dialing. This type of dialing is needed in cases where the switching domain prompts the device for more digits (by sending dial tone again or some other tone).

multistage queuing This a term typically used in Automatic Call Distributors. It is the ability to array a number of agent groups in a routing table. The notion of multiple agent groups being addressed may mean that the system may be able to "look-back" and "look forward" as it searches for a free agent in the right group to take the call presently holding.

multitasking The concurrent management of two or more distinct tasks by a computer. Although a computer with a single processing unit (as virtually all PCs are) can only execute one application's code at a given moment, a multi tasking operating system can load and manage the execution of multiple applications, allocating processing cycles to each in sequence. Because of the processing speed of computers, the apparent result is the simultaneous processing of multiple tasks. Standard mode Windows performs multi tasking only in the form of context switching. See Multi-Processing.

multithreaded See Multi-Threaded.

multiuser A computer that can support several workstations operating simultaneously.

MUMPS Massachusetts General Hospital Utility Multi-Programming System.

munge Munge is a verb meaning to render information unrecognizable. "Address munging," which also is incorrectly called "spam blocking," is the practice of monkeying around with an e-mail address so that its form is unrecognizable to computers but remains recognizable to humans. For example, you might take the e-mail address foo@foo.com and munge it into foo@foo.com.kill-spam. This munged address absolutely cannot be interpreted by a computer, but a human being can figure out the real address rather quickly. The basic reason for munging an e-mail address is to keep spammers from harvesting addresses from Usenet sites, putting you on a distribution list and flooding your mailbox with spam mail. The basic reason for not munging is that it makes it impossible for anyone to automatically respond to your e-mails. According to legend, MUNG originated at MIT as an acronym for "Mash Until No Good." See also Foo and Spam.

muni Wi-Fi See municipal Wi-Fi.

municipal Wi-Fi Citywide, city-sponsored Wi-Fi service, provided to citizens for "free" (i.e., as a tax-supported service, like road maintenance or the sewer system) or for a low cost (like water). There is a big debate over the pros and cons of municipal Wi-Fi. Personally, I think it's a perfectly reasonable thing for a municipality to do for its citizenry.

Muriel Muriel Fullam worked for me for ten years before she worked with me on this dictionary. That had to be the longest apprenticeship ever. It shows. She has now worked for me for over 20 years. She is married to Jerry Fullam who is absolutely the best husband and will love it when he sees his name here. Christine is Muriel's daughter and she now publishes this dictionary.

Murphy's Law When something can go wrong, it will. Compare with the Gotcha Law.

MURS Multi-Use Radio Service. See Personal Radio Services.

MUSE MUltiple Sub-Nyquist Samplying Encoding. The Japanese bandwidth compression algorithm for analog HDTV transmission. The Japanese began work on analog HDTV in 1968, and demonstrated it in Washington, D.C. in 1987. The Japanese system ultimately was rejected for use in the U.S., in favor of a digital standard proposed by The Grand Alliance. The Japanese government invested over US$1 billion in the project. See also The Grand Alliance.

Musepack See MPC.

mushroom All mushrooms can be eaten, some only once. Many Europeans enjoy searching for edible mushrooms in the forest. They call this activity mushrooming.

mushroom board Also called a white board or peg board. This is placed between termination blocks to support route crossing wire.

mushroom strategy A technique for dealing with dissident shareholders, recalcitrant employees and difficult families. The strategy is threefold: Keep everyone in the dark, feed them a lot of organic bovine waste (i.e. bullshit) and harvest them (i.e. bring them into the real world) when they are truly ripe, i.e. when you're ready to bring them out of their misery, or ignorance. The names derives from techniques used for growing mushrooms.

music recognition service A service that lets mobile users dial a number, hold their mobile phone toward a nearby music source for 10-15 seconds, hang up, and then receive a text message in another 10-15 seconds with the name of the song and the artist who recorded it, and information about how to purchase and download the song or a ringtone based on it.

music source An external music source such as a radio, can be connected to the Key Service Unit for Music on Hold, Background Music, or both.

music-on-camp Audio source input for use with attendant camp-on. See also MUSIC-ON-HOLD.

music-on-hold Background music heard when someone is put on hold, letting them know they are still connected. Some modern phone systems generate their own electronic synthesized music. Most phone systems have the ability to connect any sound-producing device, e.g. a radio or a cassette player. Most companies, unfortunately, devote little attention to the sound source they select. Sometimes competitors will deliberately advertise on the radio station that callers will hear on hold. Thus, Macys is now selling Gimbels. It's better to use pre-recorded music. Better yet are tapes of "specials" and other happenings around your firm. Use the "Music-on-Hold" feature as another method of selling. "Ask the operator about our special on ladies' underwear." "Ask our operator to send you a copy of our latest annual report."

musical chair stocks When the music stops, the last one holding the stock is bankrupt. This Aphorism, which refers to the phenomenon of forever-rising Internet stocks, which happened in the fall of 1998, came from the best stockbroker in North America, Todd Kingsley of Smith Barney in Washington, D.C. I know he's the best, because he, modestly, told me so.

Musical Instrument Digital Interface (MIDI) A standardized communications protocol. MIDI files contain musical note information and other performance data that can "play" a MIDI instrument or sound module to produce music.

must carry signals A term from the 1992 Cable Act, it refers to a cable system's mandatory signal carriage of both commercial and noncommercial television broadcast stations that are "local" to the area served by the cable system.

mute A feature which disconnects the handset microphone or speakerphone microphone so that side conversations won't be heard.

mutual capacitance The capacitance between two conductors when all other conductors, including the shield, are short circuited to ground.

mutual fund A portfolio of stocks managed by a professional investment company for a management fee. Investors purchase shares in the fund, hoping that the professionals managing the fund will do better than the individual investors could, had they bought stocks individually. These days mutual funds tend to specialize – in stocks of telecommunications companies, for example.

mutual synchronization A timing subsystem not employing directed control, by which the frequency of the clock at a particular node is controlled by some weighted average of the timing on all signals received from neighboring nodes.

mutual telephone company A telephone company founded, built, and operated by a community or an association of individuals in the community. In some cases a mutual telephone company was founded by an association of farmers who initially established telephone service among themselves via makeshift farmer's lines. See farmer's line.

mutually synchronized network A network-synchronized arrangement in which each clock in the network exerts a degree of control on all others.

mutually exclusive If there are two or more ways in which an event can occur then each way is mutually exclusive of the other way. The probability of the event occurring is the sum of the probabilities of the two or more ways.

MUX See Multiplexer.

MV-90 A UL listing for single or multi-conductor cables rated 2001-35000 volts complying with Art. 326 of the NEC. The MV designation stands for Medium Voltage and the 90 refers to the conductor temperature rating in wet or dry locations.

MV-90 Dry A UL medium voltage power cable listing which restricts the cable to dry location use only. Usually pertains to a single conductor 5kV nonshielded power cable.

MV-90 Wet or Dry UL medium voltage listing for power cable suitable for wet and dry locations with a maximum continuous conductor temperature of 900xAFC.

MVIP Multi-Vendor Integration Protocol. Pronounced M-VIP. MVIP is a family of standards designed to let computer telephony products from different vendors inter-operate within a computer or group of computers. MVIP started in 1990 with a computer telephony bus for use inside a single computer. Picture a printed circuit card that fits into an empty slot in a personal computer. The slot carries information to and from the computer. This is called the data bus. Printed circuit cards that do voice processing typically have a second "bus" – the voice bus. That "bus" is actually a ribbon cable which connects one voice processing card to another. The ribbon cable is typically connected to the top of the printed circuit card, while the data bus is at the bottom. As of this writing, there are at least four such buses defined. Several (AEB, PEB, SCSA) have been defined by Dialogic Corporation, Parsippany, NJ for connecting devices with Dialogic products and Dialogic compatible devices. There is also an organization called the ECTF which helps promulgate buses. The MVIP Bus was defined by Natural MicroSystems, Natick, MA with assistance from Mitel, Promptus Communications and Rhetorex as a vendor-independent means of connecting telephony devices within a computer chassis. MVIP was introduced by a seven-company group in 1990 and has been distributed by Natural MicroSystems, Mitel and NTT International (part of the Japanese telephone company). By July 1996, MVIP had over 170 companies manufacturing over 300 board-level MVIP products including telephone line interfaces, voice boards, FAX boards, video codecs, data multiplexers and LAN/WAN interfaces. MVIP now has its own trade association, the Global Organization for MVIP (see GO-MVIP), which develops extensions such as higher-level APIs and multi-chassis switching. Here is a write-up on the original MVIP Bus from Mitel's Communicating Objects Division:

"The MVIP Bus consists of communications hardware and software that allows printed circuit cards from multiple vendors to exchange information in a standardized digital format. The MVIP bus consists of eight 2 megabyte serial highways and clock signals that are routed from one card to another over a ribbon cable. Each of these highways is partitioned into 32 channels for a total capacity of 256 full-duplex voice channels on the MVIP bus.

These serial link from one card to another. They are electronically compatible with Mitel's ST_BUS specification for inter-chip communications. By letting expansion cards exchange data directly, the MVIP bus opens the PC architecture to voice/data applications that would otherwise overburden the PC processor with data transfers. The MVIP bus is equivalent to an extra backplane that is capable of routing circuit switched data.

"MVIP systems generally have two types of cards ; network cards and resource cards. They differ by the switching they provide and in the way they are wired to the bus. Network cards almost always provide more flexible switching and can drive either the input or the output side of the bus, although they usually drive the output side of the bus. Resource cards usually provide very little switching and are only able to drive the input side of the bus. Resource cards usually rely on the network cards to do most or all of the switching on the MVIP bus."

According to Brough Turner, chairman of the GO-MVIP Technical Committee, MVIP Evangelist and vice president of Natural MicroSystems, "The original MVIP bus distributes switching elements but only requires switches on telephone network interface cards. This has simplified access to MVIP for many communications service providers and has permitted easy inter-connection of pre-existing and proprietary technology to the common bus. More recently, the Flexible MVIP Interface Circuit (see FMIC) has reduced the complete MVIP interface, including clocks and switching, to one chip, further simplifying MVIP connections.

"MVIP's distributed switching architecture provides software controlled digital switching within the PC chassis. MVIP software driver standards simplify access for application developers and support the integration of components from multiple vendors. As a result of work by more than a dozen companies active in the GO-MVIP Technical Committees, MVIP now addresses multi-chassis connections and higher-level APIs. Multi-Chassis MVIP (MC-MVIP) provides telephony connections and distributed digital switching between MVIP-based PCs and supports interconnection of MVIP PCs with proprietary telephony equipment including PBXs, ACDs, and voice processing systems. MC-MVIP includes redundant clocks and specifically permits plugging and unplugging connections while a multi-chassis system is operating. This allows application developers to construct fault-tolerant systems."

Since the original MVIP standard (now known as MVIP-90), the GO-MVIP Technical Committee has developed several standards which offer significant improvement. H-MVIP (High-Capacity MVIP) was approved in 1995 as a single-chassis standard which increases the call capacity of MVIP from 512 to 3,072 time slots. Additionally, four draft standards have been developed for MC-MVIP (Multi-Chassis MVIP). Those standards address operation and synchronization of MVIP in a multi-chassis environment over twisted pair and fiber (FDDI-II and SDH/SONET), with MC4 addressing ATM networking involving both SVCs and PVCs. See FMIC, GO-MVIP, NMS, NTT, PEB and SCSA.

MVL Multiple Virtual Line, one of several modulation schemes being tried for DSL circuits. See Digital Subscriber Line.

MVNE MVNE is short for MVNO (mobile virtual network operator) Enabler, or a company that "enables" MVNOs. In order to "enable" an MVNO, the enabler builds a service platform, which provides billing, care, logistics, and carrier interfaces. By contrast, Richard Branson's Virgin Mobile, probably the most successful MVNO to date, has built out their own infrastructure, at a cost estimated to be $40-50 million. It took them more than two years to get to market by going this route. An enabler, or MVNE, builds out a shared platform, which is offered to multiple MVNOs, thus allowing them to avoid the huge capital cost of building their own infrastructure and at the same time, getting them to market faster. An MVNO enabled by an MVNE can get to market in as little as three or four months at a cost of a few million, depending on how complex and feature-laden their offerings are.

MVNO A mobile virtual network operator (MVNO) is an organization that buys minutes and services wholesale from an existing cell phone carrier (or carriers) and resells them under its own brand. Virgin Mobile is a famous MVNO. Under GSM, for example, an MVNO may issue its own SIM card. A MVNO is basically a reseller. It does not have radio frequency (spectrum). It doesn't have its own communications plant. It usually (but not always) has its own billing system. It usually targets a speciality segment of the market – teenagers, Pakistanis, Indians, etc. Globally, there are now more than 250 of these targeted cell ventures, including 42 in the United States, according to a new book by Christian Dippon and Aniruddha Banerjee of NERA Economic Consulting, "Mobile Virtual Network Operators: Blessing or Curse?" See GSM, MVNE and SIM Card.

MVPD Multichannel Video Program Distribution. It is provided by cable and satellite. See IB docket No. 95-168 or PP docket No. 93-253 In the Matter of Revision of Rules and Policies for the Direct Broadcast Satellite Service before the FCC. Also multichannel video programming distributor.

MVS Multiple Virtual Storage.

mW Abbreviation for Milliwatt.

MW Megawatt. One million watts of power.

MWI Message Waiting Indication. A fancy way of saying a light on a phone, a buzzing sound when you pick up the phone (also called a stutter dial tone), a message on a screen on your phone or a message on your PC telling you have a voice mail. In short, some way of telling you that you received a message while you were out or on the phone.

MX Records Mail Exchange Records. All machines (i.e., host computers) that are connected directly to the Internet have IP (Internet Protocol) addresses. The DNS (Domain Name Servers) translate the "names" of the machines from URLs (Uniform Resource Locators), such as Harry_Newton@TechnologyInvestor.com, into the dotted decimal notation characteristic of the IP addressing scheme. Those machines that are directly connected to the Internet have both their URLs and their IP addresses stored in the DNS as "A Records" (Address Records). Machines that are not directly connected, have their address translation information stored as "MX Records." The MX Records point to the IP address of the mail host (i.e., the machine that is accepting mail for that target machine). Actually, multiple mail hosts usually are identified, in order of precedence (i.e., priority); that way, mail still can reach the target machine in the event that the primary mail host is out of service for some reason. The mail hosts commonly are in the form of ISP (Internet Service Provider) e-mail servers that receive your mail, store it, and download it to you when you dial in via modem or turn on your client computer connected to the ISP via DSL or a cable modem. See also A Records, CNAME Records, DNS, and URL.

MX3 A designation for a multiplex which interfaces between any of the following circuit combinations: 28 DSIs to one DS3 (M13), 14 DSICs to one DS3 (MC3), or 7 DS2s to one DS3 (M23).

MXR Multiplexer.

MySpace The most popular online social network in the United States, with a market share of over 80%, measured in terms of site visits, as of August 2006. MySpace is owned by Rupert Murdoch's News Corporation. Murdoch was famously quoted in a WIRED magazine article as saying, "God knows what we're going to do with MySpace" as he considered possibilities for monetizing the social network's huge user base. Several of my friends report using MySpace to set up weekend orgies. I kid you not.

MZI Mach Zehnder Interferometer.

N Any digit 2 through 9. X is any digit 0 through 9. This is telephony nomenclature, not computer nomenclature. Accept it, whether it's logical or not.

N carrier N-Carrier and O-Carrier – Frequency division multiplexed analog carrier over long haul copper.

N port Node Port. A Fibre Channel term, referring to the link control facility which connects across a link to the F Port (Fabric Port) at the Fabric (switch). The node can be a mainframe, storage device, workstation, peripheral, or any other attached device. See Fibre Channel.

N series terminators Most local area networks are bus configurations. This means one long piece of cable (coaxial or fiber) with workstations connected along the way, typically with "T" connectors. For a network to work properly, you need to place resistance at the end of the bus, to terminate it. A thin wire Ethernet typically requires a 50 ohm resistance at either end of the bus. Thin wire Ethernet terminators are commonly called N series terminators. They may also be used with grounding wires to ground the network. Networks don't work well without resistance at the end of their buses.

N type connector The screw on connector frequently used with fifty ohm coaxial cable.

N-1 Pronounced N minus One. This is a term used in central office (also called exchange) switching. It refers to the central office switch just before the last one, i.e. the penultimate switch. It's an important term because that second last switch through which your incoming call passed will determine the signaling your central office switch received. See also N-1 Network.

N-1 network A telephone company AIN term. The telecommunications network in the call path just prior to the network containing the ported-from switch. If there are only two networks in the call path, then the N-1 network is the originating network. See also N-1.

N-ary code A code used to encode real-world characters that use n different code elements. Examples are binary (two states) and tertiary (three states).

N-ISDN Narrowband ISDN (Integrated Services Digital Network). Narrowband ISDN is an unkind name for the present form of ISDN presently implemented. In short, anything under an ISDN PRI. According to Professor Michael L. Dertouzos of MIT, narrowband ISDN suffers from the same constraints as classical voice telephony. For example, N-ISDN can carry reasonably be altered to accommodate the great information only in fixed chunks of 64 kilobits per second. This would be like a road system closed to everything except motorcycle. An 18-wheel truck carrying produce or a heating fuel tanker would be "welcome" to use this road as long as it could repackage its cargo into (and out of) motorcycle size chunks. See also B-ISDN.

n-play A generic term referring to a double-play, triple-play, quadruple-play or larger bundle of services offered by telephone and cable TV companies. Some of the services include telephone, cable TV and cell phone. Some industry observers expect n-play bundles to emerge where the number and variety services in a bundle are personalized and packaged according to each customer's unique and changing needs – translation: whatever can be most easily sold and cuts expensive customer churn down as much as possible.

N-type semiconductor See Semiconductor.

N+1 sparing Pronounced N plus One Sparing. A method of achieving redundancy, and therefore resiliency, in mission-critical systems and networks. N+1 Sparing involves the provisioning of one additional element above the number anticipated to be required under full load. See also Sparing.

   N+K sparing Pronounced N plus K Sparing. A method of achieving redundancy, and therefore resiliency, in mission-critical systems and networks. N+K Sparing involves the provisioning of some number K (where 1<\o>K<\o>N) additional elements above the number anticipated to be required under full load. See also N+1 Sparing.

N/A Not Available.

N&SS Network and Systems Support.

N1 The first short-haul multiplex carrier in the U.S. by the Bell System was Western Electric's N1. The N1 transmitted 12 voice frequency channels over separate transmit and receive pairs in a single cable. N carriers have been progressively improving.

N11 Service codes, or telephone dialing patterns, used to provide three-digit dialing to access special services. The FCC recognizes only 311, 711, and 911 as nationally assigned. Other codes with traditional uses include 411, 611, and 811. See those numbers for more information.

N2 In packet data networking technology, parameter used to specify the allowable number of retransmissions before disconnection.

NA 1. Night Answer.
   2. Network Administration.

NAB National Association of Broadcasters. A U.S.-based association which fosters and promotes radio and television broadcasting, represent those industries before the government, works to strengthen the abilities of its members to serve the public, and assists its members in operational matters. www.nab.org.

NAC 1. Network Access Center.
   2. Numbering Advisory Committee. Committee established under the Telecommunications Act 1997 to advise the Australian Communications Authority on numbering matters.

NACN The North American Cellular Network. Various operators nationwide have linked up to offer seamless roaming services so that roaming subscribers can make calls anywhere nationally without speaking to an operator. See North American Cellular Network for much more detail.

NACS 1. Network Analysis Control Surveillance.
   2. Network Analysis Control System.

NADF North American Directory Plan. An association of electronic mail providers who are figuring standards and ways of sending mail between their subscribers. They plan to use X.500 Recommendations.

Nagle's Algorithm A pair of algorithms used to control congestion in TCP networks by reducing the sending window and limiting small datagrams.

nagware Every time I turn on or reboot by PC, it nags me into doing something – updating my virus software, paying money for the real version, registering my software. You need to go into regedit and turn off some of the software that starts when you turn your machine one. See this dictionary introduction for how to do that. Nagware is like turning your operating system into your mom.

nailed connection A permanent circuit of a previously switched circuit/s. See also Nailed Data Connection and Nailed-up Circuit.

nailed up circuit A private line. A circuit permanently established through a circuit switching facility for point-to-point connectivity. Originally, private lines were, in fact, dedicated circuits which literally could be physically traced through the network. They also were known as "nailed-up circuits," as telephone company technicians hung the circuits on nails driven into the walls of the central offices. Today, private lines are actually dedicated channel capacity typically now provided over high-capacity, multi-channel transmission facilities. See Private Line.

nailing To make a circuit permanent. See Nailed Up Circuit and Nailed Connection.

nailing jelly to a tree A task thought to be impossible, especially when the difficulty comes from poor specifications or sloppiness.

NAK 1. Negative AcKnowledgement. NAK is a control character in ASCII that means a packet arrived with the check digits in error. It is sent from the computer receiving the packets to the sender, implying that the packet should be retransmitted so that all bits will arrive intact in the next go-round. The binary code is 0101001. The hex code is 51. See Check Digits and ACK.

2. No Acknowledgement.

naked call An incoming call that is routed into an ACD queue without getting call menus or flexible routing.

naked DSL In the United States, most phone companies charge both for DSL service and the phone line the DSL is being carried on. That adds significant money to the cost of your DSL service. In my case, my DSL line costs about $49 a month. The phone line costs another $17 a month. "Naked" DSL service would simply charge for the DSL service – not for the phone line. That would save me around $17 a month. The "logic" for charging the extra $17 is that, theoretically, with a filter, I could in fact make voice and data calls on the same line – simultaneously, no less. In fact, the reality is that it's often impossible to have satisfactory phone calls on the same line.

naked technology Forrester's Research's talented boss, George F. Colony defines "naked technology" as injecting technology into a company without process and organizational change which in turn creates waste and chaos. Colony writes, "Deploying technology without changing process and organization will create little impact – and it often brings negative consequences. Naked technology wipes out productivity improvements, hurts return on investment, and dulls the bright edge of well-conceived strategies."

NAM 1. Number Assignment Module. An electronic or module in a cellular phone which associates the MIN (Mobile Identification Number) with the ESN (Electronic Serial Number). Phones with dual or Multi-NAM features offer the user the option of having a cellular phone with more than one phone number.

2. National Account Manager.

NAMAS NAtional Measurement Accreditation Service.

name A name, as opposed to an address, is a location. Independent description of an end-station or node on a network (LAN or WAN) that contains no information about where the name entity is located. Certain protocols, such as IBM NetBIOS, make extensive use of a naming scheme.

name devolution A process by which a DNS resolver appends one or more domain names to an unqualified domain name, making it a fully qualified domain name, and then submits the fully qualified domain name to a DNS server.

name registration A Windows 95 definition. The method by which a computer registers its unique name with a name server on the network. In a Microsoft network, a WINS server can provide name registration services.

name resolution Name resolution is the act of translating a name from a difficult to remember number to something easier. Most people find it easier to remember www.HarryNewton.com than 208.222.46.156. When you choose to go to a Web site (e.g. HarryNewton.com) or send an email to someone (e.g. BillG@Microsoft.com), your computer network (which may be the Internet) determines the appropriate IP address. This is done using a name server and/or a host table file. Name resolution is the process of mapping a name to the corresponding address. It is the process used on a network for resolving a computer address as a computer name, to support the process of finding and connecting to other computers on the network. See Name Server.

name server 1. An AIN (Advanced Intelligent Network) term. A directory service located in the SLEE (Service Logic Execution Environment) that provides a mapping between a resource's global name and its physical location in the network.

2. An electronic messaging term. A program which provides information about network objects, such as domains and hosts within a domain, by answering queries. See Name Resolution.

3. Software that runs on network hosts charged with translating (or resolving) text-style names into numeric IP addresses.

named pipe A connection used to transfer data between separate processes, usually on separate computers. Named pipes are the foundation of interprocess communications (IPC). An administrator can set permissions on named pipes, but only LAN Manager and network applications can create them. See also Named Pipes.

named pipes A technique used for communications between applications operating on the same computer or across the local area network. It includes an applications programming interface, providing application programmers with a way to create interprogram communications using routines similar to disk-file opening, reading, and writing. In Microsoft's words, named pipes allow two or more processes to communicate with each other. Any process that knows the name of a named pipe can access it (subject to security checks). See also Named Pipe.

naming authority An authority responsible for the allocation of names.

naming service A service, such as that provided by WINS or DNS, that allows friendly names to be resolved to an address, or other specially defined resource data used to locate network resources of various types and purposes.

NAMPS Narrowband Analog Mobile Phone Service. A proposed new standard for cellular radio. NAMPS combines current voice processing with digital signaling. According to Motorola, NAMPS triples the capacity of today's cellular AMPS system, reduces the number of dropped calls and offers a range of new performance enhancements and digital messaging services. The other cellular standards include E-TDMA (Extended Time Division Multiple Access), TDMA (Time Division Multiple Access) and CDMA (Code Division Multiple Access).

NAN Neighborhood Area Networks. A NAN is a bunch of 802.11 wireless networks connected together.

NANC North American Numbering Council, pronounced "nancy." An industry council chartered by the FCC in October 1995 to assume administration of the NANP (North American Numbering Plan) from Bellcore, as well as to select LNP (Local Number Portability) administrators. The impartial council comprises 32 voting members from the carrier, manufacturer and end user communities. Another four non-voting members were selected, including representatives from Bellcore, ATIS, the U.S. NTIA, and the U.S. State Department. Ex-officio participants are selected from Canada, the Caribbean, and Bermuda. In October 1997, Lockheed Martin was selected as the primary administrator of NANP, formally replacing Bellcore. For that specific purpose, Lockheed Martin formed NeuStar as an independent business unit in 1996. Neustar subsequently was spun off as a separate business entity, and continues as NANPA (North American Numbering Plan Administrator). See also Bellcore, NANP and LNP. www.NeuStar.com.

NAND Also known as flash memory. Think of those little SanDisk memory sticks. Think of MP3 players.

NAND is a form of non-volatile computer memory that can be erased and reprogrammed. It is a technology that is primarily used in memory cards. Unlike EEPROM, it is erased and programmed in blocks consisting of multiple locations (in early flash the entire chip had to be erased at once). Flash memory costs far less than EEPROM and therefore has become the dominant technology wherever a significant amount of non-volatile, solid-state storage is needed. Examples of applications include digital audio players, digital cameras and mobile phones. Flash memory is also used in USB flash drives (thumb drives, handy drive), which are used for general storage and transfer of data between computers. It has also gained some popularity in the game console market, where it is often used instead of EEPROMs or battery-powered SRAM for game save data.

Nancy Hanks A radio code word used by US Navy ships in convoy, signifying that ship-to-ship infrared signaling will follow.

NAND NAND memory is a form of rewritable memory chip that, unlike a Random Access Memory (RAM) chip, holds its content without the need of a power supply. It is used in memory cards, USB flash drives, MP3 players, digital cameras and mobile phones.

NAND Gate An AND Gate with a digital inverter at its output.

NANOG North American Network Operators Group. A not-for-profit group for Internet network service providers of various descriptions. NANOG provides a forum for the exchange of technical information, promotes discussion of implementation issues that require community cooperation, and promotes and coordinates interconnection of networks within North America and to other continents. www.nanog.org. See also IOPS, Internet Society (ISOC) and IOPS.

nanny phone A cell phone designed for young children. Typically the handset allows parental management of incoming and outgoing calls, and has a small number of keys, in some cases as few as five, for example, one button to speed-dial mom, another button to speed-dial dad, and another button to display a directory of names. The phone probably will also have lights, sounds, and animated graphics on its LCD.

nanometer One billionth of a meter. Written nm. The nanometer is a convenient unit for describing the wavelength of light. The light spectrum extends from 750 nm (near infrared) to 390 nm (lowest energy ultraviolet). A nanometer is equal to 10 angstroms. A nanometer is also a millimicron.

nanoparticles A new class of magnetic materials made of iron and platinum. These particles boost storage densities to as high as 150 gigabits per square inch from the current standard (Fall of 2001) of 35 gigabits per square inch.

nanosecond 1. One billionth of a second. Written nsec. It's ten to the minus 9. One nanosecond – a billionth of a second – is the speed at which transistors in today's computers turn on and off to represent the ones and zeros of binary logic and arithmetic. It is a time-duration so short that light, which can speed seven times around Earth in the second between our heartbeats, travels only one foot. See Picosecond.

2. The amount of time to say "NO" when your wife asks, "Do I look fat?" – from comedian Don McMillan.

nanotechnology Nanotechnology comes from nanos, the Greek word for dwarf. Nanotechnology describes many types of research where the characteristic dimensions are less than about 1,000 nanometers. Continued improvements in lithography have resulted in line widths that are less than one micron. This work is often called "nanotechnology." Sub-micron lithography is clearly very valuable (ask anyone who uses a computer!) but it is equally clear that lithography will not let us build semiconductor devices in which individual dopant atoms are located at specific lattice sites. Many of the exponentially improving trends in computer hardware capability have remained steady for the last 50 years. There is fairly widespread confidence that these trends are likely to continue for at least another ten years, but then lithography starts to reach its fundamental limits. If we are to continue these trends we will have to develop a new "post-lithographic" manufacturing technology which will let us inexpensively build computer systems with mole quantities of logic elements that are molecular in both size and precision and are interconnected in complex and highly idiosyncratic patterns. Nanotechnology will let us do this. See Nanotube and Nanowire.

nanotube A nanotube resembles a rolled-up tube of chicken wire. It is about one hundred- thousandth the thickness of a human hair. Its thinness, only about 10 atoms wide, makes it a promising candidate for circuits in future faster and smaller computer chips. It takes its name from nanometer, a unit of measurement one-billionth of a meter long, which is a convenient length for specifying molecular dimensions. In the summer of 2001, I.B.M. scientists announced they have built a computer circuit out of a single strand of carbon. The I.B.M. circuit performs only a single, simple operation – flipping a "true" to "false" and vice versa – but it marks the first time that a device made of carbon strands known as nanotubes has been able to carry out any sort of logic. It is also the first logic circuit made of a single molecule. Though it will take more years of research, the fact that the researchers were able to build the circuit raises hopes that nanotubes could eventually be used for computer processors that pack up to 10,000 times more transistors in the same amount of space. See Not Statement.

nanowire Nanowires are long, thin, and tiny – perhaps one-ten-thousandth the width of a human hair. Researchers can now manipulate the wires' diameters (from five to several hundred nanometers) and lengths (up to hundreds of micrometers). Wires have been made out of such materials as the ubiquitous semiconductor silicon, chemically sensitive tin oxide, and light-emitting semiconductors like gallium nitride. The wires can be fashioned – ultimately – into lasers, transistors, memory arrays, perhaps even chemical-sensing structures akin to a bloodhound's famously sensitive sniffer.

NANP North American Numbering Plan. Invented in 1947 by AT&T and Bell Telephone Laboratories (now Lucent). The NANP assigns area codes and sets rules for calls to be routed across North America (i.e. the US and Canada). The new one, put into effect in January, 1995 has one major change: The middle number in a North American area code no longer is required to be a 1 or a 0 (one or zero); rather, it can range between 0 and 9. NANP numbers are 10 digits in length, in the format NXX-NXX-XXXX. The first three are the

NPA code (i.e., area code). The second three are the central office code or central office prefix, and the last four are the line number. NANP numbers conform to E.164, which is the ITU-T international standard for numbering plans. NANP administration was shifted from Bell Labs to Bellcore, when it was formed in 1986. Due to Bellcore's obvious conflict of interest, responsibility was shifted to NANC (North American Numbering Council) in 1995; it was shifted again in 1997 to Lockheed Martin. In November, 1999, if was shifted to NeuStar Inc., when it was discovered that Lockheed Martin had a conflict of interest. NeuStar originally was an independent business unit of Lockheed Martin, but was spun off in order to resolve the conflict. See North American Area Codes, North American Numbering Plan and NANC. www.nanpa.com.

NANPA North American Numbering Plan Administration. See also NANP and NANC.

NAOM National Accounts Operations Manager.

NAP 1. Network Action Point. The switching point through which a call is processed. The NAP switches the call based on routing instructions received from the NCP. A telephone company AIN term.

2. Network Access Point. A point of access into the Internet used by ISPs and providers of Internet regional and local subnets. NAPs operate at Layer 2 (Link Layer) of the OSI Reference Model, providing meet points where ISPs exchange traffic and routes. Similar to the original concept of the CIX (Commercial Internet eXchange), the NAPs provide a means of direct connection to the Internet, rather than serving solely as an intermediate point of exchanging commercial traffic. The initial NAPs were located in San Francisco under the operation of PacBell; Chicago, Bellcore and Ameritech, and New York (actually, Pennsauken, New Jersey), SprintLink. A fourth was awarded for MAE-East (MERIT Access Exchange) in Washington, DC, and is operated by MFS (Metropolitan Fiber Systems), which now is a business unit of Worldcom. On April 30, 1995, the NSFNet backbone was essentially shut down, and the NAP architecture effectively became the Internet. See also GigaPOP, FIX and MAE. In a nutshell, a NAP is a facility where backbone providers, carriers, ISPs go to peer (exchange Internet traffic). Tier 1 refers to the size/reach of the backbone providers, carriers, ISPs, etc. So, a Tier-1 NAP is a peering facility that has as its customers a majority of national and global telecom companies (ie Global Crossing, France Telecom, AOL/Time Warner, etc.)

3. Network Access Point, an AIN term. See Network Access Point.

4. Network Access Provider. The NAP provides a transit network service permitting connection of service subscribers to NSPs. The NAP is typically the network provider that has access to the copper twisted pairs over which the DSL-based service operates.

NAPI Numbering/Addressing Plan Identifier.

NAPLPS North American Presentation-Level Protocol Syntax. A protocol for videotex text graphics and screen formats, developed by AT&T and since standardized within ANSI, based on Canada's Telidon videographics protocol.

Napster Napster is a web-based service that allowed its "members" to download music from the hard disks of the PCs belonging to its "members." Napster's brilliance was creating the software that allowed "peer to peer" (P2P) computing. Napster's sin was making copyrighted material available for free to anyone who signed up for the service. Eventually the music industry hawled Napster into court and sought injunctions to stop it. The final injunction, interestingly, didn't hold Napster solely responsible for stopping all unauthorized trading of copyrighted files. Rather, the responsibility for policing the network is shared between Napster and the record companies that filed suit. Napster only must show that it is making every effort, within the limits of its technology, to stop trading specific recordings on lists provided by the plaintiffs. As a result a number of companies have cropped up to provide sophisticated technology to identify music files. The companies use digital signal processing to measure essential parts of the music itself. That information provides a small fingerprint, on the order of a few hundred to a few thousand bytes, which cannot be altered without significantly changing the sound of the song. One company claimed each identification takes, on average, less than half a second. See also Bit Torrent and P2P.

NAR 1. Network Access Register. Centrex term describing a Central Office register which is required in order to complete a call involving access to the network outside the confines of that Centrex CO. NARs may be incoming, outgoing or two-way. NARs may be defined in support of local, intraLATA or interLATA traffic. The specifics of NAR implementation vary by Centrex provider.

2. Nothing Added Reseller. In contrast to a VAR, which is a Value Added Reseller. Clearly nothing added reseller is a lam joke.

3. National Accounts Representative.

Narcipost A shamelessly egocentric blog post that's of no interest to anyone except the person who wrote and posted it.

narrative traffic Messages normally prepared in accordance with standardized procedures for transmission via optical character recognition equipment or teletypewriter. In contrast to data pattern traffic, narrative messages must contain additional message format lines.

narrowband 1. An imprecise term. Some people think it's sub-voice grade channels capable of only carrying 100 to 200 bits per second. Others think it means lines or circuits able to carry data up to 2400 bits per second. So as lines get broader, narrowband gets broader. The latest definition of narrowband is up to and including T-1 – or 1.544 megabits per second. See also Bandwidth, Wideband, Broadband, N-ISDN and B-ISDN.

2. In cellular radio terminology, narrowband refers to the methodology of gaining more channels (and hence more capacity) by splitting FM channels into channels that are narrower in bandwidth. See NAMPS and NTACS.

3. PCS. Mobile or portable radio services which can be used to provide services to both individuals and businesses such as acknowledgement and voice paging and data services.

4. A WDM (Wavelength Division Term). Once upon a time, digital optical fiber transmission systems transmitted pulses of white light. Pulse on for a binary "1" bit, and pulse off for a binary "0" bit. White light is a combination of all of the light frequencies in the light spectrum. If you transmit white light, one optical fiber can carry only one signal. Now, they still pulse on and off. Increasingly, however, the laser diodes that create the light pulses are tunable so that they emit light pulses that are narrowly defined to specific frequencies that we would relate, at the most basic level, to specific colors of light such as ruby red or violet. Each specific frequency, or wavelength, is known as a "lambda" in the optical domain. An optical fiber simultaneously can support a vast range of narrowly defined lambdas, which do not interfere with each other as they are separated by frequency. While each lambda operates at broadband speed (i.e., equal to or greater than 45 Mbps), in the context of a WDM transmission system each lambda is referred to as "narrowband."

Narrowband Advanced Mobile Phone Service NAMPS. Narrowband AMPS. NAMPS triples the capacity of AMPS, by compressing three 10 KHz analog FM channels into a signal 30 kHz analog FM AMPs channel, along with improved signaling. Pronounced "N-AMPS."

Narrowband FM Narrowband FM is an FM signal with a bandwidth approximately equal to that of an AM signal modulated with the same audio information. Narrowband FM is used on many emergency bands because it conserves bandwidth while being clear and free from static.

Narrowband ISDN Any ISDN speed up to 1.544 Mbps, which is called PRI or PRA. But this definition is imprecise. And as speeds get faster, so the definition of narrowband ISDN means faster and faster. See N-ISDN and B-ISDN.

narrowband signal Any analog signal or analog representation of a digital signal whose essential spectral content is limited to that which can be contained within a voice channel of nominal 4-kHz bandwidth.

narrowband TACS N-TACS. The narrowband version of TACS from Motorola which doubles the capacity of TACS by splitting the 25 kHz TACs channel into two 2.5 kHz channels.

narrowcasting First, there was broadcasting. One signal went to many people. Radio and TV are the classic concepts of broadcasting. One signal – the same signal – to many people. Then came the idea of narrowcasting. One signal to a select number of people – maybe only those people who subscribed to the service and had the equipment to receive it. Then there came pointcasting. This is a fancy name for sending someone a collection of customized information – snippets of stuff that they chose from a palette of information offerings.

NARS 1. Network Audio Response System.

2. A Nortel switching term for Network Automatic Route Selection.

NARTE National Association of Radio and Telecommunications Engineers. A worldwide, non-profit, professional organization which certifies engineers and technicians in the areas of telecommunications and electromagnetic compatibility (EMC). NARTE was founded in 1983 to address the professional testing and certification void created when the FCC reduced its role in that regard. www.narte.org.

NARUC National Association of Regulatory Utility Commissioners. Members are commissioners of utility regulatory agencies of the states, the federal government and U.S. territories (i.e., the District of Columbia, Puerto Rico, and the Virgin Islands). Objectives are the advancement of uniform regulation, coordinated action, and protection of the common interests of the people with respect to utility regulation. www.erols.com/naruc.

NAS 1. NetWare Access Server. See Remote Access Server.

2. Network-Attached Storage. NAS devices, which include iSCSI devices, are simply storage devices that include NICs (network interface cards). This makes the devices accessible via a standard Ethernet network connection. See Network Attached Storage and Storage

Area Network (SAN) for longer explanations.

NASA The National Aeronautics and Space Administration, founded in 1958 as the successor to the National Advisory Committee for Aeronautics.

NASC Number Administration and Service Center. Provides centralized administration of the Service Management System (SMS) database of 800 numbers. The NASC keeps track of the 800 numbers that are in use, or available for use, by new 800 users.

Nasdaq An exchange where stocks are traded electronically across vast computer networks. This contrasts, for example, with the New York Stock Exchange, which is a physical place (i.e. a building) and stocks are traded by people at that building. Nasdaq stands for The National Association of Securities Dealers Automated Quotation System.

Nasdaq Composite Index An index representing the movement of stocks traded by the National Association of Securities Dealers (NASD).

NAT Network Address Translation. An Internet standard that enables a local area network (LAN) to use one set of IP addresses for internal traffic and a second set of addresses for external traffic. This allows a company to shield internal addresses from the public Internet. According to Cisco, NAT has several applications. You want to connect to the Internet, but not all your hosts have globally unique IP addresses. NAT enables private IP internetworks that use nonregistered IP addresses to connect to the Internet. NAT is configured on the router at the border of a stub domain (referred to as the inside network) and a public network such as the Internet (referred to as the outside network). NAT translates the internal local addresses to globally unique IP addresses before sending packets to the outside network. You must change your internal addresses. Instead of changing them, which can be a considerable amount of work, you can translate them by using NAT. You want to do basic load sharing of TCP traffic. You can map a single global IP address to many local IP addresses by using the TCP load distribution feature. As a solution to the connectivity problem, NAT is practical only when relatively few hosts in a stub domain communicate outside of the domain at the same time. When this is the case, only a small subset of the IP addresses in the domain must be translated into globally unique IP addresses when outside communication is necessary, and these addresses can be reused when no longer in use.

A significant advantage of NAT, according to Cisco, is that it can be configured without requiring changes to hosts or routers other than those few routers on which NAT will be configured. NAT may not be practical if large numbers of hosts in the stub domain communicate outside of the domain. Furthermore, some applications use embedded IP addresses in such a way that it is impractical for a NAT device to translate. These applications may not work transparently or at all through a NAT device. NAT also hides the identity of hosts, which may be an advantage or a disadvantage. A router configured with NAT will have at least one interface to the inside and one to the outside. In a typical environment, NAT is configured at the exit router between a stub domain and backbone. When a packet is leaving the domain, NAT translates the locally significant source address into a globally unique address. When a packet is entering the domain, NAT translates the globally unique destination address into a local address. If more than one exit point exists, each NAT must have the same translation table. If the software cannot allocate an address because it has run out of addresses, it drops the packet and sends an ICMP Host Unreachable packet. A router configured with NAT must not advertise the local networks to the outside. However, routing information that NAT receives from the outside can be advertised in the stub domain as usual.

NATA 1. North American Telecommunications Association. A trade association of manufacturers and distributors of telephone equipment that was formed in 1970, two years after the FCC's Carterfone decision, which said that the Carterfone and other customer phone devices could be connected to the nation's phone network – if they were "privately beneficial, but not publicly harmful." The Carterfone decision was a landmark. It allowed the connection of non-telephone company equipment to the public telephone network. This decision marked the beginning of the telephone interconnect business as we know it today. And NATA's mission was to fight all the restrictive rules and regulations which the telephone companies subsequently threw up to make connection of customer-owned equipment difficult and expensive. At that time the phone companies (especially AT&T and GTE) were among the largest manufacturers of phone equipment. Eventually NATA won the legal fight and lost its mission, its reason for being. In the early 1990s, NATA collapsed. The remaining bits of it were re-assembled into a new trade association called the MMTA, which stood for MultiMedia Telecommunications Association. MMTA is organized around five divisions – computer telephony integration, conferencing/collaboration and messaging, LAN/WAN internetworking, Voice/Multimedia and Wireless Communications. In the fall of 2000 the MMTA was integrated into the TIA, which stands for the Telecommunications Industry Association. See TIAOnline.org.

2. National Association of Testing Authorities. Australian body that accredits services

that test equipment against standards.

NATD National Association of Telecommunications Dealers. A national association of dealers in used telecommunications equipment. www.natd.com.

NATI Kirk Forman works for a company with 3,000 employees. Turns out that the network and workstation folks are two different groups. It seems that 85% of the problem calls he in telecom get from the workstation group are NATI, Not A Telecom Issue.

National Association Of Regulatory Utility Commissioners NARUC. An organization supporting the needs of the commissioners of U.S. federal and state regulatory agencies. www.naruc.org.

National Association of Telecommunications Dealers See NATD.

National Association of Testing Authorities See NATA.

National Bureau Of Standards NBS. The U.S. government organization that helps prepare non-Department of Defense communications standards and operates a testing service to indicate conformity to existing standards. Address: Institute for Computer Sciences and Technology, National Bureau of Standards, Gaithersburg, MD 20899.

National Cable Television Association NCTA. An organization representing the major U.S. cable television operators in the United States. Founded in 1952, the NCTA's mission is to advance the public policies of the CATV industry before the legislative, judicial and regulatory bodies, as well as before the American public. www.ncta.com.

National Cell Phone Courtesy Month July. Examples of cell phone courtesy: (1) switch cell phone to "mute" in movie theaters, restaurants, places or worship, buses, checkout lines, and elevators; and (2) have your calls go directly to voicemail when you are in a meeting.

National Code, The Telecommunications National Code (of Australia). Carriers were formerly required to comply with the provisions of this code when installing or rolling out telecommunications facilities and infrastructure.

National Communications System NCS. The NCS was set up by President John F. Kennedy in 1963 after a snarl in communications during the Cuban missile crisis among the United States, The Soviet Union and NATO. The system has continued over the years to try to maintain communications among emergency service officials during times of crises. Formally, the NCS was established by Section 1(a) of Executive Order No. 12472 to assist the President, the National Security Council, the Director of the Office of Science and Technology Policy, and the Director of the Office of Management and Budget, in the discharge of their national security emergency preparedness telecommunications functions. See also Priority Access Methods.

National Convergence Alliance See NCA.

National Coordinating Center NCC. The joint telecommunications industry-Federal Government operation established by the National Communications System to assist in the initiation, coordination, restoration, and reconstitution of NS/EP telecommunication services or facilities.

National Electrical Code NEC. A nationally recognized safety standard for the design, construction, and maintenance of electrical circuits. The NEC also gives rules for the installation of electrical and telephone cabling. The NEC is developed by the NEC Committee of the American National Standards Institute (ANSI), sponsored by the National Fire Protection Association (NFPA) and identified by the description ANSI/NFPA 70-1990. This code has been adopted and enforced by many states and municipalities as law. NEC Article 800 covers telephone and telegraph electrical communications circuits, and telephone systems. Article 810 covers commercial radio and television receiving equipment, and amateur radio transmitting and receiving equipment. Article 820 covers CATV (Community Antenna Television) coaxial cable systems.

National Electrical Manufacturers Association See NEMA.

National Emergency Number Association NENA. A US non-profit organization promoting 911 as the standard emergency number, including technical support, public awareness, certification programs, and legislative representation. www.nena9-1-1.org.

National Exchange Carrier Association NECA. An association of local exchange carriers mandated by the U.S. Federal Communications Commission in 1983, in anticipation of the breakup of the Bell System. NECA's primary responsibility was to file new interstate access tariffs and to perform the "settlement" functions previously performed by AT&T and the BOCs (Bell Operating Companies). Under the direction of the FCC, NECA administers approximately $2 billion in annual revenues, based on pooled tariff rates. Through the pooling process, the LECs (Local Exchange Carriers) bill the IXCs (IntereXchange Carriers) for the use of their local exchange networks in originating and terminating long distance calls. Individual revenues and costs then are submitted to NECA, which distributes

them to the member companies on an averaged basis. NECA also administers the Universal Service Fund (USF), the Lifeline Assistance Programs, and the Telecommunications Relay Services Fund. See also Division of Revenues, Lifeline Assistance Programs, Separations and Settlements, Telecommunications Relay Services Fund, and Universal Service Fund. www.neca.org.

National Fire Protection Association NFPA. A not-for-profit association which works with the U.S. Congress and federal agencies to promote the adoption and use of fire protection codes and standards, and to promote a uniform national approach to combating the problem of fires. The NFPA has published the National Electrical Code (NEC) since 1897. The NEC, published as NFPA 70, includes all sorts of provisions for electrical wires, connectors, and outlets. Included in the NEC are specifications for the cable jackets used in telecommunications inside wire and cable systems. www.nfpa.org. See National Electric Code.

National Information Infrastructure NII. What the Clinton Administration prefers to call the Information Superhighway – basically a switched, broadband network that could, in theory, deliver everything from switched video to high-speed access to the Internet. The NII is intended to connect people, businesses, institutions and governments with one another. Its purpose is to expand the availability of a wide variety of information and communications services.

National Institute For Standards And Technology NIST. The U.S. government agency that oversees the operation of the U.S. National Bureau of Standards. The NIST is based in Gaithersburg, MD. See NIST.

National ISDN Council NIC. A council of carriers and manufacturers which determines the generic guidelines for National ISDN. It also publishes those guidelines, as they are developed. NIC is a voluntary council. Membership includes Ameritech, Bell Atlantic, BellSouth, Cincinnati Bell, Lucent, Nortel and SBC. Administration and project management are the responsibility of Bellcore. See also National ISDN-1.

National ISDN-1 NI-1. National ISDN-1 is a set of specifications for a "standard" national implementation of ISDN (BRI and PRI). Based on international ISDN standards recommendations from the ITU-T, and technical references (TRs) specified by Bellcore for the U.S., NI-1 lays the groundwork for a national ISDN infrastructure. Bellcore issued the National ISDN-1 document, SR-NWT-001937, Issue 1, in February 1991. Currently, National ISDN is the responsibility of the National ISDN Council (NIC), comprising certain carriers and manufacturers, and administered by Bellcore. The idea of National ISDN-1 is that it be a set of standards to which every manufacturer and carrier can conform. Thereby, all manufacturers can build ISDN equipment (i.e., terminal equipment, terminal adapters, PBXs and COs) which can interoperate effectively, both through and across the various carrier ISDN networks. Similarly, all carriers can interconnect through a standard set of interfaces in order to support seamless connectivity, and transparency of feature content and access. As a result, a consumer can buy an ISDN phone (one conforming to National ISDN-1) at his local Radio Shack (or other store) take it home, plug it in and know it will work, irrespective of carrier and central office manufacturer. This was not always the case. Sadly, National ISDN-1 has not been a rousing success, as ISDN generally has been a failure; also, NI-1 addressed a set of only 17 features. National ISDN-2 documents were first published in 1992, and National ISDN-3 documents in 1993. See also Bellcore Custom ISDN, ISDN and National ISDN Council.

National ISDN-2 NI-2. A set of specifications published in 1992 and building on NI-1. NI-2 defines a set of features and functional capabilities to be provided in the NI offering, and composed of those directly related to serving user applications. NI-2 also defines operations support and billing capabilities directed toward service providers. See National ISDN-1.

National ISDN-3 NI-3. A set of specifications published in 1993, building on NI-1 and NI-2. NI-3 defines a marketable and feasible set of features in consideration of market needs, difficulty of implementation and ease of use. See National ISDN-1.

national park A government-provided repository of on-line information and entertainment, such as is provided by the Smithsonian.

National Radio Quiet Zone A 13,000 square mile area in West Virginia, Virginia and Maryland that was established by the FCC in 1958, for the purpose of minimizing possible harmful radio frequency interference (RFI) to the National Radio Astronomy Observatory (NRAO) at Green Bank, West Virginia and the US Navy's space receiving facility at Sugar Grove, West Virginia. The surrounding Allegheny Mountains provide natural protection against RFI and the surrounding national forests provide protection against encroaching development. Applications for radio services within the National Radio Quiet Zone (NRQZ) are reviewed by both the FCC and NRAO for compliance with strict FCC guidelines. All approved major transmitters in the NRQZ are required to install, configure and power their antennas in compliance with NRQZ-specific guidelines, and they must coordinate their operations with the NRAO.

National Relay Service NRS. In Australia, the NRS provides people who are deaf, hearing or speech impaired with access to the standard telephone service through the relay of voice, modem or teletypewriter communications. The NRS operates as a translation service between voice and non-voice users of the standard telephone service.

National Repository Line Level Database NRLLDB. Probably this dictionary's most clumsy and least memorable acronym. As defined by the Alliance for Telecommunications Industry Solutions (ATIS), the NRLLDB is a centralized database containing technical informaton that identifies company, switch and consumer elements, elements that define each phone and wireless line. NRLLDB helps carriers with the billing done in the back office by identifying the carrier.   The NRLLDB includes data that can help carriers with provisioning, CLEC-to-CLEC migration, return code 50s, wireless number portability, and end user, direct and access billing. ATIS spent a little over 5 years defining this national database, it's history is tracked on their site: www.atis.org/atis/clc/obf/rfi.htm. ATIS origianlly intended to 'own' the repository and allow companies to bid for the devlopment and management of NRLLDB. In the end, they decided not to own it but seeing that they defined the solution to the industry problem, allow companies to pick from providers.

National Science Foundation National Science Foundation. An independent agency of the U.S. government established by the National Science Foundation Act of 1950 with the mission "to promote the progress of science; to advance the national health, prosperity, and welfare; and to secure the national defense." Among the NSF activities are the fostering of the interchange of scientific information among scientists and engineers in the U.S. and foreign countries, and the fostering of the development and use of computers. This mission and charter gave rise to NSFNET, which was in large part the impetus behind the development of the Internet. See also National Technology Grid, NCSA, NREN.

National Security Agency See NSA.

National Technical Information Service. NTIS. An non-appropriated agency of the U.S. Department of Commerce's Technology Administration, the NTIS serves as the official resource for government-sponsored U.S. and worldwide scientific, technical, engineering and business-related information. Information is acquired from over 200 U.S. government agencies, numerous international governmental departments and other international organizations, and through contracts or cooperative agreements with the private sector and other organizations. Information is available in a variety of formats. www.nist.com.

National Technology Grid An effort of the NSF (National Science Foundation) to develop a nationwide computational infrastructure. The NSF will fund the National Computational Science Alliance with up to $170 million over a period of five years, beginning in 1997. The aim of the Alliance is to enable the science and engineering community to take full advantage of rapidly improving high-performance computing and communications technologies. The alliance comprises more than 50 research partners, and is led by the National Center for Supercomputing Applications (NCSA). See also NCSA.

National Telecommunications and Information Administration NTIA. U.S. government agency responsible for, among other things, administering the use of spectrum allocated for government usage.

National Television Standards Committee NTSC. The North American standard for the generation, transmission, and reception of television communication wherein the 525-line picture is the standard. The picture information is transmitted in AM and the sound information is transmitted in FM. Compatible with CCIR Standard M. This standard is used also in Central America, a number of South American countries, and some Asian countries, including Japan. See NTSC.

nationwide paging A method of national or regional paging in which a single frequency is used throughout the nation or region for sending messages to a pager carried by a subscriber.

nationwide/statewide cost averaging A method of averaging costs to establish uniform prices for telephone service so that subscribers using more costly-to-serve, lightly trafficked routes – such as those between small communities – receive the same service for the same price as subscribers on lower-cost highly trafficked routes. The idea is that people in rural areas shouldn't pay more for phone service than those living in cities. Theoretically, they would, since theoretically it costs more to provide phone service to rural communities.

native See Native Application.

native address An address that matches one of a given node's summary addresses.

native application 1. Software that runs directly on the computer's operating system – without requiring an emulation or other program to sit between the software and the operating system. The term "native" acquired new meaning when Apple moved from its Motorola 680x0 microprocessor line of computers to the Power Macintosh line. See Native Mode.

2. An ATM term. A term given to any application written to use any communications protocol prior to ATM.

native format The original form of a file. A file created with one application can often be read by others, but a file's native format remains the format it was given by the application that created it. In most cases the specific attributes of a file (for example, fonts in a document) can only be changed when it is opened with the program that created it.

native implementation See Native Mode, definition #2.

Native LAN Services Carrier-provided services which interconnect LANs at their full speed, i.e., the native speed of the LANs.

native management policy Management policy derived from the customer requirements. Native policies represent coordination between NGOSS components and reflect the expected system behavior based on functional requirements. Native policies cover also the specification of potential interaction with an external management system. A native management policy is complementary to a native functional policy, which is also derived from the customer requirements; the latter which is commonly embedded into the functional specification.

native mode 1. Uncompressed.

2. Able to run software directly in the computer's indigenous operating system without intervening emulation software. The term became important when Apple moved from its Motorola 680x0 microprocessor line of computers (called Quadra, Macintosh IIX, Performa, etc.) to the Power Macintosh line of computers powered by the RISC microprocessors called PowerPC 601. This new line of computers runs an operating system different to what Apple had been running on its Macintoshes. The new Power Macintoshes can run software for the Macintosh 680x0 machines – but only in emulation. This makes the software run slowly. So, for the software to run at speeds the new faster machines are capable at running, application software (word processing, desktop publishing, etc.) has to be rewritten to run in native mode.

native protocol A native protocol is the information format in which communications occur within a particular service.

native signal processing Analog signals, such as voice and video, often are converted into digital signals – for transmission, for computer storage, for compression, etc. Once an analog signal becomes digital, it often has to be manipulated – compressed (to save storage space or transmission time), to be edited, to be conferenced, etc. These speciality tasks of working on digital signals that originally were analog have largely been done by a specialized microprocessor, called a digital signal processor (DSP). DSPs are used extensively in telecommunications for tasks such as echo cancellation, call progress monitoring, voice processing and for the compression of voice and video signals. Lately, as the PC's general purpose microprocessor has become more powerful, the makers of these general purpose microprocessors (especially Intel) have started talking about using some of their spare MIPS (spare processing power) to do some of the tasks previously done by DSPs. They call this new idea – Native Signal Processing.

Native X.400 A term describing a messaging system or service developed using the protocol specifications and service definitions in the X.400 recommendations. Typically used to describe full X.400 services native to a user's home mail environment.

NATO phonetic alphabet Words that represent the letters of the alphabet. Used in radio communications among NATO military forces, in order to avoid mis-hearing letters of the alphabet that have similar sounds. Also approved by the International Telecommunication Union, the International Civil Aviation Organization, and the Federal Aviation Administration. The NATO phonetic alphabet is as follows: Alpha, Bravo, Charlie, Delta, Echo, Foxtrot, Golf, Hotel, India, Juliet, Kilo, Lima, Mike, November, Oscar, Papa, Quebec, Romeo, Sierra, Tango, Uniform, Victor, Whiskey, Xray, Yankee, Zulu.   In some NATO countries the following spelling variations may be used Alfa (Alpha), Juliette (Juliet), Oskar (Oscar) and Viktor (Victor).

NATOA The National Association of Telecommunications Officers and Advisors, an affiliate of the National League of Cities. NATOA is involved in telecommunications issues that affect state and local governments. Such issues include rights-of-way, radio frequency emissions, placement of radio towers, and universal service. www.natoa.org.

natural A voice recognition term for a language as in normal spoken conversational sentences. The vocabulary would include words like fifty, sixty, hundred etc. and be used in digit recognition.

natural computing As computers get more and more powerful, they get easier to use as they are endowed with human-like senses – fluid speech, a good ear (speech recognition) and keen vision. IBM coined the term "natural computing" to describe this.

natural frequency The natural frequency of an antenna, the lowest frequency at which the antenna resonates without the addition of any inductance or capacitance.

natural language query A query written in natural language (for example, plain English) seeking information from a database.

natural monopoly A term used by economists to justify regulation. The idea is that one company can provide certain services (such as gas, water, or telecommunications) considerably cheaper than two or three. Therefore, let one company have the monopoly on the service. But substitute government regulation for free competition and this way keep prices down. How well the theory and the practice of government regulation has worked is the subject of acres of learned prose. Suffice, the theory of "natural monopoly" has evaporated in most areas it was practiced – from airlines to telephones, local and long distance. When regulation is removed, prices have usually fallen.

NAU Network Addressable Unit. SNA term for LU, PU and SSCP. Each unit in SNA has a unique address.

NAUCS Network Access Usage and Cost System.

NAUN Nearest Active Upstream Neighbor. In Token Ring or IEEE 802.5 networks, the closest upstream network device from any given device that is still active.

nautical mile 6,076 feet. 15% longer than a normal mile, which is 5,280 feet. A measure of distance equal to one minute of arc on the Earth. An international nautical mile is equal to 1,852 meters or 6,076.11549 feet.

NAV 1. Network Applications Vehicle.

2. Net Asset Value.

navaid Navigational aid. Any visual or electronic device, airborne or on the ground, that provides guidance information or position data to aircraft in flight.

NAVAREA The world's oceans are divided into 16 NAVAREAs, in which the NAVTEX maritime safety information system and the WWNMS navigational warning service operate.

navigate An Internet term. It means simply to move around by the World Wide Web by following hypertext links / paths from document to document on different computers, typically in different places.

navigation The ability to route telecommunications traffic over diverse circuit options to achieve communications continuity between the desired end customer stations.

Navigation Systems Information Dissemination Network A US Coast Guard network that provides navigation information via HTTP, FTP, LISTSERVER, and DBMS services. NSIDN provides information from LORAN and radio beacons, the Global Positioning System (GPS), and differential GPS (dGPS), and disseminates local notices to mariners (LNM).

NAVTEX The principal worldwide means of transmitting short-range maritime safety information (MSI), including navigational warnings, weather forecasts and warnings, ice reports, search and rescue (SAR) information, pilot messages and the details of changes in navigational aids. In 1979, NAVTEX was first implemented in countries bordering the Baltic Sea, and in 1987 it became part of GMDSS (the Global Maritime Distress and Safety System). International NAVTEX messages are broadcast at set times on 518 kHz in English using narrow-band direct printing (NBDP) and are effective in coastal areas up to 400 nautical miles offshore. In some sea areas, critical safety messages also may be broadcast in national languages on 518 kHz and on 4 MHz. NAVTEX is included in GMDSS.

Navy Marine Corps Intranet Known more commonly by its acronym, NMCI, the Navy Marine Corps Intranet is a single integrated network that connects all of the U.S. Navy's and Marine Corps' shore commands, and provides secure, integrated access to voice, video and data services. NMCI also provides a standard desktop platform and configuration (collectively called a "seat").

navy shipboard cable Cables for use aboard naval vessels and for shore applications. Manufactured in accordance with a Navy specification.

Nax Network fAXing; to send faxes over the internet and any other network. Can also be used as a noun. Naxing.

NBAD See network behavior anomaly detection.

NBC Network Bus Controller.

NBF NETBIOS Frame. See NetBEUI.

NBFM Narrow Band Frequency Modulation, in which the total signal bandwidth is constrained by limiting the modulation index to less than or equal to 0.5; an NBFM signal bandwidth is approximately equal to that of an amplitude modulated signal.

NBFP NetBIOS Frame Format Protocol. See NetBEUI.

NBFCP NetBios Framing Control Protocol; protocol for transporting NetBIOS traffic over a PPP connection. See NetBEUI.

NBI Some people think this once famous company's name stood for "Nothing But Initials." Others thought it stood for Nectum Bilinium Inc. It actually stood for "Nothing But Initials."

NBIC A new acronym meaning the coming together of the previously disparate fields of Nanotechnology, Biology & medicine; Information sciences; and Cognitive sciences, and the allegedly resulting synergy.

NBIF No Basis In Fact. Acronym used as a greeting in e-mail, during online chat sessions, and in newsgroup postings to save keystrokes and time.

NBMA NonBroadcast MultiAccess. A term for networks which provide access for multiple routers but do not support broadcasting, or in which broadcasting is not feasible. Examples of NBMA networks are Frame Relay, ATM and X.25. Although Ethernet is multiaccess, broadcasting is used on the network, so it is not considered NBMA. The most common NBMA technology today is Frame Relay. Frame Relay has largely replaced X.25 as a transmission technology.

NBP Name Binding Protocol. AppleTalk protocol for translating device names to addresses.

NBS National Bureau of Standards. A US government agency that produces Federal Information Processing Standards (FIPS) for all other agencies except the Department of Defense (DoD).

NBS/ICST National Bureau of Standards/Institute for Computer Sciences and Technology. The NBS directorate, based in Gaithersburg, MD, is concerned with developing computer and data communications.

NC Network Computer. Larry Ellison of Oracle's idea of a $500 (or so) PC that lacks a hard disk and may lack a monitor but can be used to browse the Internet and run applications on a server on the Internet or corporate intranet. Ellison, who is Oracle's chairman, sees the NC as a "universal digital appliance."

The New Yorker of September 8, 1997 discussed the implications of the network computer thus: Microsoft's worries about Ellison and NCs are not trivial. After a prolonged period of being in denial about the rise of the Internet, Gates and his team now understand that it is the central fact of the next phase of computing, and that it poses a real threat to Microsoft's power. In 1995, Sun Microsystems introduced an Internet-centric programming language called Java, which creates programs that can run on any operating system and is fast becoming the standard lingo of the Net. In a Java-fuelled future, the reign of the PC might be challenged by the NC which would let users "borrow" programs from the Net and would have no need for Microsoft's Windows – developments that would create enormous upheaval in many of the software markets that Gate's firm now dominates. See also Internet Terminal, NetPC, and NetStation.

NC Code The Network Channel (NC) Code is an encoded representation used to identify both switched and non-switched channel services. Included in this code set are customer options associated with individual channel services or feature groups and other switched services.

NCA 1. Number of Calls Abandoned. The number of incoming calls accepted by an ACD (Automatic Call Distributor) but abandoned by the caller before being connected to a person. The caller either dialed a wrong number or just got tired of waiting and hung up. It's usually the latter.

2. National Convergence Alliance. An association of product and service vendors in the convergence industry, whatever that is. According to the NCA, the alliance "creates industry and market focus, establishes a unified voice, enhances our ability as an industry to identify and drive technology, enables the overall acceleration of convergence adoptions and capabilities, creates a forum where equipment and service providers can discuss their customers needs, and, most importantly, educates providers and enterprises about the benefits of convergence." I don't know anything about this organization except for what they say on their website, but they win the prize for the longest compound sentence I've seen in the past decade. www.ConvergenceAlliance.com. See also Convergence.

NC&A Network Construction and Administration.

NCAS NCAS is a generic acronym for Non Call-path Associated Signaling. NCAS is out-of-band signaling used to provide emergency signaling information separate from the wireless 911 call to the Public Safety Answering Point (PSAP). This signaling information includes the phone number of the wireless phone and coding to derive a general location of the caller, and meets the Enhanced 911 Phase 1 FCC requirements. This coding can be either a p-ANI or an ESRD. Out-of-band signaling is separated from the channel carrying the voice. Several solutions are available to accomplish this in the wireless 911 calling environment. CAS is a generic acronym for Call-path Associated Signaling. This definition contributed by Glenda Drizos and Doug Puckett of Sprint PCS, Overland Park, Kansas.

NCBS New Commercial Billing System.

NCC 1. Network Control Center. A central location on a network where remote diagnostics and network management are controlled. See Network Control Center.

2. National Computer Conference, once upon a time the largest annual conference of the computer industry.

NCCF Network Communications Control Facility (IBM, SNA). Along with NPDA this program product allows for the monitoring of a complete SNA Network from a central location. An IBM Term. See Netview.

NCCS Network Control Center System.

NCEO NonCompliant End Office.

NCH Number of calls handled.

NCHD National Customer Help Desk.

NCI Network Channel Interface Code. Identifies the electrical conditions that are required for Pacific Bell when the circuit is handed off at the designated premises or Point of Termination (POT). NCI is formatted at each circuit location. It has a maximum 12 alphanumeric entry that contains such information as:

- number of wires
- signaling characteristics
- impedance levels
- protocol options
- transmission levels

NCIC National Criminal Information Center. The FBI's data warehouse of criminal histories that is accessed over phone lines by every law enforcement agency.

NCL 1. Network Control Language. A command line interface language used by Digital Equipment Corp.'s Digital Network Architecture (DNA).

2. Non-Computing Module Load.

NCO 1. Network Control Office.

2. Neutral Central Office. Also sometimes called a "Telephone Hotel." A NCO is a neutral location owned by a party other than the ILEC (Incumbent Local Exchange Carrier). The third-party owner of the NCO provides a totally neutral environment in which various carriers, both incumbent and competitive, may lease space to terminate their transmission facilities and locate their various switches, multiplexers, and other equipment. A NCO is much like a Private Peering Point, in Internet terminology. See Carrier Hotel for a full explanation. See also Private Peering Point.

NCOP Network Code Of Practice.

NCP 1. In AT&T language, Network Control Point. A routing, billing, and call control data base system for DSDC which uses the AT&T 3B20D computer as the feature processor.

2. In IBM language, Network Control Program, which is a program that controls the operations of the communication controllers, 3704 and 3705 in an IBM SNA network.

3. In Northern Telecom language, it means Network Configuration Process.

4. In Novell language, it means NetWare Core Protocol. It is NetWare's format for requesting and replying to requests for file and print services.

NCPS National Customer Product Support.

NCR Originally National Cash Register, then called NCR Corp. AT&T acquired the company in 1991, and changed the name to AT&T Global Information Solutions (GIS) in January 1994. AT&T had grand plans to become a voice and data powerhouse, in both the equipment and network domains. It didn't work out that way. AT&T split into three companies on January 1, 1997. AT&T GIS was spun off and once again became NCR Corp. NCR is trying to regain its status as a leading manufacturer of data processing systems, ATMs and electronic cash registers.

NCRA National Cellular Resellers Association. A Washington-based trade association and lobbying organization which became the NWRA (National Wireless Resellers Association), which merged into TRA (Telecommunications Resellers Association) in 1997, which changed its name to ASCENT in 2000. See ASCENT.

NCS 1. Network Control System.

2. See National Communications System.

3. Network Computing System.

4. Network-based Call Signaling has been defined by CableLabs as the standard for telephony over HFC (Hybrid Fiber Coax) CATV cable plants. The NCS specification is a CATV-specific enhancement to MGCP (Media Gateway Control Protocol), which is the de facto standard for multimedia call control between the traditional PSTN (Public Switched Telephone Network) and IP-based packet networks. NCS serves as the core of the Cable-Labs PacketCable architecture. PacketCable is a fast-track initiative aimed at developing interoperable interface specifications for delivering advanced, real-time multimedia services over IP-based packet channels carved out of two-way CATV cable plants. PacketCable is built on the infrastructure set by CableLabs Certified, previously known as DOCSIS (Data Over Cable Service Interface Specification), which sets standards for CATV modems and related

network elements supporting high-speed Internet access over CATV networks. PacketCable includes specifications for call signaling, QoS (Quality of Service) control, PSTN interconnection, security, network management, codec support, billing event messages, and network announcements. The NCS signaling occurs between the Multimedia Terminal Adapter/cable modem (MTA) and the NCS Gateway (NCSG), which works as interface between the MTA and the Call Feature Server connected to the CATV network or PSTN. See also CableLabs, CableLabs Certified, CATV, MGCP, and PacketCable.

NCSA National Center for Supercomputer Applications. A supercomputer operations and research center located at the University of Illinois at Urbana-Champaign. The NCSA was one of the national supercomputer centers funded by the NSF (National Science Foundation). The six centers were interconnected over a network known as the NSFNET, which originally operated at 56 Kbps, and which was upgraded in 1987 to 1.544 Mbps (T-1) and to 45 Mbps (T-3) in 1989. The NSFNET was essentially disbanded in 1995, in favor of commercialization of the backbone network. See also NSF.

NCSANet National Center for Supercomputer Applications Network. A regional TCP/IP network which connected users in Illinois, Wisconsin and Indiana. NCSANet was also a mid-level network in NSFNET, which has been abandoned. The NSF (National Science Foundation) decided to get out of the business of being an Internet backbone provider, in favor of allowing the commercialization of the backbone. See also NCSA and NSF.

NCTA National Cable Television Association. A trade organization representing U.S. cable television carriers.

NCTE Network Channel Terminating Equipment. The general name for equipment that provides line transmission termination and layer-1 maintenance and multiplexing, terminating a 2-wire U-interface. Another name for a CSU, a Channel Service Unit. Also called a Data Service Unit. A device to terminate a digital channel on a customer's premises. It performs certain line conditioning and equalization functions, and responds to loop back commands sent from the central office. A CSU sits between the digital line coming in from the central office and devices such as channel banks or data communications devices. FCC decisions have established that most NCTE is customer premises equipment (CPE) and may therefore, by supplied by third party vendors as well as the telephone company.

NCUG National Centrex Users Group.

NDA 1. A non-disclosure agreement is a signed formal agreement in which one party agrees to give a second party confidential information about its business or products and the second party agrees not to share this information with anyone else for some time – sometimes as long as five years. NDAs are often signed when a company wants to sell something – itself or a subsidiary and it doesn't want

NDAs tend to protect the company more than the individual signing the agreement. As a matter of practice, I've signed very very few NDAs. Most people will show me things because they trust me. Making me sign an NDA is not going to protect them much if I'm not trustworthy. I personally prefer trust and integrity to an NDA.

A non-disclosure agreement (NDA) is a signed formal agreement in which one party agrees to give a second party confidential information about its business or products and the second party agrees not to share this information with anyone else for a specified period of time.   Translation...is promising not to tell anyone else what a lousy product, idea or business plan you have before you have a chance to cash in on the idea...

2. National Directory Assistance, or Nationwide Directory Assistance. A Directory Assistance that removes the need to know the area code of the target party in order to get the telephone number. The area codes are expanding and changing so frequently that it's difficult to keep up with them. A number of telephone companies are offering NDA services to help solve the problem. It's a huge business.

NDAC National Database Administration Center.

NDC National Destination Code. A wireless telecommunications term. The second part of an MSISDN (Mobile Station ISDN) number. It is used to identify a PLMN (Public Land Mobile Network) within a country.

NDD Network Descriptive Database.

NDE Network Design Engineering.

NDIS Network Driver Interface Specification. A device driver specification co-developed by Microsoft and 3Com and used by LAN Manager and Vines and supported by many network card vendors. Besides providing hardware and protocol independence for network drivers, NDIS supports both MS-DOS and OS/2. It offers protocol multiplexing so that multiple protocol stacks can coexist in the same host. NDIS is conceptually similar to ODI.

NDIS miniport drivers A type of minidriver that interfaces network class devices to NDIS.

NDM Network Data Mover, a delayed interface using the "Connect:Direct" transmission method. I don't know much about this term, but apparently it is a one-way street, may be somewhat error prone, and is being moved away from by the long distance industry as a means of transmission. See Connect:Direct.

NDMP Network Data Management Protocol.

NDN Measured standard, no allowance.

NDO Network Design Order. See Telephone Equipment Order.

NDR See Network Design Request.

NDS Novell Directory Services. According to Novell, NDS is a multiple-platform, distributed database that stores information about the hardware and software resources available on a network. It provides network users, administrators and application developers global access to all network resources. NDS also provides a flexible directory database schema, network authentication/security, and a consistent cross-platform development environment. NDS uses objects to represent all network resources and maintains them in a hierarchical directory tree. NDS was originally NetWare Directory Services, but it was changed about 1996 when Novell started to explore putting it on platforms other than NetWare. NDS is now available on a wide variety of platforms, including NetWare, Windows NT, Sun Solaris, Linux (several variants), OS/390, AIX, and Compaq Tru64 UNIX.

NDSF Non Dispersion Shifted Fiber is the earliest form of Single Mode Fiber (SMF) developed, and was a considerable improvement over MMF (MultiMode Fiber). NDSF generally runs in the O-Band, at about 1310nm (nanometers). When NDSF was first developed, it was discovered that signals suffered from chromatic dispersion, the phenomenon by which different colors (i.e., wavelengths) of light travel at slightly different speeds. Since even the most tightly tuned laser light sources create signals composed of multiple wavelengths, each pulse of light contains components that travel at slightly different speeds. The portions of the light pulses that enter the cladding experience this phenomenon to a greater extent, as the cladding is much more pure that the core. Therefore, all signals travel at an accelerated rate through the cladding, and the impact of the phenomenon is more pronounced. The problem is exacerbated by waveguide dispersion, the phenomenon by which higher frequency (shorter wavelength) signals travel more in the cladding (and less in the core) than do lower frequency signals. NDSF runs in the 1310nm window (i.e., wavelength range), as chromatic dispersion and waveguide dispersion cancel each other at that wavelength. Dispersion Shifted Fiber (DSF) is an improvement over NDSF. See also Chromatic Dispersion, DSF, MMF, O-Band, SMF, Waveguide Dispersion, and Window.

NDT 1. No Dial Tone. Abbreviation often used on phone company repair orders by staff.
2. Direct inward dialing (DID) Trunk.

NDX NetWare Directory Services. A feature of Novell's NetWare operating system which allows users to sign on to multiple NetWare servers through one simple sign-on. The ultimate idea is that NDS will become somewhat of a front-end sign-on to other networked devices, e.g. telephone systems, voice mail, electronic mail, Lotus Notes, etc. Sign on in one place, get into many. A big user time-saving benefit. A competitor to NDS is Microsoft's ODSI Open Directory Services Interface.

NDZ PBX Service, establish trunk group and provide first group.

NE Network Element. A single piece of telecommunications equipment used to perform a function or service integral to the underlying bearer network. In ATM networks, NE is a system that supports at least NEFs and may also support Operation System Functions/Mediation Functions. An ATM NE may be realized as either a standalone device or a geographically distributed system. It cannot be further decomposed into managed elements in the context of a given management function.

NE&I Network Engineering and Implementation. An area of responsibility in a telecommunications organization that identifies business processes and conducts activities that identify, plan, design, and construct network and system resources.

near end The originating end of a trunk circuit or connecting path.

near end crosstalk NEXT. Interference from an adjacent channel. If crosstalk at the near end is great enough, it may interfere with signals received across the circuit. See NEXT for a full explanation.

near field magnetic communication A wireless communications technology for very short-range communications (up to three meters) between wireless devices, such as a headset and a belt-mounted two-way radio or CD player. Near-field magnetic communication (NFMC) does not use radio waves. Instead, it involves modulating a non-propagating, quasi-static magnetic field between devices. NFMC systems also have significantly lower power requirements than RF systems. The very short-range nature of NFMC also means that a NFMC system consumes much less power than a comparable RF system.

Near Instantaneous Companding NIC. The very fast quantizing of an analog signal into digital representation...and converting it back. See also Companding.

near line A storage term typically used as "near-line storage." A digital audio tape (DAT) might be considered near-line storage. A DAT would typically store a database of infor-mation sequentially. To find a record might take anywhere from three to 23 seconds. That's an eternity in most computer applications, but it may be adequate for finding something you rarely need to find – maybe several times a year. Thus the concept "near-line" storage.

near real-time Not quite in real-time, but nearly so. Real-time involves the immediate processing of information as the transaction occurs. Near real-time is not immediate, but nearly so. Batch processing is neither, and may involve considerable delay.

near video on demand A service with which a subscriber can watch a programmer-chosen video program at nearly any time. Usually implemented by showing a hot movie on several channels by staggering start times,. like every half-hour. Compare to Video-on-Demand.

nearest active upstream neighbor See NAUN.

NEARNET A commercial Internet access service run by Bolt Beranek and Newman, Inc., Cambridge, MA.

NEBE Near End Block Error. See FEBE (Far End Block Error) for a full explanation.

NEBS 1. New Equipment Building System.
2. Network Equipment Building Standards. NEBS defines a rigid and extensive set of performance, quality, environmental and safety requirements developed by Bellcore, the R&D and standards organization once owned by the seven regional Bell operating companies (RBOCs). NEBS compliance is often required by telecommunications service providers such as BOCs (Bell Operating Companies) and Interexchange Carriers (IEC) for equipment installed in their switching offices. NEBS defines everything from fire spread and extinguish ability test to Zone-4 earthquake tests to thermal shock, cyclic temperature, mechanical shock, and electro-static discharge. Conforming to NEBS is not inexpensive.
Here is a more detailed explanation of NEBS: NEBS, Generic Equipment Requirement, is a subset of the Family of Requirements as published by Bellcore known as LSSGR, LATA Switching System Generic Requirement. Bellcore has published a technical reference (TR), TR-NWT-000063, Issue 5, 1994, which outlines the NEBS "standard." Bellcore states that the intent of this TR is to "inform the industry of Bellcore's view of the proposed minimum generic requirements that appear to be appropriate for all new telecommunications equipment systems used in central offices (CO) and other telephone buildings of a typical BCC (Bellcore Client Company)."
The requirements defined in this TR are broken into two major categories: 1) spacial requirements and 2) environmental requirements. Spacial requirements apply to the equipment systems' cable distributing and interconnecting frames, power equipment, operations support systems and cable entrance facilities. Compliance with these requirements is intended to improve the use of space in the CO, simplify building-equipment interfaces and help make the planning and engineering of central offices simpler and more economical. In general, the spacial requirements relate to the size and location of the equipment going into a CO, how to power it and how to cable it all together.
The second category relates to environmental requirements that define the conditions under which the equipment could potentially be exposed to and still operate reliably or not cause any catastrophic situation such as fire spread. These environmental requirements include temperature and humidity, fire resistance (which seems to be at the top of many BCCs' lists), shock and vibration (which covers transportation, office and earthquake), electrostatic discharge (susceptibility and immunity) and electromagnetic compatibility (emission and immunity). Bellcore has also defined test methods in this TR which provide procedures on how to test the equipment to ensure that it meets the requirement. See Bellcore.

NEC See National Electric Code.

NEC requirements The National Electrical Code (NEC) is written and administered by the National Fire Protection Agency (NFPA). The latest 1990 version states that any equipment connected to the telecommunications networks must be listed for that purpose. Listing is acquired through Underwriters Laboratories (UL) or a similar approved lab. Listing requirements for premises wiring between the Network Interface Device and the modular jack at the work area took effect on October 1, 1990. See NFPNA 90A.

NECA National Exchange Carrier Association. An association of local exchange carriers mandated by the U.S. Federal Communications Commission in 1983, in anticipation of the breakup of the Bell System. NECA's primary responsibility was to file new interstate access tariffs and to perform the "settlement" functions previously performed by AT&T and the BOCs (Bell Operating Companies). Under the direction of the FCC, NECA administers approximately $2 billion in annual revenues, based on pooled tariff rates. Through the pooling process, the LECs (Local Exchange Carriers) bill the IXCs (IntereXchange Carriers) for the use of their local exchange networks in originating and terminating long distance calls. Individual revenues and costs then are submitted to NECA, which distributes them to the member companies on an averaged basis. NECA also administers the Universal Service Fund

(USF), the Lifeline Assistance Programs, and the Telecommunications Relay Services Fund. See also Division of Revenues, Lifeline Assistance Programs, Separations and Settlements, Telecommunications Relay Services Fund, and Universal Service Fund. www.neca.org.

NECP Network Element Control Protocol. An Internet protocol which provides methods by which network elements (NEs) can learn about the capabilities and availability of services, and which provides hints as to which data flows can and cannot be serviced. Thereby, NECP allows intelligent NEs to perform load balancing across an array of servers, interception and redirection of data flows to transparent proxy servers, and cut-through of data flows to the origin in the event that the proxy servers cannot serve the data flows. See also Cache, Client, Proxy, and Server.

NECTAR Nectar is a collaboration among Bell Atlantic/Bell of PA, Carnegie-Mellon University, the Pittsburgh Supercomputing Center and Bellcore.

NEDS Short for New Economy Depression Syndrome. The stress, fatigue, and loss of productivity caused by information overload and endless interruption. Coined by Tim Sanders, Yahoo!'s chief solutions officer.

need-to-know principle The principle that a user or process should only be able to access resources that they are authorized to access and only to complete a legitimate task at hand.

needle message Sent by the Supervisor to go from node to node with instructions for setting up a virtual circuit.

neep-neep A rapidly increasing group of people who are fascinated with computers but who do not necessarily have any scientific skills beyond getting the computer to boot up.

NEF Network Element Function: A function within an ATM entity that supports the ATM based network transport services, (e.g., multiplexing, cross-connection).

negated A signal is negated when it is in the state opposed to that which is indicated by the name of the signal. Opposite of Asserted.

negative absorption Amplification. The positive difference between stimulated and absorbed radiation.

Negative Acknowledgement NAK. A communications control character sent by a receiving station to indicate that the last message or block received was not received correctly.

negative trapping Use of a notch filter, usually in cable systems, to trap out part of a signal and thereby deny the signal to non-subscribers.

negligent Negligent is a condition in which you absentmindedly answer the door in your nightie.

negotiation 1. Think about a voice conversation. We call someone. They answer. We say "Joe." They say "Harry." We've identified ourselves. We know each other so we choose the language we speak – in this case English. This is a simple explanation of what two machines do when trying to speak with each other. When they first start talking, they query each other on speed, protocol, language, etc. etc. This is called "negotiation." For example, if one device is speaking slowly (the line is a slow speed one), the other device will drop (negotiate) its speed down. Here's a more technical explanation: Negotiation is the process whereby two end-points announce their speed (10/100 Mbps) and duplex value (Half/Full) and determine the best values for communication. For example, if a device is connected to an Internet switch port, the network card in the device and the Internet switch port negotiate their speed and duplex values. If the values are different, the lower values are used.

2. Negotiation also means getting the product or service you want at the price you want to pay. Books have been written on the art of negotiation. For me, the key is to go for the best deal. The worst that can happen is for them to say "NO." If you lose that deal, there's always a better one around the corner.

Negroponte Switch This term was coined by George Gilder after Nicholas Negroponte, founder of MIT's Media Lab. Negroponte predicted that the advent of personal wireless technologies would cause signals that historically traveled by wireless methods (broadcast television, for example) to "switch" to wireline technologies (such as CATV), and signals that historically traveled by wireline methods (telephone, for example) would switch to wireless (such as cell phones). Negroponte's prediction was based on considerations of efficiency – cell-based wireless networks are much more efficient in spatial re-use of the spectrum than broadcast networks – and business opportunity. Broadcasters/content providers seeking to "narrowcast" their content in order to collect more subscription revenues. The model holds up pretty well in developed countries, but less so in other countries where wireline network infrastructure (telephone or CATV plant) may not be as well-developed.

neighbor node An ATM term. A node that is directly connected to a particular node

via a logical link.

NEL Network Element Layer: An abstraction of functions related specifically to the technology, vendor, and the network resources or network elements that provide basic communications services. The NEL is the layer of an integrated digital network whose function and capabilities include the information necessary for billing and collection, for routing or transmission of a telecommunications service.

Nelson, Lorraine Routh The lady who records the messages on Lucent's Audix voice mail system. She records her timeless messages at the Lucent Bell Labs in Denver. She lives in Oregon.

Nelson, Ted The software visionary who coined the term "hypertext" in a 1965 paper delivered to the Association for Computing Machinery. Extending the concept, Nelson proposed "zippered lists," whereby elements in one textual document would be linked to identical or related elements in other texts. Nelson also started Xanadu, a yet unsuccessful venture for the development of a system which would support the network sale of documents, or snippets of documents, with an automatic royalty on every byte. Nelson since has gone on to other things. See also Hypertext and Xanadu.

NEMA National Electrical Manufacturers Association. A trade association comprising over 500 member companies of the "electroindustry." NEMA develops and promotes positions on standards and government regulations, serves its members to acquire information on industry and market economics. NEMA publishes over 200 standards, which it offers for sale, along with certain standards originally developed by the American National Standards Institute (ANSI) and the International Electrotechnical Commission (IEC). NEMA is organized into nine divisions: Industrial Automation, Lighting Equipment, Industrial Equipment, Electronics, Building Equipment, Insulating Materials, Wire and Cable, Power Equipment, and Diagnostic Imaging and Therapy systems. The Wire and Cable division specifically includes voice and data types used for internal premises wiring, and specifically excludes exterior telephone distribution cables. See also ANSI and IEC.

NEMAS Network Management and Administration System.

NEMKO Norwegian Board for Testing and Approval of Electrical Equipment.

NENA National Emergency Number Association. A US non-profit organization promoting 911 as the standard emergency number, including technical support, public awareness, certification programs, and legislative representation. NENA's mission is to foster the technological advancement, availability, and implementation of a universal emergency telephone number system. In carrying out its mission, NENA promotes research, planning, training and education. The protection of human life, the preservation of property and the maintenance of general community security are among NENA's objectives. www.nena9-1-1.org.

Neoprene Trade name for polychloroprene, used for jacketing.

NEP 1. Ted Rose works in a NEP for a large telephone company. His employer calls NEP a Network Entry Point. Mr. Rose explains, "We are not unlike the hub of a wagon wheel where data comes in from the two computer centers at DS1 and DS3 speeds. The equipment here breaks down those high speeds to both DS0 and DS1 speeds and then we feed that low speed data to Pacific Bell offices and garages to the people who are the users of the data that are customer facing. They would be the spokes of the wagon wheel."

2. See Noise Equivalent Power.

NEPA The National Environment Policy Act of 1969. An Act of Congress which requires federal agencies to take into consideration the potential environmental effects of a particular proposal such as construction of a radio station.

NEPE Basic unit of a logarithmic SCAL used for the expression of ratios of voltages, currents, and similar quantities.

Neper Np. A method of expressing the ratio between two quantities (Q1, Q2). $Np = log(\text{to the base } e) \times Q1/Q2$. One Neper equals 8.686 decibels.

nerd Slang term coined in the 1950s for an unpleasant, unattractive, or insignificant person – a square, or loser. It would seem that the term clearly was coined by people who considered themselves to be pleasant, attractive, or significant, i.e. winners. One theory traces the term it to a passage in Dr. Seuss' lesser-known 1950 children's book "If I Ran the Zoo." The thought is that Dr. Seuss probably got it from the U.S. Navy acronym NERD, which referred to Naval Enlisted Requiring Discipline, a term commonly used for sailors locked away in the brig. Another theory considers it a variation of the 1940s put-down "nerts to you," as in "nuts to you." Since a lot of nerds are now rich Cybergeeks, the term probably needs to be redefined. See also Geek, Nerb Bird and Nerdistan.

nerd bird A slang term for the non-stop plane that flies between Austin, Texas and San Jose, California. Both these town are high tech centers, thus full of "nerds." See Nerd.

nerdistan An upscale and largely self-contained suburb or town with a large population of high-tech workers employed in nearby office parks that are dominated by high-tech industries. Also any large collection of nerds. See Nerd and Nerd Bird.

nest 1. To embed a set of instructions or a block data within another.

2. Novell Embedded Systems Technology. Software that Novell wants to put into cars, office machines, telephones and other noncomputer products. It is a variant of NetWare.

nested list A list that is contained within a member of another list. Nesting is indicated by indentation in most Web browsers. When you create one list element within another list element in FrontPage, the new list element is automatically nested.

nesting Literally putting one thing inside another. In the security business of IPSec, this refers to creating tunnels within tunnels by wrapping additional headers/trailers around network packets.

net 1. The Internet. See Internet.

2. NETwork.

3. Normes Europeans de Telecommunication (European Telecommunications Standards).

4. Network Equipment Technologies.

net 1000 A data communications network and processing service of AT&T that never got off the ground and was closed down at a cost of about $1 billion. At one stage it was called Advanced Communications Service.

net additions A cellular term. The amount of new subscribers signing up for service after adjusting for disconnects (churn).

net address The location on a network where an addressee's mail is held, usually in storage until the user logs into the system. The system delivers the message at that time. The Net Address in the header of a message gives the information required by an automated message processing or message switch, to deliver a message to an intended addressee. A net address of RA3Y#SAIL would designate that the message was addressed to the user with the "Net Name" of "RA3Y" (pronounced "Ray". The 3 is silent) located at (signified by the symbol "#") the system in the network called "SAIL", in this case, the Stanford (University) Artificial Intelligence Laboratory.

Net Card Detection Library NCD. Part of Microsoft's Windows for Workgroups. Its purpose is to determine the network adapted installed in the workstation and minimize mistakes made by the user.

net citizen An inhabitant of Cyberspace. One usually tries to be a good net.citizen, lest one be flamed, i.e. insulted through e-mail.

Net Data Throughout NDT. The rate at which data is transferred on a communications channel, normally specified in bits per second.

net income See Operating Income.

net logon service A Windows NT term. For Window NT Server, performs authentication of domain logons, and keeps the domain's database synchronized between the primary domain controller and the other backup domain controllers of the Windows NT Server domain.

net loss The signal loss encountered in a transmission facility or network, or the sum of all the losses the signal encounters on its way to its final destination.

net metering Selling your kind power utility more power than it sells you. You do this by generating your own electricity using photovoltaic cells, a windmill or whatever and not using as much as you produce. Some states don't allow you net metering.

net nanny A Net Nanny program lets parents, guardians and teachers keep children from accessing pornographic material on the 'Net, while preventing the childrens' personal information – names, addresses, telephone numbers, etc. – from being accessed.

net personality Somebody sufficiently opinionated/flaky/with plenty of time on his hands to regularly post in dozens of different Usenet newsgroups, whose presence is known to thousands of people. See Internet.

net police Derogatory term for those who would impose their standards on other users of the Internet. Often used in vigorous flame wars (in which it occasionally mutates to net.nazis).

net POP Another name for POP, or Point of Presence. See POP.

net staffing A call center term. The actual number of staff minus the required number of staff in a given period. Net staffing that is positive indicates overstaffing; net staffing that is negative, understaffing. See Net Staffing Matrix.

net staffing matrix A call center term. A report that shows the actual number of staff, required number of staff, and net staffing for each period of a given day. See Net Staffing.

net station Radio station which broadcasts its programming over the Internet.

NetBEUI There's only one thing to know about the NetBEUI protocol – install it. On any networked Windows PC, go to Networking/Properties/Install/Protocol/Microsoft NetBEUI. The reason? Windows machines networked to each other over Ethernet work better and more reliably when NetBEUI protocol is installed. This is irrespective of what Microsoft tells you about its machines happily using the TCP/IP protocol to talk to each other. IN fact they can. But TCP/IP seems to be unreliable in for communicating between PCs on a small local area network. NetBEUI stands for NetBIOS Extended User Interface. It was one of the earliest protocols available for use on networks composed of personal computers. It was designed around the Network Basic Input/Output System (NetBIOS) interface as a small, efficient protocol for use in department-sized local area networks (LANs) of 20 to 200 computers, which would not need to be routed to other subnets. Windows 2000-based NetBEUI, known as NetBIOS Frame (NBF), implements the NetBEUI version 3.0 specification. NBF is the underlying implementation of the NetBEUI protocol installed on a computer running Windows 2000. It provides compatibility with existing LANs using the NetBEUI protocol, and is compatible with the NetBEUI protocol driver included with previous versions of Microsoft networking products. In addition, NBF:

Uses the Windows 2000-based Transport Driver Interface (TDI), which provides an emulator for interpretation of NetBIOS network commands. + Uses the Windows 2000-based Network Device Interface Specification (NDIS) with improved transport support and a full 32-bit asynchronous interface. + Uses memory dynamically to provide automatic memory tuning. + Supports dial-up client communications with a remote access server. + Provides connection-oriented and connectionless communication services. + Removes the NetBIOS session number limit.

Network Basic Input/Output System (NetBIOS) is software developed by IBM. It provides the interface between a computer's operating system, the I/O bus, and the network; a de-facto network standard. NetBIOS Extended User Interface (NetBEUI) is the original personal computer networking protocol and interface designed by IBM for their LAN Manager server. NetBEUI implements the OSI LLC2 protocol and has a limitation of 254 session connections. Windows 2000 NetBEUI 3.0, also known as NetBIOS Frame Format Protocol (NBFP), is the Microsoft implementation of the IBM NetBEUI protocol. It eliminates the previous NetBEUI limitation of 254 sessions to a server and uses the Microsoft TDI layer as an interface to NetBIOS. See NETBIOS.

NETBIOS Network Basic Input/Output System. A layer of software originally developed by IBM and Sytek to link a network operating system with specific hardware. Originally designed as the network controller for IBM's PC Network LAN, NetBIOS has now been extended to allow programs written using the NetBIOS interface to operate on the IBM Token ring. NetBIOS has been adopted as something of an industry standard and now it's common to refer to Netbios-compatible LANs. It offers LAN applications, a variety of "hooks" to carry out inter-application communications and data transfer. Essentially, NetBIOS is a way for application programs to talk to the network. Other applications interfaces are also being used these days, such as IBM's APPC. To run an application that works with NetBIOS, a non-IBM network operating system or network interface card must offer a NetBIOS emulator. More and more hardware and software vendors offer these emulators. They aren't always perfectly compatible, though. Today, many vendors either provide a version of NetBIOS to interface with their hardware or emulate its transport layer communications services in their network products. See NetBEUI.

NETBIOS Emulator An emulator program provided with NetWare that allows workstations to run applications that support IBM's NetBIOS calls.

NetBIOS Extended User Interface See NetBEUI.

NetBIOS Frame Format Protocol NBFP. See NetBEUI.

NetBSD NetBSD is a free, secure, and highly portable UNIX-like operating system available for many platforms, from 64-bit AlphaServers and desktop systems to handheld and embedded devices. Its clean design and advanced features make it excellent in both production and research environments, and it is user-supported with complete source. Apple's operating system OS X is a variation of Unix. It was built on NetBSD, which is different than Linux. The original SunOS from Sun Microsystems was build from an earlier version of NetBSD as well.

NetBT NBT. NetBios over TCP/IP. A protocol supporting NetBIOS services in a TCP/IP environment, defined by RFCs 1001 and 1002.

netcasting A method used to deliver Web content automatically to the desktop. Netcasting is referred to as push technology because content is pushed from a Web site to those users who requested receipt of the content. Content can include weather forecasts, stock market quotes, or software updates.

NETCOM A system that provides information on Customer Premises Equipment (CPE), location equipment, and contracts.

NETDDE Network Dynamic Data Exchange, a feature of Microsoft Windows.

Netfind A white pages service which enables a person to query one service and have that service search other databases for addresses matching the originally entered query.

NetFX Consortium NetFX consortium is a group of companies that have come together to provide solutions for delivering real-time multimedia over the Internet's World Wide Web. The members are Diamond Multimedia Systems www.diamondmm.com, NET-COM www.netcom.com, SAIC http://merkury.saic.com/demo), Template Graphics Software www.sd.tgs.com/~template), and Xing Technology Corp www.xingtech.com).

Netgod Also spelled NET.GOD. A person very visible on a network and who played an important role in the development of the network.

nethead Also spelled net.head. A nethead, who believes that the PSTN is a relic, the Internet is the future of all telecommunications and networking, and Bellheads – members of the telecom establishment – should not be allowed to control the Internet.

Nethopper A wonderfully cute name for a device that organizes data to hop from one local area network to another. Nethoppers are also called bridges and routers.

NetHub The home of the RIME network located in Bethesda, MD. All Super-Regional Hubs in the RIME network call the NetHub to exchange mail packets.

NETI Network Information Table.

NETIPC Network Interprocess Communications. Permits inter-CPU program sharing and allows programs running on different systems to exchange data through a set of programmatic calls. This peer-to-peer service is important for developing distributed applications.

netiquette A pun on "etiquette". Contraction for "Network Etiquette". It means proper behavior on Internet. Because the recipients of text-only messages cannot see your face or hear the inflections of your voice, special care must be taken to avoid misunderstanding and to convey the "flavor" that you intended your messages to have. Some common techniques are the use of smileys (also called emoticons): punctuation that suggests sideways faces, bracketed abbreviations like <\><>g> for grin and <\><>G> for Big Grin, and careful capitalization as in I AM NOW SHOUTING AT YOU!

netizen A citizen of the Internet.

netkeeper See IPTC.

netmail A private message transmission capability on FidoNet in which nodes directly communicate with one another on a point-to point (in FidoNet) or hub-routed (in PCRelay). NetMail was originally developed for use by SYSOPS to communicate with one another and is available on some BBs for regular users. See Echo.

netmask A colloquial way of saying IP Address Mask. See IP Address Mask.

NetMeeting Microsoft's NetMeeting is a conferencing and collaboration tool designed for the Internet or intranet. I'm giving a lot of space to it in this edition of my dictionary, because I believe NetMeeting is an important product that will have a major impact on the way we communicate (voice, video, data, images, etc.) over the PSTN, over private corporate networks and over IP networks, such as the Internet and private IP networks, known as intranets.

NetMeeting, according to Microsoft, has the following features: H.323 standards-based voice and video conferencing. Real-time, point-to- point audio conferencing over the Internet or corporate intranet enables a user to make voice calls to associates and organizations around the world. NetMeeting voice conferencing offers many features, including half-duplex and full-duplex audio support for real-time conversations, automatic microphone sensitivity level setting to ensure that meeting participants hear each other clearly, and microphone muting, which lets users control the audio signal sent during a call. This voice conferencing supports network TCP/IP connections.

Support for the H.323 protocol enables interoperability between NetMeeting and other H.323-compatible voice clients. The H.323 protocol supports the ITU G.711 and G.723 audio standards and Internet Engineering Task Force (IETF) RTP and RTCP specifications for controlling audio flow to improve voice quality. On MMX-enabled computers, NetMeeting uses the MMX-enabled voice codecs to improve performance for voice compression and decompression algorithms. This will result in lower CPU use and improved voice quality during a call.

With NetMeeting, a user can send and receive real-time visual images with another conference participant using any video for Windows-compatible equipment. They can share ideas and information face-to-face, and use the camera to instantly view items, such as hardware or devices, that the user chooses to display in front of the lens. Combined with the video and data capabilities of NetMeeting, a user can both see and hear the other conference participant, as well as share information and applications. This H.323 standards-based video technology is also compliant with the H.261 and H.263 video codecs.

Multipoint data conferencing using T.120. Two or more users can communicate and collaborate as a group in real time. Participants can share applications, exchange information through a shared clipboard, transfer files, collaborate on a shared whiteboard, and use a text-based chat feature. Also, support for the T.120 data conferencing standard enables interoperability with other T.120-based products and services. The following features comprise multipoint data conferencing:

Application sharing: A user can share a program running on one computer with other participants in the conference. Participants can review the same data or information, and see the actions as the person sharing the application works on the program (for example, editing content or scrolling through information.) Participants can share Windows-based applications transparently without any special knowledge of the application capabilities. The person sharing the application can choose to collaborate with other conference participants, and they can take turns editing or controlling the application. Only the person sharing the program needs to have the given application installed on their computer.

Shared Clipboard: The shared clipboard enables a user to exchange its contents with other participants in a conference using familiar cut, copy, and paste operations. For example, a participant can copy information from a local document and paste the contents into a shared application as part of a group collaboration. File Transfer: With the file transfer capability, a user can send a file in the background to one or all of the conference participants. When one user drags a file into the main window, the file is automatically sent to each person in the conference; they can then accept or decline receipt. This file transfer capability is fully compliant with the T.127 standard.

Whiteboard: Multiple users can simultaneously collaborate using the whiteboard to review, create, and update graphic information. The whiteboard is object-oriented (versus pixel-oriented), enabling participants to manipulate the contents by clicking and dragging with the mouse. In addition, they can use a remote pointer or highlighting tool to point out specific contents or sections of shared pages.

Chat: A user can type text messages to share common ideas or topics with other conference participants, or record meeting notes and action items as part of a collaborative process. Also, participants in a conference can use chat to communicate in the absence of audio support. A "whisper" feature lets a user have a separate, private conversation with another person during a group chat session.

TAPI 3.0 and NetMeeting both support core IP Telephony capabilities. Each platform offers unique benefits: TAPI 3.0 integrates traditional telephony with IP Telephony, providing a COM-based, protocol-independent call-control and data streaming infrastructure. NetMeeting SDK supports T.120 conferencing and application sharing in addition to IP Telephony. Applications using TAPI 3.0 and the NetMeeting API interoperate using H.323 audio and video conferencing.

Because TAPI 3.0 and NetMeeting both support core IP Telephony capabilities (including support for H.323), developers may want to consider the following guidelines when choosing an API for their IP Telephony applications: TAPI 3.0. This is the API to use if you are doing IP Telephony in your application. TAPI 3.0 is especially valuable in the world of client/server computer telephony integration, for combining IP Telephony with traditional telephony, and for IP multicast of voice and video.

NetMeeting API. This is the API to use if you are doing real-time collaboration and want to integrate voice, video, and data conferencing into your application. The NetMeeting API is useful for applications that want to integrate application sharing, whiteboard functionality, and multipoint file transfer with voice and video sessions. www.microsoft.com/netmeeting.

netopath The most extreme, ugly and deranged form of Net abuser.

NETOPS Network Operations System.

NetPC See Network Computer and Thin Client.

Netscape Netscape was once the most famous Web Browser in the world. It owes its genesis to a software program called Mosaic, which Marc Andreessen and a small team of student programmers working for $6.85 an hour in 1993 wrote at the National Center for Supercomputing Applications at the University of Illinois at Urbana-Champaign. A year later, after becoming annoyed at the way the University had taken over his Mosaic creation, Mr. Andreessen proposed a "Mosaic Killer" – a new and improved version of his own creation. The team was back at work by April 1994 in a company called Netscape. And by October, they had created a new version of Mosaic, called Netscape Navigator. Netscape went public in August of 1995 in one of the most successful IPOs (Initial Public Offerings) ever. When Microsoft started giving away its Internet Explorer Browser for free, Netscape fell on hard times. And in late 1998, America OnLine (AOL) bought it for a gigantic amount of money. See Mozilla.

Netserv A file server used for distributing files directly related to the BITNET network.

Netsite The term Netscape Navigator uses to refer to a URL or WWW address.

Netsploitation Flick Any one of the Hollywood films about the Internet.

Netstat A utility program used to show server connections running over TCP/IP (Transmission Control Protocol/Internet Protocol) and statistics, including current connections, failed connection attempts, reset connections, segments received, segments sent, and segments retransmitted.

netstation In an Internet scenario, thin clients are known as NetPCs or Netstations.

The NetStation is reliant on the server, which is provided by your company or a service provider (e.g., America OnLine, CompuServe, or your ISP). In addition to providing some combination of content and Internet access, the service provider's server will provide your client NetPC with access to all necessary applications (e.g., word processing and spreadsheet), will store all your personal files, will provide all significant processing power, and so on. In this Internet example, the Netstation differs from the standard thin client by virtue of the fact that it does contain a modem, a communications port and communications software, all of which are required for Internet access. See also Client, Client/Server, Client/Server Model, Fat Client, Mainframe Server, Media Server, and Thin Client.

Netview An IBM product for management of heterogeneous networks that integrates the functions of three formerly separate Communications Network Management (CNM) software programs: 1. NCCF. Network Communication Control Facility. 2. NLDM. Network Logical Data Manager, which uses functions from NCCF and helps pinpoint problems along the logical connection/path of an SNA session. 3. NPDA. Network Problem Determination Application, which displays various alerts using IBM equipment located at strategic points in the network and allows diagnostic information to be displayed. Also, NetView incorporates some of the functions from two other programs: VNCA (Virtual Telecommunications Access Method/Node Control Application) which monitors the status and current activity of all resources in a domain, and NMPF (Network Management Productivity Facility) which helps the network operator to install, learn and use many network management products. See also Network and Network Management.

NetWare NetWare used to be a popular and good operating system for a local area network from Novell, Orem, UT. NetWare is actually its own operating system. This means it was the link between machine hardware (file servers, printers, modems, etc.) and people who want it use that hardware. NetWare is neither DOS, nor OS/2 nor Windows though it can be made to look and act like them. That's part (a small part) of its charm. Unfortunately Windows killed it. Or more precisely, Novell's once arrogant management killed it. See Netware MHS, Netware Workstation Files, NETx.COM and Novell.

NetWare Bindery Centralized authentication database for NetWare 3.xx LANs.

NetWare Directory Services. See NDS.

NetWare Global MHS Novell's implementation of MHS as a NetWare Loadable Module (NLM), providing powerful integration with NetWare services. This supports additional modules to connect to X.400, SNA and SMTP systems.

NetWare Loadable Module NLM. An driver that runs in a server on a local area network under Novell's NetWare operating system and can be loaded or unloaded on the fly as it's needed. In other networks, such applications could require dedicated PCs. A telephony NLM might allow a workstation on a LAN to control a PBX attached to a NetWare file server. It might also allow the workstation to control one or more voice processing cards sitting on in a NetWare server. In early 1993, AT&T became the first PBX maker to ink a deal with Novell, the creator of NetWare, to put telephony onto Novell LANs. AT&T created a PC-card resident in a Novell File server. The card connects to the ASAI (Adjunct Switch Applications Interface) BRI port on the AT&T Definity PBX. Anyone with a PC on the Novell network and an AT&T phone on their desk can use telephone features, such as auto-dialing, conference calling and message management (a new term for integrating voice, fax and e-mail on your desktop PC via your LAN). The Novell/AT&T deal intends to create open Application Programming Interfaces (APIs) that third party developers can work with. A Novell/AT&T example of what could be developed: A user could select names from a directory on his PC. He could tell the Definity PBX through the PC over the LAN to place a conference call to those names. At the same time, a program running under NetWare would automatically send an e-mail to the people, alerting them to the conference call and giving them the agenda. All participants would have access to both the document and the conference call simultaneously. See Telephony Services.

NetWare MHS Netware MHS, which is software that provides store-and-forward capability. Fax and E-mail systems that support MHS format their message transmissions according to MHS specifications. MHS reads compatible transmissions, determines the intended recipient and his location, and then sends the message to that location, regardless of the type of fax or E-mail system at the different ends. See MHS.

NetWare Telephony Services See Telephony Services.

network Networks are common in our lives. Think about trains and phones. A networks ties things together. Computer networks connect all types of computers and computer related things – terminals, printers, modems, door entry sensors, temperature monitors, etc. The networks we're most familiar with are long distance ones, like phones and trains. But there are also Local Area Networks (LANs) which exist within a limited geographic area – like the few hundred feet of a small office, an entire building or even a "campus," such as a university or industrial park. There are also Metropolitan Area Networks (MANs). See also LAN and MAN.

network access control Electronic circuitry that determines which workstation may transmit next or when a particular workstation may transmit.

network access line NAL. A communications channel between a customer's premises and the central office.

network access point NAP. See NAP. A telephone company AIN term. Software within a switch capable of recognizing a call that requires processing by AIN logic which, upon recognizing such a call, routes the call to an SSP or ASC switch.

network access server See access server.

network accounting A system or application software module that monitors and reports on packet-switched data network traffic, generally focusing on IP (Internet Protocol) traffic. Network accounting software captures data packets as they transverse the LAN, compresses them, and stores them on a centralized data repository to which the network administrator, cost center managers, and privileged others can gain access to run various reports. Much like a call accounting system in the circuit-switched voice domain, a network accounting system captures data output from a switch or router. The SDR (Session Detail Record) output is much like CDR (Call Detail Record) output from a voice PBX. Much like PBX CDR information identifies the originating and terminating extension/telephone number, a network accounting system captures the originating and terminating IP address, and can translate that into both the MAC address of the LAN-attached user workstation, and the URL (Uniform Resource Locator) of the Website the user has visited. Much like a call accounting system keeps track of the duration and time of day of a voice call, a network accounting system keeps track of the duration and time of day of a data network user's session. Network accounting systems are highly effective monitoring systems used by large corporations to ensure that expensive network resources are neither abused or misused. Such resources include both LAN resources (e.g., hubs, switches, servers, and routers) and high-speed (e.g., T-1 and T-3) circuits connecting to the Internet. See also Call Accounting, Electronic Communication Privacy Act, and Keystroke Monitoring.

network ACD Network ACD allows ACD agent groups, at different locations (nodes), to service calls over the network independent of where the call first entered the network. NACD uses ISDN D-channel messaging to exchange information between nodes.

network address Every cardEvery node on an Ethernet network has one or more addresses associated with it, including at least one fixed hardware address such as "ae-34-2c-1d-69-f1" assigned by the device's manufacturer. Most nodes also have protocol specific addresses assigned by a network manager.

network address translation NAT. Network Address Translation. A Cisco version of Port Address Translation (PAT), NAT enables a local area network (LAN) to use one set of IP addresses for internal traffic and a second set of addresses for external traffic. This allows a company to shield internal addresses from the public Internet. According to Cisco, NAT has several applications. You want to connect to the Internet, but not all your hosts have globally unique IP addresses. NAT enables private IP internetworks (i.e., Intranets) that use nonregistered IP addresses to connect to the Internet, or another public IP-based network. NAT is configured on the router at the border of a stub domain (referred to as the inside network) and a public network such as the Internet (referred to as the outside network). NAT translates the internal local addresses to globally unique IP addresses before sending packets to the outside network. You must change your internal addresses. Instead of changing them, which can be a considerable amount of work, you can translate them by using NAT. You want to do basic load sharing of TCP traffic. You can map a single global IP address to many local IP addresses by using the TCP load distribution feature. As a solution to the connectivity problem, NAT is practical only when relatively few hosts in a stub domain communicate outside of the domain at the same time. When this is the case, only a small subset of the IP addresses in the domain must be translated into globally unique IP addresses when outside communication is necessary, and these addresses can be reused when no longer in use.

A significant advantage of NAT, according to Cisco, is that it can be configured without requiring changes to hosts or routers other than those few routers on which NAT will be configured. NAT may not be practical if large numbers of hosts in the stub domain communicate outside of the domain. Furthermore, some applications use embedded IP addresses in such a way that it is impractical for a NAT device to translate. These applications may not work transparently or at all through a NAT device. NAT also hides the identity of hosts, which may be an advantage or a disadvantage. A router configured with NAT will have at least one interface to the inside and one to the outside. In a typical environment, NAT is configured at the exit router between a stub domain and backbone. When a packet is leaving the domain, NAT translates the locally significant source address into a globally unique address. When a packet is entering the domain, NAT translates the globally unique destination address into a local address. If more than one exit point exists, each NAT must have the same translation table. If the software cannot allocate an address because it has run out of addresses, it drops the packet and sends an ICMP Host Unreachable packet. A router configured with NAT must not advertise the local networks to the outside. However, routing information that NAT receives from the outside can be advertised in the stub domain as usual. See also NAT and Port Address Translation.

network addressable unit NAU. In IBM's SNA, a logical unit (LU), physical unit (PU) or system services control point (SSCP), which is host-based, that is the origin or destination of information transmitted by the path control portion of an SNA network.

network agent A network agent is a device, such as a workstation or a router, that is equipped to gather network performance information to send to the network management agent. See Network Management Agent.

network agnostic Able to work with any type of network, any generation of network, and any size of network. AGNOSTIC, FREETHINKER, and ATHEIST can all apply to one who does not take an orthodox religious position. AGNOSTIC is the most neutral; it usually implies only an unwillingness on available evidence to affirm or deny the existence of God or subscribe to tenets that presuppose such existence. FREETHINKER is broader; it can apply to one of no determinable religious position or to one who feels truth is made more available by not committing oneself to any orthodoxy, especially a belief in God's existence. Often it can suggest a reprehensible and dangerous license of opinion. ATHEIST can apply strictly and neutrally to one who denies the existence of God or tenets presupposing it. More frequently than FREETHINKER, however, it has carried ideas of reprehensible license of opinion and menacing godlessness.

network analyzer A microwave test system that characterizes devices in terms of their complex small-signal scattering parameters (S-parameters). Measurements involve determining the ratio of magnitude and phase of input and output signals at the various ports of a network with the other ports terminated in the specified characteristic impedance (generally 50 ohms).

network and services integration forum See NSIF.

network application architecture A generalized architecture allowing interoperability at the application level. Examples are Digital Equipment Corp.'s Network Application Support (NAS) and IBM Corp.'s Systems Application Architecture (SAA).

network application support Digital Equipment Corporation's set of open software which allegedly allows its customers to integrate, port and distribute applications across different computer systems, including VMS, UNIX, MS-DOS, OS/2 and Apple MacIntosh.

network architecture The philosophy and organizational concept for enabling communications between multiple locations and multiple organizational units. Network architecture is a structured statement of the terminal devices, switching elements and the protocols and procedures to be used for the establishment effective telecommunications. See Architecture.

network attached storage NAS. Network Attached Storage is simply one or more storage devices (e.g., disk arrays) associated with a single server which exists as a node on a LAN (Local Area Network). The server assumes the responsibility for all data storage, offloading that responsibility from application servers, and making the data available to all users on the network. Besides basic storage and file sharing NAS devices can also be configured to provide services. For instance, a NAS device could be an Internet cache to increase the speed of surfing certain web sites, they could provide multimedia content across the network, or backup services for important content. As with all storage devices that are made available to the general network security is a concern. To protect sensitive information contained on NAS devices several security features can be put in place such as access control lists (ACL), encryption, antivirus, and packet filtering to regulate who has access to the stored information. Currently, NAS devices are relatively inexpensive compared to other available technologies and quality ranges from home-use units to full-featured commercial models. NAS devices available for the enterprise network have survivability options such as backup power and RAID support. However, they are still limited in scalability and performance when compared to storage area network (SAN) technologies. Advances in NAS technology are overcoming these hurdles by using Gigabit Ethernet networks and newer storage technologies. At some time, the gaps between SAN and NAS technologies will disappear. A Storage Area Network (SAN), by contrast, is a much more complex, high-speed dedicated sub-network. See also SAN.

network balancing 1. Lumped circuit elements (inductances, capacitances and resistances) connected so as to simulate the impedance of a uniform cable or open-wire circuit over a band of frequencies.

2. Moving circuits around in a multi-node switching network such the switching loads on

each of similar switching modules are roughly equal.

network basic input/output system NETBIOS. Within the context of the MS-DOS operating system, the software or software and firmware services that implement the interface between applications and a network adaptor, such as a CSMACD or token-ring adaptor.

network behavior anomaly detection NBAD. NBAD systems monitor network traffic for deviations that might indicate worm outbreaks, zero-day attacks and other forms of harmful network misuse. NBAD systems use passive sensors at key network intersections. They create profiles of normal traffic and tend to get alarmed (i.e. trigger alarms) when anomalies are detected. NBAD have their good points. They let you know when something awful may be happening to your network. And they're a lot better than perimeter security systems. They have their bad points. They still need people to check into what's causing the bump in traffic – maybe it's the half-yearly sale? They can also be expensive.

network board 1. A circuit board installed in each network station to allow stations to communicate with each other and with the file server.

2. An SCSA term. A board device designed to act as an interface between a computer-based signal processing system and a telephone network.

network border switch The latest "modern" way to move and switch phone calls is via packet switched IP (Internet Telephony). The problem is that standards aren't universal. Most carriers have their proprietary technology, which is fine if you stay on their network. Carriers trying to connect their IP voice networks with those of other carriers face a string of challenges. Maintaining quality of service: Carriers must understand markers used by their peers to designate voice-quality service and tag packets accordingly. Keeping call records: Details of which customers make use of the networks must be gathered, maintained and shared in a format compatible with billing software. Resolving addresses: The networks must support network address translation to deliver traffic to   devices   with private IP addresses that sit behind firewalls. Handling signaling differences: Calls initiated using H.323 signaling must be able to cross networks where Session Initiation Protocol (SIP) signaling is used. Minimizing delay: Protocol conversions and re-forming of packets can't be allowed to introduce enough delay to hurt voice quality. In early 2004, Sonus Network introduced software that lets its switching gear connect phone calls between IP networks run by different carriers even if the networks use different protocols. In doing so, Sonus is naming a new category of network device – the network border switch – that it says is needed to link carrier IP-voice networks. Sonus says its new software can translate signaling from a network that uses the H.323 standard and one that uses Session Initiation Protocol. This is necessary if the two networks are to set up and control calls. Carriers must resolve private IP addresses of IP phones with those of the firewall that protects them, something some firewalls can handle and that can be resolved by session controllers, which are separate devices. Using network border switches can avoid the need for traditional voice switches at the junction of carrier networks. Typically, a carrier using IP to carry voice on its network converts the traffic to TDM in a tandem switch that connects to a similar TDM tandem in the other carrier's network. Carriers can accomplish the same ends, but it takes multiple devices, Sonus says. Putting the features of many such devices on one piece of hardware simplifies networks and reduces maintenance costs, along with the need for power and space. It also can reduce provisioning time because there are fewer devices to configure. See also Media Gateway.

network byte order The Internet standard way of ordering of bytes corresponding to numeric values.

Network Channel Interface See NCI.

network channel terminating equipment NCTE. A device or devices at the user's premises used to amplify, match impedance or match network signaling to the customer's equipment connected to the network. Basically, network channel terminating equipment is a general name for equipment linking the network to a customer's premises. When NCTE connects to digital circuits, it typically consists of DSUs and CSUs. They are used for balancing of signals and providing for loop-back testing.

network computer NC. Larry Ellison of Oracle's idea of a $500 (or so) PC that lacks a hard disk and may lack a monitor but can be used to browse the Internet and run applications on a server on the Internet or corporate intranet. Ellison, who is Oracle's chairman, sees the NC as a "universal digital appliance."

The New Yorker of September 8, 1997 discussed the implications of the network computer thus: Microsoft's worries about Ellison and NCs are not trivial. After a prolonged period of being in denial about the rise of the Internet, Gates and his team now understand that it is the central fact of the next phase of computing, and that it poses a real threat to Microsoft's power. In 1995, Sun Microsystems introduced an Internet-centric programming language called Java, which creates programs that can run on any operating system and is fast becoming the standard lingo of the Net. In a Java-fuelled future, the reign of the PC might be challenged by the NC which would let users "borrow" programs from the Net and would have no need for Microsoft's Windows – developments that would create enormous upheaval in many of the software markets that Gate's firm now dominates. See also Internet Terminal, NetPC and NetStation.

network computing system NCS. A RPC (Remote Procedure Call) system developed by Apollo, and used in DEC and Hewlett-Packard computer systems. The NCS protocol later was adopted by the Open Software Foundation (OSF). See also OSF.

network control center A physical point within a network where various management and control functions are implemented.

network control program NCP. An IBM Systems Network Architecture (SNA) term. This is the program that switches the virtual circuit connections into place, implements path control, and operates the Synchronous Data Link Control (SDLC) link. The Network Control Program is normally resident in the communications controller or the host processor.

network control signaling The transmission of signals used in the telecommunications system which perform functions such as supervision, address signaling and audible tone signals to control the operation of switching machines in the telecommunication system.

network control signaling unit A telephone set that controls the transmission of signals into the telephone system which will perform supervision, number identification and control of the switching machines.

network controller A powerful microprocessor device designed to perform communications protocol translations between various terminals and computers and an X.25 packet switching network.

network data management protocol NDMP. An Internet draft specification from the IETF (Internet Engineering Task Force), NDMP is an open protocol for enterprise-wide, network-based data backup. NDMP is a secure backup technique which makes use of the TCP/IP protocol, running on networked file servers.

Network DDE Service A Windows NT definition. The Network DDE (dynamic data exchange) service manages shared DDE conversations. It is used by the Network DDE service.

network demarcation point The network demarcation point is the point of interconnection between the local exchange carrier's facilities and the wiring and equipment at the end user's facilities. The demarcation point is located on the subscriber's side of the telephone company's protector.

network design and optimization Network design and optimization is a process which balances network performance (availability) against cost. There are two fundamental tools in network design and optimization: a traffic usage recorder and software to interpret the results and make recommendations. A traffic usage recorder (TUR) is a device which connects to a network element in order to capture and record traffic statistics. Most network elements (e.g., PBXs, ACDs, data switches and routers) have special ports to which such a device can connect, usually via a RS-232 cable. As traffic flows through the network element, various information about that traffic is sent to the TUR in real time. The TUR holds that raw data in buffer memory until such time as it is polled by a centralized computer and the data is downloaded to that centralized computer. Later, the data is processed and reports are generated by traffic analysis software. That software will help you figure out which circuits you need, what speeds, to where, etc.

network design order NDO. See Telephone Equipment Order.

network design request NDR. Process required to establish the scope of a network project and to develop preliminary timeframes for providing service to a CLEC .

network device driver Software that coordinates communication between the network adapter card and the computer's hardware and other software, controlling the physical function of the local area network adapter cards.

network distribution The physical facilities of a communications carrier which are located outside the carrier's buildings. It includes cables, pole lines, conduit, etc.

network diversity A simply concept that says if you have a network which is important you need to have multiple ways of moving information around that network. Some of those ways should be provided by circuits from one vendor and some should be provided by circuits from several vendors. In other words, there should be diversity in both circuits and vendors. As an example of the necessity of taking network diversity very seriously, consider the year 1991. In 1991, nearly 30 million residential and business customers, more than one hundred thousand airline passengers and an alphabet soup of state and federal agencies were temporarily crippled by nine major telephone network

outages. Also in a 12-month period, the FAA (Federal Aviation Commission) reported a total of 114 phone service failures that disrupted air traffic facilities around the nation. Between June 10 and July 1, 1991, six major outages in three states comprised the largest series of network outages in history.

network drive A disk drive that is available to multiple users and computers on a network. Network drives often store data files for many people in a work group.

network economy The network economy is a term economists use when they apply the law of increasing returns to our entire economy. The idea is that one fax machine is worth nothing. Connect a hundred of them together and they're now worth something. Connect one million of them and suddenly you have something truly valuable. By the end of 1998, there were 350 million email addresses worldwide. Bingo, the network economy.

network effects As the name implies, these refer to the implications of being networked. A product that is part of a network has the property that its value to the user is a function of the number of other users of that product. A telephone or cellular phone becomes more valuable as the number of people who have one increases. The concept is also known as a positive feedback loop or self-reinforcing virtuous circle. It helps explain the super-fast growth that many products or services experience over a segment of their life cycle.

network element As defined by the Telecommunications Act of 1996, the The term 'network element' means a facility or equipment used in the provision of a telecommunications service. Such term also includes features, functions, and capabilities that are provided by means of such facility or equipment, including subscriber numbers, databases, signaling systems, and information sufficient for billing and collection or used in the transmission, routing, or other provision of a telecommunications service. See also Network Elements.

network element control protocol See NECP.

network elements NE. Processor controlled entities of the telecommunications network that primarily provide switching and transport network functions and contain network operations functions. Examples are: Non-AIN switching systems, digital cross-connect systems, AIN switching systems, and Signaling Transfer Points (STPs). In SONET, five basic network elements are: add/drop multiplexer; broadband digital cross-connect; wideband digital cross-connect; digital loop carrier; and switch interface.

Network Entity Identifier NEI. A cellular radio term. An Internet protocol address, or ConnectionLess Networking Protocol (CLNP) address, or any other protocol addressing used by the provider in transmission and receipt of Cellular Digital Packet Data (CDPD) services. The address that identifies that a user is authorized to use the service.

network equalizing A device connected to a transmission path in order to alter the characteristics of that path in a specified way. Often used to equalize the frequency response characteristic of a circuit for data transmission.

network extension unit NEU. An AT&T thing which sits in a telephone satellite closet (their words) which links up to 11 Starlan daisy chained clusters of PCs, etc. in a star configuration through standard telephone wiring, modular cords and plugs.

network externalities First seen in an article by John Cassidy in The New Yorker Jan 12, 1998 issue. In plain English, "network externalities" means that the value of a product increases along with the number of other people using it. Few people care about how many people are buying corn flakes or pizzas. But it usually applies to high-tech goods for two reasons. They have to be compatible with one and other (how useless is a Betamax videocassette today?). Second, the buyers are often linked to a network. The more people on the network, the more valuable the product becomes. Clearly, one person and one phone has no value. Two people and two phones have more value. 10,000 people and 10,000 phones have even more value, etc.

network fax server An network fax server is typically a PC on LAN. The PC has one or more fax/modem cards. Its job is to send and receive faxes for everyone on the LAN. As a result it is connected to dial out phone lines. See Server.

network file system NFS. A method developed by Sun Microsystems Inc. for distributing files within a heterogeneous network. NFS has become a de facto standard. It requires special software drivers to suit specific vendor hardware and operating system software. NFS hardware and software gateway products are also available.

network harm The reasoning behind the Bell System's insistence that the public could not hook its own equipment to the telephone line for fear it would produce "Irreparable Harm To The Network". A Bell System publication quoted the National Academy of Sciences as identifying four areas of potential harm. They were: Excessive signal power, hazardous voltage, improper network control signaling and line imbalance. Since the Carterfone Decision in the summer of 1968, the FCC has seen fit to allow devices, which pass an FCC Registration program, to be connected to the network, and while the quality of the stuff is occasionally poor, the fears of Network Harm have proven groundless. The fears were, of course, raised to preserve the Bell System's erstwhile almost monopoly on the manufacture of telephone equipment in the U.S.

Network Information Center NIC. Centers providing user assistance, document service, training on the Internet.

network information management system NIMS. Provides customers with browsing capability in their 800 and VNET databases.

network interface The point of interconnection between Telephone Company communications facilities and terminal equipment, protective apparatus or wiring at a subscriber's premises. The network interface or demarcation point shall be located on the subscriber's side of the Telephone Company's protector, or the equivalent thereof in cases where a protector is not employed, as provided under the local telephone company's reasonable and nondiscriminatory standard operating practices.

network interface card Also called a NIC card. A printed circuit board comprising electronic circuitry for the purpose of connecting a workstation to a LAN. A NIC usually is in the form of a card that fits into one of the expansion slots inside a PC. Alternatively, it can fit into a slot of a MAU (Multistation Access Unit), which serves multiple LAN-attached devices, such as workstations and printers. In the context of IEEE standards, NICs operate at the MAC (Medium Access Control) layer. In the context of the OSI Reference Model, NICs operate at Layers 1 (Physical Layer) and 2 (Data Link Layer). The basic job of the NIC is to take data from the transmitting workstation, form it into the specific packet format demanded by the LAN protocol you are running (e.g., Ethernet or Token Ring), and present it to the shared medium (usually a cable). On the receiving end, the process is reversed, of course. Hard-coded into the NIC at the time of manufacture is a MAC address, unique in all the world to that NIC card; the MAC address effectively identifies the LAN-attached device with which it is associated. A NIC works with the network software and computer operating system to transmit and receive messages on the network.

network interface controller Same as a Network Interface Card. See above definition.

network interface device 1. NID. A device wired between a telephone protector and the inside wiring to isolate the customer's equipment from the network.

2. A device that performs, functions such as code and protocol conversion, and buffering required for communications to and from a network.

3. A device used primarily within a local area network to allow a number of independent devices, with varying protocols, to communicate with each other. This communication is accomplished by converting each device protocol into a common transmission protocol.

network interface module Electronic circuitry connecting a system (typically a PC) to the telephone network. Network interface modules come in as many versions as there are ways of connecting to the telephone network. They range from a simple loop start telephone line to complex ISDN PRI circuits. Usually the network interface modules slides into one of the expansion slots inside a PC. The card transmits and receives messages from the resource modules and provides access to the telephone network.

network interface unit 1. A Network Interface Card (NIC) or Multistation Access Unit (MAU) that attaches to a LAN. It implements the local network protocols and provides an interface for device attachment. See also NIC and MAU for much more detail.

2. NIU. A semi-intelligent device that serves as the point of physical and logical demarcation between the LEC (Local Exchange Carrier) and your residential or small business premise. The NIU includes a silicon-based protector that trips the circuit in the event of a lightening strike or some other form of aberrant voltage that otherwise would fry your telephone and burn down your house. The NIU also has enough intelligence to allow the carrier to conduct an automated loopback test, which tests the integrity of the electrically-based, twisted-pair, local loop from the central office (CO) to your premise...and back. See also Loopback and NIU.

network interworking A Frame Relay/ATM term. Network internetworking is a method of connecting two frame relay devices over an ATM backbone network. This approach leaves the entire frame relay frame intact, including header, payload and trailer; the ATM devices act to set up the connection on a cut-through basis through the use of a tunneling protocol. Network internetworking is defined in the FRF.5 specification from the Frame Relay Forum, and is recognized by the ATM Forum. Contrast with Service Interworking.

network inward dialing NID. A service feature of an automatically switched telephone network that allows a calling user to dial directly to an extension number at the called user facility without operator intervention.

Network Intrusion Protection NIP. See NIP.

network job entry An application layer protocol developed by IBM for transmitting jobs, data sets, operator commands and messages, and job accounting information from one computing system to another. Network Job Entry (NJE) enables work and data to be transferred throughout a network of computers for further processing, or for delivery to a user or network-attached device. NJE is an extension of remote job entry (RJE), which was also developed by IBM. NJE has been around for years and supported the largest computer networks in existence prior to the widespread use of the Internet. BITNET is an example of an NJE network.

network layer Third layer of the OSI model of data communications, sometimes called the packet layer. Involves routing data messages through the network on alternate routes. See OSI Standards.

network layer address NLA. An address appended to the LAN packet that, unlike a MAC name, indicates exactly where a computing device is located within an internetwork. TCP/IP, DECNet and IPX support network layer addressing and each has its own unique NLA format. Protocol dependent routers use the NLA to make routing decisions.

network layer protocol identifier NLPI. An identifier allowing entities providing different Network Layer protocols to be distinguished from each other.

network location In a URL, the unique name that identifies an Internet server. A network location has two or more parts, separated by periods, as in my.network.location. Also called host name and Internet address.

network logon Logging onto a machine via the network. Typically, a user first interactively logs onto a local machine, then provides logon credentials to another machine on the network, such as a server, that they are authorized to use.

network management Procedures, software, equipment and techniques designed to keep a network operating near maximum efficiency. Several standards bodies have established frameworks for network management. The ISO (International Organization for Standardization), for example, has provided a framework known as FCAPS (Fault, Configuration, Accounting, Performance, and Security). FCAPS and other network management frameworks generally comprise five broad management areas:

1. Fault Management addresses the identification, isolation and correction of network faults (i.e., hard, or catastrophic, failures. Once the corrective action has been taken, Fault Management supports the testing and acceptance of the action, and the creation and maintenance of a database containing the fault history of a given network, subnetwork, or network element. Such tools allow the network management staff to discover failure conditions in real-time or near real-time, provide them with the necessary information to correct the failure, and to work around the failure during the restoration process. Fault location and management tools have strong error and alarm characteristics.

2. Configuration Management deals with installing, initializing, "boot" loading, modifying, and tracking the configuration parameters of network hardware and software. Configuration Management supports the identification and tracking of both physical and logical elements of a network, which process is central to the ongoing management of any network.

3. Accounting Management applications take a number of different forms, including traffic and usage analysis, capital and operations costs, and cost allocation. Accounting Management addresses all physical and logical network resources, including hardware, circuits, and software.

4. Performance Management Tools support the monitoring and improvement of the performance of networks, sub-networks, and the network elements that comprise them. Closely related to Fault Management, Performance Management deals with performance issues, rather than hard failure conditions. Through regular and intensive network monitoring, Performance Management develops real-time or near real-time information of the performance of the network, and develops and maintains historical databases that are used for constant trending and analysis of overall network operation. Such tools show, for example, the number of packets being transmitted over a given circuit at any given moment, the level of bandwidth being consumed, the number of users logged into a specific server and application over specific ports, the level of error performance over the circuit, and the levels of congestion being experienced. Corrective techniques also may be suggested to the network management personnel.

5. Security Management Tools allow the network manager to manage access to network resources, maintain confidentiality, ensure data integrity, and provide auditability of usage. Password protection typically is supplemented with authentication and encryption mechanisms, and firewalls commonly are employed.

network management control center NMCC. A central place from which the network is maintained and changed and from where statistical information is collected.

Network Management Forum NMF. Now known as the TeleManagement Forum (TMF). An international consortium of service providers and suppliers working toward the development of "practical solutions for cost-effective integration of support systems for improved management of services and networks...through a common, service-based approach." NMF is not a standards organization. Rather, it picks up where the standards bodies leave off through highly-focused activities, oriented toward standards implementation. www.nmf.org. See Network Management and TeleManagement Forum.

network management software The software that manages and controls all network functions within a network. Read the definition above for network management and you get a good idea what network management software is meant to do. Writing about its network management software, Neon said, "Network administrators and IT managers must make sure their networks operate smoothly 24/7. They also must document software licensing to reduce audit risks." Neon's product LANsurveyor (which won an Editor's Choice by Network Computing Magazine) helps administrators solve these problems:

Automatically build complete network maps to visualize physical network.

- Continuously scan your network to monitor for intruders.
- Monitor network systems to pinpoint the source of network problems and identify potential issues before they become problems.
- Remotely identify and solve network and desktop problems.
- Automatically generate software meter and software installation reports to help manage software licensing and protect the organization from piracy.
- Systematically quantify important data assets so administrators can plan, implement, and monitor a disaster recovery strategy before disaster strikes.

network management station NMS. A network management station is a dedicated workstation that gathers and stores network performance data, obtaining that data from network nodes (computers) running network agent software that enables them to collect the data. The NMS runs network management software that enables it to compile the information and perform network management functions. See NMS.

network management system A comprehensive system of equipment used in monitoring, controlling and managing a data communications network. Usually consists of testing devices, CRT displays and printers, patch panels and circuitry for diagnostics and reconfiguration of channels, generally housed together in an operator console unit. See NMS.

network mask A filter that selectively includes or excludes certain values. For example, when defining a database field, it is possible to assign a mask that indicates what sort of value the field should hold. Values that do not conform to the mask cannot be entered.

network monitor agent A service that can be installed on an NT workstation, NT server or Windows 95 that lets that computer collect statistics about its network performance, such as the number of packets sent from and received. It gathers the information through its NIC about NetBEUI and NWLink (IPX/SPX) traffic that passes through it.

network name automatic routing A system for getting network-delivered faxes delivered to individual users. The fax network assigns each user a name (usually the user's network login name), and the sender uses this name. The receiving fax system detects this name and routes the fax to the intended recipient.

network news transfer protocol The protocol used to distribute network news messages to NNTP servers and to NNTP clients (news readers) on the Internet. NNTP provides for the distribution, inquiry, retrieval, and posting of news articles by using a reliable stream-based transmission of news on the Internet. NNTP is designed so that news articles are stored on a server in a central database, thus users can select specific items to read. Indexing, cross-referencing, and expiration of aged messages are also provided. Defined in RFC 977.

network neutrality The proposed principle that a network must nondiscriminatorily deliver packets, with no awareness of what specific application, device, or end-user generated them. The issue of network neutrality regulation has become a bone of contention between pipe-owners (i.e., bandwidth providers), on the one hand, and application service providers and content providers, on the other hand, whose services and/or content consume bandwidth and may be in competition with the pipe-owners, as in the case of VoIP. Proponents of network neutrality regulation want the Federal Communications Commission to forbid pipe-owners from discriminating against 3rd-party services, applications, and content.

network node See Node.

network number The part of an Internet address that designates the network to which the addressed node belongs.

Network Numbering Exchange See NNX Codes.

network operating system NOS. See NOS.

network operations Functions that provide, maintain and administer services supported by the network systems. These functions reside in network systems or network operations applications that interface directly to network systems and include memory administration, surveillance, testing, network traffic management and network data collection. Definition from Bellcore in reference to its concept of the Advanced Intelligent Network.

network operations center A name for the central place which monitors the status of a corporate network and sends out instructions to repair bits and pieces of the network when they break. In more formal terms, monitoring of network status, supervision and coordination of network maintenance, accumulation of accounting and usage data and user support.

network outward dialing NOD. A service feature of an automatically switched telephone network that allows a calling user to dial directly all network user numbers without operator intervention.

network penetration testing Network penetration testing uses tools and techniques to check how vulnerable your network is to being hacked, to being broken into, to being compromised. Network penetration testing helps refine a company's security policies, identify vulnerabilities and ensures that the security implementation (what the company has in place) actually provides the protection the company needs.

network printer A printer shared by multiple computers over a network. See also local printer.

Network Problem Determination Application NPDA. A host resident IBM program product that aids a network operator in interactively identifying network problems from a central point.

network processor A network processor is a specialized piece of silicon

Network Professional Association NPA. A self-regulating, not-for-profit organization with the mission of advancing the network computing profession by educating and providing resources for its members. Membership is approximately 7,000 in 100 chapters, worldwide. NPA runs a program called CNP, which stands for Certified Network Professional. That program is designed for individuals whose career is to design, integrate, manage and maintain networked computing environments. www.npa.org.

network protection A term used to describe an array of strategies to protect your network from crashing around your ears should a disaster happen. The ultimate network protection means duplicating every item in the network – including the people who operate it. Obviously anything else (which is what we are all forced to do) is a compromise. The typical network protection these days tends to focus on alternate routing and duplication of network lines, including local loops and long distance lines.

network protection device NPD. A device which provides isolation between PBX circuits and CO trunks or tie lines.

network protocol data unit NPDU. Network Layer Protocol Data Unit. The NPDU comprises the Network Layer Service Data Unit and the Network Layer protocol Control Information.

network redlining If a telecommunications carrier deploys network services in certain areas and discriminatorily does not deploy them in other areas, basing its decisions on the low average household income or other demographics of the two areas, this is known as "network red-lining."

network redundancy Network redundancy means that the network topology has been constructed such that a failure of a network component can be automatically and/or rapidly recovered by using identical components engineered into the network for recovery purposes. For example, IXC network redundancy may take the form of spare network capacity on geographically diverse IXC circuits so that applications can be recovered to an alternative network path in the event of a failure on the primary path.

network relay A device which allows interconnection of dissimilar networks.

network reliability council A committee comprising senior-level officials from a cross-section of telecommunications service providers and user organizations. Organized under the provisions of the Federal Advisory Committee Act, the council provides recommendations to both the FCC and the telecommunications industry to enhance the reliability of the nations telecom networks.

network resource A facility or device that supports call processing.

network security There is no perfect way of defending a network against attacks. Network security is a way of designing your network and properly implementing hardware and software security measures. To defend against the next breed of network attacks and the changing nature of enterprise networks, the protection paradigm must change from securing one or two connections to a pervasive network wide strategy where security functions are broken down into constituent parts or layers. See also Intrusion Detection.

network security administrator A users who manages network and information security. A network security administrator should implement a security plan that addresses network security threats.

network server A powerful computer on a LAN designed to serve the needs of all the people on the LAN with everything from database, email, fax, images, voice messages, communications, etc. See also Email Server and Fax Server.

network services In IBM's SNA, the services within network addressable units (NAUs) that control network operations via sessions to and from the SSCP.

network slip In a digital circuit, such as a T-1, if you lose timing (because the clocks are not operating correctly) you start to lose information. The smallest loss of information is a network slip. A slip is defined as a one frame (193 bits) shift in time difference between the two signals in question. This time difference is equal to 125 microseconds. What's the penalty to a network slip? Does the connected equipment stop working? Not usually. Voice equipment tends to re-acquire frame synchronization quickly, resulting in a pop or click, which is not usually a problem. Data circuits lose some number of bits depending on the data rate being transmitted, and on whether or not forward error correction is being used. Some multiplex equipment that provides add and drop services interrupt all output trunks while a new source of synchronization is acquired. Such interruptions, if due to circuit noise, may render a network temporarily useless, as the slip causes further slips downstream (called error or slip multiplication). A decent clock system provides a stable frequency source during circuit impairments. www.laruscorp.com/tchap03.htm.

network supervisor A network supervisor is responsible for monitoring network-wide call activity.

network synchronization Within a data transmission network, especially one using multiplexers, the need for network synchronization is two-fold:

1. The network must transmit data to the receiving DTE device at the same rate it is being received from the transmitting DTE; and

2. The data originating from the low-speed channels must be capable of being inserted into the composite link (and vice versa) without the loss of any information (channel signaling rate too low) or the creation of unwanted information (channel signaling too high). The need for network synchronization is necessary with networks with time division multiplexing and synchronous data. It does not arise with time division multiplexing in which only asynchronous data is transmitted, because asynchronous data contains "start" and "stop" bits and therefore doesn't need the synchronization of a single master clock. When time division multiplexers are to multiplex synchronous channels, it is essential that the composite links and the low-speed channels are strictly synchronized, i.e. that their clocks operate at precise multiples of the basic clock rate. A single clock, therefore, must be in overall control of the whole network. See Synchronous.

network synchronization unit See Network Synchronization and NSU.

network systems A telephone company AIN term. Processor controlled entities of the telecommunications network that provide ancillary network functions and contain network operations functions. Examples are: Service Control Points (SCPs), Adjuncts, Service Nodes (SNs), and Intelligent Peripherals (IPs).

Network Terminal Number NTN. The number assigned to a data terminal under the Data Network Identification Code system. In the ITU-T International X.121 format, the sets of digits that comprise the complete address of the data terminal end point. For an NTN that is not part of a national integrated numbering format, the NTN is the 10 digits of the ITU-T X.25 14-digit address that follow the Data Network Identification Code (DNIC). When part of a national integrated numbering format, the NTN is the 11 digits of the ITU-T X.25 14-digit address that follow the DNIC.

network terminal option NTO. An IBM program product that enables an SNA network to accommodate a select group of non-SNA asynchronous and bisynch devices via the NCP-driven communications controller.

Network Terminating Equipment NTE. See NTE and Network Terminating Interface.

Network Terminating Interface 1. NTI. The point where the network service provider's responsibilities for service begin or end.

2. NTI. The interface between DCE and its connected DTE.

Network Terminating Unit NTU. The part of the network equipment which connects directly to the data terminal equipment.

Network Termination See NT.

Network Termination Type 1 NT1 or NT-1. An ISDN term. The NT1 is a device which provides the interface between an ISDN local loop and the customer premises. The NT1 provides the functions of physical and electrical connection to the ISDN local loop, thereby operating at Layer 1 of the OSI Reference Model. In a BRI implementation, the NT1 is in the form of a special NIU (Network Interface Unit). Many TAs (Terminal Adapters) also perform NT1 functions; some ISDN-compatible terminal devices (e.g., ISDN phones and PCs) also include NT1s. The basic NT1 functions are:

Line transmission termination. The ISDN local loop is a multichannel, full duplex loop which may be presented physically in the form of either a two-wire or four-wire connection. In a BRI implementation, this function typically is provided by the NIU; in a PRI implementation, by the DSU (Data Service Unit), which generally is embedded in the intelligent switching or concentrating device.

Line maintenance functions and performance monitoring. Through the D Channel, the ISDN local loop is supervised and monitored from the central office to the user premises in order to ensure that it performs at a satisfactory level. Loopback testing is included in this functional grouping. In a BRI implementation, this function typically is provided by the NIU; in a PRI implementation, by the DSU (Data Service Unit).

Multiplexing. This function allows multiple devices to share access to the multichannel digital local loop. In a BRI implementation, this function typically is provided by the TA; in a PRI implementation, by the PBX, ACD, or other CPE switching device.

Interface termination, including multi drop termination and contention resolution. At the physical level, the devices which access to an ISDN loop may connect through 2, 4 or more wires; the ISDN loop is in the form of a physical circuit of either two or four wires. Physical interconnectivity is resolved through the NT1 functional grouping, generally in the form of a TA for a BRI implementation; a PBX, ACD, router, or other intelligent switching or concentrating device performs this function in a PRI implementation. Contention resolution is the process of resolving access issues as multiple physical devices contend for a limited number of ISDN logical channels. Contention resolution generally is performed by a TA in a BRI implementation; a PBX, ACD, router, or other intelligent switching or concentrating device performs this function in a PRI implementation.

See also ISDN and Network Termination Type 2.

Network Termination Type 2 NT2 or NT-2. An ISDN term for a functional grouping embodied in an intelligent switching device such as a PBX, ACD, router or concentrator. Depending on the requirements of the specific device, NT2 functions fall into Layer 1 (Physical Layer), Layer 2 (Data Link Layer) and Layer 3 (Network Layer) of the OSI Reference Model. A full description of Layer 1 functions is described immediately above in the definition of Network Termination type 1. NT2 functions, again depending on the requirements of the specific device, also include the following:

• Protocol handling at Layers 2 and 3.
• Multiplexing at Layers 2 and 3.
• Switching at Layer 2.
• Concentration at Layer 2.

See also ISDN and Network Termination type 1.

network time protocol NTP. A protocol built on top of TCP that assures accurate local time keeping with reference to radio and atomic clocks located on the Internet. This protocol is capable of synchronizing distributed clocks within milliseconds over long time periods.

network to node interface A set of ATM Forum-developed specifications for the interface between two ATM nodes in the same network. Two variations are being developed: an interface between nodes in a public network and an interface between nodes in a private network.

network topology The geometric arrangement of links and nodes of a network. The geography of a network. Networks are typically either a star, ring, tree or bus topology, or some hybrid combination.

network traffic The number, size, and frequency of packets transmitted on the network in a given amount of time.

network traffic management NTM. Functionality that maximizes the traffic throughput of the network during times of overload or failure and minimizes the impact of one service on the performance of others. This is done through centralized surveillance of maintenance conditions and traffic, and centralized control of traffic volumes being originated in network systems. Definition from Bellcore (now Telcordia) in reference to its concept of the Advanced Intelligent Network.

network trunks Circuits connecting switching centers.

network vehicle A network vehicle is basically a car equipped with PC-like gadgetry. Such a car, according to the Economist Magazine, "could tap into a regional road system not only to get directions but also to map out a route around rush-hour snags. Drivers and passengers will be able to send and receive email, track the latest sports scores or stock quotes, surf the Web and even play video games. Apparently one of Detroit's largest suppliers of electrical and electronic hardware to the automotive industry, United Technologies Automotive (UTA), has licensed Microsoft's CE operating system. Writing about the deal, the Economist commented, "Any driver familiar with the flakiness of some early releases of Microsoft's products need not fear, however: his car will not crash each time the operating system does. The UTA/Microsoft system will, according to a spokesman, be rigorously tested; and to start with, it will only be put in charge of office-type tastes and certain "comfort and convenience" fripperies such as a garage door open, or more extravagantly, temperature-controlled cup-holders.

network virtual terminal A communications concept wherein a variety of DTEs, with different data rates, protocols, codes and formats, are accommodated in the same network. This is done as a result of network processing, where each device's data is converted into a network standard format and then converted into the format of the receiving device at the destination end.

networked multimedia This definition courtesy Intel: Just as standalone PCs bring you rich audio, video, animation and three dimensional graphics, so networked multimedia brings these same capabilities to PCs connected to networks and the Internet. You could create an electronic letter with an embedded video clip of yourself in the salutation. Include hotlinks to your company's web site for more information on new products. Allow the receiver to have the entire letter read aloud by a real human voice. Hold a virtual conference in shared interactive spaces, with people in different locations virtually meeting and interacting. Videoconferencing and broadcasting over the Internet will become commonplace. With innovations such as real-time video streaming and multicasting, your company could design a live instructional video that covers their problem. If they wanted more help, they might click and initiate an interactive video conference with a customer supper engineer. In the personal sphere, networked multimedia will create many new applications, from real-time multi-player games between people scattered all over the world, to interactive chat sessions complete with voice-enabled and video images of participants, to virtual amusements.

Networked PBXs Two or more telephone systems that communicate to each other through an Hub.

neugents Neural network agents. Neugents are nifty pieces of programming that monitor systems and applications to detect behavior patterns – be they a computer network, an online shopper or a manufacturing process. Once installed on a system, Neugents begin crunching numbers to come up with a basic model of how that system normally behaves. Neugents begin looking for abnormalities to predict when a problem may occur, and then can trigger an action to deal with the abnormality. The longer a neugent has been installed on a system, the smarter it gets. See Neural Network.

neural network A neural network, or ANN (Artificial Neural Network), is designed to take a pattern of data and generalize from it, much as would the human brain, even if the data are "noisy" or incomplete. Thus, if the data are daily temperatures in New York City over two years, the neural network should emerge with a simple undulating curve that describes the way temperature rises in summer and falls in winter. It does this in effect by a sophisticated form of trial and error, or, in the jargon, varying the strengths of connections between individual processing elements (analogous to neurons in the human brain) until the input yields the right output. The two essential features of neural network technology are that it improves its performance on a particular task by trial and error (neural networkers prefer to say that "it learns"), and that it can be a "black box." Neural networks commonly comprise multiple processors residing on multiple host computers, and tackle a problem through an advanced form of parallel processing. The first ANN was Perceptron, invented in 1958 by psychologist Frank Rosenblatt to model how the human brain processes visual data and learns to recognize objects. Thanks to the Economist Magazine for help on this definition. There are two methods for training a neural network. A self-organizing ANN, or Kohonen after its inventor, discovers patterns and relationships in a large set of data. Such an approach often is used when dealing with experimental data. A back-propagation ANN is trained by humans to perform specific tasks. During the training period, neural weightings are reinforced if the output is correct, and diminished if the output is incorrect. This approach is used in cognitive research and for problem-solving applications. ANNs are used in bomb detectors in airports. Large financial institutions use them to improve performance in areas such as bond rating, credit scoring, and detection of likely fraud in credit card transactions. See Neural Network Computer.

neural network agent See Neugent.

neural network computer A very different kind of computer. Neural network computers are built from webs of randomly connected electronic neurons. These machines are designed to be trained, not programmed. In design and in function, they are meant to closely resemble the human brain. As in the brain, the neurons send signals to one another through thousands of adjustable connections, or synapses. As the machine learns, the settings of these controls are automatically turned up or down. And the chaos of connections evolves into a finely tuned machine, one that can read handwritten letters or recognize the spoken words. With a neural network computer, a trainer simply speaks words to the machine, rewarding it if it acts correctly and punishing it when it acts incorrectly. There is hope that neural network computers may perform better than traditional computers in running artificial intelligence software. See Neural Network.

neuromarketing The use of functional MRI (Magnetic Response Imagery) to scan consumers' brains for clues to subconscious responses to advertising.

neuromorphic electronics A class of electronic circuits and systems whose forms are based on pieces of biological nervous systems. The nervous systems of even the simplest of animals efficiently extract and process information in an unconstrained environment in a way that puts to shame the best of our algorithms running on our most advanced computers. The idea underlying the enterprise of neuromorphic systems engineering is that by building electronic systems organized along similar lines to those of nervous systems, we can build artificial systems that perform similar tasks as those done easily by nervous systems with similar levels of efficiency.

NeuStar See LNP.

neutral The ac power system conductor that is intentionally grounded on the supply side of the service disconnect. It is the low potential (white) side of a single phase ac circuit or the low potential fourth wire of a three-phase Wye distribution system. The neutral provides a current return path for ac power currents whereas the safety ground (green) conductor should not, except during fault conditions.

Neutral Central Office See NCO.

neutral ground An intentional ground applied to the neutral conductor or neutral point of a circuit, transformer, machine, apparatus, or system.

neutral hosting company Imagine you're a big convention center. Imagine that you run different events – trade shows, rock concerts, political conferences, etc. At various times, various telecom carriers working for their various clients want to get into your facility in order to film the event, to rent out cell phone usage, to set up Internet access, etc. So you employ a company which exclusively brings in fiber, equipment, antennas, wires up your building, and then rents out telecom facilities to everyone else, then does the billing and remits you (the Convention Center) a share of the action. For now, we call this company a neutral hosting company.

neutral transmission Unipolar transmission. A form of signaling which employs two distinct states, one of which represents the existence of a space as well as the absence of current.

Nevada Bell The Nevada Bell Operating Company (BOC) which, along with Pacific Bell (California), formed Pacific Telesis (PacTel). PacTel was one of the seven original Regional Bell Operating Companies (RBOCs) formed in 1984 by the Modified Final Judgment (MFJ). As PacTel was acquired by SBC Corporation (Southwestern Bell) in 1996, Nevada Bell is now a subsidiary of SBC.

Never-Busy Fax A fax service offered by LECs and IXCs. Here's how it works. You identify the telephone line dedicated to your fax machine, only. If someone tries to send you a fax, but your fax line is busy, the carrier redirects the fax call to it own fax server, which receives the fax and stores it. The fax server then periodically tries to forward the fax to you. After a predetermined number of tries or after a predetermined period of time has elapsed, an exception report is created so that the carrier personnel can call you on your voice line to advise you of the waiting fax.

new A term used in the secondary telecom equipment business. Generally defined as being sold by authorized vendor of the OEM and carrying the OEM's standard warranty. See Unused.

new economy See Internet Time.

New Line ASCII character 10, abbreviated "NL". This character replaced the abbreviation "LF" which meant Line Feed. Newer terminals accept the NL character to mean both CR (Carriage Return) and LF functions for "end of line" sequences. The function of transmitting an "end of line" sequence to a printer or remote computer system is to designate to a computer to print what follows on the next line of a print out. This usually consists of transmitting the characters CR, LF (Carriage Return, followed by Line Feed). On older printing terminals, these two characters would be followed by a number of null characters whose job would be to waste time until the print head got back to the left margin before printing the first letter of the next line on the page. Otherwise, that character would be in the middle of the page, and the second character would start the line. Telex machines usually send the sequence CR, LF, CR as a New Line sequence to give the head enough time to get back.

New Media Old media is magazines and TV, etc. New media is Internet and web "magazines." The term became important when America Online (new media) announced it had agreed to buy Time Warner (old media) in January, 2000.

New York Stock Exchange On May 17, 1792, a group of New York brokers who had been buying and selling securities under an old buttonwood tree met to formalize rules and methods for securities trading. A agreement was composed and signed by the local brokers. These were the humble beginnings of the New York Stock Exchange.

New York Telephone A phone company that once was part of the Bell System. When AT&T was broken up in 1984, it was lumped with New England Telephone to become Nynex. Then in January of 1994, Nynex decided to eliminate the names New York Telephone and New England Telephone and call them both Nynex. A public relations spokesman told me the reason for this name change had something to do with consistency of marketing image. Frankly, I never understood the necessity for the name change, nor its cost – which included painting the 16,700 trucks the two companies owned. Then Nynex got bought by Bell Atlantic, which changed its name to Verizon after it bought GTE. The trucks have been painted many times over the years.

newbie A newcomer to cyberspace; usually applied condescendingly, the way sophomores talk about freshmen. Somebody new to the Net. Sometimes used derogatorily by net.veterans who have forgotten that, they, too, were once newbies who did not know the answer to everything. "Clueless newbie" is always derogatory.

newco When someone thinks of a new idea for a new product, they often call the new company "newco." My friend, David Epstein, creates newcos more often than I change underwear. (I wonder if he'll ever see this definition?)

newsfeed ISPs (Internet Service Providers) get their newsgroups from different newsfeeds, or news sources, by transferring them over the Internet or other networks. See Newsgroup.

newsgroup A USENET newsgroup is a place on the Internet where people can have conversations about a well-defined topic. Physically it is made of the computer files that contain the conversation elements to the discussions currently in progress about the agreed-upon topic. See Newsfeed.

newsgroup distribution software A set of computer programs used to manage the distribution of USENET news groups and articles.

newsgroup feed One remote computer receiving USENET newsgroups from another remote computer.

newsgroup management software A set of computer programs used to manage USENET newsgroups and articles.

newspaper telegraphese Once reporters were charged by the word for sending their stories in. Reporters started joining words together to save money. This new language was known as newspaper telegraphese. The old United Press, which wanted to hold down its costs, used to wire its overseas correspondents, "Downhold expenses," thereby saving the cost of a word. U.P. reporters called their favorite bar "the Downhold club."

newsreader Software for reading and posting articles to newsgroups on the Internet. A newsreader is a program that organizes conversations in a sensible and presentable manner. The news reader allows the person using it to read and/or participate in those conversations.

Newton 1. The name of Apple's personal communicator, or PDA (Personal Digital Assistant). In its day, the Newton was truly revolutionary. Sadly, Apple Computer Inc. announced on February 27, 1998 the demise of the Newton operating system and Newton OS-based products, including the MessagePad 2100 and eMate 300.

2. The author of this dictionary, Harry Newton, has reached a pinnacle of achievement, of sorts. According to Susan, his wife of over 21 years, he has become a sex symbol for women who no longer care. The photo on the back cover (in case you hadn't already figured) is years old. Maybe decades old.

Newton, Isaac English cleric and scientist, 1642 - 1727, discovered the classical laws of motion and gravity. The story with the apple is probably apocryphal.

Newton's Law of Universal Gravitation From Sir Isaac Newton. Two bodies attract each other with equal and opposite forces; The magnitude of this force is proportional to the product of the two masses and is also proportional to the inverse square of the distance between the centers of mass of the two bodies.

NexGen Next Generation. NexGen is used to describe emerging technologies. NexGen switching, for example, can refer to Gigabit Ethernet (GE) in the LAN domain, and Asynchronous Transfer Mode (ATM) or Internet Protocol (IP) in the LAN and WAN domains. NexGen network access can refer to Digital Subscriber Line (DSL) technologies such as ADSL and RADSL. Hopefully, these NexGen technologies will be widely available, and even affordable, before the NexGen of people.

NEXT Near End Crosstalk. A type of crosstalk which occurs when signals transmitted on one pair of wires are fed back into another pair. Since at this point on the link the transmitted signal is at maximum strength and the receive signal has been attenuated, it may be difficult to maintain an acceptable ACR (Attenuation-to-Crosstalk Ratio). NEXT is particularly troublesome when a number of high-speed transmission services (e.g., ADSL, HDSL and T-carrier) are supported within a single copper cable system. Shielded or screened cable

systems are more desirable in addressing this problem than are unshielded varieties. See also ACR and FEXT.

Next Generation Network 1. A next generation network is a fancy (basically meaningless) name for a phone company that, instead of using circuit switching to move phone calls, uses packets and voice over Internet protocol (VoIP) technology. One would hope that a next generation network would also have a zillion new features – like never busies, automatic searching. But alas no. Most "next generation" networks are marketed the way old generation networks are marketed – on lower price and often on lower service. See also Softswitch.

2. A next-generation news portal focusing on technology for businesses, ZCast. tv uniquely uses streaming media as a primary means of content delivery. "Streaming media," a term describing broadcasting on the Internet, brings news stories to life with sound and video and is available on demand. This all-powerful broadcast medium allows businesspeople to view and listen to newsmakers rather than just reading about them. Or they can listen at their desks while performing other tasks.

NF Noise Figure.

NFAS Non-Facility-Associated Signaling. Another term for out-of-band signaling, such as SS7 (Signaling System 7), which is a Common Channel Signaling technique. ISDN is dependent on the use of SS7 for non-intrusive signaling and control. In an ISDN PRI (Primary Rate Interface) application, a PRI delivers 23 B (Bearer) channels of 64 Kbps each for information transfer, and a D (Data) channel of 64 Kbps primarily for signaling and control purposes between the CPE (Customer Premises Equipment) switch (e.g., PBX, ACD, or router) and the LEC (Local Exchange Carrier) CO (Central Office) switch. However, the original ISDN standards specify that as many as five PRIs can be supported with a single D channel, as the signaling and control activities of a B channel aren't excessively bandwidth-intensive. North American NI-2 (National ISDN 2) standards allow as many as 20 PRIs to be supported by a combination of a primary D channel on the first PRI in a trunk group, and a backup D channel on the second PRI in a trunk group. In the cellular domain, NFAS is known as NCAS (Non Callpath Associated Signaling).See also ISDN, NCAS, PRI, and D Channel.

NFPA National Fire Protection Association. That association responsible as the Administrative Sponsor of the National Electrical Code. Also identified as "ANSI Standards Committee CI".

National Fire Protection Association. www.npfa.org. See NEC Requirements and the next definition.

NFPNA 90A Writing in an issue of Cabling Business Magazine, Andrew Bushelman and Xiaomei Fang talked about "The issue with plenum ceiling cabling installations is that the large amounts of cable may impose an unacceptably large fuel load hazard for plenum spaces. Plenums are a special fire safety concern because they connect occupied areas and contain moving air. Plenums have been involved in serious fires in the past, and they can convey smoke and flame, and supply air to fires. Building codes rely on the NEC (National Electric Code) for safety requirements, including those for plenum cable. For plenum spaces, the NEC is based on NFPA 90A, which requires that materials exposed to airflow be non-combustible or limited combustible, and also meet stringent requirements for low smoke generation. Combustibility is measured by the NFPA 259 fuel load test and smoke generation and flame spread by the NFPA 255 Steiner tunnel test. Under an Exception to NFPA 90A adopted in 1975, plenum cables need not meet these requirements. Instead, they must pass the less stringent NFPA 262 Steiner tunnel test for smoke generation and flame spread. There is no requirement for fire load in the Exception. In response to requests from AHJs and end users, the cable industry has developed a new type of plenum cable that meets the full requirements of NFPA 90A for limited combustibility and low smoke generation. Called limited combustible (to be proposed as CMP-50 rated), the cable is listed by Underwriters Laboratories (UL) and Intertek Testing Services/ETL Semko and is available from leading manufacturers. CMP-50 cable can be used today to provide the best available fire performance in the cable industry. It is well suited for densely populated buildings, laboratories and facilities with extensive computer installations and other electronic equipment. A proposal to recognize CMP-50 cable as a new class of plenum cable has been developed by the NEC Code Panel 16 for approval at the NFPA 2001 national meeting. Such recognition would enable AHJs and building owners to specify CMP-50 in high-risk situations. Under this proposal, CMP-50 is viewed as a cabling option for enhanced fire safety, not an across-the-board replacement for CMP (Communications Plenum) cable."

NFS Network File System. One of many distributed-file-system protocols that allow a computer on a network to use the files and peripherals of another networked computer as if they were local. This protocol was developed by Sun Microsystems and adopted by other vendors. Network File System is an industry standard for remote file access across a common network. It allows workstations to share file systems in a multivendor network of machines and operating systems. Includes RFS, RPC, XDR, and YP.

NFT Network File Transfer. Copies files between any two nodes on a network, either interactively or programmatically. Allows user to 1) copy remote files, 2) translate file attributes, and 3) access remote accounts.

ngDLC next generation Digital Loop Carrier. A DLC that supports a hybrid fiber/copper transmission system, and both voice and data. The hybrid transmission system supports embedded UTP (Unshielded Twisted Pair) local loops to the customer premises, and uses optical fiber to the Central Office (CO). This configuration takes advantage of the UTP over short distances within a neighborhood, aggregates large volumes of traffic from multiple premises, and uses high-bandwidth optical fiber between the DLC and the edge of the carrier network. A ngDLC also supports both voice and packet data traffic, in combination with xDSL (generic Digital Subscriber Line) technology. xDSL supports both voice and packet data over embedded UTP local loops. Those UTP loops terminate in the ngDLC, which splits the voice and data channels based on Frequency Division Multiplexing (FDM). The voice conversations are sent over the optical fiber medium in voice-grade channels which terminate in the circuit-switched Central Office (CO) for presentation to the PSTN (Public Switched Telephone Network). The packet data traffic is concentrated in a DSLAM (DSL Access Multiplexer) and sent over the optical fiber transmission system in data-grade channels which terminate in a packet switch or router for presentation to the Internet or other packet data network. See also DLC, DSLAM, FDM, PSTN, UTP, and xDSL.

NGI Next Generation Internet. An initiative of the Clinton administration announced in February 1997, NGI is "a second generation of the Internet so that our leading universities and national laboratories can communicate in speeds 1,000 times faster than today." A large part of NGI is Internet2, a project of the University Corporation for Advanced Internet Development (UCAID). See also Internet2.

NGN Next Generation Network. I think that has something to do with voice delivered by IP and allied data services also delivered by IP on the same IP circuit. Basically, next generation network is another marketing term.

NGNP Non-Geographic Number Portability. Toll-free numbers (e.g., 800, 888, 877 and 866 in the U.S.) are non-geographic. The ability to change service providers while retaining the same telephone number is known as number portability. The ability to change service providers while retaining the same toll-free number is known as NGNP.

NGOSS New Generation Operations Services and Software. A program of the TeleManagement Forum to produce a framework that will help design new generation OSS/BSS solutions, and a repository of documentation, models and code to support these developments. The goal of NGOSS is "to facilitate the rapid development of flexible, low cost of ownership, OSS/BSS solutions to meet the business needs of the Internet enabled economy". See TeleManagement Forum.

NGSO Satellite Non-Geostationary Orbit Satellite. NGSO Satellites do exactly what you'd think: They move. However, they must do so without interfering with GSO (GeoStationary Orbit) Satellites, especially when they come between the crowded geostationary orbital arc and GSO earth stations. They do this by "arc avoidance," that is, shutting down transmissions just before crossing between them. This requires complex programming, and the International Telecommunications Union (ITU) is currently working on developing standards for NGSO. Negotiations at the ITU between GSO and NGSO operators, as to the margin NGSO satellites will give GSO satellites before passing in front of them, are proceeding slowly. See GSO and Geostationary Satellite.

NH National Host.

NHRP Next Hop Resolution Protocol. A protocol suggested by the IETF (Internet Engineering Task Force) as a means of extending the issue of address resolution beyond the borders of individual IP subnets. The central issue is one of address resolution between IP addresses and ATM addresses when one is attempting to send IP packets over an ATM network, i.e. Classical IP over ATM. NHRP makes use of a NHS (Next Hop Server) to advise routers and other network clients where the next hop toward a destination resides. NHRP is a critical protocol for NBMA (NonBroadcast MultiAccess) networks like ATM and X.25. NHRP is likely to be integrated into the Multiprotocol Over ATM (MPOA) architecture. See also Classical IP over ATM and MPOA.

NHS Next Hop Server. See NHRP.

NI 1. National ISDN.

2. Network Interface. Demarcation point between PSTN and CPE.

3. Network Implementation.

NI-1 See National ISDN-1.

NI-2 See National ISDN-2.

NI-3 See National ISDN-3.

Ni-MH Nickel Hydride is a new technology used in batteries for portable devices, such

as cell phones and laptops, etc. Allegedly, these batteries do not have a "memory effect." Today, Ni-MH batteries cost about 30% more for a 20% or so gain in capacity. The average life of Ni-MH batteries is about 500 cycles. See Nickel Metal Hydride.

NIA Network Interface Adapter. An IBM hardware device that with certain software, will allow an SNA device to communicate over a packet switching network.

nibble Four bits. Computer coding schemes usually, but not always, use eight-bit bytes. In other words, they encode a number, a letter or a character into 8-bits. Sometimes, they make use of smaller "bytes", like nibbles. Sometimes, but rarely, they make use of larger bytes. A nibble sometimes is referred to as a "quadbit." See also Byte and Hexadecimal.

nibble interleaving/multiplexing A technique where four bit nibbles (one at a time) from each lower speed input a channel are used to build the higher speed frame output of a multiplexer.

nibble mode The Nibble mode is the most common way to get reverse channel data from a printer or peripheral. This mode is usually combined with the compatibility mode or a proprietary forward channel mode to create a complete bi-directional channel. All of the standard parallel ports provide five lines from the peripheral to the PC to be used for external status indications. Using these lines, a peripheral can send a byte of data (8-bits) by sending two nibbles (4-bits) of information to the PC in two data transfer cycles. Unfortunately, since the nACK line is generally used to provide a peripheral interrupt, the bits used to transfer a nibble are not conveniently packed into the byte defined by the Status register. For this reason, the software must read the status byte and then manipulate the bits in order to get a correct byte. www.fapo.com/nibble.htm.

NIC 1. Network Interface Card. The device that connects a device to a LAN. Usually in the form of a PC expansion board, the NIC executes the code needed by the connected device to share a cable or some other media with other stations. See Network Interface Card for a full explanation.

2. Near Instantaneous Companding. This describes the very fast, essentially real-time, process of quantizing an analog signal into digital symbols..and converting it back into its native form. See Companding.

3. Network Information Center; any organization that's responsible for supplying information about any network.

4. The InterNIC, which plays an important role in overall Internet coordination. See InterNIC.

5. The DDN NIC (Defense Data Network Network Information Center), which provides support for the defense community, much as does the InterNIC for all other Internet user communities.

6. National ISDN Council. A council of carriers and manufacturers which determines the generic guidelines for National ISDN. It also publishes those guidelines, as they are developed. NIC is a voluntary council. Membership includes Ameritech, Bell Atlantic, BellSouth, Cincinnati Bell, Lucent, Nortel and SBC. Administration and project management are the responsibility of Bellcore. See also National ISDN-1.

nickel cadmium battery NiCad: Nickel Cadmium is the most popular and durable type of rechargeable battery. It is quick to charge, lasts about 700 charge and discharge cycles, and works well in extreme temperatures. Unfortunately, NiCads suffer from "memory effect" if they are not completely discharged during each cycle. The memory effect reduces the overall capacity and run time of the battery.

Nickel Metal Hydride (NiMH) batteries do not suffer from memory effect. Compared to a NiCad battery of equal size, NiMH batteries run for 30% longer on each charge. They are also made from non-toxic metals so they are environmentally friendly. The downside to NiMH technology is overall battery life. These batteries last for 400 charge and discharge cycles.

Lithium Ion (LiON) is the latest development in portable battery technology. These batteries do not suffer from memory effect. Compared to a NiMH of equal size, a LiON will deliver twice the run time from each charge. Unfortunately, these batteries are only available for a limited number of models and are expensive. Similar to NiMH technology, LiON batteries have a life expectancy of 400 charge and discharge cycles.

See Nickel Metal Hydride.

nickel metal hydride battery Ni-MH or NiMH. A rechargeable battery, now being used in laptop computers. Most of today's rechargeable batteries are nickel cadmium. These suffer from several problems: 1. They have a "memory." Which means they must be fully discharged once a month or they won't deliver their full potential. 2. They are hard to dispose of. Nickel hydride is a new technology with some benefits: It's easier to dispose of. It can hold at least 1.25 times as much power per unit as a standard nickel cadmium battery. It doesn't have a "memory." (Neither do car or telephone batteries, which are typically lead acid.) But there are some downsides: Nickel hydride batteries lose their charge faster than nickel cadmium. Nickel hydride batteries are ABOUT 30% more expensive than nickel cadmium. The Nickel hydride battery also can not put out power as fast as a nickel cadmium one can, so it probably won't be suitable for power tools and appliances that drain batteries quickly. But it's great for "slow-burning" items, like laptops and cellular phones. See Nickel Cadmuim Battery.

Nicname A LAN protocol specified in RFC-812. It requests information about a specific user or hostname from the Network Information Center (or NIC) name database service.

NID 1. Network Inward Dialing. A service feature of an automatically switched telephone network that allows a calling user to dial directly to an extension number without operator intervention.

2. Network Interface Device. An electronic device that connects the telephone line and the POTS splitter to the local loop.

NIF Network Interface Function.

night answer Incoming calls to a switchboard during evening and weekend hours are automatically rerouted to ring only at designated night answering phones such as the security desk. See the next few definitions.

night answer -- Assigned Night answer going to specific, assigned telephones.

night answer -- Offsite Phones being answered after hours by people and machinery located offsite – someplace else.

night answer -- Universal Anybody and everybody can answer the incoming calls from any phone. In TELECONNECT, if the phone rings after hours, we simply touch "6" on any phone and we can answer that incoming call – whether it's a local, normal DDD or incoming WATS.

night audit This feature provides automatic printout of message registration data for all quest rooms at the front desk console.

night chime An auxiliary ringer, usually wall mounted, used to indicate a ringing trunk during night operations or used as a "phantom" extension for overflow applications.

night console position Provides an alternate attendant position which can be used at night in lieu of the regular console. Usually cheaper than buying a second normal console.

night patch Assigned Night Answer, Fixed Night Service, Programmable Night Connections. Provides arrangements (which are prewired into the system) to route incoming central office calls normally answered at the attendant position, to preselected stations within the PBX system when the attendant is not on duty.

night service Your operator goes home, puts your phone system into "night service." A call now comes in. The night bell rings. You hear it. You go to a phone and hit a code (such as #6) or hit the flashing "Nigh Answer" button. And you answer the call. Not all phone systems need to be placed in Night Service for you to answer an incoming call from any phone. And you can program the night bell to ring all day.

night service automatic switching Should the attendant neglect to place the console in the night answering mode, after a certain period of timed ringing from an incoming central office call, the entire system will automatically jump to the night service.

night service -- Expanded Service Routes calls normally directed to the attendant to preselected station lines within the system when it is arranged for night service. Calls to specific exchange trunks can be arranged to route to specific station lines and can be assigned on a flexible basis. Trunk Answer From Any Station capability is provided for calls which are handled by assigned night stations.

night service -- Fixed Service Provides arrangements to route calls normally directed to the attendant to preselected station lines within the PBX system when regular attendant positions are not in use. In addition, calls to specific trunks can be arranged to ring on specific phones. The receiving phone can then transfer the call.

night station The phone assigned to automatically handle incoming calls after the main switchboard has shut down for the day.

NIH 1. Not Invented Here. The tendency of organizations to reject ideas and inventions which they didn't think of. A major and continuing problem in the telephone industry.

2. National Information Highway. A term coined by Al Gore when he was Vice President of the United States under President Clinton. See also NII.

NII National Information Infrastructure. A clumsy name for the Information SuperHighway. NII is a term that came from a paper which the Gore-Clinton Administration released, called "National Information Infrastructure; Agenda for Action." According to one engineer, NII is "multiple, interconnected interoperable networks."

NIMBY Not In My BackYard. I love cellular telephones, although the coverage and transmission quality isn't always as good as I'd like. More cell sites would help, but NIMBY. The same goes for toxic waste dumps, nuclear reactors, public fills, and so on.

NIMTO Not In My Term of Office.

NiMH Nickel Metal Hydride. See Nickel Metal Hydride.

NIMQ (pronounced "nihm-kyoo") Acronym for "Not in My Queue." Said in response to suggestions to take on more work when you're already overwhelmed. Similar to the more common "It's not my job."

NIMS Network Information Management System.

nine yards 1. There are many theories on where this term comes from. The theory I like the most is that it came from World War II fighter pilots battling in the Pacific. Laid on the ground, before loading, the .50-caliber machine gun ammo belts measured exactly 27 feet. If a pilot fired all his ammunition at a target, it got "the whole nine yards."

2. The Federal Reserve's boardroom table in Washington is nine yards long. I don't know why.

Ninkasi In ancient Egypt, Ninkasi was the goddess of beer.

NIP Network Intrusion Prevention. NIP systems are devices that are meant to protect your network from the bad guys getting and wreaking havoc on your servers and computers. In the old days such devices were called firewalls. But our industry has a tendency to create new terms so that vendors have something new to sell. This view is probably a little cynical. But you get the point. There are two broad types of NIP systems: signature based IPSs (Intrusion Protection Systems), which match packets or flows to known signatures and traffic anomaly IPSs, which learn normal flow behavior for a network and alert you (the IP manager) to statisticaly significant deviant events.

Nipper The name of the RCA dog, once used by RAC as its logo.

NIPRNET The network formerly known as the Non-secure Internet Protocol Router Network, but now known as the Unclassified but Sensitive Internet Protocol Router Network. Created by the Defense Information Systems Agency (DISA), NIPRNET is used for the exchange of unclassified but sensitive information between DoD users and to provide them with access to the Internet.

NIOD Network Inward/Outward Dialing.

NIOSH National Institute of Occupational Safety and Health.

NIS 1. Network Information Service. A way of centralizing user configuration files in a big distributed computing environment. NIS was created by Sun Microsystems. Sun named it Yellow Pages (YP for short) but they had to rename it because yellow pages apparently is a registered trademark of British Telecom. Although officially renamed, the UNIX commands for NIS still start with "yp".

2. Network Imaging Server. A local area network based server largely devoted to storing, retrieving and possibly manipulating images. See also Server.

NISDN-1 See National ISDN-1.

NIST The National Institute of Standards and Technology, which was formed in 1901 as the National Bureau of Standards, is part of the U.S. Department of Commerce. NIST works with industry and government to advance measurement science and to develop standards in support of industry, commerce, scientific institutions, and all branches of government. NIST comprises four major programs, according to their literature. The Advanced Technology Program (ATP) provides cost-shared awards to companies and consortia for project geared toward the development of high-risk, enabling technologies during pre-product phases of research and development. The Manufacturing Extension Partnership (MEP) is a network of extension centers, co-funded by state and local governments, to provide small and medium-sized manufacturers with technical assistance toward improved performance and competitiveness. Laboratory research and services focus on development and delivery of measurement techniques, test methods and standards. NIST manages the Baldrige National Quality Award program, including providing U.S. industry with comprehensive guides to quality improvement. www.nist.gov.

NIU Network Interface Unit. Also known as a NID (Network Interface Device). The NIU serves as the point of demarcation between the local exchange carrier network and the customer premise. As required by the Modified Final Judgment (MFJ), which broke up the Bell System, the NIU currently is positioned outside the main body of the premise, generally on an exterior wall, inside the garage, or just inside the point of cable entrance. NIUs are multi-function devices, including a protector block which serves to protect premise equipment and inside wire from high-voltage surges. The NIU also typically includes electronics which allow the carrier to initiate a loop-back test from the central office for purposes of testing the integrity of the local loop.

NIUF North American ISDN Users' Forum.

NJE See network job entry.

NKT Nederlands Keuringsinstituut voor Telecommunicatieapparatuur (Private laboratory for regulatory testing, The Netherlands).

NL See New Line.

NL-port Port that connects a node to an FC-AL. See Fibre Channel.

NLA Network Layer Address. An address appended to the LAN packet that, unlike a MAC name, indicates exactly where a computing device is located within an internetwork. TCP/IP, DECNet and IPX support network layer addressing and each has its own unique NLA format. Protocol dependent routers use the NLA to make routing decisions.

NLDM Network Logical Data Manager. An IBM term. See also Netview.

NLETS The National Law Enforcement Telecommuncation System.

NLM NetWare Loadable Modules. NLMs are applications and drivers that run in a server on a local area network under Novell's NetWare operating system and can be loaded or unloaded on the fly. The category of applications ranges from complex voice mail/auto attendant programs and electronic mail gateways to simple programs like drivers for PCI cards. In some other networks, such applications could require dedicated PCs. See Netware Loadable Module for a longer explanation.

NLOS Non Line of Sight. See also WiMAX.

NLPI Network Layer Protocol Identifier.

NLS 1. Name Lookup Service. An electronic directory service which is designed to respond to external queries for general information about a large group of users. Once installed, NLS can be accessed via finger or whois.

2. Non-Linear Schrodinger equation. Refers to the propagation of an optical pulse in a dispersive medium (in the presence of Kerr nonlinearity). Note: there should be an umlaut over the "o" in Schrodinger.

NLSP NetWare Link Services Protocol. A link state protocol from Novell to improve performance of IPX traffic in large internetworks.

nm 1. nm. Nanometer. A billionth of a meter. Often used as a measure of frequency in a fiber optic link.

2. Network Management. A business process in a telecommunications organization that constantly monitors the state of the network and invokes changes to the network when outages or problems occur. In an ATM network, NM is the body of software in a switching system that provides the ability to manage the PNNI protocol. NM interacts with the PNNI protocol through the MIB.

3. Non-metallic sheathed cable, plastic covered. For dry use, 600xAFC.

NM Forum Network Management Forum. International forum for network management.

NMC 1. Network Management Center. A centralized location at the network management layer used to consolidate input form various network elements to monitor, control, and manage the state of a network in a telecommunications organization.

2 Non-metallic sheathed cable, plastic covered. For wet or dry use 600xAFC. ceiling.

NMCCC Network Management Command and Control Center.

NMCI See Navy Marine Corps Intranet.

NMCM Network Management Communications Manager.

NMCN Network Management Communications Network.

NMEA National Marine Electronics Association, US industrial organization whose activities include stipulating standards for marine navigation systems. www.nmea.org

NMF Network Management Forum. An international forum of service providers and suppliers developing network management solutions. Now known as the TeleManagement Forum. See Network Management Forum and TeleManagement Forum. www.nmf.org.

NMI No Middle Initial.

NML Network Management Layer. The layer in the Telecommunications Management Network (TMN) standard addressing network functions including monitoring, management, and control. The network management layer is an abstraction of the functions provided by systems which manage network elements on a collective basis, so as to monitor and control the network end-to-end.

NMOS N-channel Metal Oxide Semiconductor.

NMP Network Management Protocol. An AT&T-developed set of protocols designed to exchange information with and control the devices that govern various components of a network, including modems and T-1 multiplexers.

NMS 1. Network Management Station or Network Management System. The system responsible for managing a portion of a network. The NMS talks to network management agents, which reside in the managed nodes, via a network management protocol. The NMS is the entity that implements functions at the Network Management Layer. It may also include Element Management Layer functions. A Network Management System may manage one or more other Network Management Systems.

2. Network Management Support.

NMSE Network Management Systems Engineering.

NMT Nordic Mobile Telephone. One of the earliest 1G cellular network developed jointly in Denmark, Finland, Iceland, Norway and Sweden. Originally operated in the 450 MHz band.

Later the 900 MHz was used as well.

NMVT Network Management Vector Transport. A network management protocol which provides alert problem determination statistics and other network management data within SNA management services.

NN Network Node. In frame relay, a network node is typically the frame relay service port connection and its associated virtual circuits.

NN National Number.

NNC National Control Center.

NNI 1. Network Node Interface. An Asynchronous Transfer Mode (ATM) term. The interface between two public network pieces of equipment (contrast that to UNI, which stands for User Network Interface).

2. Network to Network Interface. A protocol defined by the Frame Relay Forum and the ATM forum to govern how ATM switches establish connection and how ATM signaling requests are routed through an ATM network. Equivalent to a routing protocol in a router environment.

3. Network to Network Interface. There are two types of network interfaces specified by frame relay standards. The first is a user-to-network (UNI) interface and the second is a network-to-network interface (NNI). The NNI describes the connection between two public frame relay services, and includes elements such as bidirectional polling, to assist the network services providers with gaining information on the status of the public networks being interconnected.

NNM National Network Management.

NNMC National Network Management Center.

NNTP Network News Transport Protocol (NNTP), an extension of the TCP/IP protocol suite, is the standard for the exchange of Usenet messages over the Internet. Specified in the IETF's RFC 977, NNTP provides for the distribution, inquiry, retrieval, and posting of news articles by using a reliable stream-based transmission of news on the Internet. NNTP is designed so that news articles are stored on a server in a central database, thus users can select specific items to read. Indexing, cross-referencing, and expiration of aged messages are also provided. See also UUCP.

NNS National Network Support NNS: Network Software Support.

NNSC National Network Surveillance Center.

NNTM National Network Trouble Management.

NNX The first three digits of a North American local telephone number, NNX once upon a time was used to identify the local central office exchange, or CO prefix. This prefix code breaks down as follows: N is a specific digit (i.e., 2- 9) and X is any digit (i.e., 0-9). Originally, only NNX codes were used to identify and number local central offices. Now all subscribers dial 1+ when making a direct distance dialed long distance call. NNX has been changed to NXX, which allows local central offices to have numbers which look like area codes, e.g. 206, 210, etc. This gives us more central office numbers. We needed more central office prefixes for several reasons. First, CO prefixes are unique within each area code, or NPA, and we assign both NPAs and CO prefixes on a very wasteful basis. Second, the popularity of fax machines, pagers and cell phones has resulted in demand for millions of telephone numbers. See Local Number Portability, NPA and NXX.

NNX Codes The 3-digit code used historically for local telephone exchange prefixes. In the electromechanical era, limitations applied to the range of numbers usable in each position of the prefix. The current numbering plan allows for more variation in assigning Exchange Codes, and under it exchange codes are commonly referred to as "NNXs." See NNX for a full explanation.

NO 1. No is the hardest word in the English language to say. Saying it can win you long-term friends, though it may cost you something short-term. Mostly, it's better to say No politely and not give a reason for saying No. If you do give a reason, you might find yourself in an argument about the "logic" or otherwise of your reason. And that's usually a no-win argument.

2. Network Object.

3. Number.

no answer transfer A service provided by a cellular carrier that automatically transfers an incoming cellular call to another phone number if the cellular subscriber is unable to answer. Most no-answer transfer systems will automatically transfer an incoming cellular call to another phone number if the cellular subscriber is unable to answer (it's not turned on) or if it's not answered after the third or fourth ring if the phone is turned on.

no attendant option CBX systems with Direct Inward Dialing may be designed (configured) without an attendant console.

no bill phone A name for a cellular phone from which you can make free phone calls – local or long distance. The name "no bill" was given by crooks, largely in Southern

California in the Spring of 1992, to phones they had modified to emulate other legitimate cellular phones. Thus all calls made by "no bill" phones are billed to legitimate users, who happen to have other cellular phones with identical codes and now have huge cellular bills. Sorry about that.

no busy test A circuit used to connect to a busy subscriber's line number.

no hold conference/transfer In the event that a call to 911 is being transferred to a secondary or what they call a downstream PSAP (Public Service Answering Position), it is important that the caller never be left in disconcerting silence. After all it is an emergency call. "No hold" features allow the conference or transfer to be done while the PSAP 911 agent is in full uninterrupted communication with the caller.

no incoming calls A cell phone carrier restriction that prevents incoming calls to the assigned cellular number. Only outgoing calls are permitted. This is an optional feature that some cellular phone users subscribe to because it saves them money, In North America, calls coming into a cell phone typically cost the user a per minute charge that is equal to the cost of an outgoing call. This not the case in most GSM cell phone systems in Europe, Japan and Australia, etc.

no line preference Requires the user to manually select (i.e. punch down) a line for each call.

no op Instruction that does nothing. It is used to hold the place for future insertion of a machine instruction.

no worries Australian for no problem. "Will my phone be fixed by 5 PM?" "Yes, no worries."

NOAA National Oceanic and Atmospheric Administration. A US government agency which runs satellites used, inter alia, to track wildlife movements. Two satellites, called NOAA-11 and NOAA-12, orbit the earth via the poles every 100 minutes. As the Earth revolves beneath them, they are able to scan every point on its surface. The satellites work by listening for the "Doppler Shift" in the signal being transmitted by a collar round an animal's neck. The shift is a change in the perceived frequency of the radio signal – similar to what a pedestrian hears happen to the pitch of a police siren as the police car speeds by. Using this technique, you can track animals to with an accuracy of a little more than half a mile, or one kilometer.

NOBIS Network Operations Business Information System.

NOC Network Operations Center, a group which is responsible for the day-to-day care and feeding of a network. Each service provider usually has a separate NOC, so you need to know which one to call when you have problems. Also called NCC for Network Control Center. See also NOCC.

NOCC Network Operations Control Center. See also NOC.

NOD Network Outward Dialing. A service feature of an automatically switched telephone network that allows a calling user to dial directly all user numbers on the network without operator intervention. See also Direct Outward Dialing.

nodal architectures Also called Hub architectures. Nodal network architectures means that network traffic from many locations are connected into a single network site for consolidation and aggregation to higher speed circuits. The traffic is then transmitted from this hub site to a central network site, or to other hub sites. Nodal architectures save money, but also increase the risk of a single network failure affecting multiple network locations.

nodal attribute A nodal state parameter that is considered individually to determine whether a given node is acceptable and/or desirable for carrying a given connection.

nodal clock The principal clock or alternate clock located at a particular node that provides the timing reference for all major functions at that node.

nodal constraint A restriction on the use of nodes for path selection for a specific connection.

nodal ID Another name for a MAC address. Every Network Interface Card (NIC) has a MAC address hardcoded into it. That address is unique in all the world to that NIC card. The MAC address effectively identifies the LAN-attached device with which it is associated. A NIC works with the network software and computer operating system to transmit and receive messages on the network. I first saw the use of the term Nodal ID in instructions to instal a PCMCIA network card into a laptop running Windows NT. Those instructions asked you for the card's nodal ID, which happened (though I didn't know it) to be printed on the card itself. I had never run into this problem because Windows for Workgroups, Windows 95 and Windows 98 automatically recognize the MAC address. See Network Interface Card.

nodal metric A nodal parameter that requires the values of the parameter for all nodes along a given path to be combined to determine whether the path is acceptable and/or desirable for carrying a given connection.

nodal state parameter Information that captures an aspect or property of a node.

node 1. A point of connection into a network. In multipoint networks, it means it's a unit that's polled. In LANs, it's a device on the ring. In packet switched networks, it's one of the many packet switches which form the network's backbone.

2. An SCSA term. An independent SCSA unit in a distributed processing SCSA network, consisting of one or more resource and/or network boards, and one or more SCxbus adapter boards. Communication between nodes take place via the SCxbus. From a device programming point of view, a node is simply an addressable system unit which contains boards connected by an SCbus. See S.100.

node address The unique identifier used to describe a specific node. See Node Number.

node number A node number identifies a network board on a local area network. Every station on a network must contain at least one network board. Each network board must have a unique node number to distinguish it from all the other network boards in that network. In a file server with more than one network board, the node number in LAN A is designated for all traffic addressed to that server. Node numbers can be set in a variety of ways, depending on which network board you use: (1) with jumpers or switches on boards such as Arcnet, (2) at the factory for Token-Ring and Ethernet boards, or (3) with software.

node type In IBM's SNA, the classification of a network device based on the protocols it supports and the network addressable units (NAUs) it can contain. Type 1 and Type 2 nodes are peripheral nodes. Type 4 and 5 nodes are subarea nodes.

NoFA NoFA (pronounced as one word, /NO-fuh/). Notice of Field Activity. A notification created by engineers to notify others (users, other engineers, etc...) of scheduled service outages. Used at Broadwing and other fine companies. Contributed by Harvey Madison in customer care planning (whatever that is). This mention, Harvey claims, will get him free beer from his co-workers. He claims it's a legitimate word.

NOI Notice Of Inquiry. The first public notification that the FCC is about to hold a public inquiry into a particular subject.

NOIS Microsoft has coined the term, NOIS, for the group of four Net-centric companies that tend to believe that what happens on the Internet might be so strong as to alleviate Microsoft's hegemony over the PC industry. The four members are Netscape, Oracle, IBM and Sun.

noise 1. Unwanted electrical signals introduced into telephone lines by circuit components or natural disturbances which tend to degrade the performance of the line. Also known as Line Noise.

2. Noise in stock market parlance is market rumors, guesses, and faulty analysis. It contrasts with information, which consists of solid verifiable information. The concept of noise was first introduced by Fischer Black in his 1986 presidential address to the American Finance Association. His speech, called "Noise", contains this wonderful paragraph, "Noise makes financial markets possible, but it also makes them imperfect. If there is no noise trading, there will be very little trading in individual assets. People will hold individual assets, directly or indirectly, but they will rarely trade them.... The whole structure of financial markets depends on relatively liquid markets in the shares of individual firms. Noise trading provides the missing ingredient."

noise canceling Headset manufactures have long sought to reduce the background noise transmitted via headsets. One approach is the use of noise canceling microphones. These microphones consist of two separate microphones, one directed at the headset user's mouth, the other in the opposite direction. The room side element will pick up ambient room noise along with some ambient user sound. The microphone directed at the user will receive the same amount of ambient room noise as the other microphone, but a much greater amplitude of the user's voice. Both signals are then transmitted to the amplifier. At this point, signals common to both microphones are cancelled out. What remains is the extra voice signals received by the user side microphone. This signal is then amplified and transmitted to the party on the receiving end of the call. This approach has one drawback. It demands perfect microphone positioning, because without it, the headset user's voice is cancelled. The technology works well with highly-trained people such as pilots, astronauts, and military personnel, but can be difficult to implement in the office environment where less skilled personnel struggle to properly position sensitive microphones. Headset manufacturers compromised by using noise canceling microphones with more limited capabilities but that were easier to use.

A second approach to noise reduction is the use of voice switching technology. This technique only allows the microphone to transmit when volume reaches a predetermined level. When the headset user is not talking, or is pausing during the conversation, no sound is transmitted. When the headset user speaks at a normal level, the microphone is "live" and will transmit in a normal fashion. This approach also has it drawbacks. When the microphone is "live" it picks up not only the voice of the person using the headset, but any and all background noise. Voice switching helps the headset user hear what is being said more clearly, but does little to help the person to whom they are talking.

As a solution, some headset manufacturers have merged the two technologies. By using a noise canceling microphone and voice switching, they achieve near perfect noise reduction. Each manufacturer offers noise canceling technology on some of their headsets.

Noise canceling is important in a telephone call center. In a large center, as room noise rises, agents speak louder. For those employees, noise is more than just an inconvenience, or a black spot on a professional image, it directly affects productivity. When conversations must be repeated, call durations increase. Multiply this by enough calls, and staffing and equipment must also be increased. The above information from headset distributor, CommuniTech.

noise equivalent power At a given data-signaling rate or modulation frequency, operating wavelength, and effective noise bandwidth, the radiant power that produces a signal-to-noise ratio of unity at the output of a given optical detector. Information Gatekeepers defines NEP as a measurement in fiber optics that at a given modulation frequency, wavelength, and for a given effective noise bandwidth, the radiant power that produces a signal-to-noise ratio of 1 at the output of a given detector. Some manufacturers and authors, according to Information Gatekeepers, define NEP as the minimum detectable power per root unit bandwidth; when defined in this way, NEP has the units of watts/(hertz) 1/2. Therefore, the term is a misnomer, because the units of power per watts. Some manufacturers define NEP as the radiant power that produces a signal-to-dark current noise ratio of unity. This is misleading when dark-current noise does not dominate, as is often true in fiber systems.

Noise Figure NF. The ratio (in dB) between the signal-to-noise ratio applied to the input of the microwave component and the signal-to-noise ratio measured at its output. It is an indication of the amount of noise added to a signal by the component during normal operation. Lower noise figures mean less degradation and better performance.

noise floor The lowest input signal power level which will produce a detectable output signal from a microwave component, determined by the thermal noise generated within the microwave component itself. The noise floor limits the ultimate sensitivity to the weak signals of the microwave system, since any signal below the noise floor will result in an output signal with a signal-to-noise ratio of less than one and will be more difficult to recover.

noise generator Let's say you're designing wireless local area networking transmitter/receivers. In your lab everything is peaceful and tranquil. But the real world is different. There's the guy next door with his vacuum cleaner and the elevator in his building is old, creaky and noisy. And there's a bunch of cordless phones in the neighborhood. So now what do you do in your pristine perfect lab? You get yourself a noise generator. Basically this gadget produces what's called white Gaussian noise. For example, a noise generator to test wireless LAN receivers might generate white Gaussian noise from 2.0 through 6.0 GHz. The output might be controllable in 1 dB steps up to -15 dBm. You might also want features on this gadget, like high peak to average noise ratio, and some software for programmable routines such as automatic steps in noise or signal as a function of time.

noise measurement units A series of terms used to express circuit noise. These units include: -dB RN – Decibel rated noise -dBrnC – C message weighting refers to the noise measured at 1000 Hz.

noise suppressor Filtering or digital signal-processing circuitry in a receiver or transmitter that automatically eliminates or reduces noise.

noise temperature An amplifier noise rating based upon temperature measured in degrees.

noise voltage In optical communication, an rms component of the optical detector electrical output voltage which is incoherent with the signal radiant power.

noise weighting A method of assigning a specific value to the transmission impairment due to the noise encountered by an average user operating a particular telephone. Noise weightings generally in use have been established by regulatory agencies concerned with public telephone service. They are tightened as technology improves.

NOL NOL is shorthand for net operating loss carried forward for income tax purposes and is an indication of how much future earnings are potentially able to sheltered from taxes, i.e. how much you can earn and not pay taxes on.

Nom De Ligne For those of you who don't speak French, particularly Old French, "Nom de Ligne" translates to "Line Name." It's a pseudonym, or "handle" that you use when on-line-in an Internet Chat Room, for instance. You adopt a Nom de Ligne to give yourself anonymity for whatever reasons you find convenient. Nom de ligne is called a "handle" in some circles. Other French terms for pseudonyms include "Nom de Guerre" (war name) and "Nom de Plume" (pen name). I don't have a "Nom de Ligne" or a "Nom de Plume." My "Nom de Guerre" is "Harry The Terrible."

NOM112 Mexican flavor of C7, which is the ITU flavor (i.e. version) of Signaling System 7, also called SS7. See Signaling System 7.

nomadic broadband Nomadic broadband means broadband that can be accessed anywhere within the service area. The CPE/modem is connected to the desktop/laptop via USB or Ethernet port and can be moved to get broadband access anywhere...in your house, office, local cafe', etc.

nomadic computing A fancy name for allowing laptop and home computer users to dail over the public telephone network, from hotel rooms, from client offices, from home or from airports to access information on the corporate LAN.

nomadic node See Mobile IP.

NOMDA National Office Machine Dealers Association. NOMDA merged with LANDA (Local Area Network Dealers Association) and became NOMDA/LANDA. It's now known as BTA (Business Technology Association). See BTA.

nominal Approximate. For example, the nominal signaling rate of T-1 is 1.5 Mbps, and the actual rate is 1.544 Mbps. The nominal signaling rate of T-3 is 45 Mbps, and the actual rate is 44.736 Mbps. In conversational terms we use the nominal rates of 1.5 Mbps and 45 Mbps just because they're easier to say. We assume that the other party in the conversation understands that the nominal rate is not the actual rate. Assuming, of course, can cause big problems. The devil is in the details, as the old saying goes.

nominal bit stuffing rate The approximate rate at which stuffing bits are inserted (or deleted) when both the input and output bit rates are at their nominal values.

nominal linewidth In facsimile systems, the average separation between centers of adjacent scanning or recording lines.

nominal range An RFID definition. The read range at which the tag can be read reliably.

nominal velocity of propagation The approximate speed at which a signal moves through a cable, expressed as a percentage or fraction of the speed of light in a vacuum. Some cable testers use this speed, along with the time it takes for a signal to return to the testing device, to calculate cable lengths.

non-adaptive routing A routing method that cannot adapt to or accommodate changes in a network.

non-analytical positive Coined by the US Anti-Doping Agency as a euphemism for guilty of using performance-enhancing drugs based solely on circumstantial evidence. Usage: "He won't submit to our voluntary drug testing? Then he must be non-analytical positive."

non-blocking A device which can support a full traffic load without experiencing congestion. The term typically is applied to a switch, such as a PBX or an ATM switch. A PBX can be characterized as fully non-blocking if the internal switching matrix is capable of supporting connectivity between all ports, simultaneously. In other words, the total number of transmission paths is equal to the total number of ports. In a purely voice application, this means that every user with a telephone set can be talking at the same time-this is a highly unlikely scenario involving a ridiculously expensive PBX. It is especially expensive if the PBX is designed to provide a non-blocked path between every PBX station set and the wide area network, as trunk cards and trunks both are very expensive-it's one thing to equip a system so that all station users can talk to each other, but quite another to expect that they all would be connected to the outside world at the same time. ACDs, on the other hand, typically are effectively non-blocking, as productivity objectives demand that all agents in a call center be active as much as possible. Non-blocking switch architectures can be very important in the data world, as holding times are very long in comparison to voice calls. ATM backbone switches commonly are virtually non-blocking, as traffic loads are very heavy and as Quality of Service (QoS) requirements can be extreme.

non-blocking switch A switching system where a connection path always exists for attached device. See Non-Blocking.

non-busy season A telephone company definition. The nine months not selected as part of the Busy Season.

non-coherent interference Any unintelligible television interference such as sparklies, heavy snow, lines in the picture, tearing, or herringbone.

non-concur IBM-speak for to withhold approval, as in I non-concur with this proposal.

non-connectable mode A Bluetooth term. A device that does not responds to paging (an attempt to establish a communication link) is said to be in non-connectable mode. The opposite of connectable mode is connectable mode.

non-consecutive hunting Often referred to as "jump hunting." NON-CONSECUTIVE lines, trunks or extensions can be accessed or "searched" by the switching equipment upon dialing the initial number in the hunting group to find a connection to the first non-busy phone. NON-CONSECUTIVE hunting can be used on incoming and outgoing lines. For example, you could order four trunks from your local phone company which are to "hunt" on. Should the first be busy, the call will go to the next. If the next is busy, then it will hunt to the next. TELECONNECT Magazine has "consecutive" numbering in its first four phone numbers – 212-691-8215, 8216, 8217 and 8218. But let's say we only started with one number and then, because of growth, we needed three more. It's possible consecutive numbers might be taken. Therefore to get the hunting feature, the phone company might have assigned us NON-CONSECUTIVE hunting numbers like 691-8220, 8256, 8678. Sometimes the phone company will also assign us "coded" trunks as part of our trunk group. These are trunks which have no dialable number associated with them. It's best to get real numbers so you can test them by calling them individually. You cannot test coded trunks.

noncontainer object An object that cannot logically contain other objects. For example, a file is a noncontainer object.

non-contaminating PVC A polyvinylchloride formulation, which does not produce electrical contamination.

non-critical technical load That part of the technical load not required for synchronous operation.

non-data bit Bit with encoding violating normal format: used for special control purposes.

non-dedicated server A node on which user applications are available while network resource maintenance applications execute in the background.

non-deterministic A term which refers to the inability of being able to predict the performance or delay on a network where collision is possible.

non-deterministic network A network where access to the transmission medium within some specified period, cannot be guaranteed.

non-dialable toll point Six-digit telephone numbers which follow the pattern 88X-NXX, for routing and billing of calls to extremely isolated stations. Such stations are either beyond the reach of the wireline public switched telephone network (PSTN) or are too low (e.g., in a canyon) for satellite communications. Operator intervention is required to reach these remote areas by radio or some other means. Note that each non-dialable toll point consumes a full NNX, often referred to as a Central Office prefix, of 10,000 telephone numbers. This fact adds to the strain on the North American Numbering Plan (NANP).

non-directional antenna An antenna that transmits and receives equally well in all directions, usually on one plane; also called omnidirectional antenna.

non-disclosure agreement See NDA.

non-discoverable mode A Bluetooth term. A device that cannot respond to an inquiry is said to be in non-discoverable mode. The device will not enter the INQUIRY_RE-SPONSE state in this mode. See also discoverable mode. non-pairable mode A device that does not accepts pairing. is said to be in non-pairable mode. The opposite of non-pairing mode is pairable mode.

non-dispersion shifted fiber A type of Single Mode Fiber (SMF) used in fiber optic transmission systems. See NDSF.

non-DIV Non DIVersity.

non-dominant carrier Non-dominant carriers were identified by the Federal Communications Commission as small long distance carriers. These carriers received a special dispensation from the FCC – namely they didn't have to make public the long distance prices they charged.

non-duplication rules Restrictions placed on cable television systems prohibiting them from importing distant programming that is simultaneously available locally.

non-email Stockbrokers are subject to all sorts of rules and regulations by agencies of the Federal Government. Every phone call they make or receive is recorded. Every correspondence they send – from faxes to handwritten notes, from xeroxes to business cards, from birthday cards to requests for golf time– is checked by a compliance officer. Every email they send is monitored and checked. It's sort of like having on-staff your own censor. This can be stifling. As a result, some brokers have subscribed to their own email service, e.g. with AOL, hotmail or MSN. They refer to this email as "non-email." This way they can say things to their clients which their firm would not otherwise want them to – like recommending the sale of securities, which their firm officially likes.

non-erasable A switch where a through traffic path always exists for each attached phone. Generically, a switch or switching environment designed never to experience a busy condition due to call volume. See Non-Blocking, which is a better term.

Non-Geographic Number Portability See NGNP.

non-impact printer Refers to printers that do not strike a hammer to a platen as typewriters do. Usually a heat sensitive paper or laser printing technology is involved.

non-intelligible crosstalk Crosstalk which is not of sufficient level to be un-

derstood by a listener but which is more annoying than other crosstalk because you think it's intelligible. Or think it should be intelligible.

non-interlaced Non-interlaced refers to monitors whose electron gun scans the entire screen without skipping any scan lines. There's a good definition on interlacing and non-interlacing under Interlacing.

non-ionizing emissions Radio waves, infrared rays and visible light rays, none of which can affect an atom's electrical balance.

non-ISDN line Any connection from a CPE to a central office switch that is not served by D-channel signaling.

non-ISDN trunk In ISDN language, a non-ISDN trunk is any trunk not served by either SS7 or D-channel signaling.

non-linear distortion Amplitude distortion of a signal in which the output signal does not have a linear relationship to the input signal.

non-loaded lines Cable pairs or transmission lines with no added inductive loading coils. In short, straight, raw copper pairs. See Load Coils.

non-maskable interrupt An interrupt on a PC that cannot be disabled by another interrupt.

non-persistent In local area networking technology, describes a carrier sense multiple access (CSMA) local area network (LAN) in which the stations involved in a collision do not try to retransmit immediately – even if the network is quiet. Compare with persistent and p-persistent.

non-printing character A character in a transmission code which performs a control function but is not reproduced when the transmission is printed.

non-progressive display See Interlaced GIF.

non-proprietary LAN A Local Area Network that can connect the equipment of many vendors. See Proprietary LAN.

non-published There are various interpretations of what constitutes an "non-published," "unpublished" and "unlisted" phone number in North America. Some phone companies use these words interchangeably. Some don't. In California, Pacific Bell offers unpublished phone service. Your phone number is not listed in the paper phone directories, but is listed with dial up "Directory Assistance." Pacific Bell also has a more expensive service called "Unlisted Service." Here, your phone number is not included in the paper phone directories or given out to callers to Directory Assistance. Verizon (and I presume other phone companies) has a service whereby you can leave a message for the owner of an unlisted number. "Please call me. You've won the lottery." The owner of the unlisted number then has the choice to return the call or not. He doesn't pay to receive this message. Some telephone companies confuse the definitions and some invent new ones. For example, some phone companies use the term "non-published" number. You won't find the number in a phone book or by calling Directory Assistance. Over 25% of many private phone numbers in major metropolitan areas are unlisted, unpublished or non-published – a "service" their subscribers pay extra for. To my simple brain, it's a lot easier to simply publish your name as "Apple Plumpudding." See Unpublished.

non-receipt notification In X.400, a non-receipt notification is a report prepared by a recipient UA (User Agent) or Access Unit (upon request) and sent to the originating UA or Access Unit when a message is deemed unreceivable by a recipient.

non-repudiation A mechanism which prevents a user from denying a legitimate, billable charge. For instance, a user engaged in an Internet-based voice or video conference might later repudiate the charge, alleging that he was not a party to the conference. A non-repudiation mechanism would provide for monitoring of all endpoints (connected users or machines) during the course of the conference in order that any applicable charges might be supported in the event that they are challenged.

nonresident attribute A file attribute whose value is contained in one or more runs, or extents, outside the master file table (MFT) record and separate from the MFT.

non-routable protocols LAN protocols, such as IBM's NetBIOS, LAN Server and SNA, that use names and not Network Layer Addresses to identify devices and therefore supply no routing information. Internetworking devices must find other ways to route traffic in networks that use nonroutable protocols.

non-selective routing In the US emergency services telephone network, the routing of 911 calls to a serving public safety answering point based only on the originating Central Office. Compare to Selective Routing. See E-911 Service.

non-sent paid Utility industry term for calls made as third party billings, reversed charges or with a Calling Card.

non-session A type of synchronous access mode that can exist between an end user and a host computer in an Amdahl packet network. In non-session access, the calling line (PU) is mapped to a specific offering line (LU). The offering line is reserved for that calling line at all times.

Non-Signaling Data Protocol for Billing and Settlements
NSDP-B&S. Part of the EIA/TIA Interim Standard-124 (IS-124) for Data Message Handler (DMH), a method used in the cellular industry for exchanging non-signaling messages between service providers on a near real-time basis. DMH originally was used to facilitate call hand-offs between carriers. It was extended to serve a number of other purposes. NSDP-B&S is the DMH extension for the transfer of billing and settlement information on an electronic basis. It facilitates the exchange of roamer information between systems and countries, regardless of the air interface (e.g., AMPS, TDMA, and GSM), thereby considerably reducing clearing costs for both charges and adjustments.

Non-Signaling Data Protocol for Fraud NSDP-F. Part of the EIA/TIA Interim Standard-124 (IS-124) for Data Message Handler (DMH), a method used in the cellular industry for exchanging non-signaling messages between service providers on a near real-time basis. DMH originally was used to facilitate call hand-offs between carriers. It was extended to serve a number of other purposes. NSDP-F is the DMH extension for the detection and prevention of cellular fraud. As the carriers can communicate access and usage on an electronic basis in near real-time through DMH, it is possible to more readily detect fraud by identifying collision, measuring velocity, and profiling. Collision is the association of multiple call records occurring at overlapping times, and is indicative of multiple fraudulent callers accessing one or more cellular networks at the same time, masquerading as a single legitimate user. Velocity is a measurement of the speed of calling activity; i.e., the number of calls placed during a given period of time. If the velocity is exceptionally high, the implication is that multiple fraudulent callers might be masquerading as a single legitimate user, perhaps across multiple networks. Profiling simply involves the home service provider's maintaining a profile of calling activity over a period of time. If calling activity suddenly fails to match that profile, especially on a roaming basis, the implication is that calling activity may be fraudulent in nature. In all cases, it is the responsibility of the home service provider to detect the fraudulent activity, and to advise all other service providers to deny future calling attempts.

non-simultaneous transmission Half duplex transmission. Transmission in one direction at a time. This mode of transmission may be the result of limitations of the transmission channel or of the transmitting/receiving equipment.

non-subscriber calling See Casual Calling.

non-synchronous communications See Asynchronous Communications.

non-traditional retailing A fancy word for selling your wares via a Web page on the Internet.

non-traffic sensitive plant Telephone company facilities which are unaffected by changes in volume of telephone activity.

non-transparent mode A transmission environment, mainly of bisynch transmission, in which control characters and control-character sequences are recognized through the examination of all transmitted data. Compare with transparent mode. Also called normal mode.

non-trivial Microsoft word for "hard."

non-volatile Memory which is not lost (i.e. that does not "forget") when the power is shut off.

non-volatile random access memory Electronic circuitry that provides back-up operation of CMOS RAM and/or Flash PROM in case of a power failure.

non-volatile (Nonvolatile) storage A storage medium that does not lose its contents when power is removed.

non-wireline Also called the block "A" carrier. The "A" originally stood for "alternate." The FCC, in setting up the licensing and systems in each market, it reserved one for the local wireline telephone company, and opened the second system – the Block A system – to other interested applicants. The distinction between Block A and Block B is meaningful only during the licensing phase at the FCC. Once a system is constructed, it can be sold to anyone. Thus in some markets today, both the A and B systems are owned by a telephone company. One happens to be the local phone company, and the other is a phone company that decided to buy a cellular system outside its home territory. Non-Wireline, or Block A systems, operate on the radio frequencies from 824 to 849 Megahertz.

Non Zero Dispersion Shifted Fiber NZDF. A type of Dispersion Shifted Fiber (DSF) that is used in long haul, high speed fiber optical transmission systems. See DSF for a full explanation.

nonextended network AppleTalk Phase 2 network that supports addressing of up to 253 nodes and only 1 zone.

nonce 1) In information technology, a nonce is a parameter that varies with time. A nonce can be a time stamp, a visit counter on a Web page, or a special marker intended to

limit or prevent the unauthorized replay or reproduction of a file. Because a nonce changes with time, it is easy to tell whether or not an attempt at replay or reproduction of a file is legitimate; the current time can be compared with the nonce. If it does not exceed it or if no nonce exists, then the attempt is authorized. Otherwise, the attempt is not authorized. 2) In general usage, a nonce is a pronounceable string of characters invented and used only in a given context. The origin of the term goes back to the Middle Ages with "the anes," an expression meaning "for the immediate occasion." This evolved to "the nonce."

nonswitched A leased or hardwired connection line permanently installed between two points. Also called leased line or private line.

Noordung, Herman, Pseudonym of Captain Potocnik (1892-1929) of the old Austrian Imperial Army, who in 1929 in Berlin published a small monograph in German, ""Das Problem Der Befahrung Des Weltraums" ("The Problems of Space Travel"), in which he included text and drawings that comprise the first known proposal for communications satellites in geostationary orbit. The depression, the rise of Nazi Germany and World War II kept Noordung's work from being more widely known, though Arthur C. Clarke referenced it in his landmark paper in the October 1945 issue of Wireless World. In 1993, Noordung's book was rescued from oblivion by being reprinted by Turia & Kant of Vienna.

NOP Network Operations Protocol.

NOPS New Order Processing System.

NOR North Royalton.

NOR Gate An OR Gate with a digital inverter at its output.

NORC Network Operators Research Committee.

Norda, Ray

Nordic Mobile Telephone @ 450 MHz (NMT450) The Nordic Mobile Telephone system operating at 450 MHz; introduced in 1981. See the next definition.

Nordic Mobile Telephone @ 900 MHz (NMT900) Essentially an upgrade of the NMT450 to 900 MHz with enhancements and more channels. See previous definition.

normal My loving wife says I have no idea of the meaning of the word normal.

normal distribution (Gaussian) Bell shaped curve, with 90% of the values within 1.645 standard deviations of the mean and 98% of the values within 2.33 standard deviations of the mean. In each case, half of the values are greater than the mean and half are less than the mean. See Mean.

normal mode The AC voltage that exists between the normal current-carrying wires, that is, between neutral and hot or live.

Normal Response Mode NRM. HDLC mode for use on links with one primary and one or more secondaries. Under NRM, a secondary can transmit only after receiving a poll addressed to it by a primary. It may then send a series of responses. But after it sets the F bit in a response, it cannot transmit any more until it receives another poll.

normal tennis Tennis without the emotion, mistakes and frustration. What Patrick Bullot and Harry Newton would prefer to play, but can't.

normalize A call center term. To change an unusual call statistic reported by the ACD so as to reflect what would have been usual for that period of the day or that day of the week. Normalizing is something you do before updating historical patterns so that the historical patterns will not be distorted by the unusual data.

normalized average transfer delay Average transfer delay divided by packet transmission time at the clock rate of the medium.

normalized network throughput Network throughput in packets per second divided by maximum throughput possible at clock rate of medium. Less than one.

normalized offered traffic The average number of attempted packet transmissions per second divided by the average number of packet transmissions/second possible at the clock rate of the medium. May exceed one.

Nortel Nortel is a shorthand version of Northern Telecom's corporate name (which remains Northern Telecom Limited). According to the company's annual report, it is designed to reflect the corporation's heritage, while signaling a new direction, presenting a single face that reinforces Northern Telecom's global presence. The new logo, with its Globemark "O", creates a dynamic visual symbol for a corporation whose business knows no boundaries and whose spirit is one of leadership, innovation, dedication and excellence. In the U.S., Northern Telecom's principal subsidiary is called Northern Telecom Inc. Nortel is now commonly called Nortel Networks since its acquisition of Bay Networks in 1998. The former BNR is now integrated into Nortel Networks. www.nortel.com.

North American area codes North American area codes are a numbering plan (called NANP, North American Numbering Plan) for the public switched telephone network in the United States and its territories, Canada, Bermuda, and many Caribbean nations, including Anguilla, Antigua & Barbuda, Bahamas, Barbados, British Virgin Islands,

North American cellular network

Cayman Islands, Dominica, Dominican Republic, Grenada, Jamaica, Montserrat, St. Kitts and Nevis, St. Lucia, St. Vincent and the Grenadines, Trinidad and Tobago, and Turks & Caicos. NANP numbers are ten digits in length, and they are in the format: NXX-NXX-XXXX, where N is any digit 2-9 and X is any digit 0-9. The first three digits are called the number-ing plan area (NPA) code, often called simply the area code. The second three digits are called the central office code or prefix. The final four digits are called the telephone line number. In May of 2001 there were 344 area codes in North America. For an up-to-date collection, see www.nanpa.com. Here are two lists, one numerically by area code and one

North American Area Codes By Number

201 New Jersey	318 Louisiana	512 Texas	705 Ontario	830 Texas
202 District of Columbia	319 Iowa	513 Ohio	706 Georgia	831 California
203 Connecticut	320 Minnesota	514 Quebec	707 California	832 Texas
204 Manitoba	323 California	515 Iowa	708 Illinois	843 South Carolina
205 Alabama	330 Ohio	516 New York	709 Newfoundland	847 Illinois
206 Washington	334 Alabama	517 Michigan	710 U.S. Government	850 Florida
207 Maine	336 North Carolina	518 New York	711 TRS Access	855 Inbound toll-free calling
208 Idaho	340 US Virgin Islands	519 Ontario	712 Iowa	858 California
209 California	345 Cayman Islands	520 Arizona	713 Texas	860 Connecticut
210 Texas	352 Florida	530 California	714 California	864 South Carolina
211 Social Services	360 Washington	540 Virginia	715 Wisconsin	866 Inbound toll-free calling
212 New York		541 Oregon	716 New York	867 for NorthWest, Yukon &
213 California	401 Rhode Island	559 California	717 Pennsylvania	Nunavut Territories
214 Texas	402 Nebraska	561 Florida	718 New York	868 Trinidad and Tobago
215 Pennsylvania	403 Alberta	562 California	719 Colorado	869 St. Kitts & Nevis
216 Ohio	404 Georgia	570 Pennsylvania	720 Colorado	870 Arkansas
217 Illinois	405 Oklahoma	573 Missouri	724 Pennsylvania	876 Jamaica
218 Minnesota	406 Montana	580 Oklahoma	727 Florida	877 800 Service (also 888)
219 Indiana	407 Florida		732 New Jersey	880 PAID-800 Service
224 Illinois	408 California	600 Canada (Services)	734 Michigan	881 PAID-888 Service
225 Louisiana	409 Texas	601 Mississippi	740 Ohio	882 PAID-877 Service
228 Mississippi	410 Maryland	602 Arizona	757 Virginia	888 800 Service (also 877)
240 Maryland	411 Local Directory Assistance	603 New Hampshire	758 St. Lucia	
242 Bahamas	412 Pennsylvania	604 British Columbia	760 California	900 900 Service
246 Barbados	413 Massachusetts	605 South Dakota	765 Indiana	901 Tennessee
248 Michigan	414 Wisconsin	606 Kentucky	767 Dominica	902 Nova Scotia
250 British Columbia	415 California	607 New York	770 Georgia	903 Texas
252 North Carolina	416 Ontario	608 Wisconsin	773 Illinois	904 Florida
253 Washington	417 Missouri	609 New Jersey	775 Nevada	905 Ontario
254 Texas	418 Quebec	610 Pennsylvania	780 for Northern Alberta, Canada	906 Michigan
256 Alabama	419 Ohio	611 Repair Service	781 Massachusetts	907 Alaska
264 Anguilla	423 Tennessee	612 Minnesota	784 St. Vincent & Grenada	908 New Jersey
267 Pennsylvania	424 California	613 Ontario	785 Kansas	909 California
268 Antigua/Barbuda	425 Washington	614 Ohio	786 Florida	910 North Carolina
270 Kentucky	435 Utah	615 Tennessee	787 Puerto Rico	911 Emergency
281 Texas	440 Ohio	616 Michigan		912 Georgia
284 British Virgin Islands	441 Bermuda	617 Massachusetts	800 Inbound toll-free calling	913 Kansas
	443 Maryland	618 Illinois	801 Utah	914 New York
301 Maryland	450 Quebec	619 California	802 Vermont	915 Texas
302 Delaware	456 Inbound International	626 California	803 South Carolina	916 California
303 Colorado	469 Texas	630 Illinois	804 Virginia	917 New York
304 West Virginia	473 Grenada	649 Turks & Caicos Islands	805 California	918 Oklahoma
305 Florida	484 Pennsylvania	650 California	806 Texas	919 North Carolina
306 Saskatchewan		651 Minnesota	807 Ontario	920 Wisconsin
307 Wyoming	500 Personal Communication	660 Missouri	808 Hawaii	925 California
308 Nebraska	Services	661 California	809 Caribbean Islands	931 Tennessee
309 Illinois	501 Arkansas	664 Montserrat	810 Michigan	935 California
310 California	502 Kentucky	670 CNMI	811 Business Office	937 Ohio
311 Non-Emergency Access	503 Oregon	671 Guam	812 Indiana	940 Texas
312 Illinois	504 Louisiana	672 Antartica	813 Florida	941 Florida
313 Michigan	505 New Mexico	678 Georgia	814 Pennsylvania	949 California
314 Missouri	506 New Brunswick		815 Illinois	954 Florida
315 New York	507 Minnesota	700 IC Services	816 Missouri	956 Texas
316 Kansas	508 Massachusetts	701 North Dakota	817 Texas	970 Colorado
317 Indiana	509 Washington	702 Nevada	818 California	972 Texas
	510 California	703 Virginia	819 Quebec	973 New Jersey
	511 Road Conditions	704 North Carolina	828 North Carolina	978 Massachusetts

alphabetically by area:

North American cellular network A Craig McCaw idea to join together a bunch of cellular phone companies providers who would provide a painless, simple way for someone making or receiving a cellular call in their territory. Previously, traveling cellular users – what the industry calls "roamers" – were forced to pay heavy charges for calling from the territory of a cellular company not theirs. And roamers were, in effect, incommunicado from incoming calls. It was impossible to call someone on a cellular phone if they were not in the own territory and you didn't know where they were. Craig McCaw was the founder of

North American Area Codes By Place

800 800 Service	303 Colorado	225 Louisiana	709 Newfoundland	758 St. Lucia
888 800 Service Expansion	719 Colorado	318 Louisiana	311 Non-Emergency Access	784 St. Vincent & Grenada
877 888 Service Expansion	720 Colorado	504 Louisiana	252 North Carolina	423 Tennessee
900 900 Service	970 Colorado	207 Maine	336 North Carolina	615 Tennessee
205 Alabama	203 Connecticut	204 Manitoba	704 North Carolina	901 Tennessee
256 Alabama	860 Connecticut	240 Maryland	828 North Carolina	931 Tennessee
334 Alabama	302 Delaware	301 Maryland	910 North Carolina	210 Texas
907 Alaska	202 Dist. of Columbia	410 Maryland	919 North Carolina	214 Texas
403 Alberta	767 Dominica	443 Maryland	701 North Dakota	254 Texas
780 Alberta	911 Emergency	413 Massachusetts	902 Nova Scotia	281 Texas
264 Anguilla	305 Florida	508 Massachusetts	216 Ohio	409 Texas
672 Antartica	352 Florida	617 Massachusetts	330 Ohio	469 Texas
268 Antigua/Barbuda	407 Florida	781 Massachusetts	419 Ohio	512 Texas
520 Arizona	561 Florida	978 Massachusetts	440 Ohio	713 Texas
602 Arizona	727 Florida	231 Michigan	513 Ohio	806 Texas
501 Arkansas	786 Florida	248 Michigan	614 Ohio	817 Texas
870 Arkansas	813 Florida	313 Michigan	740 Ohio	830 Texas
242 Bahamas	850 Florida	517 Michigan	937 Ohio	832 Texas
246 Barbados	904 Florida	616 Michigan	405 Oklahoma	903 Texas
441 Bermuda	941 Florida	734 Michigan	580 Oklahoma	915 Texas
250 British Columbia	954 Florida	810 Michigan	918 Oklahoma	940 Texas
604 British Columbia	404 Georgia	906 Michigan	416 Ontario	956 Texas
284 British Virgin Islands	678 Georgia	218 Minnesota	519 Ontario	972 Texas
811 Business Office	706 Georgia	320 Minnesota	613 Ontario	868 Trinidad and Tobago
209 California	770 Georgia	507 Minnesota	647 Geographic Relief Code	711 TRS Access
213 California	912 Georgia	612 Minnesota	705 Ontario	649 Turks & Caicos Islands
310 California	473 Grenada	651 Minnesota	807 Ontario	710 U.S. Government
323 California	671 Guam	228 Mississippi	905 Ontario	340 US Virgin Islands
408 California	808 Hawaii	601 Mississippi	503 Oregon	435 Utah
415 California	700 IC Services	314 Missouri	541 Oregon	801 Utah
424 California	208 Idaho	417 Missouri	880 PAID-800 Service	802 Vermont
510 California	217 Illinois	573 Missouri	882 PAID-877 Service	540 Virginia
530 California	224 Illinois	660 Missouri	881 PAID-888 Service	703 Virginia
559 California	309 Illinois	816 Missouri	215 Pennsylvania	757 Virginia
562 California	312 Illinois	406 Montana	267 Pennsylvania	804 Virginia
619 California	618 Illinois	664 Montserrat	412 Pennsylvania	206 Washington
626 California	630 Illinois	308 Nebraska	484 Pennsylvania	253 Washington
650 California	708 Illinois	402 Nebraska	570 Pennsylvania	360 Washington
661 California	773 Illinois	702 Nevada	610 Pennsylvania	425 Washington
707 California	815 Illinois	775 Nevada	717 Pennsylvania	509 Washington
714 California	847 Illinois	506 New Brunswick	724 Pennsylvania	304 West Virginia
760 California	456 Inbound International	603 New Hampshire	814 Pennsylvania	414 Wisconsin
805 California	219 Indiana	201 New Jersey	500 Personal Communication Services	608 Wisconsin
818 California	317 Indiana	609 New Jersey	787 Puerto Rico	715 Wisconsin
831 California	765 Indiana	732 New Jersey	418 Quebec	920 Wisconsin
858 California	812 Indiana	908 New Jersey	450 Quebec	307 Wyoming
909 California	319 Iowa	973 New Jersey	514 Quebec	867 Yukon & NW Territories.
916 California	515 Iowa	505 New Mexico	819 Quebec	
925 California	712 Iowa	212 New York	611 Repair Service	
935 California	876 Jamaica	315 New York	401 Rhode Island	
949 California	316 Kansas	516 New York	306 Saskatchewan	
600 Canada (Services)	785 Kansas	518 New York	803 South Carolina	
809 Caribbean Islands	913 Kansas	607 New York	843 South Carolina	
345 Cayman Islands	270 Kentucky	716 New York	864 South Carolina	
670 CNMI	502 Kentucky	718 New York	605 South Dakota	
	606 Kentucky	914 New York	869 St. Kitts & Nevis	
	411 Local Directory Assistance	917 New York		

McCaw Communications, a large cellular phone company, now owned by AT&T.

North American Directory Forum NADF. An association of electronic mail providers who are figuring standards and ways of sending mail between their subscribers.

North American ISDN Users Forum NIUF. Here is the complete explanation of this important group, as excerpted from their brochure: The barriers to the widespread use of ISDN nationally and internationally is difficult because the technology is complex and developing rapidly. One underlying problem is the lack of standard implementations of ISDN applications. ISDN standards are currently developed by the International Telecommunications Union (ITU-T) and by accredited standards committee T-1 under the umbrella of the American National Standards Institute (ANSI). But standards designed to meet many requirements offer multiple options which are open to diverse interpretations. As a result, services and products produced by different manufacturers are incompatible.

To solve this problem, NIST (National Institute of Standards and Technology) collaborated with industry in 1988 to establish the North American ISDN Users' Forum (NIUF). NIST's Computer Systems Laboratory (CSL) serves as chair of the forum and hosts the NIUF Secretariat under the terms of a cooperative Research and Development Agreement (CRADA) established with industry in 1991. Through support of the forum, CSL advances new uses of computer and telecommunications technology in government and industry. The objectives of the NIUF are "to provide users the opportunity to influence developing ISDN technology to reflect their needs; to identify ISDN applications, develop implementation requirements, and facilitate their timely, harmonized, and interoperable introduction; and to solicit user, product provider, and service provider participation in the process." www.niuf.nist.gov.

North American Numbering Plan NANP. The method of identifying telephone trunks in the public network of North America, called World Numbering Zone 1 by the ITU-T. The Plan has three ways of identifying phone numbers in North America – a three digit area code, a three digit exchange or central office code and four digit subscriber code. Other countries have much more complicated numbering schemes. There are some countries, for example, where the length of the area code actually exceeds the subscriber code. There are many countries where there is no consistency in the length of phone numbers. Some are nine-digit. Some are 12-digit, etc. All these varying number lengths may be within 100 miles of each other.

Under the NAN P format prior to January 1, 1995, the second digit of the area code was always a one or a zero. With this system, approximately one billion telephone numbers, 152 area code combinations and 640 prefixes were available. Numbers started to run out as a result of increased use of fax, modem and cellular lines. With the new Numbering Plan introduced on January 1, 1995, the area code can be any combination of three digits. This will allow more than 6 billion telephone numbers and 792 area code combinations. See NANC and NANP.

North American Telecommunications Association NATA. Now known as the MultiMedia Telecommunications Association MMTA. The national trade association for companies providing customer premise telephone equipment. NATA represents the interests of the industry before the Congress, the FCC and in court actions. It also does market research. www.MMTA.org.

North Dakota In North Dakota, it is legal to shoot an Indian on horseback, provided you are in a covered wagon.

NOS A network operating system is software that runs a network (as if that wasn't self-evident). It typically runs in a file server and controls access to shared and non-shared files and allows users to communicate with each other and to gain access to shared facilities, like printers, data storage devices, scanners, routers, bridges, and of course. It's also the software that communicates with the LAN hardware (also called network interface card) in each device (PC, printer, file server, gateway, router, etc.) attached to the network. What makes a network operating system (NOS) different from today's computer operating system (OS)? Today, the answer is: very little. Virtually all computers now include a network operating system. However, all operating systems were not always network-aware, and just because an OS was network-aware did not necessarily qualify it as a Network OS. The first true NOS products to emerge included Novell NetWare and Banyan Vines. These network operating systems were engineered to provide network-specific functionality such as shared access to large (and expensive) disk drives, printers, multi-user applications, and provided additional security that most mere operating systems lacked. Such network operating systems were either not intended to support, or were incapable of supporting, end-user applications. Most were/are dedicated to providing network services, hence the term, "dedicated server". By this definition, since Vines is defunct, NetWare is the only remaining NOS. Some may also insist that many versions of Windows, Unix/Linux, as well as the Mac OS, are in fact network operating systems. While all current releases of these operating systems are certainly network-aware, they are by no means dedicated to supporting network services only. Of these, only Microsoft offers a server-specific release of its Windows operating system(s), however, Windows 2000 Server and Advanced Server share the same operating system kernel as the client workstation version, Windows 2000 Professional. The server versions augment this basic OS with additional networking capabilities and server-related applications and utilities. The new Windows XP Home and XP Professional operating systems are also network-aware, however there is currently no server version of Windows XP.

nose The human nose cleans, warms, and humidifies over 500 cubic feet of air every day.

not! An expression of disagreement with a previous statement.

NOT Function See Digital Inverter.

not spot A cellular dead zone.

not statement A "not" operator is one of three fundamental logic operations that underlie all computer calculations. The other two operations are "and" and "or". The "and" operator compares whether two statements are both true. The "or" operator determines if at least one statement is true. In the binary language of 1's and 0's used by computers, a "not" operator converts a "0" (the equivalent of "false") into a "1" ("true"). It also works the other way, changing a "1" into a "0." Computer calculations are all reduced to combinations of "and," "or" and "not" operators.

NoTAR No TelecomAction Required.

notch An out-of-phase impulse causing spontaneous dip in voltage. This is an under-voltage impulse, similar to a spike, but of reverse polarity to the instantaneous value of the AC sine wave. A notch normally is associated with the power company removing a generator from the power grid. See Notched Noise.

notch a frequency Block a frequency. See notch filter.

notch filter A filter that blocks or passes a specific band of frequencies. The filters are: lowpass/block, high pass/block, notch pass/block. If the filter is set in a series with a circuit, the desired frequencies pass down the line. If it's passed to ground, the desired frequencies are sent to ground. They can't get through the circuit.

notch depth In order to avoid causing radio interference in a particular geographical area, a radio system may be required to operate at a reduced power level (measured in decibels) for a certain frequency band. The amount of required power reduction is called the notch depth. See "notch filter" and "notch a frequency."

notched noise Noise in which a narrow band of frequencies has been removed. Normally used for testing devices or circuits.

notebook A laptop computer that weighs approximately five to seven pounds. Sub-notebooks clearly weigh less. See also laptop computer.

notes on the network A famous book explaining how the North American public switched network works. It's now called BOC Notes on the LEC Networks.

NOTHS Network Operations Trouble Handling System.

notwork A network in its nonworking state.

Notice of Inquiry NOI. A Federal Communications Commission term, a Notice of Inquiry (NOI) is adopted by the Commissioners primarily for fact gathering, a way to seek comments from the public or industry on a specific issue. The NOI also states where and when comments may be submitted, where and when you can review comments others have made, and how to respond to those comments. After reviewing comments, the FCC may issue a Notice of Proposed Rulemaking or it may release a Report & Order (R&O) explaining what action – or non-action – is taken.

Nouse A peripheral device that tracks the movement of the tip of your nose to control a cursor.

Novell Novell is the reincarnation of NDSI, a computer firm that almost went under in the early 1980s. The then nearly 60-year-old Ray Noorda, who had 20 years of experience in systems automation with General Electric was called in to help NDSI prepare for a trade show. He spotted it had some potentially interesting technology, and bought the ailing company in 1983. He sold most of his holdings in the 1990s. Novell was one of the earliest pioneers in PC networking. At one stage everyone who had a network of PCs ran Novell's software, called NetWare. Microsoft seriously impacted Novell's business by simply including networking software in its various incarnations of Windows.

Novell Directory Services See NDS.

Novell IPX See IPX.

Novell NLSP Novell NetWare Link Services Protocol. A link state protocol under development by Novell to improve performance of IPX traffic in large internetworks.

Novell SAP Novell Service Advertisement Protocol. See SAP.

Novell Telephony Services See Telephony Services.

Noyce, Robert A coinventor or the computer chip. Noyce helped start Fairchild, and in 1968 left Fairchild to start Intel. See Kilby, Jack. S.

NP See Number Portability.

NP3 Non-published telephone number service.

NPA 1. In IBM language, it means Network Performance Analyzer, a product for network tuning, determining performance, degradation and determining the affect of network growth.

2. Numbering Plan Area. A fancy way of saying Area Codes. There are well over 200 area codes in the United States, Canada, Bermuda, the Caribbean, Northwestern Mexico, Alaska and Hawaii. Within any of these area codes, no two telephone lines may have the same seven digit phone number. The middle number has been either "1" or "0" creating "N 1/0 x" codes. The number of codes available on this basis were nearing depletion. Bellcore (now Telcordia Technologies) then modified the plan to obtain more area codes by changing the area codes numbering scheme to NXX like the central office numbering scheme. Switching systems in the national network differentiate between the central office and area codes by recognizing the subscriber always dials 1+ or 0+ preceding an area code when direct dialing such long distance calls. The responsibility for numbering plan administration (NPA) was shifted in 1997 from Bellcore to Lockheed Martin, as Bellcore no longer was considered to be an impartial party.

Here are the special, unassigned and reserved NPAs:

200 Reserved for special services
211 Assigned to local operators
300 Assigned to special services
311 Reserved for special local services
400 Reserved for special services
500 Reserved for special services
511 Reserved for special local services
600 Reserved for special services
700 Assigned special access code for interLATA carriers/resellers
711 Reserved for special local services

See NNX and North American Area Codes.

NPA Code A unique 3-digit code in the N 0/1X series that identifies and NPA (Numbering Plan Area). An NPA code is the first three digits of the destination code for all inter-NPA toll calls. See NPA.

NPA/NXX The first six digits of a North American telephone number; the area code and exchange.

NPAC Number Portability Administration (also called Access) Center. Regional centers which will be developed to assist in the implementation of Local Number Portability (LNP). The NPAC will interact with the LSOAs (Local Service Order Administration) systems of all carriers in order to effect the transportation of the end user's telephone number from the legacy Local Exchange Carrier (LEC) to the new LEC of choice. LNP effectively decouples customer identification number (Directory Number) from Network Routing Address (Location Routing Number or LRN). The NPACs serve to synchronize the numbering databases, and to coordinate the porting process. As orders are placed to change the serving LEC, the OSSs (Operations Support Systems) of the victorious LEC communicate that request to its Local Service Order Administration (LSOA) system, which passes the request to the NPAC. The NPAC notifies the legacy carrier's LSOA of the request for purposes of confirmation. Once confirmed and ported, the NPAC passes relevant data to to all carrier LSOAs in its region, which, in turn, pass that information to its LSMSs (Local Service Management Systems), which pass the change to the LSOAs, which pass it on to the appropriate OSSs. There are seven NPACs designated in the U.S., with Lockheed Martin serving all seven regions. Originally, Perot Systems served three regions, but it pulled out. Now Lockheed Martin runs all seven. There is a single NPAC in Canada. See also LNP.

NPC 1. Network Processing Card.

2. Network Parameter Control: Network Parameter Control is defined as the set of actions taken by the network to monitor and control traffic from the NNI. Its main purpose is to protect network resources from malicious as well as unintentional misbehavior which can affect the QoS of other already established connections by detecting violations of negotiated parameters and taking appropriate actions. Refer to UPC.

NP&D Network Planning and Design.

NPD 1. See Network Protection Device.

2. Network Performance Display.

NPDA Network Problem Determination Application. A program which allows for the monitoring of an entire network from a single location, collection of statistics, and isolation of communication faults. An IBM term. See Netview.

NPDU Network Protocol Data Unit.

NPI 1. Network Product Implementation.

2. A term used by computer support personnel to calm clueless users, it stands for Not Plugged In, as in "The reason your mouse doesn't work is because of an NPI error."

NPL Non Performing Loan. A euphemism for a loan which has gone bust. The borrower is no longer paying the lender interest or principal. In the late 1990s and early 2000s, Japanese banks had oodles of NPLs on their balance sheets. They typically carried the NPLs at the full theoretical value of the loan – the full amount they'd loaned to a borrower plus the interest he was committed to pay. These loans were carried as "assets" on the banks' balance sheets and made them look financially stronger than they really were. American banks are obliged to write NPLs off their balance sheets.

NPM Network Process Monitor.

NPR DOA is Dead On Arrival. And it's a term several manufacturers use to refer to equipment which arrives at the customer's premises not working. A person who receives a DOA machine will ask the company for a NPR number – New Product Return number. This allows them to return the product and have the factory replace it with another new one.

NPRM Notice of Proposed Rule Making. A term used in regulatory agencies. The agency runs an idea up the flagpole, then typically hold hearings to find out how people react.

NPS Network Product Support.

NPSI IBM X.25 NCP Packet Switching Interface. Networking software package that allows Systems Network Architecture (SNA) 3270 traffic to be transmitted over an X.25 packet data network (PDN). See also QLLC. Contrast with DSP.

NPSP Network Products and Systems Planning.

NPU 1. Listing for the phone number is not in the printed Directory nor is it available from Directory Assistance.

2. Network Processor Unit. Typically a microprocessor that does communications functions.

3. Network Processing Unit. Typically a microprocessor that does communications functions.

NQM Megalink Channel Service.

NQP Megalink Channel Service.

NRAM Nonvolatile Random Access Memory, also called nanotube nonvolatile RAM) remembers everything even though your PC is switched off. With NRAM, your PC could turn on and off immediately, dispensing with all of its tedious booting up and shutting down. NRAM (as in nonvalilate random access memory) does not lose its memory when you turn off the computer or phone system. Many modems, for example, use nonvolatile memory to store configuration information (in place of the switches used on other modems). The command &W writes instructions to the NRAM in many PC modems.

NRC 1. Non-Recurring Charge. 2. Network Reliability Council.

3. The National Research Council of Canada is the main federal scientific and technical research organization. See www.nrc.ca.

NRCE Non-Redundant Common Equipment.

NRDA National Routing and Database Administration.

NREN The National Research and Education Network, an ultragigabit network established by legislation in 1991 by the U.S. House and Senate. Preliminary steps toward deploying this information superhighway (for talking between University computers) include some gigabit network testbeds and the cutover to a 45 megabit per second backbone for the National Science Foundation Network (NSFnet) used by the scientific research and university community nationwide.

NRLLDB See National Repository Line Level Database for an explanation of the world's most difficult-to-remember acronym.

NRM 1. NoRMal response of HDLC. See Normal Response Mode.

2. An ATM term. An ASP service parameter, Nrm is the maximum number of cells a source may send for each forward RM-cell.

NRN No Response Necessary. A Proposed e-mail convention to prevent endless back-and-forth acknowledgments: "Thanks for the info." " You're welcome ...hope it helps." "I hope so too. Thanks." By putting NRN at the bottom of your mail, you absolve the receiver from having to reply.

NS Record Name Server Record. An NS record declares that a given zone is served by a given name server. Every NS record is either a delegation record or an authority Record. If the name of the NS record is the name of the zone it appears in, it is an authority record. If the name of the NS record is that of a descendant zone, then it is a delegation record. BIND, a common Unix DNS server, uses NS records in hint files for telling resolvers where to find root servers.

NRS 1. Network Reconfiguration System.

2. Network Routing System.

3. National Relay Service.

NRT Near Real Time.

NRTC National Rural Telecommunications Cooperative. The organization that provides telecommunications services to rural electric and rural telephone cooperatives in the United States.

NRTL Pronounced "nurtle." Nationally Recognized Test Laboratory. A laboratory that is nationally recognized for the performance of certain tests on equipment or cable systems for compliance with various standards or legal requirements. UL (Underwriters Laboratories) is a NRTL. See also UL.

NRUG National Rolm Users Group.

NRZ Non-Return to Zero. A binary encoding scheme in which ones and zeroes are represented by opposite and alternating high and low voltages and where there is no return to a zero (reference) voltage between encoded bits. NRZ is now used as an encryption scheme for getting data onto and off hard disk fast. It eliminates the need for clock pulses and yields up to 18.5 kilobytes per track and high read/write speeds.

NRZI Non-Return to Zero Inverted. A binary encoding scheme that inverts the signal on a "one" and leaves the signal unchanged for a "zero". Where a change in the voltage signals a "one" bit, and the absence of a change denotes a "zero" bit value. Also called transition coding.

NS-CPE Non-Standard Customer Premise Equipment.

NS/EP Telecommunications National Security and Emergency Preparedness Telecommunications. A Federal government definition. Telecommunications services that are used to maintain a state of readiness or to respond to and manage any event or crisis (local, national, or international) that causes or could cause injury or harm to the population, damage to or loss of property, or degrade or threaten the national security or emergency preparedness of the United States.

NSA 1. Network Service Address. NSAs are unique addresses that define physical or logical locations in equipment, such as a residential telephone number (TN).

2. National Security Agency. An agency of the Federal Government, created in 1952 in a top-secret presidential order issued by Harry Truman. The agency was cloaked in secrecy until the 1982 publication of James Bamford's book, "The Puzzle Palace." NSA is the Federal agency responsible for the design and use of nonmilitary encryption technology, developing sophisticated codes to scramble data, voice or video information. In short, it is charged with signals intelligence and is widely assumed to monitor all communications traffic (phone, fax, data, video, etc.) into and out of the United States with foreign countries. It is barred from intercepting domestic communications. NSA grabbed the headlines in 1993 and 1994 when it adopted its most visible attempt to outgun cybervillains with something called the Clipper Chip. The idea is that the Clipper Chip (a microprocessor) would be installed in every phone, computer, and personal digital assistant in America would carry a device identification number or electronic "key" – a family key and unit key unique to each Clipper Chip. The device key is split into two numbers that, when combined into what's called a Law Enforcement Access Field number, can unscramble the encrypted messages. The device keys and the corresponding device numbers, according to NSA proposals, would be kept by the US government through key escrow agents. Under a plan proposed, the attorney general would deposit the two device keys in huge, separate electronic database vaults. One key would be held by the National Institute for Standards and Technology (NIST) and the other by the Automated Systems Division of the U.S. Treasury. Access to these keys would be limited to government officials with legal authorization to conduct a digital wiretap. When a law enforcement agency wants to tap into information encrypted by the Clipper Chip, they must obtain a court order and then apply to each of the escrow agents. The agents electronically send their key into to an electronic black box operated by the law enforcement agency. When these keys are electronically inserted, encrypted conversations stream into the black box and come as standard voice transmissions or as ASCII characters in the case of electronic mail. At least that's the theory. American Industry resisted the Clipper Chip and NSA backed down, only to start pumping for something Fortezza. See NSA Line Eater.

Since I wrote the above, NSA has sort of come part way of its secret shell. It's now got its own Web site, www.nsa.gov, in which it describes itself thus: The National Security Agency (NSA) was established by Presidential directive in 1952 as a separately organized agency within the Department of Defense under the direction, authority, and control of the Secretary of Defense, who acts as Executive Agent of the U.S. government for the production of communications intelligence (COMINT) information.

The Central Security Service (CSS) (which is part of NSA) was established by Presidential memorandum in 1972 in order to provide a more unified cryptologic organization within the Department of Defense. The Director, NSA, serves as chief of the CSS and exercises control over the signals intelligence activities of the military services. The resources of NSA/CSS are organized for the accomplishment of two national missions:

The information systems security or INFOSEC mission provides leadership, products, and services to protect classified and unclassified national security systems against exploitation through interception, unauthorized access, or related technical intelligence threats. This mission also supports the Director, NSA, in fulfilling his responsibilities as Executive Agent for interagency operations security training.

The foreign signals intelligence or SIGINT mission allows for an effective, unified organization and control of all the foreign signals collection and processing activities of the United States. NSA is authorized to produce SIGINT in accordance with objectives, requirements and priorities established by the Director of Central Intelligence with the advice of the National Foreign Intelligence Board. Executive Order 12333 of 4 December 1981 describes in more detail the responsibilities of the National Security Agency. See Fortezza and Sigint. www.nsa.gov.

NSA Line Eater The more paranoid Internet users believe that the National Security Agency has a super-powerful computer assigned to reading everything posted on the Net. They will jokingly refer to this line eater in their postings. Goes back to the early days of the Internet when the bottom lines of messages would sometimes disappear for no apparent reason.

NSAP Network Service Access Point. The point at which the OSI Network Service is made available to a Transport entity. The NSAPs are identified by OSI Network Addresses. The NSAP is a generic standard for a network address consisting of 20 octets. ATM has specified E.164 for public network addressing and the NSAP address structure for private network addresses.

NSAP-Selector A component of an Network Service Access Point (NSAP)-Address used to select the Network Layer service user. The NSAP-Selector is sometimes referred to as a Transport-Selector; however, a user of the Network Layer need not be a transport service.

NSAPI Netscape Server Application Programming Interface. A Netscape-only Web server application development interface, developed by Netscape Communications Corporation. NSAPI was designed as a more robust and efficient replacement for CGI.

NSC 1. Network Service Center. See Network Control Point and SDN.

2. Non-Standard facilities Command. A response to the called fax DIS response.

NSCP National Scalable Cluster Project. A consortium of universities and corporate partners dedicated to developing distributed computing and deploying high-speed networking through ATM technology. The project is intended to solve the technical hardware and software problems associated with the support of sophisticated multi-processing at widely separated locations. The meta-cluster will operate over an ATM-based WAN; cluster computing within the LAN domain is emerging as a scalable and cost effective to supercomputing and massively parallel computing. Applications are anticipated to include digital libraries and linguistic data, imaging and virtual reality, and data mining. Initial participants include the University of Pennsylvania; the University of Illinois at Chicago; and the University of Maryland, College Park.

NSE 1. Network Support Element.

2. Network Surveillance Engineer.

3. Network Systems Engineer.

NSE/OES Network Support Element/Operator Evaluation System.

NSDETP Network Systems Development Electronic Transaction Processing.

NSDP-B&S Non-Signaling Data Protocol for Billing and Settlements.

NSDP-F See Non-Signaling Data Protocol for Fraud.

NSEP Telecommunications See NS/EP Telecommunications.

NSESS Network Systems Engineering Systems Support.

NSF 1. National Science Foundation.

2. Network Specific Facilities.

NSIDN See Navigation Systems Information Dissemination Network.

NSIF Network and Services Integration Forum. A not-for-profit membership organization sponsored by ATIS (Alliance for Telecommunications Industry Solutions), NSIF intends to "provide an open industry forum for the discussion and resolution of multiple technology integration and SONET interoperability issues and to enable delivery of services across a set of networks, leading to the definition of recommendations and strategies." Members include equipment vendors, service providers, and other industry players. See also ATIS.

NSFNET National Science Foundation's TCP/IP-based NETwork, funded by the U.S. Government. Linking supercomputing centers and over 2500 academic and scientific institutions across the world, largely through the Internet, the NSFNET was founded in 1985. It was officially retired in 1995, replaced by the MERIT Network, which now has been replaced by the vBNS (Very high-speed Backbone Network Service). See also NREN, MERIT, and vBNS.

NSG National Systems Group. Division of Canadian telcos responsible for multiple accounts – large customers with service in multiple telephone company territories.

NSLOOKUP Name Server Lookup. An interactive query program of the InterNIC DNS (Domain Name Server). NSLOOKUP allows the user to contact servers to request information about a specific host or to print a list of hosts in the domain. See Internic, DNS, WHOIS, RWHOIS and DIG.

NSO 1. National Services Organization. An NSO provides the interconnecting network infrastructure, intelligent network services, network management, and roaming in a national Personal Communications Network (PCN). An NSO has ownership of resources that allow it to provide services to local operators. Operations such as marketing are centralized.

2. National Switch Operations.

NSP 1. Network Service Provider. Can include a local telephone company, ISP, or CLEC.

2. Native Signal Processing. Intel's idea to use the "spare MIPS" on its Pentium, Pentium Pro and later versions to take over the processing that once dedicated chips, like DSPs, used to do. The idea, clearly, is to sell more Intel chips and fewer chips from other makers.

3. Network Service Plan.

NSR 1. Non-Source Routed: Frame forwarding through a mechanism other than Source Route Bridging.

2. Number Portability Surcharge.

NSS 1. Non Standard Facilities Setup command, a response to an NSF frame.

2. Network and Switching Sub-system. A wireless telecommunications term. The part of the GSM (Global System for Mobile Communications) system in charge of the management of calls and the interface with other networks.

3. Network Status Subsystem.

4. Network Support System.

NSTHS Network Systems Technical Support.

NSTS/NSS Network Systems Technical Support/Network Systems Support.

NSU Network Synchronization Unit. A timing distribution system that provides synchronization signals to all electronic equipment within a wireline network node or office.

NT 1. Network Termination. Network Termination represents the termination point of a Virtual Channel, Virtual Path, or Virtual Path/Virtual Channel at the UNI. See NT-1.

2. New Technology, usually known as Windows NT. It's a new operating system from Microsoft which will let Windows run on high-end machines, such as file servers and workstations. NT has two sets of goals: to provide true multi-tasking, security, network connectivity and 32-bit power. Secondly, it's to provide a smooth upgrade path from Windows and MS-DOS. For a much larger explanation, see Windows NT and Windows NT Advanced Server.

3. Night transfer of ringing and station class of service.

NT-1 National ISDN 1. See ISDN.

NT-2 National ISDN 2. See ISDN.

NT-3 National ISDN 3. See ISDN.

NT1 Network Termination type 1. An ISDN term. The NT1 provides functions related to the physical and electrical termination of the local loop between the carrier network and the user premise. These functions, which fall into Layer 1 of the OSI Reference Model, are necessary to support multichannel digital communications over the loop, support loopback testing, and a variety of other functions. NT1 functions are required for both BRI (Basic Rate Interface), also known as 2B+D, and PRI (Primary Rate Interface), also known as 23B+D. See Network Termination type 1 for a more detailed explanation. See also BRI, ISDN, NT2, NT12, OSI Reference Model and PRI.

The NT1 is the first customer premise device on a two-wire ISDN circuit coming in from the ISDN central office. It does several things. It converts the two-wire ISDN circuit (called "U" interface) to four-wire so you can hook up several terminals – like a voice phone and a videophone. The NT1 typically has several lights on it which indicate if it's working. An ISDN central office can usually "talk" to the NT1 and do testing and maintenance by instructing the NT1 to loop signals back to the central office. An NT1 will support up to eight terminal devices, though I've never seen it work with that many. The basic NT1 functions are:

Line transmission termination.

Layer 1 line maintenance functions and performance monitoring.

Layer 1 multiplexing, and

Interface termination, including multi drop termination employing layer 1 contention resolution.

Increasingly, NT1s are being incorporated into customer premises equipment, like phones, PC ISDN data cards, etc.

See ISDN.

NT12 Network Termination type 0xB4. An ISDN term for a NT2 device which supports both NT1 and NT2 functionality. Such a device performs functions described in Layers 1 and 2, and perhaps Layer 3, of the OSI Reference Model. See also NT1 and NT2.

NT2 In ISDN, the Network Termination type 2 is an intelligent CPE (Customer Premise Equipment) switching or concentrating device (e.g., PBX, ACD, router, or concentrator). A NT2 device typically terminates PRI (Primary Rate Interface), also known as 23B+D (or 30B+D in Europe), access lines from the local ISDN CO switch. Depending on the specific nature of the NT2 device, it performs protocol handling functions described in Layers 1, 2 and 3 of the OSI Reference Model. See also BRI, ISDN, NT1, NT12, OSI Reference Model and PRI.

NTACS Narrow TACS. Cellular radio system deployed in Japan using narrow band technology to increase capacity by splitting TACs channels into two narrow channels.

NTAS Network Traffic Analysis System.

NTCA National Telecommunications Cooperative Association. A trade association representing primarily rural telephone cooperatives and other small telephone companies. It used to be called National Telephone Cooperative Association. www.ntca.org.

NTD Newton's Telecom Dictionary. What else?

NTE 1. Network Termination Equipment. The equipment on both sides of a subscriber line. A term used in the DSL (Digital Subscriber Line) world. See ADSL, ATU-C, ATU-R, and xDSL.

2. Network Transmission Elements.

NTF No Trouble Found.

NTFS Windows file system; an advanced file within the Windows operating system. It supports file system recovery, extremely large storage media. It also supports object-oriented applications by treating all files as objects with user-defined and system-defined attributes. If you're given the choice of having your Windows file system format your disk for FAT32 or NTFS, choose NTFS. NTFS has features to improve reliability, such as transaction logs to help recover from disk failures. To control access to files, you can also set permissions for directories and/or individual files. NTFS files are not accessible from other operating systems such as MS-DOS. NTFS supports spanning volumes, which means files and directories can be spread out across several physical disks. See FAT32.

NTI 1. Northern Telecom Inc. Now called Nortel Networks Limited.

2. Network Terminating Interface. A. The point where the network service provider's responsibilities for service begin or end, or B. The interface between DCE and its connected DTE.

NTIA National Telecommunications and Information Administration. An agency of the U.S. Department of Commerce concerned with spectrum management, public safety (e.g., electromagnetic fields), and communications (primarily telephony) standards. The NTIA is responsible for managing the Federal government's use of the spectrum, while the FCC is responsible for managing the use of spectrum by the private sector, and state and local governments. Clearly, the two agencies work closely in such matters. Both the FCC and the NTIA serve to represent the U.S. at the biannual World (Administrative) Radiocommunications Conference (WRC, pronounced "wark" as in "cork"), which are sponsored by the ITU (International Telecommunications Union) every two years. The NTIA also serves as the principal adviser to the President, Vice President and Secretary of Commerce on issues of domestic and international communications and information. The NTIA has been highly instrumental in furthering the concept of the NII (National Information Infrastructure). www.ntia.doc.gov See FCC, ITU, NII and WARC.

NTIS National Technical Information Service. www.ntis.gov. See National Technical Information Service.

NTLM A security package that provides authentication between clients and servers.

NTLM authentication protocol A challenge/response authentication protocol. The NTLM authentication protocol was the default for network authentication in Windows NT version 4.0 and earlier and Windows Millennium Edition and earlier. The protocol continues to be supported in Windows 2000 and Windows XP but no longer is the default.

NTN See Network Terminal Number.

NTO Network Terminal Option. An IBM program product that enables an SNA network to accommodate a select group non-SNA asynchronous and bisynchronous devices via the NCP-driven communications controller.

NTP 1. Network Termination Point.

2. The Network Time Protocol was developed to maintain a common sense of "time" among Internet hosts around the world. Many systems on the Internet run NTP, and have the same time (relative to Greenwich Mean Time), with a maximum difference of about one second.

NTRC National Telecommunications Cooperative is an organization of rougly 1,300 rural utilities. The NTRC helps its members deliver telecommunications and information technology, including broadband data and video.

NTS 1. Network Test System.

2. Non-Traffic-Sensitive commercial line costs levied on the user.

3. Number Translation Services.

NTSC National Television Systems Committee or National Television Standards Committee. It's not clear which is correct, though Systems seems to be more correct. It is the committee of Electronic Industries Association (EIA) that prepared the specifications standards approved by the Federal Communications Commissions in 1953 for commercial analog color TV broadcasting in the United States. The initials are used to describe the method of analog television transmission in the U.S., Canada, Japan, Central America and half of South America. The NTSC system uses interlaced scans and 525 horizontal lines per frame at a rate of 30 frames per second. The picture information is transmitted in AM (amplitude modulation) and the sound information is transmitted in FM (frequency modulation). NTSC is compatible with ITU-R (nee CCIR) Standard M. NTSC is not the only method of transmitting TV in the world. PAL (which stands for Phase Alternate Line) is the name of the format for color TV signals used in West Germany, England, Holland, Australia and several other countries. It uses an interlaced format with 25 frames per second and 625 lines per frame. The extra lines – 625 versus 525 – makes PAL a bit "crisper." NTSC and PAL are not compatible. You cannot view an Australian videotape on a U.S. TV, for example. The background to the NTSC signal is simple. Before NTSC there were millions of black and white TV receivers being used in the USA in the 1940s and early 1950s. As color came along, a color broadcast standard compatible with black and white sets was needed. The idea was that the signal would go out in color. If you had a black and white TV, you could see it in black and white. If you had a color TV, you could see the same signal in color. The method used was to encode the color information separately from the black and white information, and thus to broadcast two signals, called the luminance and chrominance signals. This gives good color reproduction for video, but broadcast TV suffers from phase shifts in the signal caused by reflections from buildings. This gives rise to hue changes, a solution to which was to include an auto-tint circuit in the TV set to reduce the hue fluctuations.

Another solution to the problem with the NTSC standard was adopted in some parts of Europe (and now in Brazil, Africa and Australia). The color signal reverses its phase (direction) on every second line, in order to cancel out any phase shift. This is called Phase Alternate Lines (PAL). Although PAL achieves stable hues it can reduce the color saturation, which luckily is less of a problem. The French (God bless them for being different) developed a a different color TV system called Sequential Couleur avec Memoire (SECAM). It shares with PAL the ability to display the correct hue, but it also ensures consistent saturation of color. This is still not the perfect solution since the color encoding is by frequency modulation, which can create patterning effects even on non-colored objects. It is used in France, the former Eastern Block countries (OIRT) and some Middle East countries (MESECAM - using an AM encoding). The black and white broadcast system in use had 891 lines! Unfortunately, limited bandwidth meant a color system could not broadcast this amount of information. The number of lines was reduced to 625 lines for color television. It and was therefore incompatible with the black and white TV sets (now a moot anyway, since no one watches black and white TV).

When TV engineers get together to hoist some brews, the initials for the various TV broadcast standards take on other meanings, such as: NTSC, Never Twice the Same Color. PAL, Peace At Last. PAL-M, Peace At Last – Maybe. SECAM (the French system) becomes System Essentially Contrary to the American Method.

NTSC Signal National Television Standards Committee specified signal. De-facto standard governing the format of television transmission signals in the United States. See NTSC.

NTSI Format A color television format having 525 scan lines; a field frequency of 60 Hz; a broadcast bandwidth of 4 MHz; line frequency of 15.75 KHz; frame frequency of 1/30 of a second; and a color subcarrier frequency of 3.58 MHz.

NTT The Japanese Nippon Telephone and Telegraph Cellular System. The large Japanese company that is 65% government owned. NTT has developed a number of cellular systems operating at 450 MHz and 800 MHz.

NTU Network Terminating or Termination Unit. A device which is placed at the final interconnect point between the PSTN and the customer owned equipment. NTUs allow the Carrier to isolate their facilities from those of their customers for testing, fault detection, and some service feature functionality.

NU 1. Network unit.

2. Network Unavailable. Equivalent to "All Trunks Busy". The NU tone, an European term, is in Europe is equivalent to "fast busy tone" in the USA.

NuBus (Pronounced "New Bus.") The name of the bus design for most Apple Macintosh computers. See Nubus Card.

NuBus Card An add-on card that fits inside the NuBus slots of a Modular Macintosh. Often used for video cards and modems.

nucleus An ATM term. The interior reference point of a logical node in the PNNI complex node representation.

nudist camp A nudist camp is a place where men and women meet to air their differences. I believe many go there to air out their differences.

NUI Network User Identifier. A unique alphanumeric number provided to dial-up users to identify them to packet switched networks around the world. The number is used to get onto the network and for billing.

NUID Network User Identification.

Nuisance Call Bureau Generally part of a telephone company's Security Department, the Nuisance Call Bureau handles complaints from customers who have received harassing, obscene, or threatening phone calls. If nuisance calls are determined to be locally originated, the Nuisance Call Bureau sets a trap to identify the source of the call and takes whatever action it thinks is appropriate in the circumstances. If nuisance calls are long-distance, the Nuisance Call Bureau will work with the appropriate long-distance carrier to have the latter set a trap to identify the source of the call and take whatever action is appropriate, including calling in the police.

null Having no value. A dummy letter, letter symbol, or code group inserted in an encrypted message to delay or prevent its solution, or to complete encrypted groups for transmission or transmission security purposes. See also Null Characters.

null Call_id AN ISDN term. A null call_id is the call_id used to convey information that does not pertain to a specific call between the ISDN System Adapter and the central office switch. Null call_ids are primarily used for accessing feature buttons that do not relate to a specific call.

null characters Characters transmitted to fill space, time or to "pad" something. They add nothing to the meaning of a transmitted message, but the null characters are expected by the system. On older teletype machines, for example, when the type head reaches the end of a line and the New Line sequence is transmitted, it usually includes a number of Null Characters in order to give the mechanical type head enough time to reach the left margin of the page before transmitting the next line to the terminal. In this manner, no characters are lost. MCI Mail uses five null characters at the beginning of every line – unless you tell it otherwise.

NULL encryption algorithm An algorithm that does nothing to transform plaintext data; that is, a no-op. It originated because of IPsec (a collection of IP security measures) ESP, which always specifies the use of an encryption algorithm to provide confidentiality. The NULL encryption algorithm is a convenient way to represent the option of not applying encryption in ESP (or in any other context where this is needed). See IPsec.

null fill A modification of an antenna radiation intended to fill the null areas in an antenna's usual radiation pattern. Null fill use to cover areas which would otherwise fall into an antenna's null zones.

null modem A null modem is either a shortened way of saying "Null Modem Cable" or a device directly connecting two DTE devices. See Null Modem Cable and Crossover Cable.

null modem cable Crossover or cross-pinned wiring of an RS-232 cable such that a DTE (Data Terminal Equipment) device (such as a PC) can talk to another such device without the use of a modem, hence the term "null", which means "amounting to nothing." A null modem cable allows one PC to connect directly to another PC for file transfer over maximum distances of 50-100 feet (depending on the quality of the cable) without the use of either a modem or a line driver. A null modem cable also can be used to connect one DCE (Data Communications Equipment) device to another, in what is known as a "tail circuit" configuration. Essentially, a null modem cable reverses pins 2 and 3 on an RS-232 cable. But there are no standard null modem cables. And other pins sometimes need changing and jumpering together. Null Modems also are known as Modem Eliminators.

null spot An RFID definition. Area in the reader field that doesn't receive radio waves. This is essentially the reader's blind spot. It is a phenomenon common to UHF systems.

null state A state in which there is no relationship between a call and device. Synonymous with Idle State.

null suppression A data compression technique whereby streams of null characters are identified at a transmission source and replaced by two or more control characters. The first character indicates the null suppression, and more characters indicate the number of null characters removed. The receiver uses this information to replace the removed data.

null value A parameter or field position for which no value is specified.

nulls See Leaky Coax.

NUMA Non-Uniform Memory Access. A symmetrical multiprocessing (SMP) technology. Proponents claim the technology is an improvement over traditional Intel-based SMP systems that can suffer from traffic jams on their shared-memory buses and typically cannot accommodate more than 16 or 32 processors. With NUMA, each Intel processor has its own local memory and is able to form static or dynamic connections with other chips' memories. NUMA servers can be powered by 64 or more processors.

Number Administration And Service Center NASC. Provides centralized administration of the Service Management System (SMS) database of 800 numbers. The NASC keeps track of the 800 numbers that are in use, or available for use, by new 800 customers.

number capacity A telephone company term. The network access line capacity of numbers or terminals is the maximum number of network access lines that can be working on installed numbers at the entity's derived objective percent number fill.

number cruncher A number cruncher is a supercharged computer, typically with tens of thousands of microprocessors, that performs complicated calculations. Supercomputers that astrophysicists and chaos theoreticians at universities use are usually referred to as number crunchers.

number crunching Mathematical calculations. "Joe, go do some number crunching and tell if the fundamentals on that stock are any good."

number pooling Number pooling allows local phone companies to share a "pool" (a collection) of telephone numbers with the same exchange. Number pooling is a way to allocate scarce telephone numbers more efficiently. Without pooling, a CLEC (Competitive Local Exchange Carrier) was assigned an entire 10,000 block of phone numbers (for example those on the 212-69X exchange) but might require only a portion of the block, while other phone companies in the same area may be desperate for numbers for their customers. The North American Numbering Council (NANC) developed Thousand Block Number Pooling, also known as 1K Pooling, in order to allow the sharing of numbers in increments of 1,000. For example, one Service Provider (SP) might have 212-691-xxxx block, and another the 212-692-xxxx block. Such pooling can be implemented in areas where Local Number Portability (LNP) is in place. See also CLEC, Local Number Portability, NANC, and Number Portability.

Number Portability Number Portability (NP) refers to the ability of end users to retain their geographic or non-geographic telephone number when they change any of:

1. Their service provider. This means that users can retain the same telephone number as they change from one service provider (telephone company) to another.

2. Their location. This means that users can retain the same telephone number as they move from one permanent location to another.

3. Their service. This means that users can keep the same telephone number as they move from one type of service to another (e.g., POTS to ISDN).

Number portability started with 800 toll-free numbers as that industry and then moved to local number portability (LNP) as competition developed for local telephone companies. Here's a longer explanation of how 800 number portability developed.

Once upon a time, 800 numbers belonged to the phone companies who supplied 800 service. So if you got 800 service from AT&T and you wanted to keep your number, you were stuck with AT&T. For example, we had (and still have) 1-800-LIBRARY (800-542-7279). The reason? The database of 800 numbers was maintained by Bellcore, who allocated certain "exchanges," like 800-542, to certain carriers. In the case of 542, it was AT&T. If we wanted to go to a different phone company, we had to give up our phone number. Since many companies had invested vast monies in promoting their 800 numbers, this was inconvenient and, in essence, forced companies to stay with the same long distance company. In 1993, things changed. The FCC mandated Number Portability, which allowed you to change your long distance carrier but keep your valued 800 number. The way the whole thing works now is simple to understand, although somewhat complex and expensive for the carriers to implement. You pick up the phone. You dial an 800-number. Your local central office holds the call for a moment, while it" dips" (checks) into an external, central database of 800 numbers for the routing of that call. (A number of such centralized databases are maintained by the carriers. In other words, the centralized databases are distributed throughout the network in order to balance the load on the processors and associated databases, placing them in strategic proximity to concentrations of traffic.) This database dip typically takes place via the packet-switched SS7 (Signaling System 7) network. When it receives a reply from that external database, it simply then sends the call to the appropriate long distance carrier which is providing service on that 800 number. To change the carrier of your 800 number, simply tell the database who the new carrier is. And that's 800 (actually, 800/888) number portability. See also LNP (Local Number Portability)

As defined by the Telecommunicaitons Act of 1996, the term 'number portability' means the ability of users of telecommunications services to retain, at the same location, existing telecommunications numbers without impairment of quality, reliability, or convenience when switching from one telecommunications carrier to another.

numbering To have any sort of network work, everyone on the network has to have a unique number which we can all dial. I only included this definition because I found the following definition in a glossary of LAN terms and I thought, "Maybe someone doesn't know, though if they didn't know, this definition wouldn't help them much." Here's the LAN definition: "The assignment of unique identities to a user-network interface."

Numbering Advisory Committee See NAC.

numbering plan 1. In Wide Area Networks, the method for assigning NNX codes to provide a unique telephone address for each subscriber, special line or trunk destination.

2. In PBXs, the method of assigning extension numbers and trunk designations at the local premises.

Numbering Plan Area NPA. A fancy term the Bell System came up with years ago to mean Area Codes in North America. NPA/NPA Code. A geographical division in the North American numbering plan within which no two telephones will have the same 7 digit number. "N" originally was any number between "2" and "9"; "P" was always "1" or "0"; and "A" was any number excluding "0" in the electromechanical era of switching machines. Colloquial synonym: "area code." Internationally equivalent to "City Codes" or "Routing Codes" in other nations, where various number sequences may be used.

numbers shift A character in the Baudot code which establishes that the characters following in the transmission are to be interpreted as numeric characters. See Letters Shift.

numeric key pad A separate section of a computer keyboard which contains all the numerals 0 through 9. Sometimes, some special keys are included – a plus sign, a minus sign, a multiplication sign and a division sign. The numeric key pad on a computer is the same as that found on calculators and adding machines. The top row is 789. The second top row is 456. The third top row is 123. The lowest row is typically 0, "." and "+". The numeric key pad is exactly opposite that of the touchtone telephone keypad, which was designed deliberately to be unfamiliar to users, so they may not input digits into the nation's telephone system faster than it could take them. Early touchtone central offices were very slow.

numeric user identifier According to the 1988 X.400 recommendations, a numeric user identifier is a standard attribute of an O/R (Originator/Recipient) address that consists of a unique sequence of numbers for identifying a user. (Numeric User Identifier was referred to as Unique Identifier in the 1984 recommendations.)

numerical aperture The measure of the light-acceptance angle of an optical fiber.

numeris The French name for ISDN.

numeronym A telephone number that spells a word. For example, 1-800-542-7279 is the telephone for the company which distributes this dictionary. That telephone number is advertised as 1-800-LIBRARY. It is a great number to have because it's easy to remember, although it's not so easy to dial. Numeronyms are in great demand, for obvious reasons. With the expansion of the toll-free dialing plan to include 888 (and other prefixes in the near future), it's also important (but increasingly difficult) to protect those numbers from competition.

nutmeg In the 16th and 17th centuries, nutmeg was the third most valuable commodity in the world, after silver and gold. In addition to preserving and flavoring food, nutmeg was believed to be a soporific, an emetic, a prophylactic against plague and a hallucinogen. The great naval powers of Europe at that time – the Portuguese, Dutch, English and Spanish – all competed for control of the trade in nutmeg and other spices, including pepper, cinnamon and cloves. Under the Treaty of Breda in 1667, Holland traded New Amsterdam, as the Dutch called Manhattan (i.e. New York City) to the English for Pulau Run, one of the principal nutmeg-growing islands in the Bandas, a series of Islands known as The Spice Islands, which are now part of Indonesia. That was not one of the Dutch's better deals. Nutmeg is reputed to be the secret ingredient in Coca Cola, the beverage and the largest buyer of nutmeg in the world. Most of the world's nutmegs are now not grown in the Spice Islands.

NUXI Problem The conversion of data between big-endian and little-endian data architectures is known as the NUXI Problem. If the word "UNIX" were stored in two two-byte words, a big-endian system would store it as "UNIX," while a little-endian system would store it as "NUXI." According to legend, the origin of the NUXI Problem refers to the first time that UNIX was ported to the IBM Series 1 minicomputer. The Series 1 had the same size words (two bytes) as the PDP-11, but the bytes were swapped. When the Series 1 machine started up for the first time, it printed out "NUXI" instead of "UNIX". Ever since then, the "NUXI Problem" has referred to byte-ordering problems between machines of different "byte sexes." See Big-endian for a full explanation.

NVOD Near Video On Demand. Providing a consumer a multimedia item – movie, TV program, etc. – on a rotating schedule, thus giving the appearance of an on-demand system, i.e. VOD (Video On Demand). See VOD.

NVP 1. Network Voice Protocol. Circa 1973 ARPANET protocol. Used to support real-time voice over the ARPANET. Both LPC and CVSD encoding schemes were successfully implemented by Culler-Harrison, Inc., the Information Sciences Institute, Lincoln Laboratory and Stanford Research Institute.

2. In regards to cable, NVP stands for the Nominal Velocity of Propagation. All communications cable has a spec called NVP. An electrical signal in a vacuum would travel at the speed of light. In the world of land-based communications, electrical signals travel through twisted copper pairs at a percentage of the speed of light, around 72% for good cat 5 cable. When you test cat 5 cable, the testing instrument is supposed to be calibrated to the NVP of the cable so that the device can measure the time it takes to go end to end and back (or end to fault and back). Without knowing the NVP, the test device cannot accurately locate a fault in the cable. See also Time Domain Reflectometer. This definition contributed kindly by Steven Waxman.

NVRAM NonVolatile Random Access Memory. RAM that doesn't lose its memory when you shut the electricity off to it.

NVS NonVolatile Storage is a storage device, like a disk or EPROM, that retains data when you turn off.

NWL Non-WireLine. Cellular radio licenses received from the FCC with no initial association to telephone company. Also referred to as A-Block.

NWLink Microsoft's network protocol that simulates Novell's IPX/SPX for Windows 95 and NT communications with Novell NetWare file servers and compatible devices. NWLink is an IPX/SPX-compatible transport stack that gives NetWare-compatible clients access to NT applications services.

NWRA National Wireless Resellers Association. A Washington-based trade association and lobbying organization which formerly was known as the NCRA (National Cellular Resellers Association), NWRA merged into TRA (Telecommunications Resellers Association) in 1997. TRA changed its name to ASCENT in 2000. See ASCENT.

NWS Network Services.

Nx384 N-by 384. The ITU-T's approach to creating a standard algorithm for video codec interoperability It is based on the ITU-T's H0 switched digital network standard, which was expanded into the Px64 or H.261 standard, approved in 1990.

Nx64 An ATM term. This refers to a circuit bandwidth or speed provided by the aggregation of nx64 kbps channels (where n= integer> 1). The 64K (64,000 bits per second) or DS-0 channel is the basic rate provided by the T Carrier systems. See T Carrier.

NXX 1. In a seven digit local phone number, the first three digits identify the specific telephone company central office code which serves that number. These digits are referred to as the NXX where N can be any number from 2 to 9 and X can be any number. At one stage, many moons ago, it was not permissible to have a 1 or a 0 as the second digit in an NXX and it was called an NNX. But that was before everyone had to dial a "1" before making a direct distance dialed long distance call, whether within their own area code or outside it. This little trick of forcing everyone to dial "1" for long distance allowed us to introduce telephone exchanges with the same three digits as area codes. For example, one of our company's numbers is 212-206-6660. The "206" elsewhere is an area code for Seattle and other parts of Washington state.

2. Network Exchange.

NYNEX Corporation Pronounced "nine X." One of the original seven Regional Holding Companies (RHCs) formed at Divestiture. It included New York Telephone and New England Telephone Company and sundry service and cellular radio companies. The company said its name was spelled in all capitals, thus NYNEX. The New York Times spelled it Nynex. In any event, Nynex no longer exists. The company was merged (read "acquired") by Bell Atlantic in 1997 and it now is part of Bell Atlantic. Internally, Bell Atlantic refers to Nynex as Bell Atlantic North, but we're not supposed to know that. Internally, some Bell Atlantic people tend to look down on Nynex people as inferior. After all, Bell Atlantic bought Nynex. Anyway, the saga continues. Bell Atlantic later bought GTE and changed its name on June 30, 2000 to Verizon Communications. See Verizon for a neat explanation of where the word Verizon comes from.

Nyetscape Nickname for AOL's less-than-full-featured Web browser. But this definition has fallen out of favor since AOL bought Netscape, the company, and they've introduced some better Netscape browsers, though why remains a mystery, since all browsers are free.

Nyquist Theorem In communications theory, a formula stating that two samples per cycle is sufficient to characterize an analog signal. In other words, the sampling rate must be twice the highest frequency component of the signal (i.e., sample 4 KHz analog voice channels 8000 times per second.) The Nyquist Theorem is the mathematical theorem used in the Mu-Law encoding technique used in T-carrier transmission systems. Shannon's Law, a similar but incompatible theorem, is used in E-carrier systems. See also E-Carrier, Shannon's Law, and T-Carrier.

Nyquist, Harry A distinguished electrical engineer, Harry Nyquist worked in the research department of AT&T and, later, at Bell Telephone Laboratories. In 1924, he published a paper entitled "Certain Factors Affecting Telegraph Speed," which was an analysis of the relationship between the speed of the telegraph system and the number of signal values it used. That paper was refined in 1928, when he published "Certain Topics in Telegraph Transmission Theory." That paper expressed the Nyquist Theorem, which established the principle of sampling continuous signals to convert them to digital signals. The two papers are cited in the first paragraph of Claude Shannon's classic paper, "The Mathematical Theory of Communication." Nyquist also is noted for his pioneering work on a mathematical explanation for thermal noise, sometimes known as "Nyquist noise," and "telephotography," which was AT&T's 1924 version of a fax machine. See also Nyquist Theorem; and Shannon, Claude Elwood.

NZDF Non Zero Dispersion Shifted Fiber. A type of Dispersion Shifted Fiber that is used in long haul, high-speed fiber optical transmission systems. See DSF for a full explanation.

O Used on switches to mean "OFF." The "ON" setting is "I."

O band The optical band, or window, specified by the ITU-T at a wavelength range between 1260nm and 1310nm (nanometers) for fiber optic transmission systems.

O carrier N-Carrier and O-Carrier – Frequency division multiplexed analog carrier over long haul copper.

O-band See O Band.

O&M Operations and Maintenance.

O.E.M. Original Equipment Manufacturer. See OEM.

O/E Optic to Electric conversion.

O/R Short for Originator/Recipient in the X.400 MHS (Message Handling System).

O/R name A ITU (International Telegraph and Telephone Consultative Committee) term for the set of user attributes that identifies a specific MHS (Message Handling System) user. Example components are: given name, surname, ADMD (ADministrative Management Domain), country and PRMD (PRivate Management Domain), country and PRMD (PRivate Management Domain). An example of a complete O/R Address is: G=HARRY; S=NEWTON; I=HN; A=MARK400; O=PIPELINE; OUI=COMPUTER TELEPHONY; C=US.

O2E Optical-to-electrical (O2E) converter is a converter that provides electrical current that is proportional to the optical power that it receives.

OA 1. Office Automation. Nobody knows what it means. But there are many consultants out there who will tell you for the right amount of money. Actually, the early "office automation" seemed to translate into word processing for the masses. When someone discovered that word processing didn't enhance office productivity, office automation fell into disrepute.

2. Operating Agency. An ITU (International Telecommunications Union) term. Any individual, company, corporation or governmental agency which operates a telecommunication installation intended for an international telecommunication service or capable of causing harmful interference with such a service. See also ROA.

OA&M Operations, Administration and Maintenance. See OAM.

OACSU Off Air Call Set Up. A wireless telecommunications term. A method of establishing a call to a mobile telephone where the radio channel is set up at the last possible moment.

OADM Optical Add/Drop Multiplexer. Another term for ADM, which is a SONET term. See ADM and SONET.

OADX An optical add/drop switch. It performs high-speed wavelength switching and wavelength transmission through all-optical (OOO) or optical- electrical-optical (OEO) means, and also provides the functionality of an OADM. See OADM.

OAI Open Application Interface. Basically one or many openings in a telephone system that lets you link a computer to that phone system and lets the computer command the phone system to answer, delay, switch, hold etc. calls. The term is also called PHI – as in PBX-Host-Interface. The term OAI was first used by PBX makers, NEC and InteCom. And now the term has become somewhat generic. Essentially every manufacturer of phone systems is evolving towards open application interfaces of their own. Many of these open interfaces are TSAPI-compatible and increasingly most are TAPI-compatible. See TAPI and TSAPI. According to Probe Research, there are really two separate "markets" for OAI or PHI:

First, there's Horizontal/Office Automation applications. These are applications that support business functions across organizational groups or industry verticals in inter- or intra-department business settings. Examples include voice mail, electronic mail, message centers, corporate telephone directories, automated screen-based dialing, personal productivity tools, conferencing, PBX feature enhancements, ANI interfaces, time clocks, 911 emergency service and compound image (data, text, image and voice) processing. I see these applications as all those useful, productivity-enhancing things I always wished my telephone systems would do if only they would let me program the thing. (Until OAI, all phone systems were totally closed architecture.)

Second, there's transaction applications. These are applications that support an actual business transaction – customer service, inbound telephone order taking, outbound telemarketing, market research, data gathering, inventory inquiry, account time billing, credit collections, locator services with or without transfer to the local dealer. These always require access to a computer database. These applications are generally complicated, time-sensitive and customized for each installation.

A sample OAI arrangement: In early 1993, Novell and AT&T inked a deal to put telephony onto Novell LANs. The Telephony Server NetWare Loadable Module (NLM) will be the first product. It is an AT&T PC-card sitting in the Novell File server. The card connects to the ASAI (Adjunct Switch Applications Interface) port on the AT&T Definity PBX. Anyone with a PC on the network and an AT&T phone on their desk can use telephone features, such as auto-dialing, conference calling and message management (a new term for integrating voice, fax and e-mail on your desktop PC via your LAN). The Novell/AT&T deal intends to create open Application Programming Interfaces (APIs) that third party developers can work with. A Novell/AT&T example of what could be developed: A user could select names from a directory on his PC. He could tell the Definity PBX through the PC over the LAN to place a conference call to those names. At the same time, a program running under NetWare would automatically send an e-mail to the people, alerting them to the conference call and giving them the agenda. All participants would have access to both the document and the conference call simultaneously. Here are some of the names which manufacturers of PBXs and computers have coined for their open application interfaces, also called PHIs:

ACL – Applications Connectivity Link – Siemens' protocol

ACT – Applied Computer Telephony – Hewlett Packard's generic application interface to PBXs

Application Bridge – Aspect Telecommunications' ACD to host computer link

ASAI – Lucent's Adjunct Switch Application Interface

CAM – Tandem's Call Applications Manager – the name of the Tandem software interface which provides the link between a call center switch telephone switch (either a PBX or an ACD) and all Tandem NonStop (fault tolerant) computers.

CIT – Digital Equipment Corporation's Computer Integrated Telephony (works with major PBXs)

CSA – Callpath Services Architecture – IBM's Computer to PBX link

Call Frame – Harris' PBX to computer link

Callbridge – Rolm's CBX and Siemens to IBM host or non-IBM host computer link

Callpath – IBM's announced, CICS application link to IBM's CSA, available on the AS400 in April of 1991

Callpath Host – IBM and ROLM's CICS-based integrated voice and data applications platform which links to ROLM's 9751 PBX

CompuCall – Northern Telecom's DMS central office link to computer interface

CPI – Computer to PBX Interface developed by Northern Telecom and DEC

CSTA – Computer Supported Telephony Application, RSL standard from ECMA

DECags – DEC ASAI Gateway Services. Two-directional link to AT&T's Definity

DMI – AT&T's Digital Multiplexed Interface, a T-1 PBX to computer interface

HCI – Host Command Interface. Mitel's digital PBX link to DEC computer

IG – AT&T's ISDN Gateway (one direction from the switch to the host)

ITG – AT&T's Integrated Telemarketing Gateway (two directional)

ISDN/AP – Northern Telecom's host to SL1 PBX protocol, which supports NT's Meridian Link

Meridian Link – Northern Telecom's host to PBX link available on the Meridian PBX

ONA – Open Network Architecture (for telephone central offices)

PACT – Siemens' PBX and Computer Teaming, protocols between Siemens PBXs and computers

PDI – Telenova/Lexar's Predictive Dialing Interface

PHI – PBX Host Interface (a generic term coined by Probe Research)

SAI – Stratus Computer Switch Application Interface

SCAI – Switch to Computer Application Interface, one name given by Northern Telecom to PHI

SCIL – Aristacom's Switch Computer Interface Link Transaction Link

Solid State Applications Interface Bridge – Solid's State's PBX to external computer link.

STEP – Speech and Telephony Environment for Programmers; WANG's link

Transaction Link – Rockwell's link from its Galaxy ACD to an external computer

Teleos IRX-9000 – Teleos' Intelligent Call Distribution platform

VoiceFrame – Harris Digital Telephone Systems Division Platform

OASIS The Organization for the Advancement of Structured Information Standards. This organization is the largest standards group for electronic commerce on the Web.

Oakley A key determination protocol, which defines a method for establishing security keys for use on a session basis by Internet hosts and routers. Oakley often is used in conjunction with ISAKMP (Internet Security Association and Key Management Protocol), a protocol that provides a method for authentication of communications between peers over the Internet. ISAKMP provides a framework for the management of security keys, and support for the negotiation of security attributes between associated peers. See also ISAKMP.

OAM Operations, Administration and Maintenance. This usually term refers to the specifics of managing a system or network. It is typically a group of network management functions that provide network fault indication, performance information, and data and diagnosis functions. Some switches have computers devoted to OAM.

OAM&P Operation, Administration, Maintenance and Provisioning.

Oberith, Herman Julius 1894-1989. The scientist who taught World War II rocket expert Werner von Braun, and who, in 1923, was the first to propose that satellites be used for communications, though by optical means.

OBEX Object Exchange: a set of high-level protocols allowing objects such as vCard contact information and vCalendar schedule entries to be exchanged using either IrDA (IrOBEX) or Bluetooth. Symbian OS implements IrOBEX for exchange of vCards, for example between a Nokia 9210 Communicator and an Ericsson R380 Smartphone and vCalendar.

OBF Ordering and Billing Forum. A forum of the Carrier Liaison Committee (CLC) of ATIS (Alliance for Telecommunications Industry Solutions), originally called the Exchange Carriers Standards Association (ECSA). ATIS is heavily involved in standards issues including interconnection and interoperability. According to ATIS, the OBF "provides a forum for telecommunications customers and providers to identify, discuss and resolve national issues which affect ordering, billing, provisioning and exchange of information about access services, other connectivity and related matters." The six standing committees of the OBF are heavily involved in the development of standard mechanisms by which CLECs (Competitive Local Exchange Carriers) and ILECs (Incumbent LECs) can interface effectively; such interface is required in the competitive environment fostered by the Telecommunications Act of 1996, and by various state initiatives. See also ATIS, CLEC and ILEC. www.atis.org.

OBI Open Buying on the Internet. A standard which provides a generic set of requirements, an architecture, and a technical specification for Internet purchasing solutions in the general context of Electronic Commerce. The OBI standard was developed by the OBI Consortium; membership is open to buying and selling organizations, technology companies, financial institutions, and other interested parties. See also Electronic Commerce. www.supplyworks.com/obi.

object 1. In the context of network management, an object is a numeric value that represents some aspect of a managed device. An object identifier is a sequence of numbers separated by periods, which uniquely defines the object within a MIB. See MIB.

2. In its simplest form in computing, an object is a unit of information. It can be used much more broadly, depending on the application. In X.400, an object contains both attributes and method describing how the content is to be interpreted and/or operated on.

3. In object-oriented programming, a variable comprising both routines and data that is treated as a discrete entity. An object is based on a specific model, where a client using an object's services gains access to the object's data through an interface consisting of a set of methods or related functions. The client can then call these methods to perform operations.

4. An entity or component, identifiable by the user, that may be distinguished by its properties, operations, and relationships.

See also Object Oriented Programming.

object cache scavenger The code that periodically scans the cache for objects to be discarded. It deletes from the cache files that have not been used recently and therefore are unlikely to be used again in the near future.

object code A term usually applied to the executable, machine-readable, form of a software program. Object code is instruction code in machine language produced as the output of a compiler or a assembler. The original program, or code, is called the source Code. The term is also used to refer to an intermediate state of compilation, which is different from the initial source code and the final binary object code. See Source Code.

object definition alliance A group formed to establish software standards for interactive multimedia applications that will operate uniformly over a variety of hardware, operating systems and networks.

object encapsulation Data and procedures may be encapsulated to produce a single object, thereby hiding complexity.

object handle Code that includes access control information and a pointer to the object itself.

object inheritance The transfer of characteristics down a hierarchy from one to another.

Object Linking And Embedding OLE. An enhancement to DDE protocol that allows you to embed or link data created in one Windows application in a document created in another application, and subsequently edit that data in the original application without leaving the compound document. See DDE and OLE.

object orientation Representing the latest approach to accurately model the real world in computer applications, object orientation is an umbrella concept used to describe a suite of technologies that enable software products that are highly modular and reusable. Applications, data, networks, and computing systems are treated as objects that can be mixed and matched flexibly rather than as components of a system with built-in relationships. As a result, an application need not be tied to a specific system or data to a specific application. The four central object-oriented concepts are encapsulation, message passing, inheritance, and late binding. This definition courtesy Microsoft.

object oriented file system A file system, based on object-oriented programming, that allows permanent storage of objects and associated links.

Object Oriented Programming OOP. Object oriented programming is a form of software development that models the real world through representation of "objects" or modules that contain data as well as instructions that work upon that data. These objects are the encapsulation of the attributes, relationships, and methods of software-identifiable program components. Object-oriented methodology differs from conventional software programming where functions contained in code are found within an application. Although hype about object-oriented technology is fairly recent, the approach was first introduced in the programming language Simula developed in Norway in the late 1960s. The methodology was based on the way children learn (i.e., object + action = result). The idea of object oriented programming is make the writing of complex computer software much easier. The idea is to simply combine objects together to produce a fully-written software application.

The work, goes into producing all the objects. They are the building blocks. Theoretically, libraries of objects will be worth a fortune to those companies who develop them.

Object oriented programming is not easy to understand. Here's a definition of object-oriented programming from Business Week, September 30, 1991. "Software objects are chunks of programming and data that can behave like things in the real world. An object can be a business form, an insurance policy or even an automobile axle. The axle object would include data describing its physical dimensions and programming that describes how it interacts with other parts, such as wheels and struts. A system for a human resources department would have objects called employees, which would have data about each worker and the programming needed to calculate salary raises and vacation pay, sign up dependents for benefits, and make payroll deductions. Because objects have 'intelligence,' they know what they are and what they can and can't do. Thus objects can automatically carry out tasks such as calling into another computer, perhaps to update a file when an employee is promoted. The biggest advantage of objects is they can be reused in different programs. The object in an electronic-mail program that places messages in alphabetical order can also be used to alphabetize invoices. Thus, programs can be built from prefabricated, pretested building blocks in a fraction of the time it would take to build them from scratch. Programs can be upgraded by simply adding new objects."

The key concepts of object-oriented programming, according to ComputerWorld are 1. OBJECTS. The basic building block of a program is an object. Objects are software entities. They may model something physical like a person, or they may model something virtual like a checking account. Normally an object has one or more attributes (fields) that collectively define the state of the object; behavior defined by a set of methods (procedures) that can modify those attributes; and an identity that distinguishes it from all other objects. Some objects may be transient, existing temporarily during the execution of a program, i.e., only during run time. Others may be persistent, existing on some form of permanent storage (file, database, programming library) after the program finishes.

2. ENCAPSULATION. This concept refers to the hiding of most of the details of the object. Both the attributes (data structure) and the methods (procedures) are hidden. Associated with the object is a set of operations it can perform. These are not hidden. They constitute a well-defined interface – that aspect of the object that is externally visible. The point of encapsulation is to isolate the internal workings of the object so that, if they must be modified, those changes will also be isolated and not affect any part of the program.

3. MESSAGING. One object requests another object to perform its operation through messaging. The client object sends a message to the server object consisting of the identity of the server object, the name of the operation and, in some cases, optional parameters. The names of the operations are limited to those defined for that object. For example, the operations for a checking account object may be defined to be OPEN, DEBIT, CREDIT, COMPUTE INTEREST, ISSUE STATEMENT, SCHEDULE AUDIT, AND CLOSE.

4. DATA ABSTRACTION. An object is sometimes referred to as an instance of an abstract data type or class. Abstract data types are constructed using the built-in data types supported by the underlying programming language, such as integer and date. The common characteristics (both attributes and methods) of a group of similar objects are collected to create a new data type or class. Not only is this a natural way to think about the problem domain, it is a very efficient way to write programs. Instead of individually describing several dozen instances, the programmer describes the class once. Once identified, each instance is complete with the exception of its instance variables. The instance variables are associated with each instance, i.e., each object; methods exist only with the classes.

5. INHERITANCE. Data abstraction can be carried up several levels. Classes can have super-classes and subclasses. In moving to a level of greater specificity, the application developer has the option to retain some attributes and methods of the super-class, while dropping or adding new attributes or methods. This allows greater flexibility in class definition. It is even possible in some languages to inherit from more than one parent. This is referred to as multiple inheritance.

objective percent fill -- lines A telephone company term. The objective percent line fill provides for line equipments that are administratively unusable because of their use for test, assignment restrictions such as class-of-service, and for being out on assignment lists. This percent which is a percent of the total line equipment installed less those required for trunks will vary according to the entity equipment type and its service features. The objective percent line fill should not be one value applied to all entities, but is derived for each individual entity on an empirical basis by the Network Administrator. Local OTC policy may dictate specific values to be used with each type of switching system and the procedures to be followed when changing these values. See also Objective Percent Fill – Numbers.

objective percent fill -- numbers A telephone company term. The objec-

tive percent number fill provides for numbers that are administratively unusable because of intercept requirements, PBX growth coin and official series, rate protection, and for other requirements such as being out on assignment lists. Care also needs to be taken, where applicable, to make allowance for CENTREX-CO and CENTREX-CU requirements. As in percent line fill, the quantity of numbers classified as administratively unusable will vary with an entity's characteristics. The objective percent number fill which is a percent of the total numbers installed less those required for trunks is derived on an empirical basis for each individual entity by the Network Administrator. Local OTC policy may dictate specific values to be used with each type of switching system and the procedures to be followed when changing these values. See also Objective Percent Fill – Lines.

Object Request Broker ORB. An object-oriented system consisting of middleware which manages message traffic between application software and computer/software platforms. As an application sends a message to an object, it need only identify the object by name. The ORB keeps track of the actual addresses of all such objects and, therefore, is able to route the request to the specific address space where the object resides in the system. CORBA (Common Object Request Broker Architecture), developed as an ORB standard by the OMG (Object Management Group), provides a standard interface for interoperability between object management systems residing on disparate platforms. SEE CORBA and OMG.

OBL Op0xC7rateur de Boucle Locale (Local Loop Operator) in France.

obscene Soon we'll be able to clone ourselves. If you pushed your naked clone off the top of a tall building, would it be:

A) murder,
B) suicide, or
C) merely making an obscene clone fall.

obscenity The Supreme Court of the United States has defined obscenity as speech lacking in any social value. The Court has said that obscenity is entitled for First Amendment protection. But it may be regulated to a greater or less degree, depending on the medium – TV yes, paper no. Indecency on the Internet is under the jurisdiction of the Communications Decency Act of 1996.

obsession Obsession is the most popular name for a boat in the United States.

obstruction lights Warning lights mounted on the side of an antenna tower.

OC 1. Operator Centralization.

2. Optical Carrier. A SONET optical signal. See OC-1.

OC-1 Optical Carrier-level 1. OC-1 is 51.840 million bits per second. The optical counterpart of STS-1 (Synchronous Transport Signal-1), which is the fundamental signaling rate of 51.840 Mbps on which the SONET (Synchronous Optical NETwork) hierarchy is based. OC-1 provides for the direct electrical-to-optical mapping of the STS-1 signal with frame synchronous scrambling. The STS-1 signal originates as a DS-3 (T-3) electrical signal, which operates at a raw signaling rate of 44.736 Mbps and which supports 672 voice-grade digital channels of 64 Kbps; the additional bps are attributable to optical processing overhead (signaling and control data). All higher levels are direct multiples of OC-1 (e.g., OC-3 equals three times OC-1). See also Concatenation, OC-3, OC-12, OC-48, OC-192 and OC-256.

OC-3 Optical Carrier equal to three DS-3s, which is equal to 155.52 Mbps. See also Concatenation, OC-1, OC-3c, and SONET.

OC-3c Optical Carrier-level 3 concatenated. An OC-3 signal comprising three OC-1s which are linked together to form a contiguous chunk of bandwidth of 155.52 Mbps. The first OC-1 in the OC-3c data stream identifies this concatenation so that all other devices which act on the signal subsequently can quickly and easily multiplex, switch, transport and deliver the OC-3 data stream as a single entity. Further, concatenation improves SONET efficiency, as only the first OC-1 frame in the OC-3 data stream carries the full SONET overhead (i.e., signaling and control data). Each of the subsequent two SONET frames in the OC-3c signal are relieved of Path Overhead (POH) of nine bytes (i.e., 72 bits). That fact frees up 576 Kbps of bandwidth for data payload in each of the subsequent two frames, as each OC-1 frame is sent 8,000 times per second. See also Concatenation, OC-1, OC-3, and SONET.

OC-12 Optical Carrier-level 12. SONET channel of 622.08 Mbps. See also Concatenation, OC-1, OC-12c, and SONET.

OC-12c Optical Carrier-level 12 concatenated. An OC-12 signal comprising 12 OC-1s which are linked together to form a contiguous chunk of bandwidth of 622.08 Mbps. The first OC-1 in the OC-12c data stream identifies this concatenation so that all other devices which act on the signal subsequently can quickly and easily multiplex, switch, transport and deliver the OC-12 data stream as a single entity. Further, concatenation improves SONET efficiency, as only the first OC-1 frame in the OC-12c data stream carries the full SONET overhead (i.e., signaling and control data). Each of the subsequent SONET frames in the OC-12c signal are relieved of Path Overhead (POH) of nine bytes (i.e., 72 bits). That fact frees up 576 Kbps of bandwidth for data payload for each of the subsequent 11 frames, as each OC-1 frame is sent 8,000 times per second. See also Concatenation, OC-1, OC-12, and SONET.

OC-192 Optical Carrier level 192. SONET channel of 9.953 thousand million bits per second (Gbps). How you calculate the capacity of an OC-192 is to multiply times 192 by 51.840 million bits per second and thus you get 9.953 thousand million bits per second – gigabits per second. See OC-1, OC-N and SONET.

OC-256 Optical Carrier-level 256. The digital bit rate of OC-256 is 13.271.04 thousand million (Gbps) bits per second, which will accommodate 172,032 voice circuits, which is equivalent to 7,168 DS1s (T-1s) and equivalent to 256 DS3s (T-3s). See also Concatenation, OC-1, OC-N and SONET.

OC-3 Optical Carrier level 3, SONET, equivalent to 3 DS3s, line speed equals 155.5 million bits per second.

OC-48 Optical Carrier-level 48. SONET channel of 2.488 Gbps. How you calculate OC-48 is to multiply 51.84 Mbps by 48. That gives you 2.488 thousand million bits per second, or roughly 2.5 Gbps (Gigabit per second). See also Concatenation, OC-1, OC-N and SONET.

OC-48c OC-48c is OC-48 concatenated. See OC-48 and Concatenation.

OC-768 Optical Carrier-level 768. A SONET and DWDM transmission rate of 39.812 Gbps, OC-768 currently is the highest speed level defined for SONET. OC-768 is the equivalent of sending seven CDs full of data every second. See also Concatenation, OC-1, OC-N and SONET.

OC-N Optical Carrier-number. The optical interface designed to work with the STS-n signaling rate in a Synchronous Optical Network (SONET). The optical signal that results from an optical conversion of an STS-N signal. N= 1, 3, 9, 12, 18, 24, 36, 48, 192, 256, or 768. See also Concatenation, OC-1, OC-N and SONET.

OCAP OpenCable Application Platform. It is a specification that defines the software interface for OpenCable interactive applications and services. The basic idea is that by writing to the OCAP interface, a developer only has to write an application once, and it will run on any OCAP-compliant set-top box (STB). This essentially defragments the STB market, from an application developer's perspective, thereby making the STB market a bigger and more attractive market to write software for. Cable companies hope that this will lead to more applications for STBs, thereby increasing the value and stickiness of their networks. See OpenCable.

OCC Other Common Carrier. A long distance carrier other than AT&T. When AT&T was the major long distance carrier in North America, other long distance carriers were called (guess what?) "other common carrier." Now there are many, and AT&T's share of the long distance market is much smaller, the term is no longer used much (except at AT&T). These long distance companies (including AT&T) are now called IntereXchange Carriers (IXCs, or IECs).

Occam's Razor "Pluralitas non est ponenda sine neccesitate" or "plurality should not be posited without necessity." In other words, one should not increase, beyond what is necessary, the number of entities required to explain anything. Also known as the "principle of unnecessary plurality," the "principle of parsimony" and "the principle of simplicity," Occam's Razor is a principle of logic attributed to William of Ockham (ca. 1285-1349), a medieval English philosopher and Franciscan monk. Occam's Razor essentially means that the simplest plausible explanation that fits the facts is the best. The application of Occam's Razor might lead you to first investigate the electrical circuit powering a computer system such as a PC, PBX or central office. Many system failures and performance problems are the result of dirty power, improper grounding or the buildup of corrosion or dirt on connectors. In physics, razor refers to the method by which one cuts away at metaphysical concepts, eliminating those elements of an explanation that cannot be observed by the scientific method.

occupancy 1. The time a circuit or a switch is in use, i.e. occupied. Occupancy is normally expressed as a percentage, occupancy represents the actual usage versus the maximum amount of time available during a 1-hour period.

2. A call center term. The percentage of the scheduled work time that employees are actually handling calls or after-call wrap-up work, as opposed to waiting for calls.

occupational spam A term that refers to unwanted and unnecessary email that clogs corporate mail systems, produced by excessive CCing, inappropriate use of distribution lists and the "reply all" button, the use of email for non-business purposes, and other practices that generate unnecessary email in a company. It is estimated that 30% of a company's email volume consists of occupational spam.

OCDMA Optical Code Division Multiple Access is a variation on the CDMA theme used in RF (Radio Frequency) based networks. OCDMA begins with an inexpensive incoherent light source, i.e., an LED (Light Emitting Diode) light source generating light signals comprising multiple wavelengths which are not necessarily in phase. Programmable optical filters then

are used to filter out specific wavelengths of light in the signal, creating an optical signal with a unique spectral signature, that conceptually resembles a barcode. Multiple OCDMA signals can be introduced into and can coexist on the same fiber, with each comprising a unique channel. OCDMA is proposed by some as a much less expensive alternative to DWDM (Dense Wavelength Division Multiplexing). DWDM requires multiple finely tuned lasers, one for each channel, at each point that channels are added to or dropped from the fiber. OCDMA replaces these with inexpensive passive optical filters, with OCDMA gear residing only at the entry and exit points of the network. See also CDMA, Coherent Light, DWDM, and Incoherent.

OCE See Open Collaboration Environment.

ocean region One of four areas of coverage of a single geostationary Inmarsat satellite: Atlantic Ocean East (AOE), Atlantic Ocean West (AOW), Pacific Ocean (POR) and Indian Ocean (IOR). The coverages overlap, as each satellite views more than a third of the earth's surface, from 81.3 degrees south to 81.3 degrees north in latitude. However, to achieve performance within specifications, Inmarsat stipulates a minimum antenna elevation of 5 degrees above the horizon, which corresponds to a satellite coverage from 70 degrees south to 70 degrees north N in latitude.

OCh See Optical Channel.

OCL Office Code Location.

OCN Operating Company Number. A code used in the telephone industry to identify a telephone company. Company codes assigned by NECA (National Exchange Carriers Association) may be used as OCNs. See also AOCN and NECA.

OCR See Optical Character Recognition or Outgoing Call Restriction.

OCR Automatic Routing Implemented by Optus and others, this technique also assigns a number to each user up to four digits long. The sender types this number within double parentheses anywhere on the cover sheet of a fax transmission, and the LAN-networked fax system uses optical scanning technology to read the number and route the fax to the intended recipient's workstation.

OCRI The Ottawa Centre for Research and Innovation, brings together industry, govt. and universities. See www.ocri.ca.

Octal A numbering system with the base eight.

octathorp Also spelled Octothorpe, Octotherp, or Octothorp. The character on the bottom right of your touchtone keyboard, which is also typically above the 3 on your computer keyboard. It's commonly called the pound sign, but it's also called the number sign, the crosshatch sign, the tic-tack-toe sign, the enter key, the octathorp (also spelled octothorpe) and the hash. On some phones it represents "NO." And on others it represents "YES." Most U.S. long distance companies use it as the key for making another long distance credit card call without having to punch in your authorization code again. Here is the origin of this term in telephony. One day many moons ago, AT&T wanted to make sure that every symbol on the telephone keypad had a proper name that can be used in their manuals and user's guides that is published for every product that they made. Initially, the phone keypad had two strange looking symbols one under the 7 and looked like a "star", and the other under the 9 and looked like a "diamond". These two characters were actually impressed on these two keys. The problem with these two symbols was that they were not available on any typewriter or a printing operation of the time, hence AT&T decided to change them to something that was available on a standard typewriter. So they checked all the available symbols on a typewriter and found that the closet match to the "star" was the "*" which is properly referred to as the "asterisk", and the closest match to the "diamond" was the "#", or the "number sign". The problem with the "#" symbol was that it didn't officially have a proper name. It had many descriptions as the number sign, or the pound sign, or the crosshatch sign, the tic-tac-toe sign, or the hash key, but not a proper name like the asterisk. In the English language, most printed characters have a name, for example, the dot "." carries the name of "period", the "&" sign carries the name of "ampersand", and the "!" carries the name of "exclamation mark". So, on a rainy afternoon in the summer of 1965, several employees of AT&T marketing department on Church Street in New York City after a heavy luncheon meeting, were contemplating this particular problem of a name for the "number sign". After studying the character for a while, one of the men present noted that there were eight spaces around the outside of the character and he proposed that the name of this character should begin with the Greek word "Octo" which means eight. Another fellow, in jest, suggested this should be followed by a sound which is unique in the English language and he proposed the letters "th". In reality, the sound of the letters "th" is very unique to the English language, and no other language has anything like it (I checked in over 50 languages, and found none has a sound like "th", some were close but never the same). A psychologist in the group suggested that the word should end very positively with a hard ending to have a good impact on the name. While they were contemplating what should the ending be, another chap who apparently was having some stomach problems since lunch, gave vent to a mild long belch "burrrrpppp" which one of the conspirators seized upon to make the ending of the word "erp"!! The name of the number sign immediately became "Octotherp". Later in time, as the word gaind in usage, people who were unaware of its genesis began to question the lexical background of the word. Although there was little discussion about "Octo", the rest of the word "therp" could not be found in any dictionary or encyclopedia. The nearest anyone could think of was the old Saxon word "thorp" meaning a village or hamlet, like the birthplace of Sir Isaac Newton was Woolsthorp. As a result, the spelling of the word was changed to "Octothorp" by an English language professor at McGill University in Montreal who was asked by Bell Canada to find the origin of the word "Octotherp", since they were planning to use it in their printed manual and user's guides. The professor was unable to account for the origin of the word and thought it was a misprint or a typo and "corrected" the word to "Octothorp". Thanks to Elias Zaydan for the explanation. His email address: elias@canada.com.

octet An eight-bit byte.

octets received OK The number of octets (bytes) received without error.

octopus 1. A 25-pair cable that at one end has an amphenol connector (typical of what 1A2 phone systems were connected with) and at the other has many individual 2, 4, 6 and 8 wire connectors, typically male RJ-11s. The reason it's called an octopus is that it looks a bit like an octopus – one body and many arms. It's also called a Hydra.

2. The octopus. A disparaging term that small independent telephone companies and anti-big business journalists used in the late 19th century and early 20th century when referring to AT&T. The image of AT&T as an octopus was used in cartoons in the yellow press, with AT&T's spreading telephone lines drawn as tentacles of the beast. The term was also applied by the yellow press to large railroad companies and large oil companies, with railroad lines and oil pipelines drawn as octopus tentacles.

octothorpe The character on the bottom right of your touchtone keyboard, which is also typically above the 3 on your computer keyboard. It's commonly called the pound sign, but it's also called the number sign, the crosshatch sign, the tic-tack-toe sign, the enter key, the octothorpe (also spelled octathorp) and the hash. On some phones it represents "NO." And on others it represents "YES." Most US long distance companies use it as the key for making another long distance credit card call without having to redial. My friend Ralph Carlsen who worked at Bell Labs, wrote me, "First, where did the symbols * and # come from? In about 1961 when DTMF dials were still in development, two Bell Labs guys in data communications engineering (Link Rice and Jack Soderberg) toured the USA talking to people who were thinking about telephone access to computers. They asked about possible applications, and what symbols should be used on two keys that would be used exclusively for data applications. The primary result was that the symbols should be something available on all standard typewriter keyboards. The * and # were selected as a result of this study, and people did not expect to use those keys for voice services. The Bell System in those days did not look internationally to see if this was a good choice for foreign countries. Then in the early 1960s Bell Labs developed the 101 ESS which was the first stored program controlled switching system (it was a PBX). One of the first installations was at the Mayo Clinic. This PBX had lots of modern features (Call Forwarding, Speed Calling, Directed Call Pickup, etc.), some of which were activated by using the # sign. A Bell Labs supervisor Don MacPherson went to the Mayo Clinic just before call over to train the doctors and staff on how to use the new features on this state of the art switching system. During one of his lectures he felt the need to come up with a word to describe the # symbol. Don also liked to add humor to his work. His thought process which took place while at the Mayo Clinic doing lectures was as follows:

- There are eight points on the symbol so "OCTO" should be part of the name.

- We need a few more letters or another syllable to make a noun, so what should that be? (Don MacPherson at this point in his life was active in a group that was trying to get JIM THORPE's Olympic medals returned from Sweden) The phrase THORPE would be unique, and people would not suspect he was making the word up if he called it an "OCTOTHORPE".

So Don Macpherson began using the term Octothorpe to describe the # symbol in his lectures. When he returned to Bell Labs in Holmdel NJ, he told us what he had done, and began using the term Octothorpe in memos and letters. The term was picked up by other Bell Labs people and used mostly for the fun of it. Some of the documents which used the term Octothorpe found their way to Bell Operating Companies and other public places. Over the years, Don and I have enjoyed seeing the term Octothorpe appear in documents from many different sources."

OCUDP Office Channel Unit Data Port. A channel bank unit used to interface between the channel bank and a customer's DDS CSU or DSU.

OCX Ole Custom Control. According to InfoWorld, OCXes are the core of Microsoft's plans for packaging prewritten code that can be downloaded from Web servers.

ODA Office Document Architecture. ISO's standard 8613-1/8 for document architecture and interchange format adopted by MAP/TOP 3.0, GOSIP, and standardized by ECMA as ECMA-101.

ODBC ODBC – Open DataBase Connectivity – is a way of writing software in one program so you can get access to data in another program. A simple idea: You company might be a have sales order entry system. You may wish to tie that system to a inventory system. You join the two program with ODBC. More formally, ODBC is a standard database access method developed by Microsoft (September 1992), based on the Call-Level Interface (CLI) defined by the SQL (Structured Query Language) Access Group, which now is part of The Open Group. ODBC is a standardized API (Application Programming Interface) that provides a set of functions for access to databases. Independent of database format, ODBC supports access to both relational and non-relational databases. Thereby, a wide range of database programs (e.g. Microsoft Access, Microsoft Excel, Oracle and Xbase) can be accessed by a common interface. ODBC allows one to write an application that links to an appropriate ODBC driver, which translates ODBC requests into the native format for a specific file source (i.e., database).

ODBC Compliant Your software is ODBC compliant if it can talk to and access other ODBC databases. See ODBC.

ODC See optical directional coupler.

odd ball day A telephone company term. A day experiencing an extremely heavy traffic load caused by an event that is not expected to reoccur. The data would be excluded from the engineering historical base.

odd parity One of many methods for detecting errors in transmitted data. An extra bit is added to each character sent and that bit is given a value of 0 ("zero") or 1 ("one") such that the total number of ones in the character (including the parity bit) will be odd.

ODF Optical Distribution Frame.

ODI Open Data-link Interface. A device driver standard from Novell. ODI allows you to run multiple protocols on the same network adapter card. Interconnectivity strategy that adds functionality to Novell's NetWare and network computing environments by supporting multiple protocols and drivers. ODI's benefits allow you to:

- Expand your network by using multiple protocols without adding network boards to the workstation. ODI creates a "logical network board" to send different packet types over one board and wire.
- Communicate with a variety of workstations, file servers, and mainframe computers via different protocols without rebooting your workstation.
- Configure the LAP driver for any possible hardware configuration with NET.CFG, instead of using limited SHELL.CFG choices.

ODM Original design manufacturer. From Business Week of June 17, 2002: "Few products have won Compaq Computer as much favorable buzz as the sleek iPaq Pocket PC. But Compaq, now part of Hewlett-Packard, neither designed nor made the ipaq. The handheld and its clever system of interchangeable accessory sleeves were products of a Taiwanese company called HTC (www.htc.com.tw), one of an increasingly important breed known in the trade as original design manufacturers, or ODMs. PC makers have been using contract manufacturers to build many of their products for years. Even IBM, one of the last U.S. companies to do most of its own manufacturing, has contracted out all of its desktops and some ThinkPads. What's new here is the farming out of the design work as well. The amount of involvement by the "name" company ranges from extensive modifications of the ODM's work to simply slapping a logo on an off-the-shelf product. Some ODMs, such as South Korea's Samsung and LG Electronics or Taiwan's Acer, market product lines under their own brands, too. Others, such as HTC and Quanta, also Taiwanese, exist almost entirely on contract work. If consumers can't tell who makes laptops or even who designs them, how are they to choose? In the final tally, what really matter are the intangibles, such as the warranty, the service, and the tech support provided by the dealer and the company whose brand is on the product."

ODP Open Distributed Processing.

ODSI Open Directory Services Interface. Announced by Microsoft in the summer of 1995, two years after Novell's NetWare Directory Services (NDS), it is designed to do essentially what NDS does, namely give users a way to sign on to multiple servers through one simple sign-on. The ultimate idea is that NDS or ODSI will become a single place for a networked user (i.e. one on a local area network – a LAN) to sign on to multiple networked devices, e.g. telephone systems, voice mail, electronic mail, Lotus Notes and, of course, multiple file servers (wherever they're located). Sign on in one place, get into many. A big user time-saving benefit.

OEM Original Equipment Manufacturer. This term is confusing. An OEM maker is a manufacturer of original equipment. Intel is an OEM, because it makes semiconductors, though it in turn buys raw silicon from somebody else. Compaq, Dell or Gateway are not OEMs, since they buy their components from someone else and simply assemble the components into PCs and servers. They manufacture very little. They are often called value-added resellers.

OEO Optical-Electrical-Optical. Conventional fiber optic transmission systems generally are not purely optical in nature. Rather, the signal always begins and ends life as electrical, as the connected computer systems (e.g., concentrators, multiplexers, switches, and routers) are electrically-based. So, the signal is converted from electrical to optical format in order to be sent over the transmission system, and ultimately must be converted back into electrical format at the receiving computer system. Now, over a distance, the optical signal attenuates (i.e., weakens) to the point that it must be boosted. Usually, that boosting unit is in the form of a regenerative repeater, which is an OEO device. That is to say that the incoming optical (O) signal is converted to electrical (E) format, boosted in signal strength and filtered for noise, and the outgoing signal is converted back into (O) optical format as it exits the repeater. It works just fine, but OEO repeaters are expensive, are prone to failure, require local power, and must be upgraded as network speeds increase. OOO (Optical-Optical-Optical) techniques have been developed in the forms of Erbium-Doped fiber Amplifiers (EDFAs) and Raman amplifiers, which generally are considered to be substantial improvements over OEO repeaters. See also EDFA, Raman Amplifier, and Repeater.

Ofcom Office of Communications. Ofcom is the FCC's counterpart in the UK. Ofcom has regulatory authority over communications industries in the UK, with responsibilities across television, radio, wireline telecommunications and wireless telecommunications.

OFDM Orthogonal Frequency Division Multiplexing. A modulation technique for wireless communications, OFDM was patented by Bell Labs in 1970, and initially was used in a naval communications system dubbed Catherine. Much like DMT (Discrete MultiTone), OFDM splits the datastream into multiple RF (Radio Frequency) channels, each of which is sent over a subcarrier frequency. DMT and other more conventional techniques encode data symbols for a given data stream onto one radio frequency. In an OFDM system, however, each tone (i.e., frequency) is orthogonal (i.e., independent or unrelated) to the other tones; multiple data symbols are encoded concurrently onto multiple tones in a parallel fashion. The signal-to-noise ratio of each of those very precisely defined frequencies is carefully monitored to ensure maximum performance. OFDM eliminates the requirement for guard bands to separate the frequencies and, thereby, avoid interference from adjacent RF channels. Guard bands are required only around the edges of a set of tones, i.e., RF channels. This yields greater spectral efficiency, as virtually all of the allocated RF spectrum can be used for data transmission. OFDM is based on the mathematical concept of FFT (Fast Fourier Transform), which allows individual channels to maintain their orthogonality (i.e., distance) to the adjacent channels. As OFDM makes use of many narrowband tones, frequency selective fading due to multipath propagation degrades only a small portion of the signal, and has little or no effect on the remaining RF components. As a result, OFDM can eliminate Line-of-Sight (LOS) requirement for fixed wireless systems, also known as WLL (Wireless Local Loop). "Orthogonal," by the way, is from the Greek "orthos" and "gonia," translating into "straight angle," i.e., perpendicular.

OFDM is used by the HomePlug Powerline alliance to avoid line noise in a residential power-line network. How It Works: 1. Broadband content from cable modem, DSL or satellite is fed to home gateway. 2. By plugging gateway into any electrical outlet, content is bridged to any HomePlug-enabled device. 3. OFDM takes multiple signals of different frequencies and combines them to form one signal to avoid line "noise." 4. HomePlug network monitors power lines for "noise." If noise occurs on a particular frequency, OFDM engine shifts transmission to another frequency. OFDM also is used in the IEEE 802.16 WLL specification, also known as WiMAX. See also 802.16, DMT, WiMAX and WLL.

off-air Jargon for "over the air." More correctly called "over the air" signals to avoid confusion between "on the air" and "off the air." See Over-the-Air.

off-hook When the handset is lifted from its cradle it's Off-Hook. When you lift the handset of many phones, the hookswitch is moved by a spring and alerts the central office that the user wants the phone to do something like receive an incoming call or dial an outgoing call. A dial tone is a sign saying "Give me an order." The term "off-hook" originated when the early handsets were actually suspended from a metal hook on the phone. When the handset is removed from its hook or its cradle (in modern phones), it completes the electrical loop, thus signaling the central office that it wishes dial tone. Some leased line circuits work by lifting the handset, signaling the central office at the other end which rings the phone at the other end; such circuits are known as "ring down circuits." Some phones have autodialers in them. Lifting the phone signals the phone to dial that one number. An example is a phone without a dial at an airport, which automatically dials the local taxi

company. All this by simply lifting the handset at one end-going "off-hook." See also Ring Down Circuit.

off-hook call announce A telephone system feature. A telephone has a speaker. If the person is speaking on the phone, another person (inside the building, on the same phone system) can "off-hook voice announce" you and can give you a message or speak with you. You will hear their voice coming through the speaker on your phone. (So may the person on the other end.) Depending on your phone, you may be able to put your hand over your telephone handset and whisper something back to the person who's "off-hook voice announcing" you. Otherwise you'll have to hang up or put the person on hold, and speak to the person on another line.

off-hook queue There are two types of queuing: ON-HOOK and OFF-HOOK. In On-Hook Queuing, the user dials his number, the switch tells him the outgoing trunks are busy. The user then hangs up. The switch calls him back when a trunk becomes available. In Off-Hook dialing, the user waits with his receiver screwed into his ear until a trunk comes free and the PBX connects him to the next available trunk. Off-Hook queues are usually shorter than on-hook queues. If a trunk doesn't come free quickly in off-hook queuing, the call will often flow over onto the more expensive DDD trunks. Off-hook queuing costs more but keeps the user waiting less. On-hook queuing costs less by waiting for a cheaper trunk but can be tiring and frustrating for the workers waiting for their calls to go through.

off-hook routing See Off-Hook and Ringdown.

off-hook voice announce See Off-Hook Call Announce.

off-line Any equipment not actively connected to a phone line but which can be activated to work with that system is Off-Line. This concept also applies to computer systems. For example, a modem attached to or built into a microcomputer can be plugged permanently into a phone line. But the microcomputer can be used for word processing most of the time. While it's doing word processing, it is "off-line." When the user loads the communications software and turns on the modem, the microcomputer is now said to be "on line." Off-line computer storage is a place to put stuff which a computer cannot access "on-line," like a hard-disk. Off-line storage might be microfilm or microfiche.

off-line storage Storage that is not under the control of a processing unit.

off-net calling Phone calls which are carried in part on a network but are destined for a phone not on the network, i.e. some part of the conversation's journey will be over the public switched network or over someone else's network. MCI defines off-net calls as "Billable calls to non-tariffed cities. Can be MCI Off-Net or WATS Off-Net. Classified as Tier 2 for tariff purposes."

Off-Network Access Line ONAL. A circuit in a private network which allows the user to go off the private network and complete calls on the public dial network.

off-peak The periods of time after the business day has ended during which carriers offer discounted airtime charges. Usually, OFF-PEAK rates are available for cellular calls between 7:00 p.m. and 7:00 a.m. and on weekends and holidays, but times vary among carriers. Among landline carriers, the business day usually ends at 5 p.m., after which time residential calling builds, and that ends at 11 p.m., after which little happens, except rates drop once again until they rise at the beginning of the next business day at 8 a.m. the next morning.

off-portal Of or pertaining to content and/or services that are not provisioned directly through the network operator's portal, but instead through 3rd-party websites. See carrier portal.

off-premises extension OPX. Now also called OPS for Off-Premises Station. A telephone located in a different office or building from the main phone system. The OPX is connected by a phone line dedicated to it. It acts as if it were in the same place as the main phone system and can use its full capabilities. Here's another explanation a reader sent in. OPX is the appearance of an actual telephone line (such as 212-691-8215) in two physically separate locations. For example, this line (212-691-8215) could appear and ring in my office and at my home without my home phone being a part of the office telephone system. An OPX is commonly used for answering services or for small businesses. Bell operating phone companies are sharply increasing the charge for dedicated OPS lines – often by several hundred percent per year.

off-site night answer A feature of some phone systems that allows phone calls to the main line to be forwarded to another phone line after hours.

off-the-grid Euphemism for being off the Net. "Sorry I didn't email you last week; I was off the grid in Europe." Off-the-grid also refers to someone who lives in a rural area without running water, electricity or other means.

off-the-shelf When something has already been produced and is available for immediate delivery, it is said to be available "off-the-shelf." It is presumably sitting on the shelf in a warehouse waiting for your order. Sometimes called "shrinkwrapped," referring to the plastic wrap around the box. It means you don'1t are not meant to need intervention by a programmer or integrator to make the software work.

off/on hook A modem term. Modem operations which are the equivalent of manually lifting a phone receiver (taking it off hook) and replacing it (going on hook). See ff-Hook.

offered call 1. A call that is presented to a trunk or group of trunks. See Traffic Engineering.

2. A call center term. A call that is received by the ACD. Offered calls are then either answered by an employee (handled) or abandoned.

offered load 1. The total traffic load, including load that results from retries, submitted to a system, group of servers, or the network over a circuit. See also Offered Traffic.

2. In Frame Relay terms, the total data rate presented to the network. Note that Offered Load does not translate to carried load. See also Committed Information Rate.

offered traffic The total attempts to seize a group of servers.

office automation Nobody knows what office automation means, though there are many consultants out there who will tell you what it means for an grand sum of money. In reality, it's a benign, imprecise term for data processing when it applies to self-focused (as opposed to customer-focused) white collar-type activities – accounting, word processing, communications, document management.

office busy hour OBH. Normally, the hour in which the maximum load on a switching system, desk, etc., occurs.

office characteristics A telephone company definition. The peculiarities that make one switching entity different from others. Some examples: a. Class of Service Mixture b. Trunking Configuration c. Holding Times d. Special Services e. Calling habits of subscribers and f. Size of office.

office class Functional ranking of a telephone central office switch depending on transmission requirements and hierarchical relationship to other switching centers. (Awful mouthful!) There used to be five classes of switches in the U.S. telephone network hierarchy, with the one closest to the end-subscriber being a class 5 central office. But technology and marketing is changing things and, by distributing intelligence closer to the end user, it is diffusing our traditional definitions of network hierarchies and the class of switches. See Class 5 Central Office. See Office Classification.

office classification Prior to divestiture, those numbers that were assigned to offices according to their hierarchical function in the U.S. public switched telephone network. The following class numbers are used:

    Class 1: Regional Center (RC)
    Class 2: Sectional Center (SC)
    Class 3: Primary Center (PC)
    Class 4: Toll Center (TC) if operators are present, or else Toll Point (TP)
    Class 5: End Office (EO) (local central office)

    Any one center handles traffic from one to two or more centers lower in the hierarchy. Since divestiture and with more intelligent software going into telephone switching offices, these designations have become less firm.

office code The first three digits of your seven-digit local telephone number. Also called NXX code. See also NXX.

office network A network within an office. An older term for a Local Area Network. User concern is with application sharing, file/database sharing, electronic mail, word processing and circuit switching.

Office Repeater Bay ORB. Mounting and powering arrangement for digital regenerators, such as for T-1 lines.

office user interface A special shell program which sets up windows with menus of available utilities and applications.

office window interface See Office User Interface.

offline reader An application that lets you read postings to Usenet newsgroups without having to stay connected to the Internet. The program downloads all the newsgroup postings you have not read and disconnects from your Internet Service Provider. You can then read the postings at your convenience without incurring online charges or tying up your telephone line. If you reply to any of these postings, the program will automatically upload them to the correct newsgroup the next time you connect to your service provider.

offset The offset of a port or a memory location is the difference between the address of the specific port or memory address and the address of the first port or memory address within a contiguous group of ports or a memory window. This term is used when identifying the locations of registers located with respect to the base address of the 16 contiguous I/O ports in a PCMCIA card. It is also used when identifying the location of memory mapped

registers with respect to the base address of the memory window. See also Offset Parabolic Antenna and Offset Geometry.

offset geometry Shadow-free geometry. The feeder in the primary focus is mounted so that the effect of its shadow on the secondary radiation is negligible.

offset parabolic antenna An offset antenna is a new form of satellite antenna that is taller than it is wide. According to the manufacturers, the antenna design makes for more efficient use of the antenna surface. What that means is that it captures more of the satellite signal hitting the antenna. Offset antennas are more expensive than the "normal" parabolic satellite antennas, which are called "prime focus parabolic antennas." They are more expensive because they cost more to make since they typically must be made out of one sheet of metal. Offset antennas are harder to carry around, since you can't make them out of several foldover sheets of metal.

offshoring Sending some corporate function – typically manufacturing overseas to a cheaper labor country. See Outsourcing.

offsite night answer This mode allows incoming after-hours calls to be forwarded automatically to an off site location.

OFHC Abbreviation for oxygen-free, high conductivity copper. It has no residual deoxidant, 99.95% minimum copper content and an average annealed conductivity of 101%.

OFM Operational Fixed Microwave. See IMASS.

OFNR Optical Fiber Nonconductive Riser.

OFNP Optical Fiber Nonconductive Plenum.

OFS Operational Fixed Station.

OFTEL The OFfice of TELecommunications in the United Kingdom. OFTEL is the main regulatory body over the U.K. telecommunications industry, which includes phones and cable. OFTEL was created at the time British telecommunications was sort of de-regulated in 1984 by the Telecommunications Act. On its Web site, Oftel describes itself: "OFTEL is the regulator - or "watchdog" - for the UK telecoms industry. Broadcast transmission is also part of OFTEL's remit. Our aim is for customers to get the best possible deal in terms of quality, choice and value for money. OFTEL is a government department but independent of ministerial control. It is headed by the Director General of Telecommunications, who is appointed by the Secretary of State for Trade and Industry." All telecommunications operators in the U.K. – such as BT, Mercury, local cable companies, mobile network operators and the increasing number of new operators – must have an operating licence. These set out what the operators can – or must - do or not do. For example, BT's licence contains the formula (currently RPI – 7.5%) which controls the prices of its main network services. Users of telecom services supplied by the operators also need a licence. In nearly all cases they are covered by a class licence – a licence issued to a group, not an individual, allowing certain activities. For example, the Self Provision Licence (SPL) enables customers to use telephones in their homes. What are OFTEL's functions? Under the Telecommunications Act 1984, OFTEL has a number of functions. Briefly these are:

- to ensure that licensees comply with their licence conditions.
- to initiate the modification of licence conditions either by agreement with the licensee or, failing that, by reference to the Monopolies and Mergers Commission (MMC) together with the Director General of Fair Trading to enforce competition legislation - under both the Fair Trading Act 1973 and the Competition Act 1980 - in relation to telecommunications. OFTEL expects to gain wider powers under the new 1998 Competition Bill. This will bring UK law into line with European law, and is much more flexible.
- to advise the Secretary of State for Trade and Industry on telecommunications matters and the granting of new licences.
- to obtain information and arrange for publication where this would help users.
- to consider complaints and enquiries made about telecommunications services or apparatus.
    www.oftel.gov.uk.

OFX Open Financial eXchange. A technical specification for the exchange of electronic financial data over the Internet for the purpose of Electronic Commerce. OFX was developed jointly by CheckFree, Intuit and Microsoft in concert with financial services and technology companies. The OFX specification is an Internet-oriented client/server system which provides security, features full data synchronization, and offers error recovery mechanisms to simplify and streamline the process by which financial services companies connect to transactional Web sites, thin clients and financial software. OFX supports a range of financial activities including consumer and small business banking, bill presentment, and investments. According to Open Financial Exchange, Specification 1.0.2 issued on May 30, 1997, Open Financial Exchange is a broad-based framework for exchanging financial data and instructions between customers and their financial institutions. It allows institutions to connect directly to their customers without requiring an intermediary. Open Financial Exchange is an open specification that anyone can implement: any financial institution, transaction processor, software developer, or other party. It uses widely accepted open standards for data formatting (such as SGML), connectivity (such as TCP/IP and HTTP), and security (such as SSL).

Open Financial Exchange defines the request and response messages used by each financial service as well as the common framework and infrastructure to support the communication of those messages. This specification does not describe any specific product implementation.

The following principles were used in designing Open Financial Exchange:

Broad Range of Financial Activities. Open Financial Exchange provides support for a broad range of financial activities. Open Financial Exchange 1.0.1 specifies the following services:

- Bank statement download
- Credit card statement download
- Funds transfers including recurring transfers
- Consumer payments, including recurring payments
- Business payments, including recurring payments
- Brokerage and mutual fund statement download, including transaction
- history, current holdings, and balances.

Broad Range of Financial Institutions - Open Financial Exchange supports communication with a broad range of financial institutions (FIs), including Banks, Brokerage houses, Merchants, Processors, Financial advisors, Government agencies

Platform Independent. Open Financial Exchange can be implemented on a wide variety of front-end client devices, including those running Windows 3.1, Windows 95, Windows NT, Macintosh, or UNIX. It also supports a wide variety of Web-based environments, including those using HTML, Java, JavaScript, or ActiveX. Similarly on the back-end, Open Financial Exchange can be implemented on a wide variety of server systems, including those running UNIX, Windows, or OS/2. The design of Open Financial Exchange is as a client and server system. An end-user uses a client application to communicate with a server at a financial institution. The form of communication is requests from the client to the server and responses from the server back to the client. Open Financial Exchange uses the Internet Protocol (IP) suite to provide the communication channel between a client and a server. IP protocols are the foundation of the public Internet and a private network can also use them. Clients use the HyperText Transport Protocol (HTTP) to communicate to an Open Financial Exchange server. The World Wide Web throughout uses the same HTTP protocol. In principle, a financial institution can use any off-the-shelf web server to implement its support for Open Financial Exchange. To communicate by means of Open Financial Exchange over the Internet, the client must establish an Internet connection. This connection can be a dial-up Point-to-Point Protocol (PPP) connection to an Internet Service Provider (ISP) or a connection over a local area network that has a gateway to the Internet. Clients use the HTTP POST command to send a request to the previously acquired Uniform Resource Locator (URL) for the desired financial institution. The URL presumably identifies a Common Gateway Interface (CGI) or other process on an FI server that can accept Open Financial Exchange requests and produce a response. See also Electronic Commerce. See also Electronic Commerce, GOLD and XML. www.ofx.net.

OGG Ogg Vorbis. A lossy audio format similar to MP3. It's an open-source format that can produce files with better results than MP3. Although hardware support for this codec isn't as widespread, the format is gaining popularity.

OGM OutGoing Message. The message an answering machine delivers to someone who calls. Sample, "I'm not here. Leave a message after the beep."

OGT OutGoing trunk.

OhNoSecond That tiny fraction of time in which you realize you've just made a gigantic mistake.

OHD Optical Hard Drive. A term pioneered by Pinnacle Micro, Irvine, CA. OHD technology, according to Pinnacle, combines the advantages of magneto-optical technology with speeds faster than most hard drives.

Ohm The practical unit of resistance. The resistance that will allow one ampere of current to pass at the electrical potential of one volt. Ohm's Law dictates the relation between the current, electromotive force and resistance in a circuit:

Amperes = Volts divided by Ohms Volts = product of Amperes and Ohms Ohms = Volts divided by Amperes

Ohm's Law The law that for any circuit, the electric current is directly proportional to the voltage and is inversely proportional to the resistance. The law relates current measured as Amps (I), voltage (V) and resistance measured in Ohms (R). Ohm's Law is V = I x R.

It can also be expressed as I = V/R, or R = V/I. Sometimes E is used instead of volts. E is short for EMF or Electro Motive Force, a synonym for voltage.

Ohms Measures of resistance. A resistance of one Ohm allows one Ampere to flow when a potential difference of one volt is applied to the resistance. See Ohm's LAW.

OHQ Off-Hook Queue. See Off-Hook Queuing.

OHR Optical Handwriting Recognition. Exactly what it says. Machine reading of handwriting.

OHSA Occupational Health and Safety Act. Specifically the Williams-Steiger law passed in 1970 covering all factors relating to safety in places of employment.

OHVA Off-Hook Voice Announce. A phone system feature that permits an intercom announcement to be heard through a speaker at a phone where the handset is in use on an outside call.

OIC Online Insertion and Removal is the practice of replacing or removing equipment components without powering off the system. This provides a high availability of service because the components are hot swappable. This term is commonly found in Cisco equipment documentation.

OID Object Identifier. See MIB.

OIF Optical Internetworking Forum. A coalition of manufacturers, telecom service providers, and end users that cooperate in the acceleration of the deployment of optical internetworks. Members of the OIF are engaged in the development of optical products based on specifications that ensure compatibility and interoperability. www.oiforum.com.

OIR Online Insertion and Removal. Feature that permits the addition, the replacement, or the removal of cards without interrupting the system power, entering console commands, or causing other software or interfaces to shutdown. Sometimes called hot swapping or power-on servicing. See also Hot Swappable.

OLAP On-Line Analytical Processing, also called a multidimensional database. According to PC Week, these databases can slice and dice reams of data to produce meaningful results that go far beyond what can be produced using the traditional two-dimensional query and report tools that work with most relational databases. OLAP data servers are best suited to work with data warehouses. A database warehouse consolidates information from many departments within a company. This data can either be accessed quickly by users or put on an OLAP server for more thorough analysis.

According to Microsoft, OLAP refers to a class of database-management systems and client software that arranges data in multiple dimensions for high-speed analysis. Microsoft does not currently ship its own OLAP software, but is in the process of adding OLAP functionality to its SQL Server database and some client applications. In October, 1996, Microsoft announced the acquisition of OLAP technology from Panorama Software Systems in Tel Aviv. At the time, Microsoft said it intended to use the technology to add OLAP features to SQL Server and to some of its desktop tools, such as its Excel spreadsheet and the Internet Explorer browser.

OLAP Client End-user applications, that can request slices from OLAP servers and provide two- or multidimensional displays, user modifications, selections, ranking, calculations, etc.,for visualization and navigational purposes. OLAP clients can be as simple as a spreadsheet program retrieving a slice for further work by a spreadsheet-literate user or or as high functioned as a financial-modeling or sales-analysis application.

OLAP Server An OLAP server is a high-capacity, multi-user, data manipulation engine specifically designed to support and operate on multi-dimensional data structures. A multi-dimensional is arranged so that every data item is located and accessed based on intersection of the dimension members that define the item. The design of the server and the structure of the data are optimized for rapid, ad-hoc information retrieval in any orientation, as well as for fast, flexible calculation and transformation of raw data based on formulaic relationships.

OLC See Overload Class.

Old Media Old media is magazines and TV, etc. New media is Internet and web "magazines" and web blogs. The term old media became important when America Online (new media) announced it had agreed to buy Time Warner (old media) in January, 2000. Ultimately, the old media part of Time Warner did financially better than the new media part, namely AOL. But that's a whole long boring story.

OLE Object Linking and Embedding. A Microsoft Corp. software technology that allows Windows programs to exchange information and work together. For instance, a word processing document with OLE capabilities could contain a link to a chart created in a spreadsheet. Version 2.0 of OLE was released in the Windows 95 operating system. OLE means tying one piece of information in one form into a document in another form, such that a change in one piece of information will be automatically reflected in the other document. Here's an explanation from the New York Times: Business reports may contain information in a variety of formats, including text and numbers, charts, tables, images, graphics, sound and video. Typically, these are created in separate applications programs (e.g. spreadsheet, word processing, charting, database, etc.) and are merged into a single document (i.e. the report). But when the numbers used to create a chart are changed the chart must be updated as well. The executive then has to track down all the various components of the report, call up their respective applications, make the changes and stitch everything back together. OLE promises to keep track of those links and update the various components as they change. Here's an explanation from PC Magazine: Ole is a complex specification that describes the interfaces used for such tasks as embedding objects created by one application within documents created by another, performing drag-and-drop data transfers within or between applications, creating automation servers that expose their inner functionality to other programs, extending the Windows 95 shell with custom DLLs, and much more. Version 1.0 of the specification was originally created for placing objects such as Excel spreadsheets inside documents created by other applications such as Microsoft Word for Windows. OLE 2.0 greatly expanded the scope of OLE and made the original name obsolete, but the name had achieved widespread recognition and was retained. See OLE DB.

OLE DB Abbreviation of Object Linking and Embedding (pronounced as separate letters or as "oh-leh"). OLE is a compound document standard developed by Microsoft Corporation. It enables you to create objects with one application and then link or embed them in a second application. Embedded objects retain their original format and link to the application that created them. Support for OLE is built into the Windows and Macintosh operating systems. A competing compound document standard developed jointly by IBM, Apple Computer, and other computer firms is called OpenDoc.

OLEC Originating Local Exchange Carrier. Where the call comes from. See also LEC, CLEC and TLEC.

OLED OLED stands for organic light-emitting diode. It's a technology that's brighter and more cost-effective than the current LCD – or liquid crystal display – technology. The organic chemicals in OLED displays emit their own light when electrically charged and don't rely on backlighting, which adds weight, cost and thickness to conventional (i.e. non-OLED) screens. OLED screens will be lighter and brighter and consume less power than today's displays. They will also be thin and flexible, even foldable eventually.

OLIU Optical Line Interface Unit.

OLNS Originating Line Number Screening.

OLO This is a European term that stands for "Other Licensed Operators" referring to all Public Telecommunications Operators – except for the one that was originally the PTT.

OLS 1. Originating Line Screening.
    2. OnLine Services. See Internet Access Provider.

OLT Optical line terminal or optical line termination device. A device serving to terminate an optical trunk in a carrier CO (Central Office) or headend, an OLT is one of the primary elements of a FSAN (Full Service Optical Network). See also FSAN and FTTP.

OLTP OnLine Transaction Processing. A generic concept in the computer industry to cover everything from issuing airline tickets to dispensing money out of street-corner, automated teller machines.

OM Operational Measurement.

OMAP Texas Instruments' Open Multimedia Applications Protocol. OMAP is designed to reduce the number of chips required to put together a cell phone, thereby reducing the cost. It's also designed to create new applications – such as videoconferencing – for the new faster 2.5G and 3G cellphones. See 2.5G and 3G.

OMAT Operational Measurement and Analysis Tool.

OMC Operation and Maintenance Center. Computer hardware and software assigned specifically to monitor and manage one part of a telecommunications network, usually employed in GSM (Global System for Mobile Communications).

OMC-Env Operations and Maintenance Center – Environment. An OMC dedicated to monitoring and managing the physical environment where telecommunications equipment resides.

OMC-IN Operations and Maintenance Center – Intelligent Network. A wireless telecommunications term. An OMC dedicated to monitoring and managing components of the Intelligent Network in GSM (Global System for Mobile Communications). This includes the HLR, AUC, EIR, SMSC, and VMS.

OMC-Misc Operations and Maintenance Center – Miscellaneous. An OMC that manages and monitors non-intelligent devices in a mobile telecommunications network under GSM (Global System for Mobile Communications).

OMC-R Operations and Maintenance Center – Radio. An OMC that manages and monitors the radio interface under GSM (Global System for Mobile Communications) including the BSS, BSC, and BTS.

OMC-S Operations and Maintenance Center – Switching. An OMC dedicated to monitoring and managing switches in a GSM (Global System for Mobile Communications) telecommunications network.

OMC-SS7 Operations and Maintenance Center – SS7. A wireless telecommunications term. An OMC dedicated to monitoring and managing the SS7 signaling network in a GSM (Global System for Mobile Communications) telecommunications network.

OMC-T Operations and Maintenance Center – Transmission. A wireless telecommunications term. An OMC dedicated to managing and monitoring transmission activities under a GSM (Global System for Mobile Communications) telecommunications network.

OMC-WAN Operations and Maintenance Center – Wide Area Network. A wireless telecommunications term. An OMC dedicated to managing and monitoring components of the Wide Area Network under GSM. The WAN links all or most devices in a GSM (Global System for Mobile Communications) telecommunications network together.

OMEGA A global radio navigation system that provides position information by measuring phase difference between signals radiated by a network of eight transmitting stations deployed worldwide. The transmitted signals time-share transmission on frequencies of 10.2, 11.05, 11.33, and 13.6 KHz. Since the transmissions are coordinated with UTC (Universal Time Coordinated), they also provide time reference. In the U.S., UTC is the responsibility of the USNO (U.S. Naval Observatory). See also UTC.

OMG Object Management Group. A group of major systems vendors involved in the definition of standards for object management. According the OMG, "a non-profit consortium dedicated to promoting the theory and practice of Object Technology for the development of distributed computing systems. OMG was formed to help reduce the complexity, lower the costs, and hasten the introduction of new software applications." The stated goal of the OMG is to "provide a common architectural framework for object-oriented applications based on widely available interface specifications." OMG membership currently stands at over 600 software vendors, developers and end users. www.omg.org.

OMNI Operating Missions as Nodes on the Internet. A proof-of-concept project by NASA from 1999 to 2003, to demonstrate that space satellites, Space Shuttles, high-altitude balloons, aircraft, and other airborne and remote terrestrial sites could serve as nodes on the Internet. NASA's interest in proving the concept, which it succeeded in doing, was to set the stage for it to use low-cost, commercial-off-the-shelf networking gear in an Internet-based networking scenario, instead of more expensive, custom-made, proprietary solutions.

omnidirectional A microphone with a pickup pattern essentially uniform in all directions.

omnidirectional antenna An antenna whose pattern is nondirectional in azimuth. The vertical pattern may be of any shape.

OMNIPoint A program established by the Network Management Forum (NMF) to speed the implementation of TMN (Telecommunications Management Network). OMNIPoint is a collaborative partnership between NMF and a number of other groups. Each OMNIPoint document release specifies a strategy and a comprehensive set of network management components, such as standards, de-facto standards, software development tools, and implementation and procurement guides. The intent of the document releases is to provide users with sufficient information to prepare and evaluate responses to RFPs and to guide suppliers in implementing network management products. See also Network Management Forum and TMN.

OMR Optical Mark Recognition. Refers to machine recognition of filled-in "bubbles" on reader service bingo cards.

on the horn To be "on the horn" in England or Australia is to be on the phone.

On-board processing in satellite communications, the functions in an orbiting satellite that process the signals it relays; may include regeneration such as change of access method, and switching, such as to another satellite and multiplexing such as of several user terminal signals.

on-demand connection An ISDN BRI term. The ability to automatically suspend and resume a physical connection while "spoofing" network protocols, routing and applications. The physical connection is only brought up on-demand. This ensures that users' ISDN holding time charges are proportional to their useful holding time not the total holding time.

on-demand dialing ISDN cost-savings feature that sets up, transfers, and closes a call only if the ISDN device detects a data packet that is addressed to a remote network.

on-going maintenance A term used in the secondary telecom equipment business. A manufacturer's guarantee of maintainability from one owner to the next. There are all types of conditions which must be met, including: The machine has been under the manufacturer's maintenance contract until its time of deinstallation. There hasn't been any damage in storage or transit. And there have not been any modifications made.

on-hook When the phone handset is resting in its cradle. The phone is not connected to any particular line. Only the bell is active, i.e. it will ring if a call comes in. On-Hook is thus the normal, inactive condition of a telephone system terminal device. See On-Hook Dialing and Off-Hook.

on-hook dialing Allows a caller to dial a call without lifting his handset. After dialing, the caller can listen to the progress of the call over the phone's built-in speaker. When you hear the called person answer, you can pick up the handset and speak or you can talk hands-free in the direction of your phone, if it's a speakerphone. Critical: Many phones have speakers for hands-free listening. Not all phones with speakers are speakerphones – i.e. have microphones, which allow you to speak, also.

on-line When a device is actively connected to a PBX or a computer, it is On-Line. Terminals, PCs, modems and phones are often On-Line. More and more people are spelling it online, as one word.

on-net Telephone calls which stay on a customer's private network, traveling by private line from beginning to end are said to be on-net. Here's MCI's definition: Billable calls to MCI-tariffed cities, including those cities reached via leased lines. Can be MCI On-Net or WATS On-Net. Classified as Tier 1 for tariff purposes.

on-network calling A term used to describe a call that originates and terminates within the limits and facilities of a private network.

On-Off Keying See OOK.

on-portal Of or pertaining to content and/or services that are provisioned directly through the network operator's portal.

on-ramps A way of getting onto The Information Superhighway. Such an on-ramp could be anything from a phone line to a two-way cable TV channel. Many companies are trying to create on-ramps, however they define them.

on-the-fly switching Refers to a method of Frame Switching where the switching device commences forwarding a frame after it has determined the destination port without waiting for the entire frame to have been received on the incoming port. Also known as cut-through switching.

ONA See Open Network Architecture.

ONAC Operations Network Administration Center.

ONAL Off Network Access Line.

ONC Open Network Computing. A distributed applications architecture developed by Sun Microsystems. Includes NFS, NIS and RPC. Now part of Solaris OS.

ONC+ Part of Sun Microsystems' ENOS Networking Solutions.

one armed cable locator A backhoe. Ben Kurtzer of Qwest Communications first heard this term at AT&T, but it's used throughout the company here also. It refers to the uncanny ability of the backhoe to locate buried cable, even when you clearly aren't looking for it and, in fact, are trying to avoid it.

one call 1. Fax-back system that requires you to call from your fax machine to send documents on the same line after picking from menu of verbal prompts. the other fax-back system is called a two-call system. You dial from one line and tell the fax-back machine on the other end that you want it to send your requested fax to second number, i.e. one where your fax machine will receive your requested message.

2. In Texas, "One Call" is a special system set up for all utilities with underground installations. It means that you can make one phone call to a toll-free number to get the various utilities to come out to a specific address and mark the exact locations of the underground utilities (gas, electric, cable, telephone, water, etc.). Writes Sally Hahn of the Texas PUC, "With the boom in fiber optic cable installation in and around Austin, we've seen way too many cases in which the installers failed to make the One Call, and have seen gas lines, water lines, and other fiber optic lines destroyed by some over eager (and ill-informed) backhoe operator. The telecom companies (and/or their insurers), and thus their customers, eventually end up eating these costs. I would offer a suggestion to add this additional definition of this term to your dictionary. Maybe someone with a brain will actually look it up before they send out the backhoes."

one dimensional coding A data compression scheme for fax machines that considers each scan line as being unique without referencing it to a previous scan line. One dimensional coding operates horizontally only.

one hop set A set of hosts which are one hop apart in terms of internetwork protocols.

one number calling You give someone a phone number at which they can reach you 24 hours a day. It's typically an 800 number but it could be a local number. Here's how it works: They call the number. A computer answers. That computer might ask the caller to touchtone in some digits. That will identify whom the caller is trying to call. The computer will then check its memory. What number is the person likely to be at, at this very moment? It will then dial that number. If the number answers, it will connect the caller. If the number doesn't answer, it will call another number, and, if it answers, it will connect the call. How does it know which numbers to call? The subscriber (i.e. the person who wants to be reached) might have given the computer several numbers – office number, cell phone numbers, home number, etc.

one number presence This is the consistent use of one telephone number (particularly an 800 number) across all advertising media. The long distance vendors can arrange this for both in-state and national 800 services.

one number systems Also called Follow Me Systems. Follow me systems and services are based on the premise that people are mobile (e.g., they move around a lot in and out of the office), and have many phone numbers or places they might be. A person could have an office number, a cellular number, a voice mail number, a home number, and a pager number. Which phone number will a caller be at? Follow me systems will "trackdown" the user being called no matter where they are and connect the caller to the user. The caller need only dial a single phone number. Usually, network or local switch provided call data (or data gathered by a voice response unit) is used to identify each call as being intended for a specific user. Based on options the user has selected, the caller will hear an answering prompt customized to that user, and the one number system will then automatically attempt to locate the user at one of several locations. The tracking-down process varies considerably between follow me systems. Some systems try multiple locations at once, others will try the possible destination locations sequentially. Almost all follow me services provide the caller an exit to voice mail at various points of the call.

one plus bulk restriction This is the name for a local service provided by a Northern Telecom DMS central office switch. One-Plus Bulk Restriction allows subscribers to deny or permit all one-plus (i.e. long distance) calls from their phones by dialing a special PIN (Personal Identification Number).

one plus per-call restriction This is the name for a local service provided by a Northern Telecom DMS central office switch. Subscribers can restrict one-plus toll calls from their phones by requiring that a PIN (Personal Identification Number) be dialed prior to a one-plus call. If the PIN is valid, the caller hears a second dial tone, and the one-plus number can then be dialed. When a one-plus call is attempted without the PIN or with a wrong PIN, the caller is routed to a tone or announcement and is not able to place the one-plus call.

one plus restriction A central office service. Telephone subscribers can now limit one-plus toll calls by selecting an authorization code that must be dialed before any one-plus call will be connected. If the code is valid, the caller hears a second dial tone, permitting the number to be dialed and connected.

one screen company A video content provider that provides content exclusively for a single type of display screen, such as only for a movie screen, or only for a TV screen, or only for a PC screen. These days the number of one-screen companies is shrinking, as video providers repurpose their content or develop customized content for multiple types of display screens. All of this is driven by and is driving network convergence, device convergence, and application convergence.

one thirtieth One thirtieth of a second is the time it takes human eyes to react to light. Project each frame of a home movie for one thirtieth of a second, and viewers, unable to distinguish separate frames, see continuous motion. Light, during the time one frame is projected, travels 6,200 miles. If you climb aboard a light beam in Chicago, you'll be in Tokyo in the blink of an eye.

One Time Programmable OTP. A term describing memory that can be programmed to a specific value once, and thereafter cannot be changed (or can only be revised in a limited way.) OTP EPROMs are typically ordinary EPROMs that have been packaged in such as way that ultra-violet light cannot be used to erase the contents of the EPROM. Such packaging is usually less expensive.

one way bypass This is a term we are beginning to increasingly see as international telecommunications markets are liberalized, i.e. deregulated and competitors are allowed in. One Way Bypass is an abuse of its dominant position by an incumbent operator in an unliberalised country. The incumbent insists on receiving all traffic into the country via the accounting rate system (and so gets the benefits of very high settlement rates) but takes advantage of the liberalised system in other countries (to avoid paying similarly high outgoing settlement rates) by bypassing the accounting rate system for outgoing traffic and finding other ways to terminate its traffic. Some regulators have devised rules to try to prevent this by insisting that all operators have a proportion of total incoming traffic no larger than their proportion of total outgoing traffic on any given route (known as "proportional return"). See also whipsaw.

one way encryption Irreversible transformation of plaintext to ciphertext, such that the plaintext cannot be recovered from the ciphertext by other than exhaustive procedures even if the cryptographic key is known.

one way operation See SIMPLEX.

one way splitting When the attendant is connected to an outside trunk and an internal phone, pushing a button on the console allows her to speak privately with the internal extension, thus "splitting" her off from the external trunk.

one way trade A call center term. A schedule trade in which only one employee is working the other's schedule.

one way trunk A trunk between a switch (PBX) and a central office, or between central offices, where traffic originates from only one end. You can, of course, still speak and listen on the trunk. It's just like a normal two-way trunk except that a one-way trunk can only be used for dialing out or only to receive calls.

One Wilshire A 30-story, 656,000-square foot building at the corner of South Grand Avenue and Wilshire Boulevard in downtown Los Angeles. It is the largest carrier hotel in the western United States. The building houses approximately 250 domestic and international service providers. The building has backup generators and fuel storage and pumping facilities to operate all generators on a continuous basis for up to 24 hours without refueling. See carrier hotel.

ones density The requirement for digital transmission lines in the public switched telephone network that eight consecutive zeros cannot exist in a digital data transmission. On a T-1 line, 0 means no voltage, no pulse. Too many zeroes and the repeaters lost count because they had no signal pulses to count. Ones density exists because repeaters and clocking devices within the network will lose timing after receiving eight zeros in a row. There are many techniques or algorithms used to insert a one after every seventh-consecutive zero. The question of how many consecutive zeros you can have on a digital line is changing. In the old days, the FCC actually said you could have 15 zeroes in a row. Now the FCC says you can have up to about 40 something (just like you) Os without 'harming the network." For all practical purposes, seven consecutive zeros is the maximum today. See Bit Stuffing.

ONI 1. Operator Number Identification.

2. Optical Network Interface. A device which converts photons to electrons and vice versa. It's a device which converts an optical signal into an electrical signal that non-optical telecommunications transmission and switching devices can understand and vice versa.

onion routing Onion routing and onior routers are colorful terms for network technology that has been developed (and which continues to be improved) to make it difficult for an entity to piece together Internet traffic data and Web usage data to identify and monitor specific Internet users and their online activities. Onion routing involves low-latency Internet-based connections that resist traffic analysis, eavesdropping, and other attacks both by outsiders (e.g. Internet routers) and insiders (Onion Routing servers themselves). Onion Routing prevents the transport medium from knowing who is communicating with whom – the network knows only that communication is taking place. In addition, the content of the communication is hidden from eavesdroppers up to the point where the traffic leaves the onion routing (OR) network. Onion Routing works in the following way: An application, instead of making a (socket) connection directly to a destination machine, makes a socket connection to an Onion Routing Proxy. That Onion Routing Proxy builds an anonymous connection through several other Onion Routers to the destination. Each Onion Router can only identify adjacent Onion Routers along the route. Before sending data over an anonymous connection, the first Onion Router adds a layer of encryption for each Onion Router in the route. As data moves through the anonymous connection, each Onion Router removes one layer of encryption, so it finally arrives as plaintext. This layering occurs in the reverse order for data moving back to the initiator. Data passed along the anonymous connection appears different at each Onion Router, so data cannot be tracked en route and compromised Onion Routers cannot cooperate. When the connection is broken, all information about the connection is cleared at each Onion Router. Most of the above comes from the U.S. Navy. See also www.onion-router.net/

online Available through the computer. Online may refer to information on the hard disk, such as online documentation or online help, or a connection, through a modem, to another computer.

online fallback A modem feature. It allows high speed error-control modems to monitor line quality and fall back to the next lower speed if line quality degrades. Some modems fall forward as line quality improves.

online gaming Gambling over the Internet or on a dial-up connection. It's illegal in the U.S., but not in most countries outside the U.S. Many millions of Americans gamble illegally each day via the Internet.

Online Insertion and Removal OIC. Online Insertion and Removal is the practice of replacing or removing equipment components without powering off the system. This provides a high availability of service because the components are hot swappable. This term is commonly found in Cisco equipment documentation.

online service A commercial service that gives computer users (i.e. its customers) access to a variety of online offerings such as shopping, games, and chat rooms, as well as access to the Internet. America Online and Microsoft Network (MSN) are examples of online services.

online system A communicating system that requires a terminal to be in communication with and usually under control of a central point ... as in the case of a telephone set.

Online Transaction Processing See OLTP.

ONMS Open Network Management System. Digital Communications Associates architecture for products confirming to ISO's CMIP.

ONP 1. One Night Process.

2. Open Network Provision.

ONS Object Name Service. An Auto-ID Center-designed system for looking up unique Electronic Product Codes and pointing computers to information about the item associated with the code. ONS is similar to the Domain Name Service, which points computers to sites on the Internet.

ONT Optical Network Terminal. Also known as Optical Network Unit (ONU). See ONU.

ONTC Optical Networks Technology Consortium. This research consortium was organized and coordinated by Bellcore's Optical Network Research Department with assistance from ARPA. The ONTC's research spans material and device technologies to network design and management. Other ONTC members include Case Western Reserve University, Columbia University, Hughes Research Laboratories, Northern Telecom, Bell Northern Research, Lawrence Livermore National Laboratory, Rockwell Science Center, United Technologies Photonics, and United Technologies Research Center.

ONU Optical Network Unit. Also known as ONT (Optical Network Terminal). A device serving to terminate an optical circuit at the customer premises, at the curb, or in the neighborhood, an ONU is one of the primary elements of a FSAN (Full Service Optical Network). Various hybrid networks make use of ONUs to accomplish the interface between fiber optic feeder cables and metallic cables (e.g., coaxial cable or twisted pair), converting optical signals to electrical signals and vice versa. See also FSAN and HFC.

OOCM Object-oriented call model.

OOF 1. Out Of Frame. A designation for a condition defined as either the network or the DTE equipment sensing an error in framing bits. It's declared when 2 of 4 or 2 of 5 framing bits are missed (the OOF condition existing for 2.5 seconds generally creates a local Red Alarm). See LOF.

2. Out Of Franchise. Often used by a local phone company to refer to business and other activities that are outside its local franchise boundaries. The concept often is that these OOF activities are often subject to fewer government regulations. Also referred to as OOR, or Out Of Region.

OOK On-Off Keying (OOK) is a digital signal modulation format that simply involves the representation of a 1 (one) bit by the presence of a signal and the representation of a 0 (zero) bit by the absence of a signal. OOK is used in telegraphy, with various combinations of short and long signal pulses, or dots and dashes, separated by pauses, represent alphanumeric characters. OOK is used extensively in optical systems, including both fiber optics and FSO (Free Space Optics) due to its inherent simplicity and low cost.

OOO Optical-Optical-Optical. Conventional fiber optic transmission systems generally are not purely optical in nature. Over a distance, the optical signal attenuates (i.e., weakens) to the point that it must be boosted. In conventional fiber optic transmission systems, that boosting unit generally is in the form of a regenerative repeater, which is an OEO (Optical-Electrical-Optical) device. That is to say that the incoming optical (O) signal is converted to electrical (E) format, boosted in signal strength and filtered for noise, and the outgoing signal is converted back into (O) optical format as it exits the repeater. It works just fine, but OEO repeaters are expensive, are prone to failure, require local power, and must be upgraded as network speeds increase. OOO (Optical-Optical-Optical) techniques have been developed in the forms of Erbium-Doped fiber Amplifiers (EDFAs) and Raman amplifiers, which generally are considered to be substantial improvements over OEO repeaters. See also EDFA, Raman Amplifier, and Repeater.

OOP Object-oriented programming (OOP) is simply combining objects to produce a finished software application. The real work is developing the objects that carry out tasks within the program. Programs can be produced by recycling objects from other programs and upgraded by replacing old objects.

OOPS Open Outsourcing Policy Services. A specification from the IETF for policy-based networking. OOPS defines a protocol for exchanging QoS (Quality of Service) policy information and policy-based decisions between a RSVP-capable router and a policy server.

The data are transferred between routers and servers via TCP/IP. Proposed for use in large and complex networks, OOPS is intended to allow the prioritization of traffic based on policy-level parameters established by the network administrator. Those parameters are stored in a policy server, which is queried by the client routers. This centralized approach avoids the requirement for programming each router. See also Policy-Based Networking, QoS, and RSVP.

OOR See OOF.

OPAC Outside Plant Access Cabinet.

opacification The transformation of a material from its being transparent to its being opaque. See fiber opacification.

OPB On Premise Box. A box used by broadband cable systems to house and protect splitters, filters and/or coaxial connectors against atmospheric conditions and vermin.

open Means the circuit is not complete or that the fiber is broken. There is a break in it. A break does not necessarily mean it's malfunctioning, only that it's been turned off.

open air transmission Referring to a transmission type or associated equipment, that uses no physical communications medium other than air. Most radio communications systems, including microwave, shortwave and FM radio and infrared are open-air (also called "through-the-air") transmission systems. Open air transmission is not to be confused with the OpenAir specification, an ad hoc standard promoted by the Wireless LAN Interoperability Forum (WLI Forum). See also OpenAir.

Open Application Interface See OAI.

Open Architecture See OAI.

open cable See OpenCable.

open circuit A circuit is not complete. There is no complete path for current flow. Electrical current cannot flow in the circuit. In electrical engineering, a loop or path that contains an infinite impedance. In communications, a circuit available for use.

Open Collaboration Environment O.C.E. Apple's Open Collaboration Environment extends the Macintosh operating system to provide a platform for the integration of fax, voicemail, electronic mail, directories, telephony and agents. From a user's perspective, according to Apple, O.C.E.'s functionality will be seen through:

System-wide directory services, including a desktop directory browser and electronic business cards.

A compound mailbox for mail from all sources – fax, voice mail, e-mail, pager, etc.

Application integration, with all applications having the ability to send documents.

Open Database Connectivity See ODBC.

open end The end of a Switched Access Service that transmits ringing and dial tone and receives address signaling.

open ended access Term used to describe the ability to terminate a call to any public network destination. For example, on the open end of a foreign exchange circuit, the customer may call any number in the local calling area without being charged for long distance service.

open financial exchange See OFX for a full explanation.

Open Grid Services Architecture Working Group OGSA WG of the Global Grid Forum (GGF).

open ground, neutral or hot In AC electrical power, an "open" is a break, an extremely loose or an unconnected wire in any electrical path. Dangers of an "open" GROUND include serious shock and fire hazard and are life-threatening. Caution: an "open" GROUND will not stop equipment from operating. However it will stop a fuse or circuit breaker from operating should a ground fault occur.

open line dealing This is a term used by British Telecom in its turrets. The system electronically recreates the original "pit" share trading environment in the stock exchange in which everybody could talk to everybody else. In open line dealing, the trader has the ability to program a number of parties onto speakers. Full duplex speech is achieved using the associated microphones and all lines receive a simultaneous broadcast.

open loop system A control system which does not use feedback to determine its output.

open loops The incomplete tasks and projects in your life that constantly cycle through your head, leading to anxiety, stress, and creative constipation. Popularized by David Allen's work-flow management book, Getting Things Done; The Art of Stress-Free Productivity.

open mic A type of stand-up entertainment where the microphone is open to any member of the audience who wishes to take the stage and perform. "Open mic" also refers to the situation where a public speaker or radio/TV show host or newscaster incorrectly believes that the microphone is turned off, and says something inappropriate that is picked up by the microphone and broadcast to the viewing/listening audience.

open mike See open mic.

Open Network Architecture ONA. The "network" refers to the public switched network. The FCC wants to encourage companies to get into the value-added telecom business – voice mail, electronic mail, shopping by phone, etc. These companies may be called "value added providers" or "enhanced service providers." The FCC's idea of encouraging companies to add value to phone lines is a nice idea, except that all these companies will rely on phone lines provided by local phone companies who also want to be in value-added telecommunications business. And the local phone companies would prefer that the business be a monopoly (easier to manager, higher prices, etc.) If the FCC is to allow the Bell operating companies and other local phone companies into the value-added business and encourage others in, then it must figure a way the Bell operating companies don't organize things so they have an unfair advantage. The FCC's latest idea is called ONA – Open Network Architecture. Under this concept, the telephone companies are obliged to provide a certain class of service to their own internal value-added divisions and the SAME class of service to nonaffiliated (i.e. outside) valued-added companies. The concept is that the phone company's architecture is to be "open" and that everyone and anyone can gain access to it on equal footing. ONA is only a concept at present and still needs some rigorous defining. There is not much pressure from outside entrepreneurial companies for ONA access. Thus ONA at the FCC and elsewhere drags its feet.

The March 18, 1991 issue of Telephony Magazine said that as conceptualized by the FCC, ONA "is the overall design of a carrier's basic network facilities and services to permit all users of the basic network, including the enhanced services operations of a carrier and its competitors, to interconnect to specific basic network functions on an unbundled and "equal access" "basis." Selected regional Bell holding companies ONA services would include:

1. Basic Serving arrangements (BSAs):

A BSA is the basic interconnection access arrangement which offers a customer access to the public network and provides for the selection of available Basic Service Elements (BSEs, see below). Basic serving arrangements are:

• Switched, line side connection
• Switched, trunk side connection
• Dedicated, metallic dedicated

2. Basic service elements (BSEs):

Basic Service Elements are optional basic network functions that are not required for an ESP to have a BSA, but when combined with BSEs can offer additional features and services. Most BSEs allow an ESP to offer enhanced services to their customers in a more flexible manner. BSEs fall into four general categories: Switching, where call routing, call management and processing are required; Signaling, for applications like remote alarm monitoring and meter reading; Transmission, where dedicated bandwidth or bit rate is allocated to a customer application; and Network Management, where a customer is given the ability to monitor network performance and reallocate certain capabilities. The selection of available BSEs is an ongoing process, with new arrangements being developed many times in response to customer demands. ANI, Audiotext "Dial-It" Services, and Message Waiting Notification are all examples of BSEs, which also include:

Multiline hunt group
• Uniform call distribution
• Central office announcements
• Three-way call transfer

3. Complementary network services (CNSs):

CNSs are basic services associated with end user's lines that make it easier for ESPs (Enhanced Service Providers) to offer enhanced services. Some examples of CNSs include

Call forwarding Busy/Don't Answer
• Three way calling
• Call waiting
• Virtual dial tone
• Message waiting/indicator
• Speed calling
• Warm line

4. Ancillary services.

Ancillary Services. These are options available to an ESP which support and complement the provision of enhanced services. Examples of ancillary services are protocol conversion, and DID with third number billing inhibited.

In June, 1991, according to Communications Week, the FCC established a tariff structure that will determine how much the telcos can charge enhanced-services providers – and ultimately how much end users will have to pay for those services. FCC Chairman Alfred Sikes called the agency's action "one of our most pivotal steps" in the implementation of ONA. The FCC's idea is that ONA tariffs will be filed with the FCC in November, 1991 and will take effect February 1, 1992. They didn't. At the time I wrote this, the FCC had a new chairman, with a different agenda.

See also CompuCALL, Enhanced Services, OAI, and Open Application Interface.

open office Typically, this layout places the manager's desk in the foreground, within view of all other desks and enclosed spaces. The first open office appeared around 1960 when it was introduced in Germany as the "office landscape."

Open Outcry Auction An auction in which bidders openly shout out their bids.

Open Settlement Protocol OSP is the international standard for Inter-Domain pricing, authorization and usage exchange of IP communications. The OSP standard is defined by TIPHON Working Group 3 of the European Telecommunications Standards Institute (ETSI) – www.etsi.org. The OSP standard is officially known as ETSI Technical Specification 101 321. Version 1.4.2 of this standard was ratified in December 1998. Version 4.1.1 of the OSP standard was ratified in November 2003 and may be downloaded here, www.transnexus.com/OSP%20Toolkit/ts_101321v040101p.pdf.

open skies When a government or government agency allows virtually anyone to sell satellite telecommunications service, you have "OPEN SKIES." The United States has an Open Skies satellite policy. Virtually anyone can apply to launch and operate a telecommunications satellite and, with a high degree of certainty, you'll be granted your wish. The European community is just now (summer, 1991) beginning to think of opening its skies. They have gone one small step – namely allowing anyone to buy and operate a receive-only satellite earth station not connected to the public network.

Open Software Foundation OSF. An industry organization founded in 1988 to deliver technology innovations in all areas of open computer systems, including interoperability, scalability, portability and usability. The OSF was an international coalition of vendors and users in industry, government and academia that work to provide technology solutions for a distributed computing environment. In February 1996, the OSF consolidated with X/Open Company Ltd. to form The Open Group. www.osf.org. See The Open Group www.opengroup.org.

open solutions developers A Dialogic term for companies which develop and sell end-user voice processing applications. We'd probably call these people value added resellers (VARs).

open source software Open source software is typically free. It's typically written by programmers all over the world contributing their efforts for the common good of mankind. Linux is open source software. So is something call OpenOffice. According to the website, www.OpenSource.com, open source promotes software reliability and quality by supporting independent peer review and rapid evolution of source code. To be certified as open source, the license of a program must guarantee the right to read, redistribute, modify, and use it freely. See the next definition.

open source software license An open source softare license permits users to read, access, change and reuse the source code of a software product. See also Open Source Software.

Open System Interconnection OSI. A Reference Model published by the ISO (International Standards Organization, as translated into English) that defines seven independent layers of communication protocols. Each layer enhances the communication services of the layer just below it and shields the layer above it from the implementation details of the lower layer. In theory, this allows communication systems to be built from independently developed layers, i.e. software and hardware developed by different people. See OSI.

open systems Open systems refers to that best of all possible worlds, where everyone would comply with a set of hardware and software standards. You could buy a server from company A, a client from company B, a networking system from companies C, D or E, and applications software from companies F-Z, and everything would work harmoniously. In real life, some "open" systems are more open than others. Some companies talk, without embarrassment, of their "proprietary open systems." Some things are open and closed. PBXs have many "open" ports on which you can attach things. But there are no PBXs I know of which have open backlanes, meaning you (or someone else) can design a board and plug it into the PBX's backplane. That's closed. The concept of open systems has been more popular in the computer industry.

Open Systems Interconnection See OSI.

open toolkit developers A Dialogic term for outside developers (outside of Dialogic) who provide applications generators that simply application development and work in a variety of operating systems – including MS-DOS, UNIX, OS/2, Windows, etc.

open wire A transmission facility typically consisting of pairs of bare (uninsulated) conductors supported on insulators which are mounted on poles to form an aerial (above ground) pole line. Most basic of all practical types of transmission media. Open wire may be used in both communication and power.

OpenAir A wireless LAN specification promoted by the Wireless LAN Interoperability Forum (WLI Forum), Open Air is recognized by many as an ad hoc standard. Actually an interface specification, OpenAir describes the physical and MAC layer (Layers 1 and 2 of the OSI Reference Model) interface used by WLI Forum products. Based on the RangeLAN2 protocol developed by Proxim, the specification employs Frequency Hopping Spread Spectrum (FHSS) technology in the 2.4 GHz portion of the unlicensed Industrial, Scientific and Medical (ISM) radio band. The specification includes Access Points (APs) in the form of protocol-independent wireless bridges which connect to a standard wired backbone, such as IEEE 802.3 Ethernet. Client workstations are equipped with compatible client adapters, completing the wireless circuit. Security is provided at several levels. First, the FHSS implementation includes 79 frequencies, each of which can be used in a pseudo-random fashion in any of 15 hopping sequences. Second, every station is programmed with a 20-bit security code number which must be validated by the network during the process of circuit initiation and network synchronization. Third, the security code is encrypted. See also FHSS, ISM and WLI Forum.

OpenCable OpenCable is a trademarked term for a project established by CableLabs and aimed at defining a next-generation digital device for interactive services over CATV networks. The project addresses both hardware and software specifications. See also CableLabs.

OpenOffice OpenOffice is a core reason for the increasing popularity of the Linux desktop. Based on initially proprietary software that was later made open source by Sun, it includes a word processor, a spreadsheet program, a presentation builder, and an image editor and has become one of the most popular open-source alternatives to Microsoft's productivity software. Companies such as Novell and Red Hat distribute it along with their own versions of Linux, and Sun sells an enhanced version called StarOffice. The key feature of OpenOffice is that it behaves pretty much the same way users of Windows software would expect. See Linux.

OpenView Hewlett-Packard's suite of a network-management application, a server platform, and support services. OpenView is based on HP-UX, which complies with AT&T's Unix system.

operand That which is being operated on. An operand is usually identified by the address part of an instruction.

Operating Agency See OA.

operating environment Referring to the combination of (usually IBM) host software that includes operating system, telecommunications access method, database software and user applications. Some common operating environments include MVS/CICS and MVS/TSO.

Operating Company Number OCN. A code used in the telephone industry to identify a telephone company. Company codes assigned by NECA (National Exchange Carriers Association) may be used as OCNs. See also AOCN and NECA.

operating income There are two ways American telecom companies traditionally report their earnings – operating income and net income. Operating income purports to show how much money the company earned from running its basic business – that of making and selling its primary products and services. Operating income starts with total revenues (i.e. sales from products and services) and then deducts the cost of delivering those sales, i.e. raw materials, components, supplies, packagings, etc. In short, all direct costs. Next deduction is sales and marketing expenses. The final result gives you "operating income." If you then deduct interest expense (or add in interest income), goodwill amortization, "other (unusual) income or losses" (e.g. profit or loss on the sale of a building) and state and federal income taxes, you end up with a number that is referred to as Net Income. See GAAP.

operating system An operating system is a software program which manages the basic operations of a computer system. It figures how the computer's main memory will be apportioned, how and in what order it will handle tasks assigned to it, how it will manage the flow of information into and out of the main processor, where in memory it will place material, how it will get material to the printer for printing, to the screen for viewing, how it will receive information from the keyboard, etc. In short, the operating system handles the computer's basic housekeeping. These days operating system also handle the computer's access to various flavors of networks – from local area networks (LANs) to corporate networks that span the world and, of course, to the Internet. MS-DOS, Linux, UNIX, PICK, Windows, Solaris, Symbian, etc. are operating systems. Historically, an operating system was the minimal set of software needed to manage a device's hardware capability and share it between application programs. Practically, "OS" is now used to mean all software including kernel, device drivers, communications, graphics, data management,

GUI framework, system shell application, and utility applications. See also NOS (Network Operating System).

operating time The time required for seizing the line, dialing the call and waiting for the connection to be established.

operation code The command part of a machine instruction.

Operational Data Integrator ODI. An MCI term: Combines data from the Customer Information Manager, the MCI Information Manager, and the Management Information Systems to build databases containing network information.

operational fixed station A service established by the FCC operating between 2.65 GHz and 2.68 GHz.

operational grammar A voice recognition term. A vocabulary structure where certain word sets activate other word sets.

operational load The total power requirements for communication facilities.

Operational Security OPSEC. A thorough on-site examination of an operation or activity to determine if there are vulnerabilities that would permit adversaries and exploitation of critical information during the planning, preparation, execution, and post-execution phases of any operation or activity. A Federal Government definition.

operational service period A performance measurement period, or succession of performance measurement periods, during which a telecommunication service remains in an operational service state. An operational service period begins at the beginning of the performance measurement period in which the telecommunications service enters the operational service state, and ends at the beginning of the performance measurement period in which the telecommunications service leaves the operational service state.

operationalize Our American English language has a tendency to make verbs out of nouns. An operation means an action. Therefore to operationalize means to make something happen. In military terms, it means a plan being brought into action. People using big words when they could just as easily use little ones are trying to impress others with their importance and learning. Frankly, I'm not.

Operations Applications OAs. A telephone company AIN term. A class of functions to provide provisioning, administration, maintenance, and management capabilities for network elements, network systems, software, and services (e.g., assessment of service quality over the group of systems and software that support the service). These functions usually reside in Operations Systems but may be assigned to network elements or network systems.

operations domains A telephone company AIN term. The set of operations functions residing in network elements, network systems, Operations Applications, and associated interfaces necessary to accomplish memory administration, network surveillance, network testing, network traffic management, and network data collection.

Operations Evaluation System OES. An MCI internal system, which generates daily, weekly, and monthly switch data reports; used to scan high-level switch degradation problems and to analyze specific switch problems.

Operations Support System See OSS.

operator 1. Employee of telephone company, or an individual business or institution, who aids in the completion of phone calls. Traditionally a woman's occupation, now increasingly the role of men and machines. In some countries, like Germany, the phone company doesn't have operators.

2. In PCs, an operator is a symbol that represents a mathematical action, such as a +-/ and * (plus, minus, divide and multiply). Operators can also be words like AND, OR and NOT.

operator assisted A phone call placed with the assistance of the carrier's operator. You pay more when you use an operator.

operator console Same as attendant console. See Attendant Console.

operator service system Equipment capable of processing certain kinds of traffic originating or terminating to an end office; this processing may take place either with or without an operator's assistance. Use of such equipment includes call rating and charge recording functions, operator assistance functions, coin control and collection functions, automatic or manual identification of calling line number, and verification of the busy/idle condition of subscriber lines.

Operator Services OS. Any of a variety of telephone services which need the assistance of an operator or an automated "operator" (i.e. using interactive voice response technology and speech recognition). Such services include collect calls, third party billed calls and person-to-person calls. The responsibility for operator services used to be very straightforward. The LECs had their own operators and so did the IXCs. That's no longer necessarily true. Many IXCs use an Alternative Operator Services (AOS) service bureau, which essentially is a huge incoming call center. The AOSs typically provide operator services

for a large number of small long distance companies, many of which are resellers and aggregators who don't own their own facilities. That is to say that they own no switches or transmission facilities – they simply resell long distance services which they buy at bulk wholesale rates from the larger carriers. Since the MFJ (Modified Final Judgment) broke up the Bell System in 1984, operator services have gotten more confusing still. The RBOCs can, and generally do, provide their own operator services for both local and IntraLATA calls within their home states. The Telecom Act of 1996 added to the confusion. Where the RBOCs operate outside their home states, they may provide operator services directly or they may use an AOS. The larger CLECs (Competitive LECs), many of which also are IXCs (e.g., AT&T, MCIWorldcom, and Sprint), also generally provide their own operator services; the smaller CLECs generally use an AOS. As a rule, operator services are much faster today, thanks to automation. Also as a rule, operator services today are much less personal, much, much more expensive, and much less accurate. Accuracy is a problem particularly in the case of Directory Assistance. See also Directory Assistance.

Operator Workstation OWS. The OWS is an advanced voice and data workstation (typically a PC running a flavor or Windows) that streamlines and automates many of the routine tasks of an operator, thus reducing the amount of time needed for call handling. Color screens, pop-up windows, one-touch commands, and database look-up are some of the features that simplify the operator's tasks and speed call processing.

OPEX OPerating EXpenses. See CAPEX, which stands for CApital EXpenses.

OPL A BASIC-like programming language, for rapid application development, used in Symbian OS.

OPN See VMN.

opportunity My dear wife doesn't appreciate my sense of humor. She claims that I never miss an opportunity to miss an opportunity.

OPRE Operations Order Review.

OPS 1. Operator Services.

2. Off-Premises Station. See Off-Premises Extensions.

3. Open Profiling Standard. A recent (1997) privacy initiative intended to provide guidelines for collecting information on users accessing Web sites. OPS software will allow the user to fill out a personal profile, which will be stored in a file residing on a client computer. The OPS file will conform to the vCard specification which is managed by the Internet Mail Consortium. Much like an electronic version of a business card, the OPS file will store information such as name, company, address, telephone number, fax number, and e-mail address. The OPS information will be shared only with the consent of the user – very much unlike a cookie, which is embedded by the host of the Web site in a file on the client computer. It's a privacy issue. See also Cookie and vCard.

4. What my wife calls my messy home office – my Own Personal Slum.

opt Verb meaning to choose. You can opt in, meaning you choose to join or receive something. You can opt out, meaning you choose not. In short, opt in means to give your consent. Opt out means not to give your consent.

opt-in email To sign up for an email newsletter or other regular communications. See Opt.

optical amplifier A device to amplify an optical signal without converting the signal from optical to electrical back again to optical energy. The two most common optical amplifiers are erbium-doped fiber amplifiers (EDFAs), which amplify with a laser pump diode and a section of erbium-doped fiber, and semiconductor laser amplifiers.

optical attenuator In optical communications, a device used to reduce the intensity of the optical signal. In some optical attenuators used in optical fiber systems, the amount of attenuation depends on the modal distribution of the optical signal.

optical bandpass The range of optical wavelengths that can be transmitted through a transmission medium.

optical bands The generation of optical signals over optical fibers is accomplished by lasers, which produce light waves in very narrow optical wavelengths. In telecommunications networks, the three main center wavelengths used by source lasers are 1550nm, 1600nm, and 1310nm, encompassing the C, L, and S optical bands.

C band (1530nm-1550nm). Because the atomic structure of erbium provides light amplification in this range, this is the spectrum at which the first erbium-doped fiber amplifiers (EDFAs) operated.

L band (1565nm-1610nm). Longer than the C band, the L band enlarges the usable spectrum for optical transport, thus increasing the number of DWDM channels. With higher dispersion (spreading) of the optical signal than at the C band, however, additional EDFAs are required for optical transport.

S band (1460nm-1500nm). Wavelength at which original single mode fiber was optimized. providing a large effective area for transmission of optical signals.

optical blank A casting consisting of an optical material molded into the desired geometry for grinding; polishing; or, in the case of some optical fiber manufacturing processes, drawing to the final optical/mechanical specifications.

optical cavity A region bounded by two or more cavity surfaces, referred to as mirrors or cavity mirrors, whose elements are aligned to provide multiple reflections of lightwaves. The resonator in a laser is an optical cavity.

Optical Channel OCh, or OCH. In a network based on WDM (Wavelength Division Multiplexing) or DWDM (Dense WDM), each wavelength, or lambda, also is referred to as an Optical Channel. See also DWDM and WDM.

optical channel spacing The wavelength separation between adjacent WDM channels.

optical channel width The optical wavelength range of a channel.

Optical Character Recognition OCR. Reading data using a machine that visually scans the characters in a document and converts that data into standard form which can be stored on conventional magnetic medium, e.g. floppy or hard disk. OCR is not 100% accurate and so requires manual cleanup or fuzzy logic to make your OCRed relatively clean and thus retrievable through a search engine. See Optical Scanner.

optical clocks Thanks to Eric Smalley of Technology Research News for this definition: Making the most of time requires that time be well defined. To be precise, a definition of time must involve something that happens very quickly. The current definition of a second is the duration of 9,192,631,770 oscillations of cesium atoms excited by microwaves. Today's cesium atomic clocks are accurate to within one million billionth of a second or 1 second in 30 million years. This is precise enough for a cluster of orbiting satellites to calculate the position of a stationary object to within a millimeter. For moving objects like cars and planes, the accuracy is a few meters, which is not enough to allow a global positioning system to automatically land a plane. Researchers from the National Physical Laboratory in England have made a prototype atomic clock that divides time into slices based on optical radiation, or lightwaves, rather than microwave radiation. Such clocks could eventually improve global positioning systems, and make space exploration more accurate. Because lightwaves are smaller and faster than microwaves, optical clocks have operating frequencies as many as 100,000 times higher than today's cesium microwave clocks. Today's atomic clocks measure the vibration frequency of cesium atoms to calibrate quartz crystal electronic oscillators. A laser excites the atoms to the energy level where they resonate with a microwave field that is tuned by an electronic oscillator. The microwave field cycles through a range of frequencies close to the cesium atoms' resonant frequency, and as the microwaves resonate with the atoms the atoms give off energy in the form of photons. A photodetector measures the peak amount of light and locks the microwave field on that frequency. Optical atomic clocks use lasers instead of microwaves to resonate the atoms, and atoms that have higher resonant frequencies than cesium. The researchers' prototype uses a single strontium ion that is held in an electromagnetic trap and laser cooled to near absolute zero. Another laser causes the ion to oscillate at its resonant frequency – 444,779,044,095,484.6 cycles per second. The difference between the strontium frequency and the cesium frequency is the difference between one second and 13 and a half hours. This higher frequency could lead to optical atomic clocks that are so accurate they would lose less than a second over the lifetime of the universe.

Optical Code Division Multiple Access See OCDMA.

optical combiner A passive device in which power from several input fibers is distributed among the smaller number (one or more) of output fibers.

optical computer A computer that uses photons, not electrons as in today's old-fashioned computers. Scientists think photon computers and photon switches could be a thousand times faster than present computers and switches, and, of course, totally impervious to electromagnetic interference.

optical connector variation The maximum value in dB of the difference in insertion loss between mating optical connectors (e.g. with remating, temperature cycling, etc.). Also known as Connector Variation.

optical connectors Connectors designed to terminate and connect either single or multiple optical fibers. Optical connectors are used to connect fiber cable to equipment and interconnect cables.

optical continuous wave reflectometer An instrument used to characterize a fiber optic link wherein an unmodulated signal is transmitted through the link, and the resulting light scattered and reflected back to the input is measured. Useful in estimating component reflectance and link optical return loss.

Optical Cross-Connect See OXC.

optical cross-connect panel A cross-connect unit used for circuit administration and built from modular cabinets. It provides for the connection of individual optical fibers with optical fiber patch cords.

optical directional coupler ODC. Component used to combine and separate optical power.

optical drive See Optical Storage Device.

optical disk Peripheral storage disk for programs and information. Optical disks are emerging as computer storage devices because of their tremendous storage capacities in comparison to magnetic disk. Optical disks include CD-Rs and DVDs. Originally optical disks were WORM - which stands for Write Once, Read Many Times, meaning you wrote once to the disk, and read that information many times. However re-writable CDs are available which allow overwriting of data. Emerging standards, such as Blu-ray, offer up to 27 gigabytes (GB) on a single-sided 12-centimeter disc, the same size and shape of a CD you currently buy recorded music on. See also OPTICAL STORAGE DEVICE, CD-R, Blu-Ray.

optical ethernet A new platform for extending Ethernet "natively" from the LAN into the MAN (Metropolitan Area Network), affording enterprise users unprecedented bandwidth at LAN costs. For service providers, Optical Ethernet establishes an entirely new service platform that is data-optimized, easily provisioned and as scalable as SONET/ATM solutions are– without the cost penalties or complexity of management. It is a "supercharged" Ethernet, suitable for service providers' access and transport networks. Optical Ethernet leverages the simplicity and cost-effectiveness of Ethernet and the reach and reliability of optics, overcoming traditional LAN Ethernet limitations of distance, quality of service and scalability. It provides carriers with a unified protocol for converged network services and gives enterprise users ways to access broadband applications or outsourced network resources at affordable costs. This definition contributed by Henry Brent.

optical fall time The time interval for the falling edge of an optical pulse to transition from 90% to 10% of the pulse amplitude. Alternatively, values of 80% and 20% may be used.

optical fault finder A device which measures power and distance characteristcs in fiber-optic cable. Optical Fault Finders are more expensive than Optical Power Meters but have the additional ability to make distance measurements and locate tiny fiber breaks called Microbends. See Optical Power Meter.

optical fiber A thin (approximately 125 micro-meters in diameter) silica glass cable with an outer cladding material and around 9 micro-meters diameter inner core with a slightly higher index of refraction than the cladding. A typical index of refraction is 1.443 so that light travels in a fiber at roughly 2/3 the speed of light in a vacuum. Optical fiber is made of glass or plastic, and guides light, whether or not it is used to transmit signals. Optical fiber is an almost ideal transmission medium. It has these advantages:

Transmission losses are very small.

- Bandwidth is greater than any other transmission medium we know of today. And we have no idea what the theoretical bandwidth of a strand of fiber might be.
- Fiber is immune to electromagnetic interference. This means it can operate in hostile or hazardous environments, like on the factory floor, in elevators shafts, on battleships, etc.
- Fiber does not radiate. You can't place a receiver next to it and figure out what's going on the fiber, as you can with cable.
- You can put many strands of fiber carrying much information in the same bundle and they won't interfere with each other, i.e. there won't be any significant cross-talk between the adjacent fibers.
- Its basic raw material, silica (sand) is the second most abundant element on earth. Actually, optical fiber is made from a synthetic glass produced by burning two chemicals together; the resulting soot collects on a bait rod or in a bait tube, and is baked until all moisture is removed. this is called the blank which fiber is drawn from.

The insulating properties of the glass fiber produces the only major disadvantage. Metallic conductors need to be included to power repeater amplifiers needed over long fiber runs. However, these can form the strength members that are necessary to aid the laying of fiber cables. Typical fiber applications range from use in local area network sections, where there is a high degree of electrical interference, to trans-oceanic telecommunications cables.

An optical fiber not carrying signals is typically called a "dark fiber." There are two types of optical fiber: Single Mode and Multimode. In single mode fiber, light can only take a single path through a core that measures about 10 microns in diameter. A micron is one millionth of a meter. Multimode fibers have thicker cores – typically 50 to 200 microns. Single mode fiber is more efficient. It offers low dispersion, travels great distances without repeaters and has enormous information-carrying capacity. The relatively large core of multimode fiber lightguide allows light pulses to zig-zag along many different paths. It's also ideal for light sources larger than lasers, such as LEDs (Light Emitting Diodes). Multimode fiber is not the preferred method of optical telecommunications any longer. See also Fiber,

Optical Fiber Cable and Optical Spectrum.

optical fiber cable A transmission medium consisting of a core of glass or plastic surrounded by a protective cladding, strengthening material, and outer jacket. Signals are transmitted as light pulses, introduced into the fiber by light transmitter (either a laser or light emitting diode). Low data loss, high-speed transmission, large bandwidth, small physical size, light weight, and freedom from electromagnetic interference and grounding problems are some of the advantages offered by optical fiber cable. There are five common types: single, dual, quad, stranded, and ribbon. See Optical Fiber.

optical fiber duplex adapter Mechanical media termination device designed to align and join two duplex connectors.

optical fiber duplex connection Mated assembly of two duplex connectors and a duplex adapter.

optical fiber duplex connector Mechanical media termination device designed to transfer optical power between two pairs of optical fibers.

optical fiber facility Transmission system which uses glass fibers as the transmission medium. See Optical Fiber.

optical fiber patch panel One way to terminate fiber optic cable. Fiber patch panels have a fiber splice tray with pigtails. Pigtails are fiber connectors with a piece of fiber optic connected so fiber from within a cable can be easily spliced to it. Connectors are on the front of the fiber patch panel.

optical fiber preform Optical fiber material from which an optical fiber is made, usually by drawing or rolling.

optical fiber ribbon A cable of optical fibers laminated in a flat plastic strip.

optical fiber splice A permanent joint whose purpose is to couple optical power between two fibers.

optical fibre British spelling. In an optical fibre transmission system, the data is carried by pulses of light along glass fibres. This method of transmission has a much higher bandwidth than copper cables and is less subject to distortion and interference. It is safer than copper because it provides electrical isolation and as it carries no current can be used in flammable areas. See Optical Fiber for a longer explanation.

optical filter A device capable of passing one portion of the optical spectrum while attenuating all other portions.

optical interconnection panel An interconnection unit used for circuit administration and built from modular cabinets. It provides interconnection for individual optical fibers. Unlike the optical cross-connect panel, the interconnection panel does not use patch cords.

Optical Internetworking Forum. See OIF.

optical isolator A component used to block out reflected and unwanted light. Used in laser modules, for example. Also called an isolator.

Optical Line Interface Unit OLIU. See LIU.

optical line terminal OLT. Located in a central office, an OLT is the connection point for optical fiber drops to customers' premises, specifically, to multiple single-family units (SFUs). Also known as Optical Network Unit (ONU). See FTTP and ONU.

optical link loss budget The range of optical loss over which a fiber optic link will operate and meet all specifications. The loss is relative to the transmitter output power.

optical loss test set A source and power meter combined to measure attenuation in optical fiber.

optical modulator An active optoelectronic component, typically made from lithium niobate or gallium arsenide, that turns an optical signal on and off to encode and transmit data throughout the network. Modulation can be achieved directly by turning a source laser on and off or externally by altering a continuous source laser signal to achieve a similar on/off effect. Long-distance and submarine optical networks typically use high-power lasers and external modulators, while shorter-distance optical networks are better suited for direct modulation.

optical network unit ONU. Located in the telco closet on the premises of a multi-dwelling unit (MDU) or multi-tenant unit (MTU), an ONU is the connection point for an optical fiber run to the MDU/MTU. It is also the interconnection point with subscriber lines (usually copper) to individual customers within the MDU/MTU. See also FTTP and ONU.

optical path power penalty The additional loss budget required to account for degradations due to reflections, and the combined effects of dispersion resulting form intersymbol interference, mode-partition noise, and laser chirp.

optical power meter A device which measures the light signal transmitted through fiber-optic cable.

optical pump laser A shorter wavelength laser that is used to pump a length of

fiber with energy to provide amplification at one of more longer wavelengths. See also EDFA.

optical receiver The receiver converts the optical signal back into a replica of the original electrical signal. The detector of the optical signal is either a PIN-type photodiode or avalanche-type photodiode.

optical repeater In an optical fiber communication system, an optoelectronic device or module that receives a signal, amplifies it (or, in the case of a digital signal, reshapes, retimes, or otherwise reconstructs it) and retransmits it. See also Optical Amplifier.

optical return loss Optical Return loss measurements usually fall into two distinct classes. One of these is the measurement of an installed fibre system. This is very easy to do, with a measurement in the range of 25 - 30 dB. The other class is measuring the performance of individual components, such as an isolator, and for this an extended range is required, typically around 50 dB. In short, ORL is the ratio, expressed in dB, of optical power reflected by a component or an assembly to the optical power incident on a component port when that component or assembly is introduced into a link or system. See also UPC.

optical rise time The time interval for the rising edge of an optical pulse to transition from 10% to 90% of the pulse amplitude. Alternatively, values of 20% to 80% may be used.

optical scanner A hardware device that recognizes images on paper, film and other media and converts them into digital form which can be stored in a conventional computer readable magnetic medium, such as floppy or hard disk. Optical scanners are getting better and better but are still not perfect. See OCR.

Optical Spatial Division Multiplexing See OSDM.

optical spectrum Generally, the electromagnetic spectrum within the wavelength region extending from the vacuum ultraviolet at 1 nm to the far infrared at 0.1 mm. The term was originally applied to that region of the electromagnetic spectrum visible to the normal human eye, but is now considered to include all wavelengths between the shortest wavelengths of radio and the longest of X-rays. See Optical Fiber.

optical storage device Optical storage devices use a source of coherent light – usually a semiconductor laser – to read and write the data. There are three big advantages to using a laser – size, safety and portability. Because you can focus a laser into approximately one micron in size – a far smaller area for encoding a bit of data than conventional drives – you can fit more data in.

Optical media are also more stable than metal-oxide disks. They aren't affected by light, normal temperatures or electromagnetic fields. (You can put through as many airport x-ray machines as you wish.) And best, the read/write head doesn't get as close to the recording medium as it does in conventional disk drives. Optical drives are interchangeable, also. You can remove them and store them. That makes them great for archiving.

optical time domain reflectometer OTDR. A device that measures distance to a reflection surface by measuring the time it takes for a lightwave pulse to reflect from the surface. Reflection surfaces include the ends of cables and breaks in fiber. The reflectometer measures the ratio of incident and reflected light power, or backscatter. By using this device you can figure precisely where a fiber optic link is broken. This device operates like Radar. It sends a light pulse down the cable and waits for it to return. It measures the time taken and calculates the distance based on the speed of light through the fiber optic cable. The reflectometer usually also displays the reflected waves on a time axis for precise reading of, e.g., the leading edges of the transmitted and reflected waves. The reflectometer is also capable of launching a light pulse into the fiber optic transmission medium and measuring the time required for its reflection to return by backscattering or end reflection, thus indicating the continuity, crack, fracture, break, or other anisotropic features of the medium. See also Backscatter.

Optical-to-electrical converter Optical-to-electrical (O2E) converter is a converter that provides electrical current that is proportional to the optical power that it receives.

optical telegraph This is a historical entry.It's about a clever solution for long-distance communications in the days before the electric telegraph was created. It's called the optical telegraph, and it involved a network of towers within line-of-sight of each other.Each tower was manned and the tower itself was topped with a mechanical device, often a mast whose cross arms and end pieces could be adjusted to communicate messages.There was a variety of solutions, some of which used shutters or other mechanical devices to create visual communications. The optical telegraph was invented in France by Claude Chappe, a former priest. He and his brother Ignace built an incredible optical telepgraph network in France, network diagrams of which are at http://chappe.ec-lyon.fr/carte.html. A message could travel 240 kilometers (approx. 148 miles) over the 15-node network between Paris to Lille in 9 minutes. Optical telegraphs spread throughout Europe, the US and Canada in the late 1700s and the first half of the 1800s. See also Telegraph Hill.

optical transmitter The optical transmitter converts an electrical analog or digital signal into a corresponding optical signal. The source of the optical signal can be either a light emitting diode, or a solid state laser diode. The most popular wavelengths of operation for optical transmitters are 850, 1300, or 1550 nanometers.

Optical Transport Network OTN. The Optical Transport Network is a fancy name for a dream – an all optical telecommunications network with the flexibility of the present all-electronic one. The idea is that a call – narrow or broadband – will be "dialed," then wend its way to the other end through pure optical pipes and pure optical switching. The so-called optical network of today is actually a collection of dumb wavelength-division multiplexed (WDM) point-to-point links, which are typically converted to electrons at some stage (unless the circuit is an end-to-end leased line.) Telecommunications service providers of tomorrow will provide on-demand lightpath services. For this to happen, we'll need, as a first step, dynamically controllable equipment such as configurable optical add-drop multiplexers and large port count optical cross-connects. The next crucial step is to create a standardized management and control infrastructure to allow all the various optical switching and transmission equipment to interwork with one another. There is some debate as to how much "network" should we place? Since most traffic will be dominated by data in the future, some have suggested using an IP-centric minimalist philosophy to build the OTN. On the other hand, the increasing value of data, driven by the growth of e-commerce and companies running their day-to-day business through the Internet, argues for a more sophisticated network. The debate continues.

optical unifying layer OUL. A heady concept. A theoretical way for carriers to integrate the diverse technologies of their existing network into one physical infrastructure.

optical virtual tributary group OVTG contains four DS-1 signals packaged into an optical SONET virtual tributary, or VT.

optical waveguide Technically, any structure that can guide light. Sometimes used as a synonym for optical fiber, it also can apply to planar light waveguides. See also Photoni Crystals.

optical waveguide connector A device whose purpose is to transfer optical power between two optical waveguides or bundles, and that is designed to be connected and disconnected repeatedly.

optical wireless Optical wireless is a vague term which tries to capitalize on the popularity of optical stocks. Several vendors use the term to create interest in various freespace (i.e., through-the-air) transmission systems, some of which run at microwave frequencies and others of which use infrared (Ir) light. One maker described his product as "providing optical wireless connectivity with data speeds up to 2.5 Gbps, allowing high speed access to customers where the existing last-mile infrastructure, whether based on copper or wireless radio frequency technology, provides inadequate bandwidth. This family of products can be installed and operational quickly and easily on a small rooftop space or in an office behind a window. Installation of two units at a location enables a short mesh configuration, allowing redundancy in the event of a link or equipment failure." The term "optical wireless" certainly would be more meaningful if it were limited to Ir and laser-based WLL (Wireless Local Loop) systems, but not all vendors seem to be as concerned about "meaningful" as I am. Everybody's got his own agenda. See also Agenda and WLL.

optically active material A material that can rotate the polarization of light that passes through it. An optically active material exhibits different refractive indices for left and right circular polarizations (circular birefringence).

option In the stock market, an option is a contract you buy which allows you to buy (a "call") or sell (a "put") a stock at a specified per-share price by a specified future time period (option expiration date).

An option gives its holder the right to buy a stock at a fixed, or exercise price, usually for up to 10 years after the date of the grant. By backdating the grant – or pretending it was made earlier than it really was – the employee gets an instant boost to the potential profits from the award. The practice can violate securities laws and accounting rules.

In a March 2002 proxy statement, Apple told its shareholders – wrongly – that the 2001 grant to Mr. Jobs was made at fair market value on the date of grant. On Aug. 8, 2002, Mr. Jobs personally signed a routine disclosure statement to the SEC bearing the false price, and false date of Oct. 19, 2001, for the grant. Apple has said the grant wasn't finalized until December of that year, when Apple's share price was higher. An Apple spokesman declined to comment on the matter.

optimist 1. "A pessimist," Winston Churchill once said, "sees the difficulty in every opportunity. An optimist sees the opportunity in every difficulty."

2. An optimist could also be described as an ant crawling up an elephant's leg with rape in mind. Bill McGowan told me in the very early days of MCI, when MCI had six employees and no money, and AT&T had over a million employees. Bill McGowan was the ultimate optimist. See McGowan, William G.

OPTIS Overlapped Phase Trellis-code Interlocked Spectrum. OPTIS is a modulation technique use in HDSL2 (High-bit-rate Digital Subscriber Line). See also HDSL2.

opto-electric transducer A device which converts electrical energy to optical energy and vice versa. Used as transmitters and receivers in fiber optic communications systems.

opto-electronics The range of materials and devices associated with fiber optic and infrared transmission systems. As there are no practical optical computers, all information originates as an electrical signal. Therefore, opto-electronic light sources convert the electrical signal generated to an optical signal which is transmitted to the receiving light detector for reverse conversion back to an electrical signal. In fiber optic systems, the light sources are in the form of either LEDs (Light-Emitting Diodes) or laser diodes. The light detectors are in the form of either PINs (Photo-INtrinsic diodes or APDs (Avalanche PhotoDiodes). Laser diodes and APDs are matched in high-speed, long-haul networks. LEDs commonly are used in fiber optic LANs and other short-haul, relatively low-speed applications. Optical repeaters, which also are opto-electronic devices, repeat the optical signal. Such repeaters accept the optical signal through a light detector, convert it to electrical energy, boost the signal and clean it up, and reconvert it to an optical signal for insertion to the next link of the optical fiber system. Also, each of these functions requires electrical energy to operate and depends on electronic devices to sense and control this energy. See also Optical Amplifiers, a new breed of amplifiers that doesn't take the light signal back to electrical energy.

optoelectric transducer Electronic components that turn light energy into electrical energy and electrical energy into light energy.

optronics Opt(o-Elect)ronics. See Opto-Electronics.

Optus Communications Ltd Historically, Australia's second general communications carrier, licensed by the government in 1991 to provide competition to then government-owned Telstra Corporation. Initially formed from a consortium of Mayne Nickless, AMP Society, National Mutual, AIDC Telecommunications Fund, Bell South and Cable and Wireless. Optus provides mobile and fixed network services.

OPX Off Premise Extension. An extension or phone terminating in a location other than the location of the PBX. The station uses a line circuit out of the PBX. OPX is commonly used to provide a company executive with an extension off the PBX in his home. This way he can pretend to be at the office when he's really at home. He can also make toll calls and have them easily charged to the office.

OR Gate A digital device which outputs a high state if either or both of its inputs are high.

or statement See Not Statement.

orange book The common name for the U.S. Department of Defense's Trusted Computer System Evaluation Criteria (TSEC).

ORB 1. Office Repeater Bay. Mounting and powering arrangement for digital regenerators, such as for T-1 lines.

2. Object Request Broker. See Object Request Broker and CORBA.

order entry A voice processing application which allows someone with a touchtone phone to buy something, i.e. enter their order.

order wire 1. A circuit used by telephone personnel for fixing, installing and removing phone lines. See also Order Wire Circuit.

2. Equipment and the circuit providing a telephone company with the means to establish voice contact between central office and carrier repeater locations.

3. A SONET/SDH term for a connection request, and consisting of one octet contained within the SOH (Section Overhead).

order wire circuit A voice or data circuit used by telephone company technical control and maintenance personnel for the coordination and control action relating to activation, deactivation, change, rerouting, reporting and maintenance of communication systems and services. See also Order Wire.

orderwire See Order Wire.

organic There are two ways you can grow a company's revenues. You can buy other companies and add their revenues to yours. Or you can grow the revenues in your own business by thinking up and creating new products, then marketing and selling more, adding more customers and selling more to your existing customers. Growing your present business' revenues without acquiring other businesses is called organic growth.

organic growth See Organic.

originate mode The "originate mode" sets the modem to begin a data phone call – i.e. dial the phone, listen for a carrier tone from a remote modem and connect to that modem. The modem at the receiving end must be set to "Answer" mode. In any asynchronous data conversation, one side must be set to "Originate" and the other to "Answer." Such settings are usually made in software.

originate/answer The two modes of operation for a modem. Originate and answer states define the frequencies used to transmit and receive. In a two-way communication system, one modem must be set to originate and the other to answer.

originating direction The use of Access Service for the origination of calls from an End User premise to a customer premise.

originating office The central office that serves the calling party.

originating restriction A phone line with this restriction cannot place calls at any time. Calls directed to the phone, however, will be completed normally.

origination A call that is placed by the mobile subscriber, calling either a land-line circuit or another mobile subscriber.

origination cablecasting Programming over which a cable television system operator exercises editorial control. This term includes programming produced by the operator; Non-broadcast local programming produced by other entities and carried voluntarily by the system. Example: PRISM; regional news channels; Satellite-delivered non-broadcast programming carried voluntarily by the system, such as HBO, ESPN, CNN, C-SPAN, QVC, etc.

This term does not include programming over which the operator does not exercise editorial control, including any broadcast signal, including satellite-delivered broadcast "superstations" (WGN-TV, WWOR, etc.); Any access channel designated by franchise for public, educational, or governmental use; Leased-access channels.

The cable system operator is required by Section 76.225c of the FCC Rules to maintain records, in the PIF, to verify compliance with rules governing commercial matter in children's programming carried on origination-cablecasting channels. See PIF.

originator The user that is the ultimate source of a message or probe.

orl Optical return loss. See UPC.

ORM Optically Remote Module. A type of switching module made by AT&T which connects directly to the 5ESS switch communications module via optical fibers.

orphan A member of a mirrored volume or a RAID-5 volume that has failed due to a severe cause, such as a loss of power or a complete hard-disk head failure. When this happens, the fault-tolerant driver determines that it can no longer use the orphaned member and directs all new reads and writes to the remaining members of the fault-tolerant volume.

orthogonal Having, meeting or determined at right angles.

Orthogonal Frequency Division Multiplexing See OFDM.

OS 1. Outage Seconds.

2. Operating System, as in MS-DOS (Microsoft Disk Operating System), Windows NT, Windowsn 2000, Windows XP, Solaris, Unix, Linux, Symbian or OS/2. See Operating System.

3. Operator Services. See Operator Services.

4. Operations System. Includes SCOTS, FMAS, etc.

OS/2 Operating System/2. An operating system originally developed by IBM and Microsoft for use with Intel's microprocessors and for use with IBM personal system/2 personal computers. OS/2 has pretty well died. Microsoft's various flavors of Windows survive it.

osaifu-keitai 'Osaifu' means 'wallet' and 'keitai' means 'mobile phone' in Japanese. In other words, an osaifu-keitai is a 'wallet phone,' i.e., a mobile phone that can be used as electronic money, credit card, electronic membership card, electronic ticket, house entry key, and more. Basically, it replaces a traditional wallet and its contents. Additional services can be added by downloading and installing application programs. The term 'Osaifu-Keitai,' spelled as shown, is a registered trademark of NTT DoCoMo.

Osborne Effect Once there was a personal computer company called Osborne Computer Company. One day, the president announced a revolutionary new computer. It was so good not one of his dealers wanted to (or could) sell the existing product and they sent all their inventory back. Meantime, it was six months before the company could deliver the new product. But without any sales in the meantime, Osborne had no money and it went broke. There is a lesson here for companies who are attempting to manage transition between old and new product lines. Be careful, or suffer the horrible consequences of The Osborne Effect.

Oscar Holywood gives our Oscars for great movies, performances, etc. Apparently when the first statue was cast, someone quipped, "My God. It looks like my uncle Oscar." Apparently it stuck.

oscillator 1. A device for generating an analog test signal.

2. Electronic circuit that creates a single frequency signal.

oscilloscope Electronic testing device that can display wave forms and other information on a TV-screen-like cathode ray tube. A basic fixture in sci-fi movies.

OSDM Optical Spatial Division Multiplexing is a technology developed to improve the efficiency with which SONET (Synchronous Optical NETwork) supports bursty packet data traffic such as LAN traffic. OSDM accomplishes this by dynamically allocating arbitrary levels of bandwidth to such traffic, guaranteeing minimum levels that are supplemented by higher levels of bandwidth as it becomes available. OSDM is a protocol-independent, self-contained technology that adapts to various current and developing physical layer technologies such as digital wrappers and DWDM (Dense Wavelength Division Multiplexing).

OSF Open Software Foundation. An industry organization founded in 1988 to deliver technology innovations in all areas of open computer systems, including interoperability, scalability, portability and usability. The OSF was an international coalition of vendors and users in industry, government and academia that worked to provide technology solutions for a distributed computing environment. In February 1996, the OSF consolidated with X/Open Company Ltd. to form The Open Group. See The Open Group. www.opengroup.org.

OSF/1 Version 1 of the Open Software Foundation's Unix-based operating system

OSI Open Systems Interconnection. A Reference Model developed by the ISO (International Organization for Standardization, as translated into English). The OSI Reference Model is the only internationally accepted framework of standards for communication between different systems made by different vendors. ISO's goal is to create an open systems networking environment where any vendor's computer system, connected to any network, can freely share data with any other computer system on that network or a linked network. Most of the dominant communications protocols used today have a structure based on the OSI model. Although OSI is a model and not an actively used protocol, and there are still very few pure OSI-based products on the market today, it is still important to understand

OSI Reference Model

Layer 7	Application	Semantics
Layer 6	Presentation	Syntax
Layer 5	Session	Dialog Coordination
Layer 4	Transport	Reliable Data Transfer
Layer 3	Network	Routing & Relaying
Layer 2	Data Link	Technology-Specific Transfer
Layer 1	Physical	Physical Connections

its structure. The OSI model organizes the communications process into seven different categories and places these categories in a layered sequence based on their relation to the user. Layers 7 through 4 deal with end to end communications between the message source and the message destination, while layers 3 through 1 deal with network access.

Layer 1 – The Physical Layer deals with the physical means of sending data over lines (i.e., the electrical, mechanical and functional control of data circuits). Examples include EIA-232 (RS-232), T-carrier and SONET.

Layer 2 – The Data Link Layer is concerned with procedures and protocols for operating the communications lines. It also has a way of detecting and correcting message errors. Examples include Frame Relay, PPP (Point-to-Point Protocol), and SLIP (Serial Line Internet Protocol). ATM runs at Layers 1 & 2, as do LANs.

Layer 3 – The Network Layer determines how data is transferred between computers. It also addresses routing within and between individual networks. The most visible example is IP (Internet Protocol).

Layer 4 – The Transport Layer defines the rules for information exchange and manages end-to-end delivery of information within and between networks, including error recovery and flow control. TCP (Transmission Control Protocol) is an example, as is the OSI Transport Protocol (TP), which comprises five layers of its own. Layer 4 protocols ensure end-to-end integrity of the data in a session. The X.25 packet-switching protocol operates at Layers One, Two, Three, and Four.

Layer 5 – The Session Layer is concerned with dialog management. It controls the use of the basic communications facility provided by the Transport layer. If you've ever lost your connection while Web surfing, you've likely experienced a session time-out, so you have some sense of the Session Layer.

Layer 6 – The Presentation Layer provides transparent communications services by masking the differences of varying data formats (character codes, for example) between dissimilar systems. Conversion of coding schemes (e.g., ASCII to EBCDIC to Unicode), and text compression and decompression exemplify Presentation Layer functions.

Layer 7 – The Applications layer contains functions for particular applications services, such as file transfer, remote file access and virtual terminals. TCP/IP application protocols such as FTP (File Transfer Protocol), Simple Mail Transfer Protocol (SMTP), SNMP (Simple Network Management Protocol) and TELNET (TELecommunications Network) take place at Layer 7.

See also OSI Standards, which compares Layers 1 through 2 on OSI to making a phone call on the public switched telephone network. If you're looking for an easy way to remember the seven OSI layers, remember "Please Do Not Tell Sales People Anything."

Please : Physical Layer (Layer 1)
Do : Data Link Layer (Layer 2)
Not: Network Layer (Layer 3)
Tell : Transport Layer (Layer 4)
Sales : Session Layer (Layer 5)
People : Presentation Layer (Layer 6)
Anything : Application Layer (Layer 7)

OSI Model Open Systems Interconnection Model. See OSI.

OSI Network Address The address, consisting of up to 20 octets, used to locate an OSI Transport entity. The address is formatted into an Initial Domain Part which is the responsibility of the addressing authority for that domain and a domain-specific part which is the responsibility of the addressing authority for that domain.

OSI Presentation Address The address used to locate an OSI Application entity. It consists of an OSI Network Address and up to three selectors, one each for use by the Transport, Session, and Presentation entities.

OSI Standards The International Standards Organization (ISO) has established the Open Systems Interconnection (OSI) Reference Model is to provide a standar network design framework to allow equipment from different vendors to be able to communicate. Standards allow us to buy items such as batteries and light bulbs. Many of us have learned "the hard way" that the lack of computer standards can make it impossible for computers from different vendors to talk to each other. Because a major goal of a LAN (Local Area Network) is to connect varied systems, standards have been developed to specify the set of rules networks will follow. The OSI Model is a design in which groups of protocols, or rules for communicating, are arranged in layers. Each layer performs a specific data communications function. The concept of layered protocols is analogous (but not identical) to the steps we follow in making a phone call:

Step 1 – Listen for dial tone.
Step 2 – Dial a phone number.
Step 3 – Wait for a ring.
Step 4 – Exchange greetings to check that the connection is made and we're speaking the same language.
Step 5 – Talk, i.e. communicate messages back and forth.
Step 6 – Prepare to end conversation. For example, say Goodbye.
Step 7 – Take physical action. Hang up.

Each of these steps, or OSI "layers," builds upon the one below it. Although each step must be performed in preset order, within each layer there are several options. Within the OSI model, there are seven layers. The first three are the Physical (PHY), Data Link (DLL), and Network layers, all of which are concerned with data transmission and routing. The last three – Session, Presentation and Application – focus on user applications. The fourth layer, Transport, provides an interface between the first and last three layers. The X.25 Protocol which created a standard for data transmission and routing is equivalent to the first three layers of the OSI Reference Model." See also OSI and X.25.

OSINet A test network sponsored by the National Bureau of Standards (NBS) designed to provide vendors of products based on the OSI model a forum for doing interoperability testing.

osmics The science of smells. See Snortal.

OSMINE Operations System Modifications for the Integration of Network Elements. OSMINE enables equipment used by Regional Bell Operating Companies (RBOCs) and other service providers to be managed effectively from the same software program, helping to ensure multi-vendor interoperability.

OSN Operations System Network.

OSP 1. Operator Service Provider. A new breed of long distance phone company. It handles operator-assisted calls, in particular Credit Card, Collect, Third Party Billed and Person-to-Person. Phone calls provided by OSP companies are often more expensive than phone calls provided by "normal" long distance companies, i.e. those which have their own long

distance networks and which you see advertised on TV. You normally encounter an OSP only when you're making a phone call from a hotel or hospital phone, or privately-owned payphone. It's a good idea to ask the operator what the cost of your call will be before you make it.

2. Online Service Provider. A company that provides content only to subscribers of their service. This content is not available to regular Web surfers. The idea was to build subscription and other revenues from a closed knit group of people. The problem with this idea was the Internet came along and no one any longer could afford a team to compete with the Web's exploding and varied content. So, some online service providers dropped their attempt at content altogether. Others severely limited it. But all were forced to offer (and do offer) access to the Internet. As a result the term "online service provider" has virtually become obsolete, to be replaced by the term, Internet Service Provider.

3. See Open Settlement Protocol.

OSPF Open Shortest Path First. My definition is that OSPF is a link-state routing algorithm that is used to calculate routes based on the number of routers, transmission speed, delays and route cost. Here's a longer explanation from Alcatel:

Open Shortest Path First (OSPF) as described in RFC 1245 and RFC 1583 is a routing protocol designed for larger or more complex networks than those typically supported by the Routing Information Protocol (RIP). OSPF uses link state and interior gateway protocols to create a network map on each router and then uses the Dijkstra shortest path algorithm to find the optimum path between network devices. RIP has visibility only to the next hop and uses the distance vector algorithm.

Link state protocol algorithms determine the state of, or status of, each link connected to the router. In a network each router constructs a link state advertisement (LSA) with the status of its links and transmits this to its neighbors. Each router builds a list of all routes to all destinations, based on the compilation of LSAs from each router. Each router identifies which routers and subnets are directly connected to it. Then, it distributes this information to all other routers. OSPF routers take the information and build a table of what the network looks like. Using this table, each router can identify where the sub-networks are located, what routers are in direct connection, and how to get to any specific router.

As an interior gateway protocol, OSPF distributes routing information between routers in a single autonomous system. Once all routers have constructed their databases based on the LSA information, they run the Shortest Path First Algorithm. This results in a tree structure with each router at the "root" of its own tree, and the shortest path to all other destinations mapped out. The selection of the path to these destinations is based on metrics. These metrics may be based on hop count, bandwidth, load, cost, reliability, delay, or controlled statically by the user. This provides the network manager greater control over how routing occurs in the network. Dijkstra's Shortest Path Algorithm is a mathematical process by which it is possible to find the shortest path between points. Essentially, the Dijkstra Shortest Path Algorithm calculates the cost of a path between points beginning with the closest points to the starting point and works its way outward until it reaches the desired end point. A high bandwidth link costs less because more information can be sent across it at one time. Conversely, a lower speed/smaller bandwidth connection costs more because it is not able to send information as quickly. For instance, when sending packets across a 56k point-to-point serial connection there is more delay and overhead than if the same packet was sent over a 100Mbps Ethernet connection. Therefore, it would cost more time to send a transmission over the 56k connection compared to the 100Mbps connection.

OSPF is an excellent protocol in a larger network because it can build a map of large complex networks and then navigate a path between two of the network devices with visibility of the entire network providing the most efficient routing paths possible. Because of its ability to handle large complex networks, OSPF can be complex for the network manager to configure and set up and requires greater computing power within the router. However, OSPF is often the routing protocol of choice when configuring larger networks due to its ability to quickly adapt to network changes (faster route convergence), larger network metrics, area-based topology, low traffic overhead and the ability to support complex address structures and route summarization. Such speed and efficiency means minimized bandwidth usage, faster routing compared to other comparable protocols (RIP, RIPv2), lower network latency and better overall network performance, which is especially useful in networks where bandwidth is at a premium such as in a WAN.

OSPFIGP Open Shortest-Path First Internet Gateway Protocol. An experimental replacement for RIP. It addresses some problems of RIP and is based upon principles that have been well-tested in non-internet protocols. Often referred to simply as OSPF. See OSPF.

OSPR Optical Shared Protection Ring.

OSPS An AT&T word for Operator Services Position System.

OSS Operations Support System. Methods and procedures (mechanized or not) which directly support the daily operation of the telecommunications infrastructure. The average LEC (Local Exchange Carrier) has hundreds of OSSs, including automated systems supporting order negotiation, order processing, line assignment, line testing and billing.

OSS7 Operator Services Signaling System Number 7.

OSSI Operations Support System Interface. An element of DOCSIS (Data Over Cable Service Interface Specification), a project intended to develop a set of specifications for high-speed data transfer over cable television systems. At the head-end of the network, the OSSI provides the interface between the cable modem system and the OSSs. The OSSs, according to the OSI (Open Systems Integration) model, provide for the management of faults, performance, configuration, security and accounting. See also DOCSIS and OSI.

OSTA The Optical Storage Technology Association. An international trade association dedicated to promoting the use of writable optical technology for storing computer data and images. With a membership of more than 60, OSTA helps the optical storage industry define practical implementations of standards to assure the compatibility of resulting products. www.osta.org.

OTA Over The Air. See also Preferred Roaming List.

OTA DM Over-the-air device management (OTA DM) - Provisioning, configuration and management of wireless devices remotely via wireless transmission. See also firmware over the air.

OTASP Over-The-Air Service Provisioning. The ability of a wireless carrier to provision new services over the network, rather than requiring the customer to bring the terminal device into a shop for programming.

OTC Operating Telephone Company.

OTDR Optical Time Domain Reflectometer, a test and measurement device often used to check the accuracy of fusion splices and the location of fiber optic breakers. See GR.196 and Optical Time Domain Reflectometer.

OTGR Operations Technology Generic Requirements.

Other Common Carrier OCC. In the early days of long-distance competition, providers of long distance telephone service other than with AT&T were called Other Common Carriers. All long distance carriers – including AT&T – are now called interexchange carriers.

OTIA NTIA's Office of Telecommunications and Information Applications (OTIA) assists state and local governments, educations and health care entities, libraries, public service agencies, and other groups in effectively using telecommunications and information technologies to better provide public services and advance other national goals. This is accomplished through the administration of the Telecommunications and Information Infrastructure Assistance Program (TIIAP), the Public Telecommunications Facilities Program (PTFP) and the National Endowment for Children's Educational Television (NECET). The Telecommunications and Information Infrastructure Assistance Program promotes the widespread use of advanced telecommunications and information technologies in the public and non-profit sectors. The program provides matching demonstration grants to state and local governments, health care providers, school districts, libraries, social service organizations, public safety services, and other non-profit entities to help them develop information infrastructures and services that are accessible to all citizens, in rural as well as urban areas. The program was specifically created to support the development of the National Information Infrastructure. The Public Telecommunications Facilities Program supports the expansion and improvement of public telecommunications services by providing matching grants for equipment that disseminate noncommercial educational and cultural programs to the American public. The main objective of the program is to extend the delivery of public radio and television to unserved areas of the United States. Under the program's authority, funds are also allocated to support the Pan-Pacific Educational and Cultural Experiments by Satellite (PEACESAT) project. PEACESAT provides satellite-delivered education, medical, and environmental emergency telecommunications to many small-island nations and territories in the Pacific Ocean. The National Endowment For Children's Educational Television supports the creation and production of television programming that enhances the education of children. The program provides matching grants for television productions, which are designed to supplement the current children's educational program offerings and strengthen the fundamental intellectual skills of children. In addition, a ten-member national Advisory Council on Children's Educational Television provides advice to the Secretary of Commerce on funding criteria for the program and other matters pertaining to its administration. See www.ntia.doc.gov/otiahome/otiahome.html.

Otlet, Paul A Belgian lawyer who in his 1934 Traite de Documentation conceived the idea of a Universal Network for Information and Documentation in which access would be gained through multimedia workstations that had not even been invented yet.

OTN See Optical Transport Network.

OTOH Abbreviation for "On The Other Hand;" commonly used on E-mail and BBSs (Bulletin Board Systems).

OTS Operations Technical Support. See also Office Telesystem.

OUI Organizationally Unique Identifier: The OUI is a three-octet field in the IEEE 802.1a defined SubNetwork Attachment Point (SNAP) header, identifying an organization which administers the meaning of the following two octet Protocol Identifier (PID) field in the SNAP header. Together they identify a distinct routed or bridged protocol.

out-of-band A LAN term. It refers to the capacity to deliver information via modem or other asynchronous connection.

out-of-band network management A method of managing LAN bridges and routers that uses telephone lines for communications between the network management station and the managed devices. This type of management is normally in addition to the conventional method which uses the LANs and WANs that are being connected by these devices. The principal advantage is that in the event of a system failure (which may take a LAN or a WAN down), a network supervisor can bypass the failed system and use a telephone link to reach a bridge/router to diagnose a network problem. Bridges and routers must have built-in telephone modems for this to work.

out-of-band signaling Signaling that is separated from the channel carrying the information. Also known as NFAS (Non Facilities Associated Signaling). In the cellular domain, it is known as NCAS (Non Callpath Associated Signaling). Out-Of-Band Signaling is non-intrusive, as it is carried over separate facilities or over separate frequency channels or time slots than those used to support the actual information transfer (i.e., the call). Thereby, the signaling and control information does not intrude on the information transfer. SS7 (Signaling System 7) is an example of NFAS. The signaling information includes called number, calling number, and other supervisory signals. See also In-Band Signaling, NCAS, NFAS and SS7.

out-of-frame In T-1 transmission, an OOF (Out Of Frame) error occurs when two or more of four consecutive framing bits are in error. When this condition exists for more than 2.5 seconds a RED alarm is sent by OOF detecting unit. Equipment receiving this RED alarm responds with a YELLOW alarm.

out-of-order tone A tone which indicates the phone line is broken.

out-of-paper reception The ability to receive a facsimile transmission into memory when the facsimile machine is out of paper. The facsimile paper will be printed when you put in new paper.

out-of-service Or Used. A term used in the secondary telecom equipment business. Equipment taken from service. Can be in any condition. Expected to work and be complete. May not be.

out-tasking Using a vendor to perform specific network management tasks; as opposed to "outsourcing" where the whole operation is turned over to an outside vendor. See also Outsourcing.

out-of-territory Out-of-territory equals out-of-franchise which equals out-of-region.

outage Service interrupted.

outage ratio The sum of all the outage durations divided by the time period of measurement.

outdoor jack closure Closures that protect jacks from moisture, dirt and the elements.

0x06outer coder The first of two concatenated error-correcting coders in transmission and the second of two in reception used in the digital signal processing (DSP) of bitstreams in radio communications systems.

outgassing Percentage of a gas released during the combustion of insulation or jacketing material.

outgoing access A ITU description of the ability of a device in one network to communicate with a device in another network.

outgoing calls barred A switch configuration option that blocks call origination attempts. Only incoming calls are allowed.

outgoing line restriction The ability of the system to selectively restrict any outgoing line to "incoming only."

outgoing sender Equipment used to transmit call completion information on an interoffice call.

outgoing station restriction The ability of the system to restrict any given phone from making outside calls.

outgoing trunk A line or trunk used to make calls.

outgoing trunk circuit Used to carry traffic to a connecting (distant) office. depending on the traffic in an individual office. The types of outgoing trunks used will vary depending on the traffic in an individual office.

Outgoing Trunk Queuing OTQ. Extensions can dial a busy outgoing trunk group, be automatically placed in a queue and then called back when a trunk in the group is available. This feature allows more efficient use of expensive special lines such as WATS or FX. Instead of having to redial the trunk access code until a line is free, the caller can activate OTQ. See also Off-Hook Queuing.

Outgoing WATS An outgoing WATS (OUTWATS) trunk can only be used for outgoing bulk-rate calls from a customer's phone system to a defined geographical area via the dial-up telephone network. Originally WATS lines came in only lines that could receive calls or lines that could make calls. Now, you can buy a WATS line that handles both incoming and outgoing lines. See WATS.

outlet A set of openings containing electrical contacts into which an electrical device can be plugged. See Outlet Telecommunications.

outlet box A metallic or nonmetallic box mounted within a wall, floor, or ceiling and used to hold telecommunications outlets/connectors or transition devices.

outlet cable A cable placed in a residential unit extending directly between the telecommunications outlet/connector and the distribution device.

outlet connector A connecting device in the work area on which horizontal cable terminates.

outlet telecommunications A single-piece cable termination assembly (typically on the floor or in the wall) and containing one or more modular telecom jacks. Such jacks might be RJ-11, RJ-45, coaxial terminators, etc.

outlier An ATM term. A node whose exclusion from its containing peer group would significantly improve the accuracy and simplicity of the aggregation of the remainder of the peer group topology.

outline font Font is the design of printed letters, like the ones you see on this page. The first type was produced with raised metal or wooden blocks. Put ink on the blocks. Put paper on the inked blocks. Lift paper off. Bingo you have type on paper. Blocks came in fonts – styles of type, which has neat names like Times Roman, Helvetica, Souvenir, etc. Blocks also came in various sizes – 10 point, 12 point, 14 point, 36 point, etc. "Point" is simply the name for a way of measuring the size of type, like miles measure distance. When computers came along, they simply copied this technique. You picked type and you picked the size. Printers with print cartridges still work this way. They have to. They couldn't simply take one size font and enlarge or contract it because type enlarged or contracted doesn't look "right." Then two men, John Warnock and Martin Newell, said there had to be a better way and they came up with the idea of an outline font, originally called JaM, then Interpress and now PostScript. In PostScript letters and numbers become mathematical formulas for lines, curves and which parts of the character are to be filled with ink and which parts are not. Because they are mathematical, outline fonts are resolution independent. They can be scaled up or down in size in as fine detail as the printer or typesetter is capable of producing. PostScript outline fonts contain "hints" which control how much detail is given up as the type becomes smaller. This makes smaller type faces much more readable than they otherwise would be. Before outline fonts can be printed, they have to be rasterized. This means that a description of which bits to print where on the page has to be generated. And this is one reason printing outline fonts is so consuming of computer power (whether the power is in the computer or in the printer – usually it's in both). But it's also the reason why outline fonts, of which PostScript is the most successful and the most common, look so great.

Outlook See Inbox Repair Tool.

outpulse dial A pushbutton dial which allows rotary dial users the convenience of "Touch-Tone" dialing. Pushing the buttons makes the phone pretend to be a rotary dial phone. This is necessary because touch-tones are not recognized everywhere.

outpulsing The process of transmitting address information over a trunk from one switching center to another.

output Data that flows out of a computer to any device.

output device A device by which a computer transfers its information to the outside world. For example, a monitor, a printer and a speaker.

output impedance The impedance a device presents to its load. The impedance measured at the output terminals of a transducer with the load disconnected and all impressed driving forces taken as zero.

output return loss A measure of the accuracy of the impedance match between a signal source (such as a cable) and its terminating load. An unequal impedance match causes some of the power from the source to be reflected back to the source, resulting in signal distortion. The ratio of the signal voltage at the load to that voltage reflected back to the source is defined as the return loss. This ratio is generally expressed in decibels (dB).

output to output isolation The ratio of attenuation provided by the output

stage to an interfering signal driving one output compared to a second output. The ratio is measured at the second output. A good specification protects output signals against incorrect cabling, such as accidental untermination or double termination.

outshored Outsourcing software coding jobs to countries like China and India where high tech labor is cheaper.

outside line A term used by PBX users to denote a line connected to the outside world, as opposed to a station line, which essentially is an intercom line. In the US, you usually dial "9" for an outside line to call home to check on the kids. In South Africa, you dial "0." Either way, the kids usually don't appreciate your calling to check on them.

outside link An ATM term. A link to an outside node.

outside node An ATM term. A node which is participating in PNNI routing, but which is not a member of a particular peer group.

outside plant The part of the LEC (Local Exchange Carrier) telephone network that is physically located outside of telephone company buildings. This includes cables, conduits, poles and other supporting structures, and certain equipment items such as load coils. Microwave towers, antennas, and cable-system repeaters traditionally are not considered outside plant. Outside plant includes the local loops from the LEC's switching centers to the customers' premises, and all facilities which serve to interconnect the various switches (e.g., central office and tandem) in the carrier's internal network. Dedicated outside plant comprises physical local loop facilities which are dedicated from the switching center to the customers' premises. See also Dedicated Outside Plant and Inside Plant.

outside the box As in "He thinks outside the box." This is an expression that means he is creative, whatever that means. Where the expression came from is anyone's guess.

outsourcing Outsourcing is the contracting by a company of one or more of its internal functions to an outside vendor, one based in the same country or one based in a low-wage country such as China. There's probably nothing you can't outsource Those functions might include running the company's phone systems and telecom networks and/or running the company's computer system. A company might be motivated to do this because they lack the internal resources (e.g., capital budget or staff), or feel they can bring their phone costs into line and those phone costs (or at least certain of them) might now become able to be budgeted with some precision. This has appeal to senior management, who are trying to reduce their uncertainties. This usually has no appeal to lower level management who might be fired, especially if the new corporate outsourcing manager felt they were useless. See also Out-Tasking. An more contemporary word for what we used to call "facilities management." See also Offshoring.

outtasking See managed service provider.

outward restriction Phone lines within the PBX can be denied the ability to access the exchange network without the assistance of the attendant. Restricted calls are routed to intercept tone.

outward trunk queuing A process of holding outgoing long distance calls in queue until the appropriate long distance facility is available. See Off-Hook Queuing.

OutWATS Outward Wide Area Telephone Service. See WATS.

OV Ground. See Ground Start.

over modulation Occurs when the level of a modulation signal is so high that the processing amplifiers become saturated resulting in clipping of the signal. When a transmitter is over modulated it can put out spurious interference emissions or create "harmful interference."

over-the-air Refers to radio or television signals received on a local antenna and processed for re-transmission into a television distribution system.

over-the-top video Video provisioned by 3rd parties, such as Google, Yahoo! and TV studios, and delivered directly to customers over cable companies' and telephone companies' networks, leaving the cablecos and telcos in the role as bit-haulers, and competing with these network operators' own video content.

over-the-top VoIP VoIP that is provided by a 3rd-party entity other than the customer's Internet service provider.

overarching Basically overarching in telecommunications means the same as it does in real life – namely to over-reach something. In telecom, it means that as distance increases, signal quality will decrease leading to an adverse effect on performance. Most commercial wireless transmitters and receivers attempt to combat this trend using a variety of technologies.

overbuild 1. Overbuild traditionally means building more capacity into a network than you really need. Overbuilding has several potential benefits, depending on which side of the equation you find yourself. First, it provides for anticipated growth in traffic requirements. Second, it yields greater revenues for the vendor, although at the expense of the customer.

ILECs (Incumbent Local Exchange Carriers) long have been accused of overbuilding their networks, as the total investment goes into the rate base, on which the carrier realizes a rate of return guaranteed by the regulator. Overbuilding also is known as "gold-plating." See also Overbuilder.

2. The installation of a new network on top of and/or alongside of an existing network, without ripping-and-replacing the existing network. Some of the existing network's facilities (e.g., conduits and poles) may be shared by the new network.

overbuilder An overbuilder is a fancy name for a local phone company which runs fiber to your building and on that fiber runs telephone, phone, TV, and data. The idea of an overbuilder is that it rationalizes service provision. That's the benefit to the customer. The benefit to the overbuilder is that he'll suffer far less churn – as much as 78%, according to numbers I'm hearing.

overcall Paging calls over and above the standard amount of calls allowed to a subscriber. The customer is charged for overcalls.

overclocking Setting a microprocessor to run at speeds above the rated specification.

overdrive processor Intel's name for its line of single-chip performance upgrade chips. Based on Intel486 DX2 "speed doubling" technology, the Intel overdrive processors allow users of Intel486SX systems to double the internal speed of their computer's CPU by adding a single chip, without upgrading or modifying any other system components.

overfloor duct method A distribution method that uses metal or rubber ducts to protect and conceal exposed wiring across floor surfaces.

overflow Additional traffic beyond the capacity of a specific trunking group which is then offered to another group or line. For example, overflowing calls from WATS lines to (DDD) direct distance dial lines.

overflow calls Calls sent to other answering groups, other call centers, or outside service bureaus to handle peak calling issues. There are planning, training and scheduling issues involved with handling overflow calls.

overflow load The part of an offered load that is not carried. Overflow load equals offered load minus carried load.

overflow tie-line enhancement A call center term. Using Overflow Tie-Line Enhancement, non-ISDN calls diverted to an overflow call center now convey the city-of-origin announcement prior to being connected to an agent.

overflow traffic The part of the offered traffic that is not carried, for example, overflow traffic equals offered traffic minus carried traffic.

overhead In communications, all information, such as control, routing and error-checking characters, that is in addition to user-transmitted data. Includes information that carries network status or operational instructions, network routing information, as well as retransmissions of user-data messages that are received in error.

overhead bit A bit other than one containing information. It may be an error checking bit or a framing bit. See Overhead Bits.

overhead bits Overhead bits are bits assigned at the source. They are transmitted with the information payload and are for functions associated with transporting that payload.

overlap sending The process of setting up a call by sending dialed digits one at a time, as they are dialed by the originating device. The term comes from the fact that you can key digits faster than an old pulse telephone set with a keypad can send them. In other words, your dialing overlaps the pulses, as the dialing instructions are held in buffer memory in the set until such time as they can be output in pulses. Overlap sending is the way you normally set up a voice call over an analog circuit. Overlap sending also is used in some ISDN BRI (Basic Rate Interface) implementations. A better approach for ISDN is "en-bloc" sending, in which all call setup information is gathered in a block and sent to the network all at once, as in cell phone calls.

Overlapped Phase Trellis-code Interlocked Spectrum See OPTIS.

overlay 1. Typically a piece of cut-out cardboard, which you place over several keys on a phone or console. When you punch in a certain code, the buttons become what's written on the programming overlay. Also called a Programming Overlay.

2. The ability to superimpose computer graphics over a live or recorded video signal and store the resulting video image on videotape. It is often used to add titles to videotape.

3. A Northern Telecom definition: Generally used to describe some software that is not always memory resident. It is loaded on request. In the Meridian 1 most configuration, administration and maintenance functions are done from a tty terminal using various overlays. Each overlay is designed for a specific task: For example, Overlay 10 is used to configure PBX (500/2500) sets, Overlay 11 is used to configure proprietary sets. Overlay 17 is used to configure I/O ports, Overlay 15 is used to configure customer data, Overlay 48 is used to

configure link maintenance, and so on. The Meridian 1 has some 100 overlays. An overlay is loaded from the tty by typing LD nn where is the overlay number. Overlays are exited by typing ****./

4. See Overlay Area Code.

overlay area code A new area code which overlays an existing area code. New York's Manhattan has always had 212 as an area code. When the local phone company needed more numbers, it simply added an area code, 917. It originally used that area code for cell phones and beepers, but then it ran out of new numbers and used that area code for new business and home phones. Good . Thus no one had to change their phone number and go through all the expense of printing new stationery and contacting all their customers and friends with their new phone number. In contrast, we have something called a "split" area code. The "authorities" simply say the left hand side of town will have a new area code, which means that all the people and companies on the left side of town now get a new area code, and thus a new phone number. This is a "dumb" idea since it forces a huge number of people and companies to reprint all their stationery and all their customers to change their listings. In contrast an overlay area code allows everyone to keep their old phone numbers, forcing only the new area code on new phone numbers. The problem, according to some state public service commissions is that an overlay area code forces some neighbors to dial 1+area code+a number (i.e. eleven digits) when dialing their neighbors. In today's exploding communicating world, this doesn't seem like too high a price to pay for the huge savings and lack of business disruption. Local public service commissions make the split/overlay decision. For example, a few years ago when Dallas, Texas, ran out of numbers, the local phone companies simply wanted to overlay a new area code – 972 – on top of the existing 214 area code. But the Texas PUC refused to allow it for whatever dumb reason. The serious problem with splits is that a customer of five years back may no longer be able to find you. If he dials your old number, he will now get some new company who's never heard of you and you'll lose the sale. I recommend to all my friends that when they get clobbered by an area code split they immediately subscribe to their "old" number, put an answering machine on it, directing everyone who calls it to your new number. See Split Area Code.

overlay cell A cellular/PCS term for an additional cell sector which overlays the underlay cell. Overlay cells are used when traffic regularly exceeds the capacity of an individual cell. The overlay cell handles the excess traffic.

overlay network A separate network for a particular service covering most of the same geographical locations as the basic telephone network, but operating independently. Overlay networks typically are deployed on a selective basis to address performance issues in the existing PSTN (Public Switched Telephone Network). For example, developing countries in Asia and Central/Eastern Europe have built digital microwave overlay networks to serve major businesses, educational institutions, and government offices. Those networks overlay the legacy network, bypassing it and its inherent problems resulting from old, overloaded and poorly maintained infrastructure. Such an approach serves as an immediate stop-gap measure, providing state-of-the-technology voice and data communications capabilities. Eventually, such overlay networks will be fully integrated into the PSTN at such time as the PSTN is upgraded.

Overload Class OLC. The means used to control system access by mobile stations, typically in emergency or other overload conditions. Mobile phones are assigned one or more of 16 overload classes. Which class you get assigned means your phone will or won't work in the next emergency, e.g. the day of September 11, 2001 when most peoples' mobile phones (including mine) simply ceased to be able to make calls.

overload control How a system responds to being overstressed is called Overload Control. When a system is overloaded, frequently there are so many extra events being processed that the system's actual capacity or throughput goes down. Even though it may be rated at, say 10,000 busy hour calls, when overloaded, for example, with 11,000 calls, the computer telephony system may be only able to process only 8,000 calls.

overload management An AT&T term for handling peak demands by selectively delaying, degrading, or dropping only those portions of traffic flow that are tolerant of those particular types of impairments.

overload protection An uninterruptible power supply (UPS) definition. This feature automatically shuts the unit off when the battery is overloaded in order to protect it against overload damage.

override When a circuit already in use is seized. For example, when your boss can break into your telephone conversation.

override prompts The ability of callers and users to key over system prompts.

overrun Loss of data because the receiving equipment could not accept the data at the speed at which it was being transmitted.

oversampling Time division multiplexing (TDM) technique where each bit from each channel is sampled more than once.

overscan The image fills the screen from bezel to bezel. The bezel is the metal or plastic part – in short, the frame – that surrounds a cathode ray tube – a "boob" tube. NEC's term for overscan is "full scan."

overspeed Condition in which the transmitting device runs slightly faster than the data presented for transmission. Overspeeds of 0.1% for data PABXs are typical.

oversubscription In frame relay, this refers to a situation in which a device is transmitting over a circuit at a rate greater than the target device can receive the data over a circuit. For instance, the transmitter may be pumping data at 128 Kbps over a T-1 circuit, while the receiver has only a 64 Kbps circuit connection to the network. In such an instance, the network will impose flow control, adapting to the rate of reception within the limits of its buffering capability. Once that limit is reached, the network is oversubscribed. In order to maintain its ability to support "normal" traffic requirements of users in the aggregate, the network will mark the excess frames as Discard Eligible and spill them onto the switchroom floor for the night shift to clean up. (Just kidding about the switchroom floor part.) As always, it is the responsibility of the end user equipment to detect the fact that this had occurred and to recover through retransmission of the discarded frames. See DE.

overtime period Those minutes of use of a telephone service beyond the initially defined period for which a basic charge is quoted. The initial period on many calls is one or three minutes. After that, the next minute is overtime.

overview Proteon's architecture for products conforming to SNMP.

overwrite A call center term. To replace the contents of something – a file or record, for example – with new data.

OVS Open Video Systems, the successor to Video Dialtone.

OVTG See Optical Virtual Tributary Group.

OWL Web Ontology Language. Yes, the acronym is OWL, not WOL. OWL has more connotative expressiveness than WOL, at least in English. OWL is a language developed by the W3C. It is designed for applications that need to extract and meaningfully process the information content of a web page instead of just displaying the content to humans. OWL is more powerful than XML and RDF in terms of enabling greater machine interpretability of Web page content. See Semantic Web.

own When I take charge of a project, a product, a feature or whatever I "own" it. I am now responsible for it. "Own" is Microsoft jargon. But it's moving into mainstream corporate America.

owner Microsoft jargon for person in charge. See also Own.

OWS Operator Workstation. See Operator Workstation.

OWT Operator Work Time.

OXC Optical Cross-Connect. A device that serves to physically interconnect multiple optical fibers on an automated basis, thereby allowing optical signals to be directed and redirected from any of many input fibers to any of a number of output fibers without the requirement for a technician to manually connect and reconnect the fibers. In other words, an OXC is much like an automated patch panel, or cross-connect block, for DWDM (Dense Wavelength Division Multiplexing) optical networks. An OXC can take several forms. An FXC (Fiber Switch Cross-Connect) switches all of the wavelength channels from one input fiber to one output fiber. A WSXC (Wavelength Selective Cross Connect) can switch a subset of the wavelength channels from one input fiber to an output fiber, in which case the wavelength channels must first be demultiplexed. A WIXC (Wavelength Interchanging Cross-Connect) is a WSXC with the additional ability to change the wavelength (i.e., optical frequency) of an input channel on an input fiber to another wavelength on an output channel on an output fiber. See also DWDM.

oxymoron A phrase that contains an obvious contradiction. Examples include jumbo shrimp, gourmet food, military intelligence, accurate billing and constructive criticism.

P Connector Also called a male amp connector or 25-pair male connector. The female version is called a C connector.

p-ANI pseudo-ANI. A cellular term. p-ANI is a fictitious non-dialable telephone number which comprises the seven to ten digits of the cell site or base station identifier. p-ANI information is used by the PSAP (Public Safety Answering Position) in a wireless E-911 (Enhanced-911) application in order that the 911 personnel can locate the caller. Along with p-ANI information are sent the 10 digits of the cellular calling number, also known as the MDN (Mobile Dialing Number), in order that the 911 personnel can callback for more information. The FCC has mandated that, in the future, p-ANI information will be used to identify the specific location of the caller, within a range of 125 meters. See also E-911 Service and PSAP.

P-Code The Precise or Protected Code. A very long sequence of pseudo random binary biphase modulations on the GPS (Global Positioning System) carrier at a chip rate of 10.23 MHz which repeats every 267 days. Each one week segment of this code is unique to one GPS satellite and is reset each week.

P-Frame Predictive framing, specified by the MPEG Recommendation. Pictures are coded through a process of predicting the current frame using a past frame. The picture is broken up into 16x16 pixel blocks and each block is compared to a previous frame's block that occupies the same vertical and horizontal position.

P-MAC Packet Media Access Control. An FDDI-II term. See FDDI-II.

P-Phone Enhanced Business Service (also known as P-Phone) is an analog Centrex offering provided by Northern Telecom (now called Nortel). It operates over a single-pair subscriber loop., providing normal full duplex voice conversations and a secondary 8 KHz half-duplex amplitude shift-keyed signal, which is used to transmit signaling information to and from the Northern Telecom-equipped central office.

p-to-p Peer-to-peer technology. See Peer to peer technology.

P-type Semiconductor See Semiconductor.

P/AR Peak to Average Ratio. A standard analog transmission-line test signal of varying frequencies and amplitudes, which is then compared with the received signal. Composite results are a weighted number from 1 to 100 being the maximum. The P/AR is used increasingly as a standard quick test of a telecommunications channel comparative quality. Per Bell standard, the minimal acceptable P/AR rating for medium-speed data transmission is 48.

P/E Ratio See Price Earnings Ratio.

P.139 Serial Bus The P.139 Serial Bus is promulgated by the IEEE (202-371-0101). Initially it will operate at 100 million bits per second. But it has extensions for 200 and 400 megabits per second. The cable topology allows for branching and daisychaining of the peripherals attache, with a maximum of 32 4.5 meter hops using the baseline copper cable. That cable is a specially-designed combination consisting of two shielded twisted pairs for clock and data, two wires for power and ground and an overall shield that surrounds the twisted pairs and power and ground wires. See also USB, which stands for Universal Serial Bus.

P.800 The ITU-T specification "Methods for Subjective Determination of Voice Quality." P.800 involves the subjective evaluation of preselected voice samples of voice encoding and compression algorithms by a panel of "expert listeners" comprising a mixed group of men and women under controlled conditions. The result of the evaluation is a Mean Opinion Score (MOS) in a range from 1 to 5, with 1 being "bad" and 5 being "excellent." Pulse Code Modulation (PCM), the encoding standard traditionally used in most circuit-switched PSTN (Public Switched Telephone Network) environments, rates a MOS of 4.4, with 4.0 considered to be "toll quality." See also MOS, P.861, and Toll Quality.

P.861 The ITU-T specification "Objective Quality Measurement of Telephone Band (300-3400 Hz) Speech Codecs," P.861 builds on the subjective P.800. P.861 is an intrusive automated method which rates voice quality based on comparison of a predefined speech sample before and after transmission through a codec and/or network. P.861 describes Perceptual Speech Quality Measure (PSQM) as the automated rating method. The PSQM algorithm attempts to measure distortion in an "internal psychoacoustic domain," which mimics the sound perception of humans in real-life situations. The PSQM conversion of signals in the physical domain (e.g., frequency and time) to the more meaningful psychoacoustic domain, the process provides a quantifiable, objective rating, rather than the subjective rating offered by the P.800 specification. The source signal is specified in ITU-T P.50 as an artificial voice with an active speech level of -20dBm, which includes both male and female genders, and which reproduces many of the essential characteristics of human speech. The output of the test process is a PSQM value ranging from 0 to 6.5, with 0 being perfect quality and 6.5 being the highest level of signal degradation. Note that the PSQM value of PCM (Pulse Code Modulation), the traditional method for converting voice signals from analog to digital format in PSTN applications, is 0.5. This value reflects a relatively minor difference between the analog input and the uncompressed digital output of the encoding process. See also MOS, P.800, and Toll Quality.

P0 Protocol The protocol for messaging headers used for undefined messaging in an X.400 MHS (Message Handling System). Not officially part of the X.400 standards but became the U.S. intercept solution to P35 and X.435. See POP and POP3.

P01, P.01, Pnn or P.nn The Grade of Service for a telephone system. The digits following the P, i.e. nn, indicate the number of calls per hundred that are or can be blocked by the system. It is a goal or a measure of an event. In this example, P01 (also spelled P.01) means one call in a hundred (i.e. one divided by 100) can be blocked, so the system is designed to meet this criterion. See Grade of Service and Traffic Engineering.

P1 protocol P1 is the protocol defined in the X.400 standard used between MTAs (Message Transfer Agents) for relaying messages. Defined as the envelope, P1 is a portion of an X.400 message the identifies the message originator and potential recipients, records information about the message's path through the MTS (Message Transfer Service), directs the message's subsequent movement through the MTS, and includes characteristics of the message's contents. The composition of an envelope changes as the message is submitted, relayed and delivered.

P2 format The 1984 defined message format, or protocol, used between cooperating User Agents in the IPM (Interpersonal Messaging System). The P2 heading is a component of an IP-message that indicates such items as the originator, recipients(s) and subject of the message.

P25 See Project 25.

P2P 1. Peer-to-Peer. Think of Napster or Gnutella, the music people. You sign up. Your computer and its hard disk become part of a giant network. Now you can go to anyone else's hard disk and download the music you wish. Downloading music hogged the P2P headlines, but the ideas behind the P2P networking stirred lots of imaginations. Some argued that the technology had been around for a long time – the video chat and Net-Meeting-style software from Intel, Microsoft, CU-See Me, and others are P2P – but the generalized idea of a dispersed, searchable, serverless database of files is the real story behind P2P. Napster does, of course, have a central database, but it gets out of the way once you've found what you're looking for, and lets the PCs with the music commune among themselves. Can you use technology like this, asked PC Magazine, for non-musical purposes? Sure, and it tested several new collaboration tools using P2P. which it liked. Said the magazine, "Individuals become a part of an open collaboration community, thereby overcoming the barriers of traditional closed collaboration systems." P2P is also being used to make voice calls over the Internet. See Skype.

2. Person To Person. Another web term that says one person sells something to another person. It's commonly used in the auction business. One person selling something like a used camera to another person.

P3 protocol X.400 standard protocol used between a RUA (Remote User Agent) and an MTA (Message Transfer Agent) and an MTA (Message Transfer Agent) or between a remote MS (Message Store) and an MTA. P3 may be used optically between a co-resident UA (User Agent) and MTA. See P3P.

P3P Platform for Privacy Preferences Project. P3P is a way of that Web sites you visit could capture a great deal of information about you – so long as you allow that capture. You have control over how much information also. Writing about P3P, the New York Times' John Markoff said, "the new standard establishes a complicated computerized negotiation that dramatically extends both the information-gathering potential and the privacy-protection possibilities inherent in each visit to a Web site." The P3P standard, wrote Markoff, "foreshadows new Internet technology that will make each visit to a Web site a complicated exchange of information in which a personal computer will have the opportunity to automatically disclose a range of information covering every conceivable category, from birth data to shoe size, depending on rules set by the computer user. At the same time, the Web site will be forced to disclose its policy for using the information it is gathering. hopefully giving the computer user the ability to decline to share data about himself."

P5 protocol The 1984 X.400 protocol used between a MTA (Message Transfer Agent) and a Teletex Unit. This protocol is not popular.

P6 Intel's successor to the Pentium processor. The P6 is now called the Pentium Pro.

P7 protocol The protocol used between a UA (User Agent) and a MS (Message Store). Defined in the 1988 X.400 standard.

P22 format An enchanted version of the P2 format that appeared in the 1988 X.400 standard.

P2T Pulse to Tone & Rotary Dial Recognition products.

P35 format The EDI (Electronic Data Interchange) message header enhancement to X.400 that enables EDI-specific addressing, routing and handling of EDI messaging. Defined in 1990 in the X.435 standard.

P802.11 P802.11 is a series of evolving standards in the wireless Local Area Network arena. The IEEE Program P802.11 Standards Working Group aims to define universal protocols for wireless LANs in the 900 MHz, 2.4 GHz and infrared frequency bands. Main components in the 802.11 standard are twofold: 1. The physical specifications for medium-dependent protocols. There are different physical specifications for each frequency band supported in 802.11. 2. The Medium Access Control (MAC) specifications for ad-hoc wireless networks and wireless network infrastructures. A single medium-independent MAC protocol provides a unified network interface between different wireless PHYs (Physical Specifications) and wired networks. See 802.11.

PA 1. Public Address. Loud speaker system, sometimes used for paging.

2. Pooling Administrator. See Number Pooling.

PABX Private Automatic Branch eXchange. Originally, PBX was the word for a switch inside a private business (as against one serving the public). PBX means a Private Branch Exchange. Such a "PBX" was typically a manual device, requiring operator assistance to complete a call. Then the PBX went "modern" (i.e. automatic) and no operator was needed any longer to complete outgoing calls. You could dial "9." Thus it became a "PABX." Now all PABXs are modern. And a PABX is now commonly referred to as a "PBX.

PAC Personal Activity Center. A combination IBM PC clone, alarm clock, answering machine, speakerphone, fax machine, modem, compact-disk player and AM/FM radio all rolled into one unit sitting on your desk.

Pac Bell Pacific Bell, the California Bell Operating Company (BOC) which, along with Nevada Bell, formed Pacific Telesis (PacTel). PacTel was one of the seven original Regional Bell Operating Companies (RBOCs) formed in 1984 by the Modified Final Judgment (MFJ). As PacTel was acquired by SBC Corporation (Southwestern Bell) in 1996, Pac Bell is now a subsidiary of SBC.

Pacific Telesis One of the seven, independent Regional Holding Companies (RHCs) formed by the Divestiture of AT&T of 1984. Pacific Telesis held Pacific Bell, Nevada Bell and several non-regulated subsidiaries. In 1996, SBC Corporation bought Pacific Telesis. Pacific Telesis no longer exists. Pacific Bell and Nevada Bell retained their identities. See also RBOC.

pacing Controlled rate of flow dictated by the receiving component, to prevent congestion. A method of flow control in IBM's SNA. See Pacing Group.

pacing algorithm The mathematical rules established to control the rate at which calls are placed by an automatic dialing machine, also called a predictive dialer. See Predictive Dialer.

pacing control SNA term for flow control. See Pacing Group.

pacing group In IBM's SNA, the number of data units (Path Information Units, or PIUs) that can be sent before a response is received. An IBM term for window.

pack To compress data items so they take up less space. A process used by many database programs to remove records marked for deletion.

packet 1. Generic term for a bundle of data, usually in binary form, organized in a specific way for transmission. The specific native protocol of the data network may term the packet as a packet, block, frame or cell. A packet consists of the data to be transmitted and certain control information. The three principal elements of a packet include: 1. Header – control information such as synchronizing bits, address of the destination or target device, address of originating device, length of packet, etc. 2. Text or payload – the data to be transmitted. The payload may be fixed in length (e.g., X.25 packets and ATM cells), or variable in length (e.g., Ethernet and Frame Relay frames). 3. Trailer – end of packet, and error detection and correction bits. See also Block, Cell and Frame.

2. Specific packaging of data in a packet-switched network, such as X.25, Frame Relay or ATM. A true packet-switched network such as X.25 involves packets of a specific and fixed length. In a public packet switched network, such packet payloads originally were specified as being either 128B or 256B, where B=Byte. In later versions of the standards, that packet size was increased to a maximum of 4,096B, although the packet payload size generally does not exceed either 512B or 1024B. The larger packet size traditionally is used in airline reservation systems and other applications where relatively large sets of data routinely are transmitted. A X.25 packet prepends the payload with header information including a flag of eight bits; the flag denotes the beginning of the packet and also serves to assist the network nodes (packet switches) in synchronizing on the rate of transmission. An address field of eight bits also prepends the payload, with four bits identifying the target device and four bits identifying the transmitting device. Control data of eight-to-sixteen bits comprise the last element of the header; included in control data is the packet number in order that the network nodes might identify and correct for lost or errored packets. Appending the payload is a trailer consisting of a Cyclic Redundancy Check (CRC), which is used by all packet nodes for purposes of error detection and correction. See ATM, CRC, Frame Relay, Packet Assembler/Disassembler (PAD), X.25.

Packet Assembler/Disassembler PAD. A hardware/software combination that forms the interface between an X.25 network such as PDN and an asynchronous device such as a PC. The PAD generates call request, call clear, and other information packets in addition to the ones that contain user data. The PAD is responsible for packetizing the data from the transmitting device before it is forwarded through the packet network. On the receiving end of the transmission, the PAD strips away the control information contained in the header and trailer in order to get at the original text or payload, in effect disassembling the packets and reconstituting the original set of data in its native or original form. The PAD may be in the form of a standalone DCE device on the customer premise and supporting one or more terminals, or in the form of a printed circuit board which fits into an expansion slot of the terminal. Smaller users generally rely on the carrier to provide the PAD, which is embedded in the packet switch.

packet buffer Memory set aside for storing a packet awaiting transmission or for storing a received packet. The memory may be located in the network interface controller or in the computer to which the controller is connected. See Buffer.

packet burst protocol A protocol built on top of IPX that speeds the transfer of NCP data between a workstation and a NetWare server by eliminating the need to sequence and acknowledge each packet. With packet burst, the server sends a whole set (or burst) of packets before it requires an acknowledgement.

packet controller The hub of the AT&T ISDN system. It acts as a fast packet switch providing virtual circuit services to the devices hooked to the system.

packet driver The specification developed by John Romkey at FTP Software to allow TCP/IP and other transport protocols to share a common network interface card. Packet Drivers have been written for a variety of network interface cards, and in many cases provide NetWare compatibility.

packet filtering Packet filtering is the recognition and selective transmission or blocking of individual packets based on destination addresses or other packet contents. Packet filtering can be an elementary form of firewall in that it can accept or reject packets based on predefined rules. This ability helps to control network traffic. See Packet Filtering Firewall.

packet filtering firewall A packet filtering firewall is a router or a computer running software that has been configured to block certain types of incoming and outgoing packets. A packet-filtering firewall screens packets based on information contained in the packets' TCP and IP headers, including some or all of the following: Source address; Destination address; Application or protocol; Source port number; and Destination port number. See Packet Filtering.

packet forwarding Copying the packet to another node without looking at the destination address.

packet handler function The packet switching function within an ISDN switch, for the packet mode bearer service.

packet header In network protocol communications, a specially reserved field of a defined bit length that is attached to the front of a packet for carry and transfer of control information. When the packet arrives at its destination, the field is then detached and discarded as the packet is processed and disassembled in a corresponding reverse order for each protocol layer.

packet inspection Essentially what it sounds like. Your device – typically a switch – looks at every packet it is moving around. Inspecting every packet that a switch switches gives the switch the ability to do neat things to the packets – like rate limiting, quality of service (e.g. voice and video gets moved faster) and various levels of security (porn gets dropped, selective encryption, etc.)

packet interleaving Refers to the process of multiplexing multiple incoming packets from multiple channels on to a single outgoing channel by sampling one or more packets from the first channel, then the next, and so on.

packet level In packet data networking technology, level 3 of X.25. Defines how user messages are broken into packets, how calls are established and cleared over the packet data network (PDN) and how data flows across the entire PDN. The packet level also handles missing and duplicate packets.

Packet Level Procedure PLP. A full-duplex protocol that defines the means of packet transfer between a X.25 DTE and a X.25 DCE. It supports packet sequencing, flow control (including maintenance of transmission speed), and error detection and recovery.

packet mode bearer service An ISDN term for X.25 packet data transmission over the D channel in a BRI application. Always a part of the ITU-T (nee CCITT) standards, the 16-Kbps D channel can accomplish its primary responsibilities for signaling and control while still leaving 9.6 Kbps free for end user transmission of low-speed data. Retailers use of this service for credit card authorization. See also AO/DI, BRI and ISDN.

packet over SONET A metropolitan area network (MAN) or wide area network (WAN) transport technology that carries IP packets directly over SONET transmission without any data link facility such as ATM in between. Packet over SONET is intended to transmit data at the highest rates possible, because SONET has a smaller packet header overhead than ATM (28 bytes out of an 810-byte frame compared with 5 out of a 53-byte ATM cell).

packet overhead A measure of the ratio of the total packet bits occupied by control information to the number of bits of data, usually expressed as a percent.

packet radio Packet Radio is the transmission of data over radio using a version of the international standard X.25 data communications protocol adapted to radio (AX.25). It takes your information, and breaks it up into "packets" which are each sent and ac-

knowledged separately. This assures error-free delivery from sender to receiver. A packet is a stream of characters consisting of a header, the information the user is sending, and a check sequence. The header gives the destination call sign, the call sign of the sender, and any digipeaters (digital repeater) call signs that will be used for relaying the packet. The check sequence makes certain that the data received is what was sent. AlohaNET, a packet radio network developed for a number of years ago for use at the University of Hawaii, was an early packet radio network for LAN networking among the islands and laying a foundation for subsequent packet networks, both wired and wireless. Packet radio data networks recently have been deployed by a number of carriers serving mobile and fleet applications, with such carriers including ARDIS, RAM Mobile Data and Nextel.

packet size The length of a packet, expressed in bytes (B). Packet size is of specified and fixed length in X.25 and other true packet networks. The size of the "packet" in other networks may be variable within limits, as is the case with an Ethernet frame or a Frame Relay frame.

packet sniffer Most data is now transmitted in packets. A packet sniffer is a piece of software that simply examines every packet on whichever circuit/s it's assigned to monitor. A packet sniffer could be used by a legitimate company to protect itself against unwanted intruders into its network. It could also be used by an intruder to monitor a data stream for a pattern such as a password or credit card numbers. See the Internet and IP.

packet switching Sending data in packets through a network to some remote location. The data to be sent is assembled by the PAD (Packet Assembler/Disassembler) into individual packets of data, involving a process of segmentation or subdivision of larger sets of data as specified by the native protocol of the transmitting device. Each packet has a unique identification and each packet carries its own destination address. Thereby, each packet is independent, with multiple packets in a stream of packets often traversing the network from originating to destination packet switch by different routes. Since the packets may follow different physical paths of varying lengths, they may experience varying levels of propagation delay, also known as latency. Additionally, they may encounter varying levels of delay as they are held in packet buffers awaiting the availability of a subsequent circuit. Finally, they may be acted upon by varying numbers of packet switches in their journeys through the network, with each switch accomplishing the process of error detection and correction. As a result, the packets may also arrive in a different order than they were presented to the network. The packet sequence number allows the destination node to reassemble the packet data in the proper sequence before presenting it to the target device.

Originally developed to support interactive communications between asynchronous computers for time-share applications, packet switched networks are shared networks, based on the assumption of varying levels of latency and, thereby, yielding a high level of efficiency for digital data networking. Isochronous data such as realtime voice and video, on the other hand, are stream-oriented and highly intolerant of latency. As a result, packet switched networks are considered to be inappropriate for such applications. Recent development of certain software and making use of complex compression algorithms, however, has introduced packetized voice and video to the corporate intranets and the Internet, which was the first public packet-switched data network and remains by far the most heavily used.

Here is another way of explaining packet switching: There are two basic ways of making a call. First, the one everyone's familiar with – the common phone call. You dial. Your local switch finds an unused path to the person you called and joins you. While you are speaking, the circuit is 100% all yours. It's dedicated to the conversation. This is called circuit switched. Packet switching is different. In packet switching, the "conversation" (which may be voice, video, images, data, etc.) is sliced into small packets of information. Each packet is given a unique identification and each packet carries its own destination address – i.e. where it's going. Each packet may go by a different route. The packets may also arrive in a different order than how they were shipped. The identification and sequencing information on each packet lets the data be reassembled in proper sequence. Packet switching is the way the Internet works. Circuit switching is the way the worldwide phone system works, also called the PSTN (Public Switched Telephone Network).

Packet and Circuit Switching each have their own significant advantages. Packet switching for example does a wonderful job getting oodles of data into circuits. Think about a voice conversation. When you are talking, he's listening. Therefore half the circuit is dead. There are pauses between your voice. Packet switching takes advantage of those pauses to send data. Packet switching has been used primarily for data. But with the growth of the Internet, it has been used also for voice. Because of the need to re-assemble packets and other reasons, there's up to a half second delay between talking and the person at the other end hearing anything. Packet voice on the Internet is not as clear as circuit switched voice. But that's changing as the packets come faster and the technology improves. See Internet, IP Telephony and TAPI 3.0

Packet Switching Exchange PSE. The part of a packet switching network that receives the data from a PAD Packet Assembler Disassembler) through a modem. The PSE makes and holds copies of each packet before sending them to the PSE they're addressed to. After the far-end PSE acknowledges receipt of the original, the copies are discarded.

packet switching network A network designed to carry data in the form of packets. See Packet Switching.

packet telephony Another name for Internet Telephony. Also called Voice Over the Internet. See VoIP.

packet tracing The monitoring and reporting a particular packet addresses or types for diagnostic purposes.

packet type identifier In packet data networking technology, the third octet in the packet header that identifies the packet's function and, if applicable, its sequence number.

packet-centric A growing focus in the telecom industry away from voice-dominant (circuit-centric) networks and toward IP packet networks as the future delivery system for combined data and telephony. Definition courtesy Wireless Magazine.

PacketCable PacketCable is a project managed by CableLabs on behalf of its member companies. PacketCable is a fast-track initiative aimed at developing interoperable interface specifications for delivering advanced, real-time multimedia services over IP-based packet channels carved out of two-way CATV cable plants. PacketCable is built on the infrastructure set by CableLabs Certified, previously known as DOCSIS (Data Over Cable Service Interface Specification), which sets standards for CATV modems and related network elements supporting high-speed Internet access over CATV networks. The PacketCable architecture is built around the NCS (Network-based Call Signaling) specification, which is a CATV-specific enhancement to MGCP (Media Gateway Control Protocol). MGCP is the de facto standard for multimedia call control between the traditional PSTN (Public Switched Telephone Network) and IP-based packet networks. PacketCable includes specifications for call signaling, QoS (Quality of Service) control, PSTN interconnection, security, network management, codec support, billing event messages, and network announcements. See also CableLabs, CableLabs Certified, MGCP, and NCS.

PacketCable Multimedia An architecture that enables cablecos to provide QoS for latency-sensitive IP-based multimedia services, such as video on demand, VoIP, and online gaming.

packetized video First, read the definition of "Packet Switching." Then read the definition of "Packetized Voice" just below. The concept of packetized video is basically the same as that of packetized voice. A video camera feeds the signal into a codec, which converts the native analog signal into a digital format, and segments the data into data packets. The packets are sent across a packet network as a packet stream for reassembly by a codec on the receiving end of the transmission before presentation on a monitor. While packetized video performance is improving in quality through the application of increasingly sophisticated video compression techniques, it suffers from the same intrinsic packet-switching characteristics as does packetized voice. Namely, packet latency and loss. The result often is a video image which is less than pleasing. Note that voice and video are isochronous data, meaning that they are stream-oriented. In other words, the transmitting device must have regular and reliable access to the network. Further, the network must transport and deliver the data on a regular and reliable basis in order that a stream of information reach the presentation device. Such regular and reliable ingress, transport and egress of data results in a image of consistent quality. As packet-switched networks are not designed to support isochronous data communications, they generally are considered unsuitable for voice and video communications. Additionally, video is very bandwidth-intensive, thereby placing additional stress on packet-switched networks such as the Internet, which already is overloaded.

An example might help. Let's say that you are using an inexpensive ($200 or so) videoconferencing package consisting of a camera and software. Your friend has the same package. At a pre-arranged time, you place a call over the Internet to establish a videoconference. At two fps (frames per second) the videoconference goes along pretty smoothly, although both the video and voice quality are a bit rough. At some point, your friend turns his head quickly; at the same time, the Internet bogs down. The packet which contains the image of your friend's nose gets delayed or lost in the network. The video image of your friend now is missing a nose. Funny the first time, aggravating the second, maddening thereafter. The upside is that the videoconference is cheap, if not free, depending on your cost of Internet access. See also media gateway and packet switching.

packetized voice First read the definition of "Packet Switching" just above. The idea is to digitize voice and, compress it, and then slice it up into packets and send those packets from the sender by various routes and assemble them as they get to the receiver.

Packet switching for data makes sense. Packet switching for voice has not made sense because the voice is too sensitive to latency, or delay, especially the variable delay which is part and parcel of packet-switched networking. Recently developed software and DSP hardware, which employs sophisticated compression techniques has improved the ability to conduct "reasonable" quality packet voice conversations over the Internet. See media gateway, packet switching, IP Telephony, TAPI 3.0.

PacketNet Sprint's internal X.25 Packet Network.

PACS Personal Communications Access System. PACS is a cellular system providing limited, regional mobility in a given area. It provides mobility between that of a cordless phone and a full-fledged cellular system. Originally developed by Bell Labs in the early 1980s, PACS is a comprehensive framework for the deployment of PCS and applies to both licensed and unlicensed applications. Now it is approved by the TIA and Exchange Carriers Standards Associations. Today's currently implemented versions of PCS are "up-banded" versions of the 900 MHz AMPS and GSM cellular standards.

PACT Siemens' PBX And Computer Teaming. It defines protocols between Siemens PBXs and external computers.

pad 1. A pad is a device inserted into a circuit to introduce loss, i.e., to reduce the level of a signal. "Level" is a measurement of amplitude (signal power) at a specified point in a circuit known as a Transmission Level Point (TLP). "Loss" is the measurement of the decrease in amplitude between two TLPs, and is measured in decibels (dB). A pad may consist of any combination of inductors, resistors and capacitors. Pad or padding, as a verb, means to attenuate or reduce a signal, as in "to pad down the level". For example, we might say "the signal is 'hot', give me 5db of padding."

2. Packet Assembler/Disassembler. A device that accepts characters from a terminal or host computer and puts the characters into packets that can be handled by a packet switching network. It also accepts packets from the network, and disassembles them into character streams that can be handled by the terminal or host. PADs generally are associated with X.25, an ITU-T Standards Recommendation for an access protocol used in older packet-switched networks. See also X.25.

3. A concrete slab used as the foundation for a microwave radio tower or satellite dish.

4. Portable Application Description, or PAD for short, is a data set that is used by shareware authors to disseminate information to anyone interested in their software products.

pad characters In (primarily) synchronous transmission, characters that are inserted to ensure that the first and last characters of a packet or block are received correctly. Inserted characters that aid in clock synchronization at the receiving end of a synchronous transmission link. Also called Fill Characters.

pad switching A technique of automatically cutting a transmission loss pad into and out of a transmission circuit for different operating conditions.

PAF file A British term. Post Office Address file, a publicly available data file that, when integrated with an application, links postcodes to full addresses. When using a PAF file, an agent can save time by entering only the postcode. The PAF file automatically inserts post town, street and country.

page A chunk of information, like a document or file, on the Web. A hypermedia document as viewed through a World Wide Web browser. Pages are the way you make information available on the Web. They can contain text, black and white and color photographs, audio and video.

page hits A measure of the number of Web pages accessed at a particular site, or of the number of times a single page is accessed.

page mirroring You're surfing the Internet. You come upon a Web site and find something you want to buy. It has a button that says, "To speak to an Agent, push here." You do. An agent in a distant office calls you on another phone line. You're now both speaking to each other and your agent is also seeing the screen you're seeing. As you move to different screens (to different pages), the agent sees the same screens you are looking at. This is called page mirroring. At present, the technology won't allow the agent to move the pages which you see. Only you can move the screens.

page recall A feature which allows subscribers to telephone a toll-free number to check messages at any time. Messages are stored in the IVR (Interactive Voice Response) device for a period of 99 hours.

page scan state A Bluetooth term. A mode where a device listens for page strains containing its own device access code (DAC). A mode that a RemDev enters when advertising that a service is available.

page state A Bluetooth term. A mode that a LocDev enters when searching for services. The LocDev sends out a page to notify other devices that it wants to know about the other devices and/or their services.

page train A Bluetooth term. A series of paging messages sent over the baseband.

page zone A local area in the office that can receive directed Page announcements independently of the remainder of the office.

pager A small one-way (typically) wireless receiver you carry with you. When someone wants you, they make your pager receiver alert you via a tone or a vibrator. They can activate your pager in a number of ways, including dialing your pager digits directly into a computer; calling your pager from a telephone; or the old, low-tech approach of giving your name and pager number to an operator who then punches out your numbers. Pagers have become small, cheap and very reliable. Monthly service costs have dropped and the area which you can be paged in has widened dramatically. With most pagers you can be reached in most major metropolitan areas of the US. This minor miracle is accomplished through a combination of satellite and terrestrial radio networks; if the network can't reach you to deliver the page, it will store the message until you can be reached. Some pagers also display small alphanumeric messages-like phone numbers to call and names of babies born; this capability is known as SMS (Short Message Service). Many pagers and pager networks now include an "acknowledgment" feature, which allows you to press a button to acknowledge the receipt of the page through two-way communications capability. All this has made the pager far more useful. Some multi-function devices incorporate pagers into cellular phones. See also Paging and SMS.

pager codes Forget the phone. Forget the PC. Pagers are teenagers' new communication tools of choice. Wondering what your kid is receiving? Here are the secret codes, according the September 1997 Seventeen Magazine and Danielle Cioffi.

    007 – I've got a secret.
    143 – I love you.
    07734 – Hello (upside down).
    55 – Let's go for a drive.
    2468 – Who do we appreciate? You.
    13 – I'm having a bad day.
    10-4 – Is everything ok?
    2-2 – Shall we dance?
    666 – He's a creep.
    90210 – She's a snob.
    9-5 – Time to go home.
    121 – I need to talk to you alone.
    100-2-1 – Bad odds. That's not likely.

Pages Blanches White Pages in French.
Pages Jaunes Yellow Pages in French.
Paginas Amarelas Yellow Pages in Portuguese.
Paginas Amarillas Yellow Pages in Spanish.
Paginas Blancas White Pages in Portuguese and Spanish.
Pagine Bianche White Pages in Italian.
Pagine Gialle Yellow Pages in Italian.

paging 1. To send a message to someone who is somewhere, but where we don't know. Paging can be done with a little "beeper" carried in her purse or on his belt. Paging can also be done through speakers in phones or from speakers in the ceiling. Most phone systems offer a paging channel access. You dial that number and page your party. The system comes "live" and your voice is heard everywhere. Similar to the stuff they have at airline terminals. Paging systems as an accessory to phone systems always cost extra. They're one of the most valuable features on a phone system. Don't skimp, however, on the quality of the speakers or the power of the paging system. If you do, your system will sound awful and people will not use it. See Pager.
    2. A Bluetooth term. A paged device is typically contacted by a paging device to establish a communication link. See acceptor.

paging access, rapid Your attendant and you or anybody else with an extension off your PBX can make a page (access the paging equipment) by pushbuttoning one or several digits. Sometimes you can dial one several numbers for different paging alternatives. One number gives you the tenth floor. One number gives you the fourth floor, etc. And one number to page everyone – also called an "ALL CALL" page. See also Rapid Paging Access.

paging by zone By dialing the appropriate access code, any phone is able to selectively page "groups" of pre designated phones or speakers.

Paging Channel PCH (from Paging CHannel). Specified in IS-136, PCH carries signaling information for set up and delivery of paging messages from the cell site to the user terminal equipment. PCH is a logical subchannel of SPACH (SMS (Short Message Service) point-to-point messaging, Paging, and Access response CHannel), which is a logical channel of the DCCH (Digital Control CHannel), a signaling and control channel which is employed in cellular systems based on TDMA (Time Division Multiple Access). The DCCH operates on a set of frequencies separate from those used to support cellular conversations. See also DCCH, IS-136, Paging, SPACH and TDMA.

paging code call access A feature of the ROLM Attendant Console which offers direct, one-touch access to the paging or code call features.

paging device A Bluetooth term. A paging device is typically attempting to establish a communication link with other devices. See initiator.

paging file A hidden file on the hard disk that Windows uses to hold parts of programs and data files that do not fit in memory. The paging file and physical memory, or RAM, comprise virtual memory. Windows moves data from the paging file to memory as needed and moves data from memory to the paging file to make room for new data. Paging file is also called a swap file.

paging procedure A Bluetooth term. With the paging procedure, an actual connection can be established. The paging procedure typically follows the inquiry procedure. Only the Bluetooth device address is required to set up a connection. Knowledge about the clock will accelerate the setup procedure. A unit that establishes a connection will carry out a page procedure and will automatically be the master of the connection.

paging speakers Speakers in the telephone. Also external units located in ceilings, on walls, etc.

paging total system Upon dialing the appropriate special code, any station may make a paging announcement through all the loudspeakers.

PaGP Port Aggregation Group Protocol is a Cisco proprietary technology which conforms to the IEEE 802.3ad (Link Aggregation Control Protocol-LACP), which itself is a specification for bundling multiple Ethernet links into what appears to be one spanning tree protocol (STP) link.   Before this specification was ratified, various vendors had their own proprietary mechanisms for providing this functionality, but it would not work in mixed vendor environments. The technology allows an uplink to have 8 Gbps aggregate uplink speed in the situation where 8 Gigabit Ethernet links are used between two switches. ithout the technology, seven of the links would have gone into the STP blocking state because spanning-tree protocol detected a loop in the network.

paid call The usual type of a toll telephone call automatically billed to the calling telephone number.

paid hours A call center term. The time that an employee is either on duty handling calls, doing other work, in meetings, etc., or on a paid schedule exception, such as an excused absence.

paid search A fast-growing segment of the online advertising market in which advertisers pay to be listed in Web search results. For example, I search for folding bicycles. When the results of my search come up, they also bring up the web sites for those suppliers of folding bicycles who paid to show up, should someone search for folding bicycles. This area of paid searching is a fast-growing area, since such advertising really works – or at least that's what advertisers have discovered.

paint In military terms, painting means that your aircraft have been targeted for ground-based air defense ground-to-air missiles. The term is used in Iraq where coalition forces regularly fly surveillance missions and their planes are often "painted" by the Iraqi air defense radars.

paint-on antenna A conductive coating that can be applied to a surface and tuned to emit or absorb frequencies.

paintmonkey Someone with a less-than-glamorous, entry-level computer graphics job. A paintmonkey may spend months on a nanosecond of digitized film footage, painting mattes, or doing monotonous touch-ups.

pair The two wires of a circuit. Those which make up the subscriber's loop from his office to the central office.

pair gain The multiplexing of x phone conversations over a lesser number of physical facilities. Pair gain usually refers to electronic systems used in outside plant – from the central office to the subscriber's premises. In "pair gain" you might do something as simple as take one pair of wires and carry two conversations on it. You might also take two pairs and carry 128 conversations. "Pair gain" is actually the number of conversations you get minus the number of wire pairs used by the system. Lucent Technologies has various subscriber pair gain devices called "SLC" (pronounced "slick" and standing for Subscriber Loop Carrier Systems). Other companies, like Rockwell, have comparable systems. The more circuits these devices produce, i.e. the more cable pairs they save you, the more they cost. The cost of subscriber pair gain equipment – like all electronics – has been dropping in recent years, reducing the phone company's need to install outside cable and thus making better use of the presently installed cable. T-1 is a type of subscriber pair gain equipment.

pairable mode A Bluetooth term. A device that accepts pairing. is said to be in pairable mode. The opposite of pairing mode is non-pairable mode. paired device A device with which a link key has been exchanged (either before connection establishment was requested or during connecting phase).

paired cable A cable in which all conductors are arranged in twisted pairs. This form of cable is the most common for communications.

pairing A Bluetooth term. Getting a Bluetooth device to speak to another device. Typically it's getting a headset to speak to a phone – cellphone or landline phone. In technical language, it's the creation and exchange of a link key between two devices. The devices (LocDev and RemDev) use the link key for future authentication when exchanging information. Pairing is also called an association between a LocDev and a RemDev based on a common link key. The link key is also referred to as a bond. Pairing can also establish a link by the user entering a PIN which is authenticated by the device providing the service. Getting one Bluetooth device to pair with another is not easy. You typically have to carefully read the manual for both devices and carefully watch blinking led lights.

Pairs in Metal Foil PiMF. See SSTP.

PAL 1. Public Access Line. Also called a COCOT, if you live outside of US West territory. A line that is tariffed to be attached to a pay phone. There are two basic flavors, "smart" and "basic" (or dumb). Smart lines may be connected to dumb phones, and offer features such as coin signalling. Basic lines may be connected to smart phones, and the phone does all the work. Between the phone and the line, somebody has to have some intelligence. I first heard about this from one of my customers, a pay phone provider that puts up phones around town and pays commissions to the building owners. The service is tariffed by US West as a Public Access Line. (With multiple providers such as this one, anxious to pay you a percentage on a pay phone installed at your location, it's pretty silly to pay Bell for a pay phone).

2. Programmable Array Logic.

3. Proprietary ALgorithm. A designation for a privately designed and owned intelligence-based electronic method for performing a task (such as voice compression).

4. Phase Alternate Line. PAL is the format for color TV signals used in the United Kingdom, West Germany, Holland, much of the rest of western Europe, several South American countries, some Middle East and Asian countries, several African countries, Australia, New Zealand, and other Pacific island countries. PAL inverts the phase of the color signal 180 degrees on alternate lines, hence the term Phase Alternate Line. It was invented in 1961 and is used in England and many other European countries. With its 625-line scan picture delivered at 25 frames/second (primary power 220 volts), it provides a better image and an improved color transmission compared to the US system which is called NTSC, which uses interlaced scans and 525 horizontal lines per frames at a rate of 30 frames per second. SECAM is used in France and in a modified form in Russia. SECAM uses an 819-line scan picture which provides better resolution than PAL's 625-line and NTSC's 525. All three systems are not compatible. You cannot view an Australian or English videotape on a US TV. Well, at least, you couldn't until recently when a handful of manufacturers started making PAL/NTSC NTSC/PAL Video Cassette Recorder decks which will play and record in all formats. These VCRs are now very popular in both Europe and the US. See NTSC, mPAL-M and PVR.

5. Paradox Application Language. A programming language used in PDP (Programmed Data Processor) computers developed and marketed by Digital Equipment Corporation in the 1970s. See also PDP and PDP-11.

PAL-M A modified version of the phase-alternation-by-line (PAL) television signal standard (525 lines, 50 hertz, 220 volts primary power), used in Brazil. See also NTSC, PAL, SECAM.

PALAPA A major satellite communication system in south-east Asia, based in Indonesia and operational since 1976. www.satelindo.co.id/

palette In some programs, a palette is a collection of drawing tools, brush widths, line widths and colors. In other programs it is the part of the color lookup table that determines the number and type of colors that will be displayed on the screen.

pALI See pseudo Automatic Location Identification.

PalmOS The computer operating system used with the PalmPilot and Palm series of handheld digital personal assistants, as well as other organizers including Handspring and IBM. The system comes with a number of built-in applications including dates, address book, to-do list, memo pad and calculator. In addition, PalmOS interfaces with infrared devices and TCP/IP. Other operating systems for handheld devices include Windows CE and EPOC.

PALS A standard library database interface. An Internet term.

PAM 1. Pulse Amplitude Modulation. Process of representing a continuous analog signal (a voice conversation) with a series of discrete analog samples. This concept is based on the information theory which suggests that the signal can be accurately recreated from a sufficient sample. Why bother? Sampling allows several signals to then be combined on a channel that otherwise would only carry one telephone conversation. PAM was used as part of a method of switching phones calls in several PBXs. It is not a truly "digital" switching system. PAM is the basis of PCM, Pulse Code Modulation. See PCM and T-1.

2. Presence and Availability Management. See PAM Forum.

PAM Forum. The PAM Forum is an independent nonprofit consortium dedicated to establishing and promoting presence and availability management (PAM) as an ad hoc industry standard. The focus of the PAM Forum is to develop and promote a presence and availability interface specification that enables software vendors and service providers to bring personalized, interoperable communications services to market. PAM-based services enable subscriber control of communications choices and privacy options across multiple devices and networks. www.pamforum.org.

PAMS Perceptual Analysis Measurement System is a speech quality measure developed by British Telecommunications to address the problem of objectively measuring subjective speech clarity. The specification was first issued in 1998. To perform a PAMS measurement, a sample of recorded human speech is input into a system or network. The characteristics of the input signal follow those that are used for MOS testing and are specified in P.830. Though natural speech samples may be used, PAMS is optimized for proprietary artificial-like speech samples. The output signal is recorded as it is received. The input and output signals are then input into the PAMS model. PAMS performs time alignment, level alignment, and equalization to remove the effects of delay, overall systems gain/attenuation, and analog phone filtering. Time alignment is performed in time segments, so that the negative effects of large delay variations (that cause problems for PSQM) are removed. PAMS then compares the input and output signal in the time-frequency domain, comparing time-frequency cells within time frames. This comparison is based on human perception factors. The results of the PAMS comparison are scores that range from 0-5, corresponding on the same scale as MOS testing. PAMS produces a listening quality score and a listening effort score that correspond with the ACR opinion scale in P.800 and the opinion scale in P.830. A newer newer technique is called PESQ. See PESQ.

PAMA Pulse Address Multiple Access. Where carriers are distinguished by their time and space characteristics simultaneously.

PAMR Public Access Mobile Radio. The European term for what we call Specialized Mobile Radio (SMR) or Trunked Mobile Radio (TMR) in the US. The private version for use in fleet applications is called Private Mobile Radio (PMR). See SMR and TETRA.

PAN Personal Area Network. A personal area network is a very small wireless local area network that joins your own personal communications devices such as your PDA, your PC and perhaps your phone. For more, see Bluetooth and 802.15.1.

pan and scan When a program or movie which has originally been created for theatre viewing on a 16 by 9 aspect screen is shown on a 4 by 3 aspect television screen by selecting the portion of the picture which contains the action and centering that on the display. This allows for a full screen display on a television screen that captures the action and eliminates only static parts of the original picture. At least that's the theory. It doesn't work too well if one actor one side of the screen is speaking to another actor on the other side of the screen. See also Letterbox.

PAN-PAN A maritime radio distress call when a boat or person is in jeopardy. Pronounced "pahn-pahn," from the French "pann0xC7" (broken, ruined, hardship).

pan-European Across all of Europe, including England and the countries on the continent – France, Germany, Italy, etc. . "We'll launch that magazine in a Pan-European edition."

pan-and-scan A TV term referring to the translation of widescreen movies for TV broadcast through the introduction of moves and cuts which were never intended in the original. Less than the complete frame is transmitted, and portions of the picture are left out. The technique makes the action visible in a narrower frame such as your TV set. Contrast with Letterboxing.

PanAmSat Founded in 1984, PanAmSat was the first private-sector company to provide international satellite communications services, which historically had been provided by government consortia. PanAmSat quickly gained publicity through its irreverent company motto and full page ads showing its company mascot, Spot, urinating on the word "bureaucracy." PanAmSat was acquired by Intelsat, the world's largest commercial satellite communications services provider, in July 2006.

pancake coil A type of inductance having flat spiral windings. An old radio definition.

panda eyes The term panda eyes is used in the fiber optic business. Fiber optics are tiny threads of glass that carry pulses of light, and with them, much of the world's telephone, television and computer signals. But, according to the Wall Street Journal, "you can get a real education standing in a San Jose, Calif., clean room belonging to JDS Uniphase, the leading maker of the lasers, amplifiers and the like used in the world's fiber networks. The room is brightly lit, with a steady hum from all the air-filtering equipment. Workers sit in rows at work stations, each having a sign above the worker's head. One of the stations is labeled 'Panda Eyes.' The job here is to take two fiber-optic strands and, while looking through a microscope, join them together, lengthwise, like the strings in a violin bow. When the two strands are perfectly aligned and viewed head-on under the microscope, they look just like, yes, a pair of panda's eyes. The worker then can move on to the next pair.

pane What Australians call a window in a multi-windowed computer screen. For example. an agent in a call center might see a screen with several panes, one showing the customer's order status, another showing the customer's payment schedule, etc.

panel, patch See Patch Panel.

panel antenna An antenna which consists of a dipole array built in the configuration of a panel.

panel office A very early type of central office switch.

panel system Workstation defined by thin panels that provide privacy and insulation from noise.

Pangloss, Dr. Dr. Pangloss is a character in Candide, by Voltaire. It's a farce in which all these horrible things keep happening to Candide, but Dr. Pangloss keeps telling him, "It's the best of all possible worlds."

panne fatale An equipment crash in Italian or a fatal breakdown in French.

pANI pseudo Automatic Number Identification. Pronounced "pee-annie" it is a modification of Automatic Number Identification (ANI) and is used to pass information across systems that can handle ANI traffic. Also called Routing Number. pANI is a number employed in wireless E-911 call setup that can be used to route the call to an appropriate public service answering point (PSAP). The pANI generally identifies the cell/sector from which the call originates, whereas an ANI carries the actual telephone number of a wireline caller. For example, one of the wireless E911 mandates requires that a 20 digit number be passed from the call originator to the receiving 911 Call Center when processing an emergency call. The first 10 digits is the calling number, where the second 10 digits is the pANI. This pANI is in standard numbering format (that is, (NPA)NXX-xxxx for North America), and is cross-referenced in a database that holds cell site co-ordinates. This allows conventional telephone networks to move a number (that will ultimately equate to a very rough location of a cell site) to emergency service folks. Thus, a 911 Center will get the calling number (ANI), and a rough location (using a pANI).

Panic Panic never made anyone a nickel in the stockmarket.

PANS Pretty Amazing New Services/Stuff. PANS is a term coined to describe ISDN Capabilities which should eventually replace POTS. Contrast with POTS. In May, 1998, Rodney G. Seiler, Telecommunications Engineer, QUALCOMM Incorporated, wrote me "Harry, I take issue with your definition and origin of 'PANS.' I first heard this as 'Peculiar And Novel Services' in 1978, possible even earlier. I can assure you that ISDN was not in the picture when PANS first came along. Thank you for an almost perfect (this is not faint praise) reference much used in the business."

PAP 1. Packet-Level Procedure. A protocol for the transfer of packets between an X.25 DTE and an X.25 DCE.X.25 PAP is a full-duplex protocol that supports data sequencing, flow control, accountability, and error detection and recovery.

2. Password Authentication Protocol and CHAP are widely-used authentication methods for communicating between routers, both for reaching the Internet and for securing temporary WAN connections such as a dial-backup line. CHAP uses a 3 way handshake process that, in concept, resembles a dial-back routine and uses encrypted passwords. With PAP, one router connects to the other and sends a plaintext login and password.

paper To paper is a noun that means, in certain circles, to put the agreement onto paper – a process usually done by lawyers. Usage is: "We have a deal. Let's now paper it." The goal of papering the transaction is not to spend more on legal bills than the amount of the deal. Says Michael Dubin, real estate seller extraordinaire, "Most of our deals are papered too extensively and expensively. That's why I no longer practice as a lawyer."

Paper Sizes

US	Europe and Japan
A = 8 1/2" x 11"	A3 = 11.7" x 16.5"
B = 11" by 17"	A4 = 8.3" x 11.7"
C = 18" by 24"	A5 = 5.8" x 8.3"
D = 24" by 36"	B4 = 10.1" x 14.3"
E = 34" by 44"	B5 = 7.2" x 10.1"
	B6 = 5.1" x 7.2"

paper tape A long thin paper roll on which data is stored in the form of punched holes. Usually used as input to other systems. Many old-fashioned telex machines still use paper tape as their storage medium. Punch up the message on the paper tape, rewind the paper tape, call the distant telex machine, then start the paper tape containing the message. The primary benefit of paper tape is that you save on transmission line cost. The paper tape will run through at the maximum speed of the line, while a human operator typing manually would be slower. The disadvantage of paper tape is that you can't change the message once you've typed it. Magnetic medium – floppy disks, hard disks, bubble memory – are much more flexible. They are rapidly replacing paper tape, even on telex machines, or on personal computers, which are replacing telex machines as telex data entry devices.

paper tape punch A device to physically punch holes in a roll of paper tape in order to store information.

paper tape reader A device which translates the holes in coded perforated tape into electrical signals suitable for further handling. The reader may be attached to a keyboard-printer or it may be a free standing device.

paper tiger A paper tiger is a threat or a person who appears to be outwardly powerful or dangerous but is inwardly weak and/or ineffectual. The expression comes from the paper tigers that Chinese generals used to hang on city walls in an effort to frighten attacking forces. The tactic didn't work.

paperless office The concept of the paperless office arose in the 1970s, as the possibilities of alternative media became apparent. First, microfiche replaced paper records in many offices. Computer screens were next, as dumb terminals appeared. Finally, PCs and LANs entered the scene in the 1980s and became commonplace in the 1990s. With most, if not all, information readily accessible via fiche or electronic media, why would anyone need paper? Well, it turns out that people just like to read words on paper. Rather than printing and distributing paper, we now distribute and print it. In other words, we used to print out a document, make 50 copies and distribute it around the office. Now we distribute a document over a LAN and each recipient prints out a copy. (I'm being a little cynical here.) IBM estimates that the average office employee generates as many as 13,000 pieces of paper annually. Several learned people believe the office becomes paperless the day the bathroom becomes paperless.

PAR 1. Positive Acknowledgement Retransmit.

2. See Peak-to-Average Ratio.

parabola A shape which can focus a microwave signal into one narrow beam. All satellite and microwave antennas are parabolic, not spherical. See Parabolic Antenna.

parabolic antenna The most frequently found satellite TV antenna, it takes its name from the shape of the dish described mathematically as a parabola. The function of the parabolic shape is to focus the weak microwave signal hitting the surface of the dish into a single focal point in front of the dish. It is at this point that the feedhorn is usually located.

parabolic reflector The technical name for a dish antenna shaped like a perfect parabola. See Parabolic Antenna.

paradigm An assumption about the ways things work. The word paradigm (pronounced par-a-dime) is typically used by people who want to sound a little more pompous and intellectual than you and I. A yuppie word. If you want to talk about how things are changing you can talk about a "paradigm shift." According to the Economist Magazine, Thomas Kuhn invented the notion of the paradigm shift to explain what happens in scientific revolutions. A revolution happens, his theory goes, not because of startling new facts, but because of a change in the overall way the universe is seen. After this shift, old knowledge suddenly takes on new meaning. A classic paradigm shift is the way we have changed our concepts of computing from mainframe centralized mainframe computing to distributed LAN-based computing.

paradigm shift A paradigm is the way we think about something. It's a collection of theories, laws and generalizations that cause us to think the way we do about something. When something "BIG" comes along to cause us to think differently – very differently – it's called a paradigm shift.

paradise Utility is when you have one telephone, luxury is when you have two. Opulence is when you have three. Paradise is when you have none.

parallel Classically, parallel means extending in the same direction, equidistant at all points, and never converging or diverging, e.g. parallel rows of houses. In computing, parallel means the apparent or actual performance of more than one operation at a time, by the same or different devices (distinguished from serial): Some computer systems join more than one CPU for parallel processing. In telecommunications, parallel transmission means to take one stream of data, break that stream into several and transmit all the streams simultaneously. Parallel transmission is like having a four lane highway. Serial transmission is like having a single lane highway. See Parallel Data, Parallel Transmission, Series and Serial Transmission.

parallel circuit In a parallel circuit there are at least two paths for the electric current to flow through. To find the resistance of a parallel circuit, add up the reciprocal of the resistance of each of the paths the electric current may follow. The reciprocal of the sum is the total resistance of the circuit. As components are added to a circuit in series the total resistance of the circuit increases. As components are added to a circuit in parallel, the total resistance of the circuit decreases.

Household wiring is the most common type of parallel circuitry. Every outlet is parallel with every other outlet on the same circuit-breaker. So, if a bulb blows in a light fixture on that circuit, all the other devices on that circuit will still function. Another explanation. Imagine three 1.5 volt batteries. With the batteries connected in parallel, the circuit will deliver 1.5 volts. Connected in series it will deliver 3 x 1.5 volts, or 4.5 volts. By contrast, a string of Christmas tree lights is strung in a series circuit, one long continuous circuit. Should one bulb go out, usually the rest will also. See also Parallel Data.

parallel connection A connection in which the current divides, only a part of the total current passing through each device.

parallelism Indicates that multiple paths exist between two points in a network. These paths might be of equal or unequal cost. Parallelism is often a network design goal: If one path fails, there is redundancy in the network to ensure that an alternate path to the same point exists.

parallel cut See Cutover.

parallel data The transmission of bits over multiple wires at one time. This is usually accomplished by having one wire for each bit of an eight-bit byte going from a device, usually a computer, to another device, usually a printer. Thus the word "Parallel." Data transmission in parallel is very fast, but usually happens only over short distances (typically under 500 feet) because of the need for huge amounts of cable. In contrast, the other common method of data transmission, serial transmission, takes place over one pair of wires and is usually slower than parallel transmission, but can happen over much longer distances, especially using phone lines. Parallel data transmission does not happen on phone lines. See Serial Data and see the Appendix.

parallel interfacing A method of interfacing peripherals to computers, usually printers. Not as common as RS-232-C serial interfacing.

parallel networks Parallel networks, or segregated networks, exist when a single network location supports more than one physical wide area network connection for the purpose of supporting one or more applications.

parallel optic interfaces See POI.

parallel port An output receptacle often located on the rear of a computer. Unlike serial, there is no EIA standard for parallel transmission, but most equipment adheres to a quasi-standard called the Centronics Parallel Standard. Almost every PC since the original IBM PC has come with an ordinary, 25-pin D-connector parallel port. These low-speed ports are fine for sending output to a printer (which is usually the slowest device in a computer system). But when transferring data between two PC parallel ports or using the parallel port as a method of getting to and from external hard disks, the speed is too slow. As a result, there have been a number of attempts to speed up and add intelligence to the lowly parallel port. How can you tell what type of parallel port you have? A free utility called PARA14. ZIP is available on the Internet at netlab2.usu.edu/misc, or on CompuServe in the IBMHW forum, Library2. See EPP (Enhanced Parallel Port), ECP (Extended Capabilities Port) and USB (Universal Serial Bus).

parallel processing 1. A computer technology in which several or even hundreds of low-cost microprocessors are linked and able to work on different parts of a problem simultaneously.

2. A computer performs two or more tasks simultaneously. This contrasts with multitasking in which the computer works fast and gives the impression of performing several tasks at once.

parallel sessions In IBM's SNA, two or more concurrently active sessions between the same two logical units (LUs), using different network addresses. Each session can have different transmission parameters.

parallel tasking Technology which allows LAN adapters to transmit data to the network before an entire frame has been loaded from the computer to the adapter's buffer and to transmit the data to the computer's main memory before an entire frame has been received from the network. In effect, a frame can reside on the network, the adapter and in the computer memory simultaneously, thus boosting throughput.

parallel transmission 1. Method of information transfer in which all bits of a character are sent simultaneously as opposed to serial transmission where the bits are sent

one after another. See Parallel Data.

2. Method of achieving higher system reliability through use of completely redundant transmission facilities.

parallel wiretaps A parallel wiretap is connected across the two lines of a telephone line pair in parallel with the telephone instrument. A parallel must have a high resistance, otherwise the telephone line will be closed and the central office will think that the telephone is off hook. Parallel wiretaps can use high value resistors to isolate the tap. These are easy to detect. Serious wiretaps will use capacitors to isolate the tap from the telephone line.

parameter A limit, boundary, or threshold. Software often includes user-definable parameters, which allow the user to set threshold values. If a threshold value is exceeded, an alarm is triggered, an exception report is generated, or an action takes place. For example, you can set a user-definable parameter that causes your PC's screen saver to activate after a certain number of seconds. The screen saver keeps your CRT (Cathode Ray Tube), or monitor, from being ruined.

parameterize A verb which means to control the behavior of a piece of software by supplying required data at time of execution. A basic precept of structured programming is that, wherever possible, data should be kept separate from active program code. If, for example, you're writing a function that dials a phone number, the phone number should never be "hardwired" into the function itself – but should be passed to it, at runtime, as a so-called "argument" or "parameter." This encourages the creation of flexible, bulletproof software components that are easy to re-use. Much application software is written so that its behavior can be controlled by supplying parameters at time of execution. For example, the DOS command 'delete' requires a filename (the name of a file to be deleted) or wildcard expression as a parameter. More complex application programs (e.g., voicemail systems) are configured by supplying parameters on special screens or in dialog boxes, or by modifying external parameter databases.

parameters The record in a stored program control central office's data base that specifies equipment and software and options and addresses of peripheral equipment for use in call processing. See also Parameterize.

parametric amplifier Paramp. An amplifier that (a) has a very low noise level, (b) has a main oscillator that is tuned to the received frequency, (c) has another pumping oscillator of a different frequency that periodically varies the parameters, i.e., the capacitance or inductance, of the main oscillator circuit, and (d) enables amplification of the applied signal by making use of the energy from the pumping action. Note: Paramps with a variable-capacitance main-oscillator semiconductor diode are used in radar tracking and communications Earth stations, Earth satellite stations, and deep-space stations. The noise temperature of paramps cooled to the temperature of liquid helium, about 5 K, is in the range of 20 to 30 K. Paramp gains are about 40 dB.

parametric equalizer A device for manipulating sound by boosting and cutting selected frequencies by specific amounts. Basically, a much more elaborate and precise version of the bass and treble controls found on stereo systems. See Equalization.

Paramp See Parametric Amplifier.

PARC Palo Alto Research Center. A laboratory owned by Xerox and populated by it in the 1970s with some of the most creative scientists of the day. It is legendary for having pioneered technologies ranging from the laser printer to the Ethernet local area network and the graphical user interface for PCs. Sadly, Xerox senior management didn't recognize what it had and didn't recognize the value of its inventions, and didn't exploit most of them. Fortunately, other companies did. PARC was incorporated as a subsidiary company in 2002.

parasite 1. A radio tap that takes its power from the phone line.

2. One seeking financial gain from the effort or fame of another. In the world of the World Wide Web, a parasite is much like a Cyber squatter, or squatter. A squatter simply registers (i.e., squats on) a URL (Uniform Resource Locator) in advance of the "legitimate" owner of a trademark or service mark in hopes that the "rightful owner" will pay an exorbitant sum of money to buy it. A parasite registers a similar URL, which typically is a commonly mistyped version of a well-known name, often taking advantage of telephone dialing mnemonics. For example, AT&T owns 1-800-OPERATOR, which is a collect-calling service access by dialing the telephone number associated with that combination of numbers and digits on the touchtone keypad. Some years ago, another company (which will remain unnamed herein) got the number 1-800-OPERATER, relying on the fact that a large percentage of the US population can't spell "OPERATOR." It worked, and they got a lot of business. AT&T took exception to the parasitic service, i.e., they took legal action. They won. See also Cybersquatter. Ray Horak, my contributing editor, wrote this definition. And while I'm

3. A Windows software that is quietly loaded onto your PC when you install an ordinary

program or carrier. When a PC user infected by a parasite visits an e-business site on the Internet, the parasite inserts a code that routes the commission to its carrier, even though it did nothing to refer the visitor. According to InfoWorld, this has been shown to suck countless dollars from Dell.com, Buy.com, OfficeMax.com, Staples.com and others.

parasitic grid The ad hoc network created when many 802.11b (also called Wi-Fi) users steal Internet access from their neighbors' wireless network, or, more politely, get that Internet access for free. See 802.11b and 802.1X which deals with attempts to place security on wireless LANs.

parent domain Another way of saying a top level domain. See Domain.

parent node The logical group node that represents the containing peer group of a specific node at the next higher level of the hierarchy.

parent object An object in which another object resides. For example, a folder is a parent object in which a file, or child object, resides. An object can be both a parent and a child object. For example, a subfolder that contains files is both the child of the parent folder and the parent folder of the files.

parent peer group The parent peer group of a peer group is the one containing the logical group node representing that peer group. The parent peer group of a node is the one containing the parent node of that node.

Pareto Principle Named after Vilfredo Pareto, the 19th-Century economist and sociologist, the Pareto Principle is also known as "the 80:20 rule." It says that 80 percent of an enterprise's revenue comes from 20 percent of its customers. More correctly, the Pareto Principle refers to the fact that in any population that contributes to a common effect, a relative small number of the population account for the bulk of the effect. Vilfredo Pareto observed this relationship with respect to the distribution of wealth in a human population, and developed a theory of logarithmic law of income distribution to fit the phenomenon. The Pareto Principle later became applied to the distribution of quality losses (i.e., errors) in manufacturing, and since has become a universal shorthand name for such phenomena in virtually any field. In large part, popularity of the term is attributable to J.M. Juran, who used it incorrectly to describe the principle of the "vital few and trivial many."

Paris metro pricing Andrew Odlyzko, a former researcher at Bell Labs, has proposed "Paris metro pricing" (PMP) as a price-driven way to deal with congestion on packet networks. With PMP, a packet network is partitioned into several logically separate channels, each of which treats its own packets equally on a best-effort basis. These channels (gold, silver, bronze, etc. – they can be named any way the network operator chooses) would differ only in the price paid for using them. The idea is that the more highly priced the channel, the less traffic it will attract, thereby "rewarding" customers who use it with less congestion and more bandwidth on a per-user basis. Conversely, more users would choose the lower-priced channels, which would mean that they would experience more congestion and less bandwidth on a per-user basis. With PMP pricing, price would be the primary tool of traffic management. PMP resembles, in a way, the scenario that automobile drivers face when they have a choice between a toll road and a freeway.

parity A process for detecting whether bits of data (parts of characters) have been altered during transmission of that data. Since data is transmitted as a stream of bits with values of one or zero, each character of data composed of, say seven bits has another bit added to it. The value of that bit is chosen so that either the total number of one bits is always even if Even Parity error correction is to be obeyed or always Odd if odd Parity error correction is chosen.

Here's an explanation (better, but longer) from The Black Box Corporation in Pittsburgh: Many asynchronous systems append a parity bit following the data bits for error detection. Parity bits trap errors in the following way. When the transmitting device frames a character, it counts either the number of 0s or 1s in the data bits and appends a parity bit that corresponds to whether or not the count in the data bits was even or odd. The receiving end also counts the data bit 0s or 1s as it receives them and then compares the computation to the parity bit. If an error is detected, a flag can be set and retransmission may be requested. When even parity is chosen, the parity bit is set at 0 if the number of 1's in the data bits is even and it is set at 1 if the number of 1's is odd. Conversely, odd parity sets the parity bit at 1 if the number of 1's in the data bits is even, and it is set at 0 if the number of 1's is odd. Other parity selections include mark, space or off. Mark parity always sets parity at 1. Space parity always sets parity at 0, and "off" tells the system to ignore the parity bit.

parity bit A binary bit appended to an array of bits to make the sum of all the bits always odd or always even. See ASCII and Parity.

parity check A method of error-detection in binary data transmission whereby an extra bit is added to each group of bits (usually a character of data). If parity is to be odd, then the extra or parity bit is assigned either a one or zero so the total number of ones in the character will be odd. If the parity is even, the parity bit is assigned a value so that the total

number of ones in the character is even. This way errors can be detected. See Parity.

park 1. A telephone system feature that (like many features) may mean different things depending on who created it. One definition of "park" is that I dial another extension and park the call at that extension. It doesn't ring. Then I go over to that extension and pick up the phone and I'll be speaking with whoever I parked over there. This feature is useful if I have to go to another phone to find some information the caller wants. There's another definition of the telephone system meaning of "park." You have a single line phone. You put that call on a variation of hold. Then you or anyone else can pick up any phone in that pickup group and you will have your parked call.

2. In the language of hard disks, "parking" means moving the read/write head to a safe area of the hard disk when you're ready to turn the disk off. "Parking" places the heads of a hard disk in a locked position so that the storage medium (i.e. the hard disk) will not be damaged during transit. This is useful because it keeps the head from bouncing on data areas of the disk and damaging the disk. Some hard disks have a program called "park" which you run before you turn off the machine. Others do it (self-park) automatically. All hard disks on laptops are self-parking. Most modern disks are. You ought to check. It's very important.

park timeout A PBX feature. This is the period of time before an unanswered Call Park call is redirected to the Prime Phone for the line the call is on.

parked domain A parked domain has been pointed to a generic or simple Web site on the Registrar's network. Domains are usually parked while Web sites for them are under development. This is done to avoid paying for a hosting account before a site is ready to be uploaded.

parked unit A Bluetooth term. Devices in a piconet which are synchronized but do not have a MAC addresses.

parking When two domains point to the same IP Address.

parliament A group of owls is called a parliament.

PARS 1. PARS are Purchase of Accounts Receivables. These are what the LECs (local exchange carriers) send to their long distance carriers who they have a B&C (Billing and Collections)agreement with.

2. Periodic Auction Reset Security. PARS are typically short-term securities. You can buy them from your broker.

parse In linguistics it means to divide the language into components that can be analyzed. Parsing a sentence involves dividing it into words and phrases, then identifying and naming each component. Parsing is very common in computer science. Compilers must parse source code to translate it into object code. Applications that processes complex commands must also parse the commands. Parsing is divided into lexical analysis and semantic parsing. Lexical analysis divides strings into components, called tokens, based on punctuation and other keys. Semantic parsing works to define the meaning of the string once it's been broken down into individual components.

Part 68 Requirements Specifications established by the FCC as the minimum acceptable protection communications equipment must provide the telephone network. Meeting these requirements does not certify that equipment performs any task. Part 68 is the section of Title 47 of the Code of Federal Regulations governing the direct connection of telecommunications equipment and premises wiring with the public switched telephone network and certain private line services, e.g., foreign exchange lines (customer premises end), the station end of off-premises stations associated with PBX and Centrex services, trunk-to-station tie lines (trunk end only), and switched service network station lines (common control switching arrangements); and the direct connection of all PBX (or similar) systems to private line services for tie trunk type interfaces, off-premises station lines, automatic identified outward dialing and message registration. These rules provide the technical, procedural and labeling standards under which direct electrical connection of customer-provided telephone equipment, systems, and protective apparatus may be made to the nationwide network without causing harm and without a requirement for protective circuit arrangements in the service provider's network. Form 730 Application Guide is a collection of literature you'll need to register your telephone/telecom equipment under Part 68 of Title 47 at the Federal Communications Commissions. To get this material (it's free) drop a line or call the Federal Communications Commission, Washington DC 20554. You can file Form 730 yourself, but the Form 730 Application Guide also contains a list of Part 68 Certification Laboratories, a list of technical references and a list of reference sources. www.fcc.gov

More recently, the FCC decided to privatize its Part 68 responsibilities, selecting ATIS (Alliance for Telecommunications Industry Solutions) and the TIA (Telecommunications Industry Association) as joint sponsors of the Part 68 ACTA (Administrative Council for Terminal Attachments). ACTA comprises 18 members, with two each elected from six interest segments, including Local Exchange Carriers (LECs), Interexchange Carriers (IXCs, or IECs), Terminal Equipment Manufacturers, Network Equipment Manufacturers, Testing Laboratories, and Other Interested Parties. "Invited Observers," a non-voting category, will include members approved by the council on a case-by-case basis. ACTA responsibilities include adopting and publishing technical criteria for terminal equipment submitted by ANSI-accredited standards development organizations, and operating and maintaining a database of approved terminal equipment. The first meeting of ACTA was scheduled for May 2, 2001. See also ATIS and TIA.

Part X A reference to Part 64 or 68 of the MFJ (Modified Final Judgment) given to the RBOCS by Judge Harold Green. It specified the separation of customer-owned equipment and telephone owned equipment as well as telephone company demarcation.

partial conversion A Verizon definition. This situation occurs when a Verizon end user elects to use a CLEC or Reseller billing service for some, but not all of the lines on his/her account.

partial meshed network A type of wide area network topology in which every remote location is not connected directly to every other remote location, but instead is connected directly to a small subset of locations. It is a topology with more direct connectivity than a star configuration, but less direct connectivity than a fully meshed configuration.

partial packet discard (PPD). An intelligent packet discard algorithm used to control congestion. PPD checks for any of the following conditions: a policing violation, a CLP threshold violation, or no free buffer space available. If it encounters one of these conditions, it discards all remaining cells from point of encounter (including violating cells) up to but not including the next end-of-frame cell. See PPD for a bigger explanation.

partially perforated tape Same as chadless tape. See Chadless Tape.

particle beam Beams of neutral particles such as deuterium or heavy hydrogen at very high particle energies and low currents. The atoms are accelerated through electric fields as negative ions with an extra electron attached; then the electron is stripped off in passage through a gas cell, leaving a beam of neutral atoms. The advantage of the beams as weapons is that their target penetration is so high that it is virtually impossible to shield against them. However, the beams must have very long dwell times on a target to produce lethal depositions of energy.

partition 1. As a verb, partition means to divide a network into independent segments or to divide a disk or tape drive into independent volumes. As a noun, a partition is a division of memory or hard disk. For example, the Windows operating systems can allocate hard disk space for one or more partitions, each of which behaves as a physically distinct hard disk.

2. To break down a spectrum license into two or more geographic areas.

3. Many overseas administrations (i.e., PTTs) partition their return traffic to the United States. What this means is that they apportion their international long-distance traffic to the US on a carrier-by-carrier basis, in direct proportion to each carrier's respective share of US-originated traffic that is handed off to that carrier for termination in the carrier's country. For example, if 80% of the US-originating, Japan-terminating traffic to KDD International (one of Japan's international carriers) was brought in by AT&T, then KDDI will give 80% of its Japan-originating, US-terminating traffic to AT&T. Carriers that partition their return traffic determine their partitioning percentages on inbound traffic measured, typically, over a 6-month period, and the return traffic allocation percentages that are calculated for that measurement period are put into effect for the next 6-month period. In other words, the greater a carrier's share of outbound traffic to a partitioning administration, the greater the carrier's share of the return traffic from that administration.

Partitioned Emulation Programming Extension PEP. An IBM special software package that, with the Network Control Program (NCP), allows the same communications controller to operate in split mode, controlling an SNA network while at the same time managing a number of non-SNA communications lines. It was developed by IBM to facilitate migration of users to SNA.

partitions Sections on a hard disk. You can divide your hard disk into as many as four partitions to run four operating systems. See also 32-Megabyte Barrier.

Parts Per Million See PPM.

party A particularly stupid word for the person making or receiving a phone call, as in the calling party (caller) or the called party (person called). Sometimes the phone industry calls a subscriber to their service a "party," as in four-party phone lines. Party is now now used in the airline business, as in "How many people will there be in your party?" As if traveling were fun any longer.

party identification Identifying the person who is placing a call on a party line, i.e. phone line with several people sharing it. Often found in rural locations.

party line 1. Saying what your company or boss wants you to say.

2. A telephone line with several subscribers sharing its use.

party line service Telephone service which provides for two or more phones to share the same loop circuit. Party line service, which is becoming less common, is offered in two-party, four-party and eight-party versions. Interestingly, there is a version of ISDN in which several subscribers do share the same ISDN line – but they would rarely be affected by it because of ISDN's specialized signaling and the two phone lines in its 2B+D bandwidth.

party line stations Two party phone service can be expanded to support to multi-party service.

PAS 1. PAS (Personal Access System) is an MLL-based (Mobile Local Loop) personal wireless system that uses the PHS (Personal Handyphone System) standard to offer a limited mobility cordless/portable phone solution to end-users. You can't take it from one city to the other. It's a cell phone that only works in your home city. Think of it a giant cordless phone that only works in your own city. In other words you can't roam with it. This is less a technical limitation, rather than a marketing limitation. PAS basically offers consumers the convenience of a mobile phone, with the cost advantages of a fixed-line phone. PAS enables landline telephone companies to offer limited mobility phone services to communities of up to as big as you want, based on how many cell sites. As far as I can tell, PAS is a proprietary system of a manufacturer called UTStarcom and is mainly used in China and Vietnam – other developing countries. Air interface technical specs are: Standard – RCR STD-28, Version 2; Spectrum – 1,895 to 1,918.1 MHz; Voice coding – 32 Kbps ADPCM; Data coding – PIAFS (PHS Internet Access Forum Standard). As I write this, it has 30 million subscribers in China and offers Chinese subscribers cell phone service for about one-third the price of normal GSM cell phone. PAS is probably the fastest and lowest price way of installing telephone service where there previously was none. It bolts onto the existing switchs and it looks like a loop carrier in an existing central office.

2. Profile alignment system. A system that provides an automated method by which a fiber optic fusion splicer aligns a individual fibers prior to splicing. Collimated light (i.e., light comprising parallel rays) is directed through the fibers at right angles to the axis of each. Two highly sensitive cameras produce images that the machine uses to align each fiber exactly with the other along both the X and Y axes. See also fusion splicer.

Pascal A programming language designed for general information processing and noted for its structured design. Pascal originally was specified by Niklaus Wirth, a computer scientist at the Institut fur Informatic in Zurich, Switzerland in 1968. It is named in honor of Blaise Pascal, a 17th century mathematician who developed one of the first calculating machines.

pass band A spectrum of frequencies conducted by an electronic device and usually defined by upper and lower -3 dB points. See Passband.

pass through The process of accessing one device via another device. The intermediate device that sends backs the transmitted messages for testing.

pass window The range of frequencies used in a transmission system to transmit voice or data signals. More often referred to as bandwidth. See Bandwidth.

passband The range of frequencies that can pass through a filter without being attenuated (i.e. stopped). See Pass Band.

passing the buck In card games, it was once customary to pass an item, called a buck, from player to player to indicate whose turn it was to deal. If a player did not wish to assume the responsibility, he would "pass the buck" to the next player.

passionate A friend got fired from a busted dot com. He called me to help him new employment. His criterion? He wanted a company he "could feel passionate about." He wanted a job he could love. He wanted a job he could feel strongly about. The concept of feeling passionate about your job became popular with dot com companies in the late 1990s when people who joined them felt they were creating a new world – the New Economy, as it was called. Passion thus replaced older values of security, stability and the chance to climb the corporate ladder.

passive No electronics. See Passive Backplane.

passive backplane A technology where all of the active circuitry that is normally found on an "active" PC motherboard (such as the CPU) is moved onto a plug-in card. The motherboard itself is replaced with a passive backplane that has nothing on it other than connectors and joining, etched-in wires. This is why this technology is sometimes referred to as "slot cards". The chance of a passive backplane failing is very low, since it has essentially no functioning componentry. A passive backplane has several advantages: You can swap cards in and out faster. You can upgrade your processor and change faster and easier. A computer made with a passive backplane will typically have slots – as many as 25 versus only 8 or so in a "normal" PC. Passive backplane computers are increasingly used in critical computer telephony applications.

passive branching device A device which divides an optical input into two or more optical outputs.

passive bus ISDN feature which allows up to six terminal devices and two voice devices (also called telephones) to simultaneously share the same twisted pair, each being uniquely identifiable to the switched ISDN telephone network. See ISDN.

passive components Passive components, which include resistors, capacitors and inductors, do not generate signals or energy. They adjust and regulate current, store energy and filter frequencies, etc. Passive components belong to a group called discrete components, which perform a single function, e.g. regulate current or switch signals. By contrast, integrated circuits combine the functions of multiple electronic components on one chip. Together, discrete components and integrated circuits are the building blocks of electronic devices. See Active Components.

passive contract In the software business, there are two types of contracts. One you sign and one you don't. A passive contract is the one you don't sign. A passive contract typically comes with an over-the-counter, shrink-wrapped software package and you execute it by breaking the seal on the package. The passive contract spells out terms and conditions you agree to – like not copying the software, not selling, etc.

passive coupler A coupler that divides entering light among output ports without generating new light.

passive device Electronic components that don't require external power to manipulate or react to electronic output. These include capacitors, resisters and coils (inductors). Active devices include transistors, op amps, diodes, cathode ray tubes and ICs.

passive headend A device that connects the two broadband cables of a dual-cable system. It does not provide frequency translation.

passive hub A device used in certain network topologies to split a transmission signal, allowing additional workstations to be added. A passive hub cannot amplify the signal, so it must be connected directly to a workstation or an active hub.

passive leg A telephone company AIN term. The leg to a terminating access of an SSP or ASC switch. There is no access signaling on a passive leg to directly control the progress of a call.

passive loitering A law passed in France in the fall of 2002. My friend, an American lawyer in Paris, wrote me, "It is only something that a rightwing French Interior Minister could come up with. Apparently, it is hanging around in a manner and dressed in a manner that suggests a woman is an available prostitute. It doesn't require that she apparently do anything. hence the word 'passive.'" It wouldn't last 10 seconds in a U.S. federal court."

passive optical components Components used to guide and manipulate optical wavelengths, including:

- Attenuators. Used to control signal amplitude.
- Couplers. Used to split or combine light.
- DWDM couplers. Split or combine light by wavelength.
- Optical isolators. Used to eliminate back reflections.
- Optical switches. Used to direct light to fiber.
- Tunable bandpass filter. Allows for wavelength selection.

Passive Optical Network See PON.

passive reflector A simple reflector used to change the direction of radiation from a microwave beam. For example, a reflecting surface mounted on a hill top and so positioned as to direct the energy down onto a valley receiving site. See Passive Repeater.

passive repeater A passive reflector system constructed from two reflectors that are simply coupled together with a short length of waveguide. The first reflector acts as a receiver while the second transmits but in a different direction. There's no electronics in the system.

passive side When describing a loopback test, the passive side is used to identify the device that sends back the transmitted messages for testing.

passive splicing Aligning the two ends of a fiber without monitoring its splice loss.

passive star A star-topology local network configuration in which the central switch or node is a passive device. Each station is connected to the central node by two links, one for transit and one for receive. A signal input on one of the transmit links passes through the central node where it is split equally among and output to all of the receive links. Also called a star coupler or a retransmissive star.

passive tag An RFID tag without a battery. When radio waves from the reader reach the chip's antenna, it creates a magnetic field. The tag draws power from the field and is able to send back information stored on the chip. Today, simple passive tags cost around 50 cents to several dollars. See also passive telemetry sensor.

passive telemetry sensor A passive telemetry sensor is a wireless, battery-less sensor that is used for implantable sensors. It consists of a sensor unit which does not any power source of its own, and a transceiver unit, which provides energy to the sensor. The sensor provides information back to the transceiver via the inductive link. Since the sensor is battery-less, it is used in an application where the battery of sensor cannot be changed, or it is cumbersome to connect wires to the sensor. Applications include an implantable blood pressure sensor, implantable humidity sensor, implantable pressure sensor and in civil engineering for strain measurement where is is cumbersome to connect wires to each of the sensors.

passive terminator A crude type of single-ended SCSI terminator that can't compensate for variations in terminator power or bus impedance. No longer recommended by ANSI, it's adequate for most simple SCSI-1 Applications. See also Active Terminator and Forced-Perfect Terminator.

passphrase Used much like a password, a passphrase simplifies the WEP encryption process by automatically generating the WEP encryption keys for Linksys products, according to Linksys.

passthrough Gaining access to one network through another element. Also spelled Pass Through.

password A word or string of characters recognized by automatic means permitting a user access to a place or to protected storage, files or input or output devices. In order to be somewhat secure, passwords should be at least eight characters and should include both numbers and letters, both upper and lower case. Passwords should not be obvious things like your name or initials, your wife's name or initials, your children's names or initials, your dog's name, your birthday, or your social security or employee number. "Password," is not a good password, either. Don't write your password on a sticky note and stick it on your computer. Most people's password are obvious, thus easy to crack. The harder you make a security system – of which a password is but one element – the more complaints you will get from your users. All security systems are a compromise between absolute security and useability.

Password Authentication Protocol PAP, A security protocol that establishes a simple PPP authentication method using a two-way handshake to verify the identity of the two computers or communicating devices. PAP sends passwords as text, which makes it vulnerable to hackers.

password control of changes A feature that makes it impossible to alter the performance of a piece of equipment without first entering a password.

pastoral call In Australia, a pastoral call is a call made between basic telephone services which are either: in the same extended charging zone, in the same community access zone but with one service in an extended access zone, or in the same community access zone but with one service in an standard charging zone adjacent to one or more extended charging zones. Telstra (the dominant carrier in Australia) defines extended charging zones, community access zones and standard charging zones. For your enlightenment, I did not make the above up. I copied it from an Australian government report that purports to explain the Australian phone system. Good luck.

PAT See Port Address Translation.

patch 1. A small addition to the original software code, written to bypass or correct a problem. See Y2K.

2. To connect circuits temporarily with a jack and a cable. Patching is typically done on devices called PATCH BAYS, PATCHBOARDS or PATCH PANELS. See also Patching.

patch antenna An antenna that is printed on a thin circuit board. The most common shapes are rectangular and circular, but triangular and ring-shaped patch antennas are also common. Patch antennas are used in avionics, GPS receivers, and RFID devices and in other devices where a low-profile, lightweight, or surface-conformable antenna is needed. Also called a microstrip antenna.

patch bay A collection of hardware put together in such a way that circuits appear on jacks and can be connected together for transmission, monitoring and testing. See Patch Panel.

patch cord A short length of wire or fiber cable with connectors on each end, a patch cord is used to join communication circuits at a cross connect point. A patch cord is much like an extension cord. In the context of telephony, it's much like the cords that the telephone operators in the early 1900s used to use on a manual switchboard. They would use a short cord with a plug on each end to connect to one jack for the calling party and another for the called party. Thereby, a unique physical and electrical path was established. When the call was concluded, the operator unplugged the cord from the jacks. The next call involved a repeat of the same process, and so on. Patch cords still have a very important purpose where semi-permanent and highly reliable connections must be made between links. See also Cross Connect.

patch management You run a the IT department of a large corporation. You have thousands of computers to take care of. Most are Windows machines. Every day (or so it seems) Microsoft issues yet another patch to its Windows/Office and other software. You need to test the patch – to make sure it's not buggy. Many are. Second, you need to get the patch onto each of your thousands of Windows machines. In the IT world, that's called patch management. There are automated tools that patch computers over private corporate networks and over the public Internet. I believe some of them work.

patch panel A device in which temporary connections can be made between incoming lines and outgoing lines. It is used for modifying or reconfiguring a communications system or for connecting devices such as test instruments to specific lines. A patch panel differs from a distribution frame in that the connections on a distribution frame are intended to be permanent. See also Cross-Connect and DACS. Also called a jackfield.

Patch Tuesday The second Tuesday of each month, when Microsoft releases the newest fixes for its operating systems and applications.

patchboard Same as a patch bay. See Patch Bay.

patching Means of connecting circuits via cords and connectors that can be easily disconnected and reconnected at another point. May be accomplished by using modular cords connected between jack fields or by patch cord assemblies that plug onto connecting blocks.

patchmaster A patch panel in which multiple pair lines can be interconnected as a group.

patent A patent is a grant, limited in time and technological extent, of the right to exclude others from making, using or selling the invention. The temporal extent of a U.S. patent (other countries' laws are similar) is usually at least 17 years. The technological extent of protection is defined by the claims of the patent. Claims are allowed only after examination by a technically degreed examiner of the Patent and Trademark Office to ensure that an opinion is defined that is nonobvious with respect to the predecessor technology, or "prior art." Having a patent does not confer the right to manufacture the thing claimed; others may have dominating patents. Independent invention is no defense to a charge of infringement. Rights in an invention made by an employee in performance of his normal duties belong to the employer in most states, even without a written agreement.

The claims made for the invention in applying for the patent may define a "process, machine, manufacture, or composition of matter". The great software patent controversy has largely been settled with the understanding that software is normally best claimed as a process, although sometimes claims usefully intermix hardware and software. The only software that remains per se unpatentable is that defining a process operating on pure numbers, that is, not tied to any particular end use, control process, or the like. A Fast Fourier Transform algorithm would thus not be patentable although an unobvious method of using it to (say) remove harmonic noise from a signal might be.

Patent serve several functions. A patent is a technical disclosure, forming part of the scientific literature, as the invention must be described with sufficient particularity that others can use it "without undue experimentation." The idea is that the inventor is given a limited monopoly in exchange for the benefit to the public of having inventions made and disclosed. A patent is also a legal document, in that the claims define the exact extent of protection. The third function of a patent is to market the invention.

A well-written patent explains the underlying technical problem, the deficiencies of the prior art, and the way in which the invention solves these problems, in clear and non-technical language. Only then can a federal judge or jury, or the CEO of a competitor being asked to pay damages, be expected to understand the relation of the claimed invention to the prior art and the allegedly infringing product. See also Intellectual Property, Patent Troll and Prior Art.

patent troll A company that files a broad patent, or buys a patent, often from a bankrupt firm, and then sues another company by claiming that one of its products infringes on their patent. Patent trolls have hurt companies such as Research in Motion and various companies offering voice mail service. See patent.

path The physical route a telecommunications signal follows from transmitter to receiver. The path includes all circuits and all intermediate devices, such as switches and routers.

path clearance In through-the-air microwave transmission, you must find a line-of-sight path, free of obstruction of buildings, trees, other microwave towers, etc. In microwave line-of-sight communications, the perpendicular distance from the radio-beam axis to obstructions such as trees, buildings, or terrain. The required path clearance is usually expressed, for a particular k-factor, as some fraction of the first Fresnel zone radius. That's the technical definition of path clearance.

path constraint A bound on the combined value of a topology metric along a path for a specific connection.

path control IBM Corp.'s implementation of what is normally referred to as the network layer in the International Standards Organization Open Systems Interconnect (OSI) layered network architecture.

path control layer In IBM's SNA, the network processing layer that handles primarily the routing of data units as they travel through the network and manages shared link resources.

Path Delay Value PDV. An indexed (i.e., relative) measurement of propagation delay in communications circuits. The point of reference is that of optical fiber, which has a designated PDV of 1.0. CAT 5 copper wire has a PDV of 1.11, meaning that the signal takes 111% as long to travel the same distance through the transmission medium. CAT 3 wire has a PDV of 1.14. A PDV of 1.11 may seem insignificant, since that translates into 200,000 kilometers per second, but it can be very important in the context of high-speed data transmission protocols. See also Propagation Delay.

path loss The power loss which occurs when radio waves move through space along specific paths.

path planning The process that is followed to ensure that the hardware selection and placement decisions for a wireless network will ensure transmission accuracy and reliability. Path planning involves site audits to collect GPS data for sites, paths, and obstructions; a determination of the types of antennas that will be needed, along with the gain, height, tilt angle, and installation location for each of them; the calculated fade margin for each antenna; a topological terrain map showing the network path; and a complete radio bill-of-materials.

path replacement A method by which the normal physical path set up in support of a call is replaced by another path which is more efficient. Path replacement is intended to overcome the inefficiencies of trombonboning associated with a preselected path which normally would be used to connect a call. See also Tromboning.

path switched ring A technique for providing redundancy in a SONET network. Path switched rings use 2 fibers, with both transmitting simultaneously in both directions. Through this technique, a failure in a SONET ring will not prevent devices from communicating, as they transmit and receive in both clockwise and counter-clockwise directions. All devices monitor both rings, locking in on the better signal, thereby improving on the inherently high quality of fiber optic transmission. See Line Switched Ring.

path table One of two tables contained in the volume descriptor of a CD-ROM and which comprises the file management for the disc. The path table contains the names of all directories on the disc and is the latest fastest way to access a directory that is not close to the root directory. Access to directory and path tables is handled automatically by the Virtual CD Changer Driver.

Path Vector routing protocol A Path Vector (PV) routing protocol, such as BGP (Border Gateway Protocol), is used by routers to select a path across a network. Path Vector protocols are similar to Distance Vector (DV) protocols, but with a key difference. A Distance Vector protocol selects the best path between two border routers based on the hop count (i.e., number of routers transversed). A border router (BR) running a Path Vector routing protocol advertises the destinations it can reach to its neighboring border routers. Further, a path vector protocol pairs each of those destinations with the attributes of the path to it. The attributes include the number of hops (i.e., routers transversed) and the administrative "distance." The attribute of administrative distance weights routes learned from IBGP (Internal BGP) more heavily than those learned from EBGP (External BGP). In other words, interior routes are weighted more heavily (and preferred) than are exterior routes, which cross network domains. Note, however, that BGP does not take into account factors such as link speed or network load. See also BGP, Distance Vector Routing Protocol, Link State Protocol, Media Gateway, Policy Routing Protocol, Router, and Static Routing.

pathway A facility for routing communications cables.

pattern recognition A small element of human intelligence. The ability to recognize and match visual patterns. (Auditory pattern recognition is the ability to recognize spoken words.) Pattern recognition basically works by having the computer seek out particular attributes of the character (assuming it's pattern recognition for reading words) and then having the computer compare what it finds to what's in its database of patterns. By a process of breaking down letters into curves and lines, and by a process of elimination, the computer can figure out what it's seeing. As Forbes said, "think of pattern recognition as a kind of super detective, a tireless if unimaginative collector of clues, distinguished not by brilliance, but by ceaseless legwork."

pau hana Hawaiian for "end of the work day."

pause This feature on some phone systems which, by hitting #, inserts a 1.5 second delay in a speed dialing sequence. This way you can program your phone to call a main number, wait a few seconds for the machine to answer and then punch out your person's extension.

PAX Private Automatic eXchange. Typically an intercom system not joined to the public telephone system. PAXs are more common in Europe, where is it common for business people to have two phones on their desk – one for internal intercom calls and one for external calls.

pay-as-you-grow An approach to network investments whereby a company doesn't invest up-front in a network with enough excess capacity to accommodate expected growth over the planning horizon but instead invests in additional capacity incrementally as the need for it arises. In addition to the short-term budgetary advantages of this approach, it also reduces the risk of making large investments in technologies that may soon become obsolete.

pay phone See Payphone.

Pay-Per-View PPV. A television service which allows viewers to select movies or other programming for viewing for a fee which is charged to their bill. See Impulse Pay Per View.

payload 1. From the perspective of a network service provider: of a data field, block or stream being processed or transported, the part that represents information useful to the user, as opposed to system overhead information. Payload includes user information and may include such additional information as user-requested network management and accounting information. In Sonet, the STS-1 signal is divided into a transport overhead section and an information payload section (similar to signaling and data). See SPE (Synchronous Payload Envelope) for a description of what would be found in the payload.

2. The activity carried out by a computer virus when it is activated by a triggering event. Depending on the virus, the payload may be as benign as putting a message on your screen or as destructive as erasing your hard disk or scrambling your data.

payload pointers Payload Pointers indicate the beginning of the synchronous payload envelope. See Payload.

Payload Type Indicator Field PTI. A three-bit field in the ATM cell header that indicates the type of information being carried in the payload. The PTI is used to distinguish between cells carrying user data and those carrying service information such as call set-up and call termination.

payoff A business term used to predict the amount of time it will take for an investment such as a switch, new computers, new telephones, and applications to pay for itself in increased sales, reduced costs, or a combination.

payphone Also known as a paystation, a payphone used to be just a public phone that accepted only coins. Now payphones can be coinless and can read credit cards. They also now commonly include keyboards, computer screens, and dataports for plugging in fax machines and laptop computers. Sadly, pay phones are disappearing rapidly, largely having been rendered obsolete by cell phones. Also, residents of some neighborhoods have pushed for their removal, associating them with such undesirable activities as drug-dealing. Further, they are expensive to operate and maintain, as the coins much be collected (assuming that they haven't been stolen), and the phones and phone booths are subject to acts of theft and vandalism. As of the end of 1999, only 2.5 million payphones remained in the U.S.; approximately 300,000 had been removed during the previous two years. Some of the ILECs (Incumbent Local Exchange Carriers), the primary operators of payphones, have announced their intentions to sell their remaining payphones to independent operators.

When using the first pay telephone, a caller did not deposit his coins in the machine. He gave them to an attendant who stood next to the telephone. Coin telephones did not appear until 1899. The payphone was invented by William Gray, an American whose previous inventions included the inflatable chest protector for baseball players. Mr. Gray's first phone lacked a dial. Its instructions read:

"Call Central in the usual manner. When told by the operator, drop coin in proper channel and push plunger down."

There are several different ways in which payphones can work. In today's nomenclature, Mr. Gray's original phone is known as a postpay coin phone. Postpay payphones require that you pay for the call only after it is completed. Semi-postpay payphones allow you to dial the number first, requiring that you pay with coins after the called party answers. Both postpay and semi-postpay phones were used for many years for local calling in rural areas, but now are considered to be obsolete for obvious reasons. Prepay payphones require that you pay for calls before they can be dialed. Virtually all contemporary payphones are prepay.

payphone-postpay Calls are paid for after they are completed, typically with a credit card or calling card, etc. See Payphone.

payphone-prepay At a coin phone, calls must be paid for before they can be dialed. Virtually all local calls are prepay. See Payphone.

payphone-private Referred to as Customer Owned Coin Operated Telephone Companies (COCOTs). Private payphones are installed and maintained by companies other than the LECs (Local Exchange Carriers) that traditionally provided the service. Be very careful if someone calls you with an offer to invest in private payphones. According to the North American Securities Administrators Association (NASAA), payphone scams were among the top 10 investment fraud schemes in the U.S. in 2000. Such scams generally are characterized as Ponzi Schemes. See also Ponzi Scheme and Scam.

payphone-public A coin phone installed in a "public" place. The owner of the property may get a commission for allowing the service provider to install it. See also Payphone-Semi-Public.

payphone-semi-public A coin phone installed for public use but installed in a "semi-public" place, such as a restaurant or bar. The proprietor of the establishment is obliged to guarantee that the service provider will realize a minimum amount of revenue from the phone. However, the service provider typically will not pay a commission on this type of phone. What is a "public" and what is a "semi-public" phone is a decision made by the local telephone company for whatever reason it chooses. The pay phone business is rapidly deregulating. It is now legal to own your own payphone.

paystation See Payphone.

paystation, postpay Calls are paid for after they are completed, typically with a credit or calling card, etc.

paystation, prepay Calls must be paid for before they can be dialed. Virtually all local calls are prepay.

paystation, public A coin phone installed in a "public" place. The phone company is totally responsible for its installation. The phone company will typically pay someone – the city, the bus station owner – a commission on the calls made from this phone. See Paystation, Semi-Public.

paystation, semi-public A coin phone installed for public use but installed in a "semi-private" place, such as a restaurant or bar. The proprietor of the establishment is obliged to guarantee that the phone company will receive a minimum amount of money out of the phone. The phone company will typically not pay a commission on this type of phone and takes all the money in the coin box for itself. What is a "Public" and what is a "Semi-Public" phone is a decision made by the local telephone company for whatever reasons it chooses. The pay phone business is rapidly deregulating. So the rules are changing. And it is now legal to own your own payphone.

PB Petabyte. See Petabyte.

PBC See Polarization Beam Combiner.

PBNM Policy-Based Network Management. See Policy-Based Networking.

PBS Personal Base Station. A PCS (Personal Communications System) term. A PCS subscriber might use a "High-Tier" PCS service, which effectively is cellular service using PCS frequencies. When at home, the PCS set acts as a cordless phone, establishing a wireless link to the PBS. When in close enough proximity to have sufficient signal strength, the PBS takes over from the PCS carrier's cell site. All PBS calls then are routed over the landline PSTN, thereby avoiding cellular usage charges. In a business environment using a PCS wireless office system, the PCS set and the wireless controllers establish the same relationship.

PBX Private Branch eXchange. A private (i.e. you, as against the phone company owns it), branch (meaning it is a small phone company central office), exchange (a central office was originally called a public exchange, or simply an exchange). In other words, a PBX is a small version of the phone company's larger central switching office. A PBX is also called a Private Automatic Branch Exchange, though that has now become an obsolete term. In the very old days, you called the operator to make an external call, except in Europe. Then later someone made a phone system that you simply dialed nine (or another digit – in Europe it's often zero), got a second dial tone and dialed some more digits to dial out, locally or long distance. So, the early name of Private Branch Exchange (which needed an operator) became Private AUTOMATIC Branch Exchange (which didn't need an operator). Now, all PBXs are automatic. And now they're all called PBXs, except overseas where they still have PBXs that are not automatic.

At the time of the Carterfone decision in the summer of 1968, PBXs were electro-mechanical step-by-step monsters. They were 100% the monopoly of the local phone company. AT&T was the major manufacturer with over 90% of all the PBXs in the U.S. GTE was next. But the Carterfone decision allowed anyone to make and sell a PBX. And the resulting inflow of manufacturers and outflow of innovation caused PBXs to go through five, six or seven generations – depending on which guru you listen to. (See my definition for GENERATIONS in this dictionary). Anyway, by the fall of 1991, PBXs were thoroughly digital, very reliable, and very full featured. There wasn't much you couldn't do with them. They had oodles of features. You could combine them and make your company a mini-network. And you could buy electronic phones that made getting to all the features that much easier. Sadly, by the late 1980s the manufacturers seemed to have finished innovating and were into price cutting. As a result, the secondary market in telephone systems was booming. Fortunately, that isn't the end of the story. For some of the manufacturers in the late 1980s figured that if they opened their PBXs' architecture to outside computers, their customers could realize some significant benefits. (You must remember that up until this time, PBXs were one of the last remaining special purpose computers that had totally closed architecture. No one else could program them other than their makers.) Some of the benefits customers could realize from open architecture included:

- Simultaneous voice call and data screen transfer.
- Automated dial-outs from computer databases of phone numbers and automatic transfers to idle operators.
- Transfers to experts based on responses to questions, not on phone numbers.
- And a million more benefits.

There are two alternatives to getting a PBX. You can buy the newer, open more full-featured version called a communications server. Or you can subscribe to your local telephone company's Centrex service. For a long explanation on Centrex and its benefits, see Centrex. Here are some of the benefits of a PBX versus Centrex:

1. Ownership. Once you've paid for it, you own it. There are obvious financial and tax benefits.

2. Flexibility. A PBX is a far more flexible than a central office based Centrex. A PBX has more features. You can change them faster. You can expand faster. Drop another card in, plug some phones in, do your programming and bingo you're live.

3. Centrex benefits. You can always put Centrex lines behind a PBX and get the advantages of both. In some towns, Centrex lines are cheaper than PBX lines. So buy Centrex lines and put them behind your PBX. Make sure you don't pay for Centrex features your PBX already has. (It has most.)

4. PBX phones. There are really no Centrex phones – other than a few Centrex consoles. If you want to take advantage of Centrex features, you have to punch in cumbersome, difficult-to-remember codes on typically single line phones. PBXs have electronic phones, often with screens and dedicated buttons. They're usually a lot easier to work. A lot easier to transfer a call. Conference another, etc. A lot more productive.

5. Footprint savings. Modern PBXs take up room, more than Centrex. But the space they take up is far less than it used to be. PBXs are getting smaller.

6. Voice Processing/Automated Attendants. Centrex's DID (Direct Inward Dialing) feature was always pushed as a big "plus." You saved operators. However, you can now do operator-saving things with PC-based voice processing and automated attendants you couldn't do five years ago. These things work better with on-site standalone PBXs than with distant, central office based Centrex. Moreover, virtually every PBX in existence today supports DID. You can dial directly into PBXs and reach someone at their desk just as easily as you can dial directly using Centrex.

7. Open Architecture. Most PBXs have open architecture. See OAI for the benefits. Central offices don't.

8. Good Reliability. There have been sufficient central office crashes and sufficient improvement in the reliability of PBXs that you could happily argue that the two are on a par with each other today. Both are equally reliable, or unreliable. The only caveat, of course, is that you back your PBX up with sufficient batteries that it will last a decent power outage. Of course, that assumes that your people will be prepared to hang around and answer the phones during a blackout.

9. Expansion. Central offices are big. Allegedly you can grow your lines to whatever size you want. In contrast, PBXs have finite growth. It's true about PBXs. But it's equally true about central offices. I've personally heard too many stories about central office line shortages to believe in the nonsense about "infinite Centrex" growth. Fact is central offices grow out, just like PBXs. Given the tight economy of recent years, local phone companies have not been buying the central offices they should have. And they have been filling central offices up a little too tight for my taste.

10. Technological obsolescence. Allegedly central offices are upgraded faster than PBXs and therefore are always up to date technologically. It's nonsense. The life cycle of a typical central office was 40 years until recently. It's now around 20 years. Think of what's happened to PCs in the past 10 years – the IBM PC debuted only in 1981 – and you can imagine how obsolete many of the nation's central offices are.

PBX Call Through This enables a cell phone user to make a call into a PBX and for the caller identification to be mapped to the user's network name and number associated within the PBX directory. This means that when calling across the PBX network, internal users can recognize the calling mobile party. Cellular cost savings are possible when calling into the corporate voice network, especially where the company has a an extensive private

enterprise network, perhaps one stretching overseas. Outbound calls from the cellular phone can also be placed through the PBX, offering some cost savings where companies have negotiated attractive PSTN rates. One other benefit is it that allows direct extension dialing, saving

PBX Central Office Trunk PBX central office (CO) trunks connect the PBX switch to the central office serving the PBX location. The trunks appear as station lines at the central office equipment.

PBX Driver Profiles Telephony Services (also called TSAPI) is software which AT&T (now Lucent) invented to run on NetWare servers and allow those servers to communicate with PBXs. According to a presentation made by Oliver Tavakoli of Novell in late August, 1994, "The purpose of TSAPI is to make it easier for developers to create telephony applications. Given the broad scope of TSAPI and the number of ways in which PBX vendors can package and deliver solutions, applications being developed against TSAPI are unlikely to work with a large cross-section of PBXs. This is particularly problematic for application developers wishing to add simple telephony features to a product that is not telephony centric. In order to spur the growth of CTI application development, Novell has created 'profiles' that are placed on top of TSAPI. The profiles listed in this document (i.e. the speech) are based on documents created by other organizations:

"- A NOTA document entitled PABX Driver NLM Conference Meeting and Draft Recommendations dated July 20th, 1994

"- An ECMA document entitled 'Technical report on CSTA Scenarios/Second Draft' dated July 1994.

"Each profile is made up of the following components: Functions (and matching confirmation events) included in the profile; unsolicited events that may be encountered in the event flow scenarios in the profile; event flow scenarios that attempt to describe what should occur in the common scenarios encountered when using the functions in the profile.

"Group A Profile: Using the functions and event flows in the Group A profile, an application should be able to: provide "screen pops" to the end user; make outbound calls originating at an end user's device; provide hands-free operation for answering the phone and making calls (assuming the PBX and end user's device support hands-free operation)

"Group B Profile: Using the functions and event flows in the Group B profile, an application should be able to: perform operations involving two calls on a single device (the connection to one of the calls is in the held state); perform operations involving more than two parties on the same call; obtain information and unsolicited events from a call-centric view (as opposed to a device-centric view).

"Group C Profile: Using the functions and event flows in the Group C profile, an application should be able to: perform the rest of the telephony functions usually done from a phone – pickup, group pickup, call completion and perform more sophisticated monitoring functions.

(There is apparently no Group D profile.)

"Group E Profile: Using the functions in the Group E ('E' for Environmental) profile, an application should be able to: activate and deactivate features of a device, query the state of features on a device, query the generic information about a device."

Once the profiles are finalized, they will be incorporated into Novell's PBX Driver certification process; groups supported by the driver will be included in literature describing all certified drivers. See also Netware Telephony Services.

PBX Extension A telephone phone line connected to a PBX.

PBX Fraud Same as Toll Fraud.

PBX Generations See Generations, PBX.

PBX Integration A loose term to mean joining the PBX to any number of outside computer based gadgets and services, from voice mail to call accounting. To make voice mail integrate into PBXs, you minimally need the ability to provide a message waiting indicator (light or stutter dial-tone) at the user's phone when a message is received, and to forward a call to the user's mailbox when a call is sent to the recipient and they are on the phone or do not answer (forward on busy or ring no answer). This requires PBX "integration." Most PBX integrations provide the attached voice mail system call data that includes the calling phone number, the number of the caller, why the call was presented (such as forwarded on busy, or ring no answer), to pass message waiting on or off indications. PBX integration data may be implemented in-band or out-of-band on a separate link, most often a serial link. Some PBXs "integrate" with outside equipment better than others.

PBX Profiles PBXs do things differently. To make a conference call, one PBX's phone may put the caller on hold automatically, while another may insist that you put that person on hold manually and then dial the next person to join the conference call. As Novell in the fall of 1994 attempted to get as many PBXs to conform to TSAPI, it discovered that PBX features often work very differently. So it decided to categorize PBXs and their features. This, it called, PBX profiles. The idea being that Profile A would contain the most common, easy-to-integrate-to-TSAPI features. B would contain the second most common, etc. Novell also calls PBX Profiles PBX Driver Profiles. See PBX Driver Profiles for a much bigger explanation.

PBX Station Line A transmission path extending from the station (phone instrument) location to the switching equipment.

PBX Tie Line A tie line between two PBX's, permitting extensions in one PBX to be connected to extensions in the other without having to dial through the public switched network. See also OPX and OPS, which are different and are lines between PBXs and distant extensions, not tie lines between PBXs.

PBX Trunk A circuit which connects the PBX to the local telephone company's central office switching center or other switching system center.

PC 1. Personal Computer. See Personal Computer for a bigger explanation.

2. Peg Count.

3. Point Code.

4. Printed Circuit.

5. Product Committee.

6. Protocol Control.

7. Politically Correct.

8. AT&T uses this acronym for "project coordinator".

9. Physical contact finish on a fiber optic connector. See UPC.

PC Administration Server A Sun Microsystems term, Part of Solaris' Server Suite. Automates and centralizes PC network administration.

PC As Phone See also Handset Management.

PC Card A memory or I/O card compatible with the PCMCIA PC Card Standard. In short, PC Cards are a new name for PCMCIA cards. For a much fuller definition, see PCMCIA, which stands for the Personal Computer Memory Card International Association.

PC centric There are two ways you can organize a computer to control telephone calls on an office telephone system. One way is to join a file server on a local area network to a phone system. Commands to move calls around are passed from the desktop PC over the LAN to the server and then to the phone system via the cable connection between the server and the system. A second way to get a computer to control phone calls is through a connection at the desktop. This is called PC Centric. There are two ways you can do this. The first is to join the desktop phone to the computer with a cable. This is often done via the PC's serial port connecting via cable to the phone's data communications port (if it has one – if it doesn't, you get one). The second way to be PC Centric is by simply replacing the standalone phone with a board that emulates in a phone and drop it into the PC's bus.

PC companion This generic term is used to describe a handheld PC that acts as a satellite, companion, or backup to a user's main PC. It contains communications. It is different to a HPC, Handheld PC. It is also different to a PDA. PDAs can generally be classified as stand-alone, keyboardless devices with proprietary user interfaces and applications that require pen-based entry and navigation.

PC network IBM's first LAN (Local Area Network).

PC telephony Another term for Computer Telephony. See Computer Telephony.

PC's The plural of the word PC, according to the New York Times. However, every other computer and general magazine spells them PCs. And that's the spelling which this dictionary writer prefers also.

PCA 1. Premises Cabling Association. A association in Great Britain.

2. Protective Connecting Arrangement. A device that AT&T and members of the Bell System insisted be connected between a telecommunications device (like a phone) that wasn't made and sold by AT&T and a phone line provided by a local Bell operating phone company. Many years later, the PCAs were found by the FCC to be totally unnecessary and AT&T and members of the Bell System were ordered to refund all payments received for rental of PCAs. The Bell System insisted on the PCAs as a way of protecting AT&T's effective monopoly of telecommunications equipment. See also Voice Connecting Arrangement.

PCB Printed Circuit Board.

PCC Personal Companion Computer. What other companies call a PDA (Personal Digital Assistant), Intel calls a PCC. A PCC or PDA is meant to have significant telecommunications abilities – including wired and wireless. See PDA.

PCCA Portable Computer and Communications Association. An association formed to provide a forum for exchange of information between the computer and communications industries. PCCA has developed standards for wireless and mobile computing. www.pcca.org.

PCCA AT Command Set The new PCCA AT command set for wireless modems contains well-defined commands for obtaining link status information.See PCCA and www.pcca.org.

PCD See Post Completion Discrepancy.

PCF Physical Control Fields. The AC (Access Control) and FC (Frame Control) bytes in a Token Ring header.

PCH Paging CHannel. Specified in IS-136, PCH carries signaling information for set up and delivery of paging messages from the cell site to the user terminal equipment. PCH is a logical subchannel of SPACH (SMS (Short Message Service) point-to-point messaging, Paging, and Access response CHannel), which is a logical channel of the DCCH (Digital Control CHannel), a signaling and control channel which is employed in cellular systems based on TDMA (Time Division Multiple Access). The DCCH operates on a set of frequencies separate from those used to support cellular conversations. See also DCCH, IS-136, PAGING, SPACH and TDMA.

PCI 1. Protocol Control Information. The protocol information added by an OSI entity to the service data unit passed down from the layer above, all together forming a Protocol Data Unit (PDU).

2. Peripheral Component Interconnect. A 32-bit local bus inside a PC or a Mac, PCI was designed by Intel for the PC. According to Intel's original specifications, it can transfer data between the PC's main microprocessor (its CPU) and peripherals (hard disks, video adapters, etc.) at up to 132 megabytes per second, compared to only five megabytes per second for the original ISA bus. The PCI design calls for one 64-bit bus running at 66 MHz, and additional busses are either 32-bit running at 66 MHz or 64-bit running at 33 MHz. The maximum amount of data transfer that the PCI design currently will support between the processor and peripherals is 532 MB (MegaBytes, or millions of bytes) per second. The newer PCI-X bus is much faster. PCI is one of two widely adopted local-bus standards. The other, the VL-Bus, is primarily used in 486 PCs. See also CompactPCI, PCI-X, and VLB.

PCI Express PCIe. PCI Express is a bus which connects a computer's microprocessor to its various peripheral cards which drives things like networks and video cards. PCI (Peripheral Component Interconnect) was originally developed by Intel Corporation, but is now slowly coming to an end. PCI Express is a faster version. The PCI 2.2 bus and other revisions just don't provide us with enough bandwidth to support the ever more demanding peripheral cards. Hard drive controllers and networking cards just aren't providing enough bandwidth that some hard drives have the potential to offer. The reason is because PCI only offers a throughput (maximum theoretical bandwidth) of 1.056 Gbps (gigabits per second) while Serial-ATA hard drives can offer 1.5 Gbps (3.0 Gbps with SATA II). If SATA controller can only have a transfer rate of 1.056 Gbps, the SATA drive will have a transfer rate of only 132MBps (megabytes per second), which is roughly equivalent to ATA-133 technology. If you have a Gigabit LAN cards, at one thousand megabits per second, you are using approximately 95% of your available PCI bus' bandwidth. This leaves little room for any other peripherals to use the bandwidth. Most of the time, however, you will only be using a sound card on the PCI bus, but if you have a few other bandwidth-demanding components, you will have performance loss somewhere in the PCI bus, and that may likely sacrifice data transfer rates with your LAN. While most of us actually may find PCI to be satisfying currently, we are rapidly approaching the point where PCI just won't be enough for our needs. Thus, enter PCI Express, which was approved in July 2002 and which could offer data transfer rates as fast as 4 gigabits per second in each direction, depending on how many slots you use. See also ExpressCard.

PCI-X Peripheral Component Interconnect-Extended. An extension of the original PCI design, PCI-X increases the internal bus speed from 66 MHz to 133 MHz. PCI-X calls for one 64-bit bus to run at 133 MHz, with the others running at 66 MHz. Thereby, PCI-X supports a maximum rate of data exchange of 1.06 GB (GigaBytes, or billions of bytes) per second. This level of bandwidth is critical for servers running Gigabit Ethernet, Fibre Channel, and other high-speed networking applications. PCI-X was developed by IBM, HP and Compaq, and was submitted to the PCI SIG (PCI Special Interest Group) of the (ACM) Association for Computing Machinery) in 1998. See also PCI.

PCIA Personal Communications Industry Association. A leading association of providers of wireless voice and data communications services. PCIA member companies include PCS licensees and others in the cellular, paging, ESMR, SMR, mobile data, cable, computing, manufacturing, and local and interexchange sectors of the industry. www.pcia.com.

PCIe See PCI Express.

PCL 1. Hewlett-Packard's Printer Control Language, developed by HP in 1984 as a way for the then-new PC to communicate with a new breed of laser printers – the HP LaserJet printer. HP's PCL language is now the de facto industry standard for PC printing. Most of the printers in the world today are equipped with PCL or a PCL-compatible language. PCL allows the type of sophisticated page creation generally referred to as "laser quality output." PCL supports such advanced features as fully scalable typefaces and rotation of text. PCL defines a standard set of commands enabling applications to communicate with HP or HP-compatible printers. PCL has become a de facto standard for laser and ink jet printers and is supported by virtually all printer manufacturers. On April 8, 1996 HP announced PCL 6 which it billed as "the next generation" of HP Printer Control Language). HP said that PCL 6 includes font synthesis technology for true what-you-see-is-what-you-get (WYSIWYG) printing and better document fidelity. PCL 6 commands were designed by HP to closely match Microsoft Windows GDI (Graphical Direct Interface) commands.

    2. Product Compute-Module Load.

PCLEC Packet Competitive Local Exchange Carrier. Covad invented this term for a CLEC who provides dedicated high-speed digital communications services using DSL technology to Internet Service Providers ("ISPs") and corporate enterprise customers.

PCM Pulse Code Modulation. The most common method of encoding an analog voice signal into a digital bit stream. First, the amplitude of the voice conversation is sampled. This is called PAM, Pulse Amplitude Modulation. This PAM sample is then coded (quantized) into a binary (digital) number. This digital number consists of zeros and ones. The voice signal can then be switched, transmitted and stored digitally. There are three basic advantages to PCM voice. They are the three basic advantages of digital switching and transmission. First, it is less expensive to switch and transmit a digital signal. Second, by making an analog voice signal into a digital signal, you can interleave it with other digital signals – such as those from computers or facsimile machines. Third, a voice signal that is switched and transmitted end-to-end in a digital format will usually come through "cleaner," i.e. have less noise, than one transmitted and switched in analog. The reason is simple: An electrical signal loses strength over a distance. It must then be amplified. In analog transmission, everything is amplified, including the noise and static the signal has collected along the way. In digital transmission, the signal is "regenerated," i.e. put back together again, by comparing the incoming signal to a logical question: Is it a one or a zero? Then, the signal is regenerated, amplified and sent along its way.

PCM refers to a technique of digitization. It does not refer to a universally accepted standard of digitizing voice. The most common PCM method is to sample a voice conversation at 8000 times a seconds. The theory is that if the sampling is at least twice the highest frequency on the channel, then the result sounds OK. Thus, the highest frequency on a voice phone line is 4,000 Hertz. So one must sample it at 8,000 times a second. Many PCM digital voice conversations are typically put on one communications channel. In North America, the most typical channel is called the T-1 (also spelled T1). It places 24 voice conversations on two pairs of copper wires (one for receiving and one for transmitting). It contains 8000 frames each of 8 bits of 24 voice channels plus one framing (synchronizing bit) bit which equals 1.544 Mbps, i.e. 8000 x (8 x 24 + 1) equals 1.544 megabits.

Countries outside of the United States and North America use a different scheme for multiplexing voice conversations. It is based not on 24 voice channels, but on 32. This scheme keeps two of the 32 channels for control, actually transmitting 30 voice conversations at a data rate of 2.048 Mbps. The European system is calculated as 8 bits x 32 channels x 8000 frames per second. European PCM multiplexing is not compatible with North American multiplexing. The two systems cannot be directly connected. Some PBXs in the U.S. conform to the U.S. standard only. Some (very few) conform to both. Both the European and North American T-1 "standards" have now been accepted as ISDN "standards." In addition to PCM, there are many other ways of digitally encoding voice. PCM remains the most common. See also ADPCM, DPCM, Nyquist Theorem, Shannon's Law, T-1 and Voice Compression.

PCM Upstream See V.92.

PCM Voice Transmission Synchronization There are three levels in PCM voice transmission synchronization:

    1. Bit Level Synch – operating the transmitter and receiver at the same bit rate so the bits are not lost.

    2. Frame Level Synch – phase alignment between the transmitter and receiver so the beginning of the frame can be identified.

    3. Time Slot Synch – phase alignment between the transmitter and receiver so the time slots are lined up for information retrieval.

PCM-30 Short name of international 2.048 Mbps T-1 (also known as E1) service derived from the fact that 30 channels are available for 64 Kbps digitized voice each using pulse code modulation (PCM).

PCMCIA The Personal Computer Memory Card International Association (an awful mouthful) standardizes credit-card size packages for memory and input/output (modems, LAN cards, etc.) for computers, laptops, palmtops, etc. There are three physical standards for PCMCIA cards – Type I, II, and III and undefined standard called Type IV, which only Toshiba has at this moment. The cards are 69.2 millimeters (3.37 inches) long x 51.46 millimeters (2.126 inches) wide. All three types use the same 68 female pin edge con-

nector for attachment to the computer, and differ only in thickness. The thickness for Type I, Type II and Type III are 3.3, 5.0, 10.5 millimeters respectively. Toshiba's Type IV is 16 mm. A Type I PC Card is typically used for various types of memory enhancements, including RAM, FLASH memory, one-time programmable (OTP) memory, and electronically erasable programmable read only memory (EEPROM). A Type II PC Card (the most common) is typically used for input/output such as modem, LANs, host communications and SCSI device connection. A Type III PC Card is twice the thickness of the Type II and is typically used for I/O features that require a larger size, such as rotating mass storage devices (removable hard disk drives) and radio communication devices. Since Type I, Type II and Type III Cards all use the same interface, the size of the card chosen for the application is dependent on the miniaturization of the technology to be implemented.

PCMCIA's first standards were issued in September, 1991. The idea is that small computers will use these cards for modems, fax cards, hard disks, LAN connections, Ethernet connections, SCSI device connections, etc. A PCMCIA card is, in most cases, the only way to get to a laptop's bus without attaching a docking station. In terms of performance, the bus is comparable, but not equivalent to the ISA bus on desktops. The PCMCIA bus is 8 bits wide and does not allow for direct memory access (DMA) transfers or bus mastering. This means, for example, that you can't equip a laptop with a sound blaster compatible PCMCIA card, since it requires DMA transfers. A new spec called PCMCIA v3 (now called CardBus) calls for a number of improvements, including 32-bit data paths, DMA, bus mastering and significant improvements in speed with the bus as fast as the PCI bus.

PCMCIA standards exist so others can make these cards. Some computers – like some pre-summer 1991 notebooks and laptops from Compaq and Toshiba – don't comply because these computers were released before the standards were released. The PCMCIA v2.0 standard contains a software specification for XIP, the "eXecute In Place" mechanism that maps application software stored on the PCMCIA card into the system address space. This means application software will run directly from the card, start faster and not require precious RAM from the host computer.

One key element of the PCMCIA software architecture are Socket Services and Card Services. Socket Services is a BIOS level software interface that provides a way to access the PCMCIA sockets (slots) of a computer. Socket Services identifies how many sockets are in your computer system and detects the insertion or removal of a PC Card while the system is powered on. Socket Services is part of the PCMCIA Specification and interfaces with Card Services. Card Services is a software management interface that allows the allocation of system resources (such as memory and interrupts) automatically, once the Socket Services detects that a PC Card has been added. Card Services also releases these resources when the PC Card has been removed. Card Services also provides you with an interface to higher level software to load any needed hardware drivers.

The combination of PC Card hardware, Card Services software and Socket Services software provides an almost "plug-and-play" capability in the portable computing environment. Once the software has been installed, it is theoretically possible to add and remove PC Cards without powering off the system. But this is theory. And in practice, it has worked only intermittently for this writer. It doesn't work for network cards. It's meant to be possible, for example, to insert a modem PCMCIA Card to access another computer system, download information into the portable computer's memory, remove the modem PCMCIA Card, replace it with a flash PCMCIA Card, and store the downloaded information – all while your portable computer is still powered on. Great theory.

The PCMCIA has around 300 members, including manufacturers of semiconductors, connectors, peripherals and systems, as well as BIOS and software developers and related industries. Members include Intel, IBM, Toshiba, Lotus, Epson and Fujitsu. The association is based in Sunnyvale, CA. 408-720-0107. It has an electronic bulletin board – 408-720-9388. Its standards are also recognized by the Japanese Electronic Industry Development Association (JEIDA). The Association publishes a free book listing all the manufacturers making cards which comply to their standards.

Advice in buying PCMCIA cards: PCMCIA standards are still being formed and are thus not truly standards. Check that cards you buy work in your machine. The manufacturer should have tested it in your machine. Don't assume compatibility. Check your socket and card services supports that card. It may not. Use "enabler" software instead. Avoid PCMCIA cards with "pigtails." They break. You lose them. Use PCMCIA cards with an "X-JACK." You can plug directly into a PCMCIA card with an X-JACK, or equivalent. If you have to buy a card with a pigtail, buy a second pigtail, just in case.

In 1994, the PCMCIA started calling its specs, PC Card. Some people said it was because many people claimed that PCMCIA stood for "People Can't Memorize Computer Industry Acronyms."

In the latter part of 1995, the PCMCIA came out with a new specification called CardBus which is a 32-bit bus, as against the present 16-bit cards. CardBus is an extension of the PCI bus. This means that new CardBus will support 132 Mbps, much faster than the present 8 Mbps. See CardBus for a full explanation of that new standard. See also Card Services, CardBus, PCMCIA standards, Socket Services, and Slot Sizes. www.pcmcia.org

PCMCIA Standards The complete set of all of the PCMCIA PC Card Standards. It includes the PC Card Standard Release v2.01, Socket Services Specification Release v2.0, Card Services Specification Release v2.0, ATA Specification Release v1.01, AIMS Specification Release v1.0, and the Recommended Extensions Release v1.0. Standard v3.0 has been proposed. See PCMCIA. www.pcmcia.org

PCMM See PacketCable Multimedia.

PCN Personal Communications Network. Another name for GSM 1800 (it is also known as DCS 1800). It is used in Europe and Asia Pacific. Originally PCN was a type of wireless telephone system that would use light, inexpensive handheld handsets and communicate via low-power antennas. When it was originally conceived, PCN was primarily seen as an a city communications system, with far less range than cellular. Subscribers would be able to make and receive calls while they are traveling, as they can do today with cellular radio systems, but at a low price. One idea for PCN was to locate a PCN cell site (transmitter/receiver) in a residential community. When someone wanted a new phone line, they'd simply drop down to their local phone store, pick up a PCN portable phone and, by the time, they got back home, their frequency would be "switched on" and they'd be "live." The original plans for PCN never materialized fully. However, the concept has been somewhat implemented in the forms of Personal Communications Service (PCS) and Wireless Local Loop (WLL). Now PCN is simply another name for GSM 1800. See also PCS, Personal Communications Network, Wireless Local Loop.

PCNFS Daemon PCNFSD. A program that receives requests from PCNFS clients for authentication on remote computers.

PCO 1. Point of Control and Observation: A place (point) within a testing environment where the occurrence of test events is to be controlled and observed as defined by the particular abstract test method used.

2. Private Cable (TV) Operator.

PCP 1. Post Call Processing.

2. Program Clock Reference: A timestamp that is inserted by the MPEG-2 encoder into the Transport Stream to aid the decoder in the recovering and tracking the encoder clock.

PCR An ATM term. Peak Cell Rate: The Peak Cell Rate, in cells/sec, is the cell rate which the source may never exceed.

PCS 1. the plural of PCs, i.e. PCs.

2. Personal Communications Service. A low-powered, high-frequency alternative to traditional cellular, mainly in the US. Whereas cellular typically operates in the 800-900 MHz range, PCS operates in the 1.9 GHz range. The specific technologies include CDMA, Digital AMPS, and GSM 1900. PCS is a digital system making use of relatively cheap phones. The higher frequency range limits the cell size, as 1.9 GHz signals attenuate (weaken) more quickly than signals in 800-800 MHz range. The FCC awarded PCS licenses as follows:

- "C-Block" Carrier A 30 MHz PCS carrier serving a Basic Trading Area (BTA) in the frequency block 1895-1910 MHz paired with 1975-1990 MHz.
- "D-Block" Carrier A 10 MHz PCS carrier serving a Basic Trading Area (BTA) in the frequency block 1865-1870 MHz paired with 1945-1950 MHz.
- "E-Block" Carrier A 10 MHz PCS carrier serving a Basic Trading Area (BTA) in the frequency block 1885-1890 MHz paired with 1965-1970 MHz.
- "F-Block" Carrier A 10 MHz PCS carrier serving a Basic Trading Area (BTA) in the frequency block 1890-1895 MHz paired with 1970-1975 MHz.

See Personal Communications Networks and Personal Communication Services.

3. Private Client Services. A term used in the investment banking for taking money from rich people and providing them investments that, hopefully, don't go down.

PCS 1900 PCS 1900 is a GSM system offering 148 full-duplex voice channels per cell. The system operates in the 1.9-GHz band used in the United States and is now known as GSM 1900.

PCS Over Cable You run a CATV – cable TV company. You have a wires strung all over the neighborhood. On one of your wires you attach a six foot by four foot by four box of electronics and three two feet antennae. Bingo, you're now a way station – also called a cell site – for a PCS cellular phone system. People who are PCS subscribers will talk and receive calls when they're close to your cell site. Calls come and go via your coax cable, up it to a landline connection point with the PCS carrier. You, the CATV company, get paid money for completing calls. See www.sanders.com/telecomm

PCSA Personal Computing System Architecture. A PC implementation of DECnet, that lets PCs work in a DECnet environment. PCSA is a network architecture defined and supported by Digital Equipment Corporation for the incorporation of personal computers into server-based networks.

PCT Personal Communications Technology. A security protocol developed by Microsoft for online Web commerce and financial transactions. Transparent to the user, PCT provides authentication and encryption routines that complement credit-card based commerce on the World Wide Web. Internet Explorer, Microsoft's Web browser, makes use of PCT. See also Authentication and Encryption.

PCTA Personal Computer Terminal Adapter. A printed circuit card that slips into an IBM PC or PC compatible and allows that PC to be connected to the ISDN T-interface. See Personal Computer Terminal Adapter.

PCTE Portable Common Tool Environment.

PCTS Public Cordless Telephone Service. A Canadian digital cordless telephone service for residential, business and public use. For other variations of digital cordless telephone service, see CT1, CT2, CT2Plus, CT3, and DECT.

PCWG Personal Conferencing Work Group. www.gogcwg.org/pcwg.

PCX Server Software PCX server software turns your PC into a graphics terminal front-end for Unix and X applications. Thus, your PC can display application output generated by remote X-based client applications.

PD Powered Device. See 802.3af.

PDA Personal Digital Assistant. A consumer electronics gadget that looks like a palmtop computer. Unlike personal computers, PDAs perform specific tasks – acting like an electronic diary, carry-along personal database, multimedia player, personal communicator, memo taker, calculator, alarm clock. The communications take place through wire to the PC, infrared link, modem and the phone or through wireless. Apple announced a PDA, which it has named Newton, and which now is MD (Manufacturer Discontinued). When I added this definition initially in the late fall of 1992, sales of PDAs weren't doing well and some wag in Silicon Valley called them Probably Disappointed Again. IBM prefers to call them Personal Communicators. General Magic preferred to called them PICs, Personal Intelligent Communicators, but then General Magic no longer exists. The biggest-selling PDA is the Palm Pilot, which was created by US Robotics, which subsequently was acquired by 3Com and then spun out as a separate company called Palm. See HPC.

PDAia PDA Industry Association. A not-for-profit industry organization formed to create a marketplace for PDAs by promoting the awareness of the usage of PDAs and hand-held technology, in general. Although the PDAia does not develop standards, it does act as a clearinghouse for standards being discussed. Working relationships exist with IrDA, PCCA, PCMCIA, and other associations which relate to the PDA market. www.pdaia.org.

PDAU Physical Delivery Access Unit. A gateway device that facilitates the delivery of messages (excluding probes and reports) in physical form. This is an X.400 term.

PDC 1. Personal Digital Cellular: the 2G TDMA-based protocols used in Japan, owned by NTT DoCoMo. PDC services operate in the 800 and 1500 MHz bands. PDC is the Japanese equivalent of GSM but is incompatible with other systems. It is operated by NTT DoCoMo, as well as by all the other Japanese operators, but the technology was developed by NTT DoCoMo. Previously known as PHP (Personal HandyPhone) and Japanese Digital Cellular (JDC). See PHS.

2. See Primary Domain Controller.

3. Pacific Digital Cellular.

PDF 1. Portable Document Format. Let's say you produce a document. You use Microsoft Word to write it. On a PC, that program will produce a file with a extension .doc. You now want to send someone your document. You send them your file electronically. They will need a copy of Microsoft Word or Microsoft Word Viewer to read it. What happens if they don't have it? Tough. They're out of luck. They can't read it. Of course, you say Harry, everyone has Word. What happens if you write the document in QuarkXpress? Few people have copies of that program. Here's how to solve the problem. Write the program with whatever program you wish, then use software to convert it to a PDF file, which typically ends with a .pdf. PDF is the file format for documents invented by Adobe and now in the public domain. The PDF file format was developed to allow everyone to be able to read documents – even though they don't have the program which created them. Note that PDF files are typically read-only, which also means that the reader can't change them or otherwise steal your work. All you need to view a PDF document is a free version of Acrobat Reader, which is available from www.adobe.com. In order to create a PDF file, however, you must buy a copy of Adobe Acrobat, PDF Converter or any number of software applications that make PDF documents. PDF documents can include text, colors, diagrams, and photographs. A PDF image of a document can be very good quality. You can typically read even the smallest type. And magazines now send their files electronically to their printers in PDF format – but in higher quality than is typically used for PDFs distributed in offices. Here, according to Adobe, are some highlights of how documents are stored in PDF:

- PDF represents text and graphics by using the imaging model of the PostScript language. Like a PostScript program, a PDF page description draws a page by placing "paint" on selected areas, which allows for device independence and resolution independence.

- PDF files are extremely portable across diverse hardware and operating-system environments. PDF makes use of binary as well as ASCII-encoded data.

- To reduce file size, PDF supports JPEG, CCITT Group 3, CCITT Group 4, ZIP, and LZW industry-standard compression filters.

- PDF files contain information necessary for either displaying embedded fonts or for font substitution. A PDF file contains a font descriptor for each font used in the document. The font descriptor includes the font name, character metrics, and style information. If a font used in a document is available on the computer on which the document is viewed, or if it's embedded in the PDF file, it is used. If the font is not available or is not embedded, a special serif or sans serif Multiple Master font is used to simulate the font. This solution applies to Type 1 fonts and to fonts in the TrueType format. Symbol fonts and Expert fonts are automatically embedded or converted to graphics.

- PDF font substitution does not cause documents to reformat. Substitute fonts created from serif and sans serif Multiple Master fonts retain the width and height of the original characters.

- A PDF file contains a cross-reference table that can be used to locate and directly access pages and other important objects in the file. Because it uses this cross-reference table (called xref), the time needed to view a given page can be nearly independent of the total number of pages in a document.

- PDF is designed to be extensible; that is, new features can easily be added to the file format through the plug-in architecture. (Plug-ins are software programs that add functionality to a base program such as Acrobat.)

The PDF standard is a public, open specification. Though it was invented by Adobe Systems Inc. the Portable Document Format is an open specification and has been implemented by over 1,800 hardware and software vendors. Several companies make software products which make PDF documents out of others, e.g. word processing documents.

2. Power Distribution Frame.

PDH Plesiochronous Digital Hierarchy. Developed to carry digitized voice over twisted pair cabling more efficiently. This evolved into the North American, European, and Japanese Digital Hierarchies where only a discrete set of fixed rates is available, namely, nxDS0 (DS0 is a 64 kbps rate) and then the next levels in the respective multiplex hierarchies. See also Plesiochronous. See T Carrier for hierarchy detail.

PDI A Versit term. Personal Data Interchange, a collaborative application area which involves the communication of data between people who have a business or personal relationship, but do not necessarily share a common computing infrastructure.

PDL A page description language (PDL) is a is a clever short-cut for transmitting bit-mapped images from a PC application to a printer. They save processing time, by sending only "instructions" to a printer, rather than the entire bitmapped image. They also allow the printer to print any font, any size. PDL is the generic term. Hewlett-Packard has been the major proponent of the concept of PDLs. And they include something called PCL with all their printers. PCL stands for Printer Command Language. Postscript is also a PDL, but different to PCL. HP includes PCL with all its laser printers, but has only included

What we really meant was that HP had never built PostScript into their printers. They still haven't, really. You still have to buy PostScript as either a plug-in cartridge or an add-on "SIM" chip from HP, Adobe or third parties for most of the HP line. The exception is the top-of-the-line LaserJet IIIsi, for which built-in PostScript is an option that you pay extra for.

That may change. For now, if you're using Windows applications, Windows' built-in "True Type" scalable fonts – which do work on HP and other non-PostScript printers – will satisfy most of your printing needs.

PDM 1. Pulse Duration Modulation.

2. See Personal Data Module.

PDN 1. An ISDN Term. Primary Directory Number (7 digits for 5ESS switch; 10 digits for DMS-100 switch)

2. Public Data Network. A public network for the transmission of data, particularly a network compatible with X.25 protocol. A public data network is to data what the Public Switched Voice Network is to voice. To access a public data network, you typically dial a local number, receive a carrier tone and then follow very specific instructions. Public data networks send their digital data in packets over high speed channels. The major reason to use a PDN is that it may be cheaper than dialing directly on a switched voice line. Also,

they get you into some databases and services which are hard to get into by dialing direct. There are many public data networks in the United States. The two best-known public data networks are Tymnet and Telenet. But there are probably 30 more. Every industrialized foreign country has at least one, usually owned by the local government phone company.

3. Premises Distribution Network. The electronics and cabling system which connect terminal equipment to the NTI (Network Terminating Interface). The PDN may be either point-to-point or multipoint, may be configured as either a bus or star, and may be either active or passive in nature. In an ADSL Asymmetric Digital Subscriber Line) implementation, for example, a PDN system would be used to connect the ATU-R (ADSL Termination Unit-Remote) to Service Modules which perform terminal adaptation functions. See also NTI, ADSL and ATU-R.

PDP 1. See Power Distribution Panel.

2. Plasma Display Panel.

3. Packet Data Protocol. A GPRS (General Packet Radio Service) term for a range of protocols which support the transfer of packet data over a 3G (3rd Generation) wireless cellular network. See also GPRS.

4. Programmed Data Processor. A family of minicomputers developed by Digital Equipment Corporation in the 1970s, the most popular of which was the PDP-11. PDP systems used the PAL (Paradox Application Language) programming language.

5. Policy Decision Point. A network-based server that makes decisions relative to traffic priorities in an IP network running the COPS (Common Open Policy Server) protocol. See COPS.

PDP-11 Programmed Data Processor-11. On the last day of March, 1970 Digital Equipment Corporation (DEC) shipped its first PDP-11/20 minicomputer to a customer in Tennessee. The PDP-11 family was among the first minicomputers that incorporated open standards into its operations. And the PDP-11 software platform has continued through subsequent generations of hardware. It's now obsolete. DEC got bought by Compaq and is now effectively closed down. See also PDP.

PDS 1. Premise Distribution System. Lucent's proprietary structured cable and wire system for intrabuilding application. Such a cabling system provides a number of options for deploying various combinations of UTP, STP, coax and fiber optic cables contained within a single cable sheath. Thereby, various media can be deployed in appropriate combination, considerably reducing overall deployment costs.

2. Processor Direct Slot. Some non-NuBus Apple Macintosh computers have one PDS that allows for expansion cards.

PDU Protocol Data Unit. OSI (Open Systems Interconnection) terminology for a generic "packet." A PDU is a message of a given protocol comprising payload and protocol-specific control information, typically contained in a header. PDUs pass over the protocol interfaces which exist between the layers of protocols (per OSI model). Basically, PDU is OSI (Open Systems Interconnection) terminology for "packet". A PDU is a data object exchanged by protocol machines (entities) within a given layer. PDUs consist of both data and control (protocol) information that allows the two to coordinate their interactions. The native PDU may be altered through a process of Segmentation and Reassembly (SAR) in order to achieve compatibility with a service offering. For example, an Ethernet PDU in the form of a frame becomes known as an SMDS Service Data Unit (SDU) when considered in the context of a SMDS network service. The SDU then is encapsulated with control information to become a Level 3 (SMDS) PDU, which subsequently is segmented into 48-octet payloads, each of which is preceded with another 5 octets of control data to constitute a 53-octet SMDS cell, also known as a Level 2 PDU. The cells are transported over the SMDS network, with the process of reassembly being performed on the receiving end. As a result, the receiving device is presented with the data in its native PDU format.

PDV Path Delay Value

PE 1. Processing Element.

2. PolyEthelyne. A type of plastic material used to manufacture jackets for outside plant cable systems.

peak That part of the business day in which customers expect to pay full service rates. For cellular customers, peak hours are generally 7:00 a.m. to 7:00 p.m. For business landline customers, peak hours are generally 8:00 a.m. to 5:00 p.m.

peak emission wavelength Of an optical emitter, the spectral line having the greatest power.

Peak Envelope Power PEP. The average power supplied to the antenna transmission line by a radio transmitter during one radiofrequency cycle at the crest of the modulation envelope taken under normal operating conditions.

peak hour When used with an automatic call distributor, the peak hour is when the number of calls coming into your center are at their highest level. ACDs allow you to track and report on calls by hour. Some allow you to also track peak half-hours, or peak delays of the week or months of the year.

peak load A higher than average quantity of traffic. Peak Load is usually expressed for a one-hour period, often the busiest hour of the busiest day of the year. See Busy Hour.

peak mount An antenna mounting system (tripod or ridge mount) used for erecting an antenna at the ridge of roof. See Ridge Mount and Tripod.

peak position requirements A call center term. The maximum number of base staff required in any half hour within a given date range.

peak rate The per-minute price for using a communications device in the "peak" time period. For cellular, "Peak" time generally includes hours such as evenings Monday through Friday, and all day Saturday, Sunday, and certain holidays. For normal landline phone service, peak rates will include workday days.

Peak To Average Ratio PAR. Also known as the "crest factor." The ratio between the peak and the average power level applied to a carrier frequency (the frequency which carries the information signal) over a circuit.

peak power Many phone systems use more power when more people are talking on them. You need to know peak power so you don't suddenly blow all your fuses when everybody gets on the phone.

peaked load The load that results from peaked traffic.

peaked traffic Random traffic that has a variance-to-mean ratio grater than one.

peakedness Within-the-hour or 'moment-to-moment' variations in traffic.

peaking Momentary bursts of high volume traffic which occur during the busy hour.

peanut tubes A name given to the smaller sizes of vacuum tubes.

PEB PCM Expansion Bus. A digital voice bus for sending voice across different voice processing cards and components in the same PC. PEB is from Dialogic, Parsippany, NJ. It is an open platform. Many companies make voice processing products connecting via PEB. See also Dynamic Node Access.

PEBCAK Tech support shorthand for "Problems Exist Between Chair And Keyboard." A way of indicating there's nothing wrong with the computer or the network. The user is clueless.

PECL Positive Emitter Logic.

Ped See Pedestal.

pedestal 1. A small green box sits outside and houses cables coming in, cables going out and cable splices inside the box to join cables. This way a cable coming in from the telephone central office can be joined to one going to someone's house.

2. A pedestal is also a mounting device used in pay telephone installations where the instrument is not attached to a wall.

PEDI X.435. The standard defines how EDI (Electronic Data Interchange) is handled in the X.400 world. The standard includes many advanced service features addressing security, message forwarding, etc.

peek When you store a number in memory you "poke" it there. When you want to read it back, you take a "peek." Peek and poke are thus instructions that view and alter a byte of memory by referencing a specific memory address. Peek displays the contents; poke changes it.

peer A peer is an equal in civil standing or rank. It is one's equal before the law. The word had little meaning in computing or telecommunications until PCs came along and got progressively more powerful – in fact more powerful than early mainframes. Someone thought if you joined up a bunch of PCs together on networks their combined computing and storage capability would be big, powerful and cheap. One of the first places PCs were used in peer-to-peer computing was with the short-lived Napster music-sharing operation. With Napster there was a central database as to where all the music was. But the music was stored on remote computers joined by the telecom lines of the Internet. The original Napster was closed down because the music companies objected to people getting their hands on free music and because they could find Napster. But peer-to-peer computing has become far more sophisticated. Companies are using their employee computers to divide up complex processing and storage tasks – perhaps using the machines only at night when employees aren't using them. See all the Peer and Peer-to-Peer definitions following, including Skype.

peer-assisted file delivery See file swarming.

peer entities Entities within the same communications layer.

peer-entity authentication The corroboration that a peer entity in an association is the one claimed.

peer group A set of logical nodes which are grouped for purposes of creating a routing hierarchy. PTSEs (PNNI Topology State Elements) are exchanged among all members of the group.

peer group identifier A string of bits that is used to unambiguously identify a peer group.

peer group leader A node which has been elected to perform some of the functions associated with a logical group node.

peer group level The number of significant bits in the peer group identifier of a particular peer group.

peer networking. See Peer and Peer-to-peer.

peer network entities The origin and destination of all data transmissions.

peer node A node that is a member of the same peer group as a given node.

peer protocol The set of rules defining the procedures for communication between like entities. The identical entities may be devices or software modules at specific layers in a layered network architecture implementation.

Peer-to-Peer P-2-P. Peer-to-Peer is a fancy way of saying grid computing and communications. Peer-to-peer describes communications between two entities that operate within the same protocol layer of a system. Peer-to-peer has come to mean a bunch of smaller computers helping each other work on a larger computing task. Many computers, joined by communications lines, bring their little piece of computing and storage together to solve a big task. The most famous instance of peer-to-peer has been the music sharing sites, first Napster and then Kazaa. All computers of these sites share information as to where the music is (the directory) and where its stored (the storage). Some companies are replacing their big expensive mainframes with a collection of smaller computers. The concept is that the whole system becomes intelligent enough to use the computing and storage of those machines that are temporarily idle. See also Peer and the next six definitions.

peer-peer directory propagation A way of updating user addresses in which changes in any post office on a LAN (local area network) are sent to all other post offices.

peer-to-peer flow Term used to describe communication flow in a protocol stack. It means that a layer in one protocol stack communicates with a peer layer in another protocol stack through a physical link.

peer-to-peer network 1. A network (typically a local area network) in which every node has equal access to the network and can send and receive data at any time without having to wait for permission from a control node. While peer-to-peer resource sharing is effective in small networks, security and reliability issues prevent its widespread use in larger networks.

2. A new telephony term describing the relationship between a telephone system and the external computer working with it. Picture a telephone switch acting as an automatic call distributor and an outboard computer processor. The idea is to coordinate the call and the screen at the agent. Communication must take place between the switch and the computer. See also Extended Call Management.

peer-to-peer protocols Describes the relationship between a telephone system and the external computer.

peer-to-peer resource sharing An architecture that lets any PC contribute resources to the network while still running local application programs.

peer-to-peer technology A communications model in which Peer-to-peer is a communications model in which each party has the same capabilities and either party can initiate a communication session. In recent usage, peer-to-peer has come to describe applications in which users can use the Internet to exchange files with each other directly or through a mediating server. See also Napster. There are four categories of peer-to-peer:

1. Pure peer-to-peer: Completely decentralized network characterized by a lack of a central server or central entity. Users of a pure peer-to-peer network make direct contact with one another.

2. Computational peer-to-peer: Uses peer-to-peer technology to disseminate computational tasks over multiple clients; peers do not have a direct connection to one another.

3. Datacentric peer-to-peer: Information and data residing on systems or devices that is accessible to other when users connect. It is sometimes called peer-assisted or grid-assisted delivery. Applications include distributed file and content sharing.

4. User centric/hybrid peer-to-peer: Involves clients contacting others via a central server or entity to communicate, share data or process data. Often used in collaboration applications.

peering Once upon a time America's entire phone system was run by AT&T (the Bell System) and a bunch of independents who served subscribers in areas AT&T wasn't. Essentially the whole gigantic thing was a monopoly, with monies for transmission and switching changing hands under rules run by government regulators and essentially administered by one company. The Internet is different. There are many companies providing telecom and switching service. These companies have to move traffic of the Internet between them. Hence they established something they called peering. It's a relationship established between two or more ISPs (Internet Service Providers) for the purpose of exchanging traffic directly, rather than doing so through a backbone Internet provider. The traditional Internet architecture (to the extent that there is a "traditional architecture" in the amorphous world of the Internet) calls for ISPs and regional carriers to exchange traffic at Network Access Points (NAPs), using carrier-class switches and routers, of which there currently are a dozen or so around the world. Traditionally, that traffic has been exchanged between one ISP and another at no cost, although that no longer is necessarily true. Money is raising its grubby head. In order to avoid the headaches and expenses of figuring who should get paid and for how much, many of the larger ISPs have developed peering relationships which allow them to exchange traffic directly over dedicated circuits. In some geographical regions, mostly in North America, several ISPs have formed "private peering points." These packet switching centers allow them to exchange traffic on a switched basis, once again avoiding the cost of doing so through a NAP. Peering requires the exchange and updating of router information between the peered ISPs, typically using the border gateway protocol (BGP). Initially, peering arrangements did not include an exchange of money. More recently, some larger ISPs have charged smaller ISPs for peering. Each major ISP generally develops a peering policy that states the terms and conditions under which it will peer with other networks for various types of traffic. See Internet backbone provider.

peering agreement An agreement to directly interconnect two or more networks and thus carry each other's traffic. See Peering.

peering point A point at which ISPs (Internet Service Providers) exchange traffic, a peering point is in the form of a switch or router. Public peering points are in the form of NAPs (Network Access Points), which include MAEs (Metropolitan Access Exchanges, or Merit Access Exchanges) and CIXs (Commercial Internet Exchanges), also known as just IXs. Private peering points also have been established by consortia of ISPs in order that they can exchange traffic at costs lower than those imposed by the NAPs. See Peering.

peers Equals. See also Piconet. See Peering.

PEG Public, Educational or Government access. A cable TV term to denote the local public access channel/s.

peg board Also called a mushroom board or white board. This is placed between termination blocks to support route crossing wire.

peg count A raw count of some event. Because this was originally maintained by moving pegs on a board with units of 1,10s,100, 1000s it became a peg count. A count of the number of calls placed or received at a certain point or over certain lines during a period such as an hour or day or week. A peg count simply tells you how many calls you made or received. It does not tell you how long they were or where they went or anything else. In the old days before we had accurate and relatively inexpensive call accounting equipment, we relied on Peg Counts to figure out how many circuits we needed. No more. The peg count method is too inaccurate.

PEG ratio Price to Earnings Growth ratio. A new way of comparing the price of stocks. Let's say that your favorite stock is selling at $100 and growing at 50% a year, that means its PEG ratio (actually pronounced peg) is 2.0.

PEL Picture ELement. A pel is the smallest area on a video screen that can be controlled by software. Pels are arranged on screens in a grid-like fashion. Depending on the screen mode selected by an application, a pel may be a single pixel or several pixels. (A pixel is the smallest visual element on a screen, which can be turned on or off or varied in intensity.) The larger the number of pixels in a pel, the greater the size of the palette of distinct colors. When a pel consists of multiple pixels, the number of colors can be quite large. The pel size determines the clarity of the image – called screen resolution. Larger individual pels reduce the total number of available pels, resulting in lower resolution. Smaller pels increase the number of pels that can fit on the screen, resulting in higher resolution and a clearer picture. See also Pixel, Fax and VGA.

PEM Privacy Enhanced Mail. An Internet electronic mail capability which provides confidentially and message integrity using various encryption methods. PEM was adopted by the IETF in 1985 as an Internet standard. PEM involves digital certificates which are issued by the Internet Policy Registration Authority (IPRA), which serves as the top-level "trusted authority." The IPRA signs certificates for a second level of trusted bodies, which then sign certificates for Certificate Authorities (CAs), which then issue them to the public. As PEM does not address MIME (Multipurpose Internet Mail Extension), it is seldom used. S/MIME (Secure MIME) is the preferred approach. See also Certificate Authority, Digital Signature, Encryption, MIME, and S/MIME.

pen register Also called a Dialed Number Recorder (DNR). An instrument that records telephone dial pulses as inked dashes on paper tape. A touchtone decoder performs the same thing for a touchtone telephone. Pen registers and DNRs are used by law enforcement agencies to gather information about the telephone numbers that suspected criminals are calling. They also are used by toll fraud criminals to steal calling card numbers.

pen windows A new Microsoft operating system for notebook size computers that uses a stylus instead of a keyboard.

penal colony A place where a country sends undesirables such as thieves where a couple of generations later, organizations are formed to validate their aristocratic members' claim of descent from these founding fathers and mothers. Everybody knows that Australia started off as a British penal colony. What is not generally known is that it was the United States' fault: After the American colonies seceded to form the U.S., Great Britain could no longer send its criminals here. The author of this dictionary, Harry Newton, was born in Australia.

penetration 1. The number of homes actually served by cable or telephone in a given area, expressed as a percentage of homes passed (e.g. cable penetration in November 1998 was 67.4 percent nationwide).

2. Bypassing the security mechanisms of a system.

penetration tap A connection method used in installations that allows devices to be connected to cable without interrupting network operation. Penetration taps are most commonly found used with coaxial cable. A sharp, pointed probe is used to penetrate the outer insulation and grounding shield of the coaxial cable and to make direct contact with the inner conductor.

penetration testing A test of a network's security by a trustworthy security professional. It involves looking for weaknesses and technical flaws, using manual and automated techniques that hackers use. In short, penetration testing (also called pen testing) is attempting to circumvent a system's security features in order to identify the network's security weaknesses.

Pentium 4 In coming up with names, companies often vacillate between the technical and the poetic. Intel's first microprocessor in the early 1970s was called the 4004, and Intel kept naming its chips by numbers – 286, 386 – until the courts ruled it couldn't trademark them that way. Thus was born the Pentium. The Pentium 4 chip with 42 million transistors in a 217-square-millimeter die, compared to Pentium III's count of 28 million transistors in a 106-square-millimeter die.

pentode A version of a vacuum tube. The pentode adds a suppressor grid between the screen grid and the plate. It's usually connected to ground potential in order to repel any stray electrons from bouncing back to the screen grid instead of the plate. Because the screen grid has a positive charge with respect to the cathode, it tends to attract a percentage of the electrons away from the plate, reducing the efficiency of the tube. The suppressor acts as a negatively-charged (with respect to the plate) barrier to prevent the diversion of electrons from the plate to the screen grid. This increases overall efficiency and gain of the tube by directing more of the electrons to their intended destination. See also Diode, Triode and Tetrode.

people churner A boss who drives away talented people.

people meter An electronic device on which individuals record their television viewing by touching predestinated "buttons" assigned to each individual in the household. Current people meters also have separate remote-controlled, portable handsets.

PEP 1. Protocol Extensions Protocol. A part of the JEPI (Joint Electronics Payments Initiative) specification from the World Wide Web Consortium (W3C) and CommerceNet for a universal payment platform to allow merchants and consumers to transact E-Commerce (Electronic Commerce) over the Internet. PEP is an extension layer that sits on top of HTTP (HyperText Transfer Protocol). PEP works in conjunction with Universal Payment Preamble (UPP), the negotiation protocol that identifies appropriate payment methodology. These protocols are intended to make payment negotiations automatic for end users, happening at the moment of purchase, based on browser configurations. See also Electronic Commerce and JEPI.

2. Packet Ensemble Protocol. A high-speed modulation method from Telebit Corporation for dial-up modems. Now obsolete.

3. Policy Enforcement Point. An router or IP switch that enforces the traffic priority decisions made by a PDP (Policy Decision Point) in an IP network running the COPS (Common Open Policy Server) protocol. See COPS.

4. See Peak Envelope Power.

PEPCI Protocol for Exchange of PoliCy Information. A draft protocol from the IETF intended to support the exchange of policy information between the policy server and its clients. Policy-based networks manage traffic through the establishment of priorities based on parameters such as traffic type, application type, and user. See also Policy-Based Networking.

Per Call Calling Identity Delivery Blocking Feature Allows a caller to toggle or override the value of a calling identity item's Permanent Presentation Status (PPS) for a particular call.

Percent of Interstate Usage PIU. A Verizon definition. Refers to the amount of traffic subject to FCC authority and charged under GTOC FCC #1.

Percent of Local Usage PLU. A Verizon definition. Verizon and the service provider provide this factor to each other each quarter or (as otherwise specified in their agreement) in order to identify the jurisdiction of each call type carried over their trunk groups.

Percentage ATB Percentage of All Trunks Busy. Percentage of time during a reporting period that all trunks in a group or split were busy. This may be measured in two ways, actual simultaneous busies and call length per event, or backed into statistically. Neither technique is absolutely accurate as each depends on "snap shots" in a environment of random interleaved call events.

Percentage CA Percentage of Calls Abandoned. Indicates the percentage of calls abandoned by callers after being accepted by the ACD.

Percentage HLD Percentage of total calls HeLD in queue within a reporting group.

Percentage NCO Percentage of total of Number of Calls Offered to a particular reporting group.

Percentage TUT Percentage of Trunk Utilization Time. The percentage of a time during a reporting period that a trunk is in use and not idle.

Perceptual Analysis Measurement System See PAMS.

Perceptual Evaluation of Speech Quality See PESQ.

Perceptual Speech Quality Measure See PSQM.

percussive maintenance The fine art of whacking the crap out of an electronic device to get it to work again. A joke.

perforator An instrument for the manual preparation of a perforated tape, on which telegraph signals are represented by holes punched in accordance with a predetermined code.

perforator, paper tape An electro-mechanical device which converts electrical signals into coded holes in a paper tape. See Paper Tape Punch.

performance management Measures and records resource utilization. It is one of the categories of network management defined by the ISO (international Standards Organization).

periapsis In satellite systems, the point on a satellite's orbit at which it is closest to the center of the primary body about which it is orbiting. Where Earth-based satellite systems are concerned, the term is synonymous with perigee. See also Geostationary Orbit.

perigee The point at which a satellite orbit is the least distance from the center of the gravitational field of the Earth. The point in an orbit at which the satellite is farthest from the Earth is known as the apogee. In commercial application, the terms have most significance with respect to LEOs (Low Earth Orbiting) and MEOs (Middle Earth Orbiting) satellite constellations, which travel in elliptical orbits. See also LEO, MEO, and PERIAPSIS.

perimeter firewall There are two types of perimeter firewalls: static packet filtering and dynamic firewalls. Both work at the IP address level, selectively passing or blocking data packets. Static packet filters are less flexible than dynamic firewalls. See Firewall.

perimeter protection system A field disturbance sensor which uses buried leaky cables installed around a facility to detect any unauthorized entry or exit. See also network behavior anomaly detection (NBAD).

perimeter security system See also network behavior anomaly detection.

period of a satellite The time elapsing between two consecutive passages of a satellite through a characteristic point on its orbit.

periodic postings Articles that are posted periodically to a newsgroup for the benefit of people who are new to the newsgroup. An Internet term.

periodic registration A cellular term. A MSC (Mobile Switching Center) periodically sends out a broadcast message, requesting that all active cell phones register themselves. This process helps the MSC keep track of the active devices, and determine which cell sites are serving them.

periodicity Periodicity refers to the uniformly spaced variations in the diameter of the insulation of a cable system as a result of the manufacturing process. These variations cause signal reflections when the wavelength (or multiples of the wavelength) of the transmitted signal is exactly equal to the distance between the variations in the insulation.

peripheral device See Peripheral Equipment or Applications Processor.

peripheral equipment Equipment not integral to but working with a phone system. An example might be a printer or television screen on which calling traffic statistics are displayed. It might also be a voice mail or an automated attendant system. AT&T once called PBX peripheral equipment "applications processors," because they process specific applications. Some people now call them Adjunct Processor or Outboard Processors.

Perl Practical Extraction and Report Language. An interpreted scripting programming language, first released in 1987 by Larry Wall to streamline the administration of a network of Sun and DEC VAX computers. Perl is a highly portable language widely used in writing CGI (Common Gateway Interface) scripts, which are the standard means of performing actions – like searching or running applications when the user clicks on certain buttons or on parts of Web screen pages. The form Perl is preferred for the language itself; perl is used for the interpreter for the Perl language. See CGI.

perma-bears Skeptical investors who would rather own gold, silver or water rights than own a blue-chip stock. Perma-bears believe that the stockmarket will always go down.

permalancer A permanent freelancer. A person hired on a per-project basis who lives a benefits-free existence.

permanent path Fixed communication path between fibre channel nodes that provides guaranteed bandwidth. See Dynamic Path.

Permanent Presentation Status PPS. A PPS has a public value of a "public" or "anonymous" and is used as the presentation status of a call if no per-call CIDB (Calling Line Identification Delivery Blocking) feature is active. A PPS should exist for each calling identity item. A Bellcore definition.

permanent shift type A call center term. A shift definition that the program gives priority to in creating schedules but uses only as long as no overstaffing results in any intra-day period (at which point flexible shift types are used). When scheduling is done for more than a week at a time, the permanent schedules are always identical from one week to the next.

permanent signal A sustained off-hook supervisory signal originating outside a switching system and not related to a call in progress. Permanent signals can occupy a substantial part of the capacity of a switching system.

Permanent Virtual Circuit 1. PVC. A virtual circuit that is "permanently" defined in routing tables in packet network switches or routers, at least until such time as it is "permanently" changed. The network path, therefore, is fixed in programmed logic, and can be exercised quickly. A PVC uses a fixed logical channel over a physical network. Once a PVC is defined, it requires no setup operation before data is sent and no disconnect operation after data is sent. PVCs are used in packet-switched networks such as Frame Relay and TCP/IP. See also Switched Virtual Circuit.

2. PolyVinyl Chloride. A type of plastic material used in the manufacture of flame-retardant cable jackets.

Permanent Virtual Path Tunneling See PVP Tunneling.

permissible interference Observed or predicted interference which complies with quantitative interference and sharing criteria contained in these [Radio] Regulations or in CCIR Recommendations or in special agreements as provided for in these Regulations. (RR) See also accepted interference, interference.

permission A Windows NT term. A rule tied to an object (usually a directory, file or printer) to regulate which users can have access to the object and in what manner.

permission marketing A marketing method whereby companies get their customers' permission to market products or services to them. Sign up for a email newsletter. It often asks you if it could also send you offers to sell you stuff. It asks your permission. The theory is that by consumers will pay more attention to the marketing message. These days email that is is not permissions driven is called spam. The term was coined by Seth Godin in his book, Permission Marketing.

permissions 1. A call center term. Privileges granted to each user with respect to what data that user is allowed to access and what menu options or commands he or she is allowed to use. Permissions are under control of the System Administrator.

2. A Northern Telecom Norstar definition to define specific characteristics that can be assigned to an individual telephone. Permissions includes Full Handsfree, Handsfree Answerback, Pickup Group, Page Zone, Auxiliary Ringer, Receive tones, and Priority Call.

permissive dialing There's an area code change. Many peoples' phones now have different area codes. People calling them will have to dial a different area code. There'll be a time, perhaps three months, in which a caller will be able to reach the number by dialing the old area code or the new area code. That period is known as permissive dialing. Once the period of permissive dialing is over, the period of mandatory dialing begins.

permitivity See Dielectric Constant.

perpetrator Someone who carries out an illegal or malicious act affecting information security.

persistence 1. In a CRT, the time a phosphor dot remains illuminated after being energized. Long-persistent phosphors reduce flicker, but generate ghost-like images that linger on screen for a fraction of a second.

2. The probability that when a device on a local area network which has the data to transmit senses a free transmission line it will attempt the transmission. 1.0 persistence in-

dicates that there is 100 percent probability that the device will always attempt to transmit. IEEE 802.3 and Ethernet both use 1.0 persistence. A 0.5 persistence indicates that when a device senses a free line it will only attempt to transmit 50 percent of the time.

persistency A persistent configured site will attempt to re-establish the connection in the event of an unexpected line drop. A non-persistent configured site will not.

persistent communication A dialogue between an ASC (AIN Switch Capabilities) and a SLEE (Service Logic Execution Environment) that may involve a sequence of messages. Definition from Bellcore in reference to its concept of the Advanced Intelligent Network.

persistent information Information for which a permanent data object exists. Definition from Bellcore in reference to its concept of the Advanced Intelligent Network.

person call There will be two types of calls: person calls and place calls. I make a place call when I call a phone number which ends in one designated, fixed RJ-11 jack attached to the wall or floor. I make a person call when I call a phone number which doesn't necessarily end on a fixed place. A typical person call might be a cellular or wireless phone call. It might also be a service like MCI's Personal 800 Service. The major characteristic of a person call is that I am calling a person and I don't know where that person is. As a result, the person I called might answer my call anywhere – from his car, from vacation home, from his wireless phone, etc.

person to person call The most expensive way to make a long distance call. Call the operator. Say "I want to speak to Harry Newton on 212-691-8215." The operator dials Mr. Newton's phone number, gets Mr. Newton on the phone. "Are you Mr. Newton?" Mr. Newton replies: "Yes, I am." The operator bows out of the conversation and sends you, the caller, a hefty bill for that personalized service. Until recently, person-to-person service was only offered by AT&T. Now it's offered by most phone companies. But prices, in the main, are not regulated. And some companies charge an arm and a leg. Please be careful.

personal 800 number Several long distance companies are now offering Personal 800 numbers, which are basically party line 800 numbers with call routing. The way they work is as follows: You dial a number, e.g. 800-484-1000. A machine answers with a double beep. You punch in four or five digits on your touchtone pad. A voice response unit at the other end hears the digits, says "Thank you for using MCI" and dials out your long distance number which might be 212-691-8215 (mine). The long distance carriers are charging under $5 a month and 15 to 25 cents a minute for the service. The per minute charges are more expensive than normal 800 lines. One company, MCI, is also offering FOLLOW ME 800 which allows you to change the routing of your personal 800 number instantly with one phone line.

Personal Access System See PAS.

personal agent A personal agent is a piece of software on your personal computer that does your bidding. It may answer your emails (or at least some of them). It may find things on the web you're looking for. It may schedule appointments for you. The concept is easy: No one has a secretary anymore. So why not program your computer to do your bidding? Easier said than done. A software agent is software in the making.

personal area network See 802.15.1 and PAN.

Personal Authentication Device P.A.D. A portable or fixed device which allows for precise identification and validation of each user. Used in fraud prevention and unauthorized access. Also called a "token."

personal central office trunk line Also known as a "private line." Allows a user behind a PBX to access a central office trunk line dedicated to him; also allows people to call him directly, bypassing the PBX. Such a private line can appear on a line access button on an electronic PBX phone, or can terminate on a separate phone. As the private line bypasses the PBX, the console attendants can't listen to your conversations; neither can the technicians, unless they physically tap the line. Private lines, therefore, are more secure than the typical PBX station line. They also are CO-powered; as a result, they are not susceptible to power outages which might affect the PBX system. Senior executives use them, as do venture capitalists and others who conduct highly confidential negotiations. See also Private Line.

Personal Communications Industry Association PCIA. The trade association of the new cellular phone providers. PCIA represents a wide range of interests, including Broadband PCS, paging and narrowband PCS, antenna site owners and managers, Specialized Mobile Radio (SMR), suppliers and manufacturers. See also PCS. www.pcia.com

Personal Communications Networks PCN. A new type of wireless telephone system that would use light, inexpensive handheld handsets and communicate via low-power antennas. PCN is primarily seen as a city communications system, with far less range than cellular. Subscribers would be able to make and perhaps receive calls while they

are traveling, as they can do today with cellular radio systems, but at a low price. There's talk that they'll put PCN antennas in communities and issue everyone in the community with a PCN phone that would hub off the PCN antenna. This would save the phone company wiring up each house. In this way, the PCN phone would resemble the common household cordless phone, but with a larger range. Dr. Sorin Cohn of Northern Telecom calls the new low-power, wireless, personal communications systems an "enabler of unplanned growth." Nice definition. See PCN and Personal Communications Services.

In the fall of 1992 MCI broadened the definition of PCN. It proposed a national consortium of local PCNs joined together by a long distance carrier (namely it). In its release, MCI said that "PCNs are the next generation of digital wireless communications technology. PCNs use less power and are less expensive than the current cellular technology and permit the use of inexpensive pocket telephones with much longer battery life than cellular portables. PCN phones will have many more features than today's cellular or conventional telephones, and unlike cellular phones, will be usable in most areas of the world. PCNs will operate in the same frequency band in most countries, (1850-1990 MHz) while cellular is operating in several different frequency bands in various countries, and thus is not portable from country to country." See also Personal Communications Services.

Personal Communications Services PCS are a broad range of individualized telecommunications services that let people or devices to communicate irrespective of where they are. Some of the services include:

- Personal numbers assigned to individuals rather than telephones
- Call completion regardless of location ("find me")
- Calls to the PCS customer can be paid for by the caller, or by the PCS customer
- Call management services giving the called party much greater control over incoming calls.

PCS can both find and complete a call to a person regardless of location, but give that person the choice of accepting or rejecting the call or sending it somewhere else. PCS will possibly use a new category of wireless, voice and data communications – using low power, lightweight pocket telephones and hand-held computers.

Personal Computer PC. A computer for one person's use. That's why it was originally called a personal computer – to distinguish it from other computers that existed at the time of the PC's invention, or 1981, the date IBM introduced its first PC. Those other computers were mainframes and mini-computers. Their main use was to be shared by many users. Airline reservations, being an example. Of course, as the PC got more powerful, things got more complex. In 1983, Novell introduced its first local area network software called NetWare. NetWare was originally introduced to allow a handful of personal computers to share a single hard disk, which at that stage was costly and scarce. As hard disks became more available and cheaper, NetWare evolved to allow sharing of printers and file servers. Networking developed in the 1980s and 1990s and gave PCs the shared-user power of mainframes and minicomputers. As a result, today some PCs are personal computers, used by one person standalone. Some PCs are PCs that sit on a local area network (LAN). For some reason, people call these PCs workstations. And there are some PCs which have become servers – i.e. they "serve" many PCs. They may do faxing. They may do voice mail for the company. They may run the company's entire phone system. In short, meanings change as technology changes. Not all PCs are personal computers.

Personal Computer Memory Card International Association See PCMCIA.

Personal Computing Systems Architecture PCSA. A network architecture defined and supported by Digital Equipment Corp. for the incorporation of personal computers into server-based networks.

Personal Computer Terminal Adapter PCTA. A printed circuit card that slips into an IBM PC or PC compatible and connects the PC to the ISDN T-interface. The PCTA basically turns a normal PC into an ISDN phone and ISDN terminal, ready for voice and data communications. According to Northern Telecom, one of the manufacturers of such a device, the PCTA does:

- Functional signaling call setup.
- B channel for circuit switched data.
- X.25 packet data services on the B or D channel.
- Simultaneous operation of the B and D channels.
- M5317T digital telephone to PC messaging for integrated voice and data operation.
- NetBIOS interface to applications software, such as Microsoft Networks.

For more, see Northern Telecom's brochure ISDN PC Terminal Adapter NetBIOS Interface Description (D307-1).

Personal Data Interchange PDI. A format for exchanging information, such as electronic business, cards via wired or wireless connections.

personal data module A removable module, unique to the ROLM Cypress and Cedar, which stores all information entered by the user. Such items as phone numbers, terminal profiles and log-on sequences are stored in the PDM, which also features battery backup to protect the memory in the event of a power failure. The user can take his personal data module with him, plug it into another Cypress or Cedar and get all his speed dial and other personalized programming on his phone he just moved from.

Personal Digital Assistant PDA. A consumer electronics gadget that looks like a palmtop computer. Unlike personal computers, PDAs will perform specific tasks-acting like an electronic diary, carry-along personal database, personal communicator, memo taker, calculator, alarm clock. The communications will take place through the phone, through a cable attached to your PC or through infrared. See also Windows CE.

Personal Identification Number PIN number. 1. An AT&T term meaning the last four digits of your AT&T, MCI Bell operating company Credit Card – the card you use for making long distance numbers. 2. Some banks and financial institutions issue credit cards for machine, teller-less banking. These machines, called Automated Teller Machines, ask you for a password consisting of several numbers or characters. These are not on your credit card. These numbers or characters, called PIN numbers, are designed to make sure the right person is using your card. It's not a good idea to use your birthday as your PIN number.

personal intelligent communications What General Magic calls the products and services its alliance members will create using General Magic technologies that will help people remember, communicate and know things in new and powerful ways. According to General Magic, "the alliance's shared long-term vision is to bring personal communications to people who may not use a computer today, to people whose personal technology is a car, a television set and a telephone."

personal information assistant Tandy's name for what Apple calls a Personal Digital Assistant. See PDA.

Personal Information Manager PIM. Software application which allow the user to organize personal information. similar to an appointment book but personalized and programmed into the PC. PIMs now support Caller ID. When the phone rings, the phone system passes the calling number to the PC, which does a lookup in the PIM on the phone number and throws information about the caller up on screen. See PIM.

personal IVR An Interactive Voice Response system running on your own personal PC and designed to serve the needs of only one person. See IVR.

personal line A feature which allows specific key telephones to have their own private Central Office line. Sometimes called an AUXILIARY LINE. You can typically receive and make calls on this line. No one else can answer it, since it does not appear on any other phone instrument in the office. You can give this number to your wife or girlfriend. It is not a good idea to give it to both.

Personal Locator Beacons PLB. Peronal locator beacons are used by hikers, and people in remote locations to alert search and rescue personnel of a distress situation. See Personal Radio Services.

personal name An X.400 term for a standard attribute of an O/R (Originator/Recipient) Address form that identifies a person relative to another attribute (e.g., an organization name). The personal name may include surname, given name, initials, and generation qualifier. Initials consists of the first letter of all the user's names except the user's surname.

personal productivity tool Another term for a computer. John Perry Barlow thinks the expression was created by the "druids" who run Microsoft and Apple. Mr. Barlow is a cattle rancher, computer hacker, poet, and a lyricist for the rock band The Grateful Dead.

Personal Radio Services According to the FCC, Personal radio services provide short-range, low power radio for personal communications, radio signaling, and business communications not provided for in other wireless services. The range of applications is wide, spanning from varied one- and two way voice communications systems to non-voice data transmission devices used for monitoring patients or operating equipment by radio control. Licensing and eligibility rules vary. Some personal radio services require a license grant from the FCC, while others require only that you use equipment that is properly authorized under the FCC's rules. The personal radio services are:

218-219 MHz Service - One or two way communications for transmission of information to subscribers within a specific service area.

Citizens Band (CB) Radio Service - 1-5 mile range two-way voice communication for use in personal and business activities.

Family Radio Service (FRS) - 1 mile range Citizen Band service for family use in their neighborhood or during group outings

General Mobile Radio Service (GMRS) - 5-25 mile range Citizen Band service for family use in their neighborhood or during group outings

Low Power Radio Service (LPRS) - private, one-way communications providing auditory assistance for persons with disability, language translation, and in educational settings, health care, law, and AMTS coast stations.

Medical Implant Communications Service (MICS) - for transmitting data in support of diagnostic or therapeutic functions associated with implanted medical devices.

Multi-Use Radio Service (MURS) - private, two-way, short-distance voice or datacommunications service for personal or business activities of the general public.

Personal Locator Beacons (PLB) - used by hikers, and people in remote locations to alert search and rescue personnel of a distress situation.

Radio Control Radio Service (R/C) - one-way non-voice radio service for on/off operation of devices at places distant from the operator.

Wireless Medical Telemetry Service (WMTS) - for remote monitoring of patients' health through radio technology and transporting the data via a radio link to a remote location, such as a nurses' station.

personal scanners Personal scanners are tiny devices that read bar codes off ads in magazines and newspapers and allow us to go directly to the spot on the web where there's more information about the product or service. Early personal scanners were the CueCat, the Gamut and the CS 2000. They haven't been very successful so far. I threw my three CueCats out. Some people are predicting a big future for them.

personal speed dial Simplified ways of dialing. You do them by dialing a couple of digits. Or you punch in a button at your phone. Personal Speed Dial codes are programmed for each telephone, and can only be used at the telephone on which they are programmed. System Speed Dials, in contrast, can be used from every phone in the system.

Personal System/2 PS/2. IBM's old and now no-longer-manufactured family of microcomputers some of whom sport one major difference from its predecessors, namely the existence of a 32-bit micro-channel bus. This "bus" serves the same purpose as a PBX's backplane – namely to move information from the printed circuit cards and to other printed circuit cards, which may contain their own individual microprocessors (computers on a chip) and which may communicate with the outside world through their own communications ports.

personal telephone The category of cellular telephones pioneered by Motorola's Pan American Cellular Subscriber Group with the introduction of the MicroTAC Digital Personal Communicator Telephone. Weighing less than one pound, they are so compact and lightweight, they fit comfortably into a shirt pocket or purse making them "body friendly."

personal universal controller PUC. An all-in-one remote, pioneering work on which is being done by the Pebbles Project at Carnegie Mellon University. A personal universal controller (PUC) is a hand-held computer that can be used to control many different kinds of electronic devices in the home, office, or factory. The basic concept is that when a user points the PUC at a device, such as a TV, VCR or DVR at home; or a photocopier in an office; or a machine tool in a factory; the target device will wirelessly transmit a description of its control parameters to the PUC. The PUC uses this data to create an appropriate control panel on its screen display, taking into account the properties of the controls that are needed, the properties of the specific handheld device serving as the PUC (the display type and input techniques available), and the preferences of the user (e.g., language preference, whether the user is left or right-handed, display button size, and so on). The user can then use the PUC to control the target device.

Personal Video Recorder See PVR.

Personal Wireless Telecommunications PWT. Also known as Personal Wireless Telephony. A U.S. cordless telephony standard for in-building wireless communications systems, PWT is a variant of the DECT (Digital European Cordless Telecommunications) standard. PWT is an air interface which operates in the unlicensed PCS (Personal Communications Services) 1.9 GHz radio band. PWT(E) is an enhanced version which operates in the licensed PCS bands. See also Air Interface, DECT, and PCS.

personality module A small motherboard added to a voice board to give it the "personality" of a proprietary electronic PBX telephone. "Personality" means electrical characteristics and the same button configuration and responsiveness, all of which can be recreated on the screen of a PC which has the voice board installed.

personalization technology Software that lets suppliers of information to "nonintrusively" (their word, not mine) learn the individual interests of their customers in order to deliver personalized Web content, targeted advertising, and product recommendations.

personalized ring 1. A telephone feature which allows you to select different ringing sounds for your telephone. This feature is useful if you work in a big room with lots of other people and it's hard to tell whose phone is ringing.

2. A cell phone feature which allows you to customize the ring on your cell phone. You can customize that ring with one of the various ringtones that the manufacturer gives you or which you download from a web site. You can get a bunch from www.yourmobile.com.

3. A telephone feature offered by most local phone companies which adds a new phone number to your existing line. When someone calls the new phone number the same line will ring with a different ring pattern. Many fax machines can distinguish between the different ring patterns and only answer the one you select.

PersonalJava A Java platform optimized for the requirements and constraints of mobile devices.

PERT Project Evaluation and Review Technique. A variation on the Critical Path Method of organizing the completion of projects. Projects are examined for the their worst, best, average completion times. A critical path is determined and overall standards for completion times are created. The PERT technique was created by the military. It is used for organizing complex tasks.

pervasive computing A form of computing that involves ubiquitous access to information through the use of portable computers and extensive networking.

PES 1. Packetized Elementary Stream. In MPEG-2 (Moving Pictures Experts Group) compression, once the media stream has been digitized and compressed, it is formatted into a stream of packets. The resulting PES then is multiplexed into either a Program Stream or Transport Stream. See also MPEG-2.

2. Packet over Ethernet over SONET. A technique by which IP (Internet Protocol) packets are encapsulated within Ethernet frames to transit an Ethernet LAN segment, which is connected to another Ethernet segment across an network based on a SONET optical fiber transmission system. A developing PES application is that of 10GbE (10 Gigabit Ethernet). See also Ethernet, IP, PEW, and SONET.

PESQ Perceptual Evaultion of Speech Quality, currently under review by the ITU, with expected approval as P.862 in 2001. It is widely anticipated that PESQ will eventually replace previous methods (like PSQM and PAMS) for objective speech quality testing. PESQ is directed at narrowband telephone signals. It is applicable to systems with speech coding (including low bit-rate vocoders), variable delay, filtering, packet or cell loss, time-clipping, and channel errors. PESQ scores predict listening quality scores for ACR (Absolute Category Rating) listening tests. PESQ leverages the best featurs of PAMS and PSQM99. It combines the robust time-alignment techniques of PAMS with the accurate perceptual modeling of PSQM99, and it adds new methods including transfer function equalization and a new method for averaging distortion over time.

pessimist 1. "A pessimist," Winston Churchill once said, "sees the difficulty in every opportunity. An optimist sees the opportunity in every difficulty."

2. Borrow money from pessimists. They don't expect it back.

peta P. A prefix that denotes 10 to the 15th power or one quadrillion. In computer terms, however, peta- is actually equal to 2 raised to the 50th power , or 1,125,899,906,842,624, the power of 2 that is closest to one quadrillion.

Petabits Pb. A petabit is literally equal to 2 50 bits, or 1,125,899,906,842,624 bits, but is often calculated with a base of 10, making it equal to 10 to the 15th or one quadrillion bits. See Petabyte.

Petabyte PB. A combination of the Greek "pente," meaning "five," and the English "bite," meaning "a small amount of food." A unit of measurement for physical data storage on some form of storage device – hard disk, optical disk, RAM memory etc. and equal to two raised to the 50th power, i.e. 1,125,899,906,842,600 bytes.

KB = Kilobyte (2 to the 10th power)
MB = Megabyte (2 to the 20th power)
GB = Gigabyte (2 to the 30th power)
TB = Terabyte (2 to the 40th power)
PB = Petabyte (2 to the 50th power)
EB = Exabyte (2 to the 60th power)
ZB = Zettabyte (2 to the 70th power)
YB = Yottabyte (2 to the 80th power)
One googolbyte equals 2 to the 100th power.

Peter A government which robs Peter to pay Paul can always depend on the support of Paul. - George Bernard Shaw

Peter the Great of Russia Peter the Great of Russia had his wife's lover beheaded, pickled the head in a jar of alcohol and had her keep it in her bedroom.

Petition for Reconsideration A Petition for Reconsideration is a written request submitted to the FCC for review of an action it has previously taken. Applicants have 30 days after a Report and Order is published in the Federal Register to file comments with the FCC. The agency will consider public comments, replies, and industry concerns

before finalizing its initial decision. As a result of the review process, the FCC will either issue a Memorandum Opinion and Order amending its initial decision, or deny the Petition for Reconsideration.

PEW Packet over Ethernet over WDM. A technique by which IP (Internet Protocol) packets are encapsulated within Ethernet frames to transit an Ethernet LAN segment, which is connected to another Ethernet segment across an network based on a DWDM (Dense Wavelength Division Multiplexing) optical fiber transmission system. A developing PEW application is that of 10GbE (10 Gigabit Ethernet). See also DWDM, Ethernet, IP, and PES.

PF Xfer Power Failure Transfer.

PF-E PK-E is a suffix to an FCC registration number, indicating the function of a system in accordance with its FCC registration. A PF-E is a Fully Protected PBX Telephone System, with "PF" denoting "PBX Function," and "E" denoting that the system will accept both rotary and tone signaling. See also KF-E, MF-E, PBX, and Registration Number.

Pfishing See phishing, which is the more common spelling.

PFM Pure F...ing Magic. A term to indicate that it works but the software engineer has no real idea how it works. It works by PFM. Sort of like an accidental winning shot in tennis. Looks great. The hitter won the point, but he had no idea how he did it. See PFTS.

PFOC Pending the Firm Order Commitment.

PG Peer Group A set of logical nodes which are grouped for purposes of creating a routing hierarchy. PTSEs are exchanged among all members of the group.

PGA 1. Programmable Gain Amplifier, a part of the Analog Front End.

2. Pin Grid Array. A silicon chipset in which all of the connecting pins are arranged in concentric squares. See also DIP, SIMM, and SIP.

PGL Peer Group Leader: A single real physical system which has been elected to perform some of the functions associated with a logical group node.

PGP Pretty Good Privacy. PGP is a cryptography program for computer data, electronic mail and voice conversations. PGP was written in 1991 by Philip R. Zimmermann, who gave the software away. It was posted on a public computer on the Internet, and thousands downloaded copies. Originally, PGP was just used for the transmission of computer data and electronic mail. Then it got extended to voice conversations on the phone, called PGPfone. The idea is that you use modems to dial. Then the PCs "shake hands" and jointly agree on a complex number that is plugged in a scrambling algorithm equation. This notoriously complex scrambling algorithm, called Blowfish, recalculates the digital ones and zeroes of the sampled voices into a stream of numbers unintelligible even to highly sophisticated eavesdroppers. Finally, PGPfone unscrambles the stream to provide intelligible – though not great quality – sound. In February 1993, the U.S. government notified Zimmerman that he was being investigated for violating this restriction on the export of encryption technologies. The case was dropped in 1996, leaving Zimmerman free to found PGP, Inc. The company later was acquired by Network Associates. See also Encryption.

PGPfone The name of the protocol used for voice conversations on Pretty Good Privacy. See PGP for a full explanation.

PH 1. Packet Handler, or Packet Handling function.

2. The Ph system allows you to look up directory information, usually including e-mail addresses at universities, research institutions, and some governmental agencies throughout the world. You need a program that lets you use Ph. Tell that program which Ph server to use, and then enter a name you would like to search for.

phantom circuit A circuit derived from two suitably arranged pairs of wires, called side circuits, with each pair of wires being a circuit in itself and at the same time acting as one conductor of the phantom.

phantom directory number Also called a virtual DN. A directory number with a voice mailbox, but not a phone on it. Calls are then transferred to this number. The mailbox user can dial into the system, enter the extension number, security information and retrieve their messages.

phantom load The electricity consumed by a device when it is not doing its main function and may be theoretically turned off. Studies show that about 6 per cent of the electricity used in typical homes is by devices that consume power even when turned off.

phantom phone rings Also known as ringxiety or fauxcellarm. It's what happens when you hear a sound and think it's your cell phone ringing. That sound could be anything from a passing ice cream truck to a piece of music on the radio.

phantom power A method for supplying electrical power, usually 6 to 60 volts DC, to a device such as a microphone, over the same cable that carries the signal from the device. Phantom power is defined by IEC 61938 for 12V, 24V and 48V implementations.

phantom traffic Phantom traffic is defined as all communication services using the switched network that is either un-billed or under-billed due to:

1. insufficient information to properly identify and invoice the responsible originating entity,

2. traffic delivered by a connecting company over common trunk groups without agreements or knowledge of the terminating company, or

3. traffic routed fraudulently or inadvertently to conceal the geographic origination of the traffic, including but not limited to routing over trunks with lower-priced jurisdictions.

Phantom traffic is a growing concern to small, medium and large companies that terminate traffic over their networks. Carriers that have undertaken studies frequently determine that a substantial share of traffic in studied central offices is "phantom." Phantom traffic has both direct and indirect costs including uncollected and under-collected revenues, high expenditures for quality of service network augmentation to meet increased traffic demand, "revenue assurance" personnel, consultants, hardware, and software and enforcement actions. These costs, carriers claim, undercut their ability to invest and serve their customers. Some argue there should be wholesale intercarrier compensation reform. See revenue assurance and revenue leakage.

pharming A type of man-in-the-middle attack whereby users' HTTP requests are redirected to a masquerading website. This is accomplished by corrupting a record on a DNS server so that a targeted URL is associated with the masquerading website's IP address. This redirects traffic to the IP address of the masquerading website, where transactions are mimicked and information such as logon ids and passwords are collected. Using the illicitly captured logon information, hackers can then penetrate the real website. Think of a bank and how much money you could steal if you knew peoples' ids and passwords. Here's a neat (though illegal) way of finding them out. I'm hoping that the only farms most pharmers will farm are those in prison.

phase The relationship between a signal and its horizontal axis, also called zero-crossing point. A full cycle describes a 360 degree arc. A sine wave that crosses the zero-point when another has attained its highest point is 90 degrees out of phase with the other. See Phase Shift Keying.

Phase 1 Phase 1 Environmental Site Assessment. Investigation to determine the condition of the land at the proposed site location. Soil contaminants, pollution etc. To be done at any site where equipment will be touching the ground. See SHPO – State Historic Preservation Officer.

Phase A Phase A is the first part of a fax machine's call process. It is the call establishment. It occurs when transmitting and receiving units connect over the phone line, recognizing one another as fax machines. This is the start of the handshaking procedure. See Phase B.

Phase B Phase B is the second part of a fax machine's call process. It is the premessage procedure, where the answering machine identifies itself, describing its capabilities in a burst of digital information packed in frames conforming to the HDLC standard. See Phase C.

Phase C Phase C is the third part of a fax machine's call process. It is the fax transmission portion of the operation. This step consists of two parts C1 and C2 which take place simultaneously. Phase C1 deals with synchronization, line monitoring and problem detection. Phase C2 includes data transmission. See Phase D.

Phase D Phase D is the fourth part of a fax machine's call process. This phase begins once a page has been transmitted. Both the sender and receiver revert to using HDLC packets as during Phase B. If the sender has further pages to transmit, it sends an MPS and Phase C recommences for the following page. See Phase E.

phase displacement antenna An antenna constructed from a driven element and a group of reflectors, the secondary radiation from which produces an antenna with directivity. The Yagi-Uda array is a member of this family.

phase delay See Envelope Delay.

phase distortion An unwanted modification of a transmitted signal caused by the non-uniform transmission of the different frequency components of the signal. Same as Delay Distortion.

Phase E Phase E is the fifth part of a fax machine's call process. This phase is the call release portion. The side that transmitted last sends a DCN frame and hangs up without awaiting a response.

phase hit In telephony, the unwanted and significant shifting in phase of an analog signal. As defined by AT&T: any case where the phase of a 1004 Hz test signal shifts more than 20 degrees. Also, error-causing events more severe than phase jitter, especially for data transmission equipment using PSK modulation.

phase inversion The condition whereby the output of a circuit produces a wave of the same shape and frequency but 180 degrees out of phase with the input.

phase jitter In telephony, the measurement, in degrees out of phase, that an analog signal deviates from the referenced phase of the main data-carrying signal. Phase jitter is often caused by alternating current components in a network.

phase lock The phase of a signal follows exactly the phase of a reference signal.

Phase Locked Loop PLL. Phase Locked Loop is a mechanism whereby timing information is transferred within a data stream and the receiver derives the signal element timing by locking its local clock source to the received timing information.

phase modulation One of three ways to change a sine wave (or L signal) to let it carry information. In this case, the phase of the sine wave is changed as the information to be carried is changed. See Phase Shift Keying and Modulation.

phase roll Variations in the phase of a transmitted signal and its echoed back modem verification. Phase roll is encountered most often in international systems.

phase shift A change in the time or amplitude that a signal is delayed with respect to a reference signal.

Phase Shift Keying Also known as PSK and Phase Modulation. Used by relatively sophisticated modems for transmitting digital signals over analog phone lines. Picture an electromagnetic sine wave. In its natural state, the sine wave is a continuous and uninterrupted waveform of a certain amplitude and carrier frequency. If we want to place a digital signal on it, it is necessary that we cause the signal to change in some way to reflect the presence of a "1" or the absence of a one, i.e., a zero. In phase shift keying, we simply change the phase of the carrier signal to reflect a change in value of the adjacent bits. The receiving device looks for the phase shift of the signal in absolute, rather than relative terms. Differential Phase Shift Keying (DPSK) is a more sophisticated technique, involving the comparison of the signal shift to the carrier frequency, with the latter being used as a reference. Compare with Differential Phase Shift Keying. Contrast with Amplitude Modulation and Frequency Modulation. See also BPSK and QPSK.

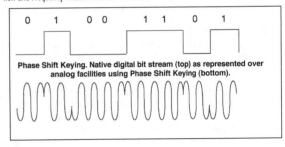

Phase Shift Keying. Native digital bit stream (top) as represented over analog facilities using Phase Shift Keying (bottom).

phase shift modulation Another way of saying Phase Shift Keying. See above.

phased array antenna A phased antenna system consists of two or more active antennas – called antenna elements – arranged (also called arrayed) so the electromagnetic fields effectively add in some directions and cancel in other directions. This produces enhanced transmission and reception in the directions where the fields add, and reduces the strength of radiated and received signals in the directions where the fields cancel. A phased array antenna is used in certain satellite, wireless TV, and WLL (Wireless Local Loop) applications. Phased Array Antennas are typically small (e.g., 4.5 inches square), flat antennas which mount on the side of your building or on your rooftop. Inside the thin flat box is an array (i.e. several) of chip-based radio receivers which lock in on the desired transmission frequency on a dynamic basis. That is to say that they have sufficient intelligence to direct and redirect their focus in order to maximize the strength of the incoming or outgoing signal. As neither the device nor its components, are physically dynamic (i.e., they cannot be repositioned physically, in the moving kinetic sense of the word), this capability is critical. These devices have the appearance of a pizza box. The term "pizza box" sometimes is applied to them. Many AM (amplitude-modulation) broadcast stations use sets of two, three, or four phased vertical antennas. This results in a directional pattern that optimizes coverage, so the station can reach the greatest possible number of listeners in its designated area. The nulls in the pattern reduce or eliminate interference with other AM broadcast stations. Some new wireless 802.11b transmission systems used phased array antenna, thus effectively getting around limitations on the strength of the signal they can emit. Using phased array antennas gives them greater range and reliability. In two-way radio communications, several vertical dipole antenna can be placed end-to-end and fed in phases. This is known as a collinear antenna and is a specialized type of phased array. At low elevation angles, the radiation and response are enhanced relative to a single vertical dipole. This gain occurs at the expense of radiation and response at higher elevation angles, increasing the range of communication for surface-to-surface communications. See also WLL.

phasing The process of ensuring that both sending and receiving facsimile machines start at the same position on a page.

phasing orbit A temporary satellite orbit which is used prior to putting the satellite into its final orbit.

phasor Temporary buffer storage that compensates for slight differences in data rate between TDM I/O ports and devices.

PHB Per-hop-behavior; also described in RFCs 2597 and 2598 (also describes "Expedited" PHB).

phenolic insulating materials A type of insulating materials, one of which is bakelite. Now no longer used.

phenom A slang use of the word phenomenon used to mean a prodigy, for example: John Elway at 18 or Serina Williams at 17.

PHF Packet Handling Function. The switching capability that processes and routes X.25 virtual calls.

PHI PBX-to-Host Interface. The same as CPI (Computer to PBX Interface) but it puts the PBX first, which is the way many telephone manufacturers prefer to see it. It refers to a connection between a telephone system and a computer, such that the computer can signal the telephone system to switch calls and the telephone system can signal the computer when it has switched them. There are major advantages in joining a telephone system to a computer. For a much greater explanation, see OAI, which stands for Open Application Interface.

Phi Phenomenon A theory developed by psychologist Hugo Munsterberg, it explains the illusion of motion created by rapid presentation of a series of still images. Munsterberg suggests that the brain hallucinates, effectively filling in the voids between the images. At 30 fps (frames per second), the brain processes the images as fully fluid motion. See Frame Rate.

phiber optik See Master of Deception

phishing Phishing (pronounced fishing) is short for password harvesting fishing. Phishing (also spelled pfishing) is an Internet email scam. Through bogus emails, it lures unsuspecting people (you and I) to bogus Web sites which appear to be legitimate business sites (e.g. Citigroup, Bank of America, Visa) and then asks us to divulge confidential information, such as our credit card and social security numbers. It does this by pretending to be a legitimate site (replete with logos) and pretending that some problem has arisen with our credit card or bank account and then asking us to update account information by resubmitting it. The website to which you are directed masquerades as a legitimate website of a financial institution or merchant, complete with logos and other official-looking graphics. A word of warning. Don't. Virtually all legitimate financial institutions do not request information in this way. If someone gets your personal information and steals your financial identity, you will spend months and months digging yourself out. I repeat: Don't give out financial information on yourself on the Internet. The only exception I make is when buying from a legitimate seller, e.g. Amazon. See also antiphishing, identify theft, smishing, spear phishing and vishing.

Phoenix MCI's ill-fated order entry system that never rose from the dead, i.e. it never worked.

phone A simpler way of saying the word "telephone." See Telephone.

phone bomb A phone bomb is a booby-trapped cellular phone. I saw the term first used on the front page of the January 6, 1996 New York Daily News. The story read: "'The engineer,' an extremist Palestinian bomb maker which was No. 1 on Israel's most-wanted list was killed on the Gaza Strip yesterday when his head was blown up by a booby-trapped cellular phone. The phone was rigged with about two ounces of explosives." No one admitted to killing the man.

phone centric users These are knowledge workers for whom the telephone plays a key role in the success of their business –, e.g. salesmen, stock brokers.

phone farm A place where you prepare cell phones for giving to the ultimate user. The theory is that a good specialist will not hand you a product that has not been activated, tested, or charged for immediate use. A reader wrote me "When I order several phones I get a surge-protected strip and connect all the chargers up to the phones and charge the phones. This also works with digital phone testing and can work with most landlines when many phones have to be tested."

phone flu Attacks that target vulnerabilities in mobile phones and VoIP systems. Symptoms include, but are not limited to, viruses, service disruptions, toll fraud, and identity theft.

phone freak See Phone Phreaks.

phone ladies Women in Bangladeshi villages who buy cheap cell phones from the country's leading provider, then sell time to phoneless fellow villagers for a profit, building a cottage industry in the process

phone parlor Often found in immigrant neighborhoods in large cities in the 1990s but less so these days, a phone parlor is a private establishment with multiple establishment-owned phones where people can come in off the street and make phone calls, generally paying for the calls in hard cash. A phone parlor caters to people who don't have a phone at home and don't want to pay the phone company's standard pay phone rates. Phone parlors with cell phones are often set up in poor countries.

phone phreaks Communication hobbyists. People, usually kids, who like to figure out how the telephone network works and sometimes make free calls on the network by figuring a way to bypass billing mechanisms. Phone Phreaks have become Computer Phreaks with the advent of PCs and the advent of out-of-band signaling, making it a lot more difficult to make long distance calls for free. See Phreak.

phone services database A Bluetooth term. The portion of the BT implementation that stores information about device services, both local services and remote services.

phone telepathy Norwich, England (Reuters), September 5, 2006 – Many people have experienced the phenomenon of receiving a telephone call from someone shortly after thinking about them – now a scientist says he has proof of what he calls telephone telepathy. Rupert Sheldrake, whose research is funded by the respected Trinity College, Cambridge, said on Tuesday he had conducted experiments that proved that such precognition existed for telephone calls and even e-mails. Each person in the trials was asked to give researchers names and phone numbers of four relatives or friends. These were then called at random and told to the subject who had to identify the caller before answering the phone. "The hit rate was 45 percent, well above the 25 percent you would have expected," he told the annual meeting of the British Association for the Advancement of Science. "The odds against this being a chance effect are 1,000 billion to one."

phone tree A list of people, along with their phone numbers, arranged to facilitate a chain of calls when information needs to be communicated to a large group of people quickly. The person at the top of the phone tree calls a small group of people, each of whom then proceeds to contact other people, and so on.

phone-In A phone-in promotion uses the telephone as a key element in its communication to and/or obtaining response from its target audience. Phone-ins can be used with most types of sales promotion. The key to success depends upon the effectiveness with which the phone-in number is communicated to the customer.

PhoneMail A ROLM term for Voice Mail. Rolm's PhoneMail is a voice messaging system that provides telephone answering (with the user's own greeting), the capability to store and forward voice messages and the capability to turn on a message waiting light or message on the recipient's phone. PhoneMail can be used positively to speed the flow of information. It can also be used negatively to allow the user to "hide behind" the system and avoid the outside world and anyone in the outside world who might actually want to buy something. See also Voice Mail.

phoneme A voice recognition term. The minimal significant structural unit in the sound system of any language that can be used to distinguish one word from another. For example, the p of pit and the b of bit are considered two separate phonemes, while the p of spin is not. These minimal sound units comprise words.

PhoneNet Farallon's twist on Apple's local area network called LocalTalk. PhoneNet uses standard one pair UTP (unshielded twisted pair) wiring for networking. PhoneNet is compatible with LocalTalk.

phonetic alphabet A list of standard words used to identify letters in a message transmitted by radio or telephone. The following are the authorized words, listed in order, for each letter in the alphabet: Alpha, Bravo, Charlie, Delta, Echo, Foxtrot, Golf, Hotel, India, Juliet, Kilo, Lima, Mike, November, Oscar, Papa, Quebec, Romeo, Sierra, Tango, Uniform, Victor, Whiskey, X-ray, Yankee, Zulu.

phosphor Substance which glows when struck by electrons. The back of a picture tube face is coated with phosphor.

photo diode Preferably spelled photodiode. The basic element that responds to light energy in a solid state imaging system. It generates an electric current that is proportional to the intensity of the light falling on it. Photo diodes are the light source or detector in a fiber optic transmission system. Light sources and light detectors are paired: LEDs (Light Emitting Diodes) as light sources, and PINs (Photo INtrinsic diodes) as light detectors in the slower systems; Laser Diodes and APDs (Avalanche PhotoDiodes) in the high-bandwidth systems such as SONET. See also APD, Laser Diode, LED, photodiode, PIN and SONET.

photo electric cell A light activated switch frequently used to turn lights on at dusk or off at sun rise. Sometimes they are also used to dim lights at night.

photo etch The process of forming a circuit pattern in metal film by light hardening a photo sensitive plastic material through a photo negative of the circuit and etching away the unprotected metal.

photoconductive effect Some non-metallic materials exhibit a marked increase in electrical conductivity when they absorb photon or light energy. This is called the photoconductive effect. The conductivity increase is due to the additional free carriers generated when photon energies are absorbed in electronic transitions. The rate at which free carriers are generated and the length of time they persist in conducting states (their lifetime) determines the amount of conductivity change.

photoconductivity The conductivity increase exhibited by some nonmetallic materials, resulting from the free carriers generated when photon (i.e. light) energy is absorbed in electronic transitions. The rate at which free carriers are generated, the mobility of the carriers, the length of time they persist in conducting states (their lifetime) are some of the factors that determine the extent of conductivity charge. See Photoconductive Effect.

photoconductor 1. Any transducer that produces a current which varies in accordance with the incident light energy. A fiber optic communications term.

2. Photoconductor is also material, available in many forms (sheets, belts, and drums), which changes in electrical conductivity when acted upon by light. Electrophotography (a form of facsimile machine printing) relies on the action of light to selectively change the potential of a charged photoconductive surface, creating areas receptive to an oppositely charged toner, thus making the latent charged-image visible.

photocurrent The current that flows through a photosensitive device (such as a photodiode) as the result of exposure to radiant power. Internal gain, such as that in an avalanche photodiode, may enhance or increase the current flow but is a distinct mechanism.

photodetector, receiver In a lightwave system, a device which turns pulses of light into bursts of electricity. For relatively fast speeds and moderate sensitivity in the 0.750xCAm to 0.950xCAm area wavelength, the silicone photodiode is most commonly used. See photodiode.

photodiode A type of photoelectric component that that detects optical signals and converts them into electrical signals, i.e., converts photons into electrons, or converts electrical signals into optical signals, i.e., converts electrons into photons. It generates an electric current that is proportional to the intensity of the light falling on it, or vice versa. (Note that diodes always work only one way, like a valve.) Photodiodes are the light sources and detectors in fiber optic transmission systems. Light sources and light detectors are paired: LEDs (Light Emitting Diodes) as light sources, and PINs (Photo INtrinsic diodes, or Positive Intrinsic Negative diodes) as light detectors in the slower systems; Laser Diodes and APDs (Avalanche PhotoDiodes) in the high-bandwidth systems such as those used in SONET. See also APD, Laser Diode, LED, PIN and SONET.

photoelectric effect An effect explained by Albert Einstein which demonstrates that light seems to be made up of particles, or photons. Light can excite electrons (called photoelectrons in this context) to be ejected from a metal. Light with a frequency below a certain threshold, at any intensity, will not cause any photoelectrons to be emitted from the metal. Above that frequency, photoelectrons are emitted in proportion to the intensity of incident light. The reason is that a photon has energy in proportion to its wavelength, and the constant of proportionality is the Planck constant. Below a certain frequency – and thus below a certain energy – the incident photons do not have enough energy to knock the photoelectrons out of the metal. Above that threshold energy, called the work function, photons will knock the photoelectrons out of the metal, in proportion to the number of photons (the intensity of the light). At higher frequencies and energies, the photoelectrons ejected obtain a kinetic energy corresponding to the difference between the photon's energy and the work function.

photomask Photomasks are typically quartz plates several inches thick and six inches across. They transfer chip designs onto silicon wafers in a process called optical lithography. An advanced mask set can include two-dozen or more plates, or reticles. Each imparts specific design elements or layers of the chip onto the silicon. In the 1990s, mask makers were able to do extraordinary things with light to etch shrinking chip designs onto wafers. Ordinary light produces wavelengths about 0.6 microns wide, or about 1/200th the width of a human hair. By the mid 1990s, when mask makers came into their own by providing a key enabling technology for the chip industry, they used ultraviolet light and then deepultraviolet light, which has a wavelength of 0.248 microns. That cleared the way for ships at the 0.25-micron node. Those mask sets cost about $100,000 initially, and the price has dropped somewhat since then. On their inexorable march toward tighter design rules, leading makers of integrated circuits soon designed chips at the 0.18-micron node, or 1/500th the thickness of a human hair. Ics at the 0.13-micron node are ramping now, with the 90-nanometer node next in the coming cycle. To keep up, mask makers have figured out how to exploit predictable behavior of light.

photon The Photon is a particle of light. For hundreds of years light was thought of solely as a wave. In 1905 Einstein discovered that under certain circumstances the energy of a light wave only came in specific amounts or quanta. These quanta are called photons.

photonic band gap See Photonic Crystals.
photonic crystal fiber See holey fiber.
photonic crystals Photonic crystals are microscopically patterned materials that might be used to produce the true optical equivalent of the electronic integrated circuit (IC) – but at much higher speeds. The idea is that when electronic ICs (integrated circuits) run out of speed, photonic crystals will take over. To produce a photonic crystal involves micromachining arrays of very small holes into a planar wafer. The diameter of each hole (they're usually circular) is actually smaller than the wavelength of light used for telecommunications. The easiest way to understand the behavior of light in a photonic crystal is to compare it to the movement of electrons and holes in a semiconductor. Take a deep breath. This stuff is complicated. In a silicon crystal, the atoms are arranged in a diamond-lattice structure, and electrons moving through this lattice experience a periodic potential as they interact with the silicon nuclei via the Coulomb force. This interaction results in the formation of allowed and forbidden energy states. For pure and perfect silicon crystals, no electrons will be found in an energy range called the forbidden energy gap or simply the band gap. However, the situation is different for real materials: electrons can have an energy within the band gap if the periodicity of the lattice is broken by a missing silicon atom or by an impurity atom occupying a silicon site, or if the material contains interstitial impurities (additional atoms located at non-lattice sites). Now consider photons moving through a block of transparent dielectric material that contains a number of tiny air holes arranged in a lattice pattern. The photons will pass through regions of high refractive index – the dielectric - interspersed with regions of low refractive index – the air holes. To a photon, this contrast in refractive index looks just like the periodic potential that an electron experiences travelling through a silicon crystal. Indeed, if there is large contrast in refractive index between the two regions, then most of the light will be confined either within the dielectric material or the air holes. This confinement results in the formation of allowed energy regions separated by a forbidden region – the so-called photonic band gap. Since the wavelength of the photons is inversely proportional to their energy, the patterned dielectric material will block light with wavelengths in the photonic band gap, while allowing other wavelengths to pass freely. It is possible to create energy levels in the photonic band gap by changing the size of a few of the air holes in the material. This is the photonic equivalent to breaking the perfect periodicity of the silicon-crystal lattice. In this case, the diameter of the air holes is a critical parameter, together with the contrast in refractive index throughout the material. With photonic crystals, it's possible to make a so-called "perfect mirror." This is formed from a regular hexagonal array of holes, giving the appearance of a honeycomb. The holes must be identical in size. In practice, the holes are quite shallow because they are drilled into the surface of a thin wafer, but this doesn't stop them from working. The perfect mirror can reflect light incident from any angle in the plane of the wafer (unlike the edge of a standard waveguide, the properties of which change with angle) for a selected band of wavelengths. Take two perfect mirrors, put them side by side with a narrow gap in between, and the result is a waveguide. Unlike today's waveguides, which leak out light when angles get tight, a photonic crystal waveguide can guide light around 900xAF bends with zero loss – at least that's the theory. But making and measuring these things has proven fairly challenging. To make this idea work in reality, the light has to be confined in the vertical direction; otherwise it gets radiated out of the top of the wafer. That's because the photonic crystal is a two-dimensional structure – it controls light in the plane of the wafer, but can't stop it from escaping out of the top or bottom.

Two teams of researchers have made breakthroughs that could lead to a big leap forward in making optical integrated circuits. One project at Sandia National Laboratories in the U.S., promises to yield optical waveguides that can guide light around tight bends with zero losses. Currently, optical waveguides rely on small differences in refractive index to channel light. To get light around a corner with no losses requires a bend radius of several centimeters in some cases - not good news for making compact optical circuits. Sandia's technology has huge potential for overcoming this limitation. In a separate but related development, scientists at Kyoto University in Japan have made the integrated equivalent of an optical add-drop multiplexer – though at the moment it only performs the drop function. The device is based on microcavities, which select a specific frequency of light from a waveguide (as a fiber Bragg grating does) and spew it out of the top surface. But unlike Bragg gratings, which need to be several centimeters in length, the microcavity is less than one micrometer wide. The number of manufacturers working on optical integration is an indicator of how important a direction this is considered to be. And there are many. However, most of these companies – for example Bookham Technologies Ltd. and the newest, Sparkolor Corp. - take components made in separate processes and then glue them onto a substrate. Intense Photonics Ltd. and a few others are attempting to make monolithic chips, but they're at a very early stage, and the number of components they can integrate is small. Making large numbers of components at the same time on the same chip will require a radically different approach. In that sense, the integrated optics industry is at the same stage that the electronics industry was 30 years ago, when people thought that six transistors was the most they would ever squeeze onto a chip. I believe I stole much of the above explanation from the Economist.

photonic ethernet A high-speed networking technology based on Polymer Optical Fiber (POF) cabling that can deliver gigabit networking speeds for a fraction of the cost of conventional fiber cable, even though the new POF has the same optical characteristics as glass. With this new technology, Gigabit Ethernet can be delivered to the desktop cost-effectively – about $200 per port (Summer of 1998) – making it potentially interesting for bandwidth-constricted workgroups.

photonic layer The lowest of four layers of Sonet capability, which specifies the kind of fiber to be used including sensitivity and laser type. See SONET.

photonic switch A switch which switches photonics, or light signals. See Lambda Switch and Photonic Cystals.

photonics The technology that uses light particles (photons) to carry information over hair-thin fibers of very pure glass. See also Photonic Cystals.

photophone In 1880 (four years after he invented the telephone) Alexander Graham Bell and his assistant, Charles Tainter, invented the photophone, which Bell felt was his greatest invention. In fact, Bell was so proud of it that he wanted to name his daughter Photophone. A cooler head prevailed, with that head sitting on his wife's shoulders. Consisting of a set of specially ground and shaped mirrors, and associated electrical gear, the photophone was capable of voice transmission over short distances, using sunlight. It was, in fact, the first optical transmission system, preceding fiber optics by nearly 100 years. It also was highly impractical, relying on fragile mirrors and sunny days. The Nazis experimented with a variation on the theme for application in W.W.II tank warfare – the results were not positive. See also Bell, Alexander Graham.

phototransistor A transistor that detects light and amplifies the resulting electrical signal. Light falling on the base-collector junction generates a current, which is amplified internally.

photovoltaics Photovoltaics is the name of the science that uses a semiconductive device to convert sunlight into electricity. The photovoltaic effect was first discovered by a French Physicists by the name of Edmond Becquerel in 1839. It is the only area of work that Einstein received a Nobel Prize in. It was commercially developed in the 1950s but did not expand into a significant industry until the 1990s.

photovoltaic effect Using light to produce electricity. Shine light on a device, typically a "cell." If the device produces electricity, that's called the photovoltaic effect.

PHP 1. Personal Handy Phone. Japan's standard for digital cordless phones.

2. Initially known as Personal Home Page Tools and now officially called PHP: Hypertext Preprocessor, PHP was developed in late 1994 by Rasmus Lerdorf. Its aim is to make it easy to write dynamic Web pages. PHP code is embedded into standard HTML Web pages. www.informationweek.com/769/devlop.htm

phreak A phone phreak (also known as a Phreaker) is to the phone network community what the original hackers were to the computer revolution. These were the hobbyists who couldn't get enough information about the telephone from reading published works or taking one apart; rather they learned how the touchtone frequencies worked to route calls. Granted, one aspect of what they did was making illegal phone calls, but the larger picture was a hunger for information that they couldn't get elsewhere. See also Cracker, Hacker, Phone Phreak, Script Kiddies, and Sneaker.

PHS Personal Handyphone System, or Personal Handyphone Service. Also known as PDC (Personal Digital Cellular), PHS previously was known as PHP (Personal HandyPhone) and Japanese Digital Cellular. PHS is the Japanese version of the U.S.'s PCS (Personal Communications Service, with two key differences. It's not as powerful as PCS. You can't use a PHS phone in a rapidly moving vehicle, since there is no cell-handoff, i.e. it won't move you from one cell to another. And thus, if you move outside your cell with PHS, you lose connection. PHS is a perfect mobile phone for pedestrians in high density cities like Tokyo, as long as they don't move around a lot during the course of a call.

PHY PHYsical, as in physical specifications. OSI Physical Layer: The physical layer provides for transmission of cells over a physical medium connecting two ATM devices. This physical layer is comprised of two sublayers: the PMD Physical Medium Dependent sublayer, and the TC Transmission Convergence sublayer. See Physical Layer, PMD and TC.

physical address 1.A number of digits which identifies the physical location of a communications channel or port within a system. In the old Rolm CBX, for instance, the number was in the form of xxyyzz, where xx=shelf, yy=slot and zz=channel.

2. The address where something physically resides. A physical addresses is translated from a logical address. Allow me to illustrate. When someone dials your telephone number, they are dialing a logical address; in other words, the series of numbers means nothing until they are translated into a physical address. The physical address is the port to which your local loop is connected to which your telephone is connected. Similarly, your postal address is a logical address. It has meaning only when translated by the post office into the plot of earth on which your house sits. A logical address, on the other hand and just to confuse you, may have no fixed physical address. For example, your e-mail address has no fixed physical address. Rather, it is translated into an IP (Internet Protocol) address which is associated with your e-mail server, which can be moved from place to place. Ultimately, your e-mail address actually is associated with you, and you and your computer can move all over the world without losing access to your e-mail. Rather, you gain access to your e-mail by going on a network connected to the Internet – e.g. dialing a telephone number (logical address) which connects you to your e-mail server which has a physical address which can change as the server is moved from one location to another.

physical change The modification of an existing circuit, dedicated access line, or port, at the request of the customer, requiring some physical change or determination.

physical channel A Bluetooth term. A synchronized Bluetooth baseband-compliant RF hoping sequence. Physical link A Baseband level association between two devices established using paging. A physical link comprises a sequence of transmission slots on a physical channel alternating between master and slave transmission slots.

physical colocation An incumbent local exchange carrier (ILEC) provides space within the building housing its central office to other phone companies (e.g. CLECs), to interconnect companies and/or to users to place their equipment. Typically it's done to connect circuits – transmission or switching – to the phone company's copper local loops. The interconnector (i.e. the company placing the equipment) installs, maintains, and repairs its own equipment, while the LEC provides power, environmental conditioning, and conduit and riser space for the interconnector's cable. See Colocation.

physical connection The full-duplex physical layer association between adjacent PHYs in an FDDI ring.

physical delivery Delivery of a message in physical form through a Physical Delivery System; for example, delivery of a letter through the U.S. Postal Service. This term is used in X.400.

physical delivery address component An X.400 address component that describes how to physically deliver a message. For example, the name and mail stop to hand deliver a message after it is printed. The concept is that the X.400 address would cause a message to be printed on a printer and an individual would complete the hand delivery.

physical delivery office name Standard attribute of a Postal O/R (Original/Recipient) Address, in the context of physical delivery, specifying the name of the city, town, etc., where the physical delivery is to be accomplished. An X.400 term.

physical delivery office number Standard attribute in a Postal O/R (Originator/Recipient) Address that distinguishes between more than one physical delivery office within a city, etc. An X.400 term.

physical delivery organization name A free form name of the addressed entity in the postal address, taking into account the specified limitations in length. An X.400 term.

physical delivery personal name In a postal address a free form name of the addressed individual containing the family name and optionally the given name(s), the initial(s), title(s) and generation qualifier, taking into account the specified limitations in length. An X.400 term.

physical delivery service The service provided by a Physical Delivery System. An X.400 term.

physical delivery service name Standard attribute of a Postal O/R (Original/Recipient) Address in the form of the name of the service in the country electronically receiving the message on behalf of the physical delivery service. An X.400 term.

physical formatting The second step in structuring a hard drive so that you may write to it. Physical formatting follows partitioning.

Physical Interface Card PIC. An intelligent line card that provides a high-speed and secure processing capability beyond the typical connectivity of first generation router cards. This robust line card basically connects a communications channel or circuit to multi-port switching equipment. This term is used by Riverstone Networks and Juniper Networks and will most likely pick up popularity when competitors like Cisco also adopt PIC as a more robust term than Network Interface Card (NIC) or line card.

physical layer 1. The OSI model defines Layer 1 as the Physical Layer and as including all electrical and mechanical aspects relating to the connection of a device to a transmission medium, such as the connection of a workstation to a LAN. Included at this layer are issues specific to the manner in which a device gains physical access to the medium and how it goes about putting bits on the wire or extracting bits from the wire. As the lowest level of network processing, below the Link Layer, the Physical Layer deals with issues such as volts, amps, and pin configurations and handshaking procedures. Communications hardware (e.g., NICs and MAUs) and software drivers are specified at the Physical Layer.

2. The ATM Physical Layer (PHY) loosely corresponds with the OSI version. In the ATM world, Physical Layer functionality is discussed in terms of the Physical Medium sublayer (PM) and the Transmission Convergence (TC) sublayer. The implementation of the ATM Physical Layer is addressed in the ATM Forum's UNI (User Network Interface) specifications.

physical layer connection An association established by the PHY (OSI Physical Layer) between two or more ATM entities. A PHY connection consists of the concatenation of PHY links in order to provide an end-to-end transfer capability to PHY SAPs.

physical layer medium dependent The Physical Layer sublayer that defines the media dependent portion of the Physical Layer in FDDI. Items defined by PMD include transmit and receive power levels, connector requirements, and fiber optic cable requirements.

physical layer protocol The Physical Layer sublayer that defines the media independent portion of the Physical Layer in FDDI. Items defined by the PMD include transmit and receive power levels, connector requirements, and fiber optic cable requirements.

physical link A real link which attaches two switching systems.

Physical Markup Language PML. An Auto-ID Center-designed method of describing products in a way computers can understand. PML is based on the widely accepted eXtensible Markup Language used to share data over the Internet in a format all computers can use.

physical media Any means in the world for transferring signals between OSI systems. Considered to be outside the OSI Model, and therefore sometimes referred to as "Layer 0." The physical connector to the media can be considered as defining the bottom interface of the Physical Layer, i.e. Layer 1 of the OSI Reference Model.

Physical Medium PM. In ATM terms, the Physical Medium sublayer is the dimension of the Physical Layer (PHY) which specifies the physical and electrical/optical interfaces with the physical media. See also Physical Layer and Transmission Convergence.

physical rendition The transformation of an MHS (Message Handling System) message to a physical message (e.g., by printing the message on paper and enclosing it in a paper envelope). An X.400 term.

physical security Ways to stop someone from gaining physical access to your stuff. Methods include locks, security personnel and guard dogs.

Physical Signaling Sublayer PLS. In a LAN or MAN system, that portion of the OSI Physical Layer that interfaces with the medium access control sublayer and performs bit symbol encoding and transmission, bit symbol reception and decoding, and optional isolation functions.

physical slots Slots that are available to cards in a card shelf.

physical topology The actual arrangement of cables and hardware that comprise a network. In other words, the actual physical appearance of a network. Typical physical topologies in the LAN world, for instance, include Bus, Ring and Star. The Physical Topology may differ significantly from the Logical Topology.

Physical Unit PU. In IBM's SNA, the component that manages and monitors the resources of a node, such as attached links and adjacent link stations. PU types follow the same classification as node types.

Physical Unit Control Point PUCP. In SNA, the component that provides a subset of the system-services control point (SSCP) within a node. Types 1, 2 and 4 nodes contain a PUCP, while a Type 5 (host) contains a SSCP.

PHz Petahertz (10 to the 15th power hertz). See also Spectrum Designation of Frequency.

PI 1. See Power Influence.

2. Presentation Indicator. A two-bit field in the Calling Party Number (CPN) subfield of the Initial Address Message (IAM). In an ISDN network, the IAM is part of the call set-up protocol. The PI indicates to the terminating switch whether it should pass to the called party the telephone number of the calling party, which telephone number is contained in the CPN. See also CPN, IAM, and ISDN.

3. pi. The constant equal to the ratio of the circumference of a circle to its diameter, which is approximately 3.141593.

PIA Personal Information Appliance. A name for a product most people call a Personal Digital Assistant (PDA).

PIAFS PIAFS is a V.110-like protocol for use over PHS, a digital cell phone network found in Japan.

PIC 1. See Primary Interexchange Carrier.

2. Plastic Insulated Conductor. A metallic cabling system in which the individual conductors are covered with an extruded coating of plastic. Virtually all cabling systems fall into this category. See also Icky PIC.

3. Also an imaging term. Picture Image Compression. Intel-DVI Technology's on-line still image compression algorithm. See DVI.

4. Personal Intelligent Communicator. A General Magic term for a product most other people called a Personal Digital Assistant. See PDA.

5. Point In Call. See Basic Call State Model.

6. Programmable Interrupt Controller. A chip or device that prioritizes interrupt requests generated by keyboards, serial ports, and other devices and passes them on to the CPU in PC in order of highest priority. See also IRQ.

PIC CARE Primary Interexchange Carrier Customer Account Record Exchange. PIC CARE records are exchanged between the LEC (Local Exchange Carrier) and the IXC (Interexchange Carrier) at the time an account is provisioned. PIC CARE information includes the identity of the IXC, and the type of customer line (e.g., residential line, business line, or PBX trunk) in use. PIC CARE records also contain a "jurisdiction field" which identifies the kind of long distance calls (i.e.; only interLATA; interLATA and/or international; or interLATA, intraLATA and international) carried by the IXC for the customer.

PIC Freeze Pre-subscribed Interexchange Carrier Freeze. A PIC Freeze is when a customer tells his local phone company (LEC – Local Exchange Carrier) that nobody is allowed to change his preference on which long distance company he will use. A PIC free is designed to prevent which is unauthorized changing of your long distance telephone carrier. See also PICC, Jamming and Slamming.

Picasso Porn The semi-scrambled transmissions from adult cable channels that can sometimes be seen and heard by nonsubscribers. This definition courtesy Wired Magazine. According to Ray Horak, my Contributing Editor, the scrambling technique used by CATV operators over analog coaxial cable systems (which they mostly are) simply involves "twisting" the audio and video frequencies; thereby, you are able to hear and view the adult channel only if you have paid for the service and, therefore, have a converter box that can unscramble the signal. However, and according to Ray, you'll notice that the picture comes in clear when the actors stop moaning and the obnoxious music stops.

PICC Primary Interexchange Carrier Charge, also known as Pre-subscribed Interexchange Carrier Charge. The FCC-mandated (May 1997) flat-rate charge which applies to pre-subscribed IXCs connecting to the end user through LEC facilities. The PICC applies first to primary lines; to the extent that PICC charges, in combination with the SLC (Subscriber Line Charge) and the monthly tariff line charge, are insufficient to provide the LEC with full recovery of the costs of the local loop, a lower PICC also may apply to non-primary (i.e., secondary) residential lines and multi-line business lines. The PICC became effective in 1998, and can either increase or decrease over time. While the LEC bills the end user directly for the SLC, it bills the IXC for the PICC. The IXCs are free to recover the PICC from end users; AT&T, for instance, has imposed a Carrier Line Charge on its end users in order to recover this cost. This charge was partly eliminated in May 2000 by the FCC. See CALLS Proposal. See also Access Charge and SLC.

Pick 1. The computer operating system of Pick Systems, Inc. Pick is a neat operating system that has only caught on in a very small way. Steve Lamb of Peet's Coffee & Tea, Inc. wrote me, "Those of us who have actually become familiar with it really do love it, and the Pick community, though small, is very much a family-like group. We occasionally ponder the question of what would have happened if Pick Systems had had the same marketing approach as someone like IBM or Microsoft." www.picksys.com.

2. A cabling term. It's the distance between two adjacent crossover points of braid filaments. The measurement in picks per inch indicates the degree of coverage.

pick-and-place The manufacturing operation in which components are selected and placed in the correct position on a substrate for the purpose of interconnection to the substrate. This is most commonly done with a programmable machine equipped with a robot arm.

pick cable Outside telephone cable with plastic insulated pairs.

PICMG PCI Industrial Computer Manufacturers Group. An international consortium of vendors of industrial computer products, PICMG was formed to develop specifications for PCI (Peripheral Component Interconnect) based systems and boards for use in industrial and telecommunications computing applications. The electrical and mechanical specifications of PICMG enable CPU boards and backplanes from different manufacturers of industrial-grade computers to be interchangeable. PICMG specifications include CompactPCI, rackmount applications and PCI for passive backplane, standard format cards. See CompactPCI and PCI. www.picmg.org

pickup Means you can answer a call from your phone. There all sorts of "pickups." The most common is GROUP PICKUP. Here you are part of a group and you can answer – from your phone – the call of anybody in that group, usually by punching a digit or a button or two. There's also NIGHT PICKUP, which typically allows anyone to answer an incoming call after hours, again by punching down a digit or a button or two. In TELECONNECT Magazine, we have one GROUP PICKUP. Everybody in the company belongs to that Group and everyone can answer everyone else's phone. We believe this simplifies things. It also allows anyone, anywhere, from any phone, to play telephone attendant, answering any and all incoming calls and transferring them through the system.

pickup group Imagine you're on a phone behind a PBX. Imagine you work in accounting, a group of five people. You can program most PBXs such that a call to your phone could be answered by anyone else in your group, and vice versa, you could answer someone else's ringing phone in your group. When you set your PBX up, you need to program which Pickup Groups which phones belong in.

pickup pattern A determination of the directions from which a microphone is sensitive to sound waves. It varies with the mike element and mike design. The two most common pickup patterns are omni- and uni-directional.

pico Prefix meaning one-trillionth, or one-millionth of a millionth. A pico is ten to the minus 12. See Atto, Femto and Nanosecond.

picocell A wireless base station with extremely low output power designed to cover an extremely small area, such as one floor of an office building.

picofarad One-trillionth of a farad. A unit of capacitance usually used to designate capacitance unbalance between pairs and capacitance unbalance of the two wires of a pair to ground.

piconet A Bluetooth term. A collection of devices connected via Bluetooth technology in an ad hoc fashion. A piconet starts with two connected devices, such as a portable PC and cellular phone, and may grow to eight connected devices. All Bluetooth devices are peer units and have identical implementations. However, when establishing a piconet, one unit will act as a master and the other(s) as slave(s) for the duration of the piconet connection. All devices hare the same physical channel defined by the master device parameters (clock and BD_ADDR). Here's another explanation: a piconet is a collection of devices connected in an ad hoc fashion via Bluetooth RF (Radio Frequency) technology. The connection may begin with any two devices, such as a wireless PC and a cellular telephone, and may grow to as many as eight devices, including PDAs (Personal Digital Assistants) and anything else you might imagine. While all piconet devices are peers (i.e., equals) in terms of connectivity, one device acts as the master, and the others as slaves, for the duration of the piconet connection. A scatternet is formed when multiple, non-synchronized piconets are linked. See also Bluetooth and Pico.

picosecond One-millionth of a millionth of a second. A picosecond is ten to the minus 12 of a second. One picosecond – a trillionth of a second – is a spot of time from the domain of molecules. Light, traveling for one picosecond, would barely make it across the period at the end of this sentence. Only with a laser that generates picosecond light pulses can scientists freeze the short-duration motion of molecules and produce images of what goes on at the molecular level. Used in this way, the picosecond laser is comparable to a strobe, which can freeze the motion of a sprinter's stride in time-lapse photography. See Nanosecond.

PICS 1. Protocol Implementation Conformance Statement: A statement made by the supplier of an implementation or system stating which capabilities have been implemented for a given protocol.

2. Product Inventory Control System.

3. Plug-In Inventory Control System. Plug-In's are circuit cards which fit into central office equipment and which control various aspects of transmission and carrier circuit functionality and usage. See also PICS/DCPR.

4. A Macintosh-specific "multimedia" format for exchanging animation sequences (developed in 1988 by Macromind and others). PICS assembles several PICT files (frames) and combines them into one file.

PICS/DCPR The Plug-in Inventory Control System/Detailed Continuing Property Records (PICS/DCPR) system maintains records of plug-in equipment. It is an OSS (Operations Support System) that was developed many years ago by the Bell System. PICS/DCPR interfaces with TIRKS (Trunks Integrated Records Keeping System) in the circuit-provisioning process. PICS/DCPR and TIRKS are still used by the RBOCs. See also TIRKS.

PICT Picture Format. Developed by Apple in 1984 as the standard format for storing and exchanging files. PICT2 (1987) supports eight-bit color and gray scale. The PICT format now is widely used among Macintosh graphics and page-layout applications as an intermediary file format for transferring files between applications. The PICT format supports RGB files with a single alpha channel, and indexed-colour, grayscale, and bitmap files without alpha channels. The PICT format is especially effective at compressing images with large areas of solid colour."

picture element See Pixel.

picturephone AT&T's trademark for a video telephone that permitted the user to see as well as talk with the person at the distant end. AT&T introduced it at the 1964 World's Fair in Flushing Meadow, Queens, New York City. The device had a camera mounted on the top and a 5.25" x 4.75" screen. Audio signals were transmitted separately from video signals and the system could not use the public switched telephone network. It needed a transmission bandwidth of 6.3 Mbps and no one wanted to pay the price for the service. It never got off the ground. AT&T picked up on the name Picturephone and came out with an offering called Picturephone Meeting Service which provided full video teleconferencing. It was available through rented rooms or through equipment sold or rented to corporations. It also didn't do too well since the service was expensive and no one wanted to spend the time traveling to the few and far between conferencing room. AT&T has abandoned Picturephone, the product, and has closed down the Picturephone Meeting Service rooms it rented to corporations.

picturephone meeting service An AT&T service once provided under experimental tariff. It combined TV techniques with voice transmission. PMS is usually only available between telephone company-located picturephone centers. Most (we think all) have now been closed down. The venture was losing too much money.

PID Protocol Identifier. A field in the Call User Data included in the Call Request Packet sent to the ISP host for POS terminal initiated calls.

PIECE Productivity, Information, Education, Creativity, Entertainment. Microsoft's trick for remembering the big five multimedia computing applications.

piepser The German word for beeper. Also the name of a small beeper made by Swatch and sold by BellSouth.

piesio The Greek prefix meaning near.

piezo-electric crystal A type of crystal which, when subjected to mechanical stress, generates current; or which, when subjected to varying electrical stresses, generates mechanical movement. Most familiar type is Rochelle Salts crystal. An old radio term.

PIF 1. Personal Communications Services Industry Forum.

2. Program Information File, a binary file, which contains information about how Windows should run an MS-DOS application, such as how much memory it needs, the path to the executable file, and whether the window in which the program is run closes automatically when the program terminates.

3. Public Inspection File. A set of documents which must be maintained by every cable television system at a convenient location in the cable community: at the system's office, or at some other convenient location such as an attorney's office or the local public library. The contents of the PIF are specified in Sections 76.302 and 75.305 of the FCC Rules. Any member of the public:

- Has the right to see the PIF on request.
- Has the right to request accommodations where the PIF can be reviewed without disturbance.
- Has the right to request photocopies of any or all documents in the file at a reasonable cost.

The Public Inspection File should be kept separate from all other files, both physically and operationally. This will reduce the chance of inadvertently releasing to the public any information which is not specifically required.

pig Coin banks are often shaped like pigs. Why? Long ago, dishes and cookware in Europe were made of a dense, orange clay called "pygg." When people saved coins in jars made of this clay, the jars became known as "pygg banks." When an English potter misunderstood the word, he made a bank that resembled a pig. And it caught on.

piggyback attack Form of active wiretapping in which the attacker gains access to a system via intervals of inactivity in another user's legitimate communication connection. Sometimes called a "between-the-lines" attack.

piggy back data slurp Imagine a data communications connection between two distant computers. Somewhere along the line, someone has attached a terminal or communicating computer and begun to capture the data as it flows across the line. In short, someone had slurped out data on a piggyback terminal or communicating computer. A piggyback data slurp could continue forever, without either party being aware their data and their conversations were being stolen – so long as the piggyback terminal didn't let its presence be known. The piggyback terminal, however, may insert itself into the conversation, pretending to send back authentic communications, thus misleading one or both of the parties. You could call this active deception or proactive espionage.

piggyback A cellular term. A microcell sometimes "piggybacks" on a larger macrocell, depending on the macrocell for some of its operational intelligence. In other words, the microcell acts as a semi-intelligent remote, relying on the more intelligent, centralized macrocell for higher-level intelligence to control its more complex operations.

piggyback board Another name for a daughterboard on a card inside a PC.

piggybacking A technique used at the data link or transport layer in a layered network architecture that allows for transmission acknowledgments to be carried in transmission frames received from the destination.

pigtail 1. Multiple pieces of short cable with single circuit connectors connected to a multi-conductor cable. See Octopus.

2. A short, permanently attached piece of optical fiber used to link the transmitter and receiver to the transmission fiber.

pigtail antenna The standard cellular antenna for a car. The term "pigtail" refers to the spring-like section in the lower third of the antenna, the phasing coil.

Pilot An in-band reference signal transmitted by an earth station and used to track a satellite.

pilot error See Cockpit Problem.

pilot installation A small installation that may precede a large, expensive planned installation, used to show the benefits in order to get a customer's commitment for a larger installation, or to identify and solve problems before installing an enterprise-wide system.

pilot number Identifies a Hunt Group or Distribution Group. See also Distribution Group.

pilot-make-busy circuit A circuit arrangement by which trunks provided over a carrier system are made busy to the switching equipment in the event of carrier system failure, or during a fade of the radio system.

PIM 1. Protocol Independent Multicast. Multicasting is the process of sending one packet to many people without having to duplicate the packet at the source for each recipient. Multicast is often used for multimedia transmissions such as streaming video or sound. Many multicast routing protocols like Distance Vector Multicast Routing Protocol (DVMRP) and Multicast Extensions to OSPF (MOSPF) were designed for a network where bandwidth is plentiful and all the multicast subscribers are in the same region of the network. However, networks in the real world have a wide variety of configurations, and more often than not, multicast subscribers are scattered across the network in small groups with a limited amount of available network resources. Since most multicast routing protocols were created based on assumed ideal network conditions, they are often inefficient with network bandwidth and resources. Protocol Independent Multicast (PIM), however, makes better use of network bandwidth and adapts to network conditions and subscriber distribution for optimized multicast session performance. PIM follows standard multicast protocol procedures and uses discovery techniques to find the optimal multicast path between the multicast server and the subscribers. PIM is also very scalable, so it operates on small or very large networks. Since it doesn't rely on any single unicast routing protocol, it operates over any layer-2 network configuration. Because PIM was designed to be a simple, effective protocol, it lacks the sophistication of more complex routing protocols like DVMRP and MOSPF. PIM has to perform more operations to accomplish the same task, which results in an increased load for router processors and packet duplication. However, when compared to other protocols, PIM is an advance in multicast technologies. PIM has two distinct modes to handle varying network conditions: First, PIM sparse mode (PIM-SM) and second, PIM dense mode. (PIM-DM) Sparse mode assumes that the multicast subscribers are far apart, are not grouped together, or have limited bandwidth, while dense mode assumes subscribers are close together on the network and have plenty of bandwidth for the transmissions. In each situation PIM uses the mode best suited to optimize its performance based on the configuration of the subscribers of the multicast stream. See also Multicast.

2. Personal Information Manager. A specialized form of software used by individuals and groups for keeping track of contacts (including addresses and phone numbers), appointments, project schedules, to-do lists, reminder notes, anniversaries, etc. PIMs are also called contact managers. Contemporary PIMs also are e-mail clients that allow such information to be shared with friends and coworkers. PIMs may be Web-based, as is the case with My Personal Diary from Lycos. Examples of PIMs include Maximizer, My Personal Diary (Lycos), TeleMagic, Organizer (Lotus), Outlook (Microsoft), and SideKick (Borland).

3. Plug In ISDN Module. A module in the form of a printed circuit board (PCB) that plugs into a data switch or router to make it ISDN-compatible.

4. Presence and Instant Messaging.

PIM Dense Mode PIM DM. Protocol Independent Multicast Dense Mode. A multi-cast protocol similar to DVMRP in that it uses Reverse Path Forwarding but does not require any particular unicast protocol. See Protocol Independent Multicast Dense Mode for a longer explanation.

PIM sparse mode A multicast protocol that works by defining a rendezvous point that is common to both sender and receiver. Sender and receiver initiate communication at the rendezvous point, and when flow begins it occurs over an optimized path. See PIM for a longer explanation.

PiMF Pairs in Metal Foil. See SSTP.

PIN 1. Procedure Interrupt Negative. A fax term.

2. Photo INtrinsic diode. Also known as PIN Diode. A type of photodetector used to sense lightwave energy and then to convert it into electrical signals. PINs are matched with LEDs (Light Emitting Diodes) in fiber optic transmission systems of relatively low capacity (i.e., less than 500 Mbps for); currently, such systems are deployed for interconnection of hubs, switches and routers in a LAN environment. See also LED. See APD and Laser Diode for descriptions of the components used in high-bandwidth fiber optic systems.

3. Personal Identification Number. A code used by a mobile telephone subscriber in conjunction with a SIM card to complete a call. A code used by a credit card user. A code used by an ATM card user, etc. In short, a code to protect you against fraud. A code that you remember, which you don't write down anywhere and which theoretically can't fall into the wrong hands.

4. A Bluetooth term. Personal Identification Number. The Bluetooth PIN is used to authenticate two devices that have not previously exchanged link key. By exchanging a PIN, the devices create a trusted relationship. The PIN is used in the pairing procedure to generate the initial link that is used for further identification. See also Piconet.

PIN code Personal Identity Number code. A code used by a mobile telephone subscriber in conjunction with a SIM card to complete a call. See PIN.

pin diode A device used to convert optical signals to electrical signals in a receiver. It is a photodiode made with an intrinsic layer of undoped material between doped P and N layers and used as a lightwave detector.

pinning To make a network file or folder available for offline use.

PIN number Personal Identification Number. A group of characters entered as a secret code to gain access to a computer system, such as the one that completes long distance calls. See Personal Identification Number.

pin photodiode An optical detector that converts light into electricity. This type is the typical diode used in a fiber optic receiver.

PIN(BB) A Bluetooth term. The PIN used on the baseband level. The PIN(BB) is used by the baseband mechanism for calculating the initialization key during the paring procedure. (128 bits)

PIN(UI) A Bluetooth term. The PIN used on the user interface level. The PIN(UI) is the character representation of the PIN that is entered on the UI level.

pincushion distortion When a video screen is distorted – with the top, bottom and sides pushing in – the screen is said to be suffering pincushion distortion.

pine Program for Internet News and Email. Pine is developed at the University of Washington (there are lots of pine trees in Washington state) as an easy to use, character-based e-mail client for use with UNIX and MS-DOS host computers. Pine was based on an older e-mail program call Elm (there also are lots of elm trees in Washington state), but since has evolved well beyond those simple origins. Pine sometimes is decoded as "Pine Is No longer Elm."

ping 1. To ping a computer basically means that you send out a small amount of information, or packet, to another computer connected over a network – the Internet or a private network or both. Then you wait for a response from the other computer. If you get a response, you know the computer exists and your network is working. Ping is a actually a software program written by Mike Muuss to test whether a particular network destination on the Internet is online (i.e., working) by bouncing a "signal" off a specified IP destination address. Every PC these days comes with ping software. Pinging is among the more useful tools I have on my computer. It allows me to pinpoint when something is wrong with my network. Let's say I'm at home and I can't get to www.Microsoft.com. Whose fault is it? The first thing I'll do is to go into MS-DOS and ping 10.1.25.1. (That's not the real number, but it's close.) If I get a response, that means my firewall is working. Then I ping the server of my local DSL provider (called Prism), which is 199.105.128.148. If I get a response, I then ping 199.105.128.1, which gets me just out of their system. If I get a response, I then ping globix.com, whose server is in my town, New York City. At some point I won't get a response. At that point, I know which part of my system or which part of Prism or which part of the Internet is down (or I know they're using a firewall, which is programmed not to respond). Try it yourself. Go to DOS from Windows. Then type ping www.HarryNewton.

com. Neat. Notice how it tells you how long it takes. Notice that it also tells you what www. HarryNewton.com's numeric domain address is. Hence if you want to find out a web site's numeric domain, ping it.

According to Mr. Muuss, the name "ping" refers to the sound of a SONAR return (used by submaries, for example). It seems as though Mr. Muuss had done a good deal of college work on the modeling of SONAR and RADAR systems, both of which are based on the concept of transmitting a signal, and seeing how long it takes that signal to complete a round trip from the transmitter to the target device and back. Some years later, Mr. David Mills apparently made an acronym out of "ping;" according to his definition, "PING" means "Packet InterNet Groper." PING is an example of a backronym, a word that often is interpreted as an acronym, although it was not so intended.

In any event, the ping utility makes use of ICMP (Internet Control Message Protocol) packets. The pinging device sends ICMP Echo Request Packets to one or more target IP (Internet Protocol) addresses. If the target address is active, it has a working IP protocol stack and is online (i.e., turned on, and networked). Assuming that is so, the target device will return an ICMP Echo Reply Packet. As the IP header contains a Time-To-Live (TTL) counter which is decremented by timers in the intermediate switches and routers, the time of the round-trip signal can be calculated, which yields some information about the performance level that can be expected for a client/server application across a network. If no ICMP packet is returned, the target device is either down or unreachable, or the performance of the networked application is so slow that the packets "timed out" and "died" before they could be accepted and returned. The term is often used as verb: "Ping host X to see if it is up!" Ping is useful for testing and debugging networks. By the way, you can ping your very own PC. Ping 127.0.0.1. See also Acronym.

2. A graphics file format ending in .png, which was developed to overcome deficiencies of .gif and .jpeg file formats, which are commonly used on the Web. Ping allows 16-, 24-, and 32-bit images, providing for better color depth. The downside is that higher-quality images require more storage and involve longer download times.

pingback A pingback is a notification letting a blog or website know that it has been referenced (or linked to) by another blog. Pingbacks usually include a short excerpt of the post containing the link, along with a link to the website. This definition from blogossary.com.

Ping of Death The Ping of Death is a denial-of-service attack that crashes servers by sending invalid IP ping packets. The first version, discovered in early 1997, affected mostly Macintosh and Unix servers, not NT servers. The problem faded after vendors posted patches to their servers, and Microsoft posted a new PING.EXE file that prevented users from generating the invalid pings. The resurrected Ping of Death, however, takes a slightly different approach: According to PC Week, it modifies the header of the ping to indicate that there is more data in the packet than there actually is. "Effectively, the server hangs because the IP stack is waiting for the rest of the data," said Mike Nash, Microsoft's director of server marketing in 1997.

ping pong 1. A method of getting full duplex data transmission over a two wire circuit by rapidly alternating the transmission direction. See Ping Ponging.

2. A disruptive phenomenon that occurs in digital cellular networks when the cell phone repeatedly reselects two cell sites of approximately equal strength. This problem is overcome through the use of a buffer area known as a "hysteresis."

3. A disruptive phenomenon that occurs in digital cellular networks when both transmission and reception take place over the same frequency channel, although in separate time slots. This problem is overcome through the use of separate frequency channels.

ping ponging Routing that causes a packet to bounce back and forth between two modes. See Ping Pong.

ping scan A ping scan looks for machines that are responding to ICMP echo requests ("pings"). **ping sweep** An exploit that sends ICMP echo requests ("pings") to a range of IP addresses, for the purpose of finding machines that can be probed for vulnerabilities.

pink Adult audiotext, i.e. dirty talking over the phone for money.

pink flamingos See Flamingos, Pink.

pink noise Noise in which power distribution is logarithmic through the spectrum, with an equal amount of power in each octave.

pink pages In Australia, at one stage, the list of businesses organized by industry, were printed on pink paper and called the Pink Pages. They are the equivalent of the North American "Yellow Pages."

pinouts Pin configurations for cabling. In other words, which pin connects to which cable. Not all pins are always connected. Not all cables always connected.

pinosecond One-trillionth of a second. One-millionth of a microsecond.

PINT PSTN and Internet Internetworking. The PINT WG (Working Group) of the IETF (Internet Engineering Task Force) addresses connection arrangements through which Internet applications can request and enrich PSTN (Public Switched Telephone Network) services. For example, PINT might include a Web-based Yellow Pages service, with the ability to launch a return call via the conventional PSTN based on a Web search-and-click process. In this example, the Internet, or other IP-based network, might be used to request the placement of the call, and the PSTN would execute the call. The call might be a return call (either voice or fax) from a call center with a Website, which might include a commercial vendor that might be requested to return a call regarding a product question or service issue, or a governmental agency such as a weather bureau that might be requested to call in the event of a serious weather alert. See also IETF, IP, and PSTN.

pip tone The tone that notifies you of a call waiting, assuming you subscribe to "call waiting." See Call Waiting.

pipe 1. A communications process within the operating system that acts as an interface between a computer's devices (keyboard, disk drives, memory, and so on) and an applications program. A pipe simplifies the development of application programs by "buffering" a program from the intricacies of the hardware or the software that controls the hardware; the application developer writes code to a single pipe, not to several individual devices. A pipe is also used for program-to-program communications and can be a connection between two processes so that the output from one immediately becomes the input for the other. Indicated by the | character.

2. A transmission facility. Pipe usually is used when discussing transmission bandwidth. For instance, fiber optics is a "big pipe," because it offers lots of bandwidth. See also Bandwidth.

3. Private Investment in Public Equities. When public companies become cheaper than private companies, some venture capital firms often invest in public companies. This is valled PIPE.

pipe fillers Content providers.

pipe holders See pipe owners.

pipe mount A mounting system used for mounting an antenna mast to a vent pipe on a roof.

pipe owners Carriers and other network operators who own and operate the telecommunications infrastructure. Also called pipe holders.

pipelining 1. Executing instructions by breaking them into component parts and processing them in parallel on separate processors. This reduces reduce cycle time and increases the computer's performance.

2. In imaging, pipelining also lets an imaging card start compressing and writing the image to disk while it is still being scanned.

3. In networking, pipelining is a technique used at the transport layer or data link layer in a layered network architecture that allows for the transmission of multiple frames without waiting to see if they are acknowledged on an individual basis. Each frame may have to be acknowledged later and in sequence, or a process of implied acknowledgment may be employed. Implied acknowledgment is a process whereby negative acknowledgment of a specific frame implies that all previously transmitted frames have been received correctly.

pipes The network infrastructure. See also dumb pipes.

PIR 1. Peak Information Rate. This word is used in context of an Ethernet interface to specify its performance in terms of peak number of frames transmitted per second.

2. Passive Infra-Red.

piracy Any impersonation, unauthorized browsing, falsification or theft of data or disruption of service or control information in a network.

PISD Planned In-Service Date. The date the your vendor quotes for installation of a new system or circuit. A due date, in other words. I think that if your vendor missed the PISD, you would be "PISD off."

PITA Abbreviation for "Pain In The Ass;" commonly used on E-mail and BBSs (Bulletin Board Systems).

pitch In flat cable, the nominal distance between the index edges of two adjacent conductors.

pitch control Variable control for increasing or decreasing the speed of a tape deck or turntable.

pistoning the movement of a connectorized fiber axially in and out of a ferrule end, often caused by changes in temperature.

PIU 1. Path Information Unit in SNA. 2. See Percent of Interstate Usage.

PIX Private Internet eXchange. It's a Cisco term for a family of their remote access routers with firewall capabilities.

pixel PIcture ELement. The smallest unit of area of a video screen image that can be turned on or off, or varied in intensity. The single point on a CRT display. The single point in a facsimile transmission. The image you see on your screen is the result of some pixels being on and others off. A pixel is the smallest part of the video screen that can be turned on or off or varied in intensity. It is one of the phosphor elements that coat the inside of a CRT tube. Pixels glow when struck by an electron beam. The number of pixels in the most common computer screen, a VGA monitor, is 640 x 480, with the first number (640) being the number of pixels in each horizontal row and the second number (480) the number of rows displayed. VGA stands for Video Graphics Array. Resolution (crispness and clarity of text and images) improves as the number of pixels displayed increases. If it's a color screen, a pixel is really three dots together, or clusters of red, green and blue – the triad of colors that, when energized, add up to white, or when the set is turned off, show as black. The phrase "picture element" was first used in 1927 in the magazine "Wireless World" writing about the mosaic of dots, or picture elements. See PEL.

pixel shim A small, usually invisible graphic used in an HTML document to create a page format. "I had to use a pixel shim to get the type to space correctly." A Wired Magazine term.

pixelation 1. Sometimes parts of a digital TV screen image become degraded, with the affected portions of the screen taking on the appearance of a jagged jumble of over-sized screen pixels. This is called pixelation, also known as macroblocking. The root cause of the problem is lost packets in the digital TV stream. Since each packet contains highly compressed data, the amount of screen information that is lost is a large multiple of the packet's payload size.

2. A technique used by cinematographers and stage managers to make human performers appear to move as it artificially animated. Using a stop-frame camera, the pixilator can distort and speed up the motion of actors.

PIXIT Protocol Implementation eXtra Information for Testing: A statement made by a supplier or implementor of an IUT (Implementation Under Testing) which contains information about the IUT and its testing environment which will enable a test laboratory to run an appropriate test suite against the IUT.

Pixrects Pixrects is the primary graphics programming interface in the SunView Window System from Sun Microsystems. It is replaced in the OpenWindows XView toolkit by the Pixwin interface, which is a thin layer on top of Xlib.

Pixwin Pixwin is the primary graphics programming interface in the XView toolkit from Sun Microsystems. Pixwin is a thin layer on top of Xlib.

pizza box 1. A wireless term referring to a phased array antenna. The device is small, square, and flat, resembling a pizza box. If you open the cover of the pizza box, you will find an array of little antenna gizmos where you normally would expect to find the pepperoni slices on a pizza. The pizza box mounts flat against the side of your building in a WLL (Wireless Local Loop) application, for instance. It also can mount on your rooftop in a satellite application. Hence, it is aesthetically pleasing. See also Phased Array and WLL.

2. A server the size of a pizza box.

PJ-327 A double RCA dipole plug for connecting a headset into a PBX console. When you buy a headset you need to specify – 4-pin modular jack (for connecting to a telephone) or two prong plug (for connecting to a PBX console).

PKE See Public Key Encryption.

PKI See Public Key Infrastructure.

PKM Programmable Key Module is an expansion module for a phone. Each expansion module adds a bunch speed dial buttons (in telecom a key is a switch or a button). Such modules typically come with 24, 48 or 654 speed dial buttons. The idea is to have zillions of speed dial buttons for your 64 closest business acquaintances. Such devices are often used in the brokerage business.

PL Private Line.

PL/1 Programming Language One (IBM).

place blog A blog whose content is focused on a neighborhood, city or region. Also called a placeblog.

place call There will be two types of calls: person calls and place calls. I make a place call when I call a phone number which ends in one designated, fixed RJ-11 jack attached to the wall or floor. I make a person call when I call a phone number which doesn't necessarily end on a fixed place. A typical person call might be a cellular or wireless phone call. It might also be a service like MCI's Personal 800 Service. The major characteristic of a person call is that I am calling a person and I don't know where that person is. As a result, the person I called might answer my call anywhere – from his car, from vacation home, from his wireless phone, etc.

placeblog A blog whose content is focused on a neighborhood, city or region. Also called a place blog.

placeshifting Ability for consumers to access their digital media, from any networked

device, anywhere in the world. This includes the home network.

Plain B Wire Connector Also called a B Connector or beans. A twisted-pair splicing connector that looks like a one-inch drinking straw. They have metal teeth inside them to pierce the vinyl insulation of the wire to make a good connection. Sometimes water-retardant jelly is sometimes place inside.

plaintext A message which is not encrypted. A message which is encrypted is called a ciphertext. See Clipper Chip.

plan A route to an end or objective usually achieved accidentally and often written in hindsight. Derived from the expression: "Most people spend more time planning their annual vacation than they do planning their careers."

plan B What happens if things don't work out as intended? Go to Plan B. No one knows where this expression came from.

plan file A file that lists anything you want others on the Internet to know about you. You place it in your home directory on your public-access site. Then, anybody who fingers (sees) you, will get to see this file.

planar array antenna A planar array antenna is designed for use at microwave frequencies. It resembles a double-sided printed circuit board. One side of the substrate carries an etched pattern of microstrip whilst the other is left completely metallic to act as a ground plane.

planar board IBM's new name for a motherboard in their new series of System/2 Personal Computers. A motherboard is the main board in a PC on which the main CPU, the main memory, the clock and sundry other things like serial and parallel ports are mounted. Other boards, i.e. graphics boards, are plugged into the motherboard. Thus the expression "motherboard." No one knows why IBM dropped the word. Maybe it was too risque? Maybe they included more on their motherboards in the System/2 series that they would no longer function as motherboards? Maybe a feminist group of mothers objected?

plane management An ATM term. As described in the ATM Protocol Reference Model, Plane Management is an element of the Management Plane. Plane Management acts on the management of the ATM switch, as a whole, without consideration of each of the various specific layers of the model. See ATM Protocol Reference Model for a graphic representation of the three-dimensional model.

planned communities Imagine a large, empty piece of land. Imagine a developer coming along and building roads, sewage, electricity, telecommunications and several hundred houses. Bingo, we have a planned community. Planned communities are important because of the communities' telecommunications needs which may be provided by a central group with exclusive rights to provide phone, cable and data services to the community. That exclusive right has value. That's why several companies are going after the right to provide telecom service to planned communities. See also MDUs.

plant A general term for all equipment used by a telephone company to provide telecommunications services. In the telecom business, plant comes in two variations – inside and outside plant. Inside is in a building. Outside is outside the building – on poles, in the ground. Several people have asked me if the phone industry invented the term. The answer is no. The term has been around since the middle of the nineteenth century. It was used to describe all the implements of production. The third edition of the Oxford English Dictionary defines plant as "The fixtures, implements, machinery, and apparatus used in carrying on any industrial process; the premises and fixtures of a business or (chiefly U.S.) of an institution; a place where an industrial process is carried on; also, a single machine or large piece of apparatus..." See Inside Plant, Outside Plant and Plant Hump.

plant hump A very friendly term for a craftsperson in a phone company – installer, splicer, underground cable guy – who works hard with outside phone equipment, often under adverse conditions and whose labors tend to be undervalued, particularly by the white collar, pencil pushers back at headquarters. In British slang, "hump" means to exert oneself. There is a badge of honor to being a "plant hump." These are the people who bring phone service to your door. The word plant hump is occasionally written as one word, i.e. Planthump. Other people who work for telephone companies are often called pencil pushers.

plant test numbers Virtually every 800 IN-WATS number has a plant test number. This is its equivalent seven digit local number. That number looks like a normal local seven digit number, with a standard three-digit central office exchange code and a four-digit extension. The purpose of plant test numbers is to allow the telephone company to test the local part of the incoming 800 number by simply dialing that number. For example, Miller Freeman, which published this dictionary has an 800 number – 800-LIBRARY (or 800-542-7279). The plant test number of the first line of that 800-LIBRARY group is 212-206-6870. The second line is 212-206-6871 and so on. It is valuable to know the plant test numbers of your incoming WATS lines so you can test the local loop part of those lines. The local loop part is the part which typically gives the most problem. It is, unfortunately, the only part of your 800 lines you can test yourself – unless you ask someone (or several people) to call you regularly on your 800 lines, just to test them. You can get plant test numbers out of your local and/or your long distance carrier. When they tell you those numbers are "not available," beg a little. They are available and you are entitled to them. Calling plant test numbers costs exactly what a normal long distance IN-WATS call on that line costs. So keep your test calls short. You should call your plant test numbers once a day.

planthump A colloquial word for a telephone company craftsperson. It derives from the term "plant," a telephone company word used to describe their "factory" – i.e. everything from their inside plant, their central office switch, to their outside plant, which includes wire strung on telephone poles. In British slang, "hump" means to exert oneself. Planthump is a term of endearment in the telephone industry. Definition courtesy, Steve Marcus, New York Telephone, now called Bell Atlantic. The term is often spelled as two words, i.e. plant hump.

PLAR Private Line, Automatic Ringdown. In telecommunications, leased voice circuit that connects two single instruments together. When either handset is lifted, the other instrument automatically rings.

plasma A low-density gas in which the individual atoms are ionized (and therefore charged), even though the total number of positive and negative charges is equal, maintaining an overall electrical neutrality. See Plasma Display.

Plasma Display Panels PDPs. Type of flat visual display device in which selected electrodes, part of a grid of crisscross electrodes in a gas-filled panel, are energized, causing the gas to be ionized and light to be emitted. Some computers use plasma displays. They're fabulous, and expensive. See Plasma Display Panels.

PDPs. Plasma gases composed of helium, neon and xenon are sandwiched into cells between two vertical glass plates. Bursts of electricity are applied between transparent electrodes attached to one pane of glass. These bursts causes the plasma gases to emit ultraviolet rays. This activates red, blue and green phosphor dots, which emit visible light and form pictures on the screen. The more common cathode ray tube technology (e.g. computer monitors) uses an electron gun to direct a beam that lights up phosphors on a screen. Directing that beam requires CRT sets to be deep, heavy and unwieldy.

Plasmatron The name Sony chose for a flat screen display it demoed at Comdex in the fall of 1995. The screen measured 20 inches diagonally. It was as bright as normal CRT screen, but was less than four inches wide. It is the beginning of flat screen entertainment screens for the home. It was spectacular.

plaster ring A metal or plastic plate that attaches to wallboard for the purpose of mounting a telecommunications outlet box.

Plastic Optic Fiber See POF.

plastic deformation A cabling term. Change in dimensions under load that is not recovered when the load is removed.

plastic roaming In Europe, GSM wireless cellphones operate on 900 MHz and 1800 MHz. By contrast, GSM operates at a frequency of 1900 MHz in the United States. This makes it impossible for Europeans to roam with their GSM phones in the United States and for American GSM users to roam with their GSM phones in Europe, unless they have a tri-band phone that supports all three frequencies. Plastic roaming, also known as SIM roaming, is one solution to the problem. Here's how it works. Upon arrival in the United States a European GSM user rents a 1900 MHz GSM mobile phone and replaces the phone's SIM card with the SIM card from his European GSM mobile phone. The user therefore keeps his European mobile number, keeps his mobile phone directory, and all of his mobile phone calls in the US are billed by his regular European GSM operator. American GSM users traveling to European can use plastic roaming too by replacing the SIM card in a rental 900 MHz or 1800 MHz phone with the SIM card from their American GSM phone. Before you do this, check the prices.

plasticizer A chemical agent added in compounding plastics to make them softer and more flexible.

Plat 1. An imaging term. When a CAD/CAM plotter prints a large drawing, it's called a plat.

2. A map or plan of a small piece of ground showing boundaries, area, remainder, ownership, access, and other pertinent information.

plate The anode in a vacuum tube, which collects the electrons emitted by the filament.

plate battery The source of E.M.F. connected in the plate circuit to give the plate element its positive charge.

plate voltage The potential applied to the plate of the vacuum tube by the plate voltage supply.

platen A cylinder in a printer or typewriter around which the paper goes and which the

printing mechanism strikes to produce an impression.

platform Platform is a loosely-defined word for a software operating system and/or open hardware, which an outsider could write software for. Windows98 is a platform. So is Windows 2000. If every phone system were a platform, then every owner of that phone system could write software for his phone system or buy outside-produced software and have his phone system work more to his liking. That's the objective of creating a "platform." See also OAI and Platform Independence.

platform agnostic Several other developers in the wireless industry have created their own proprietary de-velopment platforms, or application pipelines, leaving carriers with more non-standard platforms to evaluate than content to deploy. This has created an environment with very few developers willing to commit their resources to any specific platform. Those devel-opers that do create an application must find a way to monetize their creation, a process that usually involves a relationship with carriers. Such relationships are extremely diffi-cult to develop, as carriers often will not deal with small companies that have little or no established reputation. This difficult environment has contributed to slow consumer adoption of wireless data services, and the slow deployment of the 3G networks that require killer applications to justify the expense of upgrading a network. We believe that closed and proprietary approaches are flawed, and that profitable compa-nies will create applications that run on existing platforms - instead of creating platforms themselves. This agnostic stance is Chasma's major advantage, because Chasma's ap-plications can generate revenue immediately upon distribution and are not dependant on carriers deploying any technology that is proprietary to Chasma.

platform wars Your PDA works on Palm OS. Mine works on Windows CE. Which one will succeed in the long run is the one with the most number of useful software applications running on them. The only war to get outside software developers to write for my platform is to convince them that I've won the platform war – the war between Palm OS and Windows CE. There are similar "wars" in PCs and in cell phones.

platform for privacy See P3P.

platform independence A term from IBM and Metaphor Computer Systems. The idea, they say, is to produce a layer of software that would rest atop any operating system on any piece of hardware. The applications developers would write their software just once, rather than start from scratch each time they wanted get their software working on a different computer. If the whole idea sounds rather daunting, you're right.

platter The round magnetic disk surfaces used for read/write operations in a hard disk system.

play off In voice processing, in response to questions such as "Press one for Harry," the user touchtones buttons on his phone. Those buttons generate DTMF (Dual Tone Multi-Frequency) tones. The system has to figure out what the person "said" with his touchtones. The tricky part of DTMF detection is distinguishing between tones generated from an actual "key press" and "tones" caused by speech. Mistaking a person's speech (as in leaving a message) for DTMF is called "talk-off." Mistaking a person's recorded speech (as in playing back a message) for DTMF is called "play-off."

You can imagine the havoc poor DTMF detection can cause a voice processing system. For example, if touchtoning three means "delete this message" and while playing the message, the system incorrectly detects a portion of the message playback as the touchtones for a key press three, I'll delete the message when I had intended to listen to it. On the other hand, if I'm listening to a message and want to delete it prior to finishing the message, I want the system to detect my key press three as the real thing and go ahead and delete the message.

playback Retrieval, decoding and transmission of encoded data. It is also a multimedia term. Playback is the process of viewing multimedia materials created by an author. Play-back can include a range of activities, from viewing a single video clip to participating in a series of interactive multimedia training modules. Some playback applications (for example many training and presentation applications) are sold separately from their authoring applications. However, many developers are selling authoring and playback capabilities in a single product.

playback head The part which converts the magnetic information on the tape or disk into an electrical signal. Moving the magnetic fields on the medium (tape or disk) past the playback head generates a tiny voltage, which is picked up in a conductor (a coil) in the payback head and sent onto the electronic equipment where it is amplified or transmitted.

player An SCSA definition. A resource object that plays TVM data. The audio data can come from a voice or audio encoded file, or from text that has passed through a text-to-speech service. The output of a player can be analog audio, TDD, ADSI, etc.

playtester A person who tests computer games, either for free or for a living. I first saw the term when Microsoft was sending out emails looking for "playtesters" for its Xbox electronic game – the one meant to compete with the Sony Playstation.

PLB Personal Locator Beacons. See Personal Radio Services.

PLC 1. Planar Lightwave Circuit.

2. See PowerLine Carrier and also BPL.

PLCP Physical Layer Convergence Protocol. The part of the physical layer that adapts the transmission facility to handle DQDB functions as defined in IEEE 802.6-1990. It is used for DS-3 transmission of ATM. ATM cells are encapsulated in a 125microsecond frame defined by the PLCP which is defined inside the DS3 M-frame.

PLD Programmable logic device. PLDs used to be slow, big and expensive. Now they can be customized using a PC and their performance is close to that of the ASIC. See ASIC.

PLDS Private line data circuit. I don't know why it's an S, not a C.

pleading cycle The time period established by the FCC for third parties to submit written comments on a petition submitted by a carrier, broadcaster or other entity.

Please Do Not Tell Sales People Anything A memory aid for remembering the seven layers of the OSI Reference Model.

Please : Physical Layer (Layer 1)
Do : Data Link Layer (Layer 2)
Not: Network Layer (Layer 3)
Tell : Transport Layer (Layer 4)
Sales : Session Layer (Layer 5)
People : Presentation Layer (Layer 6)
Anything : Application Layer (Layer 7)

Please Do Not Throw Sausage Pizza Away See "Please Do Not Tell Sales People Anything."

pleasure All pleasure is sin, according to John Calvin.

plenum In some modern buildings, the ducts carrying the heat return are not metal ducts but actually are part of the ceiling. This is called a plenum ceiling. Most cities now have rules and regulations which say that if you run cabling through these plenum ceilings, you must not use cabling sheathed in PVC (polyvinyl chloride), the standard jacketing of most electrical cable. The reason is that PVC burns and emits toxic smoke ferociously. Plenum cable is low smoking so that if it catches fire it won't circulate toxic smoke through the vent system and suffocate everyone. Plenum cabling is often made of teflon. It's much more expensive than normal cabling. See also FEP, NFPNA 90A, plenum area and plenum cable.

plenum area The space between the drop ceiling and the floor above. Continuous throughout the length and width of each commercial building floor.

plenum cable Cable listed by Underwriters Laboratories for installation in plenums without the need for conduit. Cable specifically designed for use in a plenum, or air-handling space (the space above a suspended ceiling used to circulate air back to the heating or cooling system in a building) As specified by the NEC (National Electrical Code), plenum rated cable uses buffers, insulation and jackets made of low smoke, low toxicity, fire retardant material with a low flame spread index and a low potential heat (i.e., fuel load) level. Otherwise, a fire can travel along a cable, from room to room through walls, fanned by the air moving through the plenum, while giving off a vile, deadly smoke as it does so. The best plenum rated jacketing material generally is agreed to be fluorinated ethylene propylene (FEP), which Dupont markets as Teflon. Building codes now require the use of plenum rated cables in plenum spaces, and many contractors use it exclusively in plenum, riser and distribution applications. See also riser cable and distribution cable. Many buildings and many cities stipulate that only plenum cable can be installed in the plenum in the ceilings. Plenum cable has fully color coded insulated copper conductors and is available in various pair sizes.

plesiochronous Plesiochronous, based on Greek and Latin roots, roughly translates as "more together in time." Plesiochronous networks involve multiple digital synchronous circuits running at different clock rates. For instance, a Verizon T-1 circuit may meet a MCI T-1 circuit, with each taking making use of a different clocking source. Also for example, multiple MCI T-1 circuits may require multiplexing into a T-3 circuit; with the T-1's and the T-3 running at different clock speeds. In either case, the differences in clock speeds must be resolved through the use of a master clocking source such as a Stratum I clock, which relies on a highly reliable cesium clocking source. T-carrier and E-carrier networks are plesiochronous. Compare to Synchronous, Asynchronous and Isochronous. See also PDH.

plesiochronous networks Network elements that derive timing from more than one primary reference source. Network elements accommodate minor frequency differences between nodes.

PLL Phase Locked Loop: Phase Locked Loop is a mechanism whereby timing information is transferred within a data stream and the receiver derives the signal element timing by locking its local clock source to the received timing information.

PLLC Professional Limited Liability Corporation, as in a law firm.

PLM Public Land Mobile. See the next definition.

PLMN Public Land Mobile Network. A mobile telephone communications network established by a provider to facilitate mobile telecommunications services. This includes equipment, operations, and staff. A single provider may have more than one PLMN.

PLMR Private Land Mobile Radio system.

Plotter A type of computer peripheral printer that displays data in two-dimensional graphics form.

PLS Premises Lightwave System.

PLSC Private Line Service Center.

PLTS Private Line Transport Service. Non-switched communications channel from one customer location to another. May be leased from a Local Exchange Carrier or Interexchange Carrier.

PLU See Percent of Local Usage.

plug A male element of a plug/jack connector system. In the Premises Wiring System it provides the means for the user to connect his communications devices to the Communications Outlet as well as the means to disconnect his service at the Network Interface Jack when trouble analysis is required.

plug 'N play 1. Manufacturers' concept of how easy it is to install their equipment. "Why it's just plug 'n play," says the manufacturer. In reality, nothing, absolutely nothing, is plug 'n play. It's a fantasy concept. See Plug and Play.

2. Also defined as a new hire who doesn't need any training. "The new guy, Harry, is great. He's 100% plug-and-play."

plug and peer A term used by VoIP peering services to describe the advantage of using their interconnection service. The basic idea is that a VoIP provider that signs up with a VoIP peering service will enjoy instant connectivity to the networks of all other VoIP providers that use the peering service. A VoIP provider that uses a VoIP peering service only has to concern itself with establishing a connection between its own network and the peering service's network; the peering service handles the rest. This spares each individual VoIP provider the time and expense of establishing network connections and interconnection agreements with other individual VoIP providers. See VoIP peering.

plug and play This explanation comes from an Intel Technology Primer: Since add-in cards first appeared over a decade ago, they've given users a lot of different ways to improve their PCs and given them a lot of installation headaches. In this brief, we'll tell you how Intel, together with industry leaders, has spent years developing Plug and Play technology to make add-in cards both easier to use and install. Never before has the PC had as many capabilities as it does today. That's due in part to the large number of add-in cards available, like those for multimedia and faxmodems. Yet, as more cards are added to a PC, their installation can become quite complex. Installing a card can be a time-consuming and technical process, and there's no guarantee it will even work the first time. Sometimes the user must configure the card manually, which means selecting a variety of system resources for each card. These include Interrupt Requests (IRQ), I/O and memory addresses, and Direct Memory Access (DMA) channels. Every PC has a limited number of these resources available. Each card is designed to use a small group of them. Assigning these resources means opening the computer and physically setting the jumpers and DIP switches. And since no standard has been set to determine which cards can use which resources, numerous conflicts can arise between cards. Often, it's a process of trial and error to determine which resources aren't already being used by other cards. Since the ISA bus was introduced, several new bus architectures have followed to solve the resource allocation problem. For example, the MCA and the EISA bus standards both defined a mechanism where add-in cards were configured somewhat automatically. These bus architectures allocated the resources, but the process wasn't always flexible and still required some manual intervention. And they still left the current ISA cards without a solution. Plug and Play technology, co-developed by Intel and other industry partners, consists of hardware and software components that card, PC, and operating system manufacturers incorporate into their products. With this technology, the user is responsible for simply inserting the card. Plug and Play makes the card capable of identifying itself and the resources it requires. The system's software automatically sets up a suitable configuration for the card. Newly developed PCI and Plug and Play ISA cards are all built to eliminate user intervention during the installation process. See plug and play BIOS extensions.

plug and play BIOS extensions Software code added to a PC's bios which purports to automatically recognize which peripherals are in the PC and automatically configure the PC for those peripherals – without the need for fiddling with dip switches or setting interrupts, etc. Plug and Play comes from Intel. And more and more PC cards are coming Plug and Play compatible. See plug and play.

plug and pray Deploy hardware or software in a production environment without prior testing.

plug compatible Devices made by different manufacturers that are totally interchangeable. The word derives from the fact that the devices are so completely interchangeable that you can simply unplug one device and plug in another device made by different manufacturer and it will work the same, or better.

plug-in A program of data that enhances, or adds to, the operation of a (usually larger) parent program. A paint package, for example, might contain plug-in tools that create special effects, like speckling. A Web browser might have a plug-in that allows you to hear sound or view movies you download over the Internet. There are hundreds of plug-ins. See www.netscape.com/plugins/index.html

plugboard A telephone switchboard on which connections are made by a jack and an attached cord representing a trunk (the male jack and the cord) and a female plug (the telephone extension). Early plugboards needed an operator to place outside calls and connect incoming calls. All calls were completed by the operator. Plugboards were common in the days of "PBXs." Then came PABXs (Private Automated Branch Exchanges) and you could dial out without the help of an operator. Then electronic PABXs came in and you could dial directly in to many internal extensions, using a feature called DID (Direct Inward Dial). Now PABXs are called PBXs because they're all automatic. Plugboards are rapidly disappearing. They do have two great uses, however. First, operators who grew up with them, still like them. Operators who now live in nursing homes like them. Second, because the jack and plug make a pure metallic connection, they're great for data transmission and occasional data switching.

plugfest An event in a neutral and standards-based environment, where multiple vendors test and demonstrate the interoperability of their products. A plugfest creates market awareness of the standards and products being demonstrated.

plugs Circuit cards which control various aspects of transmission and carrier circuit functionality and usage.

plumber's crack Imagine a fat plumber dressed in jeans. When he leans forward, exposing the top part of his bottom. That is called plumber's crack. Longer tea shirts and higher jeans have been designed to cover plumber's crack, since the industry has found that its customers – suburban housewives – are often offended by the sight.

plumbing Slang term for an organization's network infrastructure.

plumbing software If your business is like mine, you have software programs for each business task – accounting, sales automation, order entry, inventory, etc. If you could get those pieces of software to talk to each other, and to talk sense to each other, you could save time, lower labor costs, improve your products and provide better customer service. Better yet, if you could get your internal software programs talking to software at your suppliers and customers, you could save even more money, labor and time. That's what integration software does. Every business of any size can use it to improve how their business works. There are three types of integration software: enterprise application, business-to-business, and business-to-community. What's the difference? All integration software lets two or more software applications – e.g. accounting and inventory – exchange (transport) and understand (transformation) each other's data. That's why it's often called plumbing software.

PLV An imaging term. Production Level Video. DVI Technology's highest quality motion video compression algorithm. It's about 120-1 compression. Compression is done "off-line". i.e. non-real time, and playback (decompression) is real time. Independent of the technology in use, off-line compression will produce a better image quality than real time since more time and processing power is used per frame.

PLY One layer in a composite.

PM 1. Physical Medium: Physical Medium refers to the actual physical interfaces. Several interfaces are defined including STS-1, STS-3c, STS-12c, STM-1, STM-4, DS1, E1, DS2, E3, DS3, E4, FDDI-based, Fiber Channel-based, and STP These range in speeds from 1.544Mbps through 622.08 Mbps.

2. Performance Monitoring. Gives a measure of the quality of service and identifies degrading or marginally operating systems (before an alarm would be generated). Digital signal parameters, including errored seconds and out of frame, measure the integrity of a communication channel as defined in AT&T Compatibility Bulletin 149 (CB 149).

3. Peripheral Module.

4. An ATM term for the Physical Medium sublayer. See Physical Medium.

PMA Positive Mental Attitude. Todd Kingsley has it. I know because he told me so. He said he was positively mental.

PMC Public Mobile Carrier – another name for a cellphone carrier.

PMD 1. Physical Medium Dependent. This PHY (PHYsical Layer) sublayer defines the parameters at the lowest level, such as speed of the bits on the media. See also PHY.

2. Polarization Mode Dispersion. A fiber optic term describing distortion created by irregularities in the shape of the fiber optic cable and its core; the problem is exacerbated by splicing, expansion and contraction of the cable due to variations in ambient temperature, and spooling of the cable. At high transmission speeds (e.g., SONET OC-192) digital light pulses can suffer from PMD. As the pulses travel down the fiber, the cable's physical irregularities cause delays on the at the outer edges of the core; the center portion of the pulse which travels through the "sweet spot" is unimpaired and, therefore, travels at a higher rate of speed from end-to-end. The result is that portions of an individual light pulse can arrive at slightly different time, with the delay being measured in picoseconds. The effect is one of distortion as the center portion of subsequent light pulses can overrun the outer portions of the preceding pulses. The impact is a higher bit error rate (BER). PMD is especially a problem at high transmission speeds, and particularly over older fiber optic cables deployed prior to the anticipation of high-bit rate SONET. Confused?...Consider the following explanation: If you drop a perfectly round rock into a pond of water, the waves run at the same speed toward the edges of the pond. If the pond is not perfectly round, some waves will reach the bank before other waves. If you now think of shooting the rock into a water pipe, you can see that some of the resulting compression waves will reach the other end before others. The combination of resistance at the edges of the pipe and irregularities in the shape of the pipe act to compound the problem, which varies unpredictably from pipe to pipe. See Chromatic Dispersion.

PMI Project Management Institute.

PML Physical Markup Language. An Auto-ID Center-designed method of describing products in a way computers can understand. PML is based on the widely accepted eXtensible Markup Language used to share data over the Internet in a format all computers can use.

PML Server A server that responds to requests for Physical Markup Language (PML) files related to individual Electronic Product Codes. The PML files and servers will be maintained by the manufacturer of the item.

PMMU Paged Memory Management Unit. Macintosh computers equipped with a PMMU may use virtual memory with the System 7 operating system.

PMN Indicates loss of ac power at the far-end terminal.

PMP 1. Point-to-MultiPoint. An ATM term describing the connecting circuitry between a single end point (root node) and multiple end points (leaf nodes). The root node can transmit data over a PMP connection to multiple leaf nodes, usually through a switch or router. The leaf nodes can transmit back to the root node, but do not have the ability to transmit data directly to other leaf nodes.

2. Project Management Professional as certified/designated by the Project Management Institute (PMI).

PMR 1. Poor Man's Routing. A technique used in any packet-switched networks to allow a source node to predefine the routing to the destination, bypassing the normal routing algorithm implemented at the network layer.

2. Private Mobile Radio. The European term for what in the US is called Specialized Mobile Radio (SMR) or Trunk Mobile Radio (SMR). See SMR and TETRA.

PMS 1. Picturephone Meeting Service. An AT&T service once provided under experimental tariff. It combined TV techniques with voice transmission. PMS is usually only available between telephone company-located picturephone centers. Most have now been closed down. The venture was losing too much money.

2. Property Management System, a software program and computer that controls all guest billing and guest services functions in a hotel. In short, the guts of a hotel's computer system. Some telephone systems have a PMS Interface, which allows various degrees of integration between the telephone system and the hotel's computer systems. For example, voice mail could be administered through the hotel's Property Management System.

3. The Pantone Matching System, a universal language for solid-color specification and reproduction. Colors defined by PMS receive a unique number and mixing formula. Consequently, when artists specify a PMS number they can be sure that the final printed product will match the chosen color. But, be careful, PMS colors look different when printed on different papers. The biggest perceived difference is when you print on glossy or matte paper.

PMS Interface An interface that allows telephone system functions (like voice mail) to be administered through a hotel's Property Management System.

PN Public Notice. A Public Notice is issued by the Federal Communications Commission (FCC) to notify the public of an action or an upcoming event.

pneumatic telegraph Steven Schoen writes "WhenI was at Tripler Army Medical Center (TAMC) for a couple of surgeries in 1979, the hospital had a working pneumatic telegraph network which was constantly used. Basically it consisted of a network of pipes, and canisters that contained messages that traveled through a pipe to a central hub. Gravity sent the canisters from upper floors down to the central hub. At the central hub, the canister was put in the appropriate pipe and a blast of pressurized air sent it on its way to its destination. Paris had a sophisticated pneumatic telegraph network that enabled messages to be sent around town." It's also called tubular post.

PNG Portable Network Graphics, pronounced "Ping." A graphics file format ending in .png, which was developed to overcome deficiencies of .gif and .jpeg file formats, which are commonly used on the Web. PNG, under development by the W3C, allows 16-, 24-, and 32-bit images, providing for better color depth. The downside is that higher-quality images require more storage and involve longer download times. PNG overcomes this problem through the use of better image compression technology.

PNI An ATM term. Permit Next Increase: An ABR service parameter, PNI is a flag controlling the increase of ACR upon reception of the next backward RM-cell. PNI=0 inhibits increase. The range is 0 or 1.

PNNI Private Network-to-Network Interface: PNNI is a routing protocol for ATM that allows for the exchange of routing information between ATM switches. PNNI allows those switches in the network to determine the best route to establish the type of connection needed. PNNI allows for switches from different vendors, or on different networks, to exchange addressing and routing information so those who maintain the network can set up connections with minimum fuss. In short, PNNI is a routing information protocol that enables extremely scalable, full function, dynamic multi-vendor ATM switches to be integrated in the same network. PNNI has two functions: to reliably distribute network topology information so that fast routing paths can be determined to any destination and second, to provide a signaling protocol to help set up point-to-point and point-to-multipoint connnections. The idea is enable simple network configurations while helping network personnnel efficiently manager the network's resources.

PNNI protocol entity An ATM term. The body of software in a switching system that executes the PNNI protocol and provides the routing service.

PNNI routing control channel An ATM term. VCCs used for the exchange of PNNI routing protocol messages.

PNNI routing domain An ATM term. A group of topologically contiguous systems which are running one instance of PNNI routing.

PNNI routing hierarchy An ATM term. The hierarchy of peer groups used for PNNI routing.

PNNI topology state element An ATM term. A collection of PNNI information that is flooded among all logical nodes within a peer group.

PNNI topology state packet An ATM term. A type of PNNI Routing packet that is used for flooding PTSEs among logical nodes within a peer group.

PNM Public Network Management.

PNO Public Network Operator. Usually a PTT of some sort. See PTT.

PnP Plug and Play. The technology that lets Windows 95 and soon other operating systems automatically detect and configure most of the adapters and peripherals connected to or sitting inside a PC. A fully Plug and Play-enabled PC requires three PnP pieces: a PnP BIOS, PnP adapters and peripherals, and a PnP operating system. Adding a PnP-compliant CD-ROM drive, hard disk, monitor, printer, scanner, or other device to a PnP PC means little more than making the physical connection. The operating system, together with PnP logic present in the BIOS and in the device itself, handles the IRQ settings, I/O addresses, and other technical aspects of the installation to make sure that the thing will work. The idea of PnP is to make installation of complex gadgets – such as sound cards and modems – easy, taking care of the major bane of everyone's life: That your new device now conflicts with an old device, effectively killing both devices and maybe crashing your PC at the same time. PnP is a great idea. Its success has been slow in coming, because so many devices are not PnP compatible.

PNS Personal Number Service is a new concept in telecommunications that assigns a telephone number to a person, not a location, effectively allowing a subscriber to use one number for all calls and helping them manage their incoming communications. The service does not require the user to change any existing phone numbers. The subscriber simply provides the various numbers – office, cellular, pager, fax and home – and instructions on where and when the calls should be routed, and the PNS directs the calls in the order requested by the subscriber.

PO 1. Point of Origin. It is used in relationship with a Message Transfer Agent (MTA).

2. Post Office Location: City, State, ZIP.

POC 1. Points of Contact. The person or persons identified in a record. Sometimes this information is referred to as "Person Objects." See also InterNIC.

2. Push-to-talk over cellular. See also Push to Talk.

pocket bongo Picture a group of people. Suddenly, something on someone beeps. But the someone doesn't know (or pretends to not to know) which of the many wireless devices he's carrying that is bleating. Is it the cell phone? Or the pager? Or the PCS phone? The person starts patting himself all over, with mock embarrassment. But his look screams, "I'm wired and I'm proud." His behavior is called "pocket bongo." I read about pocket bongo first in an article by Joan Hamilton in the February 15, 1999 issue of Business Week. The article was headed, "We've got a bad case of digital gizmosis."

pocket call A call made when a cell phone's buttons are depressed as it jostles around in a pocket or purse. Because many handsets speed-dial 911 at the touch of a button, the calls are an ongoing problem for emergency lines.

POCSAG Post Office Code Standardization Advisory Group. A one-way paging protocol that supports numeric and text paging at data rates of 512, 1,200 and 2,400 bits per second. Most traffic occurs at 1,200 bits per second. It is one of the communications protocols used between paging towers and mobile pagers/receivers/beepers themselves. Other protocols are GOLAY, ERMES, FLEX and REFLEX. The same paging tower equipment can transmit messages one moment in POCSAG and the next moment in ERMES, or any of the other protocols.

podcache 1. Pairing MP3 files with GPS coordinates, thereby transforming an iPod into an audio tour guide.

2. Treasure hunting with clues provided by a podcast.

podcasting Postcasting is a way of having Internet audio programs – from amateur talk shows to professional radio – delivered automatically to your MP3 player so that they be recorded there and you can thus listen to the anytime. To understand how podcasting workings, think first about RSS, on which it's built; You subscribe to an audio feed from your favorite site and its headlines are downloaded to an RSS reader on your desktop. Podcasting works the same way except that the feed carries an MP3 file (called the podcast) which can be automatically sent to your MP3 player (including your iPod) next time you connect it to your computer. A podcast aggregator such iPodder.org manages your subscriptions.

potcatcher Software designed to download and aggregate podcasts for use in an audio player. Some available podcatchers include Juice and DopplerRadio. This definition from blogossary.com.

PODP Public Office Dialing Plan.

podiumware You're presenting a great speech detailing some great new concept in hardware or software. You don't have many precise details, except your vague words. This is called podiumware. When your thinking has become more concrete, and you make slides on your new hardware or software, you have moved to slideware. Eventually when you announce your new hardware or software, you have moved to hypeware or vaporware. I was first introduced to the word "podiumware" by Bob Lewis, a columnist for InfoWorld Magazine. See also Hookeware, Hyperware, Meatware, Podiumware, Shovelware, and Vaporware.

PodShanking Transferring audio files directly from one iPod to another using the headphone prots and apps such as Podzilla.

podslurping The use of a portable device such as an iPod to directly connect to an unattended computer in an organization and illicitly download (i.e., "slurp") data from it. Podslurping is a security risk to companies and government agencies.

POE Power over Ethernet. See 802.3af.

POF Plastic Optic Fiber. A fiber optic transmission medium made from plastic, rather than glass, POF is evolving as a replacement for twisted-pair copper wire. Glass clearly (double entendre intended) performs better than plastic, as it offers less attenuation and, therefore, better transmission quality at higher speeds and over longer distances. Plastic, however, is less expensive, less susceptible to breakage, and is highly tolerant of temperature extremes. The ATM Forum has approved specifications for 155-Mbps ATM transmission over POF for distances up to 50 feet.

POFS Private Operation Fixed Systems. Microwave incumbents in the 2.0 Ghz band. Must be relocated with comparable alternative facilities funded.

POGO Post Office Goes Obsolete. When MCI Mail was originally being planned, its code name was POGO. The idea was obvious. In September of 1994, I asked MCI what "POGO" meant and they answered: "Pogo" is an internal message format used by MCI for coding purposes.

POH Path OverHead. SONET overhead assigned to and transported with the payload until the payload is demultiplexed. It is used for functions that are necessary to transport the payload; i.e., end-to-end network management. These functions include parity check and trace capability. It is not implemented in SONET Lite.

POI 1. Point Of Interface. The physical telecommunications interface between the LATA access and the interLATA functions. A POI is a demarcation point between LEC and a Wireless

Services Provider (WSP). This point establishes the technical interface, the test point(s) and the point(s) for operational division of responsibility. See also Point of Presence.

2. Parallel Optic Interfaces. POI is a relatively cheap technique for optical transmission at high speeds over VSR (Very-Short-Reach) links of 300 meters or less. The technique makes use of multiple MMF (MultiMode Fiber) links, over which an optical signal at speeds up to 10 Gbps (OC-192) is spread across multiple fibers at the same wavelength through inverse multiplexing. The technique is considered to be low cost, as the MMF is much less expensive than the SMF (Single Mode Fiber) that normally is required to support such speeds, and as the quantum-well VCSELs (Vertical-Cavity-Surface-Emitting Lasers) also are low cost. The applications are in the interconnectivity of collocated devices such as switches, optical cross-connects, routers. See also MMF, SMF, VCSEL, and VSR.

point code A SS7 term for a unique code which identifies a network node in order that the SS7 network can route calls properly. When placing a call, you dial a Global Title in the form of dialed digits (i.e., a telephone number). Those digits are translated from the Global Title to a Point Code by the STP (Signal Transfer Point) through a process known as Global Title Translation (GTT). See also Global Title, Global Title Translation, SS7, and STP.

Point In Call PIC. A representation of a sequence of activities that the ASC (AIN Switch Capabilities) performs in setting up and maintaining a basic two-party call. PICs occur in Originating and Terminating BCSMs (Basic Call State Model).

point of demarcation Physical point at which the phone company's responsibility for the wiring of the phone line ends.

Point Of Interface POI. The physical telecommunications interface between the LATA access and the interLATA functions. A POI is a demarcation point between LEC and a Wireless Services Provider (WSP). This point establishes the technical interface, the test point(s) and the point(s) for operational division of responsibility.

Point Of Presence POP. A physical place where a carrier has a presence for network access, a POP generally is in the form of a switch or router. For example, an large IXC will have a great many POPs, at which they interface with the LEC networks to accept originating traffic and deliver terminating long distance traffic. The basis on which the interface is accomplished can include switched and dedicated (leased line) connections. Similarly, providers of X.25, Frame Relay and ATM services have specialized POPs, which may be collocated with the circuit-switched POP for voice traffic. A POP also is a meet point for ISPs (Internet Service Providers), where they exchange traffic and routes. See also GIGAPOP and POP.

point of purchase politics Politically correct shopping or cause-related marketing, such as that advocated by Benetton or Ben and Jerry's.

point of sale terminal A special type of computer terminal which is used to collect and store retail sales data. This terminal may be connected to a bar code reader and it may query a central computer for the current price of that item. It may also contain a device for getting authorizations on credit cards.

Point Of Termination POT. The point of demarcation within a customer-designated premises at which the telephone company's responsibility for the provision of access service ends.

point size The height of a printed character specified in archaic units called "points." A point equals approximately 1/72 inch. Also known as font size. See also Font.

point to multipoint A circuit by which a single signal goes from one origination point to many destination points. The classic example is a TV signal (say a Home Box Office program) being broadcast from one satellite to many CATV subscribers all around the country. Not to be confused with a multi-drop circuit. See Point to Point Multipoint Connection.

point to multipoint connection A Point-to-Multipoint Connection is a collection of associated ATM VC (Virtual Channel) or VP (Virtual Path) links, with associated endpoint nodes, with the following properties:

1. One ATM link, called the Root Link, serves as the root in a simple tree topology. When the Root Node sends information, all of the remaining nodes on the connection, called Leaf Nodes, receive copies of the information.

2. Each of the Leaf Nodes on the connection can send information directly to the Root Node. The Root Node cannot distinguish which Leaf is sending information without additional (higher layer) information. (Note:UNI 4.0 does not support traffic sent from a Leaf to the Root.)

3. The Leaf Nodes cannot communicate directly to each other with this connection type. See ATM.

point-to-multipoint delivery Delivery of data from a single source to several destinations.

point-to-point A private circuit, conversation or teleconference in which there is one person at each end, usually connected by some dedicated transmission line. In short, a connection with only two endpoints. See also Point-To-Multipoint.

point-to-point connection An uninterrupted connection between one piece of equipment and another.

point-to-point delivery Delivery of data from a single source to a single destination.

Point-To-Point Protocol See PPP.

point-to-point signaling A signaling method where signals must be completely received by an intermediate station before that station can set up a call connection. See End to End Signaling.

point-to-point topology A network topology where one node connects directly to another node.

Point-To-Point Tunneling Protocol PPTP. Part of the VPN suite, a protocol by which tunnels are established and terminated over the Internet. An alternative to IPsec, L2TP, SOCKSv5 tunneling protocols. See PPTP for more detail.

pointcasting First, there was broadcasting. One signal went to many people. Radio and TV are the classic concepts of broadcasting. One signal – the same signal – to many people. Then came the idea of narrowcasting. One signal to a select number of people – maybe only those people who subscribed to the service and had the equipment to receive it. Then there came pointcasting. This is a fancy name for sending someone a collection of customized information – snippets of stuff that they chose from a palette of information offerings.

pointer processing Pointer processing accommodates frequency differences by adjusting the starting position of the payload within the frame. A pointer keeps track of the starting position of the payload.

pointing stick Alternative to an external mouse and similar to a miniature joystick. You can use the tip of your finger to move the rough textured eraser sized pointer in the direction you want the cursor or arrow to go. You find pointing sticks on superior laptops.

points of contact The person or persons identified in a record. Sometimes this information is referred to as "Person Objects." See also InterNIC.

points of failure A simple term to indicate that in a complex network there are many places things can go wrong. Those places need to be identified so that you can anticipate and plan for things to go wrong. Points of failure (with or without the hyphens) should not be confused with a "Single Point of Failure". Single Points of Failure should be eliminated by establishing proper redundancy. "Points of failure" refer to the points within a single path of communication or operation. That single path is the "Single Point of Failure". So to maximize efficiency while minimizing downtime the architecture would minimize the "Points of Failure" by eliminating process steps within a path or operation. And at the same time eliminate a "Single Point of Failure" by providing a redundant path or operation.

poison pill A poison pill is a strategy used by corporations to discourage the hostile takeover by another company by making its stock less attractive to the acquirer. There are basically two types of poison pills: First, "flip-in", which allows existing shareholders (except the potential acquirer) to buy more shares at a discount. Second, the "flip-over" allows stockholders to buy the acquirer's shares at a discounted price after the merger. By buying more shares cheaply (flip-in), investors get instant profits and more importantly, they dilute the shares held by the potential acquirer, thus making the takeover attempt more difficult and expensive. An example of a flip-over is when shareholders have the right to purchase stock of the acquirer on a 2-for-1 basis in any subsequent merger. A poison pill is often called a shareholder rights plan. Companies propose all sorts of reasons for swallowing poison pills. For example, Delphi Automotive told its shareholders, "in order for Delphi to deliver stockholder value, Delphi must have the opportunity to execute its strategic business plan without the distraction of unfair, imprudent or abusive takeover attempts, particularly in an industry that is currently undervalued. In addition, the market needs some time to understand that Delphi is not just an auto parts company but a technology company, and to value it accordingly."

poison reverse Announcement state in EIGRP when an advertisement within the EIGRP network becomes unreachable, at which time it is assigned a metric of "infinity" so routes destined for that network take alternate paths.

Poisson See Poisson Distribution.

Poisson Distribution A mathematical formula named after the French mathematician S. D. Poisson, which indicates the probability of certain events occurring. It is used in traffic engineering to design telephone networks. It is one method of figuring how many trunks you will need in the future based on measurements of past calls. Poisson distribution describes how calls react when they encounter blockage (see QUEUING THEORY for a detailed explanation of blockage). There are two main formulas used today in traffic engineering: Erlang B and Poisson. The Erlang B formula assumes all blocked calls are cleared. This means they disappear, never to reappear. The Poisson formula assumes no blocked calls disappear. The user simply redials and redials. If you use the Poisson method of prediction, you will buy more trunks than if you use Erlang B. Poisson typically overestimates the number of trunks you will need, while Erlang B typically underestimates the number of trunks you will need. There are other more complex but more accurate ways of figuring trunks – Erlang C (blocked calls delayed or queued) and computer simulation. Poisson has been used extensively by AT&T to recommend to its customers the number of trunks they needed. Since AT&T was selling the circuits and preferred its customers to have excellent service, it made sense to use the Poisson formula. As competition in long distance has heated up, as circuits have become more costly and as companies have become more economically-minded (more aware of their rising phone bills), Poisson has become widely ignored.

After I wrote the above definition, Lee Goeller, a noted traffic engineering expert contributed the following definition of Poisson Distribution: A probability distribution developed by E.C. Molina of AT&T in the early 1900s for use in solving problems in telephone traffic (see TRAFFIC ENGINEERING), although it has many other uses and is widely applied in many fields. When made aware of Poisson's prior effort (circa 1820), Molina gave him full credit and even taught himself French so he could read Poisson in the original. The Poisson distribution assumes a call is in the system for one holding time, whether it is served or not (blocked calls held); the first form of the distribution estimates the probability that exactly X calls will be in the system, while the second estimates the probability that X or more calls will be present. If there are only X trunks to serve the calls, the second form gives the probability of blocking. Although limited tabulations of the Poisson distribution had been made earlier, Molina published an extensive set of tables in 1942. The Poisson distribution slightly overstates the number of trunks needed when compared to the Erlang B distribution (see Erlang B).

Poisson Process A kind of random process based on simplified mathematical assumptions which makes the development of complex probability functions easier. In traffic theory, the arrival of telephone calls for service is considered a Poisson process. Calls arrive "individually and collectively at random," and the probability of a new call arriving in any time interval is independent of the number of calls already present. A Poisson process should not be confused with the Poisson Distribution, which gives the probability that a certain number of calls will be present if certain additional assumptions are made. See Poisson Distribution.

poke When you store a number in memory you "poke" it there. When you want to read it back, you take a "peek." Peek and poke are thus instructions that view and alter a byte of memory by referencing a specific memory address. Peek displays the contents; poke changes it.

poke-through method A distribution method that involves drilling a hole through the floor and poking cables through to terminal equipment from the ceiling space of the floor below. See also Ceiling Distribution Systems and Newton.

poke-through system Penetrations through the fire-resistive floor structure to permit the installation of horizontal telecommunications cables.

pokemon A Jamaican proctologist.

polar keying A transmission technique for digital signals in which the current flows in opposite directions for 1s and 0s or marks and spaces. It is used in telegraph signaling. It is also known as polar transmission.

polar plot A 360-degree graph measuring direction by angle and levels using concentric circles.

polar relay A relay containing a permanent magnet that centers the armature. The direction of movement of the armature is governed by the direction of current flow.

polarity Which side of an electrical circuit is the positive? Which is the negative? Polarity is the term describing which is which. Knowing polarity is not critical with rotary phones. They will work irrespective of which way the telephone circuit's polarity is. Touchtone phones, however, need correct polarity for their touchtone pads to work. How to tell? If you can receive an incoming call, can speak on the phone clearly, but can't "break" dial tone by touching a digit on your touchtone pad, then the polarity of your line is reversed. Simply reverse the red and green wires. Some electronic phones behind PBXs and key systems are also sensitive to polarity. If in doubt, simply reverse the wires. In video, reversed polarity results in a negative picture.

polarization Polarization is the direction of electric field in a radiated wave. This direction may be constant or may rotate as the wave propagates (resulting in linear, circular or elliptical polarizations). Polarization considerations apply whether a signal is transmitted (or received) in air, cable, waveguide or other transmission media. Polarization matters because it is one factor in determining how much energy an antenna receives from

an incoming signal. If the polarization of the receiving antenna matches the polarization of the incident wave, no energy lost due to polarization mismatch (such an antenna and wave are referred to as "co-polarized"). However, where the polarization of the receiving antenna is orthogonal to the polarization of an incident wave, no energy will be received by the receiving antenna (such an antenna and wave are referred to as "cross-polarized"). The interesting thing is that cross-polarization can occur between an antenna with vertical linear polarization versus a wave with horizontal linear polarization, as well as between waves/antennas with left/right handed circular polarizations. Consequently, polarization is a major consideration in antenna system design. Thanks to Paul Chandler for help on this definition.

Polarization Beam Combiner PBC. A PBC is capable of combining pump laser inputs with orthogonal polarization states. A PBC combines the signal strength of two low-powered 14XX nanometer pump lasers, producing a signal with double the power. PBCs can be used in both EDFAs and Raman amplifier modules, achieving higher output power from less costly, more readily available pump lasers. See Laser and Raman Amplifier.

Polarization Mode Dispersion See polarization and PMD.

polarmount A polarmount is a moveable dish antenna mount that allows a dish to be moved to different satellites. Azimuth and elevation are automatically adjusted as the dish moves.

Poldhu Wireless Station A wireless telegraph station in Poldhu, Cornwall, in the southwest corner of England, where the first transatlantic wireless telegraph message – the letter "S" – was transmitted to Guglielmo Marconi's kite-suspended antenna over 2,000 miles away on Signal Hill, in Newfoundland, Canada.

pole attachment Cost to cable TV, cellular provider and other telecom operators (including end users) to rent space to attach cables to telephone company and power company poles. There are charges and often significant restrictions on the attachment of your cable to their pole.

pole farm A large field that existed at Bell Labs' Chester, New Jersey facility, where rows of telephone poles, unconnected to anything, were treated with various materials designed to strengthen them and prolong their life. The treated poles were studied over the years, and compared to untreated poles and with each other, to see how the different treatments worked.

pole hug See Spurring Out.

Policy Decision Point See COPS and PDP.

Policy Enforcement Point See PEP.

policy routing protocol An extension of Vector Distance Protocols used in router networks. Used primarily in Internet routers, Policy Routing Protocols determine the route of a packet in consideration of "permissions" and reciprocal business contracts between and among backbone carriers, ISPs and Internet Access Providers. In other words, the route is determined on the basis of non-technical policy, rather than the number of hops a packet must travel. Assuming that the intercarrier policy accepts the offered traffic, the packet is routed based on technical considerations according to Vector Distance Protocols. Examples of Policy Routing Protocols include BGP (Border Gateway Protocol) and IDRP (InterDomain Routing Protocol). See also Distance Vector Protocol, Link-State Protocol, and Router.

Policy-Based Networking Also known as Policy-Based Routing (PBR) and Policy-Based Network Management (PBNM). A traffic management concept involving the establishment of priorities for network traffic based on parameters such as traffic type, application, and user ID. ATM does a great job of policy-based networking as a result of QoS (Quality of Service) levels. RSVP (Resource ReserVation Protocol) from the IETF is emerging as a solution to managing traffic priorities over the Internet. Policy-based networking can be implemented in capable switches, routers and servers. See also OOPS, RSVP and QoS.

policy-based quality of service A network service that provides the ability to prioritize different types of traffic and manage bandwidth over a network.

polishing Preparing a fiber end by moving the end over an abrasive material.

politeness The most acceptable hypocrisy. Mostly seen before the sale.

political file Records required by Section 76.207 which relate to origination cablecasts by, or on behalf of, candidates for public office. This rule requires each cable television system to keep a record, in its PIF, of all requests for cablecast time, together with detailed supporting information.

politically correct PC. The art of saying something totally bland when a good insult would be more satisfying, and more deserved.

politician John Maynard Keynes said that politicians are apt to be slaves to the ideas of long-deceased economists. John Kenneth Galbraith defined economists as people who didn't have the personality to become accountants.

politics 1. A clash of self-interests masquerading as a clash of principles.

2. The technique by which most telephone systems are bought in large corporations.

3. Poli in Latin means 'many' and 'tics' meaning 'bloodsucking creatures', which seems to accurately describe the process.

poll In data communications, an individual control message from a central controller to an individual station on a multipoint network inviting that station to send if it has any traffic to send. See Polling.

poll cycle The complete sequence in which stations are polled on a polled network.

Poll/Final Bit Bit in HDLC frame control field. If frame is a command, bit is a poll bit asking station to reply. If frame is a response, bit is a final bit identifying last frame in message.

polling Connecting to another system to check for things like mail or news. A form of data or fax network arrangement whereby a central computer or fax machine asks each remote location in turn (and very quickly) whether they want to send some information. The purpose is to give each user or each remote data terminal an opportunity to transmit and receive information on a circuit or using facilities which are being shared. Polling is typically used on a multipoint or multidrop line. Polling is done to save money on telephone lines.

polling delay Communications control procedure where a master station systematically invites tributary stations on a multipoint circuit to transmit data. Polling delay is a measure of the time to transmit and receive on a polled network versus a direct point-to-point circuit.

POLSK POLarization Shift Keying.

Polybutylene Terephthalate PBT. An insulating material used extensively for buffer tubes which surround optical fibers.

polyester Polyethylene terephthalate, which is used extensively in the production of a high-strength moisture-resistant film used as a cable core wrap.

polyethylene A family of insulating (thermoplastic) materials derived from polymerization of ethylene gas. They are basically pure hydrocarbon resins with excellent dielectric properties. Used extensively in cables.

polygon The building block of sophisticated computer graphics.

polymer A material having molecules of high molecular weight formed by polymerization of lower molecular weight molecules.

polymerization A chemical reaction in which low molecular weight molecules unite with each other to form molecules with higher molecular weights.

Polymorphic Buffer Overflow PBO is a form of computer attack that experts say will increase on the Internet. PBO works like thus: 1.The worm program containing the PBO attacks a server with a message that's longer than the server expects, causing the system to read the rest of the text as executable code, infecting the computer; 2.Once the worm has infected one server, it scrambles itself before sending itself out to other vulnerable servers. That makes it unrecognizable to the intrusion detection systems designed to look for the worm. 3. When it has infected a new server, it scrambles itself again, and sends itself out to more servers and the process continues. Experts say it will be hard to catch because no worm will look the same.

polymorphism The ability of objects to handle different types of information and different requests for actions. Components are not typically polymorphic.

polyolefin Any of the polymers and copolymers of the ethylene family of hydrocarbons.

polyphonic A music file that can create multiple tones and/or notes simultaneously. This produces a sound more like CD quality music.

polypropylene A thermoplastic similar to polyethylene but stiffer and having a higher softening point (temperature) and excellent electric properties.

Polyvinylchloride PVC. A thermoplastic material composed of polymers of vinyl chloride. A tough, water and flame-retardant thermoplastic insulation material that is commonly used in the jackets of building cables when fire retardant, but not smoke retardant properties are required. Unfortunately, it burns and gives out noxious gases which kill. PVC can't be run in air return ducts, also called plenum ducts and most towns, therefore, don't allow PVC to be run in their plenum ceilings. See Plenum.

Polyvinylidene Difluoride PVDF. A fluoropolymer material that is resistant to heat and used in the jackets of plenum cable. See also Plenum Cable.

PON Passive Optical Network is a fiber optic network without active electronics, such as repeaters you plug into the wall or a battery. A PON uses passive splitters to deliver signals to multiple terminal devices. Passive optical networking (PON) technology allows a fiber optic network to be built without the costly, active electronics found in all other types of networks. Rather, a PON network relies on inexpensive optical splitters and couplers, which are placed at each fiber "junction," or connection, throughout the network, providing a tremendous fan-out of fiber to a large number of end points. By eliminating the dependence on expensive active network elements – and the ongoing powering and maintenance costs associated with them – carriers can realize significant cost savings. (The PON is however still far more expensive than alternatives such as DSL). PON technology generally is used in the local loop to connect customer premises to an all-fiber network. A PON is a tree-like structure consisting of several branches, called Optical Distribution Networks. These run from the central office to the customer premises using a mix of passive branching components, passive optical attenuators and splices. Three active devices can be used in a PON. An Optical Line Terminal (OLT) either generates light signals on its own or takes in SONET signals from a collocated SONET crossconnect. The OLT then broadcasts this traffic to either an Optical Network Unit ONU or an Optical Network Termination, which receives the signal and converts it into an electrical signal for use in the customer premises. The speed of operation depends on whether the PON is symmetrical or asymmetrical. Symmetrical PONs operate at OC-3 speeds (155.52Mbit/sec), for asymmetrical PONs the upstream transmission is also 155.52Mbit/sec from the Optical Network Termination to the customer premises; downstream transmission can range between 155.52 to 622.08Mbit/sec. Depending on where the PON terminates, the system can be described as fiber-to-the-curb (FTTC), fiber-to-the-building (FTTB), or fiber-to-the-home (FTTH). Most PON approaches start with the specifications developed by the Full Service Access Network (FSAN) initiative. Variations on the PON theme include APON (ATM over PON) and TPON (Telephony over PON). See also APON, FSAN, TPON and WPON.

pond balls Golf balls retrieved from a pond or lake.

Pony Express Out of the summer haze bursts a horse and rider, swiftly approaching a lonely sod building on the prairie. Arriving in a cloud of dust, the rider leaps from his horse and heads for a water barrel to quench his thirst. Meanwhile, a leather sack filled with mail is whisked off the tired horse and thrown over the saddle of a fresh mount. Within two minutes, the rider is gone, galloping toward the far horizon. This young man in a hurry was one of some 200 Pony Express riders who carried the mail in a giant relay between St. Joseph, Missouri, and Sacramento, California, a distance of 1,966 miles, in ten days or less. Changing horses every ten to fifteen miles at swing stations, and switching riders at home stations after a run of 75 miles or more, the riders averaged 250 miles a day. During the short time the Pony Express was in operation – from April 1860, through October, 1861 – its rider defied hostile Indians, blazing desert heat, and bone-chilling blizzards to travel a total of 650,000 miles with 34,753 pieces of mail. To save weight the letters they carried were written on tissue-thin paper as postage cost $10 an ounce, later cut to $2. The best time ever achieved was in March 1861, when Lincoln's inaugural address was carried from Missouri to California in seven days, 17 hours.

The Pony Express was organized by stagecoach operator William Hepburn Russell, who had been convinced by a group of prominent Californians that an overland mail route to their state was feasible. Russell's business partners opposed the venture because it was not protected by a U.S. mail contract. (They had competition and de-regulation even in those days.) But Russell went ahead, building stations and purchasing 500 top quality Indian horses. In advertising for riders, he hinted at the hazardous nature of the job by asking for "small, daring young men, preferably orphans." The riders received board and keep and were paid $100 to $150 a month. Their average age was 19, but one rider, David Jay, was 13, and William F. Cody, who became famous as "Buffalo Bill," was 15. In a further effort to save weight, a rider usually carried only a pistol and a knife. He was expected to out-run the Indians, not out-fight them.

The Pony Express days of glory ended abruptly in 1861 following completion of the transcontinental telegraph. Russell's firm lost more than $200,000 in the venture, but the daring of the Pony Express riders caught the imagination of every American, and their exploits became an important part of the legend and lore of the nation. The above history copyright 1979 by Panarizon Publishing Corp.

ponytail See Suit.

Ponzi Scheme A type of scam named after Charles Ponzi, who ran such a scheme in 1919-1920. A Ponzi is somewhat like a pyramid scheme, as money owed early "investors" are paid by revenues collected from those who come later. Typically the scheme works as follows. The Ponzi scheme perpetrator advertises a 50% per year return on monies invested with him. Some monies flow in. At end of a quarter, the perpetrator pays his investors a dividend or return (or whatever he calls it) of 12.5%. Word goes out that he's paid out a handsome dividend. Soon more money flows in. He pays out more dividends. More money flows in. One day not enough money flows in. He can't pay the promised dividends. The whole thing starts to crumble. People start demanding their money back... A Ponzi scheme does not involve any manufacturing of goods, or selling of goods or services.

poof The magical successful sending of a computerized form to the next step in the corporate process after all pertinent information has been entered. This definition from a nice reader who works at Verizon.

pool A collection of things available to all for the asking or the dialing. A modem pool is a collection of modems typically attached to a PBX. Dial a special extension and you can use the modem, which answers that extension (or one of the extensions in the hunt group) to make a data call. Pooling is sharing. The purpose of having a "pool" is to avoid buying everybody one of whatever it is you're pooling. Actually, "pooling" is a fancy word for something we've been doing in the telephone business for the past 100 years – sharing. We started sharing lines, then sharing switches, then sharing voice mail devices, now we're sharing equipment, like modems.

pooling point A physical place where local and long distance carriers join their networks in order to swap bandwidth. See Bandwidth Broker.

pooling point administrator See Bandwidth Broker.

poop fiction A literary genre that uses potty humor and off-color jokes to appeal to young children.

POP 1. Point Of Presence. The IXC equivalent of a local phone company's central office. The POP is a long distance carrier's office in your local community (defined as your LATA). A POP is the place your long distance carrier, called an IntereXchange Carrier (IXC), terminates your long distance lines just before those lines are connected to your local phone company's lines or to your own direct hookup. Each IXC can have multiple POPs within one LATA. All long distance phone connections go through the POPs.

2. Point Of Presence at which ISPs (Internet Service Providers) exchange traffic and routes at Layer 2 (Link Layer) of the OSI model.

3. Short for "population." One "pop" equals one person. In the cellular industry, systems are valued financially based on the population of the market served.

4. Post Office Protocol. An e-mail server protocol used in the Internet. You use POP to get your mail and download it to your PC, using SMTP (Simple Mail Transfer Protocol). POP3 is the current version, as defined in RFC 1725. POP is increasingly being replaced by IMAP.

5. Principles of Operation is a manual for IBM Mainframe Systems such as S/360, S/370, S/390, etc. which describes how each of the assembler instructions operates. It is considered the Systems Programmer's "bible", commonly referred to as a "POP".

POP3 Post Office Protocol version 3 is pronounced "pop three." Think of POP3 as the place in the sky where your incoming email from all your friends is stored, waiting for you to come by and pick it up. All you have to do is to "knock" on your POP3 door, identify yourself and pick up your mail. Conceptually it's not much different from physically picking up mail at your local post office. POP3 is actually a protocol widely used on the Internet or other IP-based networks to retrieve electronic mail from a (typically distant) email server. You use POP3 to get your mail from the server it is sitting on and to download it to your PC. Most email software (sometimes called email clients) use the POP3 protocol. POP3 can be characterized as a store-and-forward mail protocol. It runs on a client/server basis, with your email client workstation (i.e your PC) running against an email server, both of which include POP3 software. POP3 generally makes use of SMTP (Simple Mail Transport Protocol), which is an extension of TCP/IP intended specifically for email transfer. Unlike the earlier POP2 protocol, however, POP3 does not require SMTP and, therefore, is characterized as being independent of the transport layer. POP3 is run by most Internet service providers and ISPs (Internet Service Providers). When accessing a network-based email server, you generally will access a POP3 server to download email. When uploading email, you access an SMTP server, which merely forwards your mail through the Internet after translating the email addresses into IP addresses after consulting with a DNS (Domain Name Server) server. Actually, the two servers may well be in the form of one physical server, logically partitioned into two. So, when I access my main email account at www.TechnologyInvestor. com, using my main email address of Harry_Newton@TechnologyInvestor.com, I access a local Internet POP (Point Of Presence). I can use a dial up connection through a modem, or DSL (Digital Subscriber Line), or a cable modem network to access the Internet. Anyway is fine. In any event, I run my laptop running the TCP/IP protocol stack and POP3 software against a technologyinvestor.com server running the same software. Once I connect to my server, all of my email downloads immediately, whether I want all of it or not. This all or nothing approach is a major drawback of POP3. If I set up the POP3 options correctly on the client workstation, I can choose to leave a copy of the mail on the server. That way, I can check my email from a friend's or colleague's computer when I'm away from my office, and can still view the mail from my own computer when I get back. POP3 also helps solve the problem of accessing my email when I'm on the road, and outside the local calling areas of my ISP. Let's say I'm in Singapore and I want to check my email. I jump onto the Internet, using any local Internet Service Provider (ISP). I then instruct my email client (in my case, it's Microsoft Outlook.) to go find my technoloyginvestor POP3 email server, the address of which happens to be email.technologyinvestor.com. It finds that server. The server asks me

to identify myself by user name and password. Then it starts sending me my email. POP3 is particularly useful because I can pick up my email from wherever I am in the world – just so long as I can get on the Internet in some way. IMAP (Internet Message Access Protocol) is an improvement on POP3, but is not widely implemented. When you are choosing an email provider, make sure that email provider places your email on a POP3 server and that you can retrieve your email from anywhere. Some companies claim to have "POP3" servers but, for some reason (technical, security or incompetence), they don't allow you grab your mail over the Internet and download it to your machine. Some email providers insist that you be on line to read and respond to your email. When I wrote an article on the subject of POP3 servers, many readers wrote me that there are "POP3" and "POP3" servers. Thus my warning to check. See also DNS, IMAP, IP, ISP, SMTP, and TCP/IP.

pop-under Also called pop-up advertisements. You got to a web site with your browser. You close your browser. Bingo, your taskbar shows your browser has visited another "site" and it's – surprise, surprise – an advertisement for a piece of hardware, a cheap airline trip, a lottery, a casino....usually something you don't want. It's called a pop-under because it pops under your browser. It's also very annoying. Companies pay for the privilege of buying these ads. A good piece of software which gets rid of Pop-Unders is called AdsGone and is available from www.AdsGone.com.

pop-up A call center term. A button that displays, on demand, several items from which you can choose (by clicking on it, for example). In effect, this is a list box that you don't see until you push button. Pop-ups always have double lines on their right and bottom sides.

pop-up ad A typically annoying Web-based advertising technique in which Javascript code creates a small Web browser window that suddenly 'pops up' in the foreground of the visual interface. Pop-ups can contain graphics, HTML, animation, or any combination of the three. Software to prevent pop-ups is available. Our favorite is AdsGone, available from www.AdsGone.com. See Inline ads.

pop-up electronic mail An electronic mail system that runs as a terminate-and-stay-resident program (typically within DOS) and can be popped up inside any application to send or read mail. Our office, we have a TSR electronic mail program that pops up. It is called Noteworks and we really love it.

pop-up program A memory-resident program that is loaded into memory but isn't visible until you press a certain key combination or until a certain event occurs, such as receiving a message. See also TSR.

POPS A cellular industry term for its customers or its potential customers. (It varies with usage.) POPS, short for "population" (well, sort of) refers to members of the population. According to Ron Schneiderman's book on Wireless, "if the coverage area of a cellular carrier include a popular base of one million people, it is said to have one million POPS. The financial community uses the number of potential users as measuring stick to value cellular carriers."

populate The classic definition of "to populate" is to furnish with inhabitants. To populate printed circuit board, you fill it with semiconductors, capacitors and other components. To populate fields in a form, you (or your computer) fill them in.

PORPX Federal Charge, local number port.

port 1. noun. The physical interface between a device and a circuit. The device may be a system (e.g., a mainframe, PC, or other host computer), a switch (e.g., PBX, Central Office, or ATM switch) or router, a hub or bridge, a buffer, a printer or other peripheral, or virtually any other type of device. The port and circuit may be either digital or analog, and either electrical (e.g., twisted pair or coaxial cable) or optical (e.g., optical fiber). The port and circuit connect through some sort of plug and socket arrangement. For example, your PC typically has one or more serial ports, a parallel port, a USB port and maybe a firewire port.

2. noun. The logical interface between a process or program and a communications or transmission facility. One or more logical ports (Lports) are associated with a single physical port. See also Logical Port

3. noun. A logical point of connection, most especially in the context of TCP (Transmission Control Protocol, which is part of the TCP/IP protocol suite developed for what we now know as the Internet. Port numbers are 16-bit values which range from 0 to 65,536. "Well-known ports" are numbered 0 to 1,023, and assigned by the IANA (Internet Assigned Numbers Authority) for the use of system (root) processes or by programs executed by "privileged users." Examples of well-known ports include 25 for SMTP (Simple Mail Transfer Protocol), 80 for HTTP (HyperText Transport Protocol), and 107 for Remote Telnet Service). In the Internet TCP/IP-based client/server environment the ports are assigned by the server in consideration of the application-level protocol being exercised at the client level. In Internet terms, it is the identifier (16-bit unsigned integer) used by Internet transport protocols to distinguish among multiple simultaneous connections to a single destination host. "Ephemeral ports" are short-lived ports assigned randomly to the source port of the

sender, or client. Ephemeral port numbers usually have a value between 1,024 and 5,000. Their short life is due to the fact that the client normally stops using the randomly selected port number once the transaction or session is completed. See also HTTP, IANA, SMTP, TCP, TCP/IP, and Telnet.

4. verb. To move a process, program or subroutine from one processor, controller, or operating system to another. For example, a software developer might "port" an application software system from UNIX to Windows XP.

port 25 blocking ISP use "port 25 blocking" to prevent their subscribers accessing "foreign" SMTP servers, i.e. using other SMTP email servers to send our mail. The reason? So ISPs can stop their customers spamming. The real reason? To reduce traffic on their network. See Port Services.

Port Address Translation PAT. A feature which lets you number a LAN (a local area network) with inside local addresses and filter them through one globally routable IP address. Here's an example of when you would use this. Let's say that you have subscribed at your home to a DSL line to the Internet from your local CLEC. For $80 a month you get a line and one IP address. (To get more IP addresses would cost you more money.) You would like to attach several PCs to the line – your wife's PC, your son's PC, etc. You attach a router to your "digital modem" which your DSL line is terminated on. That router has PAT. It essentially assigns IP addresses behind the DSL. You assign these separate IP addresses to each of your PCs, install cheap Ethernet hubs to joint the PCs together in a LAN and bingo you have a bunch of PCs able to access the Internet. I have exactly this configuration at my home. The device which does the Port Address Translation is a Cisco 1605R router. Companies also use PAT to use one set of IP addresses for internal traffic and a second set of addresses for external traffic. This allows a company to shield internal PCs from the outside world. None of these machines can be reached from the Internet. They can only access the Internet, surf, send email, etc. Port Address Translation is sometimes called Network Address Translation (NAT).

Port Aggregation Group Protocol PaGP is a Cisco proprietary technology which conforms to the IEEE 802.3ad (Link Aggregation Control Protocol-LACP), which itself is a specification for bundling multiple Ethernet links into what appears to be one spanning tree protocol (STP) link. Before this specification was ratified, various vendors had their own proprietary mechanisms for providing this functionality, but it would not work in mixed vendor environments. The technology allows an uplink to have 8 Gbps aggregate uplink speed in the situation where 8 Gigabit Ethernet links are used between two switches. ithout the technology, seven of the links would have gone into the STP blocking state because spanning-tree protocol detected a loop in the network.

port aliasing Imagine a switch on a local area network. It switches one port to another via a common backplane. We're having trouble and we're looking to do some diagnosis. So we grab all the information flowing through the switch and mirror it (i.e. forward it, but keep it going elsewhere in the switch) to a special port, which we can hook up equipment to and then monitor and check for problems in the resulting data flow – errors in packets, etc.

port connection The point of entry into a public frame relay network service.

port density The number of ports, physical or logical, per network device.

port group A collection of switch interfaces through which packets can be switched. Port groupings can be distinct for different types of destination addresses: multicast, broadcast, and unicast sprays.

port identifier The identifier assigned by a logical node to represent the point of attachment of a link to that node.

port knocking A new form of user authentication in which closed server ports are "knocked on" in a sequence known only to legitimate users. It eliminates the need for leaving prots open and vulnerable to attack.

Port Level VLAN VLAN based on source port ID. This is a multiple bridge configuration of a switch.

port multiplier A local area network interconnect, a concentrator providing connection to a network for multiple devices.

port per pillow A goal set by colleges seeking to install network connections in the bedrooms of every student on campus.

port replicators Low-cost docking station substitutes that provide one-step connection to multiple desktop devices.

port scanning Port scanning is the technique of attempting to find listening TCP or UDP ports on an IP device and abstracting from the listening ports as much information as possible about the device. Port scanning in and of itself is not usually harmful but it lets potential crackers fingerprint your systems, learn everything they can about your possible vulnerabilities and set themselves up for a later intrusion. For example, if a port scan shows that the device is listening on port 23, the cracker knows that Telnet is likely enabled on the device and can attempt a brute force password guessing attack later.

port selector Another name for a dataPBX. Since the advent of LANs (local area networks) these devices have been getting a bad rap. Not fair. These gadgets are really great at transmitting and switching huge number of low-speed asynchronous lines. If you put this sort of traffic on a LAN, you could severely mess up its performance. Some port selectors have data throughput in excess of 20 million bits per second.

port services Servers attached to the Internet run software for various purposes – sending mail, receiving mail, publishing web pages, etc. In Internet-speak, these software programs are called "port services." RFC's (Request for Comments) define the nature of the service and standard TCP/UDP port assignments. For example SMTP (Simple Mail Transfer Protocol) is nominally run on port 25, and POP3 (Post Office Protocol version 3) is located on port 110. Typically, organizations run these services behind a firewall. A firewall is designed to keep unwanted intruders out. Because these servers are enticing targets for unethical hackers, these people often run software across the Internet designed to see if they can penetrate into these servers and cause damage, or find secrets (like email addresses, credit card numbers). These are called "intrusive scans." Most intrusions begin with a port scan, a sequential query of available ports and services on an exposed system. Most firewalls will keep track of these attempts by hackers and capture the IP address of the system doing the scanning. A network administrator can then use this information along with a network trace utility such as traceroute to identify the originating ISP. Once that is found, the network administrator and a representative of the source ISP can typically find the hacker and suggest to him/her that she lay off. Thank you to Matt Kalas at Telephone@Work for help on this definition.

port sharing 1. A system which connects multiple lines to a single port by means of a manual or automatic line selection method.

2. In frame relay, where multiple virtual connections share the same port connection.

port sharing device A system which connects multiple lines to a single port by means of a manual or automatic line selection method.

port switching According to 3Com, port switching is merely an electronic patch panel function, not the genuine switching capability that provides a performance boost. Port switching lets administrators configure their networks, allocating any port to any backplane on their hub. Unlike true switching, it doesn't increase the bandwidth available to the network manager.

portability 1. The ability of a customer to take his telephone number from place to place and, for 800 numbers, from one long-distance company to another.

2. The ability of software designed for one computer system to be used on other systems. Little software outside MS-DOS software for IBM and IBM clone computers is portable. UNIX software is portable to an extent.

portable A one-piece, self-contained cellular telephone – easily carried in a brief case or purse. Portables normally have a built-in antenna and rechargeable battery and operate with six-tenths of one watt (0.6 watt) of power. Car cellular phones operate with three watts.

Portable Application Description See PAD.

portable cellular phone Also known as a "hand-held phone". Refers to a lightweight, compact cellular handset that incorporates a battery power supply, and can be used without any peripheral power or antenna. See Portable.

portable document format See PDF.

Portable Teletransaction Computers PTC. These are typically handheld devices used for retail (inventory), healthcare (tracking supplies), mobile field repair (reporting fixes), insurance (visiting car wrecks and other disasters), etc. The devices typically have telecommunications capabilities, sometimes wireless, sometimes landlines. And they typically include microprocessors, memories, displays, keyboards, touchscreens, character recognition software, barcode readers, printers, modems and local and/or wide area data radios.

portable NXX A Verizon definition. An NXX from which at least one number has ported or there is a pending order to port a telephone number.

portal site The classic definition of a portal is a door, gate, or entrance, especially one of imposing appearance, as to a palace. In the Internet / World Wide Web, a portal is a site, which the owner positions (through marketing and heavy promotion) as an entrance to other sites on the Internet. The concept is that he convinces visitors to the Internet to visit his site first, and savor the advertising on his site (his major way of making money). A portal typically has, at minimum, search engines, free email, instant messaging and chat, personalized home pages and Web hosting. In the past, companies like America Online and CompuServe would have been called portals. Many browsers (e.g., Netscape Navigator and Internet Explorer) point you to a Web site – their own Web site, which they are endeavoring to position as a "portal." Once there, you might then want to use a search engine (e.g., AltaVista, Excite, Infoseek, Lycos, Yahoo). You are then taken to their Web site. These sites serve as your secondary portal, or point of entry, into the Web. At each of these Web sites you are assailed with advertisements. Also, cookies are embedded in your computer so they can track your movement and activities on the Web to get a feel for your viewing and buying preferences and, ultimately, develop a profile on you for purposes of targeted on-line advertising. You can change your initial portal, if you like, perhaps pointing your browser to the Web site of favorite financial information provider (e.g. Bloomberg or Nasdaq), rather than that of Netscape or Internet Explorer. Portals are BIG business. In July 1998, Walt Disney Co. offered $900 million for 43 percent interest in Infoseek, a 4-year old startup with fewer than 200 employees and annual sales of about $35 million. At the time, Infoseek had never made a profit. See also Portal Service.

portal service First, read my definition for Portal above. Then add these words. Portal Service is a whole architectural concept surrounding and including a point of entry. Portal Service introduces the concept of service on which other Internet services could be built. Such services provide an entry point on which other applications and services can be built, customized and enhanced to suit an envisaged deployment of any new services via a library of building blocks.

ported in A Verizon definition. A telephone number is considered to be ported in when service provider A provides service in their switch with a telephone number assigned to service provider B's switch using local routing number (LRN) technology.

ported out A Verizon definition. A telephone number is considered to be ported out when a number assigned to service provider A is moved from service provider A's switch to service provider B's switch, using local routing number (LRN) technology.

portrait Most computer screens are horizontal, i.e. they are wider than they are high. In the new language of computer screens, this is called "landscape." When a computer screen is higher than it is wide, it's called "portrait." Some computer screens can actually work both ways. Some even have a small mercury switch in them that determines which way the screen is standing (portrait or landscape) and will adjust their image accordingly, though software in the computer has to reflow the image. See also Portrait Mode.

portrait mode 1. In facsimile, the mode of scanning lines across the shorter dimension of a rectangular original. ITU-T Group 1, 2 and 3 facsimile machines use portrait mode.

2. In computer graphics, the orientation of a page in which the shorter dimension is horizontal. The opposite is called landscape mode. See also Portrait.

POS 1. Point Of Service. Also called Point of Presence. See Point of Presence.

2. POS Device. A point of sale device such as a credit card scanner used for authorization when a purchase is made.

3. Packet Over SONET. A high-speed means of transmitting data over a SONET fiber optic transmission system through a direct fiber connection to a data switch or router. POS is a point-to-point, dedicated leased-line approach intended purely for high-speed data applications. Where a point-to-point approach is possible, POS offers significant advantages when compared to ATM's cell-switching approach. Specifically, POS allows a user organization to pass data in its native format, without the addition of any significant level of overhead in the form of signaling and control information. For example, POS allows Ethernet frames to be packed into STS-1 (Synchronous Transport Signal-1) frames of 810 bytes, with only 36 bytes of overhead- an overhead factor of less than 1% – and then sent over an OC-1 (Optical Carrier-1) frame at 51.84 Mbps. For higher capacity applications, SONET supports example rates of approximately 155 Mbps, 622 Mbps, 2.4 Gbps and 10 Gbps-all with the same low level of overhead, or inefficiency. ATM, on the other hand, is a cell-switching approach that segments each data frame into small cells of 53 octets, of which 5 octets are overhead-an inefficiency, or overhead, factor of approximately 11% ATM, on the other hand, will support voice, video, fax and any other form of traffic, as well as true data traffic.

4. Personal Operating Space. As defined by the IEEE 802.15 Working Group, which is engaged in the development of WPAN (Wireless Personal Area Network) standards, the POS covers a range of 10 meters. Within that space, low power, low data rate WPAN technologies can operated on a fixed, portable or mobile basis. See also 802.15.

POS-PHY Packet Over SONET-PHYsical layer. A specification for Packet over SONET, with SONET running at the Physical Layer (Layer 1) of the OSI Reference Model. See also OSI Reference Model, SONET, and POS.

POSI Promoting Conference for OSI. Consists of executives from the six major Japanese computer manufacturers and Nippon Telegraph and Telephone. They set policies and commit resources to promote OSI.

position A telephone console at a switchboard manned, oops, staffed by an attendant, or operator, or agent, or whatever the latest PC-correct word is.

position calculation method The location estimate is performed by a Posi-

tion Calculation Function (PCF) located in the network or, if the MS has the capability, in the MS itself. With the same network architecture, MS functions, LMU functions and measurement inputs, the PCF can be based on one of two possible variants of E-OTD; known as 'circular' and 'hyperbolic'. See Position Determination Technology and Location Services.

Position Determination Technology PDT. also known as Geolocation Technology. A technology used to determine the geographic coordinates of a radio-equipped mobile device, e.g., a cellular handset. Generally refers to technology that allows the device's position to be monitored remotely, such as by an emergency service operator in the case of a wireless E-911 call. "Position" is preferred in some circles to avoid confusion with the use of "location" in reference to cellular system roaming. Three general classes of PDT are as follows.

Network-based. Fixed-site network infrastructure (e.g., collocated or consolidated with the cellular base station) determines device's position based on received signal measurements. Includes Angle of Arrival and Time Difference of Arrival technologies. See Angle of Arrival.

Handset-based. The mobile device determines its own position (e.g., from an integral Global Positioning System receiver) and reports it to the network.

Handset-assisted. A hybrid approach wherein the mobile device collects some measurements from its environment that are reported to the network infrastructure which in turn uses them to derive the mobile's position.

See also Position Calculation Method and Location Services.

positive action digit A digit that must be dialed before a PBX will advance a call to a higher-cost route. The WATS lines are busy. Time on the queue is over. It's time to move the call to the more expensive direct distance dial. Before it can go that route, the caller must punch in a positive action digit. This affirms that the user knows he is now making a more expensive call. It causes him to think twice, allegedly.

positive trapping Use of a notch filter to trap out an interfering carrier inserted in the channel in order to deny service to non-subscribers.

POSIX Portable Operating System Interface uniX. A proposed universal UNIX interface to user-created application programs that would run on all vendor equipment, thereby improving system interoperability.

post 1. A method of uploading files to a Web server from a HTML-capable (HyperText Markup Language. HTTP (HyperText Transfer Protocol) specifies both POST and PUT. POST performs a permanent action, such as the uploading of an order or the response to a form, or is used which a significant amount of data must be input. POST must be handled by a program or script, which must already exist. PUT can be used to update information on the server, if only a single data item is required. PUT also can be directed at a resource (e.g., program or data element) which does not yet exist. See also HTML and HTTP.

2. To compose a message for an Internet Usenet newsgroup and then send it out for others to see.

3. Power-On Self-Test.

4. See BIOS.

Post Bubble Anxiety Disorder PBAD. Financial worries caused by the economic chaos that followed the dotcom boom. A Slate writer remarked that PBAD even afflicts characters in The Sopranos. "The fact that (Tony) has money worries – aggravated by Carmela's post-bubble anxiety disorder – only makes the problem worse."

Post Completion Discrepancy PCD. A service order request that fails the billing system edits.

Post Dial Delay PDD. The time from when the last digit is dialed to the moment the phone rings at the receiving location.

post office Any part of an email system that stores or delivers mail. the term post office needs a modifier. A local post office or host post office is the module on a network that users directly interact with to send and receive mail. A domain post office is the module that controls the mail delivery within a domain of multiple hosts on a single network. For a full explanation, see POP3.

Post Office Act In 1969, in the U.K. this act began the process of removing telecommunications from direct government control. See also the British Telecommunications Act.

Post Office Protocol See POP3.

post pay A method of coin phone operation characterized by the operation of a lever or button that causes the collection of deposits after the called party answers. This method of "A" and "Buttons" is still used on coin phones overseas, especially in Great Britain.

post production The editing process after the video footage has been shot.

post restante address A standard attribute in a postal address indicating that physical delivery at the counter is requested. It may also carry a code.

post tensioned concrete A type of reinforced concrete construction in which the steel is put under tension and the concrete under compression, after the concrete has hardened.

post-it note A scientist at the 3M laboratory, Dr Spence Silver, didn't know what to do with a failed glue he developed. The thing wouldn't stick, except for a short period of time. His colleague Art Fry, remembering how annoying it was to have little bits of paper marking pages in his music book fall out during choir practice, had a suggestion for temporary adhesive book marks. Other co-workers started using strips of paper coated with the glue to write notes. Today, Post-it notes are among the five top-selling office products.

post-mortem divorce A stipulation that one must be buried separately from one's deceased spouse.

postage Money paid for conveyance by mail. Mailing a building has been illegal in the U.S. since 1916, when a man mailed a 40,000-brick house across Utah to avoid high freight charges.

Postal Telephone and Telegraph PTT. The official government body that administers and manages the telecommunications systems in many European countries.

postalize To structure rates or prices so that they do not vary with the distance you're speaking, but depend on other factors (such as duration of a call, etc.) See also Postalized.

postalized Long distance phone calls were traditionally billed based on a costing algorithm which considered the distance between the originating and terminating points, call duration, and time of day. U.S. carriers, for the most part, currently no longer consider distance in the costing algorithm for domestic calls. In other words, long distance calls are subject to "postalized" charging – from the fact that Post Office also charges a flat rate irrespective of how far it carries the mail (within the country). Charges for circuit-switched data calls (e.g., Switched 56/64 Kbps), for the most part, also are postalized.

postalized rates Refers to the way the post office prices their delivery services, namely one price to anywhere in the United States. See Postalized.

posting An individual article sent to a USENET news group on Internet; or the act of sending an article to a USENET news group.

Postlink A program used by a RIME bulletin board system node in place of mailer software.

postmaster A postmaster could be the person responsible for taking care of mail problems, answering queries about users, and performing similar work for a given site. It could also be an alias for a mail server (i.e. computer) for routing and handling of electronic mail within an organization.

postmortem divorce A phenomenon among Japanese women who, unhappy in their marriages, secretly arrange to have themselves buried anywhere but beside their husbands.

postoffice See Post Office.

Postpay See Post Pay.

postscript 1. PostScript is a popular format used to create World Wide Web screens and send documents over the Internet.

2. PostScript by Adobe Systems Inc. is the standard page description language for desktop computer systems. It describes type, graphics and halftones as well as the placement of each on the page. The big advantage of PostScript is that it is device independent. Thus if you create a postscript image (text and/or photo and/or drawing), you can print it to a relatively cheap, low quality printer like a laser printer or a magazine quality printer like a Linotronics. PostScript is a printer language, much the same that BASIC is a computer language. By sending your PostScript printer a series of commands, you can make it do almost anything from printing text in a circle to printing foot-high letters to printing halftones. If you need PostScript, buy a printer that has built-in PostScript. If your printer doesn't have built-in PostScript, you may be able to get an external software interpreter but that interpreter will slow down your printer and tie up your computer while printing. Recent versions incorporate Adobe's Portable Document Format (PDF), a popular file format designed to allow everybody to read or view files even if they don't have the software the files were created with. The program used to read these files is called Adobe Acrobat, and is freely available on the Web, but you have to purchase the full-version Acrobat to create, modify and add notes to PDF files. You can convert Postscript files (which can be identified by their ".ps" extension) to PDF files (which have a ".pdf" extension) using Acrobat. PostScript first appeared in 1985, and grew out of a programming language called Interpress that Adobe's founders, John Warnock and Chuck Geschke developed while working at Xerox's Palo Alto Research Center (PARC). See also Acrobat, PDF, PARC.

POT 1. Techie-slang for potentiometer.

2. Point Of Termination. The point of demarcation within a customer designated premises at which the telephone company's responsibility for the provision of access service ends.

POT Bay The POT Bay, or Point-Of-Termination bay, is a device placed between a competitor's network and the natural point of connection to the local exchange carrier (LEC) network. It is located between a LEC's main distribution frame (MDF), which directs traffic to proper channels for distribution throughout the LEC network and the interconnector's colocated equipment. The POT Bay, according to most new local phone companies (i.e. ones that are not a Bell operating company), is an unnecessary obstacle that adds to the costs of interconnection, serves no necessary engineering function, can degrade quality, and is nothing more than a latter-day "protective coupling arrangement."

potato 1. Also called Aerial Service Wire Splice. A tool used to splice aerial service wire.

2. When potatoes were first brought to Europe in the 17th century, they were considered disgusting and blamed for starting outbreaks of leprosy and syphilis. As late as 1720, in America, eating potatoes was believed to shorten a person's life. Potatoes were banned in Burgundy, France in 1910. Fortunately that madness didn't last long and Burgundy now sports many McDonalds' restaurants.

Potemkin Village In 1787, Catherine the Great tours Crimea and other lands newly taken from Turks. The problem was that war had ravaged the areas. But when the Empress is in the areas, she sees only apparently prosperous villages. These villages, however, were in fact were mere facades erected by her principal adviser (and lover), Prince Grigori Aleksandrovich Potemkin. He put them up to fool her into believing in the great progress of the New Russia. Potemkin Village now means pretentious fakery.

potential The difference in voltage between one point and another. One point is usually ground.

potential revenue A call center term. The revenue value per call times the number of calls forecast for a given period.

potentiometer A variable RESISTOR, such as the ubiquitous volume control.

potting 1. In many European countries, the local regulatory authorities are very strict about voice boards that are attached to phone lines. Because phone lines have high voltage for ringing – 90 volts AC and higher – the authorities feel that the phone lines may caused an electrical short and possibly a fire. As a result, the authorities insist that the area of the voice boards which receives the high voltage be covered with some protective non-flammable material. Covering your board is called potting.

2. The sealing of a cable termination or other component with a liquid which thermosets into an elastomer.

POTS Plain Old Telephone Service. Pronounced POTS, like in pots and pans. The basic service supplying standard single line telephones, telephone lines and access to the public switched network. Nothing fancy. No added features. Just receive and place calls. Nothing like Call Waiting or Call Forwarding. They are not POTS services. All POTS lines work on loop start signaling. See also Loop Start.

POTS Splitter A device that rejects the DSL signal and allows the POTS frequencies to pass through.

POTS-C Plain Old Telephone Service-Centralized. An ADSL term for the functional interface between the PSTN (Public Switched Telephone Network) and the POTS splitter at the Centralized (i.e., Central Office) end of the network. The POTS splitter is a filter that uses FDM (Frequency Division Multiplexing) to separate the low-frequency POTS voice channel from the high-frequency ADSL data channel. See also POTS-R, ADSL, FDM and Splitter.

POTS-R Plain Old Telephone Service-Remote. An ADSL term for the functional interface between the POTS splitter at the Remote (i.e., customer premise) end of the network and the individual telephone sets on premise. The POTS splitter is a filter that uses FDM (Frequency Division Multiplexing) to separate the low-frequency POTS voice channel from the high-frequency ADSL data channel. See also POTS-C, ADSL, FDM and Splitter.

POTV Plain Old TV. A Microsoft definition. I kid you not.

pound The # on a pushbutton touchtone key pad is called the pound key. It's also called the number sign, the crosshatch sign, the tic-tack-toe sign, the enter key, the octothorpe (also spelled octathorp) and the hash. Musicians call the # sign a "sharp." See also # in the numbers section at the beginning of this dictionary.

pound key See Pound.

pound sand To pound sand means to act ineffectively, to waste time and accomplish nothing. Think about hitting sand with a baseball. Now you understand the meaning.

power 1. The term which describes the amount of work an electric current can do in a unit of time. We measure power in WATTS (note we spell it with two "T"s.) A WATT measures the amount of work done in lifting a quarter-pound weight a distance of one yard in one second. Metric WATTS are a little more powerful. They go the distance of one meter. Power is the product of the current in amperes times the voltage, i.e. P = IV. See Ohm's Law.

2. Power corrupts. Absolute power corrupts absolutely. – Lord Acton.

power budget In fiber optic cable communications, power budget is the difference between the transmitted power and the receiver sensitivity, measured in decibels. It is the minimum transmitter power and receiver sensitivity needed for a signal to be sent and received intact.

power carrot A term for a mobile phone in Japan that does nothing more than receive and transmit phone calls.

power conditioner A combination voltage regulating transformer and isolation transformer, providing smooth, regulated, noise-free, AC voltage. See Power Conditioning.

power conditioning Power conditioning is a generic concept to encompass all the methods of protecting sensitive hardware against power fluctuations. When electricity leaves a commercial power generating plant, it is very clean. In fact, most power companies make sure the power they put out is a pure sine wave. Unfortunately, nearly all devices connected to power lines – and the worst are things with motors, like elevators, air conditioners, etc. – create disturbances that pollute the sine wave. As power travels through a wire away from the power plant, it picks up more of these interferences. A pure AC power sine wave appears as a smooth wave. The height of the wave is measured in volts. The wave starts at zero volts and moves to the highest point of 120 volts. The wave then cycles through a low point of -120 volts and back to zero. The speed at which it travels through this cycle is the frequency. Normal frequency in North America is 60 cycles per second (Hz). (In other places it's often 50 cycles per second.) Anything that disrupts this wave can cause hardware or data problems and needs to be regulated.

Power disturbances can be categorized in several ways. A transient, sometimes called a spike or surge, is a very short, but extreme, burst of voltage. Noise or static is a smaller change in voltage. Brownouts and blackouts are the temporary drop in or loss of electrical power. Three types of protection against these three events are available: suppression, isolation, and regulation.

Suppression protects against transients. The most common suppression devices are surge protectors that include circuitry to prevent excess voltage. Although manufacturers originally designed surge protectors to prevent large voltage changes, most have also added circuitry to reduce noise on the line. Isolation protects against noise. Ferro-resonant isolation transformers use a transformer within the circuitry to envelop the sine wave at a slightly higher and lower voltage. Any voltage irregularity that extends beyond this envelope is clamped. Isolation transformers are usually expensive.

Regulation protects against brownouts and blackouts. Regulation modifies the power wave to conform to a nearly pure wave form. The Uninterruptible Power Supply (UPS) is the most commonly used form of regulation. A UPS comes in two varieties, on-line and off-line. An on-line UPS actively modifies the power as it moves through the unit. This is closer to true regulation than the off-line variety. If a power outage occurs, the unit is already active and continues to provide power. The on-line UPS is usually more expensive but provides a nearly constant source of energy during power outages. The off-line UPS monitors the AC line. When power drops, the UPS is activated. The drawback to this method is the slight lag before the off-line UPS jumps into action. That lag is getting shorter as electronics improves. So it's rarely a problem any longer.

Because UPS systems are expensive, most companies attach them only to the most critical devices, such as phone systems, network file servers, routers, and hard disk subsystems. Attaching a UPS to a local area network file server enables the server to properly close files and rewrite the system directory to disk. Sadly, most programs run on the workstation and data stored in their RAM is not saved during a power outage unless each workstation has its own UPS. If the UPS doesn't have its own form of surge protection, it is a good idea to install a surge protector to protect the UPS from transients. Proper use of power conditioning devices greatly reduces telephone system and network maintenance costs. Make sure that proper amperage is available for each system and that all outlets are grounded. Power conditioning devices connected to poorly-grounded outlets offer very little protection.

Studies have shown that total local area network maintenance costs are higher with line-surge suppressors and ferro-resonant isolation transformers alone, than with uninterruptible power supplies.

power conditioning systems A broad class of equipment that includes filters, isolation transformers, and voltage regulators. Generally, these types of equipment offer no protection against power outages.

power cross A situation in which AC current flows into a telephone circuit, as a result of contact with a power line.

power cycle Power cycle means the same as reboot, i.e. turn your machine off, count to 30, turn your machine on. It works for everything, including your cable modem, your DSL modem, your PC, your network hub, your phone system, even your car. The rule – before you call Support, power cycle your equipment. By doing that you'll clear your device's memory, and allow it to load its software from scratch, presumably cleaning out whatever bad software got in there. See rebooting.

power density Power per unit area normal to the direction of propagation, typically expressed in units of watts per square meter (W/m2) or microwatts per square centimeter (W/cm2).

power dialer A piece of hardware to which you feed a list of phone numbers you want called. It calls them one after another. It detects busies, no answers, fax machines, answering machines, voice mail machines, etc. When it hears these, it disconnects and dials the next number in the list. If it reaches a live human being, it will pass the call to the owner of the machine, who will presumably try and sell the fellow at the other end something, or try and collect money from him – the two main uses which power dialers are typically put to. The term "power dialer" is fairly new. I first saw it used on a single line device. It seemed to need a prompt from its owner to dial the next call, very much like a preview dialer. See also Predictive Dialing and Preview Dialer.

power distribution panel A part of the Rolm CBX power distribution system that receives voltages from the main power supply and distributes them to the cabinet shelves.

power divider A splitter which divides power from a single input into two isolated outputs. Devices with more than two outputs can also be found.

power down The sequence of things you to have to do to turn off a computer or telephone system. Not following the correct power down procedures can cause a loss of data.

power factor Because this is a complex subject, I have two definitions: The first is courtesy the American Power Company:

This is a number between 0 and 1 which represents the portion of the VA (Volt-Amps rating) delivered to the AC load which actually delivers energy to the AC load. With some equipment such as motors or computers, AMPS flow into the equipment without being usefully converted to energy. This happens if the current is distorted (has harmonics) or if the current is not in phase with the voltage applied to the equipment. Computers draw harmonic currents which cause their power factor to be less than 1. Motors draw out of phase or reactive currents that cause their power factor to be less than 1. See also VA.

The second is courtesy Michael Brady:

In AC circuits, the ratio of the total power in watts dissipated, as in heat or useful mechanical work, in an electrical circuit to the total product of volts and amperes applied to the circuit. In single-phase circuits, the power factor is equal to the cosine of the phase angle between the applied voltage and the applied current, and in balanced three-phase circuits, it is equal to the cosine of the phase angle between the phase voltage and phase current. When the applied voltage and current are in phase, the angle between them is zero degrees, and the cosine is 1. Hence, the product of the voltage and current gives the power dissipated, just as in DC circuits. When the applied voltage and current are out of phase by 90 degrees, or in quadrature, the cosine is 0, and none of the apparent power (product of volts and amps) applied results in heat or useful mechanical work. In a perfect resistor, the power factor is 1, and all power applied results in heat. In a perfect inductor or a perfect capacitor, the power factor is zero, and the apparent power applied results in no heat or useful mechanical work. In an inductor, voltage is ahead of current in the AC cycle, and in a capacitor, current is ahead of voltage. Engineering students have long relied on a mnemonic to remember this rule: "ELI the ICE man", where E is voltage, I is current, L is inductance and C is capacitance.

Power Factor Corrected Supply PFC. A recently developed type of computer power supply, which exhibits an input power factor equal to one. IEC555 may force most computers to use a power supply of this type at some point in the future.

power fail bypass A feature that allows analog trunks to be answered if your commercial AC power or your telephone system crashes.

power failure backup If your AC power fails, your telephone system can still operate by switching to a backup battery power supply, often called a UPS – Uninterruptible Power Supply.

power failure transfer When the commercial AC power fails and there is no backup power source – such as a battery or a generator – this feature switches some of the trunks connected to the phone system to several single line phones, which don't need external power and can draw their power from the phone lines.

power frequency Power frequency is the term used to describe the frequency of the alternating current wave that we use for electricity. Power-frequency fields in the US vary 60 times per second (60 Hz), and have a wavelength of 5,000 km. Power in most of the rest of the world is 50 Hz, i.e. 50 times a second.

power influence 1. Power influence refers to AC electrical pollution currents. This is when AC electrical power currents induce voltage along the wires of a pair, looking for paths to ground and because they are noise currents, they are also referred to as "Noise to Ground". These currents are residual currents that are not canceled out by a cable's bonding and grounding system. These left over currents cause noise on a pair and are measured by using a Power Influence meter. Power Influence is measured from tip and ring to ground. Typical values for power influence range from below 60dBrnc to more than 100dBrnc. The higher the power influence value, the more noise that is heard in the telephone receiver. These values are used to determine the circuit balance of a telephone circuit by using the formula: Circuit Balance = Power Influence - Circuit Noise.

2. A Verizon definition. The power of a longitudinal signal induced in a metallic OSP facility by an electromagnetic field emanating from a conductor or conductors of a power system. PI is also called longitudinal noise or noise-to-ground.

power level The measure of signal power at some point. The measure can be referenced to some power level in which case the measurement is expressed in dB (decibels). It may also be referenced to 1 milliwatt in which case the measurement is expressed in dBm.

power leveling Rapidly rising through the levels of a multiplayer game by questionable means, like paying a service to play your character around the clock.

Power Line Carrier PLC. An AC or DC power line can be made to carry high frequency radio waves, which can carry "information," which could be a voice, video, images or data call, as well as carrying power, i.e. standard electricity. Through a process of Frequency Division Multiplexing (FDM), the voice, data and video signals are transmitted over frequency channels distinct from those used in power transmission, with filters serving to minimize issues of interference between the channels. Electric utility companies have done this for many years for low speed data applications (e.g., telemetry and control) over medium voltage (MV) distribution power lines connecting power plants to substations. Rural telephone companies also have used power lines for many years to provide voice communications to extremely remote subscribers over low voltage (LV) electric drops. This latter approach was cost effective in cases where the subscriber had electric service but for whom it would have been prohibitively expensive for the telco to construct copper local loops to provide telephone service. PLC also has found some application on the premises for small business and residential intercom and telephone systems, with connectivity provided over LV indoor electrical wiring and through electrical buses housed in circuit breaker boxes. In all of these applications, the service works but the transmission quality is generally poor due to inherent noise problems. More recently, interest in PLC has been revived as technology has evolved and competitive pressures have encouraged technologists and service providers to develop means of bypassing telco and CATV local loops. Wireless Local Loop (WLL) is one alternative, but has limitations due to issues such as spectrum availability, interference and topology. Broadband over Power Line (BPL) offers another alternative not only in terms of broadband network access but also high speed in-building LANs. See BPL for a detailed explanation. See also WLL, Spread Spectrum and CEBus.

power line filter A device which prevents either radio frequency signals or power line surges from passing along a power cable into equipment.

power main surge protector A surge suppressor designed for use at the main power box of a building.

power management Methods used to efficiently direct power to different components of a system. This is particulary important in portable devices which rely on battery power. The life of a battery between charges are extended significantly by powering down components not in use.

power on See Power Up.

power open A new operating system which is planned to run on a new super-powerful PC manufactured by a joint IBM-Apple alliance. The idea of the IBM-Apple alliance is make a super-powerful PC that runs virtually every PC operating system imaginable, including MS-DOS, UNIX, Windows, OS/2, Macintosh. The new, all powerful operating system, would be called "Power Open."

Power Over Ethernet POE. See 802.3af.

power product A cellular radio term. A configurable parameter broadcast by the Mobile Data Base Station (MDBS), defining the desired relationship between received signal strength and transmitted power level at any single point.

power regulator Equipment that regulates the power delivered to a system. Designed to mitigate transients in the commercial electric power source.

power rudeness Ugly behavior enabled by the digital age, such as using beepers in theaters, taking cell calls in restaurants and firing employees by email. This definition from Wired Magazine.

power seller A person who make his living buying and selling things on eBay.

Power Spectral Density PSD. A measurement of the amount of power, measured in Watts, that is applied to the spectrum of carrier frequencies (i.e., the frequency or frequencies that carry the information signal) over a circuit in order to achieve a satisfactory level of signal strength at the receiving end of the circuit. Measured in Watts/Hertz, PSD applies to both electrical circuits and radio circuits. Clearly, every carrier frequency involved in a transmission circuit is at some power level. The PSD level is tuned to the specifics of the circuit, in consideration of the frequency or frequencies involved. In an electrical circuit, the circuit specifics can include the gauge (diameter) of the copper conductor, the number of splices, and the circuit length. Given those specifics, the carrier frequency or frequencies also must be considered. As high-frequency signals attenuate (lose power) more quickly than low-frequency signals, they often are transmitted at a higher power level in order to overcome this phenomenon. However, the combination of the higher frequency signal and the higher power level causes the signal to radiate a stronger electromagnetic field, which can have a decidedly negative impact on adjacent pairs in a multi-pair cable. The adjacent pairs absorb the radiated energy, which takes on the form of electromagnetic interference (EMI), or noise. Therefore, a "PSD mask" must be imposed in order to limit the PSD to acceptable ranges. In ADSL (Asymmetric Digital Subscriber Line), for example, PSD masks are imposed on both the upstream and downstream frequencies. Specifically, ADSL T1.413 standards specify that the upstream passband (allowable frequency range) of 25-138 kHz, with the associated PSD is -38 dBm/Hz; at frequencies above 181 kHz, the PSD is required to be at least 24 dB below -38 dBm/Hz, i.e., at a level of at least -62 dBm/Hz (dBm is decibels below 1mW, or milliwatt). See also ADSL, DB, DBM, and Decibel.

power splitter See Power Divider.

power supply Most single line phones are powered by the electricity that comes in over the phone line. That's why they'll work when there's a power outage. Single line phones that have gadgetry associated with them, all ISDN phones and all multi-line phones (like key systems and PBXs) require a supply of power, i.e. electricity, in addition to what they get over the phone line. Most phones and phone systems, like computers, these days are ultimately powered by low voltage direct current (i.e. D.C.). To convert the normal 120 or 240 volts AC power that comes in from your local utility to DC at the various voltages and frequencies needed by the components and circuits of the phone or computer system, you need something called a "power supply." That term may refer to something as simple as a $10 transformer or it may be as complicated and expensive as a $20,000 power supply with an uninterruptible power supply replete with wet cell batteries. Power supplies are usually the least reliable part of modern electronic gadgetry. This is because they take the hits from the lousy power the local utility sends in and also because many manufacturers skimp on the quality of their power supplies. A cheap power supply is not evident immediately. It may take time to break down. Whenever you're having intermittent problems with your phone system or computer, suspect the power supply. And, given a choice, buy the best quality power supply you can. See UPS.

power synthesizer Power synthesizers actually use the incoming utility power as an energy source to create a new sine wave that's free from power disturbances. They can be as much as 99% effective against power disturbances. Types of power synthesizers include magnetic synthesizers (capable of generating a sine wave of the same frequency as the incoming power - 60 Hz), motor generators (which use an electric motor to drive a generator that provides electrical power), and UPSes.

power systems A system that provides a conversion of a primary alternating current power to direct current voltages required by telecom equipment, and may generate emergency power when the primary alternating current source is interrupted.

power technology New technologies to create, distribute and clean electricity. A microturbine is a power technology.

power up The sequence of things you have to do in order to turn a computer or telephone system on. You can't cut corners starting up electronic equipment. It must be done carefully and in the correct order. Always count to ten after turning something off before turning it back on again. See also Power Down.

power vendor One who has a major chunk of a market. Some users believe that a good IS strategy is to buy from a power vendor in the belief that "you can't go wrong buying from AT&T, IBM, Northern Telecom..."fill in the name of your favorite power vendor.

power, peak In a pulsed laser, the maximum power emitted.

Powered Device PD. See 802.3af.

powerline communications Sending voice, video, data – in short telecommunications signals – over copper wire that normally carry high voltage electricity (110 volts AC and higher) for use in home and businesses for lighting, heating, etc. For a longer explanation, see PLC.

PPC Pay per call.

PPD Partial Packet Discard. A technique used in ATM networks for congestion control in support of both Classical IP over ATM and Local Area Network Emulation (LANE). Such data is transmitted in the form of packets and frames, respectively, each of which typically is a subset of a much larger set of data such as a file. In the case of Classical IP over ATM, each data packet can be variable in size, up to a maximum of 65,536 octets (e.g., bytes). As the IP data packet enters the ATM switch on the ingress side of the ATM network, it is stored in a buffer until such time as the ATM switch can segment it into cells, each with a payload of 48 octets – there can be a great many such cells for each packet – and act to set up a path and circuit to forward the stream of cells which comprise the original packet. If a given cell is dropped for some reason (e.g., there is not enough buffer space at either the incoming or the outgoing buffer within the switch, the integrity of the original packet is lost. Early implementations of Classical IP over ATM simply forwarded the remainder of the cells associated with that packet. So, the earlier cells made it to the destination device while the later cells didn't. When the cells were reassembled into the packet as they exited the ATM network, the result was an incomplete packet. The higher layer protocols then requested a retransmission of the entire packet. If the ATM network was highly congested, this occurrence was repeated many times, thereby contributing to further congestion. Partial Packet Discard (PPD) involves numbering each cell associated with a segmented packet as it enters the ATM domain through the inbound buffer of the ingress switch. If any cell is dropped, the subsequent cells associated with the packet are dropped, with the exception of the cell indicating the end of the data packet. PPP enhances the performance of the ATM network by dropping those cells, which serve no purpose as the entire packet will be transmitted in either case. PPP is an earlier, and less sophisticated, technique that largely has been replaced by Early Packet Discard (EPD). See also ATM, Classical IP over ATM, EPD, and LANE.

PPDN Public Packet Data Network.

PPI Pixels Per Inch. See Resolution.

PPL Pioneer Preference Licensees. A PCS wireless term. Three US companies were awarded licenses before the A and B band auctions began. They all adopted PCS-1900, at least in part.

PPM 1.Pulse Position Modulation. Also known as Pulse Phase Modulation. A form of pulse time modulation in which the position in time of a light pulse is varied; signal amplitude and frequency both remain constant. The typical implementations are known as 4PPM and 16PPM, as a pulse of light is emitted in one of 4 or 16 time slots, respectively. For example, 4PPM is a dibit coding scheme which allows two bits of information to be impressed on a single light pulse, with the exact bit pattern (i.e., 00, 01, 10 or 11) indicated by the specific position in time in which the pulse appears in a synchronized light stream. PPM can be used in both analog and digital transmission systems; it is particularly useful when power requirements must be kept low and when the transmission medium may be pulsed easily. PPM commonly is used in infrared (IR) transmission systems, in both wireless LAN (WLAN) and short-haul networking applications; it also is used in deep-space laser communications systems.

2. Periodic Pulse Metering.

3. Private Placement Memorandum. A document that describes a private (i.e. non-public) company's aspirations and attempts to convince some poor unsuspecting soul (like me) to invest their hard-earned money in the company. Sometimes the investment will be turn out good in the long-run. Sometimes it won't. Most times, you have no idea.

4. Parts Per Million. Also known as Parts of error Per Million, clock slack, drift, skew, slewing rate, and TIE (Time Interval Error). PPM is a measurement of the accuracy of synchronization of a clocking source associated with a switching system or a transmission system. For example, T-1 systems are characterized by clock slack of 50ppm, which translates into 50/1000000 of deviation, or five percent (5%). That means that the accuracy of the clocking source is +/-5% accurate. T-3 systems are characterized by a clock slack of 20ppm, meaning that up there can be a variation of up to 1789 bits in the signal between an incoming and outgoing T-3 facility. Clock slack exceeding this specified ppm causes an unacceptable level of data error.

PPN Processor Port Network, which is the master controller (i.e., CPU) of a proprietary Avaya system.

PPP Point-to-Point Protocol. As defined in RFC 1661, PPP is a Layer 2, or Data Link Layer (DLL), protocol that allows two peer devices (e.g., two host computers, or a host computer and a bridge or router) to transport packets over a simple link. PPP commonly is used to support TCP/IP traffic between an asynchronous PC and an access router for Internet access over a dialup serial link. This generally is the way that you connect across the PSTN from your PC at home to your ISP (Internet Service Provider). PPP is a connection-oriented protocol that encapsulates packet data using a variation on the HDLC (High level Data Link Control) protocol. PPP, which largely has replaced the less robust predecessor SLIP (Serial Line Internet Protocol), supports full-duplex data transmission, both synchronous and asynchronous. PPP includes error detection and data protection features, unlike SLIP. See also HDLC, Link Control Protocol, Multilink PPP, PPPoA, PPPoE, and SLIP.

PPPoA Point-to-Point Protocol over ATM. A DSL term, defining the use of the PPP protocol over a DSL (Digital Subscriber Line) access circuit running the ATM protocol. At the customer premises, the ATM protocol is embedded in an IAD (Integrated Access Device). At the edge of the network, ATM runs in an ATM-equipped DSLAM (DSL Access Multiplexer). See also ATM, DSL, DSLAM, IAD, PPP, and PPPoE.

PPPoE Point-to-Point Protocol over Ethernet. Defined in the RFC 2516 from the IETF (Internet Engineering Task Force), PPPoE is a means of connecting from your premises to your Internet Service Provider. Its main advantage is that it eliminates the need for the ISP to manage the allocation of IP addresses. According to www.dslreports.com, PPPoE is a method of encapsulating your data for transmission to a far point. It is very similar to PPP over ATM. In this case though, a DSL modem pumping ATM is internal to the computer, rather than being a short Ethernet cable away. Originally designed for dialup lines, PPPoE is being used by DSL providers to solve the problems they get managing an open DSL network, viz.: IP address shortages, broadcasts not meant for you appearing on your local IP address (because you are on a giant ISP centered virtual net), and other (mainly ISP-end) difficulties inherent in large bridged networks. See also Ethernet, PPP, and PPPoA.

PPS 1. Pulses Per Second.

2. Precise Positioning Service. The most accurate dynamic positioning possible with GPS (Global Positioning System), based on the dual frequency P-code.

3. Packets Per Second. A measurement of the throughput of a packet switch or router, or a packet-based network. For example, the theoretical limit of 10 Mbps Ethernet, when measured in terms of the smallest (i.e., 64-byte) packets, is 14,800 packets per second (PPS). By comparison, Token Ring is 30,000 and FDDI is 170,000 pps.

4. Path Protection Switched. A ring topology defined by Bellcore TA-496. PPS is really a fancy name for a duplicated SONET signal traveling over diverse (i.e., different) physical routes. When one route of network crashes, the other will take over. This enables it to survive service outages caused by cable cuts, earthquakes, lightning strikes and equipment failures. The PPS ring gives a SONET route a greater degree of survivability than other SONET transmission paths that don't have route diversity. Such SONET rings also are known as "self-healing."

PPSN Public Packet Switched Network

PPV Pay Per View.

PPSS Public Packet Switched Service. A connection-oriented, packet-switched data communication service that permits users to communicate with data terminals of other customers and on other packet networks.

PPTP Point-to-Point Tunneling Protocol, a new protocol that enables virtual private networking – enabling secure remote access to corporate networks over the Internet. The protocol was first demonstrated at InterOp, Spring 1996. According to the press release from U.S. Robotics and Microsoft which demonstrated the protocol, PPTP will help companies deploy remote access to employees more quickly, using fewer resources, by allowing them to take advantage of existing enterprise network infrastructures such as the Internet for remote access. U.S. Robotics developed the Windows NT PPTP driver, which will be integrated into Microsoft's Windows NT Server 4.0. Under an agreement with Microsoft, U.S. Robotics will license a variety of software components for PPTP to Microsoft. PPTP will be added as a standard feature to Microsoft's Windows NT Server and U.S. Robotics' Total Control NETServer remote access server platform. PPTP streamlines access in NT networking environments, and allows NT network clients to take full advantage of the services provided by Microsoft's RAS (Remote Access Service). For remote access, over analog or ISDN lines, PPTP creates a "tunnel" directly to the appropriate departmental NT Server on a network – even if there are hundreds of NT Servers. The PPTP specification builds on standards such as PPP and TCP/IP. PPTP 'tunnels' a remote user's PPP packets from the NETServer to a Windows NT server. By terminating the remote user's PPP connection at the NT server, rather than at the remote access hardware, PPTP allows network administrators to standardize security using the existing services and capabilities built into the Windows NT security domain. Using PPTP, network administrators can extend a virtual private network from their Windows NT server throughout the Internet and still retain control of their user passwords and accounts. NT provides its own knowledge of enterprise users, databases, allowed access and network addressing integrated into its RAS capabilities. With PPTP, users accessing their NT-based network will use these services, including DHCP and WINS, for access.

PQFP Plastic Quad Flat Pack, a format used in the design of PCMCIA devices. Another format is called TQFP, which stands for Thin Quad Flat Pack.

PRA An ISDN term used internationally to refer to what essentially is an ISDN version of an E-1 trunk from the customer premises to the edge of the ISDN service provider network. PRA, also known as 30B+D, supports 31 channels, in total. The 30 B (Bearer) channels are information-bearing channels; that is to say that each supports the transfer of actual user data. Each of these B channels is a "clear channel" running at 64 Kbps. The D (Data, or Delta) channel is used for all signaling and control purposes (e.g., on-hook and off-hook indication, ringing signals, synchronization, performance monitoring, and error control). The D channel also can be used for end user packet data transfer when not in use for signaling and control, which is its primary function. Many service providers allow multiple PRAs to share a single D channel, since the signaling and control functions are not so bandwidth intensive as to require a full 64 Kbps per PRA. The ITU-T, which sets IDSN standards, specifies that as many as five PRAs can share a single D channel, although a backup D channel is recommended on another PRA circuit. Therefore, the first and second PRAs support 30B+D, and the third-fifth PRAs support 31B+OD. The North American version of PRA is known as PRI (Primary Rate Interface), and supports 23B+D. See also PRI. See also ISDN for much more detail about ISDN, in general.

practice The technical and installation manuals often used by Bell Operating Companies. A poor use of the word. A better word would be procedure.

pragmatics The relationships among computing functionality, the users of the functionality, and the users' environment. This includes computing infrastructure issues, performance issues, availability, reliability, security, etc. Similar to Ergonomics.

prairie dogging When someone yells or drops something loudly in a cube farm (an office made up of cubicles), and people's heads pop up over the walls to see what's going on.

prayer See Telephone Man's Prayer.

PRBS 1. Pseudo Random Bit Sequence/pattern. A test pattern having the properties of random data (generally 511 or 2047 bits), but generated in such a manner that another circuit, operating independently, can synchronize on the pattern and detect individual transmission bit errors.

2. A Bluetooth term. Pseudorandom Bit Sequence. pre-paired device A device with which a link key was exchanged, and the link key is stored, before link establishment. See also paired device and un-paired device.

pre- See pre plus the word.

pre-arbitrated PA. A type of SMDS slot defined in support of isochronous traffic such as voice and video. As such traffic is stream oriented and time-sensitive, isochronous data must have regular and reliable access to time slots in order that it might be presented to, travel across, and exit the network without suffering from delay. Otherwise, the presentation of the data through the target device would be most unpleasant (e.g., "herky-jerky video").

pre-emphasis A gradual increase in the amplitude of higher frequency signals prior to modulation in order to improve signal-to-noise ratios. This process is often used in frequency modulation (FM) systems to correct for distortion and noise by emphasizing higher frequency audio signals. Pre-emphasis is also sometimes used in certain video circuits.

pre-emptive calling Have you ever had an emergency situation in which you needed to speak with someone, but his or her phone line was busy? Wouldn't it be helpful if your PBX had pre-emptive calling as one of its features? Pre-emptive calling is when a PBX disconnects a low-priority call to connect a high-priority call. When pre-emptive calling is implemented on a private branch exchange (PBX), each phone call is assigned a level of precedence or importance. There are two types of precedence: explicit and implicit. Explicit precedence occurs when a special priority prefix is input before making a call. For example, security managers may be assigned a high priority prefix enabling them to make emergency calls from any phone on the system.

An implicit precedence level can be assigned to an extension and is used automatically when a call is made. A use for implicit precedence might be to assign a high precedence to an emergency phone located in a public area. In either case, the level of precedence is established before the call is connected and is always determined based on the calling party. During the connection process, if the destination is busy, the PBX compares the precedence levels of the established call and the incoming call.

If the incoming call has higher precedence than the established call, the called party may be notified of the higher-precedence call with a display message or voice prompt. If the user does not disconnect from the current call within a set time-out period, the call is dropped automatically, and the call with the higher precedence is then connected to the called party. For cases in which the incoming call does not have a higher precedence, it is handled normally (busy signal, voicemail, auto forwarding, etc.)

Pre-emptive calling is a feature that can provide valuable internal services for enterprise-level PBXs. For instance, a company with remote sites can set up a hotline for each site to the system administrator, so that if something goes wrong the remote site can get immediate help. Other applications include security red phones, special links to business partners,

and emergency service notification (police, fire, 911).

Precedence can also be used to seize outgoing or networking trunks if needed in emergency situations. In the case where all of the trunks are in use and a high-precedence call is initiated, one of the calls at a lower precedence level will automatically be disconnected to free the required trunk resource. Thank you Alcatel for this definition.

pre-fetching 1. Imagine a call center with extensive customer records. Perhaps the company is a utility, a phone company or an insurance company. Now imagine that the phone rings. CallerID picks up the customer's name and pops a screen of information on the customer. Meantime, the database system has sent a signal to what's known as "near line" file storage. That near line file storage might be a magneto optical jukebox. It might be a tape file library. The "normal" time to retrieve a record from such systems might range from 30 seconds to two minutes. But if the signal is sent early, the records could be retrieved and loaded into a cache memory, ready if the customer asks a question, for example, about his old bills. This called pre-fetching. See also Pre-Loading.

2. Running Windows? Look on your hard disk, you'll find a folder called c:Windowsprefetch. This folder contains information which Windows loads when it starts. It loads this information (usually bits and pieces of software) because it anticipates that Windows will need this stuff at some stage during your upcoming computing session. Over time Windows sticks more and more into the prefetch folder. And over time, Windows loads more and more and slows progressively down. The solution is simple: delete everything in the prefetch folder. Over time, Windows will fill it up again. At which point, you'll delete all the stuff again. It's a never ending process. But it will make your PC run faster in the meantime.

pre-loading Or pre-fetching. Documents are manually (by the administrator) retrieved from a jukebox or optical drive on a local area network in anticipation of some special need. Pre-loading may take advantage of the regular cycles of business, such as payroll processing on the 10th and 20th, personnel reviews the first week of each quarter, year-end processing of accounting reports. The network's administrator can take advantage of regular cycles to improve the system's response to users. See also Pre-Fetching.

Pre-N Pre-N is the name for wireless routers that conform to what their manufacturers believe is the new upcoming standard called 802.11n. As I was writing this definition in the late fall of 2005, Belkin introduced a Pre-N wireless router and PC card. Writing about the package, PC Magazine commented, "Belkin is the first to market with a wireless router supporting the MIMO (multiple input multiple output) technology, which promises greater throughput and increased range over 802.11a/b/g products. And from what we saw of the Belkin Wireless Pre-N Router, the performance – 40.7 Mbps at 60 feet and over 8.9 Mbps at 160 feet-is unmatched." See also 802.11n.

pre-packaged Chapter 11 Your company is not doing well. You're about to go broke. You owe everyone too much money. The choice is Chapter 11 in which case you work out a deal with your creditors after you go broke. Or pre-packaged chapter 11 in which you work out a deal with your creditors before you go chapter 11. With pre-packaged chapter 11 you know better what the outcome will be. With normal chapter 11, the court appoints trustees you may not like or may have different ideas about how you should restructure your failed company.

pre-paid calling card A credit card size card which you buy at your local store. Each card comes with information that allows you to use a certain dollar amount of local or long distance phone service. Typically the card will give you an 800 (toll-free number) to call. You call it. You touchtone the authorization number you find on the back of your calling card and the switch at the other end dials your desired number. The switch measures how long you talk and lets you talk as long as the dollar entitlement on the card – $5, $10, $20, $50, etc. Pre-paid calling cards are typically expensive for long calls, but cheap for short calls, based on the alternative – using a credit card with one of the traditional long distance carriers, many of whom apply hefty up-front, per call surcharges. See also Prepaid Calling Card.

pre-subscription See Presubscription.

pre-survey When you want to order five or more lines from Bell Atlantic, they insist on doing a "pre-survey." This is their mumbo-jumbo way of saying they want to come to your office, and check out the cabling in your building and cabling coming to the building.

pre-WiMAX service A fixed wireless service that mimics the proposed 802.16-2004 standard, but because no equipment has been approved by the WiMAX Forum at this time, it is not yet technically a WiMAX service.

pre-wiring 1. Wiring installed before walls are enclosed or finished.

2. Wiring installed in anticipation of future use or need.

preamble 1. In satellite communications. A preamble is the initial introductory information ahead of other information in an information frame; consists of two sequential parts: an unmodulated carrier to aid carrier acquisition and a modulated sequence of alternating

1s and 0s to aid clock acquisition.

2. A synchronization mechanism used in Ethernet LANs (Local area Networks), the preamble is a set of eight octets (8-bit values) which precede the Ethernet frame. A very specific bit sequence, the preamble serves to alert each Ethernet-attached device to the fact that a data frame is traveling across the circuit. Once alerted to this fact, the preamble is used by all attached devices to synchronize on the rate of transmission of the data bits across the circuit. See Ethernet.

preamplifier An electronic circuit which maintains or establishes an audio or video signal at a predetermined signal strength, prior to that signal being amplified for reproduction through a monitor or speaker.

prebuttal Preemptive rebuttal.

precedence Precedence is Federal government parlance to mean a designation assigned to a phone call by the caller to indicate to communications personnel the relative urgency (therefore the order of handling) of the call and to the called person the order in which the message is to be noted. Autovon phones which have "precedence" have an additional four touchtone buttons. You can find frequencies for those buttons under the definition for DTMF.

precedence prosign An introductory character or set of characters which indicate how a message is to be handled by the receiving unit.

precise positioning service The most accurate dynamic positioning possible with GPS (Global Positioning System), based on the dual frequency P-code.

precision air conditioning Precision air conditioning systems are primarily designed for cooling electronic equipment, rather than people. These pre-packaged systems offer excellent reliability and typically have a high ratio of sensible-to-total cooling capacity and a high CFM/ton ratio.

prediction The promise of a paperless office was the most inaccurate prediction. According to the PaperCom Alliance, a nonprofit company that studies the future of paper-based products, electronic communication actually increases paper use in virtually all market sectors. The very upstarts that promised to take advantage of electronic efficiency, e-commerce companies, have consumed massive amounts of paper with their direct mail, catalogs, and print advertising to build brand awareness.

predictive coding Method or source coding using prediction. The prediction error resulting from the difference between the prediction value and actual sample value is transmitted.

predictive dialer An automated, computerized way (hardware and software) of making many outbound calls without people dialing the calls and then, once the called person has answered, passing the calls to a live operator. Here's the story: Imagine a bunch of operators having to call a bunch of people. Those calls may be for collections. They may be for employee callups to work. They may be for alumnae fund raising. When it's done manually, here's how it works: Before each call, operators spend time reviewing paper records or computer terminal screens, selecting the person to be called, finding the phone number, dialing the number, listening to rings, listening to phone company intercepts, busy signals and answering machines. Operators also spend time updating the records after each call. Predictive dialing automates this process, with the computer choosing the person to be called and the computer dialing the number and only passing the call to an operator when a real live human being answers. In a well run manual setup, with just one trunk per agent, then with luck you might reach 25 minutes talk time in the hour, rarely beyond this. And with progressive dialing (again one trunk per agent but cutting out setup time) you might reach as many as 35 minutes in the hour, occasionally beyond that. Productivity gains with predictive dialers come in two ways. First is the bit that commonly is labelled "call progress detection". The computerized predictive dialer does its best to intercept and screen out all calls other than live ones i.e. answering machines, busy signals, network busy signals, non-completed calls, operator intercepts etc. But it's not a perfect process and some non-live calls will creep through to be dispatched by the agent. Second, the major benefit in predictive dialing comes from the actual pace of dialing, or the numbers of calls dialed relative to waiting agents. In the days before the Feds cracked down on predictive dialing in the US, it was common to talk about dialers achieving up to 50-55 minutes talk time in the hour. But no more, except under quite exceptional circumstances. Even with the best dialers, finely-tuned to cope with the new compliance rules, getting much more than 40-45 minutes in the hour is exceptional, and most outbound shops will be under 40 minutes in the hour, if they are working under compliance. The US compliance rules for dialers according to the "Final Amended Telemarketing Sales Rule" (that also established the "Do Not Call" register) effective October 1, 2003, specify that predictive dialers may not abandon more than 3% of calls (measured per-day, per campaign). Also, telemarketers (including people using predictive dialers) have two seconds to connect you to a live person once you

pick up the phone and have finished your greeting, i.e. saying 'hello'. And they must let the phone ring for at least 15 seconds before hanging up, if there is no reply. In some activities, mainly market research and collections where you often see predictive employed when actually if you just ran in progressive mode you would get at least 35 mins, and sometimes a lot more, because of long talk times.   But in telemarketing, 35 mins for progressive would normally be a a max, under compliance. Since 2003, other countries have woken up to the need to restrict bad dialer behavior. The UK brought in similar rules in 2006 and regulatory authorities in other countries including mainland Europe and Australia are also planning action.

True predictive dialing should not be confused with automatic dialing. A properly-designed dialer uses complex mathematical algorithms that consider, in real time, the number of available telephone lines, the number of available operators, the probabilities of getting different kinds of call outcome, e.g. no answers, answering machines and live calls, and the distributions for the times that agents spend talking and wrapping up calls. Some readers of this dictionary accustomed to years of dialer hype will be wondering why the judgment on dialing performance is so bleak, compared with the highs that the industry had been used to. The answer lies in the fact that historically dialers have used a range of devices in order to improve performance such as putting live calls into hold queues. And nuisance call rates were often much greater than folks admitted to. The key to good predictive dialers is good design. And that's complicated. It is not enough to monitor all agent and telephony events and run a high speed simulation to calculate whatever. You need to get the design fundamentals right.

Unfortunately, even today, many vendors don't go there. And if they do, it is an enormously difficult task given the very limited resources available - i.e. just 3 abandoned calls per 100 live calls. If the live call rate is say 20% (1 call in 5 is answered) then your ration of abandoned calls is 6 per 1000 calls dialed.

Use that up too quickly and you back to progressive dialing. Some people don't like the term "predictive dialing", since they know it's had a "bad rap" in Washington, DC by being associated with junk phone calls, which is what it often is. As a result some people would prefer to call it Computer Aided Dialing. See also Preview Dialing.

predictive text Imagine you're typing Harr on your Blackberry and instantly Harry Newton pops up. Your Blackberry has correctly predicted what you wanted to type. Brilliant. Such a huge timesaver – especially on a Blackberry where it's painfully slow to type – because of its tiny keyboard. Some people like predictive text because of its time saving. Other people find it is often fiddly and tends to produce poor results. However, most reviewers have found that SureText (as Research In Motion calls predictive text) does an excellent job of guessing the writer's intentions. See also Blackberry and thumbing.

preemptive multi-tasking operating system A multi-tasking operating system allows more than one task to be active at the same time. Under Windows 3.x, a task is defined as a single program. For instance, if you have ever had a word processor and a spreadsheet program open at the same time, then you have used Win 3.x multitasking. Win 3.x is a cooperative multi-tasking environment. In other words, applications must cooperate for multi-tasking to work. The system cannot preempt the program that has control of the CPU. It is the responsibility of each program to share the CPU. Windows 95 and Windows NT is a preemptive multi-tasking environment. This means the CPU is in charge and can seize control from applications when necessary. This environment reduces the risk of your system freezing up.

preemptive operating system An operating system scheduling technique that allows the operating system to take control of the processor at any instant, regardless of the state of the currently running application. Preemption guarantees better response to the user and higher data throughput. Most operating systems are not preemptive multitasking, meaning that task-switching occurs asynchronously and only when an executing task relinquishes control of the processor. See Preemptive, Preemptive Multitasking, Real Time Support.

preemptive, real time support When Microsoft released its At Work operating system, it said it had a number of key features, one of which was "pre-emptive, real-time support." Here's Microsoft's description:

Pre-emptive, real-time support. Communication devices such as fax machines and phones are distinct from personal computers in that they have critical real-time needs. Consequently, the software in these devices must attend to communication hardware such as modems very frequently, so that pieces of the communication are not lost. To support this need, the operating system was designed to be able to put other processes "on hold" temporarily in order to service the communication hardware before continuing other functions. See At Work and Windows Telephony.

preference setting A set of parameters on software tools, including Web Browsers that allow the user to choose which stuff shows on his screen, which colors he will use to display text, whether a signature file should be attached to his email, etc.

preferential roaming When a mobile user roams, the user's phone will typically choose the network that has the strongest signal when the phone is turned on. Sometimes the default network is different to that which the home network operator would ideally choose; the example, Vodafone D2 would ideally like all its roamers in Italy to use OPI (part of the Vodafone group). With preferential roaming, a phone's SIM card is programmed to pick up the allied network overriding the strongest signal, thus stopping the paying away of expensive interconnection fees to other operators. Operators with large international cellular footprints such as Vodafone are keen to exploit this concept to maximize their share of traffic and prevent revenue leakage.

preferred call A local phone company services which lets you forward calls from a bunch of numbers you have pre-selected. The service uses the calling number ID as the basis for choosing which calls to forward.

preferred roaming list Your take your cell phone everywhere. Sometimes you take it to places your cell phone carrier doesn't serve. This means that your cellphone is now off your cell phone provider's main network, i.e. it's roaming. You'd like to use your cellphone when you're roaming. But when you're roaming, there are often many other cell networks you could connect to. Your phone needs a way to know which network to connect to. Inside your phone is a list of SIDs (System IDentification numbers) kept to permit roaming on other wireless networks. Your cell phone service provider will set up roaming agreements with other service providers in different geographic regions and the PRL will try to locate one of these service providers' networks first when the home service provider is unavailable. PRLs do change so it's a good idea to ask for a PRL upgrade every six months or so if you do a lot of roaming outside your home service area. SIDs are typically five digit numbers which identify the particular cellular carrier from whom you are obtaining service. This number identifies your "home" system. These days, PRL lists are frequently updated OTA (Over The Air) which makes it much easier for wireless companies to provide the most updated list of roaming partnerships to their users without making them come in to a point of sale. The PRL is a list of information that resides in the memory of a digital phone. It lists the frequency bands the phone can use in various parts of the country. (The smaller bands within Cellular or PCS.) The part of the list for each area is ordered by the bands the phone should try to use first. For example, say you had a Sprint PCS phone, and were travelling in an area with no Sprint coverage, weak Verizon coverage, and strong Qwest coverage. The PRL would tell the phone to look for towers using Sprint's band for that area. Finding none, it might tell the phone to search for towers in Verizon's band next, perhaps because Sprint's roaming agreement with Qwest was not as favorable, or none existed. The phone would use Verizon's towers, as dictated by the PRL, even though Qwest might provide a better signal.

prefetch See pre-fetch.

prefix One or several digits dialed or touchtoned in front of a phone number, usually to indicate something to the phone system. For example, dialing a zero in front of a long distance number in the United States would indicate to the phone company you wanted operator assistance on the call.

preform Optical fiber source material. Preform is glass rod formed and used as source material for drawing an optical fiber. The glass structure is a magnified version of the fiber to be drawn from it.

pregroup Now obsolete term used in certain non-Bell carrier terminal equipments to form up sets of 3 channels for placement into a basic carrier group of 12 channels.

premier As a noun, it means first in rank or performance. It also is an adjective, as in Premier Issue. See Premiere.

premiere A noun. The first in a series of events or things. The adjective is premier, as in "The premier issue."

premise A thesis. a proposition supposed or proved as a basis of argument or inference. Often misused to mean the space occupied by a customer or authorized or joint user in a building or buildings on continuous or contiguous property (except railroad rights of way, etc.) not separated by a public road or highway. See Premises.

premises The space occupied by a customer or authorized or joint user in a building or buildings on continuous or contiguous property (except railroad rights of way, etc.) not separated by a public road or highway. See also premise and several following definitions.

Premises Distribution System PDS. There are two meanings for premises distribution system – a general one and a specific one, specific to Lucent. Here, first, is the general definition. A PDS is the transmission network inside a building or group of buildings that connects various types of voice and data communications devices, switching equipment, and other information management systems to each other, as well as to outside communications networks. It includes cabling and distribution hardware components and facilities between the point where building wiring connects to the outside network lines and back to the voice and data terminals in your office or other user work location. The system consists of all the transmission media and electronics, administration points, connectors, adapters, plugs, and support hardware between the building's side of the network interface and the terminal equipment required to make the system operational.

Here is the specific definition – A multi-functional distribution system from Lucent to support voice, data, graphics and video communications on premise. PDS includes cables, adapters, electronics, eight pin universal wall jacks and protective devices, all arranged in a logically coherent and economic fashion. It uses fiber optic cable and twisted pair copper wire and is suitable for single building, multi-tenant high rise or campus environment.

premises lightwave system The fiber optic part of the Premises Distribution System (PDS) from Lucent. PDS, which can replace the coaxial cables linking IBM terminals and printers, consists of two fiber optic interface units, one at a controller-end and the optic interface units and one at a terminal-end linked by a fiber optic pair. The fiber optic interface units connect to the terminals through four-pair building wiring and balun adapters. Balun adapters also enable direct connections of the terminals to the cluster controller through building wiring.

premises wire The twisted-pair, quad or other wire installed at the user's location to provide telephone service. Includes both intra-building and inter-building wiring.

premises wiring system The entire wiring system on the user's premises, especially the supporting wiring that connects the communications outlets to the network interface jack.

premium services Individual cable TV channels such as HBO and Showtime which are available to cable customers for an additional fee, typically charged monthly in the US.

PReP PowerPC Reference Platform specification. It details the hardware, operating system and software elements necessary to build PowerPC-based systems that meet certain compatibility goals. According to P Magazine, these machines will run 32-bit operating systems and will resemble today's high-end desktop system with lots of memory, CD-ROM drives, stereo audio support, and PCI/PCMCIA expansion buses.

prepackaged Chapter 11 See pre-packaged Chapter 11.

prepaid phone cards A prepaid telephone card entitles the owner to make landline or cellular phone calls up to the maximum of value on the card. Some cards are renewable with cash or money from a credit card. Some are not. Prepaid telephone cards are sold in many denominations – typically $1, $5, $10, $20 and $50. Once the owner of the card makes phone calls, the value of the card decreases. The accounting is typically done in a remote switch which the user dials to make his calls if it's a landline call or is the mobile phone company's switch. To make landline calls, a user often dials via an 800 number, puts in his card number, then dials the number he wants. Companies sell prepaid telephone calls as a business, often owning the switch and renting the long distance lines. Sometimes companies buy cards from a supplier, print up their names on the cards and sell them or give the cards away as promotion. Giving them away allows the company to reward its clients and customers with what one seller of cards calls "a universally valuable commodity – long distance service." Prepaid cards are often given to employees by their companies. It allows the companies to set limits on phone calling, and bring their budgets more into line. Allegedly the profit in prepaid cards comes from the amount left on the cards by the user. Around 15% of the value paid for on cards is not redeemed. Sometimes some of the cards expire at the end of 90 days or a year. The expiration date is often printed in small type on the card. Some people have started collecting prepaid telephone cards the way people collect baseball cards. Surprise, surprise, the value of some telephone cards is increasing. I know someone who has a collection of 30,000 telephone prepaid cards. He tells me he is going to retire on this collection. A prepaid telephone card doesn't typically have technology on it or embedded into which represents the value of the money remaining on the debit card. Such technology might be an integrated circuit or a magnetic strip. And such technology is more commonly found on a debit card. See Breakage.

prepaid wireless Prepaid Wireless is a system that allows subscribers to pay in advance for wireless service. It is generally used for customers with little or no credit or those with a strict budget.

prepay The industry standard for coin phone operation which requires that the full cost of a call be deposited before a connection is attempted through the Central Office.

prepend The opposite of append. Prepend means added to the front of, whereas append means added to the back of. Prepend is different from precede, which just means comes before, rather than being part of the data frame. See Frame.

prepended Prepended means preceded as part of the same data unit (generic "packet"), whereas appended means succeeded as part of the same data unit. See SMDS.

preprocessor A device or information handling system which converts raw data into a form more easily processed with standard equipment.

prereq Prerequisite.

presence See Presence Management.

presence awareness The New York Times wrote "making a phone call has always been a game of chance. You never know whether the person you are calling is available. You just punch in the numbers and hope to get lucky. Imagine being able to learn without dialing a single digit whether another person's phone is in use, or in the case of a cellphone, whether it is even turned on. Now imagine being able to do the same thing with any wired or wireless device of the future – whether it is in the car, in an airplane or at the gym. Not only could you learn whether a person is available for a chat, but you could also deduce what that person might be doing at that exact moment, all without exchanging a word. That is the idea behind a programming concept called presence awareness, which is based on the realization that appliances on a network can automatically be detected by other devices."

presence management Presence management is the ability to tell your phone system via a web interface, application running on your PC, or phone interface whether I am at my desk, traveling, in meetings, etc and have the phone system find me and my connect calls. It combines one number dialing and multiple outbound calls to help connect people to me, if I want.

presentation address According to the book, "Internetwork Mobility," by Mark Taylor, William Waung and Mohsen Banan, the presentation address is the network name of an Application Entity within a Data Service. The Presentation Address takes the form of: optional Presentation Selector (SSAP-Selector + Session Selector (TSAP-Selector) + Transport Selector (NSAP-Selector) + a required NET. Presentation Addresses are stored in the Directory Service. Application Entity Titles are used to retrieve the Presentation Address from the Directory Service.

Presentation Indicator PI. An ISDN term. See PI.

presentation layer The sixth layer of the OSI model of data communications. It controls the formats of screens and files. Control codes, special graphics and character sets work in this layer. See OSI Standard.

presentation manager Presentation Manager is a look and feel specification and kernel-based toolkit development environment. It was developed for IBM by Microsoft with input from IBM. Presentation Manager is the standard graphical user interface and toolkit for the OS/2 operating system, which is a multi tasking operating system for personal computers. The screens are similar to those of Microsoft Windows.

presentation status For a particular call, an item that indicates if a calling identity item may be presented to the called party. If the presentation status is "public" presentation is allowed. If it is anonymous", presentation is restricted. Presentation status has to do with the presentation or not or calling line identification numbers.

preset The "programming" of radio station frequencies on a tuner or receiver or musical selections on a tape, for instant recall at the push of a button.

preset call forwarding Incoming calls will be re-routed to a pre-determined secondary number.

press-to-talk Telephone circuits are two way. Some circuits, such as mobile dispatch services for taxis, etc., are one-way. They use a microphone or handset with a button you must press-to-talk and release to listen. You can also buy a normal telephone handset with a press-to-talk button. Such a handset is useful in noisy places.

pressure cable Telephone cable equipped with air-pressure equipment so the phone company can determine when there's a problem with the line. When a cable is cut, the pressure drops and the company is notified of the problem. Nitrogen is often used instead of air because it's noncorrosive. Nitrogen also prevents water entering the cable when there's a break.

pressure system An intricate system of hollow tubes, copper cables and air compressors designed to force compressed air into the sheaths of some copper cables. The intent of the system is to prevent moisture from entering the cable by the escape of the compressed air through any cracks or holes which develop in the cable sheath. See also Pressurization.

pressurization Pumping inert gas into a heavy casing in which a couple of thick cables will be joined. The pressure is usually maintained at a few pounds above the surrounding atmospheric pressure. The idea is that the higher pressure inside keeps moisture out of the splice and thus improves the quality of phone service.

prestel A videotex system used only in Britain.

Prestel Terminal Emulation Prestel is a character-based graphics emulation for communicating with the Viewdata service, particularly popular in the United Kingdom.

presubscription A local Bell or local independent operating telephone company service that encourages each subscriber to select one long distance carrier he may use without having to dial a multiple digit access code. If you pre-subscribe to MCI in America, you will simply reach MCI by dialing "1" plus the 10-digit long distance number. AT&T's code is 10-10-288 (as in 1-0-ATT). Sprint's is 10-10-333.

pretend During the communist era in Russia, people, lamenting on how bad things were, used to quip, "We pretend to work and they pretend to pay us." My daughter claimed on her graduation that after thousands of dollars of school fees, she would soon pretend to work and soon pretend to get paid. I'm praying.

pretexting The obtaining of customer call records and other customer proprietary network information (CPNI) under false pretexts, for subsequent sale on the Web.

pretrip A central office malfunction that causes a phone to ring only once, and then stop, as if it had been answered. IT can be very confusing and difficult to get repaired, because most telco repair people have never heard of the problem and will insist your phone is at fault. Often fixed by replacing a faulty heat coil.

pretty good privacy A powerful encryption scheme. See PGP for a full description.

prevail An office automation UNIX software package which combines functions usually available only in individual programs, such as spreadsheet, a word processor, a database management system, communication capabilities and more.

preventive maintenance The periodic inspection, cleaning, adjusting and repair to eliminate problems before they affect service. Usually ignored.

preview dialing Preview dialing is a term used to describe an automatic dialer. Preview dialing is also called "screen dialing" or "cursor dialing." Typically the prospect's account information and/or phone number appears on the screen BEFORE the call is made. Thus the agent can "preview" the number, the screen, the customer. If the agent wants to make the call, the agent hits a key, such as "Enter" and the computer dials the number. In some preview dialing equipments, the agent must hit a key if he/she DOESN'T want the number dialed. Contrast preview dialing with Predictive Dialing where the computer makes all the dialing decisions and presents the calls to the agent only after they are connected. Predictive dialing is a lot faster than preview dialing. See Predictive Dialing.

prewiring 1. Wiring installed before walls are enclosed or finished.

2. Wiring installed in anticipation of future use or need.

PRI 1. Primary Rate Interface is an ISDN term used internationally to refer to what essentially is an ISDN version of a T-1 trunk from the customer premises to the edge of the ISDN service provider network. Also known as 23B+D, PRI supports 24 channels, in total. The 23 B (Bearer) channels are information-bearing channels; that is to say that each supports the transfer of actual user data. Each of these B channels is a "clear channel" running at 64 Kbps. The D (Data, or Delta) channel, which also runs at 64 Kbps, is used for all signaling and control purposes (e.g., on-hook and off-hook indication, ringing signals, synchronization, performance monitoring, and error control). The D channel also can be used for end user packet data transfer when not in use for its primary function of signaling and control. Many service providers allow multiple PRIs to share a single D channel, since the signaling and control functions are not so bandwidth intensive as to require a full 64 Kilobits per second per PRI. The ITU-T, which sets IDSN standards, specifies that as many as five PRIs can share a single D channel, although a backup D channel is recommended on another PRI circuit. Therefore, the first and second PRIs each support 23B+D, and the third-fifth PRIs each support 24B+0D. The international version of PRI is known as PRA (Primary Rate Access), and supports 30B+D. See also PRA. See also ISDN for much more detail about ISDN. See also T-1.

2. Product Release Instructions. One cell phone carrier explained this term to me, "We tell our handset vendors to load our company-specific information into the handsets they deliver to us. We then add the Preferred Roaming List (PRL) before we sell the phones to our customers." See PRL.

PRI-EOP A fax signal. PRocedure Interrupt-End of Page.

PRI-MPS A fax signal. PRocedure Interrupt-MultiPage Signal.

price cap The phone industry has always been regulated on the basis of the profits it earns compared to the investment it had. That was called Rate of Return Regulation. How you figure profits – since you also have to figure what are allowable expenses – has been the subject of on-going debate for over 100 years. The latest idea in regulation is to replace the rate of return regulation with something called "Price caps" which allow the price of phone company services to rise by x% at maximum – the so-called price cap.

In November, 2002, UBS Warburg wrote this about Price Cap Regulation:

"The regulatory framework wherein regulators set limits on prices operators can charge for qualified telecom services. Price caps are set based on a formula that is typically influenced by some inflation factor (e.g., CPI plus 200 basis points). This regulation provides carriers an operating efficiency incentive in that it does not explicitly limit a company's level of profits. As a result, a carrier that provides service for less than its expected cost benefits. Price cap carriers take on greater risk for a greater potential reward. Price cap regulation typically exists in urban and suburban jurisdictions and markets that are open to competition. The FCC mandated this regulatory regime at the federal level for the RBOCs and GTE in 1991. RLECs such as Citizens, Iowa Network Services, and Valor Telephone are also price cap at the federal level."

Price Earnings Ratio P/E Ratio. This ratio theoretically measures the value of a stock by dividing its current price on the stock market by its earnings per share over the last twelve months. When a stock's P/E ratio is high,it is considered by investors as pricey or overvalued. Stocks with low P/Es are considered good value. When stocks rose strongly (for example, in the first quarter of 2000), P/E ratios got very high. And investors sought ways to justify the high ratios. They started, for example, to figure P/E ratios on the basis of next year's earnings (which presumably, being higher, would bring the P/E ratio more down to earth). In his studies of blue chip stocks, the publisher of Investors Business Daily, William J. O'Neil, found the higher the P/E, the better the stock and the better its performance. The average P/E of the best winners over the last fifteen years at the initial buy point prior to their huge price increases was 31 times earnings. These P/Es went on to expand more than 100% to over 70 times earnings as the stocks significantly increased in price. P/Es are misunderstood and misused by many investors, according to Mr. O'Neil.

primary agent group An automatic call distributor term: Primary agent group for which the inbound calls are intended. Intraflow goes to secondary and tertiary groups if the primary group does not have an agent available after the time or after overflow parameters are exceeded. Different ACD systems label this process differently.

primary alias A term used in the Sun Solaris Teleservices platform. See Provider.

primary area A customer's local telephone calling area.

primary bill version A Verizon definition. This version of the bill is delivered electronically or by a cartridge tape that contains the monthly CLEC or Reseller bill in Bill Data Tape (BDT) formatted records.

primary buffer A part of a computer's memory where fast incoming or outgoing data is kept until the computer has a chance to process it.

primary carrier Customer's selection of a long distance carrier (IC) to automatically carry his or her traffic in equal access areas.

primary center A control center connecting toll centers – a Class 3 Central Office. It can also serve as a toll center for its local end offices.

primary colors In light (and in monitors, which produce light) the primary colors are the basic colors - red, green and blue - that can be added together to create any other color.

Primary Domain Controller PDC. For a Windows NT Server domain, the computer that authenticates domain logons and maintains the security policy and the master database for a domain. See Domain.

primary group A group of basic signals which are combined by multiplexing. The lowest level of the multiplexing hierarchy.

primary high-usage trunk group A high-usage trunk group that is offered first-route traffic only.

primary insulation The first layer of non-conductive material applied over a conductor to act as electrical insulation.

Primary Interexchange Carrier PIC. A Primary Interexchange Carrier is the long distance company to which traffic from a given location is automatically routed when dialing 1+ in equal access areas. The PIC is identified PIC Code (PICC) which is assigned by the local telephone company to the telephone numbers of all the subscribers to that carrier to ensure the calls are routed over the correct network. When a subscriber switches long distance carriers, it often referred to as a PIC change. A LPIC is an IntraLATA Primary Interexchange Carrier, which may be different from the InterLATA PIC.

primary InterLATA carrier The IC designated by a customer to provide interLATA service automatically without requiring the customer to dial an access code for that carrier. See Presubscription. Same as a Primary Interexchange Carrier.

primary link The active LAN connection. When it fails the LAN is switched to the Backup link.

primary partition A portion of a physical disk that can be marked for use by an operating system. Under MS-DOS, there can be up to four primary partitions (or up to three, if there is an extended partition) per physical disk. A primary partition cannot be subpartitioned.

Primary Rate Access See PRA.

primary rate interface The ISDN equivalent of a T-1 circuit. The Primary Rate Interface (that which is delivered to the customer's premises) provides 23B+D (in North America) or 30B+D (in Europe) running at 1.544 megabits per second and 2.048 megabits per second, respectively. There is another ISDN interface. It's called the Basic Rate Interface. It delivers 2B+D over either one or two pairs. In ISDN, the "B" stands for Bearer, which is 64,000 bits per second, which can carry PCM-digitized voice or data. See ISDN for a much better explanation.

primary resource An SCSA definition. The main resource around which a Group is constructed. Typically, the primary resource will be an interface to the telephone network, but it may also be a switch port.

primary routing point The switch designated as the channel point for a long haul telephone call.

primary server The SFT III Novell NetWare server that has been operating longer than its partner and is currently servicing the attached workstations. The primary server is the SFT III server that network workstations "see," and the one to which they send requests for network services. Routers on the internetwork see only the primary server and send routing packets to it. The primary server's IOEngine determines the order and type of events that are sent to the MSEngine. Only the primary server sends reply packets to network workstations. The secondary server is the SFT III NetWare server that is activated after the primary server. Either server may function as primary or secondary, depending on the state of the system. You cannot permanently designate which server is primary or secondary. System failure determines each server's role, that is, when the primary server fails, the secondary server becomes the new primary server. When the failed server is restored, it becomes the new secondary server.

primary station A network node that controls the flow of information on a communications link. Also, the station that, for some period of time, has control of information flow on a communications link (in this case primary status is temporary).

primary storage The main internal storage.

primary wire center A switching center in the AT&T/Bell system hierarchy of exchange classes. The primary center is a Class 3 exchange. It is used to connect toll offices and less frequently to connect a toll center with a local end office. Primary centers are capable of connecting toll centers through sectional centers and then to local end offices to establish communication connections when simple routing possibilities are busy.

prime focus parabolic antenna There are two types of satellite antennas – a prime focus parabolic antenna and an offset parabolic antenna. A "normal" satellite antenna is a pure parabola. It's been around forever. A newer antenna, called the offset antenna is taller than it is wide. According to the manufacturers, the offset antenna design makes for more efficient use of the antenna surface than a traditional prime focus parabolic antenna. What that means is that it captures more of the satellite signal hitting the antenna. Offset antennas are more expensive than the "normal" parabolic satellite antennas, which are called "prime focus parabolic antennas." They are more expensive because they cost more to make since they typically must be made out of one sheet of metal. Offset antennas are harder to carry around, since you can't make them out of several foldover sheets of metal.

prime line A key system feature. You can program your phone set to automatically select a certain phone line whenever your lift the receiver or press the Handsfree/Mute button. The line that appears is called the Prime Line.

prime line preference When you pick up the handset on your key system or hybrid key, you are automatically connected to your preferred line (central office or intercom), rather than having to punch down an extra line button. Some phone systems tout this as a feature. Some have it set up where you simply leave one of the line buttons depressed and it doesn't pop up when you put your handset back into its cradle. Most 1A2 phone systems have this feature. Not all electronic phone systems do. Walker breaks the features into Prime Line Outgoing and Prime Line Incoming, and allows you to program the phones separately.

prime site In a simulcast, the prime site has overall control of the simulcast. Other transmitter sites involved in the simulcast are called remote sites. See simulcast.

primitive An abstract, implementation independent, interaction between a layer service user and a layer service provider. See Primitives.

primitives Abstract representations of interactions across the service access points indicating information is passed between the service user and service provider. There are four types of primitives in the OSI Reference Model – request, indication, response and confirm.

principal First or highest in importance. An owner or part-owner of a business. A person who authorizes another, as an agent, to represent him. Often confused with principle, not that a principal necessarily has any principles. See Principle.

principle A general or fundamental truth on which others are based. A rule of conduct. Often confused with principal. See Principal.

principle headend According to the FCC's definition, the principal headend of a cable television system is: If the system has one headend, that headend is the "Principal Headend" and if the system has two or more headends, the operator may designate the "Principal Headend". However, once designated, it cannot be changed except for "good cause". The location of the Principal Headend is a factor in determining the must-carry status of certain broadcast stations.

Principles of Operation POP is a manual for IBM Mainframe Systems such as S/360, S/370, S/390, etc. which describes how each of the assembler instructions operates. It is considered the Systems Programmer's "bible", commonly referred to as a "POP".

print control character A coded control character used to instruct the receiving unit on how a message is to be formatted in hard copy. Print control characters include carriage returns, back spaces, line feeds, tabs, etc.

print server A networked computer, usually consisting of fixed-disk storage and a CPU, that controls one or more printers that can be shared by users.

print spooler An application that manages print requests or jobs so that one job can be processed while other jobs are placed in a queue until the printer has finished with previous jobs. See Print Spooling.

print spooling A technique used to schedule printing tasks to one printer and to free up computer time from the slow task of feeding a slow printer (Any printer is slow compared to the speed of a computer). A small program or program/machine called the spooler does the scheduling. A user loads the print task to the spooler and when the print task's turn comes, the job is printed. Print spooling is handled several ways: You can allocate part of the computer's main memory to become a print spooler. You can allocate part of the company's disk memory to become a print spooler. You can get an external device called a print spooler. It will have all the storage space and software necessary. There are two primary advantages to Print Spooling: 1. You can use the spooler to save your and your computer's time. Dump the report to a print spooler at thousands of bits per second. Get on with something else on the computer. 2. You can use a print spooler to schedule several users' printing requests. This is particularly good in multi-user environments – for example, where the printer is a laser printer (and therefore expensive) and is attached to a LAN (Local Area Network).

Printed Circuit Board PCB. Flat material (fiberglass/epoxy) on which electronic components are mounted. A PCB also provides electrical pathways called traces, that connect components. Printed circuit boards are what PBXs and computers are made of these days. Be careful when you're replacing PCBs. They're usually very sensitive to static electricity. Handle them only when you're attached to a static electricity strap that is properly grounded. Lay them down only on a surface you're sure is static electricity free. And don't touch the components on PCBs whatever you do.

printed wire assembly A printed wire assembly is another name for a printed circuit board (PCB) or printed wiring board (PWB) with all the components stuffed into the board.

printer A device which takes computer information and prints it on paper.

Printer Control Language PCL. See PCL.

printer driver A program that controls how your computer and printer interact. A printer driver file supplies information such as the printing interface, description of fonts, and features of the installed printer.

printer emulation A fax term for mimicking a printer-generated document. This way, the outgoing fax will look as if it has come from the printer attached to the computer. This can include full formatting, as well as letterhead, signature and graphic images.

printer farm An area in a room where a bunch of network printers are located.

printer font A font stored in your printer's memory, or soft fonts that are sent to your printer before a document is printed.

printer server A computer and/or program providing LAN (Local Area Network) users with access to a centralized printer. A person using the LAN will send a message to the printer server computer. This computer will then assign it a piece of memory or disk space to store its file while it waits to be printed. With a printer server, users can send to the printer any time. Their print jobs are usually handled in the order they are received. But "big bosses" can be given priority and can be bumped to the top of the queue. Print servers allow fewer printers to satisfy more users. Print servers are also especially useful for expensive, laser or high speed printers because they (the print servers) spread the cost of these expensive machines over many users, making them more affordable. See Print Spooling.

printer, wire A matrix printer which prints using a set of wire hammers which strike the page through a carbon ribbon to generate the matrix characters.

prior art Let's say you file a patent application for your latest, greatest invention. You get issued the patent and you now believe you're protected and will soon be rich and famous. Meantime, someone steals your idea and starts making a product that looks remarkably what you thought of. So you sue them to stop or at least pay you a royalty. In turn, they argue in court that your patent is not valid as a result of "prior art." What that means that a previous publication or patent is deemed to provide essential details of aspects of your invention and therefore your patent is invalid and you can go pound sand. There seems to be an unlimited demand for early editions of my dictionary. Lawyers want to prove that your patent is valid (because of prior art) and therefore their client can continue ripping you off. See Patent.

prioritization The process of assigning different values to network users, such that a user with higher priority will be offered access or service before a user with lower priority. Increasingly available as an added option with network operation. Any procedure where different levels of precedence exist.

prioritization parameters The hierarchical rules which a network device, such as a router, applies to incoming traffic to determine which traffic should be handled first, next, and so on.

priority A ranking given to a task which determines when it will be processed.

priority access methods When there's an emergency, say a flood or volcano explosion or the September 11, 2001 attack on the World Tr4ade Center, various government authorities need to get access to communications facilities, like cell phone bandwidth. Unfortunately, when a disaster happens everybody and their uncle jumps onto their cell phones and everybody promptly gets fast busy signals – including the important emergency people. So the concept of priority access methods is to figure a way of returning a fast busy to normal cell phones, while allowing the important people to get through, i.e. to get priority access.

priority bumping The process during a link, trunk or facility failure where lower priority user access to network services is interrupted in order to offer those services or bandwidth to a pre designated higher priority user.

priority call 1. Emergency calls to the attendant bypass the normal queue and alert the attendant with some special signal.

2. The name of a Bell Atlantic service. While all callers are important, some are more critical to your business. So give your high-priority callers a ring of their own with Priority Call. Simply use your phone to program up to six callers' numbers. When any of these people call, you'll know right away because you'll hear a different ring. If you have Call Waiting, you'll hear a Priority "beep" when you're on the phone. This way, you'll only have to interrupt your work to take priority calls.

3. This PBX feature allows an urgent intercom voice call to be made when the called telephone is busy or has Do Not Disturb activated. Priority Call should not be made available to everyone, and should be selectively programmed.

priority indicator A character or group of characters which determine the position in queue of the message in relation to the urgency of other messages. Priority indicators control the order in which messages are to be delivered.

priority ringing A name for a Pacific Bell (and possibly other local telephone companies') service which alerts you to have calls from selected numbers ring at another number.

priority transport The capability of a network for certain classes of traffic to have priority over others and thus have lower delay or otherwise better performance.

priority trunk queuing Through user-chosen trunk access level, this PBX feature places any caller with this or higher level in the class of service assignment ahead of callers waiting for the same trunk group (or Agent Group in the case of incoming ACD calls).

privacy Privacy usually means that once a caller "seizes" a line, no other user can access that same line even though it appears on his/her key set. Privacy can be automatic or selected for each call.

privacy and privacy release All other extensions of a line are unable to enter a conversation in progress unless the initiating telephone releases the feature.

privacy compromise A network security term. A scenario in which a malicious user is able to gain access to personal or confidential information about another user.

Privacy Enhanced Mail PEM. An Internet electronic mail capability which provides confidentiality and message integrity using various encryption methods.

privacy lockout Privacy automatically splits the connection whenever an attendant would otherwise be included on the call, i.e. the attendant can't listen in to a call she's just extended to someone. A tone warning is generated when the attendant bridges into a conversation in progress.

privacy override Activation of a special pushbutton allows the phone user to access a given busy line, even though the automatic exclusion facility is being used by the station on that line. This privilege of Privacy Override is usually only given to Big Bosses.

privacy tone A brief tone that sounds to signal an incoming call attempt on a phone that is in privacy mode. This is a feature of some phone systems.

private ATM address A twenty-byte address used to identify an ATM connection termination point.

Private Automatic Branch Exchange PABX. A private telephone switch for a business or an organization in which people have to dial "9" to access a local line. In the old days, private branch exchanges were manual, meaning that operators/attendants were needed to manually place calls. Then the systems improved and you were able to dial the outside world from your extension without the help (or hindrance?) of an operator. Thus they became known as private automatic branch exchanges. But then all PBXs became Automatic. So these days, PABXs are all called PBXs, except in some countries outside North America, where they're still called PABXs. See also the next definition and PBX.

Private Branch Exchange PBX. Term used now interchangeably with PABX. PBX is a private telephone switching system, usually located on a customer's premises with an attendant console. It is connected to a common group of lines from one or more central offices to provide service to a number of individual phones, such as in a hotel, business or government office. For the biggest definition, see PBX. See also PABX.

private carrier An entity licensed in private services and authorized to provide communications service to others for money.

private dial-in ports A packet network term. For customers who have many calls, the packet network operator provides dedicated, unpublished phone numbers. The idea is to give the preferred user better service.

private domain name A standard attribute of an O/R (Originator/Recipient) Address that identifies a PRMD (Private Management Domain) generally relative to an ADMD (Administrative management Domain). An X.400 term.

private equity fund A bunch of people and organizations pool their money and hire a manager or team of managers, who invest the money, buying various private businesses and listed equities. The managers typically get paid a percentage of the monies managed and perhaps a reward should certain profit goals be reached. Private equity funds typically have a life of five to seven years at which point the investments are typically sold and the monies distributed to the owners, i.e. the people and organizations who put their money in the first place. Hopefully, there's a profit in the long-term. See also Hedge Fund and Mutual Fund.

Private Exchange PX. A telephone switch serving a particular organization and having no means of connection with a public exchange. In other words, a phone system just for intercom calls.

private eye Around 1925, the Pinkerton Detective Agency adopted "We Never Sleep" as its motto. To symbolize this, the motto was shown over the picture of an open, ever-wakeful eye. The popularization of this emblem led to private detectives being called "private eyes."

private facility trunk A telephone company AIN term. A transmission facility that carries non-public switched telephone network (PST) traffic. An example of a private facility trunk is an access arrangement to a switch supporting PBXs, including the switched end of a Foreign Exchange (FX) and an Off Network Access Line (ONAL).

Private Internet eXchange PIX. It's a Cisco term for a family of their remote access routers with firewall capabilities.

private key In asymmetric key cryptography, the private key is the key that must not be divulged to others. Private key encryption requires that the decrypting key be kept secret. Also known as single-key and secret-key. See Public Key Encryption and Encryption for more detail.

private leased circuit A leased communications circuit, available 24 hours a day, 7 days a week, that connects a company's premises with a remote site.

private line 1. A direct circuit or channel specifically dedicated to the use of an end user organization for the purpose of directly connecting two or more sites in a multisite enterprise. A private line that connects two points together is known as point-to-point; a private line that connects one point to multiple points is known as point-to-multipoint. Private lines are leased from one or more carriers, which may be local or interexchange in nature. Private lines provide connectivity on a non-switched basis. As they bypass the network switches, private lines use the various switching centers (e.g., Central Offices, or COs) only as wire centers for the interconnection of circuits. Thereby, private lines provide full-time and immediate availability, eliminating dialup delays and avoiding any potential for congestion in the core of the carrier networks.

Private lines offer highly available connectivity, as they are dedicated to the use of a single organization, which may run any type or combination of traffic types over them. As private lines are priced based on distance and bandwidth, with no usage-sensitive cost element, they can be used constantly and at maximum capacity at the same cost as if they were never used at all. Therefore, they offer a highly cost-effective to usage-sensitive, switched services in environments where communications between sites is frequent and intense. Originally, private lines were, in fact, dedicated circuits which literally could be physically traced through the network. They also were known as "nailed-up circuits," as telephone company technicians hung the physically distinct circuits on nails driven into the walls of the central offices. Contemporary private lines actually involve dedicated channel capacity provided through the core of the carrier networks over high-capacity, multi-channel transmission facilities. The access portions (i.e., the local loop portions) of the private line are, of course, dedicated and physically distinct circuits.

Private lines are agnostic with respect to the form of the data, and the nature of the application. They can support voice, data, video, facsimile, or multimedia communications. They can run at rates of T-1, NxT-1, T-3, OC-3, or any other technically feasible speed. They can support any communications protocol, or combination of protocols, including TCP/IP, Frame Relay, or ATM. See also Private Network.

2. An outside telephone line, with a separate telephone number, which is separate from the PBX. The line is a standard business line which goes around the PBX. It connects the user directly with the LEC central office, rather than going through the PBX. Private line connections are considered to be very "private" by virtue of the fact that it is not possible for a third party (e.g., technician or console attendant) to listen to conversations without placing a physical tap on the circuit. Additionally, private lines are not subject to congestion in the PBX. As private lines also are not susceptible to catastrophic PBX failure, they often are used to provide fail-safe communications to key individuals with mission-critical responsibilities in data centers, network operations centers, and the like.

private line service An outside telephone number separate from the PBX, can be set up to appear on one of the buttons of a key telephone. Also called an Auxiliary Line. See also Private Line.

Private Management Domain PRMD. An X.400 electronic mail term: A private domain to which MTAs (Message Transfer Agents) send mail. PRMDs are connected to ADMDs (Administrative Management Domains) for message routing over wide area links. Under X.400 addressing, the PRMD represents a private electronic messaging system that may be connected to a Administrative Management Domain. The PRMD is usually a corporate or government agency email system connected to an ADMD.

private message A message designation which prevents that message from being given to another mailbox.

private network 1. A network built and owned by an end user organization. Some very large organizations build their own private microwave networks, rather than rely on circuits leased from carriers. This generally is the case where a number of remote sites must be networked, especially where substantial bandwidth is required. In such situations, the public carriers may be unable to provide the necessary bandwidth and network performance.

2. A network comprising dedicated circuits leased from one or more public carriers. Such circuits make use of private lines over carrier transmission facilities, bypassing the switches. Many large organizations deployed complex, dedicated T-carrier networks in the 1970s and 1980s. While such networks continue to be supplemented and while such networks continue to be deployed for data communications, VPNs (Virtual Private Networks) generally are preferred for voice communications. A variety of VPN technology alternatives also exist for data communications. See also Private Line, Private Voiceband Network, and VPN.

Private Network-to-Network Interface See PNNI.

private networks marketing A Northern Telecom term which defines their organization for making and selling all telecom switches, except central offices. These products include the Meridian 1 PBX family, residential and business telephone sets, including Norstar and data communications.

private peering point A privately-owned packet-switching center at which ISPs exchange traffic, avoiding the cost of doing so through the more traditional NAPs (Network Access Points). See Peering for a more thorough explanation.

private subscriber network A virtual private network service supported by Public Packet Switched Service (PPSS) and incorporating interLATA transmission facilities owned or leased by the customer for private traffic. A Bellcore definition.

Private System Identifier PSID. An signaling identifier sent from a private cell site associated with a Wireless Office Telecommunications System (WOTS). The cell phone display shows the name of the company, rather than the name of the carrier. The PSID is defined in IS-136. See also IS-136 and Wireless Office Telecommunications System.

private use network Two or more private line channels contracted for use by a customer and restricted for use by that customer only.

private voiceband network A network that is made up of voice band circuits, and sometimes switching arrangements, for the exclusive use of one customer. These networks can be nationwide in scope and typically serve large corporations or government agencies.

private wire A private line. Derives its name from the old telegraph days when messages were carried on wires that strung across the nation.

privilege elevation A network security term. The ability of a user to gain unauthorized privileges on a machine or network. An example of privilege elevation would be an unprivileged user who could contrive a way to be added to the Administrator's group.

privileges The access rights to a directory, file or program over a local area network. Typically read, write, delete, create and execute.

PRL See Preferred Roaming List.

PRMD PRivate Management Domain. An X.400 Message Handling System private organization mail system. Example: NASAmail.

PRNG A Bluetooth term. Pseudo Random Noise Generation.

pro forma Pro forma is a financial term. It is the presentation of data, such as a balance sheet or an income statement, where certain amounts are hypothetical. For example a pro forma balance sheet might show a debt issue that has been proposed but has not yet been consummated, or a pro forma income statement might reflect a merger that has not yet been completed. Pro forma can also mean deviating from established financial practices. Here's the history. In the late 1990s the Financial Accounting Standards Board suggested that software companies adopt an accounting practice it called 97-02. That was meant to assure that companies – some with pretty aggressive sales cultures – book revenue on shipped, or fully received wares, not just on orders. The SEC then looked at 97-02 and – with the laudable intention of providing investors with a standardized measuring stick for assessing company revenue across all industries – decided that the it was broadly applicable. So, early in 2000, the SEC propagated "Staff Accounting Bulletin 101," or SAB 101, which essentially made 97-02 the accounting standard for all publicly traded companies in the U.S. SAB 101 became mandatory as of the December quarter. But, the trouble is, some companies have very complicated business models, and many of them – along with some of their analysts and investors – simply don't believe that SAB 101 fairly reflects their business activities. Under SAB 101 accounting standards, a company has to wait until a contract is completely fulfilled before it can book proceeds from the contract as revenue. But some companies say software contracts, for instance, sometimes take years to fulfill, because they often include customization, service agreements, product updates and other over-time add-ons. Companies, given pressure to meet or beat the Street's estimates, want to give investors a more dynamic picture of their quarterly business activity, especially when they honestly view the proceeds of orders as, well, money in bank, whether it has been received or not. In short, pro forma financial numbers can be basically whatever the company wishes to report. Share buyer, beware.

proactive Taking the initiative. Doing it before someone (most likely your competition) forces you to do it. The word is currently in vogue among those people who believe the telephone companies should do all the positive, forward-looking actions before the competition does them and wins the customers and gets the public kudos. The word has no real meaning, but serves a purpose as a cry to action. The word actually is grammatically incorrect. The real word is "active." It is the opposite of "reactive."

proactive maintenance Proactive maintenance means fixing things before they're broken, or before they normally are fixed. The term is loose, but newly fashionable. I first learned this term when I read, "Through an aggressive proactive maintenance program, Cincinnati Bell Telephone (CBT) reduced its trouble reports to less than 1.5 per 1,000 customers per month, one-third less than the aggregate Regional Bell Operating Company (RBOC) average. The company has also achieved substantial cost savings by fixing cable problems in bulk before the customer notices them, rather than one at a time after the problem is reported. The company says that its proactive maintenance group cleared over 30,000 pairs last year at a cost of about $1.4 million; compared to the $5.2 million it would cost to fix them on individual dispatch runs. Proactive maintenance methods are now being used by all five RBOCs, but many industry observers point to CBT as the pacesetter in this area. CBT is one of the largest of the country's 1,200 independent operating companies (IOCs), incumbent local exchange carriers (ILECs) that are independent of the RBOCs."

probability theory Probability theory studies the possible outcomes of given events together with their relative likelihoods and distributions. Probability deals with predicting the likelihood of future events. Probability theory is important in picking stocks.

probe 1. A sensing device, typically about the size and shape of a pencil, that is used to sense various physical conditions such as temperature, humidity, current flow, speed. Usually connected to a meter or oscilloscope which displays the condition being monitored.

2. An empty message that is sent to reach a particular address to determine if an address can be reached.

probe envelope In X.400, the envelope that encloses a probe in the MTS (Message transfer System). See Probe.

problem If I knew what his problem was, I'd solve it. – Al Ross

problem tracking report PTR. A report maintained by a manufacturer in its problem tracking database that describes a specific reproducible product defect or anomaly with a product. A PTR is also used to document a request for a feature enhancement. Information includes PTR number, problem description, PTR priority, system configuration and steps for reproducing the problem.

process An operating system object that consists of an executable program, a set of virtual memory addresses, and one or more threads. When a program runs, a process is created.

process manufacturing The making of things. This contrasts with flow manufacturing which is working on something – like oil – that flows through a production process.

process throttling A method of restricting the amount of processor time a process consumes, for example, using job object functions.

processing delay In data communications, the time taken by a computer to operate on an inbound message and return a response; frequently not accounted for in complaints of telecommunications response time problems.

processing gain In a spread spectrum transmission system, the original information signal is combined with a pseudo random correlating, or spreading code. The more random and the greater the length of the code, the more robust the resulting spread spectrum signal is against interference and interception. A measure of this robustness is referred to as processing gain. The FCC requires a minimum of 10 dB processing gain for non-licensed equipment operating in the Part 15 902-928 MHz, 2400-2483 MHz, and 5725-5850 MHz frequency bands. See also CDMA.

processing, batch A method of computer operation in which a number of similar input items are accumulated and sorted for processing. Compare with On-Line or Interactive Processing.

processor 1. The intelligent central element of a computer or other information handling system. Also called the Central Processing Unit (CPU).

2. A device which uses amplifiers and filters to "boost," "clean-up" and reprocess signals. Sometimes channel converters demodulate signals in order to accomplish this function. See Channel Converter.

processor card See Smart Card.

processor occupancy The time the telephone system processor is in use. There are two typical demands on the central processor in a telephone system, moving calls around and running self-diagnostics. Be sure you factor in the second when you're trying to figure out how many calls your telephone system processor will handle before it dies.

processor power The number of computations that a computer, microprocessor, or digital signal processor can complete in a fixed time interval. May be measured in MIPS (millions of instructions per second) or MFlops. Typical low-end DSP chips provide up to 10 MFlops; high-end chips 30 or more.

procurement lead time The interval in months between the initiation of procurement action and receipt into the supply system of the production model (excludes prototypes) purchased as the result of such actions, and is composed of two elements, production lead time and administrative lead time.

procr Processor.

prod A device that resembles a pencil, but containing a metal tip in an insulated handle with a wire to connect it to a piece of test equipment, such as a VOM (volt-ohm-millimeter). the metal tip is touched to various points in an electrical circuit for measurements and trouble-shooting.

Prodigy Formed in 1984 as a joint venture of IBM and Sears Roebuck & Company, Prodigy was originally called TRINTEX. The name was changed to Prodigy in 1988, and the company was acquired by employees with the help of International Wireless in 1996. Prodigy used to offer on-line computer services. The company was one of the first to offer such services for a largely flat monthly fee. Recently, Prodigy decided to terminate the activities of 50 staffers who develop "content" for its information service, and instead to link its users to the content of Excite, a Web directory and search engine. Prodigy will now become more a pure Internet Service Provider, offering connections to the Internet.

productize This is a stupid word. Let's say you've designed something special for one of client – a piece or hardware or software or a little of both. He likes it. You figure that others will also. So you make it into a product which you can put in your catalog and which others can buy also. Productize also means to turn an idea into a product. You complete the R&D on it, finish the customer documentation, finish the packaging design, assign a name, model number and stocking number. And you make sure that your people in service are ready to help customers. See also beta.

profession George Bernard Shaw observed that every profession is a conspiracy against the laity. To confirm this, most lawyers use fear as their major sales tool, viz. "If you don't do what I tell you (and pay me much money to boot) you will go directly to jail, not pass Go and not collect anything."

profile A set of parameters defining the way a device acts. In the LAN world, a profile is often used by one or more workstations to determine the connections they will have with other devices and those devices they will offer for use by other devices. Often called a login file. Profiles and login files usually work like batch files, automatically executing a number of commands when you turn on the machine.

profile alignment system PAS. A system that provides an automated method by which a fiber optic fusion splicer aligns a individual fibers prior to splicing. Collimated light (i.e., light comprising parallel rays) is directed through the fibers at right angles to the axis of each. Two highly sensitive cameras produce images that the machine uses to align each fiber exactly with the other along both the X and Y axes. See also fusion splicer.

profiling Method of classifying individuals according to any data category. However, a profile only has significance if it is compared to a base figure (usually a population or a client database).

PROFS PRofessional OFfice System. Interactive productivity software developed by IBM that is part of the Virtual Machine (VM) Productivity System and runs under the VM/CMS mainframe system. PROFS is frequently used for electronic mail and is said to give a user an edge in productivity in three areas: business communications (including electronic mail), time management and document handling.

Prograde orbit An orbit in which the satellite moves in the same direction as the earth revolves, west to east. All geosynchronous communications satellites are in prograde orbits.

program Instructions given to a computer or automated phone system to perform certain tasks. Most vendors improve (update) their software programs continuously. It's a good idea to ask what the deal is with getting updates.

program circuit A voice circuit used for the transmission of radio program materials. It is a telephone circuit which has been equalized to handle a wider range of frequencies than are required for ordinary speech signals.

program counter A device inside a computer which keeps track of which instruction in the program is next, etc.

program evaluation reviews An activity many of us are consigned to spend our aging years doing. In the early days of our careers, we used to do things – actually do tasks hands-on. Then many of us, sadly, got "successful." Our jobs then became telling younger people what to do and checking that they do what we told them, or doing something better (hopefully). To do all this, we sit in meetings. We call these meetings "Program Evaluation Reviews."

Program Evaluation Review Technique PERT. A management tool for graphically displaying projected tasks and milestones, schedules and discrepancies between tasks.

program file A file that starts an application or program. A program file has an .EXE, .PIF, .COM, or .BAT filename extension. AKA executable.

Program Information File PIF. A file that provides information about how Windows should run a non-Windows NT application. PIFs contain such items as the name of the file, a start-up directory, and multitasking options for applications running in 836 enhanced mode. See PIF.

program log Records once kept by a broadcasting station in a public file which provided a record of programs broadcast, program type and program length. The logs also included commercial and public service spots. Broadcasters are no longer required by the FCC to maintain program logs.

program logic The particular sequence of instructions in a program.

program store In an electronic switching office, the semipermanent memory used for the controlling stored program.

programmable In telephony, the ability to change a feature or a function or the extension assigned to a telephone without rewiring.

programmable call forwarding This feature of a telephone system allows a user to instruct his phone to send all his calls to another phone. That phone might be another extension in the same phone system or it might be another phone number altogether in a different part of the country. This feature is great. You're going to a meeting but don't want to miss that one special call. You can send your calls to a person close to the meeting and ask them to interrupt you if that "special" call comes in. The problem with this feature is that often people forget their phone is on "forwarding" and when they return to their office, they sit around all afternoon waiting for that special call, which got call forwarded elsewhere. There are two ways to overcome this. Some phones have lights or messages on their screens which indicate all calls are being forwarded. Also, some telecom managers program their total phone systems so that twice a day, all call forwarding shall cease and all calls shall return to their original phone.

programmable configuration select Refers to the EEPROM setup routine which allows jumperless configuration of the system board.

programmable logic A Programmable Logic device is typically a chip that can be reconfigured. It consists of arrays of basic logic gates laid out in a grid of connecting wires. These chips offer the advantage of being able to be reconfigured in minutes rather than the months that is needed to revise the design of a custom-made chip. They are ideally suited for products where flexibility is needed, such as when industry standards are still undetermined. Disadvantages of programmable logic chips include that they are invariably larger due to their inherent inefficiency, which can make them more expensive. See also Custom Chip.

programmable memory Memory that can be both read from and be written into by the processor. Synonym for Random Access Memory– RAM.

programmable terminal A user terminal that has some limited processing power. Also, intelligent terminal.

programming language A language used by a programmer to develop instructions for the computer. It is translated into machine language by language software called assemblers, compilers and interpreters. Each programming language has its own grammar and syntax.

programming overlay Typically a piece of cutout cardboard, which you place over certain of the keys on a phone or console. When you punch in a certain code, the buttons become what's written on the programming overlay.

program sharing The ability of several users or computers to use a program simultaneously.

program store Permanent memory in a stored program control central office that contains the machine's generic software program, parameters and translations.

progressive conference A PBX feature. Allows the extension user to create conferences of more than three people using the consultation hold and add-on conference features. To create a conference, an extension user typically uses the consultation hold, dials the desired internal or external number and effects an add-on conference. The conference may then be progressively expanded, in this same fashion, to the maximum capability of the phone system offering this feature. A good question to ask before you get sold this feature is "does the conferencing have amplification and balancing?" Without these features, the conferencing conversation will simply get more and more difficult to hear on.

progressive dialing A form of predictive dialing, progressive dialing is slightly more automated than preview dialing. The customer data is not displayed until the number is dialed, giving the agent less time to review it and a shorter time between calls. See also Predictive Dialing and Preview Dialing.

progressive display See Interlaced GIF.

progressive tuning A method of painting pictures on computer monitors or TV screens in which the picture is painted line by line. It is today's most common way of painting a picture or an image on a computer screen.

Project 25 P25. Project 25 is a joint government/industry standard setting effort to develop technical standards for the next generation of two-way communications equipment for public safety communications agencies. To do their jobs well, Public safety radio users require specific functions:

- Control of group communications and dispatching, with purpose-built security, dynamic management of talk groups, emergency calls, talk-around capability (ad hoc calls between handsets without involvement of a base station),   prioritization of communications, etc.
- "Instant" connections, with voice call set-up time in the range 0.3 to 1 second; 0.5 second often is cited.
- Seamless radio coverage throughout the geographic area, including guaranteed coverage under the harsh environments of disasters.   Handsets can relay connections

when events take out network base stations.
- Ability to provision additional radio capacity during major incidents automatically, while guaranteeing capacity for rescue and law enforcement.
- Uncompromising voice quality to allow a listener to recognize the speaker, regardless of the background noise.

Initially, the Association of Public-Safety Communications Officials, International (ARCO) led the development with the ARCO Project 25, a standard for first response radio. Project 25 is now a joint effort, with the participation of local, state, and federal governments. The primary objectives are to provide high quality digital, narrowband radios. Additional objectives include optimizing radio spectrum efficiency and ensuring market competition among multiple vendors based on standards throughout the life of systems. Project 25 considered various access technologies in an attempt to make the best use of the available radio frequency spectrum.   Under Phase I of P25, upgrades moved the existing radio spectrum from analog technology with a 25 kHz bandwidth to digital technology with a 12.5 kHz bandwidth. The modulation selected for Project 25 is C4FM, which is a modified, four-level, frequency-shift keying (FSK), with a raised cosine filter to minimize inter-symbol interference.   TIA-102 requires new equipment to be "backward compatible" with the analog equipment to allow for a smooth transition.

In Phase II, more spectrum-efficient equipment using Frequency and Time Division Multiple Access (FDMA and TDMA) will need no more than 6.25 kHz per voice channel. This equipment will also be backward compatible with Phase I. Additionally, P25-II allows support compatibility with TETRA radios (originally Trans-European Trunked Radio, renamed Terrestrial Trunked Radio). In addition to the voice encoding method, the P25 specifications define the following open interfaces and equipment definitions:

-Common air interface (CAI) -RF sub-system -Inter-system interface -Telephone interconnect interface -Network management interface -Host and network data interfaces.

Implementations can take the form of software-defined radios, which means the same hardware can, in principal, upgrade from phase 1 to phase 2 and possibly later versions with new firmware.

Project Brand See HiperLAN.

Project Evaluation Review Technique PERT. A technique for managing a project – say the installation of a PBX – which produces a guess at the project's critical path (longest task to complete) and of project milestone completion dates. See PERT.

PROM Programmable Read Only Memory. A PROM is a programmable semiconductor device in which the contents are not intended to be altered during normal operation. PROM acts like nonvolatile memory. When you install an autoboot PROM on a LAN network board, the workstation can boot up from the network server. This is particularly useful for diskless workstations.

promiscuous mode Most Ethernet cards ignore all the packets on the network that aren't destined for them. But in a Remote server – one serving multiple remote users all calling in over modems – the Ethernet LAN card has to get access to all the packets and grab those that are meant for it – so it can pass them over to the remote callers. I assume it's called "promiscuous mode" because it means that the Ethernet card has to have a relationship with all the packets traveling on the local area network. Another application for promiscuous mode is if you want to attach software or hardware to your computer, monitor and analyze all the packets flying around your network. You can set some (but not all) Ethernet cards to promiscuous mode. See also Incestuous Amplification.

promotion According to various dictionaries, "promotion" means to raise in station, status, rank or honor." Once upon a time, a promotion meant you got a better title, a bigger office and a raise in pay. Today, it means that a software release or your hardware product just made a change in status from alpha test to beta test, or from beta test to general release. See also Alpha Test and Beta Test.

prompt An audible or visible signal to the system user that some process is complete or some user action is required. Also used to signify a need for further input and/or location of needed input. See also the next three definitions.

prompt tagging and encoding According to Steve Gladstone, author of the book, "Testing Computer Telephony Systems," (available from 212-691-8215) many test applications require the test system "know" which prompt is being played by the computer telephony system. This is often accomplished by prompt tagging, also known as prompt encoding. Prompt tagging has the computer telephony system play-out an audio tone that can be heard by the test system with each prompt that is played. Prompt tagging is relatively inexpensive to perform, and inexpensive to automate. Prompt tagging can be accomplished in two ways, either by inserting tones into actual user prompts ("insert" mode), or by having a special programmatic switchable test mode that may be toggled on/off by the computer telephony system that will play out a tone sequence before or after each prompt

is played ("append" mode).

prompting Visually or audibly indicating to a user of a telephony device that a call has reached (and been accepted by) the device and is capable of being answered. This is typically done by ringing the device, flashing a lamp, or presenting a message on the device display.

prompts 1. Recorded instructions delivered by voice processing units. Prompts may include MENUS or other information that is played each time you get into the system.

2. Messages from the computer instructing the user on how to use the system. See Menu and Audio Menu.

proof of concept You're at a trade show. You go to a booth. You see some great new technology. You can't buy it. It's simply a demonstration of new technology. What's called "proof of concept." It proves that the idea works. It doesn't mean there's a market for it. The idea of "proof of concept" is to excite people – customers or security analysts. Maybe someone will place a big order or buy the company's stock?

proof of relativity When you're with your wife's relatives, time slows down. – from comedian Don McMillan.

propagation delay The length of time it takes a signal to travel from transmitter to receiver across a circuit, at the most fundamental level, propagation delay is a factor of the finite speed at which electromagnetic signals can travel through a transmission medium. The basis for comparison is the velocity of light in a vacuum, which is 300,000 (actually 299,792) kilometers per second, or 186,280 miles per second. Signal propagation is approximately 222,000 kilometers per second in an optical fiber, depending on factors such as impurities in the fiber core, temperature, and the refractive index of the cladding. Signal propagation in CAT 5 copper wire is roughly 200,000 kilometers per second, depending on a variety of factors including the nature of the dielectric insulating material.

Propagation delay is a huge issue in satellite communications. Given the fact that the originating signal must travel from the earth station 22,300 miles up to the satellite and 22,300 miles back down, a roundtrip (i.e., up and down in one direction) transmission takes about .25 seconds. When you add the time imposed for signal processing at the transmitting earth station (i.e., satellite dish), the space station (i.e., satellite), and the receiving earth station, the total delay is about .32 seconds. Therefore, it takes at least .64 seconds to get a response to your query. This level of propagation delay renders satellite communications ineffective for highly interactive data communications applications, as the users get really bored. Satellite communications also is highly aggravating for voice communications.

This satellite example illustrates the fact that propagation delay is affected by not only the characteristics of the physical transmission medium, but also the nature and number of various devices associated with the circuit. Examples include terminal equipment, bridges, hubs, switches, and routers. Such devices accomplish various processes, including buffering, queueing, protocol conversion, and error detection and correction. Each of these processes takes some amount of time, which generally is sensitive to the complexity of the process. Buffering and queueing of data is a means of dealing with issues of congestion in the network, or a given network element, and the level of delay imposed on the data is variable in nature, sensitive to the level of congestion existing at a given moment in time.

Propagation delay is a factor of great significance in many data communications protocols (e.g., Synchronous Data Link Control, or SDLC), where there may be a tight timing relationship between the transmitter and the receiver across the circuit. These timing considerations are significant in the control of access to the transmission medium, and in the detection and correction of errors in transmission. See also Path Delay Value and Velocity of Light.

propagation delay skew The difference between the propagation delay on the fastest and slowest pairs in a UTP cable. When one pair is much higher or lower in delay than the others, a very high skew may result.

propagation time Time required for an electrical wave to travel between two points over a transmission circuit. See also Propagation Delay.

propagation velocity The speed at which electrons or photons travel through a transmission medium. See also Propagation Delay.

propeller head An excessively technical person, whose social skills are lacking.

properties Windows 95 treats all objects, such as windows, icons, applications, disk drives, documents, folders, modems, and printers as self-contained objects. Each object has its own properties, such as the object's name, size position on-screen, and color, among others. You can change an object's properties using the properties dialog box.

Property Management Interface PMI. A telephone system's ability to talk to a hotel's computer system.

proportional font A font in which different characters have varying widths. All magazines and newspapers are printed in proportional characters, which make reading easier. By contrast, in a monospaced font, such as one on an old typewriter, all characters have the same widths.

proportional return policy A policy whereby an international long-distance carrier in Country A partitions (i.e., apportions) its outbound international traffic to competing carriers in Country B in direct proportion to the respective inbound traffic shares of each of the latter. See partitioning.

proprietary If something is proprietary it means it will only work with one vendor's equipment. See the next three definitions.

proprietary LAN A LAN (Local Area Network) that runs the equipment of only one vendor. A proprietary LAN, for example, cannot join IBM PCs to DEC minicomputers. Some people say such LANs are more "bug-free" because they have only one vendors' wares to deal with. They also tend to be more expensive. They also tend to tie you to one vendor, although some makers are now coming out with bridges which connect proprietary LANs to non-proprietary LANs. Since Ethernet and the Internet, proprietary LANs are effectively dead.

proprietary network A network developed by a vendor that is not based on protocols approved by standards body or on standards that are "open." Typically, you won't be able to connect to a network with any equipment other than that made by the manufacturer who created it.

proportional return policy A policy whereby an international long-distance carrier in Country A partitions (i.e., apportions) its outbound international traffic to competing carriers in Country B in direct proportion to the respective inbound traffic shares of each of the latter. See partitioning.

proprietary telephone sets Proprietary telephones are feature phones that are specific to a particular make of PBX, ACD or other switching system. They may be digital or analog. As they are custom designed for that system, they have non-standard electrical interfaces and have non-standard protocols to communicate between the telephone and the switch. This has several implications: 1. You can't take a proprietary phone from one switch and expect it to run on another switch. It won't. 2. Proprietary phones are expensive, and are highly profitable to their makers. Hence the manufacturers' insistence on keeping them proprietary. 3. Signaling between proprietary phones and their switches is richer than signaling between switches and single line analog phones. As a result, it's preferable to integrate voice mail and automated attendants through proprietary phones. Sometimes the manufacturer of the switch will divulge his secret signaling scheme. Other times he won't. Most times he wont. And you, as a voice mail or auto attendant manufacturer have to reverse engineer it, which is sometimes successful. ISDN is actually the first attempt to make "proprietary" sets standard. So far, only a few manufacturers have used ISDN-like phones as their proprietary phones.

prosodics In speech recognition, prosodics refers to the parts of the sentence the speaker emphasizes. For example, "I am going to Paris" with emphasis on the "I" means that only one person is going to Paris and therefore only one ticket should be issued. See also Prosody.

prosody Intonation. In text to speech, prosody refers to how natural it sounds – the ups and downs of the sentence. See also Prosodics.

prospective The opposite of Retrospective or Retroactive. Most regulatory commission rate cases are prospective, which means they relate to prices and things in the future. Some rate cases, however, are retroactive or retrospective, which means they apply to prices and things in the past. Most of these decisions involve forcing the company to return money to its subscribers in the form of a refund. Interestingly – try this one – most retroactive commission decisions are prospectively retroactive. In other words, they only take effect some time in the future, when the decision is voted upon by all the commission members.

Prospero UNIX software which helps you search archives connected to the Internet. Prospero uses a virtual file system which enables users to transparently view directions and retrieve files. In short, Prospero is a distributed directory service and file system that allows users to construct customized views of available resources while taking advantage of the structure imposed by others.

prosumer A cross between a professional and a consumer. Imagine you make a camera that's too good and too expensive for consumers, but too cheap for professionals. Therefore the camera is a now classed as being for prosumers – whoever they are. I guess they're rich consumers, or poor professionals.

protected area A zone defined by a station's FCC license which is legally protected from interference (on the station's authorized channel or channels) by all other stations.

protected zone Either of two regions of space around satellite orbits reserved by international agreement for working satellites only. The low zone is intended to protect satellites in Low Earth Orbit (LEO) and includes orbital space up to an altitude of 2000 km (1243 miles). The Geostationary Earth Orbit (GEO) zone protects the geostationary orbit where most communications satellites operate and extends 235 km (146 miles) both above and below the geostationary orbit at an altitude of 35,780 km (22,237 miles) above the equator. See also graveyard shift and IADC.

Protected Distribution System PDS. This is a US Federal Government definition: A wireline or fiber-optics telecommunication system which includes adequate acoustic, electrical, electromagnetic, and physical safeguards to permit its use for the unencrypted transmission of classified information. A complete protected distribution system includes the subscriber and terminal equipment and the interconnecting lines.

protected memory An operating system feature that keeps one program from grabbing memory set aside for another program and corrupting data to that program.

protected memory allocation mode A mode in which the operating system (the OS) reserves memory for itself. By switching the processor to this mode, the OS can execute several programs at once, transcending the one-megabyte limit normally enforced on the processor.

protected mode A computer's operating mode that is capable of addressing extended memory directly. The operating mode for the Intel 80286 and higher processors (the 80386, 80486 and Pentium) that supports multi tasking, data security, and virtual memory. The 80286 processor can run in either of two modes: real or protected. In real mode, it emulates an 8086 (it accesses a maximum of 640KB of RAM and runs only one software application at a time). Protected mode allows the 80286 processor to access up to 16MB of memory. It uses a 24-bit address bus. Since a bit can have one of two values, raising the base number of 2 to the power of 24 is equal to 16,777,316 memory addresses. Each memory address can store one byte of information (16,777,216 bytes equals 16MB). Protected mode operation also makes it possible to run more than one application at once and to handle more processes because more memory is available. Processes can be requests from an operating system or an application to perform disk I/O, memory management, printing, or other functions. Processes are assigned priority numbers in protected mode. The processor gives priority to those with higher numbers. Operating system processes always have higher priority than application processes. See also Real Mode and Virtual 8086 Mode.

Protective Connecting Arrangement PCA. A device leased from the telephone company and placed between your own (customer-provided) telephone equipment and the lines of the telephone company. The idea was to protect their lines from your junky equipment. No instance/case was ever proven of harm occurring to the network from faulty customer-provided equipment and the PCAs were thrown out and replaced by the FCC's Part 68 Registration Program. Under this program, customer-owned equipment which passes FCC tests can be registered and connected directly to the phone network without these devices. The phone industry eventually refunded most of the fees it charged on the PCAs. NATA and many manufacturers claimed the PCAs were designed to prevent the growth of the interconnect or customer-owned phone industry. They were probably right. The question is now moot, since the charges and the devices no longer exist, except in a museum or attached to very old equipment. See also Protective Coupling Arrangement and PCA.

Protective Coupling Arrangement PCA. A device placed between the phone company's trunks and your particular telephone gadget. The objective of the PCA is to isolate the telephone company's lines from your equipment and thus protect their lines from your equipment. The device is not needed if your equipment has passed FCC approval – under Part 68 of the FCC's rules. See also Protective Connecting Arrangement, which is another term for the same thing.

protector block A device connected to an exchange access line to protect connected equipment from over-voltage and/or over-current. Hazardous voltages and currents are shunted to ground. In other words, a surge protector limits unwanted surge voltages to values which can be handled safely by the insulation on inside wire and by the electronics in the customer terminal equipment. Protectors are very important in high-lightning areas, where they (theoretically) keep wires and phones from melting, phone systems from being blown off the wall, and end users from being electrocuted.

The original protectors were based on carbon blocks which effectively blocked aberrant voltage surges. Subsequently, gas tube protectors were used. Solid state protectors were the third generation. Improvements in the speed of of reacting to incoming high voltage and high currents have been at the forefront of the improvements in technology. While all variety of protectors currently are in place, those currently being deployed are either solid-state or hybrids, which incorporate both gas tube and solid-state technology. Protectors often are an element of a multi-function NID (Network Interface Device), also known as a NIU (Network Interface Unit), which acts as the point of demarcation between the local exchange carrier and the customer premise.

protector frame A frame, usually part of the MDF, that serves as termination for loop cables. The protector frame contains electrical protection devices that normally provide conducting paths, but will break down and electrically isolate a loop from the switching equipment when an abnormally high voltage contact occurs.

PROTEL PROcedure Oriented Type Enforcing Language. Protel is a block-structured, type-enforcing, high level, software language that enables extensive type checking on the source code at compile time. It was developed at Bell Northern Research, a subsidiary of Northern Telecom. Protel is used in the DMS-100, a family of Northern Telecom central office telephone switches. Both the central control CPU and the DMS SuperNode CPU are programmed in Protel.

protn Protection. See Protector Block.

protocol Protocols define the rules by which devices talk with each other, or more formally, a protocol is a set of rules governing the format of messages that are exchanged between computers and people. Imagine making a phone call. You pick up the phone, listen for dial tone, then punch out some buttons on your phone, then listen for ringing and for an answer. The person says "Hello." You say "Hello." Then you talk... What you're doing is following a protocol to make a call. When computers make calls between themselves – to transfer data, for example – they follow a protocol. They aren't smart, like you and I. They can't distinguish between dial tone and fast busies, unless those sounds and signals are specifically defined. A protocol defines the procedure for adding order to the exchange of data (i.e. a "conversation.") A protocol is a specific set of rules, procedures or conventions relating to format and timing of data transmission between two devices. It is a standard procedure that two data devices must accept and use to be able to understand each other. The protocols for data communications cover such things as framing, error handling, transparency and line control. There are three basic types of protocol: character-oriented, byte-oriented and bit-oriented.

Protocols break a file into equal parts called blocks or packets. These packets are sent and the receiving computer checks the arriving packet and sends an acknowledgement (ACK) back to the sending computer. Because modems use phone lines to transfer data, noise or interference on the line will often mess up the block. When a block is damaged in transit, an error occurs. The purpose of a protocol is to set up a mathematical way of measuring if the block came through accurately. And if it didn't, ask the distant end to re-transmit the block until it gets it right. See PROTOCOLS for a list of the more common protocols. See the following protocol definitions. See also Handshaking and Line Discipline.

protocol analyzer A specialized computer and/or program that hooks into a LAN and analyzes its traffic. Good protocol analyzers can record and display data on all levels of traffic on a LAN cable, from the lowest media access control packets to NetBIOS commands and application data. They are excellent for diagnosing network problems, but they require some expertise, as their data output can be obscure.

protocol control Protocol control is a mechanism which a given application protocol may employ to determine or control the performance and health of the application. Example, protocol liveness may require that protocol control information be sent at some minimum rate; some applications may become intolerable to users if they are unable to send at least at some minimum rate. See MCR.

Protocol Control Information PCI. the protocol information added by an OSI (Open Systems Interconnection) entity to the service data unit passed down from the layer above, all together forming a PDU (Protocol Data Unit).

protocol conversion A data communications procedure which permits computers operating with different protocols to communicate with each other. See Protocol and Protocol Converter.

protocol converter A device which does protocol conversion. It's your classic "black box." Glasgal Communications defines a protocol converter as any device which translates a binary data stream from one format to another according to a fixed algorithm. Compare with bridge and gateway, which are different animals and may contain protocol converters...and more.

Protocol Data Unit See PDU.

protocol dependent routing Any routing method in which routing decisions are made on the basis of information provided by the specific LAN protocol used by the communicating devices. TCP/IP and DECnet routers are protocol dependent routers. So are so-called multi protocol routers, because they must support each protocol running in the network. See also Protocol Independent Routing.

protocol filtering A feature available in some network bridges which allows it to be programmed to always forward or reject transmissions associated which specified protocols.

Protocol Independent Multicast - Dense Mode PIM-DM is a multicast routing protocol designed for networks that have a large number of multicast subscribers and enough bandwidth to support the multicast sessions. PIM-DM uses a flood and prune technique to reach all parts of the network to inform routers of the multicast session. Computers that want to join the multicast session contact the nearest router using the Internet Group Management Protocol (IGMP) and ask to be added to a specific multicast group. The router adds them to the group and builds a delivery tree that links the multicast sender to the computers that have joined the multicast group. Any paths to routers that do not have computers that want to join the multicast session are removed or "pruned" from the delivery tree until only the paths to those computers who have joined the multicast group are left. PIM-DM sends multicast packets to routers in the delivery tree, which are then delivered to only the computers that have joined the multicast group. PIM-DM uses reverse path forwarding (RPF) to distribute the multicast packets to the delivery tree and to keep network loops from forming in the process. The router implementing RPF receives a multicast packet from the multicast sender and compares the destination IP address of the packet to the delivery tree. If the IP address is in the delivery tree routing table the packet is forwarded, if not it is discarded. PIM-DM is similar to other multicast routing protocols such as the Distance Vector Multicast Routing Protocol (DVMRP), but unlike other protocols PIM-DM uses any underlying unicast routing protocol such as RIP, OSPF, BGP, etc. to build a multicast delivery tree. The ability to work with multiple unicast protocols allows PIM-DM to work across mixed network infrastructures that are present in large networks. The above courtesy Alcatel.

protocol independent router A routing device that provides the functionality of protocol specific routers such as TCP/IP or DECnet routers but is independent of protocols. In addition to routing "routable" protocols like TCP/IP, DECnet or XNS, it routes IBM protocols which are not routable. The protocol independent router combines the latest in computer hardware with the new advanced routing technologies such as SPF (Shortest Path First) and IS-IS (OSI routing standard). It represents an alternative to conventional routers that use old routing technologies and are protocol dependent. Protocol independent routers provide easy-to-install-and-use enterprise-wide networks in a token ring or Ethernet environment.

protocol independent routing A routing method in which routing decisions are made without reference to the protocol being used by the communicating devices. Protocol independent routers provide the functionality of protocol specific routers such as TCP/IP or DECnet routers, but can also route non-routable protocols. See also Protocol Independent Router.

protocol mapper Protocol Mappers are employed where a logical server is delivering a proprietary CTI protocol and / or providing connections through a proprietary transport protocol. Mappers deliver one of the specified CTI Protocols alter first mapping, or translating, from the non-specified, proprietary protocol.

protocol mapper device Implementation of a Protocol Mapper as device which sits on a data connection and is transparent to the logical client. See Protocol Mapper.

protocol mapper code Implementation of a Protocol Mapper as a software component. Mapper Code components mimic transport protocol stack implementation, and, using an appropriate R/W interface, are layered above any transport protocol stack and are transparent to Client Implementations. See Protocol Mapper.

protocol stack First, read the definition on what a protocol is. Understand that a protocol is a specific set of rules, procedures or conventions relating to format and timing of transmission between two devices. Protocol defines the rules for "conversations" – voice, data, video, etc. – between two computers. A protocol stack is basically a collection of modules of software that together combine to produce the software that enables the protocol to work, i.e. to allow communications between dissimilar computer devices. It is called a stack because the software modules are piled on top of each other. The process of communicating typically starts at the bottom of the pile and works itself up. Each software module typically (not always) needs the one below it. Sometimes one big protocol stack – such as the one for H.323 – might include specific protocol standards further down the stack. The TCP/IP protocol stack includes such protocols as TCP, IP, FTP, SMTP, telnet, and so on. A protocol stack is also called a protocol family or protocol suite. See Protocol.

protocol suite A hierarchical set of related protocols.

protocol translator Network device or software that converts one protocol into another similar protocol.

protocols For an explanation of protocols, see Protocol. Here are the more common PC protocol types:
MODULATION PROTOCOLS
    Bell 103: Low Speed (300 baud)
    Bell 212: Low Speed (1200bps)
    ITU-T V.22bis: Medium (2400bps)

ITU-T V.32: Medium Speed (9600bps)
ITU-T V.32bis: High Speed (14,400bps)
ITU-T V.34: High Speed (28,800bps)
ERROR CONTROL PROTOCOLS
Microcom Network Protocol (MNP)
ITU-T V.42 (Includes LAP-M & MNP)
DATA COMPRESSION PROTOCOLS
MNP/5
ITU-T V.42bis
All data compression requires an underlying error control protocol.
FILE TRANSFER PROTOCOLS
Kermit: 7-bit data path, quotes control characters.
XMODEM: 8 bit data path, ACK/NAK protocol.
YMODEM: 8-bit data path, batch capability.
ZMODEM: 8-bit data path, quotes some control characters.

proton A Proton is a heavy subatomic particle that carries a positive charge. Protons are found in the nucleus of the atom.

prototyping The development of a model that displays the appearance and behavior (look and feel) of an application to be built. A prototype may only demonstrate the application's. It may also demonstrate navigation and user controls, or it may even accept input data that can be stored in and retrieved from a simulated database. (See also iterative development.)

provider A process that represents an interface between the Sun Solaris Teleservices platform and an installed telephone device, such as a telephone line or a fax machine. Multiple instances of a provider can be configured, each based on the same information. Each configuration is identified by a unique primary alias. A primary alias is the primary label used to prefer to a provider configuration. The primary alias is a provider's default and primary name. This definition courtesy Sun.

provider of last resort The service provider that is obligated to provide basic service to a customer in the absence of the availability or willingness of an alternative competitive service provider to do so. The term generally applies to the telecommunications industry, although it now also applies to the electric industry. In the days of monopolistic utilities, when there was only one local service provider, there was no issue. Since competition has been introduced in the local exchange, however, some service provider must serve as the provider of last resort, regardless of profitability (or the lack thereof). The provider of last resort is obligated to provide basic service (i.e., local dial tone) to customers within its franchised serving areas in the event that either no other service provider is available or no other service provider is willing to provide service. Examples include people and companies with terrible credit records, people who require lifeline service, and people who need TDDs (Telecommunications Devices for the Deaf). Providing such services is decidedly unprofitable, but some company must do it as a matter of social policy. In the telecommunications domain, the provider of last resort is the ILEC (Incumbent Local Exchange Carrier); in areas where the electric industry has been deregulated, it is the incumbent electric utility. The concept is closely tied to that of Universal Service. See also Universal Service.

provisioning The act of supplying telecommunications service to a user, including all associated transmission, wiring, and equipment. The telephone industry defines provisioning as an engineering term referring to the act of providing sufficient quantities of switching equipment to meet established service standards. In NS/EP telecommunication services, "provisioning" and "initiation" are synonymous and include altering the state of an existing priority service or capability.

proximity sensor A device that detects the presence of an object and signals another device. Proximity sensors are often used on manufacturing lines to alert robots or routing devices on a conveyor to the presence of an object.

proxy 1. A proxy is an intermediate application program that acts as both a client and a server. A proxy runs on a gateway that relays packets between a other trusted clients and an untrusted host, perhaps making protocol translations in the process. A proxy accepts requests from the trusted client for specific Internet services and then acts on behalf of this client (in other words, serves acts as proxy for this client) by establishing a connection for the requested service. The request appears to originate from the gateway running the proxy, rather directly than from the client. All application-level gateways use application-specific proxies (that is, modified versions of specific TCP/IP services). Most circuit-level gateways use pipe, or generic, proxies that offer the same forwarding service but support most TCP/IP services. A "transparent proxy" makes no modifications to a request from an origin client other than what absolutely is required to identification and authentication. A "non-transparent proxy" modifies requests from the origin client and/or responses from

the origin server in order to provide additional services. See Client, Dual-Homed Gateway, Proxy Server, and Server.

2. A software agent that acts on behalf of a user. Also, the mechanism whereby one system "fronts for" another system in responding to protocol requests. Proxy systems are used in network management to avoid having to implement full protocol stacks in simple devices, such as modems. In SNMP, a proxy is a device which performs SNMP functionality for a separate managed device. The amount of responsibility may vary. Proxy ARPing refers to address recognition for another unit with SNMP capability, while a proxy agent provides an external SNMP agent for a managed device which does not have SNMP capability.

proxy ARP The technique in which one machine, usually a router, answers ARP requests intended for another machine. By "faking" its identity, the router accepts responsibility for routing packets to the "real" destination. Proxy ARP allows a site to use a single IP address with two physical networks. Subnetting would normally be a better solution.

proxy server A proxy is an application running on a gateway that relays packets between a trusted client and an untrusted host. A proxy server is software that runs on a PC and is basically a corporate telephone system for the Internet. Here's what I mean: A telephone system's main job is to allow a large number of people access to a few number of phone lines. Example: we have 100 people in our firm. But we have only 30 outside phone lines. To grab an outside phone line you dial 9 and then dial your number. When the 31st person tries to grab an outside line, he gets a busy. If this happens a lot, we install more phone lines. The reason we have fewer phone lines than people is clearly economic. We save money that way. All phone systems work that way. A proxy server performs the same function. Let's say your company is connected to the Internet on a single high-speed digital line, e.g. a T-1. The provider of this line gives you a certain number of distinct and different IP addresses which you can use – just like our phone system gives us 30 distinct and different phone lines. Since most firms will have fewer IP addresses than they have people wanting to use the Internet, they'll need a proxy server to act like a phone system – allocating precious IP addresses as the people want them. This process is called address translation. A proxy server is typically also a firewall – that means it keeps unwanted intrusion from the Internet getting into your corporate network. Thus, the firewall's IP addresses function as a proxy addresses. Proxy servers provide extra security by replacing calls to insecure systems' subroutines. Proxy servers also allow companies to provide World Wide Web access to selected people, restricting some, allowing others through the firewall – just like a phone system restricts some people from making long distance calls, etc. Acting as behind-the-scenes directors, proxy servers can also help distribute processing load and provide an added layer of security. A proxy server could also cache some of the material from popular Web sites, saving access time and phone monies.

In short, a proxy server lets your employees access the Internet right from their desktop PCs over a shared, managed, and secure connection to the Internet. No more running modems to desktops – a slow, expensive solution. That connection to the Net can be "nailed up," like a T-1 or equivalent, or it can even be an on-demand connection. That is, if there is no traffic moving over the connection for a period of time, a proxy server can turn off the connection so your company isn't wasting dollars on an Internet connection not being used. And the proxy would then re-establish the connection immediately when a user tried to access a web site.

According to Microsoft, a proxy server has the following advantages:

1. It accelerates access to the Internet with intelligent caching – no more World Wide Wait!

2. It protects your Intranet in ways a packet filtering router can not.

3. It blocks access to undesirable sites and provides other easy-to-use management features.

4. It saves money by consolidating and making the most of your Internet connection. See also Router-based Firewall and Dual-Homed Gateway.

PRS 1. Primary Reference Source. The master clocking source in a network. Other, distributed devices derive their clocking from the PRS in order that the entire network and all associated network elements maintain synchronization. See also Clock, Stratum Level and Timing.

2. See Personal Radio Services.

PRSM Post Release Software Manager.

PRX Program.

PS Paging Systems.

PS-ACR Power Sum-Attenuation-to-Crosstalk Ratio. A measurement of the strength of the data signal in one pair compared to one or more other pairs in a common cable. PS-ACR is measured in dB (decibels). See also ACR and dB.

PS-NEXT Power Sum-Near End CrossTalk. A measurement of the extent to which one cable pair resists interference generated by one or more other pairs in a common cable. PS-NEXT is measured in dB (deciBels). See also dB and NEXT.

Ps and Qs In English pubs, ale is ordered by pints and quarts. In old England, when customers got unruly, the bartender would yell at them to mind their own pints and quarts and settle down. It's where we get the phrase "mind your Ps and Qs."

PS/ALI Public Safety Automatic Line Information for E911. See ALI.

PS/2 IBM Personal System/2 personal computer. An IBM personal computer that was popular for a while. It became extinct when IBM could not make personal computers as cheap as its competitors (like Dell and Compaq) could.

PSA Professoinal Services Automation.

PSAI AT&T's Processor-to-Switch Applications Interface. See also ASAI and SCAI.

psalmtop A growing number of people are using their hand-held organizers to help them practice their religion. The New York Times coined the word on February 7, 2002. Neat, eh?

PSAP Public Safety Answering Point. PSAPs are customarily segmented as "primary," "secondary" and so on. The primary PSAP is the first contact a 911 caller will get. Here, the PSAP operator verifies or obtains the caller's whereabouts (called locational information), determines the nature of the emergency and decides which emergency response teams should be notified. ALI (Automatic Location Information), contained in a database, provides supplemental information for purposes of locating the caller, determining if hazardous materials are located at the subject, and so on. In some instances, the primary PSAP may dispatch aid. In most cases, the caller is then conferenced or transferred to a secondary PSAP from which help will be dispatched. Secondary PSAPs might be located at fire dispatch areas, municipal police force headquarters or ambulance dispatch centers. Often the primary PSAP will answer for an entire region. See also 911, Enhanced 911 and ALI.

PSC/PUC Public Service Commission. Also known as Public Utility Commission. It's the state agency charged with regulating the local phone company utility. In reality, there are only two things the PSC can do: 1. Allow the phone company to increase its prices, and 2. Restrict competition to the phone company by creating all sorts of restrictive rules and regulations. As competition in the telecommunications industry grows – chiefly because of Federal rulings – the state PSCs are losing their power. This bothers them.

PSD See Power Spectral Density Mask.

PSD Mask See Power Spectral Density.

PSDS Public Switched Digital Service. A BOC service. AT&T Circuit Switched Digital Capability (CSDC), also known commercially as AT&T's Accunet Switched 56 service. It allows a full-duplex, dial up, 56-Kbit/s digital circuits on an end-to-end basis.

PSE Power Sourcing Equipment. See 802.3af.

PSELFEXT Power Sum Equal Level Far End Crosstalk. A calculation derived from an algebraic summation of the individual ELFEXT effects on each pair by the other three pairs. There are four PSELFEXT results for each end.

Pseudo Automatic Location Identification pALI. A telephone network database record that holds the location of a wireless cell or sector (whereas an ALI contains the location of a wireline caller). In wireless E-911, the pALI is used to provide a rough estimate of the wireless caller's location. See Position Determination Technology.

Pseudo Automatic Number Identification pANI. Also called Routing Number. A number employed in wireless E-911 call setup that can be used to route the call to an appropriate public service answering point (PSAP). The pANI generally identifies the cell/sector from which the call originates, whereas an ANI carries the actual telephone number of a wireline caller.

Pseudo Code P-CODE. Program code unrelated to the hardware of a particular computer and requiring conversion to the code used by the computer before the program can be executed or acted upon. Here's a more technical explanation. Pseudo Code is a compiled program written for a hypothetical processor and interpreted at runtime by a P-code interpreter written for a native environment. P-code has many objectives in its different implementations, most often portability and space savings.

pseudo flaw A loophole planted in an operating system as a trap for intruders.

pseudo lite A ground based differential GPS (Global Positioning System) which transmits a signal like that of an actual GPS satellite, and can be used for ranging.

pseudo random A superficially random process or series of events which follow some obscure algorithm.

pseudo random bit pattern A test pattern consisting of 511 or 2,047 bits ensuring that all possible bit combinations can pass through a network without error.

pseudo random number generator A device which generates apparently random numbers based upon some algorithm.

pseudo random test signal A pseudo random test signal is a signal consisting of a bit sequence that approximates a random signal.

pseudo range A distance measurement based on the correlation of a GPS (Global Positioning System) satellite transmitted code and the local receiver's reference code, that has not been corrected for errors in synchronization between the transmitter's clock and the receiver's clock.

pseudo ternary A term used in ISDN Basic rate interface data coding. Refers to three encoded signal levels representing two-level binary data (binary "1"s are represented by no line signal, and binary "0"s by alternating positive and negative pulses).

Pseudowire Emulation Edge-to-Edge A mechanism that generalizes the pseudowire concept introduced in the Martini Draft. Pseudowire Emulation Edge-to-Edge defines encapsulations for ATM, Frame Relay, TDM, Ethernet and other services over a packet-switched network, such as an IP network. See Martini Draft.

PSI 1. Packet Switching Interface.

2. Pounds per square inch, a unit of air pressure. Telephone cables that are pressurized with nitrogen (because it's not corrosive) are kept at a pressure of around ten to 15 PSI.

PSID Private System IDentifier.

PSK Phase Shift Keying. A method of modulating the phase of a signal to carry information. See Phase Modulation.

PSM Phase Shift Modulation, another way of saying Phase Shift Keying. See Phase Shift Keying.

PSN 1. Packet Switch Node. The contemporary term for the IMPs (Interface Message Processors) originally used in the ARPANET and MILNET, which were the predecessors to what we now call the Internet. PSNs are intelligent switching nodes, which may be in the form of either packet switches or routers.

2. Processor Serial Number. Intel created quite a stir when it released the Pentium III processor in February 1999. Each Pentium III processor chip has a PSN embedded into it during the manufacturing process. The PSN serves as a unique identifier for the processor, and the associated system of which it is a part. If enabled by the client system user, the PSN is provided to the server on request. In combination with other identifiers such as login names and passwords, the PSN provides an additional authentication mechanism and, thereby, an additional level of security. In an e-commerce application, the PSN can be matched up with other personal information as a means of ensuring that you are who you say you are, and that the transaction, therefore, is legitimate. The PSN also provides corporate IT managers with the ability to inventory and track Pentium III computers through the network, without having to track them down physically and enter serial numbers either manually or through the use of a bar code scanner. Privacy advocates created a minor furor when they suggested that the PSN was a means of tracking your activities on the World Wide Web. The furor subsided, but the issue remains.

PSNEXT (Power sum NEXT. A measurement for qualifying cabling intended to support 4-pair transmission schemes such as Gigabit Ethernet. It is an algebraic summation of the individual NEXT effects on any one pair by the other three pairs.

PSP 1. PCS Service Provider.

2. Payphone Service Provider.

3. Purchase Service Provider. A company which provides ecommerce services for a fee or a commission. I've heard fees of $1 a transaction and also 25% of the total value of the sale.

PSPDN Packet Switched Public Data Network. A PSPDN is a general purpose data network using packet transmission techniques, as opposed to circuit techniques as used for instance in the PSTN. It is used primarily for communications with or between computers.

PSQM Perceptual Speech Quality Measure. An automated method which rates voice quality based on comparison of a predefined speech sample before and after transmission through a codec and/or network. P.861 describes Perceptual Speech Quality Measure (PSQM) as the automated rating method. The PSQM algorithm attempts to measure distortion in an "internal psychoacoustic domain," which mimics the sound perception of humans in real-life situations. The PSQM conversion of signals in the physical domain (e.g., frequency and time) to the more meaningful psychoacoustic domain, the process provides a quantifiable, objective rating, rather than the subjective rating offered by the P.800 specification. The source signal is specified in ITU-T P.50 as an artificial voice with an active speech level of -20dBm, which includes both male and female genders, and which reproduces many of the essential characteristics of human speech. The output of the test process is a PSQM value ranging from 0 to 6.5, with 0 being perfect quality and 6.5 being the highest level of signal degradation. Note that the PSQM value of PCM (Pulse Code Modulation), the traditional method for converting voice signals from analog to digital format in PSTN applications, is 0.5. This value reflects a relatively minor difference between the analog input and the uncompressed digital output of the encoding process. See also P.861 and Toll Quality.

psophometer An instrument arranged to give visual indication corresponding to the aural effect of disturbing voltages of various frequencies. A psophometer usually incorporates a weighting network, the characteristics of which differ according to the type of circuit under consideration; e.g., high-quality music or commercial speech circuits.

PSR Petabit Switch Router. See also Petabyte.

PSS1 Private Signaling System number 1. The formal name for QSIG, as standardized on a worldwide basis by the ISO (International Organization for Standardization) and the IEC (International Electrotechnical Commission). PSS1 is an ISDN-based protocol for signaling between nodes of a Private Integrated Services Network (PISN). QSIG predates PSS1, and remains the name under which the standard is marketed. See QSIG for a detailed explanation.

PSTN Public Switched Telephone Network. PSTN is an abbreviation used by the ITU-T. PSTN simply refers to the local, long distance and international phone system which we use every day. In some countries it's only one phone company. In countries with competition, e.g. the United States, PSTN refers to the entire interconnected collection of local, long distance and international phone companies, which could be thousands.

PSTN modernization The conversion of the PSTN's circuit-switched architecture to a carrier-grade softswitch-based architecture. PSTN modernization is occurring in phases, starting with core transport, a transformation that has been under way since the turn of the century with facilities-based IXCs' backbone networks, and which is starting to take place with some LECs' core networks. It will take literally decades for migration to a complete end-to-end IP architecture.

PSU 1. Packet Switch Unit.
2. Power Supply Unit.

pseudo cut through A switching mechanism where a packet is transmitted from its source port to its destination port only after the first 64 bytes of the packet are in the source port and its destination port is determined.

pseudo access tandem Also called pseudo tandem. It is an end-office switch in a remote geographical area to which other end offices in the area are connected instead of their being connected to the LEC's access tandem. When a long-distance call originates in one of these distant end offices, the originating end-office switch sends it to the pseudo tandem, and then the pseudo tandem routes the call to the LEC's access tandem for routing to the IXC's POP. It's more economical to have a pseudo access tandem in such a geographical situation rather than having each distant end office directly connect to the LEC's access tandem.

pseudophone A pay phone that looks like a real Bell telephone company phone but is owned by a smaller phone company that charges exorbitant fees for long-distance calls.

psychic ANI A term created by Howard Bubb from Dialogic to designate what happens when you call someone on one line while they're calling you on the other.

PT Payload Type: Payload Type is a 3-bit field in the ATM cell header that discriminates between a cell carrying management information or one which is carrying user information.

PTC 1. Portable Teletransaction Computers. These are typically handheld devices used for retail (inventory), healthcare (tracking supplies), mobile field repair (reporting fixes), insurance (visiting car wrecks and other disasters), etc. The devices typically have telecommunications capabilities, sometimes wireless, sometimes landlines. And they typically include microprocessors, memories, displays, keyboards, touchscreens, character recognition software, barcode readers, printers, modems and local and/or wide area data radios.
2. Personal Telecommunications Center. Infocorp's name for a product most people call a PDA, Personal Digital Assistant.
3. Pacific Telecommunications Council. A not-for-profit organization open worldwide to anyone or any entity interested in the Pacific hemisphere and involved with telecommunications, broadcasting, informatics, digital media and associated fields. www.ptc.org

PTE Path Terminating Equipment. SONET network elements that multiplex and demultiplex the payload and that process the path overhead necessary to transport the payload. See also Terminating Multiplexers.

PTI An ATM term. Payload Type Indicator: Payload Type Indicator is the Payload Type field value distinguishing the various management cells and user cells. Example: Resource Management cell has PTI=110, end-to-end OAM F5 Flow cell has PTI=101.

PTMPT Point-To-Multipoint: A main source to many destination connections.

PTN Public Telecommunications Network.

PTO Public Telecommunications or Telephone Operator, first established in the U.K. as part of the British Telecommunications Act of 1981, but now refers to PTOs in all European countries. The PTO has typically evolved from the previous PTT, but other companies have also obtained PTO licenses. A PTO may specialize in certain region or city or may service the entire country.

PTR See Problem Tracking Report.

PTS Presentation Time Stamp: A timestamp that is inserted by the MPEG-2 encoder into the packetized elementary stream to allow the decoder to synchronize different elementary streams (i.e. lip sync).

PTS Public Telecommunications Systems.

PTSE An ATM term. PNNI Topology State Element: A collection of PNNI information that is flooded among all logical nodes within a peer group.

PTSP An ATM term. PNNI Topology State Packet. A type of PNNI routing packet used to exchange reachability and resource information among ATM switches to ensure that a connection request is routed to the destination along a path that has a high probability of meeting the requested QOS. Typically, PTSPs include bidirectional information about the transit behavior of particular nodes (based on entry and exit ports) and current internal state.

PTT 1. Post Telephone & Telegraph administration. The PTTs, usually controlled by their governments, provide telephone and telecommunications services in most foreign countries. In ITU-T documents, these are the Administrations referred to as Operating Administrations. The term Operating Administrations also refers to "Private Recognized Operating Agencies" which are the private companies that provide communications services in those very few countries that allow private ownership of telecommunications equipment.
2. Push To Talk. Just as it sounds. A variation on the old walkie talkie. Some cellular telephones now have a features called PTT. You push a button and you can immediately talk with people you've designated – without having to dial them. For a fuller explanation, see Push To Talk.

PTX push-to-X. A term that refers generically to services that are launched by pressing a button on a wireless handset, such as talk, view, record (a voicemail).

PTY Party.

PU Physical Unit. In IBM's SNA, the component that manages and monitors the resources of a node, such as attached links and adjacent link stations. PU types follow the same classification as node types.

PU 2.0 & 2.1 IBM protocols which allow applications written to APCC and interpreted by LU 6.2 to access the mainframe (2.0) and token ring LAN (2.1).

PU Type 2 A physical unit (PU) refers to the management services in SNA node always contains one physical unit (PU), which represents the device and its resources to the network. PU Type 2 is often referred to as a cluster controller.

PU 4 An IBM SNA front end processor.

PU 5 An IBM SNA mainframe, such as a System/370 or System/390. It runs VTAM to handle data communications.

public access network A Wi-Fi network that allows anyone within the coverage area to access the network providing they have a Wi-Fi enabled device. Some public access networks charge access fees, some offer reduced rates to bridge the 'digital divide' and others offer free access.

public access terminal A kiosk (SK series), enclosure (TK series) or a system for special enclosures and custom applications (XE Series) which provides the public access to service. Contains a color monitor and keypad for customer interaction.

public announcement trunk group A trunk group used to provide multiple types of announcements, such as the weather, time and sports results.

public asynchronous dial-in port A term used in packet switching networks referring to the local phone number of a port into the packet switched network. Some networks provide different numbers for different speeds. Some provide different speeds on the same numbers. See also Private Dial-In Ports.

public data network A network available to the public for the transmission of data, usually using packet switching under the ITU-T X.25 packet switching protocol. See Packet Switched Network.

public dial up port A port on a computer system or on a communications network which is accessible to devices operating over the public switched telephone network.

public domain Imagine you write software. You've just written a great program. You now want to sell it. You have two choices. You can take advertisements, sell it to retailers, get distributors to carry it, hire salespeople, etc. In other words, go the commercial route. This is expensive and requires a major marketing / sales budget. The other choice is to go the Public Domain (also called Shareware) route. This involves giving away your software on various bulletin boards, on many Web sites, in "shareware" direct mail catalogs. In short, putting your software in the public domain. People download the software for free and try it. If they like it, they will send you money. They will do this because you offer them an instruction manual, a new version of the software that doesn't blast "unregistered" on the splash screen when you load the software, or an upgraded version of the software, with more features, or a chit that assuages your guilt at using unpaid-for software that someone (i.e. you) worked real hard on.

public exchange A British word for Central Office. Outside North America, central offices are all called "public exchanges." In the US, a public exchange is typically a local telephone switch. TELECONNECT's phone number in North America is 212-691-8215. The 212 is our area code. The 691 designates the central office or public exchange which serves us. That public exchange belongs to Nynex Company. See also Central Office and CO.

public exchange points The public exchange points are the major intersections of the Internet. At these exchanges points (MAE-East, MAE-West, PacBell NAP, etc.), the Internet backbones (UUNET, Sprint, etc.) along with hundreds of local and regional Internet access providers, meet to pass Internet transmissions on from one network to another. At these public exchanges points, enormous amounts of data are sent to and from each and every connected network. All of this data is transferred from network to network over the same, common infrastructure. See also NAP and Peering.

public key See Public Key Encryption.

Public Key Encryption PKE. Also known as asymmetric encryption, and Diffie-Hellman encryption after its inventors (1976), Whitfield Diffie and Martin Hellman. Public Key Encryption is a form of encryption that equips each user with two keys – a private key and a public key, both of which are provided by a trusted third party known as a Certificate Authority (CA). The public key for each intended recipient, which is known by everyone with access to the key registry maintained by the CA, is used to encrypt the message. The private key, which is known only to the intended recipient and which is kept secret, is used to decrypt the message. Each public key and private key are linked in a manner such that only the public key can be used to encrypt messages to a given recipient, and only the private key held by that recipient can be used to decrypt them. PGP (Pretty Good Privacy) is perhaps the best known example of public key encryption. See also Encryption and PGP.

This longish comes from Netscape (which of course no longer exists). Here goes: "All communication over the Internet uses the Transmission Control Protocol/Internet Protocol (TCP/IP). TCP/IP allows information to be sent from one computer to another through a variety of intermediate computers and separate networks before it reaches its destination. The great flexibility of TCP/IP has led to its worldwide acceptance as the basic Internet and intranet communications protocol. At the same time, the fact that TCP/IP allows information to pass through intermediate computers makes it possible for a third party to interfere with communications in these ways:

1. Eavesdropping. Information remains intact, but its privacy is compromised. For example, someone could learn your credit card number, record a sensitive conversation, or intercept classified information.

2. Tampering. Information in transit is changed or replaced and then sent on to the recipient. For example, someone could alter an order for goods or change a person's resume.

3. Impersonation. Information passes to a person who poses as the intended recipient. Impersonation can take two forms:

a. Spoofing. A person can pretend to be someone else. For example, a person can pretend to have the email address jdoe@mozilla.com, or a computer can identify itself as a site called www.mozilla.com when it is not. This type of impersonation is known as spoofing.

b. Misrepresentation. A person or organization can misrepresent itself. For example, suppose the site www.mozilla.com pretends to be a furniture store when it is really just a site that takes credit-card payments but never sends any goods.

Normally, users of the many cooperating computers that make up the Internet or other networks don't monitor or interfere with the network traffic that continuously passes through their machines. However, many sensitive personal and business communications over the Internet require precautions that address the threats listed above. Fortunately, a set of well-established techniques and standards known as public-key cryptography make it relatively easy to take such precautions. Public-key cryptography facilitates the following tasks:

Encryption and decryption allow two communicating parties to disguise information they send to each other. The sender encrypts, or scrambles, information before sending it. The receiver decrypts, or unscrambles, the information after receiving it. While in transit, the encrypted information is unintelligible to an intruder. Tamper detection allows the recipient of information to verify that it has not been modified in transit. Any attempt to modify data or substitute a false message for a legitimate one will be detected. Authentication allows the recipient of information to determine its origin–that is, to confirm the sender's identity. Nonrepudiation prevents the sender of information from claiming at a later date that the information was never sent. The sections that follow introduce the concepts of public-key cryptography that underlie these capabilities.

The most commonly used implementations of public-key encryption are based on algorithms patented by RSA Data Security. Public-key encryption (also called asymmetric encryption) involves a pair of keys – a public key and a private key – associated with an entity that needs to authenticate its identity electronically or to sign or encrypt data. Each public key is published, and the corresponding private key is kept secret. Data encrypted with your public key can be decrypted only with your private key. In short, public key encryption uses two keys, a public key (for encrypting messages) and a private key (for decrypting messages). This allows enable users to verify each other's messages without having to securely exchange secret keys. See Encryption, Netscape and Public Key Infrastucture.

Public Key Infrastructure PKI. A means by which public keys can be managed on a secure basis for use by widely distributed users or systems. PKI also is known as PKIX, as the IETF's X.509 standard is widely accepted as the basis for such an infrastructure. X.509 defines data formats, key infrastructure components (e.g., security administrators, certificate authorities, users, and directories), and procedures for the distribution of public keys via digital certificates signed by Certificate Authorities (CAs). See also Encryption and Public Key Encryption.

public mobile radio The European term for what we call Specialized Mobile Radio (SMR) or Trunk Mobile Radio (SMR) in the US. See SMR and TETRA.

public network A network operated by common carriers or telecommunications administrations for the provision of circuit switched, packet switched and leased-line circuits to the public. Compare with private network.

Public Notice PN. A Public Notice is issued by the Federal Communications Commission (FCC) to notify the public of an action or an upcoming event.

public power utelco A public power utelco is a telecommunications company that is owned by or affiliated with an electric or gas distribution or transmission utility.

public room A video conferencing center that is arranged with transmission services. Public rooms can be rented by business customers who wish to have a video conference but do not have facility available at their own offices.

Public Safety Answering Point PSAP. A generic term for the person or group of people who answer 911 emergency phone calls.

public safety network Typically a 'private' network used by emergency services organizations, such as police, fire and emergency medical services, to prevent or respond to incidents that harm or endanger persons or property.

Public Service Commission PSC. The state regulatory authority responsible for communications regulation. Also known as Public Utility Commission, Corporate Commission and in some states, the Railway Commission.

Public Switched Digital Service PSDS. A generic name for Bell telephone companies' service offerings that provide customer switches the capability of sending data at 56 Kbps over the public circuit switched network. Also known as Switched 56 Kbps Service. See also Switched 56.

public switched network Any common carrier network that provides circuit switching between public users. The term is usually applied to the public telephone network but it could be applied more generally to other switched networks such as Telex, MCI's Execunet, etc.

public switched telephone network Usually refers to the worldwide voice telephone network accessible to all those with telephones and access privileges (i.e. In the U.S., it was formerly called the Bell System network or the AT&T long distance network).

public telephone station Coin phone. Pay phone.

public telephony A new term to describe what might be possible with a new type of public "phone" which could do more than make and receive analog phone calls. Perhaps it could do videoconferencing? Perhaps it could send faxes? Perhaps it could access remote databases? In short, it could do a bunch of multimedia things. Exactly what hasn't been defined, as yet. Having an RJ-11 plug in it so I could plug my laptop into would be a good beginning, however.

public utility commission PUC. State body charged with regulating phone companies. Also called Public Service Commissions. See Public Service Commission.

public works network A Wi-Fi network that serves local government or civic employees throughout the coverage area, providing network access to databases, or voice calls, for remote or mobile employees. May include traffic management systems, remote meter reading, or other infrastructure applications.

publish To make information public. These days you can "publish" in many ways – from paper to CD-ROM to the World Wide Web (i.e. via the Internet).

publishing Making resources like databases available to network users.

PUC 1. See public utility commission.

2. See personal universal controller.

PUC-CODE Personal Unlocking Code. Some cell phones have a PUC-CODE. If you forget your PIN code (and assuming your cellphone has a PIN code), you enter your PUC-CODE and your cellphone will come alive. PUC-Code is also spelled PUK-code.

puddle A small body of water that draws other small bodies wearing dry shoes into it.

PUK-code See PUC-CODE.

puka The Hawaiian word for "hole." It was used for translating the number zero in phone numbers in the early days of phone service in Hawaii.

pull box A box with a cover inserted in a long conduit run, particularly at a corner. It makes it easier to pull wire or cable through the conduits.

pull tension The pulling force that can be applied to a cable without effecting specified characteristics for the cable.

pull-through revenue When the sale of services or equipment to a customer creates a situation where additional sales are made possible, the incremental dollars are called pull-through revenue. For example, the sale of VoIP equipment to a customer results in pull-through revenue when the customer decides to upgrade its network to better handle the VoIP traffic. Pull-through revenue can result from complementary sales at the customer site or through additional sales at the customer's other sites. Pull-through revenue can also be generated for affiliated business units. For example, a sale by a telco's data networking unit may result in pull-through revenue for the telco's wireless division.

pullcord or pullwire or pullstring A cord or wire placed within a raceway or conduit or ceiling or wall and used to pull wire and cable through.

pulling eye A device on the end of a cable to which a pulling line is attached for pulling cable into conduit or duct liner.

pulling glass Laying fiber-optic cable.

pulling strength Expressed in lbs. The maximum force which may be applied to strength members of a cable. Pulling strength limits are specified for all Belden Fiber Optic cables in the General Line and Fiber Optic catalogs. Affects pulling methods, pulling tension and operation tension.

pulp A type of older telephone twisted-pair cable whose wood-pulp "paper" insulation is formed on the cable during manufacture.

pulp cable Outside telephone cable that uses paper insulation on the twisted coppr pairs. Pulp cable is obsolete technology.

pulpware Books and magazines. They are printed on paper, which is made of wood pulp. Despite the popularity of the World Wide Web, CD-ROM, audio tapes and other media, pulpware never has been so popular, for which fact I am ever so grateful. There is nothing quite so nice as curling up in front of the fire with a good book. Books not only are user-friendly, but they just feel good to the touch. The only problem with pulpware is that it is impossible to fix "bugs" once the book is published. There are some "bugs" in this book. Please send email if you find one; I'll fix it with the next edition.

pulse 1. Pulse is the dialing mode for outside lines that is traditionally used by rotary dial telephones.

2. A quick change in the current or voltage produced in a circuit used to operate an electrical switch or relay or which can be detected by a logic circuit.

Pulse Address Multiple Access PAMA. The ability of a communication satellite to receive signals from several Earth terminals simultaneously and to amplify, translate, and relay the signals back to Earth, based on the addressing of each station by an assignment of a unique combination of time and frequency slots. This ability may be restricted by allowing only some of the terminals access to the satellite at any given time.

Pulse Amplitude Modulation PAM. A technique for placing binary information on a carrier to transmit that information. PAM is a technique for analog multiplexing. The amplitude of the information being modulated controls the amplitude of the modulated pulses. Samples of each input voltage are placed between voltage samples from other channels. The cycle is repeated fast enough so the sampling rate of any one channel is more than twice the highest frequency transmitted. See also PAM and PCM.

pulse cable A type of coaxial cable constructed to transmit repeated high voltage pulses without degradation.

Pulse Code Modulation PCM. The most common and most important method a telephone system in North America can use to sample a voice signal and convert that sample into an equivalent digital code. PCM is a digital modulation method that encodes a Pulse Amplitude Modulated (PAM) signal into a PCM signal. See PCM and T-1.

pulse code modulation upstream See V.92.

pulse density Also known as Ones Density. In electrically-based T-Carrier systems, "0"s are represented by zero voltage (i.e., no pulse) and "1"s by alternating positive and negative voltages (pulses). Pulse density refers to the number of no pulse ("0") periods allowed before a pulse ("1") must occur. Typically, no more than 15 no pulse periods ("0"s) are allowed before a pulse ("1") must occur. Pulse density is required by older T-Carrier systems, as repeaters and other network devices rely on it in order to maintain synchronization; some devices also depend on pulse density for power. Not all T-Carrier systems require pulse density; for instance, unchannelized T-Carrier provides clear-channel transmission, without regard for pulse density. See also T-Carrier.

pulse density violation A pulse density violation occurs if a signal contains more than a specified number of zeros, or the percentage of ones in the signal is less than specified.

pulse dialing One or two types of dialing that uses rotary pulses to generate the telephone number. See Rotary Dial.

pulse dispersion A fiber optic term. Pulse Dispersion is the result of Modal Dispersion. Modal dispersion is an optical phenomenon caused by the dispersion, or spreading out, of the light signals as they propagate through the pure, clear inner core of the optical fiber. Some portions of each light signal travel more or less down the center of the core, while other portions of the signal spread out and strike the edges of the core, at which points they are reflected back into the core by a layer of "cladding," which is glass of a slightly different refractive index. Therefore, portions of the light pulses may take distinctly different paths, or "modes." The portions of the light pulses which dance around the edges of the fiber travel a longer distance than those which travel more directly down the center of the core. Over a long distance from the light source to the light detector, therefore, the pulses can overlap, creating a phenomenon known as "pulse dispersion." The end result is that the light detector may not be able to distinguish between the individual pulses, and the integrity of the data stream is compromised. MultiMode Fiber (MMF) is characterized by an inner core of relatively large diameter and, therefore, is most susceptible to both modal dispersion and pulse dispersion. Single Mode Fiber (SMF) has a thinner inner core, and is less susceptible. See also Dispersion, Fiber, Modal Dispersion, Multi-Mode Fiber, and Single Mode Fiber.

pulse Distribution Amplifier DA. A device used to replicate an input timing signal, typically providing 6 outputs, each of which is identical to the input signal. May also perform cable equalization or pulse regeneration.

Pulse Duration Modulation PDM. That form of modulation in which the duration of the pulse is varied in accordance with some characteristic of the modulating signal.

pulse link repeater A signaling set that interconnects the E and M leads of two circuits. In E & M signaling, a device that interfaces the signal paths of concatenated trunk circuits. Such a device responds to a ground on the "E" lead of one trunk by applying -48Vdc to the "M" lead of the connecting trunk, and vice versa. This function is a built-in, switch-selectable option in some commercially available carrier channel units.

pulse modulation A general method of carrying digital information on any system that uses fixed-frequency pulses of transmit information from the source to the destination. Examples are pulse amplitude modulation and pulse duration modulation.

pulse overshoot In T-1, the amount of signal voltage that can remain at the trailing end of a pulse. It can be no more than 10-30% of the pulse amplitude. Also called afterkick.

Pulse Position Modulation PPM. Pulse position modulation is a variation on frequency modulation. In FM the carrier signal's frequency is modulated by the signal. In Pulse position modulation, instead of a continuous carrier signal, Pulses of constant amplitude are transmitted at different frequencies, the frequency of the pulses being modulated by the transmission.

Pulse Repetition Frequency PRF. In radar, the number of pulses that occur each second. Not to be confused with transmission frequency which is determined by the rate at which cycles are repeated within the transmitted pulse.

pulse stuffing When timing signals on digital circuits get out of whack, some method of allowing mismatches must be provided. In time division multiplexing, this is called pulse stuffing. One stream of data has bits added to it so its final rate is the same as the master clock.

pulse to tone conversion Most of the world doesn't have touchtone service. They have rotary, make-and-break phone service and phones. Most computer telephony systems – from airline timetable audiotext machines to bank balance dispensers – require the user to punch in touch tone sounds. For people with rotary phones to get access to computer telephony systems, those systems must have a device called a pulse to tone converter – electronic circuitry which counts the clicks made by a rotary phone and converts them into touch tones. This technology is not 100% accurate and should be accompanied by programming which confirms the input. "You just entered 1034. If that is correct, please say YES or dial 1."

pulse train The resulting electronic impulses that transmit encoded information.

pulse width In T-1, refers to the width (at half amplitude) of the bipolar pulse (typically 324 + or -45 nsec).

pulse width modulation Another but not very common method of modulating a signal, in which an analog input signal's DC level controls the pulse width of the digital output pulses. See Pulse Code Modulation and Pulse Amplitude Modulation.

pulsing The method used for transmitting the phone number dialed to a telephone company switching office.

PUMA Product Upgrade Manager.

pump laser An active optical component used in optical amplifiers to amplify and regenerate light signals that lose signal strength over distance – as all signals do. Pump lasers are lasers that are built into erbium-doped fiber amplifiers (EDFAs) for the purpose of exciting the erbium in the fiber. Pump lasers are actually a type of semiconductor laser about half the size of a AA battery. Pump lasers work by exciting erbium atoms, which then add their energy to optical data signals passing through the fiber and boost the signals' power enough to carry them many miles before needing another boost. See Raman Amplifier. The history of photonics shows improving technology producing increasing distances each year between where amplification/regeneration is needed in long-haul fiber optics transmission systems. See Erbium, Fiber, Laser and Photonics.

pumped To be full of confidence, as in "He's really pumped about his new DSL circuit."

pun The five worst (i.e. best puns) ever written are: 1. Two vultures board an airplane, each carrying two dead raccoons. The stewardess looks at them and says, "I'm sorry, gentlemen, only one carrion allowed per passenger."

2. Two boll weevils grew up in South Carolina. One went to Hollywood and became a famous actor. The other stayed behind in the cotton fields and never amounted to much. The second one, naturally, became known as the lesser of two weevils.

3. Two Eskimos sitting in a kayak were chilly, but when they lit a fire in the craft, it sank, proving once again you can't have your kayak and heat it, too.

4. And then there was the Buddhist who refused Novocain during a root canal. He wanted to transcend dental medication.

5. And finally, There was a man who sent ten different puns to friends, in the hope that at least one of the puns would make them laugh. Unfortunately, no pun in ten did.

And a bonus pun: Why do Mercedes cars corner so well? Answer: Because they're Mercedes Benz.

punch 1. The process of perforating a paper tape or card in order to code information into machine readable form.

2. The process of connecting jumper interconnection wires on a distribution frame. It is called punching because of the tool which places the wire on the metal post of the frame. It is called a PUNCH and requires a heavy "punch" to make it strip its wires, then connect into the Punch-Down Block.

punch tool Punch tools are used to conduct copper wires to terminations, either jacks, blocks or patch panels. They come in two varieties: non-impact and impact. Non-impact tools are less costly, and they push a conductor into its connector. Impact punch tools have a spring mechanism that delivers a jolt of force to the conductor being punched, which helps ensure the cable is properly seated. Impact punch tools are best for most applications. There are different blades for each kind of termination, including the 110 blade, the 66 blade, the five pair 110 blade, the Krone blade and the BIX blade. See Punch Down Tool.

punch down A term used to describe the connection of twisted pair wires to an insulation displacement block. (e.g. a 66 block, or a patch panel). See Punch Down Tool.

punch down tool A device used to connect twisted pair copper wires to an insulation displacement block (e.g. a 66 block or a patch panel). The punch down tool consists of a slotted blade attached to a heavy-duty plastic handle. Loop the wire between the prongs of an insulation displacement block, slide the slotted blade over the terminal and give the tool a downward push. A spring loaded mechanism in the handle completes the job automatically. The blade spreads the prongs just enough, strips the wire, drives it between the prongs, and neatly cuts the wire with a satisfying "ka chunk." Thus you've produced a perfect "gad tight" termination, which is important since the connection might corrode and the corrosion will insulate the connection. Result: Noise (buzz, hum) on the telephone line and bad voice or data transmission. See Punch Tool.

punchdown block A device used to connect one group of wires to another. Usually each wire can be connected to several other wires in a bus or common arrangement. A 66-type block is the most common type of punchdown block. It was invented by Western Electric. Northern Telecom has one called a Bix block. There are others. These two are probably the most common. A punchdown block is also called a terminating block, a connecting block, a punch-down block, a quick connect block, a cross connect block. A punchdown block will include insulation displacement connections (IDC). In other words, with a connecting block, you don't have to remove the plastic shielding from around your wire conductor before you "punch it down."

punchdown tool A punchdown tool is used to insert cable onto cross-connects, patch panels and jacks. "Punching" a cable means forcing it into an Insulation Displacement Connector (IDC), such as a 66-block. The IDC has replaced Wire Wrap and Solder and Screw Post terminations for connecting conductors to jacks, patch panels and blocks. They pierce the cable jacket to make a connection with the conductor rather than the installer having to strip off the conductor's plastic insulation, saving time. Since IDCs are very small, they can be placed very close together, reducing the size of cross-connects. IDCs are the best termination for high speed data cabling since a gas-tight, uniform connection is made. Most punchdown tools will allow you to install wires on 66-blocks, also known as RJ-21Xs. But there are other IDC blocks. Each of the following types of IDC blocks requires a unique punchdown die for your punch-down tool: (1) 110 Connector; These are very popular connectors for new installations. 110 Patch panels are typically rated Category 5, but always double-check. (2) Krone. Krone cross-connects have a patented 45 degree angled IDC, which typically surpasses standard 66-block and 110 connectors in attainable transmission speed. (3) BIX. BIX IDCs also have superior transmission properties to 66-block and 110 connectors. See Insulation Displacement Connector.

Purchase of Accounts Receivablesb See PARS.

pure aloha A random access technique developed by the University of Hawaii in the early 1970s. In this scheme, a user wishing to transmit does so at will. Collisions are resolved by retransmitting after a random period of time. See also Aloha and Alohanet.

pure play Looking to invest in a company that makes heavy duty storage devices. You might look at EMC, because that's its main business. It's a pure play. You wouldn't look at IBM, though it also make storage devices, because it's not a "pure play" data storage device maker. It makes and sells other things.

purge Verb. To remove records from a database. To get rid of old voice mail messages.

purity An imaging term. Purity is the ability of the electron beam to hit precisely the correct phosphor color dot. If a full red page were shown on the display, impurity would result in a purple or greenish color region. This impurity can occur if the shadow mask has been damaged or if the screen has become magnetized. Degaussing the screen may fix the problem.

purple applications A term coined by VoIP promoter Jeff Pulver, purple applications are "things that one can do with an IP network that never before were practical or possible with a TDM network."

purple minutes A term coined by VoIP promoter Jeff Pulver. Purple minutes are "IP traffic that is more than just plain old telephone traffic (or 'black-and-white' traffic); it's traffic with a value-added component made possible by the network's being IP-based."

purpose built A piece of hardware that's built to do one thing and one thing only is often called "purpose built." In short, it's built for one purpose.

push There are essentially two ways of getting information from a Web or Intranet site – push or pull. In the simplest terms, "pull" means that you go into a Web site, and ask for information – typically by clicking on a button. You will see a page of information. You may ask for a file to be sent (downloaded) to you. This is typically called "pull." You are pulling the information from the web site by doing something – i.e. clicking away. "Pushing" is a new technology that involves the web site sending you specific material you had only generally asked for. An example, you're surfing the net. You have told one or more sites that you're on line and ready to receive whenever they are ready to send to you. This information might be sent to you as a bar which scrolls along the bottom of your screen – like the stockmarket quotes that scroll across CNN's financial channels – or it might be sent to you while your screen saver is on. It might replace your screen saver with live information – your stocks, news of your favorite baseball teams, etc. According to contemporary wisdom in the Spring of 1997, there are two types of "push" – Active push and Directed push. Here are two definitions I found of Active and Directed Push: Active: The server on the Web interacts with the client by sending all the content to the client upon the client's request (polling), essentially the way that a client/server application might. PointCast is an example of this. Directed: The server interacts with the push client only occasionally, providing directions (agents, modules, and so on) for how content should be handled or where content is located. The client then gets the information directly from a variety of services and processes it locally. Lanacom Headliner is a good example of this.

push to talk 1. PTT. On some phone systems you push a button to talk and stop pushing the button to listen. Typically you say "over" or "OK" to indicate it's the other person's turn to talk. Push to talk is used in two-way radio dispatch systems (also called SMR for specialized mobile radio), handheld walkie-talkie systems, CB (citizen band) radios and some cell phones. Push to talk is often used where the same communications channel is used for transmit and receive. "Push to talk" was once a symbol of low quality cheap communications. You could only talk in one direction at time. It was a pain. You couldn't interrupt someone who was speaking. They could drone on forever, though for most systems that wasn't a problem. Police, fire and other emergency services used it and their messages were short, quick and often urgent, and typically everyone on the system – all the firemen in the field – could hear everything – at least from the dispatcher. Things changed in the early 1990s when a cell phone provider called Nextel introduced the "Push to talk" feature on its "cell phones" and combined the feature with speed dial. It called the double feature "Direct Connect." The feature became wildly popular among groups of people who worked together – construction workers, electricians, cable TV service technicians, emergency teams, etc. It worked like this: Everyone in the group was given their own Nextel "phone number." It didn't look like a normal phone number. It had strange numbers. You put everybody's Nextel strange phone number into the speed dial of your Nextel "cell phone," (which were made by Motorola). When you wanted to reach someone in the group, you simply hit the button and started talking. You didn't wait for them to answer and say Hello. Your voice would come immediately out of the speaker on their Nextel phone. If they wanted to reply to your words, they would push their "push to talk" button and you'd hear their reply out of your call phone's speaker, or out of the cell phone's earpiece – depending on how you set it up. Nextel's cell phone service plans typically allowed unlimited "Direct Connect" calls – local, long distance, day, night, weekends, etc. With Direct Connect you can talk to a group of people simultaneously – as you might need to in an emergency. You can also use push to talk like instant messaging on computers: Your phone screen can show a list of contacts and indicate who is available to receive a PTT call. Because PTT is often implemented these days on a bpacket-network, the technologists can design lots of similar customized software and services, like routing specific calls to voice mail or another phone. According to the industry, it's a lot cheaper to embed a cell phone with push to talk abilities than to put in a camera. At one stage, Nextel tried to register the phrase "Push To Talk" as a trademark, claiming they originated the phrase and were the exclusive originators of the phrase. Verizon, Motorola and others thought that was a little over the top and tried to stop them. Actually the term push to talk has been used in the telecom industry since the late 1930s / early 1940s. Police radio transmitters of the time used push to talk. Verizon and Nextel ended their legal battle over Nextel's attempts to trademark Push To Talk. Nextel agreed not to trademark it. See also Citizens Band, PoC and SMR.

2. Push to talk is a protocol which one must observe to conduct a successful voice conversation over some satellite links networks or some IP telephony networks. Much like SMR and CB radio communications, you must wait a brief moment to make sure that your transmission was received and to give the other person an opportunity to respond before you start to talk again; otherwise, you will overtalk the other person, which is known as "clipping." In GEO (Geosynchronous Earth Orbiting) satellite systems, this protocol is necessary due to propagation delay, which is caused by the fact that the communications satellites are roughly 22,300 miles above the equator; even at the speed of light, it takes half a second for the signal to reach the satellite and return to earth. In IP Telephony networks (especially those on Internet) this delay, or latency, is the result of the shared nature of the network – though the latency is improving as we learn more about QoS – quality of service.

3. Push to talk is also a method of payphone operation in which a push button switch is touched by the caller when the called party answers. Once pushed, money in the phone drops into the collection box and the handset microphone is turned on. This system deprives the phone company of revenues for calls to 976 numbers, answering machines or answering services – for the simple reason that you hear but don't push to talk and pay.

push technology An Internet/WWW (World Wide Web) term describing the reversal of the traditional information gathering paradigm (I hate that word). Instead of seeking out information, information seeks you, based on your demonstrated preferences. Here's how it works: You cruise the Web (Internet or Intranet, seeking certain kinds of information from certain Web sites. You either register your interests or the Web takes note of them. Automatically, you are then presented with notification of changes in that information, as changes take place. You then can access that changed information, or ignore it, as you choose. See also RSS.

push-to-X A term that refers generically to services that are launched by pressing a button on a wireless handset, such as talk, view, record (a voicemail). Acronym: PTX.

pushback Microsoft-speak for a disagreement.

pushbutton dialing Instead of rotary dialing, buttons are pushed to generate the tones needed to place a phone call. Also called Touchtone and Touch-call. Some pushbutton phones do not produce tones, but generate the dial pulses of rotary dials. Some phones and phone systems will generate both rotary dial pulses and tone signaling. See Touchtone.

pushbutton dialing to stations A special attendant console feature in which the switching system is served by rotary dial central office trunk circuits. A ten-button keyset is provided on the console which allows fast dialing of extension numbers to complete incoming calls.

pushbutton originating register A register used to store information about originating calls with pushbutton signals.

pushing the envelope The only people not interested in pushing the envelope are postal employers, says comedian, Dennis Miller.

PUT A method of uploading files to a Web server from a HTML-capable (HyperText Markup Language). HTTP (HyperText Transfer Protocol) specifies both POST and PUT. POST performs a permanent action, such as the uploading of an order or the response to a form, or is used which a significant amount of data must be input. POST must be handled by a program or script, which must already exist. PUT can be used to update information on the server, if only a single data item is required. PUT also can be directed at a resource (e.g., program or data element) which does not yet exist. See also HTML and HTTP.

put-up Refers to the packaging of wire and cable. The term itself refers to the packaged product that is ready to be stored or shipped.

putative Reputed.

Putts Law In the Spring of 2006, many people working inside Microsoft knew that the company's implementation of its Ultra Mobile PC (UMPC) software would be a disaster. Two of the people tried to explain the disaster by quoting what they felt was Putts Law. Here are their two understandings of Putts Law:

1. Two types of people dominate technology: those who understand what they do not manage, and those who manage what they do not understand.

2, Rejection of management objectives is undesirable when you are wrong, and unforgivable when you are right.

PVC 1. Premises Visit Charge.

2. PolyVinyl Chloride, a common type of plastic used for cladding telephone cable (except that to be run in plenum ceilings). See Plenum Cable.

3. Permanent Virtual Circuit, a permanent association between two DTEs established by configuration. A PVC uses a fixed logical channel to maintain a permanent association between the DTEs. Once defined and programmed by the carrier into the network routing logic, all data transmitted between any two points across the network follows a pre-determined physical path, making use of a Virtual Circuit. PVCs are widely used in X.25 networks, and are the basis on which communications take place in a Frame Relay network. It is the circuit from end to end in a frame-relay network. See also SVC.

PVCC Permanent Virtual Channel Connection: A Virtual Channel Connection (VCC) is an ATM connection where switching is performed on the VPI/VCI fields of each cell. A Permanent VCC is one which is provisioned through some network management function and left up indefinitely.

PVDF See Polyvinylidene Diflouride.

PVN Private Virtual Network.

PVP tunneling Permanent Virtual Path tunneling. Method of linking two private ATM networks across a public network using a virtual path. The public network transparently trunks the entire collection of virtual channels in the virtual path between the two private networks.

PVPC Permanent Virtual Path Connection: A Virtual Path Connection (VPC) is an ATM connection where switching is performed on the VPI field only of each cell. A Permanent VPC is one which is provisioned through some network management function and left up indefinitely.

PVR Personal Video Recorder. Also called a digital video recorder. Think of it as a hard drive connected to a cable or satellite TV input and to a telephone line or the Internet. You use your PVR for recording and playing back TV programs. Using the telephone line or the Internet connection, the device downloads a schedule of TV programs from a distant database and saves it on its hard drive. The device comes with a remote which allows you, the user, to check off which TV shows you'd like it to record and when. The PVR knows when your show will air. It has that in its downloaded database. It changes channels itself and records your desired shows onto its hard drive. When you return home, you play the shows you wish. For example, you could instruct it to record 60 Minutes each Sunday evening. You could instruct it to record each time tennis comes on – irrespective of which channel. To my tiny brain, a PVR (also called by the brand name, TiVo) is one of the most useful consumer gadgets to come along in a long long time. I own two PVRs. I like them for three reasons. I like being able to watch my favorite TV shows when I want to – not when the program is broadcast. I also like being able to tell it to record tennis – no matter which tournament.

Third, I also like being able to download the TV show to my laptop and keep the program on my laptop's hard drive for viewing when I wish. I especially like being able to study the way my favorite tennis players do their various strokes.

PWB Printed Wire Board.

PWL PWL means password list file. It's a file Windows uses to store passwords.

PWLAN Public wireless local area network.

PWM Pulse Width Modulation. In communications, encoding information based on variations of the duration of carrier pulses. Also called Pulse Duration Modulation or PDM.

PWR Power.

PWT See Personal Wireless Telecommunications.

Px64 Pronounced P-times-sixty four. Informal name for the ITU-T family of videoconferencing interoperability standards correctly known as H.261, and addressing codecs and video formats. "P" refers to the value range of 1 through 30, with "Px64" referring to 1 through 30 64Kbps channels for videoconferencing. At a value of 30, the standard address the use of a full E-Carrier facility for video transmission. Px64, or H.261, addresses standards for codecs, as well as video formats. At the upper end, H.261 supports 352 pixels per frame, 288 lines per frame and 30 fps (frames per second). H.261 also addresses much lower levels of capability, offering the advantage of a standard for digital communication between video transmitters and receivers. As a videoconferencing transmitter and receiver go through the process of handshaking, they negotiate the communications protocol, including such issues as compression technique, frame rate, and format. H.261 provides a common standard which disparate devices (not of the same manufacturer) can use for communications. H.261 is part of a family of ITU-T standards known as H.320.

PXE Preboot Execution Environment. See WFM.

pyramid configuration A communications network in which the data link(s) of one or more multiplexers are connected to I/O ports of another multiplexer.

pyramidal horn A wave guide feed horn (i.e. antenna) in which both opposite faces are tapered.

Pythagorean theorem This theorem holds that the square of the hypotenuse of a right triangle is equal to the sum of the squares of the other two sides. In the movie, "The Wizard of Oz," the Scarecrow is a few straws short of a haystack when, near the film's end, he blurts out: "The sum of the square roots of any two sides of an isosceles triangle is equal to the square root of the remaining side. Oh, joy, rapture! I've got a brain." Actually he doesn't. His formula is totally wrong. The real Pythagorean theorem is above.

Q Queue.

Q & A Question and Answer. A teleconferencing term. During a lecture style teleconference, typically only the session sponsor can transmit audio; the other participants can listen, only. Q & A allows the other participants to signal via their touchtone pads their desire to ask a question. The session moderator/speaker can accept that request off-line, screen the question, and allow the participant to ask it on-line, as appropriate.

Q band A range of radio frequencies in the 40 GHz and 50 GHz range, also known as the V Band. The "Q" is a random, arbitrarily-assigned designation with its roots in the context of military security during World War II.

Q bit The qualifier bit in an X.25 packet that allows the DTE to indicate that it wishes to transmit data on more than one level. It is Bit 8 in the first octet of a packet header. It is used to indicate whether the packet contains control information.

Q.1541 ITU-T Recommendation. UPT Stage 2 for Service Set 1 on IN CS1 Procedures for universal personal telecommunication functional modelling and information flows.

Q.2100 ITU-T Recommendation. B-ISDN signaling ATM Adaptation Layer Overview.

Q.2110 ITU-T Recommendation. B-ISDN Adaptation Layer - Service Specific Connection Oriented Protocol.

Q.2130 ITU-T Recommendation. B-ISDN Adaptation Layer - Service Specific Connection Oriented Function for Support of signaling at the UNI.

Q.2723.6 ITU-T Recommendation. Broadband integrated services digital network (B-ISDN), Extension to the SS7 B-ISDN user Part (B-ISUP): signaling capabilities to support the indication of the statistical bit rate configuration 2.

Q.2725.1 ITU-T Recommendation. B-ISDN user Part - Support of negotiation during connection setup.

Q.2725.4 ITU-T Recommendation. Broadband integrated services digital network (B-ISDN), Extensions to the signaling system No. 7 B-ISDN user Part (B-ISUP) : Modification procedures with negotiation.

Q.2766.1 ITU-T Recommendation. Switched virtual path capability.

Q.2767.1 ITU-T Recommendation. Soft PVC capability.

Q.2931 ITU-T Recommendation. The signaling standard for ATM to support Switched Virtual Connections. This is based on the signaling standard for ISDN.

Q.2934 ITU-T ITU-T Recommendation. Broadband - Integrated services digital network (B-ISDN) digital subscriber signaling system No. 2 (DSS 2) - Switched virtual path capability.

Q.2961.6 ITU-T Recommendation. Additional signaling procedures for the support of the SBR2 and SBR3 ATM transfer capabilities.

Q.2962 ITU-T Recommendation. Digital subscriber signaling system No. 2 - Connection characteristics negotiation during call/connection establishment phase.

Q.2963.3 ITU-T Recommendation. Broadband integrated services digital network (B-ISDN) digital subscriber signaling system No. 2 (DSS 2) connection modification - ATM traffic descriptor modification with negotiation by the connection owner.

Q.2931 ITU-T Recommendation. The signaling standard for ATM to support Switched Virtual Connections. This is based on the signaling standard for ISDN.

Q.699.1 ITU-T Recommendation. Interworking between ISDN access and Non-ISDN access over ISDN user part of signaling system 7 - Support of VPN applications with PSS1 information flows.

Q.700-Q709 ITU-T Recommendation. Messaging Transfer Part (MTP) of SS7.

Q.710 ITU-T Recommendation. PBX Application part of SS7.

Q.711-Q716 ITU-T Recommendation. Signaling Connection Control Part (SCCP) part of SS7.

Q.721-Q.725 ITU-T Recommendation. Telephone User Part (TUP) part of SS7.

Q.730 ITU-T Recommendation. ISDN Supplementary Systems part of SS7.

Q.741 ITU-T Recommendation. Data User Part (DUP) part of SS7.

Q.751.4 ITU-T Recommendation. Network element information model for SCCP accounting and accounting verification.

Q.755.1 ITU-T Recommendation. MTP Protocol tester.

Q.761-Q.766 ITU-T Recommendation. ISDN User Part (ISUP) part of SS7.

Q.765 ITU-T Recommendation. Signaling System No. 7 application transport mechanism.

Q.765.1 ITU-T Recommendation. Signaling System No. 7 - Application transport mechanism - Support of VPN applications with PSS1 information flows.

Q.771-Q.775 Transaction Capabilities Application Part (TCAP) part of SS7.

Q.791-Q.795 Monitoring, Operations, and Maintenance part of SS7.

Q.780-Q.783 Test Specifications part of SS7.

Q.825 ITU-T Recommendation. Specification of TMN applications at the Q3 interface: Call detail recording.

Q.850 ITU-T Recommendation. Usage of cause and location in the digital subscriber signaling system No. 1 and the signaling system No. 7 ISDN user part.

Q.921 ITU-T Recommendation. Q.921 defines the ISDN frame format at the data link layer of the OSI/ISDN Model. It contains address information. The ITU-T/OSI Layer 2 protocol used in the D channel. It is synonymous with LAPD.

Q.922 Annex A. ITU-T Recommendation defining the structure of Frame Relay frames. All Frame Relay frames entering the network automatically conform to this frame structure.

Q.931 Q.931 is the powerful message-oriented signaling protocol in the PRI ISDN D-channel. It is also referred as ITU-T Recommendation I.451. This protocol describes what goes into a signaling packet and defines the message type and content. Specifically, Q.931 provides:

- call setup and take down.

- called party number, with type of number indication (private or public).
- calling party number information (including privacy and authenticity indicators).
- bearer capability (to distinguish, for example, voice versus data for compatibility check between terminals.
- status checking (for recovery from abnormal events, such as protocol failures or the manual busying of trunks), and
- release of B-channels and the application of tones and/or announcements in the originating switch upon encountering errors.

Q.931 makes it possible to interwork PBX features with features in the public network. In addition to offering users more access to a wider range of services, this interaction, according to Northern Telecom, will improve the revenue potential of service providers. Service provided over PRA, using Q.931, include:

- access to the public network, such as equal access, WATS, DDD, international DDD, dial-800 and other special number services and operator assisted calls.
- access to and from such private networks as Northern Telecom's Meridian Switched Network (previously call Electronic Switched Network – ESN), tandem tie networks, and extension dialing network, and
- integration of voice and circuit-switched data traffic (up to 64 Kbps).

The Q.931 protocol also enables corporations to use B-channels – that is voice and data channels – in ways currently not possible. Today, for example, a separate trunk from the PBX to the central office is often required for each different service, such as voice, data, foreign exchange, 800-service. With PRA, one common trunk between the PBX and the central office can carry multiple call types. Moreover each B-channel within the PRA trunk can be assigned dynamically to carry whatever service is needed at the moment.

Q.933 ITU-T Recommendation. The signaling standard for Frame Relay to support SVCs. This is based on the signaling standard for ISDN.

Q.Sig See QSIG.

Q-signal In the NTSC color system the Q signal represents the chrominance on the green-magenta axis.

Q1 The first quarter of the calendar year. It's typically the period from January 1 to March 30. But for companies with a different reporting year – perhaps one ending on June 30, their first quarter would be the months of July, August and September. Q2 is the second quarter, Q3 is the third quarter, etc. etc.

Q2 See Q1.

Q3 See Q1.

Q4 See Q1.

QA Quality Assurance.

QAM Quadrature Amplitude Modulation. A sophisticated modulation technique, or compression technique, using variations in signal amplitude and phase, that allows multiple bits to form a single "symbol," which then is impressed on a single sine wave. "Quadrature" refers to the fact that four (i.e., "quad") distinct amplitude levels are defined. 16 QAM creates a symbol of 4 bits through 16 distinct signal points, or variations in amplitude and phase, thereby yielding a data rate of 9600 bps over a 2400 Hz carrier. (Note: 2 to the 4th power – i.e. $2 \times 2 \times 2 \times 2$ – equals 16. Thereby, to place 4 bits on a sine wave, 16 signal points are required.) 64 QAM creates a symbol of 6 bits through 64 distinct plot points, yielding a data rate of 14.4 Kbps. (Note: 2 to the 6th power equals 64.) 128 QAM creates a symbol of 7 bits through 128 distinct plot points, yielding a data rate of 16.8 Kbps. (Note: 2 to the seventh power equals 128.) As the carrier frequency varies from 2400 Hz, which is the standard usable bandwidth for a voice-grade channel, the potential bit rate increases or decreases accordingly. See also Long Reach Ethernet and QSAM.

QAT See Quantity Above Threshold Indicator.

QBE Query By Example. A database front-end that requests the user to supply an example of the type of data he wants to retrieve. Typically, the user forms a query by filling in a table with examples of the requested information. IBM created QBE in the 1970s to simplify the process of retrieving information from mainframe databases; it was later implemented on the PC platform in such products as dBASE and Paradox. See also SQL.

QBF A test message containing the "Quick Brown Fox" text. Used to test data terminals. The text is "The Quick Brown Fox jumped over the lazy dogs." It contains every letter of the alphabet. Check it out.

QC laser Quantum Cascade laser, a new type of semiconductor laser that works like an electronic waterfall. According to Bell Labs who invented the QC laser, it is the world's first laser that can be tailored to emit light at a specific wavelength at nearly any point over a very wide range from the mid- to far- infrared spectrum. This can be done by simply varying the layer thickness of the laser, using the same combination of materials. Conventional semiconductor lasers, widely used in other applications such as lightwave communications

and compact disk players, operate at wavelengths from the near infrared to the visible. When an electric current flows through the QC laser, electrons cascade down an energy staircase. Every time they hit a step they emit an infrared photon, or light pulse. At each step, the electrons make a quantum jump between well defined energy levels. The emitted photons are reflected back and forth between built-in mirrors, stimulating other quantum jumps and the emission of other photons until the amplified pulse escapes the laser cavity. The QC laser was invented by Federico Capasso and Jerome Faist in collaboration with Debbie Sivco, Carlo Sirtori, Al Hutchinson and Al Cho, according to AT&T Bell Labs.

QCELP Qualcomm Codebook Excited Linear Prediction. A proprietary voice compression algorithm developed by Qualcomm for use in digital telephone, CDMA (Code Division Multiple Access) wireless cellular, voice storage, and speech synthesis systems. According to Qualcomm, QCELP can operate in either Fixed Rate Mode or Variable Rate Mode. In Fixed Rate Mode (FRM), speech can be encoded at rates of 4, 4.8, 8, or 9.6 Kbps. In Normal Variable Rate Mode (NVRM), the data rate automatically adjusts from 800 bps to 8 Kbps. In Enhanced Variable Rate Mode (EVRM), the data rate automatically adjusts from 800 bps to 9.6 Kbps every 20 milliseconds (ms). When in Variable Rate Mode, the vocoders (voice coders) code speech at under 7 Kbps in continuous speech applications and at under 3.6 Kbps in typical two-way telephone conversations, without degradation in speech quality. See also CELP.

QCIF Quarter Common Intermediate Format, a mandatory part of the ITU-T's H.261 standard which requires that non-interlaced video frames be sent with 144 luminance lines and 176 pixels at a rate of 30 fps (frames per second). QCIF provides approximately one quarter the resolution of CIF, but requires about one quarter the bandwidth. It works quite nicely for small-screen display devices.

QD Queuing Delay: Queuing delay refers to the delay imposed on a cell by its having to be buffered because of unavailability of resources to pass the cell onto the next network function or element. This buffering could be a result of oversubscription of a physical link, or due to a connection of higher priority or tighter service constraints getting the resource of the physical link.

QDOS In 1980 IBM showed up on Bill Gates' doorstep seeking an operating system for its upcoming personal computer. Mr. Gates did not have one. But he knew someone that had one. A little firm down the road (in Seattle) had developed QDOS – the Quick and Dirty Operating System. It looked just right for IBM's PC. Mr. Gates bought QDOS for $100,000 and renamed it MS-DOS – Microsoft Disk Operating System. According to the Economist Magazine of May 22, 1993, some jealous Microsoft rivals claim that MS-DOS now stands for Microsoft Seeks Domination Over Society.

QDU Quantizing Distortion Units. ITU-T Recommendation G.113 defines one QDU as the amount of degradation introduced into a voice channel by a single conversion from analog to PCM and back to analog (analog-PCM-analog). Where several voice channels are connected in tandem, the end-to-end QDU rating for the whole circuit is calculated by adding the number of conversions from analog to PCM and back. For example: analog - PCM - analog - PCM - analog introduces 2QDUs.

QFC Quantum Flow Control. A method of flow control for ATM proposed as an alternative to ER (Explicit Rate), the current rate-based flow control mechanism. QFC is proposed by the QFC Alliance, which consists of a group of vendors including Digital Equipment Corp., Thomson-CSF and Ascom Nexion. QFC is touted as the solution for flow control in long-haul ATM implementations where data traffic is supported in addition to widely varying levels of high-priority traffic such as voice, video and multimedia information. QFC provides assurances that the buffers in the destination switch will not overflow, which would require retransmissions of low priority data (i.e., data), while the higher-priority data (i.e., voice, video and multimedia) flows through the network without difficulty. QFC also establishes limits on the bandwidth available to any individual connection, thereby avoiding the potential for monopolization of the switch buffers. See also Flow Control. Compare with Rate-Based Flow Control and ER.

QLLC Qualified Logical Link Control. Software package that allows Systems Network Architecture (SNA) commands to be transmitted over an X.25 packet data network (PDN). See also NPSI. Contrast with DSP.

QMS Queue Management System.

QNX A UNIX-like realtime operating system that works really well for computer telephony applications. Trademark of QNX Software Systems Ltd. qnx.com

QOE Quality of Experience. How's your QOE with your cell phone? Sometimes good? Sometimes awful? Like mine. Now you see the difficulty of measuring it.

QoR Query on Release. See LNP (Local Number Portability.)

QoS Quality of Service. See Class of Service and Quality of Service.

QPAM Quadrature Phase Amplitude Modulation. Used in high speed modems to send multiple data bits per baud. This type of modulation views the electrical signal as a vector that can be placed, by combining the the signals of two amplitude modulators that are 90 degrees out of phase, on a matrix of targets (sometimes called an eye-pattern) representing numeric values. The number of bits / baud determines the number of targets that are required. It is used in a wide variety of modems from voice frequency up to microwave baseband (100 MHz - 800 Mhz).

QPSK Quaternary Phase Shift Keying. A compression technique used in modems and in wireless networks, such as CDMA (Code Division Multiple Access) and 802.11a. A simple implementation of QPSK allows the transmission of 2 bits per symbol, with each symbol being a phase range of the sine wave. In this fashion, a 2:1 compression ratio is achieved, resulting in a doubling of the efficiency with which a circuit is employed. For instance, 0-90 degrees of phase indicates a 11 bit pattern; 90-180 degrees a 01; 180-270 a 10; and 270-360 a 00. In wireless networks, two carrier signals can be used, each of which is separated by 90 degrees of phase (position). If the phase of the carrier signals were not separated, one would be indistinguishable from the other. A 90-degree phase shift provides maximum phase separation and, therefore, maximum delineation between the carrier signals. See also 802.11a and CDMA.

QPSX Queued Packet Synchronous Exchange. Medium Access Control technology developed by the University of Western Australia for use in extending the reach of LANs across a Metropolitan Area Network (MAN). The technology was licensed to QPSX, Ltd. and subsequently was standardized by the IEEE as 802.6. QPSX was commercialized by Bellcore as DQDB, which is the access technology for SMDS networks. See also DQDB and SMDS.

QR Connector See XLR connector.

QRSS Quasi-Random Sequence Signals or Quasi-Random Signal Source. An industry-standard test pattern employing a fixed bit sequence to simulate random data and used to Signals used for testing digital circuits, in particular DS-1 (i.e. T-1) circuits.

QSAM Quadrature Sideband Amplitude Modulation. A sophisticated modulation technique, using variations in signal amplitude, that allows data-encoded symbols to be represented as any of 16 or 32 different states. See also QAM and Sideband.

QSIG The name under which PSS1 (Private Signaling System number 1), an international standard established by the ISO (International Organization for Standardization) and the IEC (International Electrotechnical Commission). QSIG is a global signaling and control standard for PINX-to-PINX (Private Integrated Network eXchange) applications, intended for use in private corporate ISDN networks to link multiple vendors' PBXs while retaining feature transparency. "Q" comes from the fact that the standard is an extension of the "Q" logical reference point defined by the ITU-T in its Q.93x series of recommendations for generic functions and basic services of ISDN SIGnaling systems. The early work on QSIG was accomplished by the European Computer Manufacturers Association (ECMA), which built on ITU-T ISDN standards for public networks. As a result, and for obvious reasons, QSIG, therefore, builds on the ITU-T DSS1 (Digital Subscriber Signaling 1) standard. DSS1 defines the logical reference point for ISDN at the user equipment. The impetus for this effort was that of encouraging the harmonization of existing, proprietary private network "standards" toward the reduction of technical trade barriers in the pan-European market. Subsequently, the EC (European Commission) became involved, charging ETSI (European Telecommunications Standards Institute) with the responsibility for further development and promotion of the standard in collaboration with CENELEC (translated from French as European Electrotechnical Standards Committee). QSIG standards are submitted to the JTC1 (Joint Technical Committee 1), which is a collaboration of the ISO (International Standards Organization) and the IEC (International Electrotechnical Commission). The standards also are promoted by the IPNS (ISDN PBX Networking Specification) Forum, which comprises a number of manufacturing companies such as Alcatel, Ascom, Ericsson, Lucent, Nortel, Philips and Siemens.

QSIG is much like the public network DSS1 standard set by the ITU-T, at least at Layers 1 & 2 of the OSI Reference Model. Differences appear at Layer 3, the Network Layer, as QSIG is intended for use in private networks and is symmetrical in nature, with the user side and the network side being identical. Further, QSIG is designed for peer-to-peer operation, although the standard addresses transit node capabilities, as well. QSIG also addresses both connection-oriented and connectionless services, unlike DSS1 standards which address only the former. ECMA currently is working on B-QSIG, which will extend the QSIG protocol stack to B-ISDN (Broadband ISDN). According to the IPNS Forum, QSIG offers user benefits including vendor independence, guaranteed PBX interoperability, free-form network topology, support for an unlimited number of nodes, flexible numbering plan, flexibility of interconnecting transmission technologies (i.e., analog or digital leased lines, radio and satellite links, and public VPN services). Supplementary services offered by QSIG include name identification, call intrusion, do not disturb, path replacement, operator services, mobility services, and

call completion on no reply. As a standards recommendation, QSIG provides manufacturers the freedom to develop custom features, with QSIG providing a standard mechanism for transporting such non-standard features. See also CENELEC, EC, ECMA, ETSI, IEC, ISO, and OSI Reference Model.

QTC Quick Time Conference. Apple Computer's cross-platform, video-conferencing, collaborative computing and multimedia communications technology.

QTVR See Quicktime VR.

quad A slang term for cable conductor with four single, plastic coated, not twisted wires and contained in a single plastic covering. Quad wiring has been traditionally used inside houses and small offices. Since it will not handle data well, it is no longer being recommended for installation anywhere, except in single-line analog (never data) applications. In the old days, a quad wire would support two analog phone lines. Color coding in quad wire in North America is red-green, yellow-black. When I showed this definition to a professional installer, he told me that quad wire was generally not used anymore except by ignorant do-it-yourselfers, cheap telcos (telephone companies), irresponsible contractors, etc. See Quad Wire.

quad-band A mobile (cell) phone that can operate at 850 MHz, 900 MHz, 1800 MHz and 1900 MHz. A quad-band phone can typically operate on any GSM network.

quad block Where "quad" wiring is terminated inside a residence quad block is typically a four screw terminal mounting that has some type of modular plug.

quad cable Cables where four wires are twisted as a unit. High crosstalk may be encountered among the wires within a quad unit.

quad fiber cable A cable consisting of four single optical fiber cables placed inside a polyvinyl chloride jacket with a rip cord to peel back the jacket and gain access to each single cable.

quad inside wire Quad IW. Older phone wire. It has four solid core copper conductors – red, green, black, yellow. Line one colors are green and red, line two colors are yellow and black. Since it's often not twisted, it's susceptible to RFI.

quad LNBF A combination LNBF and multi-sat switch component for DISH 500 systems. Can accommodate up to 4 DISH Network receivers.

quad lock conduit Conduit that's designed to be buried. The four conduits let companies lease space to each other in a way that's easy to track for fiber-optic cable installers/splicers, etc.

quad shield Two foil, two braid.

quad wire A type of wire which contains four untwisted copper conductors in a plastic sheath. These four conductors are not two separate twisted pairs, although the four may have a very "slow" twist to them. Quad wiring is no longer recommended by the telephone industry for installation in other than analog single line applications. In short, quad is dead. See Quad.

quad VGA QXGA. A new standard for a display resolution, namely 2,048 x 1,536 pixels.

quadded cable A cable in which at least some of the conductors are arranged in the form of a quad.

quadpod An antenna mounting system used for mounting an antenna on a sloped roof.

Quadrature Amplitude Modulation See QAM.
Quadrature Sideband Amplitude Modulation See QSAM.

quadruple play When you and I can get four telecommunications services from the same provider, that's called a quadruple play. These four services are broadband Internet, cell phone, fixed line telephone service (either circuit or VoIP) and television. I suspect an ingenuous investment banker thought up the idea of quadruple plays in order to convince unsuspecting telecom companies to buy other unsuspecting telecom companies and thus incur huge investment banking fees. See also quintuple play.

quads See Mated Pairs.

Quaero A European multimedia search engine project led by French technology firm Thomson, and announced by French President Jacques Chirac in January 2006. Quaero supporters hope that Quaero will nudge aside Google as the leading search engine in Europe.

QUALDIR QUALification DIRective. A wireless term for changes to a VLR (Visitor Location Register), a database which contains information about legitimate roamers and which describes the features to which they have access. The response to the QUALDIR is a "qualdir" (lower case). See also VLR.

Quality Of Service QoS. Quality of Service is a measure of the telecommunications – voice, data and/or video – service quality provided to a subscriber. It's not easy to define "quality" of voice telephone service. It's very subjective. Is the call easy to hear? Is it "clear?" Is it loud enough, etc.? The state Public Service Commissions (PSCs) have attempted to define the quality of service they want the residents of their states to have. And they have created various measures to which they insist phone companies conform. They tend to be more measurable. They include the longest time someone should wait after picking up the handset before they receive dial tone (three seconds in most states).

Quality of Service is more easy to define in digital circuits, since you can assign specific error conditions and compare them. For example if you were defining QoS with respect to ATM, it would be defined on an end-to-end basis in terms of the attributes of the end-to-end ATM connection, as detailed in ITU-T Recommendation I.350. The ATM Forum extended this standard through the definition of QoS parameters and reference configurations for the User Network Interface (UNI). ATM Performance Parameters include the following:

- Cell Error Ratio (CER)
- Severely Errored Cell Block Ratio (SECBR)
- Cell Loss Ratio (CLR)Cell Misinsertion Rate (CMR)
- Cell Transfer Delay (CTD)
- Mean Cell Transfer Delay (MCTD)
- Cell Delay Variability (CDV)

ATM Quality of Service (QoS) objectives set by the carriers are defined as Class of Service 1, 2, 3, and 4. Here is an explanation of the various classes: Class 1: Equivalent to digital private lines. Class 2: Supports traffic such as audioconferencing, videoconferencing and multimedia Class 3: Addresses connection-oriented protocols such as SDLC and Frame Relay Class. 4: Supports connectionless data protocols such as SMDS.

In the middle 90s, the concept of carrying voice and video over IP (Internet Protocol) networks suddenly became very important. In a White Paper which Microsoft put out in September 1997, it discussed QoS with the following words:

"What is Quality of Service? In contrast to traditional data traffic, multimedia streams, such as those used in IP Telephony or videoconferencing, may be extremely bandwidth and delay sensitive, imposing unique quality of service (QoS) demands on the underlying networks that carry them. Unfortunately, IP, with a connectionless, "best-effort" delivery model, does not guarantee delivery of packets in order, in a timely manner, or at all. In order to deploy real-time applications over IP networks with an acceptable level of quality, certain bandwidth, latency, and jitter requirements must be guaranteed, and must be met in a fashion that allows multimedia traffic to coexist with traditional data traffic on the same network." For another explanation (this time from Cisco), go to Class of Service.

Qos is full of confusing and sometimes contradictory terms. Here's how Network computing 9.4.2003 defines them:

Quality of Service: A way to provide better or stable service for select network traffic through bandwidth or latency control.

Saturation Point: The amount of load (packet count, simultaneous sessions or bandwidth utilization) that causes a network device to start dropping an unacceptable percentage packets.

Flow: A session between two hosts (such as a TCP session)> This includes handshaking, data transfer and termination. there can be multiple simultaneous flows between two hosts.

Class: A grouping of flows based on common criteria. May include protocol, source/destination address or subnet.

Classification: Detecting, identifying and potentially marking flows.

Burst Rate vs. Maximum Rate: If a QoS device supports bursting, it can let a class or flow be configured to use more bandwidth than the maximum rate, but only if extra, unused bandwidth is available. Think of it as a second max rate: Burst will always be higher than max rate. If burst equals max rate, then bursting is effectively disabled.

Quantity Above Threshold Indicator QAT. The sum of itemized call minutes that exceeds the minimum threshold of 1500 Minutes of Usage (MOU) that is eligible for a volume discount.

quantity type indicator A Verizon definition. A code that identifies the quantity type for local calling plan charges.

quantization The converting of a native analog signal to digital format through a sampling and quantizing process. This process is accomplished in a CODEC and is necessary in order to send analog data (voice or video) over a digital network (e.g., T-carrier or ATM) or through a digital switch (e.g., PBX or central office).

In the case of a voice signal and using PCM (Pulse Code Modulation), for instance, the amplitude of the native analog signal is sampled 8,000 times per second, with the each sampled amplitude value being expressed as an 8-bit digital value (byte) consisting of a specific combination of ones and zeros. At the receiving end of the communication, the process is reversed, with the digital value being translated into an analog amplitude value. The result is an approximation of the original analog signal, as it was sampled rather than

digitized exactly. Note that the original analog signal varied continuously in terms of both amplitude and frequency. Clearly, the higher the rate of sampling, the truer the reproduced approximate signal; the lower the rate of sampling, the less accurate the reproduced signal. In other words, a low rate of sampling would yield relatively unpleasant voice or fuzzy video as a result of what is known as "quantizing noise." However, a lower rate of sampling requires less bandwidth over the network or through the switch, yielding obvious cost benefits. As is that case with many things in life, there are tradeoffs between cost and quality.

quantization noise Signal errors which result from the process of digitizing (and therefore ascribing finite quantities to) a continuously variable signal. See Quantization.

quantize The process of encoding a PAM signal (Pulse Amplitude Signal) into a PCM signal (Pulse Code Modulation). See Quantization.

quantizing The second stage of pulse code modulation (PCM), for instance. The waveform samples obtained from each communication channel are measured to obtain a discrete value of amplitude. These quantized values are converted to a binary code and transmitted to a distant location to reconstruct an approximation of the original waveform. See Quantize and Quantization.

quantizing noise Noise caused by the inability of an analog signal to be exactly replicated in digital form. Such noise is the result of the fact that the original signal was sampled, yielding an approximation (but not exact replica) of the original signal as it is reconstructed on the receiving end of the communication.

quantum In physics, quantum means a very small indivisible piece of energy. This word is widely misused by people who refer to "a quantum leap," meaning a big leap. See Quantum Computing and Quantum Leap. Also:A time slice, the maximum amount of time a thread can run before the system checks for another ready thread of the same priority to run.

quantum computing A developing computing technology that exploits the properties of atoms to create a radically different type of computer architecture through quantum physics. Quantum computing relies on the basic traits of an atom, such as the direction of its spin (i.e., left-to-right and right-to-left), to create a state, such as a "1" or "0," much as conventional computers use variations in electrical energy (i.e., positive and negative polarity). Further, quantum computing theory suggests that intermediate states can be created. Further, the entanglement of spins between atoms can enable them to function as a collective whole. Qubits (quantum bits), therefore, are more than binary "1" and "0" bits. Qubits speak paragraphs, rather than bits of letters that make up words that make up sentences that make up paragraphs. This concept shatters the bounds of binary logic, linear processing, and computing speed. Don't look for it next week, or next month, or next year. Tell your grandchildren to look for it. Tell them to buy a copy of this book, which by then will be written by my grandchildren and will be in at least its 100th edition. But then, quantum publishing may have rendered the printed word to be obsolete, and they may be reading this book by watching atoms rotate. See Quantum and Quantum Leap.

Quantum Flow Control An ATM term. See QFC.

quantum leap In physics, quantum means a very small parcel or increment of energy. Also in physics, quantum leap or quantum jump refers to the abrupt transition (of something such as an electron, atom or molecule) from one discrete energy level to another. In popular usage, the term refers to an abrupt change, dramatic advance, or sudden increase. For instance, it might be said that major system enhancements which entail "forklift upgrades" involve quantum leaps in cost. Systems which are scalable do not. See also Forklift Upgrade and Scalable.

quantum mirage effect This process uses quantum waves to transfer information from one part of a nanoprocessor to another without relying on any physical connections.

quark Physicist Murray Gell-Mann named the sub-atomic particles known as quarks for a random line in James Joyce, "Three quarks for Muster Mark!"

QuarkXpress Probably the best typesetting and layout program around for the PC and for the Mac. Adobe is trying to compete with InDesign.

quarter speed An international leased teletype line capable of transmitting one quarter of Telex speed of 16 2/3 words per minute.

quarter wave antenna An antenna, the length of which is 1/4 that of the wave length received.

quartz Code-name for a tablet-like, quarter-VGA portrait screen size, pen-based, reference design, typically used in cell phones or PDAs. See also Quartz Crystal.

quartz crystal A small piece of quartz which is cut to a precise size. When electricity is applied to the crystal, it vibrates at a specific and precise frequency. Quartz crystals are often used in watches. They vibrate quickly and make the watch far more accurate than a timing device which vibrates far more slowly, like a pendulum, for example, or a tick tock watch.

quasi-analog signal A reader berated me for not including a definition for quasi-analog. I found this definition on the Internet. It's really stupid. It defines a quasi-analog signal "as a digital signal that has been converted to a form suitable for transmission over a specified analog channel. Note: The specification of the analog channel should include frequency range, bandwidth, signal-to-noise ratio, and envelope delay distortion. When quasi-analog form of signaling is used to convey message traffic over dial-up telephone systems, it is often referred to as voice-data. A modem may be used for the conversion process." Seems to my tiny brain that it's a digital signal (like from a computer) before it's converted into an analog signal for transmission over an analog phone line. So it's not really a quasi-anything. It's either one or the other. See Analog and Digital.

Quasi-Random Signal QR. A pseudo random test signal that has artificial constraints to limit the maximum number of zeros in the bit sequence.

qubit Quantum Bit. See Quantum Computing.

quenched gap A spark gap so arranged that the spark is quenched quickly by a cooling effect. A method used to give impulse excitation. An old radio term.

query 1. In data communications, it's the process by which a master station (or mainframe or boss computer) asks a slave station to identify itself and tell its status, i.e. is it busy, alive, OK, waiting, etc.?

2. In database, a query is a request for the retrieval of data.

query indicator An indication to subsequent nodes in the call path that a Local Number Portability (LNP) query has been performed. The industry has allotted the M bit within the Forward Call Indicator (FCI) for this purpose.

query language A programming language designed to make it easier to specify what information a user wants to retrieve from a database.

query optimization New database technology for optimiziing performance, minimizing administration and querying distributed, heterogeneous information. In short, a fancy term for the same old thing. Soon there'll be high-priced consultants in this field.

queue A stream of tasks waiting to be executed, or a series of calls, messages, or packets awaiting the availability of a network resource. For example, the calls, messages or packets may be awaiting the availability of a circuit, or the availability of the computational resources of a switch or router in order that they might be processed and sent on their way. The messages or packets may be held in a buffer (i.e., temporary memory) until such time as the network resource is available. A queue commonly is associated with a buffer on an incoming port. Queues and associated buffers also may be associated with outgoing ports, or may even be internal to a complex switching matrix. See Queuing.

queue management In a network, tasks like retrieval and writes to a jukebox come randomly from all the users. These tasks vary in urgency – retrievals are higher priority than writes, for example. Queue management sorts out requests from the network by priority. Queue management also enhances the performance of a jukebox, by intelligently re-ordering requests. For example, if there are three requests for images on platter 1 and two from platter 2 and the another from platter 1, queue management means the requests from platter 1 will get handled together, then go to platter two. Sometimes it's called "elevator sorting" – responding to requests in logical order, not in the order in which they were made.

queue service interval The maximum length of time a queue will go un-sampled.

Queued Arbitrated QA. A type of SMDS time slot which supports asynchronous data traffic. As such data is not time-sensitive, it is acceptable for the data transmission to take place over time slots on an "as available" basis.

queued call A call that is waiting in a queue of telephone calls to be serviced by a system resource is a queued call. An ACD group is an example.

queued mode Calls entering an Automatic Call Distributing system wait in a queue are presented, one at a time, to the first idle trunk in the chosen group.

Queued Packet Synchronous Exchange See QPSX.

Queued Telecommunications Access Method QTAM. A program component in a computer which handles some of the communications processing tasks for an application program. QTAM is employed in data collection, message switching and many other teleprocessing applications.

queuing The act of "stacking" or holding calls to be handled by a specific person, trunk or trunk group. There are two reasons to queue telephone calls:

1. Because you simply don't have enough trunks.
2. Because you want to save money.

You can queue calls mechanically using your telephone switch or manually using a human operator or attendant. There are two ways you can queue calls – hold-on or call-back. In "hold-on" queuing, you dial, you get some queuing tone (or the operator tells you you're being queued), then you wait on-line until a line becomes free and you're connected. In "call-back" queuing, you tell the operator or the machine you want to dial a call. And you hang up. When the line becomes free you are called back and connected. There are advantages and disadvantages to both systems. In "hold-on" queuing, you waste your time but save on phone time. In call-back queuing, you waste less of your time, but more phone line time. In call-back queuing, the operator or the phone system has to grab the line you want and simultaneously call you. By then, you may have left your desk. The call may be wasted, etc. The line given to you could have been used by someone else, etc. Queuing calls as a method to save money on long distance calling makes sense ONLY:

1. IF you are out of trunks because of a temporary surge in telephone traffic – perhaps at your peak, peak busy time and it's very expensive to buy sufficient "cheap long distance" trunks to handle every conceivable peak, and

2. IF you never plan on having a queue longer than 20 seconds for a hold-on queue and 60 seconds for a call-back queue and

3. IF you are queuing calls into an expensive fixed-cost line. For example a tie line between New York and London. If you queue calls into a variable cost line, like an interstate WATS line, you will save money over throwing the call onto DDD, but the pennies you save usually won't be worth it – considering the aggravation you're going to cause your people. Queuing is a very sensitive subject in corporate telecommunications departments. People don't like to wait for telephone lines. They consider that insulting to them personally, damaging to their "productivity" and to heck with the cost. Queues do, however, make enormous sense. Even a queue as short as ten seconds can save big amounts of money. Queues of a maximum length of ten seconds are rarely noticed. These days some of the more modern PBXs will allow you to offer "selective" queuing, or levels of queuing. Upper management doesn't have to queue for the cheap long distance lines before it's bounced to the expensive ones. While lower management has to wait up to 30 seconds. And the worker bees (non-management) have to wait even longer. Queuing is also used on incoming trunks. See ACD and Queuing Theory.

queueing delay Also spelled Queuing Delay. The length of time a data packet is held in queue in a buffer (i.e., temporary memory) while awaiting either processing by a device (e.g., switch or router) on the inbound side, or availability of a circuit on the outbound side. Queueing delay is one element of overall propagation delay, which is the total time required for a signal to transit a complete circuit, from transmitting device to receiving device. A circuit commonly comprises many links and many devices, each of which imposes some level of delay. See also Propagation Delay.

queuing theory The study of the behavior of a system that uses queuing, such as a telephone system. Much of queuing theory derives from the science of Operations Research (OR). Dr. Leonard Kleinrock has written the authoritative books on the subject. He is probably a genius. His books are very difficult to understand for laymen. Here is an explanation of Queuing Theory from James Henry Green's Dow Jones-Irwin Handbook of Telecommunications (you can buy a copy from www.amazon.com or www.barnesandnoble.com):

"The most common (telephone) network design method involves modeling the (phone) network according to principles of queuing theory, which describes how customers or users behave in a queue. Three variables are considered in network design. The first is the arrival or input process that describes the way users array themselves as they arrive to request service...The second variable is the service process, which describes the way users are handled when they are taken from queue and admitted into the service providing mechanism. The third method is the queue discipline, which describes the way users behave when they encounter blockage in the network...Three reactions to blockage are possible:

- Blocked calls held (BCH). When users encounter blockage, they immediately redial and reenter the queue.
- Blocked calls cleared (BCC). When users encounter blockage, they wait for some time before redialing.
- Blocked calls delayed (BCD). When users encounter blockage, they are placed in a holding circuit until capacity to serve them is available. See QUEUE.

"Traffic engineers have different formulas or tables to apply, corresponding to the assumption about how users behave when they encounter blockage." See Poisson.

After I wrote the above definition, Lee Goeller, a noted traffic engineering expert contributed the following definition:

The study of systems in which customers wait in line for servers to become available, the "blocked calls delayed" condition in telephony (see TRAFFIC ENGINEERING). Although seldom used in designing voice networks (other techniques are usually more cost-effective), queuing is very important in the design of packet networks where speed of transmission more than offsets the delay of waiting for a transmission facility to become available, and in staffing Automatic Call Distributors.

QUICC Quad Integrated Communications Controller.

Quick Clip An electrical contact used to provide an insulation displacement connection to telecommunications cables.

QuickConnect See V.92.

QuickDraw Programming routines that allow an Apple Macintosh computer to display graphics on a screen. QuickDraw is also used for outputting text and images to printers not compatible with PostScript.

Quicksilver Mercury, an extremely poisonous chemical. Actually it's not a chemical. It's an element. See "mad as a hatter."

QuickTime A dynamic-data format developed by Apple to be used for animation. QuickTime files can be used in documents created by other applications. For instance, a QuickTime video clip can be pasted into a word-processing document. QuickTime VR is the new Apple standard for Virtual Reality. See QuickTime VR.

QuickTime VR QuickTime VR (QTVR) is a way to create and view photo-realistic environments (panoramas) and real-world objects on Mac OS and Windows computers. Users interact with QuickTime VR content with a complete 360 degree perspective and control their viewpoint through the mouse, keyboard, trackpad or trackball. Using a QTVR-enabled authoring tool, panoramas and objects are automatically 'stitched together' from digitized photographs or 3D renderings to create a realistic visual perspective. The effect is awesome. As I wrote this, over 5,000 web sites were QuickTime enabled. One site I particularly enjoyed showed a hotel room. As a you moved your cursor, it seemed as though I was turning to see all 360 degrees of the hotel room. For more info http://quicktimevr.apple.com.

quick connect block Also called a 66-block or punch-down block. It's a 18" piece of metal and plastic which allows you to connect telephone wiring coming from two remote points. The quick-connect block has multiple metal "jaws" ranging horizontally and vertically. To connect up, you "punch" (or push) a wire between the two metal teeth of the "jaws." This both holds it firm and strips the wire's insulation, thus allowing for a good electrical connection. (There are special "punch-down" tools for punching wires into 66-blocks.) On a 66-block, one horizontal row of "jaws" is always the same conductor. To connect other wires to it, you simply punch those wires down along the row. Some 66-blocks have a gap between one side of the 66-block and the other. To connect one wire on one side to the wire on the other side, you have to use a BRIDGING CLIP. This is a small metal clip about one inch long. The bridging clip has one purpose: you can slip it off easily and thus cut one side of the circuit from the other. For example, if you connected central office trunks on one side of the 66-block and a PBX on the other, by removing the bridging clips, you can tell instantly if the trouble is in the PBX or in the central office. Two conductors on a 66-block makes a circuit – a trunk or a line. Therefore, the trunk 212-691-8215 (our main number) takes up the first two horizontal rows on our 66-block. The second two horizontal rows are taken up with 212-691-8216, and so on.

It is good to learn where your main 66-block is – the one that connects you to the telephone company's central office lines. The 66-block is what the telephone company calls the "demarcation point." And they (the phone company) usually install the 66-block. On one side (the trunk side) of their block, they're responsible. On the other (the PBX, key system or phone side), you're responsible. By knowing how to test your lines at this point, you can know whose fault it is – the phone company's or your equipment's. This can avoid having to wait until the phone company arrives, discovers it's not their problem and then sends you a hefty bill. Or the interconnect company arrives, finds out it's not their problem, and sends you a hefty bill, etc.

Quick Connect Blocks or 66-blocks are found in the Main Distribution Frame – where lines coming out of the PBX are connected to the individual wires going to the phones, or to big cables going to clumps of phones in other parts of the building. They're also found in Satellite Distribution Frames where they take big cable coming in from the main distribution frame and connect it to the individual cable pairs going to the individual phones. See Connecting Block.

quick format A DOS program which deletes the file allocation table and root directory of a disk but does not scan the disk for bad areas.

quick plug A device which adapts a standard four wire telephone cord into a modular connector.

Quicktime Apple Computer's video environment (like Microsoft's Video For Windows). Quicktime video files must be converted to *.AVI format to run under Microsoft's Video For Windows. Indeo video technology is supported under MacOS.

quidnunc A quidnunc is a person who is eager to know the latest news and gossip, in short a busybody.

quiescent A fancy word for quiet. No noise. No activity. Quiescent time is the best time to write this dictionary. Sadly, it wasn't always to be.

quincunx A graphics term. An anti-aliasing technique, developed by nVidia, which uses a sampling pattern that looks like the five side of a die. This dot patter is called a quincunx.

quintuple play Voice, long distance, wireless, broadband Internet access, and TV. See quadruple play for a larger explanation.

QUIPU This term is not an acronym. This public domain X.500 directory service, developed by University College London, demonstrates X.500 feasibility on TCP/IP (Transfer Control Protocol/Internet Protocol). This pioneering software package was developed to study the OSI (Open Systems Interconnections) Directory and provide extensive pilot capabilities. ISSUE (International Organization for Standardization Development Environment) provides commercial version of this software.

quorum A family of teleconferencing products linked in a system designed to meet a customer's teleconferencing needs.

QVGA A standard of sorts for displays of 320x240 pixels. The more common VGA is 640x480 color pixels.

QWERTY The name for a computer or typewriter keyboard. It got its name from the left side, top row of letter keys which spell QWERTY. One theory for the strange design of the QWERTY keyboard has to do with typewriter keyboards which had long metal arms that physically hit the paper. To keep the arms from jamming, they designed the QWERTY keyboard which split commonly used letters – i.e. a, i, o, e – to opposite sides of the keyboard. For years, people have argued that a keyboard called the Dvorak would be a much faster and more efficient. However, the U.S. General Service Administration in the 1950s contradicted the claims made by advocates of the Dvorak keyboard. The chief advocate was the patent owner, August Dvorak. According to September, 1995 Upside Magazine, his "book on the relative merits of QWERTY versus his own keyboard has about as much objectivity as a modern infomercial found on late night TV." With computers, there is no such thing as a "standard" QWERTY keyboard. Computer keyboards typically have 20 to 30 more keys than "standard" typewriter keyboards. Many of the keys are unique – on some keyboards, not on others. Many of the keys on computer keyboards are called "function" keys. If you hit one of them, they might perform a complete function on the computer, e.g. save a file, move to the end of the file, etc. There is absolutely no such thing as a standard computer keyboard.

Qwest Qwest Communications International Inc. is a broadband communications carrier that went public with an IPO (Initial Public Offering) in 1997. Qwest built a fiber network in North America and Europe, with the European network taking the form of the KPNQwest joint venture. On June 30, 2000, Qwest announced the completion of its merger with (read acquisition of) US West. US West was one of the seven Regional Holding Companies (RHCs) formed by the divestiture of the Bell Operating Companies (BOCs) by AT&T in 1984. Qwest and its senior management ran into accounting problems and the stock pummeted. In late 2005 Qwest's former, CEO, Joseph Nacchio, was indicted on insider trading. A Bloomberg story of December 20, 2005 started "Joseph Nacchio, who oversaw a $100 billion drop in the market value of Qwest Communications International Inc. as chief executive officer, has been indicted, the U.S. attorney in Denver said. A federal grand jury in Denver indicted Nacchio, 56, said U.S. Attorney William Leone in a statement today that did not specify the charges. Leone will announce the indictment at a press conference at noon local time, the statement said. The Securities and Exchange Commission sued Nacchio in March, 2005 claiming he orchestrated a $3 billion fraud from 1999 to 2002 at Qwest, the No. 4 U.S. telephone company." See also US West.

QXGA Quad VGA. A new standard for a display resolution, namely 2,048 x 1,536 pixels.

QZ Special billing arrangement provided by your local telephone company. Before there was automatic call accounting and before there was Centrex, the phone company would give you "time and charges" on every outgoing call. This service was called "QZ" billing. It was used by engineers, lawyers, accountants, consultants and other service people who had to bill their calls back to their clients.

R 1. The symbol designation for Resistance.

2. When you see CD-R, it means that the name of the disc ending in "R" can be used only once. Discs ending in "RW" can be erased and rewritten.

R interface An ISDN term. The 2-wire physical interface which is used for termination between a TA (Terminal Adapter) and TE2 (Terminal Equipment type 2), which is non-ISDN compatible terminal equipment. The physical connection generally follows either the RS-232 or the V.35 specification in terms of its electrical, functional and physical characteristics. TE2 can be in the form of a telephone set, a PC or a fax machine; none of these devices are ISDN-compatible unless specially equipped to be so. TE1 is ISDN-compatible terminal equipment, generally at significant additional cost. See also ISDN, S Interface, T Interface, TE1, TE2, and U Interface.

R reference point An ISDN reference point between non-ISDN terminal equipment (TE2) and a terminal adapter (TA). Non-ISDN (TE2) terminal equipment connects to ISDN at the R-reference point through a terminal adapter.

R.100 Fremont, California, December 13, 1999. The Enterprise Computer Telephony Forum (ECTF) today announced the availability of R.100, a new interoperability agreement that addresses the need for easily gathering, consolidating and reporting call center data. According to the ECTF, today's call centers contain a proliferation of systems that enhance the effectiveness of the center, but each one collects separate information on individual calls. All of the data collected by these different systems is needed to intelligently modify and fine-tune call center activities, but this very proliferation of systems presents a barrier to the correlation of data for an individual call. R.100 presents standardized techniques for gathering this data in a consistent manner from all the devices in a call center. R.100 defines common terminology for elements within the call center environment, minimum data sets to be available from call center functions, a common tracking method from call initiation to conclusion regardless of number of segments, to eliminate multiple counting, a method for clock synchronization, and data structures. According to ECTF, the model presented by R.100 is applicable in environments with intelligent networks, virtual call centers, networked call centers, circuit switched networks, IP telephony networks, computer telephony integration (including ECTF C.001, TAPI, TSAPI, JTAPI, and ECMA CSTA); and/or multimedia systems and calls. www.ectf.org.

R&D Research & Development.

R&E Research & Education.

R&R Rip and replace. A type of upgrade where the old system or network is completely ripped out and replaced with a new one.

R-Y A designator used to name one of the color signals (red minus luminance) of a color difference video signal. The formula for deriving R-Y from the red, green, and blue component video signals is .70R - .59G - .11B.

R/T Internet-speak for Real Time. R/T means the time it takes to download stuff. Writing in the New York Times, Charles McGrath said it was "customary in Net-speak to make a distinction between r/t, or real time – the time in which all these delays and jam-ups occur and v/t, or virtual time, which is time on the Net: a kind of external present in which it is neither day nor night and the clock never ticks. V/t is time without urgency, without priority."

R1 The ITU-T's name for a particular North American digital trunk protocol that happens to use multi-frequency (MF) pulsing. Some Europeans refer to any North American MF signaling protocol as R1 when distinguishing it from their own R2. See R2, Multi-Frequency Pulsing.

R1.5 R1.5 is Russian inter-switch address signaling over digital trunks (E1). It is based on MF R1 signals and "pulse shuttle" rules of their exchanging.

R2 A whole series of ITU-T specs which refers to European analog and digital trunk signaling. It refers to a type of trunk found in Europe which uses compelled handshaking on every MF (multi-frequency) signaling digit.

R2D Digital version of supervisory signals in R2, based on bits A and B in E-1 (in contrast to 3825 Hz signaling over analog lines).

RA 1. Rate Area.

2. Registration Authority. An authority in a network that verifies user requests for a digital certificate and tells the certificate authority (CA) to issue it. RAs are part of a public key infrastructure (PKI), a networked system that enables companies and users to exchange information and money securely. The digital certificate contains a public key that is used to encrypt and decrypt messages and digital signatures.

RA number Same as Return Material Authorization Number, or RMA. A code number provided by the seller as a prerequisite to returning product for either repair or refund. An indispensable tracking procedure, it operates like a purchase order system. If you return computer or telephone equipment without an RMA, chances are your equipment will be lost.

rabbit food Lettuce.

RAC 1. Remote Access Concentrator. A RAS is a larger Remote Access Server. According to Mark Galvin, president of RAScom, many people are now distinguishing between RAS (Remote Access Server) and RAC (Remote Access Concentrator). There seems to be two different cut-offs for the transition from RAS to RAC. The first, I believe originally defined by Dataquest, is when the port count exceeds 12. The industry seems to be adopting a different cut-off at anything T-1 or bigger as a RAC. A remote access server or remote access concentrator is a piece of computer hardware which sits on a corporate LAN and into which employees dial on the public switched telephone network to get access to their email and to software and data on the corporate LAN (e.g. status on customer orders). Remote access servers are also used by commercial service providers, such as Internet Access Providers (ISPs) to allow their customers access into their networks. Remote Access Servers are typically measured by how many simultaneous dial-in users (on analog or digital lines) they can handle and whether they can work with cheaper digital circuits, such as T-1 and E-1 connections. See also Remote Access Concentrator and Universal Edge Server.

2. Real Application Clustering. See Database Clustering for an explanation.

RACE An association in the European Economic Community. RACE stands for Research and development for Advanced Communications in Europe.

race condition A network security term. A condition caused by the timing of events within a piece of software. Race conditions typically are associated with synchronization errors that provide a window of opportunity during which one process can interfere with another, possibly introducing a security vulnerability.

raceway Metal or plastic channel used for loosely holding electrical and telephone wires in buildings. A raceway is usually located in the floor and is encased on three or four sides by concrete. A raceway is used for interior wiring and performs the same job as a conduit but is typically larger.

raceway method A ceiling distribution method in which open or closed metal trays are suspended in false ceilings from the structural floor above. The raceway method is generally used in large buildings or for complex distribution systems that demand extra support. When closed metal trays are embedded in the floor, this distribution method is often called underfloor raceways. See also Ceiling Distribution Systems and Underfloor Duct Method.

rachet factor The rachet factor is part of CABS – Carrier Access Billing System. It used to describe the apportionment of channels on a trunk between switched and facility usage. It's a percentage. Both switched usage and leased (facility) lines can be co-resident on the same trunk. The rachet percentage refers to the percentage of the trunk dedicated to facility. Obviously, you would want to know this because switched usage is tariffed and facility usage is charged at flat contract rates.

rack 1. An equipment rack. In our industry, the standard equipment rack is 19 inches (48.26 cm) wide at the front. Much equipment is designed to fit into a standard rack. A rack is typically made of aluminum or steel, onto which equipment is mounted. A rack is typically attached to a building ceiling or wall. Cables are laid in and fastened to the rack. Sometimes a rack is called a tray. What a rack is to equipment, so a frame is to wiring. See also Distribution Frame.

2. Rack (the digits). a term which implies the storing or registering of numerical data. See Register.

Rack Unit RU. Unit of measure of vertical space in an equipment rack. One rack unit is equal to 1.75 inches (4.45 cm).

rackmount Designed to be installed in a cabinet, usually 19" wide.

RACON RAdar transponder beaCON. Short-range navigation devices that provide target images on a ship's maritime navigation radar system. The transponder beacons transmit, either automatically or in response to a predetermined received signal, a pulsed radio signal with specific characteristics. RACONs generally operate in the 9300-9500 MHz band, and are used to identify specific locations such as hazards to navigation; think of them as replacements for lighthouses and you won't be far off. Most RACONs are operated by the U.S. Coast Guard. See also Radar.

racquet technician A fancy name for someone who strings tennis racquets. Also spelled Racket Technician in America.

rad 1. The unit used to measure the absorption of ionizing radiation.

2. A British Term. Recorded Announcement Device, a device which automatically answers a line and delivers a pre-recorded message. Often used to tell a caller to a telebusiness unit that the call is in a queue and will be dealt with soon. More sophisticated RADs gather information, take messages or work in conjunction with interactive fax machines.

3. An abbreviation for Rapid Application Development. Most relate it to a quick programming environment.

4. Remote Antenna Driver.

Radar RAdio Detection And Ranging. See Radar Detector.

radar detector Picture a trooper sitting in his car aiming his radar gun down the highway. The gun emits a beam of electrons at microwave frequency. Those beams bounce off approaching vehicles and reflect back to the trooper's radar at an altered frequency (the Doppler Effect). By measuring the change in frequency, the trooper calculates the speed of the oncoming vehicle. The trouble is the radar beam fans out like a searchlight. At a distance of 1,000 feet, the beam is about as wide as the highway itself. That makes it difficult for the trooper to know which vehicle he's tracking.

Also, his reading can be thrown off by any number of operating errors or by interference from power lines, neon lights or even the fan motor in the trooper's car. According to some estimates, Esquire Magazine reported, as many as 30% of all radar-generated speeding tickets were given in error. In 1979 a Miami TV station showed a police radar clocking a house going 28 miles per hour and a banyan tree doing 86! Radar detectors are very much like FM receivers. They can pick up radar signals more than a mile from the source. At that

distance the beam is too weak to bounce all the way back to the trooper's car but strong enough to make the detector beep.

radar screen 1. A typically circular cathode ray tube (CRT) showing movement of the sweep of a swirling radar beam and the objects it hits.

2. A slang expression typically deriding something. "I'm studying the market for computers laptops, but Winbook is not on my radar screen." This typically means that Winbook, as a manufacturer, is so small they're not worth studying. To be on my radar screen means they're large enough and significant enough for me to study them.

radial acceleration The rate at which a track on an optical disc accelerates toward and away from the center, because it is not perfectly aligned or perfectly round.

radials See Ground Radials.

radiant energy Energy as measured in joules which is transferred via electromagnetic waves. There is no associated transfer of matter. And typically the giver or energy and the receiver of energy are not touching.

radiation Energy emitted in the form of waves (light) or particles (photons). See Radiation Pattern.

radiation pattern The propagation characteristics of an antenna.

radichio A forum established to promote common public key infrastructure standards for ecommerce using wireless phones.

radio RF. System of communication employing electromagnetic waves propagated through space. Because of their varying characteristics, radio waves of different lengths are employed for different purposes and are usually identified by their frequency. The shortest waves are the highest frequency, or numbers of cycles per second; the longest waves have the lowest frequency, or fewest cycles per second. In honor of the German radio pioneer Heinrich Hertz, his name has been given to the cycle per second (hertz, Hz); 1 kilohertz (Khz) is 1000 cycles per second, 1 megahertz (Mhz) is 1 million cycles per second, and 1 gigahertz (Ghz) is 1 billion cycles per second. Radio waves range from a few kilohertz to several gigahertz. Waves of visible light are much shorter. In a vacuum, all electromagnetic waves (but not audio waves) travel at a uniform speed of about 300,000 km (about 186,000 miles) per second.

Radio waves are used not only in radio broadcasting but in wireless devices, telephone transmission, television, radar, navigational systems, and communication. In the atmosphere the physical characteristics of the air cause slight variations in velocity, which are sources of error in such radio-communications systems as radar. Also, storms or electrical disturbances produce anomalous phenomena in the propagation of radio waves.

Because electromagnetic waves in a uniform atmosphere travel in straight lines and because the earth's surface is spherical, long distance radio communication is made possible by the reflection of radio waves from the ionosphere. Radio waves shorter than about 10 m (about 33 ft.) in wavelength – designated as very high, ultrahigh, and super high frequencies (VHF, UHF, and SHF) - are usually not reflected by the ionosphere; thus, in normal practice, such very short waves are received only within line-of-sight distances. Wavelengths shorter than a few centimeters are absorbed by water droplets or clouds; those shorter than 1.5 cm (0.6 in) may be absorbed selectively by the water vapor present in a clear atmosphere.

A typical radio-communication system has two main components, a transmitter and a receiver. The transmitter generates electrical oscillations at a radio frequency called the carrier frequency. Either the amplitude or the frequency itself may be modulated to vary the carrier wave. An amplitude - modulated signal consists of the carrier frequency plus two sidebands resulting from modulation. Frequency modulation produces more than one pair of sidebands for each modulation frequency. These produce the complex variations that emerge as speech or other sound in radio broadcasting, and in the alterations of light and darkness in television broadcasting.

radio access network RAN. Cellular networks essentially consist of two parts: the Radio Access Network (RAN), which controls transmission and reception of radio signals, and the Core Network, which provides switching, transport, and enhanced services for traffic emanating from and directed to the cellular network's RAN.

Radio Broadcast Data System RBDS. A new system designed to let radio stations broadcasters send text messages, such as emergency warnings and traffic alerts to radios equipped with special LCD screens. The system is designed ultimately to replace the Emergency Broadcast System.

radio button 1. A call center term. A button used for selecting from a group of options that are mutually exclusive. As with a car radio, selecting a particular button de-selects the previously selected button.

2. An World Wide Web term. Radio buttons are used in forms on Web sites to indicate a list of items. Only one button can be selected at one time.

radio cache A portable or permanent storage facilty for portable radios, batteries, and repeaters, established for use in an emergency, when there is a need to provide enhanced communications support to first-responders, second-echelon support, and other personnel who arrive on the scene later, often from other regions and jurisdictions. As an emergency situation develops and respondents arrive at a staging area, assignments are made, personnel are given portable radios from the cache which are programmed on the spot, if necessary, to ensure interoperable communications across jurisdictions and disciplines.

radio check A short radio transmission from one radio station to another, and then reciprocated, for the purpose of checking the signal strength and audio quality of transmissions between the two stations. Radio checks are commonly done by military units in the field. For example, station W54 might send the following message to station Y25 in order to perform a radio check: "Yankee 2-5. This is Whiskey 5-4. Radio check, over." If Y25 hears the message with no problem, it will respond accordingly. If W54's transmission was problematic in some way, Y25 will advise W54.

Radio Common Carrier RCC. A common carrier engaged in Public Mobile Service, which also is not the business of providing land line local exchange telephone service. These carriers were once known as Miscellaneous Common Carriers.

radio communication Any telecommunication by means of radio waves.

radio conformance testing A set of tests performed on a radio to ensure spectrum compatibility and conformance to design and manufacturing criteria. Radio conformance testing includes, but is not limited to, testing of channel allocation, modulation accuracy, transmitter power level adjustment accuracy, and adjacent channel interference.

Radio Control Radio Service Radio Control (R/C) is a one-way, short distance, non-voice radio service for on/off operation of devices at places distant from the operator. The FCC authorizes your R/C unit to transmit any non-voice emission type for the purpose of (1) the operator turning on and/or off a device at a remote location, or (2) indicating device for the operator being turned on and/or off by a sensor at a remote location. You cannot communicate voice or data in the R/C. See Personal Radio Services.

Radio France Internationale France's government-funded international radio service, whose mission is to contribute to the spread of French culture worldwide. Radio France Internationale (RFI) broadcasts in French and nineteen other languages via FM radio, short-wave radio, radio relay, satellite radio, and now also by Internet radio to over 200 countries. RFI began service in 1931, and was initially called le Poste Colonial. Over the years it went through several name changes. In 1975 the service was renamed Radio France Internationale, a name which has stuck longer than any of its predecessors.

radio frequency That group of electromagnetic energy whose wavelengths are between the audio and the light range. Electromagnetic waves transmitted usually are between 500 KHz and 300 GHz.

radio frequency flooding Radio frequency flooding turns a telephone into a room listening device by transmitting a high power radio signal down a telephone line. The high power radio frequency is able to bypass the open hookswitch in the mouthpiece circuit. Room sounds cause the carbon microphone to modulate the RF signal. Radio frequency flooding is hard to implement but can only be detected by security professionals with the right equipment.

Radio Frequency IDentity RFID. A method of identifying unique items using radio waves. Typically, a reader communicates with a tag, which holds digital information in a microchip. But there are chipless forms of RFID tags that use material to reflect back a portion of the radio waves beamed at them. For a fuller explanation, see RFID.

Radio Frequency Interface shield RFI. A metal shield enclosing the printed circuit boards of the printer or computer to prevent interference with radio and TV reception.

radio frequency interference The disruption of radio signal reception caused by any source which generates radio waves at the same frequency and along the same path as the desired wave.

Radio Frequency Interference Shield RFI Shield. A metal shield enclosing the printed circuit boards of the printer or computer to prevent radio and TV interference.

radio modem A modem that transmits and receives data via radio waves. The radio modem does this by transforming bits into modulations of radio waves, and vice-versa.

radio paging access Provides attendant and phone user dial access to customer-owned radio paging equipment to selectively tone-alert, or voice-page individuals carrying pocket radio receivers. The paged party can answer by dialing an answering code from a phone within the PBX.

radio paging access with answer back Allows access to customer-provided paging systems and provides the capability in the PBX to connect the paged party when the former answers the radio page by dialing a special code from any PBX.

Radio Regulations The internationally-accepted rules governing radio communications, as issued by the ITU. The most recent edition was published in 2001.

radio resource management A management entity or subentity concerned with the operation of the radio resources management protocol. A cellular radio term.

radio resource management entity A management entity or subentity concerned with the operation of the radio resource management protocol. A cellular radio term.

radio silence A period during which radios stop transmitting. It is most often associated with the military, where a radio transmission may give away a military unit's position or plans to the enemy.

radio wave Electromagnetic waves of frequencies between 10 KHz and 3MHz, propagated without guide in free space (air).

Radiocommunications Consultative Council See RCC.

radiogram A telegram sent by radio. Totally obsolete term, but cute.

radiometer An instrument, distinct from a photometer, to measure power, in Watts, of electromagnetic radiation.

radiophone Apparatus for transmitting and/or receiving speech or music by radio. Totally obsolete term, but cute.

radiosonde An automatic radio transmitter in the meteorological aids service usually carried on an aircraft, free balloon, kite, or parachute, and which transmits meteorological data.

radiotelegraphy The use of a radio (instead of wire) to communicate telegraphy messages over a distance. An old term, not used much any more. See also Radiotelephony.

radiotelephony The science, art, and act of transmitting speech by means of radio. Now called telecommunications.

RADIUS Remote Authentication Dial-In User Service is an authentication and accounting system used by many Internet Service Providers (ISPs) and end user organizations to provide secure Internet access, especially in a VPN (Virtual Private Networks) application. When you dial in to the ISP, for example, you must enter your username and password. This information is passed to a RADIUS server, which checks that the information is correct, and then authorizes access to the ISP systems and network. Though not an official standard, the RADIUS specification is maintained by a working group of the IETF. RADIUS is a client/server-based authentication software system that centralizes the administration of user profiles maintained in authentication databases, thereby simplifying the process of supporting multiple VPN switches. The remote access servers act as RADIUS clients which connect to the centralized authentication server. RADIUS is an open specification that can be adapted to work with legacy systems and protocols. RADIUS was developed by Steve Willins of Livingston Enterprises, Inc., which was acquired by Lucent Technologies. See also Authentication, Client/Server, and VPN.

radome A plastic cover for a microwave antenna. It protects the antenna from awful weather, but has little effect on the radiation pattern of the antenna. Here's a more technical explanation: A radome is a dielectric cover placed over an antenna to protect it from the environment. For example, the nose of an aircraft which contains the airplane's weather radar antenna. There are hundreds of other radome applications encompassing telecommunications, wireless, radars, satcom, terrestrial communications, ECM equipment and other commercial and military sensors.

RADSL Rate Adaptive Digital Subscriber Line. Transmission technology that supports both asymmetric and symmetric applications on a single twisted pair telephone line and allows adaptive data rates. RADSL employs intelligent ADSL modems which can sense the performance of the copper loop and adjust transmission speed accordingly. These devices adjust dynamically as the performance of the loop varies during a session, much as does a V.34 modem. Depending on various characteristics of the subject cable plant, ADSL accommodates downstream transmission speeds of as much as seven megabits per second, plus bidirectional transmission speeds of as high as 640 Kbps over a single UTP (Unshielded Twisted Pair). Some RADSL equipment can be manually configured See ADSL.

RAID Redundant Array of Inexpensive Disks. The idea is simple: Put several disk drives into a single housing. Then write your data over the disk drives in such a way that if you lose one or more of the drives, you won't have lost any of your data. Thus the term "redundant." At its simplest, RAID mirrors data to an equal number of disk drives, e.g. two sets of two. At its most complex, RAID writes data across a bunch of drives, so that if one goes the data can be retrieved from the remaining drives. The opposite of RAID is SLED (Single Large Expensive Disk). See also Redundant Array of Inexpensive Disks for a much more detailed explanation.

railway coverage system A distributed antenna system inside of train carriages or along railroad tracks. An in-carriage system provides wireless communications within the train. A system set up along train tracks provides communications along the rail line and between passing trains and the outside world.

railway gauge The US Standard railroad gauge (distance between the rails) is 4 feet, 8.5 inches. That's an exceedingly odd number. Why was that gauge used? Because that's the way they built them in England, and the US railroads were built by English expatriates. Why did the English people build them like that? Because the first rail lines were built by the same people who built the pre-railroad tramways, and that's the gauge they used. Why did "they" use that gauge then? Because the people who built the tramways used the same jigs and tools that they used for building wagons, which used that wheel spacing. Okay! Why did the wagons use that odd wheel spacing? Well, if they tried to use any other spacing the wagons would break on some of the old, long distance roads, because that's the spacing of the old wheel ruts. So who built these old rutted roads? The first long distance roads in Europe were built by Imperial Rome for the benefit of their legions. The roads have been used ever since. And the ruts? The initial ruts, which everyone else had to match for fear of destroying their wagons, were first made by Roman war chariots. Since the chariots were made for or by Imperial Rome they were all alike in the matter of wheel spacing. Thus, we have the answer to the original questions. The United State standard railroad gauge of 4 feet, 8.5 inches derives from the original specification for an Imperial Roman army war chariot. Specs and Bureaucracies live forever. So, the next time you are handed a specification and wonder what (expletive deleted) came up with it, you may be exactly right. Because the Imperial Roman chariots were made to be just wide enough to accommodate the back-ends of two war horses.

railway revolution The railway revolution took place from about 1825 to 1875 in Britain. It saw a great connecting of commerce and the coming of steam power. See also the Steel and Electricity Revolution.

rain attenuation When radio signals encounter a heavily moisture-laden atmosphere, the signals will lose their strength. Generally, the higher the radio frequency, the more attenuation (i.e., the greater the signal loss). Since satellite and terrestrial microwave signals for satellite are essentially line-of-sight, these types of signals are very susceptible to signal attenuation caused by rain, and especially heavy rain and snow. Modern microwave paths are engineered with weather patterns in mind. In areas where heavy rainfalls occur, microwave links may be placed closer together or more attention is paid to diverse and duplicated routing. You can't have more than one path with satellite TV signals directly to the home, e.g. DirecTV or Sky Satellite. These signals tend to fade or drop out completely during rain and snow storms. And they usually drop out at the movie's climax. Moisture also affects FSO (Free Space Optics) systems, with fog impacting the signal much more than rain. Optical fiber systems are affected by residual moisture deposited during the manufacturing process. Moisture, in general, is a bad thing in transmission systems. See also Free Space Optics and Water Peak.

rain barrel effect Signal distortion of a voice telephone line caused by the under-attenuated echoes on the return path.

rain fade The loss of signal from a satellite during a heavy rain. This happens more or less to all DBS (direct broadcast) systems. The loss of signal is usually for only a few minutes. Rain fade can occur even if it is not raining at your location. A large black thunderhead can block signal if it gets between you and the satellite. See rain attenuation and rain outage.

rain outage Loss of signal at Ku or Ka Band frequencies due to absorption and increased sky-noise temperature caused by heavy rainfall.

rainbow series According to the National Security Agency (NSA), the Rainbow Series is a six-foot tall stack of books dealing with the evaluation of "Trusted Computer Systems." The term comes from the fact that each book is a different color. Colors include orange, aqua, burgundy, lavender, venice blue, pink, peach, turquoise, and violet.

raincheck When a sports event is rained out, they give you a "raincheck," voucher for a seat when the event will be played. It's used also in personal relations. "Thanks for your invitation. I can't come to dinner tonight. Can I take a raincheck?" Which means, "Can I come at another time?"

raised floor A floor distribution method in which square, steel and wood-laminated plates resting on aluminum locking pedestals are attached to the building floor. The plates are usually covered with cork, carpet, or vinyl tiles, and each plate can be removed for easy access to the cables below. Also referred to as access floor.

rake receiver A radio receiver designed to counter the effects of multipath fading; commonly used in devices such as mobile phones. Uses several sub-receivers, each slightly delayed, to tune into the individual paths a radio wave follows (multipaths). Each compo-

nent is later combined to effectively strengthen the signal.

RAM 1. Random Access Memory. The primary memory in a computer, RAM keeps the CPU (Centralized Processing Unit) efficiently fed with data or programs from the hard drive, or other storage medium. RAM can be overwritten with new information. The "random access" part of its name comes from the fact that the next "bit" of information in RAM can be located – no matter where it is – in an equal amount of time. This means that access to and from RAM memory is extraordinarily fast. By contrast, other storage media (e.g., hard drive, magnetic tape, and CD-ROM) store information serially, one bit after another. Therefore, the computer has to search for them, with the search time depending on the distance between the current bit and the target bit. Floppy disks are faster than magnetic tape because their information is readily at hand, though the read/write head will have to search for it. Hard disks are even faster because there are multiple heads and because the disks spin faster and everything moves faster. The speed of a CD-ROM search depends on the speed of the CD-ROM drive, but it can be pretty slow. RAM memory is the fastest of all. The problem with RAM memory is that it's volatile. This means when power is turned off (or power glitches occur) RAM memory is erased. RAM memory can be protected with rechargeable batteries – just remember to charge the batteries.

All of that having been said, RAM comes in two types, at least along one dimension: buffered and unbuffered. Buffered RAM contains buffer chips that serve intermediaries between the control chips and the memory chips. The buffers help deal with the relatively large electrical load involved when a computer has a lot of memory. For example, a large server may have 16 DIMM (Dual Inline Memory Module) modules, each with as many as 36 RAM chips, yielding a total of 576 RAM chips. Since the controller cannot directly power so many chips, a buffer associated with each DIMM module assumes that responsibility. Unbuffered RAM typically is used on desktops, laptops, and smaller servers. Now for registered and unregistered RAM. Registered RAM includes a small register, or memory element, that imposes a slight delay, typically one clock cycle on the information transfer, from input to output. Large servers serving large volumes of information to large numbers of users employ registered RAM in order to ensure the integrity of the data being transferred. Now let's tie these two concepts together: buffered RAM may be registered, and unbuffered RAM is always unregistered. See also DDR-SDRAM, DRAM, EDO RAM, Flash RAM, FRAM, RDRAM, SDRAM, SRAM, and VRAM.

2. Remote Access Multiplexer. The various DSL (Digital Subscriber Line) services all involve some form of splitter or modem at the customer premises, and a matching device at the edge of the carrier network. Across the UTP (Unshielded Twisted Pair) local loop, the matching DSL units variously perform such functions as analog-to-digital conversion, data compression and packetization, ATM Adaptation, and Frequency Division Multiplexing (FDM). Generally speaking, the centralized unit at the edge of the carrier network is in the form of a DSLAM (DSL Access Multiplexer) contained within a Central Office (CO) owned by the ILEC (Incumbent Local Exchange Carrier). This approach works well for most subscribers, as they are served by "home run" UTP local loops from the CO to the premises. However, customers served through DLC (Digital Loop Carrier) systems are denied DSL service, as the DLC grooms and shapes all local signals into voice-grade channels of 64 Kbps. A ng-DLC (next generation DLC) may contain one or more RAMs, which effectively are remote DSLAMs. Unfortunately, the RAMs are about the size of a pizza box, and currently serve no more than eight DSL lines. See also DLC, DSL, DSLAM, and ngDLC.

RAM BIOS BIOS transferred to RAM so things go faster.

RAM disk A logical device made from semiconductor (i.e. chip) memory which emulates the functioning of a disk drive as closely as possible. Since most semiconductor memory (RAM) is volatile, most RAM disks are also volatile, i.e. they lose their memory when you turn off power.

ram hook/ram horn Hardware attachment that holds ASW (Aerial Service Wire) drop clamps in aerial span applications.

Rambus DRAM See RDRAM.

Rambutan A symmetric cryptographic algorithm developed by Marconi.

RAMAC RAndoM ACcess. Built in 1956 by IBM, RAMAC was the first hard drive computer memory device ever built. Consisting of 50 fixed disk platters, each approximately two feet in diameter, RAMAC could store five million characters at a rate of about 2,000 bits per square inch and at a cost of approximately $10,000 per MB (MegaByte). Contemporary (1999) hard drives store information less than a dime ($.10) per megabyte.

Raman Amplifier A means of amplification used in optical fiber transmission systems. Raman amplifiers are of two basic types: Distributed and Discrete. That was easy. The rest of this definition is hard, very hard, but some things in life are just hard. I worked hard to make this easy, but it's still hard. So, read it several times, at least.

Distributed Raman amplification uses the fiber itself as the gain medium, whereas the more traditional approaches use separate network devices such as EDFAs (Erbium Doped Fiber Amplifiers) to amplify the light signal. EDFAs accept a weakened light signal and boost it to a high power level before launching it, i.e., presenting it to the next fiber link. Over a distance, of course, the light signal weakens, and must be boosted again. Distributed Raman amplifiers front-end the EDFAs, serving as "pre-amplifiers." Diode-pumped Raman amplifiers pump an optical signal in the direction opposite the data flow (i.e., in the direction of the incoming optical signal carrying the transmitted data). Raman amplification occurs when higher energy (i.e., higher frequency, and shorter wavelength) pump photons scatter off the vibrational modes (i.e., vibrating atoms in the optical fiber), resulting in coherent (relating to electromagnetic waves that have a definite relationship to each other) stimulation (i.e., adding energy to) of the lower energy (i.e., lower frequency, and longer wavelength) photons associated with the incoming optical carrier signal (i.e., the one carrying the transmitted data). The pump photons actually can simultaneously amplify multiple carrier wavelengths in a DWDM system. At the same time, the Raman amplification process serves improve the SNR (Signal-to-Noise Ratio, expressed as Signal/Noise), which improves the quality of the incoming carrier signal and, thereby, improves error performance. As Raman amplifiers use the entire fiber as the gain medium, the signal weakens much less over a distance and the overall "launch power" level of the transmission signal need not be as high. That translates into lower power costs. Also as a result of Raman amplification, the spacing of both the EDFAs and the optical regenerators can be increased considerably. That results in lower overall network cost.

For example, a counter-propagating light signal at a wavelength of 1450 nm (nanometers, or billionths of a meter) can be pumped in the direction opposite the transmission wavelength of 1550 nm. At the point that the SNR (Signal-to-Noise Ratio) begins to drop, and the BER (Bit Error Rate) therefore becomes an issue, the Raman effect contributes gain (i.e., amplification) to the transmission signal. Raman amplification reduces the number of amplifiers required by lengthening the distance between them, say from approximately 80 km (8,000 meters) between EDFAs to approximately 100 km between Raman amplifiers. Raman amplification also allows the increasing of the bit rate at a given frequency of light, say from 10 Gbps to 40 Gbps. Raman amplification also works effectively with DWDM (Dense Wavelength Division Multiplexing), as multiple wavelengths can be amplified simultaneously.

Discrete Raman amplification occurs in a box, rather than in the optical transmission fiber. The incoming optical carrier signal enters the standalone Raman amplifier and is channeled into a length of specialty fiber, where the amplification process takes place. The likely application for this approach (it is not used commercially at the time of this writing in April 2001) will be to overcome localized loss such as that which might occur internal to a switch or multiplexer. See also DWDM, EDFA, and Raman Scattering.

Raman Scattering Also known as the Raman Effect. Chandrasekhara Venkata Raman, an Indian physicist, discovered (1928) this phenomenon of the inelastic scattering of light as it encounters physical matter. Raman was awarded the Nobel Prize in Physics in 1930 for this discovery, which confirmed the quantum theory of light, which theory is that light is made up of particles. As incident light (i.e., light falling on or striking something) is scattered as it strikes an atom of solid physical matter, most of the photons retain their incident energy, or frequency, which we would see as a color of light. This "elastic" (i.e., snaps back into its original "form") scattering effect is known as Rayleigh scattering. Raman scattering describes the fact that the incident light also causes the electrical bonds between atoms to vibrate which, in turn, causes a small fraction of the photons to experience a shift in energy, or frequency (which is the inverse of wavelength). In this "inelastic" (i.e., doesn't snap back into its original "form") scattering phenomenon, the difference in the energy of the incident photon and the Raman scattered photon is equal to the energy of the vibration of the scattering molecule.

Raman scattering clearly impacts optical fiber transmission systems. The impact is especially significant in the case of DWDM (Dense Wavelength Division Multiplexing), which involves very precise frequency division multiplexing at the optical level. As the diode lasers introduce high-energy light signals into the clear core of the optical fiber, the phenomenon of Stimulated Raman Scattering (SRS) occurs. SRS occurs when light waves interact with vibrations of atoms in a crystalline lattice (i.e., optical fiber in this context). The atoms absorb the light and re-emit a photon with energy equal to the original photon plus/minus the atomic vibration. This scattering effect travels both forwards and backwards, and can cause frequency shifts in the individual light streams, which negatively affects the integrity of the bit stream. However, a Raman Gain amplifier, also known as a Raman laser, can overcome this effect by shifting the wavelength (frequency) of the incident light to a vastly different value at the output. If the incident light signal is monochromatic and sufficiently intense to reach a certain threshold value, the signal is amplified to the point that it exhibits the

characteristics of a stimulated emission. The ultimate impact of this effect is both to clean up the signal, and to amplify it. See also DWDM, Raman Amplifier, and Rayleigh Scattering.

RAMDAC Random Access Memory Digital-to-Analog Converter. The chip on a VGA board that translates the digital representation of a pixel into the analog information needed for display on the monitor. A RAMDAC actually consists of four different components – SRAM to store the color map and three digital-to-analog converters (DACs), one for each of the monitor's red, green, and blue electron guns.

rampdown The process of reducing transmission power from the nominal power level to a level below a defined threshold.

Ramses Rameses condoms are named after the great Egyptian pharaoh, who fathered 160 children.

RAN 1. Return Authorization Number, also called RMA, Returned Merchandise Authorization. A number you need for returning busted equipment to the factory. You call the factory, tell them what you want to return and its serial number, and the factory rep gives you an RMA, which you write on the outside of the box containing the thing you're sending back. The idea is that the factory sees the number on the box and immediately logs your busted thing into its computer system. This way, when you call, it can tell you where your thing is and when you might get it back and what it might cost you. At least that's the theory. The moral of this story: Don't send stuff back to the factory without a RAN or RMA (whatever the factory calls it.)

2. Recorded trunk ANnouncements. RAN devices are devices connected on 4-wires to older central offices (public exchanges). They are used to give recorded messages to callers, e.g. "The number you have called has been changed. Please make note of the new number..."

3. Regional area network. A data network that interconnects businesses, residences and governments in a specific geographic region. RANs are larger than local area networks (LANs) and metropolitan area networks (MANs), but smaller than wide area networks (WANs). RANs are usually characterized by very high-speed connections using fiber optic cable or other digital media.

4. See radio access network.

5. See remote antenna node.

ranch radio I don't know the full derivation of this term. But a nice reader sent me this email about ranch radio, "I ran across this many years ago when I worked in the sales office for a oil rig drill bit manufacturer. The sales guys in the field would call in on the "ranch radio" or we would call an operator to connect us to them. It sounded horrible, like she was laying the receiver next to a CB."

random Scattered, unfocused, a non sequitur. A favorite expression of Bill Gates, chairman and founder of Microsoft, to dismiss ideas or strategies that lack logic (or he thinks lack logic). According to Stewart Alsop writing in the February 2, 1998 issue of Fortune Magazine, "Bill Gates is the ultimate programming machine. He believes everything can be defined, examined and reduced to essentials, and rearranged into a logical sequence that will achieve a particular goal. Anything that doesn't work this way, anything illogical is 'random.' In the world of Bill Gates, being illogical is the most serious sin." See Randomize for another Microsoft extension. See Randomness for its statistical meaning.

random access Usually refers to computer memory or storage. Random Access is the ability to reach any piece of data in the memory directly without having to pass by other pieces of data. In telephony, this means the ability to reach any other subscriber through the telco switching network. See Sequential Access.

Random Access Memory See RAM.

random noise Interference to telephone communications occurring at irregular intervals.

random number generator A device which generates random numbers without following any known algorithm. See Pseudo Random Number Generator.

Random Early Detection. RED. A congestion avoidance mechanism for TCP/IP networks. Implemented in a router at the edge of the network, RED takes advantage of TCP's inherent congestion control mechanism. In advance of periods of high network congestion, RED will drop TCP/IP packets from queues on a random basis in order to avoid buffer overflow. RED is accomplished by dropping packets on a random basis, which is determined statistically, when the mean queue depth exceeds a threshold over a period of time. This technique effectively advises the packet source router to decrease its packet rate. Once the source router has been advised by the destination device that all packets in a TCP packet stream have been received, the source can assume that the congestion condition is relieved, and can again increase the packet rate. While the queue is still relieved on the basis of FIFO (First In First Out) logic, RED is an improvement over pure FIFO, which simply drops data packets from the tail of the queue, all at once. This pure FIFO approach causes all source routers to resynchronize, and reduce their transmission rates. When the congestion condition clears, the TCP sources again increase their transmission rates, which results in wave peaks of congestion followed by troughs of network underutilization. See also FIFO and Weighted Random Early Detection.

random traffic Traffic that, over time (from moment to moment), has chance fluctuations around an average value of some measure of the traffic, such as the number of attempts arriving in a specified time interval.

randomize A Microsoft made-up word. To become distracted, as in "I got heavily randomized by other stuff going on." See also Random.

randomness The state of being random. See random. As a telephone industry assumption used in the development of blocking and delay formulas, randomness states that: All subscribers originate calls randomly, that is, without common cause such as a declaration of war, and each subscriber originates calls independently of all other subscribers.

range The difference between the greatest and least of the items being considered. A measure of dispersion.

ranging The process of acquiring the correct timing offset such that the cable modemas transmissions are synchronized with the characteristics of the CMTS.

ransom notes As sales of this dictionary dropped with the telecom bust of 2002-2003, Ray Horak, who has a great sense of humor, started ending his emails with the following quote: "Whenever I'm asked what kind of writing is the most lucrative, I have to say ransom notes." Allegedly, this was said by famous literary agent H.N. Swanson, who once represented F. Scott Fitzgerald and Ernest Hemingway.

ransomware Internet extortion software that, when downloaded, encrypts the contents of your hard drive. The parties responsible for sending the extortion software demand payment to decrypt your data. I don't know how common this is. It's never happened to me. I back my hard disk up regularly and would thus be able to tell such people to screw off if they came calling.

rat A rat can go without water longer than a camel. See camel droppings.

Rate Demarcation Point RDP. The Minimum Point of Entry (MPOE) of the property or premises where the customer's service is located as determined by Verizon. This point is where network access recurring charges and Verizon responsibility stop and beyond which customer responsibility begins. Also known as the End User Point Of Termination (EU-POT).

range extender There are two definitions. I don't know which one is correct. I thought the first one was correct. That is that a range extender is a device that increases the length of a local loop by boosting battery voltage being sent out from the telephone company central office. Bellcore, however, says a range extender is a device that permits a central office to serve a line that has resistance that exceeds the normal limit for signaling. A range extender does not extend transmission range, according to Bellcore. See Range Extender with Gain.

Range Extender With Gain REG. A unit that provides range extension in a loop for both signaling and transmission.

RAPID Reserved Alternate Path with Immediate Diversion. This is a frame relay option which does pretty much what it says on the tin. See Frame Relay.

rapid shock adjustment IT term, used mainly on laptop computers. When a laptop freezes or has a touch pad issue, as in the pointer movin on its own, if you make two fists, and bang on the plastic on either side of the touch pad, it will usually remedy the situation. A reader gave me this definition. I think I'd only try it when all else has failed and after I'd removed the laptop's hard disk. This definition provided by Jonathon Scott.

RAO Revenue Accounting Office. Identifies each LEC processing billing data through CMDS. Governed by BRADS.

RARE Reseaux Associes pour la Recherche Europeenne. European association of research networks.

rare earth doping Here is an article on rare earth doping from *The Economist* Magazine of July 6, 1991: Optical fiber is the darling of the telecommunications world. Because light waves can be superimposed on one another, fiber can carry thousands of laser generated messages at the same time, over longer and longer distances. The longest fibers were once those which doctors use to explore their patients innards. Now they can stretch 70 km (40 miles). But even that does not get you across a sea much bigger than the English Channel without the messages fading. So today's transatlantic and transpacific optical cables are interrupted about every 70 km so that the messages can be sorted out, passed through an electronic amplifier, and then, turned into light again. These amplifiers are costly. Soon, though, they may be replaced.

The key technique is called rare earth doping, which was developed not by crooked bookmakers but by scientists at Southampton University in England, and AT&T Bell labo-

ratories in New Jersey. The rare earths are a group of chemical elements with particularly restless electrons in their atoms. If these electrons are stirred up by a laser, they rise to higher energy levels inside their atoms. When they fall back again, they emit light. The frequency of the light emitted depends on the element. The trick is to pick one which emits at frequency used for telecommunications. By adding the right rare earth to a stretch of fiber, you can make it amplify signals. You can also make a laser out of the fiber itself.

In the optical amplifier developed at Southampton University, a laser is used to lift electrons in the rare earth atoms in a stretch of fiber up to higher energy levels. When a light signal comes along, it may knock one of these electrons off its perch. The falling electron gives off light, which boosts the signal. The enhanced signal then knocks down more electrons, gathering strength as it goes. Rare earths in the cable can be used for other things, as the team at Bell Labs has found. Normal light waves, even those in laser beams, spread out and dissipate as they travel.

Solitons, a special kind of wave, do not. Tidal bores, the best known form of soliton, can move up rivers for miles without losing their shapes. Light that traveled in solitons could travel much farther along an optical fiber between boosts. Solitons are created either by pumping the initial signal through an optical amplifier, or by using a laser made from doped fiber. The soliton holds its shape because the passage of light through the fiber temporarily increases the speed of light in that part of the fiber, so the back of the wave is always trying to travel faster than the front. The stronger the light, the stronger the effect.

Rare earth doping, with metals called erbium and praseodymium, has resulted in fibers which can handle billions of bits of data per second, and carry them thousands of kilometers. AT&T hopes to use erbium amplifiers in its new transoceanic cables in the 1990s. Other companies – such as British Telecom and NTT – also like praseodymium, which is harder to handle, but emits light at a more commonly used frequency.

rarefaction An element of sound. When you speak in native mode, or acoustically, you create disturbances in the molecules in the air. Those disturbances vary in terms of frequency (i.e., pitch or tone) and amplitude (i.e., volume or power), and travel in a waveform. The wave comprises the compression phase and the rarefaction phase. The compression phase is the phase of high pressure in which the molecules are packed together more tightly than normal. The rarefaction phase, or decompression phase, is the phase in which the high pressure is relaxed and the molecules snap back into position.

RARP Reverse Address Resolution Protocol. A low-level TCP/IP protocol used by a workstation (typically diskless) to query a node for purposes of obtaining its logical IP address. See Reverse Address Resolution Protocol.

RAS 1. Remote Access Server or Remote Access Services. See Remote Access Server.

2. Registration, Admissions, and Status signaling function. See H.323.

3. Sun Raster Image File image format.

raster 1. A pattern of horizontal scanning lines on a TV screen. Input data causes the beam of the TV tube to illuminate the correct dots to produce the required characters. See Raster Graphics, Raster Scanning and Rastering.

raster graphics There are two ways you can digitize a picture – use raster graphics and vector graphics. Raster graphics are line drawings, such as cartoons, logos, and line art, that computers render using mathematical logarithms. Raster graphics are created using pixels. A raster is a grid of x and y coordinates on a display space. (And for three-dimensional images, a z coordinate.) A raster image file identifies which of these coordinates to illuminate in monochrome or color values. The raster file is sometimes referred to as a bitmap because it contains information that is directly mapped to the display grid. A raster file is usually larger than a vector graphics image file. A raster file is usually difficult to modify without loss of information, although there are software tools that can convert a raster file into a vector file for refinement and changes. Examples of raster image file types are: BMP, TIFF, GIF, and JPEG files. The most popular vector graphic software program is Adobe Illustrator. The most popular raster graphic program is Adobe Photoshop. Vector graphics are saved as GIF files for the Web, using websafe colors when practical.

raster scanning The method of scanning in which the scanning spot moves along a network of parallel lines, either from side to side or top to bottom.

rastering The process by which a document image is converted to a stream of bits representing either black or white, or one of sixteen levels of gray, for each element of the image. For Group 3 faxes, there are either 98 or 196 raster lines per vertical inch, with a horizontal resolution of 203 lines per inch (yielding 1.86 or 3.72 million elements per 8 1/2 by 11 inch page). Sixteen levels of grey ("halftone " setting – requiring four bits per element, rather than the one bit required for black and white) can be specified, but are not typically used for documents containing only text and/or line drawings. The bit stream is compressed for transmission, and decompressed when received.

rasterizing See Rastering.

ratchet When Michael Marcus was a kid, he asked his father what a ratchet was. His father answered: "It's a little bigger than a mouse shit."

ratcheting A process where the monthly charges for the USOCs on a HI-CAP account are reduced in proportion to the use of the available channels by switched services. See Ratchet.

rate The price of a particular service or piece of equipment from a telephone company. Telephone companies don't use the word "price." They use the word "rate." No one knows why, except that if they didn't cultivate their own jargon, there'd be no job for telecommunications dictionary writers. God forbid!

rate adaption 1. The process of converting a digital stream of data into a different format and rate. For example, rate adaption allows a 64-Kbps data channel to interoperate with a 56-Kbps channel. In this context, rate adaption also is known as flow control. As is true for much in life, the lowest common denominator rules.

2. An ISDN term for bandwidth-on-demand. Sensitive to the application and to its underlying bandwidth requirements, the Terminal Adapter (TA) in a BRI implementation, or the PBX or router in a PRI implementation, will establish some number of 64-Kbps channels. These channels are established either automatically or by conscious selection of the user. For example, a voice session requires only one channel; a videoconferencing session requires two or more channels, depending on the quality desired. On demand and as available, the proper number of channels are selected in order to support the connection; as is the case with all connection-oriented protocols, those channels remain active during the entire session, regardless of whether they are required.

rate area A telephone company term. A geographic area within which rate treatments are the same.

rate arrangements Telephone customer prices charged by tariffs for specified telephone services.

rate averaging Telephone companies' method for establishing uniform pricing by distance rather than on the relative cost (to them) of the particular route. The theory is that some routes are more heavily trafficked, have huge transmission equipment and achieve great economies of scale. Some routes, on the other hand, have little traffic, small transmission equipment and achieve no economies of scale. Therefore, it costs more to provide calls on these less-trafficked routes. But the phone industry doesn't charge more to call small towns than big cities. The phone industry simply charges by distance, averaging its costs. This is called rate averaging.

rate base A regulated telephone company's plant and equipment which forms the dollar base upon which a specified rate of return can be earned. The total invested capital on which a regulated company is entitled to earn a reasonable rate of return.

rate based flow control A means of flow control in which devices (e.g., switched or routers) in a network control the rate of data flow from a transmitter. In an ATM network, for instance, the edge switches negotiate the rate of flow from the transmitting device in consideration of both its desired rate and the ability of the destination switch (and all intermediate switches) to handle that flow without overflowing buffers. Overflowed buffers would result in lost data and overall degradation of QoS (Quality of Service). See also Flow Control and ER. Contrast with QFC.

rate center Telephone company-designated geographic locations assigned vertical and horizontal coordinates between which airline mileages are determined for the charging of private lines. Or as defined by the telephone industry, rate center is that point within an Exchange Area defined by rate map coordinates used as the primary basis for the determination of toll rates. Rate Center may also be used for the determination of selected local rates. See Airlines Mileage and V & H.

rate chip A standard, nonvolatile memory device used to retain data base information on call pricing by Area Code and Central Office. Typically used in call accounting equipment.

rate design Utilities have a specific rate for every service provided. The rates must be approved by the PUC. In a major rate case, rates for many services will be changed in tandem. In a rate design hearing, different proposals as to rate levels are considered. The level of one rate can have an impact on what the level of another rate should be. The interrelationship between rates and the impact of demand must all be considered in "designing" a rate structure.

rate elements The pricing structure of various telecommunications service offerings usually described in tariffs.

rate of return The percentage of net profit which a telephone company is authorized (by a regulatory commission) to earn on its rate base. See Price Cap and RATE BASE.

rate of return regulation Rate of return regulation provides carriers with a guaranteed rate of return on their eligible telecommunications asset base. Current FCC guidelines set in 1990 provide for an 11.25% return on interstate assets. State rates of return on intrastate assets vary. Rate of return regulation allows carriers to charge consumers a price that, when combined with government subsidies, yields a regulated return above the cost of providing service. Companies that do not achieve the mandated rate of return due to higher costs, be they operational or capital related, may petition regulators for a price increase. 'this provides relative certainty of returnS but does not allow operators to benefit from cost efficiencies. Such regulation typically exists in rural jurisdictions where it would not be economical to serve most customers without such guaranteed subsidies.

rate period Dividing a day into various slices of time for the purpose of charging differently for long distance and local calls. There are three rate periods in force today in North America for intra-North America calls. One rate period is from 11:00 P.M. to 8:00 A.M.; one is from 8:00 A.M. to 5:00 P.M. and one is from 5:00 P.M. to 11:00 P.M. If you call outside the United States, there are different rate periods.

rate period indicator A code that denotes time periods applicable for rating purposes.

rate realignment In California's Alternative Regulatory Framework Phase III, rate realignment refers to redesigning telephone rates to reduce intraLATA toll rates and increase rates for other services to make up for the phone companies lost revenues. The Public Utility Commission (PUC) must approve all rate realignment proposals in the rate design stage of the proceeding.

rate stability plan Commit yourself to keeping a a phone service for several years and you'll pay less than if you keep it only from month to month. Some use this term. Others don't.

rate table A data base that contains the cost of calls referenced to the Area Code and/or number dialed plus time of day considerations. See Rate Period.

rate zone 1. A defined geographic division of an exchange area used as the primary basis for figuring toll rates.

2. A Verizon definition. A pricing unit for rating High Capacity Switched Access Transport and Special Access Services. Rate zones are based on the volume of traffic carried by a wire center, or the traffic density.

rated temperature The maximum temperature at which an electric component can operate for extended periods without loss of its basic properties.

rated voltage The maximum voltage at which an electric component can operate for extended periods without undue degradation or safety hazard.

rating level Standard rating levels applied to movies and other programs to help customers determine the amount of sex and violence contained in that event. Ratings include: NR (Not Rated), NR-M (Not Rated-Mature), G (General), PG (Parental Guidance), PG-13 (Parents strongly cautioned), R (Restricted), or NC-17 (No Children under 17).

ratio The quotient of two mathematical expressions. For example, a Passive Optical Network (PON) uses passive splitters that split the incoming signal into multiple outgoing signals. A splitter that splits an incoming signal into two outgoing signals is a 1:2 (spoken as "1 by 2") splitter, with each outgoing signal having a signal strength of 1/2 of the incoming signal, assuming that the splitter is perfectly balanced. See also split ratio.

ratio detector A specific type of FM detector circuit which uses principles similar to a discriminator but provides some amplitude limiting.

rats nest Terminals and connections with poor maintenance and sloppy wiring techniques.

raw bite data The data channel bit rate that includes all protocol overhead and system overhead data bits.

raw sockets An operating system with raw sockets can put out data packets with faked IP addresses. This means that a recipient of these data packets – say in a Denial of Service Attach – has no way of figuring where the data packets are coming from.

ray A beam of radiant energy. Ray is most energetic, responding to email requests from Harry for strange definitions at 2:00 AM. He also is an excellent teacher. He teaches courses on all aspects of networking all over the country and all over the world. Catch one of his seminars if you can. He often teaches a day-long seminar the day before a major trade show, like InterOp or Computer Telephony Expo. He's also a brilliant consultant. Ray's mother gave him his name and named his sisters Joy and Dawn. Ray is thus the only one not named after a dishwashing detergent. There is hidden significance in this. Ray writes the hard part of this dictionary. Blame him for all the mistakes. See Margaret, his wife.

rayleigh fading Multipath fading in a radio system, arising from an ensemble of reflected signals arriving at the receiver antenna and creating standing waves. From the transmitting antenna, even a tightly focused radio signal scatters, or spreads out. The ground and bodies of water reflect the signal back upward, and the atmosphere reflects the signal downward. At the receiver, portions of the signal arrive at different times, as the signal has taken multiple paths of differing path lengths from transmitter to receiver. Occasionally, the aggregate signal from the indirect paths can be of similar strength to the signal from the direct path. If the two signals are of opposite phase, a standing wave is created, and the signal fades in overall strength. The ultimate impact is that of increased transmission errors. See Rayleigh Scattering.

rayleigh scattering A scattering phenomenon which affects optical fiber transmission systems. As incident (i.e., falling on or striking something) light is scattered as it strikes an atom of solid physical matter, most of the photons retain their incident energy, or frequency, which we would see as a color of light. This "elastic" (no energy loss, or frequency shift) scattering effect is known as Rayleigh scattering. The scattering efficiency varies inversely with the 4th power of the light frequency. Thereby, in the visible light spectrum, high-frequency blue light is scattered more strongly by the molecules in the atmosphere than are colors of longer wavelengths (i.e., lower frequencies). This fact accounts for the blue color of the sky (absent high levels of pollution) when the sun is high in the sky. When the sun is low in the sky (particularly when there is a lot of pollution in the air), the blue light is scattered so much as to lose its intensity, and red or even green light takes over. (Strictly speaking, light scattering can be termed Rayleigh scattering only if the molecules are small compared to the wavelength of the radiation.) Relatively long wavelengths are preferred in optical fiber transmission systems, as the level of scattering is less. Rayleigh scattering also affects other forms of electromagnetic energy, such as microwave radio. The phenomenon was named after John William Strutt, Third Baron Rayleigh (1832-1919), the English physicist who discovered it. See also Raman Amplifier, Raman Scattering, and Rayleigh Fading.

razor See Occam's Razor.

RB Reverse Battery.

RBAC Role-Based Access Control. Form of identity-based access control where the system entities that are identified and controlled are functional positions in an organization or process.

RBDS Radio Broadcast Data System. A new system designed to let radio stations broadcasters send text messages, such as emergency warnings and traffic alerts to radios equipped with special LCD screens. The system is designed ultimately to replace the Emergency Broadcast System.

RBOC Regional Bell Operating Company. On January 8, 1982 AT&T signed a Consent Decree with the United States Department of Justice, stipulating that at midnight December 31, 1983, AT&T would divest itself of its 22 wholly-owned telephone operating companies, also know as regional Bell operating companies. According to the terms of the Divestiture Agreement, also known as the Modified Final Judgment (MFJ), those 22 operating Bell telephone companies would be formed into seven RHCs (regional holding companies) of roughly equal size, with Federal Judge Harold H. Greene making the final determinations as to the reorganization. The seven RHCs (and the operating companies that formed them) were Ameritech (Illinois Bell, Indiana Bell, Michigan Bell, Ohio Bell, and Wisconsin Telephone), Bell Atlantic (Bell of Pennsylvania, Diamond State Telephone, The Chesapeake and Potomac Companies, and New Jersey Bell), BellSouth (South Central Bell and Southern Bell), NYNEX (New England Telephone and New York Telephone), Pacific Telesis (Pacific Bell and Nevada Bell), Southwestern Bell (Southwestern Bell), and US West (Mountain Bell, Northwestern Bell, and Pacific Northwest Bell). By the end of 2005, 21 years later, consolidations and takeovers had reduced the original seven to four – Verizon, BellSouth, Qwest and SBC (now about to change its name to at&t (in lower case). Here's how it happened:

In October, 1994, Southwestern Bell Corporation changed its name to SBC Communications, Inc., for reasons we'll see in just a sentence or two. In April, 1996 Bell Atlantic acquired NYNEX for $22.1 billion. As a result, NYNEX lost its identity. Also in April, SBC Communications bought Pacific Telesis for $16.7 billion, with Pacific Bell and Nevada Bell continuing to operate under those names. On October 9, 1999, Ameritech was acquired by SBC, and lost its identity. On June 30, 2000, Bell Atlantic acquired GTE, and changed the name of the entire company to Verizon – a made-up name that came from nowhere. Also on June 30, 2000, US West was acquired by Qwest, an upstart long distance carrier that ended up with serious accounting problems that stemmed from misstating earnings. The whole thing now operates as Qwest. In 2005, SBC acquired what remained of AT&T – largely long distance – and changed its name to at&t (note the lower case). Of the original seven RBOCs, only BellSouth has not yet been merged, acquired or otherwise morphed.

The terms of the Divestiture also placed a number of business restrictions on AT&T and the RBOCs. Those restrictions were threefold. The RBOCs weren't allowed into long distance, equipment manufacturing, or information services. But in the years since divestiture, each of these restrictions has been eased. AT&T wasn't allowed into local telecommunications,

so it couldn't compete with the newly formed RBOCs. But it was allowed to manufacture anything it wanted, including computers. (That was effectively a lifting of a prohibition on it by the earlier 1956 Consent Decree which had not allowed it to make computers.) The federal courts overseeing divestiture have slowly relaxed the restrictions. The BOCs were allowed into information services, and AT&T into local service. This is a continuing saga. See divestiture.

RBS Robbed-Bit Signaling. See Robbed-Bit Signaling.

RC 1. Rate Center.

2. Receive Clock.

RC-4 An encryption/decryption algorithm supported in Cellular Digital Packet Data (CDPD).

RCA 1. Regional Calling Area. The geographical area covered by a telephone company.

2. Once it stood for Radio Corporation of America. Now it's just RCA.

RCA Cable Audio or phono cables used to transmit sound between two pieces of equipment.

RCAT Radio Communications Analysis Test, used in MSC sites.

RCC 1. Radio Common Carrier.

2. Radiocommunications Consultative Council. Australian council established to provide advice and feedback on radiocommunications issues, consisting of senior members of industry and consumer groups.

3. Rescue Coordination Center, a facility which may have direct connection to an Inmarsat fixed Earth station to facilitate search and rescue operations, primarily maritime.

RCDD Registered Communication Distribution Designer, a title conferred on people who have acquired certain requisite education, experience and expertise in the design, implementation and integration of telecommunications and data transport systems and infrastructure, by BICSI, the Building Industry Consulting Service International. See also www.bicsi.org/Content/Index.aspx?File=rcddoverview.htm

RCEE Resource Control Execution Environment. A term from Bellcore Advanced Intelligent Network model.

RCL 1. ReCaLl.

2. Restrictive Cabling Licence.

RCP The Berkeley UNIX remote copy program.

RD 1. An ATM term. Routing Domain: A group of topologically contiguous systems which are running one instance of routing.

2. Route Distinguisher, used in MPLS (MultiProtocol Label Switching) VPN to extend the IPv4 (Internet Protocol Version 4) address. Within an MPLS VPN, the Route Distinguisher is a 64-bit VRF-based header that's added to the beginning of the customer's IPv4 prefixes to change them into globally unique VPN-IPv4 prefixes. In summary, it enables a service provider to differentiate enterprise IP addressing schemes that may overlap.

RDBMS Relational DataBase Management System. A system that manages databases on a relational basis. The individual databases are in the form of flat files, which are two-dimensional files comprising rows and columns in the logical form of a table, or spreadsheet. Each data record in such a flat file contains all of the information about that entity, or object. For example, your listing in the telephone directory is a row comprising columns for name, address, and telephone number. The same goes for all other entities, including other people and various other legal entities, such as companies, government agencies, and not-for-profit organizations. An RDBMS is able to develop and maintain relationships between entities, such as individuals and organizations with common interests (e.g., last names, addresses, telephone number prefixes, or even other attributes such as hobbies or work-related interests). Further, a RDBMS is able to automatically update the interrelationships between the entities, as entities are added or deleted, or as attributes change over time.

RDBS Routing Data Base System (an old database from which LERG was once created).

RDC Redirect Confirm packet. Used in Cellular Digital Packet Data (CDPD) mobility packet.

RDCCH Reverse Digital Control CHannel. A digital cellular term defined by IS-136, which addresses cellular standards for networks employing TDMA (Time Division Multiple Access). The RDCCH includes all signaling and control information passed upstream from the user terminal equipment to the cell site. The RDCCH acts in conjunction with the FDCCH (Forward Digital Control CHannel), which includes all such information sent downstream from the cell site to the user terminal equipment. The FDCCH consists of the RACH (Reverse Access CHannel). See also IS-136, and TDMA.

RDF 1. Radio Direction Finding.

2. An ATM term. Rate Decrease Factor: An ABR service parameter, RDF controls the decrease in the cell transmission rate. RDF is a power of 2 from $1/32,768$ to 1.

Resource Development Framework. RDF is a language developed by the W3C for representing information about resources on the World Wide Web, such as Web pages. RDF enables, for example, a web page's, title, author, modification date, copyright, licensing information, and other properties to be machine-understandable. RDF provides a standard way for encoding resource information so it can be exchanged in a fully automated manner between applications. Use of RDF enables information to be exchanged between applications other than those for which the information was originally created, without loss of meaning. See also Semantic Web.

RDI Remote Defect Indication. An indication that a failure has occurred at the far end of an ATM network. Unlike FERF (Far-End Remote Failure), the RDI alarm indication does not identify the specific circuit in a failure condition. See FERF.

RDP See Rate Demarcation Point.

RDQ Redirect Query Packet. Used in Cellular Digital Packet Data (CDPD) mobility management.

RDR Redirect request packet. Used in Cellular Digital Packet Data (CDPD) mobility management.

RDRAM Rambus Dynamic Random Access Memory. Developed jointly by Intel and Rambus Inc., is a high-speed memory chip that runs on a 400 MHz data bus. As RDRAM supports data transfer twice per CPU (Centralized Processing Unit) clock cycle, the effective speed is 800 MHz. The RDRAM architecture also supports a 2-byte wide (i.e., 16 bits) data channel, bringing its effective data transfer rate to 1.6 GBps (GigaBytes per second). While faster than its nearest competitor, DDR-SDRAM (Double Data Rate-Synchronous DRAM), it also is more expensive. See also DDR-SDRAM, DRAM, EDO RAM, Flash RAM, FRAM, Microprocessor, RAM, SDRAM, SRAM, and VRAM.

RDS Radio Data System. A way of sending data along with a standard FM radio broadcast.

RDSI The Spanish term for ISDN.

RDT 1. Recall Dial Tone.
2. Remote Digital Terminal.

RDY ReaDY.

RE-422 A high-speed electrical interface defined by the ITU-T, supporting data rates of up to 768 Kbps over up to 300 feet of cable.

re-architect Redesign software from the bottom up. I first saw this term in a New York Times piece talking about the growth of the Mozilla Firefox Internet browser and the need for Microsoft's Internet Explorer to competer more aggressively, but that would require it to be re-architected. See Mozilla.

re-engineer A term probably invented by Michael Hammer in the July-August, 1990 issue of Harvard Business Review. In that issue, he wrote "It is time to stop paving the cowpaths. Instead of embedding outdated processes in silicon and software, we should obliterate them and start over. We should 're-engineer' our business: use the power of modern information technology to radically redesign our business processes in order to achieve dramatic improvements in their performance." The term re-engineering now seems to me mean taking tasks presently running on mainframes and making them run on file servers running on LANs – Local Area Networks. The idea is to save money on hardware and make the information more freely available to more people. More intelligent companies also redesign their organization to use the now, more-freely available information. Also called value driven re-engineering.

re-initiation time The time required for a device or system to restart (usually after a power outage).

re-installed customer An MCI term. An MCI customer who is installed again with the same customer account number after having been previously canceled either at their, MCI's, or a third party's request.

REA Rural Electrification Administration. A federal agency within the Department of Agriculture, the REA was established in 1935 to bring electricity and, later, telephone service to rural America. The REA was one of the most successful federal government programs ever. Telephone companies loved the REA, as it offered loans to telcos at a very low rate of interest (2% or less, in many cases). Once the facilities were in place, the telcos nevertheless would have suffered huge losses, as the rates for basic telephone service would not have yielded a satisfactory rate of return on investment. However, the Universal Service Fund provided very substantial additional revenues to further subsidize service in such high-cost areas through the settlements process, which established the cross-subsidy mechanism between the LECs and the IXCs. In fact, a number of independent telephone companies, such as CONTEL, thrived specifically and only because of the combination of REA money and the settlements process. REA money largely dried up some years ago, at least for this purpose, as the definition of "high-cost" changed considerably and as the restrictions on access to and use of such funding became onerous. The REA now is known as the Rural Utilities Service (RUS). See also RUS.

reach Reach, according to the FCC, refers to the availability of a service in the community. It is the number of homes to which the service is available regardless of whether or not residents choose to subscribe.

reach through Reach through is a means of extending the data accessible to the end user beyond that which is stored in the OLAP server. A reach through is performed when the OLAP server recognizes that it needs additional data and automatically queries and retrieves the data from a data warehouse or OLTP system.

reactance The opposition offered to the flow of an alternating current which is due to the presence of inductance or capacitance or both, in the circuit. Reactance is measured in Ohms. The symbol designation is X.

read 1. To glean information from a storage device, like a floppy disk. The opposite of READ is to WRITE. That's when you put information onto that storage device. Some storage devices can only be READ, but not written to. On a floppy disk that's called being "WRITE PROTECTED." See also WORM, which stands for Write Once, Read Many.
2. The process of turning radio waves from an RFID tag into bits of information that can be used by computer systems.

read after write verification A means of assuring that data written to the hard disk matches the original data still in memory. If the data from the disk matches the data in memory, the data in memory is released. If the data doesn't match, the block location is recognized as "bad," and something happens. The data is transferred again. Or in Novell's NetWare, Hot Fix redirects the data to a good block location within the Hot Fix Redirection Area.

read before write A feature of some videotape recorders that plays back the video or audio signal off of tape before it reaches the record heads, sends the signal to an external device for modification, and then applies the modified signal to the record heads so that it can be re-recorded onto the tape in its original position.

read only file A PC computer term. A read only file is a file that you can read but cannot make changes to. The read-only attribute specifies whether a file is read-only. To remove the read-only attribute, you would type the following command

ATTRIB -R FILENAME

Read Only Memory ROM.
1. A computer storage medium which allows the user to recall and use information (read) but not record or amend it (write).
2. The smaller part of a computer's memory, in which essential operating information is recorded in a form which can be recalled and used (read) but not amended or recorded (written).
ROM is memory which is programmed at the factory and whose contents thereafter cannot be altered, even by a power breakdown, or being written to, or anything else. ROM memory is also random-access, which means accessing its information is very fast. See also Microprocessor and RAM.

read only tags RFID tags that contain data that cannot be changed unless the microchip is reprogrammed electronically.

read range The distance from which a reader can communicate with an RFID tag. Active tags have a longer read range than passive tags because they use a battery to transmit signals to the reader. With passive tags, the read range is influenced by frequency, reader output power, antenna design, and method of powering up the tag. Low frequency tags use inductive coupling, which requires the tag to be within a few feet of the reader.

read rate The maximum rate at which data can be read from an RFID tag expressed in bits or bytes per second.

read write cycle Time of reading and writing data onto a memory device. See Read.

read write tags RFID tags that can store new information on its microchip. San Francisco International Airport uses a read-write tag for security. When a bag is scanned for explosives, the information on the tag is changed to indicate it has been checked. The tag is scanned again before it is loaded on a plane. Read-write tags are more expensive than read only tags, and therefore are of limited use for supply chain tracking.

readable frames The number of video frames received without error.

readable octets The number of octets (bytes) received without error.

readdressing A cellular radio term. The process whereby the serving Mobile Data Intermediate System (MD-IS) receives the encapsulated packets, de-encapsulates them, then locates the Mobile End System (M-ES) to determine the cell and channel stream associated with the M-ES. The function is also performed by the Foreign Agent in Mobile IP. This definition come from the book "Internetwork Mobility," by Mark Taylor, William Waung and Mohsen Banan.

reader 1. A device which converts information into a format recognized by a machine as input.

2. A device which interprets coded data in the process of transferring that data from one coded state of storage to another.

3. Also called an interrogator. The reader communicates with the RFID tag via radio waves and passes the information in digital form to a computer system.

reader field The areaf of coverage. RFID tags outside the reader field do not receive radio waves and can't be read.

readerboard Also called Electronic Displayboard, Electronic Wall Display or Message Display Unit (MDU). Readerboards are typically found in call centers. They are electronic displays. They are typically hooked to the call center's ACD or the PC monitoring the ACD (automatic call distributor) and they throw up information about how many people are waiting in line, how long the longest person has been in line, how well the agents are doing and, whose birthday it is. The idea is that all the agents in the call center can see the readerboards and change their behavior accordingly – speak faster if there are a lot of people in queue. Readerboards aren't TVs. They're typically large hanging electronic displays sporting red LCDs or small red lights. By lighting the correct collection of lights, you can put up a message. Some readerboards are very large with letters reaching 12 inches high.

readyline 800 A toll-free service designed for the small business. Receive "800" dialed calls over your existing telephone lines and equipment – no new lines to install, no new equipment needed. You can still use those same lines to make and receive local and long distance calls. Choose the geographic areas you want to cover – from a single area code to an entire state or the whole country. Even decide when you want your toll-free number to be available. You pay a one-time start-up charge and a low monthly fee. Calling prices are based on the market coverage you choose. There are time-of-day and day-of-week discounts, and a volume usage discount. Calls are priced on a mileage/distance-sensitive basis.

real McCoy, the In 18972, Elijah McCoy (1844-1929), an African-American (Canadian by birth), invented something that no one who ran a railroad could do without – an automatic lubricator for trains. With his device, trains could run faster and did not need to stop as often for maintenance. His invention spawned a bunch of inferior copies. When railroad engineers inspected their locomotives, they made sure it was equipped with "the real McCoy."

real mode Originally there was the first IBM PC and it was powered by an Intel 8086 chip which addressed a maximum of 1MB (megabyte of RAM). Real mode is the term that later generations of Intel chips came to call their ability to run programs written for the 8086. Real mode allows 80286, 80386, or 80486 processors to emulate an 8086 processor but perform better than the 8086 because they operate at a faster clock rate. Real mode is limited to a maximum of 1MB of addressable memory because the 8086 processor uses a memory address bus of 20 bits. This is calculated thus: Since a bit can have one of two values, raising the base number of 2 to the power of 20 is equal to 1,048,576 unique memory addresses. Each memory address can store 1 byte of information (1,048,576 bytes equal 1MB). See also Protected Mode.

real soon now A on-line term used to describe when something will happen, maybe.

real time A voice telephone conversation is conducted in real time. That is, there is no perceived delay in the transmission of the voice message or in the response to it. This concept often applies to interaction between a computer and a terminal. In data processing or data communications, real time means the data is processed the moment it enters a computer, as opposed to BATCH processing where the information enters the system, is stored and is operated on a later time. See the follow definitions beginning with real time.

real time adherence Adherence is a term used in telephone call centers to connote whether the people working in the center are doing what they're meant to be doing. Are they at work? Are they on break? Are they answering the phone? Are they at lunch? All these activities are scheduled by work force management software. If they're in line, the workers are "in adherence." If not, they're "out of adherence." Some automatic call distributors have a real time adherence data link which connects the ACD to an external computer which then tracks and displays current service rep activity measured against a predefined schedule. The idea is to give call center supervisors tools to manage the call center's work force more efficiently. Supervisors are able to define the task, the start time of each task, and the task duration. In addition, thresholds and ranges of acceptable deviations for the call center can be set for each task or service rep work state. Once the schedules have been defined and thresholds set, real-time displays inform the supervisor of discrepancies between the work schedule and actual activity. Service rep information will automatically appear should their status exceed the threshold, such as being on someone being break for too long.

real time session There are essentially two types of computer "conversations." One is called batch and the other real time. Batch is older. It typically meant you established a connection with your mainframe computer and you uploaded information to it to work on. That information might have come from a batch of 80 column paper cards or it may have come from a magnetic tape drive with oodles of information on the magnetic tape. The second type of conversation, called real time, is what we're all familiar with today. We establish a conversation with a computer (e.g. over the Internet) and we communicate back and forth with the computer. We give the computer input. It gives us a response – in real time, almost instantly.

real time capacity The capacity of the central computer processor of a stored program control telephone system to process the instructions coming at it. Real Time Capacity is probably the most important measure of the size of a telephone system relying on a single main processor.

real time chat A program allowing live conversation between individuals by typing on a computer terminal. The most common tools are Talk and IRC (International Relay Chat).

real time D-channel status display This maintenance enhancement allows you to assess active or failed status of ISDN D-Channels in real-time. This saves time since ports no longer need to be evaluated.

Real Time Streaming Protocol RTSP. RFC2326; an application-level protocol for control over the delivery of data with real-time properties. RTSP provides an extensible framework to enable controlled, on-demand delivery of real-time data, such as audio and video. Sources of data can include both live data feeds and stored clips. This protocol is intended to control multiple data delivery sessions, provide a means for choosing delivery channels such as UDP, multicast UDP and TCP, and provide a means for choosing delivery mechanisms based upon RTP (RFC 1889).

Real Time Transport Protocol RTP. Developed by the IETF (Internet Engineering Task Force) it adds a layer to the Internet protocol. It is designed to address problems caused when real-time interactive exchanges such as video are transported over LANs were designed for data. Running video on LAN means you can encounter significant end-to-end latency. RTP's approach is to give video higher priority than connectionless data. RTP resides above the IP, Datagram Protocol and ST-II protocols.

realtone A cellphone ringtone with MP3 quality sound and vocals. In short, another way to part money from unsuspecting teenagers.

RealAudio RealNetworks' RealAudio client-server software system enables Internet and on-line users equipped with conventional multimedia personal computers and voice-grade telephone lines to browse, select, and play back audio or audio-based multimedia content on demand, in real time. Several radio stations broadcast their daily fare to anyone on the Internet who's listening. RealAudio is a real breakthrough compared to typical download times encountered with delivery of audio over conventional on-line methods, in which audio is downloaded at a rate that is five times longer than the actual program; the listener must wait 25 minutes before listening to just five minutes of audio. Download RealAudio from www.realaudio.com/products/player2.0.html. For Internet radio listings (what they call NetRadio Central) go to www.netradio.net. See also www.audionet.com. RealAudio is produced by a company called RealNetworks, which had previously been called Progressive Networks.

RealMedia A term encompassing RealNetworks' RealAudio and RealVideo.

RealVideo A streaming technology developed by RealNetworks (formerly Progressive Networks) for transmitting live video over the Internet. RealVideo uses a variety of data compression techniques and works with both normal IP connections as well as IP Multicast connections.

reality check Does something make sense? Does it exist? Can it happen? Is it for real? For example, "Will we get the speed out of the circuit they're promising us? Who knows? Good news is that John is doing a reality check on it."

RealPlayer See RealAudio.

realm A term sometimes used for domain, in this case to refer to user domains established for security reasons, not Internet domains. For password-protected files, the name of the protected resource or area on the server. If the user tries to access the protected resource while browsing, the name of the realm usually appears in the dialog box that asks for a user name and password.

rearrangement A fancy word for moving phone extensions around.

reason code A three digit numeric code describing the reasons for a variety of transactions, including cash (split, partial payments), adjustments, automatic write-offs, returned checks, credit memos, etc.

reasonableness checks Tests made on information reaching a real-time system or being transmitted from it to ensure that the data lie within a given range.

reassembly 1. The process by which an IP datagram is "put back together" at the receiving host after having been fragmentation and MTU.

2. The process of combining a number of the Link Layer Service Data Unit (LSDU) into an SN-Data Protocol Data Unit (PDU) or SN-Unit-data PDU.

reasserting status An ISDN term. When the ISDN phone is being directly controlled by the application program, the set's physical status may be different from the status that has been received from the network. When direct control ends, the ISDN set reasserts the status received from the network to bring its physical condition back into conformity with the network status.

reassignment Here is an explanation by Bill Etling, a senior planner for GTE. "Under the assigned plant concept, a pair is dedicated from the central office to the subscriber home and maintained at that address, even when idle. The likelihood of such a pair being reused, thus eliminating a field visit and extra assignment work, more than makes up for lost revenue while the pair is vacant. In areas of high cable fills, such a pair, when vacant, is often used to fill an order at a different address. Reassignment quickly snowballs, generating many installation field visits and assignment changes, increasing paperwork and the chance of errors."

rebalancing Rebalancing is a new term. It means changing tariffs (the price of phone calling) to levels closer to the actual costs of providing the service. Let me explain: Tariffs are published public documents which describe the prices and conditions of buying service from regulated telephone company. Tariffs developed over a period of many years. Tariffs may apply at a local, state, national, regional, or international level. Traditionally, tariffs were created in a complex fabric of balancing the overall costs of the service against regulatory and competitive issues. For instance, many regulatory authorities put in complex cross-subsidies. These allowed highly profitable or optional services (e.g., long distance and custom calling services) to subsidize residential service, i.e. to keep its price low. Similarly, business service rates commonly were set at high levels to cross-subsidize residential service rates – the logic, at least partially, was based on the assumed ability to pay and the legislators' obsession with "universal service," i.e. giving everyone phone service. As nations move toward deregulated, competitive telecommunications, older tariffs structure put burdens on the incumbent (read regulated) carriers and put them in a potentially bad competitive position. Hence, the concept of rebalancing, which seeks to reset tariffs at levels which are representative of the actual costs of provisioning the various services. At the extreme, rebalancing eliminates cross-subsidies. Thus each service would bear its rightful share of associated costs. As it relates to international calling costs, rebalancing would eliminate the disparity in calling costs. For example, it is much more expensive to call the U.S. from Argentina than it is to call Argentina from the U.S. See also Accounting Rate System, Billing Rate, Cross Subsidization, and Tariff.

rebanding The redivision and reallocation of frequency bands in the wireless spectrum.

rebiller A rebiller, also called a switchless reseller, buys long distance service in bulk from a long distance company, such as AT&T, and resells that service to smaller users. It typically gets its monthly bill on magnetic tape, then rebills the bulk service to its customers. A rebiller owns no communications facilities – switches or transmission. It has two "assets" – a computer program to rebill the tape and sales skills to sell its services to end users. The profit it makes comes from the difference between what it pays the long distance company and what it is able to sell its services at. It's not an easy business to be in, since you are selling a long distance company's services to compete against itself.

reboot See Rebooting.

rebooting Repeating a Boot. Turning on or resetting the telephone system or the computer. The word derives from "boot-strapping." Starting from scratch. Pulling oneself up by one's own bootstraps. Booting a telephone system or a computer means starting it from scratch, usually by turning its AC power on. Rebooting a telephone system is done by simply turning it off, counting to ten and turning it back on again. Rebooting is done to clear the volatile part of the telephone system's or computer's memory and its various processing and clock chips. You reboot typically when your PC "locks" inexplicably or when your telephone system does something you can't explain logically – like ring phones randomly or give strange error messages on the console. On a computer, "Lock" means that no matter which key or combination of keys you touch on your keyboard, you can't get your computer to do anything. In addition to "unlocking" your computer, you also reboot to clear RAM or RAM-resident programs. On an IBM or an IBM clone, rebooting is done by pressing the CONTROL, ALT and DELETE keys simultaneously. You can also reboot by pressing the reset button if your computer has one. (Not all do.) You can reboot any computer by turning its power off, then turning it back on. This is usually not a good idea, since the surge of power that accompanies a computer being turned on and off will reduce the life of many of its electronic

components. Some experts recommend leaving computers running full-time, though turning their hard disks off. They also recommend turning your screen off using a screen saver after several minutes of doing nothing (inactivity).

rebuilding Imagine you have five hard disks in an array. Imagine that they are organized that data is being written to all five drives in such a way that if one drive fails, no data will be lost. That failed drive is now removed and replaced with a good drive. Immediately, the remaining four drives start writing data to the new, good, but empty drive. That process of rebuilding might take a few minutes, or an hour or two. It depends on how much data is in the system and how much activity is taking place. Typically, this rebuilding process happens in a system called RAID (which stands for Redundant Array of Inexpensive Disks). And typically RAID (which is not cheap) is found on servers on LANs. The process of rebuilding is also called reconstruction.

rebundling Rebundling is the process of putting UNEs (Unbundled Network Elements) back together by a CLEC to become part of a competitive service offering by him to a customer. See UNE.

recall The recall button on many phones provides a fresh dial tone without physically putting down and picking up the handset. Don't confuse it with REDIAL, which is a feature of a phone or phone system that allows a user to call the previously-dialed number by pressing one or a few buttons.

recall dial tone A stutter or interrupted dial tone indicating to the extension user that the hookswitch flash has been properly used to gain access to system features.

recall key Used to get dial-tone or to transfer calls on a key system installed within a PBX. See also Recall.

RECAPSS REmote CAble Pair Switching System is used to remotely handle cable transfers and related cable switching tasks by connecting a distribution cable pair to either an old cable pair or a new cable pair without interrupting service. The system accommodates both POTS and special services and the computer console operator can select one pair at a time or select thousands for sequential transfer.

receipt notification A report prepared by a recipient UA (User Agent) or Access Unit (upon request) and sent to the originating UA or Access Unit when a message is received by a recipient.

receive interruption The interruption of a transmission to a terminal to receive or send a higher priority message from the terminal.

received collect See incollect.

Receive Only RO. Describing operation of a device, usually a page printer, that can receive transmissions but cannot transmit.

received line signal detector Modem interface signal defined in RS-232-C EIA interface which indicates to the attached data terminal equipment that it is receiving a signal from the distant modem.

Received Signal Level RSL. The strength of a radio signal received at the input to a radio receiver.

received signal strength indication The measured power of a received signal.

receiver 1. Any device which receives a transmission signal.

2. Any portion of a telecommunications device which decodes an encoded signal into its desired form.

3. The earpiece portion of a telephone handset, which converts an alternating electric current into sound waves, usually through an electromagnet moving a diaphragm.

4. An electronic component capable of collecting radio frequency broadcasts and reproducing them in their original audio and/or video form, e.g. a TV or radio receiver.

receiver congestion A Token Ring error reported by any ring station that receives a frame addressed to itself, but has no room in its buffer to store the frame. The frame is then discarded, and within two seconds the station will report how many times this happened over the reporting period.

receiver jitter tolerance test A test to determine the ability of a high-speed electrical or optical receiver to correctly make sense of an imperfect incoming signal, i.e., without misinterpreting 1s for 0s, and vice-versa. Also called a stressed-eye test.

receiver multicoupler A receiver multicoupler is a device that enables several radio receivers to use a single antenna system. Typically a receiver multicoupler will consist of a bandpass preselector (filter) to determine range of receiving frequencies, and a RF amplifier with low noise figure and high gain to overcome multicoupling losses, and a balanced impedance power divider to divide the amplifier's output into the number of receiving channels required. A regulated power supply is also required for the amplifier. A received multicoupler is used extensively for cellular and trunked radio sites. Natalie Duran of the Area Transmission Engineering department of the Los Angeles Department of Water

& Power writes "We use it for the simple reason of having to run only one coaxial cable versus three."

receiver off-hook tone The loud tone sent by the central office to tell the telephone user that his/her phone is off the hook.

receiver sensitivity The magnitude of the received signal necessary to produce objective BER or channel noise performance.

receiver voting system A type of radio repeater system that attempts to improve reception by receiving and interpreting the signals from several receivers, picking the best one, and routing it to a single output.

receiving perforator Reperforator. A telegraph instrument in which the received signals cause the code of the corresponding characters or functions to be punched in a tape.

recent change Changes to line and trunk translations in a stored program control switching machine that have not been merged with the permanent data base.

recession A recession is defined by economists as two consecutive quarters in which the nation's GDP (gross domestic product) declines.

recip comp See Reciprocal Compensation.

recipient switch The switch to which a local number being ported is ported to. Sorry for the mouthful.

reciprocal agreements Also called Intercarrier Roaming Agreements. An agreement between two cellular carriers that allows the respective customers of the two carriers to use each others' systems automatically, without the necessity of registering as roamers.

reciprocal compensation Recip comp. A form of financial compensation that occurs when a local or long distance service provider terminates a call on another provider's facilities. Imagine a phone call from New York to Los Angeles. It may start with the customer of a new phone company, then proceed to a local phone company (let's say New York Telephone, now called Bell Atlantic). Then it may proceed to a long distance company before ending in Los Angeles and going through another one or two local phone companies before reaching the person dialed. Under the existing rules, all the companies carrying these phone calls have to be paid in some way for their transmission and switching services. There are programs in place such that the company doing the billing and collecting the money pays over some of those monies to the other phone companies in the chain. One such program is called "reciprocal compensation." The opposite of reciprocal compensation is called "Bill and Keep." Under this program, the company billing the call gets to keep all the money. The others in the chain (or most of the others in the chain) get nothing.

reciprocal compensation call A telephone exchange service call, completed between the end users of different carriers, which qualifies for reciprocal compensation under the terms of an interconnection agreement and any prevailing regulatory rules that may exist. Reciprocal compensation is the payment by telecommunications providers to one another for terminating each other's local exchange traffic.

reciprocal link A hyperlink or link placed on one Web site to return the favor of another site putting a link on their page.

RECO A line item Profit and Loss description for a typical networking services business signifying the four major cost classifications: Resources (People), Equipment, Circuits and Other. RECO is used by countless IBMers.

recognition assisted data entry Commonly known as Forms Processing.

Recognized Operating Agency See ROA.

Recognized Private Operating Agency RPOA. The ITU-T term for a packet interexchange carrier. The status granted to a communications entity by its national government after it pledges to abide by mandatory regulations under Article 44 of the ITU (International Telecommunications Union) convention. For example, a publicly recognized VAN (Value Added Network)

recognizer A voice recognition term. A system that attempts to classify speech (input utterances) as words from an active vocabulary.

reconfiguration A fancy word for rearranging equipment, features and options.

record In a database, a record is a group of related data items treated as one unit of information – for example, your name, address and phone number. Each Record is made up of several fields. A field is simply your last name.

record communications Any form of communication which produces a "written" record of the transmission. Teletypewriter and facsimile are examples or record communications. Companies such as RCA Globecom, ITT Worldcom, TRT and MCI, which provide international telex, are known as international record carriers. Before deregulation, that business was exceptionally profitable.

record head The electromagnetic device which magnetizes the surface of a magnetic recording – tape, disk, etc. – in proportion to an electrical signal.

record length The number of bytes in a record. See Record.

record locking Think about an airline reservation. You call up. You want to change your reservation. While the airline has your record open, your travel agent calls up to change it. You change your reservation. Your travel agent changes it. Which one ends up in the "permanent" record? Confusion reigns. Clearly it makes sense to only allow one person to access one record at once and lock everyone else out. Record locking is the most common and most sophisticated means for multi-user LAN applications to maintain data integrity. In a record locking system, users are prevented from working on the same data record at the same time. That way, users don't overwrite other users' changes and data integrity is maintained. But though it doesn't allow users into the same record at the same time, record locking does allows multiple users to work on the same file simultaneously. So multi-user access is maximized. Contrast with file locking, which only allows a single user to work on a file at a time.

recorded announcement intercept Provides a recorded message to an intercepted call indicating why the call cannot be completed, as an alternative to attendant intercept or intercept one for DID and CCSA calls to restricted or unassigned numbers.

recorded announcement service A special type of central office trunk which when dialed, will connect the caller to a prerecorded message.

Recorded Answering Device See RAD.

recorded telephone dictation Phone users can dial into centralized telephone dictation equipment. The dictation equipment is usually handled as a trunk connection or it can be wired on an extension level.

Recorded Voice Announcement See RVA.

recorder A device many large phone users use to record conversations with their callers. Recording truck dispatches can help a company gain the upper hand in customer service. Purchasing departments may use the recorder to remind vendors of their promises. The financial department can document money transfer orders and investments. Recorders come in several sizes. There are cassette recorders with standard speed and slow extended play speed. Open or reel-to-reel recorders have features similar to cassette recorders. Cassette recorders may be voice-operated (VOX) or started by a recorder coupler. Channel capacities available today include 7, 10, 14, 20, 28, 30, 40, 56 and 60 channels, depending on the manufacturer. Some recorders can search for and recall conversations recorded with an option called "autosearch."

recorder warning tone A one-half second burst of 1400 Hz applied to a telephone line every 15 seconds to indicate to the called party that the calling party is recording the conversation. This tone is required by law to be generated as an integral part of any recording device used for the purpose and is required to be not under the control of the calling party. The tone is recorded together with the conversation.

recoverability The way a computer or telephone system resumes operation after overcoming a problem with the hardware (say a power failure) or a program error. Some phone systems recover quickly by themselves. Some recover slowly by themselves. Some loose data. Some need human intervention. What causes a system to fail and how and how fast it recovers is key to understand and verify during the test process. This definition from Steve Gladstone, author of the book "Testing Computer Telephony Systems."

recoverable file system A file system that ensures that if a power outage or other catastrophic system failure occurs, the file system will not be corrupted and disk modifications will not be left incomplete. The structure of the disk volume is restored to a consistent state when the system restarts.

recovered diameter Diameter of shrinkable products after heating has caused it to return to its extruded diameter.

recovery The way a computer or telephone system resumes operation after overcoming a problem with the hardware (say a power failure) or a program error. Some phone systems recover quickly by themselves. Some recover slowly by themselves. Some need human intervention. These are the slowest. Check yours out. If your recovery is slow, and if you local power company is unreliable, you might consider backing your computer up with an uninterruptible power supply.

recovery console A command-line interface that provides a limited set of administrative commands that are useful for repairing a computer.

rectifier Rectifiers are diodes designed to be placed in an alternating current circuit. (The terms "diode" and "rectifier" often are used interchangeably. However, a diode typically is a small signal device with current in the milliamp range, while a rectifier is a power device conducting current from 1 to 1,000+ amps.) When the alternating current flows in the diode's forward direction it passes with no resistance. When the alternating current reverses direction it is blocked by the diode. Rectified current in such a circuit looks like a series of pulses which are just the positive peaks of the alternating current wave form. In short, rectifiers are used for converting Alternating Current (AC) into Direct Current (DC). AC current comes out of the commercial power supply – 120 volts, 60 Hz. DC power is what drives telephone systems and the circuits that move the transmission around. Typically that DC power ranges from 5 to 48 volts. You need rectifiers to change the AC to DC. See also Diode.

recurring charges The monthly charges to the customer for services, facilities and equipment, which continue for the agreed-upon duration of the service.

recursion The ability of a programming language to be able to call functions from within themselves.

RED Random Early Detection. A QoS (Quality of Service) mechanism for IP-based networks. See Random Early Detection for a much longer explanation.

red alarm In T-1, a red alarm is generated for a locally detected failure such as when a condition like loss of synchronization exists for 2.5 seconds, causing a CGA, (Carrier Group Alarm). See T-1.

red black concept The separation of electrical and electronic circuits, components, equipment, and systems that handle classified plain text (RED) information in electrical signal form from those that handle encrypted or unclassified (BLACK) information.

red book Another name for the CD-DA audio CD format introduced by Sony and Philips. The Red Book standard defines the number of tracks on the disc that contain digital audio data and the error correction routines that prevent data loss. The format allows 74 minutes of digital sound to be transferred at a rate of 150 kilobytes per second (K/sec).

red books The CCITT's 1984 standards recommendations were published in books with red covers, hence the term "Red Books." The CCITT is now called the ITU, as in International Telecommunications Union. See ITU.

red box A device that produces tones similar to those produced by dropping coins into a pay phone to inform the operator or automatic machinery that money has been deposited. The red box is used to defraud telephone companies. It is so named because they are usually built small enough to be placed in the "crush proof box" of a packet of Marlboro cigarettes. Red boxes are illegal.

red herring According to the magazine called "Red Herring," in the 1800s wily British fugitives discovered that rubbing a herring across their trail would divert the bloodhounds hot in pursuit. Later, in debate and in detective mysteries, red herring described any clever device used to distract people from the main issue. In the 1920s, clever American investment bankers began calling preliminary investment prospectuses red herrings as a warning to investors that the documents were not complete or final. These documents were distinguished by covers printed largely in red. Today, one Wall Street curmudgeon describes a red herring as "the one shining example of truth in advertising in the securities industry." Red herring prospectuses contain words that say the information contained within the pages has not been approved or disapproved by the SEC (the Securities and Exchange Commission). It is a warning, and is known by the euphemism "red herring," which is also the color of two-day old herring left out on the kitchen counter. It stinks something awful.

Red Horde A nickname for Novell, Inc., an erstwhile leading network operating system software company, as well as the NetWare resellers worldwide. Red is Novell's corporate color.

red light district On-line pornography.

red queen The Red Queen principle is based on a passage in Lewis Carroll's "Through the Looking Glass" in which the Red Queen tells Alice "Now, here, you see, it takes all the running you can do, to keep in the same place." I saw the use of this term in a New York Times article on how some foreign governments try to block their citizens' access to certain web sites. To counter this, people in other countries create new web sites whose purpose is to allow those citizens to pass through and get to the sites they want to – politics or sex, or whatever. To stay on top, the governments need then to collect new information constantly. To counter the governments, the services must keep one step ahead. Thus, the Red Queen principle.

redialer Interface hardware device that interconnects between a fax device and a Public Switched Telephone Network (PSTN). A redialer forwards a dialed number to another destination. Redialers contain a database of referral telephone numbers. When the user dials a specific number, the redialer collects the dialed digits and matches them to a listing in its database. If there is a match, the redialer dials the referral number (transparent to the user) and forwards the call to the referral number.

redirection In the context of message handling, a transmittal event in which an MTA (Message Transfer Agent) replaces a user among a message's immediate recipients with a user preselected for that message.

redirection and forwarding The process whereby the home Mobile Data Intermediate System (MD-IS), upon the receipt of packets encapsulates the packets with the address of the serving MD-IS and forwards them on to the serving MD-IS.

redirector Networking software that accepts input/output requests for remote files, named pipes, or mailslots and then sends (redirects) them to a network service on another computer. Redirectors (also called network clients) are implemented as file system drivers in Windows 95. A redirector is a LAN software module loaded into every network workstation. It captures application programs requests for file- and equipment-sharing services and routes them through the network for action.

reduce A Windows term. To minimize a window to an icon at the bottom of the desktop by using the Minimize button or the Minimize command. A minimized application continues running, and you can select the icon to make it the active application.

reduced slope See Chromatic Dispersion.

redundancy 1. That part of any message which can be eliminated without losing the important information.

2. Having one or more "backup" systems available in case of failure of the main system.

redundancy check A technique of error detection involving the transmission of additional data related to the basic data in such a way that the receiving terminal, by comparing the two sets of data, can determine to a high degree of probability whether there has been an error in transmission.

Redundant Array Of Inexpensive Disks RAID. The idea is simple: Put several disk drives into a single housing. Then write your data over the disk drives in such a way that if you lose one or more of the drives, you won't have lost any of your data. Thus the term "redundant." At its simplest, RAID mirrors data to an equal number of disk drives, e.g. two sets of two. At its most complex, RAID writes data across a bunch of drives, so that if one goes the data can be retrieved from the remaining drives. RAID as a concept was first defined in 1987 by Patterson, Gibson and Katz of the University of California, Berkeley. As defined, RAID has three attributes:

1. It is a set of physical disk drives viewed by the user as a single logical device. 2. The user's data is distributed across the physical set of disk drives in a defined manner. 3. Redundant disk capacity is added so that the user's data can be recovered if one (but not more than one) drive fails.

The Berkeley engineers described five levels of RAID configurations called RAID-1 through RAID-5. RAID-0 and RAID-6 have since been added by industry usage. The distinguishing features among the various RAID levels are the way data is distributed and the way redundant capacity is implemented. Each RAID level represents very different trade-offs in terms of cost, availability and performance. Here's a simple explanation of the various levels of RAID:

Level 0: Disk striping across multiple disks. No error correction or redundancy provided.

Level 1: Disk mirroring or shadowing. One disk drive and an exact backup on a second disk, i.e. All data is redundantly recorded ("mirrored") on a second disk.

Level 2: Data is striped across multiple disks, and error checking and correcting (ECC) codes are written onto additional disks for use in fault recovery.

Level 3: Data is striped byte-by-byte across multiple disks and a single additional disk dedicated to recording parity data.

Level 4: Similar to RAID-3, but stripes data in large chunks. Data is striped block-by-block across multiple disks and a single additional disk is dedicated to recording parity data.

Level 5: The most popular RAID. Data is striped block-by-block across multiple disk, and parity data is also spread out over multiple disks.

Level 6: RAID-5, plus redundant disk controllers, fans buses,etc.

A caveat: The above "levels" are overly simplistic. As Raid has appeared, most manufacturers have implemented different variations on the RAID theme. When I show them the above list, I usually get "Well, that's a beginning." And, of course, some levels are combined. The most popular RAID levels are 0/1 (Zero/One), which is an integral part of NetWare and Level 5, which is not but uses proprietary software techniques. The big difference is that level 0/1 maps the information on one drive to a second. You can always take one drive out, and read it. In Level 5, the data is spread across several drives. You can remove one drive and you won't lose any data. But you can't reconstruct your data from that removed drive. You need the others. When you replace a drive in Raid Level 5 (let's say because it is broken), the others will reconstruct the failed drive fairly quickly – often in less than an hour. Level 0/1 doesn't give you as much total storage space as Level 5.

redundant bits The extra bits included in a transmission for purposes of detecting and/or correcting errors. See Redundancy Check.

redundant link A second connection between a repeater and some other network device like a repeater or switch. One of the connections is active while the other is disabled by the repeater. If link integrity is lost on the active link, it is disabled and the redundant link is enabled so the users are not affected. See also Diversity.

reed relay Two tiny pieces of metal encapsulated in a tiny nitrogen-filled glass tube. When a current is passed through a magnet around the nitrogen-filled glass capsule, one arm of the metal reed relay moves and makes contact with the other. In this way it acts as a "switch." Reed relay switches are reliable. Because they are metal, they can carry great amounts of data. They are rapidly becoming obsolete.

re-engineer To redesign a business process. Re-engineering aims to use the power of information technology to redesign business processes to improve speed, service and quality. See Downsizing.

re-gift See Regift.

Reed-Solomon A means of accomplishing Forward Error Correction (FEC) in order to compensate for errors bursts in created in data transmission. Named for Messrs. Irving S. Reed and Gustave Solomon, staff members of MIT's Lincoln Laboratory, who published a paper entitled "Polynomial Codes over Certain Finite Fields" in the Journal of the Society for Industrial and Applied Mathematics (SIAM) in 1960. Reed-Solomon coding specifies a polynomial by plotting, or statistically sampling, a large number of points in a data block. The coding technique was a quantum leap in forward error correction (FEC) technology, as it allows recovery of data even if multiple errors occurred in a single block, and does so without the requirement for the embedding of redundant data within that block. The decoding process, however, also was challenging; Elwyn Berlekamp, a professor of electrical engineering at the University of California at Berkeley, invented an efficient algorithm for that purpose. Berlekamp's algorithm was used in the Voyager II spacecraft, and is the basis for decoding in CD players. Reed-Solomon is used in MPEG-II (Moving Pictures Experts Group) compression for digital television. The encoder examines the 187 bytes of the MPEG-II data packet (having removed the packet synchronization byte), samples them, and manipulates them as a block; thereby, the contents of the data block can be characterized and described in a 20-byte field appended to the data block. The receiver compares the 187-byte block to the 20-byte description in order to determine its validity. Should errors be detected, their exact location(s) can be identified, they can be corrected, and the original data packet can be reconstructed. As many as 10 byte errors per data packet can be corrected in this fashion.

reengineer To redesign a business process. Re-engineering aims to use the power of information technology to redesign business processes to improve speed, service and quality. See Downsizing.

refarming Refarming is an FCC initiative to promote more efficient use of the frequency bands below 512 MHz allocated to private land mobile radio services.

reference channel Continuously keyed forward-transmission Radio Frequency (RF) channel, used for signal quality assessment.

reference clock A clock of high stability and accuracy that is used to govern the frequency of a network and mutually synchronize clocks of lower stability.

reference design Let's say I'm a semiconductor chip maker who specializes in making chips for broadband communications. How many chips I sell is determined by how well the product that my chips go into sell. So I create mockups of end-user products which I think might sell. For example, I may create a CATV set-top box using my chips. I will take that box to a maker of such boxes and hope that they choose to make and market such a box. The box which I give them is called a reference design.

reference junction The junction of a thermocouple which is at a known reference temperature. Also known as the "cold" junction, it is usually located at the emf measuring device.

reference level The measure of a value used as a starting point for further measurements. In communications applications this term usually refers to a power level of a signal or a noise. A common reference level is 0 dBm, that is, 1 milliwatt.

reference line In faxing, the reference line is the first scanning line in memory. The location of each black pixel of this line is kept in memory for the next scanned line. Depending on the compression technique used, more or fewer scan lines are necessary.

reference number prompting An AT&T Enhanced Fax Mail term. Reference number prompting is an option that allows you to prompt anyone sending a fax message to your mailbox for a reference number of up to 16 digits.

reference track A special magnetic track placed on Floptical diskettes used by the drive to calibrate the optical tracking system with respect to the magnetic recording tracks.

Reference Noise RN. A reference level of noise power.

reference point "R" This ISDN reference point is used when existing interfaces such as X.21, RS-232C, V.35, etc., are used. See ISDN, R Interface, T Interface and U Interface.

reference point "S" This ISDN reference point is similar to reference point "T",

but is appropriate when NT-1 and NT-2 are used. See ISDN, R Interface, T Interface and U Interface.

reference point "T" This ISDN reference point is located either between the standard CCITT user-network interface providing access to the standalone terminal equipment or between the NT-2. See ISDN, R Interface, T Interface and U Interface.

reference point "U" This ISDN reference point is the demarcation point in the network between the customer owned NT1 and the telephone network. The "U" reference point exists only in the United States, due to the Modifed Final Judgment. See ISDN, R Interface, T Interface and U Interface.

referential integrity Refers to a database's ability to link data in two or more files, so that adding data to a record in one file automatically updates data in another file.

referral whois See RWHOIS.

refile 1. Under Mail Handler (MH), refile is a command used to move messages between folders.

2. A means of reducing long distance calling costs for calls terminating in the US, and originating in another country. Refile is a means by which calls are routed through an intermediary country in order to take advantage of lower wholesale rates or settlement rates. See also International Callback.

reflash Reprogram a device's firmware (in flash memory) to repair a software flaw and/or to add new functionality. For example, a cellphone may be reflashed to fix a bug, improve speed or reception, or improve camera image quality. Reflashing involves overwriting the existing file in flash memory with a new file. Reflashing can take place in the shop (for example, at a cell phone retail outlet) or it can take place remotely via an over-the-air (OTA) or via the Internet.

reflectance The ratio of reflected light power to incident light power. Synonym for "return loss."

reflection Radio frequency waves can reflect off hills, buildings, moving cars, the atmosphere – basically almost anything. The reflected waves may vary in phase and strength from the original wave. Reflections are what allow radio waves to reach their targets around corners, behind buildings, under bridges, in parking garages, etc. RF transmissions bend around objects as a result of reflections. See Microwave Reflection.

reflection loss The loss of signal power resulting from the reflection of the a portion of the signal due to a discontinuity in the circuit. A discontinuity is created when cable pairs are spliced together, particularly when the cable pairs are of different gauges. In addition to reflection loss, such a discontinuity causes echo. See also Echo.

REFLEX A two-way alphanumeric paging protocol with broadcast speeds of up to 25.6Kbps for pager receive and 9.6Kbps for pager response channels. The ReFLEX system allows standard paging features and also provides short message communication capability between pagers and various other e-mail enabled PCs and terminals. It is one of the communications protocols used between paging towers and the mobile pagers/receivers/beepers themselves. Other protocols are POCSAG, ERMES, GOLAY and FLEX. The same paging tower equipment can transmit messages one moment in GOLAY and the next moment in ERMES, or any of the other protocols.

reflow soldering A surface-mounting process for electronic components in which a solder paste is applied to the solder lands on the PCB and the components are properly aligned and placed on them. Upon heating the solder, it melts and forms a solder bond with the component terminals, electronically and mechanically bonding the component to the board.

reformated Google ads Some web sites have real content, which you can learn something from, be entertained or buy something. Others essentialy consist of links to other sites. Every time one of us clicks on the link, the owner of that site receives some small revenue. As we click more, so the monies add up and it's worth running the site. In the trade these sites are often known as reformated Google ad sites. Many of them actually run on servers owned by Google.

last ten yards problem - The challenge of delivering a high-quality radio signal indoors from a municipal Wi-Fi network, without the user's having to spend additional money for a repeater. The "last ten yards problem" is rooted in Wi-Fi's use of unlicensed spectrum, its required use of low-power transmitters, and the limited ability of Wi-Fi signals to bounce – all of which limit the ability of Wi-Fi signals to penetrate walls and reach inner rooms of a residence or office building.

refraction The phenomenon in which light rays bend and slightly change velocity when passing between dissimilar materials. Refraction depends on two factors: the incident angle and the refractive index, as defined by Snell's Law of Refraction. See also Index of Refraction.

refractive index The ratio of the velocity of light in one medium (e.g., a vacuum) to the velocity of light in another medium (e.g., glass). Refractive index also can refer to the ratio of the velocity of light in the core of an optical fiber and the cladding of the same fiber. Step-index fiber is distinguished by an abrupt change in the refractive index between the core and the cladding. Graded-index fiber is characterized by a gradual change in refractive index between the core and the cladding. See also Graded-Index Fiber, Index of Refraction, and Step-Index Fiber.

refresh 1. To update with new data, as in the case of a Web browser. Microsoft Internet Explorer (IE), for example, has a "Refresh" button on the tool bar. You use it if you accessed a Web site, IE indicated that the download was "done," but the screen didn't fill with data. Clicking on "Refresh" initiates a fresh download. Some other browsers, such as Netscape Navigator, call it "Reload." See also Refresh Rate.

2. The process by which an electrical charge is restored in DRAM (Dynamic Random Access Memory) cells. See also DRAM.

refresh rate 1. The speed by which memory is refreshed (i.e., updated) in RAM (Random Access Memory) during a refresh cycle in order to maintain the state of the data. Measured in lines of data, the refresh rate commonly is 1K, 2K, 4K, or 8K.

2. The interval of time by which a database is updated with new data in order to maintain its currency and, therefore, its accuracy.

3. Also called Vertical Scan Frequency or Vertical Scan Rate, the refresh rate of a video monitor is the rate at which the display is repainted. The phosphor coating on a monitor tube must be repainted or "refreshed" periodically. Typically, color displays use a low persistence phosphor that must be refreshed 60 times per second, or a rate of 60Hz to 70 Hz or more for VGA and higher resolution monitors. Generally, the faster the refresh rate, the less the flicker. Monochrome displays use a phosphor coating with longer persistence and typically are refreshed at a rate of 50 hertz; this difference accounts for the flicker sometimes seen on color monitors operating in a monochrome mode. Above 70 Hz, color monitors are considered flicker-free.

refurbished 1. A term used in the secondary telecom equipment business. Refurbishing means that telephone equipment has been cleaned, polished, resurfaced and whatever else it takes to return the equipment to a "like-new" appearance. Refurbishing usually means it has been completely tested and is ready for installation. But don't take my word for it. Get a written guarantee. "Factory Refurbished" means that the manufacturer has refurbished the equipment at its own factory. See also Used, Certified, De-install, NATD, and Remanufactured.

2. In the computer industry, refurbished refers to machines (PCs, notebooks, printers, etc.) that are actually new machines returned to the factory, checked out and then sold as refurbished, i.e. slightly cheaper than new. Such "refurbished" machines typically happen when a large customer buys a boatload of machines, e.g. for its salespeople, and then returns some as unneeded. Frankly, if I can get them, I prefer refurbished computers to "new" ones for three reasons. First, they're a little cheaper and second, they often better (they've been checked more thoroughly) and third, they typically carry the same warranty as new machines. To get refurbished equipment, you have to ask your dealer to contact his manufacturer and check if refurbished machines are available. They're often in prime demand for the reasons I mentioned above.

Reg FD See Regulation Full Disclosure.

regenerate To restore a signal to its original shape. Signals need to be restored because they become distorted and acquire noise during transmission. Analog signals cannot be regenerated because it is very hard for telecommunications equipment to distinguish between unwanted noise and wanted noise (i.e. your voice) in an analog signal. Digital signals can be more easily regenerated since they consist of "ones" and "zeros." If digital signals are flattened or distorted, a simple logic circuit – "Is it a zero or a one?" – can restore the signal to its original clean squared shape. See also Repeater.

regenerated traffic A telephone company term. Traffic caused by repeated subscriber attempts to seize blocked (busy) equipment. See also Regenerate and Regeneration.

regeneration The process of receiving and reconstructing a digital signal so that the amplitudes, waveforms, and timing of its signal elements are constrained within specified limits. See also Repeater.

regenerative repeater A device which regenerates incoming signals and re-transmits these signals on an outgoing circuit. See also Regenerate and Repeater.

regenerator A receiver and transmitter combination used to reconstruct signals for digital transmission. In an optical regenerator, the receiver converts incoming optical pulses to electrical pulses, decides whether the pulses are "1s" or "0s", generates "cleaned up" electrical pulses, and then converts these to squared off pulses for transmission. See also Repeater.

regift A verb made popular by Kathleen Thomas in New York City. Ms. Thomas is the recipient of occasional gifts from yours truly. Ms. Thomas, however, is savvy. She realizes that these gifts are, in the main, gifts I received from other people and which I was now passing onto to her, thus saving me money and making Ms. Thomas feel muchly appreciated. Thus her recent email correspondence with me:

Harry: Thanks for all the hard work and the upcoming hard work.

Kathleen: It's been a pleasure. Congrats on a great deal.

Harry: Chatkash coming. You know what Chatkash are?

Kathleen: Yes. It's all the junk you've been re-gifting to me.

region In radio communications, a region is one of the three areas of the world as designated by the International Telecommunications Union (ITU): Region 1: Europe, Africa and CIS; Region 2: The Americas; Region 3: India, Asia, Australia, Pacific. Frequency allocations for radio communications are not the same in all three Regions.

Regional Bell Operating Company RBOC. Also known as Regional Holding Company (RHC). See RBOC for a fuller explanation.

Regional Calling Area RCA. A defined area within a Local Access and Transport Area (LATA). Most telephone directories define RCAs in terms of applicable area codes and/or communities.

regional center A control center (Class 1 office) connecting sectional centers of the telephone system together.

Regional Holding Company RHC. Also called Regional Bell Operating Company (RBOC). See RBOC for a fuller explanation.

register 1. See Traffic Register.

2. A temporary-memory device used to receive, hold, and transfer data (usually a computer word) to be operated upon by a processing unit. The register holds the information for manipulation by the telephone system or a computer. In an automatic telephone system, a register receives dialed pulses or pushbutton tones and then uses that information to control the switch. Computers typically contain a variety of registers. General-purpose registers perform such functions as accumulating arithmetic results. Other registers hold the instruction being executed, the address of a storage location, or data being retrieved from or sent to storage. Other words associated with "register" include buffer, fetch protection, M-sequence, read-only storage, permanent storage, random-access memory and shift register.

register differences The difference in traffic register reading after a specified time has elapsed. See also Traffic Register.

registered access In the context of message handling services, access to the service performed by subscribers who have been registered by the service provider to use the service.

Registered Jack RJ. Any of the RJ series of jacks, described in the Code of Federal Regulations, Title 47, part 68 used to provide interface to the public telephone network. See also RJ-11, RJ-45.

registered terminal equipment Terminal equipment which is registered for connection to the telecommunications network in accordance with Subpart C of Part 68 of the FCC's Rules. If a terminal device has been properly registered it will have an identification number permanently affixed to it.

registered user A user of a Web site with a recorded name and password. In a FrontPage web, you can register users with a WebBot Registration component.

registers An ISDN term. Registers are named storage areas for numbers or strings of characters that control the operation of the ISDN set.

registrant See gTLD.

registrar 1. An entity with a direct contractual relationship with, and special access to, a registry, that inserts records on behalf of others.

2. Server that accepts register requests. A registrar typically is colocated with a proxy or a redirect server and might offer location services.

registration The process of supplying the personal information needed to establish a subscriber account and get access into a network or a server.

registration number (FCC Part 68) Approval number given to telephone equipment to certify that a particular device passes the tests defined in Part 68 of the FCC Rules. These tests certify the phone won't cause any harm to the public network. They do not attest to the commercial value of the product, nor whether it will (or won't) sell. See also KF-E, MF-E, PF-E and Registration Program.

registration program The Federal Communications Commission program and associated directives intended to assure that all connected terminal equipment and protective circuitry will not harm the public switched telephone network or certain private line services. The program requires the registering of terminal equipment and protective circuitry in accordance with Subpart C of part 68, Title 47 of the Code of Federal Regulations. This includes the assignment of identification numbers to the equipment and the testing of the equipment. The registration program contains no requirement that accepted terminal equipment be compatible with, or function with, the network. In other words, a product registered under Part 68 doesn't mean that the product will actually work – i.e. make and receive phone calls (or whatever). Part 68 simply says it won't cause any harm to the network. See Registration Number and Part 68.

registration sequence count An 8-bit counter maintained by the Mobile End System (M-ES) and incremented on each successful establishment of a data link connection with a serving Mobile Data Intermediate System (MD-IS). Used to prevent registration errors due to varying network transit delays between serving MD-IS and home MD-IS.

registration statement A statement, required by Section 76.12 of the FCC Rules, which is used to notify the FCC that one or more broadcast stations will be carried by the cable television system in a specified Community Unit.

registration timer values Time values passed from Mobile Data Intermediate System (MD-IS) to a Mobile End System (M-ES) to inform the M-ES of the period of registration. The M-ES must register again prior to expiration of the registration timer.

registry 1. A database associating DNS (Domain Naming System) information with some person, legal entity, operational entity or other referent.

2. A central hierarchical database in the Windows operating system used to store information necessary to set your computer up, for users, for applications, for attached hardware devices. The registry contains information that is constantly referenced during the computer's operation, such as profiles for each user, the applications installed on the computer; and the types of documents each can create, property sheet settings for folders and application icons, what hardware exists on the system; and which ports can be used. The registry typically tells the computer what to do on startup. A corrupted (i.e. busted) registry may prevent a Windows computer from running. In normal operation, Windows backs up the registry in several places. Most IT professionals tell their users not to mess with their registry, e.g. to start certain programs from starting. They issue these warnings for they fear their users will screw things up in a major way and make their computers unusable. Frankly, if you're careful, you can mess with your registry and live. I have.

REGNOT REGistration NOTification. A wireless term for the message sent from the VLR (Visitor Location Register) to the HLR (Home Location Register). The VLR is a SS7 database residing on the SCP (Signal Control Point) of the cellular provider in whose territory you are roaming. The REGNOT is sent to the HLR database, which resides on the SCP of your service provider of record in order to verify your legitimacy and to determine the features to which you have subscribed. Confirmation of the REGNOT is in the form of a "regnot" (lower case), sent over the SS7 network from the HLR to the VLR. See also HLR, SCP, SS7, VLR.

regression analysis A method of forecasting the future by plotting events in the past and assuming there'll be some similarity in the future. About as accurate as any other pseudo scientific method.

regression testing The selective retesting of a software system that has been modified to ensure that any bugs have been fixed and that no other previously working functions have failed as a result of the reparations and that newly added features have not created problems with previous versions of the software. Also referred to as verification testing, regression testing is initiated after a programmer has attempted to fix a recognized problem or has added source code to a program that may have inadvertently introduced errors. It is a quality control measure to ensure that the newly modified code still complies with its specified requirements and that unmodified code has not been affected by the maintenance activity.

regulated 1. Controlled for uniformity. Many aspects of telecommunications are regulated – from the input voltage powering a telecom system to the output signal of a microwave system.

2. Adhering to the rules, regulations and sundry whims of a government agency. Most aspects of the telephone business are under the control of a government agency to some degree. Their rules cover everything from certifying of expenses which may be capitalized to specifying how many seconds the subscriber can be forced to wait for dial tone (three seconds). Stripped to bare essentials, a regulatory agency can only do two things. First, it can allow the regulated entity to raise its prices to a point where nobody wants to buy anymore. Second, it can stop competitors coming into the business. The first (high price) is the reason no one (or few people, anyway) send telegrams. The second (keep out the competition) reason gets stymied because new technology – e.g. cheap local microwave – comes along to force the regulatory agency's hand. In the long run, no regulated entity survives because it has a regulated monopoly. It survives because it provides good service at a fair price.

regulated charger An uninterruptible power supply (UPS) definition. Without a regulated charger, batteries can be insufficiently charged or blistered with too much charge voltage. Either case can cause permanent damage.

regulated public utility A firm that supplies an indispensable service under essentially noncompetitive conditions with governmental regulation of prices, rate of return, and service quality. In short, a telephone company, a water company, an electricity company.

regulation See Power Conditioning.

Regulation Full Disclosure Reg FD. In October 2000, the Securities and Exchange Commission issued Regulation Full Disclosure which barred companies from giving analysts and money managers key facts about their businesses that other investors (like you and me) didn't have. Reg FD had two effects: First, it generally increased the amount of information available to investors. Companies started using the Web to post more of their financial and business information. They also opened up webcasting of investor conference calls to anyone who wanted to click on the company's web site. But it also had the unintended consequence of making company executives far more careful about answering analysts' questions, basically answering most questions with the same information as was in the company's recent financial releases – even though the world had changed and the company's management was now effectively telling lies. Those lies became evident a week or two later when the company reported dramatically different financial results.

regulatory groups Refers to local, State or Federal entities that issue orders, findings, etc. that are binding upon providers and users of telecommunications and services.

rehomes See Rehoming.

rehoming A major network change which involves moving a customer's local loop termination from one Central Office wire center to another. Rehoming generally involves the retermination of private line facilities, although it can simply involve local loop termination for purposes of access to switched services. Rehomes also can be for the purposes of the carrier, perhaps in connection with a switch upgrade or switch move/decommission.

reinforced concrete A type of construction in which steel reinforcement and concrete are combined, with the steel resisting tension and the concrete resisting compression.

reinforced sheath the outermost covering of a cable that has a cable sheath constructed in layers with a reinforcing material, usually a braided fiber, molded in place between layers.

Reis, Johann Philipp The mostly forgotten inventor of the telephone. Born on 7th January 1834 in Gelnhausen, Germany. In 1861 he presented a lecture about "Telephony by means of galvanic current" and then demonstrated the telephone he invented. For a detailed biography see www.ces-germany.de/reis/english/index.htm. See also Bell, Alexander Graham.

REJ Abbreviation for REJect.

rejection A word used in voice recognition to mean a type of recognition classification where the input utterance did not meet the criteria necessary to be classified as a word in the active vocabulary. Usually the speaker is asked to repeat the utterance.

REL RELease message. The fifth of the ISUP call set-up messages. A message sent in either direction indicating that the circuit identified in the message is being released due to the reason (cause) supplied and is ready to be put into the idle state on receipt of the Release Complete Message. See ISUP and Common Channel Signaling.

relation Synonym for table.

relational database A database that is organized and accessed according to relationships between data items. The idea of a relational databases started in 1970 when E.F. Codd, a researcher at IBM's San Jose research laboratory, published a paper "A Relational Model of Data for Large Share Data Banks." His ideas enabled the logical manipulation of data to be independent of its physical location. In its simplest conception, a relational database is actually a collection of data files that "relate" to each other through at least one common field, or "key field," that serves as a thread through the various files. A relational database consists of tables comprising rows and columns, logically similar to a spreadsheet. Each row contains a single data record, and each column contains all instances of each row of one specific piece of data. A corporate telephone directory, for example, consists of columns of names, addresses and telephone numbers. Each row is a separate listing of a given individual's name, address and telephone number. This individual listing is known as a "flat file," as it is two-dimensional and as all the data is contained in a single file. If all of the data in this file is unique, a flat file works just fine. If, however, multiple employees share the same telephone number or address (e.g., cubicle), the data in the file no longer is unique at all. The telephone number becomes a key to all of the records of the employees who share it. Perhaps one's employee number can be the common thread through several data files, such as payroll, telephone directory, and security clearance. One's employee number, therefore, might be a good way of relating all the files together in one gigantic Relational DataBase Management System (RDBMS). Such keys enable database users to search and sort multiple fields deep in a wide variety of applications, such as inventory management. One might begin the search for all station equipment with certain common attributes such as 1) manufactured by Nortel, 2) touchtone, 3) supports two-lines, and 4) has an LCD display. The search works through the problem in this specific sequence, diving through multiple levels until the search either is sufficiently deep or hits bottom, i.e., can find no record matches below a certain depth. See also Query Optimization.

Relational Database Management System See RDBMS.

relationship marketing The concept is that if you develop a lasting relationship with your customer you will sell him more. Several marketing gurus have suggested that we develop that relationship. Probably the closest we've come to relationship marketing is airline frequent flyer miles. They've done an amazing job of making airline travelers more loyal to airlines. See also CRM.

relational network modeling Relational network modeling analyzes the roles of systems on a network. The modeling system detects anomalies by recognizing when behavior deviates from the norm. Once an anomaly is detected it can dispatch software to detect and kill worms, viruses, spam attacks. It can also alert network and security administrators.

relationship routing A concept introduced by automatic call distributor manufacturer, Aspect Telecommunications, to have callers' calls routed to agents they had previously developed

relative transmission level The ratio of the test tone power at one point to the test tone power at some other point in the system chosen as a reference point.

relative URL The Internet address of a page or other World Wide Web resource with respect to the Internet address of the current page. A relative URL gives the path from the current location of the page to the location of the destination page or resource. A relative URL can optionally include a protocol. For example, the relative URL doc/harry.htm refers to the page harry.htm in the directory doc, below the current directory.

Relativity, Theory of Theories of motion developed by Albert Einstein, for which he is justifiably famous. Relativity More accurately describes the motions of bodies in strong gravitational fields or at near the speed of light than Newtonian mechanics. All experiments done to date agree with relativity's predictions to a high degree of accuracy. (Curiously, Einstein received the Nobel prize in 1921 not for Relativity but rather for his 1905 work on the photoelectric effect.)

relay 1. An electrically activated switch used to operate a circuit. It connects one set of wires to another. Usually, the relay is operated by low voltage electric current and is used to open or close another circuit, which is of much higher voltage. Older telephone switches used many relays to switch (i.e. complete) their calls. Relays come in many forms. There are hermetically-sealed relays, in which thin metal contacts are sealed in an airtight glass or metal enclosure. There are also mercury relays in which a small tube of mercury tilts and completes or breaks a circuit. See also Reed Relay.

2. A station which receives signals and rebroadcasts them either on the same frequency or on a different frequency. A satellite in the sky is a relay. It takes up signals on one frequency and relays them downwards on another. See Repeater, Signal Booster, and Translator.

relay center A common point for the relay of all messages in a system.

relay checking The Internet is a network made up of interconnected networks – probably many thousands, though no one knows the real number. Some of those networks have names you recognize – AT&T, MCI, Sprint, AOL, MSN and some will have names you won't recognize – like GoDaddy and Interland. An email starts from someone's personal computer, then goes into a mail server, then across many vendors networks (also called relay points) until it finally arrives at the recipient's computer. It's possible to trace the route an email takes – showing all the places (relays) it traversed. If you want to check if the email is spam, you might want to check incoming emails to find if they are coming from places that seem to produce a lot of spam. If you do this, you might then decide to some block these emails. Some ISPs – Internet Service Providers – are providing a form of spam blocking called relay checking. This is how it works. See also Trace Route and Tracert.

relay rack Open iron work designed to mount and support electronic equipment. A relay rack is to electronic equipment what a distribution frame is to wire. See Distribution Frame.

relayer Allows a user to open or close a solenoid via the phone system.

relaying A function of a layer by means of which a layer entity receives data from a corresponding entity and transmits it to another corresponding entity.

release 1. A call comes into a switchboard. The operator calls you to tell you it's for you. Then he/she "releases" the call to you. On most switchboards there's a button labelled "RLS." That's the release button. On some phones (not consoles) the release button is the "hang-up" button. Hitting this button means disconnecting the call. Be careful.

2. The ending of an inbound ACD call by hanging up.

3. The feature key on most ACD instruments labelled Release.

4. A term used in the secondary telecom equipment business. The relinquishing of a piece of equipment to a purchaser or user upon fulfillment or anticipated fulfillment of contractual obligations, whether written or oral.

release button The release button – found always on operator consoles and occasionally on some phones – ends a call in the same way that hanging up the receiver does.

release link capability The ability for an originating switching system, on receipt of a new destination address from the current terminating switching system, to release the transmission link to that terminating switching system and continue call processing using the new destination address. Definition from Bellcore in reference to its concept of the Advanced Intelligent Network.

release link trunk RLT. Telecommunications channel used with Centralized Attendant Service to connect attendant-seeking calls from a branch location to a main location.

release with howler If a phone stays off-hook without originating a call (or the receiver is accidentally knocked off), the system transmits a loud tone over the line and then disconnects the line and the phone. The central office effectively then ignores them (the line and the phone) until someone puts the receiver back on-hook again.

Released to manufacturing. See RTM.

reliable sequenced delivery The delivery of a set of Protocol Data Unit (PDUs) from a source to a destination with no errors in any PDU, in the order transmitted, and without gaps or duplicates.

reliable service area RSA. The area specified by the field strength contour within which the reliability of communication service is 90 percent for a mobile unit.

reliable transmission The conveying of messages from a sender to a receiver using a connection-service so as to guarantee sequenced, error-free, flow controlled reception for the duration of the connection.

Region In radio communications, one of the three areas of the world as designated by the International Telecommunications Union (ITU): Region 1: Europe, Africa and CIS; Region 2: The Americas; Region 3: India, Asia, Australia, Pacific. Frequency allocations for radio communications are not the same in all three Regions.

reliability A measure of how dependable a system is once you actually use it. Very different from MTBF (Mean Time Between Failures). And very different from availability. See MTBF.

relief Relief refers to providing additional equipment to accommodate growth in customer demand.

religious Mark Young defines religious as "we take the holidays when we work for people, but work when we're self employed."

relocatable code Machine language programs that can reside in any portion of memory.

remailer Remailers are anonymous mail drops that computer hackers have set up on the Internet, untraceable electronic mail addresses where one can send or receive encrypted data. An article in the October, 1994 issue of High Times, a drug related magazine, offered plans for a similar security system as a remailer, adding one interesting twist. By incorporating a computer virus like Viper or Decide in the system, the computer could be programmed essentially to self-destruct as soon as it detected a security breach, thus rendering it worthless as evidence.

remediation VLAN A virtual LAN that a user is confined to when it is determined during network login that the user's computer doesn't have an up-to-date antivirus signature file and/or an up-to-date patch level for the computer's operating system, firewall, antivirus software, or some other security software. Tight control is exercised over the remediation VLAN to ensure that all the user can do is get the needed remediation before being allowed to revisit the organization's normal login site.

remanufactured Equipment, parts and/or systems that have been repaired and upgraded to the latest higher revision level. The remanufacturing process makes the telecom equipment (used or new) into a finished product that is the latest release and ready for resale. Remanufactured is the term for the highest level of refurbishing equipment. See also Certified, NATD, and Refurbished.

remapping The practice of redefining the meaning of keys on the keyboard.

RemDev A Bluetooth term. Remote Device. A Bluetooth device that participates in the SDP process. A Remote Device must contain a SDP server along with a service record database. A Remote Device is typically a slave device, however, a Remote Device may not always have a slave connection with a LocDev. requestor An entity that requests information from another entity via the Bluetooth API.

remind delay The period of time from when a call is put on hold to when a reminder tone is heard and a message appears on the telephone display.

remission IBM-speak to change the mission of a product or a facility.

remodulator In a split broadband cable system, a digital device at the headend that recovers the digital data from the inbound analog signal and then retransmits the data on the outbound frequency.

remote A system or device that is separated by a distance greater than usual from a related, but more substantial system or device. Something that is remote is not local. For example, a client workstation (i.e., PC) is local to a server if it is in the same building or campus, and is connected over a LAN (Local Area Network). The client workstation is remote if the user takes it home or on the road, and connects over a dial-up modem through the PSTN (Public Switched Telephone Network) and the Internet. Another example is in the PSTN, itself, where carriers often extend the geographic reach of a CO (Central Office) through the deployment of "remotes," which can be in the form of either intelligent nodes or dumb concentrators, also known as line shelves. See also Remote Access and RAS.

remote access Sending and receiving data to and from a computer or controlling a computer with terminals or PCs connected through communications (i.e. phone) links.

remote access concentrator See RAC.

remote access device RAD. Typically, a remote access device (also called a Remote Access Server) is a piece of computer hardware which sits on a corporate LAN and into which employees dial to get access to their files and their email. Remote access devices are also used by commercial service providers, such as Internet Access Providers (ISPs) to allow their customers access into their networks. For longer explanations, see also Remote Access Server and Universal Edge Server.

remote access multiplexer See RAM.

remote access server 1. RAS. A remote access server (also called a Remote Access Device or in a bigger version, a Remote Access Concentrator) is a piece of computer hardware which sits on a corporate LAN and into which employees dial on the public switched telephone network to get access to their email and to software and data on the corporate LAN (e.g. status on customer orders). Remote access servers are also used by commercial service providers, such as Internet Access Providers (ISPs) to allow their customers access into their networks. Remote Access Servers are typically measured by how many simultaneous dial-in users (on analog or digital lines) they can handle and whether they can work with cheaper digital circuits, such as T-1 and E-1 connections. See also Remote Access Concentrator and Universal Edge Server.

2. Software that enable distant PCs and workstations to get into a Remote Access Server to get to software and data on a corporate LAN. Remote access services are provided through modems, analog telephones or digital ISDN lines. Remote access services is For a much longer explanation, see Remote Access (Ref: Hands-On Networking Essentials, M.J. Palmer, Course Technology, Cambridge, MA, 1998, p. 293)

remote access to PBX services Allows a user outside the PBX to access the PBX by dialing it over a normal phone line. You dial the number. It answers. It may or may not say anything. It may just give you dial tone. You now punch in an authorization code. If your code is acceptable, the PBX gives you another dial tone. That dial tone is effectively the one all users within the PBX get. Once you have this dial tone, you can dial another extension, jump on the company's WATS network, get into the dictation unit, access its voice mail, or whatever. Suffice, you are inside the PBX. You can do whatever anyone else inside the PBX can do.

Remote Access Trojan RAT. A Trojan horse that accesses other computer systems across a network. Malicious software of this type may be used by a hacker to gain a foothold on a compromised system.

remote adapted routing The adaptation of backbone routing techniques that take into account; slow-line communications links, intermittent connections, security, charity chatty routing protocols, management, and user ergonomics.

remote alarm indication Also known as a yellow alarm. RAI is carried in the Facilities Data Link for T-1. RAI is carried in Timeslot 0 for E-1.

remote antenna node A small, low-powered antenna that augments a conventional antenna tower as part of a distributed antenna system. It is a cost-effective way to extend wireless coverage to a targeted area, for example, an urban canyon. Since a remote antenna node (RAN) provides service in a relatively small area and uses low power, it enables greater re-use of available spectrum while avoiding signal interference. RANs are typically backhauled over a leased line to a base station, often called a base station hotel.

remote attack A network security term. An attack that targets a machine other than the one that the attacker is interactively logged onto. An example of a remote attack would be an attacker logging onto a workstation and attacking a server, whether it's on the same network or an entirely different one.

remote batch processing Processing in a computer system in which batch programs and batch data are entered from a remote terminal or a remote PC (personal computer) over phone lines.

remote bridge A bridge between two or more similar networks on remote sites. Dial up or leased lines typically require a local bridge or gateway and a remote bridge or gateway an each end, in order to network.

remote call forwarding RCF. This is a neat service. It allows a customer to have a local telephone number in a distant city. Every time someone calls that number, that call is forwarded to you in your city. Remote call forwarding is very much like call forwarding on a local residential line, except that you have no phone, no office and no physical presence in that distant city. Remote Call Forwarding exists purely in the central office. You can also think of it as measured Foreign Exchange. Companies buy Remote Call Forwarding for three reasons: 1. To encourage distant customers to call them by giving them a local number in their own city to call. (This the most obvious reason for an IN-WATS line, a FX or a RCF line); 2. They buy RCF over IN-WATS or FX lines because they don't have the volume to justify these potentially more expensive lines. 3. Companies buy RCF lines as overflow lines from IN-WATS and FX lines. They use their RCF lines when the other lines (FX and IN-WATS) get busy during peak busy periods. Remote Call Forwarding calls are typically charged at the same price as normal DDD calls (i.e. the most expensive to call). And you can't, as yet, reprogram RCF calls easily. You have to place an order with your friendly telco and wait for them to do the reprogramming.

remote concentrator See Remote Line Concentrator.

remote control Remote control software allows a remote PC to connect to the network via a PC that is on the LAN. You must use such software for working from home, for sending in your work, checking on your email, etc. See Remote Node.

remote data services A Web-based technology that brings database connectivity and corporate data publishing capabilities to Internet and intranet applications.

remote diagnostics You own a phone system. You have a service company. There's some problem with it. Instead of sending a technician out, your service company dials your PBX from a data terminal or PC and "asks" your PBX in computerese what's wrong with it. If it isn't too broken, it will come back and give you some indication. This is called remote diagnostics. Some service companies call all their customers' phone systems every morning and run routine remote diagnostics on their switch. It's like going to the doctor for a daily physical. Sometimes this test may find a problem before the user is even aware. Sometimes the problem can be repaired on-line. If not, the service company will have to dispatch a technician. Remote diagnostics is a good idea. More phone systems should have it. To do remote diagnostics on a telephone system, you will typically need a phone line dedicated to the PBX and a modem on either end.

remote digital loopback A test that checks the phone link and a remote modem's transmitter and receiver. Data entered from the keyboard is transmitted from the initiating modem, received by the remote modem's receiver, looped through its transmitter, and returned to the local screen for verification.

remote hands A recent telecom term, and service. Remote hands refer to a variety of services that give you some level of remote control and oversight in distant rented facilities. In order to get you to telehouse your routers and servers in a carrier's Internet Exchange Point, the company advertises the fact that it supports "remote hands capability," meaning that you can access information about your equipment. In some cases you can also change its programming, security codes, or operating temperature, and even look at it 24 hours a day using a remote camera. See also remote infrastructure management.

remote network management Your network is in Illinois. But the people who are monitoring how it's working are in India. They do three basic tasks: 1. Security – protecting your network against viruses and attempts by outsiders to hack in. 2. Operations management – analyzing your traffic flows and ensuring adequate capacity for those flows. 3. Network management – monitoring the overall network and reviewing performance and making recommendations about adding, expanding and dropping lines.

remote installation services RIS. Software services that allow an administrator to set up new client computers remotely, without having to visit each client. The target clients must support remote booting.

remote IP A telephone company AIN term. When an SCP/Adjunct requests a local AIN switch to make a connection to an IP to which the AIN Switch does not have a direct ISDN connection, the indicated IP is referred to as a remote IP.

remote job entry RJE. Remote Job Entry occurs in computer operations where work or input is sent in remotely over phone lines. That "work" might include the day's sales of a distant store.

remote journaling The continuous replication of journal entries over a network from a production system to a target system, in order to support fast restarts, failover, high-availability, and/or other purposes. See journaling.

remote LAN interconnection The connection of two or more LANs which are remotely located from each other so that LAN users can communicate with users and servers on any of the interconnected LANs.

remote line concentrator Also known as a remote line shelf, a remote line concentrator is a concentrator that is positioned some distance from a CO (Central Office) in the PSTN (Public Switched Telephone Network). The device simply concentrates traffic from some number of users' lines onto a lesser number of high-capacity trunks that connect to the CO. The concentrator has absolutely none of the intelligence required to switch calls or provide feature service, even within its own geographic domain. Rather, all calls are concentrated and shipped to the CO, which performs all call processing functions, and ships the call right back to the remote. A remote line concentrator is used because it's cheap and easy. See Remote Concentrator.

remote line switch A line unit mounted near a cluster of users and equipped with intracalling capability.

remote line unit A remote line concentrator without intracalling capability. See Remote Concentrator and Remote Line Switch.

Remote Live Screening See LCS.

remote management A way of managing the Satellite Gateway from a network management client that is geographically remote to the Satellite Gateway itself. A wide-area network (WAN) connection is used to support the remote management of the Satellite Gateway. The Internet Protocol (IP) is used as the basis network protocol for doing remote management.

remote maintenance facility See Remote Diagnostics.

remote monitoring A call center term. Remote Monitoring is most frequently used by service agency clients. This is the process whereby a qualified/authorized party can dial into a remote call center and monitor certain telephone calls. The process is usually administered from a specially designated room or place away from the agent's work area. The agent may or may not know that the specific call is being monitored.

remote node 1. A remote node is an device that connects to a network from a point some distance away from the central host. For example, a CO (Central Office) in the PSTN (Public Switched Telephone Network) might support a number of remote nodes. Some of the nodes are dumb line concentrators that server only to concentrate traffic over high-capacity trunks in order to reduce cabling costs. Other nodes are intelligent switching partitions that can switch basic local traffic within their own geographic domains, even though they rely on the CO for guidance in the delivery of more complex services, such as custom calling features.

2. Remote node software allows remote users to dial in to the corporate LAN and work with the applications and data on the LAN as if they were "actually in the office." By dialing in, they become nodes on the LAN. Using a PC, Mac, or UNIX workstation; a modem; and a remote access server, employees can connect from any location in the world that has an analog, a switched digital, or a wireless connection.

remote office test line ROTL. A testing device that acts in conjunction with a central controller and a responder to make two-way transmission and supervision measurements.

Remote Operations Service Element ROSE. An application layer protocol that provides the capability to perform remote operations at a remote process. Definition from Bellcore in reference to its concept of the Advanced Intelligent Network.

remote order wire An order wire is a line on which maintenance and monitoring is done. A remote order wire is an order wire that has been extended to a distant point that may be more convenient.

remote procedures call RPC. A message-passing facility that allows a distributed program to call services available on various computers in a network. Used during remote administration of computers, RPC provides a procedural view, rather than a transport-centered view, of networked operations.

remote programming Dial your phone system with your friendly personal computer, modem and a communications software package and you can change the telephone system's programming remotely. This feature is great for companies with telephone systems in many locations. They can all be run from one central point. This feature is also great if you want some changes made on your system. It's obviously a lot cheaper for your vendor to make those changes from his office rather than have to visit yours. It's also a lot faster.

See Remote Diagnostics.

remote receiver An outlying receiver that feeds its signals to the voting panel in a receiver voting system.

remote resource Any device not attached to the local node, but available through the network.

remote service unit RSU. A cable telephony term. See RSU for lots of detail.

remote site The remote site is the person or location doing the sending in a file transfer operation. An example: Sales reps in the field typically update the central database on a periodic basis. The central database location is known as the host and the sales reps in the field are doing so from remote locations.

remote site location A location for a DCE device which is not at the central or control site. A typical application would have a terminal at the remote site and the host computer at the central or control site.

remote station Any piece of equipment attached to a LAN by a telephone company supplied link. Technically, that includes all devices that aren't servers. Usually it refers to a workstation at a distant location, linked to the main LAN by a modem and connected through a serial port "gateway." See Modem

remote station lamp field For use at multi-line phones, usually manned by secretaries who answer many phone lines.

remote switch unit RSU. A portion of a digital switching system which is deployed at a remote point from a host digital switch. Remote Switch Units have time-slot interchange capability for processing calls within their serving area.

remote switching system An switch that is away from its host or control office. All or most of the central control equipment for the RSU is located in the host or control office. See also Remote Concentrator.

remote terminal A terminal connected to a computer over a phone line.

remote termination A device installed at the service user site that connects to the local loop to provide high-speed connectivity. Also referred to as the ATU-R.

remote traffic measurement Traffic and feature usage data can be transmitted by the system to a distant service technician.

Remote User Agent RUA. An X.400 standard user agent that interfaces with the MS (Message Store) for remote X.400 communications.

remote workstation A terminal or personal computer connected to the LAN (local area network) by a modem. A remote workstation can be either a standalone computer or a workstation on another network.

remotely-hosted An SCSA definition. The Client and the Server are different (i.e., the application is on a different physical box than the service provider).

removable cartridge system A high-capacity storage system that can be removed from the PC. A removable cartridge systems consists of a drive mechanism and the cartridges used to store data. The most well-known removable cartridge system is the Bernoulli Box by Iomega Corp.

removable media Diskettes or cartridges that can be removed from a computer drive. For example, a Bernoulli box uses removable cartridges.

REN Ringer Equivalency Number. Part of the FCC certification number approving a telephone terminal product for direct sale to the end user as not doing harm to the network. The REN consists of a number and a letter which indicates the frequency response of that telephone's ringer. "A" = 20 Hz or 30 Hz "B" = a range from 15.3 Hz to 68 Hz. The remaining letters represent ringers that will work on very narrow ranges such as "C" = 15.3 Hz to 17.4 Hz, etc. The number indicates the quantity of ringers which may be connected to a single telephone line and still all ring. The total of all RENs of the telephones connected to the one line must not exceed the value 5 or some or all of the ringers may not operate.

render farm A server farm (i.e., a roomful of networked servers) that render complex computer graphics.

rendezvous controls A concept introduced in TAPI 3.0. The Rendezvous Controls are a set of COM components that abstract the concept of a conference directory, providing a mechanism to advertise new multicast conferences and to discover existing ones. They provide a common schema (SDP) for conference announcement, as well as scriptable interfaces, authentication, encryption, and access control features. See TAPI 3.0.

rent-a-wreck In the 1980s, the advent of new technology in the telephone industry made telephone service much more reliable. Fewer people were needed to run phone companies. Telephone company managers soon discovered that they could increase their companies' profitability by firing these managers, which they dutifully did. Unfortunately, by the time the mid-1990s came around, telephone managers discovered they were lacking the experienced expertise necessary to run their phone company. Bingo, the answer: hire the old managers back as consultants. Around NYNEX, this program was known affectionately as "Rent-A-Wreck."

reorder An announcement, or 120 interruptions per minute tone, returned to the caller when his call is blocked in the network. See Reorder Tone.

reorder tone The Reorder tone sounds like a busy signal but is twice as fast, i.e. a reorder tone is a tone applied 120 times per minute. The tone means that all switching paths are busy, all toll trunks are busy, there are equipment blockages, the caller dialed an unassigned code, or the digits he dialed got messed up along the way. Also called Channel Busy or Fast Busy Tone.

reorg A shortened form of the word reorganization. It is used by people in companies which go through management reorganizations so often they don't have to figure what the latest organization means before the next one happens. And they certainly don't have the time to say the word "reorganization" in full. I first heard the word "reorg" from someone at Pacific Bell. He used the word as an excuse for not following up on something he had promised me.

rep Repertory dialing. Speed dialing. Some cellular phones are capable of storing 100 numbers.

REPACCS REmote cable PAir Cross-Connect System is a PC controlled, metallic, automated cross-connect system that may be applied to Automated Distribution Frames, Building Terminals, Service Area Interfaces (SAIs) or cross-connect boxes and closures/terminals. It dramatically reduces dispatches, provides 100% record accuracy, facilitates multiple line testing (MLT), and operates without local or battery power as well as keeps people out of restricted or hazardous areas.

Repair And Quick Clean RQC. A term in the industry which repairs telecom equipment. It means all equipment is repaired and fully tested with a burn-in (if required) and an operational systems test. It also includes minor cosmetic cleaning of the unit. Definition courtesy Nitsuko America. See also Like New Repair and Update and Update and Repair, Update and Refurbish.

Repair, Update And Refurbish RUR. A term in the industry which repairs telecom equipment. It means equipment is repaired and updated to current manufacturer's specifications. Also includes minor cosmetic cleaning of metal cabinets, a full diagnostic test with burn-in (if required) and an operational test. Definition courtesy Nitsuko America. See also Like New Repair and Update and Quick Clean.

repair only A term used in the secondary telecom equipment business. Equipment is repaired to original working condition, but does not include refurbishment or recycling except where required to bring equipment to working condition. See also Refurbished and Remanufactured.

Repair Service Answering RSA. Functions that support the initial handling and entry of subscriber reported troubles. They enable subscribers to request trouble verification tests, to initiate a trouble report and to obtain information on the status of an open trouble report. Definition from Bellcore in reference to its concept of the Advanced Intelligent Network.

Repair Service Bureau RSB. A centralized administrative point where the telephone company receives customer reports of trouble on their telephone circuits.

Repairman Revisit RMR. A condition that exists when a dispatched technician is unable to fix the problem on the first visit.

reparameterize a verb that means to change the current or default behavior of a software component or application by supplying new data (parameters) at time of execution.

repeat The act of a station receiving a code-bit stream (frame or token) from an upstream station and placing it onto the ring to its downstream neighbor. The repeating station may examine, copy to a buffer, or modify control bits in the code-bit stream as appropriate.

repeat call The name of a Bell Atlantic service. Dialing a busy number over and over is as time-consuming as it is frustrating. With Repeat Call, your phone will continuously monitor a busy number every 45 seconds for up to 30 minutes, without interrupting your incoming or outgoing calls. So you and your employees can do other things until your phone alerts you with a special ring when the call got through.

repeat dial Another name for Automatic Callback. See also Repeat Dialing.

repeat dialing A name for a phone company service which automatically checks a busy number and when the line is free, it rings you back and completes the call.

repeatability the amount of optical power lost due to the number of matings (dematings) a connector experiences.

repeated service deficiency When you work a delay with a service provider – a telephone or data carrier, you need to create certain definitions of service so that you can figure penalties if such levels of service are not maintained. For example, we might define service deficiency as being a service outage lasting for more than ten seconds. We

might define Repeated Service Deficiency as a service deficiency that occurs at least four times in any given 30 day period. and we might define Chronic Service Deficiency as a service deficiency that occurs more than ten times in any given 30 day period. Of course, how these terms are defined will depend on the SLA – Service Level Agreement – which you sign with your carrier.

Repeated T1/E1 Local loop T1/E1 copper plant installed using repeaters every 3,000 to 6,000 feet to restore signal quality.

repeater 1. Also known as a Regenerative Repeater and a Regenerator. A device inserted at intervals along a digital circuit to regenerate the transmitted signal. As the digital signal transverses the circuit, it loses its shape due to the combined effects of attenuation and noise. Attenuation is weakening of the signal as it transverses the circuit. Noise, or distortion, can be caused by EMI (ElectroMagnetic Interference), RFI (Radio Frequency Interference, frequency shifts internal to the circuit, and various other factors. At some point, the original signal becomes incoherent unless a repeater is placed on the circuit at specific intervals, which are sensitive to the specifics of the circuit design. The repeater is capable of reading the signal, even though it is somewhat attenuated and distorted, reshaping it into proper "ones" and "zeros," and repeating (i.e., retransmitting) it at the proper level of signal strength. Repeaters are used exclusively in digital circuits, whether they are metallic (e.g., twisted pair and coaxial), radio (e.g., cellular, microwave, and satellite), or optical (e.g., optical fiber). Analog circuits make use of amplifiers, which simply serve to boost the signal strength, and which cannot reshape it. See also amplifier and three Rs.

2. The simplest type of LAN interconnection device. A repeater moves all received packets or frames between LAN segments. The primary function of a repeater is to extend the length of the network media, i.e. the cable.

repeater coil Also called a Repeat Coil. It's really just a transformer, which converts AC power to the voltages used to charge batteries and to power various devices such as PBXs. Repeater coils also are used for impedance matching, which serves to maximize the power transfer of a signal where two electrical circuits (e.g., twisted pair) are interconnected. The power transfer is improved through the elimination of echo, which is signal reflection back towards the signal source.

repeater hop The action of a data transmission passing through a repeater in a communications circuit. IEEE 802.3 standards specify the number of repeater hops allowed for various types of repeaters. For example, Class II repeaters allow up to two repeater hops per segment.

repeater set A repeater unit plus its associated physical layers interfaces (MAUs or PHYs).

repeating coil A transformer which connects one telephone circuit with another without any DC connection between the circuits. Here's a more technical explanation: A voice-frequency transformer characterized by a closed core, a pair of identical balanced primary (line) windings, a pair of identical but not necessarily balanced secondary (drop) windings, and a low transmission loss at voice frequencies. It permits transfer of voice currents from one winding to another by magnetic induction, matches line and drop impedances, and prevents direct conduction between the line and the drop.

reperforator In teletypewriter systems, a device used to punch a tape in accordance with arriving signals, permitting reproduction of the signals for retransmission. See also Chad.

Reperforator/Transmitter RT. A teletypewriter unit consisting of a reperforator and a tape transmitter, each independent of the other.

repertory dialing Sometimes known as "memory dialing" or "speed-calling." A feature that allows you to recall from nine to 99 (or more) phone numbers from a phone's memory with the touch of just one, two or three buttons.

replica A copy. See Replication.

replication Also known as data replication. Replication is the process by which a file, a database or some other computer information in one location is updated to match a mirrored version on another computer in another location. Replication includes the process of duplicating and updating data in multiple computers on a network, some of which are permanently connected to the networks. Others, such as laptops, may only be connected at intermittent times. The idea is twofold: Everyone should have access to the same information in the database/s. Second, many people can make changes to the same record and somehow, all those changes will meld themselves into the database/s and thus, everyone will have access to the new, updated information. In the old days (i.e. pre-Lotus Notes), networked databases were stored in one place, e.g. an airline database of reservations. Everyone who wanted to access information in the database needed to be physically connected to the network through some form of phone line. That's still the case in most databases. Along came Lotus Notes whose major claim was everyone could create their own database/s and carry it with them on their laptops and everyone could put their own information in and Lotus Notes would update the central database and update everyone's database every time they logged into the network. If Lotus is confused, it sends messages out asking for clarification as to what the right database entry was. In short, replication is a far more complex process than what the traditional English language definition of replication is, namely making copies of itself. In data replication, it's the coordination, updating and reconciling of constantly-being-changed databases. That's the hard part. As I wrote this, Lotus Notes had a big lead in this process of database replication. But others, like Oracle and Microsoft, were trying to catch up. The easiest replication strategy is one-way transfer. A simple case of one-way data replication is a mobile user who needs to update the information on his laptop, but not to update any information at the corporate site. See also Synchronization.

reply 1. A transmitted message which serves as a response to an original message. (What else?)

2. An SCSA definition. An event which is a service provider's response to a synchronous or asynchronous request.

Report & Order R&O. A Federal Communications Commission term. After considering comments and reply comments to Notices of Inquiries or Notices of Proposed Rulemakings, the FCC may issue a Report & Order amending the rules or deciding not to do so. Summaries of R&Os are published in the Federal Register. Issuance of an R&O triggers a 30-day period for Petitions for Reconsideration.

report mining Coined by Gartner Group, Inc. for migrating legacy report data to a server so it can be accessed by desktop query tools and regenerated into a new report.

report program generator A computer language for processing large data files.

report-only event An event that the ASC (AIN Switch Capabilities) reports to a SLEE (Service Logic Execution Environment) but the ASC does not suspend processing events for the connection segment. Definition from Bellcore in reference to its concept of the Advanced Intelligent Network.

repository A database of information about objects and components. Synonyms include library and encyclopedia.

REpresentational State Transfer See REST.

repudiable Messages that are repudiable are messages that you can deny receiving. Messages that are non-repudiable are messages that you cannot deny receiving, i.e. the system tracks that you received the message.

repurpose Reuse a network, hardware, software, service, or content for a different purpose or in a different way than that for which it was originally intended.

request The formatted information that is sent to the switching domain as a result of a computing domain issuing a service across the service boundary.

request for comments The name of the result and the process for creating a standard on the Internet. New standards are proposed and published on line, as a "Request For Comments." The Internet Engineering Task Force (ETF) is a consensus-building body that discusses and agrees on new standards. The reference number/name for the standard retains the acronym "RFC," e.g. the official standard for e-mail is RFC 822.

REquest for service A RES is a document/record created when a piece of maintenance work that a department/group is working on is to be passed onto another department/group within the company. A RES is usually in the form of a 'ticket' (which is the official name of the document/record) which one department hands the other. RES is basically telephone company language.

Request To Send RTS. One of the control signals on a standard RS-232-C connector. It places the modem in the originate mode so it can begin to send.

requester Special software loaded onto a networked workstation to manage communications between the network and the workstation. This software may also be referred to as a shell, redirector, or client, depending on the networking system in use.

required for service date A telephone company term. This is the date beyond which service impairment may be expected to occur if equipment relief is not available. This date is used for the Timing Arrow. By this date all balancing, testing, rearrangements, and trunk relief must have been concluded.

requirement A fancy way of saying "need," which means exactly the same thing.

reradiation Energy that is induced in a cellular tower by an AM broadcast station, and is reradiated by the tower. Reradiation interferes with the proper radiation pattern of an AM station's antenna, so the FCC requires a mobile network operator with an antenna in the vicinity of an AM broadcast station to take measures to prevent reradiation.

rerouting A short-term change in the routing of telephone traffic. Rerouting may be planned and recurring or a reaction to a nonrecurring situation.

RES 1. Residential Enhanced Service.

2. See Request for Service.

resale Buying local and/or long distance phone lines in quantity at wholesale rates and then selling them to someone else, hopefully at a profit.

resale carrier A long distance company that does not own its own transmission lines. It buys lines from other carriers and then resells them to its subscribers. Some resale carriers have their own switches. Some don't. Some have a mix of their own lines and leased lines. Most long distance carriers – including AT&T, MCI and Sprint – have a mix of their own lines and leased lines.

resampling Reducing or increasing the number of pixels in an image to conform to a new size or resolution.

reseller A company which purchases a block of cellular numbers from a cellular carrier for resale to its customers. Or a company which purchases a big block of long distance calling minutes or resale in smaller blocks to its customers. See also Aggregator.

reserve power A telephone system may be equipped with storage batteries to provide primary power during a commercial power failure. No loss of service will occur during transition to battery power. All this is a long way of saying your phone system is backed by batteries, typically lead acid (the same ones used in your car).

reset To restore a device to its default or original state. To restore a counter or logic device to a known state, often a zero output. In computer lingo, to reset a computer is simply to turn its power off, wait ten seconds and then turn it on. It's also called cold booting the computer.

reset generation Young people who, when a situation becomes difficult or burdensome, quit and start over again in a different direction.

reset packet A packet that identifies error conditions on an X.25 communications circuit. The reset packet does not clear the session but rather notifies the communicating DTEs of error conditions at a known point in the data-packet transfer sequence.

resident command A command located in the personal computer's operating system itself, contained in the file COMMAND.COM.

resident program See RAM-Resident Program.

residential and light commercial wiring Refers to the wiring system and all of its appurtenances required to provide convenient and useful telephone services to residences and light commercial buildings.

residential gateway A fancy name for a cable modem with wireless LAN (local area network) and Ethernet ports. A residential gateway is basically a box that accepts incoming cable TV and broadband Internet service and doles it out to computers, TVs, and phones and allows the computers to print to one or more printers.

residential system identifier RSID. An signaling identifier sent from a cellular cell site, the RSID identifies the Personal Base Station (PBS) when the user is at home, and the cell phone is in cordless mode. The cell phone display shows "cordless, rather than the name of the carrier. When the user is out of reach of the PBS, the dual-mode phone becomes a digital cell phone, and the name of the carrier is displayed. The RSID is defined in IS-136. See also IS-136.

residual error rate The ratio of the number of bits, unit elements, characters or blocks incorrectly received but undetected or uncorrected by the error-control equipment to the total number of bits, unit elements, characters or blocks sent.

resilient packet ring RPR. The IEEE 802.17 technological specification, intended to deliver packet-based services over SONET rings. RPR is an OSI-Rm (OSI Reference Model) Layer 2 technology with framing very similar to that of Ethernet. The RPR offers the same sub 50ms (millisecond) restoration times offered by SONET without the inefficient allocation of SONET time slots. for leased line services. RPR can be oversubscribed in ways similar to that of Ethernet. (Oversubscribed means too subscribers for great service.) Quality of Service (QoS) is important in RPR networks to ensure that customer traffic within the CIR defined in the SLA is not dropped during periods of congestion. See 802.17, Ethernet and SONET.

resin A synthetic organic material formed by the union (polymerization) of one or more monomers with one or more acids. conductors.

resipiscence The noun that comes from the verb to see the error of one's ways.

resistance Resistance is the opposition to the flow of electric charge and is generally a function of the number of free electrons available to conduct the electric current. When the same amount of voltage is applied to an insulator as to a metal, less current flows through the insulator than the metal because the insulator has fewer free electrons. Resistance is a property intrinsic to a conducting material. Insulators have high resistance values while good conductors like copper have low resistance values. Other factors influence resistance as well. For example, the resistance of a conductor is directly proportional to its length: the longer the wire, the greater the resistance.

In short, any electrical conductor will resist the flow of electrical current. As it resists the flow of current, so the current becomes weaker. Resistance generates heat and occasionally light. It is technically defined as a property or a characteristic of a conductor, i.e. the metal through which the electricity flows. It is measured in Ohms.

resistance design A telephone company design technique for subscriber loop circuits. This technique is designed to employ wire which will have the smallest diameter (least amount of copper), which will ensure a loop resistance less than the signaling limit of the central office equipment serving the loop.

resistor A component made of a material (such as carbon) that has a specified resistance or opposition to the flow of electrical current. A resistor is designed to oppose but not completely obstruct the passage of electrical current.

resort An electronic destination you return to, time and again, to escape normal business.

resource 1. Any facility of a computing system or operating system required by a job or task, including memory, input/output devices, processing unit, data files, and control or processing programs.

2. A network component such as a file, printer, or serial device that is shared by other components of the network.

3. An SCSA term. A voice processing technology, such as voice store and forward, fax processing, voice recognition, or text to speech. Here's the official definition: The abstraction of a standardized vendor-independent interface of a physical device used for call processing as seen by the Application. All Resources have common methods across all implementations. Examples of Resources are voice store and forward, fax send and receive, text to speech conversion, voice recognition, etc. Resources are assumed to have at a minimum one input or output of circuit switched TDM on the internal switch fabric of the system. Resources are shared among multiple applications. Once a Resource has been allocated to an application, it is locked from the use of any other application until freed. It is assumed that applications specify at some level their resource requirements to the server prior to accessing them. This may either be through explicitly attaching them to a Group, or having the server implicitly allocate them based on usage.

4. A Windows term. A resource is a program object, such as a button, menu or dialog box, that Windows treats differently than normal programs. Resources are developed either in a special resource language or using interactive tools. They can be loaded from separate files or bound directly to the executable file.

resource characteristics An SCSA definition. Resources have a set of characteristics that define their behavior beyond that defined by the methods on their interfaces. These may vary or be optional for this Resource Type. Resource may be selected by the application on the basis of their characteristics and a client may query for the characteristics of a resource which it has claimed. See Resource.

resource class An SCSA definition. A set of methods (in object-oriented terms, a class) for controlling resource instances (or a resource). May be abbreviated as just "class." See Resource.

resource dispenser A service that provides the synchronization and management of nondurable resources within a process, providing for simple and efficient sharing by COM objects. For example, the ODBC resource dispenser manages pools of database connections. See also Open Database Connectivity.

resource group An SCSA term. A resource group is a dynamically formed group of resource units that can be made to work together as if they were a single device. See Resource.

resource manager A system service that manages durable data. Server applications use resource managers to maintain the durable state of the application, such as the record of inventory on hand, pending orders, and accounts receivable. The resource managers work in cooperation with the transaction manager to provide the application with a guarantee of atomicity and isolation (using the two-phase commit protocol).

resource module A resource module is a card that slides into a PC and does everything from text-to-speech, to fax, to voice recognition, etc. Everything except interfacing to the network, which is done by another card or another part of the resource module card called the Network Interface Module. See Network Interface Module.

Resource ReserVation Protocol RSVP. See RSVP.

resolution 1. A standard by which the sharpness of a monitor is defined. The number of pixels that are used to form an image defines it. Example: A resolution of 800x600 signifies 800 pixels running horizontally and 600 lines running vertically, thus making up a total of 480,000 pixels. Beginning with the scan processing in the transmitter and ending with the display and/or printing process in the receiver, resolution is a basic parameter of any image transmission system. It affects the design of all its subsystems. In the scanner,

the resolution is a function of the spot size which the scanner optics and associated electronics "look" at the scene and through which the system can uniquely identify the smallest distance along the scan line. Resolution is measured in terms of the density of the picture elements (pixels) and is the total number of pixels (horizontal x vertical) used to display alphanumeric characters of graphic images on the screen. High resolution images are composed of more dots per inch and appear smoother than low-resolution images. The higher the resolution, the better the display of details. See also Monitor.

2. The minimum difference between two discrete values that can be distinguished by a measuring device. High resolution does not necessarily imply high accuracy.

3. A measurement of the smallest detail that can be distinguished by a sensor system under specific conditions.

resolver DNS client programs used to look up DNS name information. Resolvers can be either a small stub (a limited set of programming routines that provide basic query functionality) or larger programs that provide additional lookup DNS client functions, such as caching. See gTLD.

resonance The condition that exists when inductive reactance equals capacitive reactance. In a series circuit it results in maximum current at the resonant frequency. In a parallel circuit it results in maximum voltage at the resonant frequency.

resonant cavity Closed metal container which has the characteristics of a parallel resonant circuit.

resp org Responsible Organization – the long distance company responsible for managing and administering the 800 subscriber's records in the 800 Service Management System (SMS/800). The SMS/800 only recognizes one RESP ORG for each 800 number. Management and record administration consists of data entry, changing records, accepting trouble reports and referring and/or clearing associated documents.

responder A test line that can make transmission and supervision measurements through its host switch under control of a remote computer.

response An answer to an inquiry. In IBM's SNA, the control information sent from a secondary station to the primary station under SDLC.

response time The time it takes a system to react to a given input. In voice recognition, response time typically refers to the amount of time required for a word (or utterance) to be recognized once the end of the word is detected. True response time is longer because silence often must occur before the end of the word can be declared. When operating a terminal connected to a computer, response time would be the time between the operator pressing the last key of a series of keys and the appearance of a response on the operator's display. In a data communications system, response time includes the transmission time, the processing time, the searching for records time and the transmission time back to the originator. Response time is very critical in applications like airline reservation systems. Here the customer is on the phone awaiting a reply. That time is critical in whether the customer perceives he's getting good or bad service. Response times of more than three seconds are not acceptable in situations where the customer is waiting on the phone to buy something. Response time is a function, inter alia, of the number of phone lines you lease or use. You can save a lot of phone line costs by cutting back on lines. But you'll extend response time. Life, as always, is a trade-off.

responsible organizations RespOrgs. Telecommunications providers that have responsibility for obtaining 800 Service numbers from the Service Management System and building and maintaining customer records. See Eighthundred Service.

responsivity The ratio of an optical detectors electrical output to its optical input. Generally expressed in amperes per watt or volts per watt of incident optical power.

REST A term coined by Roy Fielding in his Ph.D. dissertation to describe an architecture style of networked systems. REST is an acronym standing for Representational State Transfer. Here is Roy Fielding's explanation of the meaning of REpresentational State Transfer: "Representational State Transfer is intended to evoke an image of how a well-designed Web application behaves: a network of web pages (a virtual state-machine), where the user progresses through an application by selecting links (state transitions), resulting in the next page (representing the next state of the application) being transferred to the user and rendered for their use."

rest stops Electronic malls and other online diversions provided for a small fee.

restart 1. A central office word or Apple Macintosh word for resetting a PC without turning it off (also called "warm boot" or "soft reset"). To restart an MS-DOS or Windows machine while it is on, press Ctrl + Alt + Del once or twice or press the reset button. See also Boot.

2. In telephony, a system initiated action designed to restore overall service capacity.

restart packet A block of data that notifies X.25 DTEs that an irrecoverable error exists within X.25 network. Restart packets clear all existing SVCs and resynchronize all existing PVCs between X.25 DET and X.25 DCE.

restocking fee A fee for returning non-defective equipment. The fee is generally a percentage of the sales price, from 10 percent to 25 percent. Many buyers object to paying restocking fees. Check before you buy something – especially in the used telephone equipment business, where restocking fees are not uncommon.

restore 1. Typically, to put a telephone system back into full operation.

restore button A Windows term. The small button containing both an up and down arrow at the right of the title bar. The Restore button appears only after you have enlarged a window to its maximum size. Mouse users can click the Restore button to return the window to its previous size. Keyboard users can use the Restore command on the Control menu.

Restricted Cabling Licence RCL. Australian license that replaced the Domestic Premises Cabling Licence, aimed at the small business and domestic cabling markets. The licensee may perform cabling work subject to certain restrictions.

Restricted top-level domain name rTLD. A top-level domain, such as .biz, .gov, .museum, .name, and .pro, that is only available to registrants who meet certain criteria.

restriction Phone systems can disallow people or extensions from making certain calls. If they're not allowed to make long distance calls, this is called toll restriction. See Toll Restriction. There are other forms of restriction, like being able to only use the company's internal network.

restriction from outgoing calls Phone users may be restricted from placing outgoing calls. See Class of Service.

restriction override password A password which allows a caller to override restrictions when making an outside call. This password must typically be entered before the call is dialed.

Restriction services These features allow the attendant to control the restriction of phones or groups of phones. It can be very useful in hotels and motels to turn off service to room phones during the time between check out and check in quests. Here are some examples of restriction services:

Controlled outward restriction: Phones can be restricted from making dialed outgoing calls while inward calls are completed normally.

Controlled station-to-station restriction: Originating phone calls to other extensions in the system are blocked, however, normal incoming and outgoing calls can be completed.

Controlled termination restriction: Phones can complete outgoing calls normally, but incoming calls are directed to either the attendant or an intercept tone or recording.

Controlled total restriction: Restricted phone lines cannot make or receive any calls.

restrictions Preselected telephones and lines may be restricted from dialing certain telephone numbers (such as long distance, directory assistance and other toll calls).

resumania Resumania is a term coined by Mr. Robert Half, founder of the placement firm of the same name. to describe the unintentional bloopers that often appear on job candidates' resumes, job applications and cover letters. Examples:

"I perform my job with effortless efficiency, effectiveness, efficacy, and expertise."

"Seek challenges that test my mind and body, since the two are usually inseparable."

"My compensation should be at least equal to my age."

"I am very detail-oreinted."

"I can play well with others."

"Married, eight children. Prefer frequent travel."

"Objection: To utilize my skills in sales."

"My salary requirement is $34 per year."

"Served as assistant sore manager."

"Previous experience: Self-employed - a fiasco."

"I vow to fulfill the goals of the company as long as I live."

"Reason for leaving last job: Pushed aside so the vice president's girlfriend could steal my job."

resume A power management feature that restores a portable computer from a power-suspended state to full operation. It is also called stand by or sleep mode. Do not confuse resume with hibernation., which I do not recommend. I do recommend resume and stand by.

resynchronization The process of returning Novell NetWare SFT III (Novell's System Fault Tolerance) servers to a mirrored state after a failure. SFT III checks changes to all active servers and ensures that those changes are copied to the other servers. The time it takes the servers to complete resynchronization depends on the amount of memory and the disk storage in each server. Server memory synchronization is much faster than disk mirroring because disk mirroring speed is limited by the disk channel.

retail therapy My sister's term for shopping.

retard coil A coil having a large inductance which retards sudden changes of the

current flowing through its winding.

retention A telemarketing term. Refers to a marketing goal to keep current customers buying. The opposite is churn.

retractile cable A cable that returns by its own stored energy from an extended condition to its original contracted form.

retractile cord A coiled cord that springs back to its original length when you let it go. Telephone handset cords are the most common retractile cords. There are wide quality variations among retractile cords. Western Electric (oops AT&T Technologies) has set a very good standard for retractile cords. But not everyone conforms to it. If you want to quickly see the quality difference among various coil cords, take six from different manufacturers and hang them over your office door and come back in a week. You'll see a bunch touching the floor. Others will still be taut. Another way is simply to connect them to your phone, one by one, and listen to the differences. Some simply sound weaker. Cheaper ones tend to sound worse. (So what's new?)

retraining Retraining is much like Handshaking, although it occurs after modems have been successfully sending and receiving data. Retraining is a feature of some modems which adapt to changing circuit conditions, which affect the maximum speed at which the devices can communicate. Relevant circuit conditions include amplitude response, delay distortions, timing recovery, and echo. Fax machines go through the retraining process at the end of each page transmitted. The processes of handshaking and line discipline take approximately 3-7 seconds per page, and can account for a significant portion of the total time involved in a fax transmission. See also Handshaking, Line Discipline and Protocol.

retransmission A method of error control in which hosts receiving messages acknowledge the receipt of correct messages and either do not acknowledge, or acknowledge in the negative, the receipt of incorrect messages. The lack of acknowledgment, or receipt of negative acknowledgment, is an indication to the sending host that it should transmit the failed message again.

retransmissive star In optical fiber transmission, a passive component that permits the light signal on an input fiber to be retransmitted on multiple output fibers. The signal comes in on one fiber, hits a star-type connector which splays the transmission out. A retransmissive star is formed by heating together a bundle of fibers to near their melting point. Such a device is used mainly in fiber-based local networks. It's also called star coupler. When you see one you'll be surprised how crude this device looks, despite its fancy name.

retrial After failing to complete a call, a person tries again. This is called a "retrial." The term is used in traffic engineering. It's critical in figuring needed trunking capacity. See Queuing Theory, Poisson and Traffic Engineering.

retrofit kit A conversion kit which makes a standard pay phone into one which will accept credit cards.

retry In the bisynchronous protocol, the process of resending the current block of data a prescribed number of times until it is accepted.

return A carriage return. This key on some keyboards is also called "Enter." Touching the CR (Carriage Return) gives you two functions: a "line terminating function" and a "new line function", abbreviated "NL". Simply put, a Return at the end of a line, terminates that line and begins a new one.

return authorization number See RMA.

return call Return Call automatically redials the number of the last person who called your number – whether you were able to answer the phone or not. If that number is busy, Return Call continues trying to get through for up to a certain number of minutes, say 30 – without interrupting your incoming or outgoing calls. It signals you with a special ring when a connection is made. Besides making a return follow-up call quick and simple, this service also lets you call back anyone who hung up when you couldn't pick up the phone in time.

return code 50 Return code 50 is a term used a lot with billing done in the back office. Return Code 50 refers to a Carrier Services Billing/Exchange Message Interface (EMI) – EMI record type 50 is used to return EMI Billing records that error or fail.

return direction Return link communications direction from a mobile terminal via a satellite to a fixed Earth station.

return loss A measure of the similarity of the impedance of a transmission line and the impedance at its termination. Return loss is a measure of the signal reflections occurring along a channel or basic link and is related to various electrical mismatches along the cabling. It is a ratio, expressed in decibels, of the power of outgoing signal to the power of the signal reflected back from an impedance discontinuity.

Return Material Authorization Number RMA. A code number provided by the seller as a prerequisite to returning product for either repair or refund. An indispensable tracking procedure, it operates like a purchase order system. If you return computer or telephone equipment without an RMA, chances are your equipment will be lost. RMA is also called a Return Authorization (RA) number.

Return To Zero RZ. Method of transmitting binary information such that, after each encoded bit, voltage returns to the zero level.

Return-To-Zero code A code form having two information states called "zero" and "one" in which the signal returns to a rest state during a portion of the bit period.

reusability The characteristic of a component that allows it to be used in more that the application for which it was created, with or without modification.

reuse limit Customer penalty value at which point Sprint automatically stops dampening customer route announcements. The current Reuse Limit is 750. See Dampen and Dampening Limit.

reuse ratio Lines of code reused per total lines of code.

revector A euphemism for laying people off. Also means changing the course of one's business.

Revenue Accounting Office RAO. A telephone company center using mainframe computers for billing other data processing. Functions performed include receipt and processing of AMA (Automatic Message Accounting) data and preparation of the subscriber's bill. Definition from Bellcore in reference to its concept of the Advanced Intelligent Network.

revenue assurance Telephone companies send out zillions of bills every day to businesses and homes. Most of the ones sent to businesses are wrong. They're too high or too low. There's always been a big business in "auditing" (i.e. checking) business phone bills, finding bills that were too high and getting refunds from the phone companies. Many consultants I know made a handsome living from this business for many years. Anyway, many phone companies have woken up to the fact that their bills are wrong – often in the customer's favor. And they'd better start sending out accurate (i.e. higher) bills. The phone companies are forming new departments, call Revenue Assurance. One friend who just joined one such department sent me the following mouthful of a definition: Revenue assurance is a quality tracking process that identifies, and captures revenue leakage through revenue diagnostic, and revenue capture via assessment of a company's billings, sales, customer acquisition/retention, management information systems infrastructure, and financials. Its primary objective is to test the integrity of revenue generating processes, and to detect and eliminate fraud in order to maximize revenue goals. It is very common in telephone and Internet companies. The terms "usage-based" and "configuration-based" are sometimes used to classify two categories of revenue assurance. Usage-based revenue assurance involves assuring that the carrier is billing for all the usage that it's supposed to be billing for and the billing is accurate. Configuration-based revenue assurance involves assuring that the carrier is optimizing the utilization of its network assets, i.e., it has an accurate inventory of its assets and is provisioning them to customers in an optimal manner so as to maximize revenue and margins. The terms "usage-based" and "configuration-based" are sometimes used to classify two categories of revenue assurance. Usage-based revenue assurance involves assuring that the carrier is billing for all the usage that it's supposed to be billing for and the billing is accurate. Configuration-based revenue assurance involves assuring that the carrier is optimizing the utilization of its network assets, i.e., it has an accurate inventory of its assets and is provisioning them to customers in an optimal manner so as to maximize revenue and margins. See also Revenue Leakage.

revenue leakage Some telecom service providers fail to capture up to 15% of all the revenue they are supposed to get, according to a November, 2001 report on revenue assurance from Chorleywood Consulting. Revenue leakage can include unbillable revenue (for calls completed more than three months ago), unbilled features, adjustments from trouble tickets and customer care and unidentified customers. Revenue assurance refers to the actions a service provider takes to ensure that bills sent to their customers are accurate, that revenue for services provided is collected and that fraudulent use of resources is detected and eradicated. In a poll of delegates from 240 U.S. service providers, telecom consulting firm Tarifica found that more than half of those surveyed said their company leaks up to 10% of revenues.

revenue recognition How does a company report its sales? An example is a company sells a subscription or contract for goods and services to be delivered over time. How much of that deal gets booked as revenue in each quarter? And how much of the cost of pursuing and closing the deal gets booked as an expense in the current quarter as opposed to over the life of the contract? And what happens if the contract is extended or terminated? The accounting rules are a mess. The simple approach is that revenues and expenses hit the books as the actual money flows in and out. But, the accountants could tell you lots of reasons why that simplistic approach is not completely appropriate. The result is rules that can be "gamed" by management to try to "manage earnings" (and revenues). And anyone who suggests that it would be "easy" to reform the revenue recognition mess

is being misleading. Here's a simple problem not solvable by accounting rule changes: near the end of each quarter, management has discretion as to how to apply sales resources to attempt to close deals. If they really want a deal to happen this quarter to boost results, they can put extra resources on getting the deal closed. If they want to shift the revenue and income into the next quarter, they can simple hold back resources and the deal gets delayed. Sometimes, simply making or not making a critical phone call at just the right moment or expediting or delaying the signing of some documents can subtract or add weeks or a month to the time to close a deal. That's just ONE of the "tools" available to corporate management to manage the flow of revenues and earnings. The next simplest way to manage earnings is to defer or accelerate an expense like buying new computers or signing a lease.

revenue volume pricing plan AT&T's Revenue Volume Pricing Plan gives discounts based on total monthly 800 and 900 billing after all other term discounts have been taken. One of two plans used by aggregators to resell 800 services, the other is Customer Specific Term Plan.

revenue requirement How much money a regulated phone company is allowed to earn is typically determined by its rate base (depreciated value of its assets). It is allowed to earn a percentage on its rate base – just as you earn interest on your rate base (what you have deposited in the bank). So how the regulation works is (in principle) simple: Figure what the phone company's assets are. Figure what percentage you want the phone company to earn on its assets. Figure the calculation. Bingo you have the revenue requirement. Except you have to allow it to pay its expenses. So that gets added onto the revenue requirement. The formula is amount of return (rate base times rate of return – ROR) plus operations expenses. See also ROE.

reversible data-hiding A procedure to verify the integrity and authenticity of digital images, which assures users they are viewing the original image, and can notify the creator if an image is misappropriated. The technique works by extracting and compressing minor details of the image and then embedding this packet of data within the file. A digital watermark is inserted into space created by the compression process, which is detected by authorized users to assure that the image is original. The procedure was under development by Xerox Corporation and the University of Rochester, New York as of early 2003.

Reverse Address Resolution Protocol RARP. A TCP/IP protocol for determining the IP address (or logical address) of a node on a local area network connected to the Internet, when only the hardware address (or physical address) is known. Although the acronym RARP refers only to finding the IP address, and Address Resolution Protocol (ARP) technically refers to the opposite procedure, the acronym ARP is commonly used to describe both procedures.

reverse auction An auction where the low bidder wins the contract. The FCC is exploring using reverse auctions to award contracts for providing phone service in rural areas.

reverse battery signaling A type of loop signaling in which battery and ground are reversed on the tip and ring of the loop to give an "off-hook" signal when the called party answers. Some systems employ reverse battery, either for a short period or until the call is finished, to indicate that it is a toll call. In some PBXs this is used to provide toll diversion.

reverse battery supervision A way of telling the originating central office that the called telephone has been answered (i.e. it has gone off-hook). The voltage of the line at the originating end is reversed. Reverse battery supervision, which puts a signal at the user's premises, is very useful for devices like call accounting systems (knowing precisely when to begin the billing cycle) and telemarketing systems (knowing precisely when to transfer the machined-dialed call over to the operator). See also Answer Supervision.

reverse channel 1. A (typically) small-bandwidth channel used for supervisory or error-control signaling. Signals are transmitted in the opposite direction to the data that is sent.

2. The channel in a dial up telephone circuit from the called party to the calling party.

reverse charge call A collect call or a calling card call that is billed on the distant end. As an example of the latter scenario, a Japanese tourist in the United States may make a calling card call back to Japan using his KDDI calling card. The American international LD carrier that handles the call will record the call record details and the KDDI calling card number and provide them to KDDI for end-user billing.

reverse DNS Also called the in-addr.arpa name space, as in "inverse-ARPA," referring to ARPAnet, the predecessor to the Internet. Reverse DNS is the reverse of the DNS (Domain Name Service) process. A DNS allows you to find a resource (a computer) on the WWW (World Wide Web) by entering a "domain name" in the form of a URL (Uniform Resource Locator), such as www.harrynewton.com. Once entered, the URL is searched against the database of all URLs contained in a Domain Name Server, and is translated into an IP (Internet

Protocol) address. Reverse DNS allows you to enter an IP address, and map it back to the URL, using a separate but similar DNS table. It's much like looking up a person's name and address by looking up the telephone number through a "crisscross" directory. Telemarketers do it all the time. Their predictive dialing systems just dial telephone numbers at random. When you answer the phone, the system matches your telephone number to your name, which they invariably will mispronounce.

Reverse DNS is used if you want to gather statistics from the domain where a particular service is used. For instance, your company wants to track all of the activity originating from Australia. You won't get that information from the traffic statistics on your server, because it only tracks the originating IP address. Reverse DNS will look up the originating URLs from the IP addresses, so you could see all of the originating URLs with a suffix of ".au," which is the Internet country designation for Australia.

Reverse DNS also is good for security purposes. Thieves, hackers, spammers, pornography dealers and other unsavory characters will not support Reverse DNS on their web sites because they wish to mask their identities. If you are a corporate Web-Master or an IT manager, you will reject any attempted access to your internal host resources if Reverse DNS is not supported for the IP address of anyone trying to gain access. Also, if you want to restrict access to particular machines (servers) in a particular multi-server domain space, you would insist on Reverse DNS support. When you register a domain name in a DNS, it generally also is registered in a Reverse DNS. You might not want to do this for security reasons. However, and for similar security reasons, many WWW, FTP and Telnet servers on the Internet will refuse connections to any connected host that does not have both forward and reverse DNS entries. Think of it this way: I have Caller ID. When someone calls me, and I don't see a telephone number and name on my Caller ID box, I don't answer the phone. The caller deliberately has chosen to mask his identity. I don't accept calls from anyone who wants to take up my time and energy, but insists on hiding behind a mask. See also ARPAnet, DNS, FTP, IP, Telnet, URL, and WWW.

reverse engineering Reverse engineering essentially is the process of taking the program or product apart, analyzing its makeup, and then putting it back together again. This technique is used in debugging programs and in developing products that mimic or improve the original, or interface to the original.

reverse interrupt In Bisync, a control character sequence (DLE) sequence sent by a receiving station instead of ACK1 or ACKO to request premature termination of the transmission in progress.

reverse lookup You have a phone number. You want to find out to whom it belongs. That's called "reverse lookup." It's called reverse because the non-reverse, i.e. "normal" way is to have the person's name and their address and get their phone number. You can find "reverse lookup" ability on various phone number sites on the Internet. See also Reverse Matching.

reverse matching Attaching the name and address to a phone number. A It's a job usually done by a specialized service bureau. Called "reverse" matching because the service bureaus started in business by attaching phone numbers to lists containing names and their addresses. With ANI (Automatic Number Identification), we get the phone numbers of people calling us. But we don't get their names and addresses. We need to get this information for many reasons. The obvious being that getting this information on-line and fast saves asking the caller for it and typing all the stuff in. That saves time on the phone – as much as 20 seconds. And fewer questions about boring stuff like phone number, address, city, state, zip means less typing time (also called data entry time, less clerking time) and more time to explain the specials we're selling today. In short, the fewer questions we ask, the less we type and the more stuff we can sell. Reverse matching can be done instantly on-line via a direct data hookup to a distant specialized service bureau or it can be done at the end of the month when we receive our 800 phone bill containing the phone numbers of the people who called us that month. See also Reverse Lookup.

reverse multiplexing Another name for inverse multiplexer. See inverse multiplexer.

reverse operation Briefly running a shredder in reverse to clear jams.

reverse transfer An Inter-Tel term for a phone feature in which a call on common hold at any phone may be retrieved from any phone anywhere in the phone system.

reverse video A video display with all the characters reversed. Characters which are normally white on the screen appear black. And blacks appear white. Reverse video is used to emphasize or enhance things – like those characters to be printed in italics or bold.

revertive pulse Ground pulses sent back to the sender in the originating panel office from the various selector frames to control the selection process.

revertive pulsing In telephone networks, a means of controlling distant switching selections by pulsing, in which the near end receives signals from the far end.

Revest-Shamir-Adleman See RSA.

revisable-form document An electronic document with its formatting information intact, readable and modifiable.

rewritable optical disks They look like CD-ROM disks but they're not. On one side you can store 284 megabytes or 335 megabytes depending on how large you make the sectors – either 512-byte sectors or 1,024-byte, respectively. All the optical disks conform to standards set up by ISO. Compared with hard drives, they are very slow.

REX Routine Exercise.

REXX REstructured eXtended eXecutor. An interpreted script language, or procedural programming language, developed by IBM originally for users of large operating systems in a mainframe environment. REXX was designed for ease of learning and ease of use for both programmers and non-programmers. REXX offers powerful character manipulation in terms of symbolic objects (e.g., words and numbers) with which people normally deal, automatic data typing and debugging capabilities. REXX can be compared with other interpreted script languages such as Microsoft's Visual Basic, Netscape's JavaScript and Larry Wall's Perl.

RF Radio Frequency. Electromagnetic waves operating between 10 kHz and 3 MHz propagated without guide (wire or cable) in free space. If you have a home computer that lets you use your home TV set as a video display device, then the computer has an rf Generator. This means that this device is generating an rf carrier to carry the video signal information.

For the purposes of the FCC's regulation of cable television systems, this term includes any carrier, modulated or unmodulated, whether radiated over the air by an antenna or carried by a coaxial cable. This term dates from the early days of radio (hence, the name "radio" frequency) when the only uses for RF were AM broadcasting and ship-to-shore communications. The term is still in use today, even though it now includes video and control signals as well as audio.

RF Channel Number An identifier assigned to a Radio Frequency (RF) channel to distinguish it from other rf channels. A cellular radio term.

RF Channel Pair Two associated Radio Frequency (rf) channels, one forward and one reverse. The former is used to support forward transmissions from the Mobile Data Base Station (MDBS) to the Mobile End System (M-ES). The reverse channel carriers data information from an M-ES to an MDBS, and is a contention based communications channel.

RF Choke Also known as a load coil. A coil of wire that filters out high frequencies. See Load Coil and RF Leakage.

RF CMOS Radio Frequency Complementary Metal Oxide Semiconductor. RF CMOS is a developing technique for the manufacture of chips based on CMOS technology for a wide variety of RF-based devices such as cellphones, wireless LANs, and wireless peripherals such as headsets keyboards. The advantage of RF CMOS is its low cost compared to chip technology based on GaA (Gallium Arsenide) and various alumina hybrids. The problem is the high level of noise created by combining analog, digital and RF technologies on a single CMOS chip.

RF fingerprinting Network administrators traditionally have located wireless devices connecting to a WLAN (Wireless Local Area Network) through the use of closest access point (closest AP) and triangulation methods. RF (Radio Frequency) fingerprinting improves on these methods through intelligent algorithms that take into account the effects that variables such as the building, furniture or people can have on the RF signal. By considering attenuation levels, signal reflection and multipath fading, device location tracking is improved considerably. Thereby, asset management and E911 applications are improved, and the location of rogue access points and clients can be more easily and exactly pinpointed.

RF leakage RF Leakage is defined as the amount of energy which "leaks" from the connector and/or component. Although rf Leakage will vary with frequency, it is typically tested at only one frequency. Leakage, like Insertion Loss, is expressed in dB. Very large negative dB values indicate that the device does not radiate much energy.

RF line filter A device installed on a phone line next to your phone which filters out sounds at frequencies you select. Those frequencies might be caused by adjacent machinery, nearby power lines or radio towers.

RF modulator See Modulator.

RF splitter Want to put attach two (or more) TV sets behind your satellite antenna? Easy, go down to Radio Shack and buy an RF Splitter. Screw the line into one side of the splitter and your two TVs sets into the other. Bingo, your two TVs will play the signal coming in from the splitter – which might be a satellite antenna, a CATV hookup, an outdoor antenna, etc. You can buy RF splitter than will allow you to hookup almost as many TVs as you, or want to have. See also Splitter.

RF sweep The use of a transmitter detector to identify wireless emissions in a facility, for example, from a rogue wireless access point or a bugging device.

RFAC Restricted Forced Authorization Code. A forced authorization code that is valid only if it is entered from a specific telephone extension.

RFC Request For Comment. The development of TCP/IP standards, procedures and specifications is done via this mechanism. RFCs are documents that progress through several development stages, under the control of IETF, until they are finalized or discarded. RFC#### documents Internet "Request For Comment" documents (i.e., RFC822, RFC1521, etc.). The contents of an RFC may range from an official standardized protocol specification to research results or proposals. A set of papers in which the Internet's standards, proposed standards and generally agreed-upon ideas are documented and published. RFCs are the official document series of Internet Architecture Board (IAB) and are achieved permanently. They are never deleted. See the RFCs below.

RFC 1144 This RFC (Request For Comment) will provide overhead compression for the TCP/IP protocol down to 5 octets. It does this by anticipating that the next packet in a file transfer sequence will have the same address as the previous and will have the same sequence number plus one. This compression technique will be useful where SDLC encapsulation, or other bridging protocol encapsulation, is being used with low-speed PVCs (Private Virtual Circuits.) In these cases, the slight increase in processing power to perform the compression is more than balanced by the increase in application performance and throughput.

RFC 1294 This Request For Comment is Inverse ARP, which allows the automatic discovery of the addresses on the router at each end of another router's DLCIs. Right now, the RFC only applies to IP, but some equipment vendors have already expanded the protocol support to include Novell, AppleTalk, Vines, and DECnet. The benefit of the RFC is to simplify network configuration.

RFC 1315 This Request For Comment is the frame relay MIB (management information database), which standardizes what management information is made available on frame relay devices and where/how that information is accessed. This simplifies the process of integrating frame relay devices into your network monitoring and management process and programs.

RFC 1490 This Request For Comment, RFC 1294, now renumbered RFC 1490, is for multiprotocol encapsulation. The bottom line benefits are to increase interoperability between frame relay devices from different vendors. This means that you can use one vendor's routers (or other equipment type) at some locations, and a different vendor's equipment at other locations. This ability to mix and match allows you to pick the best and most cost-effective tool for the job.

RFC 1577 Under control of the IETF, Request For Comments (RFCs) are documents used to develop standards, procedures and specifications for TCP/IP. RFC 1577 is the document for classical IP.

RFC 1695 Definitions of Managed Objects for ATM Management or ATM MIB.

RFC 1918 This Request for Comment describes IP address space for building private networks, which is nowadays being used quite often for NAT. The reserved address spaces are: 10.0.0.0 - 10.255.255.255 (10/8 prefix), 172.16.0.0 - 172.31.255.255 (172.16/12 prefix) and 192.168.0.0 - 192.168.255.255 (192.168/16 prefix)

RFC 2338 This Request for Comment describes the Virtual Router Redundancy Protocol (VRRP). See VRRP for lots of detail.

RFC 822 This standard specifies a syntax for text messages that are sent among computer users on the ARPA Net (the precursor network to the Internet), within the framework of "electronic mail". The standard supersedes the one specified in ARPANET Request for Comments #733, "Standard for the Format of ARPA Network Text Messages". In this context, messages are viewed as having an envelope and contents. The envelope contains whatever information is needed to accomplish transmission and delivery. The contents compose the object to be delivered to the recipient. This standard applies only to the format and some of the semantics of message contents. It contains no specification of the information in the envelope. A distinction should be made between what the specification REQUIRES and what it ALLOWS. Messages can be made complex and rich with formally-structured components of information or can be kept small and simple, with a minimum of such information. Also, the standard simplifies the interpretation of differing visual formats in messages; only the visual aspect of a message is affected and not the interpretation of information within it. Implementors may choose to retain such visual distinctions. The formal definition is divided into four levels. The bottom level describes the meta-notation used in this document. The second level describes basic lexical analyzers that feed tokens to higher-level parsers. Next is an overall specification for messages; it permits distinguishing individual fields. Finally, there is definition of the contents of several structured fields. Messages consist of lines of text. No special provisions are made for encoding drawings, facsimile, speech, or structured text. No significant consideration has been given to questions of data compression or to transmission and storage efficiency, and the standard tends to be free with the number of bits consumed. For example, field names are specified as free text, rather than special terse

codes. A general "memo" framework is used. That is, a message consists of some information in a rigid format, followed by the main part of the message, with a format that is not specified in this document. The syntax of several fields of the rigidly-formatted ("headers") section is defined in this specification; some of these fields must be included in all messages. The syntax that distinguishes between header fields is specified separately from the internal syntax for particular fields. This separation is intended to allow simple parsers to operate on the general structure of messages, without concern for the detailed structure of individual header fields. In short, a message consists of header fields and, optionally, a body. The body is simply a sequence of lines containing ASCII characters. It is separated from the headers by a null line (i.e., a line with nothing preceding the CRLF– carriage return, life feed). www.faqs.org/rfcs/rfc822.html

RFD Request For Discussion. A period of time during which comments on a particular subject are solicited. An Internet term.

RFF Radio Frequency Fingerprinting. A process in which the radio signal information and characteristics produced by the transmitter are captured and analyzed by the receiver for purposes of detecting a cloned device from accessing the network. Bursts of control data are captured and analyzed using complex signaling techniques; the data is compared to the characteristics of the legitimate transmitter in order to determine whether access should be granted or denied. Primarily used in secure military applications, the technique has been evolving since WWII; it is being considered for application in cellular telephony.

RFI 1. Request For Information. General notification of an intended purchase of equipment or equipment and lines sent to potential suppliers to determine interest and solicit general descriptive product materials, but not prices or a formal request. See RFQ for a detailed explanation.

2. Radio Frequency Interference. All computer equipment generates radio frequency signals. The FCC regulates the amount of RFI a computing device can leak past its shielding. A Class A device is sufficient for office use. A Class B is a more stringent classification for home equipment use. See EMI and Radio Frequency Interference.

RFID Radio Frequency IDentity. RFIDs are tiny chips and wireless radio antennas that can be embedded into products and used for various identification purposes. The first application I read of was the idea of embedding them into banknotes as another protection against counterfeiting. (Other security features on bank notes include holograms, foil stripes, special threads, microprinting, special inks and watermarks.) RFID is a contactless solution that works with proximity readers. There are two high level versions – those that actually store data and those that simply store a reference key for lookup on a host system. Both have specific applications. They also come in highly secure variations as well. The actual proximity varies with the type of RFID solution in use. For example, longer distance RFID are being used on some toll highways and toll bridges and tunnels. In and around new York, they're called EZ-Pass. Exxon/Mobil uses them to help its customers buy gas faster. They're called SpeedPass. RFIDs were also used in a local marathon in ankle bracelets that were attached to every competitor and were readable at various points of the race including the finish line. Some RFIDs need batteries – and that's their weakest link. The newer RFIDs don't need batteries, which means they don't have to be replaced regularly. See RFID Tag.

RFID ink An ink whose chemical properties cause it to emit a set of radio frequencies when struck by electromagnetic waves from an RFID reader. Since each chemical in the ink emits its own unique radio frequency, varying the combination of chemicals in the ink enables a large number of frequencies to be produced. One RFID ink has been manufactured that uses 70 different chemicals, each of which is assigned its own position in a 70-digit binary number. RFID ink is used to produce chipless RFID tags that are painted onto paper, wood and other materials. One company has tested a biocompatible RFID ink to create chipless RFID tags, which the company calls RFID tattoos, on cattle and laboratory rats.

RFID tag A microchip attached to an antenna that picks up signals from and sends signals to a reader. The tag contains a unique serial number, but may have other information, such as a customer's account number. Tags come in many forms, such smart labels that are stuck on boxes; smart cards and key-chain wands for paying for things; and a box that you stick on your windshield to enable you to pay tolls without stopping. RFID tags can be active tags, passive tags or semi-passive tags.

RFID tattoo A chipless RFID tag made out of a biocompatible ink that is injected into tissue to create a unique tattoo that can be read by a non-line-of-sight RFID reader. Possible applications include tagging livestock, laboratory animals, pets, and cuts of meat. See also RFID ink.

RFID virus As a proof of concept, researchers in the Netherlands have developed a virus that is spread via RFID. The virus is encoded in the firmware of an RFID tag. When the tag is scanned by an RFID reader, the malicious code is picked up by the reader and delivered to a back-end application server or database, where it executes.

RFO Reason For Outage.

RFP Request For Proposal. A detailed document prepared by a buyer defining his requirements for service and equipment sent to one or several vendors. A vendor's response to an RFP will typically be binding on the vendor, i.e. he will be obliged to deliver what he says in his RFP at the prices and following the conditions explained in that RFP. See RFQ for a detailed explanation.

RFQ Request For Quotation. A document prepared by a buyer defining his needs for service and equipment in fairly broad terms and sent to one or several vendors. The RFQ is much less detailed than the RFP. Let's start at the beginning of the buying process. We have a buyer who wants a phone system. His first step may be to issue a formal or informal RFI – Request For Information. In effect, the RFI says "Please tell me what you have. I have a vague idea of what I want but I don't know exactly what is available to suit my needs. Please send me some information."

After a buyer gets his responses to his RFIs, he may issue an RFQ – Request For Quotation. An RFQ may include a tentative configuration of the type of phone system the user wants, plus some listing of features the buyer is interested in. In the RFQ, the buyer asks for a "ballpark" (approximation) of the possible price for such a system. Usually the "price" is within plus or minus 10% of where it will eventually be in the final configuration. In short, an RFQ's purpose is not to buy, but to find out what's out there and what it might cost. The purpose may be to allocate a budget or to put aside some money for the forthcoming purchase.

An RFP – Request For Proposal – is much more formal and definitive. Its purpose is simple. The buyer wants to buy something. The RFP contains a list of what the buyer wants, when he wants it, how it should be installed, how it should delivered, what financing may be necessary. It is now up to the vendor/s to respond with their configuration, their precise prices and their terms and conditions of sale. Whatever the vendor responds with – called a Response to an RFP – constitutes a definite offer. At this point, the buyer can negotiate the terms of the vendor's Response to his RFP. This will lead to the writing, and eventual signing of a contract. Or the buyer may simply decide to accept the Vendor's Response. Often that Response may have a line at the back of it – "I accept the terms and conditions of this response." If the buyer signs this, then the Response to the RFP becomes a valid contract.

RFS 1. Ready for Service.

2. Remote File System. The ability to mount a disk drive somewhere on a network – but it's not on your competitor.

RG Radio Guide. RG numbers are used to designate various standard types of coaxial cable, all of which "guide" a "radio" frequency signal. The term was established by the U.S. military in the 1930s. The term waveguide is used to describe the similar function accomplished by optical fibers that channel light signals, and various types of tubes used to channel radio signals in microwave systems. By the way, the RG numbers, themselves, have no significance. Rather, they are just like pages in a book. See the various RG numbers following.

RG-6 The type of coax cable recommended for digital satellite TV installations. RG-6 is a larger-size cable than the lower-grade RG-59 cable found in some homes. RG-59 has a small center conductor, a small insulating dielectric (white foam inside the cable) and typically a single outer shield. By comparison, RG-6 has a larger center conductor, a quad shield, and a much larger insulating dielectric, ensuring greater bandwidth and lower frequency loss per foot.

RG-58 Coaxial cable with 50-ohm impedance used by Thinnet.

RG-59 An older coaxial cable type often used in television distribution in homes, now largely replaced by RG-6.

RG-62 Coaxial cable with 93-ohm impedance used by ARCnet.

RG/U RG/U or RG-U is the U.S. government/military designation for coaxial cable. R=Radio Frequency, G=Government, U=Universal Specification.

RGB Red. Green. Blue. The three primary colors used in video processing, often referring to the three unencoded outputs of a color camera or VTR. A color model based on the mixing of red, green, and blue – the primary additive colors used by color monitor displays and TVs. The combination and intensities of these three colors can represent the whole spectrum. Color television signals are oriented as three separate pictures: red, green and blue. Typically, they are merged together as a composite signal but for maximum quality and for computer applications the signals are segregated.

RGB Cutoff An advanced color control that lets you set your monitor to maintain color balance across different gray scales.

RGB Gain An advanced color control that lets you adjust red, blue, and green levels individually.

RGC 1604 An acronym used in a BellSouth proposal to the State of GA. The salesman who wrote the proposal didn't know what it was. It relates to Frame Relay Service.

RGU Revenue-generating unit. This is a term used by cable TV operators. An RGU is a basic cable, digital cable, cable modem or cable telephony subscriber. A customer who subscribes to all four services equals 4 RGUs.

RH Request Header or Response Header.

RHC Regional Holding Company. Also called Regional Bell Operating Company (RBOC). See RBOC for a fuller explanation.

RHCC Rural Health Care Corporation. See that term for a full explanation.

rheostat A variable resistor.

Rhetorex A manufacturer of voice processing componentry based in Campbell, CA. Inspiration for the company's name came from the word rhetoric, which is the art of effectively using speech and language.

RHH Designation for rubber-insulated, heat-resistant building wire rated 900xAFC in dry locations. power conductors and a ground in each interstice.

rhinestone The name comes from the French caillou du Rhin. It came because the colorless, hard-glass artificial gems were originally made at Strasbourg on the Rhine.

rhombic antenna An antenna composed of wire radiators describing the sides of a rhombus. It is usually terminated and unidirectional; when unterminated, it is bidirectional.

RHW Designation for rubber-insulated building wire. Heat and moisture resistant. 750xAFC wet or dry.

RI Ring Indicator. An RS-232 control line asserted by the DCE when a call has come in for the DTE.

RIB Routing Information Base. A database of routing information. BGP (Border Gateway Protocol) routers build and maintain RIBs on the basis of the reachability of other routers across various routes. They then exchange that information with their peers, and subsequently exchange incremental updates to the RIBs. Thereby, the individual Autonomous Systems (ASs) act act on the advice and with the assistance of peer ASs to route IP (Internet Protocol) traffic between endpoints across the most logical network routes, hop-by-hop. See also BGP.

ribbon cable Multi-wire cable that is flat instead of round. In ribbon cable, the conductors are laid side by side. Ribbon cable can be more easily laid under carpeting because it is flat and thus, can extend phone and computer services to places otherwise hard to reach. There are disadvantages to ribbon cable. Because ribbon cable is flat, it's hard to twist its individual wire conductors around each other (thus humming can be a problem). It is hard to put a metal shielding around the twisted wire pairs. It is hard to put coax cable into ribbon cable. It is hard to make ribbon cable sufficiently strong to withstand thousands of high heels trampling on it. It is hard to make ribbon cable which turns a corner... But there has been enormous progress in ribbon cable. And ribbon cable is finding greater use in buildings. These days it even carries commercial A.C. power.

ribbon fiber cable A fiber optic cable comprising one or more ribbons, each of which comprises six or 12 fibers laid side by side. The ribbons are then stacked on top of each other inside of a protective cable sheath. Common configurations include 12 ribbons for a total of 144 fibers and 24 ribbons for 288 fibers, although the counts can go much higher. Ribbon cables generally are used in long haul outside plant applications, although they sometimes are used for inside cable systems where complex, high speed LANs (local area networks) must be interconnected over distances greater than 100 meters (which generally is the limitation of copper wire local area network systems).

ribbon license A private land mobile radio (PLMR) license that covers a geographical area that follows the contours of a railroad or pipeline. For example, the Association of American Railroads (AAR), which represents the U.S. railroad industry, formerly had a license for each of the 1,000+ land mobile base stations used for train control. The AAR successfully petitioned the FCC to replace the 1,000+ individual licenses with a single geographic license covering an area defined as a 70-mile zone on each side of the rights-of-way of all operating rail lines in the United States. Replacing the many individual PLMR licenses with a single ribbon license streamlined the licensing process and gave the AAR flexibility when deciding where to deploy future base stations and redeploy existing base stations.

ribbon of highway Fiber optic cable.

Rician channel A radio communications channel that is subject to both Gaussian noise and to multipath fading, such as a mobile communications channel to and from a fast-moving vehicle, ship or airplane. The spectral density of the combined direct and interfering multipath signals follows a Rician distribution, named for its originator, Steven O. Rice, who when at the Bell Telephone Laboratories (now part of Lucent Technologies) in the mid 1940s and 1950s, pioneered the theory of noise in communications systems. Rice is remembered for his landmark paper, Mathematical theory of random noise, published in 1945 in the Bell System Technical Journal and reprinted in 1954 by Dover in Selected

Papers on Noise And Stochastic Processes (ed. N. Wax).

ridge marker One or more ridges running laterally along the outer surface of an insulated wire for purposes of identification.

ridge mount An antenna mounting device used for connecting a mast to the ridge of a roof.

RIF 1. Rate Increase Factor. This factor by which the cell transmission rate may increase upon receipt of an RM-cell. See also ATM and RM-Cell.

2. Routing Information Field. In the Token Ring protocol, a optional field which is used when the transmitting frame must pass through multiple Source Routing Protocol (SRP) bridges. Within the RIF, the value of the RII, or Routing Information Indicator, (1 or 0) indicates to the bridge whether the frame should be either forwarded to another ring or confined to the local ring. See also Bridge.

3. Reduction In Force, another way of saying layoffs. A euphemism. Like saying someone passed away, when you realy mean they died.

RIFF Resource Interchange File Format. Platform-independent multimedia specification (published by Microsoft and others in 1990) that allows audio, image, animation, and other multimedia elements to be stored in a common format. See also Media Control Interface (MCI).

rig What a radio amateur calls his radio equipment.

right ascension Right ascension in the celestial coordinate system is the angle from the first point of Aries to the satellite (or other celestial body). In astronomy, the right ascension is defined as the time elapsed from the instant that the first point of Aries reaches the observer's meridian and the instant that the celestial body concerned reaches that meridian. Consequently, right ascension, abbreviated R.A. and expressed in hours and minutes, is used as one of the coordinates of celestial bodies listed in astronomical atlases.

right-hand circular polarization An elliptically or circularly polarized radio wave in which the electric field intensity vector appears to rotate to the right (clockwise), with respect to the axis, as it propagates outward from the transmitter.

right hand rule A rule for indicating the direction of magnetic effect. Grasp the wire with the right hand and with the thumb extended along the wire in the direction of current. The curved fingertips will indicate the direction of magnetic flow. Not totally relevant for including in this dictionary. But cute. See also Rule of Thumb.

right of entry A legal right to enter the premises owned by another, and, in this case, for the purpose of providing telecommunications services to the tenants of a Multiple Dwelling Unit (MDU), e.g. an apartment building. Under the terms of the Telecommunications Act of 1996, a Competitive Local Exchange Carrier (CLEC) is required to negotiate with a building owner a Right of Entry and License Agreement in order that the CLEC might provide service to the tenants. The terms of such an agreement will specify the capital improvements to be made, in the form of a demarcation point (demarc), inside cable and wire systems, and various other equipment which might include DSL Access Multiplexers (DSLAMs), switches or routers, concentrators, and distribution frames and wiring closets. If the CLEC is a wireless provider, "roof rights" must be addressed for the placement of antennas. If the CLEC is a wireline provider, Right of Way must be included in order that a trench can be dug across the property and a hole bored through an exterior wall, usually sufficient to accommodate multiple optical fibers. The CLEC will suggest that the Right of Entry should be without cost, as the communications infrastructure of the building will be improved and, as a result, the competitiveness of the property will be enhanced. The building owner generally will be highly amused by this position, and charge the CLEC an annual fee anyway. See also CLEC, DSLAM, Right of Way, and Telecommunications Act of 1996.

right of way A legal right of passage over land owned by another. Carriers and service providers (telephone companies, CATV companies, cellular providers, etc.) must obtain right-of-way to dig trenches or plant poles for cable systems, and to place wireless antennas. In return for right-of-way, the owner of the land is usually given some money. Some people don't like antennas in their backyard, no matter how much money they are being paid. When MCI was young, it needed to erect a 400 foot (or so) microwave antenna in the middle of mid-western farmland. It went looking for a place, found the ideal location. However, the lady who owned the land wouldn't allow anything as grotesque as a 400ft tower in her backyard. (Would you?) So MCI sent Jack Goeken, its president at the time, out to negotiate. Jack came back with a deal whereby the tower went up if MCI agreed to deck it with Christmas lights and light it every Christmas. Since it was visible for many miles, the thing became a navigation aid to planes, The lady got herself the largest "Christmas tree" in the world. Sadly, the tower has been obsoleted by unromantic buried fiber cable.

right thing Abba Eban once said that men and nations will always do the right thing in the end - after they exhaust every other possibility.

Right To Use See RTU.

rights-of-way easements Right of access to land; used by telecommunications providers to place their facilities.

rightsizing Another term for re-engineering. See Re-Engineering.

RII Routing Information Indicator. A Token Ring term. See RIF for an explanation.

RILD Remote ISDN Line Drawer.

RIME RelayNet International Message Exchange, a multi-tier communications network which exchanges messages among member bulletin board systems.

RIMM 1. The inline memory module used with RDRAM (Rambus Dynamic Random Access Memory) chips developed by Rambus Inc. A RIMM is similar to a DIMM (Dual In-line Memory Module). Trademarked by Rambus Inc., RIMM really isn't an acronym, although it is claimed by some to stand for Rambus In-line Memory Module. See also DIMM and RDRAM.

2. Stockmarket symbol for Research In Motion, the Canadian maker of the BlackBerry PDA/cellphone.

rimm job A bogus academic study masquerading as legitimate science. Named after Marty Rimm, author of the dubious "cyberporn" study from Carnegie Mellon University that Time magazine gullibly took as gospel.

ring 1. As in Tip and Ring. One of the two wires (the two are Tip and Ring) needed to set up a telephone connection. The ring is typically the negative wire.

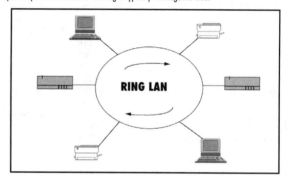

2. Also a reference to the ringing of the telephone set.

3. The design of a Local Area Network (LAN) in which the wiring loops from one workstation to another, forming a circle (thus, the term "ring"). In a ring LAN, data is sent from workstation to workstation around the loop in the same direction. Each workstation (which is usually a PC) acts as a repeater by re-sending messages to the next PC in the ring. The more PC's, the slower the LAN. Network control is distributed in a ring network. Since the message passes through each PC, loss of one PC may disable the entire network. However, most ring LANs recover very quickly should one PC die or be turned off. If it dies, you can remove it physically from the network. If it's off, the network senses that and the token ignores that machine. In some token LANs, the LAN will close around a dead workstation and join the two workstations on either side together. If you lose the PC doing the control functions, another PC will jump in and take over. This is how the IBM Token-Passing Ring works. See Topology, Bypass Cable and Token Ring.

ring again The PBX remembers the last number called by a phone and will redial it when the feature is activated.

ring back tone The sound you hear when you're calling someone else's phone. The tone you hear is generated by a device at your central office and may bear no relationship to the sound the phone at the other end is emitting – or not emitting. If your call didn't go through the first time, always call back at least once. See Ringing Tone.

ring banding A method of color coding insulated conductors by means of a small band of colored ink applied circumferentially at regular intervals along the axis of the insulated conductor.

ring battery Commonly unfiltered - 24 VDC source that supplies operating power to all local KSU components. Also called the B Battery.

ring cadence Your phone rings. In North America, it rings for two seconds and doesn't ring for four seconds. That's called ring cadence. Ring cadence is important because customer-owned telephone switches (like PBXs) expect the two second ring and silence for four seconds. If ring cadence changes to three second ring, three second silence, for example, the customer-owned phone switch may no longer answer incoming calls. See Cadence and Ring Cadence Acceptance.

ring cadence acceptance Your phone rings. In North America, it rings for two seconds and doesn't ring for four seconds. That's called ring cadence. Ring cadence acceptance is the ability of your customer-owned switching listening device to understand whatever variation of ring cadence it's presented with.

ring conductor One of the two conductors in a cable pair used to provide telephone service. This term was originally coined from its position as the second (ring) conductor of a tip-ring-sleeve switchboard plug.

ring cycle A ring cycle in North America is typically six seconds long, two of ringing, four of silence, then repeated.

ring down box Ring down boxes, also known as CO simulators or telephone line simulators, are simple devices used for generating POTS calls – without a central office. You connect a phone to both sides of the device. When one side goes off-hook, the ring down box will "ring" the other side. When both sides are off-hook, both sides are coupled together, the line is powered and the sides can talk. Ring down boxes are available with various options and configurations. These include the ability to provide dial-tone to the caller side (required to test applications with modems, faxes, or other automated out dialing devices), caller ID, and disconnect supervision. They are generally available in one to four line sizes, although special configurations may support more. Ring-down boxes are used for giving demonstrations and testing. We use them in our test labs to test drive new computer telephony systems.

ring down circuit A tie line connecting phones in which picking up one phone automatically rings the other, distant phone. In a ringdown circuit, a ringing current (AC) is sent down the line. That current may light a lamp, set off a bell, buzz a buzzer. The idea is to alert the person at the other end to the incoming call. A ringdown circuit is often used in an elevator or other emergency situation. It is also used in the financial industry, for example to allow traders to instantly access other traders and thus complete their trades as quickly as possible. Trader turrets often are loaded with ring down circuits. See also trader turrents.

ring down interface A private line two-wire interface, also called a Loop Start Trunk.

ring generator A component of virtually all phone systems, ranging from large central offices to small key systems, that supplies the power to ring the bells inside phones, typically 90 volts AC at 20 Hz.

ring group Collection of Token Ring interfaces on one or more routers that is part of a one-bridge Token Ring network.

ring indicator Modem interface signal defined in RS-232-C which indicates to the data terminal equipment that a call is coming in.

ring isolator A device placed on a telephone line to disconnect the ringer when it is an idle state. It is used for noise prevention.

ring LAN See Ring.

ring latency In a token-ring network, the time measured in bits at the data transmission rate, required for a signal to propagate once around the ring. Ring latency includes the signal propagation delay through the ring medium, including drop cables, plus the sum of propagation delays through each data station connected to the token-ring network. See also Token-Ring Network.

ring network A network that links PBXs, computers, terminals, printers and other devices in a circular communications link. See Ring.

ring out This definition contributed by Gregory Maffett. Here is the process with Ring Out: You make a cable with say 25 wires that are supposed to send signals to 25 pins at the other end. After soldering them together, you want to know if pin A on one end goes to pin A on the other end. You apply a signal to one end and put a voltmeter on the other. If the needle moves, you have success on that pin. Repeat 24 more times and you have rung out the cable successfully. Or you go back and rewire it.

ring protection switching. RPS. Nortel's "Introduction to SONET Networking" tutorial handbook (www.nt.com/broadband/reference/sonet_101.html) talks about "Automatic Healing of Failed or Degraded Optical Spans in a Two-Fiber BLSR." The handbook says "in the event of failure or degradation in an optical span, automatic ring protection switching (RPS) reroutes affected traffic away from the fault within 50 milliseconds, preventing a service outage."

ring signal The pulse ringing voltage output of the local Interrupter KSU. Typically, this signal is 105 VAC with a duty cycle of 2 seconds on and 4 seconds off.

ring splash A brief "splash" of a "ring" which announces an incoming telephone call. Ring splash is a technique used by premise-based telephone switches such as PBXs and ACDs in order to reduce the size of the "glare window," which is the length of time in which "glare" can occur. Glare is a condition in a trunk simultaneously is seized by switches at both ends. For instance, a PBX or ACD user might seize a trunk for an outgoing call at precisely (or virtually so) the same time that a central office (also called public exchange) might choose to connect an incoming call over the same trunk. In order to avoid such

embarrassment, the properly equipped PBX or ACD will send a ring splash in the form of a 500 ms (millisecond) ring that is splashed to the station immediately prior to the normal ring cadence. Thereby and in the context of the North American standard ringing cycle, the glare window might be reduced from a maximum of 4 seconds to a maximum of 200 ms (two hundred thousandths of a second, or one-fifth of a second). While the ring splash may make the beginning of the ring cadence sound slightly odd, the risk of embarrassment is reduced substantially. Have you ever picked up the phone, dialed in another person's ear, and then said something impressive like "Hello, Bob. Hello?" That's really embarrassing. Ring splash solves the problem. See also Glare and Ring Cycle.

ring tone An audible, call-progress signal connected to the calling line to indicate that the called station (telephone, fax or computer) is being rung. The industry standard in North America is a mixture of 440 Hz and 480 Hz, interrupted at the same rate, or ring cycle, as the ringing current being applied to the called station. My wife always waits for the ring tone on her incoming calls to stop, but it's not necessary. See also Ring Cycle, Ring Tones and Ringtone.

ring tones Ring tones are the name of those sounds which

Ring tones are those ubiquitous, monophonic song recordings programmed into seemingly every teenager's mobile phone. A study released in January 2003 by London-based Informa Media Group said that authors' collection societies collected $71 million in royalties from ring-tone sales in 2002, up 58 percent from the previous year. Informa's senior analyst Simon Dyson said the royalties figure – which is typically 10 percent to 15 percent of the total sales from ring tones – would suggest that the overall market is over $700 million annually, and quite possibly as high as $1 billion. The proceeds are divided between operators, labels and the artists. Ring tones started off as a promotional gimmick, with labels offering up decidedly low-fidelity renditions of new singles to Web sites and mobile phone operators as a way to keep fans humming along to their favorite artists. Despite the poor sound quality, the practice of customizing one's mobile phone with a favorite song grew with surprising speed. Now record labels regularly grant rights to mobile operators and Web sites to sell ring tones. Informa said download costs vary widely by country. For example, Russia's largest mobile phone operator, MTS, charges 30 cents per download, while Vodafone in Australia charges $1.83. Ringback tones are different to ring tones which are sounds you hear when your own cell phone rings.

ring topology A network topology in which nodes are connected to a closed loop, no terminators are required because there are no unconnected ends.

ring trip The process of stopping the AC ringing signal at the central office when the telephone being rung is answered.

ring wiring concentrator A site through which pass the links between repeaters, for all or a portion of a ring.

ring wiring configuration The same as daisy chain wiring, except the last jack is connected to jack 1, thus completing a ring.

ringback The tone heard by a calling device when, at the called-device's end, the telephone is ringing or the system is otherwise being alerted of the incoming call.

ringback tones When friends dial your phone number (typically your cell phone number) they'll hear music instead of the standard ring. You choose which music you'd like your friends to hear and, of course, you'll pay for the privilege – often on a one-time and then a per month basis. Ringback tones are different to ring tones which are sounds you hear when your own cell phone rings.

ringdown A circuit or method of signaling where the incoming signal is started by alternating current over the circuit.

ringer A bell in a telephone which indicates if a phone call is coming in. These days "ringers" are electromechanical and clunky (old-style) or small and electronic (new style). The new electronic ones are cheaper, but less interesting to listen to. Most sound like bleating sheep in heat.

ringer equivalence number REN. A number required in the U.S. for registering your telephone equipment with the phone company. Add together the RENs of all the telephones on a single line. The sum of those numbers should never exceed five otherwise none of your bells will work and you won't hear an incoming call. (Your central office simply doesn't send sufficient current down the line.) The alphabetic character after the number refers to the ringing frequency of the alternating current sent down the line to ring the bell. If the letter is "A", the ringer frequency is about 20 Hertz. Most single line phones have a Ringer Equivalence of 1.0A. If the letter is "B", the ringer will respond to any current coming down the line. Any other letter, and you are probably on a party line where the ringer frequency is used for party selection. In Canada, they use the term "Load Number" instead of Ringer Equivalence. The numbers are different, but the concept is the same. See Load Number.

ringer isolator A device in the phone which disconnects the ringer when ringing voltage is not present.

ringing Alternating Current (AC) sent out from the central office along the local loop to the subscriber. It's typically 70 to 90 volts at 17 Hz to 20 Hz. You can get a mild shock if you have your hands on a telephone circuit when ringing current comes along. The rest of the time, the lines are harmless.

ringing generator A device in a phone system that generates the AC ringer voltage. Typically, this voltage is 90 to 115 (nominally 105) VAC at 30 Hertz.

ringing key A key that sends a ringing current.

ringing signal Any AC or DC signal transmitted over a line or trunk for the purpose of alerting someone or some thing at the distant end of an incoming call.

ringing tone A low tone which is one second ON and three seconds OFF. It indicates that ringing current is being sent by the central office to the person receiving the call. Ringing tone is not produced by the calling party's central office – but by the called party's central office. Thus, it is possible for you to hear ringing tone but for the person you are calling not to hear anything. As a general rule, if the person doesn't answer, call them a second time. Often, they'll say "The phone never rang." This will not be a lie, but simply a temporary glitch in their central office.

ringing transfer A PBX feature which allows you to choose which bells in a group of phones will ring when a call is coming in for that group.

ringing voltage In addition to talk battery, a Central Office provides ringing signaling. Ring Voltage is generally 70 to 90 volts at 17 Hz to 20 Hz. See also Ringing.

ringtone In its basic sense, the ringtone (also spelled ring tone) is a tone returned by central office equipment that tells a caller that the phone at that end is ringing. (Somewhat confusingly, this meaning is also called ringback.) The tone is sent back in between the ring sequence at the receiving end. The pulsing rate is one on, two off from a 3-phase generator with each call using a single phase. The called and calling phones would not necessarily use the same phase, so if you wanted to ring someone's phone (for example, to wake them up), you would need to hear it ringing for a full cycle to make sure that the phone actually rang at the other end.

Mobile phone users use the term to mean the ring that the caller hears. The proliferation of cellular telephones in recent years has given rise to a wide variety of ringtones. These do not necesarily follow the intermittent ringdown signal. A contemporary ringtone might consist of several bars of a familiar musical tune, played by an audio oscillator through a small speaker. Such ringtones are popular because, in a crowd of people with many cellular phone sets, they make it easy to tell whose phone is calling out for attention.

ringxiety See phantom ring tones.

riogin Application that provides terminal interface between UNIX hosts using TCP/IP network protocol. Unlike Telnet, assumes remote host is (or behaves like) UNIX machine.

RIP 1. Routing Information Protocol. RIP is based on distance-vector algorithms that measure the shortest path between two points on a network, based on the addresses of the originating and destination devices. The shortest path is determined by the number of "hops" between those points. Each router maintains a routing table, or routing database, of known addresses and routes; each router periodically broadcasts the contents of its table to neighboring routers in order that the entire network can maintain a synchronized database. See also Distance Vector Protocol.

2. To record your CD or DVD's content to your PC's hard disk. You'll need to do this if you wish to transfer your CDs on a portable MP3 player – either one containing a hard disk (e.g. Apple's iPod) or simply flash memory, such as some of the MP3 players made by Sony. Software on the PC which is used to "rip" CDs does not refer to the process as ripping. They refer to it as importing or recording. As the word has developed, it has also acquired the meaning of stealing. So, to rip a DVD, which is now increasingly easy, means to copy it in contravention of copyright law. For example, you might rip a DVD you rented from

rip and replace See forklift upgrade.

rip cord A cord placed directly under the jacket of a cable in order to facilitate stripping (removal) of the jacket.

RIPE Reseaux IP Europeens, a group formed to coordinate and promote TCP/IP based networks in Europe. RIPE is responsible for management and assignment of IP (Internet Protocol) addresses in Europe, just as are ARIN and APNIC in the regions of the Americas and Asia-Pacific, respectively. RIPE holds periodic conferences to coordinate technical issues (similar to the IF) as well as running a Network Control Center (NCC) to handle operational issues such as the administration of the European domain names and routing tables. RIPE is a collaborative organization with no formal membership (over 1,000 organizations participate); all activities are performed on a voluntary basis. See also APNIC, ARIN and IP.

RIPscrip Graphics Emulation Supports RIPscrip Graphics Emulation. Popular on many bulletin board systems, RIPscrip Graphics Emulation allows users to view screens mixing text and graphics. On-screen buttons can be clicked to send commands to a remote system. Now effectively obsoleted by the World Wide Web.

RIR Regional Internet Registries. A collective term for the three non-profit organizations (ARIN, RIPE NC, and APNIC) established to register and administer IP number allocations and assignments. See also ARIN, RIPE, and APNIC.

RISC Reduced Instruction Set Computing. A microprocessor architecture that favors the speed at which individual instructions execute over the robustness of the instruction set. Computers based on RISC use an unusual high speed processing technology that uses a far simpler set of operating commands. These commands greatly speed a computer's performance, especially for calculation-intensive operations such as those performed by scientists and computer-aided design (CAD) and computer-aided manufacture (CAM) engineers. RISC is a design that achieves high performance by doing the most common computer operations very quickly. In contrast, the microprocessors used in most PCs are based on a design called CISC (Complex Instruction Set Computing). CISC does not execute instructions as quickly as RISC but it has more commands and accomplishes more with each command. Programs written for RISC are typically not compatible with those written for CISC processors. RISC is the prevailing technology for workstations today. The RISC semiconductor was an IBM baby, born in its Yorktown Heights, NY lab in 1974. But internal arguments over, and even whether the chip should be used kept IBM fiddling while Sun and other companies decisively powered ahead. IBM got its first good RISC product, the RS/6000, to market in 1990. The PowerPC architecture is one example of a RISC microprocessor design. See also CISC.

riser The conduit or path between floors of a building into which telephone and other utility cables are placed to bring service from one floor to another. Your risers should be twice the size you ever think you'll need in the next 30 years. It's expensive to build risers after the building is built. Very expensive.

riser cable High strength cables intended for use in vertical shafts (i.e., riser shafts) between floors in multi-story buildings. Riser cables must meet NEC (National Electrical Code) specifications and state and local building codes for fuel load, flame spread and smoke generation. These requirements are less stringent than those for plenum cables, but less so than those for distribution cables. Riser cables also contain strength members that support the hanging weight of the cable, thereby relieving the conductors of that stress. See also distribution cable and plenum cable.

riser closet The closet where riser cable is terminated and cross connected to either horizontal distribution cable or other riser cable. The riser closet houses cross connect facilities, and may contain auxiliary power supplies for terminal equipment located at the user work location. See also Satellite Closet.

riser shaft A riser shaft is a vertical shaft between floors in a multi-story buildings. Such riser shafts carry riser cables. See also riser cable and riser subsystem.

riser subsystem The part of a premises distribution system that includes a main cable route and facilities for supporting the cable from an equipment room (often in the building) to the upper floors, or along the same floor, where it is terminated on a cross connect in a riser closet, at the network interface, or other distribution components of the campus subsystem. The subsystem can also extend out on a floor to connect a satellite closet or other satellite location.

risk A potential liability, caused by a threat.

risk assessment The process of quantifying the potential impact on an organization from various security threats. See Ray Horak's essay at the front of this dictionary on "Disaster Recovery Planning."

risk management Process of identifying, controlling, and eliminating or minimizing uncertain events that might adversely affect system resources.

RISLU Remote Integrated Services Line Unit. One of the remoting arrangements that the Lucent 5ESS switch architecture permits. The RISLU terminates DSLs and connects to the switch DLTU via T-1.

RiSU Remote indoor Service Unit. A cable telephony term for a Remote Service Unit which interfaces the coax-based CATV system to the twisted pair inside wire and cable system in support of voice and high-speed Internet access, as well as cable TV. The RiSU is mounted indoors, and contains a battery backup, as well as an A/C (alternating current) transformer. See RSU for more detail.

RIT Rate of Information Transfer. The amount of information that can be communicated from a sender to a receiver in a given length of time.

RJ Registered Jacks. They're telephone and data plugs registered with the FCC. RJ-XX (where X is a number) are probably the most common plugs in the world. Here is a table of the most common registered jacks, courtesy the FCC. Following the table are descriptions of the most common RJ jacks.

RHCX	Single Tie Trunk, Type I or II E&M interface, 8 position.
RJ-IDC	Single-line, 4-wire, T/R, T1/R1, 6-position.
RJ-11C/W	Single-line, 2-wire, T/R, 6 position.
RJ-14C/W	Two-line, 2-wire, T/R, T(MR)/R(MR), T(OPS)/R(OPS) 6-position.
RJ-14X	Two-line, T1/R1, T2/R2, with sliding cover, 6-position.
RJ-15C	Single-line, T/R, weatherproof, 3-position.
RJ-17C	Single-line, T/R, used in hospital critical care areas, 6-position.
RJ-18C/W	Single-line, T/R, with Make Busy leads, 6 positions.
RJ-2DX	12 lines, 4 wire, T/R, T1/R1, 50 positions.
RJ-2EX	12 Tie trunks, 2-wire, T/R, E&M Type I, 50 position.
RJ-2FX	8 Tie trunks, 2-wire, T/R, E&M SG/SB Type II 50 position.
RJ-2GX	8 Tie trunks, 4-wire, T/R, T1/R1, E&M, Type I 50 position.
RJ-2HX	6 Tie trunks, 4-wire, T/R, T1/R1, E&M, SG/SB, Type II, 50 positions.
RJ-2MB	12 lines, 2-wire, T/R, Make Busy leads, 50 position.
RJ-21X	25 lines, 2-wire, T/R, 50 position.
RJ-25C	3 lines, 2-wire, T/R, T (MR)/R(MR), T(OPS)/R(OPS), 6 position.
RJ-26X	8 lines, 2-wire, T/R, FLL, or Programmed data, 50 position.
RJ-27X	8 lines, 2-wire, T/R, Programmed Data, 50 position.
RJ-4MB	Single-line, 2-wire, T/R, MB/MB1, PR/PC, with Make Busy. 8 position, keyed and programmed.
RJ-41M	Up to 8 multiple installations of FLL or Programmed Data. 8 position, keyed.
RJ-41S	Single-line, 2-wire, T/R, FLL or Programmed Data, 8 position, keyed.
RJ-45M	Up to 8 multiple installations of Programmed Data. 8 position, keyed.
RJ-45S	Single-line, 2-wire, T/R, PR/PC, programmed data, 8 position, keyed.
RJ-48C	Single-line, 4-wire, T/R, T1/R1, 1.544 Mbps, 8 position.
RJ-48H	Up to 12 lines. 4-wire, T/R, T1/R1, 1.544 Mbps, 50 position.
RJ-48M	Up to 8 lines, 4-wire, T/R, T1/R1, 1.544 Mbps, 50 position.
RJ-48S	One or two lines. T/R or T/R, T1/R1, LADC or subrate. 8-position, keyed.
RJ-48T	Up to 25 (2-wire) or 12 (4-wire)/ T/R OR T/R, T1/R1; LADC or Subrate, 50-position.
RJ-48X	Single-line, 4-wire, T/R, T1/R1, 1.544 Mbps, 8-position with shorting bar.
RJ-61X	Up to 4 lines, T/R, 8-position.
RJ-M8	Single private line, 2/4 wire, T/R or T/R, T1/R1, Non-registered service, 8-position, keyed, w/wo loopback.

RJ-11 RJ-11 is a six conductor modular jack that is typically wired for four conductors (i.e. four wires). The RJ-11 jack (also called plug) is the most common telephone jack in the world. The RJ-11 is typically used for connecting telephone instruments, modems and fax machines to a female RJ-11 jack on the wall or in the floor. That jack in turn is connected to twisted wire coming in from "the network" – which might be a PBX or the local telephone company central office. In a home installation, the red and green pair would be used for carrying the phone conversation and the black and white might be used for carrying low voltage from a plugged-in power transformer to light buttons on the phone. In many offices, the tip and ring were used for the voice conversation and the black and white were used for signaling. Increasingly, these days more and more office phone systems use only one pair, i.e. the red and green conductors. A friend always the green conductor for TIP and the red for RING. In a touchtone phone, the green and red conductors are color specific. If you have them wired the wrong way, often you can't generate touchtones. Simple solution: Reverse the wiring.

See also RJ-22 and RJ-45.

RJ-12C and RJ-12W These jacks are normally associated with one line of a key telephone system. They provide a bridged connection to the tip and ring of the telephone line and to key system A and A1 leads. The tip and ring conductors in the jack are connected ahead of the key telephone-system line circuit. The RJ-12C is surface- or flushmounted for use with desk telephone sets while the RJ-12W is for wallmounted telephone sets. Typically, these arrangements are used when registered ancillary equipment must respond to central office or PBX ringing.

RJ-13C and RJ-13W Jacks normally associated with one line of a key telephone system. They provide a bridged connection electrically behind the key-system line circuit to the tip and ring conductors and to the A and A1 leads. The RJ-13C is surface- or flush-

mounted for use with desk tele-phone sets while RJ-13W is for wallmounted telephone sets. These arrangements are generally used when the registered ancillary equipment does not require central office or PBX ringing to function properly.

RJ-14 A jack that looks and is exactly like the standard RJ-11 that you see on every single line telephone. Whereas the RJ-11 defines one line – with the two center, red and green, conductors being tip and ring, the RJ-14 defines two phone lines. One of the lines is the "normal" RJ-11 line – the red and green conductors in the center. The second line is the second set of conductors – black and yellow – on the outside. The RJ-14C is surface- or flushmounted for use with desk telephone sets while the RJ-14W is for walmounted telephone sets.

RJ-15C The RJ-15C is a weatherproof jack arranged to provide single-line bridged connection to tip and ring. Jack RJ-15C can be arranged for surface- or flushmounting depending upon customer needs.

RJ-16X A providing a single-line bridged tip and ring and is associated with -9 dBm (permissive) data arrangements that require mode indication for use with exclusion key telephone sets. The exclusion key telephone set requires a series jack, RJ-36X (described under 8-position jacks) as its normal means of connection.

RJ-17C A jack that provides a single-line bridged connection of tip and ring to special telephone sets or ancillary equipment (e.g., ECG machines) in hospital critical-care areas. Only registered equipment conforming to Article 517 of the 1978 National Electrical Code is permitted to connect to this jack arrangement. This jack differs from the RJ-11C in that tip and ring appear on pins 1 & 6 rather than 3 & 4.

RJ-18C A jack providing a bridged connection of single-line tip and ring with make-busy leads MB and MB1. When the registered equipment provides a contact closure between the MB and MB1 leads, a make-busy indication is transmitted to the network equipment busying out the line from further incoming calls. It's recommended that the busy indication (contact closure) be provided while the line is in the idle state to reduce the possibility of interfering with a call that is in the ringing or talking state. The RJ-18C is surface-or flushmounted for use with desk telephone sets.

RJ-19C A jack normally associated with one line of a key telephone system. It provides a bridged connection of single-line tip and ring behind a key-system line circuit, with A and A1 lead control, and a direct connection for MB/MB1 make-busy leads. When the modem provides a contact closure between the MB and MB1 leads, a make-busy indication is transmitted to the network equipment busying out the line from further incoming calls. It's recommended that the busy indication (contact closure) be provided while the line is in the idle state in order to reduce the possibility of interfering with a call that is in the ringing or talking state. The RJ-19C is surface or flushmounted for use with desk telephone sets.

RJ-21 Same as an RJ-21X. See RJ-21X.

RJ-21X An Amphenol connector under a different name. Here's the explanation: Amphenol is a manufacturer of electrical and electronic connectors. They make many different models, many of which are compatible with products made by other companies. Their most famous connector is probably the 25-pair connector used on 1A2 key telephones and for connecting cables to many electronic key systems and PBXs. The telephone companies call the 25-pair Amphenol connector used as a demarcation point the RJ-21X. The RJ-21X connector is made by other companies including 3M, AMP and TRW. People in the phone business often call non-amphenol 25-pair connectors, amphenol connectors. The RJ-21X is often used with Traffic-Data Recording Equipment and Multiple-Lien Communications Systems. The user must specify the connection sequence for each title appearing in the jack.

RJ-22 RJ-22 is a four position modular jack that is typically used for connecting telephone handsets to telephone instruments. It is always wired with four conductors (also called wires). It is different and slightly smaller than the more common RJ-11 which is typically used for connecting telephone instruments, modems and fax machines to a female RJ-11 jack on the wall or in the floor. That jack in turn is connected to twisted wire coming in from "the network" – which might be a PBX or the local telephone company central office. See RJ-11 and RJ-45.

RJ-22X This jack is associated with a telephone company-provided key telephone system when connection to several lines is required. It provides bridged connections of up to 12 telephone lines and their associated A and A1 leads. The tip and ring conductors in the jack are wired ahead of the lien circuit in the key telephone system. This arrangement is used when the modem must respond to central office or PBX ringing.

RJ-23X This jack is normally associated with a telephone company-provided key telephone system when connection is required to several lines. It's wired to provide bridged connections of up to 12 key-system line circuits and associated A and A1 leads. It differs from and is preferred over the RJ-22X, in that tip and ring conductors in the jack are wired behind the key-system line circuits. This arrangement is typically used when the modem doesn't require central office or PBX ringing to function properly.

RJ-24X This jack is normally associated with a telephone company-provided key telephone system. It's typically used with registered ancillary devices such as conferencing devices, music on hold, etc., and is wired to provide the same tip, ring, A, and A1 appearances as a standard five-line key telephone set.

RJ-25C RJ-25C provides for bridged connection to the tip and ring conductors of three separate telephone lines. The telephone company will wire the lines to the jack in the sequence designated by the customer. The RJ-25C is surface- or flushmounted for use with the desk telephone sets and ancillary devices.

RJ-26X An RJ-26X is a multiple-line universal data jack for up to 8 lines in a 50-position miniature ribbon connector and accommodates either fixed-loss loop (FLL) or programmed (P) types of data equipment. A switch, accessible to the customer, is provided on each line to select FLL or P type of operation. FLL equipment transmits at -4 + 1 dB with respect to one milliwatt and a pad is included in the data jack so that pad loss plus loop loss is nominally 8 dB. Programmed-type data equipment adjusts its output power in accordance with a programming resistor in the data jack. By these means, signals from either FLL or P types of registered data equipment will arrive at the local telephone company central office at a nominal -12 dB with respect to one milliwatt for optimum data transmission.

RJ-27X An RJ-27X is a multiple-line programmable data jack for up to 8 lines in a 50-position miniature ribbon connector and accommodates programmed data equipment only.

RJ-31X An RJ-31X provides a series connection to the tip and ring conductors of a telephone line. It's wired ahead of all station equipment electrically and is typically used with registered alarm-reporting devices. When there's an alarm condition, the registered device functions to cut off all station equipment wired behind it, via this jack.

RJ-32X Provides a series connection to the tip and ring conductors of a telephone line. It differs from RJ-31X in that it's wired ahead of a particular telephone set rather than ahead of all the station equipment. It's typically used with registered automatic dialers.

RJ-33X Is normally associated with a key telephone system. It provides a series connection to the tip and ring conductors of the telephone line and the key- system lien circuit A and A1 leads. The tip and ring conductors are wired ahead of the key-system line circuit. This arrangement is typically used when the modem requires central office or PBX ringing.

RJ-34X Is normally associated with a key telephone system. It's wired to provide a series connection to the key-system line circuit tip and ring conductors and it's a and A1 leads. It differs from RJ-33X in that all conductors are wired behind the key-system line circuit. This arrangement is typically used when the modem is not critical as to type of ringing signal or doesn't require central office or PBX ringing.

RJ-35X Is normally associated with a key telephone set. It's wired to provide a series connection to the tip and ring conductors of the telephone line and a bridged connection to the A and A1 leads. It differs from RJ-33X and RJ-34X in that the tip and ring leads are connected to the common wiring behind the pickup keys of the station set but ahead of the switch hook. The jack is wired to the key telephone set so that the modem functions on the line selected on the key telephone set.

RJ-36X Provides a connection for a registered telephone set equipped with an exclusion key when the telephone line is also to be used with a registered data set or registered protective circuitry. It's wired to provide a series connection to the tip and ring conductors of the telephone line and mode- indication leads MI and MIC. With this jack, the exclusion key can be used to transfer the telephone line between the modem and the telephone set. As a customer option, the exclusion key may be wired so that either the telephone set or the modem controls the line. In the former case, the exclusion key must be operated to transfer the telephone line to the modem. In the latter case, the telephone line is normally associated with the modem. Operation of the exclusion key is required to transfer the line to the telephone set. In either case, a closure on the MI and MIC leads indicates the voice mode.

RJ-37X Is used for providing two-line service with exclusion. The jack is wired to provide a bridged connection to the tip and ring conductors of two telephone lines with exclusion on line 1.

RJ-38X Provides a series connection to the tip and ring conductors of a telephone line identical to those described for RJ-31X. However, the jack also provides a continuity circuit which is used as an indication that the plug of the registered equipment is engaged with the jack. The jack is wired ahead of all station equipment electrically and is typically used with registered alarm dialers.

RJ-41M and RJ-45M Provide a multiple-mounting arrangement for mounting a number of RJ-41S or RJ-45S Single-Line Universal or Programmed data jacks. The telephone companies will terminate USOCs, RJ-41M and RJ-45M with RKM2X (which is the USOC equivalent for a mounting arrangement) and the appropriate number of RJ-41S or RJ-45S single-line data jacks as required by the user. The mounting arrangement will accommodate up to 16 single- line data jacks. In effect, this arrangement provides the features of a patch panel. The user has complete flexibility in patching the color and plug from any modem to any line. The arrangement can be mounted on a wall or on 19- or 23-inch relay racks.

RJ-41S Is a single-line universal data jack normally associated with fixed-loss loop (FLL) or programmed (P) modems. A switch, accessible to the user, is provided to select FLL to P type of operation (FLL equipment transmits at - 4 dB with respect to one milliwatt, and a pad is included in the data jack so that pad loss plus loop loss is nominally 8 dB Programmed modems adjust their output power in accordance with a programming resistor in the data jack. By these means, signals from either FLL or P types of registered modems will arrive at the local telephone company central office at a nominal -12 dB with respect to one milliwatt for optimum data transmission.) A sliding cover is provided to keep dirt and dust from entering the jack when it's not in use. The FLL/P switch selects the desired method of operation. Two matted surfaces are provided on the housing of the jack for the telephone company installer to write in the loop loss (designated LPL) and the telephone line number (designated TLN).

RJ-45 The RJ-45 is the 8-pin connector used for data transmission over standard telephone wire. That wire could be flat or twisted. And it's very important that you know what you're working with. You can easily use flat wire for serial data communications up to 19.2 Kbps. Up to that speed you're connecting with your wire to a data PBX, a modem, a printer or a printer buffer. If you wish to connect to a 10Base-T local area network, which you also do with a RJ-45, you must use twisted wire. You can typically tell the difference by looking at the cable. If it's flat grey satin (like a typical phone wire, only bigger) than it's probably untwisted. If it's circular, then it's probably twisted and therefore good for LANs. RJ-45 connectors come into two varieties – keyed and non-keyed. Keyed means that the male RJ-45 plug has a small, square bump on its end and the female RJ-45 plug is shaped to accommodate the plug. A keyed RJ-45 plug will not fit into a female, non-keyed (i.e. normal) RJ-45. See RJ-11 and RJ-22.

RJ-45S Is a single-line data jack normally associated with programmed (P) modems. This jack is the same as the universal data jack RJ-41S described above, except that the pad for fixed loss loop (FLL) equipment and the switch to select FLL or P type of operation are omitted. Its appearance is the same as RJ-41S except that RJ-45S does not have the FLL/P switch. Both jacks provide bridged connections to the tip and ring of a telephone line and provide mode-indication leads for use with exclusion key telephone sets when required. The exclusion key telephone set requires a series jack RJ-36X as its normal means of connection.

RJ-48C An 8-position keyed plug most commonly used for connecting T-1 circuits. The RJ-48C is an 8-position plug with four-wires (two for transmit, two for receive) commonly connected. When the phone company delivers T-1 to your offices, it usually terminates its T-1 circuit on a RJ-48C. And it expects you to connect that RJ-48C to your phone system or T-1 channel bank and then to your phone system.

RJ-48S Normally associated with DDS services from the telephone company, this jack is used with DDS CSU/DSUs.

RJ-71C An RJ-71C provides a multiple series arrangement of tip and ring. It's typically used with registered series devices such as toll restrictors, etc. Jack RJ-71C can accommodate up to 12 circuits per jack (i.e., one tip and ring "in" and one tip and ring "out", 4 leads per circuit). This arrangement does not currently provide restoration upon disconnection of registered equipment. Thus, a manual bridging plug is provided in order to maintain circuit continuity upon withdrawal of a registered plug.

RJ-8 A coaxial cable with a transmission impedance of 50 ohms. It's largely used for data, not video. See also RJ-58.

RJ-A1X and RJ-A3X RJ-A1X and RJ-A3X are adapters used to adapt 4- and 12- position jacks, respectively, to a 6-position miniature bridged jack. They provide bridged connections to the tip and ring of the telephone line. If A and A1 leads are already terminated in the 4- or 12-pin jack, they will appear in positions 2 and 5 in the adapter. If A and A1 leads are not involved, positions 2 and 5 are reserved for telephone company use.

RJ-A2X An RJ-A2X is an adapter that coverts a single miniature jack to two miniature jacks. It provides a bridged connection to the tip and ring conductors of the telephone line. If A and A1 leads are already terminated in an existing miniature bridged jack, they will appear in positions 2 and 5 both miniature bridged jacks in the adapters. If A and A1 leads are not provided, positions 2 and 5 are reserved for telephone company use.

RJ-A3X See RJ-A1X.

RJE Remote Job Entry. A Remote Job Entry terminal is used for the transmission of "batch" data to a remote computer system. Processed information is then returned to the printer in the terminal. This type of processing from a remote site is a standard method of data transmission. See IBM.

RJXXX Registered Jack.

RLC ReLease Complete message. The sixth ISUP call set-up message. A message sent in either direction in response to the receipt of a Release Message, or if appropriate to, a Reset Circuit Message when the circuit concerned has been brought into the idle condition. See ISUP and Common Channel Signaling.

RLCM Remote Line Concentrating Module.

RLCS Remote Live Call Screening. A Panasonic phone feature. RLCS is even better than live call screening feature. It lets you plug a single cordless phone into the External Data Port (XDP) on your Panasonic phone and monitor incoming voice mail remotely. When a call ends up in voice mail you use your cordless phone to screen your call. To take the call simply hit the same key again. This way you are always assured you will receive the important call you were waiting for – even if you're in a meeting. This feature is only available on the Panasonic Super Hybrid PBX with its integrated digital Panasonic voice mail system. But I liked the feature enough to include it in my dictionary.

RLE See Run Length Encoding.

RLEC Rural Local Exchange Carrier. Defined by the FCC as an independent phone company serving rural communities, small towns, etc but could be adjacent to a major metropolitan area. Government protects these companies by making them permitted monopolies which are extremely difficult to compete with both because of government regulation and economics. Cetain rules apply which determine whether or not an ILEC qualifies as a RLEC.

RLL 1. Radio in the Local Loop. Another term for WLL (Wireless Local Loop). RLL/WLL is a means for CLECs to deploy local loop capability rapidly, bypassing the incumbent LEC in the process. SEE WLL and CLEC for more detail.

2. Run Length Limited. A type of data coding used for disk drives. The term Run Length Limited derives from the fact that the techniques limit the distance (run length) between magnetic flux reversals on the disk platter. An RLL certified hard drive can use an MFM controller card but the storage capacity and the data transfer rate will be reduced.

Rlogin Rlogin is an application that provides a terminal interface between UNIX hosts using the TCP/IP network protocol. Unlike Telnet, Rlogin assumes the remote host is (or behaves like) a UNIX machine.

RLP Radio Link Protocol.

RLR Receive Loudness Rating.

RLS Release. You find a button labeled RLS on many phones. It means to hang up, or to end the call that you're presently on.

RLT Release Link Trunks.

RM 1. Resource Management. A mechanism used by the explicit-rate flow control scheme defined by the ATM Forum. Special control cells are used for explicit rate marking, with those cells taking up as much 3% of the network capacity, but serving to quickly convey information about network congestion back to the source (transmitting device). Explicit rate marking requires that the ATM switch mark the ABR RM (Available Bit Rate Resource Management) cells with the maximum transmission speed which can be supported over a VC (Virtual Circuit). Refer to RM-cell.

2. Reference Model, as in OSI-Rm. See OSI.

RM-Cell An ATM term. Resource Management Cell: Information about the state of the network-like bandwidth availability, state of congestion, and impending congestion, is conveyed to the source through special control cells called Resource Management Cells (RM-cells).

RMA Returned Merchandise Authorization. A code number provided by the seller as a prerequisite to returning product for either repair or refund. An indispensable tracking procedure, it operates like a purchase order system. If you return computer or telephone equipment without an RMA number, chances are that it will be lost. The manufacturer will deny they ever received it. And you will be out of pocket and blamed. RMA is also called a Return Authorization (RA) number.

RMAS Remote Memory Administration System.

RMATS Remote Maintenance and Test System. That equipment and programming used to run, maintain and test a telephone system remotely – usually by dialing in on a special phone line.

RMON Remote MONitoring specification. A simple network management protocol used to manage networks remotely. It provides multi vendor interoperability between monitoring devices and management stations. RMON, is a set of SNMP-based MIBs (Management Information Bases) that define the instrumenting, monitoring, and diagnosing of local area networks at the OSI Data-Link layer. In IETF RFC 1271, the original RMON, which is sometimes referred to as RMON-1, defines nine groups of Ethernet diagnostics. A tenth group, for Token Ring, was added later in RFC 1513. RMON uses SNMP to transport data. To be RMON-compliant, a vendor need implement only one of the nine RMON groups. See MIB,

RMON-2, RMON Probe, RMON Token Ring and SNMP.

RMON-2 The second Remote MONitoring MIB standard, called RMON-2, defines network monitoring above the Data-Link layer. It provides information and gathers statistics at the OSI Network layer and Application layer. Unlike the original RMON, RMON-2 can see across segments and through routers, and it maps network addresses (such as IP) onto MAC addresses. RMON-2 is currently a proposed standard under IETF RFC 2021. To be compliant with RMON-2, a vendor must implement all the monitoring functions for at least one protocol. RMON-2 does not include MAC-level monitoring, and thus it is not a replacement for the original RMON. See RMON.

RMON Groups The original IETF proposed standard for the RMON MIB, RFC 1271 defines nine Ethernet groups: Ethernet Statistics, Ethernet History, Alarms, Hosts, Host Top N ("N" indicates that it collects information on a number of devices), Traffic Matrix, Filters, Packet Capture, and Events. RFC 1513 extends this standard to support Token Ring. See MIB, RMON, RMON Token Ring. b

RMON Probe Sometimes called an RMON agent, an RMON probe is either firmware built into a specific network device like a router or switch, or a specific device built for network monitoring and inserted into a network segment. An RMON probe tracks and analyzes traffic and gathers statistics, which are then sent back to the monitoring software. Historically, an RMON probe was a separate piece of hardware, but now RMON firmware is embedded in high-end switches and routers. See RMON.

RMON Token Ring IETF proposed standard RFC 1513 is an extension to the original RMON MIB (RFC 1271), with support for Token Ring. Some sources refer to this standard as RMON TR, but it's generally considered a replacement for the older standard. In RFC 1513, the RFC 1271 Statistics and History monitoring groups have additional specifications for Token Ring, and a tenth group is added to monitor ring configuration and source routing. In 1994, the proposed standard became a draft standard under the designation RFC 1757; many vendors use the RFC 1513 and 1757 numbers interchangeably. See MIB, RMON.

RMP See Roving Monitor Port.

RMR See Repairman Revisit.

RMS Root Mean Square. Method of measuring amplifier power.

RMS-D1 Remote Measurement System Digital 1.

RMU Remote Mask Unit.

RNA 1. Ring No Answer: Open state in modem or Out of Band (OOB) diagnostic troubleshooting. Occurs when the remote modem is in a hung state and refuses to train (synch up) when prompted by an incoming call / request. When the modem "trains," it's establishing connectivity with the inbound RTS/CTS transaction.

2. Remote Network Access. Another term for RAS, Remote Access Server.

RNC Radio Network Controller.

RNCRN Reserved Number Charge.

RNR Abbreviation for not ready to receive.

RO 1. Receive Only.

2. An ATM term. Read-Only: Attributes which are read-only can not be written by Network Management. Only the PNNI Protocol entity may change the value of a read-only attribute. Network Management entities are restricted to only reading such read-only attributes. Read-only attributes are typically for statistical information, including reporting result of actions taken by auto-configuration.

ROA Recognized Operating Agency. An ITU (International Telecommunications Union) term. Any operating agency which operates a public correspondence or broadcasting service and upon which the certain obligations are imposed by the member state (i.e., nation) in whose territory the head office of the agency is situated, or by the member state which has authorized this operating agency to establish and operate a telecommunication service on its territory. An Operating Agency (OA) is defined as any individual, company, corporation or governmental agency which operates a telecommunication installation intended for an international telecommunication service or capable of causing harmful interference with such a service. Examples of ROAs are at AT&T, Deutsche Telekom, Telstra, and MCI Worldcom.

roach theory Coackroaches have this wonderful characteristic. Kill one. Another seems to come along. Kill that one. Another one appears. When bad news starts coming from a company you're invested in, more bad news often follows. Then bad news often follows. Bad news is not something that seems to happen in a vacuum. Wags on Wall Street have called bad news the Roach Theory. When you get a little bad news, you often get more and more, just like roaches.

road kill Road kill is an awful name for the dead animals we see along the roads and highways, killed by passing motorists. In the language of the wired world, road kill are companies that don't make it. They may be killed by their own stupidity or they may be killed by their largest competitor dropping his prices dramatically. See also Road Pizza.

road map Software that enables easier navigation to information desired. Especially helpful for Internet users.

road pizza Companies who get run over by their competitors. Also a code name for a previous new product from Apple Computer.

road warrior A businessperson who travels...a lot. The road warrior's office is an airplane seat or a hotel room, and his weapons are a laptop with a broadband access card, Wi-Fi hardware and communications software and a Treo or BlackBerry. Road warriors have a love/hate relationship with the "road." There is also a corridor warrior. They are the employees and executives, who spend their day racing from meeting to meeting, tethered to laptops so they can retrieve even the most basic of information, take notes, and remain linked to the rest of the world via e-mail. There is a second definition, namely in-house, IT service technicians who spend their days going from one office to another, fixing PC, LAN, and network problems. Corridor warriors use cell phones, BlackBerry devices or other wireless devices for messaging throughout the day as they roam the corporate corridors.

roadblocks Slang for legislation, or lack thereof, which inhibits rather than promotes the growth of the markets using interactive multimedia.

roadcasting A streaming network that, if realized, would allow motorists to broadcast in-car digital music libraries and podcasts to other vehicles within a 30-mile radius. See also podcasting.

roadhog Any company trying to dominate the information highway through control of pieces of the infrastructure.

ROADM Reconfigurable optical add/drop multiplexer. A device that lets a service provider remotely deploy, activate, reconfigure and in other ways manage wavelengths across a metropolitan optical network. See MAN.

ROADS Robust Open Architecture Distributed Switching.

roam See roaming.

roamer fraud Roamer fraud is the illegal use of another carrier's cellphone service without being charged. See Roaming.

roaming Using your cellular phone in a place besides the one you live in. Roaming usually incurs extra charges. Roaming prices may or may not be at a premium to local home area prices. See Preferred Roaming List and Roaming Agreement

roaming agreement An agreement between wireless carriers that allows their subscribers to use their phones on other wireless carriers.

roaming dial-back The ability of the Dialup Switch to dial-back the user at a roaming location.

roaming user profile A server-based user profile that is downloaded to the local computer when a user logs on and that is updated both locally and on the server when the user logs off. A roaming user profile is available from the server when logging on to a workstation or server computer. When logging on, the user can use the local user profile if it is more current than the copy on the server.

roamops The Roaming Operations working group of the IETF (Internet Engineering Task Force). Roamops is developing a standard to define an authentication method for Internet users and a billing process across ISPs. The objective is to allow travelers to use a local ISP in another city or country, much like you can roam with your cell phone for an additional charge. If you use a local ISP, you currently must either place a long distance call to your local ISP's POP (Point of Presence), where your e-mail sits in a mailbox on his mail server–much like you must place a long distance call to access your voice mail. Alternatively, you can use a national/international provider to gain access, or forward your e-mail from your local ISP to a national/international provider. All three of these alternatives are cumbersome, troublesome, and expensive, to say the least.

There also exist several roaming alliances which do the same thing, but on a proprietary basis. GRIC (Global Roaming Internet Connection) and I-Pass Alliance both act as clearing houses, much like a bank clearing house for ATM transactions, with each participating ISP setting its own rates for roaming transactions. The large telco ISPs, including both LECs and IXCs, also are establishing a proprietary roaming alliance. See also GRIC and I-PASS.

robbed-bit signaling This explanation from Gary Maier of Dianatel: ISDN is the key to future sophisticated telephone network services with its dynamic, highly configurable T-1 connection (also called PRI connection). Since T-1 is a common method of carrying 24 telephone circuits, many wonder about the uses for ISDN, especially when they learn ISDN signaling requires an entire voice channel, reducing today's T-1 from 24 voice channels to 23. But the popular signaling mechanism of "robbed bit" signaling in T-1 has serious limitations. Robbed bit signaling typically uses bits known as the A and B bits. These bits are sent by each side of a T-1 termination and are buried in the voice data of each voice channel in the T-1 circuit. Hence the term "robbed bit" as the bits are stolen from the voice data.

Since the bits are stolen so infrequently, the voice quality is not compromised by much. But the available signaling combinations are limited to ringing, hang up, wink, and pulse digit dialing. In fact, the limitations are obvious when one recognizes DNIS and ANI information are sent as DTMF tones.

This introduces a problem: time. Each DTMF tone requires at least 100 milliseconds to send, which in a DNIS and ANI situation with 20 DTMFs will take at least two full seconds. There is also a margin for error in transmission or detection, resulting in DNIS or ANI failures. With the explosion of telephone related services, the telephone companies are turning to ISDN PRI to provide the more complicated and exact signaling required for new services. ISDN employs a more robust method of signaling. ISDN uses a T-1 circuit as 23 voice channels and one signaling channel. The term 23B plus D refers to 23 bearer (voice) channels and 1 Data (signaling) channel. The data channel carries the signaling information at a rate of 64 kilobits per second. This speed is many times greater than some of the most powerful modems available. Because of this high speed, telephone calls can be placed more quickly, and because of the protocol used, DNIS or ANI transmission failures are impossible.

Additionally, since no bits are "robbed" from the voice channels, the voice quality is better than that of Robbed Bit signaling on today's T-1 circuits. Also, computer modems and high speed faxes can use the voice channel for sending digital data instead of the traditional analog bit "noise." Therefore, ISDN PRI offers the end user countless new service capabilities. One channel could be used for faxing, another for modem data, several for video, another for a LAN and the remainder for voice. Suddenly, the average T-1 circuit becomes a pipeline for all communications! Increasingly long distance carriers are using ISDN PRI to provide inbound 800 calls with ANI and DNIS and re-routing skills. See Bit Robbing.

ROBO Remote Office / Branch Office market. See also SOHO and SOMO.

robo-call An automated phone call, typically used for delivering a telemarketing pitch or a political campaign message.

robot Automaton, or humanoid machine. From the Czech "robata," meaning "drugery." The idea first was introduced by Karel Capek in his 1921 play "R.U.R.," for "Rossum's Universal Robots." Science fiction writer Isaac Asimov made them famous with his 1950 story "I, Robot," and continuing through a series of books known as the "Robot Series." Robot has been abbreviated to "bot," meaning a software agent which does the bidding of its master (i.e., you) in accomplishing Web searches and such. Also called agent, spider and crawler. See also Bot.

robot virusing hacking into an Internet-connected robot and infecting it with a virus. Wired magazine believes the proactice will spread as more businesses and homes begin using Web-controlled robot security systems.

robust A term used by telecommunications switch manufacturers to describe the alleged hefty quality of their network connections – especially their switch-to-host links. The price of their link is often directly related to the number of times the manufacturer uses the word "robust" in a customer presentation.

Robust Header Compression ROHC. There is an IETF initiative to develop generic compression schemes for packet headers on cellular network transmissions in order to improve throughput. Existing header compression schemes (RFC 1144, RFC 2508) have technical limitations and performance problems. The goal of ROHC is for generic, "future-proof" header compression schemes that perform across multiple networks and on links with high error rates. ROHC conceivably could involve multiple compression schemes, for example, some that are layer-specific. ROHC is important for reducing packet overhead for VoIP traffic in wireless environments in general, where voice payloads may be as small as 32 bytes (or even much less during periods of silence), whereas an IPv4 packet header is 40 bytes and an IPv6 header is 60 bytes. Without ROHC, that means the ratio of overhead to payload is ugly. ROHC can yield a packet header around 3 or 4 bytes in size.

Robust Security Network for WLANs. See 802.11i.

ROC Ring Operations Center. An ROC is telecommunications carriers' center that tests the circuits that that carrier is trying to turn up. The next generation after the NOC, the Network Operations Center.

rock The name of one of XM Satellite Radio's two satellites. The other is called Roll. XM has agreed to merge with its competitor, Sirius.

rock ridge format A set of CD-WO (write-once CDs) specifications to provide directory structures that may be updated as additional files are added. The specifications include: System Use Sharing Protocol (SUSP) and the Rock Ridge Interchange Protocol Specification (RRIPS). The specifications are extensions of the ISO 9660 format for CD-ROM. The SUSP extension to the ISO 9660 standard allows multiple file system extensions to coexist on one CD-ROM disc. The RRIP specification lets POSIX files and directories be recorded on CD-ROM without requiring modifications to files, such as shortening file and directory names.

rodent rubber Another term for a B connector. See B Connector.

ROE Regulatory commission authorized allowed rate of return on equity. See Revenue Requirement.

ROFL Abbreviation for "Rolling On the Floor, Laughing;" commonly used on E-mail and BBSs (Bulletin Board Systems).

Rogers, Wil Will Rogers, who died in a plane crash with Wylie Post in 1935, was probably the greatest political sage this country has ever known. Some of his statements:

Never slap a man who's chewing tobacco. + Never kick a cow chip on a hot day. + There are two theories to arguing with a woman...neither works. + Never miss a good chance to shut up. + Always drink upstream from the herd. + If you find yourself in a hole, stop digging. + The quickest way to double your money is to fold it and put it back in your pocket. + There are three kinds of men: The ones that learn by reading. The few who learn by observation. The rest of them have to pee on the electric fence and find out for themselves. + Good judgment comes from experience, and a lot of that comes from bad judgment. + If you're riding' ahead of the herd, take a look back every now and then to make sure it's still there. + Lettin' the cat outta the bag is a whole lot easier'n puttin' it back. + After eating an entire bull, a mountain lion felt so good he started roaring. He kept it up until a hunter came along and shot him.   The moral: When you're full of bull, keep your mouth shut.

rogue As an adjective, it means bad. Rogue software is bad software – software which would seriously hurt your computer. As a noun, a rogue means a scoundrel. See rogue buying.

rogue buying Rogue buying occurs when employees buy supplies for their company from unauthorized suppliers. The concept of rogue buying has enormous meaning for ecommerce. When companies set up Internet and Intranet systems to their suppliers, they do electronic ordering, delivery tracking, inventory checking, etc. They save huge amounts of money ordering this way instead of the old way of calling on the phone, sending faxes, etc. When an employee buys from an unauthorized supplier, the company loses all the dollar savings benefits of the electronic ordering over the Internet or corporate Intranet. The solution is clearly to make the rogue supplier an authorized supplier and get him into the system. There is another solution: fire the employee. But that depends on if the supplier is ready with computers and networking. By the way, it can easily cost $150 to place an order the old way and less than $10 to place an order the new way, through ecommerce. Multiply that by thousands of orders per day for big companies and you can see the savings.

rogue endpoint An unauthorized device on the network, such as a contractor's or consultant's computer, or an unauthorized computer connected to the network by an employee.

rogue IT Hardware and software initiatives by individuals, workgroups, and departments within an organization, without the permission and knowledge of the official corporate IT department. Installing Fax machines are

rogue machine An unauthorized computer on the network.

rogue software See Rogue.

rogue user An unauthorized user on the network, or an unauthorized user who has attached an unauthorized computer to the network or who has bypassed security enforcement mechanisms to gain unauthorized access to network resources.

ROH Receiver Off Hook.

ROHC See Robust Header Compression.

RoHS Directive A European Union directive that restricts the use of certain hazardous substances in new electrical and electronic equipment beyond agreed-upon levels beginning in July 1, 2006. Hazardous substances covered by this directive are lead, cadmium, mercury, hexavalent chromium, polybrominated biphenyl and polybrominated diphenyl ether. Hardware vendors such as Cisco, Nortel, Sun and IBM are in the process of reengineering their products to be RoHS-compliant. A similar regulation is expected to go into effect in California in 2007. See RoHS-compliant.

RoHS-compliant Describes an electrical or electronic component that complies with the RoHS Directive.

ROI Return On Investment. You put $100 into a new venture. It earns $10 in the first year. Your RoI is 10%. Pretty simple concept. There are more complicated ways of figuring how profitable your investment has been, like IRR – internal rate of return and present value. But this is the simplest. Many business plans talk about future ROIs – future returns on your investment. Suffice, no one in the history of mankind has ever been able to figure an accurate ROI for a new business. Yours might be the first. Cross your fingers.

RoIP Radio over IP. A Land Mobile Radio (LMR) system is a collection of portable and stationary radio units designed to communicate with each other over predefined frequencies. They are deployed wherever organizations need to have instant communication between geographically dispersed and mobile personnel. Typical LMR system users include public safety organizations such as police departments, fire departments, and medical personnel.

However, LMR systems also find use in the private sector for activities like construction, building maintenance, and site security.

In typical LMR systems, a central dispatch console or base station controls communications to the disparate handheld or mobile units in the field. The systems might also employ repeaters to extend the range of communications for the mobile users. LMR systems can be as simple as two handheld units communicating between themselves and a base station over preset channels. Or, they can be quite complex, consisting of hundreds of remote units, multiple dispatch consoles, dynamic channel allocation, and other elements.

ROLC Routing Over Large Clouds. Working group in IETF created to analyze and propose solutions to problems that arise when performing IP routing over large, shared media networks, such as ATM, Frame Relay, SMDS, and X.25. See IETF.

Role-Based Access Control See RBAC.

roll 1. Roll is the name of XM Satellite Radio's two satellites. The other is called Rock.
2. See Let's Roll.

roll a truck Send an installer or repairman to the customer's premises.

roll about A totally self-contained videoconferencing system consisting of the codec, video monitor, audio system, network interfaces and other components. These roll about systems can, in theory, be moved from room-to-room but in fact are not because they are electronic equipment that does not benefit from jostling. It's also heavy.

roll call polling A technique in which every station is interrogated sequentially by a central computer system.

roll call A teleconferencing term. In a Dial-Out (operator-initiated) teleconference, the operator will announce each of the participants as he adds them to the conference bridge.

roll off frequency Corner frequency beyond which the attenuation increases rapidly.

rollover A Web design page term. Imagine you're on a web site and you move your cursor over a piece of art, or even a blank spot. Suddenly, that piece of art changes and reveals an explanation or new menu, or some words or a new piece of art. A rollover is a piece of software which changes the appearance of objects when you roll over them. A rollover is also call a mouseover. See also Busy Out and Rollover Lines.

rollover lines You receive many incoming calls. You don't want to miss a call, so you ask your phone company to set your phone lines up to roll over, also called hunt, also called ISG (Incoming Service Group) in telephonese. You order five lines in hunt. The calls come into the first. If the first one is busy, the second rings. If it's busy, the third rings. If they're all busy, then the caller receives a busy. The commonest types of hunting are sequential and circular hunting. Sequential hunting starts at the number dialed, keeps trying one number after another in number order and ends at the last number in the group. It's typically ascending. For example, it starts at 691-8215, goes to 691-8216, then 691-8217, etc. But it can also be ascending – from 691-8217 up. Circular hunting hunts all the lines in the hunting group, regardless of the starting point. Circular hunting, according to our understanding, circles only once (though your phone company may be able to program it circle a couple of times). The differences between sequential and circular are subtle. Circular seems to work better for large groups of numbers. You don't need consecutive phone numbers to do rollovers. Nowadays you can roll lines forwards, backwards and jump around, for example most idle, least idle. Rollovers are now done in software. This also has its downside, since software fails. For example, theoretically if a rollover strikes a dead trunk, it should bounce to the next live trunk. But sometimes it hangs on the dead trunk and many of your incoming calls never get answered. They might ring and ring. They might hit a busy. My recommendation: Test your rollovers at least twice a day. In particular, test that your callers ultimately get a busy if all your lines are busy. Nothing worse your customer should receive a ring-no-answer or a constant busy when calling your company. See also Terminal Number.

ROLM A telephone equipment manufacturer based in Santa Clara, CA, at least once upon a time. ROLM was started in 1969 by four engineers to produce computers for the military. The company introduced one of the first digital PBXs in 1975. It was a great PBX. Later, they developed a line of KTSs (Key Telephone Systems) and hybrid PBX/KTS systems. They were not so good. IBM acquired ROLM in 1984 as part of their plan to integrate the worlds of computers and communications. It didn't work...at all. And IBM lost a lot of money with Rolm. In 1989, IBM sold ROLM to Siemens, at which time it became ROLM Company. In 1994, the name was changed to Siemens Rolm Communication Inc. In 1996, the name was changed to Siemens Business Communication Systems, Inc. Siemens really doesn't use the name ROLM (or Rolm) anymore, but there are a lot of ROLM systems still in service.

rolodex A Trademarked product which started life as paper card based device to keep names and address on. Now it has become more of a generic name to connote software to let you look up peoples' phone numbers and addresses. Software to do this is also called

PIM – for Personal Information Manager. If your software has some important names in it, it's called a Golden Rolodex.

ROM Read Only Memory. Computer memory which can only be read from. New data cannot be entered and the existing data is non-volatile. This means it stays there even when power is turned off. A ROM is a memory device which is programmed at the factory and whose contents thereafter cannot be altered. In contrast is the device called RAM, whose contents can be altered. See Read Only Memory and Microprocessor.

ROM Font The ROM Font is your PC's type font. It consists of a set of 256 characters which cannot be edited – unless you are running in video mode, in which case you can design your own type font.

Rom shadowing 386 and higher CPUs provide memory access on 32 & 64 bit paths. Often they will use a 16 bit data path for system ROM BIOS info. Also some adapter cards (ie. older video, network adapters etc.) with on board BIOS may use an 8 bit path to system memory. For high end computers this is a bottleneck. Like having YIELD signs out on the lanes within a freeway. ROM is very slow, 150ns-200ns. Modern RAM is 60ns or less. Therefore when the system is waiting on this data it generates wait states. For high end computers these wait states slow the entire system down. There is a system developed to transfer the contents of all the slow 8-16 bit ROM chips through out the system into 32 bit faster main memory. "This is ROM SHADOWING". This is accomplished using the MMU, the memory management unit. The MMU takes a copy of the ROM BIOS codes and places it into RAM. To the rest of the system this RAM location looks exactly like the original ROM location. This definition courtesy Charlie Irby, chasirby@foothill.net.

romaji Romaji is a system of writing Japanese using the Latin alphabet.

ROxEEntgen, Wilhelm Conrad A German scientist, (1845 - 1923), who fortuitously discovered X-rays in 1895.

roofing filter A low-pass filter used to reduce unwanted higher frequencies.

room cut-off Hotel/motel guest telephones restricted from outgoing calls when the guest room is unoccupied.

room status and selection Provides the capability to store and display the occupancy and cleaning status and the type number of each guest room. This helps house-keeping management, maid locating and room selection. Also, communications between the front desk and the housekeeper are speeded up via real-time maid activity and checkout audit printouts to indicate which rooms need cleaning next. The occupancy status is normally changed by the maid or inspector dialing from the room telephone.

root The base of a tree. The base of a hard disk. See Root Directory.

root account Privileged account on UNIX systems used exclusively by network or system administrators.

root bridge Exchanges topology information with designated bridges in a spanning-tree implementation to notify all other bridges in the network when topology changes are required. This prevents loops and provides a measure of defense against link failure.

root CA Ultimate Certificate Authority (CA), which signs the certificates of the subordinate CAs. The root CA has a self-signed certificate that contains its own public key.

root certificate Certificate for which the subject is a root. Hierarchical PKI usage: The self-signed public-key certificate at the top of a certification hierarchy.

root directory The top-level directory of a PC disk, hard or floppy. The root directory is created when you format the disk. From the root directory, you can create files and other directories.

root key Public key for which the matching private key is held by a root.

root mean square RMS. The effective value of an alternating current or voltage.

root server 1. A name server that functions at the top-level node in a network's hierarchical name space.

2. A name server that contains authoritative data approved by the Internet Assigned Numbers Authority (IANA) for the highest level of the Internet's Domain Name System (DNS). The DNS root servers are lettered A through M.

root server operator An entity that manages a root server in the Internet's Domain System. Currently the following twelve operators manage DNS' thirteen root servers (lettered A through M) worldwide: (A) VeriSign Naming and Directory Services, (B) Information Sciences Institute, (C) Cogent Communications, (D) University of Maryland, (E) NASA Ames Research Center, (F) Internet Systems Consortium, Inc., (G) U.S. DOD Network Information Center, (H) U.S. Army Research Lab, (I) Autonomica/NORDUnet, (J) VeriSign Naming and Directory Services, (K) Reseaux IP Europeens - Network Coordination Centre, (L) Internet Corporation for Assigned Names and Numbers, (M) WIDE Project.

root server system The Internet's thirteen DNS root servers, lettered A through M. It should be noted that each root server is actually a federation or cluster of multiple servers. See root server operator.

root web The FrontPage web that is provided by the server by default. To access the root web, you supply the URL of the server without specifying a page name. FrontPage is installed with a default root web named <\<>root web>. All FrontPage webs are contained by the root FrontPage web.

rope lay conductor A conductor composed of a central core surrounded by one or more layers of helically laid groups of wires.

ROSE Remote Operations Service Element. An application layer protocol that provides the capability to perform remote operations at a remote process. Definition from Bellcore (now Telcordia) in reference to its concept of the Advanced Intelligent Network.

RosettaNet A consortium of information technology, electronic components and semiconductor manufacturing companies working to create and implement open e-business process standards.

ROSI Return on security investment. Value of network security compared to the economic consequences of a security breach.

rostered staff factor RSF. A call center term. Alternatively called an Overlay, Shrink Factor or Shrinkage. RSF is a numerical factor that leads to the minimum staff needed on schedule over and above base staff required to achieve your service level and response time objectives. It is calculated after base staffing is determined and before schedules are organized, and accounts for things like breaks, absenteeism and ongoing training.

rostering A call center term. The practice of rotating employees through all existing schedules in a matrix, or roster, of schedules. This "share the grief" method is prevalent in Europe and Australia, where agents work through an entire roster.

ROT13 A way to encode things that the general Internet community can't read. Each letter in a message is replaced by the letter 13 spaces away from it in the alphabet. There are online decoders to read these. For instance, Harry Newton becomes Uneel Arjgba, which sounds a lot more exotic.

ROTA A call center term. 1. An European term for a rotating shift pattern or rotating schedule, 2. Short form for roster.

rotary dial The circular telephone dial. As it returns to its normal position (after being turned) it opens and closes the electrical loop sent by the central office. Rotary dial telephones momentarily break the DC circuit (stop current flow) to represent the digits dialed. The circuit is broken three times for the digit 3. The CO counts these evenly-spaced breaks and determines which digit has been dialed. You can hear the "clicks". The number "seven," for example consists of seven "opens and closes," or seven clicks. You can dial on a rotary phone without using the rotary dial. Simply depress the switch hook quickly, allowing pauses in between to signify that you're about to send a new digit. It's a good party trick.

rotary dial calling The telephone system will accept dialing from conventional rotary dial sets.

rotary hunt You buy several phone lines. Let's say 212-691-8215, 212-691-8216, 212-691-8217, 212-691-8218. Someone dials you on your main number – 212-691-8215. It's busy. (That's our number.) The central office slides the call over to 212-691-8216. If that number is busy, it slides it over to 212-691-8217, and so on. This is called rotary hunt. It hunts to the next line in the rotary group. In the old days, the phone lines you could rotary hunt to had to be in numerical sequence. But now with modern stored program control central offices, your lines in rotary hunt can be very different as long as they're all on the same exchange.

rotary output to central office Most central offices are equipped to provide tone dial service. In cases where the telephone company central office trunks are not designed to accept tone signaling, your on premise phone system (PBX, key system or single line phone) will translate the number entered by a phone in tones into rotary dial pulses which can be processed by the central office.

rotary splice A mechanical splice using glass capillary ferrules.

rotating cylinder (Drum) scanner A scanning technique using a drum and a photocell scan head. The original is attached to the drum, enabling the scan head to travel along the length of the document. Reflected light from the document is concentrated on the scanner photocell, which causes an analog signal.

rotating helical aperture scanner Original is illuminated by a lamp when fed onto the platen, via a mirror and lens system, the document's image is focused first through a fixed horizontal slot, then through a rotating spiral slit disk series, and finally onto a photocell to generate an analogous electrical current.

rotational latency The delay time from when a disk drive's read/write head is on-track and when the requested data rotates under it.

rotational mailboxes Information only mailboxes whose information is automatically changed on a time sensitive or usage sensitive basis.

ROTFL I'm "Rolling on the Floor, Laughing." Used in e-mail.

ROTL 1. Remote Office Test Line. Provides the capability to originate automatic inter office trunk transmission test calls under the automatic control of CAROT from a remote location.

2. A popular online abbreviation, shorthand for "Rolling On The Floor Laughing"; an appropriate typed response to a particularly amusing online remark. Other common Net acronyms include IMHO ("In My Humble Opinion") and IMNSHO ("In My Not-So-Humble Opinion").

rotor The rotating part of a motor or other electrical machines.

ROTS Rotary Out-Trunk Switches.

round cutter These are used to cut cables. The blades of the cutter are curved so that there is a space between them.

round robin This is a method of distributing incoming calls to a bunch of people. This method selects the next agent on the list following the agent that received the last call. See also Top Down and Longest Available.

Round-Trip Time See RTT.

roundtrip propagation delay Roundtrip propagation delay from a burst modem to a burst modem will be about 470 milliseconds to 570 milliseconds (About half a second). See Satellite Transmission Delay.

routable protocols Protocols, such as TCP/IP, DECnet, and XNS, that support Network Layer addressing. Packets constructed using these protocols contain information about how data should move through a network. This information, carried in the NLA (Network Layer Address) field of the packet, is used by internetworking devices to make routing decisions.

route The path that a message takes. In telephone companyese, a route is the particular trunk group or interconnected trunk groups between two reference points used to establish a path for a call. This term (or the term routing) is also used as a verb to define the act of selecting a route or routes.

Route 66 A colloquial term for the Internet, with something interesting everywhere along the way. Route 66 used to be the way we drove across America before they put in concrete highways.

route advance This feature routes outgoing calls over alternate long distance lines when the first choice trunk group is busy. The phone user selects the first choice route by dialing the corresponding access code. The phone equipment automatically advances to alternate trunks and trunk groups, based on the user's class of service. Route advance is a more primitive form of least cost routing. See Least Cost Routing.

route caching A type of load sharing in a which an application is assigned to a particular one of several parallel transmission circuits.

route control Route control technology lets a company identify the most efficient ISP and sends Internet traffic to that ISP. Border Gateway Protocol (BGP) routes are continually updated by route control devices to maintain optimal traffic paths. This is how it works: Route control devices measure real-time, end to end performance for each ISP. Traffic flows over the fastest path, as determined by the route control device. Devices update BGP routes. ISP performance and traffic results are reported by the route control device.

route daemon A program that runs under 4.2 or 4.3BSD UNIX systems (and derived operating systems) to propagate routes among machines in a local area network. Pronounced "route-dee."

route discovery Process through which a brouter can learn LAN topology by passing information about its address and the LANs it connects and receiving the same information from others.

route diversity See Diversity.

route flap Route flap describes the impact of frequent changes in state (i.e., condition and availability) of Internet routes. The changes in state are generated by routers that sense (either correctly or incorrectly) that there are problems across one or more routes that connect them to their peers. In such an event, they generate route change messages that are sent to their peers. Route flaps can be caused by events such as BGP (Border Gateway Protocol) session resets between routers, changes in state (e.g., on or off) of a router, changes in state (e.g., up or down) of a circuit, changes in router filter lists, and high error rates over a particular link or circuit interconnecting two routers. These frequent changes cause a state of confusion as packets are routed first this way and then another way, like a flag flapping in the wind.

route guide A map showing how calls are to be routed at the switch: first choice group of circuits, second choice, third choice, and fourth choice.

route hut A physical site along the a route, particularly a fiber route, where the transmission is boosted.

route indexing Provision of Interim Number Portability through direct trunks

equipped for CCIS/SS7 operation, which are provisioned between end offices of Verizon and a CLEC. Inbound traffic to a ported number is routed over these trunks.

route indicator An address or group of characters in the heading of a message defining the final circuit or terminal to which the message is to be delivered.

route length The actual length of a route, or path, between transmitter and receiver. The length of the route is one factor that determines propagation delay, i.e., the delay associated with a signal as it transverses a network. Route length is determined by measuring the actual length of a path, rather than the distance "as the crow flies."

route list A sequence of trunk groups that can be searched for a particular route. This list is comprised of trunk groups and configuration attributes (e.g. Class of Service) governing the use of a particular trunk group.

route mile Let's say that you have two sheaths of fiber, each of which contains ten fibers and runs for one mile. That is one route mile (total distance of all fibers), two sheath miles (2 sheaths running one mile), and twenty fiber miles (20 fibers running one mile).

route optimization 1. In voice communications, route optimistization is another another way of saying least cost routing, which is

2. In data communications – such as the Internet - the concept refers to devices called route optimizers which peer into the Internet and fathom efficient ways of sending their information through it. Some route optimizers use the BGP – Border Cateway Protocol. Other simpler ones take advantage of utilization thresholds, minimizing the need for BGP expertise. Route optimizers are used by webmasters especially when they maintain multiple, distant identical web servers. There's a fine art to using route optimizers on the Internet. They're often difficult to manage, but they're probably invaluable if you have a large network.

route optimizer From Network World: Organizations with branch offices are consolidating their networks to control the costs of managing their IT infrastructures. But because of latency posed by WAN connections, it's difficult to provide remote-site users with the network performance they require while ensuring the integrity of centralized data. Remote-office optimization appliances circumvent these challenges by enabling real-time access of data over WANs. This is accomplished by combining optimization techniques such as compression, dynamic caching, binary delta calculation and transaction aggregation. By deploying remote-office optimizers, organizations can offer their users real-time, synchronous access to centralized resources, ensuring the integrity and coherency of organizational data and user productivity levels at branch offices. A remote-optimizer appliance is installed at each branch and in the corporate data center. Because the solution must be bidirectional, optimizing traffic to and from a data center and remote-branch offices, devices are required in both. Once the remote-office optimizers are installed, administrators can centralize all organizational data at their data centers or headquarters. At the same time, file servers, storage and back-up resources can be removed from branches. To provide branch-office users with experiences comparable to working with local file servers at headquarters, remote-office optimizer appliances facilitate the retrieval and saving of data over WANs. When the remote user saves a file it is forwarded to the data center and saved to the central file server. The file also is saved to the remote-office optimizer's virtual server. Upon notification of the save from the central file server, the data center optimizer notifies the remote-office optimizer, which notifies the client. All messages and notices are issued by the data center optimizer, not the remote-office optimizer. When the user saves changes to the file, the remote-office optimizer compares the new file with the previously saved file. Only the changes from the previously saved file, the delta, is sent across the WAN. When other users in the branch access the same file, they are served by the remote-office optimizer. If the file was changed, the delta is sent by the device in the data center, which knows the state of the file on the branch device. The record on the remote-office optimizer is updated with the deltas as it serves the file to the user. Subsequent changes will be compared with this updated record.

route server Physical device that runs one or more network layer routing protocols and uses route query protocol to provide network layer routing forwarding descriptions to clients.

route Xpander card A board manufactured by IBM for insertion into a PC which provides the PC with a wide area interface to a frame relay network, including handling all of the necessary protocol encapsulation.

routed protocol A protocol that can be routed by a router. To do so a router must understand the logical internetwork as perceived by that routed protocol. Examples of routed protocols include DECnet, AppleTalk, and IP.

router 1. As in software, router is a system level function that directs a call to an application.

2. As in hardware, routers are the central switching offices of the Internet and corporate Intranets and WANs. Routers are bought by everybody – from backbone service providers to

local Internet Service Providers (ISPs), from corporations to Universities. The main provider of routers in the world is Cisco. It has built its gigantic business on selling routers – from small ones, connecting a simple corporate LAN to the Internet, to corporate enterprise wide networks, to huge ones connecting the largest of the largest backbone service providers. A router is, in the strictest terms, an interface between two networks.

Routers are highly intelligent devices which connect like and unlike LANs (Local Area Networks). They connect to MANs (Metropolitan Area Networks) and WANs (Wide Area Networks), such as X.25, Frame Relay and ATM. Routers are protocol-sensitive, typically supporting multiple protocols. Routers most commonly operate at the bottom 3 layers of the OSI model, using the Physical, Link and Network Layers to provide addressing and switching. Routers also may operate at Layer 4, the Transport Layer, in order to ensure end-to-end reliability of data transfer.

Routers are much more capable devices than are bridges, which operate primarily at Layer 1, and switches, which operate primarily at Layer 2. Routers send their traffic based on a high level of intelligence inside themselves. This intelligence allows them to consider the network as a whole. How they route (also called routing considerations) might include destination address, packet priority level, least-cost route, minimum route delay, minimum route distance, route congestion level, and community of interest. Routers are unique in their ability to consider an enterprise network as comprising multiple physical and logical subnets (subnetworks). Thereby, they are quite capable of confining data traffic within a subnet, on the basis of privilege as defined in a policy-based routing table. In a traditional router topology, each router port defines a physical subnet, and each subnet is a broadcast domain. Within that domain, all connected devices share broadcast traffic; devices outside of that domain can neither see that traffic, nor can they respond to it. Contemporary routers have the ability to define subnets on a logical basis, based on logical address (e.g., MAC or IP address) information contained within the packet header, and acted upon through consultation with a programmed routing table. In addition to standalone routers developed specifically for that purpose, server-based routers can be implemented. Such routers are in the form of high-performance PCs with routing software. As software will perform less effectively and efficiently than firmware, such devices generally are considered to be less than desirable for large enterprise-wide application, although they do serve well in support of smaller remote offices and less-intensive applications. Routers also are self-learning, as they can communicate their existence and can learn of the existence of new routers, nodes and LAN segments. Routers constantly monitor the condition of the network, as a whole, in order to dynamically adapt to changes in network conditions.

- Characteristics of routers can include:
- LAN Extension
- Store & Forward
- Support for Multiple Media
- Support for Multiple LAN Segments
- Support for Disparate LAN Protocols
- Filtering
- Encapsulation
- Accommodation of Various and Large Packet Sizes
- High-Speed Internal Buses (1+ Gbps)
- Self-Learning
- Routing Based on Multiple Factors
- Route Length
- Number of Hops
- Route Congestion
- Traffic Type
- Support for a Community of Interest (VLAN)
- Redundancy
- Network Management via SNMP

Router protocols include both bridging and routing protocols, as they perform both functions. Those protocols fall into 3 categories:

1. Gateway Protocols establish router-to-router connections between like routers. The gateway protocol passes routing information and keep alive packets during periods of idleness.

2. Serial Line Protocols provide for communications over serial or dial-up links connecting unlike routers. Examples include HDLC, SLIP (Serial Line Interface Protocol) and PPP (Point-to-Point Protocol).

3. Protocol Stack Routing and Bridging Protocols advise the router as to which packets should be routed and which should be bridged.

This definition courtesy of "Communications Systems & Networks," the best-selling book by Ray Horak, my Contributing Editor. To buy the book, www.amazon.com. See also

Bridges, Hubs, Internetworking and Switches.

router-based firewall A router-based firewall is a packet-filtering router. Not everyone agrees that a packet-filtering router alone is a firewall. Many people insist that only a system that includes a dual-homed gateway is a firewall. However, other people argue that a packet-filtering router is a firewall because the router meets important firewall criteria: The router is a computer through which incoming and outgoing packets must pass through which only authorized packets can pass.

router droppings The inclusions added to e-mail messages when a server or recipient cannot be found. Cryptic and foul-looking, their meaning is usually impossible to fathom. Also called "daemon droppings."

router flapping Router flapping occurs when a malfunctioning router keeps going in and out of service, forcing neighboring routers to keep updating their routing tables, until all of the processing power is being siphoned off and no traffic is being forwarded, resulting in an Internet brownout. This can occur on all types of backbones, regardless of the architecture, but routed IP networks, which deploy the most routers, are particularly vulnerable.

router protocols Router protocols figure how A formula used by routers to determine the appropriate path onto which data should be forwarded. The routing protocol also specifies how routers report changes and share information with the other routers in the network that they can reach. A routing protocol allows the network to dynamically adjust to changing conditions, otherwise all routing decisions have to be predetermined and remain static.

Open shortest path first (OSPF). A routing protocol that determines the best path for routing IP traffic over a TCP/IP network. OSPF is an interior gateway protocol (IGP) that is designed to work within an autonomous system. It is also a link state protocol that provides less router to router update traffic than the REP protocol (distance vector protocol) that it was designed to replace.

Routing information protocol (RII'). A simple routing protocol that is part of the TCP/I P protocol suite. It determines a route based on the smallest hop count between source and destination. RIP is a distance vector protocol that routinely broadcasts routing information to its neighboring routers and is known to waste bandwidth.

Border gateway protocol (BGP). A routing protocol that is used to span autonomous systems on the Internet. It is a robust and scalable protocol that was developed by the Internet Engineering Task Force (IETF). BGP4 supports the CEDR addressing scheme, which has increased the number of available IP addresses on the Internet. It is estimated that there are more than 60,000 I3GP routes currently on the Internet.

Classless interdomain routing (CIDR). A method for creating additional addresses on the Internet that are given to Internet service providers, which in turn delegate them to their customers. CIDR reduces the burden on Internet routers by aggregating routes, so that one IP P address represents thousands of addresses that are serviced by a major backbone provider. All packets sent to any of those addresses are sent to the ISP (e.g., MCI or Sprint). In 1990, there were about 2,000 routes on the Internet. Five years later, there were more than 30,000. Without CIDR, the routers would not have been able to support the increasing number of Internet sites.

Multiprotocol label switching (MPLS). A specification for Layer 3 switching from the IETF MPLS uses labels, or tags, that contain forwarding information, which are attached to packets by the initial router. The switches and routers down the road examine the label more quickly than if they had to look up a destination addresses in a routing table. When fully implemented on the Internet, MPLS is expected to deliver the quality of service required to adequately support real-time voice and video as well as service level agreements (SLAs) that guarantee bandwidth to customers.

Resource reservation protocol (RSVP). A communications protocol that signals to a router to reserve bandwidth for real-time transmission. RSVP is designed to clear a path for audio and video traffic, eliminating annoying skips and hesitations. It has been sanctioned by the I EIF, because audio and video traffic is expected to increase dramatically on the Internet.

router rip A Cisco term. This command enables the RIP (Routing Information Protocol) routing process on the router for TCP/IP.

router switches A new breed of routers that in addition to routing TCP/IP packets (Internet packets) also routes cells, frames and other types of packets. See also Router.

routine A program, or a sequence of instructions called by a program, that has some general or frequent use.

routing The process of selecting the circuit path for a message.

routing area subdomain A cellular radio term. The combined geographic area of all Mobile Data Base Stations (MDBSs) controlled by a single Mobile Data Intermediate System (MD-IS).

routing code 1. Another name for area code. See DN (as in directory number).

2. The combination of characters or digits required by the switching system to route a transmission to its desired destination.

routing computation The process of applying a mathematical algorithm to a topology database to compute routes. There are many types of routing computations that may be used. The Dijkstra algorithm is one particular example of a possible routing computation.

routing constraint A generic term that refers to either a topology constraint or a path constraint.

routing data base Distance table in DNA.

routing domain Group of end systems and intermediate systems operating under the same set of administrative rules. Within each routing domain is one or more areas, each uniquely identified by an area address.

routing flexibility The ability to send information over various network paths to avoid congestion and use portions of a total network that would otherwise be idle.

routing indicator The address or routing code in the beginning of a message which specifies to the network the final circuit or destination of the message.

Routing Information Protocol RIP is based on distance-vector algorithms that measure the shortest path between two points on a network, based on the addresses of the originating and destination devices. The shortest path is determined by the number of "hops" between those points. Each router maintains a routing table, or routing database, of known addresses and routes; each router periodically broadcasts the contents of its table to neighboring routers in order that the entire network can maintain a synchronized database. See also Distance vector Protocol.

routing label The part of a signaling message identifying its destination.

routing metric The method by which a routing algorithm determines that one route is better than another. This information is stored in routing tables. Such tables include reliability, delay bandwidth, load, MTUs, communication costs, and hop count.

routing number pseudo Automatic Number Identification (pANI). A number employed in wireless E-911 call setup that can be used to route the call to an appropriate public service answering point (PSAP). The pANI generally identifies the cell/sector from which the call originates, whereas an ANI carries the actual telephone number of a wireline caller.

routing point A location that a local exchange carrier has designated on its own network as the homing (routing) point for inbound traffic to one or more of its NPA-NXX codes. The Routing Point is also used to calculate mileage measurements for the distance-sensitive transport element charges of Switched Access Services.

routing point port This is the cornerstone term in CTI Link-based computer telephony, especially in call centers.

routing protocol A general term indicating a protocol run between routers and/or route servers in order to exchange information used to allow figuring of routes. The result of the routing computation will be one or more forwarding descriptions. In short, a protocol that accomplishes routing through the implementation of a specific routing algorithm. Examples of routing protocols include BGP, CIDR, IGRP, MPLS, RIP, OSPF and RSVP. For a bigger explanation, see Route Protocols.

routing switcher An electronic device that routes a user-supplied signal (audio, video, etc.) from any input to any user-selected output. Inputs are called sources. Outputs are called destinations.

routing table 1. Incoming Phone Calls: A routing table is a user definable list of steps which are treatment instructions for an incoming call. Ideally these steps should be addressed and the call treatment begun before the call is answered. A routing table should consist of a minimum of steps that include agent groups, voice response devices, announcements (delay and informational) music on hold, intraflow and interflow steps, route dialing (machine based call forwarding). A significant issue in the structure of routing tables is "look-back" capability, where no one previously interrogated resource is abandoned by the system (i.e. an agent group is now ignored, even though an agent is now available, because the ACD does not consider previous steps in the routing table).

2. Outgoing Phone Calls: For a specific calling site, this table lists the long distance routing choices for each location to be dialed. There may be only one choice (route) listed for some or all destinations or there may be several choices for some destinations. (It depends how many outgoing lines and how many outgoing trunk groups you have.) If there are several choices then they will be ranked by some criteria (least cost, best quality, etc.).

3. In data communications, a routing table is a table in a router or some other networking device that keeps track of routes (and, in some cases, metrics associated with those routes) to particular network destinations. See Routing Metric.

Routing Transport Number See RRN.

routing update A message sent from a router to indicate network and associated cost information. Routing updates are typically sent at regular intervals and after a change in network topology.

roving monitor port Switch feature that lets you monitor network traffic on one or more ports via a third-party LAN packet analyzer. RMP can let you change the monitoring and monitored ports via software commands instead of via hardware changes.

row In a table, a horizontal collection of cells.

RP-125 A SMPTE parallel component digital video standard.

RPC Remote Procedure Call. 1. A protocol governing the method with which an application activates processes on other nodes and retrieves the results. A popular paradigm for implementing the client-server model of distributed computing. A request is sent to a remote system to execute a designated procedure, using arguments supplied, and the result returned to the caller. There are many variations and subtleties, resulting in a variety of different RPC protocols.

2. A mechanism defined by Sun Microsystems and described in RFC-1057 that provides a standard for initiating and controlling processes on remote or distributed computer systems.

RPE A sub-switch of a Main Switch-Usually located at another site (across the street, across town, across the country, etc.), but may be located at another end of the same building to conserve cabling. Connected with T-1 or other dedicated interconnections (DS-3, DS-0 tie lines), often by T-1 over microwave (this is rapidly occurring over the Internet using IP). Allows for remote office phones to act as virtual extensions on the Main Switch.

RPG Report Program Generator. A computer language for processing large data files.

RPM Remote Packet Module.

RPN 1. Routing Recording Number. This is the number AT&T assigns to a telephone circuit, especially T-1s, DS-3s, etc. This number is also referred to as an RTN, or Routing Transport Number. It is essentially the same as a BTN (Billed Telephone Number) except that the number is not necessarily an actual telephone number. The first three digits indicate which 4ESS or 5ESS digital switch the circuit is on, as well as whether it is 4ESS or 5ESS.

The second three digits indicate the customer or trunk group the circuit is assigned to.

The last 4 digits are, usually, the DNIS for that particular circuit. If, by chance, the customer does not have a DNIS, I believe AT&T will assign a random number for that particular circuit.

2. Reverse Polish Notation. A calculator using RPN starts with the number you type in, then you hit enter, then you type in another number and the minus sign. Bingo, your screen shows the result of your calculation. In "normal" calculators you'd have to do another step, namely hit the enter button. Frankly, I prefer RPN. It's faster, easier and more logical. Most calculators don't come with RPN, sadly.

RPOA Recognized Private Operating Agency. A term used by the ITU-T to describe those companies designated as operating telephone companies – if the country's phone networks are not run by government-owned administrations, such as the PTTs in Europe. A recognized Private Operating Agency is an organization that handles internetwork communications (e.g., long distance carriers). To identify some RPOAs, you must dial a prefix before your outgoing directory number. An RPOA can also refer to one or more DNICs that will connect two X.25 endpoints. For ISDN X.25, an RPOA is usually the DNIC for the ISDN's long distance carrier. See also ITU.

RPQ Request for Price Quotation. Solicitation for pricing for a specific component, software product, service or system. See also RFQ.

RPR See Resilient Packet Ring. Emerging technology combines packet switched networks with dual rings. See Resilient Packet Ring.

RPS Ring Protection Switching. Nortel's "Introduction to SONET Networking" tutorial handbook (www.nt.com/broadband/reference/sonet_101.html) talks about "Automatic Healing of Failed or Degraded Optical Spans in a Two-Fiber BLSR." The handbook says "in the event of failure or degradation in an optical span, automatic ring protection switching (RPS) reroutes affected traffic away from the fault within 50 milliseconds, preventing a service outage."

RQC Repair and Quick Clean. A term in the industry which repairs telecom equipment. It means all equipment is repaired and fully tested with a burn-in (if required) and an operational systems test. It also includes minor cosmetic cleaning of the unit. Definition courtesy Nitsuko America. See also Like New Repair and Update and Repair, Update and Refurbish.

RR Abbreviation for Ready to Receive.

RRM Radio Resource Management.

RRME Radio Resource Management Entity.

RRN Routing Recording Number. This is the number AT&T assigns to a telephone circuit, especially T-1s, DS-3s, etc. This number is also referred to as an RTN, or Routing Transport

Number. It is essentially the same as a BTN (Billed Telephone Number) except that the number is not necessarily an actual telephone number. The first three digits indicate which 4ESS or 5ESS digital switch the circuit is on, as well as whether it is 4ESS or 5ESS. The second three digits indicate the customer or trunk group the circuit is assigned to. The last four digits are, usually, the DNIS for that particular circuit. If, by chance, the customer does not have a DNIS, I believe AT&T will assign a random number for that particular circuit.

RROCP Restricted Radio Operator's Certificate of Proficiency. Certificate issued to those who successfully complete the Australian Communication Authority's Restricted Radio Operator's examination.

RS 1. Recommended Standard, as in RS-232.

2. Record Separator, in data processing terms.

3. An ATM term. Remote single-layer (Test Method): An abstract test method in which the upper tester is within the system under test and there is a point of control and observation at the upper service boundary of the Implementation Under Test (IUT) for testing one protocol layer. Test events are specified in terms of the abstract service primitives (ASP) and/or protocol data units at the lower tester PCO.

4. Reduced Slope refers to the technical characteristics of optical fibers, specifically the dispersion slope which characterizes the chromatic dispersion of multiwavelength light in a wavelength multiplexed optical transmission path. For a much longer explanation, see Chromatic Dispersion.

RS-170 The EIA (Electronics Industries Association) standard for the combination of signals required to form NTSC monochrome (black and white) video.

RS-170A The EIA standard for the combination of signals required to form NTSC color video. It has the same base as RS-170, with the addition of color information.

RS-232 Also known as RS-232-C and in its latest version EIA/TIA-232-E. RS-232 is actually a set of standards specifying three types of interfaces – electrical, functional and mechanical. These are used for communicating between computers, terminals and modems. The RS-232-C standard, which was developed by the EIA (Electrical Industries Association), defines the mechanical and electrical characteristics for connecting DTE and DCE data communications devices. It defines what the interface does, circuit functions and their corresponding connector pin assignments. The standard applies to both synchronous and asynchronous binary data transmission. The most commonly used RS-232 interface is ideal for the data-transmission range of up to 20 Kbps for up to 50ft. It employs unbalanced signaling and is usually used with 25 pin D-shaped connectors (DB25) to interconnect DTEs (computers, controllers, etc.) and DCEs (modems, converters, etc.). Serial data exits through an RS-232 port via the Transmit Data (TD) lead and arrives at the destination device's RS-232-C port through the Receive data (RD) lead. RS-232-C is compatible with these standards: ITU V.24; ITU V.28, ISO IS2110. Most personal computers use the RS-232-C interface to attach modems. Some printers also use RS-232-C. You should be aware that despite the fact that RS-232-C is an EIA "standard," you cannot necessarily connect one RS-232-C equipped device to another one (like a printer to a computer) and expect them to work intelligently together. That's because different RS-232-C devices are often wired or pinned differently and may also use different wires for different functions. The "traditional" RS-232C plug has 25 pins. With the introduction of the IBM PC AT in the mid-1980s, most PCs and laptops switched to the "new" RS-232-C plug with only nine pins, called the DB-9. This smaller plug does essentially the same thing as its bigger cousin, but you need an adapter cable to connect one to another. They're widely available. See also interface and the RS-232-C diagram. See EIA/TIA-232-E and the APPENDIX for description of the pins and what they do. See also Crossover cable – the name for a specially-wired RS-232 cable which allows two DTE devices or two DCE devices to be connected through serial ports and transmit and receive information across the cable. The sending wire on one end is joined to the receiving wire on the other. In an RS-232 cable, this typically means that conductors 2 and 3 are reversed. Serial connections are being replaced by much faster USB and lately by Firewire. See USB and Firewire.

RS-232 Fax Server A RS-232 fax server is software which connects a network server to a fax machine via an RS-232 port attached to the fax machine. There are not many fax machines with R2-232 so you need to chose carefully. The idea of this arrangement is to let users send faxes directly from their own PC via the fax server via the attached fax machine, or directly from the fax machine. Users can also use the fax machine as a scanner.

RS-250B In telecommunications, a transmission specification for NTSC video and audio.

RS-328 October, 1966 the Electronic Industries Association issues its first fax standard: the EIA Standard RS-328, Message Facsimile Equipment for Operation on Switched Voice Facilities Using Data Communications Equipment. The Group 1 standard, as it later became known, made possible the more generalized business use of fax. Transmission was analog

and it took four to six minutes to send a page.

RS-366 An EIA interface standard for auto dialing.

RS-422 Defines a balanced interface with no accompanying physical connector. Manufacturers who adhere to this standard use many different connectors, including screw terminals, DB9, DB25 with nonstandard pinning, DB25 following RS-530, and DB37 following RS-449. RS-422 is commonly used in point-to-point communications conducted with a dual-state driver. Transmissions can run long distances at high speeds. RS is a standard operating in conjunction with RS-449 that specifies electrical characteristics for balanced circuits (circuits with their own ground leads). RS-422 (now known as EIA/TIA-422) is a balanced electrical implementation of RS-449 for high-speed data transmission. RS stands for recommended standard. The RS-422 is also known as a V11.

RS-422-A Electrical characteristics of balanced-voltage digital interface circuits.

RS-423 A standard operating in conjunction with RS-449 that specifies electrical characteristics for unbalanced circuits (circuits using common or shared grounding techniques). Another EIA standard for DTE/DCE connection which specifies interface requirements for expanded transmission speeds (up to 2 Mbps), longer cable lengths, and 10 additional functions. RS-449 applies to binary, serial, synchronous or asynchronous communications. Half- and full-duplex modes are accommodated and transmission can be over 2- or 4-wire facilities such as point-to-point or multipoint lines. The physical connection between DTE and DCE is made through a 37-contact connector; a separate 9-connector is specified to service secondary channel interchange circuits, when used.

RS-423-A Electrical characteristics of unbalanced-voltage digital interface circuits. RS-423 (now known as EIA/TIA-423) is an unbalanced electrical implementation of RS-449 for RS-232-C compatibility.

RS-449 RS-449 (now known as EIA/TIA-449) is essentially a faster (up to 2 Mbps) version of RS-232-C capable of longer cable runs. RS-449 is another "standard" data communications connector. It uses uses 37-pins and is designed for higher speed transmission. Each signal pin has its own return line, instead of a common ground return and the signal pairs (signal, return) are balanced lines rather than a signal referenced to ground. This cable typically uses twisted pairs, while a RS-232-C cable usually doesn't.

According to Black Box Corp, RS-449 defines functional/mechanical interfaces for DTEs/DCEs that employs serial binary data interchange, and is usually used with synchronous transmissions. It identifies signals (TD, RD, etc.) that correspond with the pin numbers for a balanced interface on DB37 and DB9 connectors. RS-449 was originally intended to replace RS-232-C, in order to improve data-transmission capabilities up to 2 Mbps/200ft. (60M), reduce electrical "crosstalk," and accommodate additional signal functions. RS-232-C and RS-449 were to become interoperable by using electrical interface standards RS-422 and RS-423. But right now RS-232 and RS-449 are incompatible in terms of mechanical and electrical specifications. RS-449 is technically compatible with these standards: RS-530, V.10, V.110, and ITUT X.21 bis.

RS-485 Resembles RS-422 except that associated drivers are tri-state, not dual-state. It may be used in multipoint applications where one central computer controls many different devices. Up to 64 devices may be interconnected with RS-485. RS-485 describes electrical characteristics of a balanced interface used as a bus for master/slave operation. Used in industry for the Process Field Bus, and in telco management networks.

RS-499-1 Addendum 1 to RS-449. (What else?)

RS-530 Supercedes RS-449 and complements RS-232. Based on a 25-pin connection, it works in conjunction with either electrical interface RS-422 (balanced electrical circuits) or RS-423 (unbalanced electrical circuits). RS-530 defines the mechanical/electrical interfaces between DTEs and DCEs that transmit serial binary data, whether synchronous or asynchronous. RS-530 provides a means for taking advantage of higher data rates with the same mechanical connector used for RS-232. However, RS-530 and RS-232 are not compatible. And RS-530 offers the benefits of RS-449 and the efficiency of a 25-pin design. It accommodates data transmission rates from 20 Kbps to 2 Mbps; maximum distance depends on which electrical interface is used. (RS-530 is compatible with these standards: ITU 10, V.11, X.26; MIL-188114; RS- 449.)

RSA 1. Rural Service (or Statistical) Area. The FCC designated 428 rural markets across the United States and licensed two service providers per RSA. See also MSA.

2. Rivest-Shamir-Adleman. A public key encryption algorithm invented in 1977 and named after its inventors. A large number algorithm, RSA is highly secure, as each user finds two large prime numbers ("p" and "q") which are then multiplied together ("p" x "q" = "n"). The public key is "n," and the private key is "p" and "q." RSA key sizes range from 768 to 2,048 bits. As a 2,048 bit key yields 2 to the 2,048th power possible combinations, RSA is highly immune to even the most persistent brute force security attacks. RSA Security Inc. offers a number of security tools based on the RSA core algorithm, of which

over 75 million copies have been licensed. PGP (Pretty Good Privacy) incorporates the RSA core algorithm, as does SET (Secure Electronic Commerce). RSA core technologies are part of existing and proposed standards of ANSI, IEEE, ISO, and ITU. See also Digital Certificate, Encryption, PGP, Private Key, Public Key and Set.

RSB See Repair Service Bureau.

RSC Remote Switching Center.

RSC-S Remote Switching Center-S.

RSE An ATM term. Remote Single-layer Embedded (Test Method): An abstract test method in which the upper tester is within the system under test and there is a point of control and observation at the upper service boundary of the Implementation Under Test (IUT) for testing a protocol layer or sublayer which is part of a multi-protocol IUT.

RSFG An ATM term. Route Server Functional Group: The group of functions performed to provide internetworking level functions in an MPOA System. This includes running conventional interworking Routing Protocols and providing inter-IASG destination resolution.

RSID Residential System IDentifier.

RSL Request and Status Links. A generic term for linking computers and PBXs. Every manufacturer of phone systems is evolving towards open architecture and their own "RSL." The term RSL, which is too passive, is being replaced with PHI (PBX Host Interface), a term coined by Probe Research. Manufacturer PHI names include:

  ACL – Applications Connectivity Link – Siemens' PHI link protocol
  ACT – Applied Computer Telephony – Hewlett Packard's generic application interface to PBXs
  Application Bridge – Aspect Telecommunications' ACD to host computer link
  ASAI – AT&T's Adjunct Switch Application Interface
  CIT – Digital Equipment Corporation's Computer Integrated Telephony (works with major PBXs)
  CSA – Callpath Services Architecture – IBM's Computer to PBX link
  Call Frame – Harris' PBX to computer link
  Callpath Host – IBM and ROLM's CICS-based integrated voice and data applications platform which links to ROLM's 9751
  Callpath – IBM's announced, CICS application link to IBM's CSA, available on the AS400 in April of 1991
  Callbridge – Rolm's CBX and Siemens to IBM host or non-IBM host computer link
  CompuCall – Nortel Networks' DMS central office link to computer interface
  CSP – Nabnasset's Communications Services Platform
  CSTA – Computer Supported Telephony Application, PHI standard from ECMA
  DECags – DEC ASAI Gateway Services. Two-directional link to AT&T's Definity
  DMI – AT&T's Digital Multiplexed Interface, a T-1 PBX to computer interface
  HCI – Host Command Interface. Mitel's digital PBX link to DEC computer
  IG – AT&T's ISDN Gateway (one direction from the switch to the host)
  ITG – AT&T's Integrated Telemarketing Gateway (two directional)
  ISDN/AP – NT's PHI SL1 protocol supports NT's Meridian Link PHI
  Meridian Link – NT's PHI product available on the Meridian PBX
  OAI – Open Application Interface. InteCom's and NEC's PHI
  ONA – Open Network Architecture (for telephone central offices)
  PACT – Siemens' PBX and Computer Teaming, protocols between PBXs and computers
  PDI – Telenova/Lexar's Predictive Dialing Interface
  SAI – Stratus Computer Switch Application Interface
  SCAI – Switch to Computer Application Interface, the name given by T1S1 to PHI
  SCIL – Aristacom's Switch Computer Interface Link Transaction Link
  STEP – Speech and Telephony Environment for Programmers; WANG's link
  Transaction Link – Rockwell's link from its Galaxy ACD to an external computer
  Solid State Applications Interface Bridge – Solid's State Systems' PHI
  Teleos IRX-9000 – Teleos' Intelligent Call Distribution platform
  For more information, see OAI (Open Architecture Interface.)

RSM 1. Remote Switching Module. An AT&T 5ESS switch standalone switching module that supports all line features and routes intra-RSM calls. It is either a single module or a multi module and can be situated up to 150 miles from the 5ESS switch host.

2. Radio Sub-system Management. A wireless telecommunications term. Management of radio channels including timing and frequency as well as all machines between the mobile station and the MSC.

RSRB Remote Source Route Bridging. Source-route bridging over wide-area links.

RSS Real Simple Syndication. RSS is a way of pushing small amounts of content from a web site to your computer without you actually browsing that particular site. RSS is lightweight XML-based software designed for sharing headlines. Think of it as a "What's

New" headline for that site. Originated by UserLand in 1997 and subsequently used by Netscape to fill channels for Netcenter, RSS has evolved into a popular means of sharing content between sites (including the BBC, CNET, CNN, Disney, Forbes, Motley Fool, Wired, Red Herring, Salon, Slashdot, ZDNet, and more). RSS solves myriad problems webmasters commonly face, such as increasing traffic, and gathering and distributing news. RSS can also be the basis for additional content distribution services. My friend Dan Good, who is a big RSS adherent, emails me, "RSS is a new and convenient way to bring web site information to a reader on his computer desktop. Every important site now offers this service. You can quickly skim the new items from the site and if anything interests you, you can open the full page or article with your browser." Each RSS channel can contain up to 15 items (they're all text) and is easily parsed using Perl or other open source software.

RSSI Received Signal Strength Indication.

RST An acronym that stands for readability, strength and tone. It is a multi-digit code that a radio amateur uses for describing various aspects of a received radio transmission's signal quality. The source radio could be another radio amateur's transmission, or it could be a commercial short-wave, AM or FM radio broadcast. Two of the ratings or possibly all three are done subjectively and independently of the others. If the radio receiver has an S-meter, which measures signal strength, it can be used as a basis for determining a rating for strength. If a receiver does not have an S-meter, then the strength rating, like the readability and tone ratings, is done subjectively. Readability is rated on a scale of 1 to 5, with 1 being the worst possible score and 5 being the best possible score. Strength and tone are rated on a scale of 1 to 9, with 1 being the worst possible score and 9 being the best possible score. Tone is only rated for Morse code transmissions. RST codes are a way for a radio amateur to share aspects of their radio-listening experience with other radio amateurs and/or with stations they listen to. See SINPO.

RSTP Rapid Spanning Tree Protocol.

RSU 1. Remote Service Unit. A cable telephony term for a device installed at the customer premises. The RSU functions as a standard Network Interface Device (NID) for termination of the coaxial cable network, and the interface between the twisted pair inside wire and cable system inside the premises. Thereby, the coaxial cable network can support telephone service and high-speed Internet access, in addition to CATV service. The RSU performs the RF transceiver (transmitter/receiver), modulation/demodulation of the broadband RF (Radio Frequency) signals, multiplexing/demultiplexing of the digital signals, analog-to-digital conversion (as required), signaling conversion for subscriber loop operation, and diagnostics for problem isolation. The unit provides a "virtual twisted pair" back to a local switching interface for transparent telephony operation of CLASS services and custom calling features. The RSU communicates with the Host Digital Terminal (HDT) at the head end of the CATV network. The RSU may be in the form of either an indoor or an outdoor unit, and may be powered either over the coax network or, more commonly, is powered locally powered through an A/C (Alternating Current) transformer at the customer premises. An RiSU (Remote indoor Service Unit), a variation on the theme, is locally powered and includes battery backup. A RSUM (Remote Service Unit Multiple) is used to provide service to a MDU (Multiple Dwelling Unit), such as an apartment complex, condominium, or multipurpose high-rise building. See also CATV and Network Interface Device.

2. See Remote Switch Unit.

RSVP The Resource Reservation Protocol (RSVP) is an IETF standard designed to support resource (for example, bandwidth) reservations through networks of varying topologies and media. Through RSVP, a user's quality of service requests are propagated to all routers along the data path, allowing the network to reconfigure itself (at all network levels) to meet the desired level of service. The RSVP protocol engages network resources by establishing flows throughout the network. A flow is a network path associated with one or more senders, one or more receivers, and a certain quality of service. A sending host wishing to send data that requires a certain QoS will broadcast, via an RSVP-enabled Winsock Service Provider, "path" messages toward the intended recipients. These path messages, which describe the bandwidth requirements and relevant parameters of the data to be sent, are propagated to all intermediate routers along the path. A receiving host, interested in this particular data, will confirm the flow (and the network path) by sending "reserve" messages through the network, describing the bandwidth characteristics of data it wishes to receive from the sender. As these reserve messages propagate back toward the sender, intermediate routers, based on bandwidth capacity, decide whether or not to accept the proposed reservation and commit resources. If an affirmative decision is made, the resources are committed and reserve messages are propagated to the next hop on the path from source to destination.

The idea is that for presumably a premium price, RSVP will enable certain traffic, such as videoconferences, to be delivered before e-mail. Today, all traffic on IP networks moves on a first-come-first-served basis and is charged at a flat rate. "In some ways RSVP will change what the Internet is all about, because you'll start to have different qualities of service and differential prices which are new," said Abel Weinrib, a key Internet strategist for Intel Corp. Virtually unknown among the general Internet community, RSVP has been quietly pushing ahead towards becoming acceptable and popular. It is now part of Microsoft's TAPI 3.0. It is being pushed also by Cisco Systems Inc., which makes the routers that direct most Internet traffic, and by Intel, which wants to spur demand for microprocessors by making computers and IP networks more useful for uses like phone calls and video conferencing. In an article I read, Cisco marketing manager Peter Long said RSVP technology would be included in new network software Cisco is delivering. That software controls the routers that direct Internet traffic. Cisco sells more than 80 percent of the routers used in commercial and corporate Internets. Long expects Cisco customers to start using RSVP technology to create what he calls "diamond lanes" on the Internet. "Right now, if there is congestion on the Internet, your traffic sits there, like a car stuck on an onramp," Long said. He said RSVP would act like "a big crane that picks you up and puts you over the other cars," onto these so-called diamond lanes that bypass congested parts of the Net. See TAPI 3.0.

RT 1. Reorder Tone.

2. Remote Terminal. Local loop terminates at Remote Terminal intermediate points closer to the service user to improve service reliability.

3. Remote Termination. A node at which terminates a high-capacity local distribution facility in a DLC (Digital Loop Carrier) scenario. The other end of the circuit is known as a COT (Central Office Termination). See also DLC.

RTAN Real Time ANI.

RTC RunTime Control.

RTCP 1. Real Time Conferencing Protocol. Supports real-time conferencing for large groups on the Internet. It has source identification and support for audio and video bridges/gateways. Supports multicast-to-unicast translators.

2. Real Time Control Protocol. The protocol by which VoIP (Voice over Internet Protocol) QOS (Quality of Service) information is communicated.

RTF Rich Text Format. A way of encoding documents such that the messages include boldface, italics and other limited text stylings. RTF is meant to be a cross-word processing platform such that if you send one RTF document (by email, for example) from one word processor to another word processor, that second word processor will be able to recreate the document's original format. I tend to think of RTF as a cut-down version of a document coded in Microsoft's native DOC format.

RTFM Read The Fantastic Manual. This acronym is often used when someone asks a simple or common question. The word "Fantastic" is usually replaced with an adjective much more vulgar. A friend wrote me about his new digital camera: "Downside is that it's so feature-rich, I often have to RTFM." RTFM is also used also on e-mail, newsgroups, and the Internet. See also Cockpit Problem, Idiot-Proof, and Intuitive.

rTLD Restricted top-level domain name. A top-level domain, such as .biz, .gov, .museum, .name, and .pro, that is only available to registrants who meet certain criteria.

RTLP Reference Transmission Level Point.

RTM 1. Release To Manufacture. It means that work on the software is done and it can go into wholesale and retail distribution.

2. Read The Manual.

3. Released To Manufacturing. When software has gone through its writing, testing and re-testing, it is finally ready for selling to the public. First it must go to manufacturing – sending to the factory for printing instruction manuals and duplicating CDs. Hence the acronym RTM.

RTMP Routing Table Maintenance Protocol. It is the native Appletalk routing protocol. It sends updates out every 10 seconds.

RTN See RRN.

RTNR A British term. Ring Tone No Reply, a telephone call which has not been answered. Typically a telebusiness system will automatically re-dial the number after a pre-determined period.

RTO-IS Ready To Order - In Service.

RTOS. Real-Time Operating System.

RTP 1. Realtime Transport Protocol. An IETF standard for streaming realtime multimedia over IP in packets. Supports transport of real-time data like interactive voice and video over packet switched networks. A thin protocol providing support for content identification, timing reconstruction, loss detection and security. The ARPA DARTnet transcontinental IP network experiments lead to RTPs popularity. Now championed by the Audio/Video Transport (AVT) Working Group. AVT is part of the IETF (Internet Engineering Task Force). RTP does not do resource reservation or quality of service control. It relies on resource reservation protocols like RSVP. See H.323.

2. Routing Table Protocol. Used in Banyan VINES routing with delay as a routing metric.

RTS Request To Send. One of the control signals on a standard RS-232-C connector. It places the modem in the originate mode so it can begin to send. See the Appendix.

RTSoIP Real-time session over IP. VoIP and video over IP are examples of real-time sessions over IP.

RTSE Reliable Transfer Service Element. The OSI application service element responsible for transfer of bulk-mode objects.

RTSP Real Time Streaming Protocol. See Real Time Streaming Protocol.

RTT 1. Radio Transmission Technologies. See G3.

2. Round-Trip Time. Time required for a network communication to travel from the source to the destination and back. RTT includes the time required for the destination to process the message from the source and to generate a reply. RTT is used by some routing algorithms to aid in calculating optimal routes.

RTTU Remote Trunk Test Unit.

RTU 1. Remote Termination Unit, Remote Telemetry Unit or Remote Terminal Unit. Basically an RTU is a little black box connected to some remote gadget. The RTU lets the gadget installed at one end of a analog circuit respond to commands that are sent to it over an analog, dial-up phone line. Imagine an oil pipeline. There's a leak. You have to shut the pipe off – not only at the beginning of it. You have to shut it down at the spot closest to the leak, to minimize the spill. Thus you call that switch and tell it over the phone line and using the Remote Termination Unit at the switch to shut off. It does. Obviously there's also a big electrically driven device that physically turns the handle to shut the oil down.

2. Right To Use. A term manufacturers have invented to stifle the used/secondary market in their equipment. Basically, the manufacturer says "Fine, you can sell your no-longer-needed product to some used equipment dealer. But if someone buys it from the dealer and wants to use it, they have to pay me a Right To Use fee." Without payment of this fee, the manufacturer won't contract to maintain the customer's equipment and certainly won't sell the customer software updates, etc. The right to use fee is exorbitant – typically considerably more than what the product actually sells on the used market for. A better approach for a manufacturer would be to innovate a little more and make the customer wants his new product more than his old price (despite the old product's lower price).

3. Remote Terminal (not terminating) Unit. RTUs are employed by utilities' SCADA (Supervisory Control And Data Acquisition) systems in electric substations or gas/water/steam pumping plants to monitor status/condition and/or metering data and to control operations at a remote site. SCADA systems are not limited to distribution systems; SCADA is also used to manage transmission facilities. Distribution is local delivery to end customers; transmission is backbone transport facilities. The analogies between the electrical, water, gas, and telecommunications infrastructures/networks go on and on. See SCADA.

RTV Real-Time Video. DVI software that implements quick-and-dirty, realtime video compression. Once called "edit-level video," it stores video as only 10 frames per second. Meant for use while developing DVI applications.

RTVXK 900 Blocking Option.

RTVXN 976 Blocking Option.

RU 1. Request Unit or Response Unit. A basic unit of data in SNA.

2. Receive Unit

3. Abbreviation of rack unit. See Rack Unit.

rubber bandwidth A term coined by Ascend, an inverse multiplexer manufacturer, to refer to the ability to support applications needing varying speeds. It breaks the original signal up into 56- or 64-Kbps chunks, and places these separate transmissions on the public switched digital network. See also Inverse Multiplexer.

rubber duck antenna The short, stubby rubber antenna on walkie-talkies, other handheld radios, and most wireless access points.

rubidium A silvery white alkali metal, symbol Rb, that is element 37 in the periodic table. Rubidium is used in atomic clocks for satellites and cellular base stations. See rubidium clock.

rubidium clock A low-cost atomic clock that uses properties of the element rubidium to achieve a clocking accuracy to within one second in 1,000 years. This compares favorably to the best crystal oscillators which are accurate to one second in 10 years and to cesium atomic clocks which are accurate to one second in 1.4 million years but are much more expensive than rubidium clocks. Rubidium clocks are used on board space satellites and in cellular telephone base stations. A rubidium clock also has the advantage of unfettered portability, which makes it useful for synchronizing clocks at different locations. See cesium clock.

ruby test Let's say you have a bundle of optic fibers inside a sleeve buried in the ground for miles or running inside a building. How do you figure out with great accuracy which fiber is the same at the one of the fiber as the other? Easy – the Ruby Test. Send a strong ruby coloured light down the strand. See where it comes out at the other end. Not very sophisticated, but it works.

Rue du Télégraphe French for "Telegraph Road." The name was given to various roads in France alongside of which ran telegraph lines. The various Telegraph Roads and Telegraph Avenues in the United States were so named for the same reason.

ruggedized Cable and equipment that has been engineered with enhanced impact and crush resistance and to withstand the harshest of temperatures. I don't believe there's a standard definition of "ruggedized." What is ruggedized to one manufacturer may be normal to another.

rulemaking number A number assigned to a proceeding after the appropriate Federal Communication Commission Bureau/Office has reviewed and accepted a Petition for Rulemaking, but before the Commission has taken action on the petition. The rulemaking number should appear on all appropriate documents, even those which carry a docket number.

rules based call management This a feature of some phone systems which enables the subscriber to pre-program rules specifying how they want their incoming and outgoing calls to be handled. The subscriber typically defines these rules using their browser over the Internet to speak to a server.

rules based system The most popular way to represent knowledge in an expert system. In general, a rule-based system's knowledge base contains both facts and IF..THEN production rules.

rule of thumb The phrase "rule of thumb" came from an old English law which made it illegal to beat your wife with anything wider than your thumb.

rump network The remnants of an old legacy network that is still in use and hasn't been replaced by the new network.

run To start a software program.

run/stop On a Norstar phone, this feature inserts a delay in a dialing sequence. The delay can be any length of time.

Run Length Encoding RLE. A form of data compression which is semantic-dependent in nature. Such techniques are designed to respond to specific types of local redundancy, such as image representation and processing. RLE is a common technique which involves the scanning of image elements along a scan line or row. As the device scans the image, it identifies redundant data and converts it into a code corresponding to the length of the run of such redundant data. A string of identical bits is indicated by sending only one example, preceded by a control character (e.g., *) and a shorthand description of the number of times it repeats. For example, the character string "HNBBBBBCDDDDEFFF" would be RLE Encoded as "HN*5BC*4DE*3F." The higher the level of redundancy for a particular data string, the greater the efficiency of the process. Fax machines use RLE, identifying runs of black or white dots on the page, encoding the length of the run of redundant data as they scan the document, a line at a time. The data is transmitted in compressed form, using this form of data shorthand, and the process is reversed by the receiving fax machine. The advantage, of course, is that the cost of the call is reduced considerably, as the transmission time is much less. This is particularly so in documents containing a lot of white space. RLE also is used in data processing applications to reduce the amount of processing time involved by compressing sequences of zeros or blanks in data fields. RLE also is used to compress memory-intensive files, such as bitmapped graphics. In such an application, the technique is especially useful for black-and-white or cartoon-style line graphics, as runs of the same color can be replaced with a single character. RLE files are identified by the ".rle" file extension. Commonly, .pcx files are run length encoded, as are .tiff and .bmp files; even though these files retain their own file extensions and formats. See also Compression.

run of the place In the 1981 film Raiders of the Lost Ark, one particular scene consistently brings the house down: Indiana Jones, having survived an elaborate chase through the Casbah, is confronted by a swordsman whipping through a flashy routine with a scimitar. With a look of infinite fatigue and disgust, Indy simply pulls out his gun and blows the bad guy away. That bit flowed not from the pen of a screenwriter but from Harrison Ford's desperation. His need to spend less time on this scene and more in a washroom led to an actor-inspired script change that was ultimately worked into the film. Three months' of shooting in the blazing heat in Tunisia had resulted in a terrible bout of dysentery for Ford. The original scene idea would have needed three days to shoot. Too weak to swing his whip, and not looking forward to another lengthy shoot in 130-degree heat, Ford persuaded Spielberg to try the scene this much shorter way. One could say Ford was given the run of the place.

run time The time it takes to execute a software program. See Runtime.

runt An Ethernet frame that is shorter than the valid minimum packet length, usually caused by a collision. The term is imprecise, and may indicate a collision, collision frag-

ments, a short frame with a valid FCS checksum, or a short frame with an invalid FCS checksum. In a litter of animals, the smallest and weakest animal is typically called the runt. Because the runt is weak, he/she usually ends up sucking his/her sucking hind tit – the nipple that produces the least amount of food. A person who is small and contemptible is also called a runt. See also Runt Frame and Runt Packet.

runt frame A small packet received with FCS or alignment errors. Runt frames are the result of collision occurring on connected segment or among stations connected to attached repeaters. See Runt and Runt Packet.

runt packet A data packet with a legal shorter than required by the IEEE 802.3 standard of 64 bytes or 512 bits. See also Runt and Runt Frame.

runtime 1. A computer term, usually used in the "heavy metal" world of mainframe computers. Runtime (or run time) is the amount of time it takes the CPU (Centralized Processing Unit) to execute a program or perform an operation.

2. A runtime environment is the software that plays back multimedia materials. The runtime material is created by the author. Examples of runtime applications are presentations are training, where the material cannot be edited but only viewed. The runtime software could be a slide show viewer, a software-only video playback application, or a hypermedia runtime document. See Runtime License.

Runtime Control RTC. An SCSA definition. The mechanism by which one Resource Object can influence the behavior of another. Typically used for things such as terminating conditions and speed/volume control.

runtime license A one-time or royalty-based fee paid for the inclusion of runtime code in a replicated product.

rupture in the breaking strength or tensile strength tests for cables, rupture is the point at which the material physically comes apart, as opposed to elongation yield strength, etc.

RUR Repair, Update and Refurbish. A term in the industry which repairs telecom equipment. It means equipment is repaired and updated to current manufacturer's specifications. Also includes minor cosmetic cleaning of metal cabinets, a full diagnostic test with burn-in (if required) and an operational test. Definition courtesy Nitsuko America. See also Like New Repair and Update and Quick Clean.

Rural Health Care Corporation RHCC. A not-for profit corporation formed by the Telecommunications Act of 1996, and operating as the Rural Health Care Division (RHCD) of the Universal Service Administrative Company (USAC), which also was formed by The Act. The RHCC is responsible for distributing up to $400 million per year of funds from the Universal Service Fund (USF). Those funds subsidize the cost of internal wiring, telecommunications services, and Internet access for rural health care organizations. See also Universal Service Administrative Company and Universal Service Fund.

Rural Service Area RSA. An area not included in either an MSA or a New England Country Metropolitan Area for which a common carrier may have a license to provide cellular service.

rural telephone company The Telecommunications Act of 1996 defines a rural telephone company as a local exchange carrier operating entity to the extent that such entity– (A) provides common carrier service to any local exchange carrier study area that does not include either–

(i) any incorporated place of 10,000 inhabitants or more, or any part thereof, based on the most recently available population statistics of the Bureau of the Census; or (ii) any territory, incorporated or unincorporated, included in an urbanized area, as defined by the Bureau of the Census as of August 10, 1993;

(B) provides telephone exchange service, including exchange access, to fewer than 50,000 access lines;

(C) provides telephone exchange service to any local exchange carrier study area with fewer than 100,000 access lines; or

(D) has less than 15 percent of its access lines in communities of more than 50,000 on the date of enactment of the Telecommunications Act of 1996.

RUS Rural Utilities Service. The successor to the REA (Rural Electrification Administration). The RUS is an agency of the U.S. Department of Agriculture (USDA). According to the USDA, "RUS is a vital source of financing and technical assistance for rural telecommunication systems." RUS also provides funding for electric and water programs through public/private partnerships designed to further rural infrastructure development. See also REA.

Russia In the old days of Russian communism, they had an axiom of how business was meant to work, namely, "They pretend to pay us and we pretend to work."

RVA Recorded Voice Announcement. Australian term for a recorded message advising that a number has changed.

RVVP See Revenue Volume Pricing Plan.

RW 1. An ATM term. Read-Write : Attributes which are read-write can not be written by the PNNI protocol entity. Only the Network Management Entity may change the value of a read-write attribute. The PNNI Protocol Entity is restricted to only reading such read-write attributes. Read-write attributes are typically used to provide the ability for Network Management to configure, control, and manage a PNNI Protocol Entity's behavior.

2. Referring to CDs, when you see CD-R, it means that the name of the disc ending in "R" can be used only once. Discs ending in "RW" can be erased and rewritten.

3. Rubber-insulated building wire. Moisture resistant.

RW90 Canadian Standards Association (CSA) designation for a wire or cable with rubber or rubber like insulation suitable for use in wet or dry locations at a maximum temperature of 900xAFC.

RWhois Referral Whois. An experimental distributed whois service intended to replace the centralized Whois model. Work began in April 1995, with an active test bed of RWhois servers established between each of the regional registries in September 1995. The RWhois Operational Development Working Group is a forum for coordinating the deployment, engineering and operation of the RWhois protocol. User authentication will be required and operational procedures will be established. See InterNIC, DNS and Whois.

RX Receive. See TX/RX for detail.

RZ Return to Zero. A method of transmitting binary information where voltage returns to a zero (reference) level after each encoded bit.

RZ Code Return to zero code. A code form having two information states called "zero" and "one" in which the signal returns to a rest state during a portion of the bit period.

S Designation the sleeve or control leads in electromechanical Central Offices which are used to make busy circuits, trunks and subscriber lines, as well as to test for busy conditions. It also designates the sleeve wire on a switchboard cord.

S band 1. The microwave radio band from 2 GHz to 2.4 GHz. The designation S Band applied to World War 2 radar in the range 1.55 GHz to 4.2 GHz. See also S-Band.

2. Short Wavelength Band. The optical band, or window, specified by the ITU-T at a wavelength range between 1460nm and 1530nm (nanometers) for fiber optic transmission systems. See also C-Band, E-Band, L-Band, O-Band, and U-Band.

S interface For basic rate access in ISDN, the S interface is the standard four-wire, 144-Kbps (2B+D) interface between ISDN terminals or terminal adapters and the network channel termination, which is two wires. The S interface allows a variety of terminal types and subscriber networks (e.g., PBXs, LANs, and controllers) to be connected to this type of network. At the S interface, there are 4,000 frames of 48 bits each, per second, for 192 Kbps. The user's portion is 36 bits per frame, or 144 Kbps. Out of that 144 Kbps, the user gets two B channels, each of 64 Kbps, and one D channel of 16 Kbps. The local telephone company usually needs a portion of the D channel for signaling. And often it will sell you a 9.6 Kbps packet switched service carved out of the D channel. See also T Interface, U Interface and ISDN.

S mail address Snail Mail address. Your post office address.

S port Refers to the port in an FDDI topology which connects a single attachment station or single attachment concentrator to a concentrator.

S reference point The reference point between ISDN user terminal equipment (i.e., TE1 or TA) and network termination equipment (NT2 or NT1).

S SEED Symmetric Self Electro-optic Effect Device. A switching device in which signals enter and exit as beams of light, not through electrical contacts. In 1990 AT&T Bell Labs built a general purpose digital optical processor/computer. The device contained 2,048 S-SEED chips which could be accessed simultaneously with separate beams of light. That means, that ultimately, such a computer could process huge amounts of information in parallel.

S video SVHS. This method of bringing video to a device uses two coaxial lines contained within one outer jacket with a Mini DIN 4 circular connector. One coaxial line carries the black and white information; the second coaxial line carries all the color information. This method produces a high quality video image. S-Video avoids composite video encoding, such as NTSC and the resulting loss of picture quality. Also known as Y-C Video.

S-Band Short Wavelength Band. The optical wavelength band range from 1490-1530nm (nanometers). See also C-Band, L-Band, and S Band.

S-BCCH System message-Broadcast Control CHannel. A logical channel element of the BCCH signaling and control channel used in digital cellular networks employing TDMA (Time Division Multiple Access), as defined by IS-136. See also BCCH, IS-136 and TDMA.

S-CDMA Synchronous Code Division Multiple Access (S-CDMA) is specified by for use in broadband CATV networks conforming to the Data-Over-Cable System Interface Specification (DOCSIS) standards. S-CDMA allows multiple cable modems to transmit simultaneously in the same time slot over a broadband upstream RF (Radio Frequency) channel. As many as 128 CDMA symbols can be transmitted simultaneously using 128 orthogonal codes. DOCSIS also provides for an alternative technique known as ATDMA (Advanced Time Division Multiple Access). See also ATDMA, CDMA, DOCSIS and orthogonal.

S-DMB See Digital Multimedia Broadcasting.

S-HTTP Secure Hypertext Transfer Protocol. An extension of HTTP for authentication and data encryption between a Web server and a Web browser. This is another protocol in addition to SSL for transmitting data securely over the Internet and World Wide Web. Whereas SSL creates a secure connection between a client and a server, over which any amount of data can be sent securely, S-HTTP is designed to transmit individual messages securely. SSL and S-HTTP are complementary rather than competing technologies. Both protocols have been approved by the Internet Engineering Task Force (IETF) as a standard. See also SSL.

S-meter A signal strength meter on a radio receiver.

s-video A video signal that separates the "Y" or Luma and "C" or chroma information. The s-video plug provides for better video than the old-fashioned composite signal.

S.100 A software voice processing standard established by the ECTF ((Enterprise Computer Telephony Forum). S.100, published in March, 1996 specifies a set of software interfaces that provide an effective way to develop computer telephony applications in an open environment, independent of underlying hardware. It defines a client-server model in which applications use a collection of services to allocate, configure and operate hardware resources. S.100 enables multiple vendors' applications to operate on any S.100-compliant platform. www.ectf.org. See ECTF, H.100, M.100 and S.100 Media Services API. In Communications Systems Design Magazine, Alan Percy wrote "Realizing features such as call processing, directory services, and unified messaging requires designers to coordinate a complex array of functions, services, and processes in various types of complex telephony equipment. The S.100 application programming interface (API) provides just such a toolset by brokering transactions among hardware and software components in telephony enabled products. S.100 is the result of the work done by the Enterprise Computer Telephony Forum (ECTF), which was founded in 1995 as a non-profit organization of computer telephony vendors, developers, integrators, and users from around the world. The result of this work is an architectural framework, that encompasses software and hardware, allowing a broad collection of services and telephony resources to be combined in sophisticated applications. A key component of this architecture is the S.100 API. This API fosters a development environment that opens the door to new and enhanced telephony applications such as interactive voice response (IVR), enhanced call and automated attendant services, and other useful services. The architecture provides several advantages for developers of computer telephony (CT) applications. The most important advantages include support for multiple applications, improved scaleability, and cross-vendor support. The majority of telephony application products -designed to work with servers dedicated to specific purposes -operate independently of each other, resulting in separate servers and telephone resources. Generally, if a corporation has a fax server, voice mail, and IVR system, and other applications, each resides in a separate server, uses a separate set of telephony resource cards, and requires separate telephone lines from the local PBX. A primary design goal of the S.100 architecture is flexibility, allowing multiple applications to execute and share telephony resources on a common server. S.100 accomplishes this flexibility by implementing an abstraction layer that makes the actual telephony resources and their physical location in the network transparent to the application. As a result, one telephony server can provide the required telephone interface and media processing resources to enable one or more application servers to execute a number of applications at the same time. For small-scale applications, the architecture allows the application and telephony server to be combined into a common server."

S.100 Media Services API Defines a client server model in which applications use a collection of services to allocate, configure and operate hardware resources. Independent of operating system or hardware vendor, it abstracts implementation details of call processing hardware and switch fabrics from the applications themselves. See S.100 and ECTF.

S.300 The final component of the ECTF's set of server specifications, S.300 will compliment S.100R2 released in 1998. See ECTF and S.100.

S.410 S.410 is a Java 99 language expression of the ECTF Media Services API and Architecture. S.410 is part of JTAPI 1.3, which provides an integrated suite of object-oriented APIs for telephony control and media processing. The development of S.410 was a cooperative effort of both ECTF and Sun Microsystems' Java organizations. It is the media package for the Java Telephony API (JTAPI) 1.3 replacing the media package defined in JTAPI 1.2. Closely related to S.100 (the ECTF's C language API), S.410 permits applications to use the high-level, framework-supplied services defined by the ECTF architecture. This means that Java-based S.410 applications and C-based S.100 applications can use the same servers. The complete S.410 specification may be downloaded from www.ectf.org.

S/Key Freely available authentication system, developed at Bellcore (based on paper by Leslie Lamport of DEC) that avoids many types of password attacks.

S/MIME Secure Multipurpose Internet Mail Extension. An emerging de facto security standard for securing all types of e-mail. Increasingly being used in lieu of PGP (Pretty Good Privacy) and PEM (Privacy Enhanced Mail) security techniques, S/MIME has received broad support from the vendor community. Developed by RSA Data Security in 1996, S/MIME was a proprietary security mechanism, which fact led the IETF to reject it for consideration as a standard. In November 1997, RSA announced that it would give up the trademark and other rights to the protocol and the underlying encryption algorithm, leading the IETF to reconsider it as a standard. S/MIME is built on the Public Key Encryption Standard. Digital signatures are used to ensure that the message has not been tampered with during network transit. Digital signatures also provide nonrepudiation, thereby denying senders the ability to deny that they sent a message. The message content is encrypted and enclosed in a digital envelope; the envelope can be opened and the message read only with the use of the recipient's public key, which is sent along with the message. See also Digital Signature, Encryption, MIME, PEM, PGP and Public Key.

S/N See Signal-to-Noise Ratio.

S/T Reference Point An ISDN term. In the absence of the NT2, the user-network interface is usually called the S/T reference point.

S/W Software.

S/WAN Emerging standard for secure firewall-to-firewall communication.

S&P 500 The S & P 500, or Standard and Poor's 500, is an index used to gauge the overall health of the stockmarket, similar to the Dow Jones Index in some ways. The S & P 500 consists of 500 companies selected because they are judged to be leaders in their respective businesses. The S & P has become so well respected in the market, a company's stock value may increase simply because it is added to the S & P 500. The way stocks are chosen for the S & P 500 is a bit mysterious, and not a lot is widely understood about the selection process, but the companies are predominantly United States companies operating in a wide range of businesses. Fundamental analysis plays a role in selection, as does availability of stock on the open market. www.spglobal.com.

S0 Sub zero. A European terms for BRI ISDN. Europeans call it BRA.

S10 register Hayes, the modem people, invented their "Command Set." This com-

mand set lets you control your Hayes compatible modem. In the Command Set there are "S" registers which set how the modem responds to events like answering. Should it answer on the first, second, third, etc. ring. There are 27 registers. The most important S register is S10. This register sets the time between loss of carrier and internal modem disconnect. The factory setting is 1.3 seconds. Drop carrier for 1.3 seconds and your modem will turn itself off. This is long enough for all conditions, except the awful "call waiting" signal you get at hotels and at home. There is a solution: Get your communications software to "go local." Then type ATS10=20. That will increase your S10 register to two seconds. If you have a 300 or 1200 baud you'll have to do this every time you turn on your modem. If you have a 2400 baud modem (the only one to get), you type ATS10=20&W only once. The "&W" writes it into your 2400 baud's non-volatile memory. If you want to check to see if you did it right, type ATS10? That will reply by saying 020. That means 20 tenths of a second, or two seconds. If that still doesn't work for you, increase S10 to three seconds. Other S registers control how long your modem waits for the other end to answer, how long its dialing "pause" is, how quickly it outpulses tones for dialing, etc.

SA Source Address. The address from which the message or data originated. A six octet value uniquely identifying an end point and which is sent in an IEEE LAN frame header to indicate source of frame.

SAA Systems Application Architecture. A set of specifications written by IBM describing how users should interface with applications and communications programs. The idea is to give all software "a common feel" so that training will be less burdensome. According to IBM advertising, "SAA will make it possible for everyone in an organization to access information regardless of its location. What's more, all software written to SAA specifications will provide similar screen layouts, menus and terminology." For a fuller explanation, see Systems Applications Architecture.

SAAL Signaling ATM Adaptation Layer: This resides between the ATM layer and the Q.2931 function. The SAAL provides reliable transport of Q.2931 messages between Q.2931 entities (e.g., ATM switch and host) over the ATM layer; two sublayers: common part and service specific part.

SAAS Software as a service. With SAAS you use a Web browser and the Internet to connect to the software.

SABME Set Asynchronous Balanced Mode Extended.

SAC 1. Single-Attached Concentrator.

2. Service Access Code or Special Area Code. See Service Access Code.

3. Subscriber Acquisition Cost. What it costs a telecom, wireless carrier, Internet Service Provider, satellite or cable TV or other provider (e.g. TiVo) to acquire a new customer. Usually the cost exceeds several months of revenue. SAC might include the cost of a cell phone for a wireless provider. It definitely includes the cost of the marketing and sales. See also Churn.

Sacrificial Host A computer server placed outside an organization's Internet firewall to provide a service that might otherwise hurt the local internal area network's security.

SADL Synchronous Auto Dial Language. Created by Racal Vadic, SADL is a public domain auto-dialing protocol which defines procedures in BSC, SDLC (SNA) and HDLC for PCs and larger computers that wish to control synchronous modems directly under program control. SADL does for synchronous dialing systems what the Hayes "AT" command set has done for the async PC dialing world.

saddle A device for establishing the position of the raceway or raceways within the concrete relative to the screed line, and for maintaining the spacing between the raceways.

SAF-TE SCSI Accessed Fault Tolerant Enclosures specification is a non-proprietary, standardized alert detection and status reporting system for storage subsystems which can send and receive information via a standard SCSI interface. A SAF-TE compliant enclosure is designed to monitor and provide notification to the LAN administrator on the condition of disk drives, power and cooling systems and allow for communication to server-based software agents for network notification. Under the SAF-TE specification, the enclosure is typically implemented as an assignable SCSI target using a low-cost SCSI chip and 8-bit microcontroller. The microcontroller is attached to various alarm sensors and status lights / displays on the enclosure. The enclosure target ID is periodically polled (e.g. every 10-20 seconds) by the host to detect / send changes in status. Disconnect / reconnect and asynchronous event notifications area are not used.

safe area That area in the center of a video frame which is sure to be displayed on all types of receivers and monitors. Televisions and other monitors made at different times and by different companies are slightly different in size and shape, and the outer edge of the video frame (about 10 percent) of the total picture is not produced in the same way on all sets.

safe harbor statement When companies tell you about their idea of the future, they will typically preface their forecast with a "Safe harbor Statement." This suggests that they are protecting you from the storm raging outside the harbor and everything they tell you will be just wonderful. In fact, that the opposite. The people in the safe harbor are the company. They're safe. And you're outside in the storm. Believe what they tell you with a grain of salt. Caveat emptor. In short, a safe harbor statement protects the giver of the information, not the receiver. It's a disclaimer and once you're heard it, you basically can't sue the giver of the information if something goes wrong with your investment. Here's a typical safe harbor statement:

"The Company wishes to take advantage of the Safe Harbor provisions included in the Private Securities Litigation Reform Act of 1995 ("the Act"). Statements by the Company relating to future revenues and growth, stock appreciation, plant startups, capabilities and other statements which are not historical information constitute "forward looking statements" within the meaning of the Act. All forward-looking statements are subject to risks and uncertainties which could cause actual results to differ from those projected. Factors that could cause actual results to differ materially include, but are not limited to, the following: general economic conditions; conditions in the Company's major markets; competitive factors and pricing pressures; product demand and changes in product mix; changes in pricing or availability of raw material, particularly steel; delays in construction or equipment supply; and other risks described from time to time in the Company's filings with the Securities and Exchange Commission."

Here's another one, this contained in a press release. "This press release contains forward-looking statements that involve risks and uncertainties. These forward-looking statements include the ability to attract customers for Precise's products, ability to execute as designed, acceptance of Precise's products in the market place, ability to manage existing and future strategic relationships, ability to successfully integrate the operations of W. Quinn as well as statements regarding the financial performance, strategy and plans of Precise. Precise's actual experience may differ materially from those discussed in the forward-looking statements. Factors that might cause such a difference include the size of the market; timing and acceptance of Precise's products in the market place; the future growth and acceptance of Precise's products in the market place; Precise's ability to predict and respond to market developments; the development, expansion and training of Precise's sales force; risks associated with management of growth; risks associated with existing and future strategic relationships and acquisitions, Precise being held liable for defects or errors in its products; political, economic and business fluctuations in the United States, Israel and Precise's international markets; as well as risks of downturns in economic conditions generally or as a result of recent events, and in the information technology and software industries specifically, and risks associated with competition, and competitive pricing pressures. For a more detailed description of the risk factors associated with Precise, please refer to Precise's Form 10-K Annual Report for the year ended December 31, 2000 and Registration Statement on Form S-3 on file with the Securities and Exchange Commission."

safe mode A way of starting Windows using basic files and drivers only, without networking. Safe mode is available by pressing the F8 key when prompted during startup. This allows you to start your computer when a problem prevents it from starting normally. Sometimes Windows will start in safe mode all by itself. This happens usually when there's a hardware or driver conflict or problem, or one of the hardware drivers doesn't work well. You can't run Windows usefully in safe mode. You have to fix it so it can run again in what it calls "normal" mode. There are three solutions to fixing the problem. The first is to keep rebooting Windows until it returns to normal mode all by itself, which it may do eventually. The second is to check the web site of your various hardware manufacturers and check for updated drivers. Or third, you can remove all your installed hardware, boot up in normal mode (it will easily) and then re-install everything. I had this problem of Windows booting up in safe mode constantly and refusing to go to normal mode. I fixed it by replacing my mouse driver with a new version, which my manufacturer provided.

Safebywire An internal network being developed for cars for the delivery of critical messages such as the release of airbags during a collision. See also CAN, FlexRay, LIN and MOST.

SAFENET Survivable Adaptable Fiber Network. A U.S. Navy experimental fiber-based local area network designed to survive conventional and limited nuclear battle conditions.

safety belt A thing made of leather. It's used by outside plant workers to attach themselves to and to climb utility poles. It's also called a body belt.

safety network A fieldbus that is used for safety purposes on the factory floor. Its purpose is to independently and rapidly detect degrading performance or failure on the fieldbus that controls automated manufacturing devices and on its attached devices, and

perform emergency shutdown actions to avoid danger when trouble is detected. Safety networks must meet new international safety standards, such as IEC 61508, IEC 61511, and IEC 62061.

sag 1. The downward curvature of a wire or cable due to its weight.
2. The opposite of surge. When the line voltage drops far enough to affect the operation of a phone system or computer.

sagan A large quantity. Its derivation is thought to come from Carl Sagan, the astronomer, who used the term "billions and billions" on his TV series.

SAI 1. See Serving Area Interface.
2. Stratus Computer's PBX-Switch to Stratus Computer Application Interface.

SAIC Science Applications International Corporation. According to SAIC, it is the largest employee-owned research and engineering company in the U.S. On November 21, 1996, SAIC announced its agreement to acquire Bellcore. See Bellcore. SAIC also owned Network Solutions Inc. (NSI), which currently administers the traditional Internet gTLDs (generic Top Level Domains). NSI went public in early 1998. www.saic.com. The domains under NSI control include the following: .com, which designates commercial entities; .edu, which designates institutions of higher learning; .org, which is intended to designate not-for-profit organizations, but is misused and abused.

SAID Speech Activated Intelligent Dialing.

SAID SOD Speech-Activated Intelligent Dialing Stringing of Digits.

Saint Gabriel In the Catholic religion, Saint Gabriel the Archangel first appears in the prophesies of Daniel in the Old Testament. Gabriel announced to Daniel a prophecy. He also appeared to Zachariah to announce the birth of St. John the Baptist, and he announced to Mary that she would bear a Son Who would be conceived of the Holy Spirit. Saint Gabriel was quite a communicator. He is the patron saint of telecommunications workers, and his feast day is September 29. In Geneva, Switzerland, the location of the International Telecommunications Union, there's a statue of St. Gabriel in front of the old Swiss PTT building. The feast of Gabriel is appropriately closely connected with that of the Annunciation, on March 24 in the West and on March 26 in the East. However, since 1969, the feast is kept on 29 September, together with that of Michael and All Angels. In satellite telecommunications, it is common to speak of conventional networks on land as terrestrial. The converse, networks not on the surface of the earth, is celestial, which, in addition to meaning "in the sky" (as a satellite), also means divine. How suitable, then, the choice of Gabriel/Jibril as the patron of telecommunications workers.

Saint Valentine's Day In the Middle Ages, the belief that birds chose their mates on St. Valentine's Day led to the idea that boys and girls would do the same. Up through the early 1900's, the Ozark hill people in the U.S. thought that birds and rabbits started mating on February 14, a day for them which was not only Valentine's Day but Groundhog Day as well.

Saks A young man wanted to get his beautiful blonde wife, Susie, something nice for their first wedding anniversary. So he decided to buy her a cell phone. He showed her the phone and explained to her all of its features. Susie was excited to receive the gift and simply adored her new phone. The next day Susie went shopping. Her phone rang and, to her astonishment, it was her husband on the other end. "Hi Susie," he said, "how do you like your new phone?" Susie replied, "I just love it! It's so small and your voice is clear as a bell, but there's one thing I don't understand though..." "What's that, sweetie?" asked her husband. "How did you know I was at Saks?"

salary Salt was given as monthly wages to Roman legionnaires, and was referred to as "salt money," or "salarium." Hence our word "salary."

sales agent See Aggregator.

sales automation See Sales Force Automation.

sales force automation The use of computers and computer software by salespeople to boost their sales. There are two types of sales force automation -those totally self-contained on the computers of salespeople (mostly laptops) or those which communicate with headquarters computer over phone lines. There are many purposes of the phone communication -sending orders in, finding out about back orders, getting updates on "specials," dropping letters and memos in, getting new prices, new products, new technical specs, etc. Salespeople routinely show 10% to 20% sales gains armed with a laptop PC and sales automation software (also called "personal contact") software.

salmon day One of those days where you swim upstream all day only to get screwed and die in the end.

SALT Speech Application Language Tags. SALT is an extended set of markup tags designed to add voice to Web sites. It is based on the Extensible Markup Language (XML). These markup tags are designed to add voice to application interfaces that are compatible with structured markup languages used on the Internet such as HTML, xHTML, XML, cHTML, SGML and WML. In essence, SALT allows you to add voice functions to applications built with any of the compatible languages. SALT does not change an existing language, nor does it provide all of the voice resources. It operates separately as a layer, but commands can be inserted into an existing structured language to allow access to voice functions such as speech recognition, speech synthesis, and grammar recognition that are available from a server. The main reason to use SALT is to provide speech input and output interfaces as well as the ability to control those voice functions from within a web page. There are two major applications for SALT. One is to enhance an existing application interface by adding voice, such as adding interactive voice prompts to a web page. The second is the creation of a SALT controlled, voice-only interface that could be implemented where a traditional graphical user interface (GUI) could not be used. In each case, SALT has the ability to listen for spoken input as well as accept input from a telephone touchtone pad. It then uses analyzes the input and provide the appropriate output in any combination of text, speech output or links to audio files. SALT is used in many applications. One of the most powerful is as a voice interface to web-enabled devices that are unable to adequately display a typical web interface. For instance, a web-enabled cell phone does not have a robust enough screen to properly display a web page, but through a SALT interface a user can listen to a menu of options and use vocal or keypad input to navigate the web page to find the desired information or services. SALT also enhances a typical web page. For example, an insurance company designs a Web page complete with helpful tips and online policy review to support their customers. To further enhance the page they include an interface that allows a customer to click and talk real-time with an insurance representative to handle questions, policy updates, and even claims -all from the user's computer. The effort to develop SALT technology was spearheaded by a group of companies called the SALT Forum. Their goal is to develop a platform independent, open standard that allows a variety of input and output choices, including speech recognition and speech synthesis, for real-time, web-enabled telephony applications. Thanks to Alcatel for help on this definition. See also salary.

salvo The sending of a group of commands at the same time.

SAM 1. Security Accounts Manager.

2. The Newton family's excessively spoiled cat. Also called Sammy.

See SAM Technology.

SAM technology Self Administered Maintenance. Application invented by a software engineer named Dave Tedesco. This is a technique added to large corporate web sites to allow the non-technical people to make changes to the web site without screwing up its functionality.

Samizdat Samizdat, a phenomenon which began after Khrushchev's secret speech at the Twentieth Communist Party Congress in 1956, meant the private reproduction of books, documents, letters, essays, literary works, translations, reprints from formerly published and until recent times forbidden publications, by means of typing and retyping them for dissemination by private citizens. These publications were thus completely free of censorship. Most samizdat of the pre-glasnost period is in manuscript form. Many samizdat publications found their way to the West and were published in emigre publications. There are three periods of censorship in Russian History:

1. The first period began in 1917 and ended in 1985 when Gorbachev came to power and announced a policy of glasnost. There was no freedom of speech and there was strict censorship.

2. The second period lasted from 1985 to 1 August 1990, when the new anti-censorship law (Law on the Press) came into effect. Glasnost inaugurated the collapse of the homogeneous ideological front. However, glasnost was not freedom of the press. Censorship was still in effect. Nevertheless, during this time and up to the passing of the new anti-censorship law close to 600 unofficial periodical publications appeared. These publications were uncensored and not registered in the Soviet official national bibliographies.

3. The third period, the era of freedom of speech, began on 1 August 1990 when, theoretically, after the introduction of the new law, there were no more restrictions on publishing freely. Censorship was abolished. However, the pre-coup government tried to control publishers and publishing houses by having a monopoly on paper distribution. At the beginning of this period there were still publishing houses subsidized by the government as well as publications completely independent from the government. The August 1990 Law on the Press also required the registration of printed matter with the Book chamber.

Sampling 1. Converting continuous signals, like voice or video, into discrete values, e.g. digital signals. See also Digital Signal Processing, PCM and Sampling Rate.

2. Examining a small percentage of the universe to determine makeup of the entire universe. A cook concludes that the entire pot of soup needs salt after sampling only one teaspoonful. the cook makes the assumption that the rest of the soup will taste the same as his sample.

sampling frequency The rate at which an inputted information signal is sampled to determine its instantaneous amplitude for subsequent quantization, coding and modulation to be digitally transmitted; thus 8 kilohertz becomes the sampling rate for 4 kilohertz analog speech, while 30 kilohertz is needed for 15 kilohertz music. See Pulse Code Modulation.

sampling rate 1. The number of times per second that an analog signal is measured and converted to a binary number -the purpose being to convert the analog signal each time it is sampled, to a digital byte representation. The most common digital signal -PCM -samples voice 8,000 times a second.

2. The number of times per second that a digital audio sample is taken during recording or read during playback -expressed in kilohertz (kHz). An audio CD sampled at a rate of 44kHz has 44,100 bits of information per second.

SAN 1. System Administration, Networking and Security.

2. Storage Area Network. A SAN is special purpose high-speed network designed to transport database-intensive applications, such as those used for inventory, billing, receivables, customer relationship management and supply chain. The concept is to have a dedicated network for these applications so users get fast response and don't get bogged down in the corporation's general networking traffic. Actually, a SAN generally is in the form of a sub-network that is part of a larger LAN (Local Area Network). The network protocols can include ATM (Asynchronous Transfer Mode), ESCON (Enterprise Systems Connectivity), Fast Ethernet (100 Mbps or Gigabit), FC-AL (Fibre Channel-Arbitrated Loop), or SSA (Serial Systems Architecture). The storage technology can be JBOD (Just a Bunch Of Disks), RAID (Redundant Array of Inexpensive Disks), a bunch of servers on a network, or a more complex and expensive host storage server such as a midrange or mainframe computer. SAN applications include disk mirroring, data backup and restoration, data archival and retrieval, data transfer between storage devices, and data sharing between servers. A SAN is much more complex than simple Network Attached Storage (NAS). See also NAS.

SANCHO The ITU database for terminology, short for ITU-T Sector Abbreviations and defiNitions for a teleCommunications tHesaurus Oriented database, providing access to terms, definitions, abbreviations and acronyms defined in ITU-T recommendations, available on CD-ROM. www.itu.int.

sand See Pound Sand and Sandbox.

Sand Hill Road The swank address in Menlo Park, California (i.e. Silicon Valley) for many of the world's leading venture capital firms. Firms based here have funded some of the biggest and most successful technology startups.

sandbox 1. Applications that are downloaded from a client on the Internet or an Intranet and which have the potential to damage that client. Viruses are an example of a malicious attempt to do so through damaging the hard drive, corrupting or erasing files, or perhaps damaging the operating system. Some computer programming languages or operating environments (e.g., Java) deny a distributed object access to operating system calls or to other resources. Such restricted objects are said to be "in the sandbox," according to Network magazine.

2. Many inventors look for outside financing from angels or venture capitalists. Some people look for the money to grow their company, by selling product or service and ultimately making a profit. Some people look for the money so they can continue having fun writing software, creating hardware, and doing whatever cool neat new things amuse them. These are "sandbox" companies. They will never produce a real product for their customers or a profit for their stockholders. To be a successful investor, you need to identify sandbox companies and avoid them like the plague. I get a lot of proposals from inventors seeking money. In response to one, I wrote, "The problem I have and I can smell this one is that this is a sandbox company. A bunch of incredibly bright boys are looking for money so they can keep creating cool new technologies. My job, as investor, is to provide the money, not question the use, or God Forbid, that they might actually focus their endeavors on creating a commercial product that will actually sell." See also custom software.

3. A protected area of a computer system in which programs run with limited privileges. For example Java applets may be confined to a sandbox environment which prevents them from accessing the computer system's permanent storage (e.g. hard disk) or networking services. Another example of a sandbox is an isolated network segment used for testing.

4. A network security term. A protective mechanism used in some programming environments that limits the actions that programs can take. A program normally has all the same privileges as the user who runs it. However, a sandbox restricts a program to a set of privileges and commands that make it difficult or impossible for the program to cause any damage to the user's data.

sandhog An underground worker, typically those building tunnels.

sandpit See Sandbox.

sanitize Jim Friehoff's term for forcing VARs to "get on board or get out."

sanity check A check to confirm the service capability of a switching system. This test has not been applied to the author of this dictionary.

SAP 1. Service Access Point. OSI terminology for portion of a network address that identifies the host application sending or receiving a data packet. In TCP/IP terminology, a SAP is a "port." SAPs and ports identify the specific application or service on the host computer; examples include e-mail and file server software. In the ATM Reference Model, traffic passes up and down the ATM Layers through SAPs, which are named for the specific Layers; e.g., the User Layer, ATM Adaptation Layer (AAL), ATM Layer, and Physical Layer (PHY). The term also is used in SMDS architecture, and to reference the same concept.

2. An IBM term for a logical point made available by an interface card where information can be received and transmitted.

3. Systems, Applications, and Protocols. Translated from the German Systemanalyse und Programmentwicklung. SAP AG was founded in 1972 by five former IBM employees. It has grown to be one of the largest providers of inter-enterprise software solutions in the world. Software products address applications such as Customer Relationship Management (CRM), e-commerce, supply chain management, and product lifecycle management.

4. Session Announcement Protocol. A protocol developed by the IETF as a companion to the Session Initiation Protocol (SIP). SAP is a method by which a multimedia conferencing session over an IP-based network is announced to the potential conference participants.

5. Second audio program.

SAP splitting A bandwidth conservation technique that involves sending only the accessed audio feed for a TV program to a subscriber, instead of sending all accompanying audio feeds.

SAPI Service Access Point Identifier. The SAPI identifies a logical point at which data link layer services are provided by a data link layer entity to a Layer 3 entity. ISDN jargon. See also Windows Telephony.

SAR 1. Segmentation And Reassembly. Generically speaking, a process of segmenting relatively large data packets into smaller packets for purposes of achieving compatibility with a network protocol relying on a smaller specific packet size. The process is often required in conjunction with ATM, SMDS and X.25 networks.

2. A sublayer of the ATM protocol stack, specifically of the ATM Adaptation Layer (AAL). The native Protocol Data Unit (PDU) associated with the transmitting device is segmented into 48-octet payload fields at this sublayer. At the target end of the data communication, the SAR serves to reassemble the native PDU by extracting and combining multiple 48-octet payloads from multiple ATM cells.

3. Specific Absorption Rate. Specific Absorption Rate (SAR): SAR is a measure of the rate of energy that is absorbed, or dissipated in a mass of dielectric materials, such as biological tissues (i.e. a human being, like you and me). Usually SAR is expressed in watts per kilogram (W/kg) or milliwatts per kilogram (mW/kg). Here's Motorola's explanation of SAR: Your wireless phone is a radio transmitter and receiver. It is designed and manufactured not to exceed the exposure limits for radio frequency (RF) energy set by the Federal Communications Commission (FCC) of the U.S. Government. These limits are part of comprehensive guidelines and establish permitted levels of radio frequency (RF) energy for the general population. The guidelines are based on standards that were developed by independent scientific organizations through periodic and thorough evaluation of scientific studies. The standards include a substantial safety margin designed to assure the safety of all persons, regardless of age and health. The exposure standard for wireless mobile phones employs a unit of measurement known as the Specific Absorption Rate, or SAR. The SAR limit set by the FCC is 1.6watts/kilogram (W/kg). Tests for SAR are conducted using standard operating positions reviewed by the FCC (Federal Communications Commission) with the phone transmitting at its highest certified power level in all tested frequency bands. Although the SAR is determined at the highest certified power level, the actual SAR level of the phone while operating can be well below the maximum value. This is because the phone is designed to operate at multiple power levels so as to use only the power required to reach the network. In general, the closer you are to a wireless base station antenna, the lower the power output. Before a phone model is available for sale to the public, it must be tested and certified to the FCC that it does not exceed the limit established by the government-adopted requirement for safe exposure. The tests are performed in positions and locations (for example, at the ear and worn on the body) as required by the FCC for each model. The highest SAR value for this model phone (a typical Motorola mobile phone) when tested for use at the ear is 1.51 W/kg and when worn on the body, as described in this user guide, is 0.75 W/kg. (Body-worn measurements differ among phone models, depending upon available accessories and FCC requirements.) While there may be differences between the SAR levels of various phones and at various positions, they all meet the government requirement. The FCC has granted an Equipment Authorization for this model phone with all reported SAR levels evaluated as in compliance with the FCC RF exposure guidelines. SAR information on this model phone is on file with the FCC and can be found under the Display Grant section of www.fcc.gov/oet/fccid." See also www.rfi-wireless.com/pages/press/articles/ART015.htm.

sarchasm The Washington Post's Style Invitational asked readers to take any word from the dictionary, alter it by adding, subtracting or changing one letter, and supply a new definition. This one is one of the winners. Sarchasm is the gulf between the author of sarcastic wit and the reader who doesn't get it.

sardines Canned herrings were dubbed "sardines" because the canning process was first developed in Sardinia, Italy.

SART Search and Rescue Radar Transponder for locating ships or their survival craft by generating a series of response signals upon receiving a signal from any 9 GHz ship or aircraft radar. SART is included in GMDSS -The Global Maritime Distress and Safety System. see GMDSS.

SAS 1. Simple Attachment Scheme.

2. Severely errored frame/Alarm indication Signal. A one-second period of time in which are detected multiple frame errors or an alarm indication signal over a digital circuit. See also CV, ES and SES.

3. See also SAS SCSI.

SAS SCSI The new serial-attached SCSI technology is designed to reduce storage costs of Fibre Channel devices. The SAS interface supports SAS and serial-ATA (SATA) drives, the type typically found in PCs and low-end servers. SATA drives cost less and offer less performance than Fibre Channel and SAS drives. But they are ideal candidates for near-line storage of information that is not accessed frequently; a second tier of disk-resident storage.

The compatibility of SAS and SATA drives begins with their low-level interface characteristics. In fact, SAS and SATA physical signaling is set by the same standard and is indistinguishable. The addition of the SATA Tunneling Protocol to the SAS standard paved the way for allowing a single SAS controller to support firstand second-tier storage.

Traditional storage subsystems begin with a RAID controller that must support two interfaces: one to interconnect to the host SAN (traditionally Fibre Channel) and one to support the connection of the hard-disk drives. Although the Fibre Channel loop can support up to 127 hard-disk drives, performance can be optimized by using far fewer drives per loop plus the use of Fibre Channel switches.

Although the new serial hard-disk drives share the serial nature of Fibre Channel, they do not share the ability to form loops. The SAS/SATA interfaces are strictly defined as point-to-point with unique port addresses. That's where the SAS expander comes in.

RAID controllers designed to support SAS/SATA hard-disk drives have semiconductor expander chips built into them. The Cables and Connectors Addendum to the SATA standard provides for 16 hard-disk drive ports. Implementations today usually provide for 12 to 15 hard-disk drives per enclosure. A SAS expander function is defined in the standard as a chip or printed circuit board that lets one SAS address be expanded to a number of additional ports. RAID internal expanders are chips, while external expanders are boards.

Although the Fibre Channel and SAS expander architectures appear similar, SAS offers an important advantage. The upstream SAS port can be configured as a wide port that lets multiple SAS connections (typically four) be treated as one SAS address. The result is a fat data pipe known as a SAS wide port.

For example, a connection of Fibre Channel drives in a traditional loop or point-to-point interface scenario can't exceed the Fibre Channel bandwidth, which is about 4G bit/sec. With SAS architecture and the implementation of SAS wide ports, SAS bandwidth of 3G bit/sec for a single interface can be multiplied to 12G bit/sec.

Emerging RAID controllers and storage subsystems take advantage of the bandwidth of the SAS wide port and the reduced cost. By providing a way to widen the data path while mixing SAS and SATA drives, SAS goes a long way toward meeting lower-cost, tiered-storage requirements.

SASG Special Autonomous Study Group. These ITU-T study groups are chartered to produce handbooks on basic telecommunications technical or administrative subjects for developing countries.

SASI Shugart Associates System Interface. The first SCSI interface specification defined by Shugart, a disk drive manufacturer. Later it was modified and renamed as the Small Computer System Interface (SCSI), pronounced Scuzzy. See also SCSI.

SaskTel Can I write SaskTel is the best phone company in Canada? Of course, I can't. But there are a whole bunch of fans of this dictionary who work for SaskTel, many in its headquarters in Regina, Saskatchewan who want me to write it, and -moreover -who really believe it. As I am writing this, SaskTel has over 425,000 customers and 5,200 employees across the entire huge and beautiful province of Saskatchewan. SaskTel does everything from voice, data, dial-up, high speed Internet, high-definition TV, entertainment and multimedia services, security, web hosting, text and messaging services over its neat modern digital network. It also does cellular and wireless data services, security monitoring, directory services, consulting and lots of telecom services I haven't space to write about. Take a trip to Regina. See it for yourself. Look up my dear, dear friend Heather Nord, who's gorgeous, to boot. Just don't go in winter.

SASL Simple Authentication and Security Layer. An Internet security mechanism specified in RFC 2222, SASL is a method for adding authentication support to connection-oriented protocols. SASL grew out of the work on IMAP4 (Internet Messaging Access Protocol version 4), a next-generation e-mail protocol which is likely to replace POP (Post Office Protocol) for Internet mail servers. IMAP4 includes the ability for mail clients and servers to negotiate the authentication mechanism they will use. SASL allows a client to request authentication from a server and, as an option, to negotiate the use of any authentication mechanism (e.g., Kerberos version 4, simple username/passwords, and one-time passwords such as S/Key) registered with IANA (Internet Assigned Numbers Authority). SASL mechanisms are named by strings, from 1 to 20 characters in length, and consisting of upper-case letters, digits, hyphens, and/or underscores. See also Authentication, IANA, IMAP, Kerberos, and POP.

SAT 1. Subscriber Access Termination. An SMDS term.

2. Supervisory Audio Tone. A cellular term. When a cellular call is set up, the MSC (Mobile Switching Center) sends a SAT to the cell phone. The SAT is returned by the cell phone through an automatic loopback. The MSC checks the characteristics of the SAT in order to ensure signal quality before setting up the call. See also Loopback and MSC.

SATA See Serial ATA.

SATAN Security Administrator Tool for Analyzing Networks. This tool allows a network analyst to mimic a malicious hacker (or cracker) for the purpose of identifying weaknesses in system and network security. It also provides malicious hackers a nifty tool. See Hacker.

satcaster Satcaster is a satellite broadcaster.

satcom A shortened way of saying satellite communications.

satellite 1. A microwave receiver, repeater, regenerator in orbit above the earth. Traditional communications satellites are known as GEO's, as they are in Geosynchronous Earth Orbit, which is an equatorial orbit with the satellites at high altitudes of approximately 22,300 miles. In such an orbital slot and at that altitude, they maintain their position relative to the earth's surface. More recently developed satellites are placed in Low or Middle Earth Orbits, hence the terms LEO and MEO. LEOs and MEOs vary widely in terms of orbital paths and altitudes; therefore, they are not synchronized with the earth's rotation. See GEO, MEO and LEO. See also Satellite Transmission.

2. Something distant to the main something. See Main Distribution Frame, Satellite Cabinet and Satellite Distribution Frame.

Satellite Business Systems SBS. A satellite long distance carrier originally owned jointly by IBM, Aetna Insurance and Comsat, but now owned by MCI (which acquired it in 1986). SBS started out to serve the data communications transmission marketplace but found that marketplace too small to be profitable. It then started to serve the voice transmission marketplace and did somewhat better. But everyone hated satellite voice calls because of the delay and the frequent echoes. Satellite Business Systems no longer exists as a separate entity. It has been merged into MCI. There are estimates on how much money SBS lost in its short history. They are substantial, ranging around $1 billion -a lot of money in those days. See SBS.

satellite cabinet Surface-mounted or flush-type wall cabinets for housing circuit administration hardware. Satellite cabinets, like satellite closets, supplement riser closets by providing additional facilities for connecting horizontal wiring subsystem cables from information outlets in user locations. Sometimes referred to as satellite location.

satellite closet A walk-in or shallow wall closet that supplements a backbone or riser closet by providing additional facilities for connecting riser subsystem cables to horizontal wiring subsystem cables from information outlets. Also referred to as satellite location. See also Riser Closet and Backbone Closet.

satellite communications The use of geostationary orbiting satellites to relay information.

Satellite Communications Control SCC. The earth station equipment that controls such communications functions as access, echo suppression, forward error correction and signaling.

Satellite Community See SHVERA.

satellite constellation The arrangement in space of a set of satellites.

satellite delay compensator A device that compensates for the absolute delay in a satellite circuit communicating with data terminal equipment (DTE) with the DTE's own protocol.

satellite delivered signal A television signal delivered to a cable television headend by communications satellite. This term should not be confused with the signal received from a Satellite Television Broadcast Station. Compare with Satellite Television Broadcast Station.

satellite digital audio radio service Special satellites transmitting digitally to tiny (fewer than two inches wide) antennas in the S band. The concept is that with digital technology, satellite broadcasters can stuff dozens of channels of CD-quality, interference-resistant programming into a narrow-band of frequencies. See XM for a full explanation.

satellite distribution frame An intermediate point for connecting wires running between a group of phones and the Main Distribution Frame located elsewhere in the building. A fat multi-conductor cable comes from the main distribution frame to the satellite distribution frame, where it splits into individual cables to individual phones or workstations. The satellite distribution frame is usually located in a satellite wiring closet or cabinet. These wires are ultimately connected to the telephone system. See Distribution Frame.

satellite downlink The communications path from a satellite to its ground station. Opposite to Satellite Uplink.

satellite facility A transmission system using a satellite in a geostationary orbit above the earth and a number of earth stations.

Satellite Home Viewer Improvement Act of 1999 This Act of the US Congress, for the first time, allowed satellites to carry local TV channels. The law, however, also put in place must-carry rules for satellite.

satellite link Microwave link using a satellite to receive, amplify and retransmit signals. Typically that satellite is in a geosynchronous orbit.

satellite operation A configuration of multiple PBXs or one big PBX and several smaller PBXs. The configuration gives a company with several nearby locations a unified system of centralized trunks, centralized attendants, overall call detail recording and many of the advantages of a private network. The key advantage of satellite operation is that one big centralized telephone system can contain most of the intelligence and computer smarts for the total system. This advantage is heavily economic. A variation on satellite operation is called Centralized Attendant Service (CAS).

satellite PBX A satellite PBX has no direct incoming connection from the public network. All incoming calls are routed from an associated main PBX over tie trunks. This definition places no restrictions on the handling of outgoing calls from the satellite PBX. A satellite PBX can have one-way outgoing trunks to the central office, in addition to outgoing service on trunks to the main PBX. A satellite PBX has no direct trunks to a node; however, calls to the node can be made through the main PBX.

satellite premises channel This is the cable connecting arrangement between a dedicated earth station and the Customer Provided Equipment.

satellite processor A computer with little computing power used for operations that do not require the full processing power of the main machine.

satellite radio Two companies in the U.S. have now launched satellites which broadcast 100-plus channels of high quality radio 24 hour, 7 days a week to cars or houses equipped with the special radios required to pick up the signal. At the time of writing, one service was charging $10 a month. The other was charging around $13 a month. The quality from both is excellent, but in most cases your antenna needs to have a view of the sky. I don't subscribe to either service, but friends do and they love it. You can hear snippets of what they broadcast by logging into their web sites and listening for free. Go to www. XMRadio.com and www.SiriusRadio.com.

satellite relay An active or passive repeater in geosynchronous earth orbit that amplifies the signal it receives, often shifting the radio frequency as well before retransmitting it back to earth; these functions are performed aboard the satellite in a device called a transponder. See Geosynchronous Satellite.

Satellite stable point One of two diametrically opposite points of a geosynchronous satellite orbit, located at 75 degrees east and 105 degrees west where the radius of the earth is minimum. See satellite substable point.

satellite substable point One of two diametrically opposite points of a geosynchronous satellite orbit, located at 165 degrees east and 15 degrees west, where the

65 meters (213 ft) bulge of the Earth at the equator is highest and at which satellite drift is conditionally zero, as a slight change in either direction incurs forces of drift towards a satellite stable point.

Satellite Television Broadcast Station A United States television broadcast station which: Operates pursuant to Part 73, Subpart E, of the FCC Rules. Operates at full-power levels, typically thousands of watts. Rebroadcasts all (or substantially all) of the signal of another full-power television broadcast station. Example: Stations KGMD-TV (Hilo) and KGMV (Wailuku) are satellites of station KGMB, Honolulu. Together, the three stations cover most of Hawaii. This term should not be confused with a communications satellite, or to a signal received from a communications satellite. Compare with Satellite Delivered Signal.

satellite transmission A form of transmission which sends signals to an orbiting satellite which receives them, amplifies them and returns those signals back to earth. Satellite transmission provides great clarity but suffers from delay. See Satellite Transmission Delay.

satellite transmission delay Referring to the time it takes a signal to travel from one satellite earth station to the satellite in the sky then to the satellite earth station at the other end. Since most communications satellites orbit the earth at a distance of approximately 22,300 miles, the total distance the signal travels is 44,600 miles. Since radio waves travel at the speed of light (186,000 miles per second), simple arithmetic will show a delay of approximately one-quarter of one second thus, 44,600 divided by 186,000 = 0.239 second. If you are waiting for a reply, double this time. (You double the distance.)

satellite television smart cards Satellite television smart cards are small cards roughly the size of credit cards which slide into set-top boxes. They , contain computer chips that decrypt signals from DirecTV's satellites. which slide into set-top boxes, contain computer chips that decrypt signals from DirecTV's satellites.

satellite uplink The communications path from a ground station to its satellite. Opposite to Satellite Downlink.

satellite webcasting Webcasting is defined as the real time delivery of audio, video or animation over the Internet. Satellite Webcasting is sending all that information from its source via satellite to the closest ISP serving the end user, his customer. The theory of satellite webcasting is that the signal will come through cleaner and clearer. How clean and how clear depends on the quality of the communications between the ISP and his customer. A high-speed cable modem, for example, will deliver a much better signal than a 33.6 Kbps dial-up modem.

SATPhone A telephone that works directly off a satellite. It comes with a small parabolic antenna which you aim at the satellite. You turn it on and talk. It's easy, though expensive -typically as much as $10 a minute. But for that you can talk from practically anywhere in the world, so long as you got lots of battery or easy close access to commercial power.

saturation 1. The intensity of the colors in the active picture. The voltage levels of the colors. The degree by which the eye perceives a color as departing from a gray or white scale of the same brightness. A 100% saturated color does not contain any white; adding white reduces saturation. In NTSC and PAL video signals, the color saturation at any particular instant in the picture is conveyed by the corresponding instantaneous amplitude of the active video subcarrier.

2. The point on the operational curve of an amplifier at which an increase in input amplitude will no longer result in an increase in amplitude at the output.

Saturday Vikings were not the dirty, smelly, lice-ridden lot they are depicted as having been. An English chronicler, at the height of the Viking raids on the British Isles, bitterly complained that "British womenfolk rather go with the Vikings, because they are always washing their hands and faces, combing their hair and their beards, changing their shirts, and, oh shame! every Saturday they are taking a full bath and washing all their clothes," which British men did not do because Christians of the time considered washing vanity and therefore something to avoid for the good of their souls at the expense of their bodies. The ancient Nordic word for Saturday, "Laugrdag," means "washing day."

save A telephone feature that allows the user to put a phone number into memory for future calls, by pressing one or two buttons after dialing it the first time. See also SNR.

save and repeat Another way of saying "Autodial." Electronic phones may be able to save a number so you can dial it later by simply hitting one button on the phone. This feature is similar to a "Last Number Redial" button, except that button just dials the last number called. "Save and Repeat" puts a number into temporary storage for dialing at another time. Phones should have both auto-redial and save-and-repeat buttons.

save opportunity What AOL calls it when its call center employees try to turn a disgruntled customer trying to cancel his AOL account into a great new customer for something AOL is selling.

Save Our Ship See SOS.

Save Our Souls See SOS.

SAW filter See Surface Acoustical Wave Filter.

SB Signal Battery lead, used in E&M signaling types II, III, IV. See E & M and SG.

SBAN Services oriented Building Area Network. A SBAN is a broadband infrastructure network architecture used within shared tenant facilities such as high rise buildings, office parks, malls that enables the rapid deployment of converged multimedia (voice/data/video) services. SBAN's medium could be copper or fiber. The key deliverable is to be able to converge multimedia traffic. The higher the bandwidth (i.e. fiber) the easier the task.

SBC 1. Subsequent Bill Company. A Meet Point Billing term that refers to all other LECs on a call route where multiple LECs are involved in completing the call. Opposite of IBC. See also SBC Communications Inc.

2. See Session Border Controller.

3. See SBC Communications Inc.

SBC Communications Inc. In early October, 1994, Southwestern Bell Corporation, one of the seven original regional Bell operating companies, changed its name to SBC Communications, Inc. The company's subsidiaries continue to be Southwestern Bell Telephone Co., Southwestern Bell Mobile Systems, Southwestern Bell Yellow Pages, Southwestern Bell Telecom and SBC International. SBC continues to operate as a LEC (local exchange carrier) in the states of Arkansas, Kansas, Missouri, Oklahoma and Texas. In 1997, SBC acquired Nevada Bell and Pacific Bell, at which point the Pacific Telesis (PacTel) holding company ceased to exist. In 1998 SBC bought SNET (Southern New England Telephone), which operates as a LEC in portions of Connecticut. In 1998 SBC bought Ameritech. In 2005, SBC bought what was left of AT&T -namely the long distance part. And in the fall of 2005, SBC officially changed its name to AT&T. See AT&T.

SBCA Satellite Broadcasting and Communications Association of America. A not-for-profit national trade organization founded in 1986, and representing all segments of the satellite industry. The more than 2,000 members include DBS platform providers, programmers, manufacturers, distributors, retailers, encryption vendors and software technology providers. SBCA aims to expand the use of satellite technology for the broadcast delivery of entertainment, news, information and educational programming. See also SIA. www.sbca.org.

SBM 1. DMS SuperNode Billing Manager.

2. Subnet Bandwidth Manager. The SBM provides centralized bandwidth management on shared networks. See TAPI 3.0.

SBS 1. Satellite Business Systems. A long distance satellite company that started out as a joint venture between Lockheed and MCI, was sold to IBM Aetna and Comsat and then eventually was given to MCI in exchange for shares issued to IBM. SBS never made any money. But that was irrelevant. Its job was to help IBM sell computer networks. See Satellite Business Systems.

2. Sick Building Syndrome. Phenomenon of employee discomfort and illness, or perceived illness, due mainly to a polluted indoor air supply. An office is diagnosed "sick" when more than 20% of its occupants exhibit typical symptoms, complaints persists for two weeks or more and disappear when sufferers are away from the building.

Sbus Sun Microsystems' resource sharing and expansion bus interface for the SPARC architecture. SBus expansion cards can communicate with each other through this interface. SBus competes with VME, PCI and EISA/ISA as industry standard I/O buses for computing platforms. Currently, there are several computer telephony vendors selling SBus compliant DSP and network interface cards for SBus. You can develop cards with an both SBus computer interface and a mezzanine for MVIP and SCSA busses. The Sbus specification has been adopted by the IEEE (Institute of Electrical and Electronic Engineers) as a new bus standard.

SC connector A snap-on fiber optic plastic connector. See also FC Connector, SFF Connector, and ST Connector.

SCA 1. Selective Call Acceptance.

2. Special Customer Arrangement. A billing arrangement which a carrier makes with a large customer to bundle services and charge a special, non-tariffed, discounted rate.

3. Supplemental Communications Authority. The authority granted by the Federal Communications Commission to transmit on a subcarrier.

4. Software Communications Architecture.

SCADA Supervisory Control And Data Acquisition. SCADA systems are used extensively by power, water, gas and other utility companies to monitor and manage distribution facilities. They also are used, although more sparingly, to monitor and control end user usage levels for purposes such as remote meter reading and load shedding. Traditionally, such systems made use of telephone lines for such purposes, although wireless technologies are now deployed widely. Some power utilities have deployed fiber optic transmission facilities (allegedly) for this purpose, although the small amount of bandwidth required for such an application clearly does not justify the cost of fiber. It is widely accepted that such fiber deployment is a preemptive strike against LECs and CATV providers who seek to place fiber on the power utility companies poles and in their conduits in a convergence scenario. In effect, the power utilities are laying information grids for resale to carriers which desire substantial bandwidth in competition for transmission of voice, data, video, image, TV and multimedia signals in a deregulated environment. See also Broadband Multimedia.

SCAI Switch to Computer Applications Interface. A protocol that defines how switches talk to outbound computers, i.e. computers which are external to the switch and contain such a database of customer buying information. Using SCAI, calls and data screens about a calling customer can be presented to the agent simultaneously. See Open Application Interface.

scalability Fancy way of saying size something can grow to relatively easily. See Scalable.

scalable Something that can be made larger or smaller relatively easily and painlessly. And the cost to grow is relatively straight line, rather than stair step, as in the days of "forklift upgrades." At least that was the earlier, accepted definition. Then Microsoft started referring to Windows NT as "scalable," namely that it runs on everything from Intel to RISC processors and singleto multi-processor systems. Scalable often refers to technology applications which can be made greater or smaller without quantum leaps in cost. For instance, Virtual Private Networks (e.g., Switched 56/64, X.25, Frame Relay, SMDS and ATM networks) serve as effective replacements for dedicated, leased-line networks as their capabilities are scalable, with the costs remaining in reasonable relationship to associated functionality.

scalable video Scalable video is a playback format that can determine the playback capabilities of the computer on which it is playing. Using this information, it allows video playback to take advantage of high performance computer capabilities while retaining the ability to play on a lower performance computer.

scalable typeface A set of letters, numbers, punctuation marks, and symbols that are a given design (i.e. of one font) but can be scaled to any size.

scaled point size A point size that approximates a specified point size for use on the screen. For example, text that prints at 10 point on the printer may be represented by a slightly larger font on the screen to make up for the screen's lower resolution.

scaling 1. A video compression technique which involves adjustment of the transmitted image in consideration of the presentation capabilities of the receiving device. In the case of a receiving device which is less capable in terms of resolution, for instance, the codec in the transmitting device reduces the resolution of the image prior to transmission. In this fashion, the receiving device is presented with a signal which matches its display capabilities. Additionally and more importantly, transmission bandwidth is not wasted.

2. Cost effective accommodation of traffic growth.

scam spam Junk email which contains a gross fraud, for example when you receive one of those emails from the Nigerian Minister of Finance, offering you $7,000,000 in exchange for being the custodian of millions of dollars for the Nigerian Royal Family. This wonderful definition contributed by Karen Gullett.

SCAN 1. Switched Circuit Automatic Network.

2. To examine sequentially, part by part.

3. To examine every reference or every entry in a file routinely as part of a retrieval scheme.

4. In electromagnetic or acoustic search, one complete rotation of an antenna.

5. The motion of an electronic beam through space searching for a target. Scanning is produced by the motion of the antenna or by lobe switching.

6. In imaging, a scan is the process by which an image is developed. The electron beams excites the phosphor on the monitor screen dot by dot and line by line. The faster the scanning the more stable the image.

scan time The time between two successive polls to a workstation on a data communications network.

ScanDisk ScanDisk is a piece of software that comes with all Windows computer operating systems. Its mission is to repair your computer's hard drive. ScanDisk, as its name implies, scans your hard disk for files and parts of files to make sure they're all in their proper place. If it discovers a conflict between what's actually on the hard disk and a list of what's meant to be where on your hard disk called the file allocation table, or if it determines that the table itself has been damaged, ScanDisk will try to repair the problem. The program can also look for physical imperfections on your computer's hard disk and prevent the computer

from storing data in a potentially troublesome area. Running ScanDisk on a regular basis (at least once a week) is a good maintenance habit. You should tell it to automatically fix any errors it finds. After you've run ScanDisk, run Disk Fragmenter. You can find ScanDisk on a Windows XP machine by leftclicking My Computer, rightclclicking on your C: hard drive, leftclicking Properties, hitting the Tools tab and clicking on the "Check Now" button.

Scanlation The practice of scanning Japanese manga comic books and posting them on the Web with English translations. Fans say it's not piracy because the books aren't printed in English.

scanner 1. A radio receiver which automatically skips across selected frequencies, allowing you to listen in to any of the frequencies. You can buy scanners, for example, that let you scan all police, fire, cellular frequencies and let you listen in on any conversation that is presently occurring.

2. A program on a bulletin board system which scans the message base for previously entered e-mail and pulls a copy of each message and makes them available to the BBS (Bulletin Board System) mailer program.

3. A device used to input graphic images into the computer. Scanners look at or "scan" a piece of paper and put the image's information into digital form. The information can then be recognized by the computer. Scanners come in three basic types -flat-bed, sheet-fed and as one part of a multifunction devices that prints, copies, faxes and scans. A fax machine also contains a scanner which "looks" at the original document and determines the brightness level of each pixel to be transmitted. The accuracy at which a scanner gets information from the document it is scanning and sends it to an attached computer is measured in two ways: by resolution and color information. Resolution is defined as dpi (dots per inch) or pixels, which determines the maximum size of the image. For example, a 2,400-pixel by 1,800 pixel scanned at 300 dpi (dots per inch) creates a maximum image of 8" x 6". Color information is defined by the number of bits of information per color. Today's scanners produce images with 24, 30 and 36 bits per pixel. In a 24 bit scanner, you make your color by choosing 8 bits each of red, green and blue). The more pixels and the more bits of information per color, the larger the imaged file (often going into the millions of bytes) and the more accurate the representation will be. See also Optical Character Recognition.

scanner accuracy The accuracy at which a scanner inputs information is measured in two ways: by resolution and color information. Resolution is defined as dpi (dots per inch) or pixels, which determines the maximum size of the image. For example, a 2,400-pixel by 1,800 pixel scanned by a scanner at 300 dpi creates a maximum image of 8" x 6". Color information is defined by the number of bits of information per color. Most scanners produce images with 24 bits per pixel (8 bits each of red, green and blue). The more pixels and bits of information per color, the larger the imaged file and the more accurate the representation will be.

scanning rate In video communications, the scanning rate is the rate at which the screen is refreshed. Even numbered lines are refreshed in one scan, with odd numbered lines being refreshed in the next scan. In combination, the two scans yield a frame refreshed. The process happens so quickly that it is imperceptible to the human eye/brain. The scanning rate is related directly to the frequency of the power source. For instance, the U.S. NTSC TV standard calls for 30 fps (frames per second), related to the 60 Hz power standard. The European PAL standard, on the other hand, calls for 25 fps.

SCANNS Multiplexers which perform the vital functions of monitoring and control within System 75s and 85s Automated Building Management feature. The SCANNS continuously scan sensors and send the resulting data to local control units.

SCAPI SCSA Application Programming Interface. A high-level, object oriented hardware independent, technology independent programming model that permits the design and implementation of call processing applications.

scattering A cause of lightwave signal loss in optical fiber transmission. The diffusion of a light beam caused by microscopic variations in the material density of the transmission medium. Scattering is a physical mechanism in fibers that attenuates light by changing its direction.

scattering losses In an optic fiber, power losses due to dimensional irregularities and imperfections in the fiber material.

scatternet A wireless term which is specific to the Bluetooth specification. According to the Bluetooth group, scatternet is formed when multiple, non-synchronized piconets are linked. A piconet is tiny network formed of a a collection of devices connected in an ad hoc fashion via Bluetooth RF (Radio Frequency) technology. The connection may begin with any two devices, such as a wireless PC and a cellular telephone, and may grow to as many as eight devices, including PDAs (Personal Digital Assistants) and anything else that you might imagine. While all piconet devices are peers (i.e., equals) in terms of connectivity, one device acts as the master, and the others as slaves, for the duration of the piconet connection. See also Bluetooth and Piconet.

SCbus An SCSA definition. The standard bus for communication within an SCSA node. The SCbus features a hybrid bus architecture consisting of a serial Message Bus for control and signaling, and a 16-wire TDM data bus. The SCbus is a serial time division multiplexed bus for carrying information between hardware devices in a signal processing node. The SCbus can support up to 1024 bidirectional timeslots in a PC implementation or up to 2048 timeslots in a backplane implementation. The SCbus uses 16 synchronous data lines for carrying data and a dedicated messaging channel (SCbus Message Channel) for carrying signaling information and messages between devices. See S.100 and SCbus message Channel.

SCbus Message Channel An SCSA definition. The SCbus message channel is a 2.048 Mbps serial line for carrying signaling information and messages between hardware devices in an SCSA Hardware compliant server. The SCbus message channel uses an HDLC (high level data link controller) protocol. The SCbus message channel is an optional element of the SCSA Hardware Model and provides faster system performance by allowing for direct communication of messaging information between devices at the firmware level and without consuming any data timeslots. See S.100.

SCC 1. Specialized Common Carrier. An old term for a long distance carrier in competition with AT&T. The word "Specialized" came about because these long distance carriers purported to provide "specialized" circuits for business customers. At one stage they were also known as OCCs, or Other Common Carriers (i.e. other than AT&T). These days, both terms have fallen into disrepute. All long distance carriers -including AT&T -are now called IntereXchange Carriers (IXCs).

2. Standards Council of Canada.

3. Satellite Communications Control.

4. SuperComputing Center. There are five NSF-funded supercomputing centers (SCCs): Cornell Theory Center, National Center for Atmospheric Research, National Center for Supercomputing Applications, Pittsburgh Supercomputing Center, and San Diego Supercomputing Center.

5. System Control Computer. The computer system used at CATV or MMDS headend for control of numerous technical functions. These functions include subscriber addressing, channel mapping schedules, ad insertion, encryption keys, PPV (Pay Per View), and sometimes IPPV (Impulse Pay Per View).

SCCP 1. Signaling Connection Control Part. Part of the ITU-T #7 signaling protocol. and of the SS7 protocol. It provides additional routing and management functions for transfer of messages other than call set-up between signaling points. A SS7 protocol that provides additional functions to the Message Transfer Part (MTP). It typically supports Transaction Capabilities Application Part (TCAP). See also Signaling System 7 and Common Channel Signaling.

2. Skinny Client Control Protocol. SCCP provides a means for signaling control communications to client endpoints, such as telephones. Skinny Client Control Protocol (SCCP), also referred to as "Skinny," includes a messaging set that allows communications between call control servers and endpoint clients with significant CPU and memory constraints, such as telephones and other embedded systems. It is a stimulus-based, lightweight alternative to H.323 and is proprietary to Cisco Systems. SCCP is available via license from Cisco and is widely used by Cisco telephony partners who are developing applications.

SCCS 1. Switching Control Center System

2. Switching Center Control System.

3. Specialized Common Carrier Service

SCDPI SCSA Device Programming Interface: A set of callable functions that allow SCSA application software to control SCSA hardware. The SCdpi consists of both common call processing services and technology specific modules for the application of particular resources to call processing tasks. See SCSA.

SCE Service Creation Environment. A Bellcore term used in the jargon of intelligent networks (INs) to allow outside developers to define and create new value-added (i.e., intelligent) services by connecting pre-existing blocks of code into a flow chart that describes the logical processes the service will use to handle calls. A critical and distinguishing feature of the AIN concept, the SCE comprises a toolkit for the creation of services which can be provided on a network basis. The carrier can develop a generic service which can be offered to multiple users; similarly, a third-party software developer or end user can develop such a service application. Once the application is developed, the application logic and supporting databases can be partitioned in order that multiple users can take advantage of it. In such a scenario, each user organization would have the ability to customize the application, which would draw on a customized, partitioned and secured database.

By way of example, routing logic for an ACD network might be centralized. Multiple organizations, each with multiple incoming call centers, could customize the generic call routing application in consideration of their specific Quality of Service parameters, cost issues, agent skill sets, and so on.

SCF Shared Channel Feedback. A digital wireless term defined by IS-136, the Interim Standard for digital cellular networks employing TDMA (Time Division Multiple Access). SCF is a logical channel which is part of the FDCCH (Forward Digital Control Channel) used to send signaling and control information from the cell site to the user terminal equipment. SCF information keeps all terminal devices advised of the level of network availability. SCF also provides each device with time slots for transmission and reception in order to avoid data collisions. See also FDCCH, IS-136 and TDMA.

SCFA Secondary Carrier Facility Assignment.

schedule A call center term. A record that specifies when an employee is supposed to be on duty to handle calls. The complete definition of a schedule is the days of week worked, start time, break times and durations (as well as paid/unpaid status), and stop time. See the following six terms.

schedule exception A call center term. A specific date and period when an employee cannot handle calls or is engaged in some kind of special activity. An absence, meeting, or other work assignment creates an "exception" to the employee's daily work file schedule.

schedule inflexibility A call center term. A phenomenon that tends to create overstaffing in some periods when full coverage is the objective in creating a set of schedules. This is caused by the fact that it is impractical to have extremely short schedules for covering momentary peaks in call volume. To achieve a near perfect match of staff and workload at all times would require shifts of virtually every length; for example, 2-hour shifts, 45-minute shifts, even 15-minute shifts.

schedule preference A call center term. A description of the days and hours that an employee would like to work, used by the automatic assignment process to match the employee to a suitable schedule. In some call centers, each employee can have as many as 10 schedules preferences ordered by priority.

schedule test A call center term. A variation of the scheduling process that allows you to forecast the service quality that will result from using an existing set of schedules.

schedule trade A call center term. A situation in which two employees have agreed to work each other's schedules, or an employee has agreed to work the other's schedule, on a specific date or dates.

scheduling software scheduling software makes the timetable of agent hours and shifts for your call center. The software takes into account vacation days, breaks, agent skill levels, lengths of shifts and forecasting information about when calls when arrive at your call center.

schema The set of definitions for the universe of objects that can be stored in a directory. For each object class, the schema defines which attributes an instance of the class must have, which additional attributes it can have, and which other object classes can be its parent object class.

schematic A diagram of the electrical scheme of a circuit with components represented by graphic symbols.

scheme The part of a URL that tells an HTML client, like a browser, which access method to use to retrieve the file specified in the URL. See URL.

schlepp A Yiddish word meaning to carry around, to drag around, as in "This phone system is heavy. Schlepping it is a pain." See also Chutzpah and Schlub.

schlub A crude individual lacking in social skills and blessed with insensitivity, clumsiness and no manners. In short, another great Yiddish word. Also spelled Zhlup. See also Schlepp.

schone See Harry-Proof.

Schools and Libraries Corporation SLC. A not-for profit corporation formed by the Telecommunications Act of 1996, and operating as the Schools and Libraries Division (SLD) of the Universal Service Administrative Company (USAC), which also was formed by The Act. The SLC is responsible for distributing up to $2.25 billion per year of funds from the Universal Service Fund (USF). Those funds subsidize the cost of internal wiring, telecommunications services, and Internet access for schools and libraries. See also Universal Service Administrative Company and Universal Service Fund.

SCID SONET Carrier ID. A friend wrote me "I work for a regional Bell co. and had a trouble ticket in which the problem was found to be in SCID xxxxxx."

Scientific or Industrial Organization See SIO.

SCIF Sensitive Compartmented Information Facility is a room where you keep classified documents. Cell phones don't work in such rooms. Classified documents can be read, but not removed from such rooms. We know that there are several of these rooms in Washington, D.C. Tom Kean, chairman of the commission investigating 9/11, spoke of them to the New York Times in an January 4, 2004 article.

scintillation In electromagnetic wave propagation, a random fluctuation of the received field strength about its mean value, the deviations usually being relatively small. Think of scintillation as the creation of a spark, hopefully a small one. The effects of this phenomenon become more significant as the frequency of the propagating wave increases. The discovery of scintillation must have been truly scintillating.

script A type of program consisting of a set of instructions to an application or tool program. A script usually expresses instructions by using the application's or tool's rules and syntax, combined with simple control structures such as loops and if/then expressions. "Batch program" is often used interchangeably with "script" in the Windows environment.

scriptoria Places where, for centuries, monks laboriously copied religious texts by hand -until the invention of the Guttenberg printing press around 1453.

SCM 1. Station Class Mark. A two digit number that identifies certain capabilities of your cellular phone. How the cellular network handles your call is based on these digits. The SCM tells the system if your phone transmits at standard power levels or low power levels, if it can use the full 832 channels or only the original 666 frequencies. The last attribute identified is whether your phone uses voice activated transmission (VOX).

2. Subscriber Carrier Mode.

3. See Supply Chain Management.

SCM-100 A Subscriber Carrier Module-100 Access; same as SMA.

SCN Switched Circuit Network. See Circuit Switched Network, which is the same thing.

Scope 1. A slang term for cathode ray oscilloscope.

2. An ATM term. A scope defines the level of advertisement for an address. The level is a level of a peer group in the PNNI routing hierarchy.

3. Secured Cageless Opening. A CLEC terms. It's something like co-locating at a discount because you're using only one rack at an ILEC's central office. It's in a secured Locked area. Not sure of the P and E in it the acronym.

SCOTS 1. Surveillance and COntrol of Transmission Systems. 2. Switched Circuit Ordering and Tracking System. MCI's automated tracking and order processing system for Dial up products, IMTs, and the MCI switched network.

SCP 1. Service Control Point. Also called Signal Control Point. A remote database within the System Signaling 7 network. The SCP supplies the translation and routing data needed to deliver advanced network services. The SCP translates an 800-IN-WATS number to the required routing number. It is separated from the actual switch, making it easier to introduce new services on the network. See also SSP and TCAP. For a full explanation of the Advanced Intelligent Network, see AIN.

2. Nortel Networks term for a Satellite Communications Processor.

SCPC Single Channel Per Carrier. A technique used in analog satellite and certain other radio systems. SCPC supports one transmission per frequency channel. Multiple channels can be supported through Frequency Division Multiplexing (FDM). See also FDM.

SCR 1. Abbreviation for Silicon Controlled Rectifier, a semiconductor device that allows one electric circuit to control another; often replaces electromechanical relays.

2. SCSA Call Router.

3. Sustainable Cell Rate. Parameter defined by the ATM forum for ATM traffic management. The SCR is an upper bound on the conforming average rate of an ATM connection over time scales which are long relative to those for which the PCR is defined. Enforcement of this bound by the UPC could allow the network to allocate sufficient resources, but less than those based on the PCR, and still ensure that the performance objectives (e.g., for Cell Loss Ratio) can be achieved.

scrambler A device which deliberately distorts a voice or data conversation so that only another like device can figure out the content of the message. Analog scramblers invert the frequencies of speech. Digital scramblers first convert speech to digital form and then encrypt. Both types also perform the reverse process. The sophistication (i.e. complexity) of a scrambler determines its price.

scrambling Traditionally defined in the science of cryptology as an analog method of concealing communications signals which uses the processes of heterodyne, band division, transposition, or signal inversion. Sometimes use of positive trapping is called "scrambling." See Positive Trapping. See also Scrambler.

scratch monkey A safety device based on using a scratch (a recording medium attached to a machine for testing and/or temporary use to preserve material). Before testing

or reconfiguring, always attach a "scratch monkey" to avoid losing irreplaceable data.

scratchpad A part of the random access memory of a computer or telephone system which can be used to temporarily store data. In a cellular phone system, scratch pad allows storage of phone numbers in temporary memory during a call. Silent scratch pads allows number entry into scratch pad without making beep tones. See also Register.

screen A wire mesh device or shield used for blocking radio frequency energy.

screen dump A reasonably exact copy of what's on your PC's screen printed out or saved as a file.

screen font The font that is displayed on your screen. It is, hopefully, designed to match the printer font so that documents look the same on the screen as they do when printed. You typically need a graphics interface on your company -like Windows or X-Windows -to make the font you see on your screen the same as what you see when you print it out.

screen pop Screen pop presents customer data and product and service information simultaneously with the incoming telephone call. Imagine a call center. An agent's phone is ringing. As it rings, the agent's computer screen pops up with information about the caller, what he ordered last, how much he owes, etc. This is called screen pop. The technology to make it happen typically comes from caller ID or ANI -information carried on the phone call just before the voice. It can also come from an IVR (Interactive Voice Response) system which answers the phone and asks the caller to punch in his phone or account number and then passes the phone call to the agent. In Screen Pop, the phone system typically listens for incoming digits and passes them across an attached local area network. See Caller ID.

screen refresh rate The rate at which your computer screen is re-drawn every second by a horizontal beam that scans from the top left hand corner to the bottom right hand corner. Screen refresh rates differ by the graphics standard you're running.

screen response time The time it takes to refresh a computer screen.

screen scraping This term originated in the late 1970s or early 1980s, when PC users began using their PCs and terminal emulation software to access mainframes and minicomputers. Screen scraping involved using a PC-based utility to capture character data from the screen of a mainframe or minicomputer application so that the captured data could be further processed with a PC application. The meaning of screen scraping has expanded in recent years to include screen capture and keylogging.

screen synch A colloquial term for sending an telephone service agent a phone call together with a screen of information about the incoming call, e.g. the customer's purchasing record or experiences with your product (if you're a help desk, for example).

screen tattoo A small graphical image that a user can download and install on a mobile phone to personalize the handset's screen display. Unlike wallpaper, which covers the entire handset screen, a screen tattoo covers only a small portion of the screen, and peacefully co-exists with wallpaper.

screened dual-homed gateway A screened dual-homed gateway is a dual-homed gateway that is guarded by a packet-filtering router.

screened subnet Also referred to as the demilitarized zone, a screened subnet is a collection of computers that are shielded from both the trusted network and the untrusted network by packet-filtering routers and gateways. See also DMZ.

screened transfer You are transferring a call from your phone to your boss. You dial a code for transfer, then dial your boss. The caller you're transferring is automatically put on hold. You speak to your boss, tell her who you're putting through. She okays the transfer, then you hit another digit and the call goes right through. This is called screened transfer. An unscreened transfer occurs when you simply dial your boss' office and send the call through without announcing it. Most PBXs have the ability to do both screened and unscreened calls.

Screened Twisted Pair ScTP. A type of cabling similar to UTP but ScTP has a foil shield between the conductors and the cable jacket. It also has a drain wire (a bare conductor). ScTP is used when ordinary UTP might pick up interference that would interfere with transmission. See UTP Cable.

screening router A device that, in addition to routing network traffic, is configured to reject packets which are not in keeping with the organization's policy. Screening routers are often deployed at the outer perimeter of a network and, therefore, serve as the first line of defense against network-oriented attacks; a.k.a filtering router.

screening telephone number The telephone number used by the phone company to bill, regardless of the number of phone lines associated with that number.

screensaver Before screensavers existed, phosphor would burn unwittingly into primitive computer screens. And bothered one day by the Phosphor on his computer screen Dr. Jack Eastman (physicist, geek, mathematician, software developer, and inventor of the modern-day screensaver) walked into his Berkeley, California kitchen, looked at his toaster, got inspiration, turned around, and immediately wrote the code for the Flying Toasters Screensaver. The rest is history. 80 different screensavers were developed into the "After Dark" series of classic screensavers. Today, screensavers adorn 97% of all personal computers, according to Dan Smith, General Manager of Screensavers.com. In the USA alone, there are 144 million computers with screensavers. Classic screensavers can be seen and downloaded at: www.screensavers.com/afterdark/

screw post Also called binding post. Screw posts are still used on many residential jacks. A conductor is installed on a screw post by stripping the insulation from the conductor to a half inch from the end, unscrewing the post to loosen it, wrapping the bare copper end of the conductor around the screw post between the washers and then re-tightening the screw. This doesn't make a very reliable connection, and it's easy for an installer to break the copper conductor by tightening the screw too tightly.

script A type of computer code than can be directly executed by a program that understands the language in which the script is written. Scripts do not need to be compiled into object code to be executed.

script files Some communications programs had script files that automate logging onto communications services, such as MCI Mail. The files are saved on your disk and read by your communications software when connecting to a remote service. Newer communications programs will "write" their own scripts by recording what you do in response to what questions from the remote service. This typically happens using a program feature called "Learn."

script kiddies Aspiring young hackers, usually teenagers or curious college students, who don't yet have the skill to program computers but like to pretend they do. They download ready-made scripts, languages, techniques and viruses written by more experienced crackers. They claim to have written them themselves and then set them free in an attempt to assume the role of a fearsome digital menace. Script kiddies often have only a dim idea of how the code works and little concern for how a digital plague can rage out of control. Script kiddies are also aspiring hackers who use their ready-made software to break into online distant computer sites, usually via dial-up phone lines. Script Kiddies, in the genre of computer hackers, are the lowest form of life, as they don't have the skills to develop their own techniques, and often don't understand the havoc they wreak. See also Cracker, Hacker, Phreak, and Sneaker.

script language A software language that contains English-statements for commands. A statement might be as simple as WrapPara() for wrap paragraph. Typically a script language contains commands that are specific to the type of task it's doing. For example, VOS from Parity Software in San Francisco is a script language for voice processing using Dialogic voice processing cards. A script language is more flexible than an Applications Generator, but is more difficult to program.

scripting engine A program that interprets and executes a script. See also script.

scroll bar A bar that appears at the right and/or bottom edge of a window or list box whose contents are not completely visible. Each scroll bar contains two scroll arrows and a scroll box, which enable you to scroll through the contents of the window or list box.

scrolling Browsing through information at a video terminal. Scrolling is the continuous movement of information either vertically or horizontally on a video screen as if the information were on a paper being rolled under it.

scruple The scruple was a unit of weight equal to 20 grains, used by apothecaries in olden days. Apprentices were supposed to always use these weights to measure out prescriptions. However, this was very tedious and many times the apprentice would just take a pinch of whatever was supposed to go into the prescription, without weighing it. If caught by the apothecary the apprentice was often scolded, "What is the matter with you? Have you no scruples?"

SCS Structured Cabling System. See Structured Wiring System.

SCSA Signal Computing System Architecture. SCSA is a comprehensive architecture that describes how both hardware and software building blocks work together. It has now been absorbed by S.100, but the following words still apply: It focuses on "Signal Computing" devices, which refer to any devices that are required to transmit information over the telephone network. Information can be transmitted via data modems, fax, voice or even video. SCSA defines how all these devices work together. Signal computing systems combine three major elements for call processing. Network interfaces provide for the input and output of signals transmitted and switched in telecommunications networks. Digital signal processors and software algorithms transform the signals through low-level manipulation. Application programs provide computer control of the processed signals to bring value to the end user.

SCSA is the common set of standards that telecommunication system manufacturers

and computing system manufacturers can use to create computer telephony systems. The theory is no single company today can create the total solution for all customers. SCSA represents the common ground between the two fields so that manufacturers from each area can safely develop products that will work with other manufacturers. SCSA's coverage extends from low-level bus and hardware interfaces, like the inter-board switching bus that enables boards from different suppliers to work together, to high-level application programming and software interfaces, so that software designed to work with one set of hardware products, will work with different hardware. Dialogic Corporation of Parsippany, NJ announced SCSA in the Spring of 1993. Dialogic said that SCSA was defined and created with input from a number of leading computer and switch manufacturers, call processing suppliers, and technology developers. In many cases, SCSA has drawn on existing standards, like the T.611 fax standard endorsed by the European Computer Manufacturers Association, and in other cases SCSA has extended standards to make them more useful for call processing suppliers and users.

SCSA describes all elements of the system architecture from the electrical characteristics of the SCbus and SCxbus to the high level application programming interfaces (APIs). According to TELECONNECT Magazine, this SCSA standard (and now, by extension, the S.100 standard) is remarkable for several things:

1. On the day of its announcement over 60 telecom and voice processing companies publicly endorsed SCSA. In early 1994, over 150 companies public endorsed it.

2. With SCSA -a standard for PC/LANs and VME-backplaned computers -you can build much larger telecom switches and much larger call and voice processing boxes. Previous standards, like AEB, PEB and MVIP, were basically limited to what you could do with one PC. Now PCs can be joined together. With SCSA, you can put 16 T-1 lines, or 512 voice lines in one PC and join together 16 PCs, for a total of 16 x 16 x 24 = 6,144 lines! That's a central office built out of networked PCs. A mainframe built out of a LAN. The SCSA joining is not via LAN or LAN-emulation. That would be too slow and the transmission too bursty (great for data, lousy for voice). It's via an SCbus -something that looks and works like a PBX backplane.

3. SCSA incorporates virtually every other standard in PC-based switching -including the most popular ones, Mitel's ST-Bus, MVIP, Siemens PCM Highway, AEB and PEB.

4. It's a lot faster and more reliable. All signaling is out of band. There's clock fall back and time slot bundling. It's more modular, meaning you can start with one PC and grow one at a time. That makes it more "modular" (scaleable is the new word). It's also hot pluggable. You don't have to turn off to upgrade.

5. It has applications portability. Tandem, the highly-successful fault tolerant minicomputer maker, has an SCSA application in a call center. They call it the Tandem Non-Step Call Center. It uses the Tandem 2400 VRU and the 4800 VRU.

SCSA is open, truly open. All its specs and all levels of its specs are available. To that extent, SCSA represents a remarkable gamble by its creator, Dialogic, a telecom/voice processing hardware company. It is encouraging competing manufacturers to build hardware to its specs and gambling that it won't be left in the dust, as IBM was with its PC. (Compaq, not IBM, built the first '386 PC.)

SCSA, as an idea, is revolutionary (for telecom). No one in telecom has ever promulgated an open standard everyone can adopt -hardware and software vendors. Write one application, create one applications generator, design one piece of hardware. Erector set telecom/voice processing! Build small. Build large. Just join the bits and pieces together. See also AEB, ECTF, PEB, MVIP, Signal Computing (for a differently-worded definition), S.100 and TAO.

SCSA call router An SCSA definition. A system service of SCSA which provides the basic necessities of inbound and outbound call processing and call sharing to client applications, without those applications needing to be aware of the underlying telephony interface operations. See ECTF, SCSA and TAO.

SCSA compatible An SCSA term. Able to function in an SCSA environment in its native mode.

SCSA hardware model An SCSA definition. The hardware layers of the SCSA specification. The SCSA Hardware Model defines an open architectural specification for a digital intra-node communication bus (SCbus), a switching model (SCSA Switching Model), and an multimode expansion capability (multimode Network Architecture, or MNA). The SCSA Hardware Model may be implemented independently of the SCSA Telephony Application Objects Framework. See TAO.

SCSA message protocol The open communications protocol by which entities communicate with one another in an SCSA system. The SCSA Message Protocol (SMP) is independent of the transport layers it is built upon, computer hardware, operating system, network topology (or lack thereof), and technology vendor. All SCSA-compliant AIAs will translate the functions called by client applications (via the API) into SMP messages; these are transmitted to service providers regardless of their location. Therefore, applications written to the API will be portable from one call processing environment to another. See TAO.

SCSA message protocol interface The message presentation format required by, and used by, the service provider in delivering SPM information. Contrast with Service Provider Messages. See TAO.

SCSA server A collection of service providers (objects) which in the aggregate implement the minimum set of services required for SCSA system conformance. The assumption is that these services are at a minimum provided to remote hosted client applications via common transports such as LANs, but may also be provided to client applications which are hosted on the SCSA server itself. Note that this is a logical image which may be implemented through multiple nodes (machines). See TAO.

SCSA Telephony Application Objects Framework The SCSA Telephony Application Objects (TAO) Framework originally defined the software layers of the SCSA open computer telephony specification. The SCSA TAO Framework defined a hardware-independent, open software architecture that simplifies design of distributed computer telephony systems. The SCSA TAO Framework includes a suite of interoperable, vendor-independent application programming interfaces (SCSA APIs), a set of System Services for handling various server management functions, and a set of messages and a standard transport for communication among various technology resources and system service providers (Service Provider Messages and SCSA Message Protocol). In 1995, TAO's development was taken over by a new organization ECTF -the Enterprise Computer Telephony Forum. The ECTF expanded the idea of TAO to make it the open software framework for the whole computer telephony world -to encompass hardware conforming to all major specifications, including SCSA and MVIP. Towards the end of January, 1996, ECTF promulgated TAO (now under a different name) as the software standard for the new computer telephony industry.

SCSI Small Computer System Interface. (Pronounced Scuzzie.) SCSI is a way for a devices such as magnetic hard disks, optical disk drives, tape drives, CD-ROM drives, printers and scanners to communicate with the computer's main processor. SCSI is a bus and an interface standard. SCSI has improved over the years. Here how:

Standard	Better Known As	Bus Width	Throughput	Max Devices On Chain	Cable Type
SCSI-1	Asynchronous	8-bit	4MB/sec	7	50-pin
SCSI-1	Synchronous	8-bit	5MB/sec	7	50-pinß
SCSI-2	Wide SCSI	16-bit	10MB/sec	15	68-pin
SCSI-2	Fast SCSI	8-bit	10MB/sec	7	50-pin
SCSI-2	Fast Wide	16-bit	20MB/sec	15	68-pin
SPI/SCSI-3	Ultra SCSI	8-bit	20MB/sec	7	50-pin
SPI/SCSI-3	Ultra Wide	16-bit	40MB/sec	7	68-pin
SPI-2/SCSI-3	Ultra2	8-bit	40MB/sec	7	50-pin
SPI-2/SCSI-3	Ultra2/Wide	16-bit	80MB/sec	15	68-pin
SPI-3/SCSI-3	Ultra3/Ultra160	16-bit	160MB/sec	15	68-pin

The brains of a computer is its microprocessor. That microprocessor (i.e., computer on a silicon chip) does the computer's primary work (i.e., calculations). There must be a way for information to get into and out of the microprocessor. The history of computers could be written as a continuing race to figure new, faster and more efficient ways of getting information into and out of the microprocessor. The obsession with input/output (I/O) stems from the fact that the microprocessor can work much faster than you can get information in and out and out of it. SCSI is a way for a devices such as magnetic hard disks, optical disk drives, tape drives, CD-ROM drives, printers and scanners to communicate with the computer's main processor. SCSI is a bus and an interface standard. The theory is that if you buy a SCSI device you can plug it into your computer's SCSI port and it will work, just as a parallel or serial port device will work. There are two good points about the SCSI interface, especially the newer SCSI-2 interface. First, it's fast. Second, one SCSI bus allows you can daisy chain up to seven different devices, so long as you remember to terminate the end of the chain. (In reality, you rarely have more than four devices hooked up on one SCSI link, since protocol overhead and other factors begin to degrade system performance.) Each device will work quickly and each won't siphon excessive power from the computer's main processor. That's because the SCSI bus typically has its own controller/microprocessor which takes care of the SCSI I/O workload. SCSI disk drives work faster than a "normal" IDE hard drive, which is why many new computers are coming with SCSI drives, not IDE drives. ANSI (American National Standards Institute) has set several guidelines for SCSI connection. There is SCSI-1

and SCSI-2. The SCSI specifications are available from www.ansi.org. See also iSCSI.

All Apple Macintosh computers come with built-in SCSI ports to which you can daisy chain one SCSI peripheral after another, until you have a total of seven. This is a fairly easy job, since Macintosh SCSI ports are standard and manufacturers of Macintosh SCSI peripherals will certify that their product works with the Macintosh SCSI standard. They wouldn't sell it if it didn't. One point: If you've removed the hard drive in your Macintosh and replaced it with one or more SCSI-attached drives, your Macintosh may require a hard disk terminator. Some Macintoshes require a hard disk terminator (a $5 device) if their hard disk has been removed.

To add SCSI devices to a MS-DOS machine, you must first place a SCSI adapter card in your PC's bus or your MS-DOS laptop's PCMCIA slot and connect the SCSI devices to that card. Sadly, for MS-DOS machines, SCSI is not a universal plug-n-play standard. According to Keith Comer of Toshiba, when asked why Toshiba's computers didn't come with SCSI ports as they came with parallel and serial ports, said, "I an unconvinced of SCSI's universal compatibility. It's a nontrivial task to connect SCSI devices. All devices need their own drivers. And each need to be configured for the particular SCSI card you have. Further, many of the SCSI drivers are incompatible with memory managers. In short, for us as manufacturers it would be a support nightmare."

The problem is lessening slowly. Corel (Ottawa, Canada) and others are creating "standard" SCSI Interface kits (software and/or hardware). These make connecting things less of a pain. But your desired-to-connect SCSI device (e.g., CD-ROM or magnetic optical drive) must be on the Corel list of approved devices. And -this is the Catch 22 of SCSI -if your SCSI device is new, you can be sure it will not be on Corel's list and probably will not work. In short, do not even bother trying to connect your "standard" SCSI device -unless someone has assured you that they have seen it work and it is on someone's list of approved SCSI devices. Yours truly has failed to connect many new SCSI devices using "standard" SCSI software. And when I asked the manufacturers (Corel, etc.) why they didn't work, I was told that my devices were too new and they hadn't released the necessary device driver software. But it's worse. Fingerpointing prevails. Manufacturers of SCSI deny responsibility for making SCSI device drivers to make their hardware work. And manufacturers of SCSI software and SCSCI adapter cards say they haven't been able to obtain/acquire/buy one of the new devices and figure out how to connect it. Further, there's no assurances that they will ever bother to figure out how to connect that particular device. As I write this entry, I have two SCSI devices on my desk which I cannot connect through several "standard" SCSI adapter cards I am testing. I am able to connect them through a SCSI cable connected to my computer's parallel port. The throughput is very slow, . An example: Using a parallel cable, I was able to transfer a 1-Megabyte file in 21 seconds. Using the only PCMCIA SCSI adapter card I could coax to work, I was able to transfer the same file in 12.3 seconds. See also Geoport, iSCSI, SCSI-2, SCSI Transfer Rate and USB.

SCSI-2 SCSI-2 (pronounced Scuzzie-Two) is a 16-bit implementation of the 8-bit SCSI bus. Using a superset of the SCSI commands, the SCSI-2 maintains downward compatibility with other standard SCSI devices while improving upon reliability and data throughput. SCSI-2 is capable of transferring data at rates up to 10 megabytes per second, twice as fast as SCSI-1. SCSI-2 defines more than a speed. It defines a command set and electrical characteristics. See also SCSI, SCSI-1 and SCSI Transfer Rate.

SCSI Transfer Rate SCSI transfer rate is the speed of moving data between the SCSI adapter board and the SCSI device. Host transfer rate is the speed of moving data between the adapter board and the host PC. See SCSI and SCSHI. Some hard disks come as SCSI. One way of distinguishing between these SCSI disks is to look at the pinning on the SCSI hard disks. There are three basic varieties of SCSI hard disks:

- 50-Pin. Ultra SCSI, 20 megabyte per second transfer rates, standard 50-pin cable which is backwards compatible with previous SCSI connections. Maximum cable length is 4.5ft.
- 68-Pin. Ultra Wide, 40 megabyte per second transfer rate, 68-pin Wide Cable requires Ultra Wide Controller for maximum transfer rates and optimal performance.
- 80-Pin. Ultra Wide SCA, 40 megabyte per second transfer rate, Single connector Drive designed to plug into systems with 80 pin back plane. Thus no controller Card and no Cable.

SCTE Society of Cable Telecommunications Engineers, Inc. A not-for-profit professional organization organized in 1969 to promote the sharing of operational and technical knowledge in the field of cable TV and broadband communications.

ScTP 1. Screened Twisted Pair. A type of Shielded Twisted Pair (STP) cable which employs a braided screen shield to protect the signal-carrying conductors from EMI (Electro-Magnetic Interference). See also FTP and UTP.

2. Simple Computer Telephony Protocol. SCTP is an Internet protocol authored by Brian McConnell (PhoneZone.Com) and Paul Davidson (Nortel). The protocol, modeled after other Internet application protocols (such as HTTP (worldwide web), SMTP (email), etc), creates a simple, cross-platform interface for building computer telephony applications. Unlike APIs such as TAPI and TSAPI, SCTP can be implemented on any machine which is capable of talking to TCP/IP networks. APIs, on the other hand, are operating system specific. The protocol is primarily intended for use in call control and system administration software. It is not used to create interactive voice response applications. Several vendors, such as Nexpath, a PC PBX manufacturer, have used the protocol to create cross-platform Java CTI applets which will run on virtually any operating system. SCTP is public domain, meaning the specification is public, and that anybody can use the protocol freely. www.phonezone.com/sctp

ScTP RJ-45 Plug These are used to terminate four pair ScTP patch cords. They have metal areas to connect the cable's foil shield with the equipment that it is plugged into.

scurvy Captain Cook lost 41 of his 98 crew to scurvy (a nutrition deficiency disease caused by a lack of vitamin C) on his first voyage to the South Pacific in 1768. By 1795 the importance of eating citrus was realized, and lime and lemon juice was issued on all Royal Navy ships. This is the reason British people are known as "limeys" in Australia. I was born in Australia.

SCVF Single Channel Voice Frequency.

SCWID Spontaneous Call Waiting Display.

SCxbus An SCSA term. The standard SCSA bus for communication between nodes. The SCxbus features the same architecture as the SCbus. See SCxbus Adapter.

SCxbus Adapter Inter-box expansion adapter for the SCbus.

SD 1. Starting Delimiter.

2. Secure Digital. See next definition.

SD memory card. PC Magazine called Secure Digital memory card the floppy disk of the mobile age. It is a stamp-size piece of flash memory developed by Matsushita, SanDisk and Toshiba. it lets you easily transfer data between between handheld devices such as cellphones, PDAs, MP3 players, digital cameras, digital video camcorders and laptops.

SDARS Satellite Digital Audio Receiver Services. See XM for a full explanation.

SDE 1. Synchronization Distribution Expander. 2. Secure Data Exchange as defined by the IEEE 802.10 security committee.

SDF Sub Distribution Frame. Intermediate cross connect points, usually located in wiring or utility closets. A trunk cable or LAN backbone is run from each SDF to the MDF (Main Distribution Frame).

SDH Synchronous Digital Hierarchy. A set of international fiber-optic transmission standards planned developed by the CCITT. SDH was based on the North American SONET standards, which now are considered to be a subset of SDH. See SONET for a much fuller explanation.

SDK Software Development Kit.

SDL 1. Specification and Description Language. A language defined in ITU-T Z.100 for telecommunication.

2. Signaling Data Link.

SDLC 1. Synchronous Data Link Control. A bit-oriented synchronous communications protocol developed by IBM, SDLC is at the core of IBM's SNA (System Network Architecture). Intended for high-speed data transfer between IBM devices of significance (read mainframes), SDLC forms data into packets known as frames, with as many as 128 frames being transmitted sequentially in a given data transfer. Each frame comprises a header, text and trailer. The header consists of Framing bits (F) indicating the beginning of the frame, Address information (A), and various Control data (C). The data payload, referred to as Text, consists of as many as 7 blocks of data, each of as many as 512 characters. The trailer comprises a Frame Check Sequence (FCS) for error detection and correction, and a set of Framing bits (F) indicating the end of the frame. SDLC is a protocol which supports

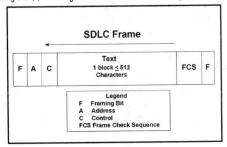

device communications generally conducted over high-speed, dedicated private line, digital circuits. SDLC can operate in either point-to-point or multipoint network configurations. See also HDLC and IBM. Contrast with Binary Synchronous Communications.

2. System development life cycle. A methodology for systematically designing and developing an information system or application.

SDLC-To-Token-Ring LLC Transformation A technique to integrate SDLC link-attached SNA devices into a LAN/WAN internet. A modified remote polling process is used to make the link-attached devices appear to be LAN-attached.

SDM 1. Subrate Data Multiplexing. A European term. In North America, it's called SRDM.

2. DMS SuperNode Data Manager.

SDMA 1. Station Detail Message Accounting. See Call Accounting.

2. Space Division Multiple Access.

SDMF Single Data Message Format. See Caller ID Message Format.

SDMI Secure Digital Music Initiative. Backed by the Recording Industry Association of America (RIAA), this initiative is working on a standard which can be built into digital music files and players to prevent illegal copies from being made.

SDN Software Defined Network. See Software Defined Network, SDN Serving Office, Virtual Network, VPN and the Appendix.

SDN Serving Office One of many AT&T-supplied switching nodes in an SDN network. See also Software Defined Network and the Appendix.

SDR 1. Session Detail Record. Data records generated by LAN-attached devices (e.g., switches and routers) that are captured, compressed, and stored on a central repository for purposes of developing various reports that are used by a network accounting system to ensure that data network resources are being used properly and effectively. Network accounting systems use SDR in the data domain much as call accounting systems use CDR (Call Detail Records) output by a PBX or Centrex system in the voice domain. See also Call Accounting and Network Accounting.

2. See Software Defined Radio.

SDRAM Synchronized Dynamic Random Access Memory. SDRAM largely has replaced DRAM (Dynamic RAM) as the most common main memory for PCs. SDRAM's memory access cycles are synchronized with the CPU clock in order to eliminate wait time associated with memory fetches between RAM and the CPU. Data bursts as high as 150 MHz are supported. PC100 and PC133 are Intel versions for SDRAM motherboards running at 100 MHz and 133 MHz, respectively, and intended for PC with faster processors running at up to 600 MHz and 1 GHz, respectively. SDRAM's leadership is being challenged by DDR-SDRAM (Double Data Rate-SDRAM) and RDRAM (Rambus DRAM), both of which are faster.

The Evolution of SDRAM				
Memory Type	Actual Clock Speed	Effective Clock Speed		Year Released
66-MHz SDRAM	66 MHz	66 MHz	0.528 GB/sec	1996
PC100 SDRAM	100 MHz	100 MHz	0.8 GB/sec	1998
PC133 SDRAM	133 MHz	133 MHz	1.064 GB/sec	1999
DDR200 SDRAM	100 MHz	200 MHz	1.6 GB/sec	2001
DDR266 SDRAM	133 MHz	266 MHz	2.1 GB/sec	2001
DDR333 SDRAM	166 MHz	333 MHz	2.7 GB/sec	2002
DDR400 SDRAM	200 MHz	400 MHz	3.2 GB/sec	2003
QBM/DDR400	200 MHz	800 MHz	6.4 GB/sec	2003
QBM/QDR/DDR400	200 MHz	1600 MHz	12.8 GB/sec	2004*
QBM/QDR/DDR400 128-bit	200 MHz	3200 MHz	25.6 GB/sec	2004*
*Estimated				

See also DDR-SDRAM, DRAM, EDO RAM, Flash RAM, FRAM, Microprocessor, RAM, RDRAM, SRAM, and VRAM.

SDRM Sub-rate Data Multiplexing. Refers to a service where a DS0 (64 Kbps) channel may contain one 56 Kbps signal, five 9.6 Kbps signals, ten 4.8 Kbps signals or twenty 2.4 Kbps signals. Although speeds may be mixed, the highest speed determines the number of signals supported.

SDP 1. Session Description Protocol. Specified in the IETF's RFC2327, SDP is intended for the description of multimedia sessions over IP-based networks. SDP is used for session announcement, session invitation, and other forms of session initiation. Session Initiation Protocol (SIP) relies on SDP. See also SIP. See TAPI 3.0 for a full description.

2. Service Delivery Point. The Minimum Point Of Entry (MPOE), where the commercial

carrier establishes a demarc (demarcation point). The demarc draws the line between the local loop, which is the responsibility or the carrier, and the inside wire and cable system, which is the responsibility of either the end user or the building owner. See also MPOE.

3. A Bluetooth term. Service Discovery Protocol.

SDP client A Bluetooth term. The SDP in a Local Device (LocDev). The SDP client requests service information from SDP servers.

SDP server A Bluetooth term. The SDP in a Remote Device (RemDev). The SDP server responds to requests made by SDP clients.

SDP session A Bluetooth term. The exchange of information between an SDP client and an SDP server. The exchange of information is referred to as an SDP transaction.

SDP transaction A Bluetooth term. The exchange of an SDP request from an SDP client to an SDP server, and the corresponding SDP response from an SDP server back to the SDP client.

SDS Short Data Service. A data transmission service for the transmission of short alphanumeric data messages in the European TETRA (TErrestrial Trunked RAdio) mobile radio system. SDS is much like SMS (Short Message Service) in cellular networks. See TETRA.

SDSAF Switched Digital Services Applications Forum, a group of manufacturers and carriers whose objective to standardize the interconnection of switched 56 kilobit and n x switched 56 channel local and long distance services. The group is based in Reston, VA. Today a switched 56 Kbps "phone" call between multiple carriers probably wouldn't get through. In short, this group is trying to bring the simplicity of the voice dial up phone system into the switched data world.

SDSL Symmetrical Digital Subscriber Line, also sometimes referred to as Single-line DSL. SDSL is a proprietary version of symmetric DSL versions such as HDSL and HDSL2. SDSL technology offers digital bandwidth of up to 2.3 Mbps both ways (that's why it's called symmetrical) over a single twisted-pair copper phone line, over distances up to about 10,000 feet on an unrepeatered basis. SDSL is aimed at the corporate and SOHO markets that require high upstream and downstream traffic rates. SDSL uses the same 2B1Q modulation scheme used in ISDN BRI. In February 2001, the ITU-T standardized on G.shdsl, which largely obsoleted SDSL. See also xDSL, ADSL, G.shdsl, HDSL, HDSL2, IDSL, RADSL, SHDSL, and VDSL.

SDT Structured Data Transfer: An AAL1 data transfer mode in which data is structured into blocks which are then segmented into cells for transfer.

SDTV Standard Definition TV. A set of standards for DTV (Digital TV) approved by the FCC in December 1996, SDTV offers about the same definition as current conventional analog TV. There really are two SDTV versions. The first calls for 480 vertical lines, 640 horizontal pixels, an aspect ratio of 4:3, and a frame rate of 24, 30, or 60 fps (frames per second). The second is much the same, but increases the number of horizontal pixels to 704, and supports aspect ratios of both 4:3 and 16:9. See also DTV and HDTV.

SDU-SMDS Data Unit The user payload in an SMDS L3PDU packet. The SDU can contain up to 9,188 bytes.

SDU An ATM term. Service Data Unit: A unit of interface information whose identity is preserved from one end of a layer connection to the other.

SDV Switched Digital Video.

SDVN Switched Digital Video Network.

SE 1. Systems Engineering

2. An ATM term. Switching Element: Switching Element refers to the device or network node which performs ATM switching functions based on the VPI or VPI/VCI pair.

SEA Self Extracting Application. You buy a piece of software. It comes delivered as one file with the extension .exe. You click on the file, bingo it executes its one main function, namely to extract the various files that go into making up the software and save those files to somewhere on your hard disk. It may then ask you if you'd like to install the software on your PC.

seagull manager A manager who flies in, makes a lot of noise, is critical of everything, then leaves. The seagull metaphor comes from the activity that seagulls are most known for, i.e. defecating. See also Albatross Manager.

SEAL Simple and Efficient Adaptation Layer: An earlier name for AAL5.

sealed case PC A fancy name for a PC that never needs to be opened up.

sealing current A designation for a powering situation that consists of a wet loop without span power.

Sealink Sealink is an error-correcting file transfer, data transmission protocol for transmitting files between PCs. It is a variant of Xmodem. It was developed to overcome the transmission delays caused by satellite relays or packet-switching networks.

seals A way of telling if a device has been tampered with.

seamless The word seamless means "perfectly smooth, without awkward transitions." In software, it means that what takes place between the user and the application or applications accessed by the user is perfectly smooth to the user and the software being used by the user will work easily with other software the user is using. On a network accessing "seamless" applications, the user doesn't perceive he's on a network because his programs run as though they were on his personal computer. In actual fact, the word "seamless" is very vague. No one has a technical definition of seamless. Originally, in years past, the word meant "without the stitches showing." This 15-century word got a boost from the phrase seamless stockings which filled a brief period between silk stockings -which had seams that always needed straightening -and most panty hose that doesn't, sadly. In short, when any vendor says they offer "seamless integration" with something else, be markedly wary. Follow the fundamental rule in this dictionary: Check.

seamless messaging Seamless messaging is a new phone service that will allow subscribers, according to Lucent, to send and respond to messages across a service provider's entire network. This will make it possible for subscribers to communicate with large groups of people such as community interest groups, athletic teams or family and friends located in different states within a service provider's region -reducing the time and effort needed to place individual calls. Single Number Retrieval will allow service providers to provide quick and easy access to voice mailboxes by assigning a single, memorable number to all of their voice mail subscribers.

seamless mobility Continuous wireless connectivity for a mobile user prior to, during, and following a move from one location to another.

search drive A drive that is automatically searched by the operating system when a requested file is not found in the current (default) directory. A search drive allows a user working in one directory to transparently access an application or data file that is located in another directory.

search engine An Internet World Wide Web term. A search engine is a program that returns a list of Web Sites (URLs) that match some user-selected criteria such as "contains the words cotton and blouse." Basically, the procedure is simple. You surf to the search engine's site. You click a couple of times and type in what you're looking for. A few seconds later you get choices. You finally make your selection and you get instantly hotlinked over to that site. Search Engines are the most useful thing to come along in years. I use them daily to find everything from information on a company I'm looking for to new definitions to fill this ever-expanding dictionary. Here are the main Internet search engines and their addresses, or more precisely the search engines I use regularly.

GOOGLE	www.google.com
CLUSTY	www.clusty.com
MSN	www.msn.com
YAHOO	www.yahoo.com
ASK JEEVES	www.askjeeves.com
TEOMA	www.Teoma.com
EXCITE	www.excite.com
HOT BOT	www.hotbot.com
LYCOS	www.lycos.com
INFOSEEK	www.go.com
OPENTEXT	http://pinstripe.opentext.com/
PROFUSION	www.profusion.com
WEBCRAWLER	www.webcrawler.com

You can find many more search engines by going to www.SearchEngineWatch.com.

Search engine optimization Google ranks Web sites based largely on the quantity and quality of other sites linking to it. Search-engine optimization refers to mining the Web for links and other page-tweaks that can help sites boost their Google rank and reel in more visitors. See also search engine.

Search Expression See query restriction.

SEAS Signaling Engineering and Administration System.

seasonality The month-to-month fluctuation in call volume that can be expected to recur each year in a call center. For example, a call center might always have peak months in the spring and early summer and slow months in the late fall and winter.

seated agents A call center term. See Base Staff.

SEB Site Event Buffer. Another name for a modem typically used for relaying error codes on PBX and voicemail systems. An SEB is primarily used for dialing into, in order to protect the PBX and voicemail with another layer of security and also so you only need one line to access both devices. We have both Nortel and Octel/Avaya equipment and this device allows for that interface.   I do not think that it is vendor specific. We also get alarms from this device if there is a power failure or if it loses connectivity to the PBX or Voicemail system. What does the SEB dial its error messages into? It connects to another modem that is serially connected to a special application that decifers the error codes that are transmitted from the SEB. The SEB has a serial interface to a TTY port on a PBX or Voicemail system that captures error codes from those devices.   It is all in ASCII.

SECABS Small Exchange Carrier Access Billing Specifications

SECAM Acronym for Systeme Electronique Couleur Avec Memoire or SEquential Couleur Avec Memoire. A television signal standard used in France, eastern European countries, Russia, the former Soviet Union and some African countries. SECAM uses an 819-line scan picture which provides a better resolution than PAL's 625-lines and NTSC's 525-lines (the US standard). All three systems are not compatible. You cannot view an Australian, English or French videotape or through-the-air broadcast on a US TV. See NTSC.

second audio program An auxiliary audio feed for a TV program that can be used for program audio in another language, such as Spanish in the United States and French in Canada, or for something completely different, such as weather information or Descriptive Video Services (DVS) for the visually impaired. The second audio program (SAP) feature is available only on TV sets that have stereo sound. To access it, use your TV remote's audio button to select the second audio program. If there is no audio button on the remote control, press the menu button then look for options like audio, audio setup, or setup and then make your audio selection.

second dialtone 1. Dialtone given to the caller, on a phone system (e.g. PBX, Centrex or hybrid), after dialing an access code (e.g. 8 or 9) to make a call out of the system (e.g. a local or long distance). Sometimes referred to as outside dialtone.

2. Dialtone returned to a caller after they've dialed a local or long distance number and reached some type of switching device. That switching device might allow you to dial into a fax machine, a modem, a phone or an answering machine. It might allow you to dial into one of several cash registers or soda machines (to check if they need being refilled). It might even allow you to dial long distance through a voice mail system or through a long distance phone company. As you dial through networks you might encounter not only second, but also third and fourth dialtones.

second level domain In the Domain Name System (DNS) of the Internet, the second level domain is the next lower level of the hierarchy underneath the top level domains. In a domain name, the second level domain is that portion of the domain name that appears immediately to the left of the top level domain. For example, the technology-investor in technologyinvestor.com. Second level domain names are often descriptive and have come to be used increasingly to represent businesses and other commercial concerns on the Internet. See also: Domain Name System. DNS.

second marriage A triumph of hope over experience.

second mover advantage The theory of second mover advantage is that market dominance hardly ever goes to the plucky startup that first releases a product or service. Rather, it does to the second company, sometimes an established company that most effectively markets and sells the product and service.

second order harmonics The second multiple of a specific frequency or a specific frequency multiplied by two.

second source When you make things that your suppliers build into other things, they prefer that they had a second place to buy things you supply. They clearly have fears that your factory will burn down, be flooded, be blown away, etc. They don't want to put out of business because something horrible happens to you. They want a second source. Some manufacturers specifically license their products to other manufacturers -so their customers will have a second source.

second time The second time you do something is always much much faster than the first time. So when your phone company tells you it'll only take you three minutes to move an extension on your PBX, they mean the second time you do it.

secondary carrier A customer can override their primary carrier selection on a call by call basis and select another long distance telephone company to carry their long distance traffic in equal access areas. See also 101XXXX and Primary Interchange Carrier.

secondary channel A low-speed transmission channel provided in many dedicated line data modems to permit simultaneous control and network management transmission to coexist with the main higher-speed data channel; actually a form of FDM (Frequency Division Multiplexing) of a voice channel for use by two modems.

secondary equipment Used telecommunications equipment. See also Used, Certified, Refurbished and Remanufactured.

secondary market The market for used business telecommunications and computer equipment.

secondary protection Primary protection is a device that sits at your building entrance between your phone line coming in from outside and your line going into and up your building. The phone company is responsible for installing primary protection. Secondary protection sits on your floor just next to your phone system. Secondary protection is designed to protect your phone equipment from spikes, surges and high electricity that might affect your phone lines between the primary protection downstairs and the secondary protection upstairs. Secondary protection typically costs $20 to $30 a line. It's worth every penny.

secondary radar A radiodetermination system based on the comparison of reference signals with radio signals retransmitted from the position to be determined.

secondary radiation Particles (such as photons, Compton recoil electrons, delta rays, secondary cosmic rays, and secondary electrons) that are produced by the action of primary radiation on matter.

secondary resource An SCSA definition. Any resource that is attached to a Group after the Group has been created around a primary resource. See SCSA.

secondary ring One of the two rings making up an FDDI ring. The secondary ring is usually used in the event of failure of the primary ring.

secondary server Under NetWare, the secondary server is the SFT III NetWare server that is activated after the primary server that receives the mirrored copy of the memory and disk from the first server activated. The secondary server mirrors the disk and memory image of the primary server. Though it cannot be used to do additional work (because it uses all of its CPU cycles keeping up with the primary server), the secondary server can act as a router for the local network segments to which it is directly attached. In addition to mirroring the primary server, the secondary server provides split seeks. Either SFT III server may function as primary or secondary, depending on the state of the system. You cannot permanently designate which server is primary or secondary. System failure determines each server's role, that is, when the primary server fails, the secondary server becomes the new primary server. When the failed server is restored, it becomes the new secondary server. See also Server Mirroring.

secondary service area The service area of a broadcast station served by the skywave and not subject to objectionable interference and in which the signal is subject to intermittent variations in strength.

secondary station In a data communication network, the secondary station responsible for performing unbalanced link-level operations, as instructed by the primary station. A secondary station interprets received commands and generates responses.

secondary winding The minor winding on a relay having two windings. The winding on a transformer that is not connected to a AC source.

secondhand speech The noise people make when they talk publicly into cellphones. Secondhand speech is the cellphone equivalent of secondhand smoke.

SECORD SEcure voice CORD board. Now extinct.

secretarial hunting The secretary's station number is programmed as the last number in one or more hunt groups. If all phones within a hunt group are busy the call will hunt to the secretary.

secretarial intercept A PBX feature. Causes calls for an executive to ring his/her secretary -even if the executive's direct extension number was dialed. The executive's phone will ring only if the secretary's phone is placed on "Do Not Disturb" or the secretary transfers the call in.

SECTEL Acronym for SEcure TELephone.

Section In Sonet, a section refers to an optical span and its equipment.

Section 251 See Telecommunications Act of 1996, U.S.

Section 254 See Telecommunications Act of 1996, U.S.

Section 271 See 271, Telecommunications Act of 1996, U.S.

Section 271 Relief The local Bell operating companies are allowed to sell long distance phone service. See 271 and 271 Hearings.

Section 706 See Telecommunications Act of 1996, U.S.

sectional center A control center connecting primary telco switching centers. A Class 2 office. The next to the highest rank (Class 2) Toll Switching Center which homes on a Regional Center (Class 1).

sectional parabolic antenna A parabolic antenna which consists of only a slice or part of a complete parabolic dish.

sector A pie-shaped portion of a hard disk. A disk is divided into tracks and sectors. Tracks are complete circuits and are divided into sectors. Under MS-DOS a sector is 512 bytes.

sector rotation Sector rotation is a stockmarket term. Investors sell out of networking stocks and they move into health care stocks. Sector is a stockmarket term for industry.

sectoring The process of dividing a mobile cellular radio cell into sectors, or smaller patterns of coverage. Traditionally, all cell antennas were omnidirectional; that is to say that they provided coverage in a 360-degree pattern. Sectoring is applied when either the capacity of the cell site is insufficient or when interference becomes a problem. Sectoring divides the number of channels assigned to the cell into smaller groups of channels, which are assigned to a sector through the use of directional antennas. Commonly, the cell antenna is divided into three sectors, each with a 120-degree coverage pattern. You probably have noticed that many contemporary cell site antennas are very tall for better coverage and triangular in shape for purposes of sectoring.

secure channel A technology that provides privacy, integrity, and authentication in point-to-point communications such as a connection on the Internet between a Web browser and a Web server. You can tell if you have a secure channel with Netscape by checking out the key on the bottom left hand side of your screen. If the key is broken, your connection is insecure. If it's together in one piece, then your Internet conversation is secure, which means it's encrypted and therefore hard for someone to break into and make sense of. See Internet Security, which details the problem of security on the Internet.

Secure Electronic Transaction SET. A system designed for electronic commerce over the Internet that promises to make stealing credit card numbers much more difficult. See Digital Cash.

secure kernel The core of a secure operating system.

Secure Hash Algorithm 1 See SHA-1.

secure phone lines There is no such animal as an absolutely secure phone line. There are only degrees of security. Think of a continuum. At one extreme is a normal analog phone line. It's totally unsecure. Anyone can attach a couple of alligator clips, join the circuit to a telephone and listen in on the call. At the other end of the continuum is a totally digital circuit (end to end digital) which is being encrypted by the topmost military encryption technology. In between, with varying degrees of security, are phone calls that start as analog and then change to digital, e.g. those that pass through a digital PBX, those that pass over the Internet or a private IP network.

secure public dial A security term. Secure public dial is dialup switching functionality that allows the service provider to offer customers the security of a private connection with the economics of public dial. Also referred to as Virtual Private Networking (VPN) allows the service provider to support multiple enterprises' dial-in requirements, securely from the same dialup switch. Each separate customer has access to their own virtual network although they may share physical ports and access paths. The service provider manages the multicustomer net with the same ease as if it were one internetwork.

secure shell protocol Protocol that provides a secure remote connection to a router through a Transmission Control Protocol (TCP) application.

Secure Sockets Layer SSL. SSL is the dominant security protocol for Internet / Web monetary transactions and communications. Information being transmitted is encrypted, and only the user's Web browser and the computer server at the other end running the Web site have the key, and thus can understand what each other is saying. But no one else can. Most reputable Web sites use SSL for credit card transactions on the Web. For a longer explanation, see SSL.

Secure Telephone Unit STU. A U.S. Government-approved telecommunication terminal designed to protect the transmission of sensitive or classified information -voice, data and fax.

secure voice Voice signals that are encoded or encrypted to prevent unauthorized listening.

Secure Voice Cord Board SECORD. A desk-mounted patch panel that provides the capability for controlling 16 wideband (50 Kbps) or narrowband (2400 bps) user lines and five narrowband trunks to AUTOVON or other DCS narrowband facilities.

Securite A maritime radio keyword that precedes a navigation or weather warning. Pronounced "seh-cure-ee-tay," from the French "securit0xC7" (safety).

security 1. See Secure Phone Lines.

2. A way of insuring data on a network is protected from unauthorized use. Network security measures can be software-based, where passwords restrict users' access to certain data files or directories. This kind of security is usually implemented by the network operating system. Audit trails are another software-based security measure, where an ongoing journal of what users did what with what files is maintained. Security can also be hardware-based, using the more traditional lock and key.

Security Accounts Manager SAM. A Windows service used during the logon process. SAM maintains user account information, including groups to which a user belongs.

Security Accounts Manager Database Also called the Directory Services Database, stores information about user accounts, groups, and access privileges on a Microsoft Windows NT server.

security blanking The ability of a switch to blank out the called digits for certain extensions so no called number detail is printed. Senior executives in serious takeover negotiations find this feature useful. There have been instances of people figuring out which company another company is about to buy based on telephone calling records. If you have this information, you can buy the company's stock before the bid is announced and make a lot of money. This feature -security blanking -is designed to avoid such occurrences.

security cabinet A cabinet, usually on casters, used to store confidential materials under lock and key prior to shredding.

security by default Security-by-default is an approach that calls for switches and routers to be in a secure configuration when sent from the factory. Out of the box, only the configuration options that are required to get the unit operational are active, while all other options and ports are turned off. From there, the administrator can turn on only what is needed without having to waste time and effort to make sure there are no open avenues of attack. For instance, there are no hidden accounts that need to be turned off or deleted such as maintenance accounts and back doors. In addition, router/switch software is designed and tested to eliminate well known coding vulnerabilities, including buffer overflow abuse, rogue format strings, exploits that bypass security, inadequate privilege checking, information leakage, and error handling, which can expose valuable information to attackers. Moreover, security-by-default switches and routers are hardened against common attacks. Types of protection include:

- ARP-ICMP hardening: A static threshold can be configured such that if ICMP/ARP levels are exceeded, the packets will be discarded and a trap is generated and sent to the network administrator that identifies the MAC and IP address of the ICMP source. Optionally, an ACL can be created so that the flooded port can be disabled.
- Self-defense against ARP poisoning: When this feature is enabled, the switch will periodically send ARP requests for the switch IP address. If a response is received from a different device, an alert will be sent to network administrator.
- Self-defense against port scan: Network administrators are able to turn off all in-band management ports, effectively hiding telnet, http, ftp etc. ports from port scanning programs.
- Valid packet attack protection: An active defense against "valid packet attacks" where the switch is flooded with packets that appear to be valid UDP or TCP packets (e.g., OSPF packets or BOOTP packets) to prevent the switch from processing valid packets.
- Source/flow learning hardening: An attacker could overflow MAC tables causing any traffic sent to the switch to be broadcast to all ports and enabling it to be sniffed. The MAC source learning process is designed to be resistant to MAC table overflow because new MAC addresses are learned only if the tables have room guarantying that active MAC addresses will not be overwritten and active Layer 2 flows will not be flooded as a result of a MAC attack.

The hardened configuration approach can also be applied outside of the router/switch environment by configuring individual systems to take the most secure approach by default. For instance, many email clients automatically open email through a preview pane or download attachments to your system and open them automatically. Newer email clients are now configured with security-by-default. One method is to first move attachments to a folder and a link to the file is inserted into the email instead of the attachment. While the attachment is in the intermediate folder it can be scanned for viruses. When the email is downloaded the attachment is not sent further limiting the risk of infection due to preview panes and automatic opening of attachments. If the user wants the attachment they can get it through the included link. Such steps make it easier for users to access email and saves time while reducing the amount of bandwidth that would be used to download garbage attachments. If the hardened configuration is the default, this is security-by-default.

The security-by-default approach reduces the number of exploitable options and eliminates the possibility of the administrator not knowing which default features need to be turned off. This makes the router or switch as secure as it can be after it is powered on for the first time while reducing the overall amount of time needed to put the router or switch into production use. This definition, courtesy Alcatel.

security by obscurity This term refers to a variety of soft, and ultimately risky, approaches toward securing a system. For example, a system administrator or network administrator may be aware of a security vulnerability but conclude that by not publicizing the vulnerability, it will go unnoticed and unexploited. Other forms of security by obscurity include the use of misleading variable names and function names in a script or program, or the use misleading filenames, or the storing of files in odd locations. All of this is done in the hope that no one will figure out how things work and discover weaknesses that can be exploited.

security code 1. A user identification code required by computer systems to protect information or information resources from unauthorized use. 2. A six-digit number used to prevent unauthorized or accidental alteration of data programmed into cellular phones. The factory default is 000000.

security dots The asterisks that appear onscreen as you type in your password.

security equivalence A security equivalence allows one user to have the same rights as another. Use security equivalence when you need to give a user temporary access to the same information or rights as another user. By using a security equivalence, you avoid having to review the whole directory structure and determine which rights need to be assigned in which directories.

security management Protects a network from invalid accesses. It is one of the management categories defined by the ISO (International Standards Organization).

security manager A Bluetooth term. The module in a Bluetooth device that controls security aspects of communications to other Bluetooth devices.

security modem A class of modem providing secure access.

security parameter index This is a number that, together with a destination IP address and a security protocol, uniquely identifies a particular security association. When using Internet Key Exchange (IKE) to establish the security associations, the SPI for each security association is a pseudo-randomly derived number. Without IKE, the SPI is specified manually for each security association.

security stud A cylindrically shaped metal finger that holds open the door to a Cash Box until the box is removed for collection.

security trapping See Positive Trapping and Negative Trapping.

SED Smoke Emitting Diode. Diodes not installed properly can become SEDs, which then can become FEDs (Fire Emitting Diodes). See also Diode.

seed money First money to invested in a startup. See also seedware.

seed router Router in an AppleTalk network that has the network number or cable range built in to its port descriptor. The seed router defines the network number or cable range for other routers in that network segment and responds to configuration queries from nonseed routers on its connected AppleTalk network, allowing those routers to confirm or modify their configurations accordingly. Each AppleTalk network must have at least one seed router.

seedware Software designed to get demand for a product or a new market segment started. Software designed to "seed" a market. Seedware is typically a less-full featured piece of software than the software you're really trying to sell. Seedware typically costs very little. It may even be free. It's also called Feedware.

seek time The time it takes to move a disk drive's read/write head to a given track. Seek time varies depending on where the head starts from and has to go it. Average seek time is a critical measure of the speed of a computer disk drive.

SEF 1. Source Explicit Forwarding. Security feature that allows transmissions only from specified stations to be forwarded by bridges.

2. Severely Errored Framing. A SONET defect which is the first indication of trouble in detecting valid signal framing patterns. Four consecutive errored framing patterns constitutes an SEF defect. If the SEF defect persists, an LOF defect is instituted, and if the LOF defect persists an LOF alarm is declared. See also LOF.

SEFS Severely Errored Framing Seconds. A count of the number of seconds during which at any point the SEF defect was present. See SEF.

segment 1. 64 characters. Use as a method of data communications billing by some overseas phone companies.

2. An electronically continuous portion of a network, usually consisting of the same wire.

3. A single ATM link or group of interconnected ATM links of an ATM connection.

segmentation Method of classifying a group of people by their purchase behavior, lifestyle and attitudes.

Segmentation and Reassembly SAR. An ATM term. See SAR.

seize To access a circuit and use it, or make it busy so that others cannot use it.

seizure signal A signal used by the calling end of a trunk or line to indicate a request for service.

SEL Selector: A subfield carried in SETUP message part of ATM endpoint address Domain specific Part (DSP) defined by ISO 10589, not used for ATM network routing, used by ATM end systems only.

select call forwarding A name for a Pacific Bell (and possibly other local telephone companies') service which allows you to have calls from selected numbers ring at another number.

selection What most people call a "block" in a word processed document, Microsoft calls a "selection." If you want to print test you've blocked, tell Windows to print a selection.

selective call acceptance Permits incoming calls only from numbers you pre-select.

selective call forwarding Forewords all calls from a pre-selected list to a specific destination.

selective call rejection Blocks incoming calls from numbers you pre-select.

selective calling The ability of the transmitting phone to specify which of several phones on the a line is to receive a message.

selective fading Fading in which the components of the received radio signal fluctuate independently.

Selective Incoming Load Control SILC. A control application available with CCIS that dynamically controls the amount and type of traffic offered to an overloaded or failed system.

selective paging to station A phone can page to individual phone instruments.

selective ringing A method of ringing only the desired party on a party line.

Selective Routing SR. In the US emergency services telephone network, selective routing is the routing of a 911 call to the appropriate public safety answering point based on the call's Emergency Service Number, i.e., the caller's location.

selective signaling A method of inband signaling used on private networks to tell switches to switch the call.

selectivity Ability of a tuner or receiver to get only a desired station while rejecting other adjacent stations. The higher the figure expressed in decibels (dB), the better the selectivity.

selector The identifier (octet string) used by an OSI entity to distinguish among multiple SAPs at which it provides services to the layer above.

selector switch The intermediate distributing switch in an (very old) step-by-step mechanical switching system. It is directly controlled by the customer dial in its vertical motion and hunts for an idle path in its angular motion.

self diagnostics Your phone system tells you when something is wrong with it by sending you a "message" via the operator console or through one of the data ports on the phone system.

self discharge When internal chemical reactions (such as the dying out of chemicals) cause the loss of useful capacity of a cell or battery in storage.

Self Electro-Optic Effect Devices SEEDs. Switches guided by light.

self extinguishing The characteristic of a material whose flame is extinguished after the igniting flame is removed.

self healing Self healing means it fixes itself, which is really a misnomer. It really doesn't fix itself. In telecom, self-healing really means 100% connectivity. The term came into telecom use with the invention of SONET fiber rings. Here's how they work: A phone company lays down several concentric rings of fiber -typically around a city or an industrial park. You, the customer, buy service from the carrier. If one ring fails, the system knows and instantly traffic is shunted in a different direction or to a different fiber. "Self-healing" is actually 100% fiber connectivity between business locations and telephone company serving wire centers. "Self-healing" means that SONET rings provide for automatic network backup with 100% redundancy so that if there is a point of failure on your fiber ring, your service continues. Self healing could also be used to describe a packet switched network (e.g. Ethernet-based) in which each node contains an intelligent router which continually senses what's going on in the network and routes traffic around bottlenecks and down circuits in real time.

There is another definition. Brian Livingston, author of books on Windows and InfoWorld columnist, says that self healing software upgrades itself by periodically calling a central phone number (via bulletin board) or a Web site and automatically downloading and installing any new components that may have become available since the last product upgrade.

Microsoft uses the term "self healing" to refer to an operating system that crashes less often than a traditional operating system. They mean that when the software encounters bad code, it goes on to find the good code and somehow replaces the bad code with good code. Several software programs, like Microsoft's Office claim to be self-healing.

A self healing grid is a proposed electrical power infrastructure that would monitor itself through an array of sensors, run simulations to identify potential glitches, and then correct itself to avert problems. See also IP VP and self restorable rings.

seal healing network See self healing.

self hosted An SCSA definition. The Client and the Server are on the same computer platform. A server with one or more self-hosted applications may be a standalone unit which is not connected to any other system.

self learning bridge See Bridge.

self replicate To re-create itself. A malicious piece of software -perhaps a Trojan Horse -that pops some

self restorable rings Optical rings can be self-restorable. Most vendors are considering the protection mechanisms (OSP and optical channel protection) for adaptation to the ring topology. Any of these mechanisms works under a failure situation. Optical channel subnetwork-connection protection (OCh-SNCP), as its name implies, protects the optical channel, while OMS shared protection ring (OMS-SPRing) protects all of the OMS. OMS-SPRing is the optical equivalent of the multiplex section shared protection ring (MS-SPRing) standardized by ITU-T in G.841. In SONET, this scheme is frequently referred to as a bidirectional line-switched ring. OCh-SNCP, meanwhile, is the optical counterpart to SDH's subnetwork connection protection, also sometimes called unidirectional line-switched ring. See also Self Healing.

self test The capability of a PBX to run programs at regular intervals to test its own operation and signal when failures have occurred or are about to occur without human intervention.

selling Selling consists of three steps: Prospecting. Qualifying and Closing.

selling steps A telemarketing term. The steps involved in a sale: A clear call objective, identify and reach the decision maker, introduction and call justification, identify needs, present solution/benefits, answer questions or objections, close, confirm the conditions, and congratulate.

Semantic Web A project of the World Wide Web Consortium (W3C) to develop technologies to enable machine-extraction of meaning from Web documents. Legacy Web technologies enable a Web document's embedded text-formatting commands and hyperlinks to be identified and processed with automated ease, but the informational content of Web documents has not been machine-discernible. The idea behind the Semantic Web is to make this information machine-extractable. The W3C has made progress toward this goal. XML, RDF and OWL are three component technologies of the Semantic Web that have already been developed by the W3C and are in use today.

semaphore 1. A message sent when a file is opened to prevent other users from opening the same file at that time. Its purpose is to preserve the integrity of data (i.e. stop it from being messed with) while you're using it.

2. An apparatus for visual signaling, such as by the use of flags. The term comes from the Greek "sema," meaning sign or signal, and "phoros," meaning carrying. Semaphores were in common use for message signaling prior to the invention of the telegraph. In fact, the British Admiralty Office on August 5, 1816 officially rejected the idea of the electric telegraph, which had been suggested by Mr. (later Sir) Ronald Edwards. It seems that the Admiralty preferred the semaphore, although it was useless during the night or when it was foggy.

semi-adjacent channel A channel separated from a specific channel by one channel.

semi-automatic message switching A network control technique whereby an operator manually switches a message to a destination according to the address information contained in the message header.

semi-passive tag An RFID definition. Similar to active tags, but the battery is used to run the microchip's circuitry but not to communicate with the reader. Some semi-passive tags sleep until they are woken up by a signal from the reader, which conserves battery life. Semi-passive tags cost a dollar or more.

semi-rigid A cable containing a flexible inner core and a relatively inflexible sheathing material, such as a metallic tube but, which can be bent for coiling or spooling and placing in duct or cable run.

semi-rigid PVC A hard semi-flexible polyvinylchloride compound with low plasticizer content.

semiconductor Semiconductors are the basic building blocks of telecom and computing. They are made into microprocessors which drive our computers, our PDAs, our communications networks, our phone systems, our watches and our cars. They are called in-

tegrated circuits, chips and a thousand other variations. In their basic form, semiconductors are materials that are midway between conductors and insulators; they have a resistance to electricity between a conductor (e.g. a copper wire) and an insulator (e.g. plastic). Hence the word "semi" conductor. Silicon and germanium are the two most commonly-used materials to make semiconductors. The flow of current in a semiconductor can be changed by light or the presence or absence of an electric or magnetic field. Semiconductors will only conduct electricity when a certain threshold voltage has been reached. Because the energy needed to "turn on" a semiconductor can be high, most semiconductors have their conductivity enhanced through a process called doping. Doping consists of adding an impurity that has either a surplus or a shortage of electrons. Semiconductors with impurities that provide a surplus of electrons are called N-type semiconductors. Semiconductors with impurities that have a shortage of electrons are called P-type semiconductors. Together N-type and P-type semiconductors are the basic building blocks of nearly all solid state electronic devices.

Making semiconductor integrated circuits can consist of more than a hundred steps, during which hundreds of copies of an integrated circuit are formed on a single wafer. Generally, the process involves the creation of eight to 20 patterned layers on and into the substrate, ultimately forming the complete integrated circuit. This layering process creates electrically active regions in and on the semiconductor wafer surface. The first step in semiconductor manufacturing begins with production of a wafer -a thin, round slice of a semiconductor material, usually silicon. In this process, purified polycrystalline silicon, created from sand, is heated to a molten liquid. A small piece of solid silicon (seed) is placed on the molten liquid, and as the seed is slowly pulled from the melt the liquid cools to form a single crystal ingot. The surface tension between the seed and molten silicon causes a small amount of the liquid to rise with the seed and cool. The crystal ingot is then ground to a uniform diameter and a diamond saw blade cuts the ingot into thin wafers. The wafer is processed through a series of machines, where it is ground smooth and chemically polished to a mirror-like luster. The wafers are then ready to be sent to the wafer fabrication area where they are used as the starting material for manufacturing integrated circuits. The heart of semiconductor manufacturing is the wafer fabrication facility where the integrated circuit is formed in and on the wafer. The fabrication process, which takes place in a clean room, typically takes from 10 to 30 days to complete. Wafers are pre-cleaned using high purity, low particle chemicals (important for high-yield products). The silicon wafers are heated and exposed to ultra-pure oxygen in the diffusion furnaces under carefully controlled conditions forming a silicon dioxide film of uniform thickness on the surface of the wafer. Masking is used to protect one area of the wafer while working on another. This process is referred to as photolithography or photo-masking. A photoresist or light-sensitive film is applied to the wafer, giving it characteristics similar to a piece of photographic paper. A photo aligner aligns the wafer to a mask and then projects an intense light through the mask and through a series of reducing lenses, exposing the photoresist with the mask pattern. Precise alignment of the wafer to the mask prior to exposure is critical. Most alignment tools are fully automatic.

The wafer is then "developed" (the exposed photoresist is removed) and baked to harden the remaining photoresist pattern. It is then exposed to a chemical solution or plasma (gas discharge) so that areas not covered by the hardened photoresist are etched away. The photoresist is removed using additional chemicals or plasma and the wafer is inspected to ensure the image transfer from the mask to the top layer is correct. Atoms with one less electron than silicon (such as boron), or one more electron than silicon (such as phosphorous), are introduced into the area exposed by the etch process to alter the electrical character of the silicon. These areas are called P-type (boron) or N-type (phosphorous) to reflect their conducting characteristics. The thermal oxidation, masking, etching and doping steps are repeated several times until the last "front end" layer is completed (all active devices have been formed). Following completion of the "front end," the individual devices are interconnected using a series of metal depositions and patterning steps of dielectric films (insulators). Current semiconductor fabrication includes as many as three metal layers separated by dielectric layers.

After the last metal layer is patterned, a final dielectric layer (passivation) is deposited to protect the circuit from damage and contamination. Openings are etched in this film to allow access to the top layer of metal by electrical probes and wire bonds. An automatic, computer-driven electrical test system then checks the functionality of each chip on the wafer. Chips that do not pass the test are marked with ink for rejection. A diamond saw typically slices the wafer into single chips. The inked chips are discarded, and the remaining chips are visually inspected under a microscope before packaging. The chip is then assembled into a package that provides the contact leads for the chip. A wire-bonding machine then attaches wires, a fraction of the width of a human hair, to the leads of the package. Encapsulated with a plastic coating for protection, the chip is tested again prior to delivery to the customer. Alternatively, the chip is assembled in a ceramic package for certain military applications. Thanks to Intersil of Irvine for description of the manufacturing of a semiconductor IC above.

semiconductor laser A laser is really an oscillator, which means an amplifier plus a feedback. In a semiconductor laser, the amplification is provided by the population inversion in the active layer obtained by injecting current in it. There are two basic types of semiconductor lasers -the Fabry-Perot and DFB lasers. They use different ways to provide the feedback. In a Fabry-Perot (FP), you a have a mirror at each end (usually obtained by cleaving the semiconductor crystal), which reflects part of the light and give the feedback. In a DFB (or Distributed FeedBack) laser, a Bragg grating is incorporated along the active layer, providing a distributed reflection of the light. Often the end facets are anti-reflection coated to avoid the Fabry-Perot effect. The grating is usually made by varying the thickness of one layer in a periodic fashion. This gives a periodic variation of the propagation constant (effective index) of the waveguide. For the right period, you get reflection of the light at a certain wavelength. The main difference between the two types is that DFBs are often single mode (only one wavelength determined by the Bragg grating), while FPs are usually multi-mode (4 to 20 different wavelengths at the same time and determined by the gain curve of the active material and the cavity length). See also Semiconductor.

semipermanent connection A connection established via a service order or via network management.

send and pray A descriptive term for a data communications protocol that provides little or no assurance that the data gets to the destination device as intended. Send and pray first was applied to Parity Checking, also known as Vertical Redundancy Checking (VRC), which is the very poor error detection mechanism used in asynchronous protocols using the ASCII coding scheme. Parity checking not only is highly uncertain in its ability to detect an error created in transmission, but also provides no mechanism for the receiving device to advise the transmitting device of a detected error in order that it might be corrected by some means. Hence, the term "send and pray"-you just send the data and pray that it gets to the other end correctly. More recently, the term has been applied to voice and data over IP networks. While IP provides a fairly robust mechanism for error detection and correction, it was intended for true data communications applications, which have time to recover from errors created in the process of transmission and switching. Voice and video, however, have no time to recover. The data is presented to the receiving device as it exits the network. Any errored, delayed or corrupted data packets are simple discarded in this stream-oriented communication mode. If you are talking over an IP-based network, therefore, you just send your voice into the phone and pray that it all gets there.

send and prey A term for the method by which the misguided amongst us seek to damage our data. These poor, misguided souls embed viruses in programs which we then download. They send them to us, and then prey upon us. This term was coined by Ray Horak, my Contributing Editor.

sender Equipment in the originating telephone system which outpulses (sends out) the routing digits and the called person's number. Senders are necessary in computerized (stored program control) switches because the switch needs to know all the digits of the numbers you are calling before it chooses and seizes a trunk.

Sendmail Sendmail is the UNIX software that handles electronic mail. It is the most common form of electronic mail software on the Internet. Sendmail provides back-end message routing and handling for Simple Mail Transfer Protocol-based electronic mail communications.

senior moment I have senior moments when I forget something my wife asks me to. those moments could also be called selective hearing loss. The buzz-phrase, "senior moment," is a joke, but it's also serious. We all fear Alzheimer's, a stroke or something that will affect us mentally as we get older.

seniority A call center term. A field in each employee record establishing that employee's seniority for the purpose of automatic schedule assignment. This is typically the employee's hire date (year, month, and day) plus a three-digit "tie breaker."

sensitivity 1. The input signal level required for a tuner, amplifier, etc., to produce a stated output. The lower the required input, the higher the sensitivity. Measured in mV (microvolts). 2. The degree to which a radio receiver responds to the wave to which it is tuned.

sensitivity analysis The process of rerunning a financial study to figure the degrees to which changing the assumptions changes the result of the analysis.

sensitivity rraining The person in history winning the prize for most in need of sensitivity training is the Roman emperor Commodus (A.D. 180-192) . His accomplish-

ment: He collected all the dwarfs, cripples, and freaks he could find in the city of Rome and had them brought to the Coliseum, where they were forced to fight each other to death with meat cleavers, much to the delight and amusement of the audience.

sensor A device that responds to a physical stimulus and produces an electronic signal. Sensors are increasingly being combined with RFID tags to detect the presence of a stimulus at an identifiable location.

sensor applications Services involving the remote monitoring of instruments including burglar alarms, fire alarms, meter reading, energy management and load shedding.

sensor glove An interface device for experiencing virtual reality with the hand. Wired with sensors, it detects changes in finger, hand and arm movements and relays them to the computer, allowing users to manipulate and move things in a virtual environment. See Virtual Reality.

sensor mote Sensor motes are typically small, coin-sized wireless devices that are designed to establish network connections on the fly with minimal power. According to Red Herring magazine, what the devices lack in intelligence, they make up for in their ability to quickly wake from hibernation and locate and establish communications with other nearby wireless motes. Using small amounts of energy, they are able to transmit short bursts of data reporting on such things as changes in temperature, light, motion, or even the presence of chemicals. Many of them can work together in concert with so-called "mesh" networks, providing precise information about movements on a battlefield or people walking through a building. Sensor motes are presently used in monitoring leaks in water pipelines, maintaining the temperature in industrial buildings and watching over an industrial process to ensure that humans are alerted before a major breakdown.

sent paid A utility industry term that describes all calls charged to the originating number or collected as coins in a pay telephone.

SEP Somebody Else's Problem. Example: "Who is going to change the DS3 termination over from BNC to fiber down to copper? SEP!"

separation 1. Extent to which two stereo channels are kept apart. Expressed in decibels, the larger the number, the better the separation and stereo effect.

2. The use of FDM (Frequency Division Multiplexing) to establish separation of upstream and downstream information channels. FDM is commonly used in cellular telephony and other radio technologies and applications. Channel separation using FDM also sometimes is used in high-speed data transmission systems in order to maintain separation between the primary data stream and other information streams such as POTS and ISDN. For example, it is proposed that FDM channel separation will be used in early versions of VDSL (Very-high-data-rate Digital Subscriber Line). See also FDM and VDSL.

separations and settlements A complex set of accounting procedures developed by the traditional telephone industry. The procedures classify telephone plant as intrastate or interstate, and return revenues from long distance phone calls to local telephone companies to compensate them for the use of local exchange facilities (e.g., switches and local loop cable plant) in the origination and termination of long distance calls. The issue is that long distance is highly profitable, while local service is not. Prior to the breakup of the Bell System, AT&T and the local telephone companies cooperated in this process. AT&T Long Lines calculated the settlement payments for the BOCs (Bell Operating Companies), which then reimbursed the independent (i.e., non-Bell) companies within their states of operation. In 1983, the FCC mandated the creation of NECA (National Exchange Carrier Association), which is charged with administering this process under Subpart G of Part 69 of the FCC's rules and regulations. The settlements pool consists of several elements. Subscribers contribute directly to the pool through the monthly Subscriber Line Charge (SLC), which is capped by the FCC at various levels for all local loops of the same type; that amount appears as a separate line item on your phone bill. (The SLC also is known variously as the Access Charge, CALC, and EUCL.) The IXCs (IntereXchange Carriers) pay into the pool a fixed amount for access trunks and termination facilities between their networks and the LEC networks; this charge is known as the CCL (Carrier Common Line Charge). The IXCs also pay per-connection and per-minute access charges into the pool. All LEC revenues from these sources are submitted to NECA, along with statements of associated costs. NECA then divides the revenues on an average-cost basis, with special consideration for "high-cost" serving areas, which are defined as having service costs in excess of 115% of the national average. In this manner, the LECs are reimbursed for the use of their networks, and "universal service" availability is assured. As deregulation and local competition have begun to alter the telecommunications landscape, the CLECs (Competitive Local Exchange Carriers) have been included in this process, as well. See also Access Charge, CLEC, NECA and Universal Service.

separator Pertaining to wire and cable, a layer of insulating material such as textile, paper, etc. placed between a conductor and its insulation, between a cable jacket and the components it covers or between various components of a multi-conductor cable. It can be utilized to improve stripping qualities and/or flexibility, or can offer additional mechanical or electrical protection to the components it separates.

SEPP Secure Electronic Payment Protocol. An open specification for secure bank card transactions over the Internet, SEPP was developed jointly by IBM, Netscape, GTE, Cyber-Cash and MasterCard. An embodiment of the iKP protocol, which uses public-key cryptography, SEPP is intended for HTTP (HyperText Transfer Protocol) transactions and is adapted to bank card payments in the general context of Electronic Commerce. SEPP messages are transmitted as MIME (Multipurpose Internet Mail Extensions) attachments, and are based on common syntax standards. See also Electronic Commerce, iKP, MIME and Public Key.

SEPT Signaling End Point Translator, part of Signaling System 7. See Signaling System 7.

SEQUEL Structured English QUEry Language. The forerunner of SQL (Structured Query Language. See SQL.

sequence A call center term. A pattern of days on and days off as defined in either a schedule preference or shift definition.

sequencing Sequencing is the process of dividing a data message into smaller pieces for transmission, where each piece has its own sequence number for reassembly of the complete message at the destination end. Sequencing is thus also the process of properly ordering the receipt of packet data at their destination, regardless of the time they have taken to travel the X.25 network. It's similar to packetizing. See Packet.

sequencing receivers All GPS (Global Positioning System) receivers must receive information from at least four satellites to calculate accurately where they (and thus you) are. Sequencing receivers use a single channel and move it from one satellite to the next to gather this data. They usually have less circuitry so they're cheaper and they consume less power than receivers which work on four satellites simultaneously. Unfortunately the sequencing can interrupt positioning and can limit their overall accuracy.

sequential Pertaining to events occurring in a specific time or code order.

sequential access The need to read data -one record after another in sequence -before getting to the information you want. Magnetic tape, for example, requires you to read the entire tape up to where your information is. This is because the computer cannot tell where on the tape your information is because records on tape files are often of variable length. Most Random Access files, usually kept on a disk drive, require records to be of a fixed length, such as 80 characters per record. Then when you seek record 23, the computer seeks character 1840 in the file (23 x 80), and takes the next 80 characters as the record you want. Using the analogy of music recorded on records and magnetic tape, a phonograph record needle has the capability of random access because the needle can be set down in the spaces between cuts on the record. With mag tape, you must fast forward past all the music you don't want before you get to the music you want to hear. Also, if the computer tape drive is fast forwarding, it cannot count characters to find the record, and must read in the data you don't want (and throw it away), before it gets to the data you need. Random access is much faster than sequential access.

sequential hunting See Rollover.

sequential logic element A device that has at least one output channel and one or more input channels, all characterized by discrete states, such that the state of each output channel is determined by the previous states of the input channels.

Sequential Packet Exchange SPX. Novell's implementation of SPP for its NetWare local area network operating system.

SER Designation for a sound Service Entrance Cable.

SerDes Serializer/Deserializer as in Serializer/Deserializer Framer Level 5 (SFI-5) is a standard electrical interface for 40-gigabit per second transponders and framer devices that promise to drive down communications cost even further. The SFI-5 Implementation Agreement was approved by the Optical Internetworking Forum in June, 2002.

serial One after another. One event after another. Serial comes from the word "series" -which is classically defined as a group or a number of related or similar things, events, etc., arranged or occurring in temporal, spatial, or other order or succession. In telecom, there are basically two types of data transmission -serial and parallel. Serial is one stream of data, one bit following the previous bit. Parallel is the same stream of data, but broken into several streams running simultaneously. The reason to go parallel is that several streams will often be faster than one stream. See Parallel Transmission and Serial Transmission.

serial acquirers Serial acquirers are people who buy things and buy more things and more. Sometimes they buy companies. Sometimes they buy material goods. Usually both. Their major characteristic is they can't stop. Some of the most colorful people in the recent history of telecommunications were serial acquirers -for example, Dennis Kozlowski of Tyco, Bernie Ebbers of WorldCom, Gary Winnick of Global Crossing and John Rigas of Adelphia.

Serial ATA Serial ATA is a slightly faster, better version of ATA (Advanced Technology Attachment). It is now available on some 2.5" hard drives which are going into new laptops. I have one on my new laptop. Frankly, I don't see the difference in speed. See also SAS SCSI.

serial bus Serial bus was the original name for Intel's standard for a type of very local, local area network that would be used for connecting peripherals to the motherboard of a PC. There'd be one plug on the back of the PC into which you'd daisy chain various peripherals, including a mouse, a keyboard, speakers, printers, a microphone and a telephone. The idea of serial bus is to clear away all the clutter on the back of the PC. In March of 1995 when the first technical specs were released, serial bus' name was changed to Universal Serial Bus (I don't know why). See USB.

serial call Telephone system feature set up by the attendant when an incoming calling party wishes to speak with more than one person internally. When the first party hangs up, the call automatically moves to the second person the outside party wants to speak with. When that person hangs up, then the call automatically goes to the third person, etc.

serial communication Networks (local and long distance) use the RS-232 serial communications standard to send information to serial printers, remote workstations, remote routers, and asynchronous communication servers. The RS-232 standard uses several parameters that must match on both systems for information to be transferred. These parameters include baud rate, character length, parity, stop bit, and XON/XOFF.

Baud rate is the signal modulation rate, or the speed at which a signal changes. Since most modems or serial printers attached to personal computers send only one bit per signaling event, baud can be thought of as bits per second. However, higher-speed modems may transfer several bits per signal change. Typical baud rates are 300, 1200, 2400, 4800, 9600 and 19,200. The higher the number, the greater the number of signal changes and, therefore, the faster the transmission.

Character length specifies the number of bits used to form a character. The standard ASCII character set (including letters, numbers, and punctuation) consists of 128 characters and requires a character length of 7 bits for transmissions. Extended character sets (containing line drawings or the foreign characters used in IBM's extended character set) contain an additional 128 characters and require a character lengths of 8 bits. Parity error checking can only be used with character lengths of 7 bits.

Parity is a method of checking for errors in transmitted data. You can set parity to odd or even, or not use parity at all. When the character length is set to 8, parity checking cannot be done because there are no "spare" bits in the byte. When the character length is 7, the eighth bit in each byte is set to 0 or 1 so that the sum of bits (0s and 1s) in the byte is odd or even (according to the parity setting). When each character is received, its parity is checked again. If it is incorrect (because a bit was changed during transmission), the communications software determines that a transmission error has occurred and can request that the data be retransmitted.

Stop bit is a special signal that indicates the end of that character. Today's modems are fast enough that the stop bit is always set to one Slower modems used to require two stop bits.

XON/XOFF is one of many methods used to prevent the sending system from transmitting data faster than the receiving system can accept the information. See also EIA/TIA-232-E, RS-232-C and serial data transmission.

serial data transmission Serial data transmission is the most common method of sending data from one DTE to another. Data is sent out in a stream, one bit at a time over one channel. When a computer is instructed to send data to another DTE, the data within the computer must pass through a serial interface to exit as serial data. Then it passes through ports, cables, and connectors that link the various devices. The boundaries (physical, functional, and electrical) shared by these devices are called interfaces. See serial communications.

serial digital Digital information that is transmitted in serial form. Often used informally to refer to serial digital television signals.

serial interface The "lowest common denominator" of data communications. A mechanism for changing the parallel arrangement of data within computers to the serial (one bit after the other) form used on data transmission lines and vice versa. At least one serial interface is usually provided on all computers for the connection of a terminal, a modem or a printer. Sometimes also called a serial port. See EIA/TIA-232-E, RS-232-C, Serial Interface Card and the Appendix.

serial interface card A printed circuit card which drops into one of the expansion slots of your computer and changes the parallel internal communications of your computer into the one-bit-at-time serial transmission for sending information to your modem or to a serial printer.

Serial Line Internet Protocol See SLIP.

serial memory Memory medium to which access is in a set sequence and not at random.

Serial Peripheral Interface See SPI.

serial over LAN A protocol for packetizing and transmitting serial data over a LAN. A typical use of serial over LAN (SOL) is to enable a remote serial console to access, over an IP network, the text-based interfaces for a server's BIOS and system utilities.

serial port An input/output port (plug) that transmits data out one bit at a time, as opposed to the parallel port which transmits data out eight bits, or one byte at a time. Most personal computers (PCs) have at least one serial and one parallel port. In a typical configuration, the serial port is used for a modem while the parallel port is used for a printer. For a diagram of a typical 25-pin RS-232-C serial port, see the Appendix at the back of this book.

serial processing Method of data processing in which only one bit is handled at a time.

serial transmission Sending pulses one after another rather than several at the same time (parallel). When transmitting data over a telephone line there is only one set of wires. Therefore, the only logical way to transmit it is to send the data in serial mode. It is possible to use eight different frequencies to transmit a character all at once (parallel), but these modems are ridiculously expensive. See Parallel, Parallel Port and Serial Port.

serialize To change from parallel-by-byte to serial-by-bit.

series A connection of electrical apparatus or circuits in which all of the current passes through each of the devices in succession or on after another. See also Parallel.

series 11000 An AT&T private line long distance tariff created in the 1970s and designed expressly to reduce MCI's chances of selling any private lines and thus of surviving. It was thrown out by the FCC and the tariff figured in MCI's and the Federal Government's antitrust against AT&T.

series circuits In a series circuit, the electric current has only one path to follow. All of the electric current flows through all the components of the circuit. To calculate the resistance of a series circuit add up the resistance of each of the components in the circuit. In contrast, see parallel circuits.

series connection A connection of electrical apparatus or circuits in which all of the current passes through each of the devices in succession or on after another. See also Parallel.

series RF tap A bugging device. It is a radio transmitter which is installed in series with one wire of the telephone circuit. Normally a parasite (i.e. takes power from the phone line). Transmits both sides of the conversation. It transmits only when the phone is off-hook. See also Series.

serve shield A type of shield used in coaxial cable systems, a serve shield, or spiral shield, is simply wound around the inner conductor. See also Coaxial Cable.

server 1. Hardware definition of server: A server is a shared computer on the local area network that can be as simple as a regular PC set aside to handle print requests to a single printer. Or, more usually, it is the fastest and brawniest PC around. It may be used as a repository and distributor of oodles of data. It may also be the gatekeeper controlling access to voice mail, electronic-mail, facsimile services. At one stage, a local area network had only one server. These days networks have multiple servers. Servers these days have multiple brains, large arrays of big disk drives (often in redundant arrays) and other powerful features. New powerful servers are called superservers. A $35,000 superserver today can match the performance of a $2 million mainframe of ten years ago. Then again, according to Peter Lewis of the New York Times, the lowliest client today has more computing power than was available to the entire Allied Army in World War II. See Downsizing for some of the benefits of running servers as against mainframes.

2. Software definition of server: A server is a program which provides some service to other (client) programs. The connection between a client program and the server program is traditionally by message passing, often over a local area or wide area network, and uses some protocol to encode the client's requests and the server's responses. Any given program may be capable of acting as both a client and a server, perhaps switching its role based on the nature of the connection. The terms "client" and "server" simply refer to the role that the software program performs during a specific connection. Similarly, any given server may function as an origin server, a proxy server, a gateway server, or a tunnel, modifying its behavior based on the specific nature of a given request from a client.

Server API A SCSA term. A communications protocol that allows a call processing application running on one computer to control SCSA hardware residing in another computer.

server appliances Little servers designed for small businesses or workgroups without supervision by a central IT department. See Server.

server application A Windows NT application that can create objects for linking or embedding into other documents.

server certificate A unique digital identification that forms the basis of a Web server's SSL security features. Server certificates are obtained from a mutually trusted, third-party organization, and provide a way for users to authenticate the identity of a Web site.

server cluster A group of server computers that are networked together both physically and with software, in order to provide cluster features such as fault tolerance or load balancing. See also Fault Tolerance and Load Balancing.

server colocation An ISP/web hoster service in which a client places their server on the Internet at an ISP's office for a monthly fee. In return, the server is, theoretically, always connected via multiple redundant high speed connections to the Internet. See also Web Hosting.

server consolidation A solution to server sprawl, whereby multiple low-end, underused servers may be replaced by a smaller number of more powerful servers. See server sprawl.

server farm Imagine a room stuffed with PCs, ranged in racks along walls, ranged in racks in lines like a library's back room. The PCs are really servers -powerful PCs containing databases and other information they are dispensing to the thousands of PCs dialing into them from afar. A server farm may be owned by one company and used by one company, or it may be owned by one company and each of the machines leased to other companies. I first heard the term when MCI described a room it had in a place called Pentagon City. There it had hundreds of servers each of which it leased to other companies who used those servers as their Web sites.

server hotel A data center where customers' servers are collocated and managed. Sometimes called a collocation facility.

Server Message Block SMB. The protocol developed by Microsoft, Intel, and IBM that defines a series of commands used to pass information between network computers. The redirector packages SMB requests into a network control block (NBC) structure that can be sent over the network to a remote device. The network provider listens for SMB messages destined for it and removes the data portion of the SMB request so that it can be processed by a local device. In short, SMB is basically a protocol to provide access to server-based files and print queues. SMB operates above the session layer, and usually works over a network using a NetBIOS application program interface. SMB is similar in nature to a remote procedure call (RPC) that is specialized for file systems.

server mirroring Server mirroring means you have two servers on your networks and each exactly what the other is doing simultaneously. It's a backup method. In Novell's NetWare, server mirroring requires two similarly configured NetWare servers. They should be evenly matched in terms of CPU speed, memory, and storage capacity. The servers are not required to be identical in terms of microprocessors type (386/486), microprocessor revision level, or clock speed. However, identical servers are recommended for NetWare SFT III 3.11. If the two servers are unequal in terms of performance, then SFT III performs at the speed of the slower server. The NetWare servers must be directly connected by a mirrored server link. SFT III servers can reside on different network segments, as long as they share a dedicated mirrored server link.

server node An individual computer in a server cluster.

Server Operating System An SCSA definition. Operating System running on the SCSA Server.

server process A process that hosts COM components. A COM component can be loaded into a surrogate server process, either on the client computer (local) or on another computer (remote). It can also be loaded into a client application process (in-process).

server push Server push is a Internet term. With server push, the Web server sends data to display on the browser display, but leaves the connection open. At some point, the server sends additional data for the browser to display. Server push is used for displaying multimedia information on the browser.

server queuing When you have too many people trying to get on your web site and they're having trouble getting on, it's called server queuing. I don't think this is a particularly profound definition.

server scriptlet A COM object that is created with Microsoft Server Scriptlet technology.

server sprawl What happens in an organization over time when it acquires server after server -some for enterprise-wide use, others department-specific, others workgroup-specific, others application-specific, and still others for other narrowly specified purposes. These servers all require physical space, electrical power, and cooling; and each server has its own warranty and service agreement. When server sprawl occurs, it is also not uncommon for some of the servers to be underused compared to other servers in the sprawl. When server sprawl occurs, an organization may look into server consolidation.

server-based network A network in which all client computers use a dedicated central server computer for network functions such as storage, security and other resources.

Severely Errored Second SES. A second during which the bit error rate over a digital circuit is greater than a specific limit. During a severely errored second, transmission performance is significantly degraded. The specific definition of SES depends on the circuit involved, e.g. T-1, T-3, OC-3 and OC-48.

servers Servers are typically ruggedized, industrial-strength PCs, i.e. they have several fans, perhaps two power supplies, perhaps two disk drives, perhaps several central processors. They are designed to work 24 hours a day, seven days a week without breaking down. They are designed to give the reliability of demanded of a business telephone system. See also Server.

serve A filament or group of filaments such as fibers or wires, wound around a central core.

Service Access Code SACs (also called Special Area Codes) are 3-digit codes in the Numbering Plan Area (NOO) format that are used as the first three digits of a 10-digit address, and that are assigned for special network uses in North America. Whereas NPA codes are normally used for identifying specific geographical areas, certain SACs have been allocated in the NANP (North American Numbering Plan) to identify generic services or provide access capability. Currently only four SACs have been assigned and are in use: 600, 700, 800, and 900. There are two general categories of NPAs: 1) Geographic NPA–Associated with a defined geographic area; all telephone numbers bearing such an NPA are associated with services provided within that geographic area.

2) Non-Geographic NPA–Associated with a specialized telecommunications service, which may be provided across multiple geographic NPA areas; for example 800, 900, 700, 500 and 888. Also known as an area code. See also SAC.

service access point A logical address that allows a system to route data between a remote device and the appropriate communications support.

service advertising protocol A protocol developed by Novell so that devices attached to a network could advertise their functionality. For instance, a file server and print server advertise different functions. An SNMP agent can also advertiser itself using SAP's. All of the Compaq manageable repeaters and switches except the 50xx switches support SAP broadcasts.

service affecting This definition courtesy Steve Gladstone, author of the book, "Testing Computer Telephony Systems." These are major bugs that significantly impact the reliability or the functionality of computer telephony systems. Comprehensive testing must uncover all service affecting problems with the goal that no computer telephony system should be installed at a live customer installation without an acceptable workaround for the service affecting problem. The goal for transition of a product from one phase to the next is that no service affecting bugs remain, and that the bug rate for new bugs be at or approaching zero.

service and equipment record A list of equipment billed to customer by type, quantity, monthly charge, location and billing dates.

service area 1. Another term for a LATA, according to some Bell operating companies.

2. The more common usage is the geographic area served by a supplier. The area in which the supplier, theoretically, stands ready to provide his service. The service area of New York Telephone (now called Bell Atlantic) is most (not all) of New York State.

service availability Service availability is what the United States Government calls its intentional degradation of GPS (Global Positioning Satellite system). The government degrades the signals in case of war or other emergencies. It does this in order to prevent enemies from taking advantages of the system. When GPS signals are degraded, they can significantly change the accuracy of position readings from GPS devices.

service boundary The boundary existing between a computing domain and a switching domain as it is established via their interconnected service boundaries over some underlying interconnection medium.

service bureau A data processing center that does work for others. There are many ways of bringing work to a service bureau, including mailing it and transmitting it over phone lines. If it comes over phone lines, the service is likely to be called "time sharing."

service charge The amount you pay each month to receive cellular service. This amount is fixed, and you pay the same fee each month regardless of how much or how little you use your cellular phone. It usually ranges from about $10 to $65 per month, depending on the carrier's tariffs and the particular plan of service you select. In addition you pay air time. Service Charge doesn't usually include any air time.

service charge detail A listing of all the telephone equipment installed as part of a specific telephone system. Usually provided by the vendor or maintenance organization.

service code A 3-digit code in general use by customers to reach telephone company service, for example, 411 (Directory Assistance), 611 (Repair Service), 811 (Business Office) and 911 (Emergency Service).

Service Control Point SCP. The local versions of the national SMS/800 number database. SCPs contain the intelligence to screen the full ten digits of an 800 number and route calls to the appropriate, customer-designated long distance carrier. Bellcore defines SCP as the network system in the Advanced Intelligent Network Release 1 architecture that contains SLEE (Service Logic Execution Environment) functionality and communicates with AIN Release 1 Switching Systems in processing AIN Release 1 calls.

service creation The set of activities that must be performed to create a new service to be offered to subscribers and the associated service-specific operations capabilities to support the new service. Definition from Bellcore in reference to its concept of the Advanced Intelligent Network.

Service Creation Environment SCE. A telephone company AIN term. The surroundings, including organizational structure and computing and communications resources, in which creation of new services takes place. See Service Creation.

service creation tool What the computer telephony industry calls an applications generator, the telephone industry calls a service creation tool. It is a software tool that, in response to your input, writes code a computer can understand. In simple terms, it is software that writes software. Service creation tools have three major benefits: 1. They save time. You can write software faster. 2. They are perfect for quickly demonstrating an application. 3. They can often be used by non-programmers. See also Application Generator.

service deficiency When you work a delay with a service provider -a telephone or data carrier, you need to create certain definitions of service so that you can figure penalties if such levels of service are not maintained. For example, we might define service deficiency as being a service outage lasting for more than ten seconds. We might define Repeated Service Deficiency as a service deficiency that occurs at least four times in any given 30 day period. and we might define Chronic Service Deficiency as a service deficiency that occurs more than ten times in any given 30 day period. Of course, how these terms are defined will depend on the SLA -Service Level Agreement -which you sign with your carrier.

Service Delivery Point See SDP.

service display Displays a specific service being presently in effect.

service enabled device A device that can communicate its role and how to interact with it to other service-enabled device (SEDs), thereby enabling such devices to discover and collaborate with each other. See Zeroconf.

service entrance The point at which network communications lines (telephone company lines) enter a building.

service flow A MAC layer service which provides unidirectional transport of upper layer packets that shapes policies and prioritizes according to QoS parameters defined for this flow.

service flow identifier. SFID. 32 bit identifier assigned by the CMTS when service flow is created.

service function A primary or secondary service function.

service identification The information uniquely identifying an NS/EP telecommunications service to the service vendor and/or service user. NS/EP (National Security and Emergency Preparedness) is federal government telecommunications services that are used to maintain a state of readiness or to respond to and manage any event or crisis (local, national, or international) that causes or could cause injury or harm to the population, damage to or loss of property, or degrade or threaten the national security or emergency preparedness of the United States. See also NS/EP Telecommunications.

service independent BAF The finite set of BAF (Bellcore Automatic Message Accounting Format) structures and modules needed to record usage of Advanced Intelligent Network (AIN) Release 1 services. The set is said to be robust because it will be designed to record future AIN Release 1 services that have not yet been identified. Definition from Bellcore in reference to its concept of the Advanced Intelligent Network. See also AIN and BAF.

service interface jack The SIJ is generally considered to be the demarcation jack for DS-1 based services in Canada. It does not necessarily imply being a "smart jack" although quite often this is the case.

service interworking A Frame Relay/ATM term. Service interworking is a means of connecting frame relay networks over an ATM backbone. The frame relay PVC (Permanent Virtual Circuit) connects to the ATM PVC through an ATM switch which accomplishes frame relay-to-ATM protocol conversion at the point of entry; at the point of exit from the ATM network, the process is reversed. This is in contrast to Network Interworking, which makes use of a tunneling protocol to provide what essentially is a cut-through path through the ATM network. Service interworking is defined in the FRF.8 specification from the Frame Relay Forum and is recognized by the ATM Forum. See Network Interworking.

Service Initiation Charge SIC. This is a charge to a client by a carrier for the initiation of a new telecom service such as the installation of a new T-1 line or a Frame Relay PVC.

service level Usually expressed as a percentage of a statistical goal. For example, if your goal is an average speed of answer of 100 seconds or less, and 80% of your calls are answered in 100 seconds or less, then your service level is 80%.

Service Level Agreement SLA. An agreement between a user and a service provider, defining the nature of the service provided and establishing a set of metrics (fancy word for measurements) to be used to measure the level of service provided measured against the agreed level of service. Such service levels might include provisioning (when the service is meant to be up and running), average availability, restoration times for outages, availability, average and maximum periods of outage, average and maximum response times, latency, delivery rates (e.g. average and minimum throughput). The SLA also typically establishes trouble-reporting procedures, escalation procedures, penalties for not meeting the level of service demanded -typically refunds to the users. An example: in May of 1998, GTE announced a new SLA that promises its "Internet Advantage dedicated Internet access customers will get a minimum packet loss guarantee from GTE. If Internet Advantage customers experience more than a 10% packet loss during any ten minute interval, they will be credited with one day of service." UUNET Technologies has a SLA that says if its network in unavailable for one, you, the user, are credited with one full day of service.

Typical SLAs include statements about:

- System/service availability
- Time to identify the cause of a customer related malfunction
- Time to repair a customer related malfunction
- Service provisioning time
- Quality of service targets

Service Level Management SLM. A suite of software tools which provide both the end user organization and the service provider a means of managing the committed service levels defined in a Service Level Agreement (SLA). SLM includes monitoring and gathering performance data, analyzing that data against committed performance levels, taking the appropriate actions to resolve discrepancies between committed and actual performance levels, and trending and reporting. SLM is a tough proposition, especially across a wide range of complex technologies (i.e., Frame Relay and ATM) in a multi-site enterprise. The SLM system must be flexible enough to reflect the service dimensions and performance metrics defined in the SLA. See also Service Level Agreement.

service line An exchange line associated with multiple data station installations to provide monitoring and testing of both customer and Telco data equipment.

Service Logic Execution Environment SLEE. A functional group residing in an SCP (Service Control Point) or Adjunct that contains the Service Logic and Control, Information Management, AMA (Automatic Message Accounting) and Operations FEs (Functional Entity). This composite set of capabilities, which includes FC routines, provides a functionally consistent interface to SLPs (Service Logic Program) independent of the underlying operating system. Definition from Bellcore in reference to its concept of the Advanced Intelligent Network.

Service Management System SMS. An operations support system used to facilitate the provisioning and administration of service data required by the SCP. Use of this term does not imply any specific technology platform. See SMS.

service measurements Measurements of the actual grade of service provided to our subscribers.

service negotiation The functionality needed to gather subscriber Advanced Intelligent Network Release 1 business needs; provide answers, AIN Release 1 service/feature descriptions, prerequisite non AIN services, availability and costs; reserve AIN Release 1 resources (e.g. 800 number and 900 number); identify required network resources; and identify available AIN Release 1 services/features by wire center. Definition from Bellcore in reference to its concept of the Advanced Intelligent Network.

service node See ATM Edge Switch.

service objectives If you're the phone company, service objectives are a statement of the quality of service that is to be provided to the customer; for example, no more than 1.5 percent of customers should have to wait more than three seconds for dial tone during the average busy hour or, the busy-hour blocking on a last choice trunk group not exceed 1 percent. According to my friends in the phone industry, service objectives are the criteria used when engineering quantities of switching equipment. Certain basic principles must be kept in mind: 1. The service must be of high quality. 2. Rates for telephone service must be reasonable. 3. When costs of operating the telephone business are subtracted from revenues produced by reasonable rates, enough profit must remain to attract new capital needed to meet increasing demands for service.

If you're designing phone systems, service objectives define the functional and performance goals for how your system will work and what will be experienced by the system's users and other systems. For example, a switch might have a service objective of providing dialtone to all users in less than two seconds. Or, of answering all incoming calls within ten seconds. Or, a VRU may have a service objective of responding to all user DTMF inputs in less than one second. When we buy or design systems, we often have many service objectives, rarely just one.

service observation 1. A generic word used by telephone companies to check the quality of the service they're providing. Some of it is done automatically with machinery. Some of it is done by senior operators who listen in on the conversations of other operators dealing with their subscribers. In short, the senior operators "observe" the service the junior operators are providing. 2. As a feature of some telephone systems, the Service Observation (SOB) command provides the capability to automatically record data about completed calls, incomplete calls and abnormal calls for the purpose of qualitative supervision of call traffic conditions.

service order A Telephone Company definition. The official form on which desired customer services are recorded for processing, e.g., new connects, changes, disconnects, etc.

Service Order Image SOI. A Verizon definition. The description of an end user's request for telephone service as represented by several Verizon North end user contact systems. The SOI contains end user information such as billing information, service needs, desired listing, etc.

Service Oriented Architecture See SOA.

service package A euphemism for a software bug fix. Also called a Service Release, Service Pack or a Second Edition, or a software release with a number that differs from the previous one because it includes a numbers after the decimal point.

service period The time during which the telephone company furnishes a circuit.

service observing period (month) A telephone company term. All business days of a month (approximately 22 days). It is recommended that the period established for measuring dial tone speed and incoming matching loss coincide as closely as possible with the Service Observing Month.

service points The points on the customer's premises where such channels or facilities are terminated in switching equipment used for communications with phones or customer-provided equipment located on the premises.

service portability A telephone company AIN term. The ability of an end user to retain the same geographic or non-geographic telephone number (NANP numbers) as he/she changes from one type of service to another. The INC Number Portability Workshop agreed that NANP numbers (e.g., 800, 500, 555, 950) should not be service portable for applications outside of their respective industry approved service definitions or guidelines, should those definitions or guidelines exist.

Service Profile Identifier See SPID.

service provider 1. In the broadest sense, a service provider is any company which provides service to anyone else. That means a service provider could be a phone company in the form of either a LEC (Local Exchange Carrier) or IXC (IntereXchange Carrier). It could be an ASP (Application Service Provider). It could be an ISP (Internet Service Provider). A service provider is thus any company which doesn't itself consume all of the service it sells.

2. A Windows Telephony Applications standard which lies between Windows Telephony and the network. It defines how the network -anything from POTS (Plain Old Telephone Service) to T-1, from a key system to a PBX -interfaces to Windows Telephony, which in turn talks to the Applications Programming Interface, which talks to the Windows telephony applications software. See also Windows Telephony.

3. An SCSA computer telephony definition. An addressable entity providing application and administrative support to the client environment by responding to client requests and

maintaining the operational integrity of the server.

4. In TAPI, a dynamic-link library (DLL) that provides an interface between an application requesting services and the controlling hardware device. TAPI supports two classes of service providers, media service providers and telephony service providers.

service provider-branded VoIP VoIP that is provided by the customer's Internet service provider, for example, the customer's LEC (local exchange provider, aka local phone company) or local cable provider.

Service Provider Interface See SPI. See Windows Telephony.

Service Provider Messages An SCSA definition. The message information required by, and provided by, the service provider to perform its functions in the environment in which it is installed. Contrast with SCSA Message Protocol Interface. See Service Provider.

Service Provider Network Identifier SPNI. An identifier for the service provider operating a particular CDPD network.

service provider portability A telephone company AIN term. The ability of an end user to retain the same geographic or non-geographic telephone number (NANP numbers) as he/she change form one service provider to another.

service provisioning tool What the computer industry calls a network manager, the telephone industry calls a service provisioning tool. It is a complex piece of software that allows telephone companies to contact their various switches and sundry computers dispersed over a wide geographic area, to log onto those machines and to upload, download and organize those machines so they are able to make different, new, updated software services for the telephone industry's customers. Telephone companies use various networks to get into their remote switches. Those networks might vary from dial-up to ISDN to packet switched networks to T-1. The better service provisioning tools allow one technician in one place to update and test multiple central offices and computers simultaneously.

service quality A call center term. A measure of how well staffing matches workload, expressed often as average delay (in answering a call).

service terminal The equipment needed to terminate the channel and connect to the phone apparatus or customer terminal.

Service Traffic Management STM. The platform functionality for detecting overloads associated with a specific service and for sending service-specific control messages to the appropriate entities. STM is the SLEE (Service Logic Execution Environment) functionality for detecting overloads associated with a specific service and for sending Automatic Code Gap messages to the appropriate entities. The SN&M (Service Negotiation and Management) OA (Operations Application) also provides STM (Service Traffic Management)-related capabilities.

service tuple Service type and level pair. For example, the service tuple data-bandwidth=45 Mbps consists of the service type data-bandwidth and the service level 45 Mbps.

Service Switching Point SSP. A telephone company AIN term. A switching system, including its remotes, that identifies calls associated with intelligent network services and initiates dialogues with the SCPs in which the logic for the services resides. See SSP.

service type A component of a service that cable providers offer subscribers. For example, devices-supported might be a service type defined for the home networking service, indicating the number of computers the subscriber can connect to the cable network from home. One or more service levels is defined for each service type.

serviceability How easy a system is to configure, service, maintain and repair.

Services Management System SMS. Administers 800 Data Base Service numbers on a national basis. Customer records for 800 Service are entered into the SCP through this system. See Eighthundred Service.

Services Node SN. A network system in the AIN architecture containing functions that enable flexible information interactions between an end user and the network.

services on demand An AT&T term for the immediate provision of almost any network service through universal ports, whenever required by a user; as opposed to provision via an expensive, time consuming, inflexible service order process.

serving area interface A serving area interface is part of a phone company's outside plant. It is a fancy name for a box on a pole, a box attached to a wall or a box in the ground that connects the phone company's feeder or subfeeder cables (those coming from the central office) to the drop wires or buried service wires that connect to the customer's premises. It's also called a cross-wire box. See also Feeder Plant and Drop Wire.

serving closet The general term used to refer to either a riser or a satellite closet; Satellite Cabinet; Satellite Closet.

Serving Mobile Data Intermediate System A cellular radio term.

The CDPD network entity that operates the Mobile Serving Function. The serving MD-IS communicates with and is the peer endpoint for the MDLP connection to the M-ES.

serving office An office of AT&T or its Connecting or Concurring Carriers, from which interstate communications services are furnished.

serving wire The term for the phone number that serves the location, referring to the phone number and terminating wire as one unit. Usually applies to a POTS number.

serving wire center The wire center from which service is provided to the customer.

servlet An applet that runs on a server. The term usually refers to a Java applet that runs within a Web server Web server environment. This is analogous to a Java applet that runs within a Web browser browser environment. Java servlets are becoming increasingly popular as an alternative to CGI programs. The biggest difference between the two is that a Java applet is persistent. This means that once it is started, it stays in memory and can fulfill multiple requests. In contrast, a CGI program disappears once it has fulfilled a request. The persistence of Java applets makes them faster because there's no wasted time in setting up and tearing down the process.

servo Short for servomechanism. Devices which constantly detect a variable, and adjust a mechanism to respond to changes. A servo might monitor optical signal strength bouncing back from a disc's surface, and adjust the position of the head to compensate.

SERVORD Service Order.

SES 1. Satellite Earth Stations.

2. Severely Errored Second. A second in which a severe number of errors are detected over a digital circuit. Each error comprises a code violation (CV), such as a bipolar violation. The specific definition of SES depends on the type of circuit involved, e.g. T-1, T-3, OC-3 and OC-48. See also CV and ES.

3. Source End Station: An ATM termination point, which is the source of ATM messages of a connection, and is used as a reference point for ABR services. See DES.

Sesame Secure European System for Applications in a Multivendor Environment. Developed by the ECMA (European Computer Manufacturers Association), it is intended for very large networks of disparate origin.

session 1. A set of transmitters and receivers, and the data streams that flow between them. In other words, an active communication, measured from beginning to end, between devices or applications over a network. Often used in reference to terminal-to-mainframe connections. Also a data conversation between two devices, say, a dumb terminal and a mainframe. It may be possible to have more than one session going between two devices simultaneously.

2. As defined under the Orange Book, a recorded segment of a compact disc which may contain one or more tracks of any type (data or audio). The session is a purely logical concept; when a multisession disc is mounted in a multisession CD-ROM player, what the user will see is one large session encompassing all the data on the disc.

Session Description Protocol See SDP.

session group Logically ordered list of sessions based on priority of the sessions. All the sessions in the session group should be configured to connect the same physical machines.

Session Initiation Protocol See SIP.

session key A digital key that is created by the client, encrypted, and sent to the server. This key is used to encrypt data sent by the client. See also Certificate, Digital Signature and Key Pair.

session layer The fifth layer -the network processing layer -in the OSI Reference Model, which sets up the conditions whereby individual nodes on the network can communicate or send data to each other. The session layer is responsible for binding and unbinding logical links between users. It manages, maintains and controls the dialogue between the users of the service. The session layer's many functions include network gateway communications.

session lead-in The data area at the beginning of a recordable compact disc that is left blank for the disc's table of contents. The session lead-in uses up 6750 blocks of space. See Track.

session lead-out The data area at the end of a session which indicates that the end of the data has been reached. When a session is closed, information about its content is written into the disc's Table of Contents, and the lead-out and the pre-gap are written to prepare the disc for a subsequent session. The lead-out and the pre-gap together take up 4650 two-kilobyte blocks (nine megabytes). See Track.

session stealing See IP splicing.

set 1. Set is another name for a telephone.

2. SET. Secure Electronic Transaction. A developing, open specification for handling credit card transactions over any sort of network, with emphasis on Internet and the World Wide Web. Rather than providing the merchant with a credit card number, you send the information to them in encoded form. The merchant can't see the encrypted credit card number, but can forward the transaction request to the credit card company or clearinghouse where the information is decrypted and verified. Only the financial institution has the key to unlock the encrypted account information. The financial institution responds to the merchant request with a digital certificate which serves to verify the authenticity of the parties and the overall legitimacy of the transaction. SET ensures that no one else (e.g., hackers) can gain access to your credit card information. It also ensures that an unscrupulous ecommerce merchant can't take advantage of your credit card number. The theory is that you don't always know with whom you are dealing in Cyberspace.

set associative mapping A caching technique where each block of main computer memory is assigned to a location in each cache set where the cache is divided into multiple sets.

set copy Set copy allows the duplication of programming settings from one telephone to another.

Set Top Box STB. The electronics box which sits on top of your TV, connecting it to your incoming CATV or MMDS (Microwave Multi-point Distribution System) TV signal and your TV's incoming coaxial cable. Set-tops, or converter boxes, vary greatly in their complexity, with older models merely translating the frequency received off the cable into a frequency suitable for the television receiver while newer models can be addressable with a unique identity much like a telephone or a node on a computer network. That identity can be addressed from the cable headend. This allows the CATV operator to turn individual channels on and off, such as pay channels. See DOCSIS and Digital Video Broadcast.

SETA SouthEastern Telecommunications Association, a user group.

SETI Search for ExtraTerrestrial Intelligence. A federally funded project which uses arrays of radiotelescopes to search the heavens for signs of intelligent life as evidenced by radio transmissions. SETI is based on rationale laid out in a "Nature" article by physicists Philip Morrison and Guiseppe Cocconi and first implemented by Frank Drake, a Cornell astronomer. As radio waves propagate infinitely at the speed of light in the pure vacuum of space, the thinking is that at least traces of intelligent life can be identified, even though they will be of millennia long past since other stars and galaxies are thousands or millions of light years away. Promoted as a far less expensive technique than that of space travel, the project is interesting but in jeopardy as results have been nil over the last 25 years or so. See Grid Computing and SETI@home.

SETI@home Search for ExtraTerrestrial Intelligence at home. A project sponsored over the Internet by the Planetary Society and the University of California at Berkley. The project harnesses home computers to sift through the billions of radio signals from the cosmos that pass the Earth each day in the hope of finding signals that have emanated from intelligent life on other planets. The PCs download the program and run it against a record of signals detected by the Aricebo radiotelescope in Puerto Rico. As each PC works on the analysis, a screen saver of sorts displays a 3-D graph charting its progress. Once the analysis is completed, the results are uploaded to UC Berkley over the Internet, and another set of data is downloaded. http://planetary.org or http://setiathome.ssl.berkley.edu. See also Grid Computing and SETI.

settlement rate See Accounting Rate System.

settop box See Set Top Box.

SEU Designation for a Flat Service Entrance cable with spirally applied copper neutral wires and thermoplastic jacket.

SF 1. Single Frequency. A method of inband signaling. Single frequency signaling typically uses the presence or absence of a single specified frequency (usually 2,600 Hz). See Signaling.

2. SuperFrame: A DS1 framing format in which 24 DS0 timeslots plus a coded framing bit are organized into a frame which is repeated 12 times to form the superframe.

SFBI Shared Frame Buffer Interconnect, a specification that makes it possible for hardware manufactures to produce a single-board video-graphics adapter for the PC.

SFC Switch Fabric Controller.

SFD Start Frame Delimiter. A binary pattern at the end of eight octets of timing information in an Ethernet frame that tells the receiving station that the timing information is over, and all subsequent signal represents an actual frame. The pattern is two 1s after a long string of alternating one, zero, one, zero, etc. The one octet SFD field is 10101011 in binary.

SFF connector Small-Form-Factor Connector. An optical fiber connector used to join single fibers together at interconnects or to connect them to optical cross connects. SFF connectors are smaller than FC and SC connectors, which makes them easier to use in high-density applications such as fiber-to-the-desktop. In fact, SFF allow two optical fibers to be connected in the same space occupied by the copper-based, eight-position modular RJ-45 footprint. Size matters! See also FC Connector, RJ-45, SC Connector, and ST Connector.

SFG Simulated Facility Group.

SFI-5 Serializer/Deserializer Framer Level 5 is a standard electrical interface for 40-gigabit per second transponders and framer devices that promise to drive down communications cost even further. The SFI-5 Implementation Agreement was approved by the Optical Internetworking Forum in June, 2002.

SFID Service Flow Identifier. 32 bit identifier assigned by the CMTS when service flow is created.

SFINX Service for French Internet Exchange. See IX.

SFP Small form-factor pluggable. A class of hot-swappable optical transceivers that can interface with small-form-factor (SFF) connectors. The Multi-source Agreement Industry Group (MSA) is finalizing a SFP standard. (See SFF).

SFQL Structured Full-Text Query Language. A proposed standard for full-text databases. The primary focus of the proposed standard is interoperability of CD-ROMs. SFQL is based on the SQL (Structured Query Language) standard for relational databases. See also SQL.

SFT System Fault Tolerance. The capability to recover from or avoid a system crash. Novell uses a Transaction Tracking System (TTS), disk mirroring, and disk duplexing as its system recovery methods. System Fault Tolerance as a Novell NetWare term means data duplication on multiple storage devices. If one storage device fails, the data is available from another device. There are several levels of hardware and software system fault tolerance. Each level of redundancy (duplication) decreases the possibility of data loss.

SFTA Scalable Fault Tolerant Architecture.

SG 1. Study Group. The ITU-T has formalized committees studying future telecommunications standards. These groups are called Study Groups.

2. Signal Ground. Ground lead used in E&M signaling types II, III, IV. See also SB and E & M.

3. Signaling Gateway. An agent which serves to resolve differences in signaling and control mechanisms between circuit-switched and packet-switched networks, SGs are defined in the overall protocol stack of MGCP (Media Gateway Control Protocol). See also MGCP.

SGCP See Simple Gateway Control Protocol.

SGML Standard Generalized Markup Language. A text-based language for describing the content and structure of digital documents SGML was developed by Tim Berners-Lee, who is generally recognized as the father of the World Wide Web. HTML, which has gained fame as the language used to create Web pages on the Internet, is a descendant of SGML. SGML documents are viewed with transformers, which render SGML data the way Web browsers render HTML data. SGML was adopted by the International Standards Organization in 1986. SGML allows organizations to structure and manage information in a cross-platform, application-independent way. It tags documents as a series of data objects rather than storing them as huge files. Theoretically, SGML can reduce errors, slice costs and speed work. SGML attempts to separate the informational content of a document from the information needed to present it, either on paper or on screen. See also HTML and XML.

SGMP Simple Gateway Monitoring Protocol. Network management protocol that was considered for Internet standardization and later evolved into SNMP.

SGSN Serving GPRS Support Node. See GPRS.

SHA-1 Secure Hash Algorithm 1. Algorithm that takes a message of less than 264 bits in length and produces a 160-bit message digest. The large message digest provides security against brute-force collision and inversion attacks. SHA-1 [NIS94c] is a revision to SHA developed by NIST that was published in 1994. See NIST.

shack What a radio amateur calls the room where his rig is. The name of the electronics retailer, Radio Shack, is derived from this term.

shadow area A dead spot in a communication area where radio communication is difficult or impossible.

Shadow BIOS ROM Shadow BIOS ROM is a concept I first found in Toshiba laptops which use Flash ROM to hold the machine's BIOS. When you start the machine, the BIOS copies itself from the flash ROM to the Shadow BIOS area. Accessing the BIOS from the Shadow BIOS is much faster than from flash ROM. I learned later that Compaq actually started what they called shadowing the BIOS. According to InfoWorld, Compaq did it because PC-compatible systems available at the time could have no more than 16 megabytes of RAM. Compaq decided to use the memory address at the top of the 15-megabyte physical address space for the shadow RAM.

shadow mask The most common type of color picture tube in which the electron beam is directed through a perforated metal mask to the desired phosphor color element.

shadow ROM A process used in many 386 machines to map ROM BIOS activities into faster 32-bit RAM memory. Shadow memory must be loaded with BIOS routines each time the computer boots. See also Shadow Bios ROM.

shag phones Prepaid cell phones used by a couple involved in an illicit affair, to keep damning evidence from showing up on caller logs, caller ID, or phone bills.

shannon A measurement of the quality of information in a message represented by one or the other of two equally probable, exclusive and exhaustive states. See Shannon's Theorem.

Shannon, Claude Elwood Professor Claude Shannon was a distinguished mathematician who spent a number of years working at Bell Telephone Laboratories. During that time, he published a paper entitled "A Mathematical Theory of Communication" in the Bell System Technical Journal. The paper showed conclusively that all information sources (e.g., telegraph keys, people speaking, and television cameras) have an associated "source rate" that can be measured in binary terms (i.e., a series of 1s and 0s) and that can be expressed in bits per second (bps). Although the contemporary technology wasn't sufficiently advanced enough to take advantage of these findings, they eventually became the basis for digital technologies as diverse as modems, magnetic storage, and all variety of digital transmission systems. Other impressive contributions in mathematics and cryptography followed. Shannon, a distant relative of Thomas Edison, also was remembered within the scientific community for his wacky inventions, such as the rocket-powered Frisbee; the gasoline-powered pogo stick; Theseus, the maze solving mechanical mouse; and THROBAC (THrifty ROman numberical BAckward-looking Computer), a calculator that performed mathematical calculations in the Roman numerical system.

Technology Review Magazine of July/August 2001 wrote of the "Reluctant Father of the Digital Age," Claude Shannon, "The entire science of information theory grew out of one electrifying paper that Shannon published in 1948, when he was a 32-year-old researcher at Bell Laboratories. Shannon showed how the once-vague notion of information could be defined and qualified with absolute precision. He demonstrated the essential unity of all information media, pointing out that text, telephone signals, radio waves, pictures, film and every other mode of communication could be encoded in the universal language of binary digits, or bits -a term that his article was the first to use in print. Shannon laid forth the idea that once information became digital, it could be transmitted without error. This was a breathtaking conceptual leap that led directly to such familiar and robust objects as CDs. Shannon had written "a blueprint for the digital age," says MIT information theorist Robert Gallager, who is still awed by the 1948 paper. Claude Elwood Shannon was born in 1916 and died on February 24, 2001. See also Shannon's Law and Shannon-Fano Coding.

Shannon-Fano Coding A statistical compression technique developed independently in 1949 by Shannon and Weaver, and Fano. Shannon-Fano coding reduces the average length of the code required to represent symbols and characters, such as the letters of an alphabet, through the use of variable-length code words. The method divides the entire set of symbols into two equal, or almost equal, subsets, based on the statistical probability of occurrence of the set of symbols in each subset. Within each subset, the symbol with the greatest frequency of occurrence is represented with the shortest code word, i.e., the fewest number of bits. The symbol with the lowest frequency of occurrence is represented with the longest code word, i.e., the number of bits increases. The codewords used to represent the symbols in the first subset all begin with a binary "0," and those in the second subset with a binary "1." With each subset, the symbol with the greatest frequency of occurrence is represented by only a "0" or "1." Additional binary digits are added to represent the less frequently occurring symbols Huffman coding is a related method. Note that this general approach was developed many years ago, as exemplified by Morse Code, in which the most frequently letters of the English alphabet are represented by the fewest number of dots and dashes. See also Compression and Huffman Coding.

Shannon-Fano Coding Developed by Claude Shannon of Bell Labs and R.M. Fano of MIT. A compression algorithm based on the ASCII character set, and supporting 256 values. The frequency of occurrence of each symbol in a set of symbols is tallied. The list is divided and subdivided, with the most frequent symbols being represented by the fewest number of bits. Shannon-Fano coding is used, along with other compression algorithms, in the implode (compression) algorithm of PKZIP, for instance. See also Compression.

Shannon's Law A theorem defining the theoretical maximum at which error-free digits can be transmitted over a bandwidth-limited channel in the presence of noise. The rough equation works out to about 10 bits per Hertz of bandwidth in practical analog circuits, making the Shannon limit about 30,000 bps for voice-grade lines. According to ZhongJin Yang, a Ph.D. engineer with Lucent Technologies, "Shannon's Law" should be

"Shannon's Theorem". Generally speaking, he wrote me, there is no law in mathematics. There are only theorems. Mr. Yang points out that mathematics is not an experimental science, and all laws are from experiments. The conclusion from logical derivation is a theorem. I stand corrected, thanks to Mr. Yang. Now comes the hard part. We've got to correct the rest of the world.

In any event, Shannon's Law is the mathematical theorem used in the A-Law encoding technique used in E-carrier transmission systems. The Nyquist Theorem, a similar but incompatible approach, is used in North American T-carrier systems. See also E-Carrier; Nyquist Theorem; Shannon, Claude; and T-Carrier.

Shannon Limit The theoretical limit of information through a channel is called the Shannon Limit (after Claude Shannon, who began information theory). You approach the Shannon Limit by adding redundant bits to the information you send, manipulating them in a standard way to arrive at the correct information, although noise distorts it. The Viterbi algorithm (symbol correction), turbo coding (bit correction), and Reed-Solomon (byte correction) are standard methods of FEC (forward-error correction).

shaping 1. The focusing of a microwave radio beam by the transmitting dish to provide maximum signal strength at the receiving dish. In most microwave systems, the transmitting antenna sits inside a concave, reflective metal dish, and transmits the signal into the dish. The dish reflects and shapes the signal towards the receiving antenna, much like the mirror inside a flashlight reflects and shapes the light signal. See also Microwave.

2. The process by which signals are modified between networks for interface purposes, at least in some instances. For example, the analog signal originating from your premises might be shaped so that it fits properly into the digital channel of a DLC (Digital Loop Carrier) system. Each channel of the DLC is 64 Kbps wide, which is voice-grade bandwidth in digital terms. The assumption is that the analog signal is 4 KHz wide, which is voice-grade in analog terms. If the analog signal exceeds 4 KHz, it is shaped down to 4 KHz before it is converted to digital format at the DLC in order that it will fit into the 64 Kbps channel. This all works just fine unless the analog signal intentionally is wider. DSL (Digital Subscriber Line) technology for high-speed Internet access is much wider than voice grade and, therefore, will not work over conventional DLC. See also DLC and ngDLC.

shaping descriptor An ATM term. N ordered pairs of GCRA parameters (I,L) used to define the negotiated traffic shape of a connection.

shaping network A network inserted in a circuit for improving or modifying the wave shape of the signals.

share To make resources, such as directories, files, printers, and ClipBook pages, available to network users.

shared clipboard A feature of Microsoft's NetMeeting. See NetMeeting.

shared data clustering Shared data clustering software allows clusters of Linux-based Intel Xeon or AMD servers to function as a single, (theoretically) easy-to-use, highly available computer system. The software theoretically provides scalable data access and sharing and a central management console for managing servers and storage as one.

shared Ethernet Ethernet configuration in which number of segments are bound together in single collision domain. Hubs produce this type of configuration where only one node can transmit at a time.

shared lock In a database a shared lock is created by non update (read) operations. Other users can read the data concurrently, but no transaction can acquire an exclusive lock on the data until all the shared locks have been released.

shared logic Simultaneous use of a single computer by multiple users.

shared loop See Line Sharing.

shared memory Portion of memory accessible to multiple processes.

shared mesh Also known as 'traditional' or 'best effort' mesh) A wireless mesh network using a single radio to communicate via mesh backhaul links to all the neighboring nodes in the mesh. Here the total available bandwidth of the radio channel is 'shared' between all the neighboring nodes in the mesh. The capacity of the channel is further consumed by traffic being forwarded from one node to the next in the mesh reducing the end to end traffic that can be passed. Because bandwidth is shared amongst all nodes in the mesh, and because every link in the mesh uses additional capacity, this type of network offers much lower end to end transmission rates than a switched mesh and degrades in capacity as nodes are added to the mesh. Wireless mesh nodes typically include both mesh backhaul links and client access. A dual radio shared mesh node uses separate access and mesh backhaul radios. Only the mesh backhaul radio is shared. In a single radio mesh node, access and mesh backhaul are collapsed onto a single radio. Now the available bandwidth is shared between both the mesh links and client access, further reducing the end to end traffic available.

shared modem pools Dial-out users share resources. Any authorized user attached to the network can dial out a port on the dial-up switch, reach a modem and go for it. Benefits: Reduced costs, improved management and security; It eliminates the need for separate modem and separate modem phone lines.

shared path A path or channel that can be used for a number of interconnecting functions; e.g., a tandem trunk that can be used to interconnect Office A, via the tandem office, with Offices B, C, or D. See Dedicated Path.

shared resource Any device, data, file or program that is used by more than one other device, program or person. For Windows shared resources refer to any resource that is made available to network users, such as directories, files, printers, and named pipes.

shared screens A multimedia concept. Shared screen applications enable two or more workstations to display the same screen simultaneously. For example, two users sharing a screen can work on the same spreadsheet. Changes made by one user can be seen by the other as they are made. Shared screens can be implemented in two ways. One way enables people to view each other's screen while one person makes changes. The other way enables people to run the same application on both screens so that both users can make changes simultaneously.

shared secrets Shared Secrets are pre-shared keys that have been allocated to the communicating parties prior to the communication process starting. Shared secrets may be used in PKI (Public Key Infrastructure), however due to the need to pre-share the keys, it has very poor scaling ability.

shared services Providing PBX-based communications and processing services to the unaffiliated tenants and/or the building manager/owner of a commercial building in a standalone or campus environment.

shared tenant services Providing centralized telecommunications services to tenants in a building or complex.

shared video memory Traditionally, PCs used a special bank of memory for the video circuitry, called Video RAM, or VRAM, that was separate from the main memory used for all the other tasks the computer performs. Having separate memory in a separate location is a good thing. However, some cheaper PCs steal part of main memory for use with their video circuitry. This can impair performance by degrading the amount of main memory the computer can use. Because of that, computer ads often reveal only in tiny type that the video memory is shared. Walt Mossberg of the Wall Street Journal warns that "a related jargon term is "integrated graphics" or "integrated video." This refers to a cheaper, less capable type of video circuitry that is bolted onto the computer's main circuit board rather than residing on a separate video card."

shared whiteboards A multimedia concept. Shared whiteboards enable you to "mark-up" a screen using a mouse or stylus input device and have the results show on other screens, often communicating over long distance telephone lines. The concept is similar to a traditional whiteboard mark-up process where everyone has a different color marking pen to circle, write, or cross out items. The background board can be a window from the workstation such as a spreadsheet, image, or blank canvas, or it can be the entire workstation screen. The shared whiteboard can be used for either real-time or store-and-forward collaboration. In the store-and-forward scenario, the mark-ups can be implemented in a time-delayed fashion so everyone can follow the entire step-by-step process.

Shared Wireless Access Protocol See SWAP.

SHARES Acronym for the SHAred RESources (SHARES) High Frequency (HF) Radio Program. The purpose of SHARES is to provide a single, interagency emergency message handling system by bringing together existing HF radio resources of federal, state and industry organizations when normal communications are destroyed or unavailable for the transmission of national security and emergency preparedness information. SHARES further implements Executive Order No. 12472, "Assignment of National Security and Emergency Preparedness Telecommunications Functions," dated April 3, 1984. As of July 2004, over 1,000 HF radio stations, representing 93 federal, state, and industry entities are resource contributors to the SHARES HF Radio Program. SHARES stations are located in every state and at 20 overseas locations. Over 150 HF frequencies have been authorized for use in SHARES. A SHARES Bulletin is published periodically to keep members updated on program activities. http://www.ncs.gov/shares/.

shareware Imagine you write software. You've just written a great program. You now want to sell it. You have two choices. You can take advertisements, sell it to retailers, get distributors to carry it, hire salespeople, etc. In other words, go the commercial route. This is expensive and requires a major marketing / sales budget. The other choice is to go the Shareware route. This involves giving away your software on various bulletin boards, on many Web sites, in "shareware" direct mail catalogs. People download the software for

free and try it. If they like it, they will send you money. They will do this because you offer them an instruction manual, a new version of the software that doesn't blast "unregistered" on the splash screen when you load the software, or an upgraded version of the software, with more features, or a chit that assuages your guilt at using unpaid-for software that someone worked real hard on.

Sharon Her last name is Kander, but could be candor. Over the years she's been a good friend, who pointed out my (many) failings and my (occasional) successes. Sometimes the failings are painful. But she's always right. Usually they relate to my total lack of sensitivity. My wife and her see eye to eye on that one. She and her wonderful family of husband, David, and seriously talented son, Matthew, live on the opposite side of the country which sadly, limits our all-too-infrequent get togethers. As I write this, I miss Sharon. Are you there?

SHARP Self Healing Alternate Route Protection. A system typically employing redundant cables (often fiber) that carry traffic between two separate local exchange carrier offices along divergent paths.

SHDSL Symmetric High-Bitrate Digital Subscriber Loop, the first multi-rate, symmetric digital subscriber loop to be standardized. SHDSL enables symmetrical data transmission of 192 kBps to 2.3 MBps on a single copper wire pair or 384 kBps to 4.6 MBps on two pairs. Hence, it supports applications previously supported by E1 and T1 ISDN and by HDSL and SDSL. The relevant recommendations on SHDSL are ITU G991.2, ETSI TS 101-524 and ANSI T1E1.4/2001-174 G:SHDSL.

shear A computer imaging term. A tool for distorting a selected area vertically or horizontally.

sheath The outer jacket (usually metal or plastic) surrounding copper and fiber cables that prevents water damage to the cables inside.

sheath miles Let's say that you have two sheaths of fiber, each of which contains ten fibers and runs for one mile. That is one route mile (total distance of all fibers), two sheath miles (two sheaths running one mile), and twenty fiber miles (20 fibers running one mile).

sheave A grooved wheel in a block or pulley that assists in the pulling of cable around a bend or corner. Used when installing cable in underground plant.

shelf life The useful life of components when not in use -such as being stored on a shelf as spare parts or in a warehouse awaiting shipment. Batteries tend to have the shortest shelf life of most telecommunications components. Today, the shelf life is less a problem of shelf decay and more a problem of technological obsolescence.

shelf registration A shelf registration is a filing by a corporation that awaits approval from the Securities and Exchange Commission with no specific dates of the offering. Essentially the SEC approves the deal with a wide latitude but no particular timeframe. A shelf registration could also give the issuer, the corporation, a wide ability to issue either debt or equity.

shelfware Software that is bought, then placed on the shelf, but never used. In the June 17, 2002 issue of Business Week, AMR Research, Inc. argued that only half the corporate software bought in the Spring of 2002 was actually ever installed and used. See also feedware, hookemware, hyperware, meatware, seedware, shovelware, smokeware slideware and vaporware.

shell An outer layer of a program that provides the user interface, or the user's way of commanding the computer. Instead of presenting the user with a bland C prompt, i.e. C:> the shell presents a list of programs that the user can choose from, making it easier, allegedly, to figure out which program to run. The problem with shells is that they often take up precious memory. That memory might better be used in actually running a program faster, or more efficiently.

shell account An Internet term. A type of interface on a dial up connection in which you log in to the host computer and use a command shell to get to the Internet. Shell accounts are typically text-based-only interfaces controlled by host servers which normally don't allow for use of graphic Web browsers.

sheriff Sheriff is a word that originated from Shire Reeve. During early years of feudal rule in England, each shire had a reeve who was the law for that shire, that is, he was the "shire reeve." When the term was brought to the U.S. it was shortened to "sheriff."

Shibboleth See Internet2.

shield A metallic layer consisting of type, braid, wire or sheath that surrounds insulated conductors in shielded cable. The shield may be the metallic sheath of the cable or the metallic layer inside a nonmetallic sheath. Shields reduce stray electrical fields and provide for safety of personnel. See Shield Effectiveness, Screen, and Microwave Absorber.

shield coverage The physical area of a cable that is actually covered by the shielding material and is expressed in percent.

shield effectiveness The relative ability of a cable shield to screen our undesirable radiation. Frequently confused with the term shield percentage, which it is not.

shielded pair Two insulated wires in a cable wrapped with metallic braid or foil to prevent the wires acting as antennas and picking up external interference (e.g. a local TV station).

shielded twisted pair A cabling system comprising wires which are separately insulated, and twisted together in a spiral manner. In addition, each pair is wrapped with metallic foil or braid, designed to insulate (i.e., shield) the pair from electromagnetic interference. See STP for a full explanation. See also SSTP.

shielding 1. The metal-backed mylar, plastic, teflon or PVC that protects a data-communications medium such as coaxial cable from Electromagnetic Interface (EMI) and Radio Frequency Interference (RFI).

2. The process by which electrical conductors are wrapped with metallic foil or braid to insulate them from interference and thus provide high quality transmission. Many devices can cause interference to cables (i.e. multiple conductors) carrying telecommunications conversations. Such things include high voltage AC power lines, machinery with motors, machines which make rays of some type (X-Ray systems, TV sets etc.). By wrapping conductors around the cable cores, these cables are less likely to be affected by these outside forces and the noise they create on telephone lines. Shielding will also lessen the chance that the information movement along the cable will interfere with signals on other, adjacent cables. The need for shielding stems from this phenomenon: If you send an electrical signal along one pair of cables, those cables will give off a small amount of electrical energy -called magnetic radiation. That radiation will cause electromagnetic interference with a cable close by. If you "shield" the pair carrying the electrical signal, you will cut down the susceptibility of those cables to interference from other cables. LANs should always be installed with the best quality shielded cable. They will run better with shielded cable. Never skimp on the quality of the cable you're installing for LANs. Most telephones don't require shielded cable unless the cable serving them is passing through some area of high electromagnetic interference.

shift 1. The movement of data to either the right or the left of an existing position in a data field. 2. The code control function of converting the characters from upper to lower case, or vice versa.

shift and shaft To shift programs to a lower level of government without providing the means with which to pay for those programs.

shift button This button acts exactly like a Shift button on a typewriter or computer. It gives the key you're touching a second meaning -either a capital letter or a second set of speed dial buttons, etc.

shift character The control character which defines the shift function.

shift definitions A call center term. A template from which the program can create schedules during a scheduling run. Each shift definition is a record in a Scenario giving more or less precise instructions on shift length, time of day, breaks, and how extensively such schedules can be used in meeting staffing requirements.

shift register 1. A register in which a clock pulse causes the stored data to move to the right or left one bit position. See Zero Stuffing.

2. Another term for a Universal Asynchronous Receiver/Transmitter (UART). See also UART.

shill bidding You're selling something on an auction on the Internet. In order to push the price up you bid against the people who are bidding. This is called "Shill bidding."

shim A shim is a piece of software. The term is associated with calls to communications protocols such as TCP or IP for communications services. The shim inserts itself into the logical space between a program asking for service and the program, such as a TCP-conforming communications program, able to provide the service. The function of a shim is to intercept calls made by higher level programs, such as applications, to translate them, and to pass them off to some other piece off software -perhaps IPX. The program requesting the service is fooled by the shim into thinking it is receiving the service from the software it addressed. This term courtesy Frank Derfler. Thank you Derf.

SHIM header MPLS term. When an IP packet (layers 2-7) is presented to the LER, it pushes the shim header between layers 2 and 3. Note that the shim header is neither a part of layer 2 or layer 3; however, it provides a means to relate both layer 2 and layer 3 information. The Shim Header consists of 32 bits in four parts twenty bits are used for the label, three bits for experimental functions, one bit for stack function, and eight bits for time to live (TTL). It allows for the marriage of ATM (a layer-2 protocol) and IP (a layer-3 protocol).

shiner Any exposed copper wire that may cause a electrical short.

Ship High In Transport See Shit.

ship to shore telephone See Marine Telephones.

ships in the night A term that describes the ability of two protocols to coexist with one neither having any knowledge of the other, nor being affected by it. For example, OSI Layer 2 (Data Link Layer) ATM switching and signaling protocols can coexist with Layer 2 and Layer 3 (Network Layer) MPLS (MultiProtocol Label Switching) protocols.

shit In the 16th and 17th centuries, everything had to be transported by ship. It was also before commercial fertilizer's invention, so large shipments of manure were common. It was shipped dry, because in dry form it weighed a lot less than wet. But once water hit it, it not only became heavier, but the process of fermentation began again, of which a by-product was methane gas. As the stuff was stored below decks in bundles you can see what could (and did) happen. Methane began to build up below decks and the first time someone came below at night with a lantern, BOOOOM! Several ships were destroyed in this manner before it was determined just what was happening. After that, the bundles of manure were always stamped with the term "Ship High In Transit." This meant to stow it high enough off the lower decks so that any water that came into the hold would not touch this volatile cargo and start making methane. Thus evolved the term "S.H.I.T." which has come down through the centuries and is in use to this very day. You probably did not know the true history of this word. Neither did I. I always thought it was a golf term. In fact, there is a site on the web called snopes.com whose sole job is to discount every neat story, like this one about how the word shit came about. Snopes says it came from a bunch of European languages which had words like shit, e.g. the Danish skide. Frankly, I'd rather believe my story. See also Camel Droppings.

shock A sudden stimulation of the nerve and convulsive contraction of the muscles caused by a discharge of electricity through the body. The severity depends on the amount and duration of the current and whether the path of the current is through a vital organ.

shock jock A radio disc jockey who uses profanity, libelous statements against minorities and sundry incitement to violence and lewd behavior. All of this is used to attract a listening audience.

shoe fetish Forty percent of women have hurled footwear at a man, according to my friend Alex Braun, West Coast/Senior Editor of Semiconductor International. No one has thrown footwear at me. But there's still time. See also shoes.

shoe hickey A mark made on your neck by someone standing on your shoulders. This definition contributed by Steve Hersee, founder of Copia International, a great fax server company. Steve does circus work and trains cheer leaders as a hobby.

shoebox A shoebox is a housing device with a power supply to support external peripherals. When the user's main computer case is filled to capacity, an external device is needed to handle the overflow. Hence a shoebox.

shoes Toward the end of the fifteenth century, men's shoes had a square tip, like a duck's beak, a fashion launched by Charles VIII of France to hide the imperfection of one of his feet, which had six toes. Most French people today have five toes per foot. See shoe fetish.

shofar The shofar is the Jewish ram's horn which is sounded on Rosh Hashanah and Yom Kippur. In olden times, the shofar was blown to announce an important event, such as the beginning of the Jewish New Year (i.e. Rosh Hashanah). There is a bad joke that explains this:

Moses and Rastus are standing outside the synagogue when they hear the shofar. Moses asks, "What's that?" Timothy replies, "Why that's the sound of the Jews blowing the shofar." Moses says, "Them's Jews really treat their hired help well."

shop-floor-to-top-floor This term describes a network solution, or policy, or anything else that extends or applies companywide, from the machine shop up to the executive suites. For example, "shop-floor-to-top-floor connectivity."

shopping agent See Shopping Bot.

shopping bot Also called a shopping agent. Bot is a shortened form of robot. A shopping bot or agent is a piece of software that prowls the Web on your behalf in search of bargains in products you specify. Want a cheap sweater, a cheap car, a cheap airline ticket, your shopping bot will go out and find it -theoretically. Of course, it ain't always that easy. Shopping bots are run by companies on the Web, who use them as ways to attract people to their sites so they can sell advertising or get a piece of your shopping bill.

shopping cart Shopping cart is the electronic equivalent of a shopping cart in a supermarket. You're on a Web site, buying something, let's say computer equipment. You choose a monitor you like. You say "buy it." It drops into

short A circuit impairment that exists when two conductors of the same pair, which normally make up an operating electrical circuit, touch or are connected.

short bus A high-speed common channel in the AT&T ISDN packet controller over which all messages between sending and receiving devices pass.

short circuit A near zero resistance connection between any two wires that disrupts transmission where two pairs are involved usually called a "cross." It disrupts transmission and may cause an excessive current flow. In AC electricity, a short circuit is an unintended connection between two supply conductors (i.e.: HOT and Neutral conductors.) A short circuit will usually cause high current flow and will operate the over current protection (fuse or breakers) to interrupt the circuit.

short event A carrier event that occurs when the activity duration is shorter than the ShortEventMaxTime (84 bits).

short haul Between a few hundred yards and 20 miles. Many people would argue with this definition.

short haul modem A data set designed for use in communicating data up to distances of 25 miles over a dedicated unloaded copper pair. Many people would argue with this definition.

Short Message Service SMS or S.M.S. A means by which short messages can be sent to and from digital cell phones, pagers and other handheld wireless devices. Alphanumeric messages up to 160 characters can be supported. That's adequate for stock quotes, short e-mail, bank account balances, buying movie tickets on line, updates on traffic conditions, answers to quizzes posed by the teacher and other really short messages. Europeans, who have relied on digital cell phones for years, love to send short text messages by tapping on their telephone dialing pads. Europeans don't use the Internet for sending email as much as Americans do. SMS is defined in IS-41C. Some cell phones announce the arrival of a new message with three short beeps, two long ones, and three short ones -Morse code for SMS. American telecommunications regulations, which encouraged different mobile operators to choose different, incompatible technologies, are also responsible for the dearth of texting in America. Only in 2003 did the largest American operators agree to pass text messages between their networks -an agreement still only patchily implemented. In addition, not all handsets sold in America support two-way texting: many older models allow only incoming messages. And texting is not included as standard in most subscription packages, but as an extra for which customers must pay a few dollars per month or rather exorbitant amounts per message. In contrast, email for most Americans is free. See also IS-41 and Wireless Application Protocol (WAP).

Short Message Service Center See SMSC.

short reach Short reach refers to optical sections of approximately 2 km or fewer in length. The sections may be interoffice or intraoffice in nature. An example application of SR is that of the interconnection of routers, cross-connects, and DWDM equipment within a CO (Central Office) or POP (Point Of Presence) over POS (Packet Over SONET) interfaces. A POS interface launches a single optical signal commonly running at SONET speeds of OC-3 (155 Mbps), OC-12 (622 Mbps), or CO-48 (2.488 Mbps). See also VSR (Very Short Reach).

Short Message Peer-to-Peer Protocol SMPP. A protocol, developed by Logica Aldiscon, that provides the capability to deliver email and voicemail between wired and wireless networks.

short sellers Investors who make bet that a company's stock will decline. They often sell the company's stock short, or they buy put options on the company's stock.

Short Tone DTMF See Short Tones.

short tones First, we invented touchtone, also called DTMF, Dual Tone Multi Frequency tones. You'd punch your number with tones, instead of dialing them. Then someone thought you could control telephone response gadgets, like voice mail, interactive voice response, etc. with touch tones. For these gadgets to work, they had to "hear" the tones you sent. No one really set standards as to the minimum length tone they would hear. But it was generally conceded that they were to be 120 milliseconds. So some manufacturers of telephone equipment started to make phone equipment that, if you pushed a touchtone button, the machine would only sent a touchtone of 120 millisecond duration. That was called a short tone. It wasn't very useful because the manufacturers quickly discovered that many pieces of equipment couldn't respond that quickly. And the manufacturers got complains that their customers couldn't call their voice mail, their bank, etc. As a result, some manufacturers of equipment brought out new hardware (replacing the old) to allow you to send "long tones," which are now defined as touchtones that last for as long as you hold down the button -just as it is (and has always been) on a normal single line, non-electronic, non-digital telephone. Isn't progress wonderful? See also DTMF for a much longer explanation of tone dialing.

shortest-path routing A routing algorithm in which paths to all network destinations are calculated. The shortest path is then determined by a cost assigned to each link.

shortwave station A broadcast station that transmits on frequencies of 6-25 megahertz. These waves are shorter than those sent out by AM stations but longer than those of the Very-High frequency FM radio and television stations.

shoulder surfing You're standing at a pay phone. You punch in your credit card numbers to make your long distance call. There's a fellow standing behind you. He's carefully watching what you're doing. He is memorizing the digits you have punched in. When you are through, he will write them down and sell them to someone else, who will use them to make fraudulent long distance phone calls. Our friend is indulging in a new "occupation." It's called "shoulder surfing."

shout-hacking The malicious use of spoken commands to trick an operating system, such as Microsoft Vista, that supports speech recognition. One way of perpetrating such an attack is to trick a computer's owner into downloading and playing an audio recording containing malicious commands; the spoken commands, sent as output to the computer's speakers, are picked up as input by the computer's microphone and are then executed by the computer's operating system.

shovelware A term used to refer to the tendency of early CD-ROM disc producers to shovel anything they could to fill up their voluminous CD-ROM discs. They did this so they could tout all the great value they were offering in their CD-ROM discs. See also Hookeware, Hyperware, Meatware, Slideware, Vaporware and Bundle Fodder.

show off A child who is more talented than yours.

showskele This is the trade name for those annoying little images you sometimes see drifting across your computer screen and which interfere with your viewing the Web page underneath. It belongs to United Virtualities, who say it was named after the middle daughter of the company founder. In April they announced another technology, Ooqa Ooqa (which daughter is that named after?), which changes your browser's toolbar in response to any Web ad you click on. This definition from Wired Magazine.

SHP Signaling Handoff Point, a type of equipment which acts as a gateway between two dissimilar Signaling System 7 networks, allowing information exchange between the two networks.

SHPO State Historic Preservation Officer. In compliance with Section 106 of the National Historic Preservation Act, this investigation reports on the "no effect/no impact or impact" of proposed telecommunication sites on Historic Properties/Communities.

shredder A device to destroy paper, plastic...in fact anything confidential that, if it fell into the wrong hands (a competitor, the public, the police, etc.), could harm the company wishing the item destroyed. The shredding business came into public renown with the Enron scandal in the winter of 2001/2002. According to the Wall Street Journal, shredding may date as far back as the age of papyrus. John Wagner, founder and chief executive of Allegheny Paper Shredders Corp., Delmont, Pa., traces its modern practice to the 1920s, when an American inventor got the idea for a document shredder from a hand-cranked Bavarian noodle cutter.

shrink tubing A cabling term. Tubing which has been extruded, crosslinked, and mechanically expanded which when reheated will return to its original diameter.

shrink wrap Software that requires no customization. So called because it comes in a package that is plastic shrink-wrapped. Of course, so does non-shrink wrap software, or software that has to be customized for your use. But this is a dictionary that explains what words mean. It doesn't explain the lack of logic behind the naming of those words.

shrinkage percentages A call center term. A group of scenario budget assumptions that define the percentage of the time employees are scheduled to work but are not available to handle calls because of absence, breaks, vacation, non-productivity, training, and other activities.

shrinkage ratio The ratio between the expanded diameter and recovered diameter of shrinkable products.

shrink temperature That temperature which effects complete recovery of a shrinkable product from the expanded state.

shroff An expert in testing coins is called a shroff.

SHT Short Hold Time.

SHTTP Secure Hypertext Transfer Protocol. An extension of HTTP for authentication and data encryption between a Web server and a Web browser.

shunt 1. A conductor joining two points in an electrical circuit so as to form a parallel or alternative path through which a portion of which the current may pass. For example, a shunt might be applied to a main circuit to connect a control circuit for purposes of regulating the amount of current passing through the main circuit.

2. A means by which traffic is switched or diverted from one network to another. For in-

sance, a number of manufacturers are developing devices intended to be positioned logically between central offices (COs) and the connected local loops. Such devices would identify data traffic destined for the Internet or an Intranet, most obviously by means of recognizing the dialed telephone number as being associated with a remote access server supporting IP (Internet Protocol) traffic. At that point, the traffic is routed over a packet network, rather than the circuit-switched Public Switched Telephone Network (PSTN). Thereby, IP traffic's significant contribution to congestion in the PSTN are mitigated. See also Circuit Switching, IP, Packet Switching and PSTN.

shunt circuit An arrangement of apparatus or circuits in which the total current is subdivided. Same as Parallel Circuit.

shuttle server A server that is typically set up on the bottom tray of a cart, and which is wheeled to wherever it is needed, for example, to support the backup, maintenance, or restoration of another server, or to set up an ad hoc network in a training room or somewhere else where a server is temporarily needed. See also crash cart.

SHVERA The Satellite Home Viewer Extension and Reauthorization Act of 2004. SHVERA, passed by Congress and enacted in December 2004, amends the Communications Act and the copyright statute. It directs the FCC to (1) publish and maintain a list of stations and communities eligible for significantly viewed status, and (2) commence a rulemaking proceeding to implement SHVERA. Significantly viewed signals have significant over-the-air non-cable viewing in a particular community. The Commission's rules relating to carriage of significantly viewed signals have been applied to the cable industry for more than 30 years. The SHVERA applies those rules to satellite providers. To implement the SHVERA, the Commission took the following actions:

- Updated Significantly Viewed List: In the Notice of Proposed Rulemaking released February 2005, the Commission published a list of stations and the communities in which they are deemed significantly viewed.

- "Significantly Viewed" Determinations: The rules will operate for satellite carriage in much the same fashion as they have for cable carriage. This means parties must petition the Commission when seeking to either add a particular station to the Significantly Viewed List for a particular community, or to restrict carriage of a listed station through application of the Commission's network non-duplication or syndicated exclusivity rules.

- Required Showing: Under our rules in effect on April 15, 1976, network affiliates demonstrate significantly viewed status by showing they have at least three percent share of viewing hours in television homes in the community and a net weekly circulation share of at least 25 percent; independent stations must show at least two percent viewing hours and a net weekly circulation of at least five percent.

- Subscriber Eligibility: The Order concludes that the SHVERA requires that a subscriber must first be receiving a specific local market network station (if it exists in the market) from the satellite carrier in order to be eligible to receive a significantly viewed station affiliated with the same network.

- Digital Significantly Viewed: The SHVERA prevents satellite carriers from retransmitting the digital signal of a local market network station in a lesser format than the digital signal of the significantly viewed station affiliated with the same network. Thus, as a condition to offering a significantly viewed digital signal, satellite carriers must comply with the "equivalent bandwidth" and "entire bandwidth" requirements, which permit satellite carriage of a significantly viewed station only if the amount of bandwidth used to carry such station is equivalent to the amount of bandwidth used to carry the signal or signals of the affiliated local network station, or the entire amount of bandwidth used by the local station. These carriage requirements for local market digital signals apply only if the satellite carrier chooses to carry an out-of-market significantly viewed digital signal, and do not affect the "carry-one, carry-all" rules in general.

- "Satellite Community" defined: Stations are deemed significantly viewed in a particular community. Satellite carriers, like cable operators, use cable communities where they exist. If there is no cable system, satellite carriers may use existing communities (e.g., cities, towns, villages, etc.) to serve as satellite communities. In the absence of a distinct community entity, a satellite community may be defined by one or more adjacent five-digit zip code areas.

- Carrier Responsibilities: At least 60 days before retransmitting a significantly viewed signal into a local market, satellite carriers must notify all television stations in such market.

SI Shift In.

Si-Ge Also written Si/Ge. Silicon Germanium (not Geranium, which is a flower). Si-GE is

a chip technology announced by IBM in the fall of 1998. Cheaper. Faster, etc.

SIA 1. Satellite Industry Association. An operating arm of SBCA (Satellite Broadcasting and Communications Association of America). Formed in 1995, the SIA promotes the role of satellite technologies and applications in the National and Global Information Infrastructure, and acts to inform policy makers, regulators, legislators, the press and the public about the capabilities and benefits of satellite communications. Members include service providers, manufacturers, launch service companies and ground equipment suppliers in the US. www.sia.org.

2. Securities Industries Association.

siamese cable A cable consisting of two individually jacketed, yet separable cables. The cables can be separated at either end at the seam where the jackets are joined, like pull-apart licorice twists. The individual cables that make up a Siamese cable can be a coax cable and CAT 5 cable, or a CAT 3 cable and a CAT 5 cable, or a video cable and a power cable, or whatever cables are needed. Cable and wire manufacturers make a wide variety of Siamese cables. Siamese cable saves time and labor when two cables need to be laid along the same path, since only one cable needs to be pulled instead of two. In addition, by virtue of its construction, Siamese cable is stronger than separate runs of the constituent cables. This means less chance of the cable's being damaged during installation. The extra bulk of Siamese cable also helps the constituent cables remain within allowable bend radiuses, insuring the integrity of the constituent cables. A Siamese cable is sometimes called a figure-of-eight cable, due to its shape when viewed from the end of the cable.

siamese pair A CATV (Community Antenna TeleVision) term. A siamese pair is a connection between a coaxial cable and a twisted pair cable in order that CATV provider can connect to devices such as telephones and PCs which do not have coax interfaces and, thereby, to provide POTS (Plain Old Telephone Service) and Internet access, as well as CATV access. A number of large CATV providers have upgraded their old analog, one-way, coax-based transmission systems in order to operate as CLECs (Competitive Local Exchange Carriers), as well as entertainment TV providers. See also CATV, CLEC, ILEC, and Sidecar.

SIBB Service Independent Building Blocks, a term coined by Bellcore and the ITU-T for the Intelligent Network. Creation of SIBBs will in theory make it easier for non-software specialists to create new services by mixing and matching SIBBs. SIBBs are like software objects -capsules of reusable software that can be combined together to create wondrous and complex new computer telephony services.

SIC Service Initiation Charge. This is a charge to a client by a carrier for the initiation of a new telecom service such as the installation of a new T-1 line or a Frame Relay PVC.

SID 1. System IDentification number. A five digit number that has been assigned to identify the particular cellular carrier from whom you are obtaining service. This number identifies your "home" system. See also preferred roaming list.

2. Security ID.

Side Band 1. A lower or higher frequency signal which is created when one signal is used to modulate another signal.

2. The band of frequencies to the right or left of a channel's center frequency. See Sideband.

side by side monitoring A call center term. The process whereby a supervisor or other qualified party listens in on the calls of an agent by sitting at their side. The supervisor is able to listen to both sides of the conversation, usually via double jacking, which is plugging a headset into the agent's phone and listening in on the line.

side circuit A metallic, single pair circuit arranged to derive a phantom circuit. The phantom circuit is derived by center tapping a repeating coil in each of two side circuits.

side hour Any hour that is not the Busy Hour. A telephone company term: An amount of time equal to one hour that is time consistent and adjacent to the CBH (Component Busy Hour). This one hour period must have average weekly usage equal to at least 90% of the CBH during the busy season and 80% of the CBH during the nonbusy season.

side lobe A minor lobe of an antenna pattern as distinguished from the main lobe(s) of an antenna pattern.

sideband The frequencies on either side of the main frequency in an RF (Radio Frequency) signal. In the early days these sideband frequencies were not used because they were too "noisy," unreliable and were not needed. Now, technology has improved and frequencies are in short supply, more and more transmission vendors are making use of their sidebands, and thus substantially broadening the throughput of their existing transmission paths. Sideband technology has made great strides especially in through-the-air microwave transmission.

sidecar An add-on module for a CATV (Community Antenna TeleVision) set-top converter box. A sidecar essentially is peripheral equipment for interactive, two-way access to the CATV provider's network. The sidecar allows you to gain access to an interactive program guide, to access the Internet through a built-in cable modem, and to access the PSTN (Public Switched Telephone Network). In order to provide a full range of capabilities, the CATV provider, of course, must have enhanced the cable system to support two-way transmission, packet-switched data access to the Internet, and circuit-switched access to the PSTN. A number of CATV providers have done so in order to position themselves as CLECs (Competitive Local Exchange Carriers), in competition against the ILECs (Incumbent LECs) in voice and data. In order to improve on the quality of transmission and to provide more bandwidth, many also have upgraded their coaxial cable transmission systems to hybrid systems, incorporating fiber optic transmission facilities in the high-capacity trunk segments of their network. See also CATV, CLEC, ILEC and Siamese Pair.

sideloading The transfer of digital content -generally music -from a PC to a cell phone.

sidereal Complete rotation of a celestial object about its axis. Literally day "according to the stars;" the time of one complete 360 degree rotation of the Earth about its axis in 86,164.091 seconds (23h, 56m, 4.091s). For a bigger explanation, see day.

sidetone A part of the design of a telephone handset which allows you to hear your own voice while speaking. The idea is to let you know that the telephone you're speaking on is working. Too much sidetone becomes an echo and is bad. Too little sidetone makes the channel unerringly quiet and people start to think it's busted. See echo and hybrid.

SIDN Security Industry Digital Network.

SIF 1. Standard Image Format. See MPEG.

2. SONET Interoperability Forum. A voluntary industry group established to define and resolve issues of SONET implementation. SIF was formed by Southwestern Bell in 1991; Bellcore and other RBOCs soon joined. SIF now is open to membership of any interested party, including vendors, service providers and end users. SIF works under the umbrella of ATIS (Alliance for Telecommunications Industry Solutions). www.atis.org/atis/sif/index.html See also ATIS.

3. Signaling Information Fields. A SS7 term. See Signaling Information Fields.

SIG 1. Special Interest Group. A SIG is an ongoing discussion group held electronically via PCs. A SIG focuses on one area of interest. Members phone in with their PCs, read messages posted, contribute their wisdom, ask questions, etc. SIGs are ways people get up-to-date accurate information on a subject. SIGs are run on most BBS (Bulletin Board Systems). SIGs are to bulletin boards what on-line services call conferences or forums.

2. Special Interest Group. In this context, a SIG is a voluntary organization which is dedicated to the advancement of a particular technology, standard or technique. Examples include the ATM Forum, CDPD Forum, Frame Relay Forum and SMDS Interest Group.

3. SMDS Interest Group. A defunct consortium of vendors and consultants who were committed to advancing worldwide SMDS as an open, interoperable solution for high-performance data connectivity. On June 16, 1997, the Board of Trustees announced that the group was disbanded, turning over all responsibilities to regional organizations. The Board declared that its mission had been fulfilled.

4. Signaling Transport. See Sigtran.

sig file Signature File. A file that automatically is appended to every e-mail message you send. The sig file commonly contains your name, title, company, telephone number, fax number, and return e-mail address. See also vCard.

Sigint A military term for signals intelligence, as opposed to humint, which is human intelligence. Sigint is a polite term for reading someone else's mail. Sigint gets information from intercepting and often decrypting voice or electronic communications. In the United States, the National Security Agency (NSA) does sigint and the Central Intelligence Agency (CIA) does humint. There is, of course, a great rivalry between the two agencies as to who can get it right. One former NSA director once scoffed, "The CIA is good at stealing a memo off a prime minister's desk. But they're not much good at anything else." See NSA.

sign-on To go through the process of beginning a working session between you, your data terminal or PC and a computer.

sign on/sign off The process of identifying oneself to a machine so as to gain access. In the case of an ACD system this process allows statistics to be kept for this person individually. It also allows for the movement of the person around the system while statistics are accumulated in one logical file.

signal 1. An electrical wave used to convey information.

2. An alert.

3. An acoustic device (e.g. a bell) or a visual device (e.g. a lamp) which calls attention. To transmit an information signal or alerting signal.

4. Bellcore defines signal as a state that is applied to operate and control the component

groups of a telecommunications circuit to cause it to perform its intended function. Generally speaking, says Bellcore, there are five basic categories of "signals" commonly used in the telecommunications network. Included are supervisory signals, information signals, address signals, control signals, and alerting signals.

signal bender See Signal Booster and Beam Bender.

signal booster By FCC definition a signal booster is a low power relay or repeater.

signal bounce When a bus topology network cable has not been properly terminated at each end of every open cable, the signal from the network will travel from one end of the cable to the other and then will continually bounce back the way it came.

signal cable A cable designed to carry current of usually less than one ampere per conductor.

signal computing Signal computing, as it has come to be, refers to the processing of analog signals for transmission over the worldwide telephone system. According to Analog Devices, the core of the Signal Computing concept is the ability to program an open, multi function chipset with easily upgraded software. The Signal Computing model gives non-traditional audio, video, speech, and communications technology providers opportunities to embrace PC channels. The term, signal computing, picked up steam in mid-1993 when Dialogic announced a new standard called SCSA (Signal Computing System Architecture). In Dialogic's words, SCSA is a comprehensive architecture that describes how both hardware and software building blocks work together. It focuses on "Signal Computing" devices, which refer to any devices that are required to transmit information over the telephone network. Information can be transmitted via data modems, fax, voice or even video. SCSA defines how all these devices work together.

Signal computing systems combine three major elements for call processing. Network interfaces provide for the input and output of signals transmitted and switched in telecommunications networks. Digital signal processors and software algorithms transform the signals through low-level manipulation. Application programs provide computer control of the processed signals to bring value to the end user.

SCSA is the common set of standards that telecommunication system manufacturers and computing system manufacturers can use to create computer telephony systems. No single company today can create the total solution for all customers. SCSA represents the common ground between the two fields so that manufacturers from each area can safely develop products that will work with other manufacturers.

SCSA's coverage extends from low-level bus and hardware interfaces, like the interboard switching bus that enables boards from different suppliers to work together, to high-level application programming and software interfaces, so that software designed to work with one set of hardware products, will work with different hardware as well. SCSA has been defined with input from leading computer and switch manufacturers, call processing suppliers, and technology developers. In many cases, SCSA has drawn on existing standards, like the T.611 fax standard endorsed by the European Computer Manufacturers Association, and in other cases SCSA has extended standards to make them more useful for call processing suppliers and users.

Signal Computing System Architecture SCSA is an open hardware and software architecture supported by over 240 companies for developing multi-technology computer telephony systems using standard interfaces. SCSA consists of two independent layers: The SCSA Hardware Model and the SCSA Telephony Application Objects (TAO) Framework. While designed for interoperability, these two layers may be implemented independently of one another. See also Signal Computing.

signal conditioning The amplification and/or modification of electrical signals to make them more appropriate for transmission over a certain medium -cable, microwave, etc.

Signal Control Point SCP. Computers that hold databases in which customer-specific information used by the "advanced intelligent network" (AIN) to route calls is stored. The AIN refers to a specific architecture -typically promulgated and created by a local or long distance phone company -that provides core capabilities in which customer-specific information held in databases within the network is used to intelligently process calls. An example of an AIN service is an 800 service that routes calls based on where the calls are coming from. The routing information is stored in the SCP, which is typically tied into the Signaling System 7 network. See SCP. For a full explanation of the Advanced Intelligent Network, see AIN.

signal converter The equipment which changes the data signal into a form suitable for the transmission medium, or the reverse. The converter can also work with DC/AC current. The signal converter comprises a modulator and/or demodulator.

signal decay The degradation of a signal's energy over time/distance, often disproportionately at different frequencies. See attenuation.

signal generator A test oscillator that can be adjusted to provide a test signal at some desired frequency, voltage, modulation, and waveform.

Signal Ground SGD. In RS-232-C signaling, Signal Ground establishes a common reference level for the voltages of all other signals (such as RXD/TXD), except Frame Ground.

Signal Hill A hill overlooking the Atlantic Ocean in Newfoundland, Canada, where wireless pioneer Guglielmo Marconi successfully received the world's first transatlantic wireless telegraph transmission in 1901. Signal Hill's name and use for signaling stations actually predated Marconi. For at least 100 years prior to Marconi, signal masts and optical telegraphs on Signal Hill were used to notify authorities and townspeople below, in the city of St. John's, of arriving ships, and also to send messages to those ships. By virtue of Signal Hill's being both high ground and the easternmost point of North America, Marconi selected it as the North American test site for his transatlantic wireless telegraphy experiments, with the European test site being in Poldhu, Cornwall, in the southwest corner of England. Many readers of this dictionary probably know of other hills named Signal Hill. Their names, too, reflect their earlier use as sites for signaling stations. For example, the town of Signal Hill, on the southern California coast, next to the city of Long Beach, traces its name to a hill there that was used for centuries by a local native tribe for signal fires. Later Spanish settlers used the hill for the same purpose.

signal leakage A measure of the degree to which signals from the nominally closed coaxial cable system are transmitted through the air.

signal level The strength of a signal, generally expressed in either absolute units of voltage or power, or in units relative to the strength of the signal at its source.

signal processing Signal processing is a combination of computer telephony call control and media processing. Call control means moving telephone calls around -answering them, hanging up on them, transferring them, conferencing them, etc. -all the stuff you do on your office phone every day. Media processing means bringing computer power to bear on the media stream, the actual voice, video or data inside the phone call. You might save it to your hard disk. You might want to have your PC software listen for key words in what you saved, for example, the caller's phone number. You might even want your software to try transcribing the conversation into text you could use in your word processor. See Call Control, Digital Signal Processing, Media Processing and Native Signal Processing.

Signal Processing Component An SCSA definition. An atomic bundle of signal processing functionality which can be allocated to a single group. It can be capable of supporting the functionality of one or more resources or a simple coder.

signal processing element An SCSA definition. That part of a Signal Processing Component which is associated with a single Resource.

signal processing platform A SCSA definition. This is a software component that supports a specific hardware package. This is typically an executable program and may control one or more instances of the vendor-specific hardware package. Each SPP may support several different types of signal processing functionality clustered into SPCs.

Signal Punch A punch made by the U.S. Army Signal Corps on special ceremonial occasions. Although a punch typically has five ingredients (the name of the beverage deriving from the Hindi word "paOxA7c" and the Sanskrit word "paOxA7chan," both meaning "five"), Signal Punch generally is made from ten or more ingredients, the exact number depending on the imagination of the master of ceremony for the punch-making ceremony. Often the ingredients are defined by colors that are significant to the Signal Corps (for example, regimental colors); at other times the ingredients are symbolic of other things that are meaningful to the Signal Corps. Often the final ingredient is a "secret" ingredient that can be almost anything -the more outrageous the better -for example, the master of ceremony's socks.

Signal Punch Ceremony The U.S. Signal Corps ritual where Signal Punch is made.

signal repeaters A signal repeater does nothing more than receive a signal and retransmit it. Repeaters are used where the original transmission is very weak, or the transmission is being sent over long distances.

signal strength indicator A display on a cellular radio that lets you know before you call about the relative strength of the cellular transmitter in your immediate area. On most cellular radios, the signal strength indicator has five bars, with five the strongest. It's best to call when you have four or five. Three is marginal. Below three, forget it. Go elsewhere and try again.

Signal Switching Point SSP. See SSP. For a full explanation of the Advanced Intelligent Network, see AIN.

signal to noise ratio 1. Often abbreviated as SNR or S/N. A measurement of the relative level of noise on a circuit and, therefore, the quality of a transmission, SNR is the ratio of the usable signal being transmitted to the noise or undesired signal. Human beings are fairly tolerant of noise, and within limits, can distinguish content from the "noise," but computer systems are not. If the level of extraneous noise is high in a data communications environment, data packets have to be re-sent. That not only slows down the data transfer, but also reduces the efficiency with which the circuit and network are used. SNR usually is measured in decibels (dB). See also dB.

2. The ratio between useful information and idle chatter to be found on an Internet Usenet newsgroup. bulletin board, or chat room. Often used derogatorily, for example: "The signal-to-noise ratio in this newsgroup is pretty low."

signal transfer point The packet switch in the Common Channel Interoffice Signaling (CCIS) system. The CCIS is a packet switched network operating at 4800 bits per second. CCIS replaces both SF (Single Frequency) and MF (Multi-frequency) by converting dialed digits to data messages. See SCP, STP and Signaling System 7. For a full explanation of the Advanced Intelligent Network, see AIN.

signalman A former rating (job specialty) in the US Navy. A signalman stood watch on a ship's signal bridge; sent and received messages by flashing light, semaphore, and flaghoist; maintained visual signal equipment; rendered passing honors to ships and boats; displayed ensigns and personal flags during salutes and during personal and national honors; performed lookout duties; sent and received visual recognition signals; repaired signal flags, pennants, and ensigns; took bearings, recognized visual navigational aids, and served as navigators' assistants. The signalman rating was disestablished on September 30, 2004. Signalmen's duties were absorbed by other ratings and formed signalmen were retrained into other ratings.

signals Signal being transmitted are information. They can be spoken words such as a telephone conversation, music, or even computer data. The simplest signal we can send is a sine wave. All sound waves are combinations of simpler sine waves at different amplitudes and frequencies to produce complicated wave forms. Similarly, computer data is a series of ones and zeros which electrically look like a square wave. Square waves, like all other types of waves, can be represented by a combination of sine waves with different frequencies and amplitudes. That a complex wave is the sum of simple sine waves was discovered by the French mathematician Jean Baptiste Joseph Fourier and is called Fourier's Theorem.

Signals Intelligence SIGINT. A federal government term. A category of intelligence information comprising, either individually or in combination, all communications intelligence, electronics intelligence, and foreign instrumentation signals intelligence, however transmitted. Signals intelligence is a polite term for reading someone else's mail. See NSA and SIGINT.

signaling In any telephone system -inside an office or across the country -some form of signaling mechanism is required to set up and tear down the calls. When you call from your office desk across the country to someone else's desk, many forms of different signaling are used. There's the signaling between your office desk phone and your office phone system. There's the signaling between your office phone system and your local telephone company central office. And there's the signaling between your local central office and the central office you're trying to reach across the country. All forms of these signaling may be different. Simple examples of signalings are ringing of your phone (someone is calling), dial tone (it's OK to dial), ringing (hopefully someone will answer), etc.

Originally, telephone systems such as POTS (Plain Old Telephone Service) used in-band signaling to carry signals. With in-band signaling, signals such as DTMF tones (touchtones), are carried in the same circuit as the talk path. Newer signaling (i.e. most of it today) is carried as out-of-band signaling, which uses a separate data network. This is much more efficient. For example, voice circuits need not be allocated for calls that do not complete. Also, this approach allows additional quantities of information to be transferred to support advanced applications such as Caller ID, roaming in wireless systems, and 800 number routing. See Signaling System 7.

Signaling Connection Control Part SCCP. Part of the SS7 protocol that provides communication between signaling nodes by adding circuit and routing information to the signaling message. The ISDN-UP (Integrated Services Digital Network User Part) and TCAP (Transaction Capabilities Application Part) use the SCCP (Signaling Connection Control Part) and the MTP (Message Transfer Part) to transport information. Definition from Bellcore in reference to its concept of the Advanced Intelligent Network.

Signaling Data Links Used to connect SS7 signaling points. In most countries, these links are 56 Kbps or 64 Kbps data.

Signaling Gateway SG. A software-based device that resolves interface issues between circuit-switched and packet-switched networks. See also MGCP and SCTP.

Signaling Information Fields SIF. In SS7 (Signaling System 7) signaling messages, the SIF is a variable length field which contains all the signaling information. Also included is any routing information in the Routing Label, which the network uses to properly connect the call. Such signaling information might include the Calling Party Number (CPN) subfield, along with the Presentation Indicator (PI). The SIF contains 2-272 octets of data. See also Calling Party Number, Presentation Indicator, and SS7.

Signaling Link Selection Code SLS. The part of a routing label that identifies the SS7 signaling link on which the message should be sent.

signaling point A node in a SS7 signaling network that either originates and receives signaling messages, or transfers signaling messages from one signaling link to another, or both. SPs are located at each switch in a Signaling System 7 network. They interface the switch with the Signal Transfer Points (STPs). See Signal Transfer Points and Signaling System 7.

signaling point code A binary code uniquely identifying a SS7 signaling point in a signaling network. This code is used, according to its position in the label, either as destination point code or as originating point code.

Signaling Point Interface SPOI. The demarcation point on the SS7 signaling link between a LEC network and a Wireless Services Provider (WSP) network. The point established the technical interface and can designate the test point and operational division of responsibility for the signaling.

Signaling System Number 1 SS1. A tone supervision system using a 500 Hertz tone modulated at a 20 Hertz rate to signal call requests between switchboards. International equivalent of Bell's 1000/20 manual ringdown signaling. Now obsolete.

Signaling System Number 2 SS2. A two-tone (600/750 Hertz) tone system for dial-pulsing selection information. Never used internationally. Closely akin to early Bell mobile radiotelephone dialing systems. Now obsolete.

Signaling System Number 3 SS3. A single-frequency (2280 Hertz) tone system used on one-way circuits only. Not intended for transit connections involving a third nation. A prime method through the late 1970s. Now obsolete.

Signaling System Number 4 SS4. A two-tone (2040 and 2400 Hertz) system for international transit and terminal traffic. The first truly global "direct dialing" signaling system. Now obsolete.

Signaling System Number 5 SS5. A two-tone (2400 and 2600 Hertz) system combined with multifrequency inter-register signaling for both terminal and transit traffic. Closest international equivalent to North American Bell "DDD trunks using SF supervision." Now obsolete.

Signaling System Number 6 SS6. A common digital data path between two switching machines to negotiate and oversee connection control on transmission facility trunks between the machines. International equivalent of Bell CCIS. Typically a 2400 bps data circuit.

Signaling System 7 SS7. SS7 typically employs a dedicated 64 kilobit data circuit to carry packetized machine language messages about each call connected between and among machines of a network to achieve connection control. International equivalent of Bell DNHR. Permits many ISDN services such as CNID or Random RCF. See below for more specific information:

All phone systems need signaling. According to James Harry Green, author of the Dow Jones-Irwin Handbook of Telecommunications, signals have three basic functions:

1. SUPERVISING. Monitoring the status of a line or circuit to see if it is busy, idle or requesting service. Supervision is a term derived from the job telephone operators perform in manually monitoring circuits on a switchboard. On switchboards, supervisory signals are shown by a lit lamp indicating a request for service on an incoming line or an on-hook condition of a switchboard cord circuit. In the network (i.e. the automated part of the network), supervisory signals are indicated by the voltage level on signaling leads, or the on-hook/off-hook status of signaling tones or bits.

2. ALERTING. Indicates the arrival of an incoming call. Alerting signals are bells, buzzers, whoofers, tones, strobes and lights.

3. ADDRESSING. Transmitting routing and destination signals over the network. Addressing signals are in the form of dial pulses, tone pulses or data pulses over loops, trunks and signaling networks.

In the old days, signaling was mostly MF (multi-frequency) and SF (single frequency) and is inband. This means that it goes along and occupies the same circuits as those which carry voice conversations. There are two problems with this. First, about 35% of all toll calls are not completed because the phone doesn't answer or is busy, or there are equipment

problems along the way. The circuit time used in signaling is substantial, expensive and wasteful. Second, inband signaling is vulnerable to fraud. So the idea of out-of-band signaling came about. It got the name of Common Channel Interoffice Signaling (CCIS) because it used a communications network totally separate from the switched voice network. In North America, CCIS started out as an AT&T packet switched network operating at 4800 bits per second. Each of the packet switches in this network (they are no longer exclusively AT&T's) are called Signal Transfer Points -STPs. CCIS has the following advantages over SF/MF signaling:

Fraud is reduced. "Talk-off" is reduced. (Talk-off occurs when your voice contains enough 2600 Hz energy to activate the tone-detecting circuits in the central office.) Signaling is faster allowing circuits and conversations to be set up and torn down (i.e. disconnected) faster. Signals can be sent in both directions simultaneously and during voice conversation if necessary. Network management information is routed over the CCIS network. For example, when trunks fail, switching systems can be told with CCIS data messages to reroute traffic around problem areas.

The older CCIS signaling has been replaced with a newer out-of-band signaling system called ITU Signaling System 7. According to an AT&T technical paper delivered at the International Switching Symposium in Spring, 1987, ITU Signaling System 7 is being required by telecommunications administrations worldwide (i.e. all the local country-owned telephone companies) for their networks. AT&T continued with the introduction of digital switches and transmission equipment with 56 Kbps and 64 Kbps transmission rates, the International Telegraph and Telephone Consultative Committee (CCITT in French) in 1980 approved the ITU 7 recommendations optimized for digital networks. This new protocol uses destination routing, octet oriented fields, variable length messages and a maximum message length allowing for 256 bytes of data. Addition of flow control, connectionless services and Integrated Services Digital Network (ISDN) capabilities were approved by ITU in 1984. A major characteristic of ITU #7 is its layered functional structure. Its transport functions are divided into four levels, three of which constitute the Message Transfer Part (MTP). The fourth consists of a common Signaling Connection Control Part (SCCP).

The SS7 protocol consists of four basic sub-protocols:

- Message Transfer Part (MTP), which provides functions for basic routing of signaling messages between signaling points.
- Signaling Connection Control Part (SCCP), which provides additional routing and management functions for transfer of messages other than call setup between signaling points.
- Integrated Services Digital Network User Part (ISUP), which provides for transfer of call setup signaling information between signaling points.
- Transaction Capabilities Application Part (TCAP), which provides for transfer of non-circuit related information between signaling points.

Signaling System 7 provides two major capabilities:

1. Fast call setup, via high-speed circuit-switched connections.

2. Transaction capabilities which deal with remote data base interactions. What this means in its simplest terms and in one simple application is that Signaling System 7 information can tell the called party who's calling and, more important, tell the called party's computer. A scenario: when you call a direct mail order business, Signaling System 7 will send a signal as to which phone is calling. The agent's CRT screen will pop the caller's name and perhaps the caller's most recent buying information. The agent may answer the phone "Good morning, Mr. Newton. Did you enjoy the three khaki pants we sent you last week?..." Signaling System 7 will be an integral part of ISDN. It will enable us to extend full PBX and Centrex-based services like call forwarding, call waiting, call screening, call transfer, etc. outside the switch to the full international network. In effect, with Signaling System 7, the entire network will acquire the "smarts" of today's smartest electronic digital PBX. See also Captain Crunch, ISUP, MTP, SCCP, Signaling System 7 Software Layers and TCAP.

Signaling System 7 Software Layers MTP (Message Transfer Part) Layers 1 through 3: These layers provide complete lower level functionality at the Physical, Data Link and Network Level. They serve as a signaling transfer point, and support multiple congestion priority, message discrimination, distribution and routing.

ISUP (Integrated Services Digital Network User Part): This layer provides the network side protocol for the signaling functions required to support voice, data, text and video services in an Integrated Services Digital Network (ISDN). Specifically, ISUP supports the call control function for the control of analog or digital circuit switched network connections carrying voice or data traffic.

SCCP (Signaling Control Connection Part): This layer supports higher protocol layers (such as TCAP and IS-634) with an array of data transfer services including connection-less and connection orientated services. SCCP supports global title translation (routing based on directory number or application title rather than point codes), and ensures reliable data transfer independent of the underlying hardware.

TCAP (Transaction Capabilities Application Part): This layer provides the signaling function for communication with network databases. TCAP is an SS7 (and ISDN) application protocol which provides noncircuit transaction based information exchange between network entities. For example, TCAP enables transaction based service applications such as enhanced dial-800 which must exchange information between a pair of signaling nodes in an SS7 network to access remote databases, referred to as Service Control Points (SCPs). Important applications which use TCAP include:

- 800 number routing.
- Automated credit card calling which queries the Line Information Database (LIDB) for calling card validation.
- Advanced Intelligent Network call processing (referred to as AIN call "triggers").

TUP (Telephony User Part): This layer provides the telephone signaling function for national and international telephone call control. TUP is primarily used outside of the U.S. in Europe, China, parts of Asia and Latin America, and can control all the various types of national and international connections used worldwide.

Higher Level Application Parts: These layers are highly application specific, which each designed to serve a particular application type. Examples include:

- GSM MAP (Mobile Application Part): This layer provides inter-system connectivity between wireless systems, and was specifically developed as part of the GSM standard.
- IS-41: This layer provides similar functionality to GSM MAP. It provides inter-system connectivity between wireless systems and is typically deployed in North American wireless networks. For example, it is widely used to provide interconnection between the analog AMPS (Advanced Mobile Phone System) cellular systems in the U.S.
- IS-634: This layer provides the interface for Mobile Switching Center (MSC) to base station communications for public 800 MHz cellular networks (i.e., for AMPS).
- INAP (Intelligent Network Application Part): This layer runs on top of TCAP and provides similar functionality to MAP but for a fixed network. Note that INAP is primarily a European standard, developed by ETSI. INAP is part of the CS1, CS2 IN Capability Set, the European equivalent to the AIN specification. While the AIN and CS specifications are similar and can be deployed using SS7 for functions such as call routing, there are some differences which the standards organizations are working to converge.
- 1129/1129+/1129A: These protocols provide a direct connection between the SCP and IP and are variants of the Bellcore 1129 and AIN 0.2 standards. In some cases, SS7 may not be required to implement the direct SCP-IP connection; in other instances, SS7 is used. Using SS7 as the underlying protocol allows any SCP to communicate directly with any IP in the SS7 network. When SS7 is used, the 1129 application layer typically runs on top of TCAP.

Thanks to Natural MicroSystems for an explanation of SS7 Software Layers. See also their white paper, "SS7 and Intelligent Networking Applications." It's on their web site, www.nmss.com

Signaling System 7 Standards SS7 standards are defined in the ITU-TS documents as noted below:

- Q.700-Q709 Messaging Transfer Part (MTP)
- Q.710 PBX Application
- Q.711-Q716 Signaling Connection Control Part (SCCP)
- Q.721-Q.725 Telephone User Part (TUP)
- Q.730 ISDN Supplementary Systems
- Q.741 Data User Part (DUP)
- Q.761-Q.766 ISDN User Part (ISUP)
- Q.771-Q.775 Transaction Capabilities Application Part (TCAP)
- Q.791-Q.795 Monitoring, Operations, and Maintenance
- Q.780-Q.783 Test Specifications

signaling system R1 International equivalent of Bell 2600 Hertz "SF signaling" used in North American DDD trunks. Largely used only in Third World nations.

signaling system R2 International equivalent of Bell out of band analog carrier system signaling. Uses a tone of 3825 Hertz placed between voice channels of a carrier system. Used primarily in Europe.

Signaling Transfer Point STP. A signaling point with the function of transfer-ring signaling messages from one signaling link to another and considered exclusively from the viewpoint of the transferrer.

signature A short text file that is automatically added to the end of your e-mail or Usenet posts. A signature file usually contains your name (or alias) and email address, and some people like to add pithy quotes; whatever your signature file contains, remember to keep it short.

significant hour Any hour that influences the sizing of a trunk group.

SIGSALY A voice-encryption device developed by Bell Labs for the US military during World War II and put into service in 1943. SIGSALY pioneered the use of pulse code modulation, companding, and enciphered telephony. Nicknamed the "Green Hornet" on account of the buzzing sound it made when operated.

Sigtran Signaling Transport. The IETF (Internet Engineering Task Force) "Sigtran" WG (Working Group) is working on a Signaling and Transport protocol to enable over the Internet, or other IP-based network, the same reliability of signaling and control that the PSTN (Public Switched Telephone Network) enjoys. Sigtran is defined in RFC 2719 as "an architecture framework for transport of message-based signaling protocols over IP networks." The work includes definition of encapsulation methods, end-to-end protocol mechanisms, and use of IP's existing capabilities in support of the functional and performance requirements of signaling transport. The ultimate yield of Sigtran is intended to be a protocol stack for the transport of SCN (Switched Circuit Network) signaling over an IP transport protocol. The protocol stack will be implemented in Signaling Gateways (SGs), which serve as the points of interface between the two networks. An SG functions to relay, translate, or terminate SS7 signaling. See also SCTP (Stream Control Transmission Protocol) and www.ietf.org/html.charters/sigtran-charter.html

silence management See Silence Suppression.

silence suppression A term used in voice compression for transmission whereby silence in the voice conversation is filled with other transmissions -e.g. data, video, imaging, etc. According to AT&T, the average voice conversation is 62% quiet and 38% not quiet (i.e. actual conversation). You can figure that for yourself: One person is speaking at a time. That's 50% of the circuit silent. The person who's speaking doesn't speak continuously. He pauses, takes a breath, thinks, etc. That's another 12%. Thus 62% in silence. A company like Micom in Simi Valley, CA makes a product called Marathon which uses the 62% silence between syllables, words and sentences to transmit data, fax and video. Micom tells me that the typical talk spurt sequence is 300 milliseconds. And it can use very small time (as short as 300 milliseconds) between talk spurts to stuff data, fax or video into and transmit.

silent alert A non-audible signal in a pager. Usually a vibrating motor that causes the pager to "shake" silently to alert you to a new message or an incoming call.

silent call See abandoned call.

silent commerce This term covers all business solutions enabled by tagging, tracking, sensing and other technologies, including RFID, which make everyday objects intelligent and interactive. When combined with continuous and pervasive Internet connectivity, they form a new infrastructure that enables companies to collect data and deliver services without human interaction.

silent device A Bluetooth term. A device that is in discoverable mode but cannot respond due to other baseband activity is said to be a silent device. The device could also be in non-discoverable mode and would also not respond to an inquiry.

silent intrusion Synonymous with Silent Participation.

silent mode Many Japanese fax machine sport a "silent mode." People in Japan who have fax machines in their houses use it so they are not awaken in the middle of the night by incoming faxes. In silent mode, your fax machine simply answers incoming calls, grabs incoming faxes into memory, but makes no noise whatsoever. In the morning when you wake up, you see a message on the machine that says "Faxes received." You hit a button and the machine starts to print the faxes it received. According to my friend, Emiko Magoshi, the shortfall of the fax machine at her house is that its memory is limited to accepting only 10 pages.

silent monitoring A call center term. The process whereby a supervisor or other qualified person listens in to the calls of an agent from a specially designated room or place away from the agent's work area. The agent may or may not be aware that his call is being monitored. Often the calling user may hear an opening message "This call may be monitored for quality assurance purposes." Typically the supervisor can hear both sides of the conversation but cannot break in on the line and say anything.

silent participation A feature that allows a third party, such as an ACD agent supervisor, to join the call. The joining party can hear the entire conversation, but cannot be heard by either originating party. The feature, sometimes called Silent Intrusion or Silent Monitoring, may provide a tone to one or both devices to indicate that they are being monitored.

Sili Code See CLLI Code.

silicon A dark gray, hard, crystalline solid. Next to oxygen, the second most abundant element in the earth's surface. Transistor chips are made from silicon, and it is the basic material for most integrated circuits and semiconductor devices. Silicon, a neutral element, is found primarily in raw form as sand. See Semiconductor.

silicon alley The area roughly corresponding to Chelsea, SOHO and NOHO in downtown Manhattan, New York. So called because so much of the web and Internet content is written by companies located in this area.

silicon avalanche diode A fast-acting surge protector which features a narrow voltage clamping range.

silicon bayou In and around New Orleans, there are a growing number of a high-tech companies. They say their nw high-tech area is called Silicon Bayou. See the next few definitions.

silicon bog Ireland.

silicon fen The vast flat and marshy lands in and around Cambridge, England, the site of the prestigious university. According to the New York Times, Saturday January 4, 1998, "the area has become home to more than 1,000 high-technology companies that employ at least 30,000 people and produce more than $3 billion a year in revenues."

silicon forest Silicon forest is the area of Portland/Beaverton/Hillsboro, where Intel has a large presence, as are Tektronix, Triquent, Lattice Semiconductors, Maximum Integrated Circuits, .. See also Silicon Bayou, Silicon Mudflats, Silicon Valley, Silicorn Valley and Sillywood.

silicon gulch Austin, Texas because of the concentration of high-tech companies that have located there. See the definitions above and below.

silicon mudflats The part of the San Francisco area that is on the Oakland side of the bay and sits between San Francisco and Oakland. That area is home to a bunch of high-tech companies. See also Silicon Bayou, Silicon Forest, Silicon Valley, Silicorn Valley and Sillywood.

silicon prairie South Dakota.

Silicon Sally Some telecommunications boxes are programmable. Some can be programmed by attaching a phone and punching in numbers in response to questions from a female voice emanating from somewhere inside the box. Many people in the telecom industry call the nice disembodied lady "Silicon Sally." I first heard this from David Epstein of BroadVoice.

Silicon Valley Silicon Valley in Santa Clara County, California, south of San Francisco Bay, is known for its microelectronics innovation. It is one of the two places (the other being Dallas, Texas) where the microchip was invented and produced. There are over 3,000 microelectronic hardware and software firms within a radius of about 30 miles. Silicon Valley owes its good fortune to:

- The fact that W. Shockley, one of the three inventors of the transistor, being a native of Palo Alto, went there in 1955 after he left Bell Laboratories, to start his own industrial research center;
- The proximity of technology universities, especially Stanford University;
- A very active financial market in California and the enterprising attitude of capitalists willing to enter into "joint ventures," often very profitable;
- The West Coast orientation to Japan.

The above information courtesy of Electronics, Computers and Telephone Switching by Robert J. Chapuis and Amos E Joel, Jr. See also Silicon Bayou, Silicon Forest, Silicon Mudflats, Silicorn Valley and Sillywood.

Silicon Valley North The Ottawa area in Canada is called Silicon Valley North by the local high-tech press. Local high-tech companies include Nortel Networks, Newbridge Networks, ObjectTime, Corel, Mitel, JDS Uniphase (formerly JDS Fitel) and Cognos. Others with with presence here include Cisco, Compaq and Siemens.

Silicone Valley A part of Silicon Valley in Northern California around San Jose which headquarters America's flourishing pornographic industry, which makes the movies (for DVD, VCR, theater and satellite distribution) and web sites. Over 80% of the "actresses" who perform in pornographic movies have had their breasts enlarged. Thus the name, Silicone Valley.

siliconize Putting all the various components of a system into microprocessor chips. For example, in the fiber business many of the pieces are individual electronic components. Getting many of them converted of them into individual microprocessors is called "silicon-

ize." Doing this saves time and money and improves the final reliability of the entire system.

silicorn valley A region in the state of Iowa that is fostering high-tech development. See also Silicon Bayou, Silicon Forest, Silicon Mudflats, Silicon Valley and Sillywood.

silly code See CLLI Code.

silly valley A silly term for Silicon Valley in California.

Sillywood The convergence of Silicon Valley and Hollywood.

SILS A Standard being formulated for Interoperating LAN Security.

silver bullet A panacea. THE solution. The one action that will solve all your problems.

silver ceiling A set of attitudes and prejudices that prevent older employees from rising to positions of power or responsibility in a workplace.

silver satin Once upon a time, phones came with cords that matched the color of the phone. This proved expensive and confusing to workers, who were color blind, who had to match the cord with the phone. So a manufacturer (we think it was AT&T) decided that the time was ripe for all phones to have a cord that matched every decor and every phone. In actuality, the color they settled upon -silver satin -matches no decor man or woman has ever created and certainly no phone has ever been produced in the silver satin color. However, the world is now stuck with every phone coming with one standard, ugly line cord, called silver satin. See also Touchtone.

silver surfers Senior citizens who surf the Web.

SIM 1. Subscriber Identity Module. A 'smart' card installed or inserted into a mobile telephone containing all subscriber-related data. This facilitates a telephone call from any valid mobile telephone since the subscriber data is used to complete the call rather than the telephone internal serial number. See GSM and SIM Card for a better description.

SIM Card Subscriber Identity Module, also called Smart Card. Every mobile phone that conforms to the GSM (global system for mobile communications) and many PCS (personal communications services) handsets have something called a SIM card. It's either postage-stamp size or credit-card sized. It has two basic functions -to make your cell phone more useful to you. For example, it may contain your personal directory of names and numbers. Second, It tells the network who you are. That way it can process your call. Basically takes a standard, off-the-shelf GSM phone (i.e. any GSM phone) and makes it yours. A GSM phone won't work without a valid SIM card, i.e. one tells the network you have enough prepaid minutes to use the network or have a valid account the network can bill your call to. Each SIM card contains a microchip that houses a microprocessor with eight kilobytes of memory. The card stores a mathematical algorithm that encrypts voice and data transmissions and makes it nearly impossible to listen in on calls. The SIM card also identifies the caller to the mobile network as being a legitimate caller. PCS and GSM cards come in two basic varieties. Some handsets work with slot-in SIM cards that are the size of conventional credit cards. Others come with smaller cards already built into the handset. GSM is the standard mobile service for Europe and Australia and most countries outside the U.S. In some countries you can buy a SIM card with a few hundred hours of prepaid minutes on it. That way you can borrow anyone's GSM cellphone, slide your SIM card in and make calls. You can also rent or buy a GSM phone and do the same thing. Some PCS operators in the U.S. have adopted the GSM standard. See GSM, SIM Toolkit and unlocking.

SIM Toolkit Software that allows cellular operators to offer enhanced value-added services on the back of their short message services. These services can include e-mail, GPS (positioning services) home zone tariffs, and information services such as stock prices and news updates. See SIM Card.

SIMD Single Instruction Multiple Data is a type of parallel processing computer, which includes dozens of processors. Each processor runs the same instructions but on different data and one chip provides central coordination. See also MIMD and MMX.

SIMM Single In-line Memory Module. Used on Macs and PCs. A form of circuit board that holds a number of silicon chips. The connectors (i.e., leads, or pins) are attached to a stiff contact strip that permits a SIMM to be inserted into a slot like an expansion adapter. On PCs, SIMM-style RAM chips have virtually replaced the dual in-line package (DIP) chips, identifiable by two rows of protruding legs, that were popular in the 1980s. The most common SIMM is the 30-pin, 9-bit wide "1 by 9", which is the standard memory upgrade for PCs. See also DIP, PGA, SIMM Socket, and SIP.

SIMM Socket The connector inside the Macintosh that holds the SIMM and connects it to the rest of the computer electronically.

SIMPLE SIP for Instant Messaging and Presence Leveraging Extensions (SIMPLE) is a developing enhancement to the SIP (Session Initiation Protocol) protocol. Developed by the SIMPLE working group of the IETF, SIMPLE adds instant messaging and presence.

Presence allows you to determine if someone on your buddy list is on-line and available to communicate via instant messaging. See also SIP.

Simple Authentication and Security Layer See SASL.

simple english What the SEC said that financial documents should be written in. Simple English is what normal people are meant to be able to read. This means that lawyers, who write most of these documents, are meant to speak. Good luck, SEC.

Simple Gateway Control Protocol SGCP is a protocol and an architecture that Bellcore has created to address the concept of a network that would combine voice and data on a single packet switched Internet Protocol (IP) network. SGCP largely operates at low level -level 2 in the OSI. So it will probably be combined with higher level concepts such as IPDC (IP Device Control) and MGCP (Media Gateway Control Protocol). According to Bellcore, the philosophy behind the Simple Gateway Control Protocol (SGCP) is that the network is dumb, the "endpoint" is simple, and services are provided by intelligent Call Agents, not in the trunking gateway (TGW) or in the residential gateway (RGW). SGCP is simple to use and easy to program, according to Bellcore, but is powerful enough to support basic telephony services and enhanced telephony services like call waiting, call transfer, and conferencing. The protocol is also flexible enough to support future IP telephony services. The SGCP, according to Bellcore, is a simple UDP-based protocol, instead of TCP-based, that allows support for managing endpoints and the connections between the endpoints. The SGCP is scalable, has support for failovers, and processes information in real-time. There is a low CPU requirement and low memory requirement for the endpoint because the SGCP is handling a small set of simple transactions at a time. This means that the endpoint can then be mass-produced cheaply. And there is no need for expensive and resource-hungry parsers. The system is text-based, has an extensible protocol, and the connection descriptions are based on SDP. When new services are introduced by the call agent, there is no need to change or update the endpoint. SGCP controls the endpoint by hooking transactions, relying on DTMF input, and by playing tones. For full details see www.bellcore.com/SGCP/SGCP-WhitePaper.rtf

Simple Mail Transfer Protocol SMTP. The TCP/IP protocol governing electronic mail transmissions and receptions. An application-level protocol which runs over TCP/IP, supporting text-oriented e-mail between devices supporting Message Handling Service (MHS). Multipurpose Internet Mail Extension (MIME) is a SMTP extension supporting compound mail, which is integrated mail, including perhaps e-mail, image, voice and video mail.

Simple MAPI Simple MAPI is a subset of MAPI that lets developers easily create "mail-aware" applications capable of exchanging messages and data files with other network clients.

Simple Network Management Protocol SNMP. The protocol governing network management and monitoring of network devices and their functions. SNMP came out of the TCP/IP environment.

Simple Network Paging Protocol See SNPP.

Simple Object Access Protocol See SOAP.

Simple Server Redundancy Protocol See SSRP.

Simplegram A Versit term. A syntax for encoding the vCard in a clear-text encoding. Simplegrams are nominally, based on the ASCII, 7-bit character set.

simplex 1. Operating a channel in one direction only with no ability to operate in the other direction. For example one side of a telephone conversation is all that could be carried by a simplex line. Obviously a simplex line is not useful for a phone conversation. See Full Duplex.

2. One-sided printing.

simplex circuit A transmission path which is capable of transmitting in only one direction. The CCITT definition differs from this more common definition. CCITT simplex is a path which can operate in either direction, but only one direction at a time. This is commonly in North America called half-duplex.

simplex loop powering In T-1, refers to the powering of the digital signal pairs that are simplex in nature (Tip or Ring) and that may have voltage applied to maintain the required 60 mA dc current to control repeater signal regeneration, loopbacks, keep alive signals and alarms.

SIMTEL20 The White Sands Missile Range used to maintain a giant collection of free and low-cost software of all kinds, which was "mirrored" to numerous other ftp sites on the Internet. In the fall of 1993, the Air Force decided it had better things to do than maintain a free software library and shut it down.

Simulcast To broadcast simultaneously on two different channels (paths).

simultaneous lobing A method of tracking in which up to four separate feed horns are used to generate simultaneous antenna beam lobes with nulls on the antenna axis, both in elevation and azimuth. The nulls provide the error signals used in tracking. Also called monopulse. See monopulse.

Simultaneous Peripheral Operations On Line SPOOL. Temporarily storing programs, data or output on magnetic tape or in RAM for later output or execution. Many PCs use a small software spooling program which accepts material to be printed very quickly, stores it in a portion of RAM, then feeds that material to the printer at a speed the printer can handle. See Spooling.

simulator A program in which a mathematical model represents an external system or process. For example, an engineer can simulate the forces that act on a building during an earthquake to find out how much damage is likely to be incurred.

simulation A technique, often involving a computer, to guess the outcome of various events in the future. Where multitudes of complex events interact, simulation may well be the only way to deal with a given problem. Simulation is often used in traffic engineering instead of or in addition to the proven formula. Many people believe that simulation should NEVER be used when standard, proven formulas (such as Poisson, Erlang B and Erlang C) are appropriate. The difficulty with simulation is actually finding out what rules to use and then programming them correctly. Once this is done, it takes time, even on a fast computer, to run thousands of simulations to get a stable statistical estimate; a single run of simulation, like a single roll of the dice, is worse than useless. Simulators were built into hardware to predict the overall behavior of AT&T's No. 5 Crossbar switch when it was developed in the late 1940s; 1ESS behavior was simulated with software in the 1960s. These were major efforts involving many people and several years, but they dealt with problems far beyond the capabilities of standard traffic equations.

sin In ancient Rome, eating the flesh of a woodpecker was considered a sin.

sin tax A tax imposed by sincere politicians on anything that might remotely be fun. In the 1970s, the Rhode Island legislature proposed the enactment of a $2.00 tax for every act of sexual intercourse. Concerns about collecting said tax finally caused the legislature, in its infinite wisdom, to set aside the idea.

SINAD A measurement of the quality of an audio signal from a communication device, calculated as the ratio of the total signal power level (wanted signal + noise + distortion, or SND) to unwanted signal power (noise + distortion, or ND). See SINAD meter.

SINAD meter A device that measures SINAD. A SINAD meter is usually built into a larger communications service monitor that provides a range of diagnostic functionality for radio repair and maintenance purposes.

sincere In ancient Rome, when a sculptor made a mistake while working on a statue -chipped the nose, for instance -he would artistically hide it with wax. The practice was so widespread that whenever someone ordered a statue, he would specify in the contract that it would be "sine cera"; that is "without wax." Hence the origin of the word "sincere."

sincerity Sincerity is everything. If you fake that, you've got it made. George Burns.

SINCGARS Single Channel Ground and Airborne Radio System. Military radios operating in the 30 MHz to 88 MHz range. They were used during the second Iraq War in 2003. SINCGARS is a family of VHF-FM military radios used for command and control in land and air combat environments. The radios have the capability to transmit and receive secure voice and data. The radios are able to operate in radio-jamming environments through the use of electronic counter-countermeasures (i.e., anti-jamming measures). SINCGARS radios are designed on a modular basis; configurations include manpack, vehicular, and airborne models. The radios conform to NATO interoperability requirements.

sine wave In an AC (Alternating Current) electrical system, the polarity (+/-) is constantly reversing, and at precise intervals of time. For example, a 120-volt, 60-Hz electrical generator has a peak positive voltage (i.e., amplitude) of 169.7 volts (+169.7 volts) and a trough negative voltage of 169.7 volts (-169.7 volts). The voltage value gradually rises to the peak positive voltage, and then gradually falls to the trough (i.e., peak negative voltage). This process takes place at a frequency of 60 times per second (60 Hz), which directly corresponds to the rotation of the electromagnet in the electrical generator which comprises the power source. Plotting the gradual rise and fall of the voltage level over time yields a curved waveform. The wave form is known as a "sine wave" because its shape is derived from plotting the sine of the angles of rotation from 0-360 degrees. The "phase" of the AC signal describes the relationship of the +/signal to the zero-voltage crossing point, as the signal alternately increases to the peak and decreases to the trough.

Now to tackle the issue of 120 volts versus 169.7 volts. In consideration of resistance in a circuit, electrical engineering principles define the current flow in a circuit as the "root mean square" (RMS) voltage, which is equal to the peak value (169.7 volts) of the sine wave divided by the square root of 2, which is approximately 1.414. Note that 169.7 /

1.414 = 120. See also Analog, Frequency, Hertz, and Wavelength.

singing An undesirable whistle or howl on a transmission circuit. Singing is usually caused by feedback, excessive gain, or unbalance of hybrid coils, or by some combination. Singing is the same effect observed when you increase the volume on a public address system until the system squeals or "sings."

singing return loss The loss at which a circuit oscillates or sings at the extreme low and high ends of the voice band.

single address message A message which is to be transmitted to only one specific terminal, as opposed to a broadcast or group message.

single attached concentrator SAC. An FDDI (or CDDI) concentrator that connects to the network by being cascaded from an M (master) port of another FDDI(or CDDI) concentrator.

single attachment concentrator SAC. A concentrator that offers one S port for connection to one ring of the FDDI network and multiple M ports for attachment of devices such as workstations. A SAC provides less reliability than does a Dual Attachment Concentrator (DAC), although at lower cost, as it connects to only one, rather than, to both fiber rings of the FDDI dual counter-rotating ring architecture. See also Single Attachment Station, Dual Attachment Concentrator, Dual Attachment Station and FDDI.

single attachment station SAS. A device such as a workstation which connects directly to only one of the FDDI dual counter-rotating rings, rather than attaching to both as does a Dual Attachment Station or gaining ring access through a Single or Dual Attachment Concentrator (DAC). See also Single Attachment Concentrator, Dual Attachment Concentrator, Dual Attachment Station and FDDI.

single channel broadband A local network scheme in which the entire spectrum of the cable is devoted to a single transmission path; frequency-division multiplexing is not used. Also known as carrierband.

Single Channel per Carrier SCPC is a radio communications channel via a satellite that is wholly allocated to carry a single carrier for the duration of a call.

single channel transmitter A transmitter used to process only one television channel.

single day assignment A call center term. A method of automatic assignment (of employees to Master File schedules) that operates only on schedules of exactly one workday in length. Unlike multi-day assignment (which operates on schedules of any length), this makes it possible for each assigned employee to be scheduled for different hours each day.

single digit dialing Provides for single-digit dialing to reach a preselected phone or group of phones.

single ended terminal device A device which terminates only one line at a given time.

single-family optical network unit SF-ONT. A SF-ONT is where a fiber drop from the central office terminates on the premises of a single-family unit (SFU).

single fiber cable A plastic-coated fiber surrounded by an extruded layer of polyvinyl chloride, encased in a synthetic strengthening material and enclosed in an outer polyvinyl chloride sheath.

Single Frequency signaling SF Signaling. The use of one tone -typically 2600 Hz -to indicate if the phone line is busy or idle (supervision) and to convey dial pulse signals from one end of a trunk or line to the other, using the presence or absence of a single specified frequency. The conversion into tones, or vice versa, is done by SF signal units. See also Frequency, Inband Signaling, Signal.

single gang receptable See Back Box.

single in line memory modules Basically memory packaged so it can be slipped into a PC or laptop PC much easier than present methods of installing memory -which typically consist of pushing memory chips with legs into printed circuit boards. The problem with memory with legs is that you're likely to bend one of the legs and thus have the installation go awry.

single line instrument A telephone set normally used to access only one line. However when used with advanced telephone systems, additional lines can be accessed by dialing specific codes rather than by depressing keys.

single master domain A relationship in a domain or amongst domains in which trusts are set up so that management control is centralized in one domain. See Trust Relationship.

single mode fiber A fiber wavelength which only one mode will propagate. See SMF.

single number service 1. Allows callers to dial a single number to reach a company with multiple local stores. For instance, a pizza chain might advertise a single number. The approximate physical location of the caller can be determined by the network by virtue of the originating number. Based on that data, the location can be compared to a network database defining the location of the closest pizza parlor. The caller, therefore need not sort through the phone book for this information, which can be especially difficult without a knowledge of the area. Additionally, all calls are routed automatically and without error to the outlet in closest proximity, ensuring that the pizza arrives quickly, as well as hot and fresh.

2. An optional feature for 800 IN-WATS Services which allows a subscriber who has or wants to have both intrastate and interstate 800 service to use the same 1-800 number for both services. If you'd like to buy another copy of this dictionary, call 1-800-LIBRARY. That phone number will be answered at our office in New York City. It will work for calls from inside and from outside New York State.

single party revertive ringing A central office feature. Single-Party Revertive Ringing provides enhanced flexibility to residential subscribers' existing telephone service by turning extension phones into intercoms. This feature creates a type of home intercom system with which subscribers can reach people in other rooms in their home and/or on a remote part of their property (i.e., barn, garage, greenhouse, etc.). To initiate a home intercom call, the subscriber dials his own number and hangs up. All extension phone rings. The intended party picks up the extension and is connected.

single point of contact SPOC. A way of handling a multi-vendor sale and implementation in which one vendor or consultant handles contacts with the other vendors and/or departments and presents a unified proposal and project management to the customer, so that the customer needs to deal with only one person or team.

single point of failure A single "point," or "element" at which a failure, or "hard fault," could bring down an entire network or subnetwork. The "point" could be a circuit if all traffic in the network or subnetwork passes through it. Similarly, the "point" could be a hub, switch or router if all traffic passes through it. The "point" also could be an application software system, if the operation of the network depends on it. Optical fiber networks are highly redundant in order to minimize the risk of catastrophic failure from "backhoe fade." Nonetheless, optical fiber networks sometimes suffer multiple cuts which render them out of service. Signaling System 7 (SS7) systems have occasionally suffered from software bugs which have resulted in the broadcasting of error messages which have overwhelmed all SS7 connected switches, causing entire networks to shut down. Similarly, Frame Relay networks have sometimes suffered from software bugs which shut them down. Paging networks have completely crashed because of software errors or because a satellite "disappeared" for several hours. Full redundancy is the only real protection against such catastrophic failures, but it's generally considered too expensive for all but the most mission-critical, time-sensitive applications or networks. ARPANET, the predecessor to the Internet, originally was designed so that there was no single point of failure. The idea with ARPANET was that every transmission would be sliced into packets. Each packet would travel whichever route was available at that time, there always being multiple routes between two routes. If one route crashed, the packet would go another way. ARPANET (Advanced Research Projects Agency Network) was a network in support of government and military research projects. It was conceived and build in the late 1960s, during the Cold War. The United States designed ARPANET to avoid the risk that a nuclear strike aimed at a Single Point of Failure might take down the entire network. See also ARPANET and Internet. See Back Hoe Fade.

single protocol router A communications device using the same mix of protocols which is designed to make decisions about which of several paths a packet of information will take. The packets are routed according to address information contained within the packet.

Single Sideband SSB. An Amplitude Modulation (AM) technique for encoding analog or digital data using either analog or digital transmission. SSB suppresses one sideband of the carrier frequency at the source. As only one sideband is used and as the carrier signal is suppressed (i.e., carries no information), less power is used and less bandwidth (one-half) is required than is the case with DSBSC (Double SideBand Suppressed Carrier). Carrier synchronization is lost. See also Amplitude Modulation, DSBSC, DSBTC, and VSB.

single sideband transmission A system of transmission which suppresses one side-band of the carrier frequency at the source. Also applied to receiving systems designed to reproduce such transmissions.

Single Sign-On SSO. An authentication and authorization mechanism that requires that a user identify himself only once. Thereafter, the user is authorized to access other connected system resources, based on his level of access privilege. The most common identification, of course, is the username / password combination. Users who access many systems often have to remember a different username/password combination for each. That results in more forgotten passwords, and more calls to the help desk. It also results in more users writing their ID information down on those little sticky notes, which they then stick to their monitors, which causes a total breakdown in security. SSO helps solve all these problems.

single slot 1. Current standard for coin phone construction that uses one slot for the deposit of all acceptable coins.

2. Cards that fit into one slot inside a Wintel (Windows/Intel) PC are called single slot cards. The significance of calling them single slot is that in the old days individual cards couldn't do very much. To build something -like an interactive voice response system -you often needed several cards. But with time, components have shrunk and cards have gotten more capabilities. Complete computer telephony systems can now be built with one single slot PC card. Thus the significance of this definition.

single use batteries Batteries that aren't rechargeable -in other words, the Duracell and Eveready batteries at your local supermarket.

single wire line A transmission path which uses a single conductor and a ground return to complete a circuit. Used a lot in rural areas.

sinister Ancient Romans always entered the home of a friend on their right foot. The left side of the body was thought to portend evil. The Latin word for "left" is sinister, from where the English word "sinister" originates, meaning anything threatening or malevolent.

sink That part of a communications system which receives information.

SINPO An acronym that stands for Signal (strength), Interference (from other stations), Noise (from the atmosphere, ionosphere or other non-radio sources), Propagation (fading), and Overall. It is a 5-digit code that a radio amateur uses for describing various aspects of a received radio transmission's signal quality. The source radio could be another radio amateur's transmission, or it could be a commercial short-wave, AM or FM radio broadcast. If the radio receiver has an S-meter, which measures signal strength, it can be used as a basis for determining a rating for signal. If the receiver does not have an S-meter, then the signal rating, like the other four ratings, is done subjectively. Each rating is done independently of the others on a scale of 1 to 5, with 1 being the worst possible evaluation and 5 being the best. The worst possible SINPO score is 11111; the best is 55555. The last rating, i.e., "overall," is a subjective rating of the overall listening experience, disregarding the actual content of the transmission. SINPO codes are a way for a radio amateur to share aspects of their radio-listening experience with other radio amateurs and/or with stations they listen to. See RST.

sintering Fusion of a spirally applied tape wrap jacket by the use of high heat to a homogenous continuum. Usually employed for fluorocarbon, non-extrudable materials.

SIO 1. Serial Input/Output. The electronic methodology used in serial data transmission.

2. Scientific or Industrial Organization. An ITU (International Telecommunications Union) term. Any organization, other than a governmental establishment or agency, which is engaged in the study of telecommunication problems or in the design or manufacture of equipment intended for telecommunication services. Examples of U.S. SIOs are ADTRAN, Advanced Micro Devices, ASCEND Communications, Cisco Systems, 3Com Corporation, and Hewlett-Packard.

SIP 1. Single Inline Package. A type of silicon chip in which all of the pins are lined up in a row. See also DIP, PGA, and SIMM.

2. SMDS Interface Protocol.

3. Session Initiation Protocol. SIP is the most important standard for setting up telephone calls, multimedia conferencing, instant messaging and other types of real-time communications on the Internet. SIP can establish sessions for features such as audio/video-conferencing, interactive gaming, and call forwarding to be deployed over IP networks, thus enabling service providers to integrate basic IP telephony services with Web, email, and chat services. SIP is much faster, more scalable and easier to implement than H.323. An array of network gear including IP phones, IP PBXs, servers, media gateways and softswitches support SIP. SIP is the Application Layer (Layer 7 of the OSI Reference Model) protocol for the establishment, modification and termination of conferencing and telephony sessions over an IP-based networks. SIP uses text-based messages, much like HTTP. SIP was developed within the IETF MMUSIC (Multiparty Multimedia Session Control) working group, and is defined in the IETF's RFC 2543. SIP is touted as being much faster, more scalable and easier to implement than H.323.SIP addressing built around either a telephone number or a Web host name. In the latter case, for example, the SIP address would be based on a URL (Uniform Resource Locator), and might look something like SIP:Harry_Newton@ TechnologyInvestor.com, which makes it very easy to guess a SIP URL based on an e-mail

address. The URL is translated into an IP address through a DNS (Domain Name Server). SIP also negotiates the features and capabilities of the session at the time the session is established. For example, a caller might wish to establish a call using G.711 audio and H.261 video. The codecs embedded in the two endpoints (i.e., originating and terminating multimedia terminals) negotiate a common set of voice and video compression algorithms (which might not include G.711 and H.261), prior to establishing the session. This advance negotiation process, which relies on the Session Description Protocol (SDP), is touted as greatly reducing the call setup time required for H.323 sessions. If the called party is not available, or does not wish to accept the call, it can be redirected (e.g., to voice mail or to an administrative assistant), with the negotiation process taking place in consideration of that endpoint. Once the session is established, the designated capabilities can be modified during the course of the call. For example, whiteboarding can be added if both terminals are capable and can negotiate a common compression algorithm. In addition to the unicast (i.e., one-to-one) session described above, SIP supports multicast (i.e., one-to-many) communications. See also CPL, H.323 and SDP.

SIP for Instant Messaging and Presence Leveraging Extensions See SIMPLE.

SIP Forum The SIP Forum is a nonprofit industry association whose mission is to promote awareness and provide information about the benefits and capabilities enabled by SIP. The SIP Forum also works to facilitate the integration of SIP with other areas of development on the Internet. The SIP Forum has become the meeting place for service providers, vendors and researchers developing commercial SIP services. SIP Forum members are offering the new services enabled by SIP, such as consumer broadband phone and IP Centrex, and are developing new technology, such as IP phones, PC clients, SIP servers and IP telephony gateways. www.SipForum.com.

SIP L2_PDU Standard Interface Protocol Level 2 Protocol Data Unit. An SMDS term. The 53-octet unit of information processed by the second level of the SIP. This process takes place once the SIP L3_PDU is presented.

SIP L3_PDU Standard Interface Protocol Level 3 Protocol Data Unit. An SMDS term. A variable-length (up to 9,220 octets long) unit of information processed by the third layer of the SIP. At this level, the SMDS SDU (Service Data Unit) of up to 9,188 octets is surrounded by a header and a trailer, before being passed to SIP Level 2 for further processing. See also SIP L2_PDU.

SIP SMDS Interface Protocol An SMDS term. The protocol defined at the interface between the SMDS network and the end user. See SIP L2_PDU.

SIPRNET Secret IP Router Network. It is the Department of Defense's largest command and control data network, supporting the Global Command and Control System (GCCS), the Defense Message System (DMS), worldwide collaborative planning, and other classified DoD applications.

SIR 1. Speaker Independent Recognition.

2. Sustained Information Rate. A SMDS term. SMDS requires that customers using access lines operating at DS-3 rates to predict the SIR, which is the level and duration of traffic flow. SIR is determined at the time of service subscription, and is enforced through a Credit Manager resident in the SMDS network switches. Five Access Classes are defined: Access Class 1 supports SIR of 4 Mbps and is intended for 4-Mbps Token Ring (802.5); Access Class 2, 10 Mbps and is intended for 10-Mbps Ethernet (802.3); Access Class 3, 16 Mbps and is intended for 16-Mbps Token Ring (802.5); Access Class 4, 25 Mbps; and Access Class 5, 34 Mbps.

Sirius Sirius is the name of the second satellite radio provider. The other one is called XM Satellite Radio. See XM for a full explanation.

SIS Single conductor with synthetic thermosetting insulation of a heat resistant, moisture resistant, flame retarding grade. Also made with chemically crosslinked polyethylene insulation. Used for switchboard wiring only.

SISO Soft Input-Soft Output. An technique used in certain Forward Error Correction (FEC) error detection and correction algorithms, such as Extended Hamming and Turbo Coding. SISO actually is used by the decoders on the receiving (Forward) side of the data transmission. FEC provides the receiving decoder with enough information to make exceptionally accurate judgments as to the integrity of a data block (e.g., packet or frame) of data. Further, FEC provides the receiving decoder with enough information to isolate the errored bit(s) in a data block, and to correct them without the requirement for a retransmission. Using Extended Hamming as an example, data to be transmitted are organized into what is viewed logically as a two-dimensional data block. The block might, for example, be 11 bits horizontal (wide) by 11 bits vertical (deep). The coding algorithm describes each 11-bit data sequence through a 5-bit parity code, or Extended Hamming code, which is appended to each 11-bit data word prior to transmission. At the receiving (Forward) decoder, the appended 5-bit descriptor is re-calculated, and the two descriptors are compared. This process takes place, both horizontally and vertically, for each 11-bit data word in a block of data, which might be 11x11 data bits, and 16x16 total bits. If there is a difference in value between the descriptors, an error has been detected. If the decoder views and evaluates each individual bit in a data block, it makes a "hard decision" on each bit. SISO is an iterative decoding technique in which the decoder not only determines the most likely transmitted sequence of data bits, but also develops a "soft decision" metric (i.e., measurement) of the confidence level (i.e., likelihood) of the value each bit in that sequence. The yield of this more involved soft decision is that each subsequently evaluated block has the advantage of the lessons learned in the previous evaluations. SISO logic is much more complex and, therefore, increases both the cost of the decoder and the time required to accomplish the process. However, SISO yields better results. See also Forward Error Correction, Hamming, and Turbo Coding.

SIT Tones 1. Standard Information Tones. These are tones sent out by a central office to a pay phone to indicate that the dialed call has been answered by the distant phone, etc. 2. Special Information Tones. These are tones for identifying network provided announcements. Here's Bellcore's explanation: Automated detection devices cannot distinguish recorded voice from live voice answer unless a machine-detectable signal is included with the recorded announcement. The ITU, which specifies signals that may be applied to international circuits, has defined Special Information Tones for identifying network provided announcement. The SIT used to precede machine-generated announcements also alerts the calling customer that a machine-generated announcement follows. Since SIT consists of a sequence of three precisely defined tones, SIT can be machine-detected, and therefore machine-generated announcements preceded by a SIT can be classified. At least four SIT encodings have been defined: Vacant Code (VC), Intercept (IC), Reorder (RO) and No Circuit (NC). With the exception of some small stored Program Control Systems (SPCSs) and some customer negotiated announcements, Bell operating companies in North America now precede appropriate announcements with encoded SITs to detect and classify announcements.

SITA Society of International Aeronautical Telecommunications. The international data communications network used by many airlines.

Sitcom SITCOMs. What yuppies turn into when they have children and one of them stops working to stay home with the kids. Stands for Single Income, Two Children, Oppressive Mortgage.

site controller An industrial grade PC located at MCI terminals and junctions. It provides the operator local visibility into alarm and performance information from the Extended Superframe Monitoring Unit and 1/0 DXC, as well as enabling the operator to interact with both devices.

Site Event Buffer See SEB.

site hosting See Hosting.

site license Companies that buy software for multiple computers typically buy one copy of the program and a license to reproduce it up to a certain number of times. This is called a site license, though it may apply to its use throughout an organization. Site licenses vary. Some require that a copy be bought for each potential user -the only purpose being to indicate the volume discount and keep tabs. Others allow for a copy to be placed on a network server but limit the number of users who can gain simultaneous access. This is called a concurrent site license. And many network administrators prefer this concurrent license, since it gives them greater control. For example, if the software is customized, it need be customized only once.

site map See sitemap.

site survey Checking the site out to see that your equipment will work there (e.g. a Wi-Fi network) and/or will fit there (e.g. a PBX). PBX suppliers often perform a site survey to determine if there is ample space in the telephone room to install their equipment.

sitekeeper See IPTC.

sitemap A web page that shows the structure and contents of a web site. Each item in a site map is a hyperlink that a user can click on to access the associated web page. Similar to a book's table of contents, a site map enables a user to find a specific page on a web site without having to explore the site's many pages on a trial-and-error basis. See sitemap protocol.

sitemap protocol A protocol for encoding a website's site map in XML, so as to inform search engines about URLs on the web site that are available for crawling. The Sitemap Protocol is an open-source protocol that was developed by Google in 2005 and adopted by Yahoo and Microsoft in late 2006. The big beneficiaries of the protocol will be the search engine companies like Google, Yahoo, and Microsoft. It will cut down the time that their "spiders" spend "crawling" a website and indexing documents on it. Search engine users (i.e. you and me) will benefit from Sitemap Protocol. The faster a search engine company's spiders and crawl through a set of websites and index them, the faster the search engine company will be able to update the indexes that users access when they run searches. The following table, from Google's website, shows the URL tags that have been developed so far for the protocol. The optional tags when used in a sitemap.xml file, are what will enable a search engine spider to save time when it visits a website.

- urlset> required. Encapsulates the file and references the current protocol standard.
- url> required. Parent tag for each URL entry. The remaining tags are children of this tag.
- loc> required. URL of the page. This URL must begin with the protocol (such as http) and end with a trailing slash, if your web server requires it. This value must be less than 2048 characters.
- lastmod> optional. The date of last modification of the file. This date should be in W3C Datetime format. This format allows you to omit the time portion, if desired, and use YYYY-MM-DD.
- changefreq> optional. How frequently the page is likely to change. This value provides general information to search engines and may not correlate exactly to how often they crawl the page. Valid values are: always, hourly, daily, weekly, monthly, yearly, never. The value "always" should be used to describe documents that change each time they are accessed. The value "never" should be used to describe archived URLs.
- priority> optional. The priority of this URL relative to other URLs on your site. Valid values range from 0.0 to 1.0. This value has no effect on your pages compared to pages on other sites, and only lets the search engines know which of your pages you deem most important so they can order the crawl of your pages in the way you would most like. The default priority of a page is 0.5. The priority you assign to a page has no influence on the position of URLs in a search engine's result pages. Search engines use this information when selecting between URLs on the same site, so a webmaster can use this tag to increase the likelihood that your more important pages are present in a search index. Google notes that "assigning a high priority to all of the URLs on your site will not help you. Since the priority is relative, it is only used to select between URLs on your site; the priority of your pages will not be compared to the priority of pages on other sites."

A hypothetical example of a sitemap that is encoded using version 0.84 of the Sitemap Protocol is the following. Such code would be included at the beginning of the first file on the web site, e.g. index.html:

```
?xml version="1.0" encoding="UTF-8"?>
urlset xmlns="http://www.newtonstelecomdictionary.net/schemas/sitemap/0.84">
url>
loc>http://www.newtonstelecomdictionary.net//loc>
changefreq>monthly/changefreq>
lastmod>2007-10-12/lastmod>
/url>
url>
loc>http://www.newtonstelecomdictionary.net/sitemap.xml/loc>
/url>
url>
loc>http://www.newtonstelecomdictionary.net/bigdic.html/loc>
/url>
/urlset>
```

SIVR Speaker Independent Voice Recognition. See Speaker Independent Voice Recognition.

six degrees of separation This theory/principle holds that any two people can find each other via people via no more than five intermediaries. The theory has been applied to areas such as search engines, power grid analysis, computer circuitry, and telecommunications networks.

six digit translation We have a long distance number 212-691-8215. The ability of a switching system to do six digit translation means that it can "look at" 212-691 and figure how to route the phone call. The criterion of choosing which way to send the call is, most often, the least expensive way. Six digit translation is often an integral part of Least Cost Routing programs within the phone system which tell the calls to go over the lines perceived by the user to be the least cost way of getting the call from point A to point B. There are typically two types of "least cost routing" translation -that which examines the first three digits of the phone number (i.e. just the area code) and the first six digits of the phone number (i.e. the area code and the three digits of the local central office). Six digit translation is preferred because it allows you more flexibility in routing, particularly to big area codes, like 213 in LA, where there are long distance calls within the area code. See also Least Cost Routing and Alternate Routing.

six in a can A technique of disguising cell antennas, this would be three dipole antennas in a tube or can that looks like the top of a flagpole, or a chimney on a building.

six nines See Five Nines.

six Webs The view, articulated by Sun Microsystems co-founder and former chief scientist Bill Joy, that the continually evolving World Wide Web has morphed into six separate but interconnected Webs. Each of the six Webs has its own purpose, organizing principles, content types and architectures, design and style principles, and user interface requirements. The 1st Web, and currently the most familiar to the average person, is the Web that is accessed by a browser from a desktop computer and is used for browsing, email and shopping. Users interact with the 1st Web using a keyboard and mouse. The 1st Web is the "near Web." The 2nd Web, also called the "far Web," and which is still in nascent form, is organized for entertainment, such as watching television and other video content. The 3rd Web contains content, applications, and services that are designed for the small screen on handheld mobile devices. The mobile device is always connected to the 3rd Web and is nearly always with the user. The 3rd Web is also known as the "here Web." The 4th Web uses the human voice for interaction and navigation, and is optimized for voice services. The 4th Web is also the Web that users will interact with in Web-connected automobiles and other situations where interactions with the Web must be hands-free and eyes-free. The first four Webs all involve human-Web interaction. Webs 5 and 6 do not. The 5th Web is an e-business Web; it is where, for example, one company's information system communicates with other companies' information systems on a fully automatic basis without human intervention, for example, when an airfare comparison service's computer polls airlines' computers for flight schedules and airfares. The 5th Web has a much higher requirement for correctness and consistency, since the machine-to-machine interactions are fully automated. The 6th Web, also known as the "device Web," involves embedded systems and sensor networks that confederate and work together to do things, such as monitor weather conditions.

sizing A telephone company term. The Network Switching Engineering activity of determining the types and quantities of equipment needed for relief. Also see Relief.

SJ Junior hard service, rubber-insulated pendant or portable cord. Same construction as type S, but 300V.

SJO Same as SJ but with an oil resistant outer jacket.

SJT Junior hard service thermoplastic or rubberinsulated conductors with overall thermoplastic jacket, 300V, 600xAFC to 1050xAFC.

skam Skype spam. An unsolicited message sent to a Skype user.

skammer A person who sends skam.

skew 1. The deviation from synchronization of two or more signals.

2. A computer imaging term. A tool that slants a selected area in any direction.

3. In parallel transmission, the difference in arrival time of bits transmitted at the same time.

4. For data recorded on multichannel magnetic tape, the difference in time of reading bits recorded in a single line.

5. In facsimile systems, the angular deviation for the received frame from rectangularity due to asynchronism between scanner and recorder. Skew is expressed numerically as the tangent of the angle of deviation.

See also Skew Ray and Skewed Distribution.

skew ray In a multimode optical fiber, any bound ray that in propagating does not intersect the fiber axis (in contrast with a meridional ray). In a straight, ideal fiber, a skew ray traverses a helical path along the fiber, not crossing the fiber axis. A skew ray is not confined to the meridian plane.

skewed distribution The mean and standard deviation describe many distributions quite well. However, a few distributions are markedly lopsided or skewed. When it is necessary to specify a distribution more precisely, the coefficient of skewness may be computed. It is sometimes expressed as the mean minus the mode divided by the standard deviation. That is, it is the spread between the mean and the mode as related to the spread or scatter of the total distribution.

skid marks American toilets have far more water in their bowls than European toilets. The reason, according to a ex-senior Kohler executive, Leo Carlin, is that "Americans don't like skid marks."

skills-based routing A call center term for routing incoming calls based on the type of service requested, assuring that calls go to agents with the skills to provide the highest quality of service to the calling customer. In other words, someone calling about a broken refrigerator should be directed to a refrigerator expert, not a vacuum cleaner expert. Skills-Based Routing takes advantage of the routing capabilities of the automatic call distributor, in consideration of the unique skills of individual agents or agent groups and the requirements of individual callers. In this manner, privileged customers with special needs can be afforded special treatment. For example, a Platinum credit card holder with a past due account balance and who prefers to conduct business in Cantonese can be directed to an agent skilled in collections involving privileged customers and who speaks fluent Cantonese. The routing process may be accomplished on the basis of a client profile stored in a database on an adjunct computer systems linked to the ACD. Prior to completing the call, the database would be queried, with the query process being initiated on the basis of the caller's touchtone entry of his account number or on the basis of Caller ID. See also Source/Destination Routing, Calendar Routing, and End-of-Shift Routing.

skimmer See Magnetic Card.

skin 1. Surface layer in a sandwich structure.

2. A custom graphical user interface that a user can select for a mobile device's screen display.

skin effect The tendency of an electrical current to pass through the outer portion, or "skin" of a metallic conductor, rather than through the center, the skin effect is a phenomenon associated with high frequency signals (i.e., over 50 kHz or so. The higher the frequency, the greater the effect.

SKIP Simple Key management for Internet Protocols. A specification that defines the way that an encryption key is exchanged between devices over an IP (Internet Protocol) network such as the public Internet. SKIP is a proprietary specification developed by Sun Microsystems. See also ISAKMP.

skip key Use "skip keys" wherever possible. Skip keys are single key-presses that take the caller to the most commonly sought places, including information message boxes. For example, a skip key is Dial 1 for Books. Dial 3 for subscriptions. Dial 4 for our fax number. Ron Acher invented this term. Or thinks he did.

skip route A control application that causes traffic to bypass a specific route and advance to another route.

skip scan See white line skip.

skip tracers People who use their knowledge of databases, motor vehicle license records, credit card records and purchases to track down people who owe money.

skip zone a ring-shaped region within the transmission range wherein signals from a transmitter are not received. It is the area between the farthest points reached by the ground wave and nearest points at which reflected sky waves come back to earth.

SKU Stock-Keeping Unit. An e-commerce term for a stock number, or inventory number. When you order something over the WWW (World Wide Web), you will reference a SKU. The SKU is unique to the product, which could be a book, a part, a complete system like a PC, or just about anything. The SKU will be compared to a database, and will match the item which you are buying, which will be pulled from stock and shipped to you. Congratulations, you have just accomplished an e-commerce transaction.

skunkworks Term for usually-secret high-pressure / high-tech research group in a company or government, often populated by people who work 24-hours a day under much pressure, don't see much sunlight, and don't have much time for washing. Hence the name, skunkworks, as in skunk, an animal which can smell big time. Original usage was from the moonshine factory in the comic strip, L'il Abner. But there are other stories. Lockheed, the maker of the stealth airplane, had a top-secret skunkworks project which it set up in a defunct airplane hangar in Burbank, California next to what is now the Burbank-Glendale-Pasadena Airport (but everyone shortens to Burbank Airport) but then it was owned by Lockheed and run by the legendary Kelly Johnson. Nearby there was a refuse processing plant which tainted the air around the airport. Anyhow, Lockheed, under Kelly Johnson's leadership, built thousands of the great B-17s and P-38 fighters in addition to dozens of secret projects including the Stealth fighter. Now, it is a vacant lot without even a plaque to commemorate its place in American aviation history. Later Ben Rich, who ran the operation for a while, wrote a great book called Skunkworks.

Sky Station Sky Station International Inc. (Washington, D.C.) intends to deliver wireless broadband services through solar-powered communications platforms measuring approximately 200 feet in both length and width. The platforms will be held aloft by tethered balloons floating at altitudes of about 14 miles. Each communications platform will serve an area of about 465,000 square miles, providing up to 150,000 channels of 64 Kbps;

the theoretical maximum, according to Sky Station, is 650,000 channels. As many as 250 of the airships are planned, with the total cost estimated at $7.5 billion. Subject to approvals from the FCC and the ITU-R, the first test is scheduled for October 1997. Commercial service is planned for the New York City area in 1999, with coverage intended for 95% of the world's population by 2005. Company officials include former U.S. Secretary of State and NATO commander Alexander M. Haig.

skyscraper ad Unlike the more common banner ads, which you see on web sites and which are bars placed across a Web page, these are advertisements that appear as vertical bars, typically down the right-hand side of the visible area, often containing snazzy Flash animations and other tricks. Sometimes also called tower ads.

Skype Skype, which rhymes with "hype" and has no particular meaning, is a service which allows free voice phone calls over the Internet between any two users who have downloaded the software to their PCs -desktops or laptops (but not Macs, so far). It is simple to use and provides amazingly clear phone calls to anyone with a broadband Internet connection. Skype uses a technology called "voice over Internet protocol," or VoIP. By routing calls over the Internet, VoIP essentially turns computers into phones. It is the core technology driving a number of small phone companies, e.g. Vonage, and is causing headaches for traditional phone companies who are trying to fend off the new VoIP rivals even as they attempt to integrate VoIP into their own systems. Skype uses Peer-to-peer ("P2P") technology which was first widely deployed and popularized by file-sharing applications such as Napster and KaZaA. In this context, P2P technology allows users to share, search for and download files and to find users who are presently online. The following are some of the techniques that Skype employs to deliver IP-based telephony (these come from Skype's web site):

- Firewall and NAT (Network Address Translation) traversal: Non-firewalled clients and clients on publicly routable IP addresses are able to help NAT'ed nodes to communicate by routing calls. This allows two clients who otherwise would not be able to communicate to speak with each other. Because the calls are encrypted end-to-end, proxies present no security or privacy risk. Likewise, only proxies with available spare resources are chosen so that the performance for these users is not affected.
- Global decentralized user directory: Most instant message or communication software requires some form of centralized directory for the purposes of establishing a connection between end users in order to associate a static username and identity with an IP number that is likely to change. This change can occur when a user relocates or reconnects to a network with a dynamic IP address. Most Internet-based communication tools track users with a central directory which logs each username and IP number and keeps track of whether users are online or not. Central directories are extremely costly when the user base scales into the millions. By decentralizing this resource-hungry infrastructure, Skype is able to focus all of our resources on developing cutting-edge functionality.
- Intelligent routing: By using every possible resource, Skype is able to intelligently route encrypted calls through the most effective path possible. Skype even keeps multiple connection paths open and dynamically chooses the one that is best suited at the time. This has the noticeable effect of reducing latency and increasing call quality throughout the network.
- Security: Skype encrypts all calls and instant messages end-to-end for privacy. Encryption was necessary since all calls are routed over the public Internet.
  See SkypeIn, SkypeMe, SkypeOut, VoIP and www.skype.com.

SkypeIn A service that allows a Skype user to receive over the Internet a phone call that originates on a regular landline or mobile phone. SkypeIn requires that the Skype user be provided with a regular phone number. When a landline and mobile phone user calls that number, an automatic call-routing arrangement routes the call via the Internet to the Skype user, no matter where he is physically located.

SkypeMe A button or other link on a web page that, when clicked, lets a user make a Skype call to the Skype user who created the link. A SkypeMe link sometimes appears in online ads and in online blogs.

SkypeOut A service that allows a Skype user to make an Internet-originated Skype call to a regular landline or mobile phone.

Skyplex Skyplex is an on board multiplexer used on DVB (Digital Video Broadcasting) satellite transmission. It means that several satellite uplink stations can transmit in SCPC (single channel per carrier) mode. On the transponder this SCPC flows have multiplexed by skyplex in TDM and then, in the downlink you have an high rate DVB stream (27.5 million symbols per second). This flow is modulated in QPSK mode. The Digital Video Broadcasting Group is a European organization which publishes its work through ETSI that

has authored many specifications for satellite and cable broadcasting of digital signals. See DVB and SCPC.

SLA See Service Level Agreement.

slab on grade Concrete floor placed directly on soil, without basement or crawlspace.

slam dunk A sure thing. Something easily done. "The customer desperately needs our product. This sale is a slam dunk." The expression comes from the basketball expression where the shooter, lacking any opposition, jumps up, grabs the bar and pushes (i.e. slams) the ball into the hoop with great force.

slamming Slamming is the practice of switching a telephone customer's long distance supplier without obtaining permission from the customer. It is illegal. A long distance company might do slamming to get itself some easy revenues. Slamming has become increasingly common in recent years as the long distance industry has seen its revenues per minute drop and dishonest telemarketing companies hired by the major long distance companies have slammed telephone users in order to get themselves some easy revenues. Often these telemarketers are paid by the long distance company on the basis of customers switched. In a "60 Minutes" TV show of December 16, 2001, Richard Blumenthal, the Connecticut Attorney General, said "Competing by cheating has become a way of life for the telecommunications industry for many of these corporations, many of the most reputable of them." According to co-host, Steve Croft, "Blumenthal has sued AT&T for billing people who aren't their customers and Qwest for a practice called slamming, stealing their competitors' customers by forging their signatures on authorization forms." Blumenthal said, "We've seen cases where AT&T continued sending people bills after they sought to terminate the service. We've seen companies like Qwest send people bills for service that was never ordered or requested."

In December, 2001 I asked Verizon for their advice to consumers to protect themselves against slamming. Here is the advice I received:

- Review your phone bill thoroughly. (That seems obvious but, as you pointed out, many people just give the bill a cursory glance.)
- Don't divulge personal information such as your telephone number, credit card number or social security number on any sweepstakes or raffle tickets. This information can be misused.
- Read the fine print on all contracts, applications and contest entry forms you sign. Avoid filling out contest entries that seem vague. (Sometimes these items contain fine print that authorizes a switch of your long distance carrier.)
- You may contact your telephone company's business office and put a PIC (primary interexchange carrier) freeze on your choice of long distance (and local and regional) carrier. That prevents future changes from being made without your authorization.
- Slamming is prohibited by state and federal laws. Victims of slamming are entitled to be switched back to the company of their choice at no charge and to have any disputed charges billed at the original company's rate.

If you want to verify your carrier, a pre-recorded message will tell you the name of your carrier. For Interexchange Carriers (IXCs), dial 1-700-555-4141. For Local Exchange Carriers (LECs), dial 1-your area code-700-4141.

SLARP Serial Line Address Resolution Protocol. A proprietary Cisco protocol used in mapping IP (Internet Protocol) addresses. SLARP is used in the initialization of new routers and servers in situations where a serial line interface is involved. The initialization process requires that the router or server discover the IP addresses of the various networked devices, and map internal IP addresses, including subnet masks, to registered IP addresses. Alternative protocols used include BOOTP and RARP. See also ARP, BOOTP, and RARP.

slashdot effect A sudden increase in load on a web server that results in either tremendous slowdown or total failure of the server. The slashdot effect occurs when a site is "backlinked" by a much more popular site. Originated from the website Slashdot (www.slashdot.org), a site with millions of visitors daily. People began to observe that once a site was linked on Slashdot, traffic would explode, often overwhelming the unprepared servers. If this happened, you were "Slashdotted". Someone wrote a paper formalizing and studying the "Slashdot Effect". Ironically, this link was posted on Slashdot and the site fell victim to the Slashdot Effect.

slave A device which operates under the control of a master -another device or system. Slave switching systems are common in rural areas. The master central office might be in town A. Twenty miles away there's a smaller town. It makes no economic sense to serve those subscribers each on single local dedicated loops from Town A. Best solution: place a "slave" central office in that distant town and drive its software, diagnostics and changes from the main central office in Town A. See also Remote Concentrator.

slave unit A Bluetooth term. All devices in a piconet that are not the master. See Bluetooth.

SLC 1. Subscriber Line Charge. A charge on the monthly bill of a phone subscriber in the United States, which produces revenues for the local telephone company. The money collected from the subscriber line charge is used to compensate the local telephone company for a part of the cost of the installation and maintenance of the telephone wire, poles and other facilities that link your home to the telephone network. These wires, poles, and other facilities are referred to as the "local loop." The SLC also is known variously as the Access Charge, CALC, and EUCL. See also Access Charge.

2. Schools and Libraries Corporation. See that term for a full explanation.

3. Subscriber Line Carrier, or Subscriber Loop Carrier. See SLC-96 and SLCC.

SLC-96 Subscriber Line Carrier 96. Pronounced "Slick 96." A short haul multiplexing device which enables up to 96 analog telephone customers to be served over a single four-wire digital circuit. SLC-96 essentially functions as a remote concentrator for up to 96 analog local loops. Here's how it works: A neighborhood of 96 residential and small business telephone customers normally would require 96 analog local loops from the LEC (Local Exchange Carrier) CO (Central Office), with each requiring a two-wire circuit all the way from the CO to the premises. That's very facilities-intensive, which translates into "expensive." SLC-96 allows the LEC to run a four-wire digital circuit to the neighborhood. That circuit is a T-2 circuit, which is a digital T-carrier circuit supporting 96 digital voice-grade channels. The SLC-96 concentrates the 96 voice-grade local loops, digitizing them in the process. SLC-96 is much more cost-effective than the original approach since it uses fewer wire pairs, even though it does require special electronics to accomplish this minor feat. On the other hand, SLC-96 limits each subscriber to a voice-grade circuit of 64 Kbps, which fact absolutely kills DSL (Digital Subscriber Line) technologies currently being deployed in support of high-speed Internet access. See also Digital Subscriber Line, T-2, and T-carrier. See SLCC.

SLCC Abbreviation for Subscriber Line (or Loop) Carrier Circuit, and pronounced "slick." It's a system that allows one pair of wires, that would normally provide one phone line, to carry multiple conversations. Various models are available, with capacity ranging from 2 to 96 lines. A SLCC is used between phone company central offices and areas where there are too many customers for the cable that is in place. It's much less expensive to install SLCCs than new cable, but the SLCC provides lower-than-normal line voltage, which may cause some phones to malfunction.

SLED Single Large Expensive Disk. The opposite of RAID (Redundant Array of Inexpensive Disks). See Raid and Redundant Array of Inexpensive Disks.

SLEE Service Logic Execution Environment. An AIN term. See AIN.

sleep mode 1. A means of increasing the battery life of portable computers. When you are using your laptop and are running on battery power, the computer will "go to sleep," if you are not actively using it for some the system "times out." When you move the mouse or touch a key, the system springs back into life. See also Hibernation.

2. A feature of digital cellular phones, allowing the phone to remain active, but not "on," when you are not engaged in a call. The phone sleeps most of the time, awakening every few milliseconds to check for incoming calls. Sleep Mode extends battery standby time by 300% or so.

sleep tight In Shakespeare's time, mattresses were secured on bed frames by ropes. When you pulled on the ropes the mattress tightened, making the bed firmer and more comfortable to sleep on. Hence the phrase, "good night, sleep tight."

sleeping policeman In Nicaragua, a sleeping policeman is a speed bump.

sleeping RSU When a remote service unit goes into a sleep mode when the tap gets too hot and is unable to reset, thus causing NDT (no dial tone). Load was taken off the RSU and then turned back on, thus resetting the RSU.

sleeves Short lengths of conduit, usually made from rigid metal pipes, used to protect cables entering a premises through a building wall or running through concrete floors between vertically aligned riser closets. Sleeves also provide for easy pulling of cable.

Sleeving A braided, extruded or woven tube.

SLI Code See CLLI Code.

SLIC Subscriber's Line Interface Circuit, a device that interfaces the SLC series of Lucent's products of local subscriber pair gain or multiplexing devices. See SLC-96.

Slideware Slideware is hardware or software whose reason for existing (eventually) has been explained in 35-mm slides, foils, charts and/or PC presentation programs. Slide is best described as "virtual vaporware." Vaporware is software which has been announced, perhaps even demonstrated, but not delivered to commercial customers. Hyperware is hardware which has been announced but has not yet been delivered. Slideware is less real than vaporware or hyperware. Classic slideware is the ISDN deployment strategies of

many of the telephone companies in the United States. See also Hookemware, Hyperware, Meatware, Podiumware, Shovelware, and Vaporware.

sliding window flow control Method of flow control in which a receiver gives the transmitter permission to transmit data until a window is full. When the window is full, the transmitter must stop transmitting until the receiver advertises a larger window. TCP, other transport protocols, and several data link layer protocols use this method of flow control.

SLIP 1. Serial Line Internet Protocol. An Internet protocol which is used to run IP datagrams over serial lines such as telephone circuits. Defined in the IETF's RFC 1055, SLIP essentially is a framing convention that identifies the end of each IP datagram, and the end of the IP session. As SLIP is not an Internet Standard protocol, specific implementations may vary. CSLIP (Compressed SLIP), defined in RFC 1144, is an enhancement that compresses the TCP/IP headers for transmission over low-speed serial lines. SLIP largely has been superseded by PPP (Point-to-Point Protocol). See IETF, Internet, IP, and PPP.

2. Slip, as in timing. See Timing.

SlipKnot A graphical browser developed by Peter Brooks and offered as restricted shareware, SlipKnot became popular among dial-up UNIX users in the early days of the Internet and World Wide Web. SlipKnot offered the advantage of not requiring SLIP (Subscriber Line Internet Protocol), PPP Point-to-Point Protocol), SLIP/PPP emulators, or TCP/IP software on the client side. All SlipKnot users required was a modem and a dial-up UNIX shell account with an ISP. SlipKnot is largely a thing of the past, as are dial-up UNIX accounts. SLIP, PPP, TCP/IP and Winsock are virtually ubiquitous today. SlipKnot now is considered to be way underpowered, as it is a 16-bit only solution that lacks the feature content embedded in contemporary browsers.

slip multiplication See Network Slip.

slip sleeve An oversized conduit that moves easily along an inner conduit and covers a gap or missing part of the smaller conduit.

slip/slip rate The loss (or rate of loss) of a data bit on a T-1 link due to a frame misalignment between the timing at a transmit node and timing at a receive node. See Timing.

slip-controlled The occurrence at the receiving terminal of a replication or deletion of the information bits in a frame.

slip-uncontrolled The loss or gain of a digit position of a set of consecutive digit positions in a digital signal resulting from an aberration of the timing processes associated with the transmission or switching of the digital signal. The magnitude or the instant of the loss or gain is not controlled.

slist Connected to a NetWare network? Type Slist in DOS on your PC and you will get a list of all the NetWare servers you are connected to.

slivercasting When entrepreneurs start TV stations on the Internet, they are aiming at small audiences of people interested in special subjects -like yachtracing or obscure tennis tournaments. It's a phenomenon that could be called slivercasting. And it only works well if you -the viewer -have a very high-speed broadband connection to the Internet.

SLM 1. System Load Module.

2. Service Level Management. See Service Level Management.

sloha Also called slotted aloha, sloha is a derivative of the aloha protocol for random multiple access satellite communications, in which the time domain is divided into time slots, each equal to the transmission time for a single packet or burst. Consequently, sloha resembles TDMA without precise time assignments.

slope 1. See Chromatic Dispersion.

2. Often refers to the slope of an amplifier curve.

sloppy clicker Someone who always manages to move Windows icons on his screen whenever he or she clicks on the icons. This makes the icons move and your computer screen look sloppy.

Slot 1 The specification that addresses the form factor (i.e., physical shape and size) and the physical and electrical interface for some of Intel's Pentium processors. Slot 1 it is the specification for one or more slots on the PC motherboard that accept a 242-contact daughterboard in the form of a microprocessor cartridge. Intel intends for Slot 1 to replace Socket 7. See also Socket 7.

slot antenna An antenna composed of a dipole across a slot shaped aperture.

slot sizes PCMCIA cards come now in three standard physical slot and card sizes: Type I: 3.3 mm thick, mostly for memory cards. Type II: 5 mm thick, fits most of the current cards on the market, including communications and networking cards. Type III: 10.5 mm, used for cards that have rotating hard disks in them. Toshiba, in its T4600, came out with a fourth size that is 16 mm thick. There is more to compatibility than just fit. The size of the slot is independent of the version of firmware that supports it. Some cards may only work with certain operating systems, BIOS, and drivers. Apple's Newton has a Type II slot but only supports cards that have been designed specifically for it. See Card Services, PCMCIA Standards, Socket Services and Slot Sizes.

slots Openings, typically rectangular, in the floor of vertically aligned riser closets that enable cable to pass through from floor to floor. A slot accommodates more cables than an individual sleeve.

slotted A Medium Access Control (MAC) protocol is slotted of attempts to transmit can only be made at times that are synchronized between contending devices. A cellular radio term.

slotted aloha An access control technique for multiple-access transmission media. The technique is the same as ALOHA, except that packets must be transmitted in well-defined time slots.

slotted ring A LAN architecture in which a constant number of fixed-length slots (packets) circulate continuously around the ring. A full/empty indicator within the slot header indicates when a workstation or PC attached to the LAN may place information into the slot. Think of a slotted ring LAN as an empty train that constantly travels in a circle, being filled and emptied at different terminals (workstations).

slotting The process of assigning a circuit to available channel capacity across the network during the circuit design process. When a circuit is slotted, it has an assigned path from one end of the network to the other.

slow switching channel A sequencing GPS (Global Positioning System) receiver channel that switches too slowly to allow the continuous recovery of the data message.

SLP Service Location Protocol. Described in the IETF's RFC 2165, SLP is a TCP/IP-based protocol which allows a computer to automatically discover and make use of network resources available over a corporate Intranet. Such resources might include printers, e-mail servers, Web servers and fax servers. It works like this. The client workstation makes use of a "user agent" to seek out the appropriate server based on its attributes, as "advertised" by "server agents." "Directory agents" serve an intermediary function between the user agents and server agents, aggregating advertisements in order to minimize the amount of time spent searching for the services across all networks and subnetworks in the enterprise Intranet.

SLR Send Loudness Rating.

SLSA Single Line Switching Apparatus.

SLT Single-Line Telephone, as opposed to a phone that has buttons to select from several lines.

SM 1. Switch module.

2. Service Module. An ADSL term for a device which performs terminal adaption functions for access to the ADSL network. Examples of terminal such devices include set-top boxes for TV sets, PC interfaces, and routers for LAN access. See ADSL.

3. See signalman.

SMG Wireless-Special Mobile Group. A standards body within ETSI that develops specifications related to mobile networking technologies, such as GSM and GPRS.

SGML Standard Generalized Markup Language. SGML is basically a standard to take ASCII text and give it meaning in order that many different computers can understand the bits and pieces of it in a standard way -i.e. all the same. Formally SGML is a document markup language. It was the beginning of a language called HTML which is now on every web page in the world. SGML is not in itself a document language, but a description of how to specify one. It is, in formal language, metadata. SGML is based on the idea that documents have structural and other semantical elements that can be described without reference to how such elements should be displayed. The actual display of such a document may vary, depending on the output medium and style preferences. Some advantages of documents based on SGML are:

They can be created by thinking in terms of document structure rather than appearance characteristics (which may change over time). They will be more portable because an SGML compiler can interpret any document by reference to its document type definition (DTD). Documents originally intended for the print medium can easily be re-adapted for other media, such as the computer display screen. The roots of SGML go back to the late 1960s when there was no GUI (graphical user interface). All computers displayed was text. After companies started using computers for document processing, it soon became obvious that a storage format should contain not only formatting codes interpreted by computer itself, but also descriptive human-legible information about the nature and role of every element in a document. The best use of SGML is generally made in big corporations and agencies that produce a lot of documents and can afford to introduce a single standard format for internal use.

XML is a simplified subset of SGML. See HTML and the World Wide Web.

S2M A European term for PRI ISDN, more commonly called PRA in Europe.

SMA 1. Subscriber Carrier Module-100A; same as SCM-100A.

2. Spectrum Management Agency. Former Australian agency established in 1993 to manage access and use of the radio spectrum. The SMA merged with AUSTEL on July 1, 1997, to form the Australian Communications Authority (ACA).

SMA 905/906 (Subminiature type A). A former microwave connector modified by Amphenol to become the "standard" fiber optic connector. The 905 version is a straight ferrule design, where as the 906 is a stepped ferrule design and uses a plastic sleeve for alignment.

Small Computer System Interface SCSI. Pronounced scuzzy, SCSI is a bus that allows computers to communicate with any peripheral device that carries embedded intelligence. The standard is covered by the American National Standards Institute (ANSI) and has developed into from SCSI-1 into SCSI-2, and now SCSI-3. Different types of device can be connected in a daisy chain via a 50-pin cable (68-pin for SCSI-3), both ends of which must be terminated. All signals on the cable are common to all devices. To avoid bus contention, each device (up to a maximum of seven) connected to the bus is given a unique "SCSI" address with each address being allocated a degree of priority. The SCSI-1 bus carries 8-bit data, 1-bit parity and 9-bit control lines, to provide a maximum synchronous data transfer rate of 5 Mbyte/s.

The maximum bus length is dependent upon the type of bus driver/receiver used. For single-ended devices, the length is restricted to 6 meters, but you wouldn't want to go that far. A desktop is about the maximum distance for an un-amplified SCSI-1 bus. The 6 meters can be extended to 25 meters through the use of differential driver/receivers. SCSI-2 has been extended to increase the data rate to 10 Mbyte/switch via either 16 or 32-bit processors. The maximum data rate is thus nearer to 40 Mbyte/s. The protocol has also been expanded to include tagged commands. This allows the execution of queued control commands according to a prescribed sequence. For a bigger explanation, see SCSI, SCSI-2 and SCSI Transfer Rate.

small fortune The only sure way to make a small fortune -in telecommunications, networking, computing, or in anything -is to start with a large one. After the Nasdaq Bubble burst in the Spring of the year 2000, many investors made small fortunes by losing their large fortunes. It was called the Tech Wreck of 2000-2002.

small vocabulary A voice recognition term. Vocabularies containing fewer than 50 words.

smart agent A software tool that acts like you. It is your agent and it acts intelligent on your behalf. For example, you might ask assign it a series of tasks to do on your behalf every day, or even throughout the day. For example, "Which orders that I sold today have been shipped today?" See Intelligent Agent and Smart Messaging.

smart antenna A fancy name for a new type of cellular base station antenna. The technology replaces conventional cell-site antennas with a multibeam antenna array that allows network operators to target the transmission and reception of calls more precisely and therefore reduce the amount of spectrum consumed and the amount of interference. The basic benefit of a "smart antenna" is to allow a cell phone provider to serve more cell phone customers without a new investment in basic cell site radio and electronics.

smart battery A type of battery that uses electronic circuitry to provide its host device (phone or notebook) with status power information so the system can conserve power intelligently.

Smart Battery Data SBD. The information accessible across the System Management Bus between the smart battery and the device.

smart border Name given to a new initiative to upgrade border surveillance technologies and procedures after terrorist traffic between Canada and the US.

smart building Also referred to as the "intelligent" building. A centrally managed structure that offers advanced technology and perhaps shared tenant services. Central to the early ideas of a smart building was the concept of one gigantic phone system shared by all the tenants. The idea was that you needed a big switch to get lots of features and all the benefits of cheap long distance. That's no longer true. Small switches offer much the same benefits. Now buildings are often smart, but frequently they're smart, using those smarts to save energy and insure safety and less to share telephone service among the tenants, who have shown a tendency to fend for themselves.

smart card A credit card-sized card which contains electronics, including a microprocessor, memory and a battery. The card can be used to store the entire repair and maintenance history on the family automobile or the health history of a member of the family. Since smart cards are tamper-resistant hardware devices that store your private keys and other sensitive information, they can be used for security applications. There is no direct contact between the "smart card" and the device which reads it. This avoids the problem of wear that afflicts traditional credit cards -both their embossed numbers on the front and their magnetic strip on the back. According to the Smart Card Industry Association, a smart card is any card with a microprocessor. Non-smart chip cards are simple memory cards and hardwired logic cards. As the smart card has developed, we see new and specialized versions of it. Memory cards store information or values, such as debit or credit information. Processor cards perform calculations, complex processing and security applications. Contact cards read information when a card is inserted into a reader. Contactless cards wirelessly read information. Combicards uses both a card reader and a wireless device. See also Satellite Television Smart Cards.

smart chip This is a chip used by GSM cell phones (also called mobiles) that stores information about the calling network. Phones that use a smart chip (also referred to as a SIM) can be easily upgraded to use different networks (such as a European GSM system). The chips also store personal information, such as your phone book, making it easy to transfer your information to a new phone. See also SIM Card.

smart display A flat screen touchscreen display with processing intelligence and communications. One idea is that you carry this device around your house and use it as a portable, fast-starting, simple PC -to look up recipes, to check your email, etc.

smart dust The Economist of February 2, 2002 wrote, "When peripatetic futurologists such as John Gage of Sun Microsystems really want to impress their audience, they talk about 'smart dust'. The concept is indeed intriguing. The dust in question is made up of tiny, wireless sensors that could be dispersed anywhere -say, over a battlefield to find out where enemy troops are, or whether chemical or biological weapons have been used. This is not science fiction. Researchers at the University of California at Berkeley led by Kris Pister, a professor of electrical engineering, are already working on a smart-dust prototype the size of a small nailhead. A klunkier version can already be bought from a Silicon Valley start-up, Crossbow Technology. It may be decades before smart dust is dispersed over real battlefields, but the tiny devices are at the forefront of an important technological trend that is often underestimated: the spread of sensors and tags. New manufacturing processes, wireless technology and intelligent software are making them ever smaller, smarter and, most important, cheaper. As with microprocessors and lasers in earlier decades, the novelty is not that these sensors exist at all, but that they have suddenly become cheap enough to be used in ordinary everyday products, says Paul Saffo, director of the Institute for the Future, a Silicon Valley think-tank."

smart home The 1990's catch phrase was "smart home," for houses with Internet-linked appliances that could, for example, tell when you ran out of milk and automatically order more, or chill the wine while you were still on the expressway. It never quite happened that way, though some early adopters have had fun testing the latest electronic gizmos. But by the early 2000s, the smart home began to center around the centralized home entertainment system, with personal video recorders, with "digital plumbing" for Internet and television connections hidden behind walls, and just enough equipment to make tapping into video, images (photos of the kids) and audio (Bob Dylan's latest and greatest). The backbone of any networked home is "structured wiring," a bundle of high-speed cables put in the walls before the drywall goes on or wireless links put in after the walls are closed. More builders are including such cables automatically. They add to the resale value of the house.

smart hub A concentrator, used in Ethernet or ARCnet networks, with certain network-management facilities built in to firmware that allow the network administrator to control and plan network configurations; also known as an intelligent hub. In token-ring networks

smart hunt groups Picture a call center with people answering the phone from customers. The people answering the phone have different skills, know different things about the company's products, and are thus better able to answer some questions better than others. Now picture that the phone system the people are using is "smart" and somewhere in it and in the databases attached to it, it knows who of the people (also called agents) is good at what. What skills they have. What they know about which products, etc. Then, by asking questions of incoming callers through a combination of an attached interactive voice response system, the incoming caller's phone number and the number the caller dialed, it is able to route the call to the exact correct person. If that correct person is not available, it routes the caller to another person. And to another, etc. There may be many smart hunt groups, each consisting of the same people. If you think of a call center with several hundred people you can easily see it's possible to have potentially thousands of "smart hunt groups." Just depends on how complex your product line is and how sophisticated you want to get. The term was coined by Rick Luhmann, editor of Computer

Telephony Magazine, New York, NY.

smart jack Industry term for the device to test integrity of T-1 circuits remotely from the central office. Installed on the customer premises in the form of a semi-intelligent demarcation point (demarc), the smart jack is completely passive until activated remotely by a digital code, typically something like "FACILITY 2," sent down the T-1. This code activates a relay that breaks the T-1 circuit and closes a receive-to-transmit loop across the T-1 at the customer end, sending the signal back to the central office (CO). This allows the CO to confirm the integrity of the loop without having to dispatch a technician to the site. Once the loopback test is completed, the relay automatically resets. The advantage of a smart jack is that the carrier can accomplish the loopback without engaging the DSU/CSU, which typically is in the CPE (Customer Premise Equipment) domain (i.e., it is customer-owned and, therefore, can possess uncertain attributes). As a result, the smart jack does not short when disconnected from the DSU/CSU; therefore, it always is available to the carrier for purposes of conducting such a loop test. The smart jack commonly has the same appearance as does a RJ48 jack; indeed, a RJ48 cable can connect the smart jack and the DSU/CSU to provide a hard-wired shorting function. The smart jack also may be rack-mounted. See also CSU, Demarc, DSU, Loopback, and RJ48C.

smart label A label that contains an RFID tag. It's considered "smart" because it can store information, such as a unique serial number, and communicate with a reader. See RFID.

smart mob Coined by cyberpundit Howard Rheingold to describe mass social movements coordinated through wireless communications. In other words protests coordinated by cell phones.

smart phone A generic term for a phone with PC-like intelligence and perhaps a slightly larger screen which could display stock price information, bank balances, etc.

smart radio Also referred to as a software radio. An advanced radio in which frequencies, tuning, and protocols are defined using software programming, rather than fixed hardware. Such a radio might be used for a wireless handset or base station, giving it the capability of utilizing various previously incompatible protocols interchangeably, such as AMPS, GSM, and CDMA. The biggest savings with software radio, according to experts, are in the areas of upgrades and maintenance.

smart retries A term used in the fax blasting industry to refer to doing different retries based on the status of the failed fax. For example, you might receive a busy and then retry 3 times 10 minutes apart, then 2 times more eight hours later. Another smart retry might be to restart sending a multi page fax at the page that the failed fax stopped at. This feature might also have an alternate cover page that can say that this fax is a continuation of a previous fax that didn't make it fully.

smart terminal A data terminal capable of operating in either a conversational or a block mode; containing a full set of local editing capabilities without reliance on a controlling external computer.

smart T-1 What some companies call metered T-1 or fractional T-1.

smart web browsers A Web Browser is a piece of software which allows you to search the World Wide Web for information. A smart Web Browser will contain a modicum of intelligence, allowing it to find what you're looking for faster, easier and less time consuming.

SmartHouse The National Society of Home Builders has registered SmartHouse. BellSouth has trademarked SuperHouse. And GTE has registered SmartPark. One day, the theory is, fiber optic will snake to everyone's house, bringing the potential of immense information services. Until that day comes, there'll be neat demonstrations at trade shows. See also Picturephone, another product that made great demonstrations.

smartphones The term came into being in 2003 when some manufacturers started joining PDAs (personal digital assistants) with cell phones. And various people called the result "smartphone." Smartphones can be used to store addresses and phone numbers, which can then, with the touch of a stylus, be dialed with the cell phone. Smartphones can play games, music and jot down short messages to transmit and receive. Some smartphones also sport tiny keyboards.

SMAS Switched Maintenance Access System.

SMATV Satellite Master Antenna Television. A distribution system that feeds satellite signals to hotels, apartments, etc. Often associated with pay-per-view.

SMB 1. Server Message Block. The protocol developed by Microsoft, Intel, and IBM that defines a series of commands used to pass information between network computers. The redirector packages SMB requests into a network control block (NBC) structure that can be sent over the network to a remote device. The network provider listens for SMB messages destined for it and removes the data portion of the SMB request so that it can be processed by a local device. In short, SMB is basically a protocol to provide access to server-based files and print queues. SMB operates above the session layer, and usually works over a network using a NetBIOS application program interface. SMB is similar in nature to a remote procedure call (RPC) that is specialized for file systems.

On June 13, 1996, Microsoft said, "The SMB protocol is an open technology widely available on UNIX, VMS and other platforms. It has been an Open Group (formerly X/Open) standard for PC and UNIX interoperability since 1992 (X/Open CAE Specification C209), and it is supported in products such as AT&T Advanced Server for UNIX, Digital's PATHWORKS, HP Advanced Server 9000, IBM Warp Connect, IBM LAN Server, Novell Enterprise Toolkit, and 3Com 3+Share, among others. SMB is also the featured file and print sharing protocol of Samba, a popular freeware network file system available for LINUX and many UNIX platforms."

The above Microsoft words came in a press release announcing something called CIFS, which is essentially an enhanced version of SMB. See CIFS.

2. Small and medium business.

SMBus System Management Bus. A two-wire bus for more intelligent handling of rechargeable batteries in portable stuff, like laptops.

SMC Standard Management Committee. Directs specialized working groups of the Architecture and Standards Steering Council.

SMDA Station Message Detail Accounting. Another name for telephone call accounting. See Call Accounting System.

SMDI Station Message Desk Interface or Simplified Message Desk Interface. The SMDI is the data link from the central office if you have ESSX, Centrex or Centron (etc.) that gives you your stutter dial tone or message waiting light. In essence, SMDI is a data line from the central office containing information and instructions to your on-premises voice mail box. With SMDI, the calling person is not required to re-enter the called phone number (or in any other way identify the called party) once the call terminates to the messaging system.

SMDR Station Message Detail Reporting. Another name for telephone call accounting. See SMDR PORT and Call Accounting System.

SMDR port Modern PBXs and some larger key systems have an Station Message Detail Recording (SMDR) electrical plug, usually an RS-232-C receptacle, into which one plugs a printer or a call accounting system. The telephone system sends information on each call made from the system to the outside world through the SMDR port. That information -who made the call, where it went, what time of day, etc. -will be printed by the printer or will be "captured" by the call accounting system on a floppy or a hard magnetic disk and later processed into meaningful management reports. See Call Accounting System.

SMDS Switched Multimegabit Data Service. A connectionless high-speed data transmission service intended for application in a Metropolitan Area Network (MAN) environment. SMDS is a public network service designed primarily to for LAN-to-LAN interconnection. An offshoot of the Distributed Queue Dual Bus (DQDB) standard defined by the IEEE 802.6 standard, SMDS owes its commercial success to Bellcore, which refined and commercialized it at the request of the RBOCs as a MAN service appropriate for RBOC offering within the confines of a LATA. SMDS offers bandwidth up to T-3 (45 Mbps) and the critical advantage of excellent congestion control, which virtually ensures that data arrives intact and as transmitted. SMDS supports asynchronous, synchronous and isochronous data. Although intended as a MAN service offering, it is possible to extend its reach over the WAN. SMDS is a cell-switched service offering, based on a 53-octet cell with a 48-octet payload similar to that of ATM. SMDS involves a process of converting data into cells before presentation to the network. Each set of data in its native form can be as much as 9,188 octets in length, constituting what is known as a Protocol Data Unit (PDU). At the point of SMDS processing, that PDU becomes known as a SMDS Service Data Unit (SDU), which is then appended with a header and trailer which contain network control information. The SDU, with header and trailer, is then segmented into units of 48 octets, each of which is prepended with a header and appended with a trailer before presentation to the SMDS network in the form of a 53-octet cell. Prepended means preceded as part of the same data unit (generic "packet"), whereas appended means succeeded as part of the same data unit. In fact, SMDS was designed with ATM in mind, providing a smooth transition path to ATM by virtue of the size of both the overall cell cell and the payload, although the specifics of the cell structure and other elements of the specific protocols are quite different. SMDS garnered a good deal of interest during the early 1990s, although broad support never developed. SMDS is not widely available and is highly unlikely to ever gain a broad level of acceptance; Frame Relay and ATM, in particular, have and will continue to overshadow SMDS. See The SMDS Interest Group.

SMDS Interest Group A defunct consortium of vendors and consultants who

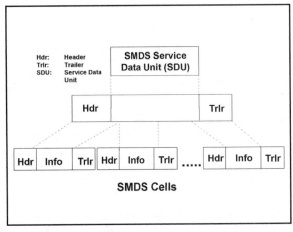

SMDS Cells

were committed to advancing worldwide SMDS as an open, interoperable solution for high-performance data connectivity. On June 16, 1997, the Board of Trustees announced that the group was disbanded, turning over all responsibilities to regional organizations. The Board declared that its mission had been fulfilled.

SMDSU Switched Multimegabit Data Service Unit. See DSU.

SME 1. Small to medium enterprises. Commonly used by EC & EDI folks.

2. Security Management Entity.

smear The undesirable blurring of edges in a compressed image, often caused by the DCT (Discrete Cosine Transform) which tends to eliminate the high-frequency portions of an image which represent sharp edges. See also Discrete Cosine Transform.

Smell-O-Vision Hans Laube was a Swiss professor of osmics, which is the science of smells. He invented a process for reproducing smells in movie theaters, which process he named "Scentavision," and which later became known as Smell-O-Vision prior to its introduction in the 1960 film "Scent of Mystery." According to articles written at the time, this development had been long awaited. Movies (moving pictures) were introduced in 1895, "talkies" (talking pictures) in 1927, and "smellies" (smelling pictures) finally in 1960. The name "Smell-O-Vision" was coined by Michael Todd, Jr. (son of Michael Todd Sr., who was one of Elizabeth Taylor's many husbands). The squirters that blasted the smells into the theater were queued from the soundtrack. Apparently, the smells were awful, and became worse as they lingered. The copycat technique of AromaRama was equally unsuccessful. Vibrating chairs were introduced for the cult classic film "The Tingler," but fared no better. Smellies were revived briefly in the film "Odorama" (1981), which relied on scratch n' sniff cards that were handed out to audience members, who were instructed to scratch 'n sniff them at special moments during the film. The concept has now been revived with the introduction of "iSmell," as in "internet Smell," introduced by Digicents. Digicents indexes various scents according their chemical compositions, and their place in the scent spectrum. The desired scent index is then described in a small file, which can be attached to any kind of Web content, interactive games, e-mail, movies, music, or other IP (Internet Protocol) file. The intended scent is recreated by your computer through personal scent synthesizers, which plug into your computer much like audio speakers. The personal scent synthesizers connect to your computer through your computer's serial port or USB (Universal Serial Bus) port, and are locally powered. They emit smells on command, such as a mouse click or a timed response triggered by a DVD or CD-ROM. The scents are recreated by an indexed combination of over 100 basic scents that are selected from a palette of scented oils, with the scent cartridges being refillable. Refills will be available from www.snortal.com, if this idea ever takes off. "Snortal," of course, is a contraction of the words "snort" and "portal." See smelltone.

smelltone Keitai KenKun, a Japanese company, at one stage was selling a cell phone accessory that releases a scent when there is an incoming call. See also Smell-O-Vision.

SMF Single Mode Fiber. Also known as monomode fiber, SMF has a much thinner inner core (8-10 microns, or so) than MultiMode Fiber (MMF). The thin inner core allows light pulses to propagate along only a single mode, or physical path, which significantly reduces the potential for various forms of dispersion. As dispersion causes signal distortion, SMF is used in applications that cover long distances at high speeds. SMF types include Non Dispersion Shifted Fiber (NDSF) and Dispersion Shifted Fiber (DSF). See also DSF, MMF, and NDSF.

SMF-PDM The ANSI X3T9.5 PDM standard which defines the requirements for the transmission of data over single mode fiber in an FDDI topology. Also refers to the ANSI working group responsible for the development and perpetuation of the standard.

SMF Single Mode Fiber: Fiber optic cable in which the signal or light propagates in a single mode or path. Since all light follows the same path or travels the same distance, a transmitted pulse is not dispersed and does not interfere with adjacent pulses. SMF fibers can support longer distances and are limited mainly by the amount of attenuation. Refer to MMF.

SMI Structure of Management Information. The set of rules and formats for defining, accessing and adding objects to the Internet MIB. SMI was elevated to full standard status in May 1990.

SMIL Synchronized Multimedia Integrated Language. Pronounced "SMILE." A recommendation from the World Wide Web Consortium (W3C) for a stylized layout language for the creation of Web-based multimedia presentations. SMIL, which is compliant with XML (eXtended Markup Language), defines the mechanism by which authors can compose a multimedia presentation, combining audio, video, text, graphics. SMIL allows the synchronization of these multimedia elements in terms of their presentation on screen, as well as the timing of their delivery to the client. The resulting SMIL file (.smi) offers the advantage of being in the form of a simple, markup text file that can be created with any text editor. Current members of the W3C Working Group developing SMIL Boston (the latest version) are key international industry players in Web multimedia, interactive television and audio/video streaming: Canon, Compaq, CSELT, CWI, France Telecom, Gateway, GLOCOM, INRIA, Intel, Macromedia, Microsoft, Oratrix, NIST, Nokia, Panasonic, Philips, RealNetworks and WGBH.

SMIL 1.0 is the W3C standard for Web-based multimedia first implemented by RealNetworks with the advent of RealSystem G2 in June of 1998. RealNetworks co-authored the SMIL 1.0 specification. SMIL 1.0 enables the delivery of long format, Web-based multimedia to a broad range of audiences, from modems to T-3 Internet connections. As an open XML-based language, SMIL enables a wide range of audio-visual presentation authoring environments,ranging from simple text editors to graphical editing tools such as RealNetworks RealSlideshow. The SMIL Boston Working Draft proposes several extensions to SMIL 1.0, such as integration with TV broadcasts, animation functionality, improved support for navigation of timed presentations, and the possibility to integrate SMILmarkup in other XML-based languages. See also W3C and XML.

smiley face :) The "smiley face is probably the most popular emoticon, from Emotional Icon. The original smiley face (the one that you see on T-shirts and buttons) was created by commercial artist Harvey Ball in 1963 as part of an in-house friendship campaign for state Mutual Life Assurance, a Massachusetts insurance company. Ball received $45 for the creation, which he failed to patent. Ball died on April 13, 2001, at the age of 79. :(See also Emoticon.

smishing Short for "SMS phishing." The use of SMS messages by con artists to trick cell phone users into revealing personal information that can later be used in an identity theft scam. Smishing takes a variety of forms, all of which begin with the sending of a fake text message to an unsuspecting cell phone user. The text message contains official-sounding instructions for the user to respond with a text message, or to visit a website, or to call a phone number to provide the requested personal information. See also phishing.

SMM System Management Mode. See APM.

smoke test Test of new or repaired equipment by turning it on. If there's smoke, it doesn't work!

Smokeware When manufacturers announce new products, sometimes the products are not ready. Smokeware derives from the expression "smoke and mirrors."

SMOP Small Matter of Programming. A little software and it will all work. Yeah!

smooth call arrival A call center term. Calls that arrive evenly across a period of time. Virtually non-existent in incoming environments.

smooth handoff See Mobile IP.

SMP 1. Symmetric MultiProcessing. The use of several CPUs (i.e. several Intel Pentium chips) inside a PC or (more typically) a server to achieve a performance boost. This is achieved by an operating system that lets CPUs run operations in parallel. This is not a "normal" PC operating system. Normal PC operating systems -MS-DOS, Windows, etc. -expect that their work will be done on one processor, not on more than one. See also Symmetric Multiprocessing.

2. SCSA Message Protocol.

3. Simple Management Protocol.

SMPI SCSA Message Protocol Interface.

SMPP Short Message Peer-to-Peer Protocol. SMPP. A protocol, developed by Logica Aldiscon, that provides the capability to deliver email and voicemail between wired and wireless networks.

SMPTE Society of Motion Picture & Television Engineers. An international society dedicated to advancing the theory and application of motion-picture technology including film, television, video, computer imaging and telecommunications. Founded in 1916 as the Society of Motion Picture Engineers, the "T" was added in 1950 in recognition of the emerging television industry. The SMPTE is an accredited ANSI Standards Developing Organization, and is recognized by both the ISO and the IEC. Current membership numbers about 8,500 in 72 countries. See SMPTE Time Code, VC-1 and www.smpte.org

SMPTE Time Code A video term. Time code that conforms to SMPTE standards. It consists of an eight-digit number specifying hours:minutes:seconds:frames. Each number identifies one frame on a videotape. SMPTE time code may be of either the drop-frame or non-drop frame type. In GVG editors, the SMPTE time code mode enables the editor to read either drop-frame or non-drop frame code from tape and perform calculations for either type (also called mixed time code). SMTPE is a standardized edit time code adopted by SMPTE, the Society of Motion Picture and Television Engineers.

SMR Specialized Mobile Radio. Also known as PAMR (Public Access Mobile Radio), PMR (Private Mobile Radio), TMR (Trunked Mobile Radio), and TRS (Trunked Radio System). A two-way radio telephony service making use of macrocells covering an area of up to 50 miles in diameter. The first SMR system was placed in service by the Detroit Police Department in 1921. The first commercial SMR service was offered in 1946 in St. Louis by AT&T. Subsequently, private SMR was widely used, and still is, in dispatch applications by truck and taxi fleets. Eventually, SMR took the form of IMTS (Improved Mobile Telephone Service), which used smaller frequency channels in support of many more conversations through improved efficiency. Many SMR operators have converted their SMR networks to digital so they can deliver both voice and data to a single device with improved efficiency and security, and thus compete with cellular and PCS radio providers. SMR systems have far less radio spectrum than cellular has, but the signal can reach 25 times farther, which means it's cheaper to build a national network. In the U.S., the frequency bands used by SMR include 220 MHz, 800 MHz, and 900 MHz. See also ESMR.

SMRP Simple Multicast Routing Protocol. Apple Computer's specialized network protocol for routing multimedia data streams on AppleTalk networks.

SMRT Single Message-unit Rate Timing. USA telephone company tariff under which local calls are timed in 5-minute increments -with a single message unit charge applied to each complete or partial increment.

SMS 1. Service Management System, a term coined by Bellcore (now called Telcordia Technologies) for the Intelligent Network. The SMS allows provision and updating of information on subscribers and services in near-real time for billing and administrative purposes. SMS also coordinates all of the national 800 telephone numbers for all the US telephone companies through service control points (SCP). See SMSC.

2. Short Message Service. A means to send or receive, short alphanumeric messages to or from mobile telephones. See Short Message Service and SMS tone.

SMS tone A downloadable tone that plays on your cell phone when there is an incoming SMS alert. These things typically cost money. See also ringtones.

SMS-R Subscriber Carrier Module-100S REmote.

SMS/800 The national database Service Management System that retains all 800 records. This database provides long distance carriers a single interface for 800 number reservations and record maintenance. Developed by Bellcore, the database has been in use by various Regional Bell Operating Companies (RBOCs) since 1988. The FCC mandated that a neutral third party administer the database after 800 portability, which occurred in May, 1993. That administration responsibility now lies with the SMS/800 Number Administration Committee (SNAC), which is part of the OBF (Ordering and Billing Forum), which operates under the auspices of the Alliance for Telecommunications Industry Solutions (ATIS). See also ATIS, OBF and SNAC.

SMSA Standard Metropolitan Statistical Area. A metropolitan area consisting of one or more cities as defined by the Office of Management and Budget and used by the FCC to allocate the cellular radio market.

SMSC Short Messaging Service Center. On a wireless network, SMSC allows short text messages to be exchanged between mobile cell telephones and other networks. Sample message: "Honey, I'll be home late tonight." It is the entity that stores and forwards Short Message Service (SMS) messages. SMSC supports message storage, point-to-point, mobile terminated short message service (SMS-PP/MT). SMSC is being enhanced to support mobile originated messages also. The product can be interfaced with multi-vendor mobile switching centers (MSCs) over an SS7 link; it can work with Gateway as well as non-Gateway MSCs. See SMS and SMSCH.

SMSCH Short Message Service CHannel. Specified in IS-136, SMSCH carries signaling information for set up and delivery of short alphanumeric messages from the cell site to the user terminal equipment. SMSCH is a logical subchannel of SPACH (SMS (Short Message Service) point-to-point messaging, Paging, and Access response CHannel), which is a logical channel of the DCCH (Digital Control CHannel), a signaling and control channel which is employed in cellular systems based on TDMA (Time Division Multiple Access). The DCCH operates on a set of frequencies separate from those used to support cellular conversations. See also DCCH, IS-136, SMS, SPACH and TDMA.

SMT 1. Surface Mounting Technology.

2. Station Management. The part of FDDI that manages stations on a ring.

3. Simultaneous MultiTasking. The idea of SMT is to mix instructions from different threads or programs (known as "processes") together in the same CPU pipeline. Normally in a microprocessor, a pipeline executes an instruction stream from a single process at a time, just as a car factory assembly line manufacturers one type of vehicle at a time.

SMTA Single-line Multi-extension Telephone Apparatus.

SMTP Simple Mail Transfer Protocol. SMTP is a TCP/IP protocol for sending e-mail between servers. Virtually all e-mail systems that send mail via the Internet use SMTP to send their messages. Typically, you will send your email via SMTP to a POP3 (Post Office Protocol) server, from where your addressee will retrieve your message. Because of SMTP and POP3, you need to specify both your POP3 server and your SMTP server when you configure your e-mail client application, e.g. Microsoft Outlook, Eudora. Your email system will ask you for your SMTP server. Mine is mail.technologyinvestor.com while my POP3 server is mail.technologyinvestor.com. SMTP is actually an application-level protocol which runs over TCP/IP, supporting text-oriented e-mail between devices supporting Message Handling Service (MHS). You can send complex attachments, however, with SMTP, by simply using MIME, which stands for Multipurpose Internet Mail Extension. MIME is an SMTP extension supporting compound mail, which is integrated mail, including perhaps e-mail, a Word document, an Excel spreadsheet, an image, a voice WAV file and perhaps also a video clip in an AVI format. The SMTP is used as the common mechanism for transporting electronic mail among different hosts within the Department of Defense Internet protocol suite. Under SMTP, a user SMTP process opens a TCP connection to a server SMTP process on a remote host and attempts to send mail across the connection. The server SMTP listens for a TCP connection on a well-known port (25), and the user SMTP process initiates a connection on that port. When the TCP connection is successful, the two processes execute a simple request/response dialogue, defined by the SMTP protocol, in which the user process transmits the mail addresses of the originator and the recipient(s) for a message. When the server process accepts these mail addresses, the user process transmits the message. The message must contain a message header and message text formatted in accordance with RFC 822. See also POP3.

SMU 1. Subscriber Carrier Module-100 URBAN.

2. System Management Unit. The card or equipment in an xDSL unit that takes care of the management of the unit.

smurf attack A category of Denial of Service Attack on a computer system. The New York Times defines a smurf attack as involving the use of many spoofed machines "but it also employs a large third party network of computers to "amplify" the data used in the attack and greatly increases the effectiveness of the assault. See Denial of Service Attack and Smurfing.

smurfing 1. A Denial of Service (DoS) attack by a hacker, usually a very young hacker hence, the origin of the term. Such an attack involves the sending of a stream of diagnostic "ping" messages to a list of IP servers, each of which forwards them to all LAN-attached workstations, each of which responds. The return address, however, is forged to reflect that of the target of the attack. The resulting stream of responses, which is magnified many times as the pinged servers and attached devices try over and over again to respond, effectively shuts down the targeted server. The targeted server might be a single server or it might be a complete Web site. See also Hacker and www.mcs.net/smurf. See also Smurf Attack.

2. Breaking down large amounts of cash which are then loaded onto various types of prepaid debit cards which can easily be transported around the world -often to avoid police.

SN An ATM term. Sequence Number: SN is a 4 octet field in a Resource Management cell defined by the ITU-T in recommendation 1.371 to sequence such cells. It is not used for

ATM Forum ABR. An ATM switch will either preserve this field or set it in accordance with 1.371.

SN Cell An ATM term. Sequence Number Cell: A cell sent periodically on each link of an AIMUX to indicate how many cells have been transmitted since the previous SN cell. These cells are used to verify the sequence of payload cells reassembled at the receiver.

SNA Systems Network Architecture. An IBM product. The most successful computer network architecture in the world. See IBM and Systems Network Architecture.

SNAC SMS/800 Number Administration Committee. SNAC is responsible for administering the Service Management System/800 (SMS/800) database for 800 and other toll-free numbers in the U.S. The SMS/800 database is the central repository for toll-free numbers, identifying the carriers to which they have been assigned, or to which they have been transferred at the request of the user. The SMS/800 database comprises all toll-free numbers, which are defined as 8NN, where NN is a set of two identical numbers (e.g., 800, 888 and 877). SNAC is a committee of the Ordering and Billing Forum (OBF), which operates under the auspices of the Alliance for Telecommunications Industry Solutions (ATIS). See also ATIS, OBF and SMS/800.

SNADS SNA Distribution Services. An IBM protocol that allows the distribution of electronic mail and attached documents through an SNA network.

SNAFU Situation Normal All Fouled Up. World War II military slang, that describes a disaster, i.e. some of the more frustrating facets of the telecom industry, like when your lines don't get installed when they're promised. Actually, GIs typically uses a much less genteel word than "Fouled." Most of us still do. See FUBAR and TARFU.

Snagless RJ-11 and RJ-45 cables have this little plastic piece sticking out. It has a nasty habit of breaking off when it gets caught in something. Some manufacturers of cables make something they call "snagless." Their snagless cables have a boot over their connector lock to keep it from getting caught on things.

Snail Each French citizen eats on average 500 snails a year.

snail mail A term used to reference delivery of messages by your local postal service. In short, mail that comes through a slot in your front door or a box mounted outside your house. See also Missile Mail.

snake A flexible strip of metal, typically 1/4 to 1/2" wide and 10 to 100' long, used to pull or push wire and cable through conduit, ceilings, walls or crawl spaces where it is difficult or impossible for a human to fit.

SNAP Subnetwork Access Protocol. A version of the IEEE local area network logical link control frame similar to the more traditional data link level transmission frame that lets you use nonstandard higher-level protocols. The Subnet Access Protocol is an Internet protocol that operates between a network entity in the subnet and a network entity in the end system and specifies a standard method of encapsulating IP datagrams and ARP messages on IEEE networks. The SNAP entity in the end system makes use of the services of the subnet and performs three key functions: data transfer, connection management, and quality of services selection.

snapshot A view of something at an instant in time, rather than over time. A snapshot of a network, for instance, might be a view of the network configuration, in both physical and logical terms, as of midnight, January 1, 2000.

snarf 1. To grab, esp. to grab a large document or file for the purpose of using it with or without the author's permission. This term was mainstream in the late 1960s, meaning 'to eat piggishly'. It may still have this connotation in context. "He's in the snarfing phase of hacking -FTPing megs of stuff a day." It also means to acquire, with little concern for legal forms or politesse (but not quite by stealing). "They were giving away samples, so I snarfed a bunch of them."

2. System Normalized And Reset Flags.

SNC Subnetwork Connection: In the context of ATM, an entity that passes ATM cells transparently, (i.e., without adding any overhead). A SNC may be either a stand-alone SNC, or a concatenation of SNCs and link connections.

SND Cellular language for SEND. You punch the digits for the phone number you want to dial into your phone. Check them on the screen. If they're fine, hit the SND button. Bingo, your call goes through.

SNDCF Subnetwork Dependent Convergence Function.

SNDCP Subnetwork Dependent Convergence Protocol. A Network Layer protocol that supports subnetwork convergence.

snd format Sound resource format. A digital audio sound resource utilized by many Macintosh applications and by the Mac OS. Double-clicking on a snd file enables playback of the sound. Also called a System 7 sound.

sneak current A low-level current that is of insufficient strength to trigger electrical surge protectors and, therefore, is able to pass through them undetected. These currents may result from contact between communications lines and AC power circuits or from power induction, and may cause equipment damage unless secondary protection is used. See Sneak Currents.

sneak currents Unwanted but steady currents which seep into a communication circuit. These low-level currents are insufficient to trigger electrical surge protectors and therefore are able to pass them undetected. They are usually too weak to cause immediate damage, but if unchecked could potentially create harmful heating effects. Sneak currents may result from contact between communications lines and AC power circuits or from power induction and may cause equipment damage due to overheating. See Sneak Current.

sneak fuse A fuse operated by a low-level current and capable of preventing sneak currents on communication lines. See also Sneak Currents.

sneaker A person who is hired to hack into (i.e., break into) systems to test the effectiveness of their security. See also Cracker, Hacker, Phreak, Script Kiddies and Sneaker Brigade.

sneaker brigade You sign a consulting contract for a big job -say a complete re-do of your website and your entire computer operations. You sign the contract with the "big cheese," the president of the consulting company. He then disappears off the scene and sends in the sneaker brigade -the young children consultants in the firm who will actually do the job. They're called the sneaker brigade because they wear sneakers.

sneakers-up A dotcom that's gone belly-up, derived from the fact that most of the employees came to work in sneakers.

sneakernet A "network" for moving files between computers. It is the oldest "network." It consists of copying a file to a floppy disk, walking to another machine, loading it on that machine. The term "sneakernet" refers to the fact that the main method of moving the disks is by feet, presumably clad in sneakers.

sneakerware A fully manual interface which relies on personnel to do the processing.

Snell's Law of Refraction See Refraction.

SNET Southern New England Telephone Corporation. Originally an independent telephone company serving a substantial portion of Connecticut, SNET stock was partially held by AT&T at the time of the Modified Final Judgment (MFJ). As SNET was not a wholly-owned subsidiary of AT&T at the time, it was not affected directly by the limitations imposed on the Bell Operating Companies (BOCs) and it was able to enter the long distance business. In January 1998, SBC Communications, Inc. (Southwestern Bell) announced an agreement to acquire SNET. According to company practice, SNET is pronounced S-N-E-T, with each letter being spelled out. See also MFJ and SBC.

SNI 1. Subscriber Network Interface. SMDS term describing generic access to a SMDS network over a dedicated circuit which can be DS-0, DS1 or DS3.

2. Systems Network Interconnection.

SNIA Storage Networking Industry Association.

snickelway A narrow footpath in a medieval city where merchants sold their wares and villagers came to shop. The company which tried

sniff and hold A term specific to the Bluetooth wireless specification. Sniff and Hold mode is the mode in which synchronized devices on a piconet (little-bitty network) can enter battery power-saving modes (i.e., "sleep" modes) during periods when device activity is lowered. See also Bluetooth and Piconet.

sniffer 1. In security terms, a sniffer is someone who is paid to twist doorknobs for a living, to see which are safely locked and which are left dangerously unsecured. So it is in the telecom world, a sniffer is a man who practices the craft of intrusion detection. You employ him/her to check out your network. Is it safe? Can he get in, using all the strange, wonderful and creative tools these people have employed for nefarious purposes?

2. Sniffer is a registered trademark owned by Network General Corporation. The Sniffer Network Analyzer is a member of the family of Network General products, that monitors traffic on a network and reports on problems on the network. The company is sensitive about the word Sniffer being used as a generic term for network monitoring. If you do use it as a generic term, their VP and General Counsel, Scott C. Neely, will write you a letter lecturing you about trademarks, etc.

3. Sadly Scott's worst nightmare has came true. Sniffer is now a generic term used to describe a piece of software which runs on a 802.11(b) equipped PC. The software snifs out (as in smells out) the existence of an 802.11(b) Wi-Fi hotspot -a small geographic area which will receive and transit data according to the 802.11(b) protocol. Starbucks coffee shops typically have Wi-Fi wireless networks. Sniffers will sniff them out.

sniglet Any word that does not appear in the dictionary, but should. A term invented by Rich Hall of the HBO Television program "Not Necessarily The News". An example of a

sniglet is the definition of "Hozone." It's obviously where socks go when they don't come back from the laundry.

sniping software Software that allows a high bid to entered in an eBay auction a millisecond or so before the auction ends -thus ensuring that you win the bidding for that glorious thing -without payhing too much and being outbid by someone else.

snips Heavy duty scissors, used for cutting metal.

SNMP 1. Signaling Network Management Protocol.

2. Simple Network Management Protocol. SNMP is the most common method by which network management applications can query a management agent using a supported MIB (Management Information Base). SNMP operates at the OSI Application layer. The IP-based SNMP is the basis of most network management software, to the extent that today the phrase "managed device" implies SNMP compliance. RMON and RMON-2 use SNMP as their method of accessing device MIB information. In 1988, the Department of Defense and commercial TCP/IP implementors designed a network management architecture for the needs of the average Internet (a collection of disparate networks joined together with bridges or routers). Although SNMP was designed as the TCP's stack network management protocol, it can now manage virtually any network type and has been extended to include non-TCP devices such as 802.1 Ethernet bridges. SNMP is widely deployed in TCP/IP (Transmission Control Protocol/Internet Protocol) networks, but actual transport independence means it is not limited to TCP/IP. SNMP has been implemented over Ethernet as well as OSI transports. SNMP became a TCP/IP standard protocol in May 1990. SNMP operates on top of the Internet Protocol, and is similar in concept to IBM's NetView and ISO's CMIP. In 1991, Microsoft started referring to SNMP as SubNetwork Access Protocol. In November of 1993 Cisco Systems announced that its internetwork routers will support version 2 of the Simple Network Management Protocol (SMNP) and it has licensed SNMP v2 developed by SNMP Research, Inc. of Knoxville, TN. See CMIP, MIB, RMON, RMON-2 SNMP-2.

SNMP-2 A major revision of the original SNMP, SNMP-2 is currently a proposed standard covered by RFC 1902 through RFC 1908. The SNMP-2 MIB -a superset of MIB-2 -addresses many performance, security, and manager-to-manager communication concerns about SNMP. For example, SNMP-2 supports encryption of management passwords. See MIB, MIB-2, SNMP.

snob In England, neighborhood lists would show -next to each name -the craft and rank of the person. This is why, next to the names of simple burghers the words "sine nobilitate" (without nobility) would appear. Often these words were abbreviated to "s.nob." which is how our modern word "snob" made its appearance.

snoop See Berkeley Snoop Protocol, Snooping and Snoopware.

snooping Looking into a packet to obtain information. See also Snoopware.

Snoopware Snoopware is software which records all the keystrokes made on your PC. Snoopware can be software which your company installs to do things like monitoring your employees' Web browsing and e-mail traffic. Snoopware can also be software which law enforcement agencies sneak into your computer to capture all your keystrokes and thus learn if you're the evil one, or not. Snoopware has wonderfully named software that includes names such Disk Tracey, Spector, LittleBrother, InternetWatchDog, NetSnitch and Snoopware. There are variations of snoopware. Spector, for example, takes snapshots of a PC's display screen at specified intervals during the day. Later, systems administrators or supervisors can replay the photo sequence and identify every web site visited, every application loaded and every e-mail chat.

Some nifty gumshoe programs like Webroot's WinGuardian also provide a search function through which suspicious bosses can fish for specific hot items -such as job application, or Curriculum Vitae -an employee might have typed into the computer during the day. Other software such as SurfCONTROL's LittleBrother and SuperScout software, even filter Internet traffic to block non-work-related instant messages and intercept attempts to access forbidden web sites. After nabbing someone goofing off on the Web at work, employers in some European countries could use this as reason to legally dismiss an employee since evidence obtained in this way is admissible in court. Sacked in 1998 for using her office computer to conduct some 150 Internet searches for her next vacation trip, an English Information Technology manager took her employer to court for unfair dismissal. The industrial tribunal that tried the case ruled in favor of the company and found the employee guilty of misconduct. In the U.S. mass dismissals of cyberslackers are already a reality: 40 Xerox employees lost their jobs last year after a snooping program called WebSense -currently used by thousands of companies worldwide -detected their office surfing activities. Some "consultants" have estimates billions of dollars lost by companies to web surfing by employees on company time.

snortal A contraction of "snort" and "portal," a snortal is a Web site from which you will be able to order refillable scent cartridges so you snort your Internet experience. In other words, you can make it smell. The concept is based on that of "Smell-O-Vision," an dubious enhancement of the moving picture experience. The snortal is the invention of Digiscents, which has introduced"iSmell," as in "internet Smell". Digiscents indexes various scents according their chemical compositions, and their place in the scent spectrum. The desired scent index is then described in a small file, which can be attached to any kind of Web content, interactive game, e-mail, movie, music, or other IP (Internet Protocol) file. The intended scent is recreated by your computer through personal scent synthesizers, which plug into your computer much like audio speakers. The personal scent synthesizers connect to your computer through your computer's serial port or USB (Universal Serial Bus) port, and are locally powered. They emit smells on command, which might take the form of a mouse click or a timed response triggered by a DVD or CD-ROM. The scents are recreated by an indexed combination of over 100 basic scents that are selected from a palette of scented oils, with the scent cartridges being refillable. Refills will be available from www.snortal.com, if this smelly idea ever takes off. See also Smell-O-Vision.

snow Random noise or interference appearing in a video picture as white specs. In short, video noise.

snowshoe A snowshoe shaped gadget that is used to maintain a minimum bend radius for installed fiber optic cable.

SNPA Subnetwork Point of Attachment.

SNPP Simple Network Paging Protocol. An IETF (Internet Engineering Task Force) protocol defined in RFCs 1645 (v2, which is one-way) and 1861 (v3, which is two-way, supporting "acknowledgement paging"), SNPP is used for sending messages (numeric and alphanumeric) between the Internet and wireless pagers through the use of a shim. The pagers must support the TAP/IXO (TAP is the Telocator Alphanumeric input Protocol, which was based on the IXO protocol, named for the company that allegedly invented it) protocol. Based on the philosophy of other Internet protocols, SNPP is based on a series of commands and replies between the client (i.e., pager) and the server. SNPP offers many advantages over SMTP (Simple Mail Transfer Protocol), including the ability to store and forward messages in the event that the client pager is temporarily out of touch. SNPP is widely used, and is based on TAP/IXO. Plans are to replace SNPP with TME (Telocator Message Entry protocol). See also IXO, Shim, and TAP.

SNR 1. Saved Number Redial. A phone system memory feature that allows the user to store a number for as long as it is useful, as opposed to other numbers that are stored more permanently. Some phones have two buttons -one for redial and one for SNR. Redial will dial the last phone number you called. Saved Number Redial will dial one you dialed earlier and chose to save because you're going to call it back, e.g. an airline that's always busy.

2. See Signal-to-Noise Ratio.

snuggling Snuggling is a method used by operators of surveillance equipment. A surveillance equipment operator, when building radio transmitters, will select a transmitter frequency close to that of a nearby high powered transmitter, usually, a commercial radio station. Most ordinary radio gear will automatically tune into the stronger of the two signals. The operator must use a specially modified receiver capable of detecting and isolating the weaker signal. "Snuggling" will be difficult to detect by low quality RF receivers when making a countermeasure sweep.

SO 1. Serving Office. Central office where IXC (IntereXchange Carrier) has POP (Point Of Presence).

2. Hard service cord, same construction as type S except oil-resistant jacket, 600V, 600xAFC to 900xAFC.

SOA Service Oriented Architecture. SOA is an architectural style whose goal is to achieve loose coupling among interacting software agents. A service is a unit of work done by a service provider to achieve desired end results for a service consumer. Both provider and consumer are roles played by software agents on behalf of their owners.

soak A means of uncovering problems in software and hardware by running them under operating conditions while they are closely supervised by their developers. See also Alarm Soaking.

soak period 0x06Alarm soaking is the allowing of an error condition to persist before action is taken. Alarm holdoff is another term for it. The term "soak period" is used for the holdoff period before some action is taken.

SOAP 1. Simple Object Access Protocol. An XML-based (Extensible Markup Language) protocol that defines a framework for passing messages between systems over the Internet. SOAP typically is used for establishing RPCs (Remote Procedure Calls) between Web servers and clients, for the purposes of initiating and controlling processes on remote or distributed computer systems. SOAP simplifies the process of packaging the application data associated with the RPC, and sending it across the Internet. XML tags the content to ensure that both sender and recipient can easily interpret message contents, and SOAP provides specific instructions that allow a network node to remotely invoke application objects and return results. The data typically is encapsulated into HTTP (HyperText Transport Protocol), although SMTP and other transport protocols also can be used. Because SOAP is based on XML, it's compatible with all programming models and allows businesses to exchange data with each other over the Internet.

Developed by Microsoft as the successor protocol to its proprietary DCOM (Distributed Component Object Model), SOAP is an open protocol that has the potential to become a de facto standard if a sufficient number of programmers adopt it. Microsoft sees it as leveling the playing field between Windows and development strategies based on Java. Instead of being forced to choose one model, companies will be free to select whichever is best suited to solving the problem at hand, Microsoft argues. At issue is the slugfest between Microsoft and its competitors over the programming models software developers use. Microsoft has its own programming model based on the Windows operating system, called the Component Object Model (COM). Its competitors support Enterprise JavaBeans (EJBs) and Common Object Request Broker Architecture (CORBA), two programming models that are tightly integrated with each other. See also COM, CORBA, DCOM, EJB, HTTP, RPC, SMTP, and XML.

2. Soap was considered a frivolous luxury of the British aristocracy from the early 1700s until 1862, and there was a tax on those who used it in England. Some wags believe it's still a luxury in England, used rarely and sparingly.

soap opera Endless daytime serials are called soap operas because their original sponsors were manufacturers of soap. In fact, in the beginning, the soap companies wrote the scripts, hired the actors and simply paid the TV station money for it broadcasting the show.

SOC System-On-a-Chip. If you can put digital signal processing, microprocessing, network, memory-and maybe even some analog-functionality on one chip, you can dramatically lower power, cost, and real estate. And increase performance. Telecosm companies like Broadcom, Texas Instruments, Analog Devices and National Semiconductor are leaders at integrating components for the cable, DSL, LAN and mobile phone markets. Altera has just introduced field programmable gate arrays with up to 114,000 logic elements, 28 DSP blocks, and 10 megabits of RAM, all on a single chip.

SoC places the contents of many integrated circuits -microprocessors, memory, logic and embedded software -onto a single semiconductor chip. In more technical language, a SoC is a silicon integrated circuit which combines generic functions (e.g., microcontrollers, UARTs, memory, FIFOs, and other analog and digital logic functions) with custom design elements to create a device that contains all major elements of a system on one integrated chip. This is one method of increasing design productivity. The SoC designer collects and integrates pre-defined (and pre-tested) components similar to the way hardware designers collect and interconnect integrated circuits on a circuit board design. The final implementation of an SOC may be in an ASIC or FPGA. See ASIC and FPGA. Thanks to Ken Coffman for help on this definition.

social Another name for your social security number, as in: "What's your social?"

social computing A term that emerged in the summer of 1993. Defined by Peter Lewis in the New York Times of September 19, 1993, social computing is a "communications-rich brew," which is "expected to create new ways for businesses and their customers to communicate, over new types of wireless as well as wired pathways, using new types of computers called personal communicators." According to Peter Lewis, "The rise of social computing is expected to shift the emphasis of computing devices away from simple number crunching and data base management to wider-ranging forms of business communications...Where client server broke away from mainframe-based systems and distributed computing power to everyone in the organization, social computing goes the next step and extends the distribution of computing power to a company's customers." See social networking.

social contract An arrangement between the local telephone company and its local regulatory authority whereby the telephone company's services are detariffed, but cannot be priced at less than cost. Quality of service standards apply.

social engineering Gaining privileged information about a computer system (such as a password) by skillful lying -usually via a phone call. Often done by impersonating an authorized user.

social networking Imagine a web site with a big database of information about people and their interests. The The goal is to connect people with like interests. One site defines itself as "an online community that connects people through a network of trusted friends. We are committed to providing an online meeting place where people can social-ize, make new acquaintances and find others who share their interests." An online social network lets people create a personal profile, describing their interests, hobbies, and activi-ties. These sites also let people post photos, blog entries, and other content, such as music, and receive posted messages from friends who visit the site. An online social network lets people control who has access to their personal space on the site, limiting it to a circle of friends or opening it up to a larger public. Some social network sites have been used for pornography, setting up orgies (I have examples) and prevailing on young

children. Most sites try to censor what can or cannot be posted. You can't join some Inter-net-based social networks. You have to be invited in. One such site defines this as we're "an organically growing network of trusted friends. That way we won't grow too large, too quickly and everyone will have at least one person to vouch for them." Examples of online social networks include MySpace, MSN Spaces, Facebook, Xanga.com, Friendster, Bebo, Hi5, and Orkut. The most popular of all online social networking sites is MySpace, which got bought by Rupert Murdoch of News Corporation, who now is trying out to figure out how to monetize his purchase, i.e. make money on his $580 million purchase.

sock puppet A pseudonym created by a usenet or user group member to second his or her own opinion. In other words another name for yourself, so you can tell everyone online what a genius you and your opinions are.

socket 1. A synonym for a PC port, a socket in is an opening or slot into which some-thing plugs. The socket serves as the physical and electrical interface between the PC and its components (e.g., a processor or motherboard), or peripherals (e.g., a PCMCIA card, a.k.a. a PC card). See also Slot 1, Socket 7, and Socket 370.

2 A technology that serves as the endpoint when computers communicate with each other.

3. An operating system abstraction which provides the capability for application pro-grams to automatically access communications protocols. Developed as part of the early work on TCP/IP.

Socket 370 The specification that addresses the form factor (i.e., physical shape and size) and the physical and electrical interface for some of Intel's motherboard processors. Socket 370 is a square socket with 370 pins, or leads. The Socket 370 is designed to accept motherboards designed in a FC-PGA (Flip Chip Pin Grid Array) configuration. See also FC-PGA.

Socket 7 The specification that addresses the form factor (i.e., physical shape and size) and the physical and electrical interface for some of Intel's Pentium processors, and com-patibles manufactured by others. Socket 7 is the specification for one or more slots on the PC motherboard that accept a daughterboard in the form of a microprocessor cartridge with leads arranged in a 37x37 or 19x19 configuration. Intel is phasing out Socket 7 in favor of Slot 1, although its competitors have announced Slot 1 enhancements. See also Slot 1.

Socket Interface The Sockets Interface, introduced in the early 1980s with the release of Berkeley UNIX, was the first consistent and well-defined application programming interface (API). It is used at the transport layer between Transmission Control Protocol (TCP) or User Datagram Protocol (UDP) and the applications on a system. Since 1980, sockets have been implemented on virtually every platform.

socket number In TCP/IP, the socket number is the joining of the sender's (or receiver's) IP address and the port numbers for the service being used. These two together uniquely identifies the connection in the Internet.

socket services The software layer directly above the hardware that provides a standardized interface to manipulate PCMCIA Cards, sockets and adapters. Socket Services is a BIOS level software interface that provides a method for accessing the PCMCIA slots of a computer, desktop or laptop (but most typically a laptop). Ideally, socket services software should be integrated into the notebook's BIOS, but few manufacturers have done so to date. For PCMCIA cards to operate correctly you also need Card Services, which is (not are) a software management interface that allows the allocation of system resources (such as memory and interrupts) automatically once the Socket Services detects that a PC Card has been inserted. You can, however, happily operate PCMCIA cards in your laptop without using socket and card services. You simply load the correct device drivers for those cards. Such drivers always come with PCMCIA cards when you buy the cards. You will, however, have to load new drivers every time you change cards and allocate the correct memory exclusions. You will have to reboot if you disconnect your network card. Theoretically, with socket and card services loaded, you do not have to reboot every time you change cards. My experience is that this works, except with network cards, which cannot be hotswapped. See PCMCIA.

sockets An application program interface (API) for communications between a user application program and TCP/IP. See Socket and Socket Number.

SOCKS A circuit-level security technology developed by David Koblas in 1990 and since made publicly available by the IETF (Internet Engineering Task Force. SOCKSv5, the current version, provides security in a client/server environment, running at the Session Layer, Layer 5 of the OSI Reference Model. SOCKSv5 supports multiple means of authentication, negotiated between client and server over a virtual circuit, on a session-by-session basis. SOCKSv5 also supports the transfer of UDP data as a stream, avoiding the need to treat each packet of UDP data as an independent message. SOCKSv5 also allows protocol filtering, which offers enhanced access control on a protocol-specific basis. For example, a network administrator can add a SMTP (Simple Mail Transfer Protocol) filter command to prevent hackers from extracting from a mail message information such as a mail alias. Reference implementations exist for most UNIX platforms, as well as Windows NT. The cross-platform nature of SOCKS offers portability to Macintosh and other operating systems and browsers. According to Network World Magazine, September 27, 1999, "the latest version of SOCKSv5 offers network managers an easier way to run videoconferencing and video and audio streaming through firewalls, which has been difficult and time-consuming. Socksv5 does this by providing a single and powerful method of authenticating users and managing security policies for all Internet applications, including multimedia." SOCKSv5 also interoperates on top of IPv4, IPsec, PPTP, L2TP and other lower-level protocols.

SOF Start Of File.

soft benefit A benefit, often a technology-enabled benefit, that is hard to quantify financially either in dollars or as a tangible return on investment. Examples of soft benefit: improved information flow, improved information sharing, improvement in employee mo-rale, reduction of clutter.

soft copy 1. A copy of a file or program which resides on magnetic medium, such as a floppy disk, or any form that is not a hard copy -which is paper.

2. Old legacy systems term reapplied to distributed computing in which reports are created on-screen from data residing within different applications.

Soft Decision See SISO.

soft ferrite Ferrite that is magnetized only while exposed to a magnetic field. Used to make cores for inductors, transformers, and other electronic components. See Barium Ferrite, Ferrite and Hard Ferrite.

soft font A font,usually provided by a font vendor, that must be installed on your computer and sent to the printer before text formatted in that font can be printed. Also known as downloadable font.

soft handoff 1. A cellular radio term. A soft handoff is a handoff between cell sites that involves first making the connection with the new cell site before breaking the con-nection with the previous cell site. A hard handoff, or "break and make" handoff, is not noticeable in a voice conversation, but has disastrous impact on a data communication. Here's a longer explanation: What happens during a Hand-Off Sequence? Hand-off occurs when a call has to be handed from one cell to another as the user moves between cells. In traditional hard hand-off, the connection to the current cell is broken and then the con-nection to the new cell is made. In CDMA technology, however, it is possible to make the connection to the new cell before leaving the current cell since all cells in CDMA use the same frequency. This is known as a "make-before-break," or "soft hand-off." Soft hand-off requires less power, which reduces interference and increases capacity.

2. A satellite term. The process of transferring a circuit from one beam or satellite to another without interruption of the call.

Soft Input-Soft Output See SISO.

soft key There are three types of keys on a telephone: hard, programmable and soft. HARD keys are those which do one thing and one thing only, e.g. the touchtone buttons 1, 2, 3, * and # etc. PROGRAMMABLE keys are those which you can program to do produce a bunch of tones. Those tones might be "dial mother." They might be "transfer this call to my home for the evening." They might be "go into data mode, dial my distant computer, log in and put in my password." SOFT keys are the most interesting. They are unmarked buttons which sit below or above on the side of a screen. They derive their meaning from what's presently on the screen. And what's on the screen will change based on where the call is at that moment -in a conference call, about to set up a conference call, about to go into voice mail, into voice mail, programming a speed dial number, etc.

soft launch An unofficial, generally by invitation-only, launch of a product, a net-work or network service before the official launch with its attendant media coverage and hoopla.

soft modem Soft modem is short for software modem. The modem consists of a small phoneline interface card and software that uses the PC's main CPU (e.g. a Pentium) for the main communications tasks. Such modem are cheaper than ones based on hardware, ones we're all used to. But they can be slower than one based mostly on hardware -depending on what else the main processor is doing. Many laptops presently have software modems. They're cheaper than a dedicated hardware-provided modem. That's why laptop makers include them now.

soft sectored A floppy disk whose sector boundaries are marked with records instead of holes. Soft-sectored disks have typically one hole. Hard sectored disks have many holes. A soft-sectored disk won't work on disk drives which use hard-sectored drives and vice versa -even though the disks might be the same size. Soft-sectored disks are now much more common.

soft selectable/soft strappable Refers to an option that is controllable through software rather than hardware.

softening of the perimeter In the days prior to wireless and remote network access, an organization was able to achieve some semblance of network security by guarding the network's perimeter with firewalls and demilitarized zones. Now, with PDAs, BlackBerries, wireless LANs, employees with VPN access from home, and traveling workers accessing the organization's network from hotel rooms and Wi-Fi hot-spots, the network perimeter is said to be softening, especially for organizations whose security policies and technologies have not kept up with the changes in network access.

softkey See Soft Key.

softkey mapping See ADSI.

softphone A softphone is a piece of software that, when installed on your PC or laptop, looks like a real telephone, with touchtone pad, line buttons, speed dial keys or directory. To use a softphone, you will need five things: A broadband connection to the Internet preferably over Ethernet, an account with a VoIP service provider, the softphone software from the VoIP service provider, a microphone and a set of speakers. The most famous softphone is from Skype. That software and Skype's service allows you to call other Skype users around the world for free and to normal telephones for a small per minute charge. I use Skype on my laptop which came, like most laptops, with its own built-in speakers and built-in microphone. Softphones are great for business travelers, especially those staying in hotels and homes with broadband connections. Softphones should not be confused with VoIP (Voice over Internet Protocol) calling. That service, which also requires a broadband Internet connection, usually comes with its own small box, one end of which is plugged into your Ethernet connnection to the Internet and the other end into a normal analog phone. To place a VoIP call is really no different to placing a normal circuit switched analog or digital call. You pick up the phone and dial. To place a softphone call, you to speak to your computer. See softswitch.

softswitch Someone described a "softswitch" as basically anything you want it to be. But it's not. See media gateway, media gateway controller and SoftSwitch Consortium.

Softswitch Consortium An international organization for global cooperation and coordination of internetworking technologies in the field of Internet-based, real-time, interactive communications and applications. This consortium was formed in May 1999 to promote the formation of interoperability standards in support of IP-based voice and multimedia communications. See media gateway and www.softswitch.org.

software The detailed instructions to operate a computer, differentiating instructions (i.e., the program) from the hardware. The term was coined by John W. Tukey, who first used it in a 1958 article for American Mathematical Monthly. See Firmware, Hardware, Program, and Tukey.

software data compression As an ISDN term, it means the ability to compress data before it arrives at the serial port of the ISDN terminal adapter. Can improve performance by as much as 400%

Software Defined Network SDN. Generically, a software defined network refers to a virtual private network. Specifically, it refers to AT&T's Software Defined Network Service, which was introduced in 1985 for AT&T's largest customers and provided only dedicated access services. In 1989, AT&T extended its SDN Network to switched access. Currently, SDN is the most commonly resold of all long distance services. The AT&T Software Defined Network Service Description of July 1986 describes SDN as a service developed for multi-location businesses which allows network managers to tailor their network to their own specific communications needs. Call processing information is stored in a database that is accessed during a call. Calls are transferred over AT&T facilities to either a location that is a dedicated part of the network (for "on-net calling") or to non-dedicated facilities that are part of the network ("off-net" calling). Any company location can become part of a network through SDN dedicated access lines to an AT&T SDN serving office. Here is further

explanation of SDN by Siemens Information Systems, whose Saturn PBX has the capability to interface to and mesh neatly with an AT&T SDN network. Here is Siemens' explanation (it's good): When a company establishes an SDN, each phone on the network has a unique seven digit number. This number may or may not be the same as the Listed Directory Number (LDN). When a call is placed at one PBX, it is sent over a dedicated access line to the long distance network. The call is received by the SDN Serving Office and digits are sent via a CCIS (Common Channel Interoffice Signaling) link to the Network Control Point (NCP) for analysis and routing. There is one NCP per SDN network. The NCP contains the unique database for the company using the SDN. The NCP analyzes the digits received against the database, determines whether it is an on-net or off-net call, and sets up the path over which the call will be rerouted on the long distance network. If it is an on-net call, the NCP translates the unique seven digit locator code to one that will be recognized by the AT&T network, sends the call over the network to another SDN Serving Office, and completes the call over dedicated lines to the PBX being called. There is a discount for any call which remains on-net throughout. If the call is off-net, the digits dialed are sent over the long distance network to a Central Office that is not part of the SDN. The call is then completed to the corporate PBX over DID (Direct Inward Dial) lines. Since the introduction of route selection features in PBXs, the caller now has the ability to dial the 10-digit LDN (Long Distance Number) and have the PBX make the decisions on the route and make the decisions on any digit translation or deletion that is necessary to route the call. The CCIS network which carries SDN call signaling is a packet switching network operating at 4,800 bits per second. It will eventually be replaced by ITU 7 Signaling, a more powerful internationally-accepted signaling system. See Signaling System 7.

Software Defined Radio SDR. As defined by the FCC (Federal Communications Commission), SDR is a generation of radio equipment that can be reprogrammed quickly to transmit and receive on any frequency within a wide range of frequencies, and using virtually any transmission format and any set of standards. The FCC began hearings on SDR in March 2000, with the thought that SDR could promote more efficient use of spectrum, expand access to broadband wireless communications, and increase competition among service providers. SDR is promoted by the Software Defined Forum. www.mmitsforum.org. See also software radio.

software engineering A broadly defined discipline that integrates the many aspects of programming, from writing code to meeting budgets, to produce affordable software that works.

software interfaces Language between programs which allows one program to call upon another for assistance in processing.

software metering Software that monitors the use of applications -such word processing, spreadsheets, databases, etc. All metering programs tend to take advantage of a software feature called "concurrency." Concurrency means that a company need only buy as many licenses to a program as it has people using the program at one time -concurrent users, in other words.

software modem Also called soft modem. Basically it's a modem which consists of a small phoneline interface card and software that uses the PC's main CPU (e.g. a Pentium) for its main communications tasks. Such modem is typically cheaper than one based on dedicated hardware (chips), i.e. ones we're all used to.

software only video playback A multimedia term. Video software playback displays a stream of video without any specialized chips or boards. The playback is done through a software application. The video is usually compressed to minimize the storage space required.

Software MPEG Playback See MPEG.

software radio See Smart Radio.

software supervision "Answer Supervision" is knowing when the person at the other end answers the phone. The main reason for wanting to know this is so that a phone company can start billing the call. There are two ways of doing answer supervision. You can get it from the nation's phone system, i.e. the distant office signals back across the country when the called person picks up the phone. Or you can fake it with "software supervision." Essentially this means there's electronics which "listens" to the call. If it "hears" voice or something like voice, it assumes the conversation has started and it's time to start billing the call. Software supervision is not accurate. But when you haven't got access to real answer supervision (for whatever reason) it's better than the previous alternative, which was "timeout." In timeout answer supervision, the carrier simply assumed the call had begun after a certain number of seconds -like 30 -had elapsed with the calling person hanging up. This meant, for example, if you called Grandma and she wasn't there, but you left it ringing, 'cause you knew she took time to answer the phone, then you'd be charged for the call

-even though she didn't answer phone!

SOH Start Of Header. A transmission control character used as the first character of the heading of an information message.

2. Section Overhead. SONET frames include 9 octets of SOH for maintenance of SONET links. SOH information includes transport status, messages and alarm indication. Without SOH data, you are SOL.

SOI See Service Order Image.

SOHO 1. Small Office Home Office. An acronym for a new market which is part work at home, part commute from home. It's getting larger as companies downsize and their workers become "consultants" or small businesspeople. It's getting larger as companies close their distant sales offices ask their salespeople to work out of home.

2. In New York City, there's an area called SOHO. It stands for South Of HOuston Street. See Silicon Alley.

SOL 1. Serial over LAN. A protocol for packetizing and transmitting serial data over a LAN. A typical use of serial over LAN (SOL) is to enable a remote serial console to access, over an IP network, the text-based interfaces for a server's BIOS and system utilities.

2. Shit Out Of Luck.

solar wind A stream of ionized hydrogen and helium that radiates outward from the Sun and exerts a forces on other heavenly bodies including orbiting satellites; the tails of comets always point away from the Sun because of the solar wind.

SOLAS Safety of Life at Sea, an international convention of the International Maritime Organization; SOLAS involves all aspects of ship safety and equipment affecting safety.

solder 1. An alloy of lead and tin having a low melting point. 2. To unite or join by solder.

solenoid A coil consisting of a number of turns in cylindrical form.

solid conductor A conductor consisting of a single wire.

solid state Any semiconductor device that controls electrons, electric fields and magnetic fields in a solid material -and typically has no moving parts.

solid state applications interface bridge Solid's State Systems' PBX to external computer link. See Open Application Interface.

solid state transfer An uninterruptible power supply (UPS) definition. Many UPSes use "mechanical relays" to switch from AC power to battery power. This technology requires 12 milliseconds switching time -slow enough to cause data loss. Solid state switching is faster and eliminates this type of problem.

solitons Solitons are light pulses that maintain their shape over long distances. The origin of the term "soliton" dates to the 19th century, when Scottish engineer John Scott Russell discovered solitary waves while he was conducting experiments to develop a more efficient design for canal boats. In early 1993, scientists at AT&T Bell Labs announced that they had transmitted error-free solitons at 10 gigabits per second over one channel and at 20 gigabits per second more than 13,000 kilometers using two channels. To accomplish this feat, they used sliding-frequency guiding filters. See also Rare Earth Doping.

On April 27, 1998, MCI (before it became Worldcom) said that it was using soliton technology to carry voice and data traffic triple the distance without costly gegeneration equipment. MCI announced a trial using solitons, a technology based on a scientific theory formulated back in 1834 and adapted for modern use by Pirelli Cables and Systems. MCI officials say the technology has the potential to reduce transmission costs by as much as 20 percent. MCI demonstrated the ability to send a single stream of data traffic at 10 gigabits per second more than 900 kilometers over existing installed fiber without regenerators -triple the distance that 10 Gbps traffic can be transported today. MCI also successfully transmitted soliton data streams using dense wavelength-division multiplexing technology, carrying four data streams at 10 Gbps each, traveling more than 450 kilometers without regenerators.

Here is MCI's explanation of how the Soliton technology works: A Soliton is a type of wave or, in the case of optical fiber, a narrow pulse of light that retains its shape as it travels long distances along the fiber. The soliton's ability to keep its shape helps to overcome the problem of lightwave dispersion, and the consequent loss of data integrity, as the data-carrying lightwave travels over long distances. Modern soliton technology is based on a phenomenon first documented in 1834 by a Scottish engineer named John Scott Russell who, while watching a boat being drawn along a canal by a pair of horses, noticed that when the boat stopped suddenly, the wave of water created by the bow continued forward at great velocity without losing speed or shape. Russell was convinced that the soliton was an important scientific discovery. But his theory wasn't fully borne out until the 1960s when scientists began to learn that many phenomena in such fields as physics, electronics

and biology can be explained by solitons. The key elements of Pirelli's 10 Gbps wavelength division multiplexed systems are the soliton converters, which transmit and receive traffic. The Soliton transmitter generates a pulse with the proper shape and power to allow for the transmission of data and voice traffic over very long distances without electronic regenerators.

solution IBM-speak to solve, as in "We've got to solution this problem if we're going to make the sale."

solution assembler Another name for a system integrator. A vendor who puts together a collection of products which purport to be the solution/s to your IS problem.

SOMO Acronym for Small Office Medium Office.

SON Service Order Number. The SON is the number issued by the local exchange carrier to confirm the order for the ISDN service. It provides a matching number for cross referencing the order to the phone company.

SONAR 1. SOund Navigation And Ranging. A system for underwater detection and location of objects through the use of acoustics. An active SONAR acoustical transmitter emits an acoustical "ping" signal at a frequency of 3500 Hz (Hertz). If the signal strikes a solid object, some of the acoustical energy is reflected in the form of an echo. Since the acoustical signal travels at approximately 1500 meters per second, the transmitter/receiver can calculate the distance of the solid object that caused the return. Further, the strength of the return ping can be calculated to provide information relative to the size and physical composition of the target object. As this acoustical echolocation technique is overt, the target object may be able to sense that it is being "pinged." Passive SONAR devices simply listen for acoustical signals, thereby remaining covert, which can be a decided advantage for an attack submarine, destroyer or other warship. Fish finders are active SONAR devices, but the fish don't seem to react to the fact that they have been "pinged." See also Ping.

2. A service that, when presented with a list of IP (Internet Protocol) addresses, attempts to order that list according to the proximity from the SONAR server. SONAR can be of value in assisting networked applications to make reasonable choices between alternative hosts in consideration of their proximity, as a "nearby" application server offers better service than a "distant" one. SONAR does not attempt to gauge the relative service levels offered by networked applications at different addresses in terms of round-trip time, hop count or available bandwidth. Rather, SONAR attempts to offer a "good" choice without consuming significant network or host resources in making that choice. Essentially, SONAR ranks host availability according to various speed-of-response service metrics, which can be affected by route distance, hop count, bandwidth availability, and application availability. While SONAR is not widely implemented at this time, it offers the advantage of avoiding the embedding of complex proximity algorithm logic in network clients. "SONAR" is intended as a pun on the "ping" utility that uses ICMP (Internet Control Message Protocol) packets to determine if an IP address has a working (i.e., installed and online) IP protocol stack. See also ICMP and Ping.

SONET Synchronous Optical NETwork. A family of fiber optic transmission rates from 51.84 million bits per second to 39.812 gigabits (billion, or thousand million) per second (and going higher, as we speak), created to provide the flexibility needed to transport many digital signals with different capacities, and to provide a design standard for manufacturers. SONET is an optical interface standard that allows interworking of transmission products from multiple vendors (i.e., mid-span meets). It defines a physical interface, optical line rates known as Optical Carrier (OC) signals, frame format and an OAM&P protocol (Operations, Administration, Maintenance, and Provisioning). The OC signals have their origins in electrical equivalents known as Synchronous Transport Signals (STSs). The base rate is 51.84 Mbps (OC-1/STS-1), which is a DS-3 (specifically, a T-3) payload of 44.736 Mbps, plus a considerable amount of overhead for network management (largely signaling and control) purposes. Higher rates are direct multiples of the base rate. Note that SONET is based in large part on T-carrier. SONET is a TDM (Time Division Multiplexed) technology, therefore, just as is T-carrier.

SONET development began at the suggestion of MCI to the Exchange Carriers Standards Association (ECSA). Bellcore then took over the project, and it ultimately came to rest at the American National Standards Institute (ANSI). Much of the development was carried out by ECSA under the auspices of ANSI. Work started on the SONET standard in the ANSI accredited T1/X1 committee in 1985, and the Phase 1 SONET standard was issued in March 1988. SONET has also been adopted by the ITU-T (International Telecommunications Union-Telecommunications Standardization Sector), previously known as the CCITT. The ITU-T version is known as SDH (Synchronous Digital Hierarchy), which varies slightly and most obviously in terms of the fact that the SDH levels begin at the OC-3 rate of 155

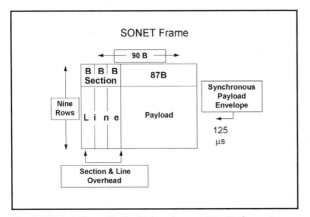

SONET Frame

Mbps. In SDH, the fundamental building blocks are known as STMs (Synchronous Transport Modules) and are equivalent in rate to three SONET STS-1s. SONET is intended to attain the following goals: Multi-vendor interworking, to be cost effective for existing services on an end-to-end basis, to create an infrastructure to support new broadband services and for enhanced operations, administration, maintenance and provisioning (OAM&P). SONET offers many advantages over asynchronous transport including: Opportunity for back-to-back multiplexing, digital cross-connect panels; Easy evolution to broadband transport; Compatibility with evolving operations standards; Enhanced performance monitoring and extension of OAM&P capabilities to end users. SONET/SDH offers the critical advantage of a standard to which manufacturers can build fiber optic gear in order to ensure interconnectivity and (at least some level of) interoperability. Thereby, carriers can safely acquire and deploy multi-vendor networks without being wed to a single manufacturer. This last point was, in fact, the primary impetus for SONET development. SONET transmission equipment interleaves frames of data in simple integer multiples to form a synchronous high speed signal known as a Synchronous Transport Signal (STS). This permits easy access to low speed signals (e.g., DS-0, DS-1, etc.) without multi-stage multiplexing and demultiplexing. The low speed signals are mapped into sub-STS-1 signals called Virtual Tributaries (VTs), or Virtual Containers (VCs) in SDH. SONET uses a 51.84 Mb/s STS-1 signal as the basic building block. Higher rate signals are multiples of STS-1 (e.g. the STS-12/OC-12 signal has a rate of 12 x 51.84 Mb/s or 622.080 Mb/s). The frame format consists of 90 x 9 bytes. The SONET frame format is divided into two main areas: Synchronous Payload Envelope (SPE) and Transport Overhead (TOH). The SPE contains the information being transported by the frame. The TOH supports the OAM&P functions of SONET, and includes a data communication channel that provides an OAM&P communication path between multiple interconnected SONET network elements. The Synchronous Payload Envelope can handle payloads in any of three ways:

1. As a continuous 50.11 Mb/s envelope for carrying asynchronous DS-3, and other payloads requiring up to 50.11 Mb/s capacity in asynchronous (byte invisible) or byte visible format;

2. In a VT (Virtual Tributary) structured envelope to accommodate DS-1, DS-1C, DS-2, European CEPT1, or future VT-based services (see chart below). These signals can have either an asynchronous or byte visible format; and

3. As concatenated payloads to accommodate services requiring more than 50.11 Mb/s capacity. For example, three STS-1 SPEs may be concatenated to transport a broadband ISDN signal of 135 Mb/s. According to AT&T, the main SONET characteristics are: A family of rates at N x 51.84 Mbps; Optical interconnect allowing mid-span meet; intraoffice mixed vendor interconnects; Overhead channels for OAM&P functions and Synchronous networking. SONET rates are

OC Level	Line Rates	Capacity
OC-1	51.84 Mbps	28 DS1s or 1 DS3
OC-3	155.52 Mbps	84 DS1s or 3 DS3s
OC-9	466.56 Mbps	252 DS1s or 9 DS3s
OC-12	622.08 Mbps	336 DS1s or 12 DS3s
OC-18	933.12 Mbps	504 DS1s or 18 DS3s
OC-24	1.244 Gbps	672 DS1s or 24 DS3s
OC-36	1.866 Gbps	1,008 DS1s or 36 DS3s
OC-48	2.488 Gbps	1,344 DS1s or 48 DS3s
OC-96	4.976 Gbps	2,688 DS1s or 96 DS3s
OC-192	9.953 Gbps	5,376 DS1s or 192 DS3s
OC-768	39.812 Gbps	21,504 DS1s or 768 DS3s

In North America, SONET rates have been limited to OC-1 plus those compatible with European SDH. Thus only OC-3, OC-12, OC-48, OC-192, and OC-768 which are equivalent to SDH-1, SDH-4, SDH-16, and SDH-64, and SDH-256, respectively; are standard.

SONET/SDH networks typically are deployed in a physical ring topology, with multiple fibers providing redundancy. In the event that a given fiber suffers a catastrophic failure, one or more other fibers are available. The rings are of two types: Line-Switched and Path-Switched. SONET also may be deployed in a physical linear topology, in which case the system operates as a logical ring.

SONET/SDH has been incredibly successful in the carrier domain, although it lately has been challenged by DWDM (Dense Wavelength Division Multiplexing). The considerable advantages of SONET have been detailed above, and at some length. The criticisms of SONET include its TDM nature, which is considered inappropriate for IP traffic; its bandwidth limitations, even at 40 Gbps; and its high level of overhead, which directly reduces user data payload, although it yields considerable network management capabilities advantages. Perhaps the greatest criticism is SONET's high cost, especially considering that an increase in bandwidth (e.g., OC-48 to OC-192) requires that the transmitting laser diode, the receiving light detector, and all intermediate optical repeaters be upgraded. DWDM is an optical transmission technique that allows multiple light signals operating at different wavelengths (i.e., frequencies of light) to share a single fiber. Thereby for example, eight or more (eight is the point at which DWM becomes DWDM) wavelengths can each operate at 10 Gbps. As a result, DWDM offers higher aggregate speeds than SONET. DWDM also is far less expensive. On the downside, DWDM does not offer the same inherent network management capabilities and does not offer the same level of standards development, which translates into lack of interconnectivity and interoperability between network elements of disparate origin. Further, each wavelength in a DWDM system is, in essence, a separate circuit. Therefore, all traffic riding over that wavelength is transported and switched as a single entity, from point of origin to point of termination. As a result, a wavelength must carry traffic of the same type (e.g., circuit-switched voice, packet voice, IP packet data, ATM, or Frame Relay), with the same QoS (Quality of Service) requirements, originating at the same place, and destined for the same place. All of that means that each wavelength must be filled to capacity, or that there must be enough available wavelengths that capacity can afford to be underutilized. The arguments over SONET vs. DWDM rage, and will continue to do so for many years. Either approach is correct, and even optimal, depending on the applications focus of a given carrier. In fact, SONET and DWDM can, and often will, coexist, with SONET-framed data riding over DWDM wavelengths. That's my view, at least. See also ADM, DWDM, Line-Switched Ring, Path Switched Ring, SONET Interface Layers, SONET Ring, STM, Stratum Level, STS, and WDM.

SONET head A device on the end of a boring machine. Such machine is used to bore holes under highways, rivers and sundry obstructions. The SONET head contains sensors which can help determine what it is about to strike as it moves ahead underground. The SONET head will signal the person operating the boring machine what lies ahead and hopefully, the operator, is sufficiently intelligent to move the boring machine up or down or sideways in order to miss the potential obstruction -which might be anything from a rock to another fibre cable to a high voltage AC power line.

SONET interface layers The SONET standards define four interface layers. Each layer requires the services of all lower-level layers to perform its functions. While conceptually similar to layering within the Open System Interconnection (OSI) reference model, SONET itself corresponds only to the OSI Physical Layer. The SONET interface layers are:

1. Physical Layer: Handles bit transport across the physical medium; primarily responsible for converting STS (electrical) signals to and from OC (optical) signals. Once the signal has been expressed optically, this layer is sometimes referred to as the photonic layer. Electro-optical devices communicate at this layer;

2. Section Layer: Transports STS-N frames and Section Overhead (SOH) across the medium; functions include framing, scrambling, and error monitoring. Section Terminating Equipment (STE) communicate at this layer;

3. Line Layer: Responsible for the reliable transport of the Synchronous Payload Envelope (SPE) (i.e., user data) and Line Overhead (LOH) across the medium; responsibilities include synchronization and multiplexing for the Path Layer and mapping the SPE and LOH into an STS-N frame. An OC-N-to-OC-M multiplexer is an example of Line Terminating Equip-

ment (LTE); and

4. Path Layer: Handles transport of services (e.g., DS-1, DS-3, E-1, or video) between Path Terminal Equipment (PTE); the main function is to map the services and Path Overhead (POH) information into the PTE includes SONET-capable switches with an interface to a non-SONET network, such as a T1-to-SONET multiplexer.

SONET Ring SONET transmission systems ideally are laid out in a physical ring for purposes of redundancy. In practice, the topology often is that of a linear ring, which is linear in its physical appearance, but which operates as a logical ring. See also SONET, Physical Topology, Logical Topology, Line-Switched Ring, and Path-Switched Ring.

SONIA A group of technology vendors which advanced a profile for NCs (Network Computers), also known as "thin clients." SONIA is derived from the names of the members of the group: Sun Microsystems, Oracle Corp., Netscape Communications Corp., IBM, and Apple Computer Inc. See also NC and Thin Client.

Sony Akio Morita, who died in October 1999, named Sony after a combination of the Latin word for "sound" and the English words "sonny boy."

Sony Mini Disc A 2 1/2 inch silvery CD (compact Disc) that can record and play 74 minutes of sounds, almost as much as its five-inch forebear. To record on this disc, a laser momentarily heats a tiny spot on the disk to 400 degrees Fahrenheit, while a magnetic head writes the signal into the heated part of the magnetic layer. To play the disk an optical pickup analyzes the polarity of the light reflected from each spot.

SOP Standard Operating Procedure.

sort To order a collection of records -for example, a telephone directory -in some specified way, say, in alphabetical order. Computers can sort in virtually any way you ask them to. Most companies don't produce sufficient "sorts" on their telephone directories.

sort scheme A call center term. A list of fields that tells the program how to sort a report or a list of records. This can be simple scheme that sorts by only one field or a complex scheme consisting of sorts within sorts.

SOS SOS is the international distress signal. SOS was officially adopted in 1908, and was transmitted on the 500kHz emergency radio wavelength. Contrary to popular belief, "SOS" does not stand for "Save Our Souls," "Save Our Ship," or anything else. Rather, it was adopted simply because of its easy radiation and unmistakable character (... –...) in Morse code telegraphy. In an emergency when sending SOS, it was defined to alternate between the S letter and the O letter continuously. It was sent like this: dididit dahdahdah dididit dahdahdah dididit dahdahdah dididit dahdahdah dididit etc..., continuously with pauses for words. Prior to the adoption of SOS, the generally accepted distress call was "CQD," which was suggested by the Marconi company. "CQ" was the signal used in England for an "all stations" general call on a landline telegraph network. "D" indicated a "distress" call. Therefore, "CQD" means "all stations–distress." Confusion reigned for some time, while the transition was made from SOS to CQD. Also, the English telegraphers favored CQD. In fact, the original distress call from the Titanic (April 15, 1912) was "CQD," sent by First Radio Officer Jack Phillips six times in rapid succession, followed by "MGY," the call letters of the Titanic. Later, and at the suggestion of Harold Bride, Second Radio Officer, Phillips interspersed his "CQD" transmissions with "SOS." That didn't help much. A few extra lifeboats would have helped a great deal more. (Bride survived, while Phillips died of hypothermia.) SOS was officially retired on February 1, 1999 by the International Maritime Organization (IMO). It was replaced by the Global Maritime Distress and Safety System (GMDSS), which uses much improved digital radio technology. See also Morse Code.

soul According to an article published in The Lancet, a respected British medical journal, a Dutch medical research team claims to have proof that humans have a soul that exists independently from the body. The team did a two-year study of near-death experiences, investigating the experiences people clinically dead at the moment claim having had. The doctors reported there were no psychological, neurophysiological, or physiological factors that could account for these experiences. Conclusion? The mind -or soul -survives death.

sound Sound is the sensation perceived by the sense of hearing. Acoustics is the science of sound. Sound is transmitted by the vibration of molecules in the air. The parameters of sound are amplitude and frequency. Amplitude, or volume, is the intensity of the acoustical signal. The louder the sound, the more intensely the molecules in the air vibrate. Frequency, or pitch or tone, is the frequency with which the molecules in the air vibrate. Recording means getting something else (e.g., a diaphragm in a microphone) to vibrate in sympathy, and turning those vibrations into electrical signals (which can then be stored as grooves on a disk, say). A loudspeaker does the reverse: electrical signals are turned into wobbles of a usually cone-shaped piece of material that batters the molecules of the air to recreate the sound. Audible sound spans a huge range of frequencies from around 20 Hertz (vibrations per second) to 20 kHz. Most loudspeakers need two to three cones of different sizes

to cover this range. Moreover, like bells and wine glasses, cones have their own natural frequencies at which they vibrate when hit. They overrespond to electrical signals close to those frequencies. As a result, sound always loses something when being electronically regurgitated, although careful loudspeaker and amplifier design can make up for a lot. See also Speakerphone.

sound files Files on PCs have their own extensions -the three letters which follow the name of the file. For example, a sound file of jungle noises might be called jungle.wav. Here are the typical extensions on sound files of various computers:

> Microsoft Windows -.wav and .wmv
> Apple -.aif and .aiff
> NeXT -.snd
> MIDI -.mid and .nni
> Sound Blaster -.voc
> Intel Indeo Video Movie clips -.avi

sound powered telephone A telephone in which the operating power is derived from the speech input only. See Sound.

sound waves The waves given off by a vibrating body, which are transmitted by an elastic material medium (such as the air) and which can be detected by the ear. See Sound and Sound Files.

source That part of a communications system which transmits information. It typically means (usually LED or laser) used to convert an electrical information-carrying signal into a corresponding optical signal for transmission by an optical waveguide.

source address The part of a message which indicates who sent the message. Just like the top left-hand address on the envelope.

source code A set of instructions, written in a programming language, that must be translated to machine instructions before the program can be run on a computer. The program which finally runs on that computer is known as the object code.

source directory The folder that contains the file or files to be copied or moved.

source explicit forwarding A feature that allows MAC-layer bridges on local area networks to forward packets from only source address specified by the network administrator.

source-quench messages Source-quench messages are Used by receiving devices to help prevent their buffers from overflowing, these messages tell the transmitting device that the receiving device is dropping segments.

source route A hierarchically complete source route. See Source Routing.

source route bridging Method of bridging originated by IBM and popular in Token Ring networks. In an SRB network, the entire route to a destination is predetermined, in real time, prior to the sending of data to the destination. See Source Routing.

source routing A method used by a bridge for moving data between two networks. Originally developed by IBM's token ring network, it relies on information contained within the token to route the packet between the two networks. Since the information in the token is supplied by the computer that sent the data packet, that computer must know on which network the destination computer is located. IBM developed a special protocol that lets computers discover that information. For source routing to work, every computer and every bridge on all networks must support this protocol. If some computers do not use this protocol, they will not receive packets from bridges that use source routing. See Bridge. Compare to Transparent Routing.

In IBM's method of routing local area network data across bridges, IBM's bridges can be configured as either single-route broadcast or all-routes broadcast. The default is single-route broadcast. Single-route broadcasting means that only one designated single-route bridge will pass the packet and only one copy of the packet will arrive at its destination. Single-route broadcast bridges can transmit both single-route and all-routes packets. All-routes broadcasting sends the packet across every possible route in the network, resulting in as many copies of the frame at the destination as there are all-routes broadcasting bridges in the network. All-routes broadcast bridges only pass all-routes broadcast packets.

source-quench messages Used by receiving devices to help prevent their buffers from overflowing, these messages tell the transmitting device that the receiving device is dropping segments. This definition is from the Cisco Certified Network Associate book.

Source Routing Protocol. SRP. See Bridge.

Source Routing Transparent SRT. See Bridge.

source traffic descriptor An ATM term. A set of traffic parameters belonging to the ATM Traffic Descriptor used during the connection set-up to capture the intrinsic traffic characteristics of the connection requested by the source.

source/destination routing A term used in call centers for routing calls based on where they originate or terminate. See also Skills-Based Routing, Calendar Routing and End-of-Shift Routing.

source/sink device A source/sink device is byte-synchronous with a byte orientation. Source devices originate; sink devices terminate.

South Sea Bubble In the 18th century, the promise of foreign trade excited Britain. A company called the South Sea Co. was given a monopoly on trade with the Spanish Empire. Shares of the South Sea Co. soared on London's stock market. According to the Wall Street Journal, trade eventually made fortunes for some people, but not for the South Sea Co. which became a disaster dubbed the South Sea Bubble and helped throw the British economy into a slump.

Southern New England Telephone Corporation SNET. Pronounced S-N-E-T, not SNET. Originally an independent telephone company serving a substantial portion of Connecticut, SNET stock was partially held by AT&T at the time of the Modified Final Judgment (MFJ). As SNET was not a wholly-owned subsidiary of AT&T at the time, it was not affected directly by the limitations imposed on the Bell Operating Companies (BOCs) and it was able to enter the long distance business. In 1998, SBC Corporation (Southwestern Bell) bought SNET.

Southwestern Bell Corporation One of the seven Regional Holding Companies formed at Divestiture. It includes Southwestern Bell Telephone Co., Southwestern Bell Mobile Systems, Southwestern Bell Yellow Pages, Southwestern Bell Telecom and SBC International. In early October, 1994, Southwestern Bell Corporation changed its name to SBC Communications Inc., apparently feeling as though the "Bell" name no longer was of value. (The old Bell Operating Companies (BOCs) fought to retain the exclusive use of the "Bell" name at divestiture. Since that time, most have abandoned it.) In 1996, SBC announced its plan to acquire Pacific Telesis, the holding company for Pacific Bell and Nevada Bell; that merger was completed in April 1997. In January 1998, SBC announced its intent to acquire SNET (Southern New England Telephone). On May 11, 1998, SBC announced its intention to acquire (Whoops, merge with) Ameritech. On January 31, 2005, SBC announced that it would buy AT&T for $16 billion+ and change its name to SBC.

SOW Statement Of Work. When you employ a consultant, he or she should give you a SOW -a list of all the things he or she is going to do for you and how much it's likely to cost you.

SP 1. Support Processor.
2. Sending Program.
3. Signal Present.
4. Signal Processor.
5. Signaling Point.
6. Service Provider.

SPA 1. Software Publishers Association. A not-for-profit organization formed in 1984, SPA is a principal software industry trade association which represents leading publishers, as well as start-ups. SPA supports "companies that develop and publish software applications, components, tools and digital content for use on the desktop, client-server networks and on-line." SPA's MPC (Multimedia PC) Working Group has published several versions of MPC standards over the years, with the current version being MPC3. SPA's headquarters is in Washington, DC; it also has offices in Paris, France. Membership totals over 1,200. www.spa.org. See also MPC3.
2. Shared Printer Access. An ISDN term for the sharing of a printer by multiple users. With a Terminal Adapter (TA) on the PC's serial port and another on the serial printer, a remote worker using ISDN BRI can send a print job over the 16 Kbps D channel. Once the transmission is complete, the call is terminated and the printer is available for another remote worker who is similarly equipped.
3. Secure Password Authentication. Microsoft uses the words in its popular Outlook email software.

space 1. In digital transmission, the space is equated to the zero (0) and the mark is equated to the one (1). In telecommunications, space is the absence of a signal. It is equivalent to a binary "0".
2. Space also stands for Service Creation and Customization.
3. What other people call marketplace, Microsoft calls space. You and I would say, "This product fits into the computer telephony marketplace." Microsoft would say "This product fits in the computer telephony space."

space brokers Companies that provide all the facilities needed to start a 976 or 900 service. Those facilities include offices, computing equipment, voice processing software, telephone lines and numbers.

space control technology A euphemism for antisatellite military weaponry.

See Hyperspectral Imaging.

space diversity Protection of a radio signal by providing a separate antenna located a few feet below the regular antenna on the same tower to assume the load when the regular transmission path on the same tower fades because of rain, a bird flying through it, etc.

space division multiplexing Each distinct signal or message travels over a separate physical path such as its own wire or wire pair within a cable.

space division switching Method for switching circuits in which each connection through the switch takes a physically separate path.

space hold A no traffic line condition where a steady space is transmitted.

space junk See Iridium.

space needle The 605-foot tower was built for the 1962 fair in Seattle.

space parity In data transmission, setting the parity bit so it is always zero.

space segment 1. The part of a satellite system that is in space.
2. This is also the imprecise term used to describe the band of frequency purchased by the satellite customer. The customer can purchase a portion of the bandwidth of a single transponder or the customer can purchase one or more entire transponder bandwidths.

spacecraft A man-made satellite.

Spacecraft Switched Time Division Multiple Access SSTDMA. A method of sharing the capacity of a communications satellite by on-board switching of signals aimed at earth stations.

SPACH SMS (Short Message Service) point-to-point messaging, Paging, and Access response CHannel. A digital wireless term defined by IS-136, the Interim Standard for digital cellular networks employing TDMA (Time Division Multiple Access). SPACH is a logical channel which is part of the FDCCH (Forward Digital Control CHannel) used to send signaling and control information downstream from the cell site to the user terminal equipment. SPACH is further subdivided into three logical subchannels: ARCH, SMSCH, and PCH. See also ARCH, FDCCH, IS-136, PCH, SMSCH and TDMA.

spade lug A metal connector attached to the end of a piece of wire, typically by soldering or by pressure. The metal spade lug is shaped like a "U." The idea is to slide the flat "U" shaped metal piece under a screw and then tighten the screw, thus making a connection. In the old days, all phones came with spade lug connectors. These days, there are other faster, more efficient ways of connecting phones -including modular jacks and punchdown tools.

SPAG 1. Europe's Standards Promotion and Application Group. 2. Standards Promotion and Application Group. A group of European OSI manufacturers which chooses option subsets and publishes these in a "Guide to the Use of Standards" (GUS).

spaghetti code A program written without thought, logic or structure. And whose "logic" is therefore very difficult to follow. Some would say this definition covers most software written today. That's unfair.

spam 1. Hormel's ever popular spiced processed pork product, consisting of leftovers from the processing of pork, plus lots of additives. Spam actually is an acronym for Shoulders, Pork And Ham. The story of SPAM luncheon meat began in 1936. Hormel Foods devised a recipe for a 12-ounce can of spiced ham. Jay C. Hormel was determined to find a brand name with a distinct identity that would set it apart from the competition. The company offered a $100 prize for the best name for the spiced ham product. The winner was Kenneth Daigneau, the actor brother of Hormel Vice President Ralph Daigneau. He created the new word by combining the "sp" from spiced ham with the "am" from ham.
2. Unsolicited commercial email. Unwanted e-mail. The term is derived from Hormel's pink, canned spiced ham that splatters messily when hurled. A milder form of spamming is called crossposting. See Dictionary Attack, Spam Filter, Spambulance Chasing and Spamming.
3. Message posted to numerous Usenet newsgroups to which it has absolutely no relevance (also a verb). See Spamming.

spam filter Software which keeps out spammed email, also called an anti-spam solution. The filter is typically based on certain criteria, like what words the email contains -like Viagra or horney women -or who the sender is. The only way to eliminate 100% of your spam email is to allow only emails from email addresses you approve -one at a time. See Spam.

spam mail See Spam and UBE.

spambulance chasing More than half the states in the U.S. now have local laws against sending spam email to people who don't want it. The laws typically say that you must have a pre-existing business relationship before you can send spam email. The laws spell out financial penalties, some severe. If you receive spam from a spammer you can find, you can hire a lawyer and go after the spammer with a class action suit and collect monies from them, either in court or settle your case "on the courtroom's steps." This business is called spambulance chasing, which is a play on the term ambulance chasing -a term applied to personal injury lawyers who represent people who have been hurt in accidents that can be traced to an organization's neglect.

spamdexing A contraction of "spamming the index." The term came from the days when the AltaVista search engine reigned supreme. Back then, people could improve their position in results by including many words many times on their web site. People who ran pornographic sites put entire dictionaries (with tens of thousands of words) on their Web pages in an attempt to increase their hits. That technique doesn't work with Google, since it employs different searching techniques.

spamming Random indiscriminate posting of items (often advertisements) on computer bulletin boards. The term is derived from a brand of pink, canned meat that splatters messily when hurled. A milder form of spamming is called crossposting.

Spain Spain literally means "the land of rabbits."

span 1. Refers to that portion of a high speed digital system than connects a C.O. (Central Office) to C.O. or terminal office to terminal office.

2. Also called a T-Span Line. A repeated outside plant four-wire, two twisted-pair transmission line.

3. A call center term. The total duration of a schedule from start time to stop time, including all breaks.

4. SPAN. Service Providers Action Network. Australian industry group of service providers.

5. A single computer file that will explode into or comes from many media -for example, several floppy disks or zip disks. See Spanned Archive.

6. In flat cables, the distance from the reference edge of the first conductor to the reference edge of the last conductor (in cables having flat conductors), or the distance between the centers of the first and last conductors (in cables having round conductors), expressed in inches or centimeters.

span line A T-1 link.

span powered In T-1, refers to the application of a varying voltage (+130V to -130V) to the digital cable pairs to maintain a 60mA DC current at each repeater and at the customer premises (this power is generally used for regeneration, loop backs, keep alive signals and alarms).

SPAN session The running of a switched port analyzer session on a switch.

spanned archive A spanned archive is typically a .ZIP or self-extracting .EXE files that was created and spans across multiple removable media, e.g. floppy disks

spanned volume A dynamic volume consisting of disk space on more than one physical disk. You can increase the size of a spanned volume by extending it onto additional dynamic disks. You can create spanned volumes only on dynamic disks. Spanned volumes are not fault tolerant and cannot be mirrored.

spanning explorer packet Follows a statically configured spanning tree when looking for paths in an SRB network. Also known as a limited-route explorer packet or a single-route explorer packet.

spanning tree Spanning Tree (802.1d) is a protocol that is resident on network bridges and switches that allows each device to communicate with all other Spanning Tree Protocol (STP) enabled devices on a port-by-port basis to detect and manage redundant links within a network.

Spanning Tree Algorithm STA. An algorithm, the original version of which was invented by Digital Equipment Corporation, used to prevent logic loops in a bridged network by creating a spanning tree. The algorithm is now documented in the IEEE 802.1d specifications, although the Digital algorithm and the IEEE 802.1d algorithm are not the same, nor are they compatible. When multiple paths exist, says PC Magazine's Frank Derfler, STA lets a bridge use only the most efficient one. If that path fails, STA automatically reconfigures the network to make another path become active, sustaining network operations. This algorithm is used mostly by local bridges; it is not economical for use over leased telephone circuits connecting remote bridges.

Spanning Tree Protocol STP. Inactivation of links between networks so that information packets are channeled along one route and will not search endlessly for a destination. See Bridge and Spanning Tree.

SPAP Secure Password Authentication Protocol.

SPARC Sun Microsystems' RISC-based (Reduced Instruction Set Computer) architecture for microprocessors. SPARC is the basis for Sun's own computer platforms.

SPARCengine Sun Microsystems' standard microprocessors supporting SBus expansion modules and I/O. 50Mhz and above speeds. Up to 512 MB memory on-board.

Sparklies Small stars in a television picture which result from interference.

spare pairs In existing distribution systems, twisted pairs that are not being used and can be used to serve new communications devices. Spare pairs are exactly what they sound like -spare pairs of cables. Best to install as many spares as you can when you initially wire up a building or office. Remember Newton's Rule: You'll always need twice as much cabling as you ever dreamed in your wildest dreams you'd need.

sparing A method of providing redundancy, or fault tolerance, in systems or networks through a design which includes one or more spare elements. Thereby, a spare element (e.g., a system processor or network circuit) is available in the event of a failure. A common approach is "N+1 Sparing," which adds one backup element to the number required at full anticipated load. While sparing is expensive, it is essential in mission-critical applications environments, where system and network resiliency demands 100% up-time. See also N+1 Sparing.

spark An arc of very short duration.

spark gap Terminals or electrodes designed to permit spark discharges to take place across a gap.

spark test A test designed to locate pin-holes in a wire's insulation by application of an electrical potential across the material for a very short period of time while the wire is drawn through an electrode field with one end of the wire grounded.

sparse network 1. A network concept describing an environment in which the intelligence of the End Offices (Central Offices) largely is stripped away in favor of the placement of relatively few centralized computer platforms which perform the majority of call processing. The dumb switches make calls to the centralized processors which consult associated databases, providing the switches with instructions. The concept of a Sparse Network is fundamental to that of the Advanced Intelligent Network (AIN).

2. A network concept involving many fewer End Offices than are currently deployed. Rather than a user gaining access to a local End Office, traffic would be concentrated at local points and shipped to a larger and more capable office serving a much larger geographic area. Advances in transmission technology, namely fiber optics, make this concept feasible as the cost of transmission bandwidth is dropping precipitously, while the cost of switches (particularly intelligent switches) is not. Hence the concentration of switches and switch intelligence.

SPATA SPeech And daTA. Watch for this expression to pick up steam once true integration of voice and data occurs. The expression does not come from the sentence: "Spata to integrate today than tomorrow."

spatial data management A technique which allows users access to information by pointing at picture symbols on the screen.

spatial diversity The use of two or more physically separated antennas for a radio to increase the probability of receiving a good signal.

spatial temporal de-interlacing A techniques that provides the smoothest possible video output by analyzing both pixel positioning and movement and filtering each pixel accordingly.

SPAX Spam Faxes, i.e. unwanted advertising faxes.

SPC 1. Stored Program Control. All phone systems these days are SPCs. There's stored software, which is the program, which controls the computer or microprocessor which in turn controls the operation of the switch. Thus switches are stored program control.

2. Signal Processing Component.

SPC Allocation Service An SCSA definition. A service which allocates SPCs (Signal Processing Components) to Groups.

SPCAS SPC Allocation Service.

SPCS Stored Program Controlled Switch. A digital switch that supports call control, routing, and supplementary services provision under software control. Pretty well switches made after 1970 in North America are SPCSs.

SPCL SPectrum CeLlular error-correction protocol.

SPE 1. Switch Processing Element or Signal Processing Element.

2. Synchronous Payload Envelope. A SONET term describing the envelope which carries the user data, or payload. The SPE comprises 783 octets, organized into 87 columns and 9 rows. Three different payload structures are defined to address different input requirements: 1. Direct-to-STS-1 line rate multiplexing takes 28 DS-1s, 14 DS-1Cs or 7 DS-2s directly into the 51.84 Mbps rate. Each is uniquely transported within the SPE; 2. Asynchronous DS-3 Multiplexing takes a complete asynchronous DS-3 bundle (the output of an M13 for example) into the SPE; 3. Synchronous DS-3 Multiplexing maps a Syntran DS-3 signal to the SPE. See also VT.

3. Semiconductor production equipment. SPE is used on Wall Street.

speak-to-dial The ability to dial a phone number without actually using a dial or keypad. The user simply speaks the digits of the telephone number or speaks a phrase such as "Call Harry."

speak-to-dial number A telephone number that is associated with a word or phrase, enabling the user to simply speak that word of phrase to dial the number.

speaker adaptive Speech recognition which improves with use. See Speech Recognition.

speaker dependent voice recognition Technology capable of recognizing speech from a given user or others who sound like this user after completion of an enrollment procedure. It is not voice verification although it is sometimes confused with this technology.

speaker identification Speaker identification is used to determine the identity of a known speaker. It is accomplished by taking spoken input and searching a database of all known system users for a match. Due to its speaker dependent recognition characteristics, you must first be enrolled as a user prior to using the system. To enroll as a user, an individual is required to speak one or more password phrases which are recorded. These phrases create a reference templates which are stored in the system user database for later use during identification sessions. When in operation, the individual using the system is prompted for a specific password or password phrase. When speaking the prompted password as input it creates a new template. This template is then compared to all reference templates in the system for that particular password. The reference template with the closest match is selected. The uniqueness of each user's voice and the finite number of users of the system makes the identification accuracy quite high. With speaker identification the speaker does not claim to be a particular individual. He or she is identified from a group of common users. For the most part, this technology is used for hands free operation of a system where messages and other information specific to that identified individual are pulled-up for use at that time.

Speaker Independent Voice Recognition SIR or SIVR. Technology capable of recognizing any user's voice without prior training or knowledge of the user. SIR converts speech to accurate and meaningful textual information (typically ASCII). SIR is used to accept input from callers to voice processors where the callers are using rotary dial phones instead of touchtone phones. SIR can substitute for the numbers on the DTMF keypad and can add the benefit of a few basic voice commands, e.g., Yes, No, Help, etc. Because computer processing demands are formidable with speaker independent recognition, accurate speaker independent products are created with limited vocabularies. In contrast, trainable or speaker dependent recognizers can feature larger vocabularies at lower prices. SIR has been slowly gaining acceptance in telephone applications. SIR is increasingly used in automated operator assistance applications. SIR will see increased use as system builders respond to pressures to provide voice processing functions to the enormous rotary phone installed base domestically and abroad.

speaker recognition Having a machine recognize human voice. This is an imprecise term.

speakerphone A telephone which has a speaker and microphone for hands free, two-way conversation. Western Electric (now Lucent Technologies) invented the loudspeaker. Western Electric was a very big name in the sound business prior to the 1956 Consent Decree. Watch the credits at the end of old movies, and you'll see "Sound by Western Electric." The original speakerphone was called the "loud-speaking telephone." Originally, a telephone loudspeaker was a peripheral device which connected to the telephone set. It wasn't until the late 1970s that they were integrated into the telephone to become speakerphones. See also Sound.

spear phishing According to Business Week, the latest twist in cyber-tricks is something called "spear-phishing." Old-style phishing e-mails, purportedly from eBay or banks, were blunt and obvious compared with spear phishing, which appear to originate within your own company. Whereas phish are blasted out to millions, spear phish are highly targeted. Spear phish often appear to be sent by reps from the human resources or info-tech department. MessageLabs provided four versions of a June 15 attack on an unnamed company, in which e-mail signed "Security Department Assistant" asked a worker to update his user name and password, or risk suspension. If successful, hackers gain access to secure networks. In another case, the MyTob virus, hidden in a spear phish, asked users to click a link. Then it dropped in spyware to steal data. In May, Israeli officials busted several companies that used e-mail with attachments disguised as vendor queries to download spyware, then gather intelligence on rivals. Workers, beware: That e-mail may not coming from inside your company.

Here is information from Microsoft: You've probably heard of phishing scams: fraudulent e-mail messages or fake Web sites designed to steal your identity. Scam artists "phish" in an attempt to persuade millions of people to disclose sensitive information. Now there's a new version of an old scam called "spear phishing," a highly targeted e-mail attack that a scammer will send only to people within a small group, such as a company. The e-mail message might appear to be genuine, but if you respond to it, you might put yourself and your employer at risk. Phishers (scammers who perpetrate phishing scams) usually take a broad approach by sending millions of e-mail messages that appear to come from popular banks, online auction houses, and other businesses. These e-mail messages, pop-up windows, and the Web sites they link to appear official enough that they deceive many people into believing that they are legitimate. Unsuspecting people often respond to these requests for credit card numbers, passwords, account information, or other personal and financial data. Spear phishing describes any highly targeted phishing attack. Spear phishers will send e-mail that appears genuine to all the employees or members within a certain company, government agency, organization, or group. The message might look like it comes from your employer, or from a colleague who might send an e-mail message to everyone in the company, such as the head of human resources or the person who manages the computer systems, and could include requests for user names or passwords.

The truth is that the e-mail sender information has been faked or "spoofed." Whereas traditional phishing scams are designed to steal information from individuals, spear phishing scams work to gain access to a company's entire computer system. If you respond with a user name or password, or if you click links or open attachments in a spear phishing e-mail, pop-up window, or Web site, you might become a victim of identity theft and you might put your employer or group at risk.

Four tips to help you avoid spear phishing scams:
- Never reveal personal or financial information in a response to an e-mail request, no matter who appears to have sent it.
- If you receive an e-mail message that appears suspicious, call the person or organization listed in the From line before you respond or open any attached files.
- Never click links in an e-mail message that requests personal or financial information. Enter the Web address into your browser window instead.
- Report any e-mail that you suspect might be a spear phishing campaign within your company.

special access The lease of private, dedicated circuits along the network of an ILEC or CAP, which run from or to the long distance carriers POP. Examples of special access services are telecommunications lines running between POPs of a single long distance carrier, from one long distance carrier POP to the POP of another long distance carrier, or from an end-user to its long distance carrier POP. Special access services do not require the use of switches.

special access code See Service Area Code.

special area code See Service Area Code

special assembly When a phone company builds or engineers something into a circuit or circuits which is not spelled out in their tariffs and which you, the customer, want, for whatever reason. They may sell you the special assembly at a high price, or at a low price. "Special assemblies" were, in the old days, ruses to avoid the strict pricing of the tariff and allow the phone company to drop its rates. These days they could be for anything.

special billing number 1) A phone number assigned to certain customers for billing purposes. It cannot be called. It may be given to an operator as the calling number on an outgoing paid call, or it may be used as a "third number billed" number. It's designed as a measure of security and accounting convenience.

2) A Verizon definition. A Customer Record Information System (CRIS) account number, in the form of a fictitious billing telephone number (BTN), which is used for non-WTN (working telephone number) specific products and services, such as unbundled loops and billable directory listings.

special characters Microsoft calls special characters that ones not found on your computer's keyboard. In Windows, these characters are accessible through Character Map, an application in the Accessories folder.

special distribution A call center term. A half-hourly or quarter-hourly call volume or average handle time distribution created for a day in which calling patterns differ significantly from those normally occurring on that day of the week.

special grade access line An AUTOVON access line specially conditioned, usually by providing amplitude and delay equalization, to give it characteristics suitable for handling special services; e.g., lower signaling rates of 600 to 2400 bits per second.

special grade network trunk A trunk specially conditioned by providing amplitude and delay equalization for the purpose of handling older special services such as "medium-speed" data (600 to 2400 bps); rapidly being made archaic by improvements in transmission facilities.

Special Information Tone SIT. A series of tones played by the telephone company at the beginning of a recorded announcement, such as indicating the telephone number dialed is no longer in service, has been changed, and so on. Automatic dialers may have the capability of recognizing the different sets of tones, allowing the user to decide whether to pass certain ones through to the agents or filter them out.

Special LATA Access An LEC (Local Exchange Company) -tariffed service offering that provides for a non-switched communications path (access link) between an long distance company Point of Presence (POP) and the premises of its end users, or to an LEC central office for centrex services. It includes all LATA access services that do not use an LEC switching system (end office or access tandem).

Special Mobile Group See SMG.

special night answering position Provides either a console or a pre-assigned single extension phone to answer all incoming night calls.

Special Routing Arrangement Service A GETS-type PSTN-based priority service to support continuity of government (COG) during a national emergency or crisis situation. GETS stands for Government Emergency Telecommunications Service.

special routing code A 3-digit code in the form 0XX and 1XX available for use within a network and used to modify routing or call-handling logic. End users are prevented from using system codes by the arrangement of the switching equipment to block all customer-dialed calls with a 0 or a 1 in the fourth digit of a 10-digit number, as well as 7-digit calls with a 0 or 1 in the first digit.

special service circuit A circuit used to provide a special service to a specific customer.

special services There are two meanings to special services in the telephone industry world, though they're related. First, special services are basically any telephone service that is different from plain old switched service. That special service may be it A variety of services that are separate from the public switched network.

Special Temporary Authority See STA.

specialized common carrier A company providing domestic long distance telecommunications services other than AT&T. See Other Common Carriers.

Specific Absorption Rate See SAR.

specific gravity The ratio of the weight of any volume of substance to a weight of an equal volume of some substance taken as a standard, usually water for liquids and hydrogen for gases.

specific inductive capacity The direct measure of the ability of a substance to store up electrical energy when used as a dielectric material in a condenser. See also Dielectric Constant.

speckle The bright and dark spots on the end face of a fiber caused by the interference of modes.

spectral bandwidth In telecommunications, the spectral bandwidth for single peak devices is the difference between the wavelengths at which the radiant intensity is 50% (or 3dB) down from the maximum value.

spectral efficiency The efficiency of a radio system in its use of the radio spectrum, usually expressed in bits per Hz for digital radios and KHz per voice channel in analog radios.

spectral purity The degree to which an optical signal is monochromatic, i.e., consists of only one wavelength.

spectrogram A basic research tool for the speech scientist which provides a three-dimensional visual representation of speech.

spectronics Technology seen from the point of view of the electromagnetic spectrum. Examples include 60 hertz power lines, 400 megahertz microprocessors, and 190 terahertz fiber optics.

spectrum A continuous range of frequencies, usually wide in extent within which waves have some specific common characteristics. See Spectrum Designation of Frequency.

spectrum analyzer Tunable RF instrument which displays a portion of the RF spectrum with amplitude of signals on the vertical axis and frequency on the horizontal axis on a screen. Used in TSCM to analyze transmissions for the characteristics of an illegitimate transmitter (radio bug).

spectrum buffer The slice of radio spectrum required by the FCC to stop adjacent radio and TV stations from interfering with each other.

spectrum cap The spectrum limit imposed by the FCC on companies to promote competition. On November 8, 2001, the FCC rule to eliminate the spectrum cap (some-

times referred to as the CMRS spectrum cap) to become effective on January 1, 2003. In the interim, the cap will become 55 MHz for all markets (up from 45 MHz in metropolitan areas). Before November 2001, the spectrum cap was determined by MSA/RSA boundaries, and was 45 MHz in urban areas (MSAs) and 55 MHz in rural areas (RSAs). For PCS to PCS overlaps or cellular to PCS overlaps, the PCS property must have at least 10% of its population in the MSA/RSA to hit the cap. For cellular properties, the cross-interest rule prohibits a company from owning one band and having more than a 20% interest in the other band. As a part of the November 2001 ruling, the FCC has changed the cross interest requirement to apply to Rural Statistical Areas (RSAs) only.

spectrum designation of frequency A method of referring to a range or band of communication frequencies. In American practice the designator is a twoor three-letter abbreviation for the name. In ITU practice, the designator is numeric. These ranges or bands are:

FREQUENCY RANGE	TYPICAL AMERICAN DESIGNATOR	ITU FREQUENCY BAND DESIGNATOR
30 - 300 Hz	ELF (Extremely Low Frequency)	2
300 - 3000 Hz	ULF (Ultra Low Frequency)	3
3 - 30 kHz	VLF (Very Low Frequency)	4
30 - 300 kHz	LF (Low Frequency)	5
300 - 3000 kHz	MF (Medium Frequency)	6
3 - 30 MHz	HF (High Frequency)	7
30 - 300 MHz	VHF (Very High Frequency)	8
300 - 3000 MHz	UHF (Ultra High Frequency)	9
3 - 30 GHz	SHF (Super High Frequency)	10
30 - 300 GHz	EHF (Extremely High Frequency)	11
300 - 3000 GHz	THF (Tremendously High Frequency)	12

FREQUENCY RANGE	ITU FREQUENCY BAND DESIGNATOR
3 - 30 THz	13
30 - 300 THz	14
300 - 3000 THz	15
3 - 30 PHz	16
30 - 300 PHz	17
300 - 3000 PHz	18
3 - 30 EHz	19
30 - 300 EHz	20
300 - 3000 EHz	21
THz = Terahertz (10 to the 12th power hertz)	
PHz = Petahertz (10 to the 15th power hertz)	
EHz = Exahertz (10 to the 18th power hertz)	

spectrum management Spectrum management is a concept you find in the cell business. It is the process of managing the radio spectrum" for purposes of imparting efficiency and intelligence to the spectrum as well as monitoring the spectrum. Spectrum management not only reduces the factors that will hinder the optimal efficiency of the allocated spectrum but also improves the overall performance of each cell and consequently the overall cellular network. Spectrum Management is based on network management principles, the difference being that instead of network elements, it is the spectrum that is being managed. Spectrum management takes into account variables like co-channel and adjacent channel interference, RSSI (Received Signal Strength Indication) values, power levels, frequencies etc. Spectrum Management involves the gathering and using such information proactively to improve network performance. The collected data can be used to generate information that is smart, concise and meaningful. This type of information will help in quicker and smarter decisions thus reducing the delay significantly and supply the carriers with an advantage that not reduces their cost but also adds value to their networks by making them smart, more reliable and proactive.

Spectrum Management Agency See SMA.

spectrum swarming The overuse of a range of radio frequencies.

speech API See MIcrosoft Speech API.

Speech Application Language Tags See SALT.

speech concatenation A term used in voice processing for economical digitized speech playback that uses independently recorded files of phrases or file segments linked together under application program control to produce a customized response in natural sounding language. For example, order status, bank balances, bus schedules or lottery results, etc. Concatenation is done for speed and economy. It lends itself to limited and structured vocabularies that are best stored in RAM (Random Access Memory) or speedily accessible from disk. Concatenation does not replace Text-To-Speech (TTS) as a method of getting the voice processor to deliver its responses. Concatenation, however, can be an excellent complement to TTS when a voice application demands broad, real time vocabulary production. See Text-to-Speech.

speech digit signaling Signaling in which digit time slots used primarily for encoded speech are periodically used for signaling (as, optionally, in ISDN). See also ISDN.

speech enabled telephony In the old days you used a telephone by rotary dialing or pushing buttons. In the new "speech enabled" telephony you talk, not dial. "Please call mother." The phone system understands you. It dials your mother. Speech enabled telephony makes heavy use of speech recognition and text-to-speech -in order to talk to you. You see these technologies embodied in "personal assistants," pseudo-assistants that understand your commands over the phone. Typically they find phone numbers, dial them, read you your voice and email messages, allow you to respond and then send your message over the Internet or over your corporate intranet. Speech enabled telephony allows your customers to call in, ask for and hear information specific for them. As speech recognition over the phone has improved by leaps and bounds in recent years, we'll see speech enabled telephony become more and more powerful and its uses broaden dramatically. It'll never be as intelligent as a full-time, intelligent secretary. But it will get close, or at least, appear to get close.

speech recognition Voice recognition is the ability of a machine to recognize your particular voice. This contrasts with speech recognition, which is different. It is the ability of a machine to understand human speech -yours and everyone else's. Voice recognition needs training. Speech recognition doesn't.

In the August/September, 1996 issue of a magazine called Speech Technology, David L. Basore talked about recent advantages which are making speech recognition systems much more natural. He wrote, "Speech recognition technology currently supports a wide range of viable applications, from voice controlled VCR remotes to sophisticated call center IVR applications. It can be separated conveniently into the following three categories:

1. Isolated word and phrase recognition in which a system is trained to recognize a discrete set of command words or phrases and to respond appropriately.

2. Connected word recognition in which a system is trained on a discrete set of vocabulary words (for example, digits), but is required to recognize fluent sequences of these words such as credit card numbers.

3. Continuous speech recognition in which a system is trained on a discrete set of subword vocabulary units (e.g., phonemes), but is required to recognize fluent speech. For more advanced applications, the vocabulary can be unlimited and the job of the recognizer is to understand the meaning of the spoken input.

A speech recognition system usually is made up of an input device, a voice board that provides analog-to-digital conversion of the speech signal, and a signal processing module that takes the digitized samples and converts them into a series of patterns. These patterns are then compared to a set of stored models that have been constructed from the knowledge of acoustics, language , and dictionaries. The technology may be speaker dependent (trained), speaker adaptive (improves with use), or fully speaker independent. In addition, features such as barge-in capability, which allow the user to speak at anytime, and key word spotting, which makes it possible to pick out key words from among a sentence of extraneous words, enable the development of more advanced applications.

This list of pros and cons is courtesy Mike Rozak, Microsoft software design engineer. What speech recognition is good for...

- Fast access to complex features. Some applications have features that are frequently used but which are difficult to present/control with a GUI. Often times the features can be more easily accessed by speech recognition.
- Magic keystrokes. Many applications have overloaded the keyboard with not only text entry features, but also commands. Some applications have so many keyboard accelerators that they distinguish between ctrl-f2, shift-f2, alt-f2, etc. These are difficult for the end-user to memorize. Using voice commands to replace these makes life easier for the user because voice commands are easier to memorize.
- Macros. Macros are related to magic keystrokes. Many applications allow users to create macros that speed up frequently-done or difficult tasks. To activate these macros, users often have to invent and memorize a magic keystroke. The keystroke can be replaced with speech, whose commands are easier to memorize and are less likely to cause an accelerator conflict.

- Global commands. At times, an application wants to have an input hook that is always active, even when the application doesn't have keyboard focus. If the application uses a keyboard hook then it's likely that the chosen key will already be used by another application, causing conflicts. Because there are so many different possible sentences, an application can provide a global command that is always active and not worry about conflict. Example: A PC-based phone can use global commands for "Call name>", so that a user can make a phone call even when the phone isn't the active application.

- Anthropomorphize the computer. If you want the user to perceive the computer as a person, then the application should use speech (synthesized or recorded) to talk to the user information, and speech recognition to get information from the user. The computer can listen for commands, or answers to questions, using speech recognition.

- Interaction with characters. Applications that have characters, especially adventure games, have a lot to gain from speech. Simulating characters is a specialized anthropomorphizing of the system.

- Form entry Applications can use speech recognition for form entry, both for selecting elements from a list, and entering numbers. Because speech can have so many commands active at once, it can be listening for all of the possible values for all of the fields at once, and infer which field the user wanted filled in. Example: If a user spoke, "Male, sixteen, one hundred and fifty pounds," the application could correctly identify which field is being referred to by "male", "sixteen", and "one hundred and fifty pounds," set the focus to the proper field for each command, and place in the proper data. The user doesn't have to worry about tabbing around.

- Sit back and relax. If you watch users sitting in front of a computer, you'll notice that whenever they type or use the mouse they hunch forward. If you want the user to relax and sit back while using your application, you should provide a speech recognition interface. Example: A consumer title might want to use speech recognition in order to make the computer seem less work-oriented because the user can sit back in his chair without touching the keyboard and mouse.

- Dictation for poor typists. Current dictation technology (which requires that users leave pauses between words) is good for people who cannot type or who are poor typists. The only reason not to include dictation is that dictation systems require about 8 megabytes of extra RAM.

- Access over the telephone. With the arrival of voice-modems in more and more PCs, applications will start providing "phone-based" UIs. Speech recognition is a much better interface than maneuvering through menus with touch-tones, especially since most Europeans don't have touch-tone phones. Example: A user will call up his computer and ask to speak to his PIM application, which then allows him to look up his schedule, address book, etc., all over the phone. The user can merely say, "Give me the phone number for John Smith

- Accessibility. Some people cannot use the keyboard or mouse effectively. The most common disability is carpel tunnel syndrome or equivalent. Adding speech recognition enables them to use the application.

- Mouse overloaded. Some applications (especially CAD systems) drag, drop, and select objects with the mouse. If the user wants to do any more complex actions, he either has to memorize a magic keystroke (which may not have the information bandwidth), or move the mouse to a menu/toolbar. Instead, the user could just speak, "Rotate this sixty two degrees."

- No keyboard/mouse available. Sometimes a keyboard or mouse is not available. This happens a lot in kiosk situations, where the kiosk has the keyboard and mouse hidden so that users don't feel intimidated, won't break the devices, and won't steal the devices. In order to get input from the users, current systems use a touch-screen. Why not use speech recognition?

- Hands busy. If a user's hands are busy, then speech is useful. Example: Some dentists' offices use speech recognition to enter charts about the patient's teeth because their hands are busy probing around the patient's mouth.

- Multiple people in front of machine. Speech recognition is good for applications that have several people sitting in front of the same computer, each participating in the application. Normally, an application which is to be used by several users would require that they pass the mouse and/or keyboard around. Since speech recognition is effective up to several feet, users would just have to sit around the computer and talk to it.

Where not to use speech recognition...

- Selecting from a large list of words. Most recognition systems break down when more than 100 words/commands are active, so keep lists below this number. For example, allowing the user to address electronic mail to any of 10,000 employees will not work well; instead, it would be better to allow any name from the 100 (or 75, or 50) employees to whom the user frequently sends mail.

- Spelling. Asking a user to spell a word does not work well because many letters such as "m" and "n" sound the same. Instead, it would be better to ask the user to type the word or offer a list of possibilities. Spelling with communications code words_"alpha," "bravo," "charlie," and the like_may be appropriate for certain vertical markets but not for general-purpose applications.

- Entering long sequences of numbers. Most engines will have a very high error rate for long series of digits that are spoken continuously. For phone numbers or other long series, either break the number into groups of four or fewer digits or have the user speak each digit as an isolated word.

- Pointing device. Do not use speech as a pointing device. Speaking "up" five times in a row is very annoying to the user.

- Action games. Because of the background noise (music and sound effects) that is typically present, speech recognition does not work well for action games. The speech-recognition engine spends time processing audio from the computer's speakers and may even recognize it as commands. Speech-recognition works for action games only if the user wears a close-talk microphone (for example, a headset). Additionally, by the time that the user finishes saying, "Fire," it's probably too late.

Speech Recognition Grammer Format See SRGF.

Speech Synthesis Markup Language See SSML.

speed There is more to life than increasing its speed. Gandhi.

speed bumps When your host system processes information faster than your network can handle and forces you to slow down.

speed calling An optional feature of a telephone system which allows the placing of calls to frequently called numbers though the use of an abbreviated number of dial digits. For example, call mom by dialing *123. Also called abbreviated dialing.

speed dial A feature that enables a PBX or PBX phone to store certain telephone numbers and dial them automatically when a code is entered. See Speed Dialing.

speed dialing Permits fast dialing of frequently used numbers. A repertory of numbers may be stored in the instrument and/or in the telephone switch. Usually a button or one, two or three digits are dialed to activate speed dialing.

speed number A one, three, or four digit number that replaces a seven or ten digit telephone number. These numbers are programmed into the switch in the carrier's office or in a PBX.

speed of light See Velocity of Light.

speed of light in a vacuum See Velocity of Light.

speed matching Feature that provides sufficient buffering capability in a destination device to allow a high-speed source to transmit data at its maximum rate, even if the destination device is a lower-speed device.

speed preview A service being tested by at least one cable broadband provider that enables its broadband customers to briefly try out the provider's higher tiered Internet access speeds, without service interruptions or the need to reboot their cable modems. The idea is that by offering customers a free and easy way to experience faster speeds, they will be able to make an informed decision about which tier of service works best for them (and sell more broadband service).

speech synthesizer A device that produces human speech sounds from input in another form.

Speech Transmission Index STI. A measure with a range from 0 to 1.0; 1.0 represents the best possible understanding of a given message. It measures how much of the message can be lost in transmission and still be understood.

speeds and feeds In the telecom world, speeds and feeds means the various technical characteristics of the various transmission services it offers. Speeds and feeds covers everything from the speed of the circuit to what media you need for it work -coaxial, copper wire, fiber, etc. The telecom industry has been widely criticized for its fixation on speeds and feeds and, consequently, paying little attention to user applications -what the speeds and feeds are used for.

Speedsync See Laplink.

spend management A new management discipline that drives enterprise-wide spending visibility and spending reductions through improved category management, procurement compliance, operational effectiveness, and supplier performance. There is now spend management software. It helps businesses aggregate data from around their entire company, analyze and source more categories more often, leverage domain expertise (whatever that is) and improve procurement. In other words, if the entire company buys paper towels, it will get a better price because it's buying more of them.

SPF Shortest Path First.

spherical semiconductor An integrated circuit, developed by BALL Semiconductor, created on the surface of a one-millimeter silicon sphere, and designed to compete with conventional integrated circuits.

SPI 1. Service Provider Interface. See Windows Telephony.

2. Serial Peripheral Interface. SPI is a Motorola specification for a full-duplex synchronous serial data link supporting high-speed connectivity at 1 Megabaud between microprocessor CPUs and other devices. The physical connection to the "Mini Board" printed circuit boards supporting SPI is in the form of a four-wire RJ-11 telephone jack. The SPI essentially is a shift register, serving to convert from parallel to serial transmission across a standard twisted-pair telephone connection over very short distances. Parallel transmission is used internal to computing systems, but generally is considered to be impractical over longer distances, even between adjacent devices, or sometimes between circuit boards housed in the same system chassis. SPI uses the each of the four wires of the RJ-11 connection for different specific purposes, with two used for data transfer and two for signaling and control. One is used for output data from the master device to the slave(s), one from a slave to the master (one slave active at a time), one for clocking in order that the transfer of the data bits is synchronized, and one for turning the individual slaves on and off. SPI is used in a number of Motorola handheld devices and other mobile platform systems. See also RJ-11 and Shift Register.

3. Stateful Packet Inspection . A firewalling technique that builds upon packet filtering technology by taking into account the state (i.e. context) of the session involved in order to decide whether to block or permit a given packet.

SPID Service Profile IDentifier. When you order an ISDN line, your phone company will give you a SPID for every terminal device (e.g., telephone port , computer port, and fax machine port) you have connected to an ISDN line for circuit-switched (not packet-switched) network access. The SPID does not relate to an ISDN line or even an ISDN B (Bearer) channel. The SPID can be the 10-digit DN (Directory Number, which is a fancy way of saying telephone number), although it usually includes a four-digit suffix or prefix. That suffix or prefix may be the same as the carrier's OCN (Operating Carrier Number), but it doesn't have to be. In any event, the SPID format is determined by the carrier, and is sensitive to the specific CO switch manufacturer, generic software load, and the carrier's local practices. Once programmed into both the customer equipment (e.g., PBX or Terminal Adapter, or TA), the SPID provides the appropriate service mode (a fancy way of saying services and features) for each device communicating over the ISDN line and B channel. By the way, your SPID changes when your area code changes. When your telephone company assigns you a SPID, you should write it down. Otherwise, you'll have a devil of a time getting them to tell you what it is in the event that you have to reprogram your TA (Terminal Adapter). Trust me, as I speak the truth. By the way, you don't have to worry so much about writing down the SPID, if you are set up for AutoSPID, which automatically negotiates the SPID between your terminal equipment and the CO.

Spider A program that prowls the Internet, attempting to locate new, publicly accessible resources such as WWW documents, files available in public FTP archives, and Gopher documents. Also called wanderers or robots (bots), spiders contribute their discoveries to a database, which Internet users can search by using an Internet-accessible search engine.

In January, 1998, I received a solicitation for Bull's Eye Gold, which bills itself as the premier email address collection tool. "This program allows you to develop targeted lists of email addresses. Doctors, florists, MLM, biz opp...Our software uses the latest in search technology called "spidering". By simply feeding the spider program a starting website it will collect for hours. The spider will go from website to targeted website providing you with thousands upon thousands of fresh targeted email addresses. When you are done collecting, the spider removes duplicates and saves the email list in a ready to send format. No longer is it necessary to send millions of ads to get a handful of responses." See also Bot.

Spidering See Spider.

SPIE Society of Photometric Industry Engineers or Society of PhotoOptical Instrumentation Engineering.

spiff A telemarketing term. An award that forms the prize for a quick motivational incentive. Spiffs can include movie tickets, pizza parties, gifts selected from a catalog and other such prizes.

spike An in-phase impulse causing spontaneous increases in voltage. Spikes are very fast impulses, less than 100 microseconds, of high-voltage electricity ranging from 400 volts to 5,600 volts superimposed on the normal 120V AC electrical sine wave. See also Metal Oxide Varistor.

spike markets A term developed by Apple. By "spike," the company means software and hardware combinations that allow Apple to rise (spike) through a noisy marketplace. By doing this, Apple hopes to grab attention in some reasonably horizontal niche markets and show people why a Mac is worth a few (hundred) extra bucks. The first of these spikes is a foray into home video editing, using Apple's new Performa 6400 and a piece of software called Avid Cinema.

spike mike Contact microphone for listening through walls.

spikes Electrical anomalies represented as short duration, instantaneous, very high voltage fluctuations on an electrical service.

spill-forward feature A service feature, in the operation of an intermediate office, that, acting on incoming trunk service treatment indications, assumes routing control of the call from the originating office. This increases the chances of completion by offering the call to more trunk groups than are available in the originating office.

spim Unsolicited commercial messages send via an instant messaging system, also known as instant messaging spam. Such SPIM can find its way onto computers and cell phones.

SPIN Service Provider Identification Number. A number that identifies the telecommunications service provider from which schools and libraries obtain discounted service through the Schools and Libraries Universal Service Program. The program, which was created by the Telecommunications Act of 1996 and which is funded by the Universal Service Fund, provides for discounted telecommunications services, Internet access and internal connections . The Schools and Libraries Corporation approves applications for such discounted rates, which are known collectively as "E-rate."

spin a board When people say "spin a board," they mean to take a given printed circuit card and do a re-layout of the components and traces, In short, to do a board revision. This is typically associated with an ECO (Engineering Change Order).

spin stabilization A method of preventing a satellite from tumbling by spinning it about its axis.

SPINA Subscriber Personal Identification Number Access. A term identified in the TIA IS-53 (Interim Standard 53), addressing security in cellular telephone networks. SPINA requires the user to enter a PIN in order to gain access to the network. Access is granted for a specified period of time or until the occurrence of some event, such as the terminal's being turned off. The PIN is transmitted "in the clear," i.e. unencrypted; therefore, it is susceptible to interception. See SPINI.

SPINI Subscriber Personal Identification Number Intercept. A term identified in the TIA IS-53 (Interim Standard 53), addressing security in cellular telephone networks. Unlike SPINA, SPINI may require that the user enter a PIN before each call in order to gain access to the network. Both SPINI and SPINA typically require a 4-digit PIN. The PIN is transmitted "in the clear," i.e. unencrypted; therefore, it is susceptible to interception. See SPINA.

spindle A spindle is metal rod serving as an axis for something that spins around it. Laptops are typically called one, two or three spindle machines. A one spindle machine would be a laptop with only a hard disk. A two spindle machine means that the laptop typically has a hard disk and a floppy disk or a hard disk and a CD or DVD drive. A three spoindle machine that the laptop an internal floppy drive, a hard disk and a CD/DVD drive, i.e. three rotating media sources.

spindle synchronization A process that coordinates all hard disks in a RAID array to use a single drive's spindle synchronization pulse.

spinner A name given to people in the United States who regularly change their long distance carrier. Several million people each year switch their long distance carriers. The problem and cost of churn has become fairly major for the long distance industry.

spinning The practice of allocating shares in hot IPOs personally to executives as an inducement to win their company's investment banking business. If they sell their newly-acquired shares the instant the company's stock starts to trade on the stock exchange, this practice is called flipping. For a longer explanation, see Flipping.

spintronics Using the spin of an electron to represent binary data (zero or one).

spiral life cycle A term used in COM development. Another term used to describe the iterative development process. Opposite of waterfall life cycle.

spiral wrap A term given to describe the helical wrap of a tape or thread over a core.

SPIRIT European consortium focused on standardizing telecom operators' procurement processes.

SPIROU Signalisation Pour l'Interconnexion des ROxC7seaux OUverts (ETSI ISUP V3 modified by ART).

SPIT Spam over Internet Telephony. Here's how SPIT works: A company interested in sending out unsolicited advertisements would harvest phone numbers from a number of different VoIP services. Any open, IP-based phone system could be a target of "spitters." The spitter then would create an audio ad, and send it to however many members he wanted. Effectively Spit is not different to those wonderful computerized auto dialers that call us regularly and tell us we're won a free trip to Florida. If no one's home, they leave an upbeat message about the wonderful opportunity and would we please call back instantly, if not sooner.

spitting dollar bills When a cell site goes on the air, it is now worth money -it spits dollar bills -as opposed to being a construction project and demanding dollar bills.

splash tone Distinctive sound used on some phone systems to indicate that a command has been received, or that something has to be done. Vaguely resembles water being splashed.

splashing A "splash" happens when an Alternate Operator Service (AOS) company, located in a city different to the one you're calling from, connects your call to the long distance carrier of your choice in the city the AOS operator is in. Splashing does not imply backhauling, but it often happens. For example, let's say you're calling from Hotel Magnificent in Chicago. You ask AT&T to handle your call. The AOS, located in Atlanta, "splashes" your call over to AT&T in Atlanta. But you're calling Los Angeles. Bingo. Your AT&T call to LA is now more expensive than it would be if you had been connected directly to AT&T from Hotel Magnificent in Chicago.

splat A slang expression for the asterisk character (*) that you can yell across a crowded room without fear of being misunderstood.

splice Verb. The joining of two or more cables together by splicing the conductors together. In copper wire telephone cables, splicing is on a mechanical basis and pair-to-pair, with the pairs organized by binder groups and color codes. In optical fiber cables, the splicing is fiber-to-fiber, with the fibers organized by ribbon or colored buffer tube and color code. Fiber optics splicing may be either mechanical splicing or fusion splicing. See also fusion splicing and mechanical splicing.

splice box A box, located in a pathway run, intended to house a cable splice.

splice closure A device used to protect a cable or wire splice.

splice tray A place where you splice fiber optic cables and then leave them in the splice tray. It's an elaborate connector.

splicing chamber An underground concrete vault in which cables may be spliced, and transmission equipment may be located.

spline A curve shape produced on a computer or video device by connecting dots or points at various intervals along the curve. In digital picture manipulators, each key frame becomes a point on a curve and the user can control how straight or curved the path of the transformed image is as it travels through the key frame points.

splint A pejorative word for Sprint, the third largest interexchange carrier. The word was created by William G. McGowan, the driving force behind MCI for so many wonderful years.

splinternet A term that refers to a portion of the Internet that splinters off, literally or figuratively, from the public Internet. One example is a darknet. Another example may be emerging in China, where China's Ministry of Information has created three of its own top-level domains, outside of the management of ICANN. A third type of splinternet is what proponents of Net Neutrality fear will happen if Congress does not pass legislation that forbids carriers from creating a tiered Internet, with each tier having its own services and pricing. See also network neutrality.

split 1. A call center term. Split is an ACD routing division that allows calls arriving on specific trunks or calls of certain transaction types to be answered by specific groups of employees. Also referred to as gate or group. Same as Group. See ACD or Automatic Call Distributor.

2. See Overlay Area Code.

split access to outgoing trunks Two separate trunk groups provided for direct outward dialing which can be accessed by dialing the same trunk access code. Controlled on class of service basis.

split area code See Overlay Area Code.

split channel modem A modem which divides a communications channel into separate send and receive channels. Most modems which use the dial-up phone network are split channel -meaning they can transmit and receive simultaneously over a two wire circuit. See also Split Stream Modem, which is another term for the same thing.

split horizon The view a router has of a wide area network interface in a partial mesh environment where an incoming packet may need to be sent out the same interface over which it is received to reach its ultimate destination. Split horizon is normally disabled to ensure that this cannot occur and that routing loops are not created. This posed a problem with frame relay because packets should be sent back out the same physical interface over which they have been received, but not the same logical interface. Router vendors have now solved this problem, enabling routers to support partially meshed frame relay networks.

split link When one multiplexer uses two links to communicate to two separate multiplexers.

split pair Something that happens in cable splicing when one wire of a pair gets spliced to the wire of an adjacent pair. It's more accurate to call it a mistake. This error cancels the crosstalk elimination characteristics of using twisted pair wiring in which the two conductors necessary for the circuit are twisted around each other. When you have a split pair, the two conductors come from two different pairs.

split ratio A ratio is the quotient of two mathematical expressions. A split ratio describes the ratio, or relationship, between an incoming signal and an outgoing signal as it passes through a splitter. For example, a Passive Optical Network (PON) uses passive splitters that split the incoming signal into multiple outgoing signals. A splitter that splits an incoming signal into two outgoing signals is a 1:2 (spoken as "1 by 2") splitter, with each outgoing signal having a signal strength of 1/2 of the incoming signal, assuming that the splitter is perfectly balanced. PONs sometimes use 1:32 splitters that split an incoming optical signal into 32 outgoing signals, each with 1/32 the power of the original incoming signal.

split resplit method Split Resplit is an inductive wiretap. Telephone wires are twisted in bundles so as to reduce crosstalk between telephone lines. The Split-resplit method involves crossing the target line pair with an unused pair of telephone lines with the goal of increasing crosstalk with the operator's line. The operator's line can then inductively pick up conversations on the target line. Because the signal levels are low in the pick up line, audio amplifiers have to be used to clearly hear intercepted audio.

split seeks A process by which Novell NetWare SFT III splits multiple read requests between the two servers' disks for simultaneous processing and faster disk rads. Disk reads are split between both servers, with only one server doing a particular read and sending the data read over the MSL (if necessary) to the other server. MSL is the Mirror Server Link.

split stream modem A modem which can handle multiple, independent channels over a single transmission path.

split system A switching system which implements the functions of more than one logical node.

split tunneling Let's start with tunneling. Tunneling typically means to secure a secure, temporary path for your communications via the Internet. For example, a telecommuter might dial into an ISP (Internet Service Provider), which would recognize the request for a high-priority, point-to-point tunnel across the Internet to a corporate gateway. The tunnel would be set up, effectively snaking its way through other, lower-priority Internet traffic. Now imagine a a corporate network that uses both the Internet and a private Intranet or a VPN (Virtual Private Network). Split tunneling simply means the ability to send material securely (that's the critical word) over both networks. You need a device to do the switching and some software. Some devices enable simultaneous connections through the VPN and to the Internet.

splitter 1. A network that supplies signals to a number of outputs which are individually matched and isolated from each other. See also Power Divider and Frequency Splitter.

2. A coaxial cable TV device. Imagine a single coaxial cable carrying TV signals. You want to connect two subscribers or two TV sets to this single cable. You insert a splitter, a small cheap (under $10) device. You screw the incoming cable in one side of the splitter. The other side has two screw terminals into which you can screw two coaxial cables -one for each subscriber, or for each TV set. Splitters are passive devices that simply split the incoming signal to create two or more identical outgoing signals, each of which travels across a tail circuit to a terminal device in the form of a TV set. Since coax splitters don't require external electricity to work, they are unaffected by power outages, although your TV set is. As non-powered passive devices, they also don't boost the power level of the signal, which means that the signal naturally loses some strength.

3. xDSL is the generic name for technology that puts several megabits of a data transmission on a local loop -from the phone company's central office to your home. A subscriber, like you or I, would use that phone line for two purposes -first, to get onto the Internet and the Web, and second, to speak on the phone. This means that at our house we'd need a xDSL box into which we'd plug our computer and our various analog phones and fax machines. The industry refers to this box as a "splitter," meaning that it splits the incoming

bit stream into voice and data. This is a stupid name for it, since the device is really a multiplexer. And typically such a device would have to be installed by a phone company technician and would replace the demarcation box outside the house. It gets worse. The phone companies have created an adjective called "splitterless" to describe a xDSL box that still splits between voice and data but doesn't require a visit from a telephone company technician. In other words, you'll be able to go down to your friendly local electronics discount store, buy a splitter box, take it home, plug one side into your phone line (your female RJ-11 jack), and into the other side you'll plug your PC and your analog phone instrument or instruments. You won't need a visit from your friendly phone company technician. In short, the concept of "splitterless" refers to whether the phone company needs to send a technician to install the box or not. Splitter means it must send a technician. Splitterless means it doesn't have to. A splitterless box typically contains the electronics for a splitter (to split out voice and data), while a splitter box would still require a DSL modem. By the way, splitterless xDSL technology is not trivial. My friend Paul Sun talks about a splitterless xDSL box as needing to contain five million transistors and have the horsepower of a 1,000 MHz Pentium PC. See also G.Lite, G.990, NIU, and xDSL.

splitterless For a detailed explanation, see Splitter immediately above. See also G.Lite and G.990.

splitterless ADSL A variation of ADSL in which a splitter is not needed at the customer's premises. Sometimes referred to as G.Lite.

splitting 1. A filter which splits or separates signals on the basis of their transmission frequency. For example, a splitter can be incorporated into an ATU-R (ADSL Termination Unit-Remote) located at the subscriber premise. The splitter would serve to separate the high-frequency data transmission from the low-frequency POTS voice transmission. The data transmission would then be delivered to a TV set or PC, while the POTS transmission would be delivered to the telephones. See also Filter.

2. A splitter is a telephone console device which permits an operator to consult privately with one party on a call without the other party's hearing. Or permits a three-party telephone conference user to consult privately with one side of the conference while the other is effectively put on hold. Jumping from one party to the other is called "Swapping."

3. See also Splitter.

splog A splog is the contraction of spam blog, a blog set up for the sole purpose of gaining links, search engine ranking, or promoting a product or website (usually pharmaceuticals, gambling, or porn), that adds no value to web content as a whole. Splogs also make it even more difficult for people to find the information they are looking for when completing searches and a significant amount of comment spam is the result of these splogs trying to promote their websites. This definition from blogossary.com.

SPM 1. Subscriber Private Meter.

2. Service Provider Messages.

3. Spectrum Peripheral Module, and is an optical interface. The capacity of SPM is much higher than DTCs (an electric interface), i.e., 2016 channels vs. 480 on DTCs. They take up less physical space and eliminate the need for echo cancellers and various other equipment. These terms are used when referring to all Nortel's products.

SPN Subscriber Premises Networks.

SPNI Service Provider Network Identifier.

SPOF Single Point of Failure. See Single Point of Failure.

SPOI Signaling Point Of Interface. The demarcation point on the SS7 signaling link between a LEC network and a Wireless Services Provider (WSP) network. The point established the technical interface and can designate the test point and operational division of responsibility for the signaling.

spoke An organization that is traditionally doing EDI (Electronic Data Interchange) as a result of a request from one of its customers. Generally has only a few trading partners.

spontaneous networking See Jini.

spoofing 1. Spoofing is another word for impersonation. Spoofing happens when someone on a network pretends to be someone else. For example, a person can pretend to have the email address HarryNewton@harryNewton.com when it's not me. Or a computer could pretend to be and identify itself as the Internet site www.HarryNewton.com. Spoofing is done for any number of reasons -lying, cheating and stealing are good beginnings. You might spoof a computer system in order to access to a computer system and find out my credit card number. You might spoof my email to convince my wife that I was going to buy her a million dollar necklace for her birthday. (She should be so lucky.) See also Public Key Encryption.

2. Spoofing is also a means of circumventing the difficulty of carrying Internet traffic via satellite: Spoofing involves speeding packet data transmission in systems that may

involve delays greater than the time-outs of the protocols involved (e.g. satellite transmission). When data are delayed or may be delayed, either the transmitter or the receiver is "fooled" into "thinking" that data are being successfully transmitted using various spoofing techniques.

SPOOL Simultaneous Peripheral Operation On Line. A program or piece of hardware that controls a buffer of data going to some output device, including a printer or a screen. A spool allows several users to send data to a device such as a printer at the same time, even when the printer is busy. The spool controls the transmission of data to the device by using a buffer or creating a temporary file in which to store the data going to the busy device. See Spooler and Spooling.

spooler A program that controls spooling. Spooling, a term mostly associated with printers, stands for Simultaneous Peripheral Operations On Line. Spooling temporarily stores programs or program outputs on magnetic tape, RAM, or disks for output or processing.

SPOOLING Simultaneous Peripheral Operations On Line. Spooling means temporarily storing programs or program outputs on magnetic tape, RAM or disks for output or processing. The word "Spooling" is mostly associated with printers. Here's an example: Pretend that a lot of people on your Local Area Network all want to send their reports to the printer today. Instead of each person having control of the printer and relinquishing it only when they're through, each user tells the print Spooler what file they want printed. The program, called the spooler, places the print request in the print queue. When your request reaches the top of the queue, your report is printed out. Using a PC as print spooler slows it down. Best not to use it for much else.

spot beam An antenna beam of a satellite that is aimed at a specific limited-area spot on the surface of the earth, the extent of which is considerably less than the total area of the earth in view.

spot beam antenna A satellite antenna capable of illuminating or focusing on a narrow portion of the earth's surface.

spot frame Single Point Of Termination Frame. See Dedicated Inside Plant.

SPOX An operating system for digital signal processors from Spectron Microsystems, Goleta, CA, now owned by Dialogic in Parsippany, NJ. Spox is a real-time, multitasking operating system that is optimized for use with fixed and floating point digital signal processors in both singleand multiprocessor systems. The SPOX environment is implemented as a library of relocatable, C-callable modules.

SPP 1. Sequenced Packet Protocol. XNS (Xerox Network Systems) protocol governing sequenced data.

2. Signal Processing Platform.

3. Standard Parallel Port.

SPRE Special PREfix code. Special digits dialed in order to access features of a telephone system, like a PBX or a Centrex. For example, in order to invoke the Call Pickup feature, a PBX user might pick up the handset, dial a three-digit SPRE code specific to the PBX, and answer the call.

spread spectrum Also called frequency hopping, spread spectrum is a modulation technique used in wireless systems. The data to be transmitted are packetized, and spread over a wider range of bandwidth than demanded by the content of the original information stream. Spread spectrum takes an input signal, mixes it with FM noise and "spreads" the signal over a broad frequency range. Spread spectrum receivers recognize a spread signal, acquire and "de-spread" it and thus return it to its initial form (the original message). A large number of transmissions can be supported over a given range of frequencies, with each transmission comprising a packet stream and with each packet in a stream being distinguished by an ID contained within the packet header. The receiver is able to distinguish each packet stream from all others by virtue of that ID, even though multiple transmissions share the same frequencies at the same time, with the potential for the overlapping of packets. Spread spectrum is highly secure. Would-be eavesdroppers hear only unintelligible blips. Attempts to jam the signal succeed only at knocking out a few small bits of it. So effective is the concept that it is now the principal antijamming device in the U.S. Government's Milstar defense communications satellite system. Spread spectrum technology also is used extensively in wireless LANs and in CDMA (Code Division Multiple Access), the access technique used in many PCS (Personal Communications Systems) cellular systems.

There are two versions of spread spectrum. Direct Sequence Spread Spectrum (DSSS) spreads the signal over a wide range of the 2.4 GHz frequency band. Frequency Hopping Spread Spectrum (FHSS) involves the transmission of short bursts of information over specific frequencies, with the frequency-hopping carefully coordinated between transmitter and receiver. See also CDMA, DSSS and FHSS.

Hedy Lamarr, the actress, created the concept of spread spectrum in 1940 and, two years later, received a U.S. patent for a "secret communication system." The patent was issued to her and George Antheil, a film-score composer, to whom Ms. Lamarr had turned for help in perfecting her idea. Spread spectrum was used extensively by the Allies during the World War II in the Pacific Theater, where it solved the problem of Japanese jamming of radio-controlled torpedoes. World War II electronics were pretty primitive, and Hedy's system used a mechanical switching system, like a piano roll, to shift frequencies faster than the Nazis or the Japanese could follow them. More recently, spread-spectrum has been combined with digital technology, for spy-proof and noise-resistant battlefield communications. In 1962, Sylvania installed it on ships sent to blockade Cuba. Ms. Lamarr never received one penny for her invention. Ms. Lamarr was quite an innovator. She delighted and shocked audiences in the 1930s by dancing in the nude in the movie "Ecstasy," which, sadly, I've never seen.

spreading loss See Free Space Loss.

Sprint The third largest IXC (IntereXchange Carrier), behind AT&T and MCI, Sprint also is a LEC (Local Exchange Carrier) of significance. Sprint began as a venture of Southern Pacific Railroad, which had the clever idea of using its right-of-way to lay a fiber optic cable network. Subsequently, Southern Pacific sold the network to GTE, at which point it became know as GTE Sprint. The company became known as US Sprint when GTE and United Telecom decided to form a (50/50) joint venture from US Telecom (United's long distance company), GTE Sprint and GTE Telnet. United Telecom bought GTE's interest, acquiring the final 19.9% in 1992. Now it's just called Sprint Corporation. Through its acquisition in 1993 of Centel, Sprint currently operates as a LEC (local exchange carrier) in at least 19 states.

sprite As used in computer graphics refers to a graphic image that can move over a background and other graphic objects in a non-destructive manner.

SPS 1. Signaling Protocols and Switching.

2. Standard Positioning Service. The normal civilian positioning accuracy obtained by using the single frequency C/A code in the GPS (Global Positioning System) system.

3. Solution Provider, also called Microsoft Solution Provider. See Microsoft Solution Provider.

4. Standby Power System. A SPS is a form of uninterruptible power supply. It monitors the power line and switches to battery power as soon as it senses a problem. The switch to battery, however, can require several milliseconds, during which time the computer (or whatever else is connected to it) does not receive any power. Standby Power Systems are sometimes called Line-interactive UPSes. See also UPS.

SPTS Single Program Transport Stream: An MPEG-2 Transport Stream that consists of only one program.

spud A special long-handled shovel used to loosen soil in a hole into which you're going to put a telephone pole.

spudger Shaped like a dental pick, it's a gadget phone technicians use to find their way through a multi-paired telephone cable in their hunt for one single pair.

spur A cable drop on a fieldbus network. What one would call a LAN drop in an office LAN environment.

spurious A term used in voice recognition. A spurious error is said to occur when a sound that is not a valid spoken input is incorrectly accepted as an input speech utterance.

spurious emission Emission on a frequency or frequencies which are outside the necessary bandwidth and the level of which may be reduced without affecting the corresponding transmission of information. Spurious emissions include harmonic emissions, parasitic emissions, intermodulation products and frequency conversion products, but exclude out-of-band emissions.

spurring out Spurring out happens when a telephone company worker is up a telephone pole working on the wires, and his or her spurs come loose from the pole. A related term is pole hug, which is what said technician does after he or she spurs out. Pole hugs usually result in nasty splinters and an awful scare. According to Scott Davis, who contributed this definition, "my impression is that spurring out is a rite of passage for new field techs.

spurs 1. The sharp metal devices on the climbers used by telephone line-persons (people who climb telephone poles). Such climbing spurs make a mess of wooden telephone poles. See Spurring Out.

2. The cowboy devices awarded by US WEST to privileged persons who have done US WEST some nice favor or are otherwise deserving of honor.

Sputnik Sputnik was the world's first artificial satellite. It was launched by the Russians on October 4, 1957. It freaked out the Americans and started the space race, which the Americans later won.

SPVC Soft Virtual Circuit. Or Smart Virtual Circuit.

spy phone A special mobile phone that looks and works like a regular mobile phone, with one exception. When you dial a special number that is programmed into the phone, the call is automatically and silently answered, without any audio or visual indication that a call had just been received and answered. The spy phone, in this scenario, becomes a bugging device, allowing the caller to hear what is going on in the vicinity of the spy phone. When a key is touched on the spy phone -for example, when the user wants to make a phone call, check the time, or send a text message -the secret connection with the remote caller is automatically broken so as to prevent suspicion. Spy phones are sold for reasons including:

- Keeping track of your wife or husband.
- See if you can trust your business partner.
- Listen in sales talks of your employees.
- Protecting your children
- To reveal secrets.

Spy phones are not cheap. See www.spyphones.com.

spyware See adware.

SPX Sequenced Packet eXchange. 1. An enhanced set of commands implemented on top of IPX to create a true transport layer interface. SPX provides more functions than IPX, including guaranteed packet delivery. 2. Novell's implementation of SPP for its NetWare local area network operating system.

SQE Signal Quality Error. The 802.3 specification defines this for signals from the MAU to the NIC. Also referred to as heartbeat, is a signal sent by transceivers after a frame is transmitted in order to verify the connection, and is also used by the transceiver to notify the station that a collision was detected. The SQE is primarily used in 10Base-5 environments as a test signal to reassure the station that the transceiver is still operating properly. Some older network devices will not operate properly unless SQE is enabled; almost all new devices do not require SQE. SQE should always be disabled when a transceiver is connected to a repeater (including a 10BASE-T hub), or if it is not required.

SQL Structured Query Language. A powerful query language for defining, maintaining, and viewing information in a relational database. The original version, invented by IBM in 1974, was known as SEQUEL (Structured English QUEry Language) and was intended for use in IBM's System R. SQL was first commercialized by Oracle in the early 1990s. SQL has evolved into a complete language for the definition and management of persistent and complex data objects. A favorite query language for DBMSs (DataBase Management Systems) running on mainframes, SQL now also widely runs on PCs and LANs. SQL been standardized by ANSI, although there also exist multiple non-standard dialects that include extensions to that standard. SQL versions have been developed by Access, Informix, Microsoft, Oracle, Sybase, and others. See also ODBC, QBE and SQL Server.

SQL Server Microsoft SQL Server. A Microsoft retail product that provides distributed database management. Multiple workstations manipulate data stored on a server, where the server coordinates operations and performs resource-intensive calculations.

square key system A square key system is one that has all telephone lines appearing on every telephone and each telephone has a separate button or "key" for each line. See Squared Key System for a longer explanation.

square operation If there are fewer than eight lines in a Merlin system, all users can access all lines. See Squared Key System.

square shoes Toward the end of the 15th century, men's shoes had a square tip, like a duck's beak, a fashion launched by Charles VIII of France to hide the imperfection of one of his feet, which had six toes.

square wave A term used to refer to a digital signal, which is binary in nature. This is in marked contrast to an analog "sine wave," which varies continuously in terms of its amplitude and frequency. In other words, digital signals involve only two values: "1" and "0." Computer systems speak digital. Every value (i.e., letter, number, punctuation mark, and control character) is expressed in terms of a specific and unique combination of 1s and 0s of a specific length according to a particular coding scheme. Not only do computers create and store information in such form, they also output information in that form and they expect to see information presented to them in that form. One advantage of digital communications in the form of a bit stream, or stream of 1 and 0 bits, is that computer communications is supported without the need for conversion to analog and back again for transmission across the network. The bit stream is transmitted in the form of a square wave, which consists of discrete values representing these 1s and 0s.

Within the CPE (Customer Premise Equipment) domain and in an electrically-based mode of operation, this representation generally involves a positive voltage for a "1" and a null voltage (0 voltage) for a "0" -this is an "on" and "off" approach. Alternatively, a "1" can be represented as a relatively high level of positive voltage such as +3.0 volts,

and a "0" as a relatively low level of positive voltage such as +1.5 volts: in this approach, the electrical circuit continues to flow energy in waves of discrete voltage values. Further still, a "1" can be in the form of a positive voltage such as +1.5 volts, and a "0" can be a negative voltage such as -1.5 volts: again, a "wave" approach. Regardless of the approach, within your own domain you can play the square wave game anyway you (and your manufacturer) choose; after all, it's your game and you pay for the privilege of setting your own rules.

In a public network, it is quite a different matter. T-carrier systems, for instance, require that 1s be represented as alternating positive and negative voltages, while 0s are null voltages. Numerous devices in the public network depend on this electrical coding scheme to maintain synchronization. Further and as the public network serves vast numbers of users, there must be uniformity in order for the network to function at all.

In a radio system, the approach is different still, with the square wave taking the form of radio waves of relatively high and relatively low amplitude.

In an optical network, several approaches can be used. One approach calls for the square wave to take the form of light waves of different levels of intensity (i.e., bright and brighter)–which essentially differences in amplitude, or power level. The second approach calls for the laser light source to pulse on and off, with the presence of light indicating a "1" and the absence of light indicating a "0."

In any event, there are two discrete values represented in the form of "on and off," or "high" and "low," or "plus" and "minus." Digital networks, which use square waves for transmission, offer clear advantages. Most especially, they're much cleaner. Any noise they picked up on their travels across the network is disregarded as the signal is received, boosted and recreated. This is in marked contrast to an analog signal, which simply is amplified, along with any noise which might be present. Square waves are also cheaper to produce. That's good. But if you send a square ringing "wave" to a device like a high-speed modem that's expecting an analog sine wave, that high-speed modem will not respond as it doesn't speak digital at that side of the connection. See also Sine Wave.

squared key system A "squared" key system is one that has all telephone lines appearing on every telephone and each telephone has a separate button or "key" for each line. No one quite knows where the word "squared" came from. So if our explanation bears no relation to the word "squared," sorry. But it goes like this (we think): In the old days there were 1A2 phone systems. These 1A2 phones had buttons on them. These buttons could correspond to trunks -any trunk. These were called non-squared systems. Then came electronic key systems. Each trunk had to "appear" (i.e. be) the same button on each phone. These electronic key systems were called squared systems. There are advantages and disadvantages. Squared systems are portrayed as having one advantage: You can go to any phone anywhere in the system and punch any button for any trunk and know it to be the same button for the same trunk. Thus less confusion. But this means you can only have as many trunks on your key system as you have trunk buttons on your key telephones. In a non-squared system -a 1A2, the newer hybrids or some of the newer programmable key systems -you can have more trunks than you have buttons on each phone. Some phones will have trunks that others don't have and vice versa. Thus you can have more trunks on your phone system than you have buttons on your phones. This means, for example, that four executives can have each have private lines and access to four trunks on a six button phone. (The other button is for Hold.)

squatter See Cybersquatting.

squeaky clean I had my bicycle cleaned by the bike shop. When it came back, the brakes squeaked. My bike was squeaky clean.

squelch A circuit function that acts to suppress the audio output of a receiver. See also squelching.

squelching Referring to the "Rerouting of Pass-Through Traffic During Node Failures", Nortel's "Introduction to SONET Networking" tutorial handbook says, "While tributaries terminating at the failed node cannot be protected, traffic passing through that node is automatically redirected. ... In an action referred to as "squelching," nodes adjacent to the failure replace non-restorable traffic with a path layer alarm indication signal (AIS) to notify the far end of the interruption in service. The squelching feature employs automatically generated squelch maps that require no manual record keeping to maintain." See also AIS.

Squid A high-performance, full-featured proxy cache server protocol developed for use in the Web. Squid supports the caching of requested Internet objects (data available via HTTP, FTP, and other protocols) on proxy servers that are more proximate to the user than the origin server. Hence, Internet network resources are used more efficiently. Squid comprises a main server Squid program, a Domain Name System (DNS) program and associated database, and management tools. Squid supports caching protocols including CARP, HTCP, and

ICP. Squid runs in a number of UNIX Operating System (OS) environments, including AIX, BSD, Linux, HP-UX, and SunOS/Solaris. Squid was funded by the National Science Foundation (NSF). It is free software copyrighted by the University of California San Diego, and is licensed under the terms of the GNU General Public License. "Squid" isn't an acronym, and it really doesn't mean anything: Harris' Lament states "All the good ones are taken."

squirt the bird To transmit a signal up to a satellite. "The crew and talent are ready; when do we squirt the bird?"

SR 1. Speech Recognition. See Speech Recognition.

2. Source Routing: A bridged method whereby the source at a data exchange determines the route that subsequent frames will use.

SRAM Static Random Access Memory. A form of RAM that retains its data without the constant refreshing that DRAM (Dynamic RAM) requires. SRAM is generally preferable to DRAM because it offers faster memory access times, and can retain data without power. However, it is more expensive to manufacture because it has more electrical components. The most common use for SRAM is to cache data traveling between the CPU and a RAM subsystem populated with DRAM. This improves your PC's performance by reducing the number of DRAM accesses needed. See also DDR-SDRAM, DRAM, EDO RAM, Flash RAM, FRAM, Microprocessor, RAM, RDRAM, SDRAM, and VRAM.

SRAS See Special Routing Arrangement Service.

SRC 1. Strategic Review Committee (ETSI).

2. Stupid Rich Customer. One who will buy anything.

SRDC SubRate Digital Cross-connect.

SRDM SubRate Data Multiplexer. The Europeans call it SDM. An SRDM typically subdivides DS-0 of 64 Kbps, into a number of circuits, each less than 64 Kbps.

SRF 1. Specifically Routed Frame: A Source Routing Bridging Frame which uses a specific route between the source and destination.

2. Special Resource Function. Another term for an IP (Intelligent Peripheral) in the context of an AIN (Advanced Intelligent Network). The term IP became confusing with the advent of network and services based on the IP (Internet Protocol). See AIN and IP for more detail.

SRGF Speech Recognition Grammer Format is a VXML (Voice eXtensible Markup Language) language for writing voice recognition grammar. See also VXML.

SRL Structural Return Loss. A measure of cable impedance uniformity relative to its own impedance.

SRM 1. Sub-Rate Multiplexing. SRM. A technique used to combine data from a number of different digital sources into a basic rate channel, efficiently using the bandwidth on the primary rate for data circuits and/or digitized voice.

2. Storage Resource Management. SRM provides a central view of a company's storage resources and their usage. SRM includes capacity management, configuration management, event and alert management and policy management.

SRMS Service Request Management System.

SRP 1. Source Routing Protocol. See Bridge.

2. Suggested Retail Price.

SRS 1. Statistics Repository System.

2. Shared Registry System. A neutral, shared, and centralized repository containing the database of Internet domain name information. In conjunction with the expansion of the Domain Naming System (DNS), the Council of Registrars (CORE) has contracted with Emergent Corporation to build, maintain and operate the SRS. SRS supports up to 90 registrars, independent organizations which are authorized to assign the new TLDs (Top Level Domains), comprising .arts, .firm, .info, .nom, .rec, .shop, and .web. See also CORE, DNS, Domain, and URL for longer explanations.

3. Stimulate Raman Scattering. An optical fiber transmission term. Stimulated Raman Scattering (SRS) results from the interaction between the optical transmission signal and the silica molecules in the fiber. SRS affects broadband optical fiber transmission, and affects the overall optical spectrum involved in a DWDM (Dense Wavelength Division Multiplexing) transmission system. The SRS phenomenon manifests itself as a transfer of power from the shorter wavelengths to the longer wavelengths, resulting in a tilt of the optical spectrum. The effect increases as the power of the signal increases, and as the width (density) of the DWDM spectrum increases. See also DWDM.

SRT 1. Station Ringing Transfer.

2. Source Routing Transparent, a token ring bridging standard that is jointly sponsored by the IEEE and IBM. It combines IBM Source Routing and Transparent Bridging (IEEE 802.1) in the same unit. This provides a way for universal bridging of token ring LANs supporting IBM and all non-IBM LAN protocols. An SRT bridge examines each data packet on the ring to

discover whether the packet is using a source routing or non-source routing protocol. It then applies the appropriate bridging method. See also Bridge, Source Routing and Transparent Routing.

3. System Response Time. It is widely recognized that system response time (SRT), the time between the user's input and the computer's response, is one of the strongest stressors during human-computer interaction. Assessments of the effects of SRT have been conducted for personal computer use in a variety of contexts. As early as 1982, researchers determined that SRT and SRT variability act in concert to increase the stress levels of some personal computer users. With increased SRTs, users rate their general well-being as lower. Also accompanying longer delays are self-reports of annoyance, frustration, and impatience.

SRTS Synchronous Residual Time Stamp: A clock recovery technique in which difference signals between source timing and a network reference timing signal are transmitted to allow reconstruction of the source timing at the destination.

SS-CDMA Spread Spectrum Code Division Multiple Access.

SS1 See Signaling System Number 1.

SS2 See Signaling System Number 2.

SS3 See Signaling System Number 3.

SS4 See Signaling System Number 4.

SS5 See Signaling System Number 5.

SS6 See Signaling System Number 6.

SS7 See Signaling System 7.

SSA In 1993, IBM, working with a committee of other major manufacturers (Conner, Western Digital, Micropolis, etc), announced an architecture, named Serial Systems Architecture (SSA). This SSA removed some of the constraints of SCSI, particularly the limitation on the attachment of storage devices. Where SCSI allows seven devices to be attached on a string, SSA attaches 127 devices on a loop. SCSI uses bulky and expensive straps where SSA employs low-cost thin cabling. Data rates are also increased from 10MB/s on SCSI to 80MB/s on SSA which is expected to increase to 160 MB/s in 1995. The use of SSA will enable faster data transfer, increase the maximum storage capacity and procedure smaller devices, all at lower costs than today's equivalents. In May of 1994, 17 companies issued a joint press release announcing their commitment to SSA. The number of companies working quietly on the development of SSA products and devices exceeds that figure. See also Fibre Channel, Firewire and SCSI.

SSADM Structured System Analysis and Design Method, a structured set of procedural, technical and documentation standards, designed by the CCTA specifically for software development, used by many European suppliers of computer equipment.

SSAP Source Service Access Point.

SSB Single SideBand. See Single Sideband.

SSB-SC Single-SideBand Suppressed Carrier.

SSCF Service Specific Coordination Function: SSCF is a function defined in Q.2130, B-ISDN Signaling ATM Adaptation Layer-Service Specific Coordination Function for Support of Signaling at the User-to-Network Interface.

SSCOP Service Specific Connection Oriented Protocol: An adaptation layer protocol defined in ITU-T Specification: Q.2110.

SSCP 1. System Services Control Point. A host based network entity in SNA that manages the network configuration, coordinates network operator and problem determination requests, maintains network address and mapping tables and provides directory support and session services.

2. Service Specific Convergence Sublayer: The portion of the convergence sublayer that is dependent upon the type of traffic that is being converted.

SSD 1. Shared Secret Data. A secret key defined in the ANSI-41 (formerly IS-41C) standard, the SSD is used in cellular networks. In conjunction with the A-key (Authentication key), the SSD provides an authentication mechanism for cell phone security. Both keys are encrypted through the CAVE algorithm, and both are known only to the cell phone and the AC (Authentication Center). See also A-key, AuC and CAVE.

2. Solid state disk (drive). The solid-state disk (SSD) uses memory chips in place of the mechanical recording system used inside today's hard drives. Such a drive has several advantages including lower power consumption, higher data rates, is stone quiet, weighs little and far fewer moving parts (i.e. none). That means no head crashes. Flash memory technology isn't new and the advantages have been known for years but such solid-state disks have never been commercially produced before because flash has one big disadvantage over hard-drive storage: it's much more expensive.

SSH Secure Shell (SSH) is a UNIX-based command interface and protocol for securely getting access to a remote computer. Network administrators use it to control Web and other kinds of servers remotely. SSH is a suite of three utilities -slogin, ssh, and scp that are secure versions of the earlier UNIX utilities, rlogin, rsh, and rcp. SSH commands are encrypted and secure in several ways. Both ends of the client/server connection are authenticated using digital certificates and passwords are protected by being encrypted. SSH uses RSA Laboratory's public key cryptography for both connection and authentication. Encryption algorithms include Blowfish, DES, and IDEA. IDEA is the default. SSH2, the latest version, is a proposed set of standards from the Internet Engineering Task Force (IETF).

SSID Short for Service Set IDdentifier, a 32-character unique identifier attached to the header of packets sent over a wireless (Wi-Fi) network that acts as a password when a computer tries to connect to the network. The SSID differentiates one WLAN from another, so all access points and all devices attempting to connect to that wireless network must use the same SSID. A device will not be permitted to join unless it can provide the unique SSID. An SSID is a short, constantly-repeating transmission from a wireless device that contains its information about itself. Turning the "Broadcast SSID" option "off" on a wireless node will prohibit Netstumbler (and anyone else) from "seeing" the access point.

SSL The Internet has opened a new world of commerce through electronic transactions. To keep unsavory characters that prowl the alleys and byways of cyberspace from stealing important, sensitive information that is sent across a network, secure socket layer (SSL) is used to provide authentication, encryption, and message integrity services. Secure socket Layer is a transport level technology for authentication and data encryption between a Web server and a Web browser, i.e. sending documents around the Internet and the Web. Developed by Netscape, SSL negotiates point-to-point security between a client and a server. SSL sends data over a "socket," a secure channel at the connection layer existing in most TCP/IP applications. Both Internet Explorer and Mozilla Firefox support SSL, and many Web sites use the protocol to obtain confidential user information, such as credit card numbers. By convention, Web pages that require an SSL connection start with https: instead of http:. You can recognize the secure nature by the S. In short, SSL is the dominant security protocol for Internet monetary transactions and communications. Information being transmitted is encrypted, and only the user's Web browser and the computer server at the other end running the Web site have the key, and thus can understand what each other is saying. No one else can. See also S-HTTP.

SSML Speech Synthesis Markup Language is a VXML (Voice eXtensible Markup Language) language for text-to-speech markup. See also VXML.

SSN Subsystem Number. The address used in the Signaling Connection Control Part (SCCP) layer of the SS7 protocol to designate an application at an end signaling point. For example, an SSN for CNAM (Calling Name) at the end office designates the CNAM application within the end office. Usually, the SSN is a three digit number. BellSouth happens to use a CNAM SSN of 232.

SSO See Single Sign-On.

SSP Service Switching Point. Also called Signal Switching Point. A PSTN switch (End Office or Tandem) that can recognize IN (Intelligent Network) calls and route and connect them under the direction of an SCP (Service ControlPoint). A computer database that holds information on IN (IntelligentNetwork) services and subscribers. The SCP is separated from the actual SCP switch, making it easier to introduce new services on the network. See SCP. For a full explanation of the Advanced Intelligent Network, see AIN.

SSRP Simple Server Redundancy Protocol. The LANE simple server redundancy feature creates fault-tolerance using standard LANE protocols and mechanisms.

SST Spread Spectrum Technology. See Spread Spectrum.

SSTP 1. Switched Services Transport Protocol. See SCTP.

2. Double-Shielded Twisted Pair, as in Shielded-Shielded Twisted Pair. Also known as PiMF (Pairs in Metal Foil). A cabling system involving multiple twisted pairs contained within the same cable sheath. Each twisted pair is separately shielded by a metallic foil. The entire group of shielded pairs is then surrounded by another shield of metallic braid or foil. See also Category 7.

SSU Session Support Utility. a DEC-proprietary protocol that allows multiple sessions to run simultaneously over a single serial cable. SSU is used to allow terminals to provide two session "windows" that can display session output simultaneously.

SSUTR2 Sous-Syst0xE4me Utilisateur T0xC7l0xC7phonie RNIS. This French national SSUT standard for the SSUT+ was defined in 1986 by the ETSI. SSUTR2 and Telephone User Part + (TUP+) are closely related. The French TUP is the interface between fixed and mobile, using SS7.

ST 1. Start signal to indicate end of outpulsing.

2. Straight Tip. A fiber-optic connector designed by AT&T which uses the bayonet style

coupling rather than screw on as the SMA uses. The ST is generally considered the eventual replacement for the SMA type connector.

3. Signaling terminal.

ST Connection An optical medium connector plug and socket.

ST Connector See Straight-Tip Connector.

ST-506/412 Interface One of several industry standard interfaces between a hard disk and hard disk controller. The "intelligence" is on the controller rather than the drive.

ST-Bus Serial Telecom Bus. Mitel Semiconductor, which makes telecom componentry, has structured its digital component product line around the ST-BUS. The ST-BUS is a high speed, synchronous serial bus for transporting information in a digital format. Whether the digital information is voice, data, or video -or a mixture of each -the ST-BUS is designed to accommodate it. The ST-BUS consists of one or several serial data streams with a framing signal and clock signals. The framing signal always has a period of 125 us, resulting in 8,000 frames per second, with the original ST-BUS clock rate of 2.048 Mbit/s (thirty two 64 kbit/s channels). The ST-BUS standard now includes higher speed modes of 4.096 or 8.192 Mbit/s ST-BUS, resulting in 64 or 128 channels of 64kbit/s, respectively. This provides the bandwidth necessary for newer multimedia applications. According to Mitel, the advantages of using the ST-BUS are:

1. Printed circuit board area devoted to information transfer between functional modules is minimized.

2. Fewer tracks, backplane connections, and intra-shelf cables are needed compared to systems that use parallel paths.

3. The ST-BUS is designed to be divided down into individual channels of 64 kbit/s, resulting in improved efficiency and lower cost when several information paths are able to share the same ST-BUS.

4. Additional glue logic is not required when using ST-BUS compatible components.

5. From an IC perspective, the ST-BUS results in lower pin counts, improved reliability, and less power consumption.

STA 1. Spanning Tree Algorithm. A technique based on an IEEE 802.1 standard that detects and eliminates logical loops in a bridged network. When multiple paths exist, STA lets a bridge use only the most efficient one. If that path fails, STA automatically reconfigures the network so that another path becomes active, sustaining network operations.

2. Special Temporary Authority. A special temporary authority granted by the FCC to operators of Public Mobile Services (e.g., cellular radio and Specialized Mobile Radio, or SMR) "upon a finding that there are extraordinary circumstances requiring operation in the public interest and that delay in the institution of such service would seriously prejudice the public interest." See also IOA and SMR.

stability factor A cabling term. The difference between the percentage power factor at 80 volts/mil and at 40 volts/mil measured on wire immersed in water at 750xAFC for a specified time.

stack A set of data storage locations that are accessed in a fixed sequence. The list of internal instructions being executed by a computer is known as a "stack," just like the one made of trays at the start of a cafeteria line. When you add an instruction, you "push" it onto the stack; you remove an instruction with a "pop," just like you're taking off a tray.

stack smashing See Buffer Overflows.

stackable A term referring to devices/system the capacity of which can be increased through connecting (daisy-chaining) the device to additional devices. Thereby, it is not necessary to increase device/system capacity through a complete replacement, often known in the telephone equipment business as a "forklift upgrade." LAN hubs and switches are often stackable. Such devices often can be interconnected (stacked) in a wiring closet, See also Daisy Chain, Forklift, and Scalable.

stackable hubs Single protocol units designed to link various hubs, thereby effectively creating one large interconnected local area network (LAN). See Stackable.

stacked antennas The use of multiple identical antennas -generally dipole antennas -to improve reception. The antennas may be stacked vertically (a broadside array), horizontally end-to-end (a collinear array), or horizontally in parallel (an echelon or end-fire array). Also called ganged antennas.

staffing basis A call center term. The basis upon which staffing requirements are calculated. Can be either desired service quality for a given day of the week or the number of staff that will product the highest net revenue.

staffing requirements forecast A call center term. A calculation of the number of employees required in each period of the day to handle the forecast call volume for that period.

stage & test A term used in the secondary telecom equipment business. The installation (stage) and diagnostic testing of a PBX switch, cabinet, part, or peripheral in a reconfiguration center facility -where a dealer tests the complete system as one entity before shipment.

stage phoning Pretending to talk on a cell phone to impress bystanders. As British researcher Sadie Plant noted, "Some mobile users tend to make a virtue of the lack of privacy of stage phoning." This definition from Wired Magazine.

stagger In facsimile systems, periodic error in the position of the recorded spot along the recorded line.

stair stepping 1. Using a low-level account to gain ever-higher levels of unauthorized access in a network.

2. Video term. Jagged raster representation of diagonals or curves; correctively called anti-aliasing.

stalker site A Web site created by an obviously obsessed fan. "Have you seen that Gillian Anderson stalker site? The guy's got like 200 pictures of her!" A Wired Magazine definition.

stakeholder Corporate stakeholders, a termed coined by SAP, include employees, customers, partners and shareholders. As business competition intensifies and planning cycles speed up, corporations will need to keep all these constituents informed and involved, Kevin McKay, CEO of SAP America Inc. said in September, 1998. "We need to communicate with every stakeholder because this will open up new opportunities," McKay said.

stand alone Any device that can perform independently of something else.

standalone ACDs These Automatic Call Distribution switching systems are designated specifically for the call center environment. Characteristics include non-blocking capacity, high-powered CPUs, comprehensive real-time and historical management reports, and programmable call routing and treatment routines. They are standalone because they are separate pieces of equipment. Another way of providing ACD service is through a telephone company central office -in which case it would not be "standalone."

stand-alone drive An online drive that is not part of a library unit. Removable Storage treats stand-alone drives as online libraries with one drive and a port.

standard Standard is a Middle English term (from Old French, and with Germanic origins) meaning something such as a specification established as a yardstick, gauge, or criterion by authority, custom, or general consent. A de jure standard is a formal standard, a de facto standard is a standard without formal authority, and a d'jour standard is one enjoying current and temporary popularity. See also De Facto, and De Jure. See also standards.

Standard A Standard A is a first generation Inmarsat Maritime telecommunication system. It has mobile to fixed, and fixed to mobile, or in other words, shore to ship or ship to shore services) voice, telex, and Data. It runs at 56 or 64 kilobites per second, i.e. not very fast in today's terms. See also FMOD.

Standard Industrial Classification SIC. The classification or segmentation of businesses that are increasingly finite based on 2, 4, 6 or more digit identifiers. Developed by the U.S. Department of Commerce in the early 1960's.

standard industry practice Terminology used to indicate normal rules used within the secondary telecom equipment business. These rules have developed over time and usage, but lack formal support by industry groups or dealers. In short, there is no clear definition as to what "standard industry practice" means in the secondary business.

standard jack The means of connecting Customer premises equipment to a circuit as specified in the FCC Registration Program.

Standard Metropolitan Statistical Area SMSA. A metropolitan area consisting of one or more cities as defined by the Office of Management and Budget and used by the FCC to allocate the cellular radio market.

standard test zone A single-frequency signal with a standardization level generally used for level alignment of single links in tandem.

standardized test tone A single frequency signal at a standardized power level.

standards An agreed-upon rule, regulation, dimension, interface and/or, technical specification. Serving as a standard of measurement, weight or value; conformed to the official standard of a unit of measure or weight.

Standards are set by standards bodies, or by individual companies or by groups of companies working together. The main purpose of standards is to expand the market for products. For example, if every

Agreed principles of protocol. Standards are set by committees working under various trade and international organizations. RS standards, such as RS-232-C are set by the EIA, the Electronics Industries Association. ANSI standards for data communications are from the X committee. Standards from ANSI would look like X3.4-1967 which is the standard for the

ASCII code. The ITU (now called the ITU-T) does not put out standards, but rather, publishes "recommendations", owing to the international egos involved. "V" series recommendations refer to data transmission over the telephone network, while "X" series recommendations, such as X.25 (properly pronounced "Eks dot twenty five"), refer to data transmission over public data networks. Notice that the ANSI standards have the year they were approved as part of the name of the standard, while ITU recommendations do not. The placement of the "dot" is another clue as to whose confusing standard belongs to whom.

When you're buying a phone system, at minimum it should conform to four standards:
- Emissions compliance according to the FCC Part 15.
- Telephone compliance according to the FCC Part 68.
- Safety standards set by the National Electric Code, OSHA and the Underwriters Laboratories 1459.
- Bellcore compliance (from the Network Equipment Building System publication and their Generic Physical Design Requirements for Telecommunications Products and Equipment publication. See Standards Bodies.

standards bodies See the Appendix at the back of this dictionary.

standby monitor In a Token Ring network, a network node that serves as a backup to the Active Monitor and can take over in the event that the Active Monitor fails.

standby processor A spare computer exists which can direct PBX operations if the primary one fails. Some standbys are just sitting there, installed but not turned on. They require someone to turn them on. Some standbys are actually running all the time, as the main one is. If the main one crashes, the standby processor is ready to take over.

standby time The amount of time you can leave your fully charged cellular portable or transportable phone turned on to receive incoming calls before the phone will completely discharge the batteries. See Talk Time.

standing wave When you look at it on an oscilloscope the pattern of the wave is perfectly flat, i.e. horizontal. It's caused by two sine waves of the same frequency moving in opposite directions. In transmission line theory the accepted definition is simply the superposition of two waves traveling in opposite directions.

Standing Wave Ratio SWR. The ratio of the amplitude of a standing wave at an anti-node to the amplitude at a node.

star 1. A topology in which all phones or workstations are wired directly to a central service unit or workstation that establishes, maintains and breaks connections between the workstations. Virtually all phone systems are stars configurations. ISDN BRI bus will be the first phone system to operate on a bus. In datacom language, the center of a star is called the hub. The advantage of a star is that it is easy to isolate a problem node. However, if the central node fails, the entire network fails. The star network we're all most familiar with is our local telephone exchange. At the center (the hub) rests the central office. Spanning out in a star are the lines going to the individual workstations (telephones) in peoples' houses and offices.

2. Advanced telecommunications for the industrially less advanced regions of the European Community.

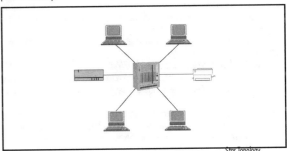
Star Topology

star button The star button on the touchtone phone is often used to mean "No" in interactive voice response or computer telephony systems.

star coupler A device that couples multiple fibers at a central point and distributes the signal from one fiber into all others simultaneously.

star key The star key is the bottom left key on your telephone's touchtone pad, i.e. the * key. See also the definition for #, which is at the front of the dictionary.

star network A computer network with peripheral nodes all connected to one or more computers at a centrally located facility.

star quadded cable Spiral-four cable. See Star Network.

star topology A LAN topology in which end points on a network are connected to a common central switch by point-to-point links. See Star.

stare and compare When a phone company installs a new billing system, it is common practice to run the old and new systems in parallel during testing. The same billing inputs are fed to both systems and each bill that is produced by the new system is visually inspected by a member of a bill validation team to see whether the bill's line items, subtotals, and totals match those on the corresponding bill produced by the old system. This type of visual bill validation process is called a "stare and compare." A "stare and compare" is also done in other parts of the business. For example, a central office technician may troubleshoot a circuit problem by doing a "stare and compare," i.e., he will visually compare the switch record for the problematic circuit with the switch record for a problem-free circuit, to see whether the former is missing field values or has incorrect field values.

StarLAN An obsolete local area network developed by AT&T using twisted pair telephone wires in a star configuration.

start bit In asynchronous data communications, characters are sent at arbitrary intervals, i.e. when the operator hits a key. In order for the computer to make heads or tails of what's coming in, each character starts its transmission with a Start Bit. This way, if the first bit of the character to be transmitted is a 1, the fact of receiving a Start Bit (always a 0) tips off the computer that the next bit is part of a transmitted character and not just part of the inter-character gap. See Stop Bit.

start element 1. The start pulse of a transmission character. It is used for synchronization of the following bits in a serial transmission process. 2. One of the input or output points in a communications system. This would include a telephone set, a data terminal, a computer communications port.

Start Of Heading SOH. A control character used in data communications that designates the beginning of the message header.

Start Of Heading Character SOH. A transmission control character used as the first character of a message heading.

Start Of Message SOM. A control character used in data communications that designates the beginning of the message.

Start Of Text STX. A control character used in data communications that designates the beginning of the text being transmitted.

start stop transmission The technique of asynchronous data transmission wherein each character is comprised of a start element at its beginning and a stop element at its end. Start-stop elements allow the receiving device to determine where the transmitted bits for one character ends and the next begins.

start time interval A call center term. A scheduling rule that governs the times at which schedules can start; for example, at 15-minute intervals as opposed to 30-minute intervals.

STAT MUX Informal for STATistical MUltipleXor.

state 1. The condition of a connection within a telephone call that reflects what the past action on that connection has been and that determines what the next set of actions may be.

2. The instantaneous properties of an object that characterize that object's current condition. See State Machine Programming.

state machine programming To control multiple telephone lines in a single voice processing program, a new program structure is required. Dialogic calls this technique state machine programming. Computer Science called state machines "Deterministic Finite State Automata."

State Public Service Commission PSC. The State legislative body responsible for among other things, regulating the operation of telephone companies and other persons involved in the furnishing of telephone service. Some states' PSCs are called Public Utilities Commissions. See the next definition.

state signal In MF pulsing, a signal used to indicate that all digits have been transmitted.

state tax Two Out Of Three Rule. When determining state tax jurisdiction for the purpose of figuring phone bills, there are three locations to consider: originating station, destination station, and the location that the bill is sent to. If two out of three are the same, then that state receives the tax.

state transition The act of moving from one state to another.

state utility commissions Each state has a utility commission responsible for the regulation of telephone service provided wholly within that state. Regulation extends to introduction of new services, their prices, who will provide them, as well as discontinuance

of existing services.

stateful Protocols that maintain information about a user's session. FTP is a stateful protocol. Stateless is the opposite.

stateful inspection firewall A stateful inspection firewall examines the contents of individual packets at all layers of the OSI model, from the network layer to the application layer. To perform this task, this firewall relies on packet-filtering algorithms to examine and compare each packet against known bit patterns of authorized packets.

Stateful Packet Inspection See SPI and Stateful Inspection Firewall.

stateless Protocols that do not maintain information about a user's session. Each transmission is considered a new session. HTTP is a stateless protocol.

statement 1. In computer programming languages, a language construct that represents a set of declarations or a step in a sequence of actions.

2. In computer programming, a meaningful expression or generalized instruction represented in a source language.

static 1. Unchanging. Fixed.

2. Interference caused by natural electric disturbances in the atmosphere, in your office, in your home. Static electricity can play havoc with telephone systems, computers, sound systems, in short anything electronic. Properly grounding your phone system to a true cold water pipe (not one that connects to a PVC plastic pipe) is the most minimal protection. Walk across a nylon carpet in your plastic shoes, touch something. You'll see a blue spark jump between you and what you touched. That blue flash can be thousands of volts. That blue flash can damage sensitive electronics. That's why factory workers always wear static electricity straps. These straps fit around their wrists and attach to a ground somewhere close by. That way any static electricity they generate will go to ground through the strap not through the valuable electronics they are touching.

static condition a cabling term. used to denote the environmental conditions of an installed cable rather than the conditions existing during cable installation.

static document A static document is one that doesn't change. Basically all web site are designed to send you static pages written in a language called HTML. If you want an updated version of that page, you have to ask your browser to refresh the page. Some web sites will feed you dynamic (i.e. changing) information but you typically have to have some software running on your computer. A simple browser won't cut it. Solutions are coming.

static IP address See Dynamic IP Address and IP Address.

static object Information that has been pasted into a document. Unlike embedded or "linked" objects, static objects cannot be changed from within the document. The only way you can change a static object is to delete it from the document, change it in the application used to create it, and paste it into the document again.

static positioning Location determination when the GPS (Global Positioning System) receiver's antenna is presumed to be stationary in the earth. This allows the use of various averaging techniques that improve the accuracy of figuring where you are by factors by over 1000.

static RAM Static Random Access Memory chips do not require a refresh cycle like Dynamic RAM chips and thus can be accessed well over twice as quickly. Static RAM chips must have power to maintain the data they are holding. Static RAM chips cost more than Dynamic RAM chips, which are also called D-RAM.

static route A route that is manually entered into a routing table. Static routes take precedence over routes chosen by all dynamic routing protocols. See Static Routing.

static routing Static routing involves the selection of a route for data traffic on the basis of routing options preset by the network administrator. Dynamic routing, on the other hand, adjusts automatically to changes in network topology or traffic. Dynamic routing automatically accomplishes load balancing, therefore optimizing the performance of the network "on the fly." Dynamic routing is more effective, but the routers are more costly and the more complex decision-making process imposes additional delays on the subject packet traffic. In addition to being less costly, static routing offers enhanced security. As there is only a single, pre-programmed route, it is impossible to bypass a security mechanism such as a firewall. See also Link State Routing, Distance Vector Routing, Path Vector Routing Protocol, and Router.

static wire A grounded wire at the very top of a telephone or utility pole intended to protect lower conductors (i.e. telephone, CATV, etc.) from lightning. See Joint Pole.

station 1. A semi-dumb word for a telephone. Also called an instrument, or a telephone instrument. An extension station is one connected "behind" a PBX or key system. In other words, the PBX or key system is between the station and the telephone central office. Howard Pena, a reader of my dictionary, wrote me that he had heard a story (several times) that the telegraph system in the US was based on a network of telegraph wires strung between railroad stations. When the telephone system reached the American frontier, the most efficient way to install it was to replace each telegraph with a telephone. So, when you placed a call, you were calling a station. If that's true, Station is not a "dumb word for a telephone." The opposite is true. It's a strong reminder that without the PBX, we'd all be standing in line to use the village telephone.

Ray Horak did "a good bit of research on this, but can't verify Howard's story." Makes sense, however. I agree with his criticism of the current definition. The dictionary defines "Station" as a stopping place. It's used in many contexts, including railroads, power transmission, and the bible (stations of the cross). Also used to describe a place established to provide a public service, e.g., fire station and police station.

For my taste, I suspect the word "station" comes from the very old days when regulation of the telephone industry was taken over by the Interstate Commerce Commission, (the ICC), which also at that time regulated the railroad industry.

2. A shortened word for workstation -a name for a PC on a LAN.

station adapters Cables and interface assemblies for connecting Dialogic network interface and switching products to telephones or analog telephone lines.

station apparatus The equipment which is installed on the customer's premises, including phones, ancillary electronics and small hardware.

station auxiliary power supply This device is used to provide power to an electronic phone that is connected more than 300 meters (or 1,000 feet) or so away from the Key Service Unit.

station battery A separate power source which provides the necessary DC power to drive a telephone system. Individual telephones are usually powered by a central source, i.e. their PBX or central office. The batteries may also power radio and telephone equipment as well as provide emergency lighting and controls for equipment. See Battery.

station busy lamps Lamps located on a station instrument, providing visual indication of each busy phone in the system. Busy Lamp Fields (BLFs) often come on key systems and sometimes on smaller PBXs. They're very handy.

station busy override Pre selected phones have the privilege and ability to preempt busy circuits and override a private conversation.

station call transfer A phone user can transfer incoming and outgoing calls to another phone without attendant assistance.

station camp-on Phones can camp-on to a busy extension. The camped-on phone will be notified of the camp-on by a special beep signal. The person at the other end may or may not hear the signal.

station clock The principal clock or alternative clock located at a particular station providing the timing reference for all major telecommunications functions at that station. A station clock may also be used to provide timing or frequency signals to other equipment.

station code The final four digits of a standard seven or 10 digit telephone number.

station conductor A wire that terminates at the equipment side of the lightning protector.

station direct station selection The phone user places a call to an extension within the PBX by pushing a single pushbutton on his phone.

station equipment Telephone instruments and associated equipment furnished to subscribers. We suspect that the word "station" came from early telephone industry which was regulated by the same government agency which regulated the railroad business.

station hunting This feature allows a calling phone which places a call to a busy phone to proceed to the next idle phone in the hunt group. This jump is done automatically. See also Rotary Hunt, which is the same thing for trunks.

station keeping The process on board a satellite for keeping it at its assigned longitude and inclination.

station line cards Station line cards sit inside telephone systems and drive a bunch of connected phones. These cards translate the software inside the phone system into electrical impulses which tell the phones at the other end what they're capable of and let them do things, like dial, transfer, conference, etc.

station line protector Circuitry that protects the telephone system from high voltage hits and lightning strikes. Such circuitry is usually on every station line card.

station load The total power requirements of the integrated station facilities.

station message detail recording Now refers to the RS-232-C "port" or plug found on the back of most modern PBXs and some larger key systems. See Call Accounting System and Call Detail Recording.

station message registers Message unit information centrally recorded on a per-station basis for each completed outgoing call.

station message waiting Special light on a phone to alert hotel/motel guests of messages waiting at the front desk.

station monitoring Selected phones can monitor (i.e. listen in on) any other phones in the system.

station override security Designated phones can be shielded against executive busy override (presumably other executives).

station protector A station protector protects phones and other phone-like devices ("stations") from lightning. A station protector is typically a gas discharge, carbon block or other device that short circuits harmful voltages to ground in the event of lightning strikes on the phone line. Sometimes it works. Sometimes it doesn't. If a bad thunderstorm is about to erupt over your house or office, it's a good idea to unplug your phones from AC and from phone lines.

station rearrangement and change Allows a user to move phones, change the features and/or restrictions assigned to phones and administer features associated with telephones.

station review A study of how people in an organization use the telephones and what communications needs are not being satisfied.

station ringer cutoff Allows the ringer on the telephone to be turned off. Not always a good idea, since calls may still come in for that phone, but no one may pick them up because they don't hear it ring.

station set Another word for a common desk telephone. Station comes from earliest days when the phone industry in the U.S. was regulated by the same agency that regulated the railroads. It made phones stations, thus easier for the government bureaucrats.

station to station call A directly dialed call. No operator is used. Most calls are now directly dialed. Some long distance companies don't even have operators to help complete calls. AT&T still does.

station tone ringing Electronic tone ringer that replaces the bell.

station transfer security If trunk call is transferred from one phone to another, and the second phone does not answer within a predetermined time, the trunk call will automatically go to the attendant.

station visual signaling Lamp on a phone which indicates flashing incoming, steady busy, and "wink" hold visual conditions associated with that phone.

station wire SW. The twisted-pair copper telephone wire used to connect stations (i.e., terminal equipment) to PBXs. Also, the telephone wire used inside the residence to connect telephone sets and computer modems to the NIU (Network Interface Unit) that provides the interface to the local loop that connects your house to the telephone company's CO (Central Office).

stationary orbit An orbit, any point on which has a period equal to the average rotational period of the Earth, is called a synchronous orbit. If the orbit is also circular and equatorial, it is called a stationary or geostationary orbit.

statistical equilibrium A telephone company definition. A state of traffic in which, over any considerable length of time, the call arrivals and departures are essentially equal. Traffic that is in statistical equilibrium has an average value of some measure of its level (such as the number of attempts arriving in a specified time interval) that does not change with time.

statistical multiplexing A multiplexing technique that differs from simple multiplexing in that the share of the available transmission bandwidth allocated to a given user varies dynamically. In other words, in statistical multiplexing, a channel is assigned only to devices (e.g., telephone, data terminal or fax machine) which are active and seeking to communicate. Static multiplexers, the original multiplexers, dedicated a channel to a device whether it was active or not. This was horribly wasteful, as devices commonly are inactive. As a result, statistical multiplexing is much more powerful than normal static FDM (Frequency Division Multiplexing) or static TDM (Time Division Multiplexing). In other words, Statistical Multiplexers act as contention devices, as well as multiplexers, making intelligent decisions about providing access to expensive bandwidth based on programmable parameters such as first-come-first-served, application priority (e.g., data vs. voice), and bandwidth reservations. See also Fast Packet Multiplexing, FDM and TDM.

statistical process control The mathematics of quality control in manufacturing.

statistics Statistics is the he only science that enables different experts using the same figures to draw different conclusions. 74.3% of statistics are made up on the spot. People will accept made-up statistics if:

1. They seem plausible. You couldn't say that 67% of people are women. It's not plausible. But 50.13% is. Simple statistics, like 75% of...are generally not believable. But 75.1% is, and

2. No one has any better statistics to contradict you.

3. If they want to believe because it suits their ends, or most importantly, if they can use your statistic to justify something they want to do.

4. If your statistics are repeated often enough and in many diverse places by many diverse sources.

Some wags also define statistics as numbers looking for an argument. My favorite statistic is that, on average, everyone has one testicle. My second favorite is: If 4 out of 5 people SUFFER from diarrhea...does that mean that one enjoys it? My third one is that marriage is the number one cause of divorce.

statistics port In network management systems, interface for reporting events and status.

STATMUX A statistical multiplexer. See Statistical Multiplexing.

status information Information about the logical state of a piece of equipment.

status signal unit Signal unit of CCS used to initiate transmission on a link or to recover from loss of transmission.

statute mile A unit of distance equal to 1.609 km, 0.869 nmi, or 5,280 ft.

stay or bail moment The precious few seconds that elapse after loading the front page of a Web site, during which one decides to either stay or leave. Defined by David Siegel in his book "Creating Killer Web Sites."

STB See Set Top Box.

STC 1. Society of Telecommunications Consultants. A professional society for telecommunications consultants. They endeavor to set standards of behavior for the consulting community, chiefly to avoid having telecom consultants recommend to their clients equipment they receive a secret commission on from the manufacturer.

2. System Time Clock: The master clock in an MPEG-2 encoder or decoder system.

3. Sound Transmission Class. This rating system provides a standard industry accepted method for comparing the sound reduction effectiveness of various soundproofing materials.

STD Subscriber Trunk Dialing. An non-North American term for direct distance dialing, i.e. dialing long distance calls directly without an operator's assistance. Pricing for long distance calls is typically done by billing a standard amount of money, e.g. a German Mark, for a length of speaking time, which shortens the further you call. For example. you might get one minute for a Mark if you're calling 50 miles. If you call 200 miles you might only get 20 seconds.

STDM Statistical Time Division Multiplexer. STDMs are TDMs (Time Division Multiplexers) with an added microprocessor that provides more intelligent data flow control and enhanced functionality, such as error control and more sophisticated user diagnostics. The major difference between TDMs and STDMs is that stat muxes dynamically allocate time slots on the link to inputting devices on an as-needed basis (rather than in round-robin fashion where all devices are polled in preordained order). Therefore, there is no idle time on the link because a device does not have information to send. Unlike TDMs, STDMs have buffers for holding data from attached devices. They can handle a combined input speed (aggregate speed) that exceeds the speed of the communications link.

STE 1. Station Terminal Equipment.

2. Section Terminating Equipment. SONET equipment that terminates a section of a link between a transmitter and repeater, repeater and repeater, or repeater and receiver. This is usually implemented in wide area facilities and not implemented by SONET Lite. STE Network elements perform section functions such as facility performance monitoring. The section is the portion of a transmission facility between a lightwave terminal and a line repeater or between two line repeaters.

3. Spanning Tree Explorer: A Source Route Bridging frame which uses the Spanning Tree algorithm in determining a route through a network. Often used in ATM networks.

steady-state condition 1. In a communication circuit, a condition in which some specified characteristic of a condition, such as value, rate, periodicity, or amplitude, exhibits only negligible change over an arbitrarily long period of time. 2. In an electrical circuit, a condition, occurring after all initial transients or fluctuating conditions have damped out, in which currents, voltages, or fields remain essentially constant or oscillate uniformly without changes in characteristics such as amplitude, frequency, or wave shape. 3. In fiber optics, synonym for equilibrium mode power distribution.

stealth tower A wireless communications antenna disguised as part of the natural or urban landscape. The antennae can appear as a flagpole, a group of rocks, a church steeple, a cactus or large tree. According to Wired news there were about 130,000 communications antennae in place across the United States in 2002, of which roughly 75 percent were standard antennae and the rest "surreptitiously stashed in scenic simulations." Many stealth towers give amusement a whole new meaning. Stealth is also a trade name for a company that makes this type of site.

stealth URL A Web site whose domain name deliberately resembles that of a popular website, and where users land when they mistype the popular site's name. Check out www.whitehouse.com which many people type when they really mean to type www.whitehouse.gov.

steel and electricity revolution, The steel and electricity revolution took place from about 1875 to 1920 in the United States and Germany. It was an age of massive engineering and the electrification of the economy.

steerable beam antenna An antenna whose main beam can be directed in various directions either by an electrical or mechanical drive system.

steganography Steganography, also known as "stego," translates from Greek as "covered writing," is a method of hiding one piece of information within another. The ancient Greek historian Herodotus described how a countryman sent a secret message warning of an invasion by carving it into the wood underneath a wax tablet. Slaves would sometimes have their heads shaved and secret messages tattooed on their scalps. After their hair grew back, they would be sent on missions. When they ultimately reached their destinations, their heads would be shaved and the messages read. Contemporary stego is the practice of encoding hidden messages into the least significant bit of other files, such as graphic, audio and HTML files, which have lots of unused space. Inexpensive, commercially available software programs are available which allow you and me to create stego files and include our very own messages. Some even allow messages to be hidden in the whitespace at the end of a line of a text file or e-mail message. Once the data is encoded, it can be decoded only with the proper password. The terrorist Osama bin Laden allegedly uses stego files to transmit maps, photos and instructions outlining future targets via pornographic bulletin boards and sports chat rooms. If the stego message is encrypted, it may be virtually impossible to decode it, assuming that it can be found. Steganography is used to watermark copyrighted digital material. See also Encryption and Watermark.

Stego See Steganoprahy.

Stella Awards The Stella Awards are named after 81-year-old Stella Liebeck who spilled coffee on herself and successfully sued McDonalds. That case inspired the Stella awards for the most frivolous successful lawsuits in the United States. Among recent candidates:

- Kathleen Robertson of Austin, Texas, was awarded $780,000 by a jury of her peers after breaking her ankle tripping over a toddler who was running inside a furniture store. The owners of the store were understandably surprised at the verdict, considering the misbehaving little toddler was Ms.Robertson's son.
- Terrence Dickson of Bristol, PA, was leaving a house he had just finished robbing by way of the garage. He was not able to get the garage door to go up since the automatic door opener was malfunctioning. He couldn't re-enter the house because the door connecting the house and garage locked when he pulled it shut. The family was on vacation and Mr.Dickson found himself locked in the garage for eight days. He subsisted on a case of Pepsi he found and a large bag of dry dog food. He sued the homeowner's insurance claiming the situation caused him undue mental anguish. The jury agreed to the tune of $500,000.
- Mr. Merv Grazinski of Oklahoma bought a brand new 32-foot Winnebago motor home. On his first trip home, having driven onto the freeway, he set the cruise control at 70 mph and calmly left the drivers seat to go into the back and make himself a cup of coffee. Not surprisingly, the RV left the freeway, crashed and overturned. Mr. Grazinski sued Winnebago for not advising him in the owner's manual that he couldn't actually do this. The jury awarded him $1,750,000 plus a new motor home. The company actually changed their manuals on the basis of this suit, just in case there were any other complete morons buying recreation vehicles.

Stentor Stentor was formed in 1992 as the long distance network of nine regional Canadian phone companies. Inspiration for the name came from the Greek poet Homer. He had immortalized Stentor, a warrior in the Trojan war, "whose voice was as powerful as the voices of 50 other men." And so the adjective "stentorian" survived to be applied to someone with a powerful voice, often a politician. The nine original members of Stentor are Bell Canada, British Columbia Telephone (BC Tel), AGT Limited, Manitoba Telephone Systems, SaskTel, Maritime Telephone, New Brunswick Telephone, Island Telephone and Newfoundland Telephone. On January 1, 1999, Stentor got changed. The members said they would concentrate their co-operative efforts in an enhanced Stentor Canadian Network Management (SCNM). to ensure continuity and seamless service for customers. Members of the alliance are now Bell Canada, British Columbia Telephone (also called BC TEL), Island Tel, Manitoba Telecom Services Inc., Maritime Tel & Tel, New Brunswick Telephone (also called NBTel), NewTel Communications, NorthwesTel, QuOxC7bec TOxC7lOxC7phone,

SaskTel and TELUS.

step 1. One movement of an electromechanical switch which typically corresponds to one impulse from a rotary dial or one impulse from a touch tone phone which has been converted to a rotary dial. See also SxS.

2. An abrupt change in the refractive index between the core and the cladding in an optical fiber. See also Refractive Index and Step-Index Fiber.

Step By Step SXS. An automatic dial-telephone system in which calls go through the switching equipment by a succession of electromechanical switches that move a step at a time, from stage to stage, each step being made in response to the dialing of a number. SXS is electromechanical switching. It was invented in the 1920s. See SxS.

step call The phone user can, upon finding that the called phone is busy, call an idle nearby phone by merely dialing an additional digit.

step down 1. To reduce the voltage. Such reduction in voltage will increase the current.

2. A feature of fax machines that makes them drop their transmission speed when the quality of the phone lines they are transmitting over begins to deteriorate. Dropping the transmission speed is the major way of getting the faxes through on "dirty" lines. All Group III fax machines have "step-down" as a built in feature, or should have.

step down transformer A transformer wound to give a lower voltage on the secondary side than that impressed on (i.e. put into) the primary. The current, however, will be stepped up. A step down (often spelled step-down or stepdown) transformer has more primary than secondary turns. See also Stepdown Transformer and Joint Pole.

step index fiber A type of MultiMode Fiber (MMF) marked by an abrupt change, or step, in the refractive index between the core and the cladding. Errant light rays that strike the core/cladding interface at various angles are reflected back into the core. However, some light rays that make up an individual digital pulse travel longer distances than others, with those traveling down the center of the core traveling the shortest distance and those bouncing around at the steepest angles traveling the longest distances. This causes a phenomenon known as modal dispersion, with a mode being the physical path taken by a light ray. Modal dispersion has the effect of causing the individual light pulses to lose their shape, i.e., smear, and run together. The longer the distance of the cable run, the greater effect. The higher the bit rate, the shorter the bit time and the greater the effect. As a result, step-index fiber is limited to relatively short haul, low speed applications. Actually, it seldom is used any longer. Rather, graded-index fiber generally is used, as it overcomes the effects of modal dispersion. See also Graded-Index Fiber, Modal Dispersion, Multimode Fiber and Refractive Index.

step index profile For an optical fiber, a refractive index profile characterized by a uniform refractive index within the core and a sharp decrease in refractive index at the core-cladding interface. See also Step Index Fiber.

step tracking A method of tracking in which the antenna beam is sequentially moved in a predetermined pattern. A maximum signal output of the receiver indicates that the beam is pointed at the satellite.

step up transformer A transformer wound to give a higher voltage on the secondary side than that impressed on the primary. The current, however, will be stepped down. It has fewer primary than secondary turns.

step-by-step See Step by Step.

stepdown transformer An oil-cooled transformer often mounted a telephone pole. Such transformer converts the primary voltage to the secondary voltage. Most step-down transformers are designed for single-phase operation; if a three-phase secondary circuit is required, three physical transformers are sometimes mounted on the same pole. See Joint Pole.

Stephanie Long Every so often, you meet someone truly wonderful and sweet, who brings great joy to one's life. For me that person is a young lady Dartmouth classmate of my son Michael. I'm not allowed to adopt Stephanie since she has a legitimate mother and father who love her. But I do, too (and so do Michael, Susan and Claire). So, for now, I'm her unofficial adopted father, if there is such a thing. There's nothing I wouldn't do for Stephanie, who is, as I write this, studying to be the world's best doctor -which she will be. Go Stephanie.

stepped index Referring to a type of optical fiber which exhibits a uniform refractive index at the core and a sharp decrease in the refractive index at the core-cladding interface.

stereophonic crosstalk An undesired signal occurring in the main channel from modulation of the stereophonic channel or that occurring in the stereophonic channel from modulation of the main channel.

stereophonic sound subcarrier A subcarrier within the FM broadcast baseband used for transmitting signals for stereophonic sound reception of the main broadcast program service.

stereophonic sound subchannel The band of frequencies from 23 kHz to 99 kHz containing sound subcarriers and their associated sidebands.

steward An entity that is responsible for maintaining the authoritative copy of a set of corporate data, as well as ensuring the semantic integrity of that set of data.

STG An imaging term. Scale To Gray. STG uses gray pixels to fill in jagged edges of document images. STG improves readability. According to a study commissioner by Cornerstone and done by Dr. Jim Sheedy, the ability to read STG images was improved between 4% and 19%, depending on the resolution, and symptoms such as headaches, tired back, blurred vision were cut way down.

STFS Standard Time and Frequency Signal.

STICI Pronounced sticky. It stands for Self Teaching Interpretive Communicating Interface. It's being touted as the "next wave of user interface" (the next wave after GUI). According to BIS Strategic Decisions, features of STICI include:

Self-Teaching: The operating system (OS) and interface use agents (special background processes) to study how the user makes use of the device. For example, agents will track exactly how a user use the OS and applications.

Interpretive: The system is able to make inferences based on the information it has collected about the user. The system does some interpretation, moving from the traditional interface approach of "Do what I say" to "Do what I mean." Unlike the static menus and dialog boxes of GUI systems, STICI systems will dynamically adapt their operation, thereby better anticipating user needs. Agents will automate common tasks, based on observed usage patterns.

Communicating: The system will be able to manage all the different communications functions offered, such as store and forward, cellular, logging on and off wireless LANs and linking to the user's desktop PC. These communications will be transparent to the user, leaving him or her free to concentrate on the task at hand.

Interface: The interface will be oriented around documents. Traditionally, interfaces have centered on applications, such as a word processor or a spreadsheet. With the STICI, the interface is centered on documents. The user creates a document, writing or drawing freely, and the various applications needed are simply tools accessed to create a chart, to write or to show numbers within that document. Users will be able to seamlessly link applications from multiple vendors, as descendent technologies evolve from OLE and Publish & Subscribe. Both Microsoft and Apple are developing document-oriented interfaces for the next generations of their respective desktop operating systems.

stick-and-click A push-pull cable connector, like those used with common audio and video plugs and sockets.

stick-and-turn A cable connector that uses a half-twist bayonet-type lock to keep the connection secure.

stickiness The New York Times calls it "the marketing buzzword of the moment." A sticky Web site keeps users glued to it, according to the Times, "either through sheer intrinsic niftiness or by piling layer upon layer of more or less related offerings, like stock quotes, weather updates or interactive whiz-bangs like sports trivia."

sticky 1, A sticky shift key lets you access the shifted functions (such as capital A) by pressing the shift key first and then pressing the second key. Sticky keys may stay down for a second or two. Or you may have to hit them again to unstick them -somewhat like the CapsLock key.

2. An adjective applied to a web site or a service where visitors hang around. For instance, a banking site that offers a financial calculator is stickier than one that does not because visitors do not have to leave to find a resource they need. When AT&T Wireless introduced voice portals to its customers one of its executives said, "When you provide a service like this it makes your relationship with customers more sticky." In other words, customers are more likely to stick with companies that offer services they like. You restrain your chuckle. See also Stickiness.

sticky note See Post-it Note.

STID Service Termination Identifier. An ISDN Service Profile term.

stimulated emission Radiation emitted when the internal energy of a quantum mechanical system drops from an excited level to a lower level when induced by the presence of radiant energy at the same frequency. An example is the radiation from an injection laser diode above lasing threshold.

STL 1. Standard Telegraph Level. 2. Studio-To-Transmitter link -typically through the air microwave.

STM Synchronous Transfer Mode. A transport and switching method that depends on information occurring in regular and fixed patterns with respect to a reference such as a frame pattern. A time division multiplex-and-switching technique to be used across the user's network interface for a broadband ISDN. It gives each user up to 50 million bits per second simultaneously -regardless of the number of users. See also ATM.

STM-1 Synchronous Transport Module 1: SDH standard for transmission over OC-3 optical fiber at 155.52 Mbps.

STM-n Synchronous Transport Module "n": (where n is an integer) SDH standards for transmission over optical fiber (OC-'n x 3) by multiplexing "n" STM-1 frames, (e.g., STM-4 at 622.08 Mbps and STM-16 at 2.488 Gbps). The SONET version is known as STS (Synchronous Transport Signal), beginning at 51.84 Mbps.

STM-nc Synchronous Transport Module "n" concatenated: (where n is an integer) SDH standards for transmission over optical fiber (OC-'n x 3) by multiplexing "n" STM-1 frames, (e.g., STM-4 at 622.08 Mbps and STM-16 at 2.488 Gbps, but treating the information fields as a single concatenated payload).

STN 1. Statens Telenamd (Swedish National Telecommunications Council).

2. Super Twist Nematic is the least expensive and most basic form of passive-matrix LCD display. It is used in low cost laptop computers. In a passive matrix color screen, like STN, the current travels along transparent electrodes printed on the glass screen. These electrodes are driven by transistors placed around the edges of the display. Horizontal and vertical electrodes form a grid-like matrix, with a pixel at every intersection. A major problem with passive technology arises when current is lost in crosstalk as the electrodes criss-cross each other. This crossing over effect greatly diminishes overall display quality. See Active Matrix, LCD and TFT.

stockbroker Someone whose services make you broker. See also Small Fortune.

stop band A spectrum of frequencies blocked by an electronic (filter) device and usually defined by upper and lower -3 dB points.

stop bit The Stop Bit is an interval at the end of each Asynchronous Character that allows the receiving computer to pause before the start of the next character. The Stop Bit is always a 1. See Start Bit.

stop element The last element of a character in asynchronous serial transmission, used to ensure recognition of the next start element.

stop order An order that becomes a market order to buy (buy stop) or a market order to sell (sell stop) only when the security trades at a specific price, known as the stop price. A buy stop order is placed above a stock's current market price and is executed if the market rises to, or through, that price. A sell stop order is placed below a stock's current market price, and is executed if the market falls to, or through, that price.

stop record signal In facsimile systems, a signal used for stopping the process of converting the electrical signal to an image on the record sheet.

stop/start transmission A method of transmission in which a group of bits are preceded by a start bit and followed by a stop bit. Also called asynchronous transmission. See Asynchronous.

Storage Area Network See SAN.

storage blade See Blade Server.

storage bytes I made this term up because of the need to explain that one million bytes can be different to one million bytes. OK. Normally, one million means 1,000,000. And it does, except inside a computer, where they measure storage in bytes. Your hard disk contains this many bytes, let's say eight gigabytes. That's fine. But they're not bytes the way we think of them in internal or external computer transmission terms. They're different and they have to do with a way computer stores material -on hard disks or in RAM. They're what I call "storage bytes." When we talk 1 Kb of storage bytes, we really mean 1,024 bytes. Which comes from the way storage is actually handled inside a computer, and calculated thus: two raised to the power of ten, thus $2 \times 2 \times 2 \times 2 \times 2 \times 2 \times 2 \times 2 \times 2 \times 2 = 1,024$. Ditto for one million, two raised to the power of twenty, thus 1,048,576 bytes. See also the introduction to this dictionary and BPs.

storage unit A device in which information can be recorded and retained for later retrieval and use.

storage virtualization In its purest form, according to InfoWorld, virtualization allows users to add storage capacity using inexpensive commodity disk and tape drives to dynamically manage those storage resources as virtual storage pools with little regard for what physically resides on the back end. InfoWorld lists the advantages of storage virtualization as follows: 1. Virtual pools of storage. Replacing the direct links between host servers and physical disks with logical volumes makes storage a plastic container that can be shared among hosts and applications. 2. Speedy Recovery. The use of logical volumes simplifies the creation of backups for quick recovery from failures or disasters and snapshots of disk images that can be used with offline applications such as data warehousing tools. 3. Huge Containers. Combining the capacities of multiple physical drives into one large logical volume simplifies the allocation of storage for very large databases. 4. Power in numbers. Data can be distributed across multiple physical devices not only to create redundant copies and reduce the impact of media failure, but also to boost the speed of data storage and retrieval. 5. Unwasted space. Because data in the storage pool can be distributed to physical devices throughout the storage network, storage is used more efficiently and available space isn't wasted.

Store And Forward S/F. In communications systems, when a message is transmitted to some intermediate relay point and stored temporarily. Later the message is sent the rest of the way. Not very convenient for voice conversations, but useful for telex type, and other one-way transmission of messages. Telephone answering machines, as well as voice mailboxes are considered forms of Store and Forward message switching.

store locator service See Single Number Dialing.

stored procedures Compiled code on a database server that reduces the processing burden on clients.

stored program A telephone company definition. The instructions which are placed in the memory of common controlled switching unit and to which it refers while processing a call. Stored programs commonly use alterable magnetic marks to record the program instruction. See also Stored Program Computer and Stored Program Control.

stored program computer A computer controlled by internally stored instructions, that can synthesize and store instructions, and that can subsequently execute those instructions. See also Stored Program Control.

Stored Program Control SPC. The routing of a phone call through a switching matrix is handled by a program stored in a computer-like device, which may well be a special-purpose computer. Before SPC switches came along, the rotary dialing of the phone caused the elements of the switch to directly "step" through their dialing path. This was slow and cumbersome, since dialing can be slow. Also subscribers can abort half way (they made a mistake) and this can mess up the switch's efficiency. Thus the move to stored program control switches was very significant. These days virtually all switches as stored program control. Nothing happens in the switching matrix until the stored program control receives all the dialing digits and decides what to do with them.

storewidth A word made up by guru George Gilder. He

stork His mother should have thrown him away and kept the stork -Mae West.

stovepipe network A network set up for a specific system. An organization with stovepipe networks has a separate network for each system, none of which communicate with the others. In such an environment there is no overall end-to-end network view. A user who needs to access multiple systems will have multiple terminals, each dedicated to a single network.

stovepiping In a call center, agents typically need access to many databases. In the past they've used dumb terminals. They log into one computer, get into one database, go further into it. When they need information out of another database, they've typically had to climb out of the previous database, the previous computer, log into another and climb down into it. This is called stovepiping, because it follows the contours of a stovepipe. These days, agents have intelligent computers as terminals. They can access several databases at once, by simply having different windows open on their screen or having a front end program that populates a screen with information from several databases, most likely using a GUI interface.

STP 1.Shielded Twisted Pair. Twisted pair (TP) wiring with a metallic shield surrounding the signal-carrying conductors in order to protect them from ambient noise in the form of EMI (ElectroMagnetic Interference). The outer shield may be in the form of a thin metallic mesh in the case of ScTP (Screened Twisted Pair). Alternatively, it may be in the form of a very thin metallic foil in the case of FTP (Foil Twisted Pair), which also is known as SSTP (Double-Shielded Twisted Pair, as in Shielded-Shielded Twisted Pair). In either case, the shield effectively serves to ensure noise-free information transfer. The shield, however, acts as an antenna, converting received noise into current flowing in the shield; it must be properly electrically grounded with a drain wire, or the shield current actually will intensify the noise problem. Any discontinuity in the shield also will result in increased noise. To function effectively, every component of a shielded cabling system must be fully shielded, and the continuity of the shield must be maintained across cable splices. See also Attenuation and STP-A.

2. Signal Transfer Point. The packet switch in the nation's emerging Common Channel Interoffice Signaling (CCIS) system. The CCIS is a packet switched network operating at

4800 bits per second. CCIS replaces both SF (Single Frequency) and MF (Multi-frequency) by converting dialed digits to data messages. It will run at 56,000 bps with the introduction of Signaling System 7. See Signaling System 7. For a full explanation of the Advanced Intelligent Network, see AIN.

3. Spanning Tree Protocol. See Bridge.

STP-A Shielded Twisted Pair-A. A modification of the original STP standard, STP-A supports increased carrier frequencies and, therefore, increased transmission speeds. STP-A makes use of the same cable, although the improved connector includes a metal shield between the two conductors as a crosstalk barrier. The new connector mates with the old. STP-A originally was tested up to 100 MHz, and now up to 300 MHz, in support of high-speed Token Ring. See also STP.

straight laced In the 1770s, ladies wore corsets which would lace up in the front. A tightly tied lace was worn by a proper and dignified lady as in "straight laced."

straight-through When wiring up phone and some data extensions, there are basically two ways of doing it -straight-through and crossover. Straight-through occurs when you wire both ends identically so the signals pass directly straight through. Crossover wiring has a reverse order of wiring. As an example, let's take a four conductor, RJ-11. In a crossover wiring (e.g. an RJ-11 two wire phone extension cord), conductor 1 on one plug would be connected to position 1 on one plug and position 4 on the other end. Conductor 2 would be connected to 3. And 3 would be connected to hole 2.

straight-tip connector ST Connector. An optical fiber connector used to join single fibers together at interconnects or to connect them to optical cross connects. See also FC Connector, SC Connector, and SFF Connector.

straightforward outward completion Operator can place an outgoing call for phone user. Also called "Through Supervision."

strain relief The connection between the cable and the termination, usually a modular plug, that bonds the cable jacket to the connector so that the individual conductors don't have to absorb tension when the cable is pulled or moved. There are two types of strain relief. The primary strain relief crimps onto the cable's outer jacket where the modular plug meets the cable, and the secondary strain relief crimps onto the rubbery insulation around each conductor inside the business end of the plug. Not all crimp dies crimp the secondary strain relief, and some crimps have a different secondary strain relief location. If the cable jacket and conductors' insulation isn't crimped, the strain of moving or pulling the cable (this is especially important at the desktop, where cables get unplugged and plugged, jostled and pulled) is all borne by your copper connection. Make sure the modular connectors (cable plugs) your technician is crimping have primary and secondary strain relief. Some dies for crimping tools don't support secondary strain relief, which anchors the insulation around the cable conductors to the plug. Strain relief is important because otherwise the fragile copper wire carrying your connection takes all of the tugging and pulling when the cable is plugged in and unplugged or moved. If the sheath of the cable is not attached to the modular plug at, (this is called primary strain relief), your connections has no strain relief at all. The cable sheath should be anchored at the end of the plug away from the connectors. Bye-bye connection.

strained silicon Strained silicon is the method developed by IBM of stretching silicon over a silicon germanium substrate layer to improve the speed of semiconductor processing.

strand 1. A single uninsulated wire.

2. Strand (as the term applies to telephone companies) is an uninsulated and unpowered stranded steel cable, installed on telephone and utility poles and similar structures to support telephone cable. Cable is lashed to the strand; other devices are fitted with clamps which attach to the strand. On joint poles, the CATV strand is usually installed below electric power facilities and above the telephone facilities.

3. Strand (as the term applies to cable television) is an uninsulated and unpowered stranded steel cable, typically 1/4" or 3/8" diameter, installed on telephone and utility poles and similar structures to support cable television distribution devices such as hard cable, amplifiers and taps. Cable is lashed to the strand; other devices are fitted with clamps which attach to the strand. On joint poles, the CATV strand is usually installed below electric power facilities and above telephone facilities. See Hard Cable, Joint Pole, and Lashing.

strand lay The distance of advance of one strand of a spirally stranded conductor, in one turn, measured axially.

stranded assets A scenario where network facilities or equipment are available for service but are not recognized as being available, or are not recognized at all, by a provisioning and/or inventory system. An asset can become stranded in a variety of ways. For example, a physical disconnection of service takes place for a customer, but the network

inventory database fails to get updated to show the availability of the asset that has been freed up by the disconnect. Stranded assets don't get reprovisioned, which means revenue is forgone. Worse, unnecessary additional investment in assets may take place, in order to replenish what is misperceived as depleted inventory.

stranded conductor A conductor composed of groups of wires twisted together, either singly or in groups.

stranded copper A type of electrical wire conductor comprised of multiple copper wires twisted together forming a single conductor and then covered with an insulating jacket. Stranded conductors perform less well than do solid-core conductors in terms of transmission quality, and are more distance-limited. Stranded conductors have greater flex strength, however; therefore, they are commonly used in applications where the cable is flexed frequently and aggressively.

stranded fiber cable A fiber optic cable in which multiple individual optical fibers contained within the same cable sheath are twisted around each other in a helix. Also twisted with the fibers are strength members, generally constructed of aramid (commonly known as Kevlar). The twisting process improves the flex strength of the cable, much as is the case with stranded copper. If the fibers were not helically stranded, each fiber essentially would stand on its own, and would be more susceptible to fatigue, which would result in the growth of surface imperfections or microcracks, and eventually fiber breakage. See also Aramid, Helix and Stranded Copper.

stranded wire Multiple small AWG strands of wire that are put together to make a flexible wire with similar electrical properties as a similar solid wire. Stranded wires are usually used in data cabling.

strap A permanent, wired connection between two more points. Older style data modems did some of the feature selection using DIP switches. They called that "straps." Even today, some software option commands have become called "software straps."

strapping The act of installing a permanent connection between a Point of Termination (POT) Bay and a collocated party's physical collocation node.

strategic alliance When a company does something with another company, both companies will announce they now have a "strategic alliance." Such strategic alliance may be anything from one company agreeing to include the other company's products in its lineup (often without a commitment to actually sell anything). Or it may simply be that each have agreed to include a hyperlink to each other on each other's web sites. In short, I've never figured what a strategic alliance really is or what its benefits are -other than an attempt to boost each other's stock by issuing more press releases to the financial community.

On August 10, 2000, the Wall Street Journal reported: SEATTLE -Amazon.com Inc. announced a strategic alliance with Toysrus.com Inc. to create a co-branded toy and video-games store. The deal could create a toy giant., combining the online unit of Toys "R" Us Inc., Montvale, N.J., with the Internet's largest retailer. Both Amazon and Toysrus.com are already leading sellers of toys online, but face a raft of competitors including eToys Inc. Under the terms of the 10-year agreement, Amazon will receive a combination of periodic fixed payments, per-unit payments and single-digit percentage of revenue. The level of the payments weren't disclosed. Amazon will also receive warrants entitling it to acquire 5% of Toysrus.com. You can figure what "strategic alliance" really means. See also Strategic Partnership.

strategic investor One who can deliver customers, managements contacts, and other help in addition to money.

strategic partnership A fancier term for strategic alliance. See Strategic Alliance for just how vague a definition can get.

stratellite A stratellite is a large balloon that hovers in a fixed position in the lower stratosphere and carries one or more repeaters for wireless microwave communications. A system of such balloons may be capable of providing cellular, mobile, fixed, and Internet communications over large geographic areas. The technological characteristics of stratellites will be similar to those of low-earth-orbit (LEO) satellite systems. In particular, the low altitude will result in low latency (minimal signal propagation delay). At an altitude of 13 miles (21 kilometers), a single stratellite will allegedly cover a roughly circular geographic area of 300,000 square miles (770,000 square kilometers). The proposed balloon radius is 100 feet (30 meters), a size that will make it possible for each balloon to carry a substantial payload. The location of each device will be maintained by ground-based remote control. So far, no one has launched a real stratellite communications system. The company talking most about stratellites is apparently called GlobeTel. You can learn more at www.Stratellite.net.

stratum level In any digital network, you need a clock -a source of timing to let transmissions know where they begin and where they end. But in any network, there are "levels" of clocks. Think of the clocks you have at home. (This analogy is stretching it. But it's a good beginning.) Somewhere at home you have your "main clock" -the one you consider to be most accurate. For me it's my digital wristwatch. When I compared my wristwatch against my computer, my wristwatch said it was two minutes later. So I called an operator at the local phone company. She gave me a different time, but she assured me that her timing was accurate, since it was on her console and derived from a distant, accurate network clock. So let's say that that's my main source of timing. That would be my level 1 timing. From there, I transfer it to my PC, my watch, etc. which would be my level two timing.

Now let's move to the digital network. An American National Standards Institute (ANSI) standard entitled "Synchronization Interface Standards for Digital Networks" (ANSI/T1.101-1987) was released in 1987. This document defines the stratum levels and minimum performance requirements. Stratum 1 is defined as a completely autonomous source of timing which has no other input, other than perhaps a yearly calibration. The usual source of Stratum 1 timing is an atomic standard or reference oscillator. The minimum adjustable range and maximum drift is defined as a fractional frequency offset delta f/f of 1 x ten raised to the minus eleven or less. At this minimum accuracy, a properly calibrated source will provide bit-stream timing that will not slip relative to an absolute or perfect standard more than once every four to five months. Atomic standards, such as cesium clocks, have far better performance. A Stratum 1 clock is an example of a Primary Reference Source (PRS) as defined in ANSI/T1.101. A PRS source can be a clock system employing direct control from Coordinated Universal Time (UTC) frequency and time services, such as Global Positioning System (GPS) navigational systems. The GPS System may be used to provide high accuracy, low cost timing of Stratum 1 quality.

A Stratum 2 clock system tracks an input under normal operating conditions, and holds to the last best estimate of the input reference frequency during impaired operating conditions. A Stratum 2 clock system requires a minimum adjustment (tracking) range of 1.6 x 10 raised to the minus eight. The drift of a Stratum 2 with no input reference is less than 1.6 x 10 raised to the minus eight in one year.

Stratum 3 is defined as a clock system which tracks an input as in Stratum 2, but over a wider range. A Stratum 3 clock system requires a minimum adjustment (tracking) range of 4.6 x 10 raised to the minus six. The short term drift of the system is less than 3.7 x 10 raised to the minus even in 24 hours. This is about 255 frame slips in 24 hours while the system is holding. Some Stratum 3 clock equipment is not adequate to time SONET network elements.

Stratum 3E is a new standard created as a result of SONET equipment requirements. Stratum 3E tracks input signals within 7.1 Hz of 1.544 MHz from a Stratum 3 or better source. The drift with no input reference is less than 1 x 10 raised to the minus eight in 24 hours.

Stratum 4 is defined as a clock system which tracks an input as in Stratum 2 or 3, except that the adjustment and drift range is 3.2 x 10 raised to the minus five. Also, a Stratum 4 clock has no holdover capability and, in the absence of a reference, free runs within the adjustment range limits.

Stratum 4E is a proposed new customer premises clock standard which allows a holdover characteristic that is not free running. This new level, intended for use by customer provided equipment in extending their networks, is not yet standardized. See Network Slip and Timing. www.laruscorp.com/tchap04.htm

stray current Current through a path other than the intended path. See also Spurious Emission.

strawman This concept is widely used in selling. The simple idea is to set up a Buyer's Checklist and tell your prospective customer that this Checklist is objective. Any product that meets all the criteria is worth buying. Of course, there's only one product that meets all the criteria. It's yours.

stream 1. A flow of digital information, such as a video stream.
2. An SCSA term. One of 16 physical data lines making up the SCbus or SCxbus Data Bus. See S.100.

streamer Streaming tape drive.

streamies This is a collective term for people who view or listen to streaming video or audio over the Internet sources such as online radio stations, webcast films and the like.

streaming An Internet term. A Web page typically consists of text and graphics (still and moving) images. The text is typically fewer bytes than the graphics which are heavy on bytes. Thus, to receive the text to your PC from their Web page typically takes much less time than receiving the graphics images. So Netscape had an idea, which they first pioneered in their browser. Let's get the text up on the user's screen fast, and paint the user's screen with the images as they came in. This allowed the user to look at a new page of text on screen as the graphics came in over the phone lines. Netscape called this streaming. That was the first use of streaming. But then a company called Real Networks came along. It had an idea. Wouldn't it be nice if we could put audio recordings and video (clips, movies, etc.) on a Web site and have people click on them and start hearing or seeing them immediately -as against (in the pre-Real Networks' days) waiting to download the entire file, then playing it. See Home Page, Internet and Streaming Audio and Streaming Media.

streaming audio see also Internet Radio.

streaming content Let's say you're a very big company with a very big website, for example Microsoft or General Motors. You have people coming to your website from all over the world at all times of the day and night. Clearly if they all home in on your one solitary web site, all sorts of horrible problems are likely to occur. You'll overload the phone lines. You'll overload the computers. You'll overload the routers. You'll overload everything. There's a solution: You go to a company who has made it its business to have servers all over the world connected to the Internet. You copy the content of your web servers to this company's geographically distributed web servers. This way when a customer of Microsoft in Australia wants content from Microsoft in Seattle, that customer doesn't have to contact Seattle, he contacts the local server in Australia. By the way, he doesn't know he's contacting the local server in Australia. He thinks he's contacting Seattle. And he really shouldn't care, since each time, Microsoft's Seattle servers are updated, the geographically distributed servers are also updated. The advantages for the customer are obvious: faster and reliable response. The advantage for Microsoft (or General Motors) is that it gives better service to its customers. The business of providing content from geographically distributed servers is called "streaming content." It is very different to streaming media. See the next definition. And it is very different to plain streaming. See the previous definition.

streaming media Streaming media is basically audio and video (i.e. a movie or a video clip) coming at you in packets over the Internet. The idea of the "stream" and the streaming software necessary to play it is that your audio and your video start playing on your screen before the entire file is downloaded to your machine. Until streaming media came along, you needed the entire audio or video file on your computer intact and in toto before you could play it -as with today's attachments to email messages. (You need those on your hard disk before you can "play" them back.) If you company starts streaming media to people -.e.g. for training -be aware that streaming video can adversely impact the performance of a network. See Streaming.

streaming media servers Software that provides multimedia support, allowing you to deliver content by using advanced streaming format over an intranet or the Internet.

streaming tape backup A device to backup files, programs and entire hard disks. A streaming tape backup looks very much like a large audio cassette. It records data sequentially, which means if you want to find specific data, it can take a long time while the tape physically moves backwards and forwards to find the data. Streaming tape backup is best used as backup for an entire computer and best used to restore a computer if all else fails. The problem with streaming tape backup, based on my limited experience, is that it's not very reliable -largely as a result of it being so mechanical. The tape mechanism breaks down more than it should.

streaming tape drive A magnetic tape unit especially designed to make a nonstop dump or restore magnetic disks without stopping at interblock gaps.

streaming video See Streaming Media.

streams An architecture introduced with Unix System V, Release 3.2 that provides for flexible and layered communication path between processes (programs) and device drivers. Many companies market applications and devices that can integrate through Streams protocols.

Streamworks The StreamWorks Player brings the power of networked audio and video to the desktop. You can play "live" and "on-demand" audio and video from StreamWorks Servers across the globe. The StreamWorks Transmitter allows for LIVE network encoding of digital audio and video over today's networks. Taking inputs from analog audio and video connections, like the ones on the back of a VCR, StreamWorks Transmitter is capable of enabling live, real-time MPEG audio and video over industry standard TCP/IP networks.

street juice Electricity supplied by the electric company.

street price The real selling price of computers, hardware, and software. Most laptop and desktop computers sell for 25 percent below list price. Software may be discounted even more.

street talk The Banyan-developed protocol for discovering and maintaining resource information distributed among the servers connected to Banyan's VINES network operating system. Also known as a global naming service.

streetsweeper A heavy duty shotgun with a revolving round magazine typically holding 18 12-gauge or 20-gauge shotgun shells. This word crept into a story the Wall Street Journal ran on cellular fraud. When the Feds rang a cellular phone store as a sting operation, one customer offered to trade his streetsweeper in on a phone. That's how dependent Detroit's drug-traffickers had become on cellular phones and beepers.

strength member The part of a fiber optic cable composed of aramid yarn, steel strands, or fiberglass filaments that increase the tensile strength of the cable.

stress puppy A person who thrives on being stressed out. Most common symptom is excessive whining.

stress test A test of a network with realistic heavy streams of traffic, so that thresholds where network performance starts to be impacted can be determined.

stressed eye test See receiver jitter tolerance test.

string A sequence of elements of the same type, such as characters, considered as a unit (a whole) by a computer. A data structure composed of a sequence of characters, usually in human-readable text.

strip force A cabling term. The force required to remove a small section of insulating material from the conductor it covers.

stripe pitch The equivalent of dot pitch on aperture grille CRTs; the distance between one stripe and the next one of the same color, expressed in millimeters. See also Resolution.

stripe set A volume that stores data in stripes on two or more physical disks. A stripe set is created by using Windows NT 4.0 or earlier. Windows XP Professional does not support stripe sets. Instead, you must create a striped volume on dynamic disks.

striping 1. A term used in data storage technology, such as RAID (Redundant Array of Inexpensive Disks), data striping is the process of dividing a large logical block of data into multiple physical blocks for storage on multiple disk drives. The resulting data partitions residing on multiple disk drives can be combined into a single logical "volume" that the operating system views as a single drive. Data striping offers the advantage of enhanced performance by enabling multiple I/O (Input/Output) operations in the volume to take place in parallel (i.e., simultaneously). Striping is used in the context of Network-Attached Storage (NAS), the most complex implementation of which is a Storage Area Network (SAN). See also NAS, RAID, and SAN.
2. A Fibre Channel definition. Striping is a way of achieving higher bandwidth using multiple N_ports in parallel to transmit a single information unit across multiple levels.

strobe A signal that triggers a data reading, transfer of information or sampling. Such a sampling might be to figure if a circuit is active and, if so, what level of activity is taking place. The sampling process might allow a carrier to bill the user correctly for circuit usage.

stroke A straight line or arc that is used as a segment of a graphic character.

stroke edge An imaging and OCR term. In character recognition, the line of discontinuity between a side of a stroke and the background, obtained by averaging, over the length of the stroke, the irregularities resulting from the printing and detecting processes.

stroke speed In facsimile systems, the number of times per minute that a fixed line perpendicular to the direction of scanning is crossed in one direction by a scanning or recording spot. In most conventional mechanical systems, this is equivalent to drum speed. In systems in which the picture signal is used while scanning in both directions, the stroke speed is twice the above figure.

stroke width In character recognition, the distance measured perpendicularly to the stroke centerline between the two stroke edges.

strong text The HTML character style used for strong emphasis. Certain browsers display this style as bold.

Strowger, Armond The man who invented the telephone dial and the earliest automatic telephone switch as a method of allowing the user to complete calls without using the Operator. In Kansas City in the late 1800's, Ol' Armond was an undertaker who wasn't getting much business. That's because the girlfriend of a rival undertaker was a telephone operator, and when she got a call asking for the local undertaker, she forwarded the calls to her boyfriend. This story may or may not be apocryphal. But it's a great story.

structure A system for organizing procedures or equipment. See all the definitions that follow.

Structured Query Language SQL. A relational database language (ANSI Standard) that consists of a set of facilities for defining, manipulating and controlling data. See SQL.

structured programming A technique for organizing and coding (computer) programs in which a hierarchy of modules is used, each having a single entry and a single exit point, and in which control is passed downward through the structure without unconditional branches to higher levels of the structure. Three types of control flow are used: sequential, test, and iteration.

structured wiring As data flows have sped up in recent years and as moves, adds and changes have proliferated, so the erstwhile idea of wiring up a building with plain old analog voice telephone wire has become no longer intelligent. The idea then came up of defining wiring standards and flexible schemes so that a user could feel comfortable about choosing a complete solution for wiring phones, workstations, PCs, LANS and other communicating devices throughout the building, the campus, the network, the company and throughout his life in the place. Consistency of design, flexible layout and logic are the keys to structured wiring systems. Typically a structured wiring system consists of two elements:

1. Manufacturer-originated standard components that link wires together in a systematic, intelligent way.

2. A set of rules for building smart wiring. The ANSI/TIA/EIA-568-A and ISO/IEC 11801 standards specify the minimum requirements for telecommunications cabling within a commercial building. The Commercial Building Telecommunications Wiring Standard is available from Global Engineering Documents, Englewood CO 314-726-0444 / 800-854-7179.

A structured cabling system will improve performance in five ways, according to Anixter, a leading supplier of structured wiring systems:

1. It eases network segmentation, the job of dividing the network into pieces to isolate and minimize traffic, and thus congestion.

2. It ensures that proper physical requirements, such as distance, capacitance, and attenuation are met.

3. It means adds, moves, and changes are easy to make without expensive and cumbersome rewiring.

4. It radically eases problem detection and isolation.

5. It allows for intelligent, easy and computerized tracking and documentation.

"Structure" brings order to what has often been an afterthought -wiring. The main pieces of a structured wiring system are:

1. Drop cable. The cable that runs from the computer to a network outlet.

2. Cable run. The cable that runs from the outlet to the wiring closet.

3. Patch panel. A board that collects all the cable runs in one place and "patches" them to different parts of the wiring concentrator. Network managers (users or their secretaries, it's that simple) change the LAN layout by plugging and unplugging "patch cables" between the patch panel and the wiring concentrators. No rewiring is necessary to move one user from one network segment to another.

4. Wiring concentrator. It makes the network connections. Some wiring concentrators are dumb, making only physical connections between network segments. Others are intelligent, making networking decisions and providing network diagnostics. A wiring concentrator can have bridges and routers that divide the network into segments. It can have the hardware necessary to change from one media, say twisted pair, to another, say fiber optic. And it can contain the hardware to change from one network type to another, say from Ethernet to Token Ring.

Here is a glossary of structured wiring words, with thanks to Anixter.

Access Method The method of "communicating" on the wire. Examples include Ethernet, Token Ring, AppleTalk, AS400 and 3270.

Cable Type (Media) The type of cable used in the system. Examples are coaxial, UTP, STP and fiber. Factors including cost, connectivity and bandwidth are important in determining cable type.

Data Speeds Different interconnect products (cables and connectors) are capable of supporting different data rates. For instance, Level 3 cable supports data rates up to 10 Mbps. (See LEVEL).

Environment Where the structured wiring system is found. The large majority of systems are located in office environments as opposed to factory or industrial environments.

Life Cycle How long the cable is physically anticipated to be in place. For example, if a customer intends to be in a large office for 10 years, fiber installation may be considered.

Methodology The physical means of getting the wiring system to the user (its distribution path). Examples include modular furniture, surface mounts, fixed wall, recessed wall, raised floor and undercarpet wiring.

Topology The way the cable is physically laid out or configured. Examples include star, ring, daisy chain and backbone. See Smart Home.

Structured Wiring System See Structured Wiring.

STS Synchronous Transport Signal. The electrical equivalent of SONET OC-level. The signal begins as electrical and is converted into optical prior to presentation to the fiber optic medium. The STS frame consists of the Synchronous Payload Envelope (SPE), Section Overhead (SOH), Line Overhead (LOH), Path Overhead (POH), and Payload. SOH and LOH comprise what is known as Transport Overhead (TOH). See also STM.

STS-1 Synchronous Transport Signal level 1. An electrical signal that is converted to or from Sonet's optically based signal; equivalent to the OC-1 signal of 51.84 Mbps.

STS-3 Synchronous Transport Signal level 2. (yes, that's right. It's STS-3, though it's level 2.) ATM Physical Layer implementation supporting 155 Mbps.

STS-n Synchronous Transport Signal "n" : (where n is an integer) SONET standards for transmission over OC-n optical fiber by multiplexing "n" STS-1 frames, (e.g., STS-3 at 155.52 Mbps STS-12 at 622.08 Mbps and STS-48 at 2.488 Gbps).

STS-nc Synchronous Transport Signal "n" concatenated: (where n is an integer) SONET standards for transmission over OC-n optical fiber by multiplexing "n" STS-1 frames, (e.g., STS-3 at 155.52 Mbps STS-12 at 622.08 Mbps and STS-48 at 2.488 Gbps but treating the information fields as a single concatenated payload).

STU Secure Telephone Unit.

STU-III The third generation of secure telephone units used by the military and its suppliers.

stub area A stub area is an OSPF-defined area of routers which only includes intra-area and inter-area routes in its tables. The ABR (area border router) for that area prevents any externally-originated routes (e.g., RIP, BGP, EIGRP, etc.) that have been redistributed into OSPF from entering the area. The ABR instead sends the area routers a default route so that they may still reach the E1/E2 destinations. A totally stubby area extends this concept further. In addition to blocking external (E1/E2) routes from entering the stub area the ABR also blocks all inter-area routes. In other words the area routers only contain intra-area routes in their tables. Again the ABR provides a default route into the area so that inter-area and external destinations can be reached.

stub domain A local domain in the context of IP networks. A stub domain commonly is a Local Area Network (LAN) which uses IP addressing for local packet data routing. Most of that IP traffic never leaves the local domain for the Internet domain, so it's just a stub, i.e., it goes nowhere. It's analogous to the stub of a finger that you got cut off in a bandsaw. It doesn't go anywhere. It isn't of any use outside of the domain of your hand. You can't even press an elevator button with it. See NAT.

stub network A network that carries packets only to and from local hosts. It does not carry traffic for other networks.

Studio-To-Transmitter Link STL. Any communication link used for transmission of broadcast material from a studio to the transmitter. It's typically microwave radio but it may also be a conventional landline link.

studpuppy The younger companion of an older woman is often called a studypuppy. He earns his vaunted title of studpuppy for his superior sexual skills. This definition contributed by Carolyn Robbs.

Study Group 15 The ITU, a United Nations agency, coordinates the development of global communications standards. Study Group 15 of the ITU Telecommunication Standardization Sector (ITU-T) is where the work on communications specifications is carried out. It is responsible for the standards development in the area of transport networks, systems and equipment. See also G.990.

Study Group 16 The ITU, a United Nations agency, coordinates the development of global communications standards. Study Group 16 of the ITU Telecommunication Standardization Sector (ITU-T) is where the work on multimedia communications specifications over the Internet is carried out. See also G.990.

stumbler An application that looks for wireless networks in the area and determines whether they are open. Net Stumbler is the name of a popular software programs that searches for available Wi-Fi networks.

STUN Simple Traversal of UDP through NAT (network address translation) devices. A protocol that lets devices behind a NAT firewall or router route packets.

stunt box A device to 1. control the nonprinting functions of a teletypewriter terminal, such as a carriage return and line feed and 2. a device to recognize line control characters.

stupid network See Dumb Network.

stutter dial tone Stutter Dial Tone is the broken-up tone a user hears on a phone when they pick up their phone to make a call and they have a message waiting in voice mail. This is used to notify users that they have a voice mail message when the phones don't or can't have a message-waiting light. They are given stutter dial tone instead of regular dial tone. See also MWI.

STV Subscriber (or Subscription) Television.

STX Start of Text. See Packet.

Stylus A pen-shaped instrument (usually made out of plastic) that is used to enter text, draw images, or point to choices on a computer -desktop or PDA. The pen is designed to make writing on the screen feel just like writing on paper. Yes!

SU 1. Subscriber Unit. A radio frequency modem used to acquire the airlink. A wireless term.

2. Service User. The end user at the customer premises.

3. Signal Unit, a group of bits forming a separately transferable entity used to convey information.

sub 1. Substitute Character. A control character used in the place of a character that has been found to be invalid or in error.

2. Subscriber.

sub-band The band of frequencies between 5 MHz and 45 MHz frequently used as a return path in CATV.

sub-band signaling A method of signaling via a band applied as modulation to a band carrying user communications.

sub-QCIF A video format that provides an image size of 128 x 96 pixels. It is used for viewing the images on a phone display.

subaddressing A name for an ISDN service which enables many different types of terminals -phones, fax machines, PCs, etc. -to be connected to the ISDN user interface and uniquely identified during a call request. See ISDN.

subcarrier 1. A carrier which modulates a main carrier so that two different modulating signals can be transmitted simultaneously, one on the main carrier and one on the subcarrier. See Sub Carrier Modulation.

2. In NTSC or PAL video, a continuous sine wave of extremely accurate frequency which constitutes a portion of the video signal. The subcarrier is phase modulated to carry picture hue information and amplitude modulated to carry color saturation information. The NTSC subcarrier frequency is 3.579545 MHz, and the PAL-I frequency is 4.43361875 MHz. A sample of the subcarrier, called color burst, is included in the video signal during horizontal blanking. Color burst serves as a phase reference against which the modulated subcarrier is compared in order to decode the color information.

subcarrier modulation Subcarrier modulation combines a signal with a single low frequency sine wave. The low frequency signal is called a sub-carrier. This combined signal is then added to a higher frequency radio signal. The resulting high frequency radio signal is very complex and the original signal is not detectable by ordinary means. To detect a signal that has been modulated by a subcarrier, it must be passed through two detector circuits, one to separate the subcarrier from the high frequency radio transmission, and a second to separate the sub-carrier from the desired information.

subchannelize Divide a wideband channel into multiple narrowband subchannels, each of which consists of one or more groups of subcarriers. One reason for subchannelization is to improve the cost to the customer, since it enables a customer to avoid paying for more bandwidth than he needs.

subchannel A subset of a wideband channel, made up of one of more groups of subcarriers.

subconference A teleconferencing term. During the course of a large teleconference, the moderator can hold an off-line subconference (i.e., a caucus, or closed meeting) with a number of participants. During the subconference, the other participants remain connected to the main conference. Once the subconference is completed, that group rejoins the main conference.

subdomain. Also called a child domain. Normal domains, such as Microsoft.com, are called parent domains.

sublayer A logical sub-division of a layer.

subloop As defined by the Telecommunications Act of 1996, a subloop is a portion of a local loop that is accessible to terminals at any point in the Incumbent Local Exchange Carrier's (ILEC's) outside plant, including inside wire. An accessible terminal is any point on the loop where technicians can access the wire or fiber within the cable without removing a splice case. Such points can include a telephone pole or pedestal, the minimum point of entry (MPOE), the single point of interconnection, the main distribution frame (MDF), and the feeder/distribution cable interface. Subloops are one of the categories of Unbundled Network Elements (UNEs), which the ILECs must make available to the Competitive LECs (CLECs). See also Local Loop, Telecommunications Act of 1996, and UNE.

submarine cable A cable designed to be laid underwater.

submarining Same as cursor submarining. When you drag your cursor across a screen and the cursor disappears as you move it. That's called cursor submarining. It happens most on monochrome LCD screens because they change slowly -much slower than active matrix screens or CRTs or VDTs (glass screens).

submission In X.400 terms, the transmission of a message or probe from an originator's UA (User Agent), MS (Message Store), o AU(Access Unit) to an MTA (Message Transfer Agent).

submodule A small circuit board that mounts on a larger module. Also called a daughterboard.

subnet A subnet is a portion of a network, which may be a physically independent network, which shares a network address with other portions of the network and is distinguished by a subnet number. A subnet is to a network what a network is to the Internet. Here's an explanation. A TCP/IP network can have a lot of traffic flowing across it at any given time. In large networks the flow of information can be too much, causing bottlenecks and congestion that essentially bogs the network down to the point it is ineffective. To alleviate this, a network will be divided into smaller networks called subnets. Subnets are created by configuring the IP addresses for all the computers in the subnet to be similar to each other, but different from other subnets. The different subnets are kept separate by using a subnet mask. A subnet mask filters IP addresses allowing computers with specific IP addresses to talk to each other directly yet other computers will not hear their broadcast traffic. By using subnets, backbones, and routers it is possible for a large network to operate efficiently without any bottlenecks or congestion. See also Subnet Mask.

subnet mask A 32-bit value that enables the recipient of IP packets to distinguish the network ID and host ID portions of the IP address. Typically, subnet masks use the format 255.x.x.x. A subnet mask is a mechanism that is used to split a network into a number of smaller subnetworks. A subnet mask can be used to reduce the traffic on each subnetwork by confining traffic to only the subnetwork(s) for which it is intended, thereby eliminating issues of associated congestion on other subnetwork(s) and reducing congestion in the network as a whole. A subnet mask also makes the entire network more manageable. In effect, each subnetwork functions as though it were an independent network, keeping local traffic local and forwarding traffic to another subnetwork only if the address of the data is external to the subnetwork. Such decisions are made on the basis of routing tables contained within the various routers, with each routing table comprising an IP (Internet Protocol) address table. (Note that the IP address may be an internal address rather than a registered IP address, with the latter being used only for access to the public Internet. For example, private IP addresses would be used for routing within a LAN, with translation to registered IP addresses taking place if the traffic is intended for the public Internet.) As a data packet is received by a router, an address lookup is accomplished, and the original IP address is translated into a subnet mask address, which also follows the IP address format. The resulting subnet mask address is divided into two parts, a subnetwork address and a host address. The IP address and subnet mask address are compared, and the router decides how to forward the data packet. See Ethernet and Local Area Network.

subnet number A part of the internet address which designates a subnet. It is ignored for the purposes internet routing.

subnetwork 1. A collection of OSI end systems and intermediate systems under the control of a single administrative domain and utilizing a single network access protocol. Examples: private X.25 networks, collection of bridged LANs.

2. A token ring LAN that is used to serve the communication needs of a department. Subnetworks are normally connected to token ring backbones via token ring bridges or routers so that they can communicate with other subnetworks via the backbone or with computers directly connected to the backbone.

3. An ATM term. A collection of managed entities grouped together from a connectivity perspective, according to their ability to transport ATM cells.

Subnetwork Access Protocol SNAP. A version of the IEEE local area network logical link control frame similar to the more traditional data link level transmission frame that lets you use nonstandard higher-level protocols.

subNMS An ATM term. Subnetwork Management System: A Network Management System that is managing one or more subnetworks and that is managed by one or more Network Management Systems.

subordination of debt A debt subordination agreement is a contract in which a junior creditor agrees that its claims against a debtor will not be paid until all senior indebtedness of the debtor is repaid. Under a general subordination agreement, a junior creditor agrees to subordinate its claim to all presently existing and future claims against the debtor. In a specific subordination agreement, a junior creditor subordinates its claim to a particular obligation of the debtor. In recent years, lenders and borrowers have used subordinated

debt, such as high yield bonds, institutional subordinated mezzanine financing, participation certificates, and subordinate trust certificates, in an ever-widening variety of transactions, including leveraged buy-outs, securitizations, and real estate financings. Businesses now commonly utilize tiered capital structures consisting of layers of common and preferred stock, management or shareholder debt, mezzanine financing, high yield bonds, and senior bank or asset-based debt. While equity has lost some of its appeal as a currency for acquisition transactions, subordinated debt has gained ground as a cost-efficient financing too.

subrate Less than the standard rate of transmission, which is defined at the voice-grade rate of 64 Kbps. The voice-grade transmission rate is defined at 64 Kbps, as that is the basic building block of the digital transmission hierarchy, based on the original, and most commonly used, PCM (Pulse Code Modulation) standard method for converting analog voice into a digital signal. Therefore, 64 Kbps is the lowest common denominator for digital transmission, across the entire PDH (Pleisiochronous Digital Hierarchy), which includes T-Carrier (North American), E-Carrier (European), and J-Carrier (Japanese). Transmissions at a rate less than 64 Kbps, therefore, are referred to as subrate.

subrate switching Through special equipment and arrangement with the public carrier(s), subrate switching (i.e., switching at less than the standard rate of 64 Kbps) can be accomplished. For example, ISDN can support subrate switching, allowing two or more transmissions to share a 64-Kbps channel over an ISDN local loop. In the carrier network, each of the transmissions is identified, and is switched separately. See also Subrate.

subroutine A functionally isolated program or sequence of instructions for a specific function that is often called by a program. A piece of software that performs a useful function that will be needed often. The code for the subroutine is stored on disk (like a letter, etc.) and dropped into a larger program as needed. A nicely-written subroutine saves you "re-inventing the wheel" and allows you to re-use your code in many programs.

subsatellite point A point on the surface of the earth directly beneath the satellite, geometrically on a direct radial line from the center of the Earth to the satellite.

subscriber A person or company who has telephone service provided by a phone company. In other industries, subscribers are called customers. Some telephone companies are beginning to call their subscribers customers. About time.

subscriber access terminal SAT. A SMDS term for DTE in the context of a SMDS network. The SAT gains access to the network through either a SNI or DXI interface.

subscriber converter See Set Top Box.

subscriber identity module See SIM.

subscriber line The telephone line connecting the local telco central office to the subscriber's telephone instrument or telephone system.

subscriber line charge SLC. A monthly charge on subscribers created by the Federal Communications Commission and paid to the local telephone company. The logic for this charge has something to do with reimbursing the local phone companies for some costs which they are allegedly not recovering elsewhere. In reality, it's just another rate increase. The SLC also is known variously as the Access Charge, CALC, and EUCL. See also Access Charge.

subscriber loop The circuit that connects the telephone company's central office to the demarcation point on the customer's premises. The circuit is most likely a pair of wires. But it could be three wires if some external signaling is being used. It could also be four wires if the circuit was a four-wire full duplex leased line.

Subscriber Loop Carrier See SLC-96.

Subscriber Network Interface See SNI

subscriber number The number that permits a user to reach a subscriber in the same local network or numbering area (same as Directory Number or DN).

subscriber plant factor A planning factor used by common carriers to allocate investment in phone equipment, subscriber lines and the non traffic sensitive portion of the central office equipment.

subscriber premises equipment Cable-related equipment located on the subscriber premises, whether owned by the subscriber or the cable system. This term includes:

Subscriber-owned consumer-electronics equipment (TV sets, VCRs, FM tuners, closed-caption decoders).

Subscriber-owned terminal devices (generic converters, digital audio tuners).

System-owned terminal devices (generic converters, converter/descramblers, digital audio tuners, special equipment to enable simultaneous reception of multiple signals). Compare with Subscriber Terminal.

subscriber self provisioning Subscriber self provisioning enables the subscriber self-service access to configure his or her own features and services using their PC and a web-based GUI (graphical user interface), e.g. a standard browser. The subscriber would be able to enable new features or modify how the existing ones work. Such features might include caller ID, call forwarding, call transfer, etc.

subscriber tap A device used to take a small portion of signal off a cable and feed it to a subscriber.

Subscriber Television STV. Also called subscription television. Television in which the subscribers pay a fee for programming as compared with commercial television where they do not.

subscriber terminal The point at which the subscriber-owned cable television equipment is connected to the cable system; typically, a 75-ohm "F"-connector or a 300-ohm balanced line. Compare to Subscriber Premises Equipment.

subscriber trunk dialing STD. The European version of direct-distance dialing. Pricing for long distance calls is typically done by billing a standard amount of money, e.g. a German Mark, for a length of speaking time, which shortens the further you call. For example. you might get one minute for a Mark if you're calling 50 miles. If you call 200 miles you might only get 20 seconds.

Subscriber Unit SU. The Radio Frequency (RF) modem used to acquire the airlink; can be an integral part of the Mobile End System (M-ES) or a separate component.

subscriber drop Wire which runs from a cable terminal or distribution point to the subscriber's premises.

subscription fraud Subscription fraud occurs when a cellular subscriber uses false information or a false identity to secure legitimate service, causing the carrier to incur softand hard-dollar losses. This type of fraud can be prevented by better credit checks and billing systems as well as through prepay offerings.

subset A contraction for Subscriber Set, or telephone set.

subsplit A method of allocating frequencies in a broadband transmission system. Transmit frequencies are in the range of 5 to 32 megahertz, and receive frequencies are in the range of 54 to 300 megahertz.

substation An additional phone which has been established as an extension to the main phone or primary line.

substitute character A transmission control character used in place of a character found to be in error.

substitution A word used in voice recognition to mean a type of error that occurs when a word within the active vocabulary is spoken correctly but classified as another word within the vocabulary. This error is usually dealt with during a verification stage in an application, i.e. " you said, 1,2,3...correct?"

subsystem A processing component addressable through SCCP. Examples include a Home Location Register (HLR) in a wireless system or an 800 number translation database.

subtend The position of a central office in the network hierarchy as it relates to another central office, normally referred to as an access tandem. A central (end) office provides the dial tone point of connection for end users and has its trunking facilities directed to the access tandem (switching point), which allows end users access to a variety of IXCs. The end office is referred to as subtending the access tandem.

subtending A network element (NE) that underlies, or has a lower position than, another network element. For example, a tandem switching system serves to interconnect subtending end offices, or central offices (COs). Similarly, a high-speed carrier-class core router or ATM switch interconnects subtending routers or switches. A long-haul backbone OC-192 SONET fiber optic ring running at 10 Gbps may serve to interconnect subtending OC-48 metropolitan rings running at 2.5 Gbps.

subtending tandem LEC tandem that would be used by a CLEC to connect to an IXC that is not directly interconnected to that CLEC.

subtree Any node within a tree, along with any selection of connected descendant nodes. The highest level of the registry (for example, HKEY_LOCAL_MACHINE).

subviral marketing Short, email-friendly video clips that appear to spoof a well-known commercial -a subversive twist on viral marketing. The clips are thought to be commissioned by the companies whose ads are parodied, such as Budweiser, Ford, Levi Strauss, and MasterCard. This definition from Wired Magazine.

subvoice-grade channel A communications channel of bandwidth narrower than a standard 3Hz voice line. A subvoice-grade channel is usually used for slow data transmission such as teletype or telemetry.

success Success is a lousy teacher. It seduces smart people into thinking they can't lose. Bill Gates

sucker trap A feature of a security firewall. Sucker traps log access attempts, separate legitimate from illegitimate users, and maintain an audit log of the illegitimate.

suckers rally In a bear market, stock prices fall over a long-term. But prices do not fall in a straight line. They occasionally bounce upwards for a few days. As prices rise, some people come to believe that the bear market is over and we're about to end a major bullmarket. But then, a week or two later, prices begin to fall again and quickly fall further than the bounce took them up. The bounce becomes known as a suckers rally, also known as a head fake.

sudden reputation death syndrome This wonderful definition from Wired Magazine: "The complete and instantaneous loss of all credibility, support and nearterm job prospects after being caught up on a high-profile scandal (e.g. Enron's Kenneth Lay and Clinton adviser, Dick Morris)."

Sufism Sufism is a mystic tradition in Islam that dates back to the eighth century and the Ottoman Empire. It is best known for its dervishes' dancing, done to achieve a higher state of awareness.

suit A pejorative term for a professional manager. The term is used by bright, hardworking folks who work for a startup company. Such people, also called ponytails, tend to believe themselves to be very creative. As a mark of their "creativity," they tend to dress casually. When the startup becomes successful, it often goes public or is bought by a large, well-established firm. In come the "suits" -professional managers who dress in fancy, expensive suits and sport arrogance to match. These professional managers then proceed to mess up the company because they try and install ponderous, "big company" practices lots of budgeting, procedures and policies -on a company that succeeded because it was light of foot. The worst of the "suits" are known as "empty suits." See Empty Suit.

suite A collection. A suite of software tools is a collection of software tools.

SuitSat-1 A satellite made, inter alia, from a surplus Russian spacesuit, which was released into orbit from the International Space Station on February 3, 2006 and re-entered the Earth's atmosphere, where it burned up, on September 7, 2006. SuitSat-1, whose formal designation was AMSAT OSCAR 54 (AO-54), was equipped with a low-power 2-meter transmitter. Its transmissions were broadcast at 145.990 MHz, enabling them to be picked up by any radio that could be tuned to that frequency. SuitSat-1's transmissions consisted of recorded greetings in several languages, followed by telemetry data in computer-generated English. SuitSat-1 generated a big following among students and teachers in science classes on Earth. SuitSat-1 was sponsored by Amateur Radio on International Space Station (ARISS), the Radio Amateur Satellite Corporation (AMSAT), the American Radio Relay League (ARRL), the Russian Space Agency and NASA.

summary address An ATM term. An address prefix that tells a node how to summarize reachability information.

summary billing Some telecom carriers will give you one monthly consolidated phone bill -no matter how many number accounts you have in your billing area. Ask.

summation check A check based on the formation of the sum of the digits of a numeral. The sum of the individual digits is usually compared with a previously computed value.

Sun Microsystems Sun Microsystems is a Californian computer manufacturer. SUN stands for Stanford University Network. See Java. www.sun.com

sun outage A complete blockage of the receiving ability of an earth station whenever the sun passes through the axis of the main beam of its antenna. Because the antenna then "sees" the noise temperature of the sun, the station's G/T is depressed below the minimum required for reception. However, in "looking" at the entire earth, the satellite "sees" only one small shiny point, of the antenna reflecting the sun, so the station may still transmit.

sun synchronous A term describing the fact that the orbits of LEOs (Low Earth Orbiting) and MEOs (Middle Earth Orbiting) satellite systems can be adjusted such that the greatest number of satellites in the constellation are positioned over geographic areas which are in the light of day. At such times, the greatest amount of traffic originates and terminates. See LEO and MEO.

sun transit outage Satellite circuit outage caused by direct radiation of the sun's rays on an earth station receiving antenna.

sundown rule A rule in voice mail which says that all messages should be returned that day, before the sun goes down.

sunlight resistant An optional UL listing that may be obtained for an insulation or jacket compound involving exposure to direct rays of the sun. An important listing for jackets of armored cables, TC Tray Cables and MV90 cables which will be used in cables trays.

sunlighting When telecommuters take on outside projects while working at home for their own employer.

SUNOS SunOS is Sun Microsystems' implementation of UNIX.

SunPC Hardware/software solutions from Sun Microsystems for MS-DOS on SPARC platforms.

sunrise industry A new industry that is rapidly expanding.

sunrise technology An emerging technology.

sunset technology A technology that is outdated and whose use is fading, and for which a discontinuation date may have been set for the technology to be phased out completely.

sunset date The date beyond which a sunset technology can no longer be used.

SunView SunView is Sun Microsystems' kernel-based window system.

SunXTL Server Part of Sun Microsystems' XTL Teleservices architecture. Provides multi-client and multi-device support. The server is the central point of contact for all tele-services services. Resource management and security are provided by the server. Communicates with the Sun XTL provider to place and receive telephone calls. An application may access the data associated with a call by acquiring a data stream from the API. www.sun.com

Super DSL A slang name for a second generation DSL line -one offering speeds as fast as eight megabits per second downstream (i.e. coming from the central office to the customer).

Super G3 Super G3 is a new unofficial "standard" for higher speed fax machines, which contain a 33.6 Kbps V.34 modem, V8 handshaking and the new ITU-T T.85 JBIG image compression. On most phone lines such a machine should get close to double the speed of the highest speed Group 3 fax machines, namely 14.4 Kbps. But, the JBIG image compression will speed faxing of gray scale images by as much as five to six times. In short, these machines will send faxes much faster -if they send to a Super G3 machine at the other end. Super G3 is compatible with and can communicate with older fax machines, Group 1, 2, 3 and 3 Enhanced.

super server A file server with more than one CPU (Central Processing Unit). At time of writing this dictionary, a high-end super server might contain four Intel Pentium chips. To take advantage of these super servers, you need an operating system capable of asymmetrical multi-processing, such as Unix and Windows NT Advanced Server.

super-JANET The latest phase in the development of JANET, the UK educational and research network run by UKERNA. It uses SMDS and ATM to provide multi-service network facilities for many new applications including Multimedia Conferencing.

supercomputing A term applied to a class of high-speed computers employing advanced technologies such as simplified instruction sets, wide data paths and pipelining.

superconductors Superconductors are materials which have no resistance to the flow of electricity. They are widely believed to have great potential for dramatically faster telecommunications switches and computers. In the past, the superconducting state -zero resistance to the flow of electricity -could be achieved only by cooling certain metal alloys to temperatures of near absolute zero, or about 460 degrees below zero. Starting in 1986, researchers discovered that ceramic materials could reach superconductivity at temperatures as high as 235 degrees below zero.

superdrive The name for Apple's 1.44 Mb floppy that can read and write MS-DOS formatted floppies and Mac formatted disks. DOS floppies require Apple File Exchange or a third party product to read the DOS format.

superframe format 1. The T-Carrier superframe transmission structure consists of 12 DS-1 frames (2316 bits). The DS-1 frame comprises 193 bit positions, the first of which is the frame overhead-bit position. Frame overhead bit positions are used for the frame and signaling phase alignment only. See also DS-1 and T-Carrier.

2. In digital cellular networks defined in IS-136, a superframe comprises 16 frames. Two superframes comprise a hyperframe. See also IS-136.

supergroup Sixty voice channels. In more technical terms: the assembly of five 12-channel groups occupying adjacent bands in the spectrum for the purpose of simultaneous modulation or demodulation.

superheterodyne A type of radio receiver operating on the heterodyne or beat principle. See Heterodyne.

superhouse The National Society of Home Builders has registered SmartHouse. Bell-South has trademarked SuperHouse. And GTE has registered SmartPark. One day, fiber optic will snake to everyone's house, bringing the potential of immense information services. Until that day comes they'll be lots of interesting demonstrations at distant trade shows.

superimposed ringing A way of stopping party line phone users from hearing each other's ring by superimposing a DC (direct current) voltage over the ringing signal and using it to alert a vacuum tube or semiconductor device in only the phone instrument that we want to ring. See also Superposed Ringing.

supermarket shopping cart Sylvan N. Goldman of Humpty Dumpty Stores and Standard Food Markets developed the shopping cart so that people could buy more in a single visit to his grocery store. He unveiled his creation in Oklahoma City on June 4, 1937.

supermastergroup SMG. Six mastergroups each comprised of 10 supergroups each comprised of five groups of 12 circuits totaling 3600 circuits carried as a unit in an analog FDM carrier system; first used in Bell's L4 coaxial cable carrier systems.

supernet See CIDR.

supernetting An IP (Internet Protocol) term. Supernetting is the linking of multiple IP network address blocks into a "supernet." In the early 1990s, the IETF (Internet Engineering Task Force) approved RFCs (Requests For Comment) 1518 and 1519, which detail the CIDR (Classless InterDomain Routing) protocol to allow the grouping of multiple Class C subnet address blocks into a single "supernet." BGP-4 (Border Gateway Protocol version 4) supports supernetting. See also BGP, CIDR, IETF, and RFC.

superposed circuit An additional channel obtained from one or more circuits, normally provided for other channels, in such a manner that all the channels can be used simultaneously without mutual interference.

superposed ringing Party-line telephone ringing in which a combination of alternating and direct currents is used, the objective being to only ring the bell in the phone of the one whose call is coming in. See also Superimposed Ringing for a better explanation.

superserver IBM's new name for a mainframe. The name is clearly an attempt to position the mainframe more competitively against the new extremely powerful PCs which are taking over the mainframe's role.

supersite Toby Corey, president of USWeb Corp., defines supersite as an Internet site combining a public Internet site, an extranet for business partners and an administrative intranet. He says a supersite has a common architecture across intranet, Internet, extranet and Web sites. He says it serves multiple audiences and it implements various levels of access control.

superstation A commercial television broadcast station which is transmitted to a cable television headend by a communications satellite and then retransmitted by the cable system to its subscribers. FCC rules provide that a cable television system may carry a superstation under the same conditions that it may carry any other television broadcast station. Examples: WGN-TV and WWOR are superstations (TBS is not a superstation, in spite of its self-proclaimed status as "Superstation TBS."). Compare Origination Cablecasting.

supertrunk A cable that carries several video channels between facilities of a cable television company. A trunk between the master and the hub headends in a hub CATV system.

supervised transfer In telephony, supervised transfer is a way to transfer the first "leg" of a call to the "second leg" or third party while keeping the first party (typically the originator of the call) "in the loop". In most cases, this is achieved by doing a conference call and then idling the transmit and receive signals to the first party. This keeps the first party in a supervising position, so the second leg can be dropped in order to establish subsequent connections with the first leg of the call. In computer telephony, supervised call transfers are necessary when the application is designed to retain supervision of the call. This means that a caller can be transferred to another system or telephone extension with the VRU (Voice Response Unit) dropping out of the call. This is especially important with some messaging systems that have automated attendant capability. In this case, the VRU will momentarily put the caller on "hold" while determining if the called party is available. This takes into account the call progress tones that are encountered when calling the destination phone. For example, if the VRU detects a busy tone, it will take the caller off of hold and indicate that the extension is busy. The caller is typically offered a chance to try again in a few moments, or to leave a message for the called party.

supervision Supervision of a phone call is detecting when a called party has picked up his phone and when that party has hung up. Supervision is used primarily for billing purposes. Not all long distance carriers have supervision capability. It depends on how "equal accessed" they have chosen to be. See Answer Supervision, Software Supervision and Signaling System 7.

supervisor The person responsible for day to day maintenance and operation of a phone system. Typically used in conjunction with an ACD -automatic call distributor.

supervisory call This service feature allows the attendant, after connecting an incoming CO line or tie line call to the wanted phone, to continuously supervise the call in progress.

supervisory control Characters or signals which automatically actuate equipment or indicators at a remote terminal.

supervisory lamp A lamp which shows the operator whether the person is speaking (off-hook) or is not speaking (on-hook). These days such lamps are called BUSY LAMP FIELDS. In some smaller key systems, all phones have them. Busy lamp fields are an operator's best friend.

supervisory program 1. A program, usually part of an operating system, that controls the execution of other computer programs and regulates the flow of work in a data processing system. 2. A computer program that allocates computer component space and schedules computer events by task queuing and system interrupts. Control of the system is returned to the supervisory program frequently enough to ensure that demands on the system are met.

supervisory relay A relay which, during a call, is controlled by the transmitter current supplied to a subscriber line to receive from the associated phone signals that control the actions of operators or switching mechanisms.

supervisory routine A routine that allocates computer component space and schedules computer events by task queuing and system interrupts. Control of the system is returned to the supervisory program frequently enough to ensure that demands on the system are met.

supervisory signal 1. Supervisory signals are the means by which a telephone user initiates a request for service; or holds or releases a connection; or flashes to recall an operator or to initiate additional features, for example, 3-way calling. Supervisory signals are also used to initiate and terminate charging on a call. A signal also indicates whether a circuit is in use, or not in use.
2. A signal used to indicate the various operating states of circuit combinations.

supplement Service Order Supplement. A term that grew out of the Telecommunications Act of 1996, which formally opened the local exchange to competition. A "supplement" is a revision to a service order or ASR (Access Service Request) of sufficient magnitude to require the creation of a new version, which is known as a supplemented service order. See also ASR.

supplementary services Telephone company talk for services above basic ability to make a phone call. Supplementary services include fast dialing, calling line ID, call waiting, call forwarding, and videoconferencing features. Here's a definitions of supplementary services as applied to ISDN service: Additional services, such as hold, conference, and call forwarding, offered to an ISDN customer. Supplementary services are always present if activated at the switch. Although the specific features and call appearances may differ between service providers, supplementary service generally provides users with the ability to connect and disconnect new calls when one (or more) call exists.

supplementary telephone service The lowest level of service in Windows Telephony Services is called Basic Telephony and provides a guaranteed set of functions that corresponds to "Plain Old Telephone Service" (POTS only make calls and receive calls). The next service level is Supplementary Telephone Service providing advanced switch features such as hold, transfer, etc. All supplementary services are optional. Finally, there is the Extended Telephony level. This API level provides numerous and well-defined API extension mechanisms that enable application developers to access service provider-specific functions not directly defined by the Telephony API. See Windows Telephony Services.

supplicant As specified in IEEE 802.1X for port-based network access control, a supplicant makes requests to an authenticator for access to system resources. The authenticator either grants or denies those requests, depending on whether or not it can authenticate (i.e., validate) the identification of the supplicant, along with its access privileges. A supplicant generally is in the form of client software. An authenticator generally is in the form of software on an authentication server, which is consulted by a router or wireless access point. The term "supplicant" is from middle French (1597) and means one who humbly and earnestly implores or begs.

supply chain All the business activities needed to satisfy the demand for products or services from the initial requirement for raw material or data to the final deliveries to the end user. See Supply Chain Management for a better explanation.

supply chain management The supply chain is the electronic link between a company and its suppliers and distributors/customers. Supply chain management seeks to apply software, hardware, networking and telecommunications links in order to get the right product to the right customer at the right time. Supply chain management software lets managers study their companies' relationships with suppliers, distributors or retailers so they can better forecast and schedule production and sales. The main objectives of supply chain management are to get the product from the supplier to the customer as cheaply and as fast as possible. Much of supply chain management is done on the Internet or private corporate Intranet.

supply chain software See Supply Chain Management.

supply side economics Supply side economics is that branch of economics which believes that tax cuts are the primary engine of economic growth.

support A really stupid word. Support is a verb that in the tech world means "This hardware, software has the following feature." An example: "This software supports 32-bit file sharing."

support hardware The racks, clamps, cabinets, brackets, trays, and other equipment that provide the physical means to hold the transmission media and connecting hardware. An AT&T definition.

support strand Called a messenger. A strength element used to carry the weight of the telecommunications cable.

suppressed carrier single-sideband emission A single-sideband emission in which the carrier is virtually suppressed and not intended to be used for demodulation.

suppressed carrier transmission A transmission technique in which only the sidebands (one or both) are transmitted and the main carrier is not transmitted and thus not used.

suppressed voltage ratings Several ranges are assigned by UL for grading transient suppression voltages. For instance, a 400 volt rating indicates a maximum peak voltage between 330 and 400 volts, These ratings appear between 300 volts and 6000 volts peak.

suppression First, there were 800 toll-free numbers in North America. Then, in the fall of 1995, the FCC introduced 888 numbers -another toll-free dialing code, since the 800 code was filling up. The FCC allowed bona fide holders of 800 numbers to request '888' replicas of their 800 numbers. 1-800-FLOWERS could request 1-888-FLOWERS, etc...for all the obvious reasons. Prior to 12/1/95, then current holders of 800 numbers could submit their request that the 888 version be suppressed, i.e. not permitted to be made available for general assignment. Many 800 owners asked for suppression of their numbers. Many asked for their use. In some cases this was successful. In others, some suppressed numbers were installed by other companies later on -i.e. a screw-up. In short, you have to be ultra-careful with suppression.

suppressors, echo Echo is controlled in long distance circuits with devices called echo suppressors. These devices automatically insert loss in the return path of a four-wire circuit. All long distance circuits are four-wire -two wires for each of the two paths (receiving and transmitting). The echo suppressor jumps back and forth between the two transmission paths. Properly adjusted, an echo suppressor puts only sufficient loss in a circuit so a listener can interrupt the talker. With very long circuits -22,300 miles -in satellites, a better way is needed. They're called echo cancellers.

surcharge A charge imposed in accordance with the Commission's Access Reconsideration decision in CC Docket 78-72, Phase 1, FCC 83-356. released August 22, 1983 and updated too many times since. The monthly charge is about $2.00 and is going up to $3.50. This charge is said to compensate the local phone company for long distance commissions (called settlements and separations) lost and now replaced with per minute access charges.

surf The word "surf" in its classic sense means to ride the crest of a wave, skimming quickly across the water underneath. The electronic world has adopted that definition of skimming quickly and included it in several definitions. For example, shoulder surfing is gazing quickly over someone's shoulder while they're making a call at a payphone and writing down the user's credit card numbers. Channel surfing is moving from one TV channel to another quickly. Surfing the Web means moving from one Web site to another, jumping around in search of knowledge or amusement, but certainly in a non-linear way. See also Internet and Web Browser.

Surface Acoustical Wave Filter SAW Filter. Specific type of filter made of piezoelectric material.

surface emitting diode An LED that emits light from its flat surface rather than its side. Simple and inexpensive, with emission spread over a wide angle.

surface mount With surface mount technology, Components sit on the surface of printed circuit boards and are soldered to conductive pads. In the "thru-the-hole" process, component leads are placed through holes in the boards and are sent through wave soldering for attachment. Surface mount technology is more cost-effective, as it allows for denser packaging on the board and components can be mounted on both sides of the surface.

surface outlet A Communications Outlet (modular jack) that is installed on the surface of the mounting location. The premises wire serving such an outlet may or may not be concealed behind the mounting surface.

Surface Resistivity The resistance of a material between two opposite sides of a unit square of its surface. It is usually expressed in ohms.

surface wave A wave that is guided along the interface between two different media or by a refractive index gradient. The field components of the wave diminish with distance from the interface. Optical energy is not converted from the surface wave field to another form of energy and the wave does not have a component directed normal to the interface surface. In optical fiber transmission, evanescent waves are surface waves. In radio transmission, ground waves are surface waves that propagate close to the surface of the Earth, the Earth having one refractive index and the atmosphere another, thus constituting an interface surface.

surfing Same as browsing. See surfing the web. See also shoulder surfing and channel surfing.

surfing the web A phrase first used by Jean Armour Polly, a former public librarian working on an article about the Internet in 1992. She wrote "At that time I was using a mouse pad from the Apple Library. The one I had pictured a surfer on a big wave. 'Information Surfer' it said. "Eureka, I said and had my metaphor."

surge A temporary large increase in line voltage that lasts longer than one cycle of the line frequency of 60Hz, the North American frequency or 50Hz in many other countries, especially those running at 240 volts.

surge protector A device which plugs between the phone system and the commercial AC power outlet. It is designed to protect the phone system from high voltage spikes (also called surges) which might damage the phone system. When a surge occurs on the power line, the surge protector sends the overload to ground. How fast it sends it to ground is a subject that could fill a book. The type of surge protector that you buy will be determined mostly by the speed you need to protect your equipment.

surge suppressor See Surge Protector.

surges The increased flow of current through an electrical device brought about by an instantaneous change in its resistance or impedance.

survivability A property of a system, subsystem, equipment, process, or procedure that provides a defined degree of assurance that the device or system will continue to work during and after a natural or man-made disturbance; e.g. nuclear attack. This term must be qualified by specifying the range of conditions over which the entity will survive, the minimum acceptable level or post-disturbance functionality, and the maximum acceptable outage duration.

Survivable Adaptable Fiber Network SAFENET. A U.S. Navy experimental fiber-based local area network designed to survive conventional and limited nuclear battle conditions.

survive We'll all survive. We have no choice. -Ray Horak, December, 2001, after a horrible year.

Susan A nice name for a wife. Everyone should have such an incredible wife. We got married in 1976. And our relationship has only gotten better since. When friends ask us how we manage such a wonderful marriage, I explain that early on in our marriage we split our responsibilities. Susan handles all the lesser decisions, while I handle all the important ones. Friends ask "What lesser decisions does Susan make?" I answer: "Decisions such as to how we bring up the kids, where the kids go to school, where we live, what we spend our money on, where and when we will vacation..." Friends then ask what decisions do I get to make? I answer it's decisions like should China be admitted to the World Trade Organization, whether we should bomb Afghanistan, whether we should invade Iraq, etc. Susan is the mother of our two children, Claire and Michael, and clearly the boss of us all. After 23 editions, she is used to seeing more of my computer screens than me. On the other hand, if she saw more of me, I'm sure our marriage would not have lasted this long. Most of our friends' marriages haven't. She's a great wife. Everyone should be so lucky. See definitions for Claire, Michael and now Ted, for the rest of the family.

susceptiveness In telephone systems, the tendency of circuits to pick up noise and low frequency induction from power systems. It depends on telephone circuit balance, transpositions, wiring spacing, and isolation from ground.

SUSE Linux Another variation of Linux, it's owned by the Novell Corporation.

suspend See Hibernation.

suspended ceiling A ceiling that creates an area or space between the ceiling material and the structure above. See also Plenum.

suspended customer An Worldcom definition. A customer who has requested service but has not yet been installed due to insufficient network capacity or some other operational/administrative constraint.

suspended on-demand connection The mode of an on-demand connection where the communications line is dropped and the connection sites are actively

spoofing.

suspension bridge When a suspension bridge was being built over the gorge near Niagara Falls, New York, there was no way a boat could carry the necessary suspension wires across the violent waters. The bridge's builders were inspired to offer $5.00 to the first boy to fly a kite from the American to the Canadian side. It worked. Once the kite string made the crossing, a succession of heavier cords and ropes tied to the kite string and each other were pulled over until the first length of cable finally spanned the river.

sustaining engineering An endeavor in which a company devotes a bunch of people to maintain the quality of engineering on existing products. Such group does not focus on new product.

SUT 1. An ATM term. System Under Test: The real open system in which the Implementation Under Test (IUT) resides.

2. Stupid User Tricks. Also called ESO, or Equipment Superior to Operator. When closing help desk tickets, it describes situations where the problem was user stupidity, such as the power cord not plugged in, the monitor unplugged, the keyboard not attached, etc.

SV Silicon Valley, California. See Silicon Valley.

SVC 1. Switched Virtual Circuit. A virtual circuit connection established across a network on an as-needed basis and lasting only for the duration of the transfer. The datacom equivalent of a dialed phone call, the specific path provided in support of the SVC is determined on a call-by-call basis and in consideration of both the end points and the level of congestion in the network. SVCs are used extensively in X.25 networks, and increasingly in Frame Relay networks. SVCs are much more complex to provision than are PVCs (Permanent Virtual Circuits), but perform much better as they effectively provide automatic and dynamic network load-balancing. In other words, SVCs are set up in consideration of the load on the network, and its subnetworks, in order that the least congested path be established and, therefore, that the data transmission receive the lowest possible level of delay. See also PVC, VC and VCC.

2. Switched Virtual Call. Basically another way of saying switched virtual circuit. See above

SVCC Switched Virtual Channel Connection: A Switched VCC is one which is established and taken down dynamically through control signaling. A Virtual Channel Connection (VCC) is an ATM connection where switching is performed on the VPI/VCI fields of each cell.

SVD Simultaneous Voice Data. In the fall of 1994, the term began to apply to several techniques for putting voice conversations and data transfers on the same analog phone line. Some of these techniques involve interrupting the voice conversation while data is transferred. Others involve transferring the data and voice simultaneously on different bandwidths.

SVG Basic A standard developed by the World Wide Web Consortium for displaying vector graphics on higher-end mobile devices, such as PDAs. See SVG Tiny.

SVG Tiny A standard developed by the World Wide Web Consortium for displaying vector graphics on mobile devices, such as cell phones, that have extremely limited CPU speed, memory and color support. See SVG Basic.

SVGA Super Video Graphics Array. An extension of the VGA video standard. SVGA enables video adapters to support resolutions of up to 800 x 600 pixels with up to 16.7 million simultaneous colors, which is known as true color because it's the number of colors someone once figured is in a Kodachrome slide. See also VGA. UGVA stands for Ultra Video Graphics Array and refers to 1024 x 768.

SVGB See SVG Basic.

SVGT See SVG Tiny.

SVHS See V Video.

SVN Subscriber Verification Number. Number issued by the long-distance carrier to confirm the order for long distance service.

SVOD subscription video on demand. A video-on-demand service offered at a flat subscription price that provides viewers with unlimited access to programs from the video libraries they subscribe to.

SVPC Switched Virtual Path Connection: A Switched Virtual Path Connection is one which is established and taken down dynamically through control signaling. A Virtual Path Connection (VPC) is an ATM connection where switching is performed on the VPI field of each cell.

SVS Switched Video Service. See IPTV.

Svyazinvest A new holding company that will control much of Russia's telecommunications industry. Svyazinvest was originally conceived both as a holding company for the state's interests in 86 local telephone firms across Russia as a competitor for Rostelcom, a state-run near-monopoly in long distance and international telephony. But, later in the planning, the state's 38% shareholding in Rostelcom was dumped into Svyazinvest. According to the Economist Magazine, Svyazinvest's bosses will therefore have power over almost all of Russia's communications systems (in many regions, the local phone companies also hold the first cellular telephone licenses.)

SW Station Wire.

SWACT Switch of Activity.

SWAG Scientific Wild Ass Guess. A means of coming up with some figures without all the facts or engineering spec being in place based on known information and past history. SWAG is more accurate than a guess and more valuable than an opinion.

SWAP 1. This came from an issue of Internet Week in the Winter of 1998. Seeking to further leverage the ability of the Web to link business partners, a group of vendors led by Netscape, Sun Microsystems and Hewlett-Packard earlier this week proposed a standard to tie together disparate workflow systems on an intranet or over the Internet. The trio, along with 20 other vendors, said they would support the Simple Workflow Access Protocol (SWAP), a proposed Internet standard that would allow disparate workflow engines to manage, monitor, initiate and control the execution of workflow processes between one another within an intranet or over the Internet.

2. Shared Wireless Access Protocol. A specification from the Home RF (Radio Frequency) Working Group intended to enable interoperability of electronic devices from a large number of manufacturers, while providing the flexibility and mobility of a wireless solution. SWAP is expected to yield a wireless home network to share voice and data communications between devices such as PCs, peripherals, PC-enhanced cordless phones, headsets, and other devices yet to be developed. SWAP also is intended to allow the sharing of a single Internet connection amongst multiple such devices. SWAP is an extension of DECT (Digital Enhanced Cordless Telephone) and WLAN (Wireless Local Area Network) technologies, and supports both TDMA (Time Division Multiple Access) for interactive voice and other real-time applications, and CSMA/CA (Carrier Sense Multiple Access/Collision Avoidance) service for high-speed packet data delivery. SWAP runs in the unlicensed ISM (Industrial Scientific and Medical) band in the 2.4 GHz range, and employs FHSS (Frequency Hopping Spread Spectrum) at 50 hops per second. Data rates are at 1 Mbps using 2FSK (Frequency Shift Keying), 2 Mbps using 4FSK, and 10 Mbps using an undefined modulation technique. SWAP will support up to 127 devices per network, and up to six full-duplex voice conversations. Data security is through the Blowfish encryption algorithm, and data compression through the LZRW3-A algorithm. See also HomeRF and HomeRF Working Group.

swap file Some operating systems and applications let you use more memory than what you have in RAM. They do this by pretending that part of your hard disk is RAM memory. They do this by creating a swap file on your hard disk and swapping memory back and forth. Some computer systems call this virtual memory. You need to be careful with swap files. Never turn your machine off when you have applications running. If you do you're likely to leave a huge swap on your hard disk, which you may not find (it's hidden) and which your system may not dispose of. To get back the space on your hard disk, you'll need to erase it separately.

swastika Between 1937 and 1945 Heinz produced a version of alphabet spaghetti especially for the German market that consisted solely of little pasta swastikas. I have no idea of how well it sold.

swatch bard In the wall paint business, the term for a person who creates the names of all the various colors of paint.

SWATS Standard Wireless AT Command Set. An extension to the Hayes AT command set to support wireless modems, such as those used in standard AMPS analog cellular phones.

SWC Serving Wire Center. The service provider's (usually telephone company) wire center (i.e., Central Office) to which you are connected.

SWEDAC Swedish Board for Technical Accreditation. They have established two standards, which effectively limit radiation emissions, MPR1 and MPR2. These standards specify maximum values for both alternating electric fields and magnetic fields and provide monitor manufacturers with guidelines in creating low emission monitors. There is, as yet, no definite proof of harm from normal computer monitors. But the argument goes that they weren't so sure about nicotine in cigarettes 30 years ago. And look at us 30 years later.

sweep acquisition A technique whereby the frequency of the local oscillator is slowly swept past the reference to assure that the pull-in range is reached.

sweep test Manufacturers use a sweep test to test the frequency response of a metallic cable as a quality control mechanism. The sweep test involves the checking of frequency response through generating an RF (Radio Frequency) voltage that is varied back and forth in a frequency range at a constant rate.

swell An increase from nominal voltage lasting one or more line cycles.

SWHK Abbreviation for SWITCH HOOK. Originally referred to an actual hook on older phones that held the receiver, and sprang upward to close a switch and activate the phone when the receiver was picked up. Today the term refers to any of various buttons and plungers that are pressed down and released when the handset is put down (physically "hung up" in the old days) and picked up.

SWIFT Society for Worldwide Interbank Financial Telecommunications.

SWIGS Special Working Interest Groups. see www.fiberchannel.com.

swim Slow, graceful, undesired movements of display elements, groups, or images about their mean position on a display surface, such as that of a monitor. Swim can be followed by the human eye, whereas jitters usually appears as a blur.

SWIP SWIP Shared Whois Project. When ARIN allocates a block of IP addresses to an ISP, they create a record of who has been assigned those addresses. If the ISP reassigns some of the IP addresses to a customer, then ARIN's records must be updated to show the new assignee. The e-mail based interface to ARIN that allows an ISP to make these changes is called SWIP. The act of making this changes is called "SWIPPing." SWIP rhymes with hip not hype.

swiped out An ATM or credit card that has been rendered useless because the magnetic strip is worn away from extensive (and expensive) use.

swiss army knife The one tool to have when you can't have more than one. Be sure it has a normal screwdriver, a phillips head screwdriver and a pair of scissors. A corkscrew also is useful.

switch A mechanical, electrical or electronic device which opens or closes circuits, completes or breaks an electrical path, or selects paths or circuits. Switches work at Layers 1 (Physical) and 2 (Data Link) of the OSI Reference Model, with emphasis on Layer 2. A switch looks at incoming data (voice data, or data data) to determine the destination address. Based on that address, a transmission path is set up through the switching matrix between the incoming and outgoing physical communications ports and links. Data switches (e.g., LAN switches and packet switches) also typically contain buffers, which can hold data packets in temporary memory until the necessary resources are available to allow the data packets to be forwarded. Voice switches, of course, don't, because you can't delay voice. Switches work link-by-link, with multiple switches typically being involved in complex networks; each switch forwards the data on a link-by-link (hop-by-hop) basis. Routers are highly intelligent data switches which are capable of setting up paths from end-to-end, perhaps in consideration of the level of privilege of the user and application. Routers commonly are used at the edges of complex data networks, where intelligence is required to set up appropriate network paths. Although such intelligent decisions impose some delay on the packet traffic, they are made only at the ingress and egress edges of the network. The routers often instruct switches in the core of the network, where speed is of the essence -switches aren't as intelligent as routers, but they are faster and less expensive. See also Ethernet Switch, OSI Reference Model, Router and Switching Fabric.

switch based resellers Switch-based resellers lease facilities from national carriers or large private line networks. They resell services provided over those facilities under their own name and provide sales, customer service, billing and technical support. Switch-based resellers own or lease their own switching equipment and, in some cases, own their transmission facilities. they typically provide originating service on a regional basis. See also Switchless Resellers.

switch busy hour The busy hour for a single switch.

switch control The technique by which a switching system responds to signals and directs the switching network.

switch domain An SCSA definition. A single instance of a particular technology-specific connection type. See S.100.

switch driver Protocol Mapper Code running on a Telephony Server that translates between a particular switches proprietary switch-server protocol and one of the specified computer telephony integration (CTI) protocols. See Protocol Mapper.

switch fabric An SCSA definition. The facility for connecting any two (or more) transmitting or receiving Service Providers.

switch fabric controller An SCSA definition. A technology-specific, replaceable ASP within the SCSA server. The SFC is designed to support both the internal connectivity within the group and the complex, multiparty call processing applications not directly addressed by the functionality of the Group.

switch feature A service provided by the switch that can be invoked by a computing domain or by manual telephone activity. "Do not disturb" is an example of a switch feature.

switch hook It is also called the Hook Switch. A switch hook or hook switch was originally an electrical "switch" connected to the "hook" on which the handset (or receiver) was placed when the telephone was not in use. The switch hook is now the little plunger at the top of most telephones which is pushed down when the handset is resting in its cradle (on-hook). When the handset is raised, the plunger pops up (the phone goes off-hook). Momentarily depressing the switch hook (under 0.8 of a second) can signal various services such as calling the attendant, conferencing or transferring calls. See also Switchhook Flash.

In ISDN, the AT&T ISDN sets have several switch hooks; one for the handset, one for the speakerphone, a "virtual" switch hook, and if an adjunct is attached, an adjunct switch hook. If all switch hooks are "on-hook" or hung up, the ISDN set is on-hook. If any switch hook is "off-hook," then the ISDN set is off-hook. If more than one switch hook is off-hook, the ISDN set uses a complex algorithm to determine whether the handset, the speakerphone, or the adjunct has precedence (only one can be used at a time).

switch hook flash A signaling technique whereby the signal is originated by momentarily depressing the switch hook. See Switch Hook.

switch interface The Ethernet MAC controller interface. In general, a switch interface on a switch is the same as a port. However, the number of interfaces does not necessarily correspond to the number of ports. For example, a MAB port on a switch may be a 4-port repeater.

switch message Information that originates in a switch. A Call-Progress Event Message is one category of switch messages. Delivered is an example of a call-progress event message.

switch network The portion of a switch that provides the connection between lines and trunks terminated in the system.

switch over When a failure occurs in the equipment, a switch may occur to an alternative piece of equipment.

switch port An SCSA definition. A resource that allows a Group to communicate with another Group. All Groups implicitly possess a Switch Port as a secondary resource, but in order to use it, the application must explicitly connect the Switch Ports of two Groups.

switch redirect A central office service which instantly, on command, redirects thousands of phone numbers to different phone numbers. Such a service has great use in a disaster.

switch room The room in which you put phone equipment. Also called the Phone Room. (What else?) The Phone Room should be large, clean and should stay at roughly seventy degrees and 50% humidity. You, the customer, are responsible for the quality and condition of your phone room. The messier it is, the hotter it is, the dirtier it is, the poorer your phone system (and its technicians) will function.

switch tag A switch tag is a series of numbers and or letters either at the beginning or end of a recording that identifies who originated the recording. A switch tag is be often used by a local or long distance phone company, wired or wireless.

switch tender In the old, old days, the switch tender was the person who took care of the switch that moved trains from one track to another. That person often stayed for many hours a day in a small hut next to the track. Based on timetables and telegraph and phone communications with central train dispatch, he would change the switch and thus move incoming trains to the right track. The job of being a switch tender is now obsolete as switches are now changed remotely by signals over phone lines. The expression "sleeping at the switch" came from the switch tender profession. Sleeping at the switch could cause train derailments and dead passengers. Sleeping was not a career-enhancing strategy. See Switch Train.

switch train In a telecom circuit (typically a step-by-step central office), the series of switching devices which a call moves through in sequence.

switchable hubs Feature pores that permit additional stackable hubs to be connected for expanded capacity. Special software is used to reassign the ports or switch the ports off and on.

switchboard The attendant position of a PBX. Most of them don't actually have "boards" (they were big), they have consoles (they're much smaller and they fit on desks). Switchboards are desks.

switchboard cable A cable used within and between the central office main frames and the switchboard.

switched 56 A switched data service which lets you dial someone else and transmit at 56 Kbps over the PSTN (Public Switched Telephone Network). Actually Switched 56/64, Switched 56 also is available at 64 Kbps, where the carrier supports non-intrusive signaling and control. It is a circuit-switched service, letting the user transmit data at 56/64 Kbps over a four-wire, digital, synchronous network. The cost of a Switched 56/64 call is calculated based on duration and time of day, with discounts for non-prime time calls; discounts

also apply according to day of week, holidays, and so on.

Provided by LECs, Switched 56/64 was the first VPN (Virtual Private Network) service offered. The term "VPN" currently is applied to IXC services, with bandwidth available at 56/64 Kbps, increments of 56/64 Kbps, 384 Kbps, and 1.544 Mbps (T1). Switched 56/64 offers a high degree of redundancy, flexibility and scalability. It often is used as an alternative to private, leased-line networks -or as a backup for them.

Applications for Switched 56 include videoconferencing, high speed data transfer, digital audio broadcasting, Group IV fax and remote LAN access for telecommuters. It also is used as an access technology for IXC-provided VPNs. While it is widely deployed, Switched 56/64 faces a great deal of pressure from ISDN BRI, which offers as much as 144 Kbps, where available. Over time, Switched 56/64 will disappear for good in North America. See also VPN.

switched access A method of obtaining test access to telecommunications circuits by using electromechanical circuitry to switch test apparatus to the circuit.

switched access line service All residential and most businesses use this type of telephone access. It refers to the connection between your phone and the long distance companies' switch (POP) when you make a regular local or long distance telephone call over standard phone lines.

switched carrier In data terms, physical line specification selection indicating a half duplex line in a bisync network.

Switched Circuit Automatic Network SCAN. A service arrangement at certain Telco premises to interconnect private line telephone service channels of a switched service network provided to certain agencies of the federal Government.

Switched Circuit Ordering And Tracking System SCOTS. MCI's automated tracking and order processing system for Dial up products, IMTs, and the MCI switched network.

switched DAL Switched Dedicated Access (Egress) Line. Dedicated trunk group (T-1, etc.) circuit(s) used to access (1+, etc.) or egress (800, etc.) through normal network switching facilities. The Switched DAL is dedicated to a particular inbound or outbound call type.

Switched Digital Services Applications Forum See SDSAF.

switched Ethernet Ethernet is the most common local area network. It is a shared network in which each piece of information being transmitted passes by each computer (or workstation) on the network. Information is sent in packets. Each packet contains an address representing which computer (or workstation) the packet is destined for. All other computers on the Ethernet network ignore the packets. If there's a lot of traffic, the time to transmit and receive slows down dramatically. Instead of getting the full 10 megabits per second, transmission speeds might get as slow as 100,000 bits per second. One way of solving this problem is switched Ethernet. Here's how it works. Imagine you have an Ethernet network with 100 users. Fifty of them are on the tenth floor. Fifty of them are on the eleventh floor. We break the cable between the two floors and insert a device known as an Ethernet Switch. Essentially we now have two separate, but joined networks. All the traffic from the tenth floor to the tenth floor stays on the tenth floor. Ditto for the eleventh floor. The duty of the Ethernet switch is to twofold: First to keep traffic from the two floors separate, i.e. no traffic from the tenth to the eleventh. Second, to send traffic from the tenth floor, which is destined for the eleventh, to the eleventh and vice versa. Thus an Ethernet switch's function is to split an Ethernet network so each piece of the network can use the full data transmission capacity. When you set up a an Ethernet switch, you want to make sure that "communities of interest" are on the same "floor" (i.e. Ethernet segment). Let's say 90% of the email, which your marketing department sends, is destined for members of your marketing department. Then it makes sense to have your marketing department on one Ethernet segment. An Ethernet switch is far more expensive than an Ethernet hub, which simply joins computers and/or workstations to an Ethernet network and makes sure that all the packets transmitted bypass all the computers and/or workstations on the Ethernet network.

switched LATA access An LEC service offering under tariff that provides for a switched communications path between an long distance company POP (Point of Presence) and the premises of its end users. It includes all LATA access services that use an LEC switching system (EO or AT).

switched line A circuit which is routed through a circuit switched network, such as the telephone or telex network.

switched local service You pick up the phone. You dial a local number. Bingo, you have switched local phone service. The reason this trivial definition is even in this dictionary is because many states in the U.S. now -finally -allow companies to offer local switched telephone service in competition with the established company, e.g. Nynex or Southern Bell. Previously, they had only allowed competition in leased lines. And then previous to that they had not allowed any competition in any area of local phone service. So things are changing, albeit very very slowly.

switched loop In telephony, a circuit that automatically releases connection from a console or switchboard, once connection has been made, to the appropriate terminal. Loop buttons or jacks are used to answer incoming listed directory number calls, dial "O" internal calls, transfer requests , and intercepted calls. The attendant can handle only one call at a time.

switched loop operation Each call requiring attendant assistance is automatically switched to one of several switched loops on an attendant position.

switched message network A network service, such as Telex or TWX, providing interconnection of message devices such as teletypewriters.

switched mesh Also known as 'high performance' mesh. A wireless mesh network using multiple radios to communicate via dedicated mesh backhaul links to each neighboring node in the mesh. Here all of the available bandwidth of each separate radio channel is dedicated to the link to the neighboring node. The total available bandwidth is the sum of the bandwidth of each of the links. Each dedicated mesh link is on a separate channel, ensuring that forwarded traffic does not use any bandwidth from any other link in the mesh. As a result, a switched mesh is capable of much higher capacities and transmission rates than a shared mesh and grows in capacity as nodes are added to the mesh.

switched multibeam A type of "smart" antenna used in Wireless Local Loop (WLL) systems. Switched multibeam antenna detect signal strength in a given connection, and select a beam between an end device and one of perhaps many WLL antennas, locking in on the strongest signal. Also in the general category of smart antenna systems is the phased array approach. See also Phased Array and WLL.

Switched Multimegabit Data Service SMDS. A 1.544 Mbps public data service with an IEEE 802.6 standard user interface. It can support Ethernet, Token Ring and FDDI (OC-3c) LAN-to-LAN connections. See SMDS and SMDS Interest Group.

switched network See PSTN.

switched port analyzer A system utility on a switch that selects network traffic to or from a port for analysis by a network analyzer. A switched port analyzer (SPAN) mirrors the targeted traffic to a destination interface for delivery to the network analyzer or other remote monitoring (RMON) probe. A switched port analyzer does not affect the normal processing of the original traffic stream(s).

switched private line network A network which results from combining point-to-point circuits with switches.

switched service network A private line network that uses scan and/or CCSA type common control switching.

switched transport A name for telephone traffic between the local exchange carriers' Central Offices and an interexchange carrier's point of presence (POP). Switched transport is generally provided on a monopoly basis as part of a LEC's network.

Switched Virtual Circuit SVC. A call which is only established for the duration of a session and is then disconnected. See SVC.

switcher Also called a production switcher. A video term. A device that allows transitions between different video pictures. May also contain special effects generators.

switchhook A synonym for hookswitch or hook switch. Also spelled switch hook. See Switch Hook.

switchhook flash A signaling technique whereby the signal is originated by momentarily depressing the switchhook. The technique is sensitive to variations in the time of depression. Too short a signal will not be recognized, and too long a signal will be interpreted as a disconnect signal.

switching Connecting the calling party to the called party. This may involve one or many physical switches.

switching arrangement A circuit component which enables a Customer to establish a communications path between two phones on a network.

switching centers There are four levels in the North American switching hierarchy run at AT&T. They are: Class 1 -Regional Center, Class 2 -Sectional Center, Class 3 -Primary Center, Class 4c -Toll Center and Class 4P -Toll Point. In addition, the local Bell operating companies run a fifth level in the hierarchy, called the Class 5 -End Office.

switching equipment Premises equipment which performs the functions of establishing and releasing connections on a per call basis between two or more circuits, services or communications systems.

switching equipment capacity A telephone company term. The capacity

of switching equipment is expressed in network access lines. these components can be grouped into four categories. For D&F Chart purposes, the four categories are: 1. Dial Tone Equipment; 2. Talking Channels; 3. Switching Control; and 4. Trunk Terminations.

switching fabric The term "switching fabric" refers to the component at the heart of a data communications switch that allows any input port to send data to any output port. Many different kinds of switching fabric have been used over the years, depending on the manufacturer, the size and type of the data communications switch, and the technology available at the time. Sometimes a switching fabric will directly connect to all ports, but usually there are a group of ports on a single card called a line card and the switching fabric connects the line cards together. There are many different types of switching fabrics available on the market today. An example of one of the most basic is the "crossbar" switching fabric, which consists of a matrix of rows and columns, where each row is connected to an input port and each column is connected to an output port. The resulting diagram looks like a fabric with threads crossing at right angles. A switch or "crosspoint" is located at each intersection between a row and a column. By closing the right crosspoints, each input port can be connected to each output port. Crossbar fabrics are very general, but expensive to create in large sizes because the number of crosspoints is equal to the number of input ports times the number of output ports. For instance, if you had a small eight port switch you would have eight potential input and output ports making a total of 64 crosspoints, but if you had a large switch with 100 ports you would need 10,000 crosspoints to allow every port to connect with each other. Other types of switching fabric use buffering, queuing, packet shaping, switching logic, and specialized application specific integrated circuits (ASIC) to enhance switching fabric performance. A well-designed switching fabric will reach switching speeds equal to the line rate of the port. For instance, a port with a theoretical speed of 100 Mbps should be able to pass packets across the switching fabric to the destination port or ports at 100Mbps, which is also known as line-rate or wire-speed switching. A poorly designed switching fabric has delays or other latency that will drop the data rate as packets travel through the switching fabric. The variety and performance of switching fabrics depend on many different variables as well as the manufacturer. However, one thing is for certain future trends in switching fabrics are hard to anticipate, but the switching fabric will always remain at the heart of the data communications switch. This definition courtesy Alcatel.

switching fee A one-time, per-line fee imposed by the LEC to reprogram their switching system to change your default long-distance carrier. Some resellers and IXCs will reimburse new subscribers for this fee.

switching hub A multiport hub that delivers the full, uncontested bandwidth between any pair of ports. An intelligent switching hub also provides bridging and multiprotocol routing capabilities.

switching office A telephone company office containing a switch.

switching point Same as end office and intermediate office.

switching system 1. An assembly of equipment arranged for establishing connections between lines, lines to trunks, or trunks to trunks.

2. An ATM term. A set of one or more systems that act together and appear as a single switch for the purposes of PNNI routing.

switchless resellers A switchless reseller buys long distance service in bulk from a long distance company, such as AT&T, and resells that service to smaller users. It typically gets its monthly bill on magnetic tape, then rebills the bulk service to its customers. A switchless reseller owns no communications facilities -switches or transmission. It has two "assets" -a computer program to rebill the tape and some sales skills to sell its services to end users. The profit it makes comes from the difference between what it pays the long distance company and what it is able to sell its services at. Switchless resellers are also called rebillers. It's not an easy business to be in, since you are selling a long distance company's services to compete against itself. See also Aggregator and Facilities Based Carrier.

switchover A changeover from one satellite channel or system to another.

swivel chair network A network or internetwork that can only be monitored and managed by using multiple terminals. The term describes the network administrator whose swivel chair enables him (or her) to switch from terminal to terminal to perform different network management tasks.

swivel chair network management How a network administrator manages a swivel chair network.

SXS Step-by-Step. The first generation of automatic dial telephone systems, SxS systems were invented by Almon B. Strowger, a Kansas City (MO) undertaker in 1891. It seems as though Mr. Strowger was increasingly frustrated with the local telephone operator, who was directing Mr. Strowger's business calls to a competing undertaker, who also just happened to be her husband. When his complaints to the telephone company had no impact (Sound familiar?), he invented a switching system that effectively made operators obsolete in the switching of local calls, at least. Strowger's original system, which was based on earlier Bell System patents, could serve 99 customers. The invention formed the foundation for his founding Automatic Electric, which later became the manufacturing arm of GTE, which much later was merged into Verizon. SxS switches are electromechanical and circuit-switching. They comprise a number of "line finders," to which groups of customers are assigned for dial tone. As the user dials the digits comprising a telephone number, originally using a dial telephone, the electrical circuit between the set and the switch is made and broken. The electrical pulses that are generated by this process cause successive mechanical SxS "line selectors" to click across contacts to set up the conversation path. Digit-by-digit the line selectors set up the call, step-by-step, hence the name. Just as SxS switches rendered obsolete manual cordboards (and the operators who ran them), they in turn were rendered obsolete by electromagnetic Xbar (Crossbar) technology, which later was rendered obsolete by ECC (Electronic Common Control) technology. A lot of SxS switches remain in service, however, most especially in developing countries. SxS switches served in various capacities, including COs (Central Offices), tandems, and PBXs. See also Cord Board, ECC, and Xbar.

Svchost Svchost stands for "Service Host". Svchost is an integral part of the Windows operating system. Many components of Windows are implemented as "services" -a fancy name for programs that run in the background, and aren't necessarily associated with whomever is logged into the Windows PC. A fair number of those services are implemented in DLLs, rather than in stand-alone executables. But a DLL is only a library of functions that can be called by running other programs. It can't be run on its own. Enter svchost. It's a standalone program whose job is to execute services that are implemented in DLLs.

Symbian Connect The PC-based Symbian Connect is a system for data synchronization, file management, printing via PC, application installation from a PC, and other utility functions allowing Symbian OS phones to integrate effectively with PC and server-based data.

Symbian Ltd. A joint venture among LM Ericsson Telephone Co., Motorola Inc., Nokia Corp. and Psion PLC to develop new operating systems based on Psion's EPOC32 platform for small mobile devices for wireless devices such as phones and handhelds.

Symbian OS An operating system for cell phones produced by a company that does not sell cell phones, just the software to make them better. The company is called Symbian. According to the company's web site, www.Symbian.com, Symbian is a software licensing company, owned by wireless industry leaders, that is the trusted supplier of the advanced, open, standard operating system -Symbian OS -for data-enabled mobile phones. Symbian was established as a private independent company in June 1998 and is owned by Ericsson, Nokia, Matsushita (Panasonic), Motorola, Psion and Sony Ericsson. Headquartered in the UK, it has offices in Japan, Sweden, UK and the USA. The world's first open Symbian OS phone became available in the first half of 2001: the Nokia 9210 Communicator.

symbol 1. An abbreviated, predetermined representation of any relationship, association or convention.

2. In digital transmission, a recognizable electrical state which is associated with a signal element, which is an electrical signal within a defined period of time. In a binary transmission, for example, a signal element is represented as one of two possible states or symbols, i.e., 1 or 0.

Symbol Rate See 802.11a.

symbolic debugger A debugger is a whollyor partly-memory-resident program that lets you closely monitor and control execution of an application under development. At the most basic level, a debugger lets you look at running machine code, and fiddle around with the contents of memory -great if you understand machine code (and are looking at machine code you've written from scratch). Not great if you don't know machine code, or are looking at machine code output by a high-level language compiler (e.g., C++ compiler). A basic symbolic debugger references the symbol table of an executable, providing readable variable names, function entry-points, etc., more or less as they appear in source. Easier for machine-language folks (because of the labels). Not much easier for high-level language folks, because you're still dealing with machine code. A source-level symbolic debugger references both the symbol table of an executable and various files produced during compilation; and lets you work with high-level language source directly, during target program execution. Fully-integrated debuggers like this are built into Microsoft's Visual/X products. Functions common to most debuggers include the ability to set "breakpoints" (i.e., run the program until you reach this step, then stop), "watch variables" (i.e., show me how the value of this variable changes -and possibly stop if it assumes a predetermined value), "single-step execution" (i.e., do this step and stop), change variable values in mid-execution, etc.

symbolic language A computer programming language used to express ad-

dresses and instructions with symbols convenient to humans rather than machines.

symbolic logic The discipline in which valid arguments and operations are dealt with using an artificial language designed to avoid the ambiguities and logical inadequacies of natural languages.

symmetric Balanced in proportion. A symmetric telecom channel has the same speed in both directions. It's important to contrast symmetric with full duplex which is transmission in two directions simultaneously, or, more technically, bidirectional, simultaneous two-way communications. For example, ISDN BRI provides full duplex, symmetric bandwidth, as each of the two B channels provides 64 Kbps in each direction and the D channel operates at 16 Kbps in each direction. Symmetric also can refer to the physical topology of the network. For example, a point-to-point circuit connects one device directly to one other device. Asymmetric, on the other hand, refers to something which is not perfectly balanced. See the next several definitions. See also Asymmetric, Bps, Byte and Full Duplex.

symmetric connection A connection with the same bandwidth (i.e. speed) in both directions. See also Bps, Byte, Full Duplex and Symmetric.

symmetric cryptography Branch of cryptography involving algorithms that use the same key for two different steps of the algorithm (such as encryption and decryption, or signature creation and signature verification).

Symmetric Multiprocessing SMP. A type of multiprocessing in which multiple processors execute the same kernel-level code at the same time, sharing a single copy of the operating system. The degree of symmetry can vary from limited, where there is very little concurrency of execution, to the theoretically ideal fully-symmetric system where any function can be executed on any processor at any time. Processors within the same system share all processes and resources, including disk I/O, network I/O and memory. Compare to Asymmetrical Multiprocessing, wherein processors in the same or different systems are dedicated to specific tasks, such as disk I/O, network I/O or memory management. They off-load these tasks from the main system CPU, which generally is responsible for running the operating system. Each processor usually has its own dedicated memory.

symmetrical channel A channel in which the send and receive directions of transmission have the same data signaling rate.

symmetrical compression A compression system which requires equal processing capability for compression and decompression of an image. This form of compression is used in applications where both compression and decompression will be utilized frequently. Examples include: still-image databasing, still-image transmission (color fax), video production, video mail, videophones, and videoconferencing.

Symmetrical Digital Subscriber Line See SDSL.

symmetrical pair A balanced transmission line in a multipair cable having equal conductor resistances per unit length, equal impedances from each conductor to earth, and equal impedances to other lines.

symocasting Sending the same audio stream out from several Web sites simultaneously. You do this because one site would not accommodate the demand for this stream.

symphony See IBOC.

syn Synchronous. SYN and syn both identify some form of synchronization between devices supporting a data session. In TCP, for example, SYN is a single bit in a field of six control bits in the the TCP header. The SYN bit is used to synchronize sequence numbers in order to ensure that every octet in a given TCP packet is received and acknowledged. In synchronous transmission, syn is a control character in character-oriented protocols used to maintain synchronization, and as a time-fill in the absence of data. The sequence of two SYN characters in succession is used to maintain synchronization following each line turnaround. Contrast with Flag.

syn flood A denial of service attack on a computer system in which the attackers hack into a large number of computers, then use those machines to bombard the victim site with requests to start an ecommerce session. The large number of requests overwhelms the victim's computers, preventing legitimate customers from gaining access to the site. See also Smurf Attack.

sync 1. Synchronization character.

2. The portion of an encoded video signal that occurs during blanking and is used to synchronize the operation of cameras, monitors, and other equipment. Horizontal sync occurs within the blanking period in each horizontal scanning line, and vertical sync occurs within the vertical blanking period.

sync bits Synchronizing bits (more properly bytes or characters) used in synchronous transmission to maintain synchronization between transmitter and receiver.

sync generator A video term. A device that generates synchronizing pulses need by video source equipment to provide proper equipment or studio timing. Pulses typically produced by a sync generator include subcarrier, burst flag, sync, blanking, H & V drives, color frame identification, and color black.

sync pulse Timing pulses added to a video signal to keep the entire video process synchronized in time.

SyncML Synchronization Markup Language is an industry-wide effort to create a single, common data synchronization protocol optimized for wireless networks. SyncML's goal is to have networked data that support synchronization with any mobile device, and mobile devices that support synchronization with any networked data. The SyncML structured data layer will use XML wherever appropriate. SyncML is intended to work on transport protocols as diverse as HTTP, WSP (part of WAP) and OBEX, and with data formats ranging from personal data (e.g. vCard & vCalendar) to relational data and XML documents. The SyncML consortium was set up by IBM, Nokia and Psion among others. Symbian is a sponsor of the SyncML consortium.

synchronet service Dedicated point to point and multipoint digital data transmission service offered by BellSouth at speeds of 2.4, 4.8, 9.6, 19.2, 56 and 64 Kbps.

synchronization 1. A networking term which means that the entire network is controlled by one master clock and transmissions arrive and depart at precise times so that information is neither lost nor jumbled. For a bigger explanation, see Network Synchronization and Synchronous.

2. An uninterruptible power supply (UPS) definition. Specially designed circuitry is "synchronized" to your AC power outlet to ensure continuity of power. Without this feature, power reversal can occur on the input.

3. A multimedia term. Synchronization is very precise real-time processing, down to the millisecond. Some forms of multimedia, such as audio and video, are time critical. Time delays that might not be noticeable in text or graphics delivery, but are unacceptable for audio and video. Workstations and networks must be capable of transmitting this kind of data in a synchronized manner. Where audio and video are combined, they must be time stamped so that they can both play back at the same time.

4. Start with a database on your server. Now, take a copy of part of it on your laptop -for example, your very own sales leads. Go traveling. Come back in a week. You want to update the database with your changes. But you don't want to destroy other peoples' changes. Some people are calling this "file synchronization." Synchronization is a critical part of what is increasingly being called "Groupware." See also Replication.

5. A Video term referring to the timing of the vertical and horizontal presentation of the multiple still images. Vertical synch prevents the picture from flipping, or scrolling unnaturally. Horizontal synch keeps the picture from twisting. If both vertical and horizontal are out of synch, the picture looks truly wretched.

synchronization bit A binary bit used to synchronize the transmission and receipt of characters in data communications.

synchronization bits Bits transmitted from source to destination for the purpose of synchronizing the clocks of the transmitting and receiving devices. The term "synchronization bit" is usually applied to digital data streams, whereas the term "synchronization pulse" is usually applied to analog signals.

synchronization code In digital systems, a sequence of digital symbols introduced into a transmission signal to achieve or maintain synchronism.

synchronization pulses Bits transmitted from source to destination for the purpose of synchronizing the clocks of the transmitting and receiving devices. The term "synchronization pulse" is usually applied to analog signals, whereas the term "synchronization bit" is usually applied to digital data streams.

synchronize The word synchronize means "to cause to match exactly." When you're synchronizing, you're causing one file on one computer to precisely match another one on another computer. Why would you want to do this? Let's say you have a database of sales contacts on a file server. One of your salesman takes a copy of his sales contacts with him on his laptop. He travels and makes changes to his contacts. Now he dials into the office via modem and wants to "synchronize" his changed database with the now-changed main database, and make them both the same, i.e. into synch. This process is far more difficult than it sounds because it means allowing for the changes made at the server and by the salesman. You have to set up elaborate rules.

In operating systems, such as Windows NT, the word "synchronize" has a narrower meaning. Windows NT instruction manual defines "synchronize" as "to replicate the domain controller to one server of the domain, or to all the servers of a domain. This is usually performed automatically by the system, but can also be invoked manually by an administrator." See also Replicate.

synchronizing Achieving and maintaining synchronism. In facsimile, achieving and maintaining predetermined speed relations between the scanning spot and the recording spot within each scanning line.

synchronizing pilot In FDM, a reference frequency used for achieving and maintaining synchronization of the oscillators of a carrier system or for comparing the frequencies or phases of the currents generated by those oscillators.

synchronous The condition that occurs when two events happen in a specific time relationship with each other and both are under control of a master clock. Synchronous transmission means there is a constant time between successive bits, characters or events. The timing is achieved by the sharing of a single clock. Each end of the transmission synchronizes itself with the use of clocks and information sent along with the transmitted data. Synchronous is the most popular communications method to and from mainframes. In synchronous transmission, characters are spaced by time, not by start and stop bits. Because you don't have to add these bits, synchronous transmission of a message will take fewer bits (and therefore less time) than asynchronous transmission. But because precise clocks and careful timing are needed in synchronous transmission, it's usually more expensive to set up synchronous transmission. Most networks are synchronous these days. See Asynchronous and Network Synchronization.

Synchronous Code Division Multiple Access See S-CDMA.

synchronous completion A computing domain issues a service request and need not wait for it to complete. If the computing domain waits for this completion, this is known as synchronous, but if it is sent off to another system entity and the computing domain goes on to other activities before the function completes (and the system later sends a message to the computing domain announcing the function's completion), that completion is known as asynchronous.

Synchronous Control Character SYN. A transmission control character used in synchronous transmission to provide a signal in the absence of any other character. SYN is defined by IBM Corp's Binary Synchronous Communications (BSC) protocol.

Synchronous Data Link Control SDLC. A data communications line protocol associated with the IBM Systems Network Architecture. See Systems Network Architecture.

synchronous data network A data network in which synchronism is achieved and maintained between data circuit-terminating equipment (DCE) and the data switching exchange (DSE), and between DSEs. The data signaling rates are controlled by timing equipment within the network. See Network Synchronization.

synchronous data transfer A physical transfer of data to or from a device that has a predictable time relationship with the execution of an I/O (Input/Output) request. See Synchronous.

Synchronous Digital Hierarchy SDH. A set of international standards developed by the CCITT (now ITU-T) for fiber optic transmission systems. The SDH standards are an international superset of the original SONET standards, which were developed for use in North America. See SONET for a full explanation.

synchronous idle character A transmission control character used in synchronous transmission systems to provide a signal from which synchronism or synchronous correction may be achieved between data terminal equipment, particularly when no other character is being transmitted.

synchronous network A network in which all the communication links are synchronized to a common clock.

Synchronous Optical NETwork. See SONET.

synchronous orbit An orbit, any point on which has a period equal to the average rotational period of the Earth. If the orbit is also circular and equatorial, it is called a stationary (geostationary) orbit. See also LEO.

synchronous payload envelope The major portion of the SONET frame format used to carry the STS-1 signal divided into an information payload section and a transport overhead system. SPE is used to address three payload structures: direct to STS-1 line rate multiplexing; asynchronous DS-3 multiplexing; and synchronous DS-3 multiplexing.

synchronous request An SCSA definition. A request where the client blocks until the completion of the request. Contrast with asynchronous request .

synchronous satellite A satellite in a synchronous orbit. See Synchronous Orbit.

synchronous TDM A multiplexing scheme in which timing is obtained from a clock that in turn controls both the multiplexer and the channel source.

Synchronous Time-Division Multiplexing STDM. A time-division multiplexing method whereby devices have access to a high-speed transmission medium at fixed time periods independent of likely load.

synchronous transfer mode A proposed transport level, a time division multiplex-and-switching technique to be used across the user's network interface for a broadband ISDN. See STM.

synchronous transmission Transmission in which the data characters and bits are transmitted at a fixed rate with the transmitter and receiver synchronized. Synchronous transmission eliminates the need for start and stop bits. See Synchronous and Asynchronous.

Synchronous Transmission Mode STM. The synchronous transmission capability of a system that is capable of both synchronous and asynchronous capabilities of Broadband Integrated Services Digital Network (B-ISDN) service.

Synchronous Transport Module 1 STM-1. SDH standard for transmission over OC-3 optical fiber at 155.52 Mbps.

Synchronous Transport Signal Level 1 STS-1. The basic signaling rate for a Synchronous Optical Network (SONET) transmission medium. The STS-1 rate is 51.84 Mbps.

Synchronous Transport Signaling Level n STS-n. A definition of the transmission speed a Synchronous Optical Network (SONET) transmission medium where n is an integer between 1 and 48 and relates to the multiplier to be applied to the basic STS-1 51.8-Mbps transmission speed. STS-48 is 48 times faster than STS-1, with a speed of 2.5 gigabits per second.

SYNDEX Syndicated exclusivity. Federal requirement that cable TV systems black out syndicated programming from distant signals (out-of-town television stations) for which a local broadcaster has exclusive contractual rights. For example, cable operators cannot import "I Love Lucy" as part of a distant TV signal if a local broadcaster has purchased the syndication rights for that program in its market. (The FCC eliminated this requirement in 1980 and subsequently reimposed it in 1990.)

syndicated exclusivity See SYNDEX.

syndrome Basic element of decoding procedure. Identifies the bits in error.

Synfuels Synthetic fuels (or "synfuels") are various fossil fuel substitutes including, gasified and liquefied coal, synthetic natural gas, oil shale, tar sands, and fermented biomass materials which are increasingly being seen as necessary to extend and improve existing fuel supplies.

syntax The rules of grammar in any language, including computer language. Specifically, it is the set of rules for using a programming language. It is the grammar used in programming statements. Also:The order in which a command must be typed and the elements that follow the command.

syntax error An error caused by incorrect programming statements according to the rules of the language being used. Sometimes the computer will throw up "SN" to indicate a syntax error.

synthesized voice Human speech approximated by a computer device that concatenates basic speech parts (or phonemes) together. Usually has a metallic, Germanic sound.

synthetic leases Synthetic leases are off-balance-sheet financings that have drawn investor concern in the wake of the collapse of Enron Corp.

synthetic operation Packets sent into the network that appear to be user data traffic but actually measure network performance. Formerly known as a probe. Also referred to as operation.

syntonization The process of setting the frequency of one oscillator equal to that of another.

syntran synchronous transmission A restructured DS-3 signal format for synchronous transmission at the 47.36 megabits per second DS-3 level of the North American Hierarchy.

SYSGEN Acronym for SYStem GENeration.

syslog System log. A file that records everything that happens on the operator's console for a mainframe or minicomputer, including, for example, job start/end times, tape mounts and backup start/end times. Syslogs now typically exist for web servers and other mission-critical host servers, for the purpose of logging server events.

SYSOP The SYStem OPerator of a PC-based electronic bulletin board/mail service or on-line computer service, such as CompuServe or America On Line. SYSOPs (pronounced sis-ops) typically put computers and modems on phone lines, then published the phone number, then invited people with computers to call them and leave them messages and interesting software programs which they had written. These programs then became "public domain," or freeware. And other callers were invited to download these programs for their

own use. Lead Sysops are called Wizops.

SYSREQ System request; the seldom used key used to get attention from another computer.

system An organized assembly of equipment, personnel, procedures and other facilities designed to perform a specific function or set of functions.

system administrator The person or persons responsible for the administrative and operational functions of a computer and a telecom system that are independent of any particular application. The system Administrator is likely to be a person with the best overview of all the applications. The System Administrator advises application designers about the data that already exists on the various services, makes recommendations about standardizing data definitions across applications, and so on.

system build This is the original manufacturer system building that occurs when the order is placed by the buyer with the vendor. The basic configuration is set up to reflect the users needs at that point in time. Thereafter, if any changes occur to reflect changes in the operating environment, the manufacturer must reconfigure the system to reflect this change. There is usually a reprogramming charge and a delay associated with the change.

system clock The clock designated as the reference for all clocking in a network of electronic devices such as a multiplexer or transmission facilities management system.

system common equipment The equipment on a premises that provides functions common to terminal devices such as telephones, data terminals, integrated work station terminals, and personal computers. Typically, the system common equipment is the PBX switch, data packet switch, or central host computer. Often called common equipment.

system connect The method by which connection is physically made to the cost computer or local area network.

system control computer The computer system used at CATV or MMDS headend for control of numerous technical functions. These functions include subscriber addressing, channel mapping schedules, ad insertion, encryption keys, PPV, and sometimes IPPV.

system coordinator This is the title assigned to the person responsible for administration programming and the training of workers on your phone system.

system disk A disk that has been formatted as a system disk. MS-DOS system disks have two hidden files and the COMMAND.COM file. You can start the computer using a system disk.

System Fault Tolerance SFT. The ability of computer to work fully regardless of component failures.

system feature A telephone switch feature that is typically available all the users.

system gain The amount of free space path loss that a radio can overcome by a combination of enhancing transmitted power and improving receiver sensitivity.

system message Messages that are not associated with a mailbox.

System On a Chip See SOC.

system redundancy The duplication of system components to protect against failure. For protection against failure, install redundant cabling, power supplies, disk storage, gateways, routers, network boards, printers, switches and other mission-critical network components.

system reload A process allowing stored data to be written from a tape into the system memory. Picture: your telephone system goes dead. For whatever reason it loses all memory of its generic programming and your specific programming (whose extension gets what, etc.). You have to quickly grab the backup (hopefully you have it on tape or magnetic disk) and load it back into your telephone system's memory. This is called system reload. Sometimes it's done automatically. Sometimes you have to do it manually.

system segment A conceptual subset of a system, usually referring to one which can be functionally replaced without damaging the capability of the system.

System Service Provider An SCSA definition. An entity that provides system wide services, such as session management and security, and the allocation and tracking of resources and groups.

system side Defines all cabling and connectors from the host computer or local area network to the cross connect field at the distribution frame.

system speed dial Simplified ways of dialing. You do them by dialing several digits. System speed dial numbers can be used by everyone on the phone system -whether they are on an electronic phone or just a simply single line phone.

system test This definition courtesy Steve Gladstone, author, "Testing Computer Telephony Systems": System test is the phase of the product life cycle that examines the entire system as a "whole" to assure it is ready to go to a true alpha or beta test. System testing is also more oriented to inter system functions as opposed to earlier phases. To pass a system test, all features and functions are expected to work correctly (function to specification) in all areas of the system -features, administration, maintenance, billing, etc. Additionally, the system must function as an "architectural whole," including all hardware and software components. Representative databases must be loaded to simulate site applications. Full load and stress testing is performed. It is in this phase that the bulk of system level testing will take place. System testing has a major focus on external load and other stimuli.

System V Interface Definition SVID. A UNIX application-to-system software interface developed and supported by AT&T. The interface is similar to POSIX.

systems analysis Analyzing an organization's activities to figure the best way of applying computer systems to its organization.

systems analyst A person who performs systems analysis and who follows through with methods, techniques and programs to meet the need.

systems integrator A systems integrator is a company that specializes in planning, coordinating, scheduling, testing, improving and sometimes maintaining a computing operation (sometimes companywide, sometimes just locally). In the old days, this was done almost exclusively by the International Business Machines Corporation. Somewhere along, companies discovered they could get more flexibility and computing power at a lower cost by shopping around. Today, hundreds of companies contribute various components -hardware, software, wiring, communications and so on -to a company's computer operation. But the added flexibility can bring stunning complexity. Systems integrators try to bring order to the disparate suppliers.

Systems Integration Interface SII. As used in the definition of the proposed multivendor integration architecture sponsored by Nippon Telegraph and Telephone (NTT) of Japan, SII specifies any set of standardized services used to connect computer based-systems.

Systems Network Architecture SNA. IBM's successful computer network architecture. At one stage the most successful computer network architecture in the world. In the days of mainframe computers, it was as successful in the computer networking world as AT&T's telephone network design was in telecommunications. The best explanation we've ever read of SNA is in James Harry Green's Dow Jones-Irwin Handbook of Telecommunications. Here is an excerpt:

"SNA is a tree-structured architecture, with a mainframe host computer acting as the network control center. The boundaries described by the host computer, front-end processors, cluster controllers and terminals are referred to as the network's domain. Unlike the switched telephone network that establishes physical paths between terminals for the duration of a session, SNA establishes a logical path between network nodes, and it routes each message with addressing information contained in the protocol. The network is therefore incompatible with any but approved protocols. SNA uses the SDLC data link protocol exclusively. Devices using asynchronous or binary synchronous can access SNA only through protocol converters...SNA works in seven layers roughly analogous to ISO's seven level OSI model. Unlike OSI, however, SNA is fully defined at each level. SNA was first announced in 1974 and is the basis for much of the OSI model, but it differs from OSI in several significant respects." For more on these differences, see page 96 in Green's Handbook.

The following is a description we received from IBM's PR department: "What is SNA?" In general, SNA is the description of the rules that enable IBM's customers to transmit and receive information through their computer networks. SNA may also be viewed as three distinct but related entities: a specification, a plan for structuring a network and a set of products. First, SNA is a specification governing the design of IBM products that are to communicate with one another in a network. It is called an architecture because it specifies the operating relationships of those products as part of system. Second, SNA provides a coherent structure that enables users to establish and manage their networks and, in response to new requirements and technologies, to change or expand them. Third, SNA may be viewed as a set of products: combinations of hardware and programming designed in accordance with the specification of SNA. In addition to a large number of computer terminals for both specific industries and general applications, IBM's SNA product line includes host processors, communication controllers, and adapters, modems and data encryption units. The SNA product line also includes a variety of programs and programming subsystems. Telecommunications access methods, network management programs, distributed applications programming and the network control program are examples.

Systems Network Interconnection SNI. A service defined by IBM that allows for the interconnection of separately defined and controlled Systems Network Architecture (SNA) networks. See Systems Network Architecture.

Systems Services Control Point SSCP. An IBM Corp. Systems Network Architecture (SNA) term for the software that manages the available connection services to be used by the Network Control Program (NCP). There is only one SSCP in an SNA network domain, and the software normally resides in the host processor, which is a member of the IBM System/370 mainframe family.

systems software A type of program used to enhance the operating systems and the computer systems they support.

Systemview The network management program that purports to let UNIX-based computers be managed along with other IBM systems.

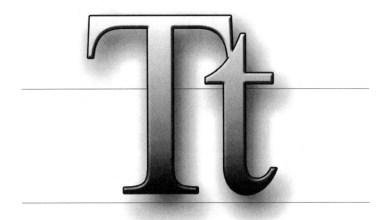

T 1. Trunk. as in T-1. See T-1.

2. Tip. See Tip & Ring.

3. Tera, which is 10 raised to the 12th power, or 1,000,000,000,000, which has 12 zeros in the number.

4. An ADSL (Asymmetric Digital Subscriber Line) term for the functional interface between the Premises Distribution Network (PDN) and the Service Module(s), both of which are installed at the user premise. The PDN is the inside cable and wire system, and associated premises electronics. The Service Modules accomplish terminal (e.g., TV, telephone, PC and router) adaption functions for access to the ADSL network. When the PDN is in the form of a point-to-point, passive wiring system, the "T" may be the same as a "T-SM" interface. The "T" may be in the form of a separate physical unit, or may be embedded in a combined ATU-R/Service Module, with the ATU-R being an Asymmetric Terminating Unit-Remote. See also ADSL, ATU-R, PDN, SM and T-SM.

5. Twisted pair. As in 10BaseT, an IEEE standard for Ethernet LANs, which run at 10Mbps (million bits per second), Baseband (single channel transmission), over Twisted pair (Category 3, 4, 5 or 6).

T 1 See T-1.

T.120 The ITU-T standard for multipoint data conferencing. T.120 provides the protocols for establishing and managing data flow, connections, and conferences. Support for T.120 enables data transfer from conferencing applications, such as file transfer and application sharing, to operate in conjunction with H.323 connections.

T Carrier T stands for trunk, meaning that the T-Carrier technology was developed for the "trunk side," or carrier side, of the network. In carrier (telco) parlance, the "line side" of the network is the end user, or local loop, side. T Carrier is a generic name for any of several digitally multiplexed carrier systems. The designators for T (Trunk) carrier in the North American digital hierarchy correspond to the broader, generic designators for the digital signal (DS) level hierarchy. To be more exact, the DS level refers to the signal before it enters the channel bank, while the T level signal refers to the signal after it leaves the channel. The T level refers, more than anything else, to the framing convention which is imposed on the information signal by the channel bank. T carrier systems were originally designed to transmit digitized voice signals over a channelized facility. Current applications also include digital data transmission, which generally is transmitted over an unchannelized facility. The table below lists the designators and rates for current T carrier levels, as well as those for E-Carrier (European) and J-Carrier (Japanese).

See T-1 below. Also see DS-, and Channel Bank.

T Connection T-shaped three-way conductor for distributing an incoming signal in two outgoing ways. Same shape as a T-connection in the road.

T Connector A T-shaped device with two female connectors and one male BNC connector used with Ethernet coaxial cable and used on local area networks.

NORTH AMERICAN HIERARCHY (T-CARRIER)		
T-1	1.544 Mbps	24 voice channels
T-1C	3.152 Mbps	48 voice channels
T-2	6.312 Mbps	96 voice channels
T-3	44.736 Mbps	672 voice channels
T-4	274.176 Mbps	4,032 voice channels
JAPANESE HIERARCHY (J-CARRIER)		
J-1	1.544 Mbps	24 voice channels
J-2	6.312 Mbps	96 voice channels
J-3	32.064 Mbps	480 voice channels
J-4	97.728 Mbps	1,440 voice channels
J-5	397.000 Mbps	5,760 voice channels
EUROPEAN HIERARCHY (E-CARRIER)		
E-1	2.048 Mbps	30 voice channels
E-2	8.448 Mbps	120 voice channels
E-3	34.368 Mbps	480 voice channels
E-4	139.264 Mbps	1,920 voice channels
E-5	565.148 Mbps	7,680 voice channels
E-6	2200.000 Mbps	30,720 voice channels

T Interface 4-wire ISDN BRI circuit. Picture this: You order an ISDN circuit from your local phone company. They deliver it on a normal phone line – one copper pair. At your offices, you plug in a small device called a network termination device. That device converts the two-wire circuit called a U interface, into a four-wire S or T interface which you'll use to plug in your ISDN terminal equipment, which might be a phone, a computer, a PBX, a videoconferencing device, etc. It may be all the above. The S or T interface is designed to allow you hook up to eight terminal devices on one ISDN line. Definitions are changing, especially in ISDN. At one stage, the T Interface (more properly the T-Reference Point) needed an NT1 rather than an NT2. These days, some people believe the T interface refers to an ISDN electrical connection to a PBX. While, the S bus refers to a connection to other devices, like phones and videoconferencing devices. There's no electrical difference between an S and a T interface. But I may be proven wrong by some phone company that changes the specs. One of ISDN's most charming features is its ability to acquire different specs and features from one provider to the next. See S Interface and U Interface and ISDN.

T Reference Point See T Interface.

T Span A telephone term for a transmission medium through which a T-carrier system is operated. Also called a Span line. See T Carrier and T-1.

T Span Line Also called a Span line. An outside plant four-wire, two twisted-pair transmission line. See T Carrier and T-1.

T Tap A passive line interface used for extracting data from a circuit. Also, for extracting optical signals from a fiber cable or electrical signals from a coaxial cable.

T&A Test and Accept. After a circuit or system is installed, you should test it before you formally accept it. Once the technician walks out the door, it's billable. (Sorry, guys. I know you were expecting another definition. Shame on you.)

T&E Technology and Evaluation Lab.

T+T Telephone and Telegraph. I found T+T engraved on a small metal closet in a hotel in Zurich, Switzerland. When I opened the closet, I found a crude telephone line crossconnect panel. The hotel ran its phone lines up a central shaft, terminated them on a crossconnect panel, then ran the lines to each guest room.

T-1 Also spelled T1, which stands for Trunk Level 1. A digital transmission link with a signaling speed of 1.544 Mbps (1,544,000 bits per second) in both directions (i.e. send and receive). T-1 is a standard for digital transmission in North America, – the United States and Canada. T-1 is part of a progression of digital transmission pipes – a hierarchy known generically as the DS (Digital Signal Level) hierarchy. (For the complete hierarchy, see the definition for T Carrier above.) In the olden days, T-1 was delivered to your business on two pairs of unshielded twisted copper wires – one pair for transmit and one pair for receive – the combination of these two simplex (unidirectional) circuits yields a full duplex symmetrical (bidirectional) circuit. These days, T-1 often is delivered on fiber optic lines, where fiber is available. If it's not available, try and get it. T-1 delivered on fiber typically works better than on copper. You can lease T-1 as a channelized service (delivered as separate voice or data channels), or as an unchannelized raw bit stream (i.e., 1.536 Mbps of transmission both ways, plus .008 Mbps framing bits) and do with the 1.526 Mbps bits as you wish – the framing bits are not under your control. North American carriers typically deliver T-1 channelized, i.e., split into 23 or 24 voice-grade channels, with each running at 56/64 Kbps (i.e., 56,000 or 64,000 bits per second), depending on the generation of the channel bank equipment involved. If you have need for a bunch of local phones, it's often cheaper to get them delivered on T-1 channels than as individual phone lines. One expensive circuit (i.e, the multi-channel T-1) is far less expensive than 24 less expensive circuits (e.g., single-channel voice circuits). While channelized T-1 was developed for and is optimized for uncompressed voice communications, it also can be used for channelized data communications. A channelized approach is required for access to the traditional PSTN, which is channelized throughout the traditional carrier networks.

On the other hand, an unchannelized approach is better for most data communications applications, and for compressed voice, video and IP telephony. The unchannelized approach provides you with 1.536 Mbps which you can split up any way you choose. If you lease a raw T-1 pipe, you could, for example, split it (i.e., multiplex it) into 12 voice grade channels to support 12 voice conversations, and use the remaining 768 Kbps for either reasonably high-speed access to the Internet or for videoconferencing with your distant office. You could also compress voice to run at speeds of perhaps 8 Kbps or less by using IP Telephony techniques and, therefore, put many more voice calls over a single T-1 pipe. Unchannelized T-1 also is commonly used for access to a frame relay or ATM network, or for Internet access. In such an application, your router or data switch or data concentrator effectively multiplexes data packets (i.e., packets, frames or cells) through the "clear" pipe. Channelization would make no sense in such an application.

In addition to use in network access applications, T-1 also can be used for private, leased line networking. In a private network, you might use channelized leased T-1 PBX tie trunks to "tie" together your voice PBXs. You might use unchannelized T-1 tie trunks to directly connect your local area network routers or data switches. Note that T-1 is medium-independent. You can run it over electrical (i.e., twisted pair or coaxial cable), optical (i.e., fiber optics or infrared) or radio (i.e., microwave or satellite) transmission media.

Outside of the United States and Canada, DS-1 is called E-1, as it was developed by the CEPT (Conference of European Postal and Telecommunications Administrations) for use in Europe. E-1 runs at a total signaling rate of 2,048,000 bits per second. Only one element remains constant between it and the North American's T-1 – the DS-0, namely the 64 Kbps channel. Most often it represents a PCM voice signal sampled at 8,000 times per second, or 64,000 bits per second. However, the form of PCM encoding, also known as companding, differs between T-1 (mu-law) and E-1 (A-law). Conversion of E-1 to T-1 involves both the compression law and the signaling format. At the higher rate of 2.048 Mbps, 32 time slots are defined at the CEPT interface, but two time slots (channels) are used for non-intrusive

signaling and control purposes. The remaining 30 channels are clear 64 Kbps channels for user information-voice, video, data, etc. T-1 is also called 23B+D. That means it can be channelized to contain 23 B channels and one D channel. 30B+2D is compatible with E-1 – namely it can be channelized to contain 30 B channels and two D channels. See also CEPT, Channel Bank, Companding, Compression, DS-1, ISDN PRI, PCM, TDM, and Time Division Multiplexing and the following five definitions.

0x05T-1 Card A modern PBX is basically an empty metal cage with a backplane, power and empty slots into which you slide various cards. One of those cards is called a T-1 card. It's used for connecting the PBX to a T-1 line. Typically the T-1 card has a built-in CSU/DSU.

T-1C Trunk Level 1 Combined. The total signaling rate is 3.152 Mbps in North America and comprises two T-1s, which are interleaved to support 48 DS-0s at 1.544 Mbps each. The additional 64 Kbps is overhead used to support additional signaling and control requirements. T-1C is seldom used, outside of limited telco applications. See T-1.

T-1 BITS interface See T-1 and BITS.

T-1 Framing Digitization and coding of analog voice signals requires 8,000 samples per second (two times the highest voice frequency of 4,000 Hz) and its coding in 8-bit words yields the fundamental T-1 building block of 64 Kbps for voice. This is termed a Level 0 Signal and is represented by DS-0 (Digital Signal at Level 0). Combining 24 such voice channels into a serial bit stream using Time Division Multiplexing (TDM) is performed on a frame-by-frame basis. A frame is a sample of all 24 channels ($24 \times 8 = 192$) plus a synchronization bit called a framing bit, which yields a block of 193 bits. Frames are transmitted at a rate of 8,000 per second (corresponding to the required sampling rate), thus creating a 1.544 Mbps ($8,000 \times 193 = 1.544$ Mbps) transmission rate, the standard North American T-1 rate. This rate is termed DS-1. See also D-4 Framing and Extended Super-Frame Format. See also T Carrier, T2, T3 and T4.

T-1 Span Line See Span.

T-2 The North American standard for DS-2 (Digital Signal Level 2). T-2 operates at a signaling rate of 6.312 Mbps, and is capable of handling 96 voice conversations, depending on the encoding scheme chosen. T-2 is four times the capacity of T-1. It generally is used only in carrier networks, although end users sometimes gain T-2 access on an ICB (Individual Case Basis). An example carrier application is that of SLC-96 (Subscriber Line Carrier 96), which supports 96 voice-grade channels of 64 Kbps over a specially conditioned four-wire circuit. See SLC-96 and T-1.

T-3 The North American standard for DS-3 (Digital Signal Level 3). T-3 operates at a signaling rate of 44.736 Mbps, equivalent to 28 T-1s. T-3 is commonly referred to as 45 megabits per second. Capable of handling 672 voice conversations each at 64 Kbps, T-3 runs on fiber optic or microwave transmission, as twisted copper pair is not capable of supporting such a high signaling rate over distances of any significance. Running on fiber, it is typically called FT-3. Both Bill Gates and George Lucas allegedly have T-3 lines coming into their houses. Sadly, I don't. See T-1.

T-4 The North American standard for DS-4. T-4 supports a signaling rate of 274.176 Mbps and is capable of handling 4,032 voice conversations. T-4 has 168 times the capacity of T-1. T-4 can run on coaxial cable, microwave radio or fiber optic transmission systems. T-4 generally is used only in carrier backbone networks, and generally is not available for end-user consumption. See also T-1.

T-Berd See T-Bird.

T-Bird also T-Berd. Colloquial term for a T-1 carrier analyzer, used by T-1 circuit technicians. Taken from the brand name of a leading device, T Berd 90A.

T-DMB See Digital Multimedia Broadcasting.

T-I Channel Either of the two external ports of a TDI or RDI which provides for transmitting or receiving eight TDM channels. Do not confuse with T-1 (as in T-one).

t-commerce Using the TV remote control to buy a product or service that is advertised on digital cable or satellite TV. T-commerce technologies currently are being tested by several cable networks and will be rolled out in 2007. T-commerce works like cable pay-per-view or video-on-demand. A shopper presses a button on the TV remote to select an item and toggles through choices such as size, color, shipping address, and payment method. Cable and satellite operators will collect fees from sellers for facilitating t-commerce transactions. I wonder if any of the items offered will be things we actually need.

T-ray Terahertz ray; electromagnetic radiation produced at terahertz (trillion cycles per second) frequencies that, like an x-ray, can penetrate solids, but that also enables the identification of certain molecules and substances. In the June 2000 issue of Technology Review, Herb Brody wrote "Just as x-ray technology came along in the 1890s – allowing doctors to peer beneath flesh to see bones and organs – another promising imaging technology is now emerging from an underused chunk of the electromagnetic spectrum: the terahertz frequencies. These so-called t-rays can, like x-rays, see through most materials. But t-rays are believed to be less harmful than x-rays. And different compounds respond to terahertz radiation differently, meaning a terahertz-based imaging system can discern a hidden object's chemical composition. Potential applications range from detecting tumors to finding plastic explosives. And since t-rays penetrate paper and clothing, a terahertz camera could detect hidden weapons." In the February 2004 issue of MIT Technology Review, the ediors wrote, "With the human eye responsive to only a narrow slice of the electromagnetic spectrum, people have long sought ways to see beyond the limits of visible light. X-rays il-luminate the ghostly shadows of bones, ultraviolet light makes certain chemicals shine, and near-infrared radiation provides night vision. Now researchers are working to open a new part of the spectrum: terahertz radiation, or t-rays. Able to easily penetrate many common materials without the medical risks of x-rays, t-rays promise to transform fields like airport security and medical imaging, revealing not only the shape but also the composition of hidden objects, from explosives to cancers."

T-SM An ADSL term for the functional interface between the ATU-R (Asymmetric Transmis-sion Unit-Remote) and the PDN (Premises Distribution Network). The ATU-R may contain one or more T-SMs (Service Modules) for terminal adaption. The T-SM may be the same as the "T" interface where the PDN is a point-to-point, passive cable and wire system. The T-SM may be a separate physical device, or may be embedded in an integrated ATU-R/SM. See also ADSL, ATU-R, PDN, SM and T.

T.120 The most important transmission protocol standard for document conferencing (viewing, changing and moving files) over transmission media ranging from analog phone lines to the Internet. T.120 is the International Telecommunications Union (ITU-T) standards suite for document conferencing via FTP (File Transfer Protocol). Virtually all major players in the document conferencing industry have announced support for this standard. Document conferencing adds a visual dimension to voice-only conference calls by allowing groups of people to share computer documents in real-time while participating in a standard voice conference call. Whatever materials would normally be distributed in a face-to-face meet-ing – graphs, spreadsheets, diagrams, or documents – can be shared on-line, in real-time. Participants can easily connect to a conference anywhere in the world, with the only require-ments being a Windows PC, a modem and a document conferencing software program. T.120 series standards provide a framework to enable multi-point data conferencing across LANs, WANs and the Internet. The T.120 architecture relies on a multilayered approach with defined protocols and service definitions between layers. Each layer presumes the existence of all layers below. The lower level layers (T.122, T.123, T.124 and T.125) specify an application-independent mechanism for providing multi-point data communications services to any application that can use these facilities. The upper level layers (T.126 and T.127) define protocols for specific conferencing applications, such as shared whiteboarding and binary file transfer. See also H.320, standards which extend data conferencing into video conferencing.

T.121. ITU standard for generic T.120 Application Template.

T.122 Multipoint Communications Service for Audiographics and Audiovisual Conferenc-ing - Service Definition. (ITU approved 1993.) See T.120 and T.125.

T.123 Protocol Stacks for Audiographics and Audiovisual Teleconference Applications. (ITU approved 1993/1994.) See T.120.

T.124 Generic Conference Control for Audiovisual Services (GCC) (ITU voted approval in 3/95.) T.125: Multipoint Communication Service - Protocol Specification (MCS). (ITU approved 1994.) See T.120 and T.122.

T.125 Multipoint Communication Service - Protocol Specification (MCS). (ITU approved 1994.) See T.120.

T.126 Provides shared whiteboard and document conferencing protocols. See T.120.

T.127 Provides multipoint binary file transfer. See T.120.

T.128 ITU standard for Audio Visual Control for Multipoint Multimedia Systems.

T.134 See T.140.

T.140 In early February, 1998, the ITU-T reported that it had just completed work on three recommendations aimed at enhancing the capability of the deaf or speech-impaired to use telecommunications. The first recommendation, V.18, describes a multi-function text telephone that bridges the gap that has, according to the ITU, between several incompatible text telephones in use today. The second (T.140) adds new facilities to enable the use of different alphabets and character sets in text communications. These include Arabic, Cyrillic and Kanjii, as well as Latin-based characters. Finally, T.134 describes how these facilities can be integrated in the multimedia communications systems defined by the ITU-T.

T.30 ITU-T standard. Fax handshake protocol. This standard describes the overall proce-dure for establishing and managing communication between two fax machines. There are five phases of operation covered: call set up, pre message procedure (selecting the com-munication mode), message transmission (including phasing and synchronization), post message procedure (end-of-message and confirmation) and call release (disconnection).

T.35 ITU-T recommendation proposing a procedure for the allocation of ITU-T members' country or area codes for non-standard facilities in telematic services.

T.37 An ITU-T Recommendation for store-and-forward fax via e-mail through the incor-poration of SMTP (Simple Mail Transfer Protocol) and MIME (Multipurpose Internet Mail Extension). SMTP is an application-layer extension of TCP/IP which governs electronic mail transmissions and receptions. MIME is a SMTP extension which supports compound mail. In this context, MIME provides for the attachment of a compressed fax image to an e-mail. Fax image documents are attached to e-mail headers and are encoded in the TIFF-F (Tagged Image File Format-Fax) compressed data format. In simple-mode, T.37 restricts fax transmission to the most popular fax machine formats (e.g., standard or fine resolution, and standard page size). This restriction is done through limitation of TIFF-F encoding to the S-profile. Simple mode provides no confirmation of delivery. Full-mode extensions include mechanisms for ensuring call completion through negotiation of capabilities between the transmit and receive devices. Full-mode also provides for delivery confirmation. This defini-tion is from "Communications Systems & Networks," Ray Horak's best-selling book. Ray is Contributing Editor of this dictionary. See also MIME, SMTP, T.30, T.38, and TIFF.

T.38 An ITU-T Recommendation for store-and-forward fax via e-mail. Derived from X.25 packet standards, T.38 addresses IP fax transmissions for IP-enabled fax devices and fax gateways, defining the translation of T.30 fax signals and Internet Fax Protocol (IFP) packets. The specific methods for various T.38 implementations include fax relay and fax spoofing. Fax relay, also known as demod/remod, addresses the demodulation of standard analog fax transmissions from originating machines equipped with modems, and their re-modulation for presentation to a matching destination device. Fax relay depends on a low latency IP network (i.e., one second or less) in order that the session between the fax machines does not time out. Fax spoofing is used for fax transmissions over IP networks characterized by longer and less predictable levels of packet latency that could cause the session with the conventional fax machines to time out. Packet transmission over such a network can result in variable levels of delay of packet receipt, packet-by-packet, which timing phenomenon is known as jitter. Compensation for both the longer level of delay and the jitter are accomplished by padding the line with occasional keep-alive packets to keep the session active, rather than allowing it to time out. The receiving device is spoofed, or fooled, into thinking that the incoming transmission is over a realtime, carefully timed voice network. Delays up to five seconds can be tolerated in this manner. T.38 provides for two transport protocols, User Datagram Protocol (UDP) and Transmission Control Protocol (TCP). UDP is the faster of the two, but the less reliable, as no error detection and correction is included at the network level. T.38 overcomes this shortcoming either through redundant transmission of the image data packets, which is inherently inefficient at the network level, or through a Forward Error Correction (FEC) technique, which is inherently inefficient at the device level. TCP includes an error correction mechanism employed at the router level, with the routers typically being positioned only at the edges of the network-switches typically are positioned in the core of the network. Although T.38 strips this process from consider-ation for the IP fax packets, the level of delay nonetheless is increased; thereby, spoofing techniques are required to maintain fax sessions. This definition is from "Communications Systems & Networks," Ray Horak's best-selling book. Ray is Contributing Editor to this dictionary. See also Fax Relay, Fax Spoofing, FEC, T.30, T.37, TCP/IP, UDP, and X.25.

T.4 ITU-T standard for Group 3 fax machines, using T.30 and various V series standards. It also describes the data compression methods MH and MR.

T.434 The concept is simple: use a fax machine equipped with a disk drive to send and receive binary files as easily as you send a fax. The system would benefit from the fax's technical sophistication and ease of use, such as calling tone and the called unit's identification of capabilities. T.434 is an evolving ITU-T recommendation which defines a format used to encode a binary file and its attributes into a set of octets. This encoded binary file can then by sent over phone lines using error-corrected T.30 fax pages. The union of these two elements (file format and T.30 ECM) is known as "Binary File Transfer." The T.434 file attribute encoding is independent of ECM's block and page segmentation. T.434 defines 27 attributes which are used to describe a file. These attributes include protocol version, filename, permitted actions, contents type, storage account, date and time of creation, date and time of last modification, date and time of last read access, identity of creator, identity of last modifier, identity of last readers, filesize, future filesize, access control, legal qualifications, private use, structure, application reference, machine, operating system, recipient, character set, compression, environment, pathname, user visible string and data file content. Fisk Communications of San Diego, CA has done extensive work in the area of T.434 and has granted an irrevocable royalty-free and compensation free license to the Telecommunications Industry Association for basically all Fisk's work on T.434. Fisk hopes to make T.434 an extensive standard so that its equipment, which sends binary file transfers, can communicate with and receive from other machines. Microsoft also has defined a binary file transfer in the fax portion of its Microsoft At Work architecture. Some observers believe the At Work binary file transfer architecture is richer and more robust than T.434, supporting password and public/private-key encryption as well as digital signature verification. As of writing, it is not clear which standard will win.

T.568A T.568A Is a wiring scheme used for terminating four pair, all category 5e cable. There are three main choices when deciding which wiring scheme to use. The T.568A can accommodate two pair voice (2 pair USOC Voice Wiring)) and also 10Base-T networks, while the T.568B can only handle one pair voice. The other wiring scheme would be USOC wiring, used only when wiring for voice (Not recommended).

T.568B See T.568A.

T.6 ITU-T recommendation for Group 3 fax machines using T.30 and various V series standards. It also describes compression methods (Modified Huffman and Modified READ).

T.611 Also known as Appli/Com. A messaging standard proposed by France and Germany defining a Programmable Communication Interface (PCI) for Group 3 fax, Group 4 fax, teletex and telex service.

T1 See T-1.

T1.5 Same as T1. Se T1.

T1.601 The ANSI specification for ISDN BRI outside wire, known as the U interface. T1.601 uses the 2B1Q line code operating at 160 Kb/s (144 Kb/s of 2B+D plus Layer 1 overhead bits). The electrical signal can tolerate a maximum loss of 42 dB at 40 KHz, which usually limits the local loop length to 18 Kft or less.

T11 A technical committee of the National Committee for Information Technology Standards , titled T11 I/O Interfaces. It is tasked with developing standards for moving data in and out of central computers.

T1 1. A committee accredited by ANSI and sponsored by the Alliance For Telecommunication Industry Solutions (ATIS). The committee's role is to establish U.S. standards for digital telephony, particularly T1. ATIS provides all the administrative and logistical support. See Alliance for Telecommunications Industry Solutions.

2. See T-1 above (as in T-ONE).

T1 Multiplexer A statistical multiplexer that divides the 1.544MbpsT1 bandwidth into 24 separate 64Kbps channels of digitized data or voice.

T1 Small Aperture Terminal TSAT. A small satellite terminal used for digital communications that can handle T1 data rates of up to 1.544Mbps.

T1 Y1 (J1) E1 Voice channels, digitized and combined into a single digital stream. T1s are 1.544 Mbps. Y1 and E1 are similar in bandwidth and are used in Japan and Europe respectively.

T1-606/T1-6ac/T1-gfr ANSI's frame relay service specifications.

T-120 See T.120

T1C 3.152 million bits per second. Capable of handling 48 voice conversations. T1C is further up the North American digital carrier hierarchy. See T CARRIER and T-1.

T1D1 This is the specific technical subcommittee within ITU-T responsible for ISDN standards.

T1E1.4 An ANSI committee studying emerging the creation of a new high bit-rate digital subscriber line standard. The technology promises to deliver symmetric 1.544 Mbps data rates (i.e. T1 speeds) to distances of up to 12,000 feet using the American National Standards Institute's T1E1.4/99-006 draft standard for a single pair T-1 transport. With single pair transmission (T-1 over copper wires has always been transported over two pairs, one pair for receive and one pair for transmit), the cost of provisioning and maintaining a T-1 line will drop dramatically. The big push to develop T1E1.4 have been North American telephone companies, who see a huge demand for high speed data. But they see their future in delivering it on fiber. Meantime, they have a huge investment in copper pairs. So let's figure a technology to use those copper wires. Bingo the big interest in T1E1.4. See HDSL and HDSL2.

T1E1 An ANSI standards sub-committee dealing with Network Interfaces.

T1M1 An ANSI standards sub-committee dealing with T-1 Inter-Network Operations, Administration and Maintenance.

T1Q1 An ANSI standards sub-committee dealing with performance.

T1S1 T1S1 is a technical subcommittee to T-1 responsible for standards related to services,architectures, and signaling.

T1X1 T1X1 is a technical subcommittee to T-1 responsible for standards pertaining to synchronous interfaces and hierarchical structures relevant to interconnection of network transport signals.

T2 6.312 million bits per second. Capable of handling at least 96 voice conversations depending on the encoding scheme chosen. T-2 is four times the capacity of T-1. T-2 is further up the North American digital carrier hierarchy. In this dictionary we have adopted the style of writing T2 as T-2. See T-1.

T2L Text-to-landline.

T3 Twenty eight (28) T-1 lines or 44.736 million bits per second. Commonly referred to as 45 megabits per second. Capable of handling 672 voice conversations. T-3 typically runs on fiber optic and is then called FT3. T-3 is further up the North American digital carrier hierarchy. In this dictionary we have adopted the style of writing T3 as T-3. Some people believe that FT3 also means fractional T3. See T-1.

T3POS Transaction Processing Protocol for Point of Sale. T3POS is a transaction switching and transport protocol designed to provide existing Point-Of-Sale (POS) equipment and future POS terminals with efficient and economical switching and transport service over an X.25 based packet network.

T4 274.176 million bits per second. Capable of handling 4032 voice conversations. T-4 has 168 times the capacity of T-1. T-4 can run on coaxial cable, waveguide, millimeter radio or fiber optic. T-4 is further up the North American digital carrier hierarchy. In this dictionary we have adopted the style of writing T4 as T-4. See also T-1.

TA 1. Terminal Adapter. A Terminal Adapter allows existing non-ISDN terminals to operate on ISDN lines. It provides conversion between a non-ISDN terminal device and the ISDN user/network interface. See Terminal Adapter.

2. Technical Advisory. These publications are documents describing Bellcore's preliminary view of proposed generic requirements for products, new technologies, services, or interfaces.

3. Termination Attempt.

4. Telecommunication Administration.

TAB Technical Application Bulletin.

tabbed browsing It's the ability to open multiple Web pages in a single window, and to switch among them by clicking on tabs at the top of each page. This allows you to quickly scan a whole bunch of Web sites at once. It's especially useful if you group bookmarks (which Microsoft calls Favorites) into a folder, and then open all the pages in the folder at the same time. Tabbed browsing is now available in most browsers, including Internet Explorer and Firefox.

table 1. A collection of data in which each item is arranged in relation to the other items. Many telephony functions use "look-up tables" to determine the routing of calls. These tables solve the problem, "If the call is going to this exchange in this area code, then use this trunk and this routing pattern." See Table Driven.

2. The major design tool of most web pages. Let's say you want to design a web page with two photos next to each other, separated by words talking about the photos. The only way to design this into a web page is to set it up with a table with three columns – two for the photos and one for your text.

table driven Describing a logical computer process, widespread in the operation of communications devices and networks, in which a user-entered variable is matched against an array of pre defined values. Frequently used in network routing, access security and modem operation. It involves a table look up that is a reference to a collection of pre defined values.

Table Mountain Radio Receiving Zone A radio quiet zone in Boulder County, Colorado that was established in 1961 to protect the Table Mountain Radio Receiving Zone Research Laboratories from radio interference. The Research Laboratories are used by the National Telecommunications and Information Administration (NTIA) for telecommunications research and engineering studies. The FCC closely reviews applications to operate a radio station in the vicinity of the Table Mountain Radio Receiving Zone, and it will reject any application to operate a station that may cause interference in the zone.

table hook-up method An information retrieval system in which the input information and the related output information are stored as a pair. When a particular input is given, the table is accessed and the output data which coincides with the input is taken out.

table stable A large one-room office that's filled with worktables but has no partitions. In short, it's a cube farm without the cubes, i.e. the cubicles.

tablet PC Imagine a laptop with a screen that swivels so its screen can sit on the outside of the laptop when "closed." The screen is touch and stylus sensitive. You can touch touch icons or handwrite on the screen – just like you do on your PDA. You also have a keyboard and some form of mouse, which you can use also. So, the tablet is basically an enhanced (and thus slightly more expensive laptop). So the question is what's it good for? Think application. I've seen a street artist use one to sketch people and have his portable printer print out the photo. I've seen salespeople walking around supermarkets tapping orders into their tablet – all the stuff the supermarket needs from that company. Some people like taking a tablet to meeting and taking handwritten notes on the tablet. That way they can save the notes as images (and read them later) or have some software do handwriting recognition and convert the handwriting into text, which the computer can understand. I don't use a tablet because they don't work the way I do. But for special applications, they're great.

TABS AT&T's Telemetry Asynchronous Block Serial protocol. A polled point-to-point or multi point "master-slave" (remote-monitored equipment) communication protocol that supports moderate data transfer rates over intra office wire pairs. The remotes send "requests" or "polls" to monitored equipment. The monitored equipment answers the request with "responses." Defines two physical interfaces for direct connection between the telemetry remote and the monitored equipment:
- R5422 Point-to-Point
- RS485 Point-to-Point or Multi-Point.

Four wire, two Tx (remote to monitored) and two to Tx (monitored to remote), 22 or 21 gauge twisted pair, max 4 kft remote-to-monitored.

TAC Test Access Controller.

TACACS Terminal Access Controller Access Control System. An IETF (RFC 1492) standard security protocol which runs between client devices on a network and against a TACACS server. TACACS is an authentication mechanism which is used to authenticate the identity of a device seeking remote access to a privileged database. Variations on the theme include TACACS+, which provides services of authentication, authorization and accounting independently. TACSAS+ supports a challenge/response system and password encryption, as well as the standard TACACS user authentication. See also Authentication.

TACS Total Access Communications Systems. The original analog cell phone system system launched in 1985 by Vodafone. Used in the U.K., China, Asia, Japan, and Italy, TACS is also called ETACS, ITACS, IETACS, NTACS, and JTACS. It operates using FDMA. TACS was short-lived. TACS is a derivative of AMPS (Advanced Mobile Phone System), the analog cellular standard developed by Motorola and widely deployed in the U.S. and other parts of the world. TACS operated in the 900 MHz band, supporting 1,000 voice grade channels. TACS gave way to GSM, which is a much better digital technology. JTAC (Japanese TACS), which operated in the 800 and 900 MHz ranges, suffered a similar fate. See also AMPS and GSM.

TACT Trend Analysis for Circuit Troubles.

Tactical Automatic Digital Switching System TADSS. A transportable store-and-forward, message-switching system designed for rapid deployment in support of tactical forces. A military definition.

tactical command and control (C2) systems The equipment, communication, procedures, and personnel essential to a commander for planning, directing, coordinating, and controlling tactical operations of assigned forces pursuant to the missions assigned. A military definition.

tactical communication A military term. A method or means of conveying information of any kind, especially orders and decisions from one command, person, or place to another within the tactical forces, normally by means of electronic equipment (including communications security equipment). Excluded from this definition are communications provided to tactical forces by DCS, to non tactical forces by DCS, to tactical forces by non tactical military commands, and to tactical forces by civil organizations.

tactical communication system A system configured by various types of fixed-size, self-contained assemblages, such as radio terminals and repeaters; switching, transmission, and terminal equipment; and interconnect and control facilities, that are used within or in support of tactical military forces. The system provides securable voice and data communications and among mobile users to facilitate command and control within, and in support of, tactical forces.

Tactical Data Information Link TADIL. A military term. A Joint-Chiefs-of-Staff-approved standardized communication link suitable for transmission of digital information. A TADIL is characterized by its standardized message formats and transmission characteristics.

Tactical Data Information Link -- A TADIL-A. A military term. A netted link in which one unit acts as a net control station and interrogates each unit by roll call. Once interrogated, that unit transmits its data to the net. This means that each unit receives all the information transmitted. This is a direct transfer of data and no relaying is involved.

Tactical Data Information Link -- B TADIL-B. A military term. A point-to-point data link between two units which provides for simultaneous transmission and reception of full duplex data.

tactical load A military term. That part of the operational load required by the host service consisting of weapons, detection, command control systems, and related functions.

TAD Telephone Answering Device.

TADIL Tactical Data Information Link.

TADSS Tactical Automatic Digital Switching System.

TAF Targeted Accessibility Fund. In the late Spring of 1998, the New York State Public Service Commission ordered the establishment of a Targeted Accessibility Fund (TAF), which, according to the NYPSC, "will fund the costs incurred by all local exchange carriers, including competitive carriers, for E-911, lifeline and telecommunications relay service. All telecommunications carriers will be required to contribute to the fund based upon their intra-state gross revenues net of payments made to underlying carriers...Under PSC guidelines, the TAF is to be administered by a ten person board."

TAFAS Trunk Answer From Any Station. The ability to answer an incoming phone call from any telephone attached to the system.

TAFKAP The Artist Formerly Known As Prince. In 1993, Prince Rogers Nelson, the rock artist known as "Prince" was engaged in a contract dispute with Warner-Chappel, his former record label. To spite them and thereby hopefully affect their sales of his records, he changed his name to an unpronounceable symbol. Therefore, he became known as "The Artist Formerly Known As Prince," i.e. TAFKAP. On December 31, 1999, his contract with Warner-Chappel formally expired. On May 16, 2000, TAFKAP announced to an anxiously-awaiting world that he was changing his name back to "Prince."

tag 1. A field in a cache that contains information that allows determination whether a word in the cache corresponds to a requested word.

2. A tag is a command that specifies how a document or portion of a document should be formatted such as headings, paragraphs, lists, hypertext links, etc. Tags are codes used for formatting HTML documents for the World Wide Web. There are both single and compound tags. For example, the single code for a line break is <\>
. Bold text requires compound tags. For example, if you want to bold the word help, you would mark it <\><b>help<\></b>, where <\><b> means to turn on bolding and <\></b> means to shut it off.

3. See RFID Tag.

4. Bus and tag cables were multi-wire copper cables used by IBM from the 1960s to the early 1990s to connect a peripheral device, such as a printer or tape drive, to a local mainframe. Bus and tag cables were always used in pairs; the "bus" carried the data, and the "tag" carried the control information.

tag and locate I first heard about this "definition" from BellSouth. Apparently a technician goes to a business and locates where multiple lines come in and tags them for identification purposes.

Tag Image File Format TIFF provides a way of storing and exchanging digital image data. Aldus Corp., Microsoft Corp., and major scanner vendors developed TIFF to help link scanned images with the popular desktop publishing applications. It is now used for many different types of software applications ranging from medical imagery to fax modem data transfers, CAD programs, and 3D graphic packages. The current TIFF specification supports three main types of image data: Black and white data, halftones or dithered data, and grayscale data. Some wags think TIFF stands for "Took It From a Fotograf." It doesn't.

tag switching A technique developed by Cisco Systems for high-performance packet forwarding through a router. A label or "tag" is assigned to destination networks or hosts. As a packet stream is presented to the tag edge switch, it analyzes the network-layer header prepended to each packet, selects a route for the packet from its internal routing tables, prepends the PDU with a tag from its Tag Information Base (TIB), and forwards the packet to the next-hop tag switch, which typically is a core tag switch. That core switch then forwards the packet solely on the basis of the tag, eliminating the need to re-analyze the header. As the packet reaches the tag edge router at the egress point of the network, the tag is stripped off and the packet is delivered to the target device. While this approach adds a small amount of overhead to each packet, the speed of packet processing is improved considerably, particularly in a complex network in which multiple routers or switches must act on each packet. Tag switching can be applied to IPX and other network protocols, as well as IP. Tag switching also allows Layer 2 switches to participate in Layer 3 routing, reducing the number of routing peers with which each edge router must deal, and, thereby, enhancing the scalability of the network. Tag switching was submitted to the IETF (Internet Engineering Task Force), where it became the basis for MPLS (MultiProtocol Label Switching). See also MPLS.

tagged traffic ATM cells that have their CLP bit set to 1. If the network is congested, tagged traffic can be dropped to ensure the delivery of higher-priority traffic. Sometimes called DE traffic. See also CLP.

TAI inTernational Atomic tIme. See International Atomic Time.

tail An echo cancellation term. The tail, measured in milliseconds, is the amount of your conversation which returns to you in the echo, as measured in milliseconds. A tail of zero milliseconds clearly means there's no echo.

tail circuit 1. A feeder circuit or an access line to a network. Typically, a connection from a satellite, microwave receiver to a user's equipment location.

2. A point-to-point circuit connecting a remote terminal to a local terminal via two modems at an intermediate site. A crossover cable connects the two modems at the intermediate site.

3. A communications line from the end of a major transmission link, such as a microwave link, satellite link, or LAN, to the end-user location. A tail circuit is a part of a user-to-user connection.

tail-circuit flooding A term that describes the effect of a distributed denial of service attack that overwhelms the communications path between a router and the attacked server. See denial of service attack.

Tail-End-Hop-Off TEHO. In a private network with several nodes (locations), TEHO occurs when a call placed from one location on the network to a location not on the network leaves the network at the node closest to its destination.

tailgating See Employee Tailgating.

tailing In facsimile systems, the excessive prolongation of the decay of the signal. Also called Hangover.

take back and transfer See Callback.

take ownership A new term for the old expression, to take charge. In this case, there's a project that needs to be done, let's say building a new factory. Someone has to be in charge. Someone has to "take ownership." In normal business, to take ownership usually involves being in charge of something smaller – for example making sure that the hot dogs are delivered to the company picnic.

TALI Tekelec Adaptation Layer Interface. Official Tekelec words: Tekelec developed the Transport Adapter Layer Interface (TALI) to fulfill an industrywide need for an open interface to ensure interoperability of convergence solutions from multiple vendors. In December 1999, Tekelec released the TALI interface source code to the telecommunications and Internet industries in an effort to promote an open network architecture for signaling over IP. With the support of Level 3, U S WEST Wireless, Lucent, Cisco, Orange PCS, and other carriers and equipment suppliers that are currently using the interface, Tekelec has submitted TALI to the Internet Engineering Task Force (IETF) for consideration as a standard.

talk A service on Internet whereby you hold conversations with others by typing into your computer. And they reply by typing into their computer.

talk battery Talk battery is the telephone company's term for the DC voltage it supplies to the subscriber's loop (i.e. your phone line). This voltage typically powers the circuitry in your phone and lets you dial, receive calls and have conversations. In North America, talk battery delivers around 48 volts DC. Of course, this is true of single line phones and most older phones. But these days virtually all phone systems (especially those with multipe lines) get their power from being plugged in locally an outlet. They need the power to do things as simple as turning on lights to things as complex as encoding your voice conversation for sending it over the Internet. Talk battery is the reason you'll still get dial tone at your house when the commercial power drops. But – and here's a tip – you better have an old-fashioned single ling phone to plug into the line. because your main multi-line phone system will be dead. Talk battery is also. also known as A BATTERY. See B Battery.

talk off In voice processing, in response to questions such as "Press one for Harry," the user touchtones buttons on his phone. Those buttons generate DTMF (Dual Tone Multi-Frequency) tones. The system has to figure out what the person "said" with his touchtones. The tricky part of DTMF detection is distinguishing between tones generated from an actual "key press" and "tones" caused by speech. Mistaking a person's speech (as in leaving a message) for DTMF is called "talk-off." Mistaking a person's recorded speech (as in playing back a message) for DTMF is called "play-off." You can imagine the havoc poor DTMF detection can cause a voice processing system. For example, if touchtoning three means "delete this message" and while playing the message, the system incorrectly detects a portion of the message playback as the touchtones for a key press three, I'll delete the message when I had intended to listen to it. On the other hand, if I'm listening to a message and want to delete it prior to finishing the message, I want the system to detect my key press three as the real thing and go ahead and delete the message.

talk path The tip and ring conductors of a telephone circuit.

talk set, optical An instrument for talking over fibers – used when installing and testing the cable.

talk time 1. The amount of time agents spend on the phone, as opposed to the time between calls spent updating records, sending out literature or going to the bathroom.

2. The length of time you can talk on your portable or transportable cellular phone from a fully charged battery without standby time. The battery capacity of a cellular portable or transportable is usually expressed in terms of so many minutes of talk time or so many hours of standby time. When you are talking, the phone draws more power from the battery.

talk-off See Talk Off.

talkdown Missed signals in the presence of speech. Commonly used to describe the performance of a DTMF receiver when it fails to recognize a valid DTMF tone due to cancellation of that tone by speech.

talkgroup A group of radio users who communicate with each other by radio as a group and share a common radio call sign, by virtue of their having a common functional responsibility and needing to communicate only among themselves.

talker Provides interaction with a Dialogic board, allowing for Voice Mail functionality.

talking battery See Battery and Talk Battery.

Talking Channel Capacity TC. The network access line capacity of talking channel equipment of an entity is the maximum number of network access lines that can be served without exceeding the percent incoming matching loss objective for that entity .

talking head That part of the person seen in the typical business videoconference or webcast; the head and shoulders. This type of image is fairly easy to capture with compressed video because there is very little motion in a talking head image.

In a TV newscast, the delay of a second or two from when a question is asked in the studio until the "talking head" responds indicates that the coverage is provided via satellite and consequently suffers propagation delay in both directions. The term talking head is often used derogatively to indicate that the TV or cable TV program has no creating programming; it only has talking heads – the cheapest way of putting a TV program together.

talkoff See Talk Off.

TAM 1. Telephone Answering Machine.

2. Telecommunications Access Method.

3. Total Available Market. A cable TV word that refers to all the houses in the neighborhood of the cable.

tamper detection In encrypted messaging, tamper detection allows the recipient of information to verify that it has not been modified in transit. Any attempt to modify data or substitute a false message for a legitimate one will be detected. See Public Key Encryption.

tandem In a telecommunications context, the term refers to switches, circuits, or other Network Elements (NEs) that serve to allow other NEs to work together. For example, tandem switches, or tandem offices, serve to interconnect other, lesser switches, i.e., Central Offices (COs) or lesser tandems. Tandem switches, in the purest sense of the term, serve no end users directly, as that is the responsibility of the COs. Rather, they strictly serve to interconnect the COs, which are at the lowest level of the switching hierarchy in the PSTN (Public Switched Telephone Network). Tandem tie trunks serve to interconnect tandem switches. See the following definitions.

tandem access trunk groups A category of trunk groups that originates at EOs (End Offices) and terminates at tandems.

tandem architecture A physical network topology where connectivity between locations is achieved by linking several locations together in a chain using private line circuits. In a tandem architecture, a packet may have to pass through several intermediate locations before reaching its final destination. A single network failure can affect connectivity between several locations, a primary weakness of the topology.

tandem call A call processed by two or more switches. Also used to designate this type of call at a switch where a connection is established from one trunk to another (tandem trunking). See Tandem Switch.

tandem center In a communication system, an installation in which switching equipment connects trunks to trunks, but not any customer loops.

tandem completing trunk groups A category of trunk groups that originates at tandems and terminates at EOs.

Tandem Connecting Trunk TC Trunk. A 1- or 2-way trunk between an end office switching system and an LATA tandem switching system.

tandem data circuit A data channel passing through more than two data circuit-terminating equipment (DCE) devices in series. At the relay point, transmitted and received data, clock and other control signals must be transposed; if not accomplished in a switching system, a cable between the two must do so, often called a "rollover cable," or "null modem." See Null Modem Cable.

Tandem InterLATA Connecting Trunk TIC Trunk. The trunk used for switched LATA access that interconnects an long distance company's Point of Presence (POP) with an local phone company's switching system. See Direct InterLATA Connecting Trunk Group.

tandem office A major phone company switching center for the switched telephone network. It serves to connect central offices when direct interoffice trunks are not available.

Tandem PBX A main PBX is one which has a Directory Number (DN) and can connect PBX stations to the public network for both incoming and outgoing calls. A main PBX can have an associated satellite PBX, and can be part of a Tandem Tie Trunk Network (TTTN). If the main PBX provides tandem switching for tie trunks, it is called a tandem PBX.

tandem point An intermediate location in a tandem architecture.

tandem queuing A telephone company term. When for example: A shortage of receivers in a terminating office will be reflected in originating offices trunking traffic into the terminating switch. The obvious effect will be a slowing down of the originating office transmitters, i.e., Tandem Queuing.

tandem switch Tandem is a telephony term meaning to "connect in series." Thus a tandem switch connects one trunk to another. A tandem switch is an intermediate switch or connection between an originating telephone call location and the final destination of the call. The tandem point passes the call along. A PBX can often handle tandem calls from other/to other locations as well as process calls to, from and within its own location. A tandem switch generally is designed primarily with trunk interfaces rather than subscriber interfaces.

tandem switching system Synonym of Tandem Tie Trunk Network. See the next definition.

Tandem Tie Trunk Network TTTN. A serving arrangement that permits sequential connection of tie trunks between PBX/CENTREX locations by utilizing tandem operation. TTTNs are large, private switched networks, generally used by the federal government, state and county governments, and the military to interconnect sites (e.g., government buildings, and military posts and bases). The tandem switches can be either totally dedicated switches, or partitions of switches also used for the PSTN (Public Switched Telephone Network). See also AUTOVON.

tandem tie trunk switching The PBX permits tie lines to "tandem" through the switch. This means an incoming tie line call from a distant PBX receives a dial tone instead of automatically connecting with the operator. The caller can then dial a connection with either a phone on the PBX or an outgoing line. The outgoing line can be a local trunk in which case the distant PBX has access to a form of foreign exchange service, or another tie line which links a third system. This system of tie lines is widely used to form a corporate communications system, allowing economical connections between distant offices. To provide tie line tandeming ability, the PBX must be able to detect when either tie line goes on-hook at the distant end so that it can break its tandem connection and allow the tie lines to be used for other calls.

tandem trunks Trunks between an end office and a tandem switching machine or between tandem switching machines. Tandem trunks can provide direct routing or alternate routing capability when direct trunks are occupied.

tangerine A stock that is trading for less than the sum of its parts. Its value can be found (perhaps) by breaking it into juicy segments.

TANDM Traffic Analysis Data Management.

tank test A voltage dielectric test in which the wire or cable test sample is submerged in water and voltage is applied between the conductor and water as ground.

TANSTAAFL There Ain't No Such Thing As A Free Lunch.

tantalum See Coltan and Tantalum Capacitor.

tantalum capacitor A tantalum capacitor is a device circuit switched equipment that processes voice calls from central offices. See Trunk Exchange.

TAO 1. The creative principle that orders the universe, according to Taoists, who follow the mystical Chinese philosophy founded by Lao-tzu (also spelled Lao-tse) in the 6th century B.C. Tao also is the path of virtuous conduct conceived by Confucians. In our context, "Tao" is both a set of principles which guide standards processes, and a set of rules of order or conduct which guide the participants in the processes. The Internet Engineering Task Force (IETF) and other standards bodies use "Tao" in order to avoid using terms like "rules," which its participants might find offensive. If you are a "Nethead" or if you know any "Netheads," you can relate to this. The IETF actually writes "Pronounced 'dow,' Tao is the basic principle behind the teachings of Lao-tse, a Chinese master. Its familiar symbol is the black and white Yin-Yang circle. Taoism conceives the universe as a single organism, and human beings as interdependent parts of a cosmic whole. Tao is sometimes translated 'the

way,' but according to Taoist philosophy the true meaning of the word cannot be expressed in words." In August, 2001 the IETF published "The Tao of IETF: A Novice's Guide to the Internet Engineering Task Force." You can read this long document at http://www.ietf.org/tao.html#8.2

2. Telephony Application Object. Part of the SCSA programming framework. See S.100 and SCSA Telephony Application Objects Framework.

tap 1. An electrical connection permitting signals to be transmitted onto or off a bus. The link between the bus and the drop cable that connects the workstation to the bus. Also a device used on CATV cables for matching impedance or connecting subscriber drops.

2. Telocator Alphanumeric Protocol, also known as IXO. A 7-bit messaging protocol which allows someone sitting at a terminal or computer to send a one-way message to a pager (also known as a beeper). TAP is ACSII-based and half-duplex. TAP provides an error detecting link from the sender to the paging service provider. See also IXO and SNPP.

3. A term in video compression referring to the number of pixels or lines considered in the process of averaging values through the filtering process. MPEG uses a 7-tap filter.

4. To draw energy from a circuit.

tap button A button found on single line phones behind a PBX or Centrex. The tap button gives a precisely measured Hookswitch flash. The purpose of this button is to signal the PBX that it is about to receive a command – typically a transfer. To transfer a call on a single line phone, you typically depress the hookswitch, then punch out the extension you want to transfer the call to, announce the call when someone answers, then hang up and the PBX or Centrex transfers the call. The problem with using a hookswitch to make this transfer is that if you depress the hookswitch for too long you will cut the call off. As a result, some manufacturers put a tap button on their single line phones. This button gives the precise hookswitch signal for the precise length of time necessary – no more, no less. The Tap Button is also called a Flash button or a Tap Key.

tap key Also called Tap Button or Flash Key. A button on a phone that accomplishes the same function as a switch hook but is not a switch hook. See Tap Button.

tap loss In a fiber optic coupler, the ratio of power at the tap port to the power at the input port.

tap port In a coupler where the splitting ratio between output ports is not equal, the output port containing the lesser power.

TAPAC Terminal Attachment Program Advisory Committee. Body which recommends telecom standards to the Canadian Federal Government.

tape bomb Troubling news that scrolls across the bottom of CNBC, Bloomberg TV and CNNfn, catching investors off guard and without time to sell (or buy).

tape wrap A spirally applied tape over an insulated or uninsulated wire.

tape drive The physical unit that holds, reads and writes magnetic tape.

tape reader A device which reads information recorded on punched paper tape or magnetic tape.

tape relay A method of retransmitting TTY traffic from one channel to another, in which messages arriving on an incoming channel are recorded in the form of perforated tape, this tape then being either fed directly and automatically into an outgoing channel, or manually transferred to an automatic transmitter for transmission on an outgoing channel.

tapenet Analogous to a sneakernet, but involving magnetic tapes and large computer systems. A tapenet involves writing one or more files to tape, usually in a mainframe or minicomputer environment, and then sending the tape by courier to the destination computing center. The high storage capacity of magnetic tape actually makes this file transfer procedure much more effective than it may sound. Network guru and computer scientist Andrew Tanenbaum does the math to prove this point in his classic textbook, Computer Networks, concluding his analysis with his famous observation, "Never underestimate the bandwidth of a station wagon full of tapes hurtling down the highway." See sneakernet.

tapeout Tapeout is a term used in the semiconductor business. When a processor or other silicon product is "taped out," all the various components that are necessary for it to function are complete. It marks the end of development of the product and the beginning of manufacturing. The term "tape out" itself is an anachronism, recalling the era when a design was actually transferred to magnetic tape in data form. The tape, in turn, was given to mask design to produce a "mask" or blueprint that was then used to manufacture the product. Today, tapeout means generating the final files for silicon masks. These masks are then shipped (or electronically transferred) to a mask making unit.

tapered fiber An optical fiber in which the cross section, i.e., cross-sectional diameter or area, varies, i.e., increases or decreases, monotonically with length.

tapewire Wire that is built into adhesive tape that can be mounted on any clean, smooth, nonporous surface. Just peel off the tape's backing and press the tapewire into place. Tapewire can be left uncovered or it can be painted, papered or carpeted over.

Tapewire is often used in the home for speaker, satellite dish, and cable TV connections. It provides an alternative to visible wires and drilling holes in walls and ceilings. Also called adhesive wire.

TAPI Telephone Application Programming Interface. Also called Microsoft/Intel Telephony API. A term that refers to the Windows Telephony API. TAPI is a changing (i.e. improving) set of functions supported by Windows that allow Windows applications (Windows 3.xx, 95 and NT) to program telephone-line-based devices such as single and multi-line phones (both digital and analog), modems and fax machines in a device-independent manner. TAPI essentially does to telephony devices what Windows printer system did to printers – make them easy to install and allow many application programs to work with many telephony devices, irrespective of who made the devices. TAPI is one of numerous high-level device interfaces that Windows offers as part of the Windows Open Services Architecture (WOSA). TAPI simplifies the process of writing a telephony application that works with a wide variety of modems and other devices supported by TAPI drivers. See also Dial String, Microsoft Fax, TAPI 2.0, TAPI 3.0, Windows 95 and Windows Telephony for fuller explanations.

TAPI 2.0 See TAPI 3.0.

TAPI 3.0 IP telephony is a set of technologies from Microsoft that enables voice, data, and video collaboration over existing IP-based LANs, WANs, and the Internet. According to Microsoft, IP telephony uses open IETF and ITU standards to move multimedia traffic over any network that uses IP. As a result, the same ubiquitous networks that carry Web, e-mail, and data traffic can used TAPI to connect to individuals, businesses, schools, and governments worldwide. TAPI has now pretty well been superceded by newer technologies SIP and VXML.

TAPTS Tenant Package.

TAR TAR is the UNIX standard program for combining and compressing files. It's like a UNIX Winzip program. I believe it stands for "Tape Archiver". The UNIX System V manual lists the command description as "tape file archiver". One no longer uses it just for tape archiving, but that was its initial usage.

TARFU TARFU is a military term used on TTY circuits to state that "things are really F...ked up". In the military there were three stages to a project, which is not going well – SNAFU, FUBAR and TARFU. See SNAFU and FUBAR.

TARGA Truevision Advanced Raster Graphics Adapter.

target A SCSI device that performs an operation requested by an initiator.

target host number The number that identifies the destination software program during the user logon process.

Target ID Address Resolution Protocol. TARP is the established multivendor standard for SONET network elements that support TL1 Operating System interfaces.

Target Token Rotation Time TTRT. An FDDI (Fiber Distributed Data Interface)token travels along the network ring from node to node. If a node does not need to transmit data, it picks up the token and sends it to the next node. If the node possessing the token does need to transmit, it can send as many frames as desired for a fixed amount of time.

tariff Documents filed by a regulated telephone company with a state public utility commission or the Federal Communications Commission. The tariff, a public document, details services, equipment and pricing offered by the telephone company (a common carrier) to all potential customers. Being a "common carrier" means it (the phone company) must offer its services to everybody at the prices and at the conditions outlined in its public tariffs. Tariffs do not carry the weight of law behind them. If you or the telephone company violate them, no one will go to jail. The worst that can happen to you, as a subscriber, is that your service will be cut off, or threatened to be cut off. Regulatory authorities do not normally approve tariffs. They accept them – until they are successfully challenged before a hearing of the regulatory body or in court (usually Federal Court). Many tariffs were accepted by regulatory commissions only to be struck down in court as unlawful, discriminatory, not cost-justified, etc. Monies collected under the tariff have been refunded and unnecessary equipment removed. In these new, competitive days, many telephone companies are violating their own tariffs by charging less money than their tariffs say they should, or bundling services together at a discount. They are also providing service and equipment on terms less onerous than outlined in their tariffs. Many users now regard tariffs as starting bargaining points, rather than ending bargaining points.

Tariff 12 A user-specific long distance tariff of AT&T. Tariff 12 gives AT&T the ability to price its long distance services for one company practically any which way it feels – giving them a mix of services at stable prices over the long term with significant volume discounts. As this dictionary was going to the printer, a federal appeals court overturned the Federal Communications Commission's April 1989 decision allowing AT&T to offer custom networks and ordered the FCC to reopen its investigation into the legality of the Tariff 12 deals. There are still some users. They are "grandfathered" until we get a final say on the tariff. And, as we go to press, AT&T can offer Tariff 12 customized services to any company – but cannot include 800 services in its Tariff 12 pricing.

Tariff 15 A user-specific long distance tariff of AT&T. Tariff 15 gives AT&T the ability to price its long distance services for one company practically any way it feels. Tariff 15 is single-customer discounting. Some of AT&T competitors claim the tariff is "illegal."

tariff rebalancing Largely an initiative of the FCC in the US, national, regional (e.g., EU) and international (ITU) regulatory authorities and policy-making bodies are considering the rebalancing of tariffs in order that they be more closely related to the costs of providing the various telecommunications products and services. While most attention is focused on the accountings rate for long distance, both domestic and international, tariff rebalancing encompasses all tariffed products (including equipment rentals) and services (e.g., local service), and across both the business and consumer domains. Historically, tariffs for individual products and services were designed in the context of the overall tariff structure, which addressed the full range of such products and services. Within this overall structure existed a complex set of cross-subsidies which generally resulted in low tariffs for basic consumer services, such as residential local service. Relatively high tariffs existed both for optional services, such as long distance, international and custom calling features. The primary justification for this arrangement was that of the desire to gain "universal service" - a phone in everybody's home. Mixed in with this was the concept of "ability to pay." Further, individual consumers vote, while companies do not. As consumer rights advocates became more vocal during the past twenty years or so, the pressure to retain these cross-subsidies increased. It generally is recognized, however, that the traditional tariff structure places incumbent carriers and service providers at a decided disadvantage in a competitive environment, as the pricing policies of the newer competitors are not constrained by tariffs, nor strange societal concepts of universal service and ability to pay. Further – and this is perhaps a major impetus – there exists a clear imbalance in the accounting rates for international long distance calls. For instance, a call from Argentina to the U.S. is much more expensive than is a call in the reverse direction, even though the costs for call origination are roughly equal to the costs of termination. This imbalance in international long distance tariffs has led to a huge imbalance of trade, to the detriment of the US. Therefore, both domestic US deregulation, and the resulting increase in competition, and the international imbalance of trade resulting from imbalance in international long distance tariffs have led to the FCC initiative. Many other nations and regions have followed the U.S. lead in terms of deregulation and competition, and are in the process of rebalancing domestic tariffs. Although the WTO (World Trade Organization) has taken the initiative in terms of the rebalancing of international long distance tariffs, many nations are resisting, citing their opinions that the WTO is unduly influenced by the U.S. Further, many developing nations rely heavily on this imbalance of trade as a major source of hard currency. See also Cross Subsidization, Separations and Settlements, Tariff and Universal Service.

TARM Telephone Answering and Recording Machines.

TARP Target ID Address Resolution Protocol. TARP is the established multivendor standard for SONET network elements that support TL1 Operating System interfaces.

TAS Telephone Answering Service.

TASC Telecommunications Alarm, Surveillance, and Control system. Expands the scope of maintenance from the traditional alarm monitoring and control functions to include performance monitoring and fault locating.

TASI Time Assignment Speech Interpolation. A money saving analog multiplexing procedure which keeps the connection to the circuit as long as someone is speaking and lets other conversations use the circuit during the intervals (measured in microseconds) when there's no speaking. Since a long distance circuit is usually only half used – one person speaking, one person listening – at least 50% of the circuit can be used by someone else. TASI is typically used by long distance companies on submarine cable across the Atlantic and the Pacific. Unfortunately, the flip-flopping around of circuit allocation by TASI means that the first tiny bit of a conversation is often lost. This can be disastrous for data. The key in data is to keep transmitting. The key in voice is to say something like "Ah" to seize the channel and then say what you want. TASI is very much like a very fast version of mobile dispatch radio. A modern version of TASI is called Digital Speech Interpolation (DSI). TASI is somewhat comparable to statistical multiplexing of data. Ray Horak, my Contributing Editor, recently had a gentleman from Cable & Wireless attending one of his public seminars. Apparently this gentleman was one of the engineers working on an early submarine cable between the US and the UK. The cable system used TASI, which was by no means perfect, but it certainly enhanced the voice traffic carrying capacity of that very limited 50-pair cable system. The system worked fine most of the time, but occasionally crashed at no particular time and for no apparent reason. The engineers were baffled. Finally, they recorded the conversations until the system crashed. It seems that the TASI approach would fail when required to support a large number of conversations between women. It seems that women have a tendency to talk and listen at the same time, and to do so quite effectively. Men, it seems, either lack this skill or prefer the "I talk and you listen, then you talk and I listen" protocol. Mystery solved! (Ladies, please don't call me to complain about this definition. Ray swears that it is true. Call him.)

task management Allocating resources and overseeing the sequence of tasks completed by the computer.

task switching You have a computer and you want to have it do several tasks at once. There are two ways. One is multi tasking. The computer will keep working on several tasks at once, though you may not see them on your screen. For example, you pay start a spreadsheet recalculating. And then you may call your electronic mail system. While you're receiving your mail, your spreadsheet is still recalculating. When you're through, you can switch back to your spreadsheet and see the final results. That's called multi tasking and MS-DOS doesn't have it. MS-DOS 5.0 through its Shell has something called task switching, whose idea is that you can load several programs into your computer and switch quickly between them. But the programs you put in background won't run. They stop the moment you put them in background. When you cycle back to them, they will start running again. Sadly, MS-DOS 5.0 Shell's task switching capabilities are very weak. Your programs will lock up and you will lose your data. I do not recommend using Task Switching in MS-DOS 5.0. Windows 3.1 is alleged to have a form of multi tasking. I don't trust it either yet. Windows has locked up on me on several occasions.

task-to-task communication The process whereby one computer program exchanges data with another. May also be called program-to-program communications.

taste tribe Specialized affinity group built around shared cultural interests and brought together through online social networks, blogs, egroups, and mailing lists

Tauzin-Dingell Bill A House of Representatives bill circa 2002 pushing to allow the Regional Bell Operating Companies (RBOCs) to offer broadband Internet services over long-distance lines without opening up their local phone service monopolies to outside competition. The Tauzin-Dingell bill, or more formally H.R. 1542, the Internet Freedom and Broadband Deployment Act, is legislation written by Representatives John D. Dingell (D-MA) and Billy Tauzin (R-LA).

tax identification number A unique identifier for business organizations that is used for reporting tax payments to the government (similar to the social security number for individuals).

tax impulse A high-frequency pulse placed on the phone line by telephone companies in Austria, Belgium, Czech Republic, Germany, Spain and Switzerland to meter a telephone call for billing purposes. These signals are inaudible to the human ear but can (and do) disrupt dial-up computer modem connections. See tax impulse filter.

tax impulse filter A small device that plugs into the phone line between the wall and a modem to filter out tax impulses. It doesn't stop the call from getting billed but it does prevent tax impulses from interfering with modem connections. See tax impulse.

TAXI Transparent Asynchronous Transmitter Receiver Interface. 100-Mbps ATM physical interface specification based on the FDDI PHY.

taxidermist The only difference between a tax man and a taxidermist is that the taxidermist leaves the skin. - Mark Twain

taxonomy The orderly classification of things according to their natural relationships. The Internet Engineering Task Force (IETF) uses this term a lot in their documents to describe the orderly interrelationships between protocols used in the Internet.

TAT TransAtlantic Telephone cable. A bunch of phone companies pool their monies and build an undersea cable connecting their various countries. They typically own the percentage of the cable which they contributed in money. In August, 2001, TAT-14 was completed. It joined Denmark, France, Germany, the Netherlands, the U.K. and the U.S. at a cost of $1.4 billion. There were 48 participating telecommunications carriers. Capacity on the cable was 640 giabits per second. That's 640,000,000,000 bits per second.

TB An ATM term. Transparent Bridging: An IETF bridging standard where bridge behavior is transparent to the data traffic. To avoid ambiguous routes or loops, a Spanning Tree algorithm is used.

TB/S TeraBits per Second. See also Tera.

TBA To Be Announced. Pricing of a product that may exist and that may, one day, be priced. You often see TBA after products which are hyperware, vaporware, mirrorware or smokeware.

TBB 1. Transnational Broadband Backbone. An international, high-speed (T-3 or faster), backbone system of transmission facilities and, perhaps, switching systems.

2. Telecommunications Bonding Backbone. A copper conductor extending from the telecommunications main grounding busbar (TMGB) to the farthest floor telecommunications grounding busbar (TGB). The idea is to eestablish an electrically conductive path for telecommunications, the metallic parts of which path are permanently joined to ensure 1) electrical continuity, 2) the capacity to conduct safely any current likely to be imposed, and 3) the ability to limit dangerous electrical potentials. In other words, a TBB is a permanently hard-wired metallic cable system, which is safely grounded and surge-protected in order to protect your equipment.

TBBIBC TBB Interconnecting Bonding Conductor. A conductor used specifically to interconnect metallic wires in a TBB (Telecommunications Bonding Backbone). See also TBB.

TBD To Be Determined.

TBE Transient Buffer Exposure: This is a negotiated number of cells that the network would like to limit the source to sending during startup periods, before the first RM-cell returns.

TBM Transport Bandwidth Manager. A TBM consists of a OC-3 or OC-12 terminal shelf. It is an optical fibre term.

TBOS Telemetry Byte Oriented Serial protocol. TBOS is a protocol for transmitting alarm, status, and control points between NE and OS. TBOS defines one physical interface for direct connection between telemetry remote and the monitored equipment. This is a point-to-point communication, RS 422A modified four wire, two to Tx (remote to monitored) and two to Tx (monitored to remote), 26 gauge, max 4 kft remote-to-monitored. Remote sees a 100 to 180 ohms resistor at monitored terminal.

Tbps Tera bit per second. One million million bits per second. See Tera.

TBR Timed BReak.

TC 1. Tc (Time committed). Also known as Measurement Interval. A Frame Relay term defining the interval of time which the carrier uses to measure data rates that burst above the CIR (Committed Information Rate). See also Committed Information Rate.

2. Transmission Convergence. The TC sublayer is a dimension of the ATM Physical Layer (PHY), working in tight formation with the Physical Medium (PM) sublayer. The TC accepts frames of data transmitted across the PM, delivering them to the ATM Layer for segmentation into cells, generates the Header Error Check (HEC), and sends idle cells when the ATM layer has none to send. On reception, the TC sublayer delineates individual cells in the received bit stream and reconstitutes the frames of data, using the HEC to detect and correct received errors.

3. Telecommunications Closet. A closet which houses telecommunications wiring and telecom wiring equipment. It contains the BHC (Backbone to Horizontal Cross-connect). It may also contain the network demarcation, or MC (Main Cross-connect). The telecommunications closet is used to connect up telecom wiring. The closet typically has a door. It's a good idea to lock the door and not put anything else in the closet, like mops, buckets and brooms.

4. Transmission Control.

5. Technical Consultant.

6. Tray Cable. A multiconductor flame-retardant cable with an overall nonmetallic jacket rated 600 volts. The cable may or may not have grounding conductors in the assembly.

7. Transmit Clock.

TD-CDMA A packet data implementation of the 3GPP Universal Mobile Telecommunications System (UMTS) standard, first launched commercially in 2003. Also known as UMTS TDD. TD-CDMA is a high-speed, low-latency, packet-based platform that uses time-division multiplexing to increase available bandwidth. TD-CDMA supports hand-offs and roaming, which enables it to be used for mobile wireless broadband.

TC-PAM Trellis-Coded Pulse Amplitude Modulation, a robust modulation format used in SHDSL. TC-PAM provides adaptability in bit rate and line reach. With SHDSL, rates of 192 kBps can be attained on line reaches of more than 20,000 feet (6 km), and higher rates, up to 2.3 MBps can be attained over line reaches of more than 10,000 feet (3 km). SHDSL is spectrally compatible with ADSL. See ADSL, PAM, SHDSL and Trellis Coding.

TCA TeleCommunications Association. A not-for-profit users association of communications management professionals. Formed in 1961, TCA recently expanded its focus to position itself as "The Information Technology and Telecommunications Association." Most of its members are West-of-the-Rockies telecommunications management professionals. For many years, TCA held a great annual conference in San Diego. The TCA conference was huge, and it was lots of fun. The vendors spent fortunes on their exhibits. They spent even more on hospitality suites and lavish dinner parties. Over time, the show declined as the conference business became for-profit, as the technologies moved ever more quickly, as the number of conferences proliferated, as the conference business became more highly focused and more fragmented, and as companies downsized and could no longer afford to subsidize associations like TCA. The TCA show moved to Reno, Nevada in 1997. That was the last show. Several of the TCA chapters still hold annual events, but they are much smaller. www.tca.org

TCAM TeleCommunications Access Method. A popular telecommunications software package to run on IBM 370 computers. See IBM.

TCAM Telecommunications Access Method. An IBM macro language for creating communications applications programs and message control.

TCAP Transactional Capabilities Application Part. Provides the signaling function for network data bases. TCAP is an ISDN application protocol. In addition to PRA and ISUP, the third major ISDN protocol in the delivery of advanced network services is TCAP, a CCS7 application protocol that provides the platform to support non-circuit related, transaction-based information exchange between network entities. This capability is required by transaction-based services that must exchange information between a pair of signaling nodes in a CCS7 network. Examples of these services include enhanced dial-800 service, automated credit card calling and virtual private networking. The TCAP protocol enables these services to access remote databases called service control points (SCPs) to process part of the call. The SCP supplies the translation and routing data needed to deliver advanced network services – like translating a dial calls into the required routing number. TCAP is useful also in coordinating some enhanced call-related services. For example, network ring again requires the connection of two users when both stations become idle. In this case, TCAP is used to coordinate between the users' switches while waiting for each line to become idle. And it can do this without tying up network trunks.

One of the major advantages of TCAP is that it provides a set of protocol building blocks for use in a variety of service definitions. The TCAP building blocks are subdivided into the transaction sublayer and the component sublayer. For more on TCAP, see the 1988-3 issue of Northern Telecom's Telesis publication.

TCAS T-Carrier Administration System. Provides mechanized support for the facility maintenance and administration center to achieve centralized administration and control of the digital network.

TCC Telephone Country Code.

TCCKT Telephone Company Circuit ID.

TCF Training Check Frame. Last step in a series of signals in a fax transmission called a training sequence, designed to let the receiver adjust to telephone line conditions.

TCH TCH means Traffic CHannel is used in combination with BCCH to set up frequency hopping on GSM/TDMA networks. The BCCH (Broadcast Control Channel) controls the traffic; the TCH carries it.

tchotchke A Yiddish word meaning trinkets, best exemplified by the giveaway junk we often pick up at telecommunications trade shows. It's pronounced choch-kuh in Yiddish. The plural is pronounced choch-key. See re-gift and Yiddish.

TCIC Trunk Circuit Identification Code. Only relevant to SS7. A number that uniquely identifies a trunk between an origination point code and a destination point code. An example would be between two telephone company switches.

TCIF TeleCommunications Industry Forum. A voluntary special interest group under ATIS (Alliance for Telecommunications Industry Solutions). TCIF addresses areas such as electronic commerce, including bar coding and EDI (Electronic Data Interchange). www.atis.org/atis/tcif/index.html See also ATIS.

TCL 1. Tool Command Language. Tcl is actually two things: a language and a library. First, Tcl is a simple textual language, intended primarily for issuing commands to interactive programs such as text editors, debuggers, illustrators, and shells. It has a simple syntax and is also programmable, so Tcl users can write command procedures to provide more powerful commands than those in the built-in set. Second, Tcl is a library package that can be embedded in application programs. The Tcl library consists of a parser for the Tcl language, routines to implement the Tcl built-in commands, and procedures that allow each application to extend Tcl with additional commands specific to that application. The application program generates Tcl commands and passes them to the Tcl parser for execution. Commands may be generated by reading characters from an input source, or by associating command strings with elements of the application's user interface, such as menu entries, buttons, or keystrokes. When the Tcl library receives commands it parses them into component fields and executes built-in commands directly. For commands implemented by the application, Tcl calls back to the application to execute the commands. In many cases commands will invoke recursive invocations of the Tcl interpreter by passing in additional strings to execute (procedures, looping commands, and conditional commands all work in this way). www.neosoft.com/tcl/default.html.

2. Tool Command Language.

TCM 1. Traveling Class Mark.

2. Trellis Coding Modulation.

3. Time Compression Multiplexing. A digital transmission technique that permits full duplex data transmission by sending compressed bursts of data in a "ping-pong" fashion.

4. Telecommunications Manager. The TCM is the manager of the department that plans, controls, and administers the telephony and telecommunications assets of the company. He ensures that the telephone and telecommunications systems are well-run and functioning smoothly. These assets may include the PBX and ISDN, T-1, local and long distance telephone lines, telephone sets, authorization codes, cable pairs, WANs, Fax machines, voice mail systems, automated attendants, interactive voice response systems, automatic call distribution, multiplexors, modem pools, etc. The internal data facilities such as LANs and routers may be under the administration of the TCM, or could be the responsibility of the Management Information Systems (MIS) department. But since the TCM has responsibility for both the inside wiring and the outside Carrier facilities, close coordination would be required if the internal data facilities are controlled by the MIS department.

The following are the functions of the TCM;

- Operating, administering, monitoring, and maintaining the existing telecommunications systems.
- Dealing with the various vendors and providers, including verifying and paying the bills.
- Preparing and managing the Telecommunications budget.
- Keeping abreast of changes in technology, services, industry structure, and rates.
- Assisting company management in developing a corporate telecommunications policy that meets business objectives.
- Developing and implementing company telephone and telecommunications procedures for efficient and cost effective use, and training company employees in these procedures.
- Upgrading, procurement, selecting, contracting, or purchasing a system, new system, equipment, or services.
- Planning and analyzing for growth, new requirements, or future functionality.

The goal of the TCM is to provide good telecommunications services for an organization and its employees at the lowest possible cost. This definition courtesy, Robert J. Perillo, Perillo@dockmaster.ncsc.mil.

TCN Collective Number Group Table.

TCNS Thomas Conrad Networking System is a 100 million bit per second proprietary networking system (LAN) based on ARCnet that can use most standard ARCnet drivers on any network operating systems.

TCO Total Cost of Ownership. A term coined by The Gartner Group to bring attention to the actual, total cost to the enterprise of owning a PC. In 1997 that figure was $29,353 for owning a standard, networked, Windows 95 PC for a period of three years. It's probably a lot higher now. The point is clear and fairly obvious – consider not only the acquisition/implementation cost of a workstation (networked or not), but also consider the total cost, including administration, maintenance, support, software upgrades and training. Of course, one problem with TCO is that it has now way to measure the actual benefits of owning that equipment. A product can also have a low cost of ownership, but not be the best product for the job. IN short, like all statistics, you can reach pretty well whatever conclusion you would like.

TCP 1. Transmission Control Protocol. ARPAnet-developed transport layer protocol corresponding to OSI Layer 4, the Transport Layer. TCP is a transport layer, connection-oriented, end-to-end protocol. It provides reliable, sequenced, and unduplicated delivery of bytes to a remote or local user. TCP provides reliable byte stream communication between pairs of processes in hosts attached to interconnected networks. It is the portion of the TCP/IP protocol suite that governs the exchange of sequential data. UDP (User Datagram Protocol) is a Layer 4 option that does not provide the same level of reliability, but does involve much less overhead. See Internet and TCP/IP for a much longer explanations.

2. An ATM term. Test Coordination Procedure: A set of rules to coordinate the test process between the lower tester and the upper tester. The purpose is to enable the lower tester to control the operation of the upper tester. These procedures may, or may not, be specified in an abstract test suite.

3. Total Call Processor.

TCP/IP According to Microsoft: Transmission Control Protocol/Internet Protocol (TCP/IP) is a networking protocol that provides communication across interconnected networks, between computers with diverse hardware architectures and various operating systems. TCP (Transmission Control Protocol) and IP (Internet Protocol) are only two protocols in the family of Internet protocols. Over time, however, "TCP/IP" has been used in industry to denote the family of common Internet protocols. The Internet protocols are a result of a Defense Advanced Research Projects Agency (DARPA) research project on network interconnection in the late 1970s. It was mandated on all United States defense long-haul networks in 1983 but was not widely accepted until the integration with 4.2 BSD (Berkeley Software Distribution) UNIX. The popularity of TCP/IP (Harry's note: it's the Internet's networking protocol) is based on:

Robust client-server framework. TCP/IP is an excellent client-server application platform, especially in wide-area network (WAN) environments.

Information sharing. Thousands of academic, defense, scientific, and commercial organizations share data, electronic mail and services on the connected Internet using TCP/IP.

General availability. Implementations of TCP/IP are available on nearly every popular computer operating system. Source code is widely available for many implementations. Additionally, bridge, router and network analyzer vendors all offer support for the TCP/IP protocol family within their products.

TCP/IP is the most complete and accepted networking protocol available. Virtually all modern operating systems offer TCP/IP support, and most large networks rely on TCP/IP for all their network traffic. Microsoft TCP/IP provides cross-platform connectivity and a client-server development framework that many software vendors and corporate developers are using to develop distributed and client-server applications in heterogeneous enterprise networks over TCP/IP.

How TCP Works: TCP is a reliable, connection-oriented protocol. Connection-oriented implies that TCP first establishes a connection between the two systems that intend to exchange data. Since most networks are built on shared media (for example, several systems sharing the same cabling), it is necessary to break chunks of data into manageable pieces so that no two communicating computers monopolize the network. These pieces are called packets. When an application sends a message to TCP for transmission, TCP breaks the message into packets, sized appropriately for the network, and sends them over the network.

Because a single message is often broken into many packets, TCP marks these packets with sequence numbers before sending them. The sequence numbers allow the receiving system to properly reassemble the packets into the original message. Being able to reassemble the original message is not enough, the accuracy of the data must also be verified. TCP does this by computing a checksum. A checksum is a simple mathematical computation applied, by the sender, to the data contained in the TCP packet. The recipient then does the same calculation on the received data and compares the result with the checksum that the sender computed. If the results match, the recipient sends an acknowledgment (ACK). If the results do not match, the recipient asks the sender to resend the packet. Finally, TCP uses port IDs to specify which application running on the system is sending or receiving data.

The port ID, checksum, and sequence number are inserted into the TCP packet in a special section called the header. The header is at the beginning of the packet containing this and other "control" information for TCP.

How IP Works: IP is the messenger protocol of TCP/IP. The IP protocol, much simpler than TCP, basically addresses and sends packets. IP relies on three pieces of information, which you provide, to receive and deliver packets successfully: IP address, subnet mask, and default gateway.

The IP address identifies your system on the TCP/IP network. IP addresses are 32-bit addresses that are globally unique on a network. They are generally represented in dotted decimal notation, which separates the four bytes of the address with periods. An IP address looks like this: 102.54.94.97

Although an IP address is a single value, it really contains two pieces of information: (a.) Your system's network ID, and (b.) Your system's host (or system) ID.

The subnet mask, also represented in dotted decimal notation, is used to extract these two values from your IP address. The value of the subnet mask is determined by setting the network ID bits of the IP address to ones and the host ID bits to zeros. The result allows TCP/IP to determine the host and network IDs of the local workstation. Here's how to understand an IP address. For example:

When the IP address is 102.54.94.97 (specified by the user) And the subnet mask is 255.255.0.0 (specified by the user) The network ID is   102.54 (IP address and subnet mask) And the host ID is 94.97 (IP address and subnet mask)

OK. the above was Microsoft's definition. Here's my definition, which covers some areas Microsoft doesn't. TCP/IP is a set of protocols developed by the Department of Defense to link dissimilar computers across many kinds of networks, including unreliable ones and ones connected to dissimilar LANs. TCP/IP is the protocol used on the Internet. It is, in essence, the glue that binds the Internet. Developed in the 1970s by the U.S. Department of Defense's Advanced Research Projects Agency (DARPA) as a military standard protocol, its assurance of multi vendor connectivity has made it popular among commercial users as well, who have adopted TCP/IP. Consequently, TCP/IP now is supported by many manufac-

turers of minicomputers, personal computers, mainframes, technical workstations and data communications equipment. It is also the protocol commonly used over many Ethernet LANs (as well as X.25) networks. It has been implemented on everything from PC LANs to minis and mainframes.

TCP/IP currently divides networking functionality into only four layers:

A Network Interface Layer that corresponds to the OSI Physical and Data Link Layers. This layer manages the exchange of data between a device and the network to which it is attached and routes data between devices on the same network.

An Internet Layer which corresponds to the OSI network layer. The Internet Protocol (IP) subset of the TCP/IP suite runs at this layer. IP provides the addressing needed to allow routers to forward packets across a multiple LAN inter network. In IEEE terms, it provides connectionless datagram service, which means it attempts to deliver every packet, but has no provision for retransmitting lost or damaged packets. IP leaves such error correction, if required, to higher level protocols, such as TCP.

IP addresses are 32 bits in length and have two parts: the Network Identifier (Net ID) and the Host Identifier (Host ID). Assigned by a central authority, the Net ID specifies the address, unique across the Internet, for each network or related group of networks. Assigned by the local network administrator, the Host ID specifies a particular host, station or node within a given network and need only be unique within that network.

A Transport Layer, which corresponds to the OSI Transport Layer. The Transmission Control Protocol (TCP) subset runs at this layer. TCP provides end-to-end connectivity between data source and destination with detection of, and recovery from, lost, duplicated, or corrupted packets – thus offering the error control lacking in lower level IP routing. In TCP, message blocks from applications are divided into smaller segments, each with a sequence number that indicates the order of the segment within the block. The destination device examines the message segments and, when a complete sequence of segments is received, sends an acknowledgement (ACK) to the source, containing the number of the next byte expected at the destination.

An Application Layer, which corresponds to the session, presentation and application layers of the OSI model. This layer manages the function required by the user programs and includes protocols for remote log-in (Telnet), file transfer (FTP), and electronic mail (SMTP).

See also OSI and Public Key Encryption.

TCPA See Telephone Consumer Protection Act of 1991.

TCR 1. Transaction Confirmation Report. A report from a fax machine listing the faxes received and transmitted. It provides details about each fax, including date, time, the remote fax's number, results, total pages.

2. An ATM term. Tagged Cell Rate: An ABR service parameter, TCR limits the rate at which a source may send out-of-rate forward RM-cells. TCR is a constant fixed at 10 cells/second.

TCS 1. Transmission Convergence Sublayer: This is part of the ATM physical layer that defines how cells will be transmitted by the actual physical layer.

2. TCS is a manufacturer of excellent call center manpower scheduling software packages. It is based in Nashville, TN.

TCS-AT A Bluetooth term. A set of AT-commands by which a mobile phone and modem can be controlled in the multiple usage models. In BT, AT-commands are based on ITU-T recommendation v.250 and ETS 300 916(GSM 07.07). In addition, the commands used for fax services are specified by the implementation. TCS-AT will also be used for dial-up networking and headset profiles.

TCS Binary A Bluetooth term. Bluetooth Telephony Control protocol Specification using bit-Oriented protocol. It is also referred to as the TCS-BIN system. TCS-BIN will be used for cordless telephony profiles.

TCU 1. Timing Control Unit.

2. Closed User Group Table.

TCV Address Conversion Table.

TD-SCDMA Time Division – Synchronous Code Division Multiple Access. A unique 3G wireless cell phone standard being developed by China for use inside China. The spectral efficiency of TD-SCDMA radio systems is three to five times higher than that of GSM, according to some Chinese government officials. The other two main 3G standards, the so-called WCDMA and CDMA2000, were developed in Europe and the U.S. Some China cell phone providers are concerned that 3G cell phone technology in China will be unique and not enjoy the continued technical advances of the other two. See 3G.

TDAS See Traffic Data Administration System.

TDC 1. Time Division Controller.

2. Dial Code Table.

TDD 1. Time Division Duplex. A method used in cellular and PCS networks employing TDMA (Time Division Multiple Access) to support full duplex communications. Each radio channel is divided into multiple time slots through TDMA, thereby supporting multiple conversations. TDD supports transmission in the forward direction (from the cell phone to the cell site) through one radio frequency channel and one time slot. Another radio channel and time slot supports transmission in the backward direction (from the cell site to the cell phone). TDD is flexible with regard to allocation of timeslots for forward and reverse, making it best suited for asymmetrical applications, such as web browsing. See also FDD.

2. IPWireless's term for UMTS TD-CDMA. They claim that their TDD implementation allows a datacom wireless service provider to transmit data on a single radio channel alternating between uplink and downlinks as opposed to a traditional Frequency Division Duplex (FDD) approach of first and second generation cellular, which require separate channels for each of the uplink and downlink channels.

2. Telecommunications Device for the Deaf. Under the Communications Act of 1934, a TDD is defined as a machine "that employs graphic communication in the transmission of coded signals through a wire or radio." TDD devices (which typically look like simple computer terminals) use the Baudot method of communications. Most TDD devices are acoustically coupled and are slow, running at 300 baud.

There is a special TDD/TTY Operator Services number. It's 800-855-1155. Users of TDDs often abbreviate commonly used words or expressions to save time. Here are some of the most frequently used:

ANS Answer	R Are
CUD Could	REC Receive
GA Go Ahead	SK Stop Keying
LTR Letter	THRU Through
MSG Message	THX Thank You
MIN Minute	U You
NITE Night	UR Your
PLS Please	WUD Would
QQ Question	XOX Hugs & Kisses

TDDRA Telephone Disclosure and Dispute Resolution Act. A US federal act passed in 1992 which required both the FCC and the FTC to prescribe regulations governing pay-per-call services. Subsequently, the FTC adopted its 900-Number Rule, which became effective November 1, 1993. Under the TDDRA, a consumer's telephone service cannot be disconnected for failure to pay charges for a 900-number call, and 900-number blocking must be made available to consumer who do not wish to have access to 900-number services. See also 900-Number Rule.

TDEL Technical Development and Evaluation Laboratory.

TDF 1. Trunk Distributing Frame.

2. Transborder Data Flows are movements of machine-readable data across international boundaries. TDF legislation began in the 1970s and has been put into effect by many countries in an attempt to protect personal privacy of citizens. This term has particular meaning as it relates to electronic commerce or EDI and is becoming more and more relevant with the use of the Internet as a means to conduct global business.

TDHS Time Domain Harmonic Scaling

TDI Transmit Division Intertie.

TDM See Time Division Multiplex.

TDMA Time Division Multiple Access. One of several technologies used to separate multiple conversation transmissions over a finite frequency allocation of through-the-air bandwidth. As with FDMA (Frequency Division Multiple Access), TDMA is used to allocate a discrete amount of frequency bandwidth to each user, in order to permit many simultaneous conversations. However, each caller is assigned a specific timeslot for transmission. A digital cellular telephone system using TDMA assigns 10 timeslots for each frequency channel, and cellular telephones send bursts, or packets, of information during each timeslot. The packets of information are reassembled by the receiving equipment into the original voice components. TDMA promises to significantly increase the efficiency of cellular telephone systems, allowing a greater number of simultaneous conversations. See also ATDMA, CDMA and FDMA.

TDMA-3 An abbreviated way of describing a TDMA system which can support 3 calls in single radio frequency "carrier." IS-54 TDMA divides a carrier frequency into 6 time slots. Initial "full-rate" vocoders will require 2 time slots per user and thus support 3 users on each on each carrier frequency. This is called TDMA-3. Future "half-rate" vocoders will require only 1 time slot per user and support 6 users per carrier frequency, yielding TDMA-6.

TDMA-6 An abbreviated way of describing a TDMA system which is capable of supporting 6 users per carrier frequency. IS-54 TDMA divides a carrier frequency into 6 time slots

per user and thus support 3 users on each carrier frequency. Future "half-rate" vocoders will require only 1 time slot per user and thus will support 6 users per carrier frequency, yielding TDMA-6.

TDMoIP TDM over IP. The delivery of PSTN services (T1, T3, ATM, Frame Relay, SS7, etc.) over IP. An IP multiplexer (IPmux) multiplexes multiple PSTN traffic streams and tunnels it over IP; the process is reversed on the distant end.   TDMoIP protects the customer's investment in legacy TDM equipment (TDMoIP supports legacy telephone services) while providing the benefits of IP.

TDMS 1. Technical Document Management Systems.

2. Time Division Multiplex System.

3. Transmission Distortion Measuring Set.

TDOA Time Difference Of Arrival. A precise method of locating a radio receiver, TDOA is being proposed to support wireless 911 services for cellular and PCS networks. Operating much like GPS (Global Positioning Satellite Systems), although in reverse, GPOA uses three cell site antennas to lock in on the signal from the cell phone. The times of signal arrival at each cell site are compared through the use of a precise master clock. Although the differences in time of signal arrival may be only microseconds, the location of the cell phone can be determined through a process of time triangulation, allowing the exact location of the device to be plotted. GPOA offers much improved location-determination than does the old method of triangulation, which relies on signal strength. See angle of arrival, 911 and Triangulation.

TDP Telocator Data Protocol. A new 8-bit protocol for sending messages and binary files (images, spreadsheets, word processing files, executables, etc.) to pagers (also known as beepers). The older (and more common) 7-bit messaging protocol now widely in use is called TAP, which stands for Telocator Alphanumeric Paging Protocol. This protocol can only send simple alphanumeric messages, like "Your shares in XYZ are now $23, up 98%."

TDR Time Domain Reflectometer.

TDRSS In 1989, NASA began a program called Tracking and Data Relay System Satellites (TDRSS), a network of geosynchronous satellites originally designed to communicate with naval vessels and submarines. Today, one of its functions is to use and produce Land Sat mapping data. Using TDRSS, seismic vessels are able to upload information acquired from TDRSS satellites at speeds in excess of 311 megabits per second. The raw data is transferred to an array of space antennas at White Sands Air Force Base in Nevada and then relayed via fiber optics to Houston for processing. Once complete, the entire upload can be put on the Internet and reviewed by anybody who has the code to view it. An entire day's seismic shoot can be shuffled to space and back within a couple of hours.

TDS 1. MCI's name for Terrestrial Data Services, i.e. services that run through on-the-ground fiber, rather than through-the-air satellite services.

2. Terrestrial Digital Service.

3. Transmission Data Service.

TDSAI Transit Delay Selection And Indication.

1. ISDN Terminal Equipment. See the next two definitions.

2. Terminal Equipment: As an ATM term, terminal equipment represents the endpoint of ATM connection(s) and termination of the various protocols within the connection(s).

TDU Tape Drive Unit.

TE 1. Terminating Equipment.

2. Telecom Enclosure. See FTTE.

TE1 Terminal equipment type 1 that supports ISDN standards and thus can connect directly to the ISDN network. TE1 could be an ISDN telephone, a personal computer capable of working with ISDN, a videophone, etc. In short, any device that can attach to and work with ISDN.

TE2 Terminal equipment that does not support ISDN standards and thus requires a Terminal Adapter. Non-ISDN terminal equipment (e.g. analog telephone) linked at the RS-232, RS-449, or V.35 interfaces.

teamware According to Internet Week, May 17, 1999, teamware is a class of web collaboration tools for sharing information over extranets while sparing IT managers much of the administrative work associated with traditional groupware. Teamware applications lack some customization features but are less complex to deploy because they normally let users perform the majority of management functions and also feature a browser interface. Teamware does not provide connectivity to back-end databases and does not not scale to handle large volumes of data like Lotus Notes, Microsoft Exchange or Novell GroupWise. However, teamware is intuitive and can be quickly implemented.

tear strength The force required to initiate or continue a tear in a material under specified conditions. or reheated and becomes firm on cooling.

teardown analysis An analysis of equipment, such as a wireless handset or laptop computer, that is performed by taking apart (tearing down) the equipment into its electronic, electro-mechanical, and mechanical components. A teardown analysis is done to study the design and manufacturing of a competitor's equipment, and to get an idea of the cost to build it. Teardown analysis can progress into reverse engineering, namely figuring how to copy the chosen piece.

teardrop attack A teardrop attack occurs when a malicious person sends instructions over a network to a server attempting to crash the server. Some implementations of the TCP/IP IP fragmentation re-assembly code do not properly handle overlapping IP fragments. Teardrop is a widely available attack tool that exploits this vulnerability. There are workarounds that will prevent your server from being attacked.

teased mexican scorpion When AOL Time Warner announced a mammoth $98.7 billion writoff in late January, 2003, an analyst on CNBC reported that Ted Turner was mad about the losses. He was, according to the analyst, as angry as "a teased Mexican scorpion."

TECK90 CSA designation for a single or multiconductor cable with interlocked armor and an inner and outer PVC jacket. Cable has 900xAFC temperature rating.

TEC Abbreviation for thermoelectric cooler.

TEC NIS Telecommunications and Electronics Consortium in the Newly Independent States. An organization based in Moscow and administered by TIA (Telecommunications Industry Association) to assist US telecommunications and telecommunications-related electronics companies with doing business in the region.

TECF Traffic Editor Control File.

Tech Wreck Technology stocks reached a peak in the Spring of 2000. Then they began a calamitous drop until the fall of 2002, when they (or a least the survivors) finally turned around and edged upwards. As I write this in the late Fall of 2005, some leaders have still not recovered to their peaks of the year 2000, e.g. Cisco, Microsoft and Intel. Some five trillion dollars in shareholder wealth was wiped out during the Tech Wreck of 2000-2002. That wealth was later more than recovered by many investors through the increased value of residential and commercial real estate owned by homeowners and investors. The appreciation in real estate was caused by the substantial drop in interest rates which Federal Reserve Bank under the leadership of Alan Greenspan engineered in order to prevent the economy from falling into recession.

techinfo A common campuswide information system developed at MIT. An Internet term.

Technical Advisory TA. A Bellcore document containing a preliminary view of proposed generic requirements for a technology, equipment, service or interface. The TA document type is being replaced by the Generic Requirements (GR) document type. See Generic Requirements.

technical control center A testing center for telecommunications circuits. The center provides test access and computer-assisted support functions to aid in circuit maintenance.

technical control facility A federal government term . A term plant, or a designated and specially configured part thereof, containing the equipment necessary for ensuring fast, reliable, and secure exchange of information. This facility typically includes distribution frames and associated panels, jacks, and switches; and monitoring, test, conditioning, and order wire equipment.

technical load A military term. The portion of the operational load required for communications, tactical operations, and ancillary equipment including necessary lighting, air conditioning, or ventilation required for full continuity of communications.

technical means A term in the spy business which means spy satellites and electronic eavesdropping stations, typically costing tens of billions of dollars.

Technical Office Protocol TOP. A seven-layer network architecture designed for office automation that uses International Standards Organization (ISO) or ITU specifications at each level. TOP was defined by Boeing Vertol Corp. and is now controlled by the MAP/TOP (Manufacturing Automation Protocol/Technical Office Protocol) Users Group.

Technical Reference TR. A Bellcore document containing the current view or performance of a technology, equipment, service or interface. The TR document type are being replaced by the Generic Requirements (GR) document type. See Generic Requirements.

technical tap When you physically hit a piece of electronic equipment in a desperate attempt to fix it, and some how manage to make it work.

technically feasible points Points at which it is technically or operationally feasible to interconnect with the Verizon network without threatening the reliability or security of the existing network.

technicolour yawn What Americans refer to as throwing up, Australians say, more politely, "to make a technicolour yawn." Some American campuses do have an expression

"to raise the technicolour flag." In dire cases (i.e. where you are very drunk), this is also know as "hugging the porcelain god." "In extremely dire cases, Chuck McDonald suggests calling Ralph On "The Porcelain Phone. "

techno-geek A geek. See Geek.

technology investor Another term for masochist, according to some of my friends. The name of a magazine I once started, ran for 11 issues and then closed it down. The magazine failed.

TED 1. Trunk Encryption Device.
 2. See Ted Maloney.

teddy bear A stuffed animal named after President "Teddy" Roosevelt, a keen hunter who once took pity on a baby bear and didn't shoot the baby. At the height of the teddy bear's huge popularity in the early 1900s, a Michigan priest publicly denounced the teddy as "an insidious weapon." He claimed that the stuffed toy would lead to "the destruction of the instincts of motherhood and eventual racial suicide."

Ted Maloney Ted has been part of our greater family after he and my daughter, Claire, fell in love during their college days in Maine in the early 2000s. And this year, in September 2007, he will marry Claire. And I will cry all the way down the aisle. I can't imagine anyone more caring or more capable that I'd rather trust my precious Claire to. I love spending time with Ted. He's bright, interesting, handsome, loveable and often laughs at my jokes – a sure way to win my heart. One day he and I (and maybe Michael, my son) will do something great in business or charity together. I'm super lucky and very happy Claire found Ted, and vice versa.

TEDIS Trade Electronic Data Interchange Systems.

Teen Buzz A ringtone that operates at a frequency (17 kHz and higher) that teenagers can hear but which most older people can't hear. According to The Merck Manual of Diagnosis and Therapy, this condition begins after the age of 20, first affecting the highest hearing frequencies (18 to 20 kilohertz). It is most common in persons over 65.

teen service A feature of some central offices which allows two telephone numbers to be assigned to a single party phone line. Each number has a distinctive ringing pattern so that the called parties can recognize which line is ringing. The inventor of this service named it after the fact that his teenage children were always receiving phone calls. And he wanted a way for them to recognize when the calls were for them and when they were for the parents. Sadly, this phenomenon now begins earlier in life, with children as young as six receiving their own calls. We speak from experience. Teen service is now used by home businesses, roommates, boarders, college dorm suite-mates, and live-in relatives.

tee coupler A three-port optical coupler.

TEF Telecommunications Entrance Facility (also called EF or Entrance Facility).

Teflon duPont's registered trademark for fluorinated ethylene propylene (FEP). In addition to working its wonders in the modern kitchen, Teflon is an exceptional insulating material for cable systems. Teflon is also coated on cables. See also FEP.

Tefzel A duPont trade name for a fluorocarbon material typically used as the insulation on wire wrap wire.

TEHO See Tail End Hop-Off (traffic engineering).

TEI Terminal Endpoint Identifier. Up to eight devices can be connected to one ISDN BRI line. The TEI defines for a given message which of the eight devices is communicating with the Central Office switch. In general, more than one of the eight may be communicating.

TEK Traffic Encryption Key.

Tekelec Adaptation Layer Interface See TALI.

tel 1. Telephone.
 2. .tel is a top-level domain, approved by ICANN. Tel enables a Web address to serve as an Internet phone number. For example, clicking on www.HarryNewton.tel would initiate an Internet phone call to Harry Newton.

telabuse A term coined by John Haugh of Telecommunications Advisors in Portland, OR to include "insider" toll fraud, waste and abuse.

telbanking Banking transactions conducted through telecommunications.

Telco 1. The local telephone company. Often a term of endearment. Americanism for local telephone company.
 2. In some LAN circles, a telco is known as a 25-pair polarized connector that is used to consolidate multiple voice or data lines. Also known as an amphenol connector.

Telco Farm A building housing many phone companies – typically one ILEC and many CLECs.

telco hotel Also known as a Carrier Hotel or Neutral Central Office (NCO), a telco hotel is a neutral location in which carriers can lease space for termination of circuits and placement of equipment. See Carrier Hotel for a full explanation.

telco splice block What some parts of the data communications industry call a

66-block, i.e. a terminating block for twisted pair voice and data cable.

telco think A term that is used to describe telephone companies' strange and wonderful way of thinking about the telephone business, including and especially regulatory stuff, and anything else that affects their business. I could write a whole book about telco think. But you'd be bored. Suffice they don't think like other businesses.

Telcoland A somewhat derogatory term that refers to LECs and IXCs, and their world view and way of doing business.

Telcordia Technologies Telcordia Technologies is the new name for Bellcore. Here's how Telcordia describes itself and its history: Bell Communications Research, or Bellcore, was created during the divestiture of the Bell System in 1984 to serve the Bell operating companies by providing a center for technological expertise and innovation. Eighty percent of the U.S. telecommunications network depends on software invented, developed, implemented, or maintained by Bellcore. We hold hundreds of patents, including key patents for broadband data communications technologies like ADSL, AIN, ATM, ISDN, Frame Relay, SMDS, SONET, and video-on-demand. We currently keep more than 100 million lines of code maintained and running, through more than 150 operations support systems. We are the world's largest provider of telecommunications training services; each year we train more than 30,000 students from 1,300 companies. In 1997, Bellcore was acquired by Science Applications International Corporation (SAIC), one of the world's largest providers of systems integration and program management. www.telcordia.com.

tele- 1. Tele comes from the Greek word meaning "far." Telecommunications is therefore communicating over a distance. Television is seeing over a distance.

2. For any word that you think should be spelled tele-word, please check the definition below, spelled without the dash.

teleaction zervice In ISDN applications, a telecommunications service using very short messages with very low data transmission rates between the user and the network.

teleadmin A means in GSM (non-North America cellular digital standard) to update a Subscriber Identity Module (SIM) card via a short message sent by the network operator using Short Message Services (SMS). In GSM, the SIM card is located in the cell phone's handset and is customized to a specific subscriber's service options. TeleAdmin is also known as remote SIM card updating.

telebusiness A British term for telemarketing. Here's a definition I found in England: Activities conducted by telephone in a planned and controlled manner. The term encompasses telesales, telemarketing, customer service and information broadcast. Telebusiness can be conducted between an organization and its customers and prospects, or conducted as in internal service.

telecenters Business centers where many companies rent space for their employees. Centers are equipped with receptionist, clerical help, e-mail, and voice mail.

telecom A shortened and perfectly acceptable way of saying the word "telecommunications." See Telecommunications.

telecom bus A byte-serial TDM (Time Division Multiplexed) bus technology, which has become something of a de facto standard.

telecom hotel See Carrier Hotel.

telecom operations map TOM, or eTOM (for enhanced Telecom Operations Map). This definition is from www.TMForum.org. Frankly I don't get it, but a kindly reader of this dictionary said that I should include a definition. So here goes: "enhanced Telecom Operations Map (eTOM): The Business Process Framework-for the Information and Communications Services Industry - GB921 v3.0 Saturday, June 01, 2002. The enhanced Telecom Operations Map (eTOM) Business Process Framework serves as the blueprint for process direction and the starting point for development and integration of Business and Operations Support Systems (BSS and OSS respectively). The eTOM Framework also helps drive TM Forum member's work to develop NGOSS solutions. For service providers, it provides a neutral reference point as they consider internal process reengineering needs, partnerships, alliances, and general working agreements with other providers. For suppliers, the eTOM Framework outlines potential boundaries of software components, and the required functions, inputs, and outputs that must be supported by products. This document consists of:

A description of the role of the eTOM Business Process Framework.

• An ebusiness context for service providers and the more complex Business Relationship Context Model required.

• A high-level business process framework and explanation of service provider enterprise processes and sub-processes that are top down, customer-centric, and end-to-end focused.

• A total enterprise framework for service providers.

• Process Decompositions of all processes from the highest conceptual view of the framework to the working level of the eTOM and many selected lower level decompositions in the Framework

• Selected process flows and descriptions of the decomposed processes that include the process purpose or description, business rules, high level information and more.

telecom server A telecom server is Intel's word for a computer that directs telephone calls and computer traffic across networks. Such servers allegedly meet rigorous tests for telecom equipment, including the ability to survive extreme heat and cold, earthquakes and fires. I think the more common word for a telecom server is a switch.

telecom trinity A "triple-play" bundle of voice, data, and video.

telecommerce A British way of saying telemarketing. See Telemarketing.

telecommunication architecture The governing plan showing the capabilities of functional elements and their interaction, including configuration, integration, standardization, life-cycle management, and definition of protocol specifications, among these elements.

telecommunication facilities The aggregate of equipment, such as telephones, teletypewriters, facsimile equipment, cables, and switches, used for various modes of transmission, such as digital data, audio signals, image and video signals.

telecommunication service Any service provided by a telecommunication provider. A specified set of user-information transfer capabilities provided to a group of users. The telecommunication service provider has the responsibility for the acceptance, transmission, and delivery of the message.

telecommunicationally challenged A politically correct term for being under-phoned, i.e. having too few phones. This definition contributed by John Warrington of Ashland University, Ashland, OH.

telecommunications 1. The art and science of "communicating" over a distance by telephone, telegraph and/or radio. The transmission, reception and the switching of signals, such as electrical or optical, by wire, fiber, or electromagnetic (i.e. through-the-air) means.

2. A fancy word for "telephony," which it replaced and which many thought meant only analog voice, but didn't.

Telecommunications Act of 1981, U.K. The Telecommunications Act of the U.K. is passed. It is the first step towards liberalizing the telecommunications market in the U.K. and has four main consequences:

• The General Post Office (the erstwhile monopoly provider of telecommunications services in the U.K.) was divided into two separate entities: The Post Office and British Telecommunications (BT), which retained the monopoly over existing telecommunications networks.

• It determined that a duopoly would be created as a first step towards the introduction of competition in telecommunications.

• The Secretary of State for Trade and Industry was empowered to license other organizations to be known as Public Telecommunications Operators (PTOs), to operate public telecommunications networks (including cellular networks) in the U.K.

• It paved the way for the gradual deregulation of equipment supply, installation and maintenance which had previously been the monopoly of the GPO.

Following the Act, Mercury Communications, majority-owned by Cable & Wireless was created to compete with British Telecommunications.

Telecommunications Act of 1984, U.K. The 1984 Telecommunications Act established British Telecommunications, now known as BT, as a public limited company which would, as such, have to apply for a PTO licence from the Secretary of State. Following the Act, BT was privatized. The Act also created Oftel, the office of telecommunications, to become a watchdog over all aspects of the telecommunications industry in the U.K. See also Telecommunications Act of 1981. www.oftel.gov.uk.

Telecommunications Act of 1996, U.S. A federal bill signed into law on February 8, 1996 "to promote competition and reduce regulation in order to secure lower prices and higher quality services for American telecommunications consumers and encourage rapid deployment of new telecommunications technologies." The Act is widely reputed to be among the worst pieces of legislation ever passed by Congress. The Act required local service providers in the 100 largest metropolitan areas of the United States, i.e. the local regional Bell operating companies, to implement Local Number Portability by the end of 1998. The Act also allowed the local regional Bell operating phone companies into long distance once they had met certain conditions about allowing competition in their local monopoly areas. The main thrust of the bill was to force the local phone companies to rent their local copper loops to new telecommunications carriers, later called CLECs (Competitive Local Exchange Carriers). The first and main service provided by the CLECs was a service called DSL (Digital Subscriber Line). Subscribers could use these line to access the Internet at 50 times faster than a traditional dial-up phone line. President Bill Clinton signed the

Telecommunications Act of 1996 into law using the very pen President Dwight D. Eisenhower used in 1957 to authorize the interstate highways. "We will help to create an open marketplace where competition and innovation can move quick as light," Clinton said. The Act lead to an explosion in the number of new phone companies – especially CLECs – and to a lesser extent, combination CLEC/long distance carriers. Wall Street raised billions of dollars for these new companies. Observers, journalist and Wall Street analysts fell in love with the new opportunities. The words "the demand is insatiable" was seen in print again and again. And the shares of many of these new companies took off for the stars. By the middle of 2001, the whole telecom boom had pretty well bust. Most of the CLECs were in Chapter 11 bankruptcy. Many DSL lines had been disconnected. And billions of dollars of telecom debt was now valueless. History will probably ascribe the collapse of the CLEC boom to the local phone companies, who pretty well did everything they could to deny their new CLEC competitors timely access to the lines the Act had made them, theoretically, entitled to. You can read the entire Act on www.fcc.gov/Reports/tcom1996.pdf. See also CLEC. Some important sections of this law are:

Section 251 Sets interconnection requirements for ILECs and CLECs.

Section 254 Defines and sets the principles of and procedures to review Universal Service.

Section 271 Sets the requirements and process for Bell long distance re-entry. Details the 14-point checklist. See 271.

Section 706 Requires the FCC and state regulatory agencies to promote the deployment of advanced telecommunications services throughout the United States.

The following are some definitions contained in the Act.

Affiliate – The term 'affiliate' means a person that (directly or indirectly) owns or controls, is owned or controlled by, or is under common ownership or control with, another person. For purposes of this paragraph, the term 'own' means to own an equity interest (or the equivalent thereof) if more than 10 percent.

AT&T Consent Decree – The term 'AT&T Consent Decree' means the order entered August 24, 1982, in the antitrust action styled United States v. Western Electric, Civil Action No. 82-0192, in the United States District Court for the District of Columbia, and includes any judgment or order with respect to such action entered on or after August 24, 1982.

Bell Operating Company – The term 'Bell operating company'

(A) means any of the following companies: Bell Telephone Company of Nevada, Illinois Bell Telephone Company, Indiana Bell Telephone Company, Incorporated, Michigan Bell Telephone Company, New England Telephone and Telegraph Company, New Jersey Bell Telephone Company, New York Telephone Company, U S West Communications Company, South Central Bell Telephone Company, Southern Bell Telephone and Telegraph Company, Southwestern Bell Telephone Company, The Bell Telephone Company of Pennsylvania, The Chesapeake and Potomac Telephone Company, The Chesapeake and Potomac Telephone Company of Maryland, The Chesapeake and Potomac Telephone Company of Virginia, The Chesapeake and Potomac Telephone Company of West Virginia, The Diamond State Telephone Company, The Ohio Bell Telephone Company, The Pacific Telephone and Telegraph Company, or Wisconsin Telephone Company; and

(B) includes any successor or assign of any such company that provides wireline telephone exchange service; but

(C) does not include an affiliate of any such company, other than an affiliate described in subparagraph (A) or (B).

Customer Premises Equipment: The term 'customer premises equipment' means equipment employed on the premises of a person (other than a carrier) to originate, route, or terminate telecommunications.

Dialing Parity: The term 'dialing parity' means that a person that is not an affiliate of a local exchange carrier is able to provide telecommunications services in such a manner that customers have the ability to route automatically, without the use of any access code, their telecommunications to the telecommunications services provider of the customer's designation from among 2 or more telecommunications services providers (including such local exchange carrier).

Exchange Access: The term 'exchange access' means the offering of access to telephone exchange services or facilities for the purpose of the origination or termination of telephone toll services.

Information Service: The term 'information service' means the offering of a capability for generating, acquiring, storing, transforming, processing, retrieving, utilizing, or making available information via telecommunications, and includes electronic publishing, but does not include any use of any such capability for the management, control, or operation of a telecommunications system or the management of a telecommunications service.

Interlata service: The term 'interLATA service' means telecommunications between a point located in a local access and transport area and a point located outside such area.

Local access and transport area: The term 'local access and transport area' or 'LATA' means a contiguous geographic area– (A) established before the date of enactment of the Telecommunications Act of 1996 by a Bell operating company such that no exchange area includes points within more than 1 metropolitan statistical area, consolidated metropolitan statistical area, or State, except as expressly permitted under the AT&T Consent Decree; or

(B) established or modified by a Bell operating company after such date of enactment and approved by the Commission.

Local exchange carrier: The term 'local exchange carrier' means any person that is engaged in the provision of telephone exchange service or exchange access. Such term does not include a person insofar as such person is engaged in the provision of a commercial mobile service under section 332(c), except to the extent that the Commission finds that such service should be included in the definition of such term.

Network element: The term 'network element' means a facility or equipment used in the provision of a telecommunications service. Such term also includes features, functions, and capabilities that are provided by means of such facility or equipment, including subscriber numbers, databases, signaling systems, and information sufficient for billing and collection or used in the transmission, routing, or other provision of a telecommunications service.

Number portability: The term 'number portability' means the ability of users of telecommunications services to retain, at the same location, existing telecommunications numbers without impairment of quality, reliability, or convenience when switching from one telecommunications carrier to another.

Rural telephone company: The term 'rural telephone company' means a local exchange carrier operating entity to the extent that such entity– (A) provides common carrier service to any local exchange carrier study area that does not include either–

(i) any incorporated place of 10,000 inhabitants or more, or any part thereof, based on the most recently available population statistics of the Bureau of the Census; or (ii) any territory, incorporated or unincorporated, included in an urbanized area, as defined by the Bureau of the Census as of August 10, 1993;

(B) provides telephone exchange service, including exchange access, to fewer than 50,000 access lines;

(C) provides telephone exchange service to any local exchange carrier study area with fewer than 100,000 access lines; or

(D) has less than 15 percent of its access lines in communities of more than 50,000 on the date of enactment of the Telecommunications Act of 1996.

Telecommunications: The term 'telecommunications' means the transmission, between or among points specified by the user, of information of the user's choosing, without change in the form or content of the information as sent and received.

Telecommunications carrier: The term 'telecommunications carrier' means any provider of telecommunications services, except that such term does not include aggregators of telecommunications services (as defined in section 226). A telecommunications carrier shall be treated as a common carrier under this Act only to the extent that it is engaged in providing telecommunications services, except that the Commission shall determine whether the provision of fixed and mobile satellite service shall be treated as common carriage.

Telecommunications equipment: The term 'telecommunications equipment' means equipment, other than customer premises equipment, used by a carrier to provide telecommunications services, and includes software integral to such equipment (including upgrades).

Telecommunications service: The term 'telecommunications service' means the offering of telecommunications for a fee directly to the public, or to such classes of users as to be effectively available directly to the public, regardless of the facilities used.

See also CLEC for an assessment of the impact of the Act.

telecommunications bonding backbone A conductor that interconnects the telecommunications main grounding busbar (TMGB) to the telecommunications grounding busbar (TGB).

telecommunications broker A person or an organization which buys telecommunications services at bulk rates and resells these services at below "normal" i.e. retail prices.

telecommunications carrier A broad term for any company which provides telecommunications transmission services. That company might be any company from a huge, established local phone company, such as Bell Atlantic or the tiniest CLEC with three subscribers. The term is about as broad as it gets. The Telecommunications Act of 1996 defines it as any provider of telecommunications services, except that such term does not include aggregators of telecommunications services (as defined in section 226). A telecommunications carrier shall be treated as a common carrier under this Act only to the extent that it is engaged in providing telecommunications services, except that the Commission shall determine whether the provision of fixed and mobile satellite service shall be treated as common carriage.

Telecommunications Closet TC. A closet which houses telecommunications wiring and telecom wiring equipment. Contains the BHC (Backbone to Horizontal Cross-connect). May also contain the Network Demarcation, or MC (Main Cross-connect). The telecommunications closet is used to connect up telecom wiring. The closet typically has a door. It's a good idea to lock the door and not put anything else in the closet, like mops, buckets and brooms.

Telecommunications Information Networking Architecture Consortium See TINA-C.

telecommunications lines Telephone and other communications lines used to transmit messages from one location to another.

telecommunications network The public switched telephone exchange network.

Telecommunications Relay Service The FCC defines Telecommunications Relay Services (TRS) as "Telephone transmission services that provide the ability for an individual who has a hearing or speech disability to engage in communication by wire or radio with a hearing individual in a manner that is functionally equivalent to the ability of an individual who does not have a hearing or speech disability to communicate using voice communication services by wire or radio." Such term includes services that enable two-way communication between an individual who uses a text telephone or other nonvoice terminal device and an individual who does not use such a device, speech-to-speech services, video relay services and non-English relay services." Under Title IV of the Americans with Disabilities Act, all telephone companies must provide free relay services either directly or through state programs throughout the 50 states, the District of Columbia, Puerto Rico, and all of the U.S. territories.

Telecommunications Resellers Association TRA. See ASCENT.

telecommunications supplier This is the term used in Europe for a company that sells, installs and supports business telephone systems. This type of company may be known as an Interconnect Company in North America.

telecommuting The process of commuting to the office through transferring information over a communications link, rather than transferring one's physical presence. In short, working at home on a telephone, a computer, a modem and maybe a facsimile machine, rather than going into the office. As the story goes, the concept of telecommuting was invented and the term was coined in 1973 by Jack Nilles, a spacecraft designer for The Aerospace Corp. Nilles was intrigued by the questions posed by an urban planner who wondered why we could not put a man on the surface of the moon but couldn't solve the problems of vehicular traffic congestion on the surface of the Earth. Eventually, Nilles left his job to become director of interdisciplinary programs at the University of Southern California, where he studied telecommuting for the next 10 years or so. There are clear benefits to telecommuting: you can live and work somewhere charming. There are disadvantages, especially accentuated if you work with others: "When you're getting data from afar, you're not in touch with the soul of the business, anymore," according to one telecommuter interviewed by the New York Times. He went on to say, "All the electronic communications are simply backup, I just hadn't factored the importance of personal loyalty and contact into the equation. And I was very wrong." Ray Horak, my Contributing Editor, telecommutes. He works on this dictionary from his SOHO in Seattle, WA, and sends me new definitions and edits old ones electronically over the Internet. The key to this successful relationship is severalfold: What's we're both doing – writing a dictionary – is a very defined, very structured and quite simple task. We both know what our goals are – to make the dictionary the best in the world. And we know how to do it. We're not debating the design of a new automobile or selling customers. These sorts of activities are far more people-oriented and are less conducive to telecommuting. In short, the nature of the task determines its success for telecommuting.

telecomputer Telecomputer appears to be the couch potato's ultimate toy. Peter Coy writing in Business Week of November 1, 1993 called telecomputer a "computerized television." He said "the idea is that you can watch anything in the (on-line) video library anytime. Your telecomputer lets you scroll through a menu of programs, click on your choice, and send an order up the line." James H. Clark, chairman of Silicon Graphics Inc., a manufacturer of computers with heavy video skills calls telecomputer a term for a combination computer/CATV controller that is being popularized by the new media industry. The idea is to use the telecomputer to do interactive games, choose a movie to play out of thousands of choices, buy things, send electronic mail, etc.

telecoms British usage. A shortened and perfectly acceptable way of saying the word "telecommunications."

Télécoms Sans Frontières Telecoms Without Borders. An international telecom and networking relief organization, modeled after Medécins Sans Frontières (Doctors Without Borders). Founded in 2005, Télécoms Sans Frontières (TSF) provides essential communications services for victims and aid workers in areas hit by a natural disaster or war. Headquartered in France, with branch offices in Asia and Latin America, TSF is staffed mostly by European telecom and networking experts. Funding is provided by the United Nations Foundation, the Vodafone Group Foundation, Inmarsat, and charitable donations from organizations and individuals. While the United Nations and other major rescue agencies have their own telecom and networking gear and staff, TSF plays an important role in the first few weeks of an emergency, before the UN and mainstream rescue groups arrive on the scene. TSF is now expanding its mission beyond emergencies, often staying at or returning to trouble spots to build more permanent telecom and networking facilities. http://www.tsfi.org/html_e/index_gb.php.

teleconference A telephone conversation with three or more people. They may be distant from each other. They may all be in the same office.

teleconferencing A term for a conference of more than two people linked by telecommunications through a conference bridge. The term is applied to voice conferencing, which also is known as audioconferencing and which can include other forms of audio, such as music. Teleconferencing, in the broader sense, also includes videoconferencing and document (data) conferencing. For years, teleconferencing has been heralded as a great coming event, and a significant replacement for travel. As corporations increasingly downsize, decentralize, and encourage telecommuting, they will continue to expect more productivity from fewer people who are geographically dispersed. Teleconferencing, clearly, is one powerful solution to this dilemma. See NetMeeting, IP Telephony, The Internet, TAPI 3.0.

Teleconnect magazine Teleconnect was a monthly magazine covering developments in telecommunications equipment in business's offices and factories. Teleconnect's goal was to help its readers choose, install and maintain their telecom equipment. I founded Teleconnect. It was my first magazine. I sold in 1997. Sadly, the new owners closed it down. I don't quite know why. I tried to buy it back but they didn't want to sell it to me. It was my first magazine baby. I'm very sad about it being no more.

telecopier A fancy word for a facsimile or fax machine.

telecosm The domains of technology unleashed by the discovery of the electromagnetic spectrum and the photon. Fiber optics, cellular telephony, and satellite communications are examples. George Gilder made this term up. It's the name of one of his books. This is his definition. www.GilderTech.com.

telecrats High-ranking telephone company executives who speak more like government bureaucrats than businesspeople. There are many of them.

teledensity A measure of the number of phone lines per 100 of population. Originally teledensity meant landlines. And between 40 and 50 lines per 100 of population indicated pretty good density. Under 10 indicated pretty bad density. Now you have to factor in landlines and mobile lines. Teledensity is a measure of a country's economic development. Over 70 means your country is pretty advanced. Some towns, like Washington, D.C., are over 100. That means there is more than one phone for every person. Whether this means Washington, D.C. is more advanced than other places is an interesting question.

Teledesic Teledesic LLC is a private company that tried to build a global, broadband "Internet-in-the-Sky." Teledesic's idea was 288 LEOs (Low-Earth-Orbiting satellites) plus spares, operating in the Ka-band of radio spectrum (28.6-29.1 GHz uplink and 18.8-19.3 GHz downlink). The system was proposed to support millions of simultaneous users, with each having asymmetric, two-way connectivity at rates of up to 2 Mbps on the uplink and up to 64 Mbps on the downlink. The user equipment will be in the form of small (laptop-size) antennas which will mount flat on a rooftop. Teledesic was founded in June of 1990 and is headquartered in Kirkland, Washington, near Seattle. The Teledesic vision was created by Craig O. McCaw, founder of McCaw Cellular Communications (now AT&T Wireless), and William H. Gates III, co-founder of Microsoft. Principal investors include McCaw, Gates, Saudi Prince Alwaleed Bin Talal ($200 million), and Boeing, which is providing the launch vehicles. Motorola and Matra Marconi Space round out the founding industrial team. Latest estimate of the required total investment-a mind-boggling $9 billion. On September 30, 2002 Teledesic announced that it had suspended work under its satellite construction contract with Italian satellite manufacturer Alenia Spazio SpA and will significantly reduce its staff as it evaluates possible alternative approaches to its business. "Teledesic has dedicated and talented employees passionate about the Teledesic vision, leading industrial partners, and some of the most astute private investors from around the world," said Teledesic Chairman and Co-CEO Craig McCaw. "We have met our regulatory milestones to date and remain financially solvent. Our decision to suspend our activities results from an unprecedented confluence of events in the telecommunications industry and financial markets. We do not presently see elements in place that would result in returns to our shareholders that are commensurate with the risk. We continue to believe that the Teledesic service would ultimately provide unique and measurable benefits to the world, and we are looking at scenarios to preserve the ability for that service to be realized." And that's the last press release on the company's web site – www.teledesic.com – as of this writing, December, 2003.

telefax 1. European term for fax.

2. A high-speed, 64 kilobit per second facsimile service that uses Group 4 fax machines and one Bearer channel of an ISDN circuit, or any other 64 Kbps circuit. Group 4 fax machines take about six seconds to transmit a page. They're fast and impressive.

telefelony Another made-up word from the people who are trying to sell you consulting services. This from Jennifer Poulsen, Consultant, High Road Communications, jpoulsen@highrd.com: "Telco fraud is a big problem, and getting bigger. In 1997, phone companies across North America lost more than $12 billion in long distance to tele-felony, a term that accurately describes one of the biggest issues facing the telephone world. The perpetrators for this crime have been dubbed tele-felons. Tele-felons hack their way into company phone systems and make lengthy and expensive long distance calls all over the world. They gain access to corporate calling card numbers for the same purpose. They even let friends make such calls from a number where they work. And the damage doesn't stop at the phone system. Once the PBX is hacked, it's merely a conduit to the computer system and a gold mine of valuable data. Here are some facts:

Fraudsters love targeting new competitive carriers first as they know (or hope) the infrastructure is not in place. An average hit by an organized fraud group costs telcos $350,000 per occurrence (two hits by an organized fraud group could wipe out a new telco's entire yearly profit). A telco's image as a quality service provider is tarnished without fraud protection. Fraud is a cause of customer churn and retaining customers is crucial to the long-term viability of CLECs."

telegaming Using communications lines – from dial-up through local area networks through WANs – to play games interactively with someone at the other end. The upcoming expected explosion of telegaming is what's driving the growth of DSVD (Digital Simultaneous Voice Data) modems and ultimately the growth of ISDN lines.

telegram Hard-copy information, in written, printed or pictorial form, routed to the general telegraph service for transmission and delivery to the addressee. Telegrams are dying due to the high cost of delivery.

telegraph A system employing the interruption of, or change in, the polarity of DC current signaling to convey coded information.

Telegraph Hill A hill in San Francisco where the famous landmark, Coit Tower, is currently located. The hill has a commanding 360-degree view of the city, San Francisco Bay, and more distant points. Telegraph Hill gets its name from the optical telegraph signal station that was built there in 1849 to signal information about arriving ships to local authorities, merchants and townspeople. The arrival of the electric telegraph in San Francisco in 1853 eventually put the signal station on Telegraph Hill out of service. That gave it a life of four years. I mention this because many people living in the 21st century believe that technology has speeded up and obsolescence happens faster. But here's a case of short obsolescence in the 19th century. Another case is the Pony Express. It started in 1860 and ended only 18 months later with the following completion of the transcontinental telegraph line. See optical telegraph.

telegraph key A type of switch for making and breaking a circuit at will for the purpose of transmitting dots and dashes.

telegraph weed A perennial that grows about 6 feet high. Its tall, erect single stem divides into branches at the top, giving it a resemblance, to those who were around in the days of the telegraph, to a telegraph pole. Also known as a telegraph plant to those who don't regard it as a noxious weed, and as a camphorweed to those who do. The telegraph weed's scientific name is Heterotheca grandiflora. It's a member of the sunflower family.

telegraphese An abbreviated style of writing used in telegraphy to minimize message length. It enabled telegraphers to send messages more quickly and conserve network bandwidth. It also saved money for customers who sent telegraph messages, since the messages were paid for on a per-character or per-word basis. Code books, which were dictionaries of telegraphese, were developed from the mid-1800s to the mid-1900s. They contained many useful phrases and sentences, each with its code word. Some code books were general-purpose. Others were trade-specific. Related to the latter were government, diplomatic, and military code books.

telegraphone Introduced at the 1900 Paris Fair, this prototype answering machine was ust slightly ahead of its time. It looked somewhat like an old-fashioned Edison cylindrical gramophone.

telegraphy Aging data transmission technique characterized by maximum data rates of 75 bits per second and signaling where the direction, or polarity, of DC current flow is reversed to indicate bit states.

teleguerilla A term coined by Telecom Australia for "the first wave of informal and unofficially sanctioned telecommuters – those who occasionally work from home with the informal approval of their immediate boss." Says Telecom Australia, "They're the ones whose bosses say, 'I don't care where you work as long as you get the project finished.'"

telehousing Offering space, security, and environmental controls for equipment owned by various carriers. This allows a telecom giant like a RBOC to rent space to friends and competitors in their Network Access Points and other similar facilities.

telemanagement A term for the application of computer systems to the management of the telephone and telecommunications expenses of a user organization. Telemanagement includes virtually every function which the contemporary corporate telecommunications manager performs. Ray Horak, my Contributing Editor, says that telemanagement comprises the management of costs, assets, processes and security. Cost management includes call accounting, cost allocation, bill consolidation, and bill reconciliation-costs include usage-sensitive network costs (e.g., long distance calls), nonrecurring costs (e.g., installation and repair), and recurring costs (e.g., circuits and maintenance agreements). Asset management includes the cradle-to-grave management of systems (e.g., PBXs and computer systems), terminals (e.g., phones and PCs), and inside cable and wire systems. Process management includes work order and trouble ticket management, traffic analysis, and network design and optimization. Security management includes the management of toll fraud and network abuse/misuse. Telemanagement systems generally are in the form of premise-based application software systems, which typically are modular. Such systems typically are either PC-based, or client/server (PC-LAN) in nature; very large user organizations make use of mainframe-based systems. Service bureaus also offer telemanagement services, although they tend to be limited to call accounting and cost allocation.

TeleManagement Forum TM Forum. The TeleManagement Forum has provided leadership, strategic guidance and practical solutions to improve the management and operation of information and communications services. The TM Forum boasts 340 members worldwide including incumbent and new-entrant service providers, equipment suppliers, software solution suppliers and Systems Integrators (SIs). The TM Forum's New Generation Operations Systems and Software (NGOSS) program is a widely accepted integrated framework for developing, procuring and deploying operational and business support systems and software. See also MTMN Version 3. www.tmforum.org.

telecommunications maintainer The European terms for a company that does not sell or install a business telephone system, but is authorized by the manufacturer to maintain it. This is more common in Europe than in North America.

Telecommunications Management Network See TMN.

telemarketing Marketing and sales conducted via the telephone. There are two sides to telemarketing – incoming and outgoing. Incoming telemarketing is largely run through 800 toll-free IN-WATS numbers and local FX (foreign exchange) lines. Outgoing telemarketing is organized over OUT-WATS lines. An expanding range of telecom gadgetry is being developed to automate telemarketing – including automated outbound dialers, voice processing technology and automatic call distributors. The tone recognition, voice detection and transaction audiotex and transaction processing capabilities of voice processing gear can be used to enhance all telemarketing applications.

Telemarketing and Consumer Fraud and Abuse Protection Act A Congressional bill passed in August 1994 with the stated purpose of combating the growth of telemarketing fraud. The bill gave law enforcement agencies new tools, and consumers new protections and guidance to help prevent the planned, fraudulent use of the telephone. See Telemarketing Sales Rule.

telemarketing sales rule This rule was adopted by the FTC December 31, 1995 pursuant to the Telemarketing and Consumer Fraud and Abuse Protection Act of 1994. Key provisions require specific disclosures, prohibit misrepresentations, set limits on the times when telemarketers may call customers, prohibit calls after a consumer requests not to be called, set payment restrictions for the use of certain goods and services, and require that specific business be kept for two years. See Telemarketing and Consumer Fraud and Abuse Protection Act.

Telematic Agent TLMA. An X.400 AU (Access Unit) serving Teletex users of other telematic services (using Teletex, Fax, etc.).

telematics 1. A generic term for a wireless network supporting the collection and dissemination of data. Static, or fixed, applications include SCADA (Supervisory Control And Data Acquisition), which is used in the power utility industry for meter reading and load control (e.g., load shedding). Mobile applications include vehicle tracking and positioning, on-line navigation, and emergency assistance. Probably the best known example is GM's OnStar system, which automatically calls for assistance if the vehicle is in an accident. These systems can also perform such functions as remote engine diagnostics, tracking stolen vehicles, providing roadside assistance, etc. www.onstar.com

2. The use of wireless data transmission to and from your car, to be used for everything from GPS (global positioning systems) to surfing the Internet.

telemedicine The provision of health-care services from a distance using networks supporting audio, video, and computer data transmissions. Telemedicine traditionally uses videoconferencing to diagnose illness and provide medical treatment over a distance. Often used to view or teach surgical procedures. Used also in rural areas where health care is not readily available and to provide medical services to prisoners. At the ICA Show in 1994, Southwestern Bell demonstrated telemedicine applications including a dermatology microscope, a video scope, an electronic stethoscope and a telepathology system that allows a pathologist to exercise computer control over a remote microscope.

telemetry A communications system for the transmission of digital or analog data which represents status information on a remote process, function or device.

Telemetry, tracking and control TT&C. These three functions control and monitor a group of satellites.

telemonkey I was given this definition by a fine gentleman, who'd just been given the job of running his company's call center. He was, as you can, quickly disgusted with the quality of the labor he had to manage. He says he didn't make this term up. It's for real. Maybe. Anyway, here goes. A pejorative word for a call center agent. Originally companies staffed their centers with highly educated, well-paid agents who were usually capable of thinking independently when dealing with a customer's inquiry, but now companies have started to replace such staff with less educated, less trained, and lower paid agents who are trained to respond to customers' inquiry by referring to a database help desk, guide book or manual. Hence the idea that monkeys could handle an agent's job.

telenet A private, commercially available network providing both packet-switched and circuit-switched service to subscribers in North America, Europe and some parts of Asia.

telenet remote login protocol A virtual terminal service specified by the U.S. Department of Defense and implemented by most versions of UNIX.

Telenet International Quotations TIQ. A market data information subscription service operated by Telrate International Co. over a network that uses proprietary protocols to enhance security and other functions.

telenomics Bandwidth economics, referring to the buying and selling of telecommunications capacity and bandwidth.

Telenor Telenor is Norways largest telecommunications group.

teleparents Parents who equip their children with pagers before allowing them to go out.

telephone 1. The invention of the devil.

2. Telephone in London cockney rhyming slag is dog 'n' bone, bone obviously rhyming with phone, In most cases of cockney the rhyming word is dropped, leaving just dog.

3. The most intrusive device ever invented.

4. The biggest time waster of all time, as in: "What did you do all day?" "Nothing. Just spent the day on the phone."

5. Many companies actually don't call them telephones. Inter-Tel calls them "endpoints." AT&T and its spinoffs Lucent and Avaya call them "voice terminals." Nortel Networks calls them "business series terminals." Panasonic sometimes calls them "handsets" or "h'sets."

6. Also a truly remarkable invention. Here are the eight things a telephone actually does.

a. When you lift the handset, it signals you with a dial tone. The dial tone actually comes from the central office, not the phone. But most people think it comes from the phone.

b. It indicates the phone system is ready for your wish by receiving a tone, called a dial tone.

c. It sends the number of the telephone to be called.

d. It indicates the progress of your call by receiving tones – ringing, busy, etc.

e. It alerts you to an incoming call.

f. It changes your speech into electrical signals for transmission to someone distant. It also changes the electrical signals it receives from the distant person to speech so you can understand them.

g. It automatically adjusts for changes in the power supplied to it.

h. When you hang up, it signals the phone system your call is finished.

And, most remarkably, most simple telephones cost under $20. See Bell and Bell, Alexander.

Telephone Account Management TAM. A telemarketing/call center term. Using the telephone channel to proactively cover an assigned group of customers with the objectives of building and retaining revenues from these customer accounts. Most often primary coverage and revenue responsibility lies with the Telephone Account Management team and individual Telephone Account Management Representatives "own" a specific set of accounts within the team. TAM coverage often requires multiple outbound and inbound contacts with assigned customers driven by information uncovered by the TAM Representatives and not by any preset campaign parameters.

telephone amplifier A device to amplify the sound of the receiver. Something no phone should be without. Some devices work strictly on line power. They can only increase volume by 10 dB, which is often not enough (especially if you're over 40). The best telephone amplifiers are powered by AC/DC adapters. Newer ones are powered by nicad batteries. They will amplify to 20 dB.

telephone answering A feature of some voice mail systems in which incoming callers are immediately directed to the called party's voice mailbox where they hear a personalized greeting in the called party's voice and are prompted to leave a detailed message.

telephone centric worker A worker whose job requires him/her to spend most of the time on the phone, such as call center agents, help desk workers, customer service reps, and receptionists.

telephone channel A transmission path suitable for carrying voice signals. Defined by its ability to transmit signals in a frequency range of about 300 to 3000 Hz.

telephone circuit 1. All telephones are made up of just three circuits, The Ringer Circuit, The Mouthpiece Circuit, and the Earpiece Circuit. The ringer circuit and the mouthpiece circuit are connected in parallel. The hookswitch keeps the mouthpiece circuit open whenever the telephone is hung up. The earpiece circuit is coupled to the mouthpiece circuit with a transformer.

The Ringer circuit is across the incoming telephone line pair at all times. The ringer circuit consists of a ringer, some sort of bell or buzzer, and a capacitor. The capacitor serves to block DC current since the telephone company's central office determines whether a phone is on or off hook by measuring the DC resistance across it's line pair. To ring the telephone's bell, a high voltage alternating current is sent down the telephone line pair.

The mouthpiece circuit is also referred to as the primary circuit, talk circuit, and DC loop. When the handset is lifted from the cradle to the off-hook position, the hookswitch closes, and creates a closed DC circuit from the central office through the line pair and to the microphone in the handset of the telephone and the primary coil of the transformer that couples the microphone circuit to the earpiece of the phone.

The DC resistance of the mouthpiece circuit is lower than the resistance of the ringer circuit. When the telephone is off hook, the central office detects the change in resistance. When the telephone is off hook, the central office will disconnect the ringing generator so as not to send high voltage down the line to a phone that is in use.

The earpiece circuit is also referred to as the secondary circuit and listen circuit. The last of the three circuits differs from the other two in that it is never directly across the incoming lines. Instead it is coupled to the primary circuit by a transformer. The current in the primary coil of the transformer is modulated by the microphones at both ends of the telephone call. This varying current induces a current in the secondary winding of the transformer. This induced current generates sound from the loudspeaker in the earpiece of the phone.

2. Electrical connection permitting the establishment of telephone communication in both directions between two telephone exchanges.

Telephone Consumer Protection Act of 1991 TCPA. Legislation passed by Congress and signed by the president in 1991. It restricts specific types of unsolicited telephone calls. Among the provisions were a prohibition on calling emergency numbers or numbers for which the recipient was charged, limiting the placement of unsolicited calls to between 8 am and 9 pm, and removing people from calling lists who request that they not be called again. The Act also makes it unlawful for any person to use a computer or other electronic device, including fax machines, to send any message unless such message clearly contains in a margin at the top or bottom of each transmitted page or on the first page of the transmission, the date and time it is sent and an identification of the business or other entity, or other individual sending the messagte and the telephone number of the sending machine or such business, other entity, or individual. The telephone number provided may not be a 900 number or any other number for which charges exceed local or long-distance transmission charges.

Telephone Disclosure and Dispute Resolution Act TDDRA. The federal law that governs the pay-per-call industry, such as 900 number services. It mandates disclosure requirements and outlines the responsibilities of service providers while providing some relief for consumers with disputed charges.

telephone drop-in mouthpiece A telephone drop-in mouthpiece used for bugging looks very much like the carbon microphone in the mouthpiece of a telephone. It is installed by unscrewing the mouthpiece, removing the old microphone, and dropping in the wiretap device. The transmitter draws power from the telephone line and only operates when the telephone is off hook. Both sides of the conversation are picked up and transmitted to a remote location. The telephone line is used as an antenna. Drop-in transmitters are simple to install and hard to detect. They require access to the telephone instrument.

Telephone Equipment Order TEO. TEOs are orders placed by Central Office Engineering for telephone apparatus and equipment. These orders may be telephone company or Vendor Engineered and are usually the downstream product of a Network Design Order (NDO).

telephone exchange A switching center for connecting and switching phone lines. A European term for what North Americans call central office.

telephone frequency Any frequency within that part of the audio frequency range essential for the transmitting speech, i.e. 300 to 3000 Hz.

telephone hotel Also known as a Neutral Central Office (NCO). A neutral location in which carriers can lease space for termination of circuits and placement of equipment. See NCO adn Telco Hotel for a full explanation.

telephone line simulator Also called ring-down box. Ring down boxes, also known as CO simulators, are simple devices used for generating calls from a POTS line to a computer telephony system (or vice versa). When one side goes offhook, the ring down box will "ring" the other side. When both sides are offhook, both sides are coupled together and the line is powered. Ring down boxes are available with various options and configurations. These include the ability to provide dialtone to the caller side (required to test applications with modems, faxes, or other automated outdialing devices), caller ID, and disconnect supervision. They are generally available in one to four line sizes, although special configurations may support more. Ring-down boxes are used for giving demonstrations and testing. We use them in our test labs to testdrive new computer telephony systems.

telephone man's prayer Now I lay me down to sleep; A reel of cable at my feet. If I should die before I wake, Blue, orange, green, brown, slate,...

telephone management system The term originally meant a system for controlling telephone costs by:

1. Automatically selecting lower-cost long distance routes for placed calls; 2. Automatically restricting certain people's abilities to make some or all long distance calls; and 3. Automatically keeping track of telephone usage by extension, time of day, number called, trunk used and sometimes by person calling and client or account to be billed for call.

These days the terms means those three functions plus a whole lot more, typically those associated with professionally managing the corporate or government telecommunications expenses, including (but not limited to):

- Computerized inventory monitoring;
- Computerized traffic engineering and network design;
- Departmental telephone bill allocation and invoicing;
- Automated telephone directory, etc.
- Project tracking;
- Automated equipment and service ordering.

In short, all the functions of professional telecommunications management that can be automated or organized in some way on a computer. The telecommunications management system thus refers to the computer hardware and the software. For more on this subject see the latest June issue of TELECONNECT Magazine. See also Call Accounting System.

telephone manager Apple's telephony API for the Macintosh world. Here is an excerpt from Apple's Web page explaining it : "Telephony is the process of managing telephones, particularly of establishing and controlling connections between telephones on a telephone network. The Telephone Manager is the part of the Macintosh system software that you can use to develop applications and other software that provide telephony capabilities. For example, you can use the Telephone Manager to place outgoing telephone calls, answer incoming telephone calls, place calls on hold or transfer them to other telephones, and accomplish many other similar tasks. The data transferred during a telephone call can be either voice, modem, or fax data, or indeed any kind of data that can be encoded for transmission across a telephone network.

"The Telephone Manager provides a set of simple but powerful programming interfaces that you can use to support telephony activities. The Telephone Manager operates independently of the particular telephone network or networks to which a user's computer is con-

nected. Accordingly, your application can provide telephony services whether the Macintosh computer on which it is executing is connected to an integrated services digital network (ISDN), to a private branch exchange (PBX), or to "plain old telephone service" (POTS).

"The Telephone Manager accesses a specific telephone network using a telephone tool, a software module that manages the connection between a network and the telephony applications or other software running on a Macintosh computer. Telephone tools control the device drivers of the telephony hardware (such as an ISDN card) installed on the user's system. Each telephone tool is designed for specific hardware. For example, the Apple ISDN Telephone Tool is designed for the Apple ISDN NB Card. For more, http://gemma.apple.com/techpubs/mac/Telephony/telephony-2.html or http://developer.apple.com/techpubs/mac/Telephony/Telephony-2.html

Telephone Pioneers of America The Telephone Pioneers of America began almost a century ago, originally consisting of the 'charter employees' of the company, or 'pioneers' in telecommunications, mainly those who served with the Bell System at its outset. As time went on, there would be fewer living or active original Pioneers, thus the TPA charter was amended to allow membership by any employee of AT&T or (as they were called) a subsidiary company who had been employed by Bell (or an independent) for at least twenty years. Membership in the Pioneers was opened to more types of telephone company people over the years (including companies that are not "Bell" or AT&T).

The Telephone Pioneers have a distinguished history of community service. Pioneers devise technical solutions to improve the lives of those with disabilities, allowing them to use telephones when this would otherwise be difficult or impossible. Pioneers also assist with general community activities such as voter registration, help those who are ill, feed those who are needy, and more. The Telephone Pioneers of America has chapters throughout the USA and Canada. At the non-Bell telcos, the same organization is known as the Independent Pioneers.

Telephone Plant Index TPI. The telecommunications indicator used to calculate inflation based on a "Market Basket of Goods."

Telephone Preference Service TPS. A service offered by the Direct Marketing Association, New York, NY. The DMA keeps a list of consumers who have requested that their names be removed from telemarketing calling lists. Telemarketing companies can have the list upon request. Use of the service does not relieve companies from their obligation under the TCPA. In Europe the TPS is called a Robinson List.

telephone receiver Telephone earpiece. Device that converts electrical energy into sound energy, designed to be held to the ear.

Telephone Relay Service TRS. A voice/data system that enables communications with the hearing impaired.

telephone number The Telephone Number is officially the 10-digit number (613-723-8231) of that subscriber, hence it includes the area code. See DN.

telephone number salary A salary that has seven digits, based on the fact that local North American phone numbers have seven digits.

Telephone Service Representative TSR. Another word for agent – the person who answers the phone on an automatic call distributor. See Agent.

telephone set A fancy name for a telephone.

Telephone Set Emulation The concept is simple: Emulate the proprietary electronic phone on a printed circuit card inside a PC. Let the PC do everything a human using the phone could do. Only the PC will do it more efficiently and the human will find it easier to use all his phone's features because the PC's screen is bigger and the PC's keyboard easier to use than the phone's keyboard. Attach the phone emulation card to voice and call processing cards, like voice synthesis, voice recognition, voice mail, touchtone generation and recognition, etc. And bingo, phone systems acquire all the benefits of integrated voice and call processing. It's powerful concept. As I wrote this, a handful of telephone phone emulation cards had appeared. Within a little while, there won't be a phone worldwide that you won't be able to emulate on a printed circuit card you can drop into a vacant slot inside your PC.

telephone set management Imagine you have a phone attached to your computer through a telephony board inside your computer. Now imagine that you pick up the phone and dial a number. If the company knows you have dialed a number and knows which number you have dialed, that feature is called handset management. It is the ability of the computer to be aware of every button pushed on the phone. The advantage of this is obvious: You really want the PC to collect those digits, so it can, for example, add a price to each call and use them for monthly billing (lawyer, accountant, etc.). You also want to be able re-dial those numbers by simply clicking on the number one you want, hitting Enter and bingo, you're redialing that number, without having to key it in again. This term, telephone set management, used to be called handset management.

Two of the early pioneers in the field of telephone set management, David Perez and Nick Nance of COM2001 Technologies in San Diego, defined telephone set management as "the ability for seamless Integration with the phone (any 2500 set) and the modem / voice processing board and /or fax machine. The hardware must notify (send a signal or command) to the software when the phone is off hook or on hook. It must also notify the software when the user presses the numeric buttons on the phone. Ultimate integration would include additional types of button support as in: volume, hold, release, redial, conference, or any button on the telephone / fax / modem etc. The reason? Telephone integration offers true Computer Telephony integration. The ability to signal the handset and feature keys allows the user to continue to use their desktop phones but take complete advantage of the Computer Telephone software on their desktop for speed dialing, transferring, conferencing, voice mail, etc.

telephone signaling device A gadget which indicates that the phone is ringing. May also be hooked up to lamps or overhead lighting to cause those lights to flash when the phone is ringing.

telephone tag I call you. But you're not there. I leave a message. You call me back. But I'm not there. You leave a message. And so on. We're now playing telephone tag.

telephone tap Telephone taps are generally defined as devices which are designed to extract audio information of intelligence from the telephone line pair. The process consists of identifying the specific telephone talk pair of interest at some accessible point, the interception of their electrical signals, and the communication of these signals to the surveillance equipment operator. Telephone companies unintentionally assist the wiretapper by installing extra telephone wires for future expansion. There are almost always extra wires that can be appropriated for use in wiretapping.

Wiretaps can be installed at the telephone company's central office if the phone company cooperates. If not, taps can be installed in splice cases or in a ready access terminal. Wiretaps should be distinguished from telephone bugs. Bugs are room audio surveillance devices that use the telephone wiring to bring the audio to the surveillance operator. Telephone bugs are used because they avoid RF interference.

telephone tone Audible tone generated by the network which provides call progress indications to the user. Different tones (e.g., ring back, busy) allow the human ear to interpret the progress of the call. On digital networks (such as PBX or ISDN), the network may send indication messages (e.g., billing, carrier, faxCNG, modemCNG) to the telephone to indicate the status of the call, and the telephone may generate certain tones locally, driven by those messages.

telephonist This is a European term referring to what North America calls a receptionist or switchboard attendant Note: the term switchboard operator is also used in Europe, but not switchboard attendant.

telephony The science of transmitting voice, data, video or image signals over a distance greater than what you can transmit by shouting. The word derives from the Greek for "far sound." For the first hundred years of the telephone industry's existence, the word telephony described the business the nation's phone companies were in. It was a generic term. In the early 1980s, the term lost fashion and many phone companies decided they were no longer in telephony, but in telecommunications – a more pompous sounding term that was meant to encompass more than just voice. The pomposity of the word may have added some value to the stock of telecommunications companies. In the early 1990s, as computer companies started entering the telecommunications industry, the word telephony was resurrected. And in a white paper on Multimedia from Sun Microsystems, the company said that telephony refers to the integration of the telephone into the workstation. For instance, making or forwarding a call will be as easy as pointing to an address book entry. Caller identification (if available from the telephone company) could be used to automatically start an application or bring up a database file. Voicemail and incoming faxes can be integrated with e-mail (electronic mail). Users can have all the features of today's telephones accessible through their workstations, plus the added benefits provided by integrating the telephone with other desktop functions. See also Computer Telephony.

Telephony Access Module See TAM.

telephony interface control A Telephony Interface Control resource is any resource that interfaces with the telephone network (public or private). This is usually claimed as the primary member of a group.

Telephony Markup Language Telephony Markup Language (TML) is a vocabulary extension to the Extensible Markup Language (XML) dealing with telephony, unified messaging, and other forms of network data exchange. Thanks to efforts by the World Wide Web Consortium (W3C), XML itself recently became a formal specification. Like HTML, XML is a simplified descendant of the enormously complex Standardized General Markup Language (SGML) that's been used for high-end, highly structured publishing ap-

plications for the last ten years. But unlike HTML, XML allows you to create your own tags (which is what "extensible" means). Whereas HMTL is concerned primarily with presentation, XML tags – which can exist on the same page as HMTL – specify content (first name, price, phone number, etc.).

The idea for TML originated with Computer Telephony Magazine editor-in-chief John Jainschigg. Two companies were instantly excited about developing and building TML into their products: SoloPoint (Los Gatos, CA) and Technology Deployment International (Santa Clara, CA). These and other companies are contributing to the TML initiative by atomizing call and messaging functions in an effort to start formulating a list of possible telephony and messaging tags. The telephony tags will be a universal way to put touchtone screens on browser-equipped handheld devices, and since XML will become a standard mechanism for exchanging data as well as documents, it's possible that voice mail, video mail, faxes and email messages from different vendors' repositories could be exchanged across the Internet and reviewed on any XML compatible device.

But TML will be more than just standardized tags for telephony and unified messaging. While XML definitely will be appearing on browsers, you can think of XML (and TML) as more of a self-describing messaging standard rather than simply some extra tags focusing on web page content. XML can be used on the Internet, intranet, extranet, VPNs, WANs, or whatever, because XML itself is not actually sending messages. XML describes the message content while other applications, components or servers actually send and interpret the messages containing XML.

So TML should not be thought of as merely a "browser client-like thing" talking to a "server-like thing." TML will be a more pervasive messaging environment running over all networks.

The closest thing to being a TML competitor is VxML or "VoxML," sponsored by some big companies such as Lucent and Motorola. VxML is a XML language for voice menu item management. It's used to represent a caller's many choices and input options when interacting with an IVR or other voice automation system. But like the TML developers, the VoxML developers (some ex-Bell Labs guys) look upon the client not just as a browser but as "audio over any phone" which makes sense for an IVR-related XML language. Still, the industry needs something more powerful and comprehensive than VxML, which is what TML's supporters hope it will be.

telephony over IP Carrier-grade VoIP and telephony services over an IP network.

telephony server A telephony server is a computer whose major function is to control, add intelligence, store, forward and manipulate the various voice, data, fax and e-mail calls flowing into and out of a computer telephony system. The traditional function of a telephony server is to move call control commands from client workstations on a LAN to an attached PBX or ACD. (This is what it does under the paradigm called "Telephony Services.") A telephony server can also be a voice response system. It can also be a fax on demand system. It can also be a conferencing device. It can also be switch. And it can be all these capabilities, which traditionally run on physically separate servers, all rolled into one machine, called generically a "telephony server." See Telephony Service Application Programming Interface And Telephony Services.

Telephony Services Application Programming Interface. TSAPI. Described by AT&T, its inventor, as "standards-based API for call control, call/device monitoring and query, call routing, device/system maintenance capabilities, and basic directory services." For a better explanation, see Telephony Server, Telephony Server NLM and Telephony Services.

Telephony Server NLM Telephony Server NetWare Loadable Module. The main part of a software product call Telephony Services announced in early 1993 by AT&T and now marketed by Lucent (but not Novell). The Telephony Server NLM is software add-on to Novell's NetWare LAN operating software. The idea is have a NetWare server equipped with the NLM, an interface card and a cable connection to an adjoining telephone system. This would mean that anyone with a PC on the network and a PBX phone on their desk will be able to use telephone features, such as auto-dialing, conference calling and multiple call handling from their desktop PC.

A Novell White Paper in Spring of 1993 said "Telephony Services for NetWare provides benefits to three main customer segments. First, applications are being developed to provide increased productivity to everyday computer desktop users. Second, call-centers take advantage of this technology as it provides a right-sizing cost-effective solution. Finally, benefits will be available to telecommunications/IS administrators by providing the ability to reduce administrative costs through easier management of user databases.

"Computer-Telephone Integration (CTI) combines telephone and computer technology to provide access and control telephone functionality from a computer terminal. It combines the easy access and usable graphical interface of the computer desktop with the features of the telephone. CTI is not a new concept. Traditionally, however, CTI has only been available in a mini and mainframe computer environments. These solutions are expensive and can be cost-justified only in large call-center applications. Consequently, the penetration of CTI solutions has been very small.

"However, providing CTI in NetWare environment brings this technology mainstream. Not only does this solution provide a more cost-effective implementation, it also allows integration with the rich set of NetWare services. In the simplest example, a Telephony Services for NetWare application allows users to make a phone call by clicking on a name from a calling list displayed on a desktop computer and having the desktop computer dial the number. Possibilities exist for applications that will allow similar functionality with the addition of conference calling capability. Instead of clicking on a single name, the user can highlight a number of names, click on a conference-call icon and have the system place the calls to all parties. The benefits which are derived from the integration of telephony with other NetWare services is far reaching. As part of continued development efforts, applications are becoming available which allow desktop video phone calls. Callers can see each other and talk on the phone, while simultaneously viewing and editing image documents.

"Other capabilities include integrating voice-mail, fax and e-mail into a single message-management application. Possibilities also exist utilizing number recognition technology to integrate computer database records with caller-id. Administrators can manage a single user database utilized by the computer network, the PBX and the voice-mail system.

"Telephony Services for NetWare takes advantages of client/server technology to provide a broad framework for creating first-party and third-party call-control applications. These applications answer the customer demand for integrated business tools and solutions. This technology provides a logical connection between the desktop computer and the telephone. The only physical connection is established between the PBX and a NetWare server. This architecture is cost-effective and efficient by utilizing a company or organization's existing equipment. The initial product deliverables include the following components:

- Client/Server API
- Telephony Server NLM
- PBX Driver
- PBX Link Hardware
- Passageway Application

"The Telephony Server NLM is the mechanism for passing information between the PBX and the NetWare server. As part of the NLM, an open PBX Driver Interface allows PBX manufacturers to write drivers which communicate with their respective PBX models. The client/server API provides support across multiple desktop operating systems. It also allows call control at either the client or the server. The Passageway application provides the user with basic autodialing and notes capability.

"Telephony Services for NetWare provides a key opportunity for developers. Open APIs which support multiple desktop operating systems provide a development platform for both traditional telecommunications developers and new or experienced NetWare developers."

Novell has effectively stopped marketing Telephony Services, but AT&T (now Lucent) continues to market it. See Telephony Services.

Telephony Service Provider TSP. A software encapsulation of all the services provided by a particular network interface device or line device. A line device may be a single POTS bearer channel or it may be several bearer channels; e.g., a single E-1 span with 30 network channels of 64 Kb/s bandwidth. The TSP is provided by the vendor who has developed a network interface device for SCSA. See S.100.

Telephony Services Telephony Services' real name is Telephony Services for NetWare, the local area network software from Novell, Orem, Utah. Telephony Services for NetWare basically consists of an addition to the NetWare operating system, called Telephony Server NLM. (See Telephony Server NLM.) That addition handles communications between a NetWare file server (a PC loaded with NetWare) and an attached telephone switch, e.g. a PBX or ACD (automatic call distributor). The concept is very simple. Picture your department's LAN. You're sitting in front of your PC which is on your department's LAN. You have a telephone which is an extension off your company's or department's PBX. You click on an icon that says "Phone." Bingo, a screen comes up with icons and pull down menus. You can now look up Joe, click on his name. Your PC sends a command to your NetWare file server, which in turn sends a command to your PBX which tells it to dial Joe from your phone. Once it's completed dialing, it might turn on your phone's speakerphone or your telephone headset. You'll then hear Joe say "Hello." Telephony Services for NetWare is basically the software in the NetWare file server which takes care of interpreting your PC commands into commands your switch can understand and respond to.

Telephony Services requires a link to your switch. Each telecom switch manufacturer has been implementing that link in a different way technically. That's fine, because Telephony Services for NetWare insulates the user and the developer. This means that computer telephony applications written for Telephony Services for NetWare will work on any switch conforming to the Telephony Services standard. As of writing, virtually every switch manufactured in North America and many made in Japan and Europe is conforming to Novell's Telephony Services.

According to a White Paper issued by Novell in March 1994 and called NetWare Telephony Services, "The three main components of Telephony Services are call control, voice processing, and speech synthesis. Call control provides the core service for PBX-to-NetWare communication and an Application Programming Interface (API) for developing client server applications. With call control, users can enjoy features, says Novell, such as making calls, transfers, or conference calling. Voice processing functions include voice mail and interactive voice response. Speech synthesis will be a key area for integrating multiple media types. Through speech/text conversion, users can access voice mail, e-mail, and fax documents through audio or text media types. The initial products Novell is delivering include:

- "Client/Server API
- "Telephony Server NetWare Loadable Module (NLM)
- "PBX Driver

"The Telephony Server NLM is the mechanism for passing information between the PBX and the NetWare server. As part of the NLM, an open PBX Driver Interface allows PBX manufacturers to write drivers that communicate with their respective PBXs. The client/server API provides support across multiple desktop operating systems and allows call control at either the client or the server."

Client server computer telephony, according to Novell, delivers ten benefits:

1. Synchronized data screen and phone call pop. Your phone rings. The call comes with the calling number attached (via Caller ID or ANI). Your PBX or ACD passes that number (via Telephony Services) to your server, which does a quick database look up to see if it can find a name and database entry. Bingo, it finds an entry. It passes the call and the database entry simultaneously to whoever is going to answer the phone: The attendant. The boss. The sales agent. The customer service desk. The help desk. All this saves asking a lot of questions. Makes customers happier.

2. Integrated messaging. Also called Unified Messaging. Voice, fax, electronic mail, image and video. All on the one screen. Here's the scenario. You arrive in the morning. Turn on your PC. Your PC logs onto your LAN and its various servers. In seconds, it gives you a screen listing all your messages – voice mail, electronic mail, fax mail, reports, compound documents .... Anything and everything that came in for you. Each is one line. Each line tells you whom it's from. What it is. How big it is. How urgent. Skip down. Click. Your PC loads up the application. Your LAN hunts down the message. Bingo, it's on screen. If it contains voice – maybe it's a voice mail or compound document with voice in it – it rings your phone (or your headset) and plays the voice to you. Or, if you have a sound card in your PC, it can play the voice through your own PC. If it's an image, it will hunt down (also called launch) imaging software which can open the image you have received, letting you see it. Ditto, if it's a video message.

Messages are deluging us. To stop them is to stop progress. But to run your eye down the list, one line per entry. Pick the key ones. Junk the junk ones. Postpone the others. That's what integrated messaging is all about. Putting some order back into your life.

3. Database transactions. Customer look ups. There are bank account balances, ticket buys, airline reservations, catalog requests, movie times, etc. Doing business over the phone is exploding. Today, the caller inputs his request by touchtone or by recognized speech. The system responds with speech and/or fax. Today's systems are limited in size and flexibility. The voice processing application and the database typically share the same processor, often a PC. Split them. Spread the processing and database access burden. Join them on a LAN (for the data) and on new, broader voice processing "LANs," like SCSA or MVIP. You've suddenly got a computer telephony system that knows no growth constraints. You could also get the system to front-end an operator or an agent. Once the caller has punched in all his information, then the call and the screen can be simultaneously passed to the agent.

4. Telephony work groups. Sales groups. Collections groups. Help desks. R&D. We work in groups. But traditional telephony doesn't. Telephony today is BIG. Telephony today is one giant phone system for the building, for the campus. Everyone shares the same automated attendant, the same voice mail, the same ubiquitous, universal, generic telephone features. But they shouldn't. The sellers need phones that grab the caller's phone number, do a look-up on what the customer bought last and quickly route the call to the appropriate (or available) salesperson. The one who sold the customer last time. The company's help desk needs a front end voice response system that asks for the customer's serial number, some indication of the problem and tries to solve the problem by instantly sending a fax or

encouraging the caller to punch his way to one of many canned solutions. "The 10 biggest problems our customers have." When all else fails, the caller can be transferred to a live human, expert at diagnosing and solving his pressing problem. A development group might need e-mails and faxes of meeting agendas sent, meeting reminder notices phoned and scheduled video conferences set up. All automatically. The accounts receivable department needs a predictive dialer to dial all our deadbeats. The telemarketing department also needs a predictive dialer, but different programming.

5. Desktop telephony. There are two important aspects. Call control and media processing services. Call control (also called call processing) is a fancy name for using your PC to get to all your phone system's features – especially those you have difficulty getting to with the forgettable commands phone makers foist on us. *39 to transfer? Or it is *79. With attractive PC screens, you point and click to easy conferencing, transferring, listening to voice mail messages, forwarding, etc. There are enormous personal productivity benefits to running your office phone from your PC: You can dial by name, not by number you can't remember. You can set up conference calls by clicking on names and have your PC call the participants and call you only when they're all on the phone. You can transfer easily. You can work your voice mail more easily on screen, instead of having to remember "Dial 3 for rewind," "Dial 2 to save," and other obscure commands. Here's a wonderful quote from Marshall R. Goldberg, Developer Relations Group at Microsoft. He says "Voice mail systems that could benefit through integration with the personal computer largely remain isolated, difficult to use, and inflexible. Browsing, storing messages in hierarchical folders, and integration of address books – functions just about everyone could use – are either unavailable or unusable."

The second benefit is media control. Media control is a fancy name for affecting the content of the call. You may wish to record the phone call you're on. You may wish to have all or part of your phone call clipped and sent to someone else – as you often today with voice mail messages. You may wish to simply file your conversations away in appropriate folders. You may wish to be able to call your PC and get it to read you back any e-mails or faxes you received in the last day or so.

6. Applying intelligence. A PC is programmable. The typical office phone isn't. A PC can be programmed to act as your personal secretary, handling different calls differently. It can be programmed to include commands, such as "If Joe calls, break into my conversation and tell me." "If Robert calls, send him to voice mail." etc.

7. The Compound Document. The typed document lacks life. But add voice, image and video clips to it and it gets life. The LAN makes the compound document easier to achieve. The Compound Document gets attention.

8. Management of phone networks. Today, phone networks are very difficult to manage. Often the PBX is managed separately from the voice mail, which is managed separately from the call accounting, etc. It's a rare day in any corporate life when the whole system is up to date, with extensions, bills and voice mail mailboxes reflecting the reality of what's actually happening. The latest generations of LAN software – NetWare 4.1 and Windows NT – have solid enterprise-wide directories and far easier management tools. Integrate these LAN management tools with telecommunications management, and potentially all you need is to make one entry (for a new employee, a change, etc.) and the whole system – telecom and computing – could update itself automatically, including even issue change orders to the MIS and telecom departments and vendors.

9. No dedicated hardware in the PC. With only one link – from the switch to the LAN – there's no need to open the desktop PC and place specialized telephony hardware in each PC that wants to take advantage of the new LAN-based telephony features.

10. Switch elimination. The ultimate potential advantage of LAN-based telephony is to eliminate the connection to the switch (PBX or ACD) by simply populating the LAN server (now called a telephony server) with specialized computer telephony cards and run the company's or department's phones off the telephony server directly.

Novell has effectively stopped marketing Telephony Services, but AT&T (now Lucent) continues to aggressively market it. and the benefits are still as valid as Novell detailed above. See Telephony Server NLM and Telephony Services Development Tools

telephony workgroup A concept that says work is done in groups and those groups need special telephony features and services. This is in contrast to most telephone installations today, where one giant phone system serves the company. Everyone shares the same automated attendant, the same voice mail, the same ubiquitous, universal, generic telephone features. But the concept of a telephony is that they shouldn't. Each group has different telephony needs: The sellers need phones that grab the caller's phone number, do a look-up on what the customer last bought and quickly route the call to the appropriate (or available) salesperson. The one who sold the customer last time.

The company's help desk needs a front end voice response system that asks for the customer's equipment serial number, some indication of the problem and tries to solve the problem by instantly sending a fax or encouraging the caller to punch his way to one of many canned solutions. "The 10 biggest problems our customers have." When all else fails, the caller can be transferred to a live human, expert at diagnosing and solving his pressing problem.

A development group might need e-mails and faxes of meeting agendas sent, meeting reminder notices phoned and scheduled video conferences set up. All automatically.

Computer telephony on a LAN can do to the workgroup what products like e-mail and Lotus Notes are doing – substantially improve productivity (or at least, the pleasure of work). Except that the telephone is still less intimidating.

There are probably as many specialized telephony workgroups features needed as there are computer workgroup features needed. And since computer workgroup features are often provided on a local area network, it makes sense to provide many telephony features for that workgroup on the same local area network.

telephotography Harry Nyquist of Bell Telephone Laboratories began work in 1918 on a method of transmitting photographs over analog telephone circuits. In 1924, the effort culminated in telephotography, a term used to describe AT&T's photographic facsimile machine. A transparency of the photograph was mounted on a spinning drum and was scanned by the transmitting machine. The machine then modulated the amplitude of electrical signals across the analog PSTN in order to represent variations in the shades and tones of the transparency. The receiving machine demodulated the signal and deposited the results onto a sheet of photographic negative film mounted on a spinning drum. The negative was developed in a darkroom to create a facsimile of the original. Many of those principles are still used in contemporary fax machines. See also Facsimile Equipment and Nyquist, Harry.

telepoint The British name for the new generation of cheap, digital mobile phones. They're also called CT2. Think of telepoint as cellular phones but using micro-cells. By having smaller cells than normal cellular cells, CT2 phones can be smaller, cheaper and lighter. The first generation of these phones didn't do well, since they weren't smaller and lighter; there weren't many micro-cells and you couldn't receive an incoming call. See CT1, CT2 and CT2+.

teleports The definition written by Gary Stix in the August 12, 1986 issue of Computer Decisions reads, "High bandwidth telecommunications distribution systems that allow major local users to obtain local, private services and long distance services. The most notable example is the New York Teleport," which is located on Staten Island. Teleports traditionally consist of two things – a fiber optic/coaxial cable network around a city and a collection of nearby satellite antennas. The cable network collects transmissions from larger customers and takes them to the antennas for shipping to and from distant offices. Teleport companies are now more successful as local communications companies than they are as long distance gateways. Which is understandable, since the cost of local calls has gone up, while the cost of long distance calls has gone down.

telepresence In 1985 a team of researchers at NASA invented the notion of telepresence – projecting yourself into someone else's virtual reality. In one version of telepresence, according to Discover Magazine, a computer prompts a robot to mimic your movements. As you manipulate objects in your virtual world a robot somewhere else does the same thing to real objects. Telepresence will be especially useful for hazardous jobs like repairing a nuclear reactor or satellite. Or going on a blind date? See Virtual Reality. In late 2006 BusinessWeek ran a story called "Are you ready for your closeup?" in which the word telepresence popped up again. The concept of this telepresence is to provide superhigh-quality "telepresence" gear, one of techdom's Holy Grails: videoconference systems intended to make participants forget they're in different places. The people captured onscreen are life-size – and lifelike: Lips move in perfect sync with the video. There's eye contact. And no audio lag. To enhance that feeling of togetherness, says BusinessWeek, each vendor installs conferencing rooms that look alike, with the same tables and wall colors. As for price, brace yourself, says BusinessWeek: Hewlett-Packard's Halo system, the priciest, costs $425,000 plus $18,000 a month per conference room for operating costs. Polycom's new ultrapremium RPX product goes for $249,000. As for Cisco's, it'll cost as much as $299,000 for the hardware and $3,500 in monthly costs.

teleprinter A teletypewriter. Also called a telex machine.

teleprocessing 1. Remote access data processing.

2. Use of data link communications to accomplish a computer-based task; distinguished from Distributed Data Processing in that an application processor is not required at each and every node as in DDP.

teleradiology A system that enables the viewing and processing of images within a hospital's nuclear medicine departments or remote image viewing from home computers or remote sites.

telesales British. Sales activities conducted by telephone in a planned and controlled manner.

teleservices 1. A generic term for services offered on phone links. Includes e-mail and facsimile features.

2. A product of SunSoft, a division of Sun MicroSystems, Mountain View, CA. According to SunSoft, "Sun's vision of the impact of widespread use of teleservices suggests that the computer workstation will become the new communications center, combining many existing communication media with new ones, while creating new paradigms for the expression and sharing of ideas. Information in the form of charts and pictures, schedules and plans, and audio and video will merge through application programs that provide a collaborative vehicle for decisions in the 1990s and beyond. The desktop will become the platform for a new set of productivity tools, seamlessly integrated into the critical business activities and methodology of today's companies, and providing a competitive edge for facing the global challenges of tomorrow. Individuals will gain new freedom in where they work and how they access information. And ideas will be communicated in more expedient and creative ways." SunSoft has developed Solaris Teleservices to provide a platform for next-generation workstation applications which leverage the benefits and capabilities of the telephone network and the commonplace use of it. Teleservices applications, according to SunSoft, include:

Desktop Teleservices. Workstation based telephone and answering machine applications allow users to efficiently plant, receive and manage telephone calls.

Remote Access. Users can place calls for their workstation from any telephone and access applications and data through DTMF signaling, or perhaps through speech, using a workstation's speech recognition capabilities.

Wide Area Networking. The ubiquity of telephone networks allow for the complete connectivity of all computers. Network links can be brought up or taken down on demand merely by placing or tearing down a telephone call.

The Solaris Teleservices Platform is called XTEL, which is a multilayered software architecture based on client server computing model. XTEL consists of four key components:

A client side library. Providing a high-level, object oriented application programming interface (API) to application programmers. Using the XTEL API, an application can place and retrieve telephone calls. The API library consists of a collection of C++ objects which is linked to applications that wish to use the systems teleservices resources.

A server. Providing multi-client and multi-device support, the server is the central point of contact for all teleservices, resource management and security are provided by the server. Communication between the XTEL API and the server occurs through the XTEL Server Protocol (XTELS).

One or more providers. Manages each telecommunication device connected to the system. The Teleservices server communicates with an XTEL provider using the XTEL provider protocol (XTELP).

A data stream multiplexer (Sun's spelling). The universal multiplexor (Umux) provides a uniform means for applications to access and share data channels associated with a telephone call. Umux is a streams pseudo-device driver used to connect data channels to applications. An XTEL application is linked with the XTEL API library, through which communicates with the server using the XTELS protocol. XTELS is a synchronous, symmetric messaging protocol using the Solaris loopback transport mechanism. The XTELS protocol is essentially the XTEL provider protocol with extensions to support multiple clients and multiple XTEL providers. For more information, see a document called Solaris Teleservices Architectural Overview, available from SunSoft, Mountain View, CA. See Teleservices API.

Teletel terminal emulation Teletel is a popular character-based graphics emulation for communicating with the Minitel service, found primarily in France.

Teletex An ITU-T standard for text and message transmission which is replacing Telex. Teletex operates at 2400 baud, about 50 times faster than telex. Teletex uses ASCII to encode its characters for transmission.

teletext A data communications information service used to transmit information from remote data banks to viewers. It was transmitted over the air in the vertical blanking interval of the TV signal of the BBS (British Broadcasting Service). Teletext was originally designed for public consumption. It gave out weather information, sports results, headlines, etc. Teletext is proving somewhat more successful among corporations for the internal dissemination of information.

teletraffic optimizer program Derives data by processing actual calls instead of using an analytical model based on estimates or summaries.

teletraining Education and training through telecommunications.

teletype A specific type of teletypewriter.

Teletyper Input Method Teletyper Input Method (TIM) is specially designed for using pushbutton phone or mobile phone to input text, command and instruct your PC through the public telephone network. Teletyper Telephone-Input Method (TIM) and Teletyper Plus use three base keys on the phone pad with the combination of other keys to form alphabets, symbols, utility and control functions in order to command your PC remotely. Information could then be sent to fax, pager or voice output.

Teletyper Input Method (TIM)

*1 = A	#4 = L	08 = W
01 = B	*5 = M	#8 = X
#1 = C	05 = N	*9 = Y
*2 = D	#5 = O	09 = Z
02 = E	*6 = P	#9 = space
#2 = F	06 = Q	** = Enter
*3 = G	#6 = R	00 = Zero
03 = H	*7 = S	## = Backspace
#3 = I	07 = T	1 to 9 press 1 to 9
*4 = J	#7 = U	
04 = K	*8 = V	

Teletypewriter TTY, as in TeleTYpewriter. A telegraph device capable of transmitting and receiving alphanumeric information over communications channels. It may also contain a keyboard similar to that of a typewriter or computer but usually with fewer keys. See Teletype.

Teletypewriter Control Unit TCU. A device that serves as the control and coordination unit between teletypewriter devices and a message switching center when controlling teletypewriter operations.

Teletypewriter Exchange Service TWX. A switched teletypewriter service in which suitably arranged teletypewriter stations are provided with lines to a central office for access to other such stations. TWX and Telex are commercial teletypewriter exchange services. They are currently both owned by AT&T. These days their revenues are in decline. A computer with a modem is a lot faster than TWX or telex.

teletypewriter signal distortion The shifting of signal-pulse transitions from their proper positions relative to the beginning of the start pulse. The magnitude of the distortion is expressed in percent of a perfect unit pulse length.

television 1. TV. Translated from Greek and Latin, "far-off sight," television traditionally is thought of as broadcast TV, transmitted over the airwaves using radio broadcast frequencies. Most of us today think of TV as being provided by a CATV (Community Access TV) provider via coaxial cable. Increasingly, as many as 5 million U.S. viewers think of TV as being delivered directly via satellite. A standard analog TV channel today fits into a frequency bandwidth of 6 MHz.

2. "Chewing gum for the eyes." Frank Lloyd Wright

3. "Why should people go out and pay money to see bad movies when they can stay at home and see bad television for nothing?" Samuel Goldwyn, film producer.

televoting A telephony-based public opinion polling service that compiles call-by-call polling statistics. Typically, the televoting customer will be a governing body, political organization, research organization, or a commercial firm. The customer presents the target audience with the issue, the available response options, and the associated phone number for each option, using mass media or some other means to deliver this information. Members of the target audience make their desired choice by calling the corresponding televoting phone number, which is then registered by the service provider. Summary reports of the number of calls to each televoting phone number reveal public opinions on various issues. The number of options that can be presented to the target audience is limited only by the number of televoting lines purchased; one televoting line is equivalent to one option. Televoting can be implemented in a variety of ways. For example, it can include an acknowledgement announcement; it can capture voting dates and times; it can capture phone numbers (except where the calling party number is marked private, in which case the first 3 digits of the phone number are captured and the last 4 are replaced with filler); and it can provide a variety of summary reports that include and exclude duplicate calls from the same calling party number. A popular show, American Idol, uses televoting to record audience votes.

telework The combination of computer and telecommunications technology which enables office workers to work at home or away from the main office on a part-time or full-time basis. See also Telecommuting.

teleworker A person who works from his home or some place distant from his company's office. A teleworker may send his completed work in and pick his new work up via a modem in his PC. A teleworker may also be on the phone at home answering calls on behalf of his company and entering the results of those calls (i.e. reservations on airlines, orders for catalogs) in on a PC connected by phone lines to his company. He may use one phone line, like an ISDN BRI line or he may simply use two analog phone lines – one for talking on and one for PC's data. Or he may simply use one analog phone line and use a protocol such as VoiceView.

Telex A worldwide switched message service. Telex service is offered in the United States by the Western Union Telegraph Company, MCI, ITT, RCA, FTCC and TRT. Telex has one gigantic advantage: Overseas it's very popular and widely used. Contacting overseas businesses by telex is often far more reliable and faster than contacting them by telephone. Telex is good for overseas time zone differences because you can send a message to an unattended telex machine. It also delivers a printed record. Telex is relatively inexpensive usually costing a little less than a phone call. Telex has one disadvantage: It's very slow and not very accurate with virtually no data communications error checking procedures. Telex is being rapidly displaced by faster, more accurate forms of data communications, including the public packet switched networks and the various electronic mail services, and most recently by massive competition from low-cost facsimile machines. See Teletex.

Telex Access Unit TLXAU. An X.400 AU (Access Unit) serving Telex users.

tell tales Directory publishing industry jargon for the letters or words that appear at the top of every page of the white pages and yellow pages in the telephone directory, and which identify the first and last entries on the page.

tellyphone A mobile phone that is also designed for mobile TV service.

Telnet A program that lets you connect to other computers on the Internet. The process by which a person using one computer can sign on to a computer in another city, state or country. Telnet is the terminal-remote host protocol developed for ARPAnet. Using Telnet, you can work from your PC as if it were a terminal attached to another machine by a hard-wired line. The format of the telenet command is telnet address.domain or telnet address. domain port #. These days, most users are insulated from TELNET by GUI browsers, such as Netscape or Internet Explorer.

Telnet Port The port address on a computer which supports remote telnet access. Normally port 23 is the default telnet port.

Telocator Alphanumeric Protocol See TAP.

Telocator Message Entry See TME.

Telpak A discontinued AT&T service that gave large customers discounts on purchases of multiple analog private lines. Telpak was to discourage users from building their own private microwave systems. Users who bought Telpak, however, were not allowed to resell any of the circuits, though they were allowed to share them. The FCC ruled Telpak as being discriminatory against competition. (It was too cheap.) It may still exist on an intrastate basis in some states. Telpak typically came in bundles of 12, 24, 60 and 240 voice lines. The bigger the bundle, the cheaper the per circuit cost.

TELR Talker Echo Loudness Rating. A score, in dB (decibels), based on ratings by test subjects of the audibility of an echo in a communications channel. Echo in telecom channels is distracting, annoying and the reason satellites aren't used much for voice any longer. The most common source of echo is at two-wire to four-wire hybrids, such as at the line card of a digital switch. Echo can also be caused by poor handset design (receiver to transmitter coupling) or from speaker phones (speaker to microphone coupling). It turns out that the objectionability (new, invented word) of an echo is a function of the delay between when a speaker utters a word and the reception of the echo back at his/her handset, as well as the level (i.e. the volume). Delay can be introduced at a number of places. The long distance network is the most significant source. An intracontinental call will have a delay less than 30 msec, whereas intercontinental calls may have as much as 100 msec. Satellite calls have nearly 600 msec. Delay can also be introduced by digital customer premise equipment, as well as small delays due to encoding and switching in the local central office switches. VOIP (Voice over IP) networks tend to introduce significantly more delay than circuit switched networks. Therefore the need for echo cancellers is also greater with VOIP. Considerable reasearch has been done on this topic using test subject ratings. Both the level of echo and the delay affect the objectionability of an echo. The ITU-T standard G.131 discusses this relationship as "Talker Echo Loudness Rating". It demonstrates, for example, that delays of 300 msec must be suppressed by 40 dB more than an equally loud echo with a delay 5 msec to achieve the same relative level of annoyance. TELR, along with MOS (Mean Opinion Score), are two important factors that determine the subjective "quality" of a com-munications channel. The actual Talker Echo Loudness Rating is a score, in dB (decibels), based on ratings by test subjects of the audibility of an echo in a communications channel. Usually used in the context of comparing the relative audibility of a fixed echo as delay is varied. TELR is one of the factors involved in determining the MOS (mean opinion score). That makes it a hot topic. Delay is almost unavoidable in large-scale VOIP networks, and that delay makes TELR important. Contributed by Rolf Taylor, Applications Engineer, Telos Systems. See also: Echo, Echo Canceller, Echo Suppressor, Hybrid and MOS.

TELRIC Total Element Long Run Incremental Cost. There are two problems in the regu-lated monopoly phone industry today. First, how to price end-user services like local phone service and second, how to price services – such as the rental of local cable to companies such as CLECs – competitive local exchange carriers. In pricing such services, there are three questions: First, what's the real cost? (There's obviously no way of figuring it.) Second, what do we want the answer to be? Third, what do you want the answer to be for political reasons? You may want the price to be low – for example to help poor people in rural areas? In the phone industry there are as many ways of calculating prices as there are prices. Telric is a way of figuring out what the end user should pay for phone service based on the incremental cost of new equipment and new labor, not counting the embedded cost of old equipment and the labor to install that old equipment. Telric has been used to put a price cap on what telephone companies can charge for links to homes and businesses. Ac-cording to George Gilder, "the Telric cap is based on an estimate of costs that would apply in a fully competitive environment, when bandwidth is a pure commodity." This is not exactly accurate, but then neither is Telric.

TELSET Telephone Set.

Telsa, Nikola Nikola Tesla was born a subject of the Austro-Hungarian Empire in 1856 in a mountainous area of the Balkan Peninsula known as Lika. It is Serbia. Yugoslavia was a federation of several small countries and they went back to their original country name after Yugoslavia fell apart.

Telstar 1 1. Telstar 1 was the world's first active communications satellite. It was launched on July 10, 1962. (Sputnik was launched on October 4, 1957.) There is some argument about Telstar's claim to fame as the first. The engineers at the RCA Astro-Electron-ics Division, Princeton NJ (now Martin-Marietta Aerospace) claimed that they launched and successfully used a satellite to broadcast the coronation of the pope a little earlier than Telstar. But it was Telstar that got all the fame and glory, and that RCA engineers don't deny that.

2. The space-age excitement and promise of Telstar inspired British record producer and composer Joe Meek to compose an instrumental recording which he named after the satel-lite. Meek got the British instrumental rock group, The Tornados, to record the song, and afterwards Meek added space-travel sound effects and space-age sounds of an electronic keyboard to the recording. In August 1962, one month after the the first transmission of Telstar, the satellite, the instrumental recording was released, to popular acclaim. It quickly climbed to the top of the charts in England and soon thereafter climbed to the top of the music charts in the US, the first record by a British group to do so, a full year before the Beatles. The American instrumental group, The Ventures, re-recorded the song in 1963 and it, too, became a hit.

Telstra Telstra is Australia's largest local and overseas phone company. The company has its origins in 1901, when the Postmaster-General's Department (PMG) was formed to manage all domestic telephone, telegraph and postal services. The Overseas Telecom-munications Commission (OTC) was formed in 1946 to manage international telecom-munications. The Australian Telecommunications Commission, trading as Telecom Australia, was created as a separate entity in July 1975, following the breakup of the PMG. Telecom Australia and OTC merged in February 1992 to form Telstra. On July 1, 1997, Australia's telecommunications markets were opened to full competition, with no limit on the number of carriers that own transmission infrastructure who can enter the market.

Telstra Research Laboratories Telstra's research arm, founded in 1923 and closed in December 2005. It was the Australian counterpart of Bell Labs in the United States.

temperature rating The maximum temperature at which the insulating material may be used in continuous operation without loss of its basic properties.

tempest Unclassified name referring to the investigation, study, and control of com-promising emanations from electrical and electronic equipment. Devices which are tempest-secure mean they do not send emanate electromagnetic signals which can, potentially, be received by others, i.e. enemies.

template 1. A voice processing term. A pattern of information as a function of time, which is intended to represent an entire word.

2. A Norstar definition: A system wide setting assigned during System Startup. The most

important effects of a template are the number of lines assigned to the telephones, and the assignment of Line Pool Access. Templates will also assign other system wide defaults, such as Prime Line and Ringing Line assignment. It is important to understand that a template is only provided as a convenience, and that any settings effected by the template can be changed.

temporal coding Compression that is achieved by comparing frames of video over time to eliminate redundancies between frames.

Temporary Signaling Connections On August 14, 1995, AT&T announced Temporary Signaling Connections, which it billed as the first service that lets banks, retail outlets and other data-intensive businesses link their Software Defined Network locations together on demand using virtual connections created in AT&T's national signaling networks. Businesses can use Temporary Signaling Connections to verify credit card transactions, update inventory databases, exchange data with automatic cash machines. The service uses a portion of the D channel capacity of an ISDN PRI channel and passes information to other ISDN PRI locations.

temporary station disconnection Allows the attendant to completely remove selected phones from service at any time on a temporary basis.

temptation I generally avoid temptation unless I can't resist it – Mae West.

Don't worry about avoiding temptation. As you grow older, it will avoid you. – Winston Churchill.

TEN Transformer Exciting Network. A specially designed drainage reactor that provides a low impedance path-to-ground for longitudinally induced currents to flow. It bridges across one of the working circuits in an INT to serve as its "exciting" pair, thus freeing up a circuit that would have otherwise had to be grounded and unavailable for revenue-producing services. See INT.

ten-code A short numeric code beginning with the number 10, which is used in radio transmissions by the police, fire departments, truckers, and ham radio operators. The most well known of these ten-codes (also known as 10-codes) by the general public is 10-4, meaning "OK, message received."

ten-high day A traffic engineering term for a traffic study which considers the average of the traffic during the same clock hour on the ten busiest normally recurring days of the busy season of the year. See Ten-High Day Busy Hour and Traffic Engineering.

ten-high day busy hour A telephone company term. The hour, not necessarily a clock hour, which produces the highest average load for the ten highest business day loads in that hour. It may be a different hour from the ABS busy hour. With present data collection procedures, the ten high days are usually selected from the busy period.

Ten-High Day Data THD. Data collected during the ten-high day busy hour. See above.

tenant partitioning Also known a Service Bureau Capability. One computer or telephone host can provide service to many tenants in the building.

tenant service Some businesses acquire a telephone system too large for their needs so they sell parts of the service to smaller offices in their own building or in the surrounding community. There are two ways to make money on tenant service – renting phone equipment or re-selling long distance lines. There's more money on re-selling long distance lines.

tensile load Refers to the maximum load or pull force that may exerted upon a cable during installation or relocation without damage. An excessive tensile load on twisted pair cables can cause elongation or untwisting which may result in signal loss.

tensile strength A term denoting the greatest longitudinal tensile stress a substance can bear without tearing apart or rupturing.

TEO See Telephone Equipment Order.

TER Pronounced "terr." From French or Latin meaning "third," or for the third time. Used in ITU terminology to indicate the third enhancement to an existing communications standard. For example, "V.27 ter" is the third enhancement to V.27 (or a subsequent enhancement to V.27 bis").

tera- T. A million million, or a thousand giga-. A prefix that denotes 10 raised to the 12th power, or one trillion. In computer terms, however, tera- is actually equal to 2 raised to the 40th power power , or 1,099,511,627,776, the power of 2 that is closest to one trillion. See Terabyte.

terabit A terabit is literally equal to 2 raised to the 40the power, 1,099,511,627,776 bits, but is often calculated with a base of 10 (10 raised to the 12th power), making it equal to one trillion bits. In short, one million million bits. The capacity of optical fiber is now moving into the terabit per second range. In April 1999, NEC announced it had achieved three terabits per second on a single strand of cable. That's 3,000,000,000,000 bits per second – more than enough capacity to carry the entire Internet in the entire world. That

brings some urgency not to break the cable.

Terabyte A combination of the Greek "tera," meaning "monster," and the English "bite," meaning "a small amount of food." A unit of measurement for physical data storage on some form of storage device-hard disk, optical disk, RAM memory etc. and equal to two raised to the 40th power, e.g. 1,099,511,627,776 bytes . Roughly a trillion bytes. Here is the progression.

KB = Kilobyte (2 to the 10th power)
MB = Megabyte (2 to the 20th power)
GB = Gigabyte (2 to the 30th power)
TB = Terabyte (2 to the 40th power)
PB = Petabyte (2 to the 50th power)
EB = Exabyte (2 to the 60th power)
ZB = Zettabyte (2 to the 70th power)
YB = Yottabyte (2 to the 80th power)
One googolbyte equals 2 to the 100th power.

To put things in perspective (courtesy of Microsoft), a terabyte holds a 100-byte record for every person on earth, as well as an index of those records; or a JPEG-compressed pixel for every square meter of land on earth, which is plenty to create a high-resolution photograph; or 1 billion business letters, which would fill 150 miles of bookshelf space; or 10 million JPEG images, which would provide 10 days and nights of continuous video.

teraflop A trillion (10 to the 12th power) floating point instructions per second. A measure of a computer's speed. A teraflop is a trillion calculations per second. That's one million million calculations per second.

terahertz THz. A unit denoting one trillion (10 to the 12th) hertz. See TERA-

terapops A point of presence (POP) with a capacity of terabits per second. See Point of Presence.

teredo A tunneling mechanism that enables IPv6 packets to pass through IPv4 networks and IPv6-unaware NAT devices by encapsulating the IPv6 packets inside of IPv4 datagrams. The mechanism is named after a type of shipworm which, like all shipworms, bores holes into wooden ship hulls.

TERENA Trans-European Research ann Education Networking Association. Formed in 1994 by a merger of Rare and Earn.

term Short for Termination. See Termination.

term sheet A term sheet is a fancy name a piece of paper which lists the terms of an deal which everyone has agreed to. A term sheet is not the final legal agreement. It's the paper which everyone initializes as correct and then hands to the lawyers and tells them to "paper" it, in other words create a legally binding agreement which the parties will then sign later.

terminal The word terminal is very loose. Some people call any piece of communicating equipment a terminal. Some people simply refer to any wall jack or wall plug as a terminal, as in "We need to connect these wires to the terminal." Some people believe it's the point at which a telephone line ends, or is connected to other circuits of a network. There are many ways of enhancing the word terminal by adding another word. See the following definitions.

Terminal Adapter TA. A Terminal Adapter, also known as an ISDN Modem, is an interface device that essentially is a protocol converter that serves to interface non-ISDN devices (e.g., PCs, fax machines and telephone sets) to an ISDN BRI (Basic Rate Interface) circuit. In more technical terms, a TA is an interface device employed at the "R" reference point in an ISDN environment that allows connection of a non-ISDN terminal at the physical layer to communicate with an ISDN network. Typically, this adapter will support standard RJ-11 telephone connection plugs for voice and fax, and RS-232C, V.35 and RS-449 interfaces for data. See also BRI and ISDN.

terminal address Where there's a terminal to punch down, there's an address. That address will have numbers on it and will enable the technician who's responsible for fixing the circuit (when it breaks) to come, find it and fix it.

terminal block A device used to connect one group of wires to another. Usually each wire can be connected to several other wires in a bus or common arrangement. A 66-type block is the most common type of connecting block. It was invented by Western Electric (now called Lucent). Nortel Networks has a terminal block called a Bix block. Other manufacturers make their own versions, though the Lucent one is most popular. A terminating block is also called a connecting block, a punchdown block, a quick-connect block, a cross-connect block. A connecting block will include insulation displacement connections (IDC). In other words, with a connecting block, you don't have to remove the plastic shielding from around your wire conductor before you "punch it down."

terminal configuration The functional interconnection of the components of a terminal. For example, a keyboard-printer may be configured to transmit keystrokes without printing them. Printing is only performed on data retrieved from the communication line. Terminals with multiple components can be configured in a variety of ways.

terminal emulation An application that allows an intelligent computing device such as a PC to mimic or emulate the operation of a dumb terminal for communications with a mainframe or minicomputer. It does this with special printed circuit boards inserted into its motherboard and/or special software. For example, TELECONNECT uses the communications software program called Crosstalk to emulate a DEC VT-100, a Digital Equipment Corporation VT-100 terminal. We do this because emulating a DEC VT-100 works better with certain software programs we call up remotely.

terminal equipment Terminal Equipment usually refers to the telephones and other equipment at the end of telephone lines. See also CPE.

Terminal Equipment Type 1 TE1. In Integrated Services Digital Network (ISDN) technology, TE1 is a type of terminal compatible with ISDN.

Terminal Equipment Type TE2. In Integrated Services Digital Network (ISDN) technology, a type of terminal that must be connected to ISDN via a specially designated point, normally an RS 232 or RS 449 interface.

terminal hunt group Also called Terminated Hunt Group. It's another name for a "top down" hunt group. You have a bunch of phone lines. They are in a "hunt" group which means that, if one is busy, the switch sends the call to the next available line in the hunt group. In a terminal hunt group, the switch always starts at the top of the hunt group and goes down, searching for the first available line from the top. This contrasts with a circular hunt group, where the switch remembers the last line it connected and, starting there, hunts down to the next available line, searching basically in a circle. A terminated or "top down" hunt group puts more calls on the first lines in the group. A circular hunt group tends to distribute the calls evenly. There are reasons why you might choose one type of hunt group over another. You might choose to evenly distribute your calls over a bunch of humans answering incoming calls or you might choose to send the calls top down to a voice mail system and watch the usage statistics carefully to tell if you need more or fewer lines and more or fewer voice cards.

terminal impedance The impedance as measured at the unloaded output terminals of transmission equipment or a line that is otherwise in normal operating condition. The ratio of voltage to current at the output terminals of a device, including the connected load.

Terminal Interface Node TIF. Provides the interface between a terminal and the Concert Packet Services network.

terminal net loss See echo.

terminal node In IBM Corp.'s Systems Network Architecture (SNA), a network device that cannot be programmed by the user.

terminal number 1. A terminal number is one or multiple circuit numbers not identified by an individual directory telephone number (DTN) but by 001, 002, 003, etc. and is only referenced as a subset of the main trunk pilot DTN. Terminal numbers may be rented from both Incumbent Local Exchange Carriers (ILECs) or Competitive Local Exchange Carriers (CLECs), like MCI Worldcom. Terminal numbers are all organized to hunt (descending, ascending, most idle, least idle). A terminal number is another name for an auxiliary or private line that doesn't have a real number, doesn't get a listing in the phone book, but gets a monthly bill. An auxiliary number is a telephone trunk you rent from your local phone company in addition to the main number you rent. Phone systems are always set up for multiple phone lines, so that when a call comes in, it doesn't hit a "busy," but rolls over to one or more auxiliary lines. That collection of lines is called an Incoming Service Group, or ISG. For example, the publisher's main office main number is 212-691-8215. But it also has 8216, 8217, 8218 and several unmarked or coded trunks. All these numbers are auxiliary lines and don't receive their own bill or directory listing from the phone company. Costs for these lines are lumped onto the bill for the main number. See Rollover and ISG.

2. TN. The physical address of a device (such as, telephone set, a truck, and attendant) on a Nortel Networks' PBX. The TN is composed of the loop, shelf, card and unit IDs.

terminal repeater A repeater for use at the end of a trunk line.

terminal server A small, specialized, networked computer that connects many terminals to a LAN (local area network) through one network connection. Any user on the network can then connect to various network hosts. A terminal server has a single network interface and several ports for terminal connections. One advantage of a terminal server is it allows many terminals to be connected to a host via a single existing LAN cable or hub, rather than a variety of point-to-point cables. A terminal server is especially valuable where lots of asynchronous terminals require low-speed access to a host computer. Since asynchronous devices output a character at a time, and since LANs work on the basis of a frame with

minimum and maximum frame sizes, asynchronous terminal devices can be very wasteful of LAN resources – the frames must be padded with stuff bits in order to satisfy the minimum frame size. The terminal server accepts individual characters output from the terminals and the ports to which they are connected, buffers them, interleaves them, packetizes them in LAN frames, and places them on the network through a process much like Time Division Multiplexing (TDM). At the target host computer, a matching terminal server recognizes the type of traffic by virtue of the port address of the incoming traffic, and breaks down the frames to get at the individual characters generating by the originating asynchronous terminals. See also Asynchronous, LAN, Server, and TDM.

terminal services The underlying technology on that enables remote desktop, remote assistance, and terminal server.

terminal shelf A terminal shelf is an area inside a FOTS (Fibre Optic Transmission System) which is used to house the circuit cards inside the bay.

terminal table An ordered collection of information that identifies each line, phone, component or application program from which a message can be sent.

terminals The screws or soldering lugs to which an external circuit can be connected.

terminals to long distance operator Commonly known as "toll terminals", they provide special trunks directly to the long distance telephone company operators. Upon completion of long distance calls, the toll operator will ring the attendant (or hotel operator) and give them "time and charges" for the phone call just ended.

terminate 1. To connect a wire conductor to something, typically a piece of equipment.

2. To end one's telecommunications service or equipment rental.

terminated 1. The condition of a wire or cable pair which is connected to (terminated on) binding posts or a terminal block.

2. The condition of a circuit connected to a network which has the same impedance the circuit would have if it were infinitely long.

terminated line A telephone circuit with a resistance at the far end equal to the characteristic impedance of the line, so no reflections or standing waves are present when a signal is entered at the near end. Compare with bridge tap.

terminating channel The name for the circuit in a private line channel that connects a local central office with the CBX/PBX or telephone instrument at the customer's premises.

Terminating Multiplexer TM. A type of Path Terminating Equipment (PTE) used to provide access to a SONET/SDH network. The Terminating Multiplexer is equivalent to a Time Division Multiplexer (TDM) in a T/E-Carrier context. The TM also serves to perform the signal conversion process from electrical to optical on the transmit side, reversing the process on the receiving end.

terminating NPA/NXX The area code and exchange of the number dialed.

terminating office The switching center (i.e. the central office) of the person you're calling (the "called party").

terminating only traffic A type of circuit operation that provides for traffic in the terminating direction only (from the carrier to the end user).

terminating resistor A grounding resistor placed at the end of a bus, line, or cable to prevent signals from being reflected or echoed. Sometimes shortened to terminator.

termination 1. Picture a phone company with lines in the United States (but not overseas) who has a customer who dials his mother in England. At some point the U.S. phone company will transfer the call to an English phone company who will carry the call to its destination, also known as terminating the call. For international termination, the English phone company charges a fee. These fees have dropped dramatically in recent years. In 1996 it was 24 cents. In 2003 it was 1.2 cents. In 1996, it cost 74 cents a minute to terminate a call in Jamaica. In 2003, it was 7.3 cents. In 1996 it cost 49 cents a minute to terminate in Singapore. In 2003 it was 1.1 cents – a 98% reduction. For USA domestic calls, from 1996 to 2003 calls dropped from 6.3 cents to 1.2 cents. These figures courtesy Arbinet, an aritrage exchange.

2. Termination involves the placement of impedance matching circuits on a bus to prevent signals from being reflected or echoed.

3. Connecting a data signal to a network or a device.

termination restriction Prevents a user from receiving any calls on the phone line. A DID call to the restricted termination routes to an attendant, an announcement or intercept tone at customer option. All other calls route to intercept tone.

termination of service The end of service of a line or equipment. All pursuant to the regulations set forth in the tariff.

terminator Some communications facilities – e.g. local area networks – are bus configurations. This means one long piece of cable with workstations connected along the way, typically with "T" connectors. For a network to work properly, you need to place resistance at the end of the cable that serves to absorb the signal on the line. A thin wire Ethernet typically requires a 50 ohm resistance at either end of the bus. You can buy these Ethernet terminators already included in a connector.

terminus A device used to terminate an optical fiber that provides a means to locate and contain an optical fiber within a connector.

terrestrial Long distance facilities which are entirely on land and do not use satellites. This includes microwave, coaxial cable, optical fiber, normal cable, etc. There are reasons to prefer terrestrial facilities over satellite facilities:

1. No echo or delay in voice conversation. Some people find satellite conversations disturbing because of the delay; and 2. No significant reduction in data throughput. Most data communications protocols send their data in "chunks" and require an acknowledgement from the other end when one chunk has been received before the next chunk can be sent. When it takes a long time for an acknowledgement to be received (as in a satellite circuit), the effective throughput of data becomes very slow.

When you're trying to send one one-way signal to many locations (such as in broadcast television), satellites usually do a better and cheaper job.

Terrestrial Interference TI. Interference emanating from sources on the earth as distinguished from those emanating in space.

Tesla, Nikola Nikola Tesla is regarded as one of the most mysterious and least recognized scientific pioneers in modern history. He is commonly associated with high-frequency electrical devices, radio transmission, and the invention of the multi-phase alternating current system in use today. Born on July 9, 1856 in Smiljan, Lika, Croatia, Yugoslavia, Tesla was the fourth of five children and son of a reverend of the Serbian Orthodox Church. Tesla was educated at the polytechnical school at Graz, Austria (1875), where he acquired an interest in the study of electrical engineering and mathematics. Tesla was dismayed at the inefficient design of DC motors. While working as a telegraph operator in Budapest (1881) and later as a telephone engineer for Edison in Paris (1882), he developed a multi-phase alternating current power system, named the Tesla Polyphase System. Other engineers had attempted AC designs modeled after direct current systems with a single circuit, but none were ever successful. Tesla's multi-circuit system included new polyphase induction motors, dynamos, and transformers, the patents to which were purchased in 1888 by the Westinghouse Electric Company from the Tesla Electric Company.

Tesla's 60 cycle AC Polyphase System was the primary competitor to the Continental Edison Company's DC Current System, and caused Thomas Edison to personally wage a massive propaganda campaign against Tesla and the "dangers of alternating current". However, AC could travel over hundreds of miles and at much higher voltages, while DC traveled only much shorter distances and required a generator every two miles. Despite Edison's efforts, the war of the currents was won by Tesla. The Chicago World's Fair in 1893 featured the first electrically powered pavilion, designed entirely by Tesla using his polyphase system and financed by Westinghouse. This pavilion featured Tesla demonstrating many high-energy devices, including the wireless transmission of energy using a tuned circuit, a device that would come to be known as radio. Guglielmo Marconi used several of Tesla's patents in his 1901 "invention" of the wireless radio transmitter. Marconi's initial patent claims for this device were rejected based on Tesla's prior patents (645,576 and 649,621 granted in 1900) for the wireless transmission of energy. Marconi's later patent claims for a signal communications device were granted. Despite a lawsuit filed by Tesla in 1915 and other efforts, history has erroneously painted Marconi as the inventor of the radio. Some redemption to this injustice occurred when the Supreme Court ruled on June 21, 1943 in the case of Marconi Wireless Telegraph Company of America vs. the United States that Tesla's radio patents predated those owned by Marconi. In effect, this ruling declared Tesla the inventor of radio; it came five months after his death.

Tesla continued to experiment with high-frequency energy. He believed that energy in phase with the natural vibrations of thunderstorms and the Earth (7.68 Hz) could be broadcast to anyplace. Using this theory, he received backing from J.P. Morgan to build a world-wide telegraphy system. Tesla also believed that electrical power was present everywhere, in unlimited quantities, and free for the taking. Tapping into such a natural source of energy could replace all other fuels. The global wireless transmission of power, however, did not interest Morgan, who did not see the benefit in providing free electricity to humanity. Tesla created hundreds of inventions and improvements during the course of his life, including the telephone repeater, wireless communications, radio, antennas, ground connections, aerial ground circuits with inductance and capacitance, tuned circuits, emitters and receivers tuned to resonance, the electronic tube, fluorescent lighting, the electromechanical audio speaker, AND and OR logic gates, radio remote-control, robotics, radar, and diathermy.

Tesla's patented inventions includes a robotic submarine (613,809), vertical take-off and landing aircraft (1,655,114), disk (bladeless) turbine engine (1,329,559), ozone generator (568,177), electro-dynamic induction lamp (514,170), and superconduction (685,012). His more fantastic concepts include anti-gravity propulsion and a thought photography machine. The more than 700 patents Tesla was awarded during his lifetime represent only a fraction of his total number of inventions and discoveries. Tesla often didn't bother to patent many of his inventions (he received his last patent in 1928) and often failed to document his work. Tesla died on January 7, 1943 secluded in his New York City apartment at the age of 86. Tesla was intelligent, highly strung, neurotic, charismatic, germphobic, and always very well-dressed. Over his lifetime his financial backers were a diverse group including George Westinghouse, J.P. Morgan, and even Thomas Edison. His friends and supporters, an equally diverse group, included Albert Einstein, Samuel Clemens, and Eleanor Roosevelt. His manor and mystery as depicted by the press inspired the creation of evil comic book geniuses that did battle with Superman and Captain Marvel using a myriad of strange inventions and energy devices. Tesla's secret work on "death ray" energy devices fueled much of this, as did the U.S. government's confiscation of many of his inventions after his death.

test and validation Physical measurements taken to verify conclusions obtained from mathematical modeling and analysis.

test antenna An antenna of known performance characteristics used in determining transmission characteristics of equipment and associated propagation paths.

test bed A constant physical and electrical environment in which devices or programs are tested in order to measure their performance against requirements, benchmarks, or each other.

test board A switchboard equipped with testing apparatus.

test center Equipment for detecting and diagnosing faults and problems with communications lines and the equipment attached to them. If centralized, a facility where a network manager or technician can gain access to (almost) any circuit in a network for the purpose of running diagnostic testing. Also called a network control center.

test desk A desk equipped with equipment to test and repair subscriber lines. See also Test Center.

test friendly busy A test to see if a line is busy. The subscriber does not know the line is being tested. Such a test is usually performed by the operator if someone calling that number requests it. It used to be free. So did a lot of things in this world.

test intrusive Breaking a circuit in order to test its functionality. Testing intrusively will drop service on the circuit.

test repeatability This definition courtesy Steve Gladstone, author, "Testing Computer Telephony Systems," (available from 212-691-8215) says that a key component in any successful test program is the ability to repeat tests simply and quickly. If problems are found, tests must be rerun, both to help recreate and document the problem, as well as to verify the bug fix. Repeating tests can be very time consuming as many tests will change the state of either the test system or the computer telephony system under test. For example, checking that messages are correctly deleted means starting the test from a known state where messages are in a mailbox. When the test is completed usually there are fewer messages left than when the test started. Repeating the test may therefore necessitates reinitializing the computer telephony system to the state before the test started.

test set A telephone handset with extra electronics designed to test telephone circuits. Also called a butt set, since it typically hangs on the technician's tool belt near the wearer's butt... Well, that's one explanation. Another is that it's called a "butt" set because it allows the use to "butt in" to a conversation and listen to its quality, etc.

test shoe A device that is applied to a circuit at a distributing frame to gain test access to circuit conductors.

test tone A tone used to find trouble on phone lines. Also called installer's tone. A small box that runs on batteries and puts an RF tone on a pair of wires. If the technician can't find a pair of wires by color or binding post, they attach a tone at one end and use an inductive amplifier (also called a banana or probe) at the other end to find a beeping tone. A more technical explanation: A tone sent at a predetermined level and frequency through a transmission system to facilitate measurement and/or alignment of the gains and/or losses of devices in the transmission circuit.

test, friendly busy A test to see if a line is busy. The subscriber does not know the line is being tested. Such a test is usually performed if someone calling that number requests it.

testability The ability of a system to support testing and verification of the services it offers.

tetherless An all-encompassing term including the concepts of wireless mobility and intelligent network services (such as the ability of a user to roam freely among wired and wireless networks while making and receiving calls.

TETRA Terrestrial Trunked Radio. An European Telecommunications Standards Institute (ETSI) standard for the type of mobile radio known variously in the US as TMR (Trunk Mobile Radio) and SMR (Specialized Mobile Radio). The typical application for such systems is in local communications with fleets of vehicles, such as taxicabs, fire trucks, police cruisers, and emergency vehicles. TETRA divides these applications into PAMR (Public Access Mobile Radio) and PMR (Private Mobile Radio). PAMR includes applications such as fleets of taxicabs, delivery vehicles, and utility vehicles. PMR applications include fleets of public service and emergency vehicles. TETRA combines the features of mobile cellular telephony, fast data communications and the workgroup capabilities of mobile radio. A measure of the reliability and range of functions of the TETRA standard is that it has already been adopted by many public safety and emergency organizations. The technology behind TETRA creates new standards of network service and functionality at a cost-effective price. It gives you "press-to-talk" instant call set-up for individual and group calls as well as the expected added-value features found in any advanced mobile network. TETRA operates on a simplex (one-way communications) basis, making use of TDMA (Time Division Multiple Access) for digital communications over 25 kHz channels in the 400 MHz frequency band. the 400 MHz frequency band supports good signal propagation over relatively long distances, compared with the GSM band of 900 MHz. TETRA's packet data transfer operates at up to 28.8 Kbps through the concatenation of four digital time slots. This, together with the Short Data Service (text and numeric messaging), provides the fastest and a flexible mobile data transfer capability available. SDS allows the transmission of four different message types, with Type 4 supporting data transfers up to 2047 bits, which is approximately 256 bytes (i.e., characters). Advanced TETRA technology delivers great voice quality which is especially impressive in noisy industrial environments and in-car, hands-free operation. TETRA operates in both the "open channel" mode and the Direct Mode. The open channel allows all terminals to hear all communications, which are handled through a centralized dispatcher. The Direct Mode is a "walkie-talkie" mode that allows direct operation between terminals. The TETRA Memorandum of Understanding (MoU) was established in December 1994 to create a forum which could act on behalf of all interested parties, representing users, manufacturers, operators, test houses and telecom agencies. Today the TETRA MoU represents 58 organizations, from 19 countries. See also APCO25, SMR, SMS.

tetrode A four-element vacuum tube, consisting of filament (or cathode), grid, screen grid and plate.

text Transmitted characters which make up the body of a message.

text based browser A browser that cannot handle hypermedia files.

text enriched The successor to MS-DOS text/richtext, is a simple text markup language for MIME that lets you mark up the document (using commands enclosed in angle brackets) without making the text unreadable to someone without the software to interpret it. See MIME.

text file A file containing only letters, numbers, and symbols. A text file contains no formatting information (like bolding and underlining and type fonts and sizes), except possibly line feeds and carriage returns. A text file is an ASCII file. A text file can be read by every word processor and editor. A text file the lowest common denominator in the word processing world. I wrote this file with an editor called The Semware Editor, which produces only text files. I did this because this dictionary has to be sent to a Macintosh for "typesetting" and to a DEC for distribution on CD-ROM. And a text file is the form both easily recognize.

text messaging Text messaging is a simple term for what it is – sending messages (also called email) in plain text to various devices – wired or wireless. The most common implementation of text messaging is something called short message service – SMS or S.M.S. It is a way by which short messages can be sent to and from digital cell phones, pagers and PDAs (personal data assistants) other handheld wireless devices. Alphanumeric messages of up to 160 characters can be supported by SMS. That's adequate for stock quotes, filtered (abbreviated) e-mail, bank account balances, buying movie tickets on line, updates on traffic conditions, answers to quizzes posed by the teacher and other really short messages. Europeans, who have relied on digital cell phones for years, prefer to send short text messages by tapping on their telephone dialing pads. SMS is defined in IS-41C. Text messaging is about to get a big boost from 2.5G – second and a half generation digital wireless phones and systems and 3G – third generation. These new generations will seriously expand the amount of characters and the quality of images that can be sent back and forth to and from cell phones. See also IS-41 and Wireless Application Protocol (WAP).

text telephone A machine that employs graphic communication in the transmission of coded signals through a wire or radio communication system. TT supersedes the term,

"TDD" or "telecommunications device for the deaf."

text-to-landline A service that lets a cell phone user send a text message to a landline number. The text message is converted into a computer-generated voice message which gets played to the recipient or deposited in the recipient's voicemail box or answering machine. In the first scenario, the called party may have the option to record a reply or select a preset text message response. For cellcos that offer text-to-landline service, the service is automatically triggered when the number that the text message is sent to is not a mobile phone number. Standard text message rates apply.

Text-To-Speech TTS. Technologies for converting writtten text into spoken speech output. A system with text-to-speech (TTS) capabilities is able to interpret electronic text and generate audible speech from it. Text-to-speech technology can be used in many different ways including reading email, airline timetables, etc. TTS is used in computers from laptops to over the phone applications. It is used in interactive voice response (IVR) systems. IVRs provide information such as credit card balances, movie listings, etc., verbally to customers who access the system by phone. See also Drunken Swede, IVR and Speech Concatenation.

texting Another way of saying text messaging. See text messaging.

TF8 Business PBX Trunk.

TFC PBX Service, Combination Flat Rate Trunk.

TF Fixture wire, thermoplastic-covered solid or 7 strands, 600xAFC.

TFE Teflon (tetrafluoroethylene)

TFF Fixture wire, thermoplastic-covered, with flexible stranding.

TFFN Fixture wire, thermoplastic insulation and nylon sheath, with flexible stranding.

TFT-LCD Thin Film Transistor Liquid Crystal Display panels are currently the most widely used flat panel display technology. TFT-LCDs are used in notebook computers, desktop monitors, televisions, digital cameras, portable DVD players, mobile phones, portable games, and car navigation systems, among other applications. TFT-LCDs work by assigning a tiny transistor to each pixel, making it possible to control pixels independently of each other. TFT screens are very fast, have a high contrast ratio and a wide viewing area.

TFTP Trivial File Transfer Protocol. A simplified version of FTP that transfers files but does not provide password protection or user-directory capability. It is associated with the TCP/IP family of protocols. TFTP depends on the connectionless datagram delivery service, UDP.

TFTS Terrestrial Flight Telephone System.

TG 1. Trunk Group.

2. Task Group. A term used by the IEEE (Institute of Electrical and Electronics Engineers) for a group of people who work on tasks associated with larger standards recommendations that are the overall responsibility of a Working Group (WG). See also 802.15.

TGB 1. Trunk Group Busy

2. Telecommunications Grounding Busbar.

See TBB.

TGC Transmission Group Control in IBM's SNA.

TGE Terminator group - type E.

TGF Terminator group - type F.

TGW Trunk Group Warning.

TH Transmission Header, an SNA term.

THD+N Total Harmonic Distortion Plus Noise. A measure of the audio clarity of a voice system. The best measure is 0%., or close. Pretty bad is 25%. The average PC voice card is around 10%.

the full monty To strip completely naked. There's a movie and a play of the same name.

the grand alliance A consortium of seven organizations which produced a workable version of HDTV – High Definition TV. The companies are AT&T, General Instrument Corporation, The Massachusetts Institute of Technology, the David Sarnoff Research Center, Philips Consumer Electronics, Thompson Consumer Electronics and the Zenith Corporation.

The Open Group Formed in February 1996 through consolidation of the Open Software Foundation (OSF) and X/Open Company Ltd. (X/Open). The stated mission of The Open Group is to make multi-vendor open systems the preferred customer choice for the delivery of the right information to the right person at the right time. Specific goals include the following: 1) Enabling a rapid vendor response to customer requirements for open systems, 2) Innovative technology research, 3) Accelerated consensus-building around standards and technology, 4) Consensus among open systems vendors, and 5) Promoting a consistent open systems message. Examples of The Open Group's work include the X Window System specification, the Motif Toolkit API, the Distributed Computing Environment (DCE) and Network File System (NFS) specifications, and the Common Desktop Environment (CDE) specification. (www.opengroup.org)

The Organization for the Advancement of Structured In-

formation Standards Also called OASIS. This organization is the largest standards group for electronic commerce on the Web.

The Phone Company Also known by the initials TPC. In the 1966 James Coburn movie "The President's Analyst", the evil worldwide conspiracy was run by TPC, which turned out to be The Phone Company. Some people believe the movie was not fiction.

The Telephone Consumer Protection Act of 1991 See Telephone Consumer Protection Act of 1991.

theater circuit So named because reels of a single movie could be sent by bicycle from one theater to the next. Show times were sometimes cut so close that one theater was showing the first reel of a film while another theater was showing the last. See also blockbuster.

theft recovery A stolen laptop may have sensitive data as well as data the user contend

theorem In the fields of mathematics and logic, a theorem is a formula, proposition, or statement that is deduced from other formulas or propositions. In popular usage, a theorem is an idea that is accepted or proposed as a demonstrable truth that may be part of a broader general theory. See also Nyquist Theorem and Shannon's Law.

Theoretical Midpoint TMP. The theoretical halfway point that divides an international private line circuit into its respective US and foreign halves. A US records carrier is responsible for the US portion of service and a foreign records carrier assumes responsibility for service to the foreign half.

thermal ducting The phenomenon that occurs when the difference in temperature from day to night increases, specifically in the Fall, that causes radio signals (cellular) to travel further than should be possible. An example would be when a handheld cellular user is in central Nebraska, but is using a cellular tower signal in Kansas.

thermal management A fancy term for cooling.

thermal noise Noise created in an electronic circuit by movement and collision of electrons.

thermal shock A test to determine the ability of a material to withstand heat and cold by subjecting it to rapid and wide changes in temperature.

thermionic emission The emission of electrons or ions under the influence of heat, as in a vacuum tube cathode.

thermistor A resistor whose resistance varies with temperature. More technically, a thermistor is a device made from mixtures of metal oxides that exhibits large negative coefficient of resistance changes as the temperature increases.

thermocoule Two dissimilar wires joined together that generate a voltage proportional to temperature when their junction is heated.

thermocouple extension cable A cable comprised of one or more twisted thermocouple extension wires under a common sheath.

thermocouple extension wire A pair of wires of dissimilar alloys having such emf-temperature characteristics complimenting the thermocouple which is intended to be used, such that when properly connected allows the emf to be faithfully transmitted to the reference junction.

thermocouple wire (grade) A pair of wires of dissimilar alloys having emf-temperature characteristics calibrated to higher temperature levels than the extension type of thermocouple wire. Unlike the thermocouple extension wire, this wire may be employed as the thermocouple hot junction in addition to serving as the entire wire connection between hot and cold reference junctions.

thermoplastic Material that will soften and distort from its formed shape when heated above a critical temperature peculiar to the material.

thermoset A plastic material which is crosslinked by a heating process known as curing. Once cured, thermosets cannot be reshaped.

thermostat Thermostats are temperature-activated on/off switches that usually work on the 'bimetal' principle, in which the bimetal strip consists of two bonded layers of conductive metal with different coefficients of thermal expansion, thus causing the strip to bend in proportion to temperature and to make (or break) physical and electrical contact with a fixed switch contact at a specific temperature. In practice, the bimetal element may be in strip, coiled, or snap-action conical disco form, depending on the application, and the thermal 'trip' point may or may not be adjustable. Figures 8(b) and (c) show the symbols used to represent fixed and variable thermostats. A variety of thermostats are readily available, and can easily be used in automatic temperature control or danger-warning (fire or frost) application. Their main disadvantage is that they suffer from hysteresis; typically, a good quality adjusted thermostat may close when the temperature rises to (say) 21 C, but not re-open again until it falls to 19.5 C.

THF Tremendously High Frequency.

THHN 900xAFC 600V nylon jacketed building wire for use in dry locations.

THI Telephone Headset Integrator. A new form of headset manufacturer who will make headsets that do new tasks, like take a phone off hook without physically having to lift the receiver off the phone.

thick Ethernet cable Thick Ethernet cable is 0.4-inch diameter, 50-ohm coaxial cable. Thick Ethernet cable can be bought in pre-cut lengths, with standard N-Series male connectors installed on each end. It also is available in bulk cable without connectors. Any of the following types of connectors will work:
Belden 9880 or Belden 89889 Montrose CBL5688 or Montrose CBL5713 Malco 250-4315-0004 or Malco 250-4314-0003 Inmac 1784 or Inmac 1785

thick route A route with a lot of telecom traffic. A major telecom traffic route. The opposite is thin route.

thicknet Jargon used to describe thick Ethernet coaxial cable. See Thickwire.

thickwire 0.4 inch diameter, 50-ohm, Ethernet IEEE 802.3 coaxial cable. See also Thick Ethernet Cable and Thinwire.

thin AP Refers to a wireless LAN architecture where the access point handles only the network's wireless communications, while other WLAN functions, such as authentication of users, encryption of communications, secure roaming, routing, and network management, are handled by network switches and the centralized management controller. See fat AP, fit AP.

thin client Here's Citrix's definition: A low-cost computing device that works in a server-centric computing model. Thin clients typically do not need state-of-the-art, powerful processors and large amounts of RAM and ROM because they accesses applications from a central server or network. Thin clients can operate in an application server environment.

Here's my longer definition: Clients are devices and software that request information – applications or files. A client is a fancy name for a PC or workstation which is connected to a network, such as a local area network (LAN), a company's Intranet or, the Internet. The client runs on a server which houses applications and/or files. Clients come in two varieties – Fat and Thin. A "thin client" is a relatively cheap workstation akin to a dumb terminal in a mainframe environment. The thin client, according to current definition, lacks a hard drive, modem, PCMCIA slot, CD-ROM drive, floppy drive, serial port, communications port. The thin client comprises a sealed unit with often no potential for enhancement, other than adding memory. However, it does contain RAM, a limited processing power and perhaps a burned-in chip with a program or two, perhaps its user interface. The bulk of the applications and the information it needs remain on the server. Hence, the thin client is totally dependent on the server. The advantage of a thin client is low TCO (Total Cost of Operation), including costs of acquisition, maintenance and support. The downside is that the thin client is totally reliant on the server, through the network. Should either the server or the network fail, the client effectively is rendered useless until the problem is resolved. Here's a definition of Thin Client, courtesy of Oracle Corporation, writing in early 1994: "The thin client modem stores and processes more data on the server, but keeps the user interface and application functions on the client device. Example: a television with a set-top box, Apple's Newton Personal Digital Assistant, or a low-end PC."

In an Internet scenario, thin clients are known as NetPCs or Netstations. The NetPC is reliant on the server, which is provided by a service provider (e.g., America OnLine, CompuServe, or your ISP). In addition to providing some combination of content and Internet access, the service provider's server will provide your client NetPC with access to all necessary applications (e.g., word processing and spread sheet applications), will store all your personal files, will provide all significant processing power, and so on. In this Internet example, the NetPC differs from the standard thin client by virtue of the fact that it does contain a modem, a communications port and communications software, all of which are required for Internet access. See also Client, Client Server, Client Server Model, Fat Client, Mainframe Server and Media Server.

thin computing See Thin Client.

thin Ethernet A coaxial (0.2-inch, RG58A/U 50-ohm) that uses a smaller diameter coaxial cable than standard thick Ethernet. Thin Ethernet is also called "Cheapernet" due to the lower cabling cost. Thin Ethernet systems tend to have transceivers on the network interface card, rather than in external boxes. PCs connect to the Thin Ethernet bus via a coaxial "T" connector. Thin Ethernet is now the most common Ethernet coaxial cable, though twisted pair is gaining. Thin Ethernet is also referred to as ThinNet, ThinWire or Cheapernet. See also 10BASE-T.

thin route A route with little telecom traffic and limited growth potential. The opposite is thick route.

thin-film interference filter Thin-film filters control the reflection, refraction, transmission, and absorption of light waves. Filters are used in a wide variety of optical components, and the majority of today's WDM systems incorporate thin-film filter technology for multiplexing and demultiplexing. Interference filters are constructed by depositing a series of coatings with different refractive indexes on a glass substrate. This construction generates interference patterns as lightwaves pass through such that certain wavelengths are reflected while others pass through undisturbed. In DWDM applications, at lower channel counts and larger spacing between wavelengths, thin-film filters are produced in volume quantities that can process channel spacings of 200 Ghz. However, at higher channel counts and smaller channel spacings (below 100 Ghz and 50 Ghz), manufacture becomes increasingly difficult. The value proposition for competing multiplexing/demultiplexing technologies such as arrayed waveguides and fiber bragg gratings becomes increasingly compelling as channel count increases.

thingy See Dingy.

thinnet Jargon used to describe thin Ethernet coaxial cable. Referred to ThinNet, Thin-Wire or Cheapernet.

thinwire The 50-ohm coaxial cable listed in IEEE 802.3 specifications and used in some Ethernet local area network installations.

third generation wireless See 3G.

third order harmonics The third multiple of a specific frequency or a specific frequency multiplied by three.

third party call Any call charged to a number other than that of the origination or destination party. It's not a good idea to let your employees make third party calls to one or more of your phone numbers. Best to ask them to place the calls on their personal phone credit cards. This way, they will spend a modicum of time justifying their exorbitant phone calls.

third party call control A call comes into your desktop phone. You can transfer that call. When the phone call has left your desk, you can no longer control it. That is called First Party Call Control. If you were still able to control the call (and let's say, switch it elsewhere) that would be called Third Party Call Control. Some Computer Telephony links allow only first party call control. Some allow third party as well. If you control the switch – the PBX or the ACD – you will typically have Third Party Call Control. If you just control the desktop, you'll typically have only First Party Call Control. There is no such animal as Second Party Call Control.

third party cookie See Cookie.

Third Party Verification TPV. Before your Primary Interexchange Carrier (PIC), or long distance carrier, can be changed, the FCC now (1988) requires that the new carrier have a in place a means of verifying that you have authorized such a change. Business customers must execute a written Letter of Agency (LOA), which the new carrier can present to the old carrier. The veracity of a change for residential customers is ensured through Third-Party Verification (TPV), which takes the following form. Once you have concluded your conversation with the sales representative of the new carrier, the sales rep will initiate a conference call to add to the call a representative of an independent third party. The third party will verify the change, including all relevant information recorded by the sales representative. This step protects you from "slamming," which is the practice of changing your PIC without your authorization. See also LOA, PIC, and Slamming.

third party wavelength An optical wavelength created by a third-party system that is injected into and integrated with the wavelengths of an existing optical network.

third place 1. A place where a person goes and hangs out when he is not at home or not in the office. Third places are another place where users want connectivity these days, and therefore represent another market opportunity for network operators. Hence the deployment of WiFi in third places.
2. A video screen, particuarly the screen on a cell phone, that a person uses almost as often as their television and computer screens. Like Starbucks' relentless attempts to sell itself as the "third place," the wireless industry is trying to sell itself as the "third screen." The electronic notion of a "third screen" is based on the idea of a "third place," which is a place other than home or work where a person can go to relax and feel part of the community, e.g. Starbucks coffee shops. People also want connectivity – e.g. Wi-Fi connection to the Internet.

third screen Think information and entertainment. The first screen was your television set. The second screen was the PC. The third screen is now the screen on your digital cell phone. Not very powerful. But that's the latest jargon.

third wire Broadband over Power Line (BPL) sometimes is referred to as the "third wire," referring to the fact that an electrical circuit usually comprises a positive (+) conductor (wire) and a negative (-) conductor. The telecommunications circuit could be considered as the third wire, at least that is the metaphor. If you consider the ground wire, the telecommunications circuit could be considered the fourth wire. But, then, not all circuits are grounded, despite the fact that grounding is always a good idea. See also BPL.

third wire tap The activating of a telephone handset microphone by using a third wire, thus bypassing the hook switch.

thirty mile zone The area around Los Angeles which web site, TMZ.com, collects news and gossip on.

THL Trans Hybrid Loss.

THOF Navy shipboard cable with three flexible stranded conductors and with heat and oil resistant coverings. respect to ground.

thought police In Imperial Japan before World War II, members of the "thought police" – Shisou Keisatsu – spread out to suppress dangerous thinking in the populace. Such dangerous thinking was obviously different to what the imperial government wanted. The Thought Police were disbanded by General McArthur when he imposed freedom of speech after the War. The thought police was later chillingly immortalized by George Orwell in his 1949 "1984."

thousand block number pooling See Number Pooling.

thrashing A computer has a finite amount of memory and processing capability. If a process or program or user makes a request that can't be met, the OS (Operating System) may borrow resources from another process in an attempt to satisfy the request. The process from which the resources were borrowed then borrows resources from another process. And so on, and so on. The computer thrashes about looking for resources, and never getting anywhere. You get either a blank screen or a frozen screen. Your computer has just crashed. Thrashing can be caused by a number of things, including your clicking around too fast, so just slow down and take things one at a time. See also Crash.

thread 1. A thread is part of a program that can be run independently of other aspects of the program. It is a sequence of computing instructions that makes up a process or program. A program can be single-threaded or multi-threaded. A single-threaded application program insists that only a single instruction can be executed at a given time. All the instructions must be executed in an exact sequence, from beginning to end. For example, a single-threaded Internet experience might involve your accessing a Web-based server that would accept your request to establish a session, and would accept and serve your request for information. During this period of time, a tightly choreographed series of steps would take place in exact sequence, and no requests from other clients would be accepted until your request was satisfied. In other words, a single-threaded program must follow a single line of logic in a very rigid manner.

A multi-threaded process has multiple threads, each executing independently and each perhaps executing on separate processors within one or multiple computers. A multi-threaded program has multiple points of execution (one per thread) and, therefore can perform multiple tasks associated with multiple processes and multiple programs supporting multiple users at any given time. As each task associated with each task associated with each process supporting each program and each user is completed, the thread for that task is resumed at the same point it had been interrupted, much as though it had been bookmarked. As multiple instruction sets can be executed concurrently, the throughput and speed of running the program is much improved. In other words, a multi-threaded program, if running on a computer with multiple processors, will run much faster than a single-threaded program running on a single processor machine.

2. In the context of an Internet Usenet newsgroup or other interactive discussion forum, a thread essentially is a train of thought or line of logic that can be followed through the fabric of a larger subject. A thread begins with a message posting. Responses are "hung off" of that initial posting in a chronological and hierarchical fashion as comments are offered and elicit additional comments, and as questions are posed and answered and the answers elicit yet other sets of questions and answers. This hierarchical threading is very easy to follow in graphical format, such as that used in the World Wide Web. See also Hyperthreading, Threaded and Threaded Code.

thread A common theme which runs through the whole course of something, connecting successive parts. In normal language, we talk about how "The movie was so complicated that I lost the thread of the story." Threads work in many ways in our industry but they all have a common thread. For example, threaded SMS (short message service) means that messages and replies are stacked on top of each other, so you figure the thread. Some computers will process several tasks simultaneously. Using multiple threads allows concurrent operations within a process and enables one process to run different parts of its program on different processors simultaneously. In this form of computing, a thread has its own set of registers, its own kernel stack, a thread environment block, and a user stack in the address space of its process. See threaded code.

threaded 1. A method of presenting articles within a newsgroup or bits of a conversation in a way that shows which articles or "conversations" refer to which other ones. See Thread.

2. t displays the messages like a conversation.

threaded code Threaded code is also known as Threaded Pseudo Code, or threaded p-code. It was first popularized in a software language called Forth. Eventually Microsoft fixed Forth and changed it into Basic. See Thread.

threat analysis Examination of all actions and events that might adversely affect a system, a network or an operation.

threat signature Data or behavior that has characteristics of known threats.

three finger salute Ctrl Alt Delete.

three degrees above zero This refers to the average temperature of the universe, which is about three degrees above absolute zero. The discovery was made in 1965 by two radio astronomers at Bell Labs, Arno Penzias and Robert Wilson, while they were investigating sources of noise in a radio antenna at a Bell Labs facility in Crawford Hill, New Jersey.

three dog night Three Dog night, attributed to Australian Aborigines and the American Eskimos) came about because on especially cold nights these nomadic people needed three dogs (dingos, actually) to keep from freezing.

three nines 99.9%. Three nines typically refers to the reliability of a system (computer, telephone system, etc.) that works 99.9% of the time. These days the industry talks increasingly of five nines reliability, i.e. 99.999% and fo six times reliability, e.g. 99.9999%.

three phase An electricity system with at least three wires in which the AC voltages on the wires are out of step with each other by one third of a cycle, or 120 degrees in phase. Three-phase is far more efficient than single-phase for power transmission, because a single line with three wires can transmit as much power as three separate single-phase, two-wire lines. This saving is why three-phase is almost universally used in power transmission. Moreover, electrical power throughout a cycle is smoother in three phase than in single-phase, which is why heavy-duty motors and generators are built for three-phase. Most of the fundamental theories of three-phase systems were developed by Charles P. Steinmetz, who at age 24 in 1882 emigrated from his native Germany to the USA, where he worked for the General Electric Company from 1893 almost until his death in 1923. In 1897, he published "Theory and Calculation of Alternating Current Phenomena", the landmark paper that helped give the USA an early lead in high-voltage AC transmission technology.

three Rs, the Retiming, reamplification, and reshaping, i.e., the three functions that a repeater performs when it regenerates a signal.

three screen delivery Refers to the delivery of content that is designed for all three consumer electronics screens: TV, PC and mobile handset.

three slot An obsolete pay phone that is identified by three separate coin slots.

three tier A type of client/server architecture consisting of three well-defined and separate processes, each running on a different platform:

1. The user interface, which runs on the user's computer, also called the client.

2. The functional modules that process the data. This middle tier runs on a server. It is often called the application server.

3. A database management system (DBMS) that stores the data required by the middle tier. This tier runs on a second server called the database server.

The three-tier design has some advantages over traditional two-tier or single-tier designs. Its modularity makes it easier to change or replace one tier without affecting the other tiers. It's also better for load balancing.

three-watt booster Optional equipment for use with a cellular phone car-mounting kit that raises a portable phone's maximum transmission power from 0.6 watts to 3.0 watts.

three-way calling A local phone company feature that allows a phone user to add another user to an existing conversation and have a three party conference call.

three-way compare What a phone company does when its billing becomes so error-ridden that it loses confidence in the integrity of the data in its OSS (Operations Support System) and BSS (Business Support System). For example, if the bills for long-distance calls are screwed up, and yet the toll rating system's tables and logic are vetted and verified to be clean, the phone company will compare every customer PIC assignment on the end-office switches with the PICs in the CRM system (or customer master file) and the PICs in call detail records downstream in the billing system, in order to uncover discrepancies that are the root cause of the billing errors. Similarly, if DSL billing is error-ridden, the phone company will eventually reach a point where it will compare DSL provisioning on the end-office switches with customer billing records and customer records in the CRM system, in order to uncover discrepancies that are the root cause of the billing errors. The incidence and severity of billing problems seems to be related to the stability of a phone company's OSS and BSS. Generally speaking, there are fewer billing problems at a phone company whose OSS and BSS have been around for years or decades, and those systems have been appropriately maintained and evolved over time to accommodate the phone company's evolving needs. By contrast, the incidence and severity of billing problems seems to more likely afflict a phone company some or all of whose OSS and BSS have been changed out completely due to a merger, acquisition, spin-off, or outsourcing. A three-way compare is a common revenue assurance activity. The basic objective of a three-way compare is to make sure that circuits and services shown as ordered in a telco's ordering system are also shown as provisioned in the telco's provisioning system, and are also shown as being billed by the telco's billing system.

three-way conference transfer A PBX feature. By depressing the switch hook, a user can dial another extension and either hang up and transfer the call, get information from the called party and then resume the first call or bridge all parties together for a three-way conference call.

three-way-handshake The process whereby two protocol entities synchronize during connection establishment.

threshold 1. The minimum value of a signal that can be detected by the system or sensor under consideration.

2. Automatic call distributors allow the definition of several different thresholds that pertain to different objectives of your organization. For instance, thresholds can be defined for the maximum length of time a customer's call should wait in queue, how long an agent should spend on each call, and how many accepting the overflow.

3. In England in the 1500s, the poor had dirt floors (hence "dirt poor"), the wealthy had slate floors that would get slippery in the winter when wet, so they spread thresh (straw) on the floor to help keep their footing. As the winter wore on, they kept adding more thresh until, when you opened the door, it would all start slipping outside. A piece of wood was placed in the entranceway – hence, a "thresh hold."

threshold of pain 1. The present price of local telephone or broadband service. Pick whatever "evil" you want.

2. Unbearable noise.

throttling What happens when your Internet service provider decides you're getting too much bandwidth (aka speed) on your connection to the Internet. He cuts down the speed of your connection. He does this because he has other people or other services (e.g. email) he wants to provide higher speed service to. Or because he's simply short of a speedy connection to the Internet himself.

through dialing Allows the attendant on a phone system to select a trunk and pass dial tone to a restricted phone user so that user may directly dial an outside call.

throughput The actual amount of useful and non-redundant information which is transmitted or processed. Throughput is the end result of a data call. It may only be a small part of what was pumped in at the other end. The relationship of what went in one end and what came out the other is a measure of the efficiency of that communications network. Throughput is a function of bandwidth, error performance, congestion, and other factors. See also Goodput.

throw As in "throw a cable". To cut over. See Cutover.

thumb economy This term refers to the SMS (short message service) boom and popularity of mobile data services in China. China accounts for over one-third of the world's SMS messages. The prevalence of mobile phones and wireless communications in China has resulted in an always-on, always-connected "thumb economy" and "thumb culture." The demand for SMS content – news alerts, commodities prices, stock quotes, society gossip, jokes, and educational content – is so huge in China that there are now approximately 500 companies in China creating SMS content. SMS messaging has also become an advertising medium in new product launches in China.

thumb-typing How we "type" on our BlackBerrys and other small wireless devices – those with tiny keyboards.

thumbing How we "type" on our Blackberrys and other small wireless devices – those with tiny keyboards. See Blackberry and predictive text.

Thumb, Rule of The Rule of Thumb of is derived from an old English law that stated a man could not beat his wife with anything wider than a thumb.

thumbnail Describes the size of an image you frequently find on Web pages. Usually photo or picture archives will present a thumbnail version of its contents (makes the page load quicker) and when a user clicks on the small image a larger version will appear. Sometimes these links will be to a new page containing the larger graphic and other times right to the image directly.

thunking A Microsoft Windows term for the transformation between 16-bit and 32-bit formats, which is carried out by a separate layer in the VM. the fundamental idea is to make older 16-bit programs work better in newer 32-bit operating systems.

Thuraya Thuraya is the name of a commercial satellite service which delivers phone calls and other mobile communications services in parts of the Middle East, Asia, Africa and Europe. It started in in May 2001, It achieved a degree of fame when, in December, 2003, it was revealed that American intelligence penetrated Saddam Hussein's inner entourage before the 2003 Iraq war, finding one of his security aides who used a Thuraya satellite telephone, of the kind that American commanders favored. According to accounts circulating in Baghdad, Mr. Hussein personally executed the security man after the second of two pinpoint bombing strikes that nearly killed him, on March 20 and April 7, 2003. After that the use of satellite telephones by his entourage virtually stopped. Mr. Hussein was captured in December, 2003. Thuraya phones offer most of the features of normal cell phones, except that they use a constellation of satellites to provide service. Like all satellite phones, Thuraya's only work outdoors. Thuraya's "country" dialing code is +88216. Each Thuraya subscriber has an eight digit number. Thuraya phones have GPS. You can access the GPS features from the "GPS manager" submenu. After you figure where you are, you can send your coordinates to them via SMS (short message service) which the phone supports. Thuraya's carrier modulation is QPSK.

THz Terahertz (10 to the 12th power hertz). See also Spectrum Designation of Frequency.

THW Thermoplastic vinyl insulated building wire. Flame retardant, moisture and heat resistant, 750xAFC, for use in wet or dry locations.

THWN Same as THW but with nylon jacket overall.

TI Terrestrial Interference.

TIA 1. Telecommunications Industry Association. TIA represents the telecommunications industry in association with the EIA (Electronics Industry Association). TIA represents companies, which provide communications materials, products, systems, distribution services and professional services in the U.S. and around the world. Activities include government relations, market support activities such as trade shows and trade missions, and standards development. TIA began as a group of equipment suppliers in the form of a committee of the USTA (United States Telephone Association, now known as the United States Telecom Association), splitting off in 1979 to form the USTSA (United States Telecommunications Suppliers Association). In 1988, the USTSA merged with the EIA/ITG (Information and Telecommunications Technologies Group of the Electronic Industries Association). TIA now operates under the umbrella of the EIA as the TIA/EIA, and works in conjunction with the USTA. TIA is accredited by ANSI (American National Standards Institute), and contributes voluntary standards to that body. In the Fall of 2000, the MultiMedia Telecommunications Association (MMTA), the successor organization to NATA, was fully integrated into the Telecommunications Industry Association.

More recently and in connection with the privatization of its Part 68 responsibilities, the FCC selected TIA and ATIS (Alliance for Telecommunications Industry Solutions) as joint sponsors of the Administrative Council for Terminal Attachments (ACTA). ACTA responsibilities include adopting and publishing technical criteria for terminal equipment submitted by ANSI-accredited standards development organizations, and operating and maintaining a database of approved terminal equipment. See also ACTA, ACTAS, NATA, and Part 68. www. eia.org and www.tiaonline.org.

2. Thanks In Advance.

TIA 568 Commercial Building Telecommunications Wiring Standard, July 91

TIA 568A Commercial Building Telecommunications Cabling Standard, October 1995

TIA 569 Commercial Building Standard for Telecommunications Pathways and Spaces, Oct 90

TIA 569A Commercial Building Standard for Telecommunications Pathways and Spaces, Aug 97

TIA 570 Residential and Light Commercial Telecommunications Wiring Standard, May 91

TIA 606 Administration Standard for the Telecommunications Infrastructure of Commercial Buildings.

TIA 607 Commercial Building Grounding and Bonding Requirements for Telecommunications.

TIA/EIA The US Telecommunications Industries Association and Electronics Industries Association, which have merged. Now just called the Telecommunications Industry Association. See TIA for a full explanation.

TIA/EIA IS-95 See TIA/EIA IS-96.

TIA/EIA IS-96 IS-96 is the Speech Service Option Standard for Wideband Spread Spectrum Digital Cellular System, a new vocoder standard. The standard supports IS-95, the North American spread spectrum digital standard based on Code Division Multiple Access (CDMA), which was published in July, 1993. The engineering effort to produce IS-96 was done in TIA Technical Subcommittee TR-45,5, Wideband Spread Spectrum Digital Technologies Standards. The specific vocoder described in IS-96 is a variable rate implementation, chosen because of its combination of high voice quality and low average transmission rate. The IS-96 vocoder provides variable vocoder rates depending on voice activity. This variability typically results in an average transmission rate of under 4 Kbps and yet provides for reasonably quality voice transmission. The IS-96 also provides a variable noise threshold which tracks and eliminates much of the background noise from the speaker's environment. You can get copies of both standards from Global Engineering Documents, 15 Inverness Way East, Englewood, CO 80112.

TIA PN-2416 Backbone Cabling Systems for Residential and Commercial Buildings

Tiananmen Square The largest public space in the world. It is five times the size of Moscow's Red Square. See 1949.

TIC 1. An AT&T term for a digital carrier facility used to transmit a DS-1 formatted digital signal at 3.152 Mbps.

2. Token-Ring Interface Coupler. An IBM device that allows a controller or processor to attach directly to a Token-Ring network. This is an optional part of several IBM terminal cluster controllers and front-end processors. See TIC CARD.

3. See Telphony Interface Control.

TIC Card Token Ring Interface Coupler is the IBM name for a variety of token ring adapter cards used to connect IBM controllers to token ring LANs. See TIC.

TICE TICE was the mantra of PC Magazine and others in 1999 and 2000. It was the great sound echoing from the (very hollow) drums of Silicon Valley: "The Internet Changes Everything"

tick 1. A tick is an increment of time, as measured by a clock. The term refers to the audible sound of a mechanical clock or watch as it clicks away the seconds. Computer systems of all sorts have internal quartz clocks that manage and synchronize all system functions. A computer's clock speed in measured in MHz (MegaHertz, or millions of cycles per second), with each cycle also known as a tick. See also Clock Speed.

2. A tick is a clock timer interrupt that causes a computer operating system to increment the system time in the DTS (Distributed Time Service) function of DCE (Distributed Computing Environment. See also DTS and DCE.

3. A clock tick is a means of measuring time on a computer operating system. There are functions that report the number of milliseconds (ms) that have elapsed since the system was booted. The Windows operating system, for example, updates the function whenever there is a system clock tick, which is every 10ms.

4. A tick mark is a mark plotted on a continuum, such as the axis of a graph. Tick marks are at regular, precise intervals in space, just as clock ticks are regular, precise intervals in time.

5. Cisco defines ticks as the delay on a data link using IBM PC clock ticks (approximately 55 milliseconds).

tick tone Clicking noise heard on some PBX lines indicating that the digits dialed will shortly be repeated to the central office.

ticker A one-way telex machine used to typically report stock or commodity prices. The machine prints on ticker tape, which is about one inch wide and perfect for throwing out windows at passing celebrities. Thus the term "Ticker Tape Parade."

ticket A telephone industry term for a filled-out form, usually a form for billing someone for a call. There are all sorts of tickets, including ones on paper, ones on computer and ones automatically generated without human intervention.

TICL Temperature Induced Cable Loss. Pronounced "tickle." A phenomenon in which the performance of fiber optic cables is adversely affected by low temperatures. The adverse impact is a multi-dB attenuation. What's strange is that the attenuation (i.e. reduction is signal strength) is localized so that it looks like a splice on an OTDR (Optical Time Domain Reflectometer, a test and measurement device often used to check the accuracy of fusion splices and the location of fiber optic breakers.) Particularly affected are cables which have been in place for at least one summer and which operate at relatively long wavelengths. Also particularly affected are cables which do not have a strong coupling between the central members (fiber cores) and the buffer tubes (protective individual fiber sheaths). The issue is that the glass fiber (expands and) contracts with changes in temperature, and to a different degree than does the buffer tube. The result is that of changes in the geometry of the fiber. TICL is a bad thing. What's difficult is fixing it, because the attenuation loss typically goes away when temperature rises, so that diagnosis is difficult ("like chasing ghosts" to use a colorful phrase). TICL is most likely to be troublesome when upgrading an existing link from single wavelength, e.g. 1310nm traffic to DWDM 1550nm traffic.

TID Terminal Identification. Used for all National-1 ISDN services, a two digit number between 00 and 62 entered after the SPID.

TIE 1. Joining cables and/or wires together.

2. Time Interval Error.

3. Trusted Information Environment, an encryption scheme.

tie down Verb meaning to terminate a wire on a main, intermediate or satellite distribution frame.

tie line A dedicated circuit linking two points without having to dial the normal phone number. A tie line may be accessed by lifting a telephone handset or by pushing one, two or three buttons.

tie trunk A dedicated circuit linking two PBXs.

tie trunk access Allows a phone system to handle tie lines which can be accessed either by dialing a trunk group access code or through the attendant.

tied in the wood Term for wire that are placed in the wiring device but not terminated. Usually, wires are tied in the wood in preparation for coordinated termination. This refers to when wiring devices had wooden wire management strips.

Tier 1 Imagine a bunch of international Internet telephony carriers. Each one has POPs (Points of Presence) in several overseas cities. A POP consists of at least one PC containing some voice cards. When someone dials another country, their call goes across the Internet, reaches the PC at the distant POP. That PC recognizes the call as for that local city, grabs it, dials the local number and conferences the Internet call with a local dial-up call. This combination Internet/local phone call is theoretically cheaper than dialing directly across the world's telephone system. In order to provide seamless, cheap international calling over the Internet, you really need POPs in every major and minor city abroad. A number of Internet telephone companies have been banding together to create this international network. They're thinking about classifying themselves into various categories. For example a Tier 1 carrier would have over 50 POPs worldwide; have a network managed by a 7x 24 NOC; have the ability to reroute and fall back to the PSTN if there is congestion or a hardware problem; have redundancy in terminating locations and have the ability to offer several levels of quality. Tier 2 might have fewer. It all hasn't been defined yet. But here's a definition a reader, Bill Coleman sent me: A Tier 1 telephone carrier has POPs in every city where there is a NFL football team. A Tier 2 telco would be a secondary market such as a city that gets bold print and a size 10 font in a Rand-McNally road atlas. A Tier 3 city would be one that nobody has really heard of unless you were born there. See also Back to Back Peering. To further confuse you, Infonetics Research, Inc. defines Tier 2 as National ISPs and CLECs that don't have fiber and regional ones that do.

Tier 2 See Tier 1.

Tier 3 See Tier 2.

tiered service Telecommunications carriers and Internet service providers are eager to differentiate their network offerings. They sometimes do this by creating tiers of service. Tier 1 service, for example, is the best, and we all want to be Tier 1, if we can justify the additional cost. Some switches and routers allow network service providers to broaden their network service portfolios by creating multiple, differentiated QoS (Quality of Service) or GoS (Grade of Service) levels so they can offer tiered services. ATM naturally supports multiple Service Categories which differentiate between various native information streams, thereby providing each with the guaranteed QoS that he or she demands. Real-time, uncompressed voice, for example, is very demanding – therefore, it is given the highest service. LAN-to-LAN data traffic, on the other hand, expects nothing in terms of QoS – therefore, it is treated on a "best effort" basis. IP networks (e.g. the Internet) basically begin as "best effort" networks. However, there are proprietary mechanisms which can be embedded in programmed logic in the network switches and routers to differentiate between various packets, and, thereby, to deliver differentiated GoS. These mechanisms, which are contained in the IP packet headers, can be associated with individual customers, based on their port numbers, circuit IDs, or other identifiers. Tiered services let network service providers differentiate their offerings to segment existing markets (by price, for example) and create new markets. There also are standards initiatives in development at the IETF (Internet Engineering Task Force) which will support standards-based GoS differentiation. At this time, it is not anticipated that IP networks will ever offer truly guaranteed QoS – that's what sets ATM apart – but differentiated GoS may well be good enough for most applications and most end user organizations.

TIES Time Independent Escape Sequence, a feature of modems.

TIF Terminal Interface Node.

TIFF Tag Image File Format. TIFF provides a way of storing and exchanging digital image data. Aldus Corp., Microsoft Corp., and major scanner vendors developed TIFF to help link scanned images with the popular desktop publishing applications. It is now used for many different types of software applications ranging from medical imagery to fax modem data transfers, CAD programs, and 3D graphic packages. The current TIFF specification supports three main types of image data: Black and white data, halftones or dithered data, and grayscale data. Some wags think TIFF stands for "Took It From a FotograF." It doesn't.

TIFF-F Tagged Image File Format-Fax. A compression technique for sending faxes across an IP (Internet Protocol) packet data network. Fax image documents are attached to e-mail headers and are encoded in the TIFF-F compressed data format. In simple-mode, T.37 restricts fax transmission to the most popular fax machine formats, which specify standard resolution and page size. Simple-mode provides no confirmation of delivery. Full-mode provides for delivery confirmation.

tiger team A group hired by an organization to defeat its own security system to learn its weaknesses.

tiger whiskers In 16th- and 17th-century Peking, one took revenge against one's enemies by placing finely chopped tiger whiskers in their food. The whisker barbs would get caught in the victim's digestive tract and cause sores and infections.

tight buffer fiber optic cables Tight-buffered fiber optic cables have a buffer of extruded polyethylene or similar material applied directly over the acrylate coating that surrounds the glass fiber, itself. The buffer, is tightly bonded to the coating, hence the term "tight buffer." The buffer serves to protect the fiber from crushing and impact loads. However, the buffer can cause microbends in the fiber if there are significant changes in temperature, as the buffer material will expand and contract at significantly different rates than the glass fiber. The resulting microbends will increase signal attenuation (i.e., signal power loss) and, over time, can cause cracks and even breaks in the fiber. Therefore, tight-buffered cables are used only in indoor applications. Loose tube cables are used in outdoor applications. Note: The outside diameter of the glass fiber is 125 microns, or 125 millionths of a meter. The acrylate coating increases the diameter to 250 microns. The buffer increases the diameter to a total of 900 microns. See also loose tube cable. Tight-buffered fiber optic cables use aramid strength members inside the cable instead of gel filling, as is the case with loose-tube gel-filled fiber optic cables. One of the advantages of tight-buffered fiber optic cables having aramid strength members along every inch of the cable is that the cable can be hung vertically and the fibers are still protected for the entire length of the cable. This is not the case with loose-tube gel-filled fiber optic cables because, when they are hung vertically, all the gel filling settles to the bottom and the optical fibers are no longer protected. Tight-buffered fiber optic cables also have buffer coatings (up to 900 microns) over each optical fiber cladding for added environmental and mechanical protection, increased visibility, and ease of handling. Tight-buffered fiber optic cables can be used indoors and outdoors which allows one cable to be cabled instead of having to switch cable types at the building entrance. This is different from loose-tube gel-filled cables because the gel is flammable and the cable must be spliced to indoor flame-retardant cables for runs into buildings. Therefore, according to manufacturers, tight-buffered fiber optic cables reduce labor, equipment and materials cost while improving system performance and reliability. See also Aramid and Tight Jacket Buffer.

tight jacket buffer A buffer construction which uses a direct extrusion of plastic over the basic fiber coating. This construction serves to protect the fiber from crushing and impact loads and to some extent from the microbending induced during cabling operations. See also Loose Tube Buffer.

tightly coupled Describing the interrelationship of processing units that share real storage, that are controlled by the same control program and that communicate directly with each other. Compare with loosely coupled.

tightly coupled CPUs Term used to describe multiple-processor computers in which several processors share the same memory and bus.

TIIAP See OTIA.

tilde The tilde is the ~ sign, which you'll find on most keyboards. It looks like an arched eyebrow. Microsoft uses it in DOS to truncate a long Windows file name. Thus c:my documents in Windows becomes c:mydocu~1 in DOS. Other programs use it in different ways.

tile A surface segment of a furniture system panel, usually removable for access to cables or patch panels contained within the panel.

tiling An unpleasant mosaic-like effect created by block-oriented video compression techniques like DCT (Discrete Cosine Transform), used in the JPEG (Joint Photographics Expert Group) standard. See DCT and JPEG.

tilt Tilt is a factor that affects the maximum length of a metallic cable system. Tilt is a mathematical expression of the difference in attenuation (i.e., loss of signal power) between high frequency and low frequency signals over the entire length of a cable segment in a carrier band system. (Note: High frequency signals attenuate more quickly over distance than do low frequency signals.) Measured in decibels (dB), tilt is determined by $N/(A1-A2)$, where N is the maximum allowable tilt, A1 is the attenuation of the high frequency signal, and A2 is the attenuation of the low frequency signal.

TIM Teletyper Input Method. See Teleypre Input Method.

timbre The resonance quality of voice by which the ear recognizes and identifies it. The quality of tone distinctive to a particular voice.

time 1. Time is critical to all forms of telecommunications. To understand why is simple. It takes time to send something from me to you. If I send you information encoded in bits and bytes, you have to have some way of figuring what I'm saying. There are basically two ways of sending information. You can send information in packets – think paper envelopes. Those "envelopes" have my address and your address on them. And they're set up so it's clear what's the addresses and what's the message, etc. That takes extra bits. A faster way is to send the information in one gigantic fast stream. But that means that your computer and mine have to be exactly on the same time. And there's a huge business in making very precise clocks that synchronize flows of information. For more on this read the definitions for Asynchronous and Synchronous and the definitions following. Here's a fun fact on time: Most grandfather clocks with metal pendulums lose time in warm weather. This phenomenon occurs because most solids expand when heated. In the case of the clock, the higher temperature makes the metal pendulum longer, and thus slower. See also Atomic Clock and Optical Clock.

2. Time Protocol (RFC 868). Time clients obtain the current time-of-day within one-second resolution from Time servers. See also TOD (as in time of day server).

3. See WWV and WWVB.

time-based authoring tool A multimedia creation tool that uses time as a metaphor for building a project. Generally, objects are set up to happen at a certain time in a project, rather than in a certain place.

Time Assignment Speech Interpolation TASI. A voice telephone technique whereby the actual presence of a speech signal activates circuit use. The result is clipping of the first bit of the speech, but more efficient use of the transmission facility. TASI is used on expensive circuits, such as long submarine cables. See TASI.

time call hour indicator The hour at which the call was placed.

time call minute indicator The minute within the hour at which the call was placed.

time congestion The time resources (outgoing trunks) are busy.

Time Difference of Arrival TDOA. A class of Position Determination Technology in which a mobile radio unit's position is calculated based on the reception time of its transmitted signal measured at three or more receiving sites. The distance from transmitter to receiver equals the propagation delay times the speed of light. However, the absolute propagation time is rarely known, leading to the use of time differences at the receiving sites. Employed in certain wireless E-911 solutions. See also E-911 and Angle of Arrival.

time divert to attendant A system feature which automatically transfers a phone to the attendant if the phone has been left off-hook too long.

time diversity A method of transmission wherein a signal representing the same information is sent over the same channel at different times. Often used over systems subject to burst error conditions and with the spacing adjusted to be longer than an error burst.

Time Division Controller TDC. A device which commands functions, monitors status and connects channels of TDM cards.

Time Division Multiple Access 1. TDMA. A technique originated in satellite communications to interweave multiple conversations into one transponder so as to appear to get simultaneous conversations. A variation on TASI. A technique now used in cellular and other wireless communications. See TDMA.

2. An RFID definition. A method of solving the problem of the signals of two readers colliding. Algorithms are used to make sure the readers attempt to read tags at different times. See RFID.

Time Division Multiplex TDM. A technique for transmitting a number of separate data, voice and/or video signals simultaneously over one communications medium by interleaving a piece of each signal one after another. Here's our problem. We have to transport the freight of five manufacturers from Chicago to New York. Each manufacturer's freight will fit into 20 rail boxcars. We have three basic solutions. First, build five separate railway lines from Chicago to New York. Second, rent five engines and schlepp five complete trains to New York on one railway track. Or, third, join all the boxcars together into one train of 100 boxcars and run them on one track. The train might look like this: Engine, Boxcar from Producer A, Box Car from Producer B, Producer C, Producer D, Producer E, and then the order begins again...Boxcar from Producer A, Producer B...Moving one large train of 100 boxcars is likely to be cheaper and more efficient than moving five smaller trains each of 20 boxcars on five separate railway tracks. Time Division Multiplexing, thus, represents substantial savings over have five separate networks (five separate tracks) and sending five separate transmissions (five separate trains).

This is what Time Division Multiplexing is all about. And the analogy is perfect. Take one large train (fast communications channel) and interleave pieces (boxcars) from each conversation one after another. If you do this fast enough, you'll never notice you've broken the conversations apart, moved them separately, and then put them back together at the distant end. In TDM, you "sample" each voice conversation, interleave the samples, send them on their way, then reconstruct the several conversations at the other end. There are several ways to do the sampling. You can sample eight bits (one byte) of each conversation, or you can sample one bit. The former is called word interleaving; the latter bit interleaving. The basic goal of multiplexing – whether it be time division multiplexing, or any other form – is to save money, to cram more conversations (voice, data, video or facsimile) onto fewer phone lines. To substitute electronics for copper. See also the following three definitions.

Time Division Multiplexer TDM. A device which derives multiple channels on a single transmission facility by connecting bit streams one at a time at regular intervals. It interleaves bits or characters from each terminal or device using the time. See Time Division Multiplex.

time division signaling Signaling over a time division multiplex system in which all voice channels share a common signaling channel, with time division providing the separation between signaling channels. See Signaling System 7.

time division switching The connection of two circuits in a network by assigning them to the same time slot on a common time division switched bus.

Time Division - Synchronous Code Division Multiple Access See TD-SCDMA.

Time Domain Reflectometer TDR. A device that measures network cable characteristics such as distance, impedance, levels of RFI/EMI, connector and terminator problems, and the presence of opens and shorts. It uses radar-like principles to determine the location of metallic circuit faults.

time guard band A time interval left vacant on a channel to provide a margin of safety against interference in the time domain between sequential operations, such as detection, integration, differentiation, transmission, encoding, decoding, or switching.

time jitters Short-term variation or instability in the duration of a specified interval.

time marker A reference signal, often repeated periodically, enabling the correlation of specific events with a time scale. markers are used in some systems for establishing synchronization.

time multiplexed switch The space switch of which the cross point settings are changed in each time slot.

time notify See TNotify.

time of day display The time and date displays on phones. Actually, it's very useful information. Sometimes it's not displayed on the operator's console. As a result, the operator may never know that every phone in the office is showing the wrong time and date.

time of day routing 1. This feature automatically changes access to certain types of lines at times when the lines change from being expensive to cheap, or vice versa. For example, it's cheaper to use WATS lines before 8:00 AM in the morning. A company has offices in New York and Los Angeles. It might be cheaper to route calls to Chicago in the morning over the tie lines to LA and then out the LA WATS lines to Chicago, than to go directly out the New York WATS lines. This is a way to allocate bandwidth for LAN traffic over corporate T-1 Networks. By programming T-1 multiplexers, customers can allocate the amount of T-1 bandwidth that can be used by voice, data, and LAN traffic on a time of day basis. For example, during the day, most of the T-1 bandwidth can be allocated for voice. At night, after employees go home, more bandwidth can be allocated to LAN and other computer data traffic so that file transfers can be done faster. This is particularly useful in IBM mainframe environments where large amounts of data needs to be transferred form remote offices/divisions to the headquarters.

time out In telecommunications and computer networks, an event which occurs at the end of a predetermined interval of time is called Time Out. For example, if you lift the phone off the cradle and do not proceed to dial, after a certain number of seconds you will hear either a voice telling you to get on with it, a howling sound of some sort, or a fast busy signal. Data networks have the same thing. Don't do anything for x minutes and the system will knock you off the air, i.e., hang up on you. For example, your Internet session will time out at some point, if a device in the network (e.g., a router or a server) senses that no activity is taking place (i.e., no data packets are flowing in either direction). In more technical terms, time out is the amount of time that hardware or software waits for an expected event before taking corrective action by terminating the connection or session. This corrective action conserves network resources, which always are limited. See also Answer Supervision.

time sensitive traffic Network traffic where latency (a fancy word of delay) must be minimized, such as voice traffic, digital TV streams, and control instructions for real-time systems.

time sharing A mode of operation that provides for the interleaving of two or more independent processes on one functional unit. Its most common use is the interleaved use of time on a computing system enabling two or more users to execute computer programs concurrently. Time sharing of computer resources is now relatively obsolete. See also Time-sharing below.

time sharing computer system A computer system permitting usage by a number of subscribers, usually through data-communication subsystems. This is usually the case where the users have only dumb terminals that cannot process data by themselves the way a stand alone computer can. Computers are being joined together to deliver more computing power where it is most needed.

time shifting A TV show is aired at 8 PM. You record it on your PVR (also called your TiVo) and play the show at 10 PM. You have time shifted your TV show by two hours. See also PVR.

time sink A consumer of our time. What happens to our personal time when things we do consume far too much of our time – for example, listening to stupid recordings of distant automatic telephone systems present us mindless requests: Dial 1 if you live in Northern Illinois. Dial 2 if you live in Southern Illinois, etc.

time slice In a multi tasking environment, each task is allotted a portion of the CPU's overall processing power. This portion is called a time-slice. And it's usually measured in milliseconds. The CPU switches between tasks, and those with higher priority receive more time-slices than lower-priority tasks. See Time Slicing.

time slicing The term used to describe the dividing of a computer resource so multiple applications or tasks requesting the resource are allocated some amount of the resource's time. See Time Slice.

time slot 1. In time division multiplexing (TDM) or switching, the slot (brief moment in time) committed to a voice, data or video conversation. It can be occupied with conversation or left blank. But the slot is always present. You can tell the capacity of the switch or the transmission channel by figuring how many slots are present. See also TDM.

2. An SCSA term. The smallest switchable data unit on the SCbus or SCxbus Data Bus. A time slot consists of eight consecutive bits of data. One time slot is equivalent to a data path with a bandwidth of 64 Kbps. See S.100 and SCSA.

Time Slot Assignment TSA. The assignment of a time slot in a forward time division multiplexed (TDM) facility in order to accommodate traffic from a tributary TDM facility, or in reverse. TDM-based transmission requires that time slots be committed across the network, from end-to-end. Therefore, it is essential that time slots be assigned by the various multiplexers that interconnect TDM circuits. Time-slot assignment enables traffic to be added to any circuit from any tributary, or to be dropped from any circuit to any tributary. The term is most commonly used in the SONET domain. See also SONET, TDM, Time Slot, and Time Slot Interchange.

Time Slot Interchange TSI. The interchanging of time slot between TDM-based links. If the timeslot committed to a given transmission (i.e., call) on an incoming tributary link is already assigned to another transmission on the outgoing link to which it connects, another time slot is selected and assigned. The term is most commonly used in the SONET domain. See also SONET, TDM, Time Slot, and Time Slot Assignment.

Time Space Time System TST. The most common form of switching matrix for small digital telephone exchanges in which a space switch is sandwiched between two time switches.

time switch A device incorporating a clock which arranges to switch equipment on or off at predetermined times.

Time T December 31, 1996, 2359 hours UTC (Universal Time Coordinated). The exact time when the maximum digit length allowed in international dialing was increased from 12 to 15 digits. It seems silly to be so precise about such a thing, but the Time T deadline marked the beginning of the expansion of the number of digits within the numbering plans of the various countries around the world. All of the switches in the networks had to be reprogrammed to understand the lengthened dialing plan, or else the new numbers couldn't be processed. Some switches were reprogrammed, but lots were not. We needed a lengthened dialing plan because we are running out of telephone numbers, and for a bunch of reasons. Blame it on fax machines, cell phones, and pagers. For that matter, blame it on me; my family of four has 18 separate telephone numbers, including fax lines, modem lines, pager (beeper) lines and cell phones, For that matter, blame it on your family; they probably have as many as I have. See also NPA and UTC.

Time to Live TTL A timer value included in packets sent over TCP/IP-based networks that tells the recipients how long to hold or use the packet or any of its included data before expiring and discarding the packet or data. For DNS, TTL values are used in resource records within a zone to determine how long requesting clients should cache and use this information when it appears in a query response answered by a DNS server for the zone.

Time To Live TTL. A mechanism used in the IP protocol, the TTL is an eight-bit field in the IP header. TTL begins at 255 (2 raised to the power eight minus one) seconds, as the TTL field in the IP header is eight bits wide, and as the value of "00000000" is the TTD (Time To Die). As an IP packet is accepted in the buffer of a switch or router, the TTL is decremented until it exits that device. This happens again and again, until either the packet reaches its destination, or until the TTL is decremented to the "00000000" value and it is killed. Without the TTL mechanism, errant packets would circle forever in a "closed loop" and the Internet (or other IP-based network) would be brought to its knees.

time varying media An SCSA definition. Time-varying media, such as audio data (as opposed to space-varying media, such as image data). See S.100.

time zone calling The ability of a dialing system to start and stop calling at predetermined times to different time zones.

timecode Any of several addressing standards used to interlock and sequence audio and video information.

timed detection As a substitute for answer supervision, some long distance phone companies use call timing and estimate that a call is completed if the caller remains off-hook for 30 seconds or more. This is not necessarily accurate, of course. The caller might be holding, thinking the person is in the shower, out in the garden, etc. Little does the caller know he is now being charged to listen to ringing signals. A long distance phone company that is "equal accessed" doesn't have this problem. A long distance company that isn't equal accessed – one that you have to dial directly with a local call – might well have this problem. Rule: When in doubt, don't wait too long on the phone listening to endless ringing. Hang up. Count to ten. Then redial.

timed purge A feature of interactive voice response systems, especially fax-back systems. If the document isn't requested for x number of days or weeks or if the document ages to a certain point, the system automatically deletes the document.

timed recall Your PBX can be instructed to place a call at a designated time. When the time comes, your PBX rings your phone. When you answer your phone, the PBX places the call.

timed reminders At 20-second intervals, timed reminders will alert an attendant that a call is still waiting, a called line has not yet been answered or a call is still on hold. Timed reminders can be made longer or shorter. They can alert attendants to all sorts of events and non-events.

timeout Two computers are "talking" on a network. One (for any reason) fails to respond. The other computer will keep on trying to communicate with the other computer for a certain amount of time, but will eventually "give up." This is called timeout. A timeout also happens in a single standalone computer. If a device (e.g. a printer) is not performing a task or responding, the computer will wait before figuring that something wrong has happened. That time period is called timeout.

times T A new, expanded dialing plan developed by the ITU-T. Times T increases the maximum number of dialed digits from the current 12 to 15, plus the three-digit international access code (country code).

timesharing The use of one computer by many users at one time. Each user is typically sitting in front of a data terminal and connected to the master computer through communications lines – local or long distance. The user asks the computer to work on his task, whether it be a simple as looking up some stock prices, checking an airline reservation or doing some accounting calculations. It appears to each user as if he/she has a computer dedicated to his own task, but the computer is large and powerful, and is moving rapidly from one user's task to the next. Timesharing's advantages are twofold: 1. The user may find it cheaper to time share a computer than to buy his own. 2. The computer may have valuable and extensive information in it, which would be virtually impossible to duplicate or handle in many stand-alone computers. Timesharing was more popular when computers were more expensive.

Timeslot Management Channel TMC. A dedicated channel for sending control messages used to set up and tear down calls in a T-1 frame. In a GR-303 interface group, the primary TMC is usually in channel 24 of the first DS1, while a redundant TMC (if used) would be located in a different DS-1.

timestamp A mark placed on a data or voice transaction used for throughput and processing calculations. Can be used to determine total work time by placing one at the beginning and one at the end of a transaction. Timestamps are used in productivity measurement, in call accounting and traffic analysis systems, and a wide variety of other applications. Timestamps also are used to synchronize various network devices, such as IRC (Internet Relay Chat) servers.

timing In the beginning, telephone systems were very simple, circuit-switched animals. When I called you, we used the entire bandwidth on the wires for our conversation. If something untoward happened, one of us would simply ask the other to repeat what he said. No sweat. Quickly it became apparent to the phone industry that devoting an entire circuit to one conversation was wasteful. So various methods to put more than one conversation on a circuit was devised. These were initially called multiplexing techniques. The early ones were typically analog, with different conversations occupying different frequencies. Filters could easily pull the various conversations out. But then came the digital revolution, which made it suddenly cheap to represent phone calls by bits and mush many conversations into one large stream of bits. (The original digital channels are originally thought of purely as "pair gain.") How to pull the various conversations out of that one gigantic bit stream? Think T-1 with a stream of 1.544 million bits per second, or 24 conversations each encoded at 64,000 bits per second. How to figure out where one conversation started and ended? You could add information to the flow. Call it "framing" information. That information would frame the data. If you know what the frame looked like, you could pluck the information out of it. The T-1 trunk's 24 channels of 64,000 bits per second, each carrying 8,000 8-bit bytes per second. Each byte represented one sample of analog information. The remaining 8,000 bits per second were framing bits (24 x 64 = 1,536,000 bits per second plus 8,000 bits per second = 1,544,000 bits per second. In the beginning each T-1 was an entity unto itself and it typically started and ended in analog loops at either end. Initially, T-1 was an asynchronous system. Each pair of end terminals ran at their own clock rate, and each terminal used its receive timing to demultiplex the incoming signal. The transmit and receive sides were independent of one another. Unfortunately, the voice quality of these analog-end and often analog-middle networks proved to be less than desirable due to the low bandwidth and losses of the analog channels. And the networks of lonely pair gain digital systems suddenly started to get more and more complex and more intertwined. It was increasingly evident that the bits and pieces of a digital networks had to become one gigantic synchronized network. So follow this evolution: When digital channel units were introduced (to bring T-1 channels right to the user's site without the disadvantage of convergence to analog), one end terminal was designated as the master and had its own timing reference. The other end terminal was a slave, and derived timing for its transmit side from the data being received (looped timing). This arrangement worked as long as the end terminals were no more complex than a channel bank. But T-1 (DS-1) was only the beginning. Soon T-3 (DS-3) came along, with speeds of 45 million bits per second (672 voice conversations). And switches became digital, plucking straight digital bit streams right out of the T-1 bit stream. Digital switches with DS-1 port interfaces exemplified the shortcomings of an asynchronous system. If the two switch clocks were not at the same frequency (i.e. their clocks were out of whack), the data would slip at a rate dependent on the difference in clock frequencies. A slip is defined as a one frame (193 bits) shift in time difference between the two signals in question. This time difference is equal to 125 microseconds. Slips were not considered a major impairment to trunks carrying voice circuits. The lost frames and temporary loss of frame synchronization resulted in occasional pops and clicks being heard during the call in progress. However, with advances in DS-1 connectivity, these impairments tended to spread throughout the network. To minimize these problems (and to allow for efficient – i.e. error-free transmission of data), a hierarchical clock scheme was developed, whose function was to produce a primary reference for distribution to switching centers in order to time the toll switches. Local switching in that early era was primarily analog, so that synchronization was not required at the end offices. Later, digital switches and direct digital services or networks (DDS or DDN) became common at the end offices (i.e. those directly serving the customer and providing digital services to customers. This meant timing had to be distributed to local levels.

The resulting hierarchy evolved into four levels. Level 1, known as Stratum 1, is the primary reference. It was known originally as the Bell System Reference Frequency or BSRF. The second level, Stratum 2, is used at toll switches. Stratum 3 is used at local switches. Channel banks and end terminals that use simple crystal oscillators are known as Stratum 4 devices. Recently, SONET (Synchronous Optical Network) networks have created the need for a clock stratum level better than Stratum 3, which is called Stratum 3E.

In 1984, when the Bell System broke up into the local service providers and the long distance carriers, the timing hierarchy became less well defined. Now each local company could no longer take its timing from the long distance carrier, but had to engineer a system, either a hierarchy or otherwise, to distribute timing to their offices. This made everything more difficult because failures in the transmission systems could cause "islands" or areas without a reference to Stratum 1. For more on timing, see www.laruscorp.com/timtut.htm.

See also Timing Advance.

timing advance See Location Services.

timing jitter Deviation of clock recovery that can occur when a receiver attempts to recover clocking as well as data from the received signal. The clock recovery will deviate in a random fashion from the transitions of the received signal.

timing recovery The derivation of a timing signal from a received signal.

timing signal The output of a clock. A signal used to synchronize connected equipment.

timing slip A sudden timing delay change during high-speed digital transmission often caused by using T-1 carriers from different suppliers.

TIMS Transmission Impairment Measurement Set. It is a basic analog testing set. In common usage it refers to the suite of fundamental tests required to ensure that a telecom circuit will support analog traffic.

tin and string Processors and network bandwidth. Used in the expression "throw tin and string at a problem," i.e., solve a performance problem on a network by upgrading servers and/or bandwidth. Sometimes the phrase "tin and wet string" is used instead.

TINA Telecommunications Information Networking Architecture. A developing standard which is intended to resolve issues of integration between TMN (Telecommunications Management Network) and IN (Intelligent Network) standards and concepts. TINA focuses on the definition and validation of an open architecture for worldwide telecom services through a flexible software architecture for both end-user and network management services. See TINA-C, TMN and IN.

TINA-C Telecommunications Information Networking Architecture Consortium. An international, voluntary, not-for-profit organization of vendors and others for the purpose of promoting an open network architecture for the delivery and management of sophisticated services.

tine Also known as a finger. An individual digital channel of a wireless rake receiver. A rake receiver can support a number of tines, which can be combined to form a stronger received signal.

tingus The dongle that fits between a PCMCIA card inside your laptop and the RJ-45 cable that is typically used to connect to your local area network.

tinned copper tin coating added to copper to aid in soldering and inhibit corrosion.

tinned wire Copper wire coated with tin to make soldering easier.

Tinsel Town Hollywood. The nickname "Tinsel Town" was coined by Oscar Levant, the pianist, and composer who observed: "Strip the phoney tinsel off Hollywood, and you'll find the real tinsel underneath."

tinsel wire A component of some phone line cord conductors. Tinsel wire is made by rolling copper into very thin, narrow rolls and then winding several strands of tinsel around a non-metallic core (a string) and then placing an insulating cover over the resulting conductor. A cord is then built up of two or more conductors encased in a plastic jacket. The essential reason for this type of construction is to obtain good cord flexibility and long life.

tint Another name for hue.

tip 1. The first wire in a pair of phone wires. The second wire is called the "ring" wire. The tip is the conductor in a telephone cable pair which is usually connected to positive side of a battery at the telephone company's central office. It is the phone industry's equivalent of Ground in a normal electrical circuit. See Tip & Ring.

2. TIP. The Transaction Internet Protocol protocol ensures that multivendor transaction monitors will work with one another to complete transactions over the Internet (RFC 2371). TIP came from a joint Microsoft/Tandem effort. See Transaction Internet Protocol for a much longer explanation.

tip & ring An old fashioned way of saying "plus" and "minus," or ground and positive in electrical circuits. Tip and Ring are telephony terms. They derive their names from the operator's cordboard plug. The tip wire (the positive wire) was connected to the tip of the plug, and the ring wire (the negative wire) was connected to the slip ring around the jack. A third conductor on some jacks was called the sleeve. See Tip, Ring & Ground.

tip cable A small cable connecting terminals on a distributing frame to cable pairs in the cable vault.

tip conductor The conductor of a pair that is grounded at the central office when the line is idle. This term was originally coined from its position as the first (tip) conductor of a tip-ring-sleeve switchboard plug. See Tip & Ring.

tip jar You go to a web site and you see something that pleases you – a video left by a visitor, an audio clip of your favorite speaker – you're asked to donate some money into the "tip jar," the electronic equivalent of something you might see at a take-out restaurant. The proceeds of the tip jar are often split between the contributor of the pleasing item and the operator of the site itself. I think this idea of the tip jar started with online communities.

tip side That conductor of a circuit which is associated with the tip of a plug, or of a telephone circuit. See Tip & Ring.

tip, ring, ground The conductive paths between a central office and a phone. The tip and ring leads constitute the circuit that carries the speech or data signal. The ground path in combination with the conductor is used occasionally for signaling.

tip-to-tip 1. Refers to how the physical separation between two antennas is measured, namely, between their tips.

2. Synonymous with "end-to-end" when describing a network infrastructure whose end points are telephone jacks, wiring closets, or LAN cable drops on the customers'/users' premises.

TIPHON Telecommunications and Internet Protocol Harmonization Over Networks (TIPHON). An ETSI (European Telecommunications Standards Institute) project to define the interactions between emerging VoIP (Voice over Internet Protocol) packet technologies and traditional circuit-switched voice networks. TIPHON is intended to ensure that VoIP networks will interface smoothly with the PSTN (Public Switched Telephone Network), as well as the GSM and other wireless networks. TIPHON is trademarked by ETSI.

TIPI Telephone Industry Price Index.

tipping point The point when everyday things reach epidemic proportions. That dramatic moment when something unique becomes common. The concept probably comes from epidemiology, where small changes will have little or no effect until a critical mass is reached. Then a further small change "tips" the system and a large effect happens. Web site entrepreneurs talk about the tipping point as the point their web site is no longer a curiosity but becomes a significant site with zillions of visitors and a positive, growing cash flow.

TIQ Telrate International Quotations.

TIRKS Trunks Integrated Records Keeping System. An Operations Support System (OSS) developed many years ago by the Bell System to mechanize circuit provisioning functions including circuit order control, circuit design, selection and assignment of equipment and facilities, work order generation and distribution, and circuit inventory control. TIRKS maintains inventory information on all assignable components for trunks and special-service circuits. The Plug-in Inventory Control System/Detailed Continuing Property Records (PICS/DCPR) system maintains records of plug-in equipment, and interfaces with TIRKS in the circuit-provisioning process. TIRKS is still used by the RBOCs (regional bell operating companies).

TIS Technical Information Sheets.

Titanic On December 21, 1993 Vice President, Al Gore, told the National Press Club in Washington, "There is a lot of romance surrounding the sinking of the Titanic 81 years ago. But when you strip the romance away, a tragic story emerges that tells us a lot about human beings – and telecommunications. Why did the ship that couldn't be sunk steam full speed into an ice field? For in the last few hours before the Titanic collided, other ships were sending messages like this one from the Mesaba: "Lat42N to 41.25 Long 49W to Long 50.30W. Saw much heavy pack ice and great number large icebergs also field ice." And why, when the Titanic operators sent distress signal after distress signal did so few ships respond?

The answer is that – as the investigations proved – the wireless business then was just that, a business. Operators had no obligation to remain on duty. They were to do what was profitable. When the day's work was done – often the lucrative transmissions from wealthy passengers – operators shut off their sets and went to sleep. In fact, when the last ice warnings were sent, the Titanic operators were too involved sending those private messages from wealthy passengers to take them. And when they sent the distress signals, operators on the other ships were in bed."

titanium Titanium resists corrosion for 100 years. It's as light as aluminum and as a strong as steel. It's also expensive and it's now "way cool." The titanium-clad Guggenheim Museum in Bilbao, Spain made the town a major tourist attraction. When I saw it in 2001, I flipped.

TISOC The Bell Atlantic Telecom Industry Services Operating Center. It provides standards, methods and procedures and services to the full spectrum of CLECs from full service providers to wholesale providers.

titles In the language of multimedia, when an author sells what he or she has created, it is called a title. The encyclopedias, dictionaries, musical works, and games available on CD are all "titles." Someone authors the material, and sells it to users who can play it back but not change the content.

tittles The dot on top of the letter "i" is called a "tittle." Tittle is Latin for something very small.

TiVo The pioneer of digital video recorders. Its success has been stymied by cable and satellite providers who supply their own cheaper DVRs to their customers. I include TiVo in this dictionary because, like Google, it's become a verb, as in "I'll TiVo that episode of Friends." As the New York Times described TiVo, as "letting you choose shows to record

from a list (without having to know their broadcast time or channel), pausing or rewinding live TV, zipping past commercials, recording the same show every week automatically, and generally bending the broadcasters' schedules to your whim." See PVR.

TiVo-Proofing The practice of embedding products and advertising (think: pop-ups) within TV shows so that owners of digital video recorders can't skip the commercials.

TJB Touch-tone central office (CO) trunk.

TJC Touch tone.

TJF Test Jack Frame.

TJN Journal Recovery Table.

TL 1. Tie Line.

2. Transmission Level.

TL 9000 TL 9000 is a set of telecommunications industry-specific quality management system standards based on ISO 9001:2000. It defines quality management system requirements for the design, development, production, delivery, installation and maintenance of telecom products and services.

TL1 Transaction Language 1. A machine to machine communications language which is a subset of ITU-T's man machine language.

TLA Three Letter Acronym. A form and usage common to our acronym-happy industry.

TLB Test LoopBack. A CSU (Channel Service Unit) operating mode that loops the telco's T-1 transmission facility back towards itself at itself at the same time it loops the CPE back toward itself.

TLD Top Level Domain. See Domain and gTLD.

TLDN A Temporary Local Directory Number is assigned by a visited wireless network's Mobile Switching Center to support call delivery to an idle roaming subscriber. This TLDN is used by the originating MSC (Mobile Switching Center) to establish a voice path to the serving MSC via existing interconnection protocols (i.e. SS7).

TLEC Terminating Local Exchange Carrier.

TLF 1. Trunk Link Frame.

2. Resource Management Table.

TLI Transport Layer Interface. TLI. An application program interface provided with UNIX System V Release 3.

TLL Logical Line Table.

TLP 1. Transmission Level Point. A physical point in a circuit at which the signal level, or amplitude, is measured. See also dBrn, Level, Loss, and Pad.

2. Logical Line Pair Table.

TLS 1. Transparent LAN Service. You have a LAN in one office. Across town you have another office with another LAN. You go to your friendly service provider and say "I want a telecommunications service that will let me send messages, mail and files, etc. between my two LANs." Bingo, they provide you a Transparent LAN Service (TLS), which is a high speed VPN (Virtual Private Network) service that hides the complexity associated with the WAN (Wide Area Network). With TLS, a service provider interconnects a corporation's LANs in such a way that the wide area is transparent to the end user. It is as though the physically separate LANs were all physically connected in the same physical location. Actually, they are logically interconnected across the WAN.

A loosely-defined high speed VPN (Virtual Private Network) service offering of various LECs (Local Exchange Carriers), IXCs IntereXchange Carriers) and MSPs (Metropolitan Service Providers), TLS provides for the interconnection of LANs over the MAN or WAN public data network (PDN). In other words, a TLS customer can establish direct Ethernet-to-Ethernet or Token Ring-to-Token Ring connectivity through a PDN without either the trouble or expense associated with a private leased-line network, or even a Frame Relay or ATM network. Internally, the carrier may provision the network through a variety of methods. Transmission facilities may be in the form of optical fiber or xDSL (e.g., ADSL, HDSL, or SDSL) for access, and will be optical fiber in the backbone. At Layer 1, the optical fiber may run the SONET protocol, although it may be DWDM. At Layer 2 and 3, various combinations of GbE (Gigabit Ethernet) and 10GbE, IP, Frame Relay, ATM, MPLS, and other protocols may be employed. In any event, all of these protocol issues are transparent to the end user organization. TLS generally is provisioned as a managed service, with the service provider retaining full ownership of and taking full responsibility for all of the technical issues, including the CPE (Customer Premises Equipment). It looks like straight Ethernet to you, although generally at a slower speed. See 10GbE, ADSL ATM, CAP, CPE, DWDM, Ethernet, Frame Relay, GbE, HDSL, IP, IXC, LEC, MPLS, MSP, SDSL, SONET, Token Ring, VPN, and xDSL.

2. Transport Layer Security. Specified by the IETF (Internet Engineering Task Force) as RFC 2246 and based on SSL (Secure Sockets Layer), TLS is a standard security protocol. TLS is modular, comprising two layers. The TLS Record Protocol ensures that the connection is private by using symmetric encryption. The TLS Record Protocol also is used to encapsu-

late higher-level protocols. The TLS Handshake Protocol is used for authentication between the server and the client through digital certificates, and for the negotiation of the encryption algorithm and cryptographic keys prior to the establishment of data communications. TLS is independent of higher-layer protocols, including applications-layer protocols such as HTTP and SMTP. See also encryption and SSL.

TM 1. Trouble Management. The responsibilities associated with receiving any network events that impact customer service whether they are generated via customer contact or from internal network elements. Trouble Management tracks all problems, groups them together (if possible), and relays trouble tickets for problem resolution. A mobile phone term.

2. Traffic Management. As an ATM term, traffic Management is the aspect of the traffic control and congestion control procedures for ATM. ATM layer traffic control refers to the set of actions taken by the network to avoid congestion conditions. ATM layer congestion control refers to the set of actions taken by the network to minimize the intensity, spread and duration of congestion. The following functions form a framework for managing and controlling traffic and congestion in ATM networks and may be used in appropriate combinations.

- Connection Admission Control
- Feedback Control
- Usage Parameter Control
- Priority Control
- Traffic Shaping
- Network Resource Management
- Frame Discard
- ABR Flow Control

TM Terminal Multiplexer.

TMA Telecommunication Managers Association.

TMAP A piece of software announced by Northern Telecom (now called Nortel Networks) in the summer of 1994 and designed to map Windows Telephony commands to Novell Telephony Services (TSAPI) commands. Tmap runs on the local workstation (also called client) and translates TAPI commands into TSAPI commands that are sent through the local area network to the telephony file server and thence to the PBX, causing the PBX to dial, or conference or ring a phone, etc. The idea is that computer telephony software running on the user's desktop PC could be implemented as easily through a card in the PC or through the company PBX attached to the LAN – and the whole process would be seamless to the user.

TMC Timeslot Management Channel. A dedicated channel for sending control messages used to set up and tear down calls in a T-1 frame. In a GR-303 interface group, the primary TMC is usually in channel 24 of the first DS1, while a redundant TMC (if used) would be located in a different DS-1.

TME Telocator Message Entry. A client/server protocol proposed for use in communications between wireless pagers and the Internet. TME is proposed as a replacement for Simple Network Paging Protocol (SNPP). See also SNPP.

TMF See TeleManagement Forum.

TMGB Telecommunications Main Grounding Busbar. A busbar placed in a convenient and accessible location and bonded by a means of the bonding conductor for telecommunications to the service equipment (power) ground. | **TMGT** Telemanagement.

TML 1. Multilink Control Table.

2. See Telephony Markup Language.

TMN Telecommunications Management Network. A network management model defined in ITU-T recommendation M.30 and related recommendations, and intended to form a standard basis for management of advanced networks such as SDH (Synchronous Digital Hierarchy) for fiber optics in land lines and GSM (Global System for Mobile Communications) in the cellular world. TMN specifies a set of standard functions with standard interfaces, and makes use of a management network which is separate and distinct from the information transmission network. Further, standard network protocols such as the OSI CMIP (Open Systems Integration Common Management Information Protocol) are specified. Implementation of this concept involves the linking of all subject device elements to OMCs (Operation and Maintenance Centers) which, in turn, are linked together over a separate network. A centralization occurs to facilitate control, monitoring, and management of all devices in the communications network, which can include legacy systems as well as newer technologies. Operation systems functions include the full range of functions defined in the OSI model: Performance Management (PM), Fault Management (FM), Configuration Management (CM), Accounting Management (AM), and Security Management (SM).

A gentleman called James Keil who wrote his master's thesis at the University of Boul-

der, Interdisciplinary Telecommunications Program, on TMN compliant equipment, says that a "quick and dirty definition of TMN" would be "A network management standard which seeks to provide IT, business and network service management in a multi-domain environments (i.e. VPN, RBOC, Cellular providers)." Mr. Keil also says TMN fully implemented can retrieve resources from disparate networks like SNMP, through the use of managed objects or ANSI.1. TMN has much more functionality than SNMP. See also TINA.

TMP Test Management Protocol: As an ATM term, it is a protocol which is used in the test coordination procedures for a particular test suite.

TMR Trunked Mobile Radio. Another name for SMR (Specialized Mobile Radio). See SMR.

TMRS Traffic Measurement and Recording System.

TMS 1. Time Multiplexed Switch. In the AT&T 5ESS switch CM, the TM provides switch paths between switching modules and passes control messages to and from the message switch, and functions as the hub for clock distribution to the switching modules.

2. TOPS Message Switch.

3. Trouble Management System.

4. Transmission Measuring Set.

TMSI Temporary Mobile Station Identifier A mobile station identifier (MSID) sent over the air interface and is assigned dynamically by the network to the mobile station.

TMU Terminal MakeUp. Refers to the electrical configuration (resistance including bridged tap) of the terminal. It is a function of the linear distance (usually measured in feet) from the Central Office and copper cable gauge (or gauges) and bridge tap.

TMUX Transmultiplexer.

TN 1. Telephone Number.

2. Twisted Nematic. Most used display technology for calculators, watches and measuring equipment. TN uses liquid crystals sandwiched between two plates of glass with integrated transparent electrodes which can be made transparent and non-transparent by applying an electric current to them. See LCD.

TN3270 A PC or Macintosh application that enables the computer to emulate an IBM 3270 terminal and remotely connect to an IBM mainframe. Also, delivery of a 3270 data stream via Telnet, provided as part of the TCP/IP protocol suite.

TN5250 A version of the Telnet protocol that supports the IBM 5250 terminal. See 5250.

TNA Tits and Ass – an irreverent name for a TV channel called the Spike Channel, which was once called TNN.

TNC A small connector used on coaxial cable, commonly used for cellular antennas, and some data and test equipment.

TNL Terminal Net Loss.

TNotify Time Notify. Specifies how often SMT initiates neighbor notification broadcasts. See also SMT.

TNPP A protocol used to send paging messages from terminal to terminal on LANs and WANs over a wire circuit.

TNS 1. Transit Network Selection. As an ATM term, it is a signaling element that identifies a public carrier to which a connection setup should be routed.

2. Transient Network Signaling.

3. Network Status Table.

TNSS Non-Synchronous test line provides for rapid testing of ringing, tripping and supervisory functions of toll completing trunks. This test line provides an operation test which is not as complete as the Synchronous test but which can be made more rapidly.

TNT 1. Test and Turnup. This is the end phase of the circuit installation and provisioning process.

2. Network I/O Device Table.

TO Transmit Only.

TOA/NPI Type Of Address/Numbering Plan Identifier.

TOC Technical Operating Center.

TOD Time Of Day. See also Time.

Toddisms Todd Kingsley, the king of stockbrokers, but the peasant of tennis players (he denies this), has advice for would-be investors. They're called Toddisms. They include such gems as:

"When in doubt, stay out." "Even a broken clock is right twice a day." "Even a blind squirrel finds an acorn every now and again." "Don't catch a falling knife." "Don't fight the tape." "Surely, you are smart enough to understand the value of flattery?" "Better to be wrong and out, than wrong and in." "Pigs gets fat. Hogs get slaughtered." "Life's too short to dance with ugly women." "Wall Street analysts suck. I don't need someone to tell me it rained yesterday. I need someone to tell me to take my umbrella tomorrow." "Sell on

Rosh Hashanah. Buy on Yom Kippur." "Capital preservation is the name of the game, not needless gambling."

TOF Time Out Factor: As an ATM term, it is an ABR service parameter, TOF controls the maximum time permitted between sending forward RM-cells before a rate decrease is required. It is signaled as TOFF where TOF=TOFF+1. TOFF is a power of 2 in the range: 1/8 to 4,096.

TOFF Time Out Factor: See TOF.

Toggle 1. A flip-flop switch that changes for every input pulse.

2. Any simple two-position switch.

TOH Transport Overhead. A SONET term describing an element of signaling and control. TOH includes Section Overhead (SOH) and Line Overhead (LOH).

ToIP Telephony over IP. See telephony over IP and VoIP.

TOK Test Okay.

token 1. In networking, a unique combination of bits used to confer transmit privileges to a computer on a local area network. It also carries important information for routing messages over the network, such as source and destination addresses, access control information, route control information, and date checking information. When a LAN-attached computer receives a token, it has been given permission to transmit. On a token ring network, the token is 24 bits long. See Token Passing and Token Ring.

2. Here is a Rolm definition: The floating master message which coordinates use of the CBX control packet network among the nodes connected to it.

token bus A local network access mechanism and topology in which all phones or workstations attached to the bus listen for a broadcast token or supervisory frame. That token confers on them the right to communicate over the share channel, the token bus. An example of a Token-Bus is IEEE 802.4. See Token Passing.

token latency The time it takes for a token to be passed around the local area network ring.

token passing A method whereby each device on a local area network receives and passes the right to use the single channel on the LAN. The key to remember is that a token passing, or token ring LAN has only one channel. It's a high-speed channel. It can move a lot of data. But it can only move one "conversation" at a time. The Token acts like a traffic cop. It confers the privilege to send a transmission. Tokens are special bit patterns or packets, usually several bits in length, which circulate from node to node when there is no message traffic. Possession of the token gives exclusive access to the network for transmission of a message. The token is generated by one device on the network. If that device is turned off or fails, another device will assume the token creation task. When the package of token and message reaches its destination, the computer copies the message. The package is then put back on the network where it continues to circulate until it returns to the source computer. The source computer then releases the token for the next computer in the sequence.

With token passing it is possible to give some computers more access to the token than others. Usually one device on the network is designated the token manager. It generates the token. If that device is turned off or fails, another device will assume management of the token. There is a complicated sequence of events that result in the generation of a token and that deal with the eventuality of token loss or destruction. The logic for this process is built into token ring cards that fit inside computers. In some manufacturers' products, the logic is slightly different and can cause incompatibilities. See Token, Token Ring and Token Ring Packet.

token ring A ring type of local area network (LAN) in which a supervisory frame, or token, must be received by an attached terminal or workstation before that terminal or workstation can start transmitting. The workstation with the token then transmits and uses the entire bandwidth of whatever communications media the token ring network is using. A token ring is a baseband network. Token ring is the technique used by IBM, Arcnet, and others. A token ring LAN can be wired as a circle or a star, with all workstations wired to a central wiring center, or to multiple wiring centers. The most common wiring scheme is called a star-wired ring. In this configuration, each computer is wired directly to a device called a Multi-station Access Unit (MAU). These are usually grouped together in a wiring closet for convenience. The MAU is wired in such a way as to create a ring between the computers. If one of the computers is turned off or breaks or its cable to the MAU is broken, the MAU automatically recreates the ring without that computer. This gives token ring networks great flexibility, reliability, and ease of configuration and maintenance.

Despite the wiring, a token ring LAN always works logically as a circle, with the token passing around the circle from one workstation to another. The advantage of token ring LANs is that media faults (broken cable) can be fixed easily. It's easy to isolate them. Token rings are typically installed in centralized closets, with loops snaking to served workstations. Some other LANs require your going up in the ceiling or into walls and finding coax taps. All

the work on a token ring can be done on one or several panels. These panels allow you to isolate workstations, and thus isolate faults.

Token Ring LANs can operate at transmission rates of either 4M bits per second or 16M bits per second. The number of computers that can be connected to a single Token Ring LAN is limited to 256. The typical installation is usually less than 100. Large installations connect multiple token ring LANS with bridges. The theoretical limit of Ethernet, measured in 64 byte packets, is 14,800 packets per second (PPS). By comparison, Token Ring is 30,000 and FDDI is 170,000. See FDDI-II and FDDI TERMS. Help on this definition courtesy Tad Witkowicz of Crosscomm, Marlboro, MA, Tim Becker, Lanquest Group, Santa Clara, CA and Elaine Jones, VP Marketing, Coral Network Corporation, Marlborough, MA. See also Bridge, IBM Token Ring, MAU, Token Passing, Token Ring, Token Ring Card and Token Ring Packet.

token ring card Name given to the circuit board inserted into a computer device for connection to a token ring LAN. This board provides the physical connection to the LAN. It also participates in the collective management of the token by sending various messages to other token ring cards. Usually, one token ring card on the network is designated the token manager. It automatically generates a token as soon as it discovers one is missing, often with the help of other token ring cards. The sending of messages between token ring cards can be used to gather information about what is taking place on the network. Statistics may be collected. These may indicated that the network should be altered in some way to improve performance. This management capability is a distinct advantage of token ring LANs. One possible drawback is that various manufacturers' token ring cards may differ slightly in how they implement token management, thereby making them incompatible in certain management features. Virtually all token ring cards will work together in basic token passing.

token ring lan service unit The ATM TLSU provides a powerful tool for offering internetworking services over ATM networks. Emulated token rings consist of up to 64 TLSU token ring ports located anywhere in the ATM network, interconnected with PVCs. These emulated token ring networks can be completely isolated form one another to ensure security and fairness among the attached LANs. The TLSUs are designed for flexible deployment, either local to an ATM switch or at a remote site. See ATM Ethernet LAN Service Unit.

token ring packet Packets on a token ring network are made up of nine fields: starting delimiter, access control, frame control, destination address, source address, routing information, the data, frame check sequence, and ending delimiter.

Starting Delimiter (SD): This is an 8-bit binary (1s and 0s) sequence which marks the beginning of a data packet.

Access Control (AC) and Frame Control (AC): These are two 8-bit sequences that are used by the computers for maintenance purposes.

Destination Address (DA): This is a 48-bit sequence that uniquely identifies the physical name of the computer to which the data packet is being transmitted. Each computer on a ring examines this field to determine if the packet is for it.

Source Address (SA): This is a 48-bit sequence that uniquely identifies the physical name of the computer that send the data packet. This is used by the receiving computer to formulate its acknowledgement.

Routing Information (RI): This is a variable-length sequence used if the data packet is being sent to a computer located on another token ring LAN. (This information can make it impossible for some bridges to route some packets. See Bridge.)

Data: This is a variable-length sequence (up to 17,800 bytes) that is the actual data being sent from source to destination.

Frame Check Sequence (FCS): This 32-bit sequence is used to protect the contents of the packet from being corrupted during transmission. See Frame Check Sequence.

Ending Delimiter (ED): This is an 8-bit sequence that signals the end of a packet.

token tree LAN A type of local area network with a topology in the form of branches interconnected via active hubs. Using a token-passing scheme, the active hubs grant nodes access to the medium. See Token Packet.

Tokentalk The original Apple Macintosh implementation of the Token Ring local area network.

toll block A restriction that is placed on a subscriber line or trunk to prevent it from making certain types of toll calls, such as domestic long distance calls, international long distance calls, calls to 900 numbers, and interstate or international directory assistance.

toll booth model When there's no front-end fee to sign up and you pay every time you use — just like on a highway. You don't pay an "installation" fee every time you hit the highway. You just pay tolls as you use the highway. Thus, some software companies would prefer to charge according to the "toll booth model."

toll bypass When you dial a long distance phone call the traditional way – on the PSTN – the public switched telephone network, you're making a toll call for which you're charged a per minute price that often works by distance also. When you make that call some other way – e.g. over your firm's private leased line network or over the Internet (e.g. with Skype or Vonage), you're bypassing the toll system, thus doing what's known as "toll bypass." The concept of toll bypass tends to raise red flags with traditional switched phone companies.

toll call Any call, local or long distance, that incurs a fee. In the old days a toll call was basically any phone call that incurred a charge. Since most local calls were free (in the old days), toll calls tended to be long distance. But now some local calls cost money per call. And some long distance calls are free. So, this definition has become simple: any call that incurs a free from your telephone company provider is a toll call.

toll center 1. A central office where operators (human or mechanical) are present to assist in completing incoming toll calls.

2. Name of a Class 4 switching center in the original Bell DDD hierarchy of long distance switching centers, providing links to the local exchanges of a city or metropolitan area; the point of interconnection between local networks and intercity networks.

toll connecting trunk A trunk used to connect a Class 5 office (local central office) to the direct distance dialing network.

toll denial Permits phone user to make local calls but denies completion of toll calls or calls to the toll operator without the assistance of the attendant. See Toll Restriction.

toll diversion A system service feature by which users are denied the ability to place toll calls without the assistance of a human attendant. Toll diversion affects the entire switching system instead of discriminating between individual extensions.

toll file guide According to Bell Atlantic, a Toll File Guide is a number that was 'inherited' by a long distance carrier after divestiture. The RBOC retains a reference to the number and will show the number on a phone bill; however any billing associated with this number comes from the LD carrier. A toll file guide exists for purposes of number portability.

toll fraud Theft of long distance service. Today's most common forms of toll fraud are DISA, voice mail and shoulder surfing. According to John Haugh of Telecommunications Advisors in Portland, OR, there are three distinct varieties of toll fraud:

"First Party" Toll Fraud, which is helped along by a member of the management or staff of a user. An example would be the telecommunications manager at the Human Resources department of New York City (an "insider") who sold his agency's internal code to the thieves, who in turn ran up unauthorized long distance charges exceeding $500,000.

"Second Party" Toll Fraud, which is facilitated by a staff member or subcontractor of a long distance carrier IXCs, vendor or local exchange telephone company selling the information to the actual thieves, or their "middlemen." An example would be a "back office clerk" working for one of these concerns who sells the codes to others.

"Third Party" Toll Fraud is facilitated by unrelated "strangers" who, though various artifices, either "hack" into a user's equipment and learn the codes and procedures, or obtain the needed information through some other source, to commit Toll Fraud.

toll free See toll-free.

toll grade This is a strange one. "Toll grade" describes what a circuit-switched, long distance phone call sounds like. In particular it means that the call doesn't have a delay or an echo. When the telephony world started going to IP – meaning it became packet switched – the first calls suffered delays and echoes. As IP calls got better they suffered fewer delays and fewer echoes, i.e. they began to sound more and more "toll grade."

toll grooming See Grooming.

toll office A central office used primarily for supervising and switching toll traffic.

toll plant The facilities that connect toll offices throughout the country.

toll quality Imagine you get a good long distance telephone voice connection. You can hear them and they can hear you. That's a "toll quality" phone call. Most "toll quality" phone calls are made on a circuit-switched basis over the PSTN (Public Switched Telephone Network). When you make a phone call over the packet-switched public Internet, your conversation sounds pretty awful. It's not "toll quality." Some of the newer packet-switched networks, however, support voice that sounds pretty good – even toll quality. Technically speaking, toll quality is defined in the ITU-T P.800 specification "Methods for Subjective Determination of Voice Quality." P.800 involves the subjective evaluation of preselected voice samples of voice encoding and compression algorithms. The evaluation is conducted by a panel of "expert listeners" comprising a mixed group of men and women under controlled conditions. The result of the evaluation is a Mean Opinion Score (MOS) in a range from 1 to 5, with 1 being "bad" and 5 being "excellent." P.800 defines toll quality as an MOS of 4.0. See also MOS, P.800, and P.861.

toll restriction To curb a telephone user's ability to make long distance calls. Toll restriction capability on modern PBXs and key telephone systems has been increasing in sophistication. Some PBXs now allow selective restriction based on specific extensions, users or geography. In other words, Joe Smith, the president, could call everywhere. John Doe in accounting might only be allowed to call Chicago and Houston, where our two factories are located. Mary Johnson, the seller for the western U.S., might only be allowed to call Denver and points west. There's considerable debate as to how useful toll restriction really is.

toll saver feature Many answering machines – both PC based and stand alone machines – allow you, the owner, to dial in and remotely retrieve your messages. Because it makes no sense to incur toll call costs if there are no messages, many machines have "a toll saver feature." They will only answer the first message on the fourth ring. They answer each additional one on the second ring. This means that if you're calling remotely, you can count the rings. If you get to three rings and the machine hasn't answered, you know that there are no new messages (i.e. ones you haven't heard) and you can safely hang up without incurring any toll costs.

toll station A Telco phone from which established long distance message rates are charged for all messages sent over company lines.

toll switching trunk A trunk connecting one or more end offices to a toll center as the first stage of concentration for intertoll traffic. Operator assistance or participation may be an optional function. In U.S. common carrier telephony service, a toll center designated "Class 4C" is an office where assistance in completing incoming calls is provided in addition to other traffic; a toll center designated "Class 4P" is an office where operators handle only outbound calls, or where switching is performed without operator assistance.

toll testboard Manual test position at which toll circuits are tested and repaired.

toll terminal A phone only furnished with long distance service.

toll terminal access Allows hotel/motel guest phones to access toll calling trunks.

toll ticket Ticket is the telephone company term for a bill. A toll ticket is a bill containing the calling number, called number, time of day, date and call duration. Some phone systems generate their own bills automatically. Some still need an operator. It depends on the equipment and the type of call.

toll trunk A communications channel between a toll office and a local central office.

toll-free call See 800 Service.

TOM See Telecom Operations Map.

Tombstone 1. A deadbug should be mounted on a circuit board with its lengthwise face parallel to and touching the laminate of the circuit board and its leads pointing away from the board. If, instead, the deadbug's end face is mounted on the board, with the component's lengthwise portion sticking out at a perpendicular angle to the circuit board, it is called a tombstone termination, because the deadbug looks like a tombstone sticking out of the ground. Tombstoning a deadbug is an industry worst-practice, because it puts too much stress on the narrow place where the deadbug is soldered to the circuit board. See deadbug.

2. In southern Arizona, the city of Tombstone is probably the most famous and most glamorized mining town in all of North America. According to legend, prospectors Ed Schieffelin and his brother Al were warned not to venture into the Apache-inhabited Mule Mountains because they would only "find their own tombstones." Thus, with a touch of the macabre, the Schieffelins named their first silver strike claim Tombstone, and it became the name of the town.

Tomlinson, Allan Allan Tomlinson won the Australian Grand Prix in motor car racing in 1939. He is now 90 and my favorite father-in-law. He's the father of my wife, Susan. He lives in Perth, Western Australia. He is a truly wonderful person.

Tomlinson, Ray Ray Tomlinson was the inventor of email. He also invented the use of the @ sign in a person's email address.

tone An audio signal consisting of one or more superimposed amplitude modulated frequencies with a distinct cadence and duration. See Tone Set and Tones.

tone alternator A motor-driven AC generator that produces audio-frequency tones.

tone dial What the Australians call tone dial, Americans call touchtone. Tone dial or touchtone dial makes a different sound (in fact, a combination of two tones) for each number pushed. The correct name for tone dial is "Dual Tone MultiFrequency" (DTMF). This is because each button generates two tones, one from a "high" group of frequencies – 1209, 1136, 1477 and 1633 Hz – and one from a "low" group of frequencies – 697, 770, 852 and 841 Hz. The frequencies and the keyboard, or tone dial, layout have been internationally standardized, but the tolerances on individual frequencies vary between countries. This makes it more difficult to take a touchtone phone overseas than a rotary phone.

You can "dial" a number faster on a tone dial than on a rotary dial, but you make more mistakes on a tone dial and have to redial more often. Some people actually find rotary dials to be, on average, faster for them. The design of all tone dials is stupid. Deliberately so. They were deliberately designed to be the exact opposite (i.e. upside down) of the standard calculator pad, now incorporated into virtually all computer keyboards. The reason for the dumb phone design was to slow the user's dialing down to the speed Bell central offices of early touch tone vintage could take. Today, central offices can accept tone dialing at high speed. But sadly, no one in North America makes a phone with a sensible, calculator pad or computer keyboard dial. On some telephone/computer workstations you can dial using the calculator pad on the keyboard. This is a breakthrough. It is a lot faster to use this pad. The keys are larger, more sensibly laid out and can actually be touch-typed (like touch-typing on a keyboard.) Nobody, but nobody can "touch-type" a conventional telephone tone pad. A tone dial on a telephone can provide access to various special services and features – from ordering your groceries over the phone to inquiring into the prices of your (hopefully) rising stocks.

tone disabling A method of controlling the operation of communications equipment by transmitting a certain tone over the phone line.

tone diversity A method of Voice Frequency Telegraph (VFTG) Transmission wherein two channels of a 16-channel VFTG carry the same information. This is commonly achieved by twinning the channels of a 16-channel VFTG to provide eight channels with dual diversity.

tone generator A handheld device which puts a tone on a cable. The tone is picked up with an inductive amplifier at connection points or the other end of the cable. Slang for the tool is Toner. See Inductive Amplifier and Tone Probe.

tone probe A testing device used to detect signals from a tone generator to identify phone circuits, often the size of a fat pencil or skinny banana. Some models contain speakers; others must be used with a headset or a butt set. See also Tone Generator.

tone ringing Either a steady or oscillating electronic tone at the phone to tell you someone is calling.

tone sender 1. A printed circuit card in Rolm CBX which supplies the data bus with the digital representations of the following tones: dial, ring, busy, error, howler (off-hook timeout) and pulse (after flashing).

2. A printed circuit card which generates the following tones: dial, ring, busy, error, howler (off-hook timeout) and pulse (after flashing).

tone set A collection of tones which are customarily used as a set for the purposes of call setup and teardown (e.g., DTMF, R1 MF, R2 MF). In the case of DTMF, the tone set can also be used by the client application during the conversation portion of a call.

tone signaling The transmission of supervisory, address and alerting signals over a telephone circuit by means of tones. Typically inband. See also Signaling System 7.

tone to dial pulse conversion Converts DTMF (Dual Tone Multiple Frequency) signals to dial pulse signals when trunks going to carry outgoing calls are not equipped to receive tone signals. A lot of electronic phones with touchtone dials have a sliding switch that allows you to choose whether the phone will outpulse in rotary, or whether it will touchtone out. You choose whichever your trunk line will accept.

tone/pulse switchable Most phones in North America come with a pushbutton dial. Many of these phones have a switch that says "Tone/Pulse." By sliding the switch one way, the pushbutton pad will dial by sending out touchtones. By sliding the switch the other way, the pushbutton pad will dial by rotary pulses. See Rotary Dial.

toner Tone generator used for identifying cable pairs.

tones There are four basic tones which you will hear as you use the telephone. These tones are used to indicate what's going on. 1. Dial tone (also called dialing tone in Europe) is typically a continuous low frequency tone of around 33 Hz depending upon the telephone company. It indicates that the line is ready to receive dialing. 2. Busy Tone when the line or equipment is in use, engaged or occupied. This is typically 400 Hz 0.75 sec on and 0.75 sec off. 3. Ring Tone is typically 133 Hz make and break 0.4 sec On: 0.2 sec Off: Indicates called line is ringing out (17 Hz intermittent applied at called end to operate the telephone bell or buzzer). 4. Number Unobtainable continuous at 400 Hz indicates out of service or temporarily suspended. Tones vary considerably from country to country and between telephone companies.

tonnage The unit of measurement used in air conditioning systems to describe the heating or cooling capacity of a system. One ton of heat represents the amount of heat needed to melt one ton (2000 lbs.) of ice in one hour. 12,000 Btu/hr equals one ton of heat. My office is on a 5,000 square foot floor. We use a ten ton air conditioner. It works most days. I wouldn't put more in. It would be a waste.

tool In some computer languages, a small program executed as a shell command. In other computer languages, such as BASIC, it is called a "utility."

toolbar A series of shortcut buttons providing quick access to commands. Usually located directly below the menu bar. Not all windows have a toolbar.

toolkit A Dialogic word for an Applications Generator.

toolkit developer program A strategic alignment by Dialogic with suppliers of voice processing applications development software to provide high-level application development tools.

tone dialing Same as touchtone dialing. See Touchtone.

TOP 1. See Technical Office Protocol.

2. Task Oriented Practice/Processes. Step-by-step procedure for engineering, ordering, installing, provisioning, operating, maintaining, testing and repairing Materiel and Licensed Software. The flow-chart system organizes information to permit the completion of a specified task and leads the user through the task in a step-by-step fashion. A TOP leads the user from an initial stimulus to all operations required to correct the problem. This procedure allows subscribers at all levels of expertise to progress at their own pace in the performance of routine and acceptance type tasks. According to my friend, Larry Morey, "This is a telecom or software user guide for call center workers, operations centers tech/engineers and on site support personnel. It is similar to a Method of Procedure (MOP). We (Qwest) ask our suppliers to provide us this documentation so we can test it in the lab then pass it on to our technicians and OPS folks. The uses of TOP – Patch installations, card/circuit packet change-outs, how to provision a customer on a device, how to perform/install routing tables into a device, how to turn-up a device for the first time, testing/certifying a new software load/patch, testing and certifying a new device in the network."

top down This is a method of distributing incoming calls to a bunch of people. It always starts at the top of a list of agents and proceeds down the list looking for an available agent. See also Round Robin and Longest Available.

top level domain A certain segment of a network in the Transmission Control Protocol/Internet (TCP/IP) UNIX environment. A network is segmented into a hierarchy of domains or groupings. In the Internet in the United States, there are six top-level domains: com (commercial organizations), edu (education organizations), gov (government agencies), mil (Military milnet hosts), net (networking organizations), and org (nonprofit organizations). The next lower level relates to specific companies, and the level below to devices within a company.

topology Network Topology. The configuration of a communication network. The physical topology is the way the network looks. LAN physical topologies include bus, ring and star. WAN physical topology may be meshed, with each network node directly connected to every other network node, or partially meshed. The logical topology describes the way the network works. For example, a 10Base-T LAN looks like a star, but works like a bus.

topology aggregation The process of summarizing and compressing topology information at a hierarchical level to be advertised at the level above.

topology attribute A generic term that refers to either a link attribute or a nodal attribute.

topology constraint An ATM term. A topology constraint is a generic term that refers to either a link constraint or a nodal constraint.

topology database As an ATM term, it is the database that describes the topology of the entire PNNI routing domain as seen by a node.

topology metric A generic term that refers to either a link metric or a nodal metric.

topology state parameter A generic term that refers to either a link parameter or a nodal parameter.

topper Refers to a proposed Ku-band or Ka-band orbit for direct broadcast satellites (DBS), i.e., above existing DBS orbits.

TOPS 1. Traffic Operator Position System. A specialized console designed for telephone company operators to help them complete toll calls.

2. A computer operating system, which originally stood for the transcendental operating system.

3. The operating system used by Digital Equipment Corp.'s DECSYSTEM-10 and DECSYSTEM-20 computers. These computers have been discontinued, but many are still in use.

TOPS MPX Nortel Networks' Traffic Operator Position System designed on a token ring for interface between operator positions and the IBM Directory Assistance system database.

torn tape relay An antiquated tape relay system in which the perforated tape is manually transferred by an operator to the appropriate outgoing transmitter. In short, it's a torn tape relay is a store and forward message switching system which uses punched paper as the storage medium.

TOS Type of Service.

total available market See TAM.

Total Cost of Ownership. See TCO.

total harmonic distortion The ratio of the sum of the powers of all harmonic frequency signals (other than the fundamental) to the power of the fundamental frequency signal. This ratio is measured at the output of a device under specified conditions and is expressed in decibels.

total internal reflection The reflection that occurs when light strikes an interface at an angle of incidence (with respect to the normal) greater than the critical angle.

Total Network Data System. TNDS. A telephone company term. The Total Network Data System is the overall data system for all types of switching equipment.

Total Service Resale TSR. The complete resale, on a wholesale basis, of an ILEC's network and services. This allows a competitor to enter a market without deploying network infrastructure. Prices charged by the ILEC for TSR are based on avoided costs, the costs incumbents avoid in selling on a wholesale versus retail basis.

total transaction call processing A Rockwell term. Rockwell's philosophy. It guides their approach to call centers. It involves managing the success of a call center, not merely supplying the ACD (Automatic Call Distributor). It could include software development, CTI integration, network management, consulting services, IVR and voice processing systems. Rockwell says it will act as the prime contractor or as a single provider for a call center solution.

touchtone Touchtone is not a trademark of AT&T, despite what editions one through six of Newton's Telecom Dictionary said. It is a generic term for pushbutton telephones and pushbutton telecommunications services and the term "touchtone" may be used by anyone. At one stage it was a trademark of AT&T. At divestiture in 1984, AT&T gave it to the public. And that's who owns it now – you and me. For a full explanation of touchtone, see DTMF, which stands for Dual Tone Multi Frequency signaling, i.e. touchtone.

touchtone adaptor A device that can be connected to a rotary dial telephone to allow for DTMF signaling.

touchtone signal to dial pulse conversion Converts touchtone dial signals to dial pulse (rotary) signals when the serving central office of the distant end of outgoing trunks is not equipped to receive touchtone signaling. External conversion equipment is not needed for this feature. Obsolete term. All central offices will accept touchtones these days.

touchtone type ahead Also known as DTMF Cut-Through. Touchtone Typed Ahead is the ability of a voice response system to receive DTMF tones while the voice synthesizer is delivering information, i.e. during speech playback. This capability of DTMF cut-through saves the user waiting until the machine has played the whole message (which typically is a menu with options). The user can simply touchtone his response anytime during the message – when he first hears his selection number, when the message first starts, etc. When the voice processor hears the touchtoned selection (i.e. the DTMF cut-through), it stops speaking and jumps to the chosen selection. For example, the machine starts to say, "If you know the person you're calling, touchtone his extension in now." But before you hear the "If you know" you push button in 230, which you know is Joe's extension. Bingo, the message stops and Joe's extension starts ringing.

tourists People who take training classes just to get a vacation from their jobs. "We have about three serious students in the class. The rest are tourists."

tower 1. A name for a PC in a vertical or upright case. Tower PCs (if they're correctly designed) have a big benefit. Heat rises and escapes more easily than in traditional horizontal machines. Heat and power surges are the most damaging threats to PC.

2. A structure used for mounting antennas. Towers may be classified ways: by (1) mounting system (self supported or guyed); (2) shape (triangular, square, or pole); (3) erection method (crank-up, tilt-over, or erected sectionals); and (4) weight and height (heavy or light).

tower ads See Skyscraper Ads.

tower farm An area with a lot of tall antennas.

tower marking Tower lighting or painting with specific patterns so that the tower can be easily seen by aircraft pilots.

towerco Tower company. A company that builds and/or manages communications towers.

TP 1. Abbreviation for Transport Protocol or Twisted Pair.

2. Test Point.

3. Transition Point. A location in the horizontal cabling subsystem where flat undercarpet cabling connects to round cabling.

TP1, TP2, TP3, TP4, TP5 The various service levels of the ISO IS 8073 Transport Protocol. TP4 is the most popular service level for information system networks and is specified in the U.S. government GOSIP architecture. TP4 stands for OSI Transport Protocol Class 4 (Error Detection and Recovery Class). This is the most powerful OSI Transport Protocol, useful on top of any type of network. TP4 is the OSI equivalent to TCP.

TP-4 Transport Protocol 4. An OSI layer-4 protocol developed by the National Bureau of Standards. See TP1.

TP-4/IP A term given to the ISO protocol suite that closely resembles TCP/IP.

TP-MIC Twisted-Pair Media Interface Connector: This refers to the connector jack at the end user or network equipment that receives the twisted pair plug.

TP-PMD Twisted-Pair Physical Media Dependent, Technology under review by the ANSI X3T9.5 working group that allow 100 Mbps transmission over twisted-pair cable. Also referred to as CDDI or TPDDI.

TPAD Terminal Packet Assembler/Disassembler linked to a cluster controller or terminal device, taking native protocol input and converting it to X.25 for transmission over a packet network.

TPC 1. TOPS Position Controller.

2. Transmission Power Control. See 802.11a.

TPCA The Trusted Computer Platform alliance. The TPCA ho;es to hammer out a specification that would include secure PC operating systems using a hardware security chip it has already specified.

TPCC Third Party Call Control: As an ATM term, it is a connection setup and management function that is executed from a third party that is not involved in the data flow.

TPDDI Twisted Pair Distributed Data Interface. Also known as ANSI X3T9.5.-TPDDI. TPDDI is a new technology that allows users to run the FDDI standard 100 Mbps transmission speed over twisted-pair wiring. Unshielded twisted-pair has been tested for distances over 50 meters (164 ft.). TPDDI is designed to help users make an earlier transition to 100 Mbps at the workstation. Also known as CDDI, Copper Distributed Data Interface.

TPDU Abbreviation for Transport Protocol Data Unit.

TPE See TransPonder Equivalent.

TPF Twists Per Foot.

TPI Tracks Per Inch. A measurement of how much data can be stored on a disk.

TPM Terminating Point Masterfile. A LEC system that tracks RAOs, NPAs, NXXs, among other things.

TPON 1. Telephony over Passive Optical Network. A passive (i.e., with no active electronics) optical local loop which connects the subscriber premises to an all-fiber telecommunications network. See also APON and PON.

2. OSI Transport Protocol Class 0 (Simple Class). This is the simplest OSI Transport Protocol, useful only on top of an X.25 network (or other network that does not lose or damage data).

TPV 1. PVC Table.

2. Third Party Verification. This term relates to a new FCC regulation (to prevent slamming) that requires an LOA or third party verification for all PIC changes on ANIs for a company's commercial customers, and TPV for all residential customers. Compliance was required as of 4/29/99.

TQFP Thin Quad Flat Pack, a format used in the design of PCMCIA devices. Another format is called PQFP, which stands for Plastic Quad Flat Pack.

TQM Total Quality Management. Doing what management should have been doing all along.

TR 1. Trouble Report.

2. Technical Reference.

3. Telecommunications Room. See 606-A.

4. Technical Requirement. These publications are the standard form of Bellcore-created technical documents representing Bellcore's view of proposed generic requirements and standards for products, new technologies, services, or interfaces. What's the difference between TRs and GRs? I asked Irvin Bingham, IBingham@carrieraccess.com. He replied: In the good old days, AT&T and Bellcore issued Technical Requirements (TRs) to dictate the way things would work in "their" phone system. After the Telecommunications Act of 1996, TRs relating to competitive products and services became General Requirements (GRs). So in 1996, TR-303 became GR-303. Bellcore also set up GR "interest groups" to elicit industry participation in defining the specifications. Although Bellcore will listen to outside suggestions and opinions, I don't think Bellcore is obligated to act on any of them. Bellcore still has absolute control over their equipment interfaces to the outside world, but they are now required to publish those specifications and make them available to their competitors-for whatever price Bellcore wants to charge. An introduction to GR-303 is available online at http://www.bellcore.com/GR/gr303.html. Unfortunately, if you want a copy of any specification, you have to pay for it. After all, when did Bellcore ever give away anything to

its competitors? Bellcore even charges exorbitant membership fees (called industry funding) for the privilege of participating in a GR interest group. See GR-303, TR-303 and also ISDN.

TR-008 A Bellcore (now Telcordia) standard describing a digital interface between the SLC-96 digital loop carrier system and a local digital switch.

TR-303 A de facto standard published by Bellcore, now Telcordia. It amounts to an industry standard high level control interface to dumb switches. It also applies to Fiber In The Loop (FITL). See GR-303 for the full explanation.

TR-444 A de facto standard published by Bellcore, now Telcordia, which spells out how the Bell regional operating phone companies want long distance companies to connect to the Bell regionals' local networks. Several observers compare the TR-444 specs to simple direct dial long distance voice phone service.

TR-57 A Bellcore (now Telcordia) standard describing a customer interface on a Digital Loop Carrier (DLC) or channel bank system and its relationship to the local digital switch. Recently it has been primarily associated with GR-303 or TR-008, and allows a standard POTS line to ring without interfering with derived voice (voice over broadband) lines.

An example of implementation is standard POTS lines co-existing peacefully with derived voice lines on an ADSL modem. The 0-4 Khz frequency is used to carry standard POTS, while the remaining frequency band is used to support VoDSL as well as broadband data services.

TRA Telecommunications Resellers Association. See ASCENT.

TRAC Technical Recommendations Approval Committee.

trace agent This is a command used in the Infoswitch product line to report all the events and transactions an agent has been involved in over a defined period of time.

trace block See Trailer.

trace packet A special kind of packet in a packet-switching network which functions as a normal packet but causes a report of each stage of its progress to be sent to the network control center.

trace program A computer program that performs a check on another computer program by showing the sequence in which the instructions are executed and usually the results of executing the instructions.

trace route A software utility that traces a data packet from your computer to a distant Internet server. After you've sent the packet to the distant host, you get a report on your screen, which shows how many hops from router to router the packet requires to reach the host and how long each hop takes. If you're visiting a website and pages are appearing slowly, you can use trace route software to figure out where the longest delays are occurring, or worse, where the bottleneck is. The original trace route is a UNIX utility, but nearly all platforms have something similar. Windows includes a utility called "tracert." In Windows 95/98, you can run this utility by going Start>Run and then entering "tracert" followed by a space and then the domain name of the host. For example: tracert www.amazon.com. Trace route utilities work by sending packets with low time-to-live (TTL) fields. The TTL value specifies how many hops the packet is allowed before it is returned. When a packet can't reach its destination because the TTL value is too low, the last host returns the packet and identifies itself. By sending a series of packet and incrementing the TTL value with each successive packet (starting with one), a trace route finds out who all the intermediary hosts are.

tracer stripe When more than one color coding stripe is required, the first or widest stripe is the base stripe. The other, usually narrower stripes are the tracer stripes.

Traceroute Traceroute is software to help you figure out what's happening on your Internet connection. Traceroute is used to evaluate the hops taken from one end of a link to the other on a TCP/IP network, such as the Internet. Traceroute shows the full connection path between your site and another Internet address. It shows how many hops a packet requires to reach the host with the time required for the packet to get to each intermediate host or router. Traceroute is a more useful superset of PING.

Tracert Trace Route. A utility – Tracert.exe – used on Windows machines and used on TCP/IP networks to trace the route that information take between your computer and the computer it's trying to reach. Tracert.exe is basically an MS-DOS program. You type Tracert www.HarryNewton.com and it will send out a packet of information to www.HarryNewton.com. It will then draw a table showing how long it takes to traverse each hop and which company is in charge of each hop, by name.Tracert is a very useful tool in identifying network trouble spots. See Relay Checking and Trace Route.

track 1. A storage channel on a disk or tape which can be magnetically encoded.

2. On a data medium, a path associated with a single read/write head as data move past the head.

3. Every time you write to a CD, you will create at least one track, which is preceded by a pre-gap and followed by a post-gap. Any session may contain one or more tracks, and the tracks within a session may be of the same of or different types (for example, a mixed-mode disc contains data and audio tracks).

4. A process by which a receiver follows, or "locks onto" a received carrier frequency or clock rate.

track access time The time it takes to move the pickup head on a disk drive from one track to another.

track and trace infrastructure Networking technology, diagnostic tools, policy support, legal support, and international agreements for cooperation and collaboration that collectively enable the tracking and tracing of cyber-attacks to their source. Such an infrastructure is needed by organizations who use the Internet, since the Internet has no built-in security. As a result, cyber attackers (the bad guys) are easily able to conceal their identity and physical location when they perpetrate their attacks, thus making it difficult to find them. Basically, the Internet is a transport mechanism that simply transports what it's given. It doesn't check the validity of anything it sends. And that's why corporations need a track and trace infrastructure.

track density The number of tracks per unit length, measured in a direction perpendicular to the tracks.

track speed The maximum speed which a train can travel over a section of railway tracks.

trackball An upside-down MOUSE; a rotatable ball in a housing used to position the cursor and move images on a computer screen. A mouse needs desktop room to work, a trackball stays in one place, and can even be part of a keyboard or built into a laptop computer. It's hard to see why anyone uses a mouse instead of a trackball. This dictionary was typeset by a fine lady called Jennifer Cooper-Farrow, who used a trackball and a Macintosh computer.

tracking 1. Figuring where a satellite is and keeping track of it. This is not an easy job, given the vastness of space.

2. The effect created in compressed video when the speed of the transmission is not great enough to keep up with the speed of the action. Tracking creates a tearing effect on the video picture.

3. A call center term. A software feature that models actual events and activities in your call center to aid you in short-term planning and evaluation of employee and call center performance. The tracking functions include employee information scheduling assignment, daily activity, and intra-day performance.

Tracking and Data Relay System See TDRSS.

Tracon Terminal Radar Approach CONtrol. An installation in an airport or close to an airport from which approaching and departing aircraft are directed by people called controllers which sit in front of giant screens which show the movement of close aircraft. These controllers speak to the pilots in the planes instructing them where to move in order to avoid collisions and to land and takeoff safely.

tractor feeder A device which attaches to a computer printer and allows the printer to use continuous, sprocket-fed, paper. Such paper has a row of evenly spaced holes on both sides. Those holes coincide with the pins on the tractor feeder. In all tractor-fed printers, the tractor moves the paper, not the printer's platen.

trade secret A trade secret can be any information, knowledge, data, or the like which is useful in business and not commonly known. A trade secret is anything from a customer list to the formula for Coke syrup. Enforcing a trade secret, for example, in order to enjoin a former employee from working for a competitor normally requires proof that the secret allegedly taken was suitably identified as such, that the employee was subject to a written contract including an obligation of confidentiality, and that physical access to the secret was suitably restricted. See Intellectual Property.

trademark A trademark can be any word, symbol, slogan, design, musical jingle, or the like capable of differentiating one party's goods or services from another's. The question is whether a member of the relevant segment of the public would be misled as to the source of the goods. Thus, a descriptive mark ("Frigidaire") is less powerful than a coined mark ("Xerox"), and the same mark can be used by different parties, if on differing goods ("Cadillac" for dog food versus "Cadillac" for automobiles.) The r symbol indicates that a mark has been registered by the Federal government, while the tm or sm symbols merely indicate that the user does not intend to waive his rights in the mark. It is not legally necessary to use the statements commonly seen that certain trademarks are the property of their owners, or to use the r symbol in text, but it does prevent any accusation of misappropriation. See Intellectual Property.

trader turret A very large key telephone used by traders of commodities, securities, etc. Turrets typically have many line buttons. Each one corresponds to a trunk, an autodial or tie-line circuit to another trader or a financial institution. The objective of turrets is to allow the trader to be in instant communication with others who might want to buy or sell whatever he is trying to sell or buy at that moment. See also ring down circuit.

trading turret See trader turret.

traffic Bellcore's definition: A flow of attempts, calls, and messages. My definition: The amount of activity during a given period of time over a circuit, line or group of lines, or the number of messages handled by a communications switch. There are many measures of "traffic." Typically it's so many minutes of voice conversation, or so many bits of data conversation. Note that Bellcore includes attempts in its definition of traffic. I don't. The decision is yours. But you should be aware of what you include in your calculations. See also Traffic Engineering and Queuing Theory.

traffic analysis Inference of information from observable characteristics of data flow(s), even when the data is encrypted or otherwise not directly available. Such characteristics include the identities and locations of the source(s) and destination(s), and the presence, amount, frequency, and duration of occurrence.

traffic cap A limit on the amount of network usage that is covered by the subscriber's monthly recurring charge (MRC), beyond which usage-sensitive rates kick in. Depending on the type of network service, the traffic cap may be measured in units of time (e.g., minutes) or units of throughput (e.g., bytes).

traffic capacity The number of CCS (hundred call seconds) of conversation a switching system is designed to handle in one hour. This is the simple definition. See Traffic Engineering.

traffic carried See Traffic Offered and Carried.

traffic channel TCH. See TCH.

traffic characteristic A basic customer or network induced property of traffic that influences a load-service relationship. Peaked traffic, Poisson traffic, and smooth traffic are examples of traffic characteristics.

traffic concentration The average ratio of the traffic during the busy hour to the total traffic during the day.

Traffic Data Administration System TDAS. A telephone company term. The TDAS program merges the data from various data acquisition systems and performs the following functions: a. The establishment of schedules for data collection; b. Maintenance of assignments records for all data collection devices; c. The acceptance of measurement data for any time interval. d. The reporting of measurement data to downstream processes via a set of standard interfaces. e. Performance of quality control reports specifically designed to permit effective management of the data collection effort. f. Adjustment and validation of measurement data.

traffic data to customer The owner of a call accounting system can poll his PBXs daily or hourly and get traffic measurements, including peg counts, usage and overflow data. Summary reports, exception reports and complete traffic register outputs can be obtained.

traffic engineering The science of figuring how many trunks, how much switching equipment, how many phones, how much communications equipment you'll need to handle the telephone, voice, data, image and video traffic you're estimating. Traffic engineering suffers from several problems:

1. You are basing your future needs on past traffic.

2. Most traffic engineering is based on one or more mathematical formulas, all of which approach but never quite match the real world situation of an actual operating phone system. Computer simulation is the best method of predicting one's needs, but it's expensive in both computer and people time.

3. Many people in the telecommunications industry do not understand traffic engineering, have not worked with it sufficiently and make dumb and costly mistakes.

4. Since there are now several hundred long distance companies in the United States, and several thousand differently-priced ways of dialing between major cities, traffic engineering has become very complex.

After I wrote the above definition, Lee Goeller, disagreed with me and contributed this definition.

Traffic Engineering: The application of probability theory to estimating the number of servers required to meet the needs of an anticipated number of customers. In telephone work, the servers are often trunks, and the customers are telephone calls, assumed to arrive at random (see POISSON Process). Then arriving calls, upon finding all trunks busy, vanish, a "blocked call cleared" situation obtains (see ERLANG B). When a call stays in the system for a given length of time, whether it gets a trunk or not, "blocked calls held" applies (see Poisson Distribution). If a call simply waits around until a trunk becomes available and then uses the trunk for a full holding time, the correct term is "blocked calls delayed" (See ERLANG C and Queuing Theory). Like any form of predicting the future on the basis of past behavior, traffic engineering has its limitations; however, when used by those who have taken the trouble to learn how it works, its track record is surprisingly good, and vastly better than most forms of simulation (see Simulation).

traffic engineering tunnel A label-switched tunnel that is used for traffic engineering. Such a tunnel is set up through means other than normal Layer 3 routing; it is used to direct traffic over a path different from the one that Layer 3 routing could cause the tunnel to take. See also Traffic Engineering.

traffic intensity A measure of the average occupancy of a facility during a period of time, normally a busy hour, measured in traffic units (erlangs) and defined as the ratio of the time during which a facility is occupied continuously or cumulatively) to the time this facility is available. A traffic intensity of one traffic unit (one erlang) means continuous occupancy of a facility during the time period under consideration, regardless of whether or not information is transmitted. See also Traffic Engineering.

traffic load Total traffic carried by a trunk during a certain time interval.

traffic measurement Memory and other software in a telephone system which collect telephone traffic data such as number of attempted calls, number of completed calls and number of calls encountering a busy. The objective of traffic measurement is to enter the results into traffic engineering and so arrange one's incoming and outgoing trunks to get the best possible service. See Traffic Engineering.

Traffic Measurement and Recording Systems TMRS. A computer generated report showing usage information of telephone systems. Usually this includes trunk utilization, outages, queuing time, and the need for additional common equipment.

traffic monitor PBX feature that provides basic statistics on the amount of traffic handled by the system.

traffic offered and carried People pick up the phone and try to place their calls. This is "Traffic Offered" to the switch. The calls that get through the switch and onto lines is called "Traffic Carried." The difference between traffic offered and carried is the traffic that was lost or delayed because of congestion. There are two basic ways of measuring traffic – erlangs and CCS (or hundred call seconds).

Traffic Order TO. A telephone company term. These are requests originated by the Network Switching Engineering organization. The requests cover new systems or additions, removals and rearrangements to existing systems. The traffic order recommends types, quantities, and arrangements of local and toll equipment in accordance with the latest forecasts of trunks, network access lines and traffic studies.

traffic overflow Occurs when traffic flow exceeds the capacity of a particular trunk group and flows over to another trunk group.

traffic path A path over which individual communications pass in sequence.

traffic policing Process used to measure the actual traffic flow across a given connection and compare it to the total admissible traffic flow for that connection. Traffic outside of the agreed upon flow can be tagged (where the CLP bit is set to 1) and can be discarded en route if congestion develops. Traffic policing is used in ATM, Frame Relay, and other types of networks. Also known as admission control, permit processing, rate enforcement, and UPC. See also tagged traffic.

traffic prioritization Imagine your job is to run a University's data network. Everything is running smoothly. Your professors are checking research with their colleagues in other universities. Your students are submitting their papers and checking their email. Then suddenly your students discover Napster. And they start downloading zillions of bytes of music. The music traffic brings your network to its knees. Your solution? Traffic prioritization. Install some hardware and software which figures out which is the important traffic and let the important traffic through and hold back the unimportant traffic until the network is free. Like a very smart traffic cop. This equipment can get pretty complex. Here are some words from CheckPoint explaining their traffic prioritization product: "Rules are established for traffic control via a combination of traffic classifications and bandwidth control criteria. Network managers can classify traffic on the basis of Internet service or application (HTTP, FTP, Telnet, BackWeb), source, destination, group of users, groups of Internet services, Internet resource (ex. URL), and traffic direction – inbound or outbound...Control criteria categories include:

"Weights – Allocates bandwidth for users and Internet services based on designated merit or importance. The weight assigned to a particular class of traffic is proportionate to the weights of all other managed traffic."

"Guarantees – Provides guaranteed bandwidth for critical applications or designated users and groups."

"Limits – Sets bandwidth restrictions for discretionary network services or user applications which are not time sensitive."

traffic radar A RADAR (RAdio Detecting and Ranging) device bounces a radio signal off of a moving object, such as a car. The reflected signal is picked up by a receiver. Traffic radar receivers measure the frequency difference between the original and reflected signals. This frequency difference is converted into a speed, which appears on the receiver's display. Radar signals, like other types of radio signals, travel in straight lines until they hit an object that either absorbs, reflects, or refracts the signal. Radar receivers cannot see around curves or over hills, so a vehicle must be in the receiver's line of sight for traffic radar to get a speed measurement. There are different radar speed detection systems: 1. Continuous Wave (CW). This traffic radar system transmits constantly. The detector alerts you up to several miles from the radar source in optimum conditions. 2. Triggered CW. Stationary Mode (also known as Instant-On, Laser Pulse, or Hawk) This system transmits radar signals in bursts, and requires less than one second to determine speed. The detector senses the burst and sounds a special signal up to several miles from the source. However, since the radar gun only transmits signals when the operator triggers it, the alert range depends on how often the operator triggers the gun. 3. Triggered CW, Moving Mode, This system uses pulses to determine the police vehicle's speed. Then when the operator triggers the system, it transmits a signal burst to determine the speed of oncoming traffic. The detector senses both the police vehicle speed pulses and the triggered signal.

traffic recorder A device which measures traffic activity on a transmission channel. It's a recorder, not a processor. It's dumb.

traffic register A software area which records occurrences within a central office, such as peg count, overflow, all trunks busy, etc.. The types of occurrences measured vary widely according to the type of system.

traffic sensitive A telephone company term. Applies to equipment whose ability to provide a specific level of service varies as the calling load varies.

traffic separations Dave Holland send me this email: Dear Harry, I am a young central office technician for a rural independent Telco. A new assignment that was given to me is traffic separations. Until we got into the process quite a way I did not know what it was. I now understand it as the mapping of all the traffic through a switching device with incoming traffic being mapped to outgoing traffic and the place that they meet on the matrix is assigned a register. The register allows you to gather the information about all calls made between the two points. I wanted to look up the "official definition for Traffic Separations (TSEP) and looked to your book. To my utmost surprise I did not find it. I think it would be a good addition to your book because it is something that all telephone companies do or should be doing. It is not the same as just a peg count it is far more detailed and requires allot of work to get set up and then administer later on.

So I asked Dave to send me a definition. He send me one from a company called Network Services Group, www.networkservicesgrp.com. It reads:

The Traffic Separations measurements are used to identify the proportion of jurisdictional usage for support of division of revenue studies. There are also secondary objectives of fulfilling federal and state regulatory requirements, as well as ownership and inter-/intra-company settlements. Usage is apportioned on the basis of relative minutes of use. These studies are normally on a monthly basis and seven consecutive days in duration. Data is collected and reported on an hourly and daily basis. In a Traffic Separations study the measurement data is collected on all switched traffic utilizing a separations matrix which correlates the "calling" and "called" party of each call. The mapping of a call on to the matrix is done with the INSEP value assigned to the "calling" party and the DESEP value assigned to the "called" party. The intersection of each INSEP/DESEP pair on the matrix is known as a cell. Each cell within the matrix is uniquely assigned to a register Each register consists of a peg counter and usage counter. Within the limits of allowed quantities of DESEPs, and INSEPs, line "class of service", incoming trunk group, outgoing route appearance, and terminating treatment call type can be uniquely assigned an INSEP or DESEP as appropriate. With the proper planning of INSEP/DESEP assignments and grouping of cell(s) to registers this can result in the data being collected on a "call type" basis. The value ranges for separations parameters are The type(s) of data to be obtained from a separation study is determined by a separation study administrator (a person or group of people in the operating company) who plan, setup and verify each study. In order to administer these "flexible" assignments all separation study parameters are administered using recent change commands. The capability to make and change assignments as well as display and validate assignments are part of the traffic separation administration process. The end product of the traffic separations feature is a set of reports for telcos which are used for statutory division of revenue purposes and for spot traffic studies.

Traffic Service Position System TSPS. A toll switchboard position configured as a push button console.

traffic shaping Traffic shaping is a phrase that describes a technique to control the rate of specific traffic types that will be allowed onto the network. Traffic shaping is a generalized term for a congestion control management procedure in which data traffic is regulated in order that it conform to a specified, desirable behavior pattern. This becomes important at locations in the network that present bottlenecks. Take the analogy of Atlanta airport. If other airports allowed planes to come to Atlanta faster than they could land, they would have to be put in holding "Queues" awaiting their turn. Some may get low on fuel and have to return to another airport, only to return to Atlanta again later adding to the congestion. If the air traffic that was permitted to arrive in the Atlanta airspace was slowed or delayed such that it did not exceed the landing capacity, then the air traffic would have been "shaped" to fit the landing capacity. This is essentially the technique of traffic shaping. But then there are more important and less important planes arriving. Using traffic shaping rules, network managers can allocate bandwidth to mission-critical user applications to ensure those applications receive the bandwidth required for efficient operation. VoIP is one application where traffic shaping is particularly useful. Network managers can identify, prioritize, and control traffic on their Frame Relay networks on a per application and per PVC basis. Shaping is accomplished by controlling the source of the traffic, not by just using queues. Traffic shaping may include reduction or elimination of excessive traffic bursts from a LAN as it is presented to a Frame Relay WAN through a router. Such bursts may exceed the CIR (Committed Information Rate) and, therefore, be marked DE (Discard Eligible). During periods of Frame Relay WAN congestion, such bursts may result in discarded frames, which require retransmission. Should the excessively long bursts be transmitted successfully across the Frame Relay WAN, surcharges may apply (such surcharges are unusual for U.S. carriers). All things considered, traffic shaping may be the best approach in such a scenario. In an ATM LAN environment, traffic shaping responsibility can be accomplished by the ATM switch, which actively would alter the traffic characteristics of a cell stream on a VCC (Virtual Channel Connection) or VPC (Virtual Path Connection). This procedure may serve to reduce the peak cell rate, limit the burst length, or minimize the cell delay variation by re-spacing the cells in time in order that traffic flow not congest the switch. This can be particularly important when dealing with long bursts of high priority traffic, as such traffic literally can bring the rest of the user traffic flow to its knees. See also ATM, Committed Information Rate, Discard Eligible, Frame Relay, VCC and CPC.

traffic table A computer database into which a PBX enters a count of feature activity. Certain detected operating errors are also entered in the traffic table.

traffic theory The branch of probability theory used to predict how many telephone lines you need for how much traffic you are likely to put on the lines.

traffic usage Total occupancy of a network. This is calculated a the product of holding time and calling rate and can be expressed as call-hours. Traffic usage may be made up of many short calls or few long calls – it doesn't matter.

traffic usage recorder A device for measuring and recording the amount of telephone traffic carried by a group, or several groups, of switches or trunks.

traffic use code A telephone company definition. A system standard two character alpha code designating the type of traffic offered to a trunk group. Traffic Use Codes are listed and defined in Section 795-400-100 (Common Language Circuit Identification – Message Trunks).

trail As an ATM term, it is an entity that transfers information provided by a client layer network between access points in a server layer network. The transported information is monitored at the termination points.

trailer 1. A nonstandard way of standard way of sending data. Trailers are used on some networks by 4BSD UNIX and some of its derivatives.

2. A block of controlling information transmitted at the end of a message to trace error impacts and missing blocks. Also referred to as a trace block.

train 1. When a modem "trains," it's establishing connectivity with the inbound RTS/CTS transaction. See also RNA.

2. The creation of word reference data by presenting words to a recognizer. A voice recognition term.

training A feature of some modems which adjust to the conditions including amplitude response, delay distortions, timing recovery, and echo characteristic, of a particular telecommunications connection by a receiving modem. See Training Up.

training up A technique that adjusts modems to current telephone line conditions. The transmitting modem sends a special training sequence to the receiving modem, which makes necessary adjustments for line conditions.

transaction 1. It is a completed event that can be assembled in chronological sequence for an audit trail.

2. An entry or an update in a database.

transaction capabilities Function that controls non-circuit-related information transfer between two or more nodes via a SS7 signaling network.

Transaction Capabilities Application Part TCAP. The application layer protocol of SS7. Transaction capabilities in the SS7 protocol are functions that control non-circuit related information transferred between two or more signaling nodes. Definition from Bellcore in reference to its concept of the Advanced Intelligent Network.

transaction detail The detail of a transaction record.

transaction engines If you sell on your site, you need an application that allows the customer to configure an order and pay by credit card or other means. These systems let you manage product and buyer information, and usually link to third parties that process the credit-card transactions. These are called transaction engines.

Transaction Internet Protocol TIP. The Transaction Internet Protocol protocol ensures that multivendor transaction monitors will work with one another to complete transactions over the Internet (RFC 2371). TIP came from a joint Microsoft/Tandem effort. I excerpted the following from a Microsoft Market Bulletin.

Two companies (Microsoft and Tandem) team have combined to publish a specification for a two-phase commit protocol to make it easier for businesses to do transaction processing across the Internet. Two-phase commit is the commonly-used application protocol used by high-end system software – including Transaction Processing (TP) Monitors and databases – to coordinate the work of multiple applications on different computers as a single unit, or transaction. Businesses want to link existing transaction processing systems together across the Internet using two-phase commit protocols, but existing implementations of two-phase commit are too complex for use on the Internet. TIP is designed to solve this problem, defining a simple protocol that existing vendors of TP Monitors and databases can easily implement into their products, solving the problem of transaction coordination across the Internet. Microsoft will implement TIP in the Distributed Transaction Coordinator (DTC), Microsoft's transaction manager that first shipped with SQL Server 6.5. DTC currently supports other open two-phase commit protocols, including OLE Transactions, the X/Open's XA protocol, and has future plans to support SNA LU 6.2 Sync Level 2. Windows NT Server 5.0 will provide native support for TIP. Tandem will support TIP in its NonStop systems. Both the reference implementation and the TIP specification can be downloaded directly from www.microsoft.com/pdc or www.tandem.com/menu_pgs/svwr_pgs/svwrnews.htm. Microsoft and Tandem have submitted the TIP specification to the Internet Engineering Task Force, who have published it at http://ds.internic.net/internet-drafts/draft-lyon-itp-nodes-00.txt.

transaction file A collection of transaction records. A transaction data entry program allows for the creation of new transaction files used to update the data base.

transaction link Rockwell's link from its Galaxy ACD to an external computer. See Open Application Interface.

transaction tracking Your software keeps track of each transaction as it happens. And if a component of your network fails, your transaction tracking software backs out of the incomplete transaction. This allows you to maintain your database's integrity. You may, however, lose the single transaction you were working on when your network got sick.

transaction processing A processing method in which transactions are executed immediately when they are received by the system, rather than at some later time as in batch-processing systems. Airline reservation databases and automatic teller machines are examples of transaction-processing systems.

transactional integrity A term that describes how your computing/telecom system handles making sure that the transaction you just made is solid and clean and that the next time you want to get to the results of the transaction you can. "Transactional integrity" becomes critical when you're storing bits and pieces of your transactions on different media, in different places. For example, you might want to store your data on a magnetic hard drive and your associated images on a separate optical drive.

Transborder Data Flow TDF. Transborder data flows are movements of machine-readable data across international boundaries. TDF legislation began in the 1970s and has been put into effect by many countries in an attempt to protect personal privacy of citizens. This term has particular meaning as it relates to electronic commerce or EDI and is becoming more and more relevant with the use of the Internet as a means to conduct global business.

transceiver 1. Any device that transmits and receives. In sending and receiving information, it often provides data packet collision detection as well.

2. In IEEE 802.3 networks, the attachment hardware connecting the controller interface to the transmission cable. The transceiver contains the carrier-sense logic, the transmit/receive logic, and the collision-detect logic.

3. A device to connect workstations to standard thick Ethernet-style (IEEE 802.3).

transceiver cable In local area networks, a cable that connects a network device

such as a computer to a physical medium such as an Ethernet network. A transceiver cable is also called drop cable because it runs from a network node to a transceiver (a transmit / receiver) attached to the trunk cable. See Transceiver.

transcoder A device that combines two 1.544 megabit per second bit streams into a single 1.544 megabit per second bit stream to enable transmission of 44 or 48 voice conversations over a DS-1 medium.

transcoding A procedure for modifying a stream of data carried so that it may be carried via a different type of network. For example, transcoding allows H.320 video encoding, carried via circuit switched TDM systems to be converted to H.323 so that it can connect with and be transmitted across packet switched ethernet LAN.

transcriptionist A person who listens to a tape recording and types the words he hears. The word, transcriptionist, derives from the verb to transcribe. The most common employment of transcribers is in the medical industry, where busy doctors talk into tape recorders telling good and bad news of their patients. And even busier transcriptionists type those words into the patient's medical records, or whatever.

transducer A device which converts one form of energy into another. The diaphragm in the telephone receiver and the carbon microphone in the transmitter are transducers. They change variations in sound pressure (your voice) to variations in electricity, and vice versa. Another transducer is the interface between a computer, which produces electron-based signals, and a fiber-optic transmission medium, which handles photon-based signals.

transfer A telephone system feature which provides the ability to move a call from one extension to another. It is probably the most commonly used and misused feature on a PBX. Before you buy a PBX, check out how easy it is to transfer a call. If you have a single line phone, you should simply hit the touch hook, hear a dial tone and then dial the chosen extension number and hang up. This sounds easy in principle, but many people find it difficult since they associate the touch hook with hanging up the phone. Some companies have gotten around this by putting a "hook flash" button on the phone itself. Such a button is like having an autodial button which just makes the exact short tone you make when you quickly hit the hook flash button. An even better solution is an electronic phone with a button specially marked "transfer," or a button next to a screen which lights up "transfer." Failing to efficiently transfer a call is the easiest way to give your customers the wrong impression of your firm. Think of how many times have you called a company only to be told it wasn't the fellow's job and he will transfer the call, but "If we get cut off, please call Joe back on extension 2358." There are typically four types of Transfer: Transfer using Hold, Transfer using Conference, and Transfer with and without Announcement.

transfer callback A phone system feature. After a specified number of rings, an unanswered transferred call will return to the telephone which originally made the transfer.

transfer delay A characteristic of system performance that expresses the time delay in processing information through a data transmission system.

transfer impedance A measure of shield effectiveness.

transfer mode A fundamental element of a communications protocol, transfer mode refers to the functioning arrangement between transmitting and receiving devices across a network. There are two basic transfer modes: connection-oriented and connectionless. Connection-oriented network protocols require that a call be set up before the data transmission begins, and that the call subsequently be torn down. Further, all data are considered to be part of a data stream. Examples of connection-oriented protocols include analog circuit-switched voice and data, ISDN, X.25 and ATM.

Connectionless protocols, on the other hand, do not depend on such a process. Rather, the transmitting device gains access to the transmission medium and begins to transmit data address to the receiver, without setting up a logical connection across the physical network. LANs (e.g. Ethernet and Token Ring) make use of connectionless protocols, as does SMDS, which actually is an extension of the LAN concept across a MAN (Metropolitan Area Network). For more detail, see Connection Oriented and Connectionless Mode Transmission.

transfer protocols Protocols are all of the packaging" that surround actual user data to tell the network devices where to send the data, who it comes from, and how to tell if it arrived. Transfer protocols are designed for the efficient moving of larger chunks of user data.

transfer rate The speed of data transfer – in bits, bytes or characters per second – between devices.

transfer switch Usually a switch which reverses two input-output combinations.

transfer time A power backup term. Transfer time can refer to either the speed to which an off-line UPS transfers from utility power to battery power, or to the speed with which an on-line UPS switches from the inverter to utility power in the event of an inverter failure. In either case the time involved must be shorter than the length of time that the computer's switching power supply has enough energy to maintain adequate output voltage. this hold-up time may range from eight to 16 milliseconds, depending on the point in the power supply's recharging cycle that the power outage occurs, and the amount of energy storage capacitance within the power supply. A transfer time of 4ms is most desirable , however, it should be noted that an oversensitive unit may make unnecessary power transfers.

transformer Transformers are devices that change electrical current from one voltage to another. A step-up transformer increases the voltage and a step-down transformer decreases voltage. The power of an electric current must be conserved so just as voltage is increased, current is decreased. Transformers work by feeding an alternating current into a primary coil. The primary coil induces a magnetic field in a secondary coil which is connected to an energy using load. The difference between the number of coils in the primary coil versus the secondary coil determines whether the voltage will be stepped up or down. One reason for using a transformer is that commercial power is typically 120 or 240 volts while many phone systems (and other computer-type "things") work best on 48, 24 or lower voltage.

Transformer Exciting Network See TEN.

transhybrid loss The transmission loss between opposite ports of a hybrid network, that is between the two ports of the four-wire connection.

transient Any high-speed, short duration increase or decrease impairment that is superimposed on a circuit. Transients can interrupt or halt data exchange on a network. See HIT.

transient mobile unit A mobile unit communicating through a foreign base station.

transistor The transistor was invented in 1947 by John Bardeen, Walter H. Brattain and William Shockley of Bell Laboratories. The first transistor comprised a paper clip, two slivers of gold, and a piece of germanium on a crystal plate. Here is an explanation of how a transistor works, taken from "Signals, The Science of Telecommunications" by John Pierce and Michael Noll:

"To understand how a transistor works, we must look at the laws of quantum mechanics. We commonly picture an atom as a positive nucleus surrounded by orbiting electrons ... Vacuum tubes rely on the ability of electrons to travel freely with any energy through a vacuum. Transistors rely on the free travel of electrons through crystalline solids called semiconductors ... Semiconductors (such as silicon or gallium arsenide) differ from pure conductors, such as metals, in how full of electrons are the energy bands that allow free travel." Depending on their design, transistors can act as amplifiers or switches. See also 1947, Transistor Milestones and transistor radio.

Transistor Milestones Point-contact transistor 1948

Single-crystal Germanium	1950
Grown junction transistor	1951
Alloy junction transistor	1952
Zone melting and refining	1952
Single-crystal Silicon	1952
Diffused-base transistor	1955
Oxide masking	1957
Planar transistor	1960
MOS transistor	1960
Epitaxial transistor	1960
Integrated circuits	1961

transistor radio Sony unveiled the first transistor radio in 1955. See Sony.

transit delay 1. In ISDN, the elapsed time between the moment that the first bit of a unit of data (such as a frame) passes a given point and the moment that bit passes another given point plus the transmission time of that data unit.

2. As an ATM term, it is the time difference between the instant at which the first bit of a PDU crosses one designated boundary and the instant at which the last bit of the same PDU crosses a second designated boundary.

transit exchange The European equivalent of a tandem exchange.

transit point A place through which phone calls pass on the way to someplace else. The last time Steve Schoen checked, he noticed that some of the calls to and from bases in Antarctica went through New Zealand. Australia was the only other transit point at the time.

transit timing A method of eliminating looping between nodes used in the network layer of some packet-switched systems. This method is used in the Internet Protocol (IP) portion of Transmission Control Protocol/Internet Protocol (TCP/IP).

transit traffic Every day telecommunications traffic moves from one carrier to another. That traffic often moves via a third carrier. When it does, it's called transit traffic. (Think transit air travel.) The formal definition is: The traffic that originates with a carrier (ILEC, CLEC or IXC) or end user and passes through a tandem switching office to be terminated to another carrier (ILEC, CLEC or IXC) or end user. Often large carriers have peering agreements with each other which often involve them agreeing to accept and carry each other's traffic for free. That has changed now and now carriers typically charge for transit traffic, even if they're only carrying the transit traffic for 100 feet. An ILEC is an incumbent local exchange carrier (i.e. a big one, like one of the old Bells.) A CLEC is a competitive local exchange carrier – a small phone company attempting to compete (usually unsuccessfully) with the ILEC. An IXC is an interexchange carrier, i.e. one like the old MCI or Global Crossing. These carriers carried telecom traffic between local phone companies, also called ILEC and CLEC. See peering.

transition point TP. A location in the horizontal cabling subsystem where flat undercarpet cabling connects to round cabling.

transition probabilities Probabilities of moving from one state to another.

transition zone The zone between the far end of the near-field region and the near end of the far-field region. The transition is gradual.

translate To change the digits dialed on your phone into digits necessary for routing the call across the country. See Translations.

translating bridge A special bridge that interconnects different LAN types using different protocols at the physical and data link layers, such as Ethernet and Token Ring. A translating bridge supports the physical and data link protocols of both LAN types. When they forward packets from one LAN to another, they manipulate the packet envelope to conform to the physical and data link protocols of the destination LAN. For a longer explanation, see Bridge.

translation The interpretation by a switching system of all or part of a destination code to determine the routing of a call. See Translations.

translations Here is a definition from Bellcore, who works with the telephone industry: Translations is the changing of information from one form to another. Example: In common control switching systems employing digit storage devices and decoding devices, the dialed digits are stored in a receiver or a tone decoder. The receiver/decoder translates the dialed digits data appropriate for the completion of the call and passes to a processor. With the advent of stored program control, as exemplified in a IA ESS, 5ESS-2000, DMS-100 systems, the translation function has been greatly expanded. When a customer originates a call, for example, the system needs to know if the line is denied outgoing service, if the line is being observed, what the line class is, what special equipment features it has, etc. The line equipment number is given to the translation program as an input. The translation program performs a translation and returns the answers to these questions in a coded form suitable for use by the central processor. The important thing to remember in considering the translation function in the stored program switches is the translation function is employed many times throughout the process of a call and the interplay between the translation programs and other programs is frequent.

Here's my definition: Translations are changes made by the network to dialed telephone numbers to allow the call to progress through the network. Sometimes the translations are made automatically. Take one series of dialed numbers; convert them to another. Sometimes, translations are done with the help of "look up" tables, also called databases. Here's an example of translations done with the help of a database. TELECONNECT Magazine has a WATS line, 1-800-LIBRARY. If you dial it on the phone, you'll see it is really 1-800-542-7279. But this is not its real number. When someone in California dials 1-800-LIBRARY, MCI's long distance network recognizes the "1-800" portion of the call and sends it to a special central office somewhere out west. When the call arrives, a computer looks up the number 800-542-7279 in its database and translates that to 1-212-691-8215 and puts the call back into the network. Within seconds, that number in New York, 212-691-8215 rings.

translator 1. A communications device that receives signals in one form, normally in analog form at a specific frequency, and retransmits them in a different form.

2. A device that converts information from one system into equivalent information in another system.

3. In telephone equipment, it is the device that converts dialed digits into call-routing information.

4. In computers, it is a program that translates from one language into another language and in particular from one programming language into another programming language.

5. In FM and TV broadcasting, it's a repeater station that receives a primary station's signal, amplifies it, shifts it in frequency, and rebroadcasts it.

transliterate To convert the characters of one alphabet to the corresponding characters of another alphabet.

transmission Sending electrical signals carrying information over a line to a destination. Bellcore says that transmission has the following definitions: (a) Designates a field work, such as equipment development, system design, planning, or engineering, in which electrical communication technology is used to create systems to carry information over a distance. (b) Refers to the process of sending information from one point to another. (c) Used with a modifier to describe the quality of a telephone connection: good, fair, or poor transmission. (d) refers to the transfer characteristic of a channel or network in general or, more specifically, to the amplitude transfer characteristic. You may sometimes hear the phrase, "transmission as a function of frequency."

transmission block A group of bits or characters transmitted as a unit, with an encoding procedure for error control purposes.

transmission channel All of the transmission facilities between the input (to the channel) from an initiating node and the output (from the channel) to a terminating node. In telephony, transmission channels may be of various bandwidths: e.g. nominal 3-kHz, nominal 4-kHz, or nominal 48-kHz (group). "Transmission channel" should not be confused with the more general term "channel."

transmission code A code by which information is sent and received on a transmission system.

transmission coefficient The ratio of the transmitted field strength to the incident field strength when an electromagnetic wave is incident upon an interface surface between media with two different refractive indices. In a transmission line, the ratio of the complex amplitude of the transmitted wave to that of the incident wave at a discontinuity in the line. A number indicating the probable performance of a portion of a transmission circuit. The value of a transmission coefficient is inversely related to the quality of the link or circuit.

transmission control Category of control characters intended to control or help transmission of information over telecommunication networks. See TCP.

transmission control characters A group of characters used to facilitate or control data transmission. Examples are NAK (Not acknowledge) and EOT (end of transmission).

Transmission Control Protocol TCP. A specification for software that bundles outgoing data into packets (and bundles incoming data), manages the transmission of packets on a network, and checks for errors. TCP is the portion of the TCP/IP protocol suite that governs the exchange of sequential data. In more technical terms, Transmission Control Protocol is ARPAnet-developed transport layer protocol. Corresponds to OSI layer 4, the transport layer. TCP is a connection-oriented, end-to-end protocol. It provides reliable, sequenced, and unduplicated delivery of bytes to a remote or local user. TCP provides reliable byte stream communication between pairs of processes in hosts attached to interconnected networks. It is the portion of the TCP/IP protocol suite that governs the exchange of sequential data. See TCP/IP for a much longer explanation.

Transmission Convergence TC. Transmission Convergence Sublayer, a dimension of the ATM Physical Layer (PHY). See TC.

transmission electronics Any of the various devices used in conjunction with different transmission media to convert from one transmission method to another. Transmission electronics devices typically include multiplexing equipment and Asynchronous Data Units.

transmission facility A piece of a telecommunications system through which information is transmitted, for example, a multi pair cable, a fiber optic cable, a coaxial cable, or a microwave radio.

transmission frame A data structure, beginning and ending with delimiters, that consists of fields predetermined by a protocol for the transmission of user and control data.

transmission level The power of a transmission signal at a specific point on a transmission facility. See Decibel.

Transmission Level Point TLP. A designated physical point on a circuit where the transmission level, or amplitude, is measured. Referencing this point in relation to others in the network can determine the performance of the network. See also Level, Loss, and Pad.

transmission limit The wavelengths above and below which the fiber ceases to be transparent and therefore, can no longer transmit information.

transmission line A signal-carrying circuit with controlled electrical characteristics used to transmit high frequency or narrow pulse signals.

transmission loss The decrease or loss in power during transmission of energy from one point to another, usually expressed in decibels.

transmission media Anything, such as wire, coaxial cable, fiber optics, air or vacuum, that is used to carry an electrical signal which has information. Transmission media usually refers to the various types of wire and optical fiber cable used for transmitting voice or data signals. Typically, wire cable includes twisted pair, coaxial, and twinaxial. Optical fiber cable includes single, dual, quad, stranded, and ribbon.

transmission objectives A stated set of desired performance characteristics for a transmission system. Characteristics for which objectives are stated include loss, noise, echo, crosstalk, frequency shift, attenuation distortion, envelope delay distortion, etc.

transmission pattern See Radiation Pattern.

transmission payload The interface bit rate minus the overhead bits.

transmission protocol 0 TP0. OSI (Open Systems Interconnection) Transmission Protocol Class 0 (Simple Class). This is the simplest OSI Transmission Protocol, useful only on top of an X.25 network (or other network that does not lose or damage data).

transmission protocol 4 TP4. OSI (Open Systems Interconnections) Transmission Protocol Class 4 (error detection and recover class). This is the most powerful OSI Transmission Protocol, useful on top any type of network. TP4 is the OSI equivalent to TCP (Transmission Control Protocol).

Transmission Security Key TSK. A key that is used in the control of transmission security processes such as frequency hopping and spread spectrum.

transmission speed Number of pulses or bits transmitted in a given period of time, expressed variably in Bits Per Second (BPS), Words Per Minute (WPM), Characters Per Second (CPS), an occasionally as Lines Per Minute (LPM) in printer transmission. Skilled technologists can translate one to the other.

transmissive The way many LCD (liquid crystal display) screens on laptops reflect light.

transmit bus In AT&T's Information Systems Network (ISN), the circuit on the backplane of the packet controller that transports message packets from sending device interface modules to the switch module.

Transmit Digital Intertie TDI. A 16-channel serial converter which converts the TDM Data Bus from parallel format to serial format for transmission between nodes.

transmittance The ratio of transmitted power to incident power. In optics, frequently expressed as optical density or percent; in communications applications, generally expressed in decibels.

transmitter The device in the telephone handset which converts speech into electrical impulses for transmission.

transmitter distributor A device in a teletypewriter system which converts the information from the parallel form in which it is used in the keyboard-printer to and from the serial form which it is transmitted on the transmission line.

transmitter start code A coded control character or code sequence transmitted to a remote terminal instructing that terminal to begin sending information.

transmobile The transmobile (not to be confused with a TRANSPORTABLE) is another type of cellular phone. It is essentially a standard 3-watt mobile unit – without an external battery pack – that can be quickly and easily moved from one vehicle to another. It draws its power from the vehicle's battery via a cigarette lighter plug. See Bag Phone.

transmultiplexer A device that takes a bunch of voice analog phone conversations and converts them directly into a T-1 1.544 megabit per second bit stream – without the need for de-multiplexing the bunches down to individual conversations, then digitizing them, then bundling them up into a T-1 digital bit stream. A transmultiplexer does it all in one go.

transparency 1. A data communications mode that allows equipment to send and receive bit patterns of virtually any form. The user is unaware that he is transmitting to a machine that receives faster or slower, or transmits to him faster or slower, or in a different bit pattern. All the translations are done somewhere in the network. He is unaware of the changes occurring – they are transparent. ISDN is planned to be transparent.

transparency "Transparent Communications" 1. A basic objective of telecommunications systems, to make the transportation of information invisible to the user.

2. In data communications, a suspension of control character recognition in certain systems while information transfer is in progress.

transparency/opacity An imaging term. A setting available in many image-processing functions that allows part of the underlying image to show through. 80 percent opacity is equivalent to 20 percent transparency.

transparent Fine or sheer enough to be seen through. Something that is transparent exists for some reason, but is invisible, or nearly so. In other words, it does not impair or affect the users' operation of the system or feature. In fact, the user need not interact with the transparent feature, and generally is totally unaware that it exists. Think of a pane of glass that serves to protect the interior of a building and its occupants from the elements, but does not affect the users' ability to see through it.

When applied to telephone communications, the term is used to characterize the provision of a feature or service such as Automatic Route Selection in a such a way that the user is unaware of it and it has no affect on the way he uses the telephone. It's "transparent" to him. Translations, for example, are transparent to the telephone user. Similarly, protocol conversions are transparent. See also Translations, Transparency, and Virtual.

transparent bridging Transparent bridging is so called because the intelligence necessary to make relaying decisions exists in the bridge itself and is thus "transparent" to the communicating workstations. It involves frame forwarding, learning workstation addresses and ensuring no topology loops exist.

transparent GIF Transparent GIFs are useful because they appear to blend in smoothly with the user's display, even if the user has set a background color that differs from that the developer expected. They do this by assigning one color to be transparent – if the Web browser supports transparency, that color will be replaced by the browser's background color, whatever it may be.

transparent image An image that has had one color, usually the background, designated as 'transparent,' so that when the image is displayed in a browser, the image's background is colored with the browser's background color. The effect is an image that does not have a visible rectangular background.

transparent LAN service See TLS.

Transport Layer Security See TLS.

transparent mode 1. The operation of a digital transmission facility during which the user has complete and free use of the available bandwidth and is unaware of any intermediate processing. Generally implies out-of-brand signaling (also called Clear Channel).

2. In BSC data transmission, the suppression of recognition of control characters, to allow transmission of raw binary data without fear of misinterpretation.

3. An operational mode supported by the T3POS PAD which enables the use of existing credit authorization and data capture link level protocols. This mode requires minimal modifications to the POS (Point Of Sale) terminal, and no modification to the ISP/Credit Card Association (CCA) host system software.

Transparent Networking Transport TNT. A service for transporting of LAN data across WANs in which all responsibility for the WAN transport is assumed by the WAN and is therefore invisible to the LAN.

transparent routing A method used by a bridge for moving data between two networks. With this type of routing, the bridge learns which computers are operating on which network. It then uses this information to route packets between networks. It does not rely on the sending computers for its decision-making routine. A special kind of bridge combines the practice of transparent routing with source routing. It is called a source routing transparent (SRT) bridge. It examines each packet that comes by to see if it is using IBM's special source routing protocol. If so, this protocol is used to forward the packet. If not, the transparent method is used. Thus, the SRT bridge will support both IBM and non-IBM network protocols. See also Bridge and SRT. Compare with Source Routing.

transponder 1. A transponder is a fancy name for radio relay equipment on board a communications satellite. Just like its domestic microwave counterpart (which you see along highways), a transponder will receive a signal, amplify it, change its frequency and then send it back to earth. On a satellite transponder that uses frequency modulation, the bandwdith required for an analog tv signal is 27Mhz. Since satellites are power limited, FM is the analog modulation of choice. In exchange for wide bandwidth and poor spectral efficiency, FM offers improved signal-to-noise ratio. On a terrestrial TV station or cable TV network where power is not an issue, amplitude modulation is used which offers better spectral efficency so the bandwidth neeeded for an analog TV signal is only 6Mhz.

2. A transponder on an airline is a slightly different kettle of fish. When a radar signal strikes a airline, it activates an electronic transmitter called a transponder. The transponder sends out a coded signal to the ground radar. The code appears next to the radar image of the plane, allowing the controller to identify each plane under his control. Newer aircraft have automatic collision avoidance systems that will change the flight path of two or more planes if they appear to the systems as though they're going to crash.

transponder equivalent TPE. A measurement of a communication satellite's total transmission capacity. One transponder equivalent (TPE) is defined as 36 MHz of capacity, i.e., the amount of bandwidth needed to broadcast one analog video channel. A satellite with two 72 MHz transponders, for example, would have four TPEs of capacity. Since satellites vary in the number of transponders that they carry, and since transponders vary in bandwidth, TPE is a useful unit of measurement for calculating satellite capacity for

a given region and/or for a satellite operator. It is also used to measure demand, utilization, and market share.

transport driver A network device driver that implements a protocol for communicating between Lan Manager and one or more media access control drivers. The transport driver transfers Lan Manager events between computers on the local area network.

transport efficiency An AT&T term for the ability to carry information through a network using no more resources than necessary. Transport efficiency is achieved, for example, by statistical transport, which removes silent intervals from voice, data or other traffic and carries only the bursts of meaningful user information.

transport layer Layer 4 in the Open Systems Interconnection (OSI) data communications reference model that, along with the underlying network, data link and physical layers, is responsible for the end-to-end control of transmitted information and the optimized use of network resources. Layer 4 defines the protocols governing message structure and portions of the network's error-checking capabilities. Also serves the session layer. Software in the transport layer checks the integrity of and formats the data carried by the physical layer (layer 1, the network wiring and interface hardware), managed by the data link layer (layer 2) and possibly routed by the network layer (layer 1, which has the rules determining the path to be taken by data flowing through a network). See OSI.

transport level Level 4 of the Open System Interconnection (OSI) model. The Transport level allows end users to communicate oblivious to network constraints imposed by the lower levels. Passes data from the Session level on to the Network Level and ensures that the data reaches the other end. Level 4 also provides for flow management.

transport medium The actual medium over which transmission takes place. Transport media include copper wire, fiber optics, microwave and satellites.

transport overhead 1.728 MB/s of bandwidth allocated within each SONET STS-1 channel to carry alarm indications, status information, and message signaling channels for the preventive and reactive maintenance of SONET transmission (Transport) links.

transport protocol A protocol that provides end-to-end data integrity and service quality on a network. Windows 95 Resource Kit defines transport protocol as how data should be presented to the next receiving layer in the networking model and packages the data accordingly. It passes data to the network adapter driver through the NDIS interface. See also Transport Protocol Class Four.

Transport Protocol Class Four TP4. An International Standards Organization (ISO) transport layer protocol designated as ISO IS 8073 Class Four Service. TP4 has been adopted by the U.S. Department of Defense and specified in the U.S. Government OSI Profile (GOSIP).

transportable cellular phone The transportable cellular phone is a standard 3-watt mobile phone that can be removed from the car and used by itself with an attached battery pack. The entire unit is generally mounted or built into a custom carrying case to make it easy to carry on your shoulder. Although technically "portable," the transportable should not be confused with the true portable one-piece cellular phone. Also known as a "bag phone" or "briefcase phone"; refers to a cellular handset that is packaged with a larger carrying case containing a full-scale power supply.

transposed pair A wiring error in a twisted-pair cabling where a twisted pair is connected to a completely different set of pins at both ends (instead of pin 1 to pin 1, and pin 2 to pin 2, the cable is incorrectly wired pin 1 to pin 8, and pin 2 to pin 7, for example).

transposition Interchanging the relative position of conductors at regular intervals to reduce crosstalk. In data transmission, a transmission defect in which, during one character period, one or more signal elements are changed from one significant condition to the other, and an equal number of elements are changed in the opposite sense.

transverse interferometry The method used to measure the index profile of an optical fiber by placing it in an interferometer and illuminating the fiber transversely to its axis. Generally, a computer is required to interpret the interference pattern.

transverse parity check Type of parity error checking performed on a group of bits in a transverse direction for each frame. See Parity Check.

transverse scattering The method for measuring the index profile of an optical fiber or preform by illuminating the fiber or preform coherently and transversely to its axis, and examining the farfield irradiance pattern. A computer is required to interpret the pattern of the scattered light.

trap 1. See Trap and Trace.

2. A programming term. A programmer sets a trap for something to happen when something else happens. You might say "Wait for the mouse to come by, when it does, close the trap." A trap might be sprung when a phone rings or when someone hangs up. In network management, a trap is a mechanism permitting a device to automatically send an alarm for certain network events to a management station. Typically, network management

information is gained by polling network nodes on a regular basis. This strategy can be modified when a trap is set from a network node. With traps, a node alerts the management station of a catastrophic problem. The management station can then immediately initiate a polling sequence to the node to determine the cause of the problem. This strategy is often called trap-directed polling.

3. A video term. A circuit often called a filter, which is used to attenuate undesired signals while not affecting desired signals. Typically a signal channel trap to remove a single premium service which the subscriber is not paying for. See Notch Filter, Positive Trapping, and Negative Trapping.

4. TRransmission Alarm Processor

transrate To convert a video or audio stream from one bit rate to another.

trap and trace A telephone company term. Trap and Trace is the term for equipment and procedures for determining the source of an incoming call (typically a harassing call). The phone company uses traps to trace the source of the incoming call. There are two types of traps – the Terminating Trap and the Originating Trap. A terminating trap sits on the receiving phone line. In the old days, a terminating trap was a physical piece of equipment. These days, with electronic central offices, it's basically a command to the computer running the central office to keep track of all information about the source of all incoming calls. That information might be the originating telephone number. It might be the trunk number on which the call came in on. Such trunk number might look like TGN701. Or it might be the CLLI code – which stands for the Common Language Location Identifier. The CLLI code (pronounced "silly") consists of 11 characters. A sample CLLI code is "nycmny18dso." That says the call is coming in from New York City, Manhattan from a central office called 18DSO (which I happen to know is an AT&T 5E central office located on West 18th Street). Once the terminating trap identifies the possible direction /source / incoming trunk of the offending phone calls, the phone company will work it back towards the originating line. It will attach an Originating Trap to the offending trunk, then to the offending tandem office, then to the local central office. This can be a tedious and time consuming business. With the advent of Caller ID – both local and nationwide – trapping and tracing is getting faster and easier. Now if you receive an harassing call, you simply hit *57 the moment the call is over (GTE uses *69). This "tags" the incoming call's number and other information in your central office's records. You, as subscriber, can't get access to that information. But a law enforcement agency (i.e. one investigating your annoying calls) can. See Annoyance Call Bureau and Wire Tap.

trap door Hidden software or hardware mechanism that, when triggered, allows system-protection mechanisms to be circumvented.

Trash-80 Pejorative term for the TRS-80 (Tandy Radio Shack-80), an early PC sold by the Tandy Corporation through its Radio Shack retail stores. See TRS-80.

trashing Also referred to as dumpster diving, a term used by hackers for going through trash in an effort to get information that will facilitate breaking into computers. People often write passwords on paper, then put the paper in the trash. Be careful.

trashware Software that is so poorly designed that it winds up in the garbage can.

TRAU Transcoder and Rate Adapter Unit. A transmission function of the BSS that converts speech from the user of a mobile station into digital representation needed for an ISDN, wireless network.

travel card Another name for a telephone credit calling card. Travel card calls that are placed against a travel card number issued by the service provider, typically a phone company. As each call is completed, the long distance switch increases that card's account balance by the amount of each call. During the processing of a call, if the travel card is invalid or if the caller does not respond to a system prompt, the serving switch will typically ask the caller to hold the line for a live operator, and transfers the call to an Operator Workstation. When the operator answers, the OWS screen shows call information, including card number (if already entered), destination number (if already entered), trunk identification, and a failure code.

Traveling Class Mark TCM. A code that accompanies a long distance call. When Automatic Route Selection (ARS) or Uniform Numbering/Automatic Alternate Routing (UN/AAR) selects a tie trunk to a distant tandem PBX, the traveling class mark (TCM) is sent over the tie trunk. It is then used by the distant system to determine the best available long distance line consistent with the user's calling privileges. The TCM indicates the restriction level to be used based on the phone, trunk or attendant originating the call or the authorization code, if dialed.

tray A cable tray system is a unit or assembly of units or sections, and associated fittings, made of noncombustible materials forming a rigid structural system used to support cables. Cable tray systems include ladders, troughs, channels, solid bottom trays, and similar structures. See Rack.

tray cable TC. A multiconductor flame-retardant cable with an overall nonmetallic jacket rated 600 volts. The cable may or may not have grounding conductors in the assembly.

TRC Transit Routing Control Table.

TRCO Trouble Reporting Central Office.

treatment A billing and collections term. The specific steps of the collection process to which an account is subject. The treatment level may begin with a "courtesy" call which may go something like "Mr. Newton, this is Mrs. Horak with your friendly telephone company. We've noticed that your account is past due. In fact, you have not paid your telephone bill for three months. When might we expect payment?" At this point, Mrs. Horak verifies employment, which is a standard step. Now the conversation takes a turn for the worse. "Mr. Newton, do I understand correctly that you no longer work for Flatiron Publishing, and that you expect me to believe that you now work for Harry Newton Enterprises? Really Mr. Newton! I must request immediate payment by cash, cashier's check or money order! Failure to comply with this demand by the end of the business day will result in the disconnection of your service. Oh, did I mention that we will require a security deposit of $5,000? That, too, will have to be paid by the end of the business day. Yes, Mr. Newton, I am fully aware that it is 4:59PM. Mr. Newton, Mr. Newton." (Aside: "Ray, those guys in the switchroom are really good! They cut Harry's service off at exactly 5:00. That'll teach him to pay his bills on time!") Note: This scenario actually is very inaccurate- the guys in the switchroom aren't nearly that good. Actually, treatment levels are highly sensitive to the size of the bill, the age of the receivable, the history of the account, and other factors. Treatment levels may begin with a courtesy call, progress through several calls of a firmer tone, a formal letter or two of successively firmer tone, suspension of service, and disconnection. Restoral of service and reconnection entail service fees and generally involve a security deposit. If you don't pay your final bill quickly, you'll be dealing with a collection agency. Pay your telephone bill on time.

treatment level Treatment level is a term used in some telephone companies' billing and collections processes. The phrase is used to help a telephone company identify where a particular customer is in the collections/overdue billing process and proper protocol in treating the customer. See Treatment.

Treaty of Breda See Nutmeg.

tree 1. A network topology shaped like a branching tree. (What else?) It is characterized by the existence of only one route between any two network nodes. Most CATV distribution networks are tree networks.

2. In MS-DOS, a tree describes the organization of directories, subdirectories, and files on a disk.

tree hugger IBM-speak for an employee who resists a move or any other change.

tree mailbox A special function mailbox that provides the caller with a menu and allows selections from the menu using single digit commands.

tree network A network configuration in which there is only one path between any two nodes.

tree search In a tree structure, a search in which it is possible to decide, at each step, which part of the tree may be rejected without further search.

tree stand Aerial Cross Box. A cross box on a pole. Used where vandals live or when there's a narrow easement.

tree structure Describes the organization of directories, subdirectories, and files on a disk.

tree topology A network cabling architecture in which nodes are connected by cables to a central, or trunk, cable with a central retransmission capability.

treeware Slang for documentation or other printed material. Paper documentation, such as hardware and software manuals, made from trees.

trellis code See Trellis Coding.

trellis coding A method of forward error correction used in certain high-speed modems where each signal element is assigned a coded binary value representing that element's phase and amplitude. It allows the receiving modem to determine, based on the value of the preceding signal, whether or not a given signal element is received in error. See V.32 and V.32 bis. In QPAM, trellis coding adds extra bits (the trellis code) to data transmitted over a modem. The extra bits are fed to a mathematical algorithm at the receiver that lowers the number of possible choices in a QPAM eye-pattern. Helps modems do "on the fly" error detection and correction. See QPAM.

Trellis Coding Modulation TCM. A modem modulation technique in which sophisticated mathematics are used to predict the best fit between the incoming signal and a large set of possible combinations of amplitude and phase changes. TCM provides for transmission speeds of 14,400 bps and above on single voice grade phone lines. See

V.32 and V.32 bis.

tremendously high frequency Frequencies from 300 GHz to 3000 GHz.

trenching Excavation from ground level to the required depth underground in order to install, maintain, or inspect a conduit or cable. The trench is then backfilled and the surface reinstated.

TREX Transmission Expert.

TRFR TRansFeR.

TRG Technical Review Group.

TRI-CWDM On August 30, 1995, MCI Communications announced the deployment of a technology that will enable it to increase the capacity of its network by 50 percent without any additional fiber optic lines. The technology, known as Tri-Color Wave Division Multiplexing (Tri-CWDM), allows existing fiber to accommodate three light signals instead of two, by routing them at different light wavelengths, through the combined use of narrow and wide band wave division multiplexing. With this method, lightwaves are transmitted at 1557 nanometers (nm) and 1553 nm to a wide band WDM device, where a 1310 nm signal is added. Once combined, the three signals are routed through a single fiber to the next site where they are separated and sent to the receivers. Transmitting three signals in each direction allows for three different transmit pairs on just two fibers, effectively increasing the total network capacity from 5 gigabits to 7.5 gigabits. MCI officials say the technology will be particularly valuable in major metropolitan areas, where the company is enjoying outstanding growth in voice and data traffic. Essentially TRI-CWDM is now obsolete, replaced by Dense Wave Division Division Multiplexing. See DWDM.

tri-mode Tri-Mode describes a cell phone that operates in North America on both digital bands – 800 Mhz and 1900 Mhz – along with analog AMPS in the 800 Mhz band. The reason you'd want such a phone is simple: Digital service is often cheaper better in areas you can get it. But you can't get it everywhere. If you travel you need a cell phone you can use everywhere. Thus the idea of carrying a three band cell phone and subscribing to a service that gets you access to all three. Tri-mode can also apply to other parts of the world but I am not familiar with the different band/mode interactions. The U.S. doesn't have a specific wireless carrier that provides all three modes but Canada does. Commonly, wireless carriers have agreements that allow handsets to receive a competitor's service when roaming or if the primary service contains areas of poor service which would otherwise cause dropped calls. So a carrier may provide only 1900 Mhz PCS but when necessary allow the phone to operate in 800 Mhz AMPS, offered by another carrier, so that calls are not dropped. This concept applies to the Dual Band phones, as well. See Dual Band.

tri-watch A marine radio that is able to monitor three radio frequencies at one time, generally channel 16 (the "hail and distress channel"), a weather information channel, and another channel.

triangulation A method of locating the source of a radio signal through the use of three receivers, each of which focuses on the direction of maximum signal strength. Through the use of three receivers, it easily is possible to plot the general location of the transmitter, even though radio signals bounce off and are absorbed by physical obstructions such as buildings, trees and cars. This process, also known as Angle of Arrival, now can be accomplished by two, or even a single, receivers employing much more sophisticated, smart-antenna technology.

triaxial cable A cable construction having three coincident axes, such as a conductor, first shield and second shield all insulated from one another.

tribit transmission A transmission technique used by some modems in which three bits are transmitted simultaneously.

tributary The lower rate signal input to a multiplexer for combination (multiplexing) with other low rate signals to form an aggregate higher rate signal.

tributary circuit A circuit connecting an individual phone to a switching center.

tributary office A local office, located outside the exchange in which a toll center is located, that has a different rate center from its toll center.

tributary PBX An exchange within the main PBX configuration but with its own listed number. The only difference between a satellite and a tributary PBX is that the tributary PBX has a direct incoming connection from the public network. See Satellite PBX.

tributary station In a data network, a station other than the control station. On a multi point connection or a point-to-point connection using basic mode link control, any data station other than the control station.

tributary unit The SDH equivalent of a Virtual Channel in SONET terminology. A Tributary Unit might comprise a voice channel within a Virtual Tributary, which might take the form of a T-1 frame.

trichromatic The technical name for RGB representation of color to create all the colors in the spectrum.

trickle The name of a BITNET mail server package which provides access to anonymous FTP archive sites via e-mail.

trickle charge The continuous charging of an electrical battery. It keeps the batteries continuously charged, which is a good thing. That way, the phones still work when the lights go out – at least until the batteries run down.

trickle down economics Some economists believe that if the government reduces taxes, rich people will spend their money, buying and building things. And the money they spend will go to other people and other companies and that they will cause the economy to boom. This is, of course, nonsense. The economy booms much better when middle class people are given given tax breaks, because they spend all the money. Basically trickle down economics seems to me to be an excuse to give tax breaks to rich people. There is another explanation of trickle down economics, which refinforces my point that it's more productive to give poor people money. At the Roosevelt Island Tennis, my locker is below that of a very rich man, Skip Hartman. Before he plays tennis, Skip takes money out of his pocket and puts it in his locker. Usually, several of Skips coins fall through the crack of my locked locker door into my locker. I am most appreciative of the free money, the ultimate manna from heaven. I spend Skip's money with consummate glee. Skip's money falling into my locker is the ultimate trickle down economics, proving conclusively that trickle down economics works with poor people, like me, especially if the money comes from rich people like Skip. See also trickle down ergonomics.

trickle down ergonomics The practice of stealing (or being given) an Herman Miller Aeron chair, desk, computer, a monitor, or other tools of the trade after you've been laid off. See Trickle Down Economics.

trigger An application-specific process invoked by a database management system as a result of a request to add, change, delete, or retrieve a data element. For example, Local Number Portability (LNP) currently typically involves the use of two telephone numbers when a customer ports from one carrier to another. If a caller dials the old telephone number, the Central Office (CO) of the previous carrier recognizes that the old number is no longer active. That old 10-digit number triggers the CO to consult the SCP (Signaling Control Point) of the supporting AIN (Advanced Intelligent Network). The SCP dips into a database, extracts the new 10-digit telephone number and the CIC (Carrier Identification Code) of the CLEC (Competitive Local Exchange Carrier) to which the service has been ported, and provides the CO with that information, in order that the call can be handed off to the new carrier and eventually terminated. See also AIN, Basic Call State Model, CIC, CLEC, LNP, and SCP.

trigger dataset A dataset, typically on an IBM mainframe, that triggers the execution of a program. For example, when a D.176 dataset (i.e., a file containing international collect call data, formatted in accordance with ITU Recommendation D.176) is transmitted from international carrier A to international carrier B, the scheduling software on carrier B's mainframe, when it detects the newly arrived D.176 dataset, will launch an application to process the file for end-user billing.

triggering The process of detecting a word (or utterance) and capturing the speech data associated with that word (or utterance) for subsequent processing. See also Trigger.

triggers Uncompiled code residing on an intelligent database server. See also Trigger.

TrIM A new system called "transmission imaging modulator," or TrIM, which is to be heart of a "light engine" for a projector or television, that translates the signals that come into the system to the image seen by the viewer. The TrIM consists of two microchip devices and a tiny optical component. The technology comes from a tiny Vancouver, Washington startup called Steridian Inc. I first got wind of the company in early December, 2004.

trinkets A low to modestly priced item. Typically these are T-shirts, hats, cups and pens used for promotional or motivational purposes. Also used to bribe and cajole software developers into working even more excessive hours. When the telex business was in full bloom, trinkets were used to motivate telex operators into preferring one supplier over another. Trinkets were necessary because the price for service and equipment was identical, since it was heavily regulated.

trinkets and trash A new term for trinkets. See Trinkets.

triode A combination of a heated cathode, a relatively cold anode, and a third electrode for controlling the current flowing between the other two; the whole enclosed in an evacuated bulb. Variously called, audion, pliotron, radiotron, oscillion, audiotron, aerotron, electron tube, vacuum tube, etc.

TRIP Telephony Routing over IP protocol. TRIP was engineered by a working group of the IETF as a tool for inter-domain exchange of telephone routing information. It can also be used as a means for gateways and soft switches to export their routing information to a Location Server (LS), which may be co-resident with a proxy or gatekeeper. This LS can then manage those gateway resources. TRIP will give VoIP (Voice over Internet Protocol) providers the ability to dynamically exchange routing and line propagation information between one another helping to create the global Public Internet Telephone Network.

Triple DES A security enhancement of single-DES encryption that employs three successive single-DES block operations. Different versions use either two or three unique DES keys. This enhancement is considered to increase resistance to known cryptographic attacks by increasing resistance to known cryptographic attacks by increasing the effective key length.

Triple LNB dish This definition applies specifically to DirecTV's satellite operation. It's an 18" x 20" dish with three LNBs and four outputs. This dish looks at the 1010xAF, 1100xAF, and 1190xAF satellites. It is required for HD customers, Spanish language services, and locals in some markets because these services are not all available from the 1010xAF satellite.

triple mode A combined analog and digital mobile phone. Allows operation of the phone in the existing analog system frequency (800 MHz) and in both digital frequencies (800 MHz and 1900 MHz).

triple play In telecommunications, triple play refers to the delivery of voice, high speed Internet access and video over a single broadband network. The term generally applies to a closed, QoS-enabled (Quality of Service) broadband network such as a PON (Passive Optical Network) local loop or CATV network. Triple play services typically require 25 to 50 megabits per second of bandwidth per subscriber – the problem being that most local loops are copper cable, whereas for these bandwidths fiber is needed. Hence, many telecom companies are investing billions in "wiring" their subscribers with fiber.

triple-tap An input method for entering text messages or search queries on a cell phone where, because the 26 letters of the alphabet are distributed 3 to a key, the user has to press a key up to 3 times to select a letter. For example, the 2 key on a cell phone keypad has ABC on it. If a user wants to select the letter A, he presses the key once; if he wants to select the letter B, he presses the key twice; and if he wants to select the letter C, he presses the key 3 times, because it is the third letter in the ABC string. The 7 and 9 key each have 4 letters, PQRS and WXYZ, respectively, which means the user has to press the appropriate number 4 times to select either S or Z. The 1 key on a cell phone keypad has 14 special characters associated with it, which means the user has to press the key as many as 14 times to select a character. See the definition of "predictive text" to see how the tedium of triple-tap can be lessened.

triple witching Third Friday in the last month of the quarter. Those months are March, June, September, and December. Equity Options, index options and options on futures (i.e. futures contracts) all expire simultaneously. also the day that options on futures expire. Historically called triple witching. More volatility. Everyone expired at Friday's close. Triple witching volatility has gone. Now lower – Some expire at the open . Some expire at the close. Users roll out their beforehand.

triple-wrap In S/MIME version 3, to add a layer of security to a message by digitally signing it, encrypting it, and then digitally signing it again. When a user signs and encrypts a message with Outlook Web Access with the S/MIME control, the message is automatically triple-wrapped. Outlook and Outlook Express do not triple-wrap messages, but they can read them. See S/MIME.

triplex cable a triplex cable is a cable composed of three single conductor cables twisted together with no overall covering.

tripod A three-legged mounting system used for mounting antenna masts to different structures, including ridged roofs and flat roofs.

Tristan da Cunha A British Overseas Territory in the South Atlantic, and recognized by the Guinness Book of World Records as being "the remotest inhabited island in the world" on account of its being over 1300 miles from the nearest inhabited island (St. Helena) and over 1700 miles away from the nearest continent (South America). In June 2006 the territory received its first broadband connection to the outside world, via a 3.4-meter dish antenna and a 256 Kbps satellite link connecting the island to the U.K.'s Foreign and Commonwealth Office's global network. By August the island had set up its first Internet cafe to enable the approximately 270 inhabitants to communicate with faraway relatives, friends, suppliers, and customers.

trivia The word "trivia" is Latin, meaning three (tri) way (via). The term historically was used to describe a three-way crossroads, or intersection. According to etymologists (i.e., those who study the origins of linguistic forms), it was at such intersections that people, in days gone by, often stopped to exchange in small talk about unimportant matters. These unimportant matters eventually became known as trivia. See also Arcane, Draconian, and Eccentricity.

Trivial File Transfer Protocol TFTP. A UNIX-based file protocol. TFTP is a simplification of the earlier Simple File Transfer Protocol (SFTP).

TRL Transistor Resistor Logic.

TRO Temporary Restraining Order.

Trojan Horse A Trojan Horse is a piece of software that appears to do something useful, but which actually performs hidden, usually damaging, action on your computer. For example, a Trojan Horse might be a game program which plays a neat game but at the same time deliberately erases files on your computer. Such software might be distributed by being posted on a web site or it may be sent via e-mail, claiming that it is a product upgrade from a software vendor – like Microsoft (which never sends out software updates by email). A Trojan Horse is dangerous software. The simple rule is never use software from someone you don't know and haven't verified isn't real. See also Trojan Horse Attack. Malicious Trojan Horse software has evolved to a point where it may sit silently on your computer until some command from the outside (it assumes you're connected to the Internet full-time) tells your computer to do something, e.g. allow a distant pornographer to make their identity and location by using your computer as a relay station. You, of course, have no idea that you're being used to send pornography or spam.

Trojan Horse attack A network security term. An attack carried out via software that purports to be useful and benign, but which actually performs some destructive purpose (like erasing all the files on your hard disk) when run. See also Trojan Horse.

troll Someone who leaves comments on blogs designed to antagonize the author.

trolley dolly A slang word for an airline cabin attendant.

tromboning 1. A condition when a call comes in a trunk and is transferred back out on a trunk over the same physical path on which it arrived. Suppose a call is placed from location A through the telephone network and arrives at location B. If the call is forwarded to location C, the call turns on itself and returns to the telephone network on its way to location C. The call is said to be "tromboned". The name derives from the suggestion of the curve in the picture similar to the U-shaped slide on a trombone musical instrument. The trombone condition results in two trunks being tied up when the optimal connection would use no trunks at location B.   The cure for the tromboning condition is "anti-tromboning". In the situation described by the figure, the edge telephony switch at location B detects that transferring the call to location C will result in a tromboned trunk, and signals the transfer to the telephony switch in the telephone network to transfer the call.   The call will then be set up over a more optimal path, and the trunk at location B will be released.

There are at least two common signaling methods for anti-tromboning 1) Centrex flash hook and 2) QSIG path replacement. Trombboning sometimes is known as hairpinning. See also Hairpinning, Path Replacement, and QSIG.

2. A form of regulatory arbitrage used in telecommunications, and particularly in the cellular world, tromboning is used by some carriers to increase revenues or decrease costs. Tromboning involves a call from country A to country B first going through an international gateway in country C. The termination costs for traffic from C to B may be much less than from A to B. Stupid regulations developed by stupid regulators cause carriers to have to jump through such hoops. The term "tromboning" comes from the unique U-shaped section of the slide trombone. As the slide is pushed out and back in, it creates the tones between the fundamentals and the harmonics. See also Arbitrage and Broker.

troposcattering Tropscattering is a way of transmitting telecommunications signals between two places by bouncing multiple beams off the atmosphere. See Trospheric Scatter, which is its real name.

troposphere The lower layers of the earth's atmosphere. You can bounce certain frequency radio signals off it and use it as an elementary transmission reflector. The troposphere is the region where clouds form, convection is active, and mixing is continuous and more or less complete. The lower layers of the Earth's atmosphere, between the surface and the stratosphere, in which about 80 percent of the total mass of air is located and in which temperature normally decreases with altitude. The thickness of the troposphere varies with season and latitude; it is usually 16 km to 18km over topical regions and 10 km or less over the poles. See Tropospheric Scater and Tropospheric Wave.

tropospheric scatter The propagation of radio waves by scattering as a result of irregularities or discontinuities in the physical properties of the troposphere. The propagation of electromagnetic waves by scattering as a result of irregularities or discontinuities in the physical properties of the troposphere. A method of transhorizon communications using frequencies from approximately 350 MHz to approximately 8400 MHz. The propagation mechanism is still not fully understood, though it includes several distinguishable but changeable mechanisms such as propagation by means of random reflections and scattering from irregularities in the dielectric gradient density of the troposphere, smooth-Earth diffraction, and diffraction over isolated obstacles (knife-edge diffraction).

tropospheric wave A radio wave that is propagated by reflection from a place of abrupt change in the dielectric constant or its gradient in the troposphere. In some cases, the ground wave may be so altered that new components appear to arise from reflection in regions of rapidly changing dielectric constant. When these components are distinguishable from the other components, they are called "tropospheric waves."

trouble number display The operator will know what the trouble is with the phone system by seeing a number pop up on her/his console. That number may pop up automatically or the operator may have to hit the ALM (for ALARM) or similar button.

Trouble Reporting Central Office TRCO. Office where circuit troubles from OCCs are reported for repair and restoral.

trouble ticket Form used to report problems. Often incorrectly filled-in. Check.

Trouble Ticket Modify TTMOD. A Verizon definition. The transaction a CLEC or Reseller uses to add, change or delete a previously created Verizon trouble ticket (for example, for rescheduling repair activities). A CLEC or Reseller may submit a modification for any open trouble ticket, but not for a closed trouble ticket.

trouble unit A weighting figure applied to telephone circuit or circuits to indicate expected performance in a given period.

troubles per hundred Troubles per hundred is a criterion for acceptable customer service which telephone companies and public utility commissions have agreed upon. It's measured in terms of the number of complaints received per hundred telephones in one month. Six complaints per hundred is considered the maximum for acceptable service. See Quality of Service.

troy A pound of feathers weighs more than a pound of gold. Feathers are weighed by "avoirdupois" weight measure, which has 16 ounces to a pound, while gold is weighed in "troy" measure, which only has 12 ounces to a pound. Just in case you're interested, "avoirdupois" is from Middle English usage of the Old French "avoir de pois," which means "goods of weight," as it originated in commerce. "Troy" also is from Middle English usage, referring to Troyes, France. "Troy" is based on an ounce of 20 pennyweights, or 480 grains, as it originated in finance.

TRP RPOA Conversion Table.

TRS 1. Report Steering Table.

2. Telecommunications Relay Service. TRS is a service for the hard of hearing. See Telecommunications Relay Service.

3. Trunked Radio System.

TRS-80 Tandy Radio Shack-80. One of the early PCs. It was introduced by Tandy Corporation through its Radio Shack stores. It was based on the Zilog Z80 chip and began selling in the late 1970s. Along with Apple and Commodore it proved the viability of personal computers. The TRS-80 helped ensure the success of later generations of PCs by introducing a spreadsheet called Visicalc, word processing such as Electric Pencil, WordStar and databases such as dBASE. It also was one of Microsoft's earliest customers for their Basic language package. There were several TRS-80 models including the original Model 1, Model 2, Xenix (UNIX), Model 3 with integrated drives and Model 4. Some of the models could run CP/M or TRSDOS. Tandy later also introduced the Model 100 which was, arguably the first Laptop/Notebook, and for which Bill Gates is alleged to have written much of the software code. Detractors of the TRS-80 referred to it as the "Trash-80." The TRS-80 Model 2 actually had a cage designed for the specific purpose of accepting printed circuit cards. Sadly, Radio Shack never released the technical specs on the cage. No one (including Radio Shack itself) produced cards and the machine was quickly superseded by the IBM PC. In short, Radio Shack once had the market for PCs right at its fingertips. But it blew an incredible opportunity. Sad.

TRU Inside a telephone system, the TRU (the Tone Receiver Unit) is used to retrieve and interpret touch-tone data received. Those tones might be sent by a TSU – Tone Sender Unit.

truck roll Phone company terminology for physically sending a technician in a service truck into the field to diagnose a problem. Phone companies strive to reduce their truck rolls as there is a real cost involved with maintaining trucks, technicians, handheld test equipment, etc. Phone companies are setting so more and more of their installations can be done by their customers. See Splitter.

true north North based upon the earth's axis, which points to the star Polaris.

true-Up A "true-up" essentially is a reconciliation of actual experience against plans. For example, one might take out a software license, the cost of which is sensitive to the number of "seats" (i.e., clients, or desktop computers) running the software against one or more servers, or to the number of simultaneous users, or even to the number of times or total length of time that the software is actually used. The server may keep track of that usage. At some point, the contract may call for a true-up, at which point the actual usage is compared to the contract terms. Any additional usage costs are then due and payable. Similarly, a utility may be granted a tariff involving certain rates that are based on certain cost assumptions. At some point in time, a true-up takes place, at which point the actual costs are compared with those on which the rates are based. If the costs are higher, the

rates may go up. If the costs are lower, the rates may go down.

truely The wrong way to spell truly, as in yours truly.

TrueSpeech TrueSpeech is a low-bandwidth method of digitizing speech, which was created by a company called DSP Group, Inc. Santa Clara, CA. TrueSpeech uses compression to drop one minute of voice down to 62 kilobytes with remarkably little degradation. It is used in many digital telephone answering devices for storing and reproducing voice. TrueSpeech's compression is not meant for high fidelity music. But it is more than acceptable for such business applications as voice mail, voice annotation, dictation, and education and training. The small file size means that it can be transferred more easily to other users by using either a corporate network. The DSP Group describe TrueSpeech an enabling technology for speech compression in personal computers and future personal communications devices. Speech compression is key technology to the effective convergence of personal computers and telephony. TrueSpeech compression is a technology based on complex mathematical algorithms which are derived from the way airflow from our lungs is shaped by the throat, mouth, and tongue when we speak. This shaping is what our ear finally hears. TrueSpeech is 5 to 15 times more efficient than other methods of digital speech compression. For example, a one minute long speech file which uses other PC audio technology would consume as much as 960 kilobytes. With TrueSpeech, the same file would be just over 60 kilobytes. TrueSpeech is used in the Microsoft Sound System, which also lets you choose the voice sampling you wish when you're recording material. Here is Microsoft Sound System's recording options:

Truetype A Windows feature. Fonts that are scalable and sometimes generated as bitmaps or soft fonts, depending on the capabilities of your printer. TrueType fonts can be sized to any height, and they print exactly as they appear on the screen. Using TrueType, you'll be able to create documents that retain their format and fonts on any Windows machine.

TrueUp See True-Up.

Truevoice 1. In the fall of 1993, AT&T announced that it was introducing new voice quality throughout its long distance network. And that it was calling that quality "true voice." AT&T set up a demo line. Some people thought they could notice an improvement. Some thought they couldn't. I personally thought true voice sounded pretty good.

2. The trademark name of Centigram's text-to-speech product, which they acquired from SpeechPlus.

Trumpet Winsock A once-popular Windows 3.xx communications program and TCP/IP stack which allowed people to dial into the Internet and use browsers to surf the Internet. I never liked the program and had great difficulty with it. Fortunately the program has effectively been killed by dial up networking capabilities now part of every Windows 95.

truncated binary exponential back off Another name for exponential back off used in IEEE 802.3 local area networks. In an exponential back-off process, the time delay between successive attempts to transmit a specific frame is increased exponentially.

truncation In data processing, the deletion or omission of a leading or a trailing portion of a string in accordance with specified criteria.

trunk 1. A communication line between two switching systems. The term switching systems typically includes equipment in a central office (the telephone company) and PBXs. A tie trunk connects PBXs. Central office trunks connect a PBX to the switching system at the central office. See also Trunk Side.

2. The term that is used for the communication bus that connects devices on a fieldbus network.

trunk access number 1. The number of the trunk over which a call is to be routed.

2. The number that needs to be dialed in order to gain access to an outbound trunk. This applies to both local and long distance trunks, as the access number can be different.

trunk answer A phone system feature. This feature allows a ringing call to be answered from any telephone in the system. Typically the feature must be activated in phone system programming.

trunk answer from any phone A phone system feature. When a call comes in, something rings. You can now answer the incoming call from any phone. To do so, you must dial a special code or hit a special feature button on your phone. When my office phone system bells ring, all we have to do is to touch "6" on any phone and we can answer the incoming call. Typically the feature must be activated in phone system programming.

trunk circuit An assemblage of electronic elements located in the switching machine.

trunk conditioning Electrical treatment of transmission lines to improve their performance for specific uses such as data transmission. The "tuning" and/or addition of equipment to improve the transmission characteristics of a leased voice-grade line so that it meets the specifications for higher-speed data transmission. Voice-grade lines often have too much "noise" on them. By altering the equipment at both ends of the line, this noise on the line can be overcome. This allows transmission of higher-speed data, which is much more sensitive to noise than voice. See also Conditioning.

Trunk Data Module TDM. Provides the interface between the DCP signal and a modem or Digital Service Unit (DSU).

trunk direct termination An option on switchboards which terminates a trunk group on one key (or button) on the console.

Trunk Encryption Device TED. A bulk encryption device used to provide secure communication over a wideband digital transmission link. It is usually located between the output of a trunk group multiplexer and a wideband radio or cable facility.

trunk exchange A telephone exchange dedicated primarily to interconnecting trunks.

	Digital Encoding Rates	Sampling Rate	Technology
TrueSpeech	62K per minute	8 KHz	Proprietary
Voice	234K per minute	8 KHz	ADPCM
Radio	322K per minute	11 KHz	ADPCM
Tape	1,291K per minute	22 KHz	PCM
CD	5,176K per minute	44 Khz	PCM

The above is for mono recordings. For stereo, double the amount of space.

trunk group A group of essentially like trunks that go between the same two geographical points. They have similar electrical characteristics. A trunk group performs the same function as a single trunk, except that on a trunk group you can carry multiple conversations. You use a trunk group when your traffic demands it. Typically, the trunks in a trunk group are accessed the same way. You dial your Band 5 WATS trunk group by dialing 62, for example. If the first trunk of that group is busy, you choose the second, then the third, etc. See Trunk Hunting.

trunk group alternate route The alternate route for a high-usage trunk group. A trunk group alternate route consists of all the trunk groups in tandem that lead to the distant terminal of the high-usage trunk group.

Trunk Group Multiplexer TGM. A time division multiplexer whose function is to combine individual digital trunk groups into a higher rate bit stream for transmission over wideband digital communication links.

trunk group warning Alerts the attendant when a preset number of trunks in a group are busy. See Trunk Group.

trunk holding time The length of time a caller is connected with a voice processing system. Defined from the time when the system goes off-hook to the time the port (i.e. the trunk) is placed back on hook.

trunk hunting Switching incoming calls to the next consecutive number if the first called number is busy.

trunk make busy A fancy name for saying that, by punching a few buttons on the console, you can make any trunks in your PBX or key system busy, effectively putting the trunk out of service. You may want to do this if your trunk is acting up. By busying it out at the console, you are effectively denying its use to anyone in the company. Thus you are protecting yourself from further complaints. Hopefully, it will be repaired promptly.

trunk monitoring Feature which allows individual trunk testing to verify supervision and transmission. You dial an access code and then the specific trunk number from the attendant console. You want the ability to test a specific trunk because normally you might be only accessing a trunk group when you dial an access code. Thus, each time you dial into the trunk group, you might end up on another individual trunk. Some PBXs have a variation of trunk monitoring, whereby if a user encounters a bad trunk, he can dial a specific code, then hang up. The PBX recognizes these digits and makes a trouble report on that specific trunk, possibly reporting it to the operator, keeping it in memory for later analysis or dialing a remote diagnostic center and reporting its agony.

trunk number display The specific trunk number of an incoming call can be displayed on the attendant console, enabling your attendant to instantly identify the origin of certain calls. For example, if you have several tie lines to branch offices, your attendant

knows immediately which office is calling. Many newer PBXs have displays on individual telephones, which show the actual trunk being used for outgoing and incoming calls. This provides an additional measure of control. You might, for example, speak faster if you knew the call was coming in on your IN-WATS line. You might also answer the call differently if you know what trunk it's coming in on. For example, you might be running several, totally-separate businesses from the same console. Each business has a different number. The only way you know what to answer – Joe's Bakery or Mary's Real Estate – is by the trunk.

trunk occupancy The percentage of time (normally an hour) that trunks are in use. Trunk occupancy may also be expressed as the carried CCS per trunk.

trunk order A document (or data system equivalent) used in an operating telephone company to request a change to a trunk group.

trunk queuing A feature whereby your phone system automatically stacks requests for outgoing circuits and processes those requests on, typically, a first-in/first-out basis. See Queuing Theory.

trunk reservation The attendant can hold a single trunk in a group and then extend it to a specific phone. This means, for example, that a WATS line can be held for someone special – a heavy caller, the president of the firm, etc.

trunk restriction Some people may not be allowed to use certain trunks at certain times. The sophistication of trunk restriction depends on the switch and the way it's programmed.

trunk segment The main segment of cable in an Ethernet network is called the trunk segment.

trunk side The portion of a communicating device (phone system, data communications equipment etc.) that is connected to external, i.e., outside plant, facilities such as trunks, local loops, and channels. See also Trunk Side Connection.

trunk side connection A carrier term. Trunk side connections are within the carrier network. InterMachine Trunks (IMTs) connect carrier switches to other carrier switches. Such switches include circuit switches such as Central Offices (COs) and Tandem switches, Frame Relay switches and routers, packet switches, and ATM switches. End user organizations can lease local loops with trunk side connections, as well; such a loop would appear to the carrier network as being a part of it, and would be used for access to ANI (Automatic Number Identification) information. Compare with Line Side Connection.

Trunk Type TT. Trunks that use the same type of equipment going to the same terminating location.

Trunk Type Master File TTMF. An MCI definition. A comprehensive listing of all trunk assignments on the MCI network for shared and dedicated services, necessary for processing and billing MCI customer calls.

trunk to tie trunk connections The ability of the switching system to provide the attendant with the capability of extending an incoming trunk call to a tie trunk terminating some place else.

trunk to trunk by station A PBX feature which permits the user who established a three-way conference involving himself and two trunks to drop from the call without disconnecting the trunk-to-trunk connection.

trunk to trunk connections The attendant can establish connections between two outside parties on separate trunks. Call your office on your IN-WATS. Ask the operator to extend that call to the VP who happens to be at his home. The operator must place an outside call to the VP on an outside trunk and join that call to the incoming call. Sometimes it works.

trunk to trunk consultations Allows a phone connected to an outside trunk circuit to gain access to a second outside trunk for "outside" consultation. No conference capability is available with this feature.

trunk tracking The capability of specially designed radios and scanners to automatically follow radio traffic that jumps from channel to channel on a trunked radio system. One receiver on the trunk tracking device monitors the trunked radio system's control channel, and captures the instructions and status messages sent on that channel for the call group being monitored. Whenever an instruction is received to change the frequency of the call group's traffic channel, the trunk tracking device adjusts the frequency of its own traffic channel accordingly, so as to follow transmissions in an orderly fashion.

trunk transfer by station Permits the user who established a three-way conference involving two lines to drop from the call without disconnecting the trunk-to-trunk connection.

trunk up-down See TUD.

Trunk Utilization Report TUR. A computer printout detailing the traffic use of a trunk.

trunk verification by customer Provides the attendant or phone user access to individual lines in a trunk group to check their condition. See also Trunk Monitoring.

trunk verification by station Provides a warning tone if a phone user enters a busy trunk.

trunked radio A system in which users share or pool a number of radio channels. Frequencies are distributed by the system according to demand and traffic levels. Trunking can enhance spectrum efficiency in some circumstances.

trunked radio system Another name for SMR (Specialized Mobile Radio). See SMR.

trunks in service The number of trunks in a group in use or available to carry calls. Trunk in service equals total trunks minus the trunks broken or made busy for any reason.

trunks required The number of trunks that result from interpreting a given offered load against a specified service or economic criterion.

trust relationship The trust relationship is the link between two domains (e.g. two servers on a network) that enables a user with an account in one domain to have access to resources on another domain. The trusting domain is allowing the trusted domain to return to the trusting domain a list of global groups and other information about users who are authenticated in the trusted domain. In the MIS world, a domain is "the part of a computer network in which the data processing resources are under common control." In the Internet, a domain is a place you can visit with your browser – i.e. a World Wide Web site. See domain and inter-domain trust relationships.

trusted See Class of Service.

trusted source marketing See viral marketing.

truth The first casualty of war.

truth is on the wire In networking, this is the principle that the only way a network administrator can have complete visibility into what is happening on the network is by capturing and studying packets on the network. Packet analysis provides insights into types of internal, outbound, and inbound traffic on a network; network applications and their effects on the network; network protocols and their effects on the network; end-user activities; and traffic being introduced onto the network from client machines and devices. Similarly, packet analysis is a way of getting at the truth in a way that available documentation, if any, invariably falls short on delivering.

truth table An operation table for a logic operation. A table that describes a logic function by listing all possible combinations of input values and indicating, for each combination, the output value.

TRX Transmitter and Receiver (Transceiver). A wireless telecommunications term. A function of the radio channel device for receiving and transmitting signals or information on the radio channel.

TRXXX Technical Reference number. Technical References are issued by Telcordia Technologies (nee Bellcore). See Telcordia.

TS 1. Transport Stream. As an ATM term, it is one of two types of streams produced by the MPEG-2 Systems layer. The Transport Stream consists of 188 byte packets and can contain multiple programs.

2. Traffic Shaping: Traffic shaping in an ATM network is a mechanism that alters the traffic characteristics of a stream of cells on a connection to achieve better network efficiency, while meeting the QoS (Quality of Service) objectives, or to ensure conformance at a subsequent interface. Traffic shaping must maintain cell sequence integrity on a connection. Shaping modifies traffic characteristics of a cell flow with the consequence of increasing the mean Cell Transfer Delay.

3. Time Stamp: As an ATM term, Time Stamping is used on OAM cells to compare time of entry of cell to time of exit of cell to be used to determine the cell transfer delay of the connection. See also Timestamp.

4. Transaction Server.

TS3 TimeStamp version 3. TS3 is a protocol run on some IRC (Internet Relay Chat) servers for purposes of maintaining their mutual synchronization in support of real-time chats among users on an IRC channel.

TSA See Time-Slot Assignment.

TSAC Time Slot Assigner Circuit; a circuit that determines when a CODEC will put its eight bits of data on a RCM bit stream.

TSAP Abbreviation for Transport Service Access Point in the OSI transport protocol layer.

TSAPI Telephony Server Application Programming Interface. Described by AT&T, its inventor, as "standards-based API for call control, call/device monitoring and query, call routing, device/system maintenance capabilities, and basic directory services." For a better explanation, see Telephony Services.

TSB Telecommunications System Bulletin. Interim changes to an interim standard. Not very interesting, but there it is. See IS.

TSB67 Part of the EIA/TIA-568-A standard. TSB67 describes the requirements for field testing an installed Category 3,4 or 5 twisted pair network cable.

TSC 1. Two-Six Code. A trunk group reference number. The first two characters are alphabet (a-z) and the last six characters are numeric digits.

2. Transmission Systems Construction.

3. Technical Service Centers.

TSCM Technical Surveillance CounterMeasures. Commonly called debugging, sweeps or electronic sweeping.

TSEP See Traffic Separations.

TServer Telephony Server. Name of NetWare Telephony Services LAN server which is joined physically (by wire) to an adjacent PBX. See Telephony Services.

TSI 1. Time Slot Interchange or Interchanger. A way of temporarily storing data bytes so they can be sent in a different order than they were received. Time Slot Interchange is a way to switch calls. See Time Slot Interchange.

2. Transmitting Subscriber Information. A frame that may be sent by the caller, with the caller's phone number, which may be used to screen calls, etc.

3. Telecommunication System Integration. A fancy name for joining many things together in a telephone system. For example, one part of TSI might be installing Internet Protocol (IP) phone systems into call centers and medium-sized businesses.

TSIC Time Slot Interchange Circuit; a device that switches digital highways in PCM based switching systems. In short, a digital crosspoint switch.

TSIU Time Slot Interchange Unit. Switching module hardware unit that provides the digital time switching function.

TSK Transmission Security Key.

TSM Terminal Server Manager (TSM), a program that allows terminal servers on a network to be remotely managed from another node. It is supported on VMS systems running the LAT protocol (and is incompatible with TCP/IP-only networks).

TSO 1. Time Share Operation.

2. Time Sharing Option. An archaic IBM environment for implementing time-shared use of a mainframe computer. A system that enables users to interact with an IBM mainframe in real time.

3. Transmission System Operation.

4. Technical Support Operations.

TSP 1. Telecommunications Service Priority. The TSP System is the regulatory, administrative, and operational system authorizing and providing for priority treatment to provision (initiate) and restore NS/EP (National Security and Emergency Preparedness) telecommunications services. Under the rules of the TSP System, telecommunications companies are authorized and required to provision and restore services with TSP assignments before services without such assignments. 2. Terminal Service Profile.

3. Telecommunications Service Provider. A new term for an Internet Service Provider.

4. Technical Support Planning.

5. Speed Conversion Table.

6. Ticket Service Provider. A new term for a web-based company that handles

TSPS Traffic Service Position System permits operator positions serving public phones and HOBIC operations to be located remotely from the CO which services the pay phone or the hotel, or the hospital, etc.

TSR 1. Telephone Service Representative. See also Agent.

2. Terminate and Stay Resident. A term for loading a software program in an MS-DOS computer in which the program loads into memory and is always ready for running at the touch of a combination of keys, e.g. Alt M, or Ctrl ESC. Here's some information from Jackie Fox writing in PC Today: You can't load TSRs willy-nilly and expect them to work with each other. Some will get along with each other. Others won't. When you install a TSR, it goes to a location in RAM (Random Access Memory) called the Interrupt Vector Table.

The interrupt vector table is like a hotel lobby, and TSRs are like guests waiting for messages. The TSR watches every incoming keystroke to see if it's the special hot key combination (message), the TSR is waiting for. If it isn't, the TSR passes it back to the regular program. What if you have four or five TSRs loaded? The one you loaded last has seniority. It checks the incoming keystrokes first. If the TSR recognizes the keystroke combination as its own hot-key combination, it takes over. If not, it passes it along to the next TSR. This process is called interrupt handler chaining.

If none of the TSRs recognize that particular combination, they pass it along to DOS so it can process it as a regular keystroke combination. Not all TSRs pass instructions along the way they should. Some TSRs intercept keystrokes and never pass them on. Some TSRs never restore their original addresses. Sometimes two TSRs fight over the same hot key combination. Then you end up with a frozen keyboard. The basic problem is there are no rules for loading and running TSRs.

   3. Terminal Service Representative.

   4. Technical Service Representative.

   5. System Resources Table.

   6. Terabit Switch Router.

   7. Total Service Resale. See Total Service Resale.

TSRM Telecommunication Standards Reference Manual.

TSS The Telecommunications Standards Section (TSS) is one of four organs of the ITU. Any specification with an ITU-T or ITU-TSS designation refers to the TSS organ. See ITU.

TST 1. Time-Space-Time system.

   2. Alarm Steering Table.

TSTN Triple Super Twisted Nematic. A display technology often used on laptop computers which uses three layers of crystal to give better contrast and more grey scales.

TSTS Transaction Switching and Transport Services. In 1992, the Regional Bell Operating Companies (RBOCs) agreed to provide uniform transaction processing capabilities under a banner called Transaction Switching and Transport Services.

TSU Tone Sender Unit. A device inside a telephone system. The TSU passes along touch-tone digits to telephone extension cards within the phone system. See also TRU (the Tone Receiver Unit) which is used to retrieve and interpret touch-tone data received.

TSV Network Terminal Control Table.

TSYN The SYNCHronous test line provides for testing of ringing, tripping and supervisory functions of toll completing trunks. See TNSS.

TT 1. Trunk Type.

   2. Touch Tone

   3. Transaction Time.

   4. Terminal Timing.

TTB 1. Talking Total Bollocks. This is a European (specifically UK) term coined by telecom experts when noticing a particularly enthusiastic sales guy getting over excited about some new piece of technology. Chris Hall contributed this dubious definition.

   2. Terminal Barring Table.

   3. Touch-tone service (usually for business).

TTC The Telecommunications Technology Committee, a Japanese standards committee.

TTCN Tree and Tabular Combined Notation: The internationally standardized test script notation for specifying abstract test suites. TTCN provides a notation which is independent of test methods, layers and protocol.

TTCP TestTCP is a test that reports the amount of data transferred, the transfer time, and the approximate throughput. By comparing the actual throughput with the theoretical bandwidth between the transmitter and receiver, you can tell whether the network is operating as expected. To use TTCP, you start a copy of TTCP in receive mode at one place within the network, then start a second copy in transmit mode at another place within the network. The results of the transfer of data from the transmitter to the receiver indicate the approximate performance of the path between the source and destination. By selecting the source and destination at various points with the network, you can analyze critical portions of the path. TTCP has a real advantage over tools like FTP. If you have a high performance network, it is difficult for any single computer system to transfer data to or from disk at rates which are sufficient for real network testing. TTCP achieves high performance by filling a memory buffer with data, then repeatedly transmitting this data. Since everything is running from memory, you have a traffic transmitter and receiver that can operate at true network speeds. Cisco has implemented a copy of TTCP in IOS 11.2 and later, currently as an undocumented command. Since it is undocumented, you will not find it by using the interactive help function. Instead, just type the command ttcp, then press RETURN. If the router model you are using supports TTCP, it will respond with a series of questions for the TTCP parameters. Because TTCP can create enormous amounts of network traffic, it is a privileged command. This gives us a more widely available resource within our networks for generating traffic when performing network analysis and tuning. Because this capability is in the router, we no longer have to install special traffic generators in the network. You'll find that Cisco routers, because they are very efficient at moving IP data, are very good traffic transmitters and receivers. Now you can perform high-speed network traffic analysis without the need for extra equipment. Here's a great explanation (from where I got much of the above explanation): www.ccci.com/product/network_mon/tnm31/ttcp.htm

TTD Temporary Text Delay. The TTD control sequence (STX ENQ) is sent by a sending station in message transfer state when it wants to retain the line but is not ready to transmit.

TTI Transmit Terminal Identification. A fax machine's stupid term for its telephone number and the name of its owner. When you receive a fax from someone, the top line of the fax typically will have a phone number and a name on it. That phone number and name does NOT come from the phone or the phone company. It comes from what the person who owns the machine programmed into his machine. He typically did that by punching buttons on his fax machine. He'll do that if he can understand the instruction booklet which came with his fax (which he probably won't). The point of all this is twofold: First, don't forget to put your name and phone number into your fax machine. Second, don't assume that what you read at the top of any fax you receive is accurate.

TTIA Telecommunications Technology Investment Act of 1993.

TTL 1. Transistor Transistor Logic. An internal transfer standard for electronics devices in which a 1 state is +5 Volts and a zero state is 0 Volts; communications systems are sometimes expected to interface to this and provide transmission converters to telecommunications standards.

   2. Time to Live, which is used in IP protocol. It is a time, typically in seconds, after which the fragment can be deleted by any device on the network. Typically this would be used if a router developed an error resulting in a packet that would otherwise circulate for ever. See also Trace Route.

TT&P Technical Training and Publications.

TTMF Trunk Type Master File.

TTMOD See Trouble Ticket Modify.

TTR Touch Tone Receiver. A device used to decode touchtones dialed from single-line telephones or Remote Access telephones.

TTRT Target Token Rotation Time. An FDDI (Fiber Distributed Data Interface)token travels along the network ring from node to node. If a node does not need to transmit data, it picks up the token and sends it to the next node. If the node possessing the token does need to transmit, it can send as many frames as desired for a fixed amount of time.

TTS Text-To-Speech. A term used in voice processing. See Text-to-Speech.

TTTN Tandem Tie Trunk Network.

TTY TeleTYpewriter. Typewriter-style device for communicating alphanumeric information over telecom networks. TTY is the most widely used type of emulation for PC computer communications.

TTY/TDD A unique Telecommunication Device for the Deaf, using TTY principles.

TU 1. Transmit Unit. Term used in a DS-3 channel bank.

   2. Tributary Unit in SDH terminology. Equivalent to a Virtual Channel.

TUA Telecommunications Users Association (UK). The TUA says it aims to support the development of UK businesses through the application of telecommunication and information technologies. It also strives to bring about a fair and competitive market within the UK. TUA holds an annual trade show in the U.K, typically around November or December. www.tua.co.uk.

TUANZ Telecommunications Users Association of New Zealand. A non-profit society of over 500 telecommunications users including major NZ corporations, small to large businesses, government departments, educational institutions and interested individuals. By the way, New Zealand is about as deregulated as you can get; hence, it is a technology testbed. All the really neat technologies that we now see being introduced in the US have been trialed in New Zealand for years. www.tuanz.gen.nz

TUBA TCP and UDP with Bigger Address. One of the three IPng candidates.

tubing A tube of extruded non-supported plastic or metallic material.

tubular post See pneumatic telegraph.

TUD Trunk Up-Down. Protocol used in ATM networks that monitors trunks and detects when one goes down or comes up. ATM switches send regular test messages from each trunk port to test trunk line quality. If a trunk misses a given number of these messages, TUD declares the trunk down. When a trunk comes back up, TUD recognizes that the trunk is up, declares the trunk up, and returns it to service. See also Trunk.

TUG Telecommunication User Group.

TUI Telephony User Interface, or Telephone User Interface. A TUI is much like a GUI (Graphical User Interface), except that you use the telephone to get to information in a computer. A TUI makes use of the touchtone keypad to make selections from a menu presented by a voice processor. For example, you might press "1" to speak to an agent about making a domestic flight reservation, "2" for an international reservation, "3" to confirm a reservation, or "4" to order a pizza while you wait for an agent to answer your call. See also GUI.

Tukey, John W. John W. Tukey was a professor of statistics at Princeton University. Previously, he was a statistician at Bell Telephone Laboratories. Tukey coined the terms "bit" and "software." See also bit and software.

tumbling A form of cellular fraud first appearing in late 1990. The crook alters a cellular telephone so that it "tumbles" through a series of ESNs in order to make the caller to appear to be another new customer each time a call is made. By the time the cellular phone network operator has checked with the network where the bogus telephone supposedly is registered and discovered the fraud, the crook has tumbled the telephone, or changed its electronic serial number (typically by one digit) and is ready to make more free calls. As the carriers have moved to IS-41 pre-call validation, this form of fraud has all but been eliminated. See also Clone Fraud.

tunable laser A tunable laser is a component used in fiber-optic systems that can send many different wavelengths of light using fewer parts than conventional lasers. Tunable lasers can tune into more than one wavelength. This allows a single laser to replace the role of up to 16 lasers on a DWDM system. This greatly streamlines DWDM systems and reduces the level of required inventory for back-up purposes (i.e., sparing). Also called Programmable Laser or Selectable Laser.

tunable operating system parameters Tuning an operating system is the same as optimizing it, in that you rewrite commands and programs so they operate faster and more efficiently. Any new operating system needs to be tuned to the specific machine on which it is running.

tuned absorber shield See Microwave Absorber.

tungsten A metallic element used in ceramic IC packaging to provide the traces within the package that connect the device circuitry to the external terminals pads or leads.

tuning Adjusting the parameters and components of a circuit so that it resonates at a particular frequency or so that the current or voltage is either maximized or minimized at a specific point in the circuit. Tuning is usually accomplished by adjusting the capacitance or the inductance, or both, of elements that are connected to or in the circuit.

tunnel According to the IETF (Internet Engineering Task Force), a tunnel is "An intermediary program which is acting as a blind relay between to connections. Once active, a tunnel is not considered a party to the HTTP communication, though the tunnel may have been initiated by an HTTP request. The tunnel ceases to exist when both ends of the connections are closed." A tunnel is a secure path for communications between clients and servers over an inherently insecure IP-based network. See also Tunneling.

tunnel coverage system A distributed antenna system inside of a tunnel, set up in order to provide wireless communications within the tunnel and between the tunnel and the outside world.

tunnel diode A tunnel diode conducts electricity very well in both directions. However, their resistance to current flowing in the forward direction is very unusual. As the voltage is increased, the current carried by the diode also increases until it reaches a peak. Increasing the voltage beyond the peak value causes the amount of current passing through the diode to decrease! The current will continue to decrease until it reaches a minimum value and then rise again with increasing voltage.

tunneling 1. As a local area network term, tunneling means to temporarily change the destination of a packet in order to traverse one or more routers that are incapable of routing to the real destination. For example, to route through a backbone whose internal routers don't contain entries for external destinations, the entry border router must "tunnel" to the exit border router.

2. As an Internet term, tunneling means to provide a secure, temporary path over the Internet, or other IP-based network, in a VPN (Virtual Private Network) scenario. In this context, tunneling is the process of encapsulating an encrypted data packet in an IP packet for secure transmission across an inherently insecure IP network, such as the Internet. The leading tunneling protocols currently are IP Security (IPsec), Layer 2 Tunneling Protocol (L2TP), Point-to-Point Tunneling Protocol (PPTP), and SOCKSv5. In a typical VPN application, a telecommuter might dial into an ISP (Internet Service Provider). The ISP's router would recognize the request for a high-priority, secure tunnel across the Internet to a corporate gateway router for purposes of access the corporate Intranet. The tunnel would be set up through all the intermediate routers, effectively snaking its way through other, lower-priority Internet traffic. This definition is largely from Ray Horak's book, "Communications Systems & Networks." It's a great book, and a perfect companion to this dictionary. (I wrote one of the forewords for Ray's book. I said pretty much the same thing there, so I guess I'm stuck with that opinion.) See also IP Security, Layer 2 Tunneling Protocol, Point-to-Point Tunneling Protocol, Router, SOCKSv5, Split Tunneling, and Virtual Private Network.

tunneling ray Leaky ray.

TUP Telephone User Part. An SS7 term for the predecessor to ISUP (Integrated Services User Part). TUP was employed for call control purposes within and between national networks, both wired and wireless. ISUP adds support for data, advanced ISDN, and IN (Intelligent Networks). See also ISUP.

tuple address In the Frame Relay (FR) network, packages are sent from switch A to switch B. Each package contains a header section. In the header section of the package, there is information on the DLCI (Data Link Connection Identifier). At switch B, the traffic

information is written to a file (based on the Bay Network FR, it is called DATA) every two hours. The billing adjunct processor (AP) will TFTP or FTP the DATA files after it is generated. In each DATA file, it contains the size of the package, number of packages, switch A's IP address (e.g., 156.52.245.1), switch A's DLCI (e.g., 1000), switch B's IP address (e.g., 166.100.221.25), and switch B's DLCI (e.g., 502). The database of the billing adjunct processor contains the tuple address (switch A's IP & DLCI and switch B's IP & DLCI information) of the circuit that the customer rents. The billing AP converts the DATA files' format to a standard format file and sends it to the billing company which then bills the customer who rents that tuple address. This is very much like our phone bill we receive monthly from a telephone company based on where we called.

TUR Traffic Usage Recorder. A device which connects to a network element in order to capture and record traffic statistics. Most network elements (e.g., PBXs, ACDs, data switches and routers) have special ports to which such a device can connect, usually via a RS-232 cable. As traffic flows through the network element, various information about that traffic is output to the TUR in real time. The TUR holds that raw data in buffer memory until such time as it is polled by a centralized computer system and the data is downloaded. Subsequently, the data is processed and reports are generated by a traffic analysis application software system. This process of traffic analysis of historical data is essential to the processes of network design and optimization, which balance network performance (availability) against network costs.

turf technician Another term for field technician.

turing machine A mathematical model of a device that changes its internal state and reads from, writes on, and moves a potentially infinite tape, all in accordance with its present state, thereby constituting a model for computer-like behavior. This is the same Alan Turing, British mathematician, who coined the Turing test which I mention under artificial intelligence.

turbo button An actual or figurative button on a broadband provider's portal that a customer can click on to buy extra bandwidth from one moment to another. See turbo button service.

turbo button service Another term for bandwidth on demand.

turbo coding A complex data encoding/decoding technique first introduced in 1993 by Messrs. Berrou, Glaviewx and Thitimajshima as an improved means of Forward Error Correction (FEC). Turbo Coding can dramatically improve the BER (Bit Error Rate) of data transmission through an iterative coding/decoding technique, combined with the interleaving/deinterleaving of data blocks. As each block of data is transmitted, it is encoded by two separate convolutional (intertwining) encoders which are concatenated (linked). Commonly, each of the encoders uses the Extended Hamming Code. In each case, the data to be transmitted are organized into data blocks which, for example, are viewed logically as 11 bits horizontal (wide) and 11 bits vertical (deep). To each 11-bit data word, both horizontal and vertical, is appended a 5-bit descriptor before the block is transmitted. This process takes place in each of the two encoders, and the separate results of the processes are intertwined through an "interleaver." At the receiving (Forward) device, two alternating decoders reverse the interleaving process in order to view the separately encoded data blocks. The decoders then recalculate the descriptors, and compare those descriptors with the ones that were calculated and appended by the encoders. Each decoder then is in a position to make a decision about the integrity of each of the bits contained within each of the blocks. Using a technique known as Soft Input-Soft Output (SISO), the decoder also calculates a confidence level (i.e., likelihood) for each of the decisions (i.e., errored versus unerrored) it has made relative to each of the received bits in each of the blocks. This soft decision-making process is iterative, as the evaluation of each block benefits from the evaluation of each preceding block in a stream of blocks. Think of it as learning from past experience, which must of us consider to be a good thing. While this FEC technique is complex and expensive to implement, it is highly reliable. As the cost of implementing it in silicon (i.e., at the chip level) comes down, it will have increasing application in data networks which are error-prone and bandwidth-limited. For example, Turbo Coding is anticipated to have significant application in deep-space communications and wireless data communications (e.g., CDMA 2000 and UMTS). In both cases, the probability of an error is high. Further, the process of error correction through retransmission is inefficient as bandwidth is limited. In the case of deep space communications, FEC is the only logical alternative, as the time required to retransmit errored data would be excessive, to say the least. See also CDMA, Forward Error Correction, Hamming, SISO, and UMTS.

turbo FAT Turbo FAT is an index NetWare v2.2 creates to group all the FAT (File Allocation Table) entries corresponding to a file larger than 262,144KB. The first entry in the turbo FAT index table consists of the first FAT number of the file. The second entry consists of the second FAT number of the file, etc. The turbo FAT enables a large file to be accessed quickly.

turbocharging A phone fraud term. Turbocharging is the practice of increasing phone charges by adding on extra time onto each call in the hope that the customer won't notice.

TURN The Utilities Reform Network, formerly known as Toward Utility Rate Normalization. A non-profit consumer advocacy that represents the small customer in California, TURN characterizes itself as the only independent, statewide, consumer utility watchdog group. TURN represents residential and small business consumers on utility (e.g., telecommunications and electric power) before the California Public Utilities Commission (CPUC), the state legislature, and the courts. www.turn.org

turn-key See Turnkey System.

turnaround time The actual time required to reverse the direction of transmission from sender to receiver or vice versa when using a half-duplex circuit. The turnaround time is needed for line propagation effects, modem timing and computer reaction.

turnbuckle A device used for tightening the tension on guy wires.

turnkey system An entire phone or computer system with hardware and software assembled and installed by a vendor and sold as a total package. The term "turnkey" means the buyer is presented with the key to the thing he has just bought. He turns the key and the system will do everything it is supposed to do, including work. Most telephone systems and some computer systems are purchased turnkey. An integral part of a contract to buy a turnkey phone system is the terms and conditions for the acceptance of the system. Someone has to define what it means for the thing to work, what you expect from it – so you, the buyer, can formally accept the system and thus incur an obligation to pay for it. Defining Acceptance Conditions is no small task on bigger phone systems and more complex computer systems. My advice: always hold some money back as long as you can. This way your contractor will have some incentive to come back and fix what later turns out is not working.

turnover British and Australian term for sales – typically annual sales.

turnup When a circuit becomes live, it is turned up and working. Turnup is result of completing the installation of a circuit and making it available to the customer who requested it.

turret A very large key system for financial traders, emergency teams at nuclear power stations and others who need single phone button access to hundreds of people. By simply pushing one button, the user can dial one of hundreds of people. These buttons may be connected to tie lines, foreign exchange lines. They may even be DDD lines with autodial capability. Like all good key systems, the buttons have a lamping display which shows if the particular line is idle, busy, ringing, on hold, etc.

TUV Technischer Uberwachungs-Verein. TUV. A German electrical testing and certification organization similar to Underwriters Laboratories (UL). TUV certifies products to European safety standards.

Tuvalu This is an interesting story about the commercialization of the Internet. Tuvalu is a small island nation (constitutional monarchy) comprising nine remote coral atolls with a total land area of 10 square miles, north of Fiji, in the southwest Pacific. The highest point in Tuvalu is 4 meters above sea level. Tuvalu has a population of about 9,000. Its main crops are coconut, taro, pandanus fruit, and bananas. Its exports are postage stamps and copra (dried coconut meat). Virtually everything is imported. It is a very poor nation, with virtually no telephone service, virtually no utility service (e.g., power, server, water), virtually no medical care, and certainly no Internet access. Tuvalu is a very poor country, and it is expected to be totally under water within the next 50 years, due to the effects of global warming. Tuvalu has absolutely no need for its Internet TLD (Top Level Domain) of .tv. So, the government of Tuvalu on August 11, 1998, sold the rights to ".tv" to the .TV Corporation, a Canadian marketing firm which has established a master domain on the WWW (World Wide Web). Anybody can register a domain name under the .tv domain – for a price. The total sales price is based on the success of this domain, but revenues to Tuvalu are pegged at a minimum of US$50 million. How's that for e-commerce? See also TLD and Federated States of Micronesia.

TV Television. See HDTV and Television.

TVC See Trunk Verification by Customer.

TLVA TV Linux Alliance. The TVLA is a consortium formed in 2001 by Sun Microsystems, Motorola, Lineo and Liberate.

TVoIP Television over IP. Same as IPTV. See also channelthink.

TVOR Terminal VHF Omni Range.

TVRO Television Receive Only Earth Station. Earth station equipment that receives video signals from satellite or MDS-type transmissions. Such stations have only receive capability and need not be licensed by the FCC unless the owner wants protection from interference.

Authority for reception and use of material transmitted must be given by the sender.

TV/PC Bill Gates' new name for a TV with PC smarts.

TVM See Time-Varying Media.

TVM Object An SCSA definition. An encapsulation of an atomic piece of time-varying media. This encapsulation may be the data itself or a reference to the data.

tuxedo The tailless dinner jacket was invented in Tuxedo Park, New York. Thus it is called the "tuxedo dinner jacket" and is named after the town...not the other way around. So there.

TVS See Trunk Verification by Station.

TW Thermoplastic insulated building wire that is flame retardant and moisture resistant, 600xAFC in wet and dry locations.

TWAIN Technology Without An Interesting Name. A protocol for communication between software and image-acquisition devices, such as cameras and scanners. It is, in essence, a cross-platform application interface standard for image capturing. It allows you to bring images into imaging programs (like HiJaak Pro, Hotshot Graphics, DocuWare, Documagic) from your graphics hardware – for example, desktop scanners, hand-held scanners, slide scanners, frame grabbers and digital cameras. If your hardware is TWAIN compliant and if you have installed the correct driver, you should be able to use that imaging hardware with any TWAIN complaint application. The TWAIN protocol is the most popular protocol for imaging sources and has become an industry standard. Any application that supports TWAIN can communicate with any TWAIN-compliant imaging device. TWAIN is spearheaded by Hewlett-Packard, Logitech, Eastman Kodak, Aldus, Caere and other imaging hardware and software vendors. It was previously known as CLASP and "Direct Connect" during its development stage. Apparently the term TWAIN comes from "Toolkit Without An Interesting Name." Despite its silly name, it is a very serious standard. There are several key elements to TWAIN including:

- Application Layer This is the application that controls and uses the TWAIN resource.
- Protocol Layer This contains the TWAIN Source Manager (the code that communicates between the application and the Source).
- Acquisition Layer This is software that controls the image acquisition device. The application layer is developed by the device manufacturer. It can be thought of as a hardware device driver.
- Device Layer This is the physical device, such as a scanner.

TWAIN Working Group An industry organization dedicated to developing and advancing software standards for the imaging world. www.twain.org. See TWAIN.

TWAMP Two-Way Active Measurement Protocol, a protocol for measuring round-trip IP performance between any two TWAMP-enabled devices on a network.

Tweak Freak A computer techie obsessed with finding the root of all tech problems, regardless of the relevance. A tweak freak might spend hours trying to track down something that could instantly be fixed by reinstalling the software.

tween Someone between 8 and 14 in age.

tweener Refers to a proposed direct broadcast satellite (DBS) orbital slot that is 4.5 degrees from an existing DBS orbital location, instead of the 7 degrees of separation that is required today.

Twenty-three Skiddoo The famed New York expression, "Twenty-three skiddoo" came to be because the wind drafts created by the height of the skyscraper raised women petticoats, and constables had to "skiddoo" the men who came to peek! See Flatiron Building.

Twilight Zone Fans of the Twilight Zone (CBS 1959-64) think writer Rod Serling invented the term "Twilight Zone." As a matter of fact, so did Serling. He'd not heard anyone use it before, so he assumed he'd created it. However, after the hit TV show debuted in 1959, Serling was informed that Air Force pilots used the phrase to describe "a moment when a plane is coming down on approach and it cannot see the horizon."

twin cable A cable composed of two insulated conductors laid parallel and either attached to each other by the insulation or bound together with a common covering.

Twinax Twinaxial Cable made up of two central conducting leads of coaxial cable. See Twinaxial Cable.

Twinaxial Cable Two insulated conductors inside a common insulator, covered by a metallic shield, and enclosed in a cable sheath. Because it carries high frequencies, twinaxial cable is often used for data transmission and video applications, especially for cable television.

twinning (pronounced twin-ning.) The act of paralleling systems to work together. An example is connecting a wireless system to the same CO line as a key system, so the user can use either instrument to access the trunk.

twinplex A frequency-shift-keyed, carrier telegraphy system in which four unique tones (two pairs of tones) are transmitted over a single transmission channel (such as one twisted pair). One tone of each tone pair represents a "mark," and the other, a "space."

twist 1. A change, as a function of temperature, in the response characteristic of a transmission line.

2. The amplitude ratio of a pair of DTMF tones. Because of transmission and equipment variations, a pair of tones that originated equal in amplitude may arrive with a considerable difference in amplitude. In short, signals at different frequencies are transmitted with differing response by the transmission system. Twist usually refers to distortion of DTMF signals.

twisted pair 1. Two insulated copper wires twisted around each other to reduce induction (thus interference) from one wire to the other. The twists, or lays, are varied in length to reduce the potential for signal interference between pairs. Several sets of twisted pair wires may be enclosed in a single cable. In cables greater than 25 pairs, the twisted pairs are grouped and bound together in a common cable sheath. Twisted pair cable is the most common type of transmission media. It is the normal cabling from a central office to your home or office, or from your PBX to your office phone. Twisted pair wiring comes in various thicknesses. As a general rule, the thicker the cable is, the better the quality of the conversation and the longer cable can be and still get acceptable conversation quality. However, the thicker it is, the more it costs. Here's a historical and technical explanation from Ray Horak's best-selling book, Communications Systems & Networks:

Metallic wires were used almost exclusively in telecommunications networks for the first 80 years, certainly until the development of microwave and satellite radio communications systems. Initially, uninsulated iron telegraph wires were leased from Western Union for this purpose, although copper was soon found to be a much more appropriate medium. The early metallic electrical circuits were one-wire, supporting two-way communications with each telephone connected to ground in order to complete the circuit. In 1881, John J. Carty, a young American Bell technician and one of the original operators, suggested the use of a second wire to complete the circuit and, thereby, to avoid the emanation of electrical noise from the earth ground. This second conductor also supports common Central Office battery; as a result, your phone stills works when the lights go out. Twisted pair involves two copper conductors, which generally are solid core, although stranded wire is used occasionally in some applications. Each conductor is separately insulated by polyethylene, polyvinyl chloride, flouropolymer resin, Teflon, or some other low-smoke, fire retardant substance. The insulation separates the conductors, thereby avoiding shorting the electrical circuit which is accomplished by virtue of the two conductors, and serves to reduce electromagnetic emissions. Both conductors serve for signal transmission and reception. As each conductor carries a similar electrical signal, twisted pair is considered to be a (electrically) "balanced" medium. The twisting process involves the separately insulated conductors being twisted 900xAF at routine, specified intervals, hence the term twisted pair. This twisting process serves to improve the performance of the medium by containing the electromagnetic field within the pair. Thereby, the radiation of electromagnetic energy is reduced and the strength of the signal within the wire is improved over a distance. Clearly, this reduction of radiated energy also serves to minimize the impact on adjacent pairs in a multi-pair cable configuration, as the other conductors absorb that radiated electromagnetic energy much as an antenna would absorb a radio signal. This is especially important in high-bandwidth applications, as higher frequency signals tend to attenuate (lose power) more rapidly over distance. Additionally, the radiated electromagnetic field tends to be greater at higher frequencies, thereby impacting adjacent pairs to a greater extent. Generally speaking, the more twists/ft., the better the performance of the wire. Several sets of twisted pair wires may be enclosed in a single cable. In cables greater than 25 pairs, the twisted pairs are grouped into "binder groups," which are contained within a common cable sheath. Twisted pair cable is the most common type of transmission media. It is the normal cabling from a central office to your home or office, or from your PBX to your office phone. Twisted pair wiring comes in various thicknesses, or gauges. As a general rule, the thicker the conductor, the better the quality of the conversation and the longer cable can be and still get acceptable conversation quality. However, the thicker it is, the more it costs. Most twisted pair circuits are Unshielded Twisted Pair (UTP). UTP involves no special shielding-just simple insulation. Shielded Twisted Pair (STP) sometimes is used in high-noise environments, where the cable must be run in proximity to electric motors or other sources of ambient electromagnetic interference which can distort the signal. STP looks much like a coaxial cable, as the central conductors are insulated and then surrounded by an outer conductor (shield) of steel, copper alloy, or some other metal. The STP shield absorbs the ambient noise, and conducts it to ground, thereby protecting the center conductor. See also Attenuation, Cat 1-5, STP, and UTP.

2. Harry Newton and Gerry Friesen are commonly referred to as the twisted pair because their brains don't quite work the way they should.

Twisted-Pair Physical Media Dependent TP-PMD. Technology under review by the ANSI X3T9.5 working group that allows 100 Mb/s transmission over twisted-pair cable. Also referred to as CDDI or TPDDI.

Twisted-Pair Distributed Data Interface TP-DDI. Trademark of 3COM Corporation. See Twisted-pair Physical Media Dependent.

Twists Per Foot TPF. The number of times per foot that the two conductors are twisted around each other in a twisted pair cable system. The twist length, or lay length, is the distance between the twists. For example, a lay length of 3 inches refers to a pair with 4 twists per foot. The more twists per foot, the better the circuit performs, as the twisting process reduces the strength of the electromagnetic field radiated from the circuit. The less energy radiated from the circuit, the more energy remains within it, and the farther the signal will travel without requiring amplification (analog) or regeneration (digital) in order to remain intelligible.

twit A pregnant goldfish is called a twit. Stupid people are often called twits. I don't know why.

twitch game A computer or arcade game that's all hand-eye coordination and little brain. Similar to "thumb candy."

twitcher Twitchers are birdwatchers in England. In the U.S., they are called birders.

two dimensional coding A data compression scheme in facsimile transmission that uses the previous scan line as a reference when scanning a subsequent line. Because an image has a high degree of correlation vertically as well as horizontally, two-dimensional coding schemes work only with variable increments between one line and the next, permitting higher data compression. See One Dimensional Coding.

two electrode vacuum tube A vacuum tube having a hot cathode and a relatively cold anode, i.e., one with filament and plate only.

two hots in outlets In AC electrical power, more than one HOT conductor has been incorrectly connected to the terminals in the outlet being tested. Dangers include extreme fire hazard and/or major damage to equipment plugged into the outlet.

two out of five code A decimal code system in which each decimal digit is represented by five binary bits, two of which are ones and three are zeroes.

two out of three rule When determining state tax jurisdiction for the purpose of figuring phone bills, there are three locations to consider: originating station, destination station, and the location that the bill is sent to. If two out of three are the same, then that state receives the tax.

two party hold on console Allows an attendant to hold a call with both a calling and a called phone (or trunk) connected. Such a feature is required for activation of Attendant Lockout, Serial Call and Trunk-to-Trunk connections features.

two party station service PBX system with two internal phones, each with selective ringing. Resembles rural two party service of old.

two pilot regulation In FDM systems, the use of two pilot frequencies within a band so that the change in attenuation due to twist can be detected and compensated for by a regulator.

two pronged vampire Black transformer boxes plugged into AC outlets and attached to things like radios, and laptops are often called two pronged vampires because they often continuously draw power, remaining warm to the touch even when their device – the radio and the laptop – are turned off. According to the New York Times, such devices waste about 5% of the power in the U.S., and as much as 10% to 12% in Japan.

two stage shutter release A term used in digital photography. A 2-stage shutter release is commonly employed on current electronic cameras. When pressed half-way, the release activates the autofocus and the light meter of the camera, setting them so as to achieve correct focus and exposure. Holding the release at mid-course maintains the focusing point and the exposure parameters (AE Lock) and allows for re-composition if desired. A further press on the shutter release takes the picture.

two tier pricing A complex and now largely obsolete AT&T pricing plan which imposed two monthly "rate elements" on every hardware piece of an AT&T (now Lucent) telephone system. Tier A was a fixed rate, not subject to rate increases. It was fixed for a certain number of months, say 60. It was, allegedly, to pay for the system. At the end of the 60 months, Tier A disappeared, as though it were a full-payout lease and you now owned the equipment (which you do it.) Tier B is the second element in this pricing scheme. It covers maintenance, and it is subject to rate increases. Neither AT&T nor Lucent offer two-tier pricing any longer. Many two-tier contracts are now finding their Tier A payments ceasing. Once Tier A payments cease, the equipment still belongs to Lucent.

two tone key Same as frequency shift keying.

two tone keying In telegraphy systems, a system employing a transmission path composed of two channels in the same direction, one for transmitting the "space" binary modulation, the other for transmitting the "mark" of the same modulation; or that form of keying in which the modulating wave causes the carrier to be modulated with a single tone for the "marking" condition and modulated with a different single tone for the "spacing" condition.

two way alternate operation Transmission in one direction or the other but not in both simultaneously. Most often referred to as half-duplex transmission.

two way simultaneous operation Transmission and reception at the same time. More often referred to as full-duplex transmission.

two way splitting PBX feature. Allows a telephone user to jump back and forth between two calls. Try this: Someone calls you. You both decide you want to speak to a third person. You call that person and conference the three of you together. Then you decide you want to consult with one of the people confidentially. So you "split" one from the other and you speak to one. Then you swap back and forth between the two, speaking to one and then the other in complete privacy. It's easier to do this sort of complicated phone transaction on a phone with an LCD screen. Fortunately, these are becoming more common these days.

two way trade A call center term. A schedule trade in which both employees are working each other's schedules.

two way traffic A type of circuit operation that provides for both originating and terminating traffic.

two way trunk A trunk which can be seized from either end. Can be used to carry conversations into or out of a telephone system, i.e. most trunks. Some trunks are set up as one-way only. A classic one-way trunk is a IN-WATS line. It is designed to only receive calls.

two wire circuit A transmission circuit composed of two wires – signal and ground – used to both send and receive information. In contrast, a four wire circuit consists of two pairs. One pair is used to send. One pair is used to receive. All trunk circuits – long distance circuits – are four wire. A four wire circuit costs more but delivers better reception. All local loop circuits – those coming from a Class 5 central office to the subscriber's phone system – are two wire, unless you ask for a four-wire circuit and pay a little more.

two-phase commit A method used in transaction processing to ensure data is posted to shared databases correctly by dividing the writing of data into two steps. Each of the steps must receive a verification of completeness from the shared databases; otherwise, the transaction-processing system rolls back the transaction and tries again.

two-tier Internet The proposal by some telecommunications carriers that two tiers of service be offered on the Internet, namely "regular" and "premium."

Content/service providers and customers that want to use the higher-speed "premium" Internet would pay extra for this premium service, whereas content/service providers and customers on the "regular" Internet would pay only their usual Internet access rate. The proposed two-tier scenario has grown out of frustration by some telephone carriers (also called telephone companies) that Web-economy companies like Google and Yahoo! are using the telcos' networks to make piles of money, while the telcos, themselves, have been relegated to the low-margin role of mere bit-haulers, forced to upgrade their networks at their own expense to accommodate increasing traffic to/from Web-economy companies' websites. Some content providers are allegedly already paying the higher fees. It remains in doubt as to actually what they get for their extra monies.

TWP X.3 User ID Conversion Table.

TWT Traveling Wave Tube.

TWTA Traveling Wave Tube Amplifier.

TWX (Pronounced TWIX.) Teletype Writer eXchange Service. An automatic teletypewriter (i.e. telex-like) switching service where subscribers may dial any other subscriber and send and receive a message. Formerly owned by AT&T and sold to Western Union in 1972. It differed from Telex in that TWX used AT&T's normal long distance phone network, was thus more ubiquitous, was faster than Telex and was incompatible with Telex, which Western Union owned. However, Western Union, in a major accomplishment, got them to talk to each other.

TX 1. The designation of a copper RJ-45 connection for Fast Ethernet.
2. Transmit. See TX/RX for detail.

TX/RX Transmit/Receive. TX/RX is used to indicate the direction of traffic from the perspective of a device such as a microwave antenna or a router. As an example, a network accounting tool can provide you with traffic statistics over a dial-up modem connection between a network-based router and your client workstation. TX data would be downstream data transmitted from the network-based router to your client, and RX data would be upstream data received by the network-based router from your client. Specific TX/RX traffic measurements might include transmission rates as measured in bps (bits per second), total

frame/packet counts, and data (as opposed to signaling and control) frame/packet counts. TX/RX information might also include levels of signal loss, as measured in dB (deciBels), in each direction.

tycoon Tycoon comes from the Japanese for "Great Lord."

TYM2 Gateway Interface The gateway interface between two networks.

TYM2 Protocol The proprietary protocol used by the TYMNET network.

Tymnet Tymnet has been billed as one of the first public X.25 packet switched networks. Actually, it is not a true X.25 network but rather an X.25 compliant network using a proprietary OS (operating system). Protocol conversion between the Tymnet OS and X.25 is done at the edge of the network. The name "Tymnet" comes from the fact that this network, like all X.25 packet networks, was established to support timeshare applications. Tymnet was created by Tymshare, spun off as a separate company, then purchased by McDonnell Douglas in 1984. BT (British Telecom) acquired the network in November, 1989; MCI subsequently acquired BT North America, including the North American portion of Tymnet. And MCI was later acquired by Worldcom, which changed its name to MCI Worldcom, which then changed its name back to Worldcom. The X.25 services are still sold by WorldCom domestically in the United States as a bundled service named "Xstream" and sold internationally by Concert Communications (which was a combination of British Telecom and MCI, but then became BT and AT&T) as "Concert Packet Services." If any employee stayed through the various changes over the years, they deserve a major bravery medal. See also Time Sharing.

Type 1 Cable The IBM Cabling System specification for two-pair, 22 gauge, solid conductor cable protected with a braided wire shield. Tested to 16 Mbps, Type 1 is used between Token Ring MAUs and from the MAU to the wallplate.

Type 1 CLEC When CLECs have their own network in place, it is referred to as Type 1 Service. In areas where CLECs don't have their own network in place, they lease facilities from the ILEC in order to provide service; this is referred to as Type 2 Service. See CLEC.

Type 2 Cable The IBM Cabling System specification for a six-pair, 22 gauge shielded cable for voice transmission application. The six-pair version of Type 1 Cable, Type 2 also is tested to 16 Mbps. Typical application is for the two shielded pairs to be used for Token Ring or 10Base-T LANs, with the remaining four pairs being outside the shield and being used for voice transmission. Type 2 is a six-pair equivalent of Category 3 (Cat 3) cable.

Type 3 Cable IBM's term for telephone wire, Type 3 is single-pair UTP (Unshielded Twisted Pair) wire of 22 or 24 gauge, and involving a minimum of 2 twists per foot. It is used in applications such as 4 Mbps Token-Ring networks.

Type 5 Cable The IBM Cabling System specification for 62.5/125 micron multimode fiber optic cable in a two-pair configuration, which is the de facto standard for FDDI (Fiber Distributed Data Interface).

Type 6 Cable The IBM Cabling System specification for two-pair, stranded 26 gauge wire used in patch cable applications, as well for connecting LAN station adapters to wall plates. Type 6 is limited to a distance of 30 meters.

Type 8 Cable The IBM Cabling System specification for untwisted, shielded two-pair, 26 gauge wire. This flat, ribbon cable commonly is used under carpets.

Type 9 Cable The IBM Cabling System specification for two-pair, shielded, 26 gauge wire, which can be either stranded or solid core. Type 9 Cable accepts RJ-45 termination, and typically is used in connecting from the wall plate to the LAN station adapter.

Type 66 Punchdown Block A standard, solderless, punchdown terminal wiring block used today. Invented by Western Electric, now Lucent.

Type A Intelligent Network term describing IN (Intelligent Network) services invoked by, and affecting, a single user. Most of them can only be invoked during call setup or teardown.

type ahead Imagine a voice processing service. It says "punch in your zip code at the beep." If you are able to punch in your zip code before you hear the beep or before the talking stops, you have "type ahead." If you are unable to punch in your zip code before you hear the beep, you don't have "type ahead." Better interactive voice response systems have "type ahead."

type approval A concept in which a design is approved by an agency and all devices subsequently manufactured according to that design are automatically approved. An administrative procedure of technical tests and vetting applied to items of telecommunication equipment before they can be sold or connected to public network.

Type B Intelligent Network term describing IN (Intelligent Network) services invoked at any point by, and affecting directly, several users.

Type I PC Card The thinnest PCMCIA Card from factor at 3.3 mm thick. The Type I format is typically used for various memory enhancements, including RAM, Flash, OTP, SRAM, and EEPROM.

Type II PC Card A PCMCIA Card which is 5 mm thick. This card is typically used for I/O such as modem, LAN, and host communications.

Type III PC Card The thickest PCMCIA Card type at 10.5 mm thick, the Type III Format is primarily used for memory enhancements or I/O capabilities that require more space, such as rotating media and wireless communication devices.

Type of Service TOS. The header of an IPv4 (Internet Protocol version 4), the version currently deployed most widely, contains an eight-bit TOS field. That field can be used to identify to the various packet switches and routers in an IP-based network those packets which would like preferential treatment on a Class of Service (COS) basis. Unfortunately, most switches and routers, and consequently most IP-based networks, currently are unable to support differential levels of service.

typebar Linear type element in a printer containing the printable symbols.

typing reperforator Same as receive only typing reperforator.

typosquatting Operating a website whose domain name is a misspelled version of a well-known domain name, in order to exploit mistakes that users make when they mistype the latter. That exploitation might include selling products at cheaper than the main site. Or it may include enticing the visitor to buy something else. The administration's official web site is www.whitehouse.gov. There is a web site at www.whitehouse.com, which seems to be an glorified search engine for sponsored (i.e. paid) links. There is also a site called www.whitehouse.org that sells posters, t-shirts and other paraphenalia that make fun of the administration.

U U is 1.75" inches. Us (i.e. the plural of U) are used as a measurement for servers in standard 19 inch racks. So a 1U server is 1.75" high and 19" wide. A 2U server is 3.5 inches high. Most servers are 1U.

U band The optical band, or window, specified by the ITU-T at a wavelength between 1625nm and 1675nm (nanometers) for fiber optic transmission systems. See also C-Band, E-Band, L-Band, O-Band, and S-Band.

U interface An ISDN term. The reference point for a BRI (Basic Rate Interface) connection between a telephone company local loop and a customer premises. BRI is intended for consumer, SOHO (Small Office Home Office), and small business applications. The U Interface specifies a single-pair loop (physical two-wire twisted pair) over which a logical four-wire circuit is derived. The resulting digital local loop supports three full duplex channels – two B (Bearer) channels, and one D (Delta, or Data) channel. Each of the B channels provides 64 Kbps of bandwidth, and is designed to support actual user data payload. The D channel provides 16 Kbps of bandwidth, 9.6 Kbps of which can be used to support X.25 packet data traffic, and the balance of which is reserved for signaling and control purposes. There are two line coding techniques used in the U interface: 2B1Q is used in North America, and 4B3T in Europe. The U Interface is designed to work over a maximum distance of 18,000 feet, which addresses the vast majority of US local loops, with loss up to 42 dB (decibels). You screw the two "U" wires (local loop pair) coming in from your local ISDN CO into a black box about the size of desk printing calculator, called an NT-1. Out the side of the black box comes four wires, which are called the "S Bus." Onto these four wires you can attach, in a loop configuration (also called single bus), as many as eight ISDN terminals, telephones, fax machines, etc. See 2B1Q, 4B3T, and ISDN.

U Law Actually it's Mu Law, but the "Mu" symbol isn't available on conventional keyboards. Mu Law is a voice amplitude compression/expansion quasi-logarithmic curve, based on the approximation with 15 linear segments. Used for PCM encoding/decoding in North America. See Mu Law and PCM.

U plane The user plane within the ISDN protocol architecture; these protocols are for the transfer of information between user applications, such as digitized voice, video and data; user plane information may be carried transparently by the network or may be processed or manipulated (e.g. A- to u-law conversion).

U-Band See U Band.

U reference point In the U.S., the point that defines the line of demarcation between user-owned and supplier-owned Integrated Services Digital Network (ISDN) facilities.

U-C An ADSL term for the functional interface between the "U" (standard, two-wire, twisted pair local loop) and the POTS splitter at the Centralized (i.e., Central Office) end of the network. The functional equivalent of the U-C at the premise end of the network is known as the "U-R." The asymmetric nature of this technology requires that the "C" and "R" interfaces be distinguished. See also ADSL, Splitter and U-R.

u-Japan Stands for "Ubiquitous-net Japan." It is a government and industrial next-generation networking initiative in Japan to enable a society where everyone/everything is easily connected to the network anytime, anywhere and everywhere, thereby improving the quality of life and favorably impacting the economy. u-Japan builds on e-Japan. Government policymakers in Japan hope to achieve u-Japan by 2010.

U-NII Unlicensed-National Information Infrastructure. A group of three frequency bands, each of 100 MHz in the 5 GHz band, set aside by the FCC in January 1997 for support of a projected family of high-speed, low-power, wireless voice and data devices. Band 1: 5.15 - 5.25 GHz/200 mW EIRP max/Indoor use only; Band 2: 5.25-5.35 GHz/1 W EIRP max/campus applications; and Band 3: 5.725 - 5.825 GHz/4W EIRP max/local access. The U-NII band is being used now for wireless last mile access (as is ISM band) in a point-to-point and point-to-multipoint fashion. Multiple vendors are working to commercialize equipment for this purpose.

U-R An ADSL term for the functional interface between the "U" (standard, two-wire, twisted pair local loop) and the POTS splitter at the Remote (i.e., customer premise) end of the network. The functional equivalent of the U-R at the premise end of the network is known as the "U-C." The asymmetric nature of this technology requires that the "C" and "R" interfaces be distinguished. See also ADSL, Splitter and U-C.

U-space See rack unit.

UA 1. User Agent. An OSI application process that represents a human user or organization in the X.400 Message Handling System. A US creates, submits, and takes delivery of messages on the user's behalf. See also Browser Sniffing.

2. User Agent. An end system, or endpoint, that acts on behalf of a user. In SIP (Session Initiation Protocol), there are two types of UA: User Agent Clients (UACs), and User Agent Servers (UASs). The UAC initiates a request which is sent to a UAS, which then returns responses. See also SIP.

UAC User Agent Client. See UA.

UADSL Universal Asymmetric Digital Subscriber Line. A new standard from the telephone companies in order to provide faster access to the Internet for their subscribers and to compete against cable modems. See also Cable Modem and DOCSIS.

UART Universal Asynchronous Receiver/Transmitter. Also called a Shift Register. PCs have a serial port, which is used for bringing data into and out of the computer. The serial port is used for data movement on a channel which requires that one bit be sent (or received) after another, i.e. serially. The UART is a device, usually an integrated circuit chip that performs the parallel-to-serial conversion of digital data to be transmitted and the serial-to-parallel conversion of digital data that has been transmitted. The UART converts the incoming serial data from a modem (or whatever else is connected to the serial port) into the parallel form which your computer handles. UART also does the opposite. It converts the computer's parallel data into serial data suitable for asynchronous transmission on phone lines. UART chips control the serial port/s on personal computers. Now read the next definition. See also 16550, Interrupt, Interrupt Latency, Interrupt Overhead, Interrupt Request and UART Overrun.

UAS User Agent Server. See UA.

UASs A measure of performance. The duration is seconds for which the resource was unavailable.

UART Overrun UART overruns are errors received from a universal asynchronous receiver/transmitter when receiving equipment cannot match transmission speed. UART overrun occurs when the UART's receive buffer is not serviced quickly enough by the CPU, and the next incoming byte of data crashes into the previous byte. The previous byte is then lost, forcing the communications driver to report and error. Your communications software must then ask for a retransmission of the lost data. High interrupt overhead is the most common cause of a UART overrun. The easiest way of solving UART overrun is to get yourself a UART with a 16 byte buffer (like the 16550), not one of the old more typical one byte buffer UART. See 16550 and UART.

UAS UnAvailable Seconds. A count of the number of seconds that a circuit or path is unavailable.

UAT User Acceptance Testing.

UAWG The Universal ADSL Working Group (UAWG), was composed of about 250 leading companies in the PC, networking, and telecommunications industries, all of which worked to develop a set of contributions building on the ANSI T1.413 standard intended to create quick deployment and adoption of G.Lite ADSL. The group aimed to accelerate both full-rate and G.Lite ADSL deployments. The group foresaw G.Lite ADSL modems being a preferred PC modem technology by the year 2000. The effort was successful. On June 8, 1999 the UAWG disbanded, handing over its remaining work to the ADSL Forum. See ADSL and ADSL Forum.

UBE Unsolicited Bulk E-mail. See Spam Mail.

Ubiquity According to Alcatel, ubiquity is the name of a feature that allows a mobile worker's cellular phone to be integrated seamlessly with the corporate voice network. Ubiquity allows employees to have a single number and voice mailbox, increasing accessibility and eliminating the frustrating experience of callers having to try multiple numbers (desk phone, cell phone, etc.) to reach an employee. One-number ubiquity services increase convenience and save time in the following ways:

When users are off-site, they have access to the same features as if they were at the corporate location. Ubiquity allows mobile users' cell phones to be used as an extension to the corporate network, so they can receive calls and access all features from anywhere. For instance, a traveling employee can take advantage of the company's dial-by-name directory from his cell phone keypad. Ubiquity gives callers choices about how to reach an employee who is away from his desk. For example, an auto attendant allows callers to leave a voice message, have their call automatically forwarded to the employee's cell phone, or speak with a live attendant. The caller does not have to hang up and try another number, and the employee does not have to give out his direct cellular phone number - he decides when and how he can be reached. Ubiquity services are fully configurable based on user needs. For example, if a user does not want to be disturbed by incoming calls to her cell phone, she can forward calls to her voicemail or an attendant. If a user is out of range or not available, an auto attendant allows callers to chose whether they'd like to leave a message or talk to an operator. Another convenience provided by ubiquity services is a single voice mailbox on the corporate network. In other words, if a call is routed to an employee's cell phone but is unanswered, the call is "pulled back" to the corporate network so the caller can record a voicemail message. Employees have access to all their messages in one place.

UBR Undefined Bit Rate or Unspecified Bit Rate. Traffic class defined by the ATM Forum. UBR is an ATM service category which does not specify traffic related service guarantees. Specifically, UBR does not include the notion of a per-connection negotiated bandwidth. No numerical commitments are made with respect to the cell loss ratio experienced by a UBR connection, or as to the cell transfer delay experienced by cells on the connection.

UC 1. Universal Controller.

2. Unit Controller.

3. Unified Communications. See also Unified Messaging.

UCA Utility Communications Architecture. An architecture for networks used to monitor and control electric power distribution systems.

UCAID University Corporation for Advanced Internet Development. Here's the explanation: Internet2 is the next generation Internet, replacing the current Internet exclusively for the use of member universities, Internet2 is a UCAID project. As a result of what they saw as the deteriorating performance of the Internet, 34 U.S. universities announced in October, 1996 the formation of Internet 2. Subsequently, the central goals of the project were

adopted as part of the Clinton administration's Next Generation Initiative (NGI). This second version of the Internet is a collaboration of the National Science Foundation (NSF), the U.S. Department of Energy, over 110 research universities, and a small number of private businesses. Each participating university has committed at least $500,000 to fund the project. Intended to serve as a private Internet for the exclusive use of its member organizations, it will be separate from the traditional Internet. The network eventually will operate over fiber optic transmission facilities at speeds of up to 2.4 Gbps (SONET OC-48), although current speeds of connection are at 155 Mbps (OC-3) and 622 Mbps (OC-12), but going higher to OC-48. Internet2 will connect through gigiPOPs, switches with throughput in the range of billions of packets per second, and will run the IPv6 protocol. www.internet2.edu. See also Internet.

UCC Uniform Commercial Code.

UCC Request Upstream Channel Change Request. A request transmitted by a CMTS to cause a CM to change the upstream channel on which it is transmitting.

UCD 1. Uniform Call Distributor. A device for allocating incoming calls to a bunch of people. Less full-featured than an Automatic Call Distributor. For a bigger explanation see Uniform Call Distributor.

2. Upstream channel descriptor. The MAC management message used to communicate the characteristics of the upstream physical layer to the CMs.

3. Urine Collection Device. Initials used in the original Mercury Space Program. NASA didn't have one ready for Alan Shepard's first suborbital flight - with predictable results.

UCS 1. Uplink Control System. Software used to support the secure delivery of digitally compressed services.

2. Universal Character Set. A standard coding scheme developed in 1993, jointly by the ISO and IEC, and specified in ISO 10646. UCS-2, also known as BMP (Basic Multilingual Plane) is a 16-bit scheme which merged with Unicode to form one standard character set. UCS-4 is a 32-bit variant, which is conceptual only. See also BMP and Unicode.

UCT Universal Coordinated Time. See ZULU Time.

UDC Connector These connectors are used to terminate 2-pair STP cable. UDC connectors form a hermaphroditic connection, meaning that there is no jack (female end) or plug (male end).

UDDI Universal Description, Discovery and Integration. UDDI is one of the key technical underpinnings of e-commerce. UDDI, which was developed by Ariba, IBM, Intel, Microsoft and SAP, is a set of specifications established for building online registries of companies and the goods and services they provide. UDDI is a kind of electronic directory for businesses that want to locate customers and suppliers through the Internet. The registries should include not just product data and contact information, but the technical details necessary to connect to a given supplier's own e-commerce systems. See Business to Business and Enterprise Application Integration.

UDF Universal Disk Format. See OSTA.

UDI Unrestricted Digital Information.

UDK A dumb GTE abbreviation, for Universal Dialing Keyset, a key pad that is switchable for either TONE or PULSE dialing. Outside GTE's private world, a keyset would mean a KEY TELEPHONE, not part of a phone.

UDLC 1. Universal Digital Loop Carrier

2. Univac Data Link Control. Sperry-Univac (now UNISYS) version of a bit-oriented computer protocol based on the ITU-T's HDLC.

UDMA Ultra Direct Memory Access. A high speed version of DMA. See DMA.

UDOP The ultimate dumb, open programmable (UDOP) switch built from multi-vendor SC-based products. A term coined by Dialogic.

UDP User Datagram Protocol. User Data Protocol is part of the TCP/IP protocol suite. It was created to provide a way for applications to access the connectionless features of IP. UDP provides for exchange of datagrams without acknowledgements or guaranteed delivery. This protocol is normally bundled with IP-layer software. UDP is a transport layer (layer 4 of the OSI reference model), connectionless mode protocol, providing a (potentially unreliable, unsequenced, and/or duplicated) datagram mode of communication for delivery of packets to a remote or local user. See also CLTP.

UDP/IP User Datagram Protocol/Internet Protocol. See UDP.

UDPU Universal Data Patch Unit.

UDSL Unidirectional HDSL (High-bit-rate Digital Subscriber Line). A variation on the HDSL theme proposed by a small group of companies in Europe. See HDSL and xDSL.

UDWDM Ultra-Dense Wavelength Division Multiplexing. Identified as the next generation of DWDM, UDWDM is planned to support as many as 400 wavelengths per optical fiber. See DWDM and WDM for detailed discussions of the technology.

UE User equipment, such as a mobile phone.

UEM Universal Equipment Module. A Nortel Networks' acronym. The basic unit of Meridian 1 PBX modular packaging. A UEM is a self-contained hardware cabinet housing a card cage, with a power supply, backplane, and circuit cards. If the UEM has the card cage for an AEM installed, it functions as an AEM.

UF Thermoplastic underground feeder and branch circuit cable.

UFD A USB Flash Drive. Usually the thing is half half the length of Sharpie. It has a male USB plug and flash memory. The USB flash drive is typically used for copying data, photos, images or video from one PC or laptop to another. Often the user of the Flash Drive puts it into his or her PC before traveling, and copies the data he or she will need, e.g. a PowerPoint presentation. If you're traveling, the easiest way to get your data printed is via a UFD at a local Kinkos or Staples. You'll plug it into one of their computers, wait a few minutes, pay them an extraordinary amount of money and, bingo, you have your documents printed out.

UFGATE A program which enables a FIDO compatible bulletin board system to exchange UUCP mail with UUCP sites.

UFSS Upstream Failed Signal State.

UG UnderGround.

UHF Ultra High Frequency. Portion of the electromagnetic spectrum ranging from about 300 MHz to about 3 GHz. The frequency ban includes television and cellular radio frequencies.

UI 1. User Interface, as in GUI, or Graphical User Interface.

2. UNIX International is a consortium of computer hardware and software vendors which is interested in the development of open software standards for the UNIX industry. Prominent members include AT&T, Sun, UNISYS and Fujitsu.

UIFN Universal International Freephone Number. In early June 1996, the ITU-T released the E.169 standard, along with the revision of E.152. This standard allows International Freephone Service (IFS) customers to be allocated a unique Universal International Freephone Number (UIFN) which will remain the same throughout the world, regardless of country or telecommunications carrier. "Freephone," the generic name for what we call "InWATS" in North America, is a service which permits the cost of a telephone call to be charged to the called party, rather than the calling party. In North America, "Freephone" numbers are in the area codes 800, 888, and 877 (in order of introduction). According to the ITU-T, a UIFN is composed of an international prefix (e.g., 011 in the U.S.), a three-digit country code (800) for global service application, and an 8-digit Global Subscriber Number (GSN). The resulting eleven-digit fixed format is "+ 800 XXXX XXXX," with "+" representing the international access code, and "X" being any number 0-9. If available, companies can choose the digits they wish and embed existing Freephone numbers into the available number space. For example, the North American InWATS number 1-800-HNEWTON might translate into the UIFN "+ 800 HNEWTON1." (These are not my real numbers, so please don't dial them. You'll cost someone else some time and money.) See also 800 Service, IFS, and ITFS.

UIPRN Universal International Premium Rate Numbers. The ITU (International Telecommunications Union) describes UIPRN as "The International Premium Rate Service (IPRS) enables an Information Service Provider (ISP) in a country to be assigned one or more Universal International Premium Rate Numbers (UIPRN), which allow IPRS callers to access information and other services provided by the ISP. For these calls, callers are charged at a premium rate. Detailed charging and accounting principles are defined in the D-Series Recommendations. In some cases, callers may need prior subscription to IPRS with the IPRS originating ROA and/or with the ISP. ITU-T Recommendation E.155 defines the service description of, and the procedures for, the implementation, operation and management of the International Premium Rate Service (IPRS), which is provided on a managed basis by a Recognized Operating Agency (ROA) in the country of an information service provider, in conjunction with an ROA(s) in the country of the caller. Through the availability of IPRS, a wide range of products offered by information service providers in one country can be made available to callers in another country. Examples of such products could include:

Access to recorded information services (speech, facsimile or data);
Access to interactive services (speech, facsimile or data);
Access to promotions, competitions and opinion surveys.

The Telecommunication Standardization Bureau (TSB) of the ITU has been requested to perform the task of Registrar for UIPRNs, responsible for processing registration requests and assignment of the SN (Subscriber Number) portion of the UIPRN in accordance with ITU-T (new) Recommendation E.169.2 and Recommendation E.155 (International Premium Rate Service)." Who can apply for a UIPRN, according to the ITU? Their answer: "Only an international telecommunication Recognized Operating Agency (ROA), as defined in the ITU Constitution, can submit an application on behalf of the IPRS (International Premium Rate

Service) customer in accordance with the Recommendations E.169.2 and E.155. UIPRNs will only be assigned to IPRS customers who will use the IPRS service between two or more countries. In other words IPRS customers offering a service that is only accessed from within a single national, or integrated numbering plan, will not be considered eligible. To qualify for a UIPRN number, the number has to be in-service between two or more countries within 180 days from the date of its reservation with the UIPRN Registrar or it will be cancelled. A UIPRN is composed of a three-digit country code for a global service application (979), followed by a single-digit charging/accounting indicator (CI), and an eight-digit Subscriber Number (SN), resulting in a twelve-digit fixed format (CC+CI+SN). For example, an IPRS customer's UIPRN could be +979 1 12345678; where + is the international prefix, 979 is the "country code" for a UIPRN, 1 is the charging/accounting indicator, and 12345678 is the IPRS customer's SN. For more, www.itu.int/ITU-T/universalnumbers/uiprn/.

UIS Universal Information Services. AT&T's vision of a single fully-integrated, user-defined digital network with a universal port of entry. Very similar to ISDN, now aggressively adopted by AT&T.

UKERNA UK Education and Research Networking Association. See JANET, Super-JANET.

UL Underwriters Laboratories, a privately-owned company that charges manufacturers a stiff fee to make sure their products meet various safety standards, some of which UL itself develops. A UL label on a product has a very specific message. It says the product conforms to the safety standards – nothing more. It does not affirm that the product will work. Among other things, UL tests inside wire and cable products to ensure that they conform to the National Electric Code (NEC). Specifically, such tests are conducted to determine compliance with UL 444, which addresses flame test procedures. Other organizations, such as the ETL Testing Laboratories, also test for NEC compliance, and also use UL 444. www.ul.com. See also National Electric Code, UL Approved and UL Cable Certification Program.

UL 1449 A method of rating and approving surge suppressors. This Underwriters Laboratories measurement is important as it tells if you're buying a true surge suppressor or just an extension cord. A UL label on a product has to be in-service between two or more countries within measures how much voltage actually reaches the attached equipment after going through the surge suppressor. It's on a scale from about 330 volts to 6,000 volts. The lower the rating, the greater the protection. Decent surge suppressors tend to be rated around 400 volts for the basic units and 340 for the advanced and superior models. In short, check for UL 1449 rating on your surge arrestor before you buy it.

UL 1459 Effective 7/1/91, telephone equipment manufacturers will be required to provide protection from current overloads and power line crosses on equipment systems. Equipment systems covered under this listing requirement include single- and multi-line telephones, PBXs, key systems and central office switches. In general, the UL 1459 requirements apply to any location where wires enter a building from the public network, as well as in most IROB (In Range Out of Building) situations. See also NEC Requirements and Underwriters Laboratories.

UL 1863 This requirement covers miscellaneous accessories intended to be electrically connected to the telecommunications network. The listing requirement applies to components that comprise the premises communications wiring system from the point of demarcation up to and including the final outlet providing modular plug and jack connection (or equivalent). Requirements are listed under Communication Circuit Accessories, UL 1863. Listing equipment for all other equipment will be covered under UL 1459, effective July 1, 1991. See also NEC Requirements and Underwriters Laboratories..

UL 444 The Underwriters Laboratories flame test procedure that is used to test inside wire and cable systems to ensure that they are in compliance with the National Electric Code (NEC). The NEC contains provisions that require that voice and data cable systems are low-smoke and fire-retardant. See also National Electric Code.

UL 497 & 497A According to the National Electrical Code, primary and secondary protection systems that will be used on a telephone circuit must be listed for that purpose. The listing requirements are UL 497 for primary protection systems and UL 497A for secondary protection systems. See also NEC Requirements and Underwriters Laboratories.

UL Approved Tested and approved by the Underwriters Laboratories. The Underwriters Laboratories, Inc. was established by the National Board of Fire Underwriters to test equipment affecting insurance risks of fire and safety. Most phone systems are tested and approved. Most of the testing focuses on the power supply feeding the phone system. The power supply is that little black box that plugs into the AC wall outlet at one end, takes 120 volt AC and converts it to low voltage DC power that the phone system typically runs on. If the power supply tests OK, then that's usually sufficient UL testing. For it is the power supply – and what happens to the commercial AC power that feeds into the power supply – that determines the potential fire hazard of your phone system. After many fire deaths in recent years, most local communities are a lot more concerned about UL Approval of installed telephone equipment. Fire departments have been known to zealously enforce these rules. In addition to the UL approval, the other major fire concern is the use of proper wire in new building construction, with especial emphasis on teflon-covered cable in plenum ceilings. See also UL, an entry which talks about UL's expanding certification business. See also UL Cable Certification Program.

UL cable certification program Underwriters Laboratories, in conjunction with companies such as Anixter, has developed a Data-Transmission Performance-Level Marking Program that covers UL Listed communications cable or power-limited circuit cable. The UL program identifies five levels of performance. UL evaluates cable samples to all of the tests required for each level. Only Levels II through V require testing.

LEVEL I: Level I cable performance is intended for basic communications and power-limited circuit cable. There are no performance criteria for cable at this level.

LEVEL II: Level II cable performance requirements are similar to those for Type 3 cable (multi-pair communications cable) of the IBM Cabling System Technical Interface Specification (GA27-3773-1). These requirements apply to both shielded and unshielded cable constructions. Level II covers cable with two to 25 pair twisted pairs of conductors.

LEVEL III: Level III data cable complies with the transmission requirements in the Electrical Industries Association/Telecommunications Wiring Standard for Horizontal Unshielded Twisted-Pair (UTP) Cable and with the requirements for Category 3 in the proposed EIA/TIA Technical Systems Bulletin PN-2841. These requirements apply to both shielded and unshielded cables.

LEVEL IV: Level IV cable complies with the requirements in the proposed National Manufacturer Association (NEMA) Standard for Low-Loss Premises Telecommunications Cable. Level IV requirements are similar to Category 4 requirements of the proposed Electronic Industries Association / Telecommunication Industry Association (EIA/TIA) Technical Systems Bulletin PN-2841. These requirements apply to both shielded and unshielded cable constructions.

LEVEL V: Level V cable complies with the requirements in the proposed National Electrical Manufacturers Association (NEMA) Standard for Low-Loss Extended-Frequency Premises Telecommunications Cable. Level V requirements are similar to Category 5 requirements of the proposed Electronic Industries Association/Telecommunication Industry Association (EIA/TIA) Technical Systems Bulletin PN-2841. These requirements apply to both shielded and unshielded cable constructions.

UL evaluates communications and data transmission cable to one of two UL Safety Standards: UL 444, the Standard for Safety for Communications Cable; and UL 13, the Standard for Safety for Power-Limited Circuit Cable.

ULA User-based LAN access control (ULA) is a new technology that redefines network admission and access. Made possible by a new breed of high-performance ASICs, emerging ULA-capable LAN security systems sit in a network at the user-access layer or at an aggregation layer, and inspect every packet on every port for security policy compliance and malware.

The technology lets an administrator identify who is using a network, where and how he logged on, what resources he can access, and whether the LAN is still secure and malware-free once the user is admitted. It also provides automatic quarantine mechanisms to isolate problem users immediately, and to dynamically change from normal to quarantine policy when malware is detected. In effect, it works to create a personal DMZ for every user on every port.

User-based LAN access control operates transparently to end users, while providing powerful security safeguards for network or security administrators. ULA-capable systems are flexible enough to offer several mechanisms for authentication, and smart enough to understand the concepts of user identity and security policies associated with each user. For example, when a user plugs his laptop in to a network, he authenticates via 802.1X, or a captive portal Web logon page, and the system immediately applies that user's security policies to all applications and network services he accesses.

This security technology also integrates with existing authentication databases to identify user-group memberships. A system matches group memberships from an existing RADIUS or LDAP database to security policies that will be applied on a LAN access port. This group-based approach guarantees scalability across a corporation, because policies are defined one time and all group members automatically inherit the policies at logon. When a user is transient (say, a contractor working on the latest SAP upgrade), policies travel with him wherever he connects to the network.

When malware, such as worms, or other inappropriate behavior is identified, the ULA system automatically applies quarantine policies to that user only. Before the availability of user-based LAN access control in LANs, the only way to protect against malware was to assign users to a quarantined virtual LAN. This is akin to throwing influenza sufferers in with malaria patients. With user-based LAN access control, a device completely isolates infected users with fully stateful firewall policies, while allowing access, for example, only to remediation servers.

Simultaneously, the device alerts the network administrator about the incident. Event details include who is responsible, what they did, where they are located and what's been done about it. Compare this with today's practice of combing through router and switch logs, or Address Resolution Protocol tables, looking for which media access control address caused the problem and to which port they are connected.

Finally, ULA systems offer robust security audit benefits. When a network understands user identity and what resources people have access to, and it has the capacity to log this network activity, compliance audits become much simpler.

User-based LAN access control enables companies to implement simple, identity-based security policy provisioning, rapid security incident resolution and complete compliance audit trails.

ULANA Unified Local Area Network Architecture. An ongoing U.S. Air Force project aimed at creating a series of interconnected local area networks using Transmission Control Protocol/Internet Protocol (TCP/IP) as the unifying transport layer.

ULH Ultra-Long-Haul. While the term is imprecise, ULH refers to a fiber optic circuit that extends a very long distance through the use of various repeater and/or amplifier technologies.

ULP Upper Layer Protocol. In the context of the OSI Reference Model, a ULP is an application-level protocol which may reside at a higher layer than something like ATM (Layers 1 and 2) or TCP (Layer 3).

ULS User Location Service (ULS) provides a mechanism for users of Microsoft's NetMeeting to locate other people on the Internet, even if their Internet addresses change. A sample of the ULS can be found at http://uls.microsoft.com/.

ULSI Ultra Large Scale Integration, the technique of putting millions of transistors on a single integrated circuit. Compare with LSI (Large Scale Integration) and VLSI (Very Large Scale Integration).

Ultimedia IBM's product in multimedia – combining sound, motion video, photographic imagery, graphics, text and touch into a unified, natural interface representing, in IBM's words, the ultimate in multimedia solutions. Ultimedia supports both Ultimotion and Indeo video. Ultimedia was coined in the Spring of 1992.

Ultimotion IBM's video compression algorithm. Although IBM supports Indeo video technology in OS/2 and Windows systems, IBM feels several OS/2 vertical applications are adequately served by the Ultimotion algorithm. Ultimotion does not offer software scalable playback or single step compression. See Ultimedia.

utelco A telecommunications company that is owned by or affiliated with an electric or gas distribution or transmission utility. A utelco may deliver services such as dial-up, broadband Internet services, satellite TV, etc. It may also provide web hosting, email and paging.

UTF-8 UCS Transformation Format 8. UCS (Universal Character Set) is a standard coding scheme developed in 1993, jointly by the ISO and IEC. UCS-2, also known as BMP (Basic Multilingual Plane) is a 16-bit scheme which merged with Unicode to form one standard character set. UTF-8 is a transformation format which is multibyte Unicode, and ASCII-compatible. UTF-8 resolves issues of Unicode coding, which may contain odd character strings that are unreadable by certain UNIX tools. UTF-8 is defined in the IETF's RFC 2279. See also ASCII, UCS, and Unicode.

UTM Unified Threat Management. A new industry created to sell seminars on.

UTRA UMTS Terrestrial Radio Access. See also UMTS.

Ultra Direct Memory Access UDMA is a high speed version of DMA. See DMA.

Ultra-dense Wavelength Division Multiplexing See UDWDM.

Ultra Hi-Res Ultra high resolution. Properly speaking, the term should be for monitors with resolutions of 1,200 x 800, 1,024 x 1024 or better, but it is sometimes used to describe monitors with 800 x 600 resolution and above.

Ultra High Frequency Frequencies from 300 MHz to 3000 MHz.

Ultra High Vacuum/Chemical Vapor Deposition UHV/CVD. The process, developed by an IBM research team led by Bernard Meyerson, by which germanium is added to silicon to form silicon germanium (SiGe).

Ultra PC Ultra Physical Contact is a term used to refer to connectors which provide back reflections (also referred to as Return Loss) of greater than 50 dB and typical insertion losses of 0.2 dB. Ultra PC connectors are available in singlemode D4-SC-FC and STr connectors.

ultrawideband UWB, also called digital pulse is a wireless technology for transmitting digital data over a wide swath of the radio frequency spectrum with very low power. It can carry signals through doors and other obstacles that tend to reflect signals. It is a less expensive alternative to so-called third-generation wireless more limited bandwidths using higher power. It can carry large amounts of data and is used for ground-penetrating radar and radio locations systems. Previously known variously as "baseband," "carrier-free," or "impulse," UWB transmission systems are typically centered within the 200 MHz to 4 GHz band and emit an average radiated power of approximately 125uW (i.e. very little). According to my techie friends, an ultrawideband system is a radiator with intentional emissions that have a fractional bandwidth greater than or equal to 25%, with the fractional bandwidth being the 20 dB bandwidth divided by the center frequency. UWB systems are characterized by their low probability of intercept and detection (LPI/D), multipath immunity, high data throughput, and precision ranging and localization. UWB is used for penetrating thick bodies, such as the ground or the walls of a building. UWB technology allows a radar system to detect buried objects such as plastic gas pipes or reveal hidden flaws in roads, bridges, or airport runways. UWB will provide law enforcement officers with a means of covert communication and to provide radar systems that will enable fire and rescue personnel to find persons inside damaged, burning, or smoke filled buildings. UWB also is used in specialized ground-penetrating radar systems to find human bodies buried in shallow graves under parking lots. UWB radios have been developed that achieve non line-of-sight communications through the use of surface or ground wave propagation. Operating at frequencies well below 100 MHz, these radios support voice and data at rates up to 128 Kbps with an operational range of 1-5 miles with intervening foliage, buildings and hills. Also based on UWB technology, electronic license plates have been developed with built-in collision avoidance radar. Much of the early work on UWB was performed under classified U.S. government restrictions, and much of that body of knowledge remains classified. For a comparison of wireless standards, see Wireless.

ultrasonic bonding The use of ultrasonic energy and pressure to join two materials.

ultraviolet That portion of the electromagnetic spectrum in which the wavelength is just below the visible spectrum, extending from approximately 4 nanometers to approximately 400 nanometers. Some scientists place the lower limit at values between 1 and 40 nanometers, 1 nm being the upper wavelength limit of X-rays. The 400-nm limit is the lowest visible frequency, namely violet. "Light" in the ultraviolet spectrum is used for erasing EPROMS.

ultraviolet fiber Special fiber which extends the usable range into the UV region of the spectrum.

Ultrawide Band Radio UWB. Also known as Digital Pulse Wireless. The new technology of ultrawide band radio uses a digital transmission consisting of small on-off bursts of energy at extremely low power but over the entire radio spectrum. According to the New York Times, "by precisely timing the pulses within accuracies of up to a trillionth of a second, the designers of ultrawide band radio systems are able to create low-power communications systems that are almost impossible to jam, tend to penetrate physical obstacles easily and are almost invulnerable to eavesdropping. Police officers could use such a system "to see through" walls and doors to detect the location of people. According to the New York Times, the most promising application for ultrawide band radio might eventually be an alternative to today's wireless office network technologies that are limited in speed. "Because of its design, ultrawide band advocates," according to the Times, say the technology has the potential to deliver vastly higher amounts of data because a large number of transmitters could broadcast simultaneously in close proximity without interfering with one another. See also Bluetooth.

UM 1. Micron – one millionth of a meter.
2. Unified Messaging. See Unified Messaging.
3. Unit Manager.

UMA 1. An acronym for Upper Memory Area. See Upper Memory Area.
2. Unlicensed Mobile Access. A set of mobile technologies that enable a cellular subscriber with a dual-mode mobile phone to seamlessly roam between a wireless LAN or Wi-Fi network and a cellular network. In the local wireless network environment, the dual-mode handset uses the network's unlicensed spectrum and a tunneling arrangement via the network's Internet connection to connect to the cellular network. This dual-mode approach improves mobility, enhances coverage, and lowers service costs. Sometimes called generic access.

UMB An acronym for Upper Memory Block, an area of upper memory (the area between 640KB and 1MB of RAM) in an MS-DOS PC that has been remapped with usable RAM. This allows device drivers and TSRs to be loaded high, into the UMB and out of conventional memory. See Upper Memory Area.

umbrella An absolute must for Outside Plant Technicians, as many jobs require sitting in the hot sun for hours on end. Also, not a bad idea for those nasty rainy days either. Commercial grade umbrellas work best. This definition contributed by a long-term, talented outside plant technician.

umbilicoplasty Plastic surgery performed on the navel, usually for cosmetic reasons.

UME UNI Management Entity: The software residing in the ATM devices at each end of the UNI circuit that implements the management interface to the ATM network.

UMIG Universal Messaging Interoperability Group. This group is an offshoot of AMIS (Audio Messaging Interchange Specification). The UMIG is being technically facilitated (their words) by the Information Industry Association in Washington, DC. The UMIG has two top priorities: First, to foster the development of "universal messaging," which entails integrating platforms supporting different types of messaging, such as voice processing, electronic mail and facsimile messaging to allow users to move easily among the three media. The second priority entails working towards standardized addressing and directory schemes that make it easy and intuitive for users to message one another.

UML Unified Modeling Language is the lingua franca for software architects who design complex computing projects.

UMPC See Putts Law.

UMTS Universal Mobile Telecommunication System. A set of third-generation (3G) access technologies for cellular networks that supports packet-switched services; circuit-switched services; improved frequency utilization; improved quality of service; faster speeds; and always-on voice, video and data for mobile users. UMTS builds on and provides full backwards-compatibility with GSM (Global System for Mobile Communication) and GPRS (General Packet Radio Service). UMTS supports data transfer rates of up 2 Mbps for users traveling slower than 6 miles per hour (10 kilometers per hour), 384 Kbps for users traveling over 6 MPH (10 KPH), and 144 Kbps for users traveling over 75 MPH (120 KPH). The slower speeds for faster traveling users are due to the overhead associated with maintaining mobile connectivity. The plan is that UMTS, once fully implemented, will allow mobile voice and data users to maintain constant connectivity, regardless where they travel. As of 2006, UMTS was deployed by 147 mobile network operators in 64 countries. See also 3G, GSM, and IMT-2000.

UMTS TD-CDMA The TD-CDMA standard is the TD (Time Division) side (the other is FDD- Frequency Division Duplex) of the Universal Mobile Telecommunication System Standard (UMTS), a key member of the "global family" of third generation (3G) mobile technologies identified by the International Telecommunications Union (ITU). Among other things, UMTS is an evolutionary step for operators of GSM networks, currently representing a customer base of more than 850 million end users in 195 countries and representing over 70% of today's digital wireless market. TD-CDMA uses time division (TD) verses frequency division duplex (FDD) to separate the uplink and downlink traffic, allowing the standard to be deployed in a single band of spectrum as well as better match the asymmetric nature of IP traffic.

UMTS Terrestrial Radio Access UTRA. UMTS terrestrial (in contrast to non-terrestrial, or satellite) radio access standards address radio parameters including radio frequency, channel spacing, modulation techniques, and protocols which form the communications link between the mobile station and the base station. The access techniques include FDD (Frequency Division Duplex) and TDD (Time Division Duplex). See also FDD, TDD, and UMTS.

UN Industry jargon for UNreachable, an unsuccessful call where the agent is unable to speak to the contact or decision maker.

UN/EDIFACT The worldwide organization responsible for defining, coordinating and integrating EDI (Electronic Data Interchange) standards under the auspices of the United Nations.

unassigned cell An ATM filler cell used to occupy available bandwidth when there are no assigned (user-generated) cells to send. Unassigned cells carry a PCI/VCI (Protocol Control Information/Virtual Channel Identifier) value of 0/0. Unassigned cells are discarded at the ATM Layer. They can be replaced by assigned cells, as required, during the process of cell multiplexing. See also ATM Layer, ATM Protocol Reference Model, PCI, VCI, and Idle Cell.

unattended Equipment working without a human attendant or operator. There are pros and cons to operators. On the pro side, they offer a personalized service that's absolutely critical to customer goodwill. On the con side, they can be slow and cumbersome. They can be very irksome when you know you could do that task yourself, but have to wait for the operator. Some companies have only one main number. Some companies use a main number and DID – on their Centrex and their PBX. Some companies use an automated attendant and an operator. There's more flexibility with DID and a main number, or Centrex

DID and a main number. Customers without knowledge dial the main number. Customers with knowledge can dial direct DID numbers. See also Automated Attendants.

unattended call Calls placed by a computerized dialing system in anticipation of an agent being available to answer the call. A called party is detected answering the phone and no agent is available to serve the call. The system hangs up on the party so as not to create any greater nuisance than has already occurred. The telemarketing industry does not believe that an unattended call can be queued for the next available agent.

unattended operation Transmission automatically controlled; not required a human operator to function.

unbalanced Unbalanced refers to the lack of electrical balance between conductors that comprise a circuit. Twisted pair circuits are balanced, which is to say that the signals across both conductors are electrically similar, although they intentionally are out of phase in order to minimize crosstalk. Coaxial cables, on the other hand, are intentionally unbalanced, as the inner conductor carries positive and negative voltages while the outer conductor is maintained at zero voltage. See also Unbalanced Line.

unbalanced line An unbalanced line is an electrical circuit comprising two conductors that do not carry equivalent electrical charges with respect to ground. Some lines are intentionally unbalanced, while others must be balanced to work correctly. Coaxial cables, for example, are intentionally unbalanced. The center coax conductor carries the information to be transferred, and carries an electrical charge; the outer conductor, or shield, is maintained at zero voltage. Twisted pair circuits, on the other hand, must be balanced. In a balanced line, both conductors carry equivalent electrical charges, which are exactly opposite in polarity. That minimizes the potential for crosstalk.

A telephone circuit in which the voltages on the two conductors are not equal with respect to ground. Unbalanced lines give poor phone service. Lines can become unbalanced when they come from the central office or when they are in the PBX or the on site phone system. Problems can and should be repaired for decent quality results. Coaxial cables are unbalanced, as the center conductor carries an electrical charge, but the outer conductor, or shield, is maintained at zero voltage.

unbundled 1. Services, programs, software and training sold separately from the hardware.

2. Services and products leased by local phone companies as a result of the Telecommunications Act of 1996. For a much better explanation, see Unbundled Network Element.

unbundled access ability Refers to the ability of a CLEC to access and use components of the ILEC's network (called network elements) to fill in the CLEC's networks.

Unbundled Network Element UNE (pronounced you nee). The Telecommunications Act of 1996 requires that the ILECs (Incumbent Local Exchange Carriers) unbundle their NEs (Network Elements) and make them available to the CLECs (Competitive LECs) on the basis of incremental cost. UNEs are defined as physical and functional elements of the network, e.g., NIDs (Network Interface Devices), local loops and subloops (portions of local loops), circuit-switching and switch ports, interoffice transmission facilities, signaling and call-related databases, OSSs (Operations Support Systems), operator services and directory assistance, and packet or data switching. When combined into a complete set in order to provide an end-to-end circuit, the UNEs constitute a UNE-P (UNE-Platform). Unbundled Network Elements is a term used in negotiations between a CLEC (Competitive local Exchange Carrier) and the ILEC (Incumbent Local Exchange Carrier) to describe the various network components that will be used or leased by the CLEC from the ILEC. These components include such things as the actual copper wire to the customers, fiber strands, and local switching. The CLEC will lease these UNEs with pricing based on the previously-signed Interconnection Agreement between the CLEC and the ILEC. Typically, a CLEC will colocate a switch at the ILEC's wire center, then pay for the "unbundled" local loop to make a connection to the customer. Alternately, a CLEC might lease both an unbundled local loop and an unbundled switch, and make a connection to their network at the LEC's switch. See CLEC, ILEC, the Telecommunications Act of 1996, UNE Rate and UNE-P.

unbundled services A CFRS service option in which the customer provides and services some of their own equipment at each site. Less expensive than Bundled service.

unbundling Here's an FCC defintion: Unbundling is the term used to describe the access provided by local exchange carriers so that other service providers may buy or lease portions of its network elements, such as interconnection loops to serve subscribers.

UNC Universal naming convention. See also UNC NAMES.

UNC Names Filenames or other resources names that begin with the string \, indicating that they exist on a remote computer.

uncontrolled terminal A user terminal that is on line at all times and that does not contain the logic that would allow it to be polled, called, or otherwise controlled by the device to which it is connected.

under mouse arrest Getting busted for violating an online service's rules of conduct. "Sorry I couldn't get back to you. AOL put me under mouse arrest."

underfill A condition for launching light into a fiber in which not all the modes that the fiber can support are excited (i.e. turned on).

underfloor duct method A floor distribution method using a series of metal distribution channels, often embedded in concrete, for placing cables. This method uses one or two levels depending on the complexity of the system. Sometimes referred to as underfloor raceways. See also Raceways Method.

underflow 1. In computing, a condition occurring when a machine calculation produces a non-zero result that is smaller than the smallest non-zero quantity that the machine's storage unit is capable of storing or representing.

2. In the transmission sense, when the input signal is operating slightly slower than the synchronized output and extra data must be periodically inserted. Generally, the last frame is repeated.

underground Cable installed in buried conduit. Does not typically include cables buried directly in the ground.

underground plant A term used to describe the network of splicing chambers, connecting sections of conduit and the cables which run though them.

underlap In facsimile, a defect that occurs when the width of the scanning line is less than the scanning pitch.

underload syndrome When you're bored, have few challenges or stmulus at work, you'll likely to get sick and/or depressed. This is now called underload syndrome.

underlying carrier A common carrier providing facilities to another common carrier which then provides services to end users.

underrun A network error indicating that buffer checks show the buffer as empty. Underruns shouldn't happen in a well managed network. An underrun is often a synchronization problem.

understaffing limit A call center term. The percentage by which you'll allow the scheduling process to fall short of the required staffing level in any period. This typically provides more economical coverage during the least-busy periods of the day.

Underwriters Laboratories, Inc. A non-profit laboratory which examines and tests devices, materials and systems for safety, not for satisfactory operation. See UL for a longer explanation.

undesired signal Any signal that tends to produce degradation in the operation of equipment or systems.

undetected error ratio The ratio of the number of bits, unit elements, characters, or blocks incorrectly received and undetected, to the total number of bits, unit elements, characters, or blocks sent.

undirected pickup A phone system feature. Undirected Pickup lets you pickup any call ringing at any extension in the pickup group in which your extension is a member. The pickup groups are pre-programmed in the switch.

undisturbed day A day in which the sunspot activity or ionospheric disturbance does not interfere with radio communications.

UNE (pronounced you nee). Unbundled Network Element. The Telecommunications Act of 1996 requires that the ILECs (Incumbent Local Exchange Carriers) unbundle their NEs (Network Elements), which must be made available to the CLECs (Competitive LECs) on the basis of incremental cost. This means that CLECs will pay the additional costs the ILECs incur in making these facilities available. the words "incremental cost" are meant to signal to the ILECs that they are not to inflate the price of these facilities by adding overhead costs (e.g. the salary of the ILEC's people in charge of investor relations). UNEs are defined as physical and functional elements of the network, e.g., NIDs (Network Interface Devices), local loops, switch ports, and dedicated and common transport facilities. When combined into a complete set in order to provide an end-to-end circuit, the UNEs constitute a UNE-P (UNE-Platform). Unbundled Network Elements is a term used in negotiations between a CLEC (Competitive local Exchange Carrier) and the ILEC (Incumbent Local Exchange Carrier) to describe the various network components that will be used or leased by the CLEC from the ILEC. These components include such things as the actual copper wire to the customers, fiber strands, and local switching. The CLEC will lease these UNEs with pricing based on the previously-signed Interconnection Agreement between the CLEC and the ILEC. Typically, a CLEC will colocate a switch at the ILEC's wirecenter, then pay for the "unbundled" local loop to make a connection to the customer. Alternately, a CLEC might lease both an unbundled local loop and an unbundled switch, and make a connection to their network at the LEC's switch. See CLEC, ILEC, the Telecommunications Act of 1996, UNE Rate and UNE-P.

UNE Rate The fee, set by state regulators, that an ILEC charges a CLEC to unbundle network elements as part of making the local exchange market competitive. Rebundling is the process of putting UNEs back together by a CLEC to become part of a competitive service offering by him to a customer.

UNE-L A common strategy used by facilities-based CLECs. A CLEC owns the local switch and leases the local loop from the ILEC. This is more capital intensive than UNE-P.

UNE-P Unbundled Network Element-Platform. See UNE.

Unequal Access Refers to long distance phone companies who do not take advantage of Judge Harold Greene's Equal Access divestiture provisions. Rather than a carrier selection code, unequal access carriers require you to dial a local seven digit number and punch in an authorization code. If the carrier elected to pay for Equal Access, you would just dial directly the same 10 digits you do today, and your local telephone company would give your billing number to your long distance company.

unerase A command for getting back files you've accidentally erased. See MS-DOS.

unfinished business See Last Piece of Unfinished Business.

ungrounded Not connected to ground. PBXs, key systems and other phone systems will not work well when not connected to a solid ground because they have no place to send high voltage spikes (static electricity, lightning strikes, etc.) Improper grounding is probably the most common cause of phone system faults. Our feeling: the better the ground, the better the phone system performance. One way of grounding is the third wire of an electrical outlet. This may be OK if you check where that wire is ultimately connected to. You can ground to the metal cold water pipe. But that may connect to a plastic PVC pipe one floor below. Best to check. A ground ultimately ending firmly routed a dozen feet below the ground is best.

UNH IOL University of New Hampshire Interoperability Lab. A testing organization affiliated with the Research Computing Center of the University of New Hampshire which tests FDDI products for vendor interoperability.

unhave Unhave is a verb which 19th century telegraphers used to say/send to mean that they didn't have something. This definition contributed by Jim Seymour.

UNI User Network Interface. Specifications for the procedures and protocols between user equipment and either an ATM or Frame Relay network. The UNI is the physical, electrical and functional demarcation point between the user and the public network service provider. By way of example, the Frame Relay UNI involves both the user's FRAD (Frame Relay Access Device) and the carrier's FRND (Frame Relay Network Device) across a dedicated link. The ATM (Asynchronous Transfer Mode) UNI was developed and is promoted by the ATM Forum; the Frame Relay UNI, by the Frame Relay Forum.

UNI A User Network Interface A. A B-ISDN term for a SONET OC-3 link from the network to the premise, operating at 155 Mbps.

UNI B User Network Interface B. A B-ISDN term for a SONET OC-12 link from the network to the premise, operating at 622 Mbps.

UNI Interface See UNI.

uni-minutes According top Faith Popcorn writing in "Dictionary of the Future," telecommunications companies will create "universal billing minutes" that are bought in advance, and can be used for landline or wireless calls.

UNIBOL A UNIX version of COBOL.

Unicast The communication from one device to another device over a network. In other words, a point-to-point communication. When you're Web browsing on the Internet or sending and receiving email, you are unicasting. Unicast communications (also called point-to-point communications) are sent between one network endpoint to another. An example of unicast communication is an email message. When a user sends an email to one recipient, his or her email client addresses and sends one message. If a user sends the same email to 10 recipients at once, her email client sends a separate copy of the message to each recipient. Unicasting is efficient for certain types of communications such as email and Web browsing; however, when multiple destinations require the same data, unicasting can be resource and bandwidth intensive. Imagine the following scenario: five end stations request a particular video stream. To unicast this data, the source creates five separate video streams. The transmission uses five times the amount of bandwidth required by one stream to traverse the network backbone. Five viewers may not use too much bandwidth, but consider the bandwidth used to unicast video to 500 or even 5,000 recipients. Unicasting (one to one communication) is often defined in comparison to its alternatives, multicasting (one to many), broadcasting (one to all). See Multicast and Broadcast.

unicasting 1. Communicating from one device to another. In contrast, multicasting sends one stream of information to many. See Unicast.

2. As an ATM term, it is the transmit operation of a single PDU by a source interface where the PDU reaches a single destination.

unicode Unicode is a 16 bit system for encoding letters and characters of all the world's languages. At 16 bits it can encode 65,536 characters. That's two raised to the 16th

power. Work it out: Multiply 2 x 2 x 2 x 2 x 2 x 2 x 2 x 2 x 2 x 2 x 2 x 2 x 2 x 2 x 2 x 2. Sixteen-bit characters (like Unicode) are also called Wide Characters. The first 128 codes of Unicode are identical to ASCII. Just add another zero byte to each ASCII character to convert to Unicode. Unicode contains over 20,000 Han characters, which are used to represent whole words or concepts in Chinese, Japanese and Korean. Unicode was developed by the Unicode Inc. consortium as a standard to replace the various proprietary 16-bit coding techniques which comprised two 8-bit bytes linked together. At the same time that Unicode was being developed, another standard was being developed jointly by the ISO and IEC. In 1992, the two coding schemes were linked to become what is known as both Unicode and BMP. See also BMP.

unidirectional The transmission of information in one direction only.

unidirectional bus A distribution conductor or set of conductors that can transfer information in one direction only.

Unidirectional Path Switched Ring UPSR. A SONET transport method in which working traffic is transmitted in one direction. UPSR is preferred for interconnected rings with numerous signals crossing the rings.

unified messaging Also called Integrated Messaging, universal messaging and unified communications (UC). You walk into your office in the morning. You turn on your PC and load up your messaging software, e.g. Microsoft Outlook. That's the software you typically use to receive and send emails. Only today, you notice that instead of seeing only emails awaiting your reading pleasure, you also see faxes and voice mails received by your telephone system. You can seem them all in one list. You can sort them by when you received them, or whom they're from or how big they are. You can click on your email and fax messages and read them on screen. You can click on your voice mail messages and hear them through your computer's speakers. Some unified messaging systems also allow you to call in and have your phone system read you your email messages, using text to speech, and, of course, listen to your voice mail messages over the phone or dialing in from afar. What's happened is that your company has acquired a server (big computer) whose job is to collect all your mail from its various places and consolidate them into one inbox. It may collect your email from various POP3 email servers (some distant and some local), from your fax server and from your voice mail server, which will be attached to your company's PBX telephone system. Once collected, it simply "serves" these messages up to you when you log in. See Integrated Messaging.

unified text messaging Under GSM cell phone systems (the ones common throughout the world) it's easy to send a message from one phone to another, irrespective of which carrier is providing wireless service – so long as it's standard GSM. In the United States, where nothing is standard, most wireless carriers let their subscribers send short messages to and from phones served by that carrier. But sending messages from my phone to a phone on a different network is not easy nor even possible in most situations. If there comes a time when the U.S. carriers set up a system to allow short message service, instant messaging, or just plain messaging between cell phones from different carriers in the U.S., it will be known a unified text messaging.

unified voice Unified voice is a bundled service that is provided via a T-1 line. It is designed to provide line side business telephone features similar to a LEC (Local Exchange Carrier) – Hunting, call forwarding, voice mail, call waiting, call blocking and conferencing. The typical UV customer will not have a PBX but may have a key system at his offices. In the old days, they used to call unified voice Centrex, with the difference that unified voice also uses the Internet, where Centrex never did.

Uniform Call Distributor A device for distributing many incoming calls uniformly among a group of people (typically called "agents" because of the early use of these machines by the airline, hotel and car reservation industry). These days the term Uniform Call Distributor is falling into disrepute as the newer term, Automatic Call Distributor comes in. According to incoming call experts, a Uniform Call Distributor is generally less "intelligent," and therefore less costly than an ACD. A UCD will distribute calls following a predetermined logic, for example "top down" or "round robin." It will not typically pay any heed to real-time traffic load, or which agent has been busiest or idle the longest. Also, a UCD's management reports tend to be rudimentary, consisting of simple pegs counts, as opposed to an ACD, which can produce reports on the productivity of agents.

Uniform Call Distributor UCD. A device located at the telephone office or in a PABX that distributes incoming calls evenly among individuals; called a "call sequencer" in some non-Bell LECs.

uniform encoding An analog-to-digital conversion process in which, except for the highest and lowest quantization steps, all of the quantization subrange values are equal.

uniform linear array An antenna composed of a relatively large number of usually identical elements arranged in a single line or in a plane with uniform spacing and usually with a uniform feed system.

uniform numbering plan A uniform seven-digit number assignment made to each phone in a private corporate network. Such a plan allows routing of calls to distant phones from any on-net telephone without any differences in the dialed number. Without a uniform numbering plan, you would dial your boss in New York differently if you were in the company's Chicago office and differently again if you were in your company's San Francisco office. With a uniform numbering plan, it would be the same from all locations. The nation's long distance network has, obviously, a uniform numbering plan.

Uniform Resource Identifier URIs have been known by many names: WWW addresses, Universal Document Identifiers, Universal Resource Identifiers, and finally the combination of Uniform Resource Locators (URL) and Names (URN). As far as HTTP is concerned, Uniform Resource Identifiers are simply formatted strings which identify – via name, location, or any other characteristic – a resource on the Internet.

Uniform Resource Locator URL. An Internet term. A standardized way of accessing various resources on the World Wide Web. In more technical terms, a URL is a string expression that can represent any resource on the Internet or local TC/IP system. The standard convention for a URL is as follows:

   method://host_spec {port} {path} {file} {misc}

   Here's an example of a URL. http://www.harrynewton.com. Typing those letters into your browser brings you to the opening screen – or home page – of my web site. In general http:// can be safely omitted with most browsers and you'll still get to the site. See URL for a detailed explanation. See also Uniform Resource Identifier.

Uniform Service Order Code See USOC.

Uniform System Of Accounts USOA. Part 3 of the FCC rules and regulations which prescribes names and numbers of accounts and describes the content of each account and gives rules for keeping records.

uniform-spectrum random noise The laboratory name for "white noise," a test signal made of noise that is constant in its power for every unit of bandwidth; used to test the crosstalk characteristics of multichannel analog transmission systems.

UNII See U-NII above.

Unimodem Unimodem, the "Universal Modem Driver" for Windows 95 and now Windows NT Server 4.0 and Windows NT Workstation 4.0, is both a TAPI service provider and a VCOMM device driver. It translates TAPI (Windows Telephony API) function calls into AT commands to configure, dial, and answer modems. See AT COMMAND SET and UNIMODEM V. See the following for Unimodem specifics: http://207.68.137.34/ntserver/communications/unimodem.htm

Unimodem V Unimodem stands for Universal Modem Driver. Unimodem V is Unimodem updated for voice. The V stands for voice, not five. It now replaces Unimodem. Unimodem stands for Universal Modem Driver. It is part of Windows 95 and Windows NT Server 4.0 and Windows NT Workstation 4.0. It is both a TAPI service provider and a VCOMM device driver. It translates TAPI (Windows Telephony API) function calls into AT commands to configure, dial, and answer modems. Unimodem V is the universal modem driver and telephony service provider for the Windows operating system. Included in Unimodem V are the features requested most often by users to support voice modems, including wave playback and record to/from the phone line, wave playback and record to/from the handset, and support for speakerphones, caller I.D., distinctive ringing and call forwarding. Unimodem now supports the most popular voice modems on the market. For Unimodem/V specifics: http://207.68.137.34/corpinfo/press/1995/95dec/unimdmpr.htm

uninstalled Euphemism for being fired. Heard on the voicemail of a Vice President at a downsizing computer firm: "You have reached the number of an uninstalled Vice President. Please dial our main number and ask the operator for assistance." See also Decruitment.

uninsured traffic Traffic within the excess rate (the difference between the insured rate and the maximum rate) for an ATM VCC. This traffic can be dropped by the network if congestion occurs.

unintelligent crosstalk Crosstalk giving rise to unintelligent signals.

Uninterruptible Power Supply UPS. A device providing a steady source of electric energy to a piece of equipment. A continuous on-line UPS is one in which the load is continually drawing power through the batteries, battery charger and invertor and not directly from the AC supply. A steady off-line UPS normally has the load connected to the AC supply. When the line is weak or down, it transfers the load without any user intervention. UPS are typically used to provide continuous power in case you lose commercial power. An UPS is typically a bank of wet cell batteries (similar to automobile batteries, but often much, much larger) engineered to power a phone system up to eight hours without any recharging. A UPS system can also include a gasoline-powered generator. And if the generator

works (make sure it has gas), you can power your phone system for much longer. According to Bell Labs, however, over 90% of all power outages last less than five minutes. The cost of Uninterruptible Power Supplies is typically a direct function of how large the battery/batteries are. The larger the batteries, the higher the cost. Many file servers on local area networks are also backed by UPSes. Many NetWare file servers, which are protected by a UPS, often are attached a printed circuit card inside the server. This card acts as an early warning system. When AC power drops, and the UPS takes over, it signals the file server through the card what has happened. The file server then will send a message to all the workstations on the network that the file server has lost AC power, is running on battery power, is running out and would everyone kindly log off the server. This protects the network.

unipolar signal A two-state signal where one of the states is represented by voltage or current and the other state is represented by no voltage or no current. The current flow can be in either direction.

unique addressing The addressing of a node by using the software-programmable address assigned to each one upon system initialization. For example, TELECONNECT's LAN has a "unique" addressing scheme. Each workstation is known by the operator's first name.

unique visitors See Hit and Hits.

unique word UW. A code word comprising a sequence of ones and zeros, used to establish frame synchronization in formatted digital transmission channels. Upon reception, a UW is sent to a UW correlator where it is correlated with a stored pattern of itself. The spike output of the UW correlator then indicates the occurrence of a unique word, which references the time of occurrence of the burst and marks the symbol time reference for decoding information in the message part of the burst.

uniserv See Universal Service.

Unisource A European provider of Virtual Network Services (VNS). Unisource was created in 1994 by three European carriers: PTT Telecom Netherlands, Swiss Telecom PTT, and Telia of Sweden. Unisource provides a wide range of voice, data, and Internet services in a number of countries through its equity partners, as well as through a group of distributors. Distributors included AT&T-Unisource Communications Services, WorldPartners, and Infonet. The three equity partners merged their international networks in June 1997 into AT&T-Unisource Carrier Services (AUCS), which operates a fiber optic backbone running ATM. AT&T (except for AT&T UK) pulled out of Unisource in 1999, citing conflicts between AUCS and its Concert venture with British Telecom, but the name AUCS stuck. See also VNS.

unit interval In a system using isochronous transmission, that interval of time such that the theoretical durations of the significant intervals of a signal are all whole multiples of this interval. The unit interval is the shortest time interval between two consecutive significant instants.

United States Telecom Association USTA. See the next definition.

United States Telephone Association USTA. Now called the United States Telecom Association. The largest trade association of telephone companies, with membership of over 1,200. USTA has its roots in the National Telephone Association, formed in 1897 to unite independent (non-Bell) telephone companies. Subsequently, the organization changed its name to the USITA (United States Independent Telephone Association). After the break-up of AT&T in 1984, the RBOCs (regional Bell Operating Companies) were admitted as members, and the name was changed to USTA, which now stands for United States Telecom Association. The organization, based in Washington, lobbies the FCC, Congress and other regulatory, legislative and judicial bodies to ensure that no regulations or legislation are passed to the detriment of its members. USTA also provides a little education for its members. USTA had a sister organization, the United States Telephone Suppliers Association, which is now merged with the EIA, forming the Telecom Association. www.usta.org. See the next definition.

United States Telephone Suppliers Association USTSA. An association of suppliers – manufacturers and wholesalers – which originally was a committee of the United States Telephone Association (USTA). USTSA merged with EIA/ITG (Electronic Industries Association/Information and Telecommunications Technologies Group) in 1988 to form the TIA (Telecommunications Industry Association), which operates under the umbrella of the EIA. www.tiaonline.org and www.eia.org See also EIA, TIA and USTA.

unity gain Refers to the balance between signal loss on a broadband network and signal gain through amplifiers.

universal access number A single number dialed from anywhere in the country which will route a customer to one or several locations for service, advice, etc. The definition varies depending on whose networking scheme you're dealing with.

universal addressing The addressing of a node by the use of the universal addresses which all nodes recognize.

universal messaging See Unified Messaging.

Universal ADSL See ADSL Lite.

Universal ADSL Working Group See UAWG.

universal agent A telephone agent who answers incoming calls and also makes outgoing calls. This duality feature may not seem worth of its inclusion in this dictionary. But the fact is that agents have largely been just "inbound" or just "outbound" – because managers felt that most agents were not capable of doing both. The skills were, allegedly, too different. Now the idea is to "empower" the agent with more flexibility and make them "universal," i.e. capable of being used for both inbound and outbound.

Universal Asynchronous Receiver-Transmitter UART. A device that converts outgoing parallel data from your computer to serial transmission and converts incoming serial data to parallel for reception. See UART for a bigger explanation.

Universal Character Set UCS. A standard coding scheme developed in 1993, jointly by the ISO and IEC. See UCS for more detail.

Universal Circuit Card See Universal Trunk Card.

universal device A SCSA device. A call processing device which has every conceivable resource for the handling of calls. The SCSA programming applies resources from many different physical devices to a call processing task. These then act as if they were a single universal device.

universal digital loop carrier A digital loop carrier system whose T-1 lines are powered by a digital channel bank, known as a central office terminal. The central office terminal interfaces with the switch (any type) through analog lines.

Universal Edge Server MediaGate, San Jose, CA, defines a "universal edge server" as a new breed of Remote Access Server (RAS) that offers telephony functionality combined with traditional data remote access capabilities. Such Universal Edge Server allows user to combine voice, email and pager communications into a single, secure message box, accessible via phone, fax, web browser or email client.

universal group A security or distribution group that can contain users, groups, and computers from any domain in its forest as members. Universal security groups can be granted rights and permissions on resources in any domain in the forest. Universal security groups are available only in native mode domains.

Universal International Freephone Number See UIFN

Universal International Premium Rate Numbers See UIPRN.

universal mailbox Allows a user of unified messaging services to have single access to all messages from internal and external electronic mail systems, fax systems and voice mail systems. A really neat idea given today's lack of standardization among electronic mail services.

universal name space The set of all unique object identifiers in a domain, network, enterprise, etc. Object naming standards and methods for locating and sending messages to mobile objects are required in large-scale object-oriented distributed-computing systems.

universal night answer A feature of telephone systems that permits any phone to pick up any incoming trunk call when the Attendant's console is unmanned (unpersonned?) and the phone system is set up (typically at the console) for "Night Answer."

universal pay phone Description for a coin-and-credit-card phone.

Universal Personal Telecommunications UPT. The ITU-T (formerly CCITT) term for the wired network architecture and capabilities to support PCS.

universal plug and play Microsoft's answer to Sun's Jini. See Jini.

universal ports A modern telephone system is typically an empty cabinet into which you slide printed circuit cards. Those cards have an edge connector and they slide into a connector at the rear of the cabinet. That connector connects via wires to other connectors in what is typically called the phone system's bus. In the old days, phone systems had dedicated slots – meaning you could only slide one type of printed circuit card into that particular slot. As phone systems got more advanced, they acquired "universal ports." Our definition of a universal port is that all the slots are totally flexible – namely that you can slide any trunk or phone card (either electronic or single line phone) into any slot in the phone system. The advantage of this is obviously a far more flexible phone system, able to accommodate lots of phones and few trunks or vice versa.

universal power supply A power supply which you can plug into electricity ranging from 100 volts to 240 volts AC. With a universal power supply (now standard with many laptops) you can travel the world, plugging yourself into virtually any power outlet and have your device work perfectly, without the need for a transformer. What you'll need, however, is a plug that converts the plug you have into the necessary plug for that country. Such a converter plug shouldn't cost you more than $2.

universal resource locator See Uniform Resource Locator.

universal sender Allows the dialed number to be sent out by the user.

Universal Serial Bus See USB.

universal service Milton Mueller of Syracuse University observes that universal service policy has gone through two generations: 1. First generation (1907 - 1965), 2. Second generation (1965 - present). The first generation was about connecting competing networks into "one system, one policy, universal service." This was the Theodore Vail vision. Vail was the first president of AT&T. The second generation started after World War II. As a response to political pressures, regulators decided to keep local rates low using the surplus generated by long distance. This system of cross-subsidies was threatened by the rise of long distance competition in the 1960s and early '70s. That was a shock to telephone monopolies because it meant that long distance rates were about to go down and therefore the subsidies were about to decline. Telephone companies tried to defend their monopoly privileges by claiming that cross subsidies were essential to the preservation of widespread household telephone penetration. This way, the term "universal service" was dusted off by the monopolies and got a new meaning: a telephone in every home (universal service as we understand it today).

Now for some history: The Communications Act of 1934 defined the nation's telecom goal as "To make available, so far as possible, to all the people of the United States a rapid, efficient Nationwide, and worldwide wire and radio communication service with adequate facilities at reasonable charges." The same act created the FCC (Federal Communications Commission), charging it with the responsibility to carry out this policy, as well to regulate the telecommunications industry, in general. Prior to the breakup of the Bell System in early 1984, AT&T and the BOCs (Bell Operating Companies) administered a fund through the "settlements" process, which essentially reimbursed the LECs (Local Exchange Carriers) for the use of their local networks in originating and terminating long distance calls. "High cost" (i.e., rural) LECs were compensated at very high levels, in recognition of the universal service policy. Since 1983, NECA (National Exchange Carrier Association) has been charged with this responsibility.

The Telecommunications Act of 1996 considerably expanded the definition of "universal service" to include "access to advanced telecommunications and information service...in all regions of the Nation, including low-income consumers and those in rural, insular, and high cost areas...reasonably comparable to those services...and those rates...for similar services in urban areas." The Act goes on to provide for discounts to elementary and secondary schools and classrooms, health care providers, and libraries. The Telecommunications Act of 1996 also directed that a special Universal Service Joint Board comprised of federal and state regulators and a consumer advocate develop recommendations for the FCC identifying services that will be supported by a federal universal service funding mechanism. Relying heavily on the Joint Board's November 8, 1996, Recommended Decision, the FCC released a Report and Order that undertakes to modernize universal service policy in an increasingly competitive marketplace and to fundamentally expand its applicability. See also FCC, High Cost, NECA, Separations and Settlements and Universal Service Fund. See also www.fcc. gov/ccb/universal_service/welcome.html and www.ntia.doc.gov/opadhome/uniserve/univweb.htm

Universal Service Administrative Company USAC. The not-for-profit corporation set up and operated by NECA (National Exchange Carriers Association) to administer the Universal Service Fund (USF). USAC accepts the Universal Service funds collected by the LECs (Local Exchange Carriers) and IXCs (IntereXchange Carriers), and disburses them in support of Universal Service (US). Universal Service first was stated as a national United States objective in the Communications Act of 1934, with the idea being that all U.S. citizens should have access to basic telephone service of good quality and at reasonable cost, regardless of where they live, and regardless of the underlying cost to the LEC of providing such service. That translated into a complex set of cross-subsidies administered by the FCC (Federal Communications Commission), which was formed by that very same act. Universal Service finally was codified (made into law) with the Telecommunications Act of 1996 (The Act), and the fund was formalized to support universal service in high-cost areas. The Act also led to the formation of USAC and its three divisions, The High Cost and Low Income Division, The Schools and Libraries Division (SLD) and the Rural Health Care Division (RHCD). The first division is responsible for disbursement of funds in connection with the original stated purposes of the USF. The last two divisions are responsible for the disbursement of funds to ensure that funds are made available to their respective constituencies to subsidize the cost of inside wiring, telecommunications services, and Internet access, with the ultimate objective being that of ensuring that all US citizens can participate in the Information Age. See also FCC, Telecommunications Act of 1996, Universal Service, and Universal Service Fund.

universal service charge A federal subsidy program that is apparently apprently being widely abused. The idea is that phone companies would add a figure to their bill, of apparently 6.9 percent. But some of the long distance companies are charging up to 12 percent. And the FCC seems to have little clue, according to testimony, where the extra money is going – though it's obviously adding to the carriers' income. See Universal Service Fund.

Universal Service Fund USF. Under the direction of the FCC, the National Exchange Carriers Association (NECA) administers the USF, which is a cost allocation mechanism designed to keep local exchange rates at reasonable levels, especially in "high cost" (i.e., rural) areas. The primary support mechanism that applies to telecom carriers is the high cost Support mechanism, of which there are five main components:

1) High-Cost Loop Support, Compensates rural operators for the cost to build local loops in areas where the cost exceeds 115% of the national average.

2) Local Switching Support. Interstate support to cover switching costs for companies that serve fewer than 50,000 customers.

3) Long-Term Support. Distribution of interstate access charges for rate-of-return regulated operators. A new universal support mechanism, Interstate Common Line Support, was created under the MAG plan (discussed below) to convert implicit subsidies to explicit subsidies.

4) High Cost Support for Non-Rural Carriers. Compensates rural operators for the cost to build the local loop in areas where the cost exceeds 135% of the national average.

5) Interstate Access Support. Distribution of interstate access charges for price cap regulated operators. The CALLS plan, which was passed in 2001, established an explicit support mechanism capped at $650 million for such charges.

NECA administers the program by collecting USF data, determining LEC eligibility, billing the IXCs (long distance phone companies), and distributing the payments. The original goal of the Universal Service Fund was to provide at least one access line for basic telephone service to every household in the U.S., and at a reasonable, subsidized cost. The fund gets money from a surcharge on phone lines, and uses those funds to offset operating costs of telcos in high-cost areas. While the concept of Universal Service was first stated in the Telecommunications Act of 1934, and while the USF has been in place for a great many years, it wasn't codified (enacted into law) until the Telecommunications Act of 1996 (The Act). The Act changed the underlying subsidy mechanism, authorizing the carriers to add a surcharge of up to five percent onto every telephone bill. The Act also established two new not-for-profit corporations, which now operate as divisions of the Universal Service Administrative Company (USAC), a not-for-profit corporation operated by NECA. The Schools and Libraries Corporation (SLC), which operates as the Schools and Libraries Division (SLD), is funded to the tune of up to $2.25 billion per year. The Rural Health Care Company (RHCC), which operates as the Rural Health Care Division (RHCD), is funded to the tune of $400 million per year. The funds are parceled out by these divisions to help fund necessary inside wiring, telecommunications services, and Internet access. See also FCC, NECA and Universal Service.

Universal Service Obligation See USO.

Universal Service Order Code USOC. The information in coded form for billing purposes use by the local telephone company pertaining to information on service and equipment (S&E) records. USOCs are not truly "universal," as operating companies can have wide differences in terminology. For example, a flat-rate single-party residence line is known variously in USOC terminology as 1FR and FR1.

Universal Service Plan See USP.

universal trunk cards Most PBXs have different circuit boards (or circuit cards - same thing) for combination trunks and for DID (direct inward dial) trunks. A Universal Circuit Board enables you to use the ports on the board for either combination trunks OR DID trunks. This capability makes the PBX more flexible. If you have spare ports on a trunk circuit board you may use them for either a combination or a DID type of trunk.

universal turret A very large key system for financial traders, emergency teams at nuclear power stations and others who need single phone button access to hundreds of people. By simply pushing one of the button in front of them, the user can dial one of hundreds of people. These buttons may be connected to tie lines, foreign exchange lines. They may even be DDD lines with autodial capability. Like all good key systems, the buttons have a lamping display which shows if the particular line is idle, busy, ringing, on hold, etc.

universal wall jack There's really no such animal. Every manufacturer of installation gadgetry is trying to propagate the idea that their jack is universal, when it really isn't. The "universal" wall jack we installed in our new offices is actually four jacks – 1. Four pairs for two PBX voice lines (one electronic two-pair phone and one tip and ring phone) and one spare. 2. One RS-232-C 12-conductor shielded cable for connecting to centralized printers, for connecting to a dataPBX and for permanent null-modem connection of computers. 3.

One for connecting to our high-speed, one megabit per second LAN, and 4. One spare twisted, shielded, stranded pair for a second LAN, or whatever comes along.

Universal Wireless Communications Consortium UWCC. A consortium of over 100 wireless telecommunications carriers and vendors dedicated to promoting and supporting TDMA (Time Division Multiple Access) cellular phone service. TDMA competes with FDMA (Frequency Division Multiple Access), the analog cellular approach, and CDMA (Code Division Multiple Access), another digital approach. Specifically, the UWCC promotes TDMA as a platform for developing and delivering enhanced personal communications features through the TIA's (Telecommunications Industry Association's) IS-136 (Interim Standard-136) for the TDMA digital air interface, and the IS-41 Wireless Intelligent Network (WIN) internetwork standard. The attempt is to bring together the three primary versions of TDMA RF (Radio Frequency) technology. GSM (Global System for Mobile Communications), is the European version. PHS (Personal Handyphone Service), also known as PDC (Personal Digital Cellular), is the Japanese version. IS-136 is the North American version, which allows AMPS (Advanced Mobile Phone System) to coexist with TDMA on the same network, sharing frequency bands and channels. GSM, PHS, and IS-136 are incompatible, meaning your phone which works on one of these networks won't work on the other networks. The UWCC also is involved heavily in the promotion of 3G (3rd Generation) TDMA-based wireless standards, including EDGE (Enhanced Data Service for GSM Evolution), GPRS (General Packet Radio Service), and UMTS (Universal Mobile Telecommunications System). See also CDMA, Cellular, EDGE, FDMA, GPRS, GSM, IS-41, IS-136, PHS, TDMA, UMTS, and WIN. www.uwcc.org.

universe A call center term. The total number of names to be attempted on an outbound call program.

UNIX An immensely powerful and complex operating system for computers for running data processing and for running telephone systems. UNIX was developed in 1969 by Ken Thompson, Dennis Ritchie and a few other programmers at Bell Telephone Laboratories. They had been working on the MULTICS (MULTiplexed Information and Computing System) project, a multiuser time-share system development project sponsored by MIT, Bell Labs and General Electric. Thompson then began to develop UNICS (a single-user variant on MULTICS) on an old DEC PDP-7 minicomputer he scrounged. According to legend, Thompson wanted to play Space War, an early computer game. In 1973, Thompson and Dennis Ritchie rewrote UNICS in the high-level C programming language, which is extremely portable, offering the incredible advantage of allowing programs to transition smoothly to newer computing platforms over time. At that point, UNICS became known as UNIX. The "X" in "UNIX" signifies that it can run on just about any computing platform. (Note: "X" is the generic "whatever.")

UNIX provides multi-tasking, multi-user capabilities that allow both multiple programs to be run simultaneously and multiple users to use a single computer. On a single-user system, such as MS-DOS, only one person at a time, on an individual task basis, can use a computer's files. programs, and other resources. Later Windows versions of MS-DOS added multi-tasking capabilities, of course. Today, the UNIX operating system is available on a wide range of hardware, from small personal computers to the most powerful mainframes, from a multitude of hardware and software vendors. UNIX is also a trademark of UNIX Systems Laboratories, Inc. which used to be owned by Novell Inc. but then was sold to SCO, Santa Cruz Operations. UNIX has given rise to a number of variants, including AIX from IBM, HP-UX from Hewlett-Lackard, LINUX, POSIX from the US government, Solaris from Sun, and ULTRIX from DEC. See also LINUX and MULTICS.

UNIX-To-UNIX Copy Program UUCP. A standard UNIX utility for exchanging information between two UNIX-based machines in a network. UUCP may also be referred to as the UNIX-TO-UNIX Communications Protocol and is widely used for electronic mail transfer.

unknown device A Bluetooth term. A device that is currently not connected with the (LocDev and the LocDev has not paired with it in the past. Also called a new device. No information about the device is stored (e.g., BD_ADDR, link key, or other information).

unlicensed bands There are two types of wireless communications devices. Those that require a licence from the Federal Communications Commission. And those that don't. Those that require a license run in a licensed communications band, a specific frequency. Those that don't require run in unlicensed communications bands can be plugged in and run – so long as they meet FCC rules for that communications band, i.e. that frequency. The FCC's rules loosely prohibit "harmful interference" of unlicensed devices, but devices that run in an unlicensed band are not guaranteed protection from interference.

unlicensed PCS Unlicensed PCS is the name for wireless frequency in the PCS band, which in the United States is 1.920 GHz - 1.930 GHz. The advantage of unlicensed PCS is that you can install a wireless telephone system in your company in this band without having to secure licenses from the Federal Communications Commission. Such systems are often called "business wireless." Such systems are typically all digital and often hang off a PBX. And the wireless phones often will have most of the features that an electronic phone wired to the PBX would have.

unlisted number There are various interpretations of what constitutes an "unlisted, an "unpublished" or a "non-published" phone number in North America. Some phone companies use these words interchangeably. Some don't. In California, Pacific Bell offers unpublished phone service. Your phone number is not listed in the paper phone directories, but is listed with dial up "Directory Assistance." Pacific Bell also has a more expensive service called "Unlisted Service." Here, your phone number is not included in the paper phone directories or given out to callers to Directory Assistance. Telephone companies have a service whereby you can leave a message for the owner of an unlisted number. "Please call me. You've won the lottery." The owner of the unlisted number then has the choice to return the call or not. He doesn't pay to receive this message. Some telephone companies confuse the definitions and some invent new ones. For example, some phone companies use the term "non-published" number. You won't find the number in a phone book or by calling Directory Assistance. Over 25% of many private phone numbers in major metropolitan areas are unlisted, unpublished or non-published – a "service" their subscribers pay extra for. To my simple brain, it's a lot easier to simply publish your name as "Apple Plumpudding." See Unpublished.

unloaded line A telephone line with its loading coils removed to increase the distance and speed with which data may be transmitted over the line. A fee is usually charged for removing the coils.

unlock See unlocking.

unlock code This is a three-digit number required to unlock a cellular phone when you have electronically locked it to prevent unauthorized use. You might lock it when you park your car with a built-in cellphone in a hotel. The factory default is 123. See also unlocking.

unlocking All GSM cell phones have a small metal card which you drop into it to get the phone working. The card is called a SIM card. GSM cell phone carriers often sell phones at loss, but make their money back by selling a year or two's service or a bundle of prepaid minutes. To make sure that you don't switch to another carrier, they lock the phone. They make the phone unusable to their competitors by encoding a SIM card restriction (subsidy lock) on the phone – which is a fancy name for a change in the phone's software. Unlocking is permanent removal (decoding) of the subsidy lock to enable the phone to accept any GSM provider's SIM card. I bought a GSM phone in Australia from Telstra, a local cellphone carrier for $99. It included about $10 of expensive pre-paid minutes. If I wanted to use a different carrier and insert their SIM card (say I wanted to use the phone in France and use a local carrier there), I would need to get the phone "unlocked." I was told that would cost me $50+ from Telstra. But any number of local independent cellphone retail stores could unlock it for $20. See SIM card.

UNMA Unified Network Management Architecture. AT&T's proprietary architecture for network management.

UNMR Universal Network Management Record.

unnumbered command In a data transmission, a command that does not contain sequence numbers in the control field.

unnumbered information frame U1 frame. A transmission frame generated by the High-Level Data Link Control (HDLC) data link protocol, where no flow control and no error control are implemented.

UnPBX An UnPBX is a server on a LAN, dedicated to communications. The UnPBX is comprised of four elements. First, it is one or several joined PCs; Second, it has boards inside the PC and software to run them; Third, it's joined to the same phone lines a PBX is joined to – analog POTS lines to digital T-1 lines; And fourth, it's joined to a local area network. Some of the cards that drop into an UnPBX are proprietary. Most are not. Most UnPBXs run on familiar operating systems – NT and Unix, etc. Most UnPBXs are open. They offer far more programming "hooks" than any other telephony device, including the "open" PBX. Most UnPBXs drop a phone on your desk via a tip and ring line. You speak on the phone. You can dial on the phone, or dial via the screen on your PC. Some UnPBXs zing your voice over ATM, Ethernet or the Internet. Some use the multimedia-equipped PC to do the dialing, speaking and phoning. You typically do everything else – from checking your messages to setting up conference calls – via software on your LAN-attached PC. Some UnPBXs have their own desktop software. Some use a browser – Netscape or Internet Explorer. Some people call the UnPBX a "total communications server." Here's what the UnPBX does:

1. Typically, the UnPBX is an office's phone system. It is a PBX in a PC. You dial your mother from your desk through it. It will switch incoming calls to your desktop. You can

call the guy in the next office on it. In short, an UnPBX has all the basic PBX functions. The UnPBX is your office's auto attendant / voice mail system. It answers incoming calls, gives out a message, listens for tones and transfers the call to you. It will alert you to the incoming call, prompt you for what it should do with the call – hold it, transfer it, conference it, dump it in voice mail, etc. If not answered, it will probably put the call in voice mail. It could also dial half a dozen outside numbers and chase you down. It can usually do solid IVR. Punch 1 to find out when we sent your order. Punch 2 to find out how much it cost. Give out and collect information from your customers at 3:00 AM. It is your office's fax and email server. It handles incoming faxes and emails. It tells you that you have mail. It lets you view your fax or email mail or listen to your voice mail. It is your one place for all your mail – fax, email, voice mail, video mail, etc. One screen showing all your mail. We call it unified messaging. The UnPBX makes you super organized. It is an ACD (automatic call distributor). It's not as sophisticated as what BIG airlines have. But it can come very close. An UnPBX's ACD can make your customers very, very happy. The UnPBX can be a predictive dialer. It can pump out telemarketing and dunning calls with the best of them.

Being open, being standards-based and being programmable, an UnPBX may be more or less than the list above. For example, it may also page you when you have a message. Or it may only page you when you have one from your largest customer.

We haven't scratched what's possible with an UnPBX. Today's UnPBX delivers many benefits. Among them:

1. It's cheaper. Don't be fooled by the sticker prices you see in Ed's Roundup. One communications system is a lot cheaper than today's multiple, cobbled-together systems.

2. It's easier to manage. It has one database. One place you assign phone extensions, email addresses, fax locations, email boxes, etc. One place to do billing for all these services. You'll pay the UnPBX off manyfold with just the administration savings.

3. It's much easier to use. You see the words, the icons, the images. Click. Drag and Drop. You now can use telephony features today's horrid telephones deny you.

4. You feel in control. You see your messages. They're all in one place. You can join voice mail messages to Excel spreadsheets and forward the multimedia message to your client. It sells.

5. You can have it your way. Write some software for your UnPBX. Get your reseller or system integrator to do it. Customize your phone system, your IVR system, your fax system, your ACD... all in one box. Gain a major competitive advantage.

unpublished phone number There are various interpretations of what constitutes an "unpublished" phone number in North America. Some phone companies use these words interchangeably. Some don't. In California, Pacific Bell offers unpublished phone service. Your phone number is not listed in the paper phone directories, but is listed with dial up "Directory Assistance." Pacific Bell also has a more expensive service called "Unlisted Service." Here, your phone number is not included in the paper phone directories or given out to callers to Directory Assistance. Some phone companies have a service whereby you can leave a message for the owner of an unlisted number. "Please call me. You've won the lottery." The owner of the unlisted number then has the choice to return the call or not. He doesn't pay to receive this message. Some telephone companies confuse the definitions and some invent new ones. For example, some phone companies use the term "non-published" number. You won't find the number in a phone book or by calling Directory Assistance. Over 25% of many private phone numbers in major metropolitan areas are unlisted, unpublished or non-published – a "service" their subscribers pay extra for. To my simple brain, it's a lot easier to simply publish your name as "Apple Plumpudding." See Unpublished.

unrecognized media pool A repository of blank media and media that are not recognized by Removable Storage.

unreliable transmission The conveying of messages from a sender to one or more receivers using connectionless-service. An unreliable service is a best effort to deliver a packet data unit (PDU), meaning that messages may lost, duplicated or received out of order.

unrestricted digital information An ISDN term. An information sequence of bits is transferred at its specified bit rate without alteration.

unshielded Wiring not protected by a metal sheathing from electromagnetic and radio frequency interference, but covered with plastic and/or PVC.

Unshielded Twisted Pair UTP. A transmission medium consisting of a pair of copper conductors which are electrically balanced. Each conductor is separately insulated (typically with plastic) in order to prevent the conductors from "shorting." The conductors are twisted around each other at routine intervals in order to confine the electromagnetic field within the conductors and, thereby, to 1) maximize signal strength over a distance, and 2) minimize interference between adjacent pairs in a multi-pair cable. UTP conductors come in various gauges and various numbers of twists per foot. The thicker the conductor, the less resistance and the better the performance; the more twists per foot, the better the performance. UTP comes in various configurations; in large cable systems, multiple pairs are combined in binder groups of 25 pairs, and multiple binder groups are combined in a single insulated cable sheath. In small configurations, such as desktop applications requiring one to four pairs, UTP is relatively inexpensive to acquire and to deploy when compared to coaxial cable and fiber optic cable; hence, its increasing popularity in both voice and data applications. For longer explanations, see also UTP Cable, Category 1 through 6, and STP.

unsolicited event Events in switching that happen without control of a program that allegedly is controlling your phone and phone system. Such an unsolicited event might be a user picking up his phone and hanging it up or simply pushing a random button on the phone. Or it may be that the switch actually does something you or your controlling computer don't expect it to do.

Unstructured Supplementary Services Data USSD. A method of transmitting data and instructions over a GSM cellular network in a two-way, interactive mode. While both USSD and SMS (Short Message Service) use the GSM network's signaling path, USSD session-oriented, rather than being store-and-forward in nature. Thereby, and once a session is established, the radio connection remains active until such time it is released by the user, the application, or a time-out condition. USSD messages can be up to 182 characters in length, compared to the maximum of 160 characters for SMS. USSD is much faster than SMS in a two-way, interactive mode, since it is session-oriented, rather than being store-and-forward. USSD also works on a roaming basis, as USSD commands are sent back to the HLR (Home Location Register), which is associated with the subscriber's home service provider. On the other hand, USSD command strings are complex and difficult to remember. See also GSM, HLR, and SMS.

unsuccessful call A call attempt that does not result in the establishment of a connection.

unsupervised transfer Someone transfers a call to someone else without telling the person who's calling. Also called Blind Transfer.

untrusted See Class of Service.

unusually heavy call volume "We are experiencing unusually heavy call volume," the machine told me. Fine, I thought, the Mercedes operator and her machine are probably snugly tucked away is a distant warm clime. Not me, I'm sitting on the side of the road with my wife's Mercedes with a crapped-out battery/electrical system. It's minus one fahrenheit. My cell phone is running out of charge. And I'm running out of patience. Why am I including this stupid definition? Because I'm supremely annoyed at Mercedes "service" (or lack thereof) and wanted someplace to mouth off against them.

unused A term used in the secondary telecom equipment business. Equipment never used and still in O.E.M. original packaging with all appropriate documents and user guides. Such equipment may or may not carry the O.E.M. standard warranty. In other words, manufacturers don't want to see independent (unauthorized) dealers advertising their equipment as new, but "unused" is acceptable.

up sell To sell a higher value product to an existing customer. For example to lease a more sophisticated photocopier to an existing customer. In contrast, cross selling is when you buy a shirt from me. I sell you a tie. You buy a car from me. I sell you a mobile phone for your car. Up selling is when I sell you a more expensive shirt or a more expensive car.

up-converter A device for performing frequency translation in such a manner that the output frequencies are higher than the input frequencies.

up-sampling A technique used for recreating an approximation of data compressed by down-sampling. Up-sampling generally doubles the data at each iteration of the signal by inserting a value between the adjacent compressed values in the case of "down-sampling by two." This operation is fundamental in the Fast Packet Algorithm used in Wavelet Transforms, which are commonly employed in image compression. See Down-Sampling, Fast Packet Algorithm and Wavelet Transform.

UPB A proprietary powerline communication technology developed by Powerline Control Systems.

UPC 1. Usage Parameter Control: As an ATM term, Usage Parameter Control is defined as the set of actions taken by the network to monitor and control traffic, in terms of traffic offered and validity of the ATM connection, at the end-system access. Its main purpose is to protect network resources from malicious as well as unintentional misbehavior, which can affect the QoS of other already established connections, by detecting violations of negotiated parameters and taking appropriate actions.

2. Universal Product Code is the US term for the numbering and bar coding system for product identification of consumer items, typically scanned at the point of sale. The UPC is a predecessor and now a subset of the European Article Number(EAN).

3. There are three different finishes you can have on the end of a fiber optic connector

– PC, APC and UPC. Whenever a connector is installed on the end of fiber, loss is incurred – also called attenuation. Some of this light loss is reflected directly back down the fiber towards the light source that generated it. These back reflections, or Optical Return Loss (ORL), will damage the laser's light source and also disrupt the transmitted signal. To reduce back reflections, we can polish connector ferrules to different finishes. A typical hand polished connector will measure at -30dB. This polish is referred to as a PC or physical contact polish, which for some systems is considered too high of an ORL measurement. To reduce the back reflection of a connector, a manufacturer can machine polish it to SPC (Super Physical Contact) polish or UPC (Ultra Physical Contact) polish. Industry standard is a minimum of -40dB for SPC back reflection measurement and -50dB for UPC back reflection measurement. If even less back reflection is required, an APC, or Angled Physical Contact polish, might be necessary. An APC connector has an 8 degree angle cut into the ferrule. These connectors are identifiable by their green color. An APC polished connector has an Industry Standard Minimum of -60dB ORL measurement. Most CATV and telephone companies require the use of these low back reflection connectors. See opical return loss.

UPCS Unlicensed Personal Communications Services. The FCC designated 1890-1930 MHz to UPCS. 1920-1930 MHz is presently being assigned by UTAM, primarily for short range, wireless PBX applications. Here are the specific designations: 1890-1910 MHz isochronous, 1910-1920 MHz asynchronous, 1920-1930 MHz isochronous. Isochronous communication is good for voice and asynchronous communication is for bursty data.

upgrade A call center term. A technique to increase the revenue of an order that is quality rated. It means getting the customer to by a better quality, more expensive version of the item sold. See also Up Sell.

uplink 1. In satellites, it's the link from the earth station up to the satellite. The link from the satellite down to the earth station is called the downlink. The uplink and downlink operate on different frequencies so that they don't interfere with each other. The uplink is at a higher frequency. For example, C-Band satellites use the 6 GHz frequency range on the uplink and 4 GHz on the downlink. International customers often buy uplinks and downlinks from different suppliers, as each nation typically awards one or more national agencies exclusive franchise rights. See DOWNLINK.

2. In data transmission, an uplink is from a data station to the headend or mainframe.

3. As an ATM term, it represents the connectivity from a border node to an upnode.

uplink time difference of arrival One of a number of technologies available for determining a caller's location after initiating a 911 call from a mobile handset. See also angle of arrival.

upload To transmit a data file from your computer to another computer. The opposite of download, which is receiving a file on your computer from another computer. Upload means the same as TRANSMIT, while DOWNLOAD means the same as receive. Before you upload or download, check at least three times you're going the direction you want. It's very easy to erase files (weeks of work) if you make a mistake and confuse uploading and downloading. (Don't laugh. We've done it several times. Dumb!)

upnode As an ATM term, it is the node that represents a border node's outside neighbor in the common peer group. The upnode must be a neighboring peer of one of the border node's ancestors.

UPP Universal Payment Preamble (UPP). The negotiation protocol that identifies appropriate payment methodology in the context of Joint Electronics Payments Initiative (JEPI), a specification from the World Wide Web Consortium (W3C) and CommerceNet for a universal payment platform to allow merchants and consumers to transact E-Commerce (Electronic Commerce) over the Internet. UPP works in conjunction with Protocol Extensions Protocol (PEP), an extension layer that sits on top of HTTP (HyperText Transfer Protocol). These protocols are intended to make payment negotiations automatic for end users, happening at the moment of purchase, based on browser configurations See also Electronic Commerce and JEPI.

upper case Upper- and lower-case letters are named "upper" and "lower," because in the time when printers set pages using loose type, the "upper-case" letters were stored in the case on top of the case that stored the smaller, "lower-case" letters.

upper crust In the 1500s in England, bread was divided according to status. Workers got the burnt bottom of the loaf, the family got the middle, and guests got the top, or the "upper crust."

upper memory area Obsolete: In PC, upper memory is the area between 640KB and 1MB of RAM. This area is made up of Upper Memory Blocks (UMBs) of various sizes. Access to this area was possible in the old days only with a special memory drive such as MS-DOS's EMM386.EXE.

upper memory blocks See Upper Memory Area.

UPS Uninterruptible Power Supply. A UPS is a device which keeps clean power going to a phone system or a computer when the power from the power company shuts down or gets flakey. Phone systems and computers are driven these days by microprocessors. These are the different electrical problems that can cause microprocessor-based systems to misbehave:

- Blackouts - resulting in the total loss of power
- Brown Out - for example, voltage dropping when an air conditioning unit switches on
- Spikes - lightning, or increased voltage for short periods of time
- Surges - occuring when incoming AC voltage goes beyond the normal 120VAC for short periods of time
- EMI (Electromagnetic Interference) - motors switching on and off
- RFI (Radio frequency interference) - radio transmitters / television station
- frequencies, etc.

UPSes can protect against all these problems, depending on how they are specified. There are basically two types of UPSes. The first has an electronic circuit which senses the change (usually drop) in power and switches over. This takes a few milliseconds, during which time you hope the computer or phone system will continue working. This is typically called a standby power system, or an off-line UPS. The second type of UPS is one that sits permanently between the outside power (your local and the device (computer or phone system). It takes power in from outside and cleans it. If power from outside is dropped or gets flakey, the UPS keep powering the device with clean power. The first type of UPS is called "off line." The second type is called "in line." UPSes span a broad array of capabilities and prices. The variables range include: (1) how quickly the UPS switches in when power drops. Is it quick enough to fool the computer or phone system that it's still getting power? (2) how long the UPS will power the computer or phone system. Does it include a small battery or a big battery? Some UPSes actually include a generator – in which case you'd better make sure that there's gas in the tank when the thing is meant to start. In general, in-line UPSs are much more expensive than SPSs.

UPS monitoring UPS monitoring allows a local area network file server to monitor an attached Uninterruptible Power Supply (UPS). When a power failure occurs, NetWare notifies users. After a time out specified with SERVER.CFG and ROUTER.CFG, the server logs out any remaining users, closes any open files, and shuts itself down. If you install a Novell-approved UPS, you must also install a printed circuit board in the file server to monitor the UPS. If you have a file server with a microchannel bus (as compared to the more common AT bus), the UPS is monitored through the mouse port and does not require a board. See UPS.

upsell See UP Sell.

UPSR Unidirectional Path-Switched Ring. A SONET term. Path-switched rings employ redundant fiber optic transmission facilities in a pair configuration, with one fiber transmitting in one direction and with the backup fiber transmitting in the other. If the primary ring fails, the backup takes over. See also PAath Switched Ring and SONET.

upspeak According to Newsweek, upspeak is the annoying way teenagers speak.

upstream In a communications circuit, there are two circuits – coming to you and going away from you. Upstream is another term for the name of the channel going away from you. In a broadband TV network, the definition of the upstream channel or signal is different. It is the channel from the transmitting stations to the CATV headend. See Upstream Channel.

upstream channel In a communications circuit, there are two circuits – coming to you and going away from you. Upstream is another term for the name of the channel going away from you. In a broadband TV network, the definition of the upstream channel or signal is different. It is the channel from the transmitting stations to the CATV headend. In yet another definition, in the cable TV industry, the upstream channel is a collection of frequencies on a CATV channel reserved for transmission from the terminal next to the user's TV set to (upstream to) the CATV company's computer. Such signals might be requests for pay movies. See Upstream.

Upstream Channel ID The identifier of the upstream channel to which the CM is to switch for upstream transmissions. This is an 8-bit field.

upstream operations Functions that provide a BCC (Bellcore Client Company) control of features and service configurations and subject to BCC control, some service management capabilities for subscribers. These functions include Service Negotiation and Management, Service Provisioning and Repair Service Answering/Work Force Administration. Definition from Bellcore in reference to its concept of the Advanced Intelligent Network.

UPT Universal Personal Telecommunications. According to L.M. Ericsson, Swedish telecom manufacturer, UPT is a "new service concept in the field of telecommunications which aims at making telecommunications both universal and personal. instead of calling a telephone line or a mobile terminal, you call the person you wish to get in touch with and leave it to the network to locate the line or terminal where he/she can be reached." There was an article on UTP in the 1993 No.4 issue of the Ericsson Review. An article in the June, 1996 issue of IEEE Communications Magazine described UPT as a service that enables users to access various services through personal mobility. It enables each UPT user to participate in a user-defined set of subscribed services, and to initiate and receive calls on the basis of a personal, network transparent UPT number across multiple networks on any fixed or mobile terminal, irrespective of geographical location. This service is limited only by terminal and network capabilities and restrictions imposed by the network operator. In short, UPT is still not totally defined and is under discussion by the world's major standards bodies. For more information on UPT, see ITU-T Recommendation F.850, Principles for Universal Personal Telecommunications, Geneva, 1993.

uptime Colloquial expression for the uninterrupted amount of time that network or computer resources are working and available to a user. In short, time between failures or periods of nonavailability (as for maintenance).

upward compatible Any device that can be easily organized, fixed or configured to work in either a different, expanded operating environment or some enhanced mode. Software is said to be upward compatible if a computer larger than the one for which it was written can run the program.

urban canyon An urban canyon is an urban environment characterized by many buildings that are in close proximity to each other, thereby hampering line-of-sight communications, for example, with satellites.

urban legend A story, which might start with a grain of truth, gets retold a million times, gets mutated in its various appearances on the Internet and finally is accepted as "fact." The growth of traffic on the Internet - namely that it was doubling every three months in 1999 - was an urban legend, which was, like many urban legends, wrong.

urban service Any of the grades of service regularly furnished inside base or locality rate areas, or outside base or locality rate areas at base or locality rates plus zone connection charges or incremental rates. Another way of saying expanded metropolitan phone service.

URI See Uniform Resource Identifier.

URL Universal Resource Locator. An Internet term. A URL is a fancy name for an Internet address. A URL is an address that can lead you to a file on any computer connected to the Internet anywhere in the world. Thus, its name – Universal Resource Locator.

In more technical terms, a URL is a string expression that can represent any resource on the Internet or local TC/IP system. The standard convention for a URL is as follows:

first, the method of protocol to be used (e.g. http)://the host's name/folder or directory on host/name of file or document

Here's an example of a URL: http://www.harrynewton.com/fantasy/happy.html

Let's see what it all means. The http stands for HyperText Transport Protocol. That tells your browser (e.g. Netscape or Internet Explorer) to use that protocol when searching for the address. Http is the "default" protocol of the Internet. But it's not the only one. There are other protocols, including ftp (file transfer protocol), news (for Usenet news groups), and "mailto" (to send email to a specific address).

The www.harrynewton.com is simply the name of my computer. All Web addresses start as numbers. This one is no different. You can reach my home page by giving your browser the following command http://209.94.129.207/. The Web's own lookup tables do an instantaneous translation to that number from www.harrynewton.com when you type in www.harrynewton.com. The translation mechanism is very much like the translation the phone industry does when you dial any 800 number, e.g. 1-800-LIBRARY – translate it to a real number, i.e. 212-691-8215.

The /fantasy/ means that there's a folder or subdirectory on my web site's computer disk called fantasy and inside that folder there's a document called happy.html. And that's what we're looking for. See also Web address.

URL equity The value of a URL, measured in terms of the number of links to it on other websites, the extent to which it is indexed and highly ranked by search engines and not blacklisted by them, the number of publications and documents that include it in their references, and the number of users who have bookmarked it.

URM User request manager.

URPS Unit Eruptible Power Supply.

U.S. Naval Observatory USNO. Established in 1830 and currently located in Washington, D.C., the mission of the U.S. Naval Observatory includes determining the positions and motions of the Earth, Sun, Moon, planets, stars and other celestial objects; providing astronomical data; determining precise time; measuring the Earth's rotation; and maintaining the Master Clock for the United States. The astronomical and timing data support navigation and communications on Earth and in space. See Coordinated Universal Time (UTC). You can the official time by going to www.time.gov.

US West One of the seven Regional Holding Companies (RHCs) formed by the divestiture of the Bell Operating Companies (BOCs) by AT&T. US West's 14-state LEC (Local Exchange Carrier) business originally comprised Mountain Telephone, Northwestern Bell and Pacific Northwest Bell, although those entities later lost their identity as they were folded into the holding company. On June 30, 2000, US West merged with (read was acquired by) Qwest Communications, and now is just part of Qwest. See also US Worst and Qwest.

US worst A derogatory term for US West. See US West.

USA Underground Service Alert. One-call System serving California, Nevada, and Hawaii providing a free Call Before You Dig service to all excavators(contractors, homeowners, and others). All USA Members will be notified who may have underground facilities at or near the location site. More information on USA can be found at:http://usanorth.org/ or 1-800-227-2600.

USAC Universal Service Administrative Company.

USACII USA Standard Code of Information Interchange. The original name of the North American version of ASCII code. Differences from ITU-T International Telegraph Alphabet #5 are so minor as to be insignificant to most applications. The name change was a result of the name change of the standards organization. When the name changed again to ANSI, most people simply reverted to ASCII.

usage A measurement of the load carried by a server or group of servers, usually expressed in CCS. Usage may also be expressed in erlangs.

usage based Usage-Based refers to a rate or price for telephone service based on usage rather than a flat, fixed monthly fee. Until a few years ago, most local phone service in the United States was charged on a flat rate basis. Increasingly, phone companies are switching their local charging over to usage-based. Flat-rate calling will probably disappear within a few years. Allegedly, usage based phone service pricing is fairer on those phone subscribers who don't use their phone much. Usage based pricing is not consistent throughout the U.S. Typically, you get charged for each call. And the charging is very much like that for long distance – by length of call, by time of day and by distance called. See also Flat Rate.

usage sensitive A form of Measured Rate Service. See Usage Based

USAN United States Advanced Network. The USAN platform is an integrated system that interfaces with the MCI network for call access and egress. The USAN platform provides automated voice prompts, billing verification, etc. for a variety of MCI products and services.

USART Universal Synchronous/Asynchronous Receiver/Transmitter. An integrated circuit chip that handles the I/O (input/output) functions of a computer port. It converts data coming in parallel form from the CPU into serial form suitable for transmission, and vice versa.

USB Universal Serial Bus is today's most commonly used way of connecting devices to personal computers. You can use USB to connect a mouse, a keyboard, a game controller, a scanner, several printers, several digital cameras, a hard disk , a DVD drive. few. With USB hubs, you can connect up to 127 devices to your computer's USB port and (theoretically) use them all at once. USB is faster than older ways of connecting devices – serial, parallel and PS/2 ports. USB presently comes in two flavors: USB 1.0 /1.1 and USB 2.0. USB 1.0 and USB 1.1 supports data rates of 1.5 million bits per second (for low speed devices like mice and keyboards) and 12 million bits per second for higher speed devices, like hard disc drives. The current USB version 2.0 supports data transfer runs up to 480 million bits per second, with the data rate between the computer and a single peripheral reaching speeds of approximately 900,000 bits per second. USB is cabled in daisy chain fashion, with one device plugging into the next and so on. Eventually USB will be overtaken by Firewire, which is much faster and used by such popular devices as Apple's iPod. Many computers come with connections for both USB and Firewire. USB is meant to be downward compatible, such that USB 1.0 and 1.1 devices are meant to work on USB 2.0 lines. Sadly, they don't always. For example, my laptop docking station's USB port won't run my older Dymo label printers, because the docking station contains a 2.0 hub. To drive my Dymo label printers I need to plug them directly into my laptop which will drive both older and newer USB devices. USB supports plug-and-play connectivity to suitable peripherals, meaning that devices are detected by the computer's operating system and configured automatically as soon they are attached – or at least that's the theory. The reality is that operating systems such as Windows often have old drivers and for the USB devices to work perfectly, you need to go to the manufacturer's web site and download the latest drivers. USB also supports "hot attach/detach," which allows adding and removing devices at any time, without powering down or rebooting. USB cables can be up to 30 meters long, and include built-in power distribution for low-power devices. The length of the cable connection can be extended through a simple bridging device. Multiple devices can be connected through

daisy-chaining in a multi-drop tiered star topology. USB is sophisticated in that it will handle certain "important" data streams – e.g. voice and video – with preference. USB supports three basic types of data transfer:

- Isochronous (i.e., streaming) real time data which occupies a prenegotiated amount of bandwidth with a prenegotiated latency. Examples include voice and video.
- Asynchronous interactive data such as characters or coordinates with few human perceptible echo or feedback responsible characteristics. Tele-gaming is an example.
- Asynchronous block transfer data which is generated or consumed in relatively large and bursty amounts and has wide dynamic latitude in transmission constraints. File storage is an example.

See also 1394 (commonly called Firewire).

USD Universal Synchronous Data.

USDLA United States Distance Learning Association. Their mission: To promote the development and application of distance learning for education and training. Constituents include K-12 education, higher education, continuing education, corporate training, and military and government training. www.usdla.org

used Equipment which was previously in service (i.e. used someplace else) and may not have been tested, refurbished or remanufactured before you bought it. Used simply means it's no longer new. No more, no less. See also Certified, NATD, and Refurbished.

used or out of service A term used in the secondary telecom equipment business. Equipment taken from service. Can be in any condition. Generally expected to work and be complete.

USENET The USENET is an informal, rather anarchic, group of computer systems that exchange "news." News is essentially similar to "bulletin boards" on other networks. USENET actually predates the Internet network. These days, the Internet is used to transfer much of the USENET's traffic. You can find newsgroups covering every conceivable subject from nude sunbathing to molecular physics. See Internet MUDS and USENET Newsgroup Organization.

USENET Newsgroup Organization USENET conversations are organized in hierarchical newsgroup trees. There are seven core newsgroup hierarchies or trees: comp (computers), misc (miscellaneous topics), news (newsgroup information), rec (recreation), sci (science), soc (society), talk (conversation). Each tree branches into different levels of newsgroup sub-topics.

user accessible tables There are many tables (databases) inside a phone system. They include the extensions with privileges and long-distance dialing selections (see LEAST COST ROUTING). In the old days, most PBX and phone system tables were not accessible to the user, on the assumptions that 1. The user would screw the tables up, and/or 2. Really didn't care about getting access. Things have changed. Users now want faster and greater control over their own destiny. So, many manufacturers are making their tables user accessible.

user account Each user has a user account that is part of local area network security and controls the user environment. Some account features are assigned to each user automatically, some must be assigned, some are optional.

user agent 1. Generally refers to the windows and menus used to make interfacing to UNIX easier.

2. UA. An OSI (Open Systems Interconnection) process that represents a human user, or organization, or application in the X.400 MHS (Message Handling System). The user creates, submits, and takes delivery of messages on the user's behalf, and in some cases, can even create the message. UA is thus an X.400 electronic mail term: It is software that prepares the message for transmission to the Message Transfer Agent. The user can be an individual or a distribution list. Users are known by their originator/recipient (O/R) addresses.

user context A user session created by an operating system in response to a logon request, and typically characterized by privilege sets that strictly define the user's authority to access system resources and information on a LAN. Contexts restrict unauthorized access to facilities and data and protect the system itself from user and applications interference, accidental or otherwise. Contexts are a feature of most multi user operating systems, usually integrated with the security system.

User Datagram Protocol UDP. A packet format included in the Transmission Control Protocol/Internet Protocol (TCP/IP) suite and used for short user messages and control messages. The transmission of UDPs is unacknowledged.

user data X.25 call control field used to transfer information concerning layers above X.25 between the originating and terminating DTEs.

user error See Cockpit Problem.

user event Input from the user (for example, clicking the mouse, pressing a key, etc.) that causes a multimedia project to perform a specific function (for example, play a movie, change pages, etc).

user friendly Computer programs or systems which are designed for simple operation by non-technical users. At least that's the theory.

user ID 1. Persistent information in an ASC (AIN Switch Capabilities) that the ASC communicates to the SLEE (Service Logic Execution Environment) as a parameter in a message to the SLEE. The SLEE uses this parameter to identify the set of information related to the user (e.g., customer record) that service logic needs to perform its task. If a user does not subscribe to any Advanced Intelligent Network Release 1 feature but invokes an AIN Release 1 feature, the user ID in the ASC may correlate in the SLEE to a set of default information to be used by service logic.

2. A compression of "user identification"; the unique account signature of an Internet user; that which precedes the @ (at) sign in an E-mail address.

user loop A 2- or 4-wire circuit connecting a user to a PBX or other phone system.

user message Part of a CPN message directing a destination node to accomplish some task.

user parameter control Traffic policing to ensure that the defined peak traffic rate is not exceeded in the ATM switch. See UPC.

user plane An ATM term referring to the functions which address flow control and error control. The User Plane cuts through all 4 layers of the ATM Protocol Reference Model.

user segment The part of a satellite system that includes the receivers on the users' premises.

sser to user messaging An ISDN service enabling voice and computer data to be transmitted simultaneously – for example, enabling one person to transmit a spreadsheet file to another so both can examine it on their individual screens, and for each to have the ability to change the spreadsheet and have the changes appear instantly on the other person's screen, and then discuss the changes. Also called User to User Signaling, or UUS. With UUS, the ISDN user can send a short message (up to 40 characters) from the ISDN terminal at the time of call setup. This will be displayed at the called terminal even before answering.

user's set Apparatus located on the premises of a communications user. Designed to work with other parts of his system.

UserID A compression of "user identification"; the unique account signature of an Internet user; that which precedes the "at" sign in an E-mail address.

username The name by which you or someone else is known by on the Internet. Used when logging into an access provider or when entering a member's only area on the Web.

USF 1. Universal Single Frequency.

2. Universal Service Fund. See Universal Service Fund.

USITA United States Independent Telephone Association, the old name for the United States Telephone Association. See USTA.

USO Universal Service Obligation. The obligation under the Australian Telecommunications Act 1997 to ensure that standard telephone services, payphones and prescribed carriage services are reasonably accessible to all Australians on an equitable basis, wherever they reside or carry on business. Telstra is the current universal service provider in Australia.

USOC Uniform Service Order Code (pronounced "U-Sock") is a structured language that allows for the development of software to support service order systems in the telephone industry. The service order process utilizes the USOC, along with field identifiers (FIDs), to provision, bill and maintain services and equipment. USOCs can be either three or five alpha/numeric characters. A plus (+) sign indicates a variable suffix position. Suffixes define options of the USOC i.e. color, jurisdiction, speed. To prevent confusion the letter "O" is used and zero is not; the number "1" is used and the letter "l" is not. USOCs are designed for tariffed services, official company services, coin services, equipment, detariffed services, etc. The Bell operating companies in the United States and many independent telephone companies use USOCs to communicate both within their company and between companies. Many new companies in the industry are using the USOC information to interpret incumbent telephone company records when they are supplying service to a new customer. The different companies may have different names for the same services, but the USOC name is generic and therefore becomes a common naming device between companies.

USOP User Service Order Profile.

USOS Universal Operations Services. A software application that supports UIS by providing traditional network operations functions.

USP 1. Usage Sensitive Pricing. A tariff for local service under which the subscriber only pays for the telephone service he uses. This is done for gas and electric service.

2. Universal Service Plan. A plan in Australia to make basic local telephone services available at an affordable price to all consumers.

USPID User Service Profile Identifier. An ISDN term.

USPTO United States Patent and Trademark Office.

USRT Universal Synchronous Receiver/Transmitter. Integrated circuit that performs conversion of parallel data to serial for transmission over a synchronous data channel.

USSD Unstructured Supplementary Services Data. Look there for a full explanation.

USSI Universal Synchronous Serial Interface. Noted in Adtran, Inc marketing literature a USSI interface is one which readily runs RS-530A,RS-449, RS-232,V.36, X.21.

USTA United States Telecom Association. USTA. The largest trade association of telecommunications companies, USTA has a membership of over 1,200 companies. USTA has its roots in the National Telephone Association, formed in 1897 to unite independent (i.e., non-Bell) telephone companies. Subsequently, the organization changed its name to USITA (United States Independent Telephone Association). After the break-up of AT&T in 1984, the RBOCs (Regional Bell Operating Companies) were admitted as members, and the name was shortened to USTA (United States Telephone Association). After the Telecommunications Act of 1996, CLECs (Competitive Local Exchange Carriers), ISPs (Internet Service Providers), wireless and cable companies were admitted and the organization changed its full name to United States Telecom Association, I suppose to reflect the convergence of voice and data. Membership is limited to facilities-based carriers. The organization, headquartered in Washington, lobbies the FCC, Congress and other regulatory, legislative and judicial bodies to ensure that no regulations are enacted or laws passed that will act to the detriment of its members. USTA also provides a little education for its members. See also USITA. www.usta.org.

USTSA See United States Suppliers Association.

UT 1. Universal Time.

2. Upper Tester. An ATM term. The representation in ISO/IEC 9646 of the means of providing, during test execution, control and observation of the upper service boundary of the IUT, as defined by the chosen Abstract Test Method.

3. User terminal.

UTAM Unlicensed Transition and Management for Microwave. Relocation in the 2.0 Ghz Band. UTAM Inc. is an open industry organization pledged to relocate incumbents (POFS) presently operating in the 1890 to 1930 MHz band. Designated by the FCC as the frequency coordinator for the UPCS spectrum. "Memorandum and Opinion and Order - June 1994," FCC Docket 90-314.

UTC 1. Coordinated Universal Time is the basis for legal time, worldwide. UTC is calculated based on International Atomic Time (TAI), which is calculated from the coordinated readings of more than 200 atomic clocks located in meteorology institutes and observatories around the world, including the U.S. Naval Observatory. UTC varies from TAI in that it is adjusted by the addition of occasional leap seconds, as the rotation of the Earth is not as precise as the rotation of electrons. UTC replaced Greenwich Mean Time (GMT), which is calculated solely on solar time. See also Greenwich Mean Time. See also the U.S. Naval Observatory.

2. UTC, The Telecommunications Association. UTC, formerly known as the Utilities Telecommunications Council primarily represents the interests of the utility industries (electric, gas, steam, and water) worldwide, but is open to vendors, manufacturers, and other interests. It is worthy to note that the utility industry is second only to the "telephone" industry in the procurement of telecommunications products and services. www.utc.org

UTDR Universal Trunk Data Record.

utility 1. In some computer languages, a small program executed as a shell command is called a "tool." In other computer languages, such as BASIC, it is called a "utility."

2. A "public utility," such as an electric, gas, sewer, telephone, or water company. In telecommunications parlance, a "utility" generally refers to an electric company. Some electric utilities also are in the telephone business. Some of them have built fiber optic networks, allegedly for remote meter-reading and SCADA (Supervisory Control And Data Acquisition) purposes such as load shedding. These applications don't require nearly the bandwidth supported by fiber optics. The electric utilities built to capital cost of these networks into the rate base, with the blessings of the various state regulators. Their thinly disguised plans were to build information grids, much like their power grids and along the same rights-of-way, and then to deny others (e.g., telephone and CATV companies) the rights to use the same power poles to string their own fibers. It was a rip-off.

Utility Communications Architecture UCA. An architecture for networks used to monitor and control electric power distribution systems.

utility computing IBM's idea that you'll buy computing power over the Internet. So you (the corporation) won't have to own huge computers which are only used part of the time. The concept is working yet and several observers have called it."futility computing."

utility nodes Nodes running software needed to control, monitor and diagnose network activities. May have a database connected to the node.

utility program A computer program in general support of the processes of a computer; for example, a diagnostic program.

utility routine A routine in general support of the operation of a computer, including input/output, diagnostic, tracing or monitoring.

utilization The extent to which a circuit, link, path, switch, or other network element is being used, either at a given point in time or over a period of time. Utilization generally is expressed as a percentage. For example a T-1 offers a total signaling rate of 1.544 Mbps, and supports a total transmission rate of up to 1.536 Mbps. If, on the average over a period of a month, it is used to support an offered load of only 768 Kbps, it can be said to be 50% utilized. Utilization of 100% is rare, as the network element would suffer congestion some percentage of the time – unless you're on a packet network, like the Internet. On such a network, you have alternative routes, so when one is 100% utilized, the traffic switches automatically to another route. See also Congestion and Utilize.

utilize An absolutely awful word created the people who believe that speaking or writing in big words is a demonstration of their superior intelligence.

UTOPIA 1. Universal Test and Operations Interface for ATM. Refers to an electrical interface between the TC (Transmission Convergence) and PMD (Physical Medium Dependent) sublayers of the PHY (Physical) layer, which is the bottom layer of the ATM Protocol Reference Model. Utopia is the interface for devices connecting to an ATM network.

2. Utah Telecommunication Open Infrastructure Agency (UTOPIA) is a Utah interlocal agency formed in 2002 to deploy a publicly owned advanced fiber optic last mile (i.e., local loop) telecommunications network within founding 18 cities, which comprise approximately one-third of the state's population. The FTTP (Fiber-To-The-Premises) project is intended ultimately to connect as many as 15,000 homes and businesses at speeds of 100/1000 Mbps to the backbone, which will run at 5.6 Terabits per second. A terabit is a thousand gigabits.

UTP Unshielded Twisted Pair. A pair of wires that is twisted, so as to minimize crosstalk with other pairs of wires in the same cable (which are each twisted at a slightly different rate) but not shielded. See Unshielded Twisted Pair.

UTP cable Unshielded Twisted Pair Cable. Most UTP cables have eight conductors. They are organized into four pairs. Each pair has a ring conductor and a tip conductor. The tip is colored white, usually with colored stripes. The ring is a solid color, usually with white stripes. The conductors of each pair are twisted around each other at a constant rate. However, each pair has different twist lengths. These exact lengths vary between manufacturers and types of cables. See Category of Performance. The conductors in most UTP and ScTP cables are 24 gauge copper wire. (Some manufacturers use 22 gauge. Conductor size doesn't matter if the cable is rated at the Category of Performance you've specified.) UTP cable can have solid copper or stranded conductors. Solid is less expensive than stranded, but stranded conductors are more flexible because they're made of tiny individual strands of copper. See Unshield Twisted Pair.

UTP RJ-45 Plug These are used to terminate four pair UTP patch cords. A clip on the plug holds it into the jack. See Unshield Twisted Pair.

UTR Universal Tone Receiver.

UTRAN UMTS Radio Access Network; used in 3G wireless; comprised of NodeB and RNC

UTS Universal Telephone Service.

utterance A word used in voice recognition to mean a vocalized sound that is typically a word.

UUCP UNIX-to-UNIX Copy Protocol (UUCP) is a standard UNIX utility and protocol for exchanging information such as files, email and news between two UNIX-based machines in a network not associated with the Internet proper. UUCP largely has been replaced by protocols such as FTP (File Transfer Protocol), NNTP (Network News Transfer Protocol) and SMTP (Simple Mail Transfer Protocol). See also FTP, NNTP and SMTP.

UUD Short for UUDecode, which is a software utility used to decode UUEncoded files. See UUEncode.

UUE Short for UUEncode, which is a de facto encoding protocol used to transfer binary files across the Internet and on-line services. See UUENCODE.

UUEncode UNIX-to-UNIX encoding. Software that allows you to take a binary computer file, e.g. a Word for Windows document or a PowerPoint presentation, convert it to ASCII for sending the binary file across the Internet or some other e-mail service and then, once received, converting it back to binary. In its Uuencoded form, the file is visible as ASCII, but it makes absolutely no sense. Direct e-mail services, like America On Line, MCI Mail or CompuServe, allow you to "attach" a binary file to a message – without ever affecting the message. The person at the other end receives it as a binary file. However, once the message passes across an X.400 gateway, from one e-mail service to another (including or not including the Internet), you need Uuencoding. Also called MIME encapsulation. See Multipurpose Internet Mail Extension for an example of UUencoding.

UUI User-to-User Information. An ISDN term. UUI comprises information of end-to-end significance sent over the ISDN D (Data) channel in the context of UUS (User-to-User Signaling), as defined by the ITU-T and ETSI. UUI services, which remain to be defined completely, must be subscribed by the sending party; the target party side of the connection has no acceptance or rejection procedure. UUS falls into three categories. UUS1 provides for the transmission and reception of UUI during call set-up and termination, through ISUP (ISDN User Part) call-control messages. UUS2 provides for the transmission and reception of UUI subsequent to call set-up, but prior to the establishment of a connection. UUS3 provides for the transmission and reception of UUI only while the connection is established, i.e., during the active phase of the circuit-switched call. UUS1/2/3 messages comprise packets of 128 bytes, with maximum numbers of packets established for each UUS category. UUI information is sent within standard Q.931 signaling messages, as separate signaling messages which are either associated with an existing call or are sent as non-call messages. UUI information sent as an associated call message might include the caller's name or account number to be transmitted to the destination. Detailed information can be found in ITU Q.931, or related ITU documents I.257.1 and D.231. See also ISUP.

UUID Universally Unique IDentifier. Also known as Globally Unique IDentifier (GUID). See Globally Unique Identifier.

UUS See User to User Signaling.

UV 1. Ultraviolet.

2. A microvolt.

3. Unified Voice.

UGVA UVGA stands for Ultra Video Graphics Array and refers to 1024 pixels by 768 pixels. See SVGA and VTGA.

UW See Unique Word.

UWB See Ultra Wideband.

UWCC Universal Wireless Communications Consortium. See that definition for detail.

UXTMN Local surcharge for emergency reporting services.

V 1. Abbreviation for Volt.

2. The International Morse Code representation of the letter V is ..._, which we hear as didididah. That refrain is reminiscent of the opening of Beethoven's Fifth Symphony, which is the reason that the British Army adopted that work as its unofficial theme music during WWII (World War, II for you youngsters). The connection, of course, is the V for Victory sign that Winston Churchill (British Prime Minister at the time) always made. Allied GI's who didn't know Schumann from shinola knew this was Beethoven and relished the irony of a German's music galvanizing the Allied effort to defeat Hitler.

V & H V&H stands for Vertical and Horizontal. Below are vertical and horizontal coordinates of major continental US cities as presented in charts published by long distance carriers in North America. The monthly charge for many leased circuits provided by either an IXC (Inter-Exchange Carrier) or a LEC (Local Exchange Carrier) is billed on the basis of "airline mileage" between the two points. The two points for an IXC (long distance carrier) private line circuit are IXC POP to IXC POP. The monthly charge for the IXC circuit is based on the mileage between the two POPs. The two points for a LEC (Local exchange carrier) are based on the customer premise SWC/CO to customer premise SWC/CO or IXC POP SWC/CO. When one end is an IXC POP, it's often called a dedicated access loop, The monthly charge is based on the mileage between them. To get your monthly charge for a dedicated circuit across the country, you typically add the IXC charge to the LEC charge. You may not need to pay an LEC charge if you're located in the same building or same complex as your IXC (long distance carrier) and can directly cable in.

Though it sounds as if the IXC charge is the distance a crow would fly directly between the two points, in reality, it is the distance in mileage between two Rate Centers whose position is laid down according to industry standards, originally created in 1956 by Jay K. Donald of AT&T Bell Telephone Laboratories. Mr. Donald referenced V&H projections in his paper "Map Projections – A Working Model," in which he described the concept as "an ellipsoidal adaptation of the two-point equidistant." Under this strange system, the entire U.S. is divided into a vertical and horizontal grid, based on a complex algorithm that projects the curvature of the earth onto a flat plane. This projection algorithm uses latitude and longitude, plus other factors, to develop a 10,000 by 10,000 V&H grid. The vertical and horizontal coordinates of each rate center are defined and applied to a square root formula which yields the "airline distance" between the two points. Think back to school. Think about a right-angled triangle. At the top is one Rate Center. At the side is the other Rate Center. The horizontal is the horizontal coordinate, in effect, the "airline mileage." The vertical is the vertical coordinate. The formula is simple: Square the vertical distance. Square the horizontal distance. Add the two together. Then take their square root. That will give you the distance across the hypotenuse – the side opposite the right angle in the triangle – i.e. the airline mileage. Here is the actual formula for calculating airline mileage. The airline mileage is the square root of ((V1 - V2) x (V1 - V2) + (H1 - H2) x (H1 - H2)) / 10. Dividing by ten is necessary in this case to allow for the way the coordinates are presented in North American mileage charts. This formula (without the division by ten) is called the Pythagorean theorem and is known to every schoolboy (and schoolgirl – a reader already pointed out my misdemeanor). It is named for its inventor, Pythagoras, Greek philosopher, mathematician, and religious reformer, who lived 582-500 B.C. Telcordia Technologies (previously Bellcore) and other independent companies develop and publish V&H data, and provide simple tables for conversion between V&H and Latitude/Longitude. As new central offices and POPS (points of presence) are installed and/or consolidated, the V&H coordinates table must be expanded and updated by the various publishers of the material. Thank you to Andrew Funk Manager, Access Pricing Strategy, Qwest Communications International for his help on this definition.

See the next page for the Vertical and Horizontal coordinates of major North American cities.

V Band A range of radio frequencies in the 40 GHz and 50 GHz range, also known as the Q Band. The "V" is a random, arbitrarily-assigned designation with its roots in the context of military security during World War II.

V Chip Violence Chip. A type of TV filter required in the Telecommunications Act of 1996. The Act requires the FCC to prescribe regulations, in conjunction with the electronic manufacturing industry, requiring that television sets manufactured after February 1998 include "features designed to enable viewers to block display of all programs with a common rating. This is commonly referred to as the 'V Chip.'" Similar capability is available through filtering agents for screening Internet content. The V Chip and Internet filtering agents primarily are intended to allow parents to filter content which they consider offensive and inappropriate for their children. These content filtering technologies can be overridden with the proper password. See also Filtering Agent.

V Commerce See V-Commerce below.

V Disk Digital Voice Storage cards.

V Fast A new higher speed over-normal-phone-line modem called V.Fast Class (V.FC) for 28,800 bits per second speed. See V.34.

V Interface The 2-wire ISDN physical interface used for single-customer termination from a remote terminal. See ISDN and U Interface.

V Reference Point The proposed interface point in an ISDN environment between the line termination and the exchange termination.

V series recommendations ITU-T standards dealing with data communications operation over the telephone network. The idea of standards is simple. If you have them and if every manufacturer conforms, then every modem can talk to every other one. That's the idea. But it's not always that simple. Sometimes you have to conform to several standards. For example, in the higher speed modems, for example those at 9,600 bps, you have to conform to speed. That's one standard. You have to conform to error control.

	V	H
ALABAMA		
Birmingham	7518	2304
Huntsville	7267	2535
Mobile	8167	2367
Montgomery	7692	2247
ARIZONA		
Flagstaff	8746	6760
Phoenix	9135	6748
Tucson	9345	6485
Yuma	9385	7171
ARKANSAS		
Fayetteville	7600	3872
Hot Springs	7827	3554
Pine Bluff	7803	3358
CALIFORNIA		
Anaheim	9250	7810
Bakersfield	8689	8060
Fresno	8669	8239
Long Beach	9217	7856
Los Angeles	9213	7878
Oakland	8486	8695
Redwood City	8556	8682
Sacramento	8304	8580
San Bernardino	9172	7710
San Diego	9468	7629
San Francisco	8492	8719
San Jose	8583	8619
Santa Monica	9227	7920
Santa Rosa	8354	8787
Sunnyvale	8576	8643
Van Nuys	9197	7919
COLORADO		
Denver	7501	5899
Fort Collins	7331	5965
Grand Junction	7804	6438
Greeley	7345	5895
Pueblo	7787	5742
CONNECTICUT		
Bridgeport	4841	1360
Hartford	4687	1373
New Haven	4792	1342
New London	4700	1242
Stamford	4897	1388
DELAWARE		
Wilmington	5326	1485
DISTRICT OF COLUMBIA (D.C.)		
Washington	5622	1583
FLORIDA		
Clearwater	8203	1206
Daytona Beach	7791	1052
Fort Lauderdale	8282	0557
Jacksonville	7649	1276
Miami	8351	0527
Orlando	7954	1031
Tallahassee	7877	1716
Tampa	8173	1147
GEORGIA		
Atlanta	7260	2083
Augusta	7089	1674
Macon	7364	1865
Savannah	7266	1379
HAWAII		
Hilo	12121	15075
Honolulu	11592	15609
Kaunakakai	11693	15443
Lanai City	11762	15429
Lihue	11340	15895
Wailuku	11770	15341
IDAHO		
Boise	7096	7869
Pocatello	7146	7250
ILLINOIS		
Chicago	5986	3426
Joliet	6088	3454
Peoria	6362	3592
Rock Island	6276	3816
Springfield	6539	3518
INDIANA		
Bloomington	6417	2984
Fort Wayne	5942	2982
Indianapolis	6272	2992
Muncie	6130	2925
South Bend	5918	3206
Terre Haute	6428	3145
IOWA		
Burlington	6449	3829
Cedar Rapids	6261	4021
Des Moines	6471	4275
Dubuque	6088	3925
Iowa City	6313	3972
Sioux City	6468	4768
KANSAS		
Dodge City	7640	4958
Topeka	7110	4369
Wichita	7489	4520
KENTUCKY		
Danville	6558	2561
Frankfort	6462	2634
Madisonville	6845	2942
Paducah	6982	3088
Winchester	6441	2509
LOUISIANA		
Baton Rouge	8476	2874
New Orleans	8483	2638
Shreveport	8272	3495
MAINE		
Augusta	3961	1870
Lewiston	4042	1391
Portland	4121	1384
MARYLAND		
Baltimore	5510	1575
MASSACHUSETTS		
Boston	4422	1249
Framingham	4472	1284
Springfield	4620	1408
Worcester	4513	1330
MICHIGAN		
Detroit	5536	2828
Flint	5461	2993
Grand Rapids	5628	3261
Kalamazoo	5749	3177
Lansing	5584	3081
MINNESOTA		
Duluth	5352	4530
Minneapolis	5777	4513
St. Paul	5776	4498
MISSISSIPPI		
Biloxi	8296	2481
Jackson	8035	2880
Meridian	7899	2639
MISSOURI		
Joplin	7421	4015
Kansas City	7027	4203
St. Joseph	6913	4301
Springfield	7310	3836
MONTANA		
Billings	6391	6790
Helena	6336	7348
Missoula	6336	7650
NEBRASKA		
Grand Island	6901	4936
Omaha	6687	4595
NEVADA		
Carson City	8139	8306
Las Vegas	8665	7411
Reno	8064	8323
NEW HAMPSHIRE		
Concord	4326	1426
Manchester	4354	1388
Nashua	4394	1356
NEW JERSEY		
Atlantic City	5284	1284
Camden	5249	1453
Hackensack	4976	1432
Morristown	5035	1478
Newark	5015	1430
New Brunswick	5085	1434
Trenton	5164	1440
NEW MEXICO		
Albuquerque	8549	5887
Las Cruces	9132	5742
Santa Fe	8389	5804
NEW YORK		
Albany	4639	1629
Binghamton	4943	1837
Buffalo	5076	2326
Nassau	4961	1355
New York City	4977	1406
Poughkeepsie	4821	1526
Rochester	4913	2195
Syracuse	4798	1990
Troy	4616	1633
Westchester	4912	1330
NORTH CAROLINA		
Asheville	6749	2001
Charlotte	6657	1698
Fayetteville	6501	1385
Raleigh	6344	1436
Winston-Salem	6440	1710
NORTH DAKOTA		
Bismarck	5840	5736
Fargo	5615	5182
Grand Forks	5420	5300

	V	H		V	H		V	H
OHIO			Spartanburg	6811	1833	**VIRGINIA**		
Akron	5637	2472	**SOUTH DAKOTA**			Blacksburg	6247	1867
Canton	5676	2419	Aberdeen	5992	5308	Leesburg	5634	1685
Cincinnati	6263	2679	Huron	6201	5183	Lynchburg	6093	1703
Cleveland	5574	2543	Sioux Falls	6279	4900	Norfolk	5918	1223
Columbus	5872	2555	**TENNESSEE**			Richmond	5906	1472
Dayton	6113	2705	Chattanooga	7098	2366	Roanoke	6196	1801
Toledo	5704	2820	Johnson City	6595	2050	**WASHINGTON**		
OKLAHOMA			Knoxville	6801	2251	Bellingham	6087	8933
Lawton	8178	4451	Memphis	7471	3125	Kennewick	6595	8391
Oklahoma City	7947	4373	Nashville	7010	2710	North Bend	6354	8815
Tulsa	7707	4173	**TEXAS**			Seattle	6336	8896
OREGON			Amarillo	8266	5076	Spokane	6247	8180
Medford	7503	8892	Austin	9005	3996	Yakima	6533	8607
Pendleton	6707	8326	Corpus Christi	9475	3739	**WEST VIRGINIA**		
Portland	6799	8914	Dallas	8436	4034	Clarksburg	5865	2095
PENNSYLVANIA			El Paso	9231	5655	Morgantown	5764	2083
Allentown	5166	1585	Fort Worth	8479	4122	Wheeling	5755	2241
Altoona	5460	1972	Houston	8938	3563	**WISCONSIN**		
Harrisburg	5363	1733	Laredo	9681	4099	Appleton	5589	3776
Philadelphia	5257	1501	Lubbock	8596	4962	Eau Claire	5698	4261
Pittsburgh	5621	2185	San Antonio	9225	4062	Green Bay	5512	3747
Reading	5258	1612	**UTAH**			La Crosse	5874	4133
Scranton	5042	1715	Logan	7367	7102	Madison	5887	3796
RHODE ISLAND			Ogden	7480	7100	Milwaukee	5788	3589
Providence	4550	1219	Provo	7680	7006	Racine	5837	3535
SOUTH CAROLINA			Salt Lake City	7576	7065	**WYOMING**		
Charleston	7021	1281	**VERMONT**			Casper	6918	6297
Columbia	6901	1589	Burlington	4270	1808	Cheyenne	7203	5958

That's another standard. And you also have to conform to data compression – if you are using data compression. ISDN terminal adapters are V series recommendations, too. ITU-T uses the term "bis" to designate the second in a family of related standards and "ter" designates the third in a family.

V-Commerce Voice Commerce. Imagine that you dial a phone number. Your bank answers. "What's your password, please?" You say, "Harry Newton." It answers, "Thank you." You have three bills to pay, but only enough money to pay two. Here are the three..." You answer "Pay the Visa bill." You get the message. Voice commerce is a fancy name for interactive voice response using speech recognition as the input mechanism, not the buttons on a touchtone phone. With touchtone buttons, you often have to keep pressing buttons to progress through zillions of menus to get what you want. Speech recognition, if it works, is faster since it gets you to where you want to go a lot faster.

V.110 Terminal rate adaptation protocols for the ISDN B channel with a V-type interface. Includes V.120.

V.120 Terminal rate adaptation protocols for the ISDN B channel with a V-type interface. Includes V.110.

V.13 ITU-T standard for simulated carrier control. Allows a full-duplex modem to be used to emulate a half-duplex modem with interchange circuits changing at appropriate times.

V.14 ITU-T standard for asynchronous-to-synchronous conversion without error control. Allows a modem that is actually synchronous to be used to carry start/stop (async) characters. If a V.42 modem connects with another modem that doesn't have error-control, it falls back to V.14 operation to work without error-control.

V.17 ITU-T standard for simplex (one-way transmission) modulation technique for use in extended Group 3 Facsimile applications only. Provides 7200, 9600, 12000, and 14400 bps trellis-coded modulation (the modulation scheme is similar to V.33), MMR (Modified Modified Read) compression and error-correction mode (ECM).

V.18 An ITU recommendation aimed at enhancing the capability of the deaf or speech-impaired to use telecommunications. The V.18 recommendation describes a multi-function text telephone that, according to the ITU, "bridges the gap that has existed between several incompatible text telephones in use today." See T.140.

V.21 ITU-T standard for 300 bit per second duplex modems for use on the switched telephone network. V.21 modulation is used in a half-duplex mode for Group 3 fax negotiation and control procedures (ITU-T T.30). Modems made in the U.S. or Canada follow the Bell 103 standard. However, the modem can be set to answer V.21 calls from overseas.

V.21 CH 2 ITU-T standard for 300 bps modem, describing the operation of modems at 300 bps, and used for critical control and handshaking functions. This low speed is highly tolerant of noise and impairments on the phone line. Fax machines use only Channel 2 of the V.21 recommendations (half duplex channel).

V.21 Fax An ITU-T standard for facsimile operations at 300 bps.

V.22 ITU-T standard for 1,200 bit per second duplex modems for use on the switched telephone network and on leased circuits. V.22 is compatible with the Bell 212A standard observed in the U.S. and Canada. See V.22 bis.

V.22 bis ITU-T standard for 2,400 bit per second duplex modems for use on the switched telephone network. "Bis: is used by the ITU-T to designate the second in a family of related standards. "ter" designates the third in a family. The standard includes an automatic link negotiation fallback to 1200 bps and compatibility with Bell 212A/V.22 modems.

The principal characteristics of the V.22bis modems are:

Duplex mode of operation on the PSTN and point-to-point 2-wire leased circuits, Channel separation by frequency division, Quadrature amplitude modulation for each channel with synchronous line transmission at 600 baud, An adaptive equalizer and a compromise equalizer, Test facilities, Data signaling rates of 1200bps and 2400bps. V.22bis is compatible with the V.22 modem and includes automatic bit rate recognition. The V.22bis modem technology is used for applications that require transactions at a rate of 2400 bps or slower. Some such applications include set-top boxes, credit card transactions, fax relay systems, satellite receivers, utility meters, and network control.

V.23 V.23 is the standard for a modem with a 600 bps or 1200 bps "forward channel" and a 75 bps "reverse" channel for use on the switched telephone network.

V.24 ITU-T definitions for interchange circuits between data terminal equipment (DTE) and data communications equipment (DCE) equipment. In data communications, V.24 is a set of standards specifying the characteristics for interfaces. Those standards include descriptions of the various functions provided by each of the pins. This standard is similar (but not identical) to the RS-232-C as established by the American TIA/EIA – Telecommunications Industry Association / Electronics Industries Association. V.24 is a ITU-T data communications hardware interface standard for communications at speeds up to 19.2 kbps.

V.25 Automatic calling and/or answering equipment on the general switched telephone network, including disabling of echo suppressors on manually established calls. Among other things, V.25 specifies an answer tone different from the Bell answer tone. Many modems, including U.S. Robotics modems, can be set with the BO command so that they use the V.25 2100 Hz tone when answering overseas calls.

V.25 bis An ITU-T standard for synchronous communications between the mainframe or host and the modem using the HDLC or character-oriented protocol. Modulation depends on the serial port rate and setting of the transmitting clock source.

V.26 V.26 is the ITU-T standard for 2400 bps modem for use on 4-wire leased lines.

V.26 bis ITU-T standard for 1.2/2.4 Kbps modem. It is important to note that V.26 bis is a half-duplex modem (1200 or 2400 bps in only one direction at a time); it provides an optional 75 bps reverse channel as well.

V.26 ter V.26 ter is a FULL DUPLEX 2400 bps modem, like V.22 bis. The difference is that V.26 ter uses echo cancellation (like V.32) instead of frequency division (like V.22 bis), making it more expensive than V.22 bis. It was intended to serve as a fallback mode from V.32, but most manufacturers ignored it and provide V.22 bis as a fallback instead (V.26 ter is used only in a few installations in France, as far as we know).

V.27 ITU-T standard for 4,800 bits per second modem with manual equalizer for use on leased telephone-type circuits. May be full-duplex on four wire leased lines, or half-duplex on two wire lines.

V.27 bis ITU-T standard for 2,400 / 4,800 bits per second modem with automatic equalizer for use on leased telephone-type circuits. 2.4 Kbps modem for 4-wire leased circuits. Either speed (2,400 is a fallback) can be used on either 4-wire leased lines (full duplex) or 2-wire leased lines (half-duplex). It also provides an optional 75 bps reverse channel.

V.27 ter ITU-T standard for 2,400 / 4,800 bits per second modem for use on the switched telephone network. Half-Duplex only. V.27 ter is the modulation scheme used in Group 3 Facsimile for image transfer at 2400 and 4800 bps. 4800 bps is a common "fallback" speed.

V.28 V.28, entitled "Electrical Characteristics for Unbalanced Double-Current Interchange Circuits" provides the ITU-T equivalent of the electrical characteristics defined in EIA-232.

V.29 ITU-T standard for 9,600 bits per second modem for use on point-to-point leased circuits. Virtually all 9,600 bps leased line modems adhere to this standard. V.29 uses a carrier frequency of 1700 Hz which is varied in both phase and amplitude. V.29 also provides fallback rates of 4800 and 7200 bps. V.29 can be full-duplex on 4-wire leased circuits, or half-duplex on two wire and dial up circuits. V.29 is the modulation technique used in Group 3 fax for image transfer at 7200 and 9600bps.

V.3 ITU-T specification that describes communications control procedures implemented in 7-bit ASCII code. An ITU-T data communication recommendation that defines the 7-bit code for the alphanumeric and control characters in a character-oriented application, such as ASCII for character coding.

V.32 ITU-T standard for 9,600 bit per second two wire full duplex modem operating on regular dial up lines or 2-wire leased lines. If you're buying a 9,600 bps modem for use on the normal dial up switched phone lines, make sure it conforms to V.32. If your modem also conforms to V.42 bis, you should be able to transmit and receive at up to 38,400 bps with other modems that conform to these two specifications. I personally use a number of V.32/V.42 bis modem and they work wonderfully fast. V.32 also provides fallback operation at 4,800 bps. See also V.32 bis, V.42 bis Error Correction and V.42 bis Data Compression and Modulation Protocols.

V.32 bis New higher speed ITU-T standard for full-duplex transmission on two wire leased and dial up lines at 4,800, 7,200, 9,600, 12,000, and 14,400 bps. Provides backward compatibility with V.32. Modems running at V.32 bis at its highest speed of 14,400 bps are actually transmitting that many bits per seconds. They do not rely on compression to achieve that high speed. However, with data compression – such as V.42 and V.42 bis – they can achieve higher speeds. The V.32 bis standard also includes "rapid rate renegotiation" feature to allow quick and smooth rate changes when line conditions change. See Modulation Protocols V.42 and V.42 bis.

V.32 terbo Modulation scheme that extends the V.32 connection range: 4800, 7200, 9600, 12K and 14.4K bps. V.32 bis terbo modems fall back to the next lower speed when line quality is impaired, and fall back further as necessary. They fall forward to the next higher speed when line quality improves.

V.33 ITU-T standard for 14,400 and 12,000 bps modem for use on four wire leased lines.

V.34 V.34 is the international standard for dial up modems of up to 28,800 bits per second. Since the standard suggests speeds twice as fast as the top standard they replace, they carry the nickname "V.Fast." New V.34 modems have a feature called line probing that will allows them to identify the capacities and quality of the specific phone line and

adjust themselves to allow, for each individual connection, for maximum throughput. The standard also supports a half-duplex mode of operation for fax applications. The new V.34 technology includes an optional auxiliary channel with a synchronous data signaling rate of 200 bits/second. Data conveyed on this channel consists of modem control data. V.34 modems contain multidimensional trellis coding, which is used to gain higher immunity to noise and other phone line impairments. V.34 modems are the first modems to identify themselves to telephone network equipment (handshaking). V.34 technology has been long in coming and has had to overcome many obstacles. At one point, members of the modem manufacturing industry became so impatient, that some of them began shipping their own proprietary versions of what they thought V.34/V.Fast/28,800 bps modems would be. Many of these modems are only compatible, at higher than 14,400 bps speeds, with themselves. See V.34bis.

V.34 bis Also known as V.34+. A faster version of the data communications standard, V.34, which supports up to 28,800 bps. V.34 bis adds two higher data rates to V.34. These speeds are 31,200 bits per second and 33,600 bits per second. V.34 bis is now the most common standard for PC data communications over dial-up phone lines.

V.35 ITU-T standard for trunk interface between a network access device and a packet network that defines signaling for data rates greater than 19.2 Kbps. It is an international standard termed "data transmission up to 1.544 Mbps" (i.e. T-1). It's typically used for DTE or DCE equipment that interface to a high-speed digital carrier. The physical interface is a 34-pin connector, which can't connect, either physically or electrically, to any other interface without a special converter. See V.36.

V.35 connector

V.36 ITU-T recommendation for 4-wire communications at speeds greater than 48 Kbps. It is intended to replace V.35. See V.35.

V.42 Error Correction ITU-T error-correction standard specifying both MNP4 and LAP-M. The ITU-T title says "Error-correcting procedures for DCEs using Asynchronous-to-Synchronous Conversion". It also notes in the text that it applies only to full-duplex devices. The ITU-T modulation schemes with which V.42 may be used are V.22, V.22 bis, V.26 ter, and V.32, and V.32 bis. LAPM, based on HDLC, is the "primary" protocol, on which all future extensions will be based. The Alternative Protocol specified in Annex A of the Recommendation is for backward compatibility with the "installed base" of error-correcting modems. See V.42 bis and V.44.

V.42bis An ITU-T data compression standard, with "bis" being the French term for "second" or "encore." It is used by the ITU/ITU to designate the second in a family of related standards. ("ter" designates the third in a family.) V.42bis compresses files "on the fly" at an average ratio of 3.5:1 (3.5 to 1) and can yield file transfer speeds of up to 9,600 bps on a 2,400 bps modem, 38,400 bits per second with a 9,600 bps modem, 57,600 bps with a 14,400 bps modem, or 115,600 bit/s on a 28,800 bps modem. On-the-fly data compression only has value if you use it to transfer and receive material that is not already compressed. Compressing stuff a second time yields no significant improvement in speed (assuming your compression technique worked the first time around). So the decision to buy a V.42 bis modem depends on the material you're working with and your pocketbook.

V.42bis modems are more expensive than predecessor modems.

V.42bis was approved by the ITU-T because of its technical merits. Existing data compression methods (MNP 5 for example) only provided up to two-to-one compression. Also, V.42bis provides for built-in "feedback" mechanisms, so that the modem can monitor its own compression performance. If the DTE starts send pre-compressed or otherwise uncompressible data, V.42bis can automatically suspend its operation to avoid expansion of the data. It continues to monitor performance even when sending data "in the clear," and when a performance improvement can be gained by reactivating compression, it will do so automatically.

V.42bis was selected because it would work with a wide variety of different implementations – different amounts of memory, different processor speeds, etc. Because of this, there are differences between various manufacturers' products in terms of throughput performance (although they will all properly compress and decompress, some will do it faster than others). If maximum throughput is important, you should check published benchmark tests to find the modem that provides the best performance.

This chart, courtesy Hayes, shows the speedup that's possible. It includes information on a modem called the Hayes Optima 288, which includes a proprietary (i.e. not compatible with anyone else) Hayes enhanced implementation of V.42 bis.

V.32	9,600 + data compression	= 38,400 bit/s
V.32bis	14,400 + data compression	= 57,600 bit/s
V.34	28,800 + data compression	=115,600 bit/s
Hayes Optima 288	28,800 + Hayes V.42 bis	=230,400 bit/s

See also V.44.

V.44 A compression standard finalized on June 30, 2000 by the ITU-T. V.44 makes use of the LZJH (Lempel-Ziv-Jeff.Heath) compression algorithm developed by Jeff Heath of Hughes Network Systems for use over satellite links. Supplanting the earlier V.42bis modem compression technology, V.44 provides 6:1 (i.e., 6 to 1) compression performance. V.44 is intended for use in V.92 modems, the successor to V.90. In combination with V.92 modems, V.44 will have a significant effect on the speed of data transmission, as did the earlier move from V.34 to V.90 modems. In the case of V.90, transmission speed is determined primarily   by the quality of the local loop. In the case of V.44, performance is dynamic (i.e., variable) and depends on the data content, as well as the quality of the local loop, which can vary during the course of a data call. When compared to V.42bis running in V.90 modems, V.44 running in V.92 modems can yield speed improvements of 20% to 60%, up to as much as 200% for certain types of highly compressible data. Taken together, V.90 and V.44 can yield effective downstream throughput of more than 300 Kbps (6:1 compression times 56 Kbps), under optimum conditions. This compares to the maximum rate of 150-200 Kbps supported by V.90 modems running V.42bis. Like the predecessor combination V.90 and V.42bis, the combination of V.92 and V.44 provides asymmetric bandwidth. While V.90 modems can support upstream speed up to 33.6 Kbps, V.92 will support up to 48 Kbps in the upstream direction. Both standards support downstream transmission rates of up to 56 Kbps, theoretically and under optimum conditions. Note that the difference between raw transmission speed and throughput is a result of the compression of the data prior to its being put on the circuit. See also LZJH, V.42bis, V.90, and V.92.

V.5 See V5

V.54 ITU-T standard for loop test devices in modems, DCEs (Data Communications Equipments) and DTEs (Data Terminal Equipment). Defines local and remote loopbacks. There are four basic tests – a local digital loopback test that is used to test the DTE's send and receive circuits; a local analog loopback test that is used to test the local modem's operation; a remote analog loopback test that is used to test the communication link to the remote modem; and a remote digital loopback test that is used to test the remote modem's operation. If a modem has V.54 capability (most V.32 and V.32 bis modems do), its manual should include documentation on performing the various tests. Version 7 of the Norton Utilities (from Symantec) also includes a local digital loopback test for your PC's COM ports, for which you will need the optional jumper plug offered with the software. Where a modem supports local digital loopback testing, it simulates the jumper plug and does not, therefore, need to be disconnected.

V.61 ITU-T V.61 is a 14.4 kbps V.32bis analog multiplexing technology standard developed by AT&T Paradyne and marketed as VoiceSpan. The data rate is reduced to 4800 bps during simultaneous voice and data. This analog Simultaneous voice and data (ASVD) standard has now been effectively obsoleted by V.70.

V.70 ITU-T standard for Digital Simultaneous Voice and Data (DSVD) modems. DSVD allows the simultaneous transmission of data and digitally-encoded voice signals over a single dial-up analog phone line. DSVD modems use for V.34 modulation (up to 33.6 kilobits per second), but may also use V.32 bis modulation (14,400 kilobits per second). The DSVD

voice coder is a modified version of an existing specification and is defined as G.729 Annex A. The DSVD voice/data multiplexing scheme is an extension of the V.42 error correction protocol widely used in modems today. DSVD also specifies fallbacks that enable DSVD modems to communicated with standard data modems (i.e. V.34, V.32 bis, V.32 and V.22). See DSVD for a bigger explanation.

V.75 ITU-T recommendations which specify DSVD control procedures. See V.70.

V.76 ITU-T recommendations which define V.70 multiplexing procedures.

V.8 A way V.34 modems negotiate connection features and options.

V.8 bis New start-up sequence for multimedia modems.

V.80 V.80 is the application interface defined in the H.324 ITU video conferencing standard. A V.80 provides a standard method for H.324 applications to communicate over modems. A V.80 modem provides three main functions:

1. Converts synchronous H.324 streams to run on asynchronous modem connections. That is, they accept and send data in synch with a timing device ("clock"). Serial ports and modems are asynchronous, meaning they accept and receive data independent of any clocking device. V.80 converts the synchronous data stream of an H.324 application so that it can communicate through an asynchronous modem connection.

2. Allows for rate adjustments based on line conditions. Modems adjust to different line conditions throughout a call. Under bad conditions a modem will slow down. When conditions clear, a modem will resume at top speed. A V.80 modem alerts an H.324 video phone of its rate adjustments thereby allowing the application to adjust the rate at which it sends video and audio.

3. Communicates lost packets to the H.324 application. During transmission, data can be lost due to buffer overflows, phone line errors and a number of other issues. Under these conditions, a V.80 modem communicates lost data information to the H.324 application, helping it to keep real-time audio and video flowing to both sides of the call.

See H.324.

V.90 An ITU-T standard for Pulse Code Modulation (PCM) modems running at speeds to 56 Kbps. It informally and variously was known as V.PCM and V.fast until the numeric designation, V.90, was assigned when the formal standard was approved on February 6, 1998. V.90 modems support transmission on an asymmetric basis. They support speeds of up to 56 Kbps in the downstream direction, from the central site equipment to the end user. The upstream "back channel" from the end user to the central site remains limited to 33.6 Kbps (i.e., V.34+ speeds). Actually and currently, the maximum downstream rate is 53.5 Kbps, as the standard exceeds the maximum amplitude levels supported over copper local loops. This restriction is expected to ease into the future, thereby allowing V.90 signaling rates to reach the full potential of 56 Kbps. See 56 Kbps Modem (for a longer technical explanation), V.91, V.92 and V.PCM.

V.91 A developing standard from the ITU-T, V.91 is the all-digital extension to V.90. V.91 will allow modem signals to be transmitted over digital circuits, such as ISDN BRI (Basic Rate Interface) local loops in consumer and home office applications, and PRI (Primary Rate Interface) local loops that connect to corporate PBXs. Thereby, end users can achieve connectivity to ISPs and others supporting V.90 modem access. In the absence of V.91, such users can connect to an ISP, for example, only if the ISP also supports IDSN. V.91 is intended to operate at signaling rates up to 64 Kbps, and will make use of both 4-wire circuit-switched connections and leased point-to-point 4-wire digital connections. See also V.90, V.92 and V.PCM.

V.92 A dial-up modem standard finalized by the ITU-T in late 2000, as the successor to V.90. V.92 improves on V.90 in a number of ways. First, while V.92 runs at the same maximum 56-Kbps downstream signaling rate as V.90, it runs at a maximum of 48 Kbps upstream, compared with V.90 upstream speed of 33.6 Kbps. This improvement in upstream signaling speed is due to the use of a modulation standard known as PCM upstream, with PCM meaning Pulse Code Modulation. PCM upstream improves the speed of the upstream channel by making use of the same clocking source for synchronization purposes as does the downstream channel. Second, V.92 makes use of the V.44 compression standard, finalized by the ITU-T in November, 2000. V.44 offers considerable improvement over the V.42bis standard used in V.90 modems. Specifically, V.44 offers a compression ratio of 6:1 (6 to 1), while V.42bis was limited to 3.5:1. Depending on the compressibility of the subject data, V.44 therefore improves throughput up to 200% or more over V.42bis. As a result, V.92 modems offer downstream throughput performance of as much as 300 Kbps, compared with the 150-200 Kbps supported by V.90 modems. Third, the V.92 standard includes a feature known as QuickConnect, which reduces the time consumed in the handshaking process between two modems. QuickConnect accomplishes this by remembering and reusing information gained during previous handshaking processes; as the same connection typically is used repeatedly, there often is no reason for the characteristics of the connection to be re-learned every time. QuickConnect can shave 10 to 20 seconds off the 20 to 30 seconds required to establish a dial-up connection required by V.90 modems. Fourth, V.92 modems can both recognize and respond to a call waiting tone, and place the data call "on hold" while the voice call is answered. This feature can eliminate the need for a second modem/fax line. As the server pauses the data session, it is easily and quickly resumed without the need for establishing another dialup session. Lastly, there is hope V.90 modems will be upgradable to V.92 modems through a software download. See also V.42bis, V.44, V.90, and V.91.

V.ASVD Analog Simultaneous Voice and Data modem.

V.AVD Alternating Voice and Data. This is the same function as provided by VoiceView products.

V.DSVD Digital Simultaneous Voice and Data.

V.Fast V.FC. An interim modem standard to support speeds to 28,800 bits per second for uncompressed data transmission rates over regular dial up, voice-grade lines. V.FAST stands for Very Fast. V.Fast was a "standard" that only a few manufacturers of modems adopted. These manufacturers adopted V.Fast because they were impatient with the ITU's slowness. Eventually, however the ITU did adopt a new standard, called V.34. See V.34 and V.34bis.

V.FC Version Fast Class. It is an interim standard that was developed for use until the ITU-T ratified V.Fast, i.e. V.34, which is the speed that a V.34 modem communicates at – namely at 28,800 bits per second. V.FC was eventually obsoleted by V.34, which the ITU-T eventually adopted. See V.34.

V.GMUX The multiplexer for V.DSVD.

V.PCM All the makers of 56 Kbps modems had been promising an interoperable 56 Kbps specification, and the International Telecommunication Union (the ITU). They got tired of waiting for a standard, and developed pre-standard versions known as x2 and K56flex. In February of 1998, ITU-T finally obliged with a new standard, called V.PCM (later termed V.90), which is an amalgamation of the pre-standard solutions from Lucent, 3Com, and Motorola. See also x2, K56flex, V.90, V.91, and V.92.

V.Standards Standards recommended by the ITU-T. See above.

V/T Internet-speak for Virtual Time. R/T means the time it takes to download stuff. Writing in the New York Times, Charles McGrath said it was "customary in Net-speak to make a distinction between r/t, or real time – the time in which all these delays and jam-ups occur and v/t, or virtual time, which is time on the Net: a kind of external present in which it is neither day nor night and the clock never ticks. V/t is time without urgency, without priority."

V11 The V11 interface is the same as the RS-422 interface., as well as supporting more than 32 remote stations, and authorizing mode changes from HDX to FDX. The ELV11 provides galvanic isolation between the two V11 interfaces via its internal opto coupler circuits. V11 I/Os use two twisted pairs, transmit data and receive data with a common signal. Specifications

V2oIP Voice and video over IP. V2oIP is the use of IP technologies for videoconferencing, visual telephony, and online collaboration. Protocol stacks that support V2oIP include H.223, H.320, H.321, H.324, and others.

V5 A standard approved by ETSI (European Telecommunications Standards Institute) in 1997 for the interface between the access network and the carrier switch for basic telephony, ISDN and semi-permanent leased lines. The V5 standard effectively provides for open access to both wired and wireless networks, thereby encouraging competition in a deregulated environment. V5 is European Telecommunications Standards Institute's (ETSI's) open standard interface between an Access Node (AN) and a Local Exchange (LE) for supporting PSTN (Public Switched Telephone Network) and ISDN (Integrated Services Digital Network). Examples of Access Nodes include Digital Loop Carrier (DLC) systems, wireless loop carrier system, and Hybrid Fiber Coax (HFC) systems. The V5 series includes V5.1, which is a non-concentrating 2.048 Mbps (i.e., E-1) Subscriber Network Interface (SNI), and V5.2, which is a concentrating SNI supporting as many as 16 2.048 Mbps physical interfaces. See also E-1 and ETSI.

VA This is a form of power measurement called "Volt-Amps". A VA rating is the Volts rating multiplied by the Amps (current) rating. The VA rating can be used to indicate the output capacity of a UPS (Uninterruptible Power Supply) or other power source, or it can be used to indicate the input power requirement of a computer or other AC load. For loads, the VA rating multiplied by the Power Factor is equal to the Watts rating. The VA rating of a load must always be greater than or equal to the Watts rating because Power Factor cannot be greater than 1. This definition courtesy American Power Company.

VAB Value Added Business partner. A term which Hewlett-Packard uses for developers which write software for its computers. HP helps its VABs sell software. Clearly, by doing so, it helps sell more HP computers.

VAC Voice Activity Compression.

vAC Volts, Alternating Current

vacant code An unassigned area code, central office or station code.

vacant code intercept Routes all calls made to an unassigned "level" (first digit dialed) to the attendant, a busy signal, a "reorder" signal or to a recorded announcement.

vacant number intercept Routes all calls of unassigned numbers to the attendant, a busy signal or a prerecorded announcement.

vacation message See Auto Responder.

vacation service A service offered by local telephone companies to subscribers who will be away. A live operator or a machine intercepts the calls and delivers a message. When you come back, you get your old number. But in the meantime, while you're away, you pay less money per month than you would for normal phone service. Also known as Absent Subscriber Service.

VACC Value Added Common Carrier. A common carrier that provides some network service other than simple end-to-end data transmission. Services include least-cost routing, accounting data, and delivery clarification.

vacuum tubes Before there were solid state devices there were vacuum tubes. A vacuum tube is an air-evacuated glass bulb with at least two electrodes: a cathode and an anode. The cathode is heated causing the electrons to "boil off." If a voltage is placed across the cathode and the anode, the electrons will be attracted to the anode completing the electric circuit. A tube with just a cathode and anode is called a diode. If a grid is placed between the anode and cathode, a small current placed on the grid can control the much larger cathode-anode current. This type of tube is called a triode. As vacuum tubes evolved, additional grids were inserted between the cathode and anode to produce tetrodes, pentodes, etc. Today, transistors have replaced vacuum tubes in all except a few specialized applications.

The first electronic computers, Eniac and Univac, built in the wartime secrecy of the 1940s, employed vacuum tubes. They had an average life span of about 20 hours, but with thousands of hot glowing tubes in a single machine, some computers shut down every seven to twelve minutes. Vacuum tubes imposed a limit on the size and power of planned next generations of computers. The second generation of computers, never used vacuum tubes. It used transistors, which were invented in 1947.

VAD 1. Value Added Dealer. Another term for Value Added Reseller (VAR). Essentially, VARs or VADs are companies who buy equipment from computer or telephone manufacturers, add some of their own software and possibly some peripheral hardware to it, then resell the whole computer or telephone system to end users, typically corporations.

2. Voice Activated Dialing.

3. Voice Activity Detection. When enabled on a voice port or a dial peer, silence is not transmitted over the network, only audible speech. When VAD is enabled, the sound quality is slightly degraded but the connection monopolizes much less bandwidth.

VADSL Very-high-speed ADSL. A variation on the theme of VDSL (Very-high-data-rate Digital Subscriber Line), which likely will support symmetric, bidirectional transmission. See also ADSL, VDSL and xDSL.

Vail, Theodore N. Theodore N. Vail began his career with the Bell System as general manager of the Bell Telephone Company in 1878. He later became the first president of the American Telephone & Telegraph Company in 1885. He left AT&T two years later. After pursuing other interests for 20 years, he returned as president of AT&T in 1907, retiring in 1919 as chairman of the board. Vail believed in "One policy, one system, universal service." He regarded telephony as a natural monopoly. He saw the necessity for regulation and welcomed it.

VAIVR Voice Activated Interactive Voice Response.

validation 1. Generally, all long distance carriers, operator service providers and private pay phone companies will not put a call through unless they can "validate," the caller's telephone company calling card, home/business phone number or credit card. Until the advent of US West's Billing Validation Service and other similar databases in 1987, the companies who needed to validate their callers' billing requests had to turn back the caller or accept the call on faith. Validating a user's calling card is, simply, a Yes-No. If the card number is validated, it is Yes. Getting the validation involves a data call from the provider to the owner of the database. There are many ways of doing this, including a dedicated trunk and an port through an X.25 network. Here's an explanation from material put out by Harris, maker long distance switches, including the P2000V: "Validation processing starts with a check of the P2000V's own internal database of invalid 'billed to' numbers. This database contains numbers that the system administrator wishes to temporarily block. If a call's 'billed to' number does not appear in the database, the P2000V then queries the external validation service. The P2000V directly accesses external validation services via an X.25 modem connected to a leased line. The P2000V can also access Line Information Database (LIDB) through LIDB service bureau providers."

2. Tests to determine whether an implemented system fulfills its requirements. The checking of data for correctness or for compliance with applicable standards, rules, and conventions. The portion of the development of specialized security test and evaluation, procedures, tools, and equipment needed to establish acceptance for joint usage of an automated information system by one or more departments or agencies and their contractors.

3. A telephone company term. The determination of the degree of validity of a measuring device. The validation checks that can be made using output data are of six general types: 1. Compare related sets of registers; 2. Compare like groups of equipments; 3. Compare past and present data; 4. Compare usage and peg count; 5. Compare usage against grade of service.

validity check Any check designed to insure the quality of transmission.

value added 1. Refers to a voice or data network service that uses available transmission facilities and then adds some other service or services to increase the value of the transmission. For example, a value added service might be a "never busy" fax service, where call-forward-busy calls are sent to a PC, which holds the incoming fax for transmission back to the fax machine when the line is free. Value-added tends to mean the addition of some computer or smart switch to the network.

2. Programmers who work overtime for free are often considered "value added."

Value Added Carrier VAC. A voice or data common carrier that adds special service features, usually computer related, to services purchased from other carriers and then sells the package of service and features.

Value Added Common Carrier VACC. A common carrier that provides some network service other than simple end-to-end data transmission. Services include least-cost routing, accounting data, and delivery clarification.

Value Added Network VAN. A data communications network in which some form of processing of a signal takes place, or information is added by the network. No one knows, however, exactly what a VAN is. The general idea is that a VAN buys "basic" transmission and sometimes switching services from local and long distance phone companies and adds something else – typically an interactive computer with a database, a computer and massive storage. In this way, the VAN adds value to basic communications services. Dial up stock market quoting services are VANs. Electronic mail providers are VANs. But VANs can also simply be basic X.25 packet switching networks which are open to the public. Such a network will use X.25 packet switching to provide error correction, redundancy, and other forms of network reliability. Private organizations (companies, universities, etc.) may set up their own value-added networks, or – as in the case of PDNs (Public Data Networks) – another fancy name for a VAN that offers its services to the public. The classic VAN is a packet-switched operation like Tymnet, GTE Telenet, MCI Mail or AT&T Mail.

A VAN can also be communication network that provides features other than transmission of information, such as translation of one type of computer signal to another type of computer signal, called protocol conversion. VAN sometimes refers to packet-switched networks with protocol conversion. The value added is referred to as dissimilar system interface capability.

Value Added Network Service VANS. A data transmission network routing transmissions according to available paths, assures that the message will be received as it was sent, provides for user security, high speed transmission and conferencing among terminals. Closely akin to courier services or shipping forwarders in physical commerce.

Value Added Reseller See OEM and VAR.

value added service A communications facility using common carrier networks for transmission and providing extra data features with separate equipment. Store and forward message switching, terminal interfacing and host interfacing features are common extras. See also Value Added Network.

value chain The value chain is the entire collection of companies that contribute to a finished product that's ultimately sold to a consumer – from the guys who make the steel to the guys who sell you the car at the local dealer. When you get to optimizing the value chain, you get something called value chain Management. See Value Chain Analysis and Value Chain Management.

value chain analysis A financial tool for identifying and quantifying cost-reduction opportunities within the supply chain. See Value Chain Management for a longer explanation.

value chain management The value chain is the entire collection of companies that contribute to a finished product that's ultimately sold to a consumer. According to Industry Week Management, "value chain management enables the synchronized flow of product, information, processes and cash – from raw material to end customers. It optimizes the profitability and productivity of the entire value chain, from the supplier's supplier to the customer's customer, and of the individual business constituents that make up the chain... value chain management binds all corporate facets – the plan, develop, buy, make, sell and move activities – into unified inter-company relationships, enabling companies to target larger than corporate goals and achieve larger than corporate benefits." Benefits from value chain management include:

- New product development cycle at companies within high-performance chains is nearly twice as fast as.
- Firms with highly effective value chain strategies are nearly four times as like to have their suppliers heavily involved in new product development.
- Manufacturers with highly effective value chain strategies turn total inventory 40% faster.
- 94% of companies within high performance chains report that increased sales are a benefit of sharing information with value chain partners.

value driven re-engineering A fancy term for Re-Engineering, which is a term probably invented by Michael Hammer in the July-August, 1990 issue of Harvard Business Review. In that issue, he wrote "It is time to stop paving the cowpaths. Instead of embedding outdated processes in silicon and software, we should obliterate them and start over. We should 're-engineer' our business: use the power of modern information technology to radically redesign our business processes to achieve dramatic improvements in their performance." The term re-engineering now seems to me mean taking tasks presently running on mainframes and making them run on file servers running on LANs – Local Area Networks. The idea is to save money on hardware and make the information more freely available to more people. More intelligent companies also redesign their organization to use the now, more-freely available information. See Re-Engineering.

value proposition You decide to introduce a new product or service. You need to figure out why people will buy it. Will it save them money? Will it save them time? Will it make them handsome? The answer to these questions is your value proposition. This is a fancy way of figuring why people should buy your fancy new thing.

valve The original British word for an electron tube.

VAM Value Added Modules which are fiber cross connect/ distribution panels for various optic carriers.

vampire tap A cable tap that penetrates through the outer shield to make connection to the inner conductor of a coax cable. The name comes from the fact that the connector pierces the insulation and outer shield by means of one or more sharp "teeth" in order to access the communications artery, much like a vampire's teeth pierce might pierce your jugular in order to drink your blood.

VAN Value Added Network. A public data communications network that provides basic transmission facilities (generally leased by the van vendor from a common carrier) plus additional, "enhanced" services such as computerized switching, temporary data storage, protocol conversion, error detection and correction, electronic mail service, etc.

Van Allen Belts Two layers of charged particles emitted from the sun that are trapped within the earth's magnetic influence. These are named after the discoverer, J. Van Allen. The inner layer exists from about 2,400 to 5,600 km altitude above the earth's surface and consists of secondary charged particles. The outer layer lies between about 13,000 and 19,000 kilometers and is thought to consist of the original particles released from the sun's surface.

Van Eck detection kit A receiver that monitors the electromagnetic radiation given off by a computer screen, allowing an eavesdropper to monitor the contents of a victim's screen from a distance (say in the bushes outside a company).

vanity phone number Also called a gold or golden number. A telephone number, often an 800 toll-free one, which spells something. By way of example, 1-800-542-7279 is advertised as 1-800-LIBRARY, the toll-free number of the publisher of this dictionary. Clearly, there is only a single set of digits which spell "LIBRARY;" therefore, such numbers can be of great value, largely due to the advantage of spontaneity of recall. The recent expansion of the North American 800 number dialing scheme to include 888 numbers has created a storm of controversy, as the publisher must now protect that vanity number from duplication via the 888 prefix. The introduction of UIFN (Universal International Freefone Number) services further promises to infringe on the uniqueness of that number. Most touchtone keypads in North America display letters as well as numbers – often not the case overseas. Hotel phones, fax machines and cellular phones generally do display letters as well as numbers. See also 800 Service and UIFN.

VANS Value Added Network Service. See Value Added Network Service.

VAPD Voice Activated Premier Dialing.

VAPN Voice Access to Private Network.

vapor seal A vapor seal is an essential infiltration of a critical space, such as a data processing center or other room that contains sensitive electronic instrumentation. Essentially, a vapor seal is a barrier that prevents air, moisture, and containments from migrating through tiny cracks or pores in the walls, floor, and ceiling into the critical space. Vapor barriers may be created using plastic film, vapor-retardant paint, vinyl wall coverings and vinyl floor systems, in combination with careful sealing of all openings into the room.

vaporware A semi-affectionate slang term for software which has been announced, perhaps even demonstrated, but not delivered to commercial customers. Hyperware is hardware which has been announced but has not yet been delivered. Slideware is hardware or software whose reason for existing (eventually) has been explained in 35-mm slides, foils, charts and/or PC presentation programs. Slideware is usually less real than vaporware or hyperware, though some people would argue with this. Allegedly the term vaporware cam as a result of the many delays in releasing Windows after Bill Gates of Microsoft announced it at the Fall, 1983 Comdex show in Las Vegas. See also Hookeware Hyperware Meatware Slideware and Sovelware.

VAR Value Added Reseller. Typically VARs are organizations that package standard products with software solutions for a specific industry. VARs include business partners ranging in size from providers of specialty turn-key solutions to larger system integrators.

variable A symbol or string of symbols whose value changes. Macros, equate statements, Configuration File parameter can have a variable value.

Variable Bit Rate Service VBR. A telecommunications service in which the bit rate is allowed to vary within defined limits. Instead of a fixed rate, the service bit rate is specified by statistically expressed parameters.

variable call forwarding An optional feature of AT&T's 800 IN-WATS service. It allows the subscriber to route calls to certain locations based on time of day or day of week.

variable format message A message in which the page format of the output is controlled by format characters embedded in the message itself. The alternative is to have the format determined by prior agreement between the origin and the destination.

variable length buffer A buffer into which data may be entered at one rate and removed at another, without changing the data sequence. Most first-in, first-out (FIFO) storage devices serve this purpose in that the input rate may be variable while the output rate is constant or the output rate may be variable while the input rate is fixed. Various clocking and control systems are used to allow control of underflow or overflow conditions.

variable length record A file in a database containing records not of uniform length and in which the distinctions between fields are made with commas, tabs or spaces. Records become uniform in length either because they are uniform to start with or they are "padded" with special characters.

Variable Quantizing Level VQL. A speech-encoding technique that quantizes and encodes an analog voice conversation for transmission at 32,000 bits per second.

variable resistor A resistance element which may be varied to afford various values.

Variable Term Pricing Plan VTPP. A rate plan developed by AT&T to replace two-tier pricing. VTPP used to provide for two, four, five or six year contracts, over which period the customer is promised stable prices for some – not all – of the equipment and/or tariffed services he uses. Generally, under VTPP, the customer does not end up owning any of the equipment. VTPP has now been replaced by more normal ways of doing commercial business – outright sale, leasing, etc.

variable timing parameter Timing durations for features such as hold recall, camp-on recall, off-hook duration, and many other programmable telephone system services.

variance The average squared deviation. To calculate the variance, determine the difference of each item from the group mean. Then square (multiply by itself) each of the differences (the deviations). Next, average the squares of the deviations to determine average square which is the variance (for samples, divide by one less than the number of items in the sample).

variolosser A device with a variable level of attenuation which is controlled by an external signal. Often this signal is the level of the signal being attenuated, that is the higher the level of the signal the more it is attenuated.

varistor A voltage controlled resistor, a voltage-limiting device used in telephony and surge protection equipment.

VARTI Value Added Reseller Telephone Integrator. A term coined at Telecom Developers '92. It refers to the VARs and interconnects of the 90s that are combining telephony and personal computers to offer products that tie the telephone network to personal computer applications.

VASCAR VASCAR is an acronym for Visual Average Speed Computer and Recorder. VAS-CAR is little more than a combination stopwatch and measuring device. In its simplest application, the police officer uses the VASCAR to measure a section of the road. The officer then starts the VASCAR when a vehicle enters the section and stops the VASCAR when it leaves the section. The VASCAR displays the vehicle's average speed over the section of road.

vay iz meer A Yiddish expression translates to "Woe is Me." It is cried out by Jewish mothers every 15 minutes. An anthem of true suffering and self-flagellation.

VAX 1. A line of minicomputers made by Digital Equipment Corporation (DEC), now part of Compaq.

2. Virtual Address Extension.

VBD Voice Band Data.

VBI Vertical Blanking Interval. The vertical blanking interval is the portion of the television signal which carries no visual information and appears as a horizontal black bar between the pictures when a TV set needs vertical tuning. The VBI is used for carrying close-captioned signals for the hearing impaired. Digitized data can also be inserted into the VBI for transmission at rates greater than 100,000 bps. Information services such as stock market quotations and news offerings are now available via the VBI of a CATV signal. The data embedded in the VBI signal is retrieved from a standard cable or satellite receiver wall outlet by a receiver set, which connects to a RS-232 port on a microcomputer. Software packages then allow subscribers instant access to the information, which may be displayed in a number of formats.

vBNS Very high-speed Backbone Network Service. A high-speed SONET fiber optic backbone network being developed by MCI (now MCI Worldcom) for the National Science Foundation (NSF), vBNS will serve as the backbone transport network for Internet2. Initially, vBNS will runs at a speed of 155 Mbps (OC-3); ultimately, the network will run at 2.4 Gbps (OC-48). The first deployment of vBNS connects five NSF-funded supercomputing centers (SCCs): Cornell Theory Center, National Center for Atmospheric Research, National Center for Supercomputing Applications, Pittsburgh Supercomputing Center, and San Diego Supercomputing Center. Also connected are the NSF-funded Network Access Points (NAPs) at Hayward, CA; Chicago, IL; Pennsauken, NJ; and Washington, DC. vBNS replaces the old NSFNET, which was decommissioned in 1995. See also Internet2, NSFNET, and SONET.

VBR Variable Bit Rate. A voice service over a an ATM switch. Voice conversations receive only as much bandwidth as they need, the remaining bandwidth is dynamically allocated to other services that may need it more at any given moment. Nortel Networks refers to this approach as "making bandwidth elastic." VBR also refers to networking processes such as LANs which generate messages in a random, bursty manner rather than continuously.

VC 1. Virtual Channel. A SONET term. Existing with a Virtual Tributary (VT), a VC is virtually to a traditional TDM channel. Note that a TDM channel is either set aside for or prioritized for a particular transmission; in other words, it's a dedicated channel. A Virtual Channel, on the other hand is not set in such a rigid environment; rather, such channels float in a SONET frame, available only when needed in a particular transmission, and not necessarily found in the same place in time. All things considered, "virtual channel" is just a tense term for a channel, but maintaining the "virtual" nomenclature of SONET-Virtually Aggravating at times, isn't it! Also known as a Tributary Unit in SDH (Synchronous Digital Hierarchy) terminology. See also SDH, SONET, TDM and Tributary Unit.

2. Virtual Channel. An ATM term. According to the ITU-T, a virtual channel is a unidirectional communication capability for the transport of ATM cells." A Virtual Channel Identifier (VCI) in the header of the ATM cell is assigned or removed, respectively, to either originate or terminate a Virtual Channel Link (VCL). VCLs are concatenated to form a Virtual Channel Connection (VCC), which is and end-to-end VP (Virtual Path) at the ATM layer. Once again, "virtual" is the operative word. Channels are "virtual" in the ATM world, as the extent to which they are made available depends on the priority level of the traffic, as defined in the cell header. High-priority traffic gets lot of VCs through the ATM switch, while low-priority traffic gets fewer. VCs also are defined in the UNI 3.0 specification. See also ATM, Concatenation, UNI, VCC, VCI, VCL, and Virtual.

3. Virtual Circuit. In packet switching, network facilities that give the appearance to the user of an actual end-to-end circuit. VCs define the physical path that all packets in a packet stream will follow during a session between two or more computing systems. Virtual circuits allow many users to share switches and transmission facilities, while each can enjoy the advantages of a Virtual Private Network (VPN). VCs can be provisioned as Permanent Virtual Circuits (PVCs) or Switched Virtual Circuits (SVCs). X.25 Packet Switched networks, Frame Relay and ATM all make use of VCs. See also ATM, Frame Relay, Packet Switching, PVC, SVC and X.25. EXPAND SLIGHTLY, XREF: VCC Virtual Channel Connection. As an ATM term, it is a concatenation of VCLs (Virtual Channel Links) that extend between the points where the ATM service users access the ATM layer. The points at which the ATM cell payload is passed to, or received from, the users of the ATM Layer (i.e., a higher layer or ATM-entity)

for processing signify the endpoints of a VCC. VCCs are unidirectional. See also Concatenation and VC.

4. Virtual Container. SDH defines a number of "containers". The container and the path overhead from a "Virtual Container" (VC) in Europe or "Virtual Tributary" (VT) in North America (ref: ITU G.709).

5. Videoconferencing

6. Venture Capital.

VC-1 VC-1 is the proposed Society of Motion Picture and Television Engineers (SMPTE) video codec standard based on Microsoft Windows Media Video 9. Once VC-1 is finalized, companies will be able to build interoperable solutions using this next generation video compression technology, allowing them the choice of either licensing it directly from Microsoft or through MPEG LA, an independent licensing body. VC-1 has already been selected as a mandatory codec in both of the leading next-generation high definition DVD specifications, HD-DVD and Blu-ray Disc. In addition to SMPTE, VC-1 is moving forward in other standards and industry organizations including DVB and ATSC, and completion of the work with SMPTE will allow interested parties to make a simple reference to the SMPTE standard in order to standardize within their group. Content created in VC-1 will be compatible with the Windows Media Video 9-based PC, CE device and TV experiences available today.

VCA See Voice Connecting Arrangement.

vCalendar A virtual calendar specification by the IMC (Internet Mail Consortium) as an electronic exchange format for personal scheduling information. vCalendar is an open specification based on industry standards including the X/Open and XAPIA Calendaring and Scheduling API, the ISO 8601 international date and time standard, and the related MIME e-mail standards. Adoption of the standard allows software products to exchange calendaring and scheduling information in an easy, automated and consistent manner. According to the IMC, it can work like this. You are at a business meeting with representatives from various companies. Every imaginable portable computing device is present, from PDAs, to hand-held organizers, to laptops and notebook PCs. The chairperson of the meeting communicates from his device to all other devices via infrared beam. All attendees then do the same. Thus, the next meeting is scheduled electronically, with conflicts identified. At the same time, personal information is passed around via vCard technology. vCalendar is being enhanced by a new specification called iCalendar, which is designed specifically for the Internet. See iCalendar, IMC and vCard.

vCard vCard is a tiny file (with the extension .vcf) that contains all the information on your business card – your street address, your phone numbers, your email address, etc. You attach this file to an email you send to someone. They receive it as an attachment. They click on it. It opens up as entry in the electronic address book or PIM (Personal Information Manager). They then click "save." And bingo, your information is added to their address book. vCards can also be shared wirelessly between PDAs by beaming cards over infrared links. The benefits are obvious: huge time savings, huge savings in accuracy, etc. Most browsers and most email clients now support the vCard specification. And most allow you to choose that your vCard file is attached to each and every email you send. My recommendation to everyone is simple: organize so that you attach your vCard to each of your outgoing emails.

vCard was a Versit idea. They defined it as an electronic, virtual information card that can be transferred between computers, PDAs, or other electronic devices through telephone lines, or e-mail networks, or infrared links. With vCard, according to Versit, individuals can consistently identify themselves without restating or rekeying their information. vCards include data such as name, address, phone number, e-mail user ID, with multimedia support for photographs, sound clips and company logos. vCard is the result of a collaborative industry effort between Versit and multiple vendors. The vCard specification is at the Internet Mail Consortium (IMC) (www.imc.org), which developed the specification in cooperation with leading producers of desktop software, hand-held organizers, Internet web clients, and others. The specification is open and based on industry standards, including ITU-T X.500 directory services.

VCC 1. Virtual Channel Connection. As an ATM term, it is a concatenation of VCLs that extends between the points where the ATM service users access the ATM layer. The points at which the ATM cell payload is passed to, or received from, the users of the ATM Layer (i.e., a higher layer or ATM-entity) for processing signify the endpoints of a VCC. VCCs are unidirectional.

2. Voice call continuity.

VCEP Video Compression/Expansion Processor chip.

vcf Virtual Card File. A vcf is a standard way of sending someone your personal information. It stands for virtual card file.See vCard.

VCI Virtual Channel Identifier. An ATM term. The address or label of a VC (Virtual Chan-

nel). The VCI is a unique numerical tag, defined by a 16 bit field in the ATM cell header, that identifies a VC over which a stream of cells is to travel during the course of a session between devices. See also VC.

VCL Virtual Channel Link. An ATM term. A means of unidirectional transport of ATM cells between the point where a VCI (Virtual Channel Identifier) value is assigned as it is presented to the ATM network, and the point where that value is translated or removed as it exits the ATM network. VCLs are concatenated to form VCCs (Virtual Channel Connections). See also VCC and VCI.

VCN Virtual Corporate Network. Stentor's name for a service it later changed to Advantage VNet. It's similar to MCI's VNet.

VCO Voltage Controlled Oscillator: An oscillator whose clock frequency is determined by the magnitude of the voltage presented at its input. The frequency changes when the voltage changes.

VCOMM Win32 (virtual) communications device driver. It protect-modes services and lets Windows apps and drivers use ports and modems. To conserve system resources, comm drivers are loaded into memory only when in use ba an application. VCOMM also uses new Plug and Play services that began in Windows 95 to help configure and install comm devices. The Win32 communications APIs provide an interface for using modems and comm devices in a device-independent fashion. Applications call the Win 32 APIs to configure modems and perform data I/O through them. Through TAPI and Unimodem, apps can control modems or other telephony devices with VCOMM. Unimodem routes TAPI/Unimodem AT +V commands through VCOMM. In addition, data from the Win 32 Comm API and Audio from the Multimedia Wave API are routed through VCOMM. The virtual device driver in turn communicates through port drivers and serial virtual device drivers to the hardware.

Microsoft defines VCOMM as: VCOMM is a static VxD always loaded at boot time in Windows 95. It functions as the Plug and Play "device loader" for installed devices of class "ports" or "modem". This means VCOMM is called to load these devices when they are enumerated by the Plug and Play Configuration Manager. This is usually at boot time for COM ports, LPT ports, and internal and external modems. However, PCMCIA ports and modems can be enumerated "on the fly" when inserted at any time during a Windows 95 session. When the Plug and Play Configuration Manager calls VCOMM to load an enumerated device, it passes a handle to the "devnode" data structure for the device. The devnode contains information about the device, including its resource allocation and registry key locations. When called to load a device, VCOMM does the following:

- Remembers the PortName and FriendlyName strings specified in the device's hardware key. Later, when called to open the device by one of these names, VCOMM will match it to the software key for this device to determine the port driver VxD to load. Note that the port driver is not loaded immediately when the device is enumerated. Rather, it is only loaded when opened by a VCOMM client such as Remote Access or a communications application. This speeds boot time, and prevents unnecessary consumption of system resources until the device is actually opened.

- Assigns a "PortName" to the device in its hardware key, if its devnode contains system resources. For COM ports (PortSubClass="01") and PCMCIA modems (PortSubClass="02"), PortName is assigned according to the base I/O port address, as shown in the following. Devices with non-standard base addresses receive port names starting at COM5 and higher.

VCOS Visible Caching Operating System. VCOS is a realtime multitasking DSP operating system for the AT&T DSP3210 Digital Signal Processor. Visible Caching means the programmer caches the program and the data onchip, in contrast to logic caching where state machines (implemented in silicon) perform all caching.

VCPI An acronym for the Virtual Control Program Interface, a standard developed by Quarterdeck and Phar Lap Software for running multiple programs and controlling the Virtual-86 mode of 386 microprocessors. A program that's VCPI-compatible and can run in the protected mode under DOS without conflicting with other programs in the system.

VCR VideoCassette Recorder (or Player). See also PVR.

VCSEL Vertical-Cavity Surface-Emitting Laser. A VCSEL (pronounced "VIXel") is a tiny laser (10 x 10 x 2 microns) that consists of two mirrors that sandwich an active region. The laser emits a cylindrical beam of light vertically from its surface. The light signal travels through a cavity that has been etched through a semiconductor to the active region of the chip. The mirrors reflect the light back and forth, resulting in a stimulated emission at a single wavelength. VCSELs offer several significant advantages over edge-emitting lasers. VCSELs are smaller (in the case of CD lasers, they're 100 times smaller; are cheaper to manufacture, as they use the same fabrication techniques as chips (so Moore's Law now applies to lasers); consume greatly reduced levels of power (requiring only 1-2 milliwatts per gigabit per second); and can be packed closely together in two-dimensional arrays, yielding a theoretically unlimited aggregate data transfer rate, depending on how many are stacked. VCSELs and fiber-optic cabling are at the heart of Gigabit Ethernet (GE) technology. VCSELs are used in Very Short Reach (VSR) transmission systems, where they typically fire at a wavelength of 850 nm (nanometers, or billionths of a meter), although any wavelength in the 830-860 range is acceptable. See also Gigabit Ethernet and VSR.

VD Virtual Destination. See VS/VD.

vDC Volts Direct Current. For example, 12 vDC is the voltage which powers automobiles in most parts of the world, including North America.

VDDD VNET International Direct Dialing Forced On-Net. VDDD is a feature enhancement for International VNET (I-VNET) customers that allows incorporation of 7-14 digit IDDD numbers into their dialing plans.

VDE Verband Deutscher Elektrotechniker. Federation of German Electrical Engineers similar in form to the IEEE.

VDECK An 8 millimeter cassette recorder developed by Sony Corporation for use as a computer peripheral.

VDI Video Device Interface. A software driver interface that improves video quality by increasing playback frame rates and enhancing motion smoothness and picture sharpness. VDI was developed by Intel and will be broadly licensed to the industry.

VDISK Virtual DISK. Part of the computer's Random Access Memory assigned to simulate a disk. VDISK is a feature of the MS-DOS operating system.

VDM Voice Data Multiplexer.

VDO A technology that enables Internet video broadcasting and desktop video conferencing on the Internet and over regular telephone lines and private networks. VDOPhone which provides the ability to have private point to point audio/video contact is currently only available for Windows95 and requires a Pentium processor. The VDOLive player however is available for Windows and Power Macs and provides the ability as a Netscape plugin for viewing and hearing live Internet Broadcasts. To download the VDOLive Player go to www.vdo.net/download. See also Realaudio.

VDPS Virtual Private Data Service.

VDRV Variable Data Rate Video. In digital systems, the ability to vary the amount of data processed per frame to match image quality and transmission bandwidth requirements. DVI symmetrical and asymmetrical systems can compress video at variable data rates.

VDS Vocabulary Development System.

VDSL Very-high-data-rate Digital Subscriber Line. A technology in the very early stages of definition, Initial VDSL implementation likely will be in asymmetric form, essentially being very high speed variations on the ADSL theme. Goals are stated in terms of submultiples of the SONET and SDH principal speed of 155 Mbps. Specifically, target downstream performance is 51.84 Mbps over UTP local loops of 1,000 feet (300 meters), 25.92 Mbps at 3,000 feet (1,000 meters), and 12.95 Mbps at 4,500 feet (1,500 meters). Upstream data rates are anticipated to fall into three ranges: 1.6-2.3 Mbps, 19.2 Mbps, and a rate equal to the upstream rate.

The application for VDSL is in a hybrid local loop scenario, with FTTN (Fiber-To-The-Neighborhood) providing distribution from the CO to the neighborhood, and with VDSL over UTP (Unshielded Twisted Pair) carrying the signal the last leg to the residential premise. Clearly, the specific application is for highly bandwidth-intensive information streams such as are required for support of HDTV and Video on Demand. Early work on VDSL has begun in standards bodies including ANSI T1E1.4, ETSI, DAVIC, The ATM Forum and The ADSL Forum. See also ADSL, ADSL Forum, HDTV, IDSL and SDSL.

VDSL2 Second-generation VDSL, designed to support simultaneous bandwidth-intensive and latency-sensitive voice, video, data, TV over IP, and interactive gaming. VDSL2 supports symmetric data rates of up to 100 Mbps over twisted pair within 100 meters of the central office, a speed that gradually falls to about 10 Mbps at a distance of three kilometers (1.86 miles) from the central office.

VDT 1. Video Display Terminal. A data terminal with a TV screen. Another name for computer monitor. VDT is the term you hear in Europe.

2. Video Dial Tone. The new concept of getting home entertainment, information and interactive services to residences over some form of new broadband network stretching into the nation's homes. Video Dial Tone is a term used by traditional telephone companies. They're the ones allegedly building this broadband network to provide "Video Dial Tone."

3. Visual Display Terminal.

VDU 1. Visual Display Unit.

2. Video Display Unit.

vector A quantity in the visual (video) telecommunications industry that describes the magnitude and direction of an object's movement – for example, a head moving to the right. See Vector Images.

vector graphics Images defined by sets of straight lines and defined by the locations of the end points. See Vector Images.

vector images Images based on lines drawn between specific coordinates. A vector image is based on the specific mathematics of lines. In contrast, a raster image is a bit-mapped (i.e. bit-drawn) image. A vector engineering image is more useful for engineering, since it can be changed easier than a bit-mapped image. A vector image can easily be converted to a raster image. But it's much more difficult to go from a raster image to a vector image. Some storage systems now store images as combination raster/vector.

vector processor Array Processor.

Vector Sum Excited Predictive Coding VSELP. A method of enhancing Linear Predictive Coders using a combination of a limited number of different excitations (sums of excitation vectors). VSELP is a sub-class of the broader class of CELP coders. VSELP is a vocoding (voice encoding) technique for encoding analog voice for transmission over a digital wireless network. VESLP is used in the IS-54 US TDMA digital cellular standard, and in Motorola's iDEN network. See also CELP, iDEN, and IS-54.

Vegetarian Vegetarian - that's an old Indian word meaning "lousy hunter," according to Andy Rooney.

Velcro Swiss mountaineer Georges de Mestral was walking in the woods in 1948. He noticed hundreds of cocklebur weeds clinging to his trousers. Hoping to oust the zipper as the 20th century's fastener of choice, he replicated this phenomenon using a loom to create a material with hundreds of hooks that grip wooly surfaces, creating Velcro brand fasteners.

velocity of light The nominal speed of light in a vacuum is 186,282.0344 miles per second, or 299,792.458 kilometers per second. The speed of light is very important because today we can measure time more accurately than length. In effect, we define the meter as the time traveled by light in 0.000000003335640952 of a second as measured by the cesium clock. The speed of light also is important, because it affects the speed of signal propagation. The speed of light in air is slower as the molecules (e.g., oxygen, carbon dioxide, smog, and water in the form of such things as rain and humidity) in the atmosphere impede the progress of the signal. For example, one common index of refraction for air is 1.0003. Do the math, that means the speed of light in air comes to 299,702.547 kilometers per second. The speed of light in an optical fiber is slower, at about 205,000 kilometers per second, and depends on the purity of the fiber and the temperature of the fiber, as the temperature affects the density of the glass. Additionally, the speed of light in the core of the fiber generally is slower than in the cladding surrounding the inner core. See also Velocity of Propagation.

velocity of propagation Vp. The speed at which a signal travels through a transmission medium. Here are some examples: The nominal Vp of light in a vaccum is 299,792.458 kilometers a second, although different wavelengths of light (and frequencies of other electromagnetic waveforms) actually travel at slightly different speeds. The nominal Vp of a radio signal also is 299,792.458 kilometers per second (or 186,282.0344 miles per second) in a vacuum. The nominal Vp of electricity in a copper cable (e.g., coax and twisted pair) ranges from 180,000 to 240,000 kilometers per second. The nominal Vp of light in an optical fiber is 205,000 kilometers per second. Signals actually travel faster through a copper twisted pair than they do through an optical fiber. Bandwidth – in other words how much informaiton you can send through the air or through a cable) has little to do with the speed of the signal (i.e., Vp). Rather, it has to do with how many pulses you can send per second and that has to do with modulation techniques. See also modulation and velocity of light.

velocity of sound The velocity of sound varies with the medium carrying it. In air at 0 degrees centigrade, it's 331 meters per second. In glass at 20 degrees centigrade, its 5485 meters per second.

Velveeta An Internet Usenet posting, often commercial in nature, excessively cross-posted to a large number of newsgroups. Similar to Spam, although that term is often used to describe an identical post that's been loaded onto lots of inappropriate newsgroups, one group at a time (rather than cross-posted). This definition courtesy Wired Magazine.

VEN Virtual Ethernet Network.

vendor code Software written by the same company that manufactured the computer system on which it is running (or not running...).

vendor ID A Plug and Play term. Vendor ID is the 32-bit vendor ID that indicates the manufacturer, specific model, and version of a device. It is this number which helps Plug and Play configure the PC to run the device.

vendor financing When the manufacturer lends a customer the money to buy the manufacturer's products, it's called vendor financing. This is a good idea if the customer eventually pays for the stuff. If it doesn't, it can be a real problem. And that depends on how the "sale" was handled in the first place. If the "sale" was booked as a real-live cash sale, then the company will have to re-state its sales and earnings for that period or take a whopping writeoff at some time. Meantime, the company's balance sheet might look bloated with "assets" of monies owed by companies. Those "assets" might be good or bad, depending on how likely the companies are to succeed. If they fail, as many will – after all, they didn't have the money to pay for the vendor's product, then things will look bad for the vendor, their stock price and the executives, who could end up in jail. It all depends on how the vendor financing is reported in the company's public financial reports.

vendor independent Hardware or software that will work with hardware and software manufactured by other vendors. The opposite of proprietary.

vendor independent messaging group A group of software and software companies who are trying to create non-proprietary, standard programming interfaces to help software and corporate developers write messaging and mail-enabled applications. Ultimately, end users should be able to work together more effectively and be able to exchange information from within desktop applications in a work group environment regardless of vendor platform. Members of the group include Apple, Borland, IBM, Lotus, Novell and WordPerfect.

Vendors ISDN VIA. In June 1996, according to a story in InfoWorld, thirteen major networking vendors united behind the banner of simple, standardized ISDN access with the announcement of the Vendors ISDN Association (VIA) at the recent ISDN World trade show in Los Angeles. The association will provide forums for discussing technical issues involved in standardizing ISDN service and products across the United States. Initial members include Cisco Systems Inc., Bay Networks Inc., 3Com Corp., and Ascend Communications Inc., as well as Microsoft Corp. and Intel Corp. The association grew out of the ISDN Forum, created in January. It is affiliated with the National ISDN Users Forum and the National ISDN Council, and it will work on standards with those organizations. The group's first focus will be on automating configuration of ISDN devices. Currently, users need to configure their devices manually by entering a Service Profile Identifier. The VIA is pushing for implementation of noninitializing terminals (NITs) by manufacturers of access devices. NITs will automatically send the configuration information, said Rob Rank, an ISDN product manager at Intel and vice president of the VIA. Apparently VIA no longer exists.

vendorfare When a vendor buys you meals to get your business.

vendorware Trash and trinkets provided by vendors to get your business.

vengeance billing A term that originated with expensive restaurants in Paris and received new meaning when long distance carriers in the U.S. started billing international calls at high prices on their poor unsuspecting customers. Term contributed by Jeddy Lieber of Paris.

VENUS-P A ITU X.25 packet-switched network operated in Japan by Nippon Telegraph and Telephone (NTT) Co.

verbal Able to whine in words.

verbal contract "A verbal contract isn't worth the paper it's written on." Samuel Goldwyn, film producer.

verbose In English, verbose means too many words. The speaker spoke verbosely means he used lots of words to say the little he had to say. In computerese, verbose means you get to see on screen what's going on. If you add the switch /v to the command line of a DOS program command, the program most likely will, as it loads, give you information on screen that shows what it's doing as it loads. If you don't add that switch, it will load or try and load and you won't be any the wiser as to what went on. If the program works well, there's no reason to turn on /v. If it doesn't, turn it on and see what happens. Not all programs have the /v switch.

verification A service of a phone company operator who dials into a busy or otherwise impossible-to-reach line and checks that line and reports on that check to the caller. Phone companies are beginning to charge for this service. As of writing, AT&T, for example, was charging 40 cents to verify the line was busy and 70 cents additional for the operator to interrupt the conversation and say another call was coming in.

verification trunk A trunk to which an operator has access and which will switch through to a called line even if the line is busy.

verified off-hook In telephone systems, a service provided by a unit that is inserted on each of a transmission circuit for the purpose of verifying supervisory signals on the circuit. Off-hook service is a priority telephone service for key personnel, affording a connection from caller to receiver by the simple expedient of removing the phone from its cradle or hook.

verifier A device that checks the correctness of transcribed data, usually by comparing with a second transcription of the same data or by comparing a retranscription with the original data.

Verizon Verizon Communications. The company created on June 30, 2000, by the merger of Bell Atlantic and GTE. According to Verizon Communications, Verizon (pronounced Vurr-EYE-zon) comes from the Latin words "veritas," and "horizon." "Veritas" means truth, and also connotes certainty and reliability. "Horizon" refers to the apparent junction between earth and sky, and signifies the endless possibilities ahead. See also Bell Atlantic and GTE.

VERONICA Very Easy Rodent Oriented Netwide Index to Computerized Archives. An Internet service that allows users to search Gopher systems for documents.

versit Versit is a loose association between Apple, AT&T, IBM and Siemens Rolm. Its "vision"? To "enable diverse communication and computing devices, applications and services from competing vendors to interoperate in all environments. Communicate and collaborate with anyone, any time, anywhere.." The products include PDAs, notebooks, phones, servers and "collaboration products." One early thrust: standardize on call control within Novell/AT&T's Telephony Services. Background: Call control among PBXs "conforming" to Telephony Services is not standard. PBXs often do the same things differently. Example: Conference a call on one PBX, the PBX may put one call automatically on hold as the other is dialed. Another PBX may expect it to be done manually. The good news: IBM has agreed to pass all its CallPath call control standards over to Versit. Versit's members (including Novell coopted for this task) are now working on making Telephony Services call control more standard. Upshot: Developers don't have to test their "standard" telephony services software on each and every PBX. What works on one will work on the others. That's the goal. In late July 1995, Versit effectively merged all its activities into another association, called ECTF. (www.versit.com) Versit@cup.portal.com. See ECTF and PBX Driver Profiles.

VersitCard A vCard. See vCard.

vertex shader A graphics term. A graphics function, much like a pixel shader, that runs on the GPU and operates on vertices (interesting lines or curves) and hence polygons. It lets programmers add complex special effects to objects in a 3-D world. Game developers can design custom animation effects using programmable vertex shaders.

vertical Descriptive of the "vertical side" of a North American wire distributing frame, on which terminal blocks for cables are mounted vertically, as opposed to the horizontally-mounted blocks on the equipment side of the frame. Frames of vertical-only blocks are not uncommon at intermediate cable connection points between main frames, hence the term Vertical Intermediate Distributing Frame, or VIDF.

vertical and horizontal coordinates V & H Coordinates. For purposes of determining airline mileage between locations, vertical and horizontal coordinates have been established across the United States. These V&H coordinates are derived from geographic latitude and longitude coordinates. See V & H.

vertical beamwidth See Elevation Beamwidth.

vertical blanking interval The interval between television frames in which the picture is blanked to enable the trace (which "paints" the screen) to return to the upper left hand corner of the screen, from where the trace starts, once again, to paint a new screen. Several companies are eyeing the vertical blanking interval as a place to send digital data, including news and weather information. The vertical blanking interval was the basis of teletext, a 1970s technology that, with the help of a decoder, displays printed information on the TV screen. Teletext has never caught on in the U.S. in part because the amount of data that could be transmitted comfortably was small. Currently the vertical blanking interval is used to transmit closed-captioned text in television broadcasts. See Closed Captioning.

Vertical Cavity Surface Emitting Laser See VCSEL (pronounced "VIX-els").

vertical integration A firm is vertically-integrated when it owns or controls a firm in an upstream or downstream market. For example, a coal-fired power station which owns a coal mine is vertically-integrated.

vertical interval The portion of the video signal that occurs between the end of one field and the beginning of the next. During this time, the electron beams in the cameras and monitors are turned off (invisible) so that they can return from the bottom of the screen to the top to begin another scan.

vertical linearity A video term. A control that allows you to set spacing consistently across the monitor. Thus a shape intended to have a 1-inch diameter will have a 1-inch diameter wherever it appears.

vertical market application An application that is industry-specific and typically very task-specific.

vertical portal A website that caters to one industry and purports to have everything that someone interested in that industry would want in their daily life.

Vertical Redundancy Check VRC. A relatively poor method of error control used in asynchronous transmission in support of the ASCII coding scheme. A check bit, or parity bit added to each ASCII character in a message such that the number of bits in each character, including the parity bit, is odd (odd parity), or even (even parity). The term comes from the fact that the bits representing each character of data conceptually is viewed in a vertical fashion. For instance, the word "CONTEXT" consists of 7 letters, each of which consists of 7 bits, viewed as follows:

BIT/VALUE	C O N T E X T
1*	1 1 0 0 1 0 0
2*	1 1 1 0 0 0 0
3*	0 1 1 1 1 0 1
4*	0 1 1 0 0 1 0
5*	0 0 0 1 0 1 1
6*	0 0 0 0 0 0 0
7*	1 1 1 1 1 1 1
8**	0 0 1 0 0 0 0

  *INFORMATION BIT
  ** PARITY BIT

The transmitting machine sums the bit values for each character, beginning with "nothing," which is an even value in mathematical terms. In the case of the letter "C," for instance, the next bit is a "1" bit, which creates an odd value. The next bit is a "1" bit, which creates an even value. The next four bits are "0" bits, which do not change the even value. The seventh bit is a "1" bit, which creates an odd value, once again. Assuming that the device is set for odd parity, which is the default, it will insert a "0" bit in the eighth bit position, thereby retaining the odd value. (Should the device be set for even parity, a "1" bit would have been inserted in the eighth bit position.) Should the value of the 7 information bits be an even value, the device appends a "1" bit in order to create an odd value. After the data, character-by-character, has been formatted in this fashion, each bit sequence is transmitted across the network to the target device, which also is set for odd (or even) parity. The receiving device goes through exactly the same process, examining each character for parity. If the parity does not match the expectation of the receiving device, the subject character is flagged as errored, although no remedial action is taken. As there is reasonable likelihood that two bits in a given character can be errored in the process of transmission, that the parity of the character therefore would not be affected, and that the receiving device would not detect the fact that the character was errored, this technique is known as "send and pray." Any remedial action must be accomplished on a man-to-machine basis. LRC (Longitudinal Redundancy Checking) often is used in conjunction with VRC to improve the likelihood of detecting an error. See also Longitudinal Redundancy Checking and Parity.

vertical scanning frequency An imaging term. The rate at which the electron beam traces across vertical phosphor dots on the CRT.

vertical service Options that the customer can add to his basic service such as touchtone, conference calling, speed dialing, etc. No one can explain why it's called "vertical" service.

Vertical Service Code VSC. Customer-dialed codes that provide access to certain features and services provided by Local Exchange Carriers (LECs), Interexchange Carriers (IXCs), Commercial Mobile Radio Service (CMRS) providers, and others. Such services include call forwarding, automatic callback, and customer-originated call trace. The VSC format is *XX or *2XX for touchtone access, and 11XX or 112XX for rotary dial access. For example, call forwarding is invoked by dialing *72 for touchtone, or 1172 for rotary dial. Vertical service code assignments are available at: www.nanpa.com/number_resource_info/vsc_assignments.html.

Very High Data Rate Digital Subscriber Line See VDSL.

Very High Frequency VHF. Frequencies from 30 MHz to 300 MHz.

Very Large Scale Integration VLSI. Semiconductor chip with several thousand active elements or logic gates – the equivalent of several thousand transistors on a single chip. VLSI is the technique for making the micro chip, the so-called "computer on a chip."

very long event A local area network term. A very long event is the condition that occurs when the repeater is forced to go into a jabber protection mode because of the excessive number of times a port receives a packet.

Very Low Frequency VLF. Frequencies from 3 KHz to 300 KHz.

Very Short Reach See VSR.

vertical refresh rate The number of times the monitor redraws its screen every second. A too-low refresh rate can result in flicker, causing eyestrain.

VESA Video Electronics Standards Association, San Jose, CA. Along with eight leading video board manufacturers, NEC Home Electronics founded VESA in the late 1980s. The

association's main goal is to standardize the electrical, timing, and programming issues surrounding 800 x 600 pixel resolution video displays, commonly known as Super VGA. VESA has also issued a standard called "local bus," a new high-speed bus for the PC designed to move video between the CPU and the screen a lot faster than the conventional AT bus.

VESDA Very Early Smoke Detection System Alarm. Used in data colocation centers to alert the Network Operation Center's technical staff of a possible fire before the sprinklers go off and destroy all the equipment.

vestal virgin A celibate woman who tended the sacred fire in the temple of Vesta in ancient Rome. There were originally four, and eventually six vestal virgins. They stayed on the job for 30 years by which time they'd probably lost all interest in having sex. It's hard to imagine which you need six people to tend one little fire.

vestigial sideband A partially suppressed sideband.

Vestigial Sideband Transmission VSB. A modified sideband transmission technique in which one sideband and the carrier are suppressed, and only a portion of the remaining sideband is transmitted. Reduced power requirements is an advantage. See also Amplitude Modulation, DSBSC, DSBTC, and SSB.

VF 1. Voice Frequency.

  2. Variance Factor. An ATM term. VF is a relative measure of cell rate margin normalized by the variance of the aggregate cell rate on the link.

VFast An earlier name for the 56Kbps modem spec. See 56 Kbps Modem.

VFAT Virtual File Allocation Table. A fat file system is a file system based on a file allocation table, maintained by the operating system, to keep track of the status of various segments of disk space used for file storage. The 32-bit implementation in Windows 95 is called the Virtual File Allocation Table (VFAT). An extension of the FAT file system in DOS and Windows 3.xx, VFAT supports long filenames while retaining some compatibility with FAT volumes. See also FAT.

VFDN Voice Frequency Directory Number. A Northern Tom term.

VFG Virtual Facilities Group. A way of limiting inbound calls to a specific number on a PRI. See Virtual Facilities Group.

VFO Variable Frequency Oscillator.

VFTG Voice Frequency TeleGraph.

VFRAD Voice Frame Relay Access Device. A CPE (Customer Premises Equipment) device used in Voice over Frame Relay (VoFR), a VFRAD is used for access to a Frame Relay network in support of voice communications. There must be matching FRADs at all locations on the network, as there must be matching logic in place for compression and decompression. See also FRAD, Frame Relay and VoFR.

VG Voice Grade. A term commonly applied to describe a circuit or channel of sufficient bandwidth to support voice communications. An analog voice grade circuit or channel provides bandwidth of 4 KHz. A digital voice grade circuit or channel provides bandwidth of 64 Kbps under traditional PCM encoding, but can be less using IP telephony.

VGA Video Graphics Array. A graphics standard developed by IBM for the IBM PC. VGA allows the PC's screen to generate any of four levels of resolution – with one of the sharpest being 640 horizontal picture elements, known as pels or pixels, by 480 pels vertically with 16 colors. VGA is superior to earlier graphics standards, such as CGA and EGA. VGA is barely adequate for CAD-CAE. See Monitor for all the numbers on pixels in various screens. VGA was the graphics standard introduced for IBM PS/2 line and quickly adopted by PC compatibles; supports analog monitors with a 31.5 Hz horizontal scan rate. Today VGA is about the lowest graphics resolution most computers support.

VGE Voice Grade Equivalent. A term commonly applied to a level of digital bandwidth sufficient to support a voice conversation using standard encoding techniques-i.e., 56/64 Kbps using PCM (Pulse Code Modulation). Voice Grade (VG) generally refers to an analog local loop circuit which provides 4 KHz bandwidth. See also VG.

VGF Voice Grade Facility.

VGPL Voice Grade Private Line.

VHD Very High Density. Techniques of recording 20 megabytes and more on a 3 1/2" magnetic disk.

VHF Very High Frequency. Portion of the electromagnetic spectrum with frequencies between about 30 MHz and 300 MHz. Operating band for radio and television channels. See VHF Drop-Ins.

VHF Drop-Ins Full power VHF TV stations that may be squeezed into locations that do not comply with the FCC's spacing requirements.

VHI Virtual Host Interface.

VHS Video Home System using half-inch tape introduced by Matsushita/JVC in 1975 and now the most popular form of video tape. There is also a VHS at 3/4". It's often used inside ad agencies for previewing work in progress. Industrial video tape – the stuff the TV stations use – is one inch. And it shows a much better quality picture than half-inch VHS.

VI Architecture See Virtual Interface Architecture

VIA See Vendors Industry Association.

Via Net Loss VNL. A planning factor used in allocating the attenuation losses of trunks in a transmission network. A specified value for this loss is selected to obtain a satisfactory balance between loss and talker echo performance. The lowest loss in dB at which it is desirable to operate a trunk facility considering limitations of echo, crosstalk, noise and signing. See echo.

Viagra Viagra was originally called Sildenafil. It was first tested on humans in 1991, but didn't prove effective for its initial indicator, namely angina or chest pain. After patients reported erections as a side effect, Pfizer began testing the compound for erectile dysfunction. In 1998, Viagra became the first drug to treat the condition, and the blockbuster drug has been a household name ever since. While visiting Australia in June of 2006, I noticed a sign at a party store: "Our (helium) ballons have Viagra. They stay up longer. "

vibratory plow A plow that rips open the ground by vibrating a plow share.

vibration isolation High precision technique for the alignment of tiny objects. Allows micropositioning to reliably occur.

VIC Voice Interface Card. Connects a system to either the PSTN or to a PBX.

video From Latin, translated as "I see," video adds the element of sight to communications. While some of us are visually-oriented, and others of us are more oriented kinesthetically-oriented (learn by doing, as in with muscles and tendons and energy and sweat), we all find a communication to be enhanced through pictures...especially motion pictures. Motion pictures are the essence of video, and video is the essence of true and full communications. Visual communications, by the way, are also extremely bandwidth-intensive. See Multimedia.

video capture Video Capture means converting an analog video signal into a digital format that can be saved onto a hard disk or optical storage device and manipulated with graphics software. This is accomplished with a device internal in a computer called a "frame grabber" or video capture board. Images thus captured are digitized, and can be dropped into a document or database record and may be transmitted locally on a LAN or long distance over a WAN. See Video Capture Board.

video capture board To capture a single frame of motion video successfully, you need a board inside your PC that can capture the two fields comprising a single video frame. The best source of single frame video images is a laser disk player which can pause and display a perfect frame of video without noise or jitter. Video cameras and camcorders aimed at a static, non-moving image also work well. VCR, which produces a jittery image when the tape is paused, are the poorest source. See also Frame Grabber.

video clamping A bandwidth-management technique that is used to convert a variable bit rate (VBR) video feed (for example from a satellite) to a constant bit rate (CBR) video stream, so that the video can be packetized for delivery to networks and devices with little-to-no tolerance for variable bit rates, such as wireless networks and DSL lines. Video clamping works by applying a rate reduction algorithm to bit rates in the source feed that are above a certain threshold in order to make them compliant with the bit rate requirements of the destination CBR device or network. Null padding is used to pad parts of the source feed whose bit rates are below the threshold to make them conform to the bit rate requirements of the destination CBR device or network.

video codec The device that converts an analog video signal into digital code.

video compression A method of transmitting analog television signals over a narrow digital channel by processing the signal digitally. You can compress an analog TV signal into one T-1 signal of 1.544 megabits per second. More advanced compression techniques will enable video signals to be compressed into fewer bits per second. One increasingly common method allows a full-color reasonably full-motion video to be compressed into two 56 Kbps channels.

video conference See Videoconference.

video dial tone Video dial tone in telco-speak means the phone company, in competition with the cable TV business, provides video to houses and offices. It does not affect the content of that video signal in any way. Thus the term video dial tone, which is like voice dial tone, whose content the phone company also does not affect or change in any way, shape or form. Video Dial Tone was named by the FCC, for a service in which subscribers can dial up any video program they wish across a network. The FCC proposed rules for telephone companies to provide Video Dial Tone in 1992, and finalized them in 1994. Included were the rules for common-carrier transport (level two Video Dial Tone). Telephone companies must apply (with "section 214" applications) to the FCC for approval to provide the Video Dial Tone service. The FCC reviews these applications on a case-by-case basis.

video driver A piece of software which translate instructions from the software you

are running into thousands of colored dots, or pixels, that appear on your video monitor. A video driver is also called a display Driver. Symptoms of a video driver giving trouble can range from colors that don't look right, to horizontal flashing lines to simply a black screen. In the Macintosh world, Apple rigidly defined video drivers. Windows, in contrast, is a free-for-all. Windows 3.1 defined the lowest common denominator of displays – namely 16 colors at 640 x 480 pixels. But most multimedia programs and many games won't run with only 16 colors. They require at least 256 colors.

Video Electronics Standards Association See VESA.

video mail Electronic mail that includes moving or still images.

video memory See Shared Video Memory.

video modulator A device used to place video information on higher frequency signals for transmission.

video monitor A high quality television set (without RF circuits) that accepts video baseband inputs directly from a TV camera, videotape recorder, etc. See video monitoring.

video monitoring A new term for when a computer analyzes video (real-time or recorded) to determine if something fishy is happening. For example, a security camera might be trained on the locked entrance to a building. The video monitoring software "watches" the entrance, the moment it detects motion it does something – perhaps contact a human being, a guard, perhaps sound an alarm, etc.

video networking I first saw the term video networking in a white paper written by L. David Passmore of Decisys, Inc. The white paper was in a press kit from Madge Networks, announcing new video hardware architecture. David wrote "When people typically think of video networking, they think of it as an application, namely video conferencing. However, video networking is really an architecture that supports a range of business applications featuring video communications. These applications can be deployed over the video network to span the LAN and WAN environments. Furthermore, if the video network is deployed properly, it can provide a consolidated (data, voice and video) WAN access solution across the enterprise."

Video On Demand VOD. Punch some buttons. Order up Gone With the Wind to start playing at your house at 8:26 P.M. or right this instant. Bingo, you have video on demand. It's a great concept with a potentially huge demand, and potentially major implications for businesses like Blockbusters and Netflix. There are problems. First, you need heavy duty bandwidth – lots of it. There's nothing more demanding of bandwidth than streaming a movie. I'm guessing you'll need six million bits per second. It will need to be even faster than that if you're shipping high definition movies. Second, the video on demand server equipment to provide the service is complex and expensive. Think about the video on demand server. Think of a movie DVD. It's typically about 4 gigabytes. VOD appears to be a technology whose time has arrived and some telephone companies and cable companies are field-testing it. The concept with some VOD services is to allow a subscriber to watch any video program at any time, with pause, resume, forward and possibly re-wind control. Compare to Near-Video-on-Demand. See NVOD, Video Dial Tone and Video Server.

video path The electronic path within the device that routes and processes the video signals. Video path length refers to the amount of time required for a signal to travel from input to output.

video pill A camera about the size of a vitamin pill that, when ingested, transmits images from a person's stomach and intestinal tract. Gives new meaning to the term "The Last Mile."

video processing amplifier A device that stabilizes the composite video signal, regenerates the synchronizing signals, and allows other adjustments to the video signal parameters.

video ringtone A video clip that plays on a mobile handset when there is an incoming call. For example, the video ringtone could be a 15-second snippet of a popular rock video.

video scaler Video scalers are used to convert television-video signals to computer-video signals. Until recently, devices called line doublers were used for this, but they cause images processed through projectors to become distorted, and are limited to a 4:3 aspect ratio and a refresh rate of 60Hz. A video scaler's processing algorithms can manipulate the signal to best fit the display or projector, and allow you to choose the refresh rate and aspect ratio.

video server A device that could store hundreds, if not thousands of movies, ready for watching by subscribers at their individual whim. A video server could be jukebox like device that would stack several hundred movies. Or it could be a powerful, large computer with several large hard disks and/or optical disk drives. The device would be used in conjunction with the local telephone companies' service called video dial tone – providing movies over normal phone lines to their subscribers or it could be used with the CATV industry's

Video On Demand service. A video server could also simply sit on a home local area network and feed movies to various TVs in various parts of your suburban mansion.

video signal Transmission of moving frames or pictures of information requiring frequencies of 1 to 6 Megahertz. A commercial quality full-color, full-motion TV signal requires 6 MHz.

video switcher Device that accepts inputs from a variety of video sources and allows the operator to select a particular source to be sent to the switcher's output(s). May also include circuits for video mixing, wiping, keying, and other special effects.

video teleconferencing Also called Videoconferencing. The real-time, and usually two-way, transmission of digitized video images between two or more locations. Transmitted images may be freeze-frame (where television screen is repainted every few seconds to every 20 seconds) or full motion. Bandwidth requirements for two-way videoconferencing range from 6 MHz for analog, full-motion, full-color, commercial grade TV to two 56 Kbps lines for digitally-encoded reasonably full motion, full color, to 384 Kbps for even better video transmission to 1,544 Mbit/s for very good quality, full-color, full motion TV. See also Videoconferencing.

video telephony Real time video call similar to a voice call.

video wall Multi-screen video system where a large number of video monitors (typically 16 monitors arrayed in a 4 x 4 matrix) or back projection modules together produce one very large image or combinations of images. Video walls come with their own software, which lets you program the video effects you want. Typically, you can feed a video wall everything from VGA computer output to moving TV (NTSC) signals. Video walls are used for exhibitions and trade shows. They're not cheap. But, when programmed properly, they ARE spectacular.

video windows A Bellcore invention which is basically a large, high capacity video conferencing device. Bellcore's Video Windows are connected by two optical links, each carrying 45 million bits of information per second. Though impressive, Bellcore's Video Windows is not considered "high definition" TV. For that to happen, you'd probably need 100 to 150 million bits being transmitted in both directions each second.

VideoCipher II VC. An encryption system belonging to M/A COM.

videoconference Videoconference is to communicate with others using video and audio software and hardware to see and hear each other. Audio can be provided through specialized videoconferencing equipment, through the telephone, or through the computer. Videoconferencing has traditionally been done with dedicated video equipment. But, increasingly personal computers communicating over switched digital lines are being used for videoconferencing. See also Videoconferencing.

videoconferencing Video and audio communication between two or more people via a videocodec (coder/decoder) at either end and linked by digital circuits. Formerly needing at least T-1 speeds (1.54 megabits per second), systems are now available offering acceptable quality for general use at 128 Kbit/s and reasonable 7 KHz audio. Factors influencing the growth of videoconferencing are improved compression technology, reduced cost through VLSI chip technology, lower-cost switched digital networks – particularly T-1, fractional T-1, and ISDN – and the emergence of standards. See Videoconferencing Standards.

videoconferencing standards ITU-T H.261 was the standards watershed. Announced in November 1990, it relates to the decoding process used when decompressing videoconferencing pictures, providing a uniform process for codecs to read the incoming signals. Originally defined by Compression Labs Inc. Other important standards are H.221: communications framing; H.230 control and indication signals and H.242d: call setup and disconnect. Encryption, still-frame graphics coding and data transmission standards are still being developed.

VideoCrypt A French encryption system created by the Thomson Company.

Videophone 2500 In January, 1992, AT&T introduced a product called Videophone 2500, which transmitted moving (albeit slowly-moving) color pictures over normal analog phone lines. The phone carried a price tag $1,500 a piece. It was not compatible with one MCI later introduced, made for it by GEC-Marconi of England and costing only $750 retail. Videophone 2500 relies on video compression from Compression Labs, Inc. of San Jose, CA. According to the New York Times, the phone took two years, about $10 million and 30 full-time people at AT&T to develop. The January 3, 1993 New York Times carried a quote from John F. Hanley, group VP for AT&T consumer products division, "We could make an AT&T phone talk to an MCI phone. It would be in both of our interests." The AT&T phone and the MCI phone are now effectively dead. See H.323.

videotape formats Videotape formats are, in general, classified by the width of magnetic tape used.

-1": Used for professional or "broadcast quality" video recording and editing. Comes in large, open reels.

-3/4": U-matic (Sony). Most industrial video uses this format, stored in inch-thick cassettes.

-1/2":Cassette based, primarily consumer format. VHS - the most popular home videotape format - is 1/2", as is Sony's Beta format. Their higher-quality counterparts (Super-VHS and Super Beta, respectively) are also in the 1/2" format.

-8mm: New consumer format that provides high-quality recording in tiny tape format. Popularly used in hand-held camera-recorders (camcorders).

videotape recorder A device which permits audio and video signals to be recorded on magnetic tape.

videotex Two-way interactive electronic data transmission or home information retrieval system using the telephone network. Videotex has not been successful because of its (erstwhile) need for expensive, proprietary (i.e. dedicated) equipment and lack of variety in information offered. There are various forms of videotex. The "classic" European version of interactive videotex typically works at 75 baud going out from the terminal and 1200 baud coming in from the central office. Some American versions ape the European system. Some have 1200 baud both ways. In interactive videotex, you can do everything from sending serious electronic mail to your business suppliers to holding raunchy conversations with perverts in distant cities. As long as you pay your bills, no European PTT seems to care about what you transmit or receive. In France, videotex is called Minitel. And it's a success because the French phone company funds it. It has also retarded the development of the Internet in France.

videotex Interactive version of teletext, with a return channel using variable forms of low-speed data. English "Prestel" typifies one realization of videotex.

VIDF Vertical Intermediate Distributing Frame. Distributing Frame in which cables terminate in vertically mounted blocks. Compare with Horizontal.

vidicon camera An image sensing device that uses an electron gun to scan a photosensitive target on which a scene is imaged.

view 1. In satellite communications, the ability of a satellite to "see" a satellite earthstation, aimed sufficiently above the horizon and clear of other obstructions so that it is within a free line of sight. A pair of satellite earthstations has a satellite in "mutual" view when both enjoy unobstructed line-of-sight contact with the satellite simultaneously.

2. An alternative way of looking at the data in one or more database tables. A view is usually created as a subset of columns from one or more tables.

viewable area An imaging term. The actual size of the live area of a CRT. For example, a 17" CRT size monitor can have a viewable area ranging from 15.8" to 16.1".

viewdata An information retrieval system that uses a remote database accessible through the public telephone network. Video display of the data is on a monitor or television receiver. Another name for Videotex, the original English (UK) name for it. See Videotex.

Vikings Vikings were not the dirty, smelly, lice-ridden lot they are depicted as having been. An English chronicler, at the height of the Viking raids on the British Isles, bitterly complained that "British womenfolk rather go with the Vikings, because they are always washing their hands and faces, combing hair and beard, changing their shirts, and, oh shame! every Saturday they are taking a full bath and washing all their clothes," which British men did not do because Christians of the time considered washing vanity and therefore something to avoid for the good of their souls at the expense of their bodies. In fact, the ancient Nordic word for Saturday, "Laugrdag," means "washing day."

VIM Vendor Independent Messaging. A new E-mail protocol developed by Lotus, Apple, Novell and Borland to provide a common layer where dissimilar messaging programs can share data and back-end services. A group called the Vendor Independent Messaging Group will is intent on developing an open, industry-standard interface that will allow e-mail features to be built into a variety of software products. See also MAPI, which is the E-mail protocol developed by Microsoft.

vining Another term for social swap nets – or the sharing of videotapes, bootleg CDs, and other tangible media through a "grapevine" of users, often done online over the Internet.

violent agreement We are definitely in agreement.

VIPA Virtual Image Phase Array .

Viper A deadly computer virus.

VIPR 1. Virtual IP Routing extends private route tables and address spaces from the enterprise into the service provider's routing/switching infrastructure. VIPR is essentially a logical partitioning of a physical IP router owned and operated by the service provider.

2. Voice over IP Router. See IP.

VIR A video term. Abbreviation of vertical interval reference. Reference signal inserted into the vertical interval of source video. This signal is used further down the video chain to verify parameters and to automatically adjust gains and phase.

viral marketing Viral comes from the word virus, which means infectious disease. Viral marketing is to have word of your product

Viral marketing is the Internet version of word-of-mouth marketing. Through e-mailing messages and instant messaging friends, images and entertaining advertisements are becoming more and more infectious. So infectious in fact that viewers want to pass them along to everyone.

And that is exactly what Kontraband.com does, as one of the top viral marketing companies in the world! It is beginning to expand beyond the traditional viral marketing techniques, and has begun to build relations with top corporations which will in fact bring viral marketing to new heights.

Your customers are your marketing tool. They show how your product or service makes them. Thus they become your greatest selling tool. For example, you sell a product called call screener. You call Harry's number. Your call screener goes out to find Harry. In the meantime, it's telling you about this product and asking if you'd like to learn more. Rick Bennett of Salt Lake City, Utah, is the world's leading expert on viral marketing. I know this because he told me so.

Virgil A guide to the Internet. Someone who's been there before.

virgule See Backslash.

virtual Something that has the essence of, or creates the effect or has the appearance of, something else. In the context of communications, there are Virtual Channels (VCs), Virtual Circuits (VCs), and Virtual Private Networks (VPNs), to name just a few. A Virtual Circuit, for example, is a pre-defined path through the network across which all the data flow, from edge-to-edge, in a connection-oriented environment. It is not a dedicated private-line circuit, dedicated to one user organization. Rather, it is shared amongst perhaps a large number of user organizations. Therefore, it is not guaranteed to be immediately available to support packet data transfer for any individual user organization, as it might be suffering some level of congestion if others are using it at that particular moment. However, the Virtual Circuit behaves something like a circuit in that the path is defined through the network, and all data packets travel that same path during the course of any given data call. Therefore and within limits, the Virtual Circuit has something of the essence of, or creates the effect or has the appearance of, a true circuit, but without being one. The advantage of a Virtual Circuit lies in its shared nature. While the Virtual Circuit is not always immediately available and while, therefore, individual data packets comprising a data stream associated with a given data call will suffer variable and unpredictable levels of latency (i.e., delay), it is much less expensive than a dedicated circuit. See also Transparent, Virtual Channel, Virtual Circuit, and Virtual Private Network.

Virtual 8086 Mode Virtual 8086 mode allows the Intel 80386 and beyond microprocessors to emulate multiple real mode processors and still switch to and from protected mode. The processor can load and execute real mode applications (in virtual 8086 mode), then switch to protected mode and load and execute another application that requires access to the full extended memory available. The microprocessor, together with a control program like Microsoft Windows or OS/2 assumes the responsibility of protecting applications from one another. See Real Mode and Protected Mode.

virtual ACD Virtual Automatic Call Distributor. In simple terms, an ACD directs incoming calls (typically toll-free ones) to an appropriate person who can answer the call and handle the caller's needs. An ACD traditionally has been both hardware and software, in short a switch designed to handle incoming calls. A Virtual ACD is a term that has come to mean a software-only application that performs call routing at an enterprise level through interfaces to carrier networks and legacy ACD equipment. Moreover, the Virtual ACD can perform queuing and agent-management functions in lieu of the traditional ACD platform. The application can reside in a carrier network or on the customer's premises. It's basically designed to switch calls before they get to a site-specific ACD. There are huge advantages to a Virtual ACD. The main one is that the application can instruct the network to adjust which calls end up on which of the customer's nodal ACDs – we're talking about companies who have many ACDs, often in different parts of the world. The Virtual ACD can be fed constant information on the busy/free status of each site-specific ACD, such as which ones are manned and which are closed (for weather or time, etc.).

virtual banding 1. In WATS services, virtual banding is the ability of trunks to carry traffic to all WATS bands, with billing based on the end points of the call instead of the band over which the traffic went. 2. MCI's definition: Allows customers of MCI's, PRISM, Hotel WATS, and University WATS to call nationwide while only paying for the distance to the actual area. For example, if a customer calls to a Band 1 area, Band 1 pricing is used. Similarly, if a call is placed to a Band 4 area, Band 4 pricing is used.

virtual bypass Virtual bypass is a way smaller users can fill the unused portion of

local T-1 dedicated loops going from a user site to a local office of a long distance company, called a POP (Point of Presence).

virtual call capability 1. Provides setup and clearing on a per call basis. Each call placed appears to have a dedicated connection for the duration of the call.

2. A data communications packet network service feature in which a call setup procedure and a call-clearing procedure will determine a period of communication between two DTEs. This service requires end-to-end transfer control of packets within a network. Data may be delivered to the network before the call setup has been completed but it will not be delivered to the destination address if the call setup is not successful. The user's data are delivered from the network in the same order in which they are received by the network. See also Virtual Circuit.

virtual call center Several Groups of agents, usually in geographically separate locations, that are treated as a single center from a management, scheduling and call-handling perspective.

virtual call service Virtual Call Service is a packet switching capability that allows a customer to establish a virtual circuit between two data terminals for the duration of a call.

virtual CD image Created by dragging and dropping files into into the main window of many CD authoring programs. Can be used to write directly to CD on-the-fly, or to master a real ISO 9660 image to hard disk.

virtual CD player A virtual CD player which the related device driver fools the operating system into believing is a real one connected to your system. It is used to stimulate CD performance from a real ISO image residing on hard disk.

virtual cell A call, established over a network, that uses the capabilities of either a real or virtual circuit by sharing all or any part of the resources of the circuit for the duration of the call.

virtual channel A single connection across a UNI or NNI allowing the switching of different ATM cells in a virtual path to different destinations.

Virtual Channel Identifier VCI. A 16-bit field in the ATM cell header identifying the Virtual Circuit which the data will travel from transmitting device to target device. The Virtual Channel is contained within a Virtual Path.

virtual channel switch An ATM term. A network element that connects VCLs. It terminates VPCs and translates VCI values. It is directed by Control Plane functions and relays the cells of a VC.

virtual circuit A communications link – voice or data – that appears to the user to be a dedicated point-to-point circuit. Virtual circuits are generally set up on a per-call basis and disconnected when the call is ended. The concept of a virtual circuit was first used in data communications with packet switching. A packetized data call may send packets over different physical paths through a network to its destination, but is considered to have a single virtual circuit. Virtual circuits have become more common in ultra-high speed applications, like frame relay or SMDS. There the connection might be permanently connected like a LAN. When the user wants to transmit he simply transmits. There's no dialing in the conventional sense, just the addition of an address field on the information being transmitted. A virtual circuit is referred to as a logical, rather than physical path for a call. A virtual voice circuit is anything from as simple as a phone with an auto dialer in it to a high-speed link in which voice calls are digitized and sent on the equivalent of a ultra high-speed, wide-area equivalent of a local area network. There are two basic reasons people buy virtual circuits. They're cheaper and faster. See Permanent Virtual Circuit.

virtual circuit capability A network service feature providing a user with a virtual circuit. This feature is not necessarily limited to packet mode transmission. e.g., an analog signal may be converted at its network node to a digital form, which may then be routed over the network via any available route. See Virtual Circuit.

virtual colocation First, you can spell it collocation or colocation. I prefer the latter. There are two definitions of this evolving term. Adjacent/Physical and Virtual. First: Imagine that you're a CLEC, a Competitive Local Exchange Carrier. Your idea is to put your switching, transmission and/or Internet equipment in the central office of an ILEC (Incumbent Local Exchange Carrier) and rent some of the ILEC's raw copper circuits out to your customers. You are now able to legally do this. And your idea is to put high-speed data (e.g. DSL) on the raw copper circuits. If you put your equipment inside the ILEC's central office (inside a cage, for example) you have physical/adjacent colocation of your equipment with theirs. Some of the central offices, however, are not large enough to accommodate all the equipment that the various new CLECs are trying to locate in their central office. So the ILECs have figured a new deal. It's called virtual colocation. The CLEC puts his equipment in the ILEC's central office. But the ILEC installs it, configures it, maintains it, fixes it, and does everything necessary to keep it running. The CLEC can remotely monitor and remotely control his equipment as much as possible. But he can't physically go near it. Obviously, the CLEC has to train the ILEC's people and trust them to do the right thing. See also Colocation.

virtual computing A new term for software that shapes computing hardware into hardware that never was. Virtual computing uses FPGAs – Field Programmable Gate Arrays. See FPGAs.

virtual connection A logical connection that is made to a virtual circuit.

virtual container See VC and SONET.

virtual device A device that software can refer to but that doesn't physically exist.

virtual disk A portion of RAM (Random Access Memory) assigned to simulate a disk drive. Also called a ram disk. See RAM DISK.

virtual enterprise network Network World of July 31, 2000 wrote IT executives gathered last week at the annual Catalyst Conference put on by consulting firm, The Burton Group, to share their concerns. They focused on how to integrate directories internally and with partners' corporate systems to help manage e-commerce and online groups of suppliers and partners, a concept The Burton Group refers to as a virtual enterprise network.

virtual facilities group A European term. A traffic control method where virtual PRIs can be created with fewer than the usual 24 or 30 B channels. This limits the amount of inbound traffic. For example, if a virtual facility were created with 10 channels, and 10 calls were in use, the 11th call would be rejected by the network with a Cause 34 (No circuit/channel available). Multiple virtual facilities can be associated with different numbers on the same PRI. The end result is to limit the number of inbound call setup messages (calls) for specific numbers on a PRI. See PRI.

virtual fax A device consisting of a personal computer and an image scanner that can duplicate the functions of a facsimile machine.

virtual fiber A wavelength channel on a single optical fiber. The "virtual fiber" concept is related to wave division multiplexing. By allocating individual wavelength channels on a single fiber to different users, a carrier can efficiently use a single fiber to serve multiple users, rather than giving each user a dedicated fiber. In other words, each wavelength channel acts like a "virtual fiber."

Virtual File Allocation Table VFAT. A fat file systems is a file system based on a file allocation table, maintained by the operating system, to keep track of the status of various segments of disk space used for file storage. The 32-bit implementation in Windows 95 is called the Virtual File Allocation Table (VFAT). See FAT.

virtual hard drive memory factor The available space on a hard drive partition that Windows can address as physical memory.

Virtual Interface Architecture April 16, 1997 - Compaq Computer Corp., Intel Corp., Microsoft Corp. and other industry leaders today announced an initiative to define high-speed communication interfaces for clusters of servers and workstations. Called the Virtual Interface (VI) Architecture specification, the initiative will enable a new class of scalable cluster products offering high performance, low total cost of ownership and broad applicability. More than 40 companies will participate in the process to complete the draft technical specification before its public release.

A cluster is a group of computers and storage devices that function as a single system. Businesses use clusters in place of individual computers for higher availability and enterprise-class scalability. It is possible to use standard local area network (LAN) and wide area network (WAN) technology to connect the machines in a cluster. However, large clusters and high-performance applications require lower latency, higher bandwidth and additional features not offered by standard LAN and WAN technology. A system area network (SAN) is a specialized network optimized for the reliability and performance requirements of clusters.

The VI Architecture specification provides standard hardware and software interfaces for cluster communications. This will spur innovation in SAN technology and make the LAN, WAN and SAN differences transparent to the applications. The VI Architecture specification will support reliable, high-performance SANs, helping clusters achieve their full potential as cost-efficient platforms for large-scale, mission-critical applications.

"Information technology industry leaders continue to lower the cost of information processing on all fronts while enabling advanced customer solutions by bringing value-added technology to the mass market," said Britt Mayo, director of information technology at Pennzoil Company. "Their efforts to drive the creation of an industry standard for the VI Architecture will make multisystem solutions widely available at new levels of price/performance."

The VI Architecture specification will be media, processor and operating system independent. The software interface will support a variety of efficient programming models to simplify development and ensure performance. The hardware interface will be compatible with standard networks such as ATM, Ethernet and Fiber Channel as well as specialized SAN products available from a variety of vendors.

virtual internet A virtual Internet is the service of a corporation or other entity that provides private-label dial-up Internet access for its customers by fully outsourcing their dial-up infrastructure and support functions. In other words, the provider is not a provider, but a virtual provider.

virtual ISDN This is an alternative way for a customer to get ISDN service. A customer can be serviced out of a nearby central office which has ISDN capabilities but not charged the extra mileage charges as they would with a foreign exchange. The phone company does not add on charges because the costs are recouped from the large volume of customers serviced out of the CO. A customer will usually have to change phone numbers if the CO where they receive their POTS service becomes ISDN capable.

virtual LAN A logical grouping of users regardless of their physical locations on the network. Racal-Datacom defines a virtual LAN as "a LAN extended beyond its geographical limit and flexibly configured to add or remove locations." LANs are typically extended beyond their geographical limits (i.e. several thousand feet within a building or campus) by using telephone company facilities, like T-1, T-3, Sonet, etc.

virtual machine A virtual machine is part of a computer's hard disk that thinks it's another computer. The virtual machine thinks it's a complete computer; it doesn't know about the "real" computer except in terms of what the software creating the virtual machine chooses to share with it (like ports or networking).

virtual machine facility VM/370. An IBM system control program, essentially an operating system that controls the concurrent execution of multiple virtual machines on a single System/370 mainframe.

Virtual Machine VM. Software that mimics the performance of a hardware device. For Intel 80386 and higher processors, a virtual machine is protected memory space that is created through the processor's hardware capabilities.

virtual memory Your computer needs memory to store the work its processing unit (its central microprcessor chip) is working on. There are two places your computer can find memory. First, semiconductor chips, also called RAM, Random Access Memory. Second, empty parts of your hard drive. In computerese, the empty parts of your hard drive are typically called virtual memory. Your computer uses virtual memory when it runs out of space on its RAM chips. The benefit of using virtual memory is that you can run more applications at one time than your system's physical RAM memory would otherwise allow. The drawbacks are the disk space required for the virtual-memory swap file and the slower execution speed of your computer when swapping back and forth to the hard disk.

virtual memory manager Virtual memory manager is a software-only approach to Expanded Memory. These work almost identically to the EMS emulators, except that they use your hard disk rather than extended memory as the storage medium for blocks of memory copied out of your program. As you can imagine, this is painfully s-l-o-w. Use this approach only as a last resort.

virtual network A network that is programmed, not hard-wired, to meet a customer's specifications. Created on as-needed basis. Also called Software Defined Network by AT&T. See Software Defined Network and Virtual Private Network.

Virtual Network Operator VNO. A service provider that offers telecom services by piggybacking on the networks of facilities-based operators.

virtual office Employees who are constantly on the move carry their offices with them. Laptops and various telecommunications services allow mobile workers to connect to the central office from virtually any location.

virtual path Contains virtual circuits that are to be switched together to a common destination such as an Interexchange Carrier.

Virtual Path Identifier VPI. An 8-bit field in the ATM header, identifying the Virtual Path (i.e. Virtual Circuit) over which the transmitted data will flow from the transmitting device to the target device.

virtual path switch An ATM term. A network element that connects VPLs. It translates VPI (not VCI) values and is directed by Control Plane functions. It relays the cell of the VP.

virtual phone number A VoIP-enabled service whereby a secondary number is established that is only able to receive incoming calls, and when such a call comes it, the call is automatically forwarded over an IP network to the customer's primary phone number. When a virtual phone number is established for another area code, it enables long-distance charges to be avoided for calls to the virtual phone number that originate in that area code. Similarly, when a virtual phone number is established for another country code, it means avoiding international charges.

virtual printer memory In a PostScript printer, virtual printer memory is a part of memory that stores font information. The memory in PostScript printers is divided into banded memory and virtual memory. Banded memory contains graphics and page-layout information needed to print your documents. Virtual memory contains any font information that is sent to your printer either when you print a document or when you download fonts.

Virtual Printer Technology VPT. Virtual Printer Technology is the enterprise network printer architecture developed by Dataproducts Corporation that enables a printer to become an intelligent node in a networked computing environment and provide printing services to other network nodes through a Client/Server type relationship.

Virtual Private Network VPN. See VPN.

virtual private office I found this definition of Virtual Private Office in the September 13, 1997 issue of the Economist, one of my favorite magazines. They had a roundup of telecommunications. Here's how they started their roundup:

In Anderson Consulting's smart new offices in Wellsley, just outside Boston, Mark Greenberg is entitled as a senior partner to three filing-cabinet drawers of storage space. In one, he keeps a bubble-wrapped package, containing the sort of personal mementoes_family photographs, shields and so on_with which businessmen like to decorate their offices, together with a diagram to show how they should be arranged. On the rare days when Mr Greenberg is not visiting a client or jetting around the world, he reserves an office. When he arrives, his treasures are neatly laid out on the desk for him to make him feel at home.

But this is, in effect, a virtual private office, his just for the day. Struck by the waste involved in maintaining expensive permanent offices for people with itinerant lives, the partners in the world's largest management consultancy have created something that feels like a cross between a hotel and a luxurious club. The Wellesley office is staffed by the cream of Boston's hotels: people who understand the arduous providing services for important and self-important people. The reception desk looks like a hotel foyer; each floor has lots of little "huddle rooms" with comfortable armchairs, as well as brainstorming rooms with less comfortable ones; and there are open spaces for coffee and conversation with colleagues.

Virtual Reality VR. The publisher of Virtual Reality Report says, "Virtual reality is a way of enabling people to participate directly in real-time, 3-D environments generated by computers." Virtual reality involves the user's immersion in and interaction with a graphic screen/s. Using 3-D goggles and sensor-laden gloves, people "enter" computer-generated environments and interact with the images displayed there. Says Business Week, "Imagine the difference between viewing fish swimming in an aquarium and donning scuba gear to swim around them. That's the sensory leap between regular computer graphics and virtual reality. There are three kinds of VR (Virtual Reality) immersion. First, the toe in the water experience of beginners who stand outside the imaginary world and communicate by computer with characters inside it. next, wading up to the hips, are the "through the window" users, who use a "flying mouse" to project themselves into the virtual, or artificial, world. Then there are the hold-the-nose plungers: "first persona interaction within the computer-generated world via the use of head-mounted stereoscopic display, gloves, bodysuits and audio systems providing binaural sound. The trick with virtual reality is not only to simulate another world but to interact with it – pouring in data affecting its plots, changing its characters and introducing real-world unpredictability into this "mirror world." Once virtual reality was called artificial reality. But artificial means "fake," while virtual means "almost." The father of virtual reality is Joran Lanier. A term close to virtual reality is telepresence. See Telepresence.

virtual route Virtual circuit in IBM's SNA. See Systems Network Architecture.

virtual route pacing control A congestion control at the path control level. See Systems Network Architecture.

Virtual Router Redundancy Protocol See VRRP.

virtual server A server that resides on a software-based partition on a physical server. With this approach, a single physical server can have multiple virtual servers. It is a way of improving utilization of a physical server and reducing server sprawl. See Virtualization.

virtual shredding The deletion of e-mail, instant messages, and other electronic documents generated by company employees.

virtual storage Storage space that may be viewed as addressable main storage to a computer user, but is actually auxiliary storage (usually peripheral mass storage) mapped into real addresses. The amount of virtual storage is limited by the addressing scheme of the computer.

Virtual Telecommunication Access Method VTAM (Pronounced "Vee-Tam.") A program component in an IBM computer which handles some of the communications processing tasks for an application program. VTAM also provides resource sharing, a technique for efficiently using a network to reduce transmission costs.

Virtual Terminal VT. A universal terminal. The ISO virtual terminal (VT) protocol is designed to describe the operation of a so-called universal terminal so any terminal can talk with any host computer.

Virtual Terminal Protocol VTP. Virtual Terminal Protocol enables computers to communicate with various types of terminals by interpreting and translating the instructions for both the computer and the terminal.

Virtual Tributary VT. A structure designed for transport and switching of SONET payloads of less than OC-1 (Optical Carrier level 1), which runs at a total signaling rate of 51.84 Mbps. Prior to conversion to an optical format, the SONET frame is defined at the electrical level in a the form of STS (Synchronous Transport Signal). It is at this point that the Synchronous Payload Envelope (SPE) is defined at a total rate of DS-3, or more exactly, T-3. The SPE can be further defined at subrate levels as follows: 1) VT1.5, which equals a T-1 frame running at 1.544 Mbps; 2) VT2, which equals and E-1 frame at 2.048 Mbps; 3) VT3, which equals a T-1c frame at 3 Mbps; and 4) VT6, which equals a T-2 frame at 6 Mbps. In applications requiring that VTs be mixed within the same STS-1 frame, the SPE can be divided into as many as seven VT Groups. A given VT Group most commonly supports four VT1.5 signals, or 3 VT2 signals. See also SONET, SPE, STS, Virtual Tributary Group, VT1.5 and VT2.

virtual tributary group A SONET/SDH Term. A SONET frame can comprise as many as seven Virtual Tributary Group, each of which contains signals of the same type, such as T-1 or E-1 frames. See also SONET and Virtual Tributary.

Virtual Wavelength Path See VWP.

virtualization You're trying to give your customers or employees access to computer service. You don't want them to have to worry about the design of the equipment and network serving them. Virtualization puts a bunch of computing and networking facilities together in a such a way that it looks like a virtual computer to a user or a customer. Virtualization hides the reality from the users, but keeps them happy with the service. There are many implementations of virtualization and over time its meaning has changed. One popular meaning is running more than one operating system on a single computer at the same time. On the Intel Duo Core Mac (i.e. Apple PC), under OS X, virtualization is done with something called Parallels. It works by creating virtual machines (VMs) that use the hardware on your Intel Mac to run Windows, Windows Server, Linux, Solaris, or even OS/2 in their own sessions. You get near-native speed, and you don't have to reboot to use it. Just start up the VM, and off you go. You can have multiple VMs and run them one at a time or simultaneously.

virtuous cycle A situation where one good thing leads to another. For example, a virtuous cycle for a vendor exists when customers who upgrade one part of their system tend to go ahead and upgrade other parts of their system, in order to maximize the benefits of the first upgrade.

virus A software program capable of replicating itself and usually capable of wreaking great harm on a computer. The term was first used by David Gerrold, a science fiction writer who wrote a story about malignant code that spread from one computer to another. The story was entitled "Virus." See also Virus Hoax.

Viscount of Vapor During the 1980s, Bill Gates, chairman and co-founder of Microsoft, became known as the Viscount of Vapor, because so many of his announced new products failed to materialize, or when they did materialize, appeared much later than he said they would.

vishing Short for "voice phishing." The use of a telephone by con artists to trick people into revealing personal information that can later be used in an identity theft scam. Vishing takes several forms. It may involve a fake PayPal or bank email, with instructions to call a phone number in the email to "confirm" account information. An automated answering service at the other end asks the caller for his account number, PIN, and other information. Vishing can also take place through a simple official-sounding phone call to an unsuspecting person. Vishing is aided by voice over IP, which enables cheap and anonymous Internet calling, and makes it easy to spoof a fake calling party number on caller ID boxes.

visibility Visibility means the ability to be seen. In the Spring and Summer of 2001, many technology companies were telling their shareholders that they had no "visibility." What they meant was that they could not predict future sales and earnings. Their customers were simply not telling them of their plans and were simply not buying until the last moment, or buying sporadically.

visible light Wavelengths between 400 and 700 nanometers and frequencies between 430 and 750 THz comprise the portion of the electromagnetic spectrum that is visible to the human eye. Constituting a mere 1/10,000,000,000,000,000,000,000,000th of the total electromagnetic spectrum, it's a very small part, but very important to us. Visible light is in a shorter wavelength (i.e., higher frequency) range that Infrared (Ir) light, and in a longer wavelength (i.e., lower frequency) range than ultraviolet (UV) light.

The spectrum of visible light can be subdivided by color. Beginning at the longest wavelength and ending at the shortest wavelength, the colors are red, orange, yellow, green, blue, and violet. An easy way to remember their order (as if you would really want to do such a thing) is to make a word, or words, out of it, just like you used to do in elementary school when you studied for a test. How about naming an imaginary person Roy G. Bv? That's not much of a name. It used to make more sense when the area between blue and violet was designated indigo. Then the key was Roy G. Biv, which is still a pretty odd name. Indigo is no longer considered to be a separate color in the visible spectrum. That's your trivia lesson for the day.

Visitors' Location Register VLR. A wireless telecommunications term. A local database maintained by the cellular provider in whose territory you are roaming. When you place a call in a roaming scenario, the local provider queries the HLR (Home Location Register) over the SS7 network through the use of a REGNOT (REGistration NOTification). The HLR is maintained by your cellular provider of record in order to verify your legitimacy and to secure your profile of features. The HLR responds to the REGNOT with a "regnot" (lower case), and transfers the necessary data. This information is maintained by the local provider as long as you remain an active roamer within that area of coverage. This process of query and download is accomplished via SS7 links between SCPs (Signal Control Points). SCPs typically are associated with MSCs (Mobile services Switching Centers), also known as MTSOs (Mobile Traffic Switching Offices) for registering visiting mobile station users. VLRs and HLRs are employed in a variety of cellular networks, including AMPS, GSM and PCS. See also AMPS, GSM, HLR, MSC and PCS.

Vista 1. Microsoft's latest operating system, introduced in early 2007. Allegedly a more reliable and prettier operating system. It didn't get rave reviews.

2. A videotext service offered in Canada.

visual area networking What one manufacturer calls its new collaborative graphics technology, whch allows remote PC users to access and manipulate complex 3-D images stored on a centralized supercomputer.

visual basic A version of the programming language BASIC written by Microsoft Corporation for Windows. The new program promises to make it much easier for businesses to develop customized Windows applications. Some programmers are calling the software a major breakthrough in ease of programming. When I wrote this, Microsoft had sold over one million copies of Visual Basic.

visual carrier The portion of a television signal which carries the video portion of the picture.

Visual Display Unit VDU. Another term for a computer monitor. VDU is preferred in Europe.

visual fault locator A device used to locate breaks and discontinuities in optical cabling, using red laser light. Fiber breaks typically show up as a spot of continuous or blinking red light. The device can also be used to locate particular cables in bundles.

Visual Message Waiting Indicator VMWI. You are talking to someone on the phone, or perhaps you went to lunch, or on a business trip, or got to work late. Someone else called and left a message on your voice mailbox. You get a Visual Message Waiting Indicator – a little lamp lights on your display phone says "Message Waiting." That's VMWI. Big fancy name for a very simple concept – a light or message on your phone that tells you someone called.

visual voice mail An application displaying and controlling voice messages on a desktop computer. Usually associated with unified messaging.

visual voice messaging A term created by Microsoft as part of its At Work announcement in June of 1993. There'll be At Work-based visual voice messaging servers sitting on a LAN. Messages for PC users on the LAN will be able to be displayed in a list, much like electronic mail, including the caller's name or number, the time he or she called and the length of the call. This information would let the user browse all messages and select the order for listening to the messages. Administrative options, such as creating a new greeting, will be accessed with a single button. Operations that are difficult today, such as forwarding a voice message to multiple people, will be dramatically simplified, according to Microsoft. One will simply select the recipients from the phone book and broadcast the message. Using visual voice messaging, users will be able to bypass today's inconsistent, time consuming and confusing audio menus and access their voice messages with the push of a button or the click of a mouse on a Windows type icon. Messages will be able to be retrieved in any order and even delivered to a single mailbox along with other messages such as e-mail and faxes. These visual voice messaging servers will, according to Microsoft, provide applications beyond basic voice messaging, such as supporting voice annotation of PC documents or reading electronic mail over the phone to a traveler.

visualization A combination of computerized graphics and imaging technology that provides high-resolution, video-like results on the workstation or personal computer's screen.

visually impaired attendant service Visually impaired attendant service capability is achieved by augmenting the normal visual signals provided on a standard attendant position with special tactile devices and/or audible signals which enable a visually impaired person to operate the position.

VITA VME International Trade Association. A widely supported industry trade group in Scottsdale, AZ. VITA is chartered to promote the growth and technical excellence of the VME bus and Futurebus-based microcomputer board market. VITA is chartered to submit standards for ANSI registration. See VME. www.vita.com

visual carrier The portion of a television signal which carries the video portion of the picture.

VITC Vertical Interval Time Code. Contains the same information as the SMPTE time code. It is superimposed onto the vertical blanking interval, so that the correct time code can be read even when a helical scanning VCR is in the pause or slow (DT) mode.

viterbi decoder A maximum likelihood technique for decoding convolutional codes first proposed in 1967 by Andrew J. Viterbi, then a professor at the University of California, Los Angeles (UCLA) and subsequently in 1985 the co-founder of QUALCOMM, Inc. Most digital satellite communications systems employ Viterbi decoding. v **Viton** DuPont trademark for a series of fluoro-elastomers based on the copolymer of unylidene fluoride and hexafluoropropylene. Viton is used in the making of cables.

vitreous silica Glass consisting of almost pure silicon dioxide.

VL-Bus A new PC bus from VESA – the Video Electronics Standards Association. The VL bus is up to 20 times as fast as an ISA bus, the most common PC bus and the one common to the original PC, the PC XT and the PC AT and clones. VL-Bus was popular on 486 PCs. Pentium-based machines now largely use the newer, PCI bus. As a result, VL-Bus is pretty obsolete. See ISA, Microchannel and PCI.

VLAN Imagine the Internet. Everybody and his uncle uses it. There is no security. Now imagine taking bits of the Internet, private connections to the Internet and a collection of special security and switching devices. Bingo, you have a VLAN – essentially a private closed network within a larger network. VLANs allow an administrator to create networks based on parameters beyond the network address, hence the name "virtual." The virtual aspect of VLANs refers to the fact that devices in a VLAN behave as though they are on the same wire, even though they may be physically located on different segments of the LAN. In fact, a VLAN can even extend across a WAN, a wide area network. The reason to create a VLAN is to segment a large subnet. This simplifies user mobility and provides broadcast controls. Switches are configured with policies (or rules) that limit which device can access which VLAN. These rules are set on a switch port or range of ports. The most secure rules combine two or more characteristics of the connecting device, such as some combination of port, MAC address, IP address, and/or routing protocol. These rules can be statically configured or dynamically learned. If a rule is violated, that device is not allowed to access the network and alarms and/or trouble logs are generated. This provides an extra level of security for non-mobile devices such as printers and servers, especially when they are deployed in semi-public spaces. See VLAN Assignment Methods and VPN (Virtual Private Network).

VLAN Assignment Methods Policy-based VLANs allow various methods for users to be assigned to VLANs independent of their physical attachment to the network, which is important for maintaining user mobility. VLANs can be formed based on a variety of characteristics including: Switch port, MAC address, IP address, Protocol type, Multicast-aware, DHCP aware, 802.1Q tags, User identity (through authentication). These characteristics can be deployed as standalone rules or combined. See VLAN.

VLC Variable length coding.

VLD Vehicle Location Tracking Device. See Location Tracking.

VLF Very Low Frequency. That portion of the electromagnetic spectrum having continuous frequencies ranging from about 3 Hz to 30 kHz.

VLM Novell Virtual Loadable Module network client architecture uses packet burst technology, so ample packets are sent without waiting for packet acknowledgement. VLM support (compared with the old IPX) improves transfer times, especially for compressible files, since there is no waiting for acknowledgement.

vloggers See vlogoshere.

vlogoshere From the New York Times, "Can you vlog a dead horse? Only if you make a video of it and post it on the Web. After blogging came photo blogging and then, suddenly last year (2004), video blogging. Video bloggers, also known as vloggers, are people who regularly post videos on the Internet, creating primitive shows for anyone who cares to watch. Some vlogs are cooking shows, some are minidocumentaries, some are mock news programs and some are almost art films. Most simply are records of ordinary life. The Das Vlog recently demonstrated the virtues of urinating in the bathroom sink. Village Girl has posted a video of her 2-year-old dancing with a friend. Josh Leo taped himself browsing through his old baby pictures and art projects. ... Fat Girl From Ohio is a man blogging largely about his wife's pregnancy. As the video blog Reality Sandwich recently put it in a video of vegetable shopping, quoting a mantra of the vlogosphere: "Hey ... mundane is the new punk." Already, though, it's beginning to look a lot like television, at least in spots. Some vlogs even share television's worries, chief among them the burden of coming up with fresh programming on a regular basis. ... One of the most winning vlogs is the 05 Project, the work of an 18-year-old in Keynes, England, Ian Mills, who has promised to post a video a day all year. He begins almost every short video by moving close to the camera and addressing the audience with a sweet formality, "Okay, so today. ..." In January, 2005 he showed the inside of his closet to prove he doesn't have just one set of clothes, but two. In February, he filmed a stuffed kangaroo seeking directions from a stuffed teddy bear sitting in front of a microwave oven...."

VLR Visitors' Location Register. A wireless telecommunications term. A local database maintained by the cellular provider in whose territory you are roaming. When you place a call in a roaming scenario, the local provider queries the HLR (Home Location Register) over the SS7 network through the use of a REGNOT (REGistration NOTification). The HLR is maintained by your cellular provider of record in order to verify your legitimacy and to secure your profile of features. The HLR responds to the REGNOT with a "regnot" (lower case), and transfers the necessary data. This information is maintained by the local provider as long as you remain an active roamer within that area of coverage. This process of query and download is accomplished via SS7 links between SCPs (Signal Control Points). SCPs typically are associated with MSCs (Mobile services Switching Centers), also known as MTSOs (Mobile Traffic Switching Offices) for registering visiting mobile station users. VLRs and HLRs are employed in a variety of cellular networks, including AMPS, GSM and PCS. See also AMPS, GSM, HLR, MSC and PCS.

VLSI Very Large Scale Integration. The art of putting hundreds of thousands of transistors onto a single quarter-inch square integrated circuit. Compare with LSI and ULSI.

VM 1. Voice Mail, Voice Messaging or Virtual memory. See Virtual Storage and Voice Mail.

2. Virtual Machine. IBM's mainframe operating system.

3. Virtual Machine. See Java Virtual Machine.

VME Acronym for "VersaModule-Eurocard." A one through 21 slot, mechanical and electrical bus standard originally developed by the Munich, Germany division of Motorola in the late 70s. VME uses most of the bus structure from then current Motorola's VersaBus board standard along with the newly developed DIN 41612 standard pin-in-socket connector for enhanced reliability. After years of work, VME was finally adopted by the ANSI/IEEE in 1987 (as ANSI/IEEE-1014). VME is known in Europe as the IEC 821 bus. This makes it an open standard. The VME backplane runs at 80 Mbytes per second. It is the most common bus on big open computers (i.e. ones larger than the PC). As of writing, there were over 300 vendors offering more than 3,000 off-the-shelf VME products. The IEEE standard is soon to lapse and be replaced by an extended VME64 specification, now in ANSI ballot being conducted by VITA. See also VMEBus and VXI.

VME64 An enhanced VME bus standard which includes multiplexed address and data cycles with 40 and 64 bit address modes and 64 bit data transfer modes allowing up to 80 MB/s transfer speed. This standard is under the ANSI ballot process conducted by VITA. See VME.

VME64 Extensions A VITA draft standard that provides extra functionality to VME64 including 5 row J1/P1 and J2/P2 connectors that support live insertion on both 3U and 6U VME boards. Other features: 3.3V power, more grounds, ETL (slew rate) drivers, geographic addressing (slot ID) as well as support for parity, a serial diagnostic bus, JTAG test support and lots of user I/O. Some mechanical features: locking extractors, RFI gasketing, and ESD chassis discharge strips. See VME.

VMEBus VersaModule-Eurocard BUS. A 32-bit bus developed by Motorola, Signetics, Mostek and Thompson CSF. Used widely in industrial, commercial and military applications with over 300 manufacturers of VMEbus products worldwide. VME64 is an expanded version that provides 64-bit data transfer and addressing.

VMEC Voice Messaging Educational Committee. An organization formed by voice messaging manufacturers and service providers to promote a better understanding of voice mail and its business benefits, and to help business implement voice mail systems in ways that meet the needs of callers and mailbox owners alike. See VME.

VMF Validation Message Fraud

VMI Voice Messaging Interface.

VMF Validation Message Fraud

VMN David Hester wrote to me: "I work for an ISP. We buy dial up service from providers in areas where we do not have our own POPs. We call these POPs Vendor Managed Networks or VMNs. We used to call them Other Peoples' Networks, or OPNs, but that term has been depricated by VMN. I think that the term VMN is a more descriptive term and could be useful to others in the telecom field."

VMR Violation Monitoring and Removal. The process of removing a violations which are detected, so that violations do not propagate beyond the maintenance span.

VMS Virtual Memory System

VMS OSI Transport Services VOTS. A Digital Equipment Corp. software product that modifies Digital's DECnet transport layer to conform to the International Standards Organization (ISO) Transport Protocol Class Four (TP4).

VMUF Voice Messaging User Interface Forum. A standards body formed by voice messaging end users, service providers and manufacturers to define a minimum set of common human interface specifications for voice messaging systems.

VMX Voice Message Exchange. One day in 1979, Gordon Matthews came back from lunch and noticed that he had received the usual half dozen messages that had been randomly written down semi-correctly by a harried receptionist. Already a noted inventor, Matthews saw an opportunity to build an adjunct device to the company phone system which would allow these messages to be recorded by the caller without an intermediary (i.e. a harried secretary) and would allow the recipient to electronically store these messages, forward them to others or to directly reply to them if they were generated from another internal extension. He called this device the Voice Message Exchange, and the company later became VMX, which later got bought by Octel, which later got bought by Lucent, which hasn't been bought by anyone, yet.

VNET Virtual private NETwork. An MCI (now MCI Worldcom) term for a service it offers to customers who want to join geographically dispersed switches (typically PBXs). Instead of private lines joining the PBX, Vnet uses fast switched lines.

VNL Via Net Loss. A loss objective for trunks, the value of which has been selected to obtain a satisfactory balance between two data terminals for the duration of the call. Loss value in db assigned to a circuit to compensate for it's added propagation delay, terminal delay or loss variability. Example: VNL for Satellite channels is 4db; VNL for Microwave channels is 0.0015 x route miles; All other terrestrial is +0.4db. See echo.

VNN Voice News Network.

VNO See Virtual Network Operator.

VNS Virtual Network Services. A VPN (Virtual Private Network) term, referring specifically to a range of international VPN services, including voice, data, Internet and multimedia. Providers of traditional VNS include AT&T, MCI, Qwest and Sprint. During the recent past, groups of international carriers have formed multilateral affiliations to provide VNS seamlessly, providing identical feature content and functionality across national borders, regardless of the affiliated country of origination or termination. Some of these ventures have failed.

VO Verification Office.

VOA Variable Optical Attenuator.

VOATM Voice over ATM. Voice Over ATM. Voice over ATM enables an ATM switch to carry voice traffic (for example, voice telephone calls and faxes) over an ATM network. When sending voice traffic over ATM, the voice traffic is encapsulated using AAL1/AAL2 ATM packets.

VOB Voice Over Broadband.

VoBB Voice Over BroadBand.

VOC In February, 1996, Investor's Business Daily, ran a story entitled "Is Your Office at Home Making You Sick?" It said that more than 20 million American workers are telecommuting, with another 20 million owning home based businesses. "Experts are finding that home based offices outfitted with the latest fax machines, photocopiers, laser printers and personal computers often foster unhealthy environments. Office machinery emits air pollutants called volatile organic compounds, or VOCs, which can make your head hurt and irritate your eyes, nose and throat. VOCs are also known to cause more serious health problems, including kidney and liver damage, experts say."

vocabulary development Development of specific word sets to be used for speaker independent recognition applications.

VoCable Voice over Cable. Cable operators are busy optimizing the bandwidth in their networks to deliver high-speed Internet access and to deliver voice service over the same cable spectrum, also called VoIP, or Voice over Internet Protocol. I make virtually all my phone calls using VoIP over my cable TV connection. See VoIP for a bigger explanation.

vocoder Voice coder. A device that synthesizes speech. Vocoders use a speech analyzer to convert analog waveforms into narrowband digital signals. They are used in digital cellular phones as well as in the entertainment business (e.g., the voice of Darth Vader in Star Wars). Vocoders are an early type of voice coder, consisting of a speech analyzer and a speech synthesizer. The analyzer circuitry converts analog speech waveforms into digital signals. The synthesizer converts the digital signals into artificial speech sounds. For COMSEC purposes, a vocoder may be used in conjunction with a key generator and a modulator-demodulator device to transmit digitally encrypted speech signals over normal narrowband voice communication channels. These devices are used to reduce the bandwidth requirements for transmitting digitized speech signals. There are analog vocoders that move incoming signals from one portion of the spectrum to another portion.

VOD Video on Demand. For bigger explanation see NVOD and Video on Demand.

VODAS Voice Operated Device Anti-Sing. A device used to prevent the overall voice frequency singing of a two-way telephone circuit by ensuring that transmission can occur in only one direction at any given instant.

vodka Experts recommend using vodka as a cleaning solution for diamond jewelry.

VoDSL Voice over a DSL line – Digital Subscriber Line. Voice phone calls over DSL. It's making a voice phone call over a digital subscriber line. See DSL.

VoFi VoIP (Voice over Internet Protocol) over Wi-Fi. The idea is to have a phone that works on Wi-Fi. You might walk into Starbucks and instead of using their Wi-Fi to "hook" your portable computer to, you take out your VoFi phone, punch in some numbers to tell it who you are and then punch out a phone number and speak. See Wi-Fi and VoIP.

VoFR Voice over Frame Relay is the transmission of voice over a Frame Relay network, VoFR is one of a family of "Voice over Packet" technologies that includes VoATM (Voice over ATM) and VoIP (Voice over Internet Protocol). Frame Relay was developed specifically for LAN-to-LAN internetworking across a WAN (Wide Area Network), and works quite well in support of such traffic. LAN traffic has no expectations in terms of QoS (Quality of Service) – meaning it may be delayed by more important voice calls – and Frame Relay inherently offers none. Voice, on the other hand, is quite demanding in terms of QoS, and that poses a problem. In order to provide some reasonable level of performance for voice. Therefore, VoFR employs several mechanisms.

First, manufacturers have developed routers that allow carriers to offer various levels of service through non-standard priority queuing mechanisms. VoFR frames can be marked as high priority and, therefore, have priority in the queuing buffers on both the inbound and outbound router ports. Frames carrying traffic of lower priority (e.g., LAN and SDLC data traffic) would be marked as such, and would have lower priority in the queuing buffers. While these mechanisms can be fairly effective, they are "best effort" in nature, offering no QoS guarantees.

Second, VoFR employs voice compression algorithms in order to improve performance. Voice, of course, always is analog in its native form. To support voice over a digital network of any sort, including Frame Relay, it first is converted to PCM (Pulse Code Modulation) format, according to the ITU-T G.711 standard. PCM takes several forms. In North America, u Law (actually mu Law, but my text editor won't support the symbol for "mu"), which is based on the Nyquist Theorem, is the choice. In Europe and internationally, A Law, which is based on Shannon's Law, is used. The two approaches are similar in concept, but different enough to be incompatible. Nonetheless, each is used to encode an analog voice stream of 4 KHz into a digital format requiring bandwidth of 64 Kbps. The analog voice signal is sampled 8000 times a second, with each sample coded into an 8-bit byte (8,000 x 8 = 64,000) at very precise intervals of 125 microseconds (millionths of a second). As each byte is formed, it is sent in a time slot across a channelized digital circuit in the form of T-carrier (North America) or E-carrier (Europe and international). On the receiving end, the process is reversed and the analog signal is reconstructed. This approach works perfectly, but requires that circuit-switches are in place to support perfectly committed connectivity. As a packet data technology, Frame Relay will not support the precise pace demanded by PCM-encoded voice. Even if each voice byte were encapsulated into a frame and the frames were sent into the network 8000 times a second at the precise pace of every 125 microseconds, the Frame Relay network couldn't provide any guarantees that they would make it through the network and to the other end at the same pace. In fact, I can guarantee you that they wouldn't. A good compression algorithm addresses the problem in several ways. First, compression reduces the pressure on bandwidth through the various switches and routers, and across the circuits that interconnect them. Second, the delay imposed on the voice traffic by the very process of decompressing a voice frame provides the receiving VFRAD (Voice Frame Relay Access Device) with a few milliseconds (thousandths of seconds) of time to seize the next voice frame and blend them together. The compression techniques employed also include various predictive techniques that enable them to predict the essence of the voice data contained in subsequent frames based on the voice data contained in previous frames. This capability is based on the fact that voice data flows in a fairly smooth pattern of am-

plitude (i.e., volume) and frequency (i.e., pitch) that is highly predictable. That allows the stretching and blending of voice data in the event that a subsequent frame is lost, errored or excessively delayed as it transits the network. While proprietary compression algorithms were developed by some manufacturers and are used by some carriers, the Frame Relay Forum specified several standard approaches in its FRF.11 IA (Implementation Agreement). Early implementations of VoFR were based on PCM, which clearly is an unsatisfactory approach. Subsequently, ADPCM (Adaptive Differential Pulse Code Modulation), standardized by the ITU-T as G.726 and included in FRF.11, gained favor. ADPCM is a popular and well understood technique that is used in some traditional PSTNs (Public Switched Telephone Networks) to yield 2:1 (two-to-one) compression, thereby requiring only 32 Kbps for "toll quality" voice. In VoFR implementations it variously compresses PCM voice to 40, 32, 24, and even 16 Kbps. As the compression rate increases, the bandwidth required decreases proportionally, but the reconstructed voice quality suffers. ADPCM does not include any predictive capability. The predictive algorithm included in FRF.11 is CS-ACELP (Conjugate Structure-Algebraic Code Excited Linear Predication), standardized by the ITU-T as G.729. CS-ACELP yields an 8:1 compression rate, which requires only 8 Kbps for voice, with some degradation in the quality of the reconstructed voice signal. Although they are not included in FRF.11, other standard compression algorithms also are used in support of VoFR. Those algorithms include Dual Rate Speech Coder for Multimedia Communication, which runs at rates of 6.3 and 5.3 Kbps, and is specified in G.723; and LD-CELP (Low Delay-Code Excited Linear Prediction), which runs at 16 Kbps, and is specified in G.728.

Third, various mechanisms have been developed to select different paths through the network. Most significant among these is MPLS (MultiProtocol Label Switching), which was standardized by the IETF (Internet Engineering Task Force) and which is based on Cisco's proprietary Tag Switching protocol. MPLS is capable of selecting a high bandwidth, low latency path through the network in support of VoFR. In addition to priority queuing and compression mechanisms, VoFR benefits from plenty of bandwidth in the Frame Relay network core. This responsibility is entirely the carrier's. If the switches or circuits along a particular VoFR path in the network core are overloaded or undersized, VoFR performance will suffer considerably. VoFR also depends on the CIR (Committed Information Rate), which is a contractual bandwidth commitment between the end user and the carrier. If the CIR is undersized by the end user organization in consideration of economics, VoFR performance will suffer. Finally, VoFR frames must be prioritized by the transmitting FRAD in terms of Discard Eligibility (DE). Clearly, the VoFR frames must not be marked as DE. If they are so marked, the network will discard them first in the event of congestion. Now, you must be asking yourself why anyone would even consider VoFR. I asked that question. Long distance rates over the circuit-switched PSTN are today in the range of 2-7 cents per minute, depending on how big your company is and how good a negotiator you are. VoIP is getting all of the attention. VoFR seems like a lot of trouble for voice of highly uncertain quality. Well, the reason people use it for voice is that it's free or can be free – if you've got a Frame Relay network in place, there is some capacity available at the moment within the CIR, and the network is not advising you of any congestion at the moment, you can compress some voice and send it between two corporate sites on the Frame Relay network for free. Frame Relay is priced on the basis of port speed, local loop speed, CIR speed, and occasionally some other elements. There is no usage-sensitive element to the pricing algorithm. The quality usually is pretty good, although it sometimes gets to be pretty bad during the course of a call. You would never install a Frame Relay network specifically for voice. You install it for data, and voice rides for free from time-to-time. That's the whole story. See also ACELP, ADPCM, CELP, CIR, Compression, CS-ACELP, DE, Dual Rate Speech Coder for Multimedia Communication, FRAD, Frame Relay, LD-CELP, MPLS, PCM, Tag Switching, VoATM, and VoIP.

VOGAD Voice-Operated Gain Adjusting Device. A voice-operated compressor circuit that is designed to provide a near-constant level of output signal from a range of input amplitudes. Such a circuit has a fast attack time with a relatively slow release time to avoid excess volume compression at the system output.

VOHDLC Voice over HDLC

Voice 2.0 The view that basic voice service will be a commoditized VoIP application on the Web, legacy and next-generation voice services will all be Web-based, Web-based voice will enable the development of new classes of applications, and the PSTN will disappear. See purple minutes and purple applications.

voice activated dialing A feature that permits you to dial a number by calling that number out to your cellular phone, instead of punching it in yourself. See Voice Activated Video.

voice activated switching Used in multipoint video conferencing so all sites automatically see the video of the person speaking.

voice activated video A microphone/camera that is activated in response to voice. Imagine you're watching a videoconference going on in four locations. You can hear what everyone is saying. What you need is to be able to see the person who is speaking the loudest, and therefore, presumably the principal speaker – the person whose attention everyone should be focused on. In voice activated video, the videoconferencing system senses who's speaking the loudest and throws that person's face up on everyone's screen.

Voice Activity Compression VAC. A method of conserving transmission capacity by not transmitting pauses in speech.

Voice Activity Detection See VAD.

voice application network A third-party provider of call-center integration. These networks typically offer clients network and voice recognition infrastructure, charging for time the application is used. Further, they allow organizations to offer their customers self-service access to voice-activated applications. For example, insurance companies can use a Voice Application Network to allow agents and customers to easily access the status of a bill payment or a claim. See also Voice Portal, VXML.

voice applications program System software providing the necessary logic to carry out the functions requested by telephone system users. It is responsible for actual call processing, making the various voice connections and providing user features, such as Call Forwarding, Speed Dialing, Conference, etc.

voice board Also called a voice card or speech card. A Voice Board is an IBM PC- or AT-compatible expansion card which can perform voice processing functions. A voice board has several important characteristics: It has a computer bus connection. It has a telephone line interface. It typically has a voice bus connection. And it supports one of several operating systems, e.g. MS-DOS, UNIX. At a minimum, a voice board will usually include support for going on and off-hook (answering, initiating and terminating a call); notification of call termination (hang-up detection); sending flash hook; and dialing digits (touchtone and rotary). See Voice Bus and VRU.

voice body part An X.400 term. A body part sent or forwarded from an originator to a recipient which conveys voice encoded data and related information. The related information consists of parameters which are used to assist in the processing of voice data. These parameters include information detailing the duration of the voice data, the voice encoding algorithm used to encode the voice data, and supplementary information.

voice browser Pick up a phone, call a VoxML-enabled Web site, ask it questions using your voice and what's known as "natural voice commands" (such as "When is my plane leaving?") and hear responses to your questions. This uses an upcoming technology called Voice Markup Language (VoxML), which several manufacturers plan to develop as an open platform and submit to the World Wide Web Consortium for standards approval. The idea is to VoxML-enable your Web site so it can respond to voice recognition.

voice bulletin boards These are voice mailboxes which contain pre-recorded information that can be updated as frequently as the provider of the mailboxes desires and can be accessed by the public 24 hours a day. Voice bulletin boards can be used by city or county departments which receive a large number of calls asking for routine information, e.g., summer programs for kids as listed by a parks and recreation department; jobs currently open in the city as listed by the personnel departments; etc.

voice bus Picture an open PC. Peer down into it. At the bottom of the PC, you'll see a printed circuit board containing chips and empty connectors. That board is called the motherboard. Fatherboards are inserted into the connectors on the motherboard. These fatherboards do things on the PC – like pump out video to your screen or material to your printer or your local area network. The motherboard controls which device does what WHEN by sending signals along the motherboard's data bus – basically a circuit that connects all the various fatherboards through their connectors. That data bus was not designed for voice. For voice you need another bus. Several voice processing manufacturers have addressed that need by creating a voice bus at the top of their PC-based voice processing cards. They have tiny pins sticking out of their cards. You attach a ribbon cable from one set of pins on one voice processing card to the next set on the adjacent card and then the next. There are several voice bus "standards." Two come from Dialogic. One is called AEB, Analog Expansion Bus. And one is called PEB, PC Expansion Bus (a digital version). One comes from a consortium of companies and is called MVIP. There are many advantages to having a voice bus. It gives you enormous flexibility to mix and match voice processing boards, like voice recognition, voice synthesis, switching, voice storage, etc. You can build really powerful voice processing systems inside today's fast '386 and '486 PCs with the great variety of voice processing now available. For more information on this exciting field, read TELECONNECT Magazine. 212-691-8215. See MVIP.

voice call A telephone call established for the purpose of transmitting voice, rather than data.

voice call continuity Seamless call handover between two networks, for example, between a mobile network and a Wi-Fi network, or between a mobile network and the PSTN. The term is most often used to describe the ability to seamlessly move an active voice session between an IP Multimedia Subsystem domain and a circuit-switched cellular network. Voice call continuity is a requirement for fixed-mobile convergence.

voice calling One manufacturer describes this as allowing a phone user to have calls automatically answered and connected to his phone's loudspeaker. Nortel defines voice calling as somewwhat differently. It says this feature allows a voice announcement to be made, or a conversation to begin, through the speaker of another telephone in the system.

Voice Carry Over VCO. A reduced form of TRS (Telecommunication Relay Service) where the person with the hearing disability speaks directly to the other end users. The Communications Assistant then types the response back to the person with the hearing disability. The Communications Assistant does not voice the conversation.

voice channel A channel suitable for transmission of speech, analog or digital data, or facsimile.

voice circuit A circuit able to carry one telephone conversation or its equivalent, i.e. the typical analog telephone channel coming into your house or office. It's the standard subunit in which telecommunication capacity is counted. It has a bandwidth between 300 Hz and 3000 Hz. The U.S. analog equivalent is 3 KHz. The digital equivalent is 56 Kbps in North American and 64 Kbps in Europe. This is not sufficient for high fidelity voice transmission. You'd probably need at least 10,000 Hz. But it's sufficient to recognize and understand the person on the other end.

voice coil The element in a dynamic microphone which vibrates when sound waves strike it. The coil of wire in a loudspeaker through which audio frequency current is sent to produce vibrations of the cone and reproduction of sound.

voice commerce See V-Commerce.

voice compression Process of reducing a voice signal to use less bandwidth during transmission. That's the obvious meaning. In telecommunications where voice was originally encoded digitally at 64 Kbps using PCM, voice compression now means to compress a voice channel to obtain a channel of 32 Kbps or fewer, nowadays to under 10 Kbps. See VoIP.

Voice Connecting Arrangement VCA. A device that, once upon a time, was necessary for connecting your own phone system to the nation's switched telephone network. Most phones now meet FCC (and other) safety standards, so VCAs are no longer necessary. Most phone systems (as opposed to phones) do have internal protection circuitry, as shown by the "F" (for fully protected) in their FCC registration number. The VCA was once called a Protective Connecting Arrangement (PCA).

voice coupler An interface arrangement once provided by the telephone company to permit direct electrical connection of customer-provided voice terminal equipment to the national telephone network. No longer needed because of the FCC's Registration Program.

voice data An SCSA definition. Encoded audio data.

voice dialing The ability to tell your phone to dial by talking to it. Say, "Call Police" and it will automatically dial the police. This feature has enormous benefits for handicapped people. It will have greater benefits for normal people when the technology of voice recognition improves.

voice digitization The conversion of an analog voice signal into binary (digital) bits for storage or transmission.

voice driver A Dialogic product that comes for MS-DOS, OS/2 and UNIX. In MS-DOS, it is a terminate and stay resident (TSR) program which acts as a central server for MS-DOS based applications. It provides all of the services required to support installable device drivers for each hardware component and for the application. See also Device Driver.

Voice DTMF Forms Applications This Voice DTMF (DUAL TONE MULTIPLE FREQUENCY) application allows a use of a voice mail system to take specific information from its customers 24 hours a day. By prompting callers to respond by speaking or pressing the keys of their touchtone phones, a city department, for example, could plan service calls, building inspections or send out appropriate forms.

Voice Extensible Markup Language See VXML.

voice frame See VFRAD and FRF11.

Voice Frequency VF. Any of the frequencies in the band 300-3,400 Hz that are transmitted in telephony systems to reproduce human speech voice with reasonable fidelity. Some Oriental languages have less than satisfactory results with this narrow a band, and emerging ISDN implementations are increasing it for them. See Voice Frequencies.

Voice Frequencies VF. Those frequencies lying within that part of the audio range that is employed for the transmission of speech. In telephony, the usable voice frequency band ranges from a nominal 300 Hz to 3400 Hz. In telephony, the bandwidth allocated for a single voice frequency transmission channel is usually 4 KHz, including guard bands.

voice frequency telegraph system A telegraph system permitting use of up to 20 channels on a single voice circuit by frequency division multiplexing.

voice grade A communications channel which can transmit and receive voice conversation in the range of 300 Hertz to 3000 Hertz.

Voice Grade Channel/Voice Grade Facility VGF. A line suitable for voice, low-speed data, facsimile, or telegraph service. Generally, it has a frequency range of about 300-3000 Hz.

voice grade wiring The term generally refers to analog lines with the bandwidth required to transmit human voice, typically about four thousand Hertz (4KHz).

voice hogging See Voice Switched.

voice integration Allows computer fax solutions to be store and forward hubs for both image as well as voice communication. Many of these products work on PC-based systems and offer all the capabilities of a message center.

Voice Interface Card See VIC.

voice logger A device that companies use to record their employees' phone calls. A voice logger is typically used in a brokerage firm or a call center. It attaches to your phone system and allows you to find calls in many ways – from which person, which extension, which incoming trunk, what time of day, what 800 line, etc. See also DID.

voice jail A poorly designed voicemail system that has so many submenus one gets lost and has to hang up and call back. Also called Voice Mail Jail.

voice mail Voice Mail allows you to receive, edit and forward messages to one or more voice mailboxes in your company or in your universe of friends. With voice mail, employees can have their own private mailboxes. Here's an explanation of how it works: You call a number. A machine answers. "Sorry. I'm not in. Leave me a message and I'll call you back." It could be a $50 answering machine. Or it could be a $200,000 voice mail "system." The primary purpose is the same – to leave someone a message. After that, the differences become profound. a voice mail system lets you handle a voice message as you would a paper message. You can copy it, store it, send it to one or many people, with or without your own comments. When voice mail helps business, it has enormous benefits. When it's abused – such as when people "hide" behind it and never return their messages – it's useless. Some people hate voice mail. Some people love it. It's clearly here to stay.

In the fall of 1991, the Wall Street Journal carried a story negative on voice mail. Les Lesniak, Rolm's Senior VP Marketing disagreed. His reply published in the Journal is one of the finest explanations of voice mail's virtues:

"The writer's observations ignore the way today's voice communication technology is making communication between people easier and more convenient, and is elevating the level of service savvy companies provide their customers. Manufacturers use it to take orders after hours and on weekends. Financial services companies use it to provide account information to customers on a 24-hour basis. Colleges use it to register students. A retail executive uses it to broadcast messages to her staff. And a lawyer uses it to respond to calls when traveling.

"Voice messaging keeps calls confidential, simplifies decision making, saves time and money, eliminates inaccurate messages and "telephone tag," allows people to use their time more productively. In short, it keeps communication crisp, clear and constant. The writer's line of thinking would demand that people remain at their desk 24 hours a day. If they don't, the phone goes unanswered, a receptionist answers the phone and takes a message, or an answering machine records the message and cuts off the caller at will. None of these scenarios is ideal.

"To be successful, voice mail technology must be understood by users and supported by top management, And it must meet the needs of the customer. Training for all employees must be mandatory and the system must be administered and managed properly. 'Must answer' lines and greetings that are changed daily are only two ideas that make voice mail not just helpful, but essential to customer service and an enhanced company image.

"Contrary to the writer's view, voice mail contributes to effective business communication and is far superior to an unanswered phone call, a misplaced message or an answering machine."

Here are some statistics which add weight to voice mail's logic:

- 75% of all business calls are not completed on the first attempt.
- This can easily waste $50 to $150 per employee per month in toll charges.
- Half of the calls are for one-way transfers of information.
- Two thirds of all-phone calls are less important than the work they interrupt.
- The average length of a voice mail message is 43 seconds. The average long distance call is 3.4 minutes. Voice mail is 80% faster.

Here are the standard benefits of voice mail:

1. No more "telephone tag." Voice mail improves communications. It lets people communicate in non-real time.

2. Shorter calls. When you leave messages on voice mail, your calls are invariably shorter. You get right to the point. Live communications encourage "chit chat" - wasting time and money.

3. No more time zone/business hour dilemma. No more waiting till noon (or rising at 6 A.M.) to call bi-coastally or across continents.

4. Reduce labor costs, Instead of answering phones and taking messages, employees are free to do more vital tasks.

5. Fewer callbacks. In some cases, as many as 50%.

6. Improved message content. Voice mail is much more accurate and private than pink slips. Messages are in your own voice, with all the original intonations and inflections.

7. Less paging and shorter holding times.

8. Less peakload traffic.

9. 24-hour availability.

10. Better customer service.

11. Voice mail allows work groups to stay in contact - morning, noon and night.

12. Voice mail reduces unwanted interruptions.

See also Voice Mail Jail and Voice Mail System.

voice mail jail What happens when you reach a voice mail message and you try and reach a human by punching "0" (zero) and you get transferred to another voice mail box and you try again by punching "0" or some other number you're told to punch...and you never reach a human. You're stuck forever inside the bowels of a voice mail machine, being instructed to go from one box to another, never reaching a real human. You're in voice mail jail.

voice mail system A device to record, store and retrieve voice messages. There are two types of voice mail devices – those which are "stand alone" and those which profess some integration with the user's phone system. A stand alone voice mail is not dissimilar to a collection of single person answering machines, with several added features. You can instruct the machines (voice mail boxes) to forward messages among themselves. You can organize to allocate your friends and business acquaintances their own mail boxes so they can dial, leave messages, pick up messages from you, pass messages to you, etc. You can also edit messages, add comments and deliver messages to a mailbox at a pre-arranged time. Messages can be tagged "urgent" or "non-urgent" or stored for future listening. The range of voice mail options varies among manufacturers.

An integrated voice mail system includes two additional features. First, it will tell you if you have any messages. It does this by lighting a light on your phone and/or putting a message on your phone's alphanumeric display. Second, if your phone rings for a certain number of rings (you set the number), the phone will transfer your caller automatically to your voice mail box, which will answer the phone, deliver a little "I am away" message and then receive and record the caller's message.

There are other levels of integration. You might have a phone which has "soft" buttons and an alphanumeric display. That display might label your phone's soft buttons like those on a cassette recorder – forward, reverse, slow, fast, stop, etc. so you can go through your messages any way you like. Telenova has such a phone. It's very impressive.

There are pros and cons to voice mail systems. Some employees will hide behind them, forwarding calls from their customers into voice mail boxes and never returning them. Some employees will make good use of them. They dial in for their messages, research what the customer wants and return the voice mail calls quickly. Many voice mail systems are being combined with automated attendants. Many are being combined with interactive voice processing systems, including sophisticated tie-ins to mainframe databases. Some people hate voice mail systems. Others love them. It all depends on how the system is used, managed and sold. See also Voice Mail, Audiotex, Automated Attendants, Information Center Mailboxes, Enhanced Call Processing and Voice Processing.

voice management Voice Management is a fancy term for managing the health and performance of a corporation's voice network. Real-time processing of voice call data enables these systems to actively calculate performance metrics such as trunk occupancy, port capacity on phone items like the corporation's main phone system, its IVR (interactive voice response), its voicemail, its ACD (automatic call distributor), etc. Voice management system administrators establish acceptable thresholds for performance and are notified (via page, email etc.) when their voice network does not meet these standards. Voice Management systems often provide a customized browser interface with one or more display panels to let you see how your network is performing. Views include:

1)   Trunk Performance: Displays the percent of circuits in a trunk group that are currently being utilized so technicians can determine if the group is under or over utilized, and optimize accordingly.

2) Route Analysis: A panel that displays how inbound and outbound calls route through the network. For instance by watching how inbound calls are transferred from auto attendant ports, a technician can determine if certain toll-free numbers are correctly reaching the final appropriate extension. Outbound route analysis helps optimize Least-Cost Routing Systems by making sure that each call goes out over the appropriate route.

3) Voice Over IP: By evaluating specific calling patterns, managers can determine the economic viability of implementing VoIP. Information provided also allows them to determine appropriate bandwidth requirements, based on volumes of traffic.

4) Call Stats: A department manager can see a real-time display of call activity within their department, even from multiple remote sites. Sales Managers particularly like this functionality.

5) System Alarms: Specific calls (i.e., 911) can be displayed at a receptionist's desk to direct emergency personnel where the call was originated. Technicians can be alerted to high trunk or port capacity conditions and quickly reroute calls to avoid an all-trunks-busy condition.

A Voice Management platform also manages the financial performance of a voice network as well. Budget warnings: Voice Management systems allows CFOs to allocate various telecom expenses across the enterprise and forecast upcoming year's budget requirements. Scheduled delivery of reports show variances of actual expenses compared to budgeted amounts.

Carrier tariff analysis: By analyzing existing carrier costs by specific call type (i.e. intralata, interstate), Voice Management systems enable users to compare alternative carrier plans to evaluate cost saving strategies.

Total Cost Of Ownership: The voice network now includes cell phones, pagers, DSL etc. Voice Management systems consolidate and report these expenses by cost centers so that executives can evaluate the total cost of ownership of their network

Jon A. Giberson of www.callaccounting.com helped compile this definition.

voice markup language See Voice Browser.

voice merging The oral tradition of African American preachers using another's words. See also Plagiarism.

voice message service A leased service typically over dial up phone lines which provides the ability for a phone user to access a voice mail system and leave a message for a particular phone user. See Voice Mail System

voice message exchange See VMX.

voice messaging Recording, storing, playing and distributing phone messages. Essentially voice messaging takes the benefits of voice mail (such as bulk messaging) beyond the immediate office to almost any phone destination you select. Voice messaging is often done through service bureaus. At one point, Nynex (now called Verizon) saw voice messaging as four distinct areas: 1. Voice Mail, where messages can be retrieved and played back at any time from a user's "voice mailbox"; 2. Call Answering, which routes calls made to a busy/no answer extension into a voice mailbox; 3. Call Processing, which lets callers route themselves among voice mailboxes via their touchtone phones; and 4. Information Mailbox, which stores general recorded information for callers to hear.

voice modem A new type of modem which handles both voice and data over standard analog phone lines. A voice modem is the classic computer telephony device, since it applies intelligence to the making and receiving of normal analog phone calls. Such voice modem might be a full-duplex speakerphone and an answering machine / voice mail device. Such modem might be able to detect incoming and outgoing touchtone and other signals, such as Caller ID. Such modem might also include music on hold, pager dialing, bong and SIT tone detect, line break detect, local phone on / off detect, extension off hook detect, remote ring back detection and VoiceView. The thrust towards voice modems is coming from chip manufacturers, including Sierra Semiconductor, Rockwell and Cirrus Logic. Some standards bodies are working on voice modems. Two standards are emerging – IS-101 and PN-3131.

voice of god You're about to give a speech. Someone has to introduce you. Suddenly, over the speakers you hear "Please welcome, all the way from New York, Mr. Harry Newton." That introduction is what's called, in that business, "the Voice of God."

voice on the net coalition An organization formed to stop regulatory attempts to stifle the growth of voice on the Internet. See VON Coalition. 802-878-9884 and www.von.org

voice operated relay (VOX) circuit A voice-operated relay circuit that permits the equivalent of push-to-talk operation of a transmitter by the operator.

voice over A feature on a phone system – namely that while you are speaking to someone on the phone, your operator can talk to you "over" the conversation you're having. What happens is that you hear your operator in your telephone's handset receiver, but the person you're speaking with can't. You can reply to the operator (telling him/her you'll be one minute, please call back, etc.) by hitting a DND/MIC (Do Not Disturb/Microphone) button on your phone. Voice Over has major benefits. It saves on long distance calls you don't have to return. It closes deals that can't wait. And it gives customers immediate answers. In short, it improves corporate efficiency and customer satisfaction.

Voice over Frame Relay See VoFR.

Voice over IP VoIP. A VoIP (Voice over Internet Protocol) phone call is transmitted over a data network, such as The Internet. The "Internet Protocol" is a catch all for the protocols and technology of encoding a voice call that allow the voice call to be slotted in between data calls on a data network. Such data network may be the public Internet, a corporate Intranet, or a managed network used by long, long distance and international traditional providers. VoIP phone calls, if properly engineered, sound just as good as a circuit switched TDM phone call – the ones we make and receive every day. There are three main benefits to VoIP phone calls:

First, they may potentially be cheaper. Since the data network is typically charged on a flat rate and thus the marginal cost of making a VoIP is zero, how cheaper depends on 1. The cost of terminating the VoIP call into the traditional phone network. Figure a penny a minute. 2. The price of a standard circuit switched TDM call. They've been getting cheaper over the years. 3. How much tax is levied on both. Taxes are horrendous on traditional circuit switched long distance calls. They aren't so big, yet, on VoIP calls, which are classified by some regulatory agencies as "information services," not voice phone calls. And therefore they escape most taxes.

Second, you may achieve benefits of managing a voice and data network as one network. If you have IP phones, moves, adds and changes will be easier and cheaper. IP phones are basically networked computers. They have individual numbers, with memories, user profiles. Their software upgrades are typically centrally managed using standard computing systems. In short, they're "user friendly" to manage and can largely be managed remotely.

Third, – and the key attraction of IP telephony – is added (and integrated) new services, including integrated messaging, voice emails, number portability, caller ID with name, call waiting, call forwarding, take your area code with you, plug into the Internet anywhere and make free calls from anywhere in the world. And best of all you can typically manage your phone via a Web site on the Internet, which will tell you which calls you made and received, etc. A VoIP phone is typically a much better animal than today's circuit switched phone.

I tried to switch all my personal and family telephone calling over IP, but the quality (at least for me) has been lacking. Often I get hangups and serious fades. The quality of your VoIP calls depends on your vendor and the technology he's using. If you VoIP phone calls goes at any point over the Internet, you will suffer occasional degradation. If your calls go over your own lines, you probably have a better shot. See Internet, Internet Protocol, Network Border Switch, Packet Switching, Skype and Vonage.

Voice over Multiservice Broadband Network See VoMBN

voice paging access Gives attendants and phone users the ability to dial loudspeaker paging equipment throughout the building. An unbelievably useful feature, if your people are prone to wander.

voice portal Call a phone number, have an interactive voice response system answer you, respond to your words with speech recognition, read your emails or the news with text-to-speech skills, perhaps even allow you to "surf" the Web. The classic definition of a portal is a door, gate, or entrance, especially one of imposing appearance, as to a palace. In the Internet / World Wide Web business, a portal is a site, which the owner positions (through marketing) as an entrance to other sites on the Internet. There are two types of portals – the conventional PC-based, browser based – and the newer one using the telephone. For a bigger explanation of a portal, see Portal.

voice print See Voiceprint.

voice processing Think of voice processing as a voice computer. Where a computer has a keyboard for entering information, a voice processing system recognizes touchtones from remote telephones. It may also recognize spoken words. Where a computer has a screen for showing results, a voice processing system uses a digitized synthesized voice to "read" the screen to the distant caller. Whatever a computer can do, a voice processing system can too, from looking up train timetables to moving calls around a business (auto attendant) to taking messages (voice mail). The only limitation on a voice processing system is that you can't present as many alternatives on a phone as you can on a screen. The caller's brain simply can't remember more than a few. With voice processing, you have to present the menus in smaller chunks. Voice processing is the broad term made up of two narrower terms – call processing and content processing. Call processing consists of physically moving the call around. Think of call processing as switching. Content consists of actually doing something to the call's content, like digitizing it and storing it on a hard disk, or editing it, or recognizing it (voice recognition) or some purpose (e.g. using it as input into a computer program.) See Voice Board, Voice Response Unit and Voice Server.

Voice Profile for Internet Messaging See VPIM.

voice recognition The ability of a machine to recognize your particular voice. This contrasts with speech recognition, which is different. Speech recognition is the ability of a machine to understand human speech – yours and most everyone else's. Voice recognition needs training. Speech recognition doesn't. See Speaker Dependent and Speaker Independent Voice Recognition.

Voice Response Unit VRU. Think of a Voice Response Unit (also called Interactive Voice Response Unit) as a voice computer. Where a computer has a keyboard for entering information, an IVR uses remote touchtone telephones. Where a computer has a screen for showing the results, an IVR uses a digitized synthesized voice to "read" the screen to the distant caller. An IVR can do whatever a computer can, from looking up train timetables to moving calls around an automatic call distributor (ACD). The only limitation on an IVR is that you can't present as many alternatives on a phone as you can on a screen. The caller's brain simply won't remember more than a few. With IVR, you have to present the menus in smaller chunks. See IVR and Voice Board.

voice retrieval Message system that stores verbal messages (from callers or operator) for automatic retrieval at the customer's convenience.

voice ring Multiple Digital Intertie Buses connected in series to all nodes. Provides extra channels for voice data transmission when direct link (DI) channels are busy.

voice ringer A voice recording used as a ringtone.

voice ringtone Voice ringer.

voice security In military and diplomatic circles, voice security refers to radiotelephone and radio conversations that are scrambled (encrypted) or protected in some other way to make them inaccessible or unintelligible.

voice server A PC sitting on a LAN (Local Area Network) and containing voice files which are accessible by the PCs on the LAN. Such voice files may be transmitted on the LAN or over phone lines under the control of the PCs on the LAN. A voice server might contain voice mail. It might contain voice annotated electronic mail. Its primary function is to store voice in such a way that it's accessible easily. Voice servers are typically faster, have more disk capacity and more backup provisions than normal PCs. According to a letter I received in early May, 1993 from the lawyers for a company called Digital Sound Corporation, that company owns federal trademark registration number 1,324,258 for the mark Voiceserver, spelled as one word, not two.

voice service personality A new name for dial tone.

voice SMS Also known as SMS voice messaging. A service whereby a wireless phone subscriber can send a voice message, typically at the standard text-messaging rate, to another wireless phone or to an email address without making a phone call. See also SMS.

In India, a country with a large number of cell phone users who are illiterate, text messaging is impossible. Several mobile operators in the country have come up with a solution: voice SMS. The caller enters a numeric shortcode to invoke the service, then enters the cell phone number of the recipient, then leaves a short voice message. The recipient receives an SMS message displaying the number of the person who sent the voice message. The recipient can listen to the message then or at a later time. While voice SMS sounds a lot like ordinary voicemail, the difference is in the billing. Unlike voicemail, which bills via a monthly recurring charge, the voice SMS user pays on a per-message basis, not via a MRC (monthly recurring charge).

voice stop Voice stop is a means for callers to interrupt a menu prompt or other instruction on a voice processing system by merely speaking into the phone. This is a capability that is similar to pressing a digit on the touch-tone pad in order to stop a recording or a prompt from continuing. Unlike Voice Cut-Through, Voice Stop does not actually analyze the word being spoken. voice Stop technology senses energy on the telephone line and stops execution based on that energy. For example, a train whistle or over head loudspeaker could be transmitted over the phone and have the same effect.

voice store and forward Voice mail. A PBX service that allows voice messages to be stored digitally in secondary storage and retrieved remotely by dialing access and identification codes. See Voice Mail System.

voice switched A device which responds to voice. When the device hears a voice, it turns on and transmits it, muting the receive side. The most common voice-switched device is the desk speakerphone. With voice switching, it's easy to hog a circuit. Just keep making a noise. Watch out for voice hogging. If you're calling someone and waiting for them by listening in on your speakerphone, mute your speakerphone. This way you'll hear them when they answer.

voice switching Equipment used in voice and video conferences. The equipment is activated by sounds of sufficient amplitude; hopefully speech, but also loud noises. Fast switching activates microphones so that only one conference participant can speak at a time. See also Voice Activated Video.

voice terminal A pretentious AT&T term for a Telephone.

voice verification The process of verifying one's claimed identity through analyzing voice patterns.

voiceband A transmission service with a bandwidth considered suitable for transmission of audio signals. The frequency range generally is 300 or 500 hertz to 3,000 or 3,400 hertz – the frequency range the common analog home phone service is made at.

voiceprint A voice recognition term. A voiceprint is a speech template used to recognize and verify callers. For example, Home shopping Network. When a voiceprint system is operating, the user's speech is compared to the stored voiceprints. If they match, the system recognizes the word and executes the command.

VoiceXML VoiceXML (VXML) is a platform independent structured language created using the extensible markup language (XML) specification to deliver voice content through several different media like the web and phone systems. It has a format similar to other structured languages like HTML, however it is entirely defined within the XML standard specification. Voice XML provides a uniform development environment that allows a business to build on its web investments for voice including application integration code, business rules, and personalized software. Voice and web channels can share the same back end integrated databases facilitating a complete view for the customer, regardless of how they choose to interact with the enterprise. VoiceXML provides the framework for:

- Delivering synthesized or digitized sound
- Recognizing user input
- Recording user input
- Controlling call flow
- Transferring and disconnecting callers

A VoiceXML document is composed of text elements and tags that instruct the system to provide the user with information and recognize user input as well as additional functions such as recording and transferring. Some VoiceXML tags that can be used to create content are:

- Tag Function
  <assign> Assign a value to a variable
  <audio> Play an audio clip
  <block> A container of executable code
  <catch> Catch an event

The structure of a VXML document is very similar to an HTML document, even though the tags and syntax of the languages are different. However, the basic similarities of HTML and VXML make it easy for a savvy web developer to learn the syntax of VXML quickly and begin to create simple or complex voice interfaces without having to know the technical details of the voice system.

Like the traditional web server VXML follows the client / server model where a user can request information from a document server, the server responds with the appropriate content and an interpreter reads the document and presents the information to the client providing the advantages of web-based development and content delivery to interactive voice response applications.

For example, a person who wants to check their bank account balance to make sure they have enough money to write a check could call their bank and access their automated customer service department. If the phone system is configured with an interactive voice response system, it is able to present and guide the caller to the point where they can check their available balance. After entering their account number and password the caller is presented with a menu of options. The caller presses the keypad number, which sends a request for their available balance. That input is interpreted and a request for the available balance is sent to the document server. The document server accepts the request, looks up the account and authenticates the user from the previously entered information. Once the document server has authenticated the request it generates a VXML document containing the account balance. The VXML document is sent to the interpreter and processed. The information in the VXML document is then interpreted and used to create a synthesized voice response so the customer can hear how much money they have in the account. Using VXML the information presented to the caller is the same interface and database the caller would use if they were retrieving the info from their Web site. This ensures a consistent view regardless of the media used by the customer.

VoiceXML Forum The VoiceXML Forum is an industry consortium engaged in educational and marketing activities in support of the VoiceXML (Voice eXtensible Markup

Language) specifications and standards being developed by the W3C (World Wide Web Consortium). VXML will make the Web accessible and browsable via voice and audio (e.g., touchtone) commands. The VoiceXML Forum was established by AT&T, IBM, Lucent Technologies, and Motorola, and now includes a large number of industry members. www. voicexml.org. See also VXML.

VoIM Voice over instant messaging, a subset of voice over IP.

VoIP Voice over Internet Protocol. The technology used to transmit voice conversations over a data network using the Internet Protocol. Such data network may be the Internet or a corporate Intranet. For much longer explanations, see VoFR, Voice over IP, VoIP peering, VoIP Forum, Skype and Vonage.

VoIP Forum Voice over Internet Protocol. The Voice over IP Forum was formed in 1996 by Cisco Systems, VocalTec, Dialogic, 3Com, Netspeak and others as a working group of the International Multimedia Teleconferencing Consortium (IMTC), which promotes the implementation of the ITU-T H.323 standard. The VoIP Forum is focused on extending the ITU-T standards to provide implementation recommendations as a means of supporting Voice over IP in order that devices of disparate manufacture can support voice communications over packet networks such as the Internet. By way of example, the VoIP Forum intends to establish directory services standards in order that Internet voice users can find each other. They also plan to port touch-tone signals to the Internet to allow the use of ACDs and voice mail systems. See also VON Coalition.

VoIP peering The connection of VoIP (voice over Internet protocol) networks – private and public – to each other without touching the PSTN (Public Switched Telephone Network) so as to avoid PSTN charges. For example, two long distance and international carriers such Level 3 and Global Crossing decide to meet at a neutral point and trade traffic destined for each others' network – without passing it through the public switched telephone network (PSTN) and thus not incurring any third party costs. If done in this manner it should be free for each carrier. Two VoIP providers like BroadVoice and Lingo could do this in a way similarly to that outlined above. BroadVoice would typically be carriers of the above carriers and thus could avoid paying them to connect our calls when they are from the subscriber of one provider to the subscriber of the other. In short, VoIP peering is an arrangement whereby cable companies and VoIP providers route VoIP traffic between their respective customers directly over each other's IP networks, thereby bypassing traditional telephone networks and avoiding PSTN (public switched telephone company) interconnection fees. VoIP peering is evolving. CableLabs, a a nonprofit research and development arm of the cable industry has announced that it "is seeking information regarding the development of a production-grade VoIP Peering infrastructure to allow VoIP traffic exchange among MSOs (multiple service operators – cable companies with more than one location), and, between MSOs and other VoIP partners. Through this RFI, CableLabs is seeking information on the technical scope, requirements, protocols, architecture, management and operations considerations to support the CableLabs VoIP Peering architecture. ... The MSO VoIP Peering project is a CableLabs-led initiative aimed at developing technical requirements, architecture and interoperable protocol specifications for delivering end-to-end VoIP and other real-time multimedia communication exchanges between service operators. The VoIP peering initiative addresses requirements for establishing a production platform to exchange VoIP traffic between operators." See plug and peer.

VoIP peering service An interconnection service that enables VoIP traffic to pass from one VoIP service provider's network to another without the individual providers' having to establish agreements with each other and directly connect their networks. See carrier hotel and VoIP peering.

VoIP substitution When a customer dumps his telephone company landline and goes with a VoIP solution (for example, over a cable or DSL connection) instead.

VolanoMark VolanoMark is a popular Java benchmark for measuring server throughput. It measures messages per second.

volatile storage Computer storage that is erased when power is turned off. RAM is volatile storage.

volleyball People in nudist colonies play volleyball more than any other sport. Personally I think tennis is more fun to play. Watching naked volleyball could be interesting.

Volser An MCI term used to denote a volume of calls. Based on the words "Volume Serial." The term "Volser" can be applied to the manual collection of calls from a switch on a switch tape or through call data transmitted via NEMAS.

volt The unit of measurement of electromotive force. Voltage is always expressed as the potential difference in available energy between two points. One volt is the force required to produce a current of one ampere through a resistance or impedance of one ohm.

volt meter An instrument for measuring voltages, resistance and current.

voltage Electricity is a essentially a flow of electrons. They're pushed into a gadget

– toaster, computer, phone – on one wire and they sucked out on the other wire. For this movement of electrons to occur there must be "pressure," just as there must be pressure in the flow of water. The pressure under which a flow of electrons moves through a gadget is called the electric voltage. Voltage doesn't indicate anything about quantity, just the pressure. The amount of electricity moving through a wire is called its current and is measured in amps. You figure the power in an electron flow (i.e. in electricity) by multiplying the flow's current by the voltage under which it flows.

voltage drop The voltage differential across a component or conductor due to current flow through the resistance or impedance of the component or conductor.

voltage rating The highest voltage that may be continuously applied to a conductor in conformance with standards or specifications.

voltage regulator A circuit used for controlling and maintaining a voltage at a constant level.

voltage spike An extremely high voltage increase on an electrical circuit that lasts only a fraction of a second, but can damage sensitive electronic equipment like telephone systems or can cause it to act "funny." If your phone system starts acting "funny," one "cure" is to shut it off, count to ten, and then turn it on again. This sometimes clears the problem.

Voltage Standing Wave Ratio VSWR. The ratio of the maximum effective voltage to the minimum effective voltage measured along the length of mis-matched radio frequency transmission line.

Voltage Tuned Oscillator VTO. The ratio of voltage to reflected voltage in a radio frequency device.

voltmeter A device for measuring the difference of potential in volts.

volume 1. A volume is a partition or collection of partitions that have been formatted for use by a computer system. A Windows NT volume can be assigned a drive letter and used to organize directories and files. In NetWare a volume is a physical amount of hard disk storage space. Its size is specified during installation. NetWare v2.2 volumes, for example, are limited to 255MB and one hard disk, but one hard disk can contain several volumes. A NetWare volume is the highest level in the NetWare directory structure (on the same level as a DOS root directory). A NetWare file server supports up to 32 volumes. NetWare volumes can be subdivided into directories by network supervisors or by users who have been assigned the appropriate rights.

2. Under ISO 9660, a single CD-Rom disc.

volume label A name you can assign to a floppy or hard disk in MS-DOS. The name can be up to 11 characters in length. You can assign a label when you format a disk or, at a later time, using the LABEL command.

volume serial number A number assigned to a disk by MS-DOS. The FORMAT command creates the serial number on a disk.

volume resistivity The electrical resistance between opposite faces of one centimeter cube of insulating material, commonly expressed in ohms-centimeter.

Volume Unit VU. The unit of measurement for electrical speech power in communications work. VUs are measured in decibels above 1 milliwatt. The measuring device is called a VU meter.

VOM Abbreviation for VOLT-OHM-MILLIAMETER, probably the most common form of electronic test equipment. It measures voltage, resistance and current, and may have either a digital or analog meter readout. Some VOMs have other test functions such as audible continuity signals and special tests for semiconductors.

VoMBN Voice over Multiservice Broadband Network. VoMBN is a means of supporting voice services (e.g., signaling services, and custom calling and Centrex services, as well as voice trunking) over DSL (Digital Subscriber Line) without the involvement of a Class 5 Central Office (CO) switch. VoMBN involves various signaling and processing servers and gateway devices at the network edge in support of VoIP (Voice over Internet Protocol) or VoATM (Voice over ATM) packet voice traffic presented to the network edge over a DSL local loop. VoMBN appeals to newer carriers that have built backbone networks based on either ATM or IP. Incumbent voice carriers generally prefer the approach of BLES (Broadband Loop Emulation Service), which accomplished much the same thing through an interface with a Class 5 switch based on traditional circuit-switching technology. See also BLES.

vomit 1. Once inside a company's firewalls or even on a consumer's computer, hackers can use software to scan files in a server looking for Internet phone packets. One such program called Vomit, which stands for voice over misconfigured Internet telephony, reassembles voice packets to allow people to listen in on conversations. . Tapping conversations carried over commercial Internet calling services that are provided by companies like Vonage and AT&T is harder because those providers have their own security controls.

2. Rats are unable to vomit, which is one of the main reasons why poison is so effective

against them.

vomit comet A plane used to simulate zero-G for astronaut flight training. Trainers often get motion sickness inside. Portuguese wine bottled in 1811 is called "comet wine." Its excellent quality is believed to be due to the Great Comet of that year. The term "comet wine" is often used for any wine made in the year of an important comet.

VON Voice On the Net (Internet), involving packetized voice. A recent development, initiating a VON call typically requires a multimedia PC or Mac computer with special software which matches that on the receiving device. More recently, Internet servers have been equipped with such software, although appropriate client (workstation) software must be installed to take advantage of this approach. More recently still, VON has been demonstrated from workstation to telephone, telephone to workstation, and telephone to telephone. Additionally, new compression techniques and new DSPs have dramatically improved the quality of VON transmission, mitigating the impacts of packet delay. See Internet Telephony for a detailed explanation. See also VON Coalition and Packetized Voice.

VON Coalition The "Voice on the Net" (VON) Coalition is an Internet organization devoted to "educating consumers and the media by monitoring and supporting present and new developed telephony, video and audio technologies that are specifically designed and manufactured for the Internet community." It was formed, inter alia, to provide a forum against the ACTA (America's Carriers Telecommunications Association) petition to the FCC which sought to ban VON as a threat to the integrity of the PSTN and the concept of Universal Service. VON Coalition, www.von.org. See Universal Service Fund.

Vonage One of the first companies to sell phone calls over the Internet (VoIP). They are now a public company. I don't think they have ever made a profit.

VORTAC VOR (VHF Omnidirectional Range) collocated and/or combined with TACAN (Tactical Air Navigation Equipment).

Vote ACK Also known as Mass ACK; in Usenet, the posting of the e-mail address of each person that voted for or against a newsgroup proposal.

voting A teleconferencing term. Also known as Polling. In a large, event-style teleconference, the participants can vote on an issue via the touchtone keypad. The teleconference service provider tabulates the electronic votes and advises the conference sponsor of the results.

voting panel The equipment in a receiver voting system that receives and interprets the signals from remote receivers, picks the best signal, and routes it to a single output.

voting receivers A group of mobile base phone receivers operating on the same frequency as a control unit to pick the best signal from among them.

VOTS VMS OSI Transport Services. A Digital Equipment Corp. software product that modified Digital's DECnet transport layer to conform to the International Standards Organization (ISO) Transport Protocol Class Four (TP4). As a company, DEC no longer exists.

VOX 1. Latin for "Voice." Some people put "Vox" on their business cards to distinguish their voice telephone number from their fax telephone number. It's a trifle pretentious.

2. Voice Operated eXchange. Your voice starts it. When you stop speaking, it stops. Tape recorders use it to figure when to start recording and when to stop. There are pros and cons to VOX. With VOX you often miss the beginning of the conversation. And the tape goes on for 3 or 4 seconds after you've stopped talking. Also if ambient noise is high, VOX might mistake it for speaking and turn the recorder on and keep it running. Cellular phones also use VOX to save battery. A cellular phone without VOX is continuously transmitting a carrier back to the cell cite the entire time your call is in progress. The VOX operation used in smaller phones allows the phone to transmit only when you're actually talking. This reduces battery drain and enables handheld phones to operate longer on a smaller battery.

VoxML Voice Markup Language. See Voice Browser.

VoWLAN Voice over Wireless LAN (Local Area Network). Typically this means you're sitting in a Wi-Fi-enabled area – let's say a Starbucks coffee shop. You turn your laptop on and bingo you're on the Internet. Now imagine you walk into Starbucks with a Wi-Fi phone. Turn it on. Hear dial tone. Dial a phone number. Bingo, you're connnected. That's it. Nothing sophisticated – except it works wirelessly, it works digitally on the equivalent of Ethernet, it works exactly as VoIP (Voice over Internet Protocol) works. You probably will need a screen on the phone which allow you to respond to questions as to who you are and give some information to allow billing on the call. See VoIP.

Voxel VOlume piXEL. The 3D equivalent of a pixel, or picture element. a voxel is the smallest distinguishable element of a three-dimensional (3D) image. The process of voxelization involves the stacking of slices, which are cross-sectional pixel images in two-dimensional format. In order that the underlying pixels are not obscured by darker, more opaque outside-layer pixels, the process also involves opacity transformation. Voxel images are used extensively in X-Rays, CAT (Computed Axial Tomography) Scans, and MRI (Magnetic Resonance Imaging) technologies. Voxel imaging also is used in some computer games.

See also Pixel.

voycall An early key system manufacturer, which made a combination 1A2 handsfree intercom telephone system. It was wood grained, inlaid into black plastic. An impressive phone system. Sadly, no more.

VP 1. Virtual Path. A SONET term for an end-to-end route between 2 points. Many Virtual Paths may share a common physical path. Each Virtual Path consists of Virtual Tributaries which, in turn, consist of Virtual Channels. In the ITU-T SDH terminology, a Virtual Path is known as a Virtual Container.

2. An ATM term. Virtual Path is a unidirectional logical association or bundle of VCs, which are communications channels that provide for the sequential unidirectional transport of ATM cells.

3. Vp. Velocity of Propagation. See Velocity of Propagation.

VPA Voice Port Adaptor.

VPAR Voice Port Adaptor Rack.

VPC An ATM term. Virtual Path Connection: A concatenation of VPLs (Virtual Path Links) between Virtual Path Terminators (VPTs). VPCs are unidirectional.

VPDN Virtual Private Data Network. A private data communications network built on public switching and transport facilities rather than dedicated leased facilities such as T1s.

VPDS Virtual Private Data Services. MCI's equivalent of Vnet for data.

VPDS 1.1 Virtual Private Data Service1.1. VPDS 1.1 provides Switched T1 and Switched T3 over a platform of Digital switch Corporation's ECS1 (DXC 3/1) and ECS3 (DXC 3/3) Cross-Connects, respectively. This is a switched data service and is intended for customers who do not want to pay a fixed price for private line services such as TDS1.5 and TDS45. VPDS 1.1 requires a customer to have dedicated access and egress, which is priced at a fixed rate. The network portion is priced on a usage basis and is the "switched" portion of the end-to-end circuit.

VPI Virtual Path Identifier. An ATM term. Virtual Path Identifier is an eight- bit field in the ATM cell header which indicates the Virtual Path (VP) over which the cell should be routed. See VPI and VCI.

VPIM Voice Profile for Internet Messaging, a proposed Internet messaging protocol to allow disparate voice messaging systems to automatically exchange voice mail over the Internet. VPIM also will allow a voice messaging system to communicate with other such systems outside the organization. VPIM works like this: You record a message and enter the target telephone number of the intended recipient. Your voice processing system does a directory look-up to a public electronic directory, using LDAP (Lightweight Directory Access Protocol) to find the e-mail address assigned for voice messages for that individual. Your system converts the voice message to a MIME (Multipurpose Internet Mail Extension) attachment, and routes the message through the Internet using SMTP (Simple Mail Transfer Protocol). The message is delivered to the voice messaging system supporting the target telephone number, where it is converted back into a voice message and stored in the recipient's voice mail box. The recipient can respond in the same fashion. Now let's take it a step further. As the messages are converted to MIME attachments, and as it uses SMTP over the Internet, VPIM has the potential to support compound mail consisting perhaps of voice mail, audio mail, e-mail, and video mail. See also LDAP, MIME, SMTP, VPIM Work Group and www.ema.org/vpimdir/index.htm

VPIM Work Group The goals of the Voice Profile for Internet Mail Work Group include establishing an internationally accepted standard profile of ESMTP/MIME to allow the interexchange of voice and fax messages between voice messaging systems; ensuring that this profile also allows interexchange with non-voice messaging MIME compatible email systems, establishing a directory service to support lookup of the routable address, and establishing a defined mapping specification with other voice messaging. The Group hosted a concept demo at EMA'96, a product demonstration at EMA'97, an info booth at CT Expo '98, and at the Fall 98 VMA Meeting in Athens. VPIM vendors are currently testing products for compatibility with the VPIM specification. The VPIM Specification, version 2 has been approved by the IETF as a Proposed Standard. After a long wait for its references to be published, VPIM v2 was published as RFC 2421 in September 1998. See also VPIM.

VPL An ATM term. Virtual Path Link is a means of unidirectional transport of ATM cells between the point where a VPI value is assigned and the point where that value is translated or removed.

VPLS VPLS is a class of VPN (virtual private network) that allows the connection of multiple sites in a single bridged domain over a service provider managed MPLS network. It's commonly called a Layer 2 VPN. From a customer's perspective, it looks as if all sites are connected to a single switched VLAN. To service providers, VPLS enables the MPLS infrastructure to offer secure any-to-any switched multiple services and to expand the portfolio beyond IP VPN. Customers access the service provider VPLS service using Ethernet.

VPLS enhances the value of WAN Ethernet services by enabling Ethernet to flow natively across service providers' MPLS networks while providing strict security. When combined with Ethernet access, VPLS transforms the WAN into a large Ethernet switch, even across national and global distances. The above explanation courtesy the Yankee Group.

VPN Virtual Private Network. With a VPN, employees can log into a distant corporate local area network, server or corporate intranet over the Internet. A VPN has the look and feel of a private network to a user. But it's really part of the Internet with heavy security – so no one on the Internet can see what's going on in the VPN. There are several definitions for VPN, and we'll go through them in some detail. But first, we need to explain the overall concept. A VPN is not a private network, but is virtually so, which means it's almost so. That is to say that it exhibits at least some of the characteristics of a private network, even though it uses the resources of a public switched network. True private networks absolutely guarantee access to network resources, and security is perfect – after all, the network is a private one, comprising dedicated leased lines. Those lines (or, more commonly today, the equivalent bandwidth) have been taken out of shared public use and dedicated to the private use of an end user organization on the basis of a lease arrangement. Those dedicated leased lines often go through various switching centers (e.g., COs or POPs), but go around, rather than through, the switches. As far as the private network is concerned, it's a wire center, rather than a switching center. The dedicated leased lines most commonly are T-carrier or even SONET in nature, directly interconnect two or more end user sites, and can be used for any purposes the end user desires. The end user can run any higher-layer protocol it chooses – after all, it's a private network. Sounds great, doesn't it? Sure, it does, but the costs are high, and the complexities of designing and implementing such a network can be way out of proportion to the benefits. Virtual Private Networks don't exhibit exactly the same characteristics and, therefore, don't perform as well as true private networks, but can come pretty close...and at much lower cost. For example, a VPN might offer priority access to bandwidth and other network resources, whereas a true private network offers guaranteed access at all times. A VPN might offer relatively tight security mechanisms, whereas a private network is totally secure. Now, let's examine the specific definitions.

1. The first VPN was developed for voice networking, but subsequently was developed for use in data networking, as well. Also known in AT&T terminology as a Software-Defined Network (SDN), these original VPNs remain in wide use on both a domestic and an international basis. Currently, they largely are used in support of voice, as Frame Relay and other packet network technologies have proved to be more effective in support of data applications. They are a public service offered by IXCs (IntereXchange Carriers) and making use of the circuit-switched PSTN (Public Switched Telephone Network). Originally known as Switched 56, the current usage of the term "VPN" distinguishes data services offered by AT&T, MCI and Sprint from Switched 56/64 Kbps services offered by the LECs (local exchange carriers, i.e.. local phone companies). Although the specifics vary by IXC, VPNs offer bandwidth options of 56/64 Kbps, increments of 56/64 Kbps, 384 Kbps and 1.544 Mbps (T-1). The last two options are designed with videoconferencing in mind. VPNs provide transmission characteristics and services similar to those of private lines, including network testing, priority access, and security. Access to a circuit-switched VPN is provided over T-carrier (e.g., T-1 or Fractional T-1) local loops, which are full-duplex, four wire, digital circuits. As VPN services are dial-up services provided over the PSTN, they offer the same inherent any-to-any connectivity provided for voice calls, with the added feature of security through a Closed User Group (CUG). In other words, any location on your VPN can dial any other location on your VPN, but can't dial any number outside the CUG and can't be dialed by any number outside the CUG. VPNs also offer the advantage of the high level of PSTN redundancy, which translates into a high level of network resiliency. This network resiliency compares favorably to private, leased-line networks, which are highly susceptible to catastrophic failure. In fact, VPNs often are deployed as a backup to leased-line networks. VPNs also are extremely effective in support of enterprise data networking in organizations with large numbers of small sites. Small locations with relatively modest communications requirements often cannot be cost-effectively connected to long-haul, leased-line networks. VPNs offer the advantages of flexibility and scalability, as sites can be added or deleted relatively easily, with costs maintaining a fairly reasonable relationship to enterprise network functionality. The processes of network configuration (design) and reconfiguration are greatly simplified as compared to a leased-line network. Provisioning time is also greatly reduced, thanks to the flexibility of the circuit-switched network core – the only dedicated portion of the VPN is the local loop, which is always dedicated, regardless of the network service accessed. Compared to a private network, the greatest disadvantage of VPNs is that all calls are priced based on a usage-sensitive algorithm much like that of a typical call over the PSTN. In other words, costs are calculated by duration and time of day, with prime-time calls being priced at a premium. Day-of-week and other special discounts also apply. Some

carriers also consider distance in the pricing of VPN calls. Note, however, that the usage-sensitive costs of a VPN typically are a lot less than the cost-per-minute of a normal dial-up call over the PSTN, sensitive to factors including the number of sites connected, usage volume commitments, and contract length. Purely from a cost standpoint, leased-lines are preferred for networking large sites with intensive communications needs. Leased line networks also can support not only data and video transmission, but also voice, thereby offering the advantage of integration of all communications needs over a single network. Access to a VPN POP (Point of Presence) can be gained directly from the IXC (Inter-eXchange Carrier), from a CAP (Competitive Access Provider), or from the LEC (Local Exchange Carrier). Appropriate access technologies include leased lines, Switched 56/64, and ISDN. See also Switched 56 and Private Line.

2. The second definition of VPN is a fairly generic one, referring to a packet data network service offering with some of the characteristics of a private network. Any packet data network can be used as the foundation for such a VPN, including X.25, TCP/IP, Frame Relay, and ATM networks. Each of these foundation networks is very different in terms of specifics, but they all are highly shared in terms of their basic nature. In order to provide services that emulate, or at least approximate, a private network over a highly shared network core, it is necessary to provide some additional features and mechanisms. One such feature is priority access to bandwidth, which can be accomplished through a variety of mechanisms which variously are intrinsic to the fundamental packet protocol (e.g., ATM) or through supplemental protocols (e.g., MPLS, or MultiProtocol Label Switching, which often is used in Frame Relay and TCP/IP networks). Security is a critical feature, which variously can be imposed through mechanisms such as a Closed User Group (e.g., Frame Relay) or tunneling (e.g., TCP/IP).

3. In contemporary usage, VPN most commonly refers to an IP (Internet Protocol) VPN running over the public Internet. While the ubiquitous nature of the Internet is a huge advantage for data networking, the Internet is inherently both insecure and subject to variable levels of congestion. In order to create a VPN over the Internet, security issues are mitigated through the use of a combination of authentication, encryption, and tunneling. Authentication is a means of access control the confirms the identity of users through password protection or intelligent tokens, thereby reducing the possibility that unauthorized users might gain access to privileged internal computing or network resources. Authentication commonly is the responsibility of an access server running the RADIUS (Remote Access Dial-In User Service) protocol, connected to an access router with embedded firewall software. Encryption is the process of encoding, or scrambling, of the data payload prior to transmission in order to secure it; the decryption process depends on the receiver's possession of the correct key to unlock the safety mechanism. The key is known only to the transmitting and receiving devices. Tunneling is the process of encapsulating the encrypted payload in an IP packet for secure transmission. Tunneling protocols include SOCKv5, PPTP (Point-to-Point Tunneling Protocol), L2TP (Layer 2 Tunneling Protocol), and IPSec (IP Security).

The applications scenarios for IP VPNs include remote access, intranets, and extranets. Remote access VPNs are highly effective in support of telecommuters, mobile workers, and virtual employees. Intranets are used to link branch, regional, and corporate offices. Extranets link vendors, affiliates, distributors, agents, affiliates, and strategic partners into the main corporate office, with the level of access afforded being sensitive to the level of privilege indicated by a combination of password and user ID, as properly authenticated. This definition is courtesy of Ray Horak's excellent book, "Communications Systems and Networks." See also Authentication, Encryption, Extranet, Firewall, Internet, Intranet, IP VPN, Tunneling and VPN concentrator.

VPN concentrator A physical gadget (also called an appliance) optimized to terminate Virtual Private Network-encrypted tunnels. Enables IP traffic to travel securely over a public TCP/IP network by encrypting all traffic from one network to another.

VPOTS Very Plain Old Telephone Service. No automated switching.

VPT Virtual Private Trunking. VPT - (as it pertains to VPN) - appears as a Frame Relay or ATM service to the enterprise, but uses VPN technology to deliver high-availability services, while enabling service providers to fully optimize trunk bandwidth. VPT accesses the flexible, high QoS capabilities of the Frame Relay and/or ATM services, and is suitable for both IP as well as non-IP requirements. For example, Switched Virtual Circuits (SVCs) could be used to link the various nodes in an Extranet, while Permanent Virtual Circuits (PVCs) could be used to link all the sites in an Intranet. PVCs provide as much security as a dedicated, circuit-based leased-line network. Enterprises concerned about using the Internet as their network backbone will find VPT's enhanced performance, QoS and security assuring. See VPN.

VPU 1. Virtual Physical Unit.
2. Voice Processing Unit.

VQL Variable Quantizing Level. Speech-encoding technique that quantizes and encodes an analog voice conversation for transmission, nominally at 32 Kbps.

VR 1. Voice Recognition. See Voice Recognition 2. Virtual Reality. See Virtual Reality.

VRAM Video RAM. Memory used to buffer an image and transfer it onto the display. It is a form of DRAM specially suited for video. VRAM differs from common DRAM in that it has two data paths – a technique known as dual porting – rather than the single path of traditional RAM; thus, it can move data in and out simultaneously. Two devices can access it at once. The CRT controller, which converts bits and bytes in video memory to pixels on the screen, and the CPU, which manipulates the contents of video memory, can access VRAM simultaneously. Conventional DRAM chips allow one read or write operation at a time. Video RAM supports simultaneous read/write, read/read and write/write operations. It's most often used in graphic accelerators. In video boards fitted with the less expensive DRAM, performance suffers somewhat because the CRT controller and the CPU must takes turns getting to the video buffer held in VRAM. See also DDR-SDRAM, DRAM, EDO RAM, Flash RAM, FRAM, Microprocessor, RAM, RDRAM, SDRAM, SRAM.

VRC Vertical Redundancy Check. Synonymous with Parity Checking.

VRD Virtual Ring Down.

VREPAIR A Novell NetWare program somewhat analogous to MS-DOS's CHKDSK program or Windows95's Scandisk. VREPAIR fixes FAT (File Allocation Table) and DIR (Directory) Tables. It's a most useful program. Highly recommended.

VRID Virtual Router IDentifier. An eight-bit identifier in the header of a VRRP (Virtual Router Redundancy Protocol) packet, the VRID identifies the virtual router for which the packet is reporting the status. See also VRRP.

VRML Virtual Reality Modeling Language. A language for writing 3D HTML applications. VRML, according to PC Magazine, is an open standard for 3-D imaging on the World-Wide Web that paves the way for virtual reality on the Internet. The way VRML code describes a 3-D scene is analogous to four points describing a square, or a center point and radius describing a sphere. VRML viewers, similar to HTML Web browsers, interpret VRML data downloaded from the Web and render it on your computer. This allows the bulk of the processing to be performed locally and drastically reduces the volume of information that must be transmitted from the Web – a key consideration if rendering is to be performed in real time. See VRML Consortium.

VRPRS Virtual Route Pacing Response in SNA.

VRRP Virtual Router Redundancy Protocol. Specified in the IETF RFC 2338, VRRP is an election (i.e., optional) protocol that allows several first-hop virtual routers on a multiaccess LAN to dynamically share a single IPv4 (Internet Protocol version 4) address. One of the virtual routers, each of which actually may be in the physical form of a VPN (Virtual Private Network) concentrator, is designated as the Master, and the other as the Backup. Should the Master fail, the Backup automatically senses the failure, and assumes responsibility for forwarding LAN packets IP-addressed to it. As a dynamic and automatic approach to virtual router redundancy, VRRP offers considerable improvements in network resiliency as compared to running a dynamic routing protocol such as RIP (Routing Information Protocol) or OSPF (Open Shortest Path First), running an ICMP (Internet Control Message Protocol) discovery client, or using a route that is defined statically. Such dynamic routing protocols as RIP and OSPF, and running client discovery mechanisms through ICMP packets, commonly involve the active participation of all hosts on the network, which fact creates security issues and involves a good deal of administrative overhead, thereby lowering the effective throughput of the network. Statically configured default routes address these issues, but necessarily involve a single point of failure and, therefore, expose the network to catastrophic failure. This latter approach is commonly used in DHCP (Dynamic Host Configuration Protocol). VRRP addresses all of these issues, dynamically assigning first-hop router responsibility and providing redundancy and resiliency in the process, all without the excessive administrative overhead demanded of host-based router discovery protocols. All of that having been said, VRRP packets are sent periodically by the Master router to all VRRP routers, communicating the state of the Master. The VRRP packets are sent via encapsulation in IP packets, using the IP multicast address 224.0.0.18 and the IP protocol number 112, as assigned by the IANA (Internet Assigned Numbers Authority). Note that there is a variation on the above theme, in which two virtual routers divide first-hop Master router responsibilities, with each serving as the Backup for the other. VRRP performs similar functions to several proprietary protocols, Cisco's HSRP (Hot Standby Router Protocol) and DEC's IP Standby Protocol. See also DHCP, HSRP, IANA, IETF, IPv4, OSPF, RIP, and VPN.

VRU See Voice Board and Voice Response Unit.

VS 1. Virtual Scheduling. As an ATM term, it is a method to determine the conformance of an arriving cell. The virtual scheduling algorithm updates a Theoretical Arrival Time (TAT), which is the "nominal" arrival time of the cell assuming that the active source sends equally spaced cells. If the actual arrival time of a cell is not "too" early relative to the TAT, then the cell is conforming. Otherwise the cell is non-conforming.

2. Virtual Source. Refer to VS/VD.

3. See Virtual Storage.

VS&F Voice Store and Forward. Voice is digitally encoded, sent to large storage devices and later forwarded to the recipient. See Voice Mail.

VS/VD Virtual Source/Virtual Destination. An ATM term, a VS/VD is an ABR connection may be divided into two or more separately controlled ABR segments. Each ABR control segment, except the first, is sourced by a virtual source. A virtual source implements the behavior of an ABR source endpoint. Backwards RM-cells received by a virtual source are removed from the connection. Each ABR control segment, except the last, is terminated by a virtual destination. A virtual destination assumes the behavior of an ABR destination endpoint. Forward RM-cells received by a virtual destination are turned around and not forwarded to the next segment of the connection.

VSAC Very Small Aperature Check.

VSAT Very Small Aperture Terminal. A relatively small satellite antenna, typically 1.5 to 3.0 meters in diameter, used for satellite-based point-to-multipoint data communications applications. While VSAT earth stations traditionally supported data rates of as much as 56 Kbps, contemporary systems can operate at rates of 1.544 Mbps. You see VSATs on top of retail stores which use them for transmitting the day's receipts and receiving instructions for sales, etc.

Consider the VSAT dishes you see on the roofs of gas stations. Large numbers of gas stations share access to a single satellite which, in turn, provides connection to a centralized data processing center. At those gas stations are intelligent gas pumps equipped with credit card readers, monitors, and limited computer memory. You swipe your credit card through the card reader, with the credit card number being transmitted through the VAST dish to the satellite to the data processing center. Once the credit is verified (i.e., the card has not been reported lost or stolen, and the balance is not overdue), the transaction is authorized in return. Once the desired amount of gas has been pumped, that information is transmitted to the data processing center, with the transaction being noted in the accounts receivable system for billing purposes. Additionally, the level of inventory (i.e., gas in the tank) is noted as having been decreased. In other words, the VSAT network supports credit verification, transaction authorization, billing and inventory management.

VSB Vestigial Sideband. A form of Amplitude Modulation (AM) that compresses required bandwidth and is commonly used for video. VSB is modulation technique used to send data over a coaxial cable network. NTSC video standardizes a VSB technique. Another (16-level, digital) VSB technique has been chosen for HDTV systems and is under study for other digital video systems. VSB is also used by hybrid networks for upstream digital transmissions, VSB is faster than the more commonly used QPSK, but it's also more susceptible to noise. See also 64QAM, Amplitude Modulation, DSBSC, DSBTC, SSB, Vestigial Sideband and QPSK.

VSC See Vertical Service Code.

VSE 1. A British Term. Voice Services Equipment, a generic term for voice response unit, interactive voice response, voice processing unit and so on.

2. Virtual Storage Extended.

VSELP 1. Vector Sum Exited Linear Prediction. A speech coding technique used in U.S. and proposed Japanese DMR standards. Second generation European DMR will probably use some version of VSELP.

2. See Vector Sum Excited Predictive Coding.

VSI Virtual Switch Interface.

VSI master A VSI master process implementing the master side of the VSI protocol in a VSI controller. Sometimes the whole VSI controller might be referred to as a VSI Master but this is not strictly correct. A device that controls a VSI switch, for example, a VSI label switch controller.

VSLAM Video Subscriber Line Access Multiplexer. A VSLAM is the device at the central office which enables VDSL – Video DSL – Video Digital Subscriber Line Service.

VSNET Virtual SS7 Network

VSP VoIP Service Provider. See VoIP.

VSR Very Short Reach. A physical-layer specification for a parallel fiber optic connection operating at up to 10 Gbps over very short distances of 300 meters or less. The reason for VSR is twofold. First, lasers that work over short distances are far less expensive than lasers that work over longer distances. So VSR systems can be very cheap, since the laser's cost is a major part of the total cost of the system. Second, there are an enormous number of applications for VSR – inside a central office, between one rack of equipment and another, between one router and another, etc. The VSR protocol uses an array of lasers typically firing at a wavelength of 850 nanometers, although any wavelength in the 830-860 nanometer range is acceptable. VSR makes use of VCSEL (Vertical-Cavity Surface-Emitting Lasers) lasers. Each VCSEL fires over a separate, parallel multimode fiber in a ribbon cable. At the near end of the connection, the input data stream is inverse multiplexed, striped over multiple fibers in the VSR transmission system, and recombined at the far end of the connection. VSR is highly cost-effective for intraoffice connectivity over very short distances between devices (e.g., routers, optical cross-connects, and long-haul optical transport gear) collocated within a CO (Central Office) or POP (Point Of Presence). The combination of multiple relatively low-speed lasers and multimode optical fiber compare quite favorably to the cost of a single high-speed laser operating over a single monomode fiber and connecting to a single high-speed port. In either case, the aggregate bandwidth supported currently is 10 Gbps, or OC-192 in SONET terms. VSR is touted as scaling well, providing the foundation for more affordable intra-office interconnections at even higher speeds of 40 and 80 Gbps. See also Gigabit Ethernet, SONET, SR, and VCSEL.

VSS Voice Server System.

VSWR Voltage Standing Wave Ratio. The ratio of the maximum effective voltage to the minimum effective voltage measured along the length of mis-matched radio frequency transmission line. Explanation: When impedance mismatches exist, some of the energy transmitted through will be reflected back to the source. Different amounts of energy will be reflected back depending on the frequency of the energy. VSWR (Voltage Standing Wave Ratio) is a unitless ratio ranging from 1 to infinity, expressing the amount of reflected energy. A value of one indicates that all of the energy will pass through, while any higher value indicates that a portion of the energy will be reflected.

VT 1. Virtual Tributary. A SONET structure designed for transport and switching of subrate DS-3 payloads. VT1.5 equals 1.544 Mbps (T-1); VT2 equals 2.048 Mbps (E-1); VT3 equals 3.456 Mbps T-1c); and VT6 (T-2) equals 6.912 Mbps. See also SONET and Virtual Tributary.

2. Virtual Tributary Group. A nine-row, 12-column structure (108 bytes) that carries one or more VTs of the same size. Seven VT groups can fit into one SRS-1 playload.

VT Pointer Virtual Tributary Pointer. Locates the VT Synchronous Payload Envelope (VT SPE) for a floating mode Virtual Tributary (where the timing is not locked in frequency nor phase to the timing of the STS-1, but is allowed to float). See also SONET and Virtual Tributary.

VT SPE VT Synchronous Payload Envelope.

VT100 Video Terminal 100. An incredibly capable CRT (Cathode Ray Terminal) developed by DEC in the early 1980s. The popularity of the VT 100 lead to its becoming an ad hoc standard, which still forms the basis for the xterm (X Terminal) terminal emulator programs for the X Windows system. VT102 is a newer version. See also Cathode Ray Tube and X Windows.

VT1.5 Virtual Tributary 1.5. A SONET term. A subrate channel with the payload equivalent of a T-1 frame at 1.544 Mbps. With SONET overhead, the total signaling rate is 1.728 Mbps. See also SONET, T-1, and Virtual Tributary.

VT2 Virtual Tributary 2. A SONET term. A subrate channel with the payload equivalent of an E-1 frame at 2.048 Mbps. With SONET overhead, the total signaling rate is 2.304 Mbps. See also E-1, SONET, and Virtual Tributary.

VT3 Virtual Tributary 3. A SONET term. A subrate channel with the payload equivalent of a T-1c frame at 3.152 Mbps. With SONET overhead, the total signaling rate is 3.456 Mbps. See also SONET, T-1c, and Virtual Tributary.

VT6 Virtual Tributary 6. A SONET term. A subrate channel of 6.912 Mbps with the payload equivalent of a T-2 at 6.312. With SONET overhead, the total signaling rate is 6.912 Mbps. See also SONET, T-2, and Virtual Tributary.

VTA Virtual Trunk Agent.

VTAC Vermont Telecommunications Applications Center. See www.vtac.org

VTAM Virtual Telecommunications Access Method. A program component in an IBM computer which handles some of the communications processing tasks for an application program. In an IBM 370 or compatible, VTAM is a method to give users at remote terminals access to applications in the main computer. VTAM resides in the host. It performs addressing and path control functions in an SNA network that allows a terminal or an application to communicate and transfer data to another application along some sort of transmission medium. VTAM also provides resource sharing, a technique for efficiently using a network to reduce transmission costs. See Systems Network Architecture.

VTC Video TeleConference, a term invented by the U.S. Air Force.

VTG Virtual Tributary Group

VTM Vendor Technical Management.

VTN Vendor Type Number.

VTNS Virtual Telecommunications Network Services.

VTO Voltage Tuned Oscillator.

VToA Voice Traffic (or Transport) over ATM.

VTOH Virtual Tributary Overhead.

VTP Virtual Terminal Protocol. An International Standards Organization (ISO) standard for virtual terminal service.

VTTH Video To The Home. The general ability to provide interactive multimedia services to people in their homes.

VU meter VU is the unit of measurement for electrical speech power in communications work. VUs are measured in decibels above 1 milliwatt. The measuring device is called a VU meter, which is an abbreviation of volume-unit meter, a type of meter used to indicate average audio amplitude.

VUI 1. Voice user interface, i.e., a speech-enabled user interface to an application or system.

2. First came the CLI (Command-Line Interface). Then came the GUI (Graphical User Interface). Get ready for the VUI: the Video User Interface. Actually, you don't need to get ready for it any time soon, but you might start wondering how to use it.

Vulcan Nerve Pinch The taxing hand positions required to reach all the appropriate keys for certain commands. For instance, the warm re-boot for a Mac II computer involves simultaneously pressing the Control Key, the Command key, the Return key and the Power On key. See also Three Finger Salute.

vulcanize To cure a thermoset insulation or jacket.

vulnerability assessment See Intrusion Detection.

vulnerability scanner A network vulnerability tester which sends data to various IP ports on a host to determine which ones are responsive; vulnerability scanners can be used by hackers to find exploitable vulnerabilities or by security specialists to identify weaknesses needing to be strengthened.

vulnerability testing I love this term. It sounds like some strange sexual perversion. In fact, it means nothing more than using tools and techniques to check how vulnerable your network is to being hacked, to being broken into, to being compromised. See also Network Penetration Testing.

VXI VME Extension for Instrumentation. An extension of the standard VME bus design, VXI is intended for high-performance instrumentation applications such as test systems, laboratory automation systems, and industrial control systems. See also VME.

VXML VXML is software designed to let you talk to the web sites, to have them answer you, i.e. give you the information you want – from stock prices, to restaurant menus, to driving instructions, to sending and hearing emails. Voice Extensible Markup Language (Voice XML) is a Web-based markup language much like HTML (HyperText Markup Language) and XML, but designed for voice-based, rather than typing, graphical or textual interaction. Like HTML, VXML relies on HTTP (HyperText Transport Protocol) for TCP/IP transport. VXML supports telephone access to Web services, and supports Web browsing and Website interaction through voice and audio (e.g., touchtone and speech recognition), rather than traditional point-and-click mouse and keyboard commands. Once fully standardized and widely implemented, VXML will support four general applications categories, according to the VoiceXML Forum. Those are information retrieval, electronic commerce, telephony services (e.g., voice-activated dialing and conference calling, and one number, find-me services) and unified communications (i.e., unified messaging). The VoiceXML Forum develops the educational and marketing aspects of the VXML standard under development by the World Wide Web Consortium (W3C). At the time of this writing, VXML is in version 1.0, a specification from the VoiceXML Forum, which is being standardized by the W3C. Version 2.0, which is under development by the W3C, will feature the Speech Recognition Grammar Format (SRGF), an XML language for writing voice recognition grammer, and Synthesis Markup Language (SSML), an XML language for text-to-speech markup. See also CCXML, HTML, HTTP, VoiceXML Forum, W3C, and XML.

VxWorks VxWorks was developed by Wind River Systems and is a development and execution environment for real-time and embedded applications on a wide variety of target microprocessors (i.e. single chip computers).

VW-1 A test used by Underwriters Laboratories to classify wires and cables by their resistance to burning. (Formerly designated as FR-1.)

VWP Virtual Wavelength Path. A VWP is a group of one or more channels between source and destination nodes. The term virtual indicates that the signal path can actually travel on different physical wavelengths throughout the network. All channels of the VWP transit the same path through the network.

Vyvx Vyvx is a division of Williams Communications and provides broadcast quality video transmission and advertising services to both television and internet broadcasters. When stations, networks, and internet companies want to get live video programming to their main broadcast facility in real-time, they either use their own satellite equipment or lease Vyvx lines. It has become common slang for any non-broadcaster owned transmission lines to be called Vyvx. Vyvx the company uses teleports, satellite trucks, or its vast network to transmit the video programming. Its services are used for the global distribution of news, sports and special events. Vyvx was the first company to provide switched, broadcast-quality fiber-optic transmission services for the broadcast television industry on a national scale in the early 1990s. Their network connects to every major news and media center in the US, to more than 100 professional sports venues in North America and to 450 television stations in the top 100 domestic markets.

W 1. Abbreviation for WATT.

2 The Hayes AT Command Set describes a standard language for sending commands to asynchronous modems. One of the commands is "W." If you embed a W in your dialing string, i.e. 212-691-8215-W-10045, the modem will dial 212-691-8215 and wait until it hears dial tone. When it hears dial tone, it will dial out 10045. That is the standard Hayes command set interpretation of W. There is another. When using some of the communications software products from Crosstalk (now a subsidiary of DCA) you can place a [W] in your dialing string. If you do, your modem will dial the number until it encounters a [W]. It will then wait until you hit any button on your keyboard. The purpose of W commands is to allow you to dial through private networks (your own), through public networks (MCI, Sprint, etc.), through fax/modem/telephone switches and through any other device or network.

3. Heavy duty portable power cable, one to six conductors, 600V, 900xAFC.

W-CDMA Wideband Code Division Multiple Access. A proposal for a 3G (Third Generation) wireless system based on CDMA technology. W-CDMA would offer bandwidth in excess of narrowband, which is commonly considered to be voice-grade bandwidth, i.e., 56/64 Kbps. W-CDMA proposes to support data rates of up to 384 Kbps initially, and up to 2 Mbps eventually. See also 3G, CDMA, and IMT-2000.

W-DCS Wideband Digital Cross-connect System. W-DCS is an electronic digital cross-connect system capable of cross-connecting signals below the DS3 rate.

W3 An abbreviation for the Internet's World Wide Web. See World Wide Web.

W3C World Wide Web Consortium. A consortium jointly hosted by the Massachusetts Institute of Technology (MIT), the Institut National de Recherche en Informatique et en Automatique (INRIA), and Keio University (Japan) Initially, W3C was established in collaboration with CERN, where the WWW originated, with support from DARPA and the European Commission. Tim Berners-Lee, inventor of the WWW, acts as Director. The W3C works to produce "free, interoperable specifications and sample code. Focus is on the domains of user interface, technology and society, and architecture. In some ways, its ambitions are not that different from those of the IETF (Internet Engineering Task Force), except that its members have commercial interests. The IETF's members are volunteers and tend to come from academia. W3C also focuses on the narrower Web, whereas the IETF focuses on the broader Internet. www.w3.org

WAAS Wide Area Augmentation System is an enhancement to the global positioning system (see GPS). Standard GPS methods produce a position accurate to about 10 meters (32.8 feet). The US Federal Aviation Administration desired to increase this accuracy and allow aviators to use GPS for more precise guidance. As a result, they developed a system similar to DGPS (Differential Global Positioning System) but on a much larger scale. This is the WAAS (Wide Area Augmentation System). Although still in its early roll-out stage, this system can improve a GPS receivers accuracy to about 3 Meters (9.8 feet) or less. How

it works. As with DGPS, the WAAS system uses base stations at known reference points to calculate the accuracy of the GPS signal. This is accomplished at each of the 25 ground reference stations (Currently only in the US) receiving a standard GPS signal. A set of correction data determined from the difference between the GPS calculated position and the known position is transferred to one of two ground control stations that then uplink the data to the WAAS satellite. The WAAS (Inmarsat) satellites then transmit this information back down to the GPS user using a GPS-like signal complete with the correction information. The GPS receiver then decodes this information and applies it to its calculated position to significantly improve the accuracy. Currently the system is only accurate in North America and primarily in the United States. However, the signal can be received over half of the world on the Inmarsat AOR-W and POR satellites. This means that in parts of the world not covered by the base station corrections, you will get a WAAS signal, but the corrections will place you well off your mark in Australia, South America and Europe, for example. This is why it is best to turn off the WAAS reception outside North America.

Wabi Sun Microsystems software for running Microsoft Windows applications on Solaris. Runs on Intel and SPARC.

WabiServer Wabi is Sun Microsystems software for running Microsoft Windows applications on Solaris. Runs on Intel and SPARC. A WabiServer allows multiple and simultaneous users to run Wabi on Intel and SPARC.

WACK 1. Wack a T-1. Here's what it means. Imagine you're an ISP – an Internet Service Provider. Your business is answering inbound calls from customers with PCs who want to send email, surf the Internet, have fun in chat rooms, etc. You are receiving your calls on digital circuits (e.g. T-1s) from a local phone company – CLEC or ILEC. Your inbound T-1 line may have as many as 24 separate phone lines, which your customers could call on. Sometimes your phone company sends you a call over one or more of those 24 lines and it doesn't get through. Its switching equipment assumes your equipment on that particular phone line is broken. So its switching equipment automatically takes that phone line out of service. It sends no more calls over that line. Later in the day, it may run a diagnostic program – called FISO in some instances – and that software program may try sending calls over the lines that it had earlier taken out of service. But between taking the lines out of service and running that program as much as six or seven hours may elapse. With these lines out of service, an ISP's customers will now enjoy lousy service. So the ISP's people (after bombarded with customer complaint calls) calls the phone company and says "wack my T-1, please." What happens then is that a real live technician at the phone company then goes into each line and manually tests it, often with a technician from the ISP on the other end of the phone. The advantage of wacking the T-1 is that the circuits out of service are put back into service much faster than if they were left to the machine to do it automatically later on.

2. Wait before transmitting positive ACKnowledgement. In Bisynch, this DLE sequence

1001

is sent by a receiving station to indicate it is temporarily not ready to receive.

3. Do not confuse "wack" with "whack," which has now become a colloquial expression for killing someone.

wafer A thin disk of a purified crystalline semiconductor, typically silicon, that is cut into chips after processing. Typically, a wafer is about one fiftieth of an inch thick and four or five inches in diameter. See Semiconductor.

wafer fabs Wafer fabs are a slang term for ultraclean factories that fabricate chips on silicon wafers.

WAG Wild Ass Guess.

WAGS Wireless Assisted Global Positioning System. See GPS.

WAIS Wide-Area Information Servers. A very powerful system for looking up information in databases (or libraries) across the Internet. WAIS allows you to perform a keyword search. WAIS is like an index, whereas Gopher, which is sometimes used as a complement to WAIS, is like a table of contents.

wait on busy An English term for the American term "Camp On" or "Call Waiting." A service allowing the subscriber to make a call to a busy phone line, wait until the call is over, then be connected automatically.

wait state A period of time when the processor does nothing; it simply waits. A wait state is used to synchronize circuitry or devices operating at different speeds. Wait states are introduced into computers to compensate for the fact that the central microprocessor might be faster than the memory chips next to it. For example, wait states used in memory access slow down the CPU so that all components seem to be running at the same speed. A wait state is a "missed beat" in the cycle of information to and from the CPU that is necessary for a memory transaction to be completed.

wake In England in the 1500s, lead cups were used to drink ale or whisky. The combination would sometimes knock people out for a couple of days. Someone walking along the road would take them for dead and prepare them for burial. They were laid out on the kitchen table for a couple of days and the family would gather around and eat and drink and wait and see if they would wake up – hence the custom of "holding a wake."

Wake On LAN This standard allows a PC to be powered on by a network server, so the server can perform routine tasks. See WFM.

Wake-Up Mail Imagine your PC receiving an important email, then waking you and reading you your email. It's bizarre. But someone is predicting it.

WAL WATS Access Line. A direct dial WATS line, as compared to a WATS line connected via a T-1 line.

walk through 1. In Computer System Development, a peer review of a system design, code, etc. The goal is to identify errors as early as possible and learn from other people's experience. Managers and people who prepare performance reviews should NOT be in the room. The concept is to invite "egoless" constructive criticism and to nurture team-oriented validation and debug responsibility.

2. In telecommunications, a formal tour and accompanying verbal description of the work that is to be done by the customer and the vendors.

walk time The time required to transfer permission to poll from one station to another.

walkaways People who walk away from coin phones though they owe extra money. You can tell a phone that has just been visited by a walkaway: It's typically ringing. And when you answer it, the operator will ask you to deposit some additional coins.

walkie-talkie Hand-held radio transmitter and receiver – like the police carry. Probably the best named device in telecommunications. You walkie, you talkie. The walkie-talkie was invented in 1938 by Al Gross, who also pioneered CB Citizen's Band radio and patented the first telephone pager device. Nextel rediscovered walkie-talkie technology, which made it available within individual coverage areas, incorporating the feature into its handsets. Nextel later made walkie-talkie communications available nationwide using Motorola-invented cell phone technology. Later other cell phone carriers copied Nextel. For the entire story, see push to talk.

wall field The layout of horizontal and vertical wiring on a designated wall in the MDF or IDF of a customer premises. Probably in the telco as well.

wall mount A mounting system used for attaching an antenna mast to a building wall.

wall outlet A phone outlet positioned at shoulder height to accept a wall telephone set. The typical installation includes a special modular jack containing two mounting bosses that insert into key-hole slots in the base of the telephone set. Electrical connection is made by a short cord or a lug element that is integral to the telephone set base.

wall phone A phone that is mounted on the wall. Where else would a wall phone be mounted? Some new phones – especially some key systems – come so you can use them on the desk or mount them on a wall, without extra hardware. Some desk phones cannot be mounted on a wall. This is a disadvantage when you run out of space on your desk, as you will with all the computers and workstations you'll be putting there.

wall thickness A term expressing the thickness of a layer of applied insulation or jacket.

wall wart Term for the common ac to dc adaptor that plugs into a wall socket. Often called a wall wart charger.

walla walla In movie-making, a "walla walla scene" is one where extras pretend to be talking in the background-they are not, they are just repeating "walla walla" over and over again. But when they say "walla walla" it looks like they are actually holding conversations. (Sounds like Monday morning meetings at work.)

walled garden This definition varies. One definition is that Web technology and Web content are used in providing content, but users cannot access the Web directly. Another definition has a walled garden referring to a service or content that is exclusive to subscribers who can venture beyond the wall into other services and content on the Web.

wallflower A stock that has fallen out of favor with investors.

wallpaper The area of your Windows or cell phone desktop behind and around your Windows and icons. The color and pattern you put on it through the desktop manager is called wallpaper.

WAM Wireless SpAM. Junk emails you receive on your wireless device – cellphone or PDA – personal digital assistant.

WAN Wide Area Network. A public voice or data network that extends beyond the metropolitan area. A LAN (Local Area Network) generally is confined to a building or campus environment, and is owned by the end user organization. A MAN (Metropolitan Area Network) is a public network that covers a metropolitan area, which may extend beyond the official city limits. A WAN extends farther, perhaps even internationally. Some people use the term GAN (Global Area Network) to refer to international networks, but I think that's a silly word.

WAN accelerator A device which is used to improve throughput and application response time over a wide area network. WAN accelerators work in pairs, with one device on one end of a WAN link and its counterpart on the other end of the link. The devices improve throughput by using a combination of file compression, compression of non-file bit streams, caching, and a customized version of TCP that reduces slowdowns when congestion occurs. A WAN accelerator is also called a WAN optimizer.

WAN emulator An arrangement of WAN links and WAN access equipment such as routers, bridges and modems, that are installed in a test environment between two segments on a local area network, in order to emulate an entire WAN infrastructure. This lets network administrators test network applications and equipment in a laboratory under a wide range of conditions. Network equipment and applications that would be used on opposite ends of the WAN link are installed on each side of the emulator. WAN parameters such as bandwidth, latency, packet loss rate, and bit error rate are varied, so that testers can assess their impact on end-to-end network performance.

WAN Interface Card WIC. A card (i.e., printed circuit board) that fits into a CPE (Customer Premises Equipment) device or system in order to effect the physical interface to an access circuit connecting to the Wide Area Network. See also WAN.

WAN optimizer See WAN accelerator.

WAN router A carrier-class router designed for Internet Service Providers to process up to several terabits of data per second. WAN routers support optical networking and electrical transport technologies.

WANabee An Information Systems employee who wants to work in the Wide Area Networking (WAN) Department of a company.

wander Long-term random variations of the significant instants of a digital signal from their ideal position in time. Wander is a matter of synchronization errors in digital networks. If a clocking source fails or degrades, it is essential that the signals be re-clocked at the point of the next network element. Otherwise, wander will increase through cascading elements, perhaps affecting the data payload. It is especially important to control wander in very speed networks, such as SONET, where even slight synchronization failures can be catastrophic. Wander variations are usually considered to be those that occur over a period greater than 1 second. See also Diurnal Wander.

WANMC West Area Network Management Center WANS: Western Area Network Service Center WARC: World Administrative Radio Conference. An international conference called by the ITU-R (formerly CCIR), focused on international agreements concerning Spectrum Allocation. The most recent meeting was February, 1992, in Spain.

WAP Wireless Application Protocol is "baby" browsing. It lets users get to information with handheld wireless devices such as mobile phones, pagers, two-way radios, smartphones and communicators. WAP will work on most wireless networks, including CDPD, CDMA, GSM, PDC, PHS, TDMA, FLEX, ReFLEX, iDEN, TETRA, DECT, DataTAC, and Mobitex. WAPs that use displays and access the Internet run what are called microbrowsers – browsers with small file sizes that can accommodate the low memory constraints of handheld devices and the low-bandwidth constraints of a wireless-handheld network. Although WAP supports HTML and XML, the WML language (an XML application) is really designed for small screens and one-hand navigation without a keyboard. WML is scalable from two-line text displays up through graphic screens found on items such as smart phones and communicators. WAP also supports WMLScript. It is similar to JavaScript, but makes minimal demands on memory and CPU power because it does not contain many of the unnecessary functions found in other scripting languages. is a text- based infotainment browsing and messaging offering, and turns it into a graphically rich, interactive, 'Internet -like' user experience. For instance, a richer user experience is achieved via color applications, pull down menus, radio buttons, and photo images. In addition to an improved GUI, WAP 2.0 offers other key advantages in terms of 'end-to-end security', capability of using XHTML and 10% faster browsing times vis 0xD6 vis WAP 1.X. A summary of WAP 2.0's technical capabilities vs. WAP 1.X is provided as follows: http://www.vzwdevelopers.com/aims/public/WapLanding.jsp

Wireless Application Protocol is a text- based infotainment browsing and messaging offering used on cell phones ,is a carrier-independent, transaction-oriented protocol for wireless data networks, designed for all type networks, but initially to be implemented on GSM networks. WAP version 1.1 was released in June, 1999. WAP will be included in a new generation of cell phones from major manufacturers such as Nokia, Ericsson and Motorola. Such phones will sport larger screens and a rolling mouse (like a normal computer mouse, except it only scrolls up and down, not side to side). The best way to think of WAP is a visual interactive computer telephony voice response system. Essentially you dial a distant WAP server (i.e. computer) through the mobile network. When you reach the server you log on and "do your business." That "do your business" might consist of figuring out how much money you have in the bank, checking the price of your stocks, booking an airline reservation, etc. You can get this information in several ways – through the expanded screen on your WAP cell phone, through a voice response unit from the other end (punch 1 for this, hear the response, etc.) or a combination of the two. The WAP protocol contains security, transaction handling, byte coding and encryption. What this means is that your "conversation" with the WAP server is secure. WAP users typically must register with their WAP server and be authenticated when they log on. What it also means is that the session the server holds with the user is a dedicated point-to-point session, unlike the Internet where sessions are shared. WAP will also get you to the Internet through a "WAP gateway." But the session doesn't typically have the security that a WAP session does.

The original thinking on WAP was not only to let users access data electronically and voice mail, make stock trades, conduct banking transactions and view miniature Web pages on a wireless terminal's LCD screen, but also to make it easier for mobile users to view shrunken Web pages using Unwired Planet's Handheld Device Markup Language (HDML). According to Unwired Planet (now called Phone.com), HDML "lets Web sites tailor the information format to fit the screen of the phone. We don't try to display the graphical Web pages on such a small device," rather, Webmasters could create smaller versions of sites more suitable for viewing on such units. With WAP, those modifications and optimizations would only have to be made once in order to be viewed on an Ericsson, Motorola or Nokia terminal, Unwired Planet's technology, called UP.Link, is used by AT&T's PocketNet and GTE Wireless services. Bell Atlantic also offers cellular digital packet data-based services combined with UP.Link. Al Haase, director of sales, GSM, at Ericsson's North American headquarters in Richardson, Texas, acknowledged that although the demand for receiving this kind of information in a wireless format is not great today, agreement on a standard may nudge wireless access deeper into the mainstream. "The systems were not designed to support data, and thus data transmission speeds are fairly low, " Hasse said. "With the new systems coming online, data access becomes more of a reality because we are more able to link the mobile user with the Net in a timely way."

The success of WAP will depend, on my opinion, on two things: First, the speed at which data can be made to pass across a GSM network. Second, the ability of companies such as Sweden's Nocom to evangelize the standard to WAP server providers. Note that DoCoMo has introduced a highly successful i-Mode (internet-Mode) service in Japan. i-Mode is a proprietary system that is not compatible with WAP. See also CDMA2000, i-Mode, WTLS and Wireless Application Protocol Forum Ltd.

WAP forum The WAP Forum has consolidated into the Open Mobile Alliance (OMA) and no longer exists as an independent organization. OMA seems to want to be the one body that does everything value-added with wireless devices, including WAP, location finding, mobile gaming, and Internet browsing. See also WAP.

WAPvert An advertisement displayed on a WAP wireless device. Such WAPvert could be a text tag or a live alert.

WAR 1. Wireless Application Environment - defines the programming interface for applications and consists of WML and WML-script - a new, Java-like local scripting language

2. The Sanskrit word for war means "desire for more cows."

war dialer Usually a computer program that can be configured to automatically dial a range of telephone numbers and make note of which ones are answered by a computer; the IT security staff can use such a tool to discover previously unknown entry points into their network or hackers can use it to identify attack points.

war chalking The marking of walls or sidewalks with chalk marks to indicate the presence of a nearby open wireless network. Also spelled warchalking.

war driving Attempting to connect to a wireless network while strolling or driving. See war tuning. Also spelled wardriving.

war plugging Using a service set identifier (SSID) – a tag of up to 32 characters – to advertise a wireless network to anyone who comes across it. Examples: "Mars network, open for all" and "please_bring_pizza." It's the latest twist on wardriving, or cruising a city in search of open wireless access points. See also 802.11b.

war room Also called a "solutions room." This is an enclosed area with a large table used for decision- or strategy-making.

war tuning Using a Wi-Fi card and a laptop to drive around, sniff out, and listen to other people's iTunes libraries. Derived from war driving - the practice of cruising for open Wi-Fi hot spots.

warble tone A tone changing in frequency at a slow enough rate to give the effect of warbling. A warble tone is the sound of an electronic ringer, according to many people.

WARC World Administrative Radio Conference. Sets international frequencies. Just before Telecom '87, WARC allocated important new frequencies for satellite-based land mobile (satellite to truck, etc.) and radio determination navigation services (electronic maps for your car). WARC is part of the 154-member International Telecommunication Union. ITU-T is part of the ITU. See ITU-T.

warchalking The marking of walls or sidewalks with chalk marks to indicate the presence of a nearby open wireless network.

wardriving Cruising around in a car with a wireless-enabled laptop computer on, looking for open wireless networks. The locations found may be recorded and posted on the Web.

wardrobe malfunction During the half-time show of the 2004 Super Bowl, Justin Timberlake and Janet Jackson were singing and performing. During one dance number, Mr. Timberlake ripped off a piece of fabric covering Ms. Jackson's right breast, revealing a large breast, whose nipple was surrounded by a metal ring shaped like a sun, with the center being not the sun, but her nipple. The incident caused a huge uproar and ultimately the FCC fined the offending broadcasting network, CBS, $500,000. Mr. Timberlake claimed the incident was a "wardrobe malfunction." MTV, which had produced the segment, said the incident was "unrehearsed, unplanned, and completely unintentional." And if you believe MTV or Mr. Timberlake, you probably believe in fairies and goblins. About a year later, the president of the FCC, Michael Powell – the man who had levied the heavy fine – resigned from the FCC, saying, in essence, he'd had enough. I have watched the incident several times. Personally I found it not offensive, but basically tasteless. It was yet another incident of pushing the envelope – a trend we've seen in entertainment and pornography.

warez Warez (pronounced "wares") is commercial software that has been pirated and made available to the public via a Bulletin Board or a Web site on the Internet. Typically, the pirate has figured out a way to remove the copy-protection or registration scheme used by the software. The use and distribution of warez software is illegal. People who create warez sites sometimes call them "warez sitez" and use "z" in their pluralizations.

Warhol Worm Really mean software that doesn't exist for now but that could theoretically infect every server on the Internet within 15 minutes. The term, of course, derives from the fact that Andy Warhol claimed that everyone in the world would be famous for 15 minutes in their life.

warm standby See Data Center.

warm start Restarting or resetting a computer without turning it off (also called "soft boot"); press Ctrl + Alt + Del on an IBM or IBM compatible.

warm swap When a system component (e.g., a computer disk drive) fails, it may be replaced without turning the system off. During this period, the system's activity is suspended, however. Also known as a Hot Plug, it is unlike a Hot Swap, during which the system remains active. See HOT SWAP and RAID.

warm zone The coverage area of a WiMAX base station.

warranty Span of time that equipment will be repaired or replaced due to failure. Usually does not include reimbursement of engineer's fees required for replacement. May not include equipment failure due to abuse or destruction by either intentional or unintentional means. Lightning, floods, and other Acts of God are not covered under warranty.

WAS 1. Wireless Access System. Another term for Wireless Local Loop (WLL). See WLL.

2. Whistling Axe Syndrome. Used to define the circumstance of a telecom employee unwillingly removed from the workforce. "That's where Bill WAS." In the two years following the telecom bust of 2000, over 500,000 employees of the telecom industry were unwillingly removed from employment.

Washington D.C. Eight square miles surrounded by reality. A comment variously attributed to John F. Kennedy, J. Edgar Hoover and sundry other luminaries.

WASI Wide Area Service Identifier

WASO Western Area Switch Operations.

WASP Wireless Application Service Provider, or wireless ASP create, deploy, run and maintain applications that let employees, partners and customers use handheld devices and wireless links to connect with corporate databases, applications, or intranets. See Application Service Provider.

WAT Date A term telephone companies use to indicate a date a day or two in advance of the promised delivery date, by which the circuit should be "up" and available for testing, so any problems can be worked out before the technician is dispatched to the customer's premises.

watch commands Watch Commands are found in programming. They allow you to "watch" the value of selected application variables while the application is executing (e.g., see the last-entered touchtone digits from a caller).

watchdog packet Used to ensure that a client is still connected to a NetWare server. If the server has not received a packet from a client for a certain period of time, it sends that client a series of watchdog packets. If the station fails to respond to a predefined number of watchdog packets, the server concludes that the station is no longer connected and clears the connection for that station.

watchdog timer There is a watchdog circuit (also called special function register) in later versions of Intel Pentium chip. When things go awry with the computer the Pentium is running, it sends out a signal to an external device. That device can then take action – for example reboot the PC. According to Intel's Web site, "The Watchdog Timer is a very useful peripheral that safeguards against software failures. The Watchdog Timer (WDT) is a 16 bit ripple counter that will reset the 8XC196KC/KD if the counter overflows. When the WDT is enabled, this peripheral monitors the execution of a program. This is useful when expensive hardware is being controlled. If we operate in an electrically noisy environment, a noise spike can be potentially fatal to the operation of the microcontroller. We can maintain control in critical applications by resetting the microcontroller when a runaway software process hangs the systems. There is only one Special Function Register to concern ourselves with in this discussion. That register is called WATCHDOG."

watching timer A circuit used in ETHERNET transceivers to ensure that transmission frames are never longer than the specified maximum length.

water I never drink water because of the disgusting things that fish do in it. – WC Fields

water absorption Water by percent weight absorbed by a material after a given immersion period.

water bore A device which bores holes underground. Pipes are then placed in the holes and cables (fiber, coax, etc.) are then pulled through. A water bore uses high-speed, pressurized water to bore through the underground.

water peak The presence of moisture in optical fibers results in significant levels of attenuation (i.e. power loss) at 1383 nm (nanometers). Depending on the magnitude of that "water peak" loss, wavelengths within +/- 50 nm also can be affected. This water peak is caused by residual moisture deposited during the manufacturing process. Specifically, hydrogen atoms in the water molecules cause the phenomenon. The hydrogen atoms readily diffuse through the glass matrix of an optical fiber and are trapped at points of defect in the glass structure. Low water peak fiber resolves this problem through special manufacturing processes. See also Low Water Peak Fiber and Rain Attenuation.

water pipe ground A water pipe to which connection is made for the ground.

waterfall life cycle The conventional software development process, consisting of a series of steps commonly defined as analysis design, construction, testing and implementation. The underlying assumption is that each phase does not begin until the preceding phase is complete.

watermark Traditionally, a watermark is an identifying mark on a piece of paper, created during the production process by varying the thickness of the paper by pressure from a mold, or by lightly applying various inks. Watermarks, which can be seen plainly only when

the paper is held up to the light, are used variously to identify the manufacturer, establish authenticity, build brand loyalty, and prevent counterfeiting. Digital watermarks are small sets of programmed code hidden within computer files or software programs, and can be either visible or invisible, or either audible or inaudible. Visible watermarks are designed to display copyright notices in order to discourage copyright infringement. Invisible watermarks also may be used to discourage copyright infringement, but by catching those who infringe. Watermarks also can be used to verify the authenticity of an image or other file or program, and to establish ownership. See also Steganography.

WATS Wide Area Telecommunications Service. Basically, a discounted toll service provided by all long distance and local phone companies. AT&T started WATS but forgot to trademark the name, so now every supplier uses it as a generic name. There are two types of WATS services – in and out WATS, i.e. those WATS lines that allow you to dial out and those on which you receive incoming calls (the typical 800 line service). You subscribe to in- and out-WATS services separately. In the old days you needed separate in and out lines to handle the in and out WATS services. But these days you can choose to have in- and out-WATS on the same line. This is not particularly brilliant traffic engineering, since you can't receive an incoming 800 call if you're making an outgoing call. But I do know someone who has an 800 line on his cellular phone!

Many users inside companies think their company's WATS lines (and thus their WATS calls) are free, so they speak longer. This can kill the idea of buying WATS lines to save money. In the old days, interstate WATS was charged at effectively a flat rate and thus, there was some reason to believe that marginal WATS calls were 'free.' These days EVERY WATS call costs money. EVERY one! Without exception. See 800 Service and Plant Test Number.

WATS aggregator A brokering billing agent having an arrangement to obtain volume discounts from common carriers for the totality of WATS services used by multiple firms; need not be the operator of a switching system.

Watson, Thomas Chairman of IBM in the 1940s. He is immortalized by his 1943 quote, "I think there is a world market for maybe five computers." It's a good thing for IBM (and the rest of us) that he was dead wrong.

watt The unit of electricity consumption and representing the product (i.e. the two multiplied together) of amperage and voltage. The power requirement of a device is listed in watts, you can convert to amps by dividing the wattage by the voltage (e.g., 1,200 watts divided by 120 volts, equal 10 amps). See OHM's LAW. Don't confuse WATTS (the measure of electricity) with WATS, which stands for Wide Area Telecommunications Service. See WATS.

wave See Wave File.

wave antenna See beverage antenna.

wave audio Also called "waveform audio," is a digital representation of actual sound waves. Wave audio "samples" the sound waveforms at regular intervals. The three standard sampling frequencies are 11.025 Kbps, 22.05 Kbps, and 44.1 kbps. Higher sampling frequencies yield higher fidelity sound.

wave file A Wave File is a Microsoft Windows proprietary format for encoding sound. You'll see the file typically as "mysounds.wav." Click on the file, Windows will launch a sound player and you'll be able to hear what's in the file. Wave files aren't use much in telecommunications because they encode voice at higher rates than are necessary for straight voice conversation. The wave format is often used to encode music. See Wave Audio.

wave length The distance between peaks of an electromagnetic (or other) wave. The distance traveled by a wave during one complete cycle. See also Wavelength.

wave wrapper Microsoft's standard DLL for modems using a separate audio port for voice communication (as opposed to just the serial port for modem and voice). This is used with Unimodem to synchronize audio data with standard modem functions. The Voice application in this case calls multimedia system functions, which in turn calls the Wave Wrapper DLL (wavewrap.drv). The Wrapper tells Unimodem to send AT + V commands to the modem to place it into voice transfer mode. The Wrapper makes subsequent calls to the multimedia system DLL to play the audio from the device. In this case, the device vendor supplies a modem (hardware) wave device interface and does not communicate directly with the serial port.

waveform The characteristic shape of a signal usually shown as a plot of amplitude over a period of time.

waveform coding Electrical techniques used to convey binary signals.

waveform editor A word processor for sound. You record something. Then you "play" it back on your PC's screen. Your PC screen now looks like an oscilloscope. Then you use this wave form editor to edit (i.e. change, replace. amplify, echo, fade in or out, cut out noise, cut/paste from other files, or generally muck with) the sound. A wave form editor is used in voice processing.

waveform monitor A device used to examine the video signal and synchronizing pulses. An oscilloscope designed especially for viewing the waveform of a video signal.

wavefront coding An imaging technique to improve the quality of digital imaging products such as surgical endoscopes, borescopes (cameras which look inside jet engines) and the machine-vision circuitry that lets robots see. Wavefront Coding brings objects of different sizes and at different distances into sharp focus in dim lighting. The focusing is done by a conventional lens coupled with a light detector. Compared with systems using traditional optics, Wavefront Coding extends the depth of field by a factor of tenfold or more, increasing throughput while minimizing the complexity of the system and sample preparation. Wavefront Coding was the result of applying mathematical techniques used in radar to optical applications. The imaging burden is shared between jointly optimized optical components and signal processing algorithms. The imaging burden is shared between jointly optimized optical components and signal processing algorithms.

Waveguide A conducting or dielectric structure able to support and propagate one or more modes. More specifically, a waveguide is a hollow, finely-engineered metallic tube used to transmit microwave radio signals from the microwave antenna to the radio and vice versa. Waveguides comes in various shapes – rectangular, elliptical or circular. They are very sensitive and should be handled very gently. Waveguides may contain a solid or gaseous dielectric material. In optical, a waveguide used as a long transmission line consists of a solid dielectric filament (optical fiber), usually circular. In integrated optical circuits an optical waveguide may consist of a thin dielectric film.

waveguide adapter A device for adapting signals on a coaxial cable to waveguide or vice versa.

waveguide bridge A roof like structure mounted over a transmission line between a transmitter enclosure and a tower. It keeps snow and ice from forming on the transmission line and protects the line from falling ice.

waveguide dispersion Waveguide dispersion is a source of noise in fiber optic transmission systems. Each pulse of light comprises some number of wavelength components, as even the more capable tunable lasers cannot create a signal comprising only one individual wavelength. Now, as the optical signals propagate through the fiber waveguide, the higher frequency (i.e., shorter wavelength) components of the signals tend to disperse (i.e., spread out) more than the lower frequency components. As they disperse from the center of the fiber core, some of them enter the cladding, where they increase in speed due to the purer nature of the glass that comprises the cladding. See also Chromatic Dispersion.

waveguide flange A mechanical "connector" for joining waveguide sections.

waveguide scattering Scattering (other than material scattering) that is attributable to variations of geometry and refractive index profile of an optical fiber.

waveguide switch A switch which uses a shutter to open or block an aperture in order to block an aperture in order to block or pass radio frequency energy.

wavelength The length of a wave measured from any point on one wave, to the corresponding point on the next wave, such as from crest-to-crest, or trough-to-trough. In other words, a wavelength is the distance an electromagnetic wave travels in the time it takes to oscillate through a complete cycle. There is a direct proportion between the wavelength of a radio signal and its frequency. For example and very close to the bottom of the radio spectrum, a Low Frequency (LF) radio signal is defined as having a frequency range of 30 kHz-300 kHz, with a corresponding wavelength of 10 Km-1Km. LF radio is used in navigation and maritime communications. At the high end of the radio spectrum is Extremely High Frequency (EHF) radio, which is defined at having a frequency range of 30 GHz-300 GHz, and a corresponding wavelength of 1 cm-.1 cm. The higher the frequency of the signal, the more Hz (Hertz, or sine waves) are sent per second, and the shorter the corresponding wavelength.

wavelength assignment See Wavelength Scheduling.

wavelength blocker A component of a reconfigurable optical add/drop multiplexer (ROADM), a wavelength blocker blocks selected wavelengths from passing through a node.

wavelength collision What happens when two lightpaths with the same wavelength are put on the same fiber in an optical network. A wavelength collision can be avoided if the wavelength of one of the lightpaths is converted to a different wavelength or if one of the lightpaths is rerouted.

Wavelength Division Multiplexing WDM. A way of increasing the capacity of an optical fiber by simultaneously sending more than one ray of light down the fiber. See WDW for a more complete explanation.

Wavelength Interchanging Cross-Connect WIXC. A type of OXC (Optical Cross-Connect). See OXC.

wavelength managed networks As optical fiber gets more and more wavelengths of light pumped down it, so the idea emerges of splitting and switching wavelengths to go in different directions down different fibers. The blue might go to Detroit. The red might go to Chicago. The concept is to do switching without actually having to drop down to electricity. The whole concept is called wavelength managed networks. See also Wavelength Division Multiplexing.

wavelength scheduling Wavelength scheduling is the art, or science, of letting distinct transmitters and receivers use wavelengths (frequency slots) to communicate over an optical network. The optical network can either be broadcast-and-select type (much like Ethernet or Token ring but in the frequency domain) or wavelength routing type. For wavelength routing networks this is often called wavelength assignment and DWDM or CWDM (Coarse WDM) channels are assumed. Wavelength scheluding can also be used for reducing transmission impairments, e.g. fiber non-linearities, as proposed by a patent held by Telia, the Swedish incumbent phone company. Under Telia's patent, wavelength channels are given permission to transmit information in scheduled time slots. The time slots are allocated so that each channel is only given access to the fiber when no more than one of its nearest neighbors is transmitting. As for most wavelength sheduling algorithms, and there are many, little pratical use have been found to date for wavelength scheduling.

Wavelength Selective Cross Connect WSXC. A type of Optical Cross-Connect (OXC). See OXC.

wavelength ripple The difference in amplitude from highest to lowest in a DWDM channel transmission. Measured in dBm. Help on this definition from Brett Wright and Ken Wilson.

wavelength tilt The difference in amplitude from each end of a DWDM channel transmission. Measured in dBm. Help on this definition from Brett Wright and Ken Wilson.

wavelet Wavelets are an economical way of compressing images, both still and moving, yielding compression ratios of 100:1 or better. For example, wavelet compression can allow a full-length movie to be stored on a conventional five-inch CD disc. Wavelets are mathematical functions that cut up data into different frequency components, and then study each component with a resolution matched to its scale. They have advantages over traditional Fourier methods in analyzing physical situations where the signal contains discontinuities and sharp spikes. Wavelets were developed independently in the fields of mathematics, quantum physics, electrical engineering, and seismic geology. Interchanges between these fields during the last ten years have led to many new wavelet applications such as image compression, turbulence, human vision, radar, and earthquake prediction. See Wavelet Packet.

wavelet packet According to the IEEE, class of time-frequency waveforms with a location (position), a scale (duration), and an oscillation (frequency). Wavelet packets are used for analyzing data with natural oscillations, such as audio signals and fingerprint images. See Wavelet.

wavelet transform A compression technique which involves the representation of a discrete signal or image through wavelet functions. Wavelet transform is computed by the fast pyramid algorithm, which involves a series of linear filtering operations in combination with down-sampling by a factor of two. See Down-Sampling, Fast Pyramid Algorithm and Wavelet.

WAW Waiter-Actor-Webmaster. Used to describe fly-by-night graphic designers and Web consultants trying to cash in on the Web boom. "Can you believe they hired that clueless WAW for $60K a year?!" This definition courtesy Wired Magazine.

way operated circuit A circuit shared by three or more phones on a party line basis. One of the phones usually operates as the control point.

way station One of the phones, other than the central controller, on a way operated circuit. See Way Operated Circuit.

waypoints Reference points used by a GPS system to use on your route.

WBA Wireless Broadcast Access.

WBAN Wireless body area network. Recent technological advances in integrated circuits, wireless communications, and physiological sensing allow miniature, lightweight, ultra-low power, intelligent monitoring devices. A number of these devices can be integrated into a Wireless Body Area Network (WBAN), a new enabling technology for health monitoring. Three researchers at the University of Alabama reported in early 2005 that using off-the-shelf wireless sensors we designed a prototype WBAN which features a standard ZigBee compliant radio and a common set of physiological, kinetic, and environmental sensors. We introduce a multi-tier telemedicine system and describe how we optimized our prototype WBAN implementation for computer-assisted physical rehabilitation applications and ambulatory monitoring. The system performs real-time analysis of sensors' data, provides guidance and feedback to the user, and can generate warnings based on the user's state, level of activity, and environmental conditions. In addition, all recorded information can be transferred to medical servers via the Internet and seamlessly integrated into the user's electronic medical record and research databases." The researchers concluded, "WBANs promise inexpensive, unobtrusive, and unsupervised ambulatory monitoring during normal daily activities for prolonged periods of time. To make this technology ubiquitous and affordable, a number of challenging issues should be resolved, such as system design, configuration and customization, seamless integration, standardization, further utilization of common off-the-shelf components, security and privacy, and social issues."

WBC Wide Band Channel. An FDDI-II term. See FDDI II.

WBEM Web-Based Enterprise Management. A wide-ranging blueprint for unified administration of network, systems and software resources (established my Microsoft, Intel, Compaq, Cisco, BMC Software, and others)....a schema that incorporates three new protocols and four current Internet standards to allow users to manage distributed systems and to access network resources using any Web browser.

WBFH Wide Band Frequency Hopping. WBFH, approved by the FCC in August 2000, is a spread spectrum technique that widens the channel bandwidth to 3 MHz and 5 MHz, as compared to the original 1 MHz. At a rate of roughly 2 Mbps per 1 MHz, WBFH considerably increases bandwidth in HomeRF systems and other wireless systems using Frequency Hopping Spread Spectrum (FHSS). See also FHSS, HomeRF, and Spread Spectrum.

WCAPS Wireless Competitive Access Providers. See Fixed Wireless Local Loop.

WCBC West Coast Billing Center.

WCCP Web Cache Control Protocol. A cache control protocol which runs between a router serving and a transparent proxy cache server. The protocol allows multiple caching proxies to register with a single router to receive intercepted and redirected Web traffic. This process allows the distribution of Web-based information on proxy serves in proximity to the client agent requesting the data, and it does so transparently. When you access a web site, you may be in touch with a proxy server, rather than the origin server. Some Internet purists are really concerned about this, as it violates the basic principles of the Internet, as well as the connection-oriented TCP protocol. See also Cache, Cache Engine, and Proxy.

WCCP V.2 Cisco's communication protocol for cache servers.

WCDMA Wideband CDMA. A high-speed 3G mobile wireless technology officially known as UMTS, (Universal Mobile Telecommunications System), which is also referred to as 3GPP. The technology works by digitizing and transmitting the input signals in a coded, spread spectrum mode over a range of frequencies. WCDMA has the ability to spread its transmissions over a 5MHz carrier. It supports (i.e. it will carry) images, mobile/portable voice, data and video communications at up to 2 Mbps for local area access or 384 Kbps for wide area access. See UMTS.

WCS Wireless Communications Service. Cellular, PCS and the like.

WCV See Weighted Call Value.

WDF Wireless Data Forum. www.wirelessdata.org.

WDM Wavelength Division Multiplexing. A means of increasing the data-carrying capacity of an optical fiber by simultaneously operating at more than one wavelength. WDM is similar to Frequency Division Multiplexing (FDM) in the analog worlds of electrical and radio transmission systems. In optical fiber communications, WDM is any technique by which two or more optical signals having different wavelengths may be simultaneously transmitted in the same direction over one strand of fiber, and then be separated by wavelength at the distant end. Each wavelength is a "virtual channel," which effectively is a separate "light pipe", that can support a given signaling rate, such as OC-48 at 2.4 Gbps or OC-192 at 10 Gbps. Although WDM technology been known since the 1980s, it was restricted to two widely separated "wideband" frequencies. (Note: frequency=speed/wavelength, where the speed of light in glass fiber is essentially constant, so wavelength and frequency are used interchangeably in describing the multiple channels of WDM). The number of distinct wavelengths supported has increased rapidly since WDM became "narrowband" capable in the early 1990s. Initial systems operated at two or four wavelengths, and the term "WDM" is usually used to refer to these low channel-count systems. Beyond WDM is WWDM (Wide WDM), operating at four channels in applications such as 10GbE (10 Gbps Ethernet). Beyond that is DWDM (Dense WDM), which generally is described as beginning at 10 channels. (Note: The terms do not have absolute differentiating definitions and are used somewhat interchangeably.) DWDM systems in commercial applications use at least eight-channel or sixteen-channel multiplexers, and standards-based systems as dense as 32-channels have been released and implemented in live carrier networks. The capacity is steadily increasing, both by ever-expanding channel counts and faster supported TDM rates within the individual wavelengths. Generally speaking, existing systems trade-off channel count against maximum supported rate: the current maximum channel count of 32 is limited to OC-48 (almost 80 Gbps net throughput), with systems supporting higher signal rates (e.g.,

OC-192) supporting fewer than half as many channels. At OC-192, a 32-channel system would yield an incredible 320 Gbps, rounded up. Within the next few years it is reasonable to expect to see systems supporting on the order of 100 wavelengths of OC-192 each, yielding bandwidth of almost 1 Tbps per fiber! The advantage of WDM is that of increasing network capacity without deploying additional fiber. To install enough new actual physical fiber to provides the same bandwidth as the equivalent virtual fiber would be enormously more expensive and often a physical impossibility. WDM also compares very favorably to exclusively upgrading SONET equipment to operate at the higher OC-n layers without WDM, especially since the rate multiplier limit of TDM is hit sooner than the channel multiplier limit of WDM. The ITU-T currently has defined 32 standard wavelengths from which WDM equipment manufacturers can choose (expressed as frequencies, the table is centered at 194.10 THz, with 100 GHz spacing, providing as many as 41 center frequencies from 192.10 to 196.10 THz). ITU-T recommendation G.692 implements an additional table which reduces the spacing to 50 GHz, making 40 additional frequencies available. Further, G.692 makes it plain that the table's end-points are not absolute, and future systems are fully anticipated to include further wavelengths. It is not unreasonable to expect that as the technology continues to advance, the channel spacings may also be reduced, although not as soon as the end-points are expanded. Overall, the contemporary wavelength range typically is between 1530nm and 1560nm, with the minimum and maximum wavelengths being restricted by the wavelength-dependent gain profiles of the optical amplifiers. As is common, the standards organizations often are outpaced by technology advances. While most initial fiber optic traffic involved SONET-formatted signals, WDM and SONET are not necessarily linked. Increasingly, carriers are deploying WDM networks in support of packet traffic (e.g., IP and Ethernet) without employing the SONET digital wrapper. SONET is derided for its overhead intensity, complexity, and cost. Note that WDM and DWDM are physical data transport terms, and not specific to SONET, or any other format standard. Long-haul amplification of WDM signals is achieved with optical EDFAs (Erbium-Doped Fiber Amplifiers), rather than optical repeaters. The EDFAs are spaced up to 100 kilometers apart, simultaneously boosting the intensity of all the multiple light channels through a "pump laser." The EDFAs operate more effectively than do conventional optical repeaters, and are much less costly to acquire and operate. See also DWDM, EDFA, SONET, and WWDM.

WDMA Wavelength Division Multiple Access. A technique which is used to provide access to multiple channels carried on different wavelengths on the same fiber-optic cable. Each optical input operates on a different wavelength of light. The multiple inputs are multiplexed over a common long-haul SONET fiber optic link through WDM (Wavelength Division Multiplexing) equipment. See WDM for much more detailed explanation.

WDME Wavelength Division Multiple Emulator. A WDM emulator is used to enhance testing ability and reduce testing time by simultaneously presenting an array of calibrated wavelengths across the spectral range of interest.

WDU Winchester Drive Unit.

WDT See WatchDog Timer.

weaponize A verb meaning to place weapons of destruction in a new environment. I first saw the term in a New York Times article relating to the weaponizing of space.

weather trunk group A trunk group used to provide customers with weather information.

weathermaster method A distribution method where the unused wall space inside heating and cooling units beneath windows is used for satellite location. Cables are fed from a riser or other serving closet to the location through baseboards, conduit, or underfloor system.

web An abbreviation for the Internet's World Wide Web. See Berners-Lee, World Wide Web and Internet.

Web 2.0 Web 2.0 doesn't formally exist. It's the name given to so-called second generation Internet services that emphasize social networking, online collaboration, wikis, podcast, blogging, content tagging and mash-ups – sites that use information from multiple sites to present unified multimedia information to the user. Web 2.0 is a concept that implies the Internet is evolving from a collection of static pages into a vehicle for delivering software services, especially those that foster self-publishing, participation, business software and project collaboration. Web 2.0 websites look and feel more like applications than web pages. The primary technology behind Web 2.0 is AJAX, which culls data without the need for a page to refresh. Web 2.0 technologies also include XML, VXML and JavaScript, which enable users to create media-rich, interactive websites. For example, Google Maps, a popular Web 2.0 website, lets a user manipulate maps, display landmarks on them, zoom in and out, and interact with maps in other ways.

Web 3.0 Like Web 2.0, Web 3.0 doesn't formally exist either. But it's an increasingly popular term for what some people are calling the Semantic Web. This means that Web

3.0 seeks to give computers the ability to understand content on the World Wide Web, like you and I would understand it, and make conclusions based on the content. much-read article in the New York Times in November 2006 defined Web 3.0 as a set of technologies that offer efficient new ways to help computers organize and draw conclusions from online data. The operative characteristics of Web 3.0, to my mind, are computer intelligence, huge computer server processing capability, gigantic fast storage, massive databases and broad, broadband high-speed access to the web. The speed of access should be indistinguishable from the speed of a high-speed desktop machine. Hence, the Web is no longer an extension of your own computer. In Web 3.0, it becomes an integral part of your computer. Its services, software and data are indistinguishable from those of your machine. **web accelerator** It was only a few years ago that the World Wide Web (WWW) was maligned as the World Wide Wait, for it was painfully slow. In those days, access for most of us was via V.34 modems running at 28.8 kbps – 56 kbps modems hadn't been invented, much less Digital Subscriber Line (DSL) or cable modems. The Internet backbone ran at T1 speeds of 1.544 Mbps – a typical speed rating for contemporary DSL. The Internet in those days was largely used for text e-mail. Photos and graphic-intensive web pages loaded so slowly that a typical web surfing experience was about as much fun as watching your fingernails grow. Well, the technology has changed to support much greater bandwidth at all levels. The applications have developed to take advantage of that fact and users once again complain about slow web site access times. Web accelerators speed up downloads in several ways. They may take the form of compression mechanisms that compress the data at the ISP server side and use decompression software at the user client side to restore that data to its original form. This approach can considerably reduce the bandwidth required. Another approach involves caching either at the ISP server or at geographically distributed web site servers or proxy servers that are closer to the user. These servers cache recently or frequently accessed content, which they may refresh periodically to ensure its timeliness. They also may prefetch content that they determine likely to be requested, based on search history. They also may filter out advertisements and other not critical content. They may compress data and may even maintain persistent TCP connections to the client. Each of these techniques serves to accelerate, or speed up, the user's access to web data across the Internet.

web address Here's a typical address of an Internet Worldwide Web site:

http://www.InSearchOfThePerfectInvestment.com/images/Harryatdesk.jpg

Let's look at what it all means:

http is Hypertext Transfer Protocol." http lets your browser know to expect a web page (as opposed to an FTP or gopher site).

:// is what several writers have referred to as random punctuation abuse. Basically, it's text to separate the next part of the address, namely the "Sub-domain," which is www. World Wide Web servers typically use "www," but you may see other names like "web3" or "w3." The next part of the address is the "High-Level domain," which tells you either the type or location of an organization. Common high level domains: .com = commercial .edu = university .gov = government .uk = England .fi = Finland

/images/ tells the browser to look in a subfolder called images. The final Harryatdesk. jpg is the file we're looking for. In this case, it's actually a photo of me sitting at my desk with a cheesy stupid grin. The photo was taken in early November, 2005 with a Canon PowerShot SD-450. See also FTP and URL.

web advertising Advertising on the Web is expanding rapidly. According to the experts, there are two types of advertising on the Web:

1. Brand advertising, which is intended to generate awareness of and create a specific image for a particular company, product or service; and

2. Direct response advertising, which is intended to generate a specific response or action from the consumer after exposure to an advertisement. Direct response advertising solutions are measured on the short-term benefit from the advertisement and are designed to maximize the number of responses per advertising dollar.

web analytics Analysis of the behavior of website visitors.

web application server Web application servers are special purpose, powerful PCs which sit between a Web server and a corporate database. They offload processing tasks form the repository and cache frequently requested information. They offer much greater growability, since multiple connections can be run to the database – instead of just one, as in the case when the Web server is linked directly. In short, Web application servers speed up getting answers to requests for information out into Web site users' hands.

web art A definition for the artwork that you are beginning to see proliferating the Internet, especially on home pages.

web beacon A way that your activity on the Internet is secretly tracked. I first learned about web beacons in reference to Yahoo! in an article written by Michael Raw of the Daily Press, Newport News. V. Here's how: When you create an account with Yahoo! (like many

people, I've had one for several years), you're asked to click on a button signifying that you agree to Yahoo!'s terms for use of the service. Somewhere within their agreement, it says that unless you specifically tell Yahoo! not to, it'll track and record ALL your Internet activity, not just that associated with Yahoo!. It does this by placing software on your hard drive that logs all the links you visit and then sends the data back to Yahoo!. What then happens to that data is unclear. Yahoo! itself says "Yahoo!! uses web beacons to conduct research on behalf of certain partners on their web sites and also for auditing purposes. Information recorded through these web beacons is used to report aggregate information about Yahoo! users to our partners. This aggregate information may include demographic and usage information." If, like me, you don't like the idea that some faceless company like Yahoo! is collecting information about your private online activities, here's allegedly how to go into the settings for your Yahoo! account and turn off the Internet browsing data-collection feature. Go to http://privacy.yahoo.com/privacy/us/. Scroll down to a section called Cookies. Inside that you'll find hyperlink for web beacons. Go there to the first hyperlink which allegedly allows you to opt out of web beacons. Note: This opt-out applies to a specific browser rather than a specific user. Therefore you will have to opt-out separately from each computer or browser that you use. Yahoo! also says that its "practice is to include web beacons in HTML-formatted email messages (messages that include graphics) that Yahoo!, or its agents, sends in order to determine which email messages were opened and to note whether a message was acted upon. In general, any file served as part of a web page, including an ad banner, can act as a web beacon. Yahoo! may also include web beacons from other companies within pages served by Yahoo! so that Yahoo!'s advertisers may receive auditing, research and reporting."

web browser A Web browser is software which allows a computer user (like you and me) to "surf" the World Wide Web. It lets us select, retrieve and interact with resources on the the Web. It lets us move easily from one World Wide Web site to another. Every time we alight on a Web Page, our Web browser moves a copy of the information on the Web to our computer. See the Internet and HTTP.

web bug A web bug is a kind of digital stalking device used for hunting down the bad guys. I don't know how it works. In a press report on the catching of a bad guy, there were these words, "A computer forensics expert embedded a Web bug, a kind of digital tracking device, in one of the e-mail messages that the good guy sent to the stalker. But the stalker screened his e-mail with decoding devices that included a hex editor, software that allows users to preview the contents of incoming files, and he uncovered the bug." Eventually, the bad guy was caught and put in jail.

Web Cache Control Protocol WCCP. Web Cache Control Protocol. A cache control protocol which runs between a router serving and a transparent proxy cache server. The protocol allows multiple caching proxies to register with a single router to receive intercepted and redirected Web traffic. This process allows the distribution of Web-based information on proxy serves in proximity to the client agent requesting the data, and it does so transparently. When you access a web site, you may be in touch with a proxy server, rather than the origin server. Some Internet purists are really concerned about this, as it violates the basic principles of the Internet, as well as the connection-oriented TCP protocol. See also Cache, Cache Engine, and Proxy.

web caching This concept is analogous to the caching currently used in computer systems. With Web caching, copies of recently accessed Web objects are stored temporarily in locations that are closer to the user, based on the high probability that some number of these objects will be accessed. As a result, the bandwidth and latency associated with fetching the original object again are eliminated. Web caching is a way to proactively store specific content closer to the users, intercepting and responding to requests for that specific content only. See Cache Engine.

web callback Let's say you're on a web site. You have a question. You want to know more than what you can find out on the site. You see a button on the site "Have us call you." That's called a web call back button. When you hit it, it will produce a small form, asking you for your name and phone number. When you fill the form in, the information is transmitted to the company's telephone automatic call distributor which dials the number when it has an agent free and ready to take the phone call and speak to the cusatomer browing the company's web site.

web casting More commonly spelled webcasting. Webcasting is defined as the real time delivery of audio, video or animation over the Internet. I first found the word in Business Week in February, 1997. Business Week described it thusly "Swamped by information on the Web? A new technology finds and delivers news for you. It also helps companies reach workers and conduct business online. Call it broadcasting, Internet-style. Call it webcasting." The basic is simple: You go to a webcasting site. You fill in a form detailing what sort of information you want to hear about. Next time you log on, information is "pushed" to you by your webcasting supplier. You might pay for this service. Or it may be advertiser-based. See also Multicasting and Satellite Webcasting.

web collaboration Imagine you're on a Web site that sells tennis racquets. You're searching for a racquet you saw last weekend. But you can't find it on this site. On the page, you find a box that says, "Ask a question." You click there. A small box opens and words appear "How may I help you?" You type in "I'm looking for a Genesis 660 tennis racquet." The answers comes, "We have several. Here's a picture of the racquet." Eventually you say you want one and you're directed to the site's order page where you can buy it. Web collaboration could be handled at the other end by a real human or software acting lie it is a human.

web crawler One of the more popular search engines on the Web. It indexes World Wide Web pages by title and URL. You can search the Internet with Webcrawler. www.webcrawler.com/cgi-bin/WebQuery

web gateway An interface between some external source of information and a Web server.

web host The computer which has your active web site on it. To host a web site, a host computer must have proper server software, connection capacity for the traffic that comes to the web site and a unique and static internet protocol (IP) address. An IP address looks like 4 sets of numbers separated by periods, i.e. "201.11.123.1" A uniform resource locator, or URL, is a unique name that has been assigned to a static IP of a specific host computer making it easier to find a web site.

web hosting A service performed by Internet Service Providers (ISPs) and Internet Access Providers (IAPs) who encourage outside companies to put their Web sites on computers owned by the ISPs. These computers are attached to communications links to the Internet – often high-speed links. For this Web hosting service, the ISPs typically charge their clients by equipment and transmission capacity used. See also Server Colocation and Web Host.

web log A web log is an online journal, usually of a personal nature, that can be updated regularly and is usually displayed in chronological order.

web master See Webmaster.

web mistress See Webmistress.

web page An HTML document on the Web, usually one of many that together make up a Web site. See World Wide Web.

web portal Same as portal site. See Portal Site.

Web Proxy Auto-Discovery Protocol See WPAD.

web search engine See search engine.

web server A web server is a powerful computer which is connected to the Internet or an Intranet. It stores documents and files – audio, video, graphics or text – and can display them to people accessing the server via hypertext transfer protocol (http). A Web server derives its name because it is part of the World Wide Web. See World Wide Web.

web service provider A vendor who provides customers with Web Pages on the vendor's computer/s. Frequently, a Web Service Provider will provide additional services such as design help and usage statistics. Often they will just provide the computer space and leave the rest to you. A Web Service Provider may or may not also be an Internet Service Provider.

web services In the networking world, the word service is often used to describe a function or set of functions provided by a computer or network device to other devices located on the same network. Services use a typical client/server relationship between network devices and can be as simple as a web server providing a web browser access to a web page, or as complicated as a virtual private network appliance providing a user with access to a private network through several layers of security protocols, encryption, and firewalls. A web service is exactly like a standard network service, except the functions are provided over the Internet. This presents different challenges from those on a private network. In a private network, the servers and clients can be tailored to use any protocol or network architecture to eliminate external problems such as network latency, low bandwidth, and network outages. In contrast, the web is a heterogeneous environment made up of a wide variety of network architectures, where almost every type of Internet device in existence is trying to try to send and try to receive data. For a web service to be useful, it must be able to talk to as many disparate systems as possible. It must be efficient enough to handle multiple requests at once without getting bogged down, and it must be able to provide nearly any service desired by a client. There are many different ways to implement a web service depending on what kind of information you want to send and receive and what structure you want to use to build the service. Some current applications include workflow automation, network management, instant messaging, client access to their accounts, updating settings on network devices (so a network manager only has to

perform one action to update all systems in the network), remote learning, chat rooms, and electronic whiteboards. To support a diverse set of users across the Internet, it is important to use a technology that is independent of a computing platform. Such a technology should be compact using little bandwidth and very portable so it can be used on many devices. There are several very powerful languages and protocols that are used between disparate systems located on the Internet, such as Extensible Markup Language (XML), Simple Object Access Protocol (SOAP) and Java. In short, web services is a mechanism for computers to talk to computers using standard Internet-based protocols, especially XML. Whereas the Internet is about how people talk to computers, XML-based web services makes programmatic interactions between computer systems much faster, cheaper and easier than previous distributed computing approaches. Web services means software on demand. Web services means information you want that's assembled specially for you. Examples of web services: An airline can link its online reservation system to that of a car-rental partner, so travelers can book a car at the same time they book a flight. An online auction company can notify bidders when they are outbid or have won an auction, or could partner with other firms to offer alternative shipping, fulfillment or payment options. As companies extend their business processes across the Internet to include customers, partners, suppliers and other constituents, the Internet must provide a new infrastructure that lets applications communicate and interoperate. Web services is being positioned as that infrastructure. See also .NET, Internet, SOAP, UDDI, XML and WSDL.

web site Any machine on the Internet that is running a Web Server to respond to requests from remote Web Browsers is a Web Site. In more common usage it refers to individual sets of Web Pages that can be visited with Web Browsers. he front page of a Web site is called its home page. It is also spelled as one word, namely website. See also Internet.

web site hosting See Hosting.

web switches Web switches are networking devices that provide high-speed switching of traffic, using information in the TCP and HTTP request header to make policy and routing decisions based on the actual content (e.g., URL) being requested. These are products that were built to provide a front end for Web server farms and cache clusters to dynamically direct specific content requests to the best site and best server at that moment. Web switches have the ability to look deep into a URL address to differentiate content requests and make additional decisions, such as determining if the request needs to go through a cache and determining the amount of bandwidth to be set aside for intense customer requests such as streaming video. Customers for these products include portals, content providers, online destinations, Web hosters, and Internet service providers.

web TV See WebTV.

webbed conductors The manufacturing process that physically binds the conductor insulation of the wire pairs of an unshielded twisted-pair cable.

webcam A webcam is a digital video camera. Train the camera on a bridge, for example. Hook the camera up to a phone line and a Web site. Now people from all over the world can visit your web site via the Internet and check out the view from your webcam. Your webcam may transmit photos every second, every minute or every day. It's your choice of how much you want to spend and how exciting the action is. People are putting up Webcams to show the view on popular sites, like stadium construction sites. Many hope to earn money by selling ads on the sites. Some do.

webcam girl A young woman who broadcasts live pictures of herself over the World Wide Web. Also called cam-girl, camgirl or cam girl.

webcasting Webcasting is the real time delivery of audio, video or animation over the Internet. In short, it's material moving from a distant web site over the Internet to your PC. If you have a high-speed line, the Internet is an ideal way to listen to distant radio stations that contain content which you can't get on your local radio stations. For example, my favorite radio station is Klassik Radio which is broadcast from Hamburg, Germany. To get to it (and a whole bunch of other radio stations) go to http://windowsmedia.com/radiotuner/default.asp. The reason I love Klassik Radio is that it plays great music without commercials. Every so often it has a few moments of news in German. Then it's back to the music.

Webinar A seminar that takes place over the web. Often a vendor might make a PowerPoint presentation, with everybody watching the PowerPoint slides on their computer and listening to the presentation over their computer's speakers. When the time comes for questions, they will often call on their phones into a special phone number, which in turn will be patched to the Internet, so everyone on the Webinar can hear the questions and answers. Webinars are much cheaper to run and much more convenient than having to schlepp to some distant city to hear a boring speaker.

Webisode A short entertainment program, usually fewer than 10 minutes, delivered on the Web. Webisodes are sitcoms for people with really short attention spans.

weblock The Internet data traffic version of gridlock.

weblog See Blog.

webmaster An Internet term. The Webmaster is the administrator responsible for the management and often design of a company's World Wide Web site. See also Webmistress.

webmistress An Internet term. The Webmistress is the female administrator responsible for the management and often design of a company's World Wide Web site. See also Webmaster.

webphony A term created by Nick Morley, an excellent salesperson who works for Computer Telephony Magazine. Webphony is a combination of the two words Web and Telephony. It means telephony-enabling your Web site. Here's a simple example: You're checking out a Web, say L. L. Bean, the direct mail catalog company. You'd like to buy a new kayak. You need to ask a question. Click on the "Reach an Operator" button in the corner of your screen. Bingo, you're speaking live to an L.L. Bean operator via your computer's sound system, or perhaps they called you on a second phone line – one for data surfing and one for voice. Or your ISDN 2B channel just got split into one for voice and one for data. There are a thousand variations on this theme of adding voice to the Internet and the Internet's World Wide Web segment. We're just beginning to explore them all.

website Some magazines are now calling web site as one word, namely website. Some call it two words. The movement to one word is a natural transition in the English language – from one word, to a hyphenated two words and then eventually two words combined into one. See Web Site.

webtone A theoretical idea pushed heavily by Nortel and others. Webtone is used to denote the idea of immediate and continuous access to the Internet. similar to a dialtorie heard when picking up a phone receiver. To have Webtone in the same way that we have dialtone, most users of the term believe that not only access is required, but also sufficient bandwidth is necessary to meet user demands as well as the same quality of service we expect currently from the telephone system. Since the telephone system and the Internet are tending to converge, some believe that eventually Webtone will include dialtone. Webtone also implies Internet access from mobile devices. supercomputers, and kitchen appliances.

webTV Well, it's finally here! Sony and Philips (Fall, 1996) have struck an alliance and licensed WebTV technology consisting of a set-top box, a wireless keyboard and a printer adapter. The set-top box costs $329 and contains a 112-MHz, 64-bit CPU. The TV set serves as the display. Internet access is provided through dial-up connection over the dial-up phone network to the WebTV Network at a monthly cost of $20 for unlimited usage. Content is provided by the WebTV Network, which also supports e-mail, help, and content lockout. Mitsubishi has taken a different approach, embedding the set-top box within its DiamondWeb TV. Sanyo and Samsung appear to be taking the same approach as Mitsubishi.

This approach to 'Net TV is a challenge to another concept of providing access through a cable modem which would be embedded in a set-top box. That box would serve as a communications controller/splitter, supporting simultaneous voice, data and TV. Regardless of the approach taken, issues abound, including lack of standards for the set-top boxes and cable modems.

webware A term coined by Ray Horak, my Consulting Editor, for misleading content on a Web site. Great graphics, some animation, and neat audio can cause you to believe that the product or service matches the description on the Web site. Basically, it's interactive brochureware. I can cite examples – I'll bet you can, too. See Brochureware.

webwench An employee given all the responsibility for a Web site without any of the authority (the opposite of a Webmaster).

webzine Magazines that are published (i.e. made public) on the World Wide Web. Typically a webzine is available for anyone to read who wants to visit the site the electronic magazine is located at. A Webzine is also called an e-zine or a Web-zine. Some webzines were published with free access for all – based on getting money from advertising. Some webzines were published as "subscription only" sites, i.e. you paid money. Few have survived.

WEC Western Electric Company, now called Lucent Technologies. See WECO.

WECA Wireless Ethernet Compatibility Alliance. WECA's stated mission is to "certify interoperability of Wi-Fi (i.e. IEEE 802.11) products and to promote Wi-Fi as the global wireless LAN standard across all market segments." www.wirelessethernet.com. See also 802.11, 802.11a and 802.11b.

WECO Western Electric COmpany. The company is now called Lucent Technologies. It used to be the equipment manufacturing arm of AT&T, but in early 1996, AT&T spun it off to the public as a separate, publicly-traded company, called Lucent Technologies. Many oldtimers are sad that about the name change and new ownership. Western Electric had a wonderful reputation and is remembered with great fondness. It has an excellent reputation

for high quality products and is still used as a brand name on some products.

wedding bouquet In England in the 1500s, most people only bathed once a year – in May. By the time they came to get married in June, the bride stunk. So brides carried a bouquet of flowers to hide their body odor. Hence the custom today of carrying a bouquet when getting married. See Bathwater.

Wei? What Chinese say when they answer the phone. It corresponds to "Hello?" or "Yes?" in English.

weight test The test that buyers once applied to proposals, as well as purchases. According to the weight test, the weight of the proposal was directly related to the level of effort that went into it, which was directly related to the size of the company that prepared it and/or to the level of interest that the proposer had in acquiring your business (i.e. making the sale).

weighted average A call center term. A method of averaging several numbers in which some numbers are increased before averaging because they have more significance relative to the other numbers.

Weighted Call Value WCV. The average handling time of a call transaction. ACD vendors count this differently. Typically, a combination of the talk time and the after-call work or wrap-up time.

Weighted Random Early Detection WRED. A congestion avoidance mechanism used in TCP/IP networks. WRED improves on RED by dropping packets on a selective basis, based on IP Precedence markings in the TCP packet header. Thereby, packets of a higher priority can be recognized by their IP Precedence values (1-7), and be handled with a higher probability of their reaching the destination device. WRED generally is implemented in TCP/IP routers in the network core, while RED generally is implemented in routers at the network edge. See also Random Early Detection.

WEP Wired Equivalent Privacy (WEP) is an optional IEEE 802.11b (aka Wi-Fi, or Wireless Fidelity) feature designed to offer privacy equivalent to that of a wired LAN. According to the standards, WEP uses the RC4 encryption algorithm with a 40-bit key, although a 128-bit key also may be used. When WEP is enabled, the network administrator assigns each wireless station (e.g., a laptop, client and access point) an encryption key string comprising a set of keys that are passed through the encryption algorithm. The key is used to scramble the data before it's transmitted over the airwaves between a mobile client or server and an access point. The secret key is used to encrypt packets before they are transmitted. The receiving access point performs an integrity check to ensure that the packets have not been modified in transit. If a station receives a packet that is not scrambled with the appropriate key, the packet will be discarded and never delivered to the host, thus ensuring that spurious material doesn't enter or leave the wireless network. WEP has proved too easy to hack and, therefore, has been replaced by WPA (Wi-Fi Protected Access). See also 802.11, Encryption, Wi-Fi and WPA.

Western Electric The telecommunications manufacturing subsidiary of AT&T, Western Electric was divested from AT&T on September 30, 1996, and renamed called Lucent Technologies. Lucent since has spun off several entities, including Avaya, which manufactures PBXs. See also AT&T, Avaya, Lucent Technologies and McGowan.

Western Union Western Union started life in 1851 as a telegraph company – sending and receiving messages for it customers. It made the dumbest decision in the history of telecom – namely not buying the patents on the telephone from Alexander Graham Bell. It did a number of "firsts" which it never followed up on, a result of some awful management. These days it sends money around the world. Here's a little of Western Union's checkered history showing how it let wonderful opportunities pass it by.

1851 A group of businessmen in Rochester, New York form The New York and Mississippi Valley Printing Telegraph Company, Western Union's predecessor company.

1856 The New York and Mississippi Valley Printing Telegraph Company changes its name to The Western Union Telegraph Company, signifying the union of "western" telegraph lines with eastern lines into one system, following acquisition of a series of competing telegraph systems.

1861 Western Union completes the first transcontinental telegraph line, providing coast-to-coast communications during the U.S. Civil War. This put the Pony Express out of business, after only 18 months in business.

1866 Western Union introduces the first stock ticker, providing brokerage firms with New York Stock Exchange quotations.

1870 Western Union launches a time service, helping to standardize time nationally. Western Union will hold the distinction as "The Nation's Timekeeper" for nearly a century.

1871 Western Union Money Transfer service was introduced and became the company's primary business.

1884 Western Union is selected as one of the original 11 stocks tracked in the first Dow Jones Average.

1914 Western Union introduces the first consumer charge card.

1923 Western Union introduces teletypewriters, joining branches and individual companies.

1933 Singing telegrams are introduced.

1935 The first inter-city facsimile service is introduced.

1943 Pioneered the first commercial inter-city microwave system.

1958 Western Union introduces Telex, a direct-dial consumer to consumer teleprinter service.

1964 Western Union inaugurates the use of a transcontinental microwave radio beam system, replacing poles and wires spanning the continent.

1970 Western Union Mailgram messages offer next-day delivery via postal service.

1974 Western Union launches Westar I, the first domestic communications satellite for America.

1982 Western Union is the first company with five satellites in orbit.

1989 Quick Collect provides creditors a service for securing fast collection of payments via flat-rate Money Transfers.

Rapid Money Transfer service becomes available outside North America.

1993 Dinero en Minutos (Money in Minutes) service is introduced, making funds sent to Mexico from the U.S. available in just minutes.

The Western Union Phone Card service is the first branded, pre-paid and disposable telephone card offered in the U.S.

1998 The Western Union Money Transfer service expands to reach 50,000 Agent locations worldwide, the world's largest money transfer network. International Regional Operating Centers open in Brussels and Costa Rica.

1999 By the end of the year, there are more than 80,000 Western Union Agent locations in over 140 countries and territories around the globe.

2000 Western Union launches westernunion.com, bringing money transfer to the Internet.

2001 Western Union celebrates its 150-year anniversary by reaching more than 100,000 Agent locations worldwide.

Western Union International WUI. Acquired by MCI in 1982 to establish MCI in the International Telex and communications market. WUI is now part of MCI International, which became part of WorldCom which went bankrupt which then emerged from bankruptcy and changed its name to MCI and then got bought by Verizon.

WestNet One of the National Science Foundation funded regional TCP/IP networks that covers the states of Arizona, Colorado, New Mexico, Utah, and Wyoming.

wet circuit A circuit carrying direct current.

wet high Also called a permanent seizure. When a phone user takes the phone off the hook for a long period of time, e.g. half a hour, the central switch simply puts phone out of order.

wet loop powering Defined as local power (non-Span provided) with use of copper pairs (power is looped at the last repeater).

wet T-1 A T-1 line with a telephony company powered interface.

wet your whistle Many years ago in England, pub frequenters had a whistle baked into the rim or handle of their ceramic cups. When they needed a refill, they used the whistle to get some service. "Wet your whistle," is the phrase inspired by this practice.

wetting agent A chemical which reduces surface tension in a liquid, motivating the liquid to spread more evenly on a surface.

wetting voltage Wetting voltage is the voltage that is present in a local circuit waiting for a contact closure to pick up a relay. It is typically DC voltage not AC.

wetware exploits 1. Human weaknesses that can be taken advantage of to hack or sabotage computer security systems. As in, "No system is immune to wetware exploits".

2. Also a type of email-borne virus in which the user himself causes the destruction. A warning message tells the user to search out and remove a virus file. The file is usually a vital part of the Windows operating system. So its removal causes system damage.

WFC Winchester Floppy Controller.

WFM Wired For Management. Intel's umbrella term for a set of management standards supported by hardware vendors. WFM communicates with network management software to help PCs send inventory data, manage power and reboot remotely. According to InfoWorld, a standard WFM-compliant PC will include a sensor to detect intruders; DMI 2.0 in the firmware; Wake on LAN 2.0 in the Ethernet chip or an Ethernet network interface card; ACPI (Advanced Configuration and Power Interface) and on the system board with software hooks to the OS, hard disk modem and monitor; and PXE (Preboot Execution Environment)

on the Ethernet chip or the NIC (Network Interface Card). The idea of WFW is to give administrators more flexibility to handle clients (i.e. PCs) remotely through a central console with protocols that are vendor independent. See also WFO.

WFO Workforce optimization is the combination of workforce management, call/interaction recording, quality management, customer surveying, coaching and e-learning, and contact center performance management/analysis. The idea is to get more from these processes, and from the underlying software, by integrating them. For example, say you're contact center gets a lot of calls from Spanish-speaking customers. You'd use quality metrics from call recording, QM, and IVR surveys to identify training opportunities for underperforming multi-lingual agents, to create e-learning modules that copy practices used by your best agents and to schedule Spanish-speaking agents and supervisors for times with high expected calls from Spanish-speaking customers.

WFQ Weighted Fair Queuing. A variation on the CBQ (Class-Based Queuing) queuing technique used in routers. As is the case with CBQ, WFQ queues traffic in separate queues, according to traffic class definition, guaranteeing each queue some portion of the total available bandwidth. As is also the case with CBQ, WFQ recognizes when a particular queue is not fully utilizing its allocated bandwidth and portions that capacity out to the other queues on a proportionate basis. WFQ takes queuing to yet another level, portioning out available bandwidth on the basis of individual information flows according to their message parameters. See also CBQ, FIFO, RED, and Router.

WFWG Windows For Workgroups. See Windows.

WG 1. An abbreviation for workstation, i.e., a computer on a desktop that isn't a server.

2. Working Group. A term used by the IEEE (Institute of Electrical and Electronics Engineers) and the IETF (Internet Engineering Task Force) for a formal group of people that work on standards recommendations. The IEEE sometimes breaks the work into tasks, which are assigned to Task Groups (TGs).

WGS-84 Worldwide Geodetic System 1984. A system of coordinates used by GPS (Global Positioning System).

WH Western Host.

Whack See Wack.

wheatstone bridge An instrument for measuring resistances.

whetstones How well does a computer work? Let's test it. The Whetstone benchmark program, developed in 1976, was designed to simulate arithmetic intensive programs used in scientific computing. It is applicable in CAD and other engineering areas where floating-point and trigonometric calculations are heavily used. The Whetstone program is completely CPU-bound and performs no I/O or system calls. The speed at which a system performs floating point operations is measured in units of Whetstones per second or floating point operations per second (flops). Whetstone I tests 32-bit, and Whetstone II tests 64-bit operations. See also Dhrystones.

While Running Backwards You'll Vomit See WRBYV.

whip cable assembly See bundled cable.

whip cable construction See bundled cable.

whipsawing Whipsawing is an abuse of its dominant position by an incumbent telephone company operator in an unliberalised country. Here the incumbent uses its market power to insist on receiving a greater amount under the accounting rate system for receiving and terminating calls from an operator in a liberalised country than it pays to that operator for the same service in the other direction. This is harmful to consumers in the liberalised country because they pay more than they should do for calls. See also "One Way Bypass".

whisper coach You're the supervisor at a call center. You need to train your agents on how to deal correctly with the customers calling in on the phone. So you listen in on several phone calls. When you hear something you don't like or when you want to help the agent with some information, you whisper. Only the agent hears your whisper. The customer doesn't hear anything. This is called whisper coaching.

whisper page Someone is on a call. You (most likely the operator) need to tell them someone important (like their boss) is calling. You call the person, while he's on the phone. But only he hears what you say to him. The person whom he is speaking to doesn't. Some PBXs and key systems have this technology. It's useful.

whisper technology A call comes into a call center. The voice response unit prompts the caller to the enter their account number. When the call is transferred to the agent, the VRU "whispers" the account number to the agent, who then manually types it into his computer. This technology is now obsolete, since VRUs can now transfer their account number directly into the agent's database and have the look up done automatically. And the call is transferred simultaneously.

whistle through A feature that allows CAMA (Centralized Automatic Message

Accounting) identification from a PBX trunk to be extended through a local central Office (public exchange) connection to an outgoing trunk to a tandem office. See also CAMA.

white board Also called a mushroom board or peg board. This is placed between termination blocks to support route crossing wire.

white box A device which is built of standard components, often customized to the requirements of the end user. The device is packaged in a white box, rather than in a fancy box with the name and logo of a manufacturer printed all over it. White boxes (i.e. custom-built PCs from local vendors) typically cost less than do those of name-brand manufacturers, although the warranty is only as good as is the local system integrator/retailer. See also Black Box and White Box Builder.

white box builder Companies of any size who build no-name or off-brand personal computers and servers to compete with the major builders of PCs – Compaq, Dell, Gateway, HP and IBM. According to John Dvorak, writing in PC Magazine, there are over 40,000 makers of PC white boxes. That's probably an exaggeration. Suffice, anyone can "build" PCs by simply buying components from all manor of manufacturers and assembling the components into full-fledged PCs, gaming machines and servers of all ilk. See White Box.

white box testing Network security testing where the evaluators conducting the tests are provided with detailed information in advance about the client's organization and network infrastructure, including, for example, network diagrams, other network documentation, and organizational information.   White box testing may be used to simulate an attack from inside the company or an external attack by ex-employees with a knowledge of the company's network and attached systems. Evaluators also test the validity and completeness of the information initially provided by the organization. See also black box testing.

white chips In poker, the blue chips are worth a lot and are thus worth having. That's why high priced shares are often called blue chips. The white chips, on the other hand, are worth very little. Most professional gamblers leave the white chips as tips.

white collar worker The term originated in the fact that, in years gone by, such folks wore white shirts with white collars as evidence of the fact that they didn't do the kind of "dirty" work that would soil a shirt. Blue collar workers, on the other hand, wore blue shirts with blue collars that didn't show dirt quite so readily. A white collar was a real status symbol, like a European luxury car. People dress much less formally these days.

white facsimile transmission In an amplitude-modulated facsimile system, that form of transmission in which the maximum transmitted power corresponds to the minimum density of the subject copy. In a frequency-modulated system, that form of transmission in which the lowest transmitted frequency corresponds to the minimum density of the subject copy.

white hats These guys are the good hackers. They don't ever cause damage to a system and they don't steal information, but instead they contact the administrator/webmaster and tell them what they did and how they did it so that steps can be taken to fix any networking or site vulnerabilities.

white label A telecom service that is provisioned by one network operator and resold and billed by another network operator under the latter's brand.

white line skip A facsimile transmission technique used to speed up the transmission time by bypassing redundant areas such as white space. (Also known as skip scan.)

white list A list of items that are meant to be spam and virus proof – hence white list. To place a name, e-mail address, Web site address, or program on a list of items that are deemed spam- or virus-free. The verb is to whitelist, namely to do the act of placing the item on the list. A white list is clearly the opposite of a black list, which is a list of people or things that are deemed unsafe or undesirable. The verb form (meaning to place on a black list) first showed up in the early 1700s. The verb to black ball comes from voting practices, most common in private clubs. A wooden box would contain two compartments. One containing many balls. These are the balls which have not been voted by the members. The other, typically a covered compartment. contains the balls voted by the members. Members vote anonymously by transferring a ball from one chamber to the other. If they approve of the proposed new member, they transfer a white ball. If they don't, they transfer a black. It only requires one black ball to be transferred and the proposed member is denied admittance. Under this method, no one ever knows who the person is who voted against admitting the applicant. See also black list and gray list.

white noise Undesirable electrical energy introduced in a communications system, principally generated by the random motion of electrons in conductors and semiconductors. The name comes from the first observation of it in the mid 1920s. Oscilloscopes of the time, which had white traces, displayed the noise as a broad band of white across the screen. In communications systems, the term is imprecise, because white noise theoretically has

an infinite bandwidth (and hence infinite power), whilst all communications systems have definite limits to their bandwidths. White noise is also called Johnson noise, in honor of the Bell Labs physicist who first observed it.

White noise results from the random motion of electrons in electronic materials as an electromagnetic signal passes through a conductor. For example, electricity passing through a copper conductor causes the electrons in the copper to vibrate. In fact, the conduction of the electrical signal really is due to that vibration. Copper often is the preferred medium, as it is distinguished by a large number of free electrons. As vibration of the electrons causes the transfer of some electromagnetic energy to heat, or thermal energy, "white noise" sometimes is called "thermal noise." White noise is uniformly distributed among all frequencies within a frequency band of interest. Seldom occurring in nature, white noise is a useful tool for theoretical research. White noise is also used less scientifically to simply mean background noise. When the first digital PBXs came out, their intercom circuits were so "clean," they spooked users out who were used to some hissing noise on the line. And some PBX manufacturers added a little "white noise" to their PBXs. White noise is also known as "average white Gaussian noise," after Karl Friedrich Gauss (1777-1855), the German mathematician who is generally recognized as the father of the mathematical theory of electricity.

white pages 1. In many countries, including the U.S., Canada and Australia, the phone company publishes two types of telephone directories. One called the "White Pages" lists all the subscribers in alphabetical order. The other, called the Yellow Pages, lists businesses by industry. On the Internet, the White Pages are the lists of Internet users that are accessible through the Internet.

white paper Imagine an 8 1/2" x 11" stapled small booklet of 16 to 32 pages. Imagine that the paper is the quality of plain-paper photocopier paper, i.e. non-glossy and weighing in at 20lb to 24lb. Imagine that the paper is printed plainly in black ink, looks "honest" (i.e. not slick and glossy) and discusses technical issues, or contains case studies of user installations. Bingo, now you have something called a White Paper. Companies write them and distribute them to prospective customers in the hopes that the "knowledge" the White Papers contain will turn the customers more favorably to the company's products and ideas. Companies issue White Papers as part of their marketing and sales programs. But it's the "subtle" part of their programs. Done correctly, White Papers can actually be useful, explaining how complex things work and what benefits they deliver. Done incorrectly, they look like bad marketing and turn potential customers off.

white peak The maximum excursion of the video signal in the white direction at the time of observation.

white signal In facsimile, the signal resulting from the scanning of a minimum-density area of the subject copy.

white space Unused, unassigned bands of spectrum between assigned bands of the electromagnetic frequency.

whiteboard A device which lets you share images, text and data simultaneously as you speak on the phone with someone else. That someone might be in the next office. Or that someone might be 3,000 miles away. The transport mechanism might be a local area network or an analog phone line running a special modem designed for whiteboarding or it might be an ISDN digital line running special PC software and hardware. The concept of whiteboarding is new; there are no standards. As a result to do whiteboarding successfully, you typically need the same equipment (hardware and software) on either end. Whiteboarding has the potential to be one of the most successful "multimedia" applications around. Whiteboarding is a document-conferencing function that lets multiple users simultaneously view and annotate a document with pens, highlighters, and drawing tools. More advanced whiteboard programs handle multi-page documents and provide tools for delivering them as presentations.

Who-Are-You Code WRU. A control character which operates the answerback unit in a terminal (typically a telex terminal) for identification of sending and receiving stations in a network.

whodunit None of the kids who live in your house.

whois 1. A command on some systems that reveals the user's name, based on that person's network username.

2. Whois is a way of looking up names in a remote database. Used initially as an aid for finding e-mail addresses for people at large institutions or companies. It is now a tool of the InterNIC DNS (Domain Name Server). Whois allows anyone to query a database of people and other Internet entities, such as domains, network, and hosts. The data includes company/individual name, address, phone number and electronic mail address. If you have your own domain and thought your personal information was hidden from view, you may be in for a shock. Don't believe me, check out www.internic.net. See also InterNIC and

RWhois.

whole nine yards The term "the whole 9 yards" came from World War fighter pilots in the Pacific. When arming their airplanes on the ground, the .50 caliber machine gun ammo belts measured exactly 27 feet, before being loaded into the fuselage. If the pilots fired all their ammo at a target, it got "the whole 9 yards."

whole person paradigm This is one of the more fascinating telecom concepts in a while. General Magic created it as some sort of psychological basis for the product/s it is producing. Here's General Magic's definition:

A psychological or behavior model of needs that all people experience. This paradigm is the design center for General Magic's personal intelligent communication products and services. It consists of three elements. 1. Remember - managing your internal agenda, such as things to do and people to see. 2. Communicate - maintaining relationships with your friends, family, and associates. 3. Know - getting information about the world.

wholesale access lines End-user access lines owned by the ILEC but served by another local carrier. Total service resale, UNE-P, and UNE-L are examples of wholesale access lines.

Wi-Fi Wireless Fidelity, A WLAN (Wireless Local Area Network) specified by the IEEE as 802.11b. Wi-Fi runs in the 2.4 GHz wireless range at speeds of up to 11 Mbps. See also 802.11a, 802.11b and Wi-Fi5.

Wi-Fi 5 A new version of Wi-Fi that's even faster, with a maximum speed of 54 megabits per second. Also called 802.11a. See also Wi-Fi.

Wi-Fi Camera A digital camera that can transmit its pictures over a Wi-Fi link.

Wi-Fi Protected Access. See WPA.

Wi-Pie In March, 2006, the Economist ran an article "Wi-Pie in the sky?" which it talked about the fact that cities across America plan to build municipal Wi-Fi networks to widen access to broadband. And the Ecoomist asked, "Will they work?" The answer was inconclusive. But the headline grabbed me and that's why it's in this dictionary. See also mesh networking.

WIA Wireless Institute of Australia. The peak body representing the amateur radiocommunications community in Australia. It is the organization to which the Australian Communications Authority (ACA) has delegated responsibility for conducting examinations for amateur radio operators.

WIC WAN Interface Card.

Wicking The longitudinal flow of a liquid in a wire or cable due to capillary action.

WID Wireless Integration/Interface Device. Also referred to as a "Proctor" box (name of the vendor), Cell Trace Box (US West's name for it), and protocol converter. One of its functions is helping convert older cellular phone systems which support only old-style 911 service (i.e. no location transmitting) to the newer E911 service which will transmit the cellphone user's location to the correct public safety people. www.proctorinc.com

Wide Area Augmentation System See WAAS.

Wide Area Network See WAN.

Wide Area Service Identifier WASI. Unique identifier for a business grouping of licensed facilities-based cellular service providers of Cellular Digital Packet Data (CDPD). It is used within CDPD for access control decisions.

Wide Area Telecommunications Service WATS. WATS permits customers to make (OUTWATS) or receive (INWATS) long-distance calls and to have them billed on a bulk rather than individual call basis. The service is provided within selected service areas, or bands, originally by means of dedicated WATS Access Lines (WAL) directly connected to the public telephone network at WATS-billing equipped central offices. The dedicated access line operation permits inward or outward service, but not both. Recent evolution permits WATS connection via regular user local PSTN dial lines.

wide area telephone service 1. A service provided by telephone companies in the United States that permits a customer to make calls to or from telephones in specific zones, with a discounted monthly charge based upon call volume.

2. See WATS and 800 Service.

wide band See Wideband.

Wide Band Frequency Hopping See WBFH.

wide center The local company's serving central office for a customer or an inter-exchange carrier.

wide characters 16-bit characters. See Unicode.

wide frequency tolerant power plant PBX power facilities are provided that will operate from AC energy sources which are not as closely regulated as commercial AC power. The wide tolerant plant will tolerate average frequency deviations of up to plus or minus 3 Hz or voltage variations of -15% to +10% as long as both of the conditions do not occur simultaneously. This feature permits operation with customer provided emergency power generating equipment.

wide SCSI A type of SCSI that uses a 16- or 32- bit bus. It can transmit twice as much information as narrow SCSI.

Wide avelength Division Multiplexing See WWDM.

wideband The original definition for a channel wider in bandwidth than a voice-grade channel. Then it became a channel wider than 12 voice channels. Now, it means a transmission facility providing capacity greater than narrowband (T-1 at 1.544 Mbps), e.g. T-3 at 45 Mbps. many rich folks in Silicon Valley now have T-1 circuits into their home. This makes surfing the Internet and accessing the Web more pleasurable. But George Lucas, the renowned filmmaker, has a T-3 in his house. He clearly is wideband. See also Bandwidth. Contrast with Narrowband and Broadband.

wideband modem A modem whose modulated output signal can have an essential frequency spectrum that is broader than that which can be wholly contained within a voice channel with a nominal 4-kHz bandwidth. A modem whose bandwidth capability is greater than that of a narrow band modem.

wideband packet transport Transmission of addressed, digitized message fragments (packets) interleaved among the addressed fragments of other messages at a rate high enough to support general purpose telecommunications services.

wideband switch Switch capable of handling channels wider in bandwidth than voice-grade lines. Radio and TV switches are examples of wideband switches.

Wi-Fi his is another name for IEEE 802.11. It is a term coined by the Wireless Ethernet Compatibility Alliance (WECA). Products certified as Wi-Fi by WECA are interoperable with each other even if they are from different manufacturers. A user with a Wi-Fi product can use any brand of Wi-Fi access point with any other brand of client hardware (e.g. a laptop PC wireless Ethernet card) that is built to the Wi-Fi standard.

Wi-Fi switching Wi-Fi switching is a new architecture for wireless local area networks that combines gigabit Ethernet switching, Wi-Fi and smart antenna design. Wi-Fi switches send and receive multiple transmissions simultaneously and extend the range of Wi-Fi from meters to kilometers.

Wi-Fi5 A Wi-Fi version specified by the IEEE as 802.11a. Wi-Fi5 runs in the 5 MHz range at speeds up to 54 Mbps. See also 802.11a and Wi-Fi.

Wi-Fi-x A generic name for 802.11a and 802.11b. When you see Wi-Fi-x chipset, you know it means a chipset that will handle both the slower and faster Wi-Fi connections. See also 802.11a and 802.11b.

Wi-Fi5 Wireless Fidelity 5, much like the term Hi-Fi (High Fidelity) is used to describe audio equipment. Wi-Fi5 is another name for a wireless locla area network (LAN) running under the 802.11a standard in the 5-GHz range. See also 802.11 and Wi-Fi.

WiFi See Wi-Fi, which is the preferred and most common spelling.

WiFly The charming name of a WiFi company in Taiwan.

Wig Queen ELizabeth I of England was completely bald. She lost her hair after suffering from smallpox at the age of 29. To disguise her loss she always wore a wig, thus creating a vogue for wigs in Europe that lasted several hundred years and giving new meaning to that incredibly bad pun about hair today, gone tomorrow.

wigwag A visual signaling system developed by Albert James Myer, a medical officer in the US Army, while he served on frontier duty in Texas in 1856. Wigwag used flags for daytime signaling and torches for nighttime signaling. The Army officially adopted Myer's wigwag system in 1860.

wiki A Wiki, in its simplest form, is a Web site that can be written on and edited by multiple users at once. That gives it pluses and minuses. On the positive side it benefits from the wisdow of many. On the negative side it suffers from the misdeeds of many – bad, unverified information, deliberately wrong information, pranks, etc. The most famous wiki is Wikipedia. To visit it, go here: http://en.wikipedia.org/wiki/Main_Page.. When visiting Wikipedia, the standard Internet caveat remains: check, check, check.

wikipedia See wiki.

wilco "Will comply." A term used in radio communications, indicating that the recipient will comply with sender's just-radioed instructions.

wild line Any incoming copper loop that is running outside of the PBX. Typically a wild line is an alarm, fax or modem line that you don't want running through the telephone switch. Many installers like to differentiate between these "wild lines" and the stuff that gets punched down and run to the phone switch.

wildcard mask A 32-bit quantity used in conjunction with an IP address to determine which bits in an IP address should be ignored when comparing that address with another IP address. A wildcard mask is specified when setting up access lists.

wildcards Special characters you use to represent one or more characters in an MS-DOS filename. An asterisk (*) represents several characters and a question mark (?) represents a single character. For example, the command

ERASE *.BAK

would erase all the files with the suffix "BAK."

The command

ERASE *.?A?

would erase all the files with "A" as the middle letter in a three-letter suffix.

wilderness protocol The recommendation that amateur radio operators (i.e., hams) in remote areas announce their presence on and monitor the national calling frequencies for five minutes beginning at the top of the hour, every three hours, from 7 AM to 7 PM. By doing so, a ham in a remote location may be able to pick up a distress call from someone in the back country and relay emergency information to another wilderness ham who has better access to a repeater. National calling frequencies (called simplex frequencies): 52.525, 146.52, 223.50, 446.00, 1294.50 MHz.

wildfeed A satellite transmission of a TV show or other broadcast that isn't meant for public viewing. Wildfeeds are raw transmissions of TV shows, sporting events or news reports sent via satellite. It's how American networks send shows to their affiliate stations and Canadian broadcasters, and how TV news reporters feed live reports home. Some of those feeds are listed on Web sites or in a satellite listings guide, but true wildfeeders prefer to go it alone: "It's got to be up there some place," says a typical wildfeeder, "If you've got a big enough dish you can find it."

wildfire The all-hearing, all-doing computer telephony slave from a company called Wildfire Communications, Lexington MA. The product uses very sophisticated voice recognition software so that its "master" (i.e. the user) can get Wildfire to take messages, find him, connect his calls, transfer his calls and act as a super intelligent on-line, computerized, 24-hour a day, never resting, all obedient secretary. Wildfire was a real breakthrough product, first introduced in the fall of 1994 and deserving of its own definition in this illustrious dictionary. Unfortunately Wildfire never caught on bigtime. It used too much processing power and consequently was too expensive. However, several computer telephony companies tried to copy it and the jury is still out. The demonstration on www.Wildfire.com is very impressive and worth listening to.

wilding Wireless hackers (aka whackers) search neighborhoods for leaky 802.11b networks to exploit. Most wireless LANs don't enable encryption, but probably should.

willful violation The act of knowingly committing a violation of the federal safety and health standards. A willful violation is the most serious finable offence.

WiLL A name Motorola uses for its Wireless Local Loop (WiLL) product, which was developed to serve the basic telephony needs of people in urban and difficult to reach rural areas. Cellular based, WiLL technology is intended to provide fixed telephony services in areas with little or no existing wireline telephone service or as a supplement to the existing wireline service. It uses very few cellular transmit/receivers – often just one at the end of the landline. The WiLL system provides three major benefits to the telecom operator looking to expand their service area: more rapid deployment of telephone service; lower cost alternative to copper wire installation, and increased flexibility in system implementation and design. A WiLL system can be operational in weeks, compared to the huge amounts of time it would take to lay and install copper wire from an end office to each of the subscriber points in a typical local loop. Although WiLL is cellular-based, the system does not require a cellular switch. This makes the WiLL system a lower cost alternative to using "typical" cellular systems for fixed telephony applications because the total system outlay costs as well as associated backhaul and maintenance costs are reduced. WiLL has three elements: the WiLL System Controller (WiSC), a Digital Loop Concentrator (DLC), and a Motorola cellular base station. It interfaces directly to the central office switch via 2-wire analog subscriber loops.

willful intercept The act of intercepting messages intended for a station experiencing a line or equipment malfunction.

Willy-Nilly The term Willy-Nilly was coined during the summer of 1914 to describe the litany of correspondence traversing Europe between Kaiser Wilhelm (Willy) of Germany and Czar Nicholas (Nilly) of Russia, as the two sought a means to avoid the collapse of Europe into a world war while maintaining their respective travel schedules on state business, and in the Kaiser's case – on vacation. Their efforts failed, along with the work of many others to find a political solution to the crisis, and the destruction that ensued eclipsed all wars prior and all wars since, including WWII.

WiMAX Worldwide Interoperability for Microwave Access (WiMAX) is a Broadband Wireless Access (BWA) solution that is based on standards recommendations from both the Institute for Electrical and Electronics Engineers (IEEE) 802.16 working group and the European Telecommunications Standards Institute (ETSI). WiMAX is promoted by the WiMAX Forum, a special interest group comprising members of the manufacturing, carrier and service pro-

vider communities. The IEEE established the 802.16 working group to standardize Local Multipoint Distribution Services (LMDS) and Multichannel Multipoint Distribution Services (MMDS). These services were highly touted in the 1990s as Wireless Local Loop (WLL) solutions that would allow competitive service providers to bypass the Incumbent Local Exchange Carrier (ILEC) copper local loops, thereby providing network access to customer premises more quickly and cost effectively. LMDS and MMDS were commercial failures due to their high cost, as well as performance issues and Line-of-Sight (LOS) requirements. The 802.16 standard from the IEEE is the primary basis for WiMAX. Officially known as the WirelessMAN Air Interface for Broadband Wireless Access (BWA), 802.16 was released in 2001 as an umbrella standard specification for WLL solutions. Specifically, 802.16 was released in 2001 as a means of standardizing LMDS and focuses on frequencies in the 10-66 GHz range and requiring LOS. Since the initial release, 802.16 has evolved considerably through the 802.16a, 802.16d and 802.16e extensions. The first BWA standard to be released by an accredited standards body, 802.16 features a protocol-independent core, supports high-bandwidth on-demand environments and hundreds of users per channel, and can handle either continuous or bursty traffic. 802.16a (2003) is based on MMDS and the European HiperMAN system. 802.16a operates in the 2-11 GHz range, which includes both licensed and license-exempt bands. It is designed for both point-to-point and point-to-multipoint topologies, and usually requires LOS (line of sight). 802.16d, also called 802.16-2004, is a compilation and modification of previous versions and amendments 802.16a, b and c, the latter two of which are considered obsolete. It operates in the 2-11 GHz range and was designed for point-to-point, point-to-multipoint and meshed topologies. 802.16d operates best with LOS, but does not require it. Included is support for indoor Customer Premises Equipment (CPE). 802.16e, at the time of this writing, is due to be finalized in October 2005. 802.16e adds hand-off capability, thereby extending the standard to include portability and mobility. This extension operates in the 2-11 GHz range and does not require LOS. The 802.16 specifications include several multiplexing options: Time Division Duplex (TDD) for half-duplex (HDX) communications and Frequency Division Duplex (FDD) for full duplex (FDX). 802.16 specification addresses both LOS and Non-Line-of-Sight (NLOS) topologies. In instances in which LOS (line of sight) can be achieved, WiMAX cell coverage can be up to 50 km (31 miles). Where LOS cannot be achieved, coverage generally is limited to approximately 8 km (5 miles). The fixed wireless standards provide for shared bandwidth up to 70 Mbps per base station, although the actual level of throughput depends on factors such as LOS, distance, air quality and interference. NLOS (non line of sght) technology addresses signal performance issues through a number of mechanisms, including Orthogonal Frequency Division Multiplexing (OFDM), sub-channelization, antenna design and error control. Specifications currently call for shared bandwidth up to 70 Mbps per base station, i.e., cell antenna. Cells typically are divided into as many as six sectors of 60 degrees, in which case each sector supports shared bandwidth of 8-11.3 Mbps, depending on output power and signal performance issues. Bandwidth is fully symmetric, i.e., the same amount of bandwidth is available to be shared upstream as well as downstream. At the customer premises, the receiver is in the form of a flat passive array antenna known affectionately as a "pizza box," by virtue of its size and shape. This antenna is installed on the side of the building or some other location that maximizes signal quality. The standards also provide for self-installed indoor CPE with no externally mounted antenna. Signal quality suffers in such a scenario and, as a result, distances are shortened.

WiMax provides for Quality of Service (QoS) through the definition of four polling schedules. Unsolicited Grant Service (UGS) is for services that periodically generate fixed units of data, such as uncompressed voice and video and services such as T1 and E-1. Real-Time Polling Service is for services that are dynamic in nature, but require periodic levels of bandwidth to meet real-time demands, with examples being compressed voice and video. Non-real-Time Polling Service supports connections that require random transmit capabilities, such as Internet access with a minimum guaranteed connection rate. Best Effort Service provides neither throughput nor delay guarantees. Security is provided through both authentication and encryption. WiMAX applications include private campus networks, T-1 and Fractional T-1 business service, rural or developing areas where cabled broadband service is not available, Wi-Fi hotspot backhaul and disaster recovery. See also HiperMAN, LMDS, LOS, OFDM, MMDS and QoS.

WiMAX Forum The WiMAX Forum is a nonprofit organization working to facilitate the deployment of broadband wireless networks based on the IEEE 802.16 standard through encouraging the compatibility and inter-operability of broadband wireless access equipment. The WiMAX Forum comprises member of the manufacturing, carrier, service provider and related communities. See WiMAX.

WiMedia Alliance The WiMedia Alliance is an industry association formed to promote personal-area range wireless connectivity and interoperability among multimedia devices in a networked environment. The Alliance develops and adopts standards-based specifications for connecting wireless multimedia devices. See also IEEE, 802.15.3.

WIMP interface Stands for Windows, Icons, Mouse and Pointing Device or Pull-down menus. A derogatory reference to GUI. Some people think WIMP is on the way out. See also Graphical User Interface.

WIN 1. Wireless In-building Network. WIN is a technology from Motorola which uses microwaves to replace local area network cabling.

2. WIN services are services that use Wireless Intelligent Network (WIN) functionalities, synonymous to AIN for the wireline services. The WIN is a standard, destined to become the successor to both IS-41 and GSM. IS-41 "Rev. D" is often used interchangeably with WIN.

WIN Services See WIN.

WIN-T Warfighter Information Network-Tactical. The U.S. Army has a strategy known as Objective Force. It is intended to make soldiers more agile, so that they can deploy more quickly, adapt more readily to changes on the battlefield and strike more lethally. WIN-T is a tactical intranet being developed by the US Army that will use commercial technologies for wired and wireless voice, data and video communications to provide networking for troops on the go. WIN-T will be mobile, secure and survivable and will integrate ground, airborne and satellite-based capabilities into a network infrastructure and will support the army's Future Combat System (FCS), which is envisioned to create an integrated battlespace, where a network of information and communications systems provide a competitive edge to soldiers in the field and commanders in the control room. With WIN-T, Army officials plan to create a mobile network environment that will enable soldiers to send and receive critical information on the fly. The Warfighter Information Network-Tactical will enable troops and their commanders to have continuous access to the information they need, even when they are in transit. WIN-T will provide planning and communications support to warfighters in fortified locations. While warfighters are en route, they will use airborne communications systems to conduct mission planning and rehearsal. WIN-T will enable commanders, staff and other users to simultaneously exchange voice, data and video information between the sustaining base and the deployed area of operation. Through the WIN-T infrastructure, warfighters will have access to specialized services such as Mobile Satellite Services, the Defense Message System, Global Broadcast Service and interfaces to joint, allied and coalition forces.

WIN XP An updated version of Windows 2000, replete with a new interface and alleged more reliability.

WIN-WIN A deal in which all parties come out better, or at least appear to, or at least feel they all came out better. Marriage, for example, is meant to be a win-win deal. And for some people, me included, it is.

WIN2000 Windows 2000, an updated version of Windows NT.

WIN32 API A 32-bit application programming interface for the Windows family of operating systems. It updates earlier versions of the Windows API with sophisticated operating system capabilities, security, and API routines for displaying text-based applications in a window.

WIN95 See Windows 95.

WIN98 See Windows 98.

WINCE An acronym for Windows CE, the portable Microsoft Windows for palm top computers.

winch A machine for pulling cable into conduit (in the street or in the building) or duct liner. A winch has a rotating drum that winds up the pulling line.

winchester disk A sealed hard disk. The Winchester magnetic storage device was pioneered by IBM for use in its 3030 disk system. It was called Winchester because "Winchester" was IBM's code name for the secret research project that led to its invention. A Winchester hard disk drive consists of several "platters" of metal stacked on top of each other. Each of the platter's surfaces is coated with magnetic material and is "read" and "written" to by "heads" which float across (but don't touch) the surface. The whole system works roughly like the old-style Wurlitzer jukebox. There are several advantages to a Winchester disk system:

1. It can store, read and write enormous quantities of information. Some Winchesters have a capacity of over 100 megabits;

2. You can access information on a Winchester faster than on most computer storage medium (RAM and ROM are obviously faster); and

3. Winchesters are reliable and relatively inexpensive. There are also disadvantages: 1. They are very sensitive to rough handling (they hate being moved); 2. They are very sensitive to the organization of their directory track (lose that and you're in big trouble); and 3. When Winchesters "crash" (i.e. the heads touch the surface of the rotating platters), you can lose an enormous amount of precious data.

wind load See dead load.

windfall The word windfall comes from an old English law prohibiting the cutting down of the King's trees, so the only firewood commoners could gather is what the wind blew down.

winding Coils of wire usually found in transformers and used to boost inductance.

window 1. A band, or range, of wavelengths at which an optical fiber is sufficiently transparent for practical use in communications applications. Each window roughly corresponds to a visible color of light in the overall light spectrum. See also DWDM, Lambda, SONET and WDM.

2. A flow-control mechanism in data communications, the size of which is equal to the number of frames, packets or messages that can be sent from a transmitter to a receiver before any reverse acknowledgment is required. It's called a pacing group in IBM's SNA.

3. A box on the CRT (cathode ray tube) of your personal computer or terminal. A software program is running inside the box. It's possible with new "windows" software to run several programs simultaneously, each accessible and visible through the "window" on your CRT.

4. A technique of displaying information on a screen in which the viewer sees what appears to be several sheets of paper much as they would appear on a desktop. The viewer can shift and shuffle the sheets on the screen. Windowing can show two files simultaneously. For example, in one window you might have a letter you're writing to someone and in another window, you might have a boilerplate letter from which you can take a paragraph or two and drop it in your present letter. Being able to see the two letters on the screen makes writing the new letter easier.

5. Video containing information or allowing information entry, keyed into the video monitor output for viewing on the monitor CRT. A window dub is a copy of a videotape with time code numbers keyed into the picture.

6. A video test signal consisting of a pulse and bar. When viewed on a monitor, the window signal produces a large white square in the center of the picture.

window control A credit or token scheme in which a limited number of messages or calls are allowed into the system.

window dressing Window dressing, a Wall Street term, happens at the end of each quarter when managers of institutions that manage other peoples' money – like mutual funds – buy and sell stocks in order to end up on the last day of quarter owning those shares that will look the best for people who might give them money to manage. The idea is to weed out the losers and fill in with winners so when the financial period's holdings are printed, that list looks and smells like roses.

window segment size A parameter used to control the flow of data across a connection. A wireless term.

window size The minimum number of data packets that can be transmitted without additional authorization from the receiver.

window treatment You take the world's most beautiful window and you screw it up with expensive stuff you affix around it. Paula Friesen invented the term.

windowing 1, A technique of running several programs simultaneously – each in running a separate window. For example, in one window you might run a word processing program. In another, you might be calculating a spreadsheet. In a third, you might be picking up your electronic mail.

2. A technique in (mostly PC) data communications protocols that permits the sender to run ahead in transmission, backing up to resend if the receiver signals an error in a recently-sent block; closely akin to "Go Back N" in IBM's SDLC.

Windows The most popular operating system for a personal computer in the world. The latest iteration is called Vista.

Windows 2000 An updated form of Windows NT.

Windows 95 Windows 95 operating system from Microsoft first shipped on August 24, 1995. Win95was the first Windows operating system designed for communications. It did for modems and phones (of all sorts – from single line analog to proprietary ISDN-phones) what Windows previously had done for printing – insulate the suffering user from the idiocies of communications and networking device drivers and implement the much-touted benefits of something called computer telephony, which had the computer controlling phone calls and building devices like voice mail boxes. It never really happened. The Internet and TCP/IP came along and usurped much of computer telephony's original ideas.

Windows 98 Windows 98 was the successor to Windows 95, which got succeeded in turn by Windows XP and then Vista.

Windows Application A term used in this document as a shorthand term to refer to an application that is designed to run with Windows and does not run without Windows. All Windows applications follow similar conventions for arrangement of menus, style of dialog boxes, and keyboard and mouse use.

Windows CE A smaller version of Microsoft's Windows operating system to be used for a range of mobile handheld communications, computing or entertainment devices. See also Microsoft At Work, an earlier attempt by Microsoft at making an operating system for devices other than PCs.

Windows character set The character set used in Windows and Windows applications. Most TrueType fonts have a set of about 220 characters.

Windows for workgroups Windows for workgroup is an obsolete local area networked version of Microsoft Windows operating system version 3.1 that offered integrated file sharing, electronic mail (Microsoft Mail) and workgroup scheduling (Schedule+), thus bringing the graphical user interface to the workgroup. Windows For Workgroups also had Network DDE, which allowed users to create compound documents that share data across network. Windows 95 superceded Windows for Workgroups.

Windows Media Audio Windows' answer to MP3. Meant primarily as a streaming format, WMA uses smaller file sizes. Some WMA files contain built-in digital rights management to check the legality of purchased music.

Windows Mobile A smaller version of Windows for PCs. Windows Mobile is a small operating system combined with a suite of basic applications for use on mobile devices (phones, PDA and smaller devices). It is designed to have the same "look and feel" of the desktop versions of Windows.

Windows MetaFile WMF. A method of encoding files. Other methods include EPS, PCX and TIFF.

Windows NT Windows New Technology, now called Windows 2000, is a 32-bit operating system from Microsoft, designed to replace Windows95 and MS-DOS. As an operating system, Windows NT is targeted at the top 10% "power" users who need the power of a big, powerful operating system. Here are the main advantages of Windows NT, as explained by Microsoft:

- Interoperability. Windows NT delivers support for open computing benefits through its protected subsystem architecture. Windows NT was also designed to be protocol independent. As such it will interoperate with all leading network systems, regardless of the native protocol of the system.

- Portability. Windows NT was designed to be portable across a variety of hardware systems. The Hardware Abstraction Layer (HAL) limits and isolates the amount of code necessary to port Windows NT to a new platform. Windows NT will run on processors other than those made by Intel. MS-DOS, for example, doesn't.

- Scalability. Windows NT scales to work on both single and multi processor computer systems. This scalability gives users the flexibility to implement their own solutions, today or over time, on machines that meet the performance needs of sophisticated client server solutions.

- System Management. Windows NT supports SubNetwork Access Protocol (SNMP) and NetView network management standards.

- Published Interfaces. The interfaces to the Windows NT operating system are fully documented and published. Software developers are free to add functionality to the system based on their interface definitions.

- Support of Industry Standards. These include POSIX.1, OSF DCE, TCP/IP and WOSA, which is Microsoft's Windows Open Services Architecture. WOSA is a standard set of interfaces to connect a variety of applications with a range of back-end devices and services, such as messaging, telephony, databases, etc. Windows Telephony is part of WOSA.

Windows NT File System NTFS. An advanced file system designed for use specifically with the Windows NT operating system. NTFS supports file system recovery and extremely large storage media. It also supports object-oriented applications by treating all files as object with user-defined and system-defined attributes.

Windows Open Services Architecture See WOSA.

Windows Telephony Introduced in the spring of 1993 jointly by Microsoft and Intel, Windows Telephony is a piece of software called a Windows Telephony DLL and two standards. The first standard is the Service Provider Interface (SPI). If a hardware manufacturer's product honors that SPI, that product can happily talk to the Windows Telephony DLL. The second standard is called the Application Programming Interface and it is directed at software developers who write applications programs. If those developers' programs adhere to the API, they can take advantage of the Windows Telephony DLL to drive whatever telephony devices or services adhere to the SPI. The Windows Telephony API is called TAPI. DLL stands for Dynamic Link Library. It is a Windows feature that allows executable code modules to be loaded on demand and linked at run time. At one stage, it was thought at Windows Telephony would bring about an explosion of shrink-wrapped

Windows based telephone software applications – from simple personal rolodexes to power dialers, to customized phone systems for banks and for bakers. It was going to bring about an explosion of new telephony hardware devices – from telephones that look more like PCs than phones, to PCs that are phones, to blackbox telephony devices that hook to laptops and transform hotel phones.

Windows Telephony removed earlier overwhelming barriers to creating PC-driven telephony applications, namely the wide enormity of telephony "network" services – from the many telephone company interfaces (POTS to T-1), to the many more proprietary interfaces behind dozens of proprietary PBXs, key systems and hybrid phone systems. Unfortunately Windows Telephony never took off, largely because the telephone industry just "didn't get it." The industry also resented that an outsider – someone from the computer industry would attempt to layer standards on the phone business. As a result, Windows Telephony has never taken off. See At Work, Fax at Work, Telephony Services, Windows Telephony Services, Windows Toolkits and WOSA.

Windows Telephony services Here is Microsoft's original definition: Windows Telephony services are provided as a WOSA (Windows Open Services Architecture) component. It consists of both an application programming interface (API) used by applications and a service provider interface (SPI) implemented by service providers. The focus of the API is to provide "personal telephony" to the Windows platform. Telephony services break down into Simple Telephony services and Full Telephony services. Simple Telephony allows telephony-enabled applications to be easily created from within these applications without these apps needing to become aware of the details of the Full Telephony services. Word processors, spreadsheets, databases, personal information managers can easily be extended to take advantage of this. Complete call control is only possible through the use of the Full Telephony services. Applications access the Full Telephony API services using a first-party call control model. This means that the application controls telephone calls as if it is an endpoint of the call. The application can make calls, be notified about inbound calls, answer inbound calls, invoke switch features such as hold, transfer, conference, pickup, park, etc., detect and generate DTMF for signaling with remote equipment. An app can also use the API to monitor call-related activities occurring in the system.

The fact the API presents a first-party call control model does not restrict its use to only first-party telephony environments. The Windows Telephony API can be meaningfully used for third-party call control. The API provides an abstraction of telephony services that is independent of the underlying telephone network and the configuration used to connect the PC to the switch and phone set. The API provides independent abstractions of the PC connections to the switch or network and the phone set. The connection may be realized in a variety of arrangements including pure client based wired or wireless connections, or client/server configurations using some sort of local area network.

The Telephony API by itself is not concerned with providing access to the information exchanged over a call. Rather, the call control provided by the API is orthogonal to the information stream management. The Telephony API can work in conjunction with other Windows services such as the Windows multimedia wave audio, MCI, or fax APIs to provide access to the information on a call. This guarantees maximum interoperability with existing audio or fax applications.

The Telephony API defines three levels of service. The lowest level of service is called Basic Telephony and provides a guaranteed set of functions that corresponds to "Plain Old Telephone Service" (POTS - only make calls and receive calls). The next service level is Supplementary Telephone Service providing advanced switch features such as hold, transfer, etc. All supplementary services are optional. Finally, there is the Extended Telephony level. This API level provides numerous and well-defined API extension mechanisms that enable application developers to access service provider-specific functions not directly defined by the Telephony API. See Windows Telephony.

Windows toolkits Windows toolkits are libraries of code that implement the graphical user interface objects that every software application uses. The toolkits save time by eliminating the need for software developers to re-implement the same code repeatedly for each application. Toolkits also have the benefit of consistent user interface implementation across all applications that use the toolkit. See also Windows.

Windows XP An updated version of Windows 2000 that has Win2000's stability and some nice new features, including the ability to dial into the machine remotely and to make phone calls. See the next definition.

Windows XP Real Time Communications When it introduced Windows XP towards the end of 2001, Microsoft announced an update of the Windows Messenger Real-Time Communications client, that will enable consumers to use their personal computers to make voice calls from their PCs to telephones anywhere in the world via a choice of participating service providers. These new PC-to-phone calling options, along with other enhancements, are included in a downloadable update to the Windows Messenger feature within Windows XP. In addition, an update to the MSN messenger client will enable this new PC-to-phone support for users of previous versions of Windows. The Microsoft Voice Services program enables carriers to participate in and benefit from this Windows Messenger Real-Time Communications update, which includes embedded Voice Over IP (VoIP) and PC-to-phone calling capability. Microsoft will host a referral service of qualified carriers who will be our partners in terminating PC to phone calls via the Messenger client. The Voice Services program will allow Microsoft, via Messenger, to refer new customers to the participating carriers in order to subscribe, and also will integrate the user online communications services with voice services from an Internet Telephony Service Provider (ITSP).

If you are an ITSP, the Voice Services program provides you with a way to build differentiating services, reduce customer churn by adding new applications and capabilities for your existing customers, find new customers (every copy of Windows XP that ships is essentially a telephone handset that needs a backbone carrier for VoIP traffic), and most importantly, increase the average revenue per customer you currently enjoy if you are not already offering VoIP services. ITSPs will be able to manage their own customers to include provisioning, call processing and termination, billing, and customer care.

Voice Services is one of the first ways an ITSP can adopt and leverage the Microsoft .NET platform. Microsoft Voice Services incorporates the Passport web service to authenticate end users securely and privately over the Internet. ITSPs also can take advantage of other Microsoft .NET Web services (Microsoft .NET My Services – formerly code-named "HailStorm") to build true Unified Messaging applications. Any company capable of terminating IP-based telephone calls to the Public Switched Telephone Network (PSTN) is a candidate for the program. In the Voice Services program, Microsoft does the following:

• Authenticates users via the Passport Internet Authentication Web Service
• Processes Session Initiation Protocol (SIP) signaling requests via our SIP proxy servers
  • In the Voice Services program, the carrier (ITSP) is responsible for the following:
Carriers need to deploy the following technical infrastructure to participate in the program:
• SIP-based proxy servers and soft switches
• VoIP trunking
• SIP signaled VoIP-to-PSTN gateways

The Referral Program works thus: Users are prompted to sign up for the new Windows Messenger when they install the Windows XP Home Edition operating system for the first time. After installing Windows Messenger, a user will be prompted to make a PC-to-telephone call. If it is the first time they have made a call, they will be taken to a referral server and shown the carriers in their area that are terminating PC to phone telephone calls. The end user selects a carrier at this time.

WinInet API The Microsoft Win32 Internet functions. These functions provide Win32 applications with access to common Internet protocols. These functions pluck out the heart of the Internet's Gopher, FTP, and HTTP protocols and turn them into an application programming interface (API). This provides a straightforward path to making applications Internet-aware.

WinISDN WinISDN is ISDN*Tek's API for talking to internal ISDN modems. It supports all of the high level functions for call setup and answering on an ISDN modem. Most of the more popular internal ISDN modems support WinISDN. One of the ways WinISDN helps increase throughput is by handling data transfers in large blocks rather than one byte at a time. The overhead on single byte transfers is much higher than handling a single block.

wink A signal sent between two telecommunications devices as part of a handshaking protocol. It is a momentary interruption in SF (Single Frequency) tone, indicating that the distant Central Office (CO) is ready to receive the digits that have just been dialed. In telephone switching systems, a single supervisory pulse. On a digital connection such as a T1 circuit attached to a Carrier Access Concentrator Access Bank II, a wink is signaled by a brief change in the A and B signaling bits. On an analog line, a wink is signaled by a brief change in polarity (electrical + and -) on the line.

wink operation A timed off-hook signal normally of 140 milliseconds, which indicates the availability of an incoming register for receiving digital information from the calling office. A control system for phone systems using address signaling.

wink pulsing Recurring pulses of a type where the off-pulse is very short with respect to the on-pulse, e.g., on key telephone instruments, the hold position (condition) of a line is often indicated by wink pulsing the associated lamp at 120 impulses per minute, 94 percent make, 6 percent break (470 ms on, 30 ms off).

wink release On most modern central offices when the person or device at the other end hangs up, your local central office will send you a single frequency tone. That tone is called wink release. Such a tone can be used to alert a data device that the device at the

other end has hung up. (Remember it can't tell by just listening – like you and me.) When a data device hears a wink release, it usually takes it as a signal to hang up also.

wink signal A short interruption of current to a busy lamp causing it to flicker. Indicates there is a line on hold.

wink start Short duration off hook signal. See Wink Operation.

WINS Windows Internet Name Service. A name resolution service that resolves Windows networking computer names to IP addresses in a routed environment. A WINS server, which is a Windows NT Server computer, handles name registrations, queries, and releases.

Winsock 2 Winsock stands for Windows Sockets. Winsocks are standard APIs between Microsoft Windows (3.1, 95 and NT) application software and TCP/IP protocol software (also called a protocol stack). Winsock 2 is a network programming interface at the transport level in the ISO reference model. It is being defined by an open, industry wide workgroup, called the Winsock Forum.

Wintel A combination of the words Windows and Intel. Wintel refers to PCs that run Microsoft Windows and use Intel microprocessors. Wintel PCs are by far the biggest-selling PCs, amounting to around 80% of all PCs sold. Observers of the PC industry refer to the "Wintel standard" to refer the phenomenon of such high market dominance by one type of PC.

wipe A transition between two video signals that takes the shape of a geometric pattern. Used also in PowerPoint.

wipeout A completely blank television picture caused by extremely strong interference.

WIPI Wireless Internet Platform for Interoperability. All three Korean cellular operators are using different operating systems (eg KTF uses BREW, SKT uses in-house developed package, LGT uses JAVA) to access the wireless internet.) WIPI is a Korean Government inspired wireless platform for accessing the internet which the Korean Government wants to compete with Qualcomm's BREW (Binary Running Environment for Wireless Internet) as part of Korea's telecoms equipment export drive. The reason is that WIPI is an open system so that content providers should be able to standardize their production based on one specification rather than having to create content for all three different operating systems, increasing consumer choice and reducing operating costs. Cellular operators' reaction to WIPI has been muted because each sees their own operating system as a competitive advantage.

WIPO World Intellectual Property Organization. An intergovernmental organization with headquarters in Geneva, Switzerland, WIPO is an agency of the United Nations. WIPO is responsible for the promotion of the protection of intellectual property throughout the world through cooperation of its member states, of which there were 161 as of February 1997. WIPO also is responsible for the administration of various multilateral treaties dealing with legal and administrative aspects of intellectual property. In 1998, WIPO convened an international process to develop recommendations concerning the intellectual property issues associated with Internet domain names, including dispute resolution. See also Domain Name, gTLD, Intellectual Property and TLD.

wire center The physical structure where the telephone company terminates subscriber outside cable plant (i.e. their local lines) with the necessary testing facilities to maintain them. Usually the same location as a class 5 central office. One or more Switching Entities which serve the plant facilities through a single main frame (or two or more main frames joined by the cables), regardless of the number of buildings involved. A wire center might have one or several class 5 central offices, also called public exchanges or simply switches. A customer could get telephone service from one, several or all of these switches without paying extra. They would all be his local switch.

wire center, multi-entity A North American telephone company term. A wire center which has two or more entities serving the plant facilities through a single main frame, (or two or more main frames joined by cables), regardless of the number of buildings involved. The wire center may include entities of various switching types and combinations, e.g., 3 entities, entity 1, IAESS, entity 2, 5ESS-2000, entity 3, DMS-100.

wire center serving area That area of an exchange served by a single wire center.

wire concentrator A conduit; a pipe within which a large number of individual wires are routed through.

wire pair Two separate conductors traveling the same route, serving as a communications channel.

wire plant The installation of all low voltage wiring, both voice and data in a facility.

wire printer A matrix printer which uses a set of wire hammers to strike the page through a carbon ribbon, generating the matrix characters.

wire rate See Wire Speed.

wire running tools Tools that help you run wire in and around a building. The most common form of wire running tools that help you fish wire through hollow drywalls.

wire speed The rate at which bits can be transmitted over a circuit or link, which generally is considered to be a wired circuit comprising one or more metallic or glass conductors, but which also may be wireless. For example, the wire speed of Ethernet is 10 Mbps, 100 Mbps, or 1 Gbps. Ethernet standards specify the medium (e.g., Cat 3 or Cat 5 UTP, optical fiber, or RF (Radio Frequency) or Ir (Infrared) wireless); the maximum and minimum allowable distances; the bit encoding scheme; and other specifics of the "wire." Wire speed generally is more closely related to signaling speed that to actual transmission speed, or throughput. Actual throughput depends on many factors, including how many devices of what specific types and what specific capabilities are associated with the circuit. All devices impose some level of delay on the signal. The more complicated the processes performed by a given device, the greater the level of delay, and oftentimes the less efficiently the circuit potential is used (i.e., the greater the difference between throughput and wire speed). To put this in context, bridges are very simple devices which accomplish very simple processes and which, therefore, typically impose very little delay. As one moves up the food chain to switches, routers and gateways, the level of complexity increases, and the level of delay increases, as well. As wire speed is a fundamental imperative of system designers, the implementation of processes in silicon and the optimization of software components is of the utmost importance in order that system throughput match wire speed as closely as possible. See also Throughput.

wire stripper A tool which takes the insulation off a wire without hurting the inside metal conductor.

wire tap The attaching to a phone line of a piece of equipment whose job is to record all conversations on that phone line. Wire taps are illegal. Law enforcement agencies use them, but must receive authorization from a court to apply the tap. Such authorizations are given if the law enforcement agency argues that applying the tap will prevent crime or help bring a suspected criminal to justice. Wire taps are not authorized lightly. See also Trap and Trace.

wire telephony The transmission of speech over wires.

wire wrap and solder Soldering and wire wrap dominated early cable connections. Some old buildings still have large boards of wire wrapped or soldered connections. Wire wrap is still used in telephone company-related applications, but solder for cross-connects is obsolete and not seen today.

Wired Equivalent Privacy See WEP.

wired for capacity The wired-for capacity represents the upper limit of capacity for a particular configuration. To bring to a phone system to its "wired for capacity," all that's necessary is to fill the empty slots in the system's metal shelving (its cage) with the appropriate printed circuit boards. "Wired-for Capacity" is a marginally useful term, giving little indication of the type of printed circuit boards – trunk, line, special electronic line, special circuit, etc. – that can be installed. And many PBXs allow only their printed circuit boards to go into assigned slots. Your PBX cabinet might, for example, have plenty of empty space for extra printed circuit boards, but it may not have any more space for boards which service electronic phones. Thus, it is effectively maxed out.

Wired For Management WFM. See WFM.

wired logic A required logic function implemented in hardware, not software.

wired love A _____ novel published in 1879 by Ella Cheever. It was about a long-distance romance between telegraph operators.

wired radio Radio programming delivered over wire (like Muzak) or cable.

wireless Without wires. Any system of transmitting and receiving information without wires. That system could be anything from your cellphone to your 802.11b-equipped laptop. It could be your wireless headset. See the following chart and also 802.11a, 802.11b, CDMA, cellular radio, GSM and the wireless definitions following.

Wireless Access Controller The first component in an in-building wireless phone system is the wireless access controller. It does many things. It provides access to the host network, be it a host PBX or the public switched telephone network (including Centrex). The access controller also manages the picocellular infrastructure of the wireless system through connections to the radio base stations. See also handoff.

Wireless Application Protocol Forum In January, 1998, Ericsson, Motorola, Nokia and Unwired Planet announced the establishment of the Wireless Application Protocol Forum Ltd. This non-profit company will administer the worldwide WAP specification process and facilitate new companies contributing to WAP specification work. According to the press release announcing the establishment of the Forum, the Wireless Application Protocol (WAP) is targeted to bring Internet content and advanced services to digital cellular phones and other wireless terminals. WAP Forum aims to create a global wireless protocol specification that works across differing wireless network technology types, for adoption by appropriate industry standards bodies. Applications using WAP will be scaleable across a va-

riety of transport options and device types. A common standard offers potential economies of scale, encouraging cellular phone and other device manufacturers to invest in developing compatible products. Cellular and other wireless network carriers and content providers will be able to develop new differentiated service offerings as a way to attract new subscribers. Consumers will benefit through more and varied choices in mobile communications applications, advanced services and Internet access. In addition to the four founding partners, new members are now welcome to join WAP Forum. Members may contribute to the current specification work, participate in driving the continuing evolution of WAP and nominate and elect additional directors to the board of WAP Forum. In order to become members of WAP Forum, interested companies need to apply to join. All the details including the application form can be found at www.wapforum.org and www.xwap.com.

Wireless Assisted Global Positioning System WAGS. See GPS.

wireless backhaul Wireline links – typically T-1 lines (E-1 lines in Europe) – that wireless operators use to connect their wireless base stations back to their mobile switching centers.

wireless cable An oxymoron which means that TV signals are broadcast by microwave to antennas on customers' homes. The former name for wireless cable was MMDS, short for Multichannel Multipoint Distribution Service.

wireless cable service An oxymoronic term if there ever was one, but it does have an industry definition: A wireless broadcast service providing cable-TV-like entertainment video channels, received at a subscriber's site by a small parabolic antenna or dish. It is an encompassing term, covering the specific services of Microwave Distribution Service (MMDS), Direct Broadcast Satellite (DBS) and Cellular Television. See also Fixed Wireless Local Loop.

wireless card Most laptops have a slot or two into which you can plug something called a PC card, or what was once called a PCMCIA card. You can plug in cards which connect you to a wireless network (local or long distance), which connect to digital RAM (e.g. from digital cameras) or cards that drive several external monitors. If the PC Card connects you to a wireless network (e.g. Wi-Fi), it's often called a wireless card.

Wireless Competitive Access Providers WCAPS. See Fixed Wireless Local Loop.

Wireless Data Forum WDF. A not-for-profit organization dedicated to publicizing successful wireless data applications and customer communities. WDF membership includes network service providers, wireless device and infrastructure equipment manufacturers and vendors, computer software and hardware developers, and information services content providers. www.wirelessdata.org.

wireless data network A radio-based network for data transmission. Cellular Digital Packet Data (CDPD) is an example.

wireless digital standards See Digital Wireless Standards.

Wireless Technologies Compared				
Formats	**Power**	**Max range**	**Max speed**	**Frequency Band**
802.11a	32mW	150-175 ft	54 Mpbs	5.15 - 5.825 GHz
802.11b	32mW	300-350 ft	11 Mbps	2.4 - 2.483 GHz
802.11g	32mW	300-350 ft	54 Mbps	2.4 - 2.483 GHz
Bluetooth	1mW	33 ft	1 Mbps	2.4 - 2.483 GHz
Ultra-wideband*	155mW	33 or 300 ft	480 Mbps	3.1 - 10.6 GHz
WiMAX*	na	up to 30 miles	to 70 Mbps	10 - 60 GHz
* Not fully defined yet				

Wireless E-911 Phase I / Phase II Refers to the technology and services mandated by FCC Report and Order 96-264 pursuant to Notice of Proposed Rulemaking (NPRM) 94-102. The FCC requirement applies to all cellular and PCS service providers, and those Specialized Mobile Radio carriers that provide public voice service with telephone network interconnection.

Phase I defines delivery of a wireless emergency 911 call with call-back number and identification of the cell and sector from which the call originated. This allows the call to be routed to an appropriate public service answering point (PSAP) based on caller's general position. Without Phase I capabilities, wireless calls are routed to some default service agency, e.g., the state highway patrol. The required Phase I availability date was April 1998, but at this time (early 1999) many public service agencies have not upgraded their equipment to accept the Phase I information and still employ default, or non-selective, call routing.

Phase II defines delivery of a wireless 911 call with Phase I requirements, plus location of the caller within 125 meters 67% of the time. In addition, the call is routed to the appropriate PSAP based on the caller's coordinates. The required Phase II availability is October 2001. This new capability has given rise to the potential for other, non-emergency, added-value location services for wireless customers, and has instigated technological development in wireless handsets, infrastructure, network signaling, and emergency service equipment. See also Position Determination Technology.

wireless edge server This definition from a company which sells wireless edge servers. A wireless edge service provides a mechanism to deliver preferential and prioritized services to high-value customers. This server can be deployed with both the traditional e-business solutions and next-generation Web solutions. The Wireless Edge Server identifies the request, prioritizes it based on the device type and connection speed, and routes the request to the available resource. It is equipped to monitor the common applications used by e-businesses. The Wireless Edge Server monitors the applications and identifies the resource that is best equipped to handle a particular request. It can also handle rich content-audio, video, and graphics-without saturating the corporate networks.

Wireless Ethernet Compatibility Alliance See WECA.

wireless fiber See Fixed Wireless Local Loop.

wireless hub A hub that works wirelessly. See hub.

Wireless Institute of Australia See WIA.

Wireless Internet Platform for Interoperability See WIPI.

wireless LANS The conventional local area network (LAN) uses wires or optical fiber as a common carrier medium. However, other possibilities exist. Low microwave frequencies (lower than about 10 GHz) can provide data rates as high 10 Mbit/s. Millimetric waves at around 60 GHz could support several 10 Mbit/s channels, while infra-red beams could support even greater data throughputs. The area covered by such a scheme would be restricted by the low allowable power radiation. The data rates of such systems tend to be restricted by walls, by interference and by multipath propagation problems that arise due to reflections within the building. Because of the wide bandwidth available, channeling can easily be provided by using spread spectrum methods and code division multiple access (CDMA), a technique that significantly improves the system security. See 802.11a and 802.11b.

Wireless Local Loop WLL. A means of provisioning a local loop facility without wires. Usually employing low power radio systems running in the microwave range, WLL also includes short-range infrared (Ir) systems, which are optical in nature. WLL allows carriers to provision local loops, with perhaps 1 Gbps or more in aggregate bandwidth per coverage area. Such systems are being deployed widely in Asia and other developing countries where they offer the advantages of rapid deployment, and rapid configuration and reconfiguration, as well as avoidance of the costs of burying wires and cables. WLL is particularly attractive where rocky or soggy terrain make cabled systems problematic. WLL also is highly attractive to CLECs (Competitive Local Exchange Carriers), who have a compelling need to deploy local loop facilities, bypassing the incumbent LECs in a deregulated, competitive environment as envisioned by the Telecommunications Act of 1996. See also ADML, Fixed Wireless Local Loop, Infrared, and LMDS.

wireless location signature One of several technologies available for determining a caller's location after initiating a 911 call from a mobile handset. Wireless Location Signature methods compare the radio signal received to a database of standard signal characteristics, such as reflections and echoes. Using this information from several cell site receivers, the caller location can be computed and sent to the PSAP. This technique works best in urban environments where lots of structures exist to provide the needed reflections. LPM works with any phone – digital, analog, TDMA, GSM, CDMA, etc.

Wireless MATV System WMATV. A coaxial distribution network which receives television programming from a wireless receiver.

Wireless Medical Telemetry Service WMTS. Wireless medical telemetry generally is the remote monitoring of a patient's health through radio technology. The use of wireless medical telemetry gives patients greater mobility and increased comfort by freeing them from the need to be connected to hospital equipment that would otherwise be required to monitor their condition. Wireless medical telemetry also serves the goal of reducing health care costs because it permits the remote monitoring of several patients simultaneously. The FCC permits all types of communications except voice and video on both a bi-directional and unidirectional basis, provided that all communications are related to the provision of medical care. See also Personal Radio Services.

wireless messaging Technology allowing the exchange of electronic messages without plugging into a wired land-based phone line. Two wireless messaging types are available: one-way, based on existing radio paging channels; and two-way, based on either radio-packet technology or cellular technology. Some people include in-room infra-red links in the term "wireless messaging." Some of the PDAs use wireless links.

Wireless Number Portability WNP. It means you can take your phone number with you as you change carriers. Some countries have it, including United Kingdom, the Netherlands, Australia and Hong Kong,. Others don't. The Telecommunications Act of 1996 mandated it for the U.S. But, as of writing, (Spring, 2003) it hadn't been implemented here. And some industry trade groups are arguing that it's a gigantic expense that could be better spent on improving service and coverage. I'm inclined to agree with them.

Wireless Office Telecommunications System WOTS. A private wireless system for use in an in-building or campus applications. A WOTS generally is in the form of an adjunct to a PBX or KTS (Key Telephone System). From the common control cabinet, wired connections are established to antennas, which are deployed in a microcell (i.e., very small cell) topology. As go off-hook, your wireless phone connects to a microcell antenna. The antenna system can accomplish handoffs, much like a cellular network, as you move about. The specific technology involved can be that of either cordless or cellular telephony.

wireless packet switching Unlike existing cellular networks, wireless packet-switched networks are designed specifically for data communications. Packet switching breaks messages into packets and sends these packets individually over the network. Here's how a message is sent over the RAM Mobile Data Wireless Networks, one of the packet radio networks in operation today:

1) After you've written a message and turned on your modem, you enter a send command in your e-mail software.

2) The modem breaks your message into packets. A typical packet has a message space for as much as 512 bytes (about 100 words). Longer messages are divided into 512 byte sections.

3) The modem then sends each packet separately over the RAM packet radio network. Each packet includes the sender's and receiver's addresses.

4) The network routes the message to the recipient.

5) The recipient's packet radio modem reassembles the individual packets into a single message.

6) Your recipient can then read the message.

A packet radio network, typically uses a hierarchical architecture to route messages. At the lowest level, base stations exchange wireless messages with nearby mobile computers. Base stations can route messages to other users who are within its service area, or the local switch to read recipients who are in other areas, on LANs, or on public e-mail services. The local switch can either route the message to a different base station or to a regional switch. Users of these packet radio networks can typically send messages anywhere in the network – regardless of the physical distance – for the same rate per message.

Wireless Personal Area Network See WPAN and 802.11a and 802.11b.

Wireless Priority Service Wireless Priority Service gives emergency personnel priority access to cellular networks when a special code is dialed. The service was established by government order following the World Trade Center attacks on September 11, 2001, when network congestion made it nearly impossible for cell phone users – civilians and emergency services personnel alike – to make cell phone calls in New York City. The White House's National Security Council thereupon ordered the creation of a nationwide Wireless Priority Service to give the Department of Homeland Security, key government officials, and civilian emergency responders first access to cellular phone systems. The Wireless Priority Service is administered by the US National Communications System, which is part of the US Department of Homeland Security. Wireless Priority Service is available only to authorized individuals whose SIM card is registered for Wireless Priority Service. Subscribers to the service are charged a one-time activation fee, a special monthly recurring charge, and a special per-minute call. Once a cellular phone has been subscribed to the Wireless Priority Service, the user dials a special code to immediately go to the head of the queue for an open cellular channel. Wireless Priority Service calls do not pre-empt other cellular calls already in progress, but being at the head of the queue for an open channel increases the probability that emergency personnel will be able to make cell phone calls during periods of cellular network congestion.

Wireless Private Branch Exchange WPBX. The WPBX offers business users the ability to make and receive calls using cordless telephones anywhere on a company's premises.

Wireless Rural Loop WRL. Wireless Rural Loop is a subset of wireless local loop (WLL) or fixed wireless local loop for a smaller number of users in a large geographic area. It is almost a polar opposite of WLL that generally serves a relatively large number of users in a localized area. See also Wireless Local Loop and Fixed Wireless Local Loop.

wireless sensor mote See sensor mote.

Wireless Service Provider WASP. A carrier authorized to provide wireless communications exchange services (for example, cellular carriers and paging services carriers).

wireless standard A group of agreed upon specifications concerned with moving data types among wireless devices. Most wireless computing standards are created by the Institute of Electrical and Electronics Engineers. The standard 802.11b is the most common wireless local area network standard. See also IEEE, WPAN, 802.11b.

wireless substitution When a customer dumps his telco landline and goes with a wireless phone instead.

Wireless Switching Center WSC. A switching system used to terminate wireless stations for purposes of interconnection to each other and to trunks interfacing with the Public Switched Telephone Network (PSTN) and other networks.

Wireless Transport Layer Security See WTLS.

WirelessReady Alliance Six companies with interests in wireless products and services came together in December, 1998 to create WirelessReady Alliance, which is aimed at creating mobile data solutions for businesses and consumers. A spokesman for the group said that it would concentrate on existing wireless technologies, including cellular packet data, GSM and CDMA for personal communications systems and cellular services. www.sierrawireless.com/alliance.

wireline 1. Telephone and data service that's provided on cables – both copper and fiber.

2. Another name for a telephone company that uses cables (copper and fiber), not radio.

3. Cellular licenses received from the FCC with initial association to telephone company. Also referred to as B-Block. See Wireline Cellular Carrier.

wireline communications Communications that require a physical connection, such as wires or cables, between users.

wireline cellular carrier Also called the Block B carrier. Under the FCC's initial cellular licensing procedures, the Block B carrier is the local telephone company. The FCC reserved one of the two systems in every market for the local telephone – or wireline company. Wireline or Block B systems operate on the frequencies 869 to 894 Megahertz. See Non-Wireline Cellular Carrier.

wiremap A cable test used to determine whether each pin has connectivity with the appropriate pin at the other end of the cable. A wiremap test is also required to test for split pairs. A wiremap test This is the most basic test that can be performed on a category-5E segment. Wiremap tests for the basic continuity between the two   devices. In 568A or B, all eight pins of each   device should be wired straight through (1 to 1, 2 to 2, 3 to 3, etc.). A wiremap (continuity) test, should also test for absence of shorts, grounding, and external voltage.

wiretap See Telephone Tap and Wiretapping.

wiretapping To listen in clandestinely to someone else's conversation. Other than scrambling, there is no known method to protect your telephone call against wiretapping, no matter what equipment you buy from companies advertising their wares nationally. Wiretapping can be accomplished without physical connection to a phone line, though technically this would be called "bugging." For all intents and purposes you should consider your telephone conversations as public and treat your conversations as such. See Telephone Tap.

wiring closet A central termination area for telephone and/or network cabling. Such wiring closet might serve the building or just a floor or just part of a floor. It's designed to accommodate the wiring needs of one organization or a department of a bigger company. A wiring closer can be a physical closet or a small room. It typically contains punch-down blocks and cross-connect panels.

wiring concentrator A wiring concentrator is an FDDI node that provides additional attachment points for stations that are not attached directly to the dual ring, or for other concentrators in a tree structure. The concentrator is the focal point of Digital's Dual Ring of Trees topology.

wiring density Refers to the number of wires that may be terminated on a connecting block in a given area. A high density block may terminate twice as many wires as a low density block, while a low density block may provide better wire management since fewer wires are being dressed into and out of the connecting block.

wiring environment A fancy term for any any building communications wiring system.

wiring grid The overall architecture of building wiring.

wiring scheme A fancy name for color coding telephone wires, so that you know how to punch down /connect up the differently colored conductors which you typically find in telephone cables. The blue goes on conductor 1, the green on conductor 2, etc.

WISeKey WISeKey operates a common root certificate service on behalf of the International Secure Electronic Transaction Organization. The common root certificate provides a service to Certificate Authorities (CAs) worldwide and for the World Trade Centres global infrastructure. WISeKey provides a high-level certification service with global recognition and trust and is open for use by any CA complying with its certification practice statement. Through its partnerships with leading companies, it also develops digital certification and electronic commerce services for a wide range of international and national organizations and promotes the use of digital certificates for both authentication and authorization.

WISP Wireless Internet Service Provider.

WIT Washington International Teleport.

witchcraft It is possible to trace the origin of witch hunting to an incident that occurred in 1242. Two of the Pope's inquisitors were staying in a house in Avignonet, in the south of France. They had traveled there to root out heretics. In the middle of the night a dozen men with axes, who belonged to a sect known as the Cathars that believed that the Old Testament God was a demon, were admitted to the house by claiming they had information about heretics. They slaughtered the two inquisitors and their servants, hacking their bodies until they were almost unrecognizable. After the massacre, the Pope became determined to stamp out heretics at all costs. A bloody crusade followed. Cathars were dragged out of their homes and burned. In 1244, 200 of them were burned on a gigantic bon-fire at Montsegur. Those who managed to avoid capture were no longer accused or heresy, but of a strange new crime: conspiring with the devil or, as it came to be known, witchcraft.

Witting To be in on a secret.

WITS Wireless Interface Telephone System.

WITSA World Information Technology and Services Alliance. A global consortium of IT industry associations, WITSA positions itself as the global voice of the IT industry. Founded in 1978 as the World Computing Services Industry Association, WITSA is an active advocate in the area of public policy, working toward increasing competition through open markets and regulatory reform, protecting intellectual property, reducing tariff and non-tariff trade barriers to IT goods and services, and safeguarding the viability and continued growth of the Internet and electronic commerce. www.witsa.org.

WIXC Wavelength Interchanging Cross-Connect. A type of OXC (Optical Cross-Connect). See OXC.

Wizard An on-line tutor that guides you through common procedures or processes, as in hardware wizard.

WIZOP A Chief Sysop (System Operator). See SYSOP.

WLAN Wireless Local Area Network. A LAN without wires. There are major benefits, the biggest being the ability to configure and reconfigure the LAN around quickly and cheaply, as wires need not be placed and moved. Groups of people use them often in temporary situations – a team of auditors, a group of firefighters, etc. Wireless LANs are often not as fast as wired LANs. Check. See also 802.11a, 802.11b and Wireless LANs.

WLANA Wireless LAN Alliance. A consortium of manufacturers of WLANs, with the purpose of "generating awareness and creating excitement about the present and future capabilities of wireless local area networks." See WLAN. www.wlana.com

WLIF Wireless LAN Interoperability Forum. A group of more than 20 mobile computing product and service suppliers formed in 1996, the WLIF promotes the OpenAir interface specification. Through a third-party laboratory, products are tested and certified for compatibility. www.wlif.com. See OpenAir.

WLL Wireless Local Loop. See Wireless Local Loop.

WLNP Wireless Local Number Portability. A government mandated project that will allow people to take their cell phone number with them when they change cell phone providers. See Local Number Portability.

WMA Windows Media Audio. Windows' answer to MP3. Meant primarily as a streaming format, WMA uses smaller file sizes. Some WMA files contain built-in digital rights management to check the legality of purchased music.

WMATV Wireless Master Antenna Television System.

WMBTOPCITBWTNTALI We May Be The Only Phone Company In Town But We Try Not To Act Like It. An advertising slogan used by Southwestern Bell Telephone Company in the early 1970's to counter public opinion. It didn't last long – subscribers didn't believe it and SWBTC employees hated it. Or so Ray Horak, my Consulting Editor, says – he should know, he was there. Southwestern Bell, by the way, no longer is the only phone company in town.

WMF Windows MetaFile. A file that stores an image as graphical objects such as lines, circles, and polygons rather than as pixels. There are two types of metafiles, standard and enhanced. Standard metafiles usually have a .wmf file name extension. Enhanced metafiles usually have a .emf file name extension. Metafiles preserve an image more accurately than pixels when the image is resized. The reason to include this definition is that wmf files are the source of my best Microsoft PowerPoint presentation tip. I use PowerPoint a lot. So here goes. You have a PowerPoint presentation. You want to bring one or several slides in from someone else's presentation. If you paste them in, PowerPoint will make the new slide look like your own presentation's format. This will screw up all the imported slides and you'll spend eons trying to fix them. The only sure way to bring in a slide and to retain its formatting is to open the presentation you want to import and save it as a Windows Metafile (.wmf) file. It will ask you if you want to save one slide or all of them? If you say all of them, it will save the presentation in as many files as there are slides. Then you simply paste the new .wmf slides you want into your presentation. One tip: You must modify the slides before you save them as .wmf. When they're in the .wmf format, they can't be messed with. They drop into your presentation as basically uneditable bitmapped images. But they retain the original PowerPoint's format and don't adopt the format of the presentation they drop into.

WMI Windows Management Instrumentation. Microsoft's WMI is the company's implementation of WBEM in Windows OSes. See WBEM.

WML WML is Wireless Markup Language, and is the WAP-proposed browser language. Analogous to, and based upon HTML, but new and different. www.wapforum.org

WMM Wi-Fi Multimedia. The Wi-Fi Alliance Multimedia extensions for wireless LANs allow prioritization of voice traffic. They are based on the IEEE 802.11e.

WMTS See Wireless Medical Telemetry and Personal Radio Services.

WNP Wireless Number Portability. See Wireless Number Portability.

wombat A harebrained idea is a wombat – a waste of money, brains and time.

woo woo tone A tone on a phone line indicating the number is unavailable. Also the words to a neat Jeffrey Osborne song, as in "Will you woo woo with me?"

wooden nickels The expression "don't take wooden nickels" came from the 19th century. Actual wooden coins were routinely "minted" as promotional gimmicks during the numerous exhibitions so popular in 19th century America. They were often were honored at "face value" by participating merchants during the run of the show. To accept a "wooden nickel" after the show had closed its gates would be financial folly (especially in the days when nickels were actually worth something).

Woofys Well Off Older Folks.

word A collection of bits the computer recognizes as a basic information unit and uses in its operation. Usually defined by the number of bits contained in it, e.g., 5, 8, 16 or 32 bits. Using DOS, the IBM PC (and compatibles) defines a word as an eight-bit byte. Such machines use the ASCII coding scheme (or a variation); such a scheme actually involves seven information bits plus a parity bit for error detection. Here's another explanation: A group of characters capable of being processed simultaneously in the processor and treated by computer circuits as an entity and capable of being stored in one storage location. See also Byte and Word Length.

word length The number of bits in a data character without parity, start or stop bits.

words per minute WPM. The speed of printing, typing or communications. 100 WPM is 600 characters per minute (six characters per average word) or 10 characters per second. In ASCII, asynchronous transmission at this rate is also 100 or 110 bits per second, depending on the number of stop bits.

work area The area where horizontal cabling is connected to the work area equipment by means of a telecommunication outlet. A station/desk which is served by a telecommunications outlet. Sometimes referred to as a work station.

work area cable A cable assembly used to connect equipment to the telecommunications outlet in the work area. Work area cables are considered to be outside the scope of cabling standards. See also Work Area.

work location wiring subsystem The part of a premises distribution system that includes the equipment and extension cords from the information outlet to the terminal device connection.

work order A term used in the secondary telecom equipment business. Internal document used by a remarketer specifying: 1. Work to be performed; 2. Machine or item on which work is to be performed; 3. Required completion date; 4. Cost of work; 5. Customer purchase order number and/or other pertinent billing information. This document is used internally to: 1. Implement required work; 2. Monitor progress; and 3. Issue final billing. A work order is implemented once a written request has been received authorizing the work to be performed.

work queue A screen listing the Forms requiring your attention.

word spotting In speech recognition over the phone, word spotting means looking for a particular phrase or word in spoken text and ignoring everything else. For example, if the word to spot was "brown," then it wouldn't matter if you said "I want the brown one," or "how about something in brown?" In short, word spotting is the process whereby specific words are recognized under specific speaking conditions (i.e. natural, unconstrained

speech). It can also refer to the ability to ignore extraneous sounds during continuous word recognition.

work station In this dictionary I spell it as one word WORKSTATION. See Workstation.

workaround A procedure or a piece of software that gets something (i.e. a computer system working). That "workaround" is typically not recommended by the manufacturer of the equipment. That manufacturer is typically surprised when your workaround actually works and often he says, "Wow, I've learned something."

workflow The way work moves around an organization. It follows a path. That path is called workflow. Here's a more technical way of defining workflow: The automation of standard procedures (e.g. records management in personnel operations) by imposing a set of sequential rules on the procedure. Each task, when finished, automatically initiates the next logical step in the process until the entire procedure is completed.

workflow management The electronic management of work processes such as forms processing (e.g. for insurance policy acceptances, college admissions, etc.) or project management using a computer network and electronic messaging as the foundation. See Workflow.

workforce management According to Jim Gordon of TCS in Nashville, Call center workforce management is the art and science of having the right number of people... agents...at the right times, in their seats, to answer an accurately forecasted volume of incoming calls at the service level you desire.

workforce optimization See WFO.

workgroup A fancy new word for a department, except that the members of the workgroup may belong to different departments. The idea is that members of the workgroup work with themselves, so they'd be perfect candidates to buy electronic mail packages that could send messages between themselves and other software packages that would allow them to share their collective wisdoms and schedule their meeting times. Typically members of the workgroup would be on the same local area network and share the same telephone system. See Workgroup Telephony.

workgroup computing An approach to the supply of computer services whereby access to computer power and information is organized on a workgroup by workgroup basis. Such systems normally consist of computers of varying capabilities connected to a local area network. See Workgroup.

workgroup manager An assistant network supervisor with rights to create and delete bindery objects (such as users, groups, or print queues) and to manage user accounts. A Workgroup Manager has supervisory privileges over a part of the bindery. When several groups share a file server, Workgroup Managers can provide autonomous control over their own users and data.

workgroup switching Method of switching that provides high-speed (100-Mbps) transparent bridging between Ethernet networks, and high-speed translational bridging between Ethernet and CDDI or FDDI.

workgroup telephony See Telephony Workgroups.

Working Group See WG.

working key A low level key that changes several times per second, also called the 'control word'. Typically this key changes every four video display frame times (133.5 msec for 525 line systems, or 160 msec for 625 line systems). The working key is used to derive keystream. For services without a video component, the working key epoch duration can be set to an appropriate interval.

workload 1. A call center term. The total duration of all calls in a given period (half hour or quarter hour), not counting any time spent in queue. This figure is equal to the number of calls times the average handle time per call.

2. Trunk workload is call volume x average trunk hold time Verification Call A call center term. The process by which a telephone sale or other disposition is verified to ensure the details are accurate, the costs quoted are precise, the delivery terms have been explained, the customer fully understands the purchase, etc. Verification may be the responsibility of a specially trained team or may be part of the role of the supervisor.

workstation In the telecom industry, a workstation is a computer and a telephone on a desk and both attached to a telecom outlet on the wall. The computer industry tends to refer to workstations as high-speed personal computers, such as Sun workstations, which are used for high-powered processing tasks like CAD/CAM, engineering, etc. A common PC – like the one you find on my desk – is not usually considered a workstation. The term workstation is vague.

world class A trendy term for cool, which was once hot. Used in sentences like "I have a world class product you ought to invest in." Ten years ago it would have been, "I have a cool product you should invest in." Twenty years ago it was "I have a hot product you should invest in." Actually the term means little, except that someone is excited about their product.

world numbering zone One of nine geographic areas used to assign a unique telephone address to each telephone subscriber. See World Zones 1-9.

world phone Another name for a quad-band phone, since it can work on any GSM network. A quad-band phone can operate at 850 MHz, 900 MHz, 1800 MHz and 1900 MHz. A quad-band phone can typically operate on any GSM network.

World Radiocommunications Conference See WRC.

World Telecommunications Day Celebrated every year on May 17, World
Telecommunications Day commemorates the founding of the International Telecommunication Union (ITU) on May 17, 1865. See ITU.

World Teleport Association WTA is headquartered in New York with website
at www.worldteleport.org.

World Trade Organization See WTO.

World Wide Web Also called WEB or W3. The World Wide Web is the universe of
accessible information available on many, many computers spread through the world and attached to that gigantic computer network called the Internet. The Web is, in essense, a sub-part of the Internet. The Web has a body of software, a set of protocols and a set of defined conventions for allowing anyone anywhere using a piece of software called a browser to get at the information on the Web. The Web uses hypertext and multimedia techniques to make the web easy for anyone to roam, browse and contribute to. The Web makes publishing information (i.e. making that information public) as easy as creating a "home page" (the first page you see when land one someone's web site) and posting it on a server somewhere in the Internet. Pick up any Web access software (e.g. Internet Explorer, Opera, Netscape, Mozilla), connect yourself to the Internet (through one of many dial-up, for-money, Internet access providers, or a broadband connection – a cable modem or a DSL line – or one of the many free terminals in Universities) and you will quickly discover an amazing diversity of information on the Web. From weather to stock reports to information on how to build nuclear bombs to the best tennis tips, it can be posted on the Web for all to read. Invented by Tim Berners-Lee at CERN, the Web is the first true "killer app" of the Internet. Ironically, Mr. Berners-Lee original originally called the Web a mesh. See Berners-Lee, Home Page, HTML and Internet.

World Zone 1 The area of the World Numbering Plan which is identified with the single-digit country code "1" and includes the territories of the United States and Canada (i.e. North America), and the following Caribbean countries: Antigua, Bahamas, Barbados, Bermuda, British Virgin Islands, Cayman Islands, Dominican Republic, Granada, Jamaica, Montserrat, Puerto Rico, St. Kitts, St. Lucia, St. Vincent, Virgin Islands.

World Zone 2 Africa. See World Numbering Zone.

World Zone 3 Europe. See World Numbering Zone.

World Zone 4 Europe. See World Numbering Zone.

World Zone 5 Central/South America. See World Numbering Zone.

World Zone 6 Pacific. See World Numbering Zone.

World Zone 7 Soviet Union, or the countries of what used to be the Soviet Union. How the world changes. See World Numbering Zone.

World Zone 8 Asia. See World Numbering Zone.

World Zone 9 Middle East. See World Numbering Zone.

WorldCom The predecessor company to MCI WorldCom. WorldCom originally was LDDS (Long Distance Discount Services), a switchless reseller of long distance services started in Jackson, Mississippi by an ex-high school basketball coach called Bernie Ebbers. LDDS grew, acquired Wiltel Communications, grew some more, acquired MFS (Metropolitan Fiber Services), grew some more, acquired Brooks Fiber and a few other companies, grew some more, and outbid British Telecom for MCI. MCI Worldcom is the company that resulted on September 13, 1998. It was an incredible success story for a time. The stock ran up to a peak of $64.51 in June 1999. At that time CEO Bernard Ebbers was listed by Forbes Magazine as one of the richest men in the US. Michael Jordan, the most popular athlete in the world, provided commercial endorsements. In October 1999, WorldCom attempted to purchase Sprint in a stock buyout for $129 billion. The deal was vetoed by the Department of Justice. At the same time, the success began to unravel with the accumulation of debt and expenses, the fall of the stock market, the long distance rates and revenue. In order to maintain its success, executives started faking up sales, moving expenses from the expense side to the income statement to the asset side of the balance. It would take two years for the extent of these problems to become public. And eventually WorldCom turned out to be the largest fraud in the history of American business. WorldCom filed Chapter 11 bankruptcy protection on July 21, 2002, nearly one month after it revealed that it had improperly booked $3.9 billion in expenses. Eventually Bernie Ebbers and the rest of his senior man-

agement left. New management came in and changed the name of the company to MCI.

WorldPartners An association of AT&T, KDD (Japan), and Singapore Telecom to provide international Virtual Network Services (VNS). VNS services are provided to over 32 countries which are major communications hubs. See also VNS.

WORM 1. Write Once, Read Many times. Refers to the new type of optical disks (similar to compact discs) which can be written to only once, but read many times. In other words, once the data is written, it cannot be erased. WORM disks typically hold around 650 plus megabytes. See also CD, CD-R, CD-RW and Erasable Optical Drive.

2. A malicious piece of software that duplicates itself repeatedly. A worm works its way onto and then through an entire computer network by duplicating itself repeatedly. A worm tends to be malicious, with intent to do damage. The Internet worm was perhaps the most famous. It successfully (and accidentally) duplicated itself on many of the systems across the Internet. A computer worm is a short computer program that spreads on its own over the Internet, causing computers to break down. A worm doesn't require a "host" file and typically spreads without any action by a computer user – making worms particularly pernicious for users who believe that they are safe as long as they don't open any suspicious e-mails. Security experts refer to both worms and viruses as "malicious code." The most ferocious threats today are "network worms," which exploit a particular flaw in a software product (often one by Microsoft). The author of Slammer, for example, noticed a flaw in Microsoft's SQL Server, an online database commonly used by businesses and governments. The Slammer worm would find an unprotected SQL server, then would fire bursts of information at it, flooding the server's data buffer, like a cup filled to the brim with water. Once its buffer was full, the server could be tricked into sending out thousands of new copies of the worm to other servers. Normally, a server should not allow an outside agent to control it that way, according to the New York Times, but Microsoft had neglected to defend against such an attack. Using that flaw, Slammer flooded the Internet with 55 million blasts of data per second and in only 10 minutes colonized almost all vulnerable machines. The attacks slowed the 911 system in Bellevue, Wash., a Seattle suburb, to such a degree that operators had to resort to a manual method of tracking calls. Unlike a virus, a worm generally does not alter or destroy data on a computer. Its danger lies in its speed: when a worm multiplies, it often generates enough traffic to brown out Internet servers, like air-conditioners bringing down the power grid on a hot summer day. The most popular worms today are mass mailers, which attack a victim's computer, swipe the addresses out of Microsoft Outlook (the world's most common e-mail program) and send a copy of the worm to everyone in the victim's address book. These days, the distinction between worm and virus is breaking down. A worm may carry a virus with it, dropping it onto the victim's hard drive to do its work, then e-mailing itself off to a new target.

3. Small, legless creeping animal. To survive, every bird must eat at least half its own weight in food each day. Young birds need even more. A young robin eats as much as 14 feet of earthworms a day.

WORN Write Once, Read Never (A joke). See also CDs.

worry Worry is interest paid on trouble before it falls due. – William Ralph Inge

worst hour of the year That hour of the year during which the median noise over any radio path is at a maximum. This hour is considered to coincide with the hour during which the greatest transmission loss occurs.

WOS 1. Wholesale Operator Services. The WOS product is a network-based service where USAN provides the automated operator services and First Data Corporation (FDC) (which provides MCI's own operator services) providers the live operator personnel.

2. Wireless Office Services, or Wireless Office System. See Wireless Office Telecommunications System.

WOSA Windows Open Services Architecture. According to Microsoft, WOSA provides a single system level interface for connecting front-end applications with back-end services. Windows Telephony, announced in May 1993, is part of WOSA. According to Microsoft, application developers and users needn't worry about conversing with numerous services, each with its own protocols and interfaces, because making these connections is the business of the operating system, not individual applications. WOSA provides an extensible framework in which Windows based applications can seamlessly access information and network resources in a distributed computing environment. WOSA accomplishes this feat by making a common set of APIs available to all applications. WOSA's idea is to act like two diplomats speaking through an interpreter. A front-end application and back-end service needn't speak each other's languages to communicate as long as they both know how to talk to the WOSA interface (e.g. Windows Telephony). As a result, WOSA allows application developers, MIS managers, and vendors of back-end services to mix and match applications and services to build enterprise solutions that shield programmers and users from the underlying complexity of the system.

This is how WOSA works: WOSA defines an abstraction layer to heterogeneous computing resources through the WOSA set of APIs. Initially, this set of APIs will include support for services such as database access, messaging (MAPI), file sharing, and printing. Because this set of APIs is extensible, new services and their corresponding APIs can be added as needed.

WOSA uses a Windows dynamic-link library (DLL) that allows software components to be linked at run time. In this way, applications are able to connect to services dynamically. An application needs to know only the definition of the interface, not its implementation. WOSA defines a system level DLL to provide common procedures that service providers would otherwise have to implement. In addition, the system DLL can support functions that operate across multiple service implementations. Applications call system APIs to access services that have been standardized in the system. The code that supports the system APIs routes those calls to the appropriate service provider and provides procedures and functions that are used in common by all providers.

The primary benefit of WOSA is its ability to provide users of Windows with relatively seamlessly connections to enterprise computing environments. Other WOSA benefits, according to Microsoft include:

- Easy upgrade paths.
- Protection of software investment.
- More cost-effective software solutions.
- Flexible integration of multiple-vendor components.
- Short development cycle for solutions.
- Extensibility to include future services and implementations.

See also ODBC, MAPI, and TAPI.

WOTS Wireless Office Telecommunications System

WPA Wi-Fi Protected Access (WPA) is an industry standard based on a subset of an early draft of the IEEE 802.11i specification, Robust Security Network for WLANs (Wireless Local Area Networks). WPA has replaced WEP (Wired Equivalent Privacy), which proved too easy to compromise. WPA replaces the WEP keying mechanism with the more robust TKIP (Temporal Key Integrity Protocol), adds a strong message integrity check and supports authentication using 802.1X. See also WEP and Wi-Fi.

WPAD Web Proxy Auto-Discovery Protocol. An Internet protocol which uses a collection of pre-existing resource discovery mechanisms to auto-discover Web proxy servers. The resource discovery mechanisms available include Dynamic Host Configuration Protocol (DHCP), Service Location Protocol (SLP), and Domain Name Service (DNS) records. See also Proxy.

WPAN Wireless Personal Area Networks. A network for interconnecting devices centered around the individual person in which the connections are wireless. This network typically consists of home multimedia devices including digital cameras and camcorders, music players, set-top boxes, game consoles and high definition television.The IEEE 802.15 Working Group (WG) is developing a set of standards for WPANs, attempting to provide for the coexistence of Bluetooth and 802.11 WLANs (Wireless Local Area Networks). In October, 2003, the IEEE Millimeter Wave Interest Group formed an interest group to explore the use of the 60 GHz band for wireless personal area networks (WPANs), which generally (according to the IEEE) have a range of 10 meters. This little-used, 7 GHz-wide portion of the radio spectrum (as defined in FCC 47 CFR 15.255) avoids interference with nearly all electronic devices, given the high attenuation of these wavelengths by walls and floors (meaning it doesn't penetrate walls and floors), and promises to allow more WPANs to occupy the same building, according to the IEEE. The IEEE 802.15.3(TM) Millimeter Wave Interest Group (mmWIG) was formed in July 2003 as part of an effort to develop a millimeter-wave-based alternative physical layer (PHY) for the IEEE high-rate WPAN standard, IEEE 802.15.3. Interest groups are the first step in the creation of a standard. See also 802.11 and Bluetooth.

WPBX Wireless private branch exchange. The WPBX offers business users the ability to make and receive calls using cordless telephones anywhere on a company's premises.

WPM See Words Per Minute.

WPON A PON (Passive Optical Network) technology that makes use of Wavelength Division Multiplexing (WDM) PONs (WPON). WPON uses wavelength splitters to deliver an individual wavelength to each premises in support of high-bandwidth applications such as MTUs (Multi-Tenant Units). WPON research focuses on wavelength splitters (which replace the power splitters found in standardized techniques and EPON). Because of the large number of possible wavelengths, they do not need to be shared. Adding a new customer means adding a laser with a different wavelength in the central office, but that has no impact on the other subscribers. See also PON.

WR-340 A specific size of rigid waveguide with a rectangular cross sectional shape.

WRAM Windows Random Access Memory. Similar to VRAM, but with added logic to accelerate common video functions such as bit-block transfers and pattern fills. See also VRAM.

WRAN Wireless Regional Area Networks. See 802.22.

WRBYV While Running Backwards You'll Vomit. The acronym is WRBYV, or White, Red, Black, Yellow, and Violet. These are the first five colors for all twisted 25-pair telephone cable in North America. The phrase makes it easier to remember the scheme. The wires are:

Tipside: (WRBYV) White, Red, Black, Yellow, Violet. Ringside:(BOGBS) Blue, Orange, Green, Brown, Slate.

The ring side color changes every pair in each group of five pairs. But the tipside only changes once for each group, as such: first pair – blue/white, second pair – orange/white, third pair green/white etc. The sixth pair would be blue/red, etc. to the last pair of a 25-pair cable, which be slate/violet. Out in the west, they usually think of BOGBS as being more important than WRBYV.

wrap 1. In data communications, to place your diagnostic and test equipment around parts of a network so you can monitor their use (i.e. do network diagnostics on them). You are, in essence, wrapping your products around theirs.

2. To make a connection between a flexible wire and a hard tag by tightly wrapping the cable around the tag. There are automatic wire wrapping tools available for this job.

3. Redundancy measure in IBM Token Ring LANS. Trunk cabling used in Token Ring TCUs contains two data paths: a main and backup normally unused). If the trunk cable is faulty, the physical disconnection of the connector at the TCU causes the signal from the main path to wrap around on to the backup path, thus maintaining the loop. The term wrap is now used on FDDI networks. If a failure occurs on one of the FDDI rings, the stations on each side of the failure reconfigure. The two rings then are combined into a single ring topology that allows all functioning stations to remain interconnected.

wrap-up Between-call work state that an ACD agent enters after releasing a caller. It's the time necessary to complete the transaction that just occurred on the phone. In wrap-up, the agent's ACD phone is removed from the hunting sequence. After wrap-up is completed, it is returned to the hunting sequence and is ready to take the next call.

wrap-up codes A call center term. Codes agents enter into the ACD to identify the types of calls they are handling. The ACD can then generate reports on call types, by handling time, time of day, etc.

wrap-up data Ad hoc data gathered by an agent in the ACD system following a call.

wrap-up time A call center term. The time an employee spends completing a transaction after the call has been disconnected. Sometimes it's a few seconds. Sometimes it can be minutes. Depends on what the caller wants.

wrapper A program that accepts parameters and passes them to another program that does the actual work. One use of wrappers is to enable legacy mainframe applications to be accessible as Web services to intranet or Internet users. In this scenario, a wrapper is a program that serves as a communications mechanism between the user's browser and the mainframe application, enabling users to interact with what appears to them to be a normal Web service, while making the users appear to be normal mainframe users to the legacy mainframe application. See also digital wrapper.

wrapping In token-ring networks, the process of bypassing cable faults without changing the logical order of the ring by using relays and additional wire circuits.

WRC World Radiocommunications Conference. AN ITU conference, usually held every two years, to review and amend the international radio regulations, allowing for the introduction of new technologies and more efficient sharing of the radiofrequency spectrum.

WREC Web REplication and Caching. A term describing the processes by which a proxy server goes about replicating data on origin Web servers and caching it in order to improve the overall efficiency with which Internet network resources are employed. See also Caching and Replication.

WRED Weighted Random Early Detection. A QoS (Quality of Service) mechanism for IP-based networks. See Weighted Random Early Detection.

wrench In the Tour de France bicycle race, a mechanic is called a wrench.

wrist candy Computers or their ilk are becoming smaller. Some manufacturers think the next big thing will be computers you wear on your wrist. They're calling them wrist candy. At least that's one explanation. The other is that wrist candy is another term for arm candy or eye candy, i.e. an attractive woman you're escorting on your arm.

write To record information on a storage device, usually disk or tape.

write head A magnetic head capable of writing only. You find write heads on everything from tape recorders to computers.

write protect Using various hardware and software techniques to prohibit the computer from recording (writing) on storage medium, like a floppy or hard disk. You can write protect a 5 1/4 diskette by simply covering the little notch with a small metal tag. The idea of "Write Protect" is to stop someone (including yourself) from changing your precious data or program. You can't write protect a hard disk easily. The easiest way to stop someone changing a file is to use the program ATTRIB.EXE. See Attributes.

write protection A scheme for protecting a diskette from accidental erasure. 5 1/4" diskettes have a notch which must be uncovered to allow data on the diskette to be modified. 3 1/2" diskettes have small window with a plastic tab which must be slid into place to cover the window to allow data on the diskette to be modified. See Write Protect and Attributes.

write protection label A removable label, the presence or absence of which on a diskette prevents writing on the diskette.

write rate An RFID definition. The rate at which information is transferred to a tag, written into the tag's memory and verified as being correct.

wrong "I am willing to admit that I may not always be right, but I am never wrong." Samuel Goldwyn.

WRSS Western Region Switch Support

WRT With respect to.

WS Web services security initiative. IBM, Microsoft and VeriSign are proposing a Web services security framework beginning with WS-Security for message integrity and confidentiality. Some of the elements are as follows:

- WS-Authorization Describes how to manage authorization data and policies. See WS-Security.
- WS-Federation Describes how to manage and broker trust relationships among unlike systems in a federated environment.
- WS-Privacy Model for stating privacy preferences and practices.
- WS-Security Describes how to attach digital signatures and encryption headers and security tokens, such as Kerberos tickets or X.509 certificates, to SOAP messages. WS-Policy: Used to express conditions and constraints of security policies.
- WS-Secure Conversation Describes how to manage and authenticate message exchanges between Web services.
- WS-Trust Framework for direct and brokered trust relationship between Web services.

WSC Wireless Switching Center. A switching center designed for wireless communications services, typically fixed wireless services including data and voice.

WSDL Web Services Description Language. Think of Web Services, a new flavor of software that makes it easier to integrate different companies' web sites (and the systems that feed them) over the Internet. This means, amongst other things, that one Web site will be able to sell visitors a combination of products and services from many web sites, making the visitors' visit easier, richer and more rewarding. Web services is based on four standards, XML, SOAP, UDDI and WSDL. Now understand that UDDI (Universal Description Discovery and Integration) is a virtual yellow pages for web services which lets software automatically figure out what web services are available and how to hook up to them. Now figure that WSDL are the few words associated with each entry into the yellow pages that describes the kind of work a web service can do for you. See UDDI and Web Services.

WSXC Wavelength Selective Cross Connect. A type of Optical Cross-Connect (OXC). See OXC.

WTAC World Telecommunications Advisory Council. WTAC is comprised of telecommunications leaders from the private and public sectors and from every region of the world. WTAC gives advice to the ITU – the International Telecommunications Union. The WTAC held its first meeting in Geneva, Switzerland, in April, 1992. In February 1993, it published a small booklet called "Telecommunications Visions of the Future."

WTLS Wireless Transport Layer Security. WTLS is based on the Transport Layer Security (TLS) protocol, a derivative of the Secure Sockets Layer (SSL) protocol. The goal of WTLS is very much like that of SSL: to provide privacy and reliability for client-server communications over a network. While SSL primarily provides security over the Internet, WTLS is specific to wireless applications using WAP. The need for WTLS over TLS and SSL is due to the restrictions present in the wireless application environment. Specifically, TLS and SSL don't offer the necessary support for the limited memory and processing capabilities of WAP-enabled phones, support for multiple transport protocol layers, or capabilities to address low-bandwidth environments.

WTN Working Telephone Number.

WTNG WaiTiNG.

WTO World Trade Organization. An organization of over 130 member countries and 30 observer countries. Formed in 1995, the WTO is a formal organization with legal status, which succeeded the ad hoc GATT (Global Agreement of Tariffs and Trade); the GATT agreement is now part of WTO agreements. According to the WTO, its main functions are assisting developing and transition economies, specialized assistance in export promotion, regional trading arrangements, cooperation in global economic policy-making, and routine notification when members introduce new trade measures or alter new ones. The WTO also assists in the settlement of trade disputes between member countries. The WTO gets involved in telecommunications in a number of ways, most recently in assisting in the development of multilateral agreements for normalization of international long distance costs. Currently, there exist incredible differences in such costs, depending on where the call originates; i.e., a 10 minute call from the U.S. to Hong Kong is much less expensive than the same call from Hong Kong to the U.S. It's more a matter of national politics and economic policy than anything else, resulting in an imbalance of trade in the magnitude of billions of dollars flowing out of the U.S. to other countries. Regardless of the direction of the call, the costs to the national carriers are roughly the same, and the revenues are split 50/50 between the two carriers. As a means around this inequity, International Callback developed. International Callback is viewed as illegal by most countries outside the U.S., as they seek to protect this significant source of hard currency. www.wto.org See also International Callback.

WTP The WAP equivalent of the reliability portion of HDTP. www.wapforum.org

WU-ATS Western Union-Advanced Transmission Systems

Wugga Wugga The sound of a computer program.

WUI 1. Web User Interface. Pronounced "wooey." A GUI (Graphical User Interface) for the WWW.

WVI Web Voice Integration. See TCP/IP and VoIP.

wVoIP Wireless VoIP. Making a VoIP phone call over a wireless connection. Typically this refers to a Wi-Fi network. The idea See VoIP.

WWDD World Wide Direct Dialing.

WWDM Wide Wavelength Division Multiplexing. WWDM is a Wavelength Division Multiplexing system employing four wavelengths. In a GbE (Gigabit Ethernet) context, a WWDM system runs each of four wavelengths at 2.5 Gbps, with all running in parallel over a single optical fiber, thereby yielding an aggregate 10 Gbps. See also GbE, DWDM, and WDM.

WWNWS The World-Wide Navigational Warning Service, established in 1979 by the International Maritime Organization (IMO) and the International Hydrographic Office (IHO) to provide ships at sea with vital non-distress information, including the malfunctioning of lights, sound signals, buoys and other aids to navigation; the location of wrecks and other hazards; and the building of new offshore structures.

WWV The callsign of the National Institute of Standards and Technology's shortwave radio time signal station near Fort Collins, Colorado. WWV continuously broadcasts official U.S. government time signals on five frequencies: 2.5 MHz, 5 MHz, 10 MHz, 15 MHz and 20 MHz. WWV's time signal can also be reached by phone at 1-303-499-7111. The time signals are derived from a set of atomic clocks located at the transmitter site. WWV is one of a few radio stations west of the Mississippi River whose call sign begins with W, an artifact of WWV's originally being located on the East Coast.

WWVB The callsign of the National Institute of Standards and Technology's time signal radio station near Fort Collins, Colorado. WWVB is the station that radio-controlled clocks throughout North America use to synchronize themselves. WWVB continuously broadcasts official U.S. government time signals at 60 MHz. WWVB is one of a few radio stations west of the Mississippi River whose call sign begins with W.

WWVH The callsign of the National Institute of Standards and Technology's shortwave radio time signal station in Kekaha, Kauai, in the state of Hawaii. WWVH continuously broadcasts official U.S. government time signals on four frequencies: 2.5 MHz, 5 MHz, 10 MHz, and 15 MHz. WWVH's time signal can also be reached by phone at 1-808-335-4363. WWVH is one of a few radio stations west of the Mississippi River whose call sign begins with W.

WWW 1. World Wide Web; a hypertext-based system for finding and accessing resources on the Internet network. For a much better explanation, see World Wide Web and Internet.

2. Also referred to as World Wide Wait, which according to some frustrated observers, is the real meaning of WWW.

WWW2 See www3.

WWW3 www3 is as an alternate name to www in a bank of web server computers, i.e. a group of computers named www1.company.com, www2.company.com, www3.company.com, etc. These servers would typically be assigned to a "round robin" DNS (Domain Name Server) and share the load for the site. To get to that site, a Web surfer could type www.company.com, www1.company.com, www2.company.com or www3.company.com. He'd get the same information and hit what appeared to him as the same server.

WX channels In maritime radio communications, these are weather channels, reserved for National Weather Service broadcasts of weather information to maritime vessels and coastal marine facilities. Ships are not permitted to broadcast on these frequencies. In North America, the following frequencies are used for weather channels.

Channel	Frequency
WX1	162.550 MHz
WX2	162.400 MHz
WX3	162.475 MHz
WX4	162.425 MHz
WX5	162.450 MHz
WX6	162.500 MHz
WX7	162.525 MHz
WX8	161.650 MHz
WX9	161.775 MHz
WX0	163.275 MHz

Different frequencies are used in other parts of the world. The channel numbers, WX1, WX2, etc., have no special significance but are often designated this way in consumer equipment. Other channel labeling schemes may be used.

WYPIWYF Acronym for "What You Print Is What You Fax," also "The Way You Print Is the Way You Fax." Coined by Intel to describe its one-step pop-up menu that makes sending faxes from the PC as easy as sending a document to a printer.

WYSIWYG (pronounced Whiz-i-wig) What You See Is What You Get. A word processing term meaning what you see on your computer screen is what you will see printed on paper. The exact typeface, the correct size, the right layout, etc. Some word processors do WYSIWYG. Others don't. You usually need a screen with graphics to get the full effect.

WZ1 World Zone One. The part of the earth covered by what used to be called The North American Numbering Plan. It includes the U.S., Canada, Alaska, Hawaii, and the Caribbean islands, but does not include Mexico or Cuba.

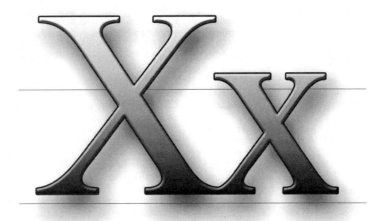

X 1. An abbreviation for the word "cross," as in crossbar (Xbar). 5XB would be the abbreviation for a No. 5 Crossbar circuit switch.

2. Generic. As in xDSL, generic Digital Subscriber Line, which is a family of access technologies including ISDN, ADSL, IDSL, HDSL, RADSL and SDSL. Also, as in 1000Base-CX, an IEEE standard for Gigabit Ethernet over some sort of Copper cable.

3. In mathematics, an unknown quantity.

4. The symbol designation for Reactance.

5. In PCs, the universal symbol for "Close me."

X Band 7 GHz and 8 GHz. Used by military satellites.

X Display Manager Control Protocol See XDMCP.

X Internet See extended Internet.

X rays Electromagnetic radiation with a wavelength of under 100 0xE8ngstroms. When Wilhelm Roentgen discovered the X-ray in 1895, some journalists were convinced that the primary user of the revealing high-frequency radiation would be the "peeping Tom." The (pardon the expression) titillating publicity led to legal and mercantile steps, resulting in New Jersey passing a law banning "X-ray opera glasses," and merchants in London selling X-ray proof feminine underwear.

X recommendations The ITU-T documents that describe data communication network standards. Well-known ones include: X.25 Packet Switching standard, X.400 Message Handling System, and X.500 Directory Services.

X series Recommendations drawn up by the ITU-T to establish communications interfaces for users' Data Terminal Equipment (DTE) and Data Circuit Terminating Equipment (DCE). They govern the attachment of data terminals to public data networks (PDNs) and the Public Switched Telephone Network (PSTN). In short, a set of rules for interfacing terminals to networks.

X terminal A networked desktop machine that displays software applications which are running on a networked server. X terminals are allegedly cheaper and easier to network than PCs or workstations. They are said to be a forerunner to thin clients, also called network computers (NCs).

X Windows X Windows is officially called X Window, or The X Window System, or X. Sometimes it is spelled with an hyphen, e.g.X-Window. It is used primarily with UNIX systems, but not exclusively. X was originally designed to be an industry-standard network windowing system, and because UNIX is widely used, X has become the windowing standard for UNIX.

X-10 protocol A proprietary technology that enables electronic devices to send and receive commands to and from each other via a home's standard electrical wiring. You can use X-10 to control conforming black boxes and devices. The protocol is a command signal that rides on the AC 60 cycle sine power curve. The signal is a series of 120 kHz pulses sent on the "zero crossing" of each cycle. The signal is in a binary fashion, transmitting the

letter code, unit code and command for a device. All receivers monitor the power line waiting for a command to respond. The limit of number of unique codes available is 256. This is derived by having 16 letter codes and 16 unit codes, hence 16x16=256 "addresses." More than one device can share an address. If you decide that every time you select "A1" on, you want all the front lights to come on, then you can place all the receivers to the same address. The system is easy to use and very flexible.

X-axis Horizontal axis on a graph or chart.

X-base A term used to describe any database application capable of generating custom programs with dBASE-compatible code.

X-dimension of recorded spot In facsimile, the center-to-center distance between two recorded spots measured in the direction of the recorded line. This term applies to facsimile equipment that responds to a constant density in the subject copy by yielding a succession of discrete recorded spots.

X-dimension of scanning spot In facsimile, the center-to-center distance between two scanning spots measured in the direction of the scanning line on the subject copy. The numerical value of this term will depend upon the type of system used.

X-Off/X-On A flow control protocol for asynchronous serial transmission. Flow control is a method of adjusting information flow. For example, in transmitting between a computer and a printer, the computer sends the information to be printed at 9600 baud. That's several times faster than the printer can print. The printer, however, has a small memory. The computer dumps to the memory, called a buffer, at 9600 baud. When it fills up, the printer signals the computer that it is full and please stop sending. When the buffer is ready to receive again, the printer (which also has a small computer in it) sends a signal to the desktop computer (the one doing the printing) to please start sending again. X-OFF means turn the transmitter off (xmit in Ham radio terms). It is the ASCII character Control-S. X-ON means turn the transmitter on. It is the ASCII character Control-Q. You can use these characters with many microcomputer functions. For example, if you do DIR in MS-DOS and you want to stop the fast rush of files, then type Control-S.

X-open An international consortium of computer vendors working to create an internationally supported vendor-independent Common Applications Environment based on industry standards.

X-ray When X-rays were discovered by Wilhelm Roentgen in 1895, some journalists were convinced that the primary user of the revealing shortwave radiation would be the "peeping Tom." The titillating publicity led to legal and mercantile steps. Two examples: A law introduced in New Jersey forbade the use of "X-ray opera glasses," and merchants in London sold X-ray-proof underwear to modest ladies.   See X-Ray Lithography.

X-ray lithography A lithographic process in which X-rays, rather than light or electron beams, are used to transfer circuit patterns to a silicon wafer. The advantage of X-rays is their shorter wavelengths, which reduce diffraction and yield greater resolution and finer line widths of features. This allows more transistors to be packed onto a chip.

X-series recommendations Set of data telecommunications protocols and interfaces defined by the ITU-T.

X-Windows The UNIX equivalent of Windows. What Windows is to MS-DOS, X-Windows is to Unix. A network-based windowing system that provides a program interface for graphic window displays. X-Windows permits graphics produced on one networked station to be displayed on another. Almost all UNIX graphical interfaces, including Motif and OpenLook, are based on X-Windows. X-Windows is a networked window system developed and specified by the MIT X Consortium. Members of the X Consortium include IBM, DEC, Hewlett-Packard and Sun Microsystems. Sun Microsystems has been contracted by the MIT X Consortium to implement PEX (PHIGS Extensions to X), which will be the standard networking protocol for sending PHIGS (Programmers Hierarchical Graphics System) graphics commands through X-Windows. Some people spell it X-Windows and some spell it X Windows.

X-Y A specific variety of electromechanical switch. Does the same things as a Stronger step-by-step switch but in a horizontal plane. It's so called because it's a two motion switch with horizontal and vertical movements. The first pulse sends the switch horizontally to the right place, then the next pulse sends it vertically up to the right place and so on, until it has switched the call through. One of the most reliable switches ever produced. Unfortunately, it's slow, space-consuming and unable to be programmed with many new customer pleasing features.

X.1 ITU-T specification that defines classes of service in a packet-switched network, such as virtual-circuit, datagram, and fast-packet services.

X.110 ITU-T recommendation that specifies routing protocol for international PDN services accessing similar switched PDNs.

X.12 The dominant EDI standard in the U.S. today. IN 1979 ANSI chartered the Accredited Standards Committee (ASC) X.12 to develop uniform standards for electronic interchange of business transactions. Pronounced "ex twelve" – the dot (X.12) was formally dropped by the standards board.

X.121 International Numbering Plan for public data networks. X.121 defines the numbering system used by data devices operating in the packet mode. X.121 is used by ITU-T X.25 packet-switched networks and has been proposed by several computer vendors as the future universal addressing scheme. X.121 comprises the ITU-T recommendations that specifies the international numbering plan for PDNs. The numbering plan uniquely identifies the world zone, country or geographical area, network, and subscriber.

X.121 Address A standard O/R (Originator/Recipient) attribute that allows Telex terminals to be identified in the context of store-and forward communications.

X.130 Call setup and clear down times for international connection to synchronous PDNs.

X.132 Grade of service over international connections to PDNs.

X.150 DTE and DCE test loops in public data networks.

X.2 ITU-T recommendation that specifies the international interface for special and optional service facilities (features) available to PDNs.

X.20 Asynchronous communications interface definitions between data terminal equipment (DTE) and data circuit terminating equipment (DCE) for start-stop transmission services on public switched telephone networks.

X.20 bis Used on public data networks of data terminal equipment (DTE) that is designed for interfacing to asynchronous duplex V-series modems.

X.200 A series of ITU (International Telegraph and Telephone Consultative Committee) recommendations that defines the type of service offered by specific layers in the OSI (Open Systems Interconnection) Model, and defines the protocol to be used at those layers. ITU is now the ITU.

X.21 Interface between Data Terminal Equipment (DTE) and DATA Circuit-Terminating Equipment (DCE) for synchronous operation on public data networks. Specifies protocols for all three lowest layers of OSI. Efficient transfer of data bit streams (unblocked) between the DTE and DCE. Only used for link establishment/disestablishment and other connection control functions. Applicable to both leased lines and circuit switching.

X.21 bis ITU-T recommendation that specifies the interface for DTE synchronous V-series modems (similar to RS-232-C). Used on public switched telephone networks of data terminal equipment (DTE) that is designed for interfacing to synchronous V-series modems.

X.21 TSS Specification for Layer 1 interface used in the X.25 packet-switching protocol and in certain types of circuit-switched data transmissions.

X.215/X.225 ITU-T Layer 5 protocol with "Call Set Up" to provide "sessions" between host computers. Access validation can be provided for security. Re-establishes failed transport connections. Synchronizes transfer of different kinds of information.

X.216/X.226, X.209 ITU-T Layer 6 Presentation Services and Protocol negotiates syntax and format (ASCII , Compressed Data, Encrypted data). X.209 Abstract Syntax Notation One to specify format or to switch from one format to another for a different session activity.

X.224 A ITU-T standard associated with the transport layer of the Open Systems Interconnect (OSI) architecture used in networks employing circuit-switched techniques.

X.225 A ITU-T standard associated with the session layer of the Open Systems Interconnect (OSI) architecture used in networks employing circuit-switched techniques.

X.226 A ITU-T standard associated with the Open Systems Interconnect (OSI) architecture that defines specific presentation layer services used with circuit-switched network services.

X.24 List of definitions for interchange circuits between data terminal equipment (DTE) and data circuit terminating equipment (DCE) on public switched telephone networks.

X.25 From its beginning as an international standards recommendation from ITU-T, the term X.25 has come to represent a common reference point by which mainframe computers, word processors, mini-computers, VDUs, microcomputers and a wide variety of specialized terminal equipment from many manufacturers can be made to work together over a type of data communications network called a packet switched network. On a packet switched data network (private or public), the data to be transmitted is cut up into fixed size blocks (i.e. packets). Each block has a header with the network address of the sender and that of the destination. As the block enters the network, the number of bits in the block are put through some mathematical functions (an algorithm) to produce a check sum. The check sum is attached as a "trailer" to the packet as it enters the network.

Packets may travel different routes through the network. But, ultimately, the packets are routed by the network to the node where the destination computer or terminal is located. At the destination, the packet is disassembled. The bits are put through the same algorithm, and if the digits computed are the same as the ones attached as the trailer, there are no detected errors. An ACK, or acknowledgement, is then sent to the transmitting end. If the check sum does not match, a NAK, or Negative Acknowledgement is sent back, and the packet is retransmitted. In this manner, high speed, low error rate information can be transmitted around the country using shared telecommunications circuits on public or private data networks. X.25 is the protocol providing devices with direct connection to a packet switched network. These devices are typically larger computers, mainframes, minicomputers, etc. Word processors, personal computers, workstations, dumb terminals, etc. do not support the X.25 packet switching protocols unless they are connected to the network via PADs – Packet Assembler/Disassemblers. A PAD converts between the protocol used by the smaller device and the X.25 protocol. This conversion is performed on both outgoing (from the network) and incoming data (to the network), so the transmission looks transparent to the terminal.

Here is a more technical definition: X.25 is the ITU-T Standards Recommendation (1976) for the protocol which provides the user interface into the original packet switched network. The basic concept of packet switching is one of a highly flexible, shared network in support of interactive computer communications across a wide area network. Previously, large numbers of users spread across a wide area and with only occasional communications requirements, had no cost-effective means of sharing access from their computer terminals to computer resources for time-share applications. In specific, the issue revolved around the fact that asynchronous communications are bursty in nature. In other words, data transmission is in bursts of keystrokes or data file transfers. Further, there is lots of idle time on the circuit between transmissions in either direction of relatively small amounts of data. Additionally, those early networks consisted of analog, twisted pair facilities, which offered very poor error performance and relatively little bandwidth.

Existing circuit-switched networks certainly offered the required flexibility, as users could dial-up the various host computers on which the desired database resided. Through a low-speed modem, which was quite expensive at the time, data could be passed over the analog network, although error performance was less than desirable. However, the cost of the connection was significant, as the calls were billed based on the entire duration of the connection, even though the circuit was idle much of the time. Dedicated circuits could address the imbalance between cost and usage, as costs are not usage-sensitive and as dedicated circuits could be shared among multiple users through a concentrator. However, dedicated circuits were expensive, especially in long-haul applications, and involved long implementation delays. Further, users tended not to be concentrated in locations where they could make effective use of dedicated circuits on a shared basis. Finally, large numbers of dedicated circuits would be required to establish connectivity to the various hosts. Packet switching solved many of those problems, in the context of the limitations of the existing networks–namely, their analog, twisted pair nature. Packet-switched networks support low-speed, asynchronous, conversational and bursty communications between computer systems. As packet-switched network usage can be billed to the user on the basis of the number of packets transmitted, such networks are very cost-effective for low-volume, interactive data communications. This cost advantage comes from the fact that the bursty nature of the such interactive applications allows large volumes of data transmissions from multiple users to be aggregated in order to share network facilities and bandwidth. Further, packet-switched networks perform the process of error detection and correction at each of the packet switches, or nodes; thereby, the integrity of the transmitted data is improved considerably.

Information is transported and switched through the network on the basis of packets, which are structured in a standard data format by a Packet Assembler/Disassembler (PAD). The PAD, which functionally sits between the user host computer or terminal and the edge of the carrier network, executes the X.25 protocol by assembling (forming) the packets on the transmit side and disassembling the packets on the receiving side of the communication. The packet format comprises a header, the payload, and a trailer. Each packet is of a fixed maximum size, typically containing 128B (B=Byte or octet) or 256B of payload (user data); packet sizes of up to 4,096B can be accommodated in some networks. The typical upper limit of a packet is 512B or 1,024B, with the latter packet size being used in many airline reservation networks. If a large file is being transmitted, the PAD fragments the file which streams out of the transmitting device, carving it into packets of the specified size. If only a few bytes are being transmitted (e.g., a few keyboard commands or mouse clicks), the PAD gathers them up over a short, pre-determined period of time, and adds the necessary number of "stuff bits" to fill the packet payload.

The payload data is encapsulated by a beginning flag (8 bits) in the header, and an ending flag (8 bits) in the trailer. These flags serve to the distinguish each packet from other packets traveling the same path. The beginning flag also serves as synchronizing bits in order that the packet nodes (intelligent switches) and the receiving terminal equipment might synchronize on the rate of transmission. This approach reduces overhead in data transmission, thereby improving the efficiency of transmission. The header also includes a packet address field of 8 bits (4 bits for the calling DTE and 4 bits for the called DTE) prepended to the data in order that the various packet nodes might route each packet on its way over an appropriate path, ultimately to the target device. Control data (8-16 bits) in the header includes the packet sequence number in order that the target node and terminal equipment are able to identify errored, corrupted or lost packets, and to resequence the packets should they arrive out of order. Additionally, the header control data includes the number of the virtual circuit (4 bits) and virtual channel (8 bits) over which the data will travel, if a path has been preordained. Finally, error control data is included in the form of a highly reliable CRC check (16 bits), which comprises the trailer. On the receiving side of the transmission, the PAD disassembles the packet, stripping away the header, trailer and stuff bits in order to get to the actual data being transmitted. See also ATM, CRC, Frame Relay, NBMA, Packet and X.25 network.

X.25 network Any network that implements the internationally accepted ITU-T standard governing the operation of packet-switching networks. The X.25 standard describes a switched communications service where call setup times are relatively fast. The standard also defines how data streams are to be assembled into packets, controlled, routed, and protected as they cross the network. See X.25.

X.28 Defines the Terminal-Pad Interface. DTE/DCE interface for start-stop-mode data terminal equipment accessing the packet assembly/disassembly facility (PAD) in a public switched telephone networks situated in the same country.

X.29 Procedures for the exchange of control information (handshaking) and user data between a packet assembly/disassembly facility (PAD) and a packet mode DTE or another PAD.

X.3 ITU-T recommendation describing the operation of a Packet Assembly/Disassembly (PAD) device or facility in a public data network. X.3 defines a set of 18 parameters that regulate basic functions performed by a PAD to control an asynchronous terminal. The setting of these parameters governs such characteristics as terminal speed, terminal display, flow control, break handling and data forwarding conditions, and so on.

X.30 Support of X.20 bis, X.21 and X.21 bis DTEs by an ISDN line.

X.300 ITU-T recommendation that specifies the essential interaction between elements of customer interfaces, interexchange signaling systems, and other network functions specifically related to the use of international user facilities and network utilities.

X.31 Support of packet mode DTEs by an ISDN system. This standard describes two modes of operation packetized data in the B-channel and packetized data over the D-channel.

X.32 ITU-T recommendation for an interface between Data Terminal Equipment (DTE) and Data Circuit Terminating Equipment (DCE) operating in a packet mode (i.e. X.25) and

accessing a packet switched public data network via a public switched telephone network or a circuit switched public data network. X.32 describes the functional and procedural aspects of the DTE/DCE interface for DTEs accessing a packet switched public data network via a public switched network.

X.38 ITU-T recommendation for the access of Group 3 facsimile equipment to the Facsimile Packet Assembly/Disassembly (FPAD) facility in public data networks situated in the same country.

X.39 ITU-T recommendation for the exchange of control information and user data between a Facsimile Packet Assembly/Disassembly (FPAD) facility and a packet mode Data Terminal Equipment (DTE) or another pad, for international networking.

X.3T9.3 The ANSI committee responsible for the creation and perpetuation of Fiber Channel standards.

X.3T9.5 The ANSI committee responsible for the creation and perpetuation of FDDI Standards.

X.4 International Alphabet No.5 for character oriented data.

X.400 X.400 is an international standard which enables disparate electronic mail systems to exchange messages. Although each e-mail system may operate internally with its own, proprietary set of protocols, the X.400 protocol acts as a translating software making communication between the electronic mail systems possible. The result is that users can now reach beyond people on their same e-mail system to the universe of users of interconnected systems. One problem with e-mail sent between X.400 networks is that the sender's name is not sent. (I kid you not.) This was one element of the protocol the committees forgot! If your message crosses an X.400 network, remember to sign your name. The X.400 standard itself is an overview which is broken down under subsequent numbers:

X.400 A standard for electronic mail exchange; developed by the ITU-T.

X.402 Overall Architecture.

X.403 Conformance testing.

X.407 Abstract service definition conventions.

X.408 Encoded information type conversion rules.

X.409 ITU-T recommendation that specifies protocol for formatting a digital message. The format specifies such things as the fields and character length (must be in bytes).

X.411 Message transfer system.

X.413 Message store.

X.419 Protocol specifications.

X.420 Interpersonal messaging system. An IPM format specification using the X.400 transfer protocol. In addition to text, it also allows CAD/CAM, graphics, Fax, and other electronic information.

X.435 An EDI (Electronic Data Interchange) format specification based on the X.400 transfer protocol. It can also allow for CAD/CAM, graphics, Fax, and other electronic information to accompany an EDI interchange.

X.440 A VM (Voice Messaging) format specification, using the X.400 transfer protocol. In addition to voice, it can also contain CAD/CAM, graphics, fax, and other electronic information.

X.445 The X.445 standard, or APS (Asynchronous Protocol Specification), lets X.400 clients and servers exchange all types of digital data over public telephone networks rather than over X.25 leased lines, which are required today. Among those backing the spec are Intel, AT&T, Microsoft Corp., Lotus and Isocor. X445 is an extension of the X.400 standard. X.400 provides an option to other messaging backbones such as System Network Architectural Distribution Services and SMTP. It supports multimedia data traffic including text, binary files, E-mail, voice images, and sound.

In its December 19, 1994 issue, PC Week said that the X.445 standard, which earlier this month gained final approval from the International Telecommunications Union, should ease messaging dramatically by allowing users to exchange X.400 data traffic over standard telephone networks.

X.50 Fundamental parameters of multiplexing scheme for the international interface between synchronous data networks.

X.500 The ITU-T international standard designation for a directory standard that permits applications such as electronic mail to access information which can either be central or distributed. The X.500 standard for directory services provides the means to consolidate e-mail directory information through central servers situated at strategic points throughout the network. These X.500 servers then exchange directory information so each server can keep all its local mail directory information current. With X.500, any e-mail user, whether on OpenVMS, Macintosh, DOS, or UNIX workstations, can be listed in a central directory that can be accessed using an X.500-compatible user agent.

X.509 A cryptography term. The part of the ITU-T X.500 Recommendation which deals with Authentication Frameworks for Directories. Within X.509 is a specification for a certificate which binds an entity's distinguished name to its public key through the use of a digital signature. Also contains the distinguished name of the certificate issuer.

X.51 Fundamental parameters of multiplexing scheme for the international interface between synchronous data networks using 10-bit envelop structure.

X.51 bis Fundamental parameters of a 48Kbit/s transmission scheme for the international interface between synchronous data networks using a 10-bit envelope structure.

X.58 Fundamental parameters of multiplexing scheme for the international interface between synchronous data networks using a 10-bit envelope structure.

X.60 Common channel signaling for circuit switched data applications.

X.61 Signaling system no. 7 – data user part.

X.70 Terminal and transit control signaling for asynchronous services on international circuits between anisochronous data networks.

X.700 Management Framework for Open Systems Interconnection (OSI) for ITU-T Applications.

X.71 Decentralized terminal and transit control signaling on international circuits between synchronous data networks.

X.75 An international standard and ITU-T recommendation for linking X.25 packet switched networks. X.75 defines the connection between public networks, i.e. for a gateway between X.25 networks. X.75 defines terminal and transit call control procedure and data transfer system between packet switched public data networks.

X.80 Interworking of inter-exchange signals for circuit switched data services.

X.92 Hypothetical reference connections for synchronous PDNs in a packet switched network.

X.95 ITU-T specification dealing with a number of internal packet-switched network parameters such as packet size limitations and service restrictions.

X.96 Call progress signals in PDNs.

X/Open A group of computer manufacturers that promotes the development of portable applications based on UNIX. Formed in 1984, X/Open is dedicated to the identification, agreement and wide-scale adoption of information technology standards which reduce issues of incompatibility and which help users realize the benefits of open information systems. In February 1996, X/Open and the Open Software Foundation (OSF) consolidated to form The Open Group (www.opengroup.org). See The Open Group.

X1 See Blind Dialing.

X10 X10 is a powerline carrier protocol that allows compatible devices throughout the home to communicate with each other via the existing 110 volt AC (alternating current) electricity wiring in the house. With X10 it is possible to control lights and virtually any other electrical device from anywhere in the house with no additional wiring. What Can I Control? According to the X10 proponents, Lighting is the most popular control category. Starter kits begin at $10 or so and are plug-and-play. You can also control Security Systems & Access Control; Home Theater & Entertainment; Phone Systems; Thermostats for adjusting room temperatures; Irrigation Have your sprinklers turn on only when it's dry. Because X10 devices don't use dedicated wiring, they can mysteriously stop working. Here's why. When X10 modules suddenly stop responding, it's usually due to something that was recently plugged into the electricity lines. Some electrical devices, like computers, televisions, and A/V gear, which have nothing to do with X10, may cause interference. Since the X10 signal goes everywhere in your home, some devices will have more effect on the signal strength. Power supplies in the following devices were designed to kill electrical noise, however, X10 signal looks like noise: Televisions, computers and monitors, battery chargers for cell phones and laptop computers, satellite receivers, surge-protection power strips, computer UPS'.

X12 ANSI (American national Standards Institute) standard that is the dominant EDI (Electronic Data Interchange) standard in the U.S. today; designed to support cross-industry exchange of business transactions. Standard specifies the vocabulary (dictionary) and format for electronic business transactions.

x2 A pre-standard 56-Kbps modem solution developed by US Robotics, x2 was one of two proprietary (read non-standard) approaches for running data over dial-up phone lines at up to 56 Kbps (actually 53.3 Kbps) one way and up to 33.6 Kbps the other way. The standard was developed for use on the Internet, with the 53.3 Kbps channel flowing downstream (i.e., to you) and the 33.6 Kbps channel running upstream (i.e., away from you). The logic is that at 53.3 Kbps, Web pages fill a lot faster on your screen. x2 was developed by US Robotics. K56flex, the competing proprietary 56-Kbps solution, was developed by Rockwell Semiconductor and Lucent Technologies. x2 was not compatible with K56flex. In other words, a K56flex modem could not talk to a x2 modem at full speed. The modems can, however, communicate at speeds up to 33.6 Kbps, which is the maximum speed of

the predecessor V.34+ modems. In February 1998, the ITU-T finalized the specification for V.90, the international standard for 56-Kbps modems. While V.90 is an amalgamation of x2 and K56flex, neither conforms to V.90, and not all x2 and K56flex modems are upgradable to V.90. See also 56 Kbps Modem (for a much longer explanation), K56flex, V.90, V.91, and V.92.

X3.15 Bit sequencing of ASCII in serial-by-bit data transmission.

X3.16 Character structure and character parity sense for serial-by-data communications in ASCII.

X3.36 Synchronous high speed data signaling rates between data terminal equipment and data circuit terminating equipment.

X3.41 Code extension techniques for use with 7-bit coded character set of ASCII.

X3.44 Determination of the performance of data communications systems.

X3.79 Determination of performance of data communications systems that use bit oriented control procedures.

X3.92 Data encryption algorithm.

X.Windows A networked GUI developed at the Massachusetts Institute of Technology (MIT) as part of Project Athena. Based on a client/server architecture it displays information from multiple networked hosts on a single workstation.

XA Transaction management protocol.

XA-SMDS Exchange Access SMDS. An access service provided by a local exchange carrier to an interexchange carrier. It enables the delivery of a customer's data over local and long distance SMDS networks.

Xanadu An idyllic place in which "did Kubla Khan a stately pleasure-dome decree," according to Samuel T. Coleridge in his poem (1798) "Kubla Khan." Jump forward to 1941, when Orson Wells released "Citizen Kane." Charles Foster Kane (allegedly William Randolph Hearst) built Xanadu, an elaborate palace (allegedly modeled on the San Simeon mansion of William Randolph Hearst). The place (Coleridge) didn't exist, and the mansion (Wells) brought only heartache. Jump forward to 1980, when Ted Nelson starts the project and the company Xanadu, a system for the network sale of documents with automatic royalty on every byte – hypertext and copyright protection (and royalty) on every byte. Nelson is a genius, but his Xanadu never (yet) was successful (more heartache). His "hypertext" concept is what we now see on the World Wide Web. See also Hypertext; Nelson, Ted; and World Wide Web.

XAPIA X.400 Application Program Interface Association. Microsoft has a new concept. It's called an Enterprise Messaging Server (EMS). The idea is to allow users to transparently access the messaging engine from within desktop applications to route messages, share files, or retrieve reference data. According to Microsoft, corporate developers will be able to add capabilities using Visual Basic and access EMS by writing either to the X.400 Application Program Interface Association's (XAPIA's) Common Mail Calls (CMC) or to Microsoft's Messaging API (MAPI). (ftp://nemo.ncsl.nist.gov/pub/olw/dssig/xapla/) See MAPI.

XAUI XAUI (pronounced "Zowie") is an interface employed by 10GbE (10 Gigabit Ethernet). "X" denotes the Roman numeral for 10, implying 10 Gbps, and "AUI" is derived from Ethernet Attachment Unit Interface. The XAUI is an interface extender for the XGMII (10 gigabit Media Independent Interface), a 74-signal wide interface comprising one 32-bit wide data path for the transmit direction and one for receive direction used to attach the Ethernet MAC (Media Access Control Layer) to the PHY (PHYsical Layer). The XAUI is a self-clocked serial bus evolved directly from the GbE 1000BASE-X PHY. The XAUI interface speed is 3.125Gbps (gigabits per second) and makes use of four serial lanes that operate in parallel over a WWDM optical fiber in order to achieve aggregate transmission speed of 10 Gbps. As is the case with 1000BASE-X, 10GbE employs the 8B10B transmission code in order to ensure signal integrity through the copper circuitry of PCBs (Printed Circuit Boards). See also 10GbE.

Xbar Crossbar. The second generation of automatic voice switching systems, Xbar switches are voice circuit switches that can be characterized as electromagnetic in nature. First installed in 1938, they quickly rendered obsolete electromechanical SxS (Step-by-Step) switches, and in turn were rendered obsolete by ECC (Electronic Common Control) systems. The first common control switches, Xbar systems also sometimes are known as Xreed (Crossreed). They rely on a common, centralized set of programmed logic to set up, maintain, and tear down conversation paths, and to provide various features to end users. An end user's request for dial tone is recognized by a "marker," which directs a "sender" to store the dialed digits. At that point, a "translator" is directed to route the call by reserving a path through a switching matrix. The switching matrix is a centrally-controlled network of electromagnetic switches which work with magnets and which interconnect horizontal and vertical paths to establish a path from port to port. Xbar switches served in various capacities, including COs (Central Offices), tandems, and PBXs. See also Switching Fabric.

XC Cross connect.

XD Dead line, same as NDT – which stands for no dial tone.

XDMCP X Display Manager Control Protocol. Protocol used to communicate between X terminals and workstations running the UNIX operating system.

XDP Ron Stadler of Panasonic dreamed this one up. It stands for eXtra Device Port. It's an analog RJ-11 equipped port on the back of a Panasonic digital telephone, which is driven by Panasonic's Digital Super Hybrid switch. The XDP is an extension line completely separate to your digital voice line. You can be speaking on the phone while receiving or sending a fax or while sending or receiving data. Or plug a cordless phone or answering machine into the XTP.

XDR External Data Representation. Standard for machine-independent data structures developed by Sun Microsystems. Similar to BER.

xDSL A generic – the letter x means generic – term for Digital Subscriber Line equipments and services, including ADSL, HDSL, IDSL, SDSL and VDSL. xDSL technologies provide extremely high bandwidth over the twisted-pair that runs from your phone company's central office to your office or home. Some xDSL lines are symmetrical (the same bandwidth in both directions). Some xDSL lines (for example, ADSL) are asymmetrical – different bandwidth (and thus speed) in both directions. Many xDSL loops are already installed. xDSL technology is being installed by local phones in order to provide faster access for their subscribers to the Internet and also (and most importantly) to counter the competition from the CATV industry with its cable modem. See ADSL, G.990, G-Lite, HDSL, IDSL, RADSL, SDSL, Splitter, Splitterless and VDSL for more detailed explanations.

xEMS xDSL Element Manager System

Xenix Microsoft name for a 16-bit microcomputer operating system derived from AT&T Bell Labs' UNIX. See also Linux.

xerographic recording Recording by action of a light spot on an electrically charged photoconductive insulating surface where the latent image is developed with a resinous powder.

Xerox Network Services XNS. An old multilayer protocol system developed by Xerox and adopted, at least in part, by Novell and other vendors. XNS is one of the many distributed-file-system protocols that allow network stations to use other computers files and peripherals as if they were local. XNS is used by some companies on Ethernet LANs. In local area networking technology, special communications protocol used between networks. XNS/ITP functions at the 3rd and 4th layer of the Open Systems Interconnection (OSI) model. Similar to transmission control protocol/internet protocol (TCP/IP), which, of course, is now much more popular, since it is the foundation of the Internet. See Internet.

Xerox Network Systems XNS. A peer-to-peer communications standard, developed by Xerox and designed for Ethernets.

xerox subsidy Euphemism for swiping free photocopies from one's workplace.

XFN X/Open Federated Naming. Used in Sun's Enterprise Server. Provides enterprise directory name Service for simplified access to and federation among multiple naming services. Works with Distributed Computing Environment (DCE), ONC, and Internet Domain Name Service (DNS).

XFN/NIS+ Part of Sun Microsystems' ENOS Networking Solutions. It's a secure repository of network information.

XFR TransFeR.

XGA eXtended Graphics Array. A new IBM level of video graphics which has a screen resolution of 1,024 dots horizontally by 768 vertically, yielding 786,432 possible bits of information on one screen, more than two and a half times what is possible with VGA. See also Monitor.

XGL Part of Sun Microsystems' Imaging and Graphics Solutions. Solaris' graphics library for 2-D/3-D applications development.

XHHW A NEC building wire with cross-linked polyethylene insulation rated 600 volts, 900xAFC dry, 750xAFC wet.

XHTML 1.0 XHTML 1.0 is a reformulation of HTML 4.0 as an XML 1.0 application. The hybrid language allows users to migrate from HTML to XML as users can create documents in HTML but also mix in XML functions. For example, complex documents can be created in HTML but can incorporate things like mathematical equations and music from XML. See XML.

XID Frame A High-Level Data Link Control (HDLC) transmission frame used to transfer operational parameters between two or more stations.

XIL Part of Sun Microsystems' Imaging and Graphics Solutions, XIL is an open Imaging API used for image enhancement, scaling, compression, color conversion, and display.

XIP eXecute-In-Place. Refers to specification for directly executing code from a PCMCIA Card without first having to load it into system memory.

XIWT Cross-Industry Working Team. About 28 companies (and growing) whose work centers on the Clinton-Gore administration's document "National Information Infrastructure – Agenda for Action." The document spells out policy initiatives required to achieve the benefits of widespread, convenient and affordable access to existing and future information resources. In short, the Information SuperHighway.

XJACK A registered trademark of modem manufacturer, MegaHertz, for one of the most innovative ideas in PCMCIA modems. XJACK is the world's tiniest female RJ-11. It's about a quarter of an inch wide and the most convenient way of attaching your laptop's modem to the phone network.

xLC xDSL Line Card, in Line Concentrating Module.

XLIU X.25/X.75/X.75' Link Interface Unit.

XLL eXtensible Linking Language. A XML linking language which provides for multidirectional linking of documents written in XML. Unlike HTML-based sites, XML sites with XLL support are capable of linkage at the object level, rather than at the page level, only. See XML.

XLP/XLPE The terms used for cross-linked polyethylene insulation, a popular polymeric type of insulation with outstanding electrical, moisture and physical properties.

XLR Connector "X"-Series "L"ockheed "R"ubber. It's a three pin connector used in all sorts of applications, including video conferencing. It was originally used by the Lockheed Corp. on fighter plane's dashboards. It's now also known as a "QG" connector or Quick Ground, because the grounding pin is longer than the other two pins making the connection grounded first.

XM XM Radio is a company which launched two satellites (one called Rock, the other called Roll) in the S-band. These satellites transmit music and talk (some commercial-free) to monthly paying subscribers on earth sitting at home or traveling in cars and equipped with the appropriate satellite antenna. The service is known as Satellite Digital Audio Receiver Services (SDARS). XM's competitor is called Sirius.

XMA eXtended Memory specifiCation. Interface that lets DOS programs cooperatively use extended memory in 80286 and higher computers. One such driver is Microsoft's HIMEM. SYS, which manages extended memory and HMA (high memory area), a 64k block just above 1Mb.

XMGII XGMII is a 74-signal wide interface comprising one 32-bit wide data path for the transmit direction and one for receive direction used to attach the 10GbE (10 Gigabit Ethernet) MAC (Media Access Control Layer) to the PHY (PHYsical Layer). The "X" is from the Roman numeral for 10, implying 10 Mbps, and the "GMII" stands for Gigabit Media Independent Interface. See 10GbE for more detail.

Xmit Transmit.

XML eXtensible Markup Language. XML, the language is a standard way of tagging data so it can be read and interpreted by a variety of Web browsers, by a variety of software, servers, and clients, regardless of how it was created. The vast bulk of the largest companies in the world use XML for electronic transactions with their customers or their suppliers, including using XML for EDI (Electronic Data Interchange). XML allows companies to automatically order from and sell to each other – without having to have a human in between physically translating between the different systems, or worse, physically entering the information again into another incompatible computer system. Developed by the W3C (World Wide Web Consortium), XML is a pared-down version of SGML, designed especially for Web documents. It enables designers to create their own customized tags to provide functionality not available with HTML. For example, XML supports links that point to multiple documents, as opposed to HTML links, which can reference just one destination each. Whether XML eventually supplants HTML as the standard web formatting specification depends a lot on whether it is supported by future Web browsers. The World Wide Web Consortium released the first spec on XML in February, 1998. According to Inter@ctive Week, a weekly newspaper covering the Internet, companies are seizing upon the ability of XML to allow structured exchanges of data between machines attached to the Web. That will enable one Web server to talk to another Web server, meaning manufacturers and merchants can begin to quickly swap data, such as pricing, stock-keeping numbers, transaction terms and product descriptions. See also HTML, OFX, SGML, Semantic Web and the three definitions that follow.

XML attributes XML attributes are used to provide more information, often meta-information, about an XML element. They are made up of the name of the attribute, an equals (=) sign, and the value of the attribute, where the value is surrounded by quotes. See XML.

XML elements XML elements are the basic building blocks of an XML document or data stream. There are three basic elements in the XML protocol: message, presence and iq. See XML.

XML namespace An XML namespace provides a simple method for qualifying element and attribute names used in XML documents by associating them with namespaces identified by a URI reference. The XML Namespace specification is currently a recommendation with the W3C. See XML.

XML schema XML Schema is a language for writing rules that constrain the kinds of elements that can appear in in documents and the ways in which they can be sequenced, grouped and nested. XML Schema is still a relatively new specification. The W3C Recommendation for XML Schema was published in May, 2001. See XML.

XML/EDI Extensible Markup Language (XML), a Web markup language, combines with scripting languages and electronic data interexchange to form XML/EDI. XML/EDI may be the means to bridge EDI into Internet Electronic Commerce (EC) by making the existing EDI (Electronic Data Interchange) knowledge base more readily usable to the Internet EC developers. See HTML.

XModem Also called "Christiansen Protocol". An error-correcting file transfer, data transmission protocol created by Ward Christiansen of Chicago for transmitting files between PCs. A file might be anything – a letter, an article, a sales call report, a Lotus 1-2-3 spreadsheet. The XModem protocol sends information in 128-byte blocks of data. Error control is provided through checksums, which are performed on each data block, and transmitted along with it. If the result does not check out at the other end, the computer at the other end sends a NAK (Negative AcKnowledgement), requesting that the transmitting device re-transmit that block. If the block checks out, the computer sends an ACK (Acknowledgement). In this way, relatively error-free transmission can be accomplished.

XModem was first used by computer hobbyists and then by business users of PCs. If you're buying a telecommunications software program for your PC – IBM, Compaq, Apple, etc. – it's a good idea to buy a program with XModem. It's among the most common data communications protocols. But it's not the fastest, just the most common. AT&T Mail supports XModem protocol. So does TELECONNECT Magazine's own E-mail InfoBoard system. MCI Mail does not support XModem protocol. We don't know why. There are many variations of XModem including XModem 1K (which uses blocks of 1,025 bytes), Modem7, YModem, Y-Modem-G, and ZModem. Most communications software packages only support (i.e. will handle) the original version of XModem (checksum) and the newer CRC variation. A study in Byte Magazine (March, 1989) showed ZModem to be a far more efficient file transfer protocol than XModem, YModem, or W/XModem. The author now tends to use ZModem more commonly. It is supported by most on-line services, such as CompuServe, etc. See also Data Compression Protocols, Error Controls Protocols, File Transfer Protocol, YModem and ZModem.

XModem-1K Xmodem-1K is an error-correcting file transfer, data transmission protocol for transmitting files between PCs. It is essentially Xmodem CRC with 1K (1024 byte) packets. On some systems and bulletin boards it may also be referred to as Ymodem.

XModem-CRC Cyclic Redundancy Checking is added to XMODEM frames for increased reliability of errors detection. See XMODEM.

XMP X/Open Management Protocol; an API and software interface specified in the Open Software Foundation's Distributed Management Environment.

XMPP eXtensible Messaging and Presence Protocol. XMPP is the part of the XML Protocol that supports Instant Messaging functionality.

XMS An acronym for eXtended Memory Specification. To run this standard, your system must have 350K of extended memory. XMS creates the HMA (High Memory Area), then governs access to and the allocation of the remainder of extended memory.

XnLC xDSL Line Card in AccessNode Express.

XNS Xerox Network System. The LAN architecture developed at the Xerox Palo Alto Research Center (Parc). It is a five-layer architecture of protocols and was the foundation of the OSI seven-layer model. It has been adopted in part by Novell and other vendors. XNS is one of the many distributed file system protocols that allow network stations to use other computers files and peripherals as if they were local. XNS is used by some companies on Ethernet LANs.

XO Crystal Oscillator.

XON/XOFF XON/XOFF are standard ASCII control characters used to tell an intelligent device to stop or resume transmitting data. In most systems typing <Ctrl>-S sends the XOFF character, i.e. to stop transmitting. Some devices understand <Ctrl>-Q as XON, i.e. start transmitting again. Others interpret the pressing of any key after <Ctrl>-S as XON.

XOpen A group of computer manufacturers that promotes the development of portable applications based on UNIX. They publish a document called the X/Open Portability Guide. X/Open is the correct spelling.

XOR gate A digital device which outputs a high state only if either of its inputs are high (but not both).

XPAD An eXternal Packet Assembler/Disassembler.

XPC A set of protocols developed by British Telecom Tymnet to allow asynchronous terminals to connect to an X.25 packet-switched network.

XPIC Cross-polarization interface cancellation (XPIC) technology allows two data streams on two polarizations of a single channel to be transmitted simultaneously, thus doubling net channel capacity. This technology is used in new broadboand wireless service.

XPM 1. Cross-phase Modulation.
2. eXtended Peripheral Module.

XPL A Private Line provided by a long distance carrier in the U.S.

XRB Transmit Reference Burst.

Xremote Tool that lets you grab the mouse and keyboard of another Windows based machine and control them with your local mouse and keyboard. It creates a form of remote control of other machines. All mouse- and keyboard-actions on the local machine are forwarded to the remote display.

XSG X.25 Service Group.

XSL eXtensible Stylesheet Language. An extension of XML which allows the automatic transformation of XML-based data into HTML and other presentation formats. Thereby, the presentation format can vary from the underlying data structure. See XML.

XT Abbreviation for crosstalk.

XTEL API The Sun Solaris Teleservices Application Programming Interface. See Teleservices for a long explanation.

XTELS The Sun Solaris Teleservices protocol. The XTELS protocol is essentially the XTEL provider protocol with extensions to support multiple clients and multiple XTEL providers. See Teleservices for a long explanation.

XTELTool A window-based tool used to configure providers in the Sun Solaris Teleservices platform. Basically a graphical user interface. See TELESERVICES for a long explanation.

XTL Sun Microsystems' "Teleservices" architecture for Solaris. XTL is the foundation library for applications using or controlling telecom data streams. It is used for computer telephony application development. The XTL subsystem and API includes call control functions. It establishes a call or connection, and data stream access methods to control the flow of data over that connection.

XTL API Part of the Sun XTL Teleservices architecture. An object-oriented interface accessed using the C++ language. The API includes XtlProvider, XtlCall, XtlCallState, and XtlMonitor objects. These base objects define the command and callback methods of XTL teleservices. There are three methods types on these objects: synchronous requests, asynchronous commands, and asynchronous events.

XTL Call Object Part of Sun Microsystems' XTL Teleservices architecture. C++ objects created by developers with the XTL API. Each XtlCall object corresponds to a telephone call. An XtlCall object has command methods to query the current state of a call or request a change in state (to put a call on hold). An XtlCall object also has callback methods for the asynchronous notification of state changes.

XTL Provider Configuration Database Part of the Sun XTL Teleservices architecture. Means of registering third-party "providers" in an XTL-based system. The database shows the existence of each provider, how to invoke it (used by the XTL system internally) and describes the specific capabilities of that provider.

XTL Provider Interface Part of Sun Microsystems' XTL Teleservices architecture. The means for third-party developers and Independent Hardware Vendors (IHVs) to integrate provider, technology or call-type-specific features into an XTL environment. Developers use the XTL Provider Library (MPI) to do this.

XTL Provider Library Part of Sun Microsystems' XTL Teleservices architecture. Isolates the provider library from the intricacies of the XTL system services. The library supplies interfaces to the provider database, the server messaging system for commands and callbacks, and the means to make device data streams accessible to applications.

XTL System Services Part of Sun Microsystems' XTL Teleservices architecture. Acts as the intermediary between the application view of a call object and the provider's implementation of the call. The XTL server, along with the application and provider libraries, handles the interprocess message passing, object identification and creation, call ownership, security, and asynchronous event notification. The XTL subsystem also manages a database of available providers and helps manage data stream routing and access.

XTLtool A Sun Microsystems term for a GUI tool supplied to browse and edit the provider configuration database in an XTL environment.

XXdecode A utility program that converts a text file created with XXencode back into its original binary file format.

XXencode A binary-to-text file conversion program, similar to UUencode. It enables binary files to be transmitted over six-bit, seven-bit and eight-bit networks, and via protocols that can only handle plain text. Text files produced by XXencode have an extension of XXE. XXencode was developed to deal with a problem that early versions of UUencode had when converting binary files to and from EBCDIC. Improvements in later versions of UUencode resolved that problem, with the result that UUencode has eclipsed XXencode.

XXXX The last four digits of a telephone number are called the "line number." A ten-digit telephone number in the U.S., for example, follows the format NXX-NXX-XXXX, where N must be a number other than "0" or "1," and X can be any number.

XWindows See X-Windows.

Y-Code A military GPS signal that is being phased out and replaced by M-Code. See GPS.

Y-dimension of recorded spot In facsimile, the center-to-center distance between two recorded spots measured perpendicular to the recorded line.

Y-dimension of scanning spot In facsimile, the center-to-center distance between two scanning spots measured perpendicular to the scanning line on the subject copy. The numerical value of this term will depend upon the type of system used.

Y/C A two channel video channel. One is for color (chrominance) and the other for black and white (luminance). This designation is used for video signals that keep separate the luminance and chrominance information, thus preventing some of the normal NTSC artifacts, E.G., cross-color and cross-luminance.

Y2K Year 2000. The millennium bug. What happened when your computer hit the year 2000. There were supposed to be three happenings. First, it would move its date to 1900 and stay there. In this case programs that used date calculations – like figuring how much money you have in your insurance policy – would screw up. For example, many old mainframe programs used the two-digit 95, instead of the four-digit 1995 in the date field. That meant that when it hit 00 or 01, the computer wouldn't be able to figure that the difference between 01 and 95 is 6. (Figure it. 2001 minus 1995 = 6). Second, it would move its date to 1900 and you'd have to manually change it back. Third, it would move its date to 2000 flawlessly. Most PCs manufactured after 1995 handled the Y2K problem flawlessly. Many mainframe and minicomputers manufactured before that wouldn't. Some phone systems would also mess up and stop working altogether. Some were expected to work in strange ways, like producing the wrong information on calls being made or being received. Billions of dollars were invested throughout the world in business and in governments rewriting computer software. As a result most of the fears of great disaster didn't materialize. Y2K was a great non-event in the history of computing.

yagi antenna An antenna system which employs a basic antenna element with parasitic reflector and director elements in order to achieve highly directional characteristics.

Yahoo! A popular search engine and directory on the Web. Yahoo! began as the bookmark lists of two Stanford University graduate students, David Filo and Jerry Yang. After putting their combined bookmark lists organized by categories on a college site, the list began to grow into an Internet phenomenon. It became the first such directory with a large following. And it earned its allegedly name – "Yet Another Hierarchical Officious Oracle." Filo and Yang postponed their graduate work and became part of a public offering for a multimillion dollar corporation. Yahoo's stock went up, then down, then up.

yard frequency The radio frequency used in a railroad yard for yard operations.

yard sale Americans love yard sales, also called garage sales. Families sell their old junk to other families who see it as treasure. The process is repeated endlessly. There are variations on this theme, as you can see from this joke Bob Dunagan sent me: A man was walking one day when he came to a big house in a nice neighborhood. Suddenly he realized there was a couple making love on the lawn. Then he noticed another couple behind a tree. Then another couple behind some bushes. He knocked on the door of the house. A well-dressed woman answered. The man asked her what kind of a place this was? "This is a brothel," replied the woman. "Well, what's all that making love on the lawn? queried the man. Replied the woman, "Oh, we're having a yard sale today."

yazam Yazam is the Hebrew word for entrepreneur. www.yazam.com is an online meeting place for entrepreneurs and investors.

yellow alarm An alarm that indicates a piece of equipment has suffered a critical (i.e., red alarm) fault. In T-1 circuits, a yellow alarm sent back toward the source of a failed transmit circuit in a DS-1 two-way transmission path. In superframe mode, the yellow alarm is generated by forcing bit two of every channel in the superframe to zero. In extended superframe mode the yellow alarm is sent via the facility data link. See also T-1. See also Yellow Signal.

yellow signal In telecommunications, a signal sent back in the direction of a failure, indicating that the input of a network element has failed. The yellow signal varies with the DS framing used. See Remote Alarm Indication and VT Path Remote Failure Indication and Yellow Alarm.

yellow cable A coaxial cable used in 10Base-5 networks. It is also referred to as "Thick" coax. This was the first cable used on many early LANs.

yellow pages A directory of telephone numbers classified by type of business. It was printed on yellow paper throughout most of the twentieth century until it was obsoleted in the late 1990s by dial up yellow page directories operated by voice processing systems and in the early 21st century by electronic directories delivered on disposable laser disks. As a concession to history, the laser disks are now painted bright yellow. Actually, yellow pages remain one of the phone companies' most lucrative sources of revenues. Advertising rates are not cheap. There is now competition. There are many "Yellow Pages" directories, since AT&T never trademarked the term "Yellow Pages." Some "yellow page" directories are better value than others. And some are more legitimate than others. Some actually never get printed or, if they are printed, are not printed in great quantity and are not distributed as widely as their sales literature implies. Many businesses have been suckered into paying money for listings and advertisements in directories that never appeared. This fictitious directory scam also has happened with "telex" and "fax" directories. This "scam" is fraud by mail and is heavily stomped upon by the US Postal Service. As a result, many fake directories (especially the telex ones) are "published" abroad.

yeppies Really silly made-up word. Yeppies is short for "young experimenting perfection seekers." Adrift in a world of conflicting opportunities, these chronically dissatisfied twentysomethings approach life as an exercise in comparison shopping, refusing to commit for fear of missing a better offer. Froogle Google is their favorite web site.

yes men "I don't want yes men around me. I want everybody to tell me the truth even if it costs them their jobs." Samuel Goldwyn, film producer.

Yiddish Yiddish is a Germanic language written in Hebrew characters, and was the native language of millions of Jews before the Holocaust. Yiddish is still very much a living language, spoken by hundreds of thousands of Jews as their first language in New York, Jerusalem, Montreal, Antwerp, and other Jewish communities. Yiddish is the language. Jewish is the religion. You don't speak Jewish. You speak Yiddish.

yield hog A person who buys very long-term bonds (20 years plus) in order to get the highest possible yield. The problem with buying such long-term bonds is that the prices of these bonds fluctuate wildly. When interest rates rise, their prices fall. When interest rates fall, their prices rise. Buying long, long-term bonds only makes sense when interest rates are falling, as they did in 2002, for example.

yield strength The minimum strength at which a material will start to physically deform.

YMMV Your Mileage May Vary. A response usually given when the answer is not precise and depends on the user's own circumstances. YMMV is an acronym used in electronic mail on the Internet to save words or to be hip, or whatever.

YModem A faster transfer variation of XMODEM. In YMODEM, XMODEM's 128-byte block grew to YMODEM's 1024 bytes (1 kilobyte). YMODEM combines the 1K block and the 128-byte block modems into the same protocols. YMODEM, or 1K as it is known, became the thrifty way to send files (i.e. it saved on phone time). Enhancements were added, such as auto-fallback to 128-byte blocks if too many errors were encountered (because of bad phone lines, etc.) See XMODEM for a much larger explanation of file transfer using X, Y and ZMODEM protocols.

YModem-G Ymodem-g is a variant of Ymodem. It is designed to be used with modems that support error control. This protocol does not provide software error correction or recovery, but expects the modem to provide it. It is a streaming protocol that sends and receives 1K packets in a continuous stream until told to stop. It does not wait for positive acknowledgement after each block is sent, but rather sends blocks in rapid succession. If any block is unsuccessfully transferred, the entire transfer is canceled. See also ZMODEM, which we prefer.

Yottabyte YB. A combination of the homonymic Greek "iota," referring to the last letter of the Latin alphabet, and the English "bite," meaning "a small amount of food." A unit of measurement for physical data storage on some form of storage device-hard disk, optical disk, RAM memory etc. and equal to two raised to the 80th power, i.e. 1,208,925 ,819,614,600,000,000,000,000 bytes. Here are the others in the chain:

KB = Kilobyte (2 to the 10th power)
MB = Megabyte (2 to the 20th power)
GB = Gigabyte (2 to the 30th power)
TB = Terabyte (2 to the 40th power)
PB = Petabyte (2 to the 50th power)
EB = Exabyte (2 to the 60th power)
ZB = Zettabyte (2 to the 70th power)
YB = Yottabyte (2 to the 80th power)

One googolbyte equals 2 to the 100th power.

Your As in "your welome." Let me be the first dictionary to include the latest abomination of the English language.

youth Youth would be an ideal state if it came a little later in life. - Herbert Henry Asquith.

YouTube YouTube is a consumer media company whose video-sharing and social-networking website, www.youtube.com, lets people upload, download, share, rate, and comment on the videos – some of which are funny, some of which are educational and some of which are just downright self-indulgent. The company makes little attempt to censor the uploaded videos for content. Nor can it. There are so many. The company was founded in February 2005 by three Silicon Valley guys in their twenties named Steve Chen, Chad Hurley and Jawed Karim. YouTube has enjoyed explosive growth. As of December 2006, its website serves over 100 million videos per day, and nearly 70,000 new original videos are uploaded to the site daily. By the time you read this, it will be many more. In addition to the millions of grassroots videos on its site, In late 2006 YouTube also began offering content from commercial video producers including Warner Music Group, Sony BMG, and CBS.

yoy Year over year. Yoy is a financial measure. It might be used thus: What earnings are like today, compared with what they were a year ago for the same period – e.g. a quarter, six months, etc.

Yuppie The phrase Yuppie came into being as a clever twist on several words and phrases. The abbreviation Y.U.P. – Young Urban Professionals or Young Upwardly mobile Professional – was used at least since the late 1970s by public opinion analysts. This abbreviation fit the patter of the first syllable of two other lifestyle words that had been around since the 1960s, hippie and preppie - so it was natural to add the '-pie' suffix to 'yup' as well. See also DINK.

Yuppie food coupons Crisp $20 bills that spew from ATM machines. Often used to split the bill after a meal. "We all owe $8, but all anybody's got is yuppie food coupons." A Wired Magazine definition.

yurt A Mongolian circular shed. Some companies are making Yurt-like sheds for installation in the back yards of telecommuters – especially those telecommuters who don't have enough room inside for a home office.

YUV A color encoding scheme for natural pictures in which luminance and chrominance are separate. The human eye is less sensitive to color variations than to intensity variations. YUV allows the encoding of luminance (Y) information at full bandwidth and chrominance (UV) information at half bandwidth.

YUV9 The color encoding scheme used in Indeo Video Technology. The YUV9 format stores information in 4x4 pixel blocks. Sixteen bytes of luminance are stored for every one byte of chrominance. For example, a 640x480 image will have 307,200 bytes of luminance and 19,200 bytes of chrominance.

Z Abbreviation for Zulu time. See Greenwich Mean Time.

Z-Wave A new low-power mesh-network protocol that can deliver two-way wireless remote control over lighting and other small appliances in the home. It uses tabletop or handheld remote controls.

Zambezi Bum My friend from Africa tells me their version of Delhi Belly and Montezuma's Revenge is Zambezi Bum.

zap To eradicate all or part of a program or database, sometimes by lightning, sometimes intentionally.

zapper message Sent by the Supervisor to teat down a virtual circuit after the transmission is complete.

zarro boogs A term meaning that all reported bugs have been taken care of, but it's understood that there probably still are undiscovered bugs. In other words, it's not "zero bugs" but it's close or (close enough). Popularized by Netscape/Mozilla developers.

ZB See Zettabyte.

ZBTSI Zero Byte Time Slot Interchange. A technique used with the T carrier extended superframe format (ESF) in which an area in the ESF frame carries information about the location of all-zero bytes (eight consecutive "O"s) within the data stream.

ZCS Zero Code Suppression.

ZDSF Zero Dispersion Shifted Fiber. A type of Dispersion Shifted Fiber (DSF) that is used in long haul, high speed fiber optical transmission systems. See DSF for a full explanation.

Zen A Japanese sect of Buddhism that stresses attaining enlightenment through intuition rather than by studying scripture. In contemporary techie lingo, you zen something if you figure it out by meditation or a sudden flash of enlightenment. Zenning is a whole lot easier that groking, as it requires much less work. See also Grok and Zen Mail.

Zen mail Email messages that arrive with no text in the message body.

zener diode A particular type of semiconductor which acts as a normal rectifier until the voltage applied to it reaches a certain point, or threshold voltage. At this point – at the Zener voltage or the avalanche voltage – the Zener diode becomes either conducting (i.e., "turns on") or non-conducting (i.e., "turns off"). These types of circuits include computer equipment (turn on), voice-activated circuits such as telephone wiretap devices (turn on), and surge protectors (turn off). As the main use of a Zener diode is to provide a reference voltage, it often is known as a "reference diode". In a RF (Radio Frequency) clamp application, the Zener diode is used to clamp (i.e., supply) a specific voltage for other, protected components, perhaps in an integrated circuit (IC). The Zener diode is the device that made it possible to make digital integrated circuits. Without Zener diode on-chip reference voltage (and, thereby, the benefit of voltage regulation and transient voltage protection), we would not be able to just "hook 'em together", as we do now. See also Diode and Rectifier.

zero The basis of Arabic numerals, including the newly invented "zero," probably originated in India around the 6th century. This new knowledge followed the trade routes to the Arab world. The shape of most of the characters were greatly modified by the Arabs and the Arabic numerals, as we know them today, were introduced into Europe around the 10th century by the Moors in Spain, although they did not come into general use for several hundred years. The symbol "zero" was the last one of the group to be accepted because most scholars believed that it was unnecessary to have something that stood for nothing.

zero beat reception Also called "homodyne" reception. A method of reception using a radio frequency current of the proper magnitude and phase relation so that the voltage impressed on the detector will be of the same nature as that of the wave. An old radio term.

zero billion dollar market In the Spring of 2004, the San Jose Mercury News wrote, "Public Wi-Fi hotspots are turning into the latest example of what are sometimes called "zero billion dollar" markets, opportunities so huge and important that no one seems to notice the lack of actual customers."

zero bit The high-order bit in a byte or a word.

Zero Byte Time Slot Interchange ZBTSI. A method of coding in which a variable address code is exchanged for any zero octet. The address information describes where, in the serial bit stream, zero octets originally occurred. It is a five-step process where data enters a buffer, zero octets are identified and removed, the nonzero bytes move to fill in the gaps, the first gap is identified, and a transparent flag bit is set in front of the message to indicate that one or more bytes originally contained zeros. See ZBTSI.

zero click access Describes a streaming video feed that is continually sent over the network to a mobile device's or PC's screen display, without the user's having to click on a button to start the feed. Examples are ticker-streams of news headlines, weather information, sports scores, and so on.

zero code suppression The insertion of a "one" bit to prevent the transmission of eight or more consecutive "zero" bits. Used primarily with digital T-1 and related telephone-company facilities which require a minimum "ones density" keep the individual sub channels of a multiplexed, high-speed facility active. Several different schemes are currently employed to accomplish this. Proposals for a standard are being evaluated by the ITU-T. See also Zero Suppression.

zero configuration networking See Zeroconf.

zero day attack A worm, virus, or other malicious, network-mediated exploit that is launched and hits users on the same day as or even before the public announcement of the system vulnerability that the attack exploits.

zero day protection A security solution that is supposed to protect a system against attacks that have never been seen before. This is more a marketing term than a reality. If zero-day protection truly existed, then security software updates would never be required.

Zero Dispersion Shifted Fiber ZDSF. A type of Dispersion Shifted Fiber (DSF) that is used in long haul, high speed fiber optical transmission systems. See DSF for a full explanation.

zero fill See Zerofill.

zero frequency The frequency (wavelength) at which the attenuation of the lightguide is at a minimum.

zero hop routing A layer 3 switch that offers cut-through services making every end station one hop away from each other

zero insertion In SDLC, the process of including a binary 0 in a transmitted data stream to avoid confusing data and SYN characters; the inserted 0 is removed at the receiving end.

zero latency A computer term describing a computer system in which there is virtually no time between the updating of an information record and its availability elsewhere in the network. See also Latency and Real Time.

zero latency enterprise The Gartner Group defines this new management buzzword as an organization that "exploits the immediate exchange of information across geographical, technical and organizational boundaries to achieve business benefit." I guess it means the place moves quickly. Nothing like saying something in big words that can't say more easily with smaller, fewer words.

zero out Someone calls one of those infernal phone machines and hears: "Thank you for calling XYZ Corporation, press 1 for sales if you're above 12th Street, press 2 for sales on 12th Street, press 3 for below 12th street, etc." When many people (like me) hear incomprehensible messages like this one, they simply hit zero and hope they'll get a real live person. In the trade, this is called a "zero out." The measure of how many "zero outs" a site has is a measure of how badly the menus are written.

zero out reset Resets the non-volatile memory in a cell phone or related device to its original factory-default state. A zero out reset obliterates user data and user-specified preferences and other settings. For security reasons, it should be done whenever you send the device into the shop for repair or when you plan to no longer keep the phone, for example, prior to exchanging it for a newer model or donating it to a charitable organization. After all, you don't want all your personal phone numbers and personal information to fall into the wrong hands. This is especially important with your PDA (BlackBerry or Treo, etc.) device, which contains serious personal stuff, like emails, contacts and appointments.

zero power modem A modem that takes its power from the phone line and therefore needs no battery or external power. Such modems are often limited in their speed and capabilities.

zero slot LAN A Local Area Network (LAN) that uses a PC's serial port to transmit and receive data. It doesn't require a network interface card to be installed in a slot in the PC, thus the name "zero-slot" LAN. RS-232 LANs usually use standard RS-232 or phone cable to link PCs. Software does the rest of the work. Due to the slow speed of serial communications on a PC, RS-232 LANs are usually restricted to speeds of around 19.2K bits per second. What they lose in speed, however, they make up in low price.

zero stuffing Get a cup of coffee right now. Synchronous data transmission is done by sending what IBM and AT&T call Frames, and what everyone else calls Packets. A frame starts off by sending a bit pattern of 01111110 (notice the six 1's in a row). Synchronous transmission is for sending a bit stream, which means that the bits may (but probably do not) have any relation to the transmission of characters. This is especially true when sending digitized voice. As the bits pass to the receiver, they go through a shift register. When the flag signifying the end of a frame goes by, the last 16 bits in the shift register are the check digits.

The receiver computes the check digits based on the data bits that have gone by. As the sender sent the data, it computed the check digit, sent it after the end of the frame, and then sent the flag. If the receiver computes the same check digit that the sender sent, then one can be reasonably assured the data came through without error. But that's not what I came to talk to you about. I came to talk about Zero Stuffing. The problem is that somewhere in the bit stream, there is the possibility of there being six 1 bits in a row. To the receiving computer, six 1's means a flag. Therefore the sending computer, if it "sees" six 1 bits, will send five 1 bits, and stuff a zero in the bit stream.

In fact, if it sees even five 1 bits, it will stuff a zero anyway, so there will be no ambiguity. The rule is, "If there are five ones in a row and it is NOT the end of a frame, stuff a zero into the bit stream." This way the receiver will know that this is in no way the end of the frame yet. Now if the receiver sees six 1's in a row, it knows without a doubt that it IS the end of a frame, and should proceed with the error checking.

zero suppression The elimination of nonsignificant zeros from a numeral. Zero suppression is the replacement of leading zeros in a number with blanks so that when the number appears, the leading zeros are gone. The data becomes more readable. For example, the number 00023 would be displayed on the monitor or printed as 23.

zero test level point A level point used as a reference in determining loss in circuits. Analogous to using sea level when defining altitude. Written as 0 TLP.

Zero Transmission Level Point ZTLP. In telephony, a reference point for measuring the signal power gain and losses of telecommunications circuit, at which a zero dBm signal level is applied.

zero transmission level reference point A point in a circuit to which all relative transmission levels are referenced. The transmission level at the transmitting switchboard is frequently taken as the zero transmission level reference point.

zero usage customer An MCI definition. An MCI customer who has not placed a call over the network, even though he/she is an active customer. Sometimes used interchangeably, but incorrectly, with the term "no usage customer."

Zeroconf The zero technology (also known as Bonjour) may just be the answer. "Zero Configuration Networking: The Definitive Guide," coauthored by technical writer Daniel Steinberg and Stuart Cheshire, original designer of Bonjour, is a must-read for hardware designers or software authors who want to use this ground-breaking technology, as well as users who want to know how Apple, and others, have built "it just works" technology, as they call it, into iTunes, SubEthaEdit, and printers.

As Cheshire and Steinberg expound, "You walk in a few minutes late to a meeting and want to know what you've missed. You open your text editor and your computer automatically discovers a shared document in which one or more attendees are taking notes. A presenter announces that anyone who wants a copy of his slides should let him know. You open your local Instant Messenger application and you see his name, even though you've never met before. A moment later he has placed a copy in your Public folder, which he has discovered in his network directory.

"This is not a fantastical glimpse of the elusive future," they add. "It is a concrete description of what is available today using Zeroconf."

Zeroconf is an open standard that was initially developed by engineers at Apple. Promoted by Apple, first under the name Rendezvous, and then Bonjour, the technologies of Zeroconf are also available for Windows, Linux, Mac OS X, BSD Unix, and other operating systems. It's a standard for building devices and applications that configure themselves: chat clients that find other people on a LAN, printing software that automatically finds printers, file-sharing software that automatically finds shared resources. Devices such as printers, cameras, PDAs, and music players also use Zeroconf so that they work "out of the box" without requiring any configuration or setup by the user.

The early chapters of the book cover the underlying components of Zeroconf: local-link addressing, Multicast DNS and DNS Service Discovery. Later chapters cover Zeroconf programming: APIs for C, Java, and Python. The Python API was developed using SWIG so that Zeroconf service discovery is available through the same API in a host of other languages including Tcl, Perl, Scheme, and PHP. Together they comprise an indispensable guide to using this technology.

Anyone interested in creating software or gadgetry that's pain-free and trouble-free for users, will want to use Zeroconf. "Zero Configuration Networking: The Definitive Guide" provides clear and detailed instructions on putting it to work.

To conclude with the words of Cheshire and Steinberg, "Zero Configuration Networking-Bonjour as Apple calls it—provides a foundation to enable hardware and software makers to produce great products."

zerofill 1. To fill unused storage locations with the character "0."

2. Here's definition from GammaLink, a fax board maker: A traditional fax device is mechanical. It must reset its printer and advance the pages as it prints each scan line it receives. If the receiving machine's printing capability is slower than the transmitting machine's data sending capability, the transmitting machine adds "fill bits" (also called Zero Fill) to pad out the span of send time, giving the slower machine the additional time it needs to reset prior to receiving the next scan line.

ZERT Zeroday Emergency Response Team. A group of security engineers with experience in reverse engineering software, firmware and hardware, who create patches for security holes. ZERT works with several Internet security organizations and has liaisons to anti-virus and networking communities, but it is not affiliated with any particular vendor. ZERT creates and releases a non-vendor patch whenever a zero-day exploit appears in the open which poses a serious risk to the public, the Internet's infrastructure or both, and for which no vendor-supplied patch exists. As ZERT's website (http://isotf.org/zert/) observes, vendor-supplied patches are preferable to non-vendor-supplied patches, but there may be times when an independent non-vendor-affiliated group such as ZERT can release a critically needed working patch before a vendor can release its own fix.

Zettabyte ZB. A combination of the Greek "zeta," the second last letter of the Greek alphabet (omega is the last letter), and the English "bite," meaning "a small amount of food." A unit of measurement for physical data storage on some form of storage device – hard disk, optical disk, RAM memory etc. and equal to two raised to the 70th power, i.e. 1,180,591,620,717,400,000,000 bytes.

KB = Kilobyte (2 to the 10th power)
MB = Megabyte (2 to the 20th power)
GB = Gigabyte (2 to the 30th power)
TB = Terabyte (2 to the 40th power)
PB = Petabyte (2 to the 50th power)
EB = Exabyte (2 to the 60th power)
ZB = Zettabyte (2 to the 70th power)
YB = Yottabyte (2 to the 80th power)
One googolbyte equals 2 to the 100th power.

Zhlup See Schlub.

ZIF Zero Insertion Force. Intel makes a bunch of math co-processor chips which are used with their 80XXX range of microprocessors. ZIF is a special device which is typically soldered to the motherboard. You place an 80387 chip on this device, move the handle down, it grabs the chip and pulls the chip down, seating it electrically. When you want to remove the chip, you simply lift the handle and up the chip comes. The device was invented by Intel because so many people were apparently breaking the legs on their math co-processor chips each time they removed them. Apparently the problem was most prevalent in the computer rental business.

ZigBee ZigBee was created to address the market need for a cost-effective, standards-based wireless networking solution that supports low data rates, low power consumption, security and reliability. The ZigBee Alliance is defining both star and mesh network topologies, a variety of data security features and interoperable application profiles. According to the Alliance, ZigBee is the only standards-based technology that addresses the unique needs of most remote monitoring and control and sensory network applications. The Alliance's members' low cost, low power solutions will enable the broad-based deployment of wireless networks that are able to run for years on standard batteries for a typical monitoring application. ZigBee-compliant products operate in the unlicensed bands worldwide, including 2.4GHz (global), 915Mhz (Americas) and 868Mhz (Europe). Raw data throughput rates of 250Kbs can be achieved at 2.4GHz (10 channels), 40Kbs at 915Mhz (6 channels) and 20Kbs at 868Mhz (1 channel). Transmission distance is expected to range from 10 to 75 meters, depending on power output and environmental characteristics. Will interference and coexistence be an issue for the ZigBee technology in the 2.4GHz band? According to the Alliance, the potential for interference exists in every band, including 2.4GHz. The IEEE 802.10 and 802.15.2 committees are addressing coexistence issues. Examples of ZigBee applications include: Wireless home security; Remote thermostats for air conditioner; Remote lighting, drape controller; Call button for elderly and disabled; Universal remote controller to TV and radio; Wireless keyboard, mouse and game pads; Wireless smoke; CO detectors; Industrial and building automation and control (lighting, etc.) According to the Alliance, the ZigBee stack is small (28Kbytes) compared to the Bluetooth stack (250K). This relates to lower cost and lower power consumption. Ultra-low power consumption is a key system design aspect of the ZigBee technology to allow long lifetime non-rechargeable battery powered devices versus rechargeable devices for Bluetooth. As an example the transition from sleep mode to data transition is much faster in ZigBee than for Bluetooth. ZigBee networking capabilities include 255 devices per network, compared to 8 for Bluetooth networks. The data rate for ZigBee technology is 250kbps compared to 1 Mbps for Bluetooth wireless technology. Range for ZigBee products is expected to be around 30 meters in a typical home, compared to around 10 meters for Bluetooth products (without power amplifier).

zinc spark gap A spark gap having zinc as the electrode.

zine An electronic "magazine" produced for distribution on the Internet.

zip 1. It all started years ago with a very popular PC program called PKZIP. You could run this program on a file on your PC and bingo PKZIP would reduce the size of the file by as much as 90%. You could then transmit the zipped file over a phone line, saving as much as 90% of the time and 90% of the cost. At the other end the recipient would run another file called PKUNZIP and bingo your file would be returned to its original size. You get the program from PKWARE, Inc. Glendale, WI 414-352-3670. www.pkware.com. After a while, the file extension ".zip" became common and more and more software converted files into zip format and unconverted them. Among the more popular ones now is WinZip. Both WinZip and Pkzip cost money but there is plenty of free software. Microsoft's Windows operating system comes with a free zip extraction program; you can extract zipped files with it.

2. The zip in zip code stands for Zone Improvement Plan. See zip code.

zip code ZIP codes are the specific codes assigned to addresses in the United States to speed up the sorting of mail and thus speed up its delivery. ZIP stands for Zoning Improvement Plan. ZIP codes were originally introduced into the United States in 1963. They were originally five digits. Then they grew to nine digits. The latest zip codes are 11 digits. Let's say you have a zip code 10036-3959-29. The 100 is the region, in this case New York City. The 36 indicates the specific post office in the region. The 3959 means the carrier route, i.e. the delivery sector in a neighborhood. The 29 is called the sequence. It indicates a specific address along a given route. Outside of the US, ZIP codes are often called postal codes. They do the same thing the American ZIP codes do. They allow machinery to sort the mail and thus get it to its destination faster. The zip in zip code stands for zone improvement plan. See also Missile Mail.

zip cord A two-wire copper cable or a two-fiber fiber optic cable, with each wire or fiber having its own jacket, and with the jackets conjoined by a thin strip of jacket material. The zip cord can be pulled apart at the seam, giving the appearance of a zipper that has been unzipped.

zip tone Short burst of dial tone to an ACD agent headset indicating a call is being connected to the agent console.

zipped See Zip.

ZModem ZMODEM is an error-correcting file transfer, data transmission protocol for transmitting files between PCs. A file might be anything – a letter, an article, a sales call report, a Lotus 1-2-3 spreadsheet. Always use ZMODEM if you can. It's the best and fastest data transmission protocol to use. This is not my sole advice. Virtually every writer in data communications recommends it. Here's an explanation, beginning with XMODEM, an older, more common and less efficient protocol.

Both XMODEM and YMODEM transmit, then receive, then transmit. The handshake (ACK or NAK) happens when the sender isn't sending. ZMODEM adds full duplex-transmission to the transfer protocol. ZMODEM does not depend on any ACK signals from the host computer. It keeps sending unless it receives a NAK, at which time it falls back to the failed block and starts to retransmit at that point. ZMODEM was written by Chuck Forsberg. According to PC Magazine (April 30, 1991) ZMODEM is the first choice of most bulletin boards. ZMODEM, according to PC Magazine, features relatively low overhead and significant reliability and speed. ZMODEM dynamically adjusts it packet size depending on line conditions and uses a very reliable 32-bit CRC error check. It has a unique file recovery feature. Let's say ZMODEM aborts a transfer because of a bad line (or whatever), it can start up again from the point it aborted the transfer. Other file transfer protocols have to start all over again. ZMODEM's ability to continue is a major benefit. ZMODEM in some communications program is a little more automated than other protocols. For example, ZMODEM will start itself when the other end gives a signal – thus saving a keystroke or two and speeding things up. See File Transfer Protocol, XMODEM and YMODEM.

zombie PC A PC computer that's been taken over by software which the owner did not put on, but came during one of the owner's visits to the Internet. Such zombie PCs can be commanded from afar to do tasks, like send spam email to hundreds of thousands of other PCs. See also bot networks.

zone 1. A telephony definition. A zone is one of a series of specified areas, beyond the base rate area of an exchange. Service is furnished in zones at rates in addition to base rates.

2. A local area network definition. A zone is part of a LAN (Local Area Network), typically defined by a router. A router will let you get into one part of someone else's network. They define what you are able to get access to. You might get to that router by an external telecommunications circuit – dial up, ISDN, Switched 56, T-1 etc.

3. A collection of all terminals, gateways, and MCUs (Multipoint Control Units) managed by a single gatekeeper in an H.323 context. See also H.323.

zone bits 1. One or two leftmost bits in a commonly used system of six bits for each character.

2. Any bit in a group of bit positions that are used to indicate a specific class of items, i.e., numbers, letters, commands.

zone cabling According to AMP, its inventor, zone cabling is the subsystem for companies on the move. Zone Cabling subsystems allow for flexible, changeable cabling of open office areas for voice, data, video and power. The open office area is divided into zones, with feeder cables running to a distribution point within each zone and short cable runs to each outlet. Zones are wired with reusable, pre-terminated cable assemblies. Office area reconfiguration is fast and easy when using plug and play assemblies not requiring retesting. Disruption and productivity loss during moves are minimal and confined to the zones you're moving. Zone cabling can be implemented in CNA or DNA architectures.

zone method A ceiling distribution method in which ceiling space is divided into sections or zones. Cables are then run to the center of each zone to serve the information outlets nearby. See also Ceiling Distribution Systems.

zone multicast address Data-link-dependent multicast address at which a node receives the NBP broadcasts directed to its zone. See also NBNS.

zone prefix A prefix that identifies the addresses to be serviced by a given gatekeeper. Zone prefixes are typically area codes and serve the same purpose as the domain names in the H.323-ID address space.

zone of silence Skip zone.

zone paging Ability to page a specific department or area in or out of a building. "Page John in the Accounting Department." Zone paging is useful for finding people who wander, as most of us do.

Zoomed Video ZV. A PC card standard that lets a computer run its operating system or application software directly from a PC card. It is a technology that allows certain streams of digital information to write directly to a laptop's screen, bypassing the CPU and its bus (ISA, PCI, EISA, etc.). Zoomed video can show full 30 frames per second movies to a laptop screen. Zoomed video used MPEG-2. To show NTSC on a laptop, you need a special ZV (Zoomed Video) CardBus PCMCIA card. It feeds NTSC video directly to the screen. Zoomed Video technology allegedly will bring full laptop screen video conferencing to laptops. See Zoom Video Port.

Zoomed Video Port ZVP. Allows specially designed PC cards to send signals directly to the computer's video adapter bypassing the system processor and data bus. This allows full screen video to run at full speed freeing up the processor for other tasks.

A PCMCIA standard which adapts the PC Card slot to allow the insertion of a ZV Port Card. Inserting a ZV Port Card establishes direct communications between the PC Card controller and the audio and video controllers, allowing large amounts of multimedia data to bypass the CPU or systems bus. The ZV Port standards makes full screen, full motion video accessible to the notebook computer user.

zulu time Military terminology for Coordinated Universal Time (UTC), which has replaced Greenwich Mean Time (GMT) as the international clock reference. "Z" (phonetically "Zulu") refers to the time at the prime meridian in Greenwich, England. The U.S. time zones are Eastern ("R", "Romeo"); Central ("S", "Sierra"); Mountain ("T", "Tango"); Pacific("U", "Uniform"); Alaska ("V", "Victor"), and Hawaii ("W", "William"). See also GMT and UCT.

ZUM Zone Usage Measured. Usage charges for calls that aren't quite local, but aren't quite long distance, either. ZUM charges are for calls to Central Office prefixes that are not within your local calling area. ZUM charges are a way for your friendly ILEC (Incumbent Local Exchange Carrier) to make extra money.

ZV See Zoomed Video. See the next definition.

ZV Port Short for zoomed video port, a port that enables data to be transferred directly from a PC Card to a VGA controller. The port is actually a connection to a zoomed video bus. This new bus was designed by the PCMCIA to enable notebook computers to connect to real-time multimedia devices such as video cameras. The first notebook computers with the ZV port arrived in late 1996.

ZZF Zentralamt fur Zulassungen im Fernmeldewessen (Approval Authority – Germany).

APPENDIX

Standards Organizations and Special Telecom Interest Groups

1OGEA
10 Gigabit Ethernet Alliance
1300 Bristol Street North
Suite 160
Newport Beach, CA 92660
Tel: 949-250-7155
Fax: 949-250-7155
www.10gea.org

3GPP
Third Generation Partnership Project
ETSI
Mobile Competence Centre
650, route des Lucioles
06921 Sophia-Antipolis
France
Tel: +33 (0)4 92 94 42 58
Fax: +33 (0)4 92 38 52 24
www.3GPP.org

ACUTA
The Association for Telecommunications Professionals in Higher Education
152 W. Zandale Drive, Suite 200
Lexington, KY 40503
Tel: 606-278-3338
Fax: 606-278-3268
www.acuta.org

AITP
Association of Information Technology Professionals
315 South Northwest Highway, Suite 200
Park Ridge, IL 60068-4278
Tel: 847-825-8124
Fax: 847-825-1693
www.aitp.org

AMTA
American Mobile Telecommunications Association
1150 18th Street, NW
Suite 250
Washington, DC 20036
Tel: 202-331-7773
Fax: 202-331-9062
www.amtausa.org

ANSI
American National Standards Institute
11 West 42nd Street
New York, NY 10036
Tel: 212.642.4900
Fax: 212.398.0023
www.ansi.org

APCO
Association of Public-Safety Communications Officials
2040 S. Ridgewood Avenue
South Daytona, Florida 32119-8437
Tel: 904-322-2500
Fax: 904-322-2501
www.apcointl.org

ARIN
American Registry for Internet Numbers
4506 Daly Drive, Suite 200
Chantilly, VA 20151
Tel: 703-227-0660
Fax: 703-227-0676
www.arin.net

ASCENT
Association of Communications Enterprises
1401 K Street, N.W., Suite 600
Washington, D.C. 20005
Tel: 202-835-9898
Fax: 202-835-9893
www.ascent.org

ATA
American Telemarketing Association
4605 Lankershim Blvd
Suite 824
North Hollywood, CA 91602-1891
Tel: 800-441-3335
Fax: 818-766-8168

ATSC
Advanced Television Systems Committee
1750 K Street, N.W., Suite 800
Washington, D.C. 20006
Tel: 202-828-3130
Fax: 202-828-3131
www.atsc.org

ATIS
Alliance for Telecommunications Industry Solutions
1200 G St. NW, Suite 500
Washington, DC 20005
Tel: 202-628-6380
www.atis.org

ATM Forum
2570 West El Camino Real, Suite 304
Mountain View, CA 94040
Tel: 415-949-6700
Fax: 415-949-6705
www.atmforum.com

Bellcore (now called Telcordia Technologies)
Bell Communications Research
8 Corporate Place
Piscataway, NJ 08854-4156
Tel: 908-699-2000
www.bellcore.com

BICSI
Building Industry Consulting Service International
8610 Hidden River Parkway
Tampa, FL 33637
Tel: 813-979-1991 or 800-242-7405
Fax: 813-971-4311
www.bicsci.org

Bluetooth Special Interest Group
www.bluetooth.com

BSI
British Standards Institution
British Standards House
389 Chiswick High Road
London W4 4AL, United Kingdom
Tel: 44(0)-181-996-9000
Fax: 44(0)-181-996-7400
www.bsi.org.uk

BTA
Business Technology Association
12411 Wornall Road
Kansas City, MO 64145
Tel: 816-941-3100
Fax: 816-941-2829 and 816-941-4838
www.btanet.org

CableLabs
Cable Television Laboratories, Inc.
400 Centennial Drive
Louisville, CO 80027-1266
Tel: 303-661-9100
Fax: 303-661-9199
www.cablelabs.com

California ISDN Users' Group
P.O. Box 27901-391
San Francisco, CA 94127
Tel: 415.241.9943
Fax: 415.753.6942
Email: info@ciug.org
Web: www.ciug.org, www.isdnworld.com

CBTA
Canadian Business Telecommunications Alliance
Canada Trust Tower
161 Bay Street, Suite 3650
P.O. Box 705
Toronto, Ontario M5J 2S1
Canada
Tel: 416-865-9993
Fax: 416-865-0859
www.cbta.ca

CCMA
Call Centre Management Association (UK)
Ranmore House
The Crescent
Leatherhead, Surrey
England KT22 8DY
Roy Bailey, Secretary
Telephone 01293 538400 Fax 01293 521313
email r.bailey@pncl.co.uk

CDG
CDMA Development Group
575 Anton Boulevard
Suite 560
Costa Mesa CA 92626
Tel: 714-545-5211
Fax: 714-545-4601
www.cdg.org

CDPD Forum
401 North Michigan Avenue
Chicago, IL 60611
Tel: 800-335-2373
Fax: 312-321-6869
www.cdpd.org

CFCA
Communications Fraud Control Association
3030 North Central Avenue, Suite 804
Phoenix, AZ 85012
Tel: 602-265-CFCA (2322)
Fax: 602-265-1015
www.cfca.org

CEMA
The Consumer Electronics Manufacturers Association
2500 Wilson Blvd
Arlington VA 22201
Tel: 703-907-7600
Fax: 703-907-7601

CEN
European Committee for Standardization
rue de Stassart 36
B-1050 Brussels, Belgium
Tel: 32-2-519-68-11
Fax: 32-2-519-68-19

CENELEC
European Committee for Electrotechnical Standards
rue de Stassart 35
B-1050 Brussels, Belgium
Tel: 32-2-51-96-871
Fax: 32-2-51-96-919

CIX Association
Commercial Internet eXchange Association
P.O. Box
Herndon, VA 20172-1726
Tel: 703-709-8200
Fax: 703-709-5249
www.cix.org

CMA
Communications Managers Association
1201 Mt. Kemble Avenue
Morristown, NJ 07960-6628
Tel: 201-425-1700
Fax: 201-425-0777
www.cma.org

CommerceNet
4005 Miranda Avenue, Suite 175
Palo Alto, CA 94304
Tel: 650-858-1930
Fax: 650-858-1936
www.commerce.net

Competitive Telephone Carriers of New York, Inc.
One Columbia Place
Albany, NY
Tel: 518-434-8112
Fax: 518-434-3232

CompTel
Competitive Telecommunications Association
1900 M Street, N.W., Suite 800
Washington, D.C. 20036
Tel: 202-296-6650
Fax: 202-296-7585
www.comptel.org

CompTIA
The Computing Technology Industry Association
450 East 22 Street, Suite 230
Lombard, IL 60148-6158
Tel: 630-268-1818
Fax: 630-268-1834
www.comptia.org

Committee T1
www.t1.org

CRTC
Canadian Radio-television and Telecommunications Commission
Les Terrasses de la Chaudiere
Central Building
1 Promenade du Portage
Hull, Quebec J8X 4B1
Tel: 819-997-0313
Fax: 819-994-0218
www.crtc.gc.ca

CSA
Canadian Standards Association
178 Rexdale Blvd.
Rexdale, Ontario M9W 1R3
Canada
Tel: 416-747-4000
www.csa.ca

CTIA
Cellular Telecommunications & Internet Association
now (2004) called CTIA-The Wireless Association
1250 Connecticut Ave., NW, Suite 800
Washington, DC 20036
Tel: 202-785-0081
www.wow-com.com and www.ctia.org

DAVIC
Digital Audio Visual Council
c/o SIA Societa' Italiana Avionice
C.P. 3176-Strada Antica di Colllegno, 253
I-10146 Torino
Italy
Tel: 39-11-7720-114
Fax: 39-11-725-679
www.davic.org

DSL Forum
39355 California Street, Suite 307
Fremont, CA 94538
Tel: 510-608-5905
Fax: 510-608-5917
www.dslforum.org

ECTF
Enterprise Computer Technology Forum
303 Vintage Park Drive
Foster City, CA 94404
Tel: 415-578-6852
www.ectf.org

ECMA
(nee European Computer Manufacturers Association)
114, Rue de Rhone CH-1204
Geneva, Switzerland
Tel: 41-22-846-60-00
www.ecma.ch

ECTA
European Competitive Telecommunications Association
3 Forest Court
Oaklands Park
Wokingham
Berks RG41 2FD
UK
Tel: 44-118-979-3282
Fax: 44-118-979-3288
info@ectaportal.com

EFF
The Electronic Frontier Foundation
1550 Bryant Street, Suite 725
San Francisco, CA 94103-4832
Tel: 415-436-9333
Fax: 415-436-9993
www.eff.org

EIA
Electronic Industries Alliance
2500 Wilson Blvd.
Arlington, VA 22201
Tel: 703-907-7500
Fax: 703-907-7501
www.eia.org

EMA
Electronic Messaging Association
1655 North Fort Myer Drive, Suite 500
Arlington, VA 22209
Tel: 703-524-5550
Fax: 703-524-5558
www.ema.org

ETSI
European Telecommunications Standards Institute
Route des Lucioles
F-06921 Sophia Antipolis Cedex - FRANCE
Tel: +33 (0)4 92 94 42 00
Fax: +33 (0)4 93 65 47 16
www.etsi.org

FCA
Fibre Channel Association
2570 West El Camino Real, Suite 304
Mountain View, CA 94040-1313
Tel: 650-949-6730
Fax: 650-949-6735
www.fibrechannel.com

FCC
Federal Communications Commission
1919 M Street, NW.
Washington DC
Tel: 202-418-0200
www.fcc.gov

Frame Relay Forum
39355 California Street, Suite 307
Fremont, CA 94538
Tel: 510-608-5920
Fax: 510-608-5917
www.frforum.com

GGF
Global Grid Forum
Argonne National Laboratory
9700 South Cass Avenue
Bldg 221/A142
Argonne, IL 60439
US Tel: 630-252-4300
EU Tel: +32-2-706-5611
US Fax: 630-252-4466
EU Fax: +32-2-706-5611
www.gridforum.org

Gigabit Ethernet Alliance
20111 Stevens Creek Boulevard, Suite 280
Cupertino, CA 95014
Tel: 408-241-8904
Fax: 408-241-8918
www.gigabit-ethernet.org

GO-MVIP, Inc.
3220 N Street, NW, Suite 360
Washington, D.C. 20007
Tel: 508-650-1388
Fax: 508-650-1375
www.mvip.org

HomePlug Powerline Alliance
2400 Camino Ramon, Suite 375
San Ramon, CA 94583
Tel: 925-275-6630
Fax: 925-886-3614
www.homeplug.org

HomePNA
Home Phoneline Networking Alliance
c/o Interprise Ventures
Bishop Ranch 2
2694 Bishop Drive, Suite 105
San Ramon, CA 94583
Tel: 925-277-8110
Fax: 925-277-8111
www.homepna.org

HRFWG
HomeRF Working Group
5440 SW Westgate Drive, Suite 217
Portland, OR 97221
Tel: 503-291-2563
Fax: 503-297-1090
www.homerf.org

ICEA
Insulated Cable Engineers Association
P.O. Box 1568
Carrollton, GA 30112
www.icea.net

ICSA
International Computer Security Association
1200 Walnut Bottom Road
Carlisle, PA 17013-7635
Tel: 717-258-1816
www.ncsa.com

ICTA
The International Computer-Telephony Association
Campus Box 350
University of Colorado
Boulder, CO 80302

IEC
International Engineering Consortium
549 W. Randolph Street
Suite 600
Chicago, IL 60661
Tel: 312-559-4100
Fax: 312-559-4111
www.iec.org

IEC
International Electrotechnical Commission
3, rue de Varembe
P.O. Box 131
1211 Geneva 20
Switzerland
Tel: 41-22-919-02-11
www.iec.ch

IEEE
Institute of Electrical and Electronics Engineers, Inc.
445 Hoes Lane
P.O. Box 1331
Piscataway, NJ 08855-1331
Tel: 908-981-0060
www.ieee.org

IETF
Internet Engineering Task Force
www.ietf.org

IMC
Internet Mail Consortium
127 Segre Place
Santa Cruz, CA 95060
Tel: 408-426-9827
Fax: 408-426-7301
www.imc.org

IMTC
International Multimedia Teleconferencing Consortium, Inc.
Bishop Ranch 2
2694 Bishop Drive, Suite 105
San Ramon, CA 94583
Tel: 925-277-1320
Fax: 925-277-8111
www.imtc.org

InterNIC
P.O. Box 1656
Herndon, VA 22070
Internet: admin@ds.internic.net
www.internic.net

IrDA
Infrared Data Association
P.O. Box 3883
Walnut Creek, CA 94598
Tel: 510-943-6546
Fax: 510-943-5600
www.irda.org

ISO
International Organization for Standardization
One Rue de Varembe CH-1211
Case Postale 56
Geneva 20, Switzerland
Tel: 41-22-749-0111
www.iso.ch

ITAA
Information Technology Association of America
1616 N. Ft. Myer Drive
Suite 1300
Arlington, VA 22209
Tel: 703-522-5055
Fax: 703-525-2279
www.itaa.org

ITIC
Information Technology Industry Council
1250 Eye Street NW Suite 200
Washington, DC 20005
Tel: 202-737-8888
Fax: 202-638-4922
www.itic.org

ITCA
International TeleConferencing Association
100 Four Falls Corporate Center, Suite 105
West Conshohocken, PA 19428
Tel: 610-941-2020
Fax: 610-941-2015
www.itca.org

ITU
International Telecommunications Union
Place des Nations
CH-1211 Geneva 20
Switzerland
Tel: +41 22 99 51 11
Fax: +41 22 33 72 56
www.itu.ch

IWTA
International Wireless Telecommunications Association
1150 18th Street, NW
Suite 250
Washington, DC 20036
Tel: 202-331-7773
Fax: 202-331-9062
www.iwta.org

Linux Phone Standards Forum
www.lipsforum.org

MMTA
See TIA.

MPLS and Frame Relay Alliance
39355 California Street #307
Fremont, CA 94538
Tel: 1-510-608-5910
Fax: 1-510-608-5917
www.frforum.com
www.mplsforum.org

NAB
National Association of Broadcasters
1771 N Street, N.W.
Washington, D.C. 20036
Tel: 202-429-5300
www.nab.org

NARTE
National Association of Radio and Telecommunications
Engineers
P.O. Box 678
Medway, MA 02053
Tel: 508-533-8333
Fax: 508-533-3815
www.narte.org

NARUC
National Association of Regulatory Utility Commissioners
1100 Pennsylvania Avenue NW, Suite 603
Post Office Box 684
Washington, D.C. 20044-0684
Tel: 202-898-2200
Fax: 202-898-2213
www.naruc.org

NATD
National Association of Telecommunications Dealers
1045 East Atlantic Avenue, Suite 206
Delray Beach, FL 33483
Tel: 561-266-9440
Fax: 561-266-9017
www.natd.com

NCA
National Convergence Alliance
133 Littleton Road
Westford, MA 01886
Tel: 978-692-3522
www.convergencealliance.com or http://63.151.41.74

NCTA
National Cable TV Association
1724 Massachusetts Avenue, N.W.
Washington, D.C. 20036
Tel: 202-775-3669
www.ncta.com

NECA
National Exchange Carrier Association
80 South Jefferson Road
Whippany, NJ 07981-1009
Tel: 973-884-8000
Fax: 973-884-8469
www.neca.org

NEMA
National Electrical Manufacturers Association
1300 North 17th Street, Suite 1847
Rosslyn, VA 22209
Tel: 703-841-3200
Fax: 703-841-3300
www.nema.org

NFPA
National Fire Protection Association
1 Batterymarch Park
P.O. Box 9101
Quincy, MA 02269-9101
Tel: 617-770-3000
Fax: 617-770-0700
www.nfpa.org

NIST
National Institute of Standards and Technology
Gaithersburg, MD 20899
Tel: 301-975-2000
www.nist.com

NIUF
North American ISDN Users' Forum
National Institute of Standards & Technology (NIST)
Building 820, Room 445
Gaithersburg, MD 20899
Tel: 301-975-2937
Fax: 301-926-9675
www.niuf.nist.gov

NMF
Network Management Forum
1201 Mt. Kemble Avenue
Morristown, NJ 07960
Tel: 973-425-1900
Fax: 973-4251515
www.nmf.org

NANOG
North American Network Operators' Group
www.nanog.org

NATOA
National Association of Telecommunications Officers and
Advisors
1650 Tysons Boulevard, Suite 200
McLean, VA 22102
Tel: 703-506-3275
Fax: 703-506-3266
www.natoa.org

NENA
National Emergency Number Association
P.O. Box 360960
Columbus, OH 43236
Tel: 614-741-2080
Fax: 614-933-0911
www.nena9-1-1.org

NPA
Network Professional Association
P.O. Box 809161
Chicago, IL 60680-9161
Tel: 801-223-9444
Fax: 801-223-9486
www.npa.org

NTIA
National Telecommunications and Information Administration
U.S. Department of Commerce
Washington, D.C. 20230
Tel: 202-377-1880
www.ntia.doc.gov

NTIS
National Technical Information Service
Technology Administration
U.S. Department of Commerce
Springfield, VA 22161
Tel: 703-605-6000
Fax: 703-321-8547
www.ntis.gov

OBI
Open Buying on the Internet Consortium
57 Bedford Street, Suite 208
Lexington, MA 02173
Tel: 781-863-5396
Fax: 781-861-1708
www.supplyworks.com/obi

OFR
Office of the Federal Register
National Archives & Records Administration
Suite 700
800 North Capitol Street NW
Washington, D.C. 20408
Tel: 202-523-3117
Fax: 202-523-6866

OFTEL
50 Ludgate Hill
London EC4M 7JJ
Tel: +44 171 834 8700
Fax: +44 171 634 8943
www.oftel.gov.uk

OIF
Optical Internetworking Forum
39355 California Street, Suite 307
Fremont, CA 94538
Tel: 510-608-5928
Fax: 510-608-5917
www.oiforum.com

OMG
Object Management Group
4902 Old Connecticut Path
Framingham, MA 01701
Tel: 508-820-4300
Fax: 508-820-4303
www.omg.org

OSTA
Optical Storage Technology Association
311 East Carrillo Street
Santa Barbara, CA 93101
Tel: 805-963-3853
Fax: 805-962-1541
www.osta.org

PCCA
Portable Computer and Communications Association
P.O. Box 2460
Boulder Creek, CA 95006
Tel: 703-793-0300
Fax: 703-836-1608
www.pca.org

PCIA
Personal Communications Industry Association
500 Montgomery Street, Suite 700
Alexandria, VA 22314
Tel: 703-739-0300
Fax: 703-836-1608
www.pcia.com

PCMCIA
Personal Computer Memory Card International Association
2635 North First Street, Suite 209
San Jose, CA 95134
Tel: 408-433-2273
Fax: 408-433-9558
www.pcmcia.org

PDAia
PDA Industry Association
815 Fairfield Road
Burlingame, CA 94010
Tel: 650-685-0842
Fax: 650-343-9753
www.padia.org

PICMG
PCI Industrial Computer Manufacturers Group
c/o Rogers Communications
401 Edgewater Place, Suite 500
Wakefield, MA 01880
Tel: 781-246-9318
Fax: 781-224-1239
www.picmg.org

PTC
Pacific Telecommunications Council
2454 S. Beretania Street, Suite 302
Honolulu, HI 96826
Tel: 808-941-3789
Fax: 808-944-4874
www.ptc.org

SBCA
Satellite Broadcasting and Communications Association of
America
225 Reinekers Lane, Suite 600
Alexandria, VA 22314
Tel: 703-549-6990
Fax: 703-549-7640
www.sbca.org

SCTE
Society of Cable Telecommunications Engineers, Inc.
140 Philips Road
Exton, PA 19341-1318
Tel: 610-363-6888, Fax: 610-363-5898
www.scte.org

SIA
Satellite Industry Association
225 Reinekers Lane, Suite 600
Alexandria, VA 22314
Tel: 703-549-8697
Fax: 703-549-9188
www.sia.org

SIF
SONET Interoperability Forum
c/o Alliance for Telecommunications Solutions
1200 G Street, N.W., Suite 500
Washington, DC 20005
Tel: 202-628-6380
Fax: 202-393-5453
www.atis.org/atis/sif/index.html

SMPTE
Society of Motion Picture & Television Engineers
595 W. Hartsdale Avenue
White Plains, NY 10607-1824
Tel: 914-761-1100
Fax: 914-761-3115
www.smpte.org

Softswitch Consortium
2694 Bishop Drive, Suite 275
San Ramon, CA 94583
Tel: 925-277-8110
www.softswitch.org

SPA
Software Publishers Association
1730 M Street, NW
Suite 700
Washington, DC 20036-4510
Tel: 202-452-1600
Fax: 202-223-8756
www.spa.org

TCA
Telecommunications Association
74 New Montgomery Street, Suite 230
San Francisco, CA 94105-3411
Tel: 415-777-4646
Fax: 415-777-5295
www.tca.org

TCIF
Telecommunications Industry Forum
c/o Alliance for Telecommunications Industry Solutions
1200 G Street, NW
Suite 500
Washington, DC 20005
Tel: 202-628-6380
Fax: 202-393-5453
www.atis.org/atis/tcif/index.html

Telecom Corridor Technology Business Council
411 Belle Grove
Richardson, TX 75080-5297
Fax: 972-680-9103

Telcordia Technologies
445 South Street
Morristown, NJ 07960-6438
Tel: 973-829-2000
Fax: 973-829-5982
www.telcordia.com

The Open Group
1010 El Camino Real
Suite 380
Menlo Park, CA 94025-4345
Tel: 650-323-7992
Fax: 650-323-8204
www.opengroup.org

TIA
Telecommunications Industry Association
2500 Wilson Boulevard
Arlington, VA 22201
Tel: 703-907-7700
Fax: 703-907-7727
www.tiaonline.org

TMF
TeleManagement Forum
1201 Mt. Kemble Avenue
Morristown, NJ 07960
Tel: 973-425-1900
Fax: 973-4251515
www.tmforum.org

TRA
Telecommunications Resellers Association
See ASCENT

The Open Group
11 Cambridge Center
Cambridge, MA 02142
Tel: 617-621-8700
Fax: 617-621-0631
www.opengroup.org

UL
Underwriters Laboratory
333 Pfingsten Road
Northbrook, IL 60062
Tel: 847-272-8800
Fax 847-272-8129
www.ul.com

Ultra-Wideband Working Group
www.uwb.org

USDLA
United States Distance Learning Association
1240 Central Boulevard, Suite A
Brentwood, CA 94513
Tel: 925-513-4253
Fax: 925-513-4255
www.usdla.org

USTA
United States Telecom Association
1401 H Street, N.W., Suite 600
Washington, D.C. 20005-2164
Tel: 202-326-7300
Fax: 202-326-7333
www.usta.org

USTTI
United States Telecommunications Training Institute
1150 Connecticut Av.., N.W. Suite 702
Washington, D.C. 20036
Tel: 202-785-7373
Fax: 202-785-1930
www.telemobile.com

U.S. Department of Commerce National Technical Information Service
5285 Port Royal Road
Springfield, VA 22161
Tel: 703-487-4600
www.ntis.gov

UTC
United Telecom Council
The Telecommunications and Information Technology Association for Utilities, Pipelines and Other Critical Infrastructure Companies
1901 Pennsylvania Avenue, NW - Fifth Floor
Washington, DC 20006
Tel: 202-872-0030
Mobile: 202-320-9790
Fax: 202-872-1331
www.utc.org

UWCC
Universal Wireless Communications Consortium
1756 114th Avenue SE, Suite 100
Bellevue, WA 98004
Tel: 425-372-8922
Fax: 425-372-8923
www.uwcc.org

VoiceXML Forum
www.voicexml.org

W3C
World Wide Web Consortium
Massachusetts Institute of Technology
Laboratory for Computer Science
545 Technology Square
Cambridge, MA 02139
Tel: 617-253-2613
Fax: 617-258-5999
www.w3.org

WAP Forum
Wireless Application Protocol Forum, Ltd.
Third Floor
Abbots House
Abbey Street
Reading
RG1 3BD
United Kingdom
Tel: 44-118-949-0000
Fax: 44-118-949-0049
www.wapforum.com

WECA
Wireless Ethernet Compatibility Alliance
www.wirelessethernet.com

WiMAX Forum
2495 Leghorn Street
Mountain View, CA 94043
Tel: 503-712-8948
www.wimax.org

WIPO
World Intellectual Property Organization
P.O. Box 18
CH-1211
Geneva 20
Switzerland
Tel: 41-22-338-9111
Fax: 41-22-733-54-28
www.wipo.org

WiMAX Forum
2495 Leghorn Street
Mountain View, CA 94043
Tel: 503-712-8948
www.wimax.org

Wireless Data Forum
1250 Connecticut Avenue, NW
Suite 800
Washington, DC 20036
Tel: 202-736-3663
Fax: 202-466-3413
wwww.wirelessdata.org

WLANA
Wireless LAN Alliance
1114 Sherman Avenue
St. Simons Island, GA 31522
www.wlana.com

WLI Forum
Wireless LAN Interoperability Forum
1111 W. El Camino Road, #109-171
Sunnyvale, CA 94087
www.wlif.com

WLANA
Wireless LAN Alliance
2723 Delaware Avenue
Redwood City, CA 94061
www.wlana.com

WSTA
Wall Street Telecommunications Association
One West Front Street
Red Bank, NJ 07701
Tel: 908-530-8808
Fax: 908-530-0020
www.wsta.org

WTO
World Trade Organization
154, rue de Lausanne
CH-1211 Geneva 21
Switzerland
Tel: 41-22-739-5111
Fax: 41-22-739-5458
www.wto.org

Publications, Computer and Telecommunications

America's Network, www.americasnetwork.com/americasnetwork/

Broadband Properties, www.BroadbandProperties.com/

Business 2.0, http://money.cnn.com/magazines/business2/

Business Communications Review, www.bcr.com/

Cabling Business Magazine, www.cablingbusiness.com/

Call Center Magazine, www.callcentermagazine.com/

CED Magazine, www.cedmagazine.com

CIO, www.cio.com/

Communications News, www.comnews.com/

Computerworld, www.computerworld.com/

Customer Inter@ction Solutions, www.tmcnet.com/call-center/

eeTimes, www.eetimes.com/

ECN, www.ecnmag.com/

EDN, www.edn.com/

Electronic Business, www.reed-electronics.com/eb-mag/

eWeek (used to be called PC WEEK), www.eweek.com/

Fast Company, www.fc.com/

Federal Computer Week, http://subscribe.101com.com/FCW/default.htm

IEEE Spectrum, www.spectrum.ieee.org/feb06/inthisissue

InfoWorld, www.infoworld.com/

Information Week, http://informationweek.com/

Intelligent Enterprise, www.intelligententerprise.com/

Internet Telephony, www.itmag.com/

IT Architect (previously Network Magazine, previously LAN Magazine), www.itarchitect.com/

Lightwave, http://lw.pennnet.com/home.cfm

MacUser, www.macuser.com/

MacWorld, www.macworld.com/

Messaging News, www.messagingnews.com/

Network Computing, www.networkcomputing.com/

Network Magazine, (see IT Architect)

Network World, www.networkworld.com/ and www.nww.com/

Optimize, www.optimizemag.com/

PC Magazine, www.pcmag.com/

PC World, www.pcworld.com/

Red Herring Magazine, www.redherring.com/

Scientific Computing, www.scimag.com/

Scientific American, www.sciam.com/

SD Times, www.sdtimes.com/

Speech Technology, 43 Danbury Road, Wilton, CT 06897

TechNet, www.technetmagazine.com/

Telecom Gear, 15400 Knoll Trail, Dallas, TX 75248

TeleManagement, www.telecomgearonline.com/memberdirectory/detail.cfm/cid/403

Telecommunications, www.telecommagazine.com/

Telephony Magazine, www.telephonyonline.com/

UnixWorld, www.networkcomputing.com/unixworld/unixhome.html/

VARBusiness, www.varbusiness.com/

0x06 Via Satellite, www.viasatellite.com/

Von Magazine, www.VonMag.com/

Wall Street & Technology, www.wallstreetandtech.com/

Wired, http://www.wired.com/

Wireless Week, www/wirelessweek.com/

xchange, www.xchangemag.com/

Home Pages of Regulatory Agencies

Regulatory Commission of Alaska www.state.ak.us/rca

Alabama Public Service Commission www.psc.state.al.us

Arizona Corporation Commission www.cc.state.az.us

Arkansas Public Service Commission www.state.ar.us/psc

California Public Utilities Commission www.cpuc.ca.gov

Colorado Public Utilities Commission www.dora.state.co.us/puc

Connecticut Department of Public Utility Control www.state.ct.us/dpuc

Delaware Public Service Commission www.state.de.us/delpsc

Florida Public Service Commission www.floridapsc.com

Georgia Public Service Commission www.psc.state.ga.us

Hawaii Public Utilities Commission www.hawaii.gov/budget/puc/

Idaho Public Utilities Commission www.puc.state.id.us

Illinois Commerce Commission www.icc.illinois.gov

Indiana Utility Regulatory Commission www.state.in.us/iurc/

Iowa Utilities Board www.state.ia.us/government/com/util/

Kansas Corporation Commission www.kcc.state.ks.us

Kentucky Public Service Commission www.psc.state.ky.us

Louisiana Public Service Commission www.lpsc.org

Maine Public Utilities Commission www.state.me.us/mpuc

Maryland Public Service Commission www.psc.state.md.us/psc/

Massachusetts Department of Telecommunications and Energy www.state.ma.us/dpu/

Michigan Public Service Commission www.michigan.gov/mpsc

Minnesota Public Utilities Commission www.puc.state.mn.us

Mississippi Public Service Commission www.psc.state.ms.us

Missouri Public Service Commission www.psc.state.mo.us

Montana Public Service Commission www.psc.state.mt.us

Nebraska Public Service Commission www.psc.state.ne.us

Public Utilities Commission of Nevada www.puc.state.nv.us

New Hampshire Public Utilities Commission www.puc.state.nh.us

New Jersey Board of Public Utilities www.bpu.state.nj.us

New Mexico Public Regulation Commission www.nmprc.state.nm.us

New York Public Service Commission www.dps.state.ny.us

North Carolina Utilities Commission www.ncuc.commerce.state.nc.us

North Dakota Public Service Commission www.psc.state.nd.us

Public Utilities Commission of Ohio www.puco.ohio.gov

Oklahoma Corporation Commission www.occ.state.ok.us

Oregon Public Utility Commission www.puc.state.or.us

Pennsylvania Public Utility Commission www.puc.state.pa.us

Rhode Island Public Utilities Commission www.ripuc.org

Public Service Commission of South Carolina www.psc.sc.gov

South Dakota Public Utilities Commission www.state.sd.us/puc/

Tennessee Regulatory Authority www.state.tn.us/tra

Public Utility Commission of Texas www.puc.state.tx.us

Public Service Commission of Utah www.psc.state.ut.us

Vermont Department of Public Service publicservice.vermont.gov

Virginia State Corporation Commission www.scc.virginia.gov

Washington Utilities & Transportation Commission www.wutc.wa.gov

Public Service Commission of West Virginia www.psc.state.wv.us

Public Service Commission of Wisconsin psc.wi.gov

Wyoming Public Service Commission psc.state.wy.us

International Calling Codes

Coountry	Code	Country	Code	Country	Code	Country	Code
Albania	355	Democratic Republic of Congo		Iran	98	New Zealand	64
Algeria	213	(formely Zaire)	243	Iraq	964	Nicaragua	505
American Samoa*	684	Denmark	45	Ireland	353	Niger	227
Andorra	376	Diego Garcia	246	Israel	972	Nigeria	234
Angola	244	Djibouti	253	Italy	39	Niue	683
Anguilla	264	Dominica	1	Ivory Coast	225	Norfolk Island	672
Antarctica	672	Dominican Republic	1	Jamaica	1	Northern Mariana Islands*	670
Antigua and Barbuda	1	East Timor	670	Japan	81	Norway	47
Argentina	54	Ecuador	593	Jordan	962	Oman	968
Armenia	374	Egypt	20	Kazakhstan	7	Pakistan	92
Aruba	297	El Salvador	503	Kenya	254	Palau	680
Australia	61	Equatorial Guinea	240	Kiribati	686	Paletinian Authority	970
Austria	43	Eritrea	291	Korea North	850	Panama	507
Azerbaijan	994	Estonia	372	Korea South	82	Papua New Guinea	675
Bahamas	1	Ethiopia	251	Kuwait	965	Paraguay	595
Bahrain	973	Falkland Islands (Malvinas)	500	Kyrgyzstan	996	Peru	51
Bangladesh	880	Faroe Islands	298	Laos	856	Philippines	63
Barbados	1	Fiji	679	Latvia	371	Pitcairn	872
Belarus	375	Finland	358	Lebanon	961	Poland	48
Belgium	32	France	33	Lesotho	266	Portugal	351
Belize	501	French Antilles	596	Liberia	231	Puerto Rico	1
Benin	229	France, Metropolitan	33	Libya	218	Qatar	974
Bermuda	1	French Guiana	594	Liechtenstein	41	Republika Srpska	387
Bhutan	975	French Polynesia	689	Lithuania	370	Reunion Island	262
Bolivia	591	Gabon	241	Luxembourg	352	Romania	40
Bosnia and Herzegovina	387	Gambia	220	Macau	853	Russian Federation	7
Botswana	267	Georgia	995	Macedonia	389	Rwanda	250
Brazil	55	Germany	49	Madagascar	261	Saint Kitts and Nevis	1
Brunei Darussalam	673	Ghana	233	Malawi	265	Saint Lucia	1
Bulgaria	359	Gibraltar	350	Malaysia	60	Saint Vincent & the Grenadines	1
Burkina Faso	226	Great Britain (UK)	44	Maldives	960	Samoa (orig. called Western)	685
Burundi	257	Greece	30	Mali	223	San Marino	378
Cambodia	855	Greenland	299	Malta	356	Sao Tome and Principe	239
Cameroon	237	Grenada	1	Mariana Islands	670	Saudi Arabia	966
Canada	1	Guadeloupe	590	Marshall Islands	692	Senegal	221
Cape Verde	238	Guam*	671	Martinique	596	Seychelles	248
Cayman Islands	1	Guatemala	502	Mauritania	222	Sierra Leone	232
Central African Republic	236	Guinea	224	Mauritius	230	Singapore	65
Chad	235	Guinea-Bissau	245	Mayotte	269	Slovak Republic	421
Chile	56	Guyana	592	Mexico	52	Slovenia	386
China	86	Haiti	509	Micronesia	691	Solomon Islands	677
Christmas Island	672	Heard and McDonald Islands	692	Moldova	373	Somalia	252
Cocos (Keeling) Islands	672	Herzegovina	387	Monaco	377	South Africa	27
Colombia	57	Honduras	504	Mongolia	976	Spain	34
Comoros	269	Hong Kong	852	Montserrat	1	Sri Lanka	94
Congo	242	Hungary	36	Morocco	212	St. Helena	290
Cook Islands	682	Iceland	354	Mozambique	258	St. Pierre and Miquelon	508
Costa Rica	506	India	91	Myanmar	95	Sudan	249
Cote D'Ivoire (Ivory Coast)	225	Inmarsat		Namibia	264	Suriname	597
Croatia (Hrvatska)	385	East Atlantic Ocean	871	Nauru	674	Svalbard & Jan Mayen Islands	378
Cuba	53	Indian Ocean	873	Nepal	977	Swaziland	268
Cyprus	357	Pacific Ocean	872	Netherlands	31	Sweden	46
Czech Republic	420	West Atlantic Ocean	874	Netherlands Antilles	599	Switzerland	41
Czechoslovakia (former)	420	Indonesia	62	New Caledonia	687	Syria	963

Country	Code
Taiwan	886
Tajikistan	992
Tanzania	255
Thailand	66
Togo	228
Tokelau	690
Tonga	676
Trinidad and Tobago	1
Tunisia	216
Turkey	90
Turkmenistan	993
Turks and Caicos Islands	649
Tuvalu	688
US Minor Outlying Islands	1
USSR (former)	7
Uganda	256
Ukraine	380
United Arab Emirates	971
United Kingdom	44
United States	1
Uruguay	598
Uzbekistan	998
Vanuatu	678
Vatican City State	396
Venezuela	58
Viet Nam	84
Virgin Islands (British)	1
Virgin Islands (U.S.)	1
Wallis and Futuna Islands	681
Western Sahara	34
WesternSamoa (now called Samoa)	685
Yemen	967
Yugoslavia	381
Zaire	243
Zambia	260
Zimbabwe	263

Standard Plugs and Connectors

5 PIN DIN (F).EPS

6 PIN MINI DIN (F).EPS

CENTRONICS (36-PIN).EPS

DB09.EPS

HD15 PIN.EPS

IEEE 488.EPS

RS-449.EPS

RS-232 INTERFACE (MALE).EPS

RS-530.EPS

V.35 ROTATED.EPS

Black Box Corporation
The World's Source for Connectivity℠

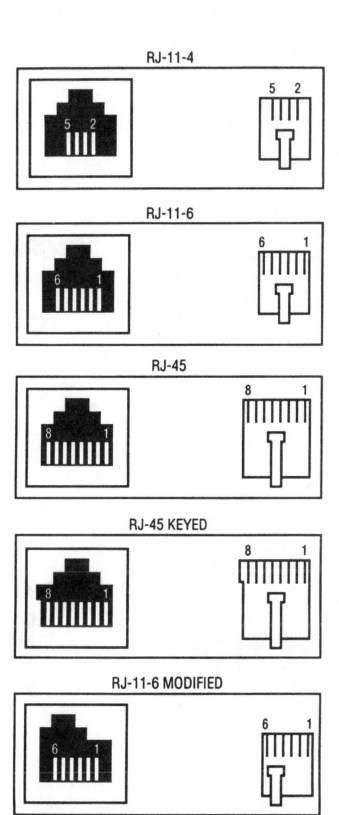

RJ-11-4

RJ-11-6

RJ-45

RJ-45 KEYED

RJ-11-6 MODIFIED